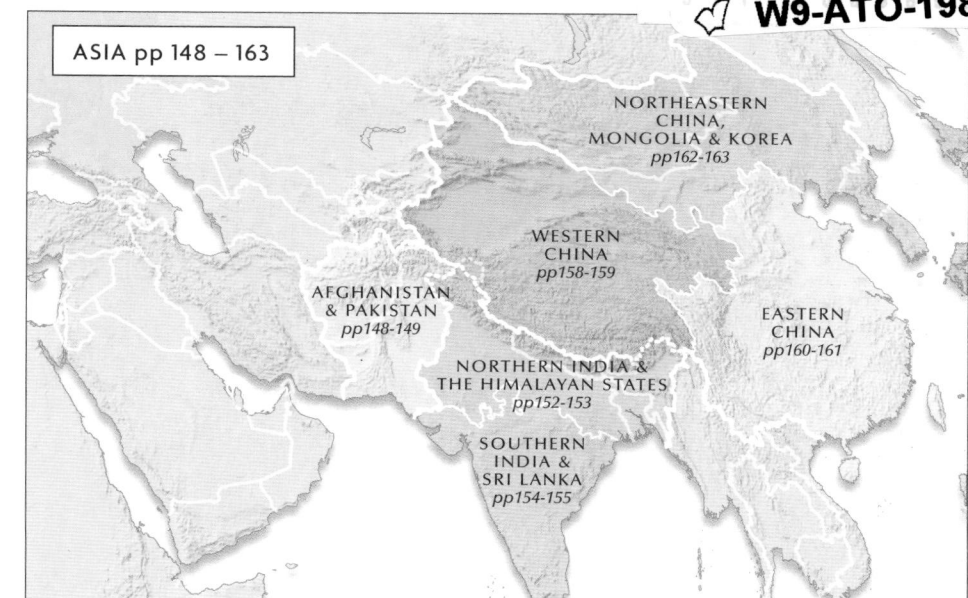

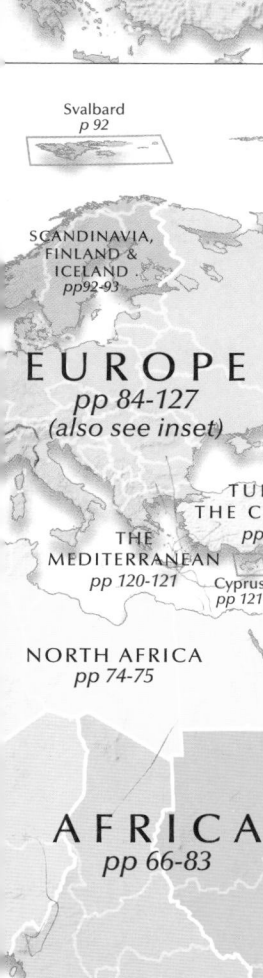

REFERENCE
WORLD
ATLAS

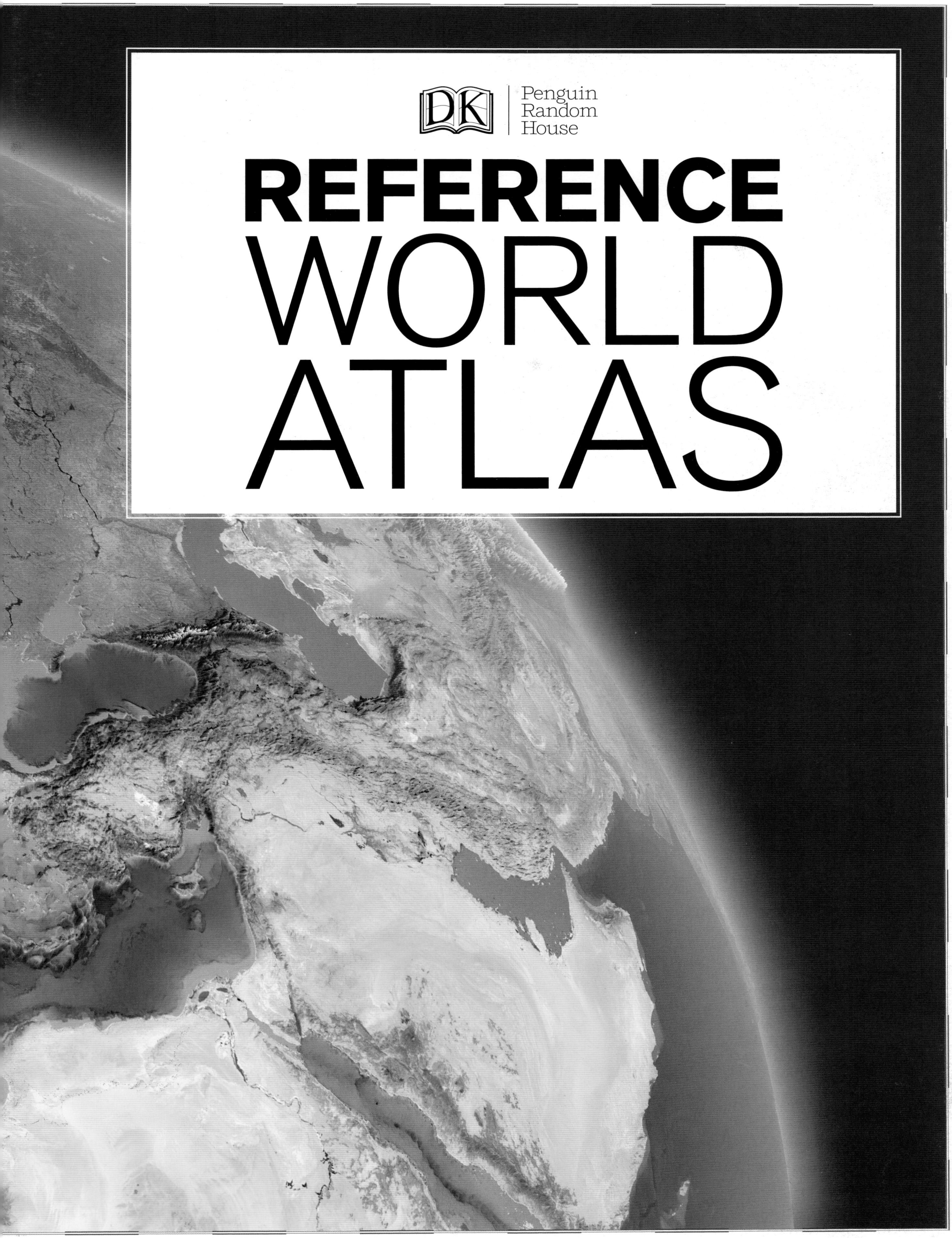

REFERENCE
WORLD
ATLAS

FOR THE TENTH EDITION

Senior Cartographic Editor Simon Mumford
Producer, Pre-Production Luca Frassinetti **Producer** Vivienne Yong
Jacket Design Development Manager Sophia MTT
Publishing Director Jonathan Metcalf **Associate Publishing Director** Liz Wheeler **Art Director** Karen Self

General Geographical Consultants

Physical Geography Denys Brunsden, Emeritus Professor, Department of Geography, King's College, London
Human Geography Professor J Malcolm Wagstaff, Department of Geography, University of Southampton
Place Names Caroline Burgess, CartoConsulting Ltd, Reading
Boundaries International Boundaries Research Unit, Mountjoy Research Centre, University of Durham

Digital Mapping Consultants

DK Cartopia developed by George Galfalvi and XMap Ltd, London
Professor Jan-Peter Muller, Department of Photogrammetry and Surveying, University College, London
Planets and information on the Solar System provided by Philip Eales and Kevin Tildsley, Planetary Visions Ltd, London

Regional Consultants

North America Dr David Green, Department of Geography, King's College, London • Jim Walsh, Head of Reference, Wessell Library, Tufts University, Medford, Massachussetts
South America Dr David Preston, School of Geography, University of Leeds **Europe** Dr Edward M Yates, formerly of the Department of Geography, King's College, London
Africa Dr Philip Amis, Development Administration Group, University of Birmingham • Dr Ieuan Ll Griffiths, Department of Geography, University of Sussex
Dr Tony Binns, Department of Geography, University of Sussex
Central Asia Dr David Turnock, Department of Geography, University of Leicester **South and East Asia** Dr Jonathan Rigg, Department of Geography, University of Durham
Australasia and Oceania Dr Robert Allison, Department of Geography, University of Durham

Acknowledgments

Digital terrain data created by Eros Data Center, Sioux Falls, South Dakota, USA. Processed by GVS Images Inc, California, USA and Planetary Visions Ltd, London, UK
Cambridge International Reference on Current Affairs (CIRCA), Cambridge, UK • Digitization by Robertson Research International, Swanley, UK • Peter Clark
British Isles maps generated from a dataset supplied by Map Marketing Ltd/European Map Graphics Ltd in combination with DK Cartopia copyright data

DORLING KINDERSLEY CARTOGRAPHY

Editor-in-Chief Andrew Heritage **Managing Cartographer** David Roberts **Senior Cartographic Editor** Roger Bullen
Editorial Direction Louise Cavanagh **Database Manager** Simon Lewis **Art Direction** Chez Picthall

Cartographers
Pamela Alford • James Anderson • Caroline Bowie • Dale Buckton • Tony Chambers • Jan Clark • Bob Croser • Martin Darlison • Damien Demaj • Claire Ellam • Sally Gable
Jeremy Hepworth • Geraldine Horner • Chris Jackson • Christine Johnston • Julia Lunn • Michael Martin • Ed Merritt • James Mills-Hicks • Simon Mumford • John Plumer
John Scott • Ann Stephenson • Gail Townsley • Julie Turner • Sarah Vaughan • Jane Voss • Scott Wallace • Iorwerth Watkins • Bryony Webb • Alan Whitaker • Peter Winfield

Digital Maps Created in DK Cartopia by
Tom Coulson • Thomas Robertshaw
Philip Rowles • Rob Stokes

Placenames Database Team
Natalie Clarkson • Ruth Duxbury • Caroline Falce • John Featherstone • Dan Gardiner
Ciárán Hynes • Margaret Hynes • Helen Rudkin • Margaret Stevenson • Annie Wilson

Managing Editor
Lisa Thomas

Senior Managing Art Editor
Philip Lord

Editors
Thomas Heath • Wim Jenkins • Jane Oliver
Siobhan Ryan • Elizabeth Wyse

Designers
Scott David • Carol Ann Davis • David Douglas • Rhonda Fisher
Karen Gregory • Nicola Liddiard • Paul Williams

Editorial Research
Helen Dangerfield • Andrew Rebeiro-Hargrave

Illustrations
Ciárán Hughes • Advanced Illustration, Congleton, UK

Additional Editorial Assistance
Debra Clapson • Robert Damon • Ailsa Heritage
Constance Novis • Jayne Parsons • Chris Whitwell

Picture Research
Melissa Albany • James Clarke • Anna Lord
Christine Rista • Sarah Moule • Louise Thomas

First American edition, 1997. This revised edition, 2016.

Published in the United States by DK Publishing, 345 Hudson Street, New York, New York 10014

16 17 18 19 20 10 9 8 7 6 5 4 3 2 1

265170 – August 2016

Published in Great Britain by Dorling Kindersley Ltd. A Division of Penguin Random House LLC.

DK Publishing books are available at special discounts when purchased in
bulk for sales promotion, premiums, fundraising, or educational use.
For details, contact:
DK Publishing Special Markets, 345 Hudson Street,
New York, New York 10014 or specialsales@dk.com

A catalog record for this book is avaiable from the Library of Congress.

ISBN 978-1-4654-5182-8

Printed and bound in Hong Kong.

A WORLD OF IDEAS:
SEE ALL THERE IS TO KNOW
www.dk.com

Introduction

EVERYTHING YOU NEED TO KNOW ABOUT OUR PLANET TODAY

For many, the outstanding legacy of the twentieth century was the way in which the Earth shrank. In the third millennium, it is increasingly important for us to have a clear vision of the world in which we live. The human population has increased fourfold since 1900. The last scraps of *terra incognita*— the polar regions and ocean depths—have been penetrated and mapped. New regions have been colonized and previously hostile realms claimed for habitation. The growth of air transportation and mass tourism allows many of us to travel farther, faster, and more frequently than ever before. In doing so we are given a bird's-eye view of the Earth's surface denied to our forebears.

At the same time, the amount of information about our world has grown enormously. Our multi-media environment hurls uninterrupted streams of data at us, on the printed page, through the airwaves, and across our television, computer, and phone screens; events from all corners of the globe reach us instantaneously and are witnessed as they unfold. Our sense of stability and certainty has been eroded; instead, we are aware that the world is in a constant state of flux and change. Natural disasters, man-made cataclysms, and conflicts between nations remind us daily of the enormity and fragility of our domain. The ongoing threat of international terrorism throws into very stark relief the difficulties that arise when trying to "know" or "understand" our planet and its many cultures.

The current crisis in our "global" culture has made the need greater than ever before for everyone to possess an atlas. DK's **REFERENCE WORLD ATLAS** has been conceived to meet this need. At its core, like all atlases, it seeks to define where places are located, to describe their main characteristics, and to map them in relation to other places. Every attempt has been made to produce information and maps that are as clear, accurate, and accessible as possible using the latest digital cartographic techniques. In addition, each page of the atlas provides a wealth of further information, bringing the maps to life. Using photographs, diagrams, at-a-glance maps, introductory texts, and captions, the atlas builds up a detailed portrait of those features—cultural, political, economic, and geomorphological—that make each region unique, and which are also the main agents of change.

This tenth edition of the **REFERENCE WORLD ATLAS** incorporates hundreds of revisions and updates, distilling the burgeoning mass of information available through modern technology into an extraordinarily detailed and reliable view of our world.

CONTENTS

THE WORLD

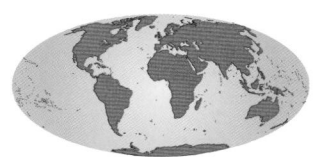

ATLAS OF THE WORLD

North America

South America

Africa

Europe

Asia

Australasia & Oceania

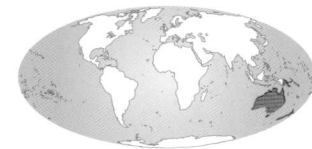

INDEX-GAZETTEER

Key to maps

Regional

Physical features

elevation

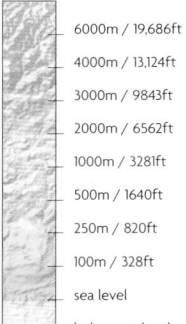

6000m / 19,686ft
4000m / 13,124ft
3000m / 9843ft
2000m / 6562ft
1000m / 3281ft
500m / 1640ft
250m / 820ft
100m / 328ft
sea level
below sea level

▲ elevation above sea level (mountain height)
▲ volcano
✕ pass
▼ elevation below sea level (depression depth)

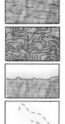

sand desert
lava flow
coastline
reef
atoll

sea depth

sea level
-250m / -820ft
-500m / -1640ft
-1000m / -3281ft
-2000m / -6562ft
-3000m / -9843ft

▲ seamount / guyot symbol
▼ undersea spot depth

Drainage features

main river
secondary river
tertiary river
minor river
main seasonal river
secondary seasonal river
canal
waterfall
rapids
dam
perennial lake
seasonal lake
perennial salt lake
seasonal salt lake
reservoir
salt flat / salt pan
marsh / salt marsh
mangrove
wadi
○ spring / well / waterhole / oasis

Ice features

ice cap / sheet
ice shelf
glacier / snowfield
• • • • summer pack ice limit
○ ○ ○ winter pack ice limit

Communications

motorway / highway
motorway / highway (under construction)
major road
minor road
→ · · · → tunnel (road)
main railroad
minor railroad
→ · · · → tunnel (railroad)
✈ international airport

Borders

full international border
undefined international border
disputed de facto border
disputed territorial claim border
indication of country extent (Pacific only)
indication of dependent territory extent (Pacific only)
demarcation / cease fire line
autonomous / federal region border
other 1st order internal administrative border
2nd order internal administrative border

Settlements

 built up area

settlement population symbols

■ more than 5 million
◙ 1 million to 5 million
◉ 500,000 to 1 million
◎ 100,000 to 500,000
⊕ 50,000 to 100,000
○ 10,000 to 50,000
○ fewer than 10,000

■ ● ● country/dependent territory capital city
■ ◉ ● autonomous / federal region / other 1st order internal administrative center
■ ◉ ⊕ 2nd order internal administrative center

Miscellaneous features

═ ═ ═ ═ ancient wall
◇ site of interest
⊙ scientific station

Graticule features

lines of latitude and longitude / Equator
Tropics / Polar circles
45° degrees of longitude / latitude

Typographic key

Physical features

landscape features ... *Namib Desert*
Massif Central
ANDES

headland .... *Nordkapp*

elevation / volcano / pass .... Mount Meru 4556 m

drainage features .... *Lake Geneva*

rivers / canals spring / well / waterhole / oasis / waterfall / rapids / dam .... *Mekong*

ice features .... *Vatnajökull*

sea features .... *Golfe de Lion*
Andaman Sea
INDIAN OCEAN

undersea features .... *Barracuda Fracture Zone*

Regions

country .... **ARMENIA**

dependent territory with parent state ..... **NIUE** (to NZ)

region outside feature area .... ANGOLA

autonomous / federal region .... MINAS GERAIS

other 1st order internal administrative region .... **MINSKAYA VOBLASTS'**

2nd order internal administrative region .... Vaucluse

cultural region .... New England

Settlements

capital city .... **BEIJING**

dependent territory capital city .... FORT-DE-FRANCE

other settlements ... **Chicago**
Adana
Tizi Ozou
Yonezawa
Farnham

Miscellaneous

sites of interest / miscellaneous .... *Valley of the Kings*

Tropics / Polar circles .... *Antarctic Circle*

How to use this Atlas

The atlas is organized by continent, moving eastward from the International Date Line. The opening section describes the world's structure, systems, and its main features. The Atlas of the World which follows, is a continent-by-continent guide to today's world, starting with a comprehensive insight into the physical, political, and economic structure of each continent, followed by integrated mapping and descriptions of each region or country.

The world

The introductory section of the Atlas deals with every aspect of the planet, from physical structure to human geography, providing an overall picture of the world we live in. Complex topics such as the landscape of the Earth, climate, oceans, population, and economic patterns are clearly explained with the aid of maps and diagrams drawn from the latest information.

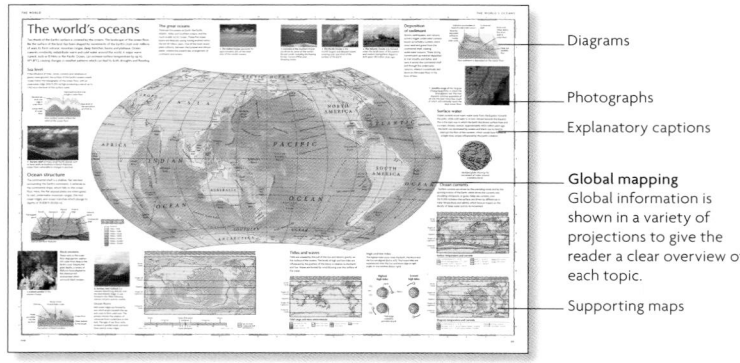

Diagrams

Photographs

Explanatory captions

Global mapping
Global information is shown in a variety of projections to give the reader a clear overview of each topic.

Supporting maps

The political continent

The political portrait of the continent is a vital reference point for every continental section, showing the position of countries relative to one another, and the relationship between human settlement and geographic location. The complex mosaic of languages spoken in each continent is mapped, as is the effect of communications networks on the pattern of settlement.

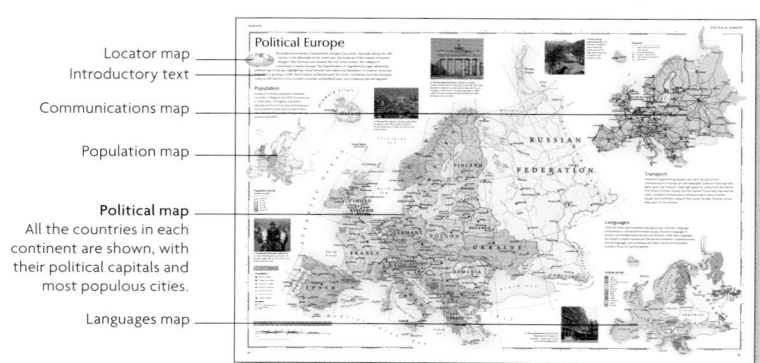

Locator map
Introductory text

Communications map

Population map

Political map
All the countries in each continent are shown, with their political capitals and most populous cities.

Languages map

Continental resources

The Earth's rich natural resources, including oil, gas, minerals, and fertile land, have played a key role in the development of society. These pages show the location of minerals and agricultural resources on each continent, and how they have been instrumental in dictating industrial growth and the varieties of economic activity across the continent.

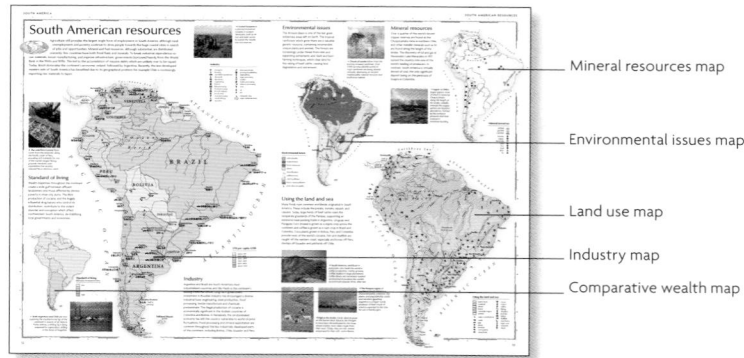

Mineral resources map

Environmental issues map

Land use map

Industry map

Comparative wealth map

The physical continent

The astonishing variety of landforms, and the dramatic forces that created and continue to shape the landscape, are explained in the continental physical spread. Cross-sections, illustrations, and terrain maps highlight the different parts of the continent, showing how nature's forces have produced the landscapes we see today.

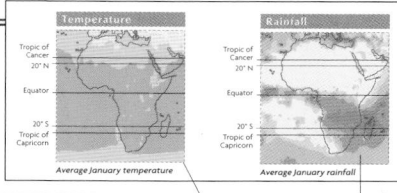

Climate charts
Rainfall and temperature charts clearly show the continental patterns of rainfall and temperature.

Average January temperature *Average January rainfall*

Climate map
Climatic regions vary across each continent. The map displays the differing climatic regions, as well as daily hours of sunshine at selected weather stations.

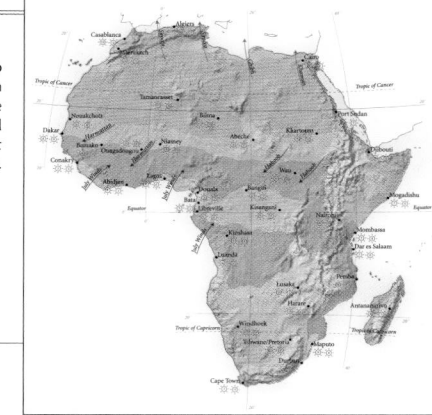

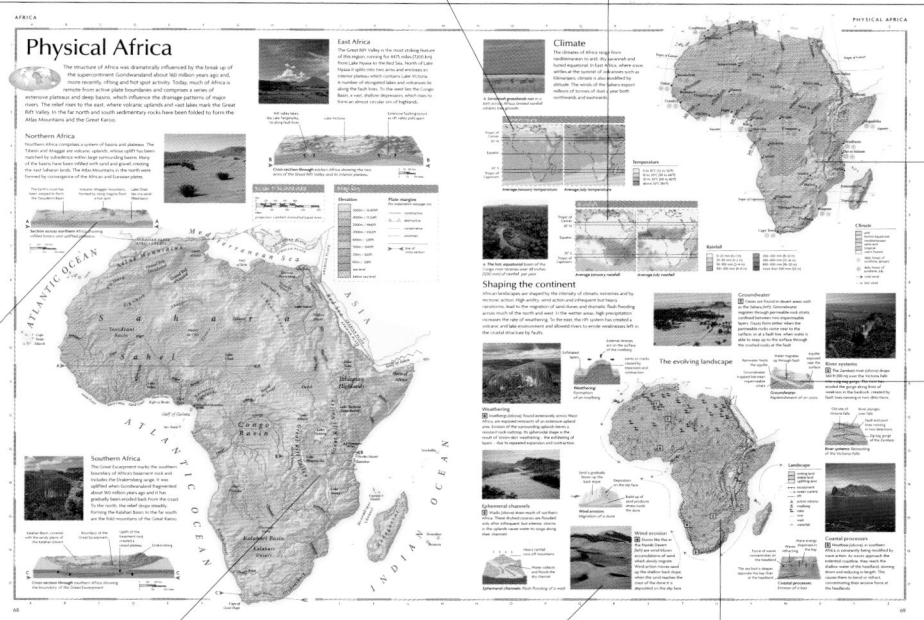

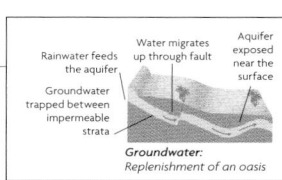

Landform diagrams
The complex formation of many typical landforms is summarized in these easy-to-understand illustrations.

Groundwater:
Replenishment of an oasis

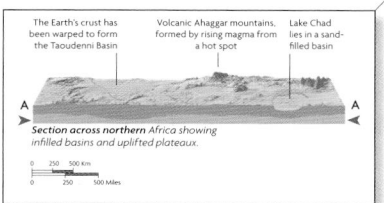

Section across northern Africa showing infilled basins and uplifted plateaux.

Cross-sections
Detailed cross-sections through selected parts of the continent show the underlying geomorphic structure.

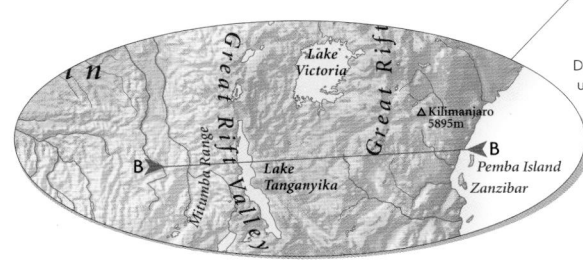

Main physical map
Detailed satellite data has been used to create an accurate and visually striking picture of the surface of the continent.

Photographs
A wide range of beautiful photographs bring the world's regions to life.

Landscape evolution map
The physical shape of each continent is affected by a variety of forces which continually sculpt and modify the landscape. This map shows the major processes which affect different parts of the continent.

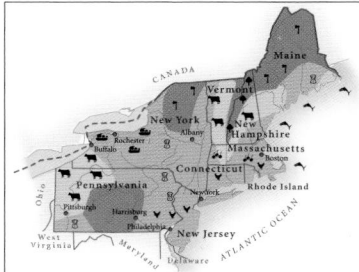

Key to transportation symbols
1 Extent of national paved road network.
2 Extent of motorways, freeways, or major national highways.
3 Extent of commercial railroad network.
4 Extent of inland waterways navigable by commercial craft.

Regional mapping

The main body of the Atlas is a unique regional map set, with detailed information on the terrain, the human geography of the region, and its infrastructure. Around the edge of the map, additional "at-a-glance" maps, give an instant picture of regional industry, land use, and agriculture. The detailed terrain map (shown in perspective), focuses on the main physical features of the region, and is enhanced by annotated illustrations, and photographs of the physical structure.

Transportation network

1	340,090 miles (544,344 km)	4813 miles (7700 km)	2
3	12,872 miles (20,592 km)	2108 miles (3389 km)	4

New York's commercial success is tied historically to its transportation connections. The Erie Canal, completed in 1825, opened up the Great Lakes and the interior to New York's markets and carried a stream of immigrants into the Midwest.

Transportation network
The differing extent of the transportation network for each region is shown here, along with key facts about the transportation system.

Regional Locator
This small map shows the location of each country in relation to its continent.

Key to main map
A key to the population symbols and land heights accompanies the main map.

World locator
This locates the continent in which the region is found on a small world map.

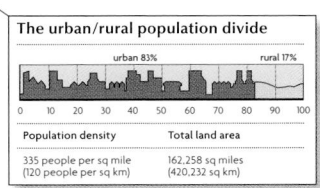

Land use map
This shows the different types of land use which characterize the region, as well as indicating the principal agricultural activities.

Map keys
Each supporting map has its own key.

Grid reference
The framing grid provides a location reference for each place listed in the Index.

The urban/rural population divide

urban 83%   rural 17%

0 10 20 30 40 50 60 70 80 90 100

Population density	Total land area
335 people per sq mile (120 people per sq km)	162,258 sq miles (420,232 sq km)

Urban/rural population divide
The proportion of people in the region who live in urban and rural areas, as well as the overall population density and land area are clearly shown in these simple graphics.

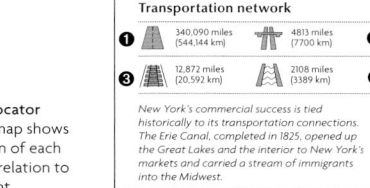

USA: NORTHEASTERN STATES

Connecticut, Maine, Massachusetts, New Hampshire, New Jersey, New York, Pennsylvania, Rhode Island, Vermont

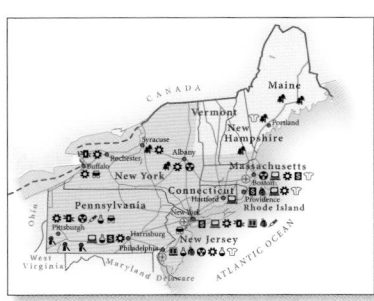

Transportation and industry map
The main industrial areas are mapped, and the most important industrial and economic activities of the region are shown.

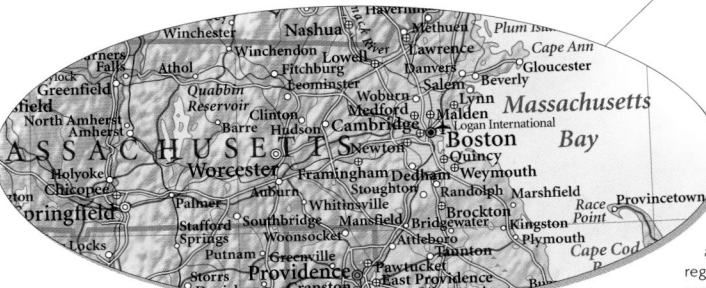

Continuation symbols
These symbols indicate where adjacent maps can be found.

Main regional map
A wealth of information is displayed on the main map, building up a rich portrait of the interaction between the physical landscape and the human and political geography of each region. The key to the regional maps can be found on page viii.

Landscape map
The computer-generated terrain model accurately portrays an oblique view of the landscape. Annotations highlight the most important geographic features of the region.

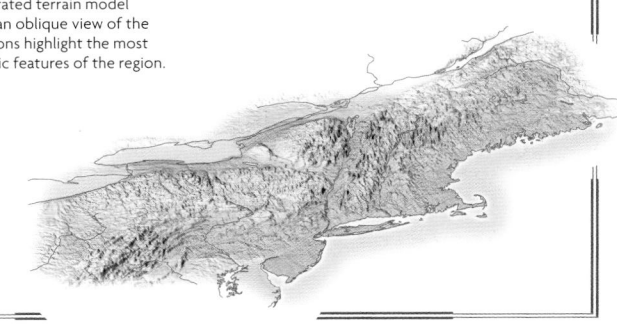

The Solar System

Nine major planets, their satellites, and countless minor planets (asteroids) orbit the Sun to form the Solar System. The Sun, our nearest star, creates energy from nuclear reactions deep within its interior, providing all the light and heat which make life on Earth possible. The Earth is unique in the Solar System in that it supports life: its size, gravitational pull and distance from the Sun have all created the optimum conditions for the evolution of life. The planetary images seen here are composites derived from actual spacecraft images (not shown to scale).

Orbits

All the Solar System's planets and dwarf planets orbit the Sun in the same direction and (apart from Pluto) roughly in the same plane. All the orbits have the shapes of ellipses (stretched circles). However, in most cases, these ellipses are close to being circular: only Pluto and Eris have very elliptical orbits. Orbital period (the time it takes an object to orbit the Sun) increases with distance from the Sun. The more remote objects not only have further to travel with each orbit, they also move more slowly.

Mercury   Venus   Earth   Mars   Ceres *(dwarf planet)*

Jupiter

The Sun

- **Diameter:** *864,948 miles (1,392,000 km)*
- **Mass:** *1990 million million million million tons*

The Sun was formed when a swirling cloud of dust and gas contracted, pulling matter into its center. When the temperature at the center rose to 1,000,000°C, nuclear fusion – the fusing of hydrogen into helium, creating energy – occurred, releasing a constant stream of heat and light.

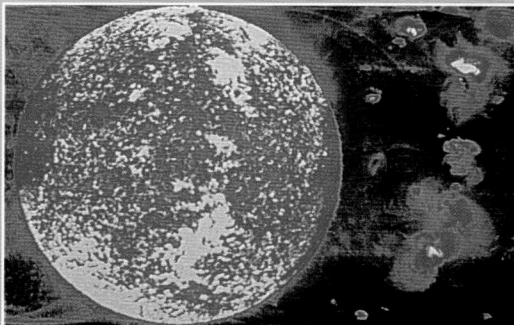

▲ **Solar flares are** *sudden bursts of energy from the Sun's surface. They can be 125,000 miles (200,000 km) long.*

The formation of the Solar System

The cloud of dust and gas thrown out by the Sun during its formation cooled to form the Solar System. The smaller planets nearest the Sun are formed of minerals and metals. The outer planets were formed at lower temperatures, and consist of swirling clouds of gases.

Solar eclipse

A solar eclipse occurs when the Moon passes between Earth and the Sun, casting its shadow on Earth's surface. During a total eclipse *(below)*, viewers along a strip of Earth's surface, called the area of totality, see the Sun totally blotted out for a short time, as the umbra (Moon's full shadow) sweeps over them. Outside this area is a larger one, where the Sun appears only partly obscured, as the penumbra (partial shadow) passes over.

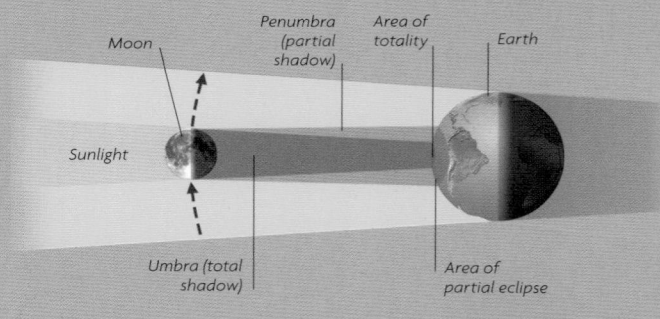

Moon Penumbra *(partial shadow)* Area of totality Earth

Sunlight

Umbra *(total shadow)* Area of *partial eclipse*

PLANETS

DWARF PLANETS

	MERCURY	VENUS	EARTH	MARS	JUPITER	SATURN	URANUS	NEPTUNE	CERES	PLUTO	ERIS
DIAMETER	3029 miles (4875 km)	7521 miles (12,104 km)	7928 miles (12,756 km)	4213 miles (6780 km)	88,846 miles (142,984 km)	74,898 miles (120,536 km)	31,763 miles (51,118 km)	30,775 miles (49,528 km)	590 miles (950 km)	1432 miles (2304 km)	1429-1553 miles (2300-2500 km)
AVERAGE DISTANCE FROM THE SUN	36 mill. miles (57.9 mill. km)	67.2 mill. miles (108.2 mill. km)	93 mill. miles (149.6 mill. km)	141.6 mill. miles (227.9 mill. km)	483.6 mill. miles (778.3 mill. km)	889.8 mill. miles (1431 mill. km)	1788 mill. miles (2877 mill. km)	2795 mill. miles (4498 mill. km)	257 mill. miles (414 mill. km)	3675 mill. miles (5915 mill. km)	6344 mill. miles (10,210 mill. km)
ROTATION PERIOD	58.6 days	243 days	23.93 hours	24.62 hours	9.93 hours	10.65 hours	17.24 hours	16.11 hours	9.1 hours	6.38 days	not known
ORBITAL PERIOD	88 days	224.7 days	365.26 days	687 days	11.86 years	29.37 years	84.1 years	164.9 years	4.6 years	248.6 years	557 years
SURFACE TEMPERATURE	-180°C to 430°C (-292°F to 806°F)	480°C (896°F)	-70°C to 55°C (-94°F to 131°F)	-120°C to 25°C (-184°F to 77 °F)	-110°C (-160°F)	-140°C (-220°F)	-200°C (-320°F)	-200°C (-320°F)	-107°C (-161°F)	-230°C (-380°F)	-243°C (-405°F)

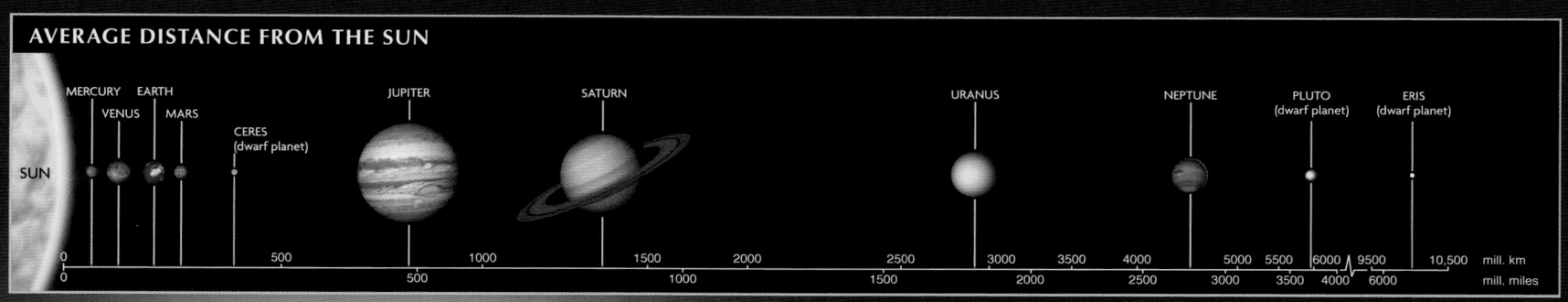

AVERAGE DISTANCE FROM THE SUN

SUN — MERCURY, VENUS, EARTH, MARS, CERES (dwarf planet), JUPITER, SATURN, URANUS, NEPTUNE, PLUTO (dwarf planet), ERIS (dwarf planet)

Saturn

Uranus

Neptune

Pluto (dwarf planet)

Eris (dwarf planet)

Space Debris

Millions of objects, remnants of planetary formation, circle the Sun in a zone lying between Mars and Jupiter: the asteroid belt. Fragments of asteroids break off to form meteoroids, which can reach the Earth's surface. Comets, composed of ice and dust, originated outside our Solar System. Their elliptical orbit brings them close to the Sun and into the inner Solar System.

▲ *Meteor Crater in* Arizona is 4200 ft (1300 m) wide and 660 ft (200 m) deep. It was formed over 10,000 years ago.

Possible and actual meteorite craters

Map key
- Possible impact craters
- Meteorite impact craters

The Earth's Atmosphere

During the early stages of the Earth's formation, ash, lava, carbon dioxide, and water vapor were discharged onto the surface of the planet by constant volcanic eruptions. The water formed the oceans, while carbon dioxide entered the atmosphere or was dissolved in the oceans. Clouds, formed of water droplets, reflected some of the Sun's radiation back into space. The Earth's temperature stabilized and early life forms began to emerge, converting carbon dioxide into life-giving oxygen.

▲ *It is thought* that the gases that make up the Earth's atmosphere originated deep within the interior, and were released many millions of years ago during intense volcanic activity, similar to this eruption at Mount St. Helens.

▲ *The orbit of* Halley's Comet brings it close to the Earth every 76 years. It last visited in 1986.

Halley's Comet

Earth's orbit

Halley's orbit

Orbit of Halley's Comet around the Sun

The physical world

The Earth's surface is constantly being transformed: it is uplifted, folded, and faulted by tectonic forces; weathered and eroded by wind, water, and ice. Sometimes change is dramatic, the spectacular results of earthquakes or floods. More often it is a slow process lasting millions of years. A physical map of the world represents a snapshot of the ever-evolving architecture of the Earth. This terrain map shows the whole surface of the Earth, both above and below the sea.

The world in section

These cross-sections around the Earth, one in the northern hemisphere; one straddling the Equator, reveal the limited areas of land above sea level in comparison with the extent of the sea floor. The greater erosive effects of weathering by wind and water limit the upward elevation of land above sea level, while the deep oceans retain their dramatic mountain and trench profiles.

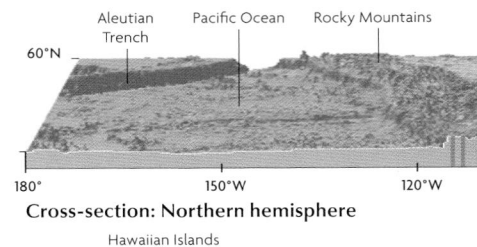
Cross-section: Northern hemisphere

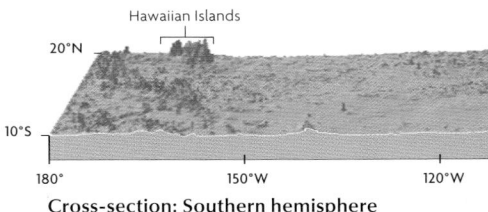
Cross-section: Southern hemisphere

Map key

Elevation
- 6000m / 19,686ft
- 4000m / 13,124ft
- 3000m / 9843ft
- 2000m / 6562ft
- 1000m / 3281ft
- 500m / 1640ft
- 250m / 820ft
- 100m / 328ft
- sea level
- below sea level

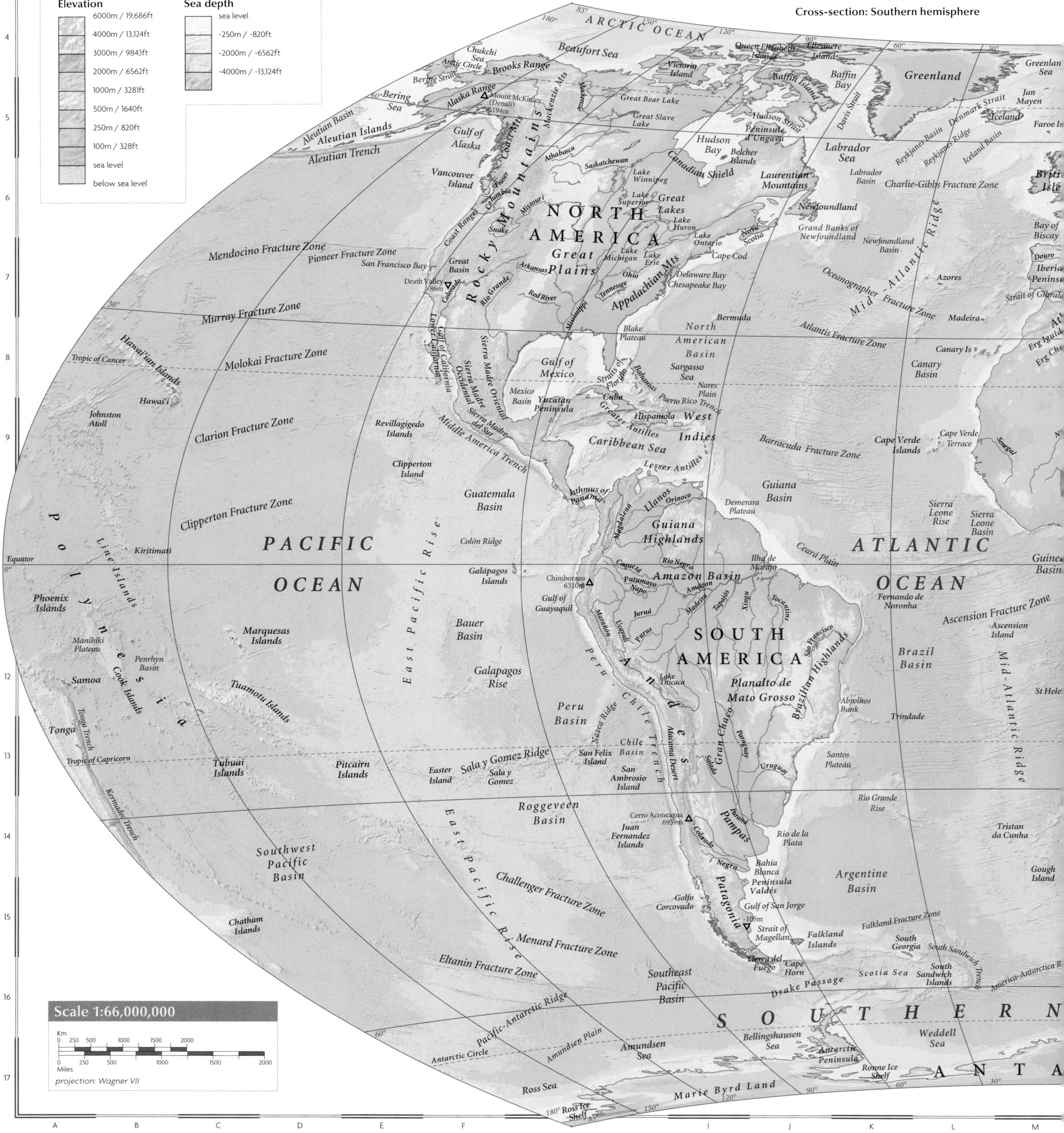

Sea depth
- sea level
- -250m / -820ft
- -2000m / -6562ft
- -4000m / -13,124ft

Scale 1:66,000,000

Km
0 250 500 1000 1500 2000

Miles
0 250 500 1000 1500 2000

projection: Wagner VII

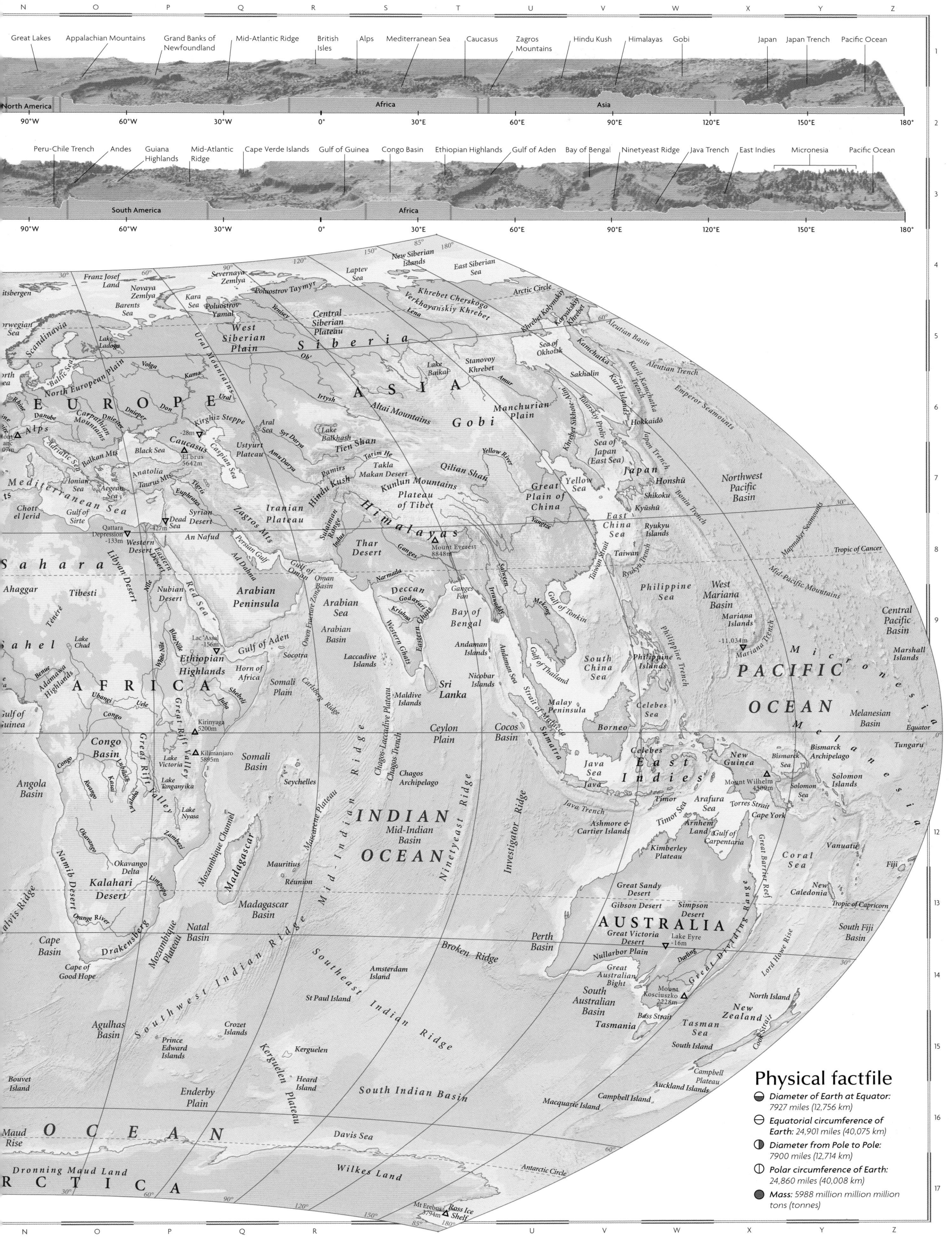

Physical factfile

- **Diameter of Earth at Equator:** 7927 miles (12,756 km)
- **Equatorial circumference of Earth:** 24,901 miles (40,075 km)
- **Diameter from Pole to Pole:** 7900 miles (12,714 km)
- **Polar circumference of Earth:** 24,860 miles (40,008 km)
- **Mass:** 5988 million million million tons (tonnes)

Structure of the Earth

The Earth as it is today is just the latest phase in a constant process of evolution which has occurred over the past 4.5 billion years. The Earth's continents are neither fixed nor stable; over the course of the Earth's history, propelled by currents rising from the intense heat at its center, the great plates on which they lie have moved, collided, joined together, and separated. These processes continue to mold and transform the surface of the Earth, causing earthquakes and volcanic eruptions and creating oceans, mountain ranges, deep ocean trenches, and island chains.

Inside the Earth

The Earth's hot inner core is made up of solid iron, while the outer core is composed of liquid iron and nickel. The mantle nearest the core is viscous, whereas the rocky upper mantle is fairly rigid. The crust is the rocky outer shell of the Earth. Together, the upper mantle and the crust form the lithosphere.

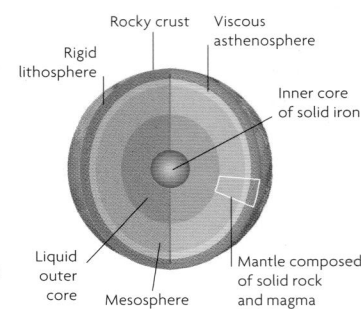

Rocky crust
Rigid lithosphere
Viscous asthenosphere
Inner core of solid iron
Liquid outer core
Mesosphere
Mantle composed of solid rock and magma

The dynamic Earth

The Earth's crust is made up of eight major (and several minor) rigid continental and oceanic tectonic plates, which fit closely together. The positions of the plates are not static. They are constantly moving relative to one another. The type of movement between plates affects the way in which they alter the structure of the Earth. The oldest parts of the plates, known as shields, are the most stable parts of the Earth and little tectonic activity occurs here.

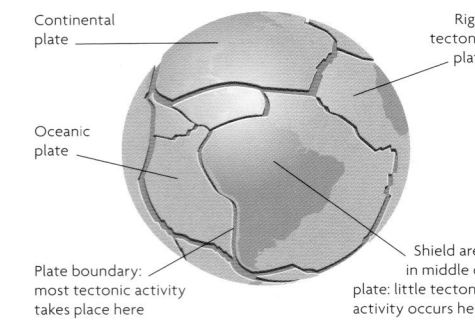

Continental plate
Rigid tectonic plate
Oceanic plate
Plate boundary: most tectonic activity takes place here
Shield area in middle of plate: little tectonic activity occurs here

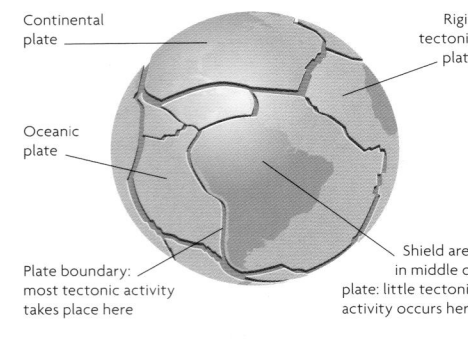

Inner core
Outer core
Subduction zone
Ocean crust
Movement of plate
Mid-ocean ridge
Lithosphere
Asthenosphere
Mesosphere
Continental crust

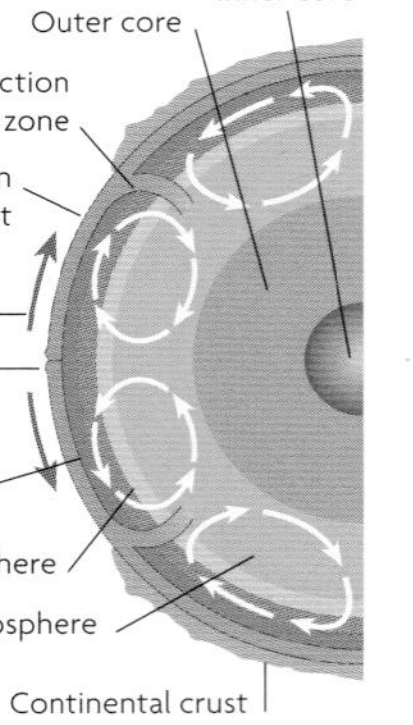

Convection currents

Deep within the Earth, at its inner core, temperatures may exceed 8,100°F (4,500°C). This heat warms rocks in the mesosphere which rise through the partially molten mantle, displacing cooler rocks just below the solid crust, which sink, and are warmed again by the heat of the mantle. This process is continuous, creating convection currents which form the moving force beneath the Earth's crust.

Plate boundaries

The boundaries between the plates are the areas where most tectonic activity takes place. Three types of movement occur at plate boundaries: the plates can either move toward each other, move apart, or slide past each other. The effect this has on the Earth's structure depends on whether the margin is between two continental plates, two oceanic plates, or an oceanic and continental plate.

▲ *The Mid-Atlantic Ridge rises above sea level in Iceland, producing geysers and volcanoes.*

Mid-ocean ridges

Mid-ocean ridges are formed when two adjacent oceanic plates pull apart, allowing magma to force its way up to the surface, which then cools to form solid rock. Vast amounts of volcanic material are discharged at these mid-ocean ridges which can reach heights of 10,000 ft (3000 m).

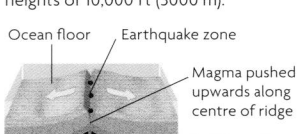

Ocean floor
Earthquake zone
Magma pushed upwards along centre of ridge
Solid mantle

Formation of a mid-ocean ridge

▲ *Mount Pinatubo is an active volcano, lying on the Pacific "Ring of Fire."*

Ocean plates meeting

△△ Oceanic crust is denser and thinner than continental crust; on average it is 3 miles (5 km) thick, while continental crust averages 18–24 miles (30–40 km). When oceanic plates of similar density meet, the crust is contorted as one plate overrides the other, forming deep sea trenches and volcanic island arcs above sea level.

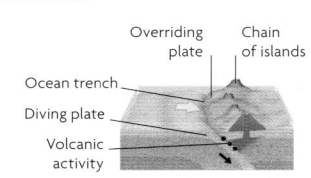

Overriding plate
Chain of islands
Ocean trench
Diving plate
Volcanic activity

Ocean plates meeting to form an island arc

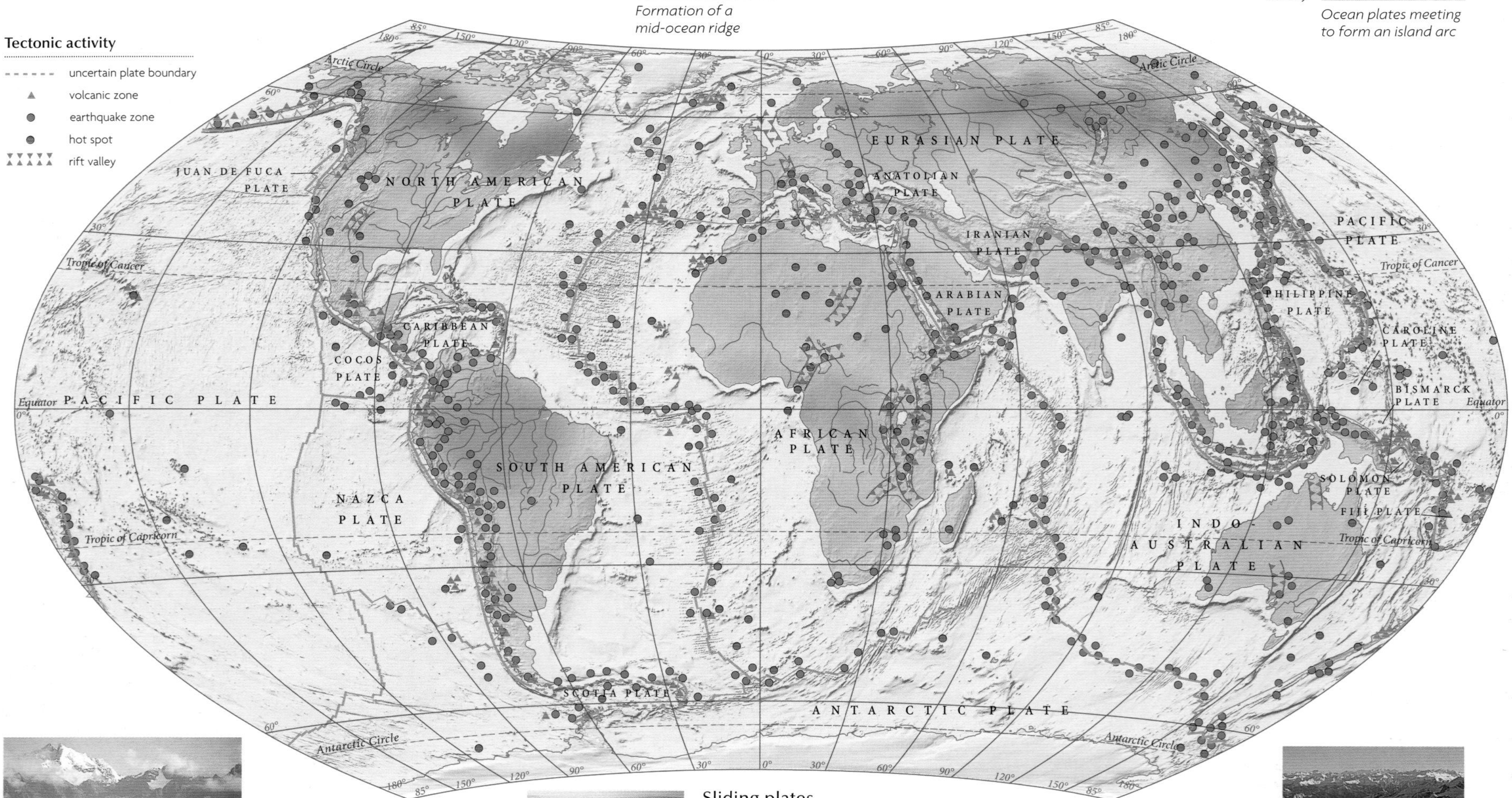

Tectonic activity

- - - - - uncertain plate boundary
▲ volcanic zone
● earthquake zone
● hot spot
ᴠᴠᴠᴠᴠ rift valley

JUAN DE FUCA PLATE
NORTH AMERICAN PLATE
EURASIAN PLATE
ANATOLIAN PLATE
IRANIAN PLATE
PACIFIC PLATE
ARABIAN PLATE
PHILIPPINE PLATE
CARIBBEAN PLATE
CAROLINE PLATE
COCOS PLATE
BISMARCK PLATE
PACIFIC PLATE
AFRICAN PLATE
SOUTH AMERICAN PLATE
SOLOMON PLATE
NAZCA PLATE
FIJI PLATE
INDO-AUSTRALIAN PLATE
SCOTIA PLATE
ANTARCTIC PLATE

Arctic Circle
Tropic of Cancer
Equator
Tropic of Capricorn
Antarctic Circle

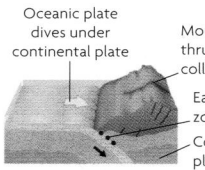

◄ *The Andean mountain chain is the typical result of the impact of a diving plate.*

Diving plates

△△ When an oceanic and a continental plate meet, the denser oceanic plate is driven underneath the continental plate, which is crumpled by the collision to form mountain ranges. As the ocean plate plunges downward, it heats up, and molten rock (magma) is forced up to the surface.

Oceanic plate dives under continental plate
Mountains thrust up by collision
Earthquake zone
Continental plate

Diving plate

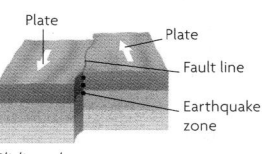

▲ *The deep fracture caused by the sliding plates of the San Andreas Fault can be clearly seen in parts of California.*

Sliding plates

When two plates slide past each other, friction is caused along the fault line which divides them. The plates do not move smoothly, and the uneven movement causes earthquakes.

Plate
Plate
Fault line
Earthquake zone

Sliding plates

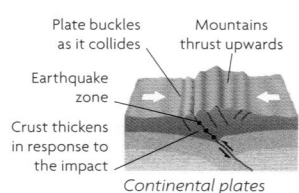

► *The Alps were formed when the African Plate collided with the Eurasian Plate, about 65 million years ago.*

Plate buckles as it collides
Mountains thrust upwards
Earthquake zone
Crust thickens in response to the impact

Continental plates colliding to form a mountain range

Colliding plates

▲▲▲ When two continental plates collide, great mountain chains are thrust upward as the crust buckles and folds under the force of the impact.

Continental drift

Although the plates which make up the Earth's crust move only a few inches in a year, over the millions of years of the Earth's history, its continents have moved many thousands of miles, to create new continents, oceans, and mountain chains

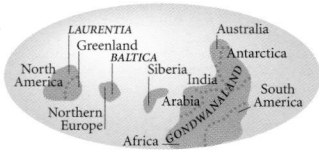

1: Cambrian period

570–510 million years ago. Most continents are in tropical latitudes. The supercontinent of Gondwanaland reaches the South Pole.

2: Devonian period

408–362 million years ago. The continents of Gondwanaland and Laurentia are drifting northward.

3: Carboniferous period

362–290 million years ago. The Earth is dominated by three continents; Laurentia, Angaraland, and Gondwanaland.

4: Triassic period

245–208 million years ago. All three major continents have joined to form the super-continent of Pangea.

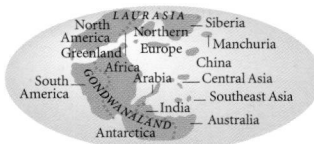

5: Jurassic period

208–145 million years ago. The super-continent of Pangea begins to break up, causing an overall rise in sea levels.

6: Cretaceous period

145–65 million years ago. Warm, shallow seas cover much of the land; sea levels are about 80 ft (25 m) above present levels.

7: Tertiary period

65–2 million years ago. Although the world's geography is becoming more recognizable, major events such as the creation of the Himalayan mountain chain, are still to occur during this period.

Continental shields

The centers of the Earth's continents, known as shields, were established between 2500 and 500 million years ago; some contain rocks over three billion years old. They were formed by a series of turbulent events: plate movements, earthquakes, and volcanic eruptions. Since the Pre-Cambrian period, over 570 million years ago, they have experienced little tectonic activity, and today, these flat, low-lying slabs of solidified molten rock form the stable centers of the continents. They are bounded or covered by successive belts of younger sedimentary rock.

The Hawai'ian island chain

A hot spot lying deep beneath the Pacific Ocean pushes a plume of magma from the Earth's mantle up through the Pacific Plate to form volcanic islands. While the hot spot remains stationary, the plate on which the islands sit is moving slowly. A long chain of islands has been created as the plate passes over the hot spot.

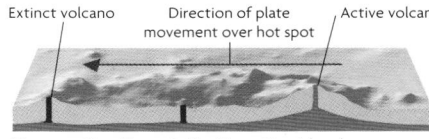

Cross-section through the Hawai'ian Islands

Evolution of the Hawai'ian Islands

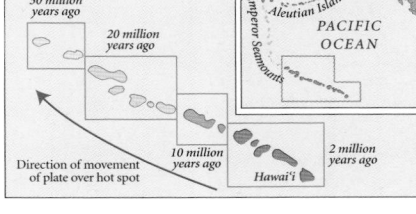

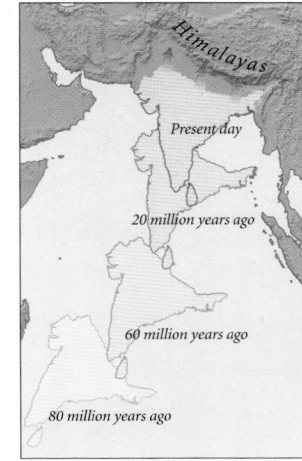

Creation of the Himalayas

Between 10 and 20 million years ago, the Indian subcontinent, part of the ancient continent of Gondwanaland, collided with the continent of Asia. The Indo-Australian Plate continued to move northward, displacing continental crust and uplifting the Himalayas, the world's highest mountain chain.

Movements of India

Cross-section through the Himalayas

▲ *The Himalayas were uplifted when the Indian subcontinent collided with Asia.*

The Earth's geology

The Earth's rocks are created in a continual cycle. Exposed rocks are weathered and eroded by wind, water, and chemicals and deposited as sediments. If they pass into the Earth's crust they will be transformed by high temperatures and pressures into metamorphic rocks or they will melt and solidify as igneous rocks.

Sandstone

[8] Sandstones are sedimentary rocks formed mainly in deserts, beaches, and deltas. Desert sandstones are formed of grains of quartz which have been well rounded by wind erosion.

▲ *Rock stacks of desert sandstone, at Bryce Canyon National Park, Utah, US.*

◀ *Extrusive igneous rocks are formed during volcanic eruptions, as here in Hawai'i.*

Andesite

[7] Andesite is an extrusive igneous rock formed from magma which has solidified on the Earth's crust after a volcanic eruption.

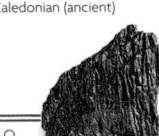

Gneiss

[1] Gneiss is a metamorphic rock made at great depth during the formation of mountain chains, when intense heat and pressure transform sedimentary or igneous rocks.

▲ *Gneiss formations in Norway's Jotunheimen Mountains.*

Basalt

[2] Basalt is an igneous rock, formed when small quantities of magma lying close to the Earth's surface cool rapidly.

◀ *Basalt columns at Giant's Causeway, Northern Ireland, UK.*

Limestone

[3] Limestone is a sedimentary rock, which is formed mainly from the calcite skeletons of marine animals which have been compressed into rock.

▲ *Limestone hills, Guilin, China.*

Coral

[4] Coral reefs are formed from the skeletons of millions of individual corals.

▲ *Great Barrier Reef, Australia.*

Geological regions

- continental shield
- sedimentary cover
- coral formation
- igneous rock types

Mountain ranges

- Alpine (new)
- Hercynian (old)
- Caledonian (ancient)

Schist

[1] Schist is a metamorphic rock formed during mountain building, when temperature and pressure are comparatively high. Both mudstones and shales reform into schist under these conditions.

▶ *Schist formations in the Atlas Mountains, northwestern Africa.*

Granite

[5] Granite is an intrusive igneous rock formed from magma which has solidified deep within the Earth's crust. The magma cools slowly, producing a coarse-grained rock.

▶ *Namibia's Namaqualand Plateau is formed of granite.*

Shaping the landscape

The basic material of the Earth's surface is solid rock: valleys, deserts, soil, and sand are all evidence of the powerful agents of weathering, erosion, and deposition which constantly shape and transform the Earth's landscapes. Water, either flowing continually in rivers or seas, or frozen and compacted into solid sheets of ice, has the most clearly visible impact on the Earth's surface. But wind can transport fragments of rock over huge distances and strip away protective layers of vegetation, exposing rock surfaces to the impact of extreme heat and cold.

Water

Less than 2% of the world's water is on the land, but it is the most powerful agent of landscape change. Water, as rainfall, groundwater, and rivers, can transform landscapes through both erosion and deposition. Eroded material carried by rivers forms the world's most fertile soils.

▲ **Waterfalls such as** the Iguaçu Falls on the border between Argentina and southern Brazil, erode the underlying rock, causing the falls to retreat.

Coastal water

The world's coastlines are constantly changing; every day, tides deposit, sift and sort sand, and gravel on the shoreline. Over longer periods, powerful wave action erodes cliffs and headlands and carves out bays.

▶ **A low, wide** sandy beach on South Africa's Cape Peninsula is continually re-shaped by the action of the Atlantic waves.

▲ **The sheer chalk** cliffs at Seven Sisters in southern England are constantly under attack from waves.

Groundwater

In regions where there are porous rocks such as chalk, water is stored underground in large quantities; these reservoirs of water are known as aquifers. Rain percolates through topsoil into the underlying bedrock, creating an underground store of water. The limit of the saturated zone is called the water table.

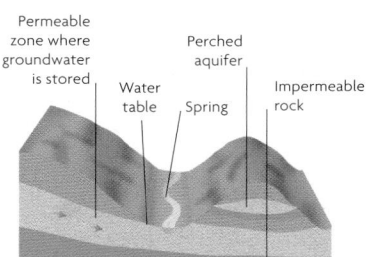

Permeable zone where groundwater is stored
Perched aquifer
Water table
Spring
Impermeable rock

Storage of groundwater in an aquifer

World river systems

drainage basin

World river systems:
Sediment deposited annually per drainage basin

tons per sq mile per year 9120
6080
1520
760
2400
1600
400
200 and less
tonnes per sq km per year

[World map showing river systems with labeled rivers: Yukon, Mackenzie, Nelson, Columbia, St. Lawrence, Colorado, Mississippi/Missouri, Rio Grande, Orinoco, Amazon, São Francisco, Paraná, Rhine, Danube, Niger, Nile, Congo, Zambezi, Orange, Ob', Yenisey, Lena, Volga, Amur, Tigris/Euphrates, Indus, Yellow River, Ganges/Brahmaputra, Yangtze, Mekong, Murray/Darling. Oceans labeled: ARCTIC OCEAN, ATLANTIC OCEAN, PACIFIC OCEAN, INDIAN OCEAN. Lines: Arctic Circle, Tropic of Cancer, Equator, Tropic of Capricorn, Antarctic Circle.]

Rivers

Rivers erode the land by grinding and dissolving rocks and stones. Most erosion occurs in the river's upper course as it flows through highland areas. Rock fragments are moved along the river bed by fast-flowing water and deposited in areas where the river slows down, such as flat plains, or where the river enters seas or lakes.

River valleys

Over long periods of time rivers erode uplands to form characteristic V-shaped valleys with smooth sides.

Resistant rock
River
Chemical erosion cuts valley in softer rock

River valley erosion

Deltas

When a river deposits its load of silt and sediment (alluvium) on entering the sea, it may form a delta. As this material accumulates, it chokes the mouth of the river, forcing it to create new channels to reach the sea.

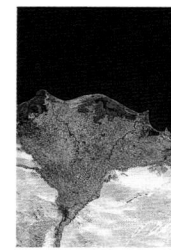

▶ **The Nile** forms a broad delta as it flows into the Mediterranean.

Drainage basins

The drainage basin is the area of land drained by a major trunk river and its smaller branch rivers or tributaries. Drainage basins are separated from one another by natural boundaries known as watersheds.

Watershed
Major trunk river
Alps
Dolomites
Apennines
Tributary river
Delta
River mouth
Po Valley

The drainage basin of the Po river, northern Italy.

Meanders

In their lower courses, rivers flow slowly. As they flow across the lowlands, they form looping bends called meanders.

▲ **The Mississippi River** forms meanders as it flows across the southern US.

▲ **The meanders of** Utah's San Juan River have become deeply incised.

Deposition

When rivers have deposited large quantities of fertile alluvium, they are forced to find new channels through the alluvium deposits, creating braided river systems.

◀ **Mud is deposited** by China's Yellow River in its lower course.

Landslides

Heavy rain and associated flooding on slopes can loosen underlying rocks, which crumble, causing the top layers of rock and soil to slip.

▶ **A huge landslide** in the Swiss Alps has left massive piles of rocks and pebbles called scree.

Gullies

In areas where soil is thin, rainwater is not effectively absorbed, and may flow overland. The water courses downhill in channels, or gullies, and may lead to rapid erosion of soil.

▲ **A deep gully** in the French Alps caused by the scouring of upper layers of turf.

Ice

During its long history, the Earth has experienced a number of glacial episodes when temperatures were considerably lower than today. During the last Ice Age, 18,000 years ago, ice covered an area three times larger than it does today. Over these periods, the ice has left a remarkable legacy of transformed landscapes.

Glaciers

Glaciers are formed by the compaction of snow into "rivers" of ice. As they move over the landscape, glaciers pick up and carry a load of rocks and boulders which erode the landscape they pass over, and are eventually deposited at the end of the glacier.

▲ *A massive glacier* advancing down a valley in southern Argentina.

Post-glacial features

When a glacial episode ends, the retreating ice leaves many features. These include depositional ridges called moraines, which may be eroded into low hills known as drumlins; sinuous ridges called eskers; kames, which are rounded hummocks; depressions known as kettle holes; and windblown loess deposits.

Glacial valleys

Glaciers can erode much more powerfully than rivers. They form steep-sided, flat-bottomed valleys with a typical U-shaped profile. Valleys created by tributary glaciers, whose floors have not been eroded to the same depth as the main glacial valley floor, are called hanging valleys

▲ *The U-shaped profile* and piles of morainic debris are characteristic of a valley once filled by a glacier.

▲ *A series of* hanging valleys high up in the Chilean Andes.

Past and present world ice-cover and glacial features

Past and present world ice cover and glacial features

extent of last Ice Age	present day ice cover
loess deposits	glacial field
post-glacial feature	
glacial feature	

Kame terrace
Retreating glacier
Kettle hole
Esker
Drumlin
Braided river
Terminal moraine
Windblown loess
Glacial till
Bedrock

Post-glacial landscape features

Ice shattering

Water drips into fissures in rocks and freezes, expanding as it does so. The pressure weakens the rock, causing it to crack, and eventually to shatter into polygonal patterns.

▲ *Irregular polygons show* through the sedge-grass tundra in the Yukon, Canada.

▲ *The profile of* the Matterhorn has been formed by three cirques lying "back-to-back."

Cirques

Cirques are basin-shaped hollows which mark the head of a glaciated valley. Where neighboring cirques meet, they are divided by sharp rock ridges called arêtes. It is these arêtes which give the Matterhorn its characteristic profile.

Fjords

Fjords are ancient glacial valleys flooded by the sea following the end of a period of glaciation. Beneath the water, the valley floor can be 4000 ft (1300 m) deep.

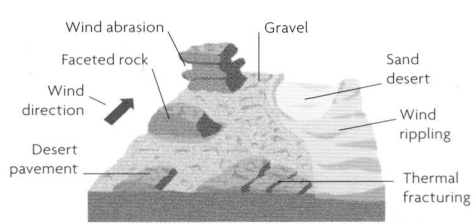

▲ *A fjord fills* a former glacial valley in southern New Zealand.

Periglaciation

Periglacial areas occur near to the edge of ice sheets. A layer of frozen ground lying just beneath the surface of the land is known as permafrost. When the surface melts in the summer, the water is unable to drain into the frozen ground, and so "creeps" downhill, a process known as solifluction.

Wind

Strong winds can transport rock fragments great distances, especially where there is little vegetation to protect the rock. In desert areas, wind picks up loose, unprotected sand particles, carrying them over great distances. This powerfully abrasive debris is blasted at the surface by the wind, eroding the landscape into dramatic shapes.

Deposition

The rocky, stony floors of the world's deserts are swept and scoured by strong winds. The smaller, finer particles of sand are shaped into surface ripples, dunes, or sand mountains, which rise to a height of 650 ft (200 m). Dunes usually form single lines, running perpendicular to the direction of the prevailing wind. These long, straight ridges can extend for over 100 miles (160 km).

Prevailing winds and dust trajectories

Prevailing winds

northeast trade
southeast trade
westerly
westerly
polar easterly
polar easterly

Dust trajectories

trajectory of aeolian dust

Hot and cold deserts

Main desert types

hot arid
semi-arid
cold polar

▲ *Barchan dunes in the* Arabian Desert.

▲ *Complex dune system in* the Sahara.

Dunes

Dunes are shaped by wind direction and sand supply. Where sand supply is limited, crescent-shaped barchan dunes are formed.

Types of dune

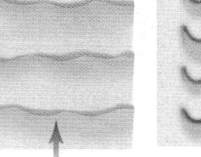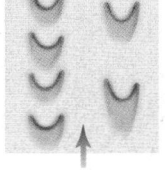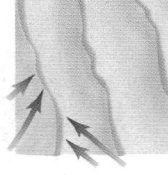

Wind direction

Transverse dune

Barchan dune

Linear dune

Star dune

Heat

Fierce sun can heat the surface of rock, causing it to expand more rapidly than the cooler, underlying layers. This creates tensions which force the rock to crack or break up. In arid regions, the evaporation of water from rock surfaces dissolves certain minerals within the water, causing salt crystals to form in small openings in the rock. The hard crystals force the openings to widen into cracks and fissures.

▲ *The cracked and* parched floor of Death Valley, California. This is one of the hottest deserts on Earth.

Temperature

Most of the world's deserts are in the tropics. The cold deserts which occur elsewhere are arid because they are a long way from the rain-giving sea. Rock in deserts is exposed because of lack of vegetation and is susceptible to changes in temperature; extremes of heat and cold can cause both cracks and fissures to appear in the rock.

Desert abrasion

Abrasion creates a wide range of desert landforms from faceted pebbles and wind ripples in the sand, to large-scale features such as yardangs (low, streamlined ridges), and scoured desert pavements.

Wind abrasion
Gravel
Faceted rock
Sand desert
Wind direction
Wind rippling
Desert pavement
Thermal fracturing

Features of a desert surface

◀ *This dry valley* at Ellesmere Island in the Canadian Arctic is an example of a cold desert. The cracked floor and scoured slopes are features also found in hot deserts.

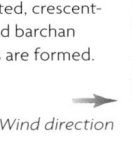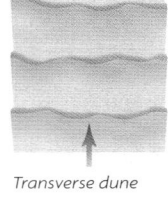

The world's oceans

Two-thirds of the Earth's surface is covered by the oceans. The landscape of the ocean floor, like the surface of the land, has been shaped by movements of the Earth's crust over millions of years to form volcanic mountain ranges, deep trenches, basins, and plateaus. Ocean currents constantly redistribute warm and cold water around the world. A major warm current, such as El Niño in the Pacific Ocean, can increase surface temperature by up to 10°F (8°C), causing changes in weather patterns which can lead to both droughts and flooding.

The great oceans

There are five oceans on Earth: the Pacific, Atlantic, Indian, and Southern oceans, and the much smaller Arctic Ocean. These five ocean basins are relatively young, having evolved within the last 80 million years. One of the most recent plate collisions, between the Eurasian and African plates, created the present-day arrangement of continents and oceans.

▲ **The Indian Ocean** accounts for approximately 20% of the total area of the world's oceans.

Sea level

If the influence of tides, winds, currents, and variations in gravity were ignored, the surface of the Earth's oceans would closely follow the topography of the ocean floor, with an underwater ridge 3000 ft (915 m) high producing a rise of up to 3 ft (1 m) in the level of the surface water.

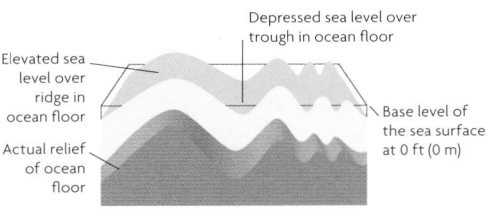
How surface waters reflect the relief of the ocean floor

▲ **The low relief** of many small Pacific islands such as these atolls at Huahine in French Polynesia makes them vulnerable to changes in sea level.

Ocean structure

The continental shelf is a shallow, flat seabed surrounding the Earth's continents. It extends to the continental slope, which falls to the ocean floor. Here, the flat abyssal plains are interrupted by vast, underwater mountain ranges, the mid-ocean ridges, and ocean trenches which plunge to depths of 35,828 ft (10,920 m).

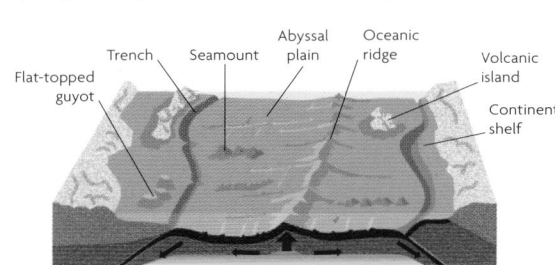
Typical sea-floor features

Ocean depth
	Sea level
	200m / 656ft
	1000m / 3281ft
	2000m / 6562ft
	3000m / 9843ft
	4000m / 13,124ft
	5000m / 16,400ft
	6000m / 19,686ft

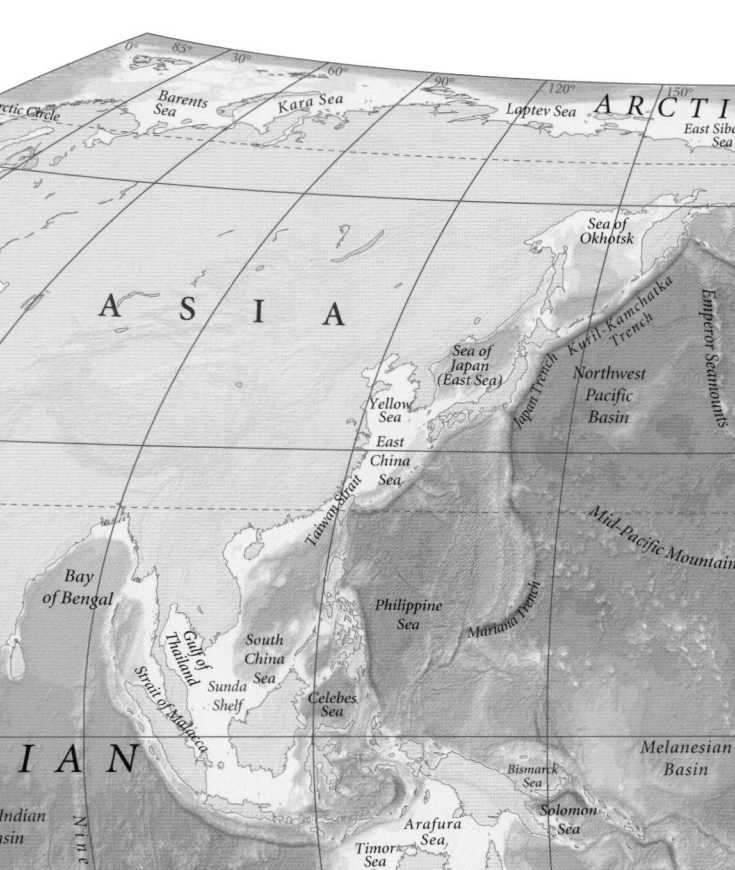

Black smokers

These vents in the ocean floor disgorge hot, sulfur-rich water from deep in the Earth's crust. Despite the great depths, a variety of lifeforms have adapted to the chemical-rich environment which surrounds black smokers.

▲ **A black smoker** in the Atlantic Ocean.

▲ **Surtsey, near Iceland,** is a volcanic island lying directly over the Mid-Atlantic Ridge. It was formed in the 1960s following intense volcanic activity nearby.

Ocean floors

Mid-ocean ridges are formed by lava which erupts beneath the sea and cools to form solid rock. This process mirrors the creation of volcanoes from cooled lava on the land. The ages of sea floor rocks increase in parallel bands outward from central ocean ridges.

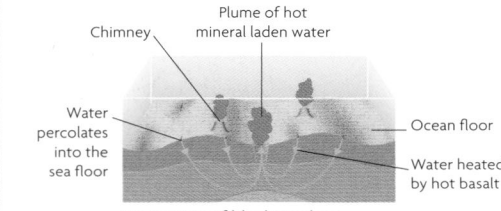
Formation of black smokers

Ages of the ocean floor

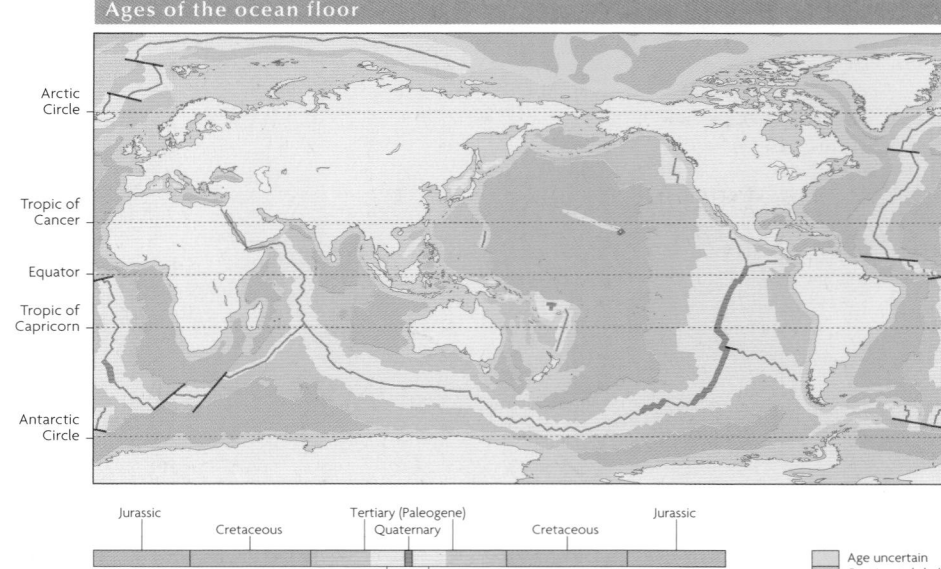

Jurassic	Cretaceous	Tertiary (Paleogene) Quaternary	Cretaceous	Jurassic
208 million years old	145	65	23	0	23	65	145	208 million years old

Tertiary (Neogene)

Age uncertain
Continental shelf and island arcs

▲ *Currents in the* Southern Ocean *are driven by some of the world's fiercest winds, including the Roaring Forties, Furious Fifties, and Shrieking Sixties.*

▲ *The Pacific Ocean is the world's largest and deepest ocean, covering over one-third of the surface of the Earth.*

▲ *The Atlantic Ocean was formed when the landmasses of the eastern and western hemispheres began to drift apart 180 million years ago.*

Deposition of sediment

Storms, earthquakes, and volcanic activity trigger underwater currents known as turbidity currents which scour sand and gravel from the continental shelf, creating underwater canyons. These strong currents pick up material deposited at river mouths and deltas, and carry it across the continental shelf and through the underwater canyons, where it is eventually laid down on the ocean floor in the form of fans.

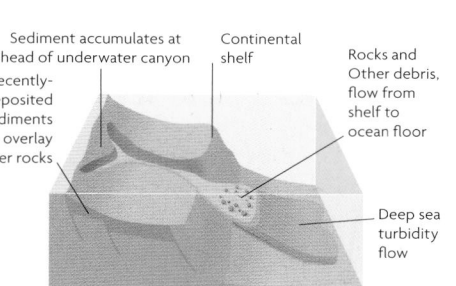

How sediment is deposited on the ocean floor

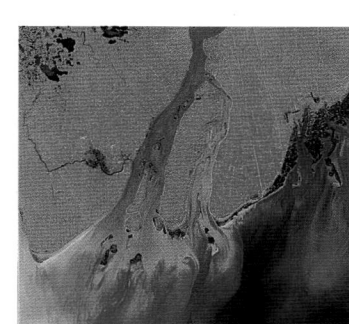

▶ *Satellite image of the Yangtze (Chang Jiang) Delta, in which the land appears red. The river deposits immense quantities of silt into the East China Sea, much of which will eventually reach the deep ocean floor.*

Surface water

Ocean currents move warm water away from the Equator toward the poles, while cold water is, in turn, moved towards the Equator. This is the main way in which the Earth distributes surface heat and is a major climatic control. Approximately 4000 million years ago, the Earth was dominated by oceans and there was no land to interrupt the flow of the currents, which would have flowed as straight lines, simply influenced by the Earth's rotation.

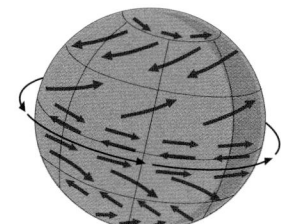

Idealized globe showing the movement of water around a landless Earth.

Ocean currents

Surface currents are driven by the prevailing winds and by the spinning motion of the Earth, which drives the currents into circulating whirlpools, or gyres. Deep sea currents, over 330 ft (100 m) below the surface, are driven by differences in water temperature and salinity, which have an impact on the density of deep water and on its movement.

Surface temperature and currents

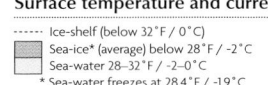

Surface temperature and currents

```
---- Ice-shelf (below 32°F / 0°C)              32–50°F / 0–10°C     → warm current
     Sea-ice* (average) below 28°F / -2°C      50–68°F / 10–20°C    → cold current
     Sea-water 28–32°F / -2–0°C               68–86°F / 20–30°C
* Sea-water freezes at 28.4°F / -1.9°C
```

Map labels

OCEAN
Chukchi Sea
Beaufort Sea
Greenland Sea
Baffin Bay
Davis Strait
Arctic Circle
Bering Sea
Hudson Strait
Hudson Bay
Labrador Sea
Gulf of Alaska
Aleutian Trench
Mendocino Fracture Zone
Murray Fracture Zone
NORTH AMERICA
Newfoundland Basin
North American Basin
Mid-Atlantic Ridge
Hawaiian Ridge
Molokai Fracture Zone
Gulf of Mexico
ATLANTIC
Canary Basin
Tropic of Cancer
Clarion Fracture Zone
Yucatan Basin
Sargasso Sea
Middle America Trench
Caribbean Sea
Barracuda Fracture Zone
PACIFIC
Clipperton Fracture Zone
Guatemala Basin
Central Pacific Basin
OCEAN
Peru Basin
Nazca Ridge
Chile Basin
SOUTH AMERICA
Brazil Basin
Equator
East Pacific Rise
Peru-Chile Trench
Sala y Gomez Ridge
OCEAN
Tonga Trench
Southwest Pacific Basin
Rio Grande Rise
Tropic of Capricorn
East Pacific Rise
Argentine Basin
Mid-Atlantic Ridge
Pacific-Antarctic Ridge
OCEAN
Southeast Pacific Basin
Scotia Sea
Ross Sea
Amundsen Sea
Bellingshausen Sea
Weddell Sea
South Sandwich Trench
Antarctic Circle

Tides and waves

Tides are created by the pull of the Sun and Moon's gravity on the surface of the oceans. The levels of high and low tides are influenced by the position of the Moon in relation to the Earth and Sun. Waves are formed by wind blowing over the surface of the water.

High and low tides

The highest tides occur when the Earth, the Moon and the Sun are aligned *(below left)*. The lowest tides are experienced when the Sun and Moon align at right angles to one another *(below right)*.

Tidal range and wave environments

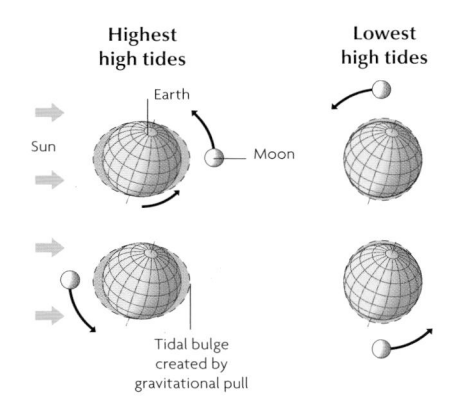

Tidal range and wave environments

```
     less than 7ft / 2m          east coast swell      tropical cyclone    ice-shelf
     7–13ft / 2–4m               west coast swell      storm wave
     greater than 13ft / 4m
```

Highest high tides

Earth
Sun
Moon

Lowest high tides

Tidal bulge created by gravitational pull

Deep sea temperature and currents

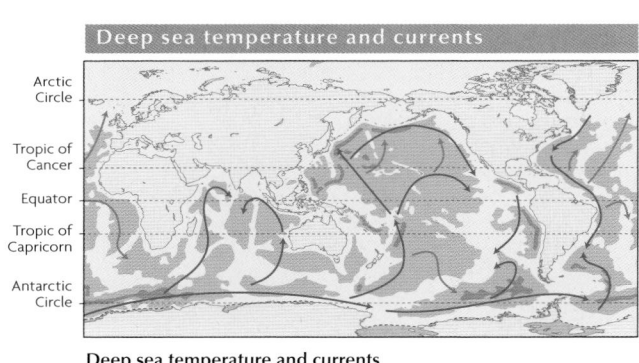

Arctic Circle
Tropic of Cancer
Equator
Tropic of Capricorn
Antarctic Circle

Deep sea temperature and currents

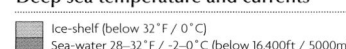

```
     Ice-shelf (below 32°F / 0°C)                                  → Primary currents
     Sea-water 28–32°F / -2–0°C (below 16,400ft / 5000m)           → Secondary currents
     Sea-water 32–41°F /0–5°C (below 13,120ft / 4000m)
```

The global climate

The Earth's climatic types consist of stable patterns of weather conditions averaged out over a long period of time. Different climates are categorized according to particular combinations of temperature and humidity. By contrast, weather consists of short-term fluctuations in wind, temperature, and humidity conditions. Different climates are determined by latitude, altitude, the prevailing wind, and circulation of ocean currents. Longer-term changes in climate, such as global warming or the onset of ice ages, are punctuated by shorter-term events which comprise the day-to-day weather of a region, such as frontal depressions, hurricanes, and blizzards.

The atmosphere, wind and weather

The Earth's atmosphere has been compared to a giant ocean of air which surrounds the planet. Its circulation patterns are similar to the currents in the oceans and are influenced by three factors; the Earth's orbit around the Sun and rotation about its axis, and variations in the amount of heat radiation received from the Sun. If both heat and moisture were not redistributed between the Equator and the poles, large areas of the Earth would be uninhabitable.

◀ **Heavy fogs, as** here in southern England, form as moisture-laden air passes over cold ground.

Temperature

The world can be divided into three major climatic zones, stretching like large belts across the latitudes: the tropics which are warm; the cold polar regions and the temperate zones which lie between them. Temperatures across the Earth range from above 86°F (30°C) in the deserts to as low as -70°F (-55°C) at the poles. Temperature is also controlled by altitude; because air becomes cooler and less dense the higher it gets, mountainous regions are typically colder than those areas which are at, or close to, sea level.

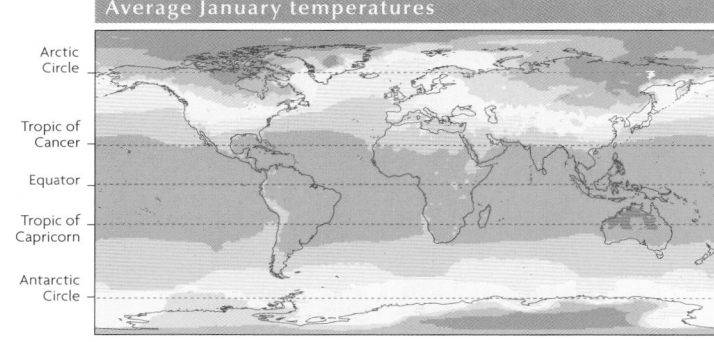
Average January temperatures

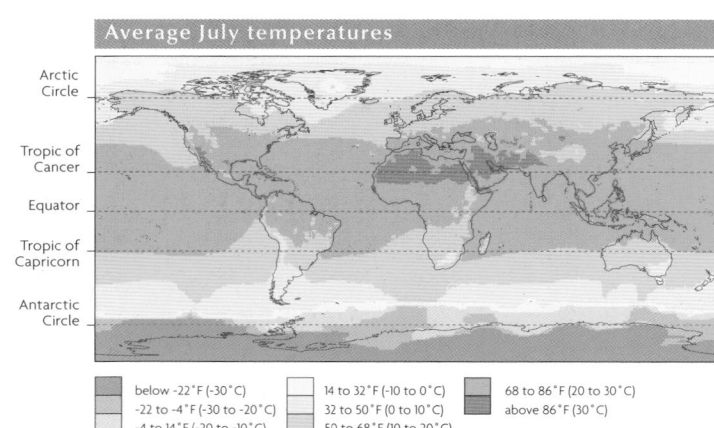
Average July temperatures

below -22°F (-30°C)	14 to 32°F (-10 to 0°C)	68 to 86°F (20 to 30°C)
-22 to -4°F (-30 to -20°C)	32 to 50°F (0 to 10°C)	above 86°F (30°C)
-4 to 14°F (-20 to -10°C)	50 to 68°F (10 to 20°C)	

Global air circulation

Air does not simply flow from the Equator to the poles, it circulates in giant cells known as Hadley and Ferrel cells. As air warms it expands, becoming less dense and rising; this creates areas of low pressure. As the air rises it cools and condenses, causing heavy rainfall over the tropics and slight snowfall over the poles. This cool air then sinks, forming high pressure belts. At surface level in the tropics these sinking currents are deflected poleward as the westerlies and toward the equator as the trade winds. At the poles they become the polar easterlies.

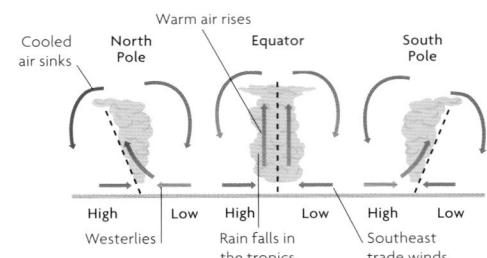

▲ **The Antarctic pack** ice expands its area by almost seven times during the winter as temperatures drop and surrounding seas freeze.

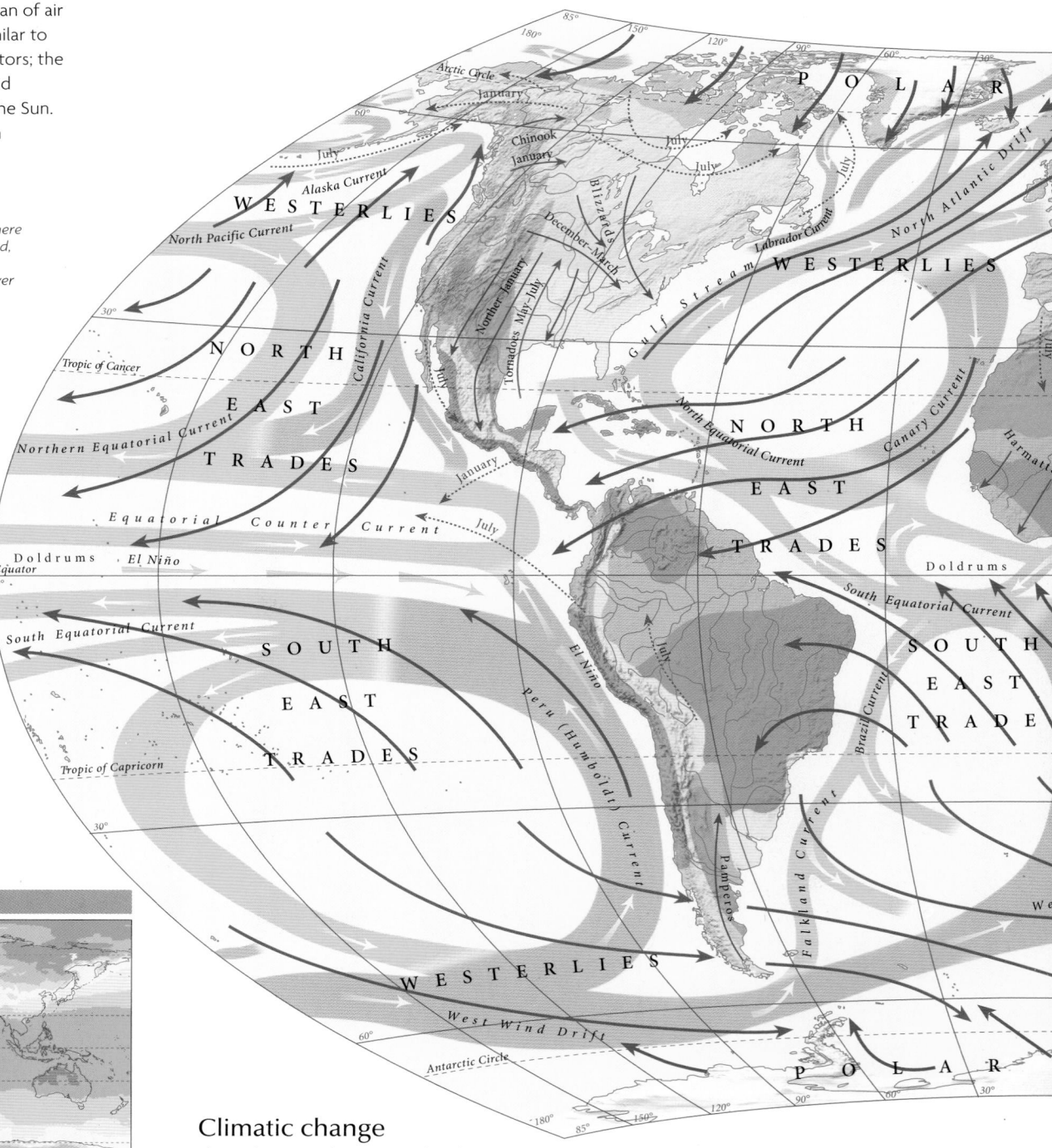

Climatic change

The Earth is currently in a warm phase between ice ages. Warmer temperatures result in higher sea levels as more of the polar ice caps melt. Most of the world's population lives near coasts, so any changes which might cause sea levels to rise, could have a potentially disastrous impact.

▲ **This ice fair,** painted by Pieter Brueghel the Younger in the 17th century, shows the Little Ice Age which peaked around 300 years ago.

The greenhouse effect

Gases such as carbon dioxide are known as "greenhouse gases" because they allow shortwave solar radiation to enter the Earth's atmosphere, but help to stop longwave radiation from escaping. This traps heat, raising the Earth's temperature. An excess of these gases, such as that which results from the burning of fossil fuels, helps trap more heat and can lead to global warming.

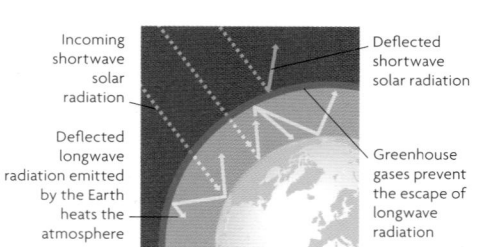

Incoming shortwave solar radiation

Deflected shortwave solar radiation

Deflected longwave radiation emitted by the Earth heats the atmosphere

Greenhouse gases prevent the escape of longwave radiation

◀ *The islands of the Caribbean, Mexico's Gulf coast and the southeastern US are often hit by hurricanes formed far out in the Atlantic.*

Oceanic water circulation

In general, ocean currents parallel the movement of winds across the Earth's surface. Incoming solar energy is greatest at the Equator and least at the poles. So, water in the oceans heats up most at the Equator and flows poleward, cooling as it moves north or south toward the Arctic or Antarctic. The flow is eventually reversed and cold water currents move back toward the Equator. These ocean currents act as a vast system for moving heat from the Equator toward the poles and are a major influence on the distribution of the Earth's climates.

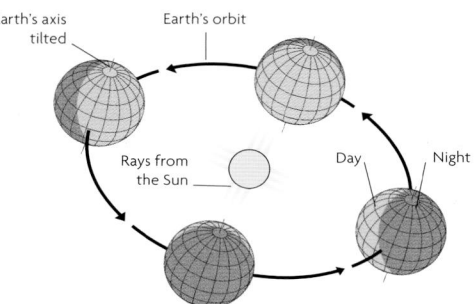

▲ *In marginal climatic zones years of drought can completely dry out the land and transform grassland to desert.*

Tilt and rotation

The tilt and rotation of the Earth during its annual orbit largely control the distribution of heat and moisture across its surface, which correspondingly controls its large-scale weather patterns. As the Earth annually rotates around the Sun, half its surface is receiving maximum radiation, creating summer and winter seasons. The angle of the Earth means that on average the tropics receive two and a half times as much heat from the Sun each day as the poles.

Earth's axis tilted
Earth's orbit
Rays from the Sun
Day
Night

The Coriolis effect

The rotation of the Earth influences atmospheric circulation by deflecting winds and ocean currents. Winds blowing in the northern hemisphere are deflected to the right and those in the southern hemisphere are deflected to the left, creating large-scale patterns of wind circulation, such as the northeast and southeast trade winds and the westerlies. This effect is greatest at the poles and least at the Equator.

Maximum deflection at North pole
Direction of Earth's rotation
Deflection to right in northern hemisphere, creates northeast trade winds
Westerlies
No deflection at Equator
Polar easterlies
Deflection to left in southern hemisphere, creates southeast trade winds
Maximum deflection at South Pole

Map key

Climate zones
ice cap
subarctic
tundra
continental
temperate
warm temperate
mediterranean
semi-arid
arid
hot humid
humid equatorial
tropical

Ocean currents
warm
cold

Prevailing winds
→ warm
→ cold

Local winds
→ warm
→ cold
→ seasonal*
* (seasonal winds which can either be warm or cold)

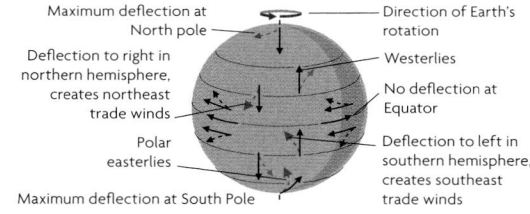
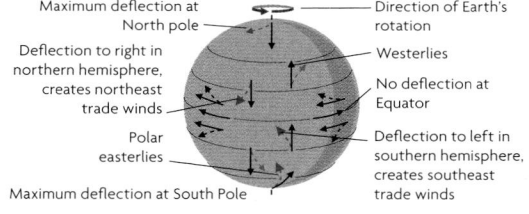

▲ *The wide range of environments found in the Andes is strongly related to their altitude, which modifies climatic influences. While the peaks are snow-capped, many protected interior valleys are semi-tropical.*

Precipitation

When warm air expands, it rises and cools, and the water vapor it carries condenses to form clouds. Heavy, regular rainfall is characteristic of the equatorial region, while the poles are cold and receive only slight snowfall. Tropical regions have marked dry and rainy seasons, while in the temperate regions rainfall is relatively unpredictable.

▲ *Monsoon rains, which affect southern Asia from May to September, are caused by sea winds blowing across the warm land.*

▲ *Heavy tropical rainstorms occur frequently in Papua New Guinea, often causing soil erosion and landslides in cultivated areas.*

Average January rainfall

Arctic Circle
Tropic of Cancer
Equator
Tropic of Capricorn
Antarctic Circle

Average July rainfall

Arctic Circle
Tropic of Cancer
Equator
Tropic of Capricorn
Antarctic Circle

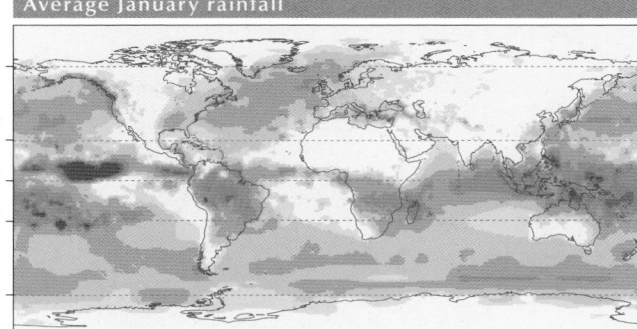

▲ *Violent thunderstorms occur along advancing cold fronts, when cold, dry air masses meet warm, moist air, which rises rapidly, its moisture condensing into thunderclouds. Rain and hail become electrically charged, causing lightning.*

▲ *The intensity of some blizzards in Canada and the northern US can give rise to snowdrifts as high as 10 ft (3 m).*

▲ *The Atacama Desert in Chile is one of the driest places on Earth, with an average rainfall of less than 2 inches (50 mm) per year.*

The rainshadow effect

When moist air is forced to rise by mountains, it cools and the water vapor falls as precipitation, either as rain or snow. Only the dry, cold air continues over the mountains, leaving inland areas with little or no rain. This is called the rainshadow effect and is one reason for the existence of the Mojave Desert in California, which lies east of the Coast Ranges.

Moist air travels inland from the sea
As air rises it cools and condenses leading to cloud
Dry air in 'shadow' of mountain

The rainshadow effect

0–1 in (0–25 mm)
1–2 in (25–50 mm)
2–4 in (50–100 mm)
4–8 in (100–200 mm)
8–12 in (200–300 mm)
12–16 in (300–400 mm)
16–20 in (400–500 mm)
above 20 in (500 mm)

Life on Earth

A unique combination of an oxygen-rich atmosphere and plentiful water is the key to life on Earth. Apart from the polar ice caps, there are few areas which have not been colonized by animals or plants over the course of the Earth's history. Plants process sunlight to provide them with their energy, and ultimately all the Earth's animals rely on plants for survival. Because of this reliance, plants are known as primary producers, and the availability of nutrients and temperature of an area is defined as its primary productivity, which affects the quantity and type of animals which are able to live there. This index is affected by climatic factors – cold and aridity restrict the quantity of life, whereas warmth and regular rainfall allow a greater diversity of species.

Biogeographical regions

The Earth can be divided into a series of biogeographical regions, or biomes, ecological communities where certain species of plant and animal coexist within particular climatic conditions. Within these broad classifications, other factors including soil richness, altitude, and human activities such as urbanization, intensive agriculture, and deforestation, affect the local distribution of living species within each biome.

Polar regions

A layer of permanent ice at the Earth's poles covers both seas and land. Very little plant and animal life can exist in these harsh regions.

Tundra

A desolate region, with long, dark freezing winters and short, cold summers. With virtually no soil and large areas of permanently frozen ground known as permafrost, the tundra is largely treeless, though it is briefly clothed by small flowering plants in the summer months.

Needleleaf forests

With milder summers than the tundra and less wind, these areas are able to support large forests of coniferous trees.

Broadleaf forests

Much of the northern hemisphere was once covered by deciduous forests, which occurred in areas with marked seasonal variations. Most deciduous forests have been cleared for human settlement.

Temperate rain forests

In warmer wetter areas, such as southern China, temperate deciduous forests are replaced by evergreen forest.

Deserts

Deserts are areas with negligible rainfall. Most hot deserts lie within the tropics; cold deserts are dry because of their distance from the moisture-providing sea.

Mediterranean

Hot, dry summers and short winters typify these areas, which were once covered by evergreen shrubs and woodland, but have now been cleared by humans for agriculture.

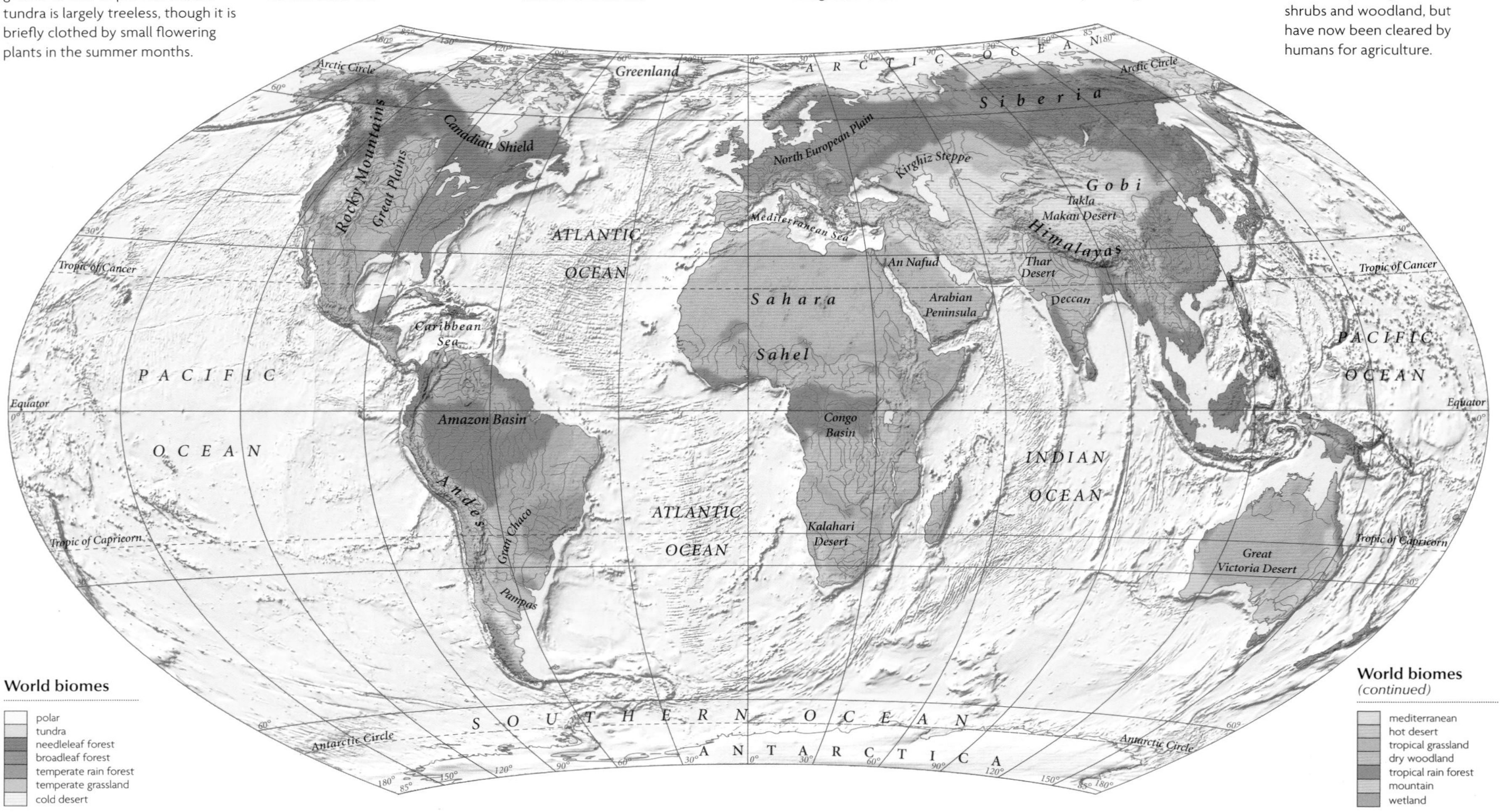

World biomes

- polar
- tundra
- needleleaf forest
- broadleaf forest
- temperate rain forest
- temperate grassland
- cold desert

World biomes
(continued)

- mediterranean
- hot desert
- tropical grassland
- dry woodland
- tropical rain forest
- mountain
- wetland

Tropical and temperate grasslands

The major grassland areas are found in the centers of the larger continental landmasses. In Africa's tropical savannah regions, seasonal rainfall alternates with drought. Temperate grasslands, also known as steppes and prairies are found in the northern hemisphere, and in South America, where they are known as the pampas.

Dry woodlands

Trees and shrubs, adapted to dry conditions, grow widely spaced from one another, interspersed by savannah grasslands.

Tropical rain forests

Characterized by year-round warmth and high rainfall, tropical rain forests contain the highest diversity of plant and animal species on Earth.

Mountains

Though the lower slopes of mountains may be thickly forested, only ground-hugging shrubs and other vegetation will grow above the tree line which varies according to both altitude and latitude.

Wetlands

Rarely lying above sea level, wetlands are marshes, swamps, and tidal flats. Some, with their moist, fertile soils, are rich feeding grounds for fish and breeding grounds for birds. Others have little soil structure and are too acidic to support much plant and animal life.

Biodiversity

The number of plant and animal species, and the range of genetic diversity within the populations of each species, make up the Earth's biodiversity. The plants and animals which are endemic to a region – that is, those which are found nowhere else in the world – are also important in determining levels of biodiversity. Human settlement and intervention have encroached on many areas of the world once rich in endemic plant and animal species. Increasing international efforts are being made to monitor and conserve the biodiversity of the Earth's remaining wild places.

Animal adaptation

The degree of an animal's adaptability to different climates and conditions is extremely important in ensuring its success as a species. Many animals, particularly the largest mammals, are becoming restricted to ever-smaller regions as human development and modern agricultural practices reduce their natural habitats. In contrast, humans have been responsible – both deliberately and accidentally – for the spread of some of the world's most successful species. Many of these introduced species are now more numerous than the indigenous animal populations.

Polar animals

The frozen wastes of the polar regions are able to support only a small range of species which derive their nutritional requirements from the sea. Animals such as the walrus *(left)* have developed insulating fat, stocky limbs, and double-layered coats to enable them to survive in the freezing conditions.

Desert animals

Many animals which live in the extreme heat and aridity of the deserts are able to survive for days and even months with very little food or water. Their bodies are adapted to lose heat quickly and to store fat and water. The Gila monster *(above)* stores fat in its tail.

Amazon rain forest

The vast Amazon Basin is home to the world's greatest variety of animal species. Animals are adapted to live at many different levels from the treetops to the tangled undergrowth which lies beneath the canopy. The sloth *(below)* hangs upside down in the branches. Its fur grows from its stomach to its back to enable water to run off quickly.

Diversity of animal species

Number of animal species per country

- more than 2000
- 1000–1999
- 700–999
- 400–699
- 200–399
- 100–199
- 0–99
- data not available

Marine biodiversity

The oceans support a huge variety of different species, from the world's largest mammals like whales and dolphins down to the tiniest plankton. The greatest diversities occur in the warmer seas of continental shelves, where plants are easily able to photosynthesize, and around coral reefs, where complex ecosystems are found. On the ocean floor, nematodes can exist at a depth of more than 10,000 ft (3000 m) below sea level.

High altitudes

Few animals exist in the rarefied atmosphere of the highest mountains. However, birds of prey such as eagles and vultures *(above)*, with their superb eyesight can soar as high as 23,000 ft (7000 m) to scan for prey below.

Urban animals

The growth of cities has reduced the amount of habitat available to many species. A number of animals are now moving closer into urban areas to scavenge from the detritus of the modern city *(left)*. Rodents, particularly rats and mice, have existed in cities for thousands of years, and many insects, especially moths, quickly develop new coloring to provide them with camouflage.

Endemic species

Isolated areas such as Australia and the island of Madagascar, have the greatest range of endemic species. In Australia, these include marsupials such as the kangaroo *(below)*, which carry their young in pouches on their bodies. Destruction of habitat, pollution, hunting, and predators introduced by humans, are threatening this unique biodiversity.

Plant adaptation

Environmental conditions, particularly climate, soil type, and the extent of competition with other organisms, influence the development of plants into a number of distinctive forms. Similar conditions in quite different parts of the world create similar adaptations in the plants, which may then be modified by other, local, factors specific to the region.

Cold conditions

In areas where temperatures rarely rise above freezing, plants such as lichens *(left)* and mosses grow densely, close to the ground.

Rain forests

Most of the world's largest and oldest plants are found in rain forests; warmth and heavy rainfall provide ideal conditions for vast plants like the world's largest flower, the rafflesia *(left)*.

Hot, dry conditions

Arid conditions lead to the development of plants whose surface area has been reduced to a minimum to reduce water loss. In cacti *(above)*, which can survive without water for months, leaves are minimal or not present at all.

Ancient plants

Some of the world's most primitive plants still exist today, including algae, cycads, and many ferns *(above)*, reflecting the success with which they have adapted to changing conditions.

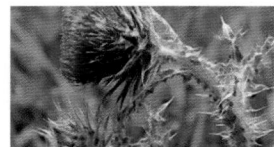

Resisting predators

A great variety of plants have developed devices including spines *(above)*, poisons, stinging hairs, and an unpleasant taste or smell to deter animal predators.

Diversity of plant species

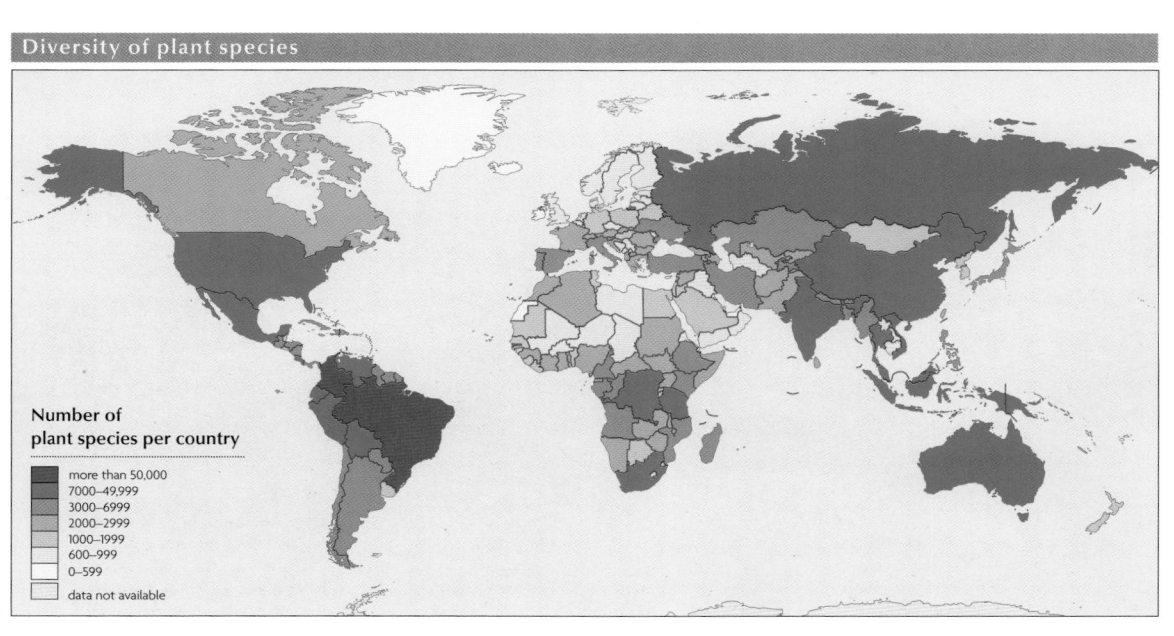

Number of plant species per country

- more than 50,000
- 7000–49,999
- 3000–6999
- 2000–2999
- 1000–1999
- 600–999
- 0–599
- data not available

Weeds

Weeds such as bindweed *(above)* are fast-growing, easily dispersed, and tolerant of a number of different environments, enabling them to quickly colonize suitable habitats. They are among the most adaptable of all plants.

Population and settlement

The Earth's population is projected to rise from its current level of about 7.2 billion to reach some 10.5 billion by 2050. The global distribution of this rapidly growing population is very uneven, and is dictated by climate, terrain, and natural and economic resources. The great majority of the Earth's people live in coastal zones, and along river valleys. Deserts cover over 20% of the Earth's surface, but support less than 5% of the world's population. It is estimated that over half of the world's population live in cities – most of them in Asia – as a result of mass migration from rural areas in search of jobs. Many of these people live in the so-called "megacities," some with populations as great as 40 million.

Patterns of settlement

The past 200 years have seen the most radical shift in world population patterns in recorded history.

Nomadic life

All the world's peoples were hunter-gatherers 10,000 years ago. Today nomads, who live by following available food resources, account for less than 0.0001% of the world's population. They are mainly pastoral herders, moving their livestock from place to place in search of grazing land.

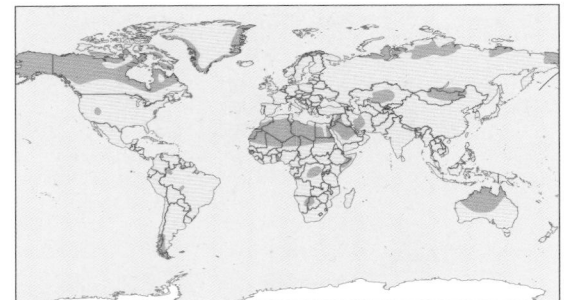

Nomadic population

Nomadic population area

The growth of cities

In 1900 there were only 14 cities in the world with populations of more than a million, mostly in the northern hemisphere. Today, as more and more people in the developing world migrate to towns and cities, there are over 70 cities whose population exceeds 5 million, and around 490 "million-cities."

Million-cities in 1900

Million-cities in 1900

• Cities over 1 million population

Million-cities in 2005

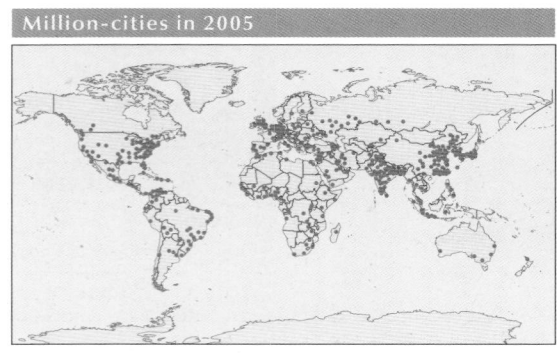

Million-cities in 2005

• Cities over 1 million population

North America

The eastern and western seaboards of the US, with huge expanses of interconnected cities, towns, and suburbs, are vast, densely-populated megalopolises. Central America and the Caribbean also have high population densities. Yet, away from the coasts and in the wildernesses of northern Canada the land is very sparsely settled.

▲ *Vancouver on Canada's* west coast, grew up as a port city. In recent years it has attracted many Asian immigrants, particularly from the Pacific Rim.

▲ *North America's central* plains, the continent's agricultural heartland, are thinly populated and highly productive.

Europe

With its temperate climate, and rich mineral and natural resources, Europe is generally very densely settled. The continent acts as a magnet for economic migrants from the developing world, and immigration is now widely restricted. Birthrates in Europe are generally low, and in some countries, such as Germany, the populations have stabilized at zero growth, with a fast-growing elderly population.

▲ *Many European cities,* like Siena, once reflected the "ideal" size for human settlements. Modern technological advances have enabled them to grow far beyond the original walls.

▲ *Within the densely-populated* Netherlands the reclamation of coastal wetlands is vital to provide much-needed land for agriculture and settlement.

Population density
(inhabitants per sq mile)

330–2600
260–330
500–1060
30–500
26–30
53–06
3–53
Less than 3

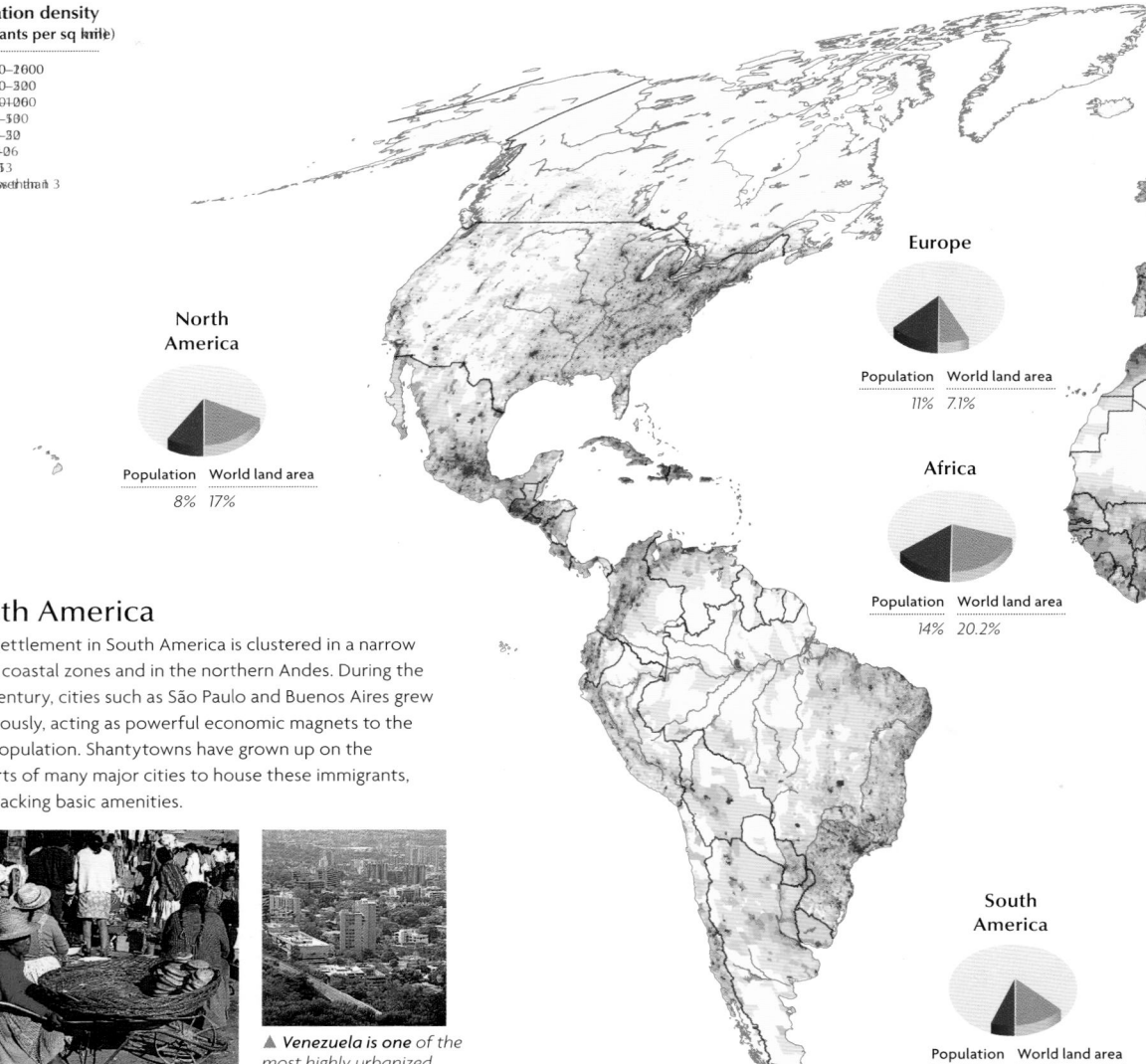

North America

Population | World land area
8% | 17%

Europe

Population | World land area
11% | 7.1%

Africa

Population | World land area
14% | 20.2%

South America

Population | World land area
6% | 11.8%

South America

Most settlement in South America is clustered in a narrow belt in coastal zones and in the northern Andes. During the 20th century, cities such as São Paulo and Buenos Aires grew enormously, acting as powerful economic magnets to the rural population. Shantytowns have grown up on the outskirts of many major cities to house these immigrants, often lacking basic amenities.

▲ *Many people in* western South America live at high altitudes in the Andes, both in cities and in villages such as this one in Bolivia.

▲ *Venezuela is one* of the most highly urbanized countries in South America, with nearly 90% of the population living in cities such as Caracas.

Africa

The arid climate of much of Africa means that settlement of the continent is sparse, focusing in coastal areas and fertile regions such as the Nile Valley. Africa still has a high proportion of nomadic agriculturalists, although many are now becoming settled, and the population is predominantly rural.

▲ *Cities such as* Nairobi (above), Cairo, and Johannesburg have grown rapidly in recent years, although only Cairo has a significant population on a global scale.

▲ *Traditional lifestyles and* homes persist across much of Africa, which has a higher proportion of rural or village-based population than any other continent.

Asia

Most Asian settlement originally centered around the great river valleys such as the Indus, the Ganges, and the Yangtze. Today, almost 60% of the world's population lives in Asia, many in burgeoning cities – particularly in the economically-buoyant Pacific Rim countries. Even rural population densities are high in many countries; practices such as terracing in Southeast Asia making the most of the available land.

▲ *Many of China's* cities are now vast urban areas with populations of more than 5 million people.

▲ *This stilt village* in Bangladesh is built to resist the regular flooding. Pressure on land, even in rural areas, forces many people to live in marginal areas.

Population structures

Population pyramids are an effective means of showing the age structures of different countries, and highlighting changing trends in population growth and decline. The typical pyramid for a country with a growing, youthful population, is broad-based *(left)*, reflecting a high birthrate and a far larger number of young rather than elderly people. In contrast, countries with populations whose numbers are stabilizing have a more balanced distribution of people in each age band, and may even have lower numbers of people in the youngest age ranges, indicating both a high life expectancy, and that the population is now barely replacing itself *(right)*. The Russian Federation *(center)* is suffering from a declining population, forcing the government to consider a number of measures, including tax incentives and immigration, in an effort to stabilize the population .

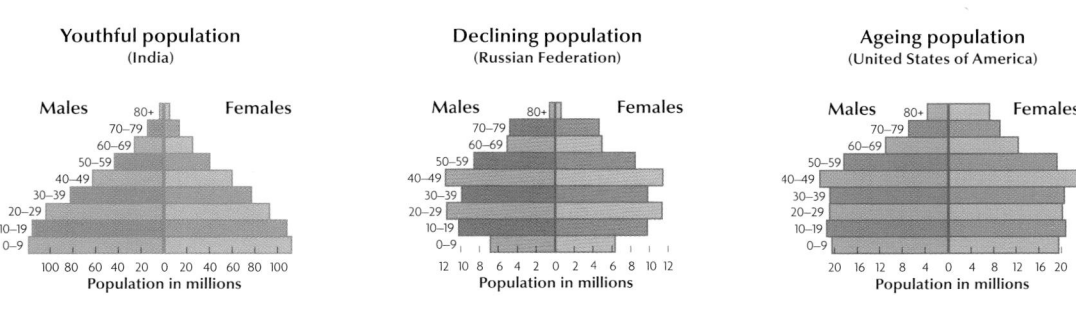

Youthful population
(India)

Declining population
(Russian Federation)

Ageing population
(United States of America)

Population growth

Improvements in food supply and advances in medicine have both played a major role in the remarkable growth in global population, which has increased five-fold over the last 150 years. Food supplies have risen with the mechanization of agriculture and improvements in crop yields. Better nutrition, together with higher standards of public health and sanitation, have led to increased longevity and higher birthrates.

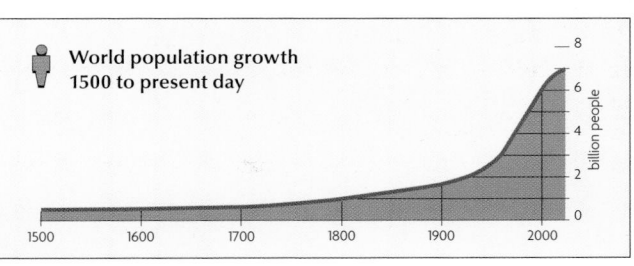

World population growth
1500 to present day

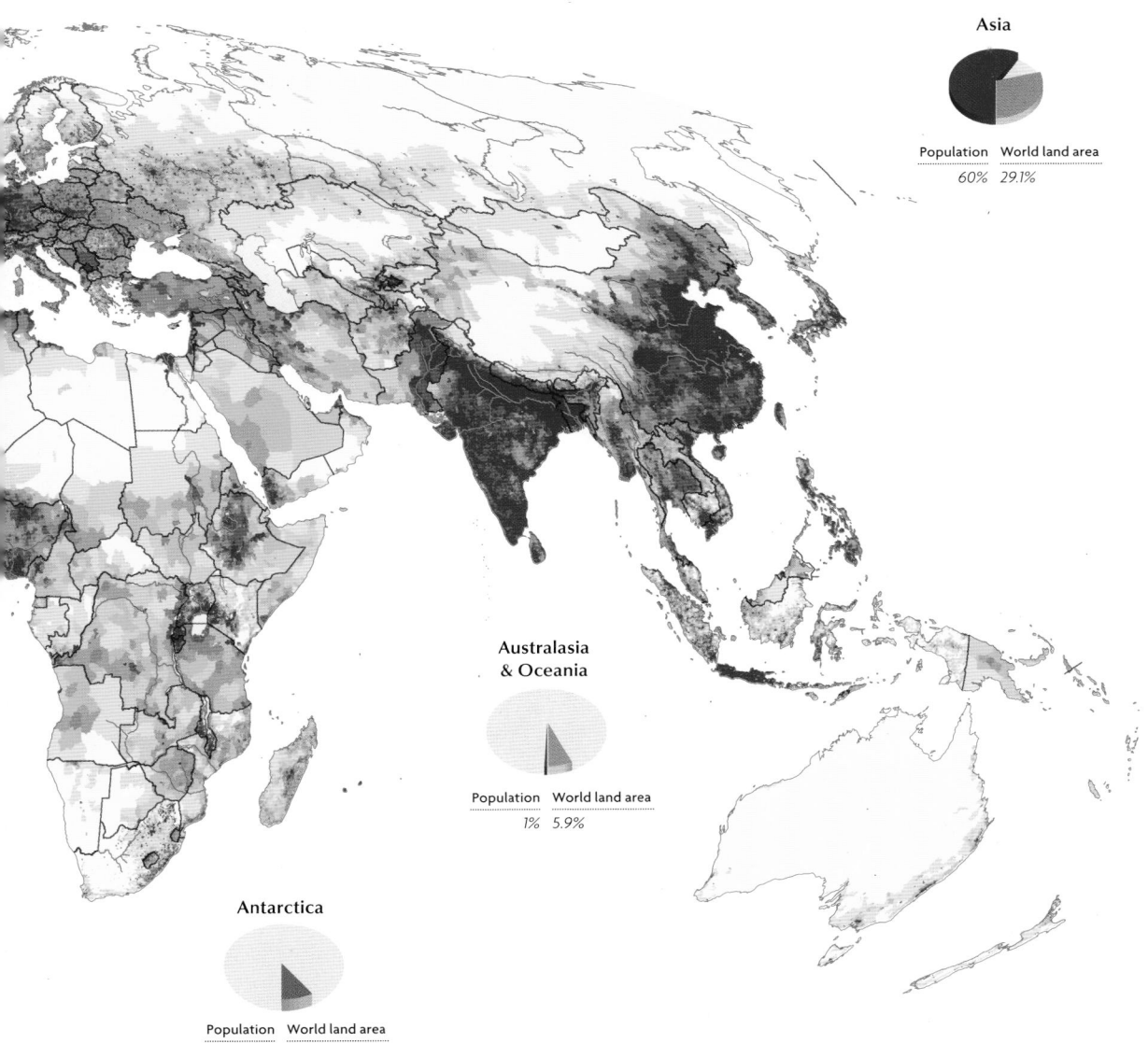

Asia
Population 60% World land area 29.1%

Australasia & Oceania
Population 1% World land area 5.9%

Antarctica
Population 0% World land area 8.9%

World nutrition

Two-thirds of the world's food supply is consumed by the industrialized nations, many of which have a daily calorific intake far higher than is necessary for their populations to maintain a healthy body weight. In contrast, in the developing world, about 800 million people do not have enough food to meet their basic nutritional needs.

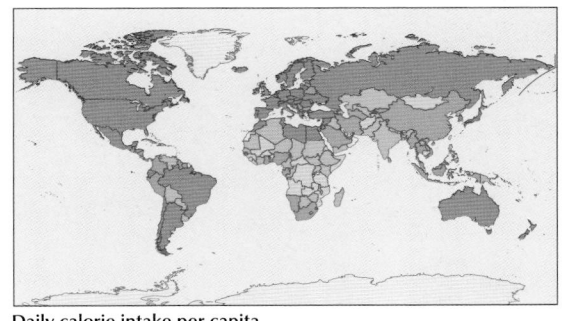

Daily calorie intake per capita

above 3000 | 2000–2499 | data not available
2500–2999 | below 2000

World life expectancy

Improved public health and living standards have greatly increased life expectancy in the developed world, where people can now expect to live twice as long as they did 100 years ago. In many of the world's poorest nations, inadequate nutrition and disease, means that the average life expectancy still does not exceed 45 years.

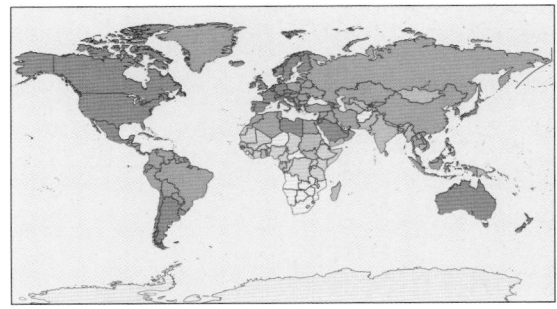

Life expectancy at birth

above 75 years | 55–64 years | below 44 years
65–74 years | 45–54 years | data not available

Australasia and Oceania

This is the world's most sparsely settled region. The peoples of Australia and New Zealand live mainly in the coastal cities, with only scattered settlements in the arid interior. The Pacific islands can only support limited populations because of their remoteness and lack of resources.

▶ *Brisbane, on Australia's* Gold Coast is the most rapidly expanding city in the country. The great majority of Australia's population lives in cities near the coasts.

◀ *The remote highlands* of Papua New Guinea are home to a wide variety of peoples, many of whom still subsist by traditional hunting and gathering.

Average world birth rates

Birthrates are much higher in Africa, Asia, and South America than in Europe and North America. Increased affluence and easy access to contraception are both factors which can lead to a significant decline in a country's birthrate.

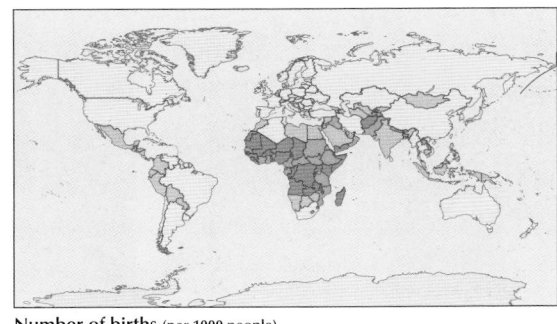

Number of births (per 1000 people)

above 40 | 20–29 | data not available
30–39 | below 20

World infant mortality

In parts of the developing world infant mortality rates are still high; access to medical services such as immunization, adequate nutrition, and the promotion of breast-feeding have been important in combating infant mortality.

World infant mortality rates (deaths per 1000 live births)

above 125 | 35–74 | below 15
75–124 | 15–34 | data not available

The economic system

The wealthy countries of the developed world, with their aggressive, market-led economies and their access to productive new technologies and international markets, dominate the world economic system. At the other extreme, many of the countries of the developing world are locked in a cycle of national debt, rising populations, and unemployment. In 2008 a major financial crisis swept the world's banking sector leading to a huge downturn in the global economy. Despite this, China overtook Japan in 2010 to become the world's second largest economy.

Trade blocs

Trade blocs

| EU | NAFTA | ASEAN | LAIA |
| CACM | SADC | ECOWAS | CEEAC |

International trade blocs are formed when groups of countries, often already enjoying close military and political ties, join together to offer mutually preferential terms of trade for both imports and exports. Increasingly, global trade is dominated by three main blocs: the EU, NAFTA, and ASEAN. They are supplanting older trade blocs such as the Commonwealth, a legacy of colonialism.

International trade flows

World trade acts as a stimulus to national economies, encouraging growth. Over the last three decades, as heavy industries have declined, services – banking, insurance, tourism, airlines, and shipping – have taken an increasingly large share of world trade. Manufactured articles now account for nearly two-thirds of world trade; raw materials and food make up less than a quarter of the total.

Shipping
Ships carry 80% of international cargo, and extensive container ports, where cargo is stored, are vital links in the international transportation network.

Multinationals
Multinational companies are increasingly penetrating inaccessible markets. The reach of many American commodities is now global.

Primary products
Many countries, particularly in the Caribbean and Africa, are still reliant on primary products such as rubber and coffee, which makes them vulnerable to fluctuating prices.

Service industries
Service industries such as banking, tourism and insurance were the fastest-growing industrial sector in the last half of the 20th century. Lloyds of London is the center of the world insurance market.

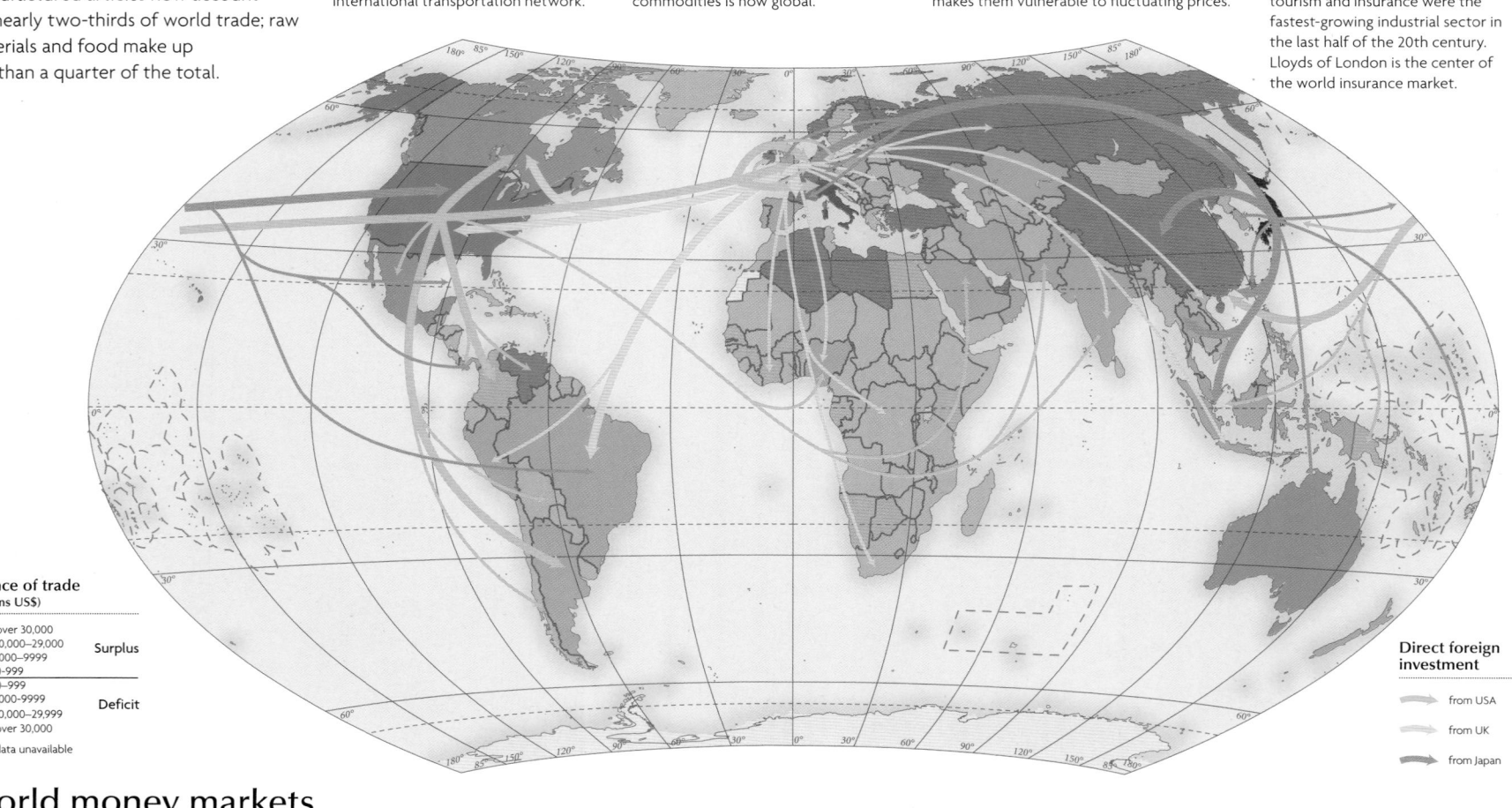

Balance of trade
(millions US$)

	over 30,000	
	10,000–29,000	
	1000–9999	Surplus
	0–999	
	0–999	
	1000–9999	
	10,000–29,999	Deficit
	over 30,000	
	data unavailable	

Direct foreign investment

from USA
from UK
from Japan

World money markets

The financial world has traditionally been dominated by three major centers – Tokyo, New York, and London, which house the headquarters of stock exchanges, multinational corporations and international banks. Their geographic location means that, at any one time in a 24-hour day, one major market is open for trading in shares, currencies, and commodities. Since the late 1980s, technological advances have enabled transactions between financial centers to occur at ever-greater speed, and new markets have sprung up throughout the world.

New stock markets

New stock markets are now opening in many parts of the world, where economies have recently emerged from state controls. In Moscow and Beijing, and several countries in eastern Europe, newly-opened stock exchanges reflect the transition to market-driven economies.

The developing world

International trade in capital and currency is dominated by the rich nations of the northern hemisphere. In parts of Africa and Asia, where exports of any sort are extremely limited, home-produced commodities are simply sold in local markets.

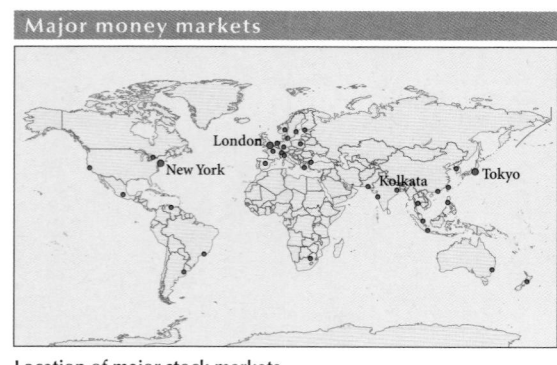

Major money markets

London
New York
Kolkata
Tokyo

Location of major stock markets

● Major stock markets

▲ *The Tokyo Stock Market* crashed in 1990, leading to a slow-down in the growth of the world's most powerful economy, and a refocusing on economic policy away from export-led growth and toward the domestic market.

▲ *Dealers at the* Kolkata Stock Market. The Indian economy has been opened up to foreign investment and many multinationals now have bases there.

▲ *Markets have thrived* in communist Vietnam since the introduction of a liberal economic policy.

World wealth disparity

A global assessment of Gross Domestic Product (GDP) by nation reveals great disparities. The developed world, with only a quarter of the world's population, has 80% of the world's manufacturing income. Civil war, conflict, and political instability further undermine the economic self-sufficiency of many of the world's poorest nations.

Urban sprawl

Cities are expanding all over the developing world, attracting economic migrants in search of work and opportunities. In cities such as Rio de Janeiro, housing has not kept pace with the population explosion, and squalid shanty towns (favelas) rub shoulders with middle-class housing.

▲ **The favelas of** Rio de Janeiro sprawl over the hills surrounding the city.

Agricultural economies

In parts of the developing world, people survive by subsistence farming – only growing enough food for themselves and their families. With no surplus product, they are unable to exchange goods for currency, the only means of escaping the poverty trap. In other countries, farmers have been encouraged to concentrate on growing a single crop for the export market. This reliance on cash crops leaves farmers vulnerable to crop failure and to changes in the market price of the crop.

Urban decay

▲ **Cities such as** Detroit have been badly hit by the decline in heavy industry.

Although the US still dominates the global economy, it faces deficits in both the federal budget and the balance of trade. Vast discrepancies in personal wealth, high levels of unemployment, and the dismantling of welfare provisions throughout the 1980s have led to severe deprivation in several of the inner cities of North America's industrial heartland.

Booming cities

Since the 1980s the Chinese government has set up special industrial zones, such as Shanghai, where foreign investment is encouraged through tax incentives. Migrants from rural China pour into these regions in search of work, creating "boomtown" economies.

◄ **Foreign investment has** encouraged new infrastructure development in cities like Shanghai.

Economic "tigers"

The economic "tigers" of the Pacific Rim – China, Singapore, and South Korea – have grown faster than Europe and the US over the last decade. Their export- and service-led economies have benefited from stable government, low labor costs, and foreign investment.

▲ **Hong Kong, with** its fine natural harbor, is one of the most important ports in Asia.

Comparative world wealth

World economies - average GDP per capita (US$)

- above 20,000
- 5000–20,000
- 2000–5000
- below 2000
- data unavailable

▲ **The Ugandan uplands** are fertile, but poor infrastructure hampers the export of cash crops.

▲ **A shopping arcade** in Paris displays a great profusion of luxury goods.

The affluent West

The capital cities of many countries in the developed world are showcases for consumer goods, reflecting the increasing importance of the service sector, and particularly the retail sector, in the world economy. The idea of shopping as a leisure activity is unique to the western world. Luxury goods and services attract visitors, who in turn generate tourist revenue.

Tourism

In 2004, there were over 940 million tourists worldwide. Tourism is now the world's biggest single industry, employing over 130 million people, though frequently in low-paid unskilled jobs. While tourists are increasingly exploring inaccessible and less-developed regions of the world, the benefits of the industry are not always felt at a local level. There are also worries about the environmental impact of tourism, as the world's last wildernesses increasingly become tourist attractions.

▲ **Botswana's Okavango Delta** is an area rich in wildlife. Tourists go on safaris to the region, but the impact of tourism is controlled.

Money flows

In 2008 a global financial crisis swept through the world's economic system. The crisis triggered the failure of several major financial institutions and lead to increased borrowing costs known as the "credit crunch". A consequent reduction in economic activity together with rising inflation forced many governments to introduce austerity measures to reduce borrowing and debt, particulary in Europe where massive "bailouts" were needed to keep some European single currency (Euro) countries solvent.

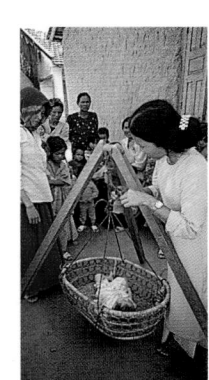

◄ **In rural Southeast Asia,** babies are given medical checks by UNICEF as part of a global aid program sponsored by the UN.

Tourist arrivals

Tourist arrivals

- over 20 million
- 10–20 million
- 5–10 million
- 2.5–5 million
- 1–2.5 million
- 700,000–999,000
- under 700,000
- data unavailable

International debt

International debt (as percentage of GNI)

- over 100%
- 70–99%
- 50–69%
- 30–49%
- 10–29%
- below 10%
- data unavailable

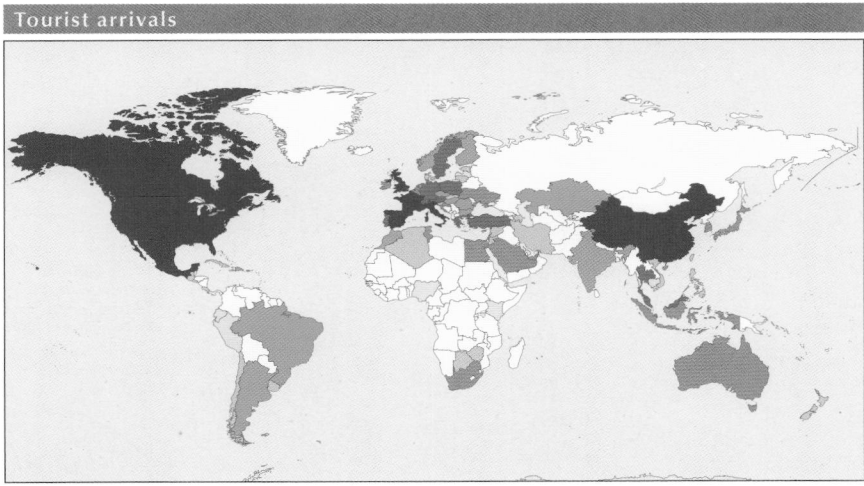

The political world

There are 196 independent countries in the world today. With the exception of Antarctica, where territorial claims have been deferred by international treaty, every land area of the Earth's surface either belongs to, or is claimed by, one country or another. The largest country in the world is the Russian Federation, the smallest is Vatican City. Some 60 overseas dependent territories remain, administered variously by France, Australia, Denmark, New Zealand, Norway, Portugal, the UK, the US, and the Netherlands.

International borders

The map shows three main types of boundary between states. Full borders represent internationally agreed and recognized territorial boundaries. Undefined borders exist where no fixed boundary between states has been demarcated; the boundaries indicated in this way show approximate areas of sovereignty. A disputed border is indicated where a *de facto* territorial boundary exists, which is not agreed or is subject to arbitration.

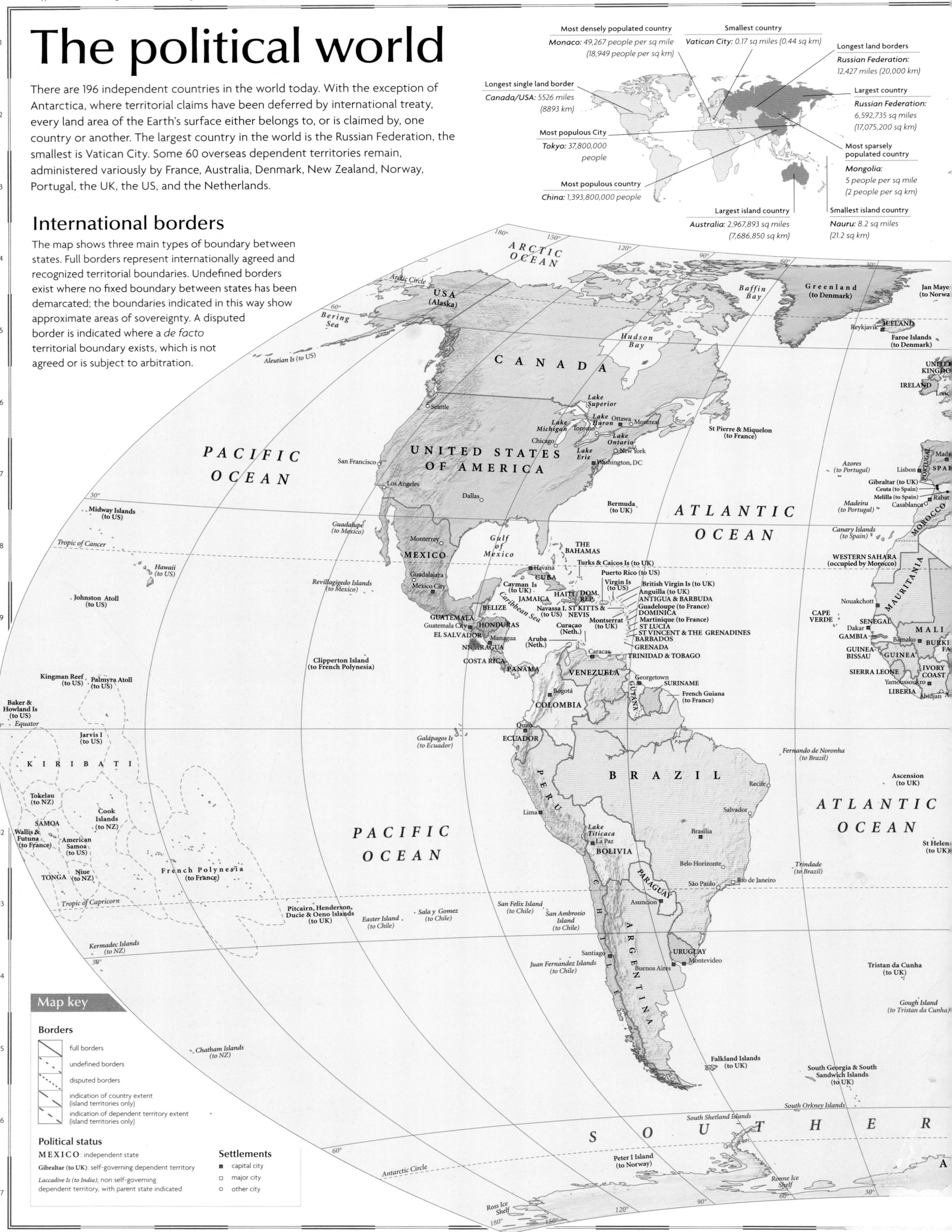

Most densely populated country
Monaco: 49,267 people per sq mile
(18,949 people per sq km)

Smallest country
Vatican City: 0.17 sq miles (0.44 sq km)

Longest land borders
Russian Federation:
12,427 miles (20,000 km)

Longest single land border
Canada/USA: 5526 miles
(8893 km)

Largest country
Russian Federation:
6,592,735 sq miles
(17,075,200 sq km)

Most populous City
Tokyo: 37,800,000
people

Most sparsely
populated country
Mongolia:
5 people per sq mile
(2 people per sq km)

Most populous country
China: 1,393,800,000 people

Largest island country
Australia: 2,967,893 sq miles
(7,686,850 sq km)

Smallest island country
Nauru: 8.2 sq miles
(21.2 sq km)

Map key

Borders

full borders

undefined borders

disputed borders

indication of country extent
(island territories only)

indication of dependent territory extent
(island territories only)

Political status

MEXICO: independent state

Gibraltar (to UK): self-governing dependent territory

Laccadive Is (to India): non self-governing
dependent territory, with parent state indicated

Settlements

■ capital city

□ major city

○ other city

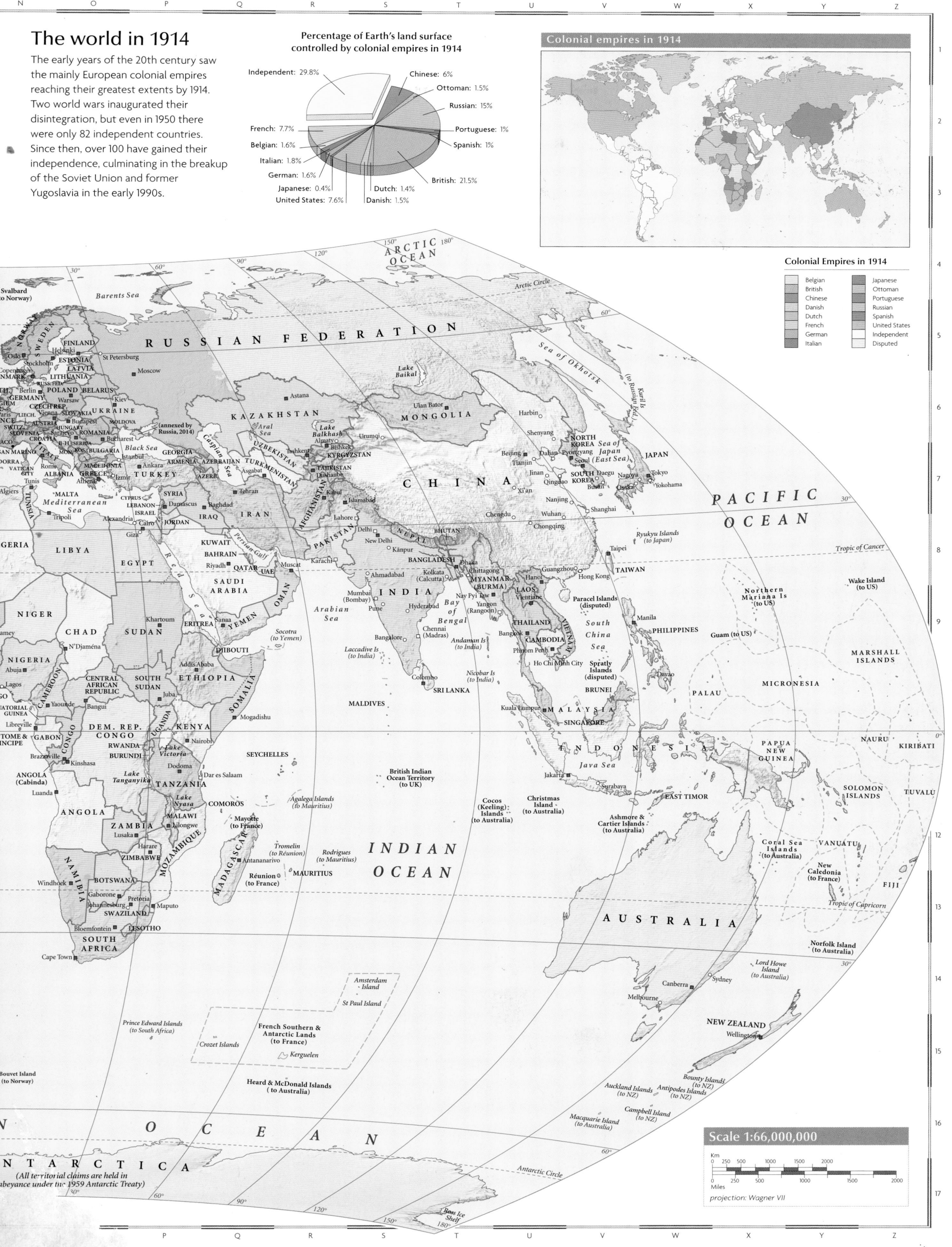

The world in 1914

The early years of the 20th century saw the mainly European colonial empires reaching their greatest extents by 1914. Two world wars inaugurated their disintegration, but even in 1950 there were only 82 independent countries. Since then, over 100 have gained their independence, culminating in the breakup of the Soviet Union and former Yugoslavia in the early 1990s.

Percentage of Earth's land surface controlled by colonial empires in 1914

- Independent: 29.8%
- Chinese: 6%
- Ottoman: 1.5%
- Russian: 15%
- Portuguese: 1%
- Spanish: 1%
- British: 21.5%
- French: 7.7%
- Belgian: 1.6%
- Italian: 1.8%
- German: 1.6%
- Japanese: 0.4%
- United States: 7.6%
- Dutch: 1.4%
- Danish: 1.5%

Colonial empires in 1914

Colonial Empires in 1914
- Belgian
- British
- Chinese
- Danish
- Dutch
- French
- German
- Italian
- Japanese
- Ottoman
- Portuguese
- Russian
- Spanish
- United States
- Independent
- Disputed

Scale 1:66,000,000

projection: Wagner VII

States and boundaries

There are almost 200 sovereign states in the world today; in 1950 there were only 82. Over the last 65 years national self-determination has been a driving force for many states with a history of colonialism and oppression. As more borders have been added to the world map, the number of international border disputes has increased.

In many cases, where the impetus toward independence has been religious or ethnic, disputes with minority groups have also caused violent internal conflict. While many newly-formed states have moved peacefully toward independence, successfully establishing government by multiparty democracy, dictatorship by military regime or individual despot is often the result of the internal power-struggles which characterize the early stages in the lives of new nations.

The nature of politics

Democracy is a broad term: it can range from the ideal of multiparty elections and fair representation to, in countries such as Singapore, a thin disguise for single-party rule. In despotic regimes, on the other hand, a single, often personal authority has total power; institutions such as parliament and the military are mere instruments of the dictator.

◀ The stars and stripes of the US flag are a potent symbol of the country's status as a federal democracy.

Types of government

- Multiparty democracy for more than 10 yrs
- Multiparty democracy within last 10 yrs
- Single-party government
- Military regime
- Theocracy
- Monarchy
- Non-party system
- Transitional regime
- ⚑ Current civil unrest

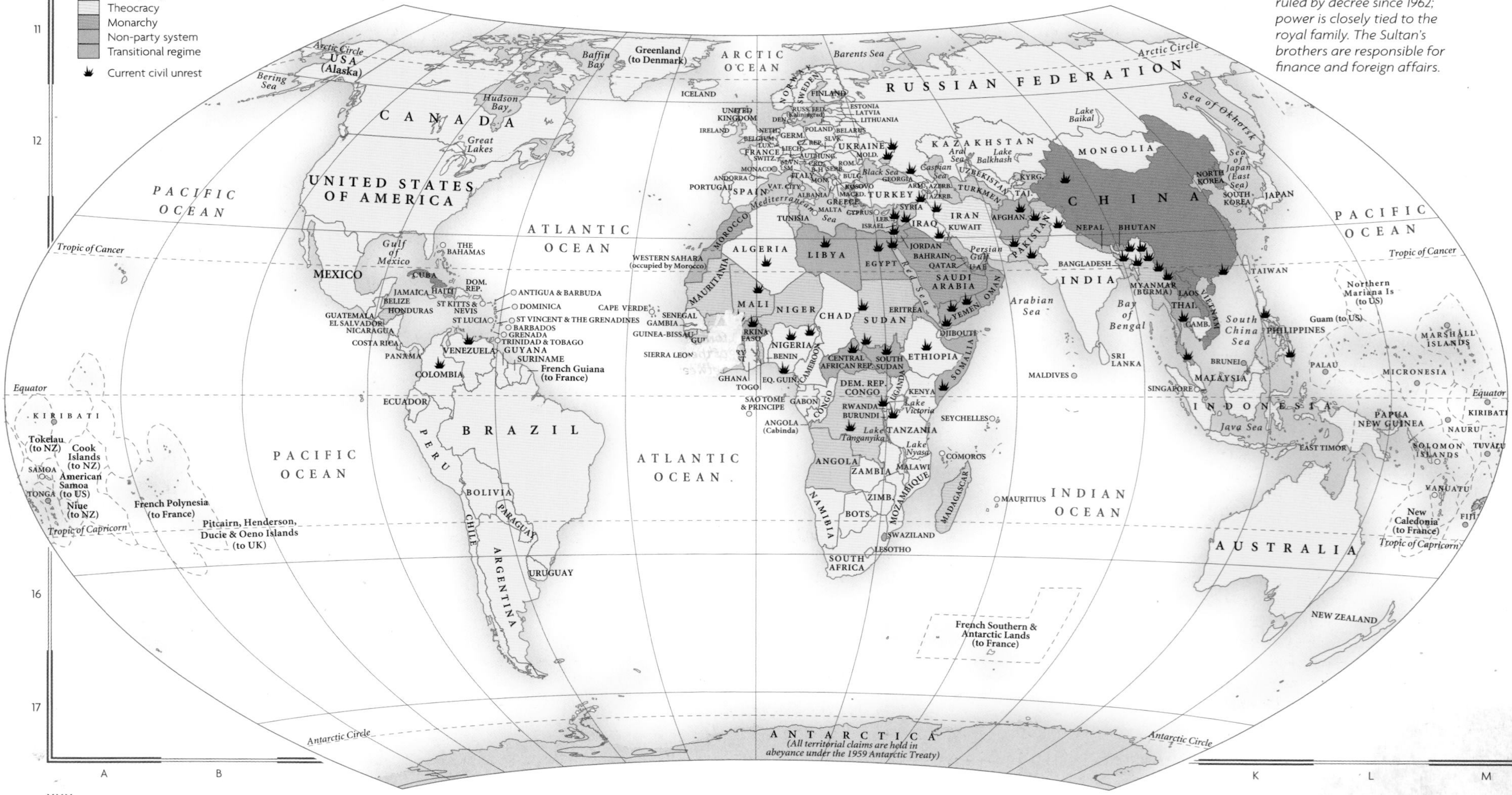

The changing world map

Decolonization

In 1950, large areas of the world remained under the control of a handful of European countries *(page xxix)*. The process of decolonization had begun in Asia, where, following the Second World War, much of southern and southeastern Asia sought and achieved self-determination. In the 1960s, a host of African states achieved independence, so that by 1965, most of the larger tracts of the European overseas empires had been substantially eroded. The final major stage in decolonization came with the breakup of the Soviet Union and the Eastern bloc after 1990. The process continues today as the last toeholds of European colonialism, often tiny island nations, press increasingly for independence.

▲ Icons of communism, including statues of former leaders such as Lenin and Stalin, were destroyed when the Soviet bloc was dismantled in 1989, creating several new nations.

▲ Iran has been one of the modern world's few true theocracies; Islam has an impact on every aspect of political life.

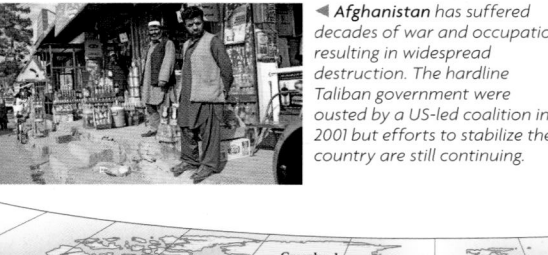

▲ Afghanistan has suffered decades of war and occupation resulting in widespread destruction. The hardline Taliban government were ousted by a US-led coalition in 2001 but efforts to stabilize the country are still continuing.

New nations 1945–1965

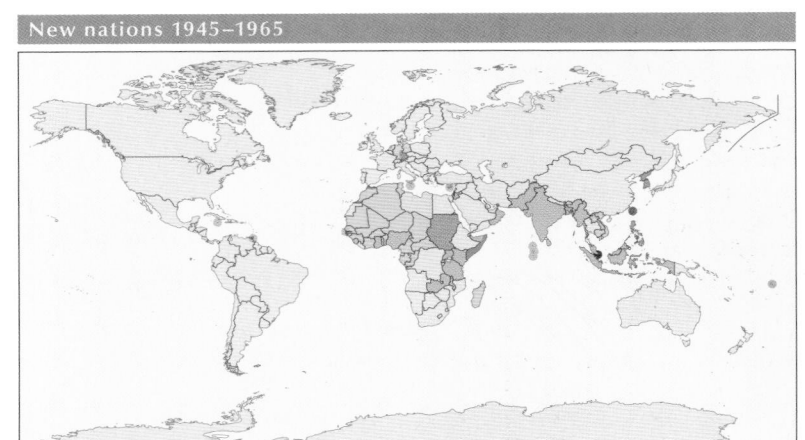

New nations 1965–present

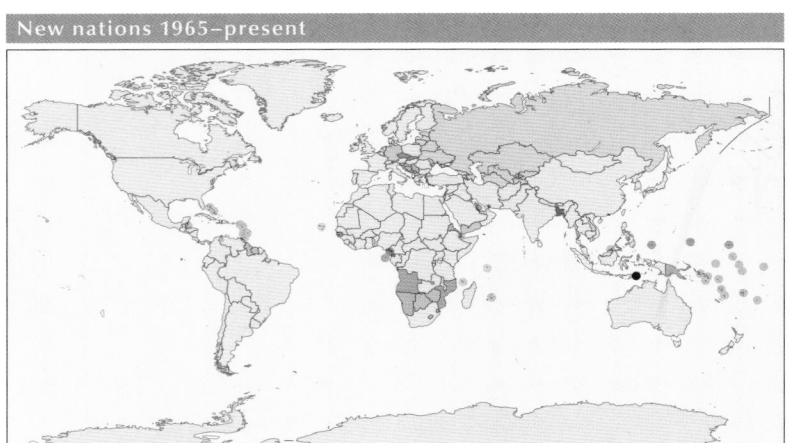

▲ North Korea is an independent communist republic. Power was transferred directly to Kim Jong-un in 2012 following the death of his father Kim Jong-il.

◀ In early 2011, Egypt underwent a revolution, part of the so called "Arab Spring," which resulted in the ousting of President Hosni Mubarak after nearly 30 years in power.

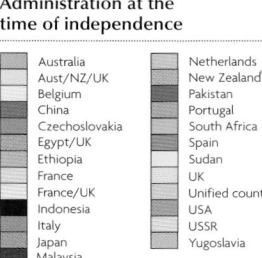

▲ In Brunei the Sultan has ruled by decree since 1962; power is closely tied to the royal family. The Sultan's brothers are responsible for finance and foreign affairs.

Administration at the time of independence

- Australia
- Aust/NZ/UK
- Belgium
- China
- Czechoslovakia
- Egypt/UK
- Ethiopia
- France
- France/UK
- Indonesia
- Italy
- Japan
- Malaysia
- Netherlands
- New Zealand
- Pakistan
- Portugal
- South Africa
- Spain
- Sudan
- UK
- Unified country
- USA
- USSR
- Yugoslavia

Lines on the map

The determination of international boundaries can use a variety of criteria. Many of the borders between older states follow physical boundaries; some mirror religious and ethnic differences; others are the legacy of complex histories of conflict and colonialism, while others have been imposed by international agreements or arbitration.

Post-colonial borders

When the European colonial empires in Africa were dismantled during the second half of the 20th century, the outlines of the new African states mirrored colonial boundaries. These boundaries had been drawn up by colonial administrators, often based on inadequate geographical knowledge. Such arbitrary boundaries were imposed on people of different languages, racial groups, religions, and customs. This confused legacy often led to civil and international war.

▲ The conflict that has plagued many African countries since independence has caused millions of people to become refugees.

Physical borders

Many of the world's countries are divided by physical borders: lakes, rivers, mountains. The demarcation of such boundaries can, however, lead to disputes. Control of waterways, water supplies, and fisheries are frequent causes of international friction.

Enclaves

The shifting political map over the course of history has frequently led to anomalous situations. Parts of national territories may become isolated by territorial agreement, forming an enclave. The West German part of the city of Berlin, which until 1989 lay a hundred miles (160km) within East German territory, was a famous example

Antarctica

When Antarctic exploration began a century ago, seven nations, Australia, Argentina, Britain, Chile, France, New Zealand, and Norway, laid claim to the new territory. In 1961 the Antarctic Treaty, now signed by 45 nations, agreed to hold all territorial claims in abeyance.

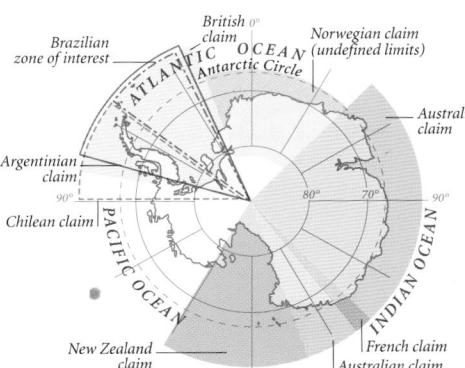

▲ Since the independence of Lithuania and Belarus, the peoples of the Russian enclave of Kaliningrad have become physically isolated.

Geometric borders

Straight lines and lines of longitude and latitude have occasionally been used to determine international boundaries; and indeed the world's second longest continuous international boundary, between Canada and the USA follows the 49th Parallel for over one-third of its course. Many Canadian, American, and Australian internal administrative boundaries are similarly determined using a geometric solution.

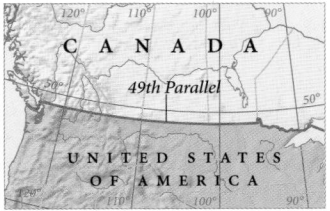

▲ Different farming techniques in Canada and the US clearly mark the course of the international boundary in this satellite map.

World boundaries

Dates from which current boundaries have existed
- 1990–present
- 1966–1989
- 1946–1965
- 1915–1945
- 1850–1914
- 1800–1849
- Pre-1800

Lake borders

Countries which lie next to lakes usually fix their borders in the middle of the lake. Unusually the Lake Nyasa border between Malawi and Tanzania runs along Tanzania's shore.

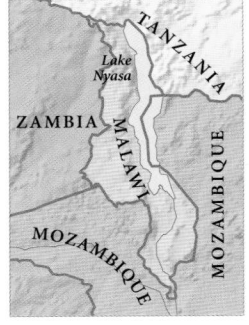

▲ Complicated agreements between colonial powers led to the awkward division of Lake Nyasa.

River borders

Rivers alone account for one-sixth of the world's borders. Many great rivers form boundaries between a number of countries. Changes in a river's course and interruptions of its natural flow can lead to disputes, particularly in areas where water is scarce. The center of the river's course is the nominal boundary line.

The Danube ...ms all or part ... border ... en nine European nations.

Mountain borders

Mountain ranges form natural barriers and are the basis for many major borders, particularly in Europe and Asia. The watershed is the conventional boundary demarcation line, but its accurate determination is often problematic.

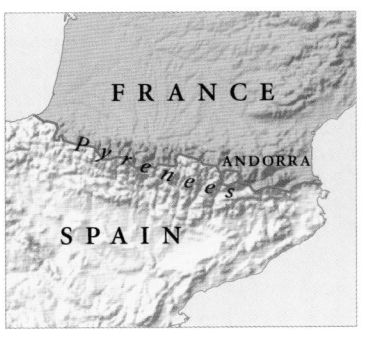

▲ The Pyrenees form a natural mountain border between France and Spain.

Shifting boundaries – Poland

Borders between countries can change dramatically over time. The nations of eastern Europe have been particularly affected by changing boundaries. Poland is an example of a country whose boundaries have changed so significantly that it has literally moved around Europe. At the start of the 16th century, Poland was the largest nation in Europe. Between 1772 and 1795, it was absorbed into Prussia, Austria, and Russia, and it effectively ceased to exist. After the First World War, Poland became an independent country once more, but its borders changed again after the Second World War following invasions by both Soviet Russia and Nazi Germany.

▲ In 1634, Poland was the largest nation in Europe, its eastern boundary reaching toward Moscow.

▲ From 1772–1795, Poland was gradually partitioned between Austria, Russia, and Prussia. Its eastern boundary receded by over 100 miles (160 km).

▲ Following the First World War, Poland was reinstated as an independent state, but it was less than half the size it had been in 1634.

▲ After the Second World War, the Baltic Sea border was extended westward, but much of the eastern territory was annexed by Russia.

International disputes

There are more than 60 disputed borders or territories in the world today. Although many of these disputes can be settled by peaceful negotiation, some areas have become a focus for international conflict. Ethnic tensions have been a major source of territorial disagreement throughout history, as has the ownership of, and access to, valuable natural resources. The turmoil of the postcolonial era in many parts of Africa is partly a result of the 19th century "carve-up" of the continent, which created potential for conflict by drawing often arbitrary lines through linguistic and cultural areas.

Jammu and Kashmir

Disputes over Jammu and Kashmir have caused three serious wars between India and Pakistan since 1947. Pakistan wishes to annex the largely Muslim territory, while India refuses to cede any territory or to hold a referendum, and also lays claim to the entire territory. Most international maps show the "line of control" agreed in 1972 as the *de facto* border. In addition, India has territorial disputes with neighboring China. The situation is further complicated by a Kashmiri independence movement, active since the late 1980s.

▲ *Indian army troops* maintain their positions in the mountainous terrain of northern Kashmir.

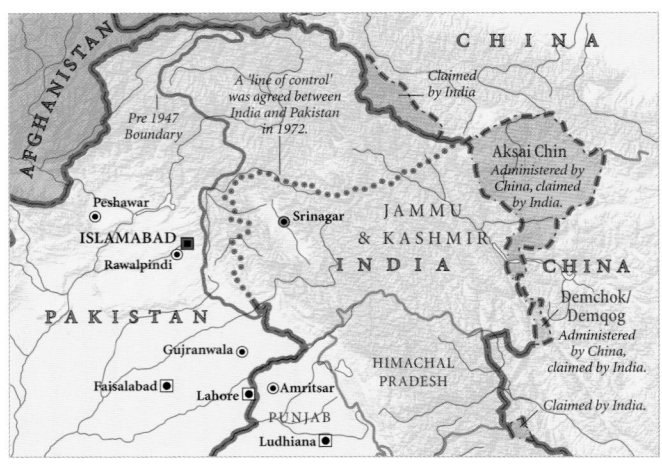

North and South Korea

Since 1953, the *de facto* border between North and South Korea has been a cease-fire line which straddles the 38th Parallel and is designated as a demilitarized zone. Both countries have heavy fortifications and troop concentrations behind this zone.

▲ *Heavy fortifications* on the border between North and South Korea.

Cyprus

Cyprus was partitioned in 1974, following an invasion by Turkish troops. The south is now the Greek Cypriot Republic of Cyprus, while the self-proclaimed Turkish Republic of Northern Cyprus is recognized only by Turkey.

▲ *The so-called "green line"* divides Cyprus into Greek and Turkish sectors.

TURKISH REPUBLIC OF NORTHERN CYPRUS (recognized only by Turkey)

Conflicts and international disputes

- UN peacekeeping missions 2005-2015
- Major active land based territorial or border disputes
- Countries involved in internal conflict
- Active land based territorial or border disputes and internal conflict

The Falkland Islands

The British dependent territory of the Falkland Islands was invaded by Argentina in 1982, sparking a full-scale war with the UK. Tensions ran high during 2012 in the build up to the thirtieth anniversary of the conflict.

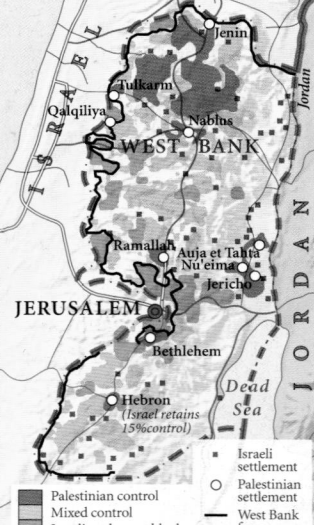

◄ *British warships* in Falkland Sound during the 1982 war with Argentina.

Israel

Israel was created in 1948 following the 1947 UN Resolution (147) on Palestine. Until 1979 Israel had no borders, only cease-fire lines from a series of wars in 1948, 1967, and 1973. Treaties with Egypt in 1979 and Jordan in 1994 led to these borders being defined and agreed. Negotiations over Israeli settlements and Palestinian self-government seen little effective progress since 2000.

- Palestinian control
- Mixed control
- Israeli settlement block
- Israeli settlement
- Palestinian settlement
- West Bank fence

▲ *Barbed-wire fences surround* a settlement in the Golan Heights.

Former Yugoslavia

Following the disintegration in 1991 of the communist state of Yugoslavia, the breakaway states of Croatia and Bosnia and Herzegovina came into conflict with the "parent" state (consisting of Serbia and Montenegro). Warfare focused on ethnic and territorial ambitions in Bosnia. The tenuous Dayton Accord of 1995 sought to recognize the post-1990 borders, whilst providing for ethnic partition and required international peace-keeping troops to maintain the terms of the peace.

- Republika Srpska
- Federacija Bosne i Hercegovine
- Brčko Distrikt

The Spratly Islands

▲ *Most claimant states* have small military garrisons on the Spratly Islands.

The site of potential oil and natural gas reserves, the Spratly Islands in the South China Sea have been claimed by China, Vietnam, Taiwan, Malaysia, and the Philippines since the Japanese gave up a wartime claim in 1951.

- Occupied by Taiwan
- Occupied by Philippines
- Occupied by Malaysia
- Occupied by China
- Occupied by Vietnam

ATLAS
OF THE WORLD

THE MAPS IN THIS ATLAS ARE ARRANGED CONTINENT BY CONTINENT, STARTING

FROM THE INTERNATIONAL DATE LINE, AND MOVING EASTWARD. THE MAPS PROVIDE

A UNIQUE VIEW OF TODAY'S WORLD, COMBINING TRADITIONAL CARTOGRAPHIC

TECHNIQUES WITH THE LATEST REMOTE-SENSED AND DIGITAL TECHNOLOGY.

North America

North America is the world's third largest continent with a total area of 9,358,340 sq miles
(24,238,000 sq km) including Greenland and the Caribbean islands.
It lies wholly within the Northern Hemisphere.

- *Greatest extent, North–South:* 4600 miles / 7400 km
- *Greatest extent, East–West:* 3500 miles / 5700 km

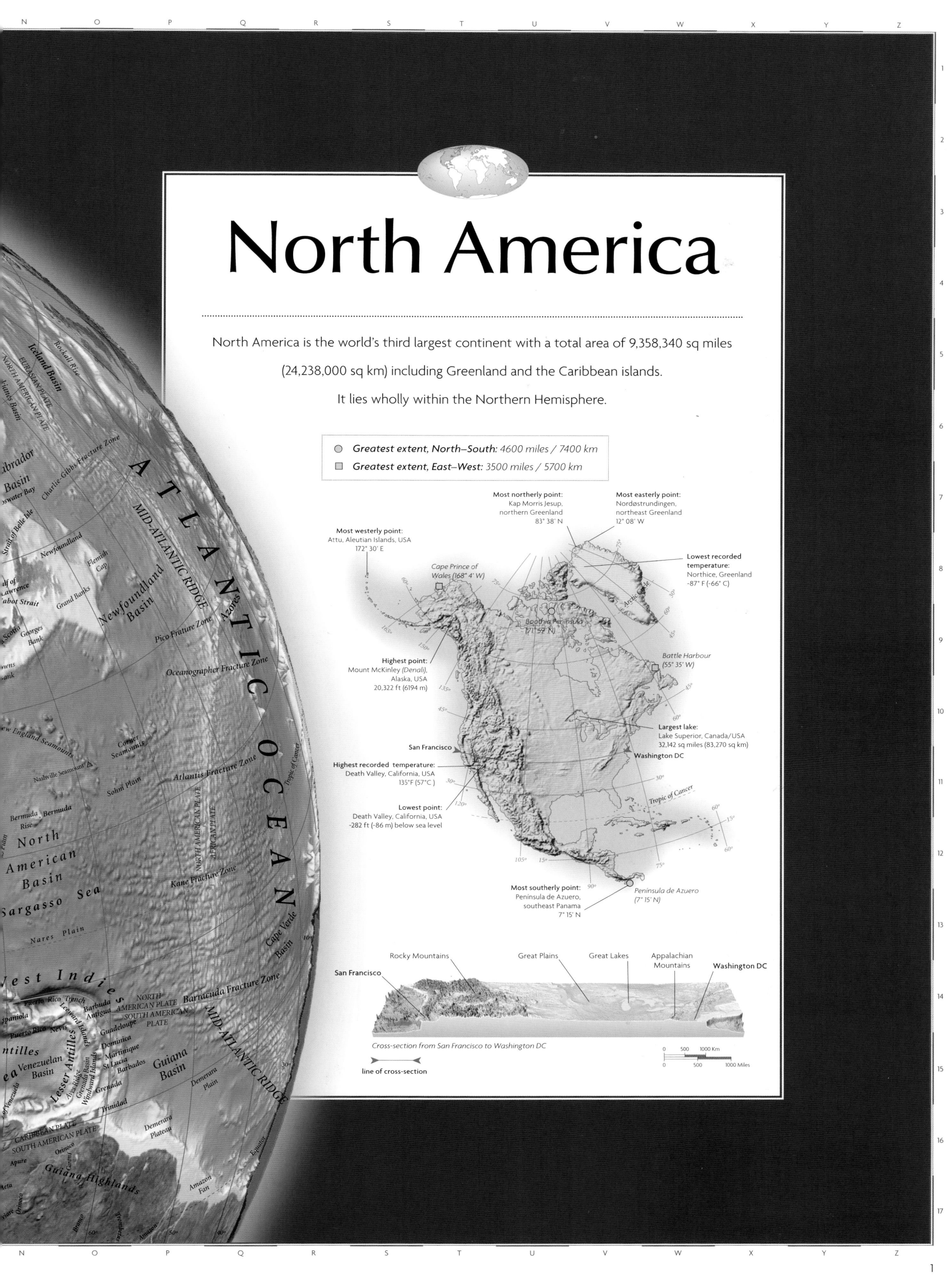

Most northerly point:
Kap Morris Jesup,
northern Greenland
83° 38′ N

Most easterly point:
Nordøstrundingen,
northeast Greenland
12° 08′ W

Most westerly point:
Attu, Aleutian Islands, USA
172° 30′ E

**Lowest recorded
temperature:**
Northice, Greenland
-87° F (-66° C)

*Cape Prince of
Wales* (168° 4′ W)

Boothia Peninsula
(71° 52′ N)

Battle Harbour
(55° 35′ W)

Highest point:
Mount McKinley (Denali),
Alaska, USA
20,322 ft (6194 m)

Largest lake:
Lake Superior, Canada/USA
32,142 sq miles (83,270 sq km)

San Francisco

Washington DC

Highest recorded temperature:
Death Valley, California, USA
135°F (57°C)

Lowest point:
Death Valley, California, USA
-282 ft (-86 m) below sea level

Most southerly point:
Peninsula de Azuero,
southeast Panama
7° 15′ N

Peninsula de Azuero
(7° 15′ N)

Tropic of Cancer

San Francisco   Rocky Mountains   Great Plains   Great Lakes   Appalachian Mountains   Washington DC

Cross-section from San Francisco to Washington DC

line of cross-section

0   500   1000 Km
0   500   1000 Miles

Atlantic Ocean features (western globe):

Iceland Basin

Rockall Rise

EURASIAN PLATE

NORTH AMERICAN PLATE

Reykjanes Ridge

Charlie-Gibbs Fracture Zone

Labrador
Basin

Hudson Bay

Strait of Belle Isle

Newfoundland

Flemish
Cap

MID-ATLANTIC RIDGE

Azores

Gulf of
St Lawrence

Cabot Strait

Grand Banks

Nova Scotia

Georges
Bank

Newfoundland
Basin

Pico Fracture Zone

Oceanographer Fracture Zone

Browns
Bank

New England Seamounts

Corner
Seamounts

Atlantis Fracture Zone

Nashville Seamount

Sohm Plain

Bermuda   Bermuda
Rise

North
American
Basin

NORTH AMERICAN PLATE

AFRICAN PLATE

Kane Fracture Zone

Sargasso Sea

Nares Plain

Cape Verde
Basin

West Indies

MID-ATLANTIC RIDGE

Puerto Rico Trench

Hispaniola

Leeward Islands

Barbuda

Antigua

Guadeloupe

NORTH
AMERICAN PLATE

SOUTH AMERICAN
PLATE

Barracuda Fracture Zone

Puerto Rico

Nevis

Dominica

Martinique

St Lucia

Lesser Antilles

Barbados

Venezuelan
Basin

Grenada
Basin

Windward Islands

Grenada

Trinidad

Guiana
Basin

Demerara
Plain

CARIBBEAN PLATE

SOUTH AMERICAN PLATE

Orinoco

Apure

Demerara
Plateau

Guiana Highlands

Amazon
Fan

Equator

Branco

Amazon

Physical North America

The North American continent can be divided into a number of major structural areas: the Western Cordillera, the Canadian Shield, the Great Plains, and Central Lowlands, and the Appalachians. Other smaller regions include the Gulf Atlantic Coastal Plain which borders the southern coast of North America from the southern Appalachians to the Great Plains. This area includes the expanding Mississippi Delta. A chain of volcanic islands, running in an arc around the margin of the Caribbean Plate, lie to the east of the Gulf of Mexico.

The Canadian Shield

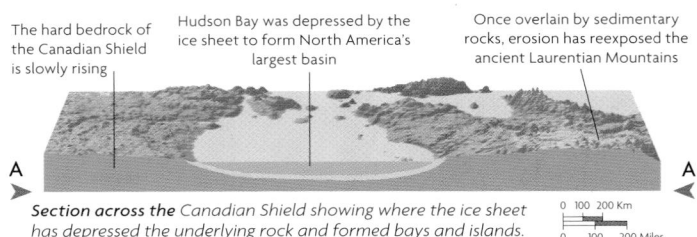

Spanning northern Canada and Greenland, this geologically stable plain forms the heart of the continent, containing rocks more than two billion years old. A long history of weathering and repeated glaciation has scoured the region, leaving flat plains, gentle hummocks, numerous small basins and lakes, and the bays and islands of the Arctic.

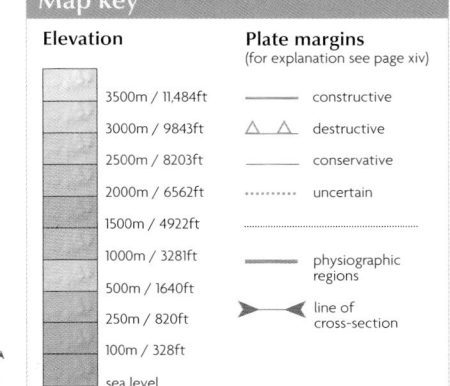

The hard bedrock of the Canadian Shield is slowly rising

Hudson Bay was depressed by the ice sheet to form North America's largest basin

Once overlain by sedimentary rocks, erosion has reexposed the ancient Laurentian Mountains

Section across the Canadian Shield showing where the ice sheet has depressed the underlying rock and formed bays and islands.

0 100 200 Km
0 100 200 Miles

The Western Cordillera

About 80 million years ago the Pacific and North American plates collided, uplifting the Western Cordillera. This consists of the Aleutian, Coast, Cascade, and Sierra Nevada mountains, and the inland Rocky Mountains. These run parallel from the Arctic to Mexico.

The weight of the ice sheet, 1.8 miles (3 km) thick, has depressed the land to 0.6 miles (1 km) below sea level

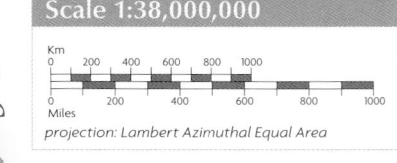

▲ *This computer-generated view* shows the ice-covered island of Greenland without its ice cap.

Strata have been thrust eastward along fault lines

Volcanic rock

The Rocky Mountain Trench is the longest linear fault on the continent

Cross-section through the Western Cordillera showing direction of mountain building.

0 50 100 Km
0 50 100 Miles

Map key

Elevation

3500m / 11,484ft
3000m / 9843ft
2500m / 8203ft
2000m / 6562ft
1500m / 4922ft
1000m / 3281ft
500m / 1640ft
250m / 820ft
100m / 328ft
sea level

Plate margins (for explanation see page xiv)

——— constructive
△ △ △ destructive
——— conservative
·········· uncertain
············
—— physiographic regions
◄——► line of cross-section

Scale 1:38,000,000

Km
0 200 400 600 800 1000
Miles
0 200 400 600 800 1000

projection: Lambert Azimuthal Equal Area

The Appalachians

The Appalachian Mountains, uplifted about 400 million years ago, are some of the oldest in the world. They have been lowered and rounded by erosion and now slope gently toward the Atlantic across a broad coastal plain.

Horizontal strata

Sedimentary strata folded and faulted into ridges and valleys

Softer strata has been crumpled against the harder basement rock

Hard basement rock

Cross-section through the Appalachians showing the numerous folds, which have subsequently been weathered to create a rounded relief.

0 25 50 Km
0 25 50 Miles

The Great Plains & Central Lowlands

Deposits left by retreating glaciers and rivers have made this vast flat area very fertile. In the north this is the result of glaciation, with deposits up to one mile (1.7 km) thick, covering the basement rock. To the south and west, the massive Missouri/Mississippi river system has for centuries deposited silt across the plains, creating broad, flat floodplains and deltas.

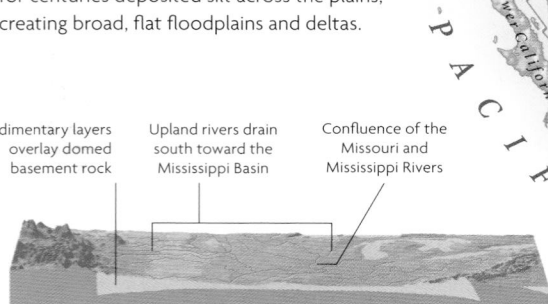

Sedimentary layers overlay basement domed rock

Upland rivers drain south toward the Mississippi Basin

Confluence of the Missouri and Mississippi Rivers

Section across the Great Plains and Central Lowlands showing river systems and structure.

0 200 400 Km
0 200 400 Miles

Map labels

ASIA
Bering Strait
Beaufort Sea
Greenland
ATLANTIC OCEAN
Baffin Bay
Baffin Island
Davis Strait
Aleutian Islands
Bering Sea
Brooks Range
Mount McKinley 6194m
Mackenzie Delta
Aleutian Range
Alaska Range
Mackenzie Mountains
Mackenzie
Great Bear Lake
Foxe Basin
Hudson Strait
Labrador Sea
Labrador
Gulf of Alaska
NORTH AMERICAN PLATE
PACIFIC PLATE
Coast Mountains
WESTERN
Great Slave Lake
Lake Athabasca
Hudson Bay
Newfoundland
Reindeer Lake
Laurentian Mountains
ROCKY MOUNTAINS
CANADIAN SHIELD
JUAN DE FUCA PLATE
CENTRAL LOWLANDS
CORDILLERA
Lake Winnipeg
Lake Manitoba
Lake Superior
St Lawrence
Nova Scotia
Cape Cod
Mount Rainier 4392m
Cascade Range
Mount St Helens 2549m
Missouri
GREAT PLAINS
Lake Huron
Lake Michigan
Lake Ontario
Great Lakes
Lake Erie
San Andreas Fault
Great Basin
Great Salt Lake
Colorado
Colorado Plateau
Ohio
APPALACHIAN MOUNTAINS
Sierra Nevada
San Joaquin Valley
Death Valley -86m
Mojave Desert
Grand Canyon
Arkansas
Mississippi
APPALACHIANS
Sonoran Desert
Lower California
GULF ATLANTIC COASTAL PLAIN
West Indies
Gulf of California
Sierra Madre Occidental
Rio Grande
Mississippi Delta
Gulf of Mexico
Greater Antilles
Lesser Antilles
PACIFIC OCEAN
Sierra Madre Oriental
Volcán Pico de Orizaba 5700m
Yucatan Peninsula
NORTH AMERICAN PLATE
CARIBBEAN PLATE
Caribbean Sea
Sierra Madre del Sur
Lake Nicaragua
Isthmus of Panama
SOUTH AMERICAN PLATE
SOUTH AMERICA

Climate

North America's climate includes extremes ranging from freezing Arctic conditions in Alaska and Greenland, to desert in the southwest, and tropical conditions in southeastern Florida, the Caribbean, and Central America. Central and southern regions are prone to severe storms including tornadoes and hurricanes.

▲ *"Tornado alley"* in the Mississippi Valley suffers frequent tornadoes.

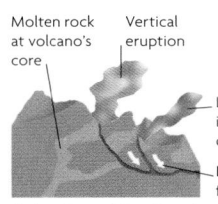

▲ *Much of the* southwest is semi-desert; receiving less than 12 inches (300 mm) of rainfall a year.

Climate

	ice cap
	tundra
	subarctic
	cool continental
	warm humid
	semiarid
	arid
	humid equatorial
	tropical

☼ daily hours of sunshine, January
☼ daily hours of sunshine, July
→ direction of hurricanes
✲ tornado zones

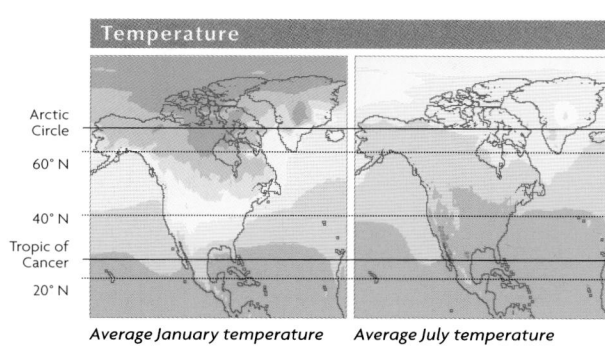

Temperature

Average January temperature *Average July temperature*

Temperature

	-22°F (below -30°C)
	-22 to -4°F (-30 to -20°C)
	-4 to 14°F (-20 to -10°C)
	14 to 32°F (-10 to 0°C)
	32 to 50°F (0 to 10°C)
	50 to 68°F (10 to 20°C)
	68 to 86°F (20 to 30°C)
	86°F (above 30°C)

Rainfall

Average January rainfall *Average July rainfall*

	0–1 in (0–25 mm)
	1–2 in (25–50 mm)
	2–4 in (50–100 mm)
	4–8 in (100–200 mm)
	8–12 in (200–300 mm)
	12–16 in (300–400 mm)
	16–20 in (400–500 mm)
	more than 20 in (500 mm)

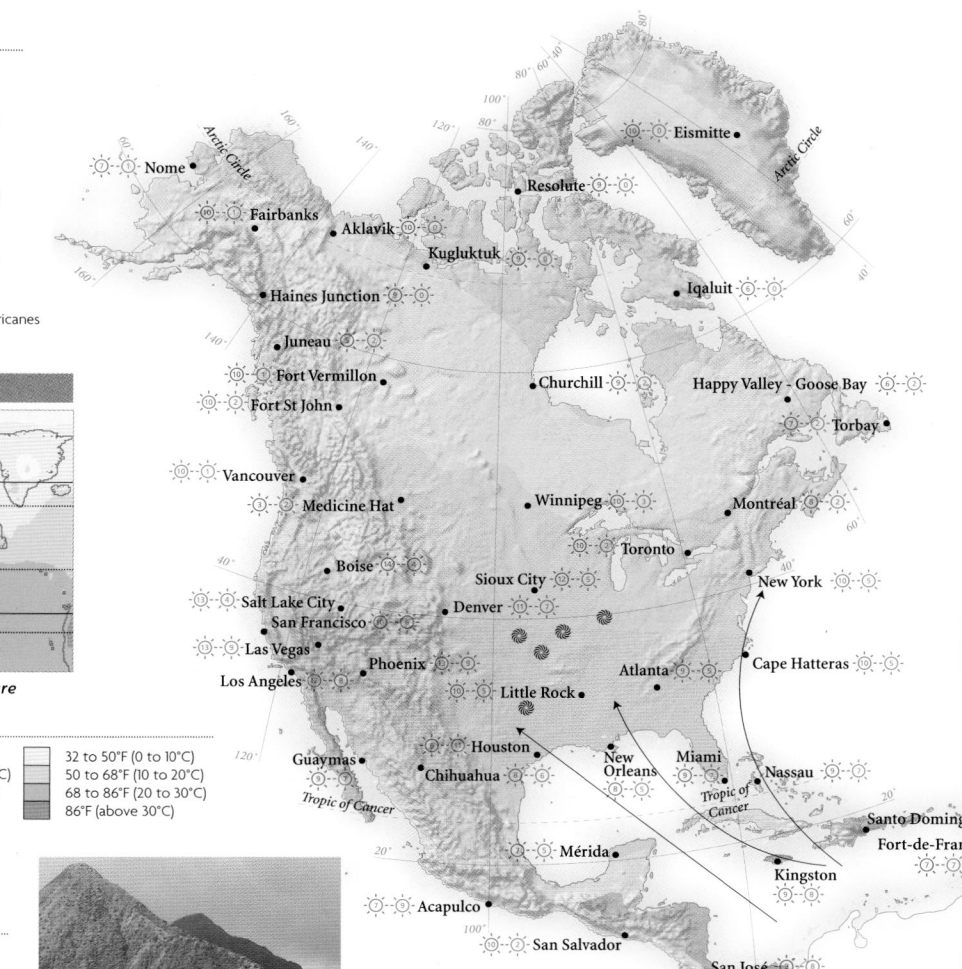

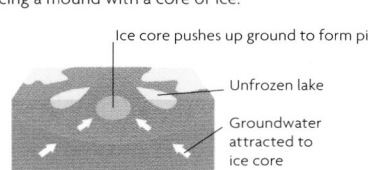

◄ *The lush, green* mountains of the Lesser Antilles receive annual rainfalls of up to 360 inches (9000 mm).

Shaping the continent

Glacial processes affect much of northern Canada, Greenland, and the Western Cordillera. Along the western coast of North America, Central America, and the Caribbean, underlying plates moving together lead to earthquakes and volcanic eruptions. The vast river systems, fed by mountain streams, constantly erode and deposit material along their paths.

Volcanic activity

1 Mount St. Helens volcano *(right)* in the Cascade Range erupted violently in May 1980, killing 57 people and leveling large areas of forest. The lateral blast filled a valley with debris for 15 miles (25 km).

Molten rock at volcano's core
Vertical eruption
Lateral explosion increases extent of damage
Landslide fills valley

Volcanic activity: Eruption of Mount St Helens

Seismic activity

5 The San Andreas Fault *(above)* places much of the North America's west coast under constant threat from earthquakes. It is caused by the Pacific Plate grinding past the North American Plate at a faster rate, though in the same direction.

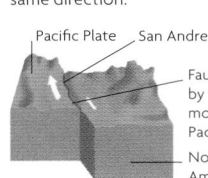

Pacific Plate San Andreas Fault
Fault is caused by faster movement of Pacific Plate
North American Plate

Seismic activity: Action of the San Andreas Fault

River erosion

6 The Grand Canyon *(above)* in the Colorado Plateau was created by the downward erosion of the Colorado River, combined with the gradual uplift of the plateau, over the past 30 million years. The contours of the canyon formed as the softer rock layers eroded into gentle slopes, and the hard rock layers into cliffs. The depth varies from 3855–6560 ft (1175–2000 m).

Soft rock is easily eroded into gentle slopes
Hard rock resists erosion
Colorado River cuts down through rock

River Erosion: Formation of the Grand Canyon

Periglaciation

2 The ground in the far north is nearly always frozen: the surface thaws only in summer. This freeze-thaw process produces features such as pingos *(left)*; formed by the freezing of groundwater. With each successive winter ice accumulates producing a mound with a core of ice.

Ice core pushes up ground to form pingo
Unfrozen lake
Groundwater attracted to ice core

Periglaciation: Formation of a pingo in the Mackenzie Delta

The evolving landscape

Landscape

	limestone region
	sinking land
	stable land
	uplifting land

▲ active volcano
⋯ area of tectonic activity
--- limit of permafrost
— maximum limit of glaciation
→ ocean current

Post-glacial lakes

3 A chain of lakes from Great Bear Lake to the Great Lakes *(above)* was created as the ice retreated northward. Glaciers scoured hollows in the softer lowland rock. Glacial deposits at the lip of the hollows, and ridges of harder rock, trapped water to form lakes.

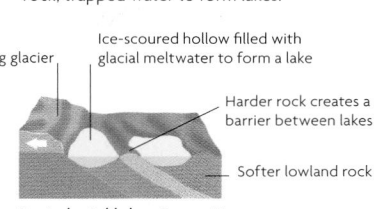

Retreating glacier
Ice-scoured hollow filled with glacial meltwater to form a lake
Harder rock creates a barrier between lakes
Softer lowland rock

Post-glacial lakes: Formation of the Great Lakes

Weathering

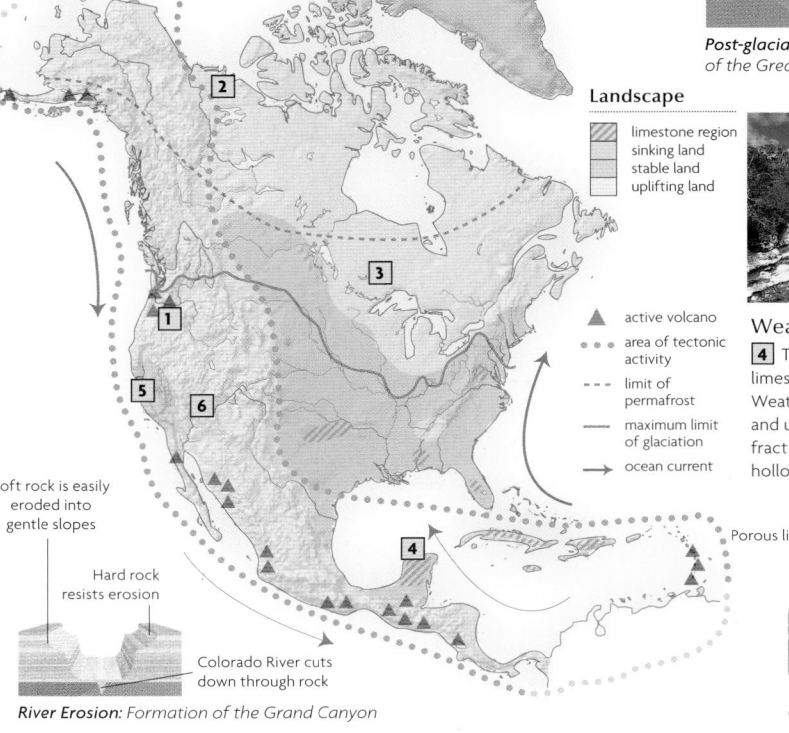

4 The Yucatan Peninsula is a vast, flat limestone plateau in southern Mexico. Weathering action from both rainwater and underground streams has enlarged fractures in the rock to form caves and hollows, called sinkholes *(above)*.

Porous limestone plateau
Rainwater erodes porous rock forming sinkholes
Sea level
Underground stream further erodes rock

Weathering: Water erosion on the Yucatan Peninsula

Eismitte
Nome
Resolute
Fairbanks
Aklavik
Kugluktuk
Haines Junction
Iqaluit
Juneau
Fort Vermilion
Fort St John
Churchill
Happy Valley - Goose Bay
Torbay
Vancouver
Medicine Hat
Winnipeg
Montréal
Boise
Toronto
Sioux City
New York
Salt Lake City
Denver
San Francisco
Atlanta
Cape Hatteras
Las Vegas
Phoenix
Los Angeles
Little Rock
Houston
Guaymas
New Orleans
Miami
Nassau
Chihuahua
Santo Domingo
Fort-de-France
Mérida
Kingston
Acapulco
San Salvador
San José

Arctic Circle Tropic of Cancer

Political North America

Democracy is well established in some parts of the continent but is a recent phenomenon in others. The economically dominant nations of Canada and the US have a long democratic tradition but elsewhere, notably in the countries of Central America, political turmoil has been more common. In Nicaragua and Haiti, harsh dictatorships have only recently been superseded by democratically elected governments. North America's largest countries, Canada, Mexico, and the US have federal state systems, sharing political power between national and state governments. The US has intervened militarily on several occasions in Central America and the Caribbean to protect its strategic interests.

Transportation

In the 19th century, railroads opened up the North American continent. Air transportation is now more common for long distance passenger travel, although railroads are still extensively used for bulk freight transportation. Waterways like the Mississippi River are important for the transportation of bulk materials, and the Panama Canal is a vital link between the Pacific and Atlantic Oceans. In the 20th century, road transportation increased massively, with the introduction of cheap, mass-produced motor cars and extensive highway construction.

◄ *This busy suburban* interchange in Los Angeles is part of the US's Interstate freeway system. Construction of the 55,000 mile (88,500 km) freeway network began in the 1950s, and it now connects most major cities, and carries one-fifth of the US's road traffic.

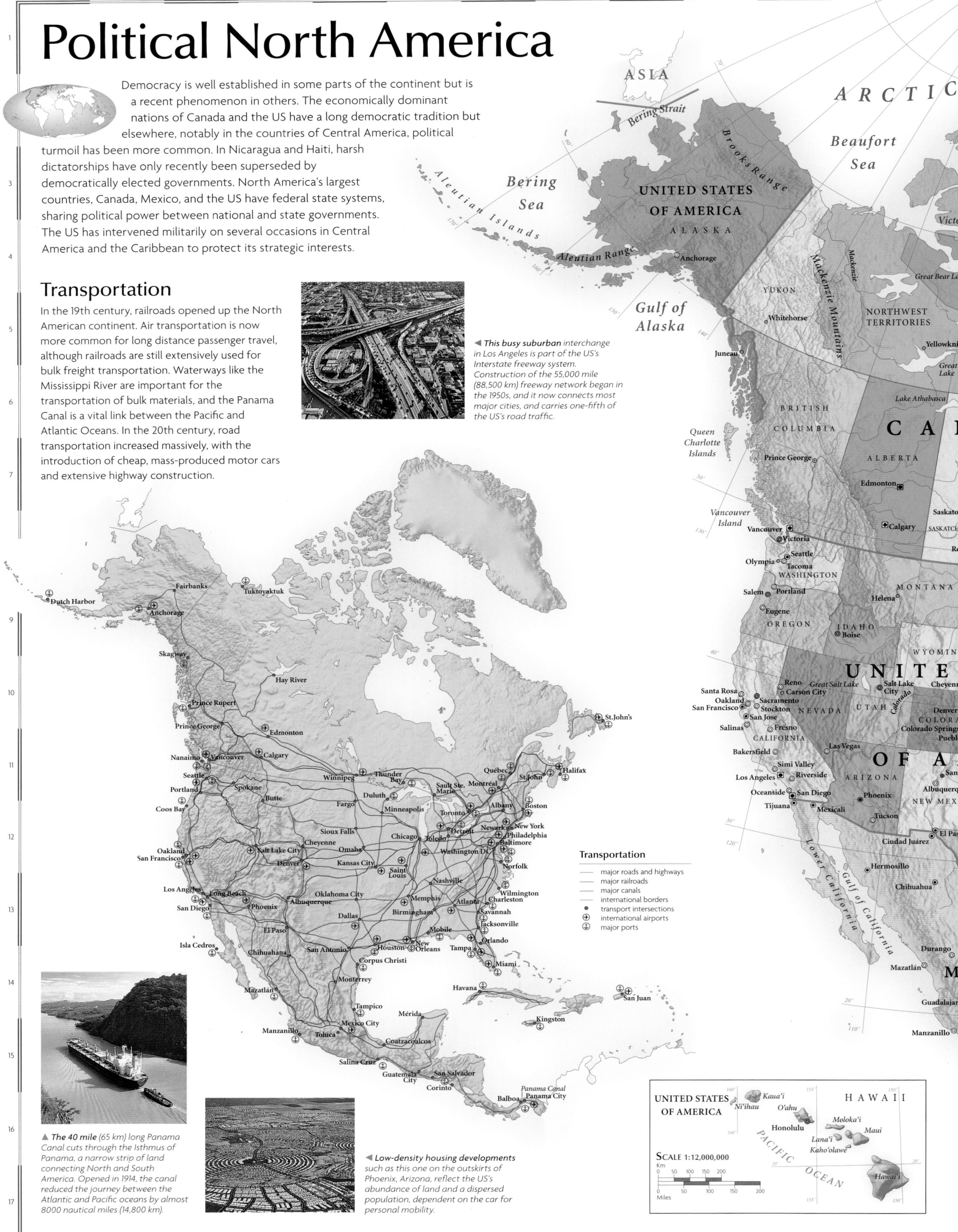

Transportation
— major roads and highways
— major railroads
— major canals
—— international borders
• transport intersections
⊕ international airports
⊕ major ports

▲ *The 40 mile* (65 km) long Panama Canal cuts through the Isthmus of Panama, a narrow strip of land connecting North and South America. Opened in 1914, the canal reduced the journey between the Atlantic and Pacific oceans by almost 8000 nautical miles (14,800 km).

◄ *Low-density housing developments* such as this one on the outskirts of Phoenix, Arizona, reflect the US's abundance of land and a dispersed population, dependent on the car for personal mobility.

UNITED STATES OF AMERICA

SCALE 1:12,000,000

HAWAI'I

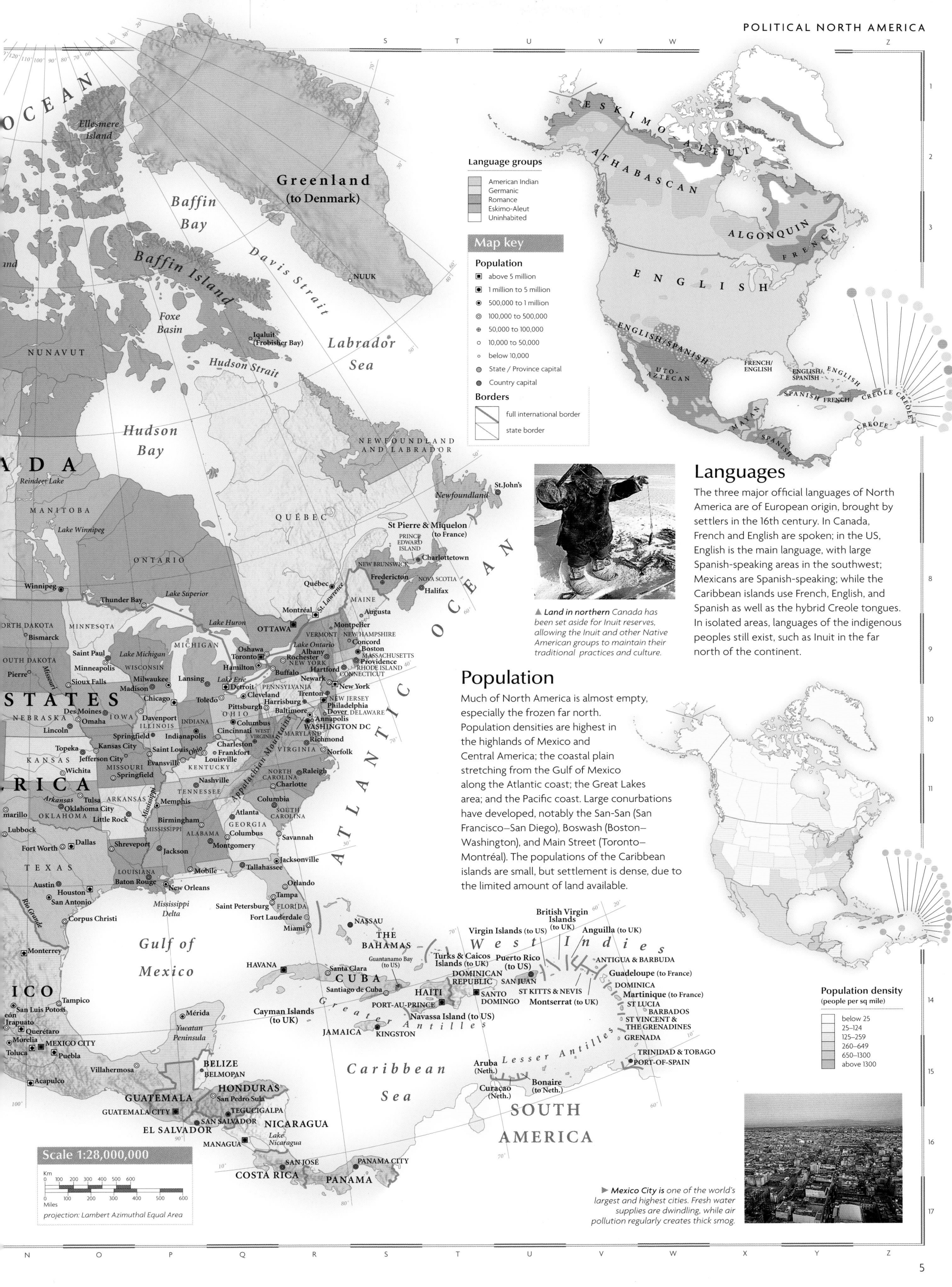

OCEAN

Ellesmere
Island

*Baffin
Bay*

Greenland
(to Denmark)

Davis Strait

Baffin Island

*Foxe
Basin*

NUUK

NUNAVUT

Iqaluit
(Frobisher Bay)

*Labrador
Sea*

Hudson Strait

*Hudson
Bay*

Reindeer Lake

MANITOBA

Lake Winnipeg

ONTARIO

Lake Superior

Winnipeg

Thunder Bay

Lake Huron

Newfoundland

St.John's

NEWFOUNDLAND
AND LABRADOR

QUÉBEC

St Pierre & Miquelon
(to France)

PRINCE
EDWARD
ISLAND

Charlottetown

NEW BRUNSWICK

NOVA SCOTIA

Québec

Fredericton

Halifax

St.Lawrence

MAINE

Augusta

ATLANTIC OCEAN

NORTH DAKOTA

MINNESOTA

Bismarck

SOUTH DAKOTA

Pierre

Saint Paul

Minneapolis

Sioux Falls

WISCONSIN

Madison

Lake Michigan

MICHIGAN

Lansing

Milwaukee

Chicago

Lake Ontario

VERMONT

NEW HAMPSHIRE

Concord

Montpelier

Boston

MASSACHUSETTS

Providence

RHODE ISLAND

CONNECTICUT

Hartford

Albany

Rochester

NEW YORK

Buffalo

Newark

New York

Oshawa

Toronto

Hamilton

OTTAWA

Montréal

Detroit

Lake Erie

Cleveland

Toledo

STATES

NEBRASKA

Lincoln

Omaha

IOWA

Des Moines

Davenport

ILLINOIS

Springfield

INDIANA

Indianapolis

Cincinnati

OHIO

Columbus

Pittsburgh

PENNSYLVANIA

Harrisburg

Philadelphia

Trenton

NEW JERSEY

Dover

DELAWARE

Baltimore

Annapolis

WASHINGTON DC

MARYLAND

WEST
VIRGINIA

Richmond

Frankfort

Louisville

KENTUCKY

VIRGINIA

Norfolk

Appalachian Mountains

RICA

KANSAS

Topeka

Kansas City

Jefferson City

Saint Louis

MISSOURI

Wichita

Springfield

Evansville

Charleston

Arkansas

Tulsa

Oklahoma City

OKLAHOMA

ARKANSAS

Little Rock

Memphis

TENNESSEE

Nashville

NORTH
CAROLINA

Charlotte

Raleigh

Columbia

SOUTH
CAROLINA

Mississippi

arillo

Lubbock

Fort Worth

Dallas

TEXAS

Austin

Houston

San Antonio

Corpus Christi

Rio Grande

LOUISIANA

Shreveport

Jackson

MISSISSIPPI

ALABAMA

Birmingham

Montgomery

GEORGIA

Columbus

Atlanta

Savannah

Baton Rouge

New Orleans

Mobile

Tallahassee

Jacksonville

Orlando

FLORIDA

Tampa

Saint Petersburg

Fort Lauderdale

Miami

*Mississippi
Delta*

*Gulf of
Mexico*

ICO

Tampico

San Luis Potosí

eón

Irapuato

Querétaro

Mérida

*Yucatan
Peninsula*

MEXICO CITY

Morelia

Toluca

Puebla

Acapulco

Villahermosa

BELIZE

BELMOPAN

GUATEMALA

GUATEMALA CITY

HONDURAS

San Pedro Sula

TEGUCIGALPA

SAN SALVADOR

EL SALVADOR

NICARAGUA

MANAGUA

*Lake
Nicaragua*

SAN JOSÉ

COSTA RICA

PANAMA CITY

PANAMA

Monterrey

NASSAU

THE
BAHAMAS

HAVANA

Santa Clara

CUBA

Santiago de Cuba

Cayman Islands
(to UK)

JAMAICA

KINGSTON

Guantanamo Bay
(to US)

Greater Antilles

HAITI

PORT-AU-PRINCE

Navassa Island (to US)

DOMINICAN
REPUBLIC

SAN JUAN

SANTO
DOMINGO

West Indies

British Virgin
Islands
(to UK)

Virgin Islands
(to US)

Anguilla
(to UK)

Turks & Caicos
Islands (to UK)

Puerto Rico
(to US)

ANTIGUA & BARBUDA

ST KITTS & NEVIS

Montserrat (to UK)

Guadeloupe (to France)

DOMINICA

Martinique (to France)

ST LUCIA

BARBADOS

ST VINCENT &
THE GRENADINES

GRENADA

Lesser Antilles

TRINIDAD & TOBAGO

PORT-OF-SPAIN

Aruba
(Neth.)

Curaçao
(Neth.)

Bonaire
(to Neth.)

*Caribbean
Sea*

SOUTH
AMERICA

Language groups

- American Indian
- Germanic
- Romance
- Eskimo-Aleut
- Uninhabited

Map key

Population

- ◼ above 5 million
- ◾ 1 million to 5 million
- ◉ 500,000 to 1 million
- ◎ 100,000 to 500,000
- ⊕ 50,000 to 100,000
- ⊙ 10,000 to 50,000
- ○ below 10,000
- ◎ State / Province capital
- ● Country capital

Borders

- full international border
- state border

ESKIMO-ALEUT

ATHABASCAN

ALGONQUIN

FRENCH

ENGLISH

ENGLISH/SPANISH

UTO-
AZTECAN

FRENCH/
ENGLISH

ENGLISH/
SPANISH

ENGLISH

SPANISH FRENCH

CREOLE

CREOLE

CREOLE

MAYAN

SPANISH

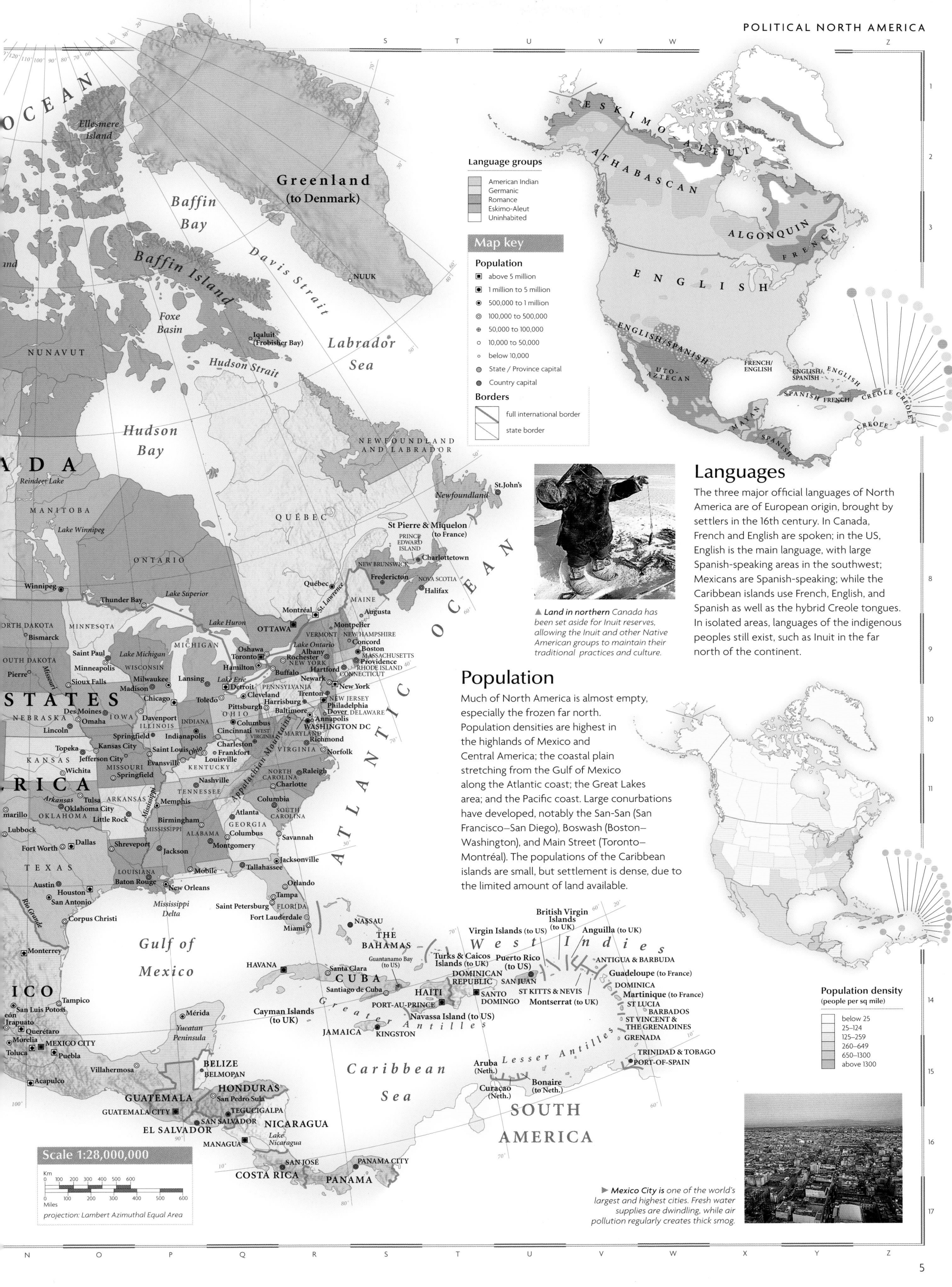

▲ Land in northern *Canada has
been set aside for Inuit reserves,
allowing the Inuit and other Native
American groups to maintain their
traditional practices and culture.*

Languages

The three major official languages of North
America are of European origin, brought by
settlers in the 16th century. In Canada,
French and English are spoken; in the US,
English is the main language, with large
Spanish-speaking areas in the southwest;
Mexicans are Spanish-speaking; while the
Caribbean islands use French, English, and
Spanish as well as the hybrid Creole tongues.
In isolated areas, languages of the indigenous
peoples still exist, such as Inuit in the far
north of the continent.

Population

Much of North America is almost empty,
especially the frozen far north.
Population densities are highest in
the highlands of Mexico and
Central America; the coastal plain
stretching from the Gulf of Mexico
along the Atlantic coast; the Great Lakes
area; and the Pacific coast. Large conurbations
have developed, notably the San-San (San
Francisco–San Diego), Boswash (Boston–
Washington), and Main Street (Toronto–
Montréal). The populations of the Caribbean
islands are small, but settlement is dense, due to
the limited amount of land available.

Population density
(people per sq mile)

- below 25
- 25–124
- 125–259
- 260–649
- 650–1300
- above 1300

▶ Mexico City is *one of the world's
largest and highest cities. Fresh water
supplies are dwindling, while air
pollution regularly creates thick smog.*

A   B   C   D   E   F   G   H   I   J   K   L   M

North American resources

The two northern countries of Canada and the US are richly endowed with natural resources that have helped to fuel economic development. The US is the world's largest economy, although today it is facing stiff competition from the Far East. Mexico has relied on oil revenues but there are hopes that the North American Free Trade Agreement (NAFTA), will encourage trade growth with Canada and the US. The poorer countries of Central America and the Caribbean depend largely on cash crops and tourism.

Industry

The modern, industrialized economies of the US and Canada contrast sharply with those of Mexico, Central America, and the Caribbean. Manufacturing is especially important in the US; vehicle production is concentrated around the Great Lakes, while electronic and hi-tech industries are increasingly found in the western and southern states. Mexico depends on oil exports and assembly work, taking advantage of cheap labor. Many Central American and Caribbean countries rely heavily on agricultural exports.

◄ *After its purchase* from Russia in 1867, Alaska's frozen lands were largely ignored by the US. Oil reserves similar in magnitude to those in eastern Texas were discovered in Prudhoe Bay, Alaska in 1968. Freezing temperatures and a fragile environment hamper oil extraction.

Standard of living

The US and Canada have one of the highest overall standards of living in the world. However, many people still live in poverty, especially in urban ghettos and some rural areas. Central America and the Caribbean are markedly poorer than their wealthier northern neighbors. Haiti is the poorest country in the western hemisphere.

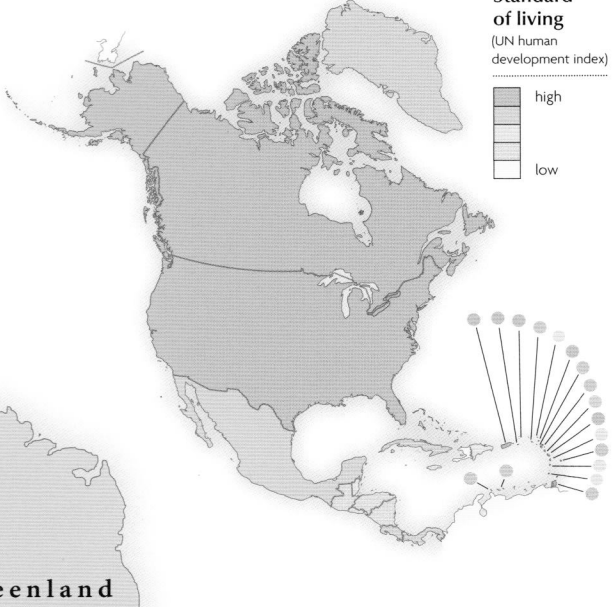

Standard of living
(UN human development index)

high

low

▲ *South of San Francisco,* "Silicon Valley" is both a national and international center for hi-tech industries, electronic industries, and research institutions.

▲ *Multinational companies rely* on cheap labor and tax benefits to facilitate the assembly of vehicle parts in Mexican factories.

▲ *Fish such as* cod, flounder, and plaice are caught in the Grand Banks, off the Newfoundland coast, and processed in many North Atlantic coastal settlements.

▲ *The health of* the Wall Street stock market in New York is the standard measure of the state of the world's economy.

ARCTIC OCEAN

RUSS. FED.

Bering Strait

Bering Sea

Beaufort Sea

Baffin Bay

Greenland (to Denmark)

Prudhoe Bay

USA

Gulf of Alaska

Hudson Strait

Labrador Sea

Hudson Bay

CANADA

Vancouver   Calgary   Winnipeg

Seattle

Portland

Montréal

Minneapolis   Toronto   Buffalo   Albany   Boston

Milwaukee   Detroit   Cleveland   New York

UNITED STATES   Chicago   Pittsburgh   Baltimore   Philadelphia

OF AMERICA   Dayton   Cincinnati

San Francisco   Denver   Kansas City   Saint Louis   Greensboro

Wichita   Nashville   Charlotte

Los Angeles   Tulsa

San Diego   Phoenix   Birmingham   Atlanta

Tijuana   Dallas

Ciudad Juárez   El Paso   Jacksonville

Houston   New Orleans   Orlando

Tampa

Monterrey   Miami

Gulf of Mexico   THE BAHAMAS

West Indies

Havana   Virgin Islands (to US)   British Virgin Islands (to UK)   Anguilla (to UK)

Turks & Caicos Islands (to UK)   ST KITTS & NEVIS

Puerto Rico (to US)   ANTIGUA & BARBUDA

CUBA   San Juan   Montserrat (to UK)   Guadeloupe (to France)

Guadalajara   DOMINICAN REPUBLIC   DOMINICA

HAITI   Santo Domingo   ST LUCIA   BARBADOS

MEXICO   Port-au-Prince   ST VINCENT & THE GRENADINES

Cayman Islands (to UK)   JAMAICA   Greater Antilles   GRENADA

Navassa Island (to US)   Lesser Antilles   TRINIDAD & TOBAGO

Mexico City   Aruba (Neth.)   Port-of-Spain

BELIZE   Caribbean   Curaçao (Neth.)   Bonaire (to Neth.)

GUATEMALA   HONDURAS   Sea   VENEZUELA

Guatemala City   Tegucigalpa

EL SALVADOR   San Salvador   NICARAGUA   COLOMBIA

Managua

San José

COSTA RICA   Panama City

PANAMA

PACIFIC OCEAN

ATLANTIC OCEAN

Industry

✈ aerospace	⬛ printing & publishing
♦ brewing	⚙ research & development
🚗 car/vehicle manufacture	⚓ shipbuilding
⚗ chemicals	⬇ sugar processing
⚙ defense	✂ textiles
🖥 electronics	🌲 timber processing
⚙ engineering	🚬 tobacco processing
🎬 film industry	
S finance	♦ coal
🍴 food processing	♦ oil
💻 hi-tech industry	♦ gas
⚒ iron & steel	• industrial cities
⚕ pharmaceuticals	▨ major industrial areas

GNI per capita (US$)

below 1999
2000–4999
5000–9999
10,000–19,999
20,000–24,999
above 25,000

Environmental issues

Many fragile environments are under threat throughout the region. In Haiti, all the primary rain forest has been destroyed, while air pollution from factories and cars in Mexico City is among the worst in the world. Elsewhere, industry and mining pose threats, particularly in the delicate arctic environment of Alaska where oil spills have polluted coastlines and decimated fish stocks.

Environmental issues
- national parks
- risk of acid rain
- tropical forest
- forest destroyed
- desert
- risk of desertification
- polluted rivers
- radioactive contamination
- marine pollution
- heavy marine pollution
- poor urban air quality

▲ Wild bison graze in Yellowstone National Park, the world's first national park. Designated in 1872, geothermal springs and boiling mud are among its natural spectacles, making it a major tourist attraction.

Mineral resources

Fossil fuels are exploited in considerable quantities throughout the continent. Coal mining in the Appalachians is declining but vast open pits exist further west in Wyoming. Oil and natural gas are found in Alaska, Texas, the Gulf of Mexico, and the Canadian West. Canada has large quantities of nickel, while Jamaica has considerable deposits of bauxite, and Mexico has large reserves of silver.

Mineral resources
- oil field
- gas field
- coal field
- bauxite
- copper
- gold
- iron
- lead
- nickel
- phosphates
- silver
- uranium

▲ In addition to fossil fuels, North America is also rich in exploitable metallic ores. This vast, mile-deep (1.6 km) pit is a copper mine in New Mexico.

▲ In agriculturally marginal areas where the soil is either too poor, or the climate too dry for crops, cattle ranching proliferates – especially in Mexico and the western reaches of the Great Plains.

Using the land and sea

Abundant land and fertile soils stretch from the Canadian prairies to Texas creating North America's agricultural heartland. Cereals and cattle ranching form the basis of the farming economy, with corn and soybeans also important. Fruit and vegetables are grown in California using irrigation, while Florida is a leading producer of citrus fruits. Caribbean and Central American countries depend on cash crops such as bananas, coffee, and sugar cane, often grown on large plantations. This reliance on a single crop can leave these countries vulnerable to fluctuating world crop prices.

Using the land and sea
- cropland
- forest
- ice cap
- mountain region
- pasture
- tundra
- wetland
- desert
- major conurbations
- cattle
- goats
- pigs
- poultry
- reindeer
- sheep
- bananas
- citrus fruits
- coffee
- corn
- cotton
- fishing
- fruit
- maple syrup
- peanuts
- rice
- shellfish
- soybeans
- sugar cane
- timber
- tobacco
- vineyards
- wheat

◄ Sugar cane is Cuba's main agricultural crop, and is grown and processed throughout the Caribbean. Fermented sugar is used to make rum.

◄ The Great Plains support large-scale arable farming throughout central North America. Corn is grown in a belt south and west of the Great Lakes, while farther west where the climate is drier, wheat is grown.

7

Canada

Canada is the second largest country in the world, and with only about one-tenth of its land area inhabited, it is one of the most sparsely populated. Canada became a confederation in 1867, though Newfoundland did not join until 1949. As a founding member of the UN and of the Commonwealth, Canada has played an important role in international affairs. A constitutional crisis, focusing on the French-speaking Québécois, and Inuit, and Native American land rights, dominated politics in the 1990s. In 1999, part of the Northwest Territories, Nunavut, became a self-governing homeland for the Inuit.

◄ *The Selwyn Mountains* in northwestern Canada form part of the Rocky Mountains. The highest point, Keele Peak, rises to 9750 ft (2972 m).

Transportation and industry

Abundant energy in the form of coal, oil, natural gas, and hydroelectric power underpins Canadian industry. Over 75% of manufacturing is concentrated in the Great Lakes–St. Lawrence region, including prospering aerospace, transportation, and hi-tech industries. Across Canada as a whole, manufacturing has developed around a diversified, high-quality resource base and a wide range of metallic and nonmetallic minerals.

◄ *Canada has one* of the world's highest rates of energy consumption per person. It is endowed with vast hydroelectric potential from which more than 60% of its electricity requirements are generated.

Major industry and infrastructure

- aerospace
- car manufacture
- chemicals
- engineering
- food processing
- hi-tech industry
- hydroelectric power
- oil & gas
- mining
- timber processing
- capital cities
- major towns
- international airports
- major roads
- major industrial areas

Transportation network

309,019 miles (497,375 km)	10,500 miles (16,900 km)
8049 miles (12,995 km)	1864 miles (3000 km)

In recent years the road network has been expanded, especially links to remote areas. Meanwhile, for long-distance travel, air transportation now supersedes the declining rail network, which focuses mainly on east–west routes.

Using the land and sea

The majority of Canada's agricultural land is found in the prairies, which cover 140 million acres (57 million ha) and support wheat and grain-fed cattle. More specialized crops, such as fruit and vegetables, are grown in pockets of agricultural land in the east and west. Of Canada's many islands, only Prince Edward Island has notable farmland. Further north, boreal forests, exploited for timber, run in an almost unbroken arc, giving way to uncultivable tundra and ice sheets in the far north.

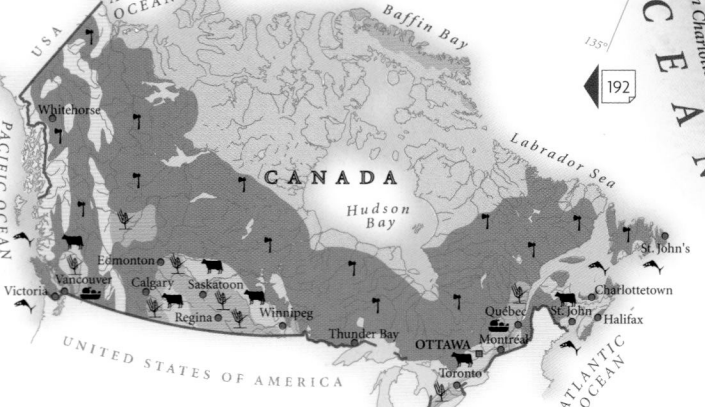

The urban/rural population divide

urban 77%   rural 23%

0 10 20 30 40 50 60 70 80 90 100

Population density	Total land area
9 people per sq mile (3 people per sq km)	3,559,294 sq miles (9,220,970 sq km)

Land use and agricultural distribution

- cattle
- cereals
- fishing
- fruit
- timber
- capital cities
- major towns

- pasture
- cropland
- forest
- wetland
- mountain region
- barren
- tundra

◄ *The climate and* topography of the prairies makes them ideally suited to farming. Long summer days, moderate temperatures, limited rainfall, and flat plains provide excellent conditions for wheat farming.

Scale 1:13,250,000

projection: Lambert Azimuthal Equal Area

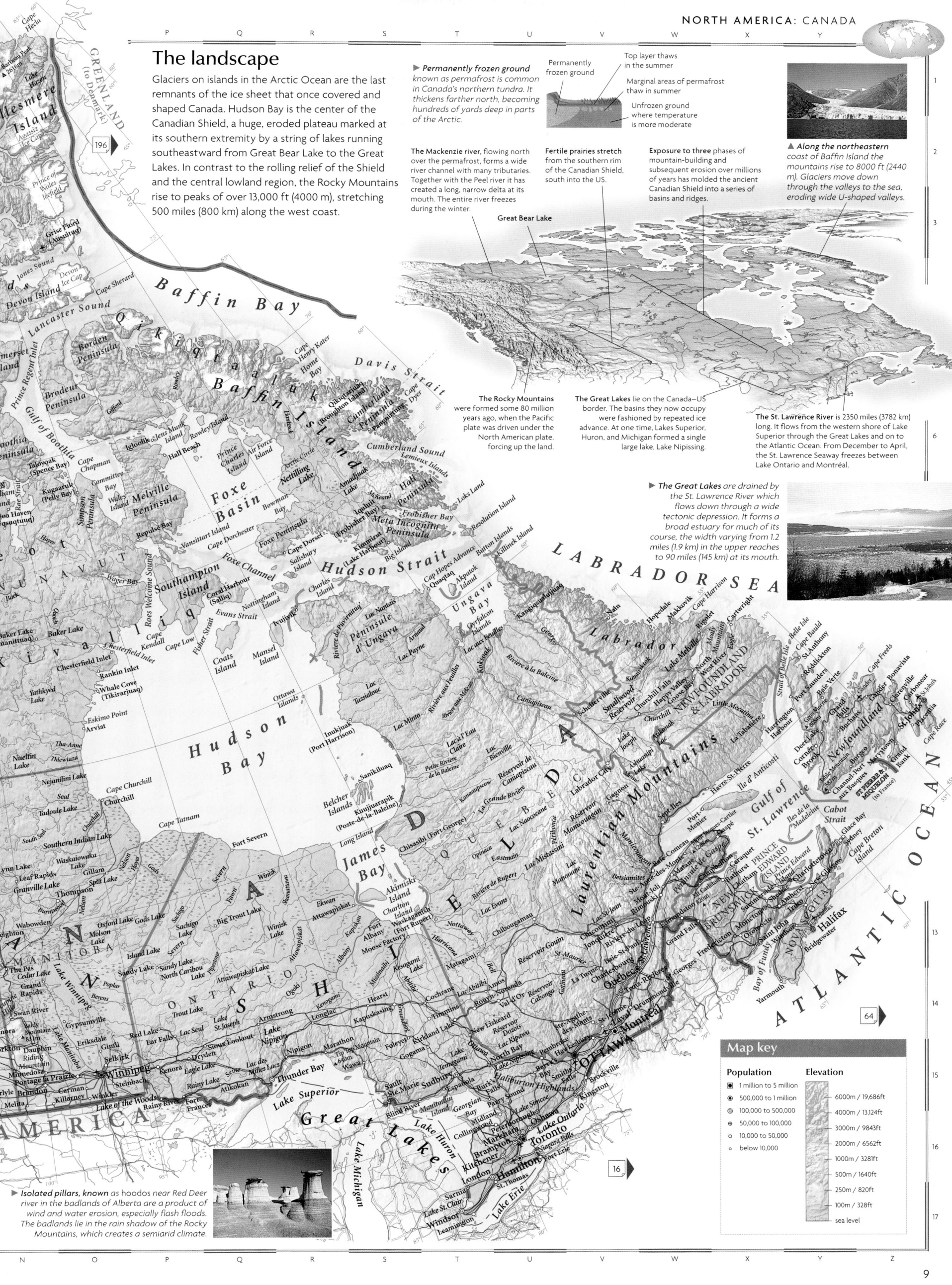

The landscape

Glaciers on islands in the Arctic Ocean are the last remnants of the ice sheet that once covered and shaped Canada. Hudson Bay is the center of the Canadian Shield, a huge, eroded plateau marked at its southern extremity by a string of lakes running southeastward from Great Bear Lake to the Great Lakes. In contrast to the rolling relief of the Shield and the central lowland region, the Rocky Mountains rise to peaks of over 13,000 ft (4000 m), stretching 500 miles (800 km) along the west coast.

▶ **Permanently frozen ground** known as permafrost is common in Canada's northern tundra. It thickens farther north, becoming hundreds of yards deep in parts of the Arctic.

Permanently frozen ground

Top layer thaws in the summer

Marginal areas of permafrost thaw in summer

Unfrozen ground where temperature is more moderate

The Mackenzie river, flowing north over the permafrost, forms a wide river channel with many tributaries. Together with the Peel river it has created a long, narrow delta at its mouth. The entire river freezes during the winter.

Fertile prairies stretch from the southern rim of the Canadian Shield, south into the US.

Exposure to three phases of mountain-building and subsequent erosion over millions of years has molded the ancient Canadian Shield into a series of basins and ridges.

▲ **Along the northeastern** coast of Baffin Island the mountains rise to 8000 ft (2440 m). Glaciers move down through the valleys to the sea, eroding wide U-shaped valleys.

The Rocky Mountains were formed some 80 million years ago, when the Pacific plate was driven under the North American plate, forcing up the land.

The Great Lakes lie on the Canada–US border. The basins they now occupy were fashioned by repeated ice advance. At one time, Lakes Superior, Huron, and Michigan formed a single large lake, Lake Nipissing.

The St. Lawrence River is 2350 miles (3782 km) long. It flows from the western shore of Lake Superior through the Great Lakes and on to the Atlantic Ocean. From December to April, the St. Lawrence Seaway freezes between Lake Ontario and Montréal.

▶ **The Great Lakes** are drained by the St. Lawrence River which flows down through a wide tectonic depression. It forms a broad estuary for much of its course, the width varying from 1.2 miles (1.9 km) in the upper reaches to 90 miles (145 km) at its mouth.

▶ **Isolated pillars,** known as hoodoos near Red Deer river in the badlands of Alberta are a product of wind and water erosion, especially flash floods. The badlands lie in the rain shadow of the Rocky Mountains, which creates a semiarid climate.

Map key

Population
- ◉ 1 million to 5 million
- ◎ 500,000 to 1 million
- ⊛ 100,000 to 500,000
- ⊕ 50,000 to 100,000
- ○ 10,000 to 50,000
- · below 10,000

Elevation
- 6000m / 19,686ft
- 4000m / 13,124ft
- 3000m / 9843ft
- 2000m / 6562ft
- 1000m / 3281ft
- 500m / 1640ft
- 250m / 820ft
- 100m / 328ft
- sea level

Canada:
WESTERN PROVINCES

Alberta, British Columbia, Manitoba, Saskatchewan, Yukon

The mountains of the west coast, incorporating British Columbia and the Yukon, descend into the vast, flat prairies of Alberta, Saskatchewan, and Manitoba. The empty lands and fertile soils of the prairie provinces attracted migrants, and the descendants of early European immigrants still make up a large proportion of the population. The mechanization of agriculture has reduced the need for labor, and rural population densities remain low. The majority of the people live within 100 miles (160 km) of the southern Canada–US border, and in British Columbia, one of the leading Canadian provinces in terms of economic wealth. The Yukon, in the far north, remains a relatively unspoiled wilderness, containing large, untapped mineral reserves. This province has a significant population of Native American people, many of whom maintain a traditional lifestyle.

Using the land and sea

Wheat farming is the economic mainstay of Alberta, Manitoba, and Saskatchewan, which contain 82% of farmland in Canada. Cattle are also raised on the prairies. Forestry and fishing are the most prominent resource-based industries in British Columbia. Despite the mountainous terrain, fruit and specialized grains can be grown in the Okanagan and Fraser valleys.

Land use and agricultural distribution
- cattle
- cereals
- fishing
- fruit
- timber
- major towns
- pasture
- cropland
- forest
- wetland
- barren
- tundra

The urban/rural population divide

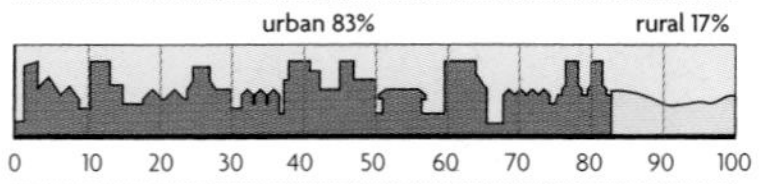

urban 83%   rural 17%

Population density	Total land area
8 people per sq mile (3 people per sq km)	1,230,547 sq miles (3,187,120 sq km)

▲ Large, highly-mechanized and often very specialized farms, requiring huge investment but little labor, characterize modern farming in the prairies.

Transportation & industry

The western provinces contain a wealth of mineral resources. Alberta holds the bulk of Canada's fossil fuels; the other provinces contain reserves of metallic ores, such as zinc, lead, and silver. Isolation from markets has slowed the development of manufacturing, restricting it to the large cities like Vancouver, Winnipeg, and Calgary. Hydroelectric power is widely exploited, although there is increasing concern about potential ecological damage.

Transportation network
- 82,438 miles (135,145 km)
- 6459 miles (10,401 km)
- 24,041 miles (38,694 km)
- None

The transportation network of the western provinces is dominated by east–west routes that weave through mountain passes and spread across the plains. Access to some northern areas is restricted to air travel.

Major industry and infrastructure
- aerospace
- chemicals
- coal
- engineering
- food processing
- hydroelectric power
- mining
- oil & gas
- timber processing
- major towns
- international airports
- major roads
- major industrial areas

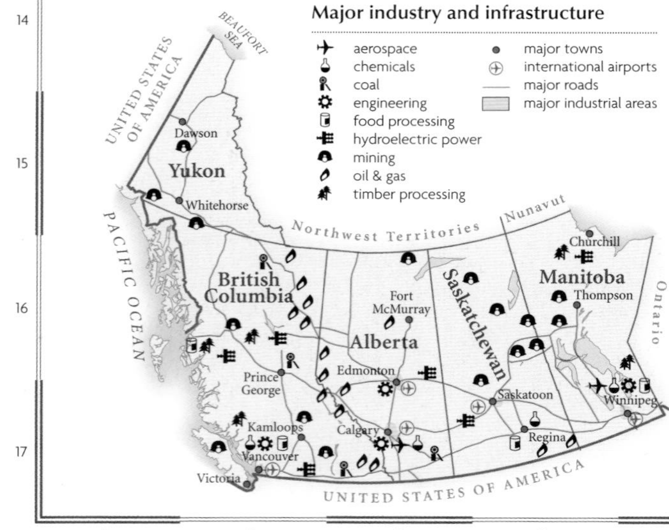

◄ Much of the Yukon is uninhabited tundra. Industry is based on the extraction of mineral resources, and to a lesser extent, on the scattered forests of the south.

▲ The Fraser River valley is a major area of settlement in British Columbia. Railroads cross the Rocky Mountains via this valley.

▲ Established in 1907, Jasper National Park lies in the heart of the Rocky Mountains. It is noted for its spectacular alpine scenery and contains part of the large Columbia Icefield.

The landscape

The massive Rocky Mountains form a continental divide between rivers flowing eastward and westward. The interior plains lie east of the mountains, stretching from the Arctic Circle south into the US. Covered with glacial deposits from the last Ice Age, these are interspersed with hilly regions and long, steep escarpments.

Map key

Population

- ◉ 500,000 to 1 million
- ◎ 100,000 to 500,000
- ⊕ 50,000 to 100,000
- ○ 10,000 to 50,000
- ∘ below 10,000

Elevation

	6000m / 19,686ft
	4000m / 13,124ft
	3000m / 9843ft
	2000m / 6562ft
	1000m / 3281ft
	500m / 1640ft
	250m / 820ft
	100m / 328ft
	sea level

Scale 1:7,500,000

Km
0 25 50 100 150 200 250

Miles
0 50 100 150 200 250

projection: Lambert Conformal Conic

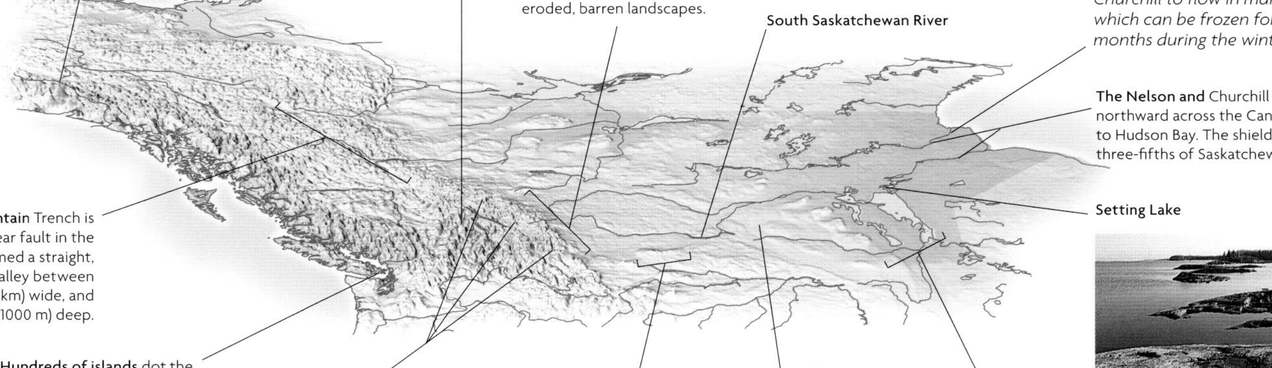

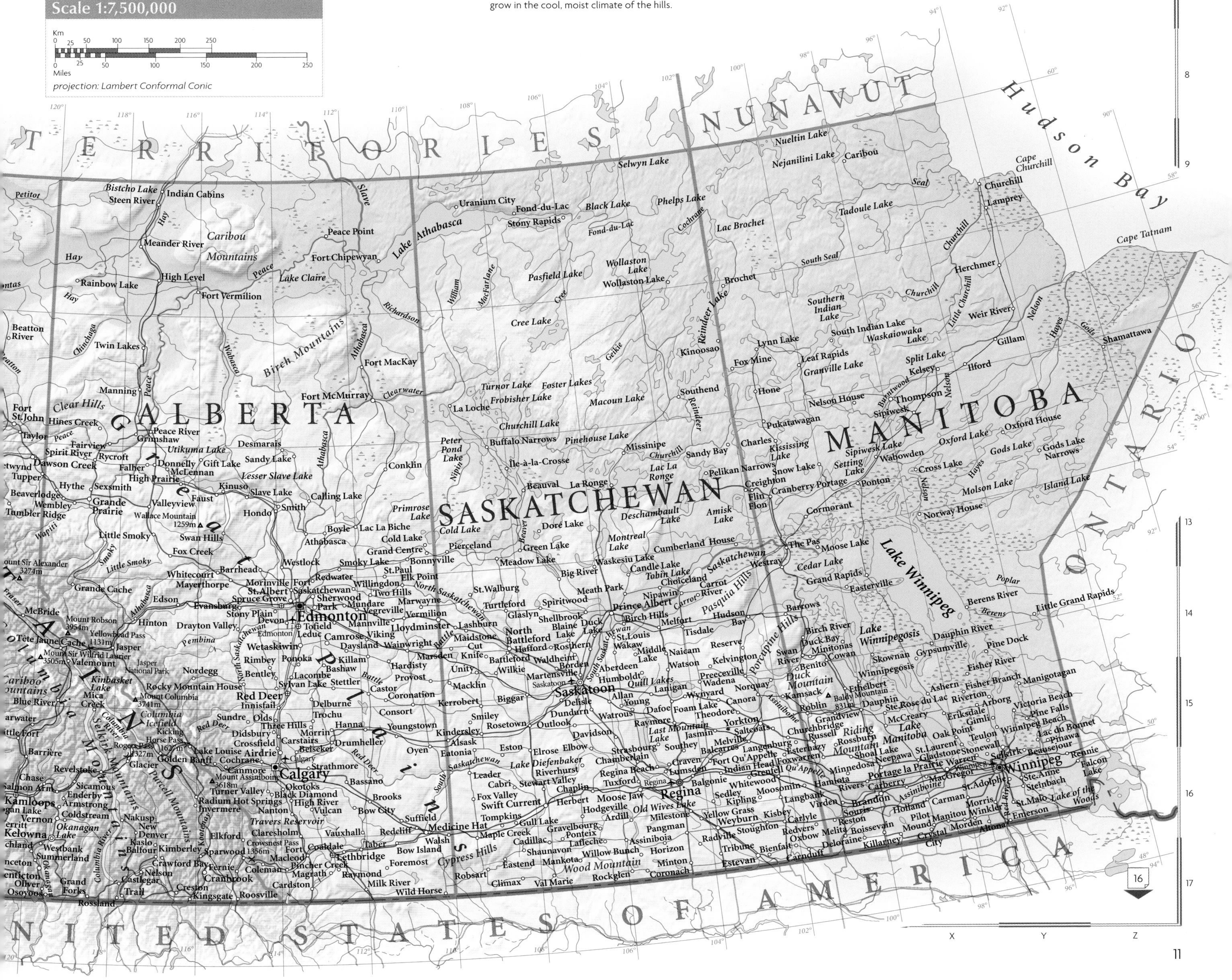

Mount Logan rises 19,551 ft (5959 m). It is the highest peak in Canada.

The Columbia Icefield in the Rocky Mountains is the source of two major rivers, the Athabasca and the North Saskatchewan.

The badlands of Alberta were created when east-flowing rivers, swollen by meltwater at the end of the last Ice Age, cut deep, wide canyons producing eroded, barren landscapes.

Vegetated island — Bar
River flow is diverted by deposited sediments — Sand flat

▲ **Braided rivers are** shallow and fast-flowing. The interlaced branches are formed when excess sediments, which can no longer be transported, are deposited. The sediments collect in the river channel forming bars and sand flats. Islands form when the bars are colonized by vegetation.

South Saskatchewan River

▲ **Across the tundra** of northern Manitoba, widespread permafrost inhibits water from permeating the soil. This causes rivers like the Churchill to flow in many channels, which can be frozen for up to six months during the winter.

The Rocky Mountain Trench is the longest linear fault in the world. It has formed a straight, flat-bottomed valley between 2–9 miles (4–15 km) wide, and up to 3280 ft (1000 m) deep.

The Nelson and Churchill rivers drain northward across the Canadian Shield to Hudson Bay. The shield covers three-fifths of Saskatchewan.

Setting Lake

Hundreds of islands dot the fjord-indented coast of British Columbia; the largest is Vancouver Island.

The Alberta and Saskatchewan plains bear strong testament to past glaciations. The Assiniboine, Saskatchewan and Qu'Appelle rivers occupy flat-bottomed, steep-sided valleys eroded during the last Ice Age by glacial meltwater.

Three major passes cut through the Rocky Mountains: Yellowhead, Kicking Horse, and Crowsnest. They are all used as transportation routes through the mountains.

The Cypress Hills rise to 4806 ft (1465 m) above the surrounding plain. Having escaped the last glaciation they contain unique plant and animal life. The silvery lupine, bunchberry, and lodgepole pine all grow in the cool, moist climate of the hills.

Ancient granite outcrops, part of the Canadian Shield, rise above the surface of Setting Lake, which was initially formed by meltwater from the last Ice Age.

The lowlands of Manitoba are a basin that once held the vast post-glacial Lake Agassiz, remnants of which include Lake Winnipeg, Lake Winnipegosis, and Lake Manitoba.

Canada: EASTERN PROVINCES

New Brunswick, Newfoundland & Labrador, Nova Scotia, Ontario,
Prince Edward Island, Québec, *St Pierre & Miquelon (to France)*

Colonized by both the English and the French during the 16th century, Canada's eastern provinces are still marked by their dual influences. They contain the last fragment of once-sizeable French territories, the islands of St. Pierre and Miquelon. French remains Canada's second official language and Québec's first language. The population of the eastern provinces is highly concentrated in the south, especially along the border with the US. A recent decline in fishing in the Atlantic provinces has encouraged a steady flow of westerly migration to more prosperous regions. The north, around Hudson Bay, remains snow-covered for most of the year and the indigenous Inuit people make up the bulk of its sparse population.

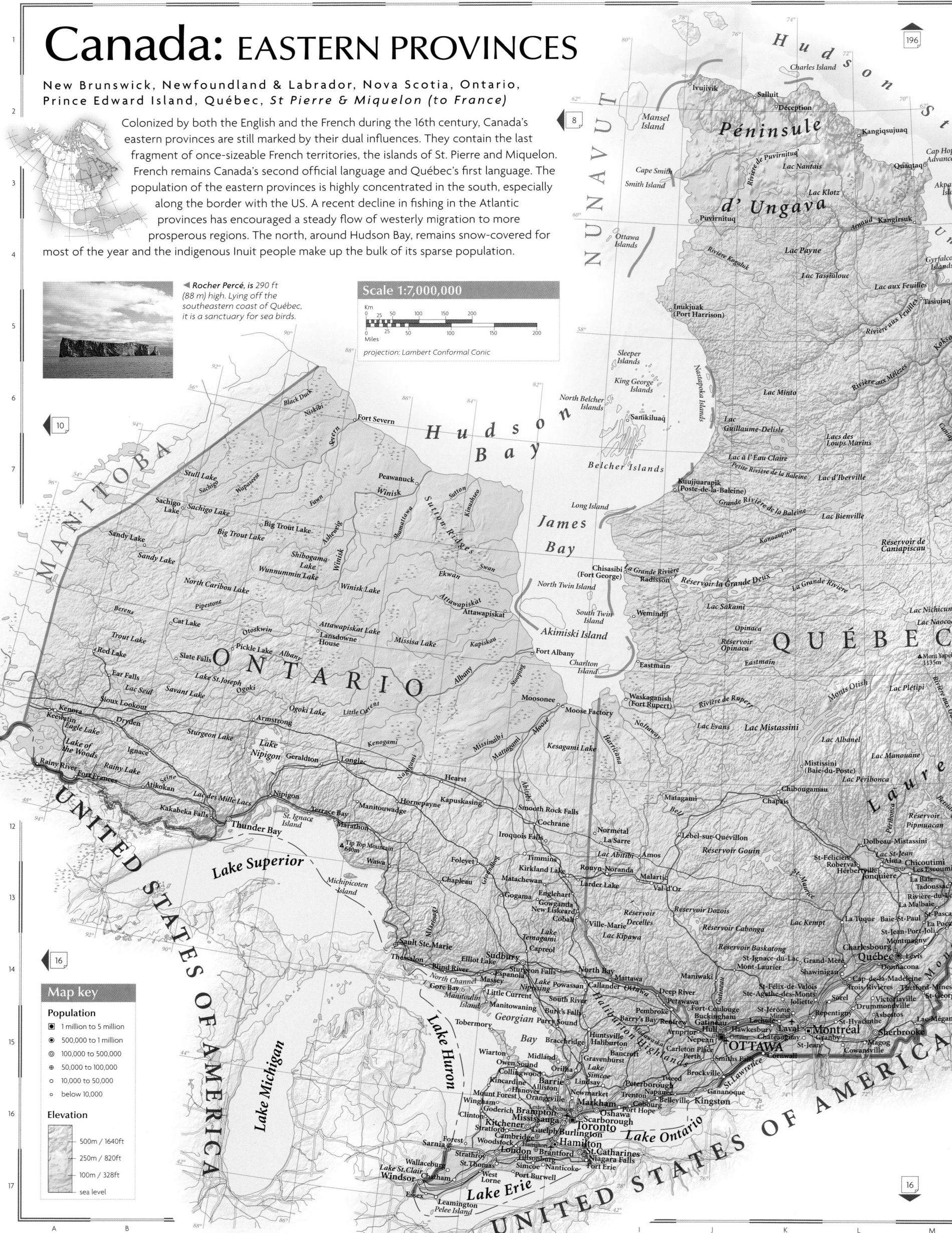

◄ *Rocher Percé, is 290 ft (88 m) high. Lying off the southeastern coast of Québec, it is a sanctuary for sea birds.*

Scale 1:7,000,000

projection: Lambert Conformal Conic

Map key

Population
- ▣ 1 million to 5 million
- ◉ 500,000 to 1 million
- ◎ 100,000 to 500,000
- ⊕ 50,000 to 100,000
- ○ 10,000 to 50,000
- ∘ below 10,000

Elevation
- 500m / 1640ft
- 250m / 820ft
- 100m / 328ft
- sea level

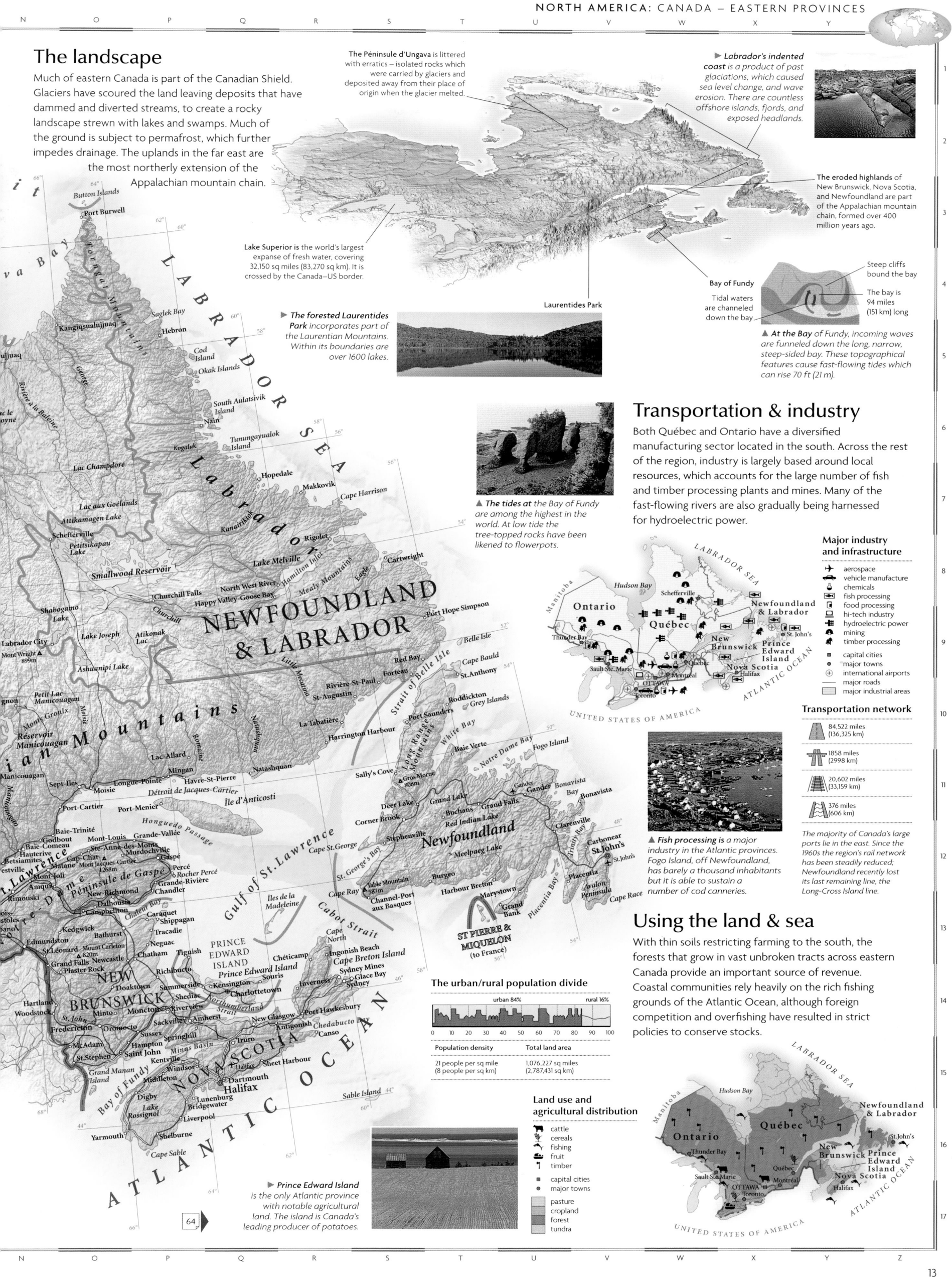

The landscape

Much of eastern Canada is part of the Canadian Shield. Glaciers have scoured the land leaving deposits that have dammed and diverted streams, to create a rocky landscape strewn with lakes and swamps. Much of the ground is subject to permafrost, which further impedes drainage. The uplands in the far east are the most northerly extension of the Appalachian mountain chain.

The Péninsule d'Ungava is littered with erratics – isolated rocks which were carried by glaciers and deposited away from their place of origin when the glacier melted.

▶ Labrador's indented coast is a product of past glaciations, which caused sea level change, and wave erosion. There are countless offshore islands, fjords, and exposed headlands.

The eroded highlands of New Brunswick, Nova Scotia, and Newfoundland are part of the Appalachian mountain chain, formed over 400 million years ago.

Lake Superior is the world's largest expanse of fresh water, covering 32,150 sq miles (83,270 sq km). It is crossed by the Canada–US border.

Laurentides Park

Bay of Fundy
Tidal waters are channeled down the bay

Steep cliffs bound the bay
The bay is 94 miles (151 km) long

▶ The forested Laurentides Park incorporates part of the Laurentian Mountains. Within its boundaries are over 1600 lakes.

▲ At the Bay of Fundy, incoming waves are funneled down the long, narrow, steep-sided bay. These topographical features cause fast-flowing tides which can rise 70 ft (21 m).

Transportation & industry

Both Québec and Ontario have a diversified manufacturing sector located in the south. Across the rest of the region, industry is largely based around local resources, which accounts for the large number of fish and timber processing plants and mines. Many of the fast-flowing rivers are also gradually being harnessed for hydroelectric power.

▲ The tides at the Bay of Fundy are among the highest in the world. At low tide the tree-topped rocks have been likened to flowerpots.

Major industry and infrastructure

- ✈ aerospace
- 🚗 vehicle manufacture
- chemicals
- 🐟 fish processing
- food processing
- hi-tech industry
- hydroelectric power
- ⛏ mining
- 🌲 timber processing
- ■ capital cities
- • major towns
- ✈ international airports
- — major roads
- major industrial areas

Transportation network

🛣	84,522 miles (136,325 km)
🛣	1858 miles (2998 km)
🚂	20,602 miles (33,159 km)
🚂	376 miles (606 km)

The majority of Canada's large ports lie in the east. Since the 1960s the region's rail network has been steadily reduced; Newfoundland recently lost its last remaining line, the Long-Cross Island line.

▲ Fish processing is a major industry in the Atlantic provinces. Fogo Island, off Newfoundland, has barely a thousand inhabitants but it is able to sustain a number of cod canneries.

Using the land & sea

With thin soils restricting farming to the south, the forests that grow in vast unbroken tracts across eastern Canada provide an important source of revenue. Coastal communities rely heavily on the rich fishing grounds of the Atlantic Ocean, although foreign competition and overfishing have resulted in strict policies to conserve stocks.

The urban/rural population divide

urban 84%   rural 16%

0 10 20 30 40 50 60 70 80 90 100

Population density	Total land area
21 people per sq mile (8 people per sq km)	1,076,227 sq miles (2,787,431 sq km)

Land use and agricultural distribution

- 🐄 cattle
- cereals
- 🎣 fishing
- fruit
- timber
- ■ capital cities
- • major towns
- pasture
- cropland
- forest
- tundra

▶ Prince Edward Island is the only Atlantic province with notable agricultural land. The island is Canada's leading producer of potatoes.

Southeastern Canada

Southern Ontario, Southern Québec

The southern parts of Québec and Ontario form the economic heart of Canada. The two provinces are divided by their language and culture; in Québec, French is the main language, whereas English is spoken in Ontario. Separatist sentiment in Québec has led to a provincial referendum on the question of a sovereignty association with Canada. The region contains Canada's capital, Ottawa, and its two largest cities: Toronto, the center of commerce, and Montréal, the cultural and administrative heart of French Canada.

▲ *The port at Montréal is situated on the St. Lawrence Seaway. A network of 16 locks allows oceangoing vessels access to routes once plied by fur-trappers and early settlers.*

Transportation & industry

The cities of southern Québec and Ontario, and their hinterlands, form the heart of Canadian manufacturing industry. Toronto is Canada's leading financial center, and Ontario's motor and aerospace industries have developed around the city. A major center for nickel mining lies to the north of Toronto. Most of Québec's industry is located in Montréal, the oldest port in North America. Chemicals, paper manufacture, and the construction of transportation equipment are leading industrial activities.

▶ *Niagara Falls lies on the border between Canada and the US. It comprises a system of two falls: American Falls, in New York, is separated from Horseshoe Falls, in Ontario, by Goat Island. Horseshoe Falls, seen here, plunges 184 ft (56 m) and is 2500 ft (762 m) wide.*

Major industry and infrastructure

- car manufacture
- chemicals
- engineering
- finance
- food processing
- hi-tech industry
- mining
- iron & steel
- textiles
- paper industry
- timber processing
- capital cities
- major towns
- international airports
- major roads
- major industrial areas

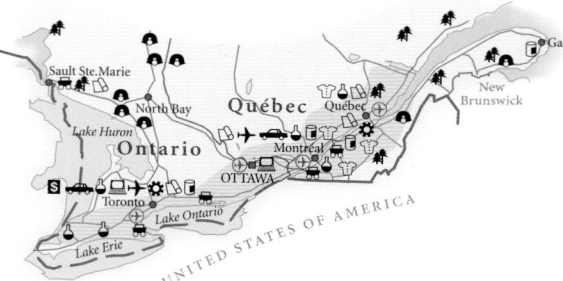

Transportation network

The opening of the St. Lawrence Seaway in 1959 finally allowed oceangoing ships (up to 24,000 tons [tonnes]) access to the interior of Canada, creating a vital trading route.

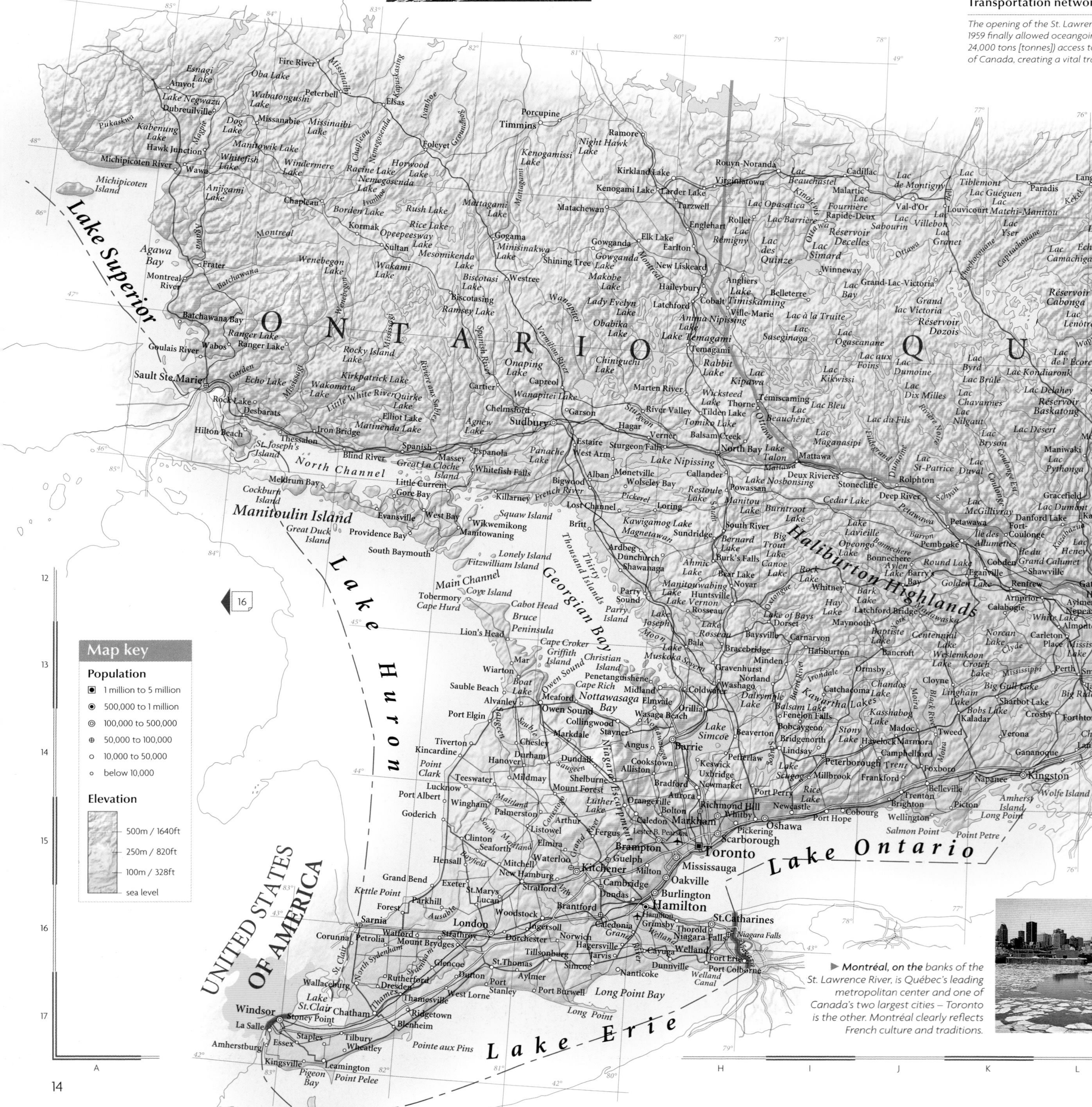

Map key

Population
- 1 million to 5 million
- 500,000 to 1 million
- 100,000 to 500,000
- 50,000 to 100,000
- 10,000 to 50,000
- below 10,000

Elevation
- 500m/1640ft
- 250m/820ft
- 100m/328ft
- sea level

▶ *Montréal, on the banks of the St. Lawrence River, is Québec's leading metropolitan center and one of Canada's two largest cities – Toronto is the other. Montréal clearly reflects French culture and traditions.*

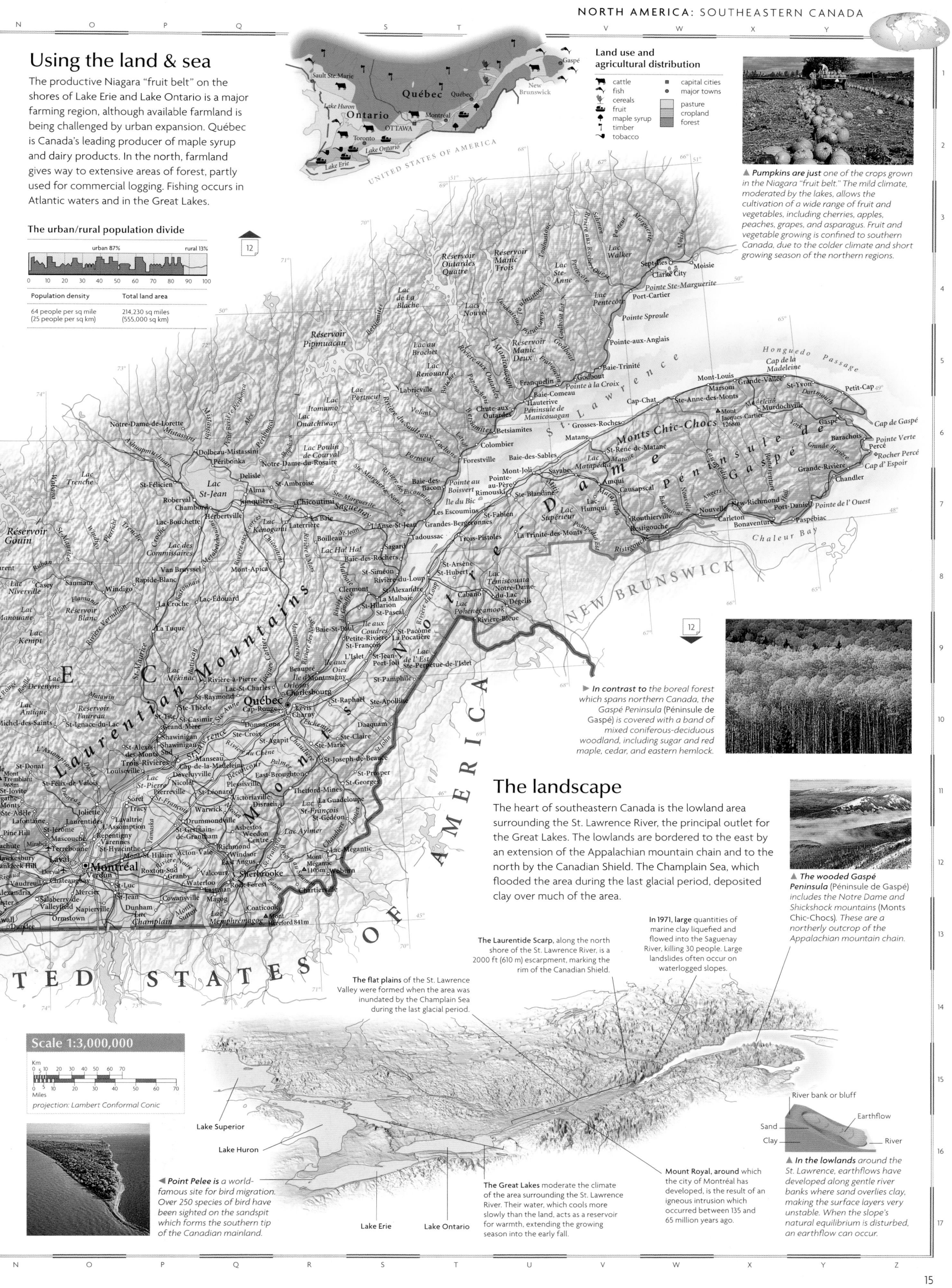

Using the land & sea

The productive Niagara "fruit belt" on the shores of Lake Erie and Lake Ontario is a major farming region, although available farmland is being challenged by urban expansion. Québec is Canada's leading producer of maple syrup and dairy products. In the north, farmland gives way to extensive areas of forest, partly used for commercial logging. Fishing occurs in Atlantic waters and in the Great Lakes.

Land use and agricultural distribution

- cattle
- fish
- cereals
- fruit
- maple syrup
- timber
- tobacco
- capital cities
- major towns
- pasture
- cropland
- forest

The urban/rural population divide

urban 87%   rural 13%

0 10 20 30 40 50 60 70 80 90 100

Population density	Total land area
64 people per sq mile (25 people per sq km)	214,230 sq miles (555,000 sq km)

▲ *Pumpkins are just* one of the crops grown in the Niagara "fruit belt." The mild climate, moderated by the lakes, allows the cultivation of a wide range of fruit and vegetables, including cherries, apples, peaches, grapes, and asparagus. Fruit and vegetable growing is confined to southern Canada, due to the colder climate and short growing season of the northern regions.

▶ *In contrast to* the boreal forest which spans northern Canada, the Gaspé Peninsula (Péninsule de Gaspé) is covered with a band of mixed coniferous-deciduous woodland, including sugar and red maple, cedar, and eastern hemlock.

The landscape

The heart of southeastern Canada is the lowland area surrounding the St. Lawrence River, the principal outlet for the Great Lakes. The lowlands are bordered to the east by an extension of the Appalachian mountain chain and to the north by the Canadian Shield. The Champlain Sea, which flooded the area during the last glacial period, deposited clay over much of the area.

▲ *The wooded Gaspé Peninsula* (Péninsule de Gaspé) includes the Notre Dame and Shickshock mountains (Monts Chic-Chocs). These are a northerly outcrop of the Appalachian mountain chain.

In 1971, large quantities of marine clay liquefied and flowed into the Saguenay River, killing 30 people. Large landslides often occur on waterlogged slopes.

The Laurentide Scarp, along the north shore of the St. Lawrence River, is a 2000 ft (610 m) escarpment, marking the rim of the Canadian Shield.

The flat plains of the St. Lawrence Valley were formed when the area was inundated by the Champlain Sea during the last glacial period.

Scale 1:3,000,000

Km
0 5 10 20 30 40 50 60 70

Miles
0 5 10 20 30 40 50 60 70

projection: Lambert Conformal Conic

◀ *Point Pelee is* a world-famous site for bird migration. Over 250 species of bird have been sighted on the sandspit which forms the southern tip of the Canadian mainland.

Lake Superior

Lake Huron

The Great Lakes moderate the climate of the area surrounding the St. Lawrence River. Their water, which cools more slowly than the land, acts as a reservoir for warmth, extending the growing season into the early fall.

Lake Erie   Lake Ontario

Mount Royal, around which the city of Montréal has developed, is the result of an igneous intrusion which occurred between 135 and 65 million years ago.

River bank or bluff

Earthflow

Sand

Clay

River

▲ *In the lowlands* around the St. Lawrence, earthflows have developed along gentle river banks where sand overlies clay, making the surface layers very unstable. When the slope's natural equilibrium is disturbed, an earthflow can occur.

The United States of America

COTERMINOUS US (FOR ALASKA AND HAWAII SEE PAGES 38-39)

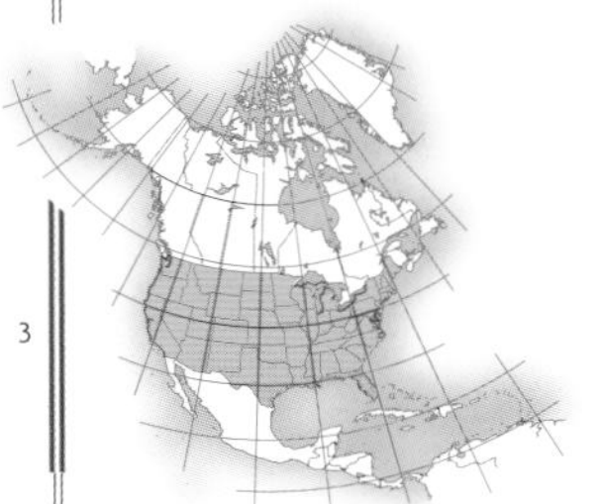

The US's progression from frontier territory to economic and political superpower has taken less than 200 years. The 48 coterminous states, along with the outlying states of Alaska and Hawaii, are part of a federal union, held together by the guiding principles of the US Constitution, which embodies the ideals of democracy and liberty for all. Abundant fertile land and a rich resource base fueled and sustained US economic development. With the spread of agriculture and the growth of trade and industry came the need for a larger workforce, which was supplied by millions of immigrants, many seeking an escape from poverty and political or religious persecution. Immigration continues today, particularly from Central America and Asia.

▲ *Washington DC was* established as the site for the nation's capital in 1790. It is home to the seat of national government, on Capitol Hill, as well as the President's official residence, the White House.

▶ *The clear waters* of Niagara Falls cascade 190 ft (58 m) into the gorge below. It is one of America's most famous spectacles and a leading tourist attraction. The falls are slowly receding and the gorge may one day stretch from Lake Ontario to Lake Erie.

▲ *Mount Rainier is* a dormant volcano in the Cascade Range, Washington. This 14,090 ft (4392 m) peak is flanked by the most extensive glacier outside Alaska.

Scale 1:11,450,000

Km 0 25 50 100 150 200 250 300 350 400
Miles 0 25 50 100 150 200 250 300 350 400

projection: Lambert Azimuthal Equal Area

Transportation & industry

The US has been the industrial powerhouse of the world since the Second World War, pioneering mass-production and the consumer lifestyle. Initially, heavy engineering and manufacturing in the northeast led the economy. Today, heavy industry has declined and the US economy is driven by service and financial industries, with the most important being defense, hi-tech, and electronics.

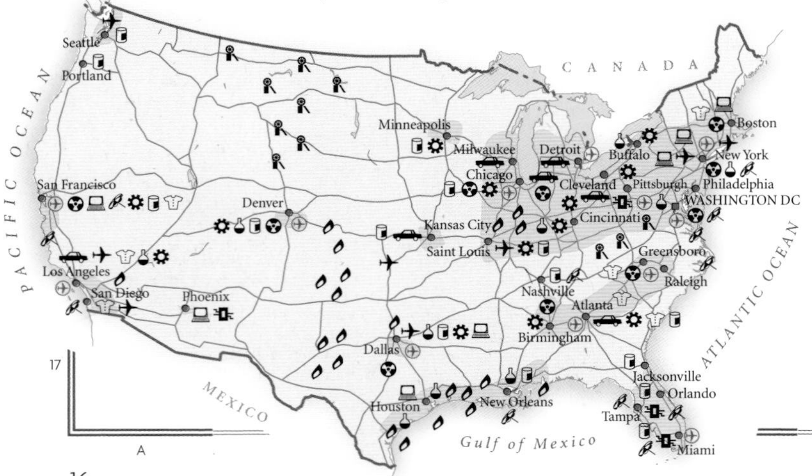

Transportation network

3,875,040 miles (6,240,000 km)	52,388 miles (84,361 km)
148,308 miles (235,238 km)	25,467 miles (41,009 km)

Transportation in the US is dominated by the car which, with the extensive Interstate Highway system, allows great personal mobility. Today, internal air flights between major cities provide the most rapid cross-country travel.

Major industry and infrastructure

- aerospace
- car manufacture
- chemicals
- coal
- electronics
- engineering
- food processing
- hi-tech industry
- oil & gas
- research & development
- textiles
- tourism
- ■ capital cities
- ● major towns
- ⊕ international airports
- — major roads
- major industrial areas

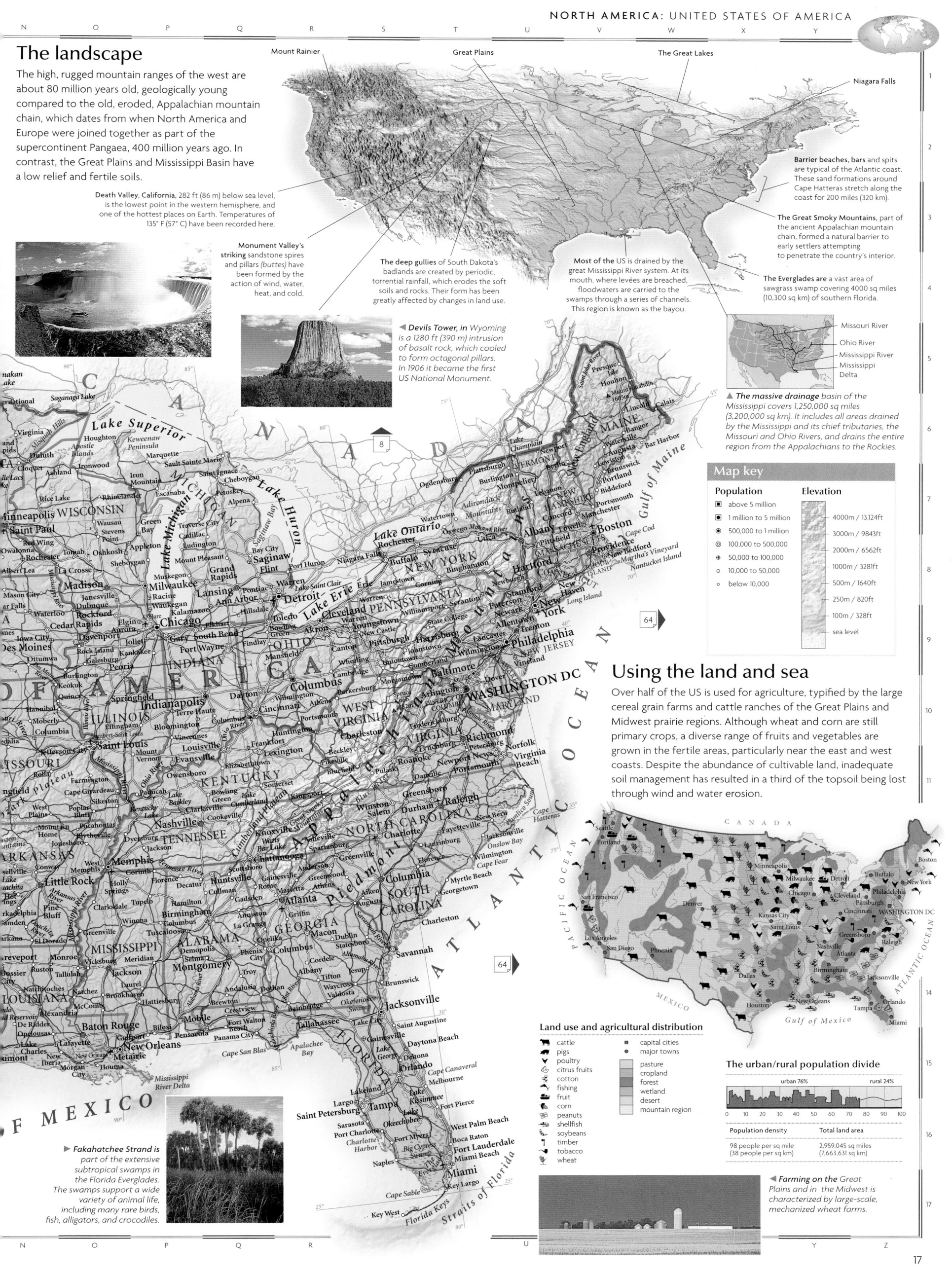

The landscape

The high, rugged mountain ranges of the west are about 80 million years old, geologically young compared to the old, eroded, Appalachian mountain chain, which dates from when North America and Europe were joined together as part of the supercontinent Pangaea, 400 million years ago. In contrast, the Great Plains and Mississippi Basin have a low relief and fertile soils.

Mount Rainier

Great Plains

The Great Lakes

Niagara Falls

Death Valley, California, 282 ft (86 m) below sea level, is the lowest point in the western hemisphere, and one of the hottest places on Earth. Temperatures of 135° F (57° C) have been recorded here.

Barrier beaches, bars and spits are typical of the Atlantic coast. These sand formations around Cape Hatteras stretch along the coast for 200 miles (320 km).

The Great Smoky Mountains, part of the ancient Appalachian mountain chain, formed a natural barrier to early settlers attempting to penetrate the country's interior.

Monument Valley's striking sandstone spires and pillars *(buttes)* have been formed by the action of wind, water, heat, and cold.

The deep gullies of South Dakota's badlands are created by periodic, torrential rainfall, which erodes the soft soils and rocks. Their form has been greatly affected by changes in land use.

Most of the US is drained by the great Mississippi River system. At its mouth, where levées are breached, floodwaters are carried to the swamps through a series of channels. This region is known as the bayou.

The Everglades are a vast area of sawgrass swamp covering 4000 sq miles (10,300 sq km) of southern Florida.

◀ **Devils Tower, in** *Wyoming is a 1280 ft (390 m) intrusion of basalt rock, which cooled to form octagonal pillars. In 1906 it became the first US National Monument.*

Missouri River
Ohio River
Mississippi River
Mississippi Delta

▲ **The massive drainage** basin of the Mississippi covers 1,250,000 sq miles (3,200,000 sq km). It includes all areas drained by the Mississippi and its chief tributaries, the Missouri and Ohio Rivers, and drains the entire region from the Appalachians to the Rockies.

Map key

Population
- ▣ above 5 million
- ■ 1 million to 5 million
- ◉ 500,000 to 1 million
- ◎ 100,000 to 500,000
- ⊕ 50,000 to 100,000
- ○ 10,000 to 50,000
- ∘ below 10,000

Elevation
- 4000m / 13,124ft
- 3000m / 9843ft
- 2000m / 6562ft
- 1000m / 3281ft
- 500m / 1640ft
- 250m / 820ft
- 100m / 328ft
- sea level

Using the land and sea

Over half of the US is used for agriculture, typified by the large cereal grain farms and cattle ranches of the Great Plains and Midwest prairie regions. Although wheat and corn are still primary crops, a diverse range of fruits and vegetables are grown in the fertile areas, particularly near the east and west coasts. Despite the abundance of cultivable land, inadequate soil management has resulted in a third of the topsoil being lost through wind and water erosion.

▶ **Fakahatchee Strand is** *part of the extensive subtropical swamps in the Florida Everglades. The swamps support a wide variety of animal life, including many rare birds, fish, alligators, and crocodiles.*

Land use and agricultural distribution

- 🐄 cattle
- 🐖 pigs
- 🦃 poultry
- 🍊 citrus fruits
- cotton
- 🐟 fishing
- fruit
- corn
- peanuts
- shellfish
- soybeans
- timber
- tobacco
- wheat

- ■ capital cities
- • major towns

- pasture
- cropland
- forest
- wetland
- desert
- mountain region

The urban/rural population divide

urban 76%   rural 24%

0 10 20 30 40 50 60 70 80 90 100

Population density	Total land area
98 people per sq mile (38 people per sq km)	2,959,045 sq miles (7,663,631 sq km)

◀ **Farming on the** *Great Plains and in the Midwest is characterized by large-scale, mechanized wheat farms.*

USA: NORTHEASTERN STATES

Connecticut, Maine, Massachusetts, New Hampshire, New Jersey, New York, Pennsylvania, Rhode Island, Vermont

The indented coast and vast woodlands of the northeastern states were the original core area for European expansion. The rustic character of New England prevails after nearly four centuries, while the great cities of the Atlantic seaboard have formed an almost continuous urban region. Over 20 million immigrants entered New York from 1855 to 1924 and the northeast became the industrial center of the US. After the decline of mining and heavy manufacturing, economic dynamism has been restored with the growth of hi-tech and service industries.

▲ *Chelsea in Vermont, surrounded by trees in their fall foliage. Tourism and agriculture dominate the economy of this self-consciously rural state, where no town exceeds 30,000 people.*

Map key

Population

- above 5 million
- 1 million to 5 million
- 500,000 to 1 million
- 100,000 to 500,000
- 50,000 to 100,000
- 10,000 to 50,000
- below 10,000

Elevation

- 1000m / 3281ft
- 500m / 1640ft
- 250m / 820ft
- 100m / 328ft
- sea level

Transportation & industry

The principal seaboard cities grew up on trade and manufacturing. They are now global centers of commerce and corporate administration, dominating the regional economy. Research and development facilities support an expanding electronics and communications sector throughout the region. Pharmaceutical and chemical industries are important in New Jersey and Pennsylvania.

CANADA
Maine
Vermont
Portland
New Hampshire
Syracuse
Albany
Rochester
Buffalo
New York
Massachusetts
Boston
Connecticut
Hartford
Providence
Rhode Island
Pennsylvania
Ohio
Pittsburgh
New York
New Jersey
Harrisburg
Philadelphia
ATLANTIC OCEAN
West Virginia
Maryland
Delaware

Major industry and infrastructure

- chemicals
- coal
- defense
- electronics
- engineering
- finance
- hi-tech industry
- iron & steel
- pharmaceuticals
- printing & publishing
- research & development
- textiles
- timber processing
- major towns
- international airports
- major roads
- major industrial area

Transportation network

340,090 miles (544,144 km)	4813 miles (7700 km)
12,872 miles (20,592 km)	2108 miles (3389 km)

New York's commercial success is tied historically to its transportation connections. The Erie Canal, completed in 1825, opened up the Great Lakes and the interior to New York's markets and carried a stream of immigrants into the Midwest.

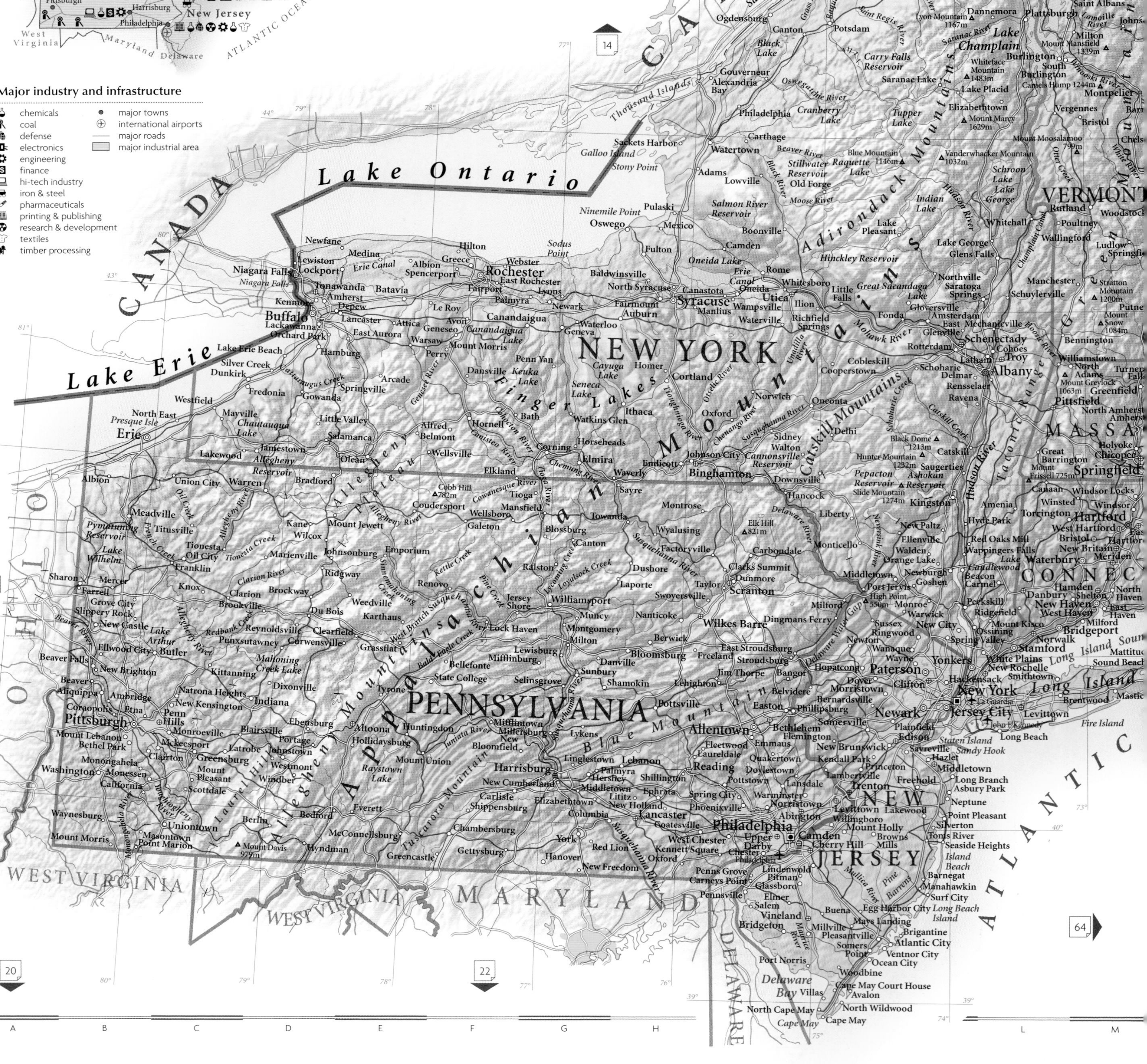

CANADA
Lake Ontario
Lake Erie
NEW YORK
PENNSYLVANIA
VERMONT
NEW JERSEY
WEST VIRGINIA
MARYLAND
DELAWARE
OHIO
ATLANTIC
Adirondack Mountains
Catskill Mountains
Appalachian Mountains
Allegheny Plateau
Finger Lakes
Blue Mountain
Saint Lawrence River
Lake Champlain
Buffalo
Rochester
Syracuse
Albany
Binghamton
Utica
Pittsburgh
Erie
Harrisburg
Philadelphia
Allentown
Scranton
Wilkes Barre
Lancaster
Reading
Trenton
Newark
Jersey City
New York
Paterson
Camden
Atlantic City
Springfield
Hartford
Stamford
Bridgeport
Yonkers
Long Island
Cape May

14
32
20
22
64

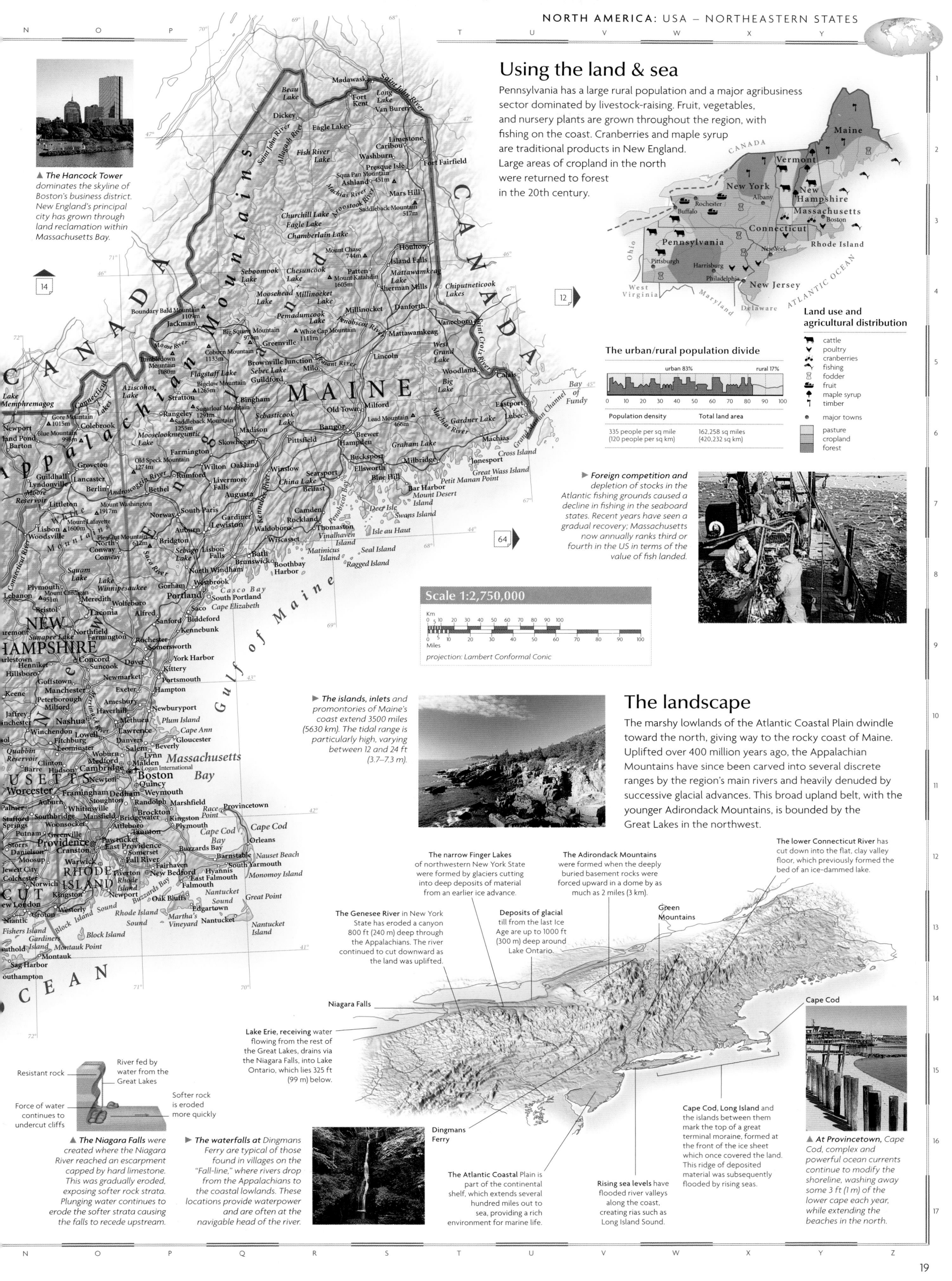

▲ *The Hancock Tower dominates the skyline of Boston's business district. New England's principal city has grown through land reclamation within Massachusetts Bay.*

Using the land & sea

Pennsylvania has a large rural population and a major agribusiness sector dominated by livestock-raising. Fruit, vegetables, and nursery plants are grown throughout the region, with fishing on the coast. Cranberries and maple syrup are traditional products in New England. Large areas of cropland in the north were returned to forest in the 20th century.

The urban/rural population divide

urban 83%   rural 17%

0 10 20 30 40 50 60 70 80 90 100

Population density	Total land area
335 people per sq mile (120 people per sq km)	162,258 sq miles (420,232 sq km)

Land use and agricultural distribution

- cattle
- poultry
- cranberries
- fishing
- fodder
- fruit
- maple syrup
- timber
- major towns
- pasture
- cropland
- forest

▶ *Foreign competition and depletion of stocks in the Atlantic fishing grounds caused a decline in fishing in the seaboard states. Recent years have seen a gradual recovery; Massachusetts now annually ranks third or fourth in the US in terms of the value of fish landed.*

Scale 1:2,750,000

Km
0 5 10 20 30 40 50 60 70 80 90 100

Miles
0 5 10 20 30 40 50 60 70 80 90 100

projection: Lambert Conformal Conic

▶ *The islands, inlets and promontories of Maine's coast extend 3500 miles (5630 km). The tidal range is particularly high, varying between 12 and 24 ft (3.7–7.3 m).*

The landscape

The marshy lowlands of the Atlantic Coastal Plain dwindle toward the north, giving way to the rocky coast of Maine. Uplifted over 400 million years ago, the Appalachian Mountains have since been carved into several discrete ranges by the region's main rivers and heavily denuded by successive glacial advances. This broad upland belt, with the younger Adirondack Mountains, is bounded by the Great Lakes in the northwest.

The narrow Finger Lakes of northwestern New York State were formed by glaciers cutting into deep deposits of material from an earlier ice advance.

The Adirondack Mountains were formed when the deeply buried basement rocks were forced upward in a dome by as much as 2 miles (3 km).

The lower Connecticut River has cut down into the flat, clay valley floor, which previously formed the bed of an ice-dammed lake.

The Genesee River in New York State has eroded a canyon 800 ft (240 m) deep through the Appalachians. The river continued to cut downward as the land was uplifted.

Deposits of glacial till from the last Ice Age are up to 1000 ft (300 m) deep around Lake Ontario.

Green Mountains

Niagara Falls

Lake Erie, receiving water flowing from the rest of the Great Lakes, drains via the Niagara Falls, into Lake Ontario, which lies 325 ft (99 m) below.

Cape Cod

Dingmans Ferry

Cape Cod, Long Island and the islands between them mark the top of a great terminal moraine, formed at the front of the ice sheet which once covered the land. This ridge of deposited material was subsequently flooded by rising seas.

▲ *At Provincetown, Cape Cod, complex and powerful ocean currents continue to modify the shoreline, washing away some 3 ft (1 m) of the lower cape each year, while extending the beaches in the north.*

Resistant rock

River fed by water from the Great Lakes

Force of water continues to undercut cliffs

Softer rock is eroded more quickly

▲ *The Niagara Falls were created where the Niagara River reached an escarpment capped by hard limestone. This was gradually eroded, exposing softer rock strata. Plunging water continues to erode the softer strata causing the falls to recede upstream.*

▶ *The waterfalls at Dingmans Ferry are typical of those found in villages on the "Fall-line," where rivers drop from the Appalachians to the coastal lowlands. These locations provide waterpower and are often at the navigable head of the river.*

The Atlantic Coastal Plain is part of the continental shelf, which extends several hundred miles out to sea, providing a rich environment for marine life.

Rising sea levels have flooded river valleys along the coast, creating rias such as Long Island Sound.

USA: MID-EASTERN STATES

Delaware, District of Columbia, Kentucky,
Maryland, North Carolina, South Carolina,
Tennessee, Virginia, West Virginia

Key events in American history took place in this diverse region, which became the front line between the North and the South during the Civil War of the 1860s. Strong regional contrasts exist between the fertile coastal plains, the isolated upcountry of the Appalachian Mountains, and the cotton-growing areas of the Mississippi lowlands to the west. While coal mining, a traditional industry in the Appalachians, has declined in recent years leaving much rural poverty, service industries elsewhere have increased, especially in Washington DC, the nation's capital.

Map key

Population
- ◉ 500,000 to 1 million
- ◎ 100,000 to 500,000
- ⊕ 50,000 to 100,000
- ○ 10,000 to 50,000
- · below 10,000

Elevation
- 6000m / 19,686ft
- 4000m / 13,124ft
- 3000m / 9843ft
- 2000m / 6562ft
- 1000m / 3281ft
- 500m / 1640ft
- 250m / 820ft
- 100m / 328ft
- sea level

Scale 1:3,000,000

Km 0 5 10 20 30 40 50 60 70 80
Miles 0 5 10 20 30 40 50 60 70 80

projection: Lambert Conformal Conic

▲ *The Bluegrass region* of Kentucky centers on the town of Lexington. This exceptionally fertile rolling plain is well known for its thoroughbred horse-breeding ranches.

Transportation & industry

In the urbanized northeast, manufacturing remains important, alongside a burgeoning service sector. North Carolina is a major center for industrial research and development. Traditional industries include Tennessee whiskey and textiles in South Carolina. The decline of open-pit coal mining in the Appalachians has been hastened by environmental controls, although adventure-tourism is a flourishing new industry.

Major industry and infrastructure
- ⌂ adventure-tourism
- 🜲 car manufacture
- coal
- ⚙ electronics
- engineering
- $ finance
- food processing
- 💻 hi-tech industry
- mining
- ⚛ research & development
- textiles
- ■ capital cities
- • major towns
- ✈ international airports
- — major roads
- major industrial areas

Transportation network

- 452,218 miles (723,548 km)
- 5737 miles (8267 km)
- 18,336 miles (29,503 km)
- 4404 miles (7081 km)

Tennessee's rivers are part of an important inland bulk transportation network. Memphis connects with New Orleans in the south, and with cities as distant as Minneapolis, Sioux City, Chicago, and Pittsburgh, via the Mississippi and its tributaries.

The landscape

The eastern tributaries of the Mississippi drain the interior lowlands. The Cumberland Plateau and the parallel ranges of the Appalachians have been successively uplifted and eroded over time, with the eastern side reduced to a series of foothills known as the Piedmont. The broad coastal plain gradually falls away into salt marshes, lagoons, and offshore bars, broken by flooded estuaries along the shores of the Atlantic.

The Mammoth Cave is part of an extensive cave system in the limestone region of southwestern Kentucky. It stretches for over 300 miles (485 km) on five different levels and contains three rivers and three lakes.

The Mississippi River and its tributary the Ohio River form the western border of the region.

Natural Bridge in eastern Kentucky is an arch 78 ft (26 m) long and 65 ft (20 m) high. It has been shaped from resistant sandstone by gradual weathering processes, which removed the softer rock lying underneath.

The Allegheny Mountains form the northwestern edge of the Appalachian mountain chain. Continuous folding has formed rich seams of bituminous coal.

Appalachian Mountains

◀ *Farmland on the* eastern shores of Chesapeake Bay is sustained by artificial drainage. The area also provides refuge for a variety of waterfowl.

The many inlets of Chesapeake Bay are the flooded tributaries of the main river valley, which have been inundated by rising sea levels.

Salt marshes such as Great Dismal Swamp, develop where the coast is sheltered. Vast areas of such marshland have been reclaimed for farmland and settlement.

Cape Hatteras is the easternmost point of an offshore barrier island, a wave-deposited sand-bar which has become permanent, establishing its own vegetation.

Barrier islands

These intertidal mudflats become submerged at high tide

Tidal inlet
Barrier island

▲ *Barrier islands are* common along the coasts of North and South Carolina. As sea levels rise, wave action builds up ridges of sand and pebbles parallel to the coast, separated by lagoons or intertidal mud flats, which are flooded at high tide.

The Cumberland Plateau is the most southwesterly part of the Appalachians. Big Black Mountain at 4180 ft (1274 m) is the highest point in the range.

The Blue Ridge mountains are a steep ridge, culminating in Mount Mitchell, the highest point in the Appalachians, at 6684 ft (2037 m).

◀ *The Great Smoky Mountains* form the western escarpment of the Appalachians. The region is heavily forested, with over 130 species of tree.

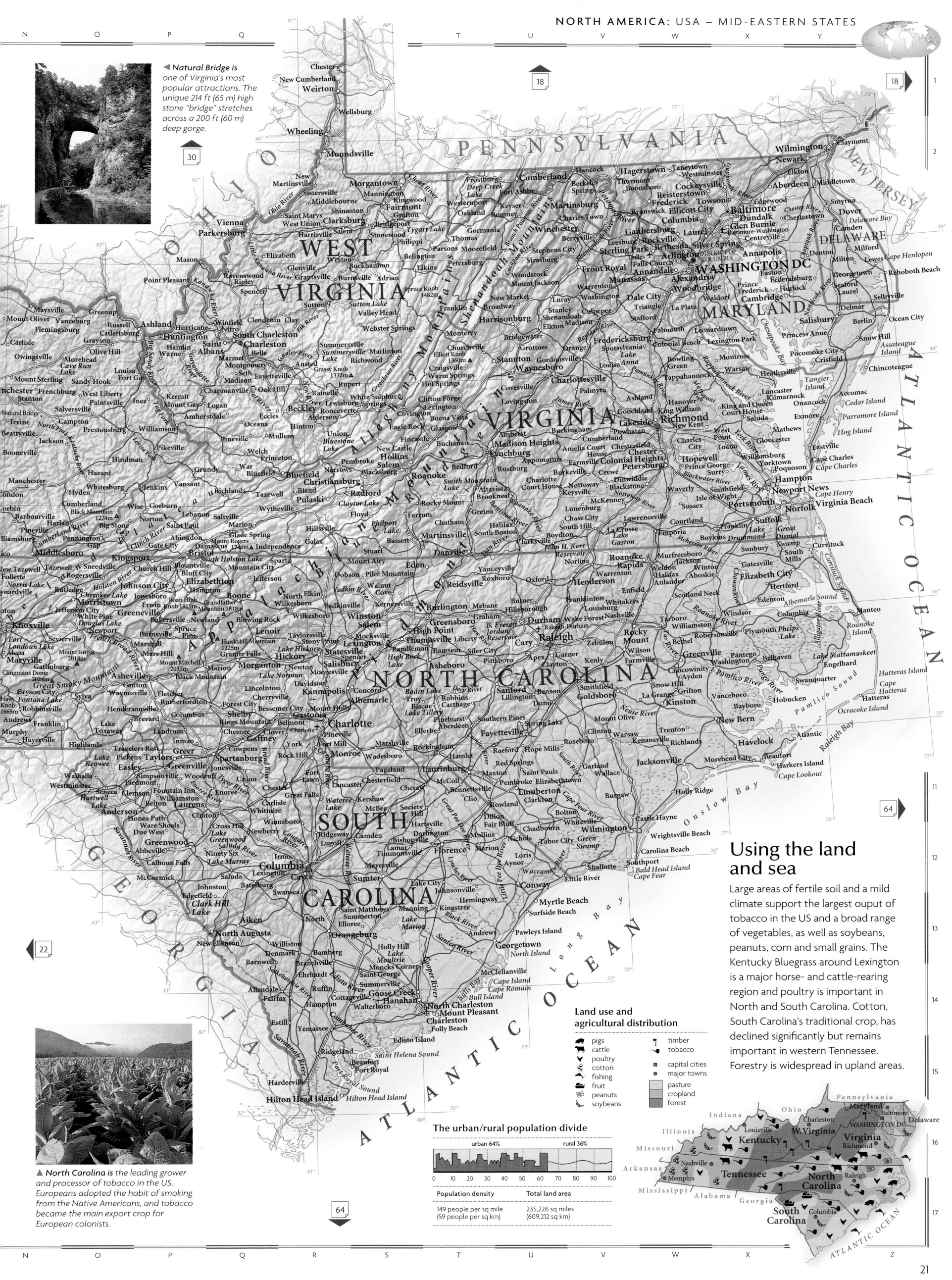

◀ *Natural Bridge is* one of Virginia's most popular attractions. The unique 214 ft (65 m) high stone "bridge" stretches across a 200 ft (60 m) deep gorge.

▲ *North Carolina is* the leading grower and processor of tobacco in the US. Europeans adopted the habit of smoking from the Native Americans, and tobacco became the main export crop for European colonists.

Using the land and sea

Large areas of fertile soil and a mild climate support the largest ouput of tobacco in the US and a broad range of vegetables, as well as soybeans, peanuts, corn and small grains. The Kentucky Bluegrass around Lexington is a major horse- and cattle-rearing region and poultry is important in North and South Carolina. Cotton, South Carolina's traditional crop, has declined significantly but remains important in western Tennessee. Forestry is widespread in upland areas.

Land use and agricultural distribution

- pigs
- cattle
- poultry
- cotton
- fishing
- fruit
- peanuts
- soybeans
- timber
- tobacco
- capital cities
- major towns
- pasture
- cropland
- forest

The urban/rural population divide

urban 64%   rural 36%

Population density	Total land area
149 people per sq mile (59 people per sq km)	235,226 sq miles (609,212 sq km)

USA: SOUTHERN STATES

Alabama, Florida, Georgia, Louisiana, Mississippi

The South has maintained a separate identity and outlook throughout the history of the US. Defeat in the Civil War (1861–65) brought chronic poverty to the former confederate states, while the subsequent liberation of four million slaves began a struggle not resolved until the 1960s, when the Civil Rights movement achieved an end to legal racial segregation. Many parts of the South have experienced rapid change. Tourism and retirement communities, together with agriculture, have fueled growth in Florida, while defense-related industries have boosted the growth of cities such as Miami and Atlanta. Many people retain a strong attachment to their history and culture, evidenced by Creole-speaking Cajuns in Louisiania and Hispanic communities in South Florida.

Transportation & industry

Florida's tourist trade is only part of a flourishing service sector, which has swelled the principal cities of the south. Petroleum and mineral extraction has made the Gulf Coast a major industrial region. Traditional textile production remains important in Georgia, while advanced new industries have grown from the NASA Space Program.

Transportation network

441,625 miles (706,600 km)	
5116 miles (8186 km)	
16,597 miles (26,555 km)	
6179 miles (9942 km)	

Atlanta's Hartsfield International airport is one of the busiest in the world. A dramatic rise in the use of regional air transportation has helped to integrate the major cities of the southern states.

◄ *The French Quarter is the traditional cultural center of New Orleans. The city, extensively damaged by Hurricane Katrina in 2005, once thrived on the cotton trade but now relies mainly on tourism and on oil from the Gulf of Mexico.*

Major industry and infrastructure

✈	aerospace	⚒ oil
🚗	car manufacture	textiles
	chemicals	✄ tourism
	coal	
	defense	● major towns
	electronics	✈ international airports
⚙	engineering	major roads
	food processing	major industrial areas

▲ *The cypress swamps of the Mississippi Delta form in the backswamps behind the levées of the river and in the multitude of subsiding delta basins.*

The landscape

The Blue Ridge mountains in the north are skirted by the gentle hills of the Piedmont, whose rivers drain south on to the great flat expanse of the coastal plain. Sandy barrier beaches and islands dominate the sea shore, tracing round the swampy limestone arm of Florida. In the west, the Mississippi meanders toward its delta, crossing the thickly mantled alluvial plain of the interior lowlands.

The Yazoo River flows parallel to the Mississippi through a common floodplain. The confluence of the rivers is deferred downstream because flood deposition has built the Mississippi channel up above the level of the Yazoo.

Cathedral Caverns near Huntsville in Alabama is a system of vast limestone caves, with a main opening 1000 ft (300 m) high and 150 ft (50 m) wide.

At De Soto Falls, Alabama, the Little River descends into the deepest canyon east of the Mississippi, with sheer cliff walls up to 700 ft (230 m) high.

Brasstown Bald in the Blue Ridge mountains of Georgia is the region's highest point, at 4784 ft (1458 m).

The Mississippi is the world's third longest river and moves over 1000 million tons (tonnes) of sediment a year, creating deep alluvial plains. Flooding is a constant threat in lowland areas.

Piedmont

▲ *In Providence Canyon, Georgia, the Chattahoochee River has cut straight down through the sandy bedrock, to leave sheer rock faces and pinnacles, which have been smoothed by subsequent weathering.*

Mississippi Delta

Atchafalaya Bay

Delta lobe

The delta of the Mississippi over 5000 years ago

Present-day delta

▲ *Over the last 5,000 years the lower course of the Mississippi has moved back and forth over great distances. These changes, caused by varying sediment loads and human modification, have resulted in a "bird's foot" delta with several lobes, each reflecting the river's different historic position*

Lake Okeechobee is actually a shallow, slow-moving river, 150 miles (240 km) long and 50 miles (80 km) wide.

The Everglades lie in a limestone hollow formed over two million years ago, which has gradually become filled with swamp deposits.

Sandbars, deposited by waves breaking offshore, form barrier beaches along much of the coastline, creating sheltered lagoons and salt marshes behind them.

Across Florida the coastal plain is mostly less than 75 ft (25 m) above sea level. The land is underlain by limestone, pitted with hollows which have been filled by over 10,000 lakes.

Florida Keys

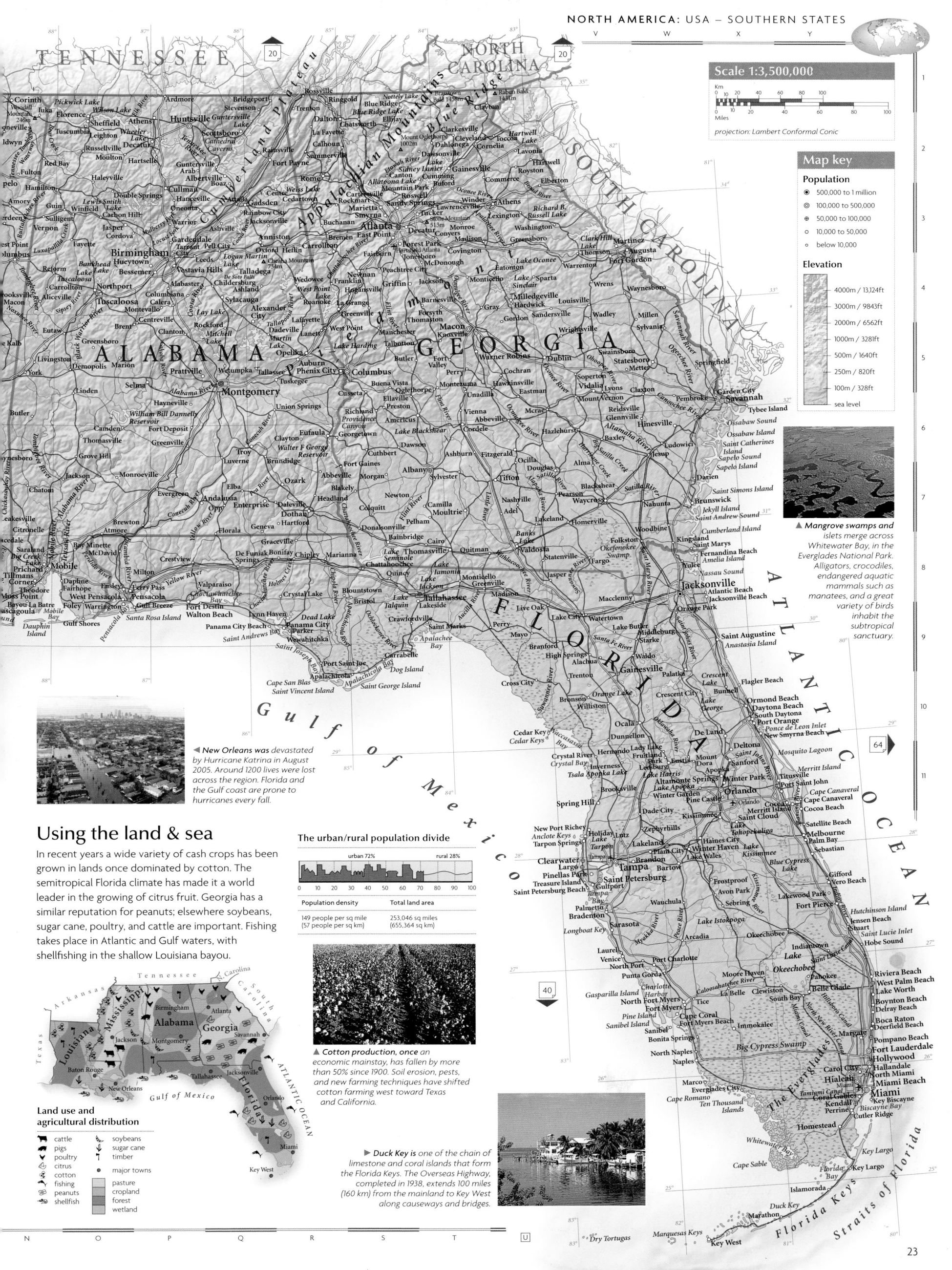

Scale 1:3,500,000

projection: Lambert Conformal Conic

Map key

Population
- 500,000 to 1 million
- 100,000 to 500,000
- 50,000 to 100,000
- 10,000 to 50,000
- below 10,000

Elevation
- 4000m / 13,124ft
- 3000m / 9843ft
- 2000m / 6562ft
- 1000m / 3281ft
- 500m / 1640ft
- 250m / 820ft
- 100m / 328ft
- sea level

▲ Mangrove swamps and islets merge across Whitewater Bay, in the Everglades National Park. Alligators, crocodiles, endangered aquatic mammals such as manatees, and a great variety of birds inhabit the subtropical sanctuary.

◄ New Orleans was devastated by Hurricane Katrina in August 2005. Around 1200 lives were lost across the region. Florida and the Gulf coast are prone to hurricanes every fall.

Using the land & sea

In recent years a wide variety of cash crops has been grown in lands once dominated by cotton. The semitropical Florida climate has made it a world leader in the growing of citrus fruit. Georgia has a similar reputation for peanuts; elsewhere soybeans, sugar cane, poultry, and cattle are important. Fishing takes place in Atlantic and Gulf waters, with shellfishing in the shallow Louisiana bayou.

The urban/rural population divide

urban 72%   rural 28%

0 10 20 30 40 50 60 70 80 90 100

Population density
149 people per sq mile
(57 people per sq km)

Total land area
253,046 sq miles
(655,364 sq km)

▲ Cotton production, once an economic mainstay, has fallen by more than 50% since 1900. Soil erosion, pests, and new farming techniques have shifted cotton farming west toward Texas and California.

Land use and agricultural distribution
- cattle
- pigs
- poultry
- citrus
- cotton
- fishing
- peanuts
- shellfish
- soybeans
- sugar cane
- timber
- major towns
- pasture
- cropland
- forest
- wetland

► Duck Key is one of the chain of limestone and coral islands that form the Florida Keys. The Overseas Highway, completed in 1938, extends 100 miles (160 km) from the mainland to Key West along causeways and bridges.

23

USA: Texas

First explored by Spaniards moving north from Mexico in search of gold, Texas was controlled by Spain and then by Mexico, before becoming an independent republic in 1836, and joining the Union of States in 1845. During the 19th century, many migrants who came to Texas raised cattle on the abundant land; in the 20th century, they were joined by prospectors attracted by the promise of oil riches. Today, although natural resources, especially oil, still form the basis of its wealth, the diversified Texan economy includes thriving hi-tech and financial industries. The major urban centers, home to 80% of the population, lie in the south and east, and include Houston, the "oil-city," and Dallas–Fort Worth. Hispanic influences remain strong, especially in southern and western Texas.

▲ *Dallas was founded* in 1841 as a prairie trading post and its development was stimulated by the arrival of railroads. Cotton and then oil funded the town's early growth. Today, the modern, high rise skyline of Dallas reflects the city's position as a leading center of banking, insurance, and the petroleum industry in the southwest.

Using the land

Cotton production and livestock-raising, particularly cattle, dominate farming, although crop failures and the demands of local markets have led to some diversification. Following the introduction of modern farming techniques, cotton production spread out from the east to the plains of western Texas. Cattle ranches are widespread, while sheep and goats are raised on the dry Edwards Plateau.

Land use and agricultural distribution

- cattle
- goats
- sheep
- cereals
- cotton
- major towns
- pasture
- cropland
- forest
- barren

The urban/rural population divide

urban 80%   rural 20%

0 10 20 30 40 50 60 70 80 90 100

Population density	Total land area
84 people per sq mile (33 people per sq km)	261,797 sq miles (678,028 sq km)

▲ *The huge cattle* ranches of Texas developed during the 19th century when land was plentiful and could be acquired cheaply. Today, more cattle and sheep are raised in Texas than in any other state.

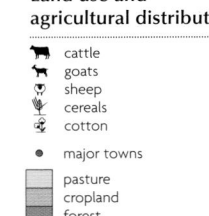

The landscape

Texas is made up of a series of massive steps descending from the mountains and high plains of the west and northwest to the coastal lowlands in the southeast. Many of the state's borders are delineated by water. The Rio Grande flows from the Rocky Mountains to the Gulf of Mexico, marking the border with Mexico.

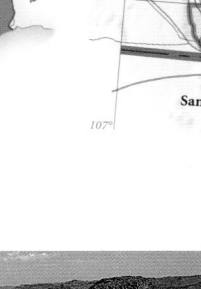

▲ *Cap Rock Escarpment* juts out from the plains, running 200 miles (320 km) from north to south. Its height varies from 300 ft (90 m) rising to sheer cliffs up to 1000 ft (300 m).

The Llano Estacado or Staked Plain in northern Texas is known for its harsh environment. In the north, freezing winds carrying ice and snow sweep down from the Rocky Mountains. To the south, sandstorms frequently blow up, scouring anything in their paths. Flash floods, in the wide, flat riverbeds that remain dry for most of the year, are another hazard.

The Guadalupe Mountains lie in the southern Rocky Mountains. They incorporate Guadalupe Peak, the highest in Texas, rising 8749 ft (2667 m).

The Rio Grande flows from the Rocky Mountains through semi-arid land, supporting sparse vegetation. The river actually shrinks along its course, losing more water through evaporation and seepage than it gains from its tributaries and rainfall.

Big Bend National Park

Edwards Plateau is a limestone outcrop. It is part of the Great Plains, bounded to the southeast by the Balcones Escarpment, which marks the southerly limit of the plains.

◄ *Flowing through* 1500 ft (450 m) high gorges, the shallow, muddy Rio Grande makes a 90° bend. This marks the southern border of Big Bend National Park, and gives it its name. The area is a mixture of forested mountains, deserts, and canyons.

The Red River flows for 1300 miles (2090 km), marking most of the northern border of Texas. A dam and reservoir along its course provide vital irrigation and hydroelectric power to the surrounding area.

Sabine River

Extensive forests of pine and cypress grow in the eastern corner of the coastal lowlands where the average rainfall is 45 inches (1145 mm) a year. This is higher than the rest of the state and over twice the average in the west.

In the coastal lowlands of southeastern Texas the Earth's crust is warping, causing the land to subside and allowing the sea to invade. Around Galveston, the rate of downward tilting is 6 inches (15 cm) per year. Erosion of the coast is also exacerbated by hurricanes.

Laguna Madre in southern Texas has been almost completely cut off from the sea by Padre Island. This sand bank was created by wave action, carrying and depositing material along the coast. The process is known as longshore drift.

Padre Island

Oil deposits

Oil accumulates beneath impermeable cap rock

Oil trapped by fault

Oil deposits migrate through reservoir rocks such as shale

Impermeable rock strata

Salt dome

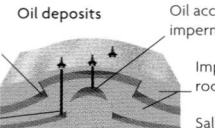

▲ *Oil deposits are* found beneath much of Texas. They collect as oil migrates upward through porous layers of rock until it is trapped, either by a cap of rock above a salt dome, or by a fault line which exposes impermeable rock through which the oil cannot rise.

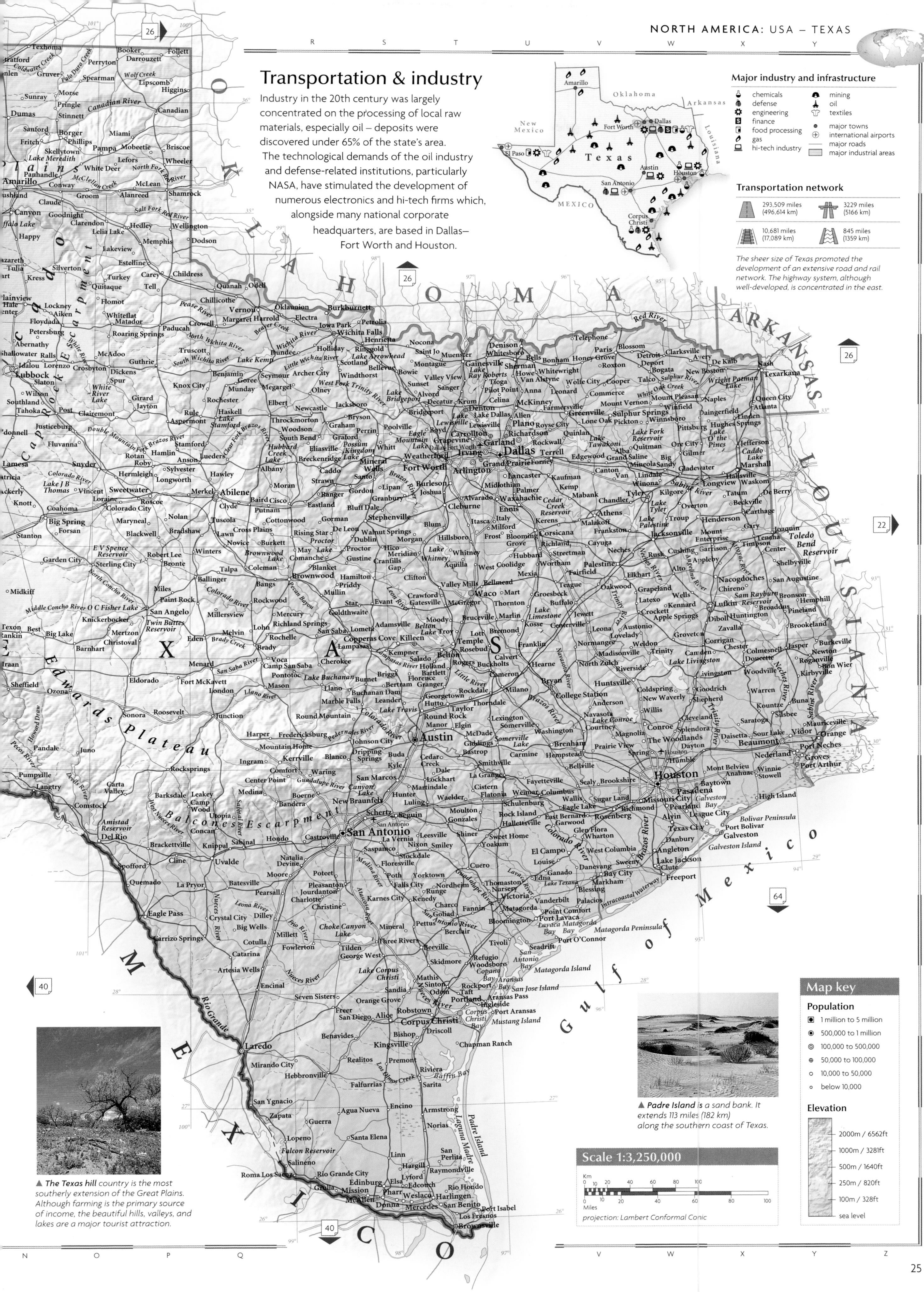

Transportation & industry

Industry in the 20th century was largely concentrated on the processing of local raw materials, especially oil – deposits were discovered under 65% of the state's area. The technological demands of the oil industry and defense-related institutions, particularly NASA, have stimulated the development of numerous electronics and hi-tech firms which, alongside many national corporate headquarters, are based in Dallas—Fort Worth and Houston.

Major industry and infrastructure

- chemicals
- defense
- engineering
- finance
- food processing
- gas
- hi-tech industry
- mining
- oil
- textiles
- major towns
- international airports
- major roads
- major industrial areas

Transportation network

293,509 miles (496,614 km)	3229 miles (5166 km)
10,681 miles (17,089 km)	845 miles (1359 km)

The sheer size of Texas promoted the development of an extensive road and rail network. The highway system, although well-developed, is concentrated in the east.

▲ *The Texas hill country is the most southerly extension of the Great Plains. Although farming is the primary source of income, the beautiful hills, valleys, and lakes are a major tourist attraction.*

▲ *Padre Island is a sand bank. It extends 113 miles (182 km) along the southern coast of Texas.*

Map key

Population

- ◉ 1 million to 5 million
- ◎ 500,000 to 1 million
- ◍ 100,000 to 500,000
- ⊕ 50,000 to 100,000
- ○ 10,000 to 50,000
- ○ below 10,000

Elevation

- 2000m / 6562ft
- 1000m / 3281ft
- 500m / 1640ft
- 250m / 820ft
- 100m / 328ft
- sea level

Scale 1:3,250,000

Km 0 10 20 40 60 80 100

Miles 0 10 20 40 60 80 100

projection: Lambert Conformal Conic

USA: SOUTH MIDWESTERN STATES

Arkansas, Kansas, Missouri, Oklahoma

The expansion of the US focused on this region in the mid-19th century. Settlers spread from the confluence of the Missouri and Mississippi rivers up onto the Great Plains. This treeless expanse, which early explorers had called the Great American Desert was turned into one of the world's richest agricultural regions. But periodic droughts, coupled with overintensive farming, led to the "dustbowl" soil erosion crisis of the 1930s, the abandonment of many farms, and a mass exodus to the west coast. The land has since recovered, although the mechanization of agriculture has led to a decline in the rural population. In recent years, suburban residential development has spread rapidly across the wooded Ozark Plateau in the east of the region.

Transportation & industry

The processing of agricultural products, such as brewing and meatpacking, has been traditionally important in these states. In Kansas and Oklahoma, diversified manufacturing now supplements income from fossil fuels; Wichita has become a world center for aeronautical engineering, an industry which also employs many people in neighboring Missouri.

Major industry and infrastructure

- ✈ aerospace
- ✿ engineering
- Ⓢ finance
- 🏭 food processing
- ◢ gas
- ⛏ mining
- ⚓ oil
- 🚗 vehicle manufacture
- • major towns
- ⊕ international airports
- — major roads
- ▨ major industrial areas

▶ *Agricultural produce from the plains is moved by barges along the Mississippi. The river now carries a far greater tonnage of freight than any other waterway system in the US.*

Transportation network

380,307 miles (608,491 km)		4068 miles (6508 km)	
16,185 miles (25,896 km)		1994 miles (3208 km)	

The Arkansas River and its tributaries allow access to over half of the US's navigable inland waterways. A system of locks and dams along the river provides Tulsa, in Oklahoma, with a navigable water route to the Gulf of Mexico.

Map key

Population

- ⊛ 100,000 to 500,000
- ⊕ 50,000 to 100,000
- ◎ 10,000 to 50,000
- ○ below 10,000

Elevation

- 1000m / 3281ft
- 500m / 1640ft
- 250m / 820ft
- 100m / 328ft
- sea level

The landscape

Most of the region consists of high, treeless plains, which gradually descend east from the Rocky Mountains. Drainage follows this slope, with rivers flowing toward the alluvial lowlands of the Mississippi in the southeast. Between the plains and the lowlands lie various ranges of wooded hills, including the deeply incised Ozark Plateau.

▲ *The Mississippi, North America's longest river, is joined by the Missouri, its main tributary, on a flood plain which spreads south to the Gulf of Mexico.*

Collapsed limestone caverns led to the formation of Big Basin in Kansas; a depression 100 ft (33 m) deep and 1 mile (1.6 km) wide.

The Great Salt Plains of northern Oklahoma cover 45 sq miles (116 sq km). The arid, white flats were left by the gradual evaporation of an ancient salt lake.

Underground water reserves

Flint Hills is the region's easternmost major escarpment. Steep, grassy uplands are interspersed with rocky, wooded ravines and outcrops of limestone and chert.

Missouri River

The Ozark Plateau is a wooded, hilly region of rivers and narrow, winding lakes. The Lake of the Ozarks was created by the damming of the Osage River in 1930.

Crowleys Ridge is a long, sandy ridge, rising from the Mississippi floodplain. It was formed over thousands of years by the deposition of sand blown eastward from the Great Plains.

Scale 1:3,000,000

Km 0 5 10 20 30 40 50 60 70
Miles 0 5 10 20 30 40 50 60 70

projection: Lambert Conformal Conic

▼ *Lake Ouachita, in Arkansas is one of a number of irregularly-shaped lakes found among the ridges of the Ouachita Mountains.*

- WY
- NE
- CO
- KS — MO
- NM — OK — AR
- TX

- Extent of the aquifer
- Kansas
- Oklahoma

Red River

Devil's Den is a dry badland area. The rugged landscape, strewn with large boulders, is the eroded remnant of a spur extending from the Arbuckle Mountains to the west.

Ouachita Mountains

Mississippi River

▲ *The Ogallala Aquifer, beneath the Great Plains, is the largest known source of underground water in the world. There is concern about the rapid depletion of this finite water supply by irrigation schemes.*

▲ *The landscape of northeast Kansas is interlaced by rivers which have cut broad wooded valleys through the gentle hills. All the rivers in Kansas form part of the massive Missouri/Mississippi drainage basin.*

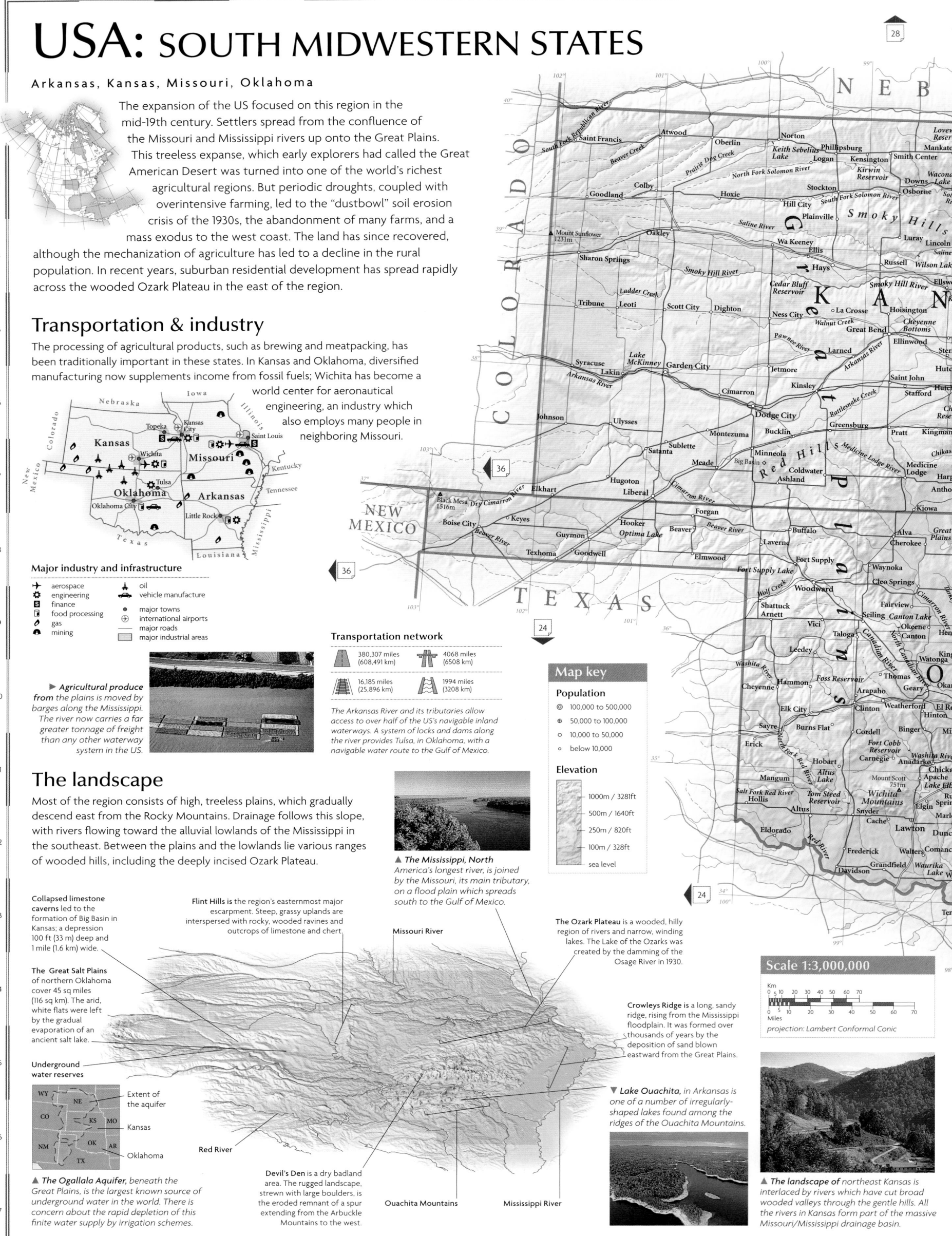

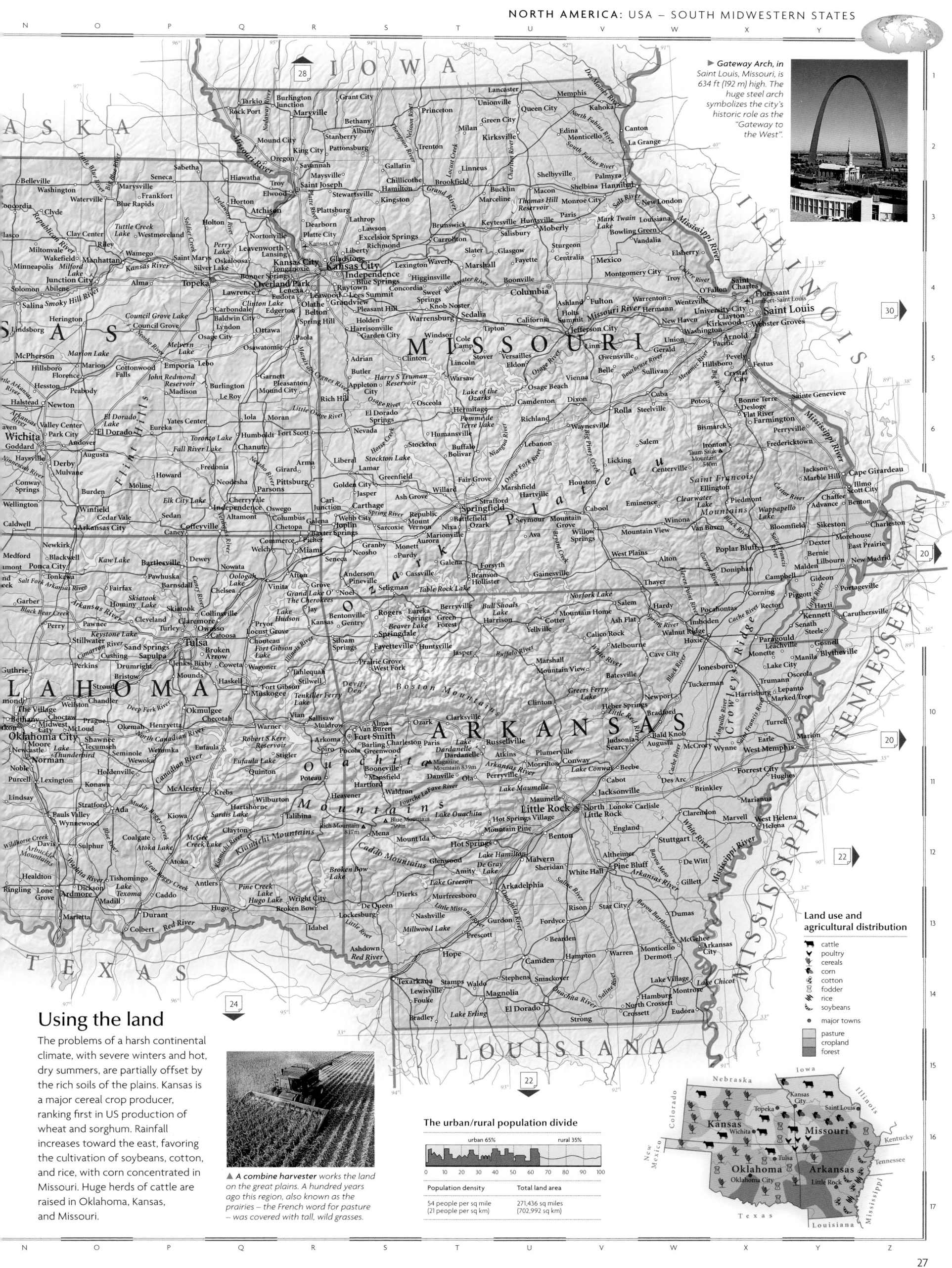

▶ *Gateway Arch,* in Saint Louis, Missouri, is 634 ft (192 m) high. The huge steel arch symbolizes the city's historic role as the "Gateway to the West".

Using the land

The problems of a harsh continental climate, with severe winters and hot, dry summers, are partially offset by the rich soils of the plains. Kansas is a major cereal crop producer, ranking first in US production of wheat and sorghum. Rainfall increases toward the east, favoring the cultivation of soybeans, cotton, and rice, with corn concentrated in Missouri. Huge herds of cattle are raised in Oklahoma, Kansas, and Missouri.

▲ *A combine harvester* works the land on the great plains. A hundred years ago this region, also known as the prairies – the French word for pasture – was covered with tall, wild grasses.

The urban/rural population divide

urban 65%   rural 35%

0 10 20 30 40 50 60 70 80 90 100

Population density	Total land area
54 people per sq mile (21 people per sq km)	271,436 sq miles (702,992 sq km)

Land use and agricultural distribution

- cattle
- poultry
- cereals
- corn
- cotton
- fodder
- rice
- soybeans
- • major towns
- pasture
- cropland
- forest

27

USA: UPPER PLAINS STATES

Iowa, Minnesota, Nebraska, North Dakota, South Dakota

Lying at the very heart of the North American continent, much of this region was acquired from France as part of the Louisiana Purchase in 1803. The area was largely bypassed by the early waves of westward migrants. When Europeans did settle, during the 19th century, they displaced the Native Americans who lived on the plains. The settlers planted arable crops and raised cattle on the immensely fertile prairie land, founding an agrarian tradition which flourishes today. Most of this region remains rural; of the five states, only in Minnesota has there been significant diversification away from agriculture and resource-based industries into the hi-tech and service sectors.

Using the land

The popular image of these states as agricultural is entirely justified; prairies stretch uninterrupted across most of the area. Croplands fall into two regions: the wheat belt of the plains, and the corn belt of the central US. Cash crops, such as soybeans, are grown to supplement incomes. Livestock, particularly pigs and cattle, are raised throughout this region.

▶ *Dark, fertile prairie soils in the southeast provide Minnesota's most productive farmland. Hot, humid summers create a long growing season for corn cultivation.*

Land use and agricultural distribution

- cattle
- pigs
- corn
- soybeans
- wheat
- • major towns
- pasture
- cropland
- forest
- wetland

The urban/rural population divide

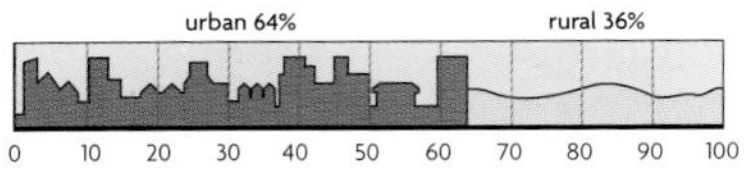

urban 64%   rural 36%

0 10 20 30 40 50 60 70 80 90 100

Population density	Total land area
31 people per sq mile (12 people per sq km)	357,212 sq miles (925,143 sq km)

Transportation & industry

Food processing and the production of farm machinery are supported by the large agricultural sector. Mineral exploitation is also an important activity: gold is mined in the ore-rich Black Hills of South Dakota, and both North Dakota and Nebraska are emerging as major petroleum producers.

▶ *Water erosion along the Little Missouri River has carried away sedimentary deposits, creating rugged landscapes known as badlands.*

Transportation network

504,522 miles (807,235 km)	3422 miles (5475 km)
16,940 miles (27,104 km)	683 miles (1098 km)

Nebraska's central location has made it an important transportation artery for east–west traffic. Minnesota's road network radiates out from the hub of the twin cities, Minneapolis–Saint Paul.

Major industry and infrastructure

- coal
- engineering
- electronics
- finance
- food processing
- oil & gas
- mining
- ⊕ major towns
- international airports
- major roads
- major industrial areas

The landscape

These states straddle the Great Plains and the lowlands of the central US, with Minnesota lying in a transition zone between the eastern forests and the prairies. The region was shaped by repeated ice advances and retreats, leaving a flat relief, broken only by the numerous lakes and broad river networks that drain the prairies.

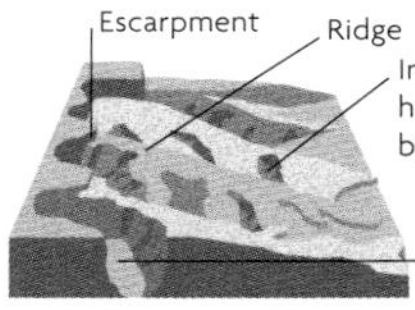

Escarpment   Ridge

In permeable strata hollows are formed by small mudslides

Water flowing into gullies erodes back the escarpment

▲ *Badlands are formed by stormwater run-off. This flows down the impermeable strata of the escarpment and saturates the permeable strata, leading to mudslides and the formation of gullies.*

North Dakota Badlands

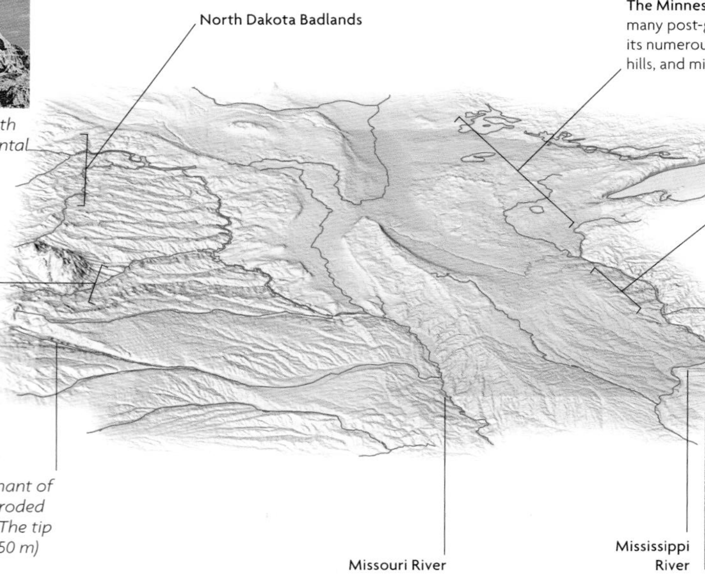

The Minnesota landscape contains many post-glacial features, including its numerous lakes, boulder-strewn hills, and mineral-rich deposits.

▲ *In the badlands of North and South Dakota, horizontal layers of sandstone have been eroded by rivers, leaving a landscape of narrow gullies, sharp crests and pinnacles.*

South Dakota Badlands

▲ *Chimney Rock is a remnant of an ancient land surface, eroded by the North Platte River. The tip of its spire stands 500 ft (150 m) above the plain.*

Missouri River

Although it escaped the last glaciation, the limestone bedrock of southeastern Minnesota has been eroded by surface and subterranean streams, leaving a network of underground caverns and steepsided valleys.

Mississippi River

◀ *In northeastern Iowa, the Mississippi and its tributaries have deeply incised the underlying bedrock creating a hilly terrain, with bluffs standing 300 ft (90 m) above the valley.*

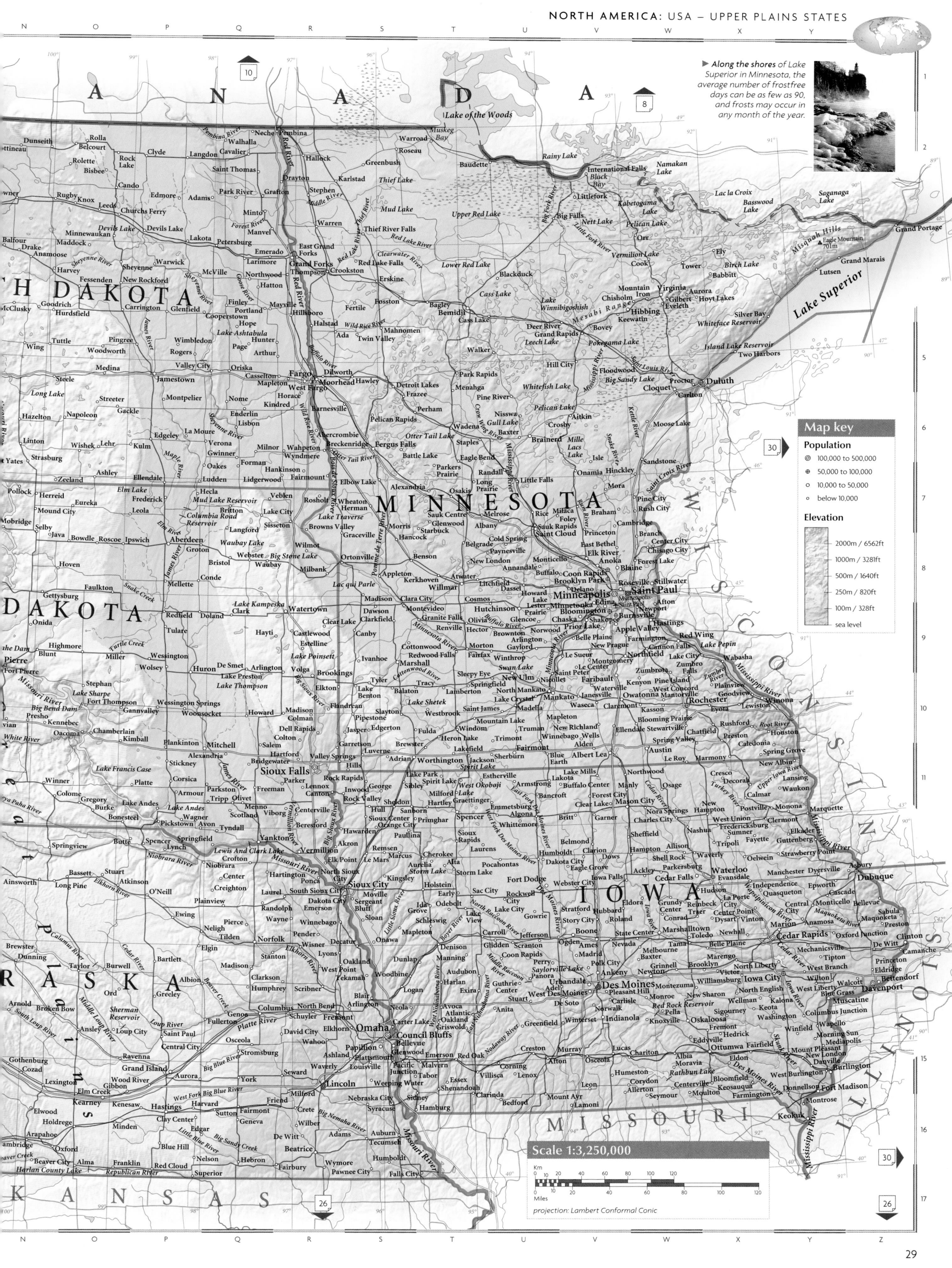

► **Along the shores** of Lake Superior in Minnesota, the average number of frostfree days can be as few as 90, and frosts may occur in any month of the year.

Map key

Population

- ◎ 100,000 to 500,000
- ⊕ 50,000 to 100,000
- ○ 10,000 to 50,000
- ○ below 10,000

Elevation

- 2000m / 6562ft
- 1000m / 3281ft
- 500m / 1640ft
- 250m / 820ft
- 100m / 328ft
- sea level

Scale 1:3,250,000

Km
0 10 20 40 60 80 100 120

Miles
0 10 20 40 60 80 100 120

projection: Lambert Conformal Conic

USA: GREAT LAKES STATES

Illinois, Indiana, Michigan, Ohio, Wisconsin

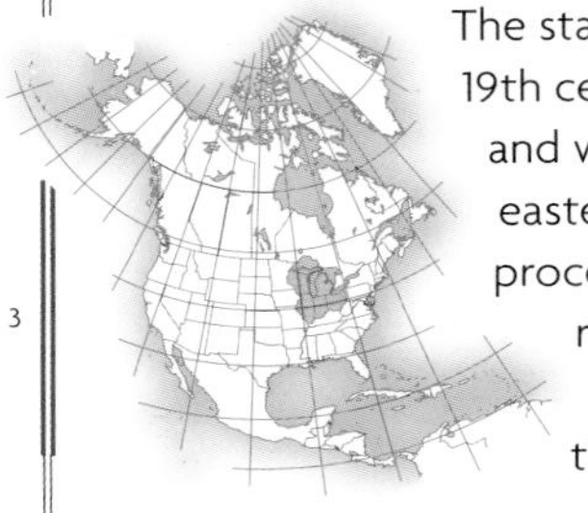

The states bordering the Great Lakes developed rapidly in the second half of the 19th century as a result of improvements in communications: railroads to the west and waterways to the south and east. Fertile land and good links with growing eastern seaboard cities encouraged the development of agriculture and food processing. Migrants from Europe and other parts of the US flooded into the region and for much of the 20th century the region's economy boomed. However, in recent years heavy industry has declined, earning the region the unwanted label the "Rustbelt."

Transportation & industry

The Great Lakes region is the center of the US car industry. Since the early part of the 20th century, its prosperity has been closely linked to the fortunes of automobile manufacturing. Iron and steel production has expanded to meet demand from this industry. In the 1970s, nationwide recession, cheaper foreign competition in the automobile sector, pollution in and around the Great Lakes, and the collapse of the meatpacking industry, centered on Chicago, forced these states to diversify their industrial base. New industries have emerged, notably electronics, service, and finance industries.

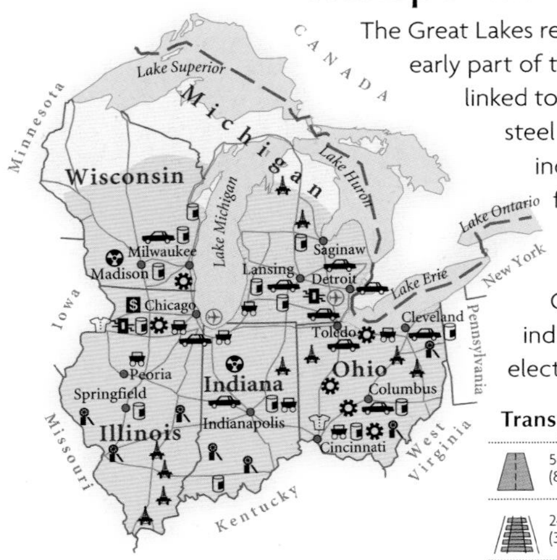

Transportation network

540,682 miles (865,091 km)		6550 miles (10,480 km)	
24,928 miles (39,884 km)		2330 miles (3748 km)	

Few areas of the US have a comparable system. Chicago is a principal transportation terminus with a dense network of roads, railroads, and Interstate freeways that radiates out from the city.

Major industry and infrastructure

- car manufacture
- coal
- electronics
- engineering
- finance
- food processing
- iron & steel
- oil
- research & development
- textiles
- major towns
- international airports
- major roads
- major industrial areas

◀ Ever since Ransom Olds and Henry Ford started mass-producing automobiles in Detroit early in the 20th century, the city's name has become synonymous with the American automotive industry.

The landscape

Much of this region shows the impact of glaciation which lasted until about 10,000 years ago, and extended as far south as Illinois and Ohio. Although the relief of the region slopes toward the Great Lakes, because the ice sheets blocked northerly drainage, most of the rivers today flow southward, forming part of the massive Mississippi/Missouri drainage basin.

◀ The dunes near Sleeping Bear Point rise 400 ft (120 m) from the banks of Lake Michigan. They are constantly being resculpted by wind action.

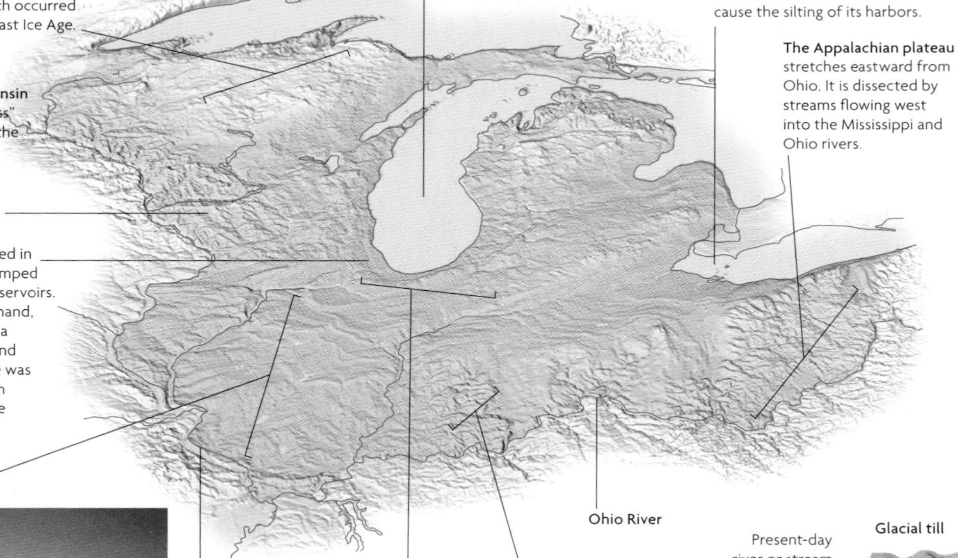

Lake Michigan

Lake Erie is the shallowest of the five Great Lakes. Its average depth is about 62 ft (19 m). Storms sweeping across from Canada erode its shores and cause the silting of its harbors.

The Appalachian plateau stretches eastward from Ohio. It is dissected by streams flowing west into the Mississippi and Ohio rivers.

The many lakes and marshes of Wisconsin and Michigan are the result of glacial erosion and deposition which occurred during the last Ice Age.

Southwestern Wisconsin is known as a "driftless" area. Unlike most of the region, low hills protected it from erosion by the advancing ice sheet.

Most of the water used in northern Illinois is pumped from underground reservoirs. Due to increased demand, many areas now face a water shortage. Around Joliet, the water table was lowered by more than 700 ft (210 m) over the last century.

Illinois plains

▲ The plains of Illinois are characteristic of drift landscapes, scoured and flattened by glacial erosion and covered with fertile glacial deposits.

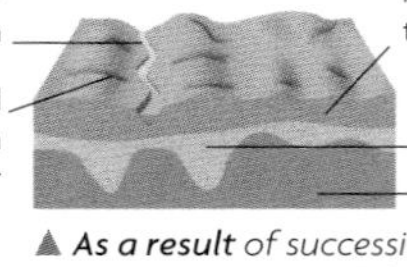

Mississippi River

Ohio River

Relic landforms from the last glaciation, such as shallow basins and ridges, cover all but the south of this region. Ridges, known as moraines, up to 300 ft (100 m) high, lie to the south of Lake Michigan.

Unlike the level prairie to the north, southern Indiana is relatively rugged. Limestone in the hills has been dissolved by water, producing features such as sinkholes and underground caves.

Glacial till

Present-day river or stream

Channels caused by outwash from melting glacier

Most recent till deposits

Older till sheet

Bedrock

▲ As a result of successive glacial depositions, the total depth of till along the former southern margin of the Laurentide ice sheet can exceed 1300 ft (400 m).

Map labels

MINNESOTA · WISCONSIN · ILLINOIS · IOWA · MISSOURI · CANADA · Michigan · Lake Superior · Lake Michigan · Lake Huron · Lake Erie · Lake Ontario · New York · Pennsylvania · West Virginia · Kentucky · Indiana · Ohio · Milwaukee · Madison · Lansing · Detroit · Cleveland · Toledo · Chicago · Peoria · Springfield · Indianapolis · Columbus · Cincinnati · Saginaw

Thunder Bay · Passage Island · Blake Point · Isle Royale · Siskiwit Ba · Eagle Point · Laurium · Hancock · Houghton · Cumberland Point · Long Point · Apostle Islands · Outer Island · Devils Island · Saint Island · Bark Point · Mount Ashwabay 437m · Madeline Island · Stockton Island · Fourteen Mile Point · Ontonagon · Porcupine Mountains · Superior · Iron River · Ashland · Washburn · Saint Croix Flowage · Big Manitou Falls · Gile Flowage · Hurley · Ironwood · Bessemer Gogebic · Wakefield · Bond Falls Flowage · Michigan · Lake Gogebic · Lac Vieux Desert · Watersmeet · Iron River · Crystal · Nelson Lake · Gile Flowage · Namekagon Lake · Turtle Flambeau Flowage · Lac Court Oreilles · Spooner · Grantsburg · Frederic · Bone Lake · Balsam Lake · Shell Lake · Long Lake · Rice Lake · Minocqua · Eagle River · Woodruff · Rhinelander · Crandon · Phillips · Willow Reservoir · Ladysmith · Jump River · Timms Hill 595m · Pelican Lake · Tomahawk · Antigo · Keshena · Chetek · Cameron · Barron · Merrill · Wausau · Wittenberg · Shawano · New Richmond · Glenwood City · Cornell · Miller Dam Flowage · Medford · Rib Mountain · Big Eau Pleine Reservoir · Mosinee · Lake Du Bay · Stevens Point · Chippewa Falls · Stratford · Spencer · Marshfield · Plover · Eau Claire · Altoona · Loyal · Wisconsin Rapids · Waupaca · New London · Kaukauna · Appleton · Menasha · Neenah · Oshkosh · Menomonie · Durand · Mondovi · Osseo · Neillsville · Nekoosa · Weyauwega · Winneconne · Chilton · Lake Pepin · Buffalo · Independence · Whitehall · Black River Falls · Yellow River · Wautoma · Berlin · Ripon · Green Lake · Alma · Arcadia · Galesville · Tomah · Castle Rock Lake · Friendship · Rush Lake · Lake Winnebago · Lac · Trempealeau · Holmen · West Salem · Sparta · Mauston · Westfield · Montello · Wisconsin Dells · Puckaway Lake · Waupun · Beaver Dam Lake · La Crosse · Westby · La Farge · Elroy · Hillsboro · Reedsburg · Portage · Columbus · Beaver Dam · Hartford · Viroqua · Richland Center · Baraboo · Sauk City · Lake Wisconsin · Juneau · Mayville · Menomonee · Prairie du Chien · Muscoda · Spring Green · Lake Mendota · Middleton · Sun Prairie · Oconomowoc · Delafield · Waukesha · Fennimore · Dodgeville · Madison · Monona · Lake Kegonsa · New Glarus · Stoughton · Whitewater · Elkhorn · Platteville · Mineral Point · Verona · Lancaster · Evansville · Delavan · Dickeyville · Hazel Green · Darlington · Monroe · Beloit · Janesville · Lake Geneva · Shullsburg · East Dubuque · Lena · Loves Park · Harvard · Woodstock · Antioch · Galena · Freeport · Rockford · Belvidere · Genoa · Arlington Heights · Savanna · Mount Carroll · Oregon · Byron · Sycamore · Schaumburg · Elgin · Mount Morris · Polo · Rochelle · De Kalb · Geneva · Wheaton · Morrison · Sterling · Rock Falls · Dixon · Prophetstown · Amboy · Sandwich · Plano · Aurora · Moline · Rock Island · East Moline · Geneseo · Princeton · Peru · La Salle · Ottawa · Marseilles · Joliet · Milan · Orion · Kewanee · Henry · Streator · Pontiac · Aledo · Galva · Toulon · Lacon · Minonk · Oquawka · Monmouth · Galesburg · Knoxville · Abingdon · Elmwood · Peoria · East Peoria · Pekin · Washington · Normal · Bloomington · Le Roy · Gibson City · Paxton · Carthage · Macomb · Canton · Lewistown · Havana · Mason City · Clinton · Champaign · Urbana · Bushnell · Tremont · Hamilton · Quincy · Rushville · Beardstown · Virginia · Petersburg · Lincoln · Clinton · Monticello · Pittsfield · Jacksonville · Springfield · Riverton · Decatur · Tuscola · Arcola · Winchester · Chatham · Auburn · Sangchris Lake · Lake Springfield · Mount Zion · Sullivan · Roodhouse · White Hall · Pawnee · Taylorville · Pana · Shelbyville · Carrollton · Virden · Girard · Nokomis · Shelbyville · Toledo · Jerseyville · Carlinville · Litchfield · Lake Lou Yaeger · Hardin · Gillespie · Mount Olive · Vandalia · Altamont · Brighton · Staunton · Greenville · Effingham · Alton · Bethalto · Wood River · Edwardsville · Carlyle · Louisville · Olney · Granite City · Glen Carbon · Highland · Carlyle Lake · East Saint Louis · Collinsville · O'Fallon · Salem · Flora · Fairfield · Albion · Cahokia · Freeburg · Centralia · Waterloo · Columbia · Mount Vernon · Red Bud · Nashville · Sparta · Rend Lake · Sesser · Steeleville · Pinckneyville · Benton · McLeansboro · Grayville · Chester · Christopher · West Frankfort · Johnston City · Eldorado · Murphysboro · Herrin · Marion · Harrisburg · Carbondale · Bald Knob · Lake of Egypt · Vienna · Jonesboro · Anna · Gwicola · Metropolis · Mound City · Cairo · Ohio River

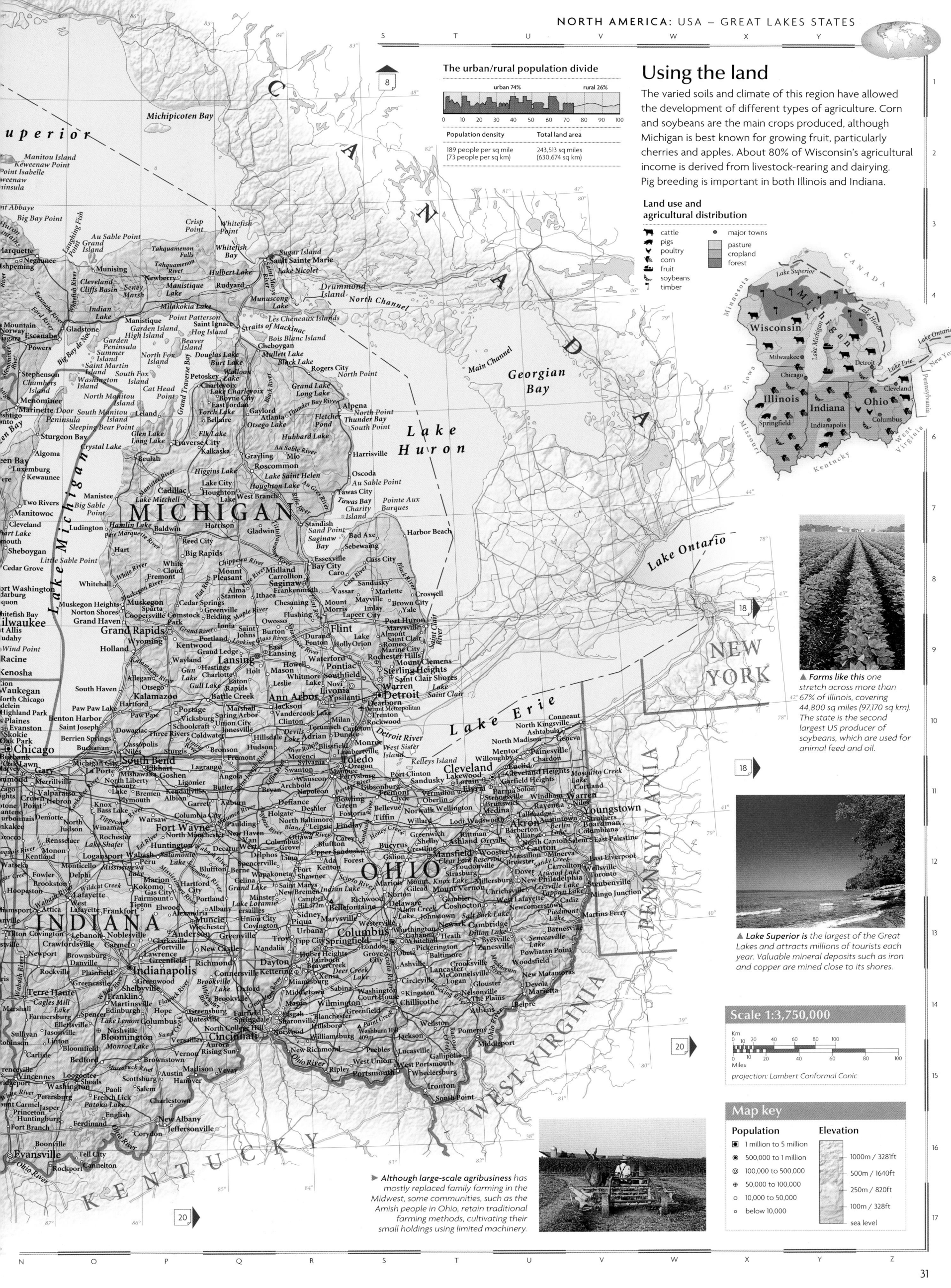

The urban/rural population divide

urban 74%   rural 26%

0 10 20 30 40 50 60 70 80 90 100

Population density	Total land area
189 people per sq mile (73 people per sq km)	243,513 sq miles (630,674 sq km)

Using the land

The varied soils and climate of this region have allowed the development of different types of agriculture. Corn and soybeans are the main crops produced, although Michigan is best known for growing fruit, particularly cherries and apples. About 80% of Wisconsin's agricultural income is derived from livestock-rearing and dairying. Pig breeding is important in both Illinois and Indiana.

Land use and agricultural distribution

- cattle
- pigs
- poultry
- corn
- fruit
- soybeans
- timber
- major towns
- pasture
- cropland
- forest

▲ *Farms like this* one stretch across more than 67% of Illinois, covering 44,800 sq miles (97,170 sq km). The state is the second largest US producer of soybeans, which are used for animal feed and oil.

▲ *Lake Superior is* the largest of the Great Lakes and attracts millions of tourists each year. Valuable mineral deposits such as iron and copper are mined close to its shores.

▶ *Although large-scale agribusiness* has mostly replaced family farming in the Midwest, some communities, such as the Amish people in Ohio, retain traditional farming methods, cultivating their small holdings using limited machinery.

Scale 1:3,750,000

Km
0 10 20   40   60   80   100

Miles
0 10 20   40   60   80   100

projection: Lambert Conformal Conic

Map key

Population
- ◼ 1 million to 5 million
- ◉ 500,000 to 1 million
- ◎ 100,000 to 500,000
- ◉ 50,000 to 100,000
- ○ 10,000 to 50,000
- ○ below 10,000

Elevation
- 1000m / 3281ft
- 500m / 1640ft
- 250m / 820ft
- 100m / 328ft
- sea level

USA: NORTH MOUNTAIN STATES

Idaho, Montana, Oregon, Washington, Wyoming

The remoteness of the northwestern states, coupled with the rugged landscape, ensured that this was one of the last areas settled by Europeans in the 19th century. Fur-trappers and gold-prospectors followed the Snake River westward as it wound its way through the Rocky Mountains. The states of the northwest have pioneered many conservationist policies, with the first US National Park opened at Yellowstone in 1872. More recently, the Cascades and Rocky Mountains have become havens for adventure tourism. The mountains still serve to isolate the western seaboard from the rest of the continent. This isolation has encouraged West Coast cities to expand their trade links with countries of the Pacific Rim.

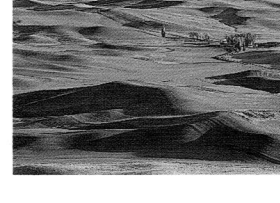

▲ *The Snake River* has cut down into the basalt of the Columbia Basin to form Hells Canyon, the deepest in the US, with cliffs up to 7900 ft (2408 m) high.

Map key

Population
- ◉ 500,000 to 1 million
- ◎ 100,000 to 500,000
- ⊕ 50,000 to 100,000
- ⊙ 10,000 to 50,000
- ○ below 10,000

Elevation
- 4000m / 13,124ft
- 3000m / 9843ft
- 2000m / 6562ft
- 1000m / 3281ft
- 500m / 1640ft
- 250m / 820ft
- 100m / 328ft
- sea level

▶ *Fine-textured, volcanic soils* in the hilly Palouse region of eastern Washington are susceptible to erosion.

Using the land

Wheat farming in the east gives way to cattle ranching as rainfall decreases. Irrigated farming in the Snake River valley produces large yields of potatoes and other vegetables. Dairying and fruit-growing take place in the wet western lowlands between the mountain ranges.

The urban/rural population divide

urban 74% | rural 26%

Population density	Total land area
26 people per sq mile (10 people per sq km)	487,970 sq miles (1,263,716 sq km)

Scale 1:3,750,000

Km 0 20 40 60 80 100
Miles 0 20 40 60 80 100

projection: Lambert Conformal Conic

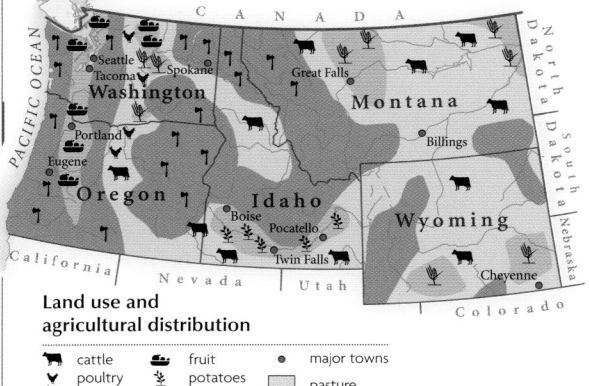

Land use and agricultural distribution
- 🐄 cattle
- 🦃 poultry
- 🌾 cereals
- 🍎 fruit
- 🥔 potatoes
- 🌲 timber
- ● major towns
- pasture
- cropland
- forest

Transportation & industry

Minerals and timber are extremely important in this region. Uranium, precious metals, copper, and coal are all mined, the latter in vast open-cast pits in Wyoming; oil and natural gas are extracted further north. Manufacturing, notably related to the aerospace and electronics industries, is important in western cities.

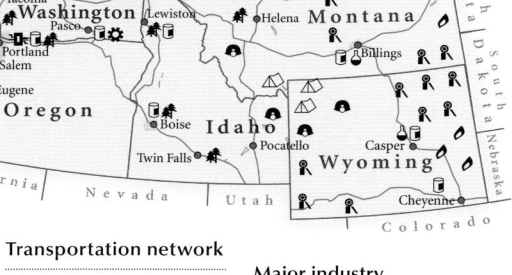

Transportation network
- 347,857 miles (556,571 km)
- 4200 miles (6720 km)
- 12,354 miles (19,766 km)
- 1108 miles (1782 km)

The Union Pacific Railroad has been in service across Wyoming since 1867. The route through the Rocky Mountains is now shared with the Interstate 80, a major east–west highway.

Major industry and infrastructure
- 🎿 adventure tourism
- ✈ aerospace
- coal
- chemicals
- electronics
- food processing
- ⛏ mining
- 🛢 oil & gas
- 🌲 timber processing
- ⊕ major towns
- international airports
- — major roads
- major industrial areas

◀ *Seattle lies in* one of Puget Sound's many inlets. The city receives oil and other resources from Alaska, and benefits from expanding trade across the Pacific.

◀ *Crater Lake, Oregon,* is 6 miles (10 km) wide and 1800 ft (600 m) deep. It marks the site of a volcanic cone, which collapsed after an eruption within the last 7000 years.

The landscape

The Rocky Mountains are flanked by lower parallel ranges, which spread onto the Great Plains in the east and surmount the broad lava plateau which extends westward. The Cascade Range divides the Columbia Basin from the coastlands, where the low areas around Puget Sound are broken by the steep, volcanic Olympic Mountains and the wooded hills of the Coast Ranges.

Molten rock cools, forming parallel columns

Surrounding strata eroded away

Molten rock wells up from the Earth's core

▲ *Devil's Tower in* Wyoming is an igneous intrusion, formed below the Earth's surface. Molten rock intruded through cracks in the overlying strata and cooled. Over time, the softer rock layers have been eroded away, leaving only the tower standing.

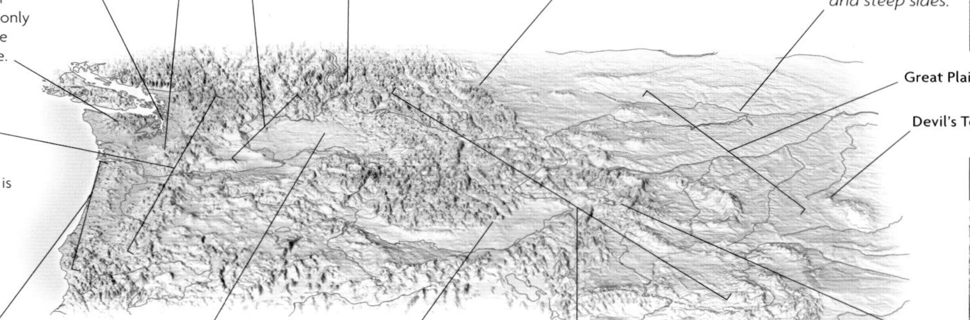

Puget Sound

Mount St. Helens erupted in 1980, killing 57 people and devastating a huge area.

Columbia Basin

Grand Coulee and the lesser *coulées* (ravines) were cut by cataclysmic floods, from the release of an ice-dammed lake, at the end of the last Ice Age.

The Continental Divide, or watershed, crosses the Lewis Range. From here, rivers flow east to Hudson Bay, south to the Gulf of Mexico and west to the Pacific Ocean.

▶ *Piney Buttes are the* remnants of an older, higher land surface gradually weathered and eroded into isolated outcrops with flat tops and steep sides.

Glacial valleys on the seaward side of the Olympic Mountains receive about 142 inches (3600 mm) of rain per year, supporting the only true rain forest of the northern hemisphere.

The Cascades are glacially scoured volcanic mountains, the highest of which is Mount Rainier, a dormant volcano at 14,409 ft (4392 m).

Coast Ranges

Great Plains

Devil's Tower

Rocky Mountains

The plateaus of the Columbia and Snake rivers represent one of the world's largest accumulations of lava. Over 5 million years ago, successive flows of molten basalt buried the existing land surface by up to 450 ft (150 m).

The contorted rock shapes at "Craters of the Moon" National Monument in Idaho were left 2000 years ago by the sporadic upwelling of viscous lava from fissures in the basalt plateau.

▲ *Water from the* hot springs in Yellowstone National Park deposits minerals as it cools in rock pools. Long periods of deposition have created these rock terraces.

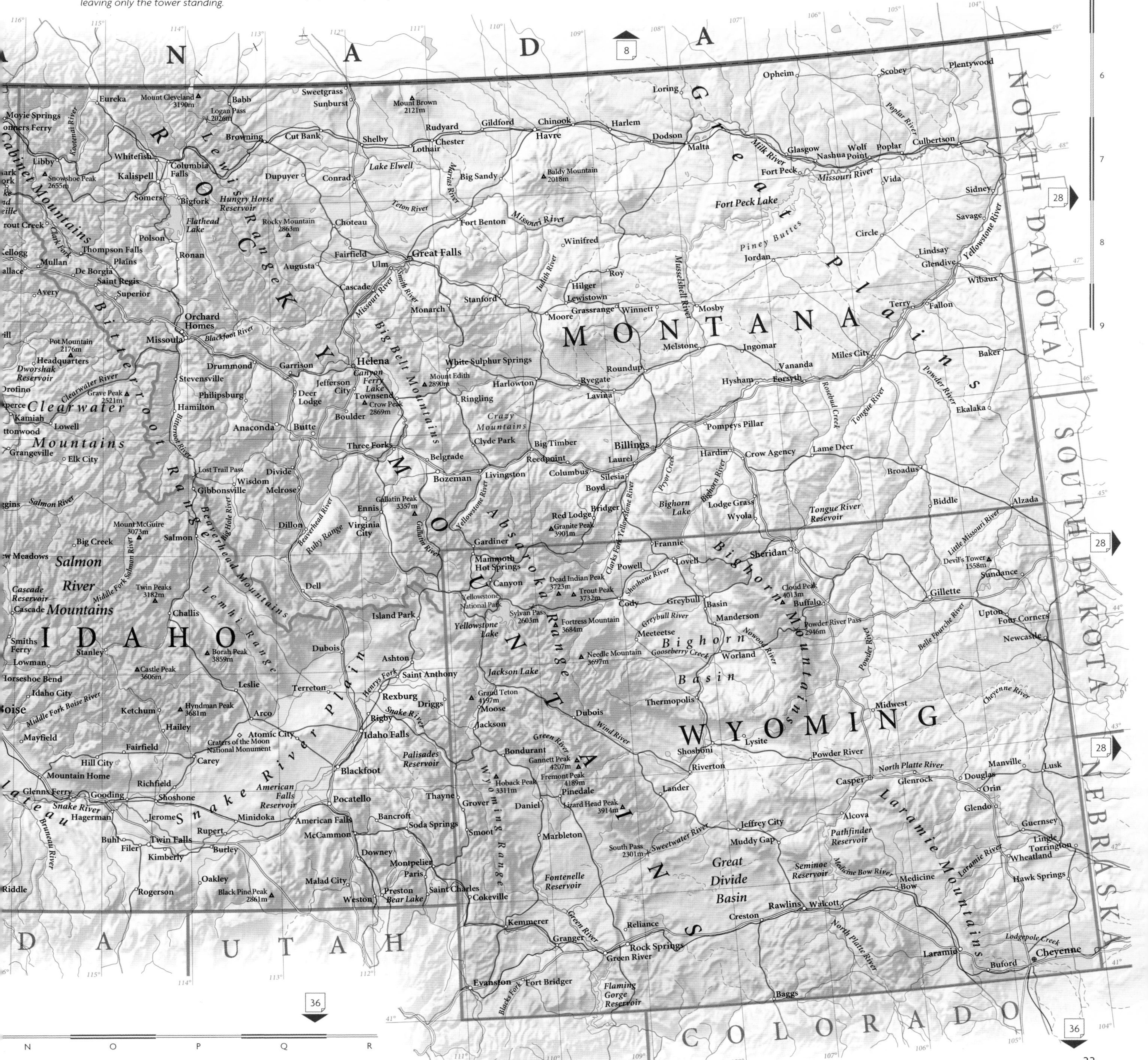

[Map of North Mountain States: Idaho, Montana, Wyoming with surrounding Canada, North Dakota, South Dakota, Nebraska, Colorado, Utah]

USA: CALIFORNIA & NEVADA

The Gold Rush of 1849 attracted the first major wave of European settlers to the West Coast. The pleasant climate, beautiful scenery, and dynamic economy continue to attract immigrants – despite the ever-present danger of earthquakes – and California has become the US's most populous state. The overwhelmingly urban population is concentrated in the vast conurbations of Los Angeles, San Francisco, and San Diego; new immigrants include people from South Korea, the Philippines, Vietnam, and Mexico. Nevada's arid lands were initially exploited for minerals; in recent years, revenue from mining has been superseded by income from the tourist and gambling centers of Las Vegas and Reno.

Map key

Population
- ◉ 1 million to 5 million
- ◎ 500,000 to 1 million
- ⊚ 100,000 to 500,000
- ⊕ 50,000 to 100,000
- ○ 10,000 to 50,000
- ○ below 10,000

Elevation
- 4000m / 13,124ft
- 3000m / 9843ft
- 2000m / 6562ft
- 1000m / 3281ft
- 500m / 1640ft
- 250m / 820ft
- 100m / 328ft
- sea level

Scale 1:3,000,000

Km
0 5 10 20 30 40 50 60 70 80

Miles
0 5 10 20 30 40 50 60 70 80

projection: Lambert Conformal Conic

Transportation & industry

Nevada's rich mineral reserves ushered in a period of mining wealth which has now been replaced by revenue generated from gambling. California supports a broad set of activities including defense-related industries and research and development facilities. "Silicon Valley," near San Francisco, is a world leading center for micro-electronics, while tourism and the Los Angeles film industry also generate large incomes.

Major industry and infrastructure
- ✈ aerospace
- 🚗 car manufacture
- defense
- $ film industry
- $ finance
- food processing
- gambling
- hi-tech industry
- mining
- pharmaceuticals
- research & development
- textiles
- tourism
- ● major towns
- ⊕ international airports
- — major roads
- major industrial areas

Transportation network

211,459 miles (338,334 km)	2944 miles (4710 km)
7822 miles (12,595 km)	190 miles (360 km)

In California, the motor vehicle is a vital part of daily life, and an extensive freeway system runs throughout the state, cementing its position as the most important mode of transport.

◀ *Gambling was legalized in Nevada in 1931. Las Vegas has since become the center of this multimillion dollar industry.*

The landscape

The broad Central Valley divides California's coastal mountains from the Sierra Nevada. The San Andreas Fault, running beneath much of the state, is the site of frequent earth tremors and sometimes more serious earthquakes. East of the Sierra Nevada, the landscape is characterized by the basin and range topography with stony deserts and many salt lakes.

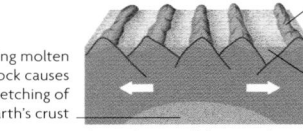

Rising molten rock causes stretching of the Earth's crust

Extensive cracking (faulting) uplifted a series of ridges

As ridges are eroded they fill intervening valleys with sediments

▲ *Molten rock (magma) welling up to form a dome in the Earth's interior, causes the brittle surface rocks to stretch and crack. Some areas were uplifted to form mountains (ranges), while others sunk to form flat valleys (basins).*

◀ *The General Sherman sequoia tree in Sequoia National Park is around 2500 years old and at 275 ft (84 m) is one of the largest living things on earth.*

Most of California's agriculture is confined to the fertile and extensively irrigated Central Valley, running between the Coast Ranges and the Sierra Nevada. It incorporates the San Joaquin and Sacramento valleys.

The dramatic granitic rock formations of Half Dome and El Capitan, and the verdant coniferous forests, attract millions of visitors annually to Yosemite National Park in the Sierra Nevada.

The Great Basin dominates most of Nevada's topography containing large open basins, punctuated by eroded features such as *buttes* and *mesas*. River flow tends to be seasonal, dependent upon spring showers and winter snow melt.

Sierra Nevada

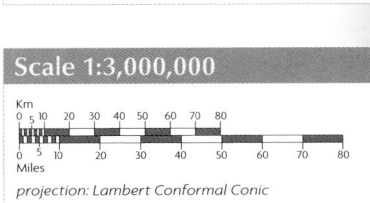

Using the land

California is the leading agricultural producer in the US, although low rainfall makes irrigation essential. The long growing season and abundant sunshine allow many crops to be grown in the fertile Central Valley including grapes, citrus fruits, vegetables, and cotton. Almost 17 million acres (6.8 million hectares) of California's forests are used commercially. Nevada's arid climate and poor soil are largely unsuitable for agriculture; 85% of its land is state owned and large areas are used for underground testing of nuclear weapons.

Wheeler Peak is home to some of the world's oldest trees, bristlecone pines, which live for up to 5000 years.

The San Andreas Fault is a transverse fault which extends for 650 miles (1050 km) through California. Major earthquakes occur when the land either side of the fault moves at different rates. San Francisco was devastated by an earthquake in 1906.

When the Hoover Dam across the Colorado River was completed in 1936, it created Lake Mead, one of the largest artificial lakes in the world, extending for 115 miles (285 km) upstream.

Amargosa Desert

Land use and agricultural distribution
- 🐄 cattle
- citrus fruits
- fruit
- irrigation
- timber
- vineyards
- ● major towns
- pasture
- cropland
- forest
- desert

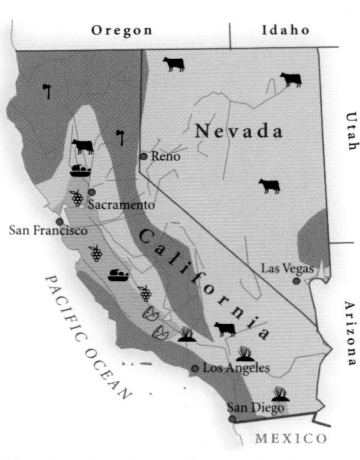

Death Valley

▶ *Named by migrating settlers in 1849, Death Valley is the driest, hottest place in North America, as well as being the lowest point on land in the western hemisphere, at 282 ft (86 m) below sea level.*

The sparsely populated Mojave Desert receives less than 8 inches (200 mm) of rainfall a year. It is used extensively for weapons-testing and military purposes.

The Salton Sea was created accidentally between 1905 and 1907 when an irrigation channel from the Colorado River broke out of its banks and formed this salty 300 sq mile (777 sq km), landlocked lake.

▲ *The Sierra Nevada create a "rainshadow," preventing rain from reaching much of Nevada. Pacific air masses, passing over the mountains, are stripped of their moisture.*

▲ *Without considerable irrigation, this fertile valley at Palm Springs would still be part of the Sonoran Desert. California's farmers account for about 80% of the state's total water usage.*

The urban/rural population divide

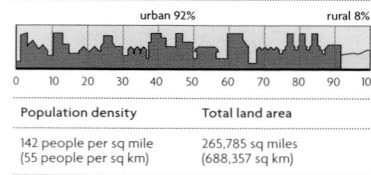

urban 92% | rural 8%

0 10 20 30 40 50 60 70 80 90 100

Population density	Total land area
142 people per sq mile (55 people per sq km)	265,785 sq miles (688,357 sq km)

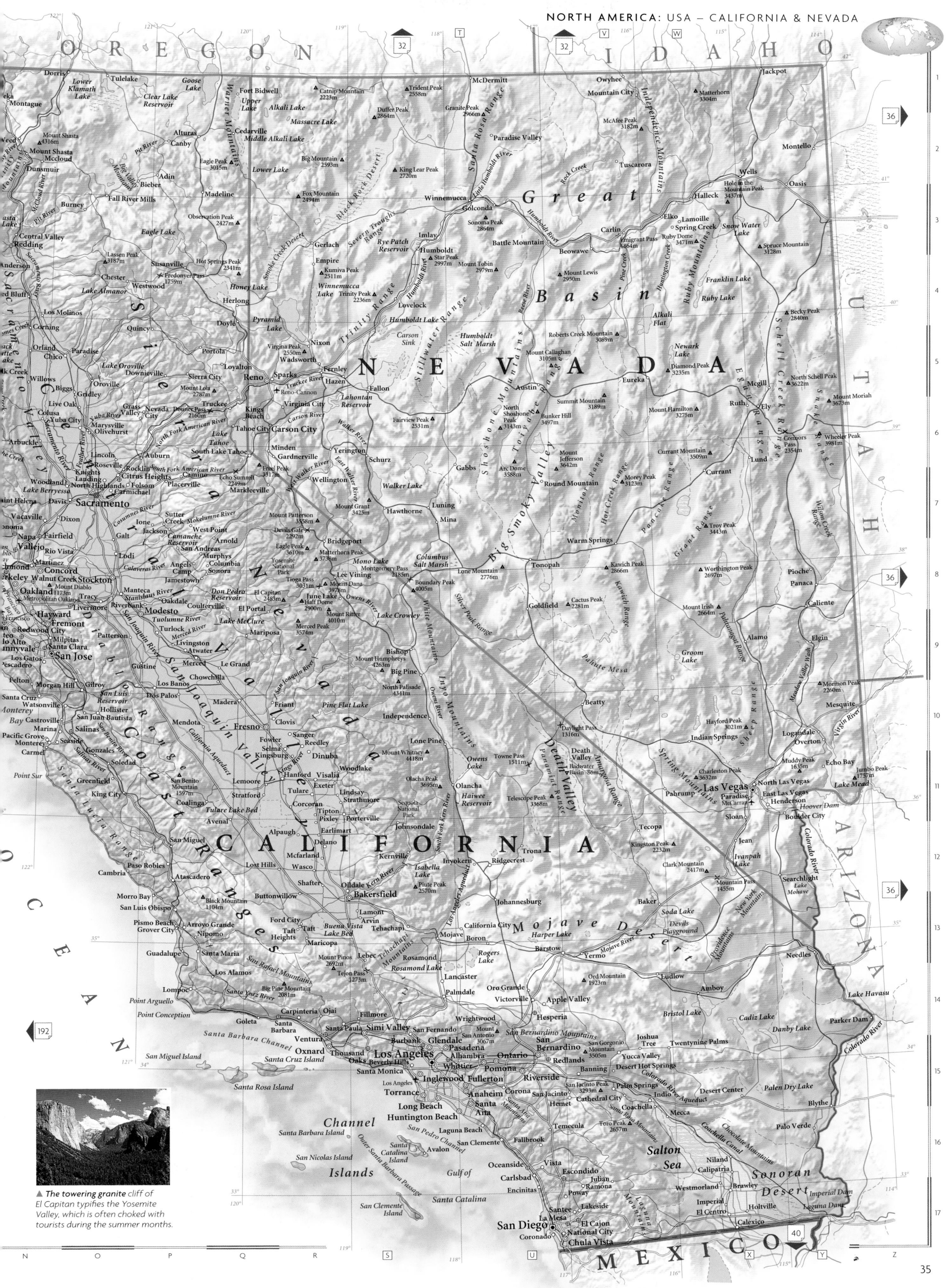

O R E G O N

I D A H O

Dorris
Tulelake
Lower Klamath Lake
Clear Lake Reservoir
Goose Lake
Fort Bidwell
Upper Lake
Alkali Lake
Catnip Mountain 2223m
Trident Peak 2558m
McDermitt
Owyhee
Mountain City
Matterhorn 3304m
Jackpot

Montague
Mount Shasta
Mccloud
Dunsmuir
Canby
Alturas
Cedarville
Middle Alkali Lake
Massacre Lake
Duffer Peak 2864m
Granite Peak 2966m
Paradise Valley
McAfee Peak 3182m
Montello

Weed
Mount Shasta 4316m
Pit River
Bieber
Adin
Madeline
Big Mountain 2593m
King Lear Peak 2720m
Winnemucca
Golconda
Elko
Lamoille
Spring Creek
Snow Water Lake
Wells
Hole in the Mountain Peak 3437m
Oasis

Burney
Fall River Mills
Lower Lake
Fox Mountain 2494m
Humboldt
Star Peak 2997m
Battle Mountain
Beowawe
Carlin
Emigrant Pass 1864m
Ruby Dome 3471m
Spruce Mountain 3128m
Halleck

Central Valley
Redding
Lassen Peak 3187m
Susanville
Hot Springs Peak 2341m
Gerlach
Empire
Kumiva Peak 2511m
Imlay
Humboldt
Mount Tobin 2979m
Sonoma Peak 2864m
Becky Peak 2840m

G R E A T

B A S I N

U T A H

Chester
Westwood
Fredonyer Pass 1759m
Honey Lake
Herlong
Winnemucca Lake
Trinity Peak 2236m
Humboldt Lake
Lovelock
Mount Lewis 2950m
Franklin Lake
Ruby Lake

N E V A D A

Red Bluff
Los Molinos
Corning
Orland
Chico
Paradise
Lake Almanor
Doyle
Portola
Pyramid Lake
Nixon
Wadsworth
Carson Sink
Humboldt Salt Marsh
Roberts Creek Mountain 3089m
Newark Lake
North Schell Peak 3622m
Mcgill

Willows
Biggs
Oroville
Quincy
Loyalton
Reno
Sparks
Fernley
Hazen
Fallon
Austin
Mount Callaghan 3105m
Diamond Peak 3235m
Eureka
Ruth
Ely
Mount Moriah 3675m

Corning
Gridley
Oroville
Lake Oroville
Downieville
Sierra City
Truckee River
Virginia City
Lahontan Reservoir
Fairview Peak 2531m
North Shoshone Peak 3143m
Bunker Hill 3497m
Mount Hamilton 3275m
Currant Mountain 3509m
Wheeler Peak 3981m

Live Oak
Colusa
Yuba City
Grass Valley
Nevada City
Mount Lola 2787m
Donner Pass 2160m
Reno-Cannon
Virginia Peak 2550m
Carson River
Yerington
Schurz
Gabbs
Arc Dome 3588m
Round Mountain
Mount Jefferson 3642m
Morey Peak 3123m
Currant
Comins Pass 2354m

Marysville
Olivehurst
Kings Beach
Tahoe City
Carson City
Minden
Gardnerville
Wellington
Walker Lake
Warm Springs
Tonopah
Kawich Peak 2866m
Worthington Peak 2697m
Pioche
Panaca

Lincoln
Roseville
Auburn
Camino
Echo Summit 2249m
Markleeville
Mount Grant 3425m
Hawthorne
Luning
Mina
Goldfield
Cactus Peak 2281m
Mount Irish 2664m
Caliente

Rocklin
Citrus Heights
Folsom
Placerville
Eagle Peak 3610m
Bridgeport
Columbus Salt Marsh
Lone Mountain 2776m
Beatty
Elgin

Sacramento
Carmichael
North Highlands
West Point
Arnold
Devils Gate
Yosemite National Park
Matterhorn Peak 3738m
Mono Lake
Montgomery Pass 2185m
Boundary Peak 4005m
Groom Lake

Davis
Jackson
San Andreas
Murphys
Columbia
Sonora
Tioga Pass
Mount Dana 3978m
Lee Vining
Daylight Pass 1316m
Indian Springs

Napa
Fairfield
Rio Vista
Lodi
Angels Camp
Jamestown
El Capitan 2483m
June Lake
Half Dome 2900m
Owens River
Silver Peak Range
Death Valley
Charleston Peak 3632m

Vallejo
Martinez
Concord
Walnut Creek
Stockton
Manteca
Riverbank
Oakdale
Coulterville
El Portal
Mount Ritter 4010m
Lake Crowley
Pahute Mesa

Berkeley
Oakland
Hayward
Fremont
Tracy
Modesto
Turlock
Lake McClure
Mariposa
Merced Peak 3574m
Bishop
Big Pine
Las Vegas
North Las Vegas
East Las Vegas
Henderson

Redwood City
Palo Alto
Milpitas
Santa Clara
San Jose
Los Gatos
Livingston
Le Grand
Chowchilla
Madera
Mount Humphreys 4263m
North Palisade 4341m
Beatty
Pahrump
Paradise
McCarran
Sloan
Lake Mead
Hoover Dam

Felton
Morgan Hill
Gilroy
Gustine
Merced
Friant
Clovis
Fresno
Sanger
Reedley
Independence
Owens River
Boulder City

Santa Cruz
Watsonville
Hollister
San Juan Bautista
Dos Palos
Mendota
Fowler
Selma
Kingsburg
Dinuba
Lone Pine
Mount Whitney 4418m
Towne Pass 1511m
Jumbo Peak 1757m
Echo Bay

Monterey Bay
Marina
Seaside
Monterey
Carmel
Gonzales
Soledad
Greenfield
Coalinga
Lemoore
Hanford
Visalia
Exeter
Tulare
Olancha Peak 3695m
Olancha
Telescope Peak 3368m
Muddy Peak 1635m
North Las Vegas

Point Sur
King City
San Miguel
Stratford
Corcoran
Pixley
Porterville
Johnsondale
Haiwee Reservoir
Tecopa
Mormon Peak 2260m
Mesquite
Logandale
Overton

Paso Robles
Atascadero
Avenal
Alpaugh
Delano
Kernville
Inyokern
Ridgecrest
Trona
Kingston Peak 2232m
Jean
Ivanpah Lake

C A L I F O R N I A

Cambria
Morro Bay
San Luis Obispo
Pismo Beach
Black Mountain 1104m
Buttonwillow
Shafter
Oildale
Bakersfield
Isabella Lake
Piute Peak 2570m
Johannesburg
Baker
Clark Mountain 2417m
Mountain Pass 1455m
Searchlight
Lake Mohave

Arroyo Grande
Grover City
Nipomo
Ford City
Taft
Lamont
Arvin
Tehachapi
California City
Mojave
Boron
Harper Lake
Mojave Desert
Soda Lake
Devils Playground
New York Mountains
Needles

Guadalupe
Santa Maria
Los Alamos
Maricopa
Buena Vista Lake Bed
Taft Heights
Mount Pinos 2692m
Rosamond
Rosamond Lake
Rogers Lake
Barstow
Yermo
Mojave River
Ludlow
Amboy

Lompoc
Tejon Pass 1273m
Lebec
Big Pine Mountain 2081m
Lancaster
Palmdale
Oro Grande
Victorville
Ord Mountain 1923m
Bristol Lake
Cadiz Lake
Danby Lake
Parker Dam

Point Arguello
Point Conception
Carpinteria
Ojai
Fillmore
Wrightwood
Apple Valley
Hesperia
Lake Havasu

Goleta
Santa Barbara
Ventura
Oxnard
Santa Paula
Simi Valley
San Fernando
Mount San Antonio 3067m
San Bernardino Mountains
San Bernardino
Joshua Tree
Twentynine Palms
Desert Center

San Miguel Island
Thousand Oaks
Burbank
Glendale
Pasadena
Alhambra
Ontario
Pomona
San Gorgonio 3505m
Yucca Valley
Palen Dry Lake
Blythe

Santa Cruz Island
Los Angeles
Beverly Hills
Santa Monica
Inglewood
Fullerton
Riverside
Redlands
Banning
Desert Hot Springs
Colorado River Aqueduct

Santa Rosa Island
Torrance
Long Beach
Huntington Beach
Anaheim
Santa Ana
Corona
Pomona
Palm Springs
Indio
Coachella
Mecca
Coachella Canal

Channel Islands
Santa Barbara Island
San Pedro Channel
Laguna Beach
San Clemente
Mount San Jacinto Peak 3293m
Palm Springs
Cathedral City
Coachella
Chocolate Mountains

San Nicolas Island
Santa Catalina Island
Avalon
Oceanside
Vista
Temecula
Hemet
Toro Peak 2657m
Salton Sea
Niland
Calipatria
Palo Verde

Santa Catalina
San Clemente Island
Carlsbad
Encinitas
Escondido
Poway
Ramona
Julian
Westmorland
Brawley
Sonoran Desert
Imperial Dam

Gulf of
Santa Barbara Passage
San Diego
La Mesa
El Cajon
Santee
Lakeside
Imperial
El Centro
Holtville
Laguna Dam
Calexico

M E X I C O

A R I Z O N A

O C E A N

▲ **The towering granite** cliff of El Capitan typifies the Yosemite Valley, which is often choked with tourists during the summer months.

USA: SOUTH MOUNTAIN STATES

Arizona, Colorado, New Mexico, Utah

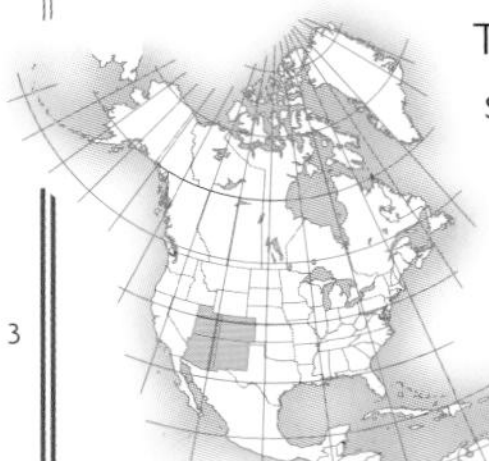

This arid region, characterized by expansive plateaus and spectacular canyons is home to several distinct peoples. The ruins of cliff dwellings built a thousand years ago by the Anasazi people still exist today, and native Americans own one-third of the land in Arizona. Spanish and Mexican conquest and settlement left a hispanic presence which is strongest in New Mexico. The Mormons, who came to the Great Salt Lake seeking religious freedom in 1847, were among the earliest Anglo-American settlers and now make up over 70% of Utah's population. The region's mineral wealth drove rapid development in the 20th century, yet the constraints of a fragile environment, including widespread water shortages, may limit prospects for growth.

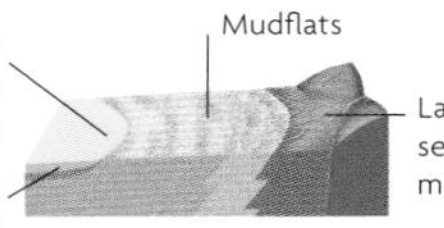

Mudflats
When water evaporates it leaves a salt pan
Lake is fed by seasonal snow melt
Water level of lake varies according to quantity of run-off received from snow melt

▲ *The Great Salt Lake is an ephemeral lake; it can remain dry for extended periods, leaving a pan of evaporated mineral salts in its center.*

The landscape

The arid, rocky expanse of the Colorado Plateau is dissected by immense canyons of the Colorado River. Desert lies to the north and south and branches of the Rocky Mountains run east and west. The Great Salt Lake and Desert lie within the Great Basin, a barren region of parallel mountain ranges that extends into Arizona.

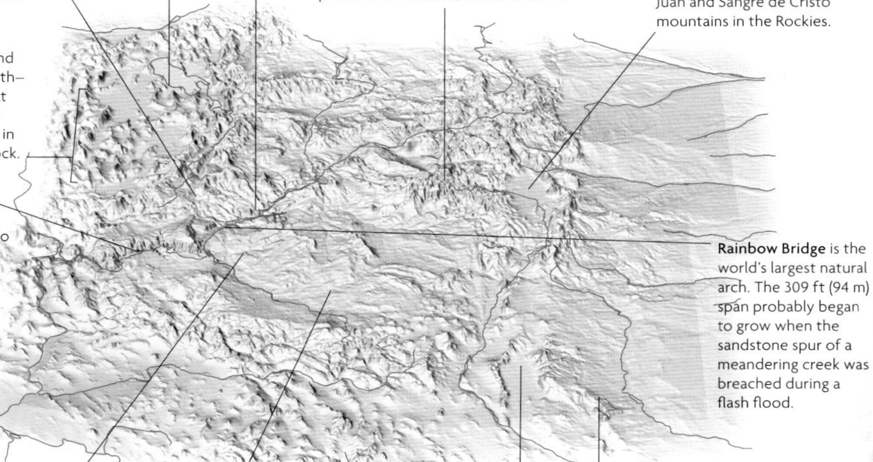

Over 13 million years of weathering has created thousands of spires and pinnacles from the alternating rock strata of Bryce Canyon.
Lake Powell
The Rio Grande has its source in several meltwater streams, which have cut deep valleys into the platform of the San Juan Mountains.
Sand dunes, 600 ft (180 m) high, have been deposited in San Luis Valley, by winds funnelled through the San Juan and Sangre de Cristo mountains in the Rockies.
The parallel basins and ridges, which run north–south along the Great Basin, reflect a major series of block-faults in the underlying bedrock.
Parts of the Grand Canyon, which cuts through the Colorado Plateau,are 16 miles (25 km) wide. The Colorado River has cut down 6262 ft (2000 m), exposing rock strata more than 2 billion years old.
Rainbow Bridge is the world's largest natural arch. The 309 ft (94 m) span probably began to grow when the sandstone spur of a meandering creek was breached during a flash flood.
The striking color effects seen in the Painted Desert come from minerals such as gypsum and haematite, combined with ambient heat and dust.
Petrified Forest
Shifting gypsum sands produce a constantly changing land surface, overwhelming plants and any other obstacles in Tularosa Valley.

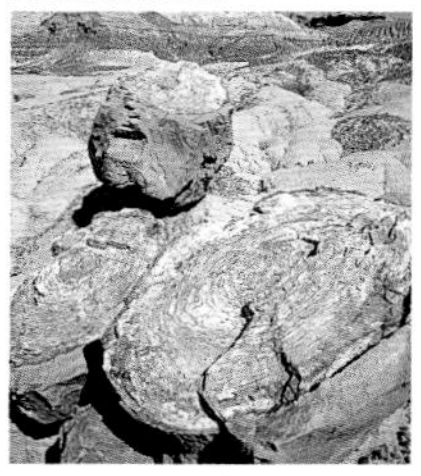

► ***In the arid** landscape of Petrified Forest National Park in Arizona, the grain of prehistoric trees has been preserved as a fossil imprint in the rocks. The bog-preserved trees were gradually turned to stone by seeping mineral-rich water.*

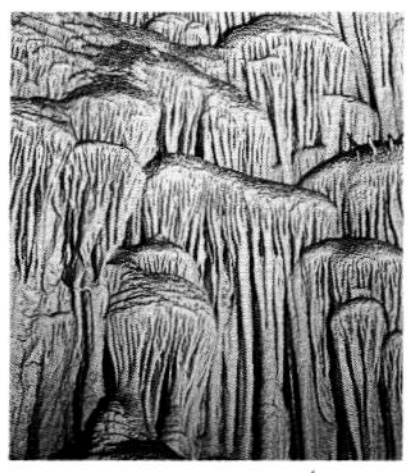

► ***The intricate stalactites** of Carlsbad Caverns have grown with the seepage of calcium-rich water over the last 100,000 years. The huge caves are home to around 100,000 Mexican freetail bats..*

34

Transportation & industry

New industries have helped reduce the region's dependence on the extraction of minerals and fossil fuels. Precision manufacture has grown rapidly, particularly in Arizona and Colorado. Salt Lake City and Denver are well-established financial centers and New Mexico, the main US producer of uranium, is a prominent region for nuclear research. Colorado is the most important US center for winter sports.

Transportation network

232,434 miles (373,986 km)	4059 miles (6515 km)
8627 miles (13,881 km)	none

The Colorado Rockies are crossed by 32 mountain passes, some as high as 12,183 ft (3713 m). The Eisenhower Tunnel west of Denver carries Interstate Highway 70 straight through the Continental Divide.

Idaho
Wyoming
Nebraska
Salt Lake City
Utah
Denver
Colorado Springs
Colorado
Kansas
Nevada
Oklahoma
Arizona
Sante Fe
Albuquerque
New Mexico
Texas
California
Phoenix
Tucson
MEXICO

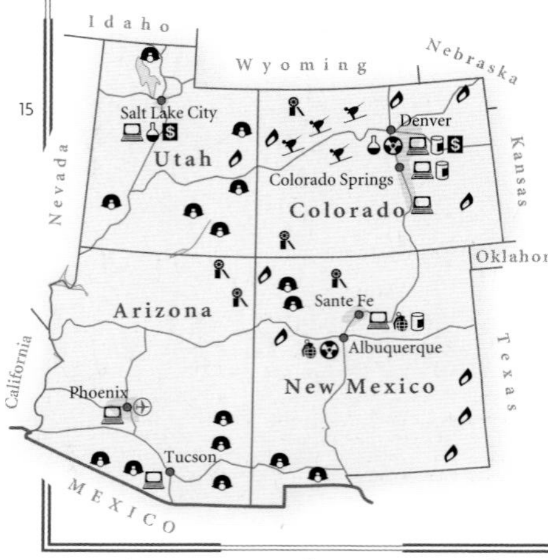

Major industry and infrastructure

- chemicals
- coal
- defense
- finance
- food processing
- hi-tech industry
- oil & gas
- mining
- research & development
- winter sports
- major towns
- international airports
- major roads
- major industrial areas

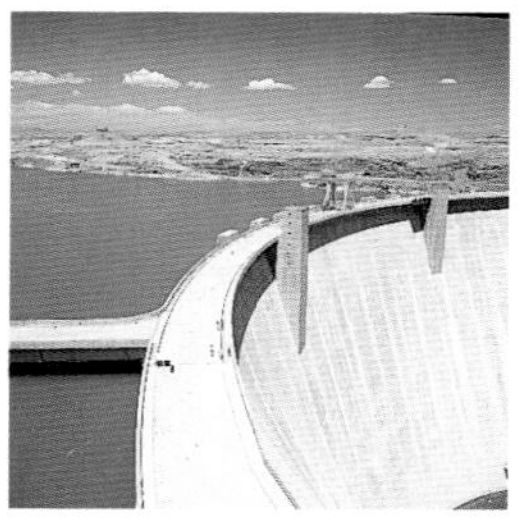

▲ ***Glen Canyon Dam** on the Colorado river was completed in 1964. it provides hydroelectric power and irrigation water as part of a long-term federal project to harness the river.*

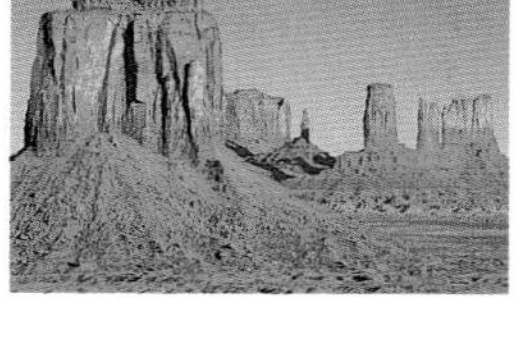

◄ ***The flat tablelands** (mesas), and the isolated pinnacles (buttes) which rise from the floor of Monument Valley are the resistant remnants of an earlier land surface, gradually cut back by erosion under arid conditions.*

IDAHO
NEVADA
UTAH
ARIZONA
CALIFORNIA
MEXICO
Great Salt Lake
Great Salt Lake Desert
Great Basin
Salt Lake City
Colorado Plateau
Grand Canyon
Lake Powell
Lake Mead
Sonoran Desert
Phoenix
Tucson
Flagstaff
Yuma
Colorado River
32
34
40

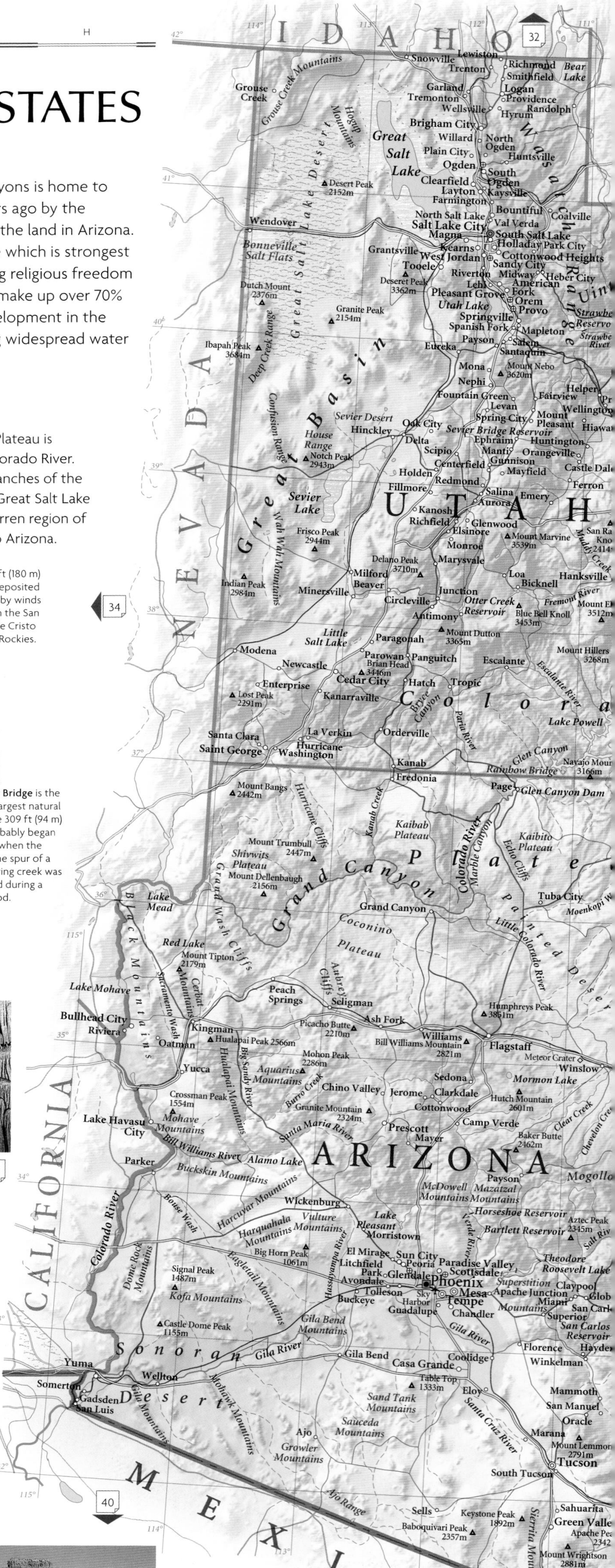

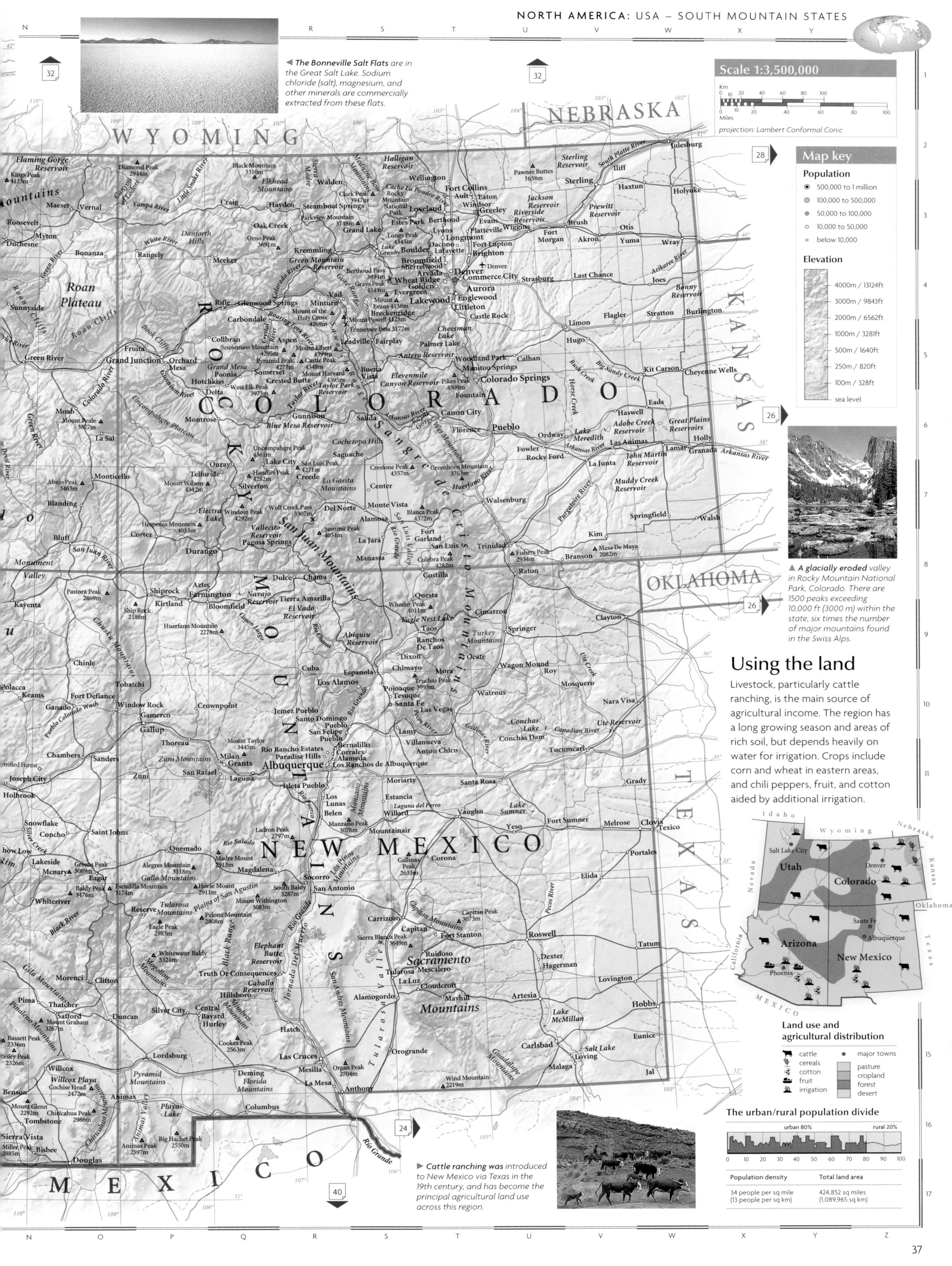

◄ *The Bonneville Salt Flats* are in the Great Salt Lake. Sodium chloride (salt), magnesium, and other minerals are commercially extracted from these flats.

Scale 1:3,500,000

Km
0   20   40   60   80   100

Miles
0   20   40   60   80   100

projection: Lambert Conformal Conic

Map key

Population

⊙ 500,000 to 1 million
◉ 100,000 to 500,000
⊕ 50,000 to 100,000
○ 10,000 to 50,000
∘ below 10,000

Elevation

4000m / 13124ft
3000m / 9843ft
2000m / 6562ft
1000m / 3281ft
500m / 1640ft
250m / 820ft
100m / 328ft
sea level

▲ *A glacially eroded* valley in Rocky Mountain National Park, Colorado. There are 1500 peaks exceeding 10,000 ft (3000 m) within the state, six times the number of major mountains found in the Swiss Alps.

Using the land

Livestock, particularly cattle ranching, is the main source of agricultural income. The region has a long growing season and areas of rich soil, but depends heavily on water for irrigation. Crops include corn and wheat in eastern areas, and chili peppers, fruit, and cotton aided by additional irrigation.

Land use and agricultural distribution

🐄 cattle   ● major towns
🌾 cereals
✿ cotton   pasture
🍎 fruit   cropland
💧 irrigation   forest
   desert

The urban/rural population divide

urban 80%   rural 20%

0 10 20 30 40 50 60 70 80 90 100

Population density	Total land area
34 people per sq mile (13 people per sq km)	424,852 sq miles (1,089,965 sq km)

▶ *Cattle ranching was* introduced to New Mexico via Texas in the 19th century, and has become the principal agricultural land use across this region.

37

USA: HAWAII

The 122 islands of the Hawai'ian archipelago – which are part of Polynesia – are the peaks of the world's largest volcanoes. They rise approximately 6 miles (9.7 km) from the floor of the Pacific Ocean. The largest, the island of Hawai'i, remains highly active. Hawaii became the US's 50th state in 1959. A tradition of receiving immigrant workers is reflected in the islands' ethnic diversity, with peoples drawn from around the rim of the Pacific. Only 2% of the current population are native Polynesians.

▲ *The island of Moloka'i is formed from volcanic rock. Mature sand dunes cover the rocks in coastal areas.*

Transportation & industry

Tourism dominates the economy, with over 90% of the population employed in services. The naval base at Pearl Harbor is also a major source of employment. Industry is concentrated on the island of O'ahu and relies mostly on imported materials, while agricultural produce is processed locally.

Transportation network

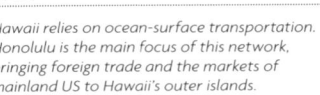

4102 miles (6600 km)		43 miles (69 km)	
none		none	

Hawaii relies on ocean-surface transportation. Honolulu is the main focus of this network, bringing foreign trade and the markets of mainland US to Hawaii's outer islands.

Major industry and infrastructure

- 🏭 food processing
- ✈ military base
- 🧵 textiles
- 🏖 tourism
- ● major towns
- ✈ international airports
- — major roads
- major industrial areas

◄ *Haleakala's extinct volcanic crater is the world's largest. The giant caldera, containing many secondary cones, is 2000 ft (600 m) deep and 20 miles (32 km) in circumference.*

Using the land & sea

The ice-free coastline of Alaska provides access to salmon fisheries and more than 129 million acres (52.2 million ha) of forest. Most of Alaska is uncultivable, and around 90% of food is imported. Barley, hay, and hothouse products are grown around Anchorage, where dairy farming is also concentrated.

The urban/rural population divide

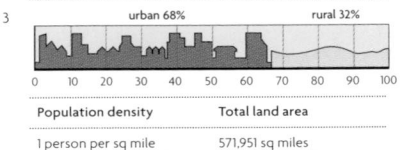

urban 68%   rural 32%

Population density	Total land area
1 person per sq mile (0.4 people per sq km)	571,951 sq miles (1,481,296 sq km)

◄ *A raft of timber from the Tongass forest is hauled by a tug, bound for the pulp mills of the Alaskan coast between Juneau and Ketchikan.*

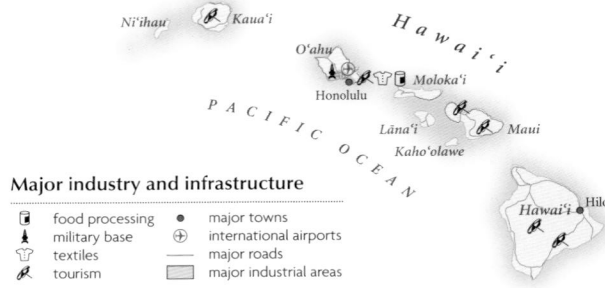

Scale 1:3,500,000

Km 0 10 20 40 60 80 100

Miles 0 10 20 40 60 80 100

projection: Lambert Conformal Conic

Map key

Population
- ◎ 100,000 to 500,000
- ⊕ 50,000 to 100,000
- ○ 10,000 to 50,000
- ○ below 10,000

Elevation
- 4000m / 13,124ft
- 3000m / 9843ft
- 2000m / 6562ft
- 1000m / 3281ft
- 500m / 1640ft
- 250m / 820ft
- 100m / 328ft
- sea level

Using the land & sea

The volcanic soils are extremely fertile and the climate hot and humid on the lower slopes, supporting large commercial plantations growing sugar cane, bananas, pineapples, and other tropical fruit, as well as nursery plants and flowers. Some land is given to pasture, particularly for beef and dairy cattle.

Land use and agricultural distribution

- 🐂 cattle
- 🐟 fishing
- 🍍 fruit
- 🌾 sugar cane
- ● major towns
- pasture
- cropland
- forest
- mountain region

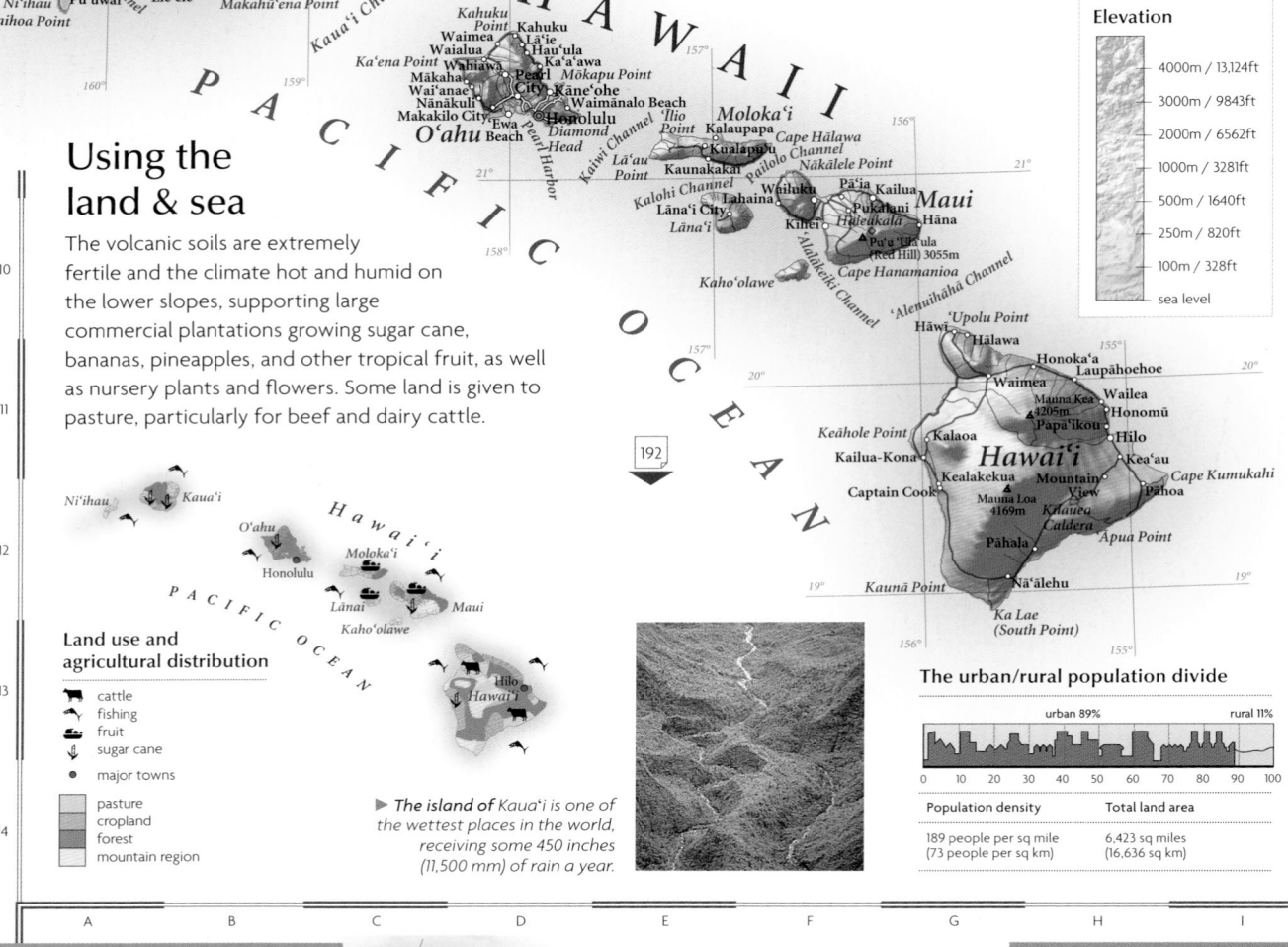

▶ *The island of Kaua'i is one of the wettest places in the world, receiving some 450 inches (11,500 mm) of rain a year.*

The urban/rural population divide

urban 89%   rural 11%

Population density	Total land area
189 people per sq mile (73 people per sq km)	6,423 sq miles (16,636 sq km)

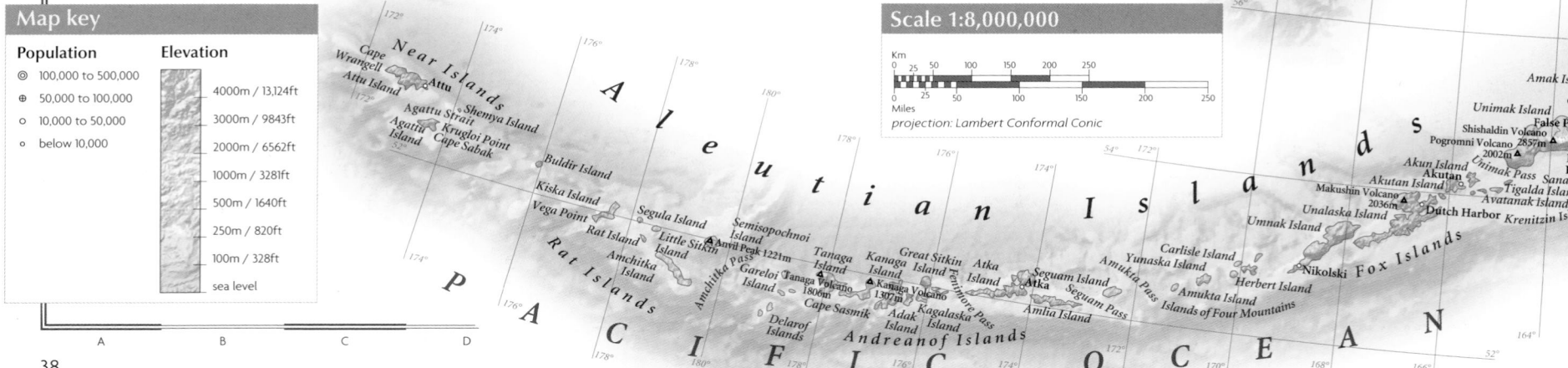

Map key

Population
- ◎ 100,000 to 500,000
- ⊕ 50,000 to 100,000
- ○ 10,000 to 50,000
- ○ below 10,000

Elevation
- 4000m / 13,124ft
- 3000m / 9843ft
- 2000m / 6562ft
- 1000m / 3281ft
- 500m / 1640ft
- 250m / 820ft
- 100m / 328ft
- sea level

Scale 1:8,000,000

Km 0 25 50 100 150 200 250

Miles 0 25 50 100 150 200 250

projection: Lambert Conformal Conic

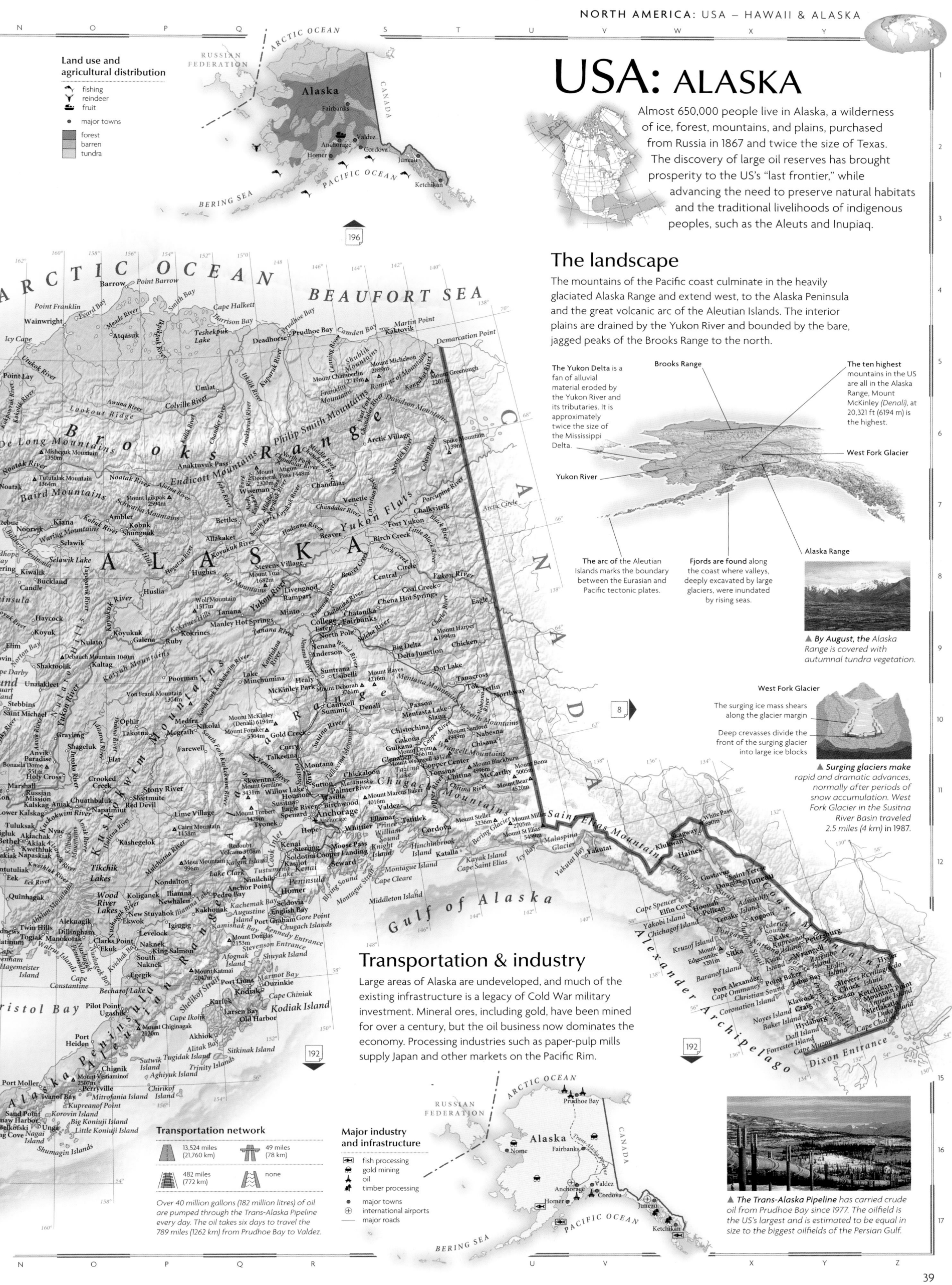

USA: ALASKA

Almost 650,000 people live in Alaska, a wilderness of ice, forest, mountains, and plains, purchased from Russia in 1867 and twice the size of Texas. The discovery of large oil reserves has brought prosperity to the US's "last frontier," while advancing the need to preserve natural habitats and the traditional livelihoods of indigenous peoples, such as the Aleuts and Inupiaq.

The landscape

The mountains of the Pacific coast culminate in the heavily glaciated Alaska Range and extend west, to the Alaska Peninsula and the great volcanic arc of the Aleutian Islands. The interior plains are drained by the Yukon River and bounded by the bare, jagged peaks of the Brooks Range to the north.

The Yukon Delta is a fan of alluvial material eroded by the Yukon River and its tributaries. It is approximately twice the size of the Mississippi Delta.

Brooks Range

The ten highest mountains in the US are all in the Alaska Range, Mount McKinley (Denali), at 20,321 ft (6194 m) is the highest.

West Fork Glacier

Yukon River

Alaska Range

The arc of the Aleutian Islands marks the boundary between the Eurasian and Pacific tectonic plates.

Fjords are found along the coast where valleys, deeply excavated by large glaciers, were inundated by rising seas.

▲ **By August, the** Alaska Range is covered with autumnal tundra vegetation.

West Fork Glacier

The surging ice mass shears along the glacier margin

Deep crevasses divide the front of the surging glacier into large ice blocks

▲ **Surging glaciers make** rapid and dramatic advances, normally after periods of snow accumulation. West Fork Glacier in the Susitna River Basin traveled 2.5 miles (4 km) in 1987.

Transportation & industry

Large areas of Alaska are undeveloped, and much of the existing infrastructure is a legacy of Cold War military investment. Mineral ores, including gold, have been mined for over a century, but the oil business now dominates the economy. Processing industries such as paper-pulp mills supply Japan and other markets on the Pacific Rim.

▲ **The Trans-Alaska Pipeline** has carried crude oil from Prudhoe Bay since 1977. The oilfield is the US's largest and is estimated to be equal in size to the biggest oilfields of the Persian Gulf.

Land use and agricultural distribution

- fishing
- reindeer
- fruit
- major towns
- forest
- barren
- tundra

Transportation network

13,524 miles (21,760 km)		49 miles (78 km)	
482 miles (772 km)		none	

Over 40 million gallons (182 million litres) of oil are pumped through the Trans-Alaska Pipeline every day. The oil takes six days to travel the 789 miles (1262 km) from Prudhoe Bay to Valdez.

Major industry and infrastructure

- fish processing
- gold mining
- oil
- timber processing
- major towns
- international airports
- major roads

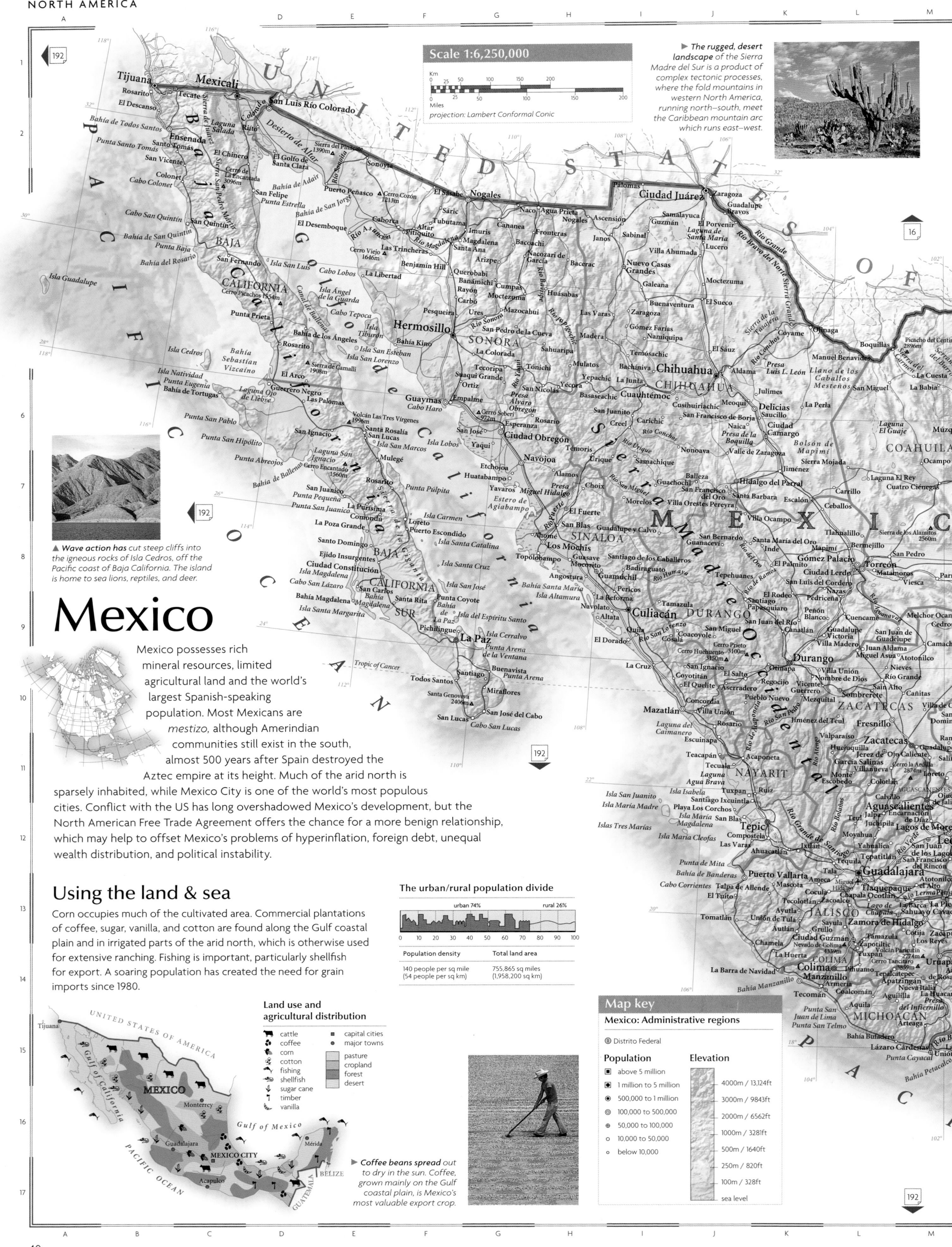

Mexico

Mexico possesses rich mineral resources, limited agricultural land and the world's largest Spanish-speaking population. Most Mexicans are *mestizo*, although Amerindian communities still exist in the south, almost 500 years after Spain destroyed the Aztec empire at its height. Much of the arid north is sparsely inhabited, while Mexico City is one of the world's most populous cities. Conflict with the US has long overshadowed Mexico's development, but the North American Free Trade Agreement offers the chance for a more benign relationship, which may help to offset Mexico's problems of hyperinflation, foreign debt, unequal wealth distribution, and political instability.

Using the land & sea

Corn occupies much of the cultivated area. Commercial plantations of coffee, sugar, vanilla, and cotton are found along the Gulf coastal plain and in irrigated parts of the arid north, which is otherwise used for extensive ranching. Fishing is important, particularly shellfish for export. A soaring population has created the need for grain imports since 1980.

Scale 1:6,250,000

projection: Lambert Conformal Conic

▶ The rugged, desert landscape of the Sierra Madre del Sur is a product of complex tectonic processes, where the fold mountains in western North America, running north–south, meet the Caribbean mountain arc which runs east–west.

▲ Wave action has cut steep cliffs into the igneous rocks of Isla Cedros, off the Pacific coast of Baja California. The island is home to sea lions, reptiles, and deer.

The urban/rural population divide

urban 74%	rural 26%

Population density	Total land area
140 people per sq mile (54 people per sq km)	755,865 sq miles (1,958,200 sq km)

Land use and agricultural distribution

- cattle
- coffee
- corn
- cotton
- fishing
- shellfish
- sugar cane
- timber
- vanilla
- capital cities
- major towns

- pasture
- cropland
- forest
- desert

▶ Coffee beans spread out to dry in the sun. Coffee, grown mainly on the Gulf coastal plain, is Mexico's most valuable export crop.

Map key

Mexico: Administrative regions

Ⓓ Distrito Federal

Population
- ▣ above 5 million
- ◙ 1 million to 5 million
- ◎ 500,000 to 1 million
- ◉ 100,000 to 500,000
- ⊕ 50,000 to 100,000
- ○ 10,000 to 50,000
- ○ below 10,000

Elevation
- 4000m / 13,124ft
- 3000m / 9843ft
- 2000m / 6562ft
- 1000m / 3281ft
- 500m / 1640ft
- 250m / 820ft
- 100m / 328ft
- sea level

The landscape

The great central plateau rises gently southward from the Rio Grande, isolated from the coastal plains by the Sierra Madre Oriental and Occidental. The two ranges converge from east and west respectively, culminating in high volcanic peaks around Mexico City. Further ranges of the Sierra Madre rise to the south of the Balsas basin, skirted by the low-lying Isthmus of Tehuantepec (Istmo de Tehuantepec) and Yucatan Peninsula.

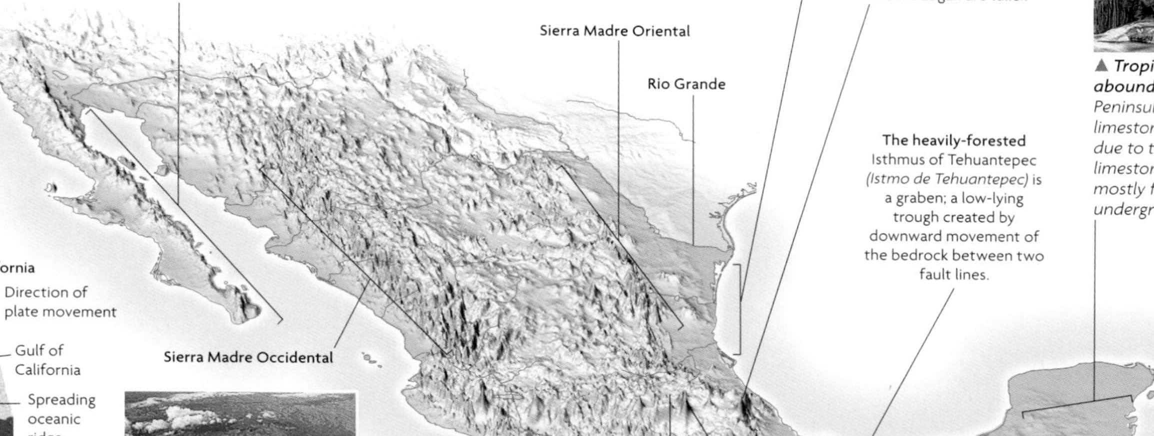

The long, narrow, extremely arid peninsula of Baja (lower) California is an elongated granite block, separated from the mainland by the flooded rift valley of the Gulf of California (Golfo de California).

Wave action has constructed sand bars which shelter lagoons along the shore of the Gulf coastal plain.

Sierra Madre Oriental

Rio Grande

The dormant cone of Volcán Pico de Orizaba is, at 18,700 ft (5700 m), the highest peak in Mexico. In North America, only Mount McKinley and Mount Logan are taller.

▲ Tropical rainforest abounds in the Yucatan Peninsula, a broad, low limestone shelf. Rivers are rare due to the porous nature of limestone, so the forest is mostly fed by streams and underground water.

Formation of the Gulf of California

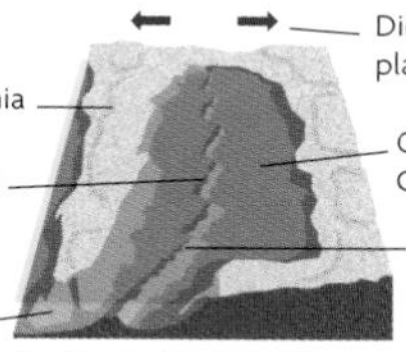

Direction of plate movement

Baja California

Transform fault

Gulf of California

Edge of continental crust

Spreading oceanic ridge

Sierra Madre Occidental

The heavily-forested Isthmus of Tehuantepec (Istmo de Tehuantepec) is a graben; a low-lying trough created by downward movement of the bedrock between two fault lines.

▲ The Gulf of California (Golfo de California) began to open out about 4 million years ago as a result of rifting and plate displacement along transform faults.

Río Balsas

▲ Popocatépetl is a dormant volcano, part of the Pacific "Ring of Fire." The crater is over half a mile (1 km) wide.

Popocatépetl

The unstable, earthquake-prone, upland basin around Mexico City was once a region of shallow lakes. Flood control measures and domestic consumption over the last four centuries have caused the virtual disappearance of this surface water.

The highlands of Chiapas are a series of horsts, blocks of land thrust upward between two fault lines. Volcanic cones have developed where lava has flowed out from the faults.

Transportation & industry

Oil and gas on the Gulf coast are Mexico's main sources of export income. Metal mining has declined but the country remains a leading global producer of silver. Manufacturing is heavily concentrated around the metropolitan area of Mexico City, while the duty-free movement of goods in the US border region, under the Maquiladora (twin plant) scheme, has created new hi-tech and service growth centers.

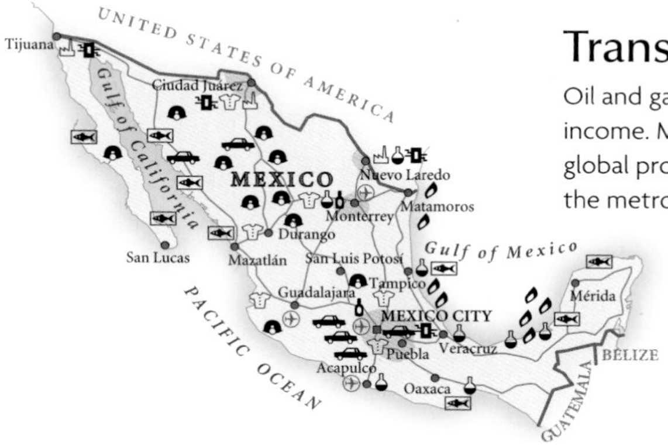

Major industry and infrastructure

- brewing
- car manufacture
- chemicals
- electronics
- fish processing
- maquiladoras
- mining
- oil & gas
- textiles
- capital cities
- major towns
- international airports
- major roads
- major industrial areas

Transportation network

67,564 miles (108,746 km)	
3994 miles (6429 km)	
16,561 miles (26,656 km)	
1801 miles (2900 km)	

Fast, modern highways or autopistas now link Mexico City with Toluca, Puebla and other satellite cities, yet distant centers like Chihuahua are still served by narrow roads and an outdated railroad network.

▲ A stone figure reclines by the Temple of Warriors, within the Mayan city of Chichén-Itzá. The Maya civilization flourished across the Yucatan Peninsula between 200 and 900 AD.

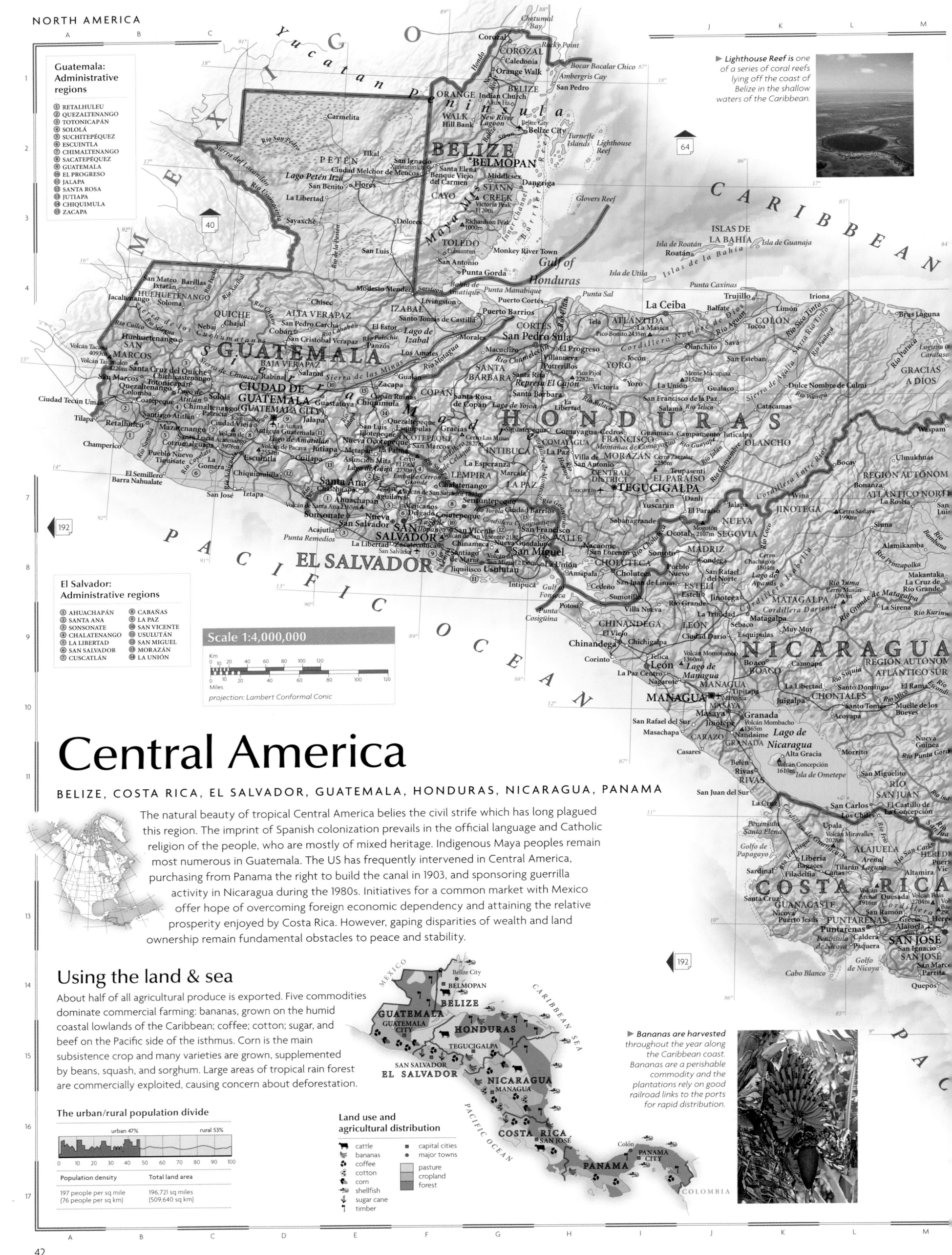

Guatemala:
Administrative regions

① RETALHULEU
② QUEZALTENANGO
③ TOTONICAPÁN
④ SOLOLÁ
⑤ SUCHITEPÉQUEZ
⑥ ESCUINTLA
⑦ CHIMALTENANGO
⑧ SACATEPÉQUEZ
⑨ GUATEMALA
⑩ EL PROGRESO
⑪ JALAPA
⑫ SANTA ROSA
⑬ JUTIAPA
⑭ CHIQUIMULA
⑮ ZACAPA

▶ *Lighthouse Reef is one of a series of coral reefs lying off the coast of Belize in the shallow waters of the Caribbean.*

El Salvador:
Administrative regions

① AHUACHAPÁN
② SANTA ANA
③ SONSONATE
④ CHALATENANGO
⑤ LA LIBERTAD
⑥ SAN SALVADOR
⑦ CUSCATLÁN
⑧ CABAÑAS
⑨ LA PAZ
⑩ SAN VICENTE
⑪ USULUTÁN
⑫ SAN MIGUEL
⑬ MORAZÁN
⑭ LA UNIÓN

Scale 1:4,000,000

```
Km
0 10 20    40    60    80   100  120
0  10   20      40      60      80     100    120
Miles
```
projection: Lambert Conformal Conic

Central America

BELIZE, COSTA RICA, EL SALVADOR, GUATEMALA, HONDURAS, NICARAGUA, PANAMA

The natural beauty of tropical Central America belies the civil strife which has long plagued this region. The imprint of Spanish colonization prevails in the official language and Catholic religion of the people, who are mostly of mixed heritage. Indigenous Maya peoples remain most numerous in Guatemala. The US has frequently intervened in Central America, purchasing from Panama the right to build the canal in 1903, and sponsoring guerrilla activity in Nicaragua during the 1980s. Initiatives for a common market with Mexico offer hope of overcoming foreign economic dependency and attaining the relative prosperity enjoyed by Costa Rica. However, gaping disparities of wealth and land ownership remain fundamental obstacles to peace and stability.

Using the land & sea

About half of all agricultural produce is exported. Five commodities dominate commercial farming: bananas, grown on the humid coastal lowlands of the Caribbean; coffee; cotton; sugar, and beef on the Pacific side of the isthmus. Corn is the main subsistence crop and many varieties are grown, supplemented by beans, squash, and sorghum. Large areas of tropical rain forest are commercially exploited, causing concern about deforestation.

The urban/rural population divide

urban 47%   rural 53%

```
0  10  20  30  40  50  60  70  80  90  100
```

Population density	Total land area
197 people per sq mile (76 people per sq km)	196,721 sq miles (509,640 sq km)

Land use and agricultural distribution

- cattle
- bananas
- coffee
- cotton
- corn
- shellfish
- sugar cane
- timber
- ▪ capital cities
- • major towns
- pasture
- cropland
- forest

▶ *Bananas are harvested throughout the year along the Caribbean coast. Bananas are a perishable commodity and the plantations rely on good railroad links to the ports for rapid distribution.*

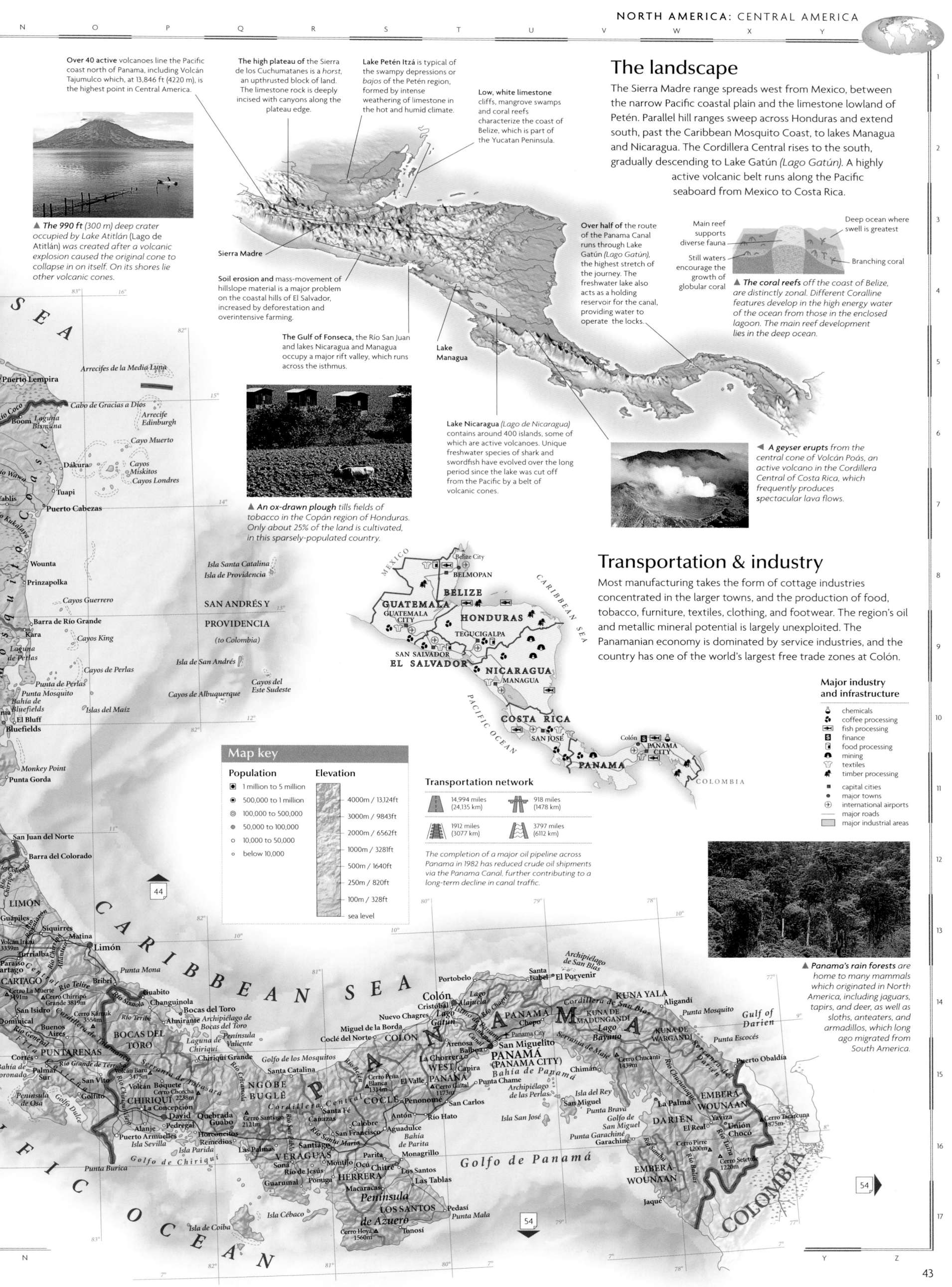

Over 40 active volcanoes line the Pacific coast north of Panama, including Volcán Tajumulco which, at 13,846 ft (4220 m), is the highest point in Central America.

The high plateau of the Sierra de los Cuchumatanes is a *horst*, an upthrusted block of land. The limestone rock is deeply incised with canyons along the plateau edge.

Lake Petén Itzá is typical of the swampy depressions or *bajos* of the Petén region, formed by intense weathering of limestone in the hot and humid climate.

Low, white limestone cliffs, mangrove swamps and coral reefs characterize the coast of Belize, which is part of the Yucatan Peninsula.

▲ *The 990 ft (300 m) deep crater occupied by Lake Atitlán (Lago de Atitlán) was created after a volcanic explosion caused the original cone to collapse in on itself. On its shores lie other volcanic cones.*

Sierra Madre

Soil erosion and mass-movement of hillslope material is a major problem on the coastal hills of El Salvador, increased by deforestation and overintensive farming.

The Gulf of Fonseca, the Río San Juan and lakes Nicaragua and Managua occupy a major rift valley, which runs across the isthmus.

Lake Managua

Over half of the route of the Panama Canal runs through Lake Gatún (*Lago Gatún*), the highest stretch of the journey. The freshwater lake also acts as a holding reservoir for the canal, providing water to operate the locks.

Main reef supports diverse fauna

Still waters encourage the growth of globular coral

Deep ocean where swell is greatest

Branching coral

▲ *The coral reefs off the coast of Belize, are distinctly zonal. Different Coralline features develop in the high energy water of the ocean from those in the enclosed lagoon. The main reef development lies in the deep ocean.*

The landscape

The Sierra Madre range spreads west from Mexico, between the narrow Pacific coastal plain and the limestone lowland of Petén. Parallel hill ranges sweep across Honduras and extend south, past the Caribbean Mosquito Coast, to lakes Managua and Nicaragua. The Cordillera Central rises to the south, gradually descending to Lake Gatún (*Lago Gatún*). A highly active volcanic belt runs along the Pacific seaboard from Mexico to Costa Rica.

Lake Nicaragua (*Lago de Nicaragua*) contains around 400 islands, some of which are active volcanoes. Unique freshwater species of shark and swordfish have evolved over the long period since the lake was cut off from the Pacific by a belt of volcanic cones.

▲ *An ox-drawn plough tills fields of tobacco in the Copán region of Honduras. Only about 25% of the land is cultivated, in this sparsely-populated country.*

◀ *A geyser erupts from the central cone of Volcán Poás, an active volcano in the Cordillera Central of Costa Rica, which frequently produces spectacular lava flows.*

Transportation & industry

Most manufacturing takes the form of cottage industries concentrated in the larger towns, and the production of food, tobacco, furniture, textiles, clothing, and footwear. The region's oil and metallic mineral potential is largely unexploited. The Panamanian economy is dominated by service industries, and the country has one of the world's largest free trade zones at Colón.

Major industry and infrastructure

- chemicals
- coffee processing
- fish processing
- finance
- food processing
- mining
- textiles
- timber processing

- capital cities
- major towns
- international airports
- major roads
- major industrial areas

Map key

Population
- 1 million to 5 million
- 500,000 to 1 million
- 100,000 to 500,000
- 50,000 to 100,000
- 10,000 to 50,000
- below 10,000

Elevation
- 4000m / 13,124ft
- 3000m / 9843ft
- 2000m / 6562ft
- 1000m / 3281ft
- 500m / 1640ft
- 250m / 820ft
- 100m / 328ft
- sea level

Transportation network

14,994 miles (24,135 km)	918 miles (1478 km)
1912 miles (3077 km)	3797 miles (6112 km)

The completion of a major oil pipeline across Panama in 1982 has reduced crude oil shipments via the Panama Canal, further contributing to a long-term decline in canal traffic.

▲ *Panama's rain forests are home to many mammals which originated in North America, including jaguars, tapirs, and deer, as well as sloths, anteaters, and armadillos, which long ago migrated from South America.*

Belize City
BELMOPAN
BELIZE
GUATEMALA
GUATEMALA CITY
HONDURAS
TEGUCIGALPA
SAN SALVADOR
EL SALVADOR
NICARAGUA
MANAGUA
COSTA RICA
SAN JOSÉ
Colón
PANAMA CITY
PANAMA
COLOMBIA
CARIBBEAN SEA
PACIFIC OCEAN

Map place names

Arrecifes de la Media Luna
Puerto Lempira
Cabo de Gracias a Dios
Río Coco
Boom
Laguna Bismuna
Arrecife Edinburgh
Cayo Muerto
Dákura
Cayos Miskitos
Cayos Londres
Yablis
Río Wawa
Río Kukalaya
Tuapi
Puerto Cabezas
Wounta
Prinzapolka
Isla Santa Catalina
Isla de Providencia
Cayos Guerrero
SAN ANDRÉS Y PROVIDENCIA
Barra de Río Grande
Cayos King
(to Colombia)
Kara
Cayos de Perlas
Laguna de Perlas
Isla de San Andrés
Cayos del Este Sudeste
Punta de Perlas
Cayos de Albuquerque
Bahía de Bluefields
Islas del Maíz
El Bluff
Bluefields
Monkey Point
Punta Gorda
San Juan del Norte
Barra del Colorado
LIMÓN
Guápiles
Siquirres
Matina
Volcán Irazú 3339m
Limón
Turrialba
Paraíso
Cartago
CARTAGO
Cerro La Muerte 3491m
Bribri
Cerro Chirripó Grande 3819m
San Isidro
Dominical
Cerro Kámuk 3554m
Buenos Aires
Cortés
Bahía de Coronado
Palmar Sur
San Vito
PUNTARENAS
Río Grande de Térraba
Peninsula de Osa
Golfo Dulce
Golfito
Punta Burica

CARIBBEAN SEA
Portobelo
Santa Isabel
Archipiélago de San Blas
El Porvenir
KUNA YALA
Aligandí
Punta Mosquito
Gulf of Darien
Colón
Lago Alajuela
Nuevo Chagres
Lago Gatún
Cristóbal
KUNA DE MADUNGANDI
Lago Bayano
Serranía de San Blas
Bocas del Toro
Changuinola
Almirante
Archipiélago de Bocas del Toro
Laguna de Chiriquí
Chiriquí Grande
Miguel de la Borda
Coclé del Norte
COLÓN
Arenosa
La Chorrera
Balboa
Panama City
PANAMA WEST
San Miguelito
PANAMA (PANAMA CITY)
Chimán
KUNA DE WARGANDÍ
Punta Escocés
Puerto Obaldía
BOCAS DEL TORO
Santa Catalina
Cerro Peña Blanca
El Valle
Capira
Punta Chame
Cerro Chucanti 1439m
NGÖBE BUGLÉ
Volcán Barú 3475m
Cerro Chorcha 2228m
Santa Fé
PANAMÁ
Bahía de Panamá
Archipiélago de las Perlas
Chiriquí
CHIRIQUÍ
La Concepción
David
Quebrada Guabo
Cerro Santiago 2121m
COCLÉ
Antón
Río Hato
Isla San José
Isla del Rey
La Palma
EMBERÁ WOUNAAN
Cerro Pirre 1200m
Alanje
Puerto Remedios
Horconcitos
San Francisco
Penonomé
San Carlos
San Miguel
Punta Brava
DARIÉN
El Real
Cerro Tacarcuna 1875m
Las Palmas
VERAGUAS
Soná
Santiago
Santa María
Aguadulce
Parita
Punta Garachiné
Garachiné
Yaviza
Unión Chocó
Golfo de Chiriquí
Guarumal
Montijo
Isla de Jesús
Pconuga
Ocú
Chitré
Los Santos
Las Tablas
Golfo de Panamá
EMBERÁ-WOUNAAN
Cerro Setetule 1200m
Isla Parida
Bahía de Parita
Macaracas
HERRERA
Peninsula de Azuero
LOS SANTOS
Pedasí
Punta Mala
Isla Cébaco
Cerro Hoya 1560m
Tonosí
Isla de Coiba
COLOMBIA
PACIFIC OCEAN

44

54

UNITED STATES OF AMERICA

GULF OF MEXICO

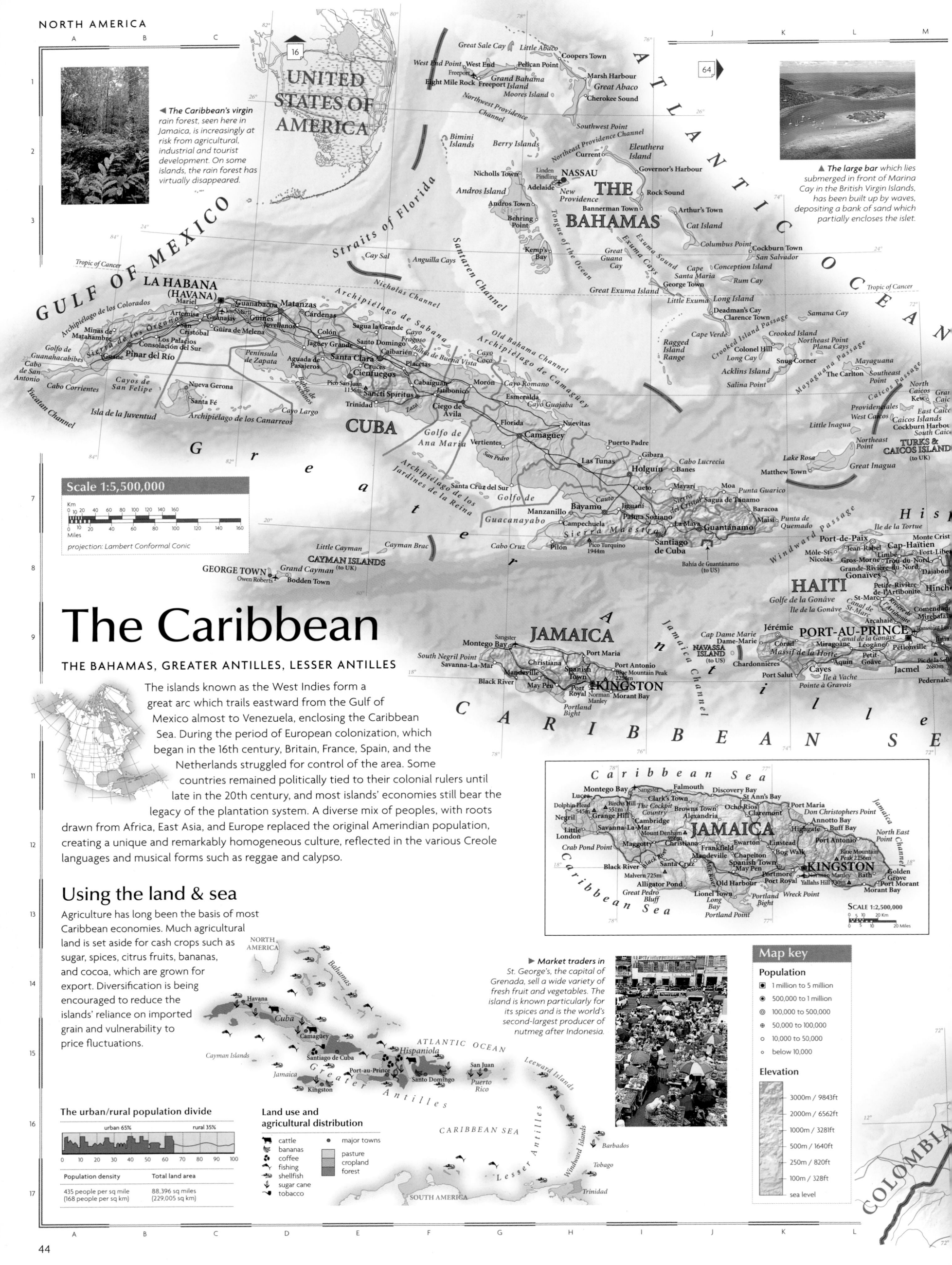

◀ *The Caribbean's virgin rain forest*, seen here in Jamaica, is increasingly at risk from agricultural, industrial and tourist development. On some islands, the rain forest has virtually disappeared.

ATLANTIC OCEAN

Great Sale Cay *Little Abaco* Coopers Town
West End Point West End Pelican Point
Freeport Grand Bahama Marsh Harbour
Eight Mile Rock Freeport Island Great Abaco
Moores Island Cherokee Sound
Northwest Providence Channel Southwest Point

Bimini Islands Berry Islands Northeast Providence Channel *Eleuthera Island* Current Governor's Harbour

Nicholls Town Linden Pindling NASSAU
Andros Island Adelaide New Providence Rock Sound
Andros Town Bannerman Town **THE BAHAMAS** Arthur's Town Cat Island
Behring Point Columbus Point Cockburn Town
Kemp's Bay Great Guana Cay Exuma Sound Exuma Cays Conception Island San Salvador
Cay Sal Anguilla Cays Cape Santa Maria George Town Rum Cay
Great Exuma Island Little Exuma Long Island
Deadman's Cay Clarence Town Samana Cay
Cape Verde Crooked Island Passage Crooked Island
Long Cay Northeast Point Plana Cays
Ragged Island Range Colonel Hill Snug Corner Acklins Island Mayaguana
Salina Point The Carlton Southeast Point
Caicos Passage North Caicos Grand Caicos
Mayaguana Passage Kew East Caic
Little Inagua Providenciales Caicos Islands
West Caicos Cockburn Harbour
South Caicos South Caicos
Lake Rosa Northeast Point **TURKS & CAICOS ISLANDS** (to UK)
Matthew Town Great Inagua

Tropic of Cancer

LA HABANA (HAVANA) Mariel
Archipiélago de los Colorados Guanabacoa Matanzas
Artemisa Guanajay Cárdenas
Minas de Matahambre San Cristóbal Güines Jovellanos
Los Palacios Consolación del Sur Güira de Melena Colón
Guane Pinar del Río Jagüey Grande Santo Domingo Sagua la Grande
Golfo de Guanahacabibes Península de Zapata Cayo Fragoso Caibarién
Cabo de San Antonio Aguada de Pasajeros Santa Clara Placetas
Cabo Corrientes Cruces Cienfuegos Bahía de Buena Vista
Cayos de San Felipe Nueva Gerona Pico San Juan 1150m Cabaiguán Cayo Coco
Santa Fé Sancti Spíritus Jatibonico Morón Cayo Romano
Isla de la Juventud Trinidad Zaza Ciego de Ávila Esmeralda Cayo Guajaba
Archipiélago de los Canarreos Cayo Largo **CUBA** Florida Nuevitas Cayo Sabinal
Golfo de Ana María Camagüey Puerto Padre Gibara
Archipiélago de Sabana Old Bahama Channel Archipiélago de Camagüey
Santaren Channel Nicholas Channel Torrent of the Ocean
Cayo Guillermo

Straits of Florida

Golfo de Guacanayabo Vertientes Las Tunas Cabo Lucrecia Banes
San Pedro Holguín Moa Punta Guarico
Archipiélago de los Jardines de la Reina Santa Cruz del Sur Cueto Mayarí Sagua de Tánamo His
Manzanillo Bayamo Jiguaní Palma Soriano Baracoa
Campechuela La Maya Guantánamo Punta de Quemado Île de la Tortue Monte Crist
Cabo Cruz Pilón Pico Turquino 1944m Santiago de Cuba Bahía de Guantánamo (to US) Windward Passage Port-de-Paix Cap-Haïtien Fort-Libe
Little Cayman Cayman Brac Sierra Maestra Môle-St-Nicolas Gros-Morne Limbé Trou-du-Nord
CAYMAN ISLANDS (to UK) Grande-Rivière-du-Nord Dajabón
GEORGE TOWN Grand Cayman Gonaïves Petite-Rivière-de-l'Artibonite Hinch
Owen Roberts Bodden Town **HAITI**

Gulf of Ana Maria

Sierra del Cristal

Scale 1:5,500,000

Km 0 20 40 60 80 100 120 140 160
Miles 0 20 40 60 80 100 120 140 160

projection: Lambert Conformal Conic

Tropic of Cancer

ATLANTIC OCEAN

The Caribbean

THE BAHAMAS, GREATER ANTILLES, LESSER ANTILLES

The islands known as the West Indies form a great arc which trails eastward from the Gulf of Mexico almost to Venezuela, enclosing the Caribbean Sea. During the period of European colonization, which began in the 16th century, Britain, France, Spain, and the Netherlands struggled for control of the area. Some countries remained politically tied to their colonial rulers until late in the 20th century, and most islands' economies still bear the legacy of the plantation system. A diverse mix of peoples, with roots drawn from Africa, East Asia, and Europe replaced the original Amerindian population, creating a unique and remarkably homogeneous culture, reflected in the various Creole languages and musical forms such as reggae and calypso.

JAMAICA
Montego Bay Sangster Port Maria
South Negril Point Christiana Port Antonio
Savanna-La-Mar Spanish Town Blue Mountain Peak 2256m
Mandeville May Pen Kingston Morant Bay
Black River Port Royal Norman Manley Portland Bight

A n t i l l e s
CARIBBEAN SEA

Golfe de la Gonâve Île de la Gonâve St-Marc Arcahaie
Cap Dame Marie Dame-Marie Jérémie Corail Miragoâne Léogâne **PORT-AU-PRINCE**
NAVASSA ISLAND (to US) Chardonnières Jima
Cayes Petit-Goâve Jacmel
Port Salut Île à Vache Pointe à Gravois Massif de la Hotte Pic de 2680m Pétionville
Pedernale

His His Hispaniola (His) I l e

Windward Passage Jamaica Channel Jamaica Channel

Using the land & sea

Agriculture has long been the basis of most Caribbean economies. Much agricultural land is set aside for cash crops such as sugar, spices, citrus fruits, bananas, and cocoa, which are grown for export. Diversification is being encouraged to reduce the islands' reliance on imported grain and vulnerability to price fluctuations.

NORTH AMERICA

Bahamas
Havana
Cuba
Cayman Islands
Camagüey
Jamaica
Santiago de Cuba
Kingston
Port-au-Prince
Hispaniola
Santo Domingo
San Juan
Puerto Rico
Greater Antilles
ATLANTIC OCEAN
Leeward Islands
CARIBBEAN SEA
Lesser Antilles
Windward Islands
Barbados
Tobago
SOUTH AMERICA
Trinidad

▶ *Market traders in St. George's*, the capital of Grenada, sell a wide variety of fresh fruit and vegetables. The island is known particularly for its spices and is the world's second-largest producer of nutmeg after Indonesia.

The urban/rural population divide

urban 65% rural 35%
0 10 20 30 40 50 60 70 80 90 100

Population density	Total land area
435 people per sq mile (168 people per sq km)	88,396 sq miles (229,005 sq km)

Land use and agricultural distribution

- cattle
- bananas
- coffee
- fishing
- shellfish
- sugar cane
- tobacco
- major towns
- pasture
- cropland
- forest

▲ *The large bar* which lies submerged in front of Marina Cay in the British Virgin Islands, has been built up by waves, depositing a bank of sand which partially encloses the islet.

JAMAICA (inset)

Caribbean Sea
Montego Bay Sangster Falmouth Discovery Bay
Lucea Clark's Town St Ann's Bay
Dolphin Head Birches Hill The Cockpit Country Browns Town Ocho Rios Port Maria Don Christophers Point
Negril 551m Grange Hill Cambridge Alexandria Claremont Annotto Bay
Little London Savanna-La-Mar Maggotty 545m Mount Denham Christiana Highgate Buff Bay North East Point
Crab Pond Point Mandeville Frankfield Ewarton Linstead Port Antonio
Black River Santa Cruz Spanish Town Bog Walk Blue Mountain Peak 2256m **KINGSTON**
Malvern 725m May Pen Chapelton Old Harbour Bath
Great Pedro Bluff Alligator Pond Lionel Town Portmore Port Royal Yallahs Hill 930m Golden Grove Port Morant
Long Bay Old Harbour Spanish Town Norman Manley Morant Bay
Portland Point Portland Bight Wreck Point

SCALE 1:2,500,000
0 5 10 20 Km
0 5 10 20 Miles

Map key

Population
- 1 million to 5 million
- 500,000 to 1 million
- 100,000 to 500,000
- 50,000 to 100,000
- 10,000 to 50,000
- below 10,000

Elevation
- 3000m / 9843ft
- 2000m / 6562ft
- 1000m / 3281ft
- 500m / 1640ft
- 250m / 820ft
- 100m / 328ft
- sea level

COLOMBIA

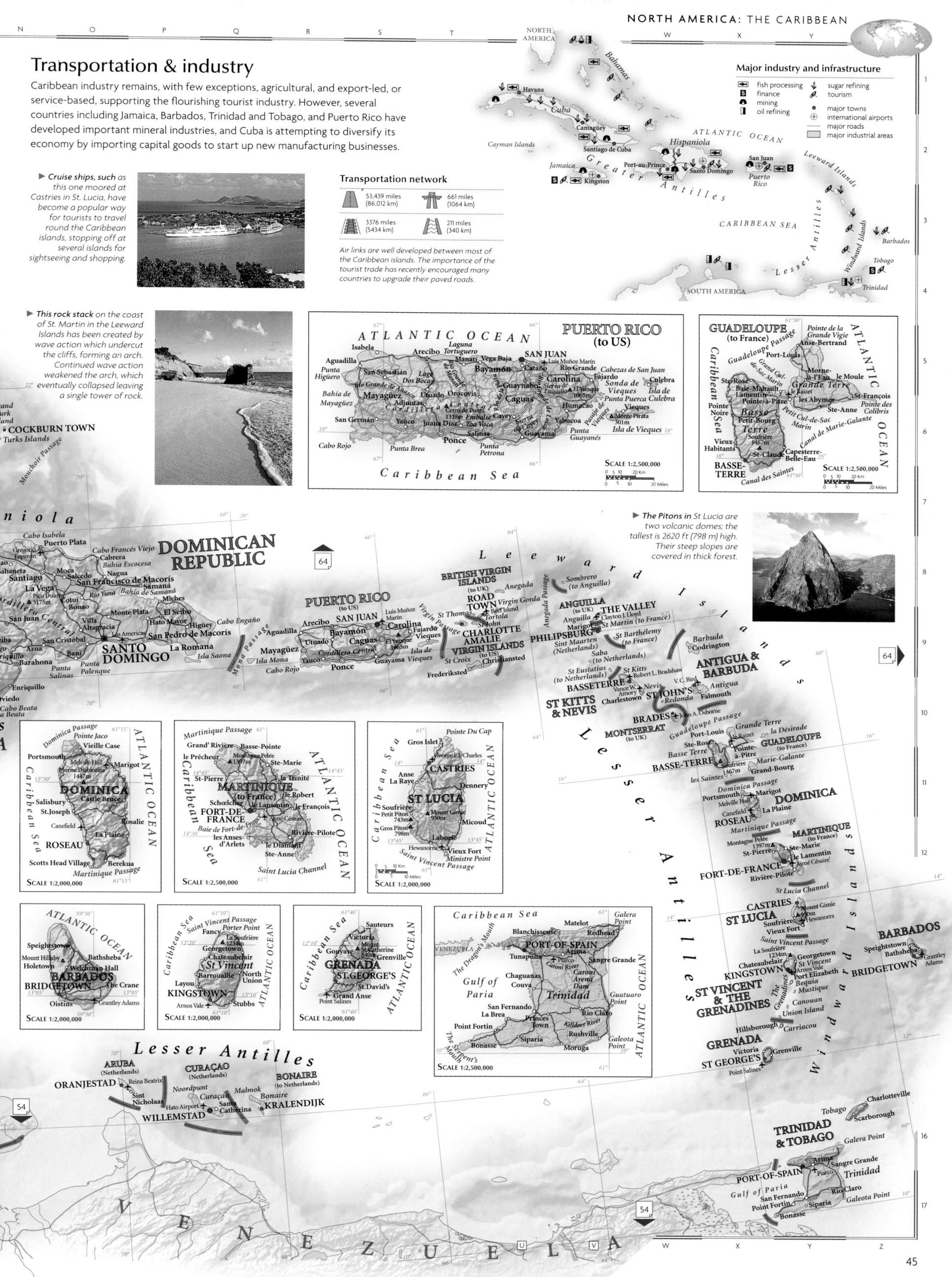

Transportation & industry

Caribbean industry remains, with few exceptions, agricultural, and export-led, or service-based, supporting the flourishing tourist industry. However, several countries including Jamaica, Barbados, Trinidad and Tobago, and Puerto Rico have developed important mineral industries, and Cuba is attempting to diversify its economy by importing capital goods to start up new manufacturing businesses.

▶ *Cruise ships*, such as this one moored at Castries in St. Lucia, have become a popular way for tourists to travel round the Caribbean islands, stopping off at several islands for sightseeing and shopping.

▶ *This rock stack* on the coast of St. Martin in the Leeward Islands has been created by wave action which undercut the cliffs, forming an arch. Continued wave action weakened the arch, which eventually collapsed leaving a single tower of rock.

Transportation network

53,439 miles (86,012 km)		661 miles (1064 km)	
3376 miles (5434 km)		211 miles (340 km)	

Air links are well developed between most of the Caribbean islands. The importance of the tourist trade has recently encouraged many countries to upgrade their paved roads.

Major industry and infrastructure

- fish processing
- finance
- mining
- oil refining
- sugar refining
- tourism
- major towns
- international airports
- major roads
- major industrial areas

▶ *The Pitons* in St Lucia are two volcanic domes; the tallest is 2620 ft (798 m) high. Their steep slopes are covered in thick forest.

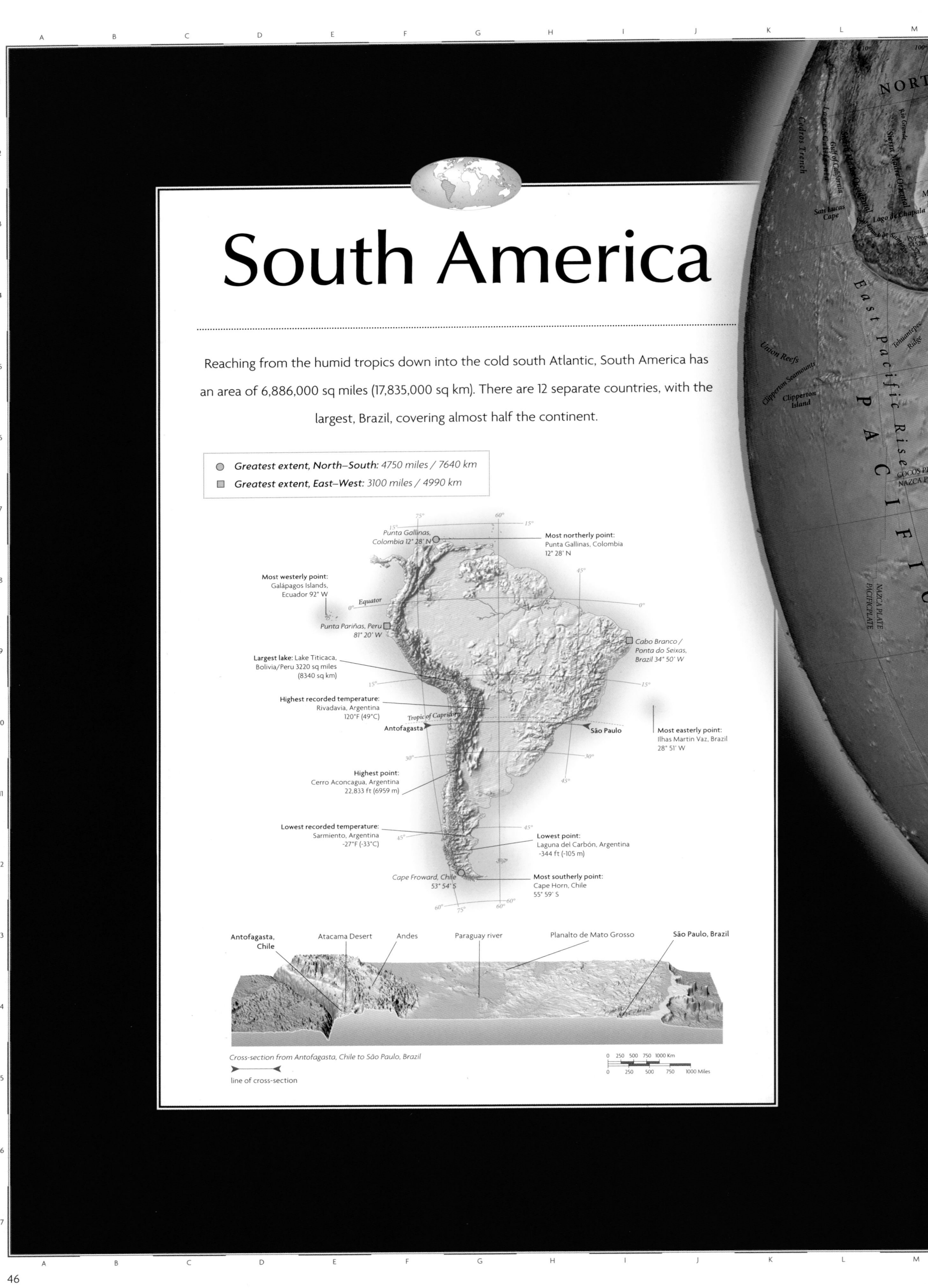

South America

Reaching from the humid tropics down into the cold south Atlantic, South America has an area of 6,886,000 sq miles (17,835,000 sq km). There are 12 separate countries, with the largest, Brazil, covering almost half the continent.

- ● *Greatest extent, North–South:* 4750 miles / 7640 km
- ■ *Greatest extent, East–West:* 3100 miles / 4990 km

Punta Gallinas, Colombia 12° 28' N

Most northerly point: Punta Gallinas, Colombia 12° 28' N

Most westerly point: Galápagos Islands, Ecuador 92° W

Equator

Punta Pariñas, Peru 81° 20' W

Cabo Branco / Ponta do Seixas, Brazil 34° 50' W

Largest lake: Lake Titicaca, Bolivia/Peru 3220 sq miles (8340 sq km)

Highest recorded temperature: Rivadavia, Argentina 120°F (49°C)

Tropic of Capricorn

Antofagasta

São Paulo

Most easterly point: Ilhas Martin Vaz, Brazil 28° 51' W

Highest point: Cerro Aconcagua, Argentina 22,833 ft (6959 m)

Lowest recorded temperature: Sarmiento, Argentina -27°F (-33°C)

Lowest point: Laguna del Carbón, Argentina -344 ft (-105 m)

Cape Froward, Chile 53° 54' S

Most southerly point: Cape Horn, Chile 55° 59' S

Antofagasta, Chile | Atacama Desert | Andes | Paraguay river | Planalto de Mato Grosso | São Paulo, Brazil

Cross-section from Antofagasta, Chile to São Paulo, Brazil

line of cross-section

0 250 500 750 1000 Km
0 250 500 750 1000 Miles

NORTH

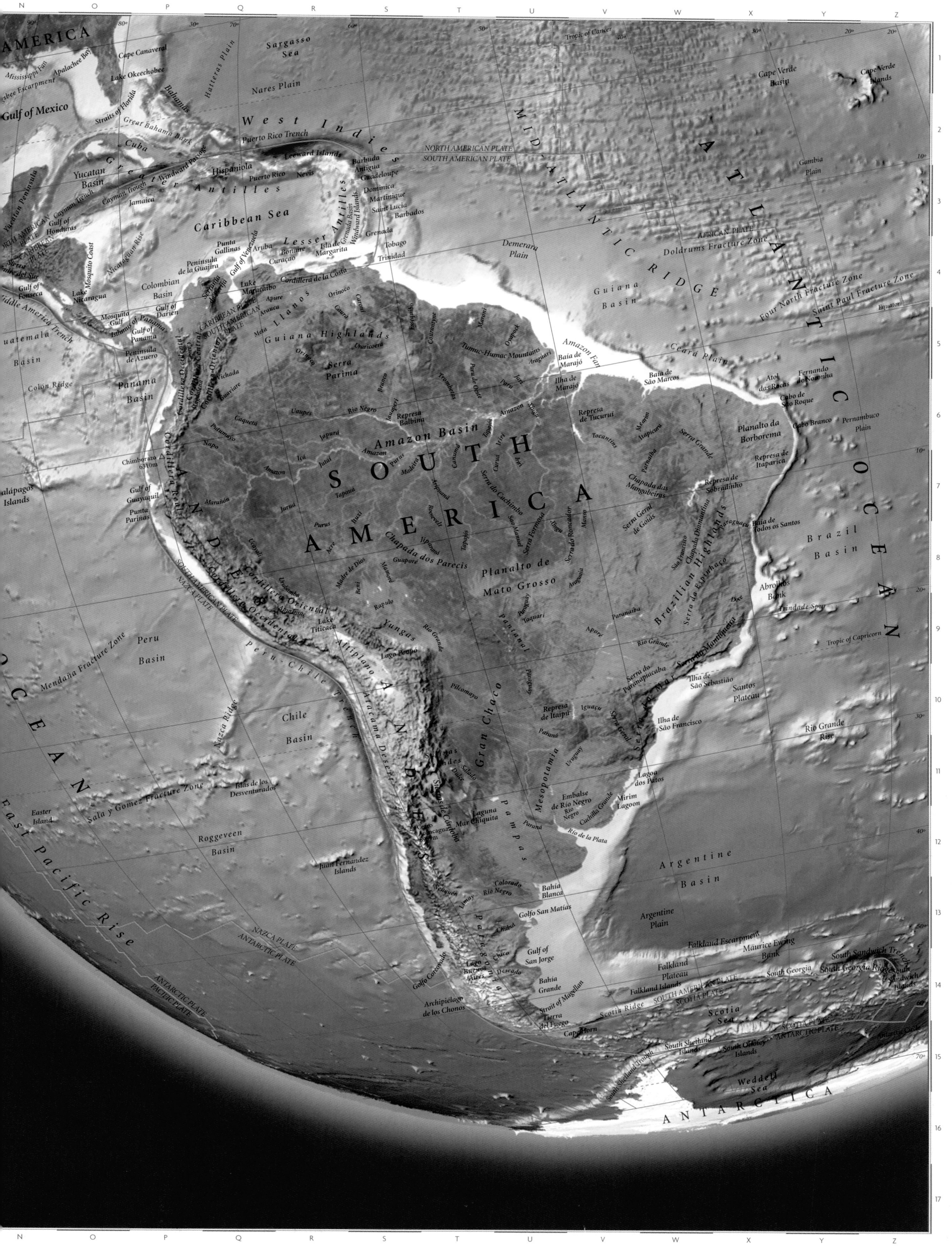

AMERICA

Cape Canaveral

Mississippi Fan

Apalachee Bay

Sigsbee Escarpment

Lake Okeechobee

Gulf of Mexico

Straits of Florida

Great Bahama Bank

Bahamas

Sargasso
Sea

Hatteras Plain

Nares Plain

Cape Verde
Basin

Cape Verde
Islands

Yucatan
Peninsula

Cuba

Greater

Cayman Trough

Windward Passage

Cayman Islands

Puerto Rico Trench

W e s t   I n d i e s

Gumbia
Plain

Yucatan
Basin

Hispaniola

Puerto Rico

Leeward Islands

Nevis

Barbuda
Antigua
Guadeloupe

NORTH AMERICAN PLATE
SOUTH AMERICAN PLATE

A
T
L
A
N
T
I
C

NORTH AMERICAN
PLATE
CARIBBEAN
PLATE

Gulf of
Honduras

Jamaica

A n t i l l e s

Dominica
Martinique
Saint Lucia
Barbados

Sierra
del Sur

Mosquito Coast

Nicaragua Rise

Caribbean Sea

Punta
Gallinas

Gulf of Venezuela

Aruba
Bonaire

Isla de
Margarita

Curaçao

L e s s e r   A n t i l l e s

Tobago

Windward Islands

Grenada

AFRICAN PLATE

Doldrums Fracture Zone

Middle America
Trench

Lake
Nicaragua

Gulf of
Fonseca

Colombian
Basin

Peninsula
de la Guajira

Lake
Maracaibo

Cordillera de la Costa

Trinidad

Demerara
Plain

G u i a n a

B a s i n

R
I
D
G
E

Guatemala
Basin

Mosquito
Gulf

Gulf of
Darién

CARIBBEAN PLATE
SOUTH AMERICAN
PLATE

Apure

Orinoco

Meta

Arauca

Casanare

Caroní

Maroni

Oyapock

Four North Fracture Zone

Saint Paul Fracture Zone

Equator

Middle America Trench

Gulf of
Panama

Peninsula
de Azuero

Isthmus of Panama

Cordillera Central

Cordillera Oriental

L l a n o s

G u i a n a   H i g h l a n d s

Vichada

Caura

Tumuc-Humac Mountains

Araguari

Amazon Fan

Ceará Plain

Colón Ridge

Panama
Basin

Inírida

Guaviare

Serra
Parima

Chiricoa

Cordillera Occidental

S O U T H

Atol
das Rocas

Fernando
de Noronha

Galápagos
Islands

Chimborazo
6310m

Napo

Uaupés

Rio Negro

Javari

Represa
Balbina

Amazon Basin

Pará de Oeste

Ilha de
Marajó

Baía de
Marajó

Baía de
São Marcos

Repres de Tucuruí

Meari

Cabo de
São Roque

Caquetá

Içá

Jutaí

A M E R I C A

Cabo Branco

Pernambuco
Plain

Gulf of
Guayaquil

Putumayo

Marañón

Amazon

Juruá

Purus

Madeira

Xingu

Tocantins

Araguaia

Tapajós

Itacaiúnas

Serra Grande

Planalto da
Borborema

Represa de
Itaparica

Punta
Parinas

Tapauá

Ipixuna

Jiparaná

Serra do Cachimbo

Chapada das
Mangabeiras

Represa de
Sobradinho

10°

Cordillera Oriental

S
O
U
T
H

Madre de Dios

Guaporé

Chapada dos Parecis

Planalto de
Mato Grosso

São Manuel

Serra Formosa

Serra Geral
de Goiás

Serra Geral
de Goiás

Serra do Espinhaço

Baía de
Todos os Santos

B r a z i l
B a s i n

Cordillera Occidental

NAZCA PLATE

Beni

Mamoré

Abuná

Rapulo

Yungas

Rio Grande

Paraguay

Iguatemi

Taquari

Paraná

Paranaíba

Paranaíba

Brazilian Highlands

Dore

Serra de Esperança

Abrolhos
Bank

Trindade-Spur

Mendaña Fracture Zone

Peru
Basin

A N D E S

SOUTH AMERICAN PLATE

Altiplano

Lake
Titicaca

Lago Poopó

Pilcomayo

Pantanal

Apa

Paraná

Rio Grande

Serra do
Paranapiacaba

Serra da Mantiqueira

Ilha de
São Sebastião

Santos
Plateau

Tropic of Capricorn

30°

Nazca Ridge

Chile
Basin

Atacama Desert

Peru-Chile Trench

Salinas Grandes

Sierra de Córdoba

Río Desaguadero

Río Grande

Represa
de Itaipú

Iguaçu

Paraná

Rio Grande

Ilha de
São Francisco

Rio Grande
Rise

Sala y Gómez Fracture Zone

Islas de los
Desventurados

Mesopotamia

Embalse
de Río Negro
Río
Negro

Cuchilla Grande

Lagoa
dos Patos

Mirim
Lagoon

Easter
Island

Roggeveen
Basin

Juan Fernández
Islands

Pampas

Uruguay

Río de la Plata

Argentine
Basin

40°

E a s t   P a c i f i c   R i s e

NAZCA PLATE

ANTARCTIC PLATE

Colorado

Rio Negro

Salinas
Grandes

Mar Chiquita

Paraná

Bahía
Blanca

Golfo San Matías

Argentine
Plain

Falkland Escarpment

Maurice Ewing
Bank

50°

ANTARCTIC
PACIFIC PLATE

Limay

Chubut

Golfo
Coronado

Gulf of
San Jorge

Patagonia

Bahía
Grande

Falkland
Plateau

Falkland Islands

South Georgia

South Sandwich Trench

South Georgia Ridge

South
Sandwich
Islands

Lago
Buenos
Aires

Deseado

SOUTH AMERICAN PLATE

Archipiélago
de los Chonos

Strait of Magellan

Tierra
del Fuego

Cape Horn

Scotia Ridge

SOUTH AMERICAN PLATE

SCOTIA PLATE

Scotia
Sea

SCOTIA PLATE

ANTARCTIC PLATE

60°

South Shetland Trough

South Shetland
Islands

South Orkney
Islands

Antarctic Circle

W e d d e l l
S e a

70°

A N T A R C T I C A

Physical South America

Three major physiographic regions characterize South America. The oldest, the ancient Brazilian Shield and the smaller Guiana and Patagonian shields, form the stable core of the continent. Stretching along the entire west coast are the younger Andean fold mountains with many summits rising to 20,000 ft (6100 m). These two diverse regions are separated by a number of sedimentary basins carrying South America's large river systems to the sea. These include the massive Amazon Basin and the basin of the Gran Chaco.

The Amazon Basin and Guiana Shield

The Amazon river occupies a large depression in the Earth's crust, formed by the uplift of the Andes. It is covered by thick volcanic deposits and layers of alluvium – these have been laid down by the Amazon's many tributaries. To the north is the smaller Guiana Shield.

Headwaters of the Amazon rise in the Andes   Thick alluvium deposits   Mouths of the Amazon

Section across northern South America showing Amazon Basin and its drainage pattern.

0   500   1000 Km
0   500   1000 Miles

Scale 1:27,500,000

Km
0   200   400   600   800
Miles
0   200   400   600   800
projection: Lambert Azimuthal Equal Area

The Andean Uplands

The Andean Uplands run along the west coast of South America. They are being uplifted as the Nazca Plate is subducted beneath the South American Plate. They contain some of the world's largest volcanoes, such as Cotopaxi, and Lake Titicaca which occupies a dormant site. The far south has many large ice-sheets and a fragmented coastline.

Nazca Plate   South American Plate   Volcanic intrusions

Cross-section through the Andes showing the subduction of the Nazca Plate beneath the South American Plate.

0   200   400 Km
0   200   400 Miles

The Brazilian Shield and Gran Chaco

The immense Brazilian Shield underlies more than one-third of South America. It is pitted with numerous volcanic intrusions, and a large basaltic plateau exists between the Paraná river and the Atlantic Ocean. The flat Gran Chaco lies to the west of the shield, covered by sedimentary deposits eroded from the Andes, and transported by South America's mighty rivers.

Young, folded Andes mountains   Volcanic intrusions   Major rivers drain to the south through the Gran Chaco   Ancient resistant shield

Section across central South America showing the flat basin of the Gran Chaco and the ancient Brazilian Shield.

0   200   400 Km
0   200   400 Miles

Map key

Elevation

6000m / 19,686ft
4000m / 13,124ft
3000m / 9843ft
2000m / 6562ft
1000m / 3281ft
500m / 1640ft
250m / 820ft
100m / 328ft
sea level

Plate margins
(for explanation see page xiv)

——— constructive
△△△ destructive
——— conservative
········ uncertain

——— physiographic regions
►◄ line of cross-section

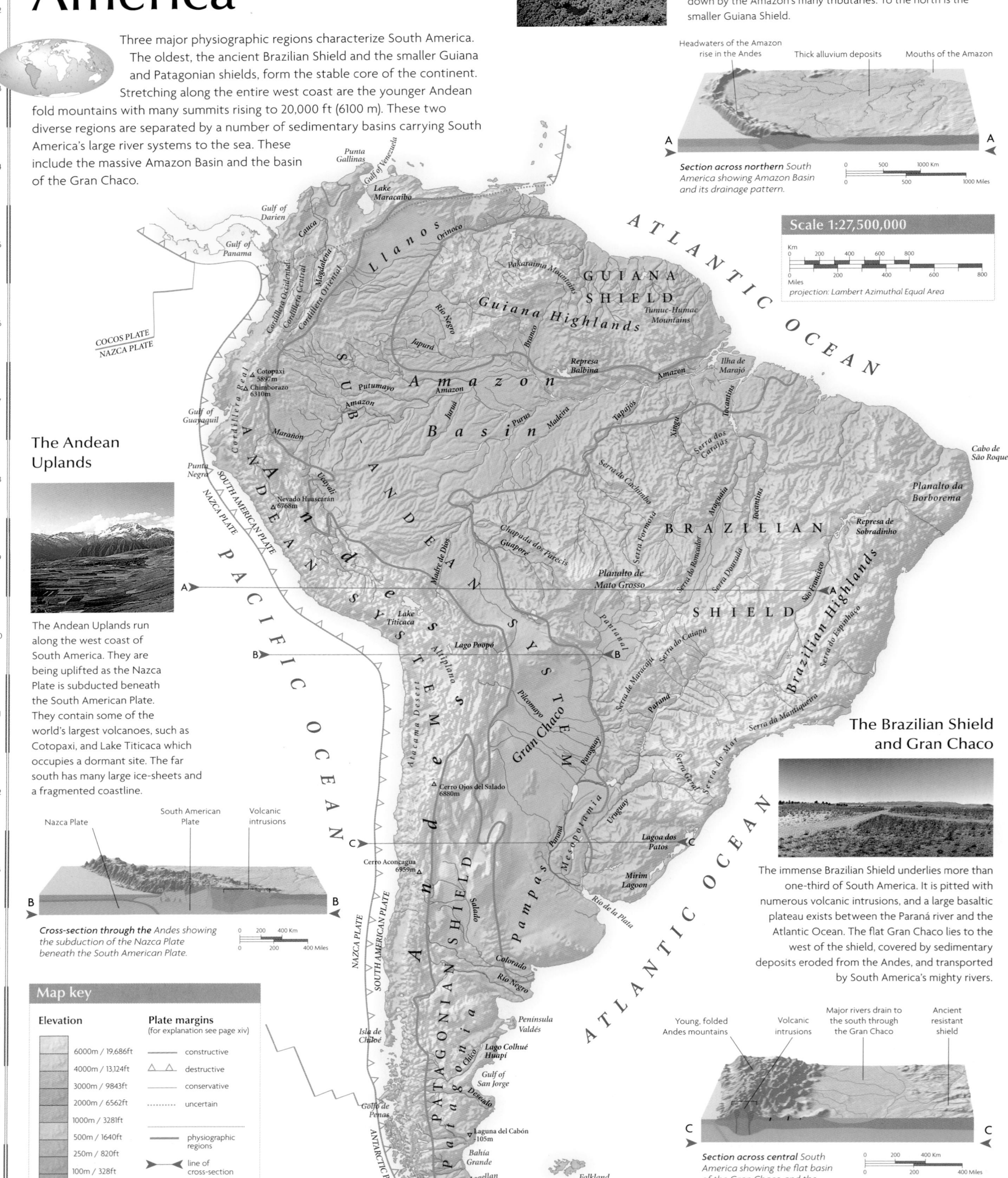

Climate

The climate of South America is influenced by three principal factors: the seasonal shift of high pressure air masses over the tropics, cold ocean currents along the western coast, affecting temperature and precipitation, and the mountain barrier produced by by the Andes, which creates a rain shadow over much of the south.

▲ *Mild winters and cool summers typify the extensive Pampas grasslands of Argentina.*

▲ *Chile's hyperarid Atacama Desert is renowned as one of the driest places on Earth.*

Climate

- tundra
- cool continental
- warm humid
- semiarid
- arid
- humid equatorial
- tropical
- ☀ daily hours of sunshine, January
- ☀ daily hours of sunshine, July
- → cold wind

Temperature

Average January temperature *Average July temperature*

Temperature

- below -22°F (-30°C)
- -22 to -4°F (-30 to -20°C)
- -4 to 14°F (-20 to -10°C)
- 14 to 32°F (-10 to 0°C)
- 32 to 50°F (0 to 10°C)
- 50°F to 68°F (10 to 20°C)
- 68 to 86°F (20 to 30°C)
- above 86°F (30°C)

Rainfall

Average January rainfall *Average July rainfall*

Rainfall

- 0–1 in (0–25 mm)
- 1–2 in (25–50 mm)
- 2–4 in (50–100 mm)
- 4–8 in (100–200 mm)
- 8–12 in (200–300 mm)
- 12–16 in (300–400 mm)
- 16–20 in (400–500 mm)
- more than 20 in (500 mm)

▲ *Tropical conditions are found across over half of South America. When both rainfall and temperatures are high, hot humid rain forests prevail.*

Shaping the continent

South America's active tectonic belt has been extensively folded over millions of years; landslides are still frequent in the mountains. The large river systems that erode the mountains flow across resistant shield areas, depositing sediment. Present-day glaciation affects the distinctive landscape of the far south.

Mass movement

6 Debris slides are common in the highlands of South America *(left)*. They occur where soil on a slope is saturated by rainwater and therefore less stable. The actual slides are often triggered by earthquakes.

- Scarp face left after soil has moved to the base of the slope
- Failure plane
- Toe of debris slide

Mass movement: *A section of a debris slide*

Chemical weathering

1 Table mountains *(left)* are the eroded remnants of an ancient upland. As water percolates along cracks in these high, flat-topped mountains it forms intricate cave systems. Chemical weathering also isolates large blocks which then collapse, accumulating as rockfalls at the foot of scarp slopes.

- Smooth summit dissected by deep gorges
- Rainfall
- Runoff surges down caverns as waterfalls

Chemical weathering: *Erosion of the Guyana Shield*

The evolving landscape

River systems

2 Along the Amazon *(above)* there is a great variation in rates of erosion. As the headwaters of the Amazon flow down from the Andes, they erode and transport vast quantities of sediment, and are known as whitewaters. Across the shield areas erosion rates are very low. These rivers, carrying rotting vegetation, are called blackwaters.

- Whitewater river
- Blackwater river
- Little erosion in shield areas
- Confluence of whitewater with blackwater

River systems: *Suspended sediments in the Amazon*

Folding

5 Folding occurs beneath the surface under high temperatures and pressures. Rocks become sufficiently malleable to flow and not fracture as tectonic plates collide. In the Valley of the Moon in Chile *(above)*, anticlines (or upfolds) and synclines (or troughs) have been exploited by erosion.

- Fold axis
- Anticline
- Syncline
- Fold axis

Folding: *Synclines and anticlines*

Deposition

4 Large alluvial fans are found extensively across South America *(above)*. Confined mountain rivers, carrying large quantities of eroded material, emerge from a mountain gorge onto the plains, where they deposit their load in huge fans.

- Confined stream in the mountains
- Subsequent fan
- Mountain front
- Fan forms as stream emerges onto the plain

Deposition: *Formation of an alluvial fan*

Landscape

- uplifting land
- stable land
- sinking land
- glacier
- ocean current
- alluvial fan
- inselberg
- river

- Unstable front in deep water, where ice is fracturing
- Original extent of glacier
- Icebergs
- Stable front
- Glacier was grounded against a shoal

Glaciation: *Retreating glacier in Patagonia*

Glaciation

3 As fjord glaciers in Patagonia *(above)* retreat, they become grounded on shoals. In deeper water the base of the glacier becomes unstable, and icebergs break off (calve) until the glacier snout grounds once more.

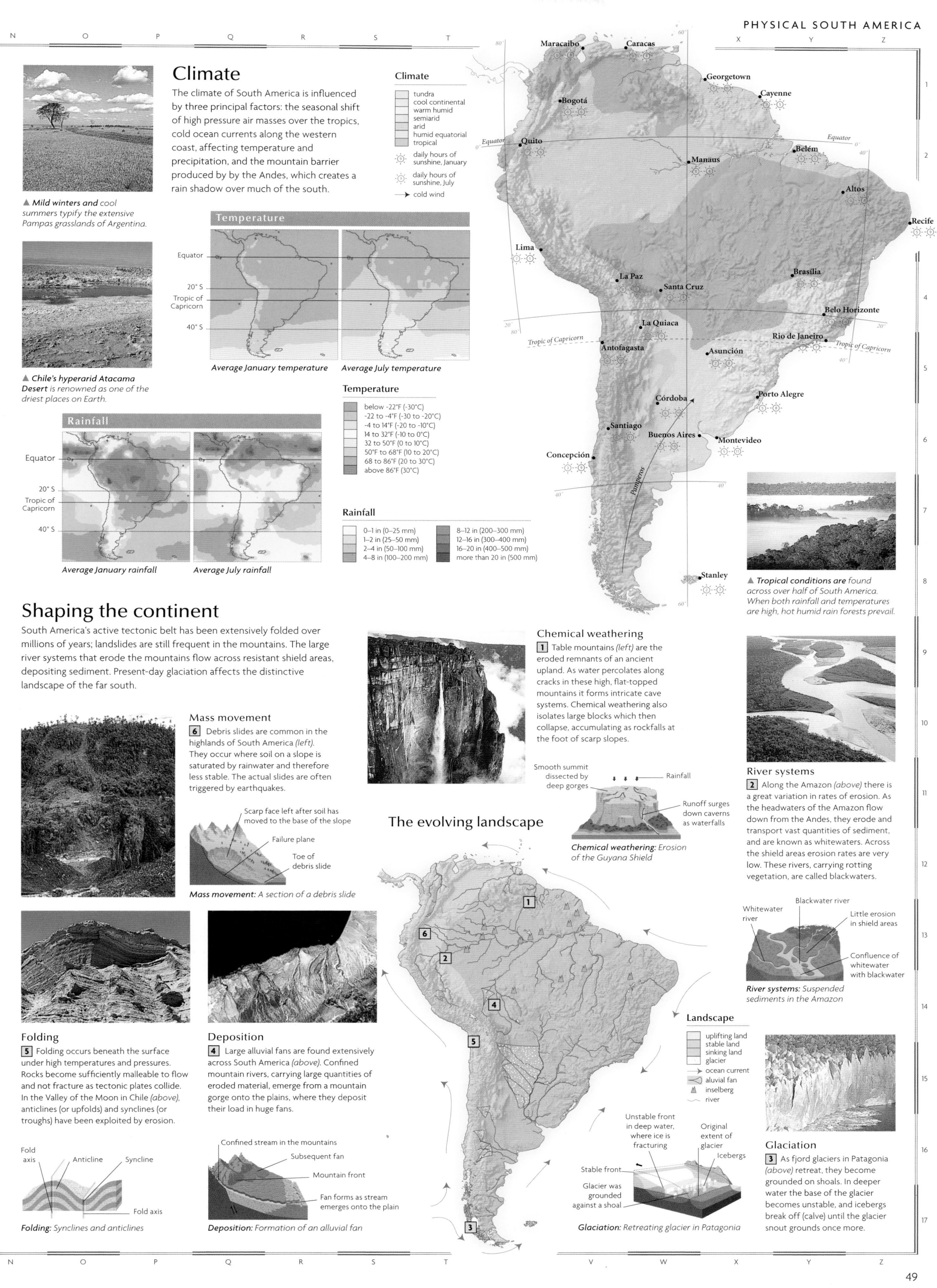

Maracaibo • Caracas • Georgetown • Cayenne • Bogotá • Quito • Belém • Altos • Recife • Lima • Manaus • La Paz • Santa Cruz • Brasília • Belo Horizonte • La Quiaca • Rio de Janeiro • Antofagasta • Asunción • Córdoba • Porto Alegre • Santiago • Buenos Aires • Montevideo • Concepción • Stanley

Equator • Tropic of Capricorn • Pamperos

Political South America

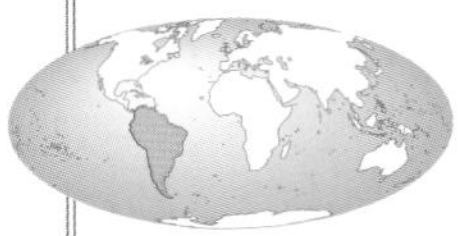

Modern South America's political boundaries have their origins in the territorial endeavors of explorers during the 16th century, who claimed almost the entire continent for Portugal and Spain. The Portuguese land in the east later evolved into the federal state of Brazil, while the Spanish vice-royalties eventually emerged as separate independent nation-states in the early 19th century. South America's growing population has become increasingly urbanized, with the growth of coastal cities into large conurbations like Rio de Janeiro and Buenos Aires. In Brazil, Argentina, Chile, and Uruguay, a succession of military dictatorships has given way to fragile, but strengthening, democracies.

◄ *Europe retains a* small foothold in South America. Kourou in French Guiana was the site chosen by the European Space Agency to launch the Ariane rocket. As a result of its status as a French overseas department, French Guiana is actually part of the European Union.

Scale 1:21,500,000

Km
0   100 200 300 400 500 600 700 800

0   100 200 300 400 500 600 700 800
Miles

projection: Lambert Azimuthal Equal Area

Transportation

Most major road and rail routes are confined to the coastal regions by the forbidding natural barriers of the Andes mountains and the Amazon Basin. Few major cross-continental routes exist, although Buenos Aires serves as a transportation center for the main rail links to La Paz and Valparaíso, while the construction of the Trans-Amazon and Pan-American Highways have made direct road travel possible from Recife to Lima and from Puerto Montt up the coast into central America. A new waterway project is proposed to transform the River Paraguay into a major shipping route, although it involves considerable wetland destruction.

► *South America's most* extensive rail network is centered on the Argentinian capital, Buenos Aires. The construction of new rail lines ouward from this important port, allowed the colonization of the Pampas lands for agriculture.

Languages

Prior to European exploration in the 16th century, a diverse range of indigenous languages were spoken across the continent. With the arrival of Iberian settlers, Spanish became the dominant language, with Portuguese spoken in Brazil, and Native American languages such as Quechua and Guaraní, becoming concentrated in the continental interior. Today this pattern persists, although successive European colonization has led to Dutch being spoken in Suriname, English in Guyana, and French in French Guiana, while in large urban areas, Japanese and Chinese are increasingly common.

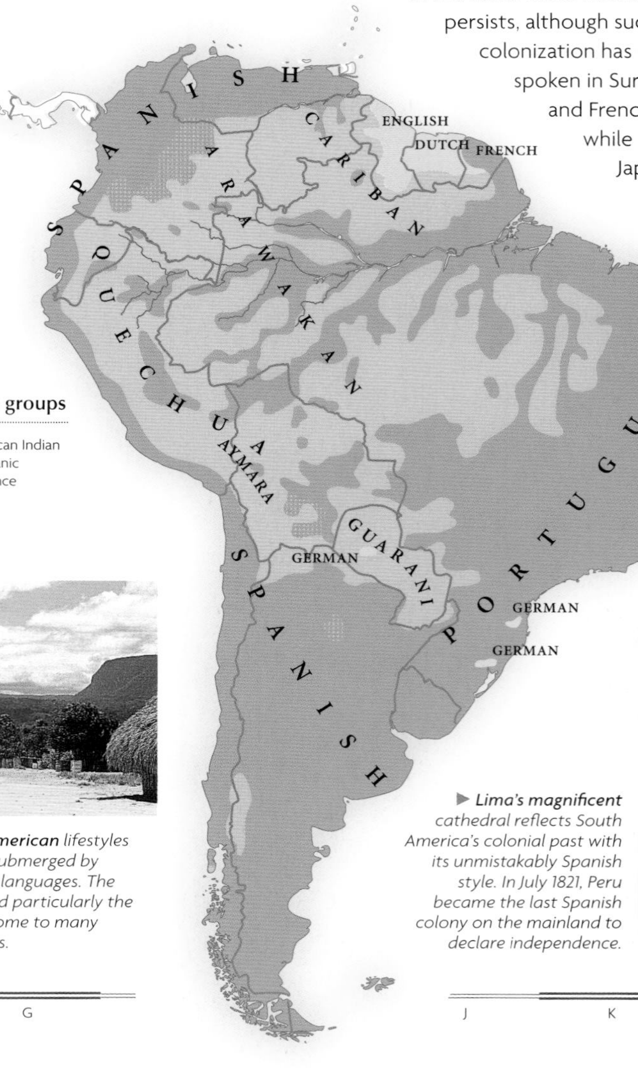

Transportation

— major roads and highways
— major railroads
— international borders
● transport intersections
⊕ international airports
⊕ major ports

Language groups

American Indian
Germanic
Romance

► *Chile's main port*, Valparaíso, is a vital national shipping center, in addition to playing a key role in the growing trade with Pacific nations. The country's awkward, elongated shape means that sea transportation is frequently used for internal travel and communications in Chile.

▲ *Indigenous South American* lifestyles have not been totally submerged by European cultures and languages. The continental interior, and particularly the Amazon Basin, is still home to many different ethnic peoples.

► *Lima's magnificent* cathedral reflects South America's colonial past with its unmistakably Spanish style. In July 1821, Peru became the last Spanish colony on the mainland to declare independence.

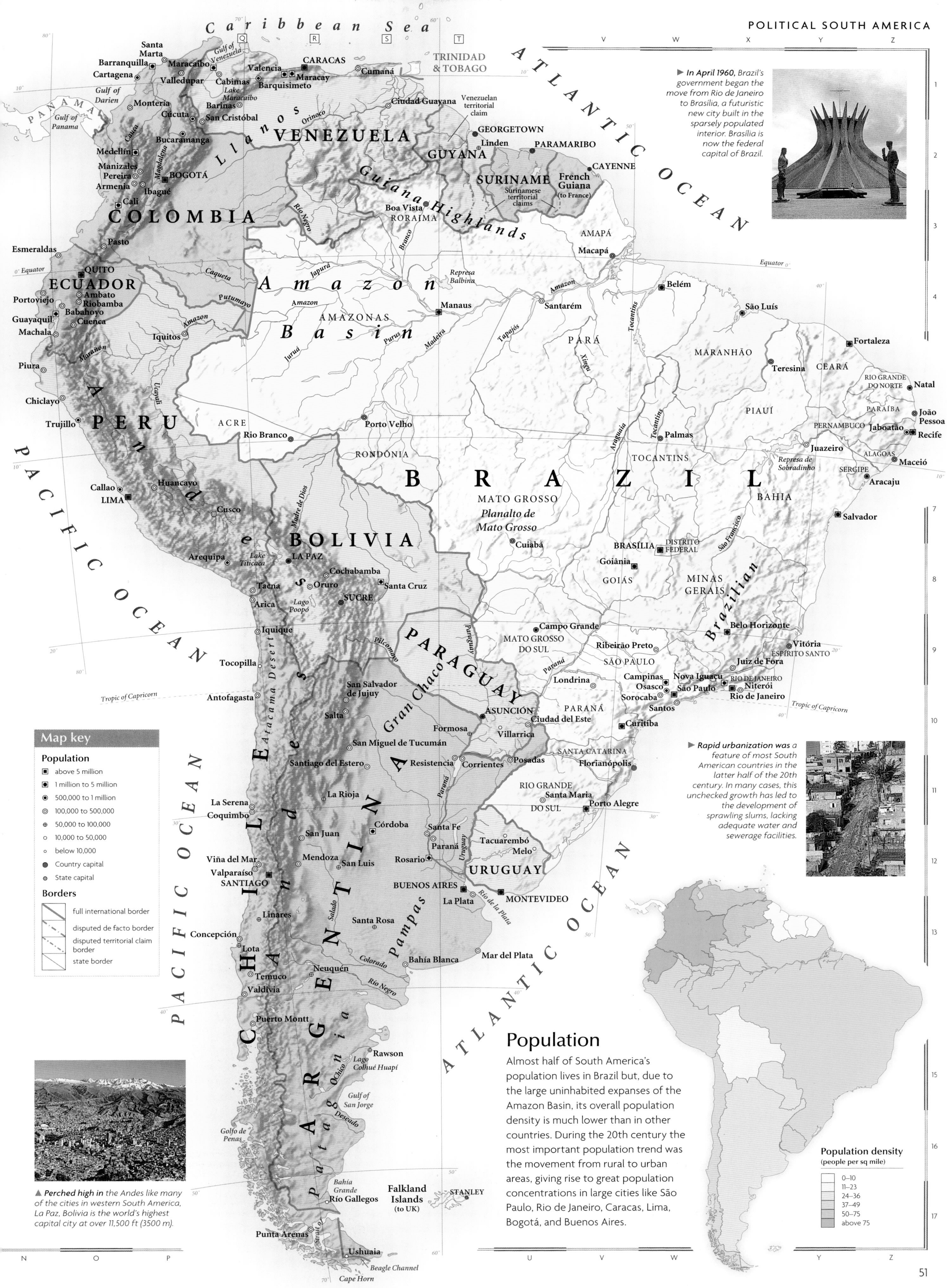

Caribbean Sea

ATLANTIC OCEAN

PACIFIC OCEAN

ATLANTIC OCEAN

Santa Marta
Barranquilla
Cartagena
Maracaibo
Valledupar
Cabimas
Valencia
CARACAS
Maracay
Cumaná
Barquisimeto
Gulf of Venezuela
TRINIDAD & TOBAGO

Gulf of Darien
Monteria
Cúcuta
Barinas
San Cristóbal
Ciudad Guayana
Venezuelan territorial claim
GEORGETOWN
Linden
PARAMARIBO
CAYENNE

PANAMA
Gulf of Panama

Bucaramanga
Medellín
Manizales
Pereira
Armenia
Ibagué
BOGOTÁ
Cali

Llanos

VENEZUELA

GUYANA

SURINAME

French Guiana (to France)

Surinamese territorial claims

COLOMBIA

Orinoco

Río Negro

Guiana Highlands

Boa Vista
RORAIMA

Branco

AMAPÁ

Macapá

Esmeraldas
QUITO
Pasto

ECUADOR

Portoviejo
Ambato
Riobamba
Babahoyo
Guayaquil
Cuenca
Machala

Equator

Caquetá

Putumayo

Amazon

Amazonas Basin

Amazon

Manaus

Santarém

Belém

São Luís

MARANHÃO

Fortaleza

Teresina

CEARÁ

PERU

Piura

Chiclayo

Trujillo

Marañón

Iquitos

Amazon

Jurúa

Purus

Madeira

Tapajós

PARÁ

Xingu

Tocantins

Araguaia

Tocantins

PIAUÍ

RIO GRANDE DO NORTE
Natal
PARAÍBA
João Pessoa
Jaboatão
Recife
PERNAMBUCO
Juazeiro
ALAGOAS
Maceió
SERGIPE
Aracaju

Callao
LIMA
Huancayo

Cusco

Arequipa

Andes

Ucayali

Madre de Dios

ACRE

Rio Branco

Porto Velho

RONDÔNIA

BRAZIL

MATO GROSSO

Planalto de Mato Grosso

Cuiabá

BRASÍLIA
DISTRITO FEDERAL

Goiânia

GOIÁS

MINAS GERAIS

Represa de Sobradinho

São Francisco

BAHIA

Salvador

Brazilian

BOLIVIA

Lake Titicaca
LA PAZ
Cochabamba
Oruro
Santa Cruz
SUCRE
Lago Poopó
Tacna
Arica

Iquique

Tocopilla

Antofagasta

Atacama Desert

Pilcomayo

Paraguay

Paraná

PARAGUAY

Gran Chaco

San Salvador de Jujuy
Salta

ASUNCIÓN

Formosa

San Miguel de Tucumán

Villarrica

Ciudad del Este

Campo Grande
MATO GROSSO DO SUL

Ribeirão Preto
SÃO PAULO
Londrina
Campinas
Osasco
Sorocaba
São Paulo
Santos
Nova Iguaçu
Niterói
Rio de Janeiro
RIO DE JANEIRO
Curitiba
PARANÁ
Belo Horizonte
Vitória
ESPÍRITO SANTO
Juiz de Fora

Santiago del Estero

La Rioja

San Juan

Córdoba

Resistencia

Corrientes

Posadas

SANTA CATARINA
Florianópolis

RIO GRANDE DO SUL
Santa Maria
Porto Alegre

La Serena
Coquimbo

Viña del Mar
Valparaíso
SANTIAGO

Mendoza
San Luis

Santa Fe
Paraná
Rosario

Tacuarembó
Melo

URUGUAY

Linares

Concepción

Lota

Temuco
Valdívia

ARGENTINA

CHILE

Andes

Patagonia

Salado

Colorado

Río Negro

Neuquén

BUENOS AIRES
La Plata
Río de la Plata
MONTEVIDEO

Pampas

Santa Rosa

Bahía Blanca

Mar del Plata

Puerto Montt

Golfo de Penas

Lago Colhué Huapí

Bahía Grande

Falkland Islands (to UK)

STANLEY

Ochoa

Río Gallegos

Punta Arenas

Ushuaia

Beagle Channel

Cape Horn

Gulf of San Jorge

Rawson

Deseado

Strait of Magellan

Tropic of Capricorn

Equator

Represa Balbina

Lago

Río de la Plata

Uruguay

▶ In April 1960, Brazil's government began the move from Rio de Janeiro to Brasília, a futuristic new city built in the sparsely populated interior. Brasília is now the federal capital of Brazil.

Map key

Population

- ▪ above 5 million
- ▣ 1 million to 5 million
- ◉ 500,000 to 1 million
- ◎ 100,000 to 500,000
- ⊕ 50,000 to 100,000
- ⊙ 10,000 to 50,000
- ∘ below 10,000
- ● Country capital
- ◉ State capital

Borders

- ⧄ full international border
- ⧄ disputed de facto border
- ⧄ disputed territorial claim border
- ⧄ state border

▶ Rapid urbanization was a feature of most South American countries in the latter half of the 20th century. In many cases, this unchecked growth has led to the development of sprawling slums, lacking adequate water and sewerage facilities.

Population

Almost half of South America's population lives in Brazil but, due to the large uninhabited expanses of the Amazon Basin, its overall population density is much lower than in other countries. During the 20th century the most important population trend was the movement from rural to urban areas, giving rise to great population concentrations in large cities like São Paulo, Rio de Janeiro, Caracas, Lima, Bogotá, and Buenos Aires.

Population density
(people per sq mile)

- 0–10
- 11–23
- 24–36
- 37–49
- 50–75
- above 75

▲ Perched high in the Andes like many of the cities in western South America, La Paz, Bolivia is the world's highest capital city at over 11,500 ft (3500 m).

South American resources

Agriculture still provides the largest single form of employment in South America, although rural unemployment and poverty continue to drive people towards the huge coastal cities in search of jobs and opportunities. Mineral and fuel resources, although substantial, are distributed unevenly; few countries have both fossil fuels and minerals. To break industrial dependence on raw materials, boost manufacturing, and improve infrastructure, governments borrowed heavily from the World Bank in the 1960s and 1970s. This led to the accumulation of massive debts which are unlikely ever to be repaid. Today, Brazil dominates the continent's economic output, followed by Argentina. Recently, the less-developed western side of South America has benefited due to its geographical position; for example Chile is increasingly exporting raw materials to Japan.

◄ *Ciudad Guayana is a planned industrial complex in eastern Venezuela, built as an iron and steel center to exploit the nearby iron ore reserves.*

Industry

✈ aerospace	✐ pharmaceuticals
🍺 brewing	📖 printing & publishing
🚗 car/vehicle manufacture	⚓ shipbuilding
chemicals	sugar processing
📺 electronics	⊤ textiles
⚙ engineering	timber processing
$ finance	tobacco processing
🐟 fish processing	wine
🍴 food processing	oil
hi-tech industry	gas
iron & steel	
meat processing	• industrial cities
△ metal refining	▨ major industrial areas
narcotics	

Standard of living

Wealth disparities throughout the continent create a wide gulf between affluent landowners and those afflicted by chronic poverty in inner city slums. The illicit production of cocaine, and the hugely influential drug barons who control its distribution, contribute to the violent disorder and corruption which affect northwestern South America, destabilizing local governments and economies.

▲ *The cold Peru Current flows north from the Antarctic along the Pacific coast of Peru, providing rich nutrients for one of the world's largest fishing grounds. However, over exploitation has severely reduced Peru's anchovy catch.*

Standard of living
(UN human development index)

low

high

▶ *Both Argentina and Chile are now exploring the southernmost tip of the continent in search of oil. Here in Punta Arenas, a drilling rig is being prepared for exploratory drilling in the Strait of Magellan.*

GNI per capita (US$)

below 999
1000–1999
2000–2999
3000–3999
4000–4999
above 5000

Industry

Argentina and Brazil are South America's most industrialized countries and São Paulo is the continent's leading industrial center. Long-term government investment in Brazilian industry has encouraged a diverse industrial base; engineering, steel production, food processing, textile manufacture, and chemicals predominate. The illegal production of cocaine is economically significant in the Andean countries of Colombia and Bolivia. In Venezuela, the oil-dominated economy has left the country vulnerable to world oil price fluctuations. Food processing and mineral exploitation are common throughout the less industrially developed parts of the continent, including Bolivia, Chile, Ecuador, and Peru.

Caribbean Sea

PANAMA

Gulf of Panama

Barranquilla
Cartagena
Maracaibo
Barquisimeto
Caracas
Valencia
VENEZUELA
Ciudad Guayana
Georgetown
Paramaribo
GUYANA
SURINAME
French Guiana
(to France)

Medellín
Bogotá
Cali
COLOMBIA

Quito
ECUADOR
Guayaquil

Iquitos

Amazon Basin

Manaus

Belém

B R A Z I L

Fortaleza

Natal

Chiclayo
Chimbote
PERU
Lima
Cusco

Recife

Maceió

Arequipa
La Paz
BOLIVIA
Santa Cruz
Sucre
Brasília

Salvador

Arica
Iquique
Chuquicamata

Belo Horizonte

Antofagasta
PARAGUAY

Asunción
Ciudad del Este
Curitiba

São Paulo
Rio de Janeiro

San Miguel de Tucumán
Corrientes

Porto Alegre

Córdoba
Santa Fe
Rosario
Rio Grande
URUGUAY

Valparaíso
Mendoza
Buenos Aires
Montevideo
Santiago
Talca
Concepción
ARGENTINA

Bahía Blanca
Neuquén
Valdivia

PACIFIC OCEAN

ATLANTIC OCEAN

Comodoro Rivadavia
Gulf of San Jorge

Falkland Islands
(to UK)

Bahía Grande

Punta Arenas

Cape Horn

Environmental issues

The Amazon Basin is one of the last great wilderness areas left on Earth. The tropical rain forests which grow there are a valuable genetic resource, containing innumerable unique plants and animals. The forests are increasingly under threat from new and expanding settlements and "slash-and-burn" farming techniques, which clear land for the raising of beef cattle, causing land degradation and soil erosion.

▲ *Clouds of smoke* billow from the burning Amazon rainforest. Over 11,500 sq miles (30,000 sq km) of virgin rainforest are being cleared annually, destroying an ancient, irreplaceable, natural resource and biodiverse habitat.

Environmental issues

- national parks
- tropical forest
- forest destroyed
- desert
- risk of desertification
- polluted rivers
- marine pollution
- heavy marine pollution
- poor urban air quality

Mineral resources

Over a quarter of the world's known copper reserves are found at the Chuquicamata mine in northern Chile, and other metallic minerals such as tin are found along the length of the Andes. The discovery of oil and gas at Venezuela's Lake Maracaibo in 1917 turned the country into one of the world's leading oil producers. In contrast, South America is virtually devoid of coal, the only significant deposit being on the peninsula of Guajira in Colombia.

▲ *Copper is Chile's* largest export, most of which is mined at Chuquicamata. Along the length of the Andes, metallic minerals like copper and tin are found in abundance, formed by the excessive pressures and heat involved in mountain-building.

Mineral resources

- oil field
- gas field
- coal field
- bauxite
- copper
- diamonds
- gold
- iron
- lead
- silver
- tin

Using the land and sea

Many foods now common worldwide originated in South America. These include the potato, tomato, squash, and cassava. Today, large herds of beef cattle roam the temperate grasslands of the Pampas, supporting an extensive meatpacking trade in Argentina, Uruguay and Paraguay. Corn is grown as a staple crop across the continent and coffee is grown as a cash crop in Brazil and Colombia. Coca plants grown in Bolivia, Peru, and Colombia provide most of the world's cocaine. Fish and shellfish are caught off the western coast, especially anchovies off Peru, shrimps off Ecuador and pilchards off Chile.

◄ *South America, and* Brazil in particular, now leads the world in coffee production, mainly growing Coffea arabica in large plantations. Coffee beans are harvested, roasted and brewed to produce the world's second most popular drink, after tea.

◄ *The Pampas region* of southeast South America is characterized by extensive, flat plains, and populated by cattle and ranchers (gauchos). Argentina is a major world producer of beef, much of which is exported to the US for use in hamburgers.

◄ *High in the Andes,* hardy alpacas graze on the barren land. Alpacas are thought to have been domesticated by the Incas, whose nobility wore robes made from their wool. Today, they are still reared and prized for their soft, warm fleeces.

Using the land and sea

- barren land
- cropland
- desert
- forest
- mountain region
- pasture
- major conurbations
- cattle
- pigs
- sheep
- bananas
- corn
- citrus fruits
- cocoa
- cotton
- coffee
- fishing
- oil palms
- peanuts
- rubber
- shellfish
- soybeans
- sugar cane
- vineyards
- wheat

Northern South America

COLOMBIA, GUYANA, SURINAME, VENEZUELA, French Guiana (to France)

Fringed by the Pacific and Atlantic oceans and the Caribbean Sea, South America's northern region has a rich range of natural resources, some exploited for centuries by colonial powers including the Spanish, French, Dutch, and British, others still to be fully explored. The prospects for further economic development in Colombia, Guyana, and Suriname are blighted by drug-related violence and political instability. Venezuela, despite huge incomes from its oil reserves, remains less developed in other industrial sectors. French Guiana is an overseas *département* of France, now seeking greater autonomy. Most of the major population centers, such as Bogotá, have grown up in the temperate conditions of the high Andes or, like Caracas, at strategic points along the Caribbean coast.

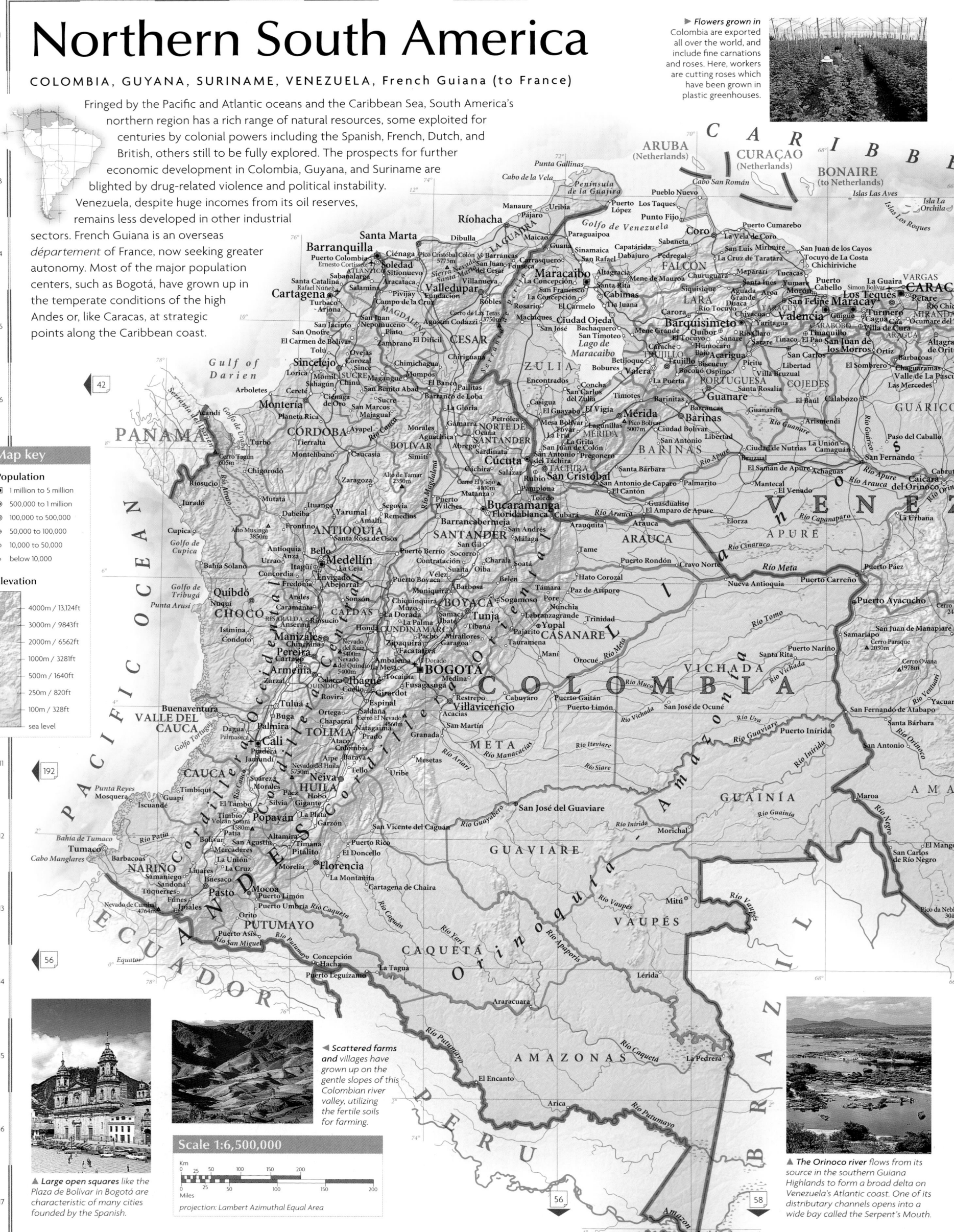

▶ *Flowers grown in* Colombia are exported all over the world, and include fine carnations and roses. Here, workers are cutting roses which have been grown in plastic greenhouses.

Map key

Population

◉ 1 million to 5 million
◉ 500,000 to 1 million
◉ 100,000 to 500,000
⊕ 50,000 to 100,000
○ 10,000 to 50,000
○ below 10,000

Elevation

	4000m / 13,124ft
	3000m / 9843ft
	2000m / 6562ft
	1000m / 3281ft
	500m / 1640ft
	250m / 820ft
	100m / 328ft
	sea level

◀ *Scattered farms and* villages have grown up on the gentle slopes of this Colombian river valley, utilizing the fertile soils for farming.

▲ *Large open squares* like the Plaza de Bolívar in Bogotá are characteristic of many cities founded by the Spanish.

Scale 1:6,500,000

Km
0 25 50   100   150   200

Miles
0 25   50   100   150   200

projection: Lambert Azimuthal Equal Area

▲ *The Orinoco river* flows from its source in the southern Guiana Highlands to form a broad delta on Venezuela's Atlantic coast. One of its distributary channels opens into a wide bay called the Serpent's Mouth.

Transportation & industry

Many mineral resources are mined in Colombia, including fuels, gold, and precious and semiprecious stones. Revenues from coffee and exports of illegal narcotics are crucial to the economy. Venezuela's major economic activity is the oil industry around Lake Maracaibo (*Lago de Maracaibo*). Sugar and bauxite are exported from Guyana and Suriname.

Transportation network

🛣	31,720 miles (51,054 km)
🛤	3411 miles (5490 km)
🚂	2448 miles (3940 km)
〰	22,429 miles (36,100 km)

Rivers are an important means of transportation in Colombia; many are extensively navigable. The Pan-American Highway runs through Colombia. In Venezuela, much infrastructure investment is linked to the oil industry.

Major industry and infrastructure

🏭 chemicals
S finance
food processing
iron & steel
narcotics
mining
oil
oil refining
pharmaceuticals
textiles
timber processing

■ capital cities
● major towns
✈ international airports
— major roads
▨ major industrial areas

▲ **Vast oil reserves** around Lake Maracaibo (*Lago de Maracaibo*) form the focus of Venezuelan industry. Incomes from oil are used to invest in other industries and in the development of infrastructure.

Using the land

The Andean basins support cereals and potatoes. Livestock graze at higher altitudes and on the drier tropical grasslands known as the *llanos;* hardy goats are reared in scrubland areas. Grown at higher elevations, coffee is an important cash crop, as is cotton, sugar cane, bananas, citrus fruits, cocoa, and rice, farmed on the Caribbean lowlands. Coca is the most widely grown narcotic plant, with heroin poppies grown in Colombia and marijuana in lowland areas throughout the region.

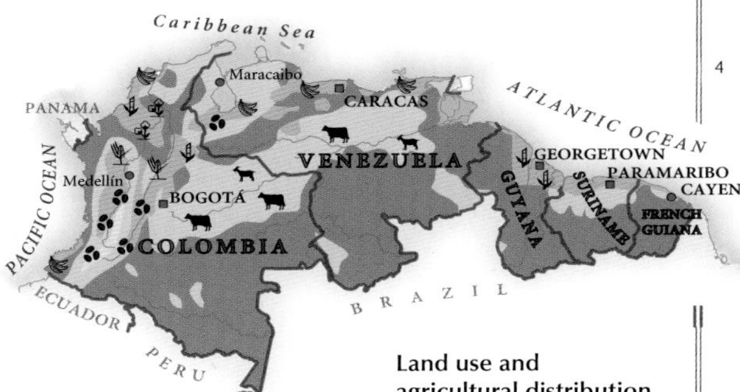

The urban/rural population divide

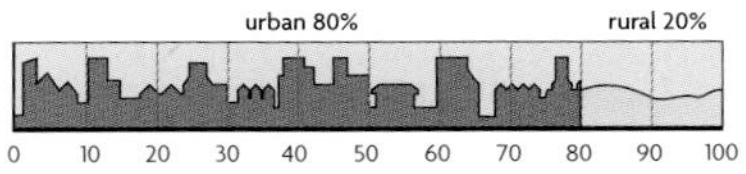

urban 80%   rural 20%

0 10 20 30 40 50 60 70 80 90 100

Population density	Total land area
78 people per sq mile (30 people per sq km)	1,111,317 sq miles (2,879,060 sq km)

Land use and agricultural distribution

cattle
goats
bananas
cereals
coffee
cotton
sugar cane

■ capital cities
● major towns

pasture
cropland
forest
wetlands
mountain region

The landscape

At its northernmost reaches, in western Colombia and Venezuela, the great Andean mountain chain splits into three distinct ranges: the Cordillera Oriental, Cordillera Central, and Cordillera Occidental, intercut by a complex series of lesser ranges and basins. The relief becomes lower toward the coast and the interior plains of the northern Amazon Basin, rising again into the tropical hills of the Guiana Highlands.

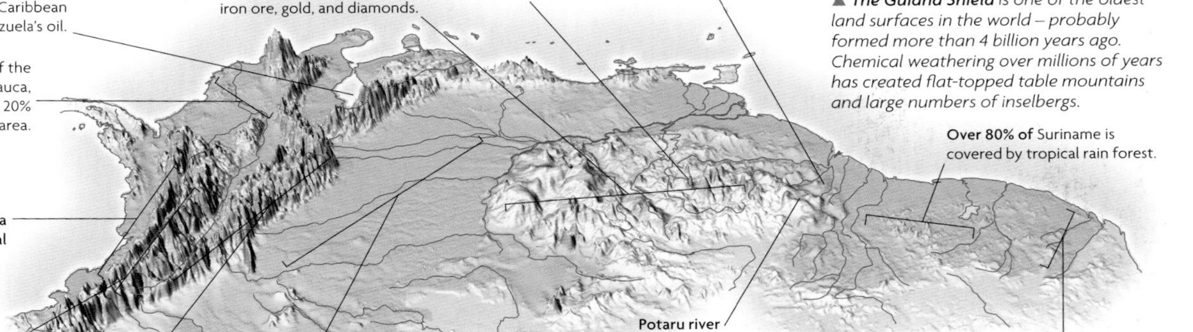

▲ **The Sierra Nevada** de Santa Marta is a granite massif which rises sharply from the Caribbean lowlands to snow-covered peaks, the tallest of which is 18,947 ft (5775 m) high.

Lake Maracaibo (*Lago de Maracaibo*) is not a true lake but a shallow inlet of the Caribbean Sea. It is the main source of Venezuela's oil.

The drainage basin of the Magdalena River and the Cauca, its main tributary, covers over 20% of Colombia's total surface area.

Cordillera Occidental

Cordillera Central

Cordillera Oriental

Colombia's eastern lowlands are known locally as *llanos*, meaning grasslands.

In the Guiana Highlands, Venezuela's most remote region, the ancient crystalline rocks contain deposits of iron ore, gold, and diamonds.

Angel Falls (*Salto Ángel*), at 3212 ft (979 m), is the world's highest waterfall.

▶ **The Potaru river** descends 741 ft (226 m) over a sandstone ledge at the Kaieteur Falls in Guyana.

Potaru river

Igneous intrusions into the crystalline plateau which forms most of central Guyana have led to the formation of the many rapids that characterize Guyana's rivers.

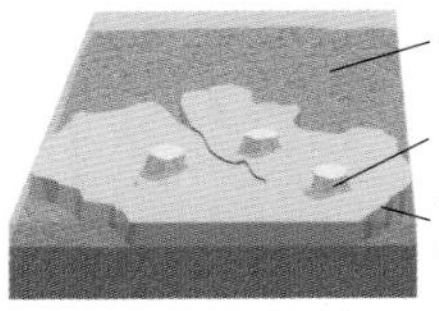

▲ **The Guiana Shield** is one of the oldest land surfaces in the world — probably formed more than 4 billion years ago. Chemical weathering over millions of years has created flat-topped table mountains and large numbers of inselbergs.

Guiana Shield

Alluvial plains
Inselbergs
Table mountains

Over 80% of Suriname is covered by tropical rain forest.

Most of the land in French Guiana is low-lying; here, the rocks of the Guiana Highlands have been eroded by rivers flowing toward the sea.

Western South America

BOLIVIA, ECUADOR, PERU

The three states of Western South America share a similar geography and recent history. Dominated by the Inca empire until Spanish conquest in the 16th century, they achieved independence from Spain in the early 19th century. The precipitous terrain of the Andes presents severe difficulties for overland transportation and continues to be a barrier to national unity and stability. Although Ecuador is now a relatively stable democracy, the military is highly influential in Peru and Bolivia, while the drug trade and associated corruption discourages external aid and economic progress. Wealth and power are still largely concentrated in the hands of a small elite of families, who attained their position during the Spanish colonial period. Energy resources and political recognition for the indigenous peoples are becoming increasingly important issues, particularly in Bolivia.

The landscape

Bolivia, Peru, and Ecuador each possess a high Andean mountain region and an eastern region consisting of tropical lowlands and the Andean slope leading down to them. Toward the south of the region, the mountains widen to form the high plateau of the Altiplano. Peru and Ecuador also have fertile, lowland coastal plains. A wide variety of environments include *selva* (tropical rain forest), *montaña* (mountain forest), and grassland.

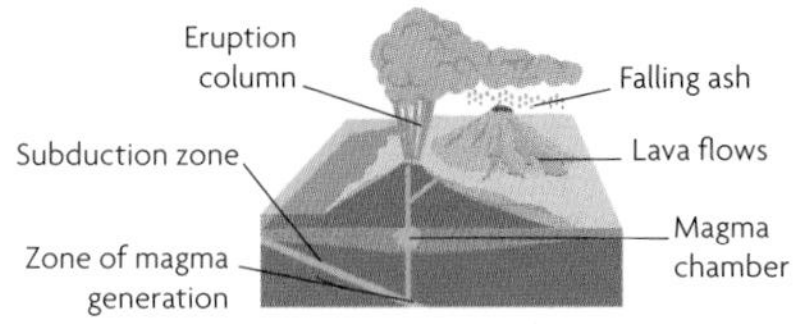

▶ **There are many large and active** volcanoes in the Andes. Magma generated in the heart of the volcano erupts in a huge cloud of ash. Ashfall deposits are common throughout the Andes and the rock produced is known as *andesite*. This is rapidly soaked by heavy rain, causing massive debris flows.

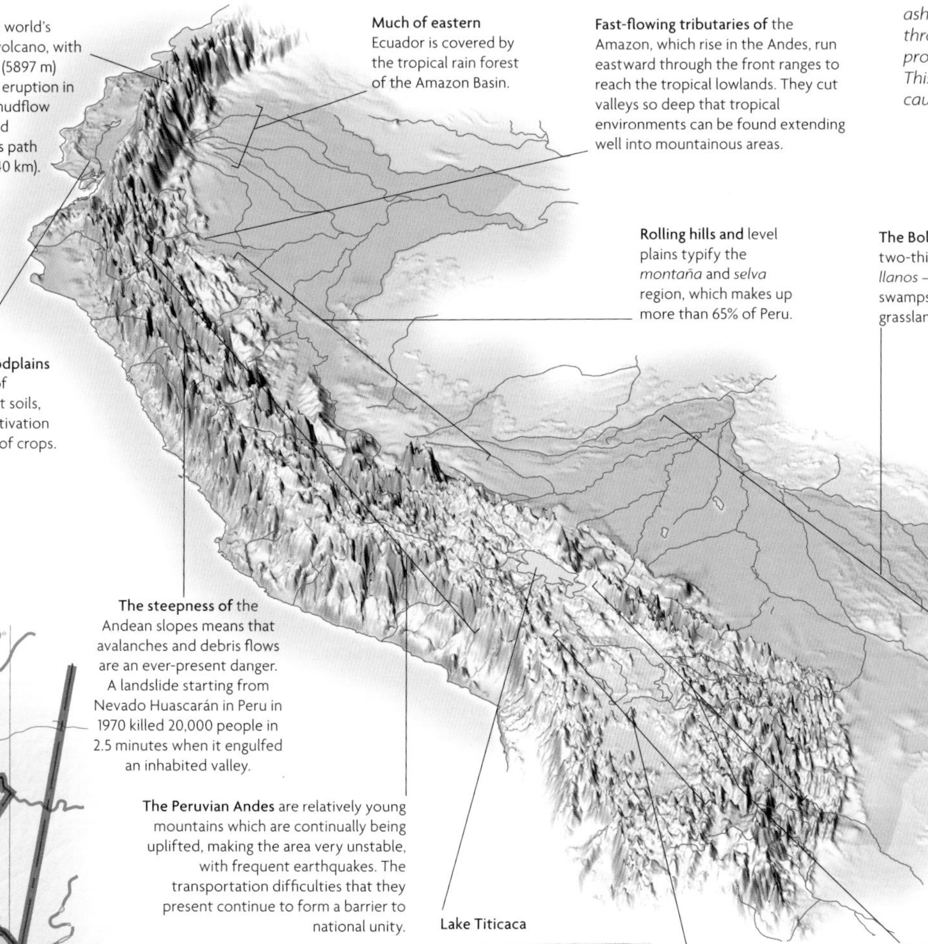

Falling ash
Lava flows
Magma chamber
Eruption column
Subduction zone
Zone of magma generation

Fast-flowing tributaries of the Amazon, which rise in the Andes, run eastward through the front ranges to reach the tropical lowlands. They cut valleys so deep that tropical environments can be found extending well into mountainous areas.

Much of eastern Ecuador is covered by the tropical rain forest of the Amazon Basin.

The Bolivian oriente covers more than two-thirds of the country. It includes *llanos* – low alluvial plains, massive swamps, flooded bottomlands, savannah grassland, and tropical forests.

Rolling hills and level plains typify the *montaña* and *selva* region, which makes up more than 65% of Peru.

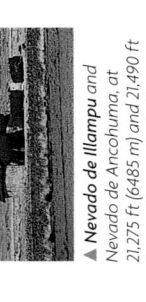
Bolivian Andes

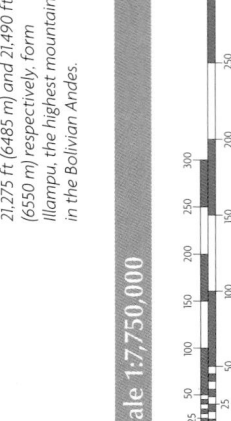
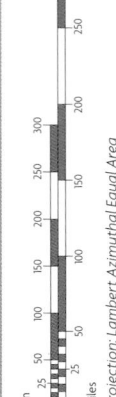

▲ **Nevado de Illampu**, at 21,275 ft (6485 m) and 21,490 ft (6550 m) respectively, form Illampu, the highest mountain in the Bolivian Andes.

Nevado de Ancohuma, at

The Altiplano is a flat, high plateau lying between the Cordillera Oriental and the Cordillera Occidental at a height of up to 12,500 ft (3800 m). At its margins lie many spurs and alluvial fans.

◀ **Lake Titicaca, which** forms part of the border between Peru and Bolivia, is the largest lake in South America and the highest significant body of water in the world at an altitude of 12,507 ft (3812 m).

Lake Titicaca

The steepness of the Andean slopes means that avalanches and debris flows are an ever-present danger. A landslide starting from Nevado Huascarán in Peru in 1970 killed 20,000 people in 2.5 minutes when it engulfed an inhabited valley.

The Peruvian Andes are relatively young mountains which are continually being uplifted, making the area very unstable, with frequent earthquakes. The transportation difficulties that they present continue to form a barrier to national unity.

Cotopaxi is the world's highest active volcano, with a peak 19,347 ft (5897 m) high. A massive eruption in 1877 caused a mudflow which destroyed everything in its path for 150 miles (240 km).

The coastal floodplains are the source of Ecuador's richest soils, enabling the cultivation of a wide range of crops.

▶ **Ecuador's capital city, Quito,** lies high in the Andes, nestling between snowcapped peaks. At 9350 ft (2850 m), Quito is the second highest capital in the world – La Paz in Bolivia is the highest.

Scale 1:7,750,000

projection: Lambert Azimuthal Equal Area

Map key

Population
- ■ above 5 million
- ◼ 1 million to 5 million
- ◉ 500,000 to 1 million
- ⊕ 100,000 to 500,000
- ⊙ 50,000 to 100,000
- ○ 10,000 to 50,000
- ○ below 10,000

Elevation
- 6000m / 19,686ft
- 4000m / 13,124ft
- 3000m / 9843ft
- 2000m / 6562ft
- 1000m / 3281ft
- 500m / 1640ft
- 250m / 820ft
- 100m / 328ft
- sea level

Ecuador: Administrative regions
- ① CARCHI
- ② TUNGURAHUA
- ③ BOLÍVAR
- ④ CHIMBORAZO
- ⑤ ZAMORA CHINCHIPE

COLOMBIA
ECUADOR
PERU
BRAZIL

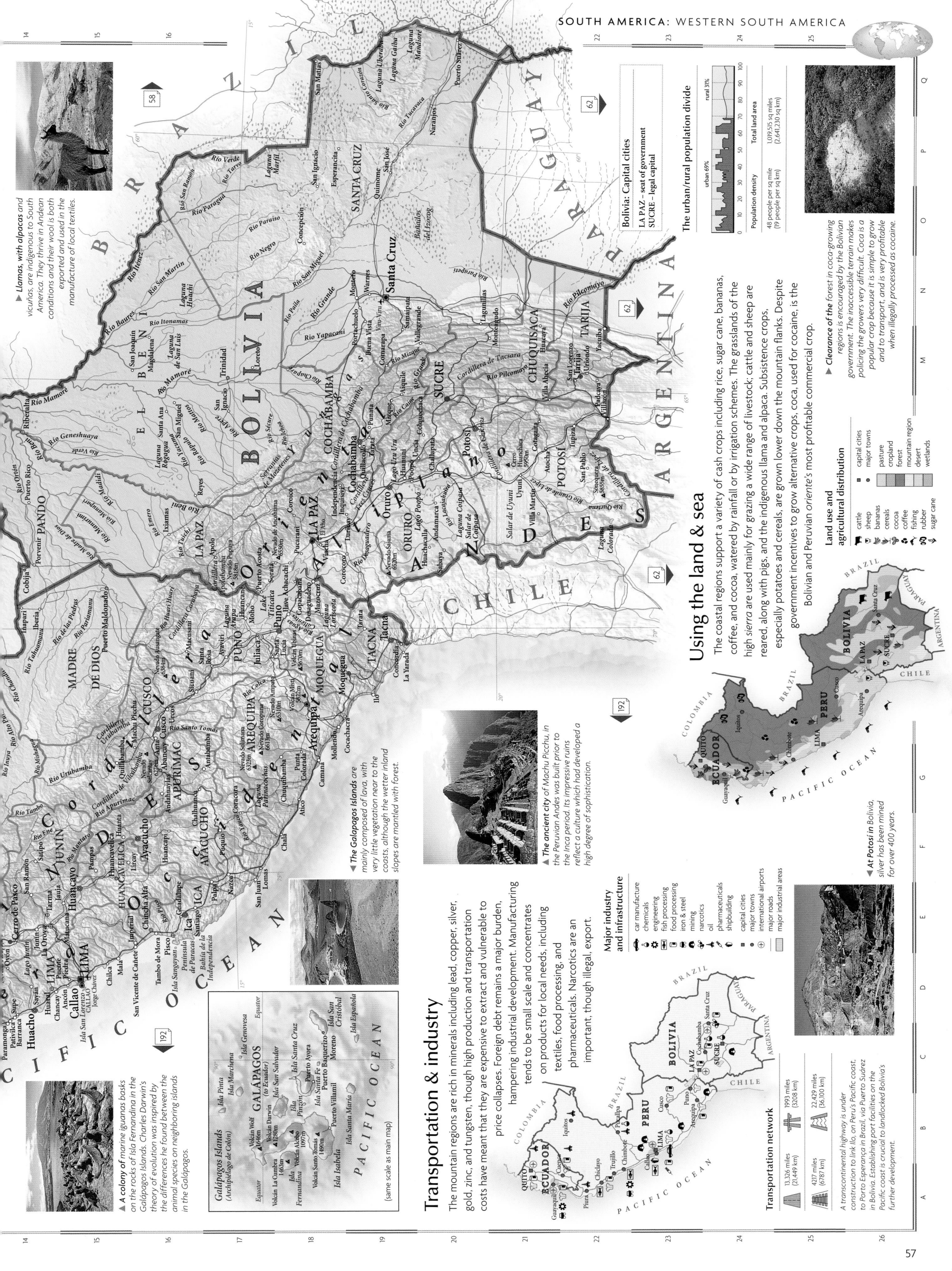

▶ *Llamas, with alpacas and vicuñas, are indigenous to South America. They thrive in Andean conditions and their wool is both exported and used in the manufacture of local textiles.*

Bolivia: Capital cities
LA PAZ – seat of government
SUCRE – legal capital

The urban/rural population divide

urban 69%   rural 31%

Total land area
1,019,515 sq miles
(2,641,230 sq km)

Population density
48 people per sq mile
(19 people per sq km)

▶ *Clearance of the forest in coca-growing regions is encouraged by the Bolivian government. The inaccessible terrain makes policing the growers very difficult. Coca is a popular crop because it is simple to grow and to transport, and is very profitable when illegally processed as cocaine.*

Using the land & sea

The coastal regions support a variety of cash crops including rice, sugar cane, bananas, coffee, and cocoa, watered by rainfall or by irrigation schemes. The grasslands of the high *sierra* are used mainly for grazing a wide range of livestock; cattle and sheep are reared, along with pigs, and the indigenous llama and alpaca. Subsistence crops, especially potatoes and cereals, are grown lower down the mountain flanks. Despite government incentives to grow alternative crops, coca, used for cocaine, is the Bolivian and Peruvian *oriente's* most profitable commercial crop.

Land use and agricultural distribution

cattle
sheep
cereals
cocoa
coffee
fishing
rubber
sugar cane

capital cities
major towns
pasture
cropland
forest
mountain region
desert
wetlands

▼ *The Galápagos Islands are mainly composed of lava, with very little vegetation near to the coasts, although the wetter inland slopes are mantled with forest.*

▲ *The ancient city of Machu Picchu, in the Peruvian Andes was built prior to the Inca period. Its impressive ruins reflect a culture which had developed a high degree of sophistication.*

▶ *At Potosí in Bolivia, silver has been mined for over 400 years.*

Transportation & industry

The mountain regions are rich in minerals including lead, copper, silver, gold, zinc, and tungsten, though high production and transportation costs have meant that they are expensive to extract and vulnerable to price collapses. Foreign debt remains a major burden, hampering industrial development. Manufacturing tends to be small scale and concentrates on products for local needs, including textiles, food processing, and pharmaceuticals. Narcotics are an important, though illegal, export.

Major industry and infrastructure

car manufacture
chemicals
engineering
fish processing
food processing
iron & steel
mining
narcotics
oil
pharmaceuticals
shipbuilding

capital cities
major towns
international airports
major roads
major industrial areas

Transportation network

13,326 miles (21,449 km)
1993 miles (3208 km)
4217 miles (6787 km)
22,429 miles (36,100 km)

A transcontinental highway is under construction to link Ilo, on Peru's Pacific coast, to Porto Esperança in Brazil, via Puerto Suárez in Bolivia. Establishing port facilities on the Pacific coast is crucial to landlocked Bolivia's further development.

Galápagos Islands
(Archipiélago de Colón)

[same scale as main map]

▲ *A colony of marine iguanas basks on the rocks of Isla Fernandina in the Galápagos Islands. Charles Darwin's theory of evolution was inspired by the differences he found between the animal species on neighboring islands in the Galápagos.*

Brazil

Brazil is the largest country in South America, with a population of 191 million – almost half the combined total of the continent. The 26 states which make up the federal republic of Brazil are administered from the purpose-built capital, Brasília. Tropical rain forest, covering more than one-third of the country, contains rich natural resources, but great tracts are sacrificed to agriculture, industry and urban expansion on a daily basis. Most of Brazil's multiethnic population now live in cities, some of which are vast areas of urban sprawl; São Paulo is one of the world's biggest conurbations, with more than 20 million inhabitants. Although prosperity is a reality for some, many people still live in great poverty, and mounting foreign debts continue to damage Brazil's prospects of economic advancement.

Using the land

Brazil has immense natural resources, including minerals and hardwoods, many of which are found in the fragile rain forest. Brazil is the world's leading coffee grower and a major producer of livestock, sugar, and orange juice concentrate. Soybeans for animal feed, particularly for poultry feed, have become the country's most significant crop.

Land use and agricultural distribution

- cattle
- pigs
- sheep
- citrus fruits
- coffee
- cotton
- soybeans
- sugar cane
- timber

- ● capital cities
- ■ major towns
- pasture
- cropland
- forest

The landscape

The Amazon Basin, containing the largest area of tropical rain forest on Earth, covers nearly half of Brazil. It is bordered by two shield areas: in the south by the Brazilian Highlands, and in the north by the Guiana Highlands. The east coast is dominated by a great escarpment which runs for 1600 miles (2565 km).

The ancient Brazilian Highlands have a varied topography. Their plateaus, hills, and deep valleys are bordered by highly-eroded mountains containing important mineral deposits. They are drained by three great river systems, the Amazon, the Paraguay-Paraná, and the São Francisco.

The São Francisco Basin has a climate unique in Brazil. Known as the "drought polygon," it has almost no rain during the dry season, leading to regular disastrous droughts.

The Amazon Basin is the largest river basin in the world. The Amazon river and over a thousand tributaries drain an area of 2,375,000 sq miles (6,150,000 sq km) and carry nearly one-fifth of the world's fresh water out to sea.

The northeastern scrublands are known as the *caatinga*, a virtually impenetrable thorny woodland, sometimes intermixed with cacti where water is scarce.

The famous Sugar Loaf Mountain (*Pão de Açúcar*) which overlooks Rio de Janeiro is a fine example of a volcanic plug a domed core of solidified lava left after the slopes of the original volcano have eroded away.

Deep natural harbors such as Baía de Guanabara were created where the steep slopes of the Serra da Mantiqueira plunge directly into the ocean.

Guiana Highlands

Brazil's highest mountain is the Pico da Neblina which was only discovered in 1962. It is 9888 ft (3014 m) high.

The floodplains which border the Amazon river are made up of a variety of different features including shallow lakes and swamps, mangrove forests in the tidal delta area, and fertile levees on river banks and point bars.

Pantanal wetlands

▲ *The Pantanal region* in the south of Brazil is an extension of the Gran Chaco plain. The swamps and marshes of this area are renowned for their beauty, and abundant and unique wildlife, including wildfowl and these caimans, a type of crocodile.

▲ *The fecundity of* parts of Brazil's rain forest results from exceptionally high levels of rainfall and the quantities of silt deposited by the Amazon river system.

▲ *The Iguaçu river* surges over the spectacular Iguaçu Falls (Saltos do Iguaçu) toward the Paraná river. Falls like these are increasingly under pressure from large-scale hydroelectric projects such as that at Itaipú.

▼ *Large-scale gullies* are common in Brazil, particularly on hillslopes from which vegetation has been removed. Gullies grow headwards (up the slope), aided by a combination of erosion through water seepage and rainwater runoff.

Hillslope gullying

- Direction of growth
- Overland water flow
- Gully
- Rainfall
- Water seeps through hillslope

Map key

Population
- ▫ above 5 million
- ◻ 1 million to 5 million
- ⊙ 500,000 to 1 million
- ⊕ 100,000 to 500,000
- ⊕ 50,000 to 100,000
- ⊕ 10,000 to 50,000
- ○ below 10,000

Elevation
- 3000m / 9843ft
- 2000m / 6562ft
- 1000m / 3281ft
- 500m / 1640ft
- 250m / 820ft
- 100m / 328ft
- sea level

The urban/rural population divide

urban 78%   rural 22%

Population density	Total land area
55 people per sq mile (21 people per sq km)	3,286,472 sq miles (8,511,970 sq km)

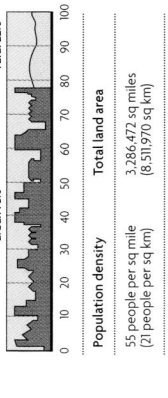

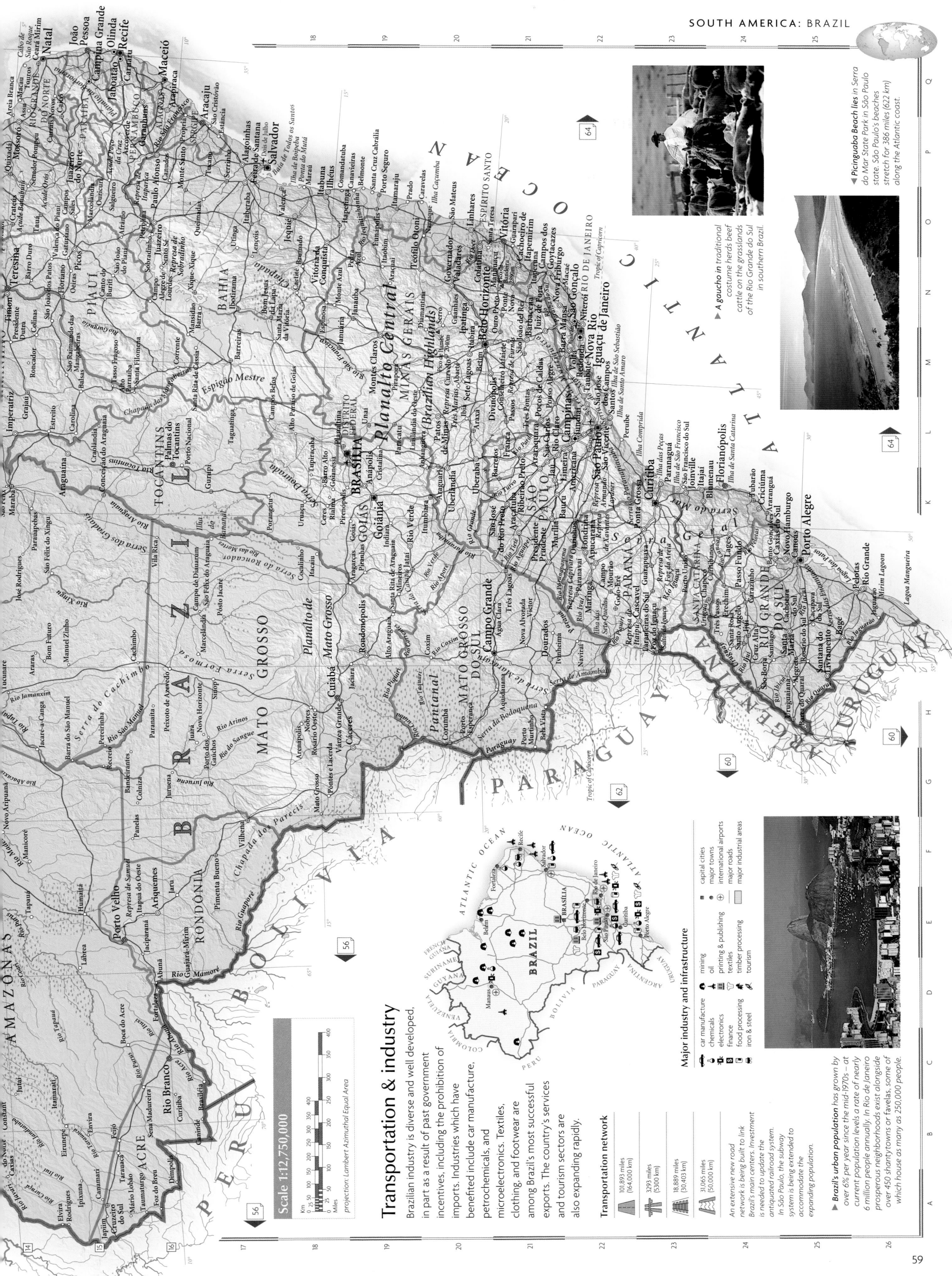

▲ *Picinguaba Beach* lies in Serra do Mar State Park in São Paulo state. São Paulo's beaches stretch for 386 miles (622 km) along the Atlantic coast.

▲ *A gaucho* in traditional costume herds beef cattle on the grasslands of the Rio Grande do Sul in southern Brazil.

Transportation & industry

Brazilian industry is diverse and well developed, in part as a result of past government incentives, including the prohibition of imports. Industries which have benefited include car manufacture, petrochemicals, and microelectronics. Textiles, clothing, and footwear are among Brazil's most successful exports. The country's services and tourism sectors are also expanding rapidly.

Scale 1:12,750,000

projection: Lambert Azimuthal Equal Area

Transportation network

101,893 miles (164,000 km)	
3293 miles (5300 km)	
18,889 miles (30,403 km)	
31,065 miles (50,000 km)	

An extensive new road network is being built to link Brazil's main centers. Investment is needed to update the antiquated railroad system. In São Paulo, the subway system is being extended to accommodate the expanding population.

▲ Brazil's urban population has grown by over 6% per year since the mid-1970s – at current population levels a rate of nearly 6 million people annually. In Rio de Janeiro prosperous neighborhoods exist alongside over 450 shantytowns or favelas, some of which house as many as 250,000 people.

Major industry and infrastructure

- car manufacture
- chemicals
- electronics
- finance
- food processing
- iron & steel
- mining
- oil
- printing & publishing
- textiles
- timber processing
- tourism
- capital cities
- major towns
- international airports
- major roads
- major industrial areas

Eastern South America

URUGUAY, NORTHEAST ARGENTINA, SOUTHEAST BRAZIL

The vast conurbations of Rio de Janeiro, São Paulo, and Buenos Aires form the core of South America's highly-urbanized eastern region. São Paulo state, with over 40 million inhabitants, is among the world's 20 most powerful economies, and São Paulo is the fastest growing city on the continent. Rio de Janeiro and Buenos Aires, transformed in the last hundred years from port cities to great metropolitan areas each with more than 10 million inhabitants, typify the unstructured growth and wealth disparities of South America's great cities. In Uruguay, over two fifths of the population lives in the capital, Montevideo, which faces Buenos Aires across the Plate River (Río de la Plata). Immigration from the countryside has created severe pressure on the urban infrastructure, particularly on available housing, leading to a profusion of crowded shanty settlements (favelas or barrios).

Transportation & industry

Southeast Brazil is home to much of the important motor and capital goods industry, largely based around São Paulo; iron and steel production is also concentrated in this region. Uruguay's economy continues to be based mainly on the export of livestock products including meat and leather goods. Buenos Aires is Argentina's chief port, and the region has a varied and sophisticated economic base including service-based industries such as finance and publishing, as well as primary processing.

Major industry and infrastructure

- car manufacture
- chemicals
- engineering
- finance
- food processing
- iron & steel
- meat processing
- printing & publishing
- shipbuilding
- textiles
- timber processing
- capital cities
- major towns
- international airports
- major roads
- major industrial areas

Transportation network

Throughout the region, road networks need to be expanded to cope with urban development. Plans are underway to build a bridge over the Plate River (Río de la Plata) to link Colonia and Buenos Aires.

▲ *The Itaipú dam on the Paraná river is one of the largest hydroelectric projects in the world, jointly financed by Brazil and Paraguay.*

Using the land

Most of Uruguay and the Pampas of northern Argentina are devoted to the rearing of livestock, especially cattle and sheep, which are central to both countries' economies. Soybeans, first produced in Brazil's Rio Grande do Sul, are now more widely grown for large-scale export, as are cereals, sugar cane, and grapes. Subsistence crops, including potatoes, corn and sugar beets, are grown on the remaining arable land.

Map key

Population
- ■ above 5 million
- ■ 1 million to 5 million
- ◉ 500,000 to 1 million
- ⊕ 100,000 to 500,000
- ○ 50,000 to 100,000
- ○ 10,000 to 50,000
- ○ below 10,000

Elevation
- 2000m / 6562ft
- 1000m / 3281ft
- 500m / 1640ft
- 250m / 820ft
- 100m / 328ft
- sea level

Scale 1:6,250,000

projection: Lambert Azimuthal Equal Area

▲ *Soybeans are harvested, pressed, and processed into soycake, which is used as animal feed. The cake is fed mainly to chickens on large-scale factory farms, and the growth in soy production has been an important factor in the expansion of the Brazilian poultry trade.*

Land use and agricultural distribution

- cattle
- sheep
- cereals
- coffee
- fruit
- soybeans
- sugar cane
- capital cities
- major towns
- pasture
- cropland
- forest
- wetlands
- barren land

▼ *The rolling grasslands of Uruguay are ideally suited to the rearing of cattle. Beef is the country's main export commodity, valued at over one billion US dollars in 2006.*

▲ *Rio de Janeiro's annual carnival, Mardi Gras, which ushers in the start of Lent, is an extravagant five-day parade through the city, characterized by fantastically decorated floats, exuberant dancing, and samba music.*

The landscape

The southern reaches of the Brazilian Highlands follow the Atlantic coast to form low, rolling hills in the northeast of Uruguay. Much of South America's mid-eastern region and all of Uruguay has a gentle relief with land rarely rising above 300 ft (100 m). Argentina's northeast comprises two main regions: a long, narrow lowland known as Mesopotamia; and part of the Pampas grasslands.

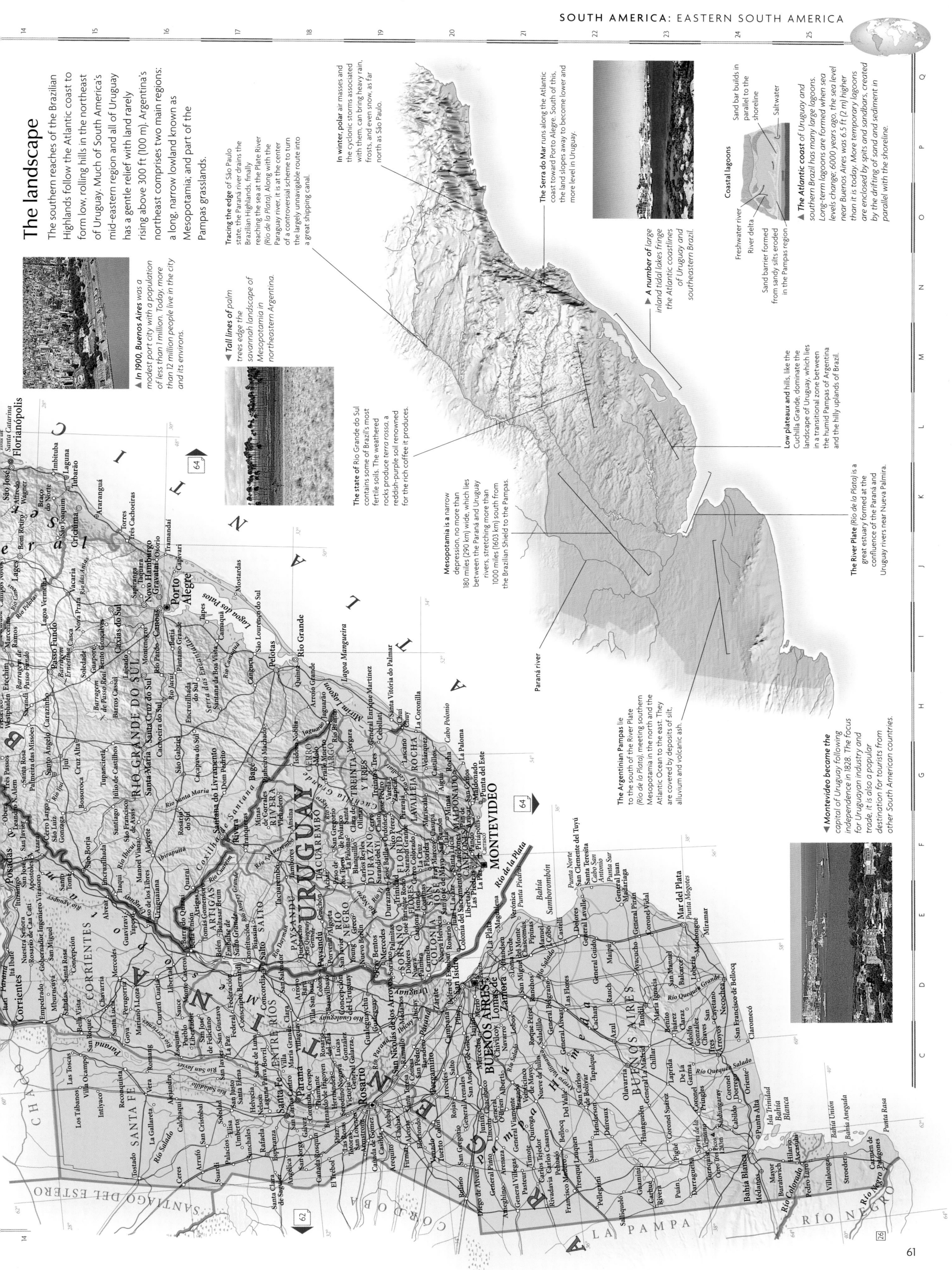

▲ *In 1900, Buenos Aires was a modest port city with a population of less than 1 million. Today, more than 12 million people live in the city and its environs.*

Tracing the edge of São Paulo state, the Paraná river drains the Brazilian Highlands, finally reaching the sea at the Plate River (*Río de la Plata*). Along with the Paraguay river, it is at the center of a controversial scheme to turn the largely unnavigable route into a great shipping canal.

▼ *Tall lines of palm trees edge the savannah landscape of Mesopotamia in northeastern Argentina.*

In winter, polar air masses and the cyclonic storms associated with them, can bring heavy rain, frosts, and even snow, as far north as São Paulo.

The Serra do Mar runs along the Atlantic coast toward Porto Alegre. South of this, the land slopes away to become lower and more level in Uruguay.

▲ *A number of* large inland tidal lakes fringe the Atlantic coastlines of Uruguay and southeastern Brazil.

Coastal lagoons

Freshwater river

River delta

Sand bar builds in parallel to the shoreline

Saltwater

Sand barrier formed from sandy silts eroded in the Pampas region

▲ *The Atlantic coast* of Uruguay and southern Brazil has many large lagoons. Long-term lagoons are formed when sea levels change. 6000 years ago, the sea level near Buenos Aires was 6.5 ft (2 m) higher than it is today. More temporary lagoons are enclosed by spits and sandbars, created by the drifting of sand and sediment in parallel with the shoreline.

The state of Rio Grande do Sul contains some of Brazil's most fertile soils. The weathered rocks produce *terra rossa*, a reddish-purple soil renowned for the rich coffee it produces.

Mesopotamia is a narrow depression, no more than 180 miles (290 km) wide, which lies between the Paraná and Uruguay rivers, stretching more than 1000 miles (1603 km) south from the Brazilian Shield to the Pampas.

Low plateaux and hills, like the Cuchilla Grande, dominate the landscape of Uruguay, which lies in a transitional zone between the humid Pampas of Argentina and the hilly uplands of Brazil.

The River Plate (*Río de la Plata*) is a great estuary formed at the confluence of the Paraná and Uruguay rivers near Nueva Palmira.

Paraná river

The Argentinian Pampas lie to the south of the River Plate (*Río de la Plata*), meeting southern Mesopotamia in the north and the Atlantic Ocean to the east. They are covered by deposits of silt, alluvium and volcanic ash.

▼ *Montevideo became the* capital of Uruguay following independence in 1828. The focus for Uruguayan industry and trade, it is also a popular destination for tourists from other South American countries.

Southern South America

ARGENTINA, CHILE, PARAGUAY

South America's cone-shaped southern region is shared by Argentina and Chile, two overwhelmingly urbanized nations whose populations live mainly in or around the capital cities, Buenos Aires and Santiago. The people are largely *mestizo* or of European origin; in the early 20th century Argentina absorbed waves of new European immigrants, many from Italy and Germany. Paraguay is far less urbanized than its neighbors, with a homogeneous population of mixed Spanish and Guaraní origin, who retain their Indian roots through the Guaraní language. Though most Paraguayans live in the southeast, near Asunción, the indigenous Indians live in the sparsely populated Gran Chaco. The Gran Chaco is also home to some of Argentina's minority indigenous peoples, who otherwise live mainly in Andean regions. Chile's estimated 800,000 Mapauche Indians live almost exclusively in the south.

Transportation & industry

Food processing and agricultural exports remain a fundamental part of Argentina's economy. The growth of manufacturing is regularly hampered by hyper-inflation and massive foreign debts. The world's most important copper producer and one of the top twenty gold producers, Chile also has a thriving wine and grape industry. Most Paraguayan exports involve primary processing, although domestic goods are produced for home markets.

▲ *Floodwaters cover the land in the Gran Chaco, partly submerging its vegetation of fan palms and hyacinths.*

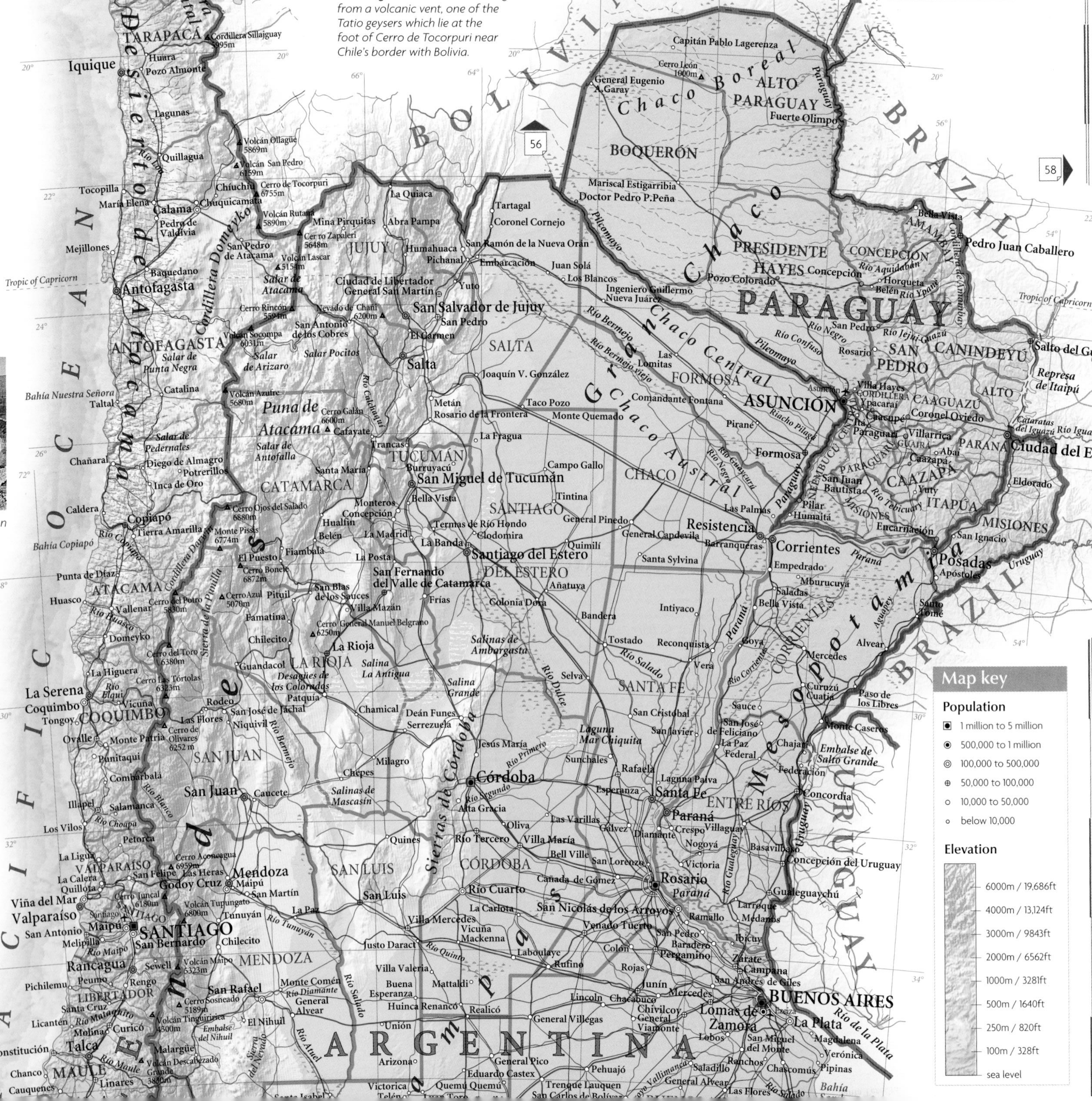

▲ *Boiling water and steam emerge from a volcanic vent, one of the Tatio geysers which lie at the foot of Cerro de Tocorpuri near Chile's border with Bolivia.*

▲ *Chuquicamata copper mine, lies on a desert plateau near Calama in the Andes of northern Chile. It is the world's largest open-pit copper mine.*

Major industry and infrastructure

- chemicals
- engineering
- food processing
- meat processing
- mining
- oil
- textiles
- timber processing

- capital cities
- major towns
- international airports
- major roads
- major industrial areas

Transportation network

55,062 miles (93,453 km)	3038 miles (4889 km)
26,811 miles (43,153 km)	9180 miles (14,775 km)

Argentina's state transportation system is under-going privatization, though the outmoded rail network requires updating. Paraguay requires foreign investment to upgrade its roads and railroads. Essential internal air routes, especially across the Andes, are well developed in all three countries.

Map key

Population
- 1 million to 5 million
- 500,000 to 1 million
- 100,000 to 500,000
- 50,000 to 100,000
- 10,000 to 50,000
- below 10,000

Elevation
6000m / 19,686ft	
4000m / 13,124ft	
3000m / 9843ft	
2000m / 6562ft	
1000m / 3280ft	
500m / 1640ft	
250m / 820ft	
100m / 328ft	
sea level	

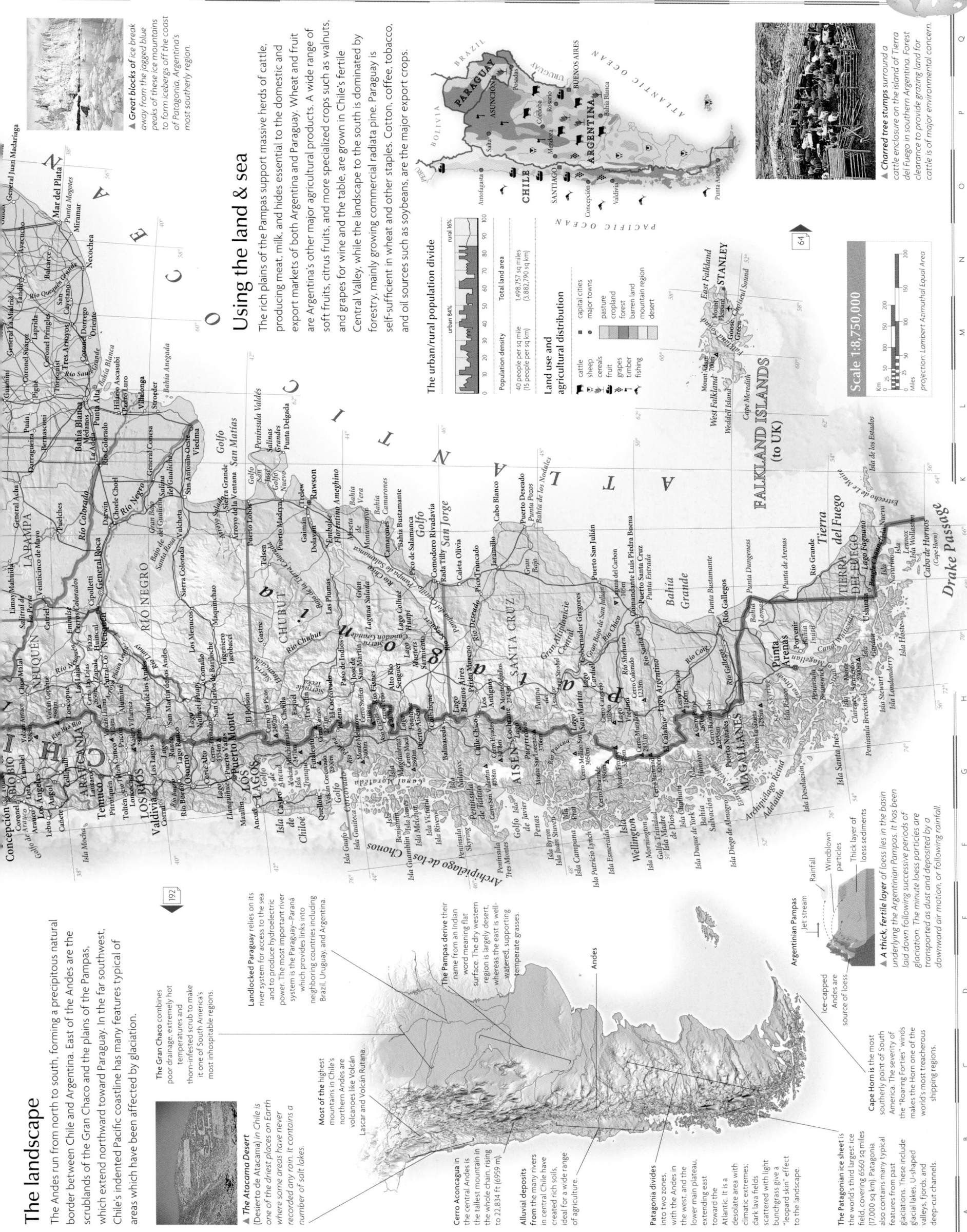

▲ *Great blocks of ice break away from the jagged blue peaks of these ice mountains to form icebergs off the coast of Patagonia. Argentina's most southerly region.*

▲ *Charred tree stumps surround a cattle enclosure on the island of Tierra del Fuego in southern Argentina. Forest clearance to provide grazing land for cattle is of major environmental concern.*

Using the land & sea

The rich plains of the Pampas support massive herds of cattle, producing meat, milk, and hides essential to the domestic and export markets of both Argentina and Paraguay. Wheat and fruit are Argentina's other major agricultural products. A wide range of soft fruits, citrus fruits, and more specialized crops such as walnuts, and grapes for wine and the table, are grown in Chile's fertile Central Valley, while the landscape to the south is dominated by forestry, mainly growing commercial radiata pine. Paraguay is self-sufficient in wheat and other staples. Cotton, coffee, tobacco, and oil sources such as soybeans, are the major export crops.

The urban/rural population divide

rural 16%
urban 84%

| Population density | 40 people per sq mile (15 people per sq km) |
| Total land area | 1,498,757 sq miles (3,882,790 sq km) |

Land use and agricultural distribution

- capital cities
- major towns
- pasture
- cropland
- forest
- barren land
- mountain region
- desert

- cattle
- sheep
- cereals
- fruit
- grapes
- timber
- fishing

Scale 1:8,750,000

projection: Lambert Azimuthal Equal Area

The landscape

The Andes run from north to south, forming a precipitous natural border between Chile and Argentina. East of the Andes are the scrublands of the Gran Chaco and the plains of the Pampas, which extend northward toward Paraguay. In the far southwest, Chile's indented Pacific coastline has many features typical of areas which have been affected by glaciation.

▲ *The Atacama Desert (Desierto de Atacama) in Chile is one of the driest places on Earth where some areas have never recorded any rain. It contains a number of salt lakes.*

The Gran Chaco combines poor drainage, extremely hot temperatures and thorn-infested scrub to make it one of South America's most inhospitable regions.

Landlocked Paraguay relies on its river system for access to the sea and to produce hydroelectric power. The most important river system is the Paraguay–Paraná which provides links into neighboring countries including Brazil, Uruguay, and Argentina.

Most of the highest mountains in Chile's northern Andes are volcanoes like Volcán Lascar and Volcán Rutana.

The Pampas derive their name from an Indian word meaning flat surface. The dry western region is largely desert, whereas the east is well-watered, supporting temperate grasses.

Cerro Aconcagua in the central Andes is the tallest mountain in the whole chain, rising to 22,834 ft (6959 m).

Alluvial deposits from the many rivers in central Chile have created rich soils, ideal for a wide range of agriculture.

Patagonia divides into two zones, with the Andes in the west, and the lower main plateau, extending east toward the Atlantic. It is a desolate area with climatic extremes; dark lava fields scattered with light bunchgrass give a "leopard skin" effect to the landscape.

The Patagonian ice sheet is the world's third largest ice field, covering 6560 sq miles (17,000 sq km) Patagonia also contains many typical features from past glaciations. These include glacial lakes, U-shaped valleys, fjords, and deep-cut channels.

Cape Horn is the most southerly point of South America. The severity of the "Roaring Forties" winds makes the Horn one of the world's most treacherous shipping regions.

Ice-capped Andes are source of loess

Andes

Argentinian Pampas

Jet stream

Rainfall
Windblown particles
Thick layer of loess sediments

▲ *A thick, fertile layer of loess lies in the basin underlying the Argentinian Pampas. It has been laid down following successive periods of glaciation. The minute loess particles are transported as dust and deposited by a downward air motion, or following rainfall.*

The Atlantic Ocean

The Atlantic is the youngest of the world's oceans, formed about 180 million years ago when the landmasses of the eastern and western hemispheres separated. Its underwater topography is dominated by the Mid-Atlantic Ridge, a huge mountain system running north to south along the center of the ocean. Although most of the ridge's peaks lie below the sea, some emerge as volcanic islands, like Iceland and the Azores. The Atlantic contains a wealth of resources, including substantial oil and gas reserves and rich fishing grounds. Until the 1950s, the north Atlantic was the world's busiest shipping route; cheaper air transportation and alternative routes have shifted patterns of world trade.

Resources

Development of the oil and gas reserves in the Atlantic began in the 1940s around the Gulf of Mexico. Since then other areas have been exploited, including the North Sea, the west coast of Africa and the area east of Newfoundland and Nova Scotia. There is also extensive mining of sand, gravel, and shell deposits by the US and UK. For centuries, the north Atlantic's fishing grounds have been utilized more heavily than other oceans, leading to a serious decline in many fish stocks.

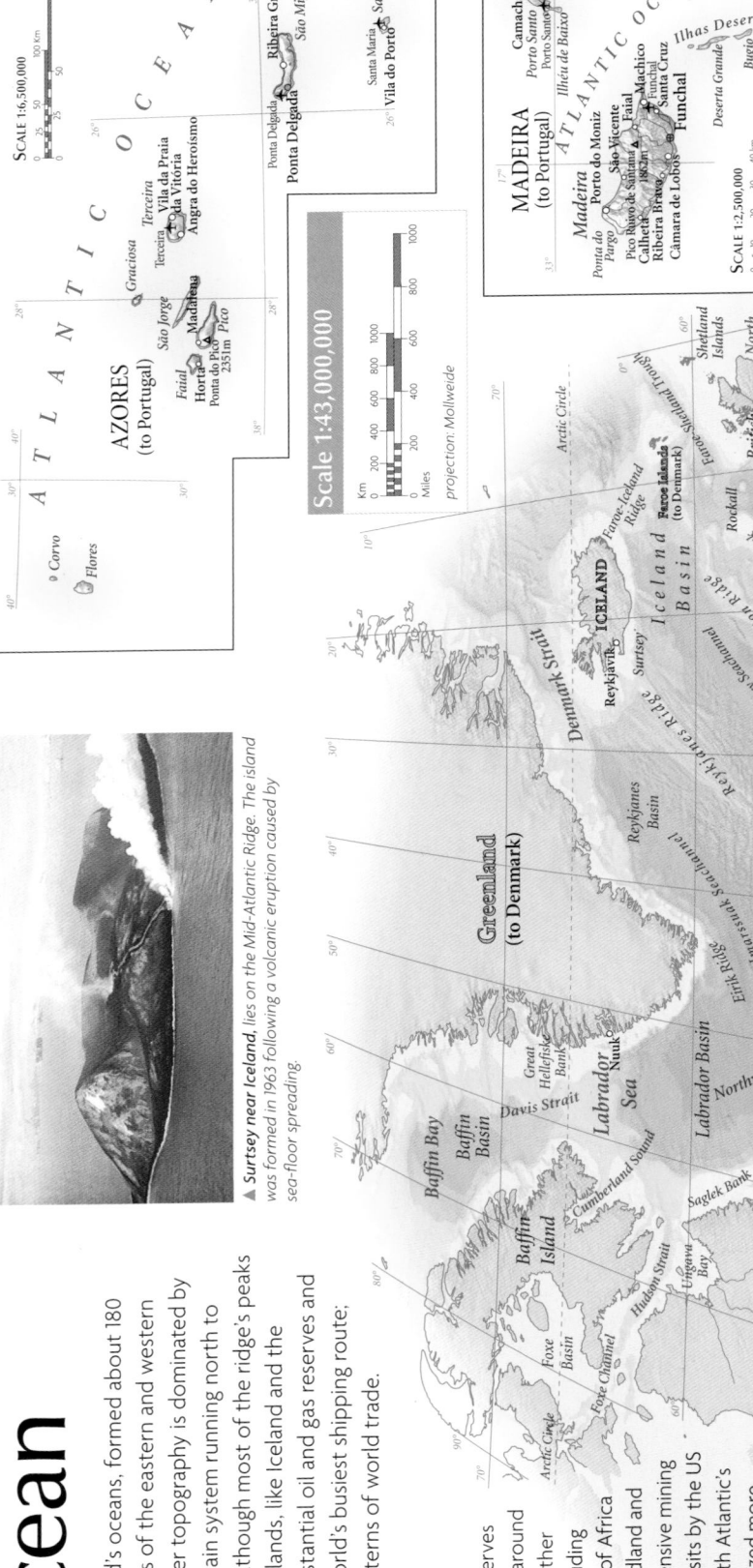

▲ **Surtsey near Iceland**, lies on the Mid-Atlantic Ridge. The island was formed in 1963 following a volcanic eruption caused by sea-floor spreading.

▲ **Fishing in the seas** around northwestern Europe dates back over 1500 years. The high nutrient content of the seas makes them ideal breeding grounds for many species of fish.

▲ **On January 5 1993**, the oil tanker Braer ran aground in the Shetland Islands, spilling 83,660 tons (85,000 tonnes) of light crude oil into the ocean, devastating the local marine ecosystem.

Resources (including wildlife)
- fish
- whales
- aggregates
- oil & gas
- major towns
- major ports

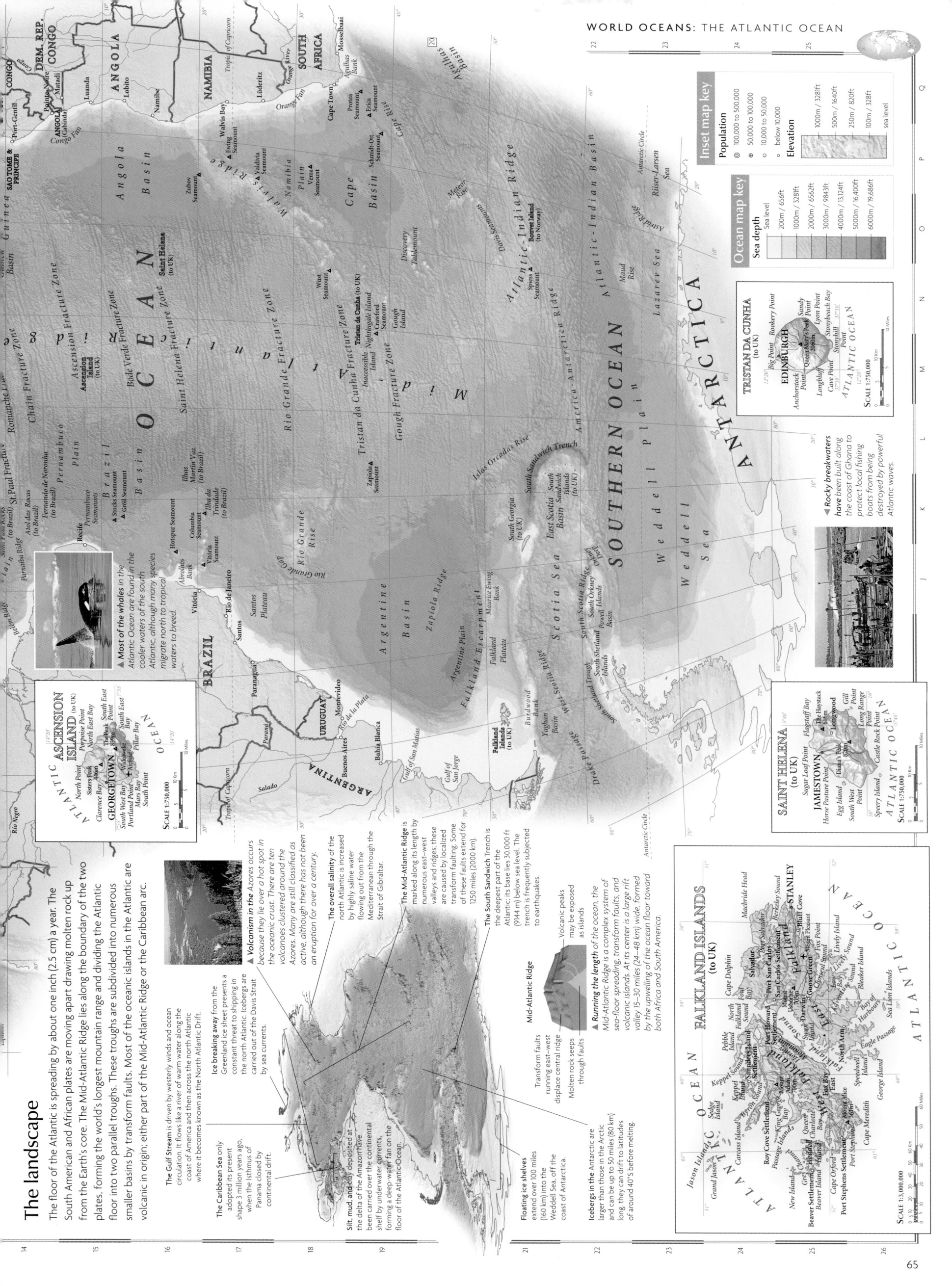

The landscape

The floor of the Atlantic is spreading by about one inch (2.5 cm) a year. The South American and African plates are moving apart drawing molten rock up from the Earth's core. The Mid-Atlantic Ridge lies along the boundary of the two plates, forming the world's longest mountain range and dividing the Atlantic floor into two parallel troughs. These troughs are subdivided into numerous smaller basins by transform faults. Most of the oceanic islands in the Atlantic are volcanic in origin; either part of the Mid-Atlantic Ridge or the Caribbean arc.

The Gulf Stream is driven by westerly winds and ocean circulation. It flows like a river of warm water along the coast of America and then across the north Atlantic where it becomes known as the North Atlantic Drift.

The Caribbean Sea only adopted its present shape 3 million years ago, when the isthmus of Panama closed by continental drift.

Ice breaking away from the Greenland ice sheet presents a constant threat to shipping in the north Atlantic. Icebergs are carried out of the Davis Strait by sea currents.

Floating ice shelves extend over 100 miles (160 km) into the Weddell Sea, off the coast of Antarctica.

Icebergs in the Antarctic are larger than those in the Arctic and can be up to 50 miles (80 km) long; they can drift to latitudes of around 40°S before melting.

Silt, mud, and clay deposited at the delta of the Amazon have been carried over the continental shelf by underwater currents, forming a deep-water fan on the floor of the Atlantic Ocean.

▲ **Volcanism in the Azores** occurs because they lie over a hot spot in the oceanic crust. There are ten volcanoes clustered around the Azores. Many are still classified as active, although there has not been an eruption for over a century.

The overall salinity of the north Atlantic is increased by highly saline water flowing out from the Mediterranean through the Strait of Gibraltar.

The Mid-Atlantic Ridge is marked along its length by numerous east–west valleys and ridges; these are caused by localized transform faulting. Some of these faults extend for 1250 miles (2000 km).

The South Sandwich Trench is the deepest part of the Atlantic; its base lies 30,000 ft (9144 m) below sea level. The trench is frequently subjected to earthquakes.

Volcanic peaks may be exposed as islands

Mid-Atlantic Ridge

Transform faults running east–west displace central ridge

Molten rock seeps through faults

▲ **Running the length** of the ocean, the Mid-Atlantic Ridge is a complex system of sea-floor spreading, transform faults, and volcanic islands. At its center is a large rift valley 15–30 miles (24–48 km) wide, formed by the upwelling of the ocean floor toward both Africa and South America.

▲ **Most of the whales in the** Atlantic Ocean are found in the cooler waters of the south Atlantic, although many species migrate north to tropical waters to breed.

▲ **Rocky breakwaters have been built** along the coast of Ghana to protect local fishing boats from being destroyed by powerful Atlantic waves.

Africa

The world's second largest continent, Africa covers an area of 11,712,434 sq miles (30,355,000 sq km). It has 54 separate countries, including Madagascar, Comoros, Mauritius, and the Seychelles in the Indian Ocean – the highest number of any continent.

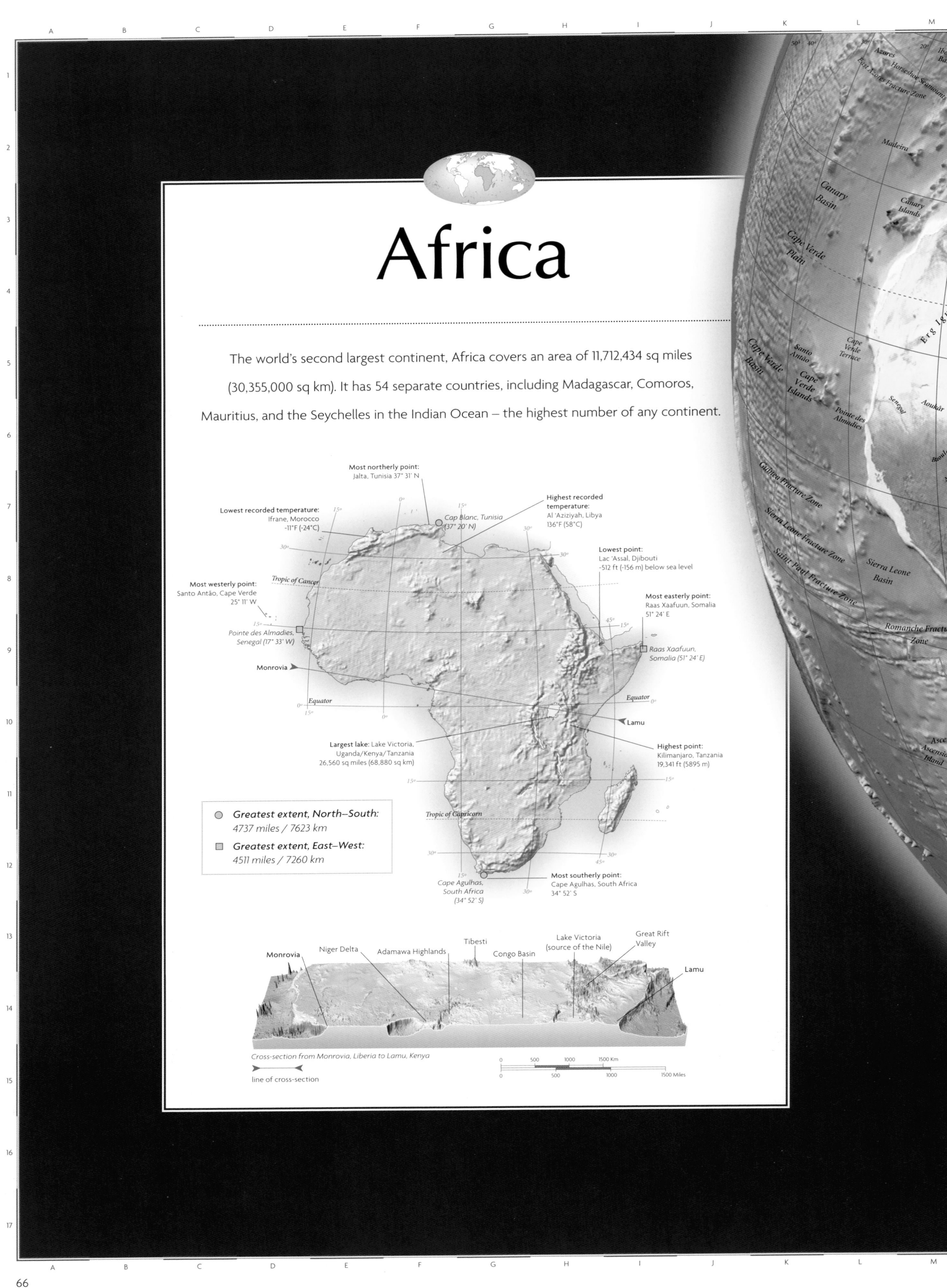

Most northerly point:
Jalta, Tunisia 37° 31' N

Lowest recorded temperature:
Ifrane, Morocco
-11°F (-24°C)

Highest recorded temperature:
Al 'Aziziyah, Libya
136°F (58°C)

Cap Blanc, Tunisia
(37° 20' N)

Lowest point:
Lac 'Assal, Djibouti
-512 ft (-156 m) below sea level

Most westerly point:
Santo Antão, Cape Verde
25° 11' W

Most easterly point:
Raas Xaafuun, Somalia
51° 24' E

Tropic of Cancer

Pointe des Almadies,
Senegal (17° 33' W)

Raas Xaafuun,
Somalia (51° 24' E)

Monrovia

Equator

Equator

Lamu

Largest lake: Lake Victoria,
Uganda/Kenya/Tanzania
26,560 sq miles (68,880 sq km)

Highest point:
Kilimanjaro, Tanzania
19,341 ft (5895 m)

● Greatest extent, North–South:
4737 miles / 7623 km

■ Greatest extent, East–West:
4511 miles / 7260 km

Tropic of Capricorn

Most southerly point:
Cape Agulhas, South Africa
34° 52' S

Cape Agulhas,
South Africa
(34° 52' S)

Monrovia
Niger Delta
Adamawa Highlands
Tibesti
Congo Basin
Lake Victoria
(source of the Nile)
Great Rift Valley
Lamu

Cross-section from Monrovia, Liberia to Lamu, Kenya

line of cross-section

0   500   1000   1500 Km
0   500   1000   1500 Miles

Azores
Iberian Basin
East Azores Fracture Zone
Horseshoe Seamounts
Madeira
Canary Basin
Canary Islands
Cape Verde Plain
Cape Verde Basin
Santo Antão
Cape Verde Terrace
Cape Verde Islands
Pointe des Almadies
Senegal
Aoukar
Bhoile
Niger
Guinea Fracture Zone
Sierra Leone Fracture Zone
Sierra Leone Basin
Saint Paul Fracture Zone
Romanche Fracture Zone
Ascension Island
Ascensi

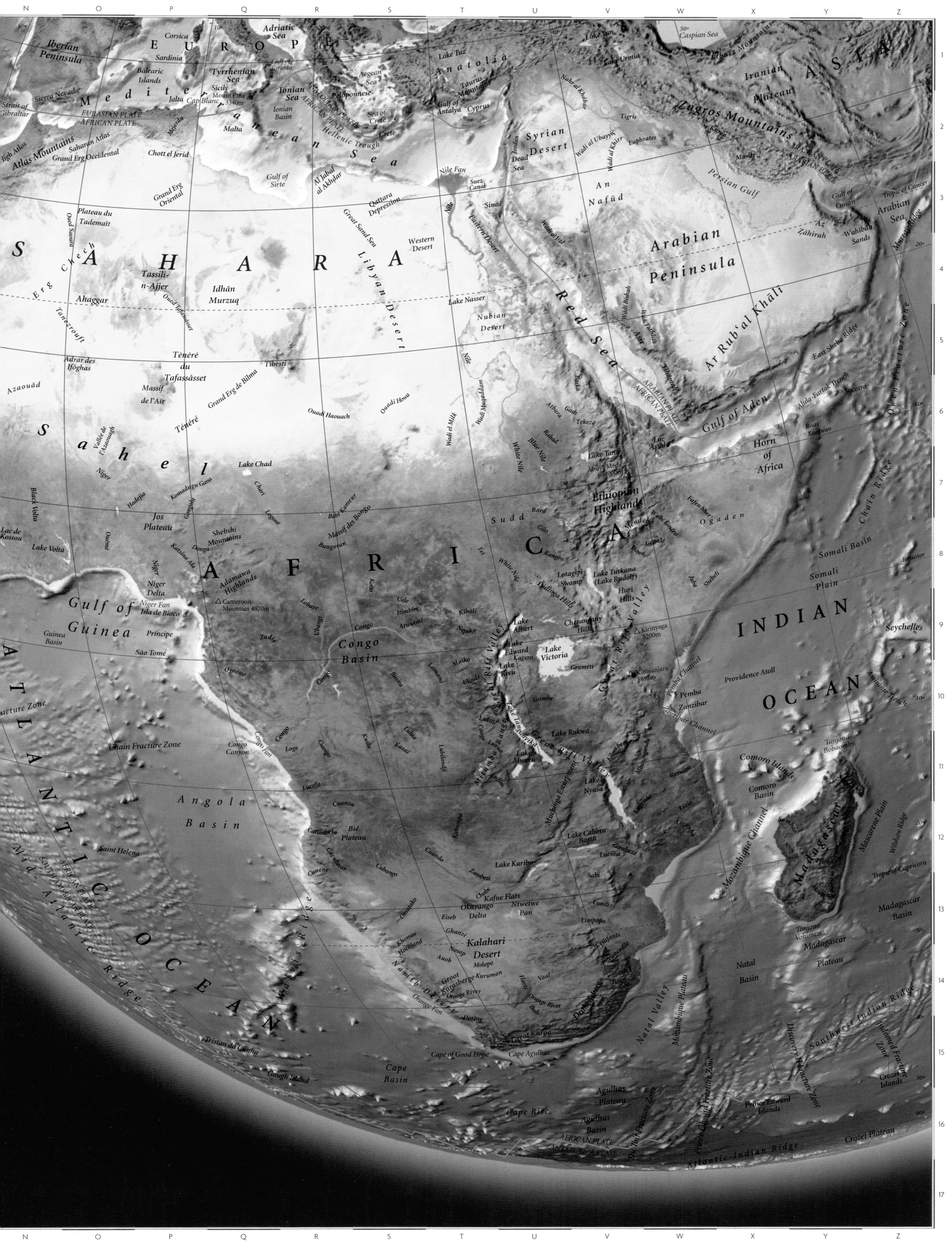

Iberian
Peninsula

EUROPE

Corsica
Sardinia
Balearic
Islands

Adriatic
Sea

Sicily
Mount Etna
3340m

Tyrrhenian
Sea

Ionian
Sea

Aegean
Sea

Poloponnese

Anatolia

Lake Tuz

Taurus
Mountains

Gulf of
Antalya

Cyprus

Elburz Mountains

Caspian Sea

Iranian
Plateau

ASIA

Zagros Mountains

Sierra Nevada

Strait of
Gibraltar

Atlas Mountains

High Atlas

Grand Erg Occidental

Saharan Atlas

EURASIAN PLATE
AFRICAN PLATE

Cap Blanc

Médraa

Malta

Gulf of
Sirte

Al Jabal
al Akhdar

Ionian
Basin

Sea of
Crete

Crete

Hellenic Trough

Mediterranean Sea

Nile Fan

Suez
Canal

Sinai

Dead
Sea

Syrian
Desert

Jordan

Wadi al Ubayyiq

Wadi as Sirr

Euphrates

Tigris

Karun

Mand

Persian Gulf

Gulf of
Oman

Arabian
Sea

Tropic of Cancer

Wahiba
Sands

Murray Ridge

Chott el Jerid

Grand Erg
Oriental

S A H A R A

Plateau du
Tademaït

Qattara
Depression

Western
Desert

An
Nafūd

Arabian

Peninsula

Az
Zāhirah

Oued Saoura

Erg Chech

Libyan Desert

Eastern Desert

Nile

Nubian
Desert

Wadi Bishah

Asir

Tihamah

Red Sea

ARABIAN PLATE
AFRICAN PLATE

Ar Rub 'al Khāli

Ahaggar

Tassili-
n-Ajjer

Idhān
Murzuq

Great Sand Sea

Lake Nasser

Azouâd

Adrar des
Ifôghas

Ténéré
du
Tafassàsset

Massif
de l'Air

Tibesti

Grand Erg de Bilma

Ouadi Howa

Wadi al Milk

Atbara

Gash

Rahad

Tekeze

Lac
Assal

Gulf of Aden

Horn
of
Africa

Alula-Fartak Trench

Socotra

East Sheba Ridge

Owen Fracture Zone

Sahel

Niger

Ténéré

Ouadi Haouach

Blue Nile

White Nile

Lake Tana

Abay Meda
4000m

Ethiopian
Highlands

Mendebo

Wabi Tertele

Fafen Sheet

Ogaden

Chain Ridge

Black Volta

Lac de
Kossou

Lake Volta

Jos
Plateau

Hadejia

Komadugu Gana

Gongola

Chari

Legome

Bahr Kameur

Massif des Bongo

Sudd

Bard

Gilo

Yei

Kangen

Genale

Dawa

Weyb

Somali Basin

Equator

Niger

Shebshi
Mountains

Katsina Ala

Donga

Bangoran

A F R I C A

White Nile

Lotagipi
Swamp

Dudinga Hills

Lake Turkana
(Lake Rudolf)

Huri
Hills

Juba

Shebeli

Somali
Plain

Niger
Delta

Adamawa
Highlands

Cameroon
Mountain 4070m

Lobaye

Uele

Itimbiri

Aruwimi

Kibali

Nzoko

Lake Albert

Charangany
Hills

Kirinyaga
5200m

INDIAN

Seychelles

Gulf of
Guinea

Guinea
Basin

Príncipe

São Tomé

Isla de Bioco

Zadié

Ubangi

Congo

Congo
Basin

Matko

Lomami

Ulindi

Lake
Edward

Lake
Kivu

Lake
Victoria

Kagera

Grumeti

Kilimanjaro
5895m

Pemba

Zanzibar

Pemba Channel

Providence Atoll

OCEAN

Fracture Zone

Ogooué

Congo

Rusizi

Kasaï

Lukuga

Lake
Tanganyika

Gombe

Lake Rukwa

Great Rift Valley

Zanzibar Channel

Chain Fracture Zone

Congo
Canyon

Congo Fan

Loge

Kwilu

Kwango

Luhlamba

Malagarasi

Lake
Mweru

Great Rift Valley

Mbarangandu

Lake
Nyasa

Ruvuma

Tanjona
Bobaomby

Comoro Islands

Amirante Trench

Lucala

Angola

Basin

Cuanza

Catumbele

Bié
Plateau

Cuango

Kabompo

Mitumba Escarpment

Mbarangandu

Comoro
Basin

Madagascar

Saint Helena

Cunene

Great Escarpment

Lake Cabora
Bassa

Luenha

Zambezi

Sabi

Mozambique Channel

Mozambique Plateau

Tropic of Capricorn

Mascarene Plain

Mid-Atlantic Ridge

South American Plate

Chuando

Okavango
Delta

Lake Karibba

Zambezi

Kafue Flats

Cubango

Ntwetwe
Pan

Limpopo

Tanjona
Vohimena

Madagascar

Madagascar
Basin

Walvis Ridge

Khomas
Highland

Ghanzi

Kalahari
Desert

Nosop

Eiseb

Molopo

Auob

Natal
Basin

Madagascar
Plateau

Namib Desert

Groot

Karasberge

Kuruman

Harts

Vaal

Orange River

Okiffants

Tugela

Wildhav Ridge

Tristan da Cunha

Orange Pan

Orange River

Fish

Great Karoo

Natal Valley

Discovery Tablemount Zone

Southwest Indian Ridge

Diamantina Fracture Zone

Gough Island

Cape of Good Hope

Cape Agulhas

Mozambique Plateau

ATLANTIC OCEAN

Cape
Basin

Cape Rise

Agulhas
Plateau

Agulhas
Basin

Prince Edward
Islands

Crozet Plateau

Crozet
Islands

AFRICAN PLATE
ANTARCTIC PLATE

Del Cano Fracture Zone

Atlantic-Indian Ridge

Prince Edward Fracture Zone

Physical Africa

The structure of Africa was dramatically influenced by the break up of the supercontinent Gondwanaland about 160 million years ago and, more recently, rifting and hot spot activity. Today, much of Africa is remote from active plate boundaries and comprises a series of extensive plateaus and deep basins, which influence the drainage patterns of major rivers. The relief rises to the east, where volcanic uplands and vast lakes mark the Great Rift Valley. In the far north and south sedimentary rocks have been folded to form the Atlas Mountains and the Great Karoo.

East Africa

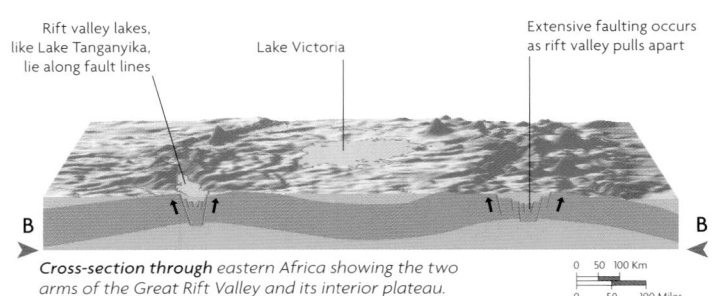

The Great Rift Valley is the most striking feature of this region, running for 4475 miles (7200 km) from Lake Nyasa to the Red Sea. North of Lake Nyasa it splits into two arms and encloses an interior plateau which contains Lake Victoria. A number of elongated lakes and volcanoes lie along the fault lines. To the west lies the Congo Basin, a vast, shallow depression, which rises to form an almost circular rim of highlands.

Northern Africa

Northern Africa comprises a system of basins and plateaus. The Tibesti and Ahaggar are volcanic uplands, whose uplift has been matched by subsidence within large surrounding basins. Many of the basins have been infilled with sand and gravel, creating the vast Saharan lands. The Atlas Mountains in the north were formed by convergence of the African and Eurasian plates.

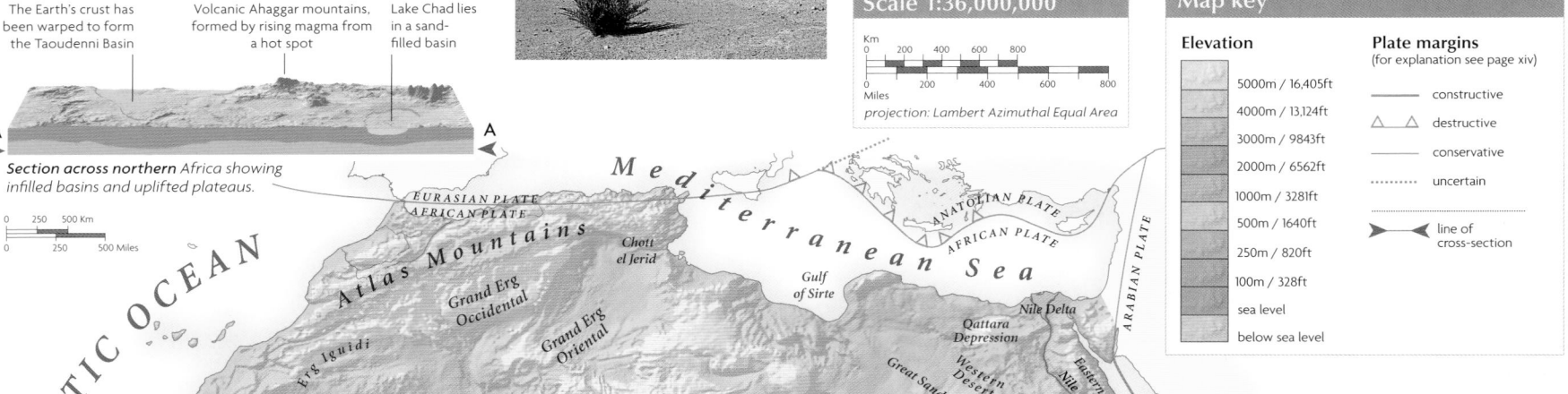

Rift valley lakes, like Lake Tanganyika, lie along fault lines

Lake Victoria

Extensive faulting occurs as rift valley pulls apart

Cross-section through eastern Africa showing the two arms of the Great Rift Valley and its interior plateau.

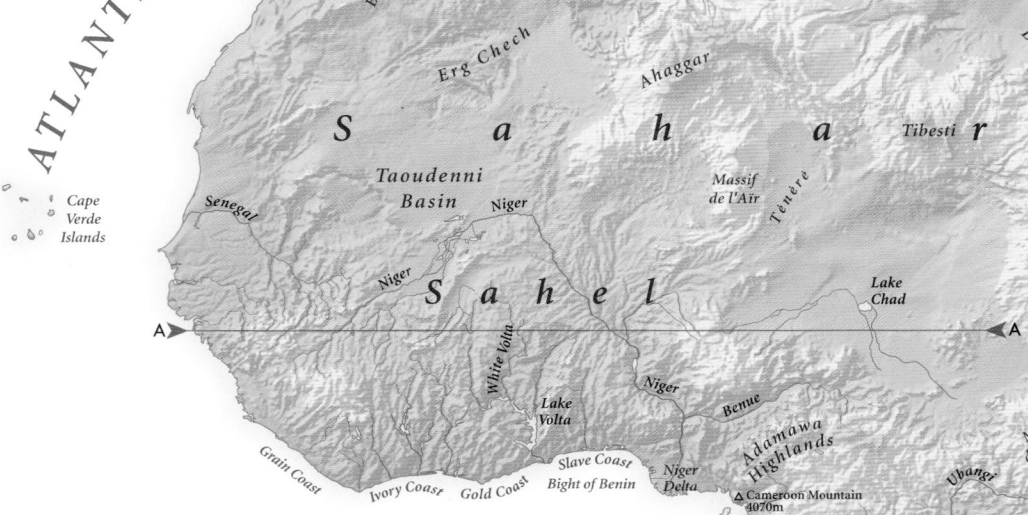

The Earth's crust has been warped to form the Taoudenni Basin

Volcanic Ahaggar mountains, formed by rising magma from a hot spot

Lake Chad lies in a sand-filled basin

Section across northern Africa showing infilled basins and uplifted plateaus.

Scale 1:36,000,000

projection: Lambert Azimuthal Equal Area

Map key

Elevation

- 5000m / 16,405ft
- 4000m / 13,124ft
- 3000m / 9843ft
- 2000m / 6562ft
- 1000m / 3281ft
- 500m / 1640ft
- 250m / 820ft
- 100m / 328ft
- sea level
- below sea level

Plate margins
(for explanation see page xiv)

- constructive
- destructive
- conservative
- uncertain
- line of cross-section

Southern Africa

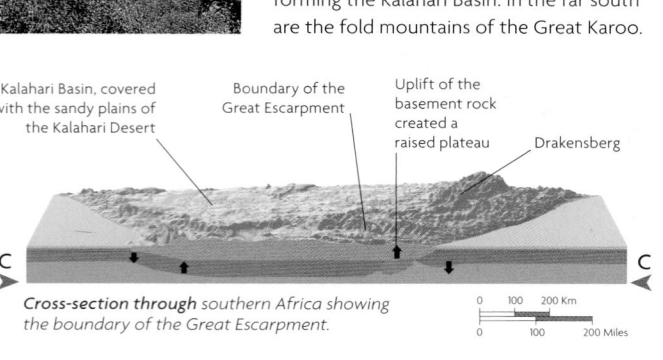

The Great Escarpment marks the southern boundary of Africa's basement rock and includes the Drakensberg range. It was uplifted when Gondwanaland fragmented about 160 million years ago and it has gradually been eroded back from the coast. To the north, the relief drops steadily, forming the Kalahari Basin. In the far south are the fold mountains of the Great Karoo.

Kalahari Basin, covered with the sandy plains of the Kalahari Desert

Boundary of the Great Escarpment

Uplift of the basement rock created a raised plateau

Drakensberg

Cross-section through southern Africa showing the boundary of the Great Escarpment.

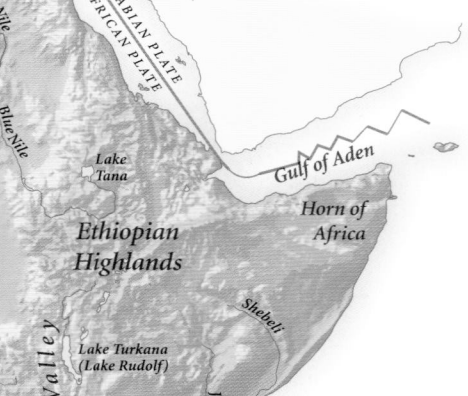

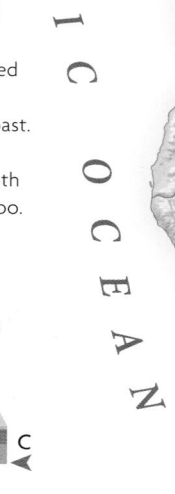

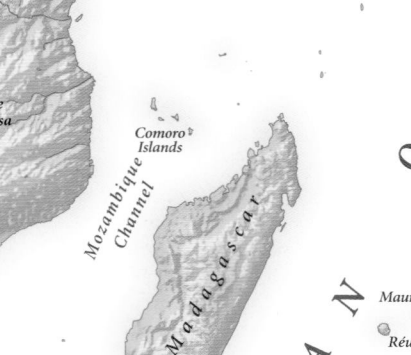

Map labels

EURASIAN PLATE
AFRICAN PLATE
Mediterranean Sea
ANATOLIAN PLATE
AFRICAN PLATE
ARABIAN PLATE

Atlas Mountains
Chott el Jerid
Gulf of Sirte
Nile Delta
Qattara Depression
Western Desert
Great Sand Sea
Libyan Desert
Eastern Desert
Red Sea
ARABIAN PLATE
AFRICAN PLATE
ASIA

Grand Erg Occidental
Grand Erg Oriental
Erg Iguidi
Erg Chech
Ahaggar
Tibesti
Nile
Lake Nasser
Nubian Desert

ATLANTIC OCEAN
Cape Verde Islands
Senegal
S a h a r a
Taoudenni Basin
Niger
Massif de l'Air
Ténéré
Blue Nile
White Nile
Lake Tana
Gulf of Aden

Sahel
Niger
Lake Chad
Ethiopian Highlands
Horn of Africa

White Volta
Lake Volta
Niger
Benue
Adamawa Highlands
Niger Delta
Cameroon Mountain 4070m
Massif des Bongo
Sudd
Shebeli
Lake Turkana (Lake Rudolf)
Juba

Grain Coast
Ivory Coast
Gold Coast
Slave Coast
Bight of Benin
Gulf of Guinea
São Tomé
Ubangi
Congo
Lake Albert
Lake Victoria
Kilimanjaro 5895m
Seychelles

ATLANTIC OCEAN
Congo Basin
Congo
Mitumba Range
Great Rift Valley
Lake Tanganyika
Pemba Island
Zanzibar

Bié Plateau
Zambezi
Lake Nyasa
Comoro Islands
INDIAN OCEAN

Namib Desert
Okavango Delta
Kalahari Basin
Kalahari Desert
Zambezi
Limpopo
Mozambique Channel
Madagascar
Mauritius
Réunion

Orange River
Drakensberg
Great Karoo
Cape of Good Hope

Climate

The climates of Africa range from mediterranean to arid, dry savannah, and humid equatorial. In East Africa, where snow settles at the summit of volcanoes such as Kilimanjaro, climate is also modified by altitude. The winds of the Sahara export millions of tonnes of dust a year both northward and eastward.

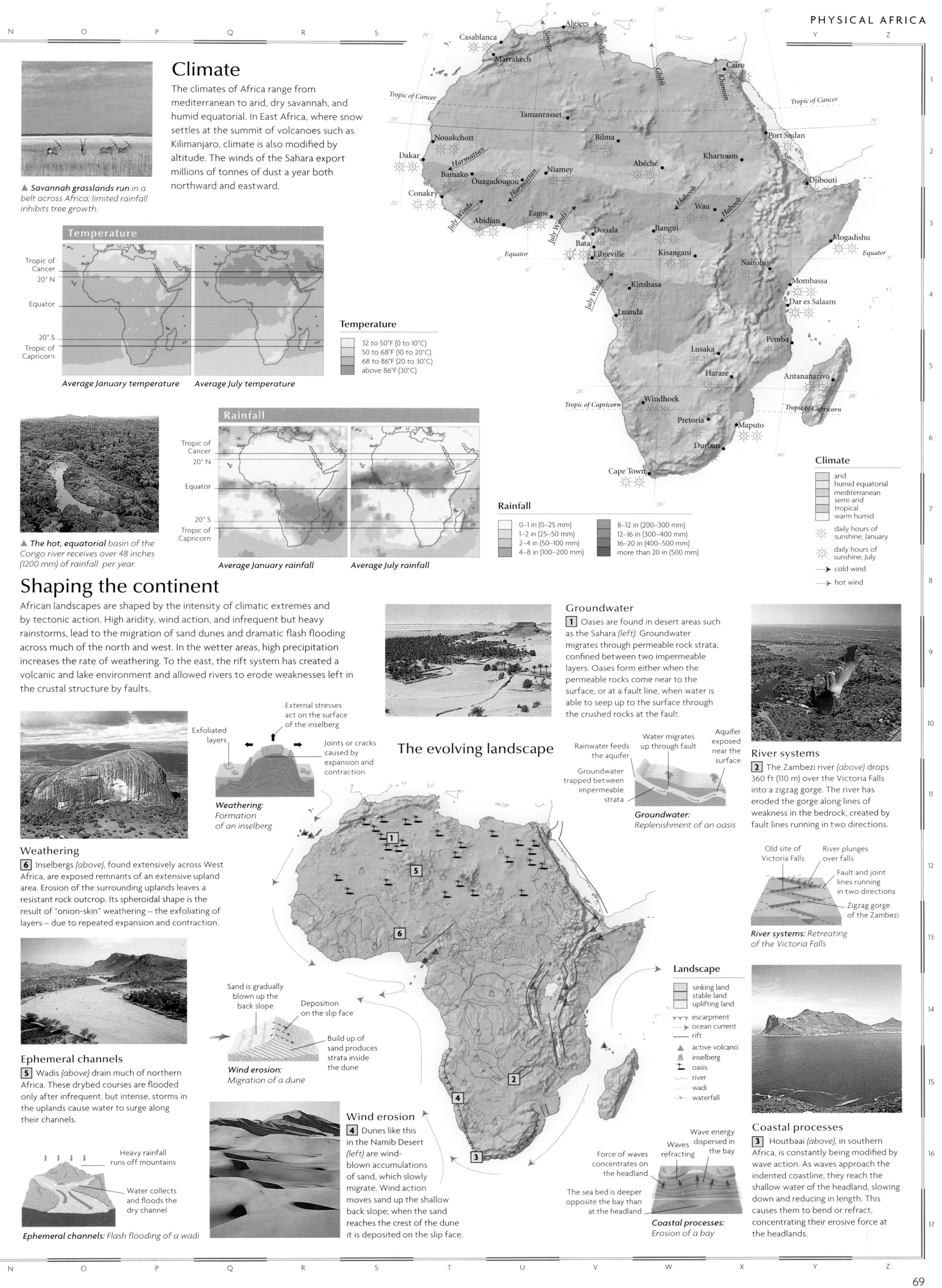

▲ *Savannah grasslands run* in a belt across Africa; limited rainfall inhibits tree growth.

Temperature

Average January temperature

Average July temperature

Temperature

- 32 to 50°F (0 to 10°C)
- 50 to 68°F (10 to 20°C)
- 68 to 86°F (20 to 30°C)
- above 86°F (30°C)

▲ *The hot, equatorial* basin of the Congo river receives over 48 inches (1200 mm) of rainfall per year.

Rainfall

Average January rainfall

Average July rainfall

Rainfall

- 0–1 in (0–25 mm)
- 1–2 in (25–50 mm)
- 2–4 in (50–100 mm)
- 4–8 in (100–200 mm)
- 8–12 in (200–300 mm)
- 12–16 in (300–400 mm)
- 16–20 in (400–500 mm)
- more than 20 in (500 mm)

Climate

- arid
- humid equatorial
- mediterranean
- semi-arid
- tropical
- warm humid
- ☀ daily hours of sunshine, January
- ☀ daily hours of sunshine, July
- → cold wind
- → hot wind

Shaping the continent

African landscapes are shaped by the intensity of climatic extremes and by tectonic action. High aridity, wind action, and infrequent but heavy rainstorms, lead to the migration of sand dunes and dramatic flash flooding across much of the north and west. In the wetter areas, high precipitation increases the rate of weathering. To the east, the rift system has created a volcanic and lake environment and allowed rivers to erode weaknesses left in the crustal structure by faults.

Groundwater

1 Oases are found in desert areas such as the Sahara (*left*). Groundwater migrates through permeable rock strata, confined between two impermeable layers. Oases form either when the permeable rocks come near to the surface, or at a fault line, when water is able to seep up to the surface through the crushed rocks at the fault.

Groundwater: Replenishment of an oasis

- Rainwater feeds the aquifer
- Water migrates up through fault
- Aquifer exposed near the surface
- Groundwater trapped between impermeable strata

The evolving landscape

Weathering: Formation of an inselberg

- External stresses act on the surface of the inselberg
- Exfoliated layers
- Joints or cracks caused by expansion and contraction

Weathering

6 Inselbergs (*above*), found extensively across West Africa, are exposed remnants of an extensive upland area. Erosion of the surrounding uplands leaves a resistant rock outcrop. Its spheroidal shape is the result of "onion-skin" weathering – the exfoliating of layers – due to repeated expansion and contraction.

River systems

2 The Zambezi river (*above*) drops 360 ft (110 m) over the Victoria Falls into a zigzag gorge. The river has eroded the gorge along lines of weakness in the bedrock, created by fault lines running in two directions.

River systems: Retreating of the Victoria Falls

- Old site of Victoria Falls
- River plunges over falls
- Fault and joint lines running in two directions
- Zigzag gorge of the Zambezi

Landscape

- sinking land
- stable land
- uplifting land
- ▽▽▽ escarpment
- → ocean current
- rift
- ▲ active volcano
- ⛰ inselberg
- oasis
- river
- wadi
- waterfall

Ephemeral channels

5 Wadis (*above*) drain much of northern Africa. These drybed courses are flooded only after infrequent, but intense, storms in the uplands cause water to surge along their channels.

Ephemeral channels: Flash flooding of a wadi

- Heavy rainfall runs off mountains
- Water collects and floods the dry channel

Wind erosion: Migration of a dune

- Sand is gradually blown up the back slope
- Deposition on the slip face
- Build up of sand produces strata inside the dune

Wind erosion

4 Dunes like this in the Namib Desert (*left*) are wind-blown accumulations of sand, which slowly migrate. Wind action moves sand up the shallow back slope; when the sand reaches the crest of the dune it is deposited on the slip face.

Coastal processes: Erosion of a bay

- Wave energy dispersed in the bay
- Waves refracting
- Force of waves concentrates on the headland
- The sea bed is deeper opposite the bay than at the headland

Coastal processes

3 Houtbaai (*above*), in southern Africa, is constantly being modified by wave action. As waves approach the indented coastline, they reach the shallow water of the headland, slowing down and reducing in length. This causes them to bend or refract, concentrating their erosive force at the headlands.

Political Africa

The political map of modern Africa only emerged following the end of the Second World War. Over the next half-century, all of the countries formerly controlled by European powers gained independence from their colonial rulers – only Liberia and Ethiopia were never colonized. The postcolonial era has not been an easy period for many countries, but there have been moves toward multiparty democracy across much of the continent. In South Africa, democratic elections replaced the internationally-condemned apartheid system only in 1994. Other countries have still to find political stability; corruption in government, and ethnic tensions are serious problems. National infrastructures, based on the colonial transportation systems built to exploit Africa's resources, are often inappropriate for independent economic development.

Languages

Three major world languages act as *lingua francas* across the African continent: Arabic in North Africa; English in southern and eastern Africa and Nigeria; and French in Central and West Africa, and in Madagascar. A huge number of African languages are spoken as well – over 2000 have been recorded, with more than 400 in Nigeria alone – reflecting the continuing importance of traditional cultures and values. In the north of the continent, the extensive use of Arabic reflects Middle Eastern influences while Bantu languages are widely-spoken across much of southern Africa.

Language groups

- Afro-Asiatic (Hamito-Semitic)
- Niger-Congo
- Nilo-Saharan
- Khoisan
- Indo-European
- Austronesian

Official African languages

- French
- English
- Arabic
- Portuguese
- Swahili
- Amharic
- Spanish
- French/English
- French/Arabic
- French/Malagasy
- English/Swahili
- Arabic/Somali

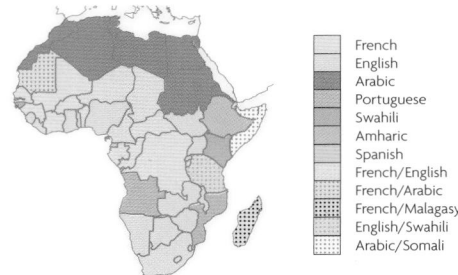

▲ *Islamic influences are* evident throughout North Africa. The Great Mosque at Kairouan, Tunisia, is Africa's holiest Islamic place.

▲ *In northeastern Nigeria,* people speak Kanuri – a dialect of the Nilo-Saharan language group.

Transportation

African railroads were built to aid the exploitation of natural resources, and most offer passage only from the interior to the coastal cities, leaving large parts of the continent untouched – five landlocked countries have no railroads at all. The Congo, Nile, and Niger river networks offer limited access to land within the continental interior, but have a number of waterfalls and cataracts which prevent navigation from the sea. Many roads were developed in the 1960s and 1970s, but economic difficulties are making the maintenance and expansion of the networks difficult.

▶ *South Africa has the* largest concentration of railroads in Africa. Over 20,000 miles (32,000 km) of routes have been built since 1870.

▲ *Traditional means of* transportation, such as the camel, are still widely used across the less accessible parts of Africa.

◀ *The Congo river,* though not suitable for river transportation along its entire length, forms a vital link for people and goods in its navigable inland reaches.

Transportation

- — major roads and highways
- — major railroads
- — major canal
- — international borders
- ● transport intersections
- ⊕ international airports
- ⊕ major ports

MOROCCO
Casablanca
Safi
Marrakech
Agadir

Canary Islands
(to Spain)

Madeira
(to Portugal)

LAÂYOUNE

Western Sahara
(Occupied by Morocco)

Tropic of Cancer

CAPE VERDE

PRAIA

MAURITANIA
NOUAKCHOTT

Senegal

SENEGAL
DAKAR
Kaolack
GAMBIA
BANJUL
GUINEA-BISSAU
BISSAU
BAMAKO
Niger

GUINEA

CONAKRY
Koidu
FREETOWN
SIERRA LEONE
YAMOUSSOUKRO
MONROVIA
LIBERIA

IVO
CO

Map labels:

Ceuta (to Spain)
Tanger
Rabat
Casablanca
Agadir
Skikda
Algiers
Oran
Tunis
Tripoli
Alexandria
Port Said
Suez Canal
Cairo
Suez
Nouâdhibou
Tamanrasset
Aswân
Wadi Halfa
Port Sudan
Nouakchott
Dakar
Banjul
Agadez
Massawa
Bamako
Niamey
Maiduguri
Khartoum
Assab
Bissau
Ouagadougou
Kano
Nyala
Djibouti
Conakry
N'Djaména
Freetown
Cotonou
Lagos
Addis Ababa
Monrovia
Accra
Lomé
Warri
Abidjan
Douala
Bangui
Malabo
Yaoundé
Mogadishu
Libreville
Kisangani
Kampala
Port-Gentil
Nairobi
Bukavu
Brazzaville
Kinshasa
Mombasa
Pointe-Noire
Kalemie
Dodoma
Matadi
Kananga
Dar es Salaam
Luanda
Mbeya
Lobito
Lubumbashi
Nampula
Namibe
Lusaka
Tsumeb
Livingstone
Harare
Antananarivo
Bulawayo
Beira
Toamasina
Walvis Bay
Windhoek
Keetmanshoop
Pretoria
Maputo
Johannesburg
Durban
Cape Town
Port Elizabeth

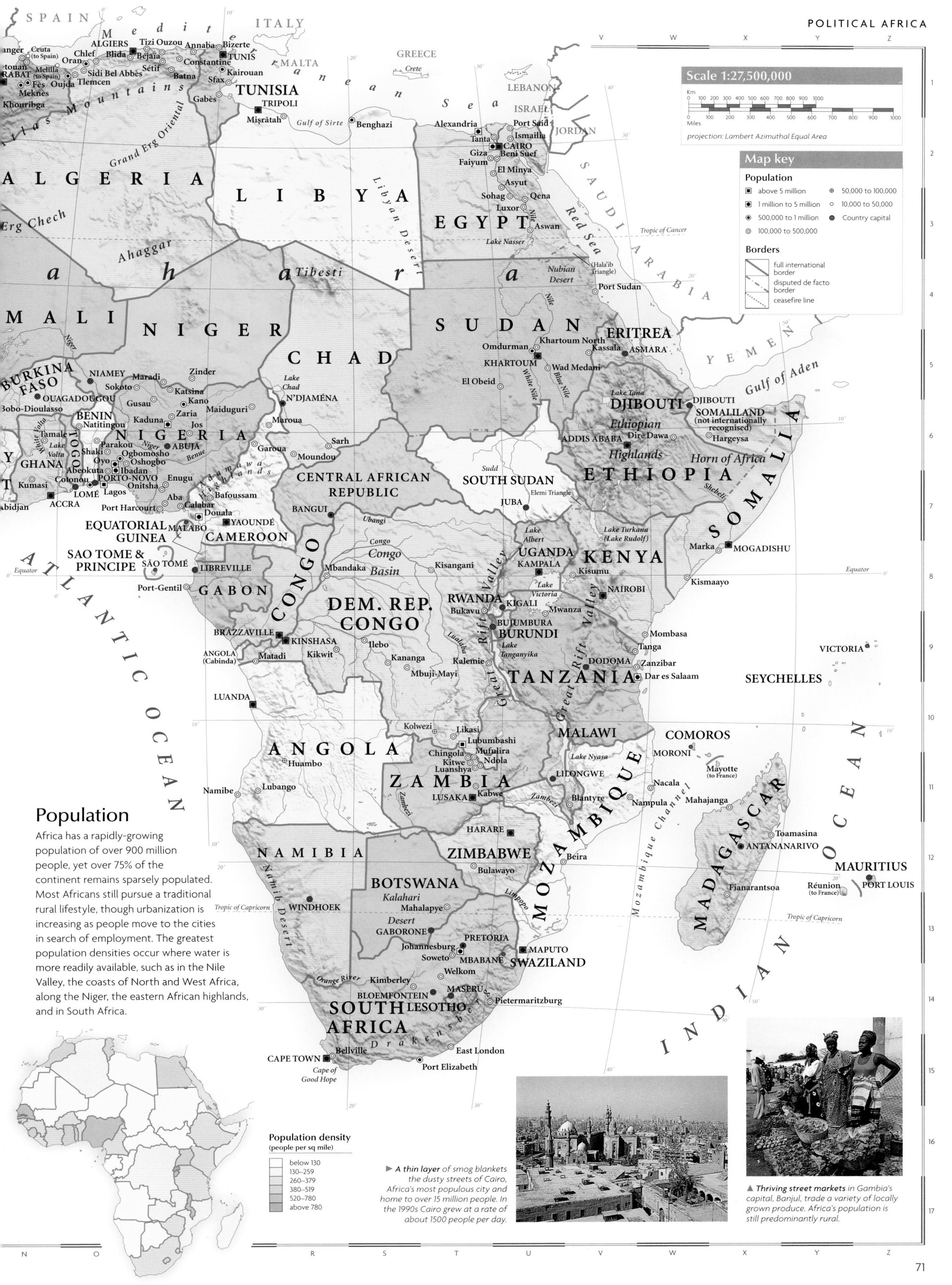

Population

Africa has a rapidly-growing population of over 900 million people, yet over 75% of the continent remains sparsely populated. Most Africans still pursue a traditional rural lifestyle, though urbanization is increasing as people move to the cities in search of employment. The greatest population densities occur where water is more readily available, such as in the Nile Valley, the coasts of North and West Africa, along the Niger, the eastern African highlands, and in South Africa.

Population density
(people per sq mile)

- below 130
- 130–259
- 260–379
- 380–519
- 520–780
- above 780

► A thin layer of smog blankets the dusty streets of Cairo, Africa's most populous city and home to over 15 million people. In the 1990s Cairo grew at a rate of about 1500 people per day.

▲ Thriving street markets in Gambia's capital, Banjul, trade a variety of locally grown produce. Africa's population is still predominantly rural.

African resources

The economies of most African countries are dominated by subsistence and cash crop agriculture, with limited industrialization. Manufacturing is largely confined to South Africa. Many countries depend on a single resource, such as copper or gold, or a cash crop, such as coffee, for export income, which can leave them vulnerable to fluctuations in world commodity prices. In order to diversify their economies and develop a wider industrial base, investment from overseas is being actively sought by many African governments.

Industry

Many African industries concentrate on the extraction and processing of raw materials. These include the oil industry, food processing, mining, and textile production. South Africa accounts for over half of the continent's industrial output with much of the remainder coming from the countries along the northern coast. Over 60% of Africa's workforce is employed in agriculture.

◄ The unspoiled natural splendor of wildlife reserves, like the Serengeti National Park in Tanzania, attract tourists to Africa from around the globe. The tourist industry in Kenya and Tanzania is particularly well developed, where it accounts for almost 10% of GNI.

Standard of living

Since the 1960s most countries in Africa have seen significant improvements in life expectancy, healthcare, and education. However, 28 of the 30 most deprived countries in the world are African, and the continent as a whole lies well behind the rest of the world in terms of meeting many basic human needs.

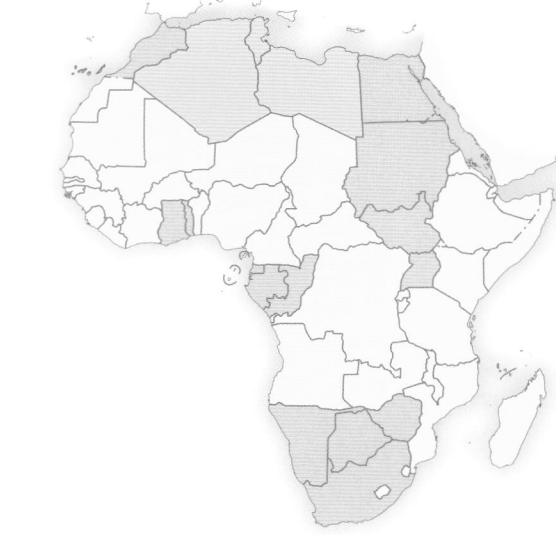

Standard of living
(UN human development index)
- high
- low

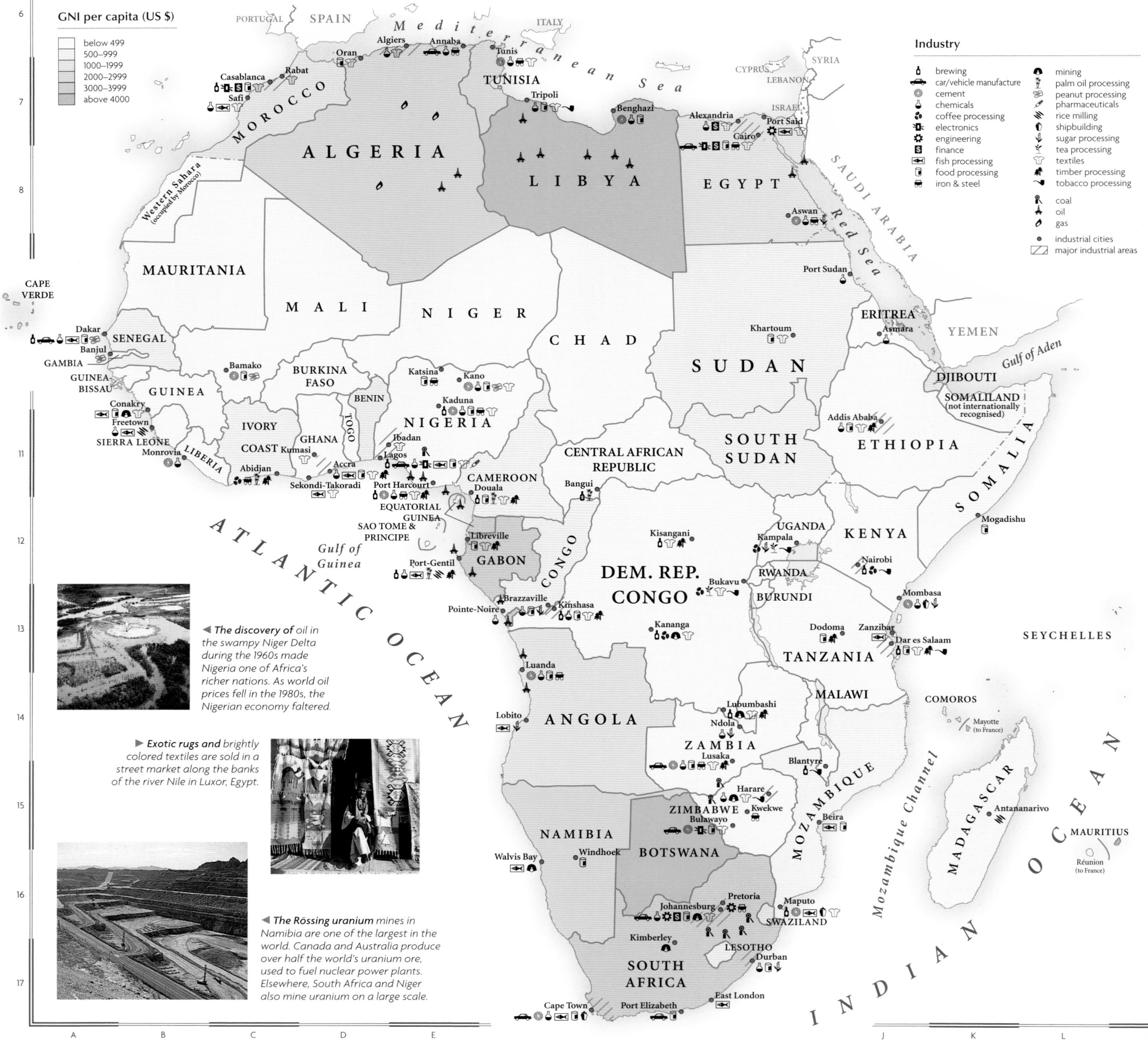

GNI per capita (US $)
- below 499
- 500–999
- 1000–1999
- 2000–2999
- 3000–3999
- above 4000

Industry
- brewing
- car/vehicle manufacture
- cement
- chemicals
- coffee processing
- electronics
- engineering
- finance
- fish processing
- food processing
- iron & steel
- mining
- palm oil processing
- peanut processing
- pharmaceuticals
- rice milling
- shipbuilding
- sugar processing
- tea processing
- textiles
- timber processing
- tobacco processing
- coal
- oil
- gas
- industrial cities
- major industrial areas

◄ The discovery of oil in the swampy Niger Delta during the 1960s made Nigeria one of Africa's richer nations. As world oil prices fell in the 1980s, the Nigerian economy faltered.

► Exotic rugs and brightly colored textiles are sold in a street market along the banks of the river Nile in Luxor, Egypt.

◄ The Rössing uranium mines in Namibia are one of the largest in the world. Canada and Australia produce over half the world's uranium ore, used to fuel nuclear power plants. Elsewhere, South Africa and Niger also mine uranium on a large scale.

PORTUGAL SPAIN ITALY
Mediterranean Sea
CYPRUS SYRIA
LEBANON
ISRAEL
Oran Algiers Annaba Tunis
Casablanca Rabat
Safi TUNISIA
Tripoli
MOROCCO Benghazi Alexandria Port Said
Cairo
Western Sahara (occupied by Morocco)
ALGERIA LIBYA EGYPT
SAUDI ARABIA
Aswan
Red Sea
MAURITANIA
CAPE VERDE
Port Sudan
MALI NIGER CHAD
Khartoum ERITREA Asmara
YEMEN
Gulf of Aden
Dakar SENEGAL
Banjul
GAMBIA SUDAN DJIBOUTI
GUINEA BISSAU
Conakry Bamako BURKINA FASO SOMALILAND (not internationally recognised)
Freetown GUINEA BENIN
Katsina Kano
SIERRA LEONE Kaduna Addis Ababa
Monrovia IVORY COAST GHANA Kumasi TOGO NIGERIA SOUTH SUDAN ETHIOPIA
LIBERIA Abidjan Accra Ibadan
Lagos CENTRAL AFRICAN REPUBLIC SOMALIA
Sekondi-Takoradi Port Harcourt Bangui
CAMEROON Douala Mogadishu
EQUATORIAL GUINEA Kisangani UGANDA KENYA
SAO TOME & PRINCIPE Libreville Kampala
ATLANTIC OCEAN Port-Gentil GABON CONGO Nairobi
Gulf of Guinea Brazzaville DEM. REP. Bukavu Mombasa
Pointe-Noire Kinshasa CONGO RWANDA
BURUNDI
Kananga Dodoma Zanzibar
Luanda Dar es Salaam SEYCHELLES
Lobito TANZANIA
ANGOLA Lubumbashi MALAWI COMOROS
Ndola ZAMBIA Mayotte (to France)
Lusaka Blantyre
MOZAMBIQUE MADAGASCAR
Harare Beira Antananarivo
Walvis Bay ZIMBABWE Kwekwe MAURITIUS
Windhoek Bulawayo Réunion (to France)
NAMIBIA BOTSWANA
Mozambique Channel
Johannesburg Pretoria Maputo
SWAZILAND INDIAN OCEAN
Kimberley LESOTHO Durban
SOUTH AFRICA East London
Cape Town Port Elizabeth

Environmental issues

One of Africa's most serious environmental problems occurs in marginal areas such as the Sahel where scrub and forest clearance, often for cooking fuel, combined with overgrazing, are causing desertification. Game reserves in southern and eastern Africa have helped to preserve many endangered animals, although the needs of growing populations have led to conflict over land use, and poaching is a serious problem.

Environmental issues
- national parks
- tropical forest
- forest destroyed
- desert
- desertification
- polluted rivers
- radioactive contamination
- marine pollution
- heavy marine pollution
- poor urban air quality

Mineral resources

Africa's ancient plateaus contain some of the world's most substantial reserves of precious stones and metals. About 15% of the world's gold is mined in South Africa; Zambia has great copper deposits; and diamonds are mined in Botswana, Dem. Rep. Congo, and South Africa. Oil has brought great economic benefits to Algeria, Libya, and Nigeria.

Mineral resources
- oil field
- gas field
- coal field
- bauxite
- copper
- diamonds
- gold
- iron
- phosphates
- tin
- uranium

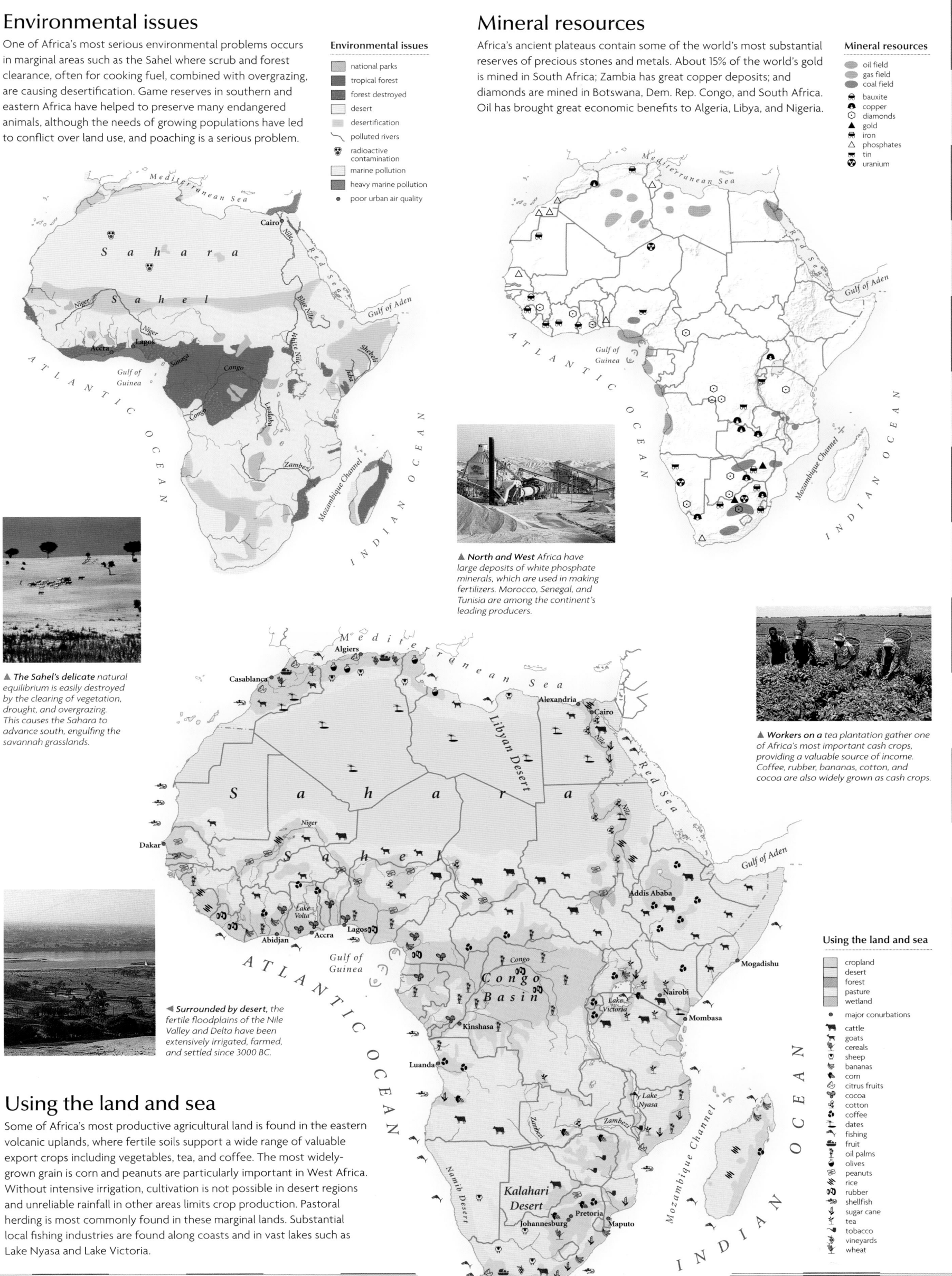

▲ **The Sahel's delicate** natural equilibrium is easily destroyed by the clearing of vegetation, drought, and overgrazing. This causes the Sahara to advance south, engulfing the savannah grasslands.

▲ **North and West** Africa have large deposits of white phosphate minerals, which are used in making fertilizers. Morocco, Senegal, and Tunisia are among the continent's leading producers.

▲ **Workers on a** tea plantation gather one of Africa's most important cash crops, providing a valuable source of income. Coffee, rubber, bananas, cotton, and cocoa are also widely grown as cash crops.

◄ **Surrounded by desert,** the fertile floodplains of the Nile Valley and Delta have been extensively irrigated, farmed, and settled since 3000 BC.

Using the land and sea

Some of Africa's most productive agricultural land is found in the eastern volcanic uplands, where fertile soils support a wide range of valuable export crops including vegetables, tea, and coffee. The most widely-grown grain is corn and peanuts are particularly important in West Africa. Without intensive irrigation, cultivation is not possible in desert regions and unreliable rainfall in other areas limits crop production. Pastoral herding is most commonly found in these marginal lands. Substantial local fishing industries are found along coasts and in vast lakes such as Lake Nyasa and Lake Victoria.

Using the land and sea
- cropland
- desert
- forest
- pasture
- wetland
- major conurbations
- cattle
- goats
- cereals
- sheep
- bananas
- corn
- citrus fruits
- cocoa
- cotton
- coffee
- dates
- fishing
- fruit
- oil palms
- olives
- peanuts
- rice
- rubber
- shellfish
- sugar cane
- tea
- tobacco
- vineyards
- wheat

North Africa

ALGERIA, EGYPT, LIBYA, MOROCCO, TUNISIA, WESTERN SAHARA

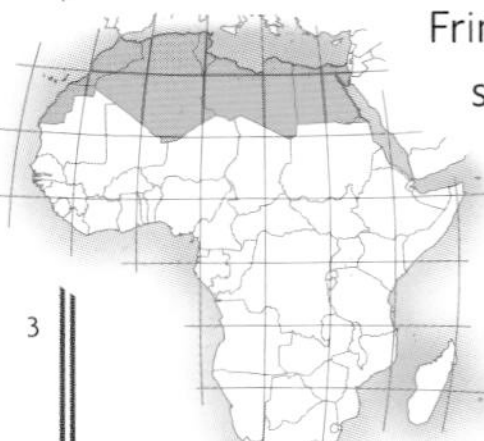

Fringed by the Mediterranean along the northern coast and by the arid Sahara in the south, North Africa reflects the influence of many invaders, both European and, most importantly, Arab, giving the region an almost universal Islamic flavor and a common Arabic language. The countries lying to the west of Egypt are often referred to as the Maghreb, an Arabic term for "west." Today, Morocco and Tunisia exploit their culture and landscape for tourism, while rich oil and gas deposits aid development in Libya and Algeria, despite political turmoil. Egypt, with its fertile, Nile-watered agricultural land and varied industrial base, is the most populous nation.

The landscape

The Atlas Mountains, which extend across much of Morocco, northern Algeria, and Tunisia, are part of the fold mountain system which also runs through much of southern Europe. They recede to the south and east, becoming a steppe landscape before meeting the Sahara desert which covers more than 90% of the region. The sediments of the Sahara overlie an ancient plateau of crystalline rock, some of which is more than four billion years old.

▲ *These rock piles in Algeria's Ahaggar mountains are the result of weathering caused by extremes of temperature. Great cracks or joints appear in the rocks, which are then worn and smoothed by the wind.*

Map key

Population
- ■ above 5 million
- ▣ 1 million to 5 million
- ◉ 500,000 to 1 million
- ◍ 100,000 to 500,000
- ⊕ 50,000 to 100,000
- ○ 10,000 to 50,000
- ○ below 10,000

Elevation
- 4000m / 13,124ft
- 3000m / 9843ft
- 2000m / 6562ft
- 1000m / 3281ft
- 500m / 1640ft
- 250m / 820ft
- 100m / 328ft
- sea level

Scale 1:11,000,000

Km 0 25 50 100 150 200 250 300

Miles 0 25 50 100 150 200 250 300

projection: Lambert Azimuthal Equal Area

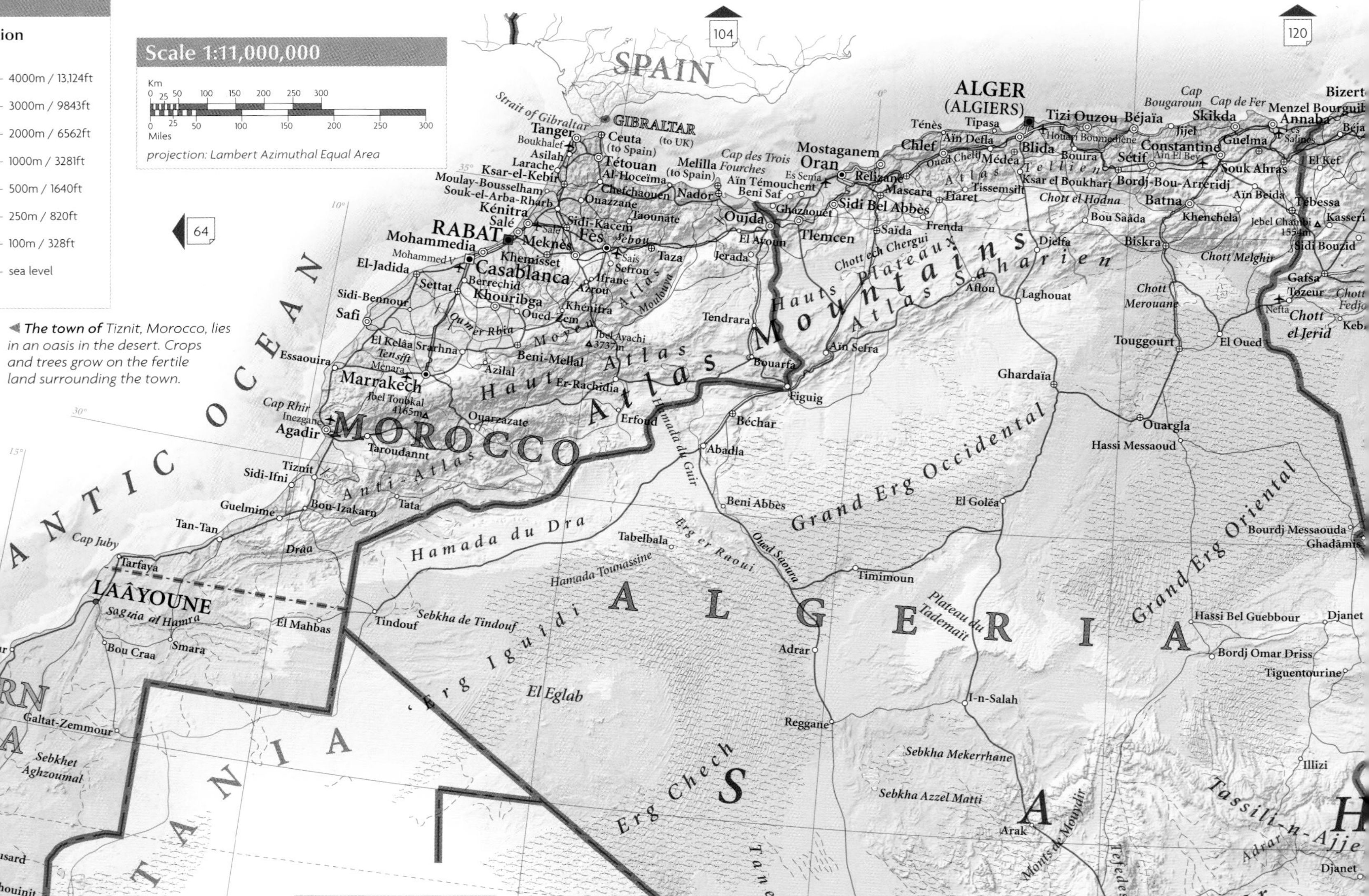

◄ *The town of Tiznit, Morocco, lies in an oasis in the desert. Crops and trees grow on the fertile land surrounding the town.*

▶ *The Grand Erg Occidental is one of Algeria's great Saharan sand seas. Wind force and direction determines the nature of landforms such as the linear or seif dunes in the foreground.*

Using the land & sea

Sheltered valleys in the Atlas Mountains, the Nile Valley and Delta, and the Mediterranean coast are the main sources of good farming land. A wide variety of valuable crops including cereals, rice, and cotton, and woods such as cedar and cork, are grown. Typical Mediterranean crops such as olives, figs, dates, and citrus fruits also thrive in these areas. The Nile Valley is particularly fertile, and most of Egypt's population lives close to the river. Elsewhere, irrigation is essential to improve crop yields on the desert margins.

The urban/rural population divide

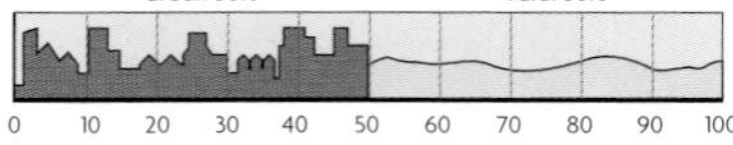

urban 50% | rural 50%

0 10 20 30 40 50 60 70 80 90 100

Population density	Total land area
65 people per sq mile (25 people per sq km)	2,215,020 sq miles (5,738,394 sq km)

Land use and agricultural distribution

- 🐐 goats
- 🐑 sheep
- 🌾 cereals
- 🍊 citrus fruits
- 🌳 cork
- 🌿 cotton
- 🌴 dates
- 🎣 fishing
- 🫒 olives
- 🍇 vineyards
- ■ capital cities
- ● major towns
- pasture
- cropland
- forest
- desert

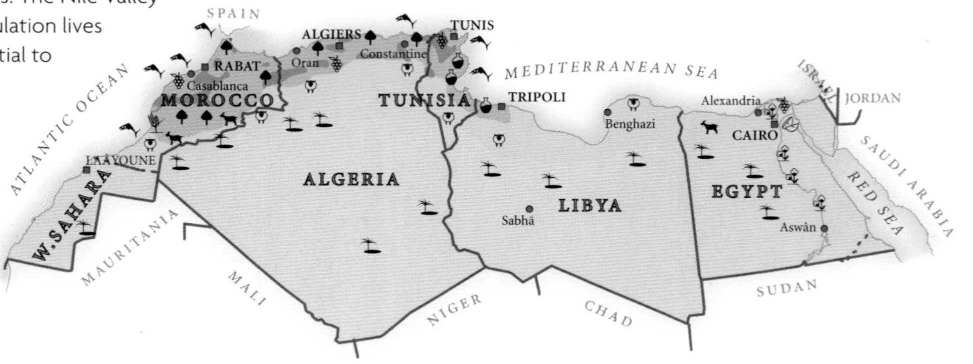

▲ *Many North African nomads, such as the Bedouin, maintain a traditional pastoral lifestyle on the desert fringes, moving their herds of sheep, goats, and camels from place to place – crossing country borders in order to find sufficient grazing land.*

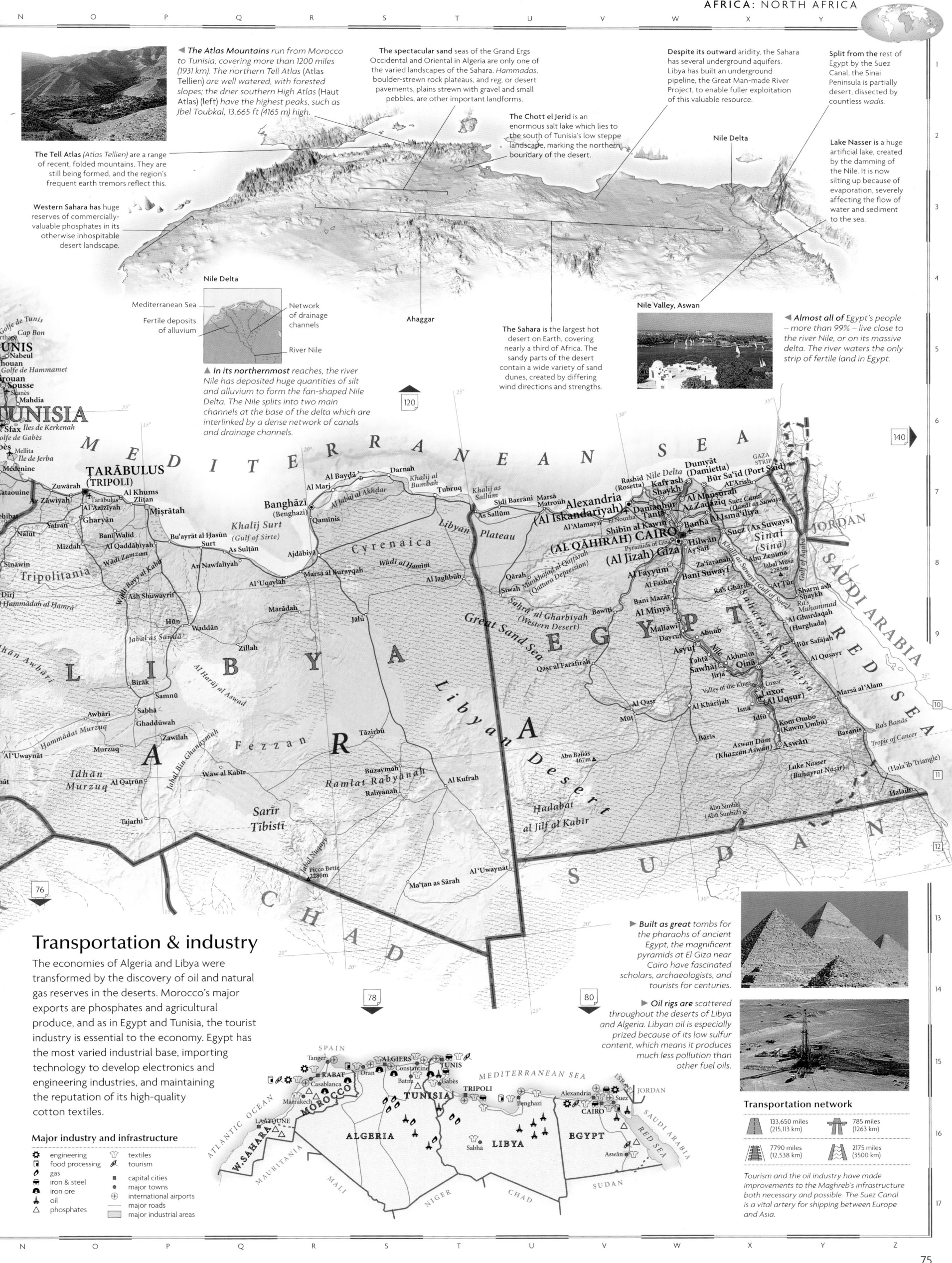

The Atlas Mountains run from Morocco to Tunisia, covering more than 1200 miles (1931 km). The northern Tell Atlas (Atlas Tellien) are well watered, with forested slopes; the drier southern High Atlas (Haut Atlas) (left) have the highest peaks, such as Jbel Toubkal, 13,665 ft (4165 m) high.

The spectacular sand seas of the Grand Ergs Occidental and Oriental in Algeria are only one of the varied landscapes of the Sahara. *Hammadas*, boulder-strewn rock plateaus, and *reg*, or desert pavements, plains strewn with gravel and small pebbles, are other important landforms.

Despite its outward aridity, the Sahara has several underground aquifers. Libya has built an underground pipeline, the Great Man-made River Project, to enable fuller exploitation of this valuable resource.

Split from the rest of Egypt by the Suez Canal, the Sinai Peninsula is partially desert, dissected by countless *wadis*.

The Tell Atlas (Atlas Tellien) are a range of recent, folded mountains. They are still being formed, and the region's frequent earth tremors reflect this.

The Chott el Jerid is an enormous salt lake which lies to the south of Tunisia's low steppe landscape, marking the northern boundary of the desert.

Nile Delta

Lake Nasser is a huge artificial lake, created by the damming of the Nile. It is now silting up because of evaporation, severely affecting the flow of water and sediment to the sea.

Western Sahara has huge reserves of commercially-valuable phosphates in its otherwise inhospitable desert landscape.

Nile Delta

Mediterranean Sea
Fertile deposits of alluvium
Network of drainage channels
River Nile

Ahaggar

The Sahara is the largest hot desert on Earth, covering nearly a third of Africa. The sandy parts of the desert contain a wide variety of sand dunes, created by differing wind directions and strengths.

Nile Valley, Aswan

Almost all of Egypt's people – more than 99% – live close to the river Nile, or on its massive delta. The river waters the only strip of fertile land in Egypt.

In its northernmost reaches, the river Nile has deposited huge quantities of silt and alluvium to form the fan-shaped Nile Delta. The Nile splits into two main channels at the base of the delta which are interlinked by a dense network of canals and drainage channels.

Transportation & industry

The economies of Algeria and Libya were transformed by the discovery of oil and natural gas reserves in the deserts. Morocco's major exports are phosphates and agricultural produce, and as in Egypt and Tunisia, the tourist industry is essential to the economy. Egypt has the most varied industrial base, importing technology to develop electronics and engineering industries, and maintaining the reputation of its high-quality cotton textiles.

Major industry and infrastructure

- engineering
- food processing
- gas
- iron & steel
- iron ore
- oil
- phosphates
- textiles
- tourism
- capital cities
- major towns
- international airports
- major roads
- major industrial areas

Built as great tombs for the pharaohs of ancient Egypt, the magnificent pyramids at El Giza near Cairo have fascinated scholars, archaeologists, and tourists for centuries.

Oil rigs are scattered throughout the deserts of Libya and Algeria. Libyan oil is especially prized because of its low sulfur content, which means it produces much less pollution than other fuel oils.

Transportation network

133,650 miles (215,113 km)		785 miles (1263 km)	
7790 miles (12,538 km)		2175 miles (3500 km)	

Tourism and the oil industry have made improvements to the Maghreb's infrastructure both necessary and possible. The Suez Canal is a vital artery for shipping between Europe and Asia.

West Africa

BENIN, BURKINA FASO, CAPE VERDE, GAMBIA, GHANA, GUINEA, GUINEA-BISSAU, IVORY COAST, LIBERIA, MALI, MAURITANIA, NIGER, NIGERIA, SENEGAL, SIERRA LEONE, TOGO

West Africa is an immensely diverse region, encompassing the desert landscapes and mainly Muslim populations of the southern Saharan countries, and the tropical rain forests of the more humid south, with a great variety of local languages and cultures. The rich natural resources and accessibility of the area were quickly exploited by Europeans; most of the Africans taken by slave traders came from this region, causing serious depopulation. The very different influences of West Africa's leading colonial powers, Britain and France, remain today, reflected in the languages and institutions of the countries they once governed.

► The dry scrub of the Sahel is only suitable for grazing herd animals like these cattle in Mali.

Transportation & industry

Abundant natural resources including oil and metallic minerals are found in much of West Africa, although investment is required for their further exploitation. Nigeria experienced an oil boom during the 1970s but subsequent growth has been sporadic. Most industry in other countries has a primary basis, including mining, logging, and food processing.

Transportation network

62,154 miles (100,038 km)		1037 miles (1669 km)	
6752 miles (10,867 km)		10,192 miles (16,405 km)	

The road and rail systems are most developed near the coasts. Some of the landlocked countries remain disadvantaged by the difficulty of access to ports, and their poor road networks.

Major industry and infrastructure

- chemicals
- cotton spinning
- food processing
- mining
- oil
- palm oil processing
- peanut processing
- textiles
- vehicle manufacture
- ■ capital cities
- ⊕ major towns
- ✈ international airports
- — major roads
- major industrial areas

Scale 1:9,000,000

Km
0 25 50 100 150 200 250
0 25 50 100 150 200 250
Miles

projection: Lambert Azimuthal Equal Area

Map key

Population
- ■ Above 5 million
- ■ 1 million to 5 million
- ◉ 500,000 to 1 million
- ◎ 100,000 to 500,000
- ⊕ 50,000 to 100,000
- ⊙ 10,000 to 50,000
- ∘ below 10,000

Elevation
- 2000m / 6562ft
- 1000m / 3281ft
- 500m / 1640ft
- 250m / 820ft
- 100m / 328ft
- sea level

CAPE VERDE

Santo Antão, Pombas, Ilhas de Barlavento
Mindelo, Ribeira Brava, Pedra Lume, Sal
São Vicente, São Nicolau, Amílcar Cabral
ATLANTIC OCEAN
Boa Vista, João Barrosa
Tarrafal, Maio
Fogo, Maio, PRAIA
São Filipe, Santiago
Ilhas de Sotavento

(same scale as main map)

◄ The southern regions of West Africa still contain great swathes of tropical rainforest, including some of the world's most prized hardwood trees, such as mahogany and iroko.

Using the land & sea

The humid southern regions are most suitable for cultivation; in these areas, cash crops such as coffee, cotton, cocoa, and rubber are grown in large quantities. Peanuts are grown throughout West Africa. In the north, advancing desertification has made the Sahel increasingly uncultivable, and pastoral farming is more common. Great herds of sheep, cattle, and goats are grazed on the savannah grasses. Fishing is important in coastal and delta areas.

▲ The Gambia, mainland Africa's smallest country, produces great quantities of peanuts. Winnowing is used to separate the nuts from their stalks.

Land use and agricultural distribution

- goats
- sheep
- cocoa
- coffee
- cotton
- oil palms
- peanuts
- rubber
- shellfish
- ■ capital cities
- major towns
- pasture
- cropland
- forest
- desert

The urban/rural population divide

urban 36%   rural 64%
0 10 20 30 40 50 60 70 80 90 100

Population density	Total land area
104 people per sq mile (40 people per sq km)	2,337,137 sq miles (6,054,760 sq km)

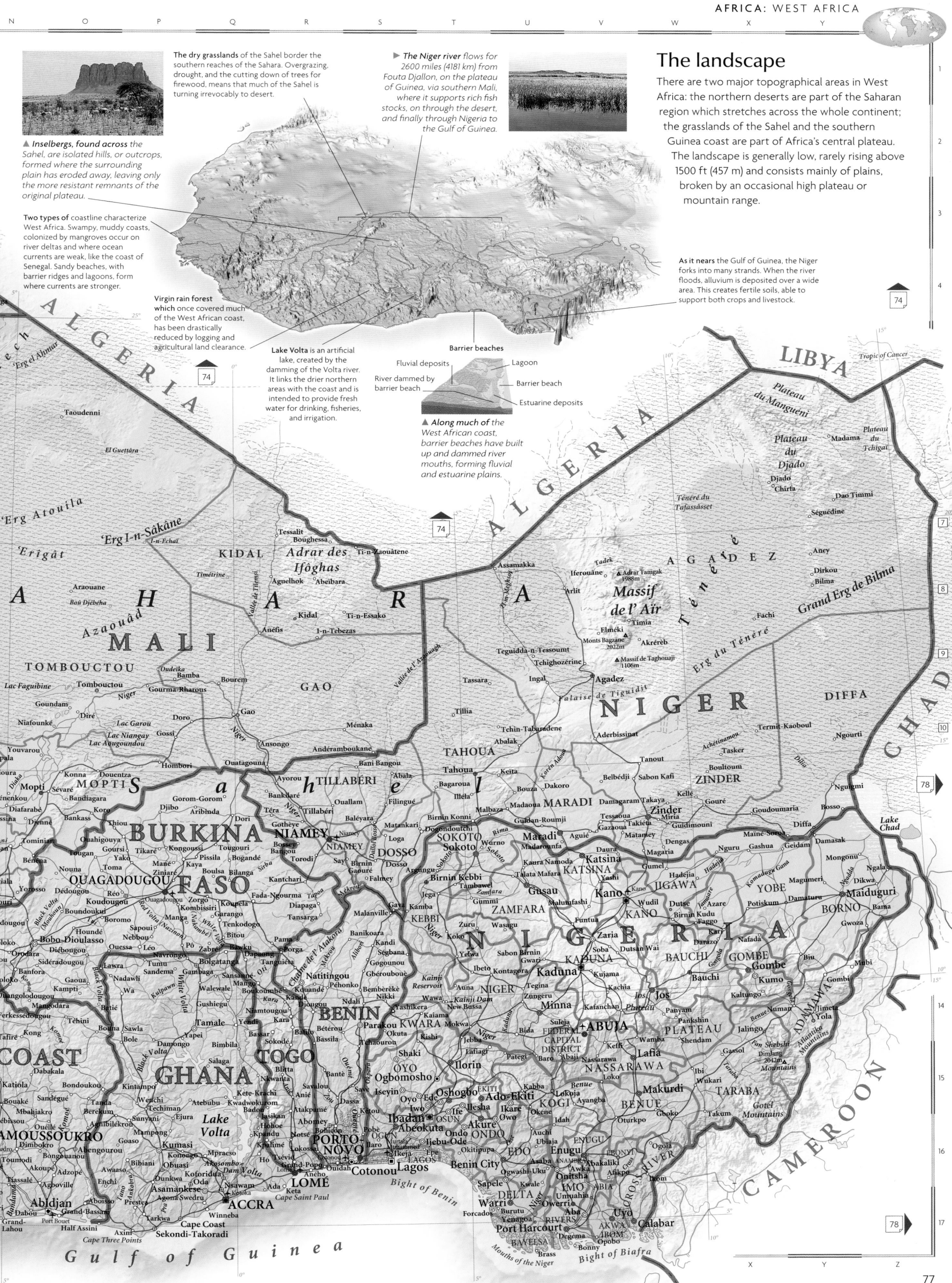

The dry grasslands of the Sahel border the southern reaches of the Sahara. Overgrazing, drought, and the cutting down of trees for firewood, means that much of the Sahel is turning irrevocably to desert.

▶ The Niger river flows for 2600 miles (4181 km) from Fouta Djallon, on the plateau of Guinea, via southern Mali, where it supports rich fish stocks, on through the desert, and finally through Nigeria to the Gulf of Guinea.

The landscape

There are two major topographical areas in West Africa: the northern deserts are part of the Saharan region which stretches across the whole continent; the grasslands of the Sahel and the southern Guinea coast are part of Africa's central plateau. The landscape is generally low, rarely rising above 1500 ft (457 m) and consists mainly of plains, broken by an occasional high plateau or mountain range.

▲ Inselbergs, found across the Sahel, are isolated hills, or outcrops, formed where the surrounding plain has eroded away, leaving only the more resistant remnants of the original plateau.

Two types of coastline characterize West Africa. Swampy, muddy coasts, colonized by mangroves occur on river deltas and where ocean currents are weak, like the coast of Senegal. Sandy beaches, with barrier ridges and lagoons, form where currents are stronger.

Virgin rain forest which once covered much of the West African coast, has been drastically reduced by logging and agricultural land clearance.

Lake Volta is an artificial lake, created by the damming of the Volta river. It links the drier northern areas with the coast and is intended to provide fresh water for drinking, fisheries, and irrigation.

As it nears the Gulf of Guinea, the Niger forks into many strands. When the river floods, alluvium is deposited over a wide area. This creates fertile soils, able to support both crops and livestock.

Barrier beaches
Fluvial deposits — Lagoon
River dammed by barrier beach — Barrier beach
— Estuarine deposits

▲ Along much of the West African coast, barrier beaches have built up and dammed river mouths, forming fluvial and estuarine plains.

Central Africa

CAMEROON, CENTRAL AFRICAN REPUBLIC, CHAD, CONGO, DEM. REP. CONGO, EQUATORIAL GUINEA, GABON, SAO TOME & PRINCIPE

The great rain forest basin of the Congo river embraces most of remote Central Africa. The interior was largely unknown to Europeans until late in the 19th century, when its tribal kingdoms were split – principally between France and Belgium – with Sao Tome and Principe the lone Portuguese territory, and Equatorial Guinea controlled by Spain. Open democracy and regional economic integration are important goals for these nations – several of which have only recently emerged from restrictive regimes – and investment is needed to improve transportation infrastructures. Many of the small, but fast-growing and increasingly urban population, speak French, the regional *lingua franca*, along with several hundred Pygmy, Bantu, and Sudanic dialects.

Transportation & industry

Large reserves of valuable minerals are found in Central Africa: copper, cobalt, zinc, and diamonds are mined in Dem. Rep. Congo and manganese in Gabon. Congo, Cameroon, Gabon, and Equatorial Guinea have oil deposits and oil has also been recently discovered in Chad. Goods such as palm oil and rubber are processed for export.

Major industry and infrastructure

- brewing
- chemicals
- cobalt
- copper
- diamonds
- food processing
- manganese
- oil
- palm oil processing
- textiles
- tin
- capital cities
- major towns
- international airports
- major roads
- major industrial areas

Transportation network

102,747 miles (165,774 km)	37 miles (60 km)
3985 miles (6414 km)	14,110 miles (22,710 km)

The Trans-Gabon railroad, which began operating in 1987, has opened up new sources of timber and manganese. Elsewhere, much investment is needed to update and improve road, rail and water transportation.

The landscape

Lake Chad lies in a desert basin bounded by the volcanic Tibesti mountains in the north, plateaus in the east and, in the south, the broad watershed of the Congo basin. The vast circular depression of the Congo is isolated from the coastal plain by the granite Massif du Chaillu. To the northwest, the volcanoes and fold mountains of the Cameroon Ridge (*Dorsale Camerounaise*) extend as islands into the Gulf of Guinea. The high fold mountains fringing the east of the Congo Basin fall steeply to the lakes of the Great Rift Valley.

▲ The Tibesti mountains are the highest in the Sahara. They were pushed up by the movement of the African Plate over a hot spot, which first formed the northern Ahaggar mountains and is now thought to lie under the Great Rift Valley.

The Congo river is second only to the Amazon in the volume of water it carries, and in the size of its drainage basin.

Lake Tanganyika, the world's second deepest lake, is the largest of a series of linear "ribbon" lakes occupying a trench within the Great Rift Valley.

Rich mineral deposits in the "Copper Belt" of Dem. Rep. Congo were formed under intense heat and pressure when the ancient African Shield was uplifted to form the region's mountains.

▲ Virgin tropical rain forest covers the Ruwenzori range on the borders of Dem. Rep. Congo and Uganda.

The lakelike expansion of the Congo river at Stanley Pool is the lowest point of the interior basin, although the river still descends more than 1000 ft (300 m) to reach the sea.

▲ A plug of resistant lava, at the southwestern end of the Cameroon Ridge (Dorsale Camerounaise), is all that remains of an eroded volcano.

The volcanic massif of Cameroon Mountain occupies an area which remains volcanically active.

Massif du Chaillu

Gulf of Guinea

Lake Chad is the remnant of an inland sea, which once occupied much of the surrounding basin. A series of droughts since the 1970s has reduced the area of this shallow freshwater lake to about 1000 sq miles (2599 sq km).

▲ The Congo river flows sluggishly through the rain forest of the interior basin. Toward the coast, the river drops steeply in a series of waterfalls and cataracts. At this point, the erosional power of the river becomes so great that it has formed a deep submarine canyon offshore.

Broad, shallow basin

Waterfalls and cataracts

Submarine canyon

▲ The vast sandflats surrounding Lake Chad were once covered by water. Changing climatic patterns caused the lake to shrink, and desert now covers much of its previous area.

▲ The ancient rocks of Dem. Rep. Congo hold immense and varied mineral reserves. This open pit copper mine is at Kolwezi in the far south.

Map key

Population
- 1 million to 5 million
- 500,000 to 1 million
- 100,000 to 500,000
- 50,000 to 100,000
- 10,000 to 50,000
- below 10,000

Elevation
- 4000m / 13124ft
- 3000m / 9843ft
- 2000m / 6562ft
- 1000m / 3281ft
- 500m / 1640ft
- 250m / 820ft
- 100m / 328ft
- sea level

Scale 1:9,500,000

projection: Lambert Azimuthal Equal Area

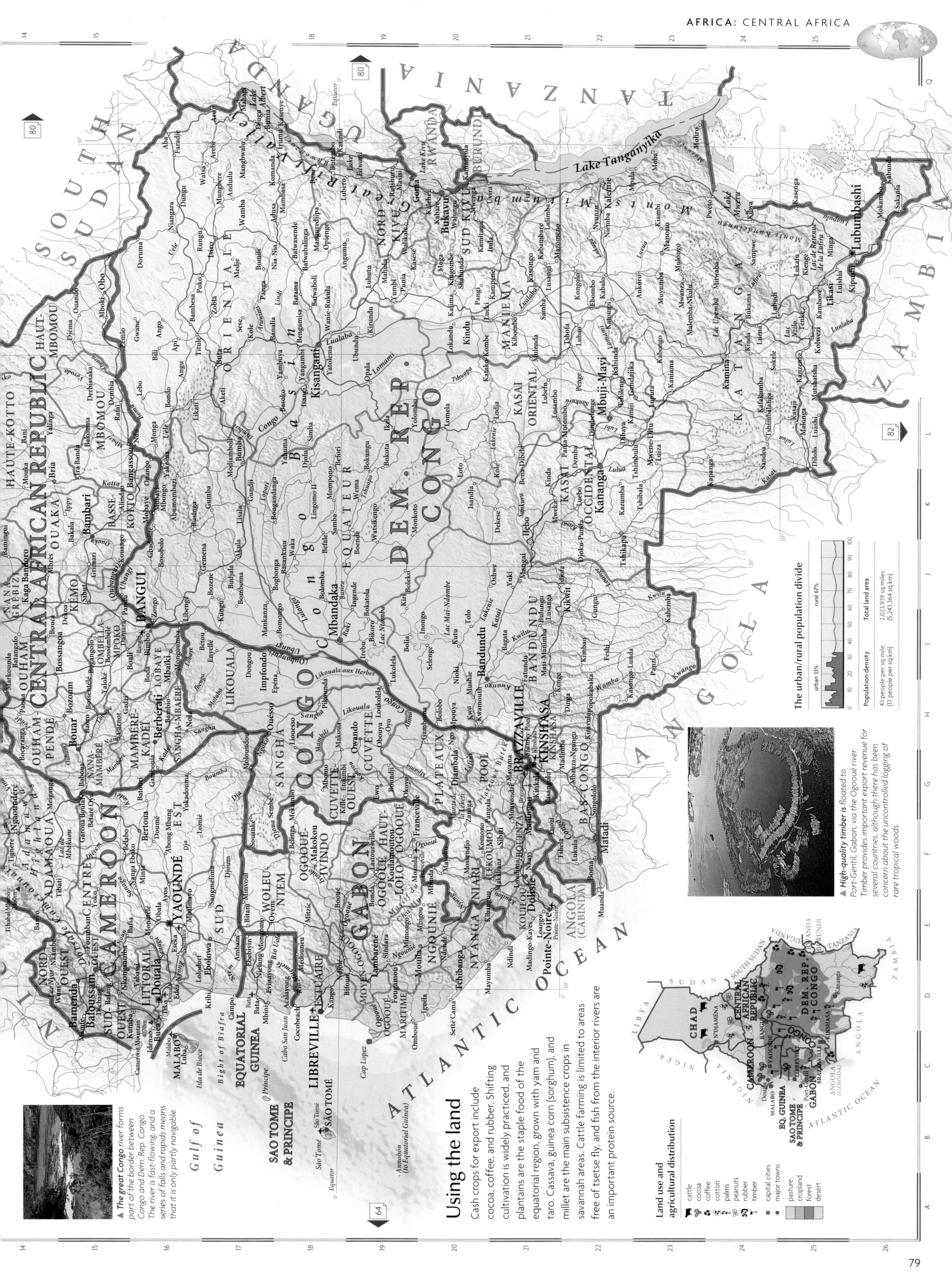

Using the land

Cash crops for export include cocoa, coffee, and rubber. Shifting cultivation is widely practiced, and plantains are the staple food of the equatorial region, grown with yam and taro. Cassava, guinea corn (sorghum), and millet are the main subsistence crops in savannah areas. Cattle farming is limited to areas free of tsetse fly, and fish from the interior rivers are an important protein source.

▲ *The great Congo river forms part of the border between Congo and Dem. Rep. Congo. The river is fast-flowing, and a series of falls and rapids means that it is only partly navigable.*

▲ *High-quality timber is floated to Port-Gentil, Gabon, via the Ogooué river. Timber provides important export revenue for several countries, although there has been concern about the uncontrolled logging of rare tropical woods.*

The urban/rural population divide

urban 33%   rural 67%

Population density	Total land area
43 people per sq mile (17 people per sq km)	2,023,939 sq miles (5,243,364 sq km)

Land use and agricultural distribution

cattle
cocoa
coffee
cotton
palms
peanuts
rubber
timber

capital cities
major towns

pasture
cropland
forest
desert

79

East Africa

BURUNDI, DJIBOUTI, ERITREA, ETHIOPIA, KENYA, RWANDA, SOMALIA, SOUTH SUDAN, SUDAN, TANZANIA, UGANDA

The countries of East Africa divide into two distinct cultural regions. Sudan and the "Horn" nations have been influenced by the Middle East; Ethiopia was the home of one of the earliest Christian civilizations, and Sudan reflects both Muslim and Christian influences. The southern countries share a closer cultural affinity with other sub-Saharan nations. Some of Africa's most densely populated countries lie in this region, and the needs of a growing number of people have put pressure on marginal lands and fragile environments. Although most East African economies remain strongly agricultural, Kenya has developed a varied industrial base.

The landscape

East Africa's most significant landscape feature is the Great Rift Valley, which formed during the most recent phase of continental movement when the rigid basement rocks cracked and buckled. Great blocks of land were raised and lowered, creating huge flat-bottomed valleys and steep escarpments, sometimes covered by volcanic extrusions in highland areas.

Ephemeral lake forms at far edge of slope

Central block slopes towards main fault

Boundary fault

▲ *The eastern arm* of the Great Rift Valley is gradually being pulled apart; however the forces on one side are greater than the other causing the land to slope. This affects regional drainage which migrates down the slope.

▼ *This dome at* Gonder, in Ethiopia, is a volcanic intrusion, formed when molten rock pushed up the surface of the Earth and then solidified, leaving an outcrop of igneous rock.

Lava flows on uplifted areas either side of the eastern branch of the Great Rift Valley gave the Ethiopian Highlands – a series of high, wide plateaus – their distinctive rounded appearance and fertile soils.

Kilimanjaro

▲ *An extinct volcano.* Kilimanjaro is Africa's highest mountain, rising 19,340 ft (5895 m). Once famed for its snow-capped peak, this has almost completely melted due to changing climatic conditions.

A vast plateau lies between the eastern and western rift valleys in Kenya, Uganda, and western Tanzania. It has been leveled by long periods of erosion to form a peneplain, but is dotted with inselbergs – outcrops of more resistant rocks.

Lake Victoria occupies a vast basin between the two arms of the Great Rift Valley. It is the world's second largest lake in terms of surface area, extending 26,560 sq miles (68,880 sq km). The lake contains numerous islands and coral reefs.

Lake Tanganyika lies 8202 ft (2500 m) above sea level. It has a depth of nearly 4700 ft (1435 m). The lake traces the valley floor for some 400 miles (644 km) of the western arm of the Great Rift Valley.

The tiny countries of Rwanda and Burundi are mainly mountainous, with large areas of inaccessible tropical rain forest.

In contrast to the desert conditions that prevail in much of Sudan to the north, annual rainfall in the tropical wetlands of the southern Sudd region in South Sudan, can sometimes exceed 40 inches (1000 mm).

▶ *The Kassala region in* eastern Sudan is watered by the Atbara River, an important tributary of the Nile. Most of the population is engaged in agriculture, growing cotton and cereals.

Scale 1:9,500,000

Km
0 25 50 100 150 200 250
Miles
0 25 50 100 150 250

projection: Lambert Azimuthal Equal Area

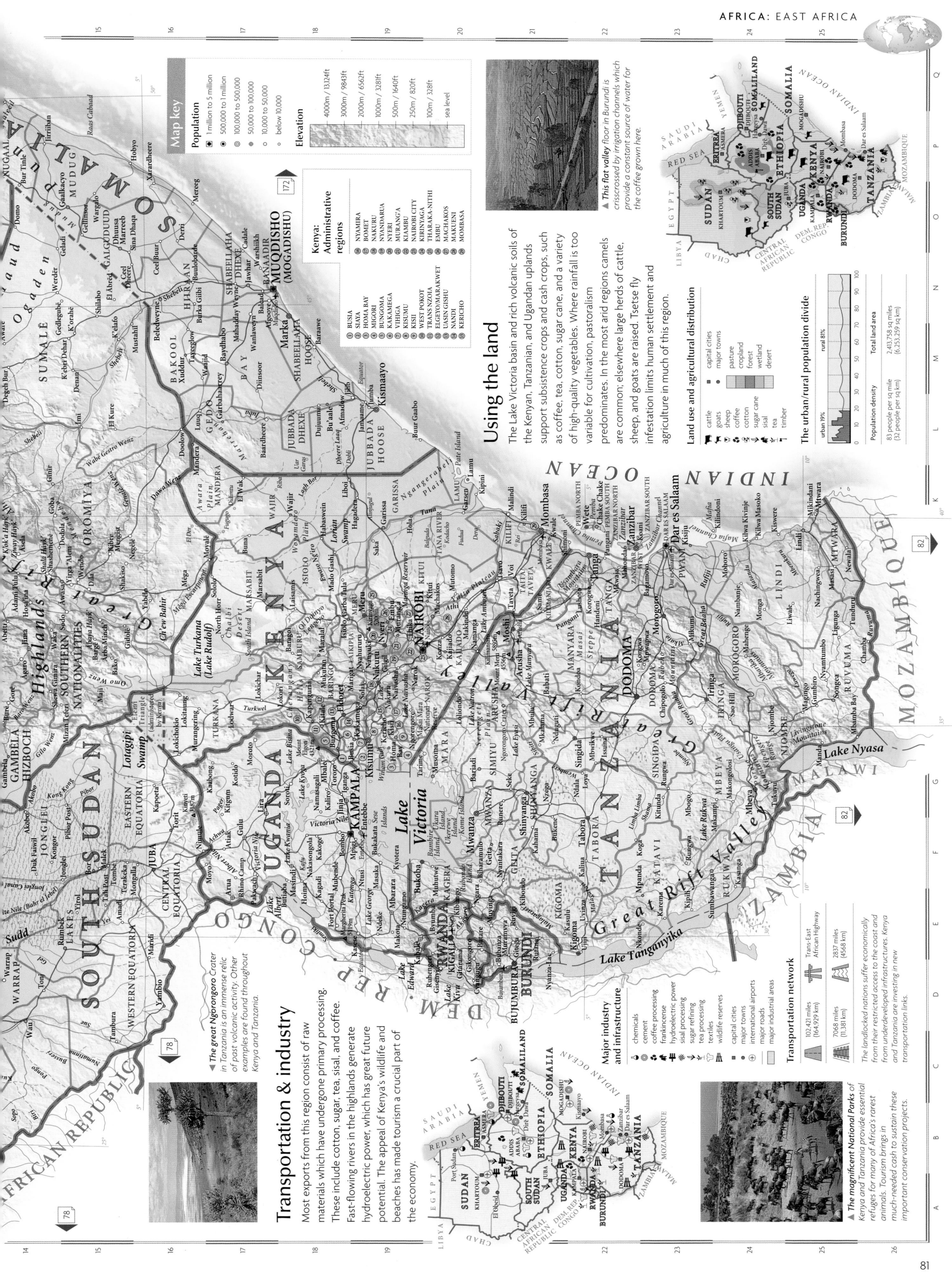

Using the land

The Lake Victoria basin and rich volcanic soils of the Kenyan, Tanzanian, and Ugandan uplands support subsistence crops and cash crops, such as coffee, tea, cotton, sugar cane, and a variety of high-quality vegetables. Where rainfall is too variable for cultivation, pastoralism predominates. In the most arid regions camels are common; elsewhere large herds of cattle, sheep, and goats are raised. Tsetse fly infestation limits human settlement and agriculture in much of this region.

Land use and agricultural distribution

cattle, goats, sheep, coffee, cotton, sugar cane, sisal, tea, timber; capital cities, major cities; pasture, cropland, forest, wetland, desert

The urban/rural population divide

urban 19%   rural 81%

Population density: 83 people per sq mile (32 people per sq km)
Total land area: 2,443,758 sq miles (6,253,259 sq km)

Transportation & industry

Most exports from this region consist of raw materials which have undergone primary processing. These include cotton, sugar, tea, sisal, and coffee. Fast-flowing rivers in the highlands generate hydroelectric power, which has great future potential. The appeal of Kenya's wildlife and beaches has made tourism a crucial part of the economy.

Major industry and infrastructure

chemicals, cement, coffee processing, frankincense, hydroelectric power, sisal processing, sugar refining, tea processing, textiles, wildlife reserves; capital cities, major towns, international airports, major roads, major industrial areas

Transportation network

102,421 miles (164,929 km)
7068 miles (11,381 km)
Trans-East African Highway
2837 miles (4568 km)

The landlocked nations suffer economically from their restricted access to the coast and from underdeveloped infrastructures. Kenya and Tanzania are investing in new transportation links.

▲ The great Ngorongoro Crater in Tanzania is an immense relic of past volcanic activity. Other examples are found throughout Kenya and Tanzania.

▲ The magnificent National Parks of Kenya and Tanzania provide essential refuges for many of Africa's rarest animals. Tourism brings in much-needed cash to sustain these important conservation projects.

▲ This flat valley floor in Burundi is crisscrossed by irrigation channels which provide a constant source of water for the coffee grown here.

Map key

Population
- 1 million to 5 million
- 500,000 to 1 million
- 100,000 to 500,000
- 50,000 to 100,000
- 10,000 to 50,000
- below 10,000

Elevation
- 4000m / 13,124ft
- 3000m / 9843ft
- 2000m / 6562ft
- 1000m / 3281ft
- 500m / 1640ft
- 250m / 820ft
- 100m / 328ft
- sea level

Kenya: Administrative regions
1 BUSIA 2 SIAYA 3 HOMA BAY 4 MIGORI 5 BUNGOMA 6 KAKAMEGA 7 VIHIGA 8 KISUMU 9 KISII 10 WEST POKOT 11 TRANS NZOIA 12 ELGEYO-MARAKWET 13 UASIN GISHU 14 KERICHO 15 NYAMIRA 16 BOMET 17 NAKURU 18 NYANDARUA 19 NYERI 20 MURANGA 21 KIAMBU 22 NAIROBI CITY 23 KIRINYAGA 24 THARAKA-NITHI 25 EMBU 26 MACHAKOS 27 MAKUENI 28 MOMBASA

81

Southern Africa

ANGOLA, BOTSWANA, LESOTHO, MALAWI, MOZAMBIQUE, NAMIBIA, SOUTH AFRICA, SWAZILAND, ZAMBIA, ZIMBABWE

Africa's vast southern plateau has been a contested homeland for disparate peoples for many centuries. The European incursion began with the slave trade and quickened in the 19th century, when the discovery of enormous mineral wealth secured South Africa's regional economic dominance. The struggle against white minority rule led to strife in Namibia, Zimbabwe, and the former Portuguese territories of Angola and Mozambique. South Africa's notorious apartheid laws, which denied basic human rights to more than 75% of the people, led to the state being internationally ostracized until 1994, when the first fully democratic elections inaugurated a new era of racial justice.

The landscape

Most of southern Africa rests on a concave plateau comprising the Kalahari basin and a mountainous fringe, skirted by a coastal plain which widens out in Mozambique. The plateau extends north, toward the Planalto de Bié in Angola, the Congo Basin and the lake-filled troughs of the Great Rift Valley. The eastern region is drained by the Zambezi and Limpopo rivers, and the Orange is the major western river.

At Victoria Falls, the Zambezi river has cut a spectacular gorge taking advantage of large joints in the basalt, which were first formed as the lava cooled and contracted

▲ *The fast-flowing Zambezi river cuts a deep, wide channel as it flows along the Zimbabwe/Zambia border.*

Lake Nyasa occupies one of the deep troughs of the Great Rift Valley, where the land has been displaced downward by as much as 3000 ft (920 m).

Great Rift Valley

Limpopo river

Bushveld intrusion

The Okavango/Cubango River flows from the Planalto de Bié to the swamplands of the Okavango Delta, one of the world's largest inland deltas, where it divides into countless distributary channels, feeding out into the desert.

Volcanic lava, over 250 million years old, caps the peaks of the Drakensberg range, which lie on the mountainous rim of southern Africa's interior plateau.

Broad, flat-topped mountains characterize the Great Karoo, which have been cut from level rock strata under extremely arid conditions.

Thousands of years of evaporating water have produced the Etosha Pan, one of the largest salt flats in the world. Lake and river sediments in the area indicate that the region was once less arid.

Planalto de Bié

Khorixas, Namibia

▲ *Finger Rock, near Khorixas, Namibia is a remnant of a former land surface, which has been denuded by erosion over the last 5 million years. These occasional stacks of partially weathered rocks interrupt the plains of the dry southern interior.*

Namib Desert

The Kalahari desert is the largest continuous sand surface in the world. Iron oxide gives a distinctive red color to the windblown sand, which, in eastern areas covers the bedrock by over 200 ft (60 m).

The mountains of the Little Karoo are composed of sedimentary rocks which have been substantially folded and faulted.

The Orange River, one of the longest rivers in Lesotho and is the only major river in the south which flows westward, rather than to the east coast.

Transportation & industry

South Africa, the world's largest exporter of gold, has a varied economy which generates about 75% of the region's income and draws migrant labor from neighboring states. Angola exports petroleum; Botswana and Namibia rely on diamond mining; and Zambia is seeking to diversify its economy to compensate for declining copper reserves.

▲ *Almost all new mining ventures in Zimbabwe are now subject to government control. This mine at Bindura in northeastern Zimbabwe produces nickel, one of the country's top three minerals in terms of economic value*

Transportation network

	84,213 miles (135,609 km)	746 miles (1202 km)
	23,208 miles (37,372 km)	3815 miles (6144 km)

Southern Africa's Cape-gauge rail network is by far the largest in the continent. About two-thirds of the 20,000 mile (32,000 km) system lies within South Africa. Lines such as the Harare–Bulawayo route have become corridors for industrial growth.

▲ *Following a series of droughts, this baobab tree in Zimbabwe now stands alone in a field once filled by sugar cane. The thick trunk and small leaves of the baobab help it to conserve water, enabling it to survive even in drought conditions.*

Major industry and infrastructure

car manufacture	gold	capital cities
coal	oil	major towns
copper	textiles	international airports
diamonds	uranium	major roads
food processing	wildlife reserves	major industrial areas

Map key

Population

- ● 1 million to 5 million
- ◉ 500,000 to 1 million
- ◎ 100,000 to 500,000
- ⊕ 50,000 to 100,000
- ○ 10,000 to 50,000
- ○ below 10,000

Elevation

- 3000m / 9843ft
- 2000m / 6562ft
- 1000m / 3281ft
- 500m / 1640ft
- 250m / 820ft
- 100m / 328ft
- sea level

Bushveld intrusion

Granite

Chromite

Gabbro and peridotite

Magnetite

Platinum minerals

▲ *The Bushveld intrusion lies on South Africa's high "veld." Molten magma intruded into the Earth's crust creating a saucer-shaped feature, more than 180 miles (300 km) across, containing regular layers of precious minerals, overlain by a dome of granite.*

South Africa: Capital cities

PRETORIA – administrative capital
CAPE TOWN – legislative capital
BLOEMFONTEIN – judicial capital

Scale 1:9,500,000

Km 0 25 50 100 150 200 250 300
Miles 0 25 50 100 150 200

projection: Lambert Azimuthal Equal Area

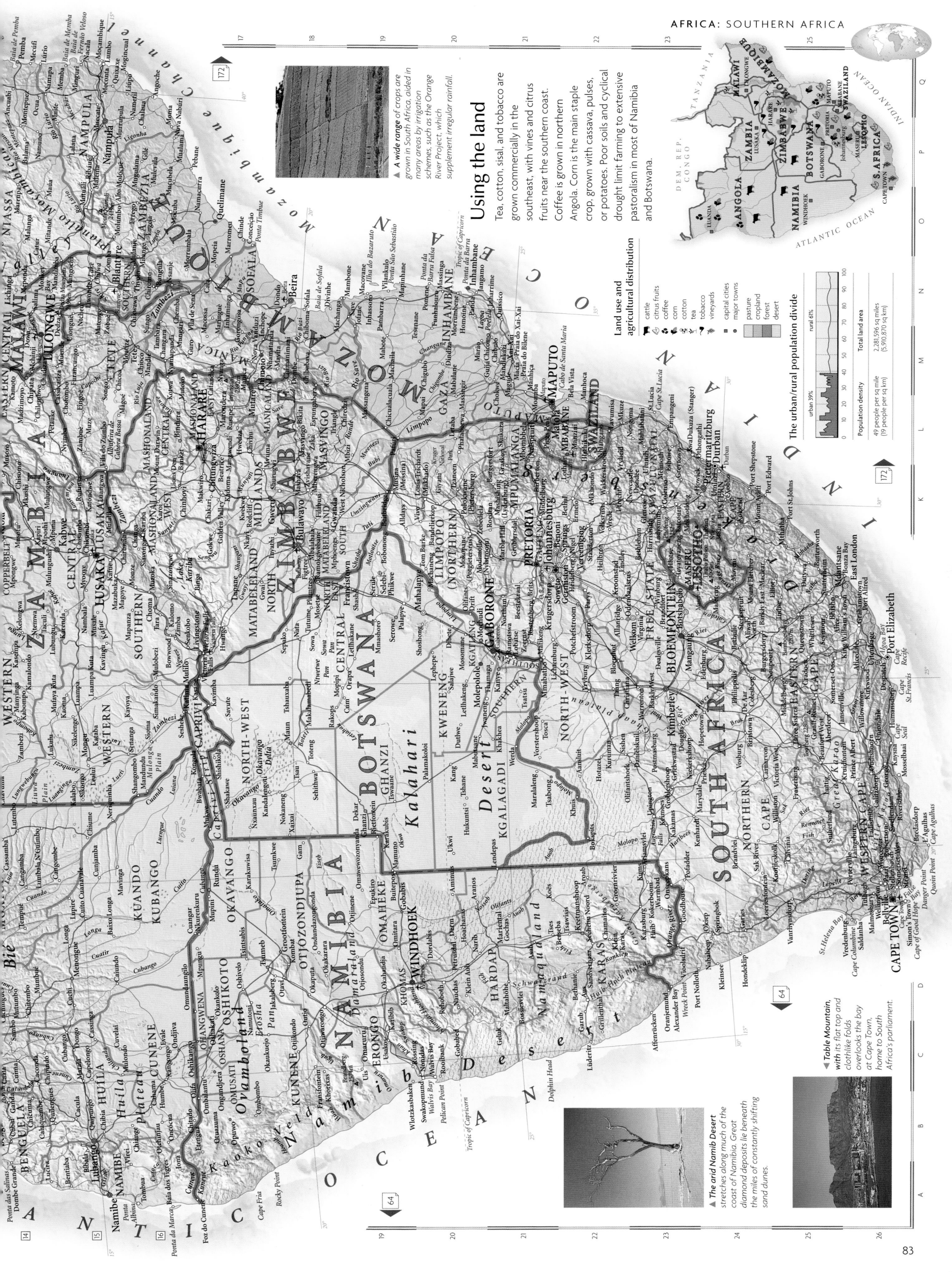

Using the land

Tea, cotton, sisal, and tobacco are grown commercially in the southeast, with vines and citrus fruits near the southern coast. Coffee is grown in northern Angola. Corn is the main staple crop, grown with cassava, pulses, or potatoes. Poor soils and cyclical drought limit farming in most of Namibia and Botswana.

▲ A wide range of crops are grown in South Africa, aided in many areas by irrigation schemes, such as the Orange River Project, which supplement irregular rainfall.

Land use and agricultural distribution

- cattle
- citrus fruits
- coffee
- corn
- cotton
- tea
- tobacco
- vineyards
- capital cities
- major towns

- pasture
- cropland
- forest
- desert

The urban/rural population divide

urban 39%
rural 61%

Population density
49 people per sq mile
(19 people per sq km)

Total land area
2,281,596 sq miles
(5,910,870 sq km)

▲ The arid Namib Desert stretches along much of the coast of Namibia. Great diamond deposits lie beneath the miles of constantly shifting sand dunes.

▲ Table Mountain, with its flat top and clothlike folds overlooks the bay at Cape Town, home to South Africa's parliament.

ARCTIC OCEAN

North Pole

Greenland

Ellesmere Island

King Frederik
VIII Land

King Christian X Land

Greenland
Sea

Spitsbergen

EURASIAN PLATE
NORTH AMERICAN PLATE

Jan Mayen Fracture Zone
Jan Mayen

Denmark Strait

Bjargtangar

Iceland
Plateau

Iceland
Yatnajökull

Reykjanes Ridge

Iceland
Basin

Hatton Ridge

Rockall
Rise

Feni Ridge

Rockall Trough

Porcupine
Plain

Charcot Seamounts

Azores-Biscay Rise

Theta Gap

Galicia
Bank

Iberian
Plain

Gorringe
Ridge

Horseshoe Seamounts

△ Ampère Seamount

Seine
Plain

△ Seine Seamount

Madeira

Dacia Seamount

Canary Islands

ATLANTIC OCEAN

Arctic Circle

Kolbeinsey Ridge

Jan Mayen Ridge

Norwegian Sea

Norwegian
Basin

Vøring Plateau

Faroe-Iceland Ridge

Bill Baileys
Bank

Faroe Islands

Faroe-Shetland Trough

Shetland
Islands

Orkney Islands

Outer Hebrides

Ben Nevis △
1343m

Grampian
Mountains

North Channel

British
Isles

Ireland
Irish Sea

Shannon

Celtic Sea

Celtic
Shelf

St. George's
Channel

Bristol Channel

Land's End

Pennines

Snowdon
△1085m

Britain

Trent

Severn

Thames

English Channel

Channel Islands

Strait of Dover

Viking Bank

Norwegian Trench

North
Sea

Jutland
Bank

Great
Fisher
Bank

Dogger
Bank

Frisian Islands

Skagerrak

Kattegat

Jutland

Sjælland

Elbe

Oder

Biscay
Plain

Bay of
Biscay

Seine

Loire

Cher

Garonne

Dordogne

Cévennes

Massif
Central

Loire

Marne

Meuse

Moselle

Ardennes

Rhine

Saône

Rhône

Vosges

Black
Forest

Danube

Lake Constance

Lake Geneva

Lake Garda

Po

Corsica

Strait of Bonifacio

Sardinia

Gulf of Lion

Ligurian
Sea

Apennines

Corno Grande
2912m

Tyrrhenian
Sea

Tyrrhenian
Basin

Strait of Sicily

Mount Etna
3340m

Sicily

Malta

Gulf of
Taranto

Ionian Sea

Adriatic Sea

Adriatic
Basin

Dinaric Alps

Balkan Mountains

Danube

Rhodope Mountains

Maritsa

Strait of Otranto

Lake
Ohrid

Lake
Prespa

Ionian Basin

Mediterranean Ridge

EUROPE

Rhône

Garonne

Iberian
Peninsula

Douro

Minho

Duero

Cordillera Cantábrica

Aragón
Ebro

Sistema Central

Sistema Ibérica

Tagus

Tagus Plain

Guadiana

Sierra Morena

Guadalquivir

Cabo
Roca

Cape
Saint Vincent

Punta de
Tarifa

Strait of
Gibraltar

Alborán Sea

Rif

Oumer Rbia

Sebou

Moulouya

Middle Atlas

High Atlas

Atlas Mountains

Tell Atlas

Saharan Atlas

Júcar

Segura

Sistemas Béticos

Gulf of
Valencia

Balearic Islands

Algerian
Basin

Mediterranean Sea

Oued Chelif

EURASIAN PLATE
AFRICAN PLATE

Chott el Jerid

Gulf of
Sirte

Grand Erg Occidental

Grand Erg Oriental

Libyan Desert

Qattara Depression
▽ -133m

Western Desert

Erg Iguidi

Erg Chech

SAHARA

AFRICA

Barents
Sea

Bjørnøya

Barents
Trough

Tromsøflaket

Vesterålen

Lofoten

North Cape

Nordkinn

Fugløya
Bank

Murmansk Rise

Kola Peninsula

Ostrov
Kolguyev

Poluostrov
Kanin

White Sea

Onega Bay

Severnaya
Zemlya

Poluostrov Taymyr

Ostrov
Rudol'fa

Franz Josef Land

Kara Sea

Mys Zhelaniya

Novaya Zemlya

Kara Strait

Pechora

Gulf of Ob

Baydaratskaya Guba

Poluostrov Yamal

Ob'

Yenisey

West Siberian
Plain

Ural Mountains

A S I A

Kemnekaise
△ 2117m

Inarijärvi

Torneälven

Kalixälven

Kemijoki

Ozero
Imandra

Timanskiy Kryazh

Mezen

Severnaya Dvina

Gaidhapiggen
△ 469m

Glåma

Scandinavia

Træna
Bank

Kölen

Inari

Oulujoki

Uncásen

Ljungan

Lusdan

Vänern

Åland

Gulf of Bothnia

Ozero
Vygozero

Lake
Ladoga

Lake
Onega

Onega

Svir

Ozero
Beloye

Rybinsk
Reservoir

Tobol

Ob'

Vättern

Gotland

Gulf of Finland

Lake Peipus

Lake Ilmen

Msta

Oka

Moskva

Kama

Votkinsk
Reservoir

Kuybyshev
Reservoir

Belaya

Ufa

Sakmara

Samara

V, Baltic Sea

Gulf of
Riga

Lake Pskov

Western Dvina

Northern Dvina

Neman

Vistula

Bug

Pripet
Marshes

Seym

Desna

Bryesina

North European Plain

Central Russian Upland

Dnieper

Don

Volga Upland

Volga

Warta

Sava

Dniester

Dnieper Lowlands

Kiev
Reservoir

Kremenchuk
Reservoir

Kryvyi

Kakhovka
Reservoir

Dniester

Podil's'ka Vysochina

Pivdennyy Buh

Dnieper

Don

Manych

Volga

Yergeni

Tisza

Tatra

Carpathian
Mountains

Bakony

Great
Hungarian
Plain

Lake Balaton

Drava

Tisza

Transylvanian Alps

Black Sea Lowland

Tsimlyansk
Reservoir

Sea of
Azov

Crimea

Kuban

Don

Kerch
Strait

Black Sea

Kirghiz Steppe

Caspian

Sea of
Marmara

EURASIAN PLATE
ANATOLIAN PLATE

Aegean Sea

Peloponnese

Mirtoan
Sea

Sea of Crete

Gökova

Anatolia

Taurus Mountains

Gulf of
Antalya

Lake Tuz

Kárpathos
Strait

Rhodes

Cyprus
Basin

Cyprus

Mediterranean Ridge

Levantine Basin

Dead
Sea

Suez Canal

Nile Fan

Nile

Gulf of Suez

Sinai

ARABIAN
PLATE

AFRICAN PLATE

Europe

Europe is the world's second smallest continent, covering 4,053,309 sq miles (10,498,000 sq km). It comprises 46 separate countries, including Turkey and the Russian Federation, although the greater parts of these nations lie in Asia.

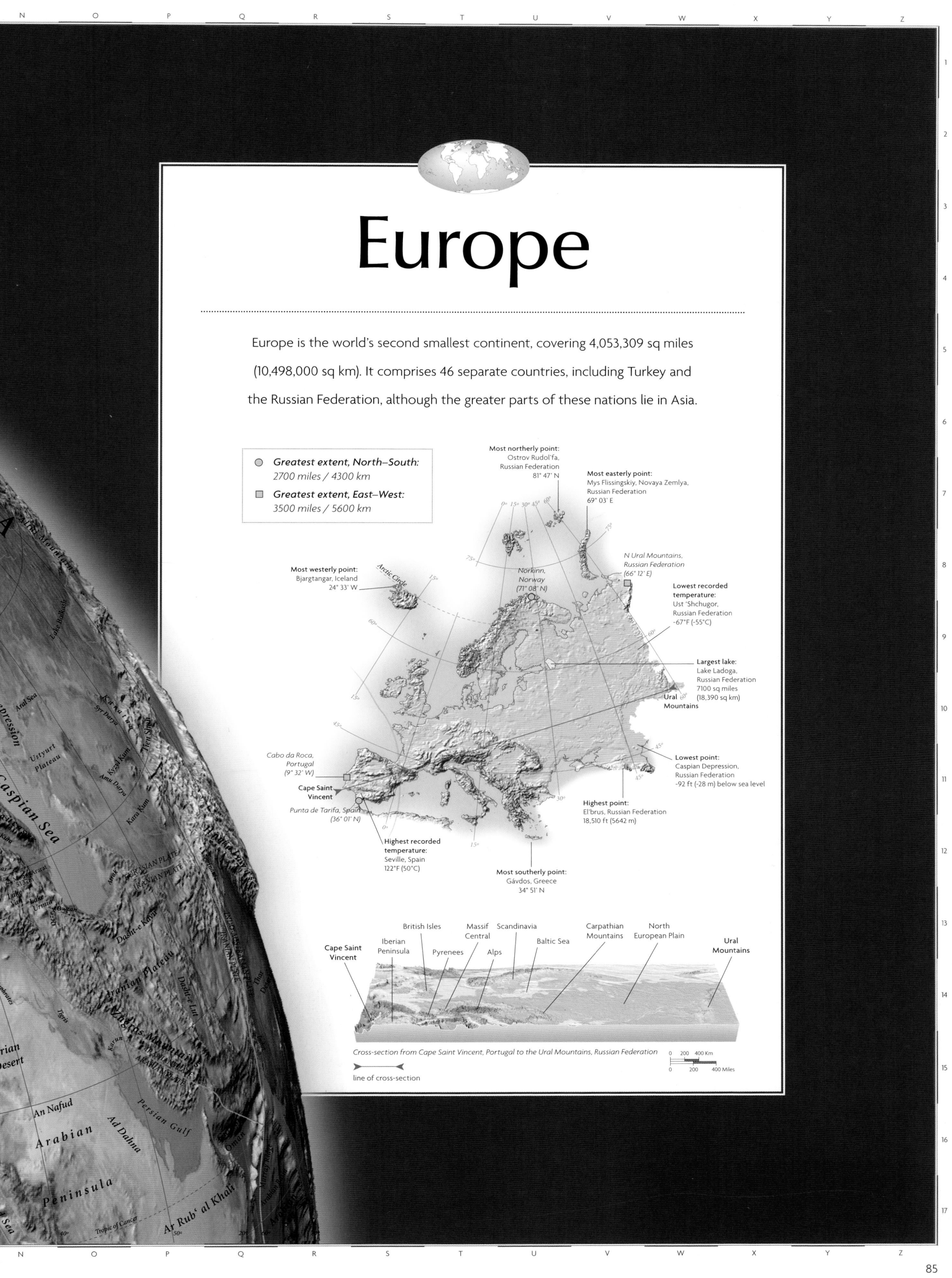

Greatest extent, North–South:
2700 miles / 4300 km

Greatest extent, East–West:
3500 miles / 5600 km

Most northerly point:
Ostrov Rudol'fa,
Russian Federation
81° 47' N

Most easterly point:
Mys Flissingskiy, Novaya Zemlya,
Russian Federation
69° 03' E

N Ural Mountains,
Russian Federation
(66° 12' E)

Most westerly point:
Bjargtangar, Iceland
24° 33' W

Arctic Circle

Nordkinn,
Norway
(71° 08' N)

Lowest recorded
temperature:
Ust 'Shchugor,
Russian Federation
-67°F (-55°C)

Largest lake:
Lake Ladoga,
Russian Federation
7100 sq miles
(18,390 sq km)

Ural
Mountains

Cabo da Roca,
Portugal
(9° 32' W)

Cape Saint
Vincent

Punta de Tarifa, Spain
(36° 01' N)

Lowest point:
Caspian Depression,
Russian Federation
-92 ft (-28 m) below sea level

Highest point:
El'brus, Russian Federation
18,510 ft (5642 m)

Highest recorded
temperature:
Seville, Spain
122°F (50°C)

Most southerly point:
Gávdos, Greece
34° 51' N

Cape Saint
Vincent

Iberian
Peninsula

British Isles

Pyrenees

Massif
Central

Alps

Scandinavia

Baltic Sea

Carpathian
Mountains

North
European Plain

Ural
Mountains

Cross-section from Cape Saint Vincent, Portugal to the Ural Mountains, Russian Federation

line of cross-section

0   200   400 Km
0   200   400 Miles

Physical Europe

The physical diversity of Europe belies its relatively small size. To the northwest and south it is enclosed by mountains. The older, rounded Atlantic Highlands of Scandinavia and the British Isles lie to the north and the younger, rugged peaks of the Alpine Uplands to the south. In between lies the North European Plain, stretching 2485 miles (4000 km) from The Fens in England to the Ural Mountains in Russia. South of the plain lies a series of gently folded sedimentary rocks separated by ancient plateaus, known as massifs.

The North European Plain

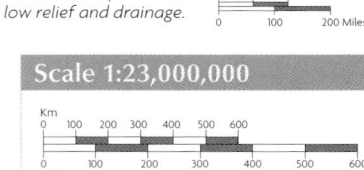

Rising less than 1000 ft (300 m) above sea level, the North European Plain strongly reflects past glaciation. Ridges of both coarse moraine and finer, windblown deposits have accumulated over much of the region. The ice sheet also diverted a number of river channels from their original courses.

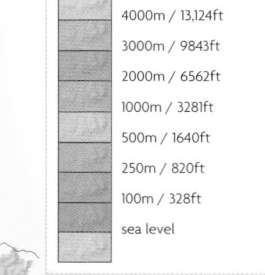

Glacial lakes · Rivers were diverted from their original course by the ice sheet · A layer of glacial sediments covers the North European Plain

Section across the North European Plain showing its low relief and drainage.

0 100 200 Km
0 100 200 Miles

The Atlantic Highlands

The Atlantic Highlands were formed by compression against the Scandinavian Shield during the Caledonian mountain-building period over 500 million years ago. The highlands were once part of a continuous mountain chain, now divided by the North Sea and a submerged rift valley.

The Atlantic Highlands continue in the British Isles · Rift valley buried by sediments · North Sea · Atlantic Highlands in Norway · Rocks affected by ancient mountain-building · Scandinavian Shield

Cross-section through northeastern Europe showing the continuous mountain chain and rift valley system.

0 100 200 Km
0 100 200 Miles

Scale 1:23,000,000

Km
0 100 200 300 400 500 600

Miles
0 100 200 300 400 500 600

projection: Lambert Azimuthal Equal Area

Map key

Elevation

- 4000m / 13,124ft
- 3000m / 9843ft
- 2000m / 6562ft
- 1000m / 3281ft
- 500m / 1640ft
- 250m / 820ft
- 100m / 328ft
- sea level

Plate margins
(for explanation see page xiv)

- constructive
- destructive
- conservative
- uncertain
- physiographic regions
- line of cross-section

The plateaus and lowlands

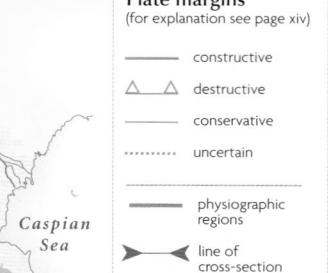

The uplifted plateaus or massifs of southern central Europe are the result of long-term erosion, later followed by uplift. They are the source areas of many of the rivers which drain Europe's lowlands. In some of the higher reaches, fractures have enabled igneous rocks from deep in the Earth to reach the surface.

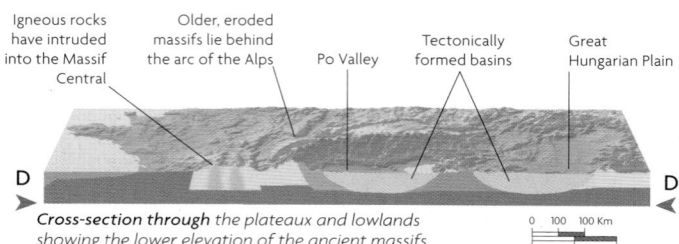

Igneous rocks have intruded into the Massif Central · Older, eroded massifs lie behind the arc of the Alps · Po Valley · Tectonically formed basins · Great Hungarian Plain

Cross-section through the plateaux and lowlands showing the lower elevation of the ancient massifs.

0 100 Km
0 100 Miles

The Alpine Uplands

The collision of the African and European continents, which began about 65 million years ago, folded and then uplifted a series of mountain ranges running across southern Europe and into Asia. Two major lines of folding can be traced: one includes the Pyrenees, the Alps, and the Carpathian Mountains; the other incorporates the Apennines and the Dinaric Alps.

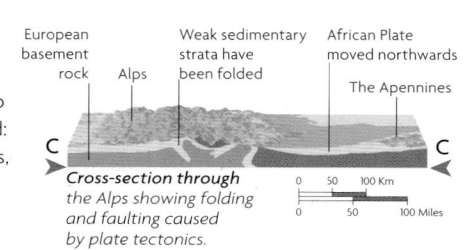

European basement rock · Alps · Weak sedimentary strata have been folded · African Plate moved northwards · The Apennines

Cross-section through the Alps showing folding and faulting caused by plate tectonics.

0 50 100 Km
0 50 100 Miles

Map labels

NORTH AMERICAN PLATE · EURASIAN PLATE · Iceland · ATLANTIC OCEAN · Faroe Islands · Shetland Islands · Outer Hebrides · British Isles · Ireland · Shannon · Britain · The Fens · Thames · English Channel · Seine · Loire · Bay of Biscay · Ardennes · Rhine · Elbe · Harz · Oder · Vistula · Danube · Garonne · Rhône · Massif Central · Mt Blanc 4807m · Po · ALPS · Pyrenees · Ebro · Iberian Peninsula · Douro · Guadalquivir · Corsica · Balearic Islands · Sardinia · Apennines · Adriatic Sea · Tyrrhenian Sea · Vesuvius 1171m · Sicily · Etna 3263m · Dinaric Alps · Ionian Sea · EURASIAN PLATE · AFRICAN PLATE · Mediterranean Sea · Norwegian Sea · Vänern · Vättern · North Sea · Jutland · Baltic Sea · Gulf of Bothnia · SCANDINAVIAN SHIELD · KÖLEN · ATLANTIC HIGHLANDS · Kola Peninsula · White Sea · Novaya Zemlya · Kara Sea · Ostrov Kolguyev · Barents Sea · Northern Dvina · Lake Onega · Lake Ladoga · Gulf of Riga · NORTH EUROPEAN PLAIN · Western Dvina · Central Russian Upland · Dnieper · Dniester · Don · Volga Uplands · Volga · Ural Mountains · Carpathian Mountains · Great Hungarian Plain · PLATEAUX AND LOWLANDS · Balkan Mountains · Sea of Azov · Crimea · Caucasus · El'brus 5642m · Caspian Sea · Black Sea · ASIA · ANATOLIAN PLATE · AFRICAN PLATE · Peloponnese · Aegean Sea · Crete

Climate

Europe experiences few extremes in either rainfall or temperature, with the exception of the far north and south. Along the west coast, the warm currents of the North Atlantic Drift moderate temperatures. Although east–west air movement is relatively unimpeded by relief, the Alpine Uplands halt the progress of north–south air masses, protecting most of the Mediterranean from cold, north winds.

▲ *Frost grips northern* and eastern Europe during the long cold winters. Lakes and rivers frequently freeze.

Temperature

Average January temperature

Average July temperature

Temperature
- below -22°F (-30°C)
- -22 to -4°F (-30 to -20°C)
- -4 to 14°F (-20 to -10°C)
- 14 to 32°F (-10 to 0°C)
- 32 to 50°F (0 to 10°C)
- 50 to 68°F (10 to 20°C)
- 68 to 86°F (20 to 30°C)
- above 86°F (30°C)

▲ *Mild temperatures and* frequent rainfall contribute to the fertile farming land found over much of northwestern Europe.

Rainfall

Average January rainfall

Average July rainfall

Rainfall
- 0–1 in (0–25 mm)
- 1–2 in (25–50 mm)
- 2–4 in (50–100 mm)
- 4–8 in (100–200 mm)
- 8–12 in (200–300 mm)
- 12–16 in (300–400 mm)
- 16–20 in (400–500 mm)
- more than 20 in (500 mm)

▶ *Dusty Sirocco winds from Africa* help create the semiarid scrubland common across the Mediterranean coastlands of southern Europe.

Climate
- tundra
- subarctic
- cool continental
- warm humid
- mediterranean
- semi-arid
- ☼ daily hours of sunshine, January
- ☼ daily hours of sunshine, July
- → cold wind
- → hot wind

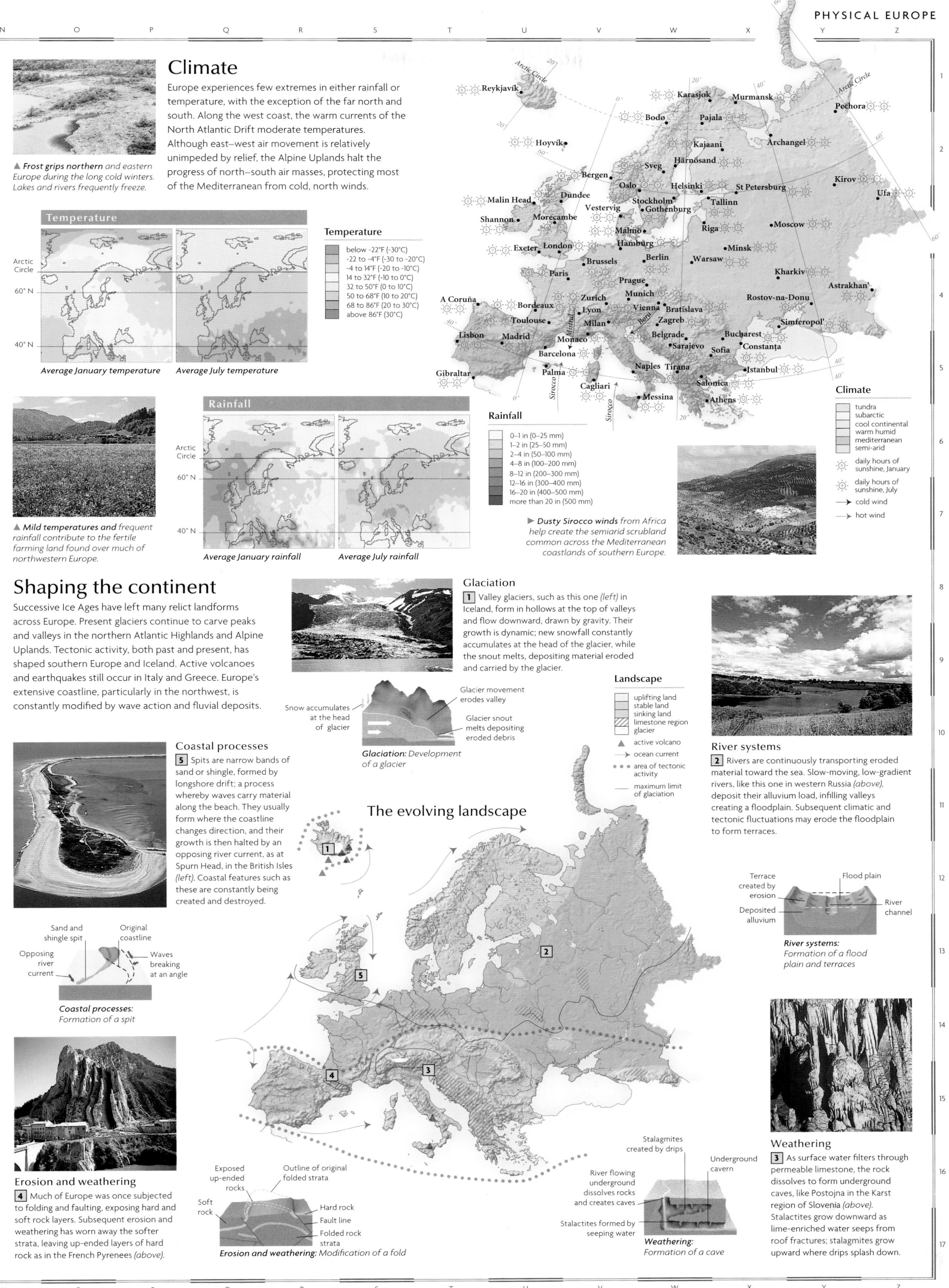

Shaping the continent

Successive Ice Ages have left many relict landforms across Europe. Present glaciers continue to carve peaks and valleys in the northern Atlantic Highlands and Alpine Uplands. Tectonic activity, both past and present, has shaped southern Europe and Iceland. Active volcanoes and earthquakes still occur in Italy and Greece. Europe's extensive coastline, particularly in the northwest, is constantly modified by wave action and fluvial deposits.

Glaciation

1 Valley glaciers, such as this one *(left)* in Iceland, form in hollows at the top of valleys and flow downward, drawn by gravity. Their growth is dynamic; new snowfall constantly accumulates at the head of the glacier, while the snout melts, depositing material eroded and carried by the glacier.

Snow accumulates at the head of glacier
Glacier movement erodes valley
Glacier snout melts depositing eroded debris

Glaciation: Development of a glacier

Landscape
- uplifting land
- stable land
- sinking land
- limestone region
- glacier
- ▲ active volcano
- ocean current
- •••• area of tectonic activity
- —— maximum limit of glaciation

River systems

2 Rivers are continuously transporting eroded material toward the sea. Slow-moving, low-gradient rivers, like this one in western Russia *(above)*, deposit their alluvium load, infilling valleys creating a floodplain. Subsequent climatic and tectonic fluctuations may erode the floodplain to form terraces.

Terrace created by erosion
Flood plain
Deposited alluvium
River channel

River systems: Formation of a flood plain and terraces

Coastal processes

5 Spits are narrow bands of sand or shingle, formed by longshore drift; a process whereby waves carry material along the beach. They usually form where the coastline changes direction, and their growth is then halted by an opposing river current, as at Spurn Head, in the British Isles *(left)*. Coastal features such as these are constantly being created and destroyed.

Sand and shingle spit
Original coastline
Opposing river current
Waves breaking at an angle

Coastal processes: Formation of a spit

The evolving landscape

Erosion and weathering

4 Much of Europe was once subjected to folding and faulting, exposing hard and soft rock layers. Subsequent erosion and weathering has worn away the softer strata, leaving up-ended layers of hard rock as in the French Pyrenees *(above)*.

Exposed up-ended rocks
Outline of original folded strata
Soft rock
Hard rock
Fault line
Folded rock strata

Erosion and weathering: Modification of a fold

Weathering

3 As surface water filters through permeable limestone, the rock dissolves to form underground caves, like Postojna in the Karst region of Slovenia *(above)*. Stalactites grow downward as lime-enriched water seeps from roof fractures; stalagmites grow upward where drips splash down.

Stalagmites created by drips
Underground cavern
River flowing underground dissolves rocks and creates caves
Stalactites formed by seeping water

Weathering: Formation of a cave

Political Europe

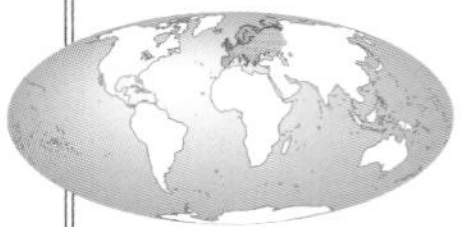

The political boundaries of Europe have changed many times, especially during the 20th century in the aftermath of two world wars, the breakup of the empires of Austria-Hungary, Nazi Germany and, toward the end of the century, the collapse of communism in eastern Europe. The fragmentation of Yugoslavia has again altered the political map of Europe, highlighting a trend toward nationalism and devolution. In contrast, economic federalism is growing. In 1958, the formation of the European Economic Community (now the European Union or EU) started a move toward economic and political union and increasing internal migration.

▲ *The Brandenburg Gate* in Berlin is a potent symbol of German reunification. From 1961, the road beneath it ended in a wall, built to stop the flow of refugees to the West. It was opened again in 1989 when the wall was destroyed and East and West Germany were reunited.

Population

Europe is a densely populated, urbanized continent; in Belgium over 90% of people live in urban areas. The highest population densities are found in an area stretching east from southern Britain and northern France, into Germany. The northern fringes are only sparsely populated.

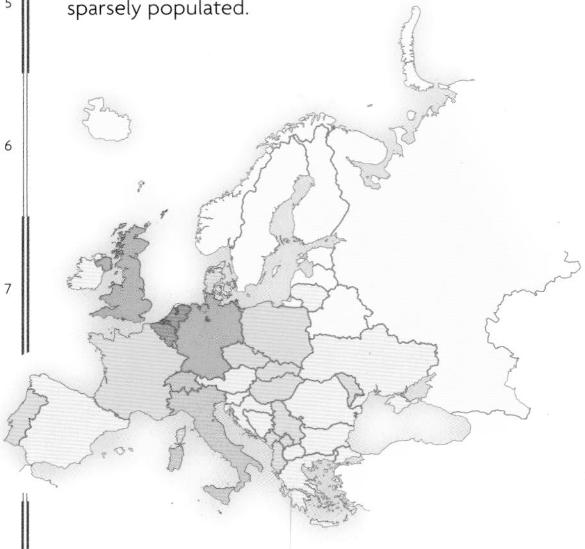

▲ *Demand for space* in densely populated European cities like London has led to the development of high-rise offices and urban sprawl.

Population density
(people per sq mile)

- below 130
- 130–259
- 260–379
- 380–519
- 520–780
- above 780

▲ *Traditional lifestyles still* persist in many remote and rural parts of Europe, especially in the south, east, and in the far north.

Map key

Population
- ■ above 5 million
- ▣ 1 million to 5 million
- ◉ 500,000 to 1 million
- ◎ 100,000 to 500,000
- ⊕ 50,000 to 100,000
- ○ 10,000 to 50,000
- ● Country capital

Borders
- ╱ full international border

Scale 1:15,500,000

Km
0 100 200 300 400 500 600 700

Miles
0 100 200 300 400 500 600 700

projection: Lambert Azimuthal Equal Area

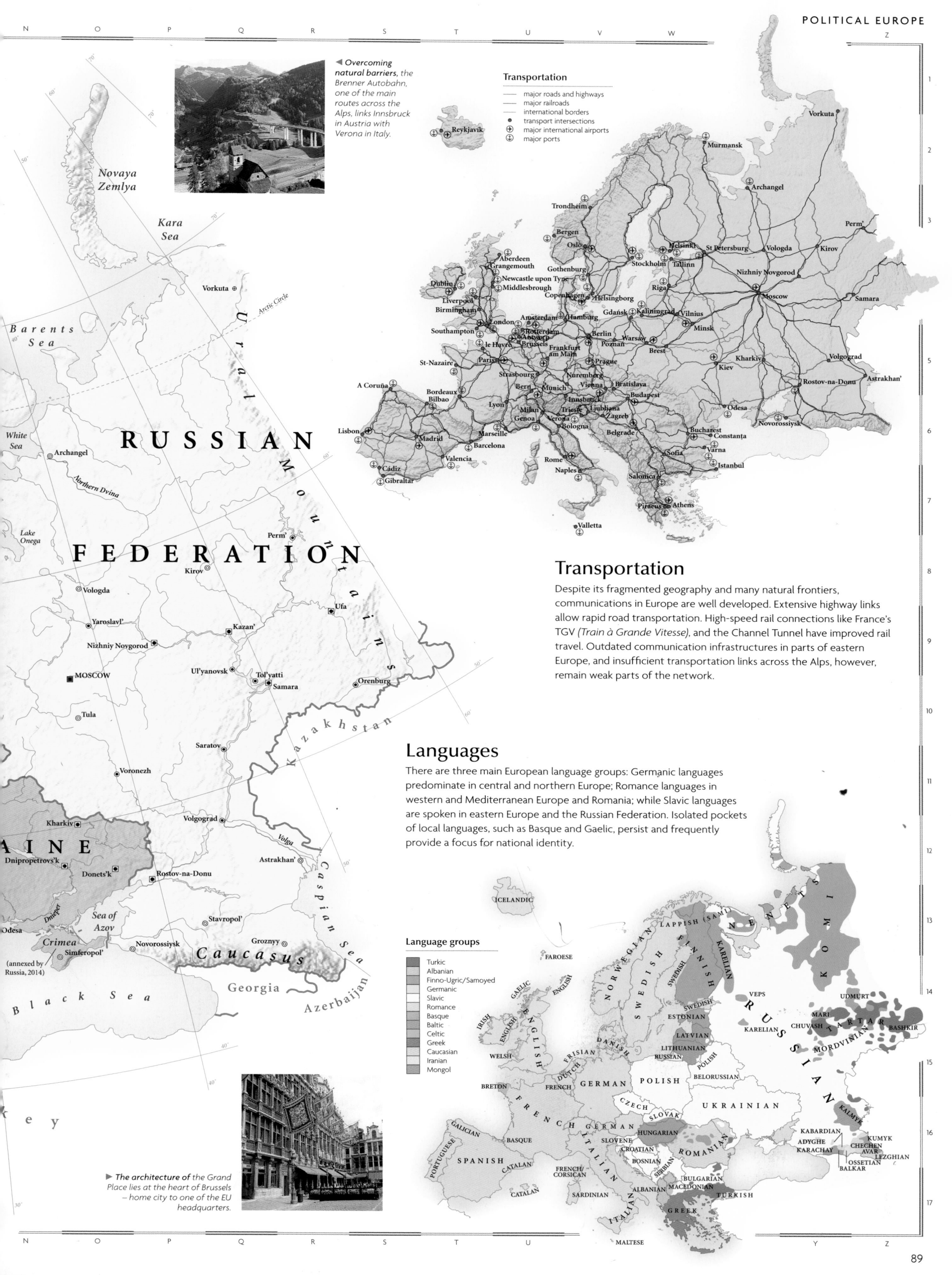

◄ Overcoming natural barriers, the Brenner Autobahn, one of the main routes across the Alps, links Innsbruck in Austria with Verona in Italy.

Transportation

Despite its fragmented geography and many natural frontiers, communications in Europe are well developed. Extensive highway links allow rapid road transportation. High-speed rail connections like France's TGV *(Train à Grande Vitesse)*, and the Channel Tunnel have improved rail travel. Outdated communication infrastructures in parts of eastern Europe, and insufficient transportation links across the Alps, however, remain weak parts of the network.

Languages

There are three main European language groups: Germanic languages predominate in central and northern Europe; Romance languages in western and Mediterranean Europe and Romania; while Slavic languages are spoken in eastern Europe and the Russian Federation. Isolated pockets of local languages, such as Basque and Gaelic, persist and frequently provide a focus for national identity.

► The architecture of the Grand Place lies at the heart of Brussels – home city to one of the EU headquarters.

European resources

Europe's large tracts of fertile, accessible land, combined with its generally
temperate climate, have allowed a greater percentage of land to be used for
agricultural purposes than in any other continent. Extensive coal and iron
ore deposits were used to create steel and manufacturing industries during
the 19th and 20th centuries. Today, although natural resources have been widely exploited,
and heavy industry is of declining importance, the growth of hi-tech and service industries
has enabled Europe to maintain its wealth.

Industry

Europe's wealth was generated by the rise of industry and colonial
exploitation during the 19th century. The mining of abundant
natural resources made Europe the industrial center of the world.
Adaptation has been essential in the changing world economy, and
a move to service-based industries has been widespread except in
eastern Europe, where heavy industry still dominates.

Standard of living

Living standards in western Europe are among the highest in the world, although
there is a growing sector of homeless, jobless people. Eastern Europeans have
lower overall standards of living – a legacy of stagnated economies.

Standard of living
(UN human development index)

- low
- high
- data not available

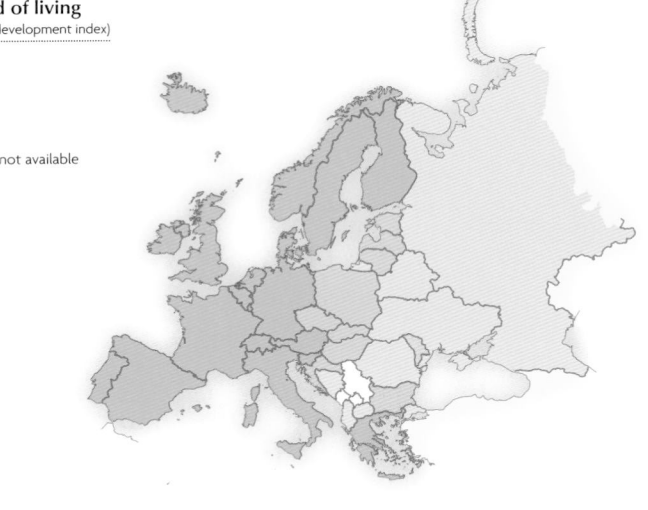

▲ *Countries like Hungary* are still
struggling to modernize inefficient
factories left over from extensive,
centrally-planned industrialization
during the communist era.

◄ *Frankfurt am Main* is an
example of a modern service-based
city. The skyline is dominated by
headquarters from the worlds of
banking and commerce.

▲ *Other power sources* are becoming more
attractive as fossil fuels run out; 16% of
Europe's electricity is now provided by
hydroelectric power.

► *Skiing brings millions*
of tourists to the slopes
each year, which means
that even unproductive,
marginal land is used to
create wealth in the
French, Swiss, Italian,
and Austrian Alps.

*Novaya
Zemlya*

*Ostrov
Kolguyev*

GNI per capita (US $)

- below 1999
- 2000–4999
- 5000–9999
- 10,000–19,999
- 20,000–24,999
- above 25,000

*Barents
Sea*

Murmansk

Archangel

RUSSIAN

FEDERATION

Perm'

Reykjavík
ICELAND

Faroe Islands
(to Denmark)

*Norwegian
Sea*

Trondheim

Bergen

Oslo

Glasgow

Belfast
IRELAND
Dublin
UNITED
Liverpool   KINGDOM
Manchester
Cardiff   Birmingham
London

Newcastle
upon Tyne

*North
Sea*

DENMARK

Gothenburg

Stockholm

Malmö

Copenhagen

Gulf of Bothnia

FINLAND

Turku

Helsinki
Tallinn
ESTONIA

St Petersburg

LATVIA

Riga

LITHUANIA

Vilnius

RUSS. FED.
(Kaliningrad)

Cherepovets

Yaroslavl'

Ivanovo

Moscow

Nizhniy Novgorod

Kazan'

Ufa

Tol'yatti

Samara

Baltic Sea

Hamburg

Gdańsk

POLAND

Poznań

Warsaw

Łódź

Minsk

BELARUS

Tula

Ryazan'

Saratov

Volgograd

KAZAKHSTAN

Amsterdam   NETH.
Rotterdam
Antwerp   Berlin
BELG.   Essen
Brussels   GERMANY
Liège   Cologne
Leipzig
Dresden

Channel
Islands

Rouen
Paris
LUX.
Metz   Frankfurt
am Main
Strasbourg
CZECH
REP.
Prague

Kraków

Katowice

Kiev

Kharkiv

Kursk

Voronezh

UKRAINE

Dnipropetrovs'k

Donets'k

Rostov-na-Donu

*Bay of
Biscay*

FRANCE

Nantes

Bordeaux

Lyon

Zürich
SWITZ.   LIECH.

Munich

Linz

AUSTRIA

Vienna

Bratislava

SLOVAKIA

Budapest

HUNGARY

ROMANIA

Kryvyy Rih

MOLDOVA

Odesa

*Caspian
Sea*

Stuttgart

A Coruña

Porto

Bilbao

Toulouse

Turin

Milan

Venice

SLVN.

Zagreb

Ploești

Bucharest

Constanța

PORTUGAL

SPAIN

Madrid

ANDORRA

Marseille

MONACO

Genoa

Bologna

ITALY

CROATIA

SAN
MARINO

BOSNIA
& HERZ.

SERBIA

Belgrade

MONT.

KOSOVO

Sofia

BULGARIA

Varna

Black Sea

GEORGIA

AZERBAIJAN

Lisbon

Seville

Gibraltar
(to UK)

Ceuta
(to Spain)

Melilla
(to Spain)

MOROCCO

Balearic Islands

Barcelona

Corsica

VATICAN
CITY
Rome

Sardinia

Adriatic Sea

MACED.

ALBANIA

GREECE

Istanbul

TURKEY

Salonica

Piraeus

Athens

*Tyrrhenian
Sea*

Naples

Taranto

Palermo

Sicily

MALTA

Mediterranean Sea

*Ionian
Sea*

*Aegean
Sea*

Crete

Industry

✈ aerospace	food processing	wine
brewing	hi-tech industry	coal
car/vehicle manufacture	iron & steel	oil
chemicals	pharmaceuticals	gas
defense	printing & publishing	
electronics	shipbuilding	• industrial cities
finance	textiles	major industrial areas
	timber processing	

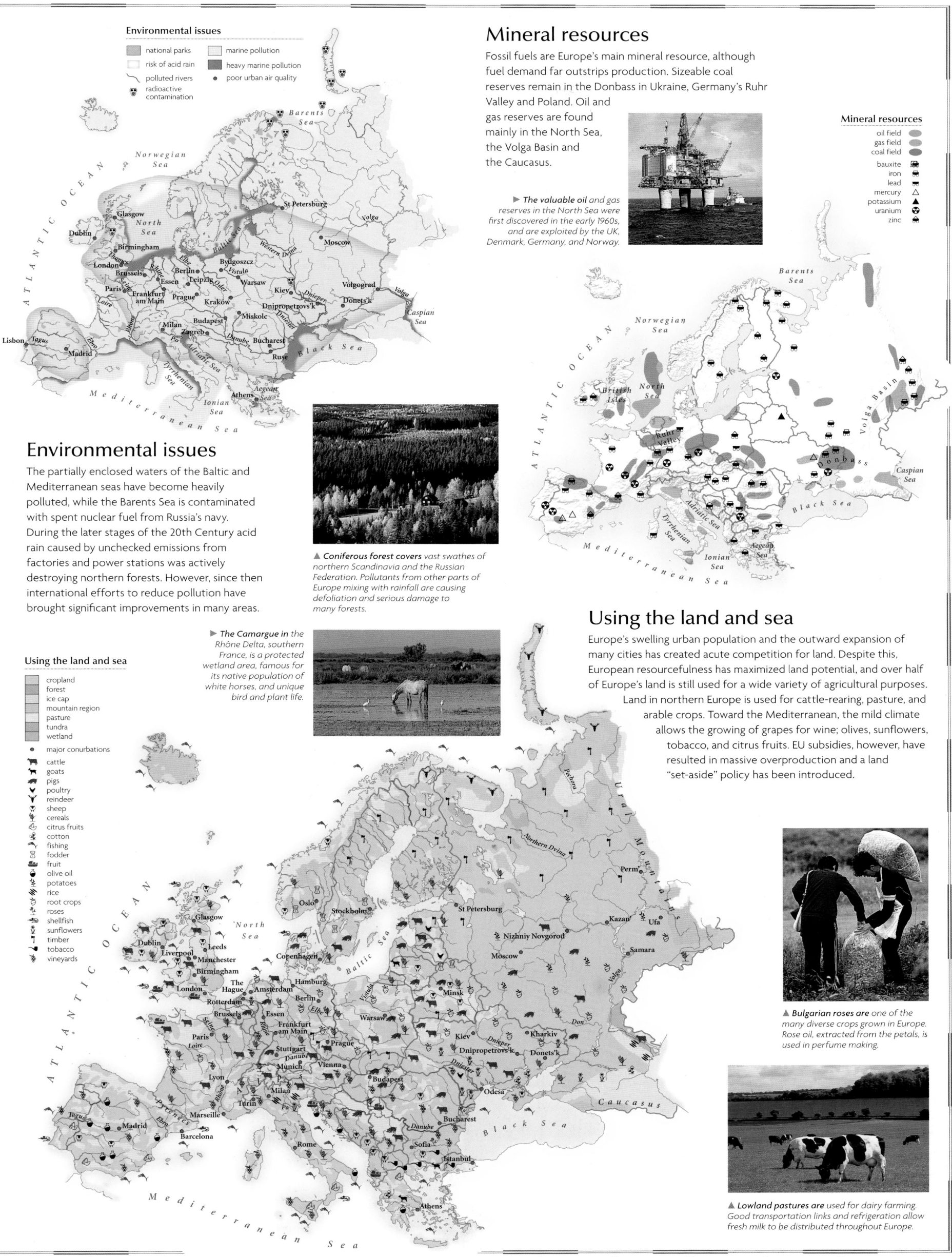

Environmental issues

Environmental issues
- national parks
- risk of acid rain
- polluted rivers
- radioactive contamination
- marine pollution
- heavy marine pollution
- poor urban air quality

Mineral resources

Fossil fuels are Europe's main mineral resource, although fuel demand far outstrips production. Sizeable coal reserves remain in the Donbass in Ukraine, Germany's Ruhr Valley and Poland. Oil and gas reserves are found mainly in the North Sea, the Volga Basin and the Caucasus.

► *The valuable oil and gas reserves in the North Sea were first discovered in the early 1960s, and are exploited by the UK, Denmark, Germany, and Norway.*

Mineral resources
- oil field
- gas field
- coal field
- bauxite
- iron
- lead
- mercury
- potassium
- uranium
- zinc

Environmental issues

The partially enclosed waters of the Baltic and Mediterranean seas have become heavily polluted, while the Barents Sea is contaminated with spent nuclear fuel from Russia's navy. During the later stages of the 20th Century acid rain caused by unchecked emissions from factories and power stations was actively destroying northern forests. However, since then international efforts to reduce pollution have brought significant improvements in many areas.

▲ *Coniferous forest covers vast swathes of northern Scandinavia and the Russian Federation. Pollutants from other parts of Europe mixing with rainfall are causing defoliation and serious damage to many forests.*

► *The Camargue in the Rhône Delta, southern France, is a protected wetland area, famous for its native population of white horses, and unique bird and plant life.*

Using the land and sea

Europe's swelling urban population and the outward expansion of many cities has created acute competition for land. Despite this, European resourcefulness has maximized land potential, and over half of Europe's land is still used for a wide variety of agricultural purposes. Land in northern Europe is used for cattle-rearing, pasture, and arable crops. Toward the Mediterranean, the mild climate allows the growing of grapes for wine; olives, sunflowers, tobacco, and citrus fruits. EU subsidies, however, have resulted in massive overproduction and a land "set-aside" policy has been introduced.

Using the land and sea
- cropland
- forest
- ice cap
- mountain region
- pasture
- tundra
- wetland
- major conurbations
- cattle
- goats
- pigs
- poultry
- reindeer
- sheep
- cereals
- citrus fruits
- cotton
- fishing
- fodder
- fruit
- olive oil
- potatoes
- rice
- root crops
- roses
- shellfish
- sunflowers
- timber
- tobacco
- vineyards

▲ *Bulgarian roses are one of the many diverse crops grown in Europe. Rose oil, extracted from the petals, is used in perfume making.*

▲ *Lowland pastures are used for dairy farming. Good transportation links and refrigeration allow fresh milk to be distributed throughout Europe.*

91

Scandinavia, Finland & Iceland

DENMARK, NORWAY, SWEDEN, FINLAND, ICELAND

Jutting into the Arctic Circle, this northern swath of Europe has some of the continent's harshest environments, but benefits from great reserves of oil, gas, and natural evergreen forests. While most early settlers came from the south, migrants to Finland came from the east, giving it a distinct language and culture. Since the late 19th century, the Scandinavian states have developed strong egalitarian traditions. Today, their welfare benefits systems are among the most extensive in the world, and standards of living are high. The Lapps, or Sami, maintain their traditional lifestyle in the northern regions of Norway, Sweden, and Finland.

The landscape

Glaciers up to 10,000 ft (3000 m) deep covered most of Scandinavia and Finland during the last Ice Age. The effects of glaciation mark the entire landscape, from the mountains to the lowlands, across the tundra landscape of Lapland, and the lake districts of Sweden and Finland.

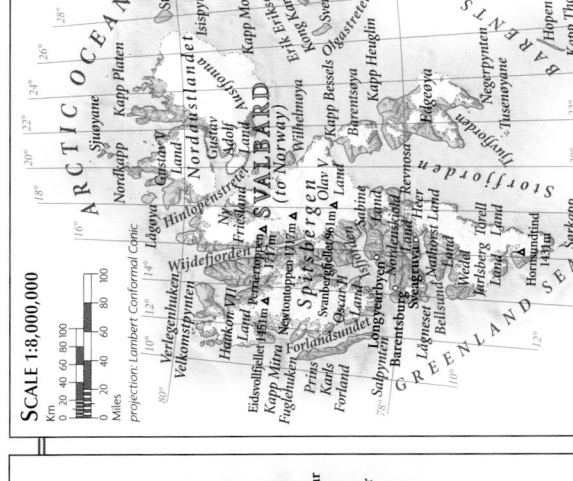

Geysers are a by-product of Iceland's volcanic activity. Geysir, Iceland's largest spring, gives them their name.

The Lofoten Islands were one of the first areas exposed as the ice sheet melted.

Halti Mountain is Finland's highest point, at 4356 ft (1328 m).

Lapland, north of the Arctic Circle, is an area of undulating fells and plains known as tundra. The subsoil is permanently frozen and therefore impermeable. There are many peat bogs. Pools reappear in the summer when the surface thaws.

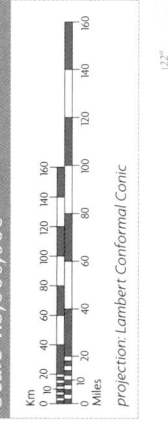

▼ Finland's landscape was fashioned by ice action. Glaciers gouged out its distinctive shallow lake basins, such as Oulujärvi, and left debris called moraines in their wake.

Oulujärvi

Area of maximum yearly uplift 0.3 in/yr (9 mm/yr)

Slower rates of uplift 0.1 in/yr (3 mm/yr)

▲ Scandinavia is still recovering from the last Ice Age, when ice depressed the land by 2000 ft (600 m). This gradual uplift is known as isostatic rebound.

Sjælland coast

▲ On the coast of Sjælland, these cliffs have been eroded by the sea, exposing layers of chalk and limestone.

Fjords

▲ The fjords on the western coast of Norway were once gentle river valleys. Their deep floors and steep sides were carved out by glaciers during the last Ice Age, and they were later flooded by the sea.

Using the land & sea

The cold climate, short growing season, poorly developed soil, steep slopes, and exposure to high winds across northern regions means that most agriculture is concentrated, with the population, in the south. Most of Finland and much of Norway and Sweden are covered by dense forests of pine, spruce, and birch, which supply the timber industries.

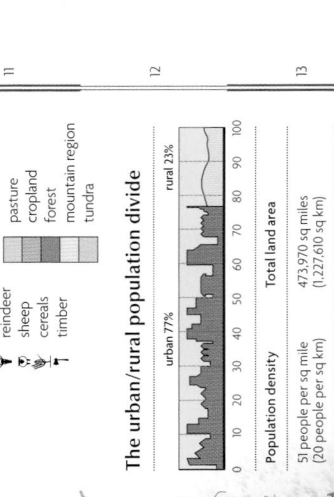

Land use and agricultural distribution

- fishing
- pigs
- reindeer
- sheep
- timber

- capital cities
- major towns

- pasture
- cropland
- forest
- mountain region
- tundra

The urban/rural population divide

urban 77%   rural 23%

Population density
51 people per sq mile
(20 people per sq km)

Total land area
473,970 sq miles
(1,227,610 sq km)

SCALE 1:8,000,000

projection: Lambert Conformal Conic

Scale 1:5,000,000

projection: Lambert Conformal Conic

(same scale as main map)

▲ **Sweden is one of** the world's largest producers of wood and wood-based products. The traditional movement of logs by floating them down rivers has now been largely replaced by the use of trucks.

Map key

Population

- ◉ 1 million to 5 million
- ◎ 500,000 to 1 million
- ⊕ 100,000 to 500,000
- ◦ 50,000 to 100,000
- ○ 10,000 to 50,000
- · below 10,000

Elevation

- 2000m / 6562ft
- 1000m / 3281ft
- 500m / 1640ft
- 250m / 820ft
- 100m / 328ft
- sea level

Transportation & industry

Norway derives its premier industry, the production of oil and gas, from the North Sea, while Denmark exploits its own oil and gas reserves. Hydroelectric power is a major industry, particularly in Sweden and Iceland. Timber processing remains significant in Finland and Sweden, but metal and engineering industries are increasingly important. In Iceland, fish products are the main source of export earnings.

Transportation network

226,735 miles (364,936 km)	
2042 miles (3286 km)	
13,704 miles (22,057 km)	
6,661 miles (10,721 km)	

Although roads now reach most areas, the railroads are markedly less developed. Much of the north is not served by rail and must rely on air and sea services for long distance travel and freight transportation.

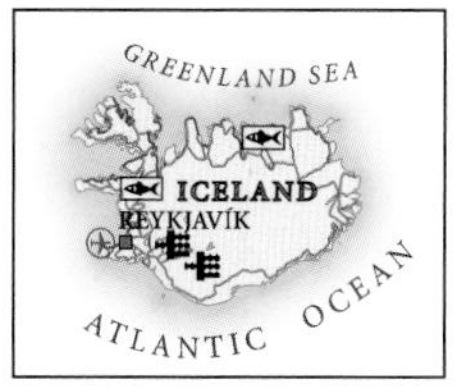

Major industry and infrastructure

- 🏭 car manufacture
- ⚙ engineering
- 🐟 fish processing
- ⚡ hydroelectric power
- ☢ nuclear power
- 🌲 timber processing
- ⛽ oil & gas
- ■ capital cities
- ● major towns
- ✈ international airports
- — major roads
- ▒ major industrial areas

▲ **The use of** geothermal power in Iceland began half a century ago. Today geothermal power stations supply 89% of the country's domestic heating requirements.

▲ **Many Lappish people**, in addition to traditional reindeer herding, now also make their living from fishing and farming, or working in cities. Tourism provides some with an extra source of income.

RUSSIAN FEDERATION

ARCTIC OCEAN

FINLAND

SWEDEN

NORWAY

ICELAND REYKJAVIK

ATLANTIC OCEAN

GREENLAND SEA

DENMARK COPENHAGEN

GERMANY

NORTH SEA

NORWEGIAN SEA

BALTIC SEA

HELSINKI

STOCKHOLM

OSLO

Gulf of Finland

HELSINKI (HELSINGFORS)

FINLAND

Tampere

Turku (Åbo)

ÅLAND

Mariehamn (Maarianhamina)

STOCKHOLM

Uppsala

Örebro

Gävle

GOTLAND

Visby

Gotland

ÖLAND

BALTIC SEA

Oulu (Uleåborg)

POHJOIS-POHJANMAA

KAINUU

Vaasa (Vasa)

Göteborg (Gothenburg)

VÄSTERBOTTEN

NORRBOTTEN

JÄMTLAND

Östersund

S W E D E N

DALARNA

VÄRMLAND

Örebro

Jönköping

Linköping

Norrköping

SKÅNE

Malmö

Helsingborg

BLEKINGE

Karlskrona

KRONOBERG

KALMAR

Kalmar

HALLAND

Halmstad

Bornholm

Trondheim

NORD-TRØNDELAG

SØR-TRØNDELAG

N O R W A Y

OPPLAND

HEDMARK

OSLO

AKERSHUS

BUSKERUD

TELEMARK

VESTFOLD

ØSTFOLD

VEST-AGDER

AUST-AGDER

ROGALAND

Stavanger

Bergen

HORDALAND

SOGN OG FJORDANE

MØRE OG ROMSDAL

Ålesund

Kristiansund

Kristiansand

N O R W E G I A N   S E A

N O R T H   S E A

Skagerrak

Kattegat

D E N M A R K

KØBENHAVN

Aalborg

Århus

Odense

GERMANY

100

93

Southern Scandinavia

SOUTHERN NORWAY, SOUTHERN SWEDEN, DENMARK

Scandinavia's economic and political hub is the more habitable and accessible southern region. Many of the area's major cities are on the southern coasts, including Oslo and Stockholm, the capitals of Norway and Sweden. In Denmark, most of the population and the capital, Copenhagen, are located on its many islands. A cultural unity links the three Scandinavian countries. Their main languages, Danish, Swedish, and Norwegian, are mutually intelligible, and they all retain their monarchies, although the parliaments have legislative control.

Using the land

Agriculture in southern Scandinavia is highly mechanized although farms are small. Denmark is the most intensively farmed country and its western pastureland is used mainly for pig farming. Cereal crops including wheat, barley, and oats, predominate in eastern Denmark and in the far south of Sweden. Southern Norway, and Sweden have large tracts of forest which are exploited for logging.

The urban/rural population divide

urban 87%   rural 13%

Population density
112 people per sq mile
(43 people per sq km)

Total land area
173,487 sq miles
(456,564 sq km)

Land use and agricultural distribution

- cattle
- pigs
- sheep
- cereals
- fodder
- root crops
- timber
- capital cities
- major towns
- pasture
- cropland
- forest
- mountain region

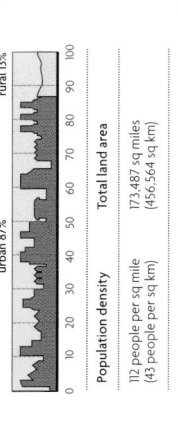

The landscape

Southern Scandinavia, with the exception of Norway, has a flatter terrain than the rest of the region. Denmark and southern Sweden are both extensions of the North European Plain. In this area, because of glacial deposition rather than erosion, the soils are deeper and more fertile.

Acid rain, caused by industrial pollution carried north from elsewhere in Europe, harms plant and animal life in Scandinavian forests and lakes. The region's surface rocks lack lime to neutralize the acid, so making the problem more serious.

▲ **In the past,** glaciers such as this one in Olden, Norway, were much larger. Today, many are retreating to yield the spectacular glacial scenery.

Distinctive low ridges, called eskers, are found across southern Sweden. They are formed from sand and gravel deposits left by retreating glaciers.

▲ **Limestone pillars eroded** by the sea dot the coast of Gotland and surrounding islands.

The lakes of southern Sweden remain from a period when the land was completely flooded. As the ice which covered the area melted, the land rose, leaving lakes in shallow, ice-scoured depressions. Sweden has over 90,000 lakes.

The peak of Glittertind in the Jotunheimen mountains is 8110 ft (2472 m) high.

Vänern in Sweden is the largest lake in Scandinavia. It covers an area of 2080 sq miles (5390 sq km).

Denmark's flat and fertile soils are formed on glacial deposits between 100–160 ft (30–50 m) deep.

When the ice retreated the valley was flooded by the sea

Old valley floor

Sea level

Erosion by glaciers deepened existing river valleys

Sognefjorden

▲ **Sognefjorden is the deepest of** Norway's many fjords. It drops to 4291 ft (1308 m) below sea level.

Map key

Population
- ■ 1 million to 5 million
- ⊙ 500,000 to 1 million
- ⊕ 100,000 to 500,000
- ◌ 50,000 to 100,000
- ○ 10,000 to 50,000
- ∘ below 10,000

Elevation
- 2000m / 6562ft
- 1000m / 3281ft
- 500m / 1640ft
- 250m / 820ft
- 100m / 328ft

Scale 1:2,900,000

projection: Lambert Conformal Conic

Km
Miles

▲ **In Norway winters** are longer and colder inland than in coastal areas, where the warm current of the North Atlantic Drift moderates the climate.

Gulf of Bothnia

NORWEGIAN SEA

NORTH SEA

BALTIC SEA

NORWAY
SWEDEN
DENMARK

STOCKHOLM
Uppsala
Örebro
OSLO
Göteborg
Trondheim
Bergen
Malmö
COPENHAGEN
Ålborg
Odense
GERMANY

VÄSTERNORRLAND
GÄVLEBORG
DALARNA
JÄMTLAND
HEDMARK
OPPLAND
NORD-TRØNDELAG
SØR-TRØNDELAG
MØRE OG ROMSDAL
SOGN OG FJORDANE
NORWAY

Trondheim
Olden

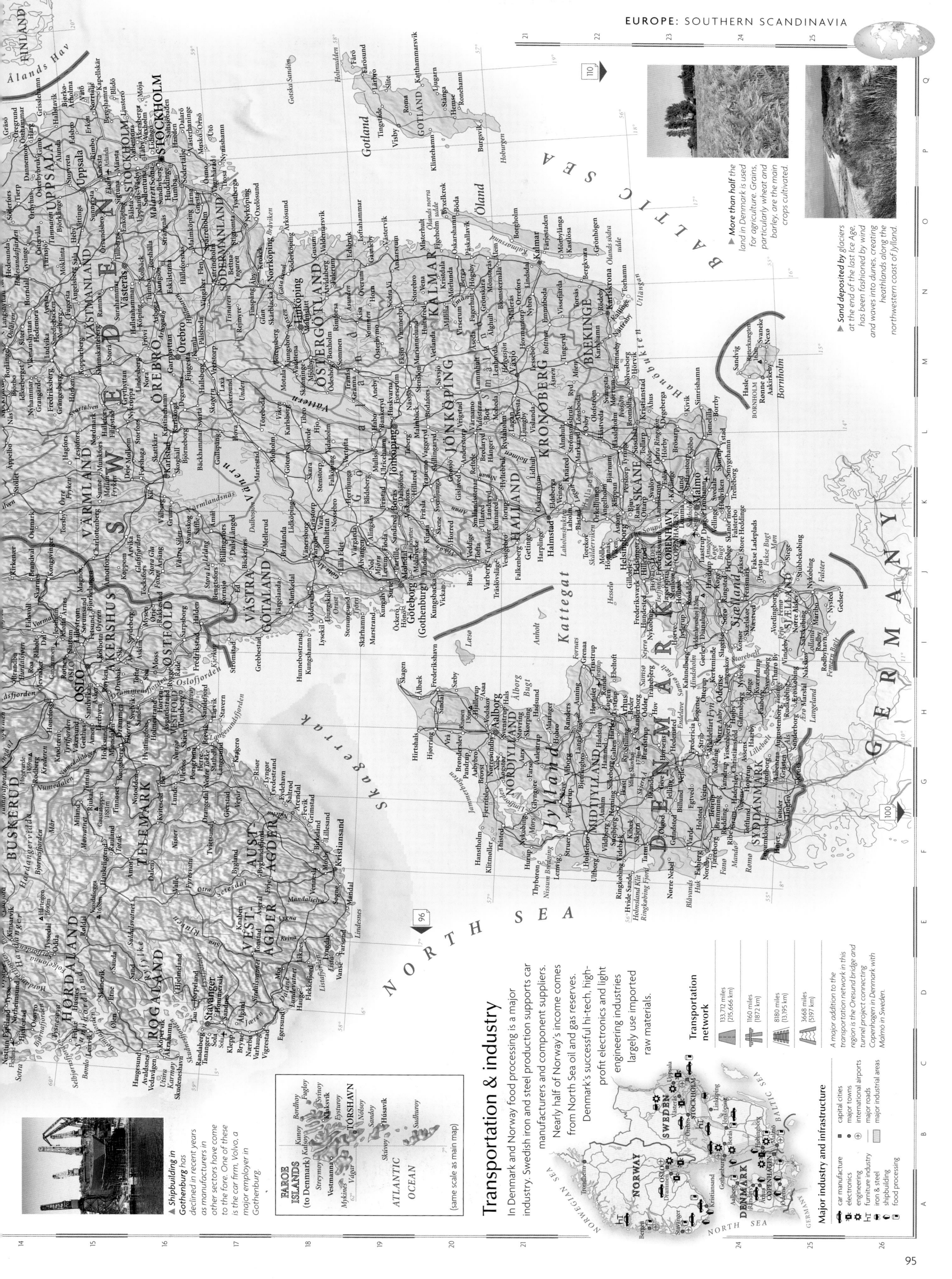

▲ More than half the land in Denmark is used for agriculture. Grains, particularly wheat and barley, are the main crops cultivated.

▲ Sand deposited by glaciers at the end of the last Ice Age, has been fashioned by wind and waves into dunes, creating heathlands along the northwestern coast of Jylland.

▲ Shipbuilding in Gothenburg has declined in recent years as manufacturers in other sectors have come to the fore. One of these is the car firm, Volvo, a major employer in Gothenburg.

Transportation & industry

In Denmark and Norway food processing is a major industry. Swedish iron and steel production supports car manufacturers and component suppliers. Nearly half of Norway's income comes from North Sea oil and gas reserves. Denmark's successful hi-tech, high-profit electronics and light engineering industries largely use imported raw materials.

Transportation network

133,712 miles (215,666 km)

1160 miles (1872 km)

8180 miles (13,195 km)

3668 miles (5197 km)

A major addition to the transportation network in this region is the Oresund bridge and tunnel project connecting Copenhagen in Denmark with Malmo in Sweden.

Major industry and infrastructure

car manufacture
electronics
engineering
furniture industry
iron & steel
shipbuilding
food processing

■ capital cities
▪ major towns
✈ international airports
major roads
major industrial areas

FAROE ISLANDS
(to Denmark)

(same scale as main map)

95

EUROPE

The British Isles

UNITED KINGDOM, IRELAND

The British Isles have for centuries played a central role in European and world history. England, Wales, Scotland, and Northern Ireland together form the United Kingdom (UK), while the southern portion of Ireland is an independent country, self-governing since 1921. Although England has tended to be the politically and economically dominant partner in the UK, the Scots, Welsh, and Irish maintain independent cultures, distinct national identities and languages. Southeastern England is the most densely populated part of this crowded region, with over eight million people living in and around the London area.

Transportation & industry

The British Isles' industrial base was founded primarily on coal, iron, and textiles, based largely in the north. Today, the most productive sectors include hi-tech industries clustered mainly in southeastern England, chemicals, finance, and the service sector, particularly tourism.

Major industry and infrastructure

- car manufacture
- chemicals
- engineering
- hi-tech industry
- iron & steel
- tourism
- capital cities
- major towns
- international airports
- major roads
- major industrial areas

Transportation network

285,947 miles (460,240 km)	2023 miles (3578 km)
11,825 miles (19,032km)	3976 miles (6400 km)

The UK's congested roads have become a major focus of environmental concern in recent years. No longer an island, the UK was finally linked to continental Europe by the Channel Tunnel in 1994.

▼ Clew Bay in western Ireland is characteristic of the heavily indented west coast, where deep wide-mouthed bays separate the mountains of Mayo, Donegal, and Kerry as they thrust out into the Atlantic Ocean.

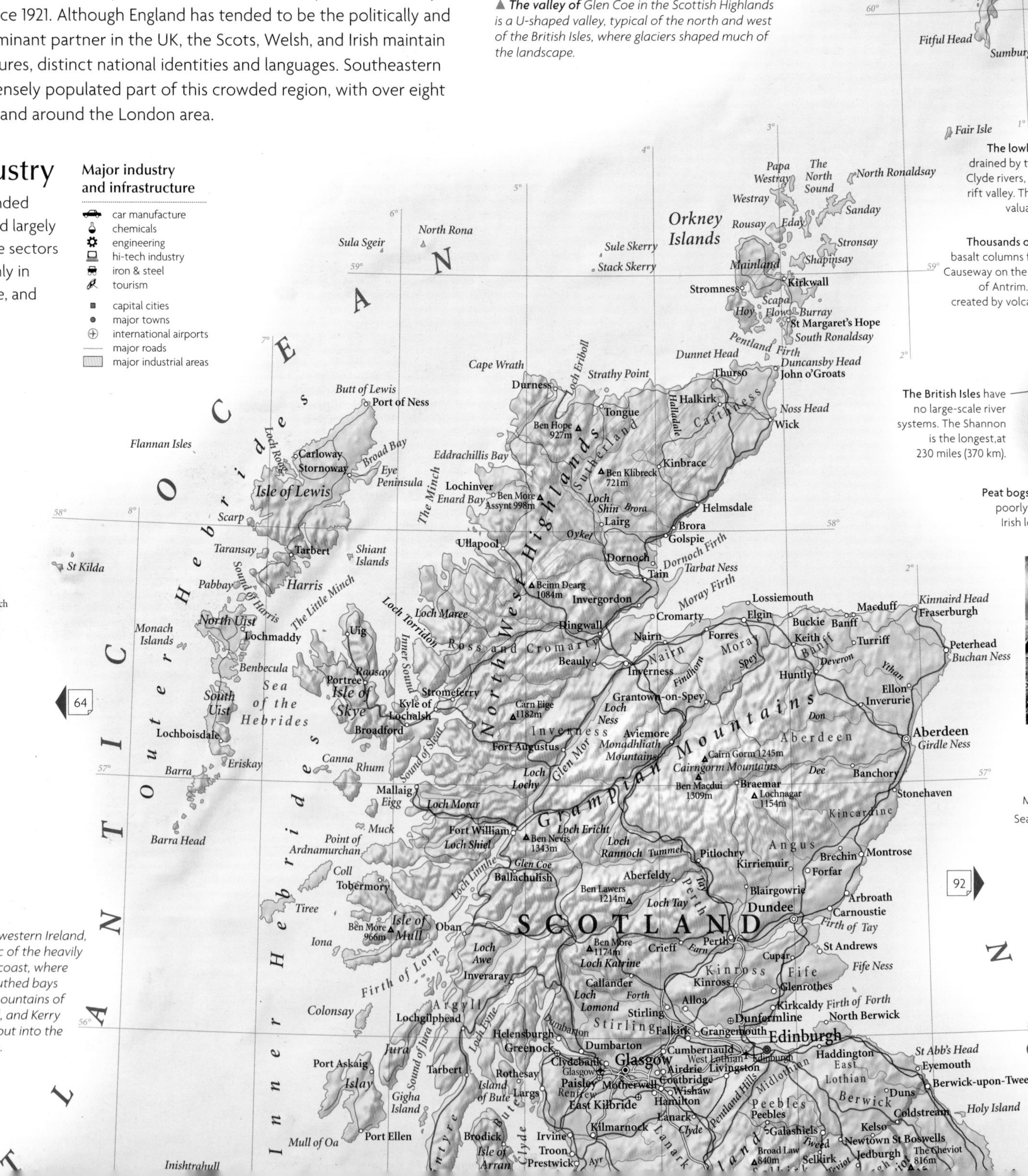

The landscape

Rugged uplands dominate the landscape of Scotland, Wales, and northern England. All the peaks in the British Isles over 4000 ft (1219 m) lie in highland Scotland. Lowland England rises in several ranges of rolling hills, including the older Mendips, and the Cotswolds and the Chilterns, which were formed at the same time as the Alps in southern Europe.

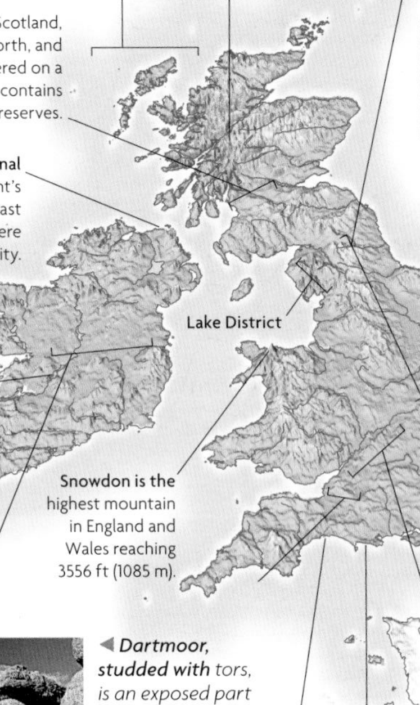

▲ The valley of Glen Coe in the Scottish Highlands is a U-shaped valley, typical of the north and west of the British Isles, where glaciers shaped much of the landscape.

▲ Ullswater in the Lake District fills a deep valley formed by glacial erosion.

The Fens are a low-lying area reclaimed from the sea.

The Cotswold Hills are characterized by a series of limestone ridges overlooking clay vales.

Chiltern Hills

The Pennines, sometimes called "the backbone of England," are formed of limestones and grits.

Ben Nevis at 4409 ft (1343 m) is the highest peak in the UK.

Over 600 islands, mostly uninhabited, lie west and north of the Scottish mainland.

Snowdon is the highest mountain in England and Wales reaching 3556 ft (1085 m).

The lowlands of Scotland, drained by the Tay, Forth, and Clyde rivers, are centered on a rift valley. The region contains valuable coal reserves.

Thousands of hexagonal basalt columns form Giant's Causeway on the north coast of Antrim. These were created by volcanic activity.

The British Isles have no large-scale river systems. The Shannon is the longest at 230 miles (370 km).

Peat bogs dot the poorly-drained Irish lowlands.

▲ Coastal erosion around the British Isles forms striking features such as this limestone arch, Durdle Door in Dorset.

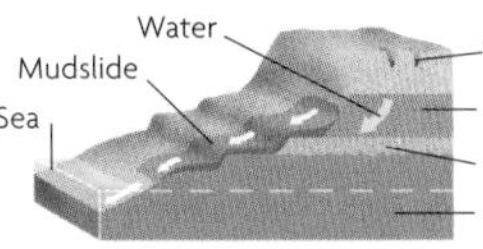

Durdle Door

▼ Dartmoor, studded with tors, is an exposed part of a vast granite dome, formed when molten rock intruded into the Earth's crust.

▲ Much of the south coast is subject to landslides. Following rain, water into the underlying, less permeable clays which then crumble and slide into the sea.

Black Ven, Lyme Regis

Cracks
Sandstone
Clay
Limestone
Water
Mudslide
Sea

Map key

Population
- above 5 million
- 1 million to 5 million
- 500,000 to 1 million
- 100,000 to 500,000
- 50,000 to 100,000
- 10,000 to 50,000
- below 10,000

Elevation
- 1000m / 328ft
- 500m / 1640ft
- 250m / 820ft
- 100m / 328ft
- sea level

96

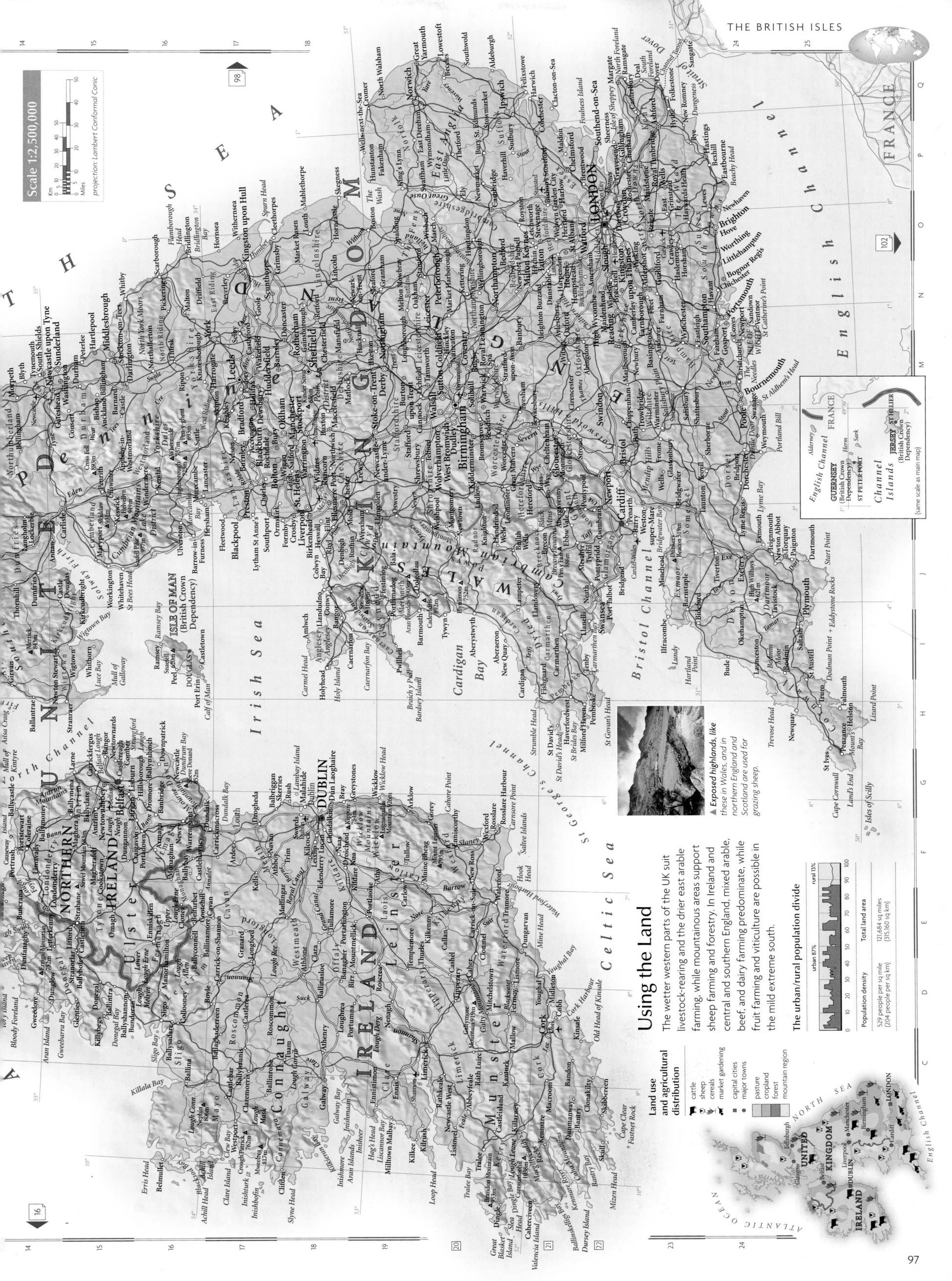

Using the Land

The wetter western parts of the UK suit livestock-rearing and the drier east arable farming, while mountainous areas support sheep farming and forestry. In Ireland and central and southern England, mixed arable, beef, and dairy farming predominate, while fruit farming and viticulture are possible in the mild extreme south.

▲ *Exposed highlands, like these in Wales, and in northern England and Scotland are used for grazing sheep.*

Land use and agricultural distribution

- cattle
- sheep
- cereals
- market gardening
- capital cities
- major towns

- pasture
- cropland
- forest
- mountain region

The urban/rural population divide

urban 87%   rural 13%

Population density	Total land area
529 people per sq mile (204 people per sq km)	121,684 sq miles (315,160 sq km)

Scale 1:2,500,000

projection: Lambert Conformal Conic

The Low Countries

BELGIUM, LUXEMBOURG, NETHERLANDS

One of northwestern Europe's strategic crossroads, the Low Countries are united by a common history in which they have often been a battleground in European wars. For over a thousand years they were ruled by foreign powers. Even after they achieved independence, the three countries maintained close links, later forming the world's first totally free labor and goods market, the Benelux Economic Union, which became the core of the European Community (now the European Union or EU). These states have remained at the forefront of wider European cooperation; Brussels, The Hague, and Luxembourg are hosts to major institutions of the EU.

The landscape

The main geographical regions of the Netherlands are the northern glacial heathlands, the low-lying lands of the Rhine and Maas/Meuse, the reclaimed polders, and the dune coast and islands. Belgium includes part of the Ardennes, together with the coalfields on its northern flanks, and the fertile Flanders plain.

Since the Middle Ages the people of the Netherlands have used ditches and drainage dikes to reclaim land from the sea. These reclaimed areas are known as polders.

▲ **Extensive sand dune** systems along the coast have prevented flooding of the land. Behind the dunes, marshy land is drained to form polders, usable land suitable for agriculture.

Sea
Polder
Dune system
Drainage ditch

The loess soils of the Flanders Plain in western Belgium provide excellent conditions for arable farming.

Sand dunes

▲ **Uplifted and folded** 220 million years ago, the Ardennes have since been reduced to relatively level plateaus, then sharply incised by rivers such as the Maas/Meuse.

Ardennes

Hautes Fagnes is the highest part of Belgium. The bogs and streams in this upland region result from high rainfall and low temperatures.

▼ **Heathlands, like these at** Schoorl, are found along the coast of the Netherlands. Much of the coast was breached by the sea in the 5th century, creating its distinctive inlets and islands.

Schoorl

▲ **One-third of the** Netherlands lies below sea level and flooding is a constant threat. Barrages have been built across the mouths of many rivers to contain floodwaters.

The parallel valleys of the Maas/Meuse and Rhine rivers were created when the Rhine was deflected from its previous course by the ice sheet which formed during the last Ice Age.

Silts and sands eroded by the Rhine throughout its course are deposited to form a delta on the west coast of the Netherlands.

Transportation & industry

In the western Netherlands, a massive, sprawling industrialized zone encompasses many new hi-tech and service industries. Belgium's central region has emerged as the country's light manufacturing and services center. Luxembourg city is home to more than 160 banks and the European headquarters of many international companies.

Transportation network

✈	2565 miles	(4129 km)
⚓	4134 miles	(6653 km)
🚂	140,588 miles	(226,281 km)
🏭	4099 miles	(6598 km)

The Low Countries hold a key position on the North Sea, containing Europe's two largest ports, Rotterdam and Antwerp, which are connected to a comprehensive system of inland waterways.

Major industry and infrastructure

- ✈ aerospace
- 💲 finance
- ⚙ engineering
- hi-tech industry
- pharmaceuticals
- textiles
- ■ capital cities
- ● major cities
- major towns
- ✈ international airports
- major roads
- major industrial areas

100

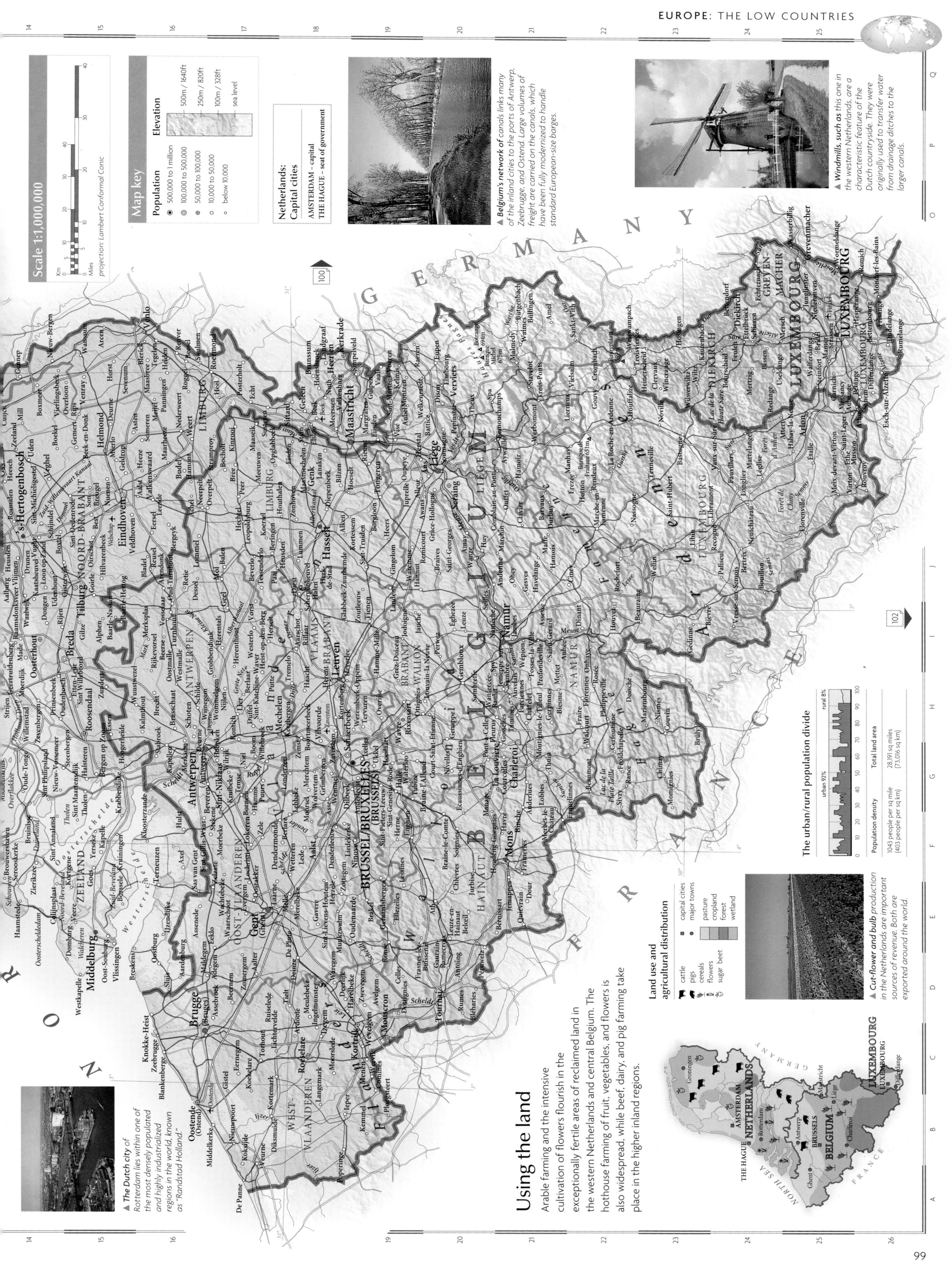

Scale 1:1,000,000

projection: Lambert Conformal Conic

Map key

Population
- ● 500,000 to 1 million
- ◉ 100,000 to 500,000
- ⊙ 50,000 to 100,000
- ○ 10,000 to 50,000
- ○ below 10,000

Elevation
- 500m / 1640ft
- 250m / 820ft
- 100m / 328ft
- sea level

Netherlands:
Capital cities

AMSTERDAM – capital
THE HAGUE – seat of government

▲ *Belgium's network of canals links many of the inland cities to the ports of Antwerp, Zeebrugge, and Ostend. Large volumes of freight are carried on the canals, which have been fully modernized to handle standard European-size barges.*

▲ *Windmills, such as this one in the western Netherlands, are a characteristic feature of the Dutch countryside. They were originally used to transfer water from drainage ditches to the larger canals.*

Using the land

Arable farming and the intensive cultivation of flowers flourish in the exceptionally fertile areas of reclaimed land in the western Netherlands and central Belgium. The hothouse farming of fruit, vegetables, and flowers is also widespread, while beef, dairy, and pig farming take place in the higher inland regions.

▲ *The Dutch city of Rotterdam lies within one of the most densely populated and highly industrialized regions in the world, known as "Randstad Holland."*

Land use and agricultural distribution
- ■ capital cities
- ● major towns
- cattle
- pigs
- cereals
- flowers
- sugar beet
- pasture
- cropland
- forest
- wetland

▲ *Cut-flower and bulb production in the Netherlands are important sources of revenue. Both are exported around the world.*

The urban/rural population divide

urban 92%   rural 8%

Population density	Total land area
1043 people per sq mile (403 people per sq km)	28,191 sq miles (73,016 sq km)

0 10 20 30 40 50 60 70 80 90 100

Germany

Despite the devastation of its industry and infrastructure during the Second World War and its separation from eastern Germany during the Cold War, West Germany made a rapid recovery in the following generation to become Europe's most formidable economic power. When the Berlin Wall was dismantled in 1989, the two halves of Germany were politically united for the first time in 40 years. Complete social and economic unity remain a longer term goal, as East German industry and society adapt to a free market. Germany has been a key player in the creation of the European Union (EU) and in moves toward a single European currency.

The landscape

The plains of northern Germany, the volcanic plateaus and mountains of the central uplands, and the Bavarian Alps are the three principal geographic regions in Germany. North to south the land rises steadily from barely 300 ft (90 m) in the plains to 6500 ft (2000 m) in the Bavarian Alps, which are a small but distinct region in the far south.

▲ *The heathlands of northern Germany are covered by glacial deposits of sandy outwash soil which makes them largely infertile. They support only sheep and solitary trees.*

Müritz lake covers 45 sq miles (117 sq km), but is only 108 ft (33 m) deep. It lies in a shallow valley formed by meltwater flowing out from a retreating ice sheet. These valleys are known as *Urstromtäler*.

Lüneburg Heath (*Lüneburger Heide*)

The Harz Mountains were formed 300 million years ago. They are block-faulted mountains, formed when a section of the Earth's crust was thrust up between two faults.

Elbe river

▼ *The Elbe flows in wide meanders across the north German plain to the North Sea. At its mouth it is 10 miles (16 km) wide.*

The Danube rises in the Black Forest (*Schwarzwald*) and flows east, across a wide valley on its course to the Black Sea.

Zugspitze, the highest peak in Germany at 9719 ft (2962 m), was formed during the Alpine mountain-building period, 30 million years ago.

Rhine Rift Valley

Much of the landscape of northern Germany has been shaped by glaciation. During the last Ice Age, the ice sheet advanced as far as the northern slopes of the central uplands.

▲ *Part of the floor of the Rhine Rift Valley was let down between two parallel faults in the Earth's crust.*

Fault lines
Rhine
Downfaulted block

The Rhine is Germany's principal waterway and one of Europe's longest rivers, flowing 820 miles (1320 km).

Scale 1:2,250,000

projection: Lambert Conformal Conic

Using the land

Germany has a large, efficient agricultural sector, and produces more than three-quarters of its own food. The major crops grown are cereals and sugar beet on the more fertile soils, and root crops, rye, oats, and fodder on the poorer soils of the northern plains and central uplands. Southern Germany is also a principal producer of high quality wines. Vineyards cover the slopes surrounding the Rhine and its tributaries.

Land use and agricultural distribution

- cattle
- pigs
- cereals
- sugar beet
- vineyards
- capital cities
- major towns
- pasture
- cropland
- forest

The urban/rural population divide

urban 87%
rural 13%

Population density	Total land area
612 people per sq mile (236 people per sq km)	137,804 sq miles (356,910 sq km)

▲ *The Moselle river flows through the Rhine State Uplands (Rheinisches Schiefergebirge). During a period of uplift, preexisting river meanders were deeply incised, to form its present dramatic contours.*

Map labels

POLAND
BALTIC SEA
Pomeranian Bay
BALTIC SEA
DENMARK
NORTH SEA
North Frisian Islands (Nordfriesische Inseln)
Helgoländer Bucht
Ostfriesische Inseln
NETHERLANDS

SCHLESWIG-HOLSTEIN
MECKLENBURG-VORPOMMERN
BRANDENBURG
NIEDERSACHSEN
BREMEN
BERLIN
Kieler Bucht
Mecklenburger Bucht

Rostock, Kiel, Lübeck, Hamburg, Bremen, Bremerhaven, Wilhelmshaven, Oldenburg, Hannover, Schwerin, Stralsund, Greifswald

92
96
98
110

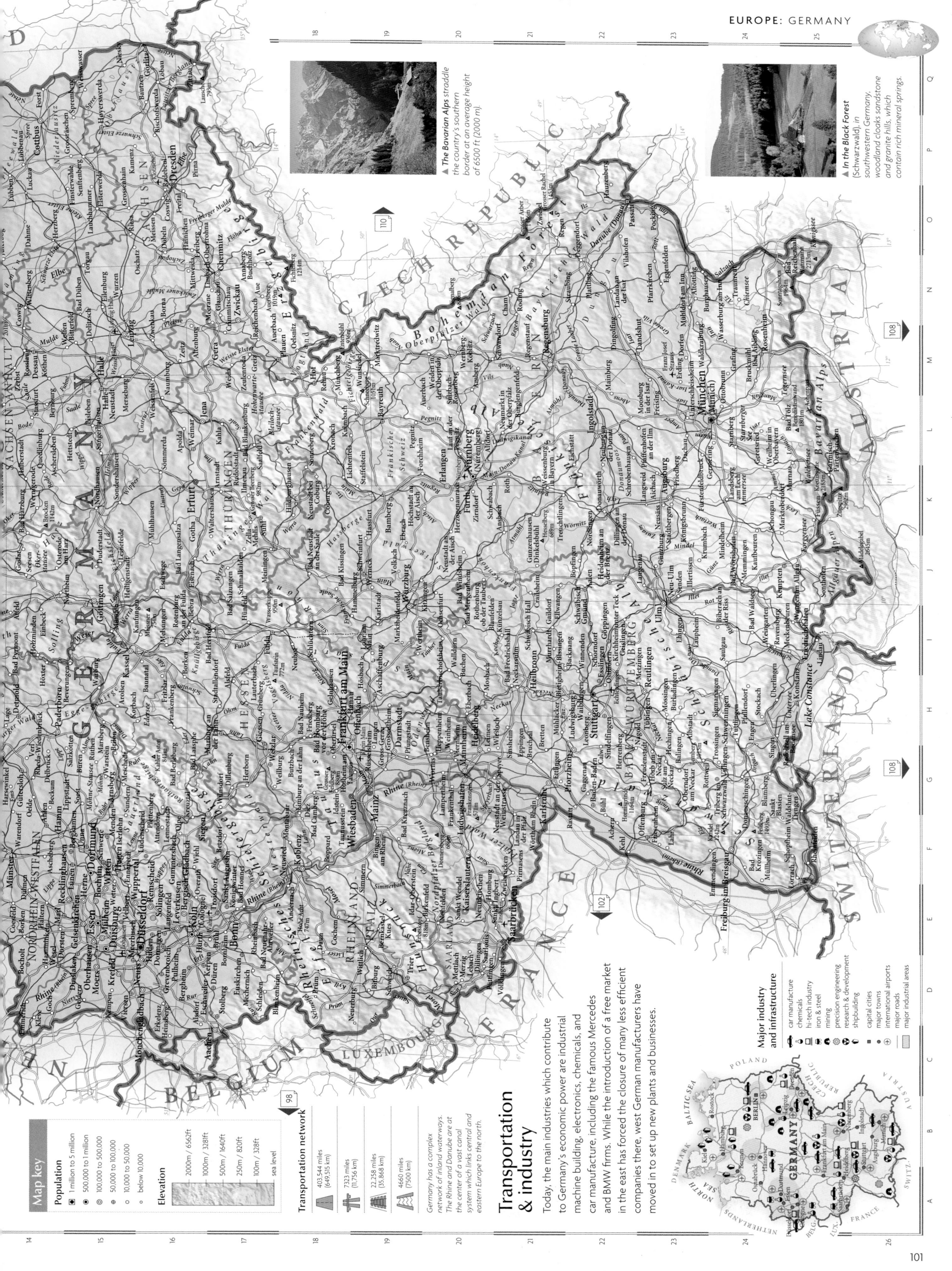

▲ The Bavarian Alps straddle the country's southern border at an average height of 6500 ft (2000m).

▲ In the Black Forest (Schwarzwald), in southwestern Germany, woodland cloaks sandstone and granite hills, which contain rich mineral springs.

Transportation & industry

Today, the main industries which contribute to Germany's economic power are industrial machine building, electronics, chemicals, and car manufacture, including the famous Mercedes and BMW firms. While the introduction of a free market in the east has forced the closure of many less efficient companies there, west German manufacturers have moved in to set up new plants and businesses.

Germany has a complex network of inland waterways. The Rhine and Danube are at the center of a vast canal system which links central and eastern Europe to the north.

Transportation network

▲	403,544 miles (649,515 km)	
✈	7323 miles (11,756 km)	
▥	22,258 miles (35,868 km)	
▲	4660 miles (7500 km)	

Map key

Population
- ◉ 1 million to 5 million
- ◎ 500,000 to 1 million
- ⊙ 100,000 to 500,000
- ⊕ 50,000 to 100,000
- ○ 10,000 to 50,000
- ∘ below 10,000

Elevation
- 2000m / 6562ft
- 1000m / 3281ft
- 500m / 1640ft
- 250m / 820ft
- 100m / 328ft
- sea level

Major industry and infrastructure

- car manufacture
- chemicals
- hi-tech industry
- iron & steel
- mining
- precision engineering
- research & development
- shipbuilding
- ▪ capital cities
- ■ major cities
- ∙ major towns
- ✈ international airports
- major roads
- major industrial areas

France

FRANCE, MONACO

Europe's second largest nation and the founder of modern Republican government, France is a major center of culture and fashion, and a leading producer of both agricultural and industrial goods. It has played a leading role in European events for centuries, and remains a key player in the push toward European unity. The Paris Basin is the most highly populated area; Île de France is home to over 11 million people. Large parts of France remain thinly populated, particularly the mountainous Massif Central, Pyrenees, and southern Alps.

◀ *The chalk cliffs of Normandy (Normandie) and southeastern England form part of a single geological region, now divided in two by the English Channel.*

The landscape

France's landscape was fashioned by two phases of mountain-building. The northwestern peninsula, the Massif Central, and the Vosges date from 220 million years ago. The complex folds of the Alps and Pyrenees, the gently-folded Jura, and the low-lying sedimentary areas of the Paris, Garonne, and Rhône basins started to form 65 million years ago.

The coast of Brittany (Bretagne) is highly indented where deep valleys in the northwestern peninsula were drowned by the sea.

The Normandy (Normandie) coastline is characterized by high chalk cliffs.

The coastline of France is 2141 miles (3427 km) long.

▲ *The Paris Basin consists of a layered sequence of sedimentary rocks. Fertile soils over much of the area make good agricultural land.*

The gently rounded summits of the Vosges are over 200 million years old.

The folded Jura form low ridges and long narrow valleys.

The Alps were forced up during several phases of mountain-building beginning 65 million years ago.

Garonne Basin

The Biscay coast, like the Mediterranean, is characterized by flat sandy beaches, interspersed with lagoons.

The Dordogne region contains spectacular examples of limestone scenery including caves and gorges.

The Pyrenees form a natural border between France and Spain.

The ancient Massif Central, disturbed by the formation of the Alps, was subject to volcanism that only ceased during the last 10,000 years.

Rhône Basin

Rhône Delta

Rhône

Delta plain

The marshes of the Camargue

Corsica's northeastern peninsula has dramatic cliffs of folded limestone.

◀ *The volcanic landscape of the Auvergne where the cones of its extinct volcanoes have worn away to leave "plugs" of lava.*

▲ *Deposition in the Rhône Delta is wave-dominated. Sea currents carry river sediments extending the delta plain westwards.*

Transportation & industry

Today the main French growth industries are hi-tech, including micro-electronics, telecommunications and aerospace. Other important sectors are the nuclear industry, only rivalled in scale by that of the US, car manufacture, dominated by the giants Renault and Peugeot, and a highly diversified tourist industry.

Major industry and infrastructure

- ✈ aerospace industry
- 🚗 car manufacture
- ⚗ chemicals
- ⚙ engineering
- 💻 hi-tech industry
- ⚛ nuclear power
- 🏖 tourism

- ■ capital cities
- ■ major towns
- ✈ international airports
- major roads
- major industrial areas

Transportation network

🛣 555,473 miles (894,050 km)		🛤 7305 miles (11,758 km)	
🚆 10,399 miles (16,737 km)		⛴ 1159 miles (1863 km)	

The French TGV (Train à Grande Vitesse) leads the world in high-speed train technology, and provides a service which can be faster, door-to-door, than air travel.

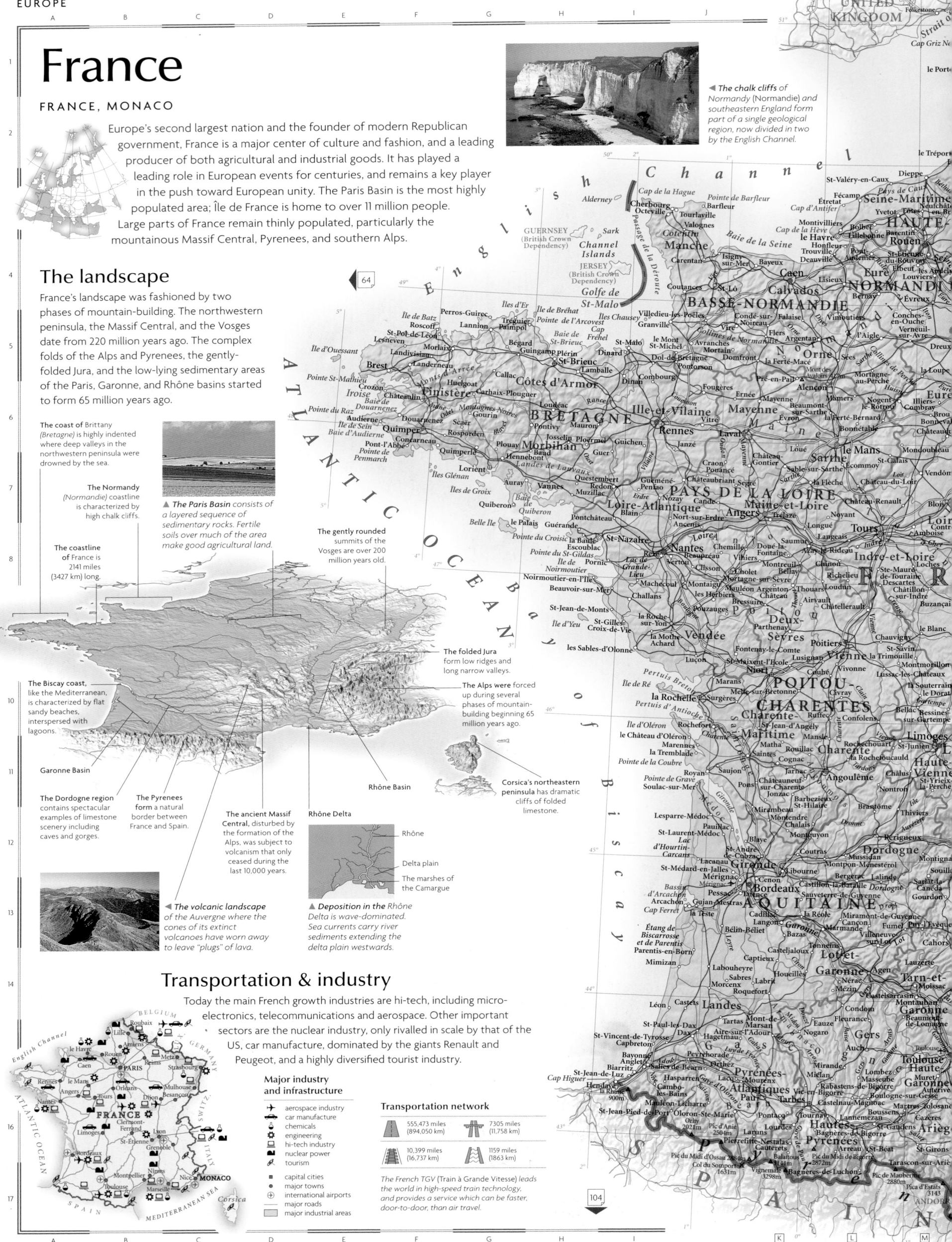

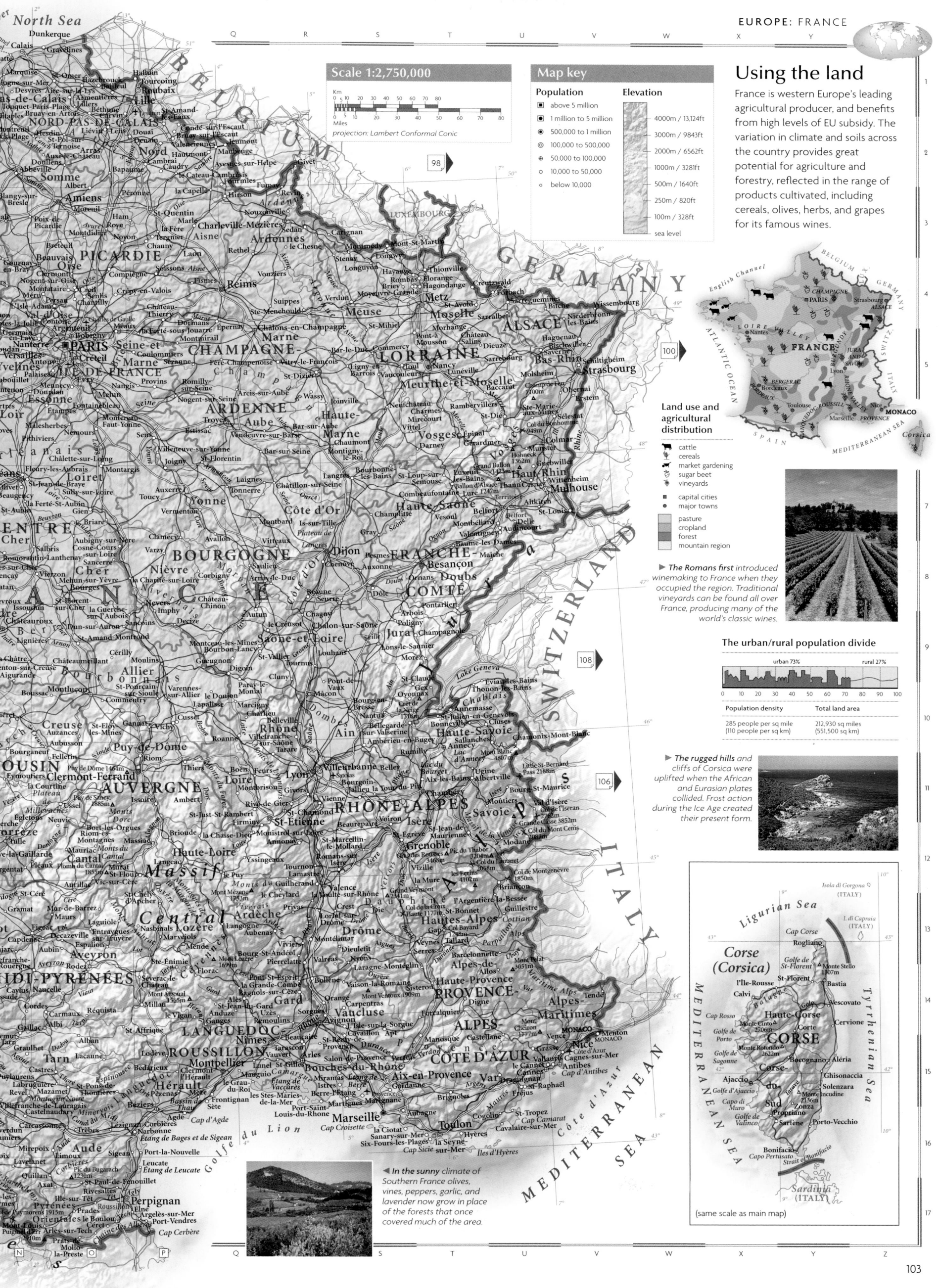

Using the land

France is western Europe's leading agricultural producer, and benefits from high levels of EU subsidy. The variation in climate and soils across the country provides great potential for agriculture and forestry, reflected in the range of products cultivated, including cereals, olives, herbs, and grapes for its famous wines.

Scale 1:2,750,000

Km
0 5 10   20   30   40   50   60   70   80
Miles
0 5 10   20   30   40   50   60   70   80

projection: Lambert Conformal Conic

Map key

Population
- ■ above 5 million
- ■ 1 million to 5 million
- ◉ 500,000 to 1 million
- ◎ 100,000 to 500,000
- ⊕ 50,000 to 100,000
- ⊙ 10,000 to 50,000
- ○ below 10,000

Elevation
- 4000m / 13,124ft
- 3000m / 9843ft
- 2000m / 6562ft
- 1000m / 3281ft
- 500m / 1640ft
- 250m / 820ft
- 100m / 328ft
- sea level

Land use and agricultural distribution

- cattle
- cereals
- market gardening
- sugar beet
- vineyards
- ■ capital cities
- ● major towns
- pasture
- cropland
- forest
- mountain region

▶ **The Romans first** introduced winemaking to France when they occupied the region. Traditional vineyards can be found all over France, producing many of the world's classic wines.

The urban/rural population divide

urban 73%   rural 27%

Population density	Total land area
285 people per sq mile (110 people per sq km)	212,930 sq miles (551,500 sq km)

▶ **The rugged hills** and cliffs of Corsica were uplifted when the African and Eurasian plates collided. Frost action during the Ice Age created their present form.

◀ **In the sunny climate of** Southern France olives, vines, peppers, garlic, and lavender now grow in place of the forests that once covered much of the area.

Corse (Corsica)

(same scale as main map)

The Iberian peninsula

ANDORRA, GIBRALTAR, PORTUGAL,
SPAIN (Azores, Canary Islands, Madeira on p.64)

The Iberian peninsula is separated from the rest of
Europe by the Pyrenees, and at its most southerly
point is only 5 miles (8 km) from North Africa.
The location of Iberia has been central to its
diverse history. The Greeks, Carthaginians, Romans,
Visigoths, and most recently the Moors, invaded
Iberia at various times. For much of the 20th century,
both Spain and Portugal were governed by right-wing
dictators. Since the establishment of democratic governments in the
mid-1970s, modernization has been rapid and both countries are now
among the most popular of European holiday destinations.

Using the land

The principal crops grown in Iberia are
cereals, especially wheat and barley. Both
countries are major wine producers, most
notably of Rioja, sherry, and port. Sheep
are kept throughout the region, and citrus
fruits thrive on the Mediterranean coast.
The successful forest industry in Iberia
produces 84% of the world's cork.

▲ The steep, terraced slopes of the
Douro Valley in northern Portugal,
are used to cultivate vines. The
grapes harvested produce
Portugal's famous port wine.

Land use and agricultural distribution

- sheep
- cereals
- citrus fruit
- olives
- vineyards
- cork
- ■ capital cities
- • major towns

- pasture
- cropland
- forest
- mountain region

The urban/rural population divide

urban 68%   rural 32%

0 10 20 30 40 50 60 70 80 90 100

Population density	Total land area
215 people per sq mile (83 people per sq km)	230,569 sq miles (597,170 sq km)

Transportation & industry

Since the 1970s, the economies of Spain and Portugal
have expanded and diversified. In both countries,
tourism has outstripped agriculture in economic
importance. Spain's resource base is varied, including
coal, iron, and the world's largest reserves of mercury.
Portugal is a leading producer of tungsten ore.

Major industry and infrastructure

- car manufacture
- chemicals
- engineering
- fish processing
- mining
- textiles
- tourism
- ■ capital cities
- • major towns
- ✈ international airports
- — major roads
- major industrial areas

Transportation network

241,720 miles (388,990 km)	1552 miles (2529 km)
11,793 miles (18,979 km)	1159 miles (1865 km)

Radiating from Madrid, the road network in
Spain dates from the 18th century, but now
includes many highways. Portugal's road
system has been completely modernized in
recent years.

◄ The eroded cliffs of the
Algarve in southern Portugal
were carved by Atlantic waves.
The numerous rocky bays and
beaches, and the region's
pleasant climate, have made it
a popular tourist destination.

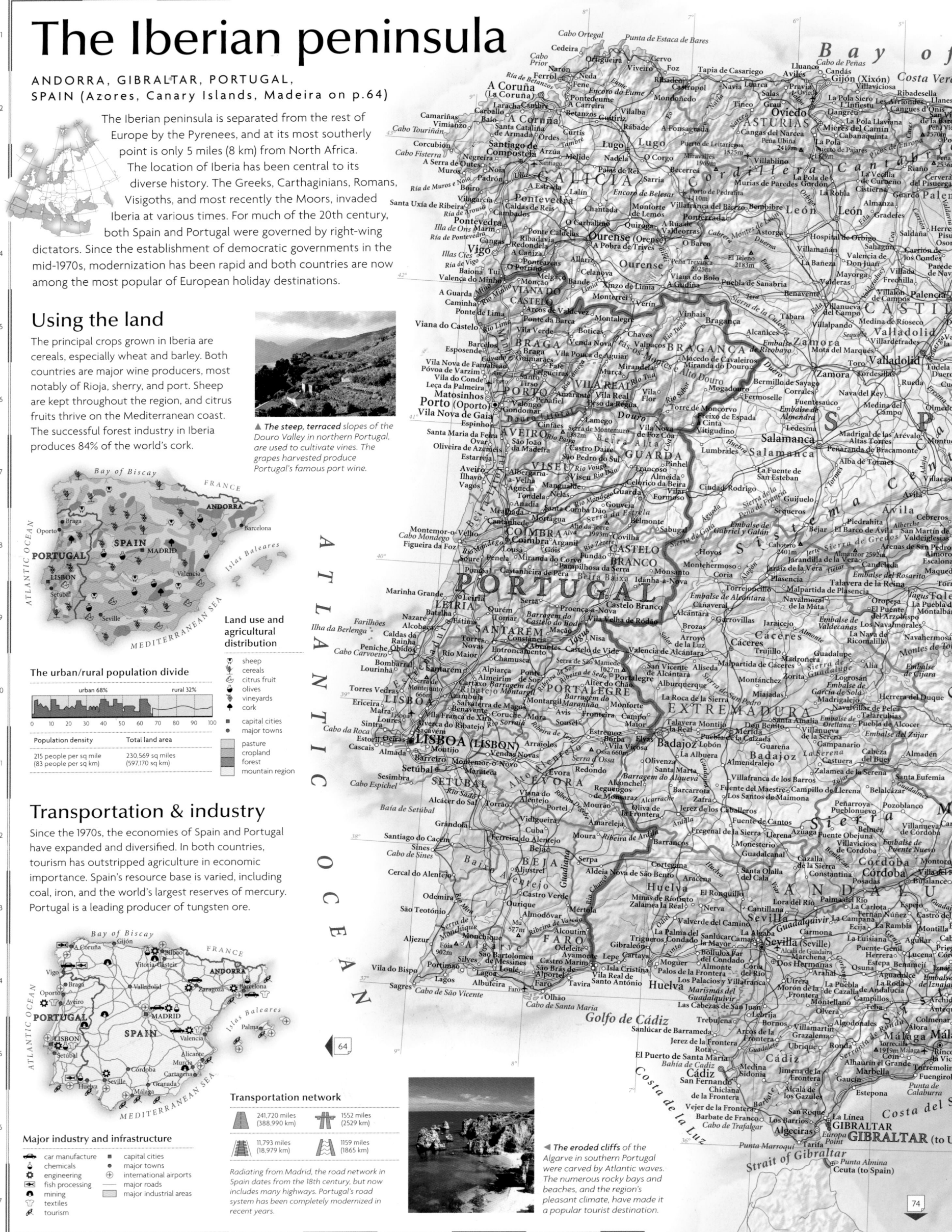

▶ The climate in northwestern Spain is milder in both summer and winter than in the rest of the country, creating a verdant environment, more commonly associated with northwestern Europe.

Map key

Population
- ▣ 1 million to 5 million
- ◉ 500,000 to 1 million
- ⊕ 100,000 to 500,000
- ⊕ 50,000 to 100,000
- ⊙ 10,000 to 50,000
- ○ below 10,000

Elevation
- 3000m / 9843ft
- 2000m / 6562ft
- 1000m / 3281ft
- 500m / 1640ft
- 250m / 820ft
- 100m / 328ft
- sea level

Scale 1:2,750,000

Km 0 5 10 20 30 40 50 60 70 80
Miles 0 10 20 30 40 50 60 70 80

projection: Lambert Conformal Conic

The landscape

A vast plateau, the Meseta dominates the centre of the peninsula, enclosed by the Cordillera Cantábrica to the north and the Sierra Morena to the south. It is drained by three major rivers, the Douro/Duero, the Tagus, and the Guadalquivir. The peninsula experiences great variations in climate and rainfall, both regionally and locally.

▲ The Pyrenees form Iberia's northeastern boundary, running for 270 miles (440 km), dividing the peninsula from the rest of Europe.

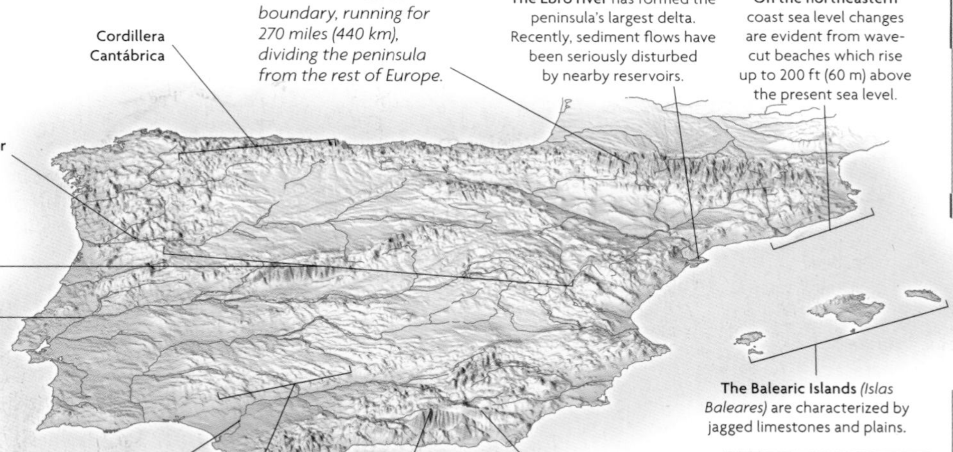

The Ebro river has formed the peninsula's largest delta. Recently, sediment flows have been seriously disturbed by nearby reservoirs.

On the northeastern coast sea level changes are evident from wave-cut beaches which rise up to 200 ft (60 m) above the present sea level.

Cordillera Cantábrica

Douro/Duero river

The Meseta plateau averages 1970 ft (600 m) in height and is now largely dry and treeless.

Tagus River

The Balearic Islands (Islas Baleares) are characterized by jagged limestones and plains.

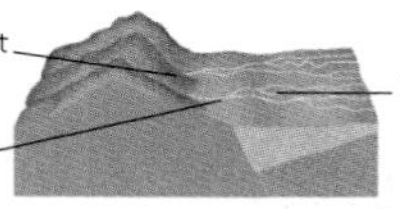

▲ Pediments are characteristic of semiarid lands across Iberia. A pediment is a flat, low-lying, eroded platform, cut into the bedrock. Weathered material is transported by streams and deposited in broad fan shapes on the pediment.

Mountain front
Weathered material
Pediment

The Guadalquivir river brings vital irrigation water to the plains, and like many of Iberia's rivers, is prone to flooding.

Sierra Morena

The Sierra Nevada in southern Spain contain Iberia's highest peak, Mulhacén, which rises 11,418 ft (3481 m).

▶ In the Sierra de los Filabres deforestation and overgrazing, which cause soil erosion, have created semidesert badlands.

Biscay

Pyrenees

MEDITERRANEAN SEA

Golfo de Valencia

Islas Baleares (Balearic Islands)

Menorca (Minorca)

Mallorca (Majorca)

ILLES BALEARS

Costa del Sol

Alboran Sea

105

The Italian peninsula

ITALY, SAN MARINO, VATICAN CITY

The Italian peninsula is a land of great contrasts. Until unification in 1861, Italy was a collection of independent states, whose competitiveness during the Renaissance resulted in the architectural and artistic magnificence of cities such as Rome, Florence, and Venice. The majority of Italy's population and economic activity is concentrated in the north, centered on the sophisticated industrial city of Milan. Southern Italy, the *Mezzogiorno*, has a harsh terrain, and remains far less developed than the north. Attempts to attract industry and investment in the south are frequently deterred by the entrenched network of organized crime and corruption.

The landscape

The mainly mountainous and hilly Italian peninsula took its present form following a collision between the African and Eurasian tectonic plates. The Alps in the northwest rise to a high point of 15,772 ft (4807 m) at Mont Blanc (*Monte Bianco*) on the French border, while the Apennines (*Appennino*) form a rugged backbone, running along the entire length of the country.

▲ *The island of Sardinia is an ancient land mass; an uplifted section of very old igneous rocks. Its rugged mountainous regions provide pasture for sheep and goats, while its valleys support some agriculture.*

Mont Blanc
(*Monte Bianco*)

Costa Smeralda

▲ *The Dolomites* (Alpi Dolomitiche) *are formed of thick limestones, overlying weaker marine strata. They have distinctive serrated peaks and many massive landslides occur.*

The distinctive square shape of the Gulf of Taranto (*Golfo di Taranto*) was defined by numerous block faults. Earthquakes are common in this region.

The Strait of Messina (*Stretto di Messina*) is between 2 and 12 miles (3–19 km) wide, and is a rich fishing ground.

Vesuvius (*Vesuvio*)

The Pontine Marshes (*Agro Pontino*) are bounded by low sand hills which prevent natural drainage.

Sicily is the largest island in the Mediterranean at 9926 sq miles (25,708 sq km).

The Apennines (*Appennino*) are the source of most of Italy's rivers. They run 823 miles (1324 km) down the length of the peninsula.

The southwestern tip of Sicily lies 95 miles (152 km) from the north African mainland and is part of the same geological region.

Sardinia is the second largest island in the Mediterranean Sea. The highest point is Punta La Marmora at 6017 ft (1834 m).

The Po Valley once formed part of the Adriatic Sea. Sediments of gravel, sand, and clay washed down from the Alps gradually filling the bay and forming a broad, cultivable plain.

Present-day crater has developed within the old crater of Monte Somma

▲ *There have been four volcanoes on the site of Vesuvius since volcanic activity began here more than 10,000 years ago.*

Vesuvius (*Vesuvio*)

Monte Somma

Old crater

Using the land

Italy produces 95% of its own food. The best farming land is in the Po Valley in northern Italy, where soft wheat and rice are grown. Irrigation is essential to agriculture in much of the south. Italy is a major producer and exporter of citrus fruits, olives, tomatoes, and wine.

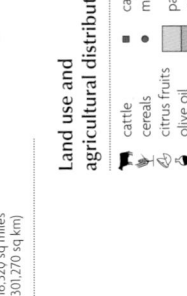

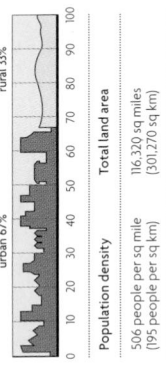
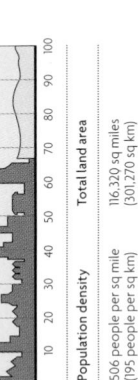

The urban/rural population divide

rural 33%

urban 67%

Population density	Total land area
506 people per sq mile (195 people per sq km)	116,320 sq miles (301,270 sq km)

Land use and agricultural distribution

- capital cities
- major towns
- pasture
- cropland
- forest
- mountain region

- cattle
- cereals
- citrus fruits
- olive oil
- rice
- vineyards

Scale 1:2,500,000

Km

Miles

projection: Lambert Conformal Conic

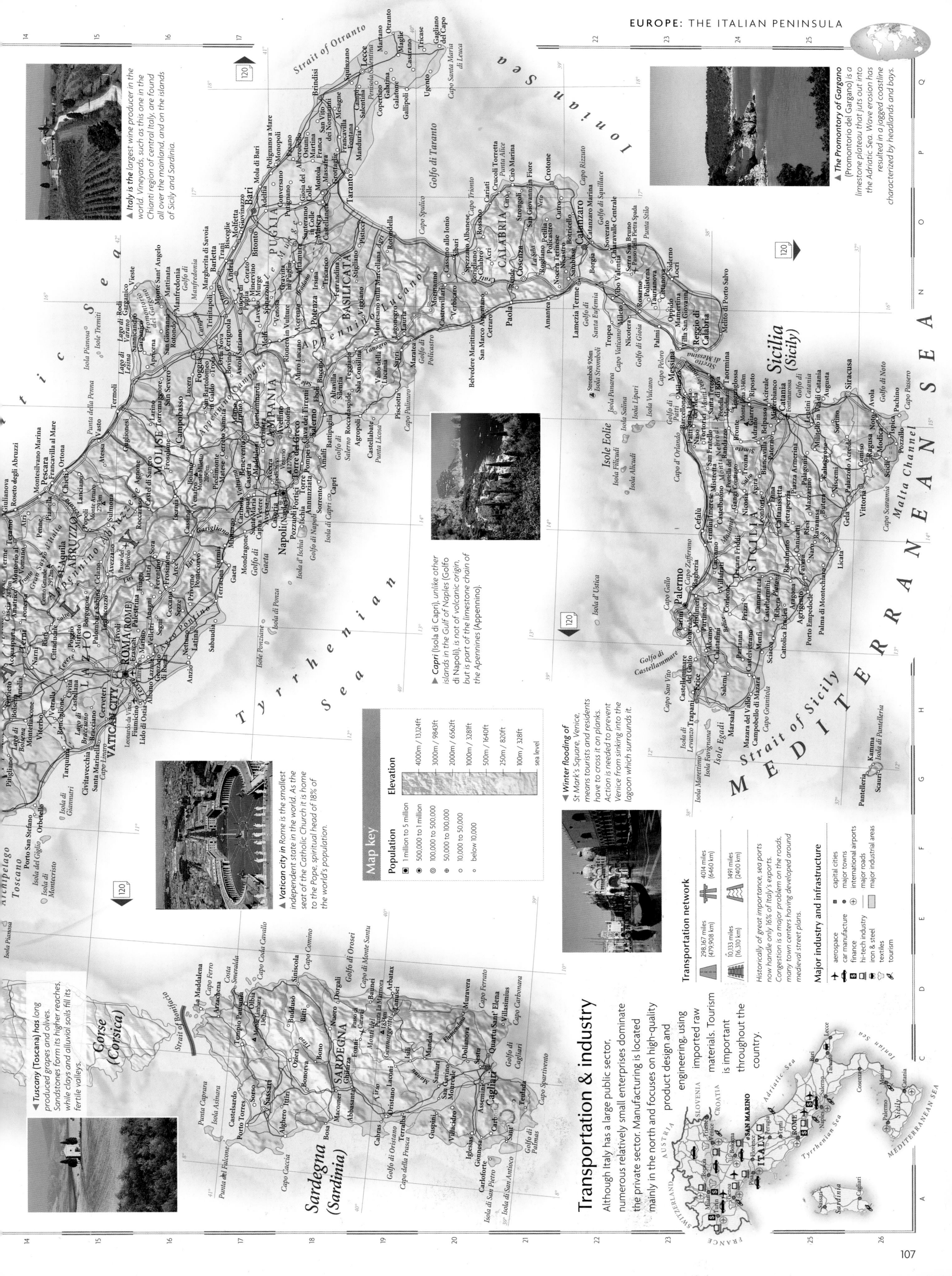

▲ **Italy is the** largest wine producer in the world. Vineyards, such as this one in the Chianti region of central Italy are found all over the mainland, and on the islands of Sicily and Sardinia.

▲ **The Promontorio del Gargano** (Promontorio del Gargano) is a limestone plateau that juts out into the Adriatic Sea. Wave erosion has resulted in a jagged coastline characterized by headlands and bays.

▲ **Capri** (Isola di Capri), unlike other islands in the Gulf of Naples (Golfo di Napoli), is not of volcanic origin, but is part of the limestone chain of the Apennines (Appennino).

▲ **Vatican city** in Rome is the smallest independent state in the world. As the seat of the Catholic Church it is home to the Pope, spiritual head of 18% of the world's population.

▼ **Winter flooding of** St Mark's Square, Venice, means tourists and residents have to cross it on planks. Action is needed to prevent Venice from sinking into the lagoon which surrounds it.

▼ **Tuscany** (Toscana) has long produced grapes and olives. Sandstones form its higher reaches, while clays and alluvial soils fill its fertile valleys.

Map key

Population

- ◉ 1 million to 5 million
- ⊙ 500,000 to 1 million
- ◎ 100,000 to 500,000
- ⊙ 50,000 to 100,000
- ○ 10,000 to 50,000
- ∘ below 10,000

Elevation

- 4000m / 13,124ft
- 3000m / 9843ft
- 2000m / 6562ft
- 1000m / 3281ft
- 500m / 1640ft
- 250m / 820ft
- 100m / 328ft
- sea level

Transportation network

298,167 miles (479,908 km)	4014 miles (6460 km)
10,133 miles (16,310 km)	149 miles (2400 km)

Historically of great importance, sea ports now handle only 16% of Italy's exports. Congestion is a major problem on the roads; many town centers having developed around medieval street plans.

Major industry and infrastructure

- ✈ aerospace
- 🚗 car manufacture
- $ finance
- hi-tech industry
- iron & steel
- textiles
- tourism
- ■ capital cities
- ■ major towns
- ⊕ international airports
- major roads
- major industrial areas

Transportation & industry

Although Italy has a large public sector, numerous relatively small enterprises dominate the private sector. Manufacturing is located mainly in the north and focuses on high-quality product design and engineering, using imported raw materials. Tourism is important throughout the country.

The Alpine states

AUSTRIA, LIECHTENSTEIN, SLOVENIA, SWITZERLAND

The Alpine countries of Austria, Switzerland, Liechtenstein, and Slovenia form a narrow strip across western Europe's geographical core, lying on the main north–south trading routes across the Alps. Switzerland, politically neutral since 1815, is an important international meeting place and houses one of the headquarters of the United Nations, it only became a member in 2002. Austria, once at the heart of the great Habsburg Empire has been a fully independent nation since 1955, and maintains a deserved reputation as an international center of culture. Slovenia declared independence from the former Yugoslavia in 1991 and despite initial economic hardship, is now starting to achieve the prosperity enjoyed by its Alpine neighbors.

Using the land

The Alpine region's mountainous terrain discourages cultivation over much of the land area. The primary agricultural activity is the raising of dairy and beef cattle on the pasture land of the lower mountain slopes. Austria is self-supporting in grains, and crops such as wheat, barley, and grapes are grown on the east Austrian lowlands. Woodlands are more prevalent in the eastern Alps; both Austria and Slovenia have large tracts of forest.

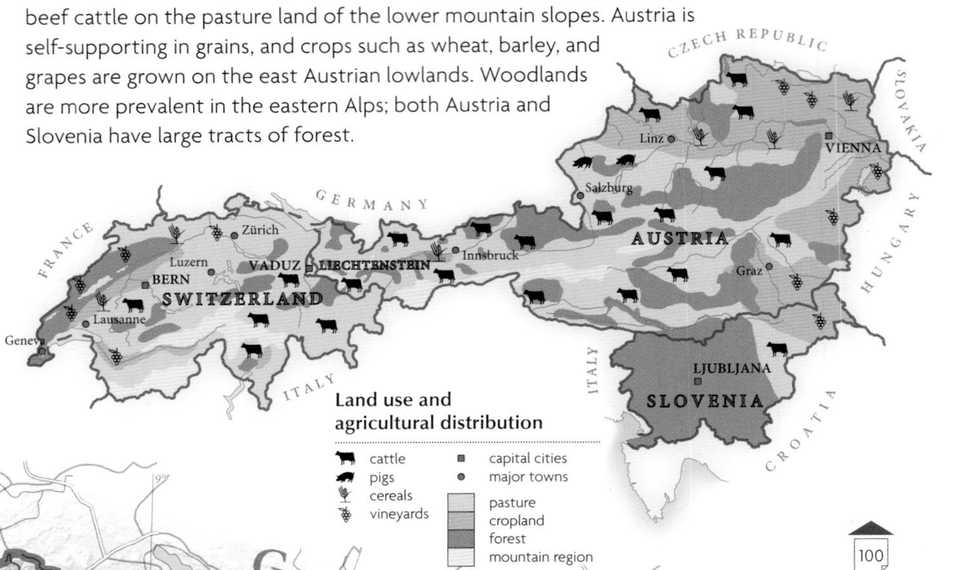

Land use and agricultural distribution

- cattle
- pigs
- cereals
- vineyards
- capital cities
- major towns
- pasture
- cropland
- forest
- mountain region

◀ *The Matterhorn*, on the Swiss-Italian border, is one of the highest mountains in the Alps, at 14,692 ft (4478 m). The term "horn" refers to its distinctive peak, formed by three glaciers eroding hollows, known as cirques, in each of its sides.

▲ *Constricted as it* cuts through ridges in the Alps, the Danube meanders across the lowlands, where uplift combined with river erosion has deepened meanders.

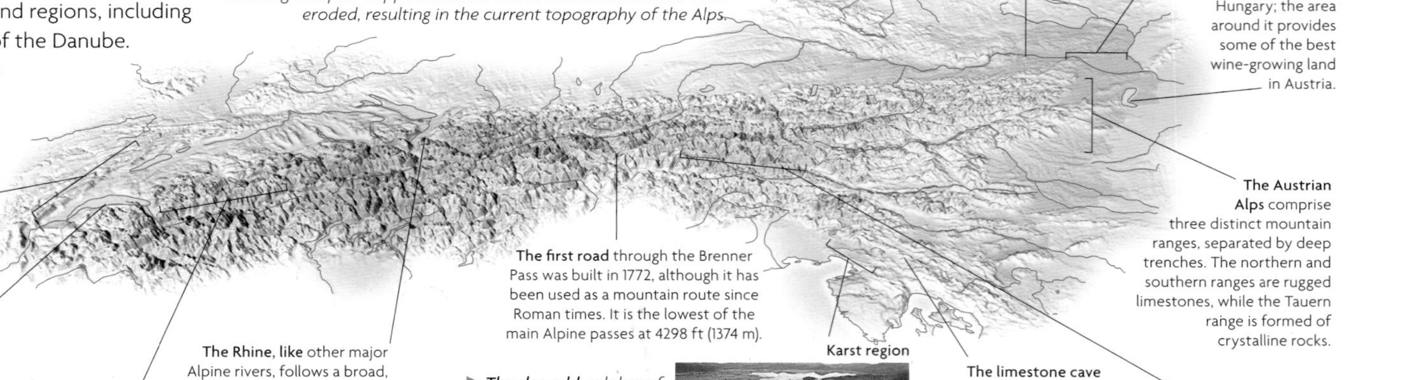

The landscape

The Alps occupy three-fifths of Switzerland, most of southern Austria and the northwest of Slovenia. They were formed by the collision of the African and Eurasian tectonic plates, which began 65 million years ago. Their complex geology is reflected in the differing heights and rock types of the various ranges. The Rhine flows along Liechtenstein's border with Switzerland, creating a broad floodplain in the north and west of Liechtenstein. In the far northeast and east are a number of lowland regions, including the Vienna Basin, Burgenland, and the plain of the Danube. Slovenia's major rivers largely flow across the lower eastern regions; in the west, the rivers flow underground through the limestone Karst region.

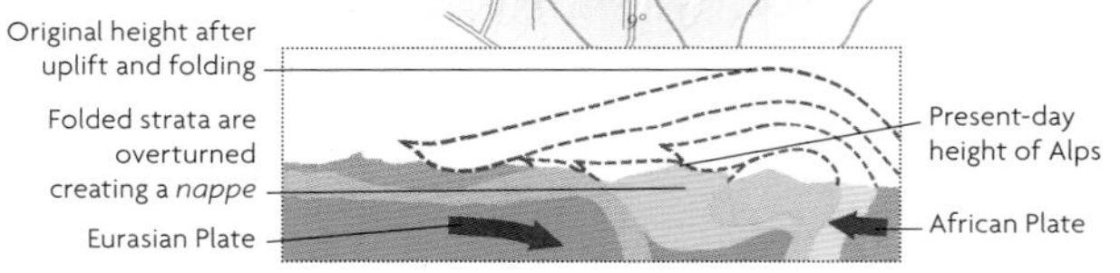

Original height after uplift and folding

Folded strata are overturned creating a *nappe*

Present-day height of Alps

Eurasian Plate

African Plate

▲ *The convergence of* the African and Eurasian plates compressed and folded huge masses of rock strata. As the plates continued to move together, the folded strata were overturned, creating complex nappes. Much of the rock strata has since been eroded, resulting in the current topography of the Alps.

The mountains of the Jura form a natural border between Switzerland and France. Their marine limestones date from over 200 million years ago. When the Alps were formed the Jura were folded into a series of parallel ridges and troughs.

Tectonic activity has resulted in dramatic changes in land height over very short distances. Lake Geneva, lying at 1221 ft (372 m) is only 43 miles (70 km) away from the 15,772 ft (4807 m) peak of Mont Blanc, on the France–Italy border.

The Bernese Alps (Berner Alpen) contain the Aletsch, which at 15 miles (24 km) is the longest Alpine glacier.

The Rhine, like other major Alpine rivers, follows a broad, flat trough between the mountains. Along part of its course, the Rhine forms the boundary between Switzerland and Liechtenstein.

▶ *The deep, blue* lakes of the Karst region are part of a drainage network which runs largely underground through this limestone area.

The first road through the Brenner Pass was built in 1772, although it has been used as a mountain route since Roman times. It is the lowest of the main Alpine passes at 4298 ft (1374 m).

Karst region

The limestone cave system at Postojna extends for more than 10 miles (16 km) and includes caverns reaching 125 ft (40 m) in height and width.

The Vienna Basin lies mainly below 390 ft (120 m). It gradually subsided and filled with sediment as the Alps were uplifted.

Neusiedler See straddles the border of Austria and Hungary; the area around it provides some of the best wine-growing land in Austria.

The Austrian Alps comprise three distinct mountain ranges, separated by deep trenches. The northern and southern ranges are rugged limestones, while the Tauern range is formed of crystalline rocks.

The Tauern range in the central Austrian Alps contains the highest mountain in Austria, the towering Grossglockner, rising 12,461 ft (3798 m).

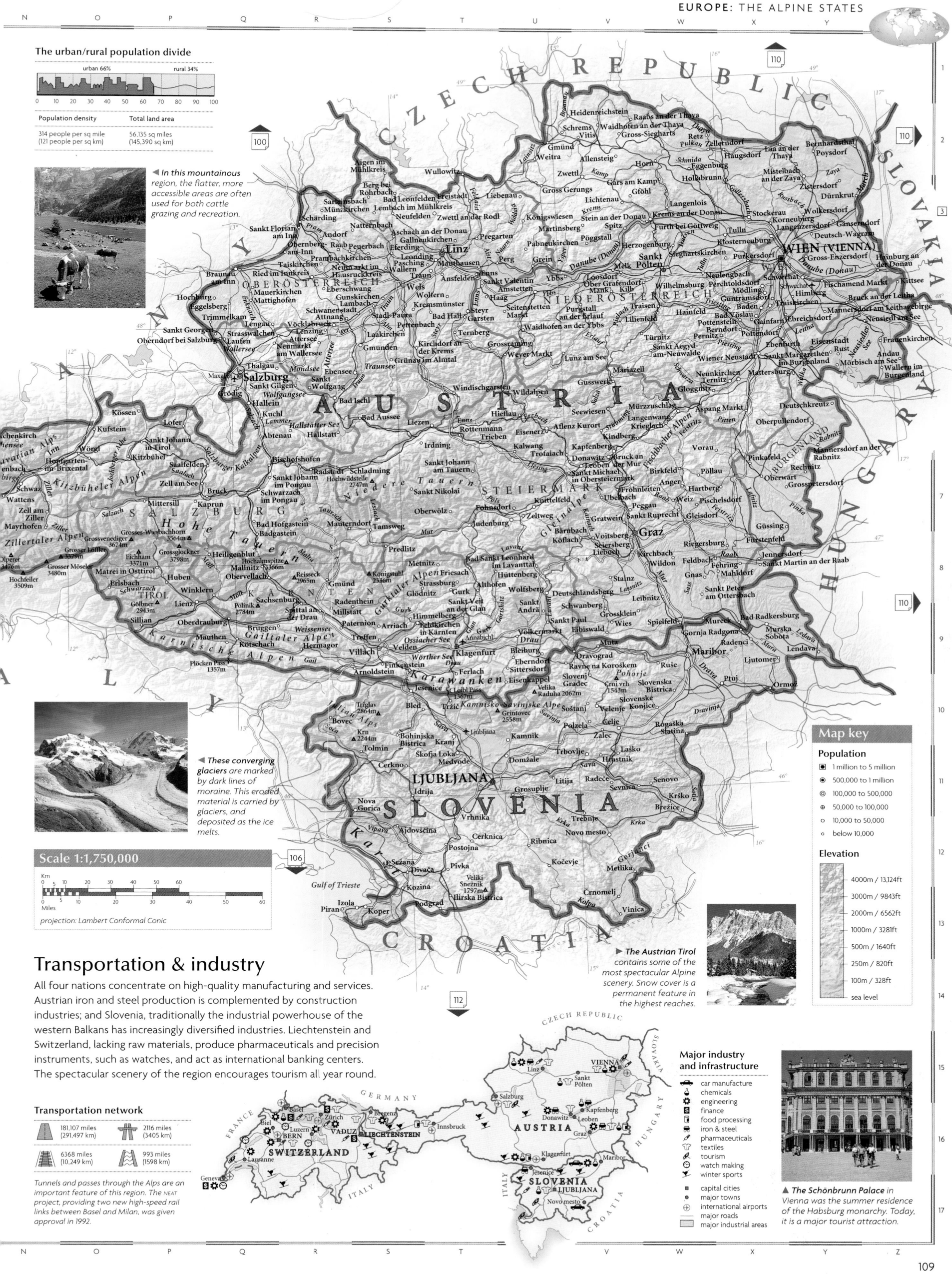

The urban/rural population divide

urban 66%   rural 34%

0 10 20 30 40 50 60 70 80 90 100

Population density	Total land area
314 people per sq mile (121 people per sq km)	56,135 sq miles (145,390 sq km)

◀ *In this mountainous region, the flatter, more accessible areas are often used for both cattle grazing and recreation.*

◀ *These converging glaciers are marked by dark lines of moraine. This eroded material is carried by glaciers, and deposited as the ice melts.*

Scale 1:1,750,000

Km
0 5 10 20 30 40 50 60

Miles
0 5 10 20 30 40 50 60

projection: Lambert Conformal Conic

Map key

Population

- ■ 1 million to 5 million
- ◉ 500,000 to 1 million
- ◎ 100,000 to 500,000
- ⊕ 50,000 to 100,000
- ○ 10,000 to 50,000
- · below 10,000

Elevation

- 4000m / 13,124ft
- 3000m / 9843ft
- 2000m / 6562ft
- 1000m / 3281ft
- 500m / 1640ft
- 250m / 820ft
- 100m / 328ft
- sea level

▶ *The Austrian Tirol contains some of the most spectacular Alpine scenery. Snow cover is a permanent feature in the highest reaches.*

Transportation & industry

All four nations concentrate on high-quality manufacturing and services. Austrian iron and steel production is complemented by construction industries; and Slovenia, traditionally the industrial powerhouse of the western Balkans has increasingly diversified industries. Liechtenstein and Switzerland, lacking raw materials, produce pharmaceuticals and precision instruments, such as watches, and act as international banking centers. The spectacular scenery of the region encourages tourism all year round.

Transportation network

181,107 miles (291,497 km)	2116 miles (3405 km)
6368 miles (10,249 km)	993 miles (1598 km)

Tunnels and passes through the Alps are an important feature of this region. The NEAT project, providing two new high-speed rail links between Basel and Milan, was given approval in 1992.

Major industry and infrastructure

- 🚗 car manufacture
- chemicals
- ⚙ engineering
- S finance
- food processing
- iron & steel
- pharmaceuticals
- textiles
- tourism
- watch making
- winter sports

- ■ capital cities
- ● major towns
- ✈ international airports
- — major roads
- ▨ major industrial areas

▲ *The Schönbrunn Palace in Vienna was the summer residence of the Habsburg monarchy. Today, it is a major tourist attraction.*

Central Europe

CZECH REPUBLIC, HUNGARY, POLAND, SLOVAKIA

When Slovakia and the Czech Republic became separate countries in 1993, they joined Hungary and Poland in a new role as independent nation states, following centuries of shifting boundaries and imperial strife. This turbulent history bequeathed the region a rich cultural heritage, shared through the works of its many great writers and composers, and celebrated in the vibrant historic capitals of Prague, Budapest, and Warsaw. Having shaken off years of Soviet domination in 1989, these states are confronting the challenge of winning commercial investment to modernize outmoded industries as they integrate their economies with those of the European Union.

Transportation & industry

Heavy industry has dominated postwar life in Central Europe. Poland has large coal reserves, having inherited the Silesian coalfield from Germany after the Second World War, allowing the export of large quantities of coal, along with other minerals. Hungary specializes in consumer goods and services, while Slovakia's industrial base is still relatively small. The Czech Republic's traditional glassworks and breweries bring some stability to its precarious Soviet-built manufacturing sector.

Major industry and infrastructure

- car manufacture
- chemicals
- engineering
- food processing
- mining
- shipbuilding
- tourism
- capital cities
- major towns
- international airports
- major roads
- major industrial areas

Transportation network

213,997 miles (344,600 km)	817 miles (1315 km)
27,479 miles (44,249 km)	3784 miles (6094 km)

▲ *The Berounka river* cuts through the precipitous wooded landscape of the Bohemian Massif, banked by a broad floodplain.

The landscape

The forested Carpathian Mountains, uplifted with the Alps, lie southeast of the older Bohemian Massif, which contains the Sudeten and Krusné Hory (*Erzgebirge*) ranges. They divide the fertile plains of the Danube to the south and the Vistula (*Wisła*), which flows north across vast expanses of glacial deposits into the Baltic Sea.

Longshore currents moving east along the Baltic coast have built a 40 mile (65 km) spit composed of material from the Vistula (*Wisła*) river.

▲ *The Biebrza river* has left meanders and oxbow lakes as it flows across low-lying ground.

Gerlachovský Štít in the Tatra Mountains, is Slovakia's highest mountain, at 8711ft (2655 m).

Carpathian Mountains

Danube river

Slip-off slope

Bluff

Direction of flow

▲ *Meanders form as rivers flow across plains at a low gradient. A steep cliff or bluff forms on the outside curve, and a gentler slip-off slope on the inside bend.*

Pomerania is a sandy coastal region of glacially-formed lakes stretching west from the Vistula (*Wisła*).

Hot mineral springs occur where geothermally heated water wells up through faults and fractures in the rocks of the Sudeten Mountains.

The Great Hungarian Plain formed by the floodplain of the Danube is a mixture of steppe and cultivated land, covering nearly half of Hungary's total area.

The Slovak Ore Mountains (*Slovenské Rudohorie*) are noted for their mineral resources, including high-grade iron ore.

Bohemian Massif

Krusné Hory (*Erzgebirge*)

▲ Budapest, the capital of Hungary, straddles the Danube. It comprises the historic towns of Buda, on the west bank, and Pest, which contains the Parliament Building, seen here on the far bank.

The huge growth of tourism and business has prompted major investment in the transportation infrastructure, with new roadbuilding schemes within and between the main cities of the region.

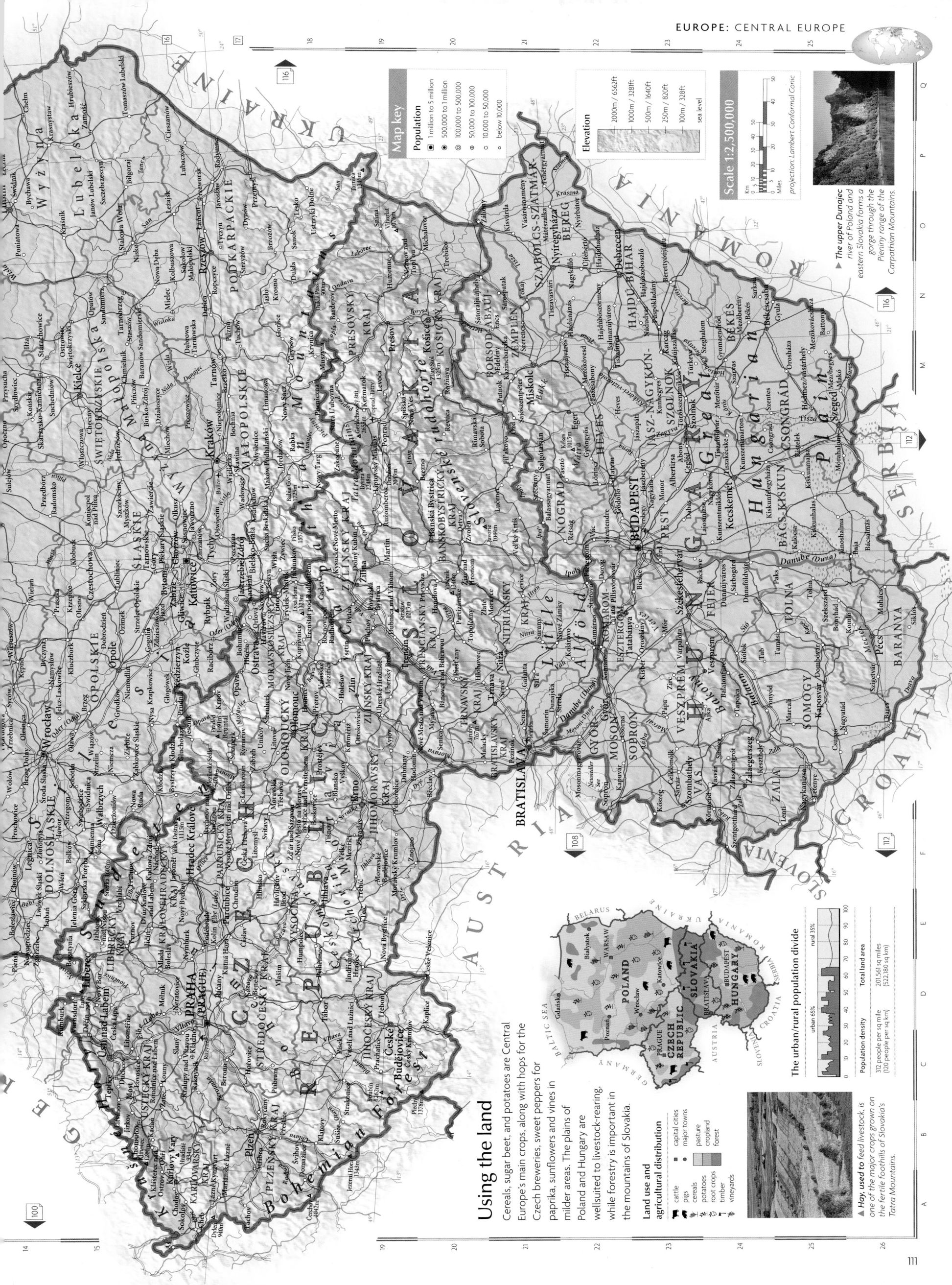

Map key

Population

- ◉ 1 million to 5 million
- ◎ 500,000 to 1 million
- ⊕ 100,000 to 500,000
- ○ 50,000 to 100,000
- ○ 10,000 to 50,000
- ○ below 10,000

Elevation

- 2000m / 6562ft
- 1000m / 3280ft
- 500m / 1640ft
- 250m / 820ft
- 100m / 328ft
- sea level

Scale 1:2,500,000

projection: Lambert Conformal Conic

▲ The upper Dunajec river of Poland and eastern Slovakia forms a gorge through the Pieniny range of the Carpathian Mountains.

Using the land

Cereals, sugar beet, and potatoes are Central Europe's main crops, along with hops for the Czech breweries, sweet peppers for paprika, sunflowers and vines in milder areas. The plains of Poland and Hungary are wellsuited to livestock-rearing, while forestry is important in the mountains of Slovakia.

Land use and agricultural distribution

- 🐄 cattle
- 🐖 pigs
- 🌾 cereals
- 🥔 potatoes
- 🌲 timber
- 🍇 vineyards
- ■ capital cities
- • major towns
- pasture
- cropland
- forest

The urban/rural population divide

urban 65%   rural 35%

Population density	Total land area
312 people per sq mile (120 people per sq km)	201,561 sq miles (522,180 sq km)

▲ Hay, used to feed livestock, is one of the major crops grown on the fertile foothills of Slovakia's Tatra Mountains.

Southeast Europe

ALBANIA, BOSNIA & HERZEGOVINA, CROATIA, KOSOVO, MACEDONIA, MONTENEGRO, SERBIA

For 46 years the federation of Yugoslavia held together the most diverse ethnic region in Europe, along the picturesque mountain hinterland of the Dalmatian coast. Economic collapse resulted in internal tensions. In the early 1990s, civil war broke out in both Croatia and Bosnia as the ethnic populations struggled to establish their own exclusive territories. Peace was only restored by the UN after NATO launched air strikes in 1995. Montenegro voted to split from Serbia in 2006. More recently, Kosovo controversially declared independence from Serbia in 2008, although this may take some time to be fully recognized. Neighboring Albania is slowly improving its fragile economy but remains one of Europe's poorest nations.

The landscape

The Tisza, Sava, and Drava Rivers drain the broad northern lowland, meeting the Danube after it crosses the Hungarian border. In the west, the Dinaric Alps divide the Adriatic Sea from the interior. Mainland valleys and elongated islands run parallel to the steep Dalmatian (Dalmacija) coastline, following alternating bands of resistant limestone.

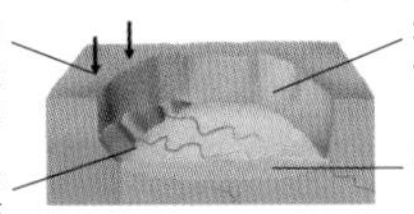

Poljes in the Kosovo region

- Sheer limestone walls enclose all sides
- Flat polje floor
- Underground drainage along joints in the rock
- Spring at foot of cliff

▲ **Rain and underground water** dissolve limestone along massive vertical joints (cracks). This creates poljes: depressions several miles across with steep walls and broad, flat floors.

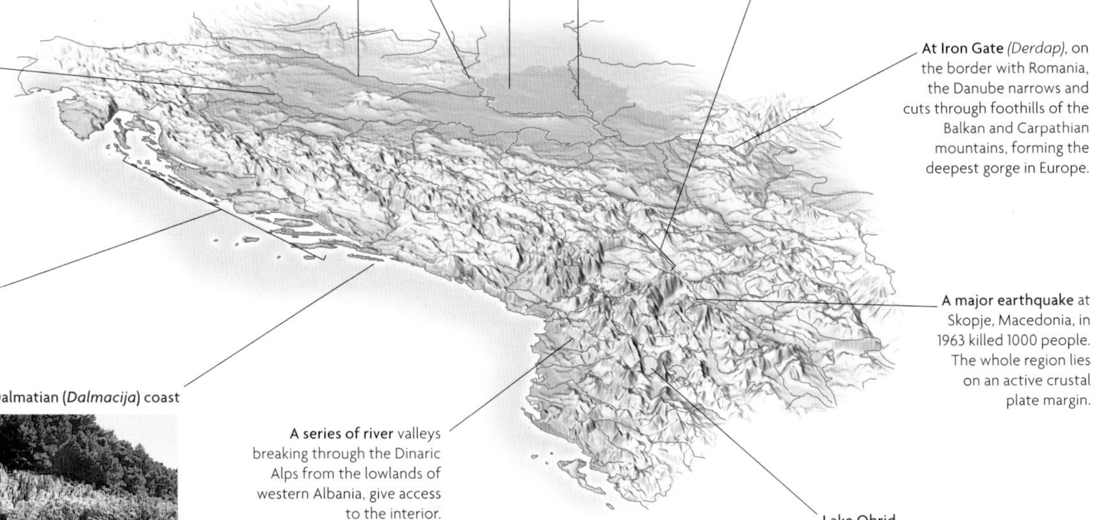

At Iron Gate (Derdap), on the border with Romania, the Danube narrows and cuts through foothills of the Balkan and Carpathian mountains, forming the deepest gorge in Europe.

A major earthquake at Skopje, Macedonia, in 1963 killed 1000 people. The whole region lies on an active crustal plate margin.

Lake Ohrid

▲ **Lake Ohrid borders** Albania and Macedonia. Ohrid is the deepest lake in the western Balkans, reaching depths of 938 ft (286 m).

The river floodplains of the Pannonian Basin are flanked by terraces of gravel and wind-blown glacial deposits known as loess.

At least 70% of the fresh water in the western Balkans drains eastward into the Black Sea, mostly via the Danube (Dunav).

Tisza river

Drava river

Sava river

A series of river valleys breaking through the Dinaric Alps from the lowlands of western Albania, give access to the interior.

Dalmatian (Dalmacija) coast

The elongated islands, promontories and straits of the Dalmatian (Dalmacija) coast were formed as the Adriatic Sea rose to flood valleys running parallel to the shore.

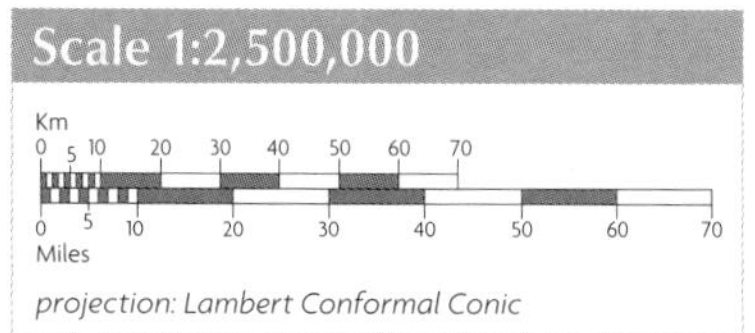

▲ **Limestone cliffs along the** Dalmatian (Dalmacija) shoreline are heavily eroded, as salt water dissolves the rock along existing horizontal cracks, or joints. This tends to form a platform of rock at the foot of the cliff.

Scale 1:2,500,000

Km
Miles

projection: Lambert Conformal Conic

▲ **Hot, dry summers and mild winters** offer excellent conditions for viticulture in Montenegro. The precipitous Dinaric Alps have kept this region relatively isolated for centuries.

108

110

114

116

120

Transportation & industry

Processing industries based on the region's wealth of mineral reserves predominate in Albania and Macedonia. In other regions, industrial plants have been commandeered, if not destroyed in the war and mineral extraction has severely declined. The fast-flowing rivers found throughout the Dinaric Alps are exploited to generate hydroelectric power.

▲ *The historic center* of Mostar in southern Bosnia, with its famous 16th-century Turkish bridge, was destroyed by shelling during 1993. The bridge was rebuilt and opened again in 2004.

In February 2008, Kosovo (a UN Protectorate within Serbia since 1999) declared independence. Although now recognized by numerous countries, this decision has proved controversial with other states wary of setting a precedent for separatist groups within their own borders. It is therefore likely to be some time before Kosovo becomes universally recognized.

Transportation network

46,996 miles (75,642 km)
685 miles (1103 km)
543 miles (8713 km)
879 miles (1415 km)

The war has resulted in the destruction or disintegration of infrastructure for transportation, communications, and power supply, though this is now in the process of recovery.

Major industry and infrastructure

aluminum refining
car manufacture
chemicals
engineering
food processing
hydroelectric power
mining
shipbuilding
textiles
timber processing
capital cities
major towns
international airports
major roads

▲ *Industrial processing plants* were established throughout Albania by the Hoxha regime, which collapsed in 1992. They remain incongruous among the villages of one of Europe's most conservative rural societies.

▲ *The ancient Croatian port* of Dubrovnik was one of the former Yugoslavia's most popular tourist resorts and an important point of access to the sea along the Dalmatian (Dalmacija) coast. Shelling of the old city by Serb forces in 1991 provoked international condemnation.

Using the land

Crops of wheat, maize, sugar beet, vegetables, and fruit are widely grown. The hilly terrain is suited to forestry and livestock farming. The mild, Mediterranean climate of the coastal regions provides ideal conditions for growing vines and olives. Albania's largely agricultural economy has been adversely affected by the recent dismantling of state farms.

▼ *Sweet red peppers are dried* in the sun, ready to make paprika. Macedonia's economy is mainly agricultural and its fertile soils support a broad range of crops.

Land use and agricultural distribution

pigs
sheep
cereals
fruit
olives
sugar beet
tobacco
vineyards
capital cities
major towns
pasture
cropland
forest
mountain region

The urban/rural population divide

urban 49%
urban 51%

Population density
240 people per sq mile
(93 people per sq km)

Total land area
95,038 sq miles
(246,278 sq km)

▲ *The Tara river* is one of Montenegro's major rivers. It flows into the Danube via the Drina and Sava rivers. Along its course the Tara has eroded spectacular gorges up to 3280 ft (1000 m) deep.

Map key

Population
1 million to 5 million
500,000 to 1 million
100,000 to 500,000
50,000 to 100,000
10,000 to 50,000
below 10,000

Elevation
2000m / 6562ft
1000m / 3281ft
500m / 1640ft
250m / 820ft
100m / 328ft
sea level

Bulgaria & Greece

Including EUROPEAN TURKEY

Greece is renowned as the original hearth of western civilization. The rugged terrain and numerous islands have profoundly affected its development, creating a strong agricultural and maritime tradition. In the past 50 years, this formerly rural society has rapidly urbanized, with one third of the population now living in the capital, Athens, and in the northern city of Salonica. Bulgaria, dominated for centuries by the Ottoman Turks, became part of the eastern bloc after the Second World War, only slowly emerging from Soviet influence in 1989. Moves toward democracy led to some instability in Bulgaria and Greece, now outweighed by the challenge of integration with the European Union.

Transportation & industry

Soviet investment introduced heavy industry into Bulgaria, and the processing of agricultural produce, such as tobacco, is important throughout the country. Both countries have substantial shipyards and Greece has one of the world's largest merchant fleets. Many small craft workshops, producing textiles and processed foods, are clustered around Greek cities. The service and construction sectors have profited from the successful tourist industry.

Major industry and infrastructure

- ⚗ chemicals
- ⚙ engineering
- 🍴 food processing
- ⚓ shipbuilding
- textiles
- 🎿 tourism
- ■ capital cities
- ● major towns
- ✈ international airports
- major roads
- major industrial areas

Transportation network

103,930 miles (167,630 km)	
345 miles (557 km)	
4346 miles (6995 km)	
294 miles (474 km)	

Bulgaria's railroads require investment to revive an outdated infrastructure. In Greece, despite a developing road network, ferry-boats remain the most effective form of transportation in many areas.

The landscape

Bulgaria's Balkan mountains divide the Danubian Plain (*Dunavska Ravnina*) and Maritsa Basin, meeting the Black Sea in the east along sandy beaches. The steep Rhodope Mountains form a natural barrier with Greece, while the younger Pindus form a rugged central spine which descends into the Aegean Sea to give a vast archipelago of over 2000 islands, the largest of which is Crete.

▲ *The Arda river cuts through the Rhodope Mountains in rugged, rocky gorges.*

The islands of Crete, Kythira, Karpathos, and Rhodes are part of an arc which bends southeastward from the Peloponnese, forming the southern boundary of the Aegean.

▲ *Layers of black volcanic ash still cover the island of Santorini. This volcano last erupted 3500 years ago, but still shows signs of volcanic activity.*

The Danube, Europe's second longest river, forms most of Bulgaria's northern border. The Danubian plain (*Dunavska Ravnina*), extending from the southern bank, is extremely fertile.

Mount Olympus is the mythical home of the Greek Gods and, at 9570 ft (2917 m), is the highest mountain in Greece.

The Peloponnese consist of several mountainous peninsulas, linked to the mainland by the isthmus of Corinth. The Corinth Canal (*Dioryga Korinthou*), built in 1893, cuts through the isthmus, linking the Aegean and Ionian Seas.

▲ *Mount Olympus is a composite of rocks formed by two major tectonic events. First the older metamorphic rocks were thrust over the limestones, then two million years ago regional warping and subsequent erosion, reexposed the limestone.*

Ancient metamorphic rock, formed miles below the surface

Mount Olympus

Limestone rocks exposed by erosion of metamorphic rocks

Younger limestones created in shallow seas

Balkan Mountains
Maritsa Basin
Pindus Mountains
Rhodes
Karpathos
Crete
Kythira
Rhodope Mountains
Corinth Canal (*Dioryga Korinthou*)

Scale 1:2,500,000

projection: Lambert Conformal Conic

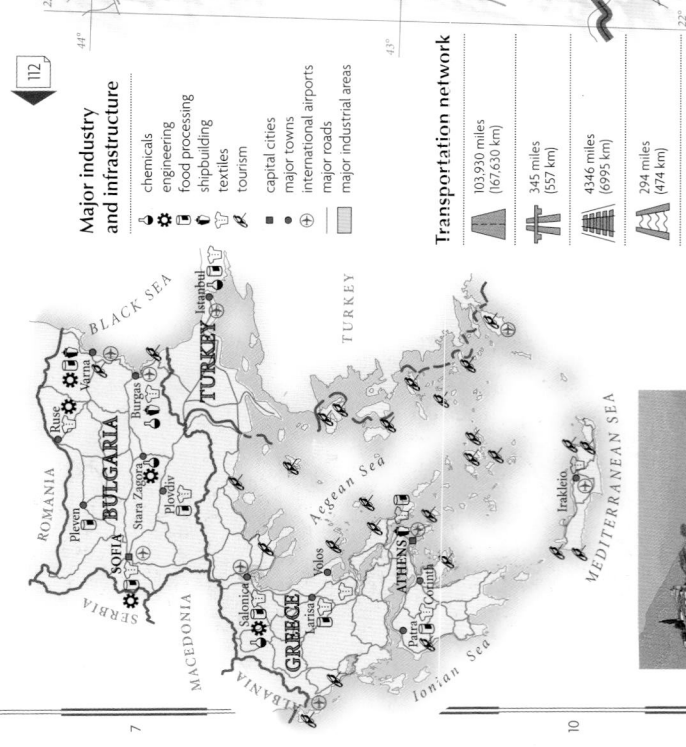

▲ *A towering pinnacle at Metéora in central Greece is home to the monastery of Roussanou. The 24 rock towers which dominate the plain of Thessalia (Thessaly) are remnants of an old plateau. Long-term weathering along fissures in the rock has worn away the rest of the plateau.*

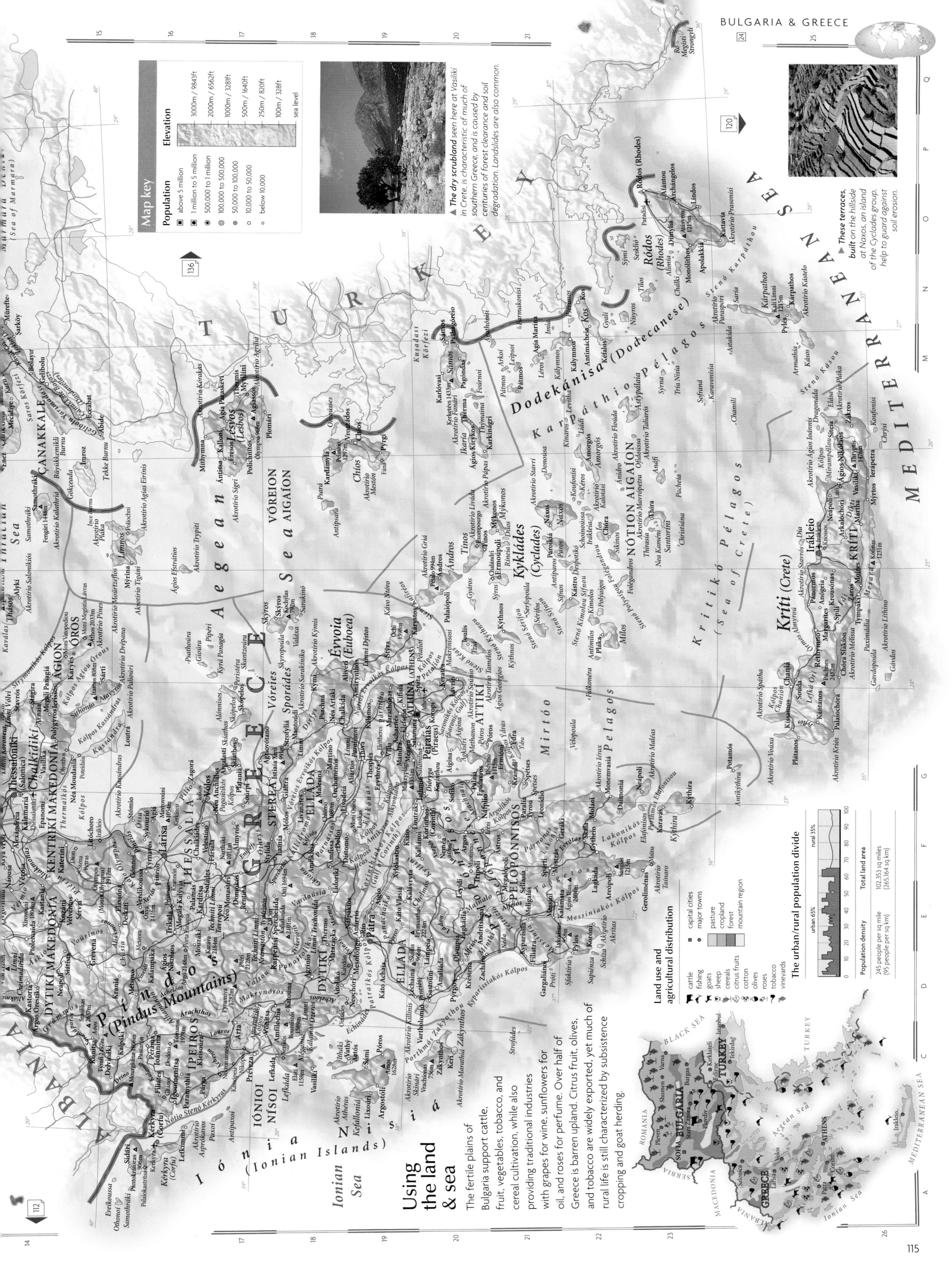

Map key

Population

■ above 5 million
■ 1 million to 5 million
◉ 500,000 to 1 million
⊕ 100,000 to 500,000
⊕ 50,000 to 100,000
○ 10,000 to 50,000
○ below 10,000

Elevation

3000m / 9843ft
2000m / 6562ft
1000m / 3281ft
500m / 1640ft
250m / 820ft
100m / 328ft
sea level

▲ The dry scrubland seen here at Vasiliki in Crete, is characteristic of much of southern Greece, and is caused by centuries of forest clearance and soil degradation. Landslides are also common.

▼ These terraces, built on the hillside at Naxos, an island of the Cyclades group, help to guard against soil erosion.

Using the land & sea

The fertile plains of Bulgaria support cattle, fruit, vegetables, tobacco, and cereal cultivation, while also providing traditional industries with grapes for wine, sunflowers for oil, and roses for perfume. Over half of Greece is barren upland. Citrus fruit, olives, and tobacco are widely exported, yet much of rural life is still characterized by subsistence cropping and goat herding.

Land use and agricultural distribution

- capital cities
- major towns
- cropland
- forest
- mountain region

- cattle
- fishing
- goats
- sheep
- cereals
- citrus fruits
- cotton
- olives
- roses
- tobacco
- vineyards

The urban/rural population divide

urban 65%   rural 35%

Population density	Total land area
245 people per sq mile (95 people per sq km)	102,353 sq miles (265,164 sq km)

Romania, Moldova & Ukraine

The industrial, social, and cultural make-up of Romania and the former Soviet states of Moldova and Ukraine still bear the imprint of their communist past. As part of the USSR, Ukraine was a leading agricultural, industrial, and energy producer. These industries, like those in Moldova and Romania, are now being reoriented more firmly toward western markets. As a result of shifting borders, and Soviet policy actively encouraging Russian immigration into other Soviet states like Ukraine and Moldova, all three countries now contain large numbers of foreign nationals. In 2014, the Russian Federation drew international condemnation by annexing the Ukrainian territory of Crimea.

Using the land

The fertile black soils of Ukraine, often called "the breadbasket of Europe," have enabled the cultivation of a variety of cereals and vegetables, which are widely exported. Romania and Moldova also grow cereals, sunflowers, and vegetables, and are noted for the quality of their wines.

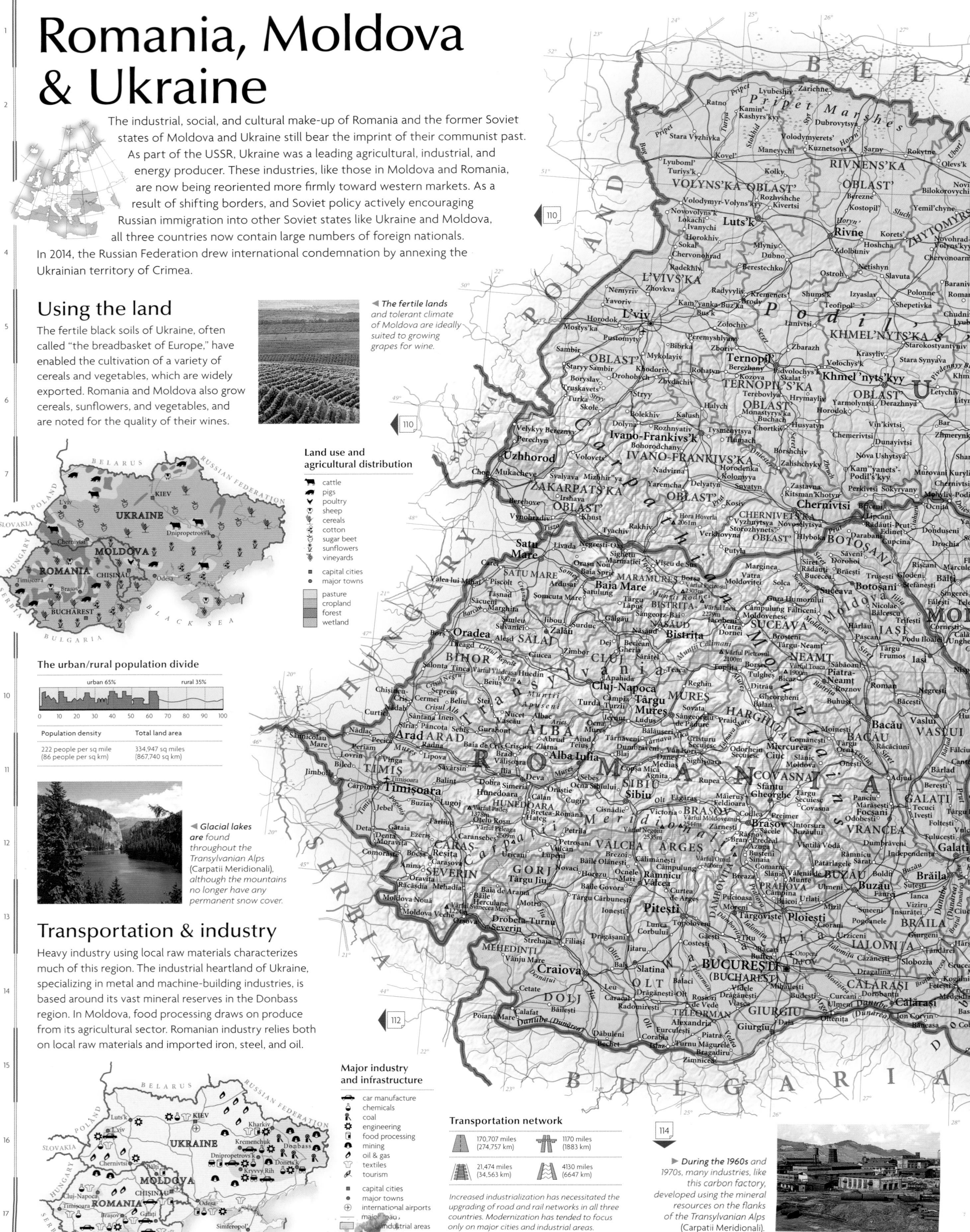

◄ The fertile lands and tolerant climate of Moldova are ideally suited to growing grapes for wine.

Land use and agricultural distribution

- cattle
- pigs
- poultry
- sheep
- cereals
- cotton
- sugar beet
- sunflowers
- vineyards
- ■ capital cities
- • major towns
- pasture
- cropland
- forest
- wetland

The urban/rural population divide

urban 65%   rural 35%

0 10 20 30 40 50 60 70 80 90 100

Population density	Total land area
222 people per sq mile (86 people per sq km)	334,947 sq miles (867,740 sq km)

◄ Glacial lakes are found throughout the Transylvanian Alps (Carpatii Meridionali), although the mountains no longer have any permanent snow cover.

Transportation & industry

Heavy industry using local raw materials characterizes much of this region. The industrial heartland of Ukraine, specializing in metal and machine-building industries, is based around its vast mineral reserves in the Donbass region. In Moldova, food processing draws on produce from its agricultural sector. Romanian industry relies both on local raw materials and imported iron, steel, and oil.

Major industry and infrastructure

- car manufacture
- chemicals
- coal
- engineering
- food processing
- mining
- oil & gas
- textiles
- tourism
- ■ capital cities
- ◉ major towns
- ⊕ international airports
- major roads
- industrial areas

Transportation network

170,707 miles (274,757 km)		1170 miles (1883 km)	
21,474 miles (34,563 km)		4130 miles (6647 km)	

Increased industrialization has necessitated the upgrading of road and rail networks in all three countries. Modernization has tended to focus only on major cities and industrial areas.

▶ During the 1960s and 1970s, many industries, like this carbon factory, developed using the mineral resources on the flanks of the Transylvanian Alps (Carpatii Meridionali).

Scale 1:3,250,000

projection: Lambert Conformal Conic

Map key

Population
- 1 million to 5 million
- 500,000 to 1 million
- 100,000 to 500,000
- 50,000 to 100,000
- 10,000 to 50,000
- below 10,000

Elevation
- 2000m / 6562ft
- 1000m / 3281ft
- 500m / 1640ft
- 250m / 820ft
- 100m / 328ft
- sea level

The landscape

Vast flat lowlands and gently rolling hills cover most of southeastern Europe. In the southwest, the Carpathian Mountains form a gentle arc. To the south of the Carpathian Mountains lies the Danube Plain, across which the Danube river flows to the Black Sea. To the north and east, the hills of Moldova level out into low plains, running east to the steppes of Ukraine.

▶ **Divided into crystalline** massifs, the southern arm of the Carpathian Mountains, the Transylvanian Alps (Carpatii Meridionali), extend 170 miles (274 km) across southwestern Romania.

▲ **The Swallow's Nest** castle at Yalta is one of many tourist resorts on the Crimean (Krym) coast, dubbed the "Russian Riviera."

The Codrii Hills dominate the landscape of central Moldova; they are intersected by deep, flat valleys and ravines.

Steppe landscape covers two-thirds of Ukraine. These flat, treeless grasslands extend from central Europe to central Asia.

Most of the major rivers in southeastern Europe, like the Danube, the Dniester, and Dnieper flow south and east to the Black Sea.

Uplifted and folded at the same time as the Alps, some 250 miles (400 km) of the eastern Carpathian Mountains contain ancient volcanic cones and craters.

The Apuseni Mountains (Muntii Apuseni) are rich in mineral deposits, including gold and iron ore.

Transylvanian Alps (Carpatii Meridionali)

The Danube forms a natural border between Romania and Bulgaria.

The three branches of the Danube Delta (Delta Dunării) form a triangle of wetlands covering some 1950 sq miles (5050 sq km).

At Kryms'ki Hory, three flat-topped, parallel limestone ridges run 80 miles (128 km) along the southern coast of the Crimean (Krym) Peninsula.

Counterclockwise currents have created the sandspits which fringe the Sea of Azov.

Old glaciated valley

Water has eroded a new post-glacial valley

▲ **Balkas are common** throughout Ukraine. They are large U-shaped valleys, formed during the last Ice Age, which contain narrower, deep valleys. These were incised by a sudden flow of water, following an icemelt.

The Baltic states & Belarus

BELARUS, ESTONIA, LATVIA, LITHUANIA, Kaliningrad

Occupying Europe's main corridor to Russia, the four distinct cultures of Estonia, Latvia, Lithuania, and Belarus share a history of struggle for nationhood against the interests of more powerful neighbors. As the first republics to declare their independence from the Soviet Union in 1990–91, the Baltic states of Estonia, Latvia, and Lithuania sought an economic role in the EU, while reaffirming their European cultural roots through the church and a strong musical tradition. Meanwhile, Belarus has shown economic and political allegiance to Russia by joining the Commonwealth of Independent States.

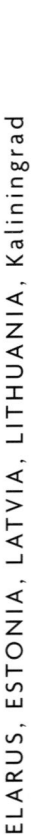

▲ *The seaport* of *Riga* is Latvia's capital and the center of economic and cultural life. With a 32% Russian minority in Latvia, language and the right to national citizenship are key issues.

Using the land

Across the four nations cattle and pig farming are widespread, together with diverse arable crops, including flax for making linen, potatoes used to produce vodka, cereals, and other vegetables. Almost a third of the land is forested; demand for timber has increased the importance of forest management.

Land use and agricultural distribution

- cattle
- pigs
- cereals
- flax
- potatoes
- timber
- capital cities
- major towns
- pasture
- cropland
- forest
- wetland

The urban/rural population divide

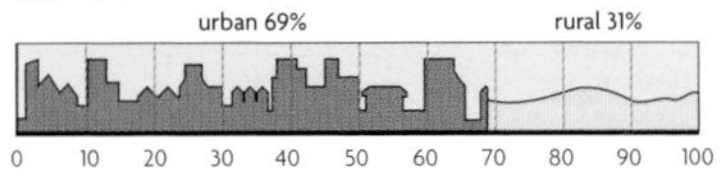

urban 69%   rural 31%

Population density
122 people per sq mile
(47 people per sq km)

Total land area
145,006 sq miles
(375,656 sq km)

▲ *A pine forest* in northern Belarus. Conifers in the north give way to hardwood forest farther south. Timber mills are supplied with logs floated along the country's many navigable waterways.

▲ *The Western Dvina* river provides hydroelectric power and, during the summer months, access to the Baltic Sea. The lower course of the river freezes from December to April.

Map key

Population
- ● 1 million to 5 million
- ◉ 500,000 to 1 million
- ⦾ 100,000 to 500,000
- ⊙ 50,000 to 100,000
- ○ 10,000 to 50,000
- ∘ below 10,000

Elevation
- 250m / 820ft
- 100m / 328ft
- sea level

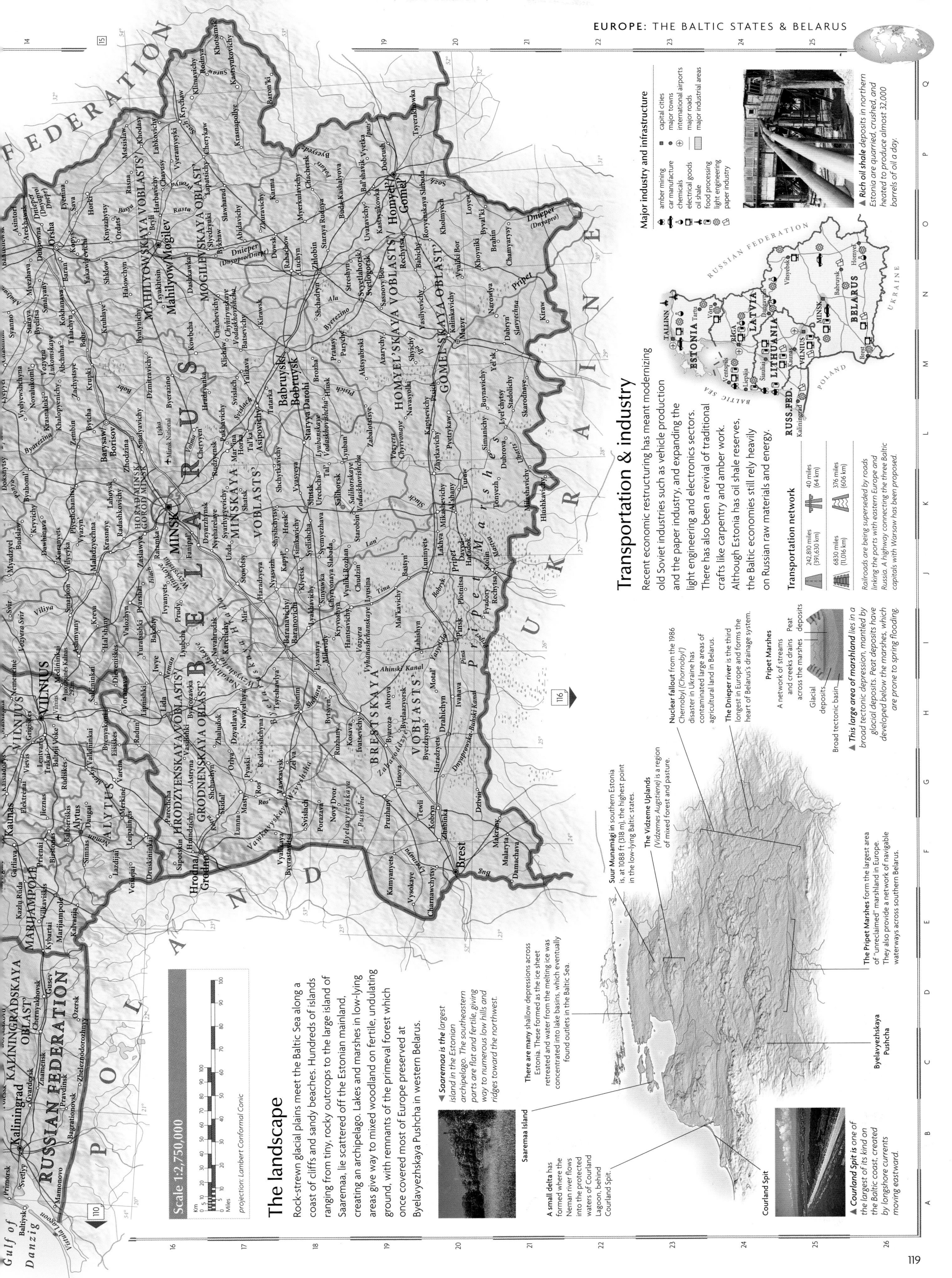

Major industry and infrastructure

- capital cities
- major towns
- international airports
- major roads
- major industrial areas

- amber mining
- car manufacture
- chemicals
- electrical goods
- oil shale
- food processing
- light engineering
- paper industry

▲ Rich oil shale deposits in northern Estonia are quarried, crushed, and heated to produce almost 32,000 barrels of oil a day.

Transportation & industry

Recent economic restructuring has meant modernizing old Soviet industries such as vehicle production and the paper industry, and expanding the light engineering and electronics sectors. There has also been a revival of traditional crafts like carpentry and amber work. Although Estonia has oil shale reserves, the Baltic economies still rely heavily on Russian raw materials and energy.

Transportation network

242,810 miles (391,630 km)	40 miles (64 km)
6830 miles (11,016 km)	376 miles (606 km)

Railroads are being superseded by roads linking the ports with eastern Europe and Russia. A highway connecting the three Baltic capitals with Warsaw has been proposed.

Nuclear fallout from the 1986 Chernobyl (Chornobyl') disaster in Ukraine has contaminated large areas of agricultural land in Belarus.

The Dnieper river is the third longest in Europe and forms the heart of Belarus's drainage system.

Pripet Marshes

A network of streams and creeks drains across the marshes

- Peat deposits
- Glacial deposits
- Broad tectonic basin

▲ This large area of marshland lies in a broad tectonic depression, mantled by glacial deposits. Peat deposits have developed below the marshes, which are prone to spring flooding.

Suur Munamägi in southern Estonia is, at 1088 ft (318 m), the highest point in the low-lying Baltic states.

The Vidzeme Uplands

(Vidzemes Augstiene) is a region of mixed forest and pasture.

The Pripet Marshes form the largest area of "unreclaimed" marshland in Europe. They also provide a network of navigable waterways across southern Belarus.

Byelavyezhskaya Pushcha

Courland Spit

▲ Courland Spit is one of the largest of its kind on the Baltic coast, created by longshore currents moving eastward.

The landscape

Rock-strewn glacial plains meet the Baltic Sea along a coast of cliffs and sandy beaches. Hundreds of islands ranging from tiny, rocky outcrops to the large island of Saaremaa, lie scattered off the Estonian mainland, creating an archipelago. Lakes and marshes in low-lying areas give way to mixed woodland on fertile, undulating ground, with remnants of the primeval forest which once covered most of Europe preserved at Byelavyezhskaya Pushcha in western Belarus.

▲ Saaremaa is the largest island in the Estonian archipelago. The southeastern parts are flat and fertile, giving way to numerous low hills and ridges toward the northwest.

Saaremaa Island

There are many shallow depressions across Estonia. These formed as the ice sheet retreated and water from the melting ice was concentrated into lake basins, which eventually found outlets in the Baltic Sea.

A small delta has formed where the Neman river flows into the protected waters of Courland Lagoon, behind Courland Spit.

Scale 1:2,750,000

projection: Lambert Conformal Conic

The Mediterranean

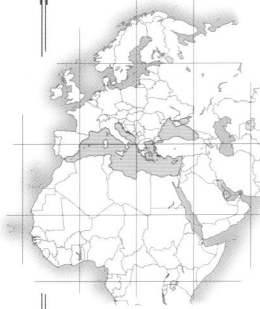

The Mediterranean Sea stretches over 2500 miles (4000 km) east to west, separating Europe from Africa. At its westernmost point it is connected to the Atlantic Ocean through the Strait of Gibraltar. In the east, the Suez canal, opened in 1869, gives passage to the Indian Ocean. In the northeast, linked by the Sea of Marmara, lies the Black Sea. The Mediterranean is bordered by almost 30 states and territories, and more than 100 million people live on its shores and islands. Throughout history, the Mediterranean has been a focal area for many great empires and civilizations, reflected in the variety of cultures found on its shores. Since the 1960s, development along the southern coast of Europe has expanded rapidly to accommodate increasing numbers of tourists and to enable the exploitation of oil and gas reserves. This has resulted in rising levels of pollution, threatening the future of the sea.

▲ *Monte Carlo is* just one of the luxurious resorts scattered along the Riviera, which stretches along the coast from Cannes in France to La Spezia in Italy. The region's mild winters and hot summers have attracted wealthy tourists since the early 19th century.

The landscape

The Mediterranean Sea is almost totally landlocked, joined to the Atlantic Ocean through the Strait of Gibraltar, which is only 8 miles (13 km) wide. Lying on an active plate margin, sea floor movements have formed a variety of basins, troughs, and ridges. A submarine ridge running from Tunisia to the island of Sicily divides the Mediterranean into two distinct basins. The western basin is characterized by broad, smooth abyssal (or ocean) plains. In contrast, the eastern basin is dominated by a large ridge system, running east to west.

The narrow Strait of Gibraltar inhibits water exchange between the Mediterranean Sea and the Atlantic Ocean, producing a high degree of salinity and a low tidal range within the Mediterranean. The lack of tides has encouraged the build-up of pollutants in many semienclosed bays.

▲ *Because the Mediterranean* is almost enclosed by land, its circulation is quite different to the oceans. There is one major current which flows in from the Atlantic and moves east. Currents flowing back to the Atlantic are denser and flow below the main current.

Industrial pollution flowing from the Dnieper and Danube rivers has destroyed a large proportion of the fish population that used to inhabit the upper layers of the Black Sea.

The Ionian Basin is the deepest in the Mediterranean, reaching depths of 16,800 ft (5121 m).

The edge of the Eurasian Plate is edged by a continental shelf. In the Mediterranean Sea this is widest at the Ebro Fan where it extends 60 miles (96 km).

Oxygen in the Black Sea is dissolved only in its upper layers; at depths below 230–300 ft (70–100 m) the sea is "dead" and can support no lifeforms other than specially adapted bacteria.

◄ *The Atlas Mountains* are a range of fold mountains that lie in Morocco and Algeria. They run parallel to the Mediterranean, forming a topographical and climatic divide between the Mediterranean coast and the western Sahara.

An arc of active submarine, island and mainland volcanoes, including Etna and Vesuvius, lie in and around southern Italy. The area is also susceptible to earthquakes and landslides.

Nutrient flows into the eastern Mediterranean, and sediment flows to the Nile Delta have been severely lowered by the building of the Aswan Dam across the Nile in Eygpt. This is causing the delta to shrink.

The Suez Canal, opened in 1869, extends 100 miles (160 km) from Port Said to the Gulf of Suez.

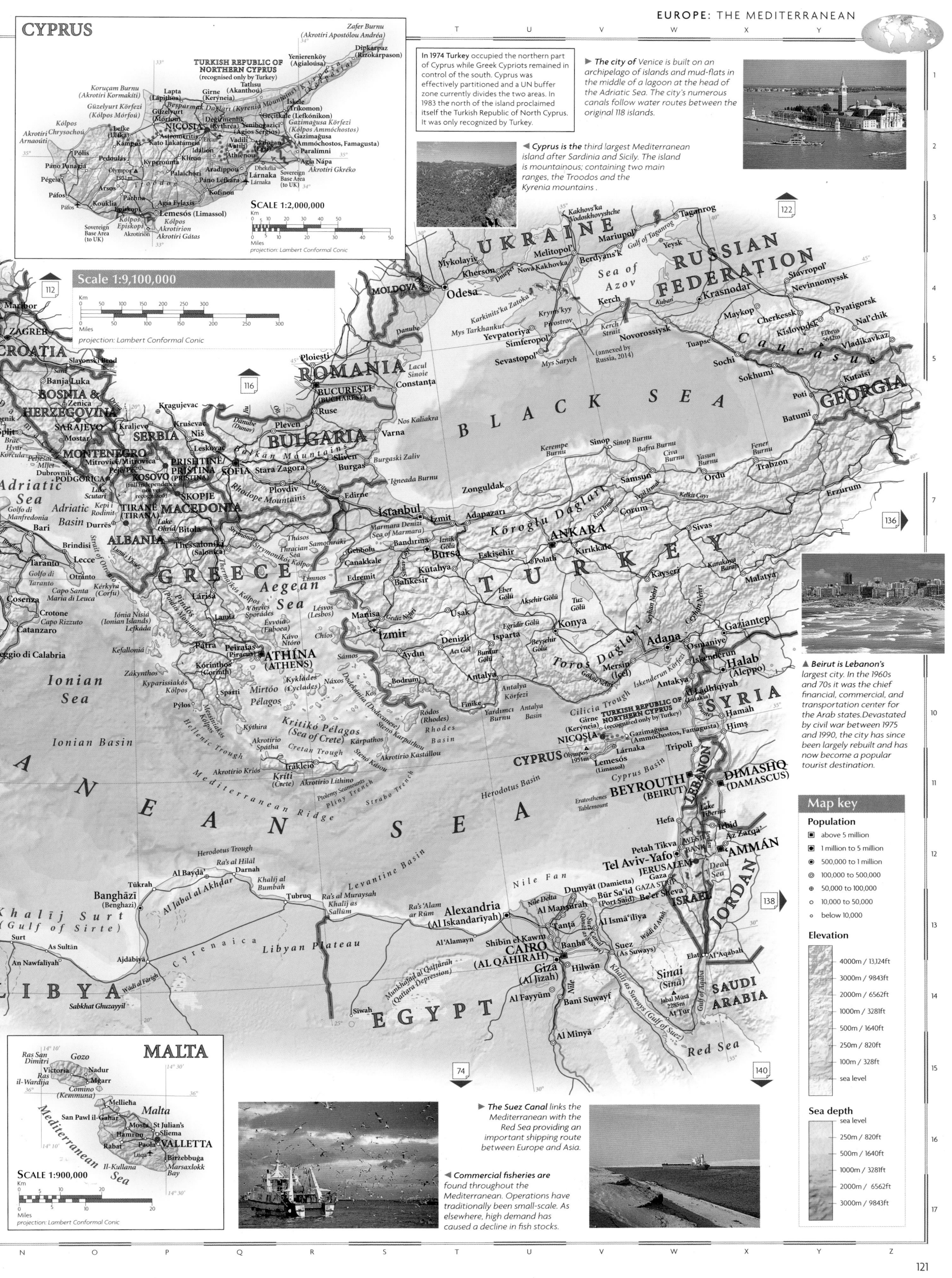

CYPRUS

SCALE 1:2,000,000

Km 0 10 20 30 40 50

Miles 0 10 20 30 40 50

projection: Lambert Conformal Conic

Scale 1:9,100,000

Km 0 50 100 150 200 250 300

Miles 0 50 100 150 200 250 300

projection: Lambert Conformal Conic

In 1974 Turkey occupied the northern part of Cyprus while Greek Cypriots remained in control of the south. Cyprus was effectively partitioned and a UN buffer zone currently divides the two areas. In 1983 the north of the island proclaimed itself the Turkish Republic of North Cyprus. It was only recognized by Turkey.

► The city of Venice is built on an archipelago of islands and mud-flats in the middle of a lagoon at the head of the Adriatic Sea. The city's numerous canals follow water routes between the original 118 islands.

◄ Cyprus is the third largest Mediterranean island after Sardinia and Sicily. The island is mountainous; containing two main ranges, the Troodos and the Kyrenia mountains.

▲ Beirut is Lebanon's largest city. In the 1960s and 70s it was the chief financial, commercial, and transportation center for the Arab states. Devastated by civil war between 1975 and 1990, the city has since been largely rebuilt and has now become a popular tourist destination.

MALTA

SCALE 1:900,000

Km 0 5 10 15 20

Miles 0 5 10 15 20

projection: Lambert Conformal Conic

► The Suez Canal links the Mediterranean with the Red Sea providing an important shipping route between Europe and Asia.

◄ Commercial fisheries are found throughout the Mediterranean. Operations have traditionally been small-scale. As elsewhere, high demand has caused a decline in fish stocks.

Map key

Population
■ above 5 million
▣ 1 million to 5 million
◉ 500,000 to 1 million
⊕ 100,000 to 500,000
⊕ 50,000 to 100,000
○ 10,000 to 50,000
○ below 10,000

Elevation
4000m / 13,124ft
3000m / 9843ft
2000m / 6562ft
1000m / 3281ft
500m / 1640ft
250m / 820ft
100m / 328ft
sea level

Sea depth
sea level
250m / 820ft
500m / 1640ft
1000m / 3281ft
2000m / 6562ft
3000m / 9843ft

The Russian Federation

The Cold War era of global relations was concluded in 1991 with the formal dissolution of the Soviet Union. The Russian Federation declared its separate sovereignty from the foundering communist empire following independence declarations from a number of former Soviet republics. As the leading member of the Commonwealth of Independent States, the Russian Federation has a central role in the development of post-Soviet Eurasia. Crossing 11 time zones, the Russian Federation is almost twice the size of the US, and with more than 150 ethnic minorities and 21 autonomous republics, regionalist dissent within its own territory remains a danger.

THE RUSSIAN FEDERATION: ADMINISTRATIVE REGIONS

124–125
126–127

The administrative area names in European Russia have been omitted west of the Ural Mountains. Please refer to pages 124–125 and 126–127 where these areas are shown at a larger scale.

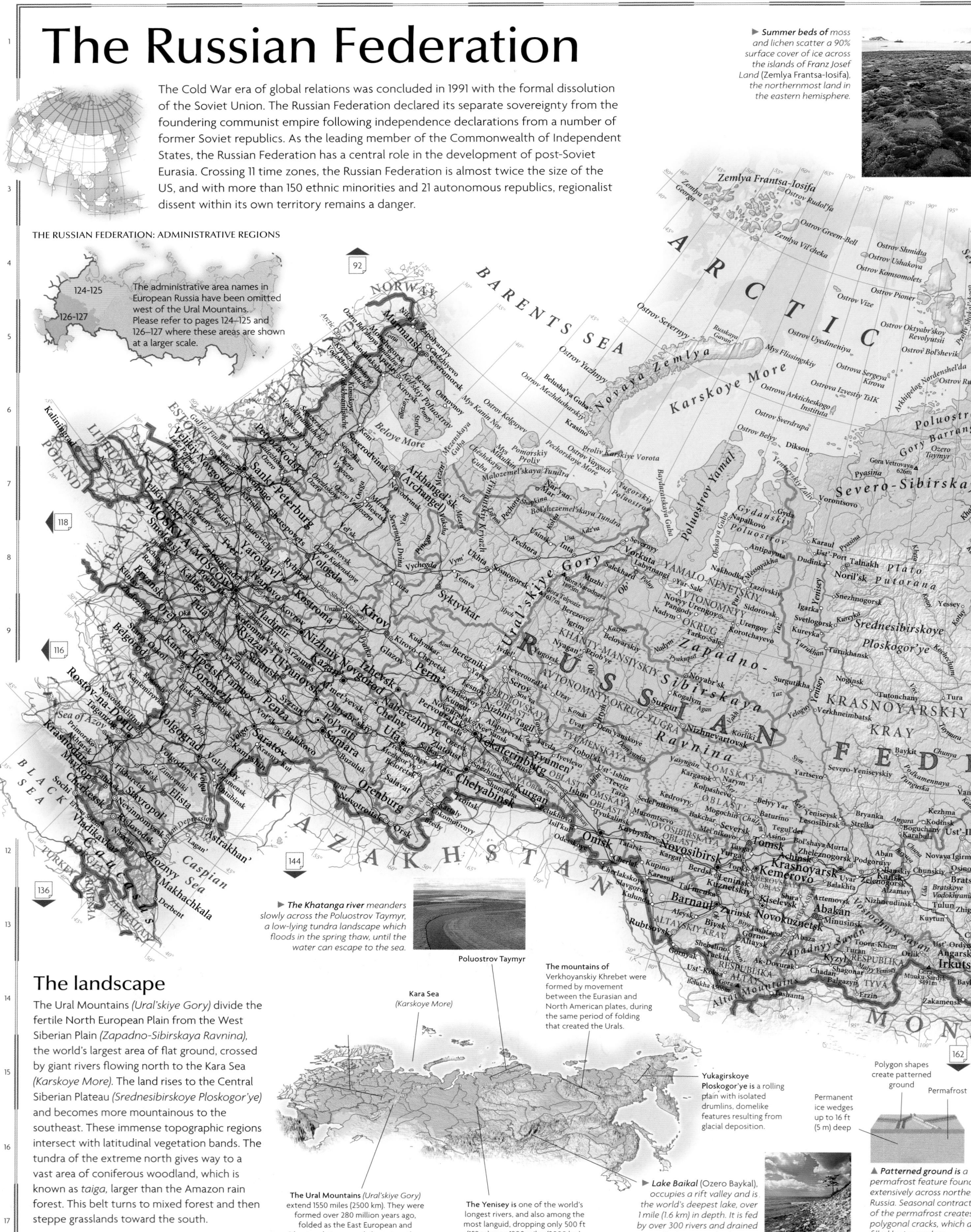

▶ *Summer beds of* moss and lichen scatter a 90% surface cover of ice across the islands of Franz Josef Land (Zemlya Frantsa-Iosifa), the northernmost land in the eastern hemisphere.

▶ *The Khatanga river* meanders slowly across the Poluostrov Taymyr, a low-lying tundra landscape which floods in the spring thaw, until the water can escape to the sea.

Poluostrov Taymyr

Kara Sea (Karskoye More)

The mountains of Verkhoyanskiy Khrebet were formed by movement between the Eurasian and North American plates, during the same period of folding that created the Urals.

The landscape

The Ural Mountains (Ural'skiye Gory) divide the fertile North European Plain from the West Siberian Plain (Zapadno-Sibirskaya Ravnina), the world's largest area of flat ground, crossed by giant rivers flowing north to the Kara Sea (Karskoye More). The land rises to the Central Siberian Plateau (Srednesibirskoye Ploskogor'ye) and becomes more mountainous to the southeast. These immense topographic regions intersect with latitudinal vegetation bands. The tundra of the extreme north gives way to a vast area of coniferous woodland, which is known as *taiga*, larger than the Amazon rain forest. This belt turns to mixed forest and then steppe grasslands toward the south.

The Ural Mountains (Ural'skiye Gory) extend 1550 miles (2500 km). They were formed over 280 million years ago, folded as the East European and Siberian plates moved closer together.

The Yenisey is one of the world's longest rivers, and also among the most languid, dropping only 500 ft (152 m) over 1200 miles (2000 km).

▶ *Lake Baikal* (Ozero Baykal), occupies a rift valley and is the world's deepest lake, over 1 mile (1.6 km) in depth. It is fed by over 300 rivers and drained by just one, the Angara.

Yukagirskoye Ploskogor'ye is a rolling plain with isolated drumlins, domelike features resulting from glacial deposition.

Permanent ice wedges up to 16 ft (5 m) deep

Polygon shapes create patterned ground

Permafrost

▲ *Patterned ground* is a permafrost feature found extensively across northern Russia. Seasonal contraction of the permafrost creates polygonal cracks, which are filled by ice wedges.

Transportation & industry

Raw materials, particularly fossil fuels, ores, and precious metals are abundant, yet often found at sites far from habitation. This inherent "friction of distance" problem was met starting in the 1930s by Soviet commitment to heavy industry and the strategic location of plants east of the Urals. It has left a pattern of isolated and often vast industrial complexes, in remote areas from Vladivostok to Murmansk, in the far north and across European Russia, with lighter manufacturing concentrated in urban areas.

Major industry and infrastructure

- aerospace
- car manufacture
- chemicals
- engineering
- gas
- iron & steel
- mining
- oil
- textiles
- timber processing
- ■ capital cities
- ● major towns
- ✈ international airports
- major roads
- major industrial areas

Transportation network

218,683 miles (351,976 km)	None
53,147 miles (85,542 km)	59,583 miles (95,900 km)

The recent growth of trade with China and East Asia has put pressure on Siberia's inadequate road and rail network, prompting increased use of the Amur river for freight transportation.

▲ *Novosibirsk was established* at the point where the Trans–Siberian railroad crosses the Ob' river. It grew as an industrial center under the Soviet Union and is now Siberia's largest city.

Map key

Population
- ■ above 5 million
- ■ 1 million to 5 million
- ◉ 500,000 to 1 million
- ◎ 100,000 to 500,000
- ⊕ 50,000 to 100,000
- ○ 10,000 to 50,000
- ○ below 10,000

Elevation
- 4000m / 13,124ft
- 3000m / 9843ft
- 2000m / 6562ft
- 1000m / 3281ft
- 500m / 1640ft
- 250m / 820ft
- 100m / 328ft
- sea level

▲ *A fishing trawler* lies at anchor in the icy waters of Karaginskiy Zaliv, at the northern end of the Kamchatka Peninsula (Poluostrov Kamchatka) in eastern Siberia. The Russian Federation's fishing fleet is the largest in the world and operates worldwide.

Using the land

The main agricultural regions follow the belt of rich, black *chernozem* soils between Ukraine and Novosibirsk, producing cereals, fodder, and a broad range of crops for industrial use. Small pockets of pastureland are also found in this region. Large areas of terrain are uncultivable, and the constraints of a severe climate force the Federation to be partly dependent on imported grain. The wilds of Siberia are given over to hunting and reindeer herding, and contain the world's largest timber reserves.

The urban/rural population divide

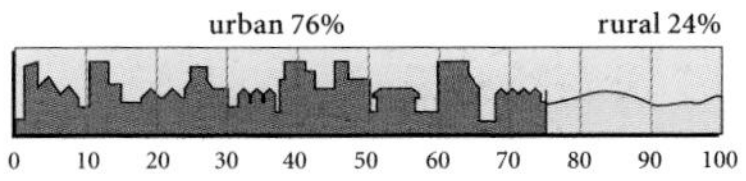

urban 76% — rural 24%

0 10 20 30 40 50 60 70 80 90 100

Population density	Total land area
22 people per sq mile (9 people per sq km)	65,592,800 sq miles (17,075,400 sq km)

Scale 1:18,750,000

Km 0 50 100 200 300 400 500 600
Miles 0 50 100 200 300 400 500 600

projection: Lambert Conformal Conic

◀ *The Kamchatka Peninsula* (Poluostrov Kamchatka) is a volcanic area on the margins of the Eurasian Plate, forming part of the Pacific "Ring of Fire." The volcano Vulkan Klyuchevskaya Sopka, at 15,585 ft (4750 m), is the highest mountain in Siberia.

Land use and agricultural distribution
- cattle
- cereals
- root crops
- timber
- ■ capital cities
- ● major towns
- pasture
- cropland
- forest
- desert
- mountain region
- barren

Northern European Russia

Reaching into the Arctic Circle, this region of lakeland, forest and tundra is historically bound to Europe by St Petersburg, the old imperial capital of Tsarist Russia and home to a third of the region's population. Communist rule from Moscow left the north politically marginalized, contributing to the present problems of outmoded industry, poor infrastructure and serious environmental neglect. However, with borders embracing Finland, Norway, the Baltic and the northern sea route to the Atlantic, the region's success in foreign trade is now of prime importance to the Russian economy.

The landscape

The ancient bedrock of the Scandinavian Shield lies exposed across the glacially scoured Khibiny Mountains of the Kola Peninsula *(Kol'skiy Poluostrov)*, becoming mantled with till toward the North European Plain. The Valdai Hills *(Valdayskaya Vozvyshennost')* form an important watershed for the plain's rivers, while thick forest veils a complicated topography of moraines, lakes, and ground disturbed by frost action. The Ural Mountains *(Ural'skiye Gory)* form a border with Asia in the east.

▲ *The Khibiny mountains were formed by volcanic intrusions into the Scandinavian Shield, over 570 million. years ago.*

Kola Peninsula *(Kol'skiy Poluostrov)*

◄ *The Kola Peninsula* (Kol'skiy Poluostrov) *is part of the Scandinavian Shield, an area of ancient bedrock underlying Scandinavia. Rocks in excess of 2500 million years old are exposed across the peninsula.*

Karst features, including sinkholes, lakes, and caverns, are found in limestone outcrops across the plain of the Severnaya Dvina and Mezen' rivers.

The low-lying plains of the Pechora, Mezen', and Severnaya Dvina rivers were flooded by the sea while the land was still isostatically depressed following the last Ice Age, a process which has hidden the landforms created by glacial deposition.

Retreating glacier | Meltwater channels

Terminal moraine

▲ *Terminal moraines are crescent-shaped ridges of glacial deposits, widely found in central Russia. Detritus is carried by the glacier and deposited at its terminus (snout) as it melts, marking the limit of the ice advance.*

Ural Mountains *(Ural'skiye Gory)*

Two of Europe's biggest rivers, the Volga and Western Dvina, rise in the swampy uplands of the Valdai Hills *(Valdayskaya Vozvyshennost'.)*

► *Lake Onega* (Onezhskoye Ozero) *is the remnant of a body of water which, 12,000 years ago, connected the White Sea* (Beloye More) *with the Gulf of Finland and the Baltic Sea.*

Using the land & sea

The cold climate confines agriculture mainly to southern and western provinces, where dairy farming predominates and arable land is given over to fodder crops as well as flax, potatoes, oats, and rye. Areas beyond the northern margins of cultivation are used for forestry, hunting, herding, and fishing, with some vegetables grown in hothouses around urban areas.

Land use and agricultural distribution

- cattle
- fishing
- reindeer
- timber
- fodder
- major towns

pasture
cropland
forest
mountain region
wetland
tundra
barren
ice

RUSSIAN FEDERATION

The urban/rural population divide

urban 80%   rural 20%

0 10 20 30 40 50 60 70 80 90 100

Population density	Total land area
26 people per sq mile (10 people per sq km)	829,398 sq miles (2,148,700 sq km)

◄ *Many rapids are found along the 175 mile (280 km) course of the Suna river.*

▶ St. Peter and Paul Fortress is the oldest building in St Petersburg, founded by Peter the Great in 1703 as a modern, European capital for Russia.

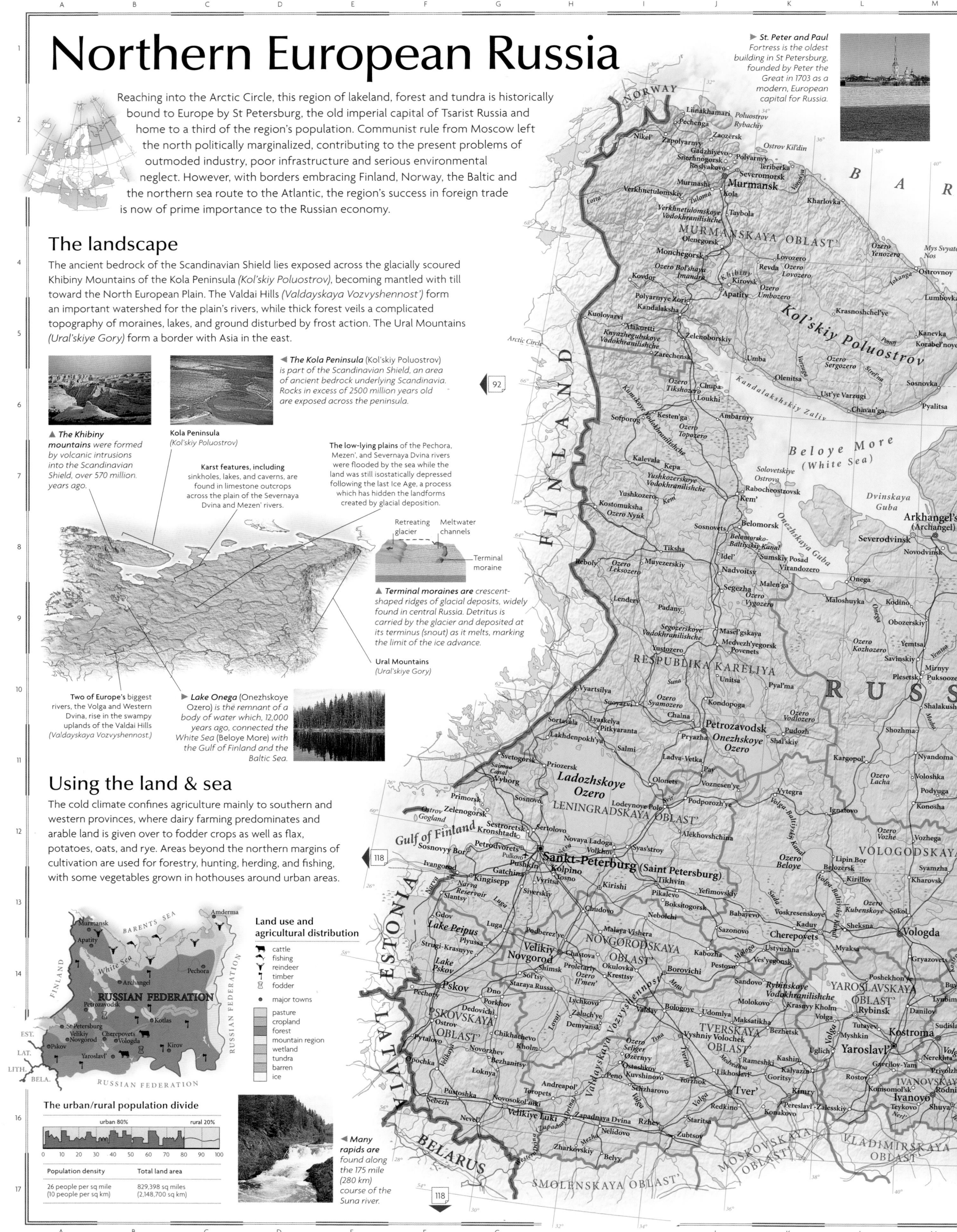

◄ *The Ural Mountains* (Ural'skiye Gory) form the traditional boundary between Europe and Asia. Elevations rarely exceed 6000 ft (1830 m). The region is extremely barren in the far northern latitudes.

Scale 1:5,500,000

Km
Miles

projection: Lambert Conformal Conic

Map key

Population
- 1 million to 5 million
- 500,000 to 1 million
- 100,000 to 500,000
- 50,000 to 100,000
- 10,000 to 50,000
- below 10,000

Elevation
- 1000m / 3281ft
- 500m / 1640ft
- 250m / 820ft
- 100m / 328ft
- sea level

Transportation & industry

The ports of St. Petersburg, Murmansk, and Archangel serve a regional economy led by large-scale resource extraction. Nickel, iron ore, and apatite are mined in the Kola Peninsula (Kol'skiy Poluostrov), and fossil fuels in the Pechora Basin. Paper production is central to Archangel's vast timber industry, while St. Petersburg, drawing on ample labor, has become a major manufacturing center.

Major industry and infrastructure
- chemicals
- coal
- defense
- engineering
- food processing
- hydroelectric power
- mining
- oil & gas
- textiles
- timber processing
- major towns
- international airports
- major roads
- major industrial areas

Transportation network
- 53,700 miles (85,920 km)
- None
- 10,300 miles (16,572 km)
- 12,500 miles (20,000 km)

Railroads linking remote industrial centers with the region's ports are the principal means of supply, although the impressive system of canals, linking natural waterways, is used for freight haulage during the summer.

► *Ice forces the port at St. Petersburg to close in winter, yet Murmansk, on the Barents Sea, remains open, its waters prevented from freezing by warmer ocean currents extending from the North Atlantic Drift.*

Southern European Russia

This region, divided from Asia by desert, seas, and mountains, has exerted a powerful influence both east and west since the 13th century. Over 70 years of Communist rule produced a highly urbanized, industrial society dominated by Moscow, which was the capital of the Soviet Union until 1991. Almost two-thirds of the Russian Federation's population live in this core area, with a relatively high per capita share of its wealth. However, the rapid growth of a market economy has caused great social upheaval, with rising crime and political instability.

The landscape

Ancient folds in the deep sedimentary strata of the North European Plain have created a sequence of high and low regions. The Central Russian Upland (Srednerusskaya Vozvyshennost') in the west is deeply incised by rivers draining into the lowland of the Oka and Don rivers. In the east the Volga, Europe's longest river, flows south to the Caspian Sea, dividing the Volga Uplands (Privolzhskaya Vozvyshennost') from the foothills of the Ural Mountains (Ural'skiye Gory). The Caucasus mountains and the Black Sea form a natural border to the southwest.

▶ Kaliningrad has been a Russian enclave since 1945. The port is an important center for the Russian Federation's Baltic fishing fleet.

◀ St Basil's Cathedral, completed in 1561, stands in Moscow's Red Square next to the Kremlin; the original fortified stronghold of the city.

▲ A plantation of Scots pine helps consolidate the loose sandy soils of the Meshchera Lowland (Meshcherskaya Nizmennost'), which lies on the bed of an old glacial lake.

The Smolensk-Moscow Upland (Smolensko-Moskovskaya Vozvyshennost') is a series of terminal moraine ridges marking the southern extent of the last glaciation.

Glacial till covers the bedrock to the north of the North European Plain, giving a gentle surface relief.

The lowland of the Oka and Don rivers lies over a broad trough, between the upfolds of the Volga Uplands (Privolzhskaya Vozvyshennost') to the east, and the Central Russian Upland (Srednerusskaya Vozvyshennost') to the west.

The southern Ural mountains (Ural'skiye Gory) consist of several parallel ranges of ancient fold mountains running from north to south.

Central Russian Upland (Srednerusskaya Vozvyshennost').

The floodplain of the Volga forms a long oasis of verdant vegetation, contrasting with the aridity of the surrounding Caspian hinterland.

The marshlands of the Volga Delta are visited by over 260 species of bird each year, migrating between South Africa and Arctic Siberia.

The Caspian Depression is a large downfold (or syncline) which became flooded, forming the Caspian Sea. The shoreline is 98 ft (30 m) below sea level.

◀ The Caucasus mountains run from the Black Sea to the Caspian Sea. They include El'brus which, at 18,511 ft (5642 m), is the highest point in Europe. It is still uplifting at a rate of 0.4 inches (10 mm) per year.

Drifting sand occupies large areas of the south, forming dunes up to 50 ft (15 m) high.

Salt dome

Salt dome is forced up and through the rock strata

Sedimentary strata

Salts are forced upwards by denser overlying strata

▲ Salt domes, rounded hills up to 500 ft (150 m) high, are produced as less dense rock salts are displaced under the extreme pressure of denser, overlying strata and forced up toward the surface creating domes. They are widespread in the Caspian Depression.

Map key

Population

- above 5 million
- 1 million to 5 million
- 500,000 to 1 million
- 100,000 to 500,000
- 50,000 to 100,000
- 10,000 to 50,000
- below 10,000

Elevation

- 4000m / 13,124ft
- 3000m / 9843ft
- 2000m / 6562ft
- 1000m / 3281ft
- 500m / 1640ft
- 250m / 820ft
- 100m / 328ft
- sea level

Using the land

In the cold, humid north and in the southern Urals (*Ural'skiye Gory*), small grains, potatoes, and flax are commonly rotated with legumes which support livestock farming. The rich *chernozem* (or black earth) areas support diverse crops such as sugar beet, hemp, sunflowers, millet, and vegetables. Further south, aridity restricts husbandry to extensive grazing, with intensive fruit and rice cultivation along the oasis of the Volga.

The urban/rural population divide

urban 71%   rural 29%

0 10 20 30 40 50 60 70 80 90 100

Population density: 119 people per sq mile (46 people per sq km)

Total land area: 705,916 sq miles (1,828,800 sq km)

Land use and agricultural distribution

- sheep
- flax
- potatoes
- rice
- sunflowers
- sugar beet
- timber
- capital cities
- major towns
- pasture
- cropland
- forest
- wetland
- mountain region
- tundra

Transportation & industry

Manufacturing is largely based around Moscow and the Volga region, which became a major industrial area during the Second World War. Both Moscow and Nizhniy Novgorod are centers of skilled labor for light manufacturing and engineering. Most of Russia's main chemical plants are located along the Volga, and one of the world's largest car factories was recently opened in Tol'yatti. Processing and machine construction plants use oil, gas, and hydroelectric power from the Volga Basin and metallic minerals from the Urals (*Ural'skiye Gory*) and Kursk.

◀ *Industrial plants are* massed along the Volga. Environmental stress from decades of unbridled industrial development has prompted widespread concern about pollution levels.

Transportation network

- 250,000 miles (402,000 km)
- None
- 28,000 miles (44,800 km)
- 16,300 miles (26,080 km)

Seventy private and national flag airlines have been created from the reorganization of the state airline Aeroflot, which maintained the world's largest fleet of aircraft during the Soviet era.

Major industry and infrastructure

- aerospace
- car manufacture
- chemicals
- defense
- electronics
- engineering
- gas
- mining
- oil
- textiles
- capital cities
- major towns
- international airports
- major roads
- major industrial areas

Asia

Asia, the world's largest continent, covers 16,838,365 sq miles (43,608,000 sq km). It comprises 49 separate countries, including 97% of Turkey and 72% of the Russian Federation. Almost 60% of the world's population lives in Asia.

● **Greatest extent, North–South:**
  4000 miles / 6440 km

■ **Greatest extent, East–West:**
  6000 miles / 9650 km

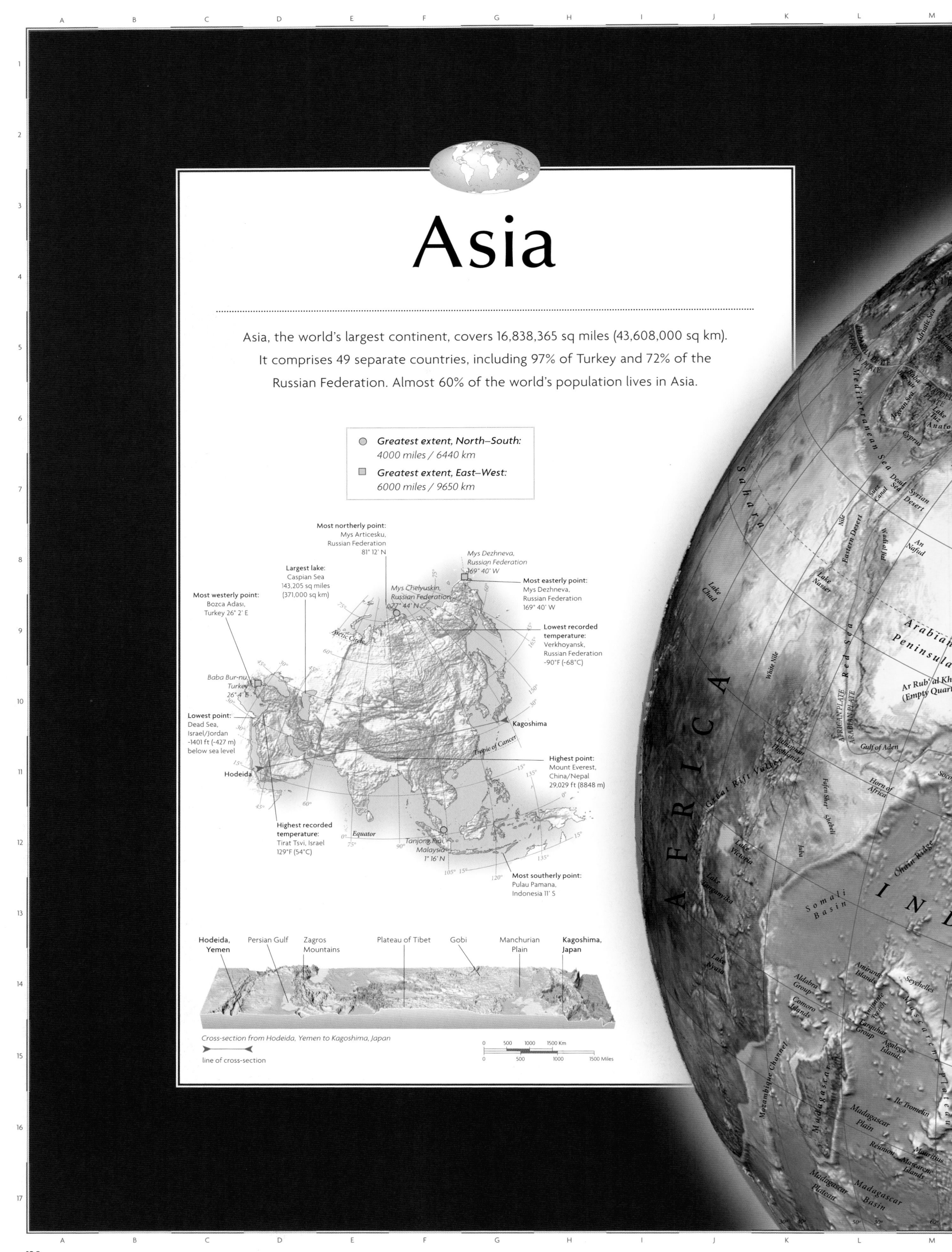

Most northerly point:
Mys Articesku,
Russian Federation
81° 12′ N

*Mys Dezhneva,
Russian Federation
169° 40′ W*

Largest lake:
Caspian Sea
143,205 sq miles
(371,000 sq km)

*Mys Chelyuskin,
Russian Federation
77° 44′ N*

Most easterly point:
Mys Dezhneva,
Russian Federation
169° 40′ W

Most westerly point:
Bozca Adası,
Turkey 26° 2′ E

Lowest recorded temperature:
Verkhoyansk,
Russian Federation
-90°F (-68°C)

*Baba Bur-nu,
Turkey
26° 4′ E*

Lowest point:
Dead Sea,
Israel/Jordan
-1401 ft (-427 m)
below sea level

Kagoshima

Tropic of Cancer

Highest point:
Mount Everest,
China/Nepal
29,029 ft (8848 m)

Hodeida

Highest recorded temperature:
Tirat Tsvi, Israel
129°F (54°C)

Equator

*Tanjong Piai,
Malaysia
1° 16′ N*

Most southerly point:
Pulau Pamana,
Indonesia 11′ S

Hodeida,
Yemen | Persian Gulf | Zagros Mountains | Plateau of Tibet | Gobi | Manchurian Plain | **Kagoshima, Japan**

Cross-section from Hodeida, Yemen to Kagoshima, Japan

→ line of cross-section

0 500 1000 1500 Km
0 500 1000 1500 Miles

ARCTIC OCEAN
North Pole
NORTH AMERICAN PLATE
EURASIAN PLATE

Norwegian Sea
Scandinavia
North Cape
Barents Sea
Novaya Zemlya
Kara Sea
Severnaya Zemlya
Mys Chelyuskin
Laptev Sea
New Siberian Islands
East Siberian Sea
Long Strait
Bering Strait
Chukot Range
Bering Sea

Franz Josef Land
Kola Peninsula
White Sea
Gulf of Bothnia
North Sea
Baltic Sea
Gulf of Finland

EUROPE
North European Plain
Central Russian Upland
Rhine
Dnieper
Dniester
Danube
Don
Volga
Oka
Kama
Pechora
Ural Mountains

Black Sea
Sea of Azov
Caspian Depression
Caspian Sea
Caucasus
Lake Van
Lake Urmia
Zagros Mountains
Iranian Plateau
Great Salt Desert
EURASIAN PLATE
IRANIAN PLATE
Elburz Mountains
Namak
Daryacheh ye
Kuh-e Baba 5143m △
Kuh-e Fuladi △
Hindu Kush
Pamirs
Karakoram Range
8611m △
Hamun
Jaz Murian
Central Makran Range
Rigestan
Sulaiman Range
Khalij Masirah
Gulf of Oman
Strait of Hormuz
Oman Basin
ARABIAN PLATE
INDO-AUSTRALIAN PLATE

West Siberian Plain
SIBERIA
North Siberian Lowland
Putorana Mountains
Kureyka
Central Siberian Plateau
Lower Tunguska
Stony Tunguska
Angara
Yenisey
Lena
Vilyuy
Markha
Olenek
Aldan
Amga
Vitim
Lake Baikal
Shilka
Chikoy
Selenga
Orkhon Gol
Ozero Tengiz
Lake Chany
Ob'
Irtysh
Tobol
Ishim
Om'
Tym
Vasyugan
Agan
Nadym
Kirghiz Steppe
Sarysu
Syr Darya
Lake Balkhash
Ili
Emba
Ural
Aral Sea
Ustyurt Plateau
Turan Lowland
Kara Kum
Amu Darya
Aydarkul
Naryn
Ozero Issyk-Kul
Ozero Alakol
Ozero Zaysan
Tien Shan
Dzungaria
Altai Mountains
Uvs Nuur
Tes-River
Hyargas Nuur
Har Us Nuur
Sayanskiy Khrebet

Verkhoyanskiy Khrebet
Indigirka
Kolyma
Khrebet Cherskogo
Koryak Range
Kamchatka
Sea of Okhotsk
Stanovoy Khrebet
Zeya Reservoir
Amur
Sablonovyy Khrebet
Sredinnyy Khrebet
Kuril–Kamchatka Trench
Kuril Islands
Hokkaido
Honshu
Sea of Japan (East Sea)
Japan Trench
PACIFIC OCEAN

ASIA

Plateau of Mongolia
Gobi
Great Khingan Range
Hulun Nur
Kerulen
Manchurian Plain
Xar Moron
Lake Khanka

Tarim He
Takla Makan Desert
Kongi He
Lop Nur
Shule He
Qarqan He
Keriya He
Yarkant He
Hotan He
Tarim Basin
Altun Shan
Nan Shan
Qilian Shan
Kunlun Mountains
Plateau of Tibet
Siling Co
Dogai Coring
Tangra Yumco
Nam Co
Yellow River
Yangtze
Ordos Desert
Wutai Shan
Qinghai Hu
Bayan Har Shan
Xiqing Shan
Bo Hai
Korea Bay
Korea Strait
Jeju-do
Yellow Sea
Great Plain of China
Huai He
Wei He
Han Shui
Hong Hu
Dongting Hu
Poyang Hu
Tai Hu
East China Sea
Ryukyu Islands
Taiwan
Okinawa Trough
Shikoku Basin

Himalayas
Annapurna 8091m △
Mount Everest 8848m △
Brahmaputra
Ganges
Ghaghara
Yamuna
Chambal
Betwa
Son
Damodar
Mahanadi
Godavari
Wainganga
Indravati
Khasi Hills
Mouths of the Ganges
Arakan Yoma
Chindwin
Irrawaddy
Sittang
Salween
Mekong
Red River
Black River
Gulf of Tonkin
Hainan
Hainan Strait
Luzon Strait
Philippine Sea
Luzon
PHILIPPINE PLATE
PHILIPPINE SEA PLATE
Mindoro
Panay
Negros
Samar
Palau
Philippine Trench
Philippine Basin
CAROLINE PLATE

Thar Desert
Lumi
Punjab Plains
Sutlej
Indus
Jhelum
Chenab
Gulf of Kachchh
Sabarmati
Mahi
Banas
Narmada
Tapti
Vindhya Range
Satpura Range
Ajanta Range
Gulf of Khambhat
Bhima
Deccan
Krishna
Tungabhadra
Penneru
Kaveri
Western Ghats
Eastern Ghats
Malabar Coast
Coromandel Coast
Cape Comorin
Gulf of Mannar
Sri Lanka
Arabian Sea
Arabian Basin
Laccadive Islands
Maldives
Laccadive Plateau
Chagos-Laccadive Plateau
Chagos Bank
Chagos Trench
Mid-Indian Basin
Mid-Indian Ridge
INDIAN OCEAN
Ceylon Plain
Ninetyeast Ridge
Nikitin Seamount △
Cocos Basin
Cocos Islands
Christmas Island
Investigator Ridge
INDO-AUSTRALIAN PLATE
AFRICAN PLATE
Owen Fracture Zone

Bay of Bengal
Gulf of Martaban
Andaman Islands
Andaman Sea
Nicobar Islands
Isthmus of Kra
Gulf of Thailand
Tônlé Sap
Chao Phraya
Mun
Chi
Mekong
Mouths of the Mekong
South China Sea
South China Basin
EURASIAN PLATE
Malay Peninsula
Strait of Malacca
Danau Toba
Tanjung Piai
Sunda
Anambas Islands
Natuna Islands
Sunda Shelf
Greater Sunda Islands
Borneo
Kapuas Sungai
Gunung Kinabalu 4094m △
Sulu Sea
Mindanao
Celebes Sea
Celebes
Makassar Strait
Buru
Seram
Molucca Sea
Halmahera
New Guinea Trench
Banda Sea
BISMARCK PLATE
New Guinea
Mentawai Ridge
Sumatra
Gunung Kerinci 3806m △
Pulau Bangka
Selat Sunda
East Indies
Java Sea
Java
Bali
Flores Sea
Lesser Sunda Islands
Sumba Islands
Timor
Arafura Sea
Torres Strait
Sunda Trough
Java Trench
Timor Trough
AUSTRALIA
Tropic of Cancer
Equator
Tropic of Capricorn

Physical Asia

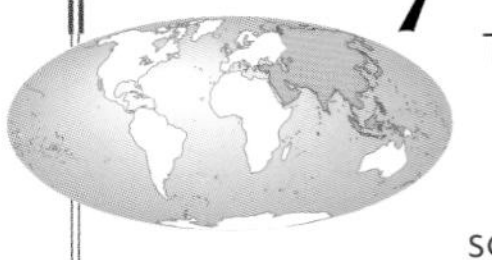

The structure of Asia can be divided into two distinct regions. The landscape of northern Asia consists of old mountain chains, shields, plateaus, and basins, like the Ural Mountains in the west and the Central Siberian Plateau to the east. To the south of this region, are a series of plateaus and basins, including the vast Plateau of Tibet and the Tarim Basin. In contrast, the landscapes of southern Asia are much younger, formed by tectonic activity beginning about 65 million years ago, leading to an almost continuous mountain chain running from Europe, across much of Asia, and culminating in the mighty Himalayan mountain belt, formed when the Indo-Australian Plate collided with the Eurasian Plate. They are still being uplifted today. North of the mountains lies a belt of deserts, including the Gobi and the Takla Makan. In the far south, tectonic activity has formed narrow island arcs, extending over 4000 miles (7000 km). To the west lies the Arabian Shield, once part of the African Plate. As it was rifted apart from Africa, the Arabian Plate collided with the Eurasian Plate, uplifting the Zagros Mountains.

Coastal Lowlands and Island Arcs

The coastal plains that fringe Southeast Asia contain many large delta systems, caused by high levels of rainfall and erosion of the Himalayas, the Plateau of Tibet, and relict loess deposits. To the south is an extensive island archipelago, lying on the drowned Sunda Shelf. Most of these islands are volcanic in origin, caused by the subduction of the Indo-Australian Plate beneath the Eurasian Plate.

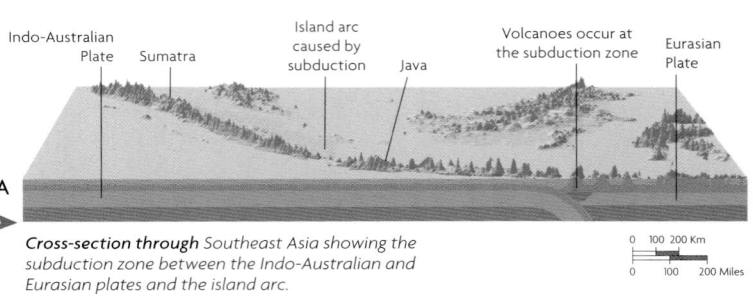

Indo-Australian Plate · Sumatra · Island arc caused by subduction · Java · Volcanoes occur at the subduction zone · Eurasian Plate

Cross-section through Southeast Asia showing the subduction zone between the Indo-Australian and Eurasian plates and the island arc.

0 100 200 Km
0 100 200 Miles

The Indian Shield and Himalayan System

The large shield area beneath the Indian subcontinent is between 2.5 and 3.5 billion years old. As the floor of the southern Indian Ocean spread, it pushed the Indian Shield north. This was eventually driven beneath the Plateau of Tibet. This process closed up the ancient Tethys Sea and uplifted the world's highest mountain chain, the Himalayas. Much of the uplifted rock strata was from the seabed of the Tethys Sea, partly accounting for the weakness of the rocks and the high levels of erosion found in the Himalayas.

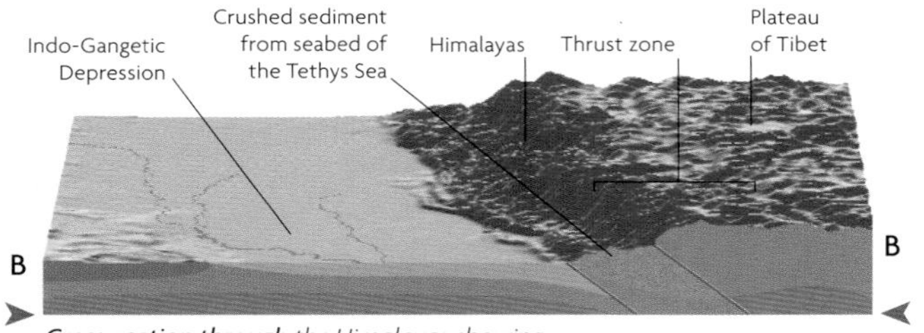

Indo-Gangetic Depression · Crushed sediment from seabed of the Tethys Sea · Himalayas · Thrust zone · Plateau of Tibet

Cross-section through the Himalayas showing thrust faulting of the rock strata.

0 50 100 Km
0 50 100 Miles

East Asian Plains and Uplands

Several, small, isolated shield areas, such as the Shandong Peninsula, are found in east Asia. Between these stable shield areas, large river systems like the Yangtze and the Yellow River have deposited thick layers of sediment, forming extensive alluvial plains. The largest of these is the Great Plain of China, the relief of which does not rise above 300 ft (100 m).

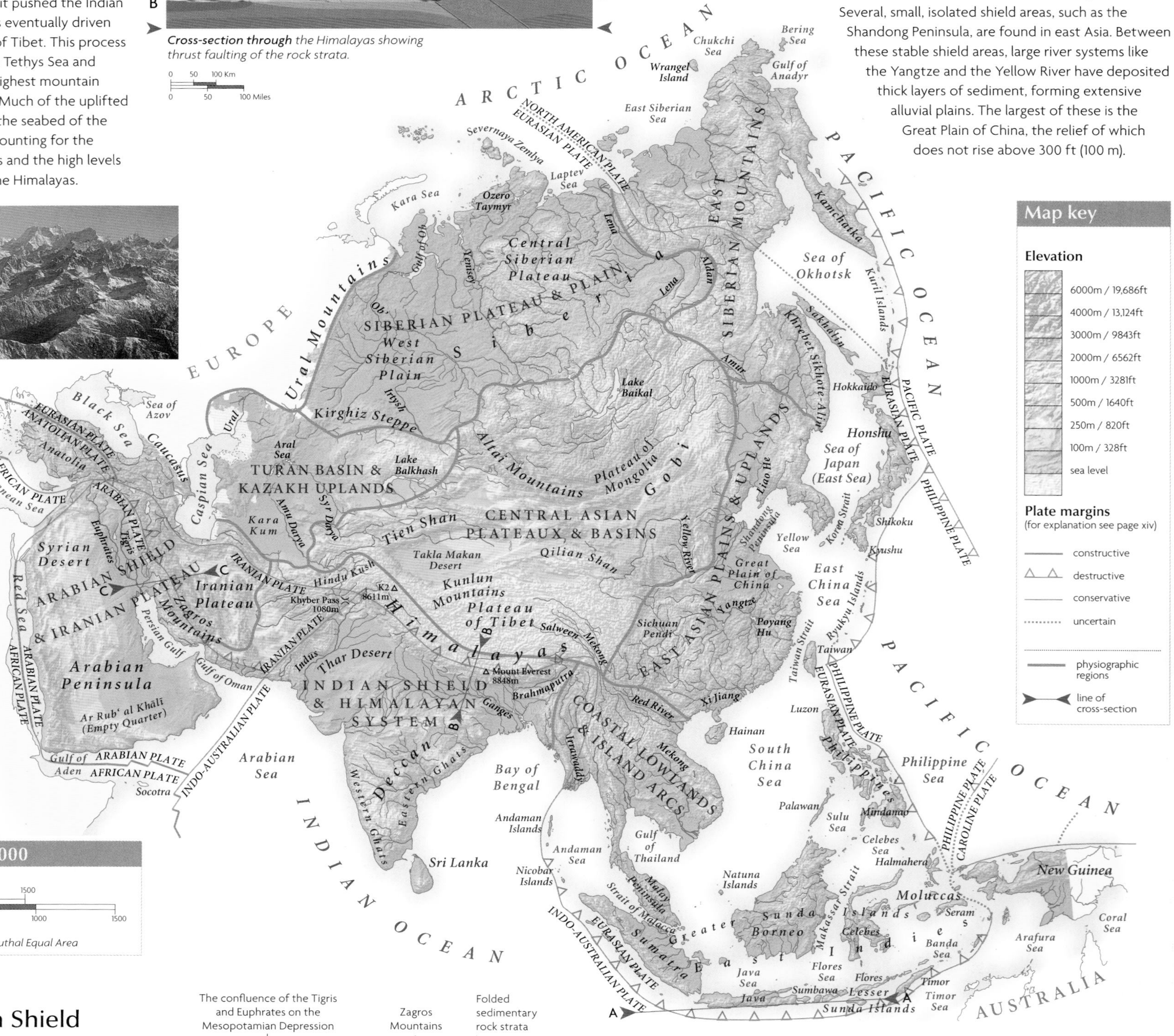

Map key

Elevation

6000m / 19,686ft
4000m / 13,124ft
3000m / 9843ft
2000m / 6562ft
1000m / 3281ft
500m / 1640ft
250m / 820ft
100m / 328ft
sea level

Plate margins
(for explanation see page xiv)

——— constructive
△ △ destructive
——— conservative
········· uncertain
——— physiographic regions
▶◀ line of cross-section

Scale 1:56,750,000

Km
0 250 500 1000 1500
Miles
0 250 500 1000 1500

projection: Lambert Azimuthal Equal Area

The Arabian Shield and Iranian Plateau

Approximately five million years ago, rifting of the continental crust split the Arabian Plate from the African Plate and flooded the Red Sea. As this rift spread, the Arabian Plate collided with the Eurasian Plate, transforming part of the Tethys seabed into the Zagros Mountains which run northwest-southeast across western Iran.

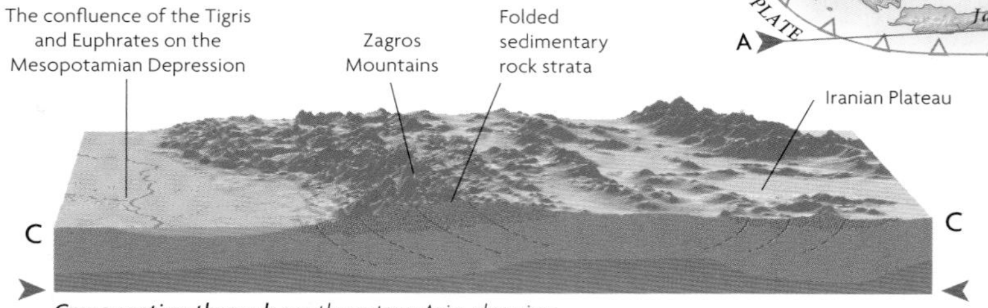

The confluence of the Tigris and Euphrates on the Mesopotamian Depression · Zagros Mountains · Folded sedimentary rock strata · Iranian Plateau

Cross-section through southwestern Asia, showing the Mesopotamian Depression, the folded Zagros Mountains, and the Iranian Plateau.

0 50 100 Km
0 50 100 Miles

Climate

The climate of Asia exhibits marked differences from region to region, with freezing polar conditions in the north, hot and cold deserts in central regions and subtropical conditions throughout the south. Much of this variation can be attributed to enormous mountain barriers and internal depressions found across the continent. Monsoon winds, which reverse semiannually, cause alternate wet and dry seasons across southern Asia. These air masses moving north from the ocean are stripped of their moisture over the Himalayas causing arid conditions across the Plateau of Tibet. Both the south and east are susceptible to tropical cyclones or typhoons.

▲ *Tropical cyclones occur* principally during late summer and early fall. The intense winds and heavy rainfall can devastate entire villages.

Temperature

Average January temperature

Average July temperature

Temperature

below -22°F (-30°C)	32 to 50°F (0 to 10°C)
-22 to -4°F (-30 to -20°C)	50 to 68°F (10 to 20°C)
-4 to 14°F (-20 to -10°C)	68 to 86°F (20 to 30°C)
14 to 32°F (-10 to 0°C)	above 86°F (30°C)

Climate

tundra	daily hours of sunshine, January
subarctic	daily hours of sunshine, July
cool continental	cyclone
warm humid	typhoon
mediterranean	cold/dry monsoon
semi-arid	warm/wet monsoon
arid	cold wind
humid equatorial	
tropical	

▶ *The Gobi Desert* experiences major extremes in climate, with winter temperatures sometimes falling below -40°C (-40°F) and summer temperatures exceeding 45°C (113°F).

Rainfall

Average January rainfall

Average July rainfall

Rainfall

0–1 in (0 –25 mm)
1–2 in (25–50 mm)
2–4 in (50–100 mm)
4–8 in (100–200 mm)
8–12 in (200–300 mm)
12–16 in (300–400 mm)
16–20 in (400–500 mm)
more than 20 in (500 mm)

◀ *Through India, the* southwest monsoon, which brings heavy rainfall from May to September, accounts for 80% of annual precipitation.

Shaping the landscape

In the north, melting of extensive permafrost leads to typical periglacial features such as thermokarst. In the arid areas wind action transports sand creating extensive dune systems. An active tectonic margin in the south causes continued uplift, and volcanic and seismic activity, but also high rates of weathering and erosion. Across the continent, huge rivers erode and transport vast quantities of sediment depositing it on the plains or forming large deltas.

River systems

[1] Vast river systems flow across Asia, many originating in the Himalayas and the Plateau of Tibet. Seasonal melting of snow and monsoon rains swell the river flow leading to flooding and erosion. The Yellow River *(right)* gets its color from the high level of eroded material from the loess plateau.

Monsoon rains
Snow melt
Yellow River dissects loess plateau
Carries large sediment load

River systems: erosion of the loess plateau by the yellow river

Chemical weathering

[2] Tower karsts are widespread across south China *(left)* and Vietnam. It is thought the karstic towers were formed under a soil cover, where small depressions in the limestone bedrock began to be weathered by soil water acids, eventually creating larger hollows. This process continued over millions of years, deepening the hollows and leaving steep-sided limestone hills.

Limestone hills
Old soil cover
Hollow being eroded by soil water acidity
Eroded hollow

Chemical weathering: formation of tower karst

Volcanic activity

[3] Volcanic eruptions occur frequently across Southeast Asia's island arcs *(below)*. Low-level eruptions occur when groundwater, superheated by underlying magma, becomes pressurized, forcing hot fluid and rocks up through cracks in the volcanic cone. This is known as aphreatic eruption.

Eruption within volcanic cone
Fluid and rocks rising under pressure
Heated groundwater
Heat rising from the magma chamber

Volcanic activity: a phreatic eruption

Sedimentation

[4] The Ganges/Brahmaputra is a tide-dominated delta *(below)*. The two rivers transport huge quantities of mountain sediment, which is deposited on the delta plain. This debris is then redistributed by tidal currents, to form extensions to the bars, beach ridges, and deltaic deposits.

Distributary channels
Ganges/Brahmaputra River
Delta plain
Redistributed sediment
Sea level at high tide

Sedimentation: the destruction of a delta

Landscape

limestone region	area of tectonic activity
sinking land	limit of permafrost
stable land	
uplifting land	ocean current
active volcano	

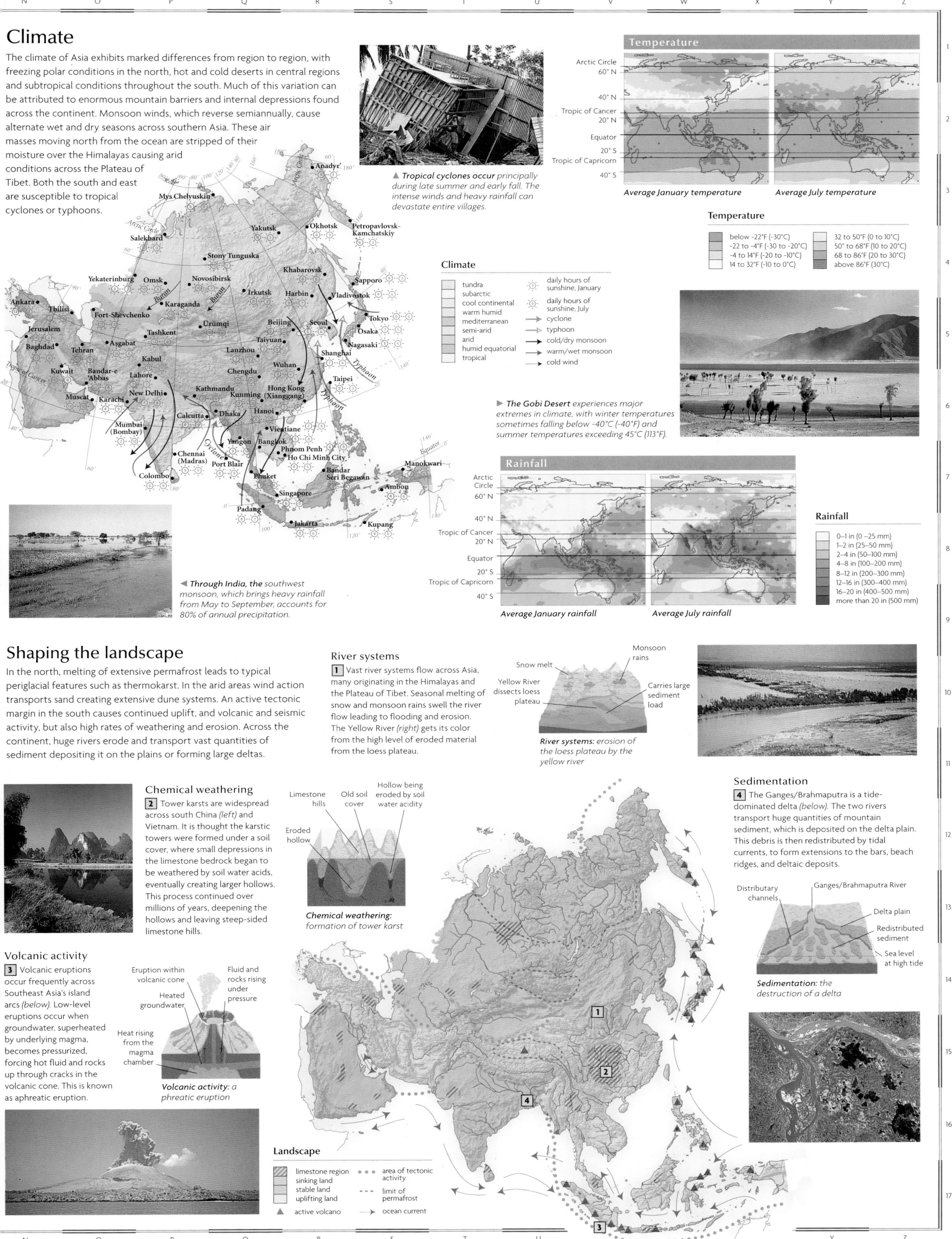

Political Asia

Asia is the world's largest continent, encompassing many different and discrete realms, from the desert Arab lands of the southwest to the subtropical archipelago of Indonesia; from the vast barren wastes of Siberia to the fertile river valleys of China and South Asia, seats of some of the world's most ancient civilizations. The collapse of the Soviet Union has fragmented the north of the continent into the Siberian portion of the Russian Federation, and the new republics of Central Asia. Strong religious traditions heavily influence the politics of South and Southwest Asia. Hindu and Muslim rivalries threaten to upset the political equilibrium in South Asia where India – in terms of population – remains the world's largest democracy. Communist China another population giant, is reasserting its position as a world and political power, while on its doorstep, the economically progressive and dynamic Pacific Rim countries, led by Japan, continue to assert their worldwide economic force.

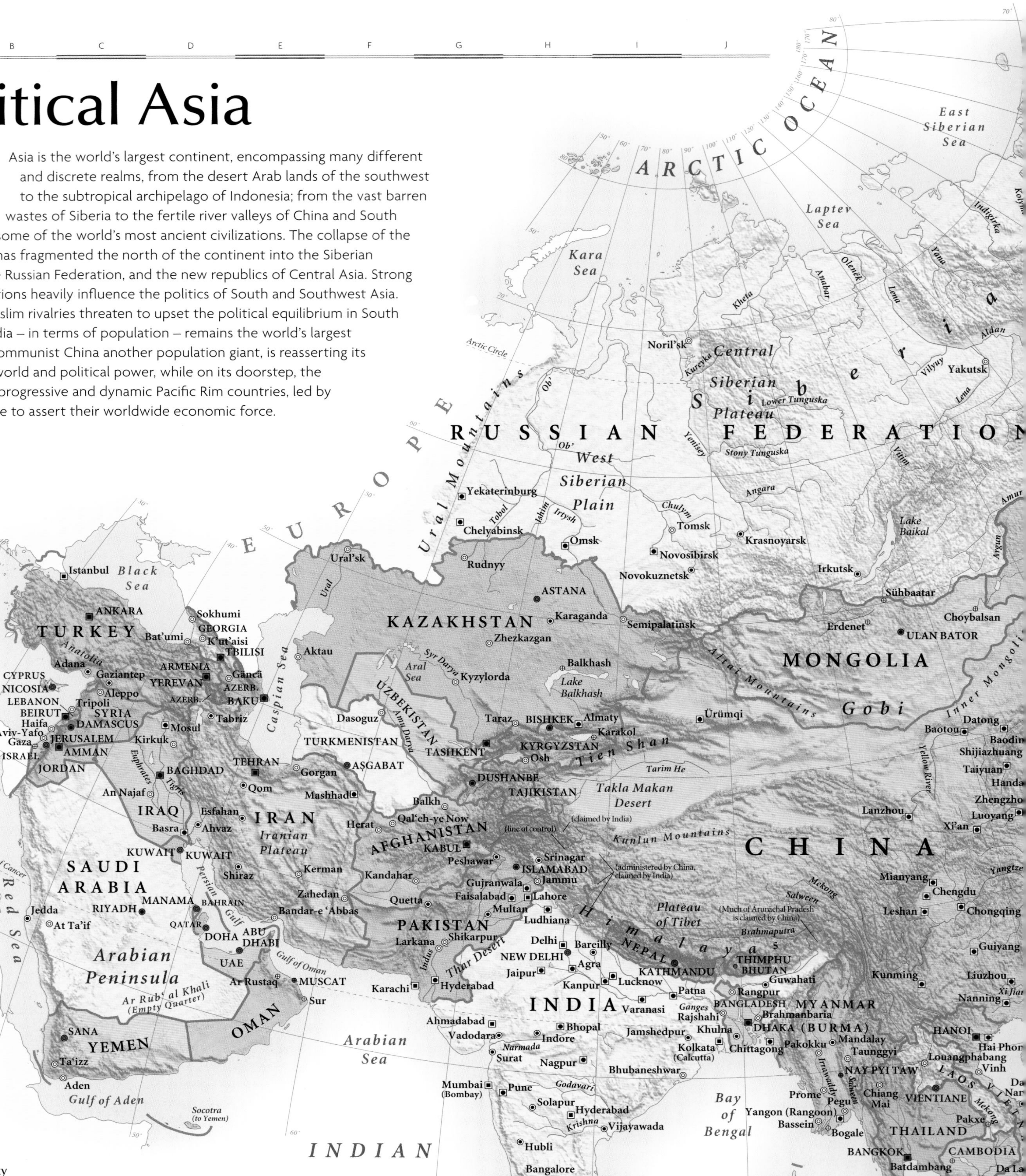

Population density
(people per sq mile)

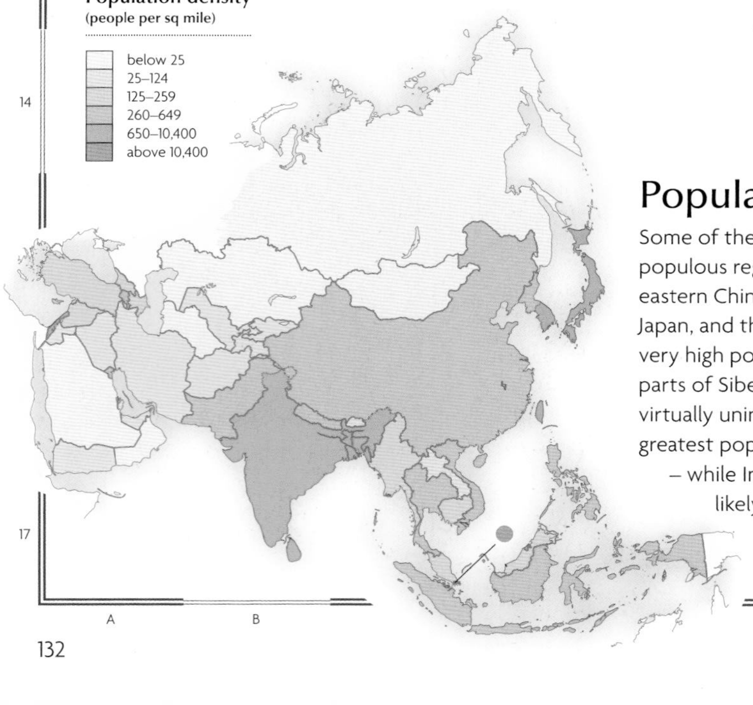

- below 25
- 25–124
- 125–259
- 260–649
- 650–10,400
- above 10,400

Population

Some of the world's most populous and least populous regions are in Asia. The plains of eastern China, the Ganges river plains in India, Japan, and the Indonesian island of Java, all have very high population densities; by contrast parts of Siberia and the Plateau of Tibet are virtually uninhabited. China has the world's greatest population – 20% of the globe's total – while India, with the second largest, is likely to overtake China within 30 years.

◀ *Over 13 million* people bustle through Kolkata's maze of crowded, narrow streets. Population densities in India's largest city reach almost 85,000 per sq mile (33,000 per sq km).

Languages

During the 19th century, Russian was introduced into Central Asia and Siberia. Under the Soviet regime, Russian-speaking became mandatory – replacing the indigenous Ural-Altaic languages in many urban areas – although today the use of Central Asian languages is being revived in the new republics. India's linguistic mosaic comprises Dravidian languages, such as Tamil, in the south, and the Indo-Aryan languages of the north such as Hindi. In China, three main languages, Mandarin Chinese, Wu Chinese, and Cantonese, share the same written form but their spoken dialects are mutually unintelligible.

▲ *Each year, Mongolians* celebrate their ancient culture at the Naadam festival of the Three Games of Men. Children aged between 7 and 12 take part in the finale; a 20 mile (32 km) cross-country horse race in full traditional dress.

Language groups

Indo-European	Dravidian
Ural-Altaic	Papuan
Sino-Tibetan	Austro-Asiatic
Hamito-Semitic	Paleo-Asiatic
Austronesian	Caucasian
Japanese and Korean	Uninhabited

Transportation

The transportation system varies enormously in extent and quality across Asia. Early trade routes included the Silk Route, from Beijing across Central Asia, and the sea routes around the coastline of southern Asia. Today, transportation networks often radiate from coastal ports, reflecting the continuing importance of sea and river travel for trade and external communications. In the interior, high mountain barriers such as the Himalayas, the Altai Mountains and the Tien Shan, deserts like the Gobi, Takla Makan, and Ar Rub' al Khali, remain virtually impenetrable to most modern terrestrial transportation. Major engineering feats are necessary to conquer these hostile frontier territories, although the success of the Trans-Siberian Railroad in overcoming the harsh Siberian landscape, proves that cross-continental transportation, if not economically viable, is physically possible.

Transportation

- —— major roads and highways
- —— major railroads
- —— international borders
- • transport intersections
- ⊕ international airports
- ⊕ major ports

Map key

Population
- ■ above 5 million
- ■ 1 million to 5 million
- ◉ 500,000 to 1 million
- ◎ 100,000 to 500,000
- ◦ 50,000 to 100,000
- ◦ 10,000 to 50,000
- ● Country capital

Borders
- full international border
- disputed de facto border
- disputed territorial claim border
- undefined border
- ceasefire line

Scale 1:29,250,000

projection: Lambert Azimuthal Equal Area

▲ *Both India and* China rely upon extensive railroad systems to transport freight and passengers. China's network is constantly expanding, in particular the link between Golmud and Lhasa, which was completed in 2006 to become the highest railroad in the world.

▲ *The Karakoram Highway* linking Mansehra in northern Pakistan with Kashi in western China was finally completed in 1978, 20 years after construction began. Regular mudslides and rockfalls necessitate continual maintenance for the road to remain open.

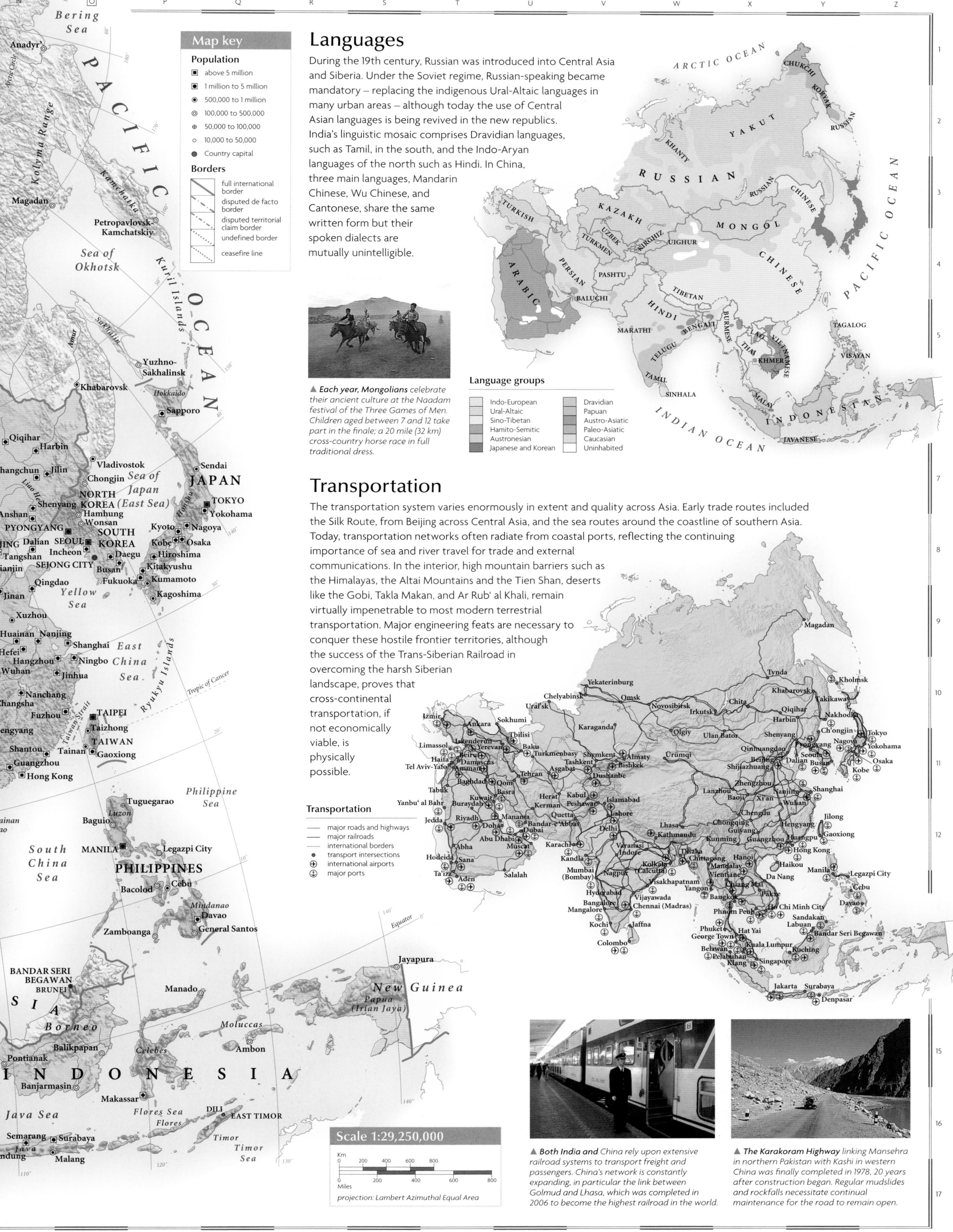

A B C D E F G H I J K L M

Asian resources

Although agriculture remains the economic mainstay of most Asian countries, the number of people employed in agriculture has steadily declined, as new industries have been developed during the past 30 years. China, Indonesia, Malaysia, Thailand, and Turkey have all experienced far-reaching structural change in their economies, while the breakup of the Soviet Union has created a new economic challenge in the Central Asian republics. The countries of The Persian Gulf illustrate the rapid transformation from rural nomadism to modern, urban society which oil wealth has brought to parts of the continent. Asia's most economically dynamic countries, Japan, Singapore, South Korea, and Taiwan, fringe the Pacific Ocean and are known as the Pacific Rim. In contrast, other Southeast Asian countries like Laos and Cambodia remain both economically and industrially underdeveloped.

Industry

East Asian industry leads the continent in both productivity and efficiency; electronics, hi-tech industries, car manufacture, and shipbuilding are important. The so-called economic "tigers" of the Pacific Rim are Japan, South Korea, and Taiwan and in recent years China has rediscovered its potential as an economic superpower. Heavy industries such as engineering, chemicals, and steel typify the industrial complexes along the corridor created by the Trans-Siberian Railroad, the Fergana Valley in Central Asia, and also much of the huge industrial plain of east China. The discovery of oil in the Persian Gulf has brought immense wealth to countries that previously relied on subsistence agriculture on marginal desert land.

Standard of living

Despite Japan's high standards of living, and Southwest Asia's oil-derived wealth, immense disparities exist across the continent. Afghanistan remains one of the world's most underdeveloped nations, as do the mountain states of Nepal and Bhutan. Further rapid population growth is exacerbating poverty and overcrowding in many parts of India and Bangladesh.

Standard of living
(UN human development index)

- low
- high

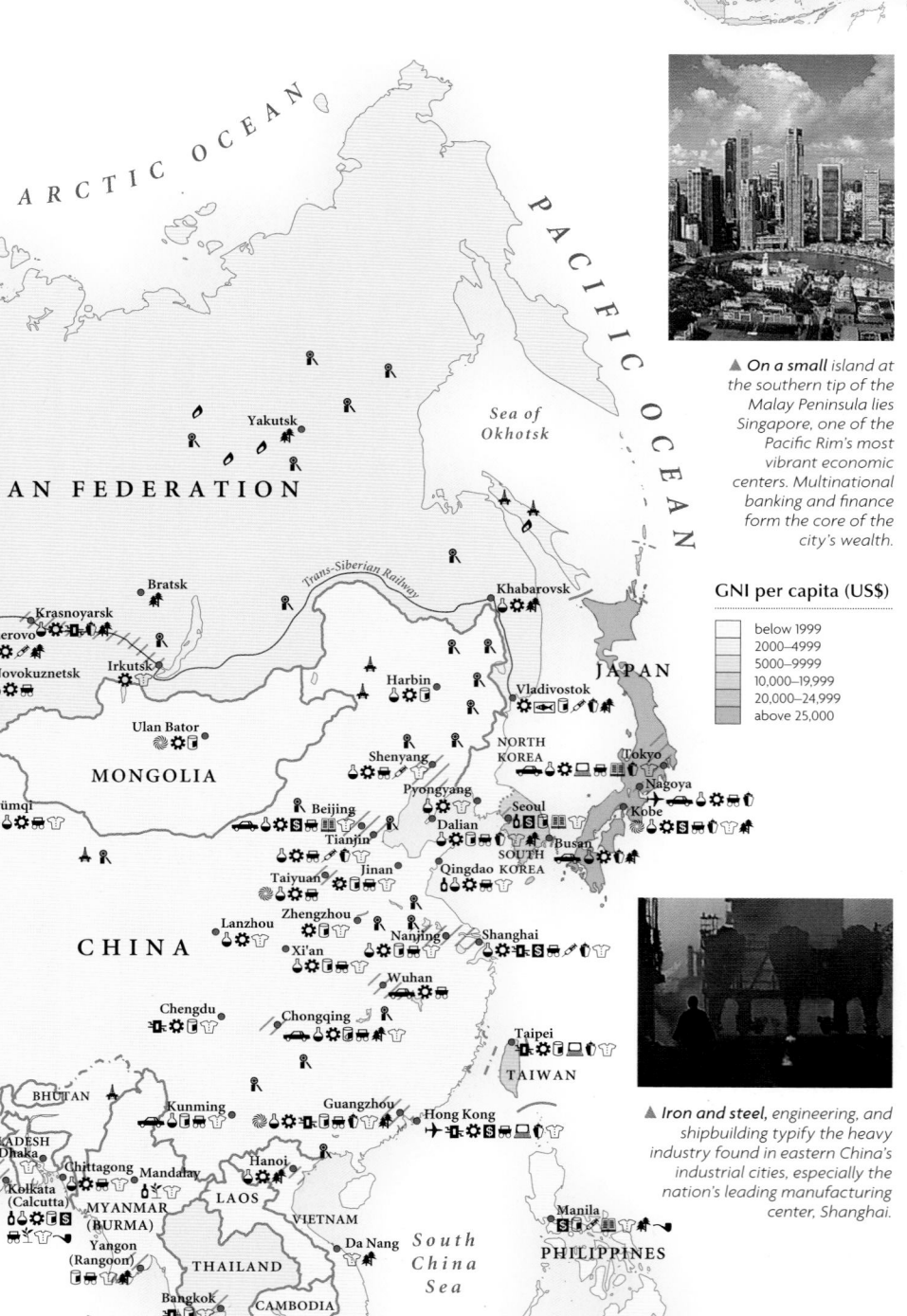

▲ On a small island at the southern tip of the Malay Peninsula lies Singapore, one of the Pacific Rim's most vibrant economic centers. Multinational banking and finance form the core of the city's wealth.

GNI per capita (US$)

- below 1999
- 2000–4999
- 5000–9999
- 10,000–19,999
- 20,000–24,999
- above 25,000

Industry

aerospace	printing & publishing
brewing	shipbuilding
car/vehicle manufacture	sugar processing
cement	tea processing
chemicals	textiles
electronics	timber processing
engineering	tobacco processing
finance	coal
fish processing	oil
food processing	gas
hi-tech industry	
iron & steel	● industrial cities
pharmaceuticals	major industrial areas

▲ Iron and steel, engineering, and shipbuilding typify the heavy industry found in eastern China's industrial cities, especially the nation's leading manufacturing center, Shanghai.

ARCTIC OCEAN

PACIFIC OCEAN

RUSSIAN FEDERATION

Sea of Okhotsk

Yakutsk

Yekaterinburg

Magnitogorsk

Chelyabinsk

Omsk

Novosibirsk

Trans-Siberian Railway

Krasnoyarsk

Kemerovo

Novokuznetsk

Irkutsk

Bratsk

Trans-Siberian Railway

Khabarovsk

JAPAN

Vladivostok

Istanbul

Izmir

Ankara

TURKEY

GEORGIA

Tbilisi

ARMENIA

Yerevan

Baku

AZERB.

CYPRUS

LEBANON

Beirut

SYRIA

Damascus

Tel Aviv-Yafo

ISRAEL

Amman

JORDAN

Kirkuk

Baghdad

IRAQ

Basra

Kuwait

KUWAIT

SAUDI ARABIA

Ad Damman

BAHRAIN

Riyadh

QATAR

Abu Dhabi

Dubai

UAE

Jedda

Red Sea

YEMEN

Gulf of Aden

OMAN

Gulf of Oman

Arabian Sea

Tehran

Isfahan

IRAN

Persian Gulf

KAZAKHSTAN

Karaganda

Aral Sea

Caspian Sea

UZBEKISTAN

TURKMENISTAN

Asgabat

Tashkent

Dushanbe

TAJIKISTAN

KYRGYZSTAN

Farghona

Almaty

Ürümqi

Ulan Bator

MONGOLIA

CHINA

Lanzhou

Xi'an

Chengdu

Chongqing

Kunming

AFGHANISTAN

Rawalpindi

Lahore

PAKISTAN

Karachi

Ahmadabad

Delhi

Kanpur

INDIA

Indore

Jamshedpur

Nagpur

Mumbai (Bombay)

NEPAL

BHUTAN

BANGLADESH

Dhaka

Chittagong

Kolkata (Calcutta)

Mandalay

MYANMAR (BURMA)

Yangon (Rangoon)

Bangalore

Chennai (Madras)

SRI LANKA

INDIAN OCEAN

Harbin

Shenyang

NORTH KOREA

Pyongyang

Beijing

Tianjin

Dalian

Seoul

Busan

SOUTH KOREA

Jinan

Qingdao

Taiyuan

Zhengzhou

Nanjing

Shanghai

Wuhan

Taipei

TAIWAN

Guangzhou

Hong Kong

Hanoi

LAOS

VIETNAM

Da Nang

THAILAND

Bangkok

CAMBODIA

Ho Chi Minh City

South China Sea

Manila

PHILIPPINES

MALAYSIA

BRUNEI

Kuala Lumpur

Singapore

SINGAPORE

INDONESIA

Jakarta

Surabaya

EAST TIMOR

Tokyo

Nagoya

Kobe

Gulf of Oman

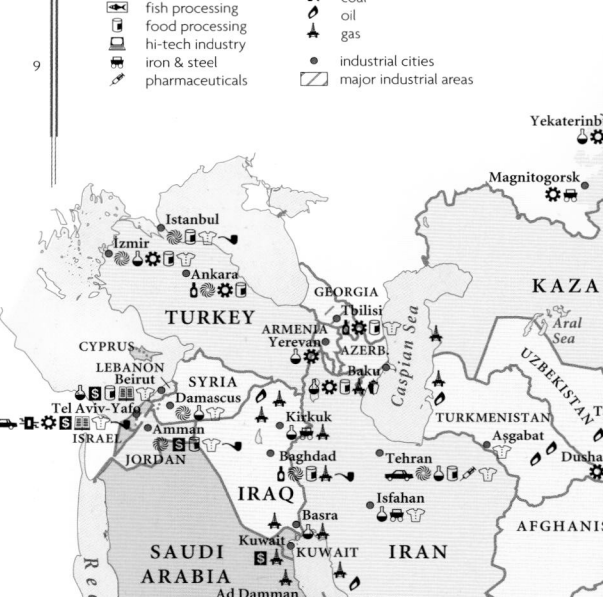

◀ Traditional industries are still crucial to many rural economies across Asia. Here, on the Vietnamese coast, salt has been extracted from seawater by evaporation and is being loaded into a van to take to market.

Environmental issues

The transformation of Uzbekistan by the former Soviet Union into the world's fifth largest producer of cotton led to the diversion of several major rivers for irrigation. Starved of this water, the Aral Sea diminished in volume by over 90% since 1960, irreversibly altering the ecology of the area. Heavy industries in eastern China have polluted coastal waters, rivers, and urban air, while in Myanmar (Burma), Malaysia, and Indonesia, ancient hardwood rainforests are felled faster than they can regenerate.

Mineral resources

At least 60% of the world's known oil and gas deposits are found in Asia; notably the vast oil fields of the Persian Gulf, and the less-exploited oil and gas fields of the Ob' basin in west Siberia. Immense coal reserves in Siberia and China have been utilized to support large steel industries. Southeast Asia has some of the world's largest deposits of tin, found in a belt running down the Malay Peninsula to Indonesia.

▲ *Although Siberia remains a quintessentially frozen, inhospitable wasteland, vast untapped mineral reserves – especially the oil and gas of the West Siberian Plain – have lured industrial development to the area since the 1950s and 1960s.*

Mineral resources

- oil field
- gas field
- coal field
- chromite
- copper
- gold
- iron
- lead
- nickel
- platinum
- tin
- wolfram

Environmental issues

- tropical forest
- forest destroyed
- desert
- desertification
- acid rain
- polluted rivers
- marine pollution
- heavy marine pollution
- radioactive contamination
- poor urban air quality

◀ *Commercial logging activities in Borneo have placed great stress on the rainforest ecosystem. Government attempts to regulate the timber companies and control illegal logging have only been partially successful.*

Using the land and sea

Vast areas of Asia remain uncultivated as a result of unsuitable climatic and soil conditions. In favourable areas such as river deltas, farming is intensive. Rice is the staple crop of most Asian countries, grown in paddy fields on waterlogged alluvial plains and terraced hillsides, and often irrigated for higher yields. Across the black earth region of the Eurasian steppe in southern Siberia and Kazakhstan, wheat farming is the dominant activity. Cash crops, like tea in Sri Lanka and dates in the Arabian Peninsula, are grown for export, and provide valuable income. The sovereignty of the rich fishing grounds in the South China Sea is disputed by China, Malaysia, Taiwan, the Philippines, and Vietnam, because of potential oil reserves.

Using the land and sea

- cropland
- desert
- forest
- mountain region
- pasture
- tundra
- wetland
- major conurbations
- cattle
- pigs
- goats
- sheep
- coconuts
- corn
- cotton
- dates
- fishing
- fruit
- jute
- peanuts
- rice
- rubber
- shellfish
- soybeans
- sugar beet
- sugar cane
- tea
- timber
- wheat

▲ *Date palms have been cultivated in oases throughout the Arabian Peninsula since antiquity. In addition to the fruit, palms are used for timber, fuel, rope, and for making vinegar, syrup and a liquor known as arrack.*

◀ *Rice terraces blanket the landscape across the small Indonesian island of Bali. The large amounts of water needed to grow rice have resulted in Balinese farmers organizing water-control co-operatives.*

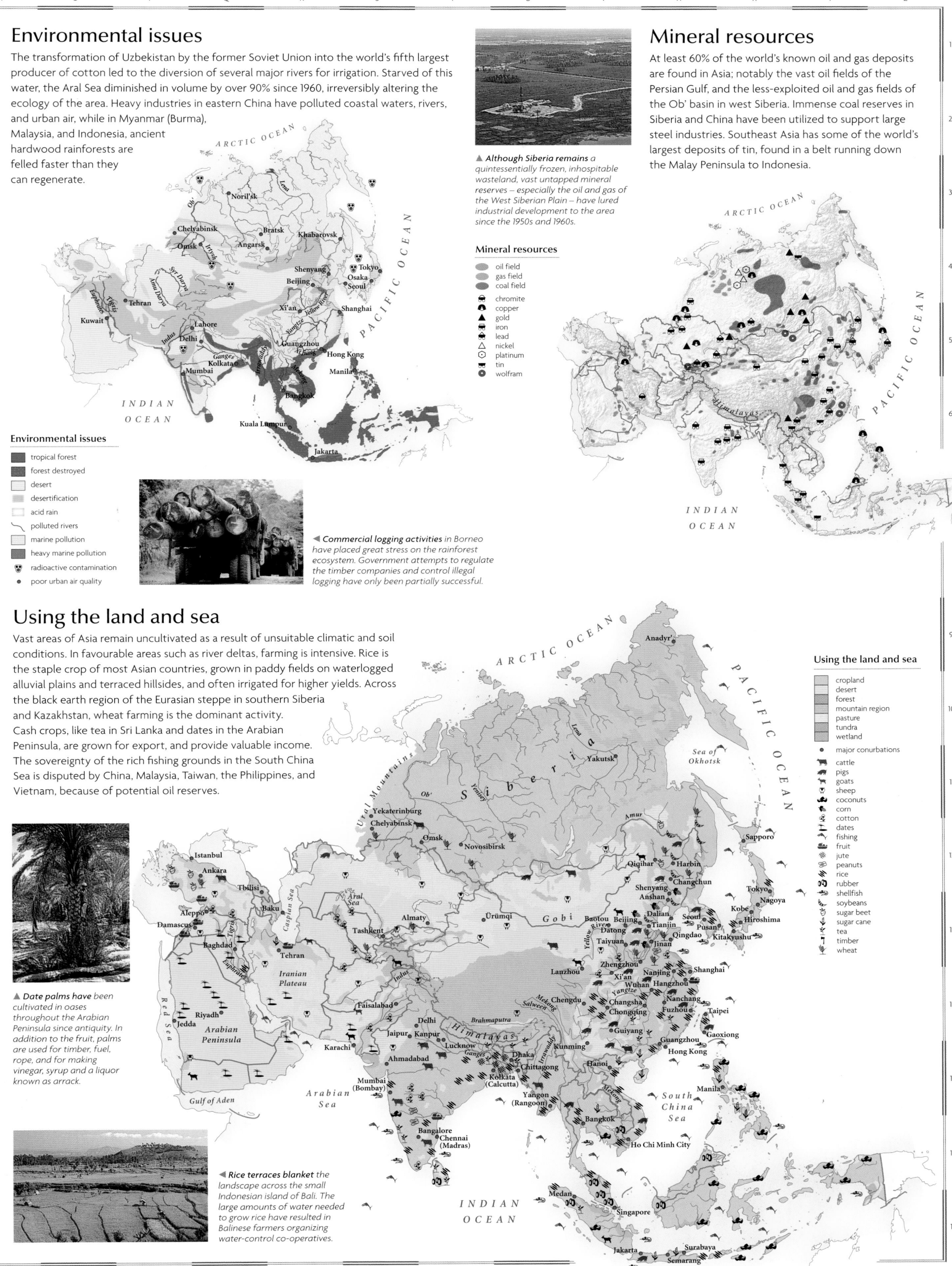

A B C D E F G H I J K L M

Turkey & the Caucasus

ARMENIA, AZERBAIJAN, GEORGIA, TURKEY

This region occupies the fragmented junction between Europe, Asia, and the Russian Federation. Sunni Islam provides a common identity for the secular state of Turkey, which the revered leader Kemal Atatürk established from the remnants of the Ottoman Empire after the First World War. Turkey has a broad resource base and expanding trade links with Europe, but the east is relatively undeveloped and strife between the state and a large Kurdish minority has yet to be resolved. Georgia is similarly challenged by ethnic separatism, while the Christian state of Armenia and the mainly Muslim and oil-rich Azerbaijan are locked in conflict over the territory of Nagorno-Karabakh.

Using the land & sea

Turkey is largely self-sufficient in food. The irrigated Black Sea coastlands have the world's highest yields of hazelnuts. Tobacco, cotton, sultanas, tea, and figs are the region's main cash crops and a great range of fruit and vegetables are grown. Wine grapes are among the labor-intensive crops which allow full use of limited agricultural land in the Caucasus. Sturgeon fishing is particularly important in Azerbaijan.

Transportation & industry

Turkey leads the region's well diversified economy. Petrochemicals, textiles, engineering, and food processing are the main industries. Azerbaijan is able to export oil, while the other states rely heavily on hydroelectric power and imported fuel. Georgia produces precision machinery. War and earthquake damage have devastated Armenia's infrastructure.

▲ **Azerbaijan has substantial** oil reserves, located in and around the Caspian Sea. They were some of the earliest oilfields in the world to be exploited.

Land use and agricultural distribution

- cattle
- goats
- cotton
- fishing
- fruit
- hazelnuts
- olives
- sugar beet
- tobacco
- vineyards

■ capital cities
● major towns

pasture
cropland
forest

The urban/rural population divide

urban 72%   rural 28%

0 10 20 30 40 50 60 70 80 90 100

Population density	Total land area
238 people per sq mile (92 people per sq km)	368,912 sq miles (955,730 sq km)

Major industry and infrastructure

- carpet weaving
- cement
- chemicals
- coal
- engineering
- food processing
- oil
- textiles
- tourism
- vehicle manufacture

■ capital cities
● major towns
✈ international airports
⊢ major roads
major industrial areas

Transportation network

114,867 miles (184,882 km)

5778 miles (9300 km)

8120 miles (13,069 km)

745 miles (1200 km)

Physical and political barriers have severely limited communications between Armenia, Georgia and Azerbaijan. Turkey has a relatively well-developed transportation network.

▲ **For many centuries,** Istanbul has held tremendous strategic importance as a crucial gateway between Europe and Asia. Founded by the Greeks as Byzantium, the city became the center of the East Roman Empire and was known as Constantinople to the Romans. From the 15th century onward the city became the center of the great Ottoman Empire.

The landscape

The deeply eroded hills and salty basins of the Anatolian Plateau are bordered by several mountain ranges along the Black Sea coast, and the limestone Taurus Mountains *(Toros Daglari)* in the south. A lowland trough divides the Caucasus and the Lesser Caucasus, which form a formidable barrier of peaks in the north.

Limestone weathering in the Anatolian Plateau

Eroded gully — High plateau

Layers of tephra — Remnant landforms

▲ **In central Turkey,** rainwater has chemically weathered away numerous layers of limestone, leaving isolated outcrops and pinnacles and deep eroded gullies.

▶ **The Caucasus are** fold mountains, which formed around the same time as the Taurus Mountains *(Toros Daglari)* around 65 million years ago and have since been modified by volcanic erruptions.

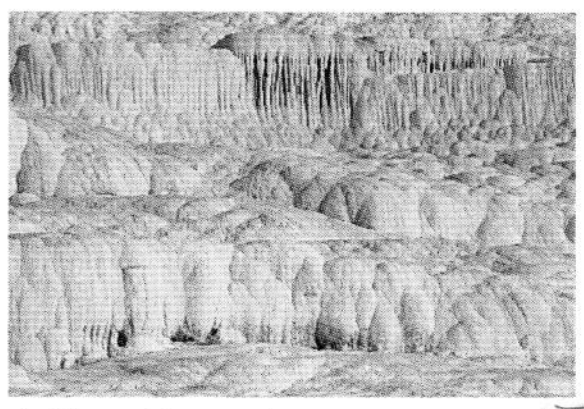

▲ **The white rock terraces** at Pamukkale in western Turkey were formed when underground water, heated by volcanic activity, dissolved minerals in the rocks. When the water reached the surface and evaporated the minerals were left behind in these extraordinary formations.

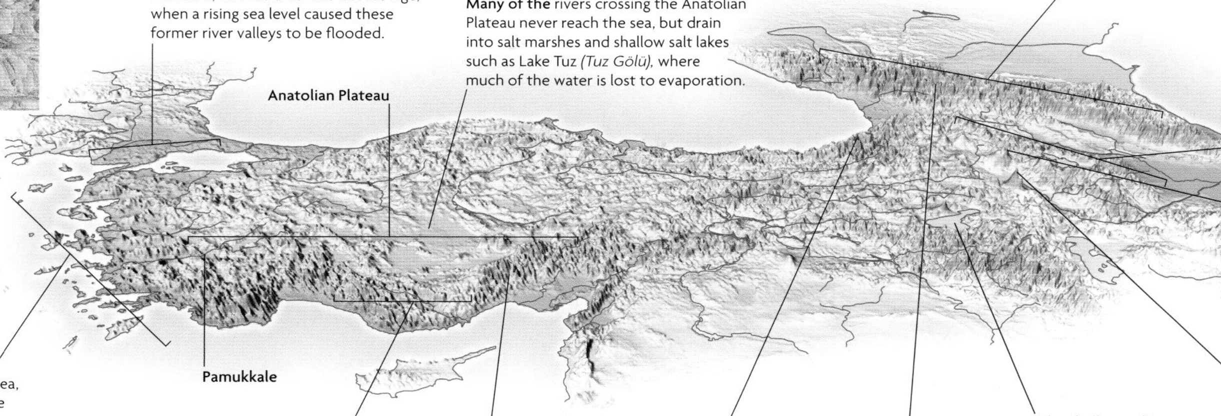

The straits of the Bosporus and the Dardanelles, respectively linking the Black and Mediterranean seas with the Sea of Marmara, formed after the last Ice Age, when a rising sea level caused these former river valleys to be flooded.

Many of the rivers crossing the Anatolian Plateau never reach the sea, but drain into salt marshes and shallow salt lakes such as Lake Tuz *(Tuz Gölü),* where much of the water is lost to evaporation.

Anatolian Plateau

Pamukkale

Lava has flowed over large areas of the Lesser Caucasus within the last five million years, producing extensive basalt plateaus.

The earthquake that struck Armenia in 1988 killed over 55,000 people and devastated the country's infrastructure.

Long, parallel mountain ranges run from east to west into the Aegean Sea, which has risen since the last Ice Age to form a drowned coastline of numerous islands and extended inlets.

The folded peaks of the Taurus Mountains *(Toros Daglari)* were formed 60–65 million years ago, at the same time as the Alps. The rock is mainly limestone, with deep caves, gorges, and underground rivers.

The Cilician Gates *(Gülek Bogazi),* a major pass through the Taurus Mountains *(Toros Daglari),* is the point where streams flow from the interior plateau onto the lowland of Adana.

Thick, temperate forest veils the seaward slopes of the Kaçkar Daglari. The southern slopes, which lie in a rainshadow, are dry and barren.

The granite massif near Surami divides the lowlands of Georgia from Azerbaijan's Kura river, which has built a large delta into the Caspian Sea.

The shallow, saline Lake Van *(Van Gölü)* is the largest lake in Turkey. Dry terraces mark a previous shoreline 181 ft (55 m) above the present water level.

The volcanic cone of Mount Ararat is the highest peak in Turkey, with an altitude of 16,853 ft (5137 m).

▶ **Since the 6th century BC,** the pinnacles and caves of east-central Anatolia have been utilized as dwellings. Many are still inhabited today.

Map key

Population

- ■ above 5 million
- ◉ 1 million to 5 million
- ◉ 500,000 to 1 million
- ◎ 100,000 to 500,000
- ⊕ 50,000 to 100,000
- ○ 10,000 to 50,000
- ∘ below 10,000

Elevation

- 4000m / 13,124ft
- 3000m / 9843ft
- 2000m / 6562ft
- 1000m / 3281ft
- 500m / 1640ft
- 250m / 820ft
- 100m / 328ft
- sea level

Scale 1:4,000,000

Km
0 10 20 40 60 80 100 120

Miles
0 10 20 40 60 80 100 120

projection: Lambert Conformal Conic

▲ **The fisheries of** Azerbaijan are noted for their hauls of sturgeon, and the Caspian Sea accounts for 80% of the world's total catch. However, stocks are now under serious threat due to overfishing.

▲ **Traditional steam baths** are found throughout the region, and are used for socializing as well as for bathing.

The Near East

IRAQ, ISRAEL, JORDAN, LEBANON, SYRIA

Some of the world's oldest civilizations developed in this region – the Fertile Crescent – which is venerated by Jews, Muslims, and Christians, but torn by competing religious, ethnic, and national claims to the land. Turkish Ottoman rule ended with the First World War and the region was divided into areas administered by Britain and France. The UN endorsed calls for a Jewish homeland in what was then Palestine and in 1948 the state of Israel was declared. Hostility towards the Jewish state led to a series of wars with its Arab neighbors. After 2000, attempts to broker peaceful resolutions with both the Palestinian population and with adjacent Arab states were hampered by a revival of Islamic militarism and conflicting international interests in the oil-rich region. This led to an Israeli retrenchment and culminated in a US-led invasion of Iraq in 2003, which toppled the Ba'athist regime of Saddam Hussein in the name of a "war on terror".

Using the land & sea

Water scarcity limits cropland to the north and to areas watered principally by the Tigris, Euphrates, and Jordan rivers. In Israel, new irrigation techniques are allowing cultivation in the arid Negev. Wheat is the chief grain and large areas of scrub support livestock herding. Commercial produce includes dates, tobacco, citrus fruits, olives, grapes, and cotton, which is Syria's main export crop. Fishing is still important in the Mediterranean.

The urban/rural population divide

urban 70%   rural 30%

0 10 20 30 40 50 60 70 80 90 100

Population density	Total land area
217 people per sq mile (84 people per sq km)	325,460 sq miles (843,160 sq km)

Land use and agricultural distribution

- sheep
- cereals
- citrus fruits
- cotton
- dates
- fishing
- rice
- tobacco
- capital cities
- major towns
- pasture
- cropland
- wetland
- desert

Transportation & industry

The petrochemical industry is well established, and central to the economies of Syria and Iraq, which was the world's second largest oil exporter before the war with Iran which began in 1980. Lebanon has traditionally been a center for commerce, while Israel has a well-diversified economy with an expanding tourist industry, despite few natural resources.

Transportation network

- 49,859 miles (80,249 km)
- 1365 miles (2197 km)
- 3826 miles (6158 km)
- 1171 miles (1885 km)

Jordan's seaport of Al 'Aqabah is connected to Damascus in Syria by road and rail. This route to the Red Sea provides for large exports of phosphate and trade with states in the Persian Gulf.

Major industry and infrastructure

- car manufacture
- cement
- chemicals
- electronics
- finance
- food processing
- iron & steel
- oil
- oil refining
- textiles
- capital cities
- major towns
- international airports
- major roads
- major industrial areas

◀ The Dome of the Rock in Jerusalem is a magnificent mosque, revered by Muslims. Close by is the Wailing Wall, the city's most sacred Jewish landmark and the Church of the Holy Sepulchre, a famous Christian place of worship.

▲ The city of Petra, carved from spectacular rose-colored limestone, lies deep within a canyon in southern Jordan. Revenues from the spice trade funded the construction of the city which was built by the Nabatean people in about 400 BC.

▶ Water and wind erosion over thousands of years have created the Canyon of the Oasis at Ein 'Avdat in the Negev Desert (HaNegev). Extreme diurnal temperature fluctuations, coupled with wind erosion, have caused layers of rock to crack and peel away.

The landscape

The Al Jazirah plateau divides the Euphrates and Tigris rivers, which cross the Mesopotamian plain to reach their confluence in the southeast. The rocky Syrian Desert extends west to the northern extremity of the Great Rift Valley, which runs from the mountains of Lebanon to the Gulf of Aqaba. The Jordan river flows south along this trough into the Dead Sea, divided from the Mediterranean coastal plain by a steep-sided plateau.

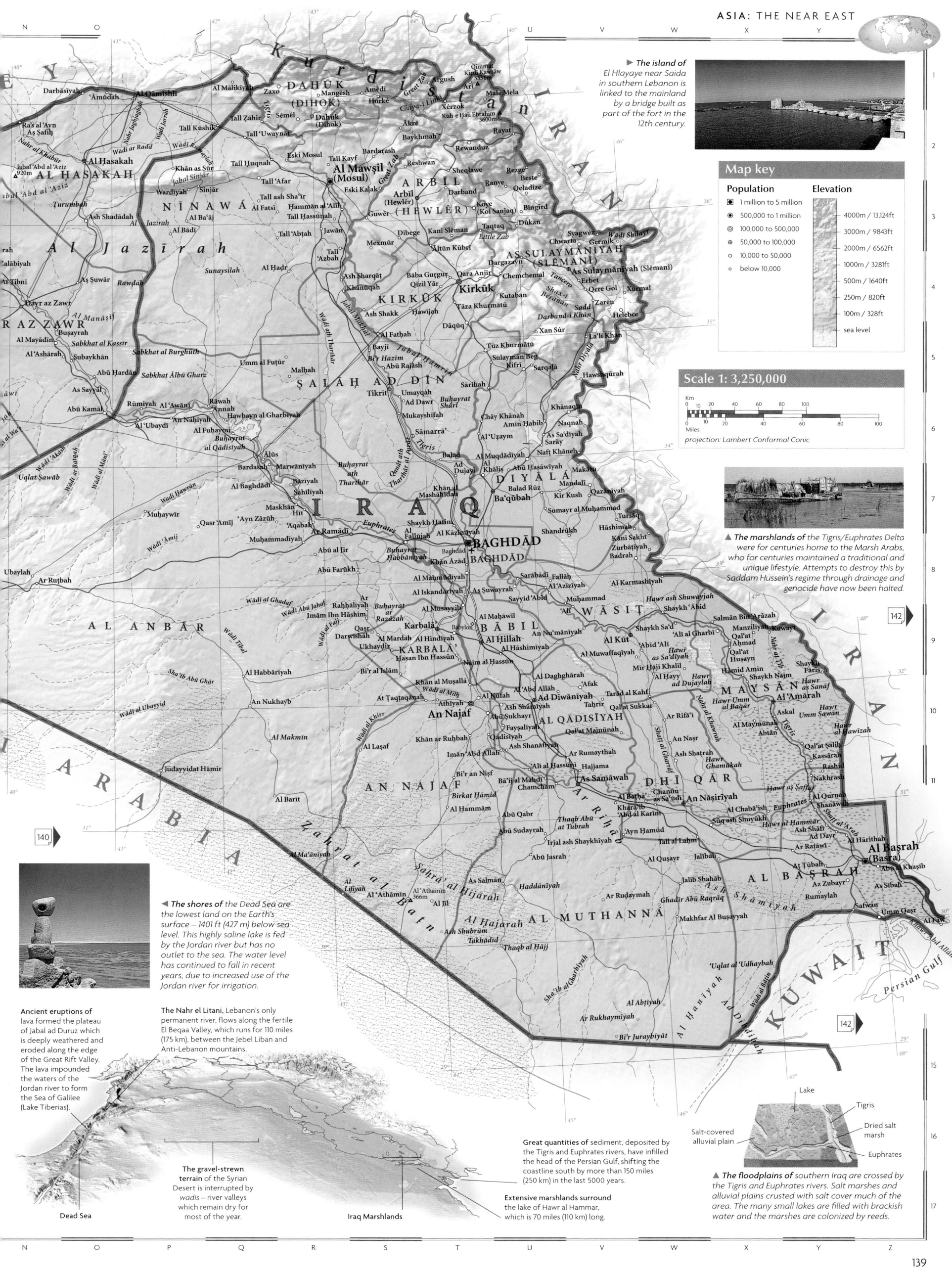

► The island of El Hlayaye near Saida in southern Lebanon is linked to the mainland by a bridge built as part of the fort in the 12th century.

Map key

Population
◉ 1 million to 5 million
◉ 500,000 to 1 million
◉ 100,000 to 500,000
⊕ 50,000 to 100,000
○ 10,000 to 50,000
○ below 10,000

Elevation
4000m / 13,124ft
3000m / 9843ft
2000m / 6562ft
1000m / 3281ft
500m / 1640ft
250m / 820ft
100m / 328ft
sea level

Scale 1: 3,250,000
Km 0 20 40 60 80 100
Miles 0 20 40 60 80 100
projection: Lambert Conformal Conic

▲ The marshlands of the Tigris/Euphrates Delta were for centuries home to the Marsh Arabs, who for centuries maintained a traditional and unique lifestyle. Attempts to destroy this by Saddam Hussein's regime through drainage and genocide have now been halted.

◄ The shores of the Dead Sea are the lowest land on the Earth's surface – 1401 ft (427 m) below sea level. This highly saline lake is fed by the Jordan river but has no outlet to the sea. The water level has continued to fall in recent years, due to increased use of the Jordan river for irrigation.

Ancient eruptions of lava formed the plateau of Jabal ad Duruz which is deeply weathered and eroded along the edge of the Great Rift Valley. The lava impounded the waters of the Jordan river to form the Sea of Galilee (Lake Tiberias).

The Nahr el Litani, Lebanon's only permanent river, flows along the fertile El Beqaa Valley, which runs for 110 miles (175 km), between the Jebel Liban and Anti-Lebanon mountains.

Dead Sea

The gravel-strewn terrain of the Syrian Desert is interrupted by wadis – river valleys which remain dry for most of the year.

Iraq Marshlands

Great quantities of sediment, deposited by the Tigris and Euphrates rivers, have infilled the head of the Persian Gulf, shifting the coastline south by more than 150 miles (250 km) in the last 5000 years.

Extensive marshlands surround the lake of Hawr al Hammar, which is 70 miles (110 km) long.

Lake
Tigris
Dried salt marsh
Salt-covered alluvial plain
Euphrates

▲ The floodplains of southern Iraq are crossed by the Tigris and Euphrates rivers. Salt marshes and alluvial plains crusted with salt cover much of the area. The many small lakes are filled with brackish water and the marshes are colonized by reeds.

139

The Arabian Peninsula

BAHRAIN, KUWAIT, OMAN, QATAR, SAUDI ARABIA,
UNITED ARAB EMIRATES (UAE), YEMEN

Huge expanses of desert cover much of the Arabian Peninsula, limiting settlement to oases, the mountains along the Red Sea, and coastal belts. The most populous area is the fertile highlands of Yemen. The Islamic faith and Arabic language give the region a cultural and religious unity, and the Saudi city of Mecca (Makkah) is Islam's most holy place, visited by over two million pilgrims each year. More than half the world's oil reserves are contained in this region, and the exploitation of oil and gas has brought great wealth, particularly to Saudi Arabia. Yemen and Oman are the least developed of the Arabian states, with large rural populations. Within Saudi Arabia over 86% of the people live in urban areas.

Using the land

Most of the Arabian Peninsula is unsuited to settled agriculture, making irrigation and land reclamation projects essential. The narrow coastal plain and isolated oases, commonly amounting to less than 1% of the land area, are used to cultivate grains, coffee, and exotic fruits. Goats, sheep, and camels are widespread throughout the region.

The urban/rural population divide

urban 64%   rural 36%

0 10 20 30 40 50 60 70 80 90 100

Population density	Total land area
50 people per sq mile (19 people per sq km)	1,147,856 sq miles (2,973,720 sq km)

Land use and agricultural distribution

- goats
- sheep
- cereals
- coffee
- dates
- fruit
- ■ capital cities
- • major towns
- pasture
- cropland
- desert

◄ *The fertile soils of Yemen have encouraged settlement of almost all of the land from sea level up to the mountains at 10,000 ft (3050 m). In the higher reaches elaborate terraces have been constructed to facilitate crop cultivation.*

The landscape

A plateau more than 2500 ft (760 m) high extends across much of the Arabian Peninsula. The plateau slopes eastward from the massive, rifted escarpment along the coast of the Red Sea, to the shallow waters of the Persian Gulf. The interior is characterized by *cuestas* and valleys, drained by a system of *wadis*. A crescent of sand and gravel deserts lies to the east.

Few areas in the Arabian Peninsula have rivers flowing through them. Most are drained by ephemeral watercourses called *wadis*.

The Hejaz (*Al Hijaz*) and Asir mountains form part of the same geological region as the highlands of Sudan and Eritrea, to which they were once joined. They were separated when faulting opened the Red Sea, over 50 million years ago.

The An Nafud Desert is covered with *barchan* dunes varying between 30–100 ft (10–30 m) high. The "horns" of the crescent-shaped dunes reflect the direction in which they are being moved by the wind.

Inselbergs are dotted over a wide area of the Najd Plateau. These resistant remnants of the ancient basement rock are left standing when the softer weathered rock has been worn away.

Evaporation / Crusted layer left behind
Storm surge flooding
Normal level of tidal range
Salt wedge penetrates inland water

▲ *A sabkha is a flat, salt-encrusted plain which occurs near the coast just above the high water mark. Flooding by sea water leads to saturation of the land with saline-rich groundwater. As this evaporates, a cracked layer of sand, cemented together with salt, gypsum, and calcium carbonate is left behind.*

Across the Najd Plateau the flat relief is broken by *mesas*; steep-sided rock plateaus and *cuestas*; ridges with one steep and one gentle slope.

▲ *Ar Rub' al Khali, also known as the Empty Quarter, is the most arid part of the Arabian Peninsula. It is the largest uninterrupted sand desert in the world. Ridges of sand up to 25 miles (40 km) long, run northeast–southwest, giving characteristic linear dunes.*

The Jabal an Nabi Shu'ayb in Yemen is the highest point on the peninsula, rising to 12,336 ft (3760 m).

The Arabian Shield underpins the west of the peninsula. It is a fragment of the ancient continent, Gondwanaland, which was separated by rifting millions of years ago.

◄ *Every Muslim must make at least one pilgrimage or hajj to Mecca (Makkah), in Saudi Arabia, during their lifetime. The cloth-covered shrine is called the Ka'bah, and is regarded by Muslims as the most sacred place on Earth.*

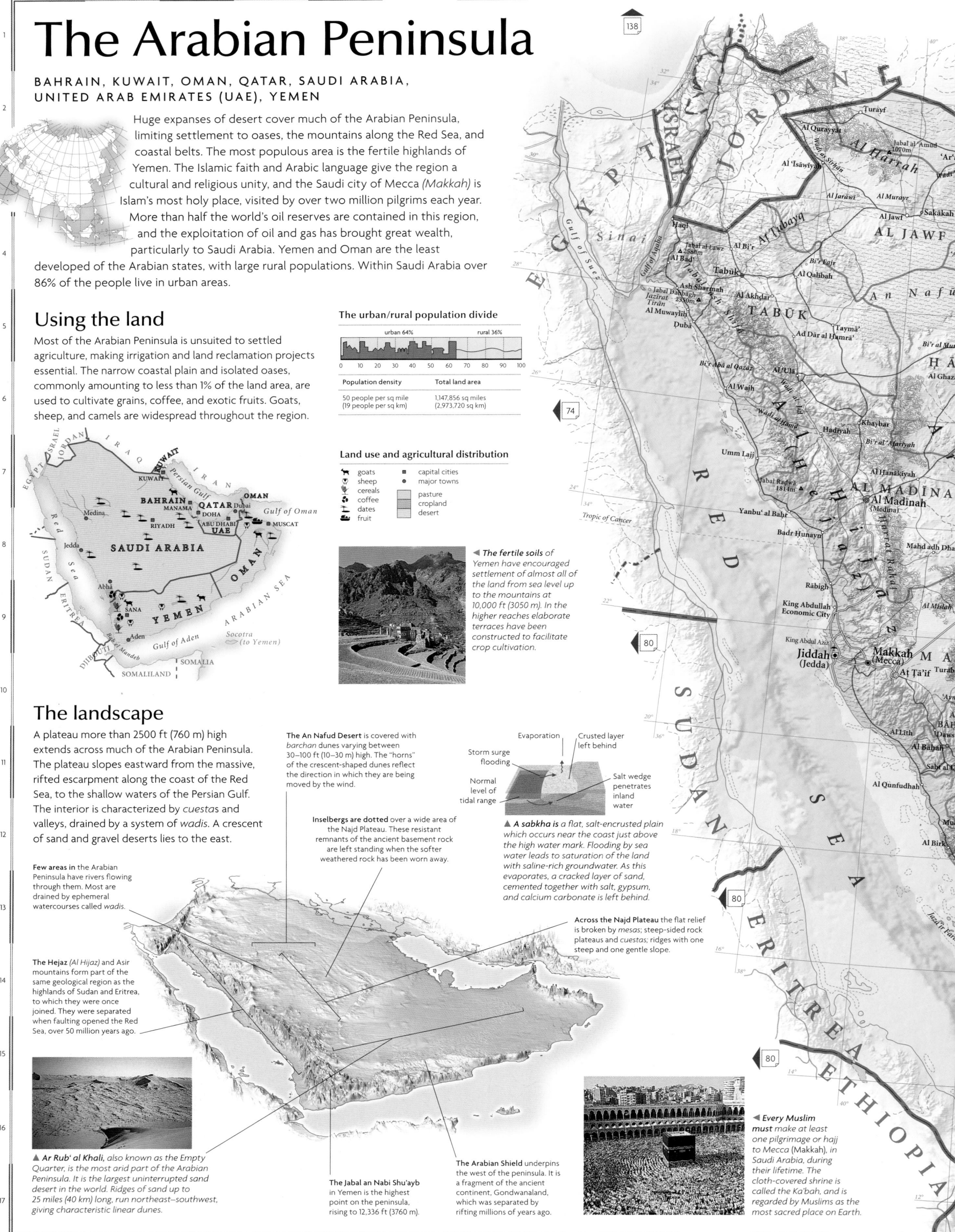

Transportation & industry

The extraction and refining of oil and gas are the major industrial activities in the Arabian Peninsula. The region also has an active construction sector, with many Arab cities reflecting the wealth generated by the oil industry. The service sector is dominated by financial and technical institutions, which, like the construction sector, mainly serve the oil industry. Traditional handicrafts such as carpet-weaving are found in rural areas.

◄ *Saudi Arabia contains the world's largest oil reserves, lying mainly along the Persian Gulf coast. Each day the region produces around 10 million barrels of oil. Here, in the desert, excess oil is being burnt off.*

Transportation network

🛣	44,832 miles (72,159 km)	🛤	673 miles (1083 km)
🚂	670 miles (1078 km)		none

Internal surface transportation is poorly developed across the peninsula. Along the coast, commercial routes have developed, but connections between bordering states rely on major airports.

Major industry and infrastructure

- cement
- chemicals
- iron & steel
- oil
- oil refining
- food processing
- capital cities
- major towns
- international airports
- major roads
- major industrial areas

Map key

Population
- 1 million to 5 million
- 500,000 to 1 million
- 100,000 to 500,000
- 50,000 to 100,000
- 10,000 to 50,000
- below 10,000

Elevation
- 3000m / 9843ft
- 2000m / 6562ft
- 1000m / 3281ft
- 500m / 1640ft
- 250m / 820ft
- 100m / 328ft
- sea level

► *Seasonal watercourses or wadis drain much of the interior of the Arabian Peninsula. Although they remain dry for much of the year, they are prone to flash floods after heavy rains.*

Scale 1:7,500,000

projection: Lambert Conformal Conic

IRAQ

KUWAIT
AL KUWAYT (KUWAIT)

Persian Gulf

IRAN

BAHRAIN
AL MANAMAH (MANAMA)

QATAR
AD DAWHAH (DOHA)

UNITED ARAB EMIRATES
ABŪ ZABY (ABU DHABI)
Dubayy (Dubai)

OMAN
MASQAT (MUSCAT)

Gulf of Oman
Strait of Hormuz

SAUDI ARABIA
AR RIYĀD (RIYADH)

AL QAŞIM
Najd (Nejd)
AR RIYĀD
ASH SHARQĪYAH
Ar Rub' al Khālī (Empty Quarter)
NAJRĀN
'ASĪR
JĪZĀN

YEMEN
SAN'Ā' (SANA)
'Adan (Aden)

ARABIAN SEA
Gulf of Aden

SOMALIA
DJIBOUTI
SOMALILAND
ERITREA
SUDAN
EGYPT
JORDAN
ISRAEL
Red Sea
Jedda

Suquţrā (Socotra) (to Yemen)

Tropic of Cancer

141

Iran & the Gulf states

BAHRAIN, IRAN, KUWAIT, QATAR, UNITED ARAB EMIRATES (UAE)

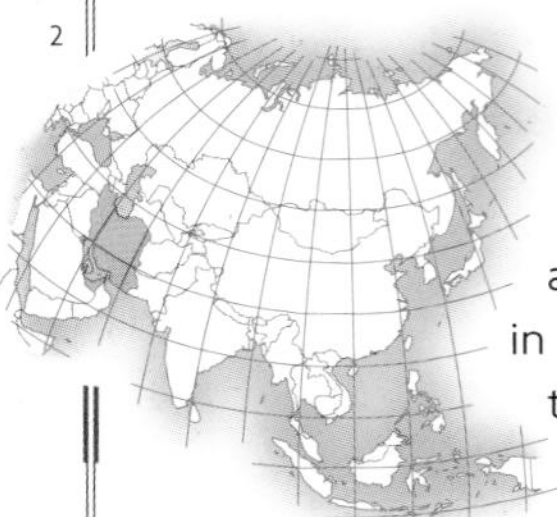

The discovery of oil in the Persian Gulf in the 1930s brought great wealth to the surrounding states. The revenue was largely used to modernize industry and infrastructure, initiating great social change in these formerly agrarian countries. Today, over 90% of the people in the Gulf states live in urban areas, and foreign nationals make up a sizeable proportion of the population in Kuwait, Qatar, and the United Arab Emirates. The importance of control of the oil reserves has led to a number of territorial disputes, including most recently the Iran–Iraq War (1980-88) and the First Gulf War (1991). Islam is practiced almost exclusively throughout the region and two distinct strands are found; Sunni Muslims in Qatar, Kuwait, and UAE, and Shi'a Muslims in Iran and Bahrain. In 1979 Iran became the world's largest theocracy.

The landscape

The land rises steeply from the fragmented coastal lowlands bordering the Persian Gulf, to reach Iran's interior plateau, bounded by heavily eroded mountain chains. An unstable plate boundary runs northwest to southeast across Iran causing frequent earthquakes. On the sandy west coast of the Persian Gulf, the relief is generally flat, with patches of salt marsh. Bahrain consists of two groups of islands, which are mostly small and rocky.

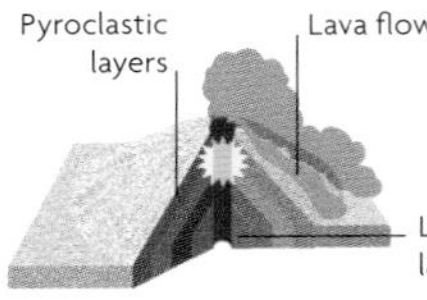

Pyroclastic layers | Lava flow

Lava flow layers

▲ *Qolleh-ye Damavand in the Elburz Mountains is a composite volcano. It comprises layers of lava and pyroclasts fragmentary rocks which accumulate on the slopes of the volcano after being ejected into the air.*

▲ *Marine sediments from deep beneath the ancient Tethys Sea have been uplifted to form the Elburz Mountains, which stretch along the shores of the Caspian Sea, northern Iran.*

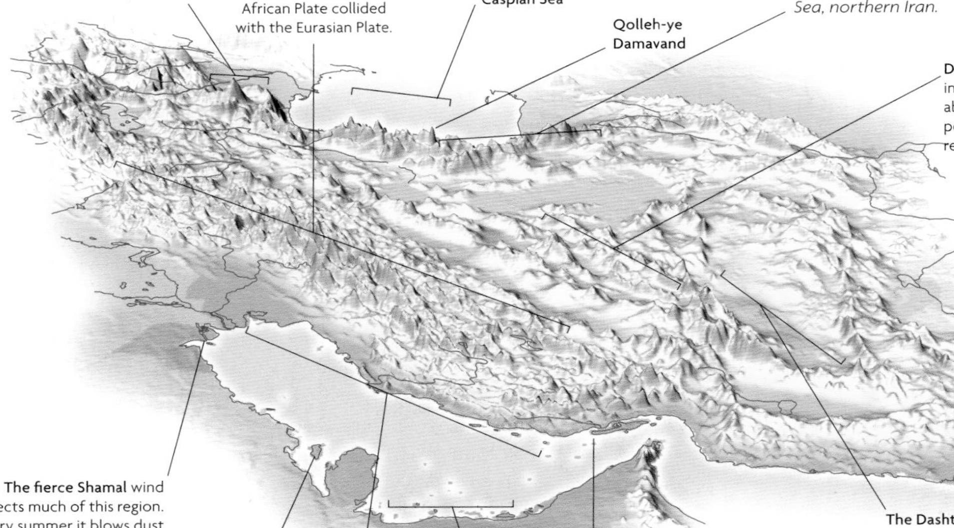

Lava and ash from previous volcanic activity covers a 200 mile (320 km) stretch from the border with Azerbaijan to the Caspian Sea.

Iran's two mountain chains, the Zagros and Elburz, were uplifted at the same time as the Alps in Europe, when the African Plate collided with the Eurasian Plate.

Caspian Sea

Qolleh-ye Damavand

Dominated by a vast, semi-arid interior plateau, most of Iran lies above 1640 ft (500 m). The region is poorly drained with many of its basins remaining dry for months at a time.

The fierce Shamal wind affects much of this region. Every summer it blows dust south from the flood plains of the Tigris and Euphrates, reducing visibility to such an extent that Kuwait International Airport is frequently forced to close.

Prolific springs tapping artesian water make cultivation possible across the north of Bahrain's main island. This provides a sharp contrast to the sandy plains in the south and west.

The oilfields of the Persian Gulf are formed from marine shale deposits lying in sedimentary basins at the margins of the Zagros Mountains.

Numerous islands lie along the southern coast of the Persian Gulf. Some of these are salt domes, created when less dense salts were displaced and forced up to the surface by denser, overlying strata.

Autumn winds blowing across the Persian Gulf can reach speeds of up to 95 mph (150 kmph) causing severe storms, squalls, and waterspouts.

The Dasht-e Lut

◄ *The Dasht-e Lut covers a large portion of eastern Iran with its dry, wind-eroded plain of scattered sandstone pillars and salty depressions. During the summer, temperatures soar, making it one of the world's hottest, driest places.*

Using the land & sea

Along the coast of the Caspian Sea, desalinated water allows fruits and vegetables to be produced, although water shortages and desert soils still limit farming. Sheep are the most important livestock raised in Iran and commercial forests cover the northwest of the country. Shrimp stocks were decimated by pollution during the Gulf War, but fishing remains important for domestic and export markets.

◄ *All of the Gulf states have commercial fishing fleets. Before the discovery of oil, fishing was the region's leading industry.*

◄ *The Kuwait Towers in the center of Kuwait are symbols of the vast wealth oil has brought to the country. Before 1960, the city had only one main street and was surrounded by a mud wall.*

Land use and agricultural distribution

- goats
- sheep
- cereals
- citrus fruits
- cotton
- dates
- fishing
- timber
- ■ capital cities
- ● major towns

- pasture
- cropland
- forest
- desert
- wetland

The urban/rural population divide

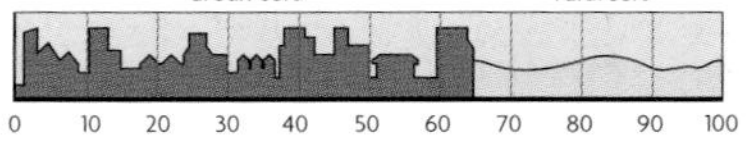

urban 65% | rural 35%

0 10 20 30 40 50 60 70 80 90 100

Population density	Total land area
112 people per sq mile (43 people per sq km)	642,883 sq miles (1,665,500 sq km)

◀ *Many volcanoes lie in Iran's 1200 mile (1930 km) volcanic belt, including the country's highest peak, the now-extinct Qolleh-ye Damavand at 18,600 ft (5671 m).*

▶ *Extensive oil and gas exploitation in the Gulf region has allowed the economic transformation of the Gulf states. Consequently, many of these states have a hugely improved per capita income compared to the 1960's.*

Transportation & industry

Both onshore and offshore oil reserves are exploited throughout the region. Kuwait not only extracts but also refines 80% of its oil. Bahrain has diversified its economy to become the main commercial and financial center in the Persian Gulf. Iran produces a wide range of products: textile mills are widespread and carpet weaving is an important export industry.

Major industry and infrastructure

- carpet manufacture
- chemicals
- S finance
- food processing
- oil
- oil refining
- textiles
- ■ capital city
- • major towns
- ✈ international airports
- — major roads
- major industrial areas

Transportation network

63,543 miles (102,274 km)		884 miles (1423 km)	
3822 miles (6151 km)		562 miles (904 km)	

Major towns and neighboring countries are linked by adequate road networks, although rural areas are less well served. Bahrain is linked to the mainland by a 15 mile (25 km) long causeway.

Map key

Population

- ■ above 5 million
- ◉ 1 million to 5 million
- ◎ 500,000 to 1 million
- ⊚ 100,000 to 500,000
- ⊕ 50,000 to 100,000
- ⊙ 10,000 to 50,000
- ○ below 10,000

Elevation

- 4000m / 13,124ft
- 3000m / 9843ft
- 2000m / 6562ft
- 1000m / 3281ft
- 500m / 1640ft
- 250m / 820ft
- 100m / 328ft
- sea level

Scale 1:5,500,000

Km
0 10 20 40 60 80 100 120 140 160 180 200
Miles
0 20 40 60 80 100 120 140 160 180 200

projection: Lambert Conformal Conic

Kazakhstan

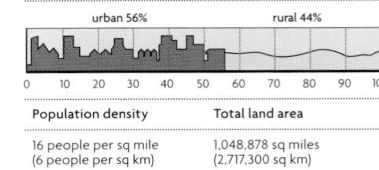

Abundant natural resources lie in the immense steppe grasslands, deserts, and central plateau of the former Soviet republic of Kazakhstan. An intensive program of industrial and agricultural development to exploit these resources during the Soviet era resulted in catastrophic industrial pollution, including fallout from nuclear testing and the shrinkage of the Aral Sea. Since independence, the government has encouraged foreign investment and liberalized the economy to promote growth. The adoption of Kazakh as the national language is intended to encourage a new sense of national identity in a state where living conditions for the majority remain harsh, both in cramped urban centers and impoverished rural areas.

Transportation & industry

The single most important industry in Kazakhstan is mining, based around extensive oil deposits near the Caspian Sea, the world's largest chromium mine, and vast reserves of iron ore. Recent foreign investment has helped to develop industries including food processing and steel manufacture, and to expand the exploitation of mineral resources. The Russian space program is still based at Baykonyr, near Kyzylorda in central Kazakhstan.

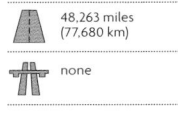

Major industry and infrastructure

- ⚗ chemicals
- ⚙ engineering
- 🐟 fish processing
- 🍴 food processing
- ⬛ iron & steel
- △ metallurgy
- ⛏ mining
- ⚓ oil

- ● capital cities
- ● major towns
- ✈ international airports
- major roads
- major industrial areas

Transportation network

🛣	48,263 miles (77,680 km)
🛤	none
🚆	8483 miles (13,660 km)
〰	3900 miles (2423 km)

Industrial areas in the north and east are well-connected to Russia. Air and rail links with Germany and China have been established through foreign investment. Better access to Baltic ports is being sought.

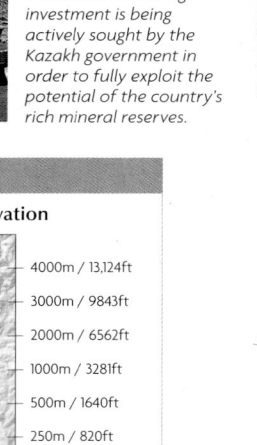

◄ *An open-pit coal mine in Kazakhstan. Foreign investment is being actively sought by the Kazakh government in order to fully exploit the potential of the country's rich mineral reserves.*

Using the land & sea

The rearing of large herds of sheep and goats on the steppe grasslands forms the core of Kazakh agriculture. Arable cultivation and cotton-growing in pasture and desert areas was encouraged during the Soviet era, but relative yields are low. The heavy use of fertilizers and the diversion of natural water sources for irrigation has degraded much of the land.

Land use and agricultural distribution

- 🐂 cattle
- 🐐 goats
- 🐑 sheep
- cotton
- 🐟 fishing
- 🌾 wheat

- ■ capital cities
- ● major towns

- pasture
- cropland
- forest
- mountain region
- desert

The urban/rural population divide

urban 56%	rural 44%

0 10 20 30 40 50 60 70 80 90 100

Population density	Total land area
16 people per sq mile (6 people per sq km)	1,048,878 sq miles (2,717,300 sq km)

◄ *The nomadic peoples who moved their herds around the steppe grasslands are now largely settled, although echoes of their traditional lifestyle, in particular their superb riding skills, remain.*

Map key

Population

- ▣ 1 million to 5 million
- ◉ 500,000 to 1 million
- ◎ 100,000 to 500,000
- ⊕ 50,000 to 100,000
- ○ 10,000 to 50,000
- ∘ below 10,000

Elevation

- 4000m / 13,124ft
- 3000m / 9843ft
- 2000m / 6562ft
- 1000m / 3281ft
- 500m / 1640ft
- 250m / 820ft
- 100m / 328ft
- sea level

Scale 1:6,250,000

Km
0 25 50 100 150 200 250

Miles
0 25 50 100 150 200 250

projection: Lambert Conformal Conic

RUSSIAN FEDERATION

KAZAKHSTAN

RUSSIA

Ural'sk · Taskala · Zachagansk · Fedorovka · Borili · Aksay · Shynggyrlau · Algabas · Shyngyrlau · Zhympity · Lubenka · Martok · Aktobe (Aktyubinsk) · Kobda · Khromtau · Akzhar · Alga · Komsomol'skoye · Karabutak

Dzhanjbek · Kaztalovka · Zhalpaktal · Chapayev · ZAPADNYY KAZAKHSTAN · Yesensay · Kaldygayty · Oyyl · Kandyagash · Zhuryn · AKTYUBINSK · Gory Mugodzhary

Khan Ordasy · Zhanakazan · Yesbol · Inderbor · Karabau · Miyaly · Oyyl · Shubarkudyk · Emba · Zharkamys · Shalkar

Ryn-Peski · ATYRAU · Makhambet · Makat · Dosbol · Sagyz · Zhem · Gory Chushkakol · Gryada Shirkala · Plato Shagyray

Ganyushkino · Akkystau · Atyrau · Komsomol · Kul'sary · Karaton · Severnyy Chink Ustyurta · Peski Bol'shiye Barsuki

Caspian Depression · Kosshagyl · Karaton · Borankul · Sarykamys · Zaliv Tushchybas · Aral'sk

Caspian Sea · Zaliv Komsomolets · Sor Mertvyy Kultuk · Beyneu · Turysh · Ostrov Vozrozhdeniya

Ostrov Kulaly · Mys Tupkaragan · Shebir · Aral Sea · KYZYLORDA

Fort-Shevchenko · Taushyk · Shetpe · Sor Kaydak · Otes · Uyaly · Aykeke Bi · Kazaly · Syr Darya · Toretam · Baykonyr · Zhosaly

Aktau · Plato Mangyshlak · Zhetybay · MANGYSTAU · Ustyurt Plateau · Maylybas Centre · Space Launching

Kuryk · Zhanaozen

TURKMENISTAN

UZBEKISTAN

KOSTANAY · Rudnyy · Tobyl · Lisakovsk · Oktyabr'sk · Denisovka · Zhitikara · KOSTANAY · Turgayskay Stolovaya Strana · Karabalyk · Toguzak · Nadezhdink · Fedorovka · Ozërnoye · Zhayylma

The landscape

Stretching more than 1250 miles (2000 km) from the Caspian Sea in the west to China in the east, more than 40% of Kazakhstan is covered by steppe grasslands which give way to barren desert in the south. The land rises eastward towards the mineral-rich central plateau, to form the Altai Mountains.

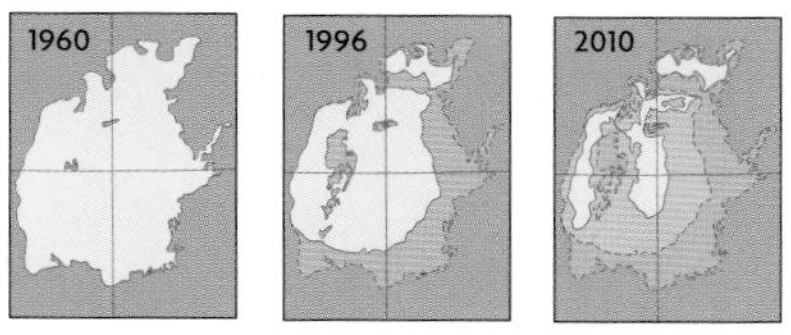

1960   1996   2010

▲ Since 1960, the Aral Sea has shrunk by 75%, become extremely saline, and lost all but five of its once-abundant fish species. Factors in this ecological disaster include the excessive use of fertilizers, defoliants and the diversion of its main source rivers for the irrigation of desert lands.

The Caspian Sea is the largest body of inland water in the world.

The desert of Peski Bol'shiye Barsuki is mainly sandy, displaying a number of classic dune formations. Groundwater supports a small amount of vegetation.

A large number of salt lakes fill depressions in the rolling uplands of central Kazakhstan.

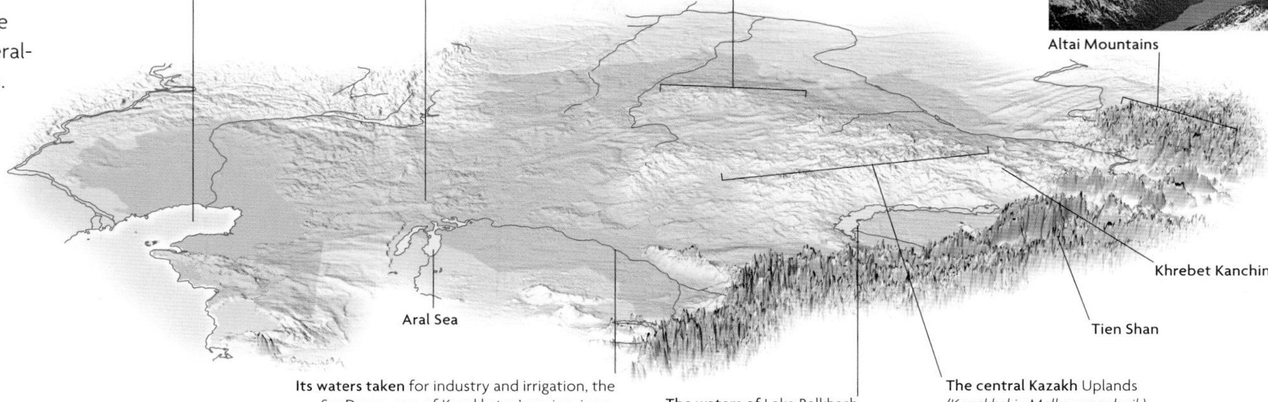

► The Altai Mountains lie on Kazakhstan's eastern borders with China and the Russian Federation. Cold and largely barren, they are the source of many of the rivers which flow across the steppe.

Altai Mountains

Aral Sea

Khrebet Kanchingiz

Tien Shan

Its waters taken for industry and irrigation, the Syr Darya, one of Kazakhstan's major rivers, now barely reaches the Aral Sea which it used to fill. Like many Kazakh rivers it has been heavily polluted with chemicals and its flow has been restricted by up to 60%.

The waters of Lake Balkhash (Ozero Balkhash), unlike those of the Aral Sea, are still able to support a fishing industry.

The central Kazakh Uplands (Kazakhskiy Melkosopochnik) contain much of the country's mineral riches. The landscape is largely flat with occasional rocky outcrops and hillocks.

► Immense stretches of steppe grasslands characterize much of the Kazakh landscape. These lowland areas have been used for arable cultivation in recent years, although problems with irrigation have meant that much of the land is being allowed to revert to its natural vegetation and pastoral usage.

▲ Rows of pine trees edge this valley near Almaty. The snow-covered slopes in the background are used for skiing.

Central Asia

KYRGYZSTAN, TAJIKISTAN, TURKMENISTAN, UZBEKISTAN

The four republics that declared independence in 1991 were created in the early years of the Soviet Union, promoting ethnic divisions in a region whose common focus, since the 8th century, has been Islam. Traditional rural, nomadic ways of life have survived the Soviet era, while the benefits of modern industry and grand irrigation schemes have resulted in severe pollution in the delicate, arid environment of the steppe, particularly in Uzbekistan. Many ethnic minority groups are scattered among the four republics, with isolated communities in the mountains of Kyrgyzstan.

The current Islamic revival has brought hope of greater regional unity, in spite of religious factionalism which, in 1992, plunged Tajikistan into civil war.

◀ **The desert of** the Kara Kum (Garagum) occupies over 70% of Turkmenistan; its wind-scoured surface of dune ridges and depressions severely limits human settlement.

▲ **The southern shoreline** of the Aral Sea has retreated over 30 miles (48 km) since 1960. A major cause is the diversion of water from the Amu Darya river for irrigation via the Kara Kum Canal (Garagum Kanaly).

Map key

Population
- ◉ 1 million to 5 million
- ◎ 500,000 to 1 million
- ⊚ 100,000 to 500,000
- ⊕ 50,000 to 100,000
- ◌ 10,000 to 50,000
- ○ below 10,000

Elevation
- 6000m / 19,686ft
- 4000m / 13,124ft
- 3000m / 9843ft
- 2000m / 6562ft
- 1000m / 3281ft
- 500m / 1640ft
- 250m / 820ft
- 100m / 328ft
- sea level

Transportation & industry

Fossil fuels are extracted and processed in all four states, with scope for further exploitation. Agriculture provides raw materials for many industries, including food and textiles processing, and the manufacture of leather goods, clothing, and carpets. Farm machinery is also produced.

Transportation network

73,658 miles (118,555 km)	87 miles (140 km)
4773 miles (7683 km)	1180 miles (1900 km)

The Kara Kum Canal (Garagum Kanaly) runs for 870 miles (1400 km) from the Amu Darya river to the Caspian Sea. The canal is principally used for irrigation but is navigable for 280 miles (450 km).

Major industry and infrastructure

- 🕸 carpet weaving
- 🧪 chemicals
- ⚙ engineering
- 🍴 food processing
- 🛢 oil & gas
- 👕 textiles

- ■ capital cities
- • major towns
- ⊕ international airports
- — major roads
- ▦ major industrial areas

The landscape

The great Tien Shan and Pamir ranges meet in a succession of high mountain chains. These mountains encircle the fertile Fergana Valley and reach west into the desert of the Kyzyl Kum, dividing the Syr Darya and Amu Darya rivers. Sandy steppeland extends to the shores of the Caspian Sea, with the desert of the Kara Kum (Garagum) in the south. The Amu Darya drains into the Aral Sea in the north.

The Amu Darya is the only river in Central Asia with a sufficient volume of water to cross the desert of the Kara Kum (Garagum) from the Pamirs to the Aral Sea, where it forms a delta largely vegetated by scrub grasses.

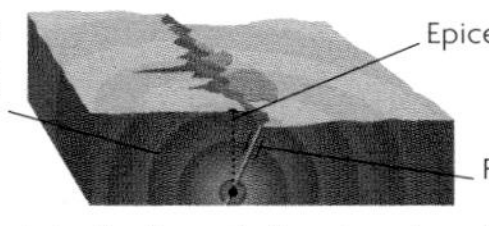

Shock waves travel through ground

Epicenter

Fault

▲ In the heavily fractured and faulted mountain region, earthquakes are common, caused by the sudden release of tension along active fault lines.

Naryn river

◀ Bare mountains provide a stark background to the croplands along the Naryn river in Kyrgyzstan. Irrigation is essential for cultivation in this dry region.

Ozero Issyk-Kul' lies at an altitude of 5193 ft (1584 m). The lake remains ice-free throughout the year, due to the slight salinity of the water.

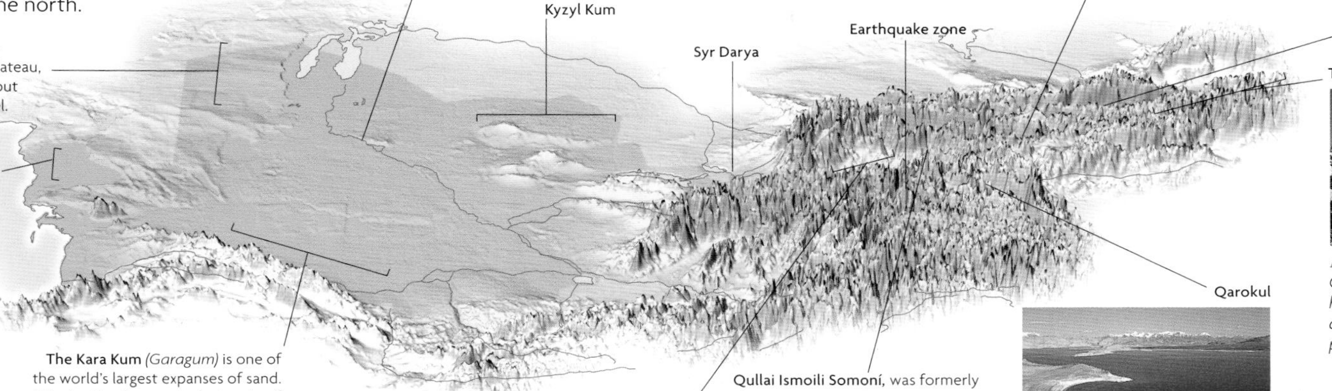

Kyzyl Kum

Syr Darya

Earthquake zone

Tien Shan

Salt marshes fill many of the depressions in the Ustyurt Plateau, a barren, rocky tableland about 650 ft (200 m) above sea level.

Some of the world's largest deposits of marine salts are found in Garabogaz Aylagy. This shallow, saline gulf has an average depth of only 33 ft (10 m), and a very high evaporation rate, producing the salty deposits.

▲ The Tien Shan extend from China in the east, reaching heights over 24,400 ft (7439 m) and branching into many parallel ranges in the west.

Qarokul

The Kara Kum (Garagum) is one of the world's largest expanses of sand. Wind action has created a terrain of shifting, crescent-shaped sand dunes known as barchans.

A series of major rock faults has created the Fergana Valley, a deep depression surrounded by high mountains. Water from the Syr Darya river and from underground sources supports intensive agriculture, despite minimal rainfall.

Qullai Ismoili Somoní, was formerly known as Mount Communism, so named because it was the highest point in the the former Soviet Union, rising to 24,590 ft (7495 m).

◀ Nestling high in the Pamir range, and fed by glacial meltwater, Qarokul is the largest of the lakes in this region.

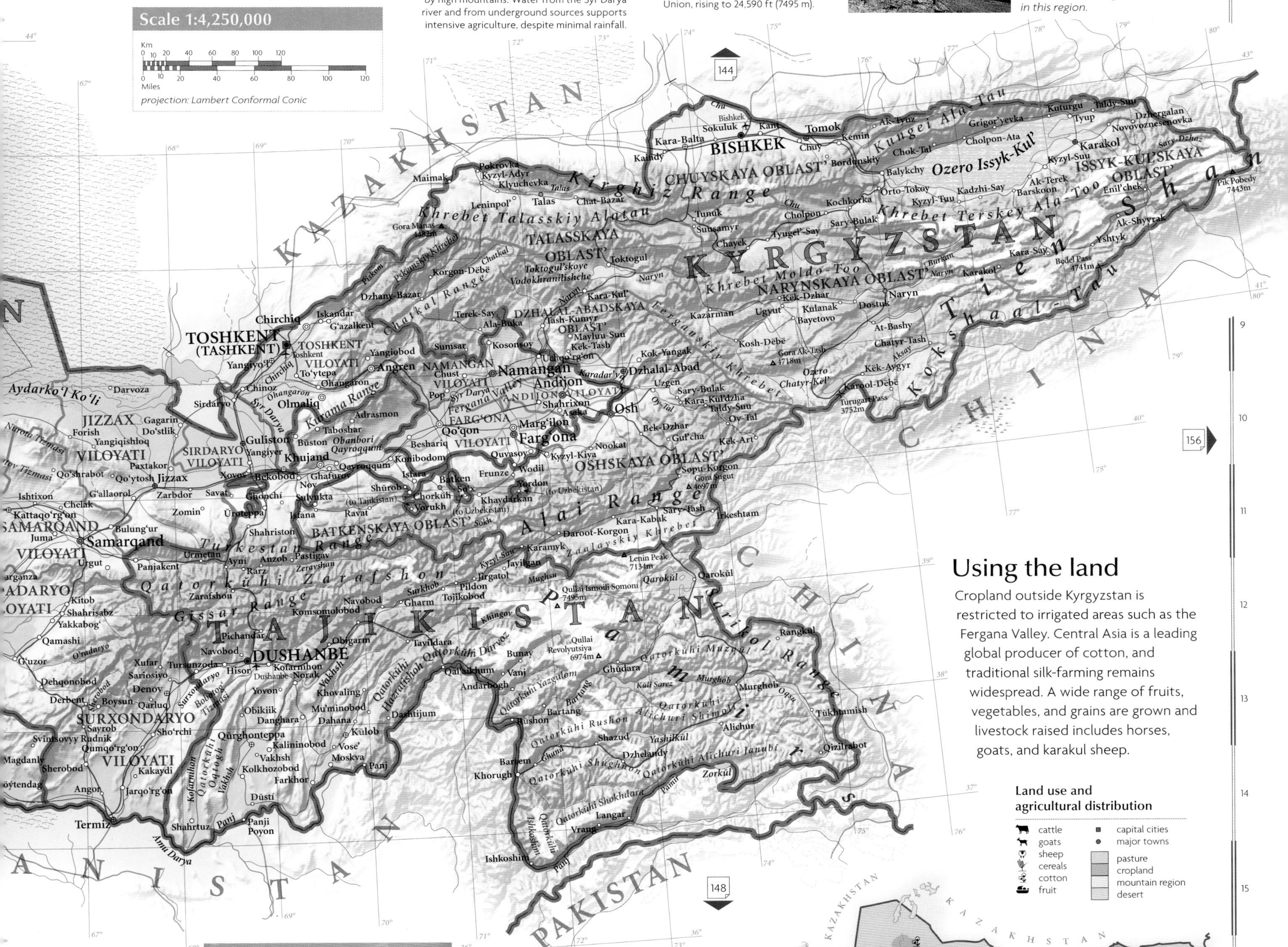

Scale 1:4,250,000

Km
0 10 20   40   60   80   120

Miles
0   20   40   60   80   100   120

projection: Lambert Conformal Conic

Using the land

Cropland outside Kyrgyzstan is restricted to irrigated areas such as the Fergana Valley. Central Asia is a leading global producer of cotton, and traditional silk-farming remains widespread. A wide range of fruits, vegetables, and grains are grown and livestock raised includes horses, goats, and karakul sheep.

Land use and agricultural distribution

- cattle
- goats
- sheep
- cereals
- cotton
- fruit
- capital cities
- major towns
- pasture
- cropland
- mountain region
- desert

▶ Plentiful sunshine, rich soils and massive irrigation schemes have made Uzbekistan the world's fifth largest cotton producer, although water shortages now prevent any further expansion of irrigated land.

The urban/rural population divide

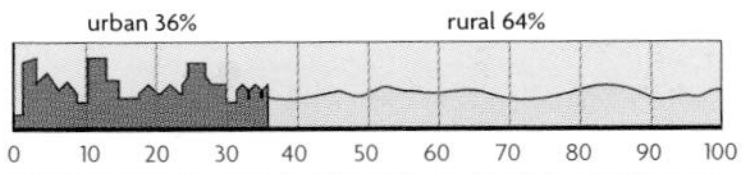

urban 36%   rural 64%

0 10 20 30 40 50 60 70 80 90 100

Population density	Total land area
88 people per sq mile (34 people per sq km)	492,961 sq miles (1,277,100 sq km)

Afghanistan & Pakistan

Pakistan was created by the partition of British India in 1947, becoming the western arm of a new Islamic state for Indian Muslims; the eastern sector, in Bengal, seceded to become the separate country of Bangladesh in 1971. Over half of Pakistan's 158 million people live in the Punjab, at the fertile head of the great Indus Basin. The river sustains a national economy based on irrigated agriculture, including cotton for the vital textiles industry. Afghanistan, a mountainous, landlocked country, with an ancient and independent culture, has been wracked by war since 1979. Factional strife escalated into an international conflict in late 2001, as US-led troops ousted the militant and fundamentally Islamist *taliban* regime as part of their "war on terror."

◄ *The town of* Bamian lies high in the Hindu Kush west of Kabul. Between the 2nd and 5th centuries two huge statues of Buddha were carved into the nearby rock, the largest of which stood 125 ft (38 m) high. The statues were destroyed by the taliban regime in March 2001.

Transportation & industry

Pakistan is highly dependent on the cotton textiles industry, although diversified manufacture is expanding around cities such as Karachi and Lahore. Afghanistan's limited industry is based mainly on the processing of agricultural raw materials and includes traditional crafts such as carpet weaving.

Major industry and infrastructure

- carpet weaving
- chemicals
- engineering
- finance
- food processing
- iron & steel
- oil & gas
- textiles
- capital cities
- major towns
- international airports
- major roads
- major industrial areas

Transportation network

96,154 miles (154,763 km)	
211 miles (340 km)	
4852 miles (7814 km)	
745 miles (1200 km)	

The Karakoram Highway was completed after 20 years of construction in 1978. It breaches the Himalayan mountain barrier providing a commercial motor route linking lowland Pakistan and China.

► *The Karakoram Highway* is one of the highest major roads in the world. It took over 24,000 workers almost 20 years to complete.

The landscape

Afghanistan's topography is dominated by the mountains of the Hindu Kush, which spread south and west into numerous mountain spurs. The dry plateau of southwestern Afghanistan extends into Pakistan and the hills which overlook the great Indus Basin. In northern Pakistan the Hindu Kush, Himalayan, and Karakoram ranges meet to form one of the world's highest mountain regions.

◄ *The Hunza river* rises in the northern Karakoram Range, running for 120 miles (193 km) before joining the Gilgit river.

Hunza river

► *The arid Hindu Kush* makes much of Afghanistan uninhabitable, with over 50% of the land lying above 6500 ft (2000 m).

The plains and foothills which extend from the northern slopes of the Hindu Kush are part of the great grassy steppe lands of Central Asia.

Hindu Kush

K2 (Mount Godwin Austen), in the Karakoram Range, is the second highest mountain in the world, at an altitude of 28,251 ft (8611 m).

Some of the largest glaciers outside the polar regions are found in the Karakoram Range, including Siachen Glacier (Siachen Muztagh), which is 40 miles (72 km) long.

Frequent earthquakes mean that mountain-building processes are continuing in this region, as the Indo-Australian Plate drifts northward, colliding with the Eurasian Plate.

Himalayas

Mountain chains running southwest from the Hindu Kush into Pakistan form a barrier to the humid winds which blow from the Indian Ocean, creating arid conditions across southern Afghanistan.

The soils of the Punjab plain are nourished by enormous quantities of sediment, carried from the Himalayas by the five tributaries of the Indus river.

The Indus Basin is part of the Indus-Ganges lowland, a vast depression which has been filled with layers of sediment over the last 50 million years. These deposits are estimated to be over 16,400 ft (5000 m) deep.

The Indus Delta is prone to heavy flooding and high levels of salinity. It remains a largely uncultivated wilderness area.

Glacis covered by coarse-grained sediment

Sediments washed down from mountains accumulate on glacis slopes

Fine sediments deposited on salt flats are removed by wind erosion

Bedrock

▲ *Glacis are gentle*, debris-covered slopes which lead into saltflats or deserts. They typically occur at the base of mountains in arid regions such as Afghanistan.

Map labels

TURKMEN
BÁDGHÍS
Kāriz-e Elyās
Towraghoudi
Qarah Bāgh
Kushk
Bālā Murghāb
Selseleh-ye Band
Eslām Qal'eh
Kūhestān
Dasht-e Hamdam Ab
Zindah Jān
Ghōriān
Qal'ah-ye Now
Qādis
Herāt
Selseleh-ye Sefíd Kūh
HERĀT
GHOR
Shahra
Namakzar
AFGHA
Shindand
Kūh-e Chehel Abdārān
Dak
Anār Darah
Dasht-e Bābūs
Farāh Rūd
FARĀH
Now Zādo
Hāmūn-e Şāberí
Hāmūn-e Pūzak
Dasht-e Khāsh
Farāh
Dilārām
Sangin
Gereshk
NĪMRŌZ
Chakhānsūr
Shelleh-ye Pūdeh Tal
Lashkar Gāh
Zaranj
Dasht-e Mārgow
Darwēshān
Kūchnay Darwēshān
Dishū
Daryā-ye Helmand
HELMAND
Dasht-e Gowd-e Zereh
Chāgai Hills
Hāmūn-i Lora
Dasht-i Tāhlāb
Nok Kundi
Yakmach
Dālbandin
Tāhlāb
Hāmūn-i Mashkel
BAL
Kamarod
Sīāhān Range
Tagas
Panjgūr
Central Makrā
Awāra
Malar
Ispikān
Nīhing
Nasīrābād
Kech
Hoshāb
Mand
Turbat
Dasht
Suntsar
Khor Kalamat
Jīwani
Gwādar West Bay
Gwādar
Gwādar East Bay
Pasni
Astola Island
Ormāra
IRAN

Afghanistan map (industry)

UZBEKISTAN
TURKMENISTAN
TAJIKISTAN
CHINA
Mazar-e Sharif
Herat
KABUL
Peshawar
ISLAMABAD
Rawalpindi
AFGHANISTAN
Kandahar
Lahore
Quetta
Faisalabad
Multan
Bahawalpur
PAKISTAN
Sukkur
Karachi
Hyderabad
IRAN
INDIA
ARABIAN SEA

Scale 1:4,500,000

Km
0 20 40 60 80 100 120 140 160

Miles
0 20 40 60 80 100 120 140 160

projection: Lambert Conformal Conic

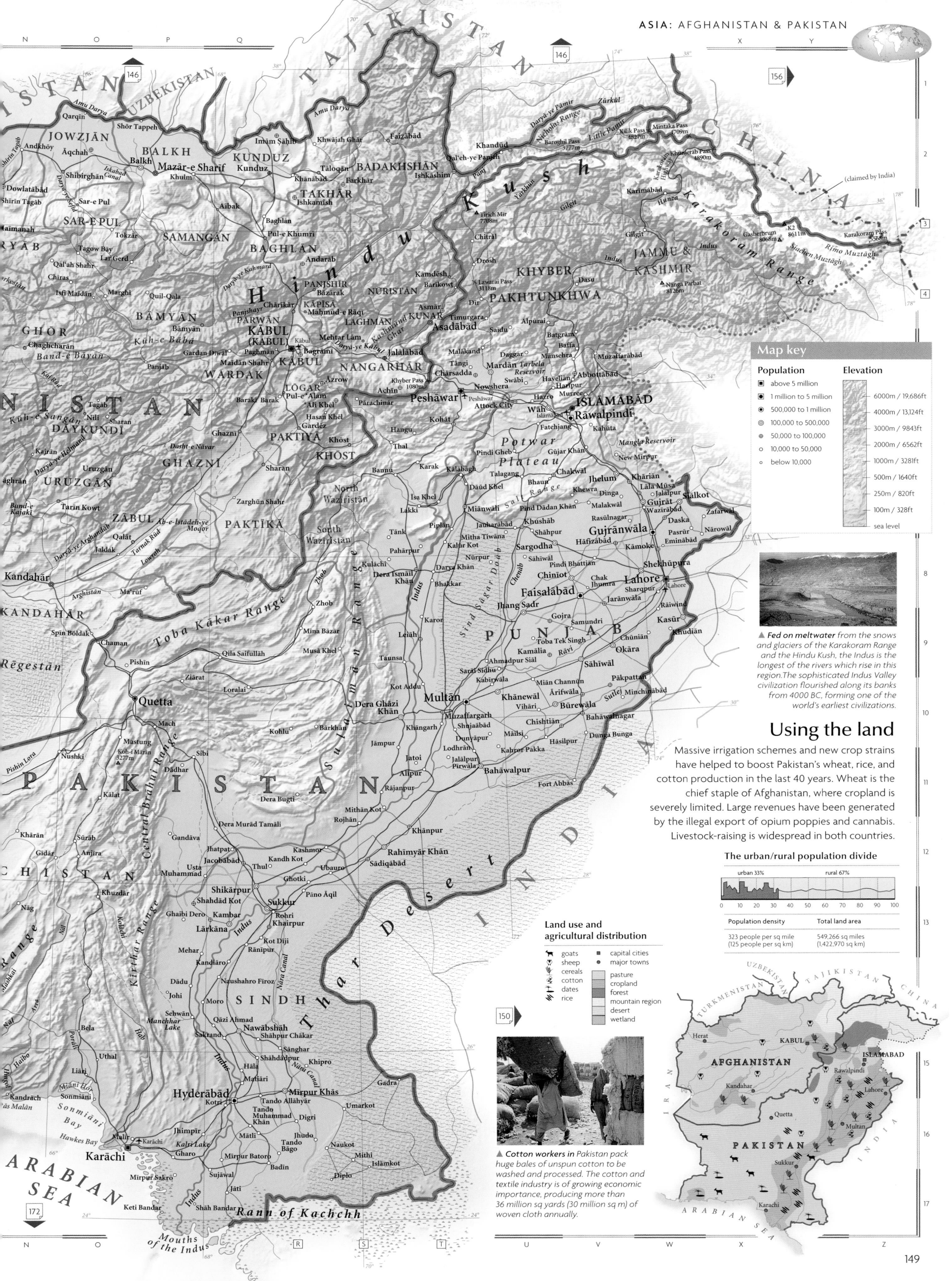

Map key

Population
- above 5 million
- 1 million to 5 million
- 500,000 to 1 million
- 100,000 to 500,000
- 50,000 to 100,000
- 10,000 to 50,000
- below 10,000

Elevation
- 6000m / 19,686ft
- 4000m / 13,124ft
- 3000m / 9843ft
- 2000m / 6562ft
- 1000m / 3281ft
- 500m / 1640ft
- 250m / 820ft
- 100m / 328ft
- sea level

▲ *Fed on meltwater* from the snows and glaciers of the Karakoram Range and the Hindu Kush, the Indus is the longest of the rivers which rise in this region. The sophisticated Indus Valley civilization flourished along its banks from 4000 BC, forming one of the world's earliest civilizations.

Using the land

Massive irrigation schemes and new crop strains have helped to boost Pakistan's wheat, rice, and cotton production in the last 40 years. Wheat is the chief staple of Afghanistan, where cropland is severely limited. Large revenues have been generated by the illegal export of opium poppies and cannabis. Livestock-raising is widespread in both countries.

The urban/rural population divide

urban 33%   rural 67%

Population density	Total land area
323 people per sq mile (125 people per sq km)	549,266 sq miles (1,422,970 sq km)

Land use and agricultural distribution
- goats
- sheep
- cereals
- cotton
- dates
- rice
- capital cities
- major towns
- pasture
- cropland
- forest
- mountain region
- desert
- wetland

▲ *Cotton workers in* Pakistan pack huge bales of unspun cotton to be washed and processed. The cotton and textile industry is of growing economic importance, producing more than 36 million sq yards (30 million sq m) of woven cloth annually.

149

South Asia

BANGLADESH, BHUTAN, INDIA, MALDIVES, NEPAL, PAKISTAN, SRI LANKA

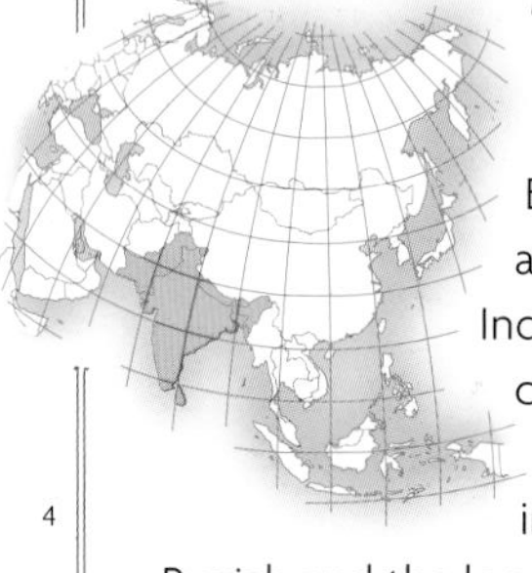

More than one-fifth of the world's population lives in the south Asian subcontinent. Great cultural diversity has come from a long succession of foreign invaders, including Hindu Aryans, Islamic Moguls, and the British, whose empire incorporated the princely states of the Maharajas and extended to the borders of Nepal and Bhutan in the Himalayas. Independent since 1947, India is the world's largest democracy, and at the current rate of growth, may overtake China as the world's most populous country during the 21st century. There are points of tension in the region over claims for independence by the Sikhs in the Indian Punjab and the long-standing dispute with Pakistan over Jammu and Kashmir in the north.

The towering Karakoram and Hindu Kush ranges, formed at the same time as the Himalayas, dominate Pakistan's northern borders. K2 on the border of northern Pakistan is the second highest mountain on Earth, at 28,251 ft (861 m).

The landscape

South Asia is effectively isolated from the rest of Asia by desert along the western flank of Pakistan, and a continuous wall of mountains, dominated by the Himalayas, to the north and east. The great basins of the Indus and Ganges separate this mountain fringe from the rolling plateau of the Indian peninsula, which is bordered by a line of coastal hills, the Eastern and Western Ghats.

The Indus river flows more than 1970 miles (3180 km) from southwestern Tibet to its mouth on the Arabian Sea. It has an estimated catchment area of 450,000 sq miles (1,165,500 sq km).

The coast of western Pakistan is a staircase of folded rock strata caused by successive periods of rapid uplift.

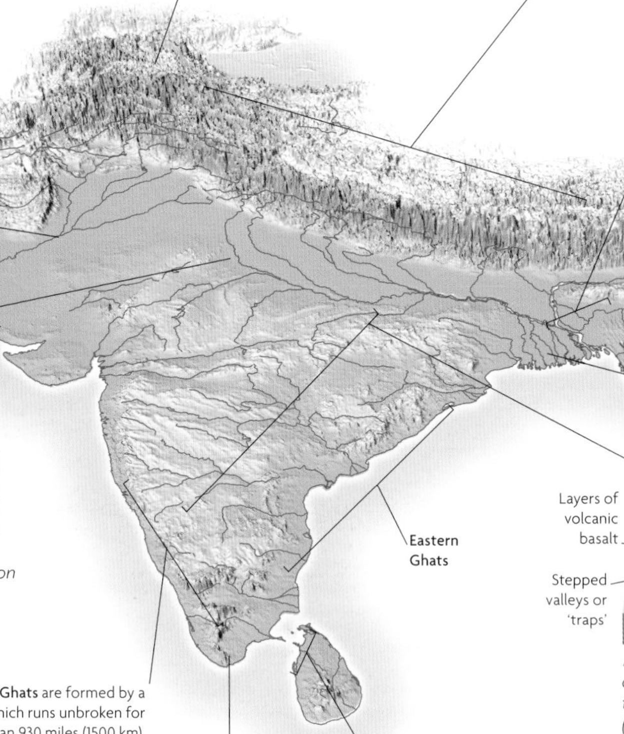

The Himalayas are the highest and most extensive mountain system in the world. They were formed when the Indo-Australian Plate collided with the Eurasian Plate about 40 million years ago, thrusting up huge masses of land and creating a "ripple" effect, which formed lesser mountain ranges in Tibet and Southeast Asia. Mount Everest is the world's tallest mountain at 29,029 ft (8848 m).

The Indus valley near Skardu in northern Pakistan has been partially infilled by great quantities of eroded sediment. Most of this is carried from the region's bare slopes by swollen rivers during the spring thaw and mass movement activity.

Almost all of Bangladesh lies in the immense delta formed by the Ganges and the Brahmaputra which merge and flow out into the Bay of Bengal.

Ganges delta

Deccan plateau

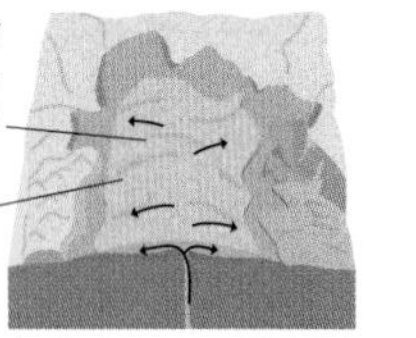

The Deccan plateau covers an area of more than 123,553 sq miles (320,000 sq km). It is formed of deep layers of volcanic basalt, reaching thicknesses of more than 9800 ft (3000 m) toward the coast. Distinctive stepped valleys cut in the basalt plateau by rivers are known as "traps."

Layers of volcanic basalt

Stepped valleys or 'traps'

Eastern Ghats

Coastal deposition has formed many typical features along the western coast of Sri Lanka. These include spits and bars, sometimes enclosing lagoons.

Trivandrum in southern India normally receives the first of the monsoon rains, which are essential to south Asian agriculture and moderate the extreme summer heat. The monsoon then moves northward over a period of about two months.

The Western Ghats are formed by a fault scarp which runs unbroken for more than 930 miles (1500 km). They reach their highest point at the southern Cardamom Hills.

Rivers flowing from the Himalayas into a broad depression in northern India have formed marshes around Bharatpur. They are now a sanctuary for numerous bird species.

Bharatpur

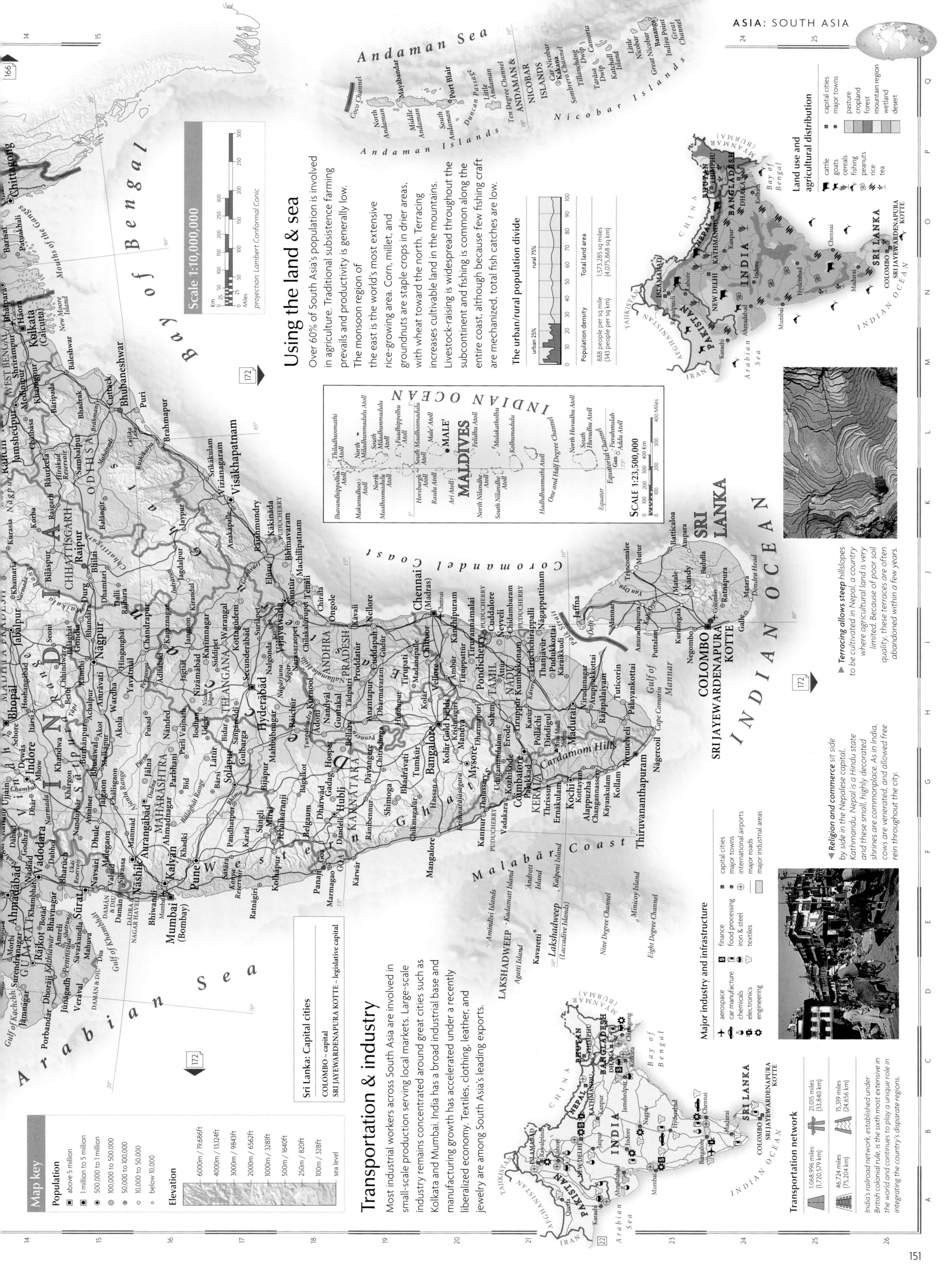

Using the land & sea

Over 60% of South Asia's population is involved in agriculture. Traditional subsistence farming prevails and productivity is generally low. The monsoon region of the east is the world's most extensive rice-growing area. Corn, millet, and groundnuts are staple crops in drier areas, with wheat toward the north. Terracing increases cultivable land in the mountains. Livestock-raising is widespread throughout the subcontinent and fishing is common along the entire coast, although because few fishing craft are mechanized, total fish catches are low.

The urban/rural population divide

urban 25%   rural 75%

	Total land area
	1,573,285 sq mile (4,075,868 sq km)

Population density
888 people per sq mile
(343 people per sq km)

Land use and agricultural distribution

- cattle
- goats
- cereals
- fishing
- peanuts
- rice
- tea
- capital cities
- major towns
- pasture
- cropland
- forest
- mountain region
- wetland
- desert

Transportation & industry

Most industrial workers across South Asia are involved in small-scale production serving local markets. Large-scale industry remains concentrated around great cities such as Kolkata and Mumbai. India has a broad industrial base and manufacturing growth has accelerated under a recently liberalized economy. Textiles, clothing, leather, and jewelry are among South Asia's leading exports.

Major industry and infrastructure

- aerospace
- car manufacture
- chemicals
- electronics
- engineering
- finance
- food processing
- iron & steel
- textiles
- capital cities
- major towns
- international airports
- major roads
- major industrial areas

Transportation network

- 1,068,996 miles (1,720,579 km)
- 44,724 miles (75,204 km)
- 21,015 miles (33,840 km)
- 15,319 miles (24,656 km)

India's railroad network, established under British colonial rule, is the sixth most extensive in the world and continues to play a unique role in integrating the country's disparate regions.

▼ *Religion and commerce sit side by side in the Nepalese capital, Kathmandu. Nepal is a Hindu state and these small, highly decorated shrines are commonplace. As in India, cows are venerated, and allowed free rein throughout the city.*

▲ *Terracing allows steep hillslopes to be cultivated in Nepal, a country where agricultural land is very limited. Because of poor soil quality, these terraces are often abandoned within a few years.*

Map key

Population
- above 5 million
- 1 million to 5 million
- 500,000 to 1 million
- 100,000 to 500,000
- 50,000 to 100,000
- 10,000 to 50,000
- below 10,000

Elevation
- 6000m / 19,686ft
- 4000m / 13,124ft
- 3000m / 9843ft
- 2000m / 6562ft
- 1000m / 3281ft
- 500m / 1640ft
- 250m / 820ft
- 100m / 328ft
- sea level

Sri Lanka: Capital cities
COLOMBO – capital
SRI JAYEWARDENAPURA KOTTE – legislative capital

Scale 1:10,000,000
projection: Lambert Conformal Conic

SCALE 1:23,500,000

Northern India & the Himalayan states

BANGLADESH, BHUTAN, NEPAL, Arunachal Pradesh, Assam, Bihar, Chandigarh, Delhi, Haryana, Himachal Pradesh, Jammu & Kashmir, Jharkhand, Manipur, Meghalaya, Mizoram, Nagaland, Punjab, Rajasthan, Sikkim, Tripura, Uttarakhand, Uttar Pradesh, West Bengal

The Ganges and Brahmaputra river basins and the massive mountain barrier of the Himalayas define this region's landscape and have served to reinforce potent cultural and religious differences among its people. Hinduism pervades most aspects of national life and is a growing political force within India, a secular country which also encompasses the center of Sikhism at Amritsar and the world's largest Muslim minority. Nepal is a crowded mountain state, which faces severe ecological problems from deforestation, while the tiny Himalayan Buddhist kingdom of Bhutan is emerging from long-term isolation, to welcome selected visitors. The Muslim state of Bangladesh, formerly East Pakistan, is one of the world's most densely populated countries and one of the poorest, with more than 145 million people living largely on the massive Ganges/Brahmaputra delta. Many Bangladeshis live under threat of repeated, catastrophic floods.

◀ *The Golden Temple* in Amritsar, the most sacred shrine of the Sikh religion, was the scene of violent clashes between Sikh separatists and government forces in 1984.

Map key

Population
- ▣ 1 million to 5 million
- ◉ 500,000 to 1 million
- ◎ 100,000 to 500,000
- ⊙ 50,000 to 100,000
- ○ 10,000 to 50,000
- ○ below 10,000

Elevation
- 6000m / 19,686ft
- 4000m / 13,124ft
- 3000m / 9843ft
- 2000m / 6562ft
- 1000m / 3281ft
- 500m / 1640ft
- 250m / 820ft
- 100m / 328ft
- sea level

Transportation & industry

Textiles, engineering, chemicals, and electronics are leading industries in north India. The plateau of Chota Nagpur provides ore for iron and steel production in the major industrial region northeast of Kolkata. Bangladesh processes jute and Nepal has a small manufacturing sector based on agricultural produce, while Bhutan's limited industry is concentrated in the southern lowland area.

Scale 1:5,750,000
projection: Lambert Conformal Conic

Major industry and infrastructure

- ⚲ adventure tourism
- ⚙ car manufacture
- ⚗ chemicals
- ⬧ coal
- ⚡ electronics
- ⚙ engineering
- $ finance
- ⬟ food processing
- ⚒ iron & steel
- ⬢ jute processing
- ⛏ oil
- ☘ tea processing
- ⊼ textiles

- ■ capital cities
- ● major towns
- ✈ international airports
- — major roads
- ▢ major industrial areas

Transportation network

Over 60% of Bangladesh's internal trade is carried by boat. The country has a very disjointed land transportation network, with no bridges over the Brahmaputra and few road crossings on the Ganges river.

152

The landscape

Most of the region is drained by the Ganges river, which meets the Brahmaputra in Bangladesh to form an immense delta before flowing into the Bay of Bengal. The Himalayas extend eastward over 1500 miles (2400 km), from the parallel ranges running through Jammu and Kashmir. The Thar Desert occupies the southwest.

The Indian Punjab lies mainly to the west of the Ganges watershed and its rivers flow into the Indus. Control of this water resource has been a source of great friction with neighboring Pakistan.

The border between India and Pakistan runs through the Thar Desert, an area of sandy seif dunes 50–100 ft (15–30 m) in height. Fossils found in the desert indicate that the dunes, stabilized by vegetation, have been in their current position for about 3000 years.

Sambhar Salt Lake in Rajasthan is India's largest lake. Unlike most of the Himalayan lakes which are glacial in origin – formed in ice-scoured basins or as the result of depositional damming – it is an ephemeral salt lake filled periodically by flash flooding.

► The Pir Panjal Range in southwestern Kashmir rises to elevations of 12,500 ft (3810 m). Despite the freezing conditions, settlements and extensive pastures are found above the tree line.

The northern ranges of the Himalayas contain the highest mountains in the world, with average heights of more than 23,000 ft (7000 m) and many peaks higher than 26,000 ft (8000 m).

In the last 40 million years, the course of the Brahmaputra has been diverted hundreds of miles to the east by the rising landmass of the Himalayas.

The Khasi Hills are an example of a *horst*, a fractured block of bedrock which has been thrust upward.

▲ The summit of Machhapuchhre rises to 22,942 ft (6993 m). It is also known as the "Fish's Tail" because of its distinctive peak.

The Ganges river, sacred to the Hindu people, drains a vast lowland area at the base of the Himalayas. The northern plains are covered by sandy deposits, broken by mud banks formed when the river floods.

The rapid deforestation of Himalayan valleys has led to acute soil erosion and increased rates of rainwater runoff, both cited as possible causes of the worsening floods downstream in the Ganges/Brahmaputra delta, although natural rates are high and may be the real cause.

Over half of the great Ganges/Brahmaputra delta floods each year during the monsoon as rivers, swollen by meltwater from the Himalayas and by excess rainwater, break their banks and fertilize the land with nutrient-rich sediment.

Debris slides in the middle Himalayas
Debris fans at base of slope
Soil blocks
Slide plain

▲ Soil loss in the middle Himalayas has largely been attributed to debris slides, where large blocks of soil are mobilized by saturation along a slide plane. Once mobile, the soil slides down the slope, gaining speed and thinning to form a fan at the base of the slope.

Using the land

Grain production dominates land use. Rice is most widely grown in the east. Irrigation and new crop strains have dramatically increased yields in the Punjab, a major wheat-producing area. River floodplains are intensively farmed and livestock herding is widespread, particularly in Bhutan. Regional crops include jute in Bangladesh, tea in Assam, cardamom in Sikkim, and saffron in Kashmir.

The urban/rural population divide
urban 23%   rural 77%
0 10 20 30 40 50 60 70 80 90 100

Population density	Total land area
993 people per sq mile (384 people per sq km)	665,104 sq miles (1,723,068 sq km)

▲ An adverse climate, steep slopes, and poor soils limit crop cultivation in Bhutan, which is a largely agrarian economy. Rice, corn, and wheat are the main staples, although orchards are being established as the soil and climate suit this type of farming.

Land use and agricultural distribution
- cattle
- goats
- sheep
- cereals
- jute
- rice
- tea
- capital cities
- major towns
- pasture
- cropland
- forest
- mountain region
- wetland
- desert

▲ Flooded streets in Dhaka, Bangladesh are a testament to the region's vulnerability to flooding. In 1988 alone, 75% of the country was flooded, leaving thousands of people dead and over 25 million homeless.

Southern India & Sri Lanka

SRI LANKA, Andhra Pradesh, Chhattisgarh, Dadra & Nagar Haveli, Daman & Diu, Goa, Gujarat, Karnataka, Kerala, Lakshadweep, Madhya Pradesh, Maharashtra, Odisha, Puducherry, Tamil Nadu, Telangana

The unique and highly independent southern states reflect the diverse and decentralized nature of southern India. The southern half of the peninsula lay beyond the reach of early invaders from the north and retained the distinct and ancient culture of Dravidian peoples such as the Tamils, whose language is spoken in preference to Hindi throughout southern India. The interior plateau of southern India is less densely populated than the coastal lowlands, where the European colonial imprint is strongest. Urban and industrial growth is accelerating, but southern India's vast population remains predominantly rural. The island of Sri Lanka has two distinct cultural groups; the mainly Buddhist Sinhalese majority, and the Tamil minority whose struggle for a homeland in the northeast led to prolonged civil war.

Using the land and sea

Rice is the main staple in the east, in Sri Lanka and along the humid Malabar Coast. Peanuts are grown on the Deccan plateau, with wheat, corn, and chickpeas, toward the north. Sri Lanka is a leading exporter of tea, coconuts and rubber. Cotton plantations supply local mills around Nagpur and Mumbai. Fishing supports many communities in Kerala and the Laccadive Islands.

The urban/rural population divide

urban 33%   rural 67%

Population density	Total land area
730 people per sq mile (282 people per sq km)	698,295 sq miles (1,809,054 sq km)

Land use and agricultural distribution

cattle, goats, cereals, cotton, fishing, peanuts
rice, rubber, tea
capital cities, major towns
pasture, cropland, forest, wetland

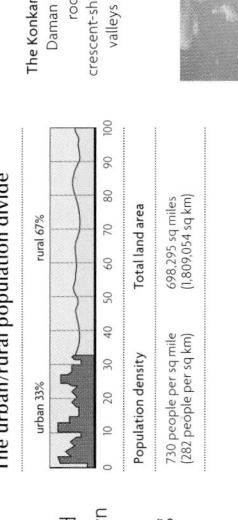

The landscape

The undulating Deccan plateau underlies most of southern India; it slopes gently down toward the east and is largely enclosed by the Ghats coastal hill ranges. The Western Ghats run continuously along the Arabian Sea coast, while the Eastern Ghats are interrupted by rivers which follow the slope of the plateau and flow across broad lowlands into the Bay of Bengal. The plateaus and basins of Sri Lanka's central highlands are surrounded by a broad plain.

Along the northern boundary of the Deccan plateau, old basement rocks are interspersed with younger sedimentary strata. This creates spectacular scarplands, cut by numerous waterfalls along the softer sedimentary strata.

The interior uplands of southern India are broadly known as the Deccan plateau. River erosion of the plateau's volcanic rock has created distinctive stepped valleys called traps.

Deep layers of river sediment have created a broad lowland plain along the eastern coast, with rivers such as the Krishna forming extensive deltas.

The island of Sri Lanka is essentially an extension of the Deccan plateau. It lies on the Indian continental shelf and is composed of the same hard, crystalline rocks.

The Rann of Kachchh tidal marshes encircle the low-lying Kachchh peninsula. For several months during the rainy season the water level of the marshes rises and Kachchh becomes an island.

The Konkan coast, which runs between Daman and Goa, is characterized by rocky headlands, and bays with crescent-shaped beaches. Flooded river valleys known as rias extend inland.

▲ The Western Ghats run north–south marking the western boundary of the Deccan plateau. Their height rises to the south where their summits reach altitudes of 8000 ft (2500m).

Adam's Bridge

Ocean currents cause sediment build up

Relict of ancient tombolo

Adam's Bridge

Sri Lanka

▲ Adam's Bridge (Rama's Bridge) is a chain of sandy shoals lying about 4 ft (1.2 m) under the sea between India and Sri Lanka. They once formed the world's longest tombolo, or land bridge, before the sea level began to rise several thousand years ago.

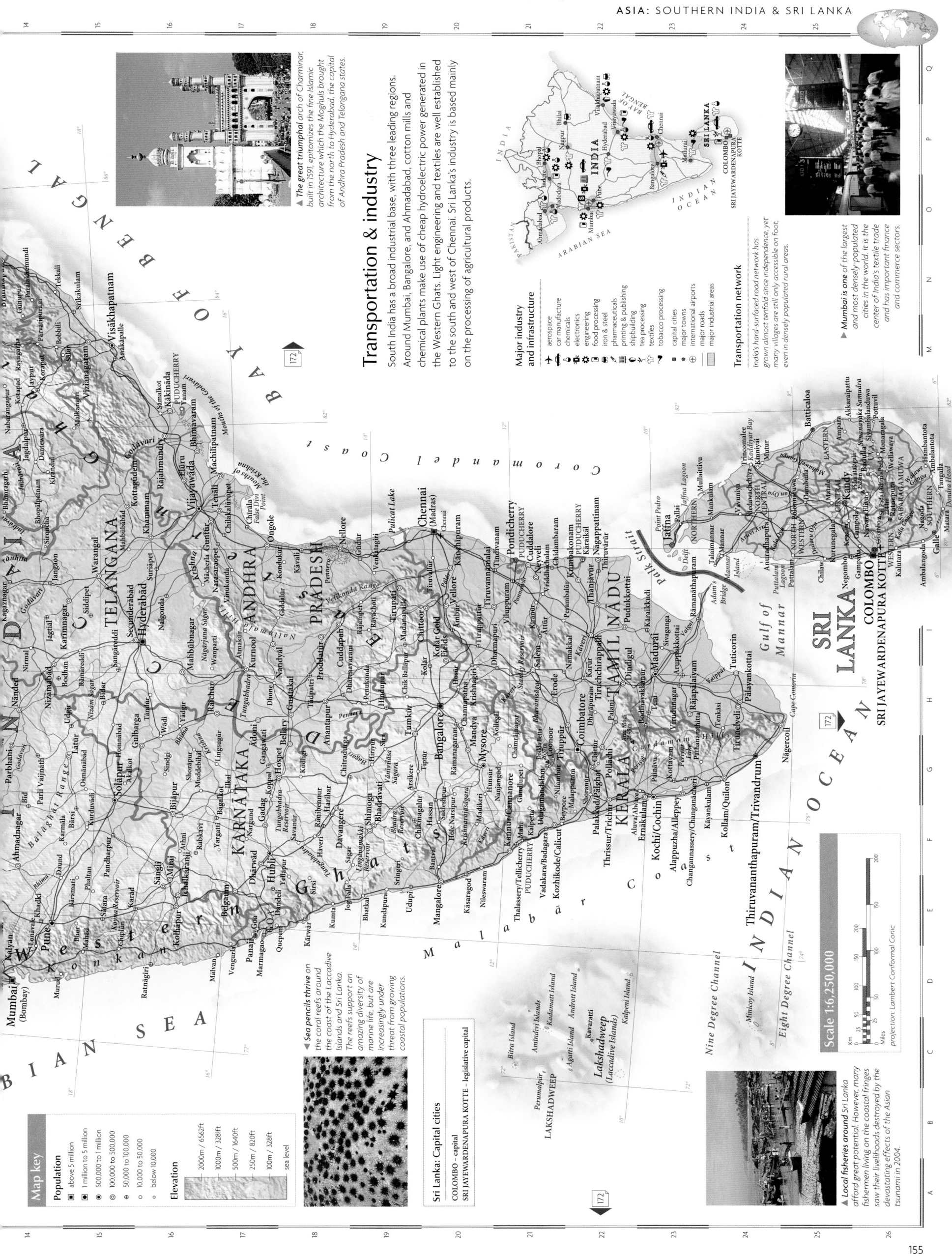

▲ *The great triumphal arch of Charminar, built in 1591, epitomizes the fine Islamic architecture which the Moghuls brought from the north to Hyderabad, the capital of Andhra Pradesh and Telangana states.*

Transportation & industry

South India has a broad industrial base, with three leading regions. Around Mumbai, Bangalore, and Ahmadabad. Light engineering and textiles are well established to the south and west of Chennai. Sri Lanka's industry is based mainly on the processing of agricultural products. The Western Ghats make use of cheap hydroelectric power generated in the chemical plants, cotton mills and

Major industry and infrastructure

- aerospace
- car manufacture
- chemicals
- electronics
- engineering
- food processing
- iron & steel
- pharmaceuticals
- printing & publishing
- shipbuilding
- tea processing
- textiles
- tobacco processing
- capital cities
- major towns
- international airports
- major roads
- major industrial areas

Transportation network

India's hard-surfaced road network has grown almost tenfold since independence, yet many villages are still only accessible on foot, even in densely populated rural areas.

▲ *Mumbai is one of the largest and most densely-populated cities in the world. It is the center of India's textile trade and has important finance and commerce sectors.*

▼ *Sea pencils thrive on the coral reefs around the coast of the Laccadive Islands and Sri Lanka. The reefs support an amazing diversity of marine life, but are increasingly under threat from growing coastal populations.*

Sri Lanka: Capital cities

COLOMBO – capital
SRI JAYEWARDENAPURA KOTTE – legislative capital

▲ *Local fisheries around Sri Lanka afford great potential. However, many fishermen living on the coastal fringes saw their livelihoods destroyed by the devastating effects of the Asian tsunami in 2004.*

Map key

Population
- above 5 million
- 1 million to 5 million
- 500,000 to 1 million
- 100,000 to 500,000
- 50,000 to 100,000
- 10,000 to 50,000
- below 10,000

Elevation
- 2000m / 6562ft
- 1000m / 3281ft
- 500m / 1640ft
- 250m / 820ft
- 100m / 328ft
- sea level

Scale 1:6,250,000

projection: Lambert Conformal Conic

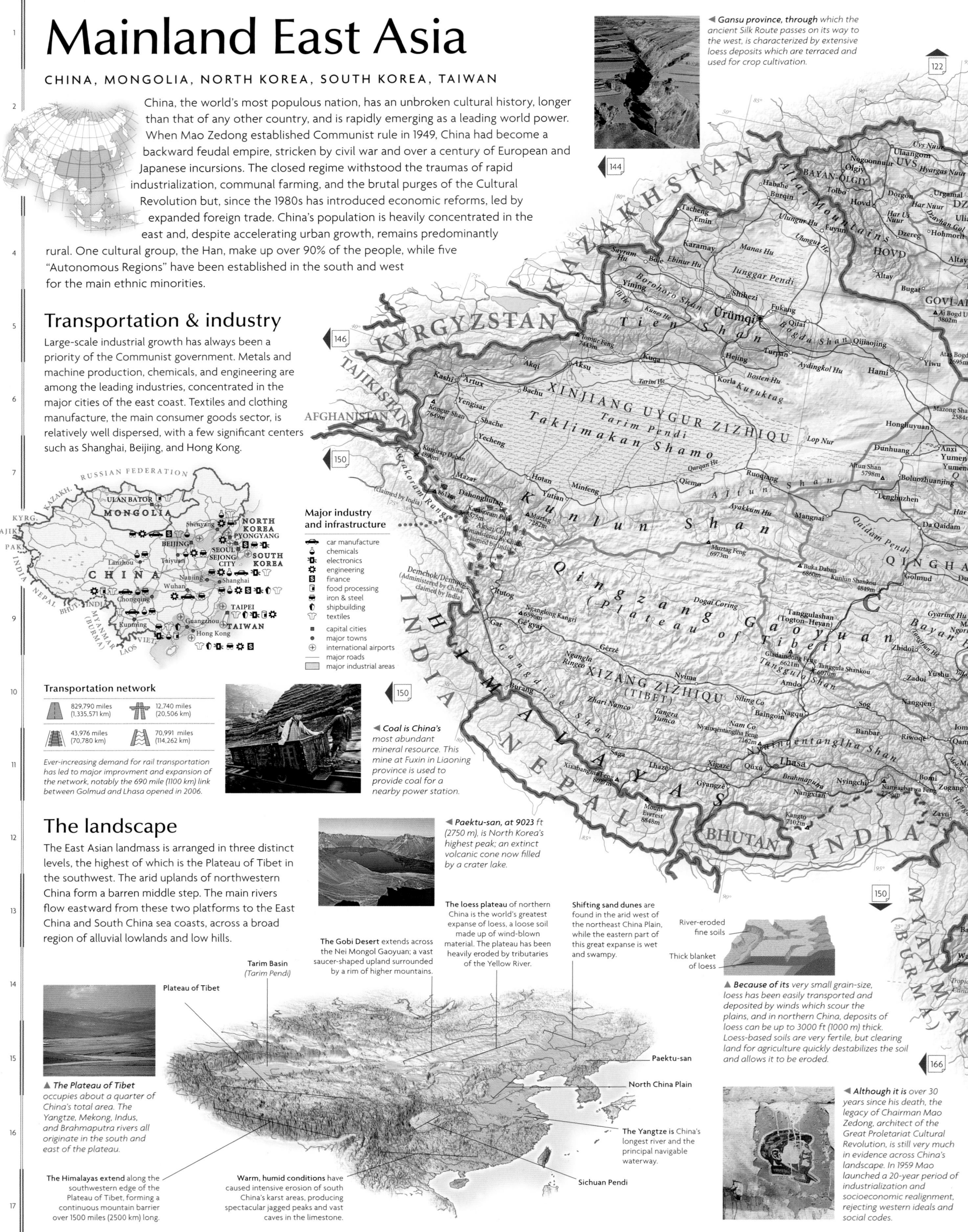

Mainland East Asia

CHINA, MONGOLIA, NORTH KOREA, SOUTH KOREA, TAIWAN

China, the world's most populous nation, has an unbroken cultural history, longer than that of any other country, and is rapidly emerging as a leading world power. When Mao Zedong established Communist rule in 1949, China had become a backward feudal empire, stricken by civil war and over a century of European and Japanese incursions. The closed regime withstood the traumas of rapid industrialization, communal farming, and the brutal purges of the Cultural Revolution but, since the 1980s has introduced economic reforms, led by expanded foreign trade. China's population is heavily concentrated in the east and, despite accelerating urban growth, remains predominantly rural. One cultural group, the Han, make up over 90% of the people, while five "Autonomous Regions" have been established in the south and west for the main ethnic minorities.

Transportation & industry

Large-scale industrial growth has always been a priority of the Communist government. Metals and machine production, chemicals, and engineering are among the leading industries, concentrated in the major cities of the east coast. Textiles and clothing manufacture, the main consumer goods sector, is relatively well dispersed, with a few significant centers such as Shanghai, Beijing, and Hong Kong.

Major industry and infrastructure

- car manufacture
- chemicals
- electronics
- engineering
- finance
- food processing
- iron & steel
- shipbuilding
- textiles
- capital cities
- major towns
- international airports
- major roads
- major industrial areas

Transportation network

829,790 miles (1,335,571 km)	12,740 miles (20,506 km)
43,976 miles (70,780 km)	70,991 miles (114,262 km)

Ever-increasing demand for rail transportation has led to major improvment and expansion of the network, notably the 690 mile (1100 km) link between Golmud and Lhasa opened in 2006.

◄ *Coal is China's most abundant mineral resource. This mine at Fuxin in Liaoning province is used to provide coal for a nearby power station.*

The landscape

The East Asian landmass is arranged in three distinct levels, the highest of which is the Plateau of Tibet in the southwest. The arid uplands of northwestern China form a barren middle step. The main rivers flow eastward from these two platforms to the East China and South China sea coasts, across a broad region of alluvial lowlands and low hills.

◄ *Gansu province, through which the ancient Silk Route passes on its way to the west, is characterized by extensive loess deposits which are terraced and used for crop cultivation.*

◄ *Paektu-san, at 9023 ft (2750 m), is North Korea's highest peak; an extinct volcanic cone now filled by a crater lake.*

The Gobi Desert extends across the Nei Mongol Gaoyuan; a vast saucer-shaped upland surrounded by a rim of higher mountains.

The loess plateau of northern China is the world's greatest expanse of loess, a loose soil made up of wind-blown material. The plateau has been heavily eroded by tributaries of the Yellow River.

Shifting sand dunes are found in the arid west of the northeast China Plain, while the eastern part of this great expanse is wet and swampy.

River-eroded fine soils

Thick blanket of loess

▲ *Because of its very small grain-size, loess has been easily transported and deposited by winds which scour the plains, and in northern China, deposits of loess can be up to 3000 ft (1000 m) thick. Loess-based soils are very fertile, but clearing land for agriculture quickly destabilizes the soil and allows it to be eroded.*

Tarim Basin (Tarim Pendi)

Plateau of Tibet

Paektu-san

North China Plain

The Yangtze is China's longest river and the principal navigable waterway.

Sichuan Pendi

▲ *The Plateau of Tibet occupies about a quarter of China's total area. The Yangtze, Mekong, Indus, and Brahmaputra rivers all originate in the south and east of the plateau.*

The Himalayas extend along the southwestern edge of the Plateau of Tibet, forming a continuous mountain barrier over 1500 miles (2500 km) long.

Warm, humid conditions have caused intensive erosion of south China's karst areas, producing spectacular jagged peaks and vast caves in the limestone.

◄ *Although it is over 30 years since his death, the legacy of Chairman Mao Zedong, architect of the Great Proletariat Cultural Revolution, is still very much in evidence across China's landscape. In 1959 Mao launched a 20-year period of industrialization and socioeconomic realignment, rejecting western ideals and social codes.*

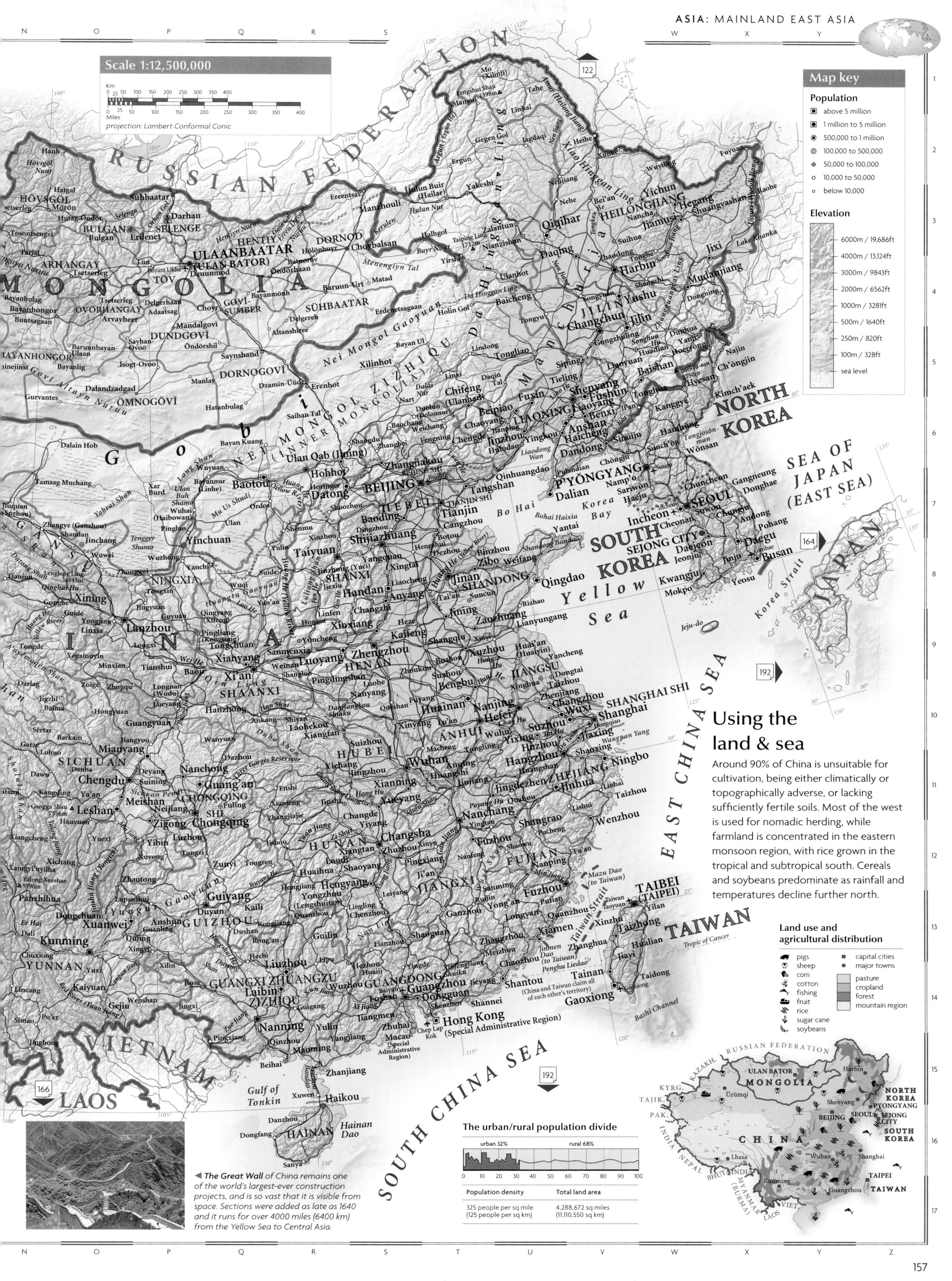

Using the land & sea

Around 90% of China is unsuitable for cultivation, being either climatically or topographically adverse, or lacking sufficiently fertile soils. Most of the west is used for nomadic herding, while farmland is concentrated in the eastern monsoon region, with rice grown in the tropical and subtropical south. Cereals and soybeans predominate as rainfall and temperatures decline further north.

▲ *The Great Wall* of China remains one of the world's largest-ever construction projects, and is so vast that it is visible from space. Sections were added as late as 1640 and it runs for over 4000 miles (6400 km) from the Yellow Sea to Central Asia.

The urban/rural population divide

urban 32%   rural 68%

Population density	Total land area
325 people per sq mile (125 people per sq km)	4,288,672 sq miles (11,110,550 sq km)

Western China

Gansu, Ningxia, Qinghai, Tibet, Xinjiang

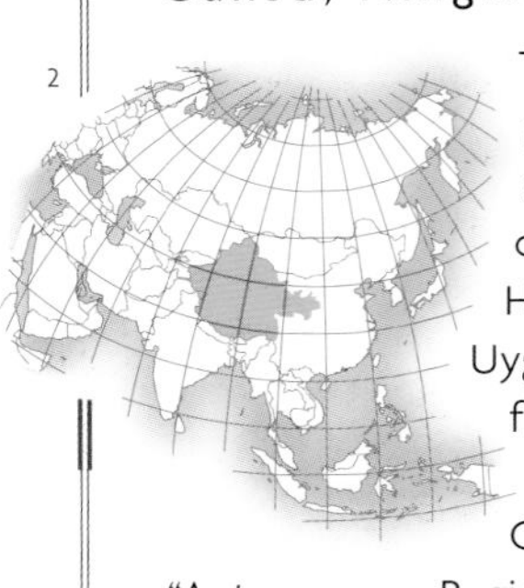

The plateaus and basins of China's dry, desolate western domain are sparsely populated and largely undeveloped, although they have rich mineral reserves; they also form a critical buffer zone for China, in a geographically important and culturally sensitive part of the Asian continent. Across most of the west, the Han Chinese are outnumbered by a range of cultural groups, including the Uygur, the largest group of the various seminomadic Muslim peoples from Central Asia. The remote, inhospitable Plateau of Tibet is the world's coldest and highest plateau. It has been occupied by the Chinese since 1950. Tibet is one of western China's five "Autonomous Regions," but its reclusive Buddhist culture has been systematically undermined by the Chinese government.

Map key

Population

- ▣ 1 million to 5 million
- ◉ 500,000 to 1 million
- ◉ 100,000 to 500,000
- ⊕ 50,000 to 100,000
- ○ 10,000 to 50,000
- ∘ below 10,000

Elevation

- 6000m / 19,686ft
- 4000m / 13,124ft
- 3000m / 9843ft
- 2000m / 6562ft
- 1000m / 3281ft
- 500m / 1640ft
- 250m / 820ft
- 100m / 328ft
- sea level

Scale 1:7,000,000

projection: Lambert Conformal Conic

▲ *The Lhasa He* is one of the many rivers that drain the vast Plateau of Tibet. From its source in the Nyainqêntanglha Shan range and fed by the spring meltwater, it eventually joins the upper Brahmaputra 40 miles (65 km) southwest of Lhasa.

Using the land

Agriculture is constrained by the cold, dry climate and lack of fertile soils in the region, although irrigation and glasshouse farming are increasing agricultural potential. Large quantities of fruit, like melons and grapes, are grown at the oases of Hami and Turpan in Xinjiang, and new irrigation schemes have greatly increased cotton and wheat production in the Tarim Basin (*Tarim Pendi*). Most of the great area of Tibet and Qinghai is devoted to pastoralism. Sheep are the principal livestock.

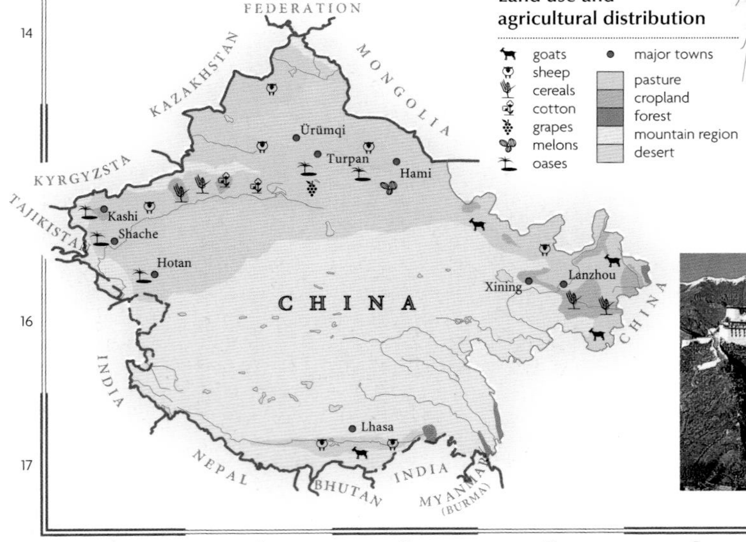

Land use and agricultural distribution

- ⚲ goats
- ⚘ sheep
- 🌾 cereals
- ⚘ cotton
- 🍇 grapes
- 🍈 melons
- oases
- ● major towns
- pasture
- cropland
- forest
- mountain region
- desert

◀ *The Potala Palace,* in Tibet's capital, Lhasa, was the former residence of the Dalai Lama, Tibetan Buddhism's spiritual leader. Tibet remains only sparsely populated; forming over 20% of China's landmass, it supports fewer than 1% of its population.

158

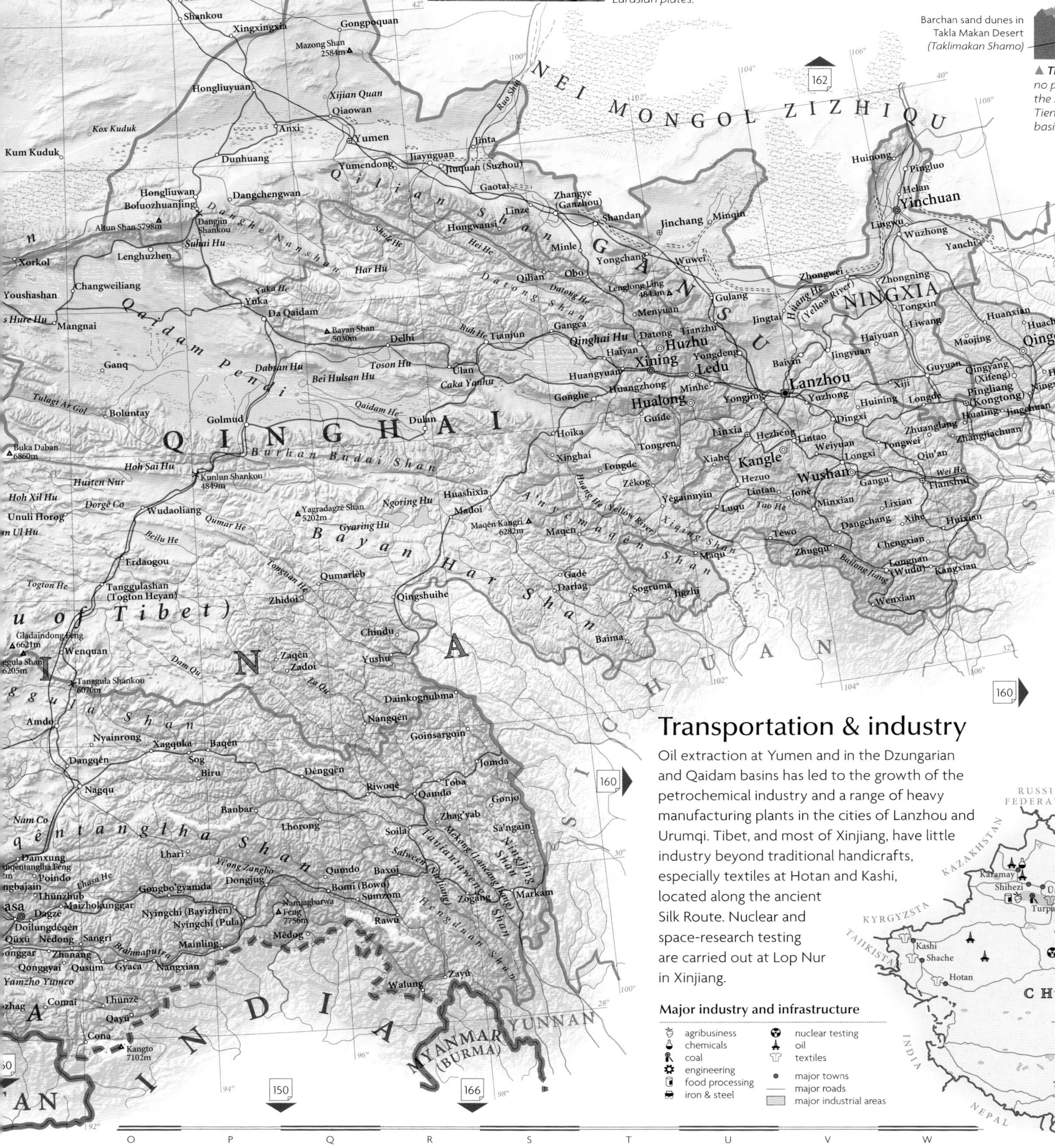

The landscape

The Himalayas mark the southwestern edge of the Plateau of Tibet, an extreme mountain wilderness which occupies nearly a quarter of China's total area. A large structural depression, the Qaidam Pendi, lies at its northeastern edge. The Kunlun mountain chain isolates the plateau from the desert to the north, where the Tien Shan range forms a spur between the Tarim Basin (Tarim Pendi) and Dzungarian Basin (Junggar Pendi).

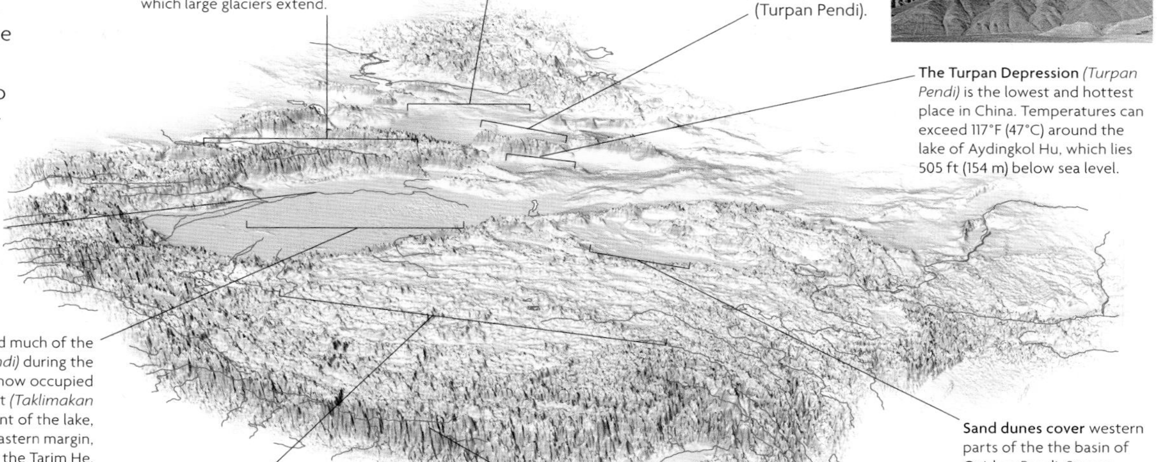

The Tien Shan reach elevations of over 24,419 ft (7435 m) and have permanent ice fields, from which large glaciers extend.

Dzungarian Basin (Junggar Pendi)

► **The Bogda Shan**, an eastward arm of the Tien Shan range, rise high above the Turpan Depression (Turpan Pendi).

The Turpan Depression (Turpan Pendi) is the lowest and hottest place in China. Temperatures can exceed 117°F (47°C) around the lake of Aydingkol Hu, which lies 505 ft (154 m) below sea level.

Northwestern China is largely a region of internal drainage. The Tarim He flows only as far as Lop Nur, where its water is lost by evapotranspiration from the lake and land surface.

A vast glacial lake filled much of the Tarim Basin (Tarim Pendi) during the last Ice Age. This area is now occupied by the Takla Makan Desert (Taklimakan Shamo). A remnant of the lake, Lop Nur, forms the eastern margin, where it is fed by the Tarim He.

◄ **The terrain of** the Plateau of Tibet consists of mountain peaks and open plateaus, dotted with brackish lakes. These are probably remnants of the Tethys Sea, which covered the area before it was uplifted following the collision of the Indo-Australian and Eurasian plates.

Mount Everest is the world's highest peak, at 29,029 ft (8848 m). The summit marks the border between China and Nepal.

Sand dunes cover western parts of the the basin of Qaidam Pendi. Strong winds frequently carry the sands east, threatening the agricultural areas around the lake of Qinghai Hu.

Tarim Basin (Tarim Pendi)

Oases at edge of basin

Barchan sand dunes in Takla Makan Desert (Taklimakan Shamo)

Lop Nur

▲ **The Tarim Basin** (Tarim Pendi) has no permanent rivers. Rainfall from the surrounding Plateau of Tibet and Tien Shan ranges drains into the basin's sand and gravel floor.

▲ **From its source**, high in eastern Qinghai, the Yellow River starts on a 3395 mile (5464 km) journey to the Yellow Sea.

Transportation & industry

Oil extraction at Yumen and in the Dzungarian and Qaidam basins has led to the growth of the petrochemical industry and a range of heavy manufacturing plants in the cities of Lanzhou and Urumqi. Tibet, and most of Xinjiang, have little industry beyond traditional handicrafts, especially textiles at Hotan and Kashi, located along the ancient Silk Route. Nuclear and space-research testing are carried out at Lop Nur in Xinjiang.

Transportation network

The construction of roads connecting Lhasa in Tibet with Sichuan, Qinghai, and Xinjiang was achieved in the 1950s, in spite of the extreme physical conditions of the Plateau of Tibet.

Major industry and infrastructure

- agribusiness
- chemicals
- coal
- engineering
- food processing
- iron & steel
- nuclear testing
- oil
- textiles
- major towns
- major roads
- major industrial areas

Eastern China

TAIWAN, Anhui, Beijing, Chongqing, Fujian, Guangdong, Guangxi, Guizhou, Hainan, Hebei, Henan, Hubei, Hunan, Jiangsu, Jiangxi, Shaanxi, Shandong, Shanghai, Shanxi, Sichuan, Tianjin, Yunnan, Zhejiang

The east is China's heartland. Massive industrial development since 1949 has transformed much of the densely populated rural landscape, in a region still prone to flooding and drought. Over 30 cities have populations of over a million, including the giant metropolis of Shanghai and the capital Beijing, which has been China's cultural and political center since the 13th century. The ethnically diverse southwest and the oil-rich interior provinces of Sichuan and Shaanxi have largely missed out on the remarkable economic growth occurring in designated free-trade areas along the coasts of the South and East China seas. The republic of Taiwan was established in 1949 by Chinese nationalists ousted from the mainland by the victorious Communist forces. Taiwan now has one of the strongest economies in the world but its sovereignty is not recognized by China. Hong Kong provides a major international trade link for China; a 99-year "lease" period of British control was concluded in 1997.

▲ North of the Qin Ling range in Shaanxi province, is an agriculturally fertile region covered with fine, wind-blown deposits and known as the loess plateau. The loose sediments are vulnerable to water erosion.

Using the land & sea

This is a region of intensive cultivation. Wheat, millet, sorghum, and cotton are the main crops of the Yellow River basin. South from Sichuan, rice becomes the principal crop, grown with wheat, corn, and cotton along the Yangtze river. Tea is produced in the hills and sugar cane along the coast of the southeast, where flat land is limited. Pigs and poultry are raised in great numbers.

Land use and agricultural distribution

- cattle
- pigs
- cereals
- corn
- cotton
- fishing
- peanuts
- rice
- sugar cane
- tea
- capital cities
- major towns
- pasture
- cropland
- forest
- mountain region

▲ On the hills above the North China Plain, slopes are terraced to utilize the rich loess soils of the Taihang Shan range.

Map key

Population
- above 5 million
- 1 million to 5 million
- 500,000 to 1 million
- 100,000 to 500,000
- 50,000 to 100,000
- 10,000 to 50,000
- below 10,000

Elevation
- 6000m / 19,686ft
- 4000m / 13,124ft
- 3000m / 9843ft
- 2000m / 6562ft
- 1000m / 3281ft
- 500m / 1640ft
- 250m / 820ft
- 100m / 328ft
- sea level

Scale 1:7,750,000

projection: Lambert Conformal Conic

◀ The former Portuguese territory of Macao, with its colonial architecture, bars and casinos, reverted to Chinese rule in 1999.

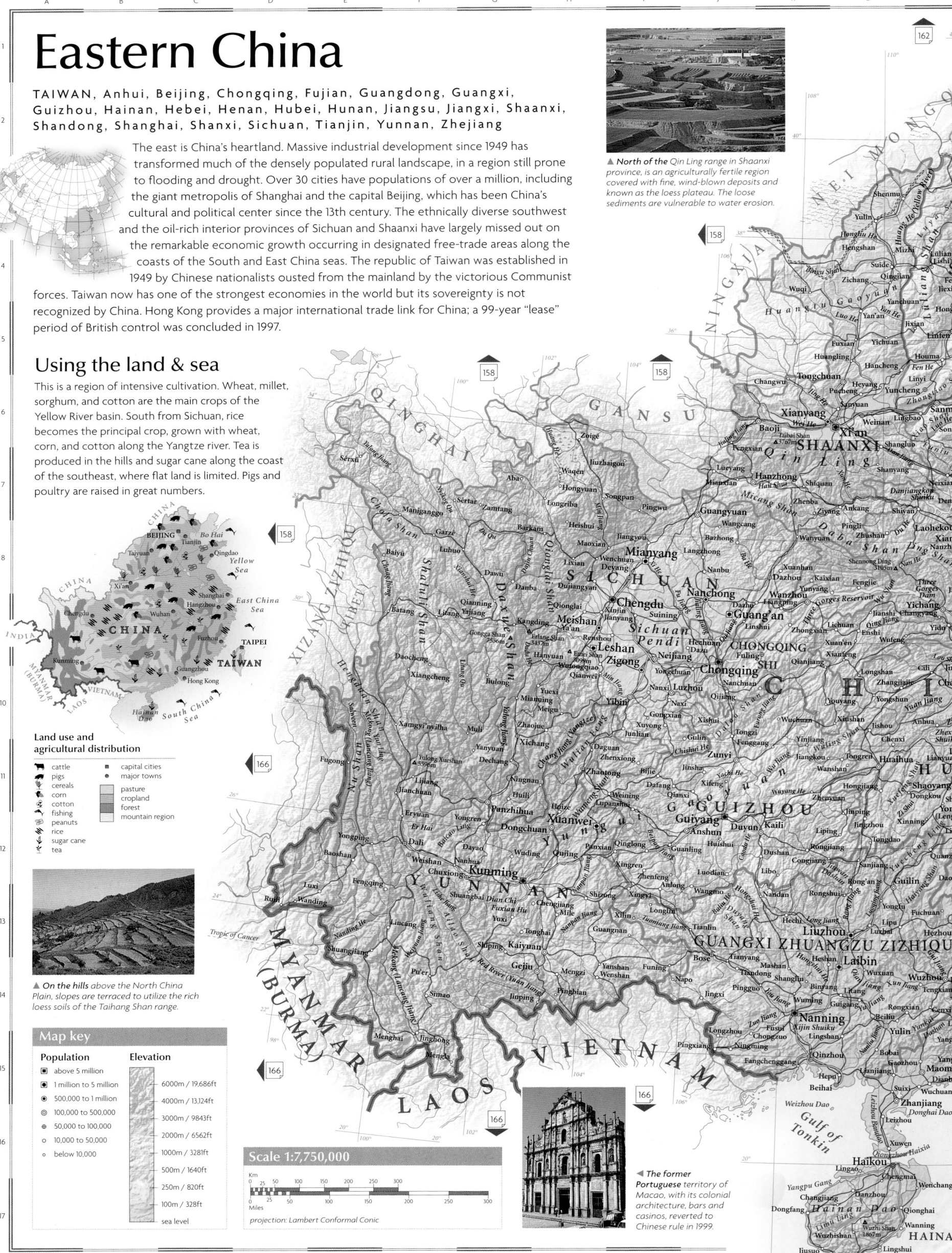

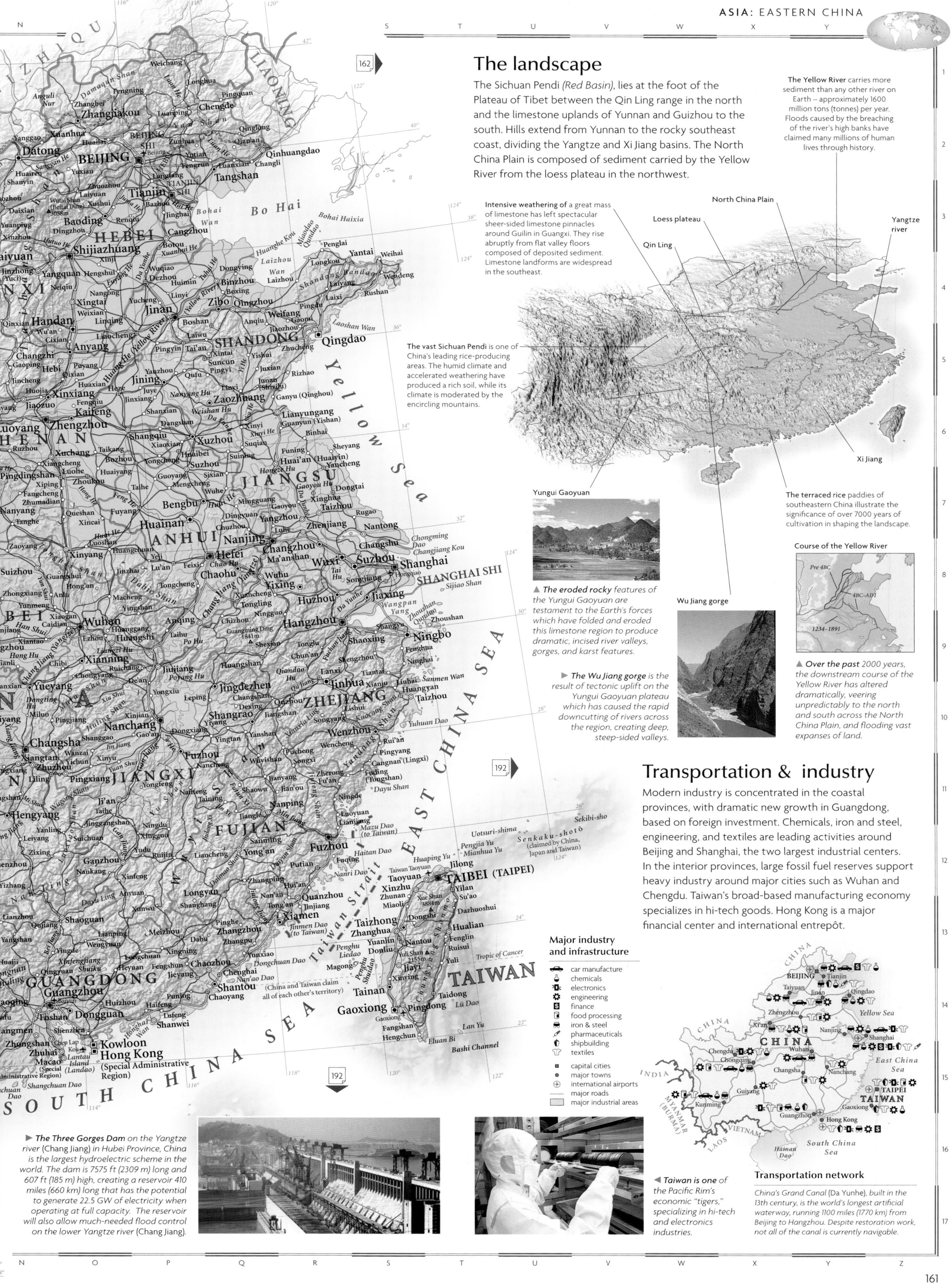

162

The landscape

The Sichuan Pendi (*Red Basin*), lies at the foot of the Plateau of Tibet between the Qin Ling range in the north and the limestone uplands of Yunnan and Guizhou to the south. Hills extend from Yunnan to the rocky southeast coast, dividing the Yangtze and Xi Jiang basins. The North China Plain is composed of sediment carried by the Yellow River from the loess plateau in the northwest.

The Yellow River carries more sediment than any other river on Earth – approximately 1600 million tons (tonnes) per year. Floods caused by the breaching of the river's high banks have claimed many millions of human lives through history.

Intensive weathering of a great mass of limestone has left spectacular sheer-sided limestone pinnacles around Guilin in Guangxi. They rise abruptly from flat valley floors composed of deposited sediment. Limestone landforms are widespread in the southeast.

Loess plateau

North China Plain

Qin Ling

Yangtze river

The vast Sichuan Pendi is one of China's leading rice-producing areas. The humid climate and accelerated weathering have produced a rich soil, while its climate is moderated by the encircling mountains.

Xi Jiang

The terraced rice paddies of southeastern China illustrate the significance of over 7000 years of cultivation in shaping the landscape.

Yungui Gaoyuan

▲ The eroded rocky features of the Yungui Gaoyuan are testament to the Earth's forces which have folded and eroded this limestone region to produce dramatic, incised river valleys, gorges, and karst features.

Wu Jiang gorge

▶ The Wu Jiang gorge is the result of tectonic uplift on the Yungui Gaoyuan plateau which has caused the rapid downcutting of rivers across the region, creating deep, steep-sided valleys.

Course of the Yellow River

Pre 4BC

4BC–AD1

1234–1891

▲ Over the past 2000 years, the downstream course of the Yellow River has altered dramatically, veering unpredictably to the north and south across the North China Plain, and flooding vast expanses of land.

Transportation & industry

Modern industry is concentrated in the coastal provinces, with dramatic new growth in Guangdong, based on foreign investment. Chemicals, iron and steel, engineering, and textiles are leading activities around Beijing and Shanghai, the two largest industrial centers. In the interior provinces, large fossil fuel reserves support heavy industry around major cities such as Wuhan and Chengdu. Taiwan's broad-based manufacturing economy specializes in hi-tech goods. Hong Kong is a major financial center and international entrepôt.

Major industry and infrastructure

- car manufacture
- chemicals
- electronics
- engineering
- finance
- food processing
- iron & steel
- pharmaceuticals
- shipbuilding
- textiles
- ■ capital cities
- • major towns
- ⊕ international airports
- major roads
- major industrial areas

▶ The Three Gorges Dam on the Yangtze river (Chang Jiang) in Hubei Province, China is the largest hydroelectric scheme in the world. The dam is 7575 ft (2309 m) long and 607 ft (185 m) high, creating a reservoir 410 miles (660 km) long that has the potential to generate 22.5 GW of electricity when operating at full capacity. The reservoir will also allow much-needed flood control on the lower Yangtze river (Chang Jiang).

◀ Taiwan is one of the Pacific Rim's economic "tigers," specializing in hi-tech and electronics industries.

Transportation network

China's Grand Canal (Da Yunhe), built in the 13th century, is the world's longest artificial waterway, running 1100 miles (1770 km) from Beijing to Hangzhou. Despite restoration work, not all of the canal is currently navigable.

Northeastern China, Mongolia & Korea

MONGOLIA, NORTH KOREA, SOUTH KOREA, Heilongjiang, Inner Mongolia, Jilin, Liaoning

This northerly region has been a domain of shifting borders and competing colonial powers for centuries. Mongolia was the heartland of Chinghiz Khan's vast Mongol empire in the 13th century, while northeastern China was home to the Manchus, China's last ruling dynasty (1644–1911). The mineral and forest wealth of the northeast helped make this China's principal region of heavy industry, although the outdated state factories now face decline. South Korea's state-led market economy has grown dramatically and Seoul is now one of the world's largest cities. The austere communist regime of North Korea has isolated itself from the expanding markets of the Pacific Rim and faces continuing economic stagnation.

▲ *The Eurasian steppe* stretches from the mouth of the Danube in Europe, to Mongolia. In Mongolia, nomadic people have lived in felt huts called yurts or gers, for thousands of years.

Map key

Population
- ▣ above 5 million
- ◉ 1 million to 5 million
- ◎ 500,000 to 1 million
- ⊕ 100,000 to 500,000
- ⊕ 50,000 to 100,000
- ○ 10,000 to 50,000
- ○ below 10,000

Elevation
- 4000m / 13,124ft
- 3000m / 9843ft
- 2000m / 6562ft
- 1000m / 3281ft
- 500m / 1640ft
- 250m / 820ft
- 100m / 328ft
- sea level

Scale 1:7,000,000

Km 0 25 50 100 150 200
Miles 0 25 50 100 150 200

projection: Lambert Conformal Conic

The landscape

The great North China Plain is largely enclosed by mountain ranges including the Great and Lesser Khingan Ranges (*Da Hinggan Ling* and *Xiao Hinggan Ling*) in the north, and the Changbai Shan, which extend south into the rugged peninsula of Korea. The broad steppeland plateau of Nei Mongol Gaoyuan borders the southeastern edge of the great cold desert of the Gobi which extends west across the southern reaches of Mongolia. In northwest Mongolia the Altai Mountains and various lesser ranges are interspersed with lakeland basins.

▲ *Much of Mongolia* and Inner Mongolia is a vast desert area. To the south and east, a semiarid region extends into China proper.

▲ *The Gobi desert* stretches from Central Asia, through Mongolia and into China. Bare rock surfaces, rather than sand dunes, typify the cold desert landscape of the Gobi.

Tributaries of the Amur river follow U-shaped valleys through the Great Khingan Range (*Da Hinggan Ling*). These were cut by ice-age glaciers between 3 and 10 million years ago.

Lesser Khingan Range (*Xiao Hinggan Ling*)

Changbai Shan

T'aebaek-sanmaek

◄ *The wooded mountain* range of T'aebaek-sanmaek forms the backbone of the Korean peninsula, running north–south along the eastern coastline.

The Altai Mountains are the highest and longest of the mountain ranges that extend into Mongolia from the northwest. These mountains provide one of the last refuges for the endangered snow leopard.

The Yellow River sweeps north around the Ordos Desert (*Mu Us Shadi*), bringing water in to an otherwise barren region.

Columns of basalt rock protrude in occasional clusters from the flat surface of the eastern Gobi. Their regular, six-sided form was produced when the rock cooled and contracted from its molten state.

Great Khingan Range (*Da Hinggan Ling*)

A crater lake occupies the 9023 ft (2750 m) snowy summit of the extinct volcano Paektu-san, the highest peak in the mountains of the Changbai Shan.

Transportation & industry

North Korea's centrally-planned economy is strongly oriented toward heavy industry, while South Korea has a broad manufacturing base which includes textiles, steel, electronics, and one of the world's largest shipbuilding industries. Mongolia and Inner Mongolia's great mineral resource potential is largely undeveloped. The heavy industrial region around Shenyang produces iron, steel, chemicals, and cement on a massive scale.

Major industry and infrastructure

- car manufacture
- chemicals
- coal
- electronics
- engineering
- finance
- food processing
- iron & steel
- pharmaceuticals
- shipbuilding
- textiles
- capital cities
- major towns
- international airports
- major roads
- major industrial areas

Transportation network

Liaoning has China's most comprehensive railroad network, the legacy of the Japanese occupation of Manchuria in the 20th century. The railroads are used primarily for freight transportation.

▲ *Ulan Bator, the Mongolian capital bears many of the hallmarks of Soviet-style central planning, the result of economic and industrial assistance from the Soviet Union following Mongolian independence in 1921.*

▶ *While North Korea has remained politically and economically isolated from the rest of the world, South Korea has enjoyed immense economic growth. It has benefited considerably from US economic aid in the aftermath of the Korean war of 1950–1953.*

South Korea: Capital cities

SEOUL – capital
SEJONG CITY – administrative capital

Using the land & sea

Mongolia and Inner Mongolia rely heavily on livestock farming, with only about 1% of the land area cultivated. Northeastern China produces wheat, corn, soybeans, and sugar beet. The cool climate limits the range of crops and large upland areas of the northeast remain forested. Rice is the staple food of North and South Korea. The latter has become a leading ocean-fishing nation.

Land use and agricultural distribution

- goats
- pigs
- sheep
- corn
- fishing
- rice
- soybeans
- sugar beet
- wheat
- capital cities
- major towns
- pasture
- cropland
- forest
- mountain region
- desert

Japan

In the years since the end of the Second World War, Japan has become the world's most dynamic industrial nation. The country comprises a string of over 4000 islands which lie in a great northeast to southwest arc in the northwest Pacific. Four major islands: Hokkaido, Honshu, Shikoku, and Kyushu are home to the great majority of Japan's population of 128 million people, although the mountainous terrain of the central region means that most cities are situated on the coast. A densely populated industrial belt stretches along much of Honshu's southern coast, including Japan's crowded capital, Tokyo. Alongside its spectacular economic growth and the increasing westernization of its cities, Japan still maintains a highly individual culture, reflected in its traditional food, formal behavioral codes, unique Shinto religion, and a deep reverence for the emperor.

Using the land & sea

Although only about 11% of Japan is suitable for cultivation, substantial government support, a favorable climate and intensive farming methods enable the country to be virtually self-sufficient in rice production. Northern Hokkaido, the largest and most productive farming region, has an open terrain and climate similar to that of the American Midwest, and produces over half of Japan's cereal requirements. Farmers are being encouraged to diversify by growing fruit, vegetables, and wheat, as well as raising livestock.

Land use and agricultural distribution

- cattle
- pigs
- fishing
- cereals
- citrus fruits
- fruit
- herbs
- rice
- root crops
- tobacco

- ■ capital cities
- • major towns

pasture
cropland
forest

The urban/rural population divide

urban 78%   rural 22%

0 10 20 30 40 50 60 70 80 90 100

Population density	Total land area
885 people per sq mile (342 people per sq km)	145,869 sq miles (377,800 sq km)

The landscape

The islands of Japan lie on the Pacific "Ring of Fire," and form a series of clearly defined arcs. The largely mountainous landscape was formed very recently in geological terms. Volcanic eruptions and earthquakes continue to reshape the terrain and shake the country's complex infrastructure. There is no single continuous mountain range; the mountains divide into many small land blocks separated by lowlands and dissected by numerous river valleys.

Sea of Japan (East Sea)
Active volcanic island
Japan Trench (subduction zone)

▲ *Japan is part* of an arc of volcanic islands, formed by the Pacific Plate diving under the Eurasian Plate. This process generates intense stress which is periodically released as earthquakes.

◄ *Mount Fuji is* Japan's highest mountain, rising 12,388 ft (3776 m) above the Kanto Plain in the central region of Honshu. The flat land below is suitable for growing crops such as tea. Like many Japanese mountains, it is revered as a sacred site.

Mount Fuji

A number of rivers which emerge from the volcanic parts of northwestern Honshu are so highly acidic that their water is unsuitable for irrigation and consumption.

▶ *Cutting terraces maximizes* the limited agricultural land, enabling Japan to produce large quantities of rice.

▶ *Trees cling to* the sheer slopes of the waterfalls on the northern island of Hokkaido. The island's climate is similar to that in northern Europe, with long, cold winters and short, warm summers.

In much of Kyushu the coast is subsiding, giving a highly indented coastline. In some places, former hilltops are barely visible above the current sea level.

There are over 60 active volcanoes – like Asahi-dake, Hokkaido's highest peak – throughout Japan. This accounts for more than 10% of the world's total.

The Inland Sea (Seto-naikai) has resulted from the depression of faulted blocks which has allowed sea water to invade the region between northern Shikoku and western Honshu.

Strong southeasterly winds blowing onshore during the winter create sand dunes which extend for miles along the eastern coasts.

Biwa-ko is the largest lake in Japan, covering 260 sq miles (673 sq km) in central Honshu. The depression in which it lies was created by recent faulting of the underlying rocks.

Rising land on the Pacific coast of Honshu leads to typical features such as raised beaches, some lying over 1000 ft (300 m) above sea level.

▼ *Autumnal trees near* Gifu, on central Honshu, create a spectacular display. Native trees on this island include camphor, pasania, Japanese evergreen oak, camellia, and holly.

▶ *The Kobe earthquake* in January 1995 highlighted Japan's vulnerability to earthquakes, despite technological advances. It shattered much of the infrastructure of this important port. More than 5000 people died as buildings and overhead highways collapsed and fires broke out.

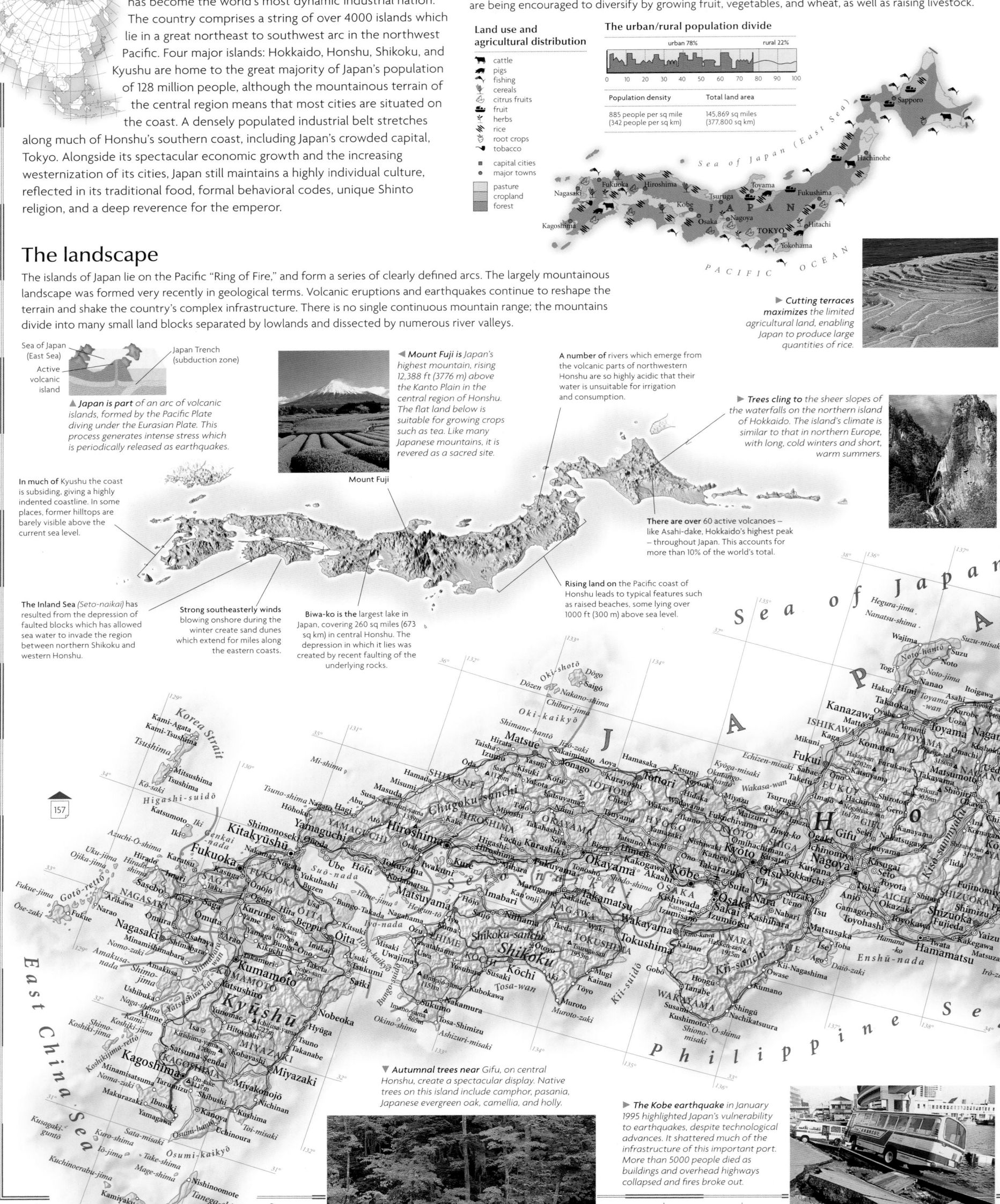

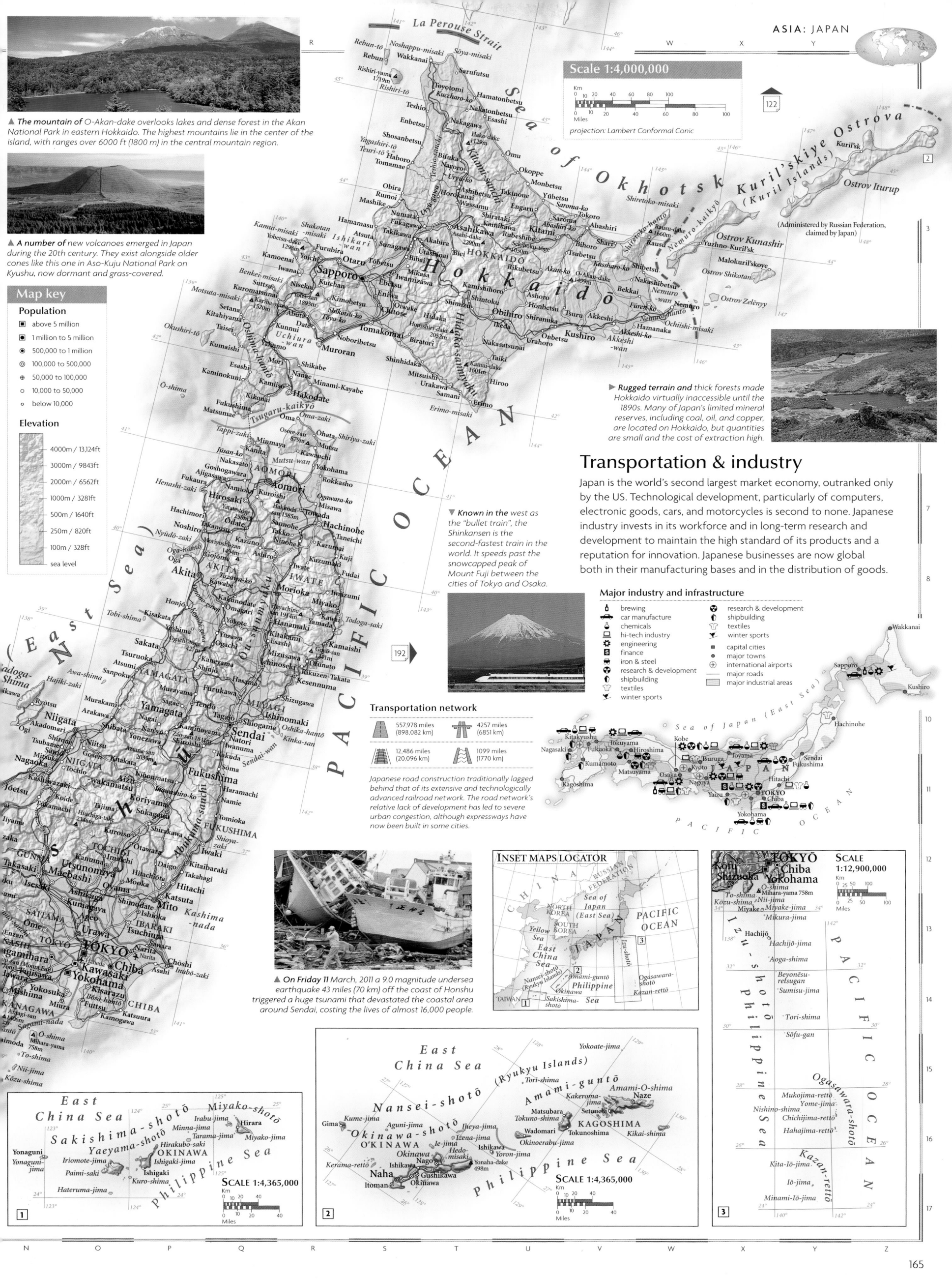

Mainland Southeast Asia

CAMBODIA, LAOS, MYANMAR (BURMA), THAILAND, VIETNAM

Thickly forested mountains, intercut by the broad valleys of five great rivers characterize the landscape of Southeast Asia's mainland countries. Agriculture remains the main activity for much of the population, which is concentrated in the river flood plains and deltas. Linked ethnic and cultural roots give the region a distinct identity. Most people on the mainland are Theravada Buddhists, and the Philippines is the only predominantly Christian country in Southeast Asia. Foreign intervention began in the 16th century with the opening of the spice trade; Cambodia, Laos and Vietnam were French colonies until the end of the Second World War, Myanmar (Burma) was under British control. Only Thailand was never colonized. Today, Thailand is poised to play a leading role in the economic development of the Pacific Rim, and Laos and Vietnam continues to mend the devastation of the Vietnam War, and to develop their economies. With ongoing political instability and a shattered infrastructure, Cambodia faces an uncertain future, while Myanmar (Burma) is seeking investment and the ending of its long isolation from the world community.

▲ *The Irrawaddy river* is Myanmar's (Burma) vital central artery, watering the ricefields and providing a rich source of fish, as well as an important transport link, particularly for local traffic.

The landscape

A series of mountain ranges runs north–south through the mainland, formed as the result of the collision between the Eurasian Plate and the Indian subcontinent, which created the Himalayas. They are interspersed by the valleys of a number of great rivers. On their passage to the sea these rivers have deposited sediment, forming huge, fertile flood plains and deltas.

The coastline of the Isthmus of Kra

Longshore drift
Eroded coastline
Spit
Lagoon
Wave attack

◀ *The east and* west coasts of the Isthmus of Kra differ greatly. The tectonically uplifting west coast is exposed to the harsh south-westerly monsoon and is heavily eroded. On the east coast, longshore currents produce depositional features such as spits and lagoons.

Hkakabo Razi is the highest point in mainland Southeast Asia. It rises 19,300 ft (5885 m) at the border between China and Myanmar (Burma).

Mountains dominate the Laotian landscape with more than 90% of the land lying more than 600 ft (180 m) above sea level. The mountains of the Chaîne Annamitique form the country's eastern border.

The Red River delta in northern Vietnam is fringed to the north by steep-sided, round-topped limestone hills, typical of karst scenery.

The Irrawaddy river runs virtually north–south, draining the plains of northern Myanmar (Burma). The Irrawaddy delta is the country's main rice-growing area.

Salween River

Isthmus of Kra

Malay Peninsula

Tonle Sap, a freshwater lake, drains into the Mekong delta via the Mekong river. It is the largest lake in Southeast Asia.

The Mekong river flows through southern China and Myanmar (Burma), then for much of its length forms the border between Laos and Thailand, flowing through Cambodia before terminating in a vast delta on the southern Vietnamese coast.

◀ *The fast-flowing waters* of the Mekong river cascade over this waterfall in Champasak province in Laos. The force of the water erodes rocks at the base of the fall.

▲ *The coast of* the Isthmus of Kra, in southeast Thailand has many small, precipitous islands like these, formed by chemical erosion on limestone, which is weathered along vertical cracks. The humidity of the climate in Southeast Asia increases the rate of weathering.

Using the land and sea

The fertile flood plains of rivers such as the Mekong and Salween, and the humid climate, enable the production of rice throughout the region. Cambodia, Laos, and Myanmar (Burma) still have substantial forests, producing hardwoods such as teak and rosewood. Cash crops include tropical fruits such as coconuts, bananas and pineapples, rubber, oil palm, sugar cane and the jute substitute, kenaf. Pigs and cattle are the main livestock raised. Large quantities of marine and freshwater fish are caught throughout the region.

▲ *Commercial logging* – still widespread in Myanmar (Burma) – has now been stopped in Thailand because of over-exploitation of the tropical rainforest.

The urban/rural population divide

urban 30%   rural 70%

0 10 20 30 40 50 60 70 80 90 100

Population density	Total land area
345 people per sq mile (133 people per sq km)	733,828 sq miles (1,901,110 sq km)

Land use and agricultural distribution

- cattle
- pigs
- bananas
- coconuts
- fishing
- oil palms
- rice
- rubber
- sugar cane
- timber

■ capital cities
● major towns

- pasture
- cropland
- forest
- wetland

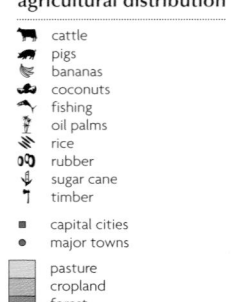

Transportation & industry

Industrial manufacturing has become increasingly important in Thailand and Vietnam in recent years. The assembling of component-based electrical and electronic goods is becoming more common throughout this region, with foreign companies benefiting from low labour costs and the upgrading of technology. The economies of Myanmar (Burma) and Cambodia are still based on agricultural produce and the processing of raw materials. Tin is the region's most important metal, and nickel, copper and chromite are also mined, although the quantities produced are not significant on a global scale. Thailand's successful tourist industry is the country's highest earner of foreign exchange.

Transportation network

82,958 miles (133,524 km)	267 miles (430 km)
7500 miles (12,071 km)	28,585 miles (46,008 km)

Transportation development has concentrated on the building of road networks. Water and sea transport remain important, although air links have improved, particularly in Thailand and the Philippines.

Major industry and infrastructure

- chemicals
- electronics
- engineering
- finance
- food processing
- iron & steel
- oil & gas
- mining
- shipbuilding
- textiles
- timber processing
- capital cities
- major towns
- international airports
- major roads
- major industrial areas

▶ **Opium poppies are** destroyed under army supervision in Thailand. This action is part of a government-sponsored initiative to reduce the trade in drugs such as heroin, which is derived from these plants. Drug trafficking is a major problem throughout the region; the area is known as the "Golden Triangle", and Laos is the third-largest producer of opium poppies in the world.

The Paracel Islands are a strategically sensitive island group, disputed by several surrounding countries. The Paracels are claimed by China, Taiwan, and Vietnam, though only China has actually occupied them.

▼ **The city of** Hue in central Vietnam was the country's capital under the 13 emperors of the Nguyen dynasty from 1802 to 1945. It is the site of a number of religious monuments, including the Thien-Mu Pagoda.

Map key

Population

- above 5 million
- 1 million to 5 million
- 500,000 to 1 million
- 100,000 to 500,000
- 50,000 to 100,000
- 10,000 to 50,000
- below 10,000

Elevation

- 4000m / 13,124ft
- 3000m / 9843ft
- 2000m / 6562ft
- 1000m / 3281ft
- 500m / 1640ft
- 250m / 820ft
- 100m / 328ft
- sea level

Scale 1:7,800,000

Km 0 25 50 100 150 200

Miles 0 50 100 150 200

projection: Lambert Conformal Conic

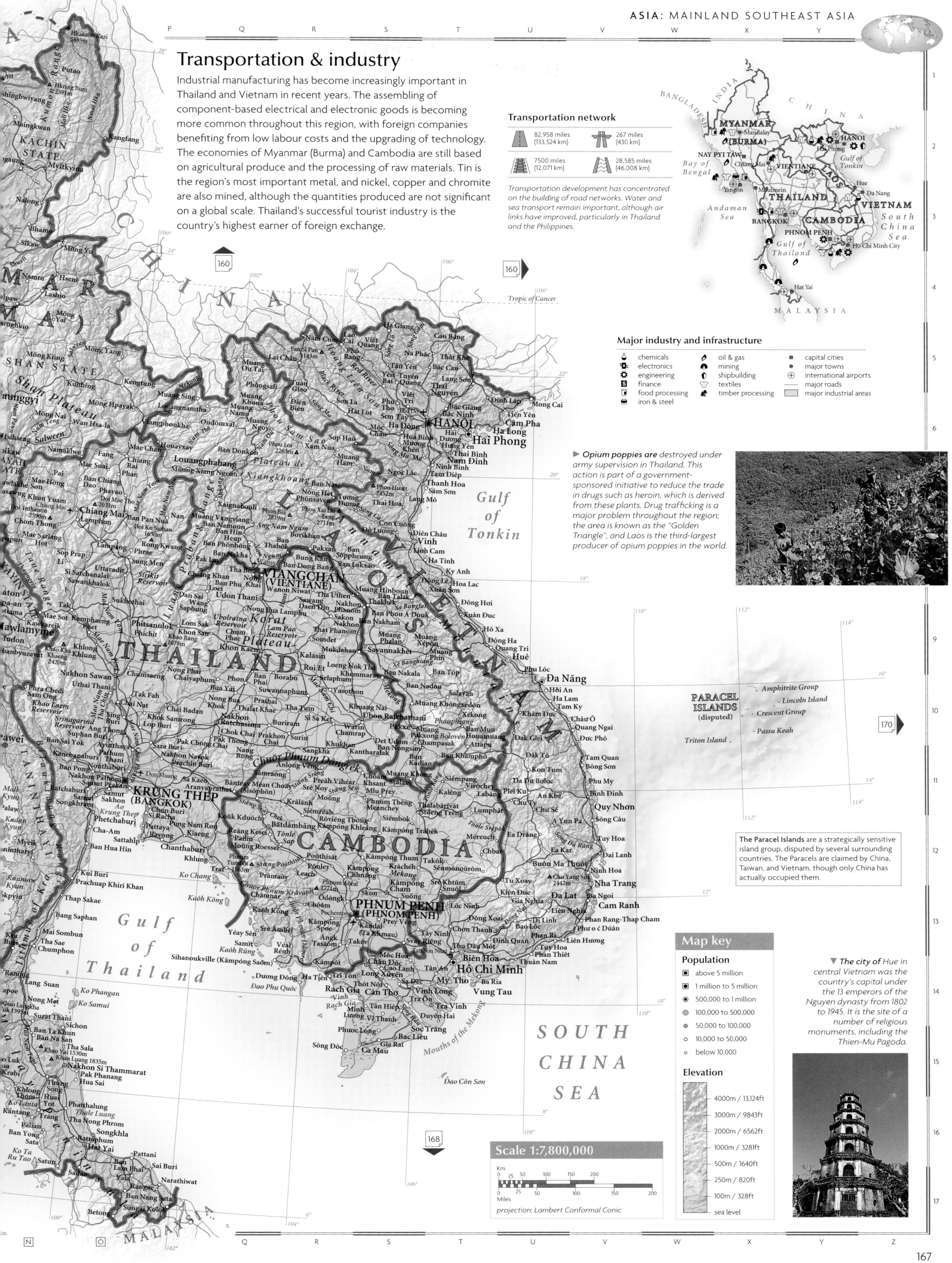

Western Maritime Southeast Asia

BRUNEI, INDONESIA, MALAYSIA, SINGAPORE

The world's largest archipelago, Indonesia's myriad islands stretch 3100 miles (5000 km) eastward across the Pacific, from the Malay Peninsula to western New Guinea. Only about 1500 of the 13,677 islands are inhabited and the huge, predominently Muslim population is unevenly distributed, with some two-thirds crowded onto the western islands of Java, Madura, and Bali. The national government is trying to resettle large numbers of people from these islands to other parts of the country to reduce population pressure there. Malaysia, split between the mainland and the east Malaysian states of Sabah and Sarawak on Borneo, has a diverse population, as well as a fast-growing economy, although the pace of its development is still far outstripped by that of Singapore. This small island nation is the financial and commercial capital of Southeast Asia. The Sultanate of Brunei in northern Borneo, one of the world's last princely states, has an extremely high standard of living, based on its oil revenues.

The landscape

Indonesia's western islands are characterized by rugged volcanic mountains cloaked with dense tropical forest, which slope down to coastal plains covered by thick alluvial swamps. The Sunda Shelf, an extension of the Eurasian Plate, lies between Java, Bali, Sumatra, and Borneo. These islands' mountains rise from a base below the sea, and they were once joined together by dry land, which has since been submerged by rising sea levels.

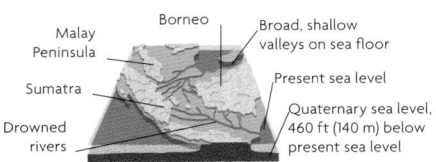

▲ **The Sunda Shelf** underlies this whole region. It is one of the largest submarine shelves in the world, covering an area of 714,285 sq miles (1,850,000 sq km). During the early Quaternary period, when sea levels were lower, the shelf was exposed.

◀ **On January 24,** 2005 a 9.2 magnitude earthquake off the coast of Sumatra triggered a devastating tsunami that was up to 90 ft (30 m) high in places. The death toll was estimated to be around 230,000 people from fourteen different countries around the Indian Ocean.

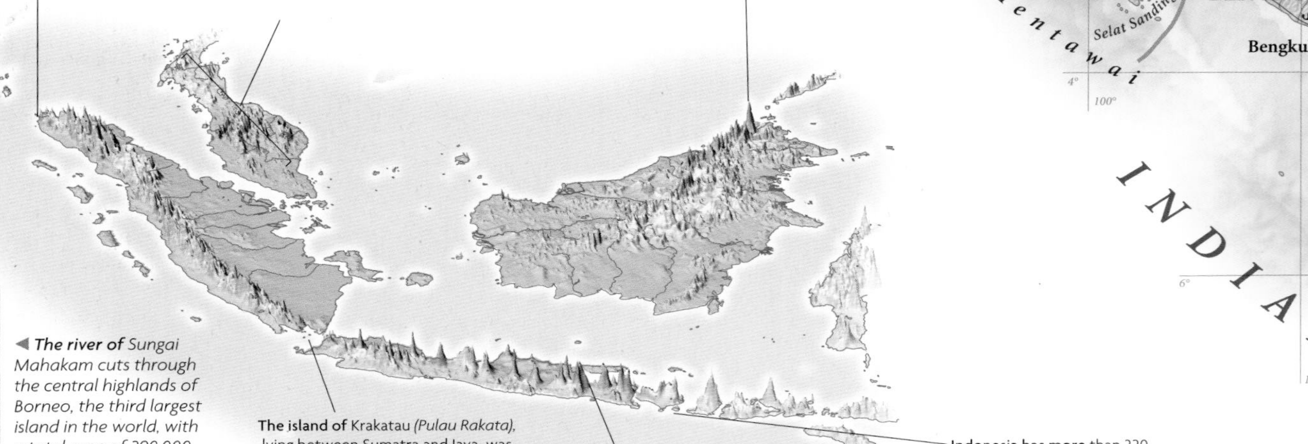

Malay Peninsula has a rugged east coast, but the west coast, fronting the Strait of Malacca, has many sheltered beaches and bays. The two coasts are divided by the Banjaran Titiwangsa, which run the length of the peninsula.

Gunung Kinabalu is the highest peak in Malaysia, rising 13,455 ft (4101 m).

◀ **The river of** Sungai Mahakam cuts through the central highlands of Borneo, the third largest island in the world, with a total area of 290,000 sq miles (757,050 sq km). Although mountainous, Borneo is one of the most stable of the Indonesian islands, with little volcanic activity.

The island of Krakatau (Pulau Rakata), lying between Sumatra and Java, was all but destroyed in 1883, when the volcano erupted. The release of gas and dust into the atmosphere disrupted cloud cover and global weather patterns for several years.

Gunung Semeru

Indonesia has more than 220 volcanoes, most of which are still active. They are strung out along the island arc from Sumatra through the Lesser Sunda Islands, into the Moluccas and Celebes.

Transportation & industry

Singapore has a thriving economy based on international trade and finance. Annual trade through the port is among the highest of any in the world. Indonesia's western islands still depend on natural resources, particularly petroleum, gas, and wood, although the economy is rapidly diversifying with manufactured exports including garments, consumer electronics, and footwear. A high-profile aircraft industry has developed in Bandung on Java. Malaysia has a fast-growing and varied manufacturing sector, although oil, gas, and timber remain important resource-based industries.

▶ **Ranks of gleaming** skyscrapers, new motorways and infrastructure construction reflect the investment which is pouring into Southeast Asian cities like the Malaysian capital, Kuala Lumpur. Traditional housing and markets still exist amidst the new developments. Many of the city's inhabitants subsist at a level far removed from the prosperity implied by its outward modernity.

Malaysia: Capital cities
KUALA LUMPUR – capital
PUTRAJAYA – administrative capital

Using the land and sea

Rice is the most important arable crop in Indonesia and Malaysia, and both countries manage to meet almost all of their domestic demand. Malaysian rubber accounts for 25% of world production and is the main cash crop, grown on plantations and small farms, along with oil palms and copra. Timber is exported from both Malaysia and Indonesia. Modern agricultural techniques enable Singapore to produce fruits and vegetables despite a shortage of suitable land.

▶ *Spiral cuts in the bark of this rubber palm show where it has been tapped. Sophisticated 'cloning' techniques mean that trees which produce consistently high quantities of rubber can be easily reproduced.*

Transportation network

	165,272 miles (266,010 km)
	958 miles (1,542 km)
	5,061 miles (8,146 km)
	18,070 miles (29,084 km)

Singapore's metro system, completed in 1991, is among the most efficient in the world. Malaysia has several fast, modern highways and most roads are paved. Indonesia's many islands make improvement of the shipping infrastructure a priority.

Major industry and infrastructure

- aerospace
- copra processing
- chemicals
- electronics
- engineering
- finance
- food processing
- iron & steel
- oil
- ship building
- timber processing
- textiles
- capital cities
- major towns
- international airports
- major roads
- major industrial areas

Land use and agricultural distribution

- coconuts
- fishing
- oil palms
- rice
- rubber
- shellfish
- sugar cane
- timber
- capital cities
- major towns
- pasture
- cropland
- forest
- wetland

The urban/rural population divide

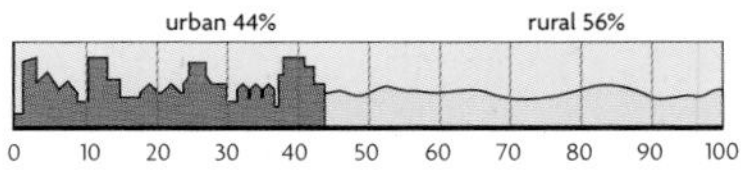

urban 44% rural 56%

0 10 20 30 40 50 60 70 80 90 100

Population density	Total land area
297 people per sq mile (115 people per sq km)	828,356 sq miles (2,146,000 sq km)

▼ *This tiny island near Kota Kinabalu, in Sabah, eastern Malaysia, is a part of a designated national park. Thickly forested, it is surrounded by broad, sandy beaches and shallow inland seas.*

▲ *The volcano of Gunung Semeru in eastern Java lies on the Pacific "Ring of Fire". It is part of the ancient Tennegger volcano and remains highly active.*

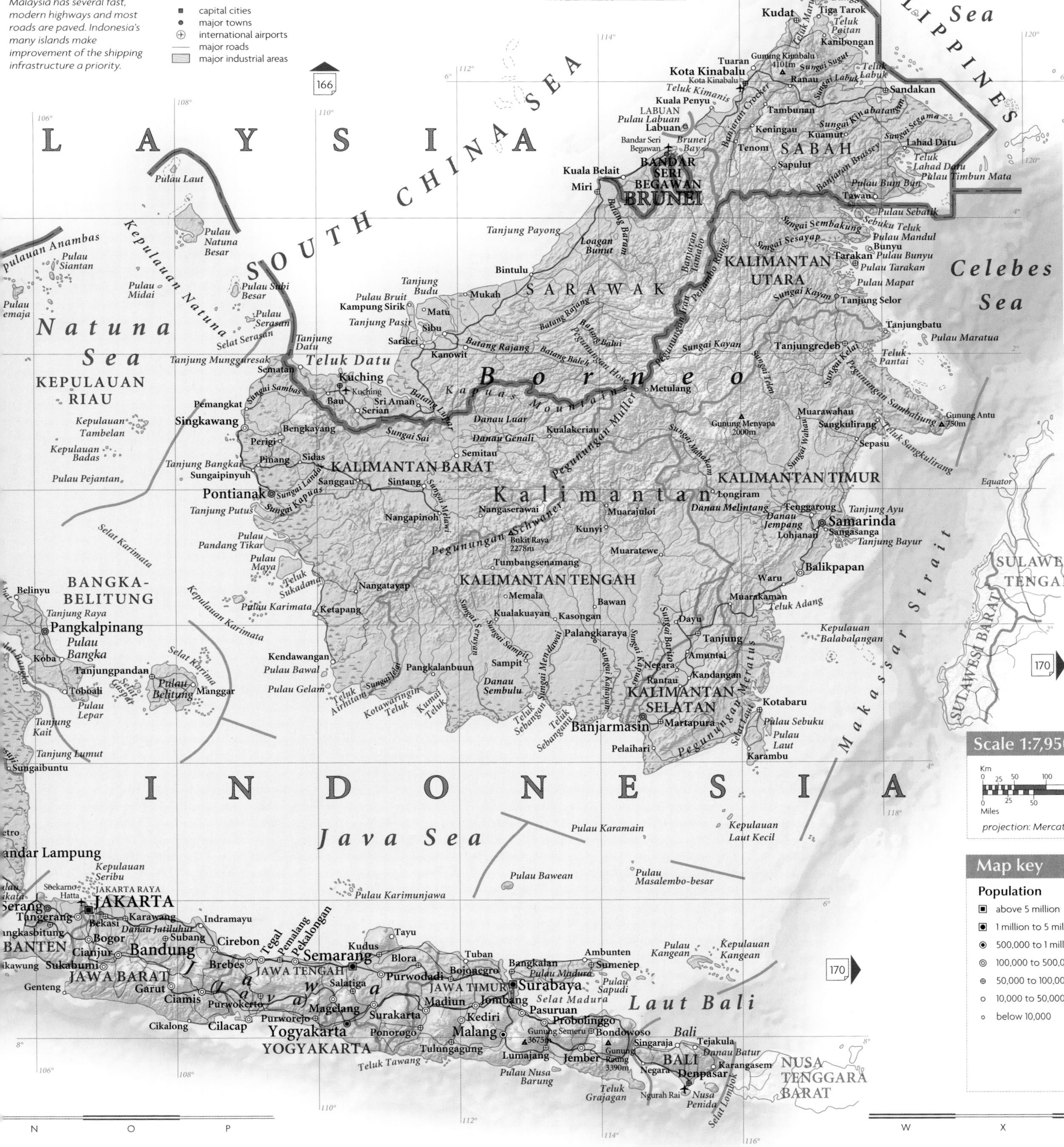

Scale 1:7,950,000

projection: Mercator

Map key

Population

- above 5 million
- 1 million to 5 million
- 500,000 to 1 million
- 100,000 to 500,000
- 50,000 to 100,000
- 10,000 to 50,000
- below 10,000

Elevation

- 4000m / 13,124ft
- 3000m / 9843ft
- 2000m / 6562ft
- 1000m / 3281ft
- 500m / 1640ft
- 250m / 820ft
- 100m / 328ft
- sea level

Eastern Maritime Southeast Asia

EAST TIMOR, INDONESIA, PHILIPPINES

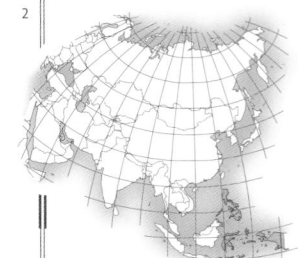

The Philippines takes its name from Philip II of Spain who was king when the islands were colonized during the 16th century. Almost 400 years of Spanish, and later US, rule have left their mark on the country's culture; English is widely spoken and over 90% of the population is Christian. The Philippines' economy is agriculturally based – inadequate infrastructure and electrical power shortages have so far hampered faster industrial growth. Indonesia's eastern islands are less economically developed than the rest of the country. Papua, which constitutes the western portion of New Guinea, is one of the world's last great wildernesses. After a long struggle, East Timor gained full autonomy from Indonesia in 2002.

▲ The traditional boat-shaped houses of the Toraja people in Sulawesi. Although now Christian, the Toraja still practice the animist traditions and rituals of their ancestors. They are famous for their elaborate funeral ceremonies and burial sites in cliffside caves.

The landscape

Located on the Pacific "Ring of Fire" the Philippines' 7100 islands are subject to frequent earthquakes and volcanic activity. Their terrain is largely mountainous, with narrow coastal plains and interior valleys and plains. Luzon and Mindanao are by far the largest islands and comprise roughly 66% of the country's area. Indonesia's eastern islands are mountainous and dotted with volcanoes, both active and dormant.

▶ Lake Taal on the Philippines island of Luzon lies within the crater of an immense volcano that erupted twice in the 20th century, first in 1911 and again in 1965, causing the deaths of more than 3200 people.

The Spratly Islands are a strategically sensitive island group, disputed by several surrounding countries. The Spratlys are claimed by China, Taiwan, Vietnam, Malaysia, and the Philippines and are particularly important as they lie on oil and gas deposits.

Mindanao has five mountain ranges many of which have large numbers of active volcanoes. Lying just west of the Philippines Trench, which forms the boundary between the colliding Philippine and Eurasian plates, the entire island chain is subject to earthquakes and volcanic activity.

The 1000 islands of the Moluccas are the fabled Spice Islands of history, whose produce attracted traders from around the globe. Most of the northern and central Moluccas have dense vegetation and rugged mountainous interiors where elevations often exceed 3000 feet (9144 m).

▲ Bohol in the southern Philippines is famous for its so-called "chocolate hills". There are more than 1000 of these regular mounds on the island. The hills are limestone in origin, the smoothed remains of an earlier cycle of erosion. Their brown appearance in the dry season gives them their name.

The four-pronged island of Celebes is the product of complex tectonic activity which ruptured and then reattached small fragments of the Earth's crust to form the island's many peninsulas.

Coral islands such as Timor in eastern Indonesia show evidence of very recent and dramatic movements of the Earth's plates. Reefs in Timor have risen by as much as 4000 ft (1300 m) in the last million years.

The Pegunungan Jayawijaya range in central Papua contains the world's highest range of limestone mountains, some with peaks more than 16,400 ft (5000 m) high. Heavy rainfall and high temperatures, which promote rapid weathering, have led to the creation of large underground caves and river systems such as the river of Sungai Baliem.

Using the land and sea

Indonesia's eastern islands are less intensively cultivated than those in the west. Coconuts, coffee and spices such as cloves and nutmeg are the major commercial crops while rice, corn and soybeans are grown for local consumption. The Philippines' rich, fertile soils support year-round production of a wide range of crops. The country is one of the world's largest producers of coconuts and a major exporter of coconut products, including one-third of the world's copra. Although much of the arable land is given over to rice and corn, the main staple food crops, tropical fruits such as bananas, pineapples and mangos, and sugar cane are also grown for export.

◀ The terracing of land to restrict soil erosion and create flat surfaces for agriculture is a common practice throughout Southeast Asia, particularly where land is scarce. These terraces are on Luzon in the Philippines.

Land use and agricultural distribution

- coconuts
- fishing
- rice
- rubber
- shellfish
- sugar cane
- ● capital cities
- ● major towns
- pasture
- cropland
- forest
- wetland

The urban/rural population divide

urban 45% · rural 55%

0 10 20 30 40 50 60 70 80 90 100

Population density	Total land area
258 people per sq mile (160 people per sq km)	654,771 sq miles (1,053,755 sq km)

▲ More than two-thirds of Papua's land area is heavily forested and the population of around 1.5 million live mainly in isolated tribal groups using more than 80 distinct languages.

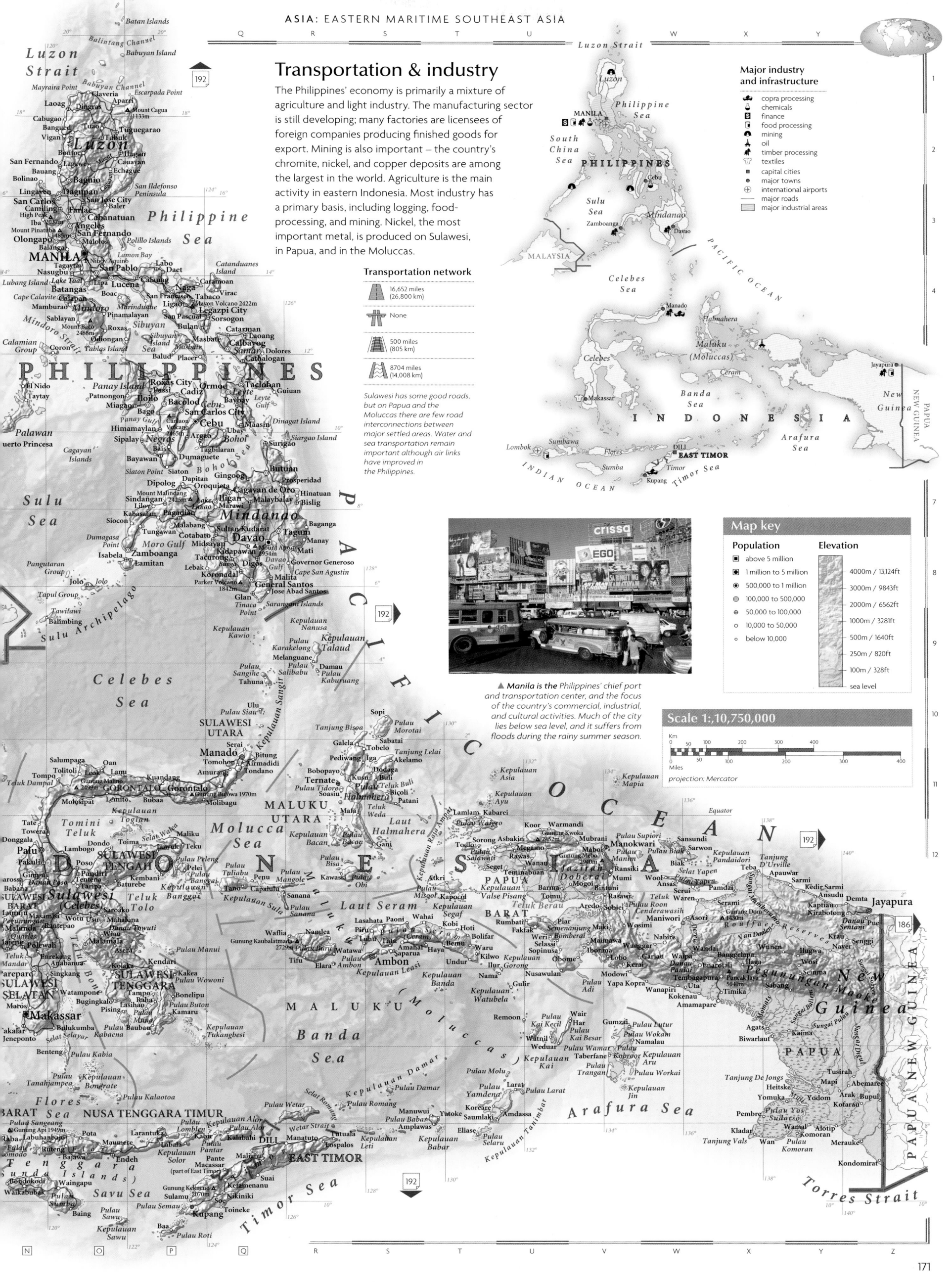

Transportation & industry

The Philippines' economy is primarily a mixture of agriculture and light industry. The manufacturing sector is still developing; many factories are licensees of foreign companies producing finished goods for export. Mining is also important – the country's chromite, nickel, and copper deposits are among the largest in the world. Agriculture is the main activity in eastern Indonesia. Most industry has a primary basis, including logging, food-processing, and mining. Nickel, the most important metal, is produced on Sulawesi, in Papua, and in the Moluccas.

Major industry and infrastructure

- copra processing
- chemicals
- finance
- food processing
- mining
- oil
- timber processing
- textiles
- ■ capital cities
- ● major towns
- ⊕ international airports
- — major roads
- ▭ major industrial areas

Transportation network

- 16,652 miles (26,800 km)
- None
- 500 miles (805 km)
- 8704 miles (14,008 km)

Sulawesi has some good roads, but on Papua and the Moluccas there are few road interconnections between major settled areas. Water and sea transportation remain important although air links have improved in the Philippines.

▲ **Manila is** the Philippines' chief port and transportation center, and the focus of the country's commercial, industrial, and cultural activities. Much of the city lies below sea level, and it suffers from floods during the rainy summer season.

Map key

Population
- ▣ above 5 million
- ◉ 1 million to 5 million
- ◎ 500,000 to 1 million
- ◉ 100,000 to 500,000
- ⊕ 50,000 to 100,000
- ○ 10,000 to 50,000
- ∘ below 10,000

Elevation
- 4000m / 13,124ft
- 3000m / 9843ft
- 2000m / 6562ft
- 1000m / 3281ft
- 500m / 1640ft
- 250m / 820ft
- 100m / 328ft
- sea level

Scale 1:10,750,000

projection: Mercator

The Indian Ocean

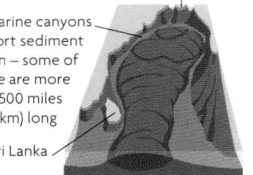

Despite being the smallest of the three major oceans, the evolution of the Indian Ocean was the most complex. The ocean basin was formed during the breakup of the supercontinent Gondwanaland, when the Indian subcontinent moved northeast, Africa moved west, and Australia separated from Antarctica. Like the Pacific Ocean, the warm waters of the Indian Ocean are punctuated by coral atolls and islands. About one-fifth of the world's population – over a billion people – live on its shores. In 2004, over 290,000 died and millions more were left homeless after a tsunami devastated large stretches of the ocean's coastline.

The landscape

The Indian Ocean began forming about 150 million years ago, but in its present form it is relatively young, only about 36 million years old. Along the three subterranean mountain chains of its mid-ocean ridge the seafloor is still spreading. The Indian Ocean has fewer trenches than other oceans and only a narrow continental shelf around most of its surrounding land.

Sediments come from Ganges/Brahmaputra river system

Submarine canyons transport sediment to fan – some of these are more than 1500 miles (2500 km) long

Sri Lanka

▲ *The Ganges Fan* is one of the world's largest submarine accumulations of sediment, extending far beyond Sri Lanka. It is fed by the Ganges/Brahmaputra river system, whose sediment is carried through a network of underwater canyons at the edge of the continental shelf.

The mid-oceanic ridge runs from the Arabian Sea. It diverges east of Madagascar. One arm runs southwest to join the Mid-Atlantic Ridge, the other branches southeast, joining the Pacific-Antarctic Ridge, southeast of Tasmania.

The Ninetyeast Ridge takes its name from the line of longitude it follows. It is the world's longest and straightest under-sea ridge.

Two of the world's largest rivers flow into the Indian Ocean; the Indus and the Ganges/Brahmaputra. Both have deposited enormous fans of sediment.

Indus River

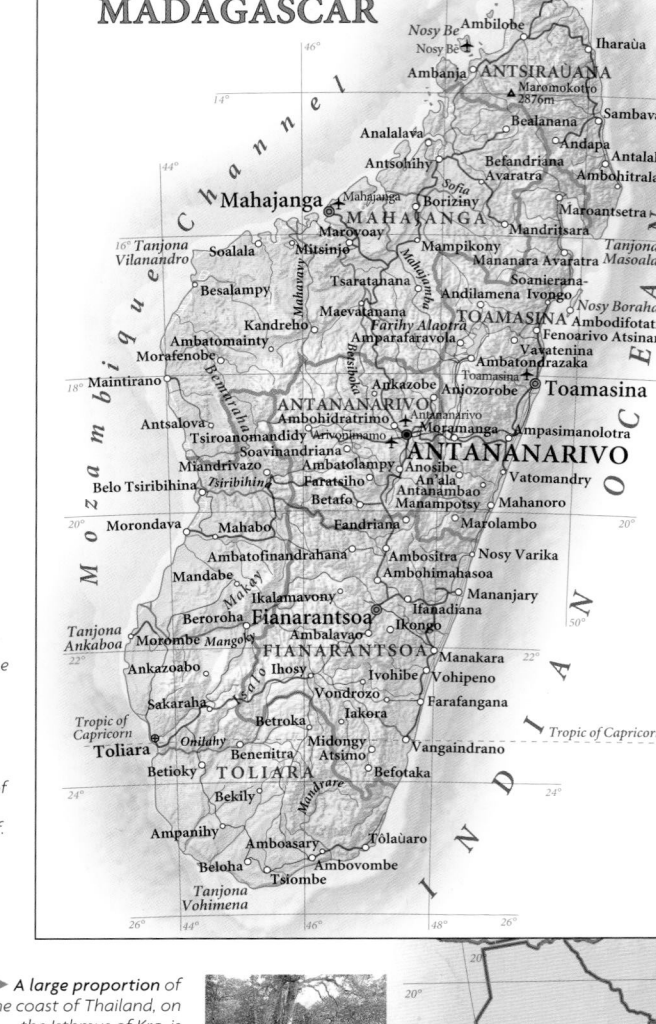

MADAGASCAR

▶ *A large proportion* of the coast of Thailand, on the Isthmus of Kra, is stabilized by mangrove thickets. They act as an important breeding ground for wildlife.

The Java Trench is the world's longest, it runs 1600 miles (2570 km) from the southwest of Java, but is only 50 miles (80 km) wide.

The relief of Madagascar rises from a low-lying coastal strip in the east, to the central plateau. The plateau is also a major watershed separating Madagascar's three main river basins.

▶ *The central group* of the Seychelles are mountainous, granite islands. They have a narrow coastal belt and lush, tropical vegetation cloaks the highlands.

The Kerguelen Islands in the Southern Ocean were created by a hot spot in the Earth's crust. The islands were formed in succession as the Antarctic Plate moved slowly over the hot spot.

The circulation in the northern Indian Ocean is controlled by the monsoon winds. Biannually these winds reverse their pattern, causing a reversal in the surface currents and alternative high and low pressure conditions over Asia and Australia.

Resources

Many of the small islands in the Indian Ocean rely exclusively on tuna-fishing and tourism to maintain their economies. Most fisheries are artisanal, although large-scale tuna-fishing does take place in the Seychelles, Mauritius and the western Indian Ocean. Other resources include oil in the Persian Gulf, pearls in the Red Sea, and tin from deposits off the shores of Myanmar, Thailand, and Indonesia.

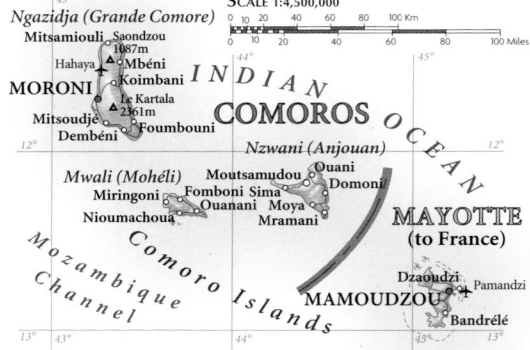

SCALE 1:4,500,000

COMOROS

MAYOTTE (to France)

▶ *The recent use* of large dragnets for tuna-fishing has not only threatened the livelihoods of many small-scale fisheries, but also caused widespread environmental concern about the potential impact on other marine species.

Resources (including wildlife)

- fish
- penguins
- shellfish
- whales
- oil & gas
- △ tin deposits
- 🏖 tourism
- major towns
- ⚓ major ports

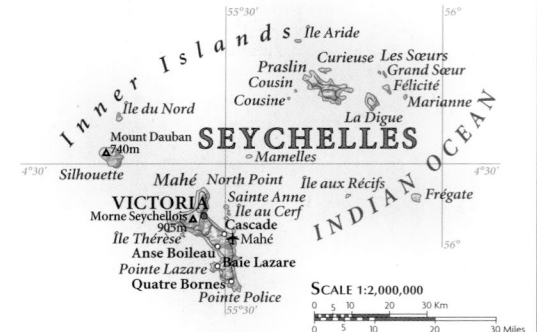

SCALE 1:2,000,000

Inner Islands

SEYCHELLES

VICTORIA

▲ *Coral reefs support* an enormous diversity of animal and plant life. Many species of tropical fish, like these squirrel fish, live and feed around the profusion of reefs and atolls in the Indian Ocean.

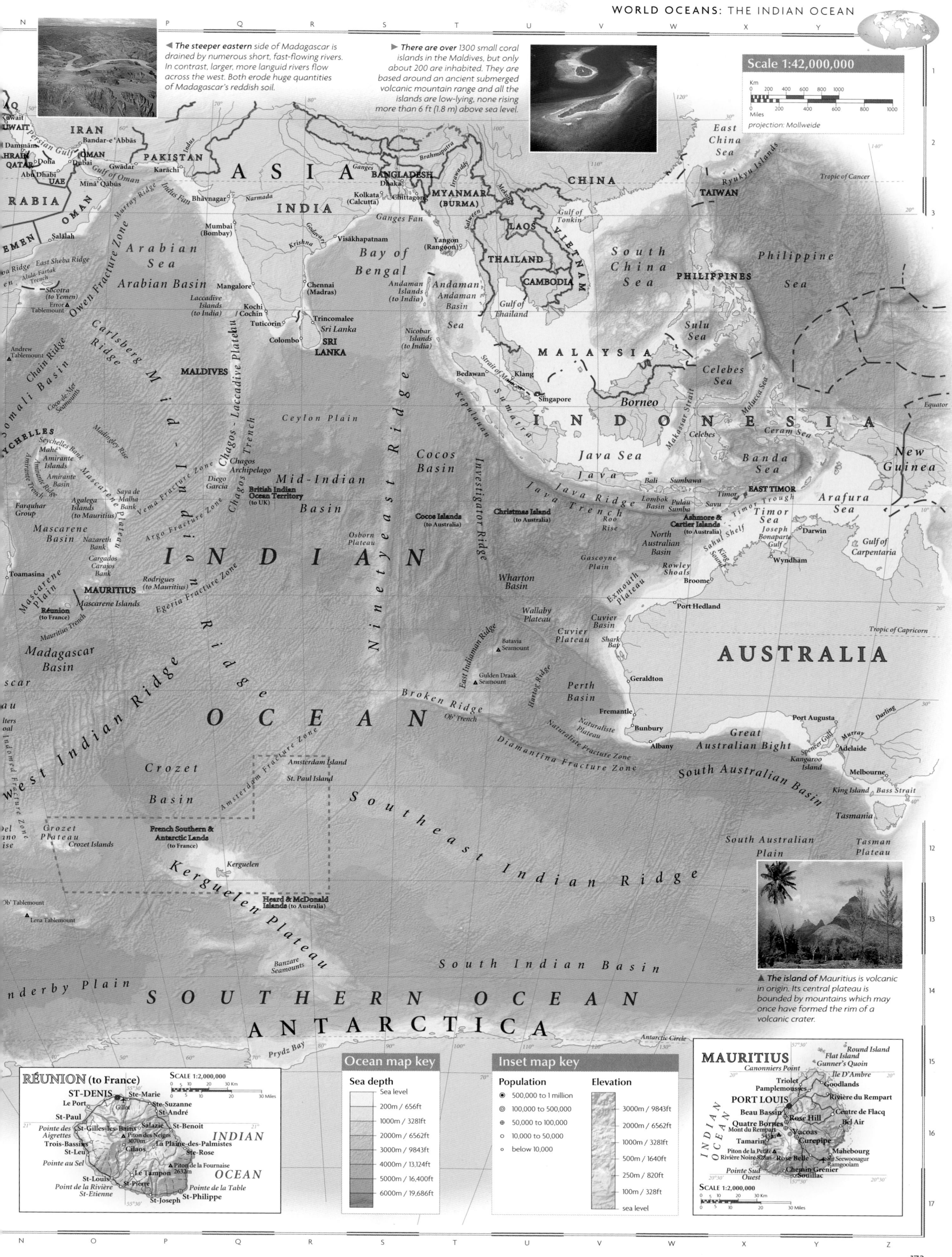

◄ *The steeper eastern* side of Madagascar is drained by numerous short, fast-flowing rivers. In contrast, larger, more languid rivers flow across the west. Both erode huge quantities of Madagascar's reddish soil.

► *There are over* 1300 small coral islands in the Maldives, but only about 200 are inhabited. They are based around an ancient submerged volcanic mountain range and all the islands are low-lying, none rising more than 6 ft (1.8 m) above sea level.

Scale 1:42,000,000

projection: Mollweide

▲ *The island of* Mauritius is volcanic in origin. Its central plateau is bounded by mountains which may once have formed the rim of a volcanic crater.

RÉUNION (to France)
SCALE 1:2,000,000

Ocean map key

Sea depth
- Sea level
- 200m / 656ft
- 1000m / 3281ft
- 2000m / 6562ft
- 3000m / 9843ft
- 4000m / 13,124ft
- 5000m / 16,400ft
- 6000m / 19,686ft

Inset map key

Population
- 500,000 to 1 million
- 100,000 to 500,000
- 50,000 to 100,000
- 10,000 to 50,000
- below 10,000

Elevation
- 3000m / 9843ft
- 2000m / 6562ft
- 1000m / 3281ft
- 500m / 1640ft
- 250m / 820ft
- 100m / 328ft
- sea level

MAURITIUS
SCALE 1:2,000,000

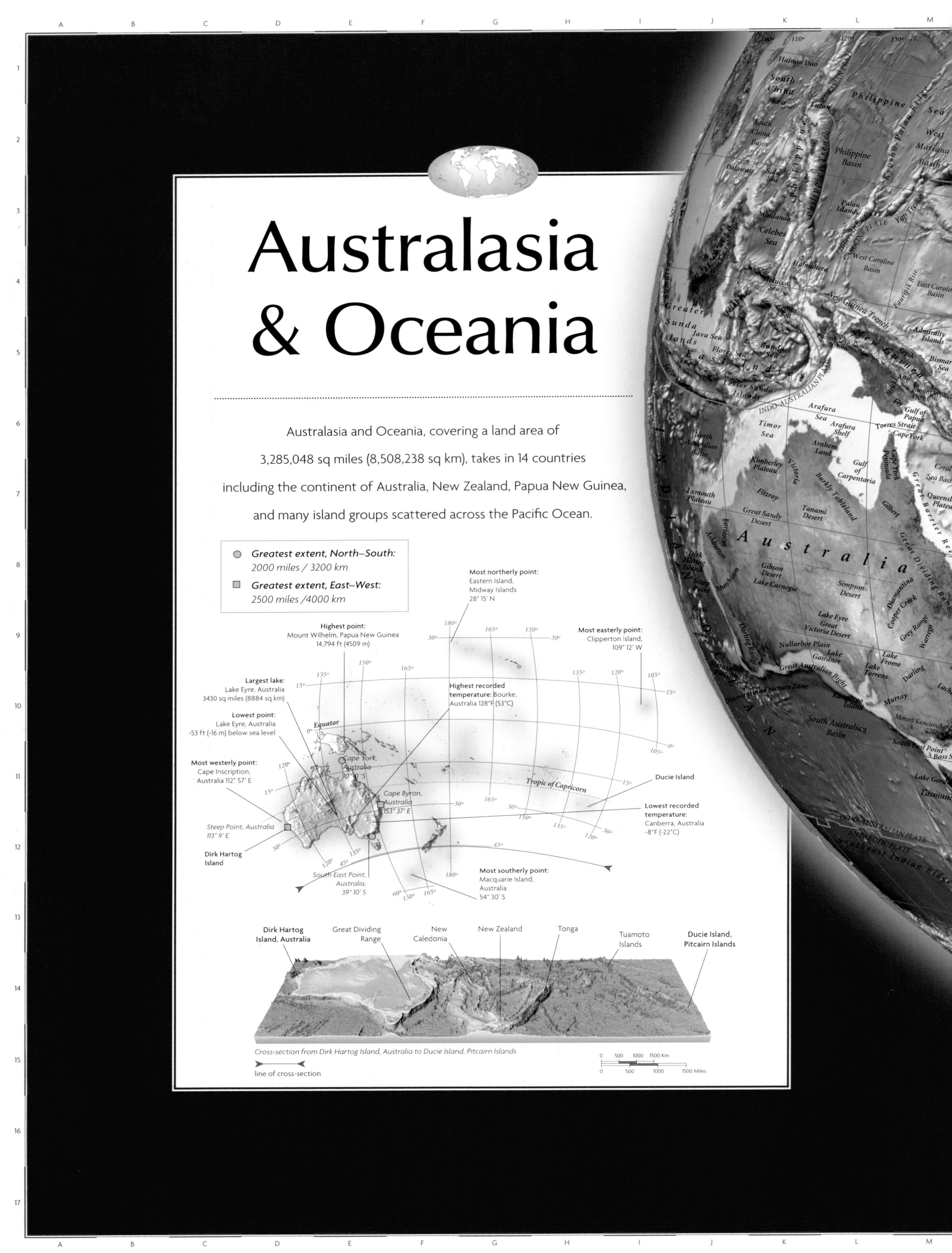

Australasia & Oceania

Australasia and Oceania, covering a land area of 3,285,048 sq miles (8,508,238 sq km), takes in 14 countries including the continent of Australia, New Zealand, Papua New Guinea, and many island groups scattered across the Pacific Ocean.

Cross-section from Dirk Hartog Island, Australia to Ducie Island, Pitcairn Islands

line of cross-section

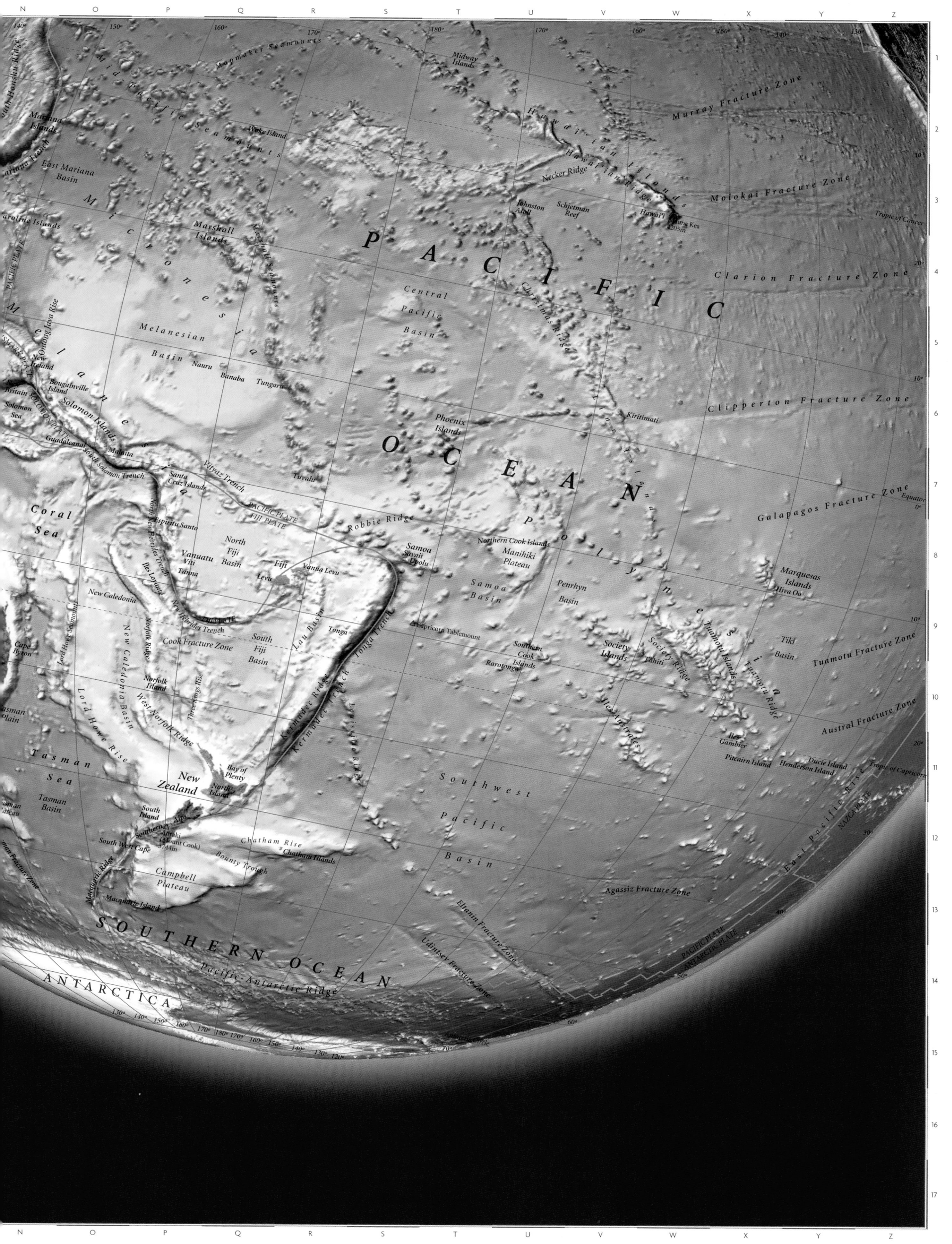

1
2
3
4
5
6
7
8
9
10
11
12
13
14
15
16
17

South Honshu Ridge

Mid-Pacific Seamounts

Mapmaker Seamounts

Midway Islands

Murray Fracture Zone

Mariana Islands

Hawaiian Islands

Hawaiian Ridge

Molokai Fracture Zone

Tropic of Cancer

Mariana Trench

East Mariana Basin

Micronesia

Necker Ridge

Johnston Atoll

Schjetman Reef

Hawai'i
Mauna Kea
4205m

Caroline Islands

PACIFIC PLATE

Marshall Islands

Marcus-Necker Ridge

Marshall Islands

P A C I F I C

Clarion Fracture Zone

Ontong Java Rise

Melanesian

Central Pacific Basin

Christmas Ridge

Melanesia

Basin   Nauru   Banaba   Tungaru

New Ireland

New Britain

Solomon Sea

Bougainville Island

Solomon Islands

Guadalcanal   Malaita

South Solomon Trench

Vityaz Trench

Santa Cruz Islands

Tuvalu

O C E A N

Kiritimati

Clipperton Fracture Zone

Galapagos Fracture Zone   Equator 0°

Coral Sea

New Hebrides Trench

Espiritu Santo

PACIFIC PLATE
FIJI PLATE

Robbie Ridge

Phoenix Islands

Samoa
Savai'i
Upolu

Northern Cook Islands
Manihiki Plateau

Marquesas Islands
Hiva Oa

North Fiji Basin

Vanuatu
Viti

Fiji   Vanua Levu

Samoa Basin

Penrhyn Basin

Polynesia

Tanna   Levu

New Caledonia

Iles Loyauté

New Hebrides Trench

Capricorn Tablemount

Society Ridge

Tiki Basin

New Caledonia   New Caledonia Basin

Cook Fracture Zone

South Fiji Basin

Lau Basin

Tonga

Southern Cook Islands
Rarotonga

Society Islands   Tahiti

Tuamotu Islands

Tuamotu Ridge

Tuamotu Fracture Zone

Norfolk Ridge

Lord Howe Seamounts

Three Kings Rise

Kermadec Ridge

Kermadec Trench

Tonga Trench

Manihiki

Australes

Austral Fracture Zone

Cape Byron

Norfolk Island

Lord Howe Rise

West Norfolk Ridge

Iles Gambier

Pitcairn Island   Henderson Island   Ducie Island   Tropic of Capricorn

Tasman Sea

Bay of Plenty

Southwest

NAZCA PLATE

Tasman Plain

Tasman Basin

New Zealand
North Island

South Pacific Basin

East Pacific Rise

Tasman Basin

South Island

Southern Alps
Aoraki
(Mount Cook)
3744m

Chatham Rise

Chatham Islands

Pacific Basin

Agassiz Fracture Zone

South West Cape

Bounty Trough

Campbell Plateau

Macquarie Ridge

Macquarie Island

Eltanin Fracture Zone

Udintsev Fracture Zone

PACIFIC PLATE
ANTARCTIC PLATE

Macquarie Fracture Zone

S O U T H E R N   O C E A N

Pacific-Antarctic Ridge

A N T A R C T I C A

Antarctic Circle

Political Australasia & Oceania

Vast expanses of ocean separate this geographically fragmented realm, characterized more by each country's isolation than by any political unity. Australia's and New Zealand's traditional ties with the United Kingdom, as members of the Commonwealth, are now being called into question as Australasian and Oceanian nations are increasingly looking to forge new relationships with neighboring Asian countries like Japan. External influences have featured strongly in the politics of the Pacific Islands; the various territories of Micronesia were largely under US control until the late 1980s, and France, New Zealand, the US, and the UK still have territories under colonial rule in Polynesia. Nuclear weapons-testing by Western superpowers was widespread during the Cold War period, but has now been discontinued.

◄ *Western Australia's mineral* wealth has transformed its state capital, Perth, into one of Australia's major cities. Perth is one of the world's most isolated cities – over 2500 miles (4000 km) from the population centers of the eastern seaboard.

Scale 1:32,000,000

projection: Lambert Azimuthal Equal Area

Population

Density of settlement in the region is generally low. Australia is one of the least densely populated countries on Earth with over 80% of its population living within 25 miles (40 km) of the coast – mostly in the southeast of the country. New Zealand, and the island groups of Melanesia, Micronesia, and Polynesia, are much more densely populated, although many of the smaller islands remain uninhabited.

Population density
(people per sq mile)

- below 10
- 10–62
- 63–130
- 131–259
- 260–519
- 520–780
- above 780

▲ *The myriad of* small coral islands that are scattered across the Pacific Ocean are often uninhabited, as they offer little shelter from the weather, often no fresh water, and only limited food supplies.

◄ *The planes of* the Australian Royal Flying Doctor Service are able to cover large expanses of barren land quickly, bringing medical treatment to the most inaccessible and far-flung places.

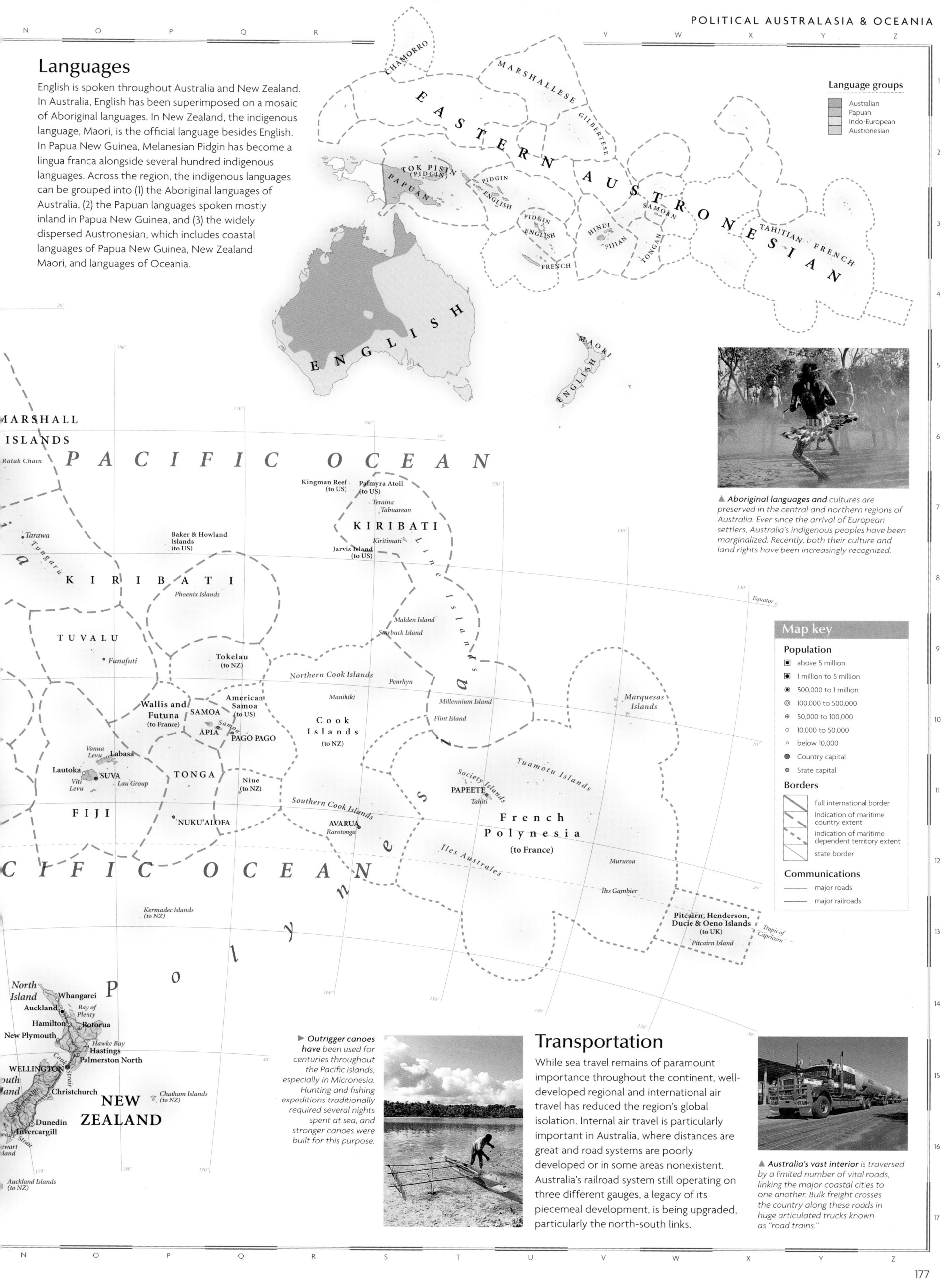

Languages

English is spoken throughout Australia and New Zealand. In Australia, English has been superimposed on a mosaic of Aboriginal languages. In New Zealand, the indigenous language, Maori, is the official language besides English. In Papua New Guinea, Melanesian Pidgin has become a lingua franca alongside several hundred indigenous languages. Across the region, the indigenous languages can be grouped into (1) the Aboriginal languages of Australia, (2) the Papuan languages spoken mostly inland in Papua New Guinea, and (3) the widely dispersed Austronesian, which includes coastal languages of Papua New Guinea, New Zealand Maori, and languages of Oceania.

Language groups
- Australian
- Papuan
- Indo-European
- Austronesian

CHAMORRO
MARSHALLESE
GILBERTESE
EASTERN AUSTRONESIAN
TOK PISIN (PIDGIN)
PAPUAN
PIDGIN ENGLISH
PIDGIN ENGLISH
SAMOAN
HINDI
FIJIAN
TONGAN
TAHITIAN FRENCH
FRENCH
ENGLISH
MAORI
ENGLISH

▲ *Aboriginal languages and cultures are preserved in the central and northern regions of Australia. Ever since the arrival of European settlers, Australia's indigenous peoples have been marginalized. Recently, both their culture and land rights have been increasingly recognized.*

MARSHALL ISLANDS
Ratak Chain
PACIFIC OCEAN
Tarawa
Tungaru
KIRIBATI
Phoenix Islands
TUVALU
Funafuti
Kingman Reef (to US)
Palmyra Atoll (to US)
Teraina
Tabuarean
KIRIBATI
Kiritimati
Jarvis Island (to US)
Baker & Howland Islands (to US)
Line Islands
Malden Island
Starbuck Island
Tokelau (to NZ)
Northern Cook Islands
Penrhyn
Manihiki
Millennium Island
Flint Island
Marquesas Islands
Wallis and Futuna (to France)
SAMOA
American Samoa (to US)
ĀPIA
PAGO PAGO
Samoa
Cook Islands (to NZ)
Vanua Levu
Labasa
Lautoka
Viti Levu
SUVA
Lau Group
TONGA
Niue (to NZ)
Society Islands
PAPEETE
Tahiti
Tuamotu Islands
FIJI
NUKU'ALOFA
Southern Cook Islands
AVARUA
Rarotonga
French Polynesia (to France)
Iles Australes
Mururoa
CIFIC OCEAN
Kermadec Islands (to NZ)
Iles Gambier
Pitcairn, Henderson, Ducie & Oeno Islands (to UK)
Tropic of Capricorn
Pitcairn Island
Equator

Polynesia

North Island
Whangarei
Auckland
Bay of Plenty
Hamilton
Rotorua
New Plymouth
Hawke Bay
Hastings
Palmerston North
WELLINGTON
Chatham Islands (to NZ)
Christchurch
NEW ZEALAND
Dunedin
Invercargill
Auckland Islands (to NZ)

Map key

Population
- ▣ above 5 million
- ◙ 1 million to 5 million
- ◉ 500,000 to 1 million
- ◎ 100,000 to 500,000
- ⊕ 50,000 to 100,000
- ⊙ 10,000 to 50,000
- ∘ below 10,000
- ■ Country capital
- • State capital

Borders
- full international border
- indication of maritime country extent
- indication of maritime dependent territory extent
- state border

Communications
- major roads
- major railroads

▶ *Outrigger canoes have been used for centuries throughout the Pacific islands, especially in Micronesia. Hunting and fishing expeditions traditionally required several nights spent at sea, and stronger canoes were built for this purpose.*

Transportation

While sea travel remains of paramount importance throughout the continent, well-developed regional and international air travel has reduced the region's global isolation. Internal air travel is particularly important in Australia, where distances are great and road systems are poorly developed or in some areas nonexistent. Australia's railroad system still operating on three different gauges, a legacy of its piecemeal development, is being upgraded, particularly the north-south links.

▲ *Australia's vast interior is traversed by a limited number of vital roads, linking the major coastal cities to one another. Bulk freight crosses the country along these roads in huge articulated trucks known as "road trains."*

Australasian & Oceanian resources

Natural resources are of major economic importance throughout Australasia and Oceania. Australia in particular is a major world exporter of raw materials such as coal, iron ore, and bauxite, while New Zealand's agricultural economy is dominated by sheep-raising. Trade with western Europe has declined significantly in the last 20 years, and the Pacific Rim countries of Southeast Asia are now the main trading partners, as well as a source of new settlers to the region. Australasia and Oceania's greatest resources are its climate and environment; tourism increasingly provides a vital source of income for the whole continent.

▲ *The largely unpolluted* waters of the Pacific Ocean support rich and varied marine life, much of which is farmed commercially. Here, oysters are gathered for market off the coast of New Zealand's South Island.

▶ *Huge flocks of* sheep are a common sight in New Zealand, where they outnumber people by 12 to 1. New Zealand is one of the world's largest exporters of wool and frozen lamb.

Standard of living

In marked contrast to its neighbor, Australia, with one of the world's highest life expectancies and standards of living, Papua New Guinea is one of the world's least developed countries. In addition, high population growth and urbanization rates throughout the Pacific islands contribute to overcrowding. In Australia and New Zealand, the Aboriginal and Maori people have been isolated, although recently their traditional land ownership rights have begun to be legally recognized in an effort to ease their social and economic isolation, and to improve living standards.

Standard of living
(UN human development index)

- low
- high
- figures unavailable

Environmental issues

The prospect of rising sea levels poses a threat to many low-lying islands in the Pacific. The testing of nuclear weapons, once common throughout the region, was finally discontinued in 1996. Australia's ecological balance has been irreversibly altered by the introduction of alien species. Although it has the world's largest underground water reserve, the Great Artesian Basin, the availability of fresh water in Australia remains critical. Periodic droughts combined with overgrazing lead to desertification and increase the risk of devastating bush fires, and occasional flash floods.

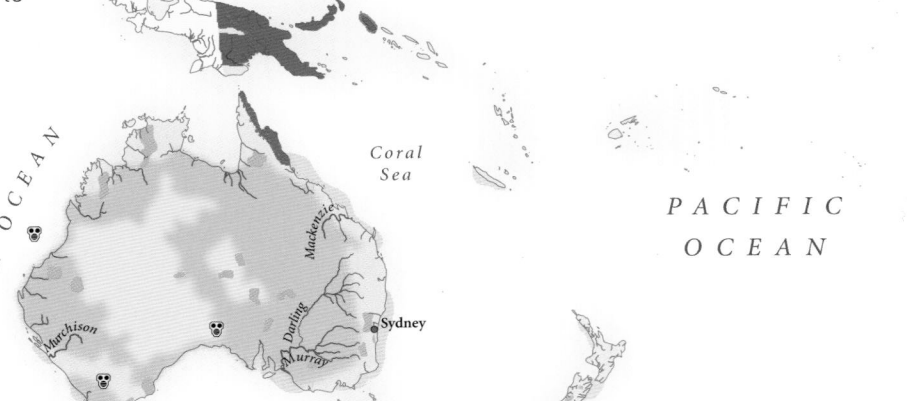

Environmental issues

- national parks
- tropical forest
- forest destroyed
- desert
- desertification
- polluted rivers
- radioactive contamination
- marine pollution
- heavy marine pollution
- poor urban air quality

▲ *In 1946 Bikini Atoll,* in the Marshall Islands, was chosen as the site for Operation Crossroads – investigating the effects of atomic bombs upon naval vessels. Further nuclear tests continued until the early 1990s. The long-term environmental effects are unknown.

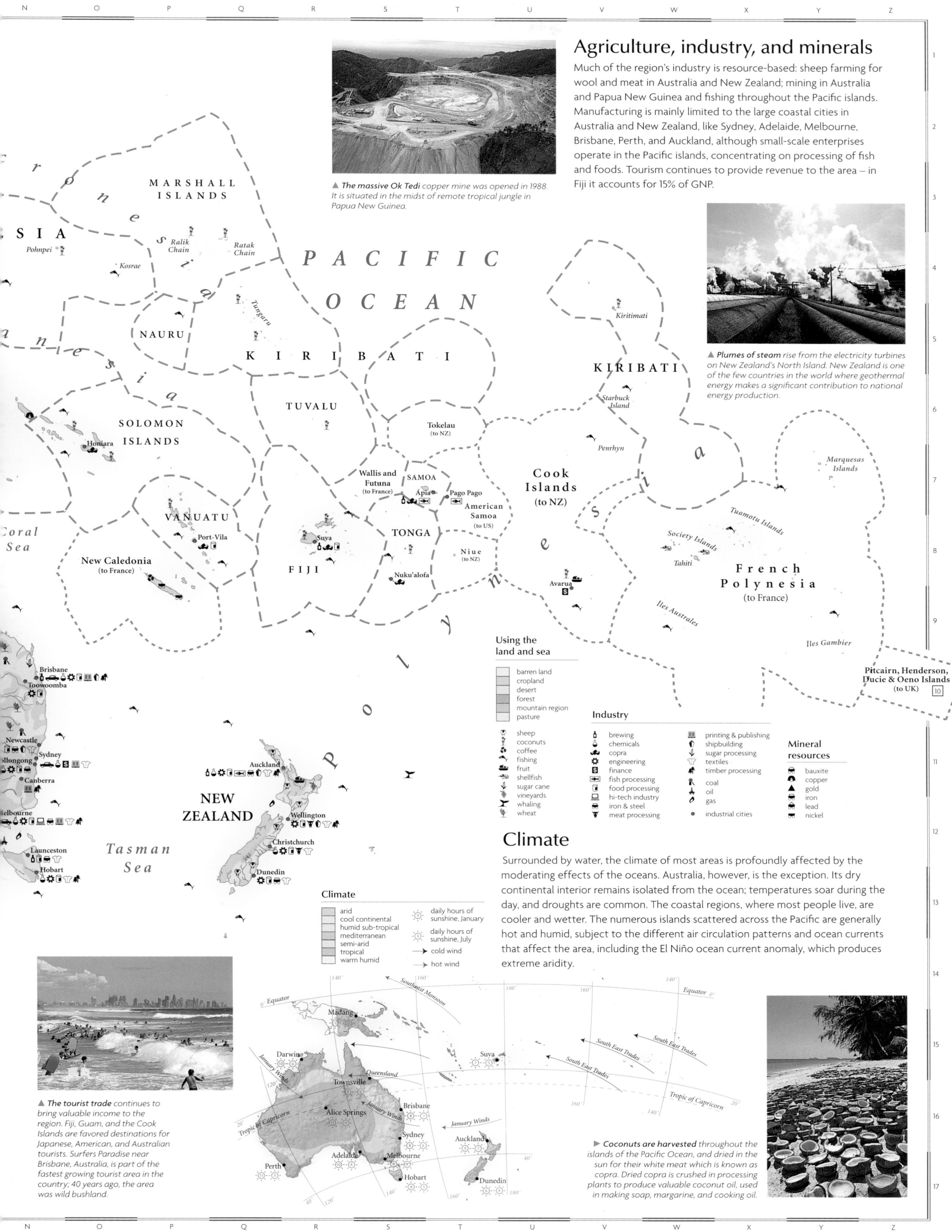

Agriculture, industry, and minerals

Much of the region's industry is resource-based: sheep farming for wool and meat in Australia and New Zealand; mining in Australia and Papua New Guinea and fishing throughout the Pacific islands. Manufacturing is mainly limited to the large coastal cities in Australia and New Zealand, like Sydney, Adelaide, Melbourne, Brisbane, Perth, and Auckland, although small-scale enterprises operate in the Pacific islands, concentrating on processing of fish and foods. Tourism continues to provide revenue to the area – in Fiji it accounts for 15% of GNP.

▲ The massive Ok Tedi copper mine was opened in 1988. It is situated in the midst of remote tropical jungle in Papua New Guinea.

▲ Plumes of steam rise from the electricity turbines on New Zealand's North Island. New Zealand is one of the few countries in the world where geothermal energy makes a significant contribution to national energy production.

Using the land and sea

- barren land
- cropland
- desert
- forest
- mountain region
- pasture

Industry

- sheep
- coconuts
- coffee
- fishing
- fruit
- shellfish
- sugar cane
- vineyards
- whaling
- wheat

- brewing
- chemicals
- copra
- engineering
- finance
- fish processing
- food processing
- hi-tech industry
- iron & steel
- meat processing

- printing & publishing
- shipbuilding
- sugar processing
- textiles
- timber processing
- coal
- oil
- gas
- industrial cities

Mineral resources

- bauxite
- copper
- gold
- iron
- lead
- nickel

Climate

Surrounded by water, the climate of most areas is profoundly affected by the moderating effects of the oceans. Australia, however, is the exception. Its dry continental interior remains isolated from the ocean; temperatures soar during the day, and droughts are common. The coastal regions, where most people live, are cooler and wetter. The numerous islands scattered across the Pacific are generally hot and humid, subject to the different air circulation patterns and ocean currents that affect the area, including the El Niño ocean current anomaly, which produces extreme aridity.

Climate

- arid
- cool continental
- humid sub-tropical
- mediterranean
- semi-arid
- tropical
- warm humid

- daily hours of sunshine, January
- daily hours of sunshine, July
- cold wind
- hot wind

▲ The tourist trade continues to bring valuable income to the region. Fiji, Guam, and the Cook Islands are favored destinations for Japanese, American, and Australian tourists. Surfers Paradise near Brisbane, Australia, is part of the fastest growing tourist area in the country; 40 years ago, the area was wild bushland.

▶ Coconuts are harvested throughout the islands of the Pacific Ocean, and dried in the sun for their white meat which is known as copra. Dried copra is crushed in processing plants to produce valuable coconut oil, used in making soap, margarine, and cooking oil.

Australia

Australia is the world's smallest continent, a stable landmass lying between the Indian and Pacific oceans. Previously home to its aboriginal peoples only, since the end of the 18th century immigration has transformed the face of the country. Initially settlers came mainly from western Europe, particularly the UK, and for years Australia remained wedded to its British colonial past. More recent immigrants have come from eastern Europe, and from Asian countries such as Japan, South Korea, and Indonesia. Australia is now forging strong trading links with these "Pacific Rim" countries and its economic future seems to lie with Asia and the Americas, rather than Europe, its traditional partner.

Using the land

Over 104 million sheep are dispersed in vast herds around the country, contributing to a major export industry. Cattle-ranching is important, particularly in the west. Wheat, and grapes for Australia's wine industry, are grown mainly in the south. Much of the country is desert, unsuitable for agriculture unless irrigation is used.

The urban/rural population divide

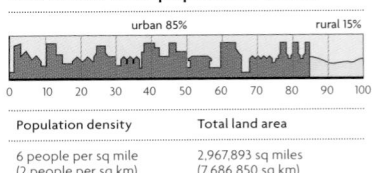

urban 85%   rural 15%

0 10 20 30 40 50 60 70 80 90 100

Population density	Total land area
6 people per sq mile (2 people per sq km)	2,967,893 sq miles (7,686,850 sq km)

Land use and agricultural distribution

- cattle
- sheep
- cereals
- sugar cane
- timber
- vineyards
- capital cities
- major towns
- pasture
- cropland
- forest
- desert
- mountain region

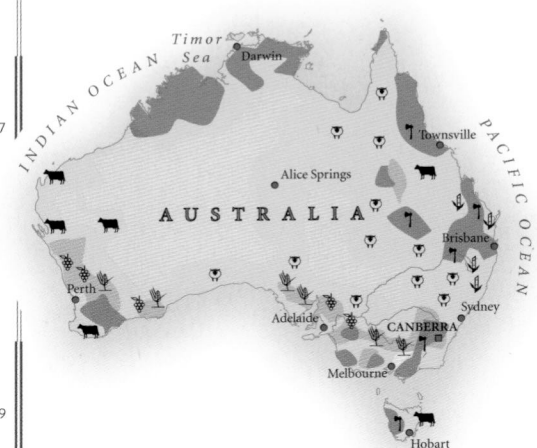

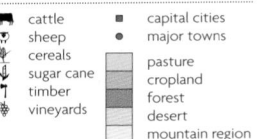

▲ *Lines of ripening* vines stretch for miles in Barossa Valley, a major wine-growing region near Adelaide.

The landscape

Australia consists of many eroded plateaus, lying firmly in the middle of the Indo-Australian Plate. It is the world's flattest continent, and the driest, after Antarctica. The coasts tend to be more hilly and fertile, especially in the east. The mountains of the Great Dividing Range form a natural barrier between the eastern coastal areas and the flat, dry plains and desert regions of the Australian "outback."

▲ *The Great Barrier Reef* is the world's largest area of coral islands and reefs. It runs for about 1240 miles (2000 km) along the Queensland coast.

▲ *The Pinnacles are* a series of rugged sandstone pillars. Their strange shapes have been formed by water and wind erosion.

The ancient Kimberley Plateau is the source of some of Australia's richest mineral deposits, including diamonds.

Uluru (Ayers Rock)

Arnhem Land

The tropical rain forest of the Cape York Peninsula contains more than 600 different varieties of tree.

Great Artesian Basin

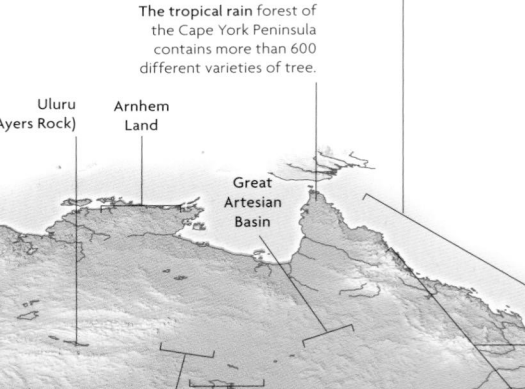

More than half of Australia rests on a uniform shield over 600 million years old. It is one of the Earth's original geological plates.

The Simpson Desert has a number of large salt pans, created by the evaporation of past rivers and now sourced by seasonal rains. Some are crusted with gypsum, but most are covered by common salt crystals.

The Nullarbor Plain is a low-lying limestone plateau which is so flat that the Trans-Australian Railway runs through it in a straight line for more than 300 miles (483 km).

The Lake Eyre basin, lying 51 ft (16 m) below sea level, is one of the largest inland drainage systems in the world, covering an area of more than 500,000 sq miles (1,300,000 sq km).

The Great Dividing Range forms a watershed between east- and west-flowing rivers. Erosion has created deep valleys, gorges, and waterfalls where rivers tumble over escarpments on their way to the sea.

Australian Alps

Tasmania has the same geological structure as the Australian Alps. During the last period of glaciation, 18,000 years ago, sea levels were some 300 ft (100 m) lower and it was joined to the mainland.

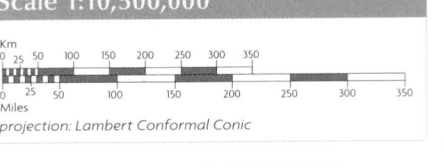

◀ *Uluru (Ayers Rock),* the world's largest free-standing rock, is a massive outcrop of red sandstone in Australia's desert center. Wind and sandstorms have ground the rock into the smooth curves seen here. Uluru is revered as a sacred site by many aboriginal peoples.

Scale 1:10,500,000

Km
0 25 50 100 150 200 250 300 350

Miles
0 25 50 100 150 200 250 300 350

projection: Lambert Conformal Conic

Map key

Population

- ▣ 1 million to 5 million
- ◉ 500,000 to 1 million
- ◎ 100,000 to 500,000
- ⊕ 50,000 to 100,000
- ○ 10,000 to 50,000
- ∘ below 10,000

Elevation

- 2000m / 6562ft
- 1000m / 3281ft
- 500m / 1640ft
- 250m / 820ft
- 100m / 328ft
- sea level

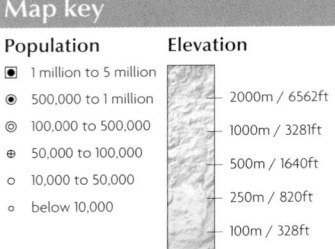
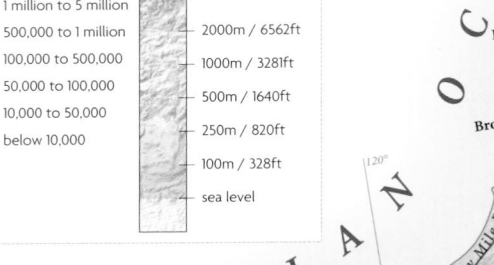

Great Artesian Basin

Rainwater replenishes aquifer

Lake Eyre

Aquifers from which artesian water is obtained

Underground water movements

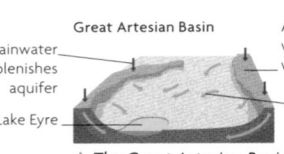

▲ *The Great Artesian Basin* underlies nearly 20% of the total area of Australia, providing a valuable store of underground water, essential to Australian agriculture. The ephemeral rivers which drain the northern part of the basin have highly braided courses and, in consequence, the area is known as "channel country."

► The Great Barrier Reef attracts thousands of tourists every year, drawn by the spectacular coral formations and exotic marine life.

▲ Lying on the border between New South Wales and Queensland, this summit is in the Great Dividing Range which splits the fertile eastern coast from the more arid interior.

Transportation & industry

Extensive mineral reserves, including coal, iron ore, gold, bauxite, and copper, once formed the heart of Australian industry, along with agricultural products. In recent years, Australia has moved from being a primary producer to a largely service-based economy, particularly the rapidly developing tourist industry.

Major industry and infrastructure

- brewing
- car manufacture
- chemicals
- coal
- electronics
- engineering
- food processing
- mining
- oil & gas
- tourism
- capital cities
- major towns
- international airports
- major roads
- major industrial areas

The Transportation network

204,470 miles (329,100 km)

11,658 miles (18,619 km)

5911 miles (9514 km)

5197 miles (8366 km)

Well-developed air transportation links, including the Royal Flying Doctor Service, connect the sparsely populated center and west. Most freight travels in massive trucks known as "road trains."

▲ Sydney Harbour is one of the world's most spectacular natural harbors. Founded in 1788, Sydney was the first major settlement in Australia.

192

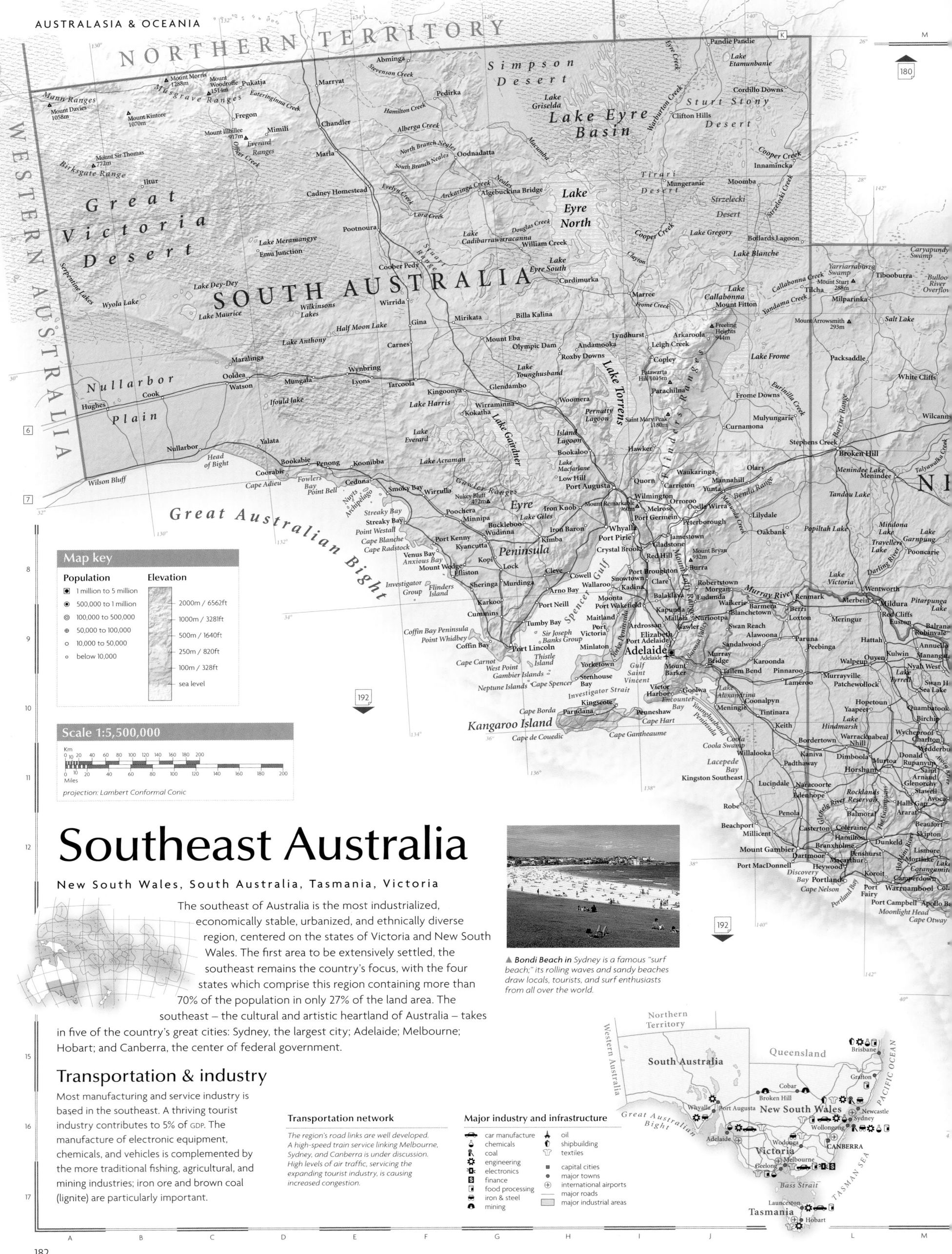

Southeast Australia

New South Wales, South Australia, Tasmania, Victoria

The southeast of Australia is the most industrialized, economically stable, urbanized, and ethnically diverse region, centered on the states of Victoria and New South Wales. The first area to be extensively settled, the southeast remains the country's focus, with the four states which comprise this region containing more than 70% of the population in only 27% of the land area. The southeast – the cultural and artistic heartland of Australia – takes in five of the country's great cities: Sydney, the largest city; Adelaide; Melbourne; Hobart; and Canberra, the center of federal government.

Transportation & industry

Most manufacturing and service industry is based in the southeast. A thriving tourist industry contributes to 5% of GDP. The manufacture of electronic equipment, chemicals, and vehicles is complemented by the more traditional fishing, agricultural, and mining industries; iron ore and brown coal (lignite) are particularly important.

Transportation network

The region's road links are well developed. A high-speed train service linking Melbourne, Sydney, and Canberra is under discussion. High levels of air traffic, servicing the expanding tourist industry, is causing increased congestion.

▲ **Bondi Beach in** Sydney is a famous "surf beach;" its rolling waves and sandy beaches draw locals, tourists, and surf enthusiasts from all over the world.

Map key

Population
- ▣ 1 million to 5 million
- ◉ 500,000 to 1 million
- ◎ 100,000 to 500,000
- ◌ 50,000 to 100,000
- ○ 10,000 to 50,000
- ∘ below 10,000

Elevation
- 2000m / 6562ft
- 1000m / 3281ft
- 500m / 1640ft
- 250m / 820ft
- 100m / 328ft
- sea level

Scale 1:5,500,000

projection: Lambert Conformal Conic

Major industry and infrastructure

- car manufacture
- chemicals
- coal
- engineering
- electronics
- finance
- food processing
- iron & steel
- mining
- oil
- shipbuilding
- textiles
- ■ capital cities
- ● major towns
- ⊕ international airports
- major roads
- major industrial areas

Using the land & sea

The western flanks of the Great Dividing Range and the northern deserts of South Australia support massive herds of sheep and cattle, while more intensive stockrearing occurs near the cities. Sugar cane is the most important industrial crop, and cereal grains including wheat, corn, barley, and sorghum are also grown. Grapes, citrus, and orchard fruits are among the wide range of fruit and vegetables cultivated in this region. Tasmania's forestry and fishing contributes to over one-third of the state's exports.

▲ The fertile Darling Downs, known as the "breadbasket of Australia," support a wide range of crops including cereals, sugar cane, and fruit.

▶ The Murray River has its source in the eastern uplands of the Great Dividing Range. Fed by melting snow, it runs for 1609 miles (2589 km), and has sufficient volume to reach the ocean southeast of Adelaide despite a minimal gradient for most of its lower reaches.

The urban/rural population divide

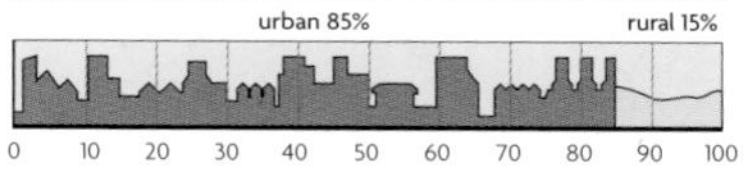

urban 85%   rural 15%

Population density	Total land area
18 people per sq mile (7 people per sq km)	778,022 sq miles (2,015,600 sq km)

Land use and agricultural distribution

- cattle
- sheep
- bananas
- fishing
- fruit
- sugar cane
- vineyards
- wheat
- capital cities
- major towns
- pasture
- cropland
- forest
- desert
- mountain region

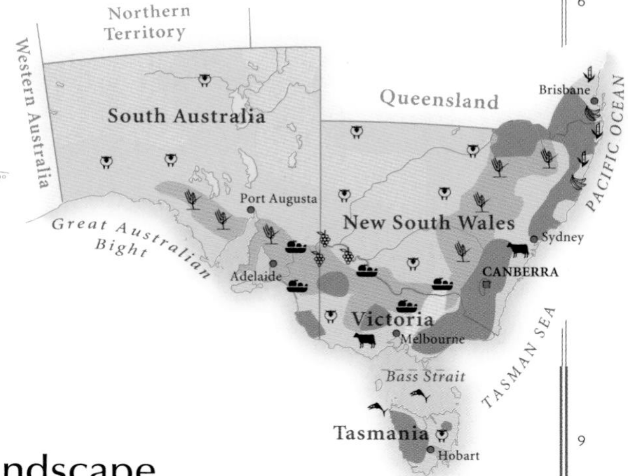

The landscape

The southern half of the Great Dividing Range runs parallel to the eastern coast of Victoria and New South Wales as far as Tasmania, which, though divided from the mainland is part of the same mountain chain. South Australia comprises the Australian shield and half of the dry, flat Nullarbor Plain. The Murray/Darling river basin is the only major river system.

◀ The heavily folded Flinders Ranges is part of an arc of sedimentary rocks reaching northward from Kangaroo Island.

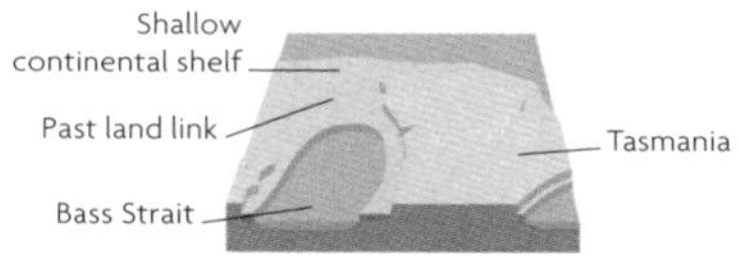

Lake Eyre is the largest of southern Australia's dry lakes. Lying -51 ft (-16 m) below sea level, it has flooded only three times in the last century.

The Musgrave and Everard ranges form bare, rounded hills made up of ancient granite and gneiss.

The Murray/Darling is Australia's longest river at 1703 miles (2739 km).

Shallow continental shelf
Past land link
Bass Strait
Tasmania

▲ Tasmania is part of Australia's eastern highlands, separated from the mainland by 155 miles (250 km) of the Bass Strait. In the recent geological past, dry land links between Tasmania and Victoria would have been possible during periods of world-wide glaciation, when the sea level was more than 180 ft (55 m) below that of present sea levels.

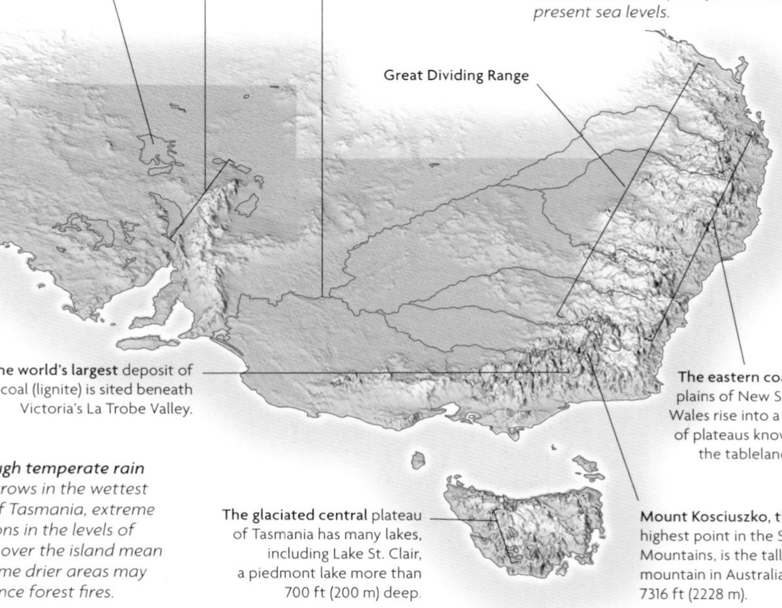

Great Dividing Range

The eastern part of the Nullarbor Plain has many sinkholes, eroded by rainwater, which run underground to form a system of long caves in the limestone rocks.

The world's largest deposit of brown coal (lignite) is sited beneath Victoria's La Trobe Valley.

◀ Though temperate rain forest grows in the wettest parts of Tasmania, extreme variations in the levels of rainfall over the island mean that some drier areas may experience forest fires.

The glaciated central plateau of Tasmania has many lakes, including Lake St. Clair, a piedmont lake more than 700 ft (200 m) deep.

The eastern coastal plains of New South Wales rise into a series of plateaus known as the tableland.

Mount Kosciuszko, the highest point in the Snowy Mountains, is the tallest mountain in Australia at 7316 ft (2228 m).

Northern Territory
Western Australia
South Australia
Queensland
New South Wales
Victoria
Tasmania
Great Australian Bight
Port Augusta
Adelaide
Brisbane
Sydney
CANBERRA
Melbourne
Hobart
PACIFIC OCEAN
TASMAN SEA

Map labels

QUEENSLAND
NEW SOUTH WALES
VICTORIA
TASMANIA
Great Dividing Range
TASMAN SEA
Bass Strait

Brisbane, Ipswich, Toowoomba, Warwick, Stanthorpe, Gold Coast, Surfers Paradise, Tweed Heads, Lismore, Casino, Ballina, Grafton, Coffs Harbour, Armidale, Tamworth, Moree, Narrabri, Gunnedah, Bourke, Cobar, Dubbo, Orange, Bathurst, Parkes, Forbes, Wagga Wagga, Albury, CANBERRA, Goulburn, Sydney, Parramatta, Campbelltown, Wollongong, Newcastle, Gosford, Penrith, Maitland, Cessnock, Port Macquarie, Kempsey, Taree, Griffith, Deniliquin, Echuca, Shepparton, Bendigo, Ballarat, Melbourne, Geelong, Sale, Bairnsdale, Warragul, Hobart, Launceston, Devonport, Burnie

Mount Kosciuszko 2228m
Mount Bogong
Lake Eyre
Murray River
Darling River
Murrumbidgee River
Lachlan River
Lake George
King Island
Flinders Island
Kangaroo Island

JERVIS BAY TERRITORY
AUSTRALIAN CAPITAL TERRITORY

New Zealand

Lying 1500 miles east-southeast of Australia, New Zealand was originally settled by the Maori people of Polynesia. It was visited by Europeans for the first time only as recently as the 1770s. The islands' rugged topography means that most settlement has been concentrated in coastal areas. People of European origin make up about 70% of the population of 4 million, following immigration which began in the 1920s. Many recent settlers have come from Asia, including India and China, and a number of the Pacific Islands. The Maori now make up a minority of less than half a million. Their ancient claims to at least half of national territory, however, are gaining increasing legal credence.

The landscape

New Zealand comprises two large islands and many scattered smaller islands. On South Island the Alpine Fault marks the boundary between the Pacific and Indo-Australian plates. Tectonic activity has strongly influenced the formation of the Southern Alps, snowcapped mountains with several peaks over 9800 ft (3000 m). North Island has a lower and less extensive mountain region, containing forested hills, a central volcanic plateau, and downlands.

▲ *The Southern Alps* have been formed by "slip" faulting. The Indo-Australian and Pacific plates run in opposite directions along the Alpine Fault. Although they slide past each other, they are also being thrust over one another, causing the continental crust of the Pacific Plate to be uplifted to form the Alps.

Mountain-building in the Southern Alps

North Island
Alpine Fault
Pacific Plate
South Island
Southern Alps
Indo-Australian Plate

The Southern Alps run for more than 300 miles (483 km) forming the backbone of South Island. They were uplifted following the collision of the Pacific and Indo-Australian plates.

Sutherland Falls

Fiordland, in the far south west, contains a large number of flooded glacial valleys.

Probable location of Alpine Fault

High levels of rainfall and a steep topography has made New Zealand's rivers swift-running. In the southern reaches of both islands, rivers such as the Mokoreta form broad, braided streams.

The Southern Alps contain more than 360 glaciers, including the Murchison, Mueller, and Godley glaciers on the eastern slopes and the Fox and Franz Josef glaciers to the west.

The Tasman Glacier, the largest glacier in New Zealand, flows for 18 miles (29 km) down the slopes of New Zealand's highest mountain, Aoraki (Mount Cook).

The coastal Canterbury Plains are the result of glacial outwash. They are the only major flat area in New Zealand.

▼ *The Rotorua and* Taupo valleys have some of the largest and most spectacular thermal springs in New Zealand. These occur when superheated groundwater rises to the surface through joints in the rocks.

Rotorua

Lake Taupo is New Zealand's largest inland lake. It occupies the crater of an extinct volcano.

Mount Taranaki, rising 8261 ft (2518 m) is an isolated, dormant volcano.

The boundary between the Indo-Australian Plate and the Pacific Plate runs through the center of North Island, leading to many typical volcanic features. The plateau which rises from the slopes of Lake Taupo contains a string of active volcanoes.

▼ *The Northland region* is characterized by many coastal inlets. These are lined by mangrove swamps, signaling the change to a subtropical climate in the far north of the island.

Northland

▲ *Clouds of steam* rise from White Island, an active, offshore volcano lying in the Bay of Plenty, off the northern coast of North Island.

Scale 1:2,750,000

projection: Lambert Conformal Conic

192

Map labels

PACIFIC OCEAN

TASMAN SEA

NEW ZEALAND

North Island

NORTHLAND
AUCKLAND
WAIKATO
BAY OF PLENTY
GISBORNE
HAWKE'S BAY
TARANAKI
MANAWATU
WANGANUI

Three Kings Islands
Cape Reinga
Cape Maria van Diemen
North Cape
Parengarenga Harbour
Te Kao
Great Exhibition Bay
Awanui
Kaitaia
Doubtless Bay
Cape Karikari
Rangaunu Bay
Mangonui
Ahipara
Ahipara Bay
Tairua Point
Ninety Mile Beach
Hokianga Harbour
Omapere
Kaikohe
Okaihau
Kerikeri
Paihia
Russell
Cape Brett
Bay of Islands
Whangaruru Harbour
Kaikohe
Kawakawa
Towai
Hikurangi
Whangarei
Whangarei Harbour
Bream Head
Bream Bay
Portland Head
Mangapere
Dargaville
Waipoua
Tutamoe Range
Kaihu
Maungaturoto
Kaipara Harbour
Wairoa
Ruawai
Paparoa
Waiotira
Cavalli Islands
Poor Knights Islands
Hen and Chickens Islands
Leigh
Warkworth
Wellsford
Kaiwaka
Helensville
Waitoki
North Head
Muriwai Beach
Whatipu
Manukau Harbour
Auckland
AUCKLAND
Papakura
Waiheke Island
Ponui Island
Waiuku
Pukekohe
Tuakau
Pokeno
Port Fitzroy
Great Barrier Island
Little Barrier Island
Kawau Island
Mokohinau Islands
Cape Rodney
Coromandel
Colville
Cape Colville
Coromandel Peninsula
Great Mercury Island
Mercury Islands
The Aldermen Islands
Mayor Island
Whitianga
Tairua
Whangamata
Thames
Paeroa
Waihi
Katikati
Tauranga
Te Puke
Matakana Island
Motiti Island
Mount Maunganui
Whakatane
Opotiki
Te Kaha
Cape Runaway
Lottin Point
Te Araroa
Hicks Bay
Raukumara
Te Araroa
Tikitiki
Ruatoria
Tolaga Bay
Tokomaru Bay
Mangahauini
Manutuke
Whangara
Gisborne
GISBORNE
Te Karaka
Poverty Bay
Young Nicks Head
Wairoa
Mahia
Mahia Peninsula
Portland Island
Table Cape
Long Point
Nuhaka
Frasertown
Putorino
Tutira
Eskdale
Napier
Havelock North
Hastings
Waipukurau
Waipawa
Onga Onga
Takapau
Porangahau
Cape Turnagain
Dannevirke
Woodville
Pahiatua
Eketahuna
Masterton
Carterton
Greytown
Martinborough
Featherston
Cape Palliser
Palmerston North
Feilding
Foxton
Otaki
Shannon
Levin
Paraparaumu
Kapiti Island
Mana Island
Porirua
Wellington
Lower Hutt
Upper Hutt
Whanganui
Wanganui
Turakina
Waiouru
Taihape
Ohakune
Raetihi
National Park
Mount Ruapehu 2797m
Mount Ngauruhoe 2291m
Tokaanu
Turangi
Mangaweka
Rangitikei
Hunterville
Marton
Bulls
Sanson
Halcombe
Bennydale
Piopio
Te Kuiti
Otorohanga
Kawhia
Kawhia Harbour
Raglan
Waitomo Caves
Hamilton
Ngaruawahia
Te Awamutu
Cambridge
Morrinsville
Matamata
Putaruru
Tokoroa
Mangakino
Lake Taupo
Taupo
Whakamaru
Atiamuri
Reporoa
Rotorua
Lake Rotorua
Tikitere
Lake Rotoiti
Kawerau
Te Teko
Matata
Waihou
Te Aroha
Waharoa
Te Poi
Paeroa
Ngatea
Kerepehi
Pukekawa
Mercer
Meremere
Huntly
Waikato
Te Kauwhata
Taupiri
Kaimai Range
Mamaku
Mount Tarawera 1111m
Rerewhakaaitu
Murupara
Waioeka
Matawai
Motu
Mount Hikurangi 1752m
Rauhini East Cape

New Plymouth
Waitara
Inglewood
Stratford
Eltham
Hawera
Patea
Waverley
Mount Taranaki (Mount Egmont) 2518m
Cape Egmont
Opunake
Manaia
Oakura
North Taranaki Bight
South Taranaki Bight
Kaponga
Okato

Farewell Spit
Cape Farewell
Golden Bay
Collingwood
Stephens Island
Cape Stephens

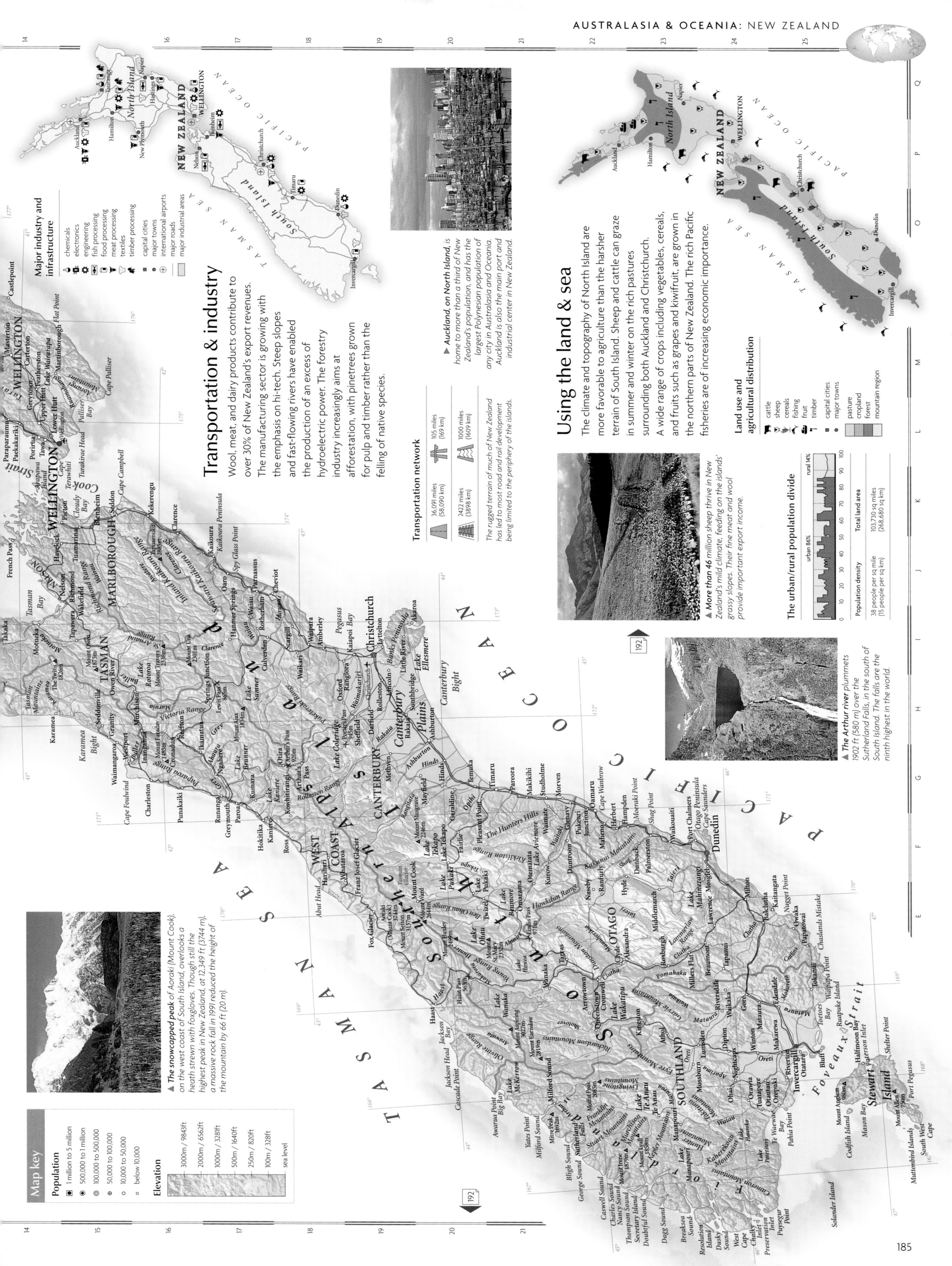

Map key

Population

- ◉ 1 million to 5 million
- ◉ 500,000 to 1 million
- ◉ 100,000 to 500,000
- ⊕ 50,000 to 100,000
- ⊙ 10,000 to 50,000
- ○ below 10,000

Elevation

- 3000m / 9843ft
- 2000m / 656ft
- 1000m / 328ft
- 500m / 1640ft
- 250m / 820ft
- 100m / 328ft
- sea level

▲ *The snowcapped peak* of Aoraki (Mount Cook), on the west coast of South Island, overlooks a heath strewn with foxgloves. Though still the highest peak in New Zealand, at 12,349 ft (3744 m), a massive rock fall in 1991 reduced the height of the mountain by 66 ft (20 m).

Transportation & industry

Wool, meat, and dairy products contribute to over 30% of New Zealand's export revenues. The manufacturing sector is growing with the emphasis on hi-tech. Steep slopes and fast-flowing rivers have enabled the production of an excess of hydroelectric power. The forestry industry increasingly aims at afforestation, with pinetrees grown for pulp and timber rather than the felling of native species.

Major industry and infrastructure

- chemicals
- electronics
- engineering
- fish processing
- food processing
- meat processing
- textiles
- timber processing
- ■ capital cities
- ■ major cities
- ⊕ international airports
- — major roads
- major industrial areas

▲ *Auckland, on North Island,* is home to more than a third of New Zealand's population, and has the largest Polynesian population of any city in Australasia and Oceania. Auckland is also the main port and industrial center in New Zealand.

Transportation network

- 36,091 miles (58,090 km)
- 105 miles (169 km)
- 2422 miles (3898 km)
- 1000 miles (609 km)

The rugged terrain of much of New Zealand has led to most road and rail development being limited to the periphery of the islands.

Using the land & sea

The climate and topography of North Island are more favorable to agriculture than the harsher terrain of South Island. Sheep and cattle can graze in summer and winter on the rich pastures surrounding both Auckland and Christchurch. A wide range of crops including vegetables, cereals, and fruits such as grapes and kiwifruit, are grown in the northern parts of New Zealand. The rich Pacific fisheries are of increasing economic importance.

Land use and agricultural distribution

- cattle
- sheep
- cereals
- fishing
- fruit
- timber
- ■ capital cities
- • major towns
- pasture
- cropland
- forest
- mountain region

▲ *More than 46 million sheep* thrive in New Zealand's mild climate, feeding on the islands' grassy slopes. Their fine meat and wool provide important export income.

The urban/rural population divide

urban 86% / rural 14%

Population density	Total land area
38 people per sq mile (15 people per sq km)	103,730 sq miles (268,680 sq km)

▲ *The Arthur river* plummets 1902 ft (580 m) over the Sutherland Falls, in the south of South Island. The falls are the ninth highest in the world.

Melanesia

FIJI, New Caledonia (to France), PAPUA NEW GUINEA, SOLOMON ISLANDS, VANUATU

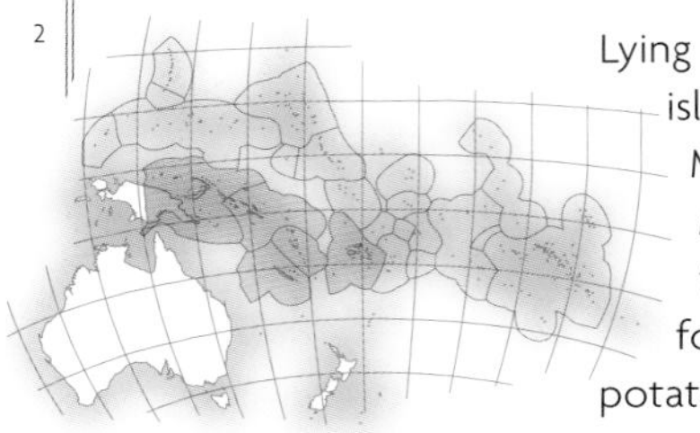

Lying in the southwest Pacific Ocean, northeast of Australia and south of the Equator, the islands of Melanesia form one of the three geographic divisions (along with Polynesia and Micronesia) of Oceania. Melanesia's name derives from the Greek melas, "black," and nesoi, "islands." Most of the larger islands are volcanic in origin. The smaller islands tend to be coral atolls and are mainly uninhabited. Rugged mountains, covered by dense rain forest, take up most of the land area. Melanesian's cultivate yams, taro, and sweet potatoes for local consumption and live in small, usually dispersed, homesteads.

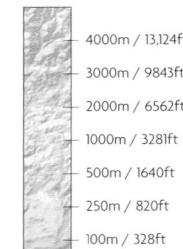
▲ *Huli tribesmen from* Southern Highlands Province in Papua New Guinea parade in ceremonial dress, their powdered wigs decorated with exotic plumage and their faces and bodies painted with colored pigments.

Map key

Population
- ⊚ 100,000 to 500,000
- ⊕ 50,000 to 100,000
- ○ 10,000 to 50,000
- ○ below 10,000

Elevation
- 4000m / 13,124ft
- 3000m / 9843ft
- 2000m / 6562ft
- 1000m / 3281ft
- 500m / 1640ft
- 250m / 820ft
- 100m / 328ft
- sea level

◀ *Lying close to* the banks of the Sepik river in northern Papua New Guinea, this building is known as the Spirit House. It is constructed from leaves and twigs, ornately woven and trimmed into geometric patterns. The house is decorated with a mask and topped by a carved statue.

▲ *On one of* Vanuatu's many islands, beach houses stand at the water's edge, surrounded by coconut palms and other tropical vegetation. The unspoilt beaches and tranquillity of its islands are drawing ever-larger numbers of tourists to Vanuatu.

Transportation & Industry

The processing of natural resources generates significant export revenue for the countries of Melanesia. The region relies mainly on copra, tuna, and timber exports, with some production of cocoa and palm oil. The islands have substantial mineral resources including the world's largest copper reserves on Bougainville Island; gold, and potential oil and natural gas. Tourism has become the fastest growing sector in most of the countries' economies.

◀ *On New Caledonia's* main island, relatively high interior plateaus descend to coastal plains. Nickel is the most important mineral resource, but the hills also harbor metallic deposits including chrome, cobalt, iron, gold, silver, and copper.

Transportation network

🛣 1236 miles (1990 km)	✈ None		
🚂 370 miles (595 km)	⛴ 6924 miles (11,143 km)		

As most of the islands of Melanesia lie off the major sea and air routes, services to and from the rest of the world are infrequent. Transportation by road on rugged terrain is difficult and expensive.

Major industry and infrastructure

- 🍶 beverages
- ☕ coffee processing
- copra processing
- 🍴 food processing
- ⛏ mining
- 👕 textiles
- 🪵 timber processing
- tourism
- ■ capital cities
- ● major towns
- ⊕ international airports
- — major roads

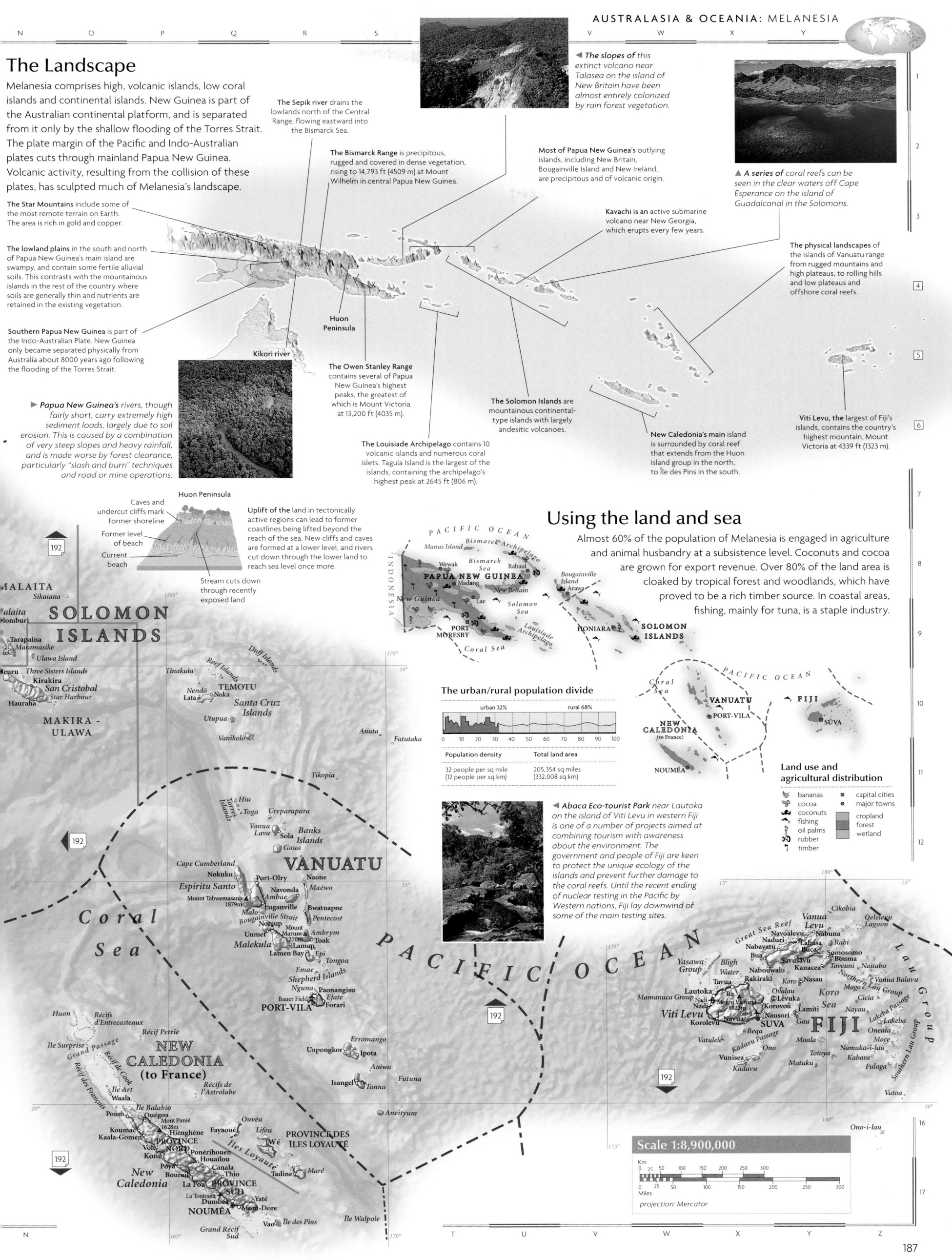

The Landscape

Melanesia comprises high, volcanic islands, low coral islands and continental islands. New Guinea is part of the Australian continental platform, and is separated from it only by the shallow flooding of the Torres Strait. The plate margin of the Pacific and Indo-Australian plates cuts through mainland Papua New Guinea. Volcanic activity, resulting from the collision of these plates, has sculpted much of Melanesia's landscape.

The Star Mountains include some of the most remote terrain on Earth. The area is rich in gold and copper.

The lowland plains in the south and north of Papua New Guinea's main island are swampy, and contain some fertile alluvial soils. This contrasts with the mountainous islands in the rest of the country where soils are generally thin and nutrients are retained in the existing vegetation.

Southern Papua New Guinea is part of the Indo-Australian Plate. New Guinea only became separated physically from Australia about 8000 years ago following the flooding of the Torres Strait.

▶ **Papua New Guinea's** rivers, though fairly short, carry extremely high sediment loads, largely due to soil erosion. This is caused by a combination of very steep slopes and heavy rainfall, and is made worse by forest clearance, particularly "slash and burn" techniques and road or mine operations.

The Sepik river drains the lowlands north of the Central Range, flowing eastward into the Bismarck Sea.

The Bismarck Range is precipitous, rugged and covered in dense vegetation, rising to 14,793 ft (4509 m) at Mount Wilhelm in central Papua New Guinea.

Most of Papua New Guinea's outlying islands, including New Britain, Bougainville Island and New Ireland, are precipitous and of volcanic origin.

◀ **The slopes of** this extinct volcano near Talasea on the island of New Britain have been almost entirely colonized by rain forest vegetation.

▲ **A series of** coral reefs can be seen in the clear waters off Cape Esperance on the island of Guadalcanal in the Solomons.

Kavachi is an active submarine volcano near New Georgia, which erupts every few years.

The physical landscapes of the islands of Vanuatu range from rugged mountains and high plateaus, to rolling hills and low plateaus and offshore coral reefs.

Huon Peninsula

Kikori river

The Owen Stanley Range contains several of Papua New Guinea's highest peaks, the greatest of which is Mount Victoria at 13,200 ft (4035 m).

The Louisiade Archipelago contains 10 volcanic islands and numerous coral islets. Tagula Island is the largest of the islands, containing the archipelago's highest peak at 2645 ft (806 m).

The Solomon Islands are mountainous continental-type islands with largely andesitic volcanoes.

New Caledonia's main island is surrounded by coral reef that extends from the Huon island group in the north, to Île des Pins in the south.

Viti Levu, the largest of Fiji's islands, contains the country's highest mountain, Mount Victoria at 4339 ft (1323 m).

Huon Peninsula

Caves and undercut cliffs mark former shoreline

Former level of beach

Current beach

Stream cuts down through recently exposed land

Uplift of the land in tectonically active regions can lead to former coastlines being lifted beyond the reach of the sea. New cliffs and caves are formed at a lower level, and rivers cut down through the lower land to reach sea level once more.

Using the land and sea

Almost 60% of the population of Melanesia is engaged in agriculture and animal husbandry at a subsistence level. Coconuts and cocoa are grown for export revenue. Over 80% of the land area is cloaked by tropical forest and woodlands, which have proved to be a rich timber source. In coastal areas, fishing, mainly for tuna, is a staple industry.

PACIFIC OCEAN
Manus Island
Bismarck Archipelago
Wewak
Bismarck Sea
Rabaul
INDONESIA
PAPUA NEW GUINEA
Madang
New Britain
Bougainville Island
Arawa
New Guinea
Lae
Solomon Sea
PORT MORESBY
Louisiade Archipelago
HONIARA
SOLOMON ISLANDS
Coral Sea

PACIFIC OCEAN
Coral Sea
VANUATU
FIJI
PORT-VILA
SUVA
NEW CALEDONIA (to France)
NOUMÉA

The urban/rural population divide

urban 32%   rural 68%
0 10 20 30 40 50 60 70 80 90 100

Population density	Total land area
32 people per sq mile (12 people per sq km)	205,354 sq miles (332,008 sq km)

◀ **Abaca Eco-tourist Park** near Lautoka on the island of Viti Levu in western Fiji is one of a number of projects aimed at combining tourism with awareness about the environment. The government and people of Fiji are keen to protect the unique ecology of the islands and prevent further damage to the coral reefs. Until the recent ending of nuclear testing in the Pacific by Western nations, Fiji lay downwind of some of the main testing sites.

Land use and agricultural distribution

- bananas
- cocoa
- coconuts
- fishing
- oil palms
- rubber
- timber
- ■ capital cities
- ● major towns
- cropland
- forest
- wetland

Map labels

SOLOMON ISLANDS
MALAITA
Sikaiana
Malaita
Olomburi
Tarapaina
Maramasike
Ulawa Island
Meuru
Three Sisters Islands
Kirakira
San Cristobal
Star Harbour
Hauraha
MAKIRA - ULAWA

Duff Islands
Reef Islands
Tinakula
Tikopia
TEMOTU
Nendō
Lata
Noka
Santa Cruz Islands
Utupua
Vanikolo
Anuta
Fatutaka

Torres Islands
Hiu
Toga
Ureparapara
Vanua Lava
Sola
Banks Islands
Gaua

Cape Cumberland
Nokuku
Port-Olry
Naone
VANUATU
Espiritu Santo
Navonda
Ambae
Maéwo
Mount Tabwemasana 1879m
Luganville
Malo
Bougainville Strait
Norsup
Bwatnapne
Pentecost
Unmet
Mount Marum 1270m
Ambrym
Malekula
Lamap
Toak
Lamen Bay
Epi
Tongoa
Emae
Shepherd Islands
Nguna
Paonangisu
Bauer Field
Efate
PORT-VILA
Forari

Coral Sea

Huon
Récifs d'Entrecasteaux
Île Surprise
Récif Petrie
Grand Passage
Récif des Français
Récifs de Cook
Île Art
Waala
Récifs de l'Astrolabe

Erromango
Unpongkor
Ipota
Aniwa
Isangel
Tanna
Futuna
Aneityum

NEW CALEDONIA (to France)
Île Balabio
Pouma
Ouégoa
Mont Panié 1628m
Hienghène
Ouvéa
Lifou
Fayaoué
PROVINCE NORD
Koumac
Kaala-Gomen
Poum
Houailou
Ponérihouen
Tadine
Maré
Poya
Canala
Bourail
Thio
La Foa
PROVINCE SUD
La Tontouta
New Caledonia
Dumbéa
Yaté
NOUMÉA
Mont-Dore
Île Walpole
Vao
Île des Pins
Grand Récif Sud
PROVINCE DES ÎLES LOYAUTÉ
Wé
Îles Loyauté

PACIFIC OCEAN

FIJI
Cikobia
Vanua Levu
Qelelevu Lagoon
Great Sea Reef
Navoalevu
Sabuna
Naduri
Labasa
Rabi
Nabavatu
Buca
Somosomo
Bouma
Yasawa Group
Bligh Water
Nabouwalu
Savusavu
Kanacea
Taveuni
Northern Lau Group
Vanua Balavu
Mago
Koro
Nasau
Cicia
Lautoka
Mamanuca Group
Ovalau
Levuka
Lamiti
Nayau
Nadi
Mount Victoria
Korovou
Nausori
Gau
Lakeba
Southern Lau Group
Viti Levu
Korolevu
Navua
SUVA
Koro Sea
Lakeba Passage
Beqa
Oneata
Moce
Vatulele
Kadavu Passage
Ono
Moala
Namuka-i-lau
Kabara
Vunisea
Totoya
Fulaga
Kadavu
Matuku
Vatoa
Ono-i-lau

Scale 1:8,900,000

Km
0 25 50 100 150 200 250 300
Miles
0 25 50 100 150 200 250 300

projection: Mercator

Micronesia

MARSHALL ISLANDS, MICRONESIA, NAURU, PALAU, Guam, Northern Mariana Islands, Wake Island

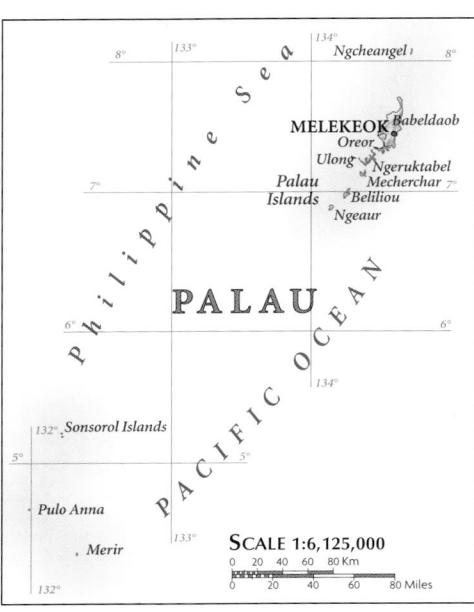

The Micronesian islands lie in the western reaches of the Pacific Ocean and are all part of the same volcanic zone. The Federated States of Micronesia is the largest group, with more than 600 atolls and forested volcanic islands in an area of more than 1120 sq miles (2900 sq km). Micronesia is a mixture of former colonies, overseas territories, and dependencies. Most of the region still relies on aid and subsidies to sustain economies limited by resources, isolation, and an emigrating population, drawn to New Zealand and Australia by the attractions of a western lifestyle.

Palau

Palau is an archipelago of over 200 islands, only eight of which are inhabited. It was the last remaining UN trust territory in the Pacific, controlled by the US until 1994, when it became independent. The economy operates on a subsistence level, with coconuts and cassava the principal crops. Fishing licenses and tourism provide foreign currency.

SCALE 1:750,000

SCALE 1:6,125,000

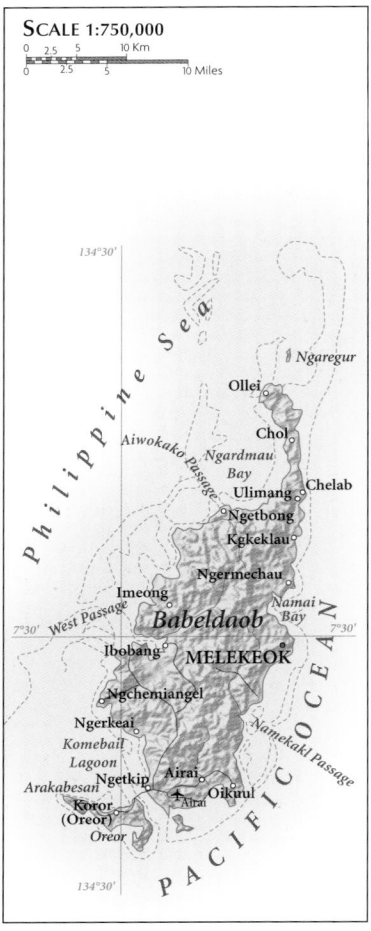

Guam (to US)

Lying at the southern end of the Mariana Islands, Guam is an important US military base and tourist destination. Social and political life is dominated by the indigenous Chamorro, who make up just under half the population, although the increasing prevalence of western culture threatens Guam's traditional social stability.

◄ The tranquility of these coastal lagoons, at Inarajan in southern Guam, belies the fact that the island lies in a region where typhoons are common.

SCALE 1:840,000

Northern Mariana Islands (to US)

A US Commonwealth territory, the Northern Marianas comprise the whole of the Mariana archipelago except for Guam. The islands retain their close links with the US and continue to receive American aid. Tourism, though bringing in much-needed revenue, has speeded the decline of the traditional subsistence economy. Most of the population lives on Saipan.

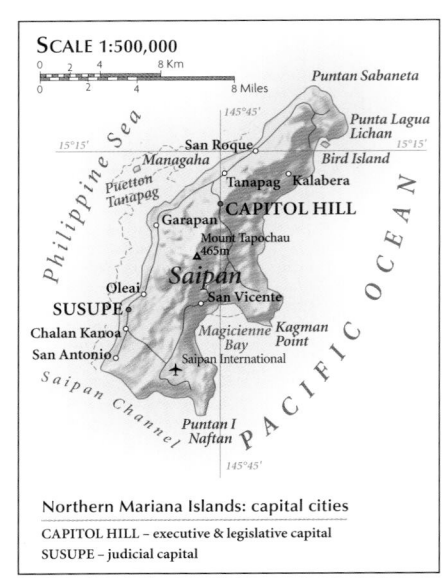

SCALE 1:500,000

Northern Mariana Islands: capital cities
CAPITOL HILL – executive & legislative capital
SUSUPE – judicial capital

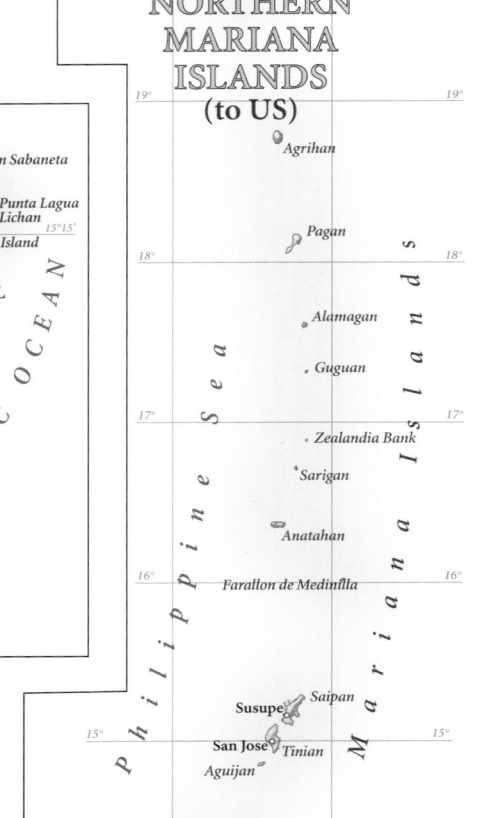

▲ The Palau Islands have numerous hidden lakes and lagoons. These sustain their own ecosystems which have developed in isolation. This has produced adaptations in the animals and plants that are often unique to each lake.

SCALE 1:5,000,000

Micronesia

A mixture of high volcanic islands and low-lying coral atolls, the Federated States of Micronesia include all the Caroline Islands except Palau. Pohnpei, Kosrae, Chuuk, and Yap are the four main island cluster states, each of which has its own language, with English remaining the official language. Nearly half the population is concentrated on Pohnpei, the largest island. Independent since 1986, the islands continue to receive considerable aid from the US which supplements an economy based primarily on fishing and copra processing.

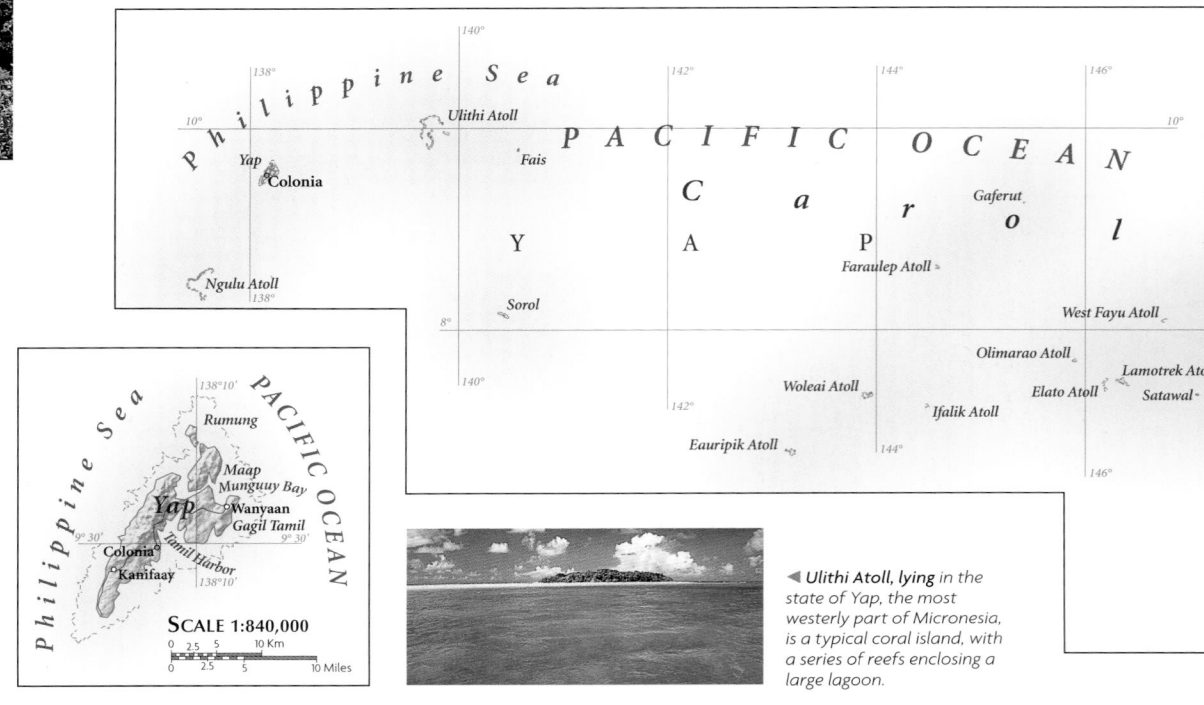

◄ Ulithi Atoll, lying in the state of Yap, the most westerly part of Micronesia, is a typical coral island, with a series of reefs enclosing a large lagoon.

SCALE 1:840,000

Marshall Islands

A group of 34 widely-scattered atolls in the central Pacific Ocean, the Marshall Islands include some of the largest atolls in the world, formed from low coral islands with sandy beaches and enclosing vast lagoons. Formerly under US protection as part of the UN Trust Territory of the Pacific Islands, and including the former US nuclear testing sites of Bikini atoll and Enewetak Atoll, the Marshall Islands became self-governing in 1979. The economy is reliant on US aid and on the rent paid by the US for its missile base on Kwajalein atoll.

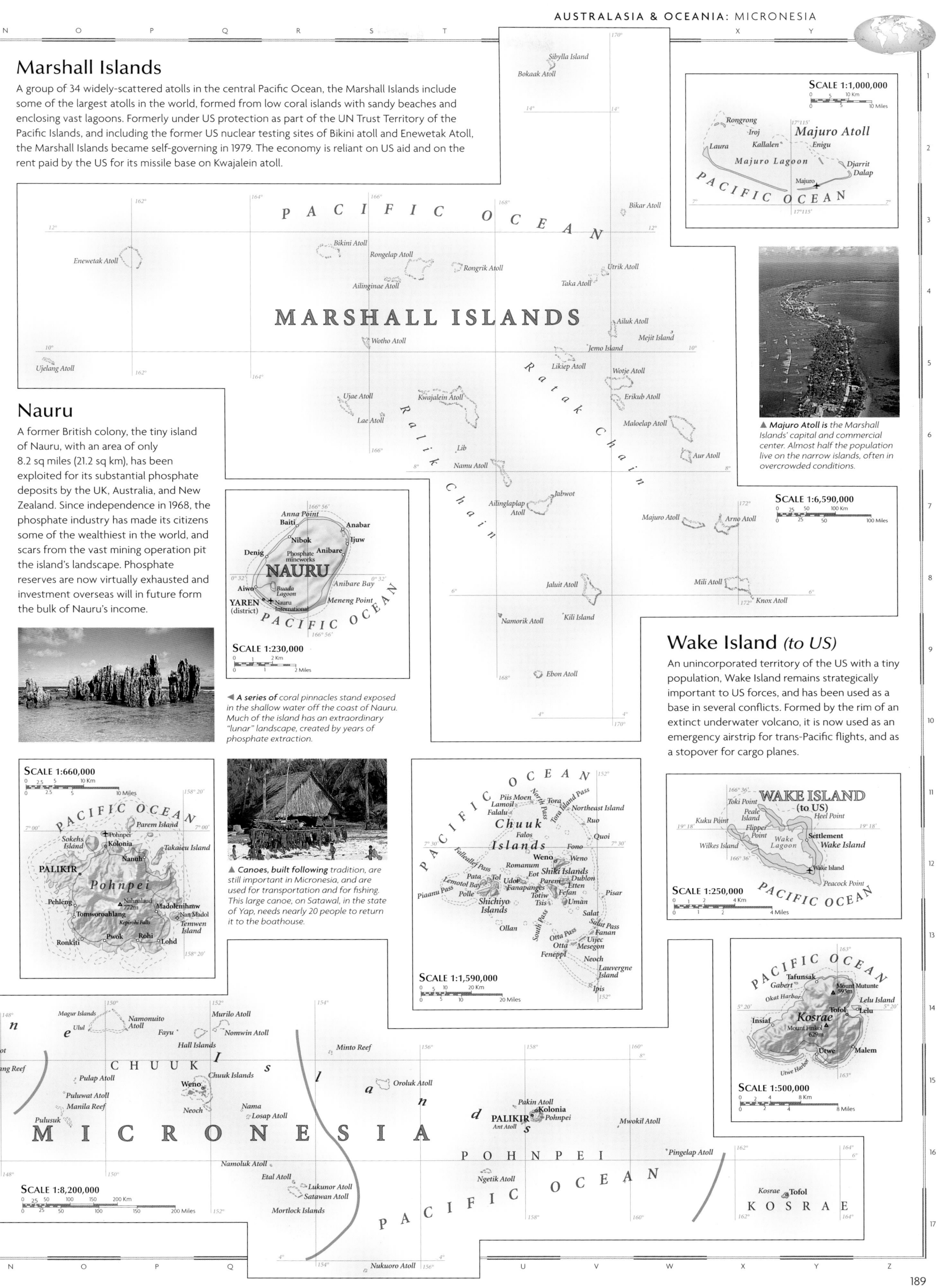

▲ *Majuro Atoll is* the Marshall Islands' capital and commercial center. Almost half the population live on the narrow islands, often in overcrowded conditions.

Nauru

A former British colony, the tiny island of Nauru, with an area of only 8.2 sq miles (21.2 sq km), has been exploited for its substantial phosphate deposits by the UK, Australia, and New Zealand. Since independence in 1968, the phosphate industry has made its citizens some of the wealthiest in the world, and scars from the vast mining operation pit the island's landscape. Phosphate reserves are now virtually exhausted and investment overseas will in future form the bulk of Nauru's income.

◄ *A series of* coral pinnacles stand exposed in the shallow water off the coast of Nauru. Much of the island has an extraordinary "lunar" landscape, created by years of phosphate extraction.

Wake Island *(to US)*

An unincorporated territory of the US with a tiny population, Wake Island remains strategically important to US forces, and has been used as a base in several conflicts. Formed by the rim of an extinct underwater volcano, it is now used as an emergency airstrip for trans-Pacific flights, and as a stopover for cargo planes.

▲ *Canoes, built following* tradition, are still important in Micronesia, and are used for transportation and for fishing. This large canoe, on Satawal, in the state of Yap, needs nearly 20 people to return it to the boathouse.

189

Polynesia

KIRIBATI, TUVALU, Cook Islands, Easter Island, French Polynesia, Niue, Pitcairn Islands, Tokelau, Wallis & Futuna

The numerous island groups of Polynesia lie to the east of Australia, scattered over a vast area in the south Pacific. The islands are a mixture of low-lying coral atolls, some of which enclose lagoons, and the tips of great underwater volcanoes. The populations on the islands are small, and most people are of Polynesian origin, as are the Maori of New Zealand. Local economies remain simple, relying mainly on subsistence crops, mineral deposits, many now exhausted, fishing, and tourism.

SCALE 1:1,000,000

Kiribati

A former British colony, Kiribati became independent in 1979. Banaba's phosphate deposits ran out in 1980, following decades of exploitation by the British. Economic development remains slow and most agriculture is at a subsistence level, though coconuts provide export income, and underwater agriculture is being developed.

► With the exception of Banaba all the islands in Kiribati's three groups are low-lying, coral atolls. This aerial view shows the sparsely vegetated islands, intercut by many small lagoons.

Tuvalu

A chain of nine coral atolls, 360 miles (579 km) long with a land area of just over 9 sq miles (23 sq km), Tuvalu is one of the world's smallest and most isolated states. As the Ellice Islands, Tuvalu was linked to the Gilbert Islands (now part of Kiribati) as a British colony until independence in 1978. Politically and socially conservative, Tuvaluans live by fishing and subsistence farming.

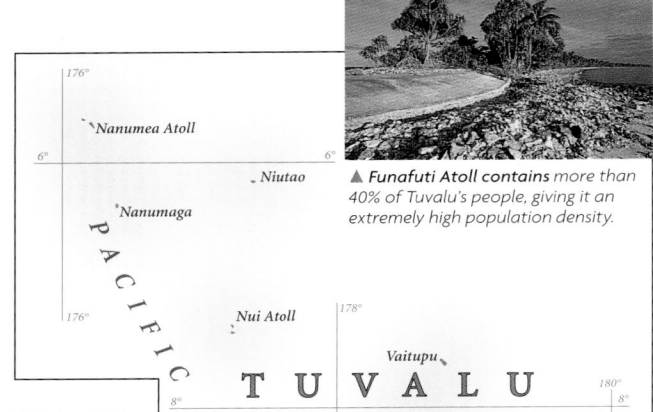

▲ Funafuti Atoll contains more than 40% of Tuvalu's people, giving it an extremely high population density.

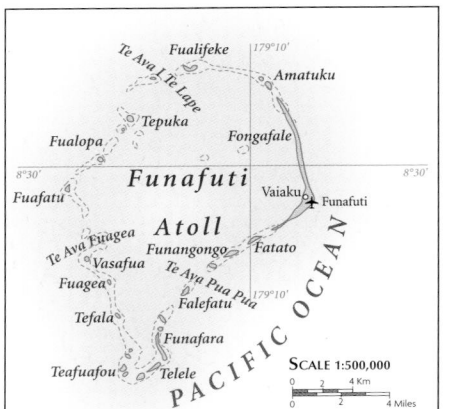

SCALE 1:500,000

SCALE 1:6,125,000

Tokelau (to New Zealand)

A low-lying coral atoll, Tokelau is a dependent territory of New Zealand with few natural resources. Although a 1990 cyclone destroyed crops and infrastructure, a tuna cannery and the sale of fishing licenses have raised revenue and a catamaran link between the islands has increased their tourism potential. Tokelau's small size and economic weakness makes independence from New Zealand unlikely.

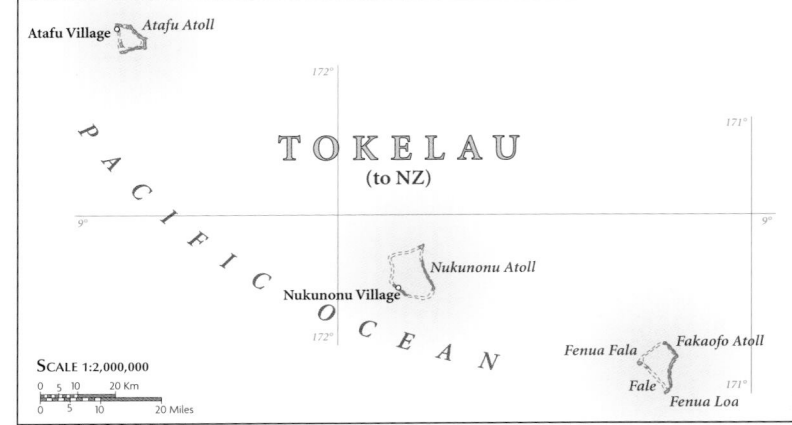

▲ Fishermen cast their nets to catch small fish in the shallow waters off Atafu Atoll, the most westerly island in Tokelau.

SCALE 1:2,000,000

Wallis & Futuna (to France)

In contrast to other French overseas territories in the south Pacific, the inhabitants of Wallis and Futuna have shown little desire for greater autonomy. A subsistence economy produces a variety of tropical crops, while foreign currency remittances come from expatriates and from the sale of licenses to Japanese and Korean fishing fleets.

SCALE 1:1,000,000

SCALE 1:1,000,000

Niue (to New Zealand)

Niue, the world's largest coral island, is self-governing but exists in free association with New Zealand. Tropical fruits are grown for local consumption; tourism and the sale of postage stamps provide foreign currency. The lack of local job prospects has led more than 10,000 Niueans to emigrate to New Zealand, which has now invested heavily in Niue's economy in the hope of reversing this trend.

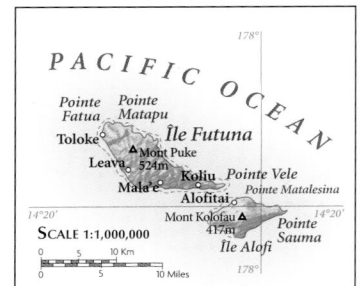

▲ Palm trees fringe the white sands of a beach on Aitutaki in the Southern Cook Islands, where tourism is of increasing economic importance.

Cook Islands (to New Zealand)

A mixture of coral atolls and volcanic peaks, the Cook Islands achieved self-government in 1965 but exist in free association with New Zealand. A diverse economy includes pearl and giant clam farming, and an ostrich farm, plus tourism and banking. A 1991 friendship treaty with France provides for French surveillance of territorial waters.

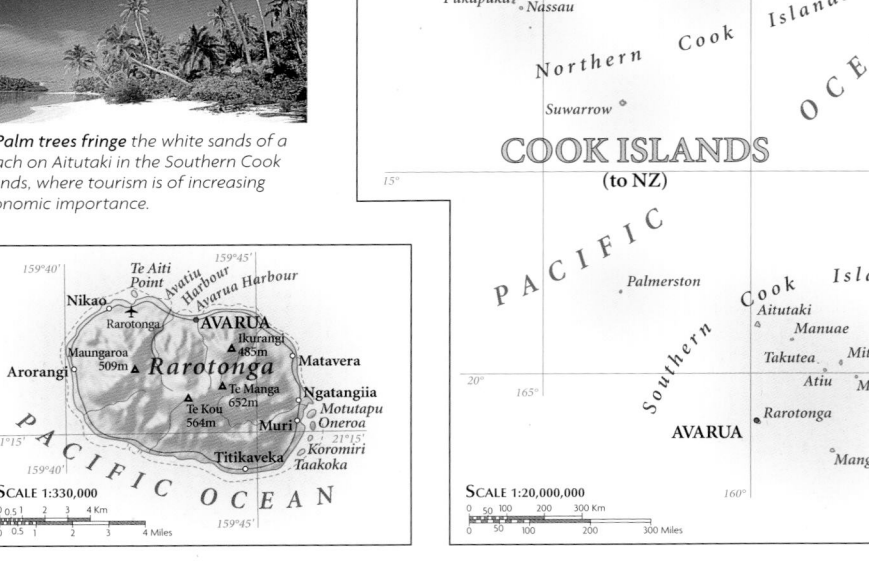

SCALE 1:20,000,000

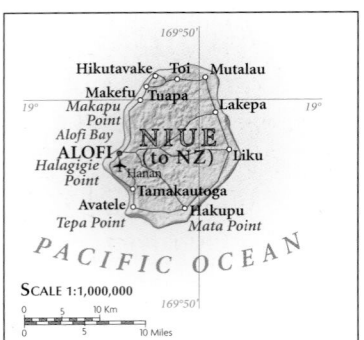

SCALE 1:1,000,000

▲ Waves have cut back the original coastline, exposing a sandy beach, near Mutalau in the northeast corner of Niue.

SCALE 1:330,000

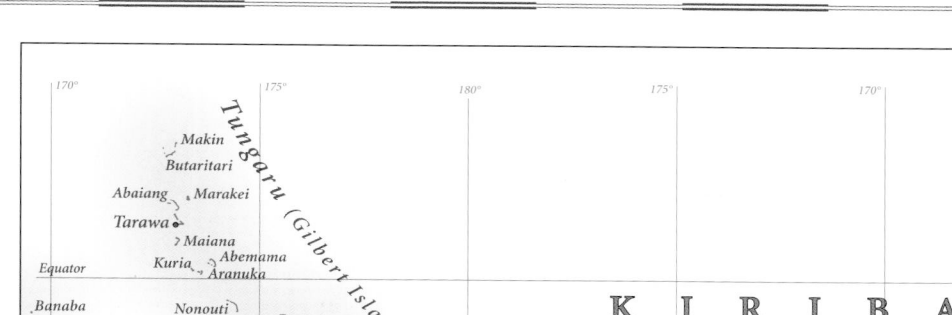

French Polynesia *(to France)*

The 130 islands of French Polynesia cover 4 million sq miles (10.5 million sq km). Nearly 75% of the people live on Tahiti. The use of Mururoa as a nuclear testing site by the French military transformed the economy, creating many jobs. The end of testing led to calls from the Polynesian majority for greater autonomy from France, the rebuilding of indigenous trade, and a reduction in tourism to stop the erosion of the islands' traditional culture.

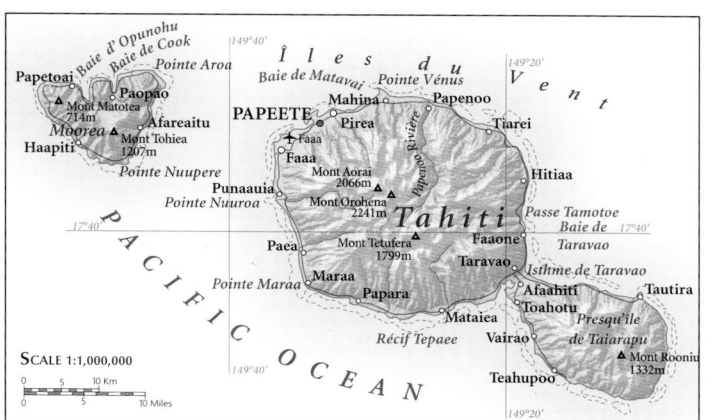

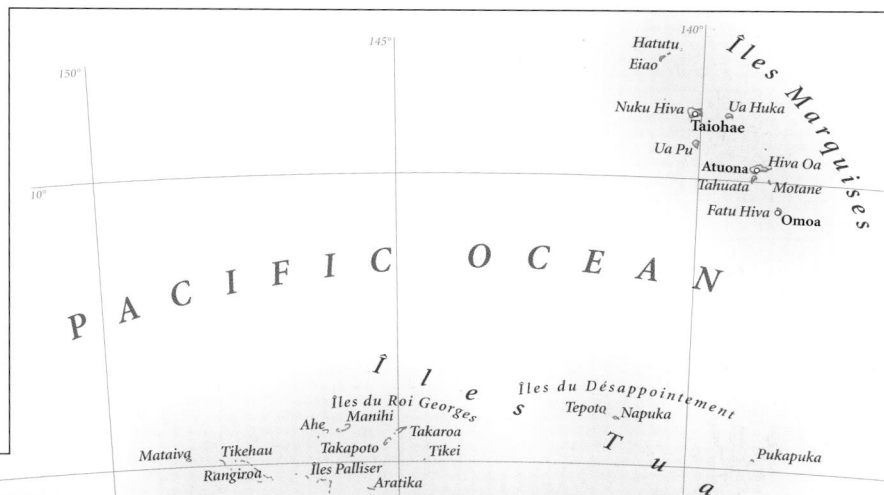

◀ *The traditional Tahitian welcome for visitors, who are greeted by parties of canoes, has become a major tourist attraction.*

Pitcairn Group of Islands *(to UK)*

Britain's most isolated dependency, Pitcairn Island was first populated by mutineers from the HMS *Bounty* in 1790. Emigration is further depleting the already limited gene pool of the island's inhabitants, with associated social and health problems. Barter, fishing and subsistence farming form the basis of the economy whilst offshore mineral exploitation may boost the economy in future.

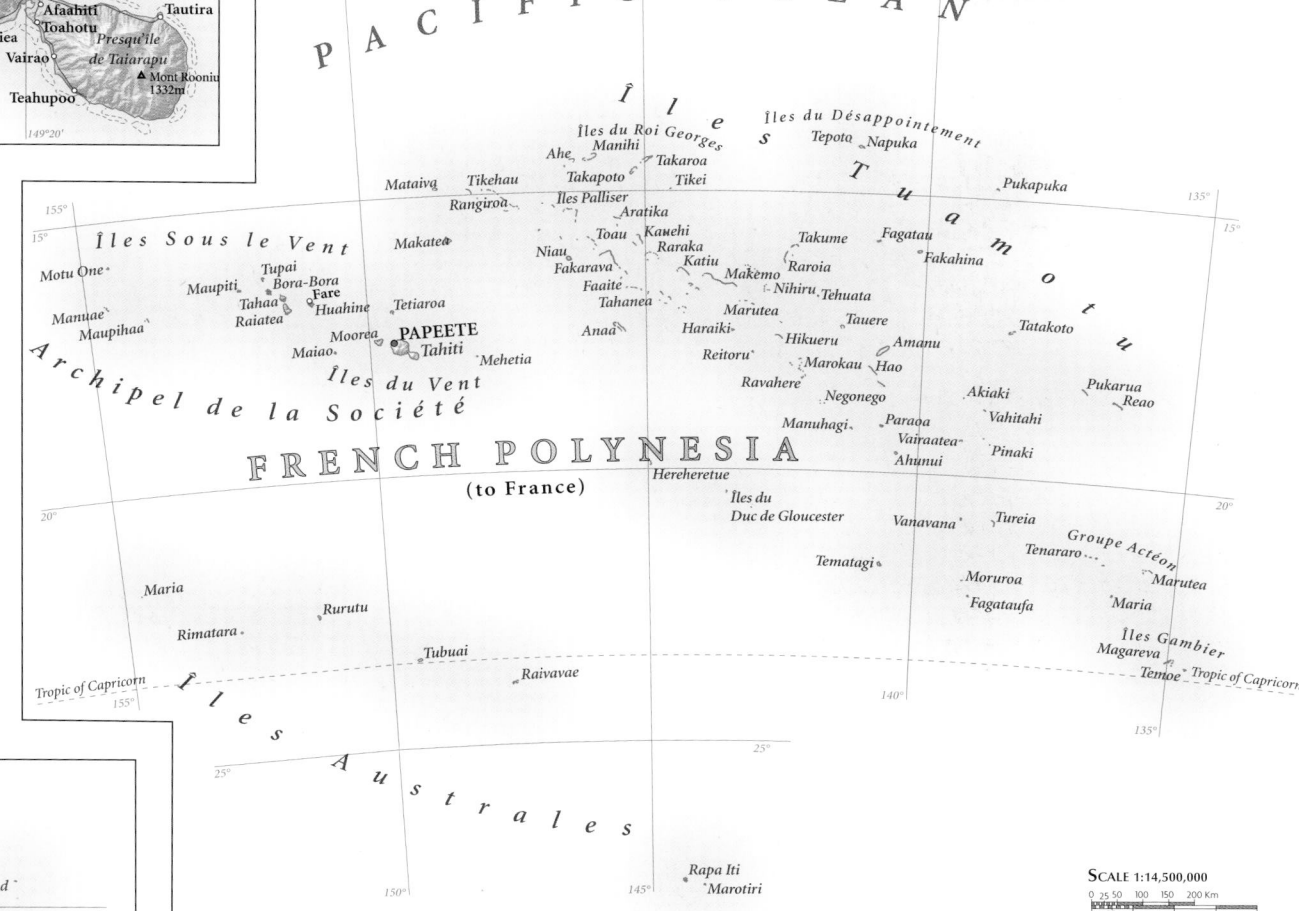

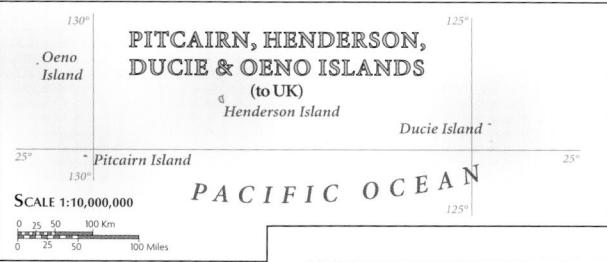

◀ *The Pitcairn Islanders rely on regular airdrops from New Zealand and periodic visits by supply vessels to provide them with basic commodities.*

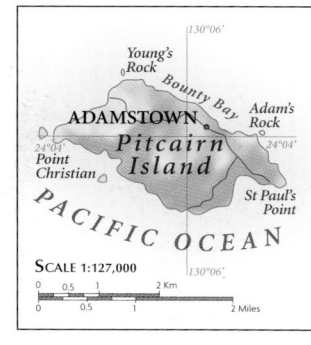

Easter Island *(to Chile)*

One of the most easterly islands in Polynesia, Easter Island *(Isla de Pascua)* – also known as Rapa Nui, is part of Chile. The mainly Polynesian inhabitants support themselves by farming, which is mainly of a subsistence nature, and includes cattle rearing and crops such as sugar cane, bananas, corn, gourds, and potatoes. In recent years, tourism has become the most important source of income and the island sustains a small commercial airport.

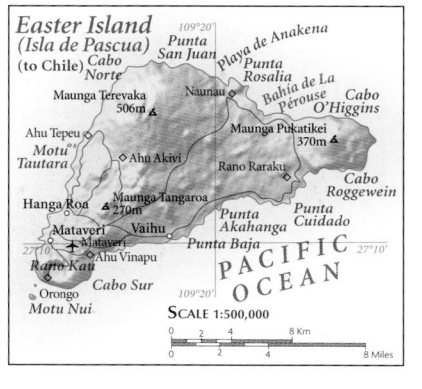

▲ *The Naunau, a series of huge stone statues overlook Playa de Anakena, on Easter Island. Carved from a soft volcanic rock, they were erected between 400 and 900 years ago.*

191

The Pacific Ocean

The Pacific is the world's largest and deepest ocean. It is nearly twice the area of the Atlantic and contains almost three times as much water. The ocean is dotted with islands and surrounded by some of the world's most populous states; over half the world's population lives on its shores. The Pacific is bordered by active plate margins known as the "Ring of Fire," causing earthquakes and tsunamis, and creating volcanic islands and subterranean mountain chains. The largest underwater mountains break the surface as island arcs. The fisheries of the Pacific are some of the most productive in the world and provide a vital resource for many of the Pacific islands. Since the Second World War there has been a shift in trading patterns, with a considerable growth in trade between the US and the countries of the Pacific Rim.

The Ring of Fire

The active plate margins surrounding the Pacific have created numerous land and island volcanoes along its border. The actual basin of the Pacific is made up of a number of separate tectonic plates which move away from each other, colliding with other plates. When they collide, the oceanic plates, being thinner, are forced beneath the thicker continental plates, forming deep ocean trenches and high ridges. These collision zones are known as subduction zones and are characterized by intense seismic and volcanic activity.

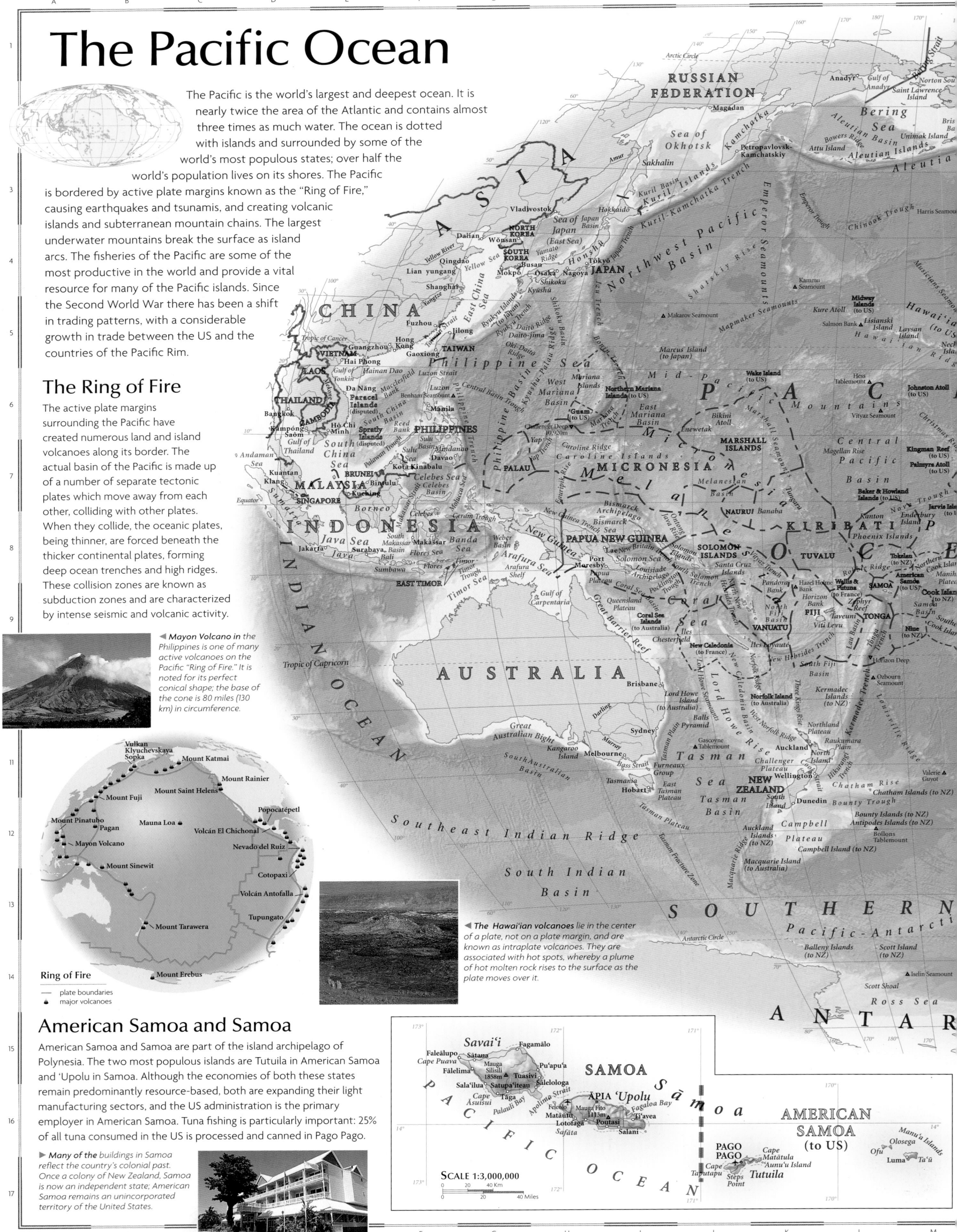

◄ *Mayon Volcano in the Philippines is one of many active volcanoes on the Pacific "Ring of Fire." It is noted for its perfect conical shape; the base of the cone is 80 miles (130 km) in circumference.*

Ring of Fire
— plate boundaries
• major volcanoes

◄ *The Hawai'ian volcanoes lie in the center of a plate, not on a plate margin, and are known as intraplate volcanoes. They are associated with hot spots, whereby a plume of hot molten rock rises to the surface as the plate moves over it.*

American Samoa and Samoa

American Samoa and Samoa are part of the island archipelago of Polynesia. The two most populous islands are Tutuila in American Samoa and 'Upolu in Samoa. Although the economies of both these states remain predominantly resource-based, both are expanding their light manufacturing sectors, and the US administration is the primary employer in American Samoa. Tuna fishing is particularly important: 25% of all tuna consumed in the US is processed and canned in Pago Pago.

▶ *Many of the buildings in Samoa reflect the country's colonial past. Once a colony of New Zealand, Samoa is now an independent state; American Samoa remains an unincorporated territory of the United States.*

SCALE 1:3,000,000

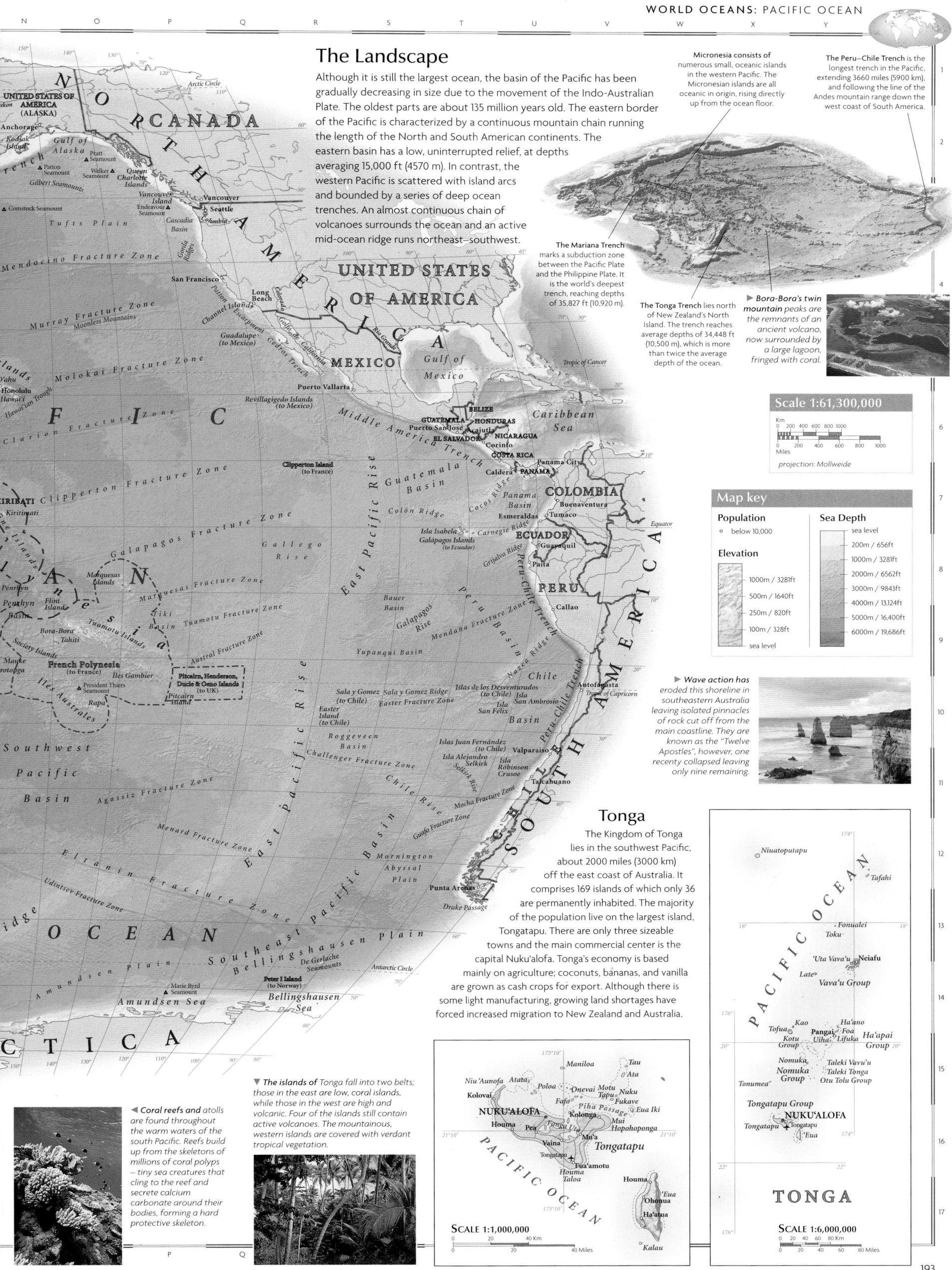

The Landscape

Although it is still the largest ocean, the basin of the Pacific has been gradually decreasing in size due to the movement of the Indo-Australian Plate. The oldest parts are about 135 million years old. The eastern border of the Pacific is characterized by a continuous mountain chain running the length of the North and South American continents. The eastern basin has a low, uninterrupted relief, at depths averaging 15,000 ft (4570 m). In contrast, the western Pacific is scattered with island arcs and bounded by a series of deep ocean trenches. An almost continuous chain of volcanoes surrounds the ocean and an active mid-ocean ridge runs northeast–southwest.

Micronesia consists of numerous small, oceanic islands in the western Pacific. The Micronesian islands are all oceanic in origin, rising directly up from the ocean floor.

The Peru–Chile Trench is the longest trench in the Pacific, extending 3660 miles (5900 km), and following the line of the Andes mountain range down the west coast of South America.

The Mariana Trench marks a subduction zone between the Pacific Plate and the Philippine Plate. It is the world's deepest trench, reaching depths of 35,827 ft (10,920 m).

The Tonga Trench lies north of New Zealand's North Island. The trench reaches average depths of 34,448 ft (10,500 m), which is more than twice the average depth of the ocean.

▶ **Bora-Bora's twin mountain** peaks are the remnants of an ancient volcano, now surrounded by a large lagoon, fringed with coral.

Scale 1:61,300,000

projection: Mollweide

Map key

Population
○ below 10,000

Elevation
1000m / 3281ft
500m / 1640ft
250m / 820ft
100m / 328ft
sea level

Sea Depth
sea level
200m / 656ft
1000m / 3281ft
2000m / 6562ft
3000m / 9843ft
4000m / 13,124ft
5000m / 16,400ft
6000m / 19,686ft

▶ **Wave action** has eroded this shoreline in southeastern Australia leaving isolated pinnacles of rock cut off from the main coastline. They are known as the "Twelve Apostles", however, one recently collapsed leaving only nine remaining.

Tonga

The Kingdom of Tonga lies in the southwest Pacific, about 2000 miles (3000 km) off the east coast of Australia. It comprises 169 islands of which only 36 are permanently inhabited. The majority of the population live on the largest island, Tongatapu. There are only three sizeable towns and the main commercial center is the capital Nuku'alofa. Tonga's economy is based mainly on agriculture; coconuts, bananas, and vanilla are grown as cash crops for export. Although there is some light manufacturing, growing land shortages have forced increased migration to New Zealand and Australia.

◀ **Coral reefs and atolls** are found throughout the warm waters of the south Pacific. Reefs build up from the skeletons of millions of coral polyps – tiny sea creatures that cling to the reef and secrete calcium carbonate around their bodies, forming a hard protective skeleton.

▼ **The islands of** Tonga fall into two belts; those in the east are low, coral islands, while those in the west are high and volcanic. Four of the islands still contain active volcanoes. The mountainous, western islands are covered with verdant tropical vegetation.

SCALE 1:1,000,000

TONGA

SCALE 1:6,000,000

Antarctica

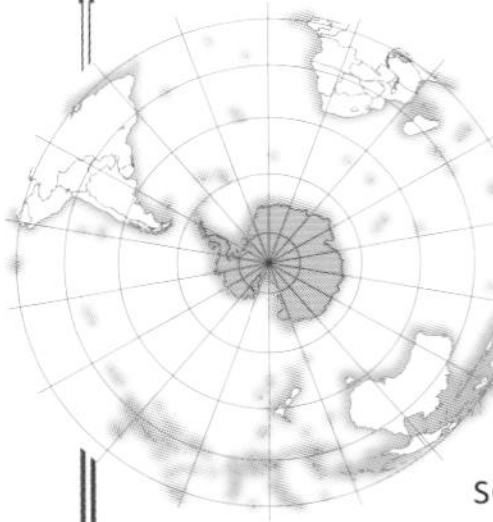

The ice-covered continent of Antarctica, which is the Earth's most southerly region, has drawn explorers and entrepreneurs seeking challenge and riches in its wintry lands for over 200 years. The extreme climate has deterred any large-scale settlement of the continent, and though commercial hunters built outposts in the past, habitation is now limited to scientific bases. The Antarctic Treaty, which came into force in 1961, provides for international governance and scientific cooperation in place of potential territorial conflict.

Resources

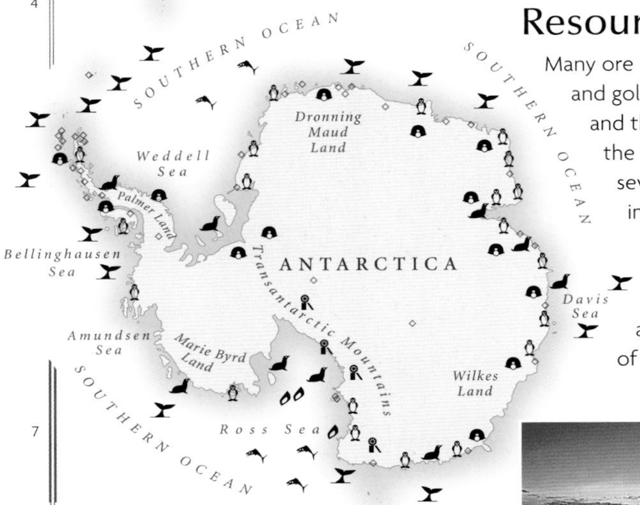

Many ore minerals, including iron and gold, are found in the Antarctic, and there are also coal reserves in the Transantarctic Mountains. The severe conditions and environmental importance of the region mean that exploitation of potential mineral resources is both uneconomic and undesirable. The unique wildlife and landscape draw a small number of tourists annually.

Resources (including wildlife)

- coal
- fish
- minerals
- oil & gas
- penguins
- seals
- whales
- polar research base

◁ Most settlements in Antarctica are research bases such as this one at Rothera on Adelaide Island, although there is a small Chilean settlement on King George Island.

The landscape

There are two distinct parts to Antarctica: West Antarctica, a series of ice-covered, mountainous islands, joined together by the ice; and the high plateau of East Antarctica. The Ross Sea and the Weddell Sea are outliers of the Southern Ocean – deep bays partially covered by thick ice shelves.

Grease ice   Pancake ice   Sea-ice sheet   Ice floe

◁ On Elephant Island, the coast is edged by glaciers, although the land is not permanently covered by ice.

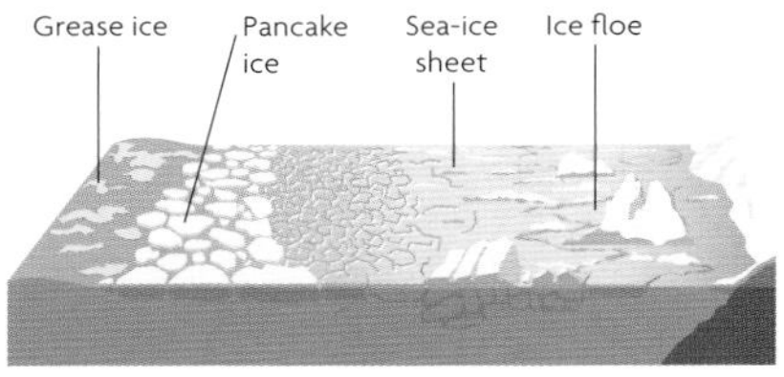

▲ Pack ice forms out at sea in freezing temperatures. At the outer limits, grease ice congeals on the surface of the ocean. This is then spun around by wind and waves into irregular "pancakes," freezing and breaking up several times before bonding together again to form sea-ice sheets, which finally cement into enormous ice floes.

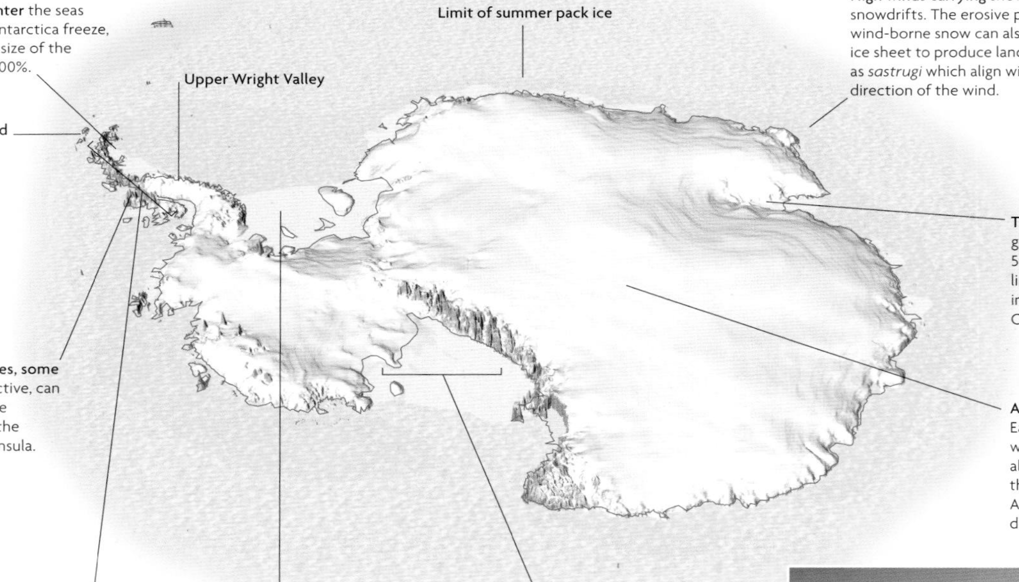

Limit of winter pack ice

Upper Wright Valley

Limit of summer pack ice

Elephant Island

During the winter the seas surrounding Antarctica freeze, increasing the size of the continent by 100%.

Many volcanoes, some of them still active, can be found in the mountains of the Antarctic Peninsula.

High winds carrying snow form huge snowdrifts. The erosive power of the wind-borne snow can also sculpt the ice sheet to produce landforms known as sastrugi which align with the direction of the wind.

The mountainous Antarctic Peninsula is formed of rocks 65–225 million years old, overlain by more recent rocks and glacial deposits. It is connected to the Andes in South America by a submarine ridge.

Nearly half – 44% – of the Antarctic coastline is bounded by ice shelves, like the Ronne Ice Shelf, which float on the Ocean. These are joined to the inland ice sheet by dome-shaped ice "rises."

More than 30% of Antarctic ice is contained in the Ross Ice Shelf.

The Lambert Glacier is the largest glacier system in the world, up to 50 miles (80 km) wide at its seaward limit, and reaching 180 miles (300 km) into the interior by way of the Prince Charles Mountains.

Antarctica is the highest continent on Earth, because of the great thickness of ice which overlays the land. In places the ice alone can each up to 15,700 ft (4800 m) thick. Much of the basement rock of west Antarctica lies below sea level, pushed down by the weight of the ice.

◁ The barren, flat-bottomed Upper Wright Valley was once filled by a glacier, but is now dry, strewn with boulders and pebbles. In some dry valleys, there has been no rain for over 2 million years.

▲ Large colonies of seabirds live in the extremely harsh Antarctic climate. The Emperor penguins seen here, the smaller Adélie penguin, the Antarctic petrel, and the South Polar skua are the only birds that breed exclusively on the continent.

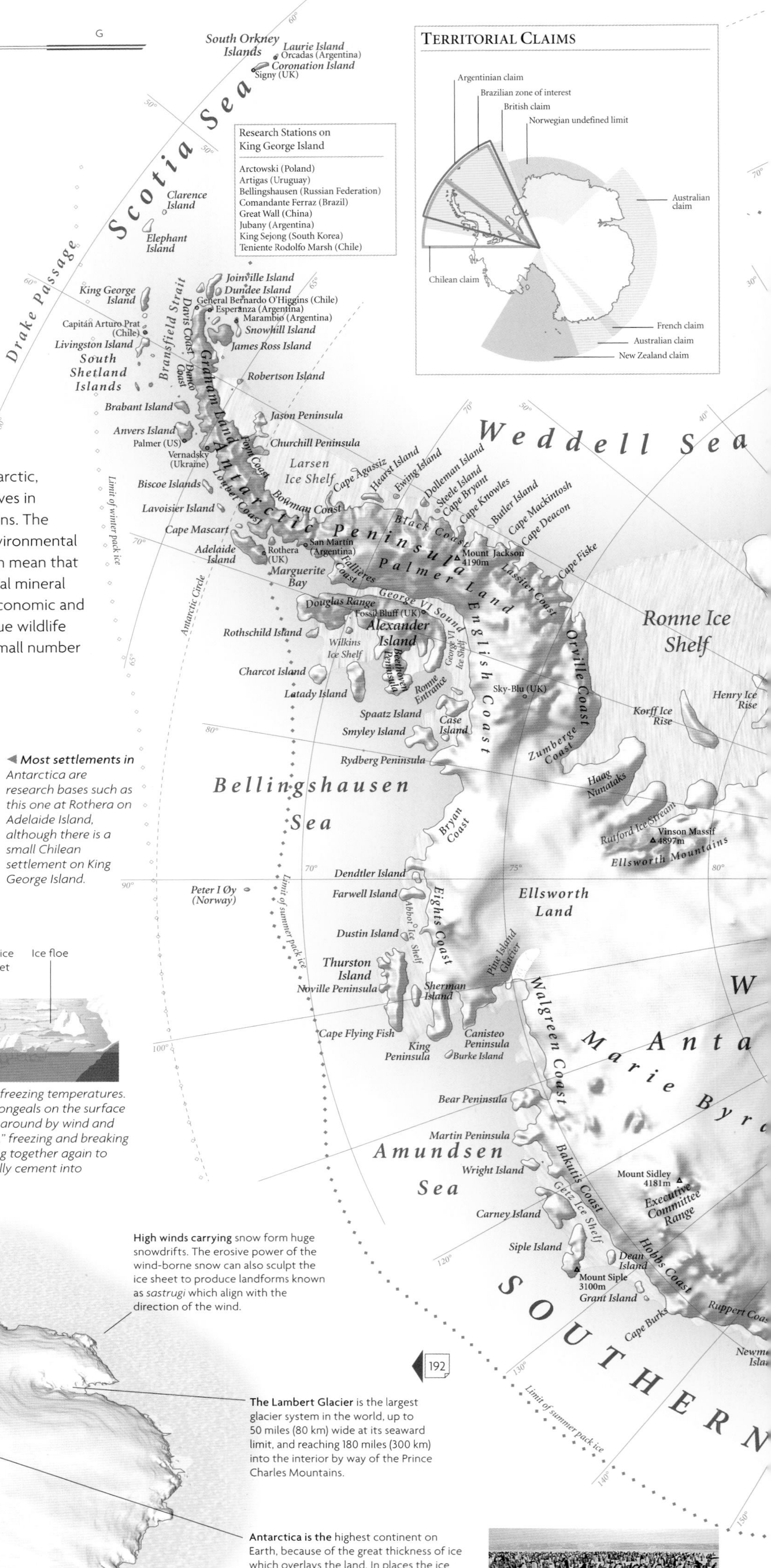

Research Stations on King George Island

Arctowski (Poland)
Artigas (Uruguay)
Bellingshausen (Russian Federation)
Comandante Ferraz (Brazil)
Great Wall (China)
Jubany (Argentina)
King Sejong (South Korea)
Teniente Rodolfo Marsh (Chile)

TERRITORIAL CLAIMS

Argentinian claim
Brazilian zone of interest
British claim
Norwegian undefined limit
Australian claim
Chilean claim
French claim
Australian claim
New Zealand claim

South Orkney Islands
Laurie Island
Orcadas (Argentina)
Coronation Island
Signy (UK)

Scotia Sea

Clarence Island

Elephant Island

Drake Passage

King George Island

Joinville Island
Dundee Island
General Bernardo O'Higgins (Chile)
Esperanza (Argentina)
Marambio (Argentina)
Snowhill Island
James Ross Island

Capitán Arturo Prat (Chile)
Livingston Island

South Shetland Islands

Robertson Island

Brabant Island

Anvers Island
Palmer (US)
Vernadsky (Ukraine)

Bransfield Strait
Graham Land
Davis Coast
Danco Coast

Larsen Ice Shelf

Jason Peninsula

Churchill Peninsula

Cape Agassiz
Hearst Island
Ewing Island
Dolleman Island
Steele Island
Cape Bryant
Black Coast
Cape Knowles
Butler Island
Cape Mackintosh
Cape Deacon

Weddell Sea

Biscoe Islands

Bowman Coast

Lavoisier Island

Cape Mascart

Rothera (UK)
San Martín (Argentina)
Marguerite Bay
Adelaide Island

Antarctic Circle

Rothschild Island

Douglas Range
Fossil Bluff (UK)
Alexander Island
Wilkins Ice Shelf

Charcot Island

Latady Island

Spaatz Island

Smyley Island

Case Island

Rydberg Peninsula

Bellingshausen Sea

Peter I Øy (Norway)

Dendtler Island

Farwell Island

Dustin Island

Thurston Island
Noville Peninsula

Cape Flying Fish

King Peninsula

Bear Peninsula

Martin Peninsula
Wright Island

Amundsen Sea

Carney Island

Siple Island

Antarctic Peninsula
Palmer Land
English Coast
George VI Sound

Ronne Entrance

Sky-Blu (UK)

Zumberge Coast

Orville Coast

Ronne Ice Shelf

Cape Fiske

Henry Ice Rise

Korff Ice Rise

Haag Nunataks

Rutford Ice Stream
Vinson Massif
4897m
Ellsworth Mountains

Ellsworth Land

Bryan Coast

Eights Coast
Abbot Ice Shelf
Canisteo Peninsula
Burke Island

Sherman Island

Walgreen Coast
Pine Island Glacier

Marie Byrd

Bakutis Coast

Mount Sidley
4181m
Executive Committee Range

Getz Ice Shelf
Dean Island
Mount Siple
3100m
Grant Island
Hobbs Coast

Cape Burks

Ruppert Coast
Newme Isla

SOUTHERN

192

Mount Jackson
4190m

ANTARCTICA

SOUTHERN OCEAN

Dronning Maud Land

Weddell Sea

Palmer Land

Bellingshausen Sea

Transantarctic Mountains

Amundsen Sea

Marie Byrd Land

Ross Sea

Wilkes Land

Davis Sea

◀ **The sun sets** over the Antarctic Peninsula for more than six months during the winter. However, there are more hours of sunshine during the brief Antarctic summer than most equatorial countries experience in a whole year.

▲ **Immense, flat-topped icebergs** are formed when blocks of ice break away from the main ice sheet. Though the exposed area is enormous, the volume of ice concealed beneath the water may be many times greater.

Scale 1:14,750,000

projection: Lambert Azimuthal Equal Area

Map key

Elevation

ice cap

ice shelf

exposed land

The Arctic

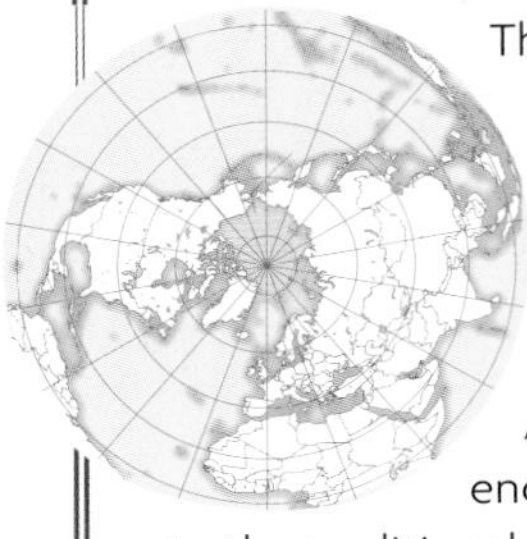

Three continents, Asia, North America, and Europe, reach into the Arctic Circle at their northernmost limits, almost entirely encircling the Arctic Ocean. Despite the region's extraordinarily harsh climate, it has been inhabited for thousands of years by peoples such as the European Lapps, the Russian Nenet, and the North American Inuit, who draw a living from fishing, herding, and hunting. More recently, particularly in the Russian Arctic, opportunities to exploit oil and other mineral reserves have encouraged immigration. Pollution of the Arctic's unique ecology and damage to the traditional lifestyles of many native peoples have been the unfortunate results of this activity, and international cooperation is needed to safeguard the future of the region.

Map key

Population

- ■ above 5 million
- ◼ 1 million to 5 million
- ◉ 500,000 to 1 million
- ◎ 100,000 to 500,000
- ⊕ 50,000 to 100,000
- ○ 10,000 to 50,000
- ∘ below 10,000

Sea depth

	Sea level
	200m / 656ft
	1000m / 3281ft
	2000m / 6562ft
	3000m / 9843ft
	4000m / 13,124ft
	5000m / 16,400ft
	6000m / 19,686ft

Scale 1:21,000,000

projection: Lambert Azimuthal Equal Area

▲ *Windblown snow etches deep patterns in the ice sheet known as sastrugi. They align with the direction of the wind*

Resources

Large quantities of coal, oil, and natural gas are to be found in the basins of the Arctic Ocean, and in northern Canada, Alaska, and the Russian Federation. The cost and difficulty of extraction and, more recently, awareness of damage to the environment, have limited exploitation to coastal regions. The unfrozen waters have stocks of fish including cod, flounder, and haddock. Quotas have now been put in place to restrict the number of fish caught annually. Reindeer are herded in large numbers by many of the native Arctic peoples. Most grain and vegetables are imported from elsewhere.

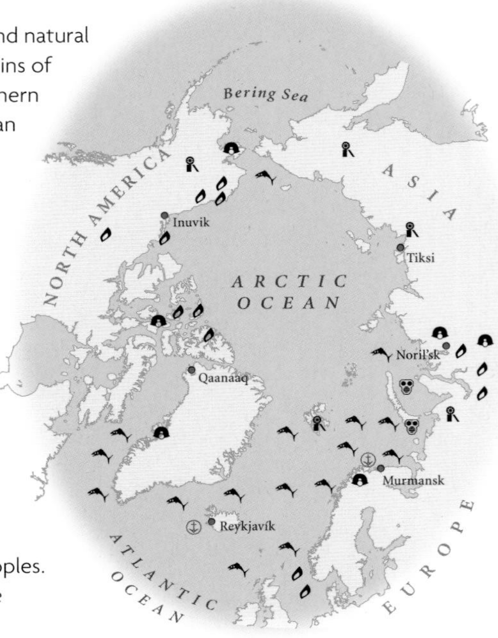

▲ *Icebreakers are ships with specially strengthened hulls, designed to break a path through the ice. They are used to keep important routes open during the winter, when falling temperatures cause much of the Arctic Ocean to freeze over.*

Resources

- ⚒ coal
- ⌐ fish
- ⚒ mining
- ◗ oil & gas
- ☢ radioactive contamination
- • major towns
- ⊕ major ports

The landscape

The Arctic Ocean comprises two large ocean basins divided by three submarine ridges, the greatest of which, the Lomonosov Ridge, is a huge underwater mountain range which has an average height of more than 10,000 ft (3000 m). The lands which encircle the Arctic Ocean are underlain by great shield areas of ancient rocks, which were heavily glaciated during the last Ice Age.

◀ *Icebergs are constantly broken up and reshaped by wind and the oceans. This flat-topped iceberg has been undercut, leaving a craggy ice cliff.*

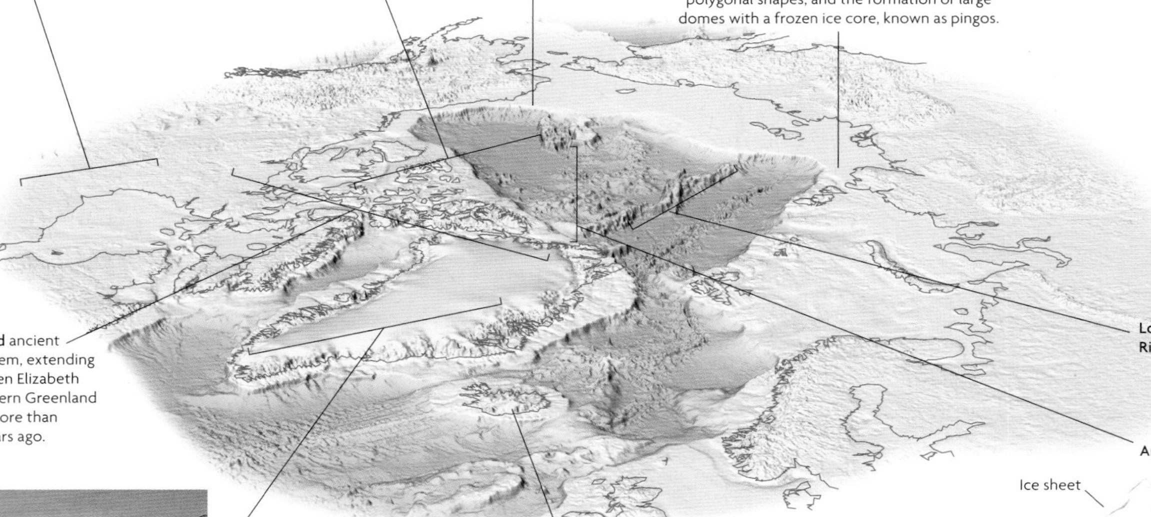

The Canadian Shield underlies almost all of the Canadian Arctic. It is a very stable plateau of ancient rock, now covered by glacial lakes and sediment, which supports tundra vegetation.

The Arctic Ocean is the world's smallest ocean with a total area of 5,440,000 sq miles (15,100,000 sq km).

At a latitude of more than 75° N, the Arctic Ocean is almost permanently covered by pack ice, though high winds and the movement of the seas may cause the ice to crack and break up.

In the more southerly reaches of the Arctic, like Siberia, much of the land is covered by permafrost. In the summer, higher temperatures warm the frozen ground, causing a number of typical phenomena. These include solifluction, the fast downhill movement of top soil layers; freeze/thaw activity, which patterns the ground into regular polygonal shapes, and the formation of large domes with a frozen ice core, known as pingos.

A complex and ancient mountain system, extending from the Queen Elizabeth Islands to eastern Greenland was formed more than 245 million years ago.

Lomonosov Ridge

Arctic ice shelf

◀ *Much of Greenland is covered by a massive ice sheet more than 650,000 sq miles (1,683,400 sq km) in extent. The weight of the ice has depressed the central land area to form a basin lying more than 1000 ft (300 m) below sea level. Only at the edges of the island is bare rock visible.*

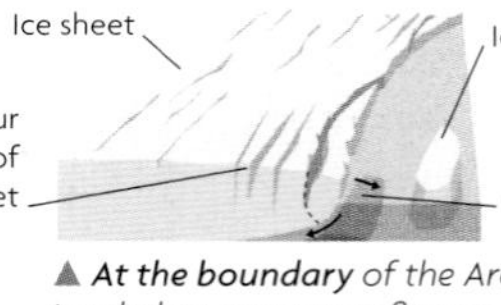

Iceland has five major glaciers, sustained by heavy snowfall. Parts of the ice cap cover active volcanoes, such as Bárdharbunga, which periodically erupt causing the melted ice to form a great lake at the glacier margins.

Ice sheet — Iceberg

Crevasses occur at the edge of the ice sheet — Sea water melts the edge of the ice sheet

▲ *At the boundary of the Arctic ice shelves, sea water flows under the ice causing melting and forming crevasses on the surface. This eventually weakens blocks of ice which break away as icebergs. This process is known as calving.*

Map labels

Bering Sea
NORTH AMERICA
ASIA
Inuvik
Tiksi
ARCTIC OCEAN
Noril'sk
Qaanaaq
Murmansk
Reykjavík
ATLANTIC OCEAN
EUROPE

CANADA
NORTH AMERICA
Mackenzie
Great Bear Lake
Great Slave Lake
Kugluktuk (Coppermine)
Coronation Gulf
Bathurst Inlet
Cambridge Bay (Ikaluktutiak)
Queen Maud Gulf
Back
King William Island
Booth Peninsula
Gulf
Churchill
Repulse Bay
Southampton Island
Melville Peninsula
Hudson Bay
Coats Island
Mansel Island
Foxe Basin
Prince Charles Island
Ivujivik
Inukjuak (Port Harrison)
Hudson Strait
Foxe Peninsula
Baffin Island
Kimmirut (Lake Harbour)
Iqaluit (Frobisher Bay)
Frobisher Bay
Cumberland Sound
Ungava Bay
Cape Chidley
Davis Strait
Maniitsoq
Nain
Labrador Sea
NUUK
Paamiut
Ivittuut
KUJALLEQ
Labrador Basin
Qaqortoq
Narsarsuaq
Nanortalik
Nunap Isua (Kap Farvel)
Eirik Ridge
ATLANTIC

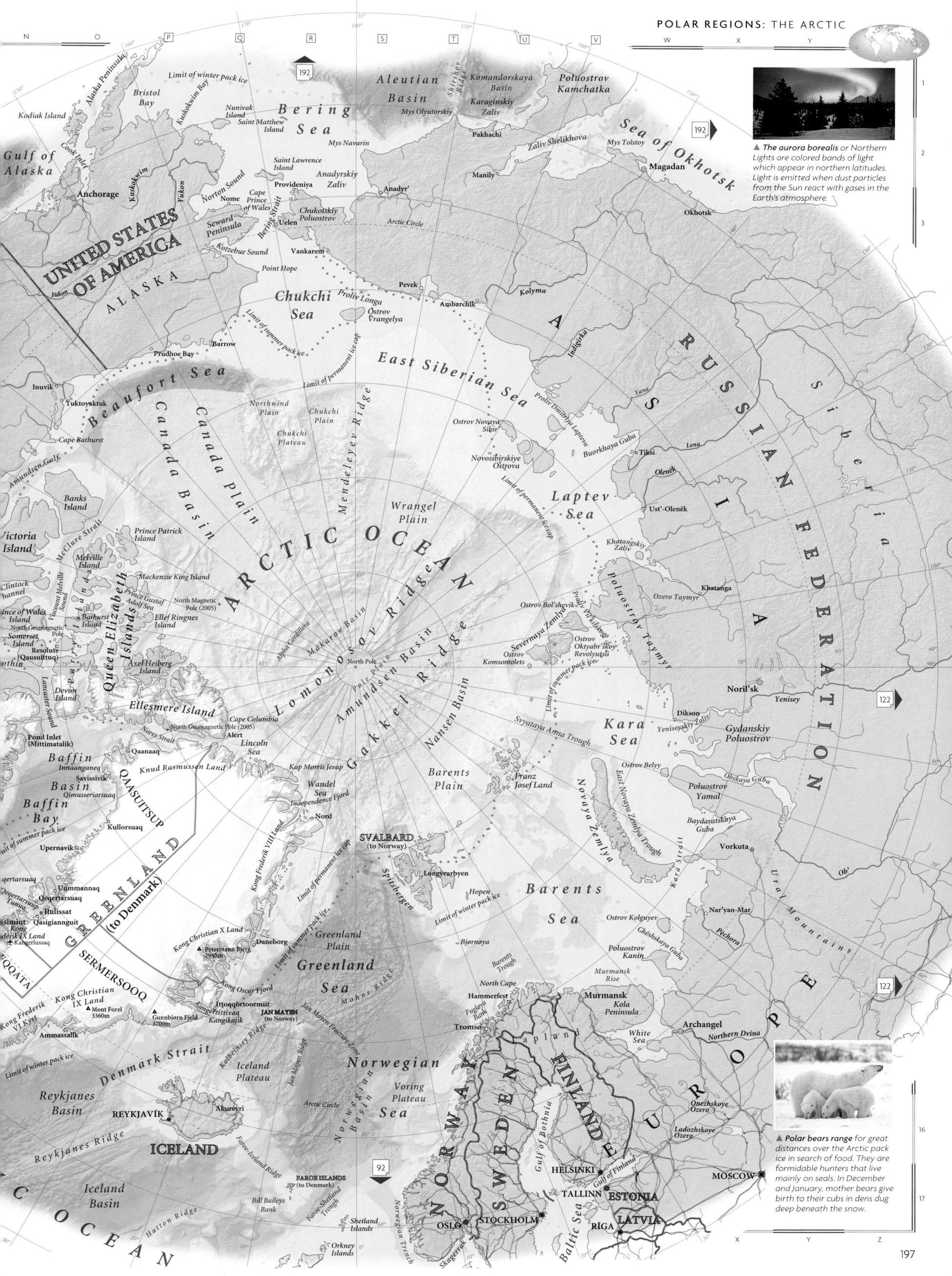

N O P Q R S T U V W X Y

The aurora borealis or Northern Lights are colored bands of light which appear in northern latitudes. Light is emitted when dust particles from the Sun react with gases in the Earth's atmosphere.

Aleutian Basin
Bering Sea
Bristol Bay
Kuskokwim Bay
Kodiak Island
Gulf of Alaska
Alaska Peninsula
Nunivak Island
Saint Matthew Island
Limit of winter pack ice

Komandorskaya Basin
Poluostrov Kamchatka
Karaginskiy Zaliv
Mys Olyutorskiy
Pakhachi
Shirshov Ridge
Sea of Okhotsk
Mys Tolstoy
Magadan

Anchorage
UNITED STATES OF AMERICA
ALASKA
Yukon
Kuskokwim
Norton Sound
Nome
Cape Prince of Wales
Seward Peninsula
Bering Strait
Uelen
Chukotskiy Poluostrov
Providenya
Anadyrskiy Zaliv
Zaliv Shelikhova
Anadyr'
Manily
Okhotsk

192
192
122
122

Arctic Circle
Kotzebue Sound
Vankarem
Point Hope
Pevek
Barrow
Prudhoe Bay
Chukchi Sea
Proliv Longa
Ostrov Vrangelya
Ambarchik
Kolyma
Indigirka
Yana
RUSSIAN FEDERATION
Siberia

Inuvik
Tuktoyaktuk
Cape Bathurst
Beaufort Sea
Amundsen Gulf
Banks Island
Victoria Island
Prince Patrick Island
Melville Island
McClure Strait
Mackenzie King Island
Canada Basin
Canada Plain
Northwind Plain
Chukchi Plain
Chukchi Plateau
Limit of summer pack ice
Limit of permanent ice cap
East Siberian Sea
Ostrov Novaya Sibir'
Novosibirskiye Ostrova
Buorkhaya Guba
Tiksi
Lena
Olenëk
Ust'-Olenëk
Laptev Sea
Proliv Dmitrya Lapteva

North Magnetic Pole (2005)
North Geomagnetic Pole
Prince Gustaf Adolf Sea
Bathurst Island
Ellef Ringnes Island
Axel Heiberg Island
Queen Elizabeth Islands
Mendeleyev Ridge
Wrangel Plain
ARCTIC OCEAN
Alpha Cordillera
Makarov Basin
North Pole
Pole Plain
Lomonosov Ridge
Amundsen Basin
Gakkel Ridge
Nansen Basin
Khatangskiy Zaliv
Khatanga
Ozero Taymyr
Poluostrov Taymyr
Ostrov Bol'shevik
Ostrov Oktyabr'skoy Revolyutsii
Ostrov Komsomolets
Severnaya Zemlya
Noril'sk
Yenisey
Dikson
Yeniseyskiy Zaliv
Svyataya Anna Trough
Kara Sea
Gydanskiy Poluostrov
Obskaya Guba

Clintock channel
Viscount Melville Sound
Somerset Island
Resolute (Qausuittuq)
Prince of Wales Island
Devon Island
Lancaster Sound
Ellesmere Island
North Geomagnetic Pole (2005)
Cape Columbia
Alert
Lincoln Sea
Qaanaaq
Kap Morris Jesup
Knud Rasmussen Land
Nares Strait
Barents Plain
Franz Josef Land
Novaya Zemlya
East Novaya Zemlya Trough
Ostrov Belyy
Poluostrov Yamal
Baydaratskaya Guba
Ob'
Vorkuta
Kara Strait
Ural Mountains

Pond Inlet (Mittimatalik)
Baffin Basin
Innaanganeq
Savissivik
Qimusseriarsuaq
Baffin Bay
Kullorsuaq
Upernavik
QAASUITSUP
Wandel Sea
Independence Fjord
Nord
Kong Frederik VIII Land
SVALBARD (to Norway)
Longyearbyen
Spitsbergen
Hopen
Limit of permanent ice cap
Limit of winter pack ice
Barents Sea
Ostrov Kolguyev
Chëshskaya Guba
Poluostrov Kanin
Nar'yan-Mar
Pechora

Qeqertarsuaq
Uummannaq
Qeqertarsuaq
Ilulissat
Qasigiannguit
Sisimiut
Kong Frederik IX Land
Kangerlussuaq
GREENLAND (to Denmark)
SERMERSOOQ
Kong Christian X Land
Daneborg
Petermann Bjerg 2940m
Kong Oscar Fjord
Greenland Plain
Limit of summer pack ice
Greenland Sea
Bjørnøya
Barents Trough
Murmansk Rise
North Cape
Hammerfest
Tromsø
Murmansk
Kola Peninsula
Archangel
Northern Dvina
White Sea
Onezhskoye Ozero
Ladozhskoye Ozero

IQQATA
Kong Frederik VI Kyst
Kong Christian IX Land
Mont Forel 3360m
Gunnbjørn Fjell 3700m
Kangerittivaq
Kangikajik
Ittoqqortoormiit
Ammassalik
JAN MAYEN (to Norway)
Jan Mayen Fracture Zone
Mohns Ridge
Kolbeinsey Ridge
Jan Mayen Ridge
Norwegian Sea
Voring Plateau
NORWAY
Lapland
SWEDEN
FINLAND
Gulf of Bothnia
HELSINKI
TALLINN
ESTONIA
RIGA
LATVIA
MOSCOW
EUROPE

Denmark Strait
Limit of winter pack ice
Reykjanes Basin
Iceland Plateau
Akureyri
REYKJAVÍK
ICELAND
Iceland Basin
Reykjanes Ridge
Arctic Circle
Norwegian Basin
Voring Plateau
FAROE ISLANDS (to Denmark)
Faroe-Iceland Ridge
Bill Bailey's Bank
Faroe-Shetland Trough
Shetland Islands
Orkney Islands
Norwegian Trench
Skagerrak
OSLO
STOCKHOLM
Baltic Sea
Gulf of Finland

92
92

OCEAN

1
2
3

16
17

Polar bears range for great distances over the Arctic pack ice in search of food. They are formidable hunters that live mainly on seals. In December and January, mother bears give birth to their cubs in dens dug deep beneath the snow.

197

Geographical comparisons

Largest countries

Russian Federation	6,592,735 sq miles	(17,075,200 sq km)
Canada	3,855,171 sq miles	(9,984,670 sq km)
USA	3,794,100 sq miles	(9,826,675 sq km)
China	3,705,386 sq miles	(9,596,960 sq km)
Brazil	3,286,470 sq miles	(8,511,965 sq km)
Australia	2,967,893 sq miles	(7,686,850 sq km)
India	1,269,339 sq miles	(3,287,590 sq km)
Argentina	1,068,296 sq miles	(2,766,890 sq km)
Kazakhstan	1,049,150 sq miles	(2,717,300 sq km)
Algeria	919,590 sq miles	(2,381,740 sq km)

Smallest countries

Vatican City	0.17 sq miles	(0.44 sq km)
Monaco	0.75 sq miles	(1.95 sq km)
Nauru	8.2 sq miles	(21.2 sq km)
Tuvalu	10 sq miles	(26 sq km)
San Marino	24 sq miles	(61 sq km)
Liechtenstein	62 sq miles	(160 sq km)
Marshall Islands	70 sq miles	(181 sq km)
St. Kitts & Nevis	101 sq miles	(261 sq km)
Maldives	116 sq miles	(300 sq km)
Malta	124 sq miles	(320 sq km)

Largest islands

	To the nearest 1000 – or 100,000 for the largest	
Greenland	849,400 sq miles	(2,200,000 sq km)
New Guinea	312,000 sq miles	(808,000 sq km)
Borneo	292,222 sq miles	(757,050 sq km)
Madagascar	229,300 sq miles	(594,000 sq km)
Sumatra	202,300 sq miles	(524,000 sq km)
Baffin Island	183,800 sq miles	(476,000 sq km)
Honshu	88,800 sq miles	(230,000 sq km)
Britain	88,700 sq miles	(229,800 sq km)
Victoria Island	81,900 sq miles	(212,000 sq km)
Ellesmere Island	75,700 sq miles	(196,000 sq km)

Richest countries

	GNI per capita, in US$
Monaco	186,950
Liechtenstein	136,770
Norway	102,610
Switzerland	90,760
Qatar	86,790
Luxembourg	69,900
Australia	65,390
Sweden	61,760
Denmark	61,680
Singapore	54,040

Poorest countries

	GNI per capita, in US$
Burundi	260
Malawi	270
Somalia	288
Central African Republic	320
Niger	400
Liberia	410
Dem. Rep. Congo	430
Madagascar	440
Guinea	460
Ethiopia	470
Eritrea	490
Gambia	500

Most populous countries

China	1,393,800,000
India	1,267,400,000
USA	322,600,000
Indonesia	252,800,000
Brazil	202,120,000
Pakistan	185,100,000
Nigeria	178,500,000
Bangladesh	159,000,000
Russian Federation	142,500,000
Japan	127,000,000

Least populous countries

Vatican City	842
Nauru	9488
Tuvalu	10,782
Palau	21,186
San Marino	32,742
Monaco	36,950
Liechtenstein	37,313
St Kitts & Nevis	51,538
Marshall Islands	70,983
Dominica	73,449
Andorra	85,458
Antigua & Barbuda	91,295

Most densely populated countries

Monaco	49,267 people per sq mile	(18,949 per sq km)
Singapore	23,305 people per sq mile	(9016 per sq km)
Vatican City	4953 people per sq mile	(1914 per sq km)
Bahrain	4762 people per sq mile	(1841 per sq km)
Maldives	3448 people per sq mile	(1333 per sq km)
Malta	3226 people per sq mile	(1250 per sq km)
Bangladesh	3066 people per sq mile	(1184 per sq km)
Taiwan	1879 people per sq mile	(725 per sq km)
Barbados	1807 people per sq mile	(698 per sq km)
Mauritius	1671 people per sq mile	(645 per sq km)

Most sparsely populated countries

Mongolia	5 people per sq mile	(2 per sq km)
Namibia	7 people per sq mile	(3 per sq km)
Australia	8 people per sq mile	(3 per sq km)
Suriname	8 people per sq mile	(3 per sq km)
Iceland	8 people per sq mile	(3 per sq km)
Botswana	9 people per sq mile	(4 per sq km)
Libya	9 people per sq mile	(4 per sq km)
Mauriania	10 people per sq mile	(4 per sq km)
Canada	10 people per sq mile	(4 per sq km)
Guyana	11 people per sq mile	(4 per sq km)

Most widely spoken languages

1. Chinese (Mandarin)	6. Arabic
2. English	7. Bengali
3. Hindi	8. Portuguese
4. Spanish	9. Malay-Indonesian
5. Russian	10. French

Largest conurbations

	Urban area population
Tokyo	37,800,000
Jakarta	30,500,000
Manila	24,100,000
Delhi	24,000,000
Karachi	23,500,000
Seoul	23,500,000
Shanghai	23,400,000
Beijing	21,000,000
New York City	20,600,000
Guangzhou	20,600,000
São Paulo	20,300,000
Mexico City	20,000,000
Mumbai	17,700,000
Osaka	17,400,000
Lagos	17,000,000
Moscow	16,100,000
Dhaka	15,700,000
Lahore	15,600,000
Los Angeles	15,000,000
Bangkok	15,000,000
Kolkatta	14,700,000
Buenos Aires	14,100,000
Tehran	13,500,000
Istanbul	13,300,000
Shenzhen	12,000,000

Countries with the most land borders

14: China	(Afghanistan, Bhutan, India, Kazakhstan, Kyrgyzstan, Laos, Mongolia, Myanmar (Burma), Nepal, North Korea, Pakistan, Russian Federation, Tajikistan, Vietnam)
14: Russian Federation	(Azerbaijan, Belarus, China, Estonia, Finland, Georgia, Kazakhstan, Latvia, Lithuania, Mongolia, North Korea, Norway, Poland, Ukraine)
10: Brazil	(Argentina, Bolivia, Colombia, French Guiana, Guyana, Paraguay, Peru, Suriname, Uruguay, Venezuela)
9: Congo, Dem. Rep.	(Angola, Burundi, Central African Republic, Congo, Rwanda, South Sudan, Tanzania, Uganda, Zambia)
9: Germany	(Austria, Belgium, Czech Republic, Denmark, France, Luxembourg, Netherlands, Poland, Switzerland)
8: Austria	(Czech Republic, Germany, Hungary, Italy, Liechtenstein, Slovakia, Slovenia, Switzerland)
8: France	(Andorra, Belgium, Germany, Italy, Luxembourg, Monaco, Spain, Switzerland)
8: Tanzania	(Burundi, Dem. Rep. Congo, Kenya, Malawi, Mozambique, Rwanda, Uganda, Zambia)
8: Turkey	(Armenia, Azerbaijan, Bulgaria, Georgia, Greece, Iran, Iraq, Syria)
8: Zambia	(Angola, Botswana, Dem. Rep.Congo, Malawi, Mozambique, Namibia, Tanzania, Zimbabwe)

Longest rivers

Nile (NE Africa)	4160 miles	(6695 km)
Amazon (South America)	4049 miles	(6516 km)
Yangtze (China)	3915 miles	(6299 km)
Mississippi/Missouri (USA)	3710 miles	(5969 km)
Ob'-Irtysh (Russian Federation)	3461 miles	(5570 km)
Yellow River (China)	3395 miles	(5464 km)
Congo (Central Africa)	2900 miles	(4667 km)
Mekong (Southeast Asia)	2749 miles	(4425 km)
Lena (Russian Federation)	2734 miles	(4400 km)
Mackenzie (Canada)	2640 miles	(4250 km)
Yenisey (Russian Federation)	2541 miles	(4090km)

Highest mountains

	Height above sea level	
Everest	29,029 ft	(8848 m)
K2	28,253 ft	(8611 m)
Kangchenjunga I	28,210 ft	(8598 m)
Makalu I	27,767 ft	(8463 m)
Cho Oyu	26,907 ft	(8201 m)
Dhaulagiri I	26,796 ft	(8167 m)
Manaslu I	26,783 ft	(8163 m)
Nanga Parbat I	26,661 ft	(8126 m)
Annapurna I	26,547 ft	(8091 m)
Gasherbrum I	26,471 ft	(8068 m)

Largest bodies of inland water

	With area and depth	
Caspian Sea	143,243 sq miles (371,000 sq km)	3215 ft (980 m)
Lake Superior	31,151 sq miles (83,270 sq km)	1289 ft (393 m)
Lake Victoria	26,828 sq miles (69,484 sq km)	328 ft (100 m)
Lake Huron	23,436 sq miles (60,700 sq km)	751 ft (229 m)
Lake Michigan	22,402 sq miles (58,020 sq km)	922 ft (281 m)
Lake Tanganyika	12,703 sq miles (32,900 sq km)	4700 ft (1435 m)
Great Bear Lake	12,274 sq miles (31,790 sq km)	1047 ft (319 m)
Lake Baikal	11,776 sq miles (30,500 sq km)	5712 ft (1741 m)
Great Slave Lake	10,981 sq miles (28,440 sq km)	459 ft (140 m)
Lake Erie	9,915 sq miles (25,680 sq km)	197 ft (60 m)

Deepest ocean features

Challenger Deep, Mariana Trench (Pacific)	35,827 ft	(10,920 m)
Vityaz III Depth, Tonga Trench (Pacific)	35,704 ft	(10,882 m)
Vityaz Depth, Kuril-Kamchatka Trench (Pacific)	34,588 ft	(10,542 m)
Cape Johnson Deep, Philippine Trench (Pacific)	34,441 ft	(10,497 m)
Kermadec Trench (Pacific)	32,964 ft	(10,047 m)
Ramapo Deep, Japan Trench (Pacific)	32,758 ft	(9984 m)
Milwaukee Deep, Puerto Rico Trench (Atlantic)	30,185 ft	(9200 m)
Argo Deep, Torres Trench (Pacific)	**30,070 ft**	**(9165 m)**
Meteor Depth, South Sandwich Trench (Atlantic)	30,000 ft	(9144 m)
Planet Deep, New Britain Trench (Pacific)	29,988 ft	(9140 m)

Greatest waterfalls

	Mean flow of water	
Boyoma (Dem. Rep. Congo)	600,400 cu. ft/sec	(17,000 cu.m/sec)
Khône (Laos/Cambodia)	410,000 cu. ft/sec	(11,600 cu.m/sec)
Niagara (USA/Canada)	195,000 cu. ft/sec	(5500 cu.m/sec)
Grande, Salto (Uruguay)	160,000 cu. ft/sec	(4500 cu.m/sec)
Paulo Afonso (Brazil)	100,000 cu. ft/sec	(2800 cu.m/sec)
Urubupungá, Salto do (Brazil)	97,000 cu. ft/sec	(2750 cu.m/sec)
Iguaçu (Argentina/Brazil)	62,000 cu. ft/sec	(1700 cu.m/sec)
Maribondo, Cachoeira do (Brazil)	53,000 cu. ft/sec	(1500 cu.m/sec)
Victoria (Zimbabwe)	39,000 cu. ft/sec	(1100 cu.m/sec)
Murchison Falls (Uganda)	42,000 cu. ft/sec	(1200 cu.m/sec)
Churchill (Canada)	35,000 cu. ft/sec	(1000 cu.m/sec)
Kaveri Falls (India)	33,000 cu. ft/sec	(900 cu.m/sec)

Highest waterfalls

	* Indicates that the total height is a single leap	
Angel (Venezuela)	3212 ft	(979 m)
Tugela (South Africa)	3110 ft	(948 m)
Utigard (Norway)	2625 ft	(800 m)
Mongefossen (Norway)	2539 ft	(774 m)
Mtarazi (Zimbabwe)	2500 ft	(762 m)
Yosemite (USA)	2425 ft	(739 m)
Ostre Mardola Foss (Norway)	2156 ft	(657 m)
Tyssestrengane (Norway)	2119 ft	(646 m)
*Cuquenan (Venezuela)	2001 ft	(610 m)
Sutherland (New Zealand)	1903 ft	(580 m)
*Kjellfossen (Norway)	1841 ft	(561 m)

Largest deserts

NB – Most of Antarctica is a polar desert, with only 50mm of precipitation annually

Sahara	3,450,000 sq miles	(9,065,000 sq km)
Gobi	500,000 sq miles	(1,295,000 sq km)
Ar Rub al Khali	289,600 sq miles	(750,000 sq km)
Great Victorian	249,800 sq miles	(647,000 sq km)
Sonoran	120,000 sq miles	(311,000 sq km)
Kalahari	120,000 sq miles	(310,800 sq km)
Kara Kum	115,800 sq miles	(300,000 sq km)
Takla Makan	100,400 sq miles	(260,000 sq km)
Namib	52,100 sq miles	(135,000 sq km)
Thar	33,670 sq miles	(130,000 sq km)

Hottest inhabited places

Djibouti (Djibouti)	86° F	(30 °C)
Tombouctou (Mali)	84.7° F	(29.3 °C)
Tirunelveli (India)		
Tuticorin (India)		
Nellore (India)	84.5° F	(29.2 °C)
Santa Marta (Colombia)		
Aden (Yemen)	84° F	(28.9 °C)
Madurai (India)		
Niamey (Niger)		
Hodeida (Yemen)	83.8° F	(28.8 °C)
Ouagadougou (Burkina Faso)		
Thanjavur (India)		
Tiruchchirappalli (India)		

Driest inhabited places

Aswân (Egypt)	0.02 in	(0.5 mm)
Luxor (Egypt)	0.03 in	(0.7 mm)
Arica (Chile)	0.04 in	(1.1 mm)
Ica (Peru)	0.1 in	(2.3 mm)
Antofagasta (Chile)	0.2 in	(4.9 mm)
Al Minya (Egypt)	0.2 in	(5.1 mm)
Asyut (Egypt)	0.2 in	(5.2 mm)
Callao (Peru)	0.5 in	(12.0 mm)
Trujillo (Peru)	0.55 in	(14.0 mm)
Al Fayyum (Egypt)	0.8 in	(19.0 mm)

Wettest inhabited places

Mawsynram (India)	467 in	(11,862 mm)
Mount Waialeale (Hawaii, USA)	460 in	(11,684 mm)
Cherrapunji (India)	450 in	(11,430 mm)
Cape Debundsha (Cameroon)	405 in	(10,290 mm)
Quibdo (Colombia)	354 in	(8892 mm)
Buenaventura (Colombia)	265 in	(6743 mm)
Monrovia (Liberia)	202 in	(5131 mm)
Pago Pago (American Samoa)	196 in	(4990 mm)
Mawlamyine (Myanmar [Burma])	191 in	(4852 mm)
Lae (Papua New Guinea)	183 in	(4645 mm)

Standard time zones

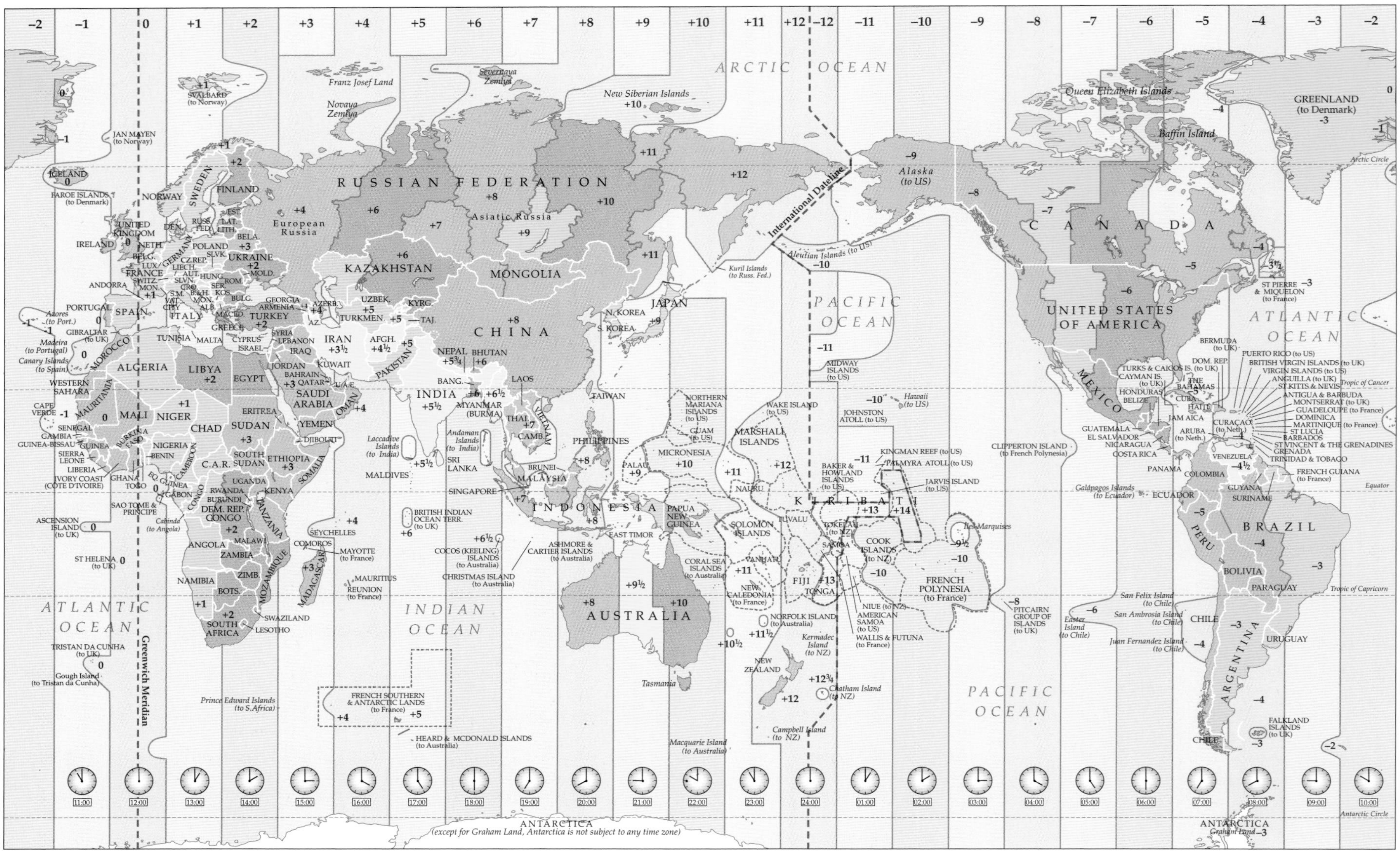

The numbers at the top of the map indicate the number of hours each time zone is ahead or behind Coordinated Universal Time (UTC).
The clocks and 24-hour times given at the bottom of the map show the time in each time zone when it is 12:00 hours noon (UTC)

Time Zones

Because Earth is a rotating sphere, the Sun shines on only half of its surface at any one time. Thus, it is simultaneously morning, evening and night time in different parts of the world (see diagram below). Because of these disparities, each country or part of a country adheres to a local time.

A region of Earth's surface within which a single local time is used is called a time zone. There are 24 one hour time zones around the world, arranged roughly in longitudinal bands.

Standard Time

Standard time is the official local time in a particular country or part of a country. It is defined by the

Day and night around the world

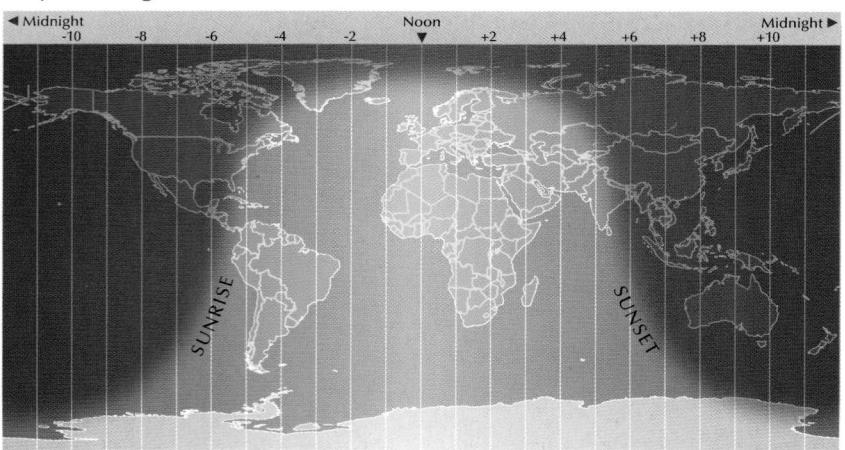

time zone or zones associated with that country or region. Although time zones are arranged roughly in longitudinal bands, in many places the borders of a zone do not fall exactly on longitudinal meridians, as can be seen on the map (above), but are determined by geographical factors or by borders between countries or parts of countries. Most countries have just one time zone and one standard time, but some large countries (such as the US, Canada, and Russia) are split between several time zones, so standard time varies across those countries. For example, the coterminous United States straddles four time zones and so has four standard times, called the Eastern, Central, Mountain, and Pacific standard times. China is unusual in that just one standard time is used for the whole country, even though it extends across 60° of longitude from west to east.

Coordinated Universal Time (UTC)

Coordinated Universal Time (UTC) is a reference by which the local time in each time zone is set. For example, Australian Western Standard Time (the local time in Western Australia) is set 8 hours ahead of UTC (it is

UTC+8) whereas Eastern Standard Time in the United States is set 5 hours behind UTC (it is UTC-5). UTC is a successor to, and closely approximates, Greenwich Mean Time (GMT). However, UTC is based on an atomic clock, whereas GMT is determined by the Sun's position in the sky relative to the 0° longitudinal meridian, which runs through Greenwich, UK.

The International Dateline

The International Dateline is an imaginary line from pole to pole that roughly corresponds to the 180° longitudinal meridian. It is an arbitrary marker between calendar days. The dateline is needed because of the use of local times around the world rather than a single universal time. When moving from west to east across the dateline, travelers have to set their watches back one day. Those traveling in the opposite direction, from east to west, must add a day.

Daylight Saving Time

Daylight saving is a summertime adjustment to the local time in a country or region, designed to cause a higher proportion of its citizens' waking hours to pass during daylight. To follow the system, timepieces are advanced by an hour on a pre-decided date in spring and reverted back in the fall. About half of the world's nations use daylight saving.

Countries of the World

There are currently 196 independent countries in the world and almost 60 dependencies. Antarctica is the only land area on Earth that is not officially part of, and does not belong to, any single country.

In 1950, the world comprised 82 countries. In the decades following, many more states came into being as they achieved independence from their former colonial rulers. Most recent additions were caused by the breakup of the former Soviet Union in 1991, and the former Yugoslavia in 1992, which swelled the ranks of independent states. In July 2011, South Sudan became the latest country to be formed after declaring independence from Sudan.

Country factfile key

Formation Date of political origin or independence/ date current borders were established
Population Total population / population density – based on total *land* area
Languages An asterisk (*) denotes the official language(s)
Calorie consumption Average number of kilocalories consumed daily per person

AFGHANISTAN
Central Asia

Official name Islamic Republic of Afghanistan
Formation 1919 / 1919
Capital Kabul
Population 31.3 million / 124 people per sq mile (48 people per sq km)
Total area 250,000 sq. miles (647,500 sq. km)
Languages Pashtu*, Tajik, Dari*, Farsi, Uzbek, Turkmen
Religions Sunni Muslim 80%, Shi'a Muslim 19%, Other 1%
Ethnic mix Pashtun 38%, Tajik 25%, Hazara 19%, Uzbek and Turkmen 15%, Other 3%
Government Nonparty system
Currency Afghani = 100 puls
Literacy rate rate 32%
Calorie consumption 2090 kilocalories

ALBANIA
Southeast Europe

Official name Republic of Albania
Formation 1912 / 1921
Capital Tirana
Population 3.2 million / 302 people per sq mile (117 people per sq km)
Total area 11,100 sq. miles (28,748 sq. km)
Languages Albanian*, Greek
Religions Sunni Muslim 70%, Albanian Orthodox 20%, Roman Catholic 10%
Ethnic mix Albanian 98%, Greek 1%, Other 1%
Government Parliamentary system
Currency Lek = 100 qindarka (qintars)
Literacy rate 97%
Calorie consumption 3023 kilocalories

ALGERIA
North Africa

Official name People's Democratic Republic of Algeria
Formation 1962 / 1962
Capital Algiers
Population 39.9 million / 43 people per sq mile (17 people per sq km)
Total area 919,590 sq. miles (2,381,740 sq. km)
Languages Arabic*, Tamazight (Kabyle, Shawia, Tamashek), French
Religions Sunni Muslim 99%, Christian and Jewish 1%
Ethnic mix Arab 75%, Berber 24%, European and Jewish 1%
Government Presidential system
Currency Algerian dinar = 100 centimes
Literacy rate 73%
Calorie consumption 3296 kilocalories

ANDORRA
Southwest Europe

Official name Principality of Andorra
Formation 1278 / 1278
Capital Andorra la Vella
Population 85,485 / 475 people per sq mile (184 people per sq km)
Total area 181 sq. miles (468 sq. km)
Languages Spanish, Catalan*, French, Portuguese
Religions Roman Catholic 94%, Other 6%
Ethnic mix Spanish 46%, Andorran 28%, Other 18%, French 8%
Government Parliamentary system
Currency Euro = 100 cents
Literacy rate 99%
Calorie consumption Not available

ANGOLA
Southern Africa

Official name Republic of Angola
Formation 1975 / 1975
Capital Luanda
Population 22.1 million / 46 people per sq mile (18 people per sq km)
Total area 481,351 sq. miles (1,246,700 sq. km)
Languages Portuguese*, Umbundu, Kimbundu, Kikongo
Religions Roman Catholic 68%, Protestant 20%, Indigenous beliefs 12%
Ethnic mix Ovimbundu 37%, Kimbundu 25%, Other 25%, Bakongo 13%
Government Presidential system
Currency Readjusted kwanza = 100 lwei
Literacy rate 71%
Calorie consumption 2473 kilocalories

ANTIGUA & BARBUDA
West Indies

Official name Antigua and Barbuda
Formation 1981 / 1981
Capital St. John's
Population 91,295 / 537 people per sq mile (207 people per sq km)
Total area 170 sq. miles (442 sq. km)
Languages English*, English patois
Religions Anglican 45%, Other Protestant 42%, Roman Catholic 10%, Other 2%, Rastafarian 1%
Ethnic mix Black African 95%, Other 5%
Government Parliamentary system
Currency East Caribbean dollar = 100 cents
Literacy rate 99%
Calorie consumption 2396 kilocalories

ARGENTINA
South America

Official name Argentine Republic
Formation 1816 / 1816
Capital Buenos Aires
Population 41.8 million / 40 people per sq mile (15 people per sq km)
Total area 1,068,296 sq. miles (2,766,890 sq. km)
Languages Spanish*, Italian, Amerindian languages
Religions Roman Catholic 70%, Other 18%, Protestant 9%, Muslim 2%, Jewish 1%
Ethnic mix Indo-European 97%, Mestizo 2%, Amerindian 1%
Government Presidential system
Currency Argentine peso = 100 centavos
Literacy rate 98%
Calorie consumption 3155 kilocalories

ARMENIA
Southwest Asia

Official name Republic of Armenia
Formation 1991 / 1991
Capital Yerevan
Population 3 million / 261 people per sq mile (101 people per sq km)
Total area 11,506 sq. miles (29,800 sq. km)
Languages Armenian*, Azeri, Russian
Religions Armenian Apostolic Church (Orthodox) 88%, Armenian Catholic Church 6%, Other 6%
Ethnic mix Armenian 98%, Other 1%, Yezidi 1%
Government Parliamentary system
Currency Dram = 100 luma
Literacy rate 99%
Calorie consumption 2809 kilocalories

AUSTRALIA
Australasia & Oceania

Official name Commonwealth of Australia
Formation 1901 / 1901
Capital Canberra
Population 23.6 million / 8 people per sq mile (3 people per sq km)
Total area 2,967,893 sq. miles (7,686,850 sq. km)
Languages English*, Italian, Cantonese, Greek, Arabic, Vietnamese, Aboriginal languages
Religions Roman Catholic 26%, Nonreligious 19%, Anglican 19%, Other 17%, Other Christian 13%, United Church 6%
Ethnic mix European origin 50%, Australian 25.5%, other 19%, Asian 5%, Aboriginal 0.5%
Government Parliamentary system
Currency Australian dollar = 100 cents
Literacy rate 99%
Calorie consumption 3265 kilocalories

AUSTRIA
Central Europe

Official name Republic of Austria
Formation 1918 / 1919
Capital Vienna
Population 8.5 million / 266 people per sq mile (103 people per sq km)
Total area 32,378 sq. miles (83,858 sq. km)
Languages German*, Croatian, Slovenian, Hungarian (Magyar)
Religions Roman Catholic 78%, Nonreligious 9%, Other (including Jewish and Muslim) 8%, Protestant 5%
Ethnic mix Austrian 93%, Croat, Slovene, and Hungarian 6%, Other 1%
Government Parliamentary system
Currency Euro = 100 cents
Literacy rate 99%
Calorie consumption 3784 kilocalories

AZERBAIJAN
Southwest Asia

Official name Republic of Azerbaijan
Formation 1991 / 1991
Capital Baku
Population 9.5 million / 284 people per sq mile (110 people per sq km)
Total area 33,436 sq. miles (86,600 sq. km)
Languages Azeri*, Russian
Religions Shi'a Muslim 68%, Sunni Muslim 26%, Russian Orthodox 3%, Armenian Apostolic Church (Orthodox) 2%, Other 1%
Ethnic mix Azeri 91%, Other 3%, Lazs 2%, Armenian 2%, Russian 2%
Government Presidential system
Currency New manat = 100 gopik
Literacy rate 99%
Calorie consumption 2952 kilocalories

THE BAHAMAS
West Indies

Official name Commonwealth of The Bahamas
Formation 1973 / 1973
Capital Nassau
Population 400,000 / 103 people per sq mile (40 people per sq km)
Total area 5382 sq. miles (13,940 sq. km)
Languages English*, English Creole, French Creole
Religions Baptist 32%, Anglican 20%, Roman Catholic 19%, Other 17%, Methodist 6%, Church of God 6%
Ethnic mix Black African 85%, European 12%, Asian and Hispanic 3%
Government Parliamentary system
Currency Bahamian dollar = 100 cents
Literacy rate 96%
Calorie consumption 2575 kilocalories

BAHRAIN
Southwest Asia

Official name Kingdom of Bahrain
Formation 1971 / 1971
Capital Manama
Population 1.3 million / 4762 people per sq mile (1841 people per sq km)
Total area 239 sq. miles (620 sq. km)
Languages Arabic*
Religions Muslim (mainly Shi'a) 99%, Other 1%
Ethnic mix Bahraini 63%, Asian 19%, Other Arab 10%, Iranian 8%
Government Mixed monarchical–parliamentary system
Currency Bahraini dinar = 1000 fils
Literacy rate 95%
Calorie consumption Not available

BANGLADESH
South Asia

Official name People's Republic of Bangladesh
Formation 1971 / 1971
Capital Dhaka
Population 159 million / 3066 people per sq mile (1184 people per sq km)
Total area 55,598 sq. miles (144,000 sq. km)
Languages Bengali*, Urdu, Chakma, Marma (Magh), Garo, Khasi, Santhali, Tripuri, Mro
Religions Muslim (mainly Sunni) 88%, Hindu 11%, Other 1%
Ethnic mix Bengali 98%, Other 2%
Government Parliamentary system
Currency Taka = 100 poisha
Literacy rate 59%
Calorie consumption 2450 kilocalories

BARBADOS
West Indies

Official name Barbados
Formation 1966 / 1966
Capital Bridgetown
Population 300,000 / 1807 people per sq mile (698 people per sq km)
Total area 166 sq. miles (430 sq. km)
Languages Bajan (Barbadian English), English*
Religions Anglican 40%, Other 24%, Nonreligious 17%, Pentecostal 8%, Methodist 7%, Roman Catholic 4%
Ethnic mix Black African 92%, White 3%, Other 3%, Mixed race 2%
Government Parliamentary system
Currency Barbados dollar = 100 cents
Literacy rate 99%
Calorie consumption 3047 kilocalories

BELARUS
Eastern Europe

Official name Republic of Belarus
Formation 1991 / 1991
Capital Minsk
Population 9.3 million / 116 people per sq mile (45 people per sq km)
Total area 80,154 sq. miles (207,600 sq. km)
Languages Belarussian*, Russian*
Religions Orthodox Christian 80%, Roman Catholic 14%, Other 4%, Protestant 2%
Ethnic mix Belarussian 81%, Russian 11%, Polish 4%, Ukrainian 2%, Other 2%
Government Presidential system
Currency Belarussian rouble = 100 kopeks
Literacy rate 99%
Calorie consumption 3253 kilocalories

BELGIUM
Northwest Europe

Official name Kingdom of Belgium
Formation 1830 / 1919
Capital Brussels
Population 11.1 million / 876 people per sq mile (338 people per sq km)
Total area 11,780 sq. miles (30,510 sq. km)
Languages Dutch*, French*, German*
Religions Roman Catholic 88%, Other 10%, Muslim 2%
Ethnic mix Fleming 58%, Walloon 33%, Other 6%, Italian 2%, Moroccan 1%
Government Parliamentary system
Currency Euro = 100 cents
Literacy rate 99%
Calorie consumption 3793 kilocalories

BELIZE
Central America

Official name Belize
Formation 1981 / 1981
Capital Belmopan
Population 300,000 / 34 people per sq mile (13 people per sq km)
Total area 8867 sq. miles (22,966 sq. km)
Languages English Creole, Spanish, English*, Mayan, Garifuna (Carib)
Religions Roman Catholic 62%, Other 13%, Anglican 12%, Methodist 6%, Mennonite 4%, Seventh-day Adventist 3%
Ethnic mix Mestizo 49%, Creole 25%, Maya 11%, Garifuna 6%, Other 6%, Asian Indian 3%
Government Parliamentary system
Currency Belizean dollar = 100 cents
Literacy rate 75%
Calorie consumption 2751 kilocalories

BENIN
West Africa

Official name Republic of Benin
Formation 1960 / 1960
Capital Porto-Novo
Population 10.6 million / 248 people per sq mile (96 people per sq km)
Total area 43,483 sq. miles (112,620 sq. km)
Languages Fon, Bariba, Yoruba, Adja, Houeda, Somba, French*
Religions Indigenous beliefs and Voodoo 50%, Christian 30%, Muslim 20%
Ethnic mix Fon 41%, Other 21%, Adja 16%, Yoruba 12%, Bariba 10%
Government Presidential system
Currency CFA franc = 100 centimes
Literacy rate 29%
Calorie consumption 2594 kilocalories

BHUTAN
South Asia

Official name Kingdom of Bhutan
Formation 1656 / 1865
Capital Thimphu
Population 800,000 / 44 people per sq mile (17 people per sq km)
Total area 18,147 sq. miles (47,000 sq. km)
Languages Dzongkha*, Nepali, Assamese
Religions Mahayana Buddhist 75%, Hindu 25%
Ethnic mix Drukpa 50%, Nepalese 35%, Other 15%
Government Mixed monarchical–parliamentary system
Currency Ngultrum = 100 chetrum
Literacy rate 53%
Calorie consumption Not available

BOLIVIA
South America

Official name Plurinational State of Bolivia
Formation 1825 / 1938
Capital La Paz (administrative); Sucre (judicial)
Population 10.8 million / 26 people per sq mile (10 people per sq km)
Total area 424,162 sq. miles (1,098,580 sq. km)
Languages Aymara*, Quechua*, Spanish*
Religions Roman Catholic 93%, Other 7%
Ethnic mix Quechua 37%, Aymara 32%, Mixed race 13%, European 10%, Other 8%
Government Presidential system
Currency Boliviano = 100 centavos
Literacy rate 94%
Calorie consumption 2254 kilocalories

BOSNIA & HERZEGOVINA
Southeast Europe

Official name Bosnia and Herzegovina
Formation 1992 / 1992
Capital Sarajevo
Population 3.8 million / 192 people per sq mile (74 people per sq km)
Total area 19,741 sq. miles (51,129 sq. km)
Languages Bosnian*, Serbian*, Croatian*
Religions Muslim (mainly Sunni) 40%, Orthodox Christian 31%, Roman Catholic 15%, Other 10%, Protestant 4%
Ethnic mix Bosniak 48%, Serb 34%, Croat 16%, Other 2%
Government Parliamentary system
Currency Marka = 100 pfeninga
Literacy rate 98%
Calorie consumption 3130 kilocalories

BOTSWANA
Southern Africa

Official name Republic of Botswana
Formation 1966 / 1966
Capital Gaborone
Population 2 million / 9 people per sq mile (4 people per sq km)
Total area 231,803 sq. miles (600,370 sq. km)
Languages Setswana, English*, Shona, San, Khoikhoi, isiNdebele
Religions Christian (mainly Protestant) 70%, Nonreligious 20%, Traditional beliefs 6%, Other (including Muslim) 4%
Ethnic mix Tswana 79%, Kalanga 11%, Other 10%
Government Presidential system
Currency Pula = 100 thebe
Literacy rate 87%
Calorie consumption 2285 kilocalories

BRAZIL
South America

Official name Federative Republic of Brazil
Formation 1822 / 1828
Capital Brasília
Population 202 million / 62 people per sq mile (24 people per sq km)
Total area 3,286,470 sq. miles (8,511,965 sq. km)
Languages Portuguese*, German, Italian, Spanish, Polish, Japanese, Amerindian languages
Religions Roman Catholic 74%, Protestant 15%, Atheist 7%, Other 3%, Afro-American Spiritist 1%
Ethnic mix White 54%, Mixed race 38%, Black 6%, Other 2%
Government Presidential system
Currency Real = 100 centavos
Literacy rate 91%
Calorie consumption 3263 kilocalories

BRUNEI
Southeast Asia

Official name Brunei Darussalam
Formation 1984 / 1984
Capital Bandar Seri Begawan
Population 400,000 / 197 people per sq mile (76 people per sq km)
Total area 2228 sq. miles (5770 sq. km)
Languages Malay*, English, Chinese
Religions Muslim (mainly Sunni) 66%, Buddhist 14%, Other 10%, Christian 10%
Ethnic mix Malay 67%, Chinese 16%, Other 11%, Indigenous 6%
Government Monarchy
Currency Brunei dollar = 100 cents
Literacy rate 95%
Calorie consumption 2949 kilocalories

BULGARIA
Southeast Europe

Official name Republic of Bulgaria
Formation 1908 / 1947
Capital Sofia
Population 7.2 million / 169 people per sq mile (65 people per sq km)
Total area 42,822 sq. miles (110,910 sq. km)
Languages Bulgarian*, Turkish, Romani
Religions Bulgarian Orthodox 83%, Muslim 12%, Other 4%, Roman Catholic 1%
Ethnic mix Bulgarian 84%, Turkish 9%, Roma 5%, Other 2%
Government Parliamentary system
Currency Lev = 100 stotinki
Literacy rate 98%
Calorie consumption 2877 kilocalories

BURKINA FASO
West Africa

Official name Burkina Faso
Formation 1960 / 1960
Capital Ouagadougou
Population 17.4 million / 165 people per sq mile (64 people per sq km)
Total area 105,869 sq. miles (274,200 sq. km)
Languages Mossi, Fulani, French*, Tuare g, Dyula, Songhai
Religions Muslim 55%, Christian 25%, Traditional beliefs 20%
Ethnic mix Mossi 48%, Other 21%, Peul 10%, Lobi 7%, Bobo 7%, Mandé 7%
Government Transitional regime
Currency CFA franc = 100 centimes
Literacy rate 29%
Calorie consumption 2720 kilocalories

BURUNDI
Central Africa

Official name Republic of Burundi
Formation 1962 / 1962
Capital Bujumbura
Population 10.5 million / 1060 people per sq mile (409 people per sq km)
Total area 10,745 sq. miles (27,830 sq. km)
Languages Kirundi*, French*, Kiswahili
Religions Roman Catholic 62%, Traditional beliefs 23%, Muslim 10%, Protestant 5%
Ethnic mix Hutu 85%, Tutsi 14%, Twa 1%
Government Presidential system
Currency Burundian franc = 100 centimes
Literacy rate 87%
Calorie consumption 1604 kilocalories

CAMBODIA
Southeast Asia

Official name Kingdom of Cambodia
Formation 1953 / 1953
Capital Phnom Penh
Population 15.4 million / 226 people per sq mile (87 people per sq km)
Total area 69,900 sq. miles (181,040 sq. km)
Languages Khmer*, French, Chinese, Vietnamese, Cham
Religions Buddhist 93%, Muslim 6%, Christian 1%
Ethnic mix Khmer 90%, Vietnamese 5%, Other 4%, Chinese 1%
Government Parliamentary system
Currency Riel = 100 sen
Literacy rate 74%
Calorie consumption 2411 kilocalories

CAMEROON
Central Africa

Official name Republic of Cameroon
Formation 1960 / 1961
Capital Yaoundé
Population 22.8 million / 127 people per sq mile (49 people per sq km)
Total area 183,567 sq. miles (475,400 sq. km)
Languages Bamileke, Fang, Fulani, French*, English*
Religions Roman Catholic 35%, Traditional beliefs 25%, Muslim 22%, Protestant 18%
Ethnic mix Cameroon highlanders 31%, Other 21%, Equatorial Bantu 19%, Kirdi 11%, Fulani 10%, Northwestern Bantu 8%
Government Presidential system
Currency CFA franc = 100 centimes
Literacy rate 71%
Calorie consumption 2586 kilocalories

CANADA
North America

Official name Canada
Formation 1867 / 1949
Capital Ottawa
Population 35.5 million / 10 people per sq mile (4 people per sq km)
Total area 3,855,171 sq. miles (9,984,670 sq. km)
Languages English*, French*, Chinese, Italian, German, Ukrainian, Portuguese, Inuktitut, Cree
Religions Roman Catholic 44%, Protestant 29%, Other and nonreligious 27%
Ethnic mix European origin 66%, other 27%, Asian 5%, Amerindian 2%
Government Parliamentary system
Currency Canadian dollar = 100 cents
Literacy rate 99%
Calorie consumption 3419 kilocalories

CAPE VERDE
Atlantic Ocean

Official name Republic of Cape Verde
Formation 1975 / 1975
Capital Praia
Population 500,000 / 321 people per sq mile (124 people per sq km)
Total area 1557 sq. miles (4033 sq. km)
Languages Portuguese Creole, Portuguese*
Religions Roman Catholic 97%, Other 2%, Protestant (Church of the Nazarene) 1%
Ethnic mix Mestiço 71%, African 28%, European 1%
Government Mixed presidential–parliamentary system
Currency Escudo = 100 centavos
Literacy rate 85%
Calorie consumption 2716 kilocalories

CENTRAL AFRICAN REPUBLIC
Central Africa

Official name Central African Republic
Formation 1960 / 1960
Capital Bangui
Population 4.7 million / 20 people per sq mile (8 people per sq km)
Total area 240,534 sq. miles (622,984 sq. km)
Languages Sango, Banda, Gbaya, French*
Religions Traditional beliefs 35%, Roman Catholic 25%, Protestant 25%, Muslim 15%
Ethnic mix Baya 33%, Banda 27%, Other 17%, Mandjia 13%, Sara 10%
Government Transitional regime
Currency CFA franc = 100 centimes
Literacy rate 37%
Calorie consumption 2154 kilocalories

CHAD
Central Africa

Official name Republic of Chad
Formation 1960 / 1960
Capital N'Djaména
Population 13.2 million / 27 people per sq mile (10 people per sq km)
Total area 495,752 sq. miles (1,284,000 sq. km)
Languages French*, Sara, Arabic*, Maba
Religions Muslim 51%, Christian 35%, Animist 7%, Traditional beliefs 7%
Ethnic mix Other 30%, Sara 28%, Mayo-Kebbi 12%, Arab 12%, Ouaddai 9%, Kanem-Bornou 9%
Government Presidential system
Currency CFA franc = 100 centimes
Literacy rate 37%
Calorie consumption 2110 kilocalories

CHILE
South America

Official name Republic of Chile
Formation 1818 / 1883
Capital Santiago
Population 17.8 million / 62 people per sq mile (24 people per sq km)
Total area 292,258 sq. miles (756,950 sq. km)
Languages Spanish*, Amerindian languages
Religions Roman Catholic 89%, Other and nonreligious 11%
Ethnic mix Mestizo and European 90%, Other Amerindian 9%, Mapuche 1%
Government Presidential system
Currency Chilean peso = 100 centavos
Literacy rate 99%
Calorie consumption 2989 kilocalories

CHINA
East Asia

Official name People's Republic of China
Formation 960 / 1999
Capital Beijing
Population 1.39 billion / 387 people per sq mile (149 people per sq km)
Total area 3,705,386 sq. miles (9,596,960 sq. km)
Languages Mandarin*, Wu, Cantonese, Hsiang, Min, Hakka, Kan
Religions Nonreligious 59%, Traditional beliefs 20%, Other 13%, Buddhist 6%, Muslim 2%
Ethnic mix Han 92%, Other 4%, Hui 1%, Miao 1%, Manchu 1%, Zhuang 1%
Government One-party state
Currency Renminbi (known as yuan) = 10 jiao = 100 fen
Literacy rate 95%
Calorie consumption 3108 kilocalories

COLOMBIA
South America

Official name Republic of Colombia
Formation 1819 / 1903
Capital Bogotá
Population 48.9 million / 122 people per sq mile (47 people per sq km)
Total area 439,733 sq. miles (1,138,910 sq. km)
Languages Spanish*, Wayuu, Páez, and other Amerindian languages
Religions Roman Catholic 95%, Other 5%
Ethnic mix Mestizo 58%, White 20%, European–African 14%, African 4%, African–Amerindian 3%, Amerindian 1%
Government Presidential system
Currency Colombian peso = 100 centavos
Literacy rate 94%
Calorie consumption 2804 kilocalories

COMOROS
Indian Ocean

Official name Union of the Comoros
Formation 1975 / 1975
Capital Moroni
Population 800,000 / 929 people per sq mile (359 people per sq km)
Total area 838 sq. miles (2170 sq. km)
Languages Arabic*, Comoran*, French*
Religions Muslim (mainly Sunni) 98%, Other 1%, Roman Catholic 1%
Ethnic mix Comoran 97%, Other 3%
Government Presidential system
Currency Comoros franc = 100 centimes
Literacy rate 76%
Calorie consumption 2139 kilocalories

CONGO
Central Africa

Official name Republic of the Congo
Formation 1960 / 1960
Capital Brazzaville
Population 4.6 million / 35 people per sq mile (13 people per sq km)
Total area 132,046 sq. miles (342,000 sq. km)
Languages Kongo, Teke, Lingala, French*
Religions Traditional beliefs 50%, Roman Catholic 35%, Protestant 13%, Muslim 2%
Ethnic mix Bakongo 51%, Teke 17%, Other 16%, Mbochi 11%, Mbédé 5%
Government Presidential system
Currency CFA franc = 100 centimes
Literacy rate 79%
Calorie consumption 2195 kilocalories

CONGO, DEM. REP.
Central Africa

Official name Democratic Republic of the Congo
Formation 1960 / 1960
Capital Kinshasa
Population 69.4 million / 79 people per sq mile (31 people per sq km)
Total area 905,563 sq. miles (2,345,410 sq. km)
Languages Kiswahili, Tshiluba, Kikongo, Lingala, French*
Religions Roman Catholic 50%, Protestant 20%, Traditional beliefs and other 10%, Muslim 10%, Kimbanguist 10%
Ethnic mix Other 55%, Mongo, Luba, Kongo, and Mangbetu-Azande 45%
Government Presidential system
Currency Congolese franc = 100 centimes
Literacy rate 61%
Calorie consumption 1585 kilocalories

COSTA RICA
Central America

Official name Republic of Costa Rica
Formation 1838 / 1838
Capital San José
Population 4.9 million / 249 people per sq mile (96 people per sq km)
Total area 19,730 sq. miles (51,100 sq. km)
Languages Spanish*, English Creole, Bribri, Cabecar
Religions Roman Catholic 71%, Evangelical 14%, Nonreligious 11%, Other 4%
Ethnic mix Mestizo and European 94%, Black 3%, Other 1%, Chinese 1%, Amerindian 1%
Government Presidential system
Currency Costa Rican colón = 100 céntimos
Literacy rate 97%
Calorie consumption 2898 kilocalories

CROATIA
Southeast Europe

Official name Republic of Croatia
Formation 1991 / 1991
Capital Zagreb
Population 4.3 million / 197 people per sq mile (76 people per sq km)
Total area 21,831 sq. miles (56,542 sq. km)
Languages Croatian*
Religions Roman Catholic 88%, Other 7%, Orthodox Christian 4%, Muslim 1%
Ethnic mix Croat 90%, Other 5%, Serb 5%
Government Parliamentary system
Currency Kuna = 100 lipa
Literacy rate 99%
Calorie consumption 3052 kilocalories

CUBA
West Indies

Official name Republic of Cuba
Formation 1902 / 1902
Capital Havana
Population 11.3 million / 264 people per sq mile (102 people per sq km)
Total area 42,803 sq. miles (110,860 sq. km)
Languages Spanish*
Religions Nonreligious 49%, Roman Catholic 40%, Atheist 6%, Other 4%, Protestant 1%
Ethnic mix Mulatto (mixed race) 51%, White 37%, Black 11%, Chinese 1%
Government One-party state
Currency Cuban peso = 100 centavos
Literacy rate 99%
Calorie consumption 3277 kilocalories

CYPRUS
Southeast Europe

Official name Republic of Cyprus
Formation 1960 / 1960
Capital Nicosia
Population 1.2 million / 336 people per sq mile (130 people per sq km)
Total area 3571 sq. miles (9250 sq. km)
Languages Greek*, Turkish
Religions Orthodox Christian 78%, Muslim 18%, Other 4%
Ethnic mix Greek 81%, Turkish 11%, Other 8%
Government Presidential system
Currency Euro = 100 cents; (TRNC: new Turkish lira = 100 kurus)
Literacy rate 99%
Calorie consumption 2661 kilocalories

CZECH REPUBLIC
Central Europe

Official name Czech Republic
Formation 1993 / 1993
Capital Prague
Population 10.7 million / 351 people per sq mile (136 people per sq km)
Total area 30,450 sq. miles (78,866 sq. km)
Languages Czech*, Slovak, Hungarian (Magyar)
Religions Roman Catholic 39%, Atheist 38%, Other 18%, Protestant 3%, Hussite 2%
Ethnic mix Czech 90%, Moravian 4%, Other 4%, Slovak 2%
Government Parliamentary system
Currency Czech koruna = 100 haleru
Literacy rate 99%
Calorie consumption 3292 kilocalories

DENMARK
Northern Europe

Official name Kingdom of Denmark
Formation 950 / 1944
Capital Copenhagen
Population 5.6 million / 342 people per sq mile (132 people per sq km)
Total area 16,639 sq. miles (43,094 sq. km)
Languages Danish*
Religions Evangelical Lutheran 95%, Roman Catholic 3%, Muslim 2%
Ethnic mix Danish 96%, Other (including Scandinavian and Turkish) 3%, Faeroese and Inuit 1%
Government Parliamentary system
Currency Danish krone = 100 øre
Literacy rate 99%
Calorie consumption 3363 kilocalories

DJIBOUTI
East Africa

Official name Republic of Djibouti
Formation 1977 / 1977
Capital Djibouti
Population 900,000 / 101 people per sq mile (39 people per sq km)
Total area 8494 sq. miles (22,000 sq. km)
Languages Somali, Afar, French*, Arabic*
Religions Muslim (mainly Sunni) 94%, Christian 6%
Ethnic mix Issa 60%, Afar 35%, Other 5%
Government Presidential system
Currency Djibouti franc = 100 centimes
Literacy rate 70%
Calorie consumption 2526 kilocalories

DOMINICA
West Indies

Official name Commonwealth of Dominica
Formation 1978 / 1978
Capital Roseau
Population 73,449 / 253 people per sq mile (98 people per sq km)
Total area 291 sq. miles (754 sq. km)
Languages French Creole, English*
Religions Roman Catholic 77%, Protestant 15%, Other 8%
Ethnic mix Black 87%, Mixed race 9%, Carib 3%, Other 1%
Government Parliamentary system
Currency East Caribbean dollar = 100 cents
Literacy rate 88%
Calorie consumption 3047 kilocalories

DOMINICAN REPUBLIC
West Indies

Official name Dominican Republic
Formation 1865 / 1865
Capital Santo Domingo
Population 10.5 million / 562 people per sq mile (217 people per sq km)
Total area 18,679 sq. miles (48,380 sq. km)
Languages Spanish*, French Creole
Religions Roman Catholic 95%, Other and nonreligious 5%
Ethnic mix Mixed race 73%, European 16%, Black African 11%
Government Presidential system
Currency Dominican Republic peso = 100 centavos
Literacy rate 91%
Calorie consumption 2614 kilocalories

EAST TIMOR
Southeast Asia

Official name Democratic Republic of Timor-Leste
Formation 2002 / 2002
Capital Dili
Population 1.2 million / 213 people per sq mile (82 people per sq km)
Total area 5756 sq. miles (14,874 sq. km)
Languages Tetum (Portuguese/Austronesian)*, Bahasa Indonesia, Portuguese*
Religions Roman Catholic 95%, Other (including Muslim and Protestant) 5%
Ethnic mix Papuan groups approx 85%, Indonesian approx 13%, Chinese 2%
Government Parliamentary system
Currency US dollar = 100 cents
Literacy rate 58%
Calorie consumption 2083 kilocalories

ECUADOR
South America

Official name Republic of Ecuador
Formation 1830 / 1942
Capital Quito
Population 16 million / 150 people per sq mile (58 people per sq km)
Total area 109,483 sq. miles (283,560 sq. km)
Languages Spanish*, Quechua, other Amerindian languages
Religions Roman Catholic 95%, Protestant, Jewish, and other 5%
Ethnic mix Mestizo 77%, White 11%, Amerindian 7%, Black African 5%
Government Presidential system
Currency US dollar = 100 cents
Literacy rate 93%
Calorie consumption 2477 kilocalories

EGYPT
North Africa

Official name Arab Republic of Egypt
Formation 1936 / 1982
Capital Cairo
Population 83.4 million / 217 people per sq mile (84 people per sq km)
Total area 386,660 sq. miles (1,001,450 sq. km)
Languages Arabic*, French, English, Berber
Religions Muslim (mainly Sunni) 90%, Coptic Christian and other 9%, Other Christian 1%
Ethnic mix Egyptian 99%, Nubian, Armenian, Greek, and Berber 1%
Government Transitional regime
Currency Egyptian pound = 100 piastres
Literacy rate 74%
Calorie consumption 3557 kilocalories

EL SALVADOR
Central America

Official name Republic of El Salvador
Formation 1841 / 1841
Capital San Salvador
Population 6.4 million / 800 people per sq mile (309 people per sq km)
Total area 8124 sq. miles (21,040 sq. km)
Languages Spanish*
Religions Roman Catholic 80%, Evangelical 18%, Other 2%
Ethnic mix Mestizo 90%, White 9%, Amerindian 1%
Government Presidential system
Currency Salvadorean colón = 100 centavos; and US dollar = 100 cents
Literacy rate 86%
Calorie consumption 2513 kilocalories

EQUATORIAL GUINEA
Central Africa

Official name Republic of Equatorial Guinea
Formation 1968 / 1968
Capital Malabo
Population 800,000 / 74 people per sq mile (29 people per sq km)
Total area 10,830 sq. miles (28,051 sq. km)
Languages Spanish*, Fang, Bubi, French*
Religions Roman Catholic 90%, Other 10%
Ethnic mix Fang 85%, Other 11%, Bubi 4%
Government Presidential system
Currency CFA franc = 100 centimes
Literacy rate 94%
Calorie consumption Not available

ERITREA
East Africa

Official name State of Eritrea
Formation 1993 / 2002
Capital Asmara
Population 6.5 million / 143 people per sq mile (55 people per sq km)
Total area 46,842 sq. miles (121,320 sq. km)
Languages Tigrinya*, English*, Tigre, Afar, Arabic*, Saho, Bilen, Kunama, Nara, Hadareb
Religions Christian 50%, Muslim 48%, Other 2%
Ethnic mix Tigray 50%, Tigre 31%, Other 9%, Afar 5%, Saho 5%
Government Mixed presidential–parliamentary system
Currency Nakfa = 100 cents
Literacy rate 70%
Calorie consumption 1640 kilocalories

ESTONIA
Northeast Europe

Official name Republic of Estonia
Formation 1991 / 1991
Capital Tallinn
Population 1.3 million / 75 people per sq mile (29 people per sq km)
Total area 17,462 sq. miles (45,226 sq. km)
Languages Estonian*, Russian
Religions Evangelical Lutheran 56%, Orthodox Christian 25%, Other 19%
Ethnic mix Estonian 69%, Russian 25%, Other 4%, Ukrainian 2%
Government Parliamentary system
Currency Euro = 100 cents
Literacy rate 99%
Calorie consumption 3214 kilocalories

ETHIOPIA
East Africa

Official name Federal Democratic Republic of Ethiopia
Formation 1896 / 2002
Capital Addis Ababa
Population 96.5 million / 225 people per sq mile (87 people per sq km)
Total area 435,184 sq. miles (1,127,127 sq. km)
Languages Amharic*, Tigrinya, Galla, Sidamo, Somali, English, Arabic
Religions Orthodox Christian 40%, Muslim 40%, Traditional beliefs 15%, Other 5%
Ethnic mix Oromo 40%, Amhara 25%, Other 13%, Sidama 9%, Tigray 7%, Somali 6%
Government Parliamentary system
Currency Birr = 100 cents
Literacy rate 39%
Calorie consumption 2131 kilocalories

FIJI
Australasia & Oceania

Official name Republic of Fiji
Formation 1970 / 1970
Capital Suva
Population 900,000 / 128 people per sq mile (49 people per sq km)
Total area 7054 sq. miles (18,270 sq. km)
Languages Fijian, English*, Hindi, Urdu, Tamil, Telugu
Religions Hindu 38%, Methodist 37%, Roman Catholic 9%, Muslim 8%, Other 8%
Ethnic mix Melanesian 51%, Indian 44%, Other 5%
Government Parliamentary system
Currency Fiji dollar = 100 cents
Literacy rate 94%
Calorie consumption 2930 kilocalories

FINLAND
Northern Europe

Official name Republic of Finland
Formation 1917 / 1947
Capital Helsinki
Population 5.4 million / 46 people per sq mile (18 people per sq km)
Total area 130,127 sq. miles (337,030 sq. km)
Languages Finnish*, Swedish*, Sámi
Religions Evangelical Lutheran 83%, Other 15%, Orthodox Christian 1%, Roman Catholic 1%
Ethnic mix Finnish 93%, Other (including Sámi) 7%
Government Parliamentary system
Currency Euro = 100 cents
Literacy rate 99%
Calorie consumption 3285 kilocalories

FRANCE
Western Europe

Official name French Republic
Formation 987 / 1919
Capital Paris
Population 64.6 million / 304 people per sq mile (117 people per sq km)
Total area 211,208 sq. miles (547,030 sq. km)
Languages French*, Provençal, German, Breton, Catalan, Basque
Religions Roman Catholic 88%, Muslim 8%, Protestant 2%, Buddhist 1%, Jewish 1%
Ethnic mix French 90%, North African (mainly Algerian) 6%, German (Alsace) 2%, Breton 1%, Other (including Corsicans) 1%
Government Mixed presidential–parliamentary system
Currency Euro = 100 cents
Literacy rate 99%
Calorie consumption 3524 kilocalories

GABON
Central Africa

Official name Gabonese Republic
Formation 1960 / 1960
Capital Libreville
Population 1.7 million / 17 people per sq mile (7 people per sq km)
Total area 103,346 sq. miles (267,667 sq. km)
Languages Fang, French*, Punu, Sira, Nzebi, Mpongwe
Religions Christian (mainly Roman Catholic) 55%, Traditional beliefs 40%, Other 4%, Muslim 1%
Ethnic mix Fang 26%, Shira-punu 24%, Other 16%, Foreign residents 15%, Nzabi-duma 11%, Mbédé-Teke 8%
Government Presidential system
Currency CFA franc = 100 centimes
Literacy rate 82%
Calorie consumption 2781 kilocalories

GAMBIA
West Africa

Official name Republic of the Gambia
Formation 1965 / 1965
Capital Banjul
Population 1.9 million / 492 people per sq mile (190 people per sq km)
Total area 4363 sq. miles (11,300 sq. km)
Languages Mandinka, Fulani, Wolof, Jola, Soninke, English*
Religions Sunni Muslim 90%, Christian 8%, Traditional beliefs 2%
Ethnic mix Mandinka 42%, Fulani 18%, Wolof 16%, Jola 10%, Serahuli 9%, Other 5%
Government Presidential system
Currency Dalasi = 100 butut
Literacy rate 52%
Calorie consumption 2849 kilocalories

GEORGIA
Southwest Asia

Official name Georgia
Formation 1991 / 1991
Capital Tbilisi
Population 4.3 million / 160 people per sq mile (62 people per sq km)
Total area 26,911 sq. miles (69,700 sq. km)
Languages Georgian*, Russian, Azeri, Armenian, Mingrelian, Ossetian, Abkhazian* (in Abkhazia)
Religions Georgian Orthodox 74%, Muslim 10%, Russian Orthodox 10%, Armenian Apostolic Church (Orthodox) 4%, Other 2%
Ethnic mix Georgian 84%, Azeri 6%, Armenian 6%, Russian 2%, Ossetian 1%, Other 1%
Government Presidential system
Currency Lari = 100 tetri
Literacy rate 99%
Calorie consumption 2731 kilocalories

GERMANY
Northern Europe

Official name Federal Republic of Germany
Formation 1871 / 1990
Capital Berlin
Population 82.7 million / 613 people per sq mile (237 people per sq km)
Total area 137,846 sq. miles (357,021 sq. km)
Languages German*, Turkish
Religions Protestant 34%, Roman Catholic 33%, Other 30%, Muslim 3%
Ethnic mix German 92%, Other European 3%, Other 3%, Turkish 2%
Government Parliamentary system
Currency Euro = 100 cents
Literacy rate 99%
Calorie consumption 3539 kilocalories

GHANA
West Africa

Official name Republic of Ghana
Formation 1957 / 1957
Capital Accra
Population 26.4 million / 297 people per sq mile (115 people per sq km)
Total area 92,100 sq. miles (238,540 sq. km)
Languages Twi, Fanti, Ewe, Ga, Adangbe, Gurma, Dagomba (Dagbani), English*
Religions Christian 69%, Muslim 16%, Traditional beliefs 9%, Other 6%
Ethnic mix Akan 49%, Mole-Dagbani 17%, Ewe 13%, Other 9%, Ga and Ga-Adangbe 8%, Guan 4%
Government Presidential system
Currency Cedi = 100 pesewas
Literacy rate 72%
Calorie consumption 3003 kilocalories

GREECE
Southeast Europe

Official name Hellenic Republic
Formation 1829 / 1947
Capital Athens
Population 11.1 million / 220 people per sq mile (85 people per sq km)
Total area 50,942 sq. miles (131,940 sq. km)
Languages Greek*, Turkish, Macedonian, Albanian
Religions Orthodox Christian 98%, Muslim 1%, Other 1%
Ethnic mix Greek 98%, Other 2%
Government Parliamentary system
Currency Euro = 100 cents
Literacy rate 97%
Calorie consumption 3433 kilocalories

GRENADA
West Indies

Official name Grenada
Formation 1974 / 1974
Capital St. George's
Population 110,152 / 841 people per sq mile (324 people per sq km)
Total area 131 sq. miles (340 sq. km)
Languages English*, English Creole
Religions Roman Catholic 68%, Anglican 17%, Other 15%
Ethnic mix Black African 82%, Mulatto (mixed race) 13%, East Indian 3%, Other 2%
Government Parliamentary system
Currency East Caribbean dollar = 100 cents
Literacy rate 96%
Calorie consumption 2453 kilocalories

GUATEMALA
Central America

Official name Republic of Guatemala
Formation 1838 / 1838
Capital Guatemala City
Population 15.9 million / 380 people per sq mile (147 people per sq km)
Total area 42,042 sq. miles (108,890 sq. km)
Languages Quiché, Mam, Cakchiquel, Kekchi, Spanish*
Religions Roman Catholic 65%, Protestant 33%, Other and nonreligious 2%
Ethnic mix Amerindian 60%, Mestizo 30%, Other 10%
Government Presidential system
Currency Quetzal = 100 centavos
Literacy rate 78%
Calorie consumption 2419 kilocalories

GUINEA
West Africa

Official name Republic of Guinea
Formation 1958 / 1958
Capital Conakry
Population 12 million / 126 people per sq mile (49 people per sq km)
Total area 94,925 sq. miles (245,857 sq. km)
Languages Pulaar, Malinké, Soussou, French*
Religions Muslim 85%, Christian 8%, Traditional beliefs 7%
Ethnic mix Peul 40%, Malinké 30%, Soussou 20%, Other 10%
Government Presidential system
Currency Guinea franc = 100 centimes
Literacy rate 25%
Calorie consumption 2553 kilocalories

GUINEA-BISSAU
West Africa

Official name Republic of Guinea-Bissau
Formation 1974 / 1974
Capital Bissau
Population 1.7 million / 157 people per sq mile (60 people per sq km)
Total area 13,946 sq. miles (36,120 sq. km)
Languages Portuguese Creole, Balante, Fulani, Malinké, Portuguese*
Religions Traditional beliefs 50%, Muslim 40%, Christian 10%
Ethnic mix Balante 30%, Fulani 20%, Other 16%, Mandyako 14%, Mandinka 13%, Papel 7%
Government Presidential system
Currency CFA franc = 100 centimes
Literacy rate 57%
Calorie consumption 2304 kilocalories

GUYANA
South America

Official name Cooperative Republic of Guyana
Formation 1966 / 1966
Capital Georgetown
Population 800,000 / 11 people per sq mile (4 people per sq km)
Total area 83,000 sq. miles (214,970 sq. km)
Languages English Creole, Hindi, Tamil, Amerindian languages, English*
Religions Christian 57%, Hindu 28%, Muslim 10%, Other 5%
Ethnic mix East Indian 43%, Black African 30%, Mixed race 17%, Amerindian 9%, Other 1%
Government Presidential system
Currency Guyanese dollar = 100 cents
Literacy rate 85%
Calorie consumption 2648 kilocalories

HAITI
West Indies

Official name Republic of Haiti
Formation 1804 / 1844
Capital Port-au-Prince
Population 10.5 million / 987 people per sq mile (381 people per sq km)
Total area 10,714 sq. miles (27,750 sq. km)
Languages French Creole*, French*
Religions Roman Catholic 55%, Protestant 28%, Other (including Voodoo) 16%, Nonreligious 1%
Ethnic mix Black African 95%, Mulatto (mixed race) and European 5%
Government Presidential system
Currency Gourde = 100 centimes
Literacy rate 49%
Calorie consumption 2091 kilocalories

HONDURAS
Central America

Official name Republic of Honduras
Formation 1838 / 1838
Capital Tegucigalpa
Population 8.3 million / 192 people per sq mile (74 people per sq km)
Total area 43,278 sq. miles (112,090 sq. km)
Languages Spanish*, Garífuna (Carib), English Creole
Religions Roman Catholic 97%, Protestant 3%
Ethnic mix Mestizo 90%, Black African 5%, Amerindian 4%, White 1%
Government Presidential system
Currency Lempira = 100 centavos
Literacy rate 85%
Calorie consumption 2651 kilocalories

HUNGARY
Central Europe

Official name Hungary
Formation 1918 / 1947
Capital Budapest
Population 9.9 million / 278 people per sq mile (107 people per sq km)
Total area 35,919 sq. miles (93,030 sq. km)
Languages Hungarian (Magyar)*
Religions Roman Catholic 52%, Calvinist 16%, Other 15%, Nonreligious 14%, Lutheran 3%
Ethnic mix Magyar 90%, Roma 4%, German 3%, Serb 2%, Other 1%
Government Parliamentary system
Currency Forint = 100 fillér
Literacy rate 99%
Calorie consumption 2968 kilocalories

ICELAND
Northwest Europe

Official name Republic of Iceland
Formation 1944 / 1944
Capital Reykjavík
Population 300,000 / 8 people per sq mile (3 people per sq km)
Total area 39,768 sq. miles (103,000 sq. km)
Languages Icelandic*
Religions Evangelical Lutheran 84%, Other (mostly Christian) 10%, Roman Catholic 3%, Nonreligious 3%
Ethnic mix Icelandic 94%, Other 5%, Danish 1%
Government Parliamentary system
Currency Icelandic króna = 100 aurar
Literacy rate 99%
Calorie consumption 3339 kilocalories

INDIA
South Asia

Official name Republic of India
Formation 1947 / 1947
Capital New Delhi
Population 1.27 billion / 1104 people per sq mile (426 people per sq km)
Total area 1,269,339 sq. miles (3,287,590 sq. km)
Languages Hindi*, English*, Urdu, Bengali, Marathi, Telugu, Tamil, Bihari, Gujarati, Kanarese
Religions Hindu 81%, Muslim 13%, Christian 2%, Sikh 2%, Buddhist 1%, Other 1%
Ethnic mix Indo-Aryan 72%, Dravidian 25%, Mongoloid and other 3%
Government Parliamentary system
Currency Indian rupee = 100 paise
Literacy rate 63%
Calorie consumption 2459 kilocalories

INDONESIA
Southeast Asia

Official name Republic of Indonesia
Formation 1949 / 1999
Capital Jakarta
Population 253 million / 364 people per sq mile (141 people per sq km)
Total area 741,096 sq. miles (1,919,440 sq. km)
Languages Javanese, Sundanese, Madurese, Bahasa Indonesia*, Dutch
Religions Sunni Muslim 86%, Protestant 6%, Roman Catholic 3%, Hindu 2%, Other 2%, Buddhist 1%
Ethnic mix Javanese 41%, Other 29%, Sundanese 15%, Coastal Malays 12%, Madurese 3%
Government Presidential system
Currency Rupiah = 100 sen
Literacy rate 93%
Calorie consumption 2777 kilocalories

IRAN
Southwest Asia

Official name Islamic Republic of Iran
Formation 1502 / 1990
Capital Tehran
Population 78.5 million / 124 people per sq mile (48 people per sq km)
Total area 636,293 sq. miles (1,648,000 sq. km)
Languages Farsi*, Azeri, Luri, Gilaki, Mazanderani, Kurdish, Turkmen, Arabic, Baluchi
Religions Shi'a Muslim 89%, Sunni Muslim 9%, Other 2%
Ethnic mix Persian 51%, Azari 24%, Other 10%, Lur and Bakhtiari 8%, Kurdish 7%
Government Islamic theocracy
Currency Iranian rial = 100 dinars
Literacy rate 84%
Calorie consumption 3058 kilocalories

IRAQ
Southwest Asia

Official name Republic of Iraq
Formation 1932 / 1990
Capital Baghdad
Population 34.8 million / 206 people per sq mile (80 people per sq km)
Total area 168,753 sq. miles (437,072 sq. km)
Languages Arabic*, Kurdish*, Turkic languages, Armenian, Assyrian
Religions Shi'a Muslim 60%, Sunni Muslim 35%, Other (including Christian) 5%
Ethnic mix Arab 80%, Kurdish 15%, Turkmen 3%, Other 2%
Government Parliamentary system
Currency New Iraqi dinar = 1000 fils
Literacy rate 79%
Calorie consumption 2489 kilocalories

IRELAND
Northwest Europe

Official name Ireland
Formation 1922 / 1922
Capital Dublin
Population 4.7 million / 177 people per sq mile (68 people per sq km)
Total area 27,135 sq. miles (70,280 sq. km)
Languages English*, Irish*
Religions Roman Catholic 87%, Other and nonreligious 10%, Anglican 3%
Ethnic mix Irish 99%, Other 1%
Government Parliamentary system
Currency Euro = 100 cents
Literacy rate 99%
Calorie consumption 3591 kilocalories

ISRAEL
Southwest Asia

Official name State of Israel
Formation 1948 / 1994
Capital Jerusalem (not internationally recognized)
Population 7.8 million / 994 people per sq mile (384 people per sq km)
Total area 8019 sq. miles (20,770 sq. km)
Languages Hebrew*, Arabic*, Yiddish, German, Russian, Polish, Romanian, Persian
Religions Jewish 76%, Muslim (mainly Sunni) 16%, Other 4%, Druze 2%, Christian 2%
Ethnic mix Jewish 76%, Arab 20%, Other 4%
Government Parliamentary system
Currency Shekel = 100 agorot
Literacy rate 98%
Calorie consumption 3619 kilocalories

ITALY
Southern Europe

Official name Italian Republic
Formation 1861 / 1947
Capital Rome
Population 61.1 million / 538 people per sq mile (208 people per sq km)
Total area 116,305 sq. miles (301,230 sq. km)
Languages Italian*, German, French, Rhaeto-Romanic, Sardinian
Religions Roman Catholic 85%, Other and nonreligious 13%, Muslim 2%
Ethnic mix Italian 94%, Other 4%, Sardinian 2%
Government Parliamentary system
Currency Euro = 100 cents
Literacy rate 99%
Calorie consumption 3539 kilocalories

IVORY COAST
West Africa

Official name Republic of Côte d'Ivoire
Formation 1960 / 1960
Capital Yamoussoukro
Population 20.8 million / 169 people per sq mile (65 people per sq km)
Total area 124,502 sq. miles (322,460 sq. km)
Languages Akan, French*, Krou, Voltaïque
Religions Muslim 38%, Traditional beliefs 25%, Roman Catholic 25%, Other 6%, Protestant 6%
Ethnic mix Akan 42%, Voltaïque 18%, Mandé du Nord 17%, Krou 11%, Mandé du Sud 10%, Other 2%
Government Presidential system
Currency CFA franc = 100 centimes
Literacy rate 41%
Calorie consumption 2799 kilocalories

JAMAICA
West Indies

Official name Jamaica
Formation 1962 / 1962
Capital Kingston
Population 2.8 million / 670 people per sq mile (259 people per sq km)
Total area 4243 sq. miles (10,990 sq. km)
Languages English Creole, English*
Religions Other and nonreligious 45%, Other Protestant 20%, Church of God 18%, Baptist 10%, Anglican 7%
Ethnic mix Black 91%, Mulatto (mixed race) 7%, European and Chinese 1%, East Indian 1%
Government Parliamentary system
Currency Jamaican dollar = 100 cents
Literacy rate 88%
Calorie consumption 2746 kilocalories

JAPAN
East Asia

Official name Japan
Formation 1590 / 1972
Capital Tokyo
Population 127 million / 874 people per sq mile (337 people per sq km)
Total area 145,882 sq. miles (377,835 sq. km)
Languages Japanese*, Korean, Chinese
Religions Shinto and Buddhist 76%, Buddhist 16%, Other (including Christian) 8%
Ethnic mix Japanese 99%, Other (mainly Korean) 1%
Government Parliamentary system
Currency Yen = 100 sen
Literacy rate 99%
Calorie consumption 2719 kilocalories

JORDAN
Southwest Asia

Official name Hashemite Kingdom of Jordan
Formation 1946 / 1967
Capital Amman
Population 7.5 million / 218 people per sq mile (84 people per sq km)
Total area 35,637 sq. miles (92,300 sq. km)
Languages Arabic*
Religions Sunni Muslim 92%, Christian 6%, Other 2%
Ethnic mix Arab 98%, Circassian 1%, Armenian 1%
Government Monarchy
Currency Jordanian dinar = 1000 fils
Literacy rate 98%
Calorie consumption 3149 kilocalories

KAZAKHSTAN
Central Asia

Official name Republic of Kazakhstan
Formation 1991 / 1991
Capital Astana
Population 16.6 million / 16 people per sq mile (6 people per sq km)
Total area 1,049,150 sq. miles (2,717,300 sq. km)
Languages Kazakh*, Russian, Ukrainian, German, Uzbek, Tatar, Uighur
Religions Muslim (mainly Sunni) 47%, Orthodox Christian 44%, Other 7%, Protestant 2%
Ethnic mix Kazakh 57%, Russian 27%, Other 8%, Uzbek 3%, Ukrainian 3%, German 2%
Government Presidential system
Currency Tenge = 100 tiyn
Literacy rate 99%
Calorie consumption 3107 kilocalories

KENYA
East Africa

Official name Republic of Kenya
Formation 1963 / 1963
Capital Nairobi
Population 45.5 million / 208 people per sq mile (80 people per sq km)
Total area 224,961 sq. miles (582,650 sq. km)
Languages Kiswahili*, English*, Kikuyu, Luo, Kalenjin, Kamba
Religions Christian 80%, Muslim 10%, Traditional beliefs 9%, Other 1%
Ethnic mix Other 28%, Kikuyu 22%, Luo 14%, Luhya 14%, Kalenjin 11%, Kamba 11%
Government Presidential system
Currency Kenya shilling = 100 cents
Literacy rate 72%
Calorie consumption 2206 kilocalories

KIRIBATI
Australasia & Oceania

Official name Republic of Kiribati
Formation 1979 / 1979
Capital Tarawa Atoll
Population 104,488 / 381 people per sq mile (147 people per sq km)
Total area 277 sq. miles (717 sq. km)
Languages English*, Kiribati
Religions Roman Catholic 55%, Kiribati Protestant Church 36%, Other 9%
Ethnic mix Micronesian 99%, Other 1%
Government Presidential system
Currency Australian dollar = 100 cents
Literacy rate 99%
Calorie consumption 3022 kilocalories

KOSOVO (not yet recognised)
Southeast Europe

Official name Republic of Kosovo
Formation 2008 / 2008
Capital Pristina
Population 1.9 million / 451 people per sq mile (174 people per sq km)
Total area 4212 sq. miles (10,908 sq. km)
Languages Albanian*, Serbian*, Bosniak, Gorani, Roma, Turkish
Religions Muslim 92%, Roman Catholic 4%, Orthodox Christian 4%
Ethnic mix Albanian 92%, Serb 4%, Bosniak and Gorani 2%, Turkish 1%, Roma 1%
Government Parliamentary system
Currency Euro = 100 cents
Literacy rate 92%
Calorie consumption Not available

KUWAIT
Southwest Asia

Official name State of Kuwait
Formation 1961 / 1961
Capital Kuwait City
Population 3.5 million / 509 people per sq mile (196 people per sq km)
Total area 6880 sq. miles (17,820 sq. km)
Languages Arabic*, English
Religions Sunni Muslim 45%, Shi'a Muslim 40%, Christian, Hindu, and other 15%
Ethnic mix Kuwaiti 45%, Other Arab 35%, South Asian 9%, Other 7%, Iranian 4%
Government Monarchy
Currency Kuwaiti dinar = 1000 fils
Literacy rate 96%
Calorie consumption 3471 kilocalories

KYRGYZSTAN
Central Asia

Official name Kyrgyz Republic
Formation 1991 / 1991
Capital Bishkek
Population 5.6 million / 73 people per sq mile (28 people per sq km)
Total area 76,641 sq. miles (198,500 sq. km)
Languages Kyrgyz*, Russian*, Uzbek, Tatar, Ukrainian
Religions Muslim (mainly Sunni) 70%, Orthodox Christian 30%
Ethnic mix Kyrgyz 69%, Uzbek 14%, Russian 9%, Other 6%, Dungan 1%, Uighur 1%
Government Presidential system
Currency Som = 100 tyiyn
Literacy rate 99%
Calorie consumption 2828 kilocalories

LAOS
Southeast Asia

Official name Lao People's Democratic Republic
Formation 1953 / 1953
Capital Vientiane
Population 6.9 million / 77 people per sq mile (30 people per sq km)
Total area 91,428 sq. miles (236,800 sq. km)
Languages Lao*, Mon-Khmer, Yao, Vietnamese, Chinese, French
Religions Buddhist 65%, Other (including animist) 34%, Christian 1%
Ethnic mix Lao Loum 66%, Lao Theung 30%, Lao Soung 2%, Other 2%
Government One-party state
Currency Kip = 100 at
Literacy rate 73%
Calorie consumption 2356 kilocalories

LATVIA
Northeast Europe

Official name Republic of Latvia
Formation 1991 / 1991
Capital Riga
Population 2 million / 80 people per sq mile (31 people per sq km)
Total area 24,938 sq. miles (64,589 sq. km)
Languages Latvian*, Russian
Religions Other 43%, Lutheran 24%, Roman Catholic 18%, Orthodox Christian 15%
Ethnic mix Latvian 62%, Russian 27%, Other 4%, Belarussian 3%, Ukrainian 2%, Polish 2%
Government Parliamentary system
Currency Euro = 100 cents
Literacy rate 99%
Calorie consumption 3293 kilocalories

LEBANON
Southwest Asia

Official name Lebanese Republic
Formation 1941 / 1941
Capital Beirut
Population 5 million / 1266 people per sq mile (489 people per sq km)
Total area 4015 sq. miles (10,400 sq. km)
Languages Arabic*, French, Armenian, Assyrian
Religions Muslim 60%, Christian 39%, Other 1%
Ethnic mix Arab 95%, Armenian 4%, Other 1%
Government Parliamentary system
Currency Lebanese pound = 100 piastres
Literacy rate 90%
Calorie consumption 3181 kilocalories

LESOTHO
Southern Africa

Official name Kingdom of Lesotho
Formation 1966 / 1966
Capital Maseru
Population 2.1 million / 179 people per sq mile (69 people per sq km)
Total area 11,720 sq. miles (30,355 sq. km)
Languages English*, Sesotho*, isiZulu
Religions Christian 90%, Traditional beliefs 10%
Ethnic mix Sotho 99%, European and Asian 1%
Government Parliamentary system
Currency Loti = 100 lisente; and South African rand = 100 cents
Literacy rate 76%
Calorie consumption 2595 kilocalories

LIBERIA
West Africa

Official name Republic of Liberia
Formation 1847 / 1847
Capital Monrovia
Population 4.4 million / 118 people per sq mile (46 people per sq km)
Total area 43,000 sq. miles (111,370 sq. km)
Languages Kpelle, Vai, Bassa, Kru, Grebo, Kissi, Gola, Loma, English*
Religions Christian 40%, Traditional beliefs 40%, Muslim 20%
Ethnic mix Indigenous tribes (12 groups) 49%, Kpellé 20%, Bassa 16%, Gio 8%, Krou 7%
Government Presidential system
Currency Liberian dollar = 100 cents
Literacy rate 43%
Calorie consumption 2251 kilocalories

LIBYA
North Africa

Official name State of Libya
Formation 1951 / 1951
Capital Tripoli
Population 6.3 million / 9 people per sq mile (4 people per sq km)
Total area 679,358 sq. miles (1,759,540 sq. km)
Languages Arabic*, Tuareg
Religions Muslim (mainly Sunni) 97%, Other 3%
Ethnic mix Arab and Berber 97%, Other 3%
Government Transitional regime
Currency Libyan dinar = 1000 dirhams
Literacy rate 90%
Calorie consumption 3211 kilocalories

LIECHTENSTEIN
Central Europe

Official name Principality of Liechtenstein
Formation 1719 / 1719
Capital Vaduz
Population 37,313 / 602 people per sq mile (233 people per sq km)
Total area 62 sq. miles (160 sq. km)
Languages German*, Alemannish dialect, Italian
Religions Roman Catholic 79%, Other 13%, Protestant 8%
Ethnic mix Liechtensteiner 66%, Other 12%, Swiss 10%, Austrian 6%, German 3%, Italian 3%
Government Parliamentary system
Currency Swiss franc = 100 rappen/centimes
Literacy rate 99%
Calorie consumption Not available

LITHUANIA
Northeast Europe

Official name Republic of Lithuania
Formation 1991 / 1991
Capital Vilnius
Population 3 million / 119 people per sq mile (46 people per sq km)
Total area 25,174 sq. miles (65,200 sq. km)
Languages Lithuanian*, Russian
Religions Roman Catholic 77%, Other 17%, Russian Orthodox 4%, Protestant 1%, Old believers 1%
Ethnic mix Lithuanian 85%, Polish 7%, Russian 6%, Belarussian 1%, Other 1%
Government Parliamentary system
Currency Euro = 100 cents
Literacy rate 99%
Calorie consumption 3463 kilocalories

LUXEMBOURG
Northwest Europe

Official name Grand Duchy of Luxembourg
Formation 1867 / 1867
Capital Luxembourg-Ville
Population 500,000 / 501 people per sq mile (193 people per sq km)
Total area 998 sq. miles (2586 sq. km)
Languages Luxembourgish*, German*, French*
Religions Roman Catholic 97%, Protestant, Orthodox Christian, and Jewish 3%
Ethnic mix Luxembourger 62%, Foreign residents 38%
Government Parliamentary system
Currency Euro = 100 cents
Literacy rate 99%
Calorie consumption 3568 kilocalories

MACEDONIA
Southeast Europe

Official name Republic of Macedonia
Formation 1991 / 1991
Capital Skopje
Population 2.1 million / 212 people per sq mile (82 people per sq km)
Total area 9781 sq. miles (25,333 sq. km)
Languages Macedonian*, Albanian*, Turkish, Romani, Serbian
Religions Orthodox Christian 65%, Muslim 29%, Roman Catholic 4%, Other 2%
Ethnic mix Macedonian 64%, Albanian 25%, Turkish 4%, Roma 3%, Serb 2%, Other 2%
Government Mixed presidential–parliamentary system
Currency Macedonian denar = 100 deni
Literacy rate 98%
Calorie consumption 2923 kilocalories

MADAGASCAR
Indian Ocean

Official name Republic of Madagascar
Formation 1960 / 1960
Capital Antananarivo
Population 23.6 million / 105 people per sq mile (41 people per sq km)
Total area 226,656 sq. miles (587,040 sq. km)
Languages Malagasy*, French*, English*
Religions Traditional beliefs 52%, Christian (mainly Roman Catholic) 41%, Muslim 7%
Ethnic mix Other Malay 46%, Merina 26%, Betsimisaraka 15%, Betsileo 12%, Other 1%
Government Mixed presidential–parliamentary system
Currency Ariary = 5 iraimbilanja
Literacy rate 64%
Calorie consumption 2052 kilocalories

MALAWI
Southern Africa

Official name Republic of Malawi
Formation 1964 / 1964
Capital Lilongwe
Population 16.8 million / 463 people per sq mile (179 people per sq km)
Total area 45,745 sq. miles (118,480 sq. km)
Languages Chewa, Lomwe, Yao, Ngoni, English*
Religions Protestant 55%, Roman Catholic 20%, Muslim 20%, Traditional beliefs 5%
Ethnic mix Bantu 99%, Other 1%
Government Presidential system
Currency Malawi kwacha = 100 tambala
Literacy rate 61%
Calorie consumption 2334 kilocalories

MALAYSIA
Southeast Asia

Official name Malaysia
Formation 1963 / 1965
Capital Kuala Lumpur; Putrajaya (administrative)
Population 30.2 million / 238 people per sq mile (92 people per sq km)
Total area 127,316 sq. miles (329,750 sq. km)
Languages Bahasa Malaysia*, Malay, Chinese, Tamil, English
Religions Muslim (mainly Sunni) 61%, Buddhist 19%, Christian 9%, Hindu 6%, Other 5%
Ethnic mix Malay 53%, Chinese 26%, Indigenous tribes 12%, Indian 8%, Other 1%
Government Parliamentary system
Currency Ringgit = 100 sen
Literacy rate 93%
Calorie consumption 2855 kilocalories

MALDIVES
Indian Ocean

Official name Republic of Maldives
Formation 1965 / 1965
Capital Male'
Population 400,000 / 3448 people per sq mile (1333 people per sq km)
Total area 116 sq. miles (300 sq. km)
Languages Dhivehi (Maldivian), Sinhala, Tamil, Arabic
Religions Sunni Muslim 100%
Ethnic mix Arab–Sinhalese–Malay 100%
Government Presidential system
Currency Rufiyaa = 100 laari
Literacy rate 98%
Calorie consumption 2722 kilocalories

MALI
West Africa

Official name Republic of Mali
Formation 1960 / 1960
Capital Bamako
Population 15.8 million / 34 people per sq mile (13 people per sq km)
Total area 478,764 sq. miles (1,240,000 sq. km)
Languages Bambara, Fulani, Senufo, Soninke, French*
Religions Muslim (mainly Sunni) 90%, Traditional beliefs 6%, Christian 4%
Ethnic mix Bambara 52%, Other 14%, Fulani 11%, Saracolé 7%, Soninka 7%, Tuareg 5%, Mianka 4%
Government Presidential system
Currency CFA franc = 100 centimes
Literacy rate 34%
Calorie consumption 2833 kilocalories

MALTA
Southern Europe

Official name Republic of Malta
Formation 1964 / 1964
Capital Valletta
Population 400,000 / 3226 people per sq mile (1250 people per sq km)
Total area 122 sq. miles (316 sq. km)
Languages Maltese*, English*
Religions Roman Catholic 98%, Other and nonreligious 2%
Ethnic mix Maltese 96%, Other 4%
Government Parliamentary system
Currency Euro = 100 cents
Literacy rate 92%
Calorie consumption 3389 kilocalories

MARSHALL ISLANDS
Australasia & Oceania

Official name Republic of the Marshall Islands
Formation 1986 / 1986
Capital Majuro
Population 70,983 / 1014 people per sq mile (392 people per sq km)
Total area 70 sq. miles (181 sq. km)
Languages Marshallese*, English*, Japanese, German
Religions Protestant 90%, Roman Catholic 8%, Other 2%
Ethnic mix Micronesian 90%, Other 10%
Government Presidential system
Currency US dollar = 100 cents
Literacy rate 91%
Calorie consumption Not available

MAURITANIA
West Africa

Official name Islamic Republic of Mauritania
Formation 1960 / 1960
Capital Nouakchott
Population 4 million / 10 people per sq mile (4 people per sq km)
Total area 397,953 sq. miles (1,030,700 sq. km)
Languages Arabic*, Hassaniyah Arabic, Wolof, French
Religions Sunni Muslim 100%
Ethnic mix Maure 81%, Wolof 7%, Tukolor 5%, Other 4%, Soninka 3%
Government Presidential system
Currency Ouguiya = 5 khoums
Literacy rate 46%
Calorie consumption 2791 kilocalories

MAURITIUS
Indian Ocean

Official name Republic of Mauritius
Formation 1968 / 1968
Capital Port Louis
Population 1.2 million / 1671 people per sq mile (645 people per sq km)
Total area 718 sq. miles (1860 sq. km)
Languages French Creole, Hindi, Urdu, Tamil, Chinese, English*, French
Religions Hindu 48%, Roman Catholic 24%, Muslim 17%, Protestant 9%, Other 2%
Ethnic mix Indo-Mauritian 68%, Creole 27%, Sino-Mauritian 3%, Franco-Mauritian 2%
Government Parliamentary system
Currency Mauritian rupee = 100 cents
Literacy rate 89%
Calorie consumption 3055 kilocalories

MEXICO
North America

Official name United Mexican States
Formation 1836 / 1848
Capital Mexico City
Population 124 million / 168 people per sq mile (65 people per sq km)
Total area 761,602 sq. miles (1,972,550 sq. km)
Languages Spanish*, Nahuatl, Mayan, Zapotec, Mixtec, Otomi, Totonac, Tzotzil, Tzeltal
Religions Roman Catholic 77%, Other 14%, Protestant 6%, Nonreligious 3%
Ethnic mix Mestizo 60%, Amerindian 30%, European 9%, Other 1%
Government Presidential system
Currency Mexican peso = 100 centavos
Literacy rate 94%
Calorie consumption 3072 kilocalories

MICRONESIA
Australasia & Oceania

Official name Federated States of Micronesia
Formation 1986 / 1986
Capital Palikir (Pohnpei Island)
Population 105,681 / 390 people per sq mile (151 people per sq km)
Total area 271 sq. miles (702 sq. km)
Languages Trukese, Pohnpeian, Kosraean, Yapese, English*
Religions Roman Catholic 50%, Protestant 47%, Other 3%
Ethnic mix Chuukese 49%, Pohnpeian 24%, Other 14%, Kosraean 6%, Yapese 5%, Asian 2%
Government Nonparty system
Currency US dollar = 100 cents
Literacy rate 81%
Calorie consumption Not available

MOLDOVA
Southeast Europe

Official name Republic of Moldova
Formation 1991 / 1991
Capital Chisinau
Population 3.5 million / 269 people per sq mile (104 people per sq km)
Total area 13,067 sq. miles (33,843 sq. km)
Languages Moldovan*, Ukrainian, Russian
Religions Orthodox Christian 93%, Other 6%, Baptist 1%
Ethnic mix Moldovan 84%, Ukrainian 7%, Gagauz 5%, Russian 2%, Bulgarian 1%, Other 1%
Government Parliamentary system
Currency Moldovan leu = 100 bani
Literacy rate 99%
Calorie consumption 2837 kilocalories

MONACO
Southern Europe

Official name Principality of Monaco
Formation 1861 / 1861
Capital Monaco-Ville
Population 36,950 / 49,267 people per sq mile (18,949 people per sq km)
Total area 0.75 sq. miles (1.95 sq. km)
Languages French*, Italian, Monégasque, English
Religions Roman Catholic 89%, Protestant 6%, Other 5%
Ethnic mix French 47%, Other 21%, Italian 16%, Monégasque 16%
Government Mixed monarchical–parliamentary system
Currency Euro = 100 cents
Literacy rate 99%
Calorie consumption Not available

MONGOLIA
East Asia

Official name Mongolia
Formation 1924 / 1924
Capital Ulan Bator
Population 2.9 million / 5 people per sq mile (2 people per sq km)
Total area 604,247 sq. miles (1,565,000 sq. km)
Languages Khalkha Mongolian, Kazakh, Chinese, Russian
Religions Tibetan Buddhist 50%, Nonreligious 40%, Shamanist and Christian 6%, Muslim 4%
Ethnic mix Khalkh 95%, Kazakh 4%, Other 1%
Government Mixed presidential–parliamentary system
Currency Tugrik (tögrög) = 100 möngö
Literacy rate 98%
Calorie consumption 2463 kilocalories

MONTENEGRO
Southeast Europe

Official name Montenegro
Formation 2006 / 2006
Capital Podgorica
Population 600,000 / 113 people per sq mile (43 people per sq km)
Total area 5332 sq. miles (13,812 sq. km)
Languages Montenegrin*, Serbian, Albanian, Bosniak, Croatian
Religions Orthodox Christian 74%, Muslim 18%, Roman Catholic 4%, Other 4%
Ethnic mix Montenegrin 43%, Serb 32%, Other 12%, Bosniak 8%, Albanian 5%
Government Parliamentary system
Currency Euro = 100 cents
Literacy rate 98%
Calorie consumption 3568 kilocalories

MOROCCO
North Africa

Official name Kingdom of Morocco
Formation 1956 / 1969
Capital Rabat
Population 35.5 million / 194 people per sq mile (75 people per sq km)
Total area 172,316 sq. miles (446,300 sq. km)
Languages Arabic*, Tamazight (Berber), French, Spanish
Religions Muslim (mainly Sunni) 99%, Other (mostly Christian) 1%
Ethnic mix Arab 70%, Berber 29%, European 1%
Government Mixed monarchical–parliamentary system
Currency Moroccan dirham = 100 centimes
Literacy rate 67%
Calorie consumption 3334 kilocalories

MOZAMBIQUE
Southern Africa

Official name Republic of Mozambique
Formation 1975 / 1975
Capital Maputo
Population 26.5 million / 88 people per sq mile (34 people per sq km)
Total area 309,494 sq. miles (801,590 sq. km)
Languages Makua, Xitsonga, Sena, Lomwe, Portuguese*
Religions Traditional beliefs 56%, Christian 30%, Muslim 14%
Ethnic mix Makua Lomwe 47%, Tsonga 23%, Malawi 12%, Shona 11%, Yao 4%, Other 3%
Government Presidential system
Currency New metical = 100 centavos
Literacy rate 51%
Calorie consumption 2283 kilocalories

MYANMAR (BURMA)
Southeast Asia

Official name Republic of the Union of Myanmar
Formation 1948 / 1948
Capital Nay Pyi Taw
Population 53.7 million / 212 people per sq mile (82 people per sq km)
Total area 261,969 sq. miles (678,500 sq. km)
Languages Myanmar (Burmese)*, Shan, Karen, Rakhine, Chin, Yangbye, Kachin, Mon
Religions Buddhist 89%, Christian 4%, Muslim 4%, Other 2%, Animist 1%
Ethnic mix Burman (Bamah) 68%, Other 12%, Shan 9%, Karen 7%, Rakhine 4%
Government Presidential system
Currency Kyat = 100 pyas
Literacy rate 93%
Calorie consumption 2571 kilocalories

NAMIBIA
Southern Africa

Official name Republic of Namibia
Formation 1990 / 1994
Capital Windhoek
Population 2.3 million / 7 people per sq mile (3 people per sq km)
Total area 318,694 sq. miles (825,418 sq. km)
Languages Ovambo, Kavango, English*, Bergdama, German, Afrikaans
Religions Christian 90%, Traditional beliefs 10%
Ethnic mix Ovambo 50%, Other tribes 22%, Kavango 9%, Damara 7%, Herero 7%, Other 5%
Government Presidential system
Currency Namibian dollar = 100 cents; and South African rand = 100 cents
Literacy rate 76%
Calorie consumption 2086 kilocalories

NAURU
Australasia & Oceania

Official name Republic of Nauru
Formation 1968 / 1968
Capital None
Population 9488 / 1171 people per sq mile (452 people per sq km)
Total area 8.1 sq. miles (21 sq. km)
Languages Nauruan*, Kiribati, Chinese, Tuvaluan, English
Religions Nauruan Congregational Church 60%, Roman Catholic 35%, Other 5%
Ethnic mix Nauruan 93%, Chinese 5%, European 1%, Other Pacific islanders 1%
Government Nonparty system
Currency Australian dollar = 100 cents
Literacy rate 95%
Calorie consumption Not available

NEPAL
South Asia

Official name Federal Democratic Republic of Nepal
Formation 1769 / 1769
Capital Kathmandu
Population 28.1 million / 532 people per sq mile (205 people per sq km)
Total area 54,363 sq. miles (140,800 sq. km)
Languages Nepali*, Maithili, Bhojpuri
Religions Hindu 81%, Buddhist 11%, Muslim 4%, Other (including Christian) 4%
Ethnic mix Other 52%, Chhetri 16%, Hill Brahman 13%, Tharu 7%, Magar 7%, Tamang 5%
Government Transitional regime
Currency Nepalese rupee = 100 paisa
Literacy rate 57%
Calorie consumption 2673 kilocalories

NETHERLANDS
Northwest Europe

Official name Kingdom of the Netherlands
Formation 1648 / 1839
Capital Amsterdam; The Hague (administrative)
Population 16.8 million / 1283 people per sq mile (495 people per sq km)
Total area 16,033 sq. miles (41,526 sq. km)
Languages Dutch*, Frisian
Religions Roman Catholic 36%, Other 34%, Protestant 27%, Muslim 3%
Ethnic mix Dutch 82%, Other 12%, Surinamese 2%, Turkish 2%, Moroccan 2%
Government Parliamentary system
Currency Euro = 100 cents
Literacy rate 99%
Calorie consumption 3147 kilocalories

NEW ZEALAND
Australasia & Oceania

Official name New Zealand
Formation 1947 / 1947
Capital Wellington
Population 4.6 million / 44 people per sq mile (17 people per sq km)
Total area 103,737 sq. miles (268,680 sq. km)
Languages English*, Maori*
Religions Anglican 24%, Other 22%, Presbyterian 18%, Nonreligious 16%, Roman Catholic 15%, Methodist 5%
Ethnic mix European 75%, Maori 15%, Other 7%, Samoan 3%
Government Parliamentary system
Currency New Zealand dollar = 100 cents
Literacy rate 99%
Calorie consumption 3170 kilocalories

NICARAGUA
Central America

Official name Republic of Nicaragua
Formation 1838 / 1838
Capital Managua
Population 6.2 million / 135 people per sq mile (52 people per sq km)
Total area 49,998 sq. miles (129,494 sq. km)
Languages Spanish*, English Creole, Miskito
Religions Roman Catholic 80%, Protestant Evangelical 17%, Other 3%
Ethnic mix Mestizo 69%, White 17%, Black 9%, Amerindian 5%
Government Presidential system
Currency Córdoba oro = 100 centavos
Literacy rate 78%
Calorie consumption 2564 kilocalories

NIGER
West Africa

Official name Republic of Niger
Formation 1960 / 1960
Capital Niamey
Population 18.5 million / 38 people per sq mile (15 people per sq km)
Total area 489,188 sq. miles (1,267,000 sq. km)
Languages Hausa, Djerma, Fulani, Tuareg, Teda, French*
Religions Muslim 99%, Other (including Christian) 1%
Ethnic mix Hausa 53%, Djerma and Songhai 21%, Tuareg 11%, Fulani 7%, Kanuri 6%, Other 2%
Government Presidential system
Currency CFA franc = 100 centimes
Literacy rate 16%
Calorie consumption 2546 kilocalories

NIGERIA
West Africa

Official name Federal Republic of Nigeria
Formation 1960 / 1961
Capital Abuja
Population 179 million / 508 people per sq mile (196 people per sq km)
Total area 356,667 sq. miles (923,768 sq. km)
Languages Hausa, English*, Yoruba, Ibo
Religions Muslim 50%, Christian 40%, Traditional beliefs 10%
Ethnic mix Other 29%, Hausa 21%, Yoruba 21%, Ibo 18%, Fulani 11%
Government Presidential system
Currency Naira = 100 kobo
Literacy rate 51%
Calorie consumption 2700 kilocalories

NORTH KOREA
East Asia

Official name Democratic People's Republic of Korea
Formation 1948 / 1953
Capital Pyongyang
Population 25 million / 538 people per sq mile (208 people per sq km)
Total area 46,540 sq. miles (120,540 sq. km)
Languages Korean*
Religions Atheist 100%
Ethnic mix Korean 100%
Government One-party state
Currency North Korean won = 100 chon
Literacy rate 99%
Calorie consumption 2094 kilocalories

NORWAY
Northern Europe

Official name Kingdom of Norway
Formation 1905 / 1905
Capital Oslo
Population 5.1 million / 43 people per sq mile (17 people per sq km)
Total area 125,181 sq. miles (324,220 sq. km)
Languages Norwegian* (Bokmål "book language" and Nynorsk "new Norsk"), Sámi
Religions Evangelical Lutheran 88%, Other and nonreligious 8%, Muslim 2%, Pentecostal 1%, Roman Catholic 1%
Ethnic mix Norwegian 93%, Other 6%, Sámi 1%
Government Parliamentary system
Currency Norwegian krone = 100 øre
Literacy rate 99%
Calorie consumption 3484 kilocalories

OMAN
Southwest Asia

Official name Sultanate of Oman
Formation 1951 / 1951
Capital Muscat
Population 3.9 million / 48 people per sq mile (18 people per sq km)
Total area 82,031 sq. miles (212,460 sq. km)
Languages Arabic*, Baluchi, Farsi, Hindi, Punjabi
Religions Ibadi Muslim 75%, Other Muslim and Hindu 25%
Ethnic mix Arab 88%, Baluchi 4%, Persian 3%, Indian and Pakistani 3%, African 2%
Government Monarchy
Currency Omani rial = 1000 baisa
Literacy rate 87%
Calorie consumption 3143 kilocalories

PAKISTAN
South Asia

Official name Islamic Republic of Pakistan
Formation 1947 / 1971
Capital Islamabad
Population 185 million / 622 people per sq mile (240 people per sq km)
Total area 310,401 sq. miles (803,940 sq. km)
Languages Punjabi, Sindhi, Pashtu, Urdu*, Baluchi, Brahui
Religions Sunni Muslim 77%, Shi'a Muslim 20%, Hindu 2%, Christian 1%
Ethnic mix Punjabi 56%, Pathan (Pashtun) 15%, Sindhi 14%, Mohajir 7%, Baluchi 4%, Other 4%
Government Parliamentary system
Currency Pakistani rupee = 100 paisa
Literacy rate 55%
Calorie consumption 2440 kilocalories

PALAU
Australasia & Oceania

Official name Republic of Palau
Formation 1994 / 1994
Capital Ngerulmud
Population 21,186 / 108 people per sq mile (42 people per sq km)
Total area 177 sq. miles (458 sq. km)
Languages Palauan*, English*, Japanese, Angaur, Tobi, Sonsorolese
Religions Christian 66%, Modekngei 34%
Ethnic mix Palauan 74%, Filipino 16%, Other 6%, Chinese and other Asian 4%
Government Nonparty system
Currency US dollar = 100 cents
Literacy rate 99%
Calorie consumption Not available

PANAMA
Central America

Official name Republic of Panama
Formation 1903 / 1903
Capital Panama City
Population 3.9 million / 133 people per sq mile (51 people per sq km)
Total area 30,193 sq. miles (78,200 sq. km)
Languages English Creole, Spanish*, Amerindian languages, Chibchan languages
Religions Roman Catholic 84%, Protestant 15%, Other 1%
Ethnic mix Mestizo 70%, Black 14%, White 10%, Amerindian 6%
Government Presidential system
Currency Balboa = 100 centésimos; and US dollar = 100 cents
Literacy rate 94%
Calorie consumption 2733 kilocalories

PAPUA NEW GUINEA
Australasia & Oceania

Official name Independent State of Papua New Guinea
Formation 1975 / 1975
Capital Port Moresby
Population 7.5 million / 43 people per sq mile (17 people per sq km)
Total area 178,703 sq. miles (462,840 sq. km)
Languages Pidgin English, Papuan, English*, Motu, 800 (est.) native languages
Religions Protestant 60%, Roman Catholic 37%, Other 3%
Ethnic mix Melanesian and mixed race 100%
Government Parliamentary system
Currency Kina = 100 toea
Literacy rate 63%
Calorie consumption 2193 kilocalories

PARAGUAY
South America

Official name Republic of Paraguay
Formation 1811 / 1938
Capital Asunción
Population 6.9 million / 45 people per sq mile (17 people per sq km)
Total area 157,046 sq. miles (406,750 sq. km)
Languages Guarani*, Spanish*, German
Religions Roman Catholic 90%, Protestant (including Mennonite) 10%
Ethnic mix Mestizo 91%, Other 7%, Amerindian 2%
Government Presidential system
Currency Guarani = 100 céntimos
Literacy rate 94%
Calorie consumption 2589 kilocalories

PERU
South America

Official name Republic of Peru
Formation 1824 / 1941
Capital Lima
Population 30.8 million / 62 people per sq mile (24 people per sq km)
Total area 496,223 sq. miles (1,285,200 sq. km)
Languages Spanish*, Quechua*, Aymara
Religions Roman Catholic 81%, Other 19%
Ethnic mix Amerindian 45%, Mestizo 37%, White 15%, Other 3%
Government Presidential system
Currency New sol = 100 céntimos
Literacy rate 94%
Calorie consumption 2700 kilocalories

PHILIPPINES
Southeast Asia

Official name Republic of the Philippines
Formation 1946 / 1946
Capital Manila
Population 100 million / 870 people per sq mile (336 people per sq km)
Total area 115,830 sq. miles (300,000 sq. km)
Languages Filipino*, English*, Tagalog, Cebuano, Ilocano, Hiligaynon, many other local languages
Religions Roman Catholic 81%, Protestant 9%, Muslim 5%, Other (including Buddhist) 5%
Ethnic mix Other 34%, Tagalog 28%, Cebuano 13%, Ilocano 9%, Hiligaynon 8%, Bisaya 8%
Government Presidential system
Currency Philippine peso = 100 centavos
Literacy rate 95%
Calorie consumption 2570 kilocalories

POLAND
Northern Europe

Official name Republic of Poland
Formation 1918 / 1945
Capital Warsaw
Population 38.2 million / 325 people per sq mile (125 people per sq km)
Total area 120,728 sq. miles (312,685 sq. km)
Languages Polish*
Religions Roman Catholic 93%, Other and nonreligious 5%, Orthodox Christian 2%
Ethnic mix Polish 98%, Other 2%
Government Parliamentary system
Currency Zloty = 100 groszy
Literacy rate 99%
Calorie consumption 3485 kilocalories

PORTUGAL
Southwest Europe

Official name Portuguese Republic
Formation 1139 / 1640
Capital Lisbon
Population 10.6 million / 299 people per sq mile (115 people per sq km)
Total area 35,672 sq. miles (92,391 sq. km)
Languages Portuguese*
Religions Roman Catholic 92%, Protestant 4%, Nonreligious 3%, Other 1%
Ethnic mix Portuguese 98%, African and other 2%
Government Parliamentary system
Currency Euro = 100 cents
Literacy rate 94%
Calorie consumption 3456 kilocalories

QATAR
Southwest Asia

Official name State of Qatar
Formation 1971 / 1971
Capital Doha
Population 2.3 million / 542 people per sq mile (209 people per sq km)
Total area 4416 sq. miles (11,437 sq. km)
Languages Arabic*
Religions Muslim (mainly Sunni) 95%, Other 5%
Ethnic mix Qatari 20%, Indian 20%, Other Arab 20%, Nepalese 13%, Filipino 10%, Other 10%, Pakistani 7%
Government Monarchy
Currency Qatar riyal = 100 dirhams
Literacy rate 97%
Calorie consumption Not available

ROMANIA
Southeast Europe

Official name Romania
Formation 1878 / 1947
Capital Bucharest
Population 21.6 million / 243 people per sq mile (94 people per sq km)
Total area 91,699 sq. miles (237,500 sq. km)
Languages Romanian*, Hungarian (Magyar), Romani, German
Religions Romanian Orthodox 87%, Protestant 5%, Roman Catholic 5%, Greek Orthodox 1%, Greek Catholic (Uniate) 1%, Other 1%
Ethnic mix Romanian 89%, Magyar 7%, Roma 3%, Other 1%
Government Presidential system
Currency New Romanian leu = 100 bani
Literacy rate 99%
Calorie consumption 3363 kilocalories

RUSSIAN FEDERATION
Europe / Asia

Official name Russian Federation
Formation 1480 / 1991
Capital Moscow
Population 143 million / 22 people per sq mile (8 people per sq km)
Total area 6,592,735 sq. miles (17,075,200 sq. km)
Languages Russian*, Tatar, Ukrainian, Chavash, various other national languages
Religions Orthodox Christian 75%, Muslim 14%, Other 11%
Ethnic mix Russian 80%, Other 12%, Tatar 4%, Ukrainian 2%, Bashkir 1%, Chavash 1%
Government Mixed Presidential–Parliamentary system
Currency Russian rouble = 100 kopeks
Literacy rate 99%
Calorie consumption 3358 kilocalories

RWANDA
Central Africa

Official name Republic of Rwanda
Formation 1962 / 1962
Capital Kigali
Population 12.1 million / 1256 people per sq mile (485 people per sq km)
Total area 10,169 sq. miles (26,338 sq. km)
Languages Kinyarwanda*, French*, Kiswahili, English*
Religions Christian 94%, Muslim 5%, Traditional beliefs 1%
Ethnic mix Hutu 85%, Tutsi 14%, Other (including Twa) 1%
Government Presidential system
Currency Rwanda franc = 100 centimes
Literacy rate 66%
Calorie consumption 2148 kilocalories

ST KITTS & NEVIS
West Indies

Official name Federation of Saint Christopher and Nevis
Formation 1983 / 1983
Capital Basseterre
Population 51,538 / 371 people per sq mile (143 people per sq km)
Total area 101 sq. miles (261 sq. km)
Languages English*, English Creole
Religions Anglican 33%, Methodist 29%, Other 22%, Moravian 9%, Roman Catholic 7%
Ethnic mix Black 95%, Mixed race 3%, White 1%, Other and Amerindian 1%
Government Parliamentary system
Currency East Caribbean dollar = 100 cents
Literacy rate 98%
Calorie consumption 2507 kilocalories

ST LUCIA
West Indies

Official name Saint Lucia
Formation 1979 / 1979
Capital Castries
Population 200,000 / 847 people per sq mile (328 people per sq km)
Total area 239 sq. miles (620 sq. km)
Languages English*, French Creole
Religions Roman Catholic 90%, Other 10%
Ethnic mix Black 83%, Mulatto (mixed race) 13%, Asian 3%, Other 1%
Government Parliamentary system
Currency East Caribbean dollar = 100 cents
Literacy rate 95%
Calorie consumption 2629 kilocalories

ST VINCENT & THE GRENADINES
West Indies

Official name Saint Vincent and the Grenadines
Formation 1979 / 1979
Capital Kingstown
Population 102,918 / 786 people per sq mile (303 people per sq km)
Total area 150 sq. miles (389 sq. km)
Languages English*, English Creole
Religions Anglican 47%, Methodist 28%, Roman Catholic 13%, Other 12%
Ethnic mix Black 66%, Mulatto (mixed race) 19%, Other 12%, Carib 2%, Asian 1%
Government Parliamentary system
Currency East Caribbean dollar = 100 cents
Literacy rate 88%
Calorie consumption 2960 kilocalories

SAMOA
Australasia & Oceania

Official name Independent State of Samoa
Formation 1962 / 1962
Capital Apia
Population 200,000 / 183 people per sq mile (71 people per sq km)
Total area 1104 sq. miles (2860 sq. km)
Languages Samoan*, English*
Religions Christian 99%, Other 1%
Ethnic mix Polynesian 91%, Euronesian 7%, Other 2%
Government Parliamentary system
Currency Tala = 100 sene
Literacy rate 99%
Calorie consumption 2872 kilocalories

SAN MARINO
Southern Europe

Official name Republic of San Marino
Formation 1631 / 1631
Capital San Marino
Population 32,742 / 1364 people per sq mile (537 people per sq km)
Total area 23.6 sq. miles (61 sq. km)
Languages Italian*
Religions Roman Catholic 93%, Other and nonreligious 7%
Ethnic mix Sammarinese 88%, Italian 10%, Other 2%
Government Parliamentary system
Currency Euro = 100 cents
Literacy rate 99%
Calorie consumption Not available

SAO TOME & PRINCIPE
West Africa

Official name Democratic Republic of Sao Tome and Principe
Formation 1975 / 1975
Capital São Tomé
Population 200,000 / 539 people per sq mile (208 people per sq km)
Total area 386 sq. miles (1001 sq. km)
Languages Portuguese Creole, Portuguese*
Religions Roman Catholic 84%, Other 16%
Ethnic mix Black 90%, Portuguese and Creole 10%
Government Presidential system
Currency Dobra = 100 céntimos
Literacy rate 70%
Calorie consumption 2676 kilocalories

SAUDI ARABIA
Southwest Asia

Official name Kingdom of Saudi Arabia
Formation 1932 / 1932
Capital Riyadh
Population 29.4 million / 36 people per sq mile (14 people per sq km)
Total area 756,981 sq. miles (1,960,582 sq. km)
Languages Arabic*
Religions Sunni Muslim 85%, Shi'a Muslim 15%
Ethnic mix Arab 72%, Foreign residents (mostly south and southeast Asian) 20%, Afro-Asian 8%
Government Monarchy
Currency Saudi riyal = 100 halalat
Literacy rate 94%
Calorie consumption 3122 kilocalories

SENEGAL
West Africa

Official name Republic of Senegal
Formation 1960 / 1960
Capital Dakar
Population 14.5 million / 195 people per sq mile (75 people per sq km)
Total area 75,749 sq. miles (196,190 sq. km)
Languages Wolof, Pulaar, Serer, Diola, Mandinka, Malinké, Soninké, French*
Religions Sunni Muslim 95%, Christian (mainly Roman Catholic) 4%, Traditional beliefs 1%
Ethnic mix Wolof 43%, Serer 15%, Peul 14%, Other 14%, Toucouleur 9%, Diola 5%
Government Presidential system
Currency CFA franc = 100 centimes
Literacy rate 52%
Calorie consumption 2426 kilocalories

SERBIA
Southeast Europe

Official name Republic of Serbia
Formation 2006 / 2008
Capital Belgrade
Population 9.5 million / 318 people per sq mile (123 people per sq km)
Total area 29,905 sq. miles (77,453 sq. km)
Languages Serbian*, Hungarian (Magyar)
Religions Orthodox Christian 85%, Roman Catholic 6%, Other 6%, Muslim 3%
Ethnic mix Serb 83%, Other 10%, Magyar 4%, Bosniak 2%, Roma 1%
Government Parliamentary system
Currency Serbian dinar = 100 para
Literacy rate 98%
Calorie consumption 2724 kilocalories

SEYCHELLES
Indian Ocean

Official name Republic of Seychelles
Formation 1976 / 1976
Capital Victoria
Population 91,650 / 881 people per sq mile (339 people per sq km)
Total area 176 sq. miles (455 sq. km)
Languages French Creole*, English*, French*
Religions Roman Catholic 82%, Anglican 6%, Other (including Muslim) 6%, Other Christian 3%, Hindu 2%, Seventh-day Adventist 1%
Ethnic mix Creole 89%, Indian 5%, Other 4%, Chinese 2%
Government Presidential system
Currency Seychelles rupee = 100 cents
Literacy rate 92%
Calorie consumption 2426 kilocalories

SIERRA LEONE
West Africa

Official name Republic of Sierra Leone
Formation 1961 / 1961
Capital Freetown
Population 6.2 million / 224 people per sq mile (87 people per sq km)
Total area 27,698 sq. miles (71,740 sq. km)
Languages Mende, Temne, Krio, English*
Religions Muslim 60%, Christian 30%, Traditional beliefs 10%
Ethnic mix Mende 35%, Temne 32%, Other 21%, Limba 8%, Kuranko 4%
Government Presidential system
Currency Leone = 100 cents
Literacy rate 44%
Calorie consumption 2333 kilocalories

SINGAPORE
Southeast Asia

Official name Republic of Singapore
Formation 1965 / 1965
Capital Singapore
Population 5.5 million / 23,305 people per sq mile (9016 people per sq km)
Total area 250 sq. miles (648 sq. km)
Languages Mandarin*, Malay*, Tamil*, English*
Religions Buddhist 55%, Taoist 22%, Muslim 16%, Hindu, Christian, and Sikh 7%
Ethnic mix Chinese 74%, Malay 14%, Indian 9%, Other 3%
Government Parliamentary system
Currency Singapore dollar = 100 cents
Literacy rate 96%
Calorie consumption Not available

SLOVAKIA
Central Europe

Official name Slovak Republic
Formation 1993 / 1993
Capital Bratislava
Population 5.5 million / 290 people per sq mile (112 people per sq km)
Total area 18,859 sq. miles (48,845 sq. km)
Languages Slovak*, Hungarian (Magyar), Czech
Religions Roman Catholic 69%, Nonreligious 13%, Other 13%, Greek Catholic (Uniate) 4%, Orthodox Christian 1%
Ethnic mix Slovak 86%, Magyar 10%, Roma 2%, Czech 1%, Other 1%
Government Parliamentary system
Currency Euro = 100 cents
Literacy rate 99%
Calorie consumption 2902 kilocalories

SLOVENIA
Central Europe

Official name Republic of Slovenia
Formation 1991 / 1991
Capital Ljubljana
Population 2.1 million / 269 people per sq mile (104 people per sq km)
Total area 7820 sq. miles (20,253 sq. km)
Languages Slovenian*
Religions Roman Catholic 58%, Other 28%, Atheist 10%, Orthodox Christian 2%, Muslim 2%
Ethnic mix Slovene 83%, Other 12%, Serb 2%, Croat 2%, Bosniak 1%
Government Parliamentary system
Currency Euro = 100 cents
Literacy rate 99%
Calorie consumption 3173 kilocalories

SOLOMON ISLANDS
Australasia & Oceania

Official name Solomon Islands
Formation 1978 / 1978
Capital Honiara
Population 600,000 / 56 people per sq mile (21 people per sq km)
Total area 10,985 sq. miles (28,450 sq. km)
Languages English*, Pidgin English, Melanesian Pidgin, 120 (est.) native languages
Religions Church of Melanesia (Anglican) 34%, Roman Catholic 19%, South Seas Evangelical Church 17%, Methodist 11%, Seventh-day Adventist 10%, Other 9%
Ethnic mix Melanesian 93%, Polynesian 4%, Micronesian 2%, Other 1%
Government Parliamentary system
Currency Solomon Islands dollar = 100 cents
Literacy rate 77%
Calorie consumption 2473 kilocalories

SOMALIA
East Africa

Official name Federal Republic of Somalia
Formation 1960 / 1960
Capital Mogadishu
Population 10.8 million / 45 people per sq mile (17 people per sq km)
Total area 246,199 sq. miles (637,657 sq. km)
Languages Somali*, Arabic*, English, Italian
Religions Sunni Muslim 99%, Christian 1%
Ethnic mix Somali 85%, Other 15%
Government Non-party system
Currency Somali shilin = 100 senti
Literacy rate 24%
Calorie consumption 1696 kilocalories

SOUTH AFRICA
Southern Africa

Official name Republic of South Africa
Formation 1934 / 1994
Capital Pretoria; Cape Town; Bloemfontein
Population 53.1 million / 113 people per sq mile (43 people per sq km)
Total area 471,008 sq. miles (1,219,912 sq. km)
Languages English, isiZulu, isiXhosa, Afrikaans, Sepedi, Setswana, Sesotho, Xitsonga, siSwati, Tshivenda, isiNdebele
Religions Christian 68%, Traditional beliefs and animist 29%, Muslim 2%, Hindu 1%
Ethnic mix Black 80%, Mixed race 9%, White 9%, Asian 2%
Government Presidential system
Currency Rand = 100 cents
Literacy rate 94%
Calorie consumption 3007 kilocalories

SOUTH KOREA
East Asia

Official name Republic of Korea
Formation 1948 / 1953
Capital Seoul; Sejong City (administrative)
Population 49.5 million / 1299 people per sq mile (501 people per sq km)
Total area 38,023 sq. miles (98,480 sq. km)
Languages Korean*
Religions Mahayana Buddhist 47%, Protestant 38%, Roman Catholic 11%, Confucianist 3%, Other 1%
Ethnic mix Korean 100%
Government Presidential system
Currency South Korean won = 100 chon
Literacy rate 99%
Calorie consumption 3329 kilocalories

SOUTH SUDAN
East Africa

Official name Republic of South Sudan
Formation 2011 / 2011
Capital Juba
Population 11.7 million / 47 people per sq mile (18 people per sq km)
Total area 248,777 sq. miles (644,329 sq. km)
Languages Arabic, Dinka, Nuer, Zande, Bari, Shilluk, Lotuko, English*
Religions Over half of the population follow Christian or traditional beliefs.
Ethnic mix Dinka 40%, Nuer 15%, Bari 10%, Shilluk/Anwak 10%, Azande 10%, Arab 10%, Other 5%
Government Transitional regime
Currency South Sudan pound = 100 piastres
Literacy rate 37%
Calorie consumption Not available

SPAIN
Southwest Europe

Official name Kingdom of Spain
Formation 1492 / 1713
Capital Madrid
Population 47.1 million / 244 people per sq mile (94 people per sq km)
Total area 194,896 sq. miles (504,782 sq. km)
Languages Spanish*, Catalan*, Galician*, Basque*
Religions Roman Catholic 96%, Other 4%
Ethnic mix Castilian Spanish 72%, Catalan 17%, Galician 6%, Basque 2%, Other 2%, Roma 1%
Government Parliamentary system
Currency Euro = 100 cents
Literacy rate 98%
Calorie consumption 3183 kilocalories

SRI LANKA
South Asia

Official name Democratic Socialist Republic of Sri Lanka
Formation 1948 / 1948
Capital Colombo; Sri Jayewardenapura Kotte
Population 21.4 million / 856 people per sq mile (331 people per sq km)
Total area 25,332 sq. miles (65,610 sq. km)
Languages Sinhala*, Tamil*, Sinhala-Tamil, English
Religions Buddhist 69%, Hindu 15%, Muslim 8%, Christian 8%
Ethnic mix Sinhalese 74%, Tamil 18%, Moor 7%, Other 1%
Government Mixed presidential–parliamentary system
Currency Sri Lanka rupee = 100 cents
Literacy rate 91%
Calorie consumption 2539 kilocalories

SUDAN
East Africa

Official name Republic of the Sudan
Formation 1956 / 2011
Capital Khartoum
Population 38.8 million / 54 people per sq mile (21 people per sq km)
Total area 718,722 sq. miles (1,861,481 sq. km)
Languages Arabic, Nubian, Beja, Fur
Religions Nearly the whole population is Muslim (mainly Sunni)
Ethnic mix Arab 60%, Other 18%, Nubian 10%, Beja 8%, Fur 3%, Zaghawa 1%
Government Presidential system
Currency New Sudanese pound = 100 piastres
Literacy rate 73%
Calorie consumption 2346 kilocalories

SURINAME
South America

Official name Republic of Suriname
Formation 1975 / 1975
Capital Paramaribo
Population 500,000 / 8 people per sq mile (3 people per sq km)
Total area 63,039 sq. miles (163,270 sq. km)
Languages Sranan (creole), Dutch*, Javanese, Sarnami Hindi, Saramaccan, Chinese, Carib
Religions Hindu 27%, Protestant 25%, Roman Catholic 23%, Muslim 20%, Traditional beliefs 5%
Ethnic mix East Indian 27%, Creole 18%, Black 15%, Javanese 15%, Mixed race 13%, Other 6%, Amerindian 4%, Chinese 2%
Government Mixed presidential–parliamentary system
Currency Surinamese dollar = 100 cents
Literacy rate 95%
Calorie consumption 2727 kilocalories

SWAZILAND
Southern Africa

Official name Kingdom of Swaziland
Formation 1968 / 1968
Capital Mbabane
Population 1.3 million / 196 people per sq mile (76 people per sq km)
Total area 6704 sq. miles (17,363 sq. km)
Languages English*, siSwati*, isiZulu, Xitsonga
Religions Traditional beliefs 40%, Other 30%, Roman Catholic 20%, Muslim 10%
Ethnic mix Swazi 97%, Other 3%
Government Monarchy
Currency Lilangeni = 100 cents
Literacy rate 83%
Calorie consumption 2275 kilocalories

SWEDEN
Northern Europe

Official name Kingdom of Sweden
Formation 1523 / 1921
Capital Stockholm
Population 9.6 million / 60 people per sq mile (23 people per sq km)
Total area 173,731 sq. miles (449,964 sq. km)
Languages Swedish*, Finnish, Sámi
Religions Evangelical Lutheran 75%, Other 13%, Muslim 5%, Other Protestant 5%, Roman Catholic 2%
Ethnic mix Swedish 86%, Foreign-born or first-generation immigrant 12%, Finnish and Sámi 2%
Government Parliamentary system
Currency Swedish krona = 100 ore
Literacy rate 99%
Calorie consumption 3160 kilocalories

SWITZERLAND
Central Europe

Official name Swiss Confederation
Formation 1291 / 1857
Capital Bern
Population 8.2 million / 534 people per sq mile (206 people per sq km)
Total area 15,942 sq. miles (41,290 sq. km)
Languages German*, Swiss-German, French*, Italian*, Romansch
Religions Roman Catholic 42%, Protestant 35%, Other and nonreligious 19%, Muslim 4%
Ethnic mix German 64%, French 20%, Other 9.5%, Italian 6%, Romansch 0.5%
Government Parliamentary system
Currency Swiss franc = 100 rappen/centimes
Literacy rate 99%
Calorie consumption 3487 kilocalories

SYRIA
Southwest Asia

Official name Syrian Arab Republic
Formation 1941 / 1967
Capital Damascus
Population 22 million / 310 people per sq mile (120 people per sq km)
Total area 71,498 sq. miles (184,180 sq. km)
Languages Arabic*, French, Kurdish, Armenian, Circassian, Turkic languages, Assyrian, Aramaic
Religions Sunni Muslim 74%, Alawi 12%, Christian 10%, Druze 3%, Other 1%
Ethnic mix Arab 90%, Kurdish 9%, Armenian, Turkmen, and Circassian 1%
Government Presidential system
Currency Syrian pound = 100 piastres
Literacy rate 85%
Calorie consumption 3106 kilocalories

TAIWAN
East Asia

Official name Republic of China (ROC)
Formation 1949 / 1949
Capital Taibei (Taipei)
Population 23.4 million / 1879 people per sq mile (725 people per sq km)
Total area 13,892 sq. miles (35,980 sq. km)
Languages Amoy Chinese, Mandarin Chinese*, Hakka Chinese
Religions Buddhist, Confucianist, and Taoist 93%, Christian 5%, Other 2%
Ethnic mix Han Chinese (pre-20th-century migration) 84%, Han Chinese (20th-century migration) 14%, Aboriginal 2%
Government Presidential system
Currency Taiwan dollar = 100 cents
Literacy rate 98%
Calorie consumption 2997 kilocalories

TAJIKISTAN
Central Asia

Official name Republic of Tajikistan
Formation 1991 / 1991
Capital Dushanbe
Population 8.4 million / 152 people per sq mile (59 people per sq km)
Total area 55,251 sq. miles (143,100 sq. km)
Languages Tajik*, Uzbek, Russian
Religions Sunni Muslim 95%, Shi'a Muslim 3%, Other 2%
Ethnic mix Tajik 80%, Uzbek 15%, Other 3%, Russian 1%, Kyrgyz 1%
Government Presidential system
Currency Somoni = 100 diram
Literacy rate 99%
Calorie consumption 2101 kilocalories

TANZANIA
East Africa

Official name United Republic of Tanzania
Formation 1964 / 1964
Capital Dodoma
Population 50.8 million / 148 people per sq mile (57 people per sq km)
Total area 364,898 sq. miles (945,087 sq. km)
Languages Kiswahili*, Sukuma, Chagga, Nyamwezi, Hehe, Makonde, Yao, Sandawe, English*
Religions Christian 63%, Muslim 35%, Other 2%
Ethnic mix Native African (over 120 tribes) 99%, European, Asian, and Arab 1%
Government Presidential system
Currency Tanzanian shilling = 100 cents
Literacy rate 68%
Calorie consumption 2208 kilocalories

THAILAND
Southeast Asia

Official name Kingdom of Thailand
Formation 1238 / 1907
Capital Bangkok
Population 67.2 million / 341 people per sq mile (132 people per sq km)
Total area 198,455 sq. miles (514,000 sq. km)
Languages Thai*, Chinese, Malay, Khmer, Mon, Karen, Miao
Religions Buddhist 95%, Muslim 4%, Other (including Christian) 1%
Ethnic mix Thai 83%, Chinese 12%, Malay 3%, Khmer and Other 2%
Government Transitional regime
Currency Baht = 100 satang
Literacy rate 96%
Calorie consumption 2784 kilocalories

TOGO
West Africa

Official name Togolese Republic
Formation 1960 / 1960
Capital Lomé
Population 7 million / 333 people per sq mile (129 people per sq km)
Total area 21,924 sq. miles (56,785 sq. km)
Languages Ewe, Kabye, Gurma, French*
Religions Christian 47%, Traditional beliefs 33%, Muslim 14%, Other 6%
Ethnic mix Ewe 46%, Other African 41%, Kabye 12%, European 1%
Government Presidential system
Currency CFA franc = 100 centimes
Literacy rate 60%
Calorie consumption 2366 kilocalories

TONGA
Australasia & Oceania

Official name Kingdom of Tonga
Formation 1970 / 1970
Capital Nuku'alofa
Population 106,440 / 383 people per sq mile (148 people per sq km)
Total area 289 sq. miles (748 sq. km)
Languages English*, Tongan*
Religions Free Wesleyan 41%, Other 17%, Roman Catholic 16%, Church of Jesus Christ of Latter-day Saints 14%, Free Church of Tonga 12%
Ethnic mix Tongan 98%, Other 2%
Government Monarchy
Currency Pa'anga (Tongan dollar) = 100 seniti
Literacy rate 99%
Calorie consumption Not available

TRINIDAD & TOBAGO
West Indies

Official name Republic of Trinidad and Tobago
Formation 1962 / 1962
Capital Port-of-Spain
Population 1.3 million / 656 people per sq mile (253 people per sq km)
Total area 1980 sq. miles (5128 sq. km)
Languages English Creole, English*, Hindi, French, Spanish
Religions Roman Catholic 26%, Hindu 23%, Other and nonreligious 23%, Anglican 8%, Baptist 7%, Pentecostal 7%, Muslim 6%
Ethnic mix East Indian 40%, Black 38%, Mixed race 20%, White and Chinese 1%, other 1%
Government Parliamentary system
Currency Trinidad and Tobago dollar = 100 cents
Literacy rate 99%
Calorie consumption 2889 kilocalories

TUNISIA
North Africa

Official name Tunisian Republic
Formation 1956 / 1956
Capital Tunis
Population 11.1 million / 185 people per sq mile (71 people per sq km)
Total area 63,169 sq. miles (163,610 sq. km)
Languages Arabic*, French
Religions Muslim (mainly Sunni) 98%, Christian 1%, Jewish 1%
Ethnic mix Arab and Berber 98%, Jewish 1%, European 1%
Government Mixed presidential–parliamentary system
Currency Tunisian dinar = 1000 millimes
Literacy rate 80%
Calorie consumption 3362 kilocalories

TURKEY
Asia / Europe

Official name Republic of Turkey
Formation 1923 / 1939
Capital Ankara
Population 75.8 million / 255 people per sq mile (98 people per sq km)
Total area 301,382 sq. miles (780,580 sq. km)
Languages Turkish*, Kurdish, Arabic, Circassian, Armenian, Greek, Georgian, Ladino
Religions Muslim (mainly Sunni) 99%, Other 1%
Ethnic mix Turkish 70%, Kurdish 20%, Other 8%, Arab 2%
Government Parliamentary system
Currency Turkish lira = 100 kurus
Literacy rate 95%
Calorie consumption 3680 kilocalories

TURKMENISTAN
Central Asia

Official name Turkmenistan
Formation 1991 / 1991
Capital Ashgabat
Population 5.3 million / 28 people per sq mile (11 people per sq km)
Total area 188,455 sq. miles (488,100 sq. km)
Languages Turkmen*, Uzbek, Russian, Kazakh, Tatar
Religions Sunni Muslim 89%, Orthodox Christian 9%, Other 2%
Ethnic mix Turkmen 85%, Other 6%, Uzbek 5%, Russian 4%
Government Presidential system
Currency New manat = 100 tenge
Literacy rate 99%
Calorie consumption 2883 kilocalories

TUVALU
Australasia & Oceania

Official name Tuvalu
Formation 1978 / 1978
Capital Funafuti Atoll
Population 10,782 / 1078 people per sq mile (415 people per sq km)
Total area 10 sq. miles (26 sq. km)
Languages Tuvaluan, Kiribati, English*
Religions Church of Tuvalu 97%, Baha'i 1%, Seventh-day Adventist 1%, Other 1%
Ethnic mix Polynesian 96%, Micronesian 4%
Government Nonparty system
Currency Australian dollar = 100 cents; and Tuvaluan dollar = 100 cents
Literacy rate 95%
Calorie consumption Not available

UGANDA
East Africa

Official name Republic of Uganda
Formation 1962 / 1962
Capital Kampala
Population 38.8 million / 504 people per sq mile (194 people per sq km)
Total area 91,135 sq. miles (236,040 sq. km)
Languages Luganda, Nkole, Chiga, Lango, Acholi, Teso, Lugbara, English*
Religions Christian 85%, Muslim (mainly Sunni) 12%, Other 3%
Ethnic mix Other 50%, Baganda 17%, Banyakole 10%, Basoga 9%, Iteso 7%, Bakiga 7%
Government Presidential system
Currency Uganda shilling = 100 cents
Literacy rate 74%
Calorie consumption 2279 kilocalories

UKRAINE
Eastern Europe

Official name Ukraine
Formation 1991 / 1991
Capital Kiev
Population 44.9 million / 193 people per sq mile (74 people per sq km)
Total area 223,089 sq. miles (603,700 sq. km)
Languages Ukrainian*, Russian, Tatar
Religions Christian (mainly Orthodox) 95%, Other 5%
Ethnic mix Ukrainian 78%, Russian 17%, Other 5%
Government Presidential system
Currency Hryvna = 100 kopiykas
Literacy rate 99%
Calorie consumption 3142 kilocalories

UNITED ARAB EMIRATES
Southwest Asia

Official name United Arab Emirates
Formation 1971 / 1972
Capital Abu Dhabi
Population 9.4 million / 291 people per sq mile (112 people per sq km)
Total area 32,000 sq. miles (82,880 sq. km)
Languages Arabic*, Farsi, Indian and Pakistani languages, English
Religions Muslim (mainly Sunni) 96%, Christian, Hindu, and other 4%
Ethnic mix Asian 60%, Emirian 25%, Other Arab 12%, European 3%
Government Monarchy
Currency UAE dirham = 100 fils
Literacy rate 90%
Calorie consumption 3215 kilocalories

UNITED KINGDOM
Northwest Europe

Official name United Kingdom of Great Britain and Northern Ireland
Formation 1707 / 1922
Capital London
Population 63.5 million / 681 people per sq mile (263 people per sq km)
Total area 94,525 sq. miles (244,820 sq. km)
Languages English*, Welsh*, Scottish Gaelic, Irish
Religions Anglican 45%, Other and nonreligious 36%, Roman Catholic 9%, Presbyterian 4%, Muslim 3%, Methodist 2%, Hindu 1%
Ethnic mix English 80%, Scottish 9%, West Indian, Asian, and other 5%, Northern Irish 3%, Welsh 3%
Government Parliamentary system
Currency Pound sterling = 100 pence
Literacy rate 99%
Calorie consumption 3414 kilocalories

UNITED STATES
North America

Official name United States of America
Formation 1776 / 1959
Capital Washington D.C.
Population 323 million / 91 people per sq mile (35 people per sq km)
Total area 3,794,100 sq. miles (9,826,675 sq. km)
Languages English*, Spanish, Chinese, French, German, Tagalog, Vietnamese, Italian, Korean, Russian, Polish
Religions Protestant 52%, Roman Catholic 25%, Other and nonreligious 20%, Jewish 2%, Muslim 1%
Ethnic mix White 60%, Hispanic 17%, Black American/African 14%, Asian 6%, American Indians & Alaksa Natives 2%, Pacific Islanders 1%
Government Presidential system
Currency US dollar = 100 cents
Literacy rate 99%
Calorie consumption 3639 kilocalories

URUGUAY
South America

Official name Oriental Republic of Uruguay
Formation 1828 / 1828
Capital Montevideo
Population 3.4 million / 50 people per sq mile (19 people per sq km)
Total area 68,039 sq. miles (176,220 sq. km)
Languages Spanish*
Religions Roman Catholic 66%, Other and nonreligious 30%, Jewish 2%, Protestant 2%
Ethnic mix White 90%, Mestizo 6%, Black 4%
Government Presidential system
Currency Uruguayan peso = 100 centésimos
Literacy rate 98%
Calorie consumption 2939 kilocalories

UZBEKISTAN
Central Asia

Official name Republic of Uzbekistan
Formation 1991 / 1991
Capital Tashkent
Population 29.3 million / 170 people per sq mile (65 people per sq km)
Total area 172,741 sq. miles (447,400 sq. km)
Languages Uzbek*, Russian, Tajik, Kazakh
Religions Sunni Muslim 88%, Orthodox Christian 9%, Other 3%
Ethnic mix Uzbek 80%, Russian 6%, Other 6%, Tajik 5%, Kazakh 3%
Government Presidential system
Currency Som = 100 tiyin
Literacy rate 99%
Calorie consumption 2675 kilocalories

VANUATU
Australasia & Oceania

Official name Republic of Vanuatu
Formation 1980 / 1980
Capital Port Vila
Population 300,000 / 64 people per sq mile (25 people per sq km)
Total area 4710 sq. miles (12,200 sq. km)
Languages Bislama (Melanesian pidgin)*, English*, French*, other indigenous languages
Religions Presbyterian 37%, Other 19%, Anglican 15%, Roman Catholic 15%, Traditional beliefs 8%, Seventh-day Adventist 6%
Ethnic mix ni-Vanuatu 94%, European 4%, Other 2%
Government Parliamentary system
Currency Vatu = 100 centimes
Literacy rate 83%
Calorie consumption 2820 kilocalories

VATICAN CITY
Southern Europe

Official name State of the Vatican City
Formation 1929 / 1929
Capital Vatican City
Population 842 / 4953 people per sq mile (1914 people per sq km)
Total area 0.17 sq. miles (0.44 sq. km)
Languages Italian*, Latin*
Religions Roman Catholic 100%
Ethnic mix The current pope is Argentinian, though most popes for the last 500 years have been Italian. Cardinals are from many nationalities, but Italians form the largest group. Most of the resident lay persons are Italian.
Government Papal state
Currency Euro = 100 cents
Literacy rate 99%
Calorie consumption Not available

VENEZUELA
South America

Official name Bolivarian Republic of Venezuela
Formation 1830 / 1830
Capital Caracas
Population 30.9 million / 91 people per sq mile (35 people per sq km)
Total area 352,143 sq. miles (912,050 sq. km)
Languages Spanish*, Amerindian languages
Religions Roman Catholic 96%, Protestant 2%, Other 2%
Ethnic mix Mestizo 69%, White 20%, Black 9%, Amerindian 2%
Government Presidential system
Currency Bolívar fuerte = 100 céntimos
Literacy rate 96%
Calorie consumption 2880 kilocalories

VIETNAM
Southeast Asia

Official name Socialist Republic of Vietnam
Formation 1976 / 1976
Capital Hanoi
Population 92.5 million / 736 people per sq mile (284 people per sq km)
Total area 127,243 sq. miles (329,560 sq. km)
Languages Vietnamese*, Chinese, Thai, Khmer, Muong, Nung, Miao, Yao, Jarai
Religions Other 74%, Buddhist 14%, Roman Catholic 7%, Cao Dai 3%, Protestant 2%
Ethnic mix Vietnamese 86%, Other 8%, Muong 2%, Tay 2%, Thai 2%
Government One-party state
Currency Dông = 10 hao = 100 xu
Literacy rate 94%
Calorie consumption 2745 kilocalories

YEMEN
Southwest Asia

Official name Republic of Yemen
Formation 1990 / 1990
Capital Sana
Population 25 million / 115 people per sq mile (44 people per sq km)
Total area 203,849 sq. miles (527,970 sq. km)
Languages Arabic*
Religions Sunni Muslim 55%, Shi'a Muslim 42%, Christian, Hindu, and Jewish 3%
Ethnic mix Arab 99%, Afro-Arab, Indian, Somali, and European 1%
Government Transitional regime
Currency Yemeni rial = 100 fils
Literacy rate 66%
Calorie consumption 2223 kilocalories

ZAMBIA
Southern Africa

Official name Republic of Zambia
Formation 1964 / 1964
Capital Lusaka
Population 15 million / 52 people per sq mile (20 people per sq km)
Total area 290,584 sq. miles (752,614 sq. km)
Languages Bemba, Tonga, Nyanja, Lozi, Lala-Bisa, Nsenga, English*
Religions Christian 63%, Traditional beliefs 36%, Muslim and Hindu 1%
Ethnic mix Bemba 34%, Other African 26%, Tonga 16%, Nyanja 14%, Lozi 9%, European 1%
Government Presidential system
Currency New Zambian kwacha = 100 ngwee
Literacy rate 61%
Calorie consumption 1930 kilocalories

ZIMBABWE
Southern Africa

Official name Republic of Zimbabwe
Formation 1980 / 1980
Capital Harare
Population 14.6 million / 98 people per sq mile (38 people per sq km)
Total area 150,803 sq. miles (390,580 sq. km)
Languages Shona, isiNdebele, English*
Religions Syncretic (Christian/traditional beliefs) 50%, Christian 25%, Traditional beliefs 24%, Other (including Muslim) 1%
Ethnic mix Shona 71%, Ndebele 16%, Other African 11%, White 1%, Asian 1%
Government Presidential system
Currency US $, South African rand, Euro, UK £, Botswana pula, Australian $, Chinese yuan, Indian rupee, and Japanese yen are legal tender
Literacy rate 84%
Calorie consumption 2110 kilocalories

GLOSSARY

This glossary lists all geographical, technical, and foreign language terms which appear in the text, followed by a brief definition of the term. Any acronyms used in the text are also listed in full. Terms in italics are for cross-reference and indicate that the word is separately defined in the glossary.

A

Aboriginal The original (*indigenous*) inhabitants of a country or continent. Especially used with reference to Australia.

Abyssal plain A broad *plain* found in the depths of the ocean, more than 10,000 ft (3,000 m) below sea level.

Acid rain Rain, sleet, snow, or mist which has absorbed waste gases from fossil-fueled power stations and vehicle exhausts, becoming more acid. It causes severe environmental damage.

Adaptation The gradual evolution of plants and animals so that they become better suited to survive and reproduce in their *environment*.

Afforestation The planting of new forest in areas that were once forested but have been cleared.

Agribusiness A term applied to activities such as the growing of crops, rearing of animals, or the manufacture of farm machinery, which eventually leads to the supply of agricultural produce at market.

Air mass A huge, homogeneous mass of air, within which horizontal patterns of temperature and *humidity* are consistent. Air masses are separated by *fronts*.

Alliance An agreement between two or more states, to work together to achieve common purposes.

Alluvial fan A large fan-shaped deposit of fine sediments deposited by a river as it emerges from a narrow, mountain valley onto a broad, open *plain*.

Alluvium Material deposited by rivers. Nowadays usually only applied to finer particles of silt and clay.

Alpine Mountain *environment*, between the *treeline* and the level of permanent snow cover.

Alpine mountains Ranges of mountains formed between 30 and 65 million years ago, by *folding*, in western and central Europe.

Amerindian A term applied to people *indigenous* to North, Central, and South America.

Animal husbandry The business of rearing animals.

Antarctic circle The parallel which lies at *latitude* of 66° 32′ S.

Anticline A geological *fold* that forms an arch shape, curving upward in the rock *strata*.

Anticyclone An area of relatively high atmospheric pressure.

Aquaculture Collective term for the farming of produce derived from the sea, including fish-farming, the cultivation of shellfish, and plants such as seaweed.

Aquifer A body of rock that can absorb water. Also applied to any rock strata that have sufficient porosity to yield *groundwater* through wells or springs.

Arable Land which has been plowed and is being used, or is suitable, for growing crops.

Archipelago A group or chain of islands.

Arctic Circle The parallel that lies at a *latitude* of 66° 32′ N.

Arête A thin, jagged mountain ridge that divides two adjacent *cirques*, found in regions where *glaciation* has occurred.

Arid Dry. An area of low rainfall, where the rate of *evaporation* may be greater than that of *precipitation*. Often defined as those areas that receive less than one inch (25 mm) of rain a year. In these areas only drought-resistant plants can survive.

Artesian well A naturally occurring source of underground water, stored in an *aquifer*.

Artisanal Small-scale, manual operation, such as fishing, using little or no machinery.

ASEAN Association of Southeast Asian Nations. Established in 1967 to promote economic, social, and cultural cooperation. Its members include Brunei, Indonesia, Malaysia, Philippines, Singapore, and Thailand.

Aseismic A region where *earthquake* activity has ceased.

Asteroid A minor planet circling the Sun, mainly between the orbits of Mars and Jupiter.

Asthenosphere A zone of hot, partially melted rock, which underlies the *lithosphere*, within the Earth's *crust*.

Atmosphere The envelope of odorless, colorless and tasteless gases surrounding the Earth, consisting of *oxygen* (23%), *nitrogen* (75%), argon (1%), *carbon dioxide* (0.03%), as well as tiny proportions of other gases.

Atmospheric pressure The pressure created by the action of gravity on the gases surrounding the Earth.

Atoll A ring-shaped island or *coral reef* often enclosing a *lagoon* of sea water.

Avalanche The rapid movement of a mass of snow and ice down a steep slope. Similar movements of other materials are described as *rock avalanches* or *landslides* and *sand avalanches*.

B

Badlands A landscape that has been heavily eroded and dissected by rainwater, and which has little or no vegetation.

Back slope The gentler windward slope of a sand *dune* or gentler slope of a *cuesta*.

Bajos An *alluvial fan* deposited by a river at the base of mountains and hills that encircle *desert* areas.

Bar, coastal An offshore strip of sand or shingle, either above or below the water. Usually parallel to the shore but sometimes crescent-shaped or at an oblique angle.

Barchan A crescent-shaped sand *dune*, formed where wind direction is very consistent. The horns of the crescent point downwind and where there is enough sand the barchan is mobile.

Barrio A Spanish term for the shantytowns – settlements of shacks – that are clustered around many South and Central American cities (*see also Favela*).

Basalt Dark, fine-grained *igneous rock* that is formed near the Earth's surface from fast-cooling *lava*.

Base level The level below which flowing water cannot erode the land.

Basement rock A mass of ancient rock often of *PreCambrian age*, covered by a layer of more recent *sedimentary rocks*. Commonly associated with *shield* areas.

Beach Lake or sea shore where waves break and there is an accumulation of loose sand, mud, gravel, or pebbles.

Bedrock Solid, consolidated and relatively unweathered rock, found on the surface of the land or just below a layer of soil or *weathered* rock.

Biodiversity The quantity of animal or plant species in a given area.

Biomass The total mass of organic matter – plants and animals – in a given area. It is usually measured in kilogrammes per square meter. Plant biomass is proportionally greater than that of animals, except in cities.

Biosphere The zone just above and below the Earth's surface, where all plants and animals live.

Blizzard A severe windstorm with snow and sleet. Visibility is often severely restricted.

Bluff The steep bank of a *meander*, formed by the erosive action of a river.

Boreal forest Tracts of mainly coniferous forest found in northern *latitudes*.

Breccia A type of rock composed of sharp fragments, cemented by a fine-grained material such as clay.

Butte An isolated, flat-topped hill with steep or vertical sides, buttes are the eroded remnants of a former land surface

C

Caatinga Portuguese (Brazilian) term for thorny woodland growing in areas of pale granitic soils.

CACM Central American Common Market. Established in 1960 to further economic ties between its members, which are Costa Rica, El Salvador, Guatemala, Honduras, and Nicaragua.

Calcite Hexagonal crystals of calcium carbonate.

Caldera A huge volcanic vent, often containing a number of smaller vents, and sometimes a crater lake.

Carbon cycle The transfer of carbon to and from the *atmosphere*. This occurs on land through *photosynthesis*. In the sea, *carbon dioxide* is absorbed, some returning to the air and some taken up into the bodies of sea creatures.

Carbon dioxide A colorless, odorless gas (CO_2) that makes up 0.03% of the *atmosphere*.

Carbonation The process whereby rocks are broken down by carbonic acid. Carbon dioxide in the air dissolves in rainwater, forming carbonic acid. *Limestone* terrain can be rapidly eaten away.

Cash crop A single crop grown specifically for export sale, rather than for local use. Typical examples include coffee, tea, and citrus fruits.

Cassava A type of grain meal, used to produce tapioca. A staple crop in many parts of Africa.

Castle kopje Hill or rock outcrop, especially in southern Africa, where steep sides, and a summit composed of blocks, give a castle-like appearance.

Cataracts A series of stepped waterfalls created as a river flows over a band of hard, resistant rock.

Causeway A raised route through marshland or a body of water.

CEEAC Economic Community of Central African States. Established in 1983 to promote regional cooperation and if possible, establish a common market between 16 Central African nations.

Chemical weathering The chemical reactions leading to the decomposition of rocks. Types of chemical weathering include *carbonation*, *hydrolysis*, and *oxidation*.

Chernozem A fertile soil, also known as "black earth" consisting of a layer of dark topsoil, rich in decaying vegetation, overlying a lighter chalky layer.

Cirque Armchair-shaped basin, found in mountain regions, with a steep back, or rear, wall and a raised rock lip, often containing a lake (or *tarn*). The cirque floor has been eroded by a *glacier*, while the back wall is eroded both by the *glacier* and by *weathering*.

Climate The average weather conditions in a given area over a period of years, sometimes defined as 30 years or more.

Cold War A period of hostile relations between the US and the Soviet Union and their allies after the Second World War.

Composite volcano Also known as a strato-volcano, the volcanic cone is composed of alternating deposits of *lava* and *pyroclastic* material.

Compound A substance made up of *elements* chemically combined in a consistent way.

Condensation The process whereby a gas changes into a liquid. For example, water vapor in the *atmosphere* condenses around tiny airborne particles to form droplets of water.

Confluence The point at which two rivers meet.

Conglomerate Rock composed of large, water-worn or rounded pebbles, held together by a natural cement.

Coniferous forest A forest type containing trees which are generally, but not necessarily, *evergreen* and have slender, needlelike leaves. Coniferous trees reproduce by means of seeds contained in a cone.

Continental drift The theory that the continents of today are fragments of one or more prehistoric *supercontinents* which have moved across the Earth's surface, creating ocean basins. The theory has been superseded by a more sophisticated one – *plate tectonics*.

Continental shelf An area of the continental crust, below sea level, which slopes gently. It is separated from the deep ocean by a much more steeply inclined *continental slope*.

Continental slope A steep slope running from the edge of the *continental shelf* to the ocean floor.

Conurbation A vast metropolitan area created by the expansion of towns and cities into a virtually continuous urban area.

Cool continental A rainy *climate* with warm summers [warmest month below 76°F (22°C)] and often severe winters [coldest month below 32°F (0°C)].

Copra The dried, white kernel of a coconut, from which coconut oil is extracted.

Coral reef An underwater barrier created by colonies of the coral polyp. Polyps secrete a protective skeleton of calcium carbonate, and reefs develop as live polyps build on the skeletons of dead generations.

Core The center of the Earth, consisting of a dense mass of iron and nickel. It is thought that the outer core is molten or liquid, and that the hot inner core is solid due to extremely high pressures.

Coriolis effect A deflecting force caused by the rotation of the Earth. In the northern hemisphere a body, such as an *air mass* or ocean current, is deflected to the right, and in the southern hemisphere to the left. This prevents winds from blowing straight from areas of high to low pressure.

Coulées A US / Canadian term for a ravine formed by river *erosion*.

Craton A large block of the Earth's *crust* which has remained stable for a long period of *geological time*. It is made up of ancient *shield* rocks.

Cretaceous A period of *geological time* beginning about 145 million years ago and lasting until about 65 million years ago.

Crevasse A deep crack in a *glacier*.

Crust The hard, thin outer shell of the Earth. The crust floats on the *mantle*, which is softer and more dense. Under the oceans (oceanic crust) the crust is 3.7–6.8 miles (6–11 km) thick. Continental crust averages 18–24 miles (30–40 km).

Crystalline rock Rocks formed when molten *magma* crystallizes (*igneous rocks*) or when heat or pressure cause re-crystallization (*metamorphic rocks*). Crystalline rocks are distinct from *sedimentary rocks*.

Cuesta A hill which rises into a steep slope on one side but has a gentler gradient on its other slope.

Cyclone An area of low *atmospheric pressure*, occurring where the air is warm and relatively low in density, causing low level winds to spiral. *Hurricanes* and *typhoons* are tropical cyclones.

D

De facto
1 Government or other activity that takes place, or exists in actuality if not by right.
2 A border, which exists in practice, but which is not officially recognized by all the countries it adjoins.

Deciduous forest A forest of trees that shed their leaves annually at a particular time or season. In *temperate* climates the fall of leaves occurs in the autumn. Some *coniferous* trees, such as the larch, are deciduous. Deciduous vegetation contrasts with *evergreen*, which keeps its leaves for more than a year.

Defoliant Chemical spray used to remove foliage (leaves) from trees.

Deforestation The act of cutting down and clearing large areas of forest for human activities, such as agricultural land or urban development.

Delta Low-lying, fan-shaped area at a river mouth, formed by the *deposition* of successive layers of *sediment*. Slowing as it enters the sea, a river deposits sediment and may, as a result, split into numerous smaller channels, known as *distributaries*.

Denudation The combined effect of *weathering*, *erosion*, and *mass movement*, which, over long periods, exposes underlying rocks.

Deposition The laying down of material that has accumulated:
(1) after being *eroded* and then transported by physical forces such as wind, ice, or water;
(2) as organic remains, such as coal and coral;
(3) as the result of *evaporation* and chemical *precipitation*.

Depression
1 In climatic terms it is a large low pressure system.
2 A complex *fold*, producing a large valley, which incorporates both a *syncline* and an *anticline*.

Desert An *arid* region of low rainfall, with little vegetation or animal life, which is adapted to the dry conditions. The term is now applied not only to hot tropical and subtropical regions, but to arid areas of the continental interiors and to the ice deserts of the *Arctic* and *Antarctic*.

Desertification The gradual extension of *desert* conditions in *arid* or *semiarid* regions, as a result of climatic change or human activity, such as over-grazing and *deforestation*.

Despot A ruler with absolute power. Despots are often associated with oppressive regimes.

Detritus Piles of rock deposited by an erosive agent such as a river or *glacier*.

Distributary A minor branch of a river, which does not rejoin the main stream, common at *deltas*.

Diurnal Daily, something that occurs each day. Diurnal temperature refers to the variation in temperature over the course of a full day and night.

Divide A US term describing the area of high ground separating two *drainage basins*.

Donga A steep-sided *gully*, resulting from *erosion* by a river or by floods.

Dormant A term used to describe a *volcano* which is not currently erupting. They differ from extinct volcanoes as dormant volcanoes are still considered likely to erupt in the future.

Drainage basin The area drained by a single river system, its boundary is marked by a *watershed* or *divide*.

Drought A long period of continuously low rainfall.

Drumlin A long, streamlined hillock composed of material deposited by a *glacier*. They often occur in groups known as swarms.

Dune A mound or ridge of sand, shaped, and often moved, by the wind. They are found in hot *deserts* and on low-lying coasts where onshore winds blow across sandy beaches.

Dyke A wall constructed in low-lying areas to contain floodwaters or protect from high tides.

E

Earthflow The rapid movement of soil and other loose surface material down a slope, when saturated by water. Similar to a mudflow but not as fast-flowing, due to a lower percentage of water.

Earthquake Sudden movements of the Earth's *crust*, causing the ground to shake. Frequently occurring at *tectonic plate* margins. The shock, or series of shocks, spreads out from an *epicenter*.

EC The European Community (*see EU*).

Ecosystem A system of living organisms – plants and animals – interacting with their *environment*.

ECOWAS Economic Community of West African States. Established in 1975, it incorporates 16 West African states and aims to promote closer regional and economic cooperation.

Element
1 A constituent of the *climate* – *precipitation*, *humidity*, temperature, *atmospheric pressure*, or wind.
2 A substance that cannot be separated into simpler substances by chemical means.

El Niño A climatic phenomenon, the El Niño effect occurs about 14 times each century and leads to major shifts in global air circulation. It is associated with unusually warm currents off the coasts of Peru, Ecuador and Chile. The anomaly can last for up to two years.

Environment The conditions created by the surroundings (both natural and artificial) within which an organism lives. In human geography the word includes the surrounding economic, cultural, and social conditions.

Eon (aeon) Traditionally a long, but indefinite, period of *geological time*.

Ephemeral A nonpermanent feature, often used in connection with seasonal rivers or lakes in dry areas.

Epicenter The point on the Earth's surface directly above the underground origin – or focus – of an *earthquake*.

Equator The line of *latitude* which lies equidistant between the North and South Poles.

Erg An extensive area of sand *dunes*, particularly in the Sahara Desert.

Erosion The processes which wear away the surface of the land. *Glaciers*, wind, rivers, waves, and currents all carry debris which causes *erosion*. Some definitions also include *mass movement* due to gravity as an agent of erosion.

Escarpment A steep slope at the margin of a level, upland surface. In a landscape created by *folding*, escarpments (or scarps) frequently lie behind a more gentle backward slope.

Esker A narrow, winding ridge of sand and gravel deposited by streams of water flowing beneath or at the edge of a *glacier*.

Erratic A rock transported by a *glacier* and deposited some distance from its place of origin.

Eustacy A world-wide fall or rise in ocean levels.

EU The European Union. Established in 1965, it was formerly known as the EEC (European Economic Community) and then the EC (European Community). Its members are Austria, Belgium, Denmark, Finland, France, Germany, Greece, Ireland, Italy, Luxembourg, Netherlands, Portugal, Spain, Sweden, and UK. It seeks to establish an integrated European common market and eventual federation.

Evaporation The process whereby a liquid or solid is turned into a gas or vapor. Also refers to the diffusion of water vapor into the *atmosphere* from exposed water surfaces such as lakes and seas.

Evapotranspiration The loss of moisture from the Earth's surface through a combination of *evaporation*, and *transpiration* from the leaves of plants.

Evergreen Plants with long-lasting leaves, which are not shed annually or seasonally.

Exfoliation A kind of *weathering* whereby scalelike flakes of rock are peeled or broken off by the development of salt crystals in water within the rocks. *Groundwater*, which contains dissolved salts, seeps to the surface and evaporates, precipitating a film of salt crystals, which expands causing fine cracks. As these grow, flakes of rock break off.

Extrusive rock *Igneous* rock formed when molten material (*magma*) pours forth at the Earth's surface and cools rapidly. It usually has a glassy texture.

F

Factionalism The actions of one or more minority political group acting against the interests of the majority government.

Fault A fracture or crack in rock, where strains (*tectonic* movement) have caused blocks to move, vertically or laterally, relative to each other.

Fauna Collective name for the animals of a particular period of time, or region.

Favela Brazilian term for the shantytowns or temporary huts that have grown up around the edge of many South and Central American cities.

Ferrel cell A component in the global pattern of air circulation, which rises in the colder *latitudes* (60° N and S) and descends in warmer *latitudes* (30° N and S). The Ferrel cell forms part of the world's three-cell air circulation pattern, with the *Hadley* and Polar cells.

Fissure A deep crack in a rock or a *glacier*.

Fjord A deep, narrow inlet, created when the sea inundates the *U-shaped valley* created by a *glacier*.

Flash flood A sudden, short-lived rise in the water level of a river or stream, or surge of water down a dry river channel, or *wadi*, caused by heavy rainfall.

Flax A plant used to make linen.

Floodplain The broad, flat part of a river valley, adjacent to the river itself, formed by *sediment* deposited during flooding.

Flora The collective name for the plants of a particular period of time or region.

Flow The movement of a river within its banks, particularly in terms of the speed and volume of water.

Fold A bend in the rock *strata* of the Earth's *crust*, resulting from compression.

Fossil The remains, or traces,of a dead organism preserved in the Earth's *crust*.

Fossil dune A *dune* formed in a once-*arid* region which is now wetter. *Dunes* normally move with the wind, but in these cases vegetation makes them stable.

Fossil fuel Fuel – coal, natural gas or oil – composed of the fossilized remains of plants and animals.

Front The boundary between two *air masses*, which contrast sharply in temperature and *humidity*.

Frontal depression An area of low pressure caused by rising warm air. They are generally 600–1,200 miles (1,000–2,000 km) in diameter. Within *depressions* there are both warm and cold fronts.

Frost shattering A form of *weathering* where water freezes in cracks, causing expansion. As temperatures fluctuate and the ice melts and refreezes, it eventually causes the rocks to shatter and fragments of rock to break off.

G

Gaucho South American term for a stock herder or cowboy who works on the grassy *plains* of Paraguay, Uruguay, and Argentina.

Geological timescale The chronology of the Earth's history as revealed in its rocks. Geological time is divided into a number of periods: *eon*, era, period, epoch, age, and chron (the shortest). These units are not of uniform length.

Geosyncline A concave fold (*syncline*) or large depression in the Earth's *crust*, extending hundreds of miles. This basin contains a deep layer of sediment, especially at its center, from the land masses around it.

Geothermal energy Heat derived from hot rocks within the Earth's *crust* and resulting in hot springs, steam, or hot rocks at the surface. The energy is generated by rock movements, and from the breakdown of radioactive elements occurring under intense pressure.

GDP Gross Domestic Product. The total value of goods and services produced by a country excluding income from foreign countries.

Geyser A jet of steam and hot water that intermittently erupts from vents in the ground in areas that are, or were, *volcanic*. Some geysers occasionally reach heights of 196 ft (60 m).

Ghetto An area of a city or region occupied by an overwhelming majority of people from one racial or religious group, who may be subject to persecution or containment.

Glaciation The growth of *glaciers* and *ice sheets*, and their impact on the landscape.

Glacier A body of ice moving downslope under the influence of gravity and consisting of compacted and frozen snow. A glacier is distinct from an *ice sheet*, which is wider and less confined by features of the landscape.

Glacio-eustacy A world-wide change in the level of the oceans, caused when the formation of *ice sheets* takes up water or when their melting returns water to the ocean. The formation of ice sheets in the *Pleistocene* epoch, for example, caused sea level to drop by about 320 ft (100-m).

Glaciofluvial To do with glacial *meltwater*, the landforms it creates and its processes; *erosion*, transportation, and *deposition*. Glaciofluvial effects are more powerful and rapid where they occur within or beneath the *glacier*, rather than beyond its edge.

Glacis A gentle slope or *pediment*.

Global warming An increase in the average temperature of the Earth. At present the *greenhouse effect* is thought to contribute to this.

GNP Gross National Product. The total value of goods and services produced by a country.

Gondwanaland The *supercontinent* thought to have existed over 200 million years ago in the southern hemisphere. Gondwanaland is believed to have comprised today's Africa, Madagascar, Australia, parts of South America, *Antarctica*, and the Indian subcontinent.

Graben A block of rock let down between two parallel *faults*. Where the graben occurs within a valley, the structure is known as a *rift valley*.

Grease ice Slicks of ice which form in *Antarctic* seas, when ice crystals are bonded together by wind and wave action.

Greenhouse effect A change in the temperature of the *atmosphere*. Short-wave solar radiation travels through the *atmosphere* unimpeded to the Earth's surface, whereas outgoing, long-wave terrestrial radiation is absorbed by materials that reradiate it back to the Earth. Radiation trapped in this way, by water vapor, carbon dioxide, and other "greenhouse gases," keeps the Earth warm. As more *carbon dioxide* is released into the atmosphere by the burning of *fossil fuels*, the greenhouse effect may cause a global increase in temperature.

Groundwater Water that has seeped into the pores, cavities, and cracks of rocks or into soil and water held in an *aquifer*.

Gully A deep, narrow channel eroded in the landscape by *ephemeral* streams.

Guyot A small, flat-topped submarine mountain, formed as a result of subsidence which occurs during *sea-floor spreading*.

Gypsum A soft mineral *compound* (hydrated calcium sulphate), used as the basis of many forms of plaster, including plaster of Paris.

H

Hadley cell A large-scale component in the global pattern of air circulation. Warm air rises over the *Equator* and blows at high altitude toward the poles, sinking in subtropical regions (30° N and 30° S) and creating high pressure. The air then flows at the surface toward the *Equator* in the form of trade winds. There is one cell in each hemisphere. Named after G. Hadley, who published his theory in 1735.

Hamada An Arabic word for a plateau of bare rock in a *desert*.

Hanging valley A tributary valley that ends suddenly, high above the bed of the main valley. The effect is found where the main valley has been more deeply eroded by a *glacier*, than has the tributary valley. A stream in a hanging valley will descend to the floor of the main valley as a waterfall or *cataract*.

Headwards The action of a river eroding back upstream, as opposed to the normal process of downstream *erosion*. Headwards erosion is often associated with *gullying*.

Hoodos Pinnacles of rock that have been worn away by *weathering* in *semiarid* regions.

Horst A block of the Earth's *crust* which has been left upstanding by the sinking of adjoining blocks along fault lines.

Hot spot A region of the Earth's *crust* where high thermal activity occurs, often leading to volcanic eruptions. Hot spots often occur far from plate boundaries, but their movement is associated with *plate tectonics*.

Humid equatorial Rainy *climate* with no winter, where the coolest month is generally above 64°F (18°C).

Humidity The relative amount of moisture held in the Earth's *atmosphere*.

Hurricane
1 A tropical *cyclone* occurring in the Caribbean and western North Atlantic.
2 A wind of more than 65 knots (75 kmph).

Hydroelectric power Energy produced by harnessing the rapid movement of water down steep mountain slopes to drive turbines to generate electricity.

Hydrolysis The chemical breakdown of rocks in reaction with water, forming new compounds.

I

Ice Age A period in the Earth's history when surface temperatures in the temperate *latitudes* were much lower and *ice sheets* expanded considerably. There have been *ice ages* from *Pre-Cambrian* times onward. The most recent began two million years ago and ended 10,000 years ago.

Ice cap A permanent dome of ice in highland areas. The term ice cap is often seen as distinct from *ice sheet*, which denotes a much wider covering of ice; and is also used refer to the very extensive polar and Greenland ice caps.

Ice floe A large, flat mass of ice floating free on the ocean surface. It is usually formed after the break-up of winter ice by heavy storms.

Ice sheet A continuous, very thick layer of ice and snow. The term is usually used of ice masses which are continental in extent.

Ice shelf A floating mass of ice attached to the edge of a coast. The seaward edge is usually a sheer cliff up to 100 ft (30-m) high.

Ice wedge Massive blocks of ice up to 6.5-ft (2-m) wide at the top and extending 32-ft (10-m) deep. They are found in cracks in *polygonally-patterned* ground in *periglacial* regions.

Iceberg A large mass of ice in a lake or a sea, which has broken off from a floating *ice sheet* (an *ice shelf*) or from a *glacier*.

Igneous rock Rock formed when molten material, *magma*, from the hot, lower layers of the Earth's *crust*, cools, solidifies, and crystallizes, either within the Earth's *crust* (*intrusive*) or on the surface (*extrusive*).

IMF International Monetary Fund. Established in 1944 as a UN agency, it contains 182 members around the world and is concerned with world monetary stability and economic development.

Incised meander A *meander* where the river, following its original course, cuts deeply into *bedrock*. This may occur when a mature, meandering river begins to erode its bed much more vigorously after the surrounding land has been uplifted.

Indigenous People, plants, or animals native to a particular region.

Infrastructure The communications and services – roads, railroads, and telecommunications – necessary for the functioning of a country or region.

Inselberg An isolated, steep-sided hill, rising from a low *plain* in *semiarid* and *savannah* landscapes. Inselbergs are usually composed of a rock, such as granite, which resists *erosion*.

Interglacial A period of global *climate*, between two *ice ages*, when temperatures rise and *ice sheets* and *glaciers* retreat.

Intraplate volcano A *volcano* which lies in the centre of one of the Earth's *tectonic plates*, rather than, as is more common, at its edge. They are thought to have been formed by a *hot spot*.

Intrusion (intrusive igneous rock) Rock formed when molten material, *magma*, penetrates existing rocks below the Earth's surface before cooling and solidifying. These rocks cool more slowly than extrusive rock and therefore tend to have coarser grains.

Irrigation The artificial supply of agricultural water to dry areas, often involving the creation of canals and the diversion of natural watercourses.

Island arc A curved chain of islands. Typically, such an arc fringes an ocean trench, formed at the margin between two *tectonic plates*. As one plate overrides another, *earthquakes* and volcanic activity are common and the islands themselves are often volcanic cones.

Isostasy The state of equilibrium that the Earth's *crust* maintains as its lighter and heavier parts float on the denser underlying mantle.

Isthmus A narrow strip of land connecting two larger landmasses or islands.

J

Jet stream A narrow belt of westerly winds in the *troposphere*, at altitudes above 39,000 ft (12,000 m). Jet streams tend to blow more strongly in winter and include: the subtropical jet stream; the *polar* front jet stream in mid-*latitudes*; the *Arctic* jet stream; and the polar-night jet stream.

Joint A crack in a rock, formed where blocks of rock have not shifted relative to each other, as is the case with a *fault*. Joints are created by *folding*; by shrinkage in *igneous rock* as it cools or *sedimentary rock* as it dries out; and by the release of pressure in a rock mass when overlying materials are removed by *erosion*.

Jute A plant fiber used to make coarse ropes, sacks, and matting.

K

Kame A mound of stratified sand and gravel with steep sides, deposited in a *crevasse* by *meltwater* running over a *glacier*. When the ice retreats, this forms an undulating terrain of hummocks.

Karst A barren *limestone* landscape created by carbonic acid in streams and rainwater, in areas where *limestone* is close to the surface. Typical features include caverns, towerlike hills, *sinkholes*, and flat limestone pavements.

Kettle hole A round hollow formed in a glacial deposit by a detached block of glacial ice, which later melted. They can fill with water to form kettle-lakes.

L

Lagoon A shallow stretch of coastal salt-water behind a partial barrier such as a sandbank or *coral reef*. Lagoon is also used to describe the water encircled by an *atoll*.

LAIA Latin American Integration Association. Established in 1980, its members are Argentina, Bolivia, Brazil, Chile, Colombia, Ecuador, Mexico, Paraguay, Peru, Uruguay, and Venezuala. It aims to promote economic cooperation between member states.

Landslide The sudden downslope movement of a mass of rock or earth on a slope, caused either by heavy rain; the impact of waves; an *earthquake* or human activity.

Laterite A hard red deposit left by *chemical weathering* in tropical conditions, and consisting mainly of oxides of iron and aluminium.

Latitude The angular distance from the *Equator*, to a given point on the Earth's surface. Imaginary lines of *latitude* running parallel to the Equator encircle the Earth, and are measured in degrees north or south of the Equator. The Equator is 0°, the poles 90° South and North respectively. Also called parallels.

Laurasia In the theory of *continental drift*, the northern part of the great *supercontinent* of *Pangaea*. Laurasia is said to consist of N America, Greenland and all of Eurasia north of the Indian subcontinent.

Lava The molten rock, *magma*, which erupts onto the Earth's surface through a *volcano*, or through a *fault* or crack in the Earth's *crust*. Lava refers to the rock both in its molten and in its later, solidified form.

Leaching The process whereby water dissolves minerals and moves them down through layers of soil or rock.

Levée A raised bank alongside the channel of a river. Levées are either human-made or formed in times of flood when the river overflows its channel, slows and deposits much of its *sediment* load.

Lichen An organism which is the symbiotic product of an algae and a fungus. Lichens form in tight crusts on stones and trees, and are resistant to extreme cold. They are often found in tundra regions.

Lignite Low-grade coal, also known as brown coal. Found in large deposits in eastern Europe.

Limestone A porous *sedimentary* rock formed from carbonate materials.

Lingua franca The language adopted as the common language between speakers whose native languages are different. This is common in former colonial states.

Lithosphere The rigid upper layer of the Earth, comprising the *crust* and the upper part of the *mantle*.

Llanos Vast grassland *plains* of northern South America.

Loess Fine-grained, yellow deposits of unstratified silts and sands. Loess is believed to be wind-carried *sediment* created in the last *Ice Age*. Some deposits may later have been redistributed by rivers. Loess-derived soils are of high quality, fertile, and easy to work.

Longitude A division of the Earth which pinpoints how far east or west a given place is from the Prime Meridian (0°) which runs through the Royal Observatory at Greenwich, England (UK). Imaginary lines of longitude are drawn around the world from pole to pole. The world is divided into 360 degrees.

Longshore drift The movement of sand and silt along the coast, carried by waves hitting the beach at an angle.

M

Magma Underground, molten rock, which is very hot and highly charged with gas. It is generated at great pressure, at depths 10 miles (16 km) or more below the Earth's surface. It can issue as *lava* at the Earth's surface or, more often, solidify below the surface as *intrusive igneous rock*.

Mantle The layer of the Earth between the *crust* and the *core*. It is about 1,800 miles (2,900-km) thick. The uppermost layer of the mantle is the soft, 125-mile (200 km) thick *asthenosphere* on which the more rigid *lithosphere* floats.

Maquiladoras Factories on the Mexico side of the Mexico/US border, that are allowed to import raw materials and components duty-free and use low-cost labor to assemble the goods, finally exporting them for sale in the US.

Market gardening The intensive growing of fruit and vegetables close to large local markets.

Mass movement Downslope movement of weathered materials such as rock, often helped by rainfall or glacial *meltwater*. Mass movement may be a gradual process or rapid, as in a *landslide* or rockfall.

Massif A single very large mountain or an area of mountains with uniform characteristics and clearly-defined boundaries.

Meander A looplike bend in a river, which is found typically in the lower, mature reaches of a river but can form wherever the valley is wide and the slope gentle.

Mediterranean climate A temperate *climate* of hot, dry summers and warm, damp winters. This is typical of the western fringes of the world's continents in the warm temperate regions between *latitudes* of 30° and 40° (north and south).

Meltwater Water resulting from the melting of a *glacier* or *ice sheet*.

Mesa A broad, flat-topped hill, characteristic of *arid* regions.

Mesosphere A layer of the Earth's *atmosphere*, between the *stratosphere* and the *thermosphere*. Extending from about 25–50 miles (40–80 km) above the surface of the Earth.

Mestizo A person of mixed *Amerindian* and European origin.

Metallurgy The refining and working of metals.

Metamorphic rocks Rocks that have been altered from their original form, in terms of texture, composition, and structure by intense heat, pressure, or by the introduction of new chemical substances – or a combination of more than one of these.

Meteor A body of rock, metal or other material, that travels through space at great speeds. Meteors are visible as they enter the Earth's *atmosphere* as shooting stars and fireballs.

Meteorite The remains of a *meteor* that has fallen to Earth.

Meteoroid A *meteor* that is still traveling in space, outside the Earth's *atmosphere*.

Mezzogiorno A term applied to the southern portion of Italy.

Milankovitch hypothesis A theory suggesting that there are a series of cycles that slightly alter the Earth's position when rotating about the Sun. The cycles identified all affect the amount of *radiation* the Earth receives at different *latitudes*. The theory is seen as a key factor in the cause of *ice ages*.

Millet A grain-crop, forming part of the staple diet in much of Africa.

Mistral A strong, dry, cold northerly or north-westerly wind, which blows from the Massif Central of France to the Mediterranean Sea. It is common in winter and its cold blasts can cause crop damage in the Rhône Delta, in France.

Mohorovicic discontinuity (Moho) The structural divide at the margin between the Earth's *crust* and the *mantle*. On average it is 20 miles (35-km) below the continents and 6-miles (10 km) below the oceans. The different densities of the *crust* and the mantle cause *earthquake* waves to accelerate at this point.

Monarchy A form of government in which the head of state is a single hereditary monarch. The monarch may be a mere figurehead, or may retain significant authority.

Monsoon A wind that changes direction biannually. The change is caused by the reversal of pressure over landmasses and the adjacent oceans. Because the inflowing moist winds bring rain, the term monsoon is also used to refer to the rains themselves. The term is derived from and most commonly refers to the seasonal winds of south and east Asia.

Montaña Mountain areas along the west coast of South America.

Moraine Debris, transported and deposited by a *glacier* or *ice sheet* in unstratified, mixed, piles of rock, boulders, pebbles, and clay.

Mountain-building The formation of *fold* mountains by tectonic activity. Also known as orogeny, mountain-building often occurs on the margin where two *tectonic plates* collide. The periods when most mountain-building occurred are known as orogenic phases and lasted many millions of years.

Mudflow An *avalanche* of mud that occurs when a mass of soil is drenched by rain or melting snow. It is a type of *mass movement*, faster than an *earthflow* because it is lubricated by water.

N

Nappe A mass of rocks which has been overfolded by repeated thrust *faulting*.

NAFTA The North American Free Trade Association. Established in 1994 between Canada, Mexico, and the US to set up a free-trade zone.

NASA The National Aeronautical and Space Administration. It is a US government agency, established in 1958 to develop manned and unmanned space programs.

NATO The North Atlantic Treaty Organization. Established in 1949 to promote mutual defense and cooperation between its members, which are Belgium, Canada, Czech Republic, Denmark, France, Germany, Greece, Iceland, Italy, Luxembourg, the Netherlands, Norway, Portugal, Poland, Spain, Turkey, UK, and US.

Nitrogen The odorless, colorless gas that makes up 78% of the atmosphere. Within the soil, it is a vital nutrient for plants.

Nomads (nomadic) Wandering communities that move around in search of suitable pasture for their herds of animals.

Nuclear fusion A technique used to create a new nucleus by the merging of two lighter ones, resulting in the release of large quantities of energy.

O

Oasis A fertile area in the midst of a *desert*, usually watered by an underground *aquifer*.

Oceanic ridge A mid-ocean ridge formed, according to the theory of *plate tectonics*, when plates drift apart and hot *magma* pours through to form new oceanic *crust*.

Oligarchy The government of a state by a small, exclusive group of people – such as an elite class or a family group.

Onion-skin weathering The *weathering* away or *exfoliation* of a rock or outcrop by the peeling off of surface layers.

Oriente A flatter region lying to the east of the Andes in South America.

Outwash plain *Glaciofluvial* material (typically clay, sand, and gravel) carried beyond an ice sheet by *meltwater* streams, forming a broad, flat deposit.

Oxbow lake A crescent-shaped lake formed on a river *floodplain* when a river erodes the outside bend of a *meander*, making the neck of the *meander* narrower until the river cuts across the neck. The meander is cut off and is dammed off with sediment, creating an oxbow lake. Also known as a cut-off or mortlake.

Oxidation A form of *chemical weathering* where *oxygen* dissolved in water reacts with minerals in rocks – particularly iron – to form oxides. Oxidation causes brown or yellow staining on rocks, and eventually leads to the break down of the rock.

Oxygen A colorless, odorless gas which is one of the main constituents of the Earth's *atmosphere* and is essential to life on Earth.

Ozone layer A layer of enriched *oxygen* (O_3) within the stratosphere, mostly between 18–50 miles (30–80 km) above the Earth's surface. It is vital to the existence of life on Earth because it absorbs harmful shortwave ultraviolet radiation, while allowing beneficial longer wave ultraviolet radiation to penetrate to the Earth's surface.

P

Pacific Rim The name given to the economically-dynamic countries bordering the Pacific Ocean.

Pack ice Ice masses more than 10 ft (3-m) thick that form on the sea surface and are not attached to a landmass.

Pancake ice Thin discs of ice, up to 8 ft (2.4 m) wide which form when slicks of *grease ice* are tossed together by winds and stormy seas.

Pangaea In the theory of continental drift, Pangaea is the original great land mass which, about 190 million years ago, began to split into Gondwanaland in the south and Laurasia in the north, separated by the Tethys Sea.

Pastoralism Grazing of livestock– usually sheep, goats, or cattle. Pastoralists in many drier areas have traditionally been *nomadic*.

Parallel *see Latitude.*

Peat Ancient, partially-decomposed vegetation found in wet, boggy conditions where there is little *oxygen*. It is the first stage in the development of coal and is often dried for use as fuel. It is also used to improve soil quality.

Pediment A gently-sloping ramp of *bedrock* below a steeper slope, often found at mountain edges in *desert* areas, but also in other climatic zones. Pediments may include depositional elements such as *alluvial fans*.

Peninsula A thin strip of land surrounded on three of its sides by water. Large examples include Florida and Korea.

Per capita Latin term meaning "for each person."

Periglacial Regions on the edges of *ice sheets* or *glaciers* or, more commonly, cold regions experiencing intense frost action, *permafrost* or both. Periglacial climates bring long, freezing winters and short, mild summers.

Permafrost Permanently frozen ground, typical of *Arctic* regions. Although a layer of soil above the permafrost melts in summer, the melted water does not drain through the permafrost.

Permeable rocks Rocks through which water can seep, because they are either porous or cracked.

Pharmaceuticals The manufacture of medicinal drugs.

Phreatic eruption A volcanic eruption which occurs when *lava* combines with *groundwater*, superheating the water and causing a sudden emission of steam at the surface.

Physical weathering (mechanical weathering) The breakdown of rocks by physical, as opposed to chemical, processes. Examples include: changes in pressure or temperature; the effect of windblown sand; the pressure of growing salt crystals in cracks within rock; and the expansion and contraction of water within rock as it freezes and thaws.

Pingo A dome of earth with a core of ice, found in *tundra* regions. Pingos are formed either when *groundwater* freezes and expands, pushing up the land surface, or when trapped, freezing water in a lake expands and pushes up lake *sediments* to form the pingo dome.

Placer A belt of mineral-bearing rock *strata* lying at or close to the Earth's surface, from which minerals can be easily extracted.

Plain A flat, level region of land, often relatively low-lying.

Plateau A highland tract of flat land.

Plate *see Tectonic plates.*

Plate tectonics The study of *tectonic plates*, that helps to explain *continental drift*, mountain formation and volcanic activity. The movement of tectonic plates may be explained by the currents of rock rising and falling from within the Earth's *mantle*, as it heats up and then cools. The boundaries of the plates are known as plate margins and most mountains, *earthquakes*, and *volcanoes* occur at these margins. Constructive margins are moving apart; destructive margins are crunching together and conservative margins are sliding past one another.

Pleistocene A period of *geological time* spanning from about 5.2 million years ago to 1.6 million years ago.

Plutonic rock *Igneous* rocks found deep below the surface. They are coarse-grained because they cooled and solidified slowly.

Polar The zones within the *Arctic* and *Antarctic* circles.

Polje A long, broad *depression* found in *karst* (*limestone*) regions.

Polygonal patterning Typical ground patterning, found in areas where the soil is subject to severe frost action, often in *periglacial* regions.

Porosity A measure of how much water can be held within a rock or a soil. Porosity is measured as the percentage of holes or pores in a material, compared to its total volume. For example, the porosity of slate is less than 1%, whereas that of gravel is 25–35%.

Prairies Originally a French word for grassy *plains* with few or no trees.

Pre-Cambrian The earliest period of *geological time* dating from over 570-million years ago.

Precipitation The fall of moisture from the *atmosphere* onto the surface of the Earth, whether as dew, hail, rain, sleet, or snow.

Pyramidal peak A steep, isolated mountain summit, formed when the back walls of three or more *cirques* are cut back and move toward each other. The cliffs around such a horned peak, or horn, are divided by sharp *arêtes*. The Matterhorn in the Swiss Alps is an example.

Pyroclasts Fragments of rock ejected during volcanic eruptions.

Q

Quaternary The current period of *geological time*, which started about 1.6-million years ago.

R

Radiation The emission of energy in the form of particles or waves. Radiation from the sun includes heat, light, ultraviolet rays, gamma rays, and X-rays. Only some of the solar energy radiated into space reaches the Earth.

Rainforest Dense forests in tropical zones with high rainfall, temperature and *humidity*. Strictly, the term applies to the equatorial rain forest in tropical lowlands with constant rainfall and no seasonal change. The Congo and Amazon basins are examples. The term is applied more loosely to lush forest in other climates. Within rain forests organic life is dense and varied: at least 40% of all plant and animal species are found here and there may be as many as 100 tree species per hectare.

Rainshadow An area which experiences low rainfall, because of its position on the leeward side of a mountain range.

Reg A large area of stony *desert*, where tightly-packed gravel lies on top of clayey sand. A reg is formed where the wind blows away the finer sand.

Remote-sensing Method of obtaining information about the *environment* using unmanned equipment, such as a satellite, that relays the information to a point where it is collected and used.

Resistance The capacity of a rock to resist *denudation*, by processes such as *weathering* and *erosion*.

Ria A flooded *V-shaped river valley* or estuary, flooded by a rise in sea level (*eustacy*) or sinking land. It is shorter than a *fjord* and gets deeper as it meets the sea.

Rift valley A long, narrow depression in the Earth's *crust*, formed by the sinking of rocks between two *faults*.

River channel The trough which contains a river and is molded by the flow of water within it.

Roche moutonée A rock found in a glaciated valley. The side facing the flow of the *glacier* has been smoothed and rounded, while the other side has been left more rugged because the *glacier*, as it flows over it, has plucked out frozen fragments and carried them away.

Runoff Water draining from a land surface by flowing across it.

S

Sabkha The floor of an isolated *depression* that occurs in an *arid environment* – usually covered by salt deposits and devoid of vegetation.

SADC Southern African Development Community. Established in 1992 to promote economic integration between its member states, which are Angola, Botswana, Lesotho, Malawi, Mauritius, Mozambique, Namibia, South Africa, Swaziland, Tanzania, Zambia, and Zimbabwe.

Salt plug A rounded hill produced by the upward doming of rock *strata* caused by the movement of salt or other evaporite deposits under intense pressure.

Sastrugi Ice ridges formed by wind action. They lie parallel to the direction of the wind.

Savannah Open grassland found between the zone of *deserts*, and that of tropical *rain forests* in the tropics and subtropics. Scattered trees and shrubs are found in some kinds of savannah. A savannah *climate* usually has wet and dry seasons.

Scarp *see Escarpment.*

Scree Piles of rock fragments beneath a cliff or rock face, caused by mechanical *weathering*, especially *frost shattering*, where the expansion and contraction of freezing and thawing water within the rock, gradually breaks it up.

Sea-floor spreading The process whereby *tectonic plates* move apart, allowing hot *magma* to erupt and solidify. This forms a new sea floor and, ultimately, widens the ocean.

Seamount An isolated, submarine mountain or hill, probably of volcanic origin.

Season A period of time linked to regular changes in the weather, especially the intensity of solar *radiation*.

Sediment Grains of rock transported and deposited by rivers, sea, ice, or wind.

Sedimentary rocks Rocks formed from the debris of preexisting rocks or of organic material. They are found in many *environments* – on the ocean floor, on beaches, rivers, and *deserts*. Organically-formed sedimentary rocks include coal and chalk. Other sedimentary rocks, such as flint, are formed by chemical processes. Most of these rocks contain *fossils*, which can be used to date them.

Seif A sand *dune* which lies parallel to the direction of the prevailing wind. Seifs form steep-sided ridges, sometimes extending for miles.

Seismic activity Movement within the Earth, such as an *earthquake* or *tremor*.

Selva A region of wet forest found in the Amazon Basin.

Semiarid, semidesert The *climate* and landscape which lies between *savannah* and *desert* or between savannah and a *mediterranean* climate. In semiarid conditions there is a little more moisture than in a true *desert*; and more patches of drought-resistant vegetation can survive.

Shale (marine shale) A compacted *sedimentary rock*, with fine-grained particles. Marine shale is formed on the seabed. Fuel such as oil may be extracted from it.

Sheetwash Water that runs downhill in thin sheets without forming channels. It can cause *sheet erosion*.

Sheet erosion The washing away of soil by a thin film or sheet of water, known as *sheetwash*.

Shield A vast stable block of the Earth's *crust*, which has experienced little or no *mountain-building*.

Sierra The Spanish word for mountains.

Sinkhole A circular *depression* in a *limestone* region. They are formed by the collapse of an underground cave system or the *chemical weathering* of the *limestone*.

Sisal A plant-fiber used to make matting.

Slash and burn A farming technique involving the cutting down and burning of scrub forest, to create agricultural land. After a number of seasons this land is abandoned and the process is repeated. This practice is common in Africa and South America.

Slip face The steep leeward side of a sand *dune* or slope. Opposite side to a *back slope*.

Soil A thin layer of rock particles mixed with the remains of dead plants and animals. This occurs naturally on the surface of the Earth and provides a medium for plants to grow.

Soil creep The very gradual downslope movement of rock debris and soil, under the influence of gravity. This is a type of *mass movement*.

Soil erosion The wearing away of soil more quickly than it is replaced by natural processes. Soil can be carried away by wind as well as by water. Human activities, such as over-grazing and the clearing of land for farming, accelerate the process in many areas.

Solar energy Energy derived from the Sun. Solar energy is converted into other forms of energy. For example, the wind and waves, as well as the creation of plant material in photosynthesis, depend on solar energy.

Solifluction A kind of *soil creep*, where water in the surface layer has saturated the soil and rock debris which slips slowly downhill. It often happens where frozen top-layer deposits thaw, leaving frozen layers below them.

Sorghum A type of grass found in South America, similar to sugar cane. When refined it is used to make molasses.

Spit A thin linear deposit of sand or shingle extending from the sea shore. Spits are formed as angled waves shift sand along the beach, eventually extending a ridge of sand beyond a change in the angle of the coast. Spits are common where the coastline bends, especially at estuaries.

Squash A type of edible gourd.

Stack A tall, isolated pillar of rock near a coastline, created as wave action erodes away the adjacent rock.

Stalactite A tapering cylinder of mineral deposit, hanging from the roof of a cave in a *karst* area. It is formed by calcium carbonate, dissolved in water, which drips through the roof of a *limestone* cavern.

Stalagmite A cone of calcium carbonate, similar to a *stalactite*, rising from the floor of a *limestone* cavern and formed when drops of water fall from the roof of a *limestone* cave. If the water has dripped from a *stalactite* above the stalagmite, the two may join to form a continuous pillar.

Staple crop The main crop on which a country is economically and or physically reliant. For example, the major crop grown for large-scale local consumption in South Asia is rice.

Steppe Large areas of dry grassland in the northern hemisphere – particularly found in southeast Europe and central Asia.

Strata The plural of stratum, a distinct, virtually horizontal layer of deposited material, lying parallel to other layers.

Stratosphere A layer of the *atmosphere*, above the *troposphere*, extending from about 7–30 miles (11–50 km) above the Earth's surface. In the lower part of the stratosphere, the temperature is relatively stable and there is little moisture.

Strike-slip fault Occurs where plates move sideways past each other and blocks of rocks move horizontally in relation to each other, not up or down as in normal *faults*.

Subduction zone A region where two *tectonic plates* collide, forcing one beneath the other. Typically, a dense oceanic plate dips below a lighter continental plate, melting in the heat of the *asthenosphere*. This is why the zone is also called a destructive margins (*see Plate tectonics*). These zones are characterized by *earthquakes*, volcanoes, *mountain–building*, and the development of oceanic trenches and *island arcs*.

Submarine canyon A steep-sided valley, that extends along the *continental shelf* to the ocean floor. Often formed by *turbidity currents*.

Submarine fan Deposits of silt and *alluvium*, carried by large rivers forming great fan-shaped deposits on the ocean floor.

Subsistence agriculture An agricultural practice in which enough food is produced to support the farmer and his dependents, but not providing any surplus to generate an income.

Subtropical A term applied loosely to *climates* which are nearly tropical or tropical for a part of the year – areas north or south of the *tropics* but outside the *temperate zone*.

Supercontinent A large continent that breaks up to form smaller continents or that forms when smaller continents merge. In the theory of *continental drift*, the supercontinents are *Pangaea*, *Gondwanaland*, and *Laurasia*.

Sustainable development An approach to development, especially applied to economies across the world which exploit natural resources without destroying them or the *environment*.

Syncline A basin-shaped downfold in rock *strata*, created when the *strata* are compressed, for example where *tectonic plates* collide.

T

Tableland A highland area with a flat or gently undulating surface.

Taiga The belt of *coniferous* forest found in the north of Asia and North America. The conifers are adapted to survive low temperatures and long periods of snowfall.

Tarn A Scottish term for a small mountain lake, usually found at the head of a *glacier*.

Tectonic plates Plates, or tectonic plates, are the rigid slabs which form the Earth's outer shell, the *lithosphere*. Eight big plates and several smaller ones have been identified.

Temperate A moderate *climate* without extremes of temperature, typical of the mid-*latitudes* between the *tropics* and the *polar* circles.

Theocracy A state governed by religious laws – today Iran is the world's largest theocracy.

Thermokarst Subsidence created by the thawing of ground ice in *periglacial* areas, creating depressions.

Thermosphere A layer of the Earth's *atmosphere* which lies above the *mesophere*, about 60–300 miles (100–500 km) above the Earth

Terraces Steps cut into steep slopes to create flat surfaces for cultivating crops. They also help reduce soil *erosion* on unconsolidated slopes. They are most common in heavily-populated parts of Southeast Asia.

Till Unstratified glacial deposits or drift left by a *glacier* or *ice sheet*. Till includes mixtures of clay, sand, gravel, and boulders.

Topography The typical shape and features of a given area such as land height and terrain.

Tombolo A large sand *spit* which attaches part of the mainland to an island.

Tornado A violent, spiraling windstorm, with a center of very low pressure. Wind speeds reach 200 mph (320 kmph) and there is often thunder and heavy rain.

Transform fault In *plate tectonics*, a *fault* of continental scale, occurring where two plates slide past each other, staying close together for example, the San Andreas Fault, USA. The jerky, uneven movement creates *earthquakes* but does not destroy or add to the Earth's *crust*

Transpiration The loss of water vapor through the pores (or stomata) of plants. The process helps to return moisture to the *atmosphere*.

Trap An area of fine-grained *igneous rock* that has been extruded and cooled on the Earth's surface in stages, forming a series of steps or terraces.

Treeline The line beyond which trees cannot grow, dependent on *latitude* and altitude, as well as local factors such as soil.

Tremor A slight *earthquake*.

Trench (oceanic trench) A long, deep trough in the ocean floor, formed, according to the theory of *plate tectonics*, when two plates collide and one dives under the other, creating a *subduction zone*.

Tropics The zone between the *Tropic of Cancer* and the *Tropic of Capricorn* where the *climate* is hot. Tropical climate is also applied to areas rather further north and south of the *Equator* where the climate is similar to that of the true tropics.

Tropic of Cancer A line of *latitude* or imaginary circle round the Earth, lying at 23° 28' N.

Tropic of Capricorn A line of *latitude* or imaginary circle round the Earth, lying at 23° 28' S.

Troposphere The lowest layer of the Earth's *atmosphere*. From the surface, it reaches a height of between 4–10 miles (7–16 km). It is the most turbulent zone of the atmosphere and accounts for the generation of most of the world's weather. The layer above it is called the *stratosphere*.

Tsunami A huge wave created by shock waves from an *earthquake* under the sea. Reaching speeds of up to 600 mph (960-kmph), the wave may increase to heights of 50 ft (15 m) on entering coastal waters; and it can cause great damage.

Tundra The treeless *plains* of the *Arctic Circle*, found south of the *polar* region of permanent ice and snow, and north of the belt of *coniferous* forests known as *taiga*. In this region of long, very cold winters, vegetation is usually limited to mosses, *lichens*, sedges, and rushes, although flowers and dwarf shrubs blossom in the brief summer.

Turbidity current An oceanic feature. A turbidity current is a mass of *sediment*-laden water thathas substantial erosive power. Turbidity currents are thought to contribute to the formation of *submarine canyons*.

Typhoon A kind of *hurricane* (or tropical cyclone) bringing violent winds and heavy rain, a typhoon can do great damage. They occur in the South China Sea, especially around the Philippines.

U

U-shaped valley A river valley that has been deepened and widened by a *glacier*. They are characteristically flat-bottomed and steep-sided and generally much deeper than river valleys.

UN United Nations. Established in 1945, it contains 188 nations and aims to maintain international peace and security, and promote cooperation over economic, social, cultural, and humanitarian problems.

UNICEF United Nations Children's Fund. A UN organization set up to promote family and child related programs.

Urstromtäler A German word used to describe *meltwater* channels that flowed along the front edge of the advancing *ice sheet* during the last Ice Age, 18,000–20,000 years ago.

V

V-shaped valley A typical valley eroded by a river in its upper course.

Virgin rain forest Tropical *rain-forest* in its original state, untouched by human activity such as logging, clearance for agriculture, settlement, or roadbuilding.

Viticulture The cultivation of grapes for wine.

Volcano An opening or vent in the Earth's *crust* where molten rock, *magma*, erupts. Volcanoes tend to be conical but may also be a crack in the Earth's surface or a hole blasted through a mountain. The magma is accompanied by other materials such as gas, steam, and fragments of rock, or *pyroclasts*. They tend to occur on destructive or constructive tectonic*plate* margins.

W-Z

Wadi The dry bed left by a torrent of water. Also classified as a *ephemeral* stream, found in *arid* and *semiarid* regions, which are subject to sudden and often severe flash flooding.

Warm humid climate A rainy climate with warm summers and mild winters.

Water cycle The continuous circulation of water between the Earth's surface and the *atmosphere*. The processes include *evaporation* and *transpiration* of moisture into the atmosphere, and its return as *precipitation*, some of which flows into lakes and oceans.

Water table The upper level of *groundwater* saturation in permeable rock *strata*.

Watershed The dividing line between one *drainage basin* – an area where all streams flow into a single river system – and another. In the US, watershed also means the whole drainage basin of a single river system – its catchment area.

Waterspout A rotating column of water in the form of cloud, mist, and spray which form on open water. Often has the appearance of a small *tornado*.

Weathering The decay and breakup of rocks at or near the Earth's surface, caused by water, wind, heat or ice, organic material, or the *atmosphere*. *Physical weathering* includes the effects of frost and temperature changes. Biological weathering includes the effects of plant roots, burrowing animals and the acids produced by animals, especially as they decay after death. *Carbonation* and *hydrolysis* are among many kinds of *chemical weathering*.

Geographical names

The following glossary lists all geographical terms occurring on the maps and in main-entry names in the Index-Gazetteer. These terms may precede, follow, or be run together with the proper element of the name; where they precede it the term is reversed for indexing purposes - thus Poluostrov Yamal is indexed as Yamal, Poluostrov.

Key

Geographical term
Language, Term

A

Å *Danish, Norwegian*, River
Āb *Persian*, River
Adrar *Berber*, Mountains
Agía, Ágios *Greek*, Saint
Air *Indonesian*, River
Akrotírio *Greek*, Cape, point
Alpen *German*, Alps
Alt- *German*, Old
Altiplanicie *Spanish*, Plateau
Älv, -älven *Swedish*, River
-ån *Swedish*, River
Anse *French*, Bay
'Aqabat *Arabic*, Pass
Archipiélago *Spanish*, Archipelago
Arcipelago *Italian*, Archipelago
Arquipélago *Portuguese*, Archipelago
Arrecife(s) *Spanish*, Reef(s)
Aru *Tamil*, River
Augstiene *Latvian*, Upland
Aukštuma *Lithuanian*, Upland
Aust- *Norwegian*, Eastern
Avtonomnyy Okrug *Russian*, Autonomous district
Āw *Kurdish*, River
'Ayn *Arabic*, Spring, well
'Ayoûn *Arabic*, Wells

B

Baelt *Danish*, Strait
Bahía *Spanish*, Bay
Baḩr *Arabic*, River
Baía *Portuguese*, Bay
Baie *French*, Bay
Bañado *Spanish*, Marshy land
Bandao *Chinese*, Peninsula
Banjaran *Malay*, Mountain range
Barajı *Turkish*, Dam
Barragem *Portuguese*, Reservoir
Bassin *French*, Basin
Batang *Malay*, Stream
Beinn, Ben *Gaelic*, Mountain
-berg *Afrikaans, Norwegian*, Mountain
Besar *Indonesian, Malay*, Big
Birkat, Birket *Arabic*, Lake, well, cistern
Boğazı *Turkish*, Strait, defile
Boka *Serbo-Croatian*, Bay
Bol'sh-aya, -iye, -oy, -oye *Russian*, Big
Botigh(i) *Uzbek*, Depression basin
-bre(en) *Norwegian*, Glacier
Bredning *Danish*, Bay
Bucht *German*, Bay
Bugt(en) *Danish*, Bay
Buḩayrat *Arabic*, Lake, reservoir
Buḩeiret *Arabic*, Lake
Bukit *Malay*, Mountain
-bukta *Norwegian*, Bay
bukten *Swedish*, Bay
Bulag *Mongolian*, Spring
Bulak *Uighur*, Spring
Burnu *Turkish*, Cape, point
Buuraha *Somali*, Mountains

C

Cabo *Portuguese*, Cape
Caka *Tibetan*, Salt lake
Canal *Spanish*, Channel
Cap *French*, Cape
Capo *Italian*, Cape, headland
Cascada *Portuguese*, Waterfall
Cayo(s) *Spanish*, Islet(s), rock(s)
Cerro *Spanish*, Hill
Chaîne *French*, Mountain range
Chapada *Portuguese*, Hills, upland
Chau *Cantonese*, Island
Chāy *Turkish*, River
Chhâk *Cambodian*, Bay
Chhu *Tibetan*, River
-chōsuji *Korean*, Reservoir
Chott *Arabic*, Depression, salt lake
Chŭli *Uzbek*, Grassland, steppe
Ch'ün-tao *Chinese*, Island group
Chuŏr Phnum *Cambodian*, Mountains
Ciudad *Spanish*, City, town
Co *Tibetan*, Lake
Colline(s) *French*, Hill(s)
Cordillera *Spanish*, Mountain range
Costa *Spanish*, Coast
Côte *French*, Coast
Coxilha *Portuguese*, Mountains
Cuchilla *Spanish*, Mountains

D

Daban *Mongolian, Uighur*, Pass
Dağı *Azerbaijani, Turkish*, Mountain
Dağları *Azerbaijani, Turkish*, Mountains
-dake *Japanese*, Peak
-dal(en) *Norwegian*, Valley
Danau *Indonesian*, Lake
Dao *Chinese*, Island
Đao *Vietnamese*, Island
Daryā *Persian*, River
Daryācheh *Persian*, Lake
Dasht *Persian*, Desert, plain
Dawḩat *Arabic*, Bay
Denizi *Turkish*, Sea
Dere *Turkish*, Stream
Desierto *Spanish*, Desert
Dili *Azerbaijani*, Spit
-do *Korean*, Island
Dooxo *Somali*, Valley
Düzü *Azerbaijani*, Steppe
-dwīp *Bengali*, Island

E

-eilanden *Dutch*, Islands
Embalse *Spanish*, Reservoir
Ensenada *Spanish*, Bay
Erg *Arabic*, Dunes
Estany *Catalan*, Lake
Estero *Spanish*, Inlet
Estrecho *Spanish*, Strait
Étang *French*, Lagoon, lake
-ey *Icelandic*, Island
Ezero *Bulgarian, Macedonian*, Lake
Ezers *Latvian*, Lake

F

Feng *Chinese*, Peak
-fjella *Norwegian*, Mountain
Fjord *Danish*, Fjord
-fjord(en) *Danish, Norwegian, Swedish*, fjord
-fjördhur *Icelandic*, Fjord
Fleuve *French*, River
Fliegu *Maltese*, Channel
-fljór *Icelandic*, River
-flói *Icelandic*, Bay
Forêt *French*, Forest

G

-gan *Japanese*, Rock
-gang *Korean*, River
Ganga *Hindi, Nepali, Sinhala*, River
Gaoyuan *Chinese*, Plateau
Garagumy *Turkmen*, Sands
-gawa *Japanese*, River
Gebel *Arabic*, Mountain
-gebirge *German*, Mountain range
Ghadīr *Arabic*, Well
Ghubbat *Arabic*, Bay
Gjiri *Albanian*, Bay
Gol *Mongolian*, River
Golfe *French*, Gulf
Golfo *Italian, Spanish*, Gulf
Göl(ü) *Turkish*, Lake
Golyam, -a *Bulgarian*, Big
Gora *Russian, Serbo-Croatian*, Mountain
Góra *Polish*, mountain
Gory *Russian*, Mountain
Gryada *Russian*, ridge
Guba *Russian*, Bay
-gundo *Korean*, island group
Gunung *Malay*, Mountain

H

Ḩadd *Arabic*, Spit
-haehyŏp *Korean*, Strait
Haff *German*, Lagoon
Hai *Chinese*, Bay, lake, sea
Haixia *Chinese*, Strait
Ḩammādah *Arabic*, Desert
Ḩammādat *Arabic*, Rocky plateau
Hāmūn *Persian*, Lake
-hantō *Japanese*, Peninsula
Har, Haré *Hebrew*, Mountain
Ḩarrat *Arabic*, Lava-field
Hav(et) *Danish, Swedish*, Sea
Hawr *Arabic*, Lake
Hāyk' *Amharic*, Lake
He *Chinese*, River
-hegység *Hungarian*, Mountain range
Heide *German*, Heath, moorland
Helodrano *Malagasy*, Bay
Higashi- *Japanese*, East(ern)
Ḩisā' *Arabic*, Well
Hka *Burmese*, River
-ho *Korean*, Lake
Holot *Hebrew*, Dunes
Hora *Belarussian, Czech*, Mountain
Hrada *Belarussian*, Mountain, ridge
Hsi *Chinese*, River
Hu *Chinese*, Lake
Huk *Danish*, Point

I

Île(s) *French*, Island(s)
Ilha(s) *Portuguese*, Island(s)
Ilhéu(s) *Portuguese*, Islet(s)
-isen *Norwegian*, Ice shelf
Imeni *Russian*, In the name of
Inish- *Gaelic*, Island
Insel(n) *German*, Island(s)
Irmağı, Irmak *Turkish*, River
Isla(s) *Spanish*, Island(s)
Isola (Isole) *Italian*, Island(s)

J

Jabal *Arabic*, Mountain
Jāl *Arabic*, Ridge
-järv *Estonian*, Lake
-järvi *Finnish*, Lake
Jazā'ir *Arabic*, Islands
Jazīrat *Arabic*, Island
Jazīreh *Persian*, Island
Jebel *Arabic*, Mountain
Jezero *Serbo-Croatian*, Lake
Jezioro *Polish*, Lake
Jiang *Chinese*, River
-jima *Japanese*, Island
Jižní *Czech*, Southern
-jõgi *Estonian*, River
-joki *Finnish*, River
-jökull *Icelandic*, Glacier
Jūn *Arabic*, Bay
Juzur *Arabic*, Islands

K

Kaikyō *Japanese*, Strait
-kaise *Lappish*, Mountain
Kali *Nepali*, River
Kalnas *Lithuanian*, Mountain
Kalns *Latvian*, Mountain
Kang *Chinese*, Harbor
Kangri *Tibetan*, Mountain(s)
Kaôh *Cambodian*, Island
Kapp *Norwegian*, Cape
Káto *Greek*, Lower
Kavīr *Persian*, Desert
K'edi *Georgian*, Mountain range
Kediet *Arabic*, Mountain
Kepi *Albanian*, Cape, point
Kepulauan *Indonesian, Malay*, Island group
Khalīg, Khalīj *Arabic*, Gulf
Khawr *Arabic*, Inlet
Khola *Nepali*, River
Khrebet *Russian*, Mountain range
Ko *Thai*, Island
-ko *Japanese*, Inlet, lake
Kólpos *Greek*, Bay
-kopf *German*, Peak
Körfäzi *Azerbaijani*, Bay
Körfezi *Turkish*, Bay
Kõrgustik *Estonian*, Upland
Kosa *Russian, Ukrainian*, Spit
Koshi *Nepali*, River
Kou *Chinese*, River-mouth
Kowtal *Persian*, Pass
Kray *Russian*, Region, territory
Kryazh *Russian*, Ridge
Kuduk *Uighur*, Well
Kūh(hā) *Persian*, Mountain(s)
-kul' *Russian*, Lake
Kŭl(i) *Tajik, Uzbek*, Lake
-kundo *Korean*, Island group
-kysten *Norwegian*, Coast
Kyun *Burmese*, Island

L

Laaq *Somali*, Watercourse
Lac *French*, Lake
Lacul *Romanian*, Lake
Lagh *Somali*, Stream
Lago *Italian, Portuguese, Spanish*, Lake
Lagoa *Portuguese*, Lagoon
Laguna *Italian, Spanish*, Lagoon, lake
Laht *Estonian*, Bay
Laut *Indonesian*, Bay
Lembalemba *Malagasy*, Plateau
Lerr *Armenian*, Mountain
Lerrnashght'a *Armenian*, Mountain range
Les *Czech*, Forest
Lich *Armenian*, Lake
Liehtao *Chinese*, Island group
Liqeni *Albanian*, Lake
Límni *Greek*, Lake
Ling *Chinese*, Mountain range
Llano *Spanish*, Plain, prairie
Lumi *Albanian*, River
Lyman *Ukrainian*, Estuary

M

Madīnat *Arabic*, City, town
Mae Nam *Thai*, River
-mägi *Estonian*, Hill
Maja *Albanian*, Mountain
Mal *Albanian*, Mountains
Mal-aya, -oye, -yy, *Russian*, Small
-man *Korean*, Bay
Mar *Spanish*, Sea
Marios *Lithuanian*, Lake
Massif *French*, Mountains
Meer *German*, Lake
-meer *Dutch*, Lake
Melkosopochnik *Russian*, Plain
-meri *Estonian*, Sea
Mifraz *Hebrew*, Bay
Minami- *Japanese*, South(ern)
-misaki *Japanese*, Cape, point
Monkhafad *Arabic*, Depression
Montagne(s) *French*, Mountain(s)
Montañas *Spanish*, Mountains
Mont(s) *French*, Mountain(s)
Monte *Italian, Portuguese*, Mountain
More *Russian*, Sea
Mörön *Mongolian*, River
Mys *Russian*, Cape, point

N

-nada *Japanese*, Open stretch of water
Nadi *Bengali*, River
Nagor'ye *Russian*, Upland
Naḩal *Hebrew*, River
Nahr *Arabic*, River
Nam *Laotian*, River
Namakzār *Persian*, Salt desert
Né-a, -on, -os *Greek*, New
Nedre- *Norwegian*, Lower
-neem *Estonian*, Cape, point
Nehri *Turkish*, River
-nes *Norwegian*, Cape, point
Nevado *Spanish*, Mountain (snow-capped)
Nieder- *German*, Lower
Nishi- *Japanese*, West(ern)
-nísi *Greek*, Island
Nisoi *Greek*, Islands
Nizhn-eye, -iy, -iye, -yaya *Russian*, Lower
Nizmennost' *Russian*, Lowland, plain
Nord *Danish, French, German*, North
Norte *Portuguese, Spanish*, North
Nos *Bulgarian*, Point, spit
Nosy *Malagasy*, Island
Nov-a, -i, *Bulgarian, Serbo-Croatian*, New
Nov-aya, -o, -oye, -yy, -yye *Russian*, New
Now-a, -e, -y *Polish*, New
Nur *Mongolian*, Lake
Nuruu *Mongolian*, Mountains
Nuur *Mongolian*, Lake
Nyzovyna *Ukrainian*, Lowland, plain

O

-ø *Danish*, Island
Ober- *German*, Upper
Oblast' *Russian*, Province
Órmos *Greek*, Bay
Orol(i) *Uzbek*, Island
Øster- *Norwegian*, Eastern
Ostrov(a) *Russian*, Island(s)
Otok *Serbo-Croatian*, Island
Oued *Arabic*, Watercourse
-oy *Faeroese*, Island
-øy(a) *Norwegian*, Island
Oya *Sinhala*, River
Ozero *Russian, Ukrainian*, Lake

P

Passo *Italian*, Pass
Pegunungan *Indonesian, Malay*, Mountain range
Pélagos *Greek*, Sea
Pendi *Chinese*, Basin
Penisola *Italian*, Peninsula
Pertuis *French*, Strait
Peski *Russian*, Sands
Phanom *Thai*, Mountain
Phou *Laotian*, Mountain
Pi *Chinese*, Point
Pic *Catalan, French*, Peak
Pico *Portuguese, Spanish*, Peak
-piggen *Danish*, Peak
Pik *Russian*, Peak
Pivostriv *Ukrainian*, Peninsula
Planalto *Portuguese*, Plateau
Planina, Planini *Bulgarian, Macedonian, Serbo-Croatian*, Mountain range
Plato *Russian*, Plateau
Ploskogor'ye *Russian*, Upland
Poluostrov *Russian*, Peninsula
Ponta *Portuguese*, Point
Porthmós *Greek*, Strait
Pótamos *Greek*, River
Presa *Spanish*, Dam
Prokhod *Bulgarian*, Pass
Proliv *Russian*, Strait
Pulau *Indonesian Malay*, Island
Pulu *Malay*, Island
Punta *Spanish*, Point
Pushcha *Belorussian*, Forest
Puszcza *Polish*, Forest

Q

Qā' *Arabic*, Depression
Qalamat *Arabic*, Well
Qatorkŭh(i) *Tajik*, Mountain
Qiuling *Chinese*, Hills
Qolleh *Persian*, Mountain
Qu *Tibetan*, Stream
Quan *Chinese*, Well
Qulla(i) *Tajik*, Peak
Qundao *Chinese*, Island group

R

Raas *Somali*, Cape
-rags *Latvian*, Cape
Ramlat *Arabic*, Sands
Ra's *Arabic*, Cape, headland, point
Ravnina *Bulgarian, Russian*, Plain
Récif *French*, Reef
Recife *Portuguese*, Reef
Reka *Bulgarian*, River
Represa (Rep.) *Portuguese, Spanish*, Reservoir
Reshteh *Persian*, Mountain range
Respublika *Russian*, Republic, first-order administrative division
Respublika(si) *Uzbek*, Republic, first-order administrative division
-retsugan *Japanese*, Chain of rocks
-rettō *Japanese*, Island chain
Riacho *Spanish*, Stream
Riban' *Malagasy*, Mountains
Rio *Portuguese*, River
Río *Spanish*, River
Riu *Catalan*, River
Rivier *Dutch*, River
Rivière *French*, River
Rowd *Pashtu*, River
Rt *Serbo-Croatian*, Point
Rūd *Persian*, River
Rūdkhāneh *Persian*, River
Rudohorie *Slovak*, Mountains
Ruisseau *French*, Stream

S

-saar *Estonian*, Island
-saari *Finnish*, Island
Sabkhat *Arabic*, Salt marsh
Sāgar(a) *Hindi*, Lake, reservoir
Şaḩrā' *Arabic*, Desert
Saint, Sainte *French*, Saint
Salar *Spanish*, Salt-pan
Salto *Portuguese, Spanish*, Waterfall
Samudra *Sinhala*, Reservoir
-san *Japanese, Korean*, Mountain
-sanchi *Japanese*, Mountains
-sandur *Icelandic*, Beach
Sankt *German, Swedish*, Saint
-sanmaek *Korean*, Mountain range
-sanmyaku *Japanese*, Mountain range
San, Santa, Santo *Italian, Portuguese, Spanish*, Saint
São *Portuguese*, Saint
Sarīr *Arabic*, Desert
Sebkha, Sebkhet *Arabic*, Depression, salt marsh
Sedlo *Czech*, Pass
See *German*, Lake
Selat *Indonesian*, Strait
Selatan *Indonesian*, Southern
-selkä *Finnish*, Lake, ridge
Selseleh *Persian*, Mountain range
Serra *Portuguese*, Mountain
Serranía *Spanish*, Mountain
-seto *Japanese*, Channel, strait
Sever-naya, -noye, -nyy, -o *Russian*, Northern
Sha'ib *Arabic*, Watercourse
Shākh *Kurdish*, Mountain
Shamo *Chinese*, Desert
Shan *Chinese*, Mountain(s)
Shankou *Chinese*, Pass
Shanmo *Chinese*, Mountain range
Shaţţ *Arabic*, Distributary
Shet' *Amharic*, River
Shi *Chinese*, Municipality
-shima *Japanese*, Island
Shiqqat *Arabic*, Depression
-shotō *Japanese*, Group of islands
Shuiku *Chinese*, Reservoir
Shŭrkhog(i) *Uzbek*, Salt marsh
Sierra *Spanish*, Mountains
Sint *Dutch*, Saint
-sjø(en) *Norwegian*, Lake
-sjön *Swedish*, Lake
Solonchak *Russian*, Salt lake
Solonchakovyye Vpadiny *Russian*, Salt basin, wetlands
Son *Vietnamese*, Mountain
Sông *Vietnamese*, River
Sør- *Norwegian*, Southern
-spitze *German*, Peak
Star-á, -é *Czech*, Old
Star-aya, -oye, -yy, -yye *Russian*, Old
Stenó *Greek*, Strait
Step' *Russian*, Steppe
Štít *Slovak*, Peak
Stœng *Cambodian*, River
Stolovaya Strana *Russian*, Plateau
Stredné *Slovak*, Middle
Střední *Czech*, Middle
Stretto *Italian*, Strait
Su Anbarı *Azerbaijani*, Reservoir
-suidō *Japanese*, Channel, strait
Sund *Swedish*, Sound, strait
Sungai *Indonesian, Malay*, River
Suu *Turkish*, River

T

Tal *Mongolian*, Plain
Tandavan' *Malagasy*, Mountain range
Tangorombohitr' *Malagasy*, Mountain massif
Tanjung *Indonesian, Malay*, Cape, point
Tao *Chinese*, Island
Ţaraq *Arabic*, Hills
Tassili *Berber*, Mountain, plateau
Tau *Russian*, Mountain(s)
Taungdan *Burmese*, Mountain range
Techníti Límni *Greek*, Reservoir
Tekojärvi *Finnish*, Reservoir
Teluk *Indonesian, Malay*, Bay
Tengah *Indonesian*, Middle
Terara *Amharic*, Mountain
Timur *Indonesian*, Eastern
-tind(an) *Norwegian*, Peak
Tizma(si) *Uzbek*, Mountain range, ridge
-tō *Japanese*, island
Tog *Somali*, Valley
-tōge *Japanese*, pass
Togh(i) *Uzbek*, mountain
Tônlé *Cambodian*, Lake
Top *Dutch*, Peak
-tunturi *Finnish*, Mountain
Ţurāq *Arabic*, hills
Tur'at *Arabic*, Channel

U

Udde(n) *Swedish*, Cape, point
'Uqlat *Arabic*, Well
Utara *Indonesian*, Northern
Uul *Mongolian*, Mountains

V

Väin *Estonian*, Strait
Vallée *French*, Valley
Varful *Romanian*, Peak
-vatn *Icelandic*, Lake
-vatnet *Norwegian*, Lake
Velayat *Turkmen*, Province
-vesi *Finnish*, Lake
Vestre- *Norwegian*, Western
-vidda *Norwegian*, Plateau
-vík *Icelandic*, Bay
-viken *Swedish*, Bay, inlet
Vinh *Vietnamese*, Bay
Víztárloló *Hungarian*, Reservoir
Vodaskhovishcha *Belarussian*, Reservoir
Vodokhranilishche (Vdkhr.) *Russian*, Reservoir
Vodoskhovyshche (Vdskh.) *Ukrainian*, Reservoir
Volcán *Spanish*, Volcano
Vostochn-o, yy *Russian*, Eastern
Vozvyshennost' *Russian*, Upland, plateau
Vozyera *Belarussian*, Lake
Vpadina *Russian*, Depression
Vrchovina *Czech*, Mountains
Vrh *Croat, Slovene*, Peak
Východné *Slovak*, Eastern
Vysochyna *Ukrainian*, Upland
Vysočina *Czech*, Upland

W

Waadi *Somali*, Watercourse
Wādī *Arabic* Watercourse
Wāḩat, Wâhat *Arabic*, Oasis
Wald *German*, Forest
Wan *Chinese*, Bay
Way *Indonesian*, River
Webi *Somali*, River
Wenz *Amharic*, River
Wiloyat(i) *Uzbek*, Province
Wyżyna *Polish*, Upland
Wzgórza *Polish*, Upland
Wzvyshsha *Belarussian*, Upland

X

Xé *Laotian*, River
Xi *Chinese*, Stream

Y

-yama *Japanese*, Mountain
Yanchi *Chinese*, Salt lake
Yanhu *Chinese*, Salt lake
Yarımadası *Azerbaijani, Turkish*, Peninsula
Yaylası *Turkish*, Plateau
Yazovir *Bulgarian*, Reservoir
Yoma *Burmese*, Mountains
Ytre- *Norwegian*, Outer
Yu *Chinese*, Islet
Yunhe *Chinese*, Canal
Yuzhn-o, -yy *Russian*, Southern

Z

-zaki *Japanese*, Cape, point
Zaliv *Bulgarian, Russian*, Bay
-zan *Japanese*, Mountain
Zangbo *Tibetan*, River
Zapadn-aya, -o, -yy *Russian*, Western
Západné *Slovak*, Western
Západní *Czech*, Western
Zatoka *Polish, Ukrainian*, Bay
-zee *Dutch*, Sea
Zemlya *Russian*, Earth, land
Zizhiqu *Chinese*, Autonomous region

100 Mile House – Ābuyē Mēda

Index

Glossary of Abbreviations

This glossary provides a comprehensive guide to the abbreviations used in this Atlas, and in the Index.

A
abbrev. abbreviated
AD Anno Domini
Afr. Afrikaans
Alb. Albanian
Amh. Amharic
anc. ancient
approx. approximately
Ar. Arabic
Arm. Armenian
ASEAN Association of South East Asian Nations
ASSR Autonomous Soviet Socialist Republic
Aust. Australian
Az. Azerbaijani
Azerb. Azerbaijan

B
Basq. Basque
BC before Christ
Bel. Belorussian
Ben. Bengali
Ber. Berber
B-H Bosnia-Herzegovina
bn billion (one thousand million)
BP British Petroleum
Bret. Breton
Brit. British
Bul. Bulgarian
Bur. Burmese

C
C central
C. Cape
°C degrees Centigrade
CACM Central America Common Market
Cam. Cambodian
Cant. Cantonese
CAR Central African Republic
Cast. Castilian
Cat. Catalan
CEEAC Central America Common Market
Chin. Chinese
CIS Commonwealth of Independent States
cm centimetre(s)
Cro. Croat
Cz. Czech
Czech Rep. Czech Republic

D
Dan. Danish
Div. Divehi
Dom. Rep. Dominican Republic
Dut. Dutch

E
E east
EC *see* EU
EEC *see* EU
ECOWAS Economic Community of West African States
ECU European Currency Unit
EMS European Monetary System
Eng. English
est estimated
Est. Estonian
EU European Union (previously European Community [EC], European Economic Community [EEC])

F
°F degrees Fahrenheit
Faer. Faeroese
Fij. Fijian
Fin. Finnish
Fr. French
Fris. Frisian
ft foot/feet
FYROM Former Yugoslav Republic of Macedonia

G
g gram(s)
Gael. Gaelic
Gal. Galician
GDP Gross Domestic Product (the total value of goods and services produced by a country excluding income from foreign countries)
Geor. Georgian
Ger. German
Gk Greek
GNP Gross National Product (the total value of goods and services produced by a country)

H
Heb. Hebrew
HEP hydro-electric power
Hind. Hindi
hist. historical
Hung. Hungarian

I
I. Island
Icel. Icelandic
in inch(es)
In. Inuit (Eskimo)
Ind. Indonesian
Intl International
Ir. Irish
Is Islands
It. Italian

J
Jap. Japanese

K
Kaz. Kazakh
kg kilogram(s)
Kir. Kirghiz
km kilometre(s)
km² square kilometre (singular)
Kor. Korean
Kurd. Kurdish

L
L. Lake
LAIA Latin American Integration Association
Lao. Laotian
Lapp. Lappish
Lat. Latin
Latv. Latvian
Liech. Liechtenstein
Lith. Lithuanian
Lus. Lusatian
Lux. Luxembourg

M
m million/metre(s)
Mac. Macedonian
Maced. Macedonia
Mal. Malay
Malg. Malagasy
Malt. Maltese
mi. mile(s)
Mong. Mongolian
Mt. Mountain
Mts Mountains

N
N north
NAFTA North American Free Trade Agreement
Nep. Nepali
Neth. Netherlands
Nic. Nicaraguan
Nor. Norwegian
NZ New Zealand

P
Pash. Pashtu
PNG Papua New Guinea
Pol. Polish
Poly. Polynesian
Port. Portuguese
prev. previously

R
Rep. Republic
Res. Reservoir
Rmsch Romansch
Rom. Romanian
Rus. Russian
Russ. Fed. Russian Federation

S
S south
SADC Southern Africa Development Community
SCr. Serbian, Croatian
Sinh. Sinhala
Slvk Slovak
Slvn. Slovene
Som. Somali
Sp. Spanish
St., St Saint
Strs Straits
Swa. Swahili
Swe. Swedish
Switz. Switzerland

T
Taj. Tajik
Th. Thai
Thai. Thailand
Tib. Tibetan
Turk. Turkish
Turkm. Turkmenistan

U
UAE United Arab Emirates
Uigh. Uighur
UK United Kingdom
Ukr. Ukrainian
UN United Nations
Urd. Urdu
US/USA United States of America
USSR Union of Soviet Socialist Republics
Uzb. Uzbek

V
var. variant
Vdkhr. Vodokhranilishche (Russian for reservoir)
Vdskh. Vodoskhovyshche (Ukrainian for reservoir)
Vtn. Vietnamese

W
W west
Wel. Welsh

This index lists all the placenames and features shown on the regional and continental maps in this Atlas. Placenames are referenced to the largest scale map on which they appear. The policy followed throughout the Atlas is to use the local spelling or local name at regional level; commonly-used English language names may occasionally be added (in parentheses) where this is an aid to identification e.g. Firenze (Florence). English names, where they exist, have been used for all international features e.g. oceans and country names; they are also used on the continental maps and in the introductory World Today section; these are then fully cross-referenced to the local names found on the regional maps. The index also contains commonly-found alternative names and variant spellings, which are also fully cross-referenced.

All main entry names are those of settlements unless otherwise indicated by the use of italicized definitions or representative symbols, which are keyed at the foot of each page.

1

10 *M16* **100 Mile House** *var.* Hundred Mile House. British Columbia, SW Canada *51°39′N 121°19′W*
25 de Mayo *see* Veinticinco de Mayo
26 Bakinskikh Komissarov *see* Häsänabad
26 Baku Komissarlary Adyndaky *see* Uzboý

A

Aa *see* Gauja
95 *G24* **Aabenraa** *var.* Åbenrå, *Ger.* Apenrade. Syddanmark, SW Denmark *55°03′N 09°26′E*
95 *G20* **Aabybro** *var.* Åbybro. Nordjylland, N Denmark *57°09′N 09°32′E*
101 *C16* **Aachen** *Dut.* Aken, *Fr.* Aix-la-Chapelle; *anc.* Aquae Grani, Aquisgranum. Nordrhein-Westfalen, W Germany *50°47′N 06°06′E*
Aaiún *see* Laâyoune
95 *M24* **Aakirkeby** *var.* Åkirkeby. Bornholm, E Denmark *55°04′N 14°56′E*
95 *G20* **Aalborg** *var.* Ålborg, Ålborg-Nørresundby; *anc.* Alburgum. Nordjylland, N Denmark *57°03′N 09°56′E*
Aalborg Bugt *see* Ålborg Bugt
101 *J21* **Aalen** Baden-Württemberg, S Germany *48°50′N 10°06′E*
95 *G21* **Aalestrup** *var.* Ålestrup. Midtjylland, NW Denmark *56°42′N 09°31′E*
98 *I11* **Aalsmeer** Noord-Holland, C Netherlands *52°17′N 04°43′E*
99 *F18* **Aalst** Oost-Vlaanderen, C Belgium *50°57′N 04°03′E*
99 *K18* **Aalst** *Fr.* Alost. Noord-Brabant, S Netherlands *51°23′N 05°29′E*
98 *O12* **Aalten** Gelderland, E Netherlands *51°56′N 06°35′E*
99 *D17* **Aalter** Oost-Vlaanderen, NW Belgium *51°05′N 03°28′E*
Aanaar *see* Inari
Aanaarjävri *see* Inarijärvi
93 *M17* **Äänekoski** Keski-Suomi, W Finland *62°34′N 25°45′E*
138 *H7* **Aanjar** *var.* ʻAnjar. C Lebanon *33°45′N 35°56′E*
83 *G21* **Aansluit** Northern Cape, N South Africa *26°41′S 22°24′E*
Aar *see* Aare
108 *F7* **Aarau** Aargau, N Switzerland *47°22′N 08°00′E*
108 *D8* **Aarberg** Bern, W Switzerland *47°19′N 07°54′E*
99 *D16* **Aardenburg** Zeeland, SW Netherlands *51°16′N 03°27′E*
108 *D8* **Aare** *var.* Aar. ~ W Switzerland
108 *F7* **Aargau** *Fr.* Argovie. ◐ *canton* N Switzerland
Aarhus *see* Århus
Aarlen *see* Arlon
95 *G21* **Aars** *var.* Års. Nordjylland, N Denmark *56°49′N 09°32′E*
99 *I17* **Aarschot** Vlaams Brabant, C Belgium *50°59′N 04°50′E*
Aassi, Nahr el *see* Orontes
Aat *see* Ath
160 *G7* **Aba** *prev.* Ngawa. Sichuan, C China *32°51′N 101°46′E*
79 *P16* **Aba** Orientale, NE Dem. Rep. Congo *03°52′N 30°14′E*
77 *V17* **Aba** Abia, S Nigeria *05°06′N 07°22′E*
140 *J6* **Abā al Qazāz, Bi'r** *well* NW Saudi Arabia
Abā as Suʻūd *see* Najrān
59 *G14* **Abacaxis, Rio** ~ NW Brazil
Abaco Island *see* Great Abaco/Little Abaco
Abaco Island *see* Great Abaco, N Bahamas
142 *K10* **Ābādān** Khūzestān, SW Iran *30°24′N 48°18′E*
146 *F13* **Abadan** *prev.* Bezmein, Büzmeýin, *Rus.* Byuzmeyin. Ahal Welaýaty, C Turkmenistan *38°08′N 57°53′E*
143 *O10* **Ābādeh** Fārs, C Iran *31°06′N 52°40′E*
74 *H8* **Abadla** W Algeria *31°04′N 02°39′W*
59 *M20* **Abaeté** Minas Gerais, SE Brazil *19°10′S 45°24′W*
62 *P7* **Abaí** Caazapá, S Paraguay *25°58′S 55°54′W*
Abai *see* Blue Nile
191 *O2* **Abaiang** *var.* Apia; *prev.* Charlotte Island. *atoll* Tungaru, W Kiribati
Abaj *see* Abay
77 *U15* **Abaji** Federal Capital District, C Nigeria *08°35′N 06°54′E*
37 *O7* **Abajo Peak** ▲ Utah, W USA *37°51′N 109°28′W*
77 *V16* **Abakaliki** Ebonyi, SE Nigeria *06°18′N 08°07′E*
122 *K13* **Abakan** Respublika Khakasiya, S Russian Federation *53°43′N 91°25′E*
77 *S11* **Abala** Tillabéri, SW Niger *14°55′N 03°27′E*
77 *U11* **Abalak** Tahoua, C Niger *15°28′N 06°18′E*
119 *N14* **Abalyanka** *Rus.* Obolyanka. ~ N Belarus
122 *L12* **Aban** Krasnoyarskiy Kray, S Russian Federation *56°41′N 96°04′E*
143 *P9* **Āb Anbār-e Kān Sorkh** Yazd, C Iran *31°22′N 53°38′E*
57 *G16* **Abancay** Apurímac, SE Peru *13°37′S 72°52′W*
190 *H2* **Abaokoro** *atoll* Tungaru, W Kiribati
Abariringa *see* Kanton
143 *P10* **Abarkūh** Yazd, C Iran *31°07′N 53°17′E*
165 *V3* **Abashiri** *var.* Abasiri. Hokkaidō, NE Japan *44°N 144°15′E*
165 *U3* **Abashiri-ko** ◎ Hokkaidō, NE Japan
Abasiri *see* Abashiri
41 *P10* **Abasolo** Tamaulipas, C Mexico *24°02′N 98°18′W*
186 *F9* **Abau** Central, S Papua New Guinea *10°04′S 148°34′E*
145 *R10* **Abay** *var.* Abaj. Karaganda, C Kazakhstan *49°38′N 72°50′E*
81 *I15* **Ābaya Hāyk'** *Eng.* Lake Margherita, *It.* Abbaia. ◎ SW Ethiopia
Ābay Wenz *see* Blue Nile
122 *K13* **Abaza** Respublika Khakasiya, S Russian Federation *52°40′N 89°58′E*
Abbaia *see* Ābaya Hāyk'
143 *Q13* **Āb Bārīk** Fārs, S Iran
107 *C18* **Abbasanta** Sardegna, Italy, C Mediterranean Sea *40°08′N 08°49′E*
Abbatis Villa *see* Abbeville
30 *M3* **Abbaye, Point** *headland* Michigan, N USA *46°58′N 88°08′W*
Abbazia *see* Opatija
Abbé, Lake *see* Abhe, Lake
103 *N2* **Abbeville** *anc.* Abbatis Villa. Somme, N France *50°06′N 01°50′E*
23 *R7* **Abbeville** Alabama, S USA *31°35′N 85°16′W*
23 *U6* **Abbeville** Georgia, SE USA *31°58′N 83°18′W*
22 *I9* **Abbeville** Louisiana, S USA *29°58′N 92°08′W*
21 *P12* **Abbeville** South Carolina, SE USA *34°10′N 82°23′W*
97 *B20* **Abbeyfeale** *Ir.* Mainistir na Féile. SW Ireland *52°24′N 09°21′W*
106 *D8* **Abbiategrasso** Lombardia, NW Italy *45°24′N 08°55′E*
93 *I14* **Abborrträsk** Norrbotten, N Sweden *65°24′N 19°33′E*
194 *J9* **Abbot Ice Shelf** *ice shelf* Antarctica
10 *M17* **Abbotsford** British Columbia, SW Canada *49°02′N 122°18′W*
30 *K6* **Abbotsford** Wisconsin, N USA *44°57′N 90°19′W*
149 *U5* **Abbottābād** Khyber Pakhtunkhwa, NW Pakistan *34°12′N 73°15′E*
119 *M14* **Abchuha** *Rus.* Obchuga. Minskaya Voblasts', NW Belarus *54°30′N 29°22′E*
98 *I10* **Abcoude** Utrecht, C Netherlands *52°17′N 04°59′E*
139 *N2* **ʻAbd al ʻAzīz, Jabal** ▲▲ NE Syria
141 *U17* **ʻAbd al Kūrī** *island* SE Yemen
127 *U6* **Abdulino** Orenburgskaya Oblast', W Russian Federation *53°37′N 53°39′E*
78 *J10* **Abéché** *var.* Abécher, Abeshr. Ouaddaï, SE Chad *13°49′N 20°49′E*
Abécher *see* Abéché
143 *S8* **Āb-e Garm va Sard** Khorāsān-e Jonūbī, E Iran
77 *R8* **Abeïbara** Kidal, NE Mali *19°07′N 01°52′E*
105 *P5* **Abejar** Castilla y León, N Spain *41°48′N 02°47′W*
54 *E9* **Abejorral** Antioquia, W Colombia *05°48′N 75°28′W*
Abela *see* Ávila
Abellinum *see* Avellino
92 *Q2* **Abeløya** *island* Kong Karls Land, E Svalbard
80 *I13* **Ābeltī** Oromīya, C Ethiopia *08°09′N 37°31′E*
191 *O2* **Abemama** *var.* Apamama; *prev.* Roger Simpson Island. *atoll* Tungaru, W Kiribati
171 *Y15* **Abemaree** *var.* Abermarre. Papua, E Indonesia *07°03′S 140°10′E*
77 *O16* **Abengourou** E Ivory Coast *06°42′N 03°27′W*
Åbenrå *see* Aabenraa
101 *L22* **Abens** ~ SE Germany
77 *S16* **Abeokuta** Ogun, SW Nigeria *07°07′N 03°21′E*
97 *I20* **Aberaeron** SW Wales, United Kingdom *52°15′N 04°15′W*
Aberbrothock *see* Arbroath
Abercorn *see* Mbala
29 *R6* **Abercrombie** North Dakota, N USA *46°25′N 96°42′W*
183 *T7* **Aberdeen** New South Wales, SE Australia *32°09′S 150°55′E*
11 *T15* **Aberdeen** Saskatchewan, S Canada *52°15′N 106°19′W*
83 *H25* **Aberdeen** Eastern Cape, S South Africa *32°30′S 24°00′E*
96 *L9* **Aberdeen** *anc.* Devana. NE Scotland, United Kingdom *57°10′N 02°04′W*
21 *X2* **Aberdeen** Maryland, NE USA *39°28′N 76°09′W*
23 *N3* **Aberdeen** Mississippi, S USA *33°49′N 88°32′W*
21 *T10* **Aberdeen** North Carolina, SE USA *35°07′N 79°25′W*
29 *P8* **Aberdeen** South Dakota, N USA *45°27′N 98°29′W*
32 *F8* **Aberdeen** Washington, NW USA *46°57′N 123°48′W*
96 *K9* **Aberdeen** *cultural region* NE Scotland, United Kingdom
8 *L8* **Aberdeen Lake** ◎ Nunavut, NE Canada
96 *J10* **Aberfeldy** C Scotland, United Kingdom *56°38′N 03°49′W*
97 *K21* **Abergavenny** *anc.* Gobannium. SE Wales, United Kingdom *51°50′N 03°W*
Abergwaun *see* Fishguard
Abermarre *see* Abemaree
25 *N5* **Abernathy** Texas, SW USA *33°49′N 101°50′W*
Abersee *see* Wolfgangsee
Abertawe *see* Swansea
Aberteifi *see* Cardigan
32 *I15* **Abert, Lake** ◎ Oregon, NW USA
97 *I20* **Aberystwyth** W Wales, United Kingdom *52°25′N 04°05′W*
Abeshr *see* Abéché
Ábeskovvu *see* Abisko
106 *F10* **Abetone** Toscana, C Italy *44°09′N 10°42′E*
125 *V5* **Abez'** Respublika Komi, NW Russian Federation *66°32′N 61°41′E*
142 *M5* **Āb Garm** Qazvin, N Iran
141 *N12* **Abhā** ʻAsīr, SW Saudi Arabia *18°16′N 42°32′E*
142 *M5* **Abhar** Zanjān, NW Iran *36°05′N 49°18′E*
Abhé Bad/Ābhē Bid Hāyk' *see* Abhe, Lake
80 *K12* **Abhe, Lake** *var.* Lake Abbé, *Amh.* Ābhē Bid Hāyk', *Som.* Abhé Bad. ◎ Djibouti/Ethiopia
77 *V17* **Abia** ◐ *state* SE Nigeria
139 *V9* **ʻAbīd ʻAlī** Wāsiṭ, E Iraq *32°20′N 45°58′E*
119 *O17* **Abidavichy** *Rus.* Obidovichi. Mahilyowskaya Voblasts', E Belarus *53°20′N 30°25′E*
115 *L15* **Abide** Çanakkale, NW Turkey *40°04′N 26°13′E*
77 *N17* **Abidjan** S Ivory Coast *05°19′N 04°01′W*
Āb-i-Istāda *see* Īstādeh-ye Moqor, Āb-e-
27 *N4* **Abilene** Kansas, C USA *38°55′N 97°14′W*
25 *Q7* **Abilene** Texas, SW USA *32°27′N 99°44′W*
Abindonia *see* Abingdon
97 *M21* **Abingdon** *anc.* Abindonia. S England, United Kingdom *51°41′N 01°17′W*
30 *K12* **Abingdon** Illinois, N USA *40°48′N 90°24′W*
21 *P8* **Abingdon** Virginia, NE USA *36°42′N 81°59′W*
Abingdon *see* Pinta, Isla
18 *J15* **Abington** Pennsylvania, NE USA *40°06′N 75°05′W*
126 *K14* **Abinsk** Krasnodarskiy Kray, SW Russian Federation *44°51′N 38°12′E*
37 *R9* **Abiquiu Reservoir** ⊠ New Mexico, SW USA
Āb-i-safed *see* Sefīd, Darya-ye
92 *I10* **Abisko** *Lapp.* Ábeskovvu. Norrbotten, N Sweden *68°21′N 18°50′E*
12 *G12* **Abitibi** ~ Ontario, S Canada
12 *H12* **Abitibi, Lac** ◎ Ontario/Québec, S Canada
80 *J10* **Ābīy Ādī** Tigray, N Ethiopia *13°40′N 38°57′E*
118 *H6* **Abja-Paluoja** Viljandimaa, S Estonia *58°08′N 25°20′E*
Abkhazia *see* Apkhazeti
182 *F1* **Abminga** South Australia *26°07′S 134°49′E*
75 *W9* **Abnūb** *var.* Abnûb. C Egypt *27°18′N 31°09′E*
Abnûb *see* Abnūb
Åbo *see* Turku
152 *G9* **Abohar** Punjab, N India *30°11′N 74°14′E*
77 *O17* **Aboisso** SE Ivory Coast *05°26′N 03°13′W*
78 *H5* **Abo, Massif d'** ▲▲ NW Chad
78 *H5* **Abo, Massif d'** ▲▲ E Chad
77 *R16* **Abomey** S Benin *07°14′N 02°00′E*
79 *F16* **Abong Mbang** Est, SE Cameroon *03°58′N 13°10′E*
111 *L23* **Abony** Pest, C Hungary *47°10′N 20°00′E*
78 *J11* **Abou-Déïa** Salamat, SE Chad *11°30′N 19°18′E*
Aboudouhour *see* Abū aḍ Ḍuhūr
Abou Kémal *see* Abū Kamāl
Abou Simbel *see* Abū Sunbul
137 *T12* **Abovyan** C Armenia *40°16′N 44°33′E*
141 *P15* **Abrād, Wādī** *seasonal river* W Yemen
Abraham Bay *see* The Carlton
104 *G10* **Abrantes** *var.* Abrántes. Santarém, C Portugal *39°28′N 08°12′W*
62 *J4* **Abra Pampa** Jujuy, N Argentina *22°47′S 65°41′W*
Abrashlare *see* Brezovo
54 *G7* **Abrego** Norte de Santander, N Colombia *08°08′N 73°14′W*
Abrene *see* Pytalovo
40 *C7* **Abreojos, Punta** *headland* NW Mexico *26°43′N 113°36′W*
65 *J16* **Abrolhos Bank** *undersea feature* W Atlantic Ocean *18°30′S 38°45′W*
119 *H19* **Abrova** *Rus.* Obrovo. Brestskaya Voblasts', SW Belarus *52°30′N 25°34′E*
116 *G11* **Abrud** *Ger.* Gross-Schlatten, *Hung.* Abrudbánya. Alba, SW Romania *46°16′N 23°05′E*
Abrudbánya *see* Abrud
118 *E6* **Abruka** *island* SW Estonia
107 *J15* **Abruzzese, Appennino** ▲▲ C Italy
107 *J14* **Abruzzo** ◐ *region* C Italy
141 *N14* **ʻAbs** *var.* Sūq ʻAbs. W Yemen *16°42′N 42°55′E*
33 *T12* **Absaroka Range** ▲▲ Montana/Wyoming, NW USA
137 *Z11* **Abşeron Yarımadası** *Rus.* Apsheronskiy Poluostrov. *peninsula* E Azerbaijan
143 *N6* **Āb Shīrīn** Eşfahān, C Iran *34°17′N 51°17′E*
139 *X10* **Abtān** Maysān, SE Iraq *31°37′N 47°06′E*
109 *R6* **Abtenau** Salzburg, NW Austria *47°33′N 13°21′E*
152 *E14* **Ābu** Rājasthān, N India *24°41′N 72°50′E*
164 *E12* **Abu** Yamaguchi, Honshū, SW Japan *34°30′N 131°26′E*
138 *I4* **Abū aḍ Ḍuhūr** *Fr.* Aboudouhour. Idlib, NW Syria *35°30′N 37°00′E*
143 *P17* **Abū al Abyaḍ** *island* C United Arab Emirates
138 *K10* **Abū al Ḥuṣayn, Khabrat** ◎ N Jordan
139 *R8* **Abū al Jīr** Al Anbār, C Iraq *33°16′N 42°55′E*
139 *Y12* **Abū al Khaṣīb** *var.* Abul Khasib. Al Başrah, SE Iraq *30°26′N 48°00′E*
139 *U12* **Abū at Tubrah, Thaqb** *well* S Iraq
Abu Balâs *see* Abu Ballās
75 *V11* **Abu Ballās** *var.* Abu Balâs. ▲ SW Egypt *24°28′N 27°36′E*
Abu Dhabi *see* Abū Ẓabī
139 *R8* **Abū Farūkh** Al Anbār, C Iraq *33°06′N 43°18′E*
80 *C12* **Abu Gabra** Eastern Darfur, W Sudan *11°02′N 26°50′E*
139 *P10* **Abū Ghār, Shaʻīb** *dry watercourse* S Iraq
80 *G7* **Abu Hamed** River Nile, N Sudan *19°32′N 33°20′E*
139 *O5* **Abū Ḥardān** *var.* Hajîne. Dayr az Zawr, E Syria *34°45′N 40°49′E*
139 *T7* **Abū Ḥasāwīyah** Diyālá, E Iraq *33°52′N 44°47′E*
138 *K10* **Abū Ḥifnah, Wādī** *dry watercourse* N Jordan
77 *V15* **Abuja** ● (Nigeria) Federal Capital District, C Nigeria *09°04′N 07°28′E*
139 *R9* **Abū Jahaf, Wādī** *dry watercourse* C Iraq
56 *F12* **Abujao, Río** ~ E Peru
139 *U12* **Abū Jasrah** Al Muthanná, S Iraq *30°43′N 44°50′E*
139 *O6* **Abū Kamāl** *Fr.* Abou Kémal. Dayr az Zawr, E Syria *34°29′N 40°56′E*
165 *P12* **Abukuma-sanchi** ▲▲ Honshū, C Japan
Abula *see* Ávila
Abul Khasib *see* Abū al Khaṣīb
79 *K16* **Abumombazi** *var.* Abumonbazi. Equateur, N Dem. Rep. Congo *03°43′N 22°06′E*
Abumonbazi *see* Abumombazi
59 *D15* **Abunã** Rondônia, W Brazil *09°41′S 65°20′W*
56 *K13* **Abunã, Rio** *var.* Río Abuná. ~ Bolivia/Brazil
138 *G10* **Abū Nuşayr** *var.* Abu Nuseir. ʻAmmān, W Jordan *32°03′N 35°58′E*
Abu Nuseir *see* Abū Nuşayr
139 *T12* **Abū Qabr** Al Muthanná, S Iraq *31°03′N 44°34′E*
138 *K5* **Abū Raḥbah, Jabal** ▲▲ C Syria
139 *S5* **Abū Rajāsh** Şalāḩ ad Dīn, N Iraq *34°47′N 43°36′E*
139 *W13* **Abū Raqrāq, Ghadīr** *well* S Iraq
152 *E14* **Ābu Road** Rājasthān, N India *24°29′N 72°47′E*
80 *I6* **Abu Shagara, Ras** *headland* NE Sudan *18°04′N 38°31′E*
Abu Simbel *see* Abū Sunbul
139 *U12* **Abū Sudayrah** Al Muthanná, S Iraq *30°55′N 44°58′E*
139 *T10* **Abū Şukhayr** Al Qādisīyah, S Iraq *31°54′N 44°27′E*
Abū Sunbul *see* Abū Sunbul
185 *E18* **Abut Head** *headland* South Island, New Zealand *43°06′S 170°16′E*
80 *E9* **Abu ʻUrug** Northern Kordofan, C Sudan *15°52′N 30°25′E*
80 *K12* **Ābuyē Mēda** ▲ C Ethiopia *10°28′N 39°44′E*

◆ Country
● Country Capital
◇ Dependent Territory
○ Dependent Territory Capital
◐ Administrative Regions
✈ International Airport
▲ Mountain
▲▲ Mountain Range
Volcano
~ River
◎ Lake
⊠ Reservoir

Abu Zabad – Aiken

80 D11 **Abu Zabad** Western Kordofan, C Sudan *12°21′N 29°16′E*
143 P16 **Abū Ẓabī** *var.* Abū Ẓabi, *Eng.* Abu Dhabi. ● (United Arab Emirates) Abū Ẓaby, C United Arab Emirates *24°30′N 54°20′E*
Abū Ẓabī *see* Abū Ẓabī
75 X8 **Abu Zenīma** E Egypt *29°01′N 33°08′E*
95 N17 **Åby** Östergötland, S Sweden *58°40′N 16°10′E*
Abyaḍ, Al Baḥr al *see* White Nile
Åbybro *see* Aabybro
80 D13 **Abyei** Southern Kordofan, S Sudan *09°35′N 28°28′E*
Abyla *see* Ávila
Abymes *see* les Abymes
Abyssinia *see* Ethiopia
Açāba *see* Assaba
54 F11 **Acacías** Meta, C Colombia *03°59′N 73°46′W*
58 L13 **Açailândia** Maranhão, E Brazil *04°51′S 47°26′W*
Acaill *see* Achill Island
42 E8 **Acajutla** Sonsonate, W El Salvador *13°34′N 89°50′W*
79 D17 **Acalayong** SW Equatorial Guinea *01°05′N 09°34′E*
41 N13 **Acámbaro** Guanajuato, C Mexico *20°01′N 100°42′W*
54 C6 **Acandí** Chocó, NW Colombia *08°32′N 77°20′W*
104 H4 **A Cañiza** *var.* La Cañiza. Galicia, NW Spain *42°13′N 08°16′W*
40 J11 **Acaponeta** Nayarit, C Mexico *22°30′N 105°21′W*
40 J11 **Acaponeta, Río de** ≈ C Mexico
41 O16 **Acapulco** *var.* Acapulco de Juárez. Guerrero, S Mexico *16°51′N 99°53′W*
Acapulco de Juárez *see* Acapulco
55 T13 **Acarai Mountains** *Sp.* Serra Acaraí. ▲▲ Brazil/Guyana
Acaraí, Serra *see* Acarai Mountains
58 O13 **Acaraú** Ceará, NE Brazil *04°35′S 37°37′W*
54 J6 **Acarigua** Portuguesa, N Venezuela *09°35′N 69°12′W*
104 H2 **A Carreira** Galicia, NW Spain *43°21′N 08°12′W*
42 C6 **Acatenango, Volcán de** ⛰ S Guatemala *14°30′N 90°52′W*
41 Q15 **Acatlán** *var.* Acatlán de Osorio. Puebla, S Mexico *18°12′N 98°02′W*
Acatlán de Osorio *see* Acatlán
41 S15 **Acayucan** *var.* Acayucán. Veracruz-Llave, E Mexico *17°59′N 94°58′W*
Accho *see* Akko
21 Y5 **Accomac** Virginia, NE USA *37°43′N 75°41′W*
77 Q17 **Accra** ● (Ghana)SE Ghana *05°33′N 00°15′W*
97 L17 **Accrington** NW England, United Kingdom *53°46′N 02°21′W*
61 B19 **Acebal** Santa Fe, C Argentina *33°14′S 60°50′W*
168 H8 **Aceh** *off.* Daerah Istimewa Aceh, *var.* Acheen, Achin, Atchin, Atjeh. ◈ *autonomous district* NW Indonesia
107 M18 **Acerenza** Basilicata, S Italy *40°46′N 15°51′E*
107 K17 **Acerra** *anc.* Acerrae. Campania, S Italy *40°56′N 14°22′E*
Acerrae *see* Acerra
57 J17 **Achacachi** La Paz, W Bolivia *16°01′S 68°44′W*
54 K7 **Achaguas** Apure, C Venezuela *07°46′N 68°14′W*
154 H12 **Achalpur** *prev.* Elichpur, Ellichpur. Mahārāshtra, C India *21°19′N 77°30′E*
61 F18 **Achar** Tacuarembó, C Uruguay *32°20′S 56°15′W*
137 R10 **Ach'ara** *prev.* Achara, *var.* Ajaria. ◈ *autonomous republic* SW Georgia
Achara *see* Ach'ara
115 H19 **Acharnés** *var.* Aharnes; *prev.* Akharnaí. Attikí, C Greece *38°09′N 23°58′E*
Ach'asar Lerr *see* Achk'asari, Mta
Acheen *see* Aceh
99 K16 **Achel** Limburg, NE Belgium *51°15′N 05°31′E*
115 D16 **Achelóos** *var.* Akhelóös, Aspropótamos; *anc.* Achelous. ≈ W Greece
Achelous *see* Achelóos
163 W8 **Acheng** Heilongjiang, NE China *45°32′N 126°56′E*
109 N6 **Achenkirch** Tirol, W Austria *47°31′N 11°42′E*
101 L24 **Achenpass** *pass* Austria/ Germany
109 N7 **Achensee** ◎ W Austria
101 F22 **Achern** Baden-Württemberg, SW Germany *48°37′N 08°04′E*
115 C16 **Acherón** ≈ W Greece
77 W11 **Achétinamou** ≈ S Niger
152 J12 **Achhnera** Uttar Pradesh, N India *27°10′N 77°45′E*
42 C7 **Achiguate, Río** ≈ S Guatemala
97 A16 **Achill Head** *Ir.* Ceann Acla. *headland* W Ireland *53°58′N 10°14′W*
97 A16 **Achill Island** *Ir.* Acaill. *island* W Ireland
100 H11 **Achim** Niedersachsen, NW Germany *53°01′N 09°01′E*
149 S5 **Achīn** Nangarhār, E Afghanistan *34°04′N 70°41′E*
Achin *see* Aceh
122 K12 **Achinsk** Krasnoyarskiy Kray, S Russian Federation *56°21′N 90°25′E*
162 E5 **Achit Nuur** ◎ NW Mongolia
137 T11 **Achk'asari, Mta** *Arm.* Ach'asar Lerr. ▲ Armenia/ Georgia *41°09′N 43°55′E*
126 K13 **Achuyevo** Krasnodarskiy Kray, SW Russian Federation *46°00′N 38°01′E*
81 F16 **Achwa** *var.* Aswa. ≈ N Uganda
136 E15 **Acıgöl** *salt lake* SW Turkey
107 L24 **Acireale** Sicilia, Italy, C Mediterranean Sea *37°36′N 15°10′E*
Aciris *see* Agri
25 N7 **Ackerly** Texas, SW USA *32°31′N 101°43′W*
22 M4 **Ackerman** Mississippi, S USA *33°18′N 89°10′W*
29 W13 **Ackley** Iowa, C USA *42°33′N 93°03′W*
44 J5 **Acklins Island** *island* SE The Bahamas
Acla, Ceann *see* Achill Head
62 H11 **Aconcagua, Cerro** ▲ W Argentina *32°36′S 69°53′W*
Açores/Açores, Arquipélago dos/Açores, Ilhas dos *see* Azores
104 H2 **A Coruña** *Cast.* La Coruña, *Eng.* Corunna; *anc.* Caronium. Galicia, NW Spain *43°22′N 08°24′W*
104 G2 **A Coruña** *Cast.* La Coruña. ◈ *province* Galicia, NW Spain
42 L10 **Acoyapa** Chontales, S Nicaragua *11°58′N 85°10′W*
106 H13 **Acquapendente** Lazio, C Italy *42°44′N 11°52′E*
106 J13 **Acquasanta Terme** Marche, C Italy *42°46′N 13°24′E*
106 I13 **Acquasparta** Lazio, C Italy *42°41′N 12°31′E*
106 C9 **Acqui Terme** Piemonte, NW Italy *44°41′N 08°28′E*
Acrae *see* Palazzolo Acreide
182 F7 **Acraman, Lake** *salt lake* South Australia
59 A15 **Acre** *off.* Estado do Acre. ◈ *state* W Brazil
Acre *see* Akko
59 C16 **Acre, Rio** ≈ W Brazil
107 N20 **Acri** Calabria, SW Italy *39°30′N 16°22′E*
Acte *see* Ágion Óros
191 Y12 **Actéon, Groupe** *island group* Îles Tuamotu, SE French Polynesia
15 P12 **Acton-Vale** Québec, SE Canada *45°39′N 72°31′W*
41 P13 **Actopan** *var.* Actopán. Hidalgo, C Mexico *20°19′N 98°59′W*
Açu *see* Assu
Acunum Acusio *see* Montélimar
77 Q17 **Ada** SE Ghana *05°47′N 00°42′E*
112 L8 **Ada** Vojvodina, N Serbia *45°48′N 20°08′E*
29 R5 **Ada** Minnesota, N USA *47°18′N 96°31′W*
31 R12 **Ada** Ohio, N USA *40°46′N 83°49′W*
27 O12 **Ada** Oklahoma, C USA *34°47′N 96°41′W*
162 L8 **Adaatsag** *var.* Tavin. Dundgovĭ, C Mongolia *46°27′N 105°43′E*
Ada Bazar *see* Adapazarı
40 D3 **Adair, Bahía de** *bay* NW Mexico
104 M7 **Adaja** ≈ N Spain
38 H17 **Adak Island** *island* Aleutian Islands, Alaska, USA
Adalia *see* Antalya
Adalia, Gulf of *see* Antalya Körfezi
141 X9 **Adam** N Oman *22°22′N 57°30′E*
Adama *see* Nazrēt
60 I8 **Adamantina** São Paulo, S Brazil *21°41′S 51°04′W*
79 E14 **Adamaoua** *Eng.* Adamawa. ◈ *province* N Cameroon
68 F11 **Adamaoua, Massif d'** *Eng.* Adamawa Highlands. *plateau* NW Cameroon
77 Y14 **Adamawa** ◈ *state* E Nigeria
Adamawa Highlands *see* Adamaoua, Massif d'
106 F6 **Adamello** ▲ N Italy *46°09′N 10°33′E*
81 J14 **Ādamī Tulu** Oromīya, C Ethiopia *07°52′N 38°39′E*
63 M23 **Adam, Mount** *var.* Monte Independencia. ▲ West Falkland, Falkland Islands *51°36′S 60°00′W*
29 R16 **Adams** Nebraska, C USA *40°25′N 96°30′W*
18 H8 **Adams** New York, NE USA *43°48′N 75°57′W*
29 Q3 **Adams** North Dakota, N USA *48°23′N 98°01′W*
155 I23 **Adam's Bridge** *chain of shoals* NW Sri Lanka
32 H10 **Adams, Mount** ▲ Washington, NW USA *46°12′N 121°29′W*
Adam's Peak *see* Sri Pada
191 R16 **Adam's Rock** *island* Pitcairn Island, Pitcairn Islands
191 P16 **Adamstown** ○ (Pitcairn Islands)Pitcairn Island, Pitcairn Islands *25°04′S 130°05′W*
20 G10 **Adamsville** Tennessee, S USA *35°14′N 88°23′W*
25 S9 **Adamsville** Texas, SW USA *31°15′N 98°09′W*
141 O17 **'Adan** *Eng.* Aden. SW Yemen *12°51′N 45°05′E*
136 K16 **Adana** *var.* Seyhan. Adana, S Turkey *37°N 35°19′E*
136 K16 **Adana** *var.* Seyhan. ◈ *province* S Turkey
Adâncata *see* Horlivka
169 V12 **Adang, Teluk** *bay* Borneo, C Indonesia
136 F11 **Adapazarı** *prev.* Ada Bazar. Sakarya, NW Turkey *40°49′N 30°24′E*
80 H8 **Adarama** River Nile, NE Sudan *17°04′N 34°57′E*
195 Q16 **Adare, Cape** *cape* Antarctica
Adāvānī *see* Ādoni
106 E6 **Adda** *anc.* Addua. ≈ N Italy
80 A13 **Adda** ≈ W South Sudan
143 Q17 **Aḍ Ḍab'īyah** Abū Ẓaby, C United Arab Emirates *24°17′N 54°08′E*
143 O18 **Aḍ Ḍafrah** *desert* S United Arab Emirates
141 Q6 **Ad Dahnā'** *desert* E Saudi Arabia
74 A11 **Ad Dakhla** *var.* Dakhla. SW Western Sahara *23°46′N 15°56′W*
Ad Dalanj *see* Dilling
Ad Damar *see* Ed Damer
Ad Damazīn *see* Ed Damazin
Ad Dāmir *see* Ed Damer
173 N2 **Ad Dammām** *desert* NE Saudi Arabia
141 R6 **Ad Dammān** *var.* Dammām. Ash Sharqīyah, NE Saudi Arabia *26°23′N 50°05′E*
Ad Dāmūr *see* Damoûr
140 K5 **Ad Dār al Ḥamrā'** Tabūk, NW Saudi Arabia *27°22′N 37°46′E*
140 M13 **Ad Darb** Jīzān, SW Saudi Arabia *17°45′N 42°15′E*
141 O8 **Ad Dawādimī** Ar Riyāḍ, C Saudi Arabia *24°32′N 44°21′E*
143 N16 **Ad Dawḥah** *Eng.* Doha. ● (Qatar) C Qatar *25°15′N 51°36′E*
143 N16 **Ad Dawḥah** *Eng.* Doha. ✈ C Qatar *25°11′N 51°37′E*
139 S6 **Ad Dawr** Ṣalāḥ ad Dīn, N Iraq *34°30′N 43°49′E*
139 Y12 **Ad Dayr** *var.* Dayr, Shahbān. Al Baṣrah, E Iraq *30°45′N 47°36′E*
Addi Arkay *see* Ādī Ārk'ay
Ad Dibakah *see* Dībege
139 X15 **Ad Dibdibah** *physical region* Iraq/Kuwait
Ad Diffah *see* Libyan Plateau
Addis Ababa *see* Ādīs Ābeba
Addison *see* Webster Springs
139 U10 **Ad Dīwānīyah** *var.* Diwaniyah. C Iraq *32°00′N 44°57′E*
Ad Dīwānīyah *see* Al Qādisīyah
Addoo Atoll *see* Addu Atoll
Addua *see* Adda
151 K22 **Addu Atoll** *var.* Addoo Atoll, Seenu Atoll. *atoll* S Maldives
Ad Dujail *see* Ad Dujayl
139 T7 **Ad Dujayl** *var.* Ad Dujail. Ṣalāḥ ad Dīn, N Iraq *33°49′N 44°16′E*
Ad Dulaym *see* Al Anbār
Ad Duwaym/Ad Duwēm *see* Ed Dueim
99 D16 **Adegem** Oost-Vlaanderen, NW Belgium *51°12′N 03°31′E*
23 U7 **Adel** Georgia, SE USA *31°08′N 83°25′W*
29 U14 **Adel** Iowa, C USA *41°36′N 94°01′W*
182 I9 **Adelaide** *state capital* South Australia *34°56′S 138°36′E*
44 H2 **Adelaide** New Providence, N The Bahamas *24°59′N 77°30′W*
182 I9 **Adelaide** ✈ South Australia *34°55′S 138°31′E*
194 H6 **Adelaide Island** *island* Antarctica
181 P2 **Adelaide River** Northern Territory, N Australia *13°12′S 131°06′E*
76 M10 **'Adel Bagrou** Hodh ech Chargui, SE Mauritania *15°33′N 07°04′W*
186 D6 **Adelbert Range** ▲▲ N Papua New Guinea
180 K3 **Adele Island** *island* Western Australia
107 O17 **Adelfia** Puglia, SE Italy *41°01′N 16°52′E*
195 V16 **Adélie Coast** *physical region* Antarctica
195 V14 **Adélie, Terre** *physical region* Antarctica
Adelnau *see* Odolanów
Adelsberg *see* Postojna
Aden *see* 'Adan
Aden *see* Khormaksar
141 Q17 **Aden, Gulf of** *gulf* SW Arabian Sea
77 V10 **Aderbissinat** Agadez, C Niger *15°30′N 07°57′E*
Adhaim *see* Al 'Uẓaym
143 R16 **Adh Dhayd** *var.* Al Dhaid. Ash Shāriqah, NE United Arab Emirates *25°19′N 55°51′E*
140 M4 **'Adhfā'** *spring/well* NW Saudi Arabia *29°15′N 41°24′E*
138 I13 **'Adhriyāt, Jabāl al** ▲ S Jordan
80 I10 **Ādī Ārk'ay** *var.* Addi Arkay. Āmara, N Ethiopia *13°18′N 37°56′E*
182 C7 **Adieu, Cape** *headland* South Australia *32°01′S 132°12′E*
106 H8 **Adige** *Ger.* Etsch. ≈ N Italy
80 J10 **Ādīgrat** Tigray, N Ethiopia *14°17′N 39°27′E*
154 I13 **Ādilābād** *var.* Ādilābād. Telangana, C India *19°40′N 78°31′E*
35 P2 **Adin** California, W USA *41°10′N 120°57′W*
171 V14 **Adi, Pulau** *island* E Indonesia
18 K8 **Adirondack Mountains** ▲▲ New York, NE USA
80 J13 **Ādīs Ābeba** *Eng.* Addis Ababa. ● (Ethiopia) Ādīs Ābeba, C Ethiopia *8°59′N 38°43′E*
80 J13 **Ādīs Ābeba** ✈ Ādīs Ābeba, C Ethiopia *08°58′N 38°53′E*
80 I11 **Ādīs Zemen** Āmara, N Ethiopia *12°00′N 37°43′E*
Adi Ugri *see* Mendefera
137 N15 **Adıyaman** Adıyaman, SE Turkey *37°46′N 38°15′E*
137 N15 **Adıyaman** ◈ *province* S Turkey
116 L11 **Adjud** Vrancea, E Romania *46°07′N 27°10′E*
45 T6 **Adjuntas** C Puerto Rico *18°10′N 66°42′W*
Adjuntas, Presa de las *see* Vicente Guerrero, Presa
Ādkup *see* Erikub Atoll
126 L15 **Adler** Krasnodarskiy Kray, SW Russian Federation *43°25′N 39°58′E*
Adler *see* Orlice
108 G7 **Adliswil** Zürich, NW Switzerland *47°19′N 08°32′E*
32 G7 **Admiralty Inlet** *inlet* Washington, NW USA
39 X13 **Admiralty Island** *island* Alexander Archipelago, Alaska, USA
186 E5 **Admiralty Islands** *island group* N Papua New Guinea
136 B14 **Adnan Menderes** ✈ (İzmir) İzmir, W Turkey *38°16′N 27°09′E*
37 V6 **Adobe Creek Reservoir** ⊞ Colorado, C USA
77 T16 **Ado-Ekiti** Ekiti, SW Nigeria *07°42′N 05°13′E*
Adola *see* Kibre Mengist
61 C23 **Adolfo González Chaues** Buenos Aires, E Argentina *38°13′S 60°05′W*
155 H17 **Ādoni** *var.* Ādavāni. Andhra Pradesh, C India *15°38′N 77°16′E*
102 K15 **Adour** *anc.* Aturus. ≈ SW France
Adowa *see* Ādwa
105 O15 **Adra** Andalucía, S Spain *36°45′N 03°01′W*
107 L24 **Adrano** Sicilia, Italy, C Mediterranean Sea *37°39′N 14°49′E*
74 I9 **Adrar** C Algeria *27°56′N 00°12′W*
76 K7 **Adrar** ◈ *region* C Mauritania
74 L11 **Adrar** ▲ SE Algeria
74 A12 **Adrar Souttouf** ▲▲ SW Western Sahara
Adrasman *see* Adrasmon
147 Q10 **Adrasmon** *Rus.* Adrasman. NW Tajikistan *40°38′N 69°56′E*
78 K10 **Adré** Ouaddaï, E Chad *13°26′N 22°14′E*
106 H9 **Adria** *anc.* Atria, Hadria, Hatria. Veneto, NE Italy *45°03′N 12°04′E*
31 R10 **Adrian** Michigan, N USA *41°54′N 84°02′W*
29 S11 **Adrian** Minnesota, N USA *43°38′N 95°55′W*
27 R5 **Adrian** Missouri, C USA *38°24′N 94°21′W*
24 M2 **Adrian** Texas, SW USA *35°16′N 102°39′W*
21 S4 **Adrian** West Virginia, NE USA *38°53′N 80°14′W*
Adrianople/Adrianopolis *see* Edirne
121 P7 **Adriatic Basin** *undersea feature* Adriatic Sea, N Mediterranean Sea *42°00′N 17°30′E*
Adriatico, Mare *see* Adriatic Sea
106 L13 **Adriatic Sea** *Alb.* Deti Adriatik, *It.* Mare Adriatico, *SCr.* Jadransko More, *Slvn.* Jadransko Morje. *sea* N Mediterranean Sea
Adriatik, Deti *see* Adriatic Sea
Adua *see* Ādwa
Aduana del Sásabe *see* El Sásabe
79 O17 **Adusa** Orientale, NE Dem. Rep. Congo *01°25′N 28°05′E*
118 J13 **Adutiškis** Vilnius, E Lithuania *55°09′N 26°34′E*
27 Y7 **Advance** Missouri, C USA *37°06′N 89°54′W*
65 D25 **Adventure Sound** *bay* East Falkland, Falkland Islands
80 J10 **Ādwa** *var.* Adowa, *It.* Adua. Tigray, N Ethiopia *14°08′N 38°51′E*
123 Q8 **Adycha** ≈ NE Russian Federation
126 L14 **Adygeya, Respublika** ◈ *autonomous republic* SW Russian Federation
Adzhikui *see* Ajyguýy
77 N17 **Adzopé** SE Ivory Coast *06°07′N 03°49′W*
125 U4 **Adz'va** ≈ NW Russian Federation
125 U5 **Adz'vavom** Respublika Komi, NW Russian Federation *66°35′N 59°13′E*
Ædua *see* Autun
115 K19 **Aegean Islands** *island group* Greece/Turkey
Aegean North *see* Vóreion Aigaíon
115 I17 **Aegean Sea** *Gk.* Aigaíon Pelagos, Aigaío Pélagos, *Turk.* Ege Denizi. *sea* NE Mediterranean Sea
Aegean South *see* Nótion Aigaíon
118 H3 **Aegviidu** *Ger.* Charlottenhof. Harjumaa, NW Estonia *59°17′N 25°37′E*
Aegyptus *see* Egypt
Aelana *see* Al 'Aqabah
Aelok *see* Ailuk Atoll
Aelōninae *see* Ailinginae Atoll
Aelōnlaplap *see* Ailinglaplap Atoll
Æmilia *see* Emilia-Romagna
Æmilianum *see* Millau
Aemona *see* Ljubljana
Aenaria *see* Ischia
Aeolian Islands *see* Eolie, Isole
191 Z3 **Aeon Point** *headland* Kiritimati, NE Kiribati *01°46′N 157°11′W*
Æsernia *see* Isernia
104 G3 **A Estrada** Galicia, NW Spain *42°41′N 08°29′W*
115 C18 **Aetós** Itháki, Iónia Nísoi, Greece, C Mediterranean Sea *38°21′N 20°40′E*
191 Q8 **Afaahiti** Tahiti, W French Polynesia *17°43′S 149°18′W*
139 U10 **'Afak** Al Qādisīyah, C Iraq *32°04′N 45°17′E*
Afanasjevo *see* Afanas'yevo
125 T14 **Afanas'yevo** *var.* Afanasjevo. Kirovskaya Oblast', NW Russian Federation *58°55′N 53°13′E*
Afándou *see* Afántou
115 O23 **Afántou** *var.* Afándou. Ródos, Dodekánisa, Greece, Aegean Sea *36°17′N 28°10′E*
80 K11 **Āfar** ◈ *region* NE Ethiopia
Afar Depression *see* Danakil Desert
191 O7 **Afareaitu** Moorea, W French Polynesia *17°33′S 149°47′W*
140 L7 **'Afarīyah, Bi'r al** *well* NW Saudi Arabia
Afars et des Issas, Territoire Français des *see* Djibouti
83 D22 **Affenrücken** Karas, SW Namibia *28°05′S 15°49′E*
Afghānestān, Dowlat-e Eslāmī-ye *see* Afghanistan
148 M6 **Afghanistan** *off.* Islamic Republic of Afghanistan, *Per.* Dowlat-e Eslāmī-ye Afghānestān; *prev.* Republic of Afghanistan. ◆ *islamic state* C Asia
Afgoi *see* Afgooye
81 N17 **Afgooye** *It.* Afgoi. Shabeellaha Hoose, S Somalia *02°09′N 45°07′E*
141 N8 **'Afīf** Ar Riyāḍ, C Saudi Arabia *23°57′N 42°57′E*
77 V17 **Afikpo** Ebonyi, S Nigeria *05°52′N 07°58′E*
Afiun Karahissar *see* Afyon
109 V6 **Aflenz Kurort** Steiermark, E Austria *47°33′N 15°14′E*
74 J6 **Aflou** N Algeria *34°09′N 02°06′E*
81 L18 **Afmadow** Jubbada Hoose, S Somalia *0°24′N 42°04′E*
39 Q14 **Afognak Island** *island* Alaska, USA
104 J2 **A Fonsagrada** Galicia, NW Spain *43°09′N 07°03′W*
186 E9 **Afore** Northern, S Papua New Guinea *09°10′S 148°30′E*
59 O15 **Afrânio** Pernambuco, E Brazil *08°32′S 40°54′W*
66-67 **Africa** *continent*
68 L11 **Africa, Horn of** *physical region* Ethiopia/Somalia
172 K11 **Africana Seamount** *undersea feature* SW Indian Ocean *37°10′S 29°10′E*
86 A14 **African Plate** *tectonic plate*
138 I2 **'Afrīn** Ḥalab, N Syria *36°31′N 36°51′E*
136 M15 **Afşin** Kahramanmaraş, C Turkey *38°14′N 36°54′E*
98 J7 **Afsluitdijk** *dam* N Netherlands
29 U15 **Afton** Iowa, C USA *41°01′N 94°12′W*
29 W9 **Afton** Minnesota, N USA *44°54′N 92°46′W*
27 R8 **Afton** Oklahoma, C USA *36°41′N 94°57′W*
136 F14 **Afyon** *prev.* Afyonkarahisar. Afyon, W Turkey *38°46′N 30°32′E*
136 F14 **Afyon** *var.* Afiun Karahissar, Afyonkarahisar. ◈ *province* W Turkey
Afyonkarahisar *see* Afyon
Agadès *see* Agadez
77 V10 **Agadez** *prev.* Agadès. Agadez, C Niger *16°57′N 07°56′E*
77 W8 **Agadez** ◈ *department* N Niger
74 E8 **Agadir** SW Morocco *30°30′N 09°37′W*
64 M9 **Agadir Canyon** *undersea feature* SE Atlantic Ocean *32°30′N 12°50′W*
145 R12 **Agadyr'** Karaganda, C Kazakhstan *48°15′N 72°55′E*
173 O7 **Agaléga Islands** *island group* N Mauritius
42 K6 **Agalta, Sierra de** ▲▲ E Honduras
122 I10 **Agan** ≈ C Russian Federation
Agana/Agaña *see* Hagåtña
188 B15 **Agana Bay** *bay* NW Guam
171 Kk13 **Agano-gawa** ≈ Honshū, C Japan
188 B17 **Aga Point** *headland* S Guam
154 G9 **Agar** Madhya Pradesh, C India *23°44′N 76°01′E*
81 I14 **Āgaro** Oromīya, C Ethiopia *07°52′N 36°36′E*
153 V15 **Agartala** *state capital* Tripura, NE India *23°49′N 91°15′E*
194 I5 **Agassiz, Cape** *headland* Antarctica *68°29′S 62°59′W*
175 V13 **Agassiz Fracture Zone** *tectonic feature* S Pacific Ocean
9 N2 **Agassiz Ice Cap** *Ice feature* Nunavut, N Canada
188 B16 **Agat** W Guam *13°20′N 144°38′E*
188 B16 **Agat Bay** *bay* W Guam
Agatha *see* Agde
115 M20 **Agathónisi** *island* Dodekánisa, Greece, Aegean Sea
171 X14 **Agats** Papua, E Indonesia *05°33′S 138°07′E*
155 C21 **Agatti Island** *island* Lakshadweep, India, N Indian Ocean
38 D16 **Agattu Island** *island* Aleutian Islands, Alaska, USA
38 D16 **Agattu Strait** *strait* Aleutian Islands, Alaska, USA
14 B8 **Agawa** ≈ Ontario, S Canada
14 B8 **Agawa Bay** *lake bay* Ontario, S Canada
77 N17 **Agboville** SE Ivory Coast *05°55′N 04°15′W*
137 V12 **Ağdam** *Rus.* Agdam. SW Azerbaijan *40°04′N 46°00′E*
Agdam *see* Ağdam
103 P16 **Agde** *anc.* Agatha. Hérault, S France *43°19′N 03°29′E*
103 P16 **Agde, Cap d'** *headland* S France *43°17′N 03°30′E*
Agedabia *see* Ajdābiyā
102 L14 **Agen** *anc.* Aginnum. Lot-et-Garonne, SW France *44°12′N 00°37′E*
Agendicum *see* Sens
165 O13 **Ageo** Saitama, Honshū, S Japan *35°58′N 139°36′E*
109 R5 **Ager** ≈ N Austria
Agere Hiywet *see* Hāgere Hiywet
108 G8 **Ägerisee** ◎ W Switzerland
142 M10 **Āghā Jārī** Khūzestān, SW Iran *30°48′N 49°45′E*
39 P15 **Aghiyuk Island** *island* Alaska, USA
74 B12 **Aghouinit** SE Western Sahara *22°14′N 13°10′W*
Aghri Dagh *see* Büyükağrı Dağı
74 B10 **Aghzoumal, Sebkhet** *var.* Sebjet Agsumal. *salt lake* E Western Sahara
115 F15 **Agiá** *var.* Ayiá. Thessalía, C Greece *39°43′N 22°45′E*
40 G7 **Agiabampo, Estero de** *estuary* NW Mexico
121 P3 **Agía Fýlaxis** *var.* Ayia Phyla. S Cyprus *34°43′N 33°02′E*
Agialoúsa *see* Yenierenköy
115 M21 **Agía Marína** Léros, Dodekánisa, Greece, Aegean Sea *37°09′N 26°51′E*
121 Q2 **Agía Nápa** *var.* Ayia Napa. E Cyprus *34°59′N 34°00′E*
115 L16 **Agía Paraskeví** Lésvos, E Greece *39°13′N 26°19′E*
115 J15 **Agías Eirínis, Akrotírio** *headland* Límnos, E Greece *39°47′N 25°21′E*
Agiasós *see* Agiassós
115 L17 **Agiassós** *var.* Agiasós, Ayiásos, Ayiássos. Lésvos, E Greece *39°05′N 26°23′E*
Aginnum *see* Agen
123 O14 **Aginskoye** Zabaykal'skiy Kray, S Russian Federation *51°10′N 114°32′E*
115 I14 **Ágion Óros** *Eng.* Mount Athos. ◈ *monastic republic* NE Greece
115 H14 **Ágion Óros** *var.* Akte, Aktí; *anc.* Acte. *peninsula* NE Greece
114 D13 **Ágios Achílleios** *religious building* Dytikí Makedonía, N Greece
115 J16 **Ágios Efstrátios** *var.* Áyios Evstratios, Hagios Evstrátios. *island* E Greece *34°00′N 25°00′E*
115 H20 **Ágios Geórgios** *island* Kykládes, Greece, Aegean Sea
Ágios Geórgios *see* Ro
115 E21 **Ágios Ilías** ▲ S Greece *36°57′N 22°19′E*
115 K25 **Ágios Ioánnis, Akrotírio** *headland* Kríti, Greece, E Mediterranean Sea *35°19′N 25°46′E*
115 L20 **Ágios Kírykos** *var.* Áyios Kírikos. Ikaría, Dodekánisa, Greece, Aegean Sea *37°34′N 26°15′E*
115 K25 **Ágios Nikólaos** *var.* Áyios Nikólaos. Kríti, Greece, E Mediterranean Sea *35°12′N 25°43′E*
115 D16 **Ágios Nikolaos** Thessalía, C Greece *39°33′N 21°21′E*
115 H14 **Agíou Órous, Kólpos** *gulf* N Greece
107 K24 **Agira** *anc.* Agyrium. Sicilia, Italy, C Mediterranean Sea *37°39′N 14°32′E*
115 G20 **Agkístri** *island* S Greece
114 G12 **Agkistro** *var.* Angistro. ▲ NE Greece *41°21′N 23°29′E*
103 O17 **Agly** ≈ S France
Agnetheln *see* Agnita
14 E10 **Agnew Lake** ◎ Ontario, S Canada
77 O16 **Agnibilékrou** E Ivory Coast *07°10′N 03°11′W*
116 I11 **Agnita** *Ger.* Agnetheln, *Hung.* Szentágota. Sibiu, SW Romania *45°59′N 24°40′E*
107 K15 **Agnone** Molise, C Italy *41°49′N 14°21′E*
164 K14 **Ago** Mie, Honshū, SW Japan *34°18′N 136°50′E*
106 C8 **Agogna** ≈ N Italy
Agoitz *see* Aoiz
Agoiz *see* Aoiz
77 P17 **Agona Swedru** *var.* Swedru. SE Ghana *05°31′N 00°42′W*
Agordat *see* Ak'ordat
Agosta *see* Augusta
103 N15 **Agout** ≈ S France
152 J12 **Āgra** Uttar Pradesh, N India *27°09′N 78°E*
Agra and Oudh, United Provinces of *see* Uttar Pradesh
Agram *see* Zagreb
105 U5 **Agramunt** Cataluña, NE Spain *41°48′N 01°07′E*
105 Q5 **Agreda** Castilla y León, N Spain *41°51′N 01°55′W*
137 S13 **Ağrı** *var.* Karaköse; *prev.* Karakılısse. Ağrı, NE Turkey *39°44′N 43°04′E*
137 S13 **Ağrı** ◈ *province* NE Turkey
107 N19 **Agri** *anc.* Aciris. ≈ S Italy
Agri Dagi *see* Büyükağrı Dağı
107 J24 **Agrigento** *Gk.* Akragas; *prev.* Girgenti. Sicilia, Italy, C Mediterranean Sea *37°19′N 13°33′E*
188 K4 **Agrihan** *island* N Northern Mariana Islands
115 L17 **Agriliá, Akrotírio** *prev.* Ákra Maléas. *cape* Lésvos, E Greece
115 D18 **Agrínio** *prev.* Agrinion. Dytikí Elláda, W Greece *38°38′N 21°25′E*
Agrinion *see* Agrínio
115 G17 **Agriovótano** Évvoia, C Greece *39°00′N 23°18′E*
107 L18 **Agropoli** Campania, S Italy *40°22′N 14°59′E*
127 T3 **Agryz** Udmurtskaya Respublika, NW Russian Federation *56°27′N 52°58′E*
137 U11 **Ağstafa** *Rus.* Akstafa. NW Azerbaijan *41°06′N 45°28′E*
137 X11 **Ağsu** *Rus.* Akhsu. C Azerbaijan
Agsumal, Sebjet *see* Aghzoumal, Sebkhet
40 J11 **Agua Brava, Laguna** *lagoon* W Mexico
54 F7 **Aguachica** Cesar, N Colombia *08°16′N 73°35′W*
59 J20 **Água Clara** Mato Grosso do Sul, SW Brazil *20°21′S 52°58′W*
44 D5 **Aguada de Pasajeros** Cienfuegos, C Cuba *22°23′N 80°51′W*
54 J5 **Aguada Grande** Lara, N Venezuela *10°38′N 69°29′W*
45 S5 **Aguadilla** W Puerto Rico *18°27′N 67°08′W*
43 S16 **Aguadulce** Coclé, S Panama *08°16′N 80°31′W*
104 L14 **Aguadulce** Andalucía, S Spain *37°15′N 04°59′W*
41 O8 **Agualeguas** Nuevo León, NE Mexico *26°17′N 99°30′W*
40 L9 **Aguanaval, Río** ≈ C Mexico
42 J5 **Aguán, Río** ≈ N Honduras
25 R16 **Agua Nueva** Texas, SW USA *26°57′N 98°34′W*
60 J8 **Aguapeí, Rio** ≈ S Brazil
61 E14 **Aguapey, Río** ≈ NE Argentina
40 G3 **Agua Prieta** Sonora, NW Mexico *31°16′N 109°33′W*
104 G5 **A Guarda** *var.* A Guardia, Laguardia, La Guardia. Galicia, NW Spain *41°54′N 08°53′W*
A Guardia *see* A Guarda
56 E6 **Aguarico, Río** ≈ Ecuador/ Peru
55 O6 **Aguasay** Monagas, NE Venezuela *09°25′N 63°44′W*
40 M12 **Aguascalientes** Aguascalientes, C Mexico *21°54′N 102°17′W*
40 L12 **Aguascalientes** ◈ *state* C Mexico
57 I18 **Aguas Calientes, Río** ≈ S Peru
105 R7 **Aguasvivas** ≈ NE Spain
60 J7 **Água Vermelha, Represa de** ⊞ S Brazil
56 E12 **Aguaytía** Ucayali, C Peru *09°02′S 75°30′W*
104 I5 **A Gudiña** *var.* La Gudiña. Galicia, NW Spain *42°04′N 07°08′W*
104 G7 **Águeda** Aveiro, N Portugal *40°34′N 08°28′W*
104 J8 **Águeda** ≈ Portugal/Spain
77 Q8 **Aguelhok** Kidal, NE Mali *19°18′N 00°50′E*
77 V12 **Aguié** Maradi, S Niger *13°28′N 07°43′E*
188 K8 **Aguijan** *island* S Northern Mariana Islands
104 M14 **Aguilar** *var.* Aguilar de la Frontera. Andalucía, S Spain *37°31′N 04°40′W*
104 M3 **Aguilar de Campóo** Castilla y León, N Spain *42°47′N 04°15′W*
Aguilar de la Frontera *see* Aguilar
42 F7 **Aguilares** San Salvador, C El Salvador *13°56′N 89°09′W*
105 Q14 **Águilas** Murcia, SE Spain *37°24′N 01°36′W*
40 L15 **Aguililla** Michoacán, SW Mexico *18°43′N 102°45′W*
Agulhas *see* L'Agulhas
172 J11 **Agulhas Bank** *undersea feature* SW Indian Ocean *35°30′S 21°00′E*
172 K11 **Agulhas Basin** *undersea feature* SW Indian Ocean *47°00′S 27°00′E*
83 F26 **Agulhas, Cape** *Afr.* Kaap Agulhas. *headland* SW South Africa *34°51′S 19°59′E*
Agulhas, Kaap *see* Agulhas, Cape
60 O9 **Agulhas Negras, Pico das** ▲ SE Brazil *22°21′S 44°39′W*
172 K11 **Agulhas Plateau** *undersea feature* SW Indian Ocean *40°00′S 26°00′E*
165 S16 **Aguni-jima** *island* Nansei-shotō, SW Japan
Agurain *see* Salvatierra
54 G5 **Agustín Codazzi** *var.* Codazzi. Cesar, N Colombia *10°02′N 73°15′W*
Agyrium *see* Agira
74 L12 **Ahaggar** *high plateau region* SE Algeria
146 E12 **Ahal Welaýaty** *Rus.* Akhalskiy Velayat. ◈ *province* C Turkmenistan
142 K2 **Ahar** Āẕarbāyjān-e Sharqī, NW Iran *38°25′N 47°07′E*
Aharnes *see* Acharnés
138 J3 **Aḥaṣ, Jabal** ▲ NW Syria
138 J3 **Aḥaṣ, Jebal** ▲ W Syria
185 G16 **Ahaura** ≈ South Island, New Zealand
100 E13 **Ahaus** Nordrhein-Westfalen, NW Germany *52°04′N 07°01′E*
191 U9 **Ahe** *atoll* Îles Tuamotu, C French Polynesia
184 N10 **Ahimanawa Range** ▲▲ North Island, New Zealand
119 I19 **Ahinski Kanal** *Rus.* Oginskiy Kanal. *canal* SW Belarus
186 G10 **Ahioma** SE Papua New Guinea *10°20′S 150°35′E*
184 I2 **Ahipara** Northland, North Island, New Zealand *35°11′S 173°07′E*
184 I2 **Ahipara Bay** *bay* SE Tasman Sea
Áhkájávrre *see* Akkajaure
Áhkká *see* Akka
39 N13 **Ahklun Mountains** ▲▲ Alaska, USA
137 R14 **Ahlat** Bitlis, E Turkey *38°45′N 42°28′E*
101 F14 **Ahlen** Nordrhein-Westfalen, W Germany *51°46′N 07°53′E*
154 D10 **Ahmadābād** *var.* Ahmedabad. Gujarāt, W India *23°03′N 72°40′E*
143 R10 **Ahmadābād** Kermān, C Iran *35°51′N 59°36′E*
Ahmadi *see* Al Aḥmadī
Ahmad Khel *see* Ḥasan Khēl
155 F14 **Ahmadnagar** *var.* Ahmednagar. Mahārāshtra, W India *19°08′N 74°48′E*
149 T9 **Ahmadpur Siāl** Punjab, E Pakistan *30°40′N 71°52′E*
77 N5 **Aḥmar, 'Erg el** *desert* N Mali
80 K13 **Ahmar Mountains** ▲▲ C Ethiopia
Ahmedabad *see* Ahmadābād
Ahmednagar *see* Ahmadnagar
114 N12 **Ahmetbey** Kırklareli, NW Turkey *41°27′N 27°35′E*
14 H12 **Ahmic Lake** ◎ Ontario, S Canada
190 G12 **Ahoa** Île Uvea, E Wallis and Futuna *13°17′S 176°12′W*
40 G8 **Ahomé** Sinaloa, C Mexico *25°55′N 109°10′W*
21 X8 **Ahoskie** North Carolina, SE USA *36°17′N 76°59′W*
101 D17 **Ahr** ≈ W Germany
143 N12 **Ahram** *var.* Ahrom. Būshehr, S Iran *28°52′N 51°18′E*
100 J9 **Ahrensburg** Schleswig-Holstein, N Germany *53°41′N 10°14′E*
Ahrom *see* Ahram
93 L17 **Ähtäri** Etelä-Pohjanmaa, W Finland *62°34′N 24°08′E*
114 N10 **Ahtopol** *var.* Akhtopol. Burgas, E Bulgaria *42°06′N 27°57′E*
40 K12 **Ahuacatlán** Nayarit, C Mexico *21°02′N 104°30′W*
42 E7 **Ahuachapán** Ahuachapán, W El Salvador *13°55′N 89°51′W*
42 A9 **Ahuachapán** ◈ *department* W El Salvador
191 V16 **Ahu Akivi** *var.* Siete Moai. *ancient monument* Easter Island, Chile, E Pacific Ocean
191 W11 **Ahunui** *atoll* Îles Tuamotu, C French Polynesia
185 E20 **Ahuriri** ≈ South Island, New Zealand
95 L22 **Åhus** Skåne, S Sweden *55°55′N 14°18′E*
Ahu Tahira *see* Ahu Vinapu
191 V16 **Ahu Tepeu** *ancient monument* Easter Island, Chile, E Pacific Ocean
191 V17 **Ahu Vinapu** *var.* Ahu Tahira. *ancient monument* Easter Island, Chile, E Pacific Ocean
142 L9 **Ahvāz** *var.* Ahwāz; *prev.* Nāsiri. Khūzestān, SW Iran *31°20′N 48°38′E*
Ahvenanmaa *see* Åland
141 Q16 **Aḥwar** SW Yemen *13°34′N 46°41′E*
Ahwāz *see* Ahvāz
94 H7 **Åi Åfjord** *var.* Åi Åfjord, Årnes. Sør-Trøndelag, C Norway *63°57′N 10°12′E* *see also* Åi Åfjord
Åi Åfjord *see* Åi Åfjord
149 P3 **Aibak** *var.* Haibak; *prev.* Āybak, Samangān. Samangān, NE Afghanistan *36°16′N 68°04′E*
101 K22 **Aichach** Bayern, SE Germany *48°26′N 11°06′E*
164 L14 **Aichi** *off.* Aichi-ken, *var.* Aiti. ◈ *prefecture* Honshū, SW Japan
Aidin *see* Aydın
Aidussina *see* Ajdovščina
Aifir, Clochán an *see* Giant's Causeway
Aigaíon Pelagos/Aigaío Pélagos *see* Aegean Sea
109 S3 **Aigen im Mülkreis** Oberösterreich, N Austria *48°39′N 13°57′E*
115 G20 **Aígina** *var.* Aíyina, Egina. Aígina, C Greece *37°45′N 23°26′E*
115 G20 **Aígina** *island* S Greece
115 E18 **Aígio** *var.* Egio; *prev.* Aíyion. Dytikí Elláda, S Greece *38°15′N 22°05′E*
108 C10 **Aigle** Vaud, SW Switzerland *46°20′N 06°58′E*
103 P14 **Aigoual, Mont** ▲ S France *44°09′N 03°34′E*
173 O16 **Aigrettes, Pointe des** *headland* W Réunion *21°02′S 55°14′E*
61 G19 **Aiguá** *var.* Aigua. Maldonado, S Uruguay *34°13′S 54°46′W*
103 S13 **Aigues** ≈ SE France
103 N10 **Aigurande** Indre, C France *46°26′N 01°49′E*
Ai-hun *see* Heihe
163 N10 **Aikawa** Niigata, Sado, C Japan *38°04′N 138°15′E*
21 Q13 **Aiken** South Carolina, SE USA *33°34′N 81°44′W*
25 N4 **Aiken** Texas, SW USA *34°06′N 101°31′W*

◆ Country ● Country Capital ◇ Dependent Territory ○ Dependent Territory Capital ◈ Administrative Regions ✈ International Airport ▲ Mountain ▲▲ Mountain Range ⛰ Volcano ≈ River ◎ Lake ⊞ Reservoir

214

◆ Country
● Country Capital
◇ Dependent Territory
○ Dependent Territory Capital
◆ Administrative Regions
✕ International Airport
▲ Mountain
▲ Mountain Range
◤ Volcano
∅ River
⊗ Lake
Reservoir

ALCAÑICES – ALMONTE

104 J5 **Alcañices** Castilla y León, N Spain 41°41′N 06°21′W
105 T7 **Alcañiz** Aragón, NE Spain 41°03′N 00°09′W
104 I9 **Alcántara** Extremadura, W Spain 39°42′N 06°54′W
104 J9 **Alcántara, Embalse de** ⊠ W Spain
105 R13 **Alcantarilla** Murcia, SE Spain 37°59′N 01°12′W
105 P11 **Alcaraz** Castilla-La Mancha, C Spain 38°40′N 02°29′W
105 P12 **Alcaraz, Sierra de** ▲▲ C Spain
104 I12 **Alcarrache** ≈ SW Spain
105 T6 **Alcarràs** Cataluña, NE Spain 41°34′N 00°31′E
105 N14 **Alcaudete** Andalucía, S Spain 37°35′N 04°05′W
Alcázar *see* Ksar-el-Kebir
105 O10 **Alcázar de San Juan** *anc.* Alce. Castilla-La Mancha, C Spain 39°24′N 03°12′W
Alcazarquivir *see* Ksar-el-Kebir
Alce *see* Alcázar de San Juan
57 B17 **Alcebo, Volcán** ⋀ Galapagos Islands, Ecuador, E Pacific Ocean 0°25′S 91°06′W
139 X12 **Al Chabā'ish** *var.* Al Kaba'ish. Dhī Qār, SE Iraq 30°58′N 47°02′E
117 Y7 **Alchevs'k** *prev.* Kommunarsk, Voroshilovsk. Luhans'ka Oblast', E Ukraine 48°29′N 38°52′E
Alcira *see* Alzira
21 N9 **Alcoa** Tennessee, S USA 35°47′N 83°58′W
104 F9 **Alcobaça** Leiria, C Portugal 39°32′N 08°59′W
105 N8 **Alcobendas** Madrid, C Spain 40°32′N 03°38′W
Alcoi *see* Alcoy
105 P7 **Alcolea del Pinar** Castilla-La Mancha, C Spain 41°02′N 02°28′W
104 I11 **Alconchel** Extremadura, W Spain 38°31′N 07°04′W
Alcora *see* L'Alcora
105 N8 **Alcorcón** Madrid, C Spain 40°20′N 03°52′W
105 S7 **Alcorisa** Aragón, NE Spain 40°53′N 00°23′W
61 B19 **Alcorta** Santa Fe, C Argentina 33°32′S 61°07′W
104 H14 **Alcoutim** Faro, S Portugal 37°28′N 07°29′W
33 W15 **Alcova** Wyoming, C USA 42°33′N 106°40′W
105 S11 **Alcoy** *Cat.* Alcoi. Valenciana, E Spain 38°42′N 00°29′W
105 Y9 **Alcúdia** Mallorca, Spain, W Mediterranean Sea 39°51′N 03°06′E
105 Y9 **Alcúdia, Badia d'** *bay* Mallorca, Spain, W Mediterranean Sea
172 M7 **Aldabra Group** *island group* SW Seychelles
139 U10 **Al Daghghārah** Bābil, C Iraq 32°10′N 44°57′E
40 J5 **Aldama** Chihuahua, N Mexico 28°50′N 105°52′W
41 P11 **Aldama** Tamaulipas, C Mexico 22°54′N 98°05′W
123 Q11 **Aldan** Respublika Sakha (Yakutiya), NE Russian Federation 58°31′N 125°15′E
123 Q10 **Aldan** ≈ NE Russian Federation
Aldar *see* Aldarhaan
al Dar al Baida *see* Rabat
162 G7 **Aldarhaan** *var.* Aldar. Dzavhan, W Mongolia 47°43′N 96°36′E
97 Q20 **Aldeburgh** E England, United Kingdom 52°12′N 01°36′E
105 P5 **Aldehuela de Calatañazor** Castilla y León, N Spain 41°42′N 02°46′W
Aldeia Nova *see* Aldeia Nova de São Bento
104 H13 **Aldeia Nova de São Bento** *var.* Aldeia Nova. Beja, S Portugal 37°55′N 07°24′W
29 V11 **Alden** Minnesota, N USA 43°40′N 93°34′W
184 N6 **Aldermen Islands, The** *island group* N New Zealand
97 L25 **Alderney** *island* Channel Islands
97 N22 **Aldershot** S England, United Kingdom 51°15′N 00°47′W
21 R6 **Alderson** West Virginia, NE USA 37°43′N 80°38′W
Al Dhaid *see* Adh Dhayd
98 L5 **Aldtsjerk** *Dutch.* Oudkerk. Fryslân, N Netherlands 53°16′N 05°52′E
30 J11 **Aledo** Illinois, N USA 41°12′N 90°45′W
76 H9 **Aleg** Brakna, SW Mauritania 17°03′N 13°53′W
64 Q10 **Alegranza** *island* Islas Canarias, Spain, NE Atlantic Ocean
37 P12 **Alegres Mountain** ▲ New Mexico, SW USA 34°09′N 108°11′W
61 F15 **Alegrete** Rio Grande do Sul, S Brazil 29°46′S 55°46′W
61 C16 **Alejandra** Santa Fe, C Argentina 29°54′S 59°50′W
193 T11 **Alejandro Selkirk, Isla** *island* Islas Juan Fernández, Chile, E Pacific Ocean
124 I12 **Alëkhovshchina** Leningradskaya Oblast', NW Russian Federation 60°22′N 33°57′E
39 O13 **Aleknagik** Alaska, USA 59°16′N 158°37′W
Aleksandriya *see* Oleksandriya
Aleksandropol' *see* Gyumri
126 L3 **Aleksandrov** Vladimirskaya Oblast', W Russian Federation 56°24′N 38°42′E
113 N14 **Aleksandrovac** Serbia, C Serbia 43°28′N 21°05′E
127 R9 **Aleksandrov Gay** Saratovskaya Oblast', W Russian Federation 50°08′N 48°34′E
127 U6 **Aleksandrovka** Orenburgskaya Oblast', W Russian Federation 52°47′N 54°14′E
Aleksandrovka *see* Oleksandrivka
125 V13 **Aleksandrovsk** Permskiy Kray, NW Russian Federation 59°12′N 57°27′E
Aleksandrovsk *see* Zaporizhzhya
127 N14 **Aleksandrovskoye** Stavropol'skiy Kray, SW Russian Federation 44°43′N 42°56′E
123 T12 **Aleksandrovsk-Sakhalinskiy** Ostrov Sakhalin, Sakhalinskaya Oblast', SE Russian Federation 50°55′N 142°12′E
110 J10 **Aleksandrów Kujawski** Kujawskio-pormorskie, C Poland 52°52′N 18°40′E
110 K12 **Aleksandrów Łódzki** Łódzkie, C Poland 51°49′N 19°19′E
114 J8 **Aleksandŭr Stamboliyski, Yazovir** ⊠ N Bulgaria
Alekseevka *see* Akkol', Akmola, Kazakhstan
Alekseevka *see* Terekty
145 P7 **Alekseyevka** *Kaz.* Alekseevka. Akmola, N Kazakhstan 53°32′N 69°30′E
126 L9 **Alekseyevka** Belgorodskaya, W Russian Federation 50°35′N 38°41′E
127 S7 **Alekseyevka** Samarskaya Oblast', W Russian Federation 52°37′N 51°20′E
Alekseyevka *see* Akkol', Akmola, Kazakhstan
Alekseyevka *see* Terekty, Vostochnyy Kazakhstan, Kazakhstan
127 R4 **Alekseyevskoye** Respublika Tatarstan, W Russian Federation 55°18′N 50°11′E
126 K5 **Aleksin** Tul'skaya Oblast', W Russian Federation 54°30′N 37°08′E
113 O14 **Aleksinac** Serbia, SE Serbia 43°33′N 21°43′E
190 G11 **Alele** Île Uvea, E Wallis and Futuna 13°14′S 176°09′W
95 N20 **Ålem** Kalmar, S Sweden 56°57′N 16°25′E
102 L6 **Alençon** Orne, N France 48°26′N 00°04′E
58 I12 **Alenquer** Pará, NE Brazil 01°58′S 54°45′W
38 G10 **'Alenuihaha Channel** *var.* Alenuihaha Channel. *channel* Hawai'i, USA, C Pacific Ocean
Alep/Aleppo *see* Ḥalab
103 Y15 **Aléria** Corse, France, C Mediterranean Sea 42°06′N 09°29′E
197 Q11 **Alert** Ellesmere Island, Nunavut, N Canada 82°28′N 62°13′W
103 Q14 **Alès** *prev.* Alais. Gard, S France 44°08′N 04°05′E
116 G9 **Aleşd** *Hung.* Élesd. Bihor, SW Romania 47°03′N 22°22′E
106 C9 **Alessandria** *Fr.* Alexandrie. Piemonte, N Italy 44°54′N 08°37′E
Ålestrup *see* Aalestrup
94 D9 **Ålesund** Møre og Romsdal, S Norway 62°28′N 06°11′E
108 E10 **Aletschhorn** ▲ SW Switzerland 46°33′N 08°01′E
197 S1 **Aleutian Basin** *undersea feature* Bering Sea 57°00′N 177°00′E
38 H17 **Aleutian Islands** *island group* Alaska, USA
39 P14 **Aleutian Range** ▲▲ Alaska, USA
0 B5 **Aleutian Trench** *undersea feature* S Bering Sea 57°00′N 177°00′W
123 T10 **Alevina, Mys** *cape* E Russian Federation
15 Q6 **Alex** ≈ Québec, SE Canada
28 J3 **Alexander** North Dakota, N USA 47°48′N 103°38′W
39 W14 **Alexander Archipelago** *island group* Alaska, USA
Alexanderbaai *see* Alexander Bay
83 D23 **Alexander Bay** *Afr.* Alexanderbaai. Northern Cape, W South Africa 28°40′S 16°30′E
23 Q5 **Alexander City** Alabama, S USA 32°56′N 85°57′W
194 J6 **Alexander Island** *island* Antarctica
Alexander Range *see* Kirghiz Range
183 O12 **Alexandra** Victoria, SE Australia 37°12′S 145°43′E
185 D22 **Alexandra** Otago, South Island, New Zealand 45°15′S 169°25′E
115 F14 **Alexándreia** *var.* Alexándria. Kentrikí Makedonía, N Greece 40°38′N 22°27′E
Alexandretta *see* İskenderun
Alexandretta, Gulf of *see* İskenderun Körfezi
15 N13 **Alexandria** Ontario, SE Canada 45°19′N 74°37′W
121 U13 **Alexandria** *Ar.* Al Iskandarīyah. N Egypt 31°07′N 29°51′E
44 J12 **Alexandria** C Jamaica 18°18′N 77°21′W
116 J15 **Alexandria** Teleorman, S Romania 43°58′N 25°18′E
31 P13 **Alexandria** Indiana, N USA 40°15′N 85°40′W
20 M4 **Alexandria** Kentucky, S USA 38°59′N 84°22′W
22 H7 **Alexandria** Louisiana, S USA 31°19′N 92°27′W
29 T7 **Alexandria** Minnesota, N USA 45°54′N 95°22′W
29 Q11 **Alexandria** South Dakota, N USA 43°39′N 97°46′W
21 W4 **Alexandria** Virginia, NE USA 38°49′N 77°06′W
Alexándria *see* Alexándreia
18 I7 **Alexandria Bay** New York, NE USA 44°20′N 75°54′W
Alexandrie *see* Alessandria
182 J10 **Alexandrina, Lake** ◎ South Australia
114 K13 **Alexandroúpoli** *var.* Alexandroúpolis, *Turk.* Dedeagaç, Dedeagach. Anatolikí Makedonía kai Thráki, NE Greece 40°52′N 25°53′E
Alexandroúpolis *see* Alexandroúpoli
10 L15 **Alexis Creek** British Columbia, SW Canada 52°06′N 123°25′W
122 I13 **Aleysk** Altayskiy Kray, S Russian Federation 52°32′N 82°46′E
139 S8 **Al Fallūjah** *var.* Falluja. Al Anbār, C Iraq 33°21′N 43°46′E
105 R8 **Alfambra** ≈ E Spain
Al Faqa *see* Faq'
141 R15 **Al Farḍah** C Yemen 14°51′N 48°33′E
105 Q4 **Alfaro** La Rioja, N Spain 42°13′N 01°45′W
105 U5 **Alfarràs** Cataluña, NE Spain 41°50′N 00°34′E
Al Fāshir *see* El Fasher
75 W8 **Al Fashn** *var.* El Fashn. C Egypt 28°49′N 30°54′E
114 M7 **Alfatar** Silistra, NE Bulgaria 43°56′N 27°17′E
139 S5 **Al Fatḥah** Şalāḥ ad Dīn, C Iraq 35°06′N 43°34′E
139 Q3 **Al Fatsī** Nīnawá, N Iraq 36°04′N 42°39′E
139 Z13 **Al Fāw** *var.* Fao. Al Başrah, SE Iraq 29°55′N 48°26′E
75 W8 **Al Fayyūm** *var.* El Faiyûm. N Egypt 29°19′N 30°50′E
115 D20 **Alfeiós** *prev.* Alfiós; *anc.* Alpheius, Alpheus. ≈ S Greece
100 I13 **Alfeld** Niedersachsen, C Germany 51°58′N 09°49′E
Alfiós *see* Alfeiós
Alföld *see* Great Hungarian Plain
94 C11 **Ålfotbreen** *glacier* S Norway
19 P9 **Alfred** Maine, NE USA 43°29′N 70°44′W
18 F11 **Alfred** New York, NE USA 42°15′N 77°47′W
61 K14 **Alfredo Wagner** Santa Catarina, S Brazil 27°40′S 49°22′W
94 M12 **Alfta** Gävleborg, C Sweden 61°20′N 16°05′E
140 K12 **Al Fuḥayḥīl** *var.* Fahaheel. SE Kuwait 29°01′N 48°05′E
139 Q6 **Al Fuḥaymī** Al Anbār, C Iraq 34°18′N 42°09′E
143 S16 **Al Fujayrah** *Eng.* Fujairah. Al Fujayrah, NE United Arab Emirates 25°09′N 56°18′E
143 S16 **Al Fujayrah** *Eng.* Fujairah. ✈ Al Fujayrah, NE United Arab Emirates 25°04′N 56°12′E
Al-Furāt *see* Euphrates
144 I10 **Alga** *Kaz.* Algha. Aktyubinsk, NW Kazakhstan 49°56′N 57°19′E
144 G9 **Algabas** *Kaz.* Alghabas. Zapadnyy Kazakhstan, NW Kazakhstan 50°43′N 52°09′E
95 C17 **Ålgård** Rogaland, S Norway 58°45′N 05°52′E
104 G14 **Algarve** *cultural region* S Portugal
182 G3 **Algebuckina Bridge** South Australia 28°03′S 135°48′E
104 K16 **Algeciras** Andalucía, SW Spain 36°08′N 05°27′W
105 S10 **Algemesí** Valenciana, E Spain 39°11′N 00°27′W
Al-Genain *see* El Geneina
120 F9 **Alger** *var.* Algiers, El Djazaïr, Al Jazair. ● (Algeria) N Algeria 36°47′N 02°58′E
74 H9 **Algeria** *off.* Democratic and Popular Republic of Algeria. ◆ *republic* N Africa
Algeria, Democratic and Popular Republic of *see* Algeria
120 J8 **Algerian Basin** *var.* Balearic Plain. *undersea feature* W Mediterranean Sea
Algha *see* Alga
138 I4 **Al Ghāb** *Valley* NW Syria
141 X10 **Al Ghābah** *var.* Ghaba. C Oman 21°22′N 57°14′E
Alghabas *see* Algabas
141 U14 **Al Ghayḍah** E Yemen 16°15′N 52°13′E
140 M6 **Al Ghazālah** Ḥā'il, NW Saudi Arabia 26°55′N 41°23′E
107 B17 **Alghero** Sardegna, Italy, C Mediterranean Sea 40°34′N 08°19′E
95 M20 **Älghult** Kronoberg, S Sweden 57°00′N 15°34′E
75 X9 **Al Ghurdaqah** *var.* Ghurdaqah, Hurghada. E Egypt 27°17′N 33°47′E
Algiers *see* Alger
105 S10 **Alginet** Valenciana, E Spain 39°16′N 00°28′W
83 I26 **Algoa Bay** *bay* S South Africa
104 L15 **Algodonales** Andalucía, S Spain 36°54′N 05°24′W
105 N9 **Algodor** ≈ C Spain
Al Golea *see* El Goléa
31 N6 **Algoma** Wisconsin, N USA 44°41′N 87°24′W
29 U12 **Algona** Iowa, C USA 43°04′N 94°13′W
20 L8 **Algood** Tennessee, S USA 36°12′N 85°27′W
105 O2 **Algorta** País Vasco, N Spain 43°20′N 03°00′W
61 E18 **Algorta** Río Negro, W Uruguay 32°26′S 57°18′W
Al Haba *see* Haba
139 Q10 **Al Habbārīyah** Al Anbār, S Iraq 32°16′N 42°12′E
Al Hadhar *see* Al Ḥaḍr
139 Q4 **Al Ḥaḍr** *var.* Al Hadhar; *anc.* Hatra. Nīnawýa, NW Iraq 35°34′N 42°44′E
139 T13 **Al Ḥajarah** *desert* S Iraq
141 W8 **Al Ḥajar al Gharbī** ▲▲ N Oman
141 Y8 **Al Ḥajar ash Sharqī** ▲▲ NE Oman
141 R15 **Al Hajarayn** C Yemen 15°29′N 48°24′E
138 L10 **Al Ḥamād** *desert* Jordan/Saudi Arabia
Al Hamad *see* Syrian Desert
75 N9 **Al Ḥamādah al Ḥamrā'** *var.* Al Ḥamrā'. *desert* NW Libya
105 N15 **Alhama de Granada** Andalucía, S Spain 37°00′N 03°59′W
105 R13 **Alhama de Murcia** Murcia, SE Spain 37°51′N 01°25′W
35 T15 **Alhambra** California, W USA 34°08′N 118°06′W
139 T12 **Al Ḥammām** An Najaf, S Iraq 31°09′N 44°04′E
141 X8 **Al Ḥamrā'** NE Oman 23°07′N 57°23′E
Al Ḥamrā' *see* Al Ḥamādah al Ḥamrā'
141 O6 **Al Ḥamūdīyah** *spring/well* N Saudi Arabia 27°05′N 44°24′E
140 M7 **Al Ḥanākīyah** Al Madīnah, W Saudi Arabia 24°55′N 40°31′E
139 W14 **Al Ḥanīyah** *escarpment* Iraq/Saudi Arabia
139 Y12 **Al Hārithah** Al Başrah, SE Iraq 30°43′N 47°44′E
140 L3 **Al Ḥarrah** *desert* NW Saudi Arabia
75 Q10 **Al Harūj al Aswad** *desert* C Libya
Al Hasaifin *see* Al Ḥusayfin
139 N2 **Al Ḥasakah** *var.* Al Hasijah, El Haseke, *Fr.* Hassetché. Al Ḥasakah, NE Syria 36°22′N 40°44′E
139 O2 **Al Ḥasakah** *off.* Muḥāfaẓat al Ḥasakah, *var.* Al Hasakah, Āl Hasakah, Hasakah, Hassakeh. ◈ *governorate* NE Syria
Al Hasakah *see* Al Ḥasakah
Al Hasakah *see* 'Āmūdah
139 T9 **Al Hāshimīyah** Bābil, C Iraq 32°24′N 44°39′E
138 G13 **Al Hāshimīyah** Ma'ān, S Jordan 30°31′N 35°46′E
Al Hasijah *see* Al Ḥasakah
104 M15 **Alhaurín el Grande** Andalucía, S Spain 36°39′N 04°41′W
141 Q16 **Al Ḥawrā** S Yemen 13°54′N 47°36′E
139 V10 **Al Ḥayy** *var.* Kut al Hai, Kūt al Ḥayy. Wāsiţ, E Iraq 32°11′N 46°03′E
141 U11 **Al Ḥibāk** *desert* E Saudi Arabia
138 H8 **Al Hījānah** *var.* Hejanah, Hijanah. Rīf Dimashq, W Syria 33°23′N 36°34′E
140 K7 **Al Ḥijāz** *Eng.* Hejaz. *physical region* NW Saudi Arabia
Al Hilbeh *see* 'Ulayyānīyah, Bi'r al
139 T9 **Al Ḥillah** *var.* Hilla. Bābil, C Iraq 32°28′N 44°29′E
139 T9 **Al Hindīyah** *var.* Hindiya. Bābil, C Iraq 32°32′N 44°14′E
138 G12 **Al Ḥisā** Aṭ Ṭafīlah, W Jordan 30°49′N 35°58′E
74 G5 **Al-Hoceïma** *var.* al Hoceima, Al-Hoceima, Alhucemas; *prev.* Villa Sanjurjo. N Morocco 35°14′N 03°56′W
Alhucemas *see* Al-Hoceïma
105 N17 **Alhucemas, Peñon de** *island group* S Spain
141 N15 **Al Ḥudaydah** *Eng.* Hodeida. W Yemen 15°N 42°50′E
141 N15 **Al Ḥudaydah** *var.* Hodeida. ✈ W Yemen 14°45′N 43°01′E
140 M4 **Al Ḥudūd ash Shamālīyah** *var.* Minṭaqat al Ḥudūd ash Shamālīyah, *Eng.* Northern Border Region. ◈ *province* N Saudi Arabia
141 S7 **Al Hufūf** *var.* Hofuf. Ash Sharqīyah, NE Saudi Arabia 25°21′N 49°34′E
al-Hurma *see* Al Khurmah
141 X7 **Al Ḥusayfin** *var.* Al Hasaifin. N Oman 24°33′N 56°33′E
138 G9 **Al Ḥuṣn** *var.* Husn. Irbid, N Jordan 32°29′N 35°53′E
139 U9 **'Alī** Wāsiţ, E Iraq 32°43′N 45°21′E
104 L10 **Alia** Extremadura, W Spain 39°25′N 05°12′W
143 P9 **'Alīābād** Yazd, C Iran 36°55′N 54°33′E
'Alīābad *see* Qā'emshahr
105 S7 **Aliaga** Aragón, NE Spain 40°40′N 00°42′W
136 B13 **Aliağa** İzmir, W Turkey 38°49′N 26°59′E
Aliákmon *see* Aliákmonas
115 F14 **Aliákmonas** *prev.* Aliákmon; *anc.* Haliacmon. ≈ N Greece
139 W9 **'Alī al Gharbī** Maysān, E Iraq 32°28′N 46°42′E
139 U11 **'Alī al Ḥassūnī** Al Qādisīyah, S Iraq 31°25′N 44°50′E
115 G18 **Alíartos** Stereá Elláda, C Greece 38°23′N 23°06′E
Äli-Bayramli *see* Şirvan
114 P12 **Alibey Barajī** ⊠ NW Turkey
77 S13 **Alibori** ≈ N Benin
112 M10 **Alibunar** Vojvodina, NE Serbia 45°06′N 20°59′E
105 S12 **Alicante** *Cat.* Alacant, *Lat.* Lucentum. Valenciana, SE Spain 38°21′N 00°29′W
105 S12 **Alicante** ◈ *province* Valenciana, SE Spain
105 S12 **Alicante** ✈ Murcia, E Spain 38°21′N 00°29′W
83 I25 **Alice** Eastern Cape, S South Africa 32°47′S 26°50′E
25 S14 **Alice** Texas, SW USA 27°45′N 98°06′W
83 I25 **Alicedale** Eastern Cape, S South Africa 33°19′S 26°05′E
65 B25 **Alice, Mount** *hill* West Falkland, Falkland Islands
107 P20 **Alice, Punta** *headland* S Italy 39°24′N 17°09′E
181 Q7 **Alice Springs** Northern Territory, C Australia 23°42′S 133°52′E
23 N4 **Aliceville** Alabama, S USA 33°07′N 88°09′W
147 U13 **Alichur** SE Tajikistan 37°49′N 73°45′E
147 U14 **Alichuri Janubí, Qatorkŭhi** *Rus.* Yuzhno-Alichurskiy Khrebet. ▲▲ SE Tajikistan
147 U13 **Alichuri Shimolí, Qatorkŭhi** *Rus.* Severo-Alichurskiy Khrebet. ▲▲ SE Tajikistan
107 K22 **Alicudi, Isola** *island* Isole Eolie, S Italy
Alifu Atoll *see* Ari Atoll
43 W14 **Aligandí** Kuna Yala, NE Panama 09°15′N 78°05′W
152 J11 **Alīgarh** Uttar Pradesh, N India 27°54′N 78°04′E
142 M7 **Alīgūdarz** Lorestān, W Iran 33°24′N 49°19′E
163 U5 **Alihe** *var.* Oroqen Zizhiqi. Nei Mongol Zizhiqu, N China 50°34′N 123°40′E
0 F12 **Alijos, Islas** *islets* California, SW USA
149 R6 **'Alī Kbel** *Pash.* 'Alī Khēl. Paktīkā, E Afghanistan 33°55′N 69°49′E
149 R6 **'Alī Khēl** *var.* Ali Khel, Jajı; *prev.* 'Alī Kheyl. Paktiyā, SE Afghanistan 33°55′N 69°46′E
'Alī Khēl *see* 'Alī Kbel, Paktīkā, Afghanistan
Ali Khel *see* 'Alī Kheyl, Paktīā, Afghanistan
'Alī Kheyl *see* 'Alī Khēl
141 V17 **Al Ikhwān** *island group* SE Yemen
Aliki *see* Alykí
79 H19 **Alima** ≈ C Congo
Al Imārāt al 'Arabīyahal Muttaḥidah *see* United Arab Emirates
115 N23 **Alimía** *island* Dodekánisa, Greece, Aegean Sea
55 V12 **Alimimuni Piek** ▲ S Suriname 02°26′N 55°46′W
79 K15 **Alindao** Basse-Kotto, S Central African Republic 04°58′N 21°16′E
95 J18 **Alingsås** Västra Götaland, S Sweden 57°55′N 12°30′E
81 K18 **Alinjugul** *spring/well* E Kenya 0°03′S 40°31′E
149 S11 **Alīpur** Punjab, E Pakistan 29°22′N 70°59′E
153 T12 **Alīpur Duār** West Bengal, NE India 26°28′N 89°25′E
18 B14 **Aliquippa** Pennsylvania, NE USA 40°36′N 80°15′W
80 L12 **'Ali Sabieh** *var.* 'Alī Sabīḥ. S Djibouti 11°07′N 42°44′E
'Alī Sabīḥ *see* 'Ali Sabieh
140 K3 **Al 'Īsāwīyah** Al Jawf, NW Saudi Arabia 30°41′N 37°58′E
104 J10 **Aliseda** Extremadura, W Spain 39°25′N 06°42′W
139 T8 **Al Iskandarīyah** Bābil, C Iraq 32°53′N 44°22′E
Al Iskandarīyah *see* Alexandria
123 T6 **Aliskerovo** Chukotskiy Avtonomnyy Okrug, NE Russian Federation 67°40′N 167°37′E
75 W7 **Al Ismā'īlīya** *var.* Ismailia, Ismâ'ilîya. N Egypt 30°32′N 32°13′E
114 H13 **Alistráti** Kentrikí Makedonía, NE Greece 41°03′N 23°58′E
39 P15 **Alitak Bay** *bay* Kodiak Island, Alaska, USA
Al Ittiḥād *see* Madīnat ash Sha'b
115 H18 **Alivéri** *var.* Alivérion. Évvoia, C Greece 38°24′N 24°02′E
Alivérion *see* Alivéri
Aliwal-Noord *see* Aliwal North
83 I24 **Aliwal North** *Afr.* Aliwal-Noord. Eastern Cape, SE South Africa 30°42′S 26°43′E
121 Q13 **Al Jabal al Akhḍar** ▲▲ NE Libya
138 H13 **Al Jafr** Ma'ān, S Jordan 30°18′N 36°13′E
75 T8 **Al Jaghbūb** NE Libya 29°45′N 24°31′E
142 K11 **Al Jahrā'** *var.* Al Jahrah, Jahra. C Kuwait 29°18′N 47°36′E
Al Jahrah *see* Al Jahrā'
Al Jamāhīrīyah al 'Arabīyah al Lībīyah ash Sha'bīyah al Ishtirākīy *see* Libya
140 K3 **Al Jarāwī** *spring/well* NW Saudi Arabia 30°12′N 38°48′E
141 X11 **Al Jawārah** *oasis* SE Oman
140 L3 **Al Jawf** *off.* Jauf. Al Jawf, NW Saudi Arabia 29°51′N 39°49′E
140 L4 **Al Jawf** *var.* Minṭaqat al Jawf. ◈ *province* N Saudi Arabia
Al Jawlān *see* Golan Heights
Al Jazair *see* Alger
139 N4 **Al Jazīrah** *physical region* Iraq/Syria
104 F14 **Aljezur** Faro, S Portugal 37°18′N 08°49′W
139 S13 **Al Jīl** An Najaf, S Iraq 30°28′N 43°57′E
138 G11 **Al Jīzah** *var.* Jiza. 'Ammān, N Jordan 31°42′N 35°57′E
Al Jīzah *see* Giza
Al Jubail *see* Al Jubayl
141 S6 **Al Jubayl** *var.* Al Jubail. Ash Sharqīyah, NE Saudi Arabia 27°N 49°36′E
141 T10 **Āl Juḥaysh, Qalamat** *well* SE Saudi Arabia
143 N15 **Al Jumaylīyah** N Qatar 25°37′N 51°05′E
Al Junaynah *see* El Geneina
104 G13 **Aljustrel** Beja, S Portugal 37°52′N 08°10′W
Al Kaba'ish *see* Al Chabā'ish
Al-Kadhimain *see* Al Kāẓimīyah
Al Kāf *see* El Kef
Alkal'a *see* Alcalá de Henares
35 W4 **Alkali Flat** *salt flat* Nevada, W USA
35 Q1 **Alkali Lake** ◎ Nevada, W USA
141 Z9 **Al Kāmil** NE Oman 22°14′N 58°15′E
138 G11 **Al Karak** *var.* El Kerak, Karak, Kerak; *anc.* Kir Moab, Kir of Moab. Al Karak, W Jordan 31°11′N 35°42′E
138 G12 **Al Karak** *off.* Muḥāfaẓat al Karak. ◈ *governorate* W Jordan
139 W8 **Al Karmashīyah** Wāsiţ, E Iraq 32°57′N 46°10′E
Al-Kashaniya *see* Al Qash'ānīyah
Al-Kasr al-Kebir *see* Ksar-el-Kebir
139 T8 **Al Kāẓimīyah** *var.* Al-Kadhimain, Kadhimain. Baghdād, C Iraq 33°22′N 44°20′E
99 J18 **Alken** Limburg, NE Belgium 50°52′N 05°19′E
141 X8 **Al Khābūrah** *var.* Khabura. N Oman 23°57′N 57°06′E
Al Khalīl *see* Hebron
139 T7 **Al Khāliş** Diyālá, C Iraq 33°51′N 44°33′E
Al Khaluf *see* Khalūf
75 W10 **Al Khārijah** *var.* El Khârga. C Egypt 25°31′N 30°36′E
141 Q8 **Al Kharj** Ar Riyāḍ, C Saudi Arabia 24°12′N 47°12′E
141 W6 **Al Khaşab** *var.* Khasab. N Oman 26°11′N 56°18′E
Al Khaur *see* Al Khawr
143 N15 **Al Khawr** *var.* Al Khaur, Al Khor. N Qatar 25°40′N 51°33′E
142 K12 **Al Khīrān** *var.* Al Khiran. SE Kuwait 28°34′N 48°21′E
141 W9 **Al Khīrān** *spring/well* NW Oman 22°31′N 55°42′E
Al Khiyām *see* El Khiyam
Al-Khobar *see* Al Khubar
Al Khor *see* Al Khawr
141 S6 **Al Khubar** *var.* Al-Khobar. Ash Sharqīyah, NE Saudi Arabia 26°00′N 50°00′E
120 M12 **Al Khums** *var.* Homs, Khoms, Khums. NW Libya 32°39′N 14°16′E
141 R15 **Al Khuraybah** C Yemen 15°05′N 48°17′E
140 M9 **Al Khurmah** *var.* al-Hurma. Makkah, W Saudi Arabia 21°59′N 42°00′E
141 V9 **Al Kidan** *desert* NE Saudi Arabia
127 V4 **Alkino-2** Respublika Bashkortostan, W Russian Federation 54°30′N 55°40′E
98 H9 **Alkmaar** Noord-Holland, NW Netherlands 52°37′N 04°45′E
139 T10 **Al Kūfah** *var.* Kufa. An Najaf, S Iraq 32°02′N 44°25′E
75 T11 **Al Kufrah** SE Libya 24°11′N 23°19′E
141 T10 **Al Kursū'** *desert* E Saudi Arabia
139 V9 **Al Kūt** *var.* Kūt al 'Amārah, Kut al Imara. Wāsiţ, E Iraq 32°30′N 45°51′E
Al-Kuwait *see* Al Kuwayt
Al Kuwayr *see* Guwēr
142 K11 **Al Kuwayt** *var.* Al-Kuwait, *Eng.* Kuwait, Kuwait City; *prev.* Qurein. ● (Kuwait) E Kuwait 29°23′N 48°00′E
142 K11 **Al Kuwayt** ✈ C Kuwait 29°13′N 47°57′E
115 G19 **Alkyonidón, Kólpos** *gulf* C Greece
141 N4 **Al Labbah** *physical region* N Saudi Arabia
138 G4 **Al Lādhiqīyah** *Eng.* Latakia, *Fr.* Lattaquié; *anc.* Laodicea, Laodicea ad Mare. Al Lādhiqīyah, W Syria 35°31′N 35°47′E
138 H4 **Al Lādhiqīyah** *off.* Muḥāfaẓat al Lādhiqīyah, *var.* Al Lathqiyah, Latakia, Lattakia. ◈ *governorate* W Syria
19 R2 **Allagash River** ≈ Maine, NE USA
152 M13 **Allahābād** Uttar Pradesh, N India 25°27′N 81°50′E
143 S3 **Allāh Dāgh, Reshteh-ye** ▲▲ NE Iran
39 Q8 **Allakaket** Alaska, USA 66°34′N 152°39′W
141 X12 **Al Lakbī** S Oman 18°27′N 56°37′E
11 T15 **Allan** Saskatchewan, S Canada 51°50′N 105°59′W
Allanmyo *see* Aunglan
83 I22 **Allanridge** Free State, C South Africa 27°45′S 26°40′E
104 H4 **Allariz** Galicia, NW Spain 42°11′N 07°48′W
139 R11 **Al Laşaf** *var.* Al Lussuf. An Najaf, S Iraq 31°38′N 43°16′E
Al Lathqiyah *see* Al Lādhiqīyah
23 S2 **Allatoona Lake** ⊠ Georgia, SE USA
83 J19 **Alldays** Limpopo, NE South Africa 22°39′S 29°04′E
Alle *see* Łyna
31 P10 **Allegan** Michigan, N USA 42°31′N 85°51′W
18 E14 **Allegheny Mountains** ▲▲ NE USA
18 E12 **Allegheny Plateau** ▲▲ New York/Pennsylvania, NE USA
18 D11 **Allegheny Reservoir** ⊠ New York/Pennsylvania, NE USA
18 E12 **Allegheny River** ≈ New York/Pennsylvania, NE USA
22 K9 **Allemands, Lac des** ◎ Louisiana, S USA
25 U6 **Allen** Texas, SW USA 33°06′N 96°40′W
21 R14 **Allendale** South Carolina, SE USA 33°01′N 81°19′W
41 N6 **Allende** Coahuila, NE Mexico 28°22′N 100°50′W
41 O9 **Allende** Nuevo León, NE Mexico 25°20′N 100°01′W
97 D16 **Allen, Lough** *Ir.* Loch Aillionn. ◎ NW Ireland
185 B26 **Allen, Mount** ▲ Stewart Island, Southland, SW New Zealand 47°05′S 167°49′E
109 V2 **Allensteig** Niederösterreich, N Austria 48°40′N 15°24′E
Allenstein *see* Olsztyn
18 I14 **Allentown** Pennsylvania, NE USA 40°37′N 75°30′W
155 G23 **Alleppey** *var.* Alappuzha. Kerala, SW India 09°30′N 76°22′E *see also* Alappuzha
100 J12 **Aller** ≈ NW Germany
29 V16 **Allerton** Iowa, C USA 40°42′N 93°22′W
99 K19 **Alleur** Liège, E Belgium 50°40′N 05°33′E
101 J25 **Allgäuer Alpen** ▲▲ Austria/Germany
28 J13 **Alliance** Nebraska, C USA 42°08′N 102°54′W
31 U12 **Alliance** Ohio, N USA 40°55′N 81°06′W
103 O10 **Allier** ◈ *department* N France
139 R13 **Al Līfīyah** *well* An Najaf, S Iraq
44 J13 **Alligator Pond** C Jamaica 17°52′N 77°34′W
21 Y9 **Alligator River** ≈ North Carolina, SE USA
29 W12 **Allison** Iowa, C USA 42°45′N 92°48′W
14 G14 **Alliston** Ontario, S Canada 44°09′N 79°51′W
140 L11 **Al Līth** Makkah, SW Saudi Arabia 21°N 41°E
Al Liwā' *see* Līwā
96 J12 **Alloa** C Scotland, United Kingdom 56°07′N 03°49′W
103 U14 **Allos** Alpes-de-Haute-Provence, SE France 44°16′N 06°37′E
108 D6 **Allschwil** Basel Landschaft, NW Switzerland 47°34′N 07°32′E
Al Lubnān *see* Lebanon
141 N14 **Al Luḥayyah** W Yemen 15°44′N 42°45′E
14 K12 **Allumettes, Île des** *island* Québec, SE Canada
Al Lussuf *see* Al Laşaf
109 S5 **Alm** ≈ N Austria
15 Q7 **Alma** Québec, SE Canada 48°32′N 71°41′W
27 S10 **Alma** Arkansas, C USA 35°28′N 94°13′W
23 V7 **Alma** Georgia, SE USA 31°32′N 82°27′W
27 P4 **Alma** Kansas, C USA 39°01′N 96°17′W
31 Q8 **Alma** Michigan, N USA 43°22′N 84°39′W
29 O17 **Alma** Nebraska, C USA 40°06′N 99°21′W
30 I7 **Alma** Wisconsin, N USA 44°21′N 91°54′W
139 R12 **Al Ma'ānīyah** *well* An Najaf, S Iraq
Alma-Ata *see* Almaty
Alma-Atinskaya Oblast' *see* Almaty
105 T5 **Almacelles** Cataluña, NE Spain 41°44′N 00°26′E
104 F11 **Almada** Setúbal, W Portugal 38°40′N 09°09′W
104 L11 **Almadén** Castilla-La Mancha, C Spain 38°47′N 04°50′W
66 L6 **Almadies, Pointe des** *headland* W Senegal 14°44′N 17°31′W
140 L7 **Al Madīnah** *Eng.* Medina. Al Madīnah, W Saudi Arabia 24°25′N 39°29′E
140 L7 **Al Madīnah** *off.* Minṭaqat al Madīnah. ◈ *province* W Saudi Arabia
138 H9 **Al Mafraq** *var.* Mafraq. Al Mafraq, N Jordan 32°20′N 36°12′E
138 J10 **Al Mafraq** *off.* Muḥāfaẓat al Mafraq. ◈ *governorate* NW Jordan
141 R15 **Al Maghārīm** C Yemen 15°00′N 47°49′E
105 N11 **Almagro** Castilla-La Mancha, C Spain 38°54′N 03°43′W
Al Maḥallah al Kubrá *see* El Maḥalla el Kubra
139 T9 **Al Maḥāwīl** *var.* Khān al Maḥāwīl. Bābil, C Iraq 32°39′N 44°28′E
Al Mahdīyah *see* Mahdia
139 T8 **Al Maḥmūdīyah** *var.* Mahmudiya. Baghdād, C Iraq 33°04′N 44°22′E
141 T14 **Al Mahrah** ▲▲ E Yemen
141 P7 **Al Majma'ah** Ar Riyāḍ, C Saudi Arabia 25°55′N 45°19′E
139 Q11 **Al Makmīn** *well* S Iraq
139 Q1 **Al Mālikīyah** *var.* Malkiye. Al Ḥasakah, N Syria 37°12′N 42°13′E
Almalyk *see* Olmaliq
Al Mamlakah *see* Morocco
Al Mamlaka al Urdunīya al Hashemīyah *see* Jordan
143 Q18 **Al Manādir** *var.* Al Manadir. *desert* Oman/United Arab Emirates
142 L15 **Al Manāmah** *Eng.* Manama. ● (Bahrain) N Bahrain 26°13′N 50°33′E
139 O5 **Al Manāşif** ▲▲ E Syria
35 O4 **Almanor, Lake** ⊠ California, W USA
105 R11 **Almansa** Castilla-La Mancha, C Spain 38°52′N 01°06′W
75 W7 **Al Manşūrah** *var.* Manşûra, El Manşûra. N Egypt 31°03′N 31°23′E
104 L3 **Almanza** Castilla y León, N Spain 42°40′N 05°01′W
104 L8 **Almanzor** ▲ W Spain 40°13′N 05°18′W
105 P14 **Almanzora** ≈ SE Spain
139 S9 **Al Mardah** Karbalā', C Iraq 32°35′N 43°30′E
Al-Mariyya *see* Almería
75 R7 **Al Marj** *var.* Barka, *It.* Barce. NE Libya 32°30′N 20°54′E
138 L2 **Al Mashrafah** Ar Raqqah, N Syria 36°25′N 39°07′E
141 X8 **Al Maşna'ah** *var.* Al Muşana'a. NE Oman 23°45′N 57°38′E
Almassora *see* Almazora
Almatinskaya Oblast' *see* Almaty
145 U15 **Almaty** *var.* Alma-Ata. Almaty, SE Kazakhstan 43°19′N 76°55′E
145 S14 **Almaty** *off.* Almatinskaya Oblast', *Kaz.* Almaty Oblysy; *prev.* Alma-Atinskaya Oblast'. ◈ *province* SE Kazakhstan
145 U15 **Almaty** ✈ Almaty, SE Kazakhstan 43°15′N 76°57′E
Almaty Oblysy *see* Almaty
al-Mawailih *see* Al Muwayliḥ
139 R3 **Al Mawşil** *Eng.* Mosul. Nīnawá, N Iraq 36°21′N 43°08′E
139 N5 **Al Mayādīn** *var.* Mayadin, *Fr.* Meyadine. Dayr az Zawr, E Syria 35°00′N 40°31′E
139 X10 **Al Maymūnah** *var.* Maimuna. Maysān, SE Iraq 31°43′N 46°55′E
141 N5 **Al Mayyāh** Ḥā'il, N Saudi Arabia 27°57′N 42°53′E
Al Ma'zam *see* Al Ma'zim
105 P6 **Almazán** Castilla y León, N Spain 41°29′N 02°31′W
141 W8 **Al Ma'zim** *var.* Al Ma'zam. NW Oman 23°22′N 56°16′E
123 N11 **Almaznyy** Respublika Sakha (Yakutiya), NE Russian Federation 62°19′N 114°14′E
105 T9 **Almazora** *Cat.* Almassora. Valenciana, E Spain 39°55′N 00°02′W
Al Mazra' *see* Al Mazra'ah
138 G11 **Al Mazra'ah** *var.* Al Mazra', Mazra'a. Al Karak, W Jordan 31°18′N 35°32′E
101 G15 **Alme** ≈ W Germany
104 I7 **Almeida** Guarda, N Portugal 40°43′N 06°53′W
104 G10 **Almeirim** Santarém, C Portugal 39°12′N 08°37′W
98 O10 **Almelo** Overijssel, E Netherlands 52°22′N 06°42′E
105 S9 **Almenara** Valenciana, E Spain 39°46′N 00°14′W
105 P12 **Almenaras** ▲ S Spain 38°31′N 02°27′W
105 P5 **Almenar de Soria** Castilla y León, N Spain 41°41′N 02°12′W
104 J6 **Almendra, Embalse de** ⊠ Castilla y León, NW Spain
104 J11 **Almendralejo** Extremadura, W Spain 38°41′N 06°25′W
98 J10 **Almere** *var.* Almere-stad. Flevoland, C Netherlands 52°22′N 05°12′E
98 J10 **Almere-Buiten** Flevoland, C Netherlands 52°24′N 05°15′E
98 J10 **Almere-Haven** Flevoland, C Netherlands 52°20′N 05°13′E
Almere-stad *see* Almere
105 P15 **Almería** *Ar.* Al-Mariyya; *anc.* Unci, *Lat.* Portus Magnus. Andalucía, S Spain 36°50′N 02°26′W
105 P14 **Almería** ◈ *province* Andalucía, S Spain
105 P15 **Almería, Golfo de** *gulf* S Spain
127 S5 **Al'met'yevsk** Respublika Tatarstan, W Russian Federation 54°53′N 52°20′E
95 L21 **Älmhult** Kronoberg, S Sweden 56°32′N 14°10′E
141 U9 **Al Miḥrāḍ** *desert* NE Saudi Arabia
Al Mīnā' *see* El Mina
104 L17 **Almina, Punta** *headland* Ceuta, Spain, N Africa 35°54′N 05°16′W
75 W9 **Al Minyā** *var.* El Minya, Minya. C Egypt 28°06′N 30°40′E
Al Miqdādīyah *see* Al Muqdādīyah
43 P14 **Almirante** Bocas del Toro, NW Panama 09°20′N 82°22′W
Almirós *see* Almyrós
140 M9 **Al Mislaḥ** *spring/well* W Saudi Arabia 22°46′N 40°47′E
Almissa *see* Omiš
104 G13 **Almodôvar** *var.* Almodóvar. Beja, S Portugal 37°31′N 08°03′W
104 M11 **Almodóvar del Campo** Castilla-La Mancha, C Spain 38°43′N 04°10′W
105 Q9 **Almodóvar del Pinar** Castilla-La Mancha, C Spain 39°44′N 01°55′W
31 S9 **Almont** Michigan, N USA 42°53′N 83°02′W
14 L13 **Almonte** Ontario, SE Canada 45°13′N 76°12′W

◆ Country
● Country Capital
◇ Dependent Territory
○ Dependent Territory Capital
◈ Administrative Regions
✈ International Airport
▲ Mountain
▲▲ Mountain Range
⋀ Volcano
≈ River
◎ Lake
⊠ Reservoir

ALMONTE – AMRAOTI

104 J14 Almonte Andalucía, S Spain 37°16′N 06°31′W
104 K9 Almonte ~ W Spain
152 K9 Almora Uttarakhand, N India 29°36′N 79°40′E
104 M8 Almorox Castilla-La Mancha, C Spain 40°13′N 04°22′W
141 S7 Al Mubarraz Ash Sharqīyah, E Saudi Arabia 25°28′N 49°34′E
Al Muḍaibī see Al Muḍaybī
138 G15 Al Mudawwarah Ma'ān, SW Jordan 29°20′N 36°E
141 Y9 Al Muḍaybī var. Al Muḍaibī. NE Oman 22°35′N 58°08′E
Almudébar see Almudévar
105 S5 Almudévar var. Almudébar. Aragón, NE Spain 42°03′N 00°34′W
141 S15 Al Mukallā var. Mukalla. SE Yemen 14°36′N 49°07′E
141 N16 Al Mukhā Eng. Mocha. SW Yemen 13°18′N 43°17′E
105 N15 Almuñécar Andalucía, S Spain 36°44′N 03°41′W
139 U7 Al Muqdādīyah var. Al Miqdādīyah. Diyālá, C Iraq 33°58′N 44°58′E
140 L3 Al Murayr spring/well NW Saudi Arabia 30°06′N 39°54′E
136 M12 Almus Tokat, N Turkey 40°22′N 36°54′E
Al Muṣana'a see Al Maṣna'ah
139 T9 Al Musayyib var. Musaiyib. Bābil, C Iraq 32°47′N 44°20′E
139 V13 Al Muthannā off. Muḥāfa at al Muthannā, var. As Samāwah. ⬖ governorate S Iraq
139 V9 Al Muwaffaqīyah Wāsiṭ, S Iraq 32°19′N 45°22′E
138 H10 Al Muwaqqar var. El Muwaqqar. 'Ammān, W Jordan 31°49′N 36°06′E
140 J5 Al Muwayliḥ var. al-Mawailih. Tabūk, NW Saudi Arabia 27°39′N 35°33′E
115 F17 Almyrós var. Almirós. Thessalía, C Greece 39°11′N 22°45′E
115 I24 Almyroú, Órmos bay Kríti, Greece, E Mediterranean Sea
Al Nūwfalīyah see An Nawfalīyah
96 L13 Alnwick N England, United Kingdom 55°27′N 01°44′W
Al Obayyid see El Obeid
Al Odaid see Al 'Udayd
190 B16 Alofi ○ (Niue) W Niue 19°01′S 169°55′E
190 A16 Alofi Bay bay W Niue, C Pacific Ocean
190 E13 Alofi, Île island S Wallis and Futuna
190 E13 Alofitai Île Alofi, W Wallis and Futuna 14°21′S 178°03′W
Aloha State see Hawai'i
118 G7 Aloja N Latvia 57°47′N 24°53′E
153 X10 Along Arunāchel Pradesh, NE India 28°15′N 94°56′E
115 H16 Alónnisos island Vóreies Sporádes, Greece, Aegean Sea
104 M15 Álora Andalucía, S Spain 36°50′N 04°43′W
171 Q16 Alor, Kepulauan island group E Indonesia
171 Q16 Alor, Pulau prev. Ombai. island Kepulauan Alor, E Indonesia
171 O16 Alor, Selat strait Flores Sea/Savu Sea
168 I7 Alor Setar var. Alor Star, Alur Setar. Kedah, Peninsular Malaysia 06°06′N 100°23′E
Alor Star see Alor Setar
Alost see Aalst
154 F9 Ālot Madhya Pradesh, C India 23°56′N 75°40′E
186 G10 Alotau Milne Bay, SE Papua New Guinea 10°20′S 150°23′E
171 Y16 Alotip Papua, E Indonesia 08°07′S 140°06′E
Al Oued see El Oued
35 R12 Alpaugh California, W USA 35°52′N 119°29′W
Alpen see Alps
31 R6 Alpena Michigan, N USA 45°04′N 83°27′W
Alpes see Alps
103 S14 Alpes-de-Haute-Provence ⬖ department SE France
103 U14 Alpes-Maritimes ⬖ department SE France
181 W8 Alpha Queensland, E Australia 23°40′S 146°38′E
197 R9 Alpha Cordillera var. Alpha Ridge. undersea feature Arctic Ocean 85°30′N 125°00′W
Alpha Ridge see Alpha Cordillera
Alpheius see Alfeiós
99 I15 Alphen Noord-Brabant, S Netherlands 51°29′N 04°57′E
Alphen see Alphen aan den Rijn
98 H11 Alphen aan den Rijn var. Alphen. Zuid-Holland, C Netherlands 52°08′N 04°40′E
Alpheus see Alfeiós
Alpi see Alps
104 G10 Alpiarça Santarém, C Portugal 39°15′N 08°35′W
24 K10 Alpine Texas, SW USA 30°22′N 103°40′W
108 F8 Alpnach Unterwalden, W Switzerland 46°56′N 08°17′E
108 D11 Alps Fr. Alpes, Ger. Alpen, It. Alpi. ▲▲ C Europe
141 W8 Al Qābil var. Qabil. N Oman 23°55′N 55°50′E
Al Qaḍārif see Gedaref
75 P8 Al Qaddāḥīyah N Libya 31°21′N 15°16′E
139 V10 Al Qādisīyah off. Muḥāfa at al Qādisīyah, var. Ad Dīwānīyah. ⬖ governorate S Iraq
Al Qāhirah see Cairo
140 K4 Al Qalībah Tabūk, NW Saudi Arabia 28°29′N 37°40′E
139 O1 Al Qāmishlī var. Kamishli, Qamishly. Al Ḥasakah, NE Syria 37°N 41°E
138 I6 Al Qaryatayn var. Qaryatayn, Fr. Qariateïne. Ḥimṣ, C Syria 34°13′N 37°13′E
142 K11 Al Qash'ānīyah var. Al-Kashaniya. NE Kuwait 29°59′N 47°42′E
141 N7 Al Qaṣīm var. Minṭaqat Qaṣīm, Qassim. ⬖ province C Saudi Arabia
75 V10 Al Qaṣr var. Al Qaṣr var. El Qaṣr. C Egypt 25°43′N 28°54′E
138 J5 Al Qaṣr Ḥimṣ, C Syria 35°06′N 37°39′E
Al Qaṣr see Al Qaṣr
Al Qaṣrayn see Kasserine
141 S6 Al Qaṭīf Ash Sharqīyah, NE Saudi Arabia 26°27′N 50°01′E
138 G11 Al Qaṭrānah var. El Qatrani, Qatrana. Al Karak, W Jordan 31°14′N 36°03′E
75 P11 Al Qaṭrūn SW Libya 24°57′N 14°40′E
Al Qayrawān see Kairouan
Al-Qsar al-Kbir see Ksar-el-Kebir
Al Qubayyāt see Qoubaïyât
Al Quds/Al Quds ash Sharīf see Jerusalem
104 H12 Alqueva, Barragem do ⊠ Portugal/Spain
138 G8 Al Qunayṭirah var. El Kuneitra, El Quneitra, Kuneitra, Qunaytra. Al Qunayṭirah, SW Syria 33°08′N 35°49′E
138 G8 Al Qunayṭirah off. Muḥāfaẓat al Qunayṭirah, var. El Q'unayṭirah, Qunaytirah, Fr. Kuneitra. ⬖ governorate SW Syria
140 M11 Al Qunfudhah Makkah, SW Saudi Arabia 19°19′N 41°03′E
140 K2 Al Qurayyāt Al Jawf, NW Saudi Arabia 31°25′N 37°26′E
139 Y11 Al Qurnah var. Kurna. Al Baṣrah, SE Iraq 31°01′N 47°27′E
75 Y10 Al Quṣayr var. Al Quṣayr var. Qusair, Quseir. E Egypt 26°05′N 34°16′E
139 V12 Al Quṣayr Al Muthanná, S Iraq 30°36′N 45°52′E
138 I6 Al Quṣayr var. El Quseir, Quṣayr, Fr. Kousseir. Ḥimṣ, W Syria 34°36′N 36°36′E
Al Quṣayr see Al Quṣayr
138 H7 Al Quṭayfah var. Quṭayfah, Qutayfe, Quteife, Fr. Kouteifé. Rīf Dimashq, W Syria 33°44′N 36°33′E
141 P8 Al Quwayīyah Ar Riyāḍ, C Saudi Arabia 24°06′N 45°18′E
Al Quwayr see Guwēr
138 F14 Al Quwayrah var. El Quweira. Al 'Aqabah, SW Jordan 29°47′E 35°18′E
Al Rayyan see Ar Rayyān
Al Ruweis see Ar Ruways
95 G24 Als Ger. Alsen. island SW Denmark
103 U5 Alsace Ger. Elsass; anc. Alsatia. ⬖ region NE France
11 R16 Alsask Saskatchewan, S Canada 51°24′N 109°55′W
Alsasua see Altsasu
Alsatia see Alsace
101 C16 Alsdorf Nordrhein-Westfalen, W Germany 50°52′N 06°09′E
10 G8 Alsek ~ Canada/USA
Alsen see Als
101 F19 Alsenz ~ W Germany
101 H17 Alsfeld Hessen, C Germany 50°45′N 09°14′E
119 K20 Al'shany Rus. Ol'shany. Brestskaya Voblasts', SW Belarus 52°05′N 27°21′E
Alsókubin see Dolný Kubín
118 C9 Alsunga W Latvia 56°59′N 21°31′E
Alt see Olt
92 K9 Alta Fin. Alattio. Finnmark, N Norway 69°58′N 23°17′E
29 T12 Alta Iowa, C USA 42°40′N 95°17′W
108 I7 Altach Vorarlberg, W Austria 47°22′N 09°39′E
92 K9 Altaelva Lapp. Álaheaieatnu. ~ N Norway
92 J8 Altafjorden fjord NE Norwegian Sea
62 K10 Alta Gracia Córdoba, C Argentina 31°42′S 64°25′W
42 K11 Alta Gracia Rivas, SW Nicaragua 11°35′N 85°38′W
54 H4 Altagracia Zulia, NW Venezuela 10°44′N 71°30′W
54 M5 Altagracia de Orituco Guárico, N Venezuela 09°54′N 66°24′W
Altai see Altai Mountains
129 T7 Altai Mountains var. Altai, Chin. Altay Shan, Rus. Altay. ▲▲ Asia/Europe
23 V6 Altamaha River ~ Georgia, SE USA
58 J13 Altamira Pará, NE Brazil 03°13′S 52°15′W
54 D12 Altamira Huila, S Colombia 02°04′N 75°47′W
42 M13 Altamira Alajuela, N Costa Rica 10°25′N 84°21′W
41 Q11 Altamira Tamaulipas, C Mexico 22°25′N 97°55′W
30 L15 Altamont Illinois, N USA 39°03′N 88°45′W
27 Q7 Altamont Kansas, C USA 37°11′N 95°18′W
32 H16 Altamont Oregon, NW USA 42°12′N 121°44′W
20 K10 Altamont Tennessee, S USA 35°28′N 85°42′W
23 X11 Altamonte Springs Florida, SE USA 28°39′N 81°22′W
107 O17 Altamura anc. Lupatia. Puglia, SE Italy 40°50′N 16°33′E
40 H9 Altamura, Isla island C Mexico
Altan see Erdenehayrhan
Altanbulag see Bayanhayrhan
163 Q7 Altan Emel var. Xin Barag Youqi. Nei Mongol Zizhiqu, N China 48°37′N 116°40′E
Altan-Ovoo see Tsenher
163 N9 Altanshiree var. Chamdmanī. Dornigovī, SE Mongolia 45°36′N 110°30′E
Altanteel see Dzereg
162 D5 Altantsögts var. Tsagaantüngi. Bayan-Ölgiy, NW Mongolia 49°06′N 90°26′E
40 F3 Altar Sonora, NW Mexico 30°41′N 111°53′W
40 D2 Altar, Desierto de var. Sonoran Desert. desert Mexico/USA see also Sonoran Desert
Altar, Desierto de see Sonoran Desert
105 Q8 Alta, Sierra ▲ N Spain 40°29′N 01°36′W
40 H9 Altata Sinaloa, C Mexico 24°40′N 107°54′W
42 D4 Alta Verapaz off. Departamento de Alta Verapaz. ⬖ department C Guatemala
Alta Verapaz, Departamento de see Alta Verapaz
107 L18 Altavilla Silentia Campania, S Italy 40°32′N 15°06′E
21 T7 Altavista Virginia, NE USA 37°06′N 79°17′W
158 L2 Altay Xinjiang Uygur Zizhiqu, NW China 47°51′N 88°06′E
162 D6 Altay var. Chihertey. Bayan-Ölgiy, W Mongolia 48°10′N 89°35′E
162 G8 Altay prev. Yösönbulag. Govĭ-Altay, W Mongolia 46°23′N 96°17′E
162 E8 Altay var. Bor-Üdzüür. Hovd, W Mongolia 45°46′N 92°13′E
Altay see Altai Mountains, Asia/Europe
Altay see Bayantes, Mongolia
122 J14 Altay, Respublika var. Gornyy Altay; prev. Gorno-Altayskaya Respublika. ⬖ autonomous republic S Russian Federation
Altay Shan see Altai Mountains
123 I13 Altayskiy Kray ⬖ territory S Russian Federation
Altbetsche see Bečej
101 L20 Altdorf Bayern, SE Germany 49°23′N 11°22′E
108 G8 Altdorf var. Altorf. Uri, C Switzerland 46°53′N 08°38′E
105 T11 Altea Valenciana, E Spain 38°37′N 00°03′W
100 L10 Alte Elde ~ N Germany
101 M16 Altenburg Thüringen, E Germany 50°59′N 12°27′E
Altenburg see București, Romania
Altenburg see Baia de Criș, Romania
100 P12 Alte Oder ~ NE Germany
104 H10 Alter do Chão Portalegre, C Portugal 39°12′N 07°40′W
92 I10 Altevatnet Lapp. Álttesjávri. ◎ N Norway
27 V12 Altheimer Arkansas, C USA 34°19′N 91°51′W
109 T9 Althofen Kärnten, S Austria 46°52′N 14°27′E
114 H7 Altimir Vratsa, NW Bulgaria 43°33′N 23°48′E
136 K11 Altınkaya Barajī ⊠ N Turkey
Altin Köprü see Āltūn Kūbrī
136 E13 Altıntaş Kütahya, W Turkey 39°05′N 30°07′E
57 K18 Altiplano physical region W South America
Altkanischa see Kanjiža
103 U7 Altkirch Haut-Rhin, NE France 47°37′N 07°14′E
Altlublau see Stará L'ubovňa
100 L12 Altmark cultural region N Germany
Altmoldowa see Moldova Veche
25 W8 Alto Texas, SW USA 31°39′N 95°04′W
104 H11 Alto Alentejo physical region S Portugal
59 I19 Alto Araguaia Mato Grosso, C Brazil 17°19′S 53°10′W
58 L12 Alto Bonito Pará, NE Brazil 01°48′S 46°18′W
83 O15 Alto Molócuè Zambézia, NE Mozambique 15°38′S 37°42′E
30 K15 Alton Illinois, N USA 38°53′N 90°10′W
27 W8 Alton Missouri, C USA 36°41′N 91°25′W
11 X17 Altona Manitoba, S Canada 49°06′N 94°35′W
18 E14 Altoona Pennsylvania, NE USA 40°32′N 78°23′W
30 J6 Altoona Wisconsin, N USA 44°49′N 91°22′W
62 N3 Alto Paraguay off. Departamento del Alto Paraguay. ⬖ department N Paraguay
Alto Paraguay, Departamento del see Alto Paraguay
59 L17 Alto Paraíso de Goiás Goiás, S Brazil 14°04′S 47°15′W
62 P6 Alto Paraná off. Departamento del Alto Paraná. ⬖ department E Paraguay
Alto Paraná see Paraná
Alto Paraná, Departamento del see Alto Paraná
59 L15 Alto Parnaíba Maranhão, E Brazil 09°08′S 45°56′W
56 H13 Alto Purús, Río ~ E Peru
Altorf see Altdorf
63 H19 Alto Río Senguer var. Alto Río Senguerr. Chubut, S Argentina 45°01′S 70°55′W
Alto Río Senguerr see Alto Río Senguer
41 Q13 Altotonga Veracruz-Llave, E Mexico 19°46′N 97°14′W
101 N23 Altötting Bayern, SE Germany 48°12′N 12°37′E
Altpasua see Stara Pazova
Altraga see Bayandzürh
105 P3 Altsasu Cast. Alsasua. Navarra, N Spain 42°54′N 02°10′W
Alt-Schwanenburg see Gulbene
Altsohl see Zvolen
108 I7 Altstätten Sankt Gallen, NE Switzerland 47°22′N 09°33′E
Álttesjávri see Altevatnet
42 G1 Altun Ha ruins Belize, N Belize
139 T4 Āltūn Kūbrî var. Altin Köprü. Kirkūk, N Iraq 35°28′N 44°26′E
158 D8 Altun Shan ▲ C China 39°19′N 93°37′E
158 L9 Altun Shan var. Altyn Tagh. ▲▲ NW China
35 P2 Alturas California, W USA 41°28′N 120°32′W
26 K12 Altus Oklahoma, C USA 34°39′N 99°21′W
26 K11 Altus Lake ⊠ Oklahoma, C USA
Altvater see Praděd
Altyn Tagh see Altun Shan
Alu see Shortland Island
al-'Ubaila see Al 'Ubaylah
139 O6 Al 'Ubaydī Al Anbār, W Iraq 34°22′N 41°15′E
141 T9 Al 'Ubaylah var. al-'Ubaila. Ash Sharqīyah, E Saudi Arabia 22°02′N 50°57′E
141 T9 Al 'Ubaylah spring/well E Saudi Arabia 22°02′N 50°56′E
Al Ubayyiḍ see El Obeid
141 T7 Al 'Udayd var. Al Odaid. Abū Ẓaby, W United Arab Emirates 24°34′N 51°27′E
118 J8 Alūksne Ger. Marienburg. NE Latvia 57°26′N 27°02′E
140 K6 Al 'Ulā Al Madīnah, NW Saudi Arabia 26°39′N 37°55′E
173 N4 Alula-Fartak Trench var. Illaue Fartak Trench. undersea feature W Indian Ocean 14°04′N 51°47′E
138 I11 Al 'Umarī 'Ammān, E Jordan 37°30′N 31°30′E
31 S13 Alum Creek Lake ⊠ Ohio, N USA
63 H15 Aluminé Neuquén, C Argentina 39°15′S 71°00′W
95 O14 Alunda Uppsala, C Sweden 60°04′N 18°04′E
117 T14 Alupka Avtonomna Respublika Krym, S Ukraine 44°24′N 34°01′E
75 P8 Al 'Uqaylah N Libya 30°13′N 19°12′E
Al Uqṣur see Luxor
Al Urdunn see Jordan
168 J9 Alur Panal bay Sumatera, W Indonesia
Alur Setar see Alor Setar
141 V10 Al 'Urūq al Mu'tariḍah salt lake SE Saudi Arabia
139 Q7 Ālūs Al Anbār, C Iraq 34°05′N 42°27′E
117 T13 Alushta Avtonomna Respublika Krym, S Ukraine 44°41′N 34°24′E
151 G22 Aluva var. Alwaye. Kerala, SW India 10°06′N 76°23′E see also Alwaye
75 N11 Al 'Uwaynāt var. Al Awaynāt. SW Libya 25°47′N 10°34′E
139 T6 Al 'Uẓaym var. Adhaim. Diyālá, E Iraq 34°12′N 44°31′E
26 L8 Alva Oklahoma, C USA 36°48′N 98°40′W
104 H8 Alva ~ N Portugal
95 J18 Älvängen Västra Götaland, S Sweden 57°56′N 12°09′E
14 F14 Alvanley Ontario, S Canada 44°33′N 81°05′W
41 S14 Alvarado Veracruz-Llave, E Mexico 18°47′N 95°45′W
25 T7 Alvarado Texas, SW USA 32°24′N 97°12′W
58 D13 Alvarães Amazonas, NW Brazil 03°13′S 64°53′W
40 G6 Álvaro Obregón, Presa ⊠ W Mexico
94 H10 Alvdal Hedmark, S Norway 62°07′N 10°39′E
94 K12 Älvdalen Dalarna, C Sweden 61°13′N 14°04′E
61 E15 Alvear Corrientes, NE Argentina 29°05′S 56°35′W
104 F10 Alverca do Ribatejo Lisboa, C Portugal 38°56′N 09°01′W
95 L20 Alvesta Kronoberg, S Sweden 56°52′N 14°34′E
25 W12 Alvin Texas, SW USA 29°25′N 95°14′W
94 O13 Älvkarleby Uppsala, C Sweden 60°34′N 17°30′E
25 S5 Alvord Texas, SW USA 33°22′N 97°39′W
93 G18 Älvros Jämtland, C Sweden 62°04′N 14°30′E
92 J13 Älvsbyn Norrbotten, N Sweden 65°41′N 21°00′E
142 K12 Al Wafrā' SE Kuwait 28°38′N 47°57′E
140 J6 Al Wajh Tabūk, NW Saudi Arabia 26°16′N 36°30′E
143 N16 Al Wakrah var. Wakra. C Qatar 25°09′N 51°36′E
138 M8 al Walaj, Sha'īb dry watercourse W Iraq
152 I11 Alwar Rājasthān, N India 27°32′N 76°35′E
141 Q5 Al Warī'ah Ash Sharqīyah, N Saudi Arabia 27°54′N 47°23′E
155 G22 Alwaye var. Aluva. Kerala, SW India 10°06′N 76°23′E see also Aluva
Alxa Zuoqi see Bayan Hot
Alx Youqi see Ehen Hudag
Al Yaman see Yemen
138 G9 Al Yarmūk Irbid, N Jordan 32°41′N 35°55′E
Alyat/Alyaty-Pristan' see Älät
115 I14 Alykí var. Aliki. Thásos, N Greece 40°36′N 24°45′E
119 F14 Alytus Pol. Olita. Alytus, S Lithuania 54°24′N 24°02′E
119 F15 Alytus ⬖ province S Lithuania
101 N23 Alz ~ SE Germany
33 Y11 Alzada Montana, NW USA 45°00′N 104°24′W
122 L12 Alzamay Irkutskaya Oblast', S Russian Federation 55°33′N 98°36′E
99 M25 Alzette ~ S Luxembourg
105 S10 Alzira var. Alcira; anc. Saetabicula, Suero. Valenciana, E Spain 39°10′N 00°27′W
Al Zubair see Az Zubayr
181 O8 Amadeus, Lake seasonal lake Northern Territory, C Australia
81 E15 Amadi Western Equatoria, SW South Sudan 05°32′N 30°20′E
9 R7 Amadjuak Lake ◎ Baffin Island, Nunavut, N Canada
95 J23 Amager island E Denmark
165 N14 Amagi-san ▲ Honshū, S Japan 34°51′N 138°57′E
171 S13 Amahai var. Masohi. Palau Seram, E Indonesia 03°19′S 128°56′E
38 M16 Amak Island island Alaska, USA
164 C15 Amakusa prev. Hondo. Kumamoto, Shimo-jima, SW Japan 32°28′N 130°12′E
164 B14 Amakusa-nada gulf SW Japan
95 J16 Åmål Västra Götaland, S Sweden 59°04′N 12°41′E
54 E8 Amalfi Antioquia, N Colombia 06°54′N 75°04′W
107 L18 Amalfi Campania, S Italy 40°37′N 14°35′E
115 D19 Amaliáda var. Amaliás. Dytikí Elláda, S Greece 37°48′N 21°21′E
Amaliás see Amaliáda
154 F12 Amalner Mahārāshtra, C India 21°03′N 75°04′E
171 W14 Amamapare Papua, E Indonesia 04°51′S 136°44′E
59 H21 Amambaí, Serra de var. Cordillera de Amambay, Serra de Amambay. ▲ Brazil/Paraguay see also Amambay, Cordillera de
62 P4 Amambay off. Departamento del Amambay. ⬖ department E Paraguay
62 P5 Amambay, Cordillera de var. Serra de Amambaí, Serra de Amambay. ▲ Brazil/Paraguay see also Amambaí, Serra de
Amambay, Departamento del see Amambay
Amambay, Serra de see Amambaí, Serra de/Amambay, Cordillera de
165 U16 Amami-guntō island group SW Japan
165 V15 Amami-Ō-shima island S Japan
186 A5 Amanab West Sepik, NW Papua New Guinea 03°38′S 141°16′E
Amānat al 'Āṣimah see Baghdād
106 J13 Amandola Marche, C Italy 42°58′N 13°22′E
107 N21 Amantea Calabria, SW Italy 39°06′N 16°05′E
191 W10 Amanu island Îles Tuamotu, C French Polynesia
58 J10 Amapá Amapá, NE Brazil 02°00′N 50°50′W
58 J11 Amapá off. Estado de Amapá; prev. Território de Amapá. ⬖ state NE Brazil
Amapá, Estado de see Amapá
42 H8 Amapala Valle, S Honduras 13°16′N 87°39′W
Amapá, Território de see Amapá
80 J12 Āmara var. Amhara. ⬖ N Ethiopia
Amara see Al 'Amārah
'Amārah, Al see Maysān
104 H6 Amarante Porto, N Portugal 41°16′N 08°05′W
166 M5 Amarapura Mandalay, C Myanmar (Burma) 21°54′N 96°01′E
Amardalay see Delgertsogt
104 I12 Amareleja Beja, S Portugal 38°12′N 07°13′W
35 V11 Amargosa Range ▲▲ California, W USA
25 N2 Amarillo Texas, SW USA 35°13′N 101°50′W
Amarinthos see Amárynthos
107 K15 Amaro, Monte ▲ C Italy 42°03′N 14°06′E
115 H18 Amárynthos var. Amarinthos. Évvoia, C Greece 38°24′N 23°53′E
Amasia see Amasya
136 K12 Amasya anc. Amasia. Amasya, N Turkey 40°37′N 35°50′E
136 K11 Amasya ⬖ province N Turkey
42 F4 Amatique, Bahía de bay Gulf of Honduras, W Caribbean Sea
42 D6 Amatitlán, Lago de ◎ S Guatemala
107 J14 Amatrice Lazio, C Italy 42°38′N 13°19′E
190 C8 Amatuku atoll C Tuvalu
99 J20 Amay Liège, E Belgium 50°33′N 05°19′E
48 F7 Amazon Sp. Amazonas. ~ Brazil/Peru
59 C14 Amazonas off. Estado do Amazonas. ⬖ state N Brazil
54 G15 Amazonas off. Comisaria del Amazonas. ⬖ province SE Colombia
56 C10 Amazonas off. Departamento de Amazonas. ⬖ department N Peru
54 M12 Amazonas off. Territorio Amazonas. ⬖ federal territory S Venezuela
Amazonas see Amazon
Amazonas, Comisaria del see Amazonas
Amazonas, Departamento de see Amazonas
Amazonas, Estado do see Amazonas
Amazonas, Territorio see Amazonas
48 F7 Amazon Basin basin N South America
47 V5 Amazon Fan undersea feature W Atlantic Ocean 05°00′N 47°30′W
58 K11 Amazon, Mouths of the delta NE Brazil
187 R13 Ambae var. Aoba, Omba. island C Vanuatu
152 I9 Ambāla Haryāna, NW India 30°19′N 76°49′E
155 J26 Ambalangoda Southern Province, SW Sri Lanka 06°14′N 80°03′E
155 K26 Ambalantota Southern Province, S Sri Lanka 06°07′N 81°01′E
172 I6 Ambalavao Fianarantsoa, C Madagascar 21°50′S 46°56′E
54 E10 Ambalema Tolima, C Colombia 04°49′N 74°48′W
79 E17 Ambam Sud, S Cameroon 02°23′N 11°17′E
172 J2 Ambanja Antsirañana, N Madagascar 13°40′S 48°27′E
123 T6 Ambarchik Respublika Sakha (Yakutiya), NE Russian Federation 69°33′N 162°08′E
62 K9 Ambargasta, Salinas de salt lake C Argentina
124 J6 Ambarnyy Respublika Kareliya, NW Russian Federation 65°53′N 33°44′E
56 C7 Ambato Tungurahua, C Ecuador 01°18′S 78°39′W
172 I6 Ambato Finandrahana Fianarantsoa, SE Madagascar
172 I5 Ambatolampy Antananarivo, C Madagascar 19°21′S 47°27′E
172 H4 Ambatomainty Mahajanga, W Madagascar 17°40′S 45°39′E
172 J4 Ambatondrazaka Toamasina, C Madagascar 17°49′S 48°28′E
101 L20 Amberg var. Amberg in der Oberpfalz. Bayern, SE Germany 49°26′N 11°52′E
Amberg in der Oberpfalz see Amberg
42 H1 Ambergris Cay island NE Belize
103 S11 Ambérieu-en-Bugey Ain, E France 45°57′N 05°21′E
185 I18 Amberley Canterbury, South Island, New Zealand 43°09′S 172°43′E
103 P11 Ambert Puy-de-Dôme, C France 45°33′N 03°45′E
Ambianum see Amiens
76 J11 Ambidédi Kayes, SW Mali 14°37′N 11°49′W
154 M10 Ambikāpur Chhattīsgarh, C India 23°09′N 83°12′E
172 J2 Ambilobe Antsirañana, N Madagascar 13°10′S 49°03′E
39 O7 Ambler Alaska, USA 67°05′N 157°51′W
Amblève see Amel
Ambo see Hāgere Hiywet
172 I8 Amboasary Toliara, S Madagascar 25°01′S 46°23′E
172 J4 Ambodifotatra var. Ambodifototra. Toamasina, E Madagascar 16°59′S 49°51′E
Amboenten see Ambunten
172 I5 Ambohidratrimo Antananarivo, C Madagascar 18°48′S 47°26′E
172 I6 Ambohimahasoa Fianarantsoa, SE Madagascar 21°07′S 47°13′E
172 K3 Ambohitralanana Antsirañana, NE Madagascar 15°13′S 50°28′E
Amboina see Ambon
102 M8 Amboise Indre-et-Loire, C France 47°25′N 01°00′E
171 S13 Ambon prev. Amboina, Amboyna. Pulau Ambon, E Indonesia 03°41′S 128°10′E
171 S13 Ambon, Pulau island E Indonesia
81 I20 Amboseli, Lake ◎ Kenya/Tanzania
172 I6 Ambositra Fianarantsoa, SE Madagascar 20°31′S 47°15′E
172 I8 Ambovombe Toliara, S Madagascar 25°10′S 46°06′E
35 W14 Amboy California, W USA 34°33′N 115°44′W
30 L11 Amboy Illinois, N USA 41°42′N 89°19′W
Amboyna see Ambon
Ambracia see Árta
Ambre, Cap d' see Bobaomby, Tanjona
18 B14 Ambridge Pennsylvania, NE USA 40°33′N 80°11′W
Ambrim see Ambrym
82 A11 Ambriz Bengo, NW Angola 07°55′S 13°11′E
Ambrizete see N'Zeto
187 R13 Ambrym var. Ambrim. island C Vanuatu
169 T16 Ambunten prev. Amboenten. Pulau Madura, E Indonesia 06°55′S 113°45′E
186 B6 Ambunti East Sepik, NW Papua New Guinea 04°12′S 142°49′E
155 I20 Āmbūr Tamil Nādu, SE India 12°48′N 78°44′E
38 E17 Amchitka Island island Aleutian Islands, Alaska, USA
38 F17 Amchitka Pass strait Aleutian Islands, Alaska, USA
141 R15 'Amd C Yemen 15°10′N 47°58′E
78 J10 Am Dam Sila, E Chad 12°46′N 20°29′E
171 U16 Amdassa Pulau Yamdena, E Indonesia 07°40′S 131°24′E
125 U1 Amderma Nenetskiy Avtonomnyy Okrug, NW Russian Federation 69°45′N 61°36′E
78 L9 Am Djarass Ennedi-Est, NE Chad 17°14′N 21°32′E
159 N14 Amdo Xizang Zizhiqu, W China 32°15′N 91°43′E
40 K13 Ameca Jalisco, SW Mexico 20°34′N 104°03′W
41 P14 Amecameca var. Amecameca de Juárez. México, C Mexico 19°08′N 98°48′W
Amecameca de Juárez see Amecameca
139 R1 Amêdî Ar. Al 'Amādiyah. Dahūk, N Iraq 37°09′N 43°27′E
61 A20 Ameghino Buenos Aires, E Argentina 34°51′S 62°28′W
99 M21 Amel Fr. Amblève. Liège, E Belgium 50°20′N 06°13′E
98 K4 Ameland Fris. It Amelân. island Waddeneilanden, N Netherlands
Amelân, It see Ameland
107 H14 Amelia Umbria, C Italy 42°33′N 12°26′E
21 V6 Amelia Court House Virginia, NE USA 37°20′N 77°59′W
23 W8 Amelia Island island Florida, SE USA
18 L12 Amenia New York, NE USA 41°51′N 73°31′W
America see United States of America
65 M21 America-Antarctica Ridge undersea feature S Atlantic Ocean
America in Miniature see Maryland
60 L9 Americana São Paulo, S Brazil 22°44′S 47°19′W
33 Q15 American Falls Idaho, NW USA 42°47′N 112°51′W
33 Q15 American Falls Reservoir ⊠ Idaho, NW USA
36 L3 American Fork Utah, W USA 40°24′N 111°47′W
192 K16 American Samoa ◇ US unincorporated territory W Polynesia
23 S6 Americus Georgia, SE USA 32°04′N 84°13′W
98 K12 Amerongen Utrecht, C Netherlands 52°00′N 05°30′E
98 K11 Amersfoort Utrecht, C Netherlands 52°09′N 05°23′E
97 N21 Amersham SE England, United Kingdom 51°40′N 00°37′W
30 I5 Amery Wisconsin, N USA 45°18′N 92°20′W
195 W6 Amery Ice Shelf ice shelf Antarctica
29 V13 Ames Iowa, C USA 42°01′N 93°37′W
19 P10 Amesbury Massachusetts, NE USA 42°51′N 70°55′W
Amestratus see Mistretta
115 F18 Amfíkleia var. Amfíklia. Stereá Elláda, C Greece 38°38′N 22°35′E
Amfíklia see Amfíkleia
115 D17 Amfilochía var. Amfilokhía. Dytikí Elláda, C Greece 38°52′N 21°09′E
Amfilokhía see Amfilochía
114 H13 Amfípoli anc. Amphipolis. site of ancient city Kentrikí Makedonía, NE Greece
115 F18 Ámfissa Stereá Elláda, C Greece 38°32′N 22°22′E
123 Q10 Amga Respublika Sakha (Yakutiya), NE Russian Federation 60°55′N 131°45′E
123 Q11 Amga ~ NE Russian Federation
163 R7 Amgalang var. Xin Barag Zuoqi. Nei Mongol Zizhiqu, N China 48°12′N 118°15′E
123 V5 Amguema ~ NE Russian Federation
123 S12 Amgun' ~ SE Russian Federation
Amhara see Āmara
13 P15 Amherst Nova Scotia, SE Canada 45°50′N 64°14′W
18 M11 Amherst Massachusetts, NE USA 42°22′N 72°31′W
18 D10 Amherst New York, NE USA 42°57′N 78°47′W
24 M4 Amherst Texas, SW USA 33°59′N 102°24′W
21 U6 Amherst Virginia, NE USA 37°35′N 79°04′W
Amherst see Kyaikkami
14 C18 Amherstburg Ontario, S Canada 42°05′N 83°06′W
21 Q6 Amherstdale West Virginia, NE USA 37°46′N 81°46′W
14 K15 Amherst Island island Ontario, SE Canada
Amida see Diyarbakır
28 J6 Amidon North Dakota, N USA 46°29′N 103°19′W
103 O3 Amiens anc. Ambianum, Samarobriva. Somme, N France 49°54′N 02°18′E
139 P8 'Āmij, Wādī var. Wadi 'Amīq. dry watercourse W Iraq
136 L17 Amik Ovasī ◎ S Turkey
76 E9 Amilcar Cabral ✈ Sal, NE Cape Verde
Amilḥayt, Wādī see Umm al Ḥayt, Wādī
Amíndaion/Amindeo see Amýntaio
155 C21 Amīndīvi Islands island group Lakshadweep, India, N Indian Ocean
139 U6 Amīn Ḥabīb Diyālá, E Iraq 34°17′N 45°10′E
83 E20 Aminuis Omaheke, E Namibia 23°43′S 19°21′E
'Amīq, Wadi see 'Āmij, Wādī
142 J7 Amīrābād Īlām, NW Iran 33°20′N 46°16′E
Amirante Bank see Amirante Ridge
173 N6 Amirante Basin undersea feature W Indian Ocean 07°00′S 54°00′E
173 N6 Amirante Islands var. Amirantes Group. island group C Seychelles
173 N7 Amirante Ridge var. Amirante Bank. undersea feature W Indian Ocean 06°00′S 53°10′E
Amirantes Group see Amirante Islands
173 N7 Amirante Trench undersea feature W Indian Ocean 08°00′S 52°30′E
11 U13 Amisk Lake ◎ Saskatchewan, C Canada
Amistad, Presa de la see Amistad Reservoir
25 O12 Amistad Reservoir var. Presa de la Amistad. ⊠ Mexico/USA
Amisus see Samsun
22 K8 Amite var. Amite City. Louisiana, S USA 30°40′N 90°30′W
Amite City see Amite
27 T12 Amity Arkansas, C USA 34°15′N 93°27′W
154 H11 Amla prev. Amulla. Madhya Pradesh, C India 21°53′N 78°10′E
38 I17 Amlia Island island Aleutian Islands, Alaska, USA
97 I18 Amlwch NW Wales, United Kingdom 53°25′N 04°23′W
Ammaia see Portalegre
138 H10 'Amman var. Amman; anc. Philadelphia, Bibl. Rabbah Ammon, Rabbath Ammon. ● (Jordan) 'Ammān, NW Jordan 31°57′N 35°56′E
138 H10 'Ammān off. Muḥāfazat 'Ammān; prev. Al 'Aṣimah. ⬖ governorate NW Jordan
'Ammān, Muḥāfazat see 'Ammān
93 N14 Ämmänsaari Kainuu, E Finland 64°51′N 28°58′E
92 H13 Ammarnäs Västerbotten, N Sweden 65°58′N 16°10′E
197 O15 Ammassalik var. Angmagssalik. Sermersooq, S Greenland 65°51′N 37°30′W
101 K24 Ammer ~ SE Germany
101 K24 Ammersee ◎ SE Germany
98 J13 Ammerzoden Gelderland, C Netherlands 51°46′N 05°07′E
Ammóchostos see Gazimağusa
Ammóchostos, Kólpos see Gazimağusa Körfezi
Amnok-kang see Yalu
Amoea see Portalegre
Amoentai see Amuntai
Amoerang see Amurang
143 O4 Āmol var. Amul. Māzandarān, N Iran 36°31′N 52°24′E
115 K21 Amorgós Amorgós, Kykládes, Greece, Aegean Sea 36°49′N 25°54′E
115 K22 Amorgós island Kykládes, Greece, Aegean Sea
23 N3 Amory Mississippi, S USA 33°58′N 88°29′W
12 I13 Amos Québec, SE Canada 48°34′N 78°08′W
95 G15 Åmot Buskerud, S Norway 59°52′N 09°55′E
95 E15 Åmot Telemark, S Norway 59°34′N 07°59′E
95 J15 Åmotfors Värmland, C Sweden 59°46′N 12°24′E
76 L10 Amourj Hodh ech Chargui, SE Mauritania 16°04′N 07°12′W
Amoy see Xiamen
172 H7 Ampanihy Toliara, SW Madagascar 24°40′S 44°45′E
155 L25 Ampara var. Amparai. Eastern Province, E Sri Lanka 07°17′N 81°41′E
172 J4 Amparafaravola Toamasina, E Madagascar 17°33′S 48°13′E
Amparai see Ampara
60 M9 Amparo São Paulo, S Brazil 22°40′S 46°49′W
172 J5 Ampasimanolotra Toamasina, E Madagascar 18°49′S 49°04′E
57 H17 Ampato, Nevado ▲ S Peru 15°52′S 71°51′W
101 L23 Amper ~ SE Germany
64 M9 Ampère Seamount undersea feature E Atlantic Ocean 35°05′N 13°00′W
Amphipolis see Amfípoli
167 X10 Amphitrite Group Chin. Xuande Qundao, Viet. Nhom An Vinh. island group N Paracel Islands
171 T16 Amplawas var. Emplawas. Pulau Babar, E Indonesia 08°01′S 129°42′E
105 U7 Amposta Cataluña, NE Spain 40°43′N 00°34′E
15 V7 Amqui Québec, SE Canada 48°28′N 67°27′W
141 O14 'Amrān W Yemen 15°39′N 43°59′E
Amraoti see Amrāvati

◆ Country ● Country Capital ◇ Dependent Territory ○ Dependent Territory Capital ⬖ Administrative Regions ✈ International Airport ▲ Mountain ▲▲ Mountain Range Volcano ~ River ◎ Lake ⊠ Reservoir

AMRĀVATI – AN TIÚR

154 H12 **Amrāvati** prev. Amraoti. Mahārāshtra, C India 20°56′N 77°45′E
154 C11 **Amreli** Gujarāt, W India 21°36′N 71°20′E
108 H6 **Amriswil** Thurgau, NE Switzerland 47°33′N 09°18′E
138 H5 **'Amrīt** ruins Tarṭūs, W Syria
152 H7 **Amritsar** Punjab, N India 31°38′N 74°55′E
152 J10 **Amroha** Uttar Pradesh, N India 28°54′N 78°29′E
100 G7 **Amrum** island NW Germany
93 I15 **Åmsele** Västerbotten, N Sweden 64°31′N 19°24′E
98 I10 **Amstelveen** Noord-Holland, C Netherlands 52°18′N 04°50′E
98 I10 **Amsterdam** ● (Netherlands) Noord-Holland, C Netherlands 52°22′N 04°54′E
18 K10 **Amsterdam** New York, NE USA 42°56′N 74°11′W
173 Q11 **Amsterdam Fracture Zone** tectonic feature S Indian Ocean
173 R11 **Amsterdam Island** island NE French Southern and Antarctic Territories
109 U4 **Amstetten** Niederösterreich, N Austria 48°08′N 14°52′E
78 J11 **Am Timan** Salamat, SE Chad 11°02′N 20°17′E
146 L12 **Amu-Buxoro Kanali** var. Aral-Bukhorskiy Kanal. canal C Uzbekistan
139 O1 **'Āmūdah** var. Amude. Al Ḥasakah, N Syria 37°06′N 40°56′E
147 O15 **Amu Darya** Rus. Amudar'ya, Taj. Dar"yoi Amu, Turkm. Amyderya, Uzb. Amudaryo; anc. Oxus. ~ C Asia
Amu-Dar'ya see Amyderýa
Amudar'ya/Amudaryo/ Amu, Dar"yoi see Amu Darya
Amude see 'Āmūdah
140 L3 **'Amūd, Jabal al** ▲ NW Saudi Arabia 30°59′N 39°17′E
38 J17 **Amukta Island** island Aleutian Islands, Alaska, USA
38 I17 **Amukta Pass** strait Aleutian Islands, Alaska, USA
Amul see Āmol
Amulla see Amla
197 S10 **Amundsen Basin** var. Fram Basin. undersea basin Arctic Ocean
195 X3 **Amundsen Bay** bay Antarctica
195 P10 **Amundsen Coast** physical region Antarctica
193 O14 **Amundsen Plain** undersea feature S Pacific Ocean
195 Q9 **Amundsen-Scott** US research station Antarctica 89°59′S 10°00′E
194 J11 **Amundsen Sea** sea S Pacific Ocean
94 M12 **Amungen** ◎ C Sweden
169 U13 **Amuntai** prev. Amoentai. Borneo, C Indonesia 02°24′S 115°14′E
129 W6 **Amur** Chin. Heilong Jiang. ~ China/Russian Federation
171 Q11 **Amurang** prev. Amoerang. Sulawesi, C Indonesia 01°12′N 124°37′E
105 O3 **Amurrio** País Vasco, N Spain 43°03′N 03°00′W
123 S13 **Amursk** Khabarovskiy Kray, SE Russian Federation 50°13′N 136°34′E
123 Q12 **Amurskaya Oblast'** ⬖ province SE Russian Federation
80 G7 **'Amur, Wadi** ~ NE Sudan
115 C17 **Amvrakikós Kólpos** gulf W Greece
Amvrosiyevka see Amvrosiyivka
117 X8 **Amvrosiyivka** Rus. Amvrosiyevka. Donets'ka Oblast', SE Ukraine 47°46′N 38°30′E
146 M14 **Amyderýa** Rus. Amu-Dar'ya. Lebap Welaýaty, NE Turkmenistan 37°58′N 65°14′E
Amyderya see Amu Darya
114 E13 **Amýntaio** var. Amindeo; prev. Amíndaion. Dytikí Makedonía, N Greece 40°42′N 21°42′E
14 B6 **Amyot** Ontario, S Canada 48°28′N 84°58′W
191 U10 **Anaa** atoll Îles Tuamotu, C French Polynesia
Anabanoea see Anabanua
171 N14 **Anabanua** prev. Anabanoea. Sulawesi, C Indonesia 03°58′S 120°07′E
189 R8 **Anabar** NE Nauru 0°30′S 166°56′E
123 N8 **Anabar** ~ NE Russian Federation
An Abhainn Mhór see Blackwater
55 O6 **Anaco** Anzoátegui, NE Venezuela 09°30′N 64°28′W
33 Q10 **Anaconda** Montana, NW USA 46°09′N 112°56′W
32 H7 **Anacortes** Washington, NW USA 48°30′N 122°36′W
26 M11 **Anadarko** Oklahoma, C USA 35°04′N 98°16′W
114 N12 **Ana Dere** ~ NW Turkey
104 G8 **Anadia** Aveiro, N Portugal 40°26′N 08°27′W
Anadolu Dağları see Doğu Karadeniz Dağları
123 V6 **Anadyr'** Chukotskiy Avtonomnyy Okrug, NE Russian Federation 64°41′N 177°22′E
123 V6 **Anadyr'** ~ NE Russian Federation
Anadyr, Gulf of see Anadyrskiy Zaliv
129 X4 **Anadyrskiy Khrebet** var. Chukot Range. ▲▲ NE Russian Federation
123 W6 **Anadyrskiy Zaliv** Eng. Gulf of Anadyr. gulf NE Russian Federation
115 K22 **Anáfi** anc. Anaphe. island Kykládes, Greece, Aegean Sea
107 J15 **Anagni** Lazio, C Italy 41°43′N 13°12′E
'Ānah see 'Annah
35 T15 **Anaheim** California, W USA 33°50′N 117°54′W
10 L15 **Anahim Lake** British Columbia, SW Canada 52°26′N 125°13′W
38 B8 **Anahola** Kaua'i, Hawai'i, USA, C Pacific Ocean 22°09′N 159°19′W
41 O7 **Anáhuac** Nuevo León, NE Mexico 27°13′N 100°09′W
25 X11 **Anahuac** Texas, SW USA 29°44′N 94°41′W
155 G22 **Anai Mudi** ▲ S India 10°16′N 77°08′E
Anaiza see 'Unayzah
155 M15 **Anakāpalle** Andhra Pradesh, E India 17°42′N 83°06′E
191 W15 **Anakena, Playa de** beach Easter Island, Chile, E Pacific Ocean
39 Q7 **Anaktuvuk Pass** Alaska, USA 68°08′N 151°44′W
39 Q6 **Anaktuvuk River** ~ Alaska, USA
172 J3 **Analalava** Mahajanga, NW Madagascar 14°38′S 47°46′E
44 F6 **Ana Maria, Golfo de** gulf N Caribbean Sea
Anambas Islands see Anambas, Kepulauan
169 N8 **Anambas, Kepulauan** var. Anambas Islands. island group W Indonesia
77 U17 **Anambra** ⬖ state SE Nigeria
29 N4 **Anamoose** North Dakota, N USA 47°50′N 100°14′W
29 Y13 **Anamosa** Iowa, C USA 42°06′N 91°17′W
136 H17 **Anamur** İçel, S Turkey 36°06′N 32°49′E
136 H17 **Anamur Burnu** headland S Turkey 36°03′N 32°49′E
154 O12 **Ānandapur** var. Anandpur. Odisha, E India 21°14′N 86°10′E
Anandpur see Ānandapur
155 H18 **Anantapur** Andhra Pradesh, S India 14°41′N 77°36′E
152 H5 **Anantnāg** var. Islamabad. Jammu and Kashmir, NW India 33°44′N 75°11′E
Ananyev see Anan'yiv
117 O9 **Anan'yiv** Rus. Ananyev. Odes'ka Oblast', SW Ukraine 47°43′N 29°51′E
126 J14 **Anapa** Krasnodarskiy Kray, SW Russian Federation 44°55′N 37°20′E
Anaphe see Anáfi
59 K18 **Anápolis** Goiás, C Brazil 16°19′S 48°58′W
143 R10 **Anār** Kermān, C Iran 30°49′N 55°18′E
Anár see Inari
143 P7 **Anārak** Eşfahān, C Iran 33°21′N 53°43′E
Anar Dara see Anār Darreh
148 J7 **Anār Darreh** var. Anar Dara. Farāh, W Afghanistan 32°45′N 61°38′E
Anárjohka see Inarijoki
23 X9 **Anastasia Island** island Florida, SE USA
188 K7 **Anatahan** island C Northern Mariana Islands
128 M6 **Anatolia** plateau C Turkey
86 F14 **Anatolian Plate** tectonic feature Asia/Europe
114 H13 **Anatolikí Makedonía kai Thráki** Eng. Macedonia East and Thrace. ⬖ region NE Greece
Anatom see Aneityum
62 L8 **Añatuya** Santiago del Estero, N Argentina 28°28′S 62°52′W
An Baile Meánach see Ballymena
Anbār, Muḥāfa at al see Al Anbār
An Bheanú see Barrow
An Bhóinn see Boyne
An Blascaod Mór see Great Blasket Island
An Cabhán see Cavan
An Caisleán Nua see Newcastle
An Caisleán Riabhach see Castlerea, Ireland
An Caisleán Riabhach see Castlereagh
56 C13 **Ancash** off. Departamento de Ancash. ⬖ department W Peru
Ancash, Departamento de see Ancash
An Cathair see Caher
102 J8 **Ancenis** Loire-Atlantique, NW France 47°23′N 01°10′W
An Chanáil Ríoga see Royal Canal
An Cheacha see Caha Mountains
39 R11 **Anchorage** Alaska, USA 61°13′N 149°52′W
39 R12 **Anchorage** ✈ Alaska, USA 61°08′N 150°00′W
39 Q13 **Anchor Point** Alaska, USA 59°46′N 151°49′W
An Chorr Chríochach see Cookstown
65 M24 **Anchorstock Point** headland W Tristan da Cunha 37°07′S 12°21′W
An Clár see Clare
An Clochán see Clifden
An Clochán Liath see Dunglow
23 U12 **Anclote Keys** island group Florida, SE USA
An Cóbh see Cobh
57 J17 **Ancohuma, Nevado de** ▲ W Bolivia 15°51′S 68°33′W
An Comar see Comber
57 D14 **Ancón** Lima, W Peru 11°45′S 77°08′W
106 J12 **Ancona** Marche, C Italy 43°38′N 13°30′E
Ancuabe see Ancuabi
82 Q13 **Ancuabi** var. Ancuabe. Cabo Delgado, NE Mozambique 13°00′S 39°50′E
63 F17 **Ancud** prev. San Carlos de Ancud. Los Lagos, S Chile 41°53′S 73°50′W
63 G17 **Ancud, Golfo de** gulf S Chile
Ancyra see Ankara
163 V8 **Anda** Heilongjiang, NE China 46°25′N 125°20′E
57 G16 **Andahuaylas** Apurímac, S Peru 13°39′S 73°24′W
An Daingean see Dingle
153 R15 **Āndāl** West Bengal, NE India 23°35′N 87°14′E
94 E9 **Åndalsnes** Møre og Romsdal, S Norway 62°33′N 07°42′E
104 K13 **Andalucía** Eng. Andalusia. ⬖ autonomous community S Spain
23 P7 **Andalusia** Alabama, S USA 31°18′N 86°29′W
Andalusia see Andalucía
151 Q21 **Andaman and Nicobar Islands** var. Andamans and Nicobars. ⬖ union territory India, NE Indian Ocean
173 T4 **Andaman Basin** undersea feature NE Indian Ocean 10°00′N 94°00′E
151 P19 **Andaman Islands** island group India, NE Indian Ocean
Andamans and Nicobars see Andaman and Nicobar Islands
173 T4 **Andaman Sea** sea NE Indian Ocean
57 K19 **Andamarca** Oruro, C Bolivia 18°46′S 67°31′W
182 H5 **Andamooka** South Australia 30°26′S 137°12′E
141 Y9 **'Andām, Wādī** seasonal river NE Oman
172 J3 **Andapa** Antsirañana, NE Madagascar 14°39′S 49°40′E
149 R4 **Andarāb** var. Banow. Baghlān, NE Afghanistan 35°36′N 69°18′E
Andarbag see Andarbogh
147 S13 **Andarbogh** Rus. Andarbag, Anderbak. S Tajikistan 37°51′N 71°45′E
109 Z5 **Andau** Burgenland, E Austria 47°47′N 17°02′E
108 I10 **Andeer** Graubünden, S Switzerland 46°36′N 09°24′E
92 H9 **Andenes** Nordland, C Norway 69°18′N 16°10′E
99 J20 **Andenne** Namur, SE Belgium 50°29′N 05°06′E
77 S11 **Andéramboukane** Gao, E Mali 15°24′N 03°03′E
Anderbak see Andarbogh
99 G18 **Anderlecht** Brussels, C Belgium 50°50′N 04°18′E
99 G21 **Anderlues** Hainaut, S Belgium 50°24′N 04°16′E
108 G9 **Andermatt** Uri, C Switzerland 46°39′N 08°36′E
101 E17 **Andernach** anc. Antunnacum. Rheinland-Pfalz, SW Germany 50°26′N 07°24′E
188 D15 **Andersen Air Force Base** air base NE Guam 13°34′N 144°55′E
39 R9 **Anderson** Alaska, USA 64°20′N 149°11′W
35 N4 **Anderson** California, W USA 40°26′N 122°21′W
31 P13 **Anderson** Indiana, N USA 40°06′N 85°40′W
27 R8 **Anderson** Missouri, C USA 36°39′N 94°26′W
21 P11 **Anderson** South Carolina, SE USA 34°30′N 82°39′W
25 V10 **Anderson** Texas, SW USA 30°29′N 96°00′W
95 K20 **Anderstorp** Jönköping, S Sweden 57°17′N 13°38′E
54 D9 **Andes** Antioquia, W Colombia 05°40′N 75°56′W
47 P7 **Andes** ▲▲ W South America
29 P12 **Andes, Lake** ◎ South Dakota, N USA
92 H9 **Andfjorden** fjord E Norwegian Sea
155 H16 **Andhra Pradesh** ⬖ state E India
98 J8 **Andijk** Noord-Holland, NW Netherlands 52°38′N 05°00′E
147 S10 **Andijon** Rus. Andizhan. Andijon Viloyati, E Uzbekistan 40°46′N 72°19′E
147 S10 **Andijon Viloyati** Rus. Andizhanskaya Oblast'. ⬖ province E Uzbekistan
Andikíthira see Antikýthira
172 J4 **Andilamena** Toamasina, C Madagascar 17°00′S 48°35′E
142 L8 **Andīmeshk** var. Andimishk; prev. Salehābād. Khūzestān, SW Iran 32°30′N 48°26′E
Andimishk see Andīmeshk
Andipáros see Antíparos
Andipaxi see Antípaxoi
Andípsara see Antípsara
136 L16 **Andırın** Kahramanmaraş, S Turkey 37°33′N 36°18′E
158 J8 **Andirlangar** Xinjiang Uygur Zizhiqu, NW China 37°38′N 83°40′E
Andírrion see Antírrio
Ándissa see Ántissa
Andizhan see Andijon
Andizhanskaya Oblast' see Andijon Viloyati
149 N2 **Andkhvōy** prev. Andkhvoy. Fāryāb, N Afghanistan 36°56′N 65°08′E
105 Q2 **Andoain** País Vasco, N Spain 43°13′N 02°02′W
163 Y15 **Andong** Jap. Antō. E South Korea 36°34′N 128°44′E
109 R4 **Andorf** Oberösterreich, N Austria 48°22′N 13°33′E
105 S7 **Andorra** Aragón, NE Spain 40°59′N 00°27′W
105 V4 **Andorra** off. Principality of Andorra, Cat. Valls d'Andorra, Fr. Vallée d'Andorre. ◆ monarchy SW Europe
Andorra see Andorra la Vella
105 V4 **Andorra la Vella** var. Andorra, Fr. Andorre la Vielle, Sp. Andorra la Vieja. ● (Andorra) C Andorra 42°30′N 01°30′E
Andorra la Vieja see Andorra la Vella
Andorra, Principality of see Andorra
Andorra, Valls d'/Andorra, Vallée d' see Andorra
Andorre la Vielle see Andorra la Vella
97 M22 **Andover** S England, United Kingdom 51°13′N 01°28′W
27 N6 **Andover** Kansas, C USA 37°42′N 97°08′W
92 G10 **Andøya** island C Norway
60 I8 **Andradina** São Paulo, S Brazil 20°54′S 51°19′W
105 X9 **Andratx** Mallorca, Spain, W Mediterranean Sea 39°35′N 02°25′E
39 N10 **Andreafsky River** ~ Alaska, USA
38 H17 **Andreanof Islands** island group Aleutian Islands, Alaska, USA
124 H16 **Andreapol'** Tverskaya Oblast', W Russian Federation 56°38′N 32°17′E
Andreas, Cape see Zafer Burnu
Andreevka see Kabanbay
21 N10 **Andrews** North Carolina, SE USA 35°19′N 84°01′W
21 T13 **Andrews** South Carolina, SE USA 33°27′N 79°33′W
24 M7 **Andrews** Texas, SW USA 32°19′N 102°34′W
173 N5 **Andrew Tablemount** var. Gora Andryu. undersea feature W Indian Ocean 06°45′N 50°30′E
Andreyevka see Kabanbay
107 N17 **Andria** Puglia, SE Italy 41°13′N 16°17′E
113 K16 **Andrijevica** E Montenegro 42°45′N 19°45′E
115 E20 **Andrítsaina** Pelopónnisos, S Greece 37°29′N 21°52′E
An Droichead Nua see Newbridge
Andropov see Rybinsk
115 J19 **Ándros** Ándros, Kykládes, Greece, Aegean Sea 37°49′N 24°54′E
115 J20 **Ándros** island Kykládes, Greece, Aegean Sea
19 O7 **Androscoggin River** ~ Maine/New Hampshire, NE USA
44 F3 **Andros Island** island NW The Bahamas
127 R7 **Androsovka** Samarskaya Oblast', W Russian Federation 52°41′N 49°34′E
44 G3 **Andros Town** Andros Island, NW The Bahamas 24°40′N 77°47′W
155 D21 **Āndrott Island** island Lakshadweep, India, N Indian Ocean
117 N5 **Andrushivka** Zhytomyrs'ka Oblast', N Ukraine 50°01′N 29°02′E
111 K17 **Andrychów** Małopolskie, S Poland 49°51′N 19°18′E
Andryu, Gora see Andrew Tablemount
92 I10 **Andselv** Troms, N Norway 69°05′N 18°30′E
79 O17 **Andudu** Orientale, NE Dem. Rep. Congo 02°26′N 28°39′E
105 N13 **Andújar** anc. Illiturgis. Andalucía, SW Spain 38°02′N 04°03′W
82 C12 **Andulo** Bié, W Angola 11°29′S 16°43′E
103 Q14 **Anduze** Gard, S France 44°03′N 03°59′E
An Earagail see Errigal Mountain
95 L19 **Aneby** Jönköping, S Sweden 57°50′N 14°45′E
Anécho see Aného
77 Q9 **Anéfis** Kidal, NE Mali 18°05′N 00°38′E
45 U8 **Anegada** island NE British Virgin Islands
61 B25 **Anegada, Bahía** bay E Argentina
45 U9 **Anegada Passage** passage Anguilla/British Virgin Islands
77 R17 **Aného** var. Anécho; prev. Petit-Popo. S Togo 06°14′N 01°36′E
197 D17 **Aneityum** var. Anatom; prev. Kéamu. island S Vanuatu
117 N10 **Anenii Noi** Rus. Novyye Aneny. C Moldova 46°52′N 29°10′E
186 F7 **Anepmete** New Britain, E Papua New Guinea 05°47′S 148°37′E
105 U4 **Aneto** ▲ NE Spain 42°36′N 00°37′E
146 F13 **Änew** Rus. Annau. Ahal Welaýaty, C Turkmenistan 37°51′N 58°22′E
Ānewetak see Enewetak Atoll
77 Y8 **Aney** Agadez, NE Niger 19°22′N 13°00′E
An Fheoir see Nore
122 L12 **Angara** ~ C Russian Federation
122 M13 **Angarsk** Irkutskaya Oblast', S Russian Federation 52°31′N 103°55′E
Angaur see Ngeaur
93 G17 **Ånge** Västernorrland, C Sweden 62°31′N 15°40′E
Angel see Úhlava
40 D4 **Ángel de la Guarda, Isla** island NW Mexico
171 O3 **Angeles** off. Angeles City. Luzon, N Philippines 15°16′N 120°37′E
Angeles City see Angeles
Angel Falls see Ángel, Salto
95 J22 **Ängelholm** Skåne, S Sweden 56°14′N 12°52′E
61 A17 **Angélica** Santa Fe, C Argentina 31°33′S 61°33′W
25 W8 **Angelina River** ~ Texas, SW USA
55 Q9 **Ángel, Salto** Eng. Angel Falls. waterfall E Venezuela
95 M15 **Ängelsberg** Västmanland, C Sweden 59°57′N 16°01′E
35 P8 **Angels Camp** California, W USA 38°03′N 120°31′W
109 W7 **Anger** Steiermark, SE Austria 47°16′N 15°41′E
Angerapp see Ozersk
Angerburg see Węgorzewo
93 H15 **Ångermanälven** ~ N Sweden
100 P11 **Angermünde** Brandenburg, NE Germany 53°02′N 13°59′E
102 K7 **Angers** anc. Juliomagus. Maine-et-Loire, NW France 47°30′N 00°33′W
15 W7 **Angers** ~ Québec, SE Canada
93 J16 **Ängesön** island N Sweden
Angistro see Ágkistro
114 H13 **Angítis** ~ NE Greece
167 R13 **Ângk Tasaôm** prev. Angtassom. Takêv, S Cambodia
185 C25 **Anglem, Mount** ▲ Stewart Island, Southland, SW New Zealand 46°44′S 167°56′E
97 I18 **Anglesey** cultural region NW Wales, United Kingdom
97 I18 **Anglesey** island NW Wales, United Kingdom
102 I15 **Anglet** Pyrénées-Atlantiques, SW France 43°29′N 01°30′W
25 W12 **Angleton** Texas, SW USA 29°10′N 95°27′W
Anglia see England
14 H9 **Angliers** Québec, SE Canada 47°33′N 79°17′W
Anglo-Egyptian Sudan see Sudan
Angmagssalik see Ammassalik
167 Q7 **Ang Nam Ngum** ◎ C Laos
79 N16 **Ango** Orientale, N Dem. Rep. Congo 04°01′N 25°52′E
83 Q15 **Angoche** Nampula, E Mozambique 16°10′S 39°58′E
63 G14 **Angol** Araucanía, C Chile 37°47′S 72°45′W
31 Q11 **Angola** Indiana, N USA 41°37′N 85°00′W
82 A9 **Angola** off. Republic of Angola; prev. People's Republic of Angola, Portuguese West Africa. ◆ republic SW Africa
65 P15 **Angola Basin** undersea feature E Atlantic Ocean 15°00′S 03°00′E
Angola, People's Republic of see Angola
Angola, Republic of see Angola
39 X13 **Angoon** Admiralty Island, Alaska, USA 57°33′N 134°30′W
147 O14 **Angor** Surkhondaryo Viloyati, S Uzbekistan 37°30′N 67°06′E
Angora see Ankara
186 C6 **Angoram** East Sepik, NW Papua New Guinea 04°04′S 144°04′E
40 H8 **Angostura** Sinaloa, C Mexico 25°18′N 108°10′W
Angostura see Ciudad Bolívar
41 U17 **Angostura, Presa de la** ▣ SE Mexico
28 J11 **Angostura Reservoir** ▣ South Dakota, N USA
102 L11 **Angoulême** anc. Iculisma. Charente, W France 45°39′N 00°10′E
102 K11 **Angoumois** cultural region W France
64 O2 **Angra do Heroísmo** Terceira, Azores, Portugal, NE Atlantic Ocean 38°40′N 27°14′W
60 O10 **Angra dos Reis** Rio de Janeiro, SE Brazil 22°59′S 44°17′W
Angra Pequena see Lüderitz
147 Q10 **Angren** Toshkent Viloyati, E Uzbekistan 41°05′N 70°18′E
Angtassom see Ângk Tasaôm
167 O10 **Ang Thong** var. Angthong. Ang Thong, C Thailand 14°35′N 100°25′E
Angthong see Ang Thong
79 M16 **Angu** Orientale, N Dem. Rep. Congo 03°38′N 24°14′E
105 S5 **Angües** Aragón, NE Spain 42°07′N 00°10′W
45 U9 **Anguilla** ◇ UK dependent territory E West Indies
45 V9 **Anguilla** island E West Indies
44 F4 **Anguilla Cays** islets SW The Bahamas
Angul see Anugul
161 N1 **Anguli Nur** ◎ E China
79 O18 **Angumu** Orientale, E Dem. Rep. Congo 0°10′S 27°42′E
14 G14 **Angus** Ontario, S Canada 44°19′N 79°52′W
96 J10 **Angus** cultural region E Scotland, United Kingdom
59 K19 **Anhangüera** Goiás, S Brazil 18°12′S 48°19′W
99 I21 **Anhée** Namur, S Belgium 50°18′N 04°52′E
95 I21 **Anholt** island C Denmark
160 M11 **Anhua** var. Dongping. Hunan, S China 28°25′N 111°10′E
161 P8 **Anhui** var. Anhui Sheng, Anhwei, Wan. ⬖ province E China
AnhuiSheng/Anhwei Wan see Anhui
39 O11 **Aniak** Alaska, USA 61°34′N 159°31′W
39 O12 **Aniak River** ~ Alaska, USA
An Iarmhí see Westmeath
189 R8 **Anibare** E Nauru 0°31′S 166°56′E
189 R8 **Anibare Bay** bay E Nauru, W Pacific Ocean
Anicium see le Puy
115 K22 **Ánidro** island Kykládes, Greece, Aegean Sea
77 R15 **Anié** C Togo 07°48′N 01°12′E
77 Q15 **Anié** ~ C Togo
102 J16 **Anie, Pic d'** ▲ SW France 42°56′N 00°44′W
127 Y7 **Anikhovka** Orenburgskaya Oblast', W Russian Federation 51°27′N 60°17′E
14 G9 **Anima Nipissing Lake** ◎ Ontario, S Canada
37 O16 **Animas** New Mexico, SW USA 31°55′N 108°49′W
37 P16 **Animas Peak** ▲ New Mexico, SW USA 31°34′N 108°46′W
37 P16 **Animas Valley** valley New Mexico, SW USA
116 F13 **Anina** Ger. Steierdorf, Hung. Stájerlakanina; prev. Ştaierdorf-Anina, Steierdorf-Anina, Steyerlak-Anina. Caraş-Severin, SW Romania 45°05′N 21°51′E
29 U14 **Anita** Iowa, C USA 41°27′N 94°45′W
123 U14 **Aniva, Mys** headland Ostrov Sakhalin, SE Russian Federation 46°02′N 143°25′E
187 S15 **Aniwa** island S Vanuatu
93 M19 **Anjalankoski** Kymenlaakso, S Finland 60°39′N 26°54′E
'Anjar see Aanjar
Anjiangying see Luanping
14 B8 **Anjigami Lake** ◎ Ontario, S Canada
149 N13 **Anjīra** Baluchistān, SW Pakistan 28°19′N 66°19′E
164 K14 **Anjō** var. Anzyô. Aichi, Honshū, SW Japan 34°56′N 137°05′E
102 J8 **Anjou** cultural region NW France
Anjouan see Nzwani
172 J4 **Anjozorobe** Antananarivo, C Madagascar 18°22′S 47°52′E
163 W13 **Anju** W North Korea 39°36′N 125°44′E
98 M5 **Anjum** Fris. Eanjum. Fryslân, N Netherlands 53°22′N 06°09′E
172 G6 **Ankaboa, Tanjona** prev./Fr. Cap Saint-Vincent. headland W Madagascar 21°57′S 43°16′E
160 L7 **Ankang** prev. Xing'an. Shaanxi, C China 32°45′N 109°00′E
136 I12 **Ankara** prev. Angora; anc. Ancyra. ● (Turkey) Ankara, C Turkey 39°55′N 32°50′E
136 H12 **Ankara** ⬖ province C Turkey
95 N19 **Ankarsrum** Kalmar, S Sweden 57°40′N 16°19′E
172 H6 **Ankazoabo** Toliara, SW Madagascar 22°18′S 44°30′E
172 I4 **Ankazobe** Antananarivo, C Madagascar 18°20′S 47°07′E
29 V14 **Ankeny** Iowa, C USA 41°43′N 93°37′W
167 V11 **An Khê** Gia Lai, C Vietnam 13°57′N 108°39′E
100 O9 **Anklam** Mecklenburg-Vorpommern, NE Germany 53°51′N 13°42′E
80 K13 **Ānkober** Āmara, N Ethiopia 09°36′N 39°44′E
77 O17 **Ankobra** ~ S Ghana
79 N22 **Ankoro** Katanga, SE Dem. Rep. Congo 06°45′S 26°58′E
99 L24 **Anlier, Forêt d'** forest SE Belgium
160 I13 **Anlong** var. Xin'an. Guizhou, S China 25°05′N 105°29′E
An Longfort see Longford
167 R11 **Ânlong Vêng** Siĕmréab, NW Cambodia 14°16′N 104°08′E
An Lorgain see Lurgan
161 N8 **Anlu** Hubei, C China 31°15′N 113°41′E
An Mhí see Meath
An Mhuir Cheilteach see Celtic Sea
An Muileann gCearr see Mullingar
93 F16 **Ånn** Jämtland, C Sweden 63°19′N 12°34′E
126 M8 **Anna** Voronezhskaya Oblast', W Russian Federation 51°31′N 40°23′E
30 L17 **Anna** Illinois, N USA 37°27′N 89°15′W
25 U5 **Anna** Texas, SW USA 33°21′N 96°33′W
74 L5 **Annaba** prev. Bône. NE Algeria 36°55′N 07°47′E
An Nabatīyah at Taḥtā see Nabatîyé
101 N17 **Annaberg-Buchholz** Sachsen, E Germany 50°35′N 13°01′E
109 T9 **Annabichl** ✈ (Klagenfurt) Kärnten, S Austria 46°39′N 14°21′E
140 M5 **An Nafūd** desert NW Saudi Arabia
139 P6 **'Annah** var. 'Ānah. Al Anbār, NW Iraq 34°50′N 42°00′E
139 P6 **An Nāḥiyah** Al Anbār, W Iraq 34°24′N 41°33′E
139 T10 **An Najaf** off. Muḥāfa at an Najaf, var. Najaf. An Najaf, S Iraq 31°59′N 44°19′E
21 V5 **Anna, Lake** ▣ Virginia, NE USA
97 F16 **Annalee** ~ N Ireland
167 S9 **Annamite Mountains** var. annamsescordillera, Fr. Chaîne Annamitique, Lao. Phou Louang. ▲▲ C Laos
Annamitique, Chaîne see Annamite Mountains
Annamses Cordillera see Annamite Mountains
97 J14 **Annan** S Scotland, United Kingdom 55°N 03°20′W
29 U8 **Annandale** Minnesota, N USA 45°15′N 94°07′W
21 W4 **Annandale** Virginia, NE USA 38°48′N 77°10′W
189 Q7 **Anna Point** headland N Nauru 0°30′S 166°56′E
21 X3 **Annapolis** state capital Maryland, NE USA 38°59′N 76°30′W
188 A10 **Anna, Pulo** island S Palau
153 O10 **Annapurṇa** ▲ C Nepal 28°30′N 83°50′E
An Nāqūrah see En Nâqoûra
31 R10 **Ann Arbor** Michigan, N USA 42°17′N 83°45′W
An Nás see Naas
139 W12 **An Nāşirīyah** var. Nasiriya. Dhī Qār, SE Iraq 31°04′N 46°17′E
139 W11 **An Naşr** Dhī Qār, E Iraq 31°34′N 46°08′E
Annau see Änew
121 O13 **An Nawfalīyah** var. Al Nūwfalīyah. N Libya 30°46′N 17°48′E
19 P10 **Ann, Cape** headland Massachusetts, NE USA 42°39′N 70°35′W
180 I10 **Annean, Lake** ◎ Western Australia
Anneciacum see Annecy
103 T11 **Annecy** anc. Anneciacum. Haute-Savoie, E France 45°53′N 06°09′E
103 T11 **Annecy, Lac d'** ◎ E France
103 T10 **Annemasse** Haute-Savoie, E France 46°10′N 06°13′E
39 Z14 **Annette Island** island Alexander Archipelago, Alaska, USA
An Nhon see Binh Đinh
An Nīl al Abyaḍ see White Nile
An Nīl al Azraq see Blue Nile
23 Q3 **Anniston** Alabama, S USA 33°39′N 85°49′W
79 A19 **Annobón** island W Equatorial Guinea
103 R12 **Annonay** Ardèche, E France 45°15′N 04°40′E
44 K12 **Annotto Bay** C Jamaica 18°16′N 76°47′W
141 R5 **An Nu'ayrīyah** var. Narıya. Ash Sharqīyah, NE Saudi Arabia 27°30′N 48°30′E
182 M9 **Annuello** Victoria, SE Australia 34°54′S 142°50′E
139 Q10 **An Nukhayb** Al Anbār, S Iraq 32°02′N 42°15′E
139 U9 **An Nu'mānīyah** Wāsiṭ, E Iraq 32°34′N 45°23′E
Áno Arkhánai see Archánes
115 J25 **Anógeia** var. Anogia, Anóyia. Kríti, Greece, E Mediterranean Sea 35°17′N 24°53′E
Anogia see Anógeia
29 V8 **Anoka** Minnesota, N USA 45°15′N 93°26′W
An Ómaigh see Omagh
172 I1 **Anorontany, Tanjona** Fr. Cap Saint-Sébastien. headland N Madagascar
172 J5 **Anosibe An'Ala** Toamasina, E Madagascar 19°24′S 48°11′E
Anóyia see Anógeia
An Pointe see Warrenpoint
161 P9 **Anqing** Anhui, E China 30°32′N 116°59′E
161 Q5 **Anqiu** Shandong, E China 36°25′N 119°10′E
An Ráth see Ráth Luirc
An Ribhéar see Kenmare River
An Ros see Rush
99 K19 **Ans** Liège, E Belgium 50°39′N 05°32′E
Anṣāb see Nişāb
171 W12 **Ansas** Papua, E Indonesia 01°44′S 135°52′E
101 J20 **Ansbach** Bayern, SE Germany 49°18′N 10°36′E
An Sciobairín see Skibbereen
An Scoil see Skull
An Seancheann see Old Head of Kinsale
45 Y5 **Anse-Bertrand** Grande Terre, N Guadeloupe 16°28′N 61°31′W
172 H17 **Anse Boileau** Mahé, NE Seychelles 04°43′S 55°29′E
45 S11 **Anse La Raye** NW Saint Lucia 13°57′N 61°01′W
54 D9 **Anserma** Caldas, W Colombia 05°15′N 75°47′W
109 T4 **Ansfelden** Oberösterreich, N Austria 48°12′N 14°17′E
163 U12 **Anshan** Liaoning, NE China 41°06′N 122°55′E
160 J12 **Anshun** Guizhou, S China 26°15′N 105°58′E
61 F17 **Ansina** Tacuarembó, C Uruguay 31°58′S 55°28′W
29 O15 **Ansley** Nebraska, C USA 41°16′N 99°22′W
25 P6 **Anson** Texas, SW USA 32°45′N 99°55′W
77 Q10 **Ansongo** Gao, E Mali 15°39′N 00°33′E
An Srath Bán see Strabane
21 R5 **Ansted** West Virginia, NE USA 38°08′N 81°06′W
171 Y13 **Ansudu** Papua, E Indonesia 02°09′S 139°19′E
57 G15 **Anta** Cusco, S Peru 13°30′S 72°08′W
57 G16 **Antabamba** Apurímac, C Peru 14°23′S 72°54′W
Antafalva see Kovačica
136 L17 **Antakya** anc. Antioch, Antiochia. Hatay, S Turkey 36°12′N 36°10′E
172 K3 **Antalaha** Antsirañana, NE Madagascar 14°53′S 50°16′E
136 F17 **Antalya** prev. Adalia; anc. Attaleia, Bibl. Attalia. Antalya, SW Turkey 36°53′N 30°42′E
136 F17 **Antalya** ⬖ province SW Turkey
136 F16 **Antalya** ✈ Antalya, SW Turkey 36°53′N 30°45′E
121 U10 **Antalya Basin** undersea feature E Mediterranean Sea
Antalya, Gulf of see Antalya Körfezi
136 F16 **Antalya Körfezi** var. Gulf of Adalia, Eng. Gulf of Antalya. gulf SW Turkey
172 J5 **Antanambao Manampotsy** Toamasina, E Madagascar 19°30′S 48°36′E
172 I5 **Antananarivo** prev. Tananarive. ● (Madagascar) Antananarivo, C Madagascar 18°52′S 47°30′E
172 I4 **Antananarivo** ⬖ province C Madagascar
172 J5 **Antananarivo** ✈ Antananarivo, C Madagascar 18°52′S 47°30′E
An tAonach see Nenagh
194-195 **Antarctica** continent
194 I5 **Antarctic Peninsula** peninsula Antarctica
61 J15 **Antas, Rio das** ~ S Brazil
189 U16 **Ant Atoll** atoll Caroline Islands, E Micronesia
An Teampall Mór see Templemore
Antep see Gaziantep
104 M15 **Antequera** anc. Anticaria, Antiquaria. Andalucía, S Spain 37°01′N 04°34′W
Antequera see Oaxaca
37 S5 **Antero Reservoir** ▣ Colorado, C USA
26 M7 **Anthony** Kansas, C USA 37°10′N 98°02′W
37 R16 **Anthony** New Mexico, SW USA 32°00′N 106°36′W
182 D5 **Anthony, Lake** salt lake South Australia
74 E8 **Anti-Atlas** ▲▲ SW Morocco
103 U15 **Antibes** anc. Antipolis. Alpes-Maritimes, SE France 43°35′N 07°07′E
103 U15 **Antibes, Cap d'** headland SE France 43°33′N 07°08′E
Anticaria see Antequera
13 Q11 **Anticosti, Île d'** Eng. Anticosti Island. island Québec, E Canada
Anticosti Island see Anticosti, Île d'
102 K3 **Antifer, Cap d'** headland N France 49°43′N 00°10′E
30 L6 **Antigo** Wisconsin, N USA 45°10′N 89°10′W
13 Q15 **Antigonish** Nova Scotia, SE Canada 45°39′N 62°00′W
64 P11 **Antigua** Fuerteventura, Islas Canarias, NE Atlantic Ocean
45 X10 **Antigua** island S Antigua and Barbuda, Leeward Islands
Antigua see Antigua Guatemala
45 W9 **Antigua and Barbuda** ◆ commonwealth republic E West Indies
42 C6 **Antigua Guatemala** var. Antigua. Sacatepéquez, SW Guatemala 14°33′N 90°42′W
41 P11 **Antiguo Morelos** var. Antiguo-Morelos. Tamaulipas, C Mexico 22°35′N 99°08′W
115 F19 **Antíkyras, Kólpos** gulf C Greece
115 G24 **Antikýthira** var. Andikíthira. island S Greece
138 I7 **Anti-Lebanon** var. Jebel esh Sharqi, Ar. Al Jabal ash Sharqī, Fr. Anti-Liban. ▲▲ Lebanon/Syria
Anti-Liban see Anti-Lebanon
115 M22 **Antimácheia** Kos, Dodekánisa, Greece 47°19′N 07°54′E
115 I22 **Antímilos** island Kykládes, Greece, Aegean Sea
36 L6 **Antimony** Utah, W USA 38°07′N 112°00′W
An tInbhear Mór see Arklow
30 M10 **Antioch** Illinois, N USA 42°28′N 88°06′W
Antioch see Antakya
102 I10 **Antioche, Pertuis d'** inlet W France
Antiochia see Antakya
54 D8 **Antioquia** Antioquia, C Colombia 06°36′N 75°53′W
54 E8 **Antioquia** off. Departamento de Antioquia. ⬖ province C Colombia
Antioquia, Departamento de see Antioquia
115 J21 **Antíparos** var. Andipáros. island Kykládes, Greece, Aegean Sea
115 B17 **Antípaxoi** var. Andipaxi. island Iónia Nísiá, Greece, C Mediterranean Sea
122 J8 **Antipayuta** Yamalo-Nenetskiy Avtonomnyy Okrug, N Russian Federation 69°08′N 76°43′E
192 L12 **Antipodes Islands** island group S New Zealand
Antipolis see Antibes
115 J18 **Antípsara** var. Andípsara. island E Greece
Antiquaria see Antequera
15 N10 **Antique, Lac** ◎ Québec, SE Canada
115 E18 **Antírrio** var. Andírrion. Dytikí Elláda, C Greece 38°20′N 21°46′E
115 K16 **Ántissa** var. Ándissa. Lésvos, E Greece 39°15′N 26°00′E
An tIúr see Newry

◆ Country
● Country Capital
◇ Dependent Territory
○ Dependent Territory Capital
⬖ Administrative Regions
✈ International Airport
▲ Mountain
▲▲ Mountain Range
Volcano
~ River
◎ Lake
▣ Reservoir

ANTIVARI – ARGUUT

Antivari see Bar
56 C6 **Antizana** ▲ N Ecuador 0°29′S 78°08′W
27 Q13 **Antlers** Oklahoma, C USA 34°15′N 95°38′W
93 J14 **Antnäs** Norrbotten, N Sweden 65°32′N 21°53′E
Antō see Andong
62 G5 **Antofagasta** Antofagasta, N Chile 23°40′S 70°23′W
62 G6 **Antofagasta** off. Región de Antofagasta. ⬖ region N Chile
Antofagasta, Región de see Antofagasta
62 I7 **Antofalla, Salar de** salt lake NW Argentina
99 D20 **Antoing** Hainaut, SW Belgium 50°34′N 03°26′E
43 S16 **Antón** Coclé, C Panama 08°23′N 80°15′W
24 M5 **Anton** Texas, SW USA 33°48′N 102°09′W
37 T11 **Anton Chico** New Mexico, SW USA 35°12′N 105°09′W
60 K12 **Antonina** Paraná, S Brazil 25°28′S 48°43′W
188 C16 **Antonio B. Won Pat International** ✈ (Agana) C Guam 13°28′N 144°48′E
103 O5 **Antony** Hauts-de-Seine, N France 48°45′N 02°17′E
Antratsit see Antratsyt
117 Y8 **Antratsyt** Rus. Antratsit. Luhans'ka Oblast', E Ukraine 48°07′N 39°05′E
97 G15 **Antrim** Ir. Aontroim. NE Northern Ireland, United Kingdom 54°43′N 06°13′W
97 G14 **Antrim** Ir. Aontroim. cultural region NE Northern Ireland, United Kingdom
97 G14 **Antrim Mountains** ▲▲ NE Northern Ireland, United Kingdom
172 H5 **Antsalova** Mahajanga, W Madagascar 18°40′S 44°37′E
Antserana see Antsirañana
An tSionainn see Shannon
172 J2 **Antsirañana** var. Antserana; prev. Antsirane, Diégo-Suarez. Antsirañana, N Madagascar 12°19′S 49°17′E
172 J2 **Antsirañana** ⬖ province N Madagascar
Antsirane see Antsirañana
An tSiúir see Suir
118 I7 **Antsla** Ger. Anzen. Võrumaa, SE Estonia 57°52′N 26°33′E
An tSláine see Slaney
172 J3 **Antsohihy** Mahajanga, NW Madagascar 14°50′S 47°58′E
63 G14 **Antuco, Volcán** Ѫ C Chile 37°29′S 71°25′W
169 W10 **Antu, Gunung** ▲ Borneo, N Indonesia 0°57′N 118°51′E
An Tullach see Tullow
An-tung see Dandong
Antunnacum see Andernach
Antwerp see Antwerpen
99 G16 **Antwerpen** Eng. Antwerp, Fr. Anvers. Antwerpen, N Belgium 51°13′N 04°25′E
99 H16 **Antwerpen** Eng. Antwerp. ⬖ province N Belgium
An Uaimh see Navan
154 N12 **Anugul** var. Angul. Odisha, E India 20°51′N 84°59′E
152 F9 **Anūpgarh** Rājasthān, NW India 29°10′N 73°14′E
154 K10 **Anūppur** Madhya Pradesh, C India 23°05′N 81°45′E
155 K24 **Anuradhapura** North Central Province, C Sri Lanka 08°20′N 80°25′E
Anvers see Antwerpen
194 G4 **Anvers Island** island Antarctica
39 N11 **Anvik** Alaska, USA 62°39′N 160°12′W
39 N10 **Anvik River** ≈ Alaska, USA
38 F17 **Anvil Peak** ▲ Semisopochnoi Island, Alaska, USA 51°59′N 179°36′E
An Vinh, Nhom see Amphitrite Group
159 P7 **Anxi** var. Yuanquan. Gansu, N China 40°32′N 95°50′E
182 F8 **Anxious Bay** bay South Australia
161 O5 **Anyang** Henan, C China 36°11′N 114°18′E
159 S11 **A'nyêmaqên Shan** ▲▲ C China
118 H12 **Anykščiai** Utena, E Lithuania 55°30′N 25°34′E
161 P13 **Anyuan** var. Xinshan. Jiangxi, S China 25°10′N 115°25′E
123 T7 **Anyuysk** Chukotskiy Avtonomnyy Okrug, NE Russian Federation 68°22′N 161°33′E
123 T7 **Anyuyskiy Khrebet** ▲▲ NE Russian Federation
54 D8 **Anzá** Antioquia, C Colombia 06°18′N 75°54′W
Anzen see Antsla
107 I16 **Anzio** Lazio, C Italy 41°28′N 12°38′E
55 O6 **Anzoátegui** off. Estado Anzoátegui. ⬖ state NE Venezuela
Anzoátegui, Estado see Anzoátegui
147 P12 **Anzob** W Tajikistan 39°24′N 68°55′E
Anzyô see Anjō
Aoba see Ambae
165 X13 **Aoga-shima** island Izu-shotō, SE Japan
Aohan Qi see Xinhui
105 R3 **Aoiz** Bas. Agoitz, var. Agoiz. Navarra, N Spain 42°47′N 01°23′W
167 O11 **Ao Krung Thep** var. Krung Thep Mahanakhon, Eng. Bangkok. ● (Thailand) Bangkok, C Thailand 13°44′N 100°30′E
186 M9 **Aola** var. Tenaghau. Guadalcanal, C Solomon Islands 09°32′S 160°28′E
166 M15 **Ao Luk Nua** Krabi, SW Thailand 08°21′N 98°43′E
Aomen Tebie Xingzhengqu see Macao
172 N8 **Aomori** Aomori, Honshū, C Japan 40°50′N 140°43′E
172 N8 **Aomori** off. Aomori-ken. ⬖ prefecture Honshū, C Japan
Aomori-ken see Aomori
Aontroim see Antrim
115 C15 **Aóos** var. Vijosa, Vijosë, Alb. Lumi i Vjosës. ≈ Albania/Greece see also Vjosës, Lumi i
Aóos see Vjosës, Lumi i
191 Q7 **Aorai, Mont** Ѫ Tahiti, W French Polynesia 17°36′S 149°29′W
185 E19 **Aoraki** prev. Aorangi, Mount Cook. ▲ South Island, New Zealand 43°38′S 170°05′E
167 R13 **Aôral, Phnum** prev. Phnom Aural. ▲ W Cambodia 12°01′N 104°10′E
Aorangi see Aoraki
185 L15 **Aorangi Mountains** ▲▲ North Island, New Zealand
184 H13 **Aorere** ≈ South Island, New Zealand
106 A7 **Aosta** anc. Augusta Praetoria. Valle d'Aosta, NW Italy 45°43′N 07°20′E
77 O11 **Aougoundou, Lac** ◎ S Mali
76 K9 **Aoukâr** var. Aouker. plateau C Mauritania
78 J13 **Aouk, Bahr** ≈ Central African Republic/Chad
Aouker see Aoukâr
74 B11 **Aousard** SE Western Sahara 22°42′N 14°22′W
164 H12 **Aoya** Tottori, Honshū, SW Japan 35°31′N 134°01′E
Aoyang see Shanggao
78 H5 **Aozou** Tibesti, N Chad 22°01′N 17°11′E
26 M11 **Apache** Oklahoma, C USA 34°57′N 98°21′W
36 L14 **Apache Junction** Arizona, SW USA 33°25′N 111°33′W
24 J9 **Apache Mountains** ▲▲ Texas, SW USA
36 M16 **Apache Peak** ▲ Arizona, SW USA 31°50′N 110°25′W
116 H10 **Apahida** Cluj, NW Romania 46°49′N 23°45′E
23 T9 **Apalachee Bay** bay Florida, SE USA
23 T3 **Apalachee River** ≈ Georgia, SE USA
23 S10 **Apalachicola** Florida, SE USA 29°43′N 84°58′W
23 S10 **Apalachicola Bay** bay Florida, SE USA
23 R9 **Apalachicola River** ≈ Florida, SE USA
Apam see Apan
Apamama see Abemama
41 P14 **Apan** var. Apam. Hidalgo, C Mexico 19°48′N 98°25′W
42 J8 **Apanás, Lago de** ◎ NW Nicaragua
54 H14 **Apaporis, Río** ≈ Brazil/Colombia
185 C23 **Aparima** ≈ South Island, New Zealand
171 O1 **Aparri** Luzon, N Philippines 18°16′N 121°42′E
112 J9 **Apatin** Vojvodina, NW Serbia 45°40′N 19°01′E
124 J4 **Apatity** Murmanskaya Oblast', NW Russian Federation 67°34′N 33°26′E
55 X9 **Apatou** NW French Guiana 05°10′N 54°22′W
40 M14 **Apatzingán** var. Apatzingán de la Constitución. Michoacán, SW Mexico 19°05′N 102°20′W
Apatzingán de la Constitución see Apatzingán
171 X12 **Apauwar** Papua, E Indonesia 01°36′S 138°10′E
Apaxtla see Apaxtla de Castrejón
41 O15 **Apaxtla de Castrejón** var. Apaxtla. Guerrero, S Mexico 18°06′N 99°55′W
118 J7 **Ape** NE Latvia 57°32′N 26°42′E
98 L11 **Apeldoorn** Gelderland, E Netherlands 52°13′N 05°57′E
Apennines see Appennino
Apenrade see Aabenraa
57 L17 **Apere, Río** ≈ C Bolivia
55 W11 **Apetina** Sipaliwini, SE Suriname 03°30′N 55°03′W
21 U9 **Apex** North Carolina, SE USA 35°43′N 78°51′W
79 M16 **Api** Orientale, N Dem. Rep. Congo 03°40′N 25°26′E
152 M9 **Api** ▲ NW Nepal 30°07′N 80°57′E
192 H16 **Āpia** ● (Samoa) Upolu, SE Samoa 13°50′S 171°47′W
Apia see Abaiang
60 K11 **Apiaí** São Paulo, S Brazil 24°29′S 48°51′W
170 M16 **Api, Gunung** ▲ Pulau Sangeang, S Indonesia 08°09′S 119°03′E
187 N9 **Apio** Maramasike Island, N Solomon Islands 09°36′S 161°25′E
41 O15 **Apipilulco** Guerrero, S Mexico 18°11′N 99°40′W
41 P14 **Apizaco** Tlaxcala, S Mexico 19°26′N 98°09′W
137 Q8 **Apkhazeti** var. Abkhazia; prev. Ap'khazet'i. ⬖ autonomous republic NW Georgia
Ap'khazet'i see Apkhazeti
104 I4 **A Pobla de Trives** Cast. Puebla de Trives. Galicia, NW Spain 42°21′N 07°16′W
55 U9 **Apoera** Sipaliwini, NW Suriname 05°12′N 57°13′W
115 O23 **Apolakkiá** Ródos, Dodekánisa, Greece, Aegean Sea 36°02′N 27°48′E
101 L16 **Apolda** Thüringen, C Germany 51°02′N 11°31′E
192 H16 **Apolima Strait** strait C Pacific Ocean
182 M13 **Apollo Bay** Victoria, SE Australia 38°40′S 143°44′E
Apollonia see Sozopol
57 J16 **Apolo** La Paz, W Bolivia 14°48′S 68°31′W
57 J16 **Apolobamba, Cordillera** ▲▲ Bolivia/Peru
171 Q8 **Apo, Mount** Ѫ Mindanao, S Philippines 06°54′N 125°16′E
23 W11 **Apopka** Florida, SE USA 28°40′N 81°30′W
23 W11 **Apopka, Lake** ◎ Florida, SE USA
59 J19 **Aporé, Rio** ≈ SW Brazil
30 K2 **Apostle Islands** island group Wisconsin, N USA
Apostolas Andreas, Cape see Zafer Burnu
61 F14 **Apóstoles** Misiones, NE Argentina 27°54′S 55°45′W
Apostólou Andréa, Akrotíri see Zafer Burnu
117 S9 **Apostolove** Rus. Apostolovo. Dnipropetrovs'ka Oblast', E Ukraine 47°40′N 33°45′E
Apostolovo see Apostolove
17 S10 **Appalachian Mountains** ▲▲ E USA
95 K14 **Äppelbo** Dalarna, C Sweden 60°30′N 14°00′E
98 N7 **Appelscha** Fris. Appelskea. Fryslân, N Netherlands 52°57′N 06°19′E
Appelskea see Appelscha
106 G11 **Appennino** Eng. Apennines. ▲▲ Italy/San Marino
107 L17 **Appennino Campano** ▲▲ C Italy
108 I7 **Appenzell** Inner-Rhoden, NW Switzerland 47°20′N 09°25′E
55 V12 **Appikalo** Sipaliwini, S Suriname 02°07′N 56°16′W
98 O5 **Appingedam** Groningen, NE Netherlands 53°18′N 06°52′E
25 X8 **Appleby** Texas, SW USA 31°43′N 94°36′W
97 L15 **Appleby-in-Westmorland** Cumbria, NW England, United Kingdom 54°35′N 02°26′W
30 K10 **Apple River** ≈ Illinois, N USA
30 I5 **Apple River** ≈ Wisconsin, N USA
25 W9 **Apple Springs** Texas, SW USA 31°13′N 94°57′W
29 S8 **Appleton** Minnesota, N USA 45°12′N 96°01′W
30 M7 **Appleton** Wisconsin, N USA 44°17′N 88°24′W
27 S5 **Appleton City** Missouri, C USA 38°11′N 94°01′W
35 U14 **Apple Valley** California, W USA 34°30′N 117°11′W
29 V9 **Apple Valley** Minnesota, N USA 44°43′N 93°13′W
21 U6 **Appomattox** Virginia, NE USA 37°21′N 78°51′W
188 B16 **Apra Harbor** harbor W Guam
188 B16 **Apra Heights** W Guam
106 F6 **Aprica, Passo dell'** pass N Italy
107 M15 **Apricena** anc. Hadria Picena. Puglia, SE Italy 41°47′N 15°27′E
114 I9 **Apriltsi** Lovech, N Bulgaria 42°50′N 24°54′E
126 L14 **Apsheronsk** Krasnodarskiy Kray, SW Russian Federation 44°27′N 39°45′E
Apsheronskiy Poluostrov see Abşeron Yarımadası
103 S15 **Apt** anc. Apta Julia. Vaucluse, SE France 43°54′N 05°24′E
Apta Julia see Apt
38 H12 **'Āpua Point** var. Apua Point. headland Hawai'i, USA, C Pacific Ocean 19°15′N 155°13′W
60 I10 **Apucarana** Paraná, S Brazil 23°34′S 51°28′W
Apulia see Puglia
54 K8 **Apure** off. Estado Apure. ⬖ state C Venezuela
Apure, Estado see Apure
54 J7 **Apure, Río** ≈ W Venezuela
57 F16 **Apurímac** off. Departamento de Apurímac. ⬖ department C Peru
Apurímac, Departamento de see Apurímac
57 F15 **Apurímac, Río** ≈ S Peru
116 G10 **Apuseni, Munţii** ▲▲ W Romania
Aqaba/'Aqaba see Al 'Aqabah
138 F15 **Aqaba, Gulf of** var. Gulf of Elat, Ar. Khalīj al 'Aqabah; anc. Sinus Aelaniticus. gulf NE Red Sea
139 R7 **'Aqabah** Al Anbār, C Iraq 33°33′N 42°55′E
'Aqabah, Khalīj al see Aqaba, Gulf of
'Aqabah, Muḥāfa at al see Al 'Aqabah
149 O2 **Āqchah** var. Āqcheh. Jowzjān, N Afghanistan 37°N 66°07′E
Āqcheh see Āqchah
Aqkengse see Akkense
Aqköl see Akkol'
Aqmola see Astana
Aqmola Oblysy see Akmola
158 L10 **Aqqikkol Hu** ◎ NW China
Aqqū see Akku
Aqqystaū see Akkystau
'Aqrah see Ākrē
Aqsay see Aksay
Aqshataū see Akshatau
Aqsū see Aksu
Aqsūat see Aksuat
Aqtas see Aktas
Aqtaū see Aktau
Aqtöbe see Aktobe
Aqtöbe Oblysy see Aktyubinsk
Aqtoghay see Aktogay
Aquae Augustae see Dax
Aquae Calidae see Bath
Aquae Flaviae see Chaves
Aquae Grani see Aachen
Aquae Panoniae see Baden
Aquae Sextiae see Aix-en-Provence
Aquae Solis see Bath
Aquae Tarbelicae see Dax
36 J11 **Aquarius Mountains** ▲▲ Arizona, SW USA
62 O5 **Aquidabán, Río** ≈ E Paraguay
59 H20 **Aquidauana** Mato Grosso do Sul, S Brazil 20°27′S 55°45′W
40 L15 **Aquila** Michoacán, S Mexico 18°36′N 103°32′W
Aquila/Aquila degli Abruzzi see L'Aquila
25 T8 **Aquilla** Texas, SW USA 31°51′N 97°13′W
44 L9 **Aquin** S Haiti 18°16′N 73°24′W
Aquisgranum see Aachen
102 J13 **Aquitaine** ⬖ region SW France
Aqzhar see Akzhar
153 P13 **Āra** prev. Arrah. Bihār, N India 25°34′N 84°40′E
105 S4 **Ara** ≈ NE Spain
23 P2 **Arab** Alabama, S USA 34°19′N 86°30′W
Araba see Álava
138 G12 **'Arabah, Wādī al** Heb. Ha'Arava. dry watercourse Israel/Jordan
117 U12 **Arabats'ka Strilka, Kosa** spit S Ukraine
117 U12 **Arabats'ka Zatoka** gulf S Ukraine
'Arab, Baḥr al see Arab, Bahr el
80 C12 **Arab, Bahr el** var. Baḥr al 'Arab. ≈ S Sudan
56 E7 **Arabela, Río** ≈ N Peru
173 O4 **Arabian Basin** undersea feature N Arabian Sea 11°30′N 65°00′E
Arabian Desert see Sahara el Sharqîya
141 N9 **Arabian Peninsula** peninsula SW Asia
85 P15 **Arabian Plate** tectonic feature Africa/Asia/Europe
141 W14 **Arabian Sea** sea NW Indian Ocean
Arabicus, Sinus see Red Sea
'Arabī, Khalīj al see Persian Gulf
Arabistan see Khūzestān
'Arabīyah as Su'ūdīyah, Al Mamlakah al see Saudi Arabia
'Arabīyah Jumhūrīyah, Miṣr al see Egypt
138 I9 **'Arab, Jabal al** ▲▲ S Syria
Arab Republic of Egypt see Egypt
139 Y12 **'Arab, Shaṭṭ al** Eng. Shatt al Arab, Per. Arvand Rūd. ≈ Iran/Iraq
136 I11 **Araç** Kastamonu, N Turkey 41°14′N 33°20′E
59 P16 **Aracaju** state capital Sergipe, E Brazil 10°45′S 37°07′W
54 F5 **Aracataca** Magdalena, N Colombia 10°38′N 74°09′W
58 P13 **Aracati** Ceará, E Brazil 04°32′S 37°45′W
60 J8 **Araçatuba** São Paulo, S Brazil 21°12′S 50°24′W
136 I11 **Araç Çayı** ≈ N Turkey
104 J13 **Aracena** Andalucía, S Spain 37°54′N 06°33′W
115 F20 **Arachnaío** ▲▲ S Greece
115 D16 **Árachthos** var. Arta, prev. Árakhthos; anc. Arachthus. ≈ W Greece
Arachthus see Árachthos
59 N19 **Araçuai** Minas Gerais, SE Brazil 16°52′S 42°03′W
138 F11 **Arad** Southern, S Israel 31°16′N 35°09′E
116 F11 **Arad** Arad, W Romania 46°12′N 21°20′E
116 F11 **Arad** ⬖ county W Romania
78 J9 **Arada** Wadi Fira, NE Chad 15°00′N 20°38′E
143 P18 **'Arādah** Abū Ẓaby, S United Arab Emirates 22°57′N 53°24′E
Aradhippou see Aradíppou
121 Q3 **Aradíppou** var. Aradhippou. SE Cyprus 34°57′N 33°37′E
174 K6 **Arafura Sea** Ind. Laut Arafuru. sea W Pacific Ocean
174 L6 **Arafura Shelf** undersea feature C Arafura Sea
Arafuru, Laut see Arafura Sea
59 J18 **Aragarças** Goiás, C Brazil 15°55′S 52°12′W
Aragats, Gora see Aragats Lerr
137 T12 **Aragats Lerr** Rus. Gora Aragats. ▲ W Armenia 40°31′N 44°06′E
32 E14 **Arago, Cape** headland Oregon, NW USA 43°17′N 124°25′W
105 R6 **Aragón** autonomous community E Spain
105 Q4 **Aragón** ≈ NE Spain
107 I24 **Aragona** Sicilia, Italy, C Mediterranean Sea 37°25′N 13°37′E
105 Q7 **Aragoncillo** ▲ C Spain 40°59′N 02°01′W
54 L5 **Aragua** off. Estado Aragua. ⬖ state N Venezuela
55 N6 **Aragua de Barcelona** Anzoátegui, NE Venezuela 09°30′N 64°51′W
55 O5 **Aragua de Maturín** Monagas, NE Venezuela 09°58′N 63°30′W
Aragua, Estado see Aragua
59 K15 **Araguaia, Río** var. Araguaya. ≈ C Brazil
59 K19 **Araguari** Minas Gerais, SE Brazil 18°38′S 48°13′W
58 J11 **Araguari, Rio** ≈ SW Brazil
Araguaya see Araguaia, Río
104 K14 **Arahal** Andalucía, S Spain 37°15′N 05°33′W
165 N11 **Arai** Niigata, Honshū, C Japan 37°02′N 138°17′E
Árainn see Inishmore
Árainn Mhór see Aran Island
Ara Jovis see Aranjuez
74 J11 **Arak** C Algeria 25°17′N 03°45′E
171 Y15 **Arak** Papua, E Indonesia 07°14′S 139°40′E
142 M7 **Arāk** prev. Sultānābād. Markazī, W Iran 34°07′N 49°39′E
188 D10 **Arakabesan** island Palau Islands, N Palau
55 S7 **Arakaka** NW Guyana 07°37′N 59°58′W
Arakan State see Rakhine State
166 K5 **Arakan Yoma** ▲▲ W Myanmar (Burma)
165 O10 **Arakawa** Niigata, Honshū, C Japan 38°06′N 139°25′E
Árakhthos see Árachthos
Araks/Arak's see Aras
158 H7 **Aral** Xinjiang Uygur Zizhiqu, NW China 40°40′N 81°19′E
Aral see Aralsk, Kazakhstan
Aral see Vose', Tajikistan
Aral-Bukhorskiy Kanal see Amu-Buxoro Kanali
137 T12 **Aralik** Iğdır, E Turkey 39°54′N 44°28′E
146 H5 **Aral Sea** Kaz. Aral Tengizi, Rus. Aral'skoye More, Uzb. Orol Dengizi. inland sea Kazakhstan/Uzbekistan
144 L13 **Aral'sk** Kaz. Aral. Kzylorda, SW Kazakhstan 46°48′N 61°40′E
Aral'skoye More/Aral Tengizi see Aral Sea
41 O10 **Aramberri** Nuevo León, NE Mexico 24°05′N 99°52′W
186 B8 **Aramia** ≈ SW Papua New Guinea
105 N5 **Aranda de Duero** Castilla y León, N Spain 41°40′N 03°41′W
112 M12 **Aranđelovac** prev. Arandjelovac. Serbia, C Serbia 44°18′N 20°32′E
Arandjelovac see Aranđelovac
97 J19 **Aran Fawddwy** ▲ NW Wales, United Kingdom 52°48′N 03°42′W
97 C14 **Aran Island** Ir. Árainn Mhór. island NW Ireland
97 A18 **Aran Islands** island group W Ireland
105 N9 **Aranjuez** anc. Ara Jovis. Madrid, C Spain 40°02′N 03°37′W
83 E20 **Aranos** Hardap, SE Namibia 24°10′S 19°08′E
25 U14 **Aransas Bay** inlet Texas, SW USA
25 T14 **Aransas Pass** Texas, SW USA 27°54′N 97°09′W
191 O3 **Aranuka** prev. Nanouki. atoll Tungaru, W Kiribati
143 N6 **Ārān-va-Bidgol** var. Golārā. Eşfahān, C Iran 34°03′N 51°30′E
167 Q11 **Aranyaprathet** Prachin Buri, S Thailand 13°42′N 102°32′E
Aranyosasztal see Zlatý Stôl
Aranyosgyéres see Câmpia Turzii
Aranyosmarót see Zlaté Moravce
164 C14 **Arao** Kumamoto, Kyūshū, SW Japan 32°58′N 130°26′E
77 O8 **Araouane** Tombouctou, N Mali 18°53′N 03°31′W
26 L10 **Arapaho** Oklahoma, C USA 35°34′N 98°57′W
29 N16 **Arapahoe** Nebraska, C USA 40°18′N 99°54′W
57 I16 **Arapa, Laguna** ◎ SE Peru
185 K14 **Arapawa Island** island C New Zealand
61 E17 **Arapey Grande, Río** ≈ N Uruguay
59 P16 **Arapiraca** Alagoas, E Brazil 09°45′S 36°40′W
140 M3 **'Ar'ar** Al Ḥudūd ash Shamālīyah, NW Saudi Arabia 31°N 41°E
54 G15 **Araracuara** Caquetá, S Colombia 0°36′S 72°24′W
61 K15 **Araranguá** Santa Catarina, S Brazil 28°56′S 49°30′W
60 L8 **Araraquara** São Paulo, S Brazil 21°46′S 48°08′W
59 O13 **Araras** Ceará, E Brazil 04°08′S 40°30′W
58 I14 **Araras** Pará, N Brazil 06°04′S 54°34′W
60 L9 **Araras** São Paulo, S Brazil 22°21′S 47°21′W
60 H11 **Araras, Serra das** ▲▲ S Brazil
137 U12 **Ararat** S Armenia 39°49′N 44°45′E
182 M12 **Ararat** Victoria, SE Australia 37°20′S 143°00′E
Ararat, Mount see Büyükağrı Dağı
140 M3 **'Ar'ar, Wādī** dry watercourse Iraq/Saudi Arabia
129 N7 **Aras** Arm. Arak's, Az. Araz Nehri, Per. Rūd-e Aras, Rus. Araks; prev. Araxes. ≈ SW Asia
Aras de Alpuente see Aras de los Olmos
105 R9 **Aras de los Olmos** prev. Aras de Alpuente. Valenciana, E Spain 39°55′N 01°09′W
137 S13 **Aras Güneyi Dağları** ▲▲ NE Turkey
Aras, Rūd-e see Aras
191 U9 **Aratika** atoll Îles Tuamotu, C French Polynesia
Aratürük see Yiwu
54 I8 **Arauca** Arauca, NE Colombia 07°03′N 70°47′W
54 I8 **Arauca** off. Intendencia de Arauca. ⬖ province NE Colombia
Arauca, Intendencia de see Arauca
63 G15 **Araucanía** off. Regíon de la Araucanía. ⬖ region C Chile
Araucanía, Regíon de la see Araucanía
54 L7 **Arauca, Río** ≈ Colombia/Venezuela
63 F14 **Arauco** Bío Bío, C Chile 37°15′S 73°22′W
63 F14 **Arauco, Golfo de** gulf S Chile
54 H8 **Arauquita** Arauca, C Colombia 06°57′N 71°19′W
Arausio see Orange
152 F13 **Arāvali Range** ▲▲ N India
186 J7 **Arawa** Bougainville, NE Papua New Guinea 06°15′S 155°35′E
185 C20 **Arawata** ≈ South Island, New Zealand
186 F7 **Arawe Islands** island group E Papua New Guinea
59 L20 **Araxá** Minas Gerais, SE Brazil 19°37′S 46°50′W
Araxes see Aras
55 O5 **Araya** Sucre, N Venezuela 10°34′N 64°15′W
Araz Nehri see Aras
105 R4 **Arba** ≈ N Spain
81 I15 **Ārba Minch'** Southern Nationalities, S Ethiopia 06°02′N 37°34′E
Arbat see Erbet
'Arbat see Erbet
107 D19 **Arbatax** Sardegna, Italy, C Mediterranean Sea 39°57′N 09°42′E
Arbe see Rab
Arbela see Arbīl
139 S3 **Arbīl** Kurd. Hewlêr, var. Erbil, Irbīl, Hawlēr; anc. Arbela. Arbīl, N Iraq 36°12′N 44°01′E
139 S3 **Arbīl** off. Muḥāfa at Arbīl, var. Irbil, off. Kurd. Parêzga-î Hewlêr, Kurd. Hewlêr. ⬖ governorate N Iraq
Arbīl, Muḥāfa at see Arbīl
95 M16 **Arboga** Västmanland, C Sweden 59°24′N 15°50′E
95 M16 **Arbogaån** ≈ C Sweden
103 S9 **Arbois** Jura, E France 46°54′N 05°45′E
54 D6 **Arboletes** Antioquia, NW Colombia 08°52′N 76°25′W
11 X15 **Arborg** Manitoba, S Canada 50°52′N 97°20′W
94 N12 **Arbrå** Gävleborg, C Sweden 61°27′N 16°21′E
96 K10 **Arbroath** anc. Aberbrothock. E Scotland, United Kingdom 56°34′N 02°35′W
35 N6 **Arbuckle** California, W USA 39°00′N 122°05′W
27 N12 **Arbuckle Mountains** ▲▲ Oklahoma, C USA
162 I5 **Arbulag** var. Mandal. Hövsgöl, N Mongolia 49°55′N 99°21′E
Arbuzinka see Arbuzynka
117 Q8 **Arbuzynka** Rus. Arbuzinka. Mykolayivs'ka Oblast', S Ukraine 47°54′N 31°19′E
103 U12 **Arc** ≈ E France
102 J13 **Arcachon** Gironde, SW France 44°40′N 01°11′W
102 J13 **Arcachon, Bassin d'** inlet SW France
18 E10 **Arcade** New York, NE USA 42°32′N 78°19′W
23 W14 **Arcadia** Florida, SE USA 27°13′N 81°51′W
22 H5 **Arcadia** Louisiana, S USA 32°33′N 92°55′W
30 J7 **Arcadia** Wisconsin, N USA 44°15′N 91°30′W
Arcae Remorum see Châlons-en-Champagne
44 L9 **Arcahaie** C Haiti 18°46′N 72°32′W
34 K3 **Arcata** California, W USA 40°51′N 124°06′W
35 U6 **Arc Dome** ▲ Nevada, W USA 38°52′N 117°20′W
107 J16 **Arce** Lazio, C Italy 41°35′N 13°34′E
41 O15 **Arcelia** Guerrero, S Mexico 18°20′N 100°16′W
99 M15 **Arcen** Limburg, SE Netherlands 51°28′N 06°10′E
115 J25 **Archánes** var. Áno Arkhánai, Epáno Archánes; prev. Epáno Arkhánai. Kríti, Greece, E Mediterranean Sea 35°12′N 25°10′E
Archangel see Arkhangel'sk
Archangel Bay see Chëshskaya Guba
115 O23 **Archángelos** var. Arhangelos, Arkhángelos. Ródos, Dodekánisa, Greece, Aegean Sea 36°13′N 28°07′E
114 F7 **Archar** ≈ NW Bulgaria
31 R11 **Archbold** Ohio, N USA 41°31′N 84°18′W
105 R12 **Archena** Murcia, SE Spain 38°07′N 01°17′W
25 R5 **Archer City** Texas, SW USA 33°36′N 98°37′W
104 M14 **Archidona** Andalucía, S Spain 37°06′N 04°23′W
65 B25 **Arch Islands** island group SW Falkland Islands
106 G13 **Arcidosso** Toscana, C Italy 42°52′N 11°30′E
103 Q5 **Arcis-sur-Aube** Aube, N France 48°32′N 04°09′E
182 F3 **Arckaringa Creek** seasonal river South Australia
106 G7 **Arco** Trentino-Alto Adige, N Italy 45°53′N 10°52′E
33 Q14 **Arco** Idaho, NW USA 43°38′N 113°18′W
30 M14 **Arcola** Illinois, N USA 39°39′N 88°19′W
105 P6 **Arcos de Jalón** Castilla y León, N Spain 41°12′N 02°13′W
104 K15 **Arcos de la Frontera** Andalucía, S Spain 36°45′N 05°49′W
104 G5 **Arcos de Valdevez** Viana do Castelo, N Portugal 41°51′N 08°25′W
59 P15 **Arcoverde** Pernambuco, E Brazil 08°23′S 37°W
102 H5 **Arcovest, Pointe de l'** headland NW France 48°49′N 02°58′W
Arctic Mid Oceanic Ridge see Gakkel Ridge
197 R8 **Arctic Ocean** ocean
8 G7 **Arctic Red River** ≈ Northwest Territories/Yukon, NW Canada
Arctic Red River see Tsiigehtchic
39 S6 **Arctic Village** Alaska, USA 68°07′N 145°32′W
194 H1 **Arctowski** Polish research station South Shetland Islands, Antarctica 61°57′S 58°23′W
114 I12 **Arda** var. Ardhas, Gk. Ardas. ≈ Bulgaria/Greece see also Ardas
142 L2 **Ardabīl** var. Ardebil. Ardabīl, NW Iran 38°15′N 48°18′E
142 L2 **Ardabīl** off. Ostān-e Ardabīl. ⬖ province NW Iran
Ardabīl, Ostān-e see Ardabīl
137 R11 **Ardahan** Ardahan, NE Turkey 41°08′N 42°41′E
137 S11 **Ardahan** ⬖ province Ardahan, Turkey
143 P8 **Ardakān** Yazd, C Iran 32°20′N 53°59′E
94 E12 **Årdalstangen** Sogn Og Fjordane, S Norway 61°14′N 07°45′E
137 R11 **Ardanuç** Artvin, NE Turkey 41°07′N 42°04′E
114 L12 **Ardas** var. Ardhas, Bul. Arda. ≈ Bulgaria/Greece see also Arda
138 I13 **Arḍ aş Şawwān** var. Ardh es Suwwān. plain S Jordan
127 P5 **Ardatov** Respublika Mordoviya, W Russian Federation 54°49′N 46°13′E
14 G12 **Ardbeg** Ontario, S Canada 45°38′N 80°05′W
Ardeal see Transylvania
Ardebil see Ardabīl
103 Q13 **Ardèche** ⬖ department E France
103 Q13 **Ardèche** ≈ C France
97 F17 **Ardee** Ir. Baile Átha Fhirdhia. Louth, NE Ireland 53°52′N 06°33′W
103 Q3 **Ardennes** ⬖ department NE France
99 J23 **Ardennes** physical region Belgium/France
137 Q11 **Ardeşen** Rize, NE Turkey 41°14′N 41°00′E
143 O7 **Ardestān** var. Ardistan. Eşfahān, C Iran 33°29′N 52°17′E
108 J9 **Ardez** Graubünden, SE Switzerland 46°47′N 10°09′E
Ardhas see Arda/Ardas
Ardh es Suwwān see Arḍ aş Şawwān
Ardila see Ardila, Ribeira de
104 I12 **Ardila, Ribeira de** Sp. Ardilla. ≈ Portugal/Spain see also Ardilla
Ardila, Ribeira de see Ardilla
11 T17 **Ardill** Saskatchewan, S Canada 49°56′N 105°49′W
104 I12 **Ardilla** Port. Ribeira de Ardila. ≈ Portugal/Spain see also Ardila, Ribeira de
40 M11 **Ardilla, Cerro la** ▲ C Mexico 22°15′N 102°33′W
114 J12 **Ardino** Kardzhali, S Bulgaria 41°38′N 25°22′E
Ardistan see Ardestān
183 P9 **Ardlethan** New South Wales, SE Australia 34°24′S 146°52′E
Ard Mhacha see Armagh
23 P1 **Ardmore** Alabama, S USA 34°59′N 86°51′W
27 N13 **Ardmore** Oklahoma, C USA 34°11′N 97°08′W
20 J10 **Ardmore** Tennessee, S USA 35°00′N 86°48′W
96 G10 **Ardnamurchan, Point of** headland N Scotland, United Kingdom 56°42′N 06°15′W
Árdni see Arnøya
99 C17 **Ardooie** West-Vlaanderen, W Belgium 50°59′N 03°10′E
182 I9 **Ardrossan** South Australia 34°27′S 137°54′E
116 H9 **Ardusat** Hung. Erdőszáda. Maramureş, N Romania 47°36′N 23°25′E
93 F16 **Åre** Jämtland, C Sweden 63°25′N 13°04′E
79 P16 **Arebi** Orientale, NE Dem. Rep. Congo 02°47′N 29°35′E
45 T5 **Arecibo** C Puerto Rico 18°29′N 66°44′W′N
171 V13 **Aredo** Papua Barat, E Indonesia 02°27′S 133°59′E
59 P14 **Areia Branca** Rio Grande do Norte, E Brazil 04°53′S 37°03′W
119 O14 **Arekhawsk** Rus. Orekhovsk. Vitsyebskaya Voblasts', N Belarus 54°42′N 30°30′E
Arel see Arlon
Arelas/Arelate see Arles
Arenal, Embalse de see Arenal Laguna
42 L12 **Arenal Laguna** var. Embalse de Arenal. ◎ NW Costa Rica
42 L13 **Arenal, Volcán** Ѫ NW Costa Rica 10°21′N 84°42′W
34 K6 **Arena, Point** headland California, W USA 38°57′N 123°44′W
59 H17 **Arenápolis** Mato Grosso, W Brazil 14°25′S 56°52′W
40 G10 **Arena, Punta** headland NW Mexico 23°28′N 109°24′W
104 L8 **Arenas de San Pedro** Castilla y León, N Spain 40°12′N 05°05′W
63 I24 **Arenas, Punta de** headland S Argentina 53°10′S 68°15′W
61 B20 **Arenaza** Buenos Aires, E Argentina 34°55′S 61°45′W
95 F17 **Arendal** Aust-Agder, S Norway 58°27′N 08°45′E
99 J16 **Arendonk** Antwerpen, N Belgium 51°18′N 05°06′E
43 T15 **Arenosa** West Panamá, N Panama 09°02′N 79°57′W
Arensburg see Kuressaare
105 W5 **Arenys de Mar** Cataluña, NE Spain 41°35′N 02°33′E
106 C9 **Arenzano** Liguria, NW Italy 44°25′N 08°43′E
115 F22 **Areópoli** prev. Areópolis. Pelopónnisos, S Greece 36°40′N 22°24′E
Areópolis see Areópoli
57 H18 **Arequipa** Arequipa, SE Peru 16°24′S 71°33′W
57 G17 **Arequipa** off. Departamento de Arequipa. ⬖ department Arequia SW Peru
Arequipa, Departamento de see Arequipa
61 B19 **Arequito** Santa Fe, C Argentina 33°09′S 61°28′W
104 M7 **Arévalo** Castilla y León, N Spain 41°04′N 04°44′W
106 H12 **Arezzo** anc. Arretium. Toscana, C Italy 43°28′N 11°50′E
105 Q4 **Arga** ≈ N Spain
Argaeus see Erciyes Dağı
115 G17 **Argalastí** Thessalía, C Greece 39°13′N 23°13′E
105 O10 **Argamasilla de Alba** Castilla-La Mancha, C Spain 39°08′N 03°05′W
158 L8 **Argan** Xinjiang Uygur Zizhiqu, NW China 40°09′N 88°16′E
105 O8 **Arganda** Madrid, C Spain 40°19′N 03°26′W
104 H8 **Arganil** Coimbra, N Portugal 40°13′N 08°03′W
171 P6 **Argao** Cebu, C Philippines 09°52′N 123°33′E
153 V15 **Argartala** Tripura, NE India
123 N9 **Arga-Sala** ≈ Respublika Sakha (Yakutiya),NE Russian Federation
103 P17 **Argelès-sur-Mer** Pyrénées-Orientales, S France 42°33′N 03°01′E
103 T15 **Argens** ≈ SE France
106 H9 **Argenta** Emilia-Romagna, N Italy 44°37′N 11°49′E
102 K5 **Argentan** Orne, N France 48°45′N 00°01′W
103 N12 **Argentat** Corrèze, C France 45°06′N 01°57′E
106 A9 **Argentera** Piemonte, NE Italy 44°25′N 06°57′E
103 N5 **Argenteuil** Val-d'Oise, N France 48°57′N 02°13′E
62 K13 **Argentina** off. Argentine Republic. ◆ republic S South America
Argentina Basin see Argentine Basin
Argentine Abyssal Plain see Argentine Plain
65 I19 **Argentine Basin** var. Argentina Basin. undersea feature SW Atlantic Ocean 45°00′S 45°00′W
65 I20 **Argentine Plain** var. Argentine Abyssal Plain. undersea feature SW Atlantic Ocean 47°31′S 50°00′W
Argentine Republic see Argentina
Argentine Rise see Falkland Plateau
63 H22 **Argentino, Lago** ◎ S Argentina
102 K8 **Argenton-Château** Deux-Sèvres, W France 46°59′N 00°22′W
102 M9 **Argenton-sur-Creuse** Indre, C France 46°34′N 01°32′E
Argentoratum see Strasbourg
116 I12 **Argeş** ⬖ county S Romania
116 K14 **Argeş** ≈ S Romania
149 O8 **Arghandāb, Daryā-ye** ≈ SE Afghanistan
Arghastān see Arghistān
Arghestān see Arghistān
149 O8 **Arghistān** Pash. Arghastān; prev. Arghestān. ≈ SE Afghanistan
Argirocastro see Gjirokastër
80 E7 **Argo** Northern, N Sudan 19°31′N 30°25′E
173 P7 **Argo Fracture Zone** tectonic feature C Indian Ocean
115 F20 **Argolikós Kólpos** gulf S Greece
103 R4 **Argonne** physical region NE France
115 F20 **Árgos** Pelopónnisos, S Greece 37°38′N 22°43′E
Argōsh see Argush
115 D14 **Árgos Orestikó** Dytikí Makedonía, N Greece 40°27′N 21°15′E
115 B19 **Argostóli** var. Argostólion. Kefallonía, Iónia Nísiá, Greece, C Mediterranean Sea 38°13′N 20°29′E
Argostólion see Argostóli
Argovie see Aargau
35 O14 **Arguello, Point** headland California, W USA 34°34′N 120°39′W
127 P16 **Argun** Chechenskaya Respublika, SW Russian Federation 43°16′N 45°53′E
157 T2 **Argun** Chin. Ergun He, Rus. Argun'. ≈ China/Russian Federation
77 T12 **Argungu** Kebbi, NW Nigeria 12°45′N 04°24′E
139 S1 **Argush** Ar. Argūsh, var. Argōsh. Dahūk, N Iraq 37°07′N 44°13′E
Argūsh see Argush
Arguut see Guchin-Us

◆ Country
● Country Capital
◇ Dependent Territory
○ Dependent Territory Capital
⬖ Administrative Regions
✈ International Airport
▲ Mountain
▲▲ Mountain Range
Ѫ Volcano
≈ River
◎ Lake
⊠ Reservoir

Argyle, Lake – As Samāwah

181 N3 **Argyle, Lake** *salt lake* Western Australia
96 G12 **Argyll** *cultural region* W Scotland, United Kingdom
Argyrokastron *see* Gjirokastër
162 I7 **Arhangay** ◐ *province* C Mongolia
Arhangelos *see* Archángelos
95 G22 **Århus** *var.* Aarhus. Midtjylland, C Denmark *56°09′N 10°11′E*
139 T1 **Ari** *Ar.* Ārī. Arbil, E Iraq *37°07′N 44°34′E*
Ārī *see* Ari
Aria *see* Herāt
83 F22 **Ariamsvlei** Karas, SE Namibia *28°08′S 19°50′E*
107 L17 **Ariano Irpino** Campania, S Italy *41°08′N 15°00′E*
54 F11 **Ariari, Río** ≈ C Colombia
151 K19 **Ari Atoll** *var.* Alifu Atoll. *atoll* C Maldives
77 P11 **Aribinda** N Burkina Faso *14°12′N 00°50′W*
62 G2 **Arica** *hist.* San Marcos de Arica. Arica y Parinacota, N Chile *18°31′S 70°18′W*
54 H16 **Arica** Amazonas, S Colombia *02°09′S 71°48′W*
62 G2 **Arica** ✈ Arica y Parinacota, N Chile *18°30′S 70°20′W*
62 H2 **Arica y Parinacota** ◐ *region* N Chile
114 E13 **Aridaía** *var.* Aridea, Aridhaía. Dytikí Makedonía, N Greece *40°59′N 22°04′E*
Aridea *see* Aridaía
172 I15 **Aride, Île** *island* Inner Islands, NE Seychelles
Aridhaía *see* Aridaía
103 N17 **Ariège** ◐ *department* S France
102 M16 **Ariège** *var.* la Riege. ≈ Andorra/France
116 H11 **Arieş** ≈ W Romania
149 U10 **Ārifwāla** Punjab, E Pakistan *30°15′N 73°08′E*
Ariguaní *see* El Difícil
138 G11 **Arīḥā** Al Karak, W Jordan *31°25′N 35°47′E*
138 I3 **Arīḥā** *var.* Arīhā. Idlib, W Syria *35°50′N 36°36′E*
Arīhā *see* Arīḥā
Arīḥā *see* Jericho
37 W4 **Arikaree River** ≈ Colorado/Nebraska, C USA
112 L13 **Arilje** Serbia, W Serbia *43°45′N 20°06′E*
45 U14 **Arima** Trinidad, Trinidad and Tobago *10°38′N 61°17′W*
Arime *see* Al 'Arīmah
Ariminum *see* Rimini
59 H16 **Arinos, Rio** ≈ W Brazil
40 M14 **Ario de Rosales** *var.* Ario de Rosáles. Michoacán, SW Mexico *19°12′N 101°42′W*
Ario de Rosáles *see* Ario de Rosales
118 F12 **Ariogala** Kaunas, C Lithuania *55°16′N 23°30′E*
47 T7 **Aripuanã** ≈ W Brazil
59 E15 **Ariquemes** Rondônia, W Brazil *09°55′S 63°06′W*
121 W13 **'Arīsh, Wādi el** ≈ NE Egypt
54 K6 **Arismendi** Barinas, C Venezuela *08°29′N 68°22′W*
10 J14 **Aristazabal Island** *island* SW Canada
60 F13 **Aristóbulo del Valle** Misiones, NE Argentina *27°09′S 54°54′W*
172 I5 **Arivonimamo** ✈ (Antananarivo) Antananarivo, C Madagascar *19°00′S 47°11′E*
Arixang *see* Wenquan
105 Q6 **Ariza** Aragón, NE Spain *41°19′N 02°03′W*
62 I6 **Arizaro, Salar de** *salt lake* NW Argentina
105 O2 **Arizgoiti** *var.* Basauri. País Vasco, N Spain *43°13′N 02°54′W*
62 K13 **Arizona** San Luis, C Argentina *35°44′S 65°16′W*
36 J12 **Arizona** *off.* State of Arizona, *also known as* Copper State, Grand Canyon State. ◐ *state* SW USA
40 G4 **Arizpe** Sonora, NW Mexico *30°20′N 110°11′W*
95 J16 **Årjäng** Värmland, C Sweden *59°24′N 12°09′E*
143 P8 **Arjenān** Yazd, C Iran *32°19′N 53°48′E*
92 I13 **Arjeplog** *Lapp.* Árjepluovve. Norrbotten, N Sweden *66°04′N 18°E*
Árjepluovve *see* Arjeplog
54 E5 **Arjona** Bolívar, N Colombia *10°14′N 75°22′W*
105 N13 **Arjona** Andalucía, S Spain *37°56′N 04°04′W*
123 S10 **Arka** Khabarovskiy Kray, E Russian Federation *60°04′N 142°17′E*
22 L2 **Arkabutla Lake** ⊠ Mississippi, S USA
127 O7 **Arkadak** Saratovskaya Oblast', W Russian Federation *51°55′N 43°29′E*
27 T13 **Arkadelphia** Arkansas, C USA *34°07′N 93°06′W*
115 J25 **Arkalochóri**; *prev.* Arkalokhórion. Kríti, Greece, E Mediterranean Sea *35°09′N 25°15′E*
Arkalohori/Arkalokhórion *see* Arkalochóri
145 O10 **Arkalyk** *Kaz.* Arqalyq. Kostanay, N Kazakhstan *50°17′N 66°51′E*
27 U10 **Arkansas** *off.* State of Arkansas, *also known as* The Land of Opportunity. ◐ *state* S USA
27 W14 **Arkansas City** Arkansas, C USA *33°36′N 91°12′W*
27 O7 **Arkansas City** Kansas, C USA *37°03′N 97°02′W*
16 K11 **Arkansas River** ≈ C USA
182 J5 **Arkaroola** South Australia *30°21′S 139°20′E*
Arkhángelos *see* Archángelos
124 L8 **Arkhangel'sk** *Eng.* Archangel. Arkhangel'skaya Oblast', NW Russian Federation *64°32′N 40°40′E*
124 L9 **Arkhangel'skaya Oblast'** ◐ *province* NW Russian Federation
127 O14 **Arkhangel'skoye** Stavropol'skiy Kray, SW Russian Federation *44°37′N 44°03′E*
123 R14 **Arkhara** Amurskaya Oblast', S Russian Federation *49°20′N 130°04′E*
97 G19 **Arklow** *Ir.* An tInbhear Mór. SE Ireland *52°48′N 06°09′W*
115 M20 **Arkoí** *island* Dodekánisa, Greece, Aegean Sea
27 R11 **Arkoma** Oklahoma, C USA *35°19′N 94°27′W*
100 O7 **Arkona, Kap** *headland* NE Germany *54°40′N 13°24′E*
95 N17 **Arkösund** Östergötland, S Sweden *58°28′N 16°55′E*
122 J6 **Arkticheskogo Instituta, Ostrova** *island* N Russian Federation
95 O15 **Arlanda** ✈ (Stockholm) Stockholm, C Sweden *59°40′N 17°58′E*
146 C11 **Arlandag** *Rus.* Gora Arlan. ▲ W Turkmenistan *39°39′N 54°28′E*
Arlan, Gora *see* Arlandag
105 O5 **Arlanza** ≈ N Spain
105 N5 **Arlanzón** ≈ N Spain
103 R15 **Arles** *var.* Arles-sur-Rhône; *anc.* Arelas, Arelate. Bouches-du-Rhône, SE France *43°41′N 04°38′E*
Arles-sur-Rhône *see* Arles
103 O17 **Arles-sur-Tech** Pyrénées-Orientales, S France *42°27′N 02°37′E*
29 U9 **Arlington** Minnesota, N USA *44°36′N 94°04′W*
29 R15 **Arlington** Nebraska, C USA *41°27′N 96°21′W*
32 J11 **Arlington** Oregon, NW USA *45°43′N 120°10′W*
29 R10 **Arlington** South Dakota, N USA *44°21′N 97°07′W*
20 E10 **Arlington** Tennessee, S USA *35°17′N 89°40′W*
25 T6 **Arlington** Texas, SW USA *32°44′N 97°05′W*
21 W4 **Arlington** Virginia, NE USA *38°54′N 77°09′W*
32 H7 **Arlington** Washington, NW USA *48°12′N 122°07′W*
30 M10 **Arlington Heights** Illinois, N USA *42°08′N 88°03′W*
77 U8 **Arlit** Agadez, C Niger *18°54′N 07°25′E*
99 L24 **Arlon** *Dut.* Aarlen, *Ger.* Arel, *Lat.* Orolaunum. Luxembourg, SE Belgium *49°41′N 05°49′E*
27 R7 **Arma** Kansas, C USA *37°32′N 94°42′W*
97 F16 **Armagh** *Ir.* Ard Mhacha. S Northern Ireland, United Kingdom *54°15′N 06°33′W*
97 F16 **Armagh** *cultural region* S Northern Ireland, United Kingdom
102 K15 **Armagnac** *cultural region* S France
103 Q7 **Armançon** ≈ C France
60 K10 **Armando Laydner, Represa** ⊠ S Brazil
115 M24 **Armathía** *island* SE Greece
137 T12 **Armavir** *prev.* Hoktemberyan, *Rus.* Oktemberyan. SW Armenia *40°09′N 43°58′E*
126 M14 **Armavir** Krasnodarskiy Kray, SW Russian Federation *44°59′N 41°07′E*
54 E10 **Armenia** Quindío, W Colombia *04°32′N 75°40′W*
137 T12 **Armenia** *off.* Republic of Armenia, *var.* Ajastan, *Arm.* Hayastani Hanrapetut'yun; *prev.* Armenian Soviet Socialist Republic. ◆ *republic* SW Asia
Armenian Soviet Socialist Republic *see* Armenia
Armenia, Republic of *see* Armenia
Armenierstadt *see* Gherla
103 O1 **Armentières** Nord, N France *50°41′N 02°53′E*
40 K14 **Armería** Colima, SW Mexico *18°55′N 103°59′W*
183 T5 **Armidale** New South Wales, SE Australia *30°32′S 151°40′E*
29 P11 **Armour** South Dakota, N USA *43°19′N 98°21′W*
61 B18 **Armstrong** Santa Fe, C Argentina *32°46′S 61°39′W*
11 N16 **Armstrong** British Columbia, SW Canada *50°27′N 119°14′W*
12 D11 **Armstrong** Ontario, S Canada *50°20′N 89°02′W*
29 U11 **Armstrong** Iowa, C USA *43°24′N 94°28′W*
25 S16 **Armstrong** Texas, SW USA *26°55′N 97°47′W*
117 S11 **Armyans'k** *Rus.* Armyansk. Avtonomna Respublika Krym, S Ukraine *46°05′N 33°43′E*
115 H14 **Arnaía** *Cont.* Arnea. Kentrikí Makedonía, N Greece *40°30′N 23°36′E*
121 N2 **Arnaoúti, Akrotíri** *var.* Arnaoútis, Cape Arnaouti. *headland* W Cyprus *35°06′N 32°16′E*
Arnaouti, Cape/Arnaoútis *see* Arnaoúti, Akrotíri
12 L4 **Arnaud** ≈ Québec, E Canada
103 Q8 **Arnay-le-Duc** Côte d'Or, C France *47°08′N 04°27′E*
Arnea *see* Arnaía
105 Q4 **Arnedo** La Rioja, N Spain *42°14′N 02°05′W*
95 I14 **Årnes** Akershus, S Norway *60°07′N 11°28′E*
Årnes *see* Åi Åfjord
26 K9 **Arnett** Oklahoma, C USA *36°08′N 99°46′W*
98 L12 **Arnhem** Gelderland, SE Netherlands *51°59′N 05°54′E*
181 Q2 **Arnhem Land** *physical region* Northern Territory, N Australia
106 F11 **Arno** ≈ C Italy
Arno *see* Arno Atoll
189 W7 **Arno Atoll** *var.* Arno. *atoll* Ratak Chain, NE Marshall Islands
182 H8 **Arno Bay** South Australia *33°55′S 136°31′E*
35 Q8 **Arnold** California, W USA *38°15′N 120°19′W*
27 X5 **Arnold** Missouri, C USA *38°25′N 90°22′W*
29 N15 **Arnold** Nebraska, C USA *41°25′N 100°11′W*
109 R10 **Arnoldstein** *Slvn.* Pod Klošter. Kärnten, S Austria *46°34′N 13°43′E*
103 N9 **Arnon** ≈ C France
45 P14 **Arnos Vale** ✈ (Kingstown) Saint Vincent, SE Saint Vincent and the Grenadines *13°08′N 61°13′W*
92 I8 **Arnøya** *Lapp.* Árdni. *island* N Norway
14 L12 **Arnprior** Ontario, SE Canada *45°31′N 76°11′W*
101 G15 **Arnsberg** Nordrhein-Westfalen, W Germany *51°24′N 08°04′E*
101 K16 **Arnstadt** Thüringen, C Germany *50°50′N 10°57′E*
Arnswalde *see* Choszczno
54 K5 **Aroa** Yaracuy, N Venezuela *10°26′N 68°54′W*
83 E21 **Aroab** Karas, SE Namibia *26°47′S 19°40′E*
Aroania *see* Chelmós
191 O6 **Aroa, Pointe** *headland* Moorea, W French Polynesia *17°27′S 149°45′W*
Aroe Islands *see* Aru, Kepulauan
101 H15 **Arolsen** Niedersachsen, C Germany *51°23′N 09°00′E*
106 C7 **Arona** Piemonte, NE Italy *45°45′N 08°33′E*
19 R3 **Aroostook River** ≈ Canada/USA
Arop Island *see* Long Island
38 M12 **Aropuk Lake** ◎ Alaska, USA
191 P4 **Arorae** *atoll* Tungaru, W Kiribati
190 G16 **Arorangi** Rarotonga, S Cook Islands *21°13′S 159°49′W*
108 I9 **Arosa** Graubünden, S Switzerland *46°48′N 09°42′E*
104 F4 **Arousa, Ría de** *estuary* E Atlantic Ocean
184 P8 **Arowhana** ▲ North Island, New Zealand *38°07′S 177°52′E*
137 V12 **Arp'a** *Az.* Arpaçay. ≈ Armenia/Azerbaijan
137 S11 **Arpaçay** Kars, NE Turkey *40°51′N 43°20′E*
Arpaçay *see* Arp'a
Arqalyq *see* Arkalyk
149 N14 **Arra** ≈ SW Pakistan
Arrabona *see* Győr
Arrah *see* Āra
Ar Rahad *see* Er Rahad
139 R9 **Ar Raḥḥālīyah** Al Anbār, C Iraq *32°53′N 43°21′E*
60 Q10 **Arraial do Cabo** Rio de Janeiro, SE Brazil *22°57′S 42°00′W*
104 H11 **Arraiolos** Évora, S Portugal *38°44′N 07°59′W*
139 R8 **Ar Ramādī** *var.* Ramadi, Rumadiya. Al Anbār, SW Iraq *33°27′N 43°19′E*
138 J6 **Ar Rāmī** Ḥimṣ, C Syria *34°32′N 37°54′E*
Ar Rams *see* Rams
138 H9 **Ar Ramthā** *var.* Ramtha. Irbid, N Jordan *32°34′N 36°00′E*
96 H13 **Arran, Isle of** *island* SW Scotland, United Kingdom
138 L3 **Ar Raqqah** *var.* Rakka; *anc.* Nicephorium. Ar Raqqah, N Syria *35°57′N 39°03′E*
138 L3 **Ar Raqqah** *off.* Muḥāfaẓat al Raqqah, *var.* Raqqah, *Fr.* Rakka. ◐ *governorate* N Syria
103 O2 **Arras** *anc.* Nemetocenna. Pas-de-Calais, N France *50°17′N 02°46′E*
105 P3 **Arrasate** *Cast.* Mondragón. País Vasco, N Spain *43°04′N 02°30′W*
138 G12 **Ar Rashādīyah** At ṭafīlah, W Jordan *30°42′N 35°38′E*
138 I5 **Ar Rastān** *var.* Rastâne. Ḥimṣ, W Syria *34°57′N 36°43′E*
139 X12 **Ar Raṭāwī** Al Baṣrah, E Iraq *30°37′N 47°12′E*
102 L15 **Arrats** ≈ S France
141 N10 **Ar Rawḍah** Makkah, S Saudi Arabia *21°19′N 42°48′E*
141 Q15 **Ar Rawḍah** S Yemen *14°26′N 47°14′E*
142 K11 **Ar Rawḍatayn** *var.* Raudhatain. N Kuwait *29°80′N 47°50′E*
143 N16 **Ar Rayyān** *var.* Al Rayyan. C Qatar *25°18′N 51°29′E*
102 L17 **Arreau** Hautes-Pyrénées, S France *42°55′N 00°21′E*
64 Q11 **Arrecife** *var.* Arrecife de Lanzarote, Puerto Arrecife. Lanzarote, Islas Canarias, NE Atlantic Ocean *28°57′N 13°33′W*
Arrecife de Lanzarote *see* Arrecife
43 P6 **Arrecife Edinburgh** *reef* NE Nicaragua
61 C19 **Arrecifes** Buenos Aires, E Argentina *34°06′S 60°09′W*
102 F6 **Arrée, Monts d'** ▲▲ NW France
Ar Refā'ī *see* Ar Rifā'ī
Arretium *see* Arezzo
Arriaca *see* Guadalajara
109 S9 **Arriach** Kärnten, S Austria *46°43′N 13°52′E*
41 T16 **Arriaga** Chiapas, SE Mexico *16°14′N 93°54′W*
41 N12 **Arriaga** San Luis Potosí, C Mexico *21°55′N 101°23′W*
139 W10 **Ar Rifā'ī** *var.* Ar Refā'ī. Dhī Qār, SE Iraq *31°47′N 46°07′E*
139 V12 **Ar Riḥāb** *salt flat* S Iraq
Arriondas *see* Les Arriondes
141 Q7 **Ar Riyāḍ** *Eng.* Riyadh. ● (Saudi Arabia) Ar Riyāḍ, C Saudi Arabia *24°50′N 46°50′E*
141 O8 **Ar Riyāḍ** *off.* Minṭaqat ar Riyāḍ. ◐ *province* C Saudi Arabia
141 S15 **Ar Riyān** S Yemen *14°43′N 49°18′E*
Arrö *see* Ærø
61 H18 **Arroio Grande** Rio Grande do Sul, S Brazil *32°15′S 53°02′W*
102 K15 **Arros** ≈ S France
103 Q9 **Arroux** ≈ C France
25 R5 **Arrowhead, Lake** ⊠ Texas, SW USA
182 L5 **Arrowsmith, Mount** *hill* New South Wales, SE Australia
185 D21 **Arrowtown** Otago, South Island, New Zealand *44°57′S 168°51′E*
61 D17 **Arroyo Barú** Entre Ríos, E Argentina *31°52′S 58°26′W*
104 J10 **Arroyo de la Luz** Extremadura, W Spain *39°28′N 06°36′W*
63 J16 **Arroyo de la Ventana** Río Negro, SE Argentina *41°41′S 66°03′W*
35 P13 **Arroyo Grande** California, W USA *35°07′N 120°35′W*
Ar Ru'ays *see* Ar Ruways
141 R11 **Ar Rub' al Khālī** *Eng.* Empty Quarter, Great Sandy Desert. *desert* SW Asia
139 V13 **Ar Rudaymah** Al Muthannā, S Iraq *30°20′N 45°26′E*
61 A16 **Arrufó** Santa Fe, C Argentina *30°15′S 61°45′W*
138 I7 **Ar Ruḥaybah** *var.* Ruhaybeh, *Fr.* Rouhaïbé. Rīf Dimashq, W Syria *33°45′N 36°40′E*
139 V15 **Ar Rukhaymīyah** *well* S Iraq
139 U11 **Ar Rumaythah** *var.* Rumaitha. Al Muthannā, S Iraq *31°31′N 45°15′E*
141 X8 **Ar Rustāq** *var.* Rostak, Rustaq. N Oman *23°34′N 57°25′E*
139 N8 **Ar Ruṭbah** *var.* Rutba. Al Anbār, SW Iraq *33°03′N 40°16′E*
140 M3 **Ar Rūthīyah** *spring/well* NW Saudi Arabia *31°18′N 41°23′E*
ar-Ruwaida *see* Ar Ruwaydah
141 O8 **Ar Ruwaydah** *var.* ar-Ruwaida. Jīzān, C Saudi Arabia *23°48′N 44°44′E*
143 N15 **Ar Ruways** *var.* Al Ruweis, Ar Ru'ays, Ruwais. N Qatar *26°08′N 51°13′E*
143 O17 **Ar Ruways** *var.* Ar Ru'ays. Abū Ẓaby, W United Arab Emirates *24°09′N 52°57′E*
Års *see* Aars
Arsanias *see* Murat Nehri
123 S15 **Arsen'yev** Primorskiy Kray, SE Russian Federation *44°09′N 133°28′E*
155 G19 **Arsikere** Karnātaka, W India *13°20′N 76°13′E*
127 R3 **Arsk** Respublika Tatarstan, W Russian Federation *56°07′N 49°54′E*
94 N10 **Årskogen** Gävleborg, C Sweden *62°07′N 17°19′E*
121 O3 **Ársos** C Cyprus *34°51′N 32°46′E*
94 N13 **Årsunda** Gävleborg, C Sweden *60°31′N 16°45′E*
115 C17 **Árta** *anc.* Ambracia. Ípeiros, W Greece *39°08′N 20°59′E*
105 Y9 **Artà** Mallorca, Spain, W Mediterranean Sea *39°42′N 03°20′E*
Arta *see* Árachthos
137 T12 **Artashat** S Armenia *39°57′N 44°34′E*
40 M15 **Arteaga** Michoacán, SW Mexico *18°22′N 102°18′W*
123 S15 **Artem** Primorskiy Kray, SE Russian Federation *43°24′N 132°20′E*
44 C4 **Artemisa** La Habana, W Cuba *22°49′N 82°47′W*
117 W7 **Artemivs'k** Donets'ka Oblast', E Ukraine *48°35′N 37°58′E*
122 K13 **Artemovsk** Krasnoyarskiy Kray, S Russian Federation *54°22′N 93°24′E*
105 U5 **Artesa de Segre** Cataluña, NE Spain *41°54′N 01°03′E*
37 U14 **Artesia** New Mexico, SW USA *32°50′N 104°24′W*
25 Q14 **Artesia Wells** Texas, SW USA *28°13′N 99°18′W*
108 G8 **Arth** Schwyz, C Switzerland *47°05′N 08°39′E*
14 F15 **Arthur** Ontario, S Canada *43°49′N 80°31′W*
30 M14 **Arthur** Illinois, N USA *39°42′N 88°28′W*
28 L14 **Arthur** Nebraska, C USA *41°35′N 101°42′W*
29 Q5 **Arthur** North Dakota, N USA *47°03′N 97°12′W*
185 B21 **Arthur** ≈ South Island, New Zealand
18 B13 **Arthur, Lake** ⊠ Pennsylvania, NE USA
183 N15 **Arthur River** ≈ Tasmania, SE Australia
185 G18 **Arthur's Pass** Canterbury, South Island, New Zealand *42°59′S 171°33′E*
185 G17 **Arthur's Pass** *pass* South Island, New Zealand
44 I3 **Arthur's Town** Cat Island, C The Bahamas *24°34′N 75°39′W*
44 M9 **Artibonite, Rivière de l'** ≈ C Haiti
61 E16 **Artigas** *prev.* San Eugenio, San Eugenio del Cuareim. Artigas, N Uruguay *30°25′S 56°28′W*
61 E16 **Artigas** ◐ *department* N Uruguay
194 H1 **Artigas** *Uruguayan research station* Antarctica *61°57′S 58°23′W*
137 T11 **Art'ik** W Armenia *40°38′N 43°58′E*
187 O16 **Art, Île** *island* Îles Belep, W New Caledonia
Artois *cultural region* N France
136 L12 **Artova** Tokat, N Turkey *40°04′N 36°17′E*
105 Y9 **Artrutx, Cap d'** *var.* Cabo Dartuch. *cape* Menorca, Spain, W Mediterranean Sea
Artsiz *see* Artsyz
117 N11 **Artsyz** *Rus.* Artsiz. Odes'ka Oblast', SW Ukraine *45°59′N 29°26′E*
158 E7 **Artux** Xinjiang Uygur Zizhiqu, NW China *39°40′N 76°10′E*
137 R11 **Artvin** Artvin, NE Turkey *41°12′N 41°48′E*
137 R11 **Artvin** ◐ *province* NE Turkey
146 G14 **Artyk** Ahal Welaýaty, C Turkmenistan *37°32′N 59°16′E*
79 Q16 **Aru** Orientale, NE Dem. Rep. Congo *02°53′N 30°50′E*
81 E17 **Arua** NW Uganda *03°02′N 30°56′E*
104 I4 **A Rúa de Valdeorras** *var.* La Rúa. Galicia, NW Spain *42°22′N 07°12′W*
Aruângua *see* Luangwa
45 O15 **Aruba** *var.* Oruba. ◇ *Dutch self-governing territory* S West Indies
47 Q4 **Aruba** *island* Aruba, Lesser Antilles
Aru Islands *see* Aru, Kepulauan
171 W15 **Aru, Kepulauan** *Eng.* Aru Islands; *prev.* Aroe Islands. *island group* E Indonesia
153 W10 **Arunāchal Pradesh** *prev.* North East Frontier Agency, North East Frontier Agency of Assam. ◐ *state* NE India
Arun Qi *see* Naji
155 H23 **Aruppukkottai** Tamil Nādu, SE India *09°31′N 78°03′E*
81 I20 **Arusha** Arusha, N Tanzania *03°23′S 36°40′E*
81 I21 **Arusha** ◐ *region* E Tanzania
81 I20 **Arusha** ✈ Arusha, N Tanzania *03°26′S 37°07′E*
54 C9 **Arusí, Punta** *headland* NW Colombia *05°36′N 77°30′W*
155 J23 **Aruvi Aru** ≈ NW Sri Lanka
79 M17 **Aruwimi** *var.* Ituri (upper course). ≈ NE Dem. Rep. Congo
Árva *see* Orava
37 T4 **Arvada** Colorado, C USA *39°48′N 105°06′W*
Arvand Rūd *see* 'Arab, Shaṭṭ al
162 J8 **Arvayheer** Övörhangay, C Mongolia *46°13′N 102°47′E*
9 O10 **Arviat** *prev.* Eskimo Point. Nunavut, C Canada *61°10′N 94°15′W*
93 I14 **Arvidsjaur** Norrbotten, N Sweden *65°34′N 19°12′E*
95 J15 **Arvika** Värmland, C Sweden *59°41′N 12°38′E*
92 J8 **Årviksand** Troms, N Norway *70°10′N 20°30′E*
35 S13 **Arvin** California, W USA *35°12′N 118°52′W*
163 S8 **Arxan** Nei Mongol Zizhiqu, N China *47.11N 119.58′E*
145 P7 **Arykbalyk** *Kaz.* Aryqbalyq. Severnyy Kazakhstan, N Kazakhstan *53°00′N 68°11′E*
Aryqbalyq *see* Arykbalyk
145 P17 **Arys'** *prev.* Arys'. Yuzhnyy Kazakhstan, S Kazakhstan *42°26′N 68°49′E*
Arys' *see* 'Arys'
Arys *see* Orzysz
Arys Köli *see* Arys, Ozero
145 O14 **Arys, Ozero** *Kaz.* Arys Köli. ◎ C Kazakhstan
107 D16 **Arzachena** Sardegna, Italy, C Mediterranean Sea *41°05′N 09°21′E*
127 O4 **Arzamas** Nizhegorodskaya Oblast', W Russian Federation *55°25′N 43°51′E*
141 V13 **Arzāt** S Oman *17°00′N 54°18′E*
104 H3 **Arzúa** Galicia, NW Spain *42°55′N 08°10′W*
111 A16 **Aš** *Ger.* Asch. Karlovarský Kraj, W Czech Republic *50°18′N 12°12′E*
95 H15 **Ås** Akershus, S Norway *59°40′N 10°50′E*
Åsa *see* Asaa
95 H20 **Asaa** *var.* Åsa. Nordjylland, N Denmark *57°07′N 10°24′E*
83 E21 **Asab** Karas, S Namibia *25°29′S 17°59′E*
77 U16 **Asaba** Delta, S Nigeria *06°10′N 06°44′E*
76 J10 **Assaba** *var.* Açâba. ◐ *region* S Mauritania
149 S4 **Asadābād** *var.* Asadābād; *prev.* Chaghasarāy. Kunar, E Afghanistan *34°52′N 71°09′E*
Asadābād *see* Asadābād
138 K3 **Asad, Buḥayrat al** *Eng.* Lake Assad. ◎ N Syria
63 H20 **Asador, Pampa del** *plain* S Argentina
165 P14 **Asahi** Chiba, Honshū, S Japan *35°43′N 140°38′E*
164 M11 **Asahi** Toyama, Honshū, SW Japan *36°56′N 137°34′E*
165 T13 **Asahi-dake** ▲ Hokkaidō, N Japan *43°42′N 142°50′E*
165 T3 **Asahikawa** Hokkaidō, N Japan *43°46′N 142°23′E*
147 S10 **Asaka** *Rus.* Assake; *prev.* Leninsk. Andijon Viloyati, E Uzbekistan *40°39′N 72°16′E*
77 P17 **Asamankese** SE Ghana *05°47′N 00°41′W*
188 B15 **Asan** W Guam *13°28′N 144°43′E*
188 B15 **Asan Point** *headland* W Guam
153 R15 **Āsānsol** West Bengal, NE India *23°40′N 86°59′E*
80 K12 **Āsayita** Āfar, NE Ethiopia *11°35′S 41°23′E*
171 T12 **Asbakin** Papua Barat, E Indonesia *0°45′S 131°40′E*
Asben *see* Aïr, Massif de l'
15 Q12 **Asbestos** Québec, SE Canada *45°46′N 71°56′W*
29 Y13 **Asbury** Iowa, C USA *42°30′N 90°45′W*
18 K15 **Asbury Park** New Jersey, NE USA *40°13′N 74°00′W*
41 Z12 **Ascención, Bahiá de la** *bay* NW Caribbean Sea
40 I3 **Ascensión** Chihuahua, N Mexico *31°07′N 107°59′W*
Ascension *see* Saint Helena, Ascension and Tristan da Cunha
65 M14 **Ascension Fracture Zone** *tectonic feature* C Atlantic Ocean
Ascension Island ◇ *dependency of St.Helena* C Atlantic Ocean
65 N16 **Ascension Island** *island* C Atlantic Ocean
Asch *see* Aš
109 S3 **Aschach an der Donau** Oberösterreich, N Austria *48°22′N 14°00′E*
101 H18 **Aschaffenburg** Bayern, SW Germany *49°58′N 09°10′E*
101 F14 **Ascheberg** Nordrhein-Westfalen, W Germany *51°46′N 07°36′E*
101 L14 **Aschersleben** Sachsen-Anhalt, C Germany *51°46′N 11°28′E*
106 G12 **Asciano** Toscana, C Italy *43°15′N 11°32′E*
106 J13 **Ascoli Piceno** *anc.* Asculum Picenum. Marche, C Italy *42°51′N 13°34′E*
107 M17 **Ascoli Satriano** *anc.* Asculub, Ausculum Apulum. Puglia, SE Italy *41°13′N 15°32′E*
108 G11 **Ascona** Ticino, S Switzerland *46°10′N 08°45′E*
Asculub *see* Ascoli Satriano
Asculum Picenum *see* Ascoli Piceno
80 L11 **'Aseb** *var.* Assab, *Amh.* Āseb. SE Eritrea *13°04′N 42°36′E*
95 M20 **Åseda** Kronoberg, S Sweden *57°10′N 15°20′E*
127 T6 **Asekeyevo** Orenburgskaya Oblast', W Russian Federation *53°36′N 52°53′E*
81 J14 **Āsela** *var.* Asella, Aselle, Asselle. Oromīya, C Ethiopia *07°55′N 39°08′E*
93 H15 **Åsele** Västerbotten, N Sweden *64°10′N 17°20′E*
Asella/Aselle *see* Āsela
98 N7 **Assen** Drenthe, NE Netherlands *53°00′N 06°34′E*
94 K12 **Åsen** Dalarna, C Sweden *61°18′N 13°49′E*
114 J11 **Asenovgrad** *prev.* Stanimaka. Plovdiv, C Bulgaria *42°00′N 24°53′E*
171 O13 **Asera** Sulawesi, C Indonesia *03°24′S 121°42′E*
95 E17 **Åseral** Vest-Agder, S Norway *58°37′N 07°27′E*
118 J3 **Aseri** *var.* Asserien, *Ger.* Asserin. Ida-Virumaa, NE Estonia *59°29′N 26°51′E*
104 G3 **A Serra de Outes** Galicia, NW Spain *42°50′N 08°54′W*
40 J10 **Aserradero** Durango, W Mexico
146 F13 **Aşgabat** *prev.* Ashgabat, Ashkhabad, Poltoratsk. ● (Turkmenistan) Ahal Welaýaty, C Turkmenistan *37°58′N 58°22′E*
146 F13 **Aşgabat** ✈ Ahal Welaýaty, C Turkmenistan *38°06′N 58°10′E*
95 H16 **Åsgårdstrand** Vestfold, S Norway *59°22′N 10°27′E*
Ashara *see* Al 'Ashārah
23 T6 **Ashburn** Georgia, SE USA *31°42′N 83°39′W*
185 G19 **Ashburton** Canterbury, South Island, New Zealand *43°55′S 171°47′E*
185 G19 **Ashburton** ≈ South Island, New Zealand
180 H8 **Ashburton River** ≈ Western Australia
145 V10 **Ashchysu** ≈ E Kazakhstan
10 M16 **Ashcroft** British Columbia, SW Canada *50°41′N 121°17′W*
138 E10 **Ashdod** *anc.* Azotos, *Lat.* Azotus. Central, W Israel *31°48′N 34°38′E*
27 S14 **Ashdown** Arkansas, C USA *33°40′N 94°09′W*
21 T9 **Asheboro** North Carolina, SE USA *35°42′N 79°50′W*
11 X15 **Ashern** Manitoba, S Canada *51°10′N 98°22′W*
21 P10 **Asheville** North Carolina, SE USA *35°36′N 82°33′W*
12 E8 **Asheweig** ≈ Ontario, C Canada
27 V9 **Ash Flat** Arkansas, C USA *36°13′N 91°36′W*
183 T4 **Ashford** New South Wales, SE Australia *29°18′S 151°09′E*
97 P22 **Ashford** SE England, United Kingdom *51°09′N 00°52′E*
36 K11 **Ash Fork** Arizona, SW USA *35°12′N 112°31′W*
Ashgabat *see* Aşgabat
27 T7 **Ash Grove** Missouri, C USA *37°19′N 93°35′W*
165 O12 **Ashikaga** *var.* Asikaga. Tochigi, Honshū, S Japan *36°21′N 139°26′E*
165 Q8 **Ashiro** Iwate, Honshū, C Japan *40°04′N 141°00′E*
164 F15 **Ashizuri-misaki** Shikoku, SW Japan
138 E10 **Ashkelon** *prev.* Ashqelon. Southern, C Israel *31°40′N 34°35′E*
Ashkhabad *see* Aşgabat
23 Q4 **Ashland** Alabama, S USA *33°16′N 85°50′W*
26 K7 **Ashland** Kansas, C USA *37°12′N 99°47′W*
21 P5 **Ashland** Kentucky, S USA *38°28′N 82°40′W*
19 S2 **Ashland** Maine, NE USA *46°36′N 68°24′W*
22 M1 **Ashland** Mississippi, S USA *34°51′N 89°10′W*
27 U4 **Ashland** Missouri, C USA *38°46′N 92°15′W*
29 S15 **Ashland** Nebraska, C USA *41°01′N 96°21′W*
31 T12 **Ashland** Ohio, N USA *40°52′N 82°19′W*
32 G15 **Ashland** Oregon, NW USA *42°11′N 122°42′W*
21 W6 **Ashland** Virginia, NE USA *37°45′N 77°28′W*
30 K3 **Ashland** Wisconsin, N USA *46°34′N 90°54′W*
20 I8 **Ashland City** Tennessee, S USA *36°16′N 87°05′W*
183 S4 **Ashley** New South Wales, SE Australia *29°21′S 149°49′E*
29 O7 **Ashley** North Dakota, N USA *46°00′N 99°22′W*
173 W7 **Ashmore and Cartier Islands** ◇ *Australian external territory* E Indian Ocean
119 I14 **Ashmyany** *Rus.* Oshmyany. Hrodzyenskaya Voblasts', W Belarus *54°24′N 25°57′E*
18 K12 **Ashokan Reservoir** ⊠ New York, NE USA
165 U4 **Ashoro** Hokkaidō, NE Japan *43°16′N 143°33′E*
Ashqelon *see* Ashkelon
Ashraf *see* Behshahr
139 O3 **Ash Shadādah** *var.* Ash Shaddādah, Jisr ash Shadadi, Shaddādi, Shedadi, Tell Shedadi. Al Ḥasakah, NE Syria *36°00′N 40°42′E*
Ash Shaddādah *see* Ash Shadādah
139 Y12 **Ash Shāfī** Al Baṣrah, E Iraq *30°49′N 47°30′E*
139 R4 **Ash Shakk** *var.* Shaykh. Ṣalāḥ ad Dīn, C Iraq *35°15′N 43°27′E*
Ash Sham/Ash Shām *see* Rīf Dimashq
139 T10 **Ash Shāmīyah** *var.* Shamiya. Al Qādisīyah, C Iraq *31°56′N 44°37′E*
139 Y13 **Ash Shāmīyah** *var.* Al Bādiyah al Janūbīyah. *desert* S Iraq
139 T11 **Ash Shanāfīyah** *var.* Ash Shināfīyah. Al Qādisīyah, S Iraq *31°35′N 44°38′E*
138 G13 **Ash Sharāh** *var.* Esh Sharā. ▲▲ W Jordan
143 R16 **Ash Shāriqah** *Eng.* Sharjah. Ash Shāriqah, NE United Arab Emirates *25°22′N 55°28′E*
143 R16 **Ash Shāriqah** *var.* Sharjah. ✈ Ash Shāriqah, NE United Arab Emirates *25°19′N 55°37′E*
140 I4 **Ash Sharmah** *var.* Šarma. Tabūk, NW Saudi Arabia *28°02′N 35°16′E*
139 R4 **Ash Sharqāṭ** Ninawýa, NW Iraq *35°31′N 43°15′E*
141 S10 **Ash Sharqīyah** *off.* Al Minṭaqah ash Sharqīyah, *Eng.* Eastern Region. ◐ *province* E Saudi Arabia
Ash Sharqīyah *see* Al 'Ubaylah
139 W11 **Ash Shatrah** *var.* Shatra. Dhī Qār, SE Iraq *31°26′N 46°10′E*
138 G13 **Ash Shawbak** Ma'ān, W Jordan *30°32′N 35°34′E*
138 L5 **Ash Shaykh Ibrāhīm** Ḥimṣ, C Syria *35°03′N 38°50′E*
141 O17 **Ash Shaykh 'Uthmān** SW Yemen *12°53′N 45°00′E*
141 S15 **Ash Shiḥr** SE Yemen *14°45′N 49°24′E*
Ash Shināfīyah *see* Ash Shanāfīyah
141 V12 **Ash Shiṣar** *var.* Shisur. SW Oman *18°13′N 53°35′E*
139 S13 **Ash Shubrūm** *well* S Iraq
141 R10 **Ash Shuqqān** *desert* E Saudi Arabia
75 O9 **Ash Shuwayrif** *var.* Ash Shwayrif. N Libya *29°54′N 14°16′E*
Ash Shwayrif *see* Ash Shuwayrif
31 U10 **Ashtabula** Ohio, N USA *41°54′N 80°46′W*
29 Q5 **Ashtabula, Lake** ⊠ North Dakota, N USA
137 T12 **Ashtarak** W Armenia *40°18′N 44°22′E*
142 M6 **Āshtiān** *var.* Āshtiyān. Markazī, W Iran *34°23′N 49°55′E*
Āshtiyān *see* Āshtiān
33 R13 **Ashton** Idaho, NW USA *44°04′N 111°27′W*
13 O10 **Ashuanipi Lake** ◎ Newfoundland and Labrador, E Canada
15 P6 **Ashuapmushuan** ≈ Québec, SE Canada
23 Q3 **Ashville** Alabama, S USA *33°50′N 86°15′W*
31 S14 **Ashville** Ohio, N USA *39°43′N 82°57′W*
30 K3 **Ashwabay, Mount** *hill* Wisconsin, N USA
128-129 **Asia** *continent*
171 T11 **Asia, Kepulauan** *island group* E Indonesia
154 N13 **Āsika** Odisha, E India *19°38′N 84°41′E*
Asikaga *see* Ashikaga
93 M18 **Asikkala** *var.* Vääksy. Päijät-Häme, S Finland *61°09′N 25°36′E*
74 G5 **Asilah** N Morocco *35°32′N 06°04′W*
'Aṣī, Nahr al *see* Orontes
107 B16 **Asinara, Isola** *island* W Italy
122 J12 **Asino** Tomskaya Oblast', C Russian Federation *56°56′N 86°02′E*
119 O14 **Asintorf** *Rus.* Osintorf. Vitsyebskaya Voblasts', N Belarus *54°43′N 30°35′E*
119 L17 **Asipovichy** *Rus.* Osipovichi. Mahilyowskaya Voblasts', C Belarus *53°18′N 28°40′E*
141 N12 **'Asīr** *off.* Minṭaqat 'Asīr. ◐ *province* SW Saudi Arabia
140 M11 **'Asīr** *Eng.* Asir. ▲▲ SW Saudi Arabia
'Asīr, Minṭaqat *see* 'Asīr
139 X10 **Askal** Maysān, E Iraq *31°45′N 47°07′E*
137 P13 **Aşkale** Erzurum, NE Turkey *39°56′N 40°39′E*
117 T11 **Askaniya-Nova** Khersons'ka Oblast', S Ukraine *46°27′N 33°54′E*
95 H15 **Asker** Akershus, S Norway *59°52′N 10°26′E*
95 L17 **Askersund** Örebro, C Sweden *58°55′N 14°55′E*
Askī Kalak *see* Eski Kaļak
95 I15 **Askim** Østfold, S Norway *59°15′N 11°10′E*
127 V3 **Askino** Respublika Bashkortostan, W Russian Federation *56°07′N 56°39′E*
115 D14 **Áskio** ▲ N Greece
152 L9 **Askot** Uttarakhand, N India *29°44′N 80°20′E*
94 C12 **Askvoll** Sogn Og Fjordane, S Norway *61°21′N 05°04′E*
136 A13 **Aslan Burnu** *headland* W Turkey *38°44′N 26°43′E*
136 L16 **Aslantaş Barajı** ⊠ S Turkey
149 S4 **Asmār** *var.* Bar Kunar. Kunar, E Afghanistan *34°59′N 71°29′E*
Asmara *see* Asmera
80 I9 **Asmera** *var.* Asmara. ● (Eritrea) C Eritrea *15°15′N 38°58′E*
95 L21 **Åsnen** ◎ S Sweden
115 F19 **Asopós** ≈ S Greece
171 W13 **Asori** Papua, E Indonesia *02°37′S 136°06′E*
80 G12 **Āsosa** Binishangul Gumuz, W Ethiopia *10°06′N 34°27′E*
32 M10 **Asotin** Washington, NW USA *46°18′N 117°03′W*
Aspadana *see* Eşfahān
Aspang *see* Aspang Markt
109 X6 **Aspang Markt** *var.* Aspang. Niederösterreich, E Austria *47°34′N 16°06′E*
105 S12 **Aspe** Valenciana, E Spain *38°21′N 00°43′W*
37 R5 **Aspen** Colorado, C USA *39°12′N 106°49′W*
25 P6 **Aspermont** Texas, SW USA *33°08′N 100°14′W*
Asphaltites, Lacus *see* Dead Sea
Aspinwall *see* Colón
185 C20 **Aspiring, Mount** ▲ South Island, New Zealand *44°21′S 168°47′E*
115 B16 **Asprókavos, Akrotírio** *headland* Kérkyra, Iónia Nísiá, Greece, C Mediterranean Sea *39°22′N 20°07′E*
Aspropotamos *see* Achelóos
Assab *see* 'Aseb
138 L4 **As Sabkhah** *var.* Sabkha. Ar Raqqah, NE Syria *35°30′N 35°54′E*
139 U6 **As Sa'dīyah** Diyālá, E Iraq *34°11′N 45°09′E*
Assad, Lake *see* Asad, Buḥayrat al
138 I8 **Aş Şafā** ▲ S Syria
138 I10 **Aş Şafāwī** Al Mafraq, N Jordan *37°12′N 37°30′E*
75 W8 **Aş Şaff** *var.* El Şaff. N Egypt *29°34′N 31°16′E*
139 N2 **Aş Şafīḥ** Al Ḥasakah, N Syria *36°42′N 40°12′E*
Aş Şaḥrā' ash Sharqīyah *see* Sahara el Sharqîya
Assake *see* Asaka
As Salamīyah *see* Salamīyah
141 Q4 **As Sālimī** *var.* Salemy. SW Kuwait *29°07′N 46°41′E*
67 W7 **'Assal, Lac** ◎ C Djibouti
75 T7 **As Sallūm** *var.* Salûm. NW Egypt *31°31′N 25°09′E*
139 T13 **As Salmān** Al Muthannā, S Iraq *30°29′N 44°34′E*
138 G10 **As Salṭ** *var.* Salt. Al Balqā', NW Jordan *32°03′N 35°44′E*
142 M16 **As Salwā** *var.* Salwa, Salwah. S Qatar *24°44′N 50°52′E*
153 V12 **Assam** ◐ *state* NE India
Assamaka *see* Assamakka
77 T8 **Assamakka** *var.* Assamaka. Agadez, NW Niger *19°24′N 05°53′E*
139 U11 **As Samāwah** *var.* Samawa. Al Muthannā, S Iraq *31°17′N 45°06′E*

◆ Country
● Country Capital
◇ Dependent Territory
○ Dependent Territory Capital
◐ Administrative Regions
✈ International Airport
▲ Mountain
▲▲ Mountain Range
Volcano
≈ River
◎ Lake
⊠ Reservoir

As Samāwah – Awakino

As Samāwah see Al Muthannā
As Saqia al Hamra see Saguia al Hamra
138 J4 Aş Şā'rān Ḩamāh, C Syria 35°15′N 37°28′E
138 G9 Aş Şarīḩ Irbid, N Jordan 32°31′N 35°54′E
21 Z5 Assateague Island island Maryland, NE USA
139 O6 As Sayyāl var. Sayyāl. Dayr az Zawr, E Syria 34°37′N 40°52′E
99 G18 Asse Vlaams Brabant, C Belgium 50°55′N 04°12′E
99 D16 Assebroek West-Vlaanderen, NW Belgium 51°12′N 03°16′E
Asselle see Āsela
107 C20 Assemini Sardegna, Italy, C Mediterranean Sea 39°16′N 08°58′E
99 E16 Assenede Oost-Vlaanderen, NW Belgium 51°15′N 03°43′E
95 G24 Assens Syddtjylland, C Denmark 55°16′N 09°54′N
Asserien/Asserin see Aseri
99 I21 Assesse Namur, SE Belgium 50°22′N 05°01′E
141 Y8 As Sīb var. Seeb. NE Oman 23°40′N 58°03′E
139 Z13 As Sībah var. Sibah. Al Başrah, SE Iraq 30°13′N 47°24′E
11 T17 Assiniboia Saskatchewan, S Canada 49°39′N 105°59′W
11 V15 Assiniboine ≈ Manitoba, S Canada
11 P16 Assiniboine, Mount ▲ Alberta/British Columbia, SW Canada 50°54′N 115°43′W
Assiout see Asyūţ
60 J9 Assis São Paulo, S Brazil 22°37′S 50°25′W
106 I13 Assisi Umbria, C Italy 43°04′N 12°36′E
Assiut see Asyūţ
Assling see Jesenice
Assouan see Aswān
59 P14 Assu var. Açu. Rio Grande do Norte, E Brazil 05°33′S 36°55′W
Assuan see Aswān
142 K12 Aş Şubayḩīyah var. Subiyah. S Kuwait 28°55′N 47°57′E
141 R16 As Sufāl S Yemen 14°06′N 48°42′E
138 L5 As Sukhnah var. Sukhne, Fr. Soukhné. Ḩimş, C Syria 34°56′N 38°52′E
139 U4 As Sulaymānīyah var. Sulaimaniya, Kurd. Slêmanî. As Sulaymānīyah, NE Iraq 35°32′N 45°27′E
139 U4 As Sulaymānīyah off. Muḩāfa at as Sulaymānīyah, off. Kurd. Parêzga-î Slêmanî, Kurd. Slêmanî. ⬖ governorate N Iraq
as Sulaymānīyah, Muḩāfa at see As Sulaymānīyah
141 P11 As Sulayyil Ar Riyāḑ, S Saudi Arabia 20°29′N 45°33′E
121 O13 Aş Şultān N Libya 31°01′N 17°21′E
141 Q5 Aş Şummān desert N Saudi Arabia
141 Q16 Aş Şurrah SW Yemen 13°56′N 46°23′E
139 N4 Aş Şuwār var. Şuwār. Dayr az Zawr, E Syria 35°31′N 40°37′E
138 H9 As Suwaydā' var. El Suweida, Es Suweida, Suweida, Fr. Soueida. As Suwaydā', SW Syria 32°43′N 36°33′E
138 H9 As Suwaydā' off. Muḩāfaẓat as Suwaydā', var. As Suwaydā, Suwaydá, Suweida. ⬖ governorate S Syria
141 Z9 As Suwayḩ NE Oman 22°07′N 59°42′E
141 X8 As Suwayq var. Suwaik. N Oman 23°49′N 57°30′E
139 T8 Aş Şuwayrah var. Suwaira. Wāsiţ, E Iraq 32°57′N 44°47′E
As Suways see Suez
Asta Colonia see Asti
Astacus see İzmit
115 M23 Astakída island SE Greece
145 Q9 Astana prev. Akmola, Akmolinsk, Tselinograd, Aqmola. ● (Kazakhstan) Akmola, N Kazakhstan 51°13′N 71°25′E
142 M3 Āstāneh var. Āstāneh-ye Ashrafīyeh. Gīlān, NW Iran 37°17′N 49°58′E
Āstāneh-ye Ashrafīyeh see Āstāneh
Asta Pompeia see Asti
137 Y14 Astara S Azerbaijan 38°28′N 48°51′E
Astarabad see Gorgān
99 L15 Asten Noord-Brabant, SE Netherlands 51°24′N 05°45′E
Asterābād see Gorgān
106 C8 Asti anc. Asta Colonia, Asta Pompeia, Hasta Colonia, Hasta Pompeia. Piemonte, NW Italy 44°54′N 08°11′E
Astigi see Ecija
Astipálaia see Astypálaia
148 L16 Astola Island island SW Pakistan
152 H4 Astor Jammu and Kashmir, NW India 35°21′N 74°52′E
104 K4 Astorga anc. Asturica Augusta. Castilla y León, N Spain 42°27′N 06°04′W
32 F10 Astoria Oregon, NW USA 46°12′N 123°50′W
0 F8 Astoria Fan undersea feature E Pacific Ocean 45°15′N 126°15′W
95 J22 Åstorp Skåne, S Sweden 56°09′N 12°57′E
Astrabad see Gorgān
127 Q13 Astrakhan' Astrakhanskaya Oblast', SW Russian Federation 46°20′N 48°01′E
Astrakhan-Bazar see Cälilabad
127 Q11 Astrakhanskaya Oblast' ⬖ province SW Russian Federation
93 J15 Åsträsk Västerbotten, N Sweden 64°38′N 20°00′E
Astrida see Butare
65 O22 Astrid Ridge undersea feature S Atlantic Ocean
187 P15 Astrolabe, Récifs de l' reef C New Caledonia
121 P2 Astromerítis N Cyprus 35°09′N 33°02′E
115 F20 Ástros Pelopónnisos, S Greece 37°24′N 22°43′E
119 G16 Astryna Rus. Ostryna. Hrodzyenskaya Voblasts', W Belarus 53°44′N 24°33′E
104 J2 Asturias ⬖ autonomous community NW Spain
Asturias see Oviedo
Asturica Augusta see Astorga
115 L22 Astypálaia var. Astipálaia, It. Stampalia. island Kykládes, Greece, Aegean Sea
192 G16 Āsuisui, Cape headland Savai'i, W Samoa 13°44′S 172°29′W
195 S2 Asuka Japanese research station Antarctica 71°49′S 23°52′E
62 O6 Asunción ● (Paraguay) Central, S Paraguay 25°17′S 57°36′W
62 O6 Asunción ✈ Central, S Paraguay 25°15′S 57°40′W
188 K3 Asuncion Island island N Northern Mariana Islands
42 E6 Asunción Mita Jutiapa, SE Guatemala 14°20′N 89°42′W
Asunción Nochixtlán see Nochixtlán
40 E3 Asunción, Río ≈ NW Mexico
95 M18 Åsunden ◎ S Sweden
118 K11 Asvyeya Rus. Osveya. Vitsyebskaya Voblasts', N Belarus 56°00′N 28°05′E
Aswa see Achwa
75 X11 Aswān var. Assouan, Assuan, Aswân; anc. Syene. SE Egypt 24°03′N 32°59′E
Aswân see Aswān
Aswan Dam see Khazzān Aswān
75 W9 Asyūţ var. Assiout, Assiut, Asyût, Siut; anc. Lycopolis. C Egypt 27°06′N 31°11′E
Asyût see Asyūţ
193 W15 Ata island Tongatapu Group, SW Tonga
62 G8 Atacama off. Región de Atacama. ⬖ region C Chile
Atacama Desert see Atacama, Desierto de
62 H4 Atacama, Desierto de Eng. Atacama Desert. desert N Chile
62 I6 Atacama, Puna de ▲ NW Argentina
Atacama, Región de see Atacama
62 I5 Atacama, Salar de salt lake N Chile
54 E11 Ataco Tolima, C Colombia 03°36′N 75°23′W
190 H8 Atafu Atoll island NW Tokelau
190 H8 Atafu Village Atafu Atoll, NW Tokelau 08°40′S 172°40′W
74 K12 Atakor ▲ SE Algeria
77 R14 Atakora, Chaîne de l' var. Atakora Mountains. ▲ N Benin
Atakora Mountains see Atakora, Chaîne de l'
77 R16 Atakpamé C Togo 07°32′N 01°08′E
146 F11 Atakui Ahal Welaýaty, C Turkmenistan 40°04′N 58°03′E
58 B13 Atalaia do Norte Amazonas, N Brazil 04°22′S 70°10′W
146 M14 Atamyrat prev. Kerki. Lebap Welaýaty, E Turkmenistan 37°52′N 65°06′E
76 I7 Atâr Adrar, W Mauritania 20°30′N 13°03′W
162 G10 Atas Bogd ▲ SW Mongolia 43°17′N 96°47′E
35 P12 Atascadero California, W USA 35°28′N 120°40′W
25 S13 Atascosa River ≈ Texas, SW USA
145 R11 Atasu Karaganda, C Kazakhstan 48°42′N 71°38′E
145 R12 Atasu ≈ Karaganda, C Kazakhstan
193 V15 Atata island Tongatapu Group, S Tonga
136 H10 Atatürk ✈ (İstanbul) İstanbul, NW Turkey 40°58′N 28°50′E
137 N16 Atatürk Barajı ⊠ S Turkey
115 O23 Atavýros prev. Attavyros. Ródos, Dodekánisa, Aegean Sea 36°10′N 27°50′E
115 O23 Atávyros prev. Attávyros. ▲ Ródos, Dodekánisa, Greece, Aegean Sea 36°10′N 27°50′E
Atax see Aude
80 G8 Atbara var. 'Aţbārah. River Nile, NE Sudan 17°42′N 34°E
80 H8 Atbara var. Nahr 'Aţbarah. ≈ Eritrea/Sudan
'Aţbārah/'Aţbarah, Nahr see Atbara
145 P9 Atbasar Akmola, N Kazakhstan 51°49′N 68°18′E
At-Bashi see At-Bashy
147 W9 At-Bashy var. At-Bashi. Narynskaya Oblast', C Kyrgyzstan 41°07′N 75°48′E
22 I10 Atchafalaya Bay bay Louisiana, S USA
22 I8 Atchafalaya River ≈ Louisiana, S USA
Atchin see Aceh
27 Q3 Atchison Kansas, C USA 39°31′N 95°07′W
77 P16 Atebubu C Ghana 07°47′N 01°00′W
105 Q6 Ateca Aragón, NE Spain 41°20′N 01°49′W
40 K11 Atengo, Río ≈ C Mexico
Aternum see Pescara
107 K15 Atessa Abruzzo, C Italy 42°03′N 14°25′E
Ateste see Este
99 E19 Ath var. Aat. Hainaut, SW Belgium 50°38′N 03°47′E
11 Q13 Athabasca Alberta, SW Canada 54°44′N 113°15′W
11 Q12 Athabasca var. Athabaska. ≈ Alberta, SW Canada
11 R10 Athabasca, Lake ◎ Alberta/ Saskatchewan, SW Canada
Athabaska see Athabasca
115 C16 Athamánon ▲ C Greece
97 F17 Athboy Ir. Baile Átha Buí. E Ireland 53°38′N 06°55′W
Athenae see Athína
97 C18 Athenry Ir. Baile Átha an Rí. W Ireland 53°19′N 08°49′W
23 P2 Athens Alabama, S USA 34°48′N 86°58′W
23 T3 Athens Georgia, SE USA 33°57′N 83°24′W
31 T14 Athens Ohio, N USA 39°20′N 82°06′W
20 M10 Athens Tennessee, S USA 35°27′N 84°38′W
25 V7 Athens Texas, SW USA 32°12′N 95°51′W
Athens see Athína
115 B18 Athéras, Akrotírio headland Kefallonía, Iónia Nísiá, Greece, C Mediterranean Sea 38°20′N 20°24′E
181 W4 Atherton Queensland, NE Australia 17°18′S 145°29′E
81 I19 Athi ≈ S Kenya
121 Q2 Athíenou SE Cyprus 35°01′N 33°31′E
115 H19 Athína Eng. Athens, prev. Athínai; anc. Athenae. ● (Greece) Attikí, C Greece 37°59′N 23°44′E
Athínai see Athína
139 S10 Athīyah An Najaf, C Iraq 32°01′N 44°04′E
97 D18 Athlone Ir. Baile Átha Luain. C Ireland 53°25′N 07°56′W
155 F16 Athni Karnātaka, W India 16°43′N 75°04′E
185 C23 Athol Southland, South Island, New Zealand 45°30′S 168°35′E
19 N11 Athol Massachusetts, NE USA 42°35′N 72°11′W
115 I15 Áthos ▲ NE Greece 40°10′N 24°21′E
Athos, Mount see Ágion Óros
Ath Thawrah see Madīnat ath Thawrah
141 P5 Ath Thumāmī spring/well N Saudi Arabia 27°56′N 45°06′E
99 L25 Athus Luxembourg, SE Belgium 49°34′N 05°50′E
97 E19 Athy Ir. Baile Átha Í. C Ireland 52°59′N 06°59′W
78 I10 Ati Batha, C Chad 13°11′N 18°20′E
81 F16 Atiak NW Uganda 03°14′N 32°05′E
57 G17 Atico Arequipa, SW Peru 16°13′S 73°13′W
105 O6 Atienza Castilla-La Mancha, C Spain 41°12′N 02°52′W
39 Q6 Atigun Pass pass Alaska, USA
12 B12 Atikokan Ontario, S Canada 48°45′N 91°38′W
13 O9 Atikonak Lac ◎ Newfoundland and Labrador, E Canada
42 C6 Atitlán, Lago de ◎ W Guatemala
190 L16 Atiu island S Cook Islands
Atjeh see Aceh
123 T9 Atka Magadanskaya Oblast', E Russian Federation 60°45′N 151°35′E
38 H17 Atka Atka Island, Alaska, USA 52°12′N 174°14′W
38 H17 Atka Island island Aleutian Islands, Alaska, USA
127 O7 Atkarsk Saratovskaya Oblast', W Russian Federation 52°15′N 43°48′E
27 U11 Atkins Arkansas, C USA 35°15′N 92°56′W
29 O13 Atkinson Nebraska, C USA 42°31′N 98°57′W
171 T12 Atkri Papua Barat, E Indonesia 01°45′S 130°04′E
41 O13 Atlacomulco var. Atlacomulco de Fabela. México, C Mexico 19°49′N 99°54′W
Atlacomulco de Fabela see Atlacomulco
23 S3 Atlanta state capital Georgia, SE USA 33°45′N 84°23′W
31 R6 Atlanta Michigan, N USA 45°01′N 84°07′W
25 X6 Atlanta Texas, SW USA 33°06′N 94°09′W
29 T15 Atlantic Iowa, C USA 41°24′N 95°00′W
21 Y10 Atlantic North Carolina, SE USA 34°52′N 76°20′W
23 W8 Atlantic Beach Florida, SE USA 30°19′N 81°24′W
18 J17 Atlantic City New Jersey, NE USA 39°23′N 74°27′W
172 L14 Atlantic-Indian Basin undersea feature SW Indian Ocean 60°00′S 15°00′E
172 K13 Atlantic-Indian Ridge undersea feature SW Indian Ocean 53°00′S 15°00′E
54 E4 Atlántico off. Departamento del Atlántico. ⬖ province NW Colombia
64-65 Atlantic Ocean ocean
Atlántico, Departamento del see Atlántico
42 K7 Atlántico Norte, Región Autónoma prev. Zelaya Norte. ⬖ autonomous region NE Nicaragua
42 L10 Atlántico Sur, Región Autónoma prev. Zelaya Sur. ⬖ autonomous region SE Nicaragua
42 I5 Atlántida ⬖ department N Honduras
77 Y15 Atlantika Mountains ▲ E Nigeria
64 J10 Atlantis Fracture Zone tectonic feature NW Atlantic Ocean
74 H7 Atlas Mountains ▲ NW Africa
123 V11 Atlasova, Ostrov island SE Russian Federation
123 V10 Atlasovo Kamchatskiy Kray, E Russian Federation 55°42′N 159°35′E
120 G11 Atlas Saharien var. Saharan Atlas. ▲ Algeria/Morocco
Atlas, Tell see Tell Atlas
120 H10 Atlas Tellien Eng. Tell Atlas. ▲ N Algeria
10 I9 Atlin British Columbia, W Canada 59°31′N 133°41′W
10 I9 Atlin Lake ◎ British Columbia, W Canada
41 P14 Atlixco Puebla, S Mexico 18°55′N 98°26′W
94 B11 Atløyna island S Norway
155 I17 Ātmakūr Andhra Pradesh, C India 15°52′N 78°42′E
23 O8 Atmore Alabama, S USA 31°01′N 87°29′W
101 J20 Atmühl ≈ S Germany
94 H11 Atna ≈ S Norway
164 E13 Atō Yamaguchi, Honshū, SW Japan 34°24′N 131°42′E
57 L21 Atocha Potosí, S Bolivia 20°55′S 66°14′W
27 P12 Atoka Oklahoma, C USA 34°22′N 96°08′W
27 O12 Atoka Lake var. Atoka Reservoir. ⊠ Oklahoma, C USA
Atoka Reservoir see Atoka Lake
33 Q14 Atomic City Idaho, NW USA 43°26′N 112°48′W
40 L10 Atotonilco Zacatecas, C Mexico 24°12′N 102°46′W
Atotonilco see Atotonilco el Alto
40 M13 Atotonilco el Alto var. Atotonilco. Jalisco, SW Mexico 20°35′N 102°30′W
77 N7 Atouila, 'Erg desert N Mali
41 N16 Atoyac var. Atoyac de Alvarez. Guerrero, S Mexico 17°12′N 100°28′W
Atoyac de Alvarez see Atoyac
41 P15 Atoyac, Río ≈ S Mexico
39 O5 Atqasuk Alaska, USA 70°28′N 157°24′W
Atrak/Atrak, Rūd-e see Etrek
95 J20 Ätran ≈ S Sweden
54 C7 Atrato, Río ≈ NW Colombia
Atrek see Etrek
107 K14 Atri Abruzzo, C Italy 42°35′N 13°59′E
Atria see Adria
165 P9 Atsumi Yamagata, Honshū, C Japan 38°38′N 139°36′E
165 S3 Atsuta Hokkaidō, NE Japan 43°28′N 141°24′E
143 Q17 Aţ Ţaff desert C United Arab Emirates
138 G12 Aţ Ţafilah var. Et Tafila, Tafila. Aţ Ţafilah, W Jordan 30°52′N 35°36′E
138 G12 Aţ Ţafilah off. Muḩāfaẓat aţ Ţafilah. ⬖ governorate W Jordan
140 L10 Aţ Ţā'if Makkah, W Saudi Arabia 21°50′N 40°50′E
Attaleia/Attalia see Antalya
23 Q3 Attalla Alabama, S USA 34°01′N 86°05′W
138 L2 At Tall al Abyaḑ var. Tall al Abyaḑ, Tell Abyad, Fr. Tell Abiad. Ar Raqqah, N Syria 36°36′N 34°00′E
At Ta'mīm see Kirkūk
138 L7 Aţ Ţanf Ḩimş, S Syria 33°29′N 38°39′E
163 N9 Attanshiree Dornogovĭ, SE Mongolia 45°36′N 110°30′E
167 T10 Attapu var. Attopeu, Samakhixai. Attapu, S Laos 14°48′N 106°51′E
139 S10 At Ţaqţaqānah An Najaf, C Iraq 32°03′N 43°54′E
Attávyros see Atávyros
12 G9 Attawapiskat Ontario, C Canada 52°55′N 82°26′W
12 F9 Attawapiskat ≈ Ontario, S Canada
12 D9 Attawapiskat Lake ◎ Ontario, C Canada
At Taybé see Ţayyibah
101 F16 Attendorn Nordrhein-Westfalen, W Germany 51°07′N 07°54′E
109 R5 Attersee Salzburg, NW Austria 47°55′N 13°31′E
109 R5 Attersee ◎ N Austria
99 L24 Attert Luxembourg, SE Belgium 49°45′N 05°47′E
138 M4 At Tibnī var. Tibnī. Dayr az Zawr, NE Syria 35°30′N 39°48′E
31 N13 Attica Indiana, N USA 40°17′N 87°15′W
18 E10 Attica New York, NE USA 42°51′N 78°13′W
Attica see Attikí
13 N7 Attikamagen Lake ◎ Newfoundland and Labrador, E Canada
115 H20 Attikí Eng. Attica. ⬖ region C Greece
19 O12 Attleboro Massachusetts, NE USA 41°55′N 71°15′W
109 R5 Attnang Oberösterreich, N Austria 48°02′N 13°44′E
149 U6 Attock City Punjab, E Pakistan 33°52′N 72°20′E
Attopeu see Attapu
25 X8 Attoyac River ≈ Texas, SW USA
38 D16 Attu Attu Island, Alaska, USA 52°53′N 173°18′E
139 Y12 Aţ Ţūbah Al Başrah, E Iraq 30°29′N 47°28′E
140 K4 Aţ Ţubayq plain Jordan/ Saudi Arabia
38 C16 Attu Island island Aleutian Islands, Alaska, USA
75 X8 Aţ Ţūr var. El Ţûr. NE Egypt 28°14′N 33°36′E
155 I21 Āttūr Tamil Nādu, SE India 11°34′N 78°39′E
141 N17 At Turbah SW Yemen 12°42′N 43°31′E
62 I12 Atuel, Río ≈ C Argentina
191 X7 Atuona Hiva Oa, NE French Polynesia 09°47′S 139°03′W
Aturus see Adour
95 M18 Åtvidaberg Östergötland, S Sweden 58°12′N 16°00′E
35 P9 Atwater California, W USA 37°19′N 120°33′W
29 T8 Atwater Minnesota, N USA 45°08′N 94°48′W
26 I2 Atwood Kansas, C USA 39°48′N 101°03′W
31 U12 Atwood Lake ◎ Ohio, N USA
127 P5 Atyashevo Respublika Mordoviya, W Russian Federation 54°36′N 46°04′E
144 F12 Atyrau prev. Gur'yev. Atyrau, W Kazakhstan 47°07′N 51°56′E
144 E11 Atyrau off. Atyrauskaya Oblast', var. Kaz. Atyraū Oblysy; prev. Gur'yevskaya Oblast'. ⬖ province W Kazakhstan
Atyraū Oblysy/ Atyrauskaya Oblast' see Atyrau
Atyrauskaya Oblast' see Atyrau
108 J7 Au Vorarlberg, NW Austria 47°19′N 10°01′E
186 B4 Aua Island island NW Papua New Guinea
103 S16 Aubagne anc. Albania. Bouches-du-Rhône, SE France 43°17′N 05°35′E
99 L25 Aubange Luxembourg, SE Belgium 49°35′N 05°49′E
103 Q6 Aube ⬖ department N France
103 R6 Aube ≈ N France
99 L19 Aubel Liège, E Belgium 50°45′N 05°49′E
103 Q13 Aubenas Ardèche, E France 44°37′N 04°24′E
103 O8 Aubigny-sur-Nère Cher, C France 47°30′N 02°27′E
103 O13 Aubin Aveyron, S France 44°30′N 02°18′E
103 O13 Aubrac, Monts d' ▲ S France
36 J10 Aubrey Cliffs cliff Arizona, SW USA
23 R5 Auburn Alabama, S USA 32°37′N 85°30′W
35 P6 Auburn California, W USA 38°53′N 121°03′W
30 K14 Auburn Illinois, N USA 39°35′N 89°45′W
31 Q11 Auburn Indiana, N USA 41°22′N 85°03′W
20 J7 Auburn Kentucky, S USA 36°52′N 86°42′W
19 P8 Auburn Maine, NE USA 44°05′N 70°15′W
19 N11 Auburn Massachusetts, NE USA 42°11′N 71°47′W
29 S16 Auburn Nebraska, C USA 40°23′N 95°50′W
18 H10 Auburn New York, NE USA 42°55′N 76°31′W
32 H8 Auburn Washington, NW USA 47°18′N 122°13′W
103 N11 Aubusson Creuse, C France 45°58′N 02°10′E
118 E10 Auce Ger. Autz. SW Latvia 56°28′N 22°54′E
102 L15 Auch Lat. Augusta Auscorum, Elimberrum. Gers, S France 43°40′N 00°37′E
77 U16 Auchi Edo, S Nigeria 07°01′N 06°17′E
23 T9 Aucilla River ≈ Florida/ Georgia, SE USA
184 L6 Auckland Auckland, North Island, New Zealand 36°53′S 174°46′E
184 K5 Auckland off. Auckland Region. ⬖ region North Island, New Zealand
184 L6 Auckland ✈ Auckland, North Island, New Zealand 37°01′S 174°49′E
192 J12 Auckland Islands island group S New Zealand
Auckland Region see Auckland
103 O16 Aude ⬖ department S France
103 N16 Aude anc. Atax. ≈ S France
Audenarde see Oudenaarde
Audern see Audru
102 E6 Audierne Finistère, NW France 48°01′N 04°30′W
102 E6 Audierne, Baie d' bay NW France
103 U7 Audincourt Doubs, E France 47°29′N 06°50′E
118 G5 Audru Ger. Audern. Pärnumaa, SW Estonia 58°24′N 24°22′E
29 T14 Audubon Iowa, C USA 41°44′N 94°56′W
101 N17 Aue Sachsen, E Germany 50°35′N 12°42′E
100 H12 Aue ≈ NW Germany
100 L9 Auerbach Bayern, SE Germany 49°41′N 11°41′E
101 M17 Auerbach Sachsen, E Germany 50°30′N 12°24′E
108 I10 Auererrhein ≈ SW Switzerland
101 N17 Auersberg ▲ E Germany 50°30′N 12°42′E
181 W9 Augathella Queensland, E Australia 25°54′S 146°38′E
31 Q12 Auglaize River ≈ Ohio, N USA
83 F22 Augrabies Falls waterfall W South Africa
31 R7 Au Gres River ≈ Michigan, N USA
Augsbourg see Augsburg
101 K22 Augsburg Fr. Augsbourg; anc. Augusta Vindelicorum. Bayern, S Germany 48°22′N 10°54′E
180 I14 Augusta Western Australia 34°18′S 115°10′E
107 L25 Augusta It. Agosta. Sicilia, Italy, C Mediterranean Sea 37°14′N 15°14′E
27 W11 Augusta Arkansas, C USA 35°16′N 91°21′W
23 V3 Augusta Georgia, SE USA 33°29′N 81°58′W
27 O6 Augusta Kansas, C USA 37°40′N 96°59′W
19 Q7 Augusta state capital Maine, NE USA 44°20′N 69°44′W
33 Q8 Augusta Montana, NW USA 47°28′N 112°23′W
Augusta see London
Augusta Auscorum see Auch
Augusta Emerita see Mérida
Augusta Praetoria see Aosta
Augusta Suessionum see Soissons
Augusta Trajana see Stara Zagora
Augusta Treverorum see Trier
Augusta Vangionum see Worms
Augusta Vindelicorum see Augsburg
95 G24 Augustenborg Ger. Augustenburg. Syddanmark, SW Denmark 54°57′N 09°53′E
Augustenburg see Augustenborg
39 Q13 Augustine Island island Alaska, USA
14 L9 Augustines, Lac des ◎ Québec, SE Canada
Augustobona Tricassium see Troyes
Augustodunum see Autun
Augustodurum see Bayeux
Augustoritum Lemovicensium see Limoges
110 O8 Augustów Rus. Avgustov. Podlaskie, NE Poland 53°52′N 22°58′E
Augustow Canal see Augustowski, Kanał
110 O8 Augustowski, Kanał Eng. Augustow Canal, Rus. Avgustovskiy Kanal. canal NE Poland
180 I9 Augustus, Mount ▲ Western Australia 24°42′S 117°42′E
Aujuittuq see Grise Fiord
186 M9 Auki Malaita, N Solomon Islands 08°48′S 160°45′E
21 W8 Aulander North Carolina, SE USA 36°15′N 77°16′W
180 L7 Auld, Lake salt lake C Western Australia
Aulie Ata/Auliye-Ata see Taraz
Äūliekōl see Auliyekol'
144 M8 Auliyekol' Kaz. Äūliekōl; prev. Semiozernoye. Kostanay, N Kazakhstan 52°22′N 64°06′E
106 E10 Aulla Toscana, C Italy 44°15′N 10°00′E
102 F6 Aulne ≈ NW France
Aulong see Ulong
37 T3 Ault Colorado, C USA 40°34′N 104°43′W
95 F22 Aulum var. avlum. Midtjylland, C Denmark 56°16′N 08°48′E
103 N3 Aumale Seine-Maritime, N France 49°45′N 01°43′E
Auminzatau, Gory see Ovminzatovo Tog'lari
77 T14 Auna Niger, W Nigeria 10°13′N 04°43′E
166 L6 Aunglan var. Allanmyo, Myaydo. Magway, C Myanmar (Burma) 19°25′N 95°13′E
95 H21 Auning Midtjylland, C Denmark 56°26′N 10°23′E
192 K17 'Aunu'u Island island W American Samoa
83 E20 Auob var. Oup. ≈ Namibia/South Africa
93 K19 Aura Varsinais-Suomi, SW Finland 60°37′N 22°35′E
109 R5 Aurach ≈ N Austria
Aural, Phnom see Aôral, Phnum
153 O14 Aurangābād Bihār, N India 24°46′N 84°23′E
154 F13 Aurangābād Mahārāshtra, C India 19°52′N 75°22′E
189 V7 Aur Atoll atoll E Marshall Islands
102 G7 Auray Morbihan, NW France 47°40′N 02°59′W
94 G13 Aurdal Oppland, S Norway 60°51′N 09°25′E
94 F8 Aure Møre og Romsdal, S Norway 63°16′N 08°31′E
29 T12 Aurelia Iowa, C USA 42°42′N 95°26′W
Aurelia Aquensis see Baden-Baden
Aurelianum see Orléans
120 J10 Aurès, Massif de l' ▲ NE Algeria
100 F10 Aurich Niedersachsen, NW Germany 53°28′N 07°28′E
103 O13 Aurillac Cantal, C France 44°56′N 02°26′E
Aurine, Alpi see Zillertaler Alpen
Aurium see Ourense
14 H15 Aurora Ontario, S Canada 44°00′N 79°26′W
55 S8 Aurora NW Guyana 06°46′N 59°45′W
37 T4 Aurora Colorado, C USA 39°42′N 104°51′W
30 M11 Aurora Illinois, N USA 41°46′N 88°19′W
31 Q15 Aurora Indiana, N USA 39°01′N 84°55′W
29 W4 Aurora Minnesota, N USA 47°31′N 92°14′W
27 S8 Aurora Missouri, C USA 36°58′N 93°43′W
29 P16 Aurora Nebraska, C USA 40°52′N 98°00′W
36 J5 Aurora Utah, W USA 38°55′N 111°55′W
Aurora see San Francisco, Philippines
Aurora see Maéwo, Vanuatu
94 F10 Aursjøen ◎ S Norway
94 I9 Aursunden ◎ S Norway
83 D21 Aus Karas, SW Namibia 26°38′S 16°19′E
Ausa see Vic
14 E16 Ausable ≈ Ontario, S Canada
31 O3 Au Sable Point headland Michigan, N USA 46°40′N 86°08′W
31 S7 Au Sable Point headland Michigan, N USA 44°19′N 83°20′W
31 R6 Au Sable River ≈ Michigan, N USA
57 H16 Ausangate, Nevado ▲ C Peru 13°47′S 71°13′W
Auschwitz see Oświęcim
Ausculum Apulum see Ascoli Satriano
105 Q4 Ausejo La Rioja, N Spain 42°21′N 02°10′W
108 H7 Ausser Rhoden ⬖ canton NE Switzerland
Aussig see Ústí nad Labem
95 F17 Aust-Agder ⬖ county S Norway
92 P2 Austfonna glacier NE Svalbard
31 P15 Austin Indiana, N USA 38°45′N 85°48′W
29 W11 Austin Minnesota, N USA 43°40′N 92°58′W
35 U5 Austin Nevada, W USA 39°30′N 117°05′W
25 S10 Austin state capital Texas, SW USA 30°16′N 97°45′W
180 J10 Austin, Lake salt lake Western Australia
31 V11 Austintown Ohio, N USA 41°06′N 80°45′W
25 V9 Austonio Texas, SW USA 31°09′N 95°39′W
Australes, Archipel des see Australes, Îles
Australes et Antarctiques Françaises, Terres see French Southern and Antarctic Lands
191 T14 Australes, Îles var. Archipel des Australes, Îles Tubuai, Tubuai Islands, Eng. Austral Islands. island group SW French Polynesia
175 Y11 Austral Fracture Zone tectonic feature S Pacific Ocean
174 M8 Australia continent
181 O7 Australia off. Commonwealth of Australia. ◆ commonwealth republic
Australia, Commonwealth of see Australia
183 Q12 Australian Alps ▲ SE Australia
183 R11 Australian Capital Territory prev. Federal Capital Territory. ⬖ territory SE Australia
Australie, Bassin Nord de l' see North Australian Basin
Austral Islands see Australes, Îles
Austrava see Ostrov
109 T6 Austria off. Republic of Austria, Ger. Österreich. ◆ republic C Europe
Austria, Republic of see Austria
92 K3 Austurland ⬖ region SE Iceland
92 G10 Austvågøya island C Norway
Ausuitoq see Grise Fiord
58 G13 Autazes Amazonas, N Brazil 03°37′S 59°08′W
102 M16 Auterive Haute-Garonne, S France 43°22′N 01°28′E
Autesiodorum see Auxerre
103 N2 Authie ≈ N France
Autissiodorum see Auxerre
40 K14 Autlán var. Autlán de Navarro. Jalisco, SW Mexico 19°48′N 104°20′W
Autlán de Navarro see Autlán
Autricum see Chartres
103 Q9 Autun anc. Ædua, Augustodunum. Saône-et-Loire, C France 46°58′N 04°18′E
Autz see Auce
99 H20 Auvelais Namur, S Belgium 50°27′N 04°38′E
103 P11 Auvergne ⬖ region C France
102 M12 Auvézère ≈ W France
103 P7 Auxerre anc. Autesiodorum, Autissiodorum. Yonne, C France 47°48′N 03°35′E
103 N2 Auxi-le-Château Pas-de-Calais, N France 50°14′N 02°06′E
103 S8 Auxonne Côte d'Or, C France 47°12′N 05°22′E
55 P9 Auyan Tebuy ▲ SE Venezuela 05°48′N 62°27′W
103 O10 Auzances Creuse, C France 46°01′N 02°29′E
27 U8 Ava Missouri, C USA 36°57′N 92°39′W
142 M5 Āvaj Qazvin, N Iran
95 C15 Avaldsnes Rogaland, S Norway 59°21′N 05°16′E
103 Q8 Avallon Yonne, C France 47°30′N 03°54′E
102 K6 Avaloirs, Mont des ▲ NW France 48°27′N 00°11′W
35 S16 Avalon Santa Catalina Island, California, W USA 33°20′N 118°19′W
18 J17 Avalon New Jersey, NE USA 39°04′N 74°42′W
13 V13 Avalon Peninsula peninsula Newfoundland and Labrador, E Canada
Avanersuaq see Avannaarsua
Avannaarsua var Avanersuaq, Dan. Nordgrønland. ⬖ province N Greenland
60 K10 Avaré São Paulo, S Brazil 23°06′S 48°57′W
Avaricum see Bourges
190 H16 Avarua ○ (Cook Islands) Rarotonga, S Cook Islands 21°12′S 159°46′E
190 H16 Avarua Harbour harbor Rarotonga, S Cook Islands
Avasfelsőfalu see Negreşti-Oaş
38 L17 Avatanak Island island Aleutian Islands, Alaska, USA
190 B16 Avatele S Niue 19°06′S 169°55′E
190 H15 Avatiu Harbour harbor Rarotonga, S Cook Islands
Avdeyevka see Avdiyivka
114 J13 Ávdira Anatolikí Makedonía kai Thráki, NE Greece 40°58′N 24°58′E
117 X8 Avdiyivka Rus. Avdeyevka. Donets'ka Oblast', SE Ukraine 48°06′N 37°46′E
Avdzaga see Gurvanbulag
104 G6 Ave ≈ N Portugal
104 G7 Aveiro anc. Talabriga. Aveiro, W Portugal 40°38′N 08°40′W
104 G7 Aveiro ⬖ district N Portugal
Avela see Ávila
99 D18 Avelgem West-Vlaanderen, W Belgium 50°46′N 03°25′E
61 D20 Avellaneda Buenos Aires, E Argentina 34°43′S 58°23′W
107 L17 Avellino anc. Abellinum. Campania, S Italy 40°54′N 14°46′E
35 Q12 Avenal California, W USA 36°00′N 120°07′W
Avenio see Avignon
94 E8 Averøya island S Norway
107 K17 Aversa Campania, S Italy 40°58′N 14°13′E
33 N9 Avery Idaho, NW USA 47°14′N 115°48′W
25 W5 Avery Texas, SW USA 33°33′N 94°46′W
Aves, Islas de see Las Aves, Islas
Avesnes see Avesnes-sur-Helpe
103 Q2 Avesnes-sur-Helpe var. Avesnes. Nord, N France 50°08′N 03°57′E
64 G12 Aves Ridge undersea feature SE Caribbean Sea 14°00′N 63°30′W
95 M14 Avesta Dalarna, C Sweden 60°09′N 16°10′E
103 O14 Aveyron ⬖ department S France
103 N14 Aveyron ≈ S France
107 J15 Avezzano Abruzzo, C Italy 42°02′N 13°26′E
115 D16 Avgó ▲ C Greece 39°31′N 21°24′E
Avgustov see Augustów
Avgustovskiy Kanal see Augustowski, Kanał
96 J9 Aviemore N Scotland, United Kingdom 57°06′N 04°01′W
185 F21 Aviemore, Lake ◎ South Island, New Zealand
103 R15 Avignon anc. Avenio. Vaucluse, SE France 43°57′N 04°49′E
104 M7 Ávila var. Avila; anc. Abela, Abula, Abyla, Avela. Castilla y León, C Spain 40°39′N 04°42′W
104 L8 Ávila ⬖ province Castilla y León, C Spain
104 K2 Avilés Asturias, NW Spain 43°33′N 05°55′W
118 J4 Avinurme Ger. Awwinorm. Ida-Virumaa, NE Estonia 58°58′N 26°53′E
104 H10 Avis Portalegre, C Portugal 39°03′N 07°53′W
Avlum see Aulum
182 M11 Avoca Victoria, SE Australia 37°09′S 143°34′E
29 T14 Avoca Iowa, C USA 41°27′N 95°20′W
182 M11 Avoca River ≈ Victoria, SE Australia
107 L25 Avola Sicilia, Italy, C Mediterranean Sea 36°54′N 15°08′E
18 F10 Avon New York, NE USA 42°53′N 77°41′W
29 P12 Avon South Dakota, N USA 43°00′N 98°03′W
97 M23 Avon ≈ S England, United Kingdom
97 L20 Avon ≈ C England, United Kingdom
36 K13 Avondale Arizona, SW USA 33°25′N 112°20′W
23 X13 Avon Park Florida, SE USA 27°36′N 81°30′W
102 J5 Avranches Manche, N France 48°42′N 01°21′W
Avveel see Ivalojoki, Finland
186 M6 Avuavu var. Kolotambu. Guadalcanal, C Solomon Islands 09°52′S 160°25′E
103 O3 Avure ≈ N France
Avveel see Ivalo, Finland
Avvil see Ivalo
77 O17 Awaaso var. Awaso. SW Ghana 06°10′N 02°18′W
141 X8 Awābī var. Al'Awābī. NE Oman 23°20′N 57°35′E
184 L9 Awakino Waikato, North Island, New Zealand 38°40′S 174°37′E

◆ Country
● Country Capital
◇ Dependent Territory
○ Dependent Territory Capital
⬖ Administrative Regions
✈ International Airport
▲ Mountain
▲ Mountain Range
△ Volcano
≈ River
◎ Lake
⊠ Reservoir

'AWĀLĪ – BAI

142 M15 **'Awālī** C Bahrain *26°07′N 50°33′E*
99 K19 **Awans** Liège, E Belgium *50°39′N 05°30′E*
184 I2 **Awanui** Northland, North Island, New Zealand *35°01′S 173°16′E*
148 M14 **Awārān** Baluchistān, SW Pakistan *26°31′N 65°10′E*
81 K16 **Awara Plain** *plain* NE Kenya
80 M13 **Āwarē** Sumalē, E Ethiopia *08°12′N 44°09′E*
138 M6 **'Awāriḍ, Wādī** *dry watercourse* E Syria
185 B20 **Awarua Point** *headland* South Island, New Zealand *44°15′S 168°03′E*
81 J14 **Āwasa** Southern Nationalities, S Ethiopia *06°54′N 38°26′E*
80 K13 **Āwash** Āfar, NE Ethiopia *08°59′N 40°16′E*
80 K12 **Āwash** *var.* Hawash. ≈ C Ethiopia
Awaso *see* Awaaso
158 H7 **Awat** Xinjiang Uygur Zizhiqu, NW China *40°36′N 80°22′E*
185 J15 **Awatere** ≈ South Island, New Zealand
75 O10 **Awbārī** SW Libya *26°35′N 12°46′E*
75 N9 **Awbāri, Idhān** *var.* Edeyen d'Oubari. *desert* Algeria/Libya
80 M12 **Awdal** *off.* Gobolka Awdal. ◈ N Somalia
80 C13 **Aweil** Northern Bahr el Ghazal, NW South Sudan *08°42′N 27°20′E*
96 H11 **Awe, Loch** ⊗ W Scotland, United Kingdom
77 U16 **Awka** Anambra, SW Nigeria *06°12′N 07°04′E*
39 O6 **Awuna River** ≈ Alaska, USA
Awwinorm *see* Avinurme
Ax *see* Dax
Axarfjördhur *see* Öxarfjörður
103 N17 **Axat** Aude, S France *42°47′N 02°14′E*
99 F16 **Axel** Zeeland, SW Netherlands *51°16′N 03°55′E*
197 P9 **Axel Heiberg Island** *var.* Axel Heiburg. *island* Nunavut, N Canada
Axel Heiburg *see* Axel Heiberg Island
77 O17 **Axim** S Ghana *04°53′N 02°14′W*
114 F13 **Axiós** *var.* Vardar. ≈ Greece/FYR Macedonia *see also* Vardar
Axiós *see* Vardar
103 N17 **Ax-les-Thermes** Ariège, S France *42°43′N 01°49′E*
120 D11 **Ayachi, Jbel** ▲ C Morocco *32°30′N 05°00′W*
61 D22 **Ayacucho** Buenos Aires, E Argentina *37°09′S 58°30′W*
57 F15 **Ayacucho** Ayacucho, S Peru *13°10′S 74°15′W*
57 E16 **Ayacucho** *off.* Departamento de Ayacucho. ◈ *department* SW Peru
Ayacucho, Departamento de *see* Ayacucho
145 W11 **Ayagoz** *var.* Ayaguz, *Kaz.* Ayaköz; *prev.* Sergiopol. Vostochnyy Kazakhstan, E Kazakhstan *47°54′N 80°25′E*
145 V12 **Ayagoz** *var.* Ayaguz, *Kaz.* Ayakoz. ≈ E Kazakhstan
Ayaguz *see* Ayagoz
Ayakagytma *see* Oyoqog'itma
Ayakkuduk *see* Oyoqquduq
158 L10 **Ayakkum Hu** ⊗ NW China
Ayaköz *see* Ayagoz
104 H14 **Ayamonte** Andalucía, S Spain *37°13′N 07°24′W*
123 S11 **Ayan** Khabarovskiy Kray, E Russian Federation *56°27′N 138°09′E*
136 J10 **Ayancık** Sinop, N Turkey *41°56′N 34°35′E*
55 S9 **Ayanganna Mountain** ▲ C Guyana *05°21′N 59°54′W*
77 U16 **Ayangba** Kogi, C Nigeria *07°36′N 07°10′E*
123 U7 **Ayanka** Krasnoyarskiy Kray, E Russian Federation *63°42′N 167°31′E*
54 E7 **Ayapel** Córdoba, NW Colombia *08°16′N 75°10′W*
136 H12 **Ayaş** Ankara, N Turkey *40°02′N 32°21′E*
57 I16 **Ayaviri** Puno, S Peru *14°53′S 70°35′W*
Āybak *see* Aibak
147 N10 **Aydarko'l Ko'li** *Rus.* Ozero Aydarkul'. ⊗ C Uzbekistan
Aydarkul', Ozero *see* Aydarko'l Ko'li
21 W10 **Ayden** North Carolina, SE USA *35°28′N 77°25′W*
136 C15 **Aydın** *var.* Aïdin; *anc.* Tralles Aydın. Aydın, SW Turkey *37°51′N 27°51′E*
136 C15 **Aydın** *var.* Aïdin. ◈ *province* SW Turkey
136 I17 **Aydıncık** İçel, S Turkey *36°08′N 33°17′E*
136 C15 **Aydın Dağları** ⛰ W Turkey
158 L6 **Aydingkol Hu** ⊗ NW China
127 X7 **Aydyrlinskiy** Orenburgskaya Oblast', W Russian Federation *52°03′N 59°54′E*
105 S4 **Ayerbe** Aragón, NE Spain *42°16′N 00°41′W*
Ayers Rock *see* Uluru
Ayeyarwady *see* Ayeyawady
Ayeyarwady *see* Irrawaddy
166 K8 **Ayeyawady** *prev.* Ayeyarwady *var.* Irrawaddy. ◈ *region* SW Myanmar (Burma)
Ayiá *see* Agiá
Ayia Napa *see* Agía Nápa
Ayia Phyla *see* Agía Fýlaxis
Ayiásos/Ayiássos *see* Agiassós
Áyios Evstrátios *see* Ágios Efstrátios
Áyios Kírikos *see* Ágios Kírykos
Áyios Nikólaos *see* Ágios Nikólaos
80 I11 **Āykel** Āmara, N Ethiopia *12°33′N 37°01′E*
123 N9 **Aykhal** Respublika Sakha (Yakutiya), NE Russian Federation *66°07′N 110°25′E*
14 J12 **Aylen Lake** ⊗ Ontario, SE Canada
97 N21 **Aylesbury** SE England, United Kingdom *51°50′N 00°50′W*
105 O6 **Ayllón** Castilla y León, N Spain *41°25′N 03°23′W*
14 F17 **Aylmer** Ontario, S Canada *42°46′N 80°57′W*
14 L12 **Aylmer** Québec, SE Canada *45°23′N 75°51′W*
15 R12 **Aylmer, Lac** ⊗ Québec, SE Canada
8 L9 **Aylmer Lake** ⊗ Northwest Territories, NW Canada
145 V14 **Aynabulak** *Kaz.* Aynabulaq. Almaty, SE Kazakhstan *44°37′N 77°59′E*
Aynabulaq *see* Aynabulak
138 K2 **'Ayn al 'Arab** *Kurd.* Kobanî. Ḥalab, N Syria *36°55′N 38°21′E*
Aynayn *see* 'Aynīn
139 V12 **'Ayn Ḥamūd** Dhī Qār, S Iraq *30°51′N 45°37′E*
147 P12 **Ayní** *prev.* Varzimanor Ayni. W Tajikistan *39°24′N 68°30′E*
140 M10 **'Aynīn** *var.* Aynayn. *spring/well* SW Saudi Arabia *20°52′N 41°41′E*
21 U12 **Aynor** South Carolina, SE USA *33°59′N 79°11′W*
139 Q7 **'Ayn Zāzūh** Al Anbār, C Iraq *33°29′N 42°34′E*
153 N12 **Ayodhya** Uttar Pradesh, N India *26°47′N 82°12′E*
123 S6 **Ayon, Ostrov** *island* NE Russian Federation
105 R11 **Ayora** Valenciana, E Spain *39°04′N 01°04′W*
77 Q11 **Ayorou** Tillabéri, W Niger *14°45′N 00°54′E*
79 E16 **Ayos** Centre, S Cameroon *03°53′N 12°31′E*
76 L5 **'Ayoûn 'Abd el Mâlek** *well* N Mauritania
76 K10 **'Ayoûn el 'Atroûs** *var.* Aïoun el Atrous, Aïoun el Atroûss. Hodh el Gharbi, SE Mauritania *16°38′N 09°36′W*
96 I13 **Ayr** W Scotland, United Kingdom *55°28′N 04°38′W*
96 I13 **Ayr** ≈ W Scotland, United Kingdom
96 I13 **Ayrshire** *cultural region* SW Scotland, United Kingdom
Aysen *see* Aisén
80 L12 **Āysha** Sumalē, E Ethiopia *10°36′N 42°31′E*
144 L14 **Ayteke Bi** *Kaz.* Zhangaqazaly; *prev.* Novokazalinsk. Kzylorda, SW Kazakhstan *45°53′N 62°10′E*
146 K8 **Aytim** Navoiy Viloyati, N Uzbekistan *42°15′N 63°25′E*
181 W4 **Ayton** Queensland, NE Australia *15°54′S 145°19′E*
114 M9 **Aytos** Burgas, E Bulgaria *42°43′N 27°14′E*
171 T11 **Ayu, Kepulauan** *island group* E Indonesia
167 U12 **A Yun Pa** *prev.* Cheo Reo. Gia Lai, S Vietnam *13°19′N 108°27′E*
169 V11 **Ayu, Tanjung** *headland* Borneo, N Indonesia *0°25′N 117°34′E*
41 P16 **Ayutlá** *var.* Ayutla de los Libres. Guerrero, S Mexico *16°51′N 99°16′W*
40 K13 **Ayutla** Jalisco, C Mexico *20°07′N 104°18′W*
Ayutla de los Libres *see* Ayutlá
167 O11 **Ayutthaya** *var.* Phra Nakhon Si Ayutthaya. Phra Nakhon Si Ayutthaya, C Thailand *14°20′N 100°35′E*
136 B13 **Ayvalık** Balıkesir, W Turkey *39°18′N 26°42′E*
99 L20 **Aywaille** Liège, E Belgium *50°28′N 05°40′E*
141 R13 **'Aywat aş Şay'ar, Wādī** *seasonal river* N Yemen
Azaffal *see* Azeffâl
105 T9 **Azahar, Costa del** *coastal region* E Spain
105 S6 **Azaila** Aragón, NE Spain *41°17′N 00°20′W*
104 F10 **Azambuja** Lisboa, C Portugal *39°04′N 08°52′W*
153 N13 **Āzamgarh** Uttar Pradesh, N India *26°03′N 83°10′E*
77 O9 **Azaouâd** *desert* C Mali
77 S10 **Azaouagh, Vallée de l'** *var.* Azaouak. ≈ W Niger
Azaouak *see* Azaouagh, Vallée de l'
61 F14 **Azara** Misiones, NE Argentina *28°03′S 55°42′W*
Azaran *see* Hashtrūd
Azärbaycan/Azärbaycan Respublikasī *see* Azerbaijan
Āžarbāyjān-e Bākhtarī *see* Āžarbāyjān-e Gharbī
142 I4 **Āžarbāyjān-e Gharbī** *off.* Ostān-e Āžarbāyjān-e Gharbī, *Eng.* West Azerbaijan; *prev.* Āžarbāyjān-e Bākhtarī. ◈ *province* NW Iran
Āžarbāyjān-e Gharbī, Ostān-e *see* Āžarbāyjān-e Gharbī
Āžarbāyjān-e Khāvarī *see* Āžarbāyjān-e Sharqī
142 J3 **Āžarbāyjān-e Sharqī** *off.* Ostān-e Āžarbāyjān-e Sharqī, *Eng.* East Azerbaijan; *prev.* Āžarbāyjān-e Khāvarī. ◈ *province* NW Iran
Āžarbāyjān-e Sharqī, Ostān-e *see* Āžarbāyjān-e Sharqī
77 W13 **Azare** Bauchi, N Nigeria *11°41′N 10°09′E*
119 M19 **Azarychy** *Rus.* Ozarichi. Homyel'skaya Voblasts', SE Belarus *52°31′N 29°19′E*
145 P13 **Azat, Gory** *hill* C Kazakhstan
102 L8 **Azay-le-Rideau** Indre-et-Loire, C France *47°16′N 00°25′E*
138 I2 **A'zāz** Ḥalab, NW Syria *36°35′N 37°03′E*
76 H7 **Azeffâl** *var.* Azaffal. *desert* Mauritania/Western Sahara
137 V12 **Azerbaijan** *off.* Azerbaijani Republic, *Az.* Azärbaycan, Azärbaycan Respublikası; *prev.* Azerbaijan SSR. ◆ *republic* SE Asia
Azerbaijani Republic *see* Azerbaijan
Azerbaijan SSR *see* Azerbaijan
74 F7 **Azilal** C Morocco *31°58′N 06°53′W*
Azimabad *see* Patna
19 O6 **Aziscohos Lake** ⊠ Maine, NE USA
Azizbekov *see* Vayk'
Azizie *see* Telish
Aziziya *see* Al 'Azīzīyah
127 T4 **Aznakayevo** Respublika Tatarstan, W Russian Federation *54°55′N 53°15′E*
56 C8 **Azogues** Cañar, S Ecuador *02°44′S 78°48′W*
64 N2 **Azores** *var.* Açores, Ilhas dos Açores, *Port.* Arquipélago dos Açores. *island group* Portugal, NE Atlantic Ocean
64 L8 **Azores-Biscay Rise** *undersea feature* E Atlantic Ocean *19°00′W 42°40′N*
Azotos/Azotus *see* Ashdod
78 K11 **Azoum, Bahr** *seasonal river* SE Chad
126 L12 **Azov** Rostovskaya Oblast', SW Russian Federation *47°07′N 39°26′E*
126 J13 **Azov, Sea of** *Rus.* Azovskoye More, *Ukr.* Azovs'ke More. *sea* NE Black Sea
Azovs'ke More/Azovskoye More *see* Azov, Sea of
138 I10 **Azraq, Wāḥat al** *oasis* N Jordan
Āzro *see* Āzrow
74 G6 **Azrou** C Morocco *33°30′N 05°12′W*
149 R5 **Āzrow** *var.* Āzro. Lōgar, E Afghanistan *34°11′N 69°39′E*
37 P8 **Aztec** New Mexico, SW USA *36°49′N 107°59′W*
36 M13 **Aztec Peak** ▲ Arizona, SW USA *33°48′N 110°54′W*
45 N9 **Azua** *var.* Azua de Compostela. S Dominican Republic *18°29′N 70°44′W*
Azua de Compostela *see* Azua
104 K12 **Azuaga** Extremadura, W Spain *38°16′N 05°40′W*
56 B8 **Azuay** ◈ *province* W Ecuador
164 C13 **Azuchi-Ō-shima** *island* SW Japan
105 O11 **Azuer** ≈ C Spain
43 S17 **Azuero, Península de** *peninsula* S Panama
62 I6 **Azufre, Volcán** *var.* Volcán Lastarria. 🌋 N Chile *25°16′S 68°35′W*
116 J12 **Azuga** Prahova, SE Romania *45°27′N 25°34′E*
61 C22 **Azul** Buenos Aires, E Argentina *36°46′S 59°50′W*
62 I8 **Azul, Cerro** ⛰ NW Argentina *28°28′S 68°43′W*
56 E12 **Azul, Cordillera** ⛰ C Peru
165 P11 **Azuma-san** ▲ Honshū, C Japan *37°44′N 140°05′E*
103 V15 **Azur, Côte d'** *coastal region* SE France
191 Z3 **Azur Lagoon** ⊗ Kiritimati, E Kiribati
'Azza *see* Gaza
Az Zāb al Kabīr *see* Great Zab
138 H7 **Az Zabdānī** *var.* Zabadani. Rīf Dimashq, W Syria *33°45′N 36°07′E*
141 W8 **Aẓ Ẓāhirah** *desert* NW Oman
141 S6 **Aẓ Ẓahrān** *Eng.* Dhahran. Ash Sharqīyah, NE Saudi Arabia *26°18′N 50°02′E*
141 R6 **Aẓ Ẓahrān al Khubar** *var.* Dhahran Al Khobar. ✈ Ash Sharqīyah, NE Saudi Arabia *26°28′N 49°42′E*
75 W7 **Az Zaqāzīq** *var.* Zagazig. N Egypt *30°36′N 31°32′E*
138 H10 **Az Zarqā'** *var.* Zarqa. Az Zarqā', NW Jordan *32°04′N 36°06′E*
138 I11 **Az Zarqā'** *off.* Muḥāfazat az Zarqā', *var.* Zarqa. ◈ *governorate* N Jordan
75 O7 **Az Zāwiyah** *var.* Zawia. NW Libya *32°45′N 12°44′E*
141 N15 **Az Zaydīyah** W Yemen *15°20′N 43°03′E*
74 I11 **Azzel Matti, Sebkha** *var.* Sebkra Azz el Matti. *salt flat* C Algeria
141 P6 **Az Zilfī** Ar Riyāḍ, N Saudi Arabia *26°17′N 44°48′E*
139 Y13 **Az Zubayr** *var.* Al Zubair. Al Başrah, SE Iraq *30°24′N 47°45′E*
Az Zuqur *see* Jabal Zuqar, Jazīrat

B

187 X15 **Ba** *prev.* Mba. Viti Levu, W Fiji *17°35′S 177°40′E*
Ba *see* Da Rāng, Sông
171 P17 **Baa** Pulau Rote, C Indonesia *10°44′S 123°06′E*
138 H7 **Baalbek** *var.* Ba'labakk; *anc.* Heliopolis. E Lebanon *34°00′N 36°15′E*
108 G8 **Baar** Zug, N Switzerland *47°12′N 08°32′E*
81 L17 **Baardheere** *var.* Bardere, *It.* Bardera. Gedo, SW Somalia *02°13′N 42°19′E*
Baargaal *see* Bargaal
99 I15 **Baarle-Hertog** Antwerpen, N Belgium *51°26′N 04°56′E*
99 I15 **Baarle-Nassau** Noord-Brabant, S Netherlands *51°27′N 04°56′E*
98 J11 **Baarn** Utrecht, C Netherlands *52°13′N 05°16′E*
162 H9 **Baatsagaan** *var.* Bayansayr. Bayanhongor, C Mongolia *45°36′N 99°27′E*
114 D13 **Baba** *var.* Buševa, *Gk.* Varnoús. ⛰ FYR Macedonia/Greece
76 H10 **Bababé** Brakna, W Mauritania *16°22′N 13°57′W*
136 G10 **Baba Burnu** *headland* NW Turkey *41°18′N 31°24′E*
117 N13 **Babadag** Tulcea, SE Romania *44°53′N 28°47′E*
137 X10 **Babadağ Dağı** ▲ NE Azerbaijan *41°02′N 48°04′E*
146 H14 **Babadayhan** *Rus.* Babadaykhan; *prev.* Kirovsk. Ahal Welaýaty, C Turkmenistan *37°39′N 60°17′E*
Babadaykhan *see* Babadayhan
146 G14 **Babadurmaz** Ahal Welaýaty, C Turkmenistan *37°39′N 59°05′E*
114 M12 **Babaeski** Kırklareli, NW Turkey *41°26′N 27°06′E*
139 T4 **Bāba Gurgur** Kirkūk, N Iraq *35°34′N 44°18′E*
56 B7 **Babahoyo** *prev.* Bodegas. Los Ríos, C Ecuador *01°53′S 79°31′W*
149 P5 **Bābā, Kūh-e** ⛰ C Afghanistan
171 N12 **Babana** Sulawesi, C Indonesia *02°03′S 119°13′E*
171 Q12 **Babar, Kepulauan** *island group* E Indonesia
171 T12 **Babar, Pulau** *island* Kepulauan Babar, E Indonesia
Bābāsar Pass *see* Babusar Pass
Babashy, Gory *see* Babaşy
146 C9 **Babaşy** *Rus.* Gory Babashy. ⛰ W Turkmenistan
168 M13 **Babat** Sumatera, W Indonesia *02°45′S 104°01′E*
Babatag, Khrebet *see* Bototog', Tizmasi
81 H21 **Babati** Manyara, NE Tanzania *04°12′S 35°45′E*
124 J13 **Babayevo** Vologodskaya Oblast', NW Russian Federation *59°23′N 35°52′E*
127 Q15 **Babayurt** Respublika Dagestan, SW Russian Federation *43°38′N 46°49′E*
33 P6 **Babb** Montana, NW USA *48°51′N 113°26′W*
29 X4 **Babbitt** Minnesota, N USA *47°42′N 91°56′W*
188 E9 **Babeldaob** *var.* Babeldaop, Babelthuap. *island* N Palau
Babeldaop *see* Babeldaob
141 N17 **Bab el Mandeb** *strait* Gulf of Aden/Red Sea
Babelthuap *see* Babeldaob
111 K17 **Babia Góra** *var.* Babia Hora. ▲ Poland/Slovakia *49°33′N 19°32′E*
Babia Hora *see* Babia Góra
Babian Jiang *see* Black River
Babichi *see* Babichy
119 N19 **Babichy** *Rus.* Babichi. Homyel'skaya Voblasts', SE Belarus *52°17′N 30°00′E*
139 U9 **Bābil** *off.* Muḥāfa at Bābil, *var.* Babylon, Al Ḥillah. ◈ *governorate* S Iraq
Bābil, Muḥāfa at al *see* Bābil
112 I10 **Babina Greda** Vukovar-Srijem, E Croatia *45°09′N 18°33′E*
10 K13 **Babine Lake** ⊗ British Columbia, SW Canada
143 O4 **Bābol** *var.* Babul, Balfrush, Barfrush; *prev.* Barfurush. Māzandarān, N Iran *36°34′N 52°39′E*
143 O4 **Bābolsar** *var.* Babulsar; *prev.* Meshed-i-Sar. Māzandarān, N Iran *36°41′N 52°39′E*
36 L16 **Baboquivari Peak** ▲ Arizona, SW USA *31°46′N 111°36′W*
79 G15 **Baboua** Nana-Mambéré, W Central African Republic *05°46′N 14°47′E*
119 M17 **Babruysk** *Rus.* Bobruysk. Mahilyowskaya Voblasts', E Belarus *53°07′N 29°13′E*
Babu *see* Hezhou
Babul *see* Bābol
Babulsar *see* Bābolsar
113 O19 **Babuna** ⛰ C FYR Macedonia
113 O19 **Babuna** ≈ C FYR Macedonia
152 G4 **Babusar Pass** *prev.* Bābāsar Pass. *pass* India/Pakistan
148 K7 **Bābūs, Dasht-e** *Pash.* Bebas, Dasht-i. ⛰ W Afghanistan
171 O1 **Babuyan Channel** *channel* N Philippines
171 O1 **Babuyan Islands** *island* N Philippines
139 T9 **Babylon** *site of ancient city* C Iraq
Babylon *see* Bābil
112 J9 **Bač** *Ger.* Batsch. Vojvodina, NW Serbia *45°24′N 19°17′E*
58 M13 **Bacabal** Maranhão, E Brazil *04°15′S 44°45′W*
41 Y14 **Bacalar** Quintana Roo, SE Mexico *18°38′N 88°17′W*
41 Y14 **Bacalar Chico, Boca** *strait* SE Mexico
171 Q12 **Bacan, Kepulauan** *island group* E Indonesia
171 S12 **Bacan, Pulau** *prev.* Batjan. *island* Maluku, E Indonesia
116 L10 **Bacău** *Hung.* Bákó. Bacău, NE Romania *46°36′N 26°56′E*
116 K11 **Bacău** ◈ *county* E Romania
Bắc Bộ, Vịnh *see* Tonkin, Gulf of
167 T5 **Bắc Can** *var.* Bach Thong. Bắc Thai, N Vietnam *22°07′N 105°50′E*
103 T5 **Baccarat** Meurthe-et-Moselle, NE France *48°27′N 06°46′E*
183 N12 **Bacchus Marsh** Victoria, SE Australia *37°41′S 144°30′E*
40 H4 **Bacerac** Sonora, NW Mexico *30°27′N 108°55′W*
116 L10 **Băcești** Vaslui, E Romania *46°50′N 27°14′E*
167 T6 **Bắc Giang** Ha Bắc, N Vietnam *21°17′N 106°12′E*
54 I5 **Bachaquero** Zulia, NW Venezuela *09°57′N 71°09′W*
Bacher *see* Pohorje
118 M13 **Bacheykava** *Rus.* Bocheykovo. Vitsyebskaya Voblasts', N Belarus *55°01′N 29°09′E*
40 I5 **Bachíniva** Chihuahua, N Mexico *28°41′N 107°13′W*
Bach Quy, Đao *see* Passu Keah
Bach Thong *see* Bắc Can
158 G8 **Bachu** Xinjiang Uygur Zizhiqu, NW China *39°46′N 78°30′E*
9 N8 **Back** ≈ Nunavut, N Canada
112 K10 **Bačka Palanka** *prev.* Palanka. Serbia, NW Serbia *44°22′N 20°57′E*
112 K8 **Bačka Topola** *Hung.* Topolya; *prev. Hung.* Bácstopolya. Vojvodina, N Serbia *45°48′N 19°39′E*
95 J17 **Bäckefors** Västra Götaland, S Sweden *58°49′N 12°07′E*
Bäckermühle Schulzenmühle *see* Żywiec
95 L16 **Bäckhammar** Värmland, C Sweden *59°09′N 14°13′E*
112 K9 **Bački Petrovac** *Hung.* Petrőcz; *prev.* Petrovac, Petrovácz. Vojvodina, NE Serbia *45°10′N 19°34′E*
101 I21 **Backnang** Baden-Württemberg, SW Germany *48°57′N 09°26′E*
167 S15 **Bac Liêu** *var.* Vinh Loi. Minh Hai, S Vietnam *09°17′N 105°44′E*
167 T6 **Bắc Ninh** Ha Bắc, N Vietnam *21°10′N 106°04′E*
40 G4 **Bacoachi** Sonora, NW Mexico *30°36′N 110°00′W*
171 P6 **Bacolod** *off.* Bacolod City. Negros, C Philippines *10°43′N 122°58′E*
Bacolod City *see* Bacolod
171 O4 **Baco, Mount** ▲ Mindoro, N Philippines *12°50′N 121°08′E*
111 K25 **Bácsalmás** Bács-Kiskun, S Hungary *46°07′N 19°20′E*
Bácsjózseffalva *see* Žednik
111 J24 **Bács-Kiskun** *off.* Bács-Kiskun Megye. ◈ *county* S Hungary
Bács-Kiskun Megye *see* Bács-Kiskun
Bácsszenttamás *see* Srbobran
Bácstopolya *see* Bačka Topola
Bactra *see* Balkh
Bada *see* Xilin
155 F21 **Badagara** *var.* Vadakara. Kerala, SW India *11°36′N 75°34′E see also* Vadakara
101 M24 **Bad Aibling** Bayern, SE Germany *47°52′N 12°00′E*
162 I13 **Badain Jaran Shamo** *desert* N China
104 I11 **Badajoz** *anc.* Pax Augusta. Extremadura, W Spain *38°53′N 06°58′W*
104 J11 **Badajoz** ◈ *province* Extremadura, W Spain
149 S2 **Badakhshān** ◈ *province* NE Afghanistan
105 W6 **Badalona** *anc.* Baetulo. Cataluña, E Spain *41°27′N 02°15′E*
154 O11 **Bādāmpāhārh** *var.* Bādāmpāhārh. Odisha, E India *22°04′N 86°06′E*
169 O10 **Badas, Kepulauan** *island group* W Indonesia
109 S6 **Bad Aussee** Salzburg, E Austria *47°35′N 13°44′E*
31 S8 **Bad Axe** Michigan, N USA *43°48′N 83°00′W*
101 G16 **Bad Berleburg** Nordrhein-Westfalen, W Germany *51°03′N 08°24′E*
101 L17 **Bad Blankenburg** Thüringen, C Germany *50°43′N 11°19′E*
Bad Borseck *see* Borsec
101 G18 **Bad Camberg** Hessen, W Germany *50°18′N 08°15′E*
100 L8 **Bad Doberan** Mecklenburg-Vorpommern, N Germany *54°06′N 11°55′E*
101 N14 **Bad Düben** Sachsen, E Germany *51°35′N 12°34′E*
109 X4 **Baden** *var.* Baden bei Wien; *anc.* Aquae Panoniae, Thermae Pannonicae. Niederösterreich, NE Austria *48°01′N 16°14′E*
108 F9 **Baden** Aargau, N Switzerland *47°28′N 08°19′E*
101 G21 **Baden-Baden** *anc.* Aurelia Aquensis. Baden-Württemberg, SW Germany *48°46′N 08°14′E*
Baden bei Wien *see* Baden
101 G22 **Baden-Württemberg** *Fr.* Bade-Wurtemberg. ◈ *state* SW Germany
112 A10 **Baderna** Istra, NW Croatia *45°12′N 13°45′E*
Bade-Wurtemberg *see* Baden-Württemberg
101 H20 **Bad Fredrichshall** Baden-Württemberg, S Germany *49°13′N 09°15′E*
100 P11 **Bad Freienwalde** Brandenburg, NE Germany *52°47′N 14°04′E*
109 Q8 **Badgastein** *var.* Gastein. Salzburg, NW Austria *47°07′N 13°09′E*
Badger State *see* Wisconsin
148 L4 **Bādghīs** ◈ *province* NW Afghanistan
109 T5 **Bad Hall** Oberösterreich, N Austria *48°03′N 14°13′E*
101 J14 **Bad Harzburg** Niedersachsen, C Germany *51°52′N 10°34′E*
101 I16 **Bad Hersfeld** Hessen, C Germany *50°52′N 09°42′E*
98 I10 **Badhoevedorp** Noord-Holland, C Netherlands *52°21′N 04°46′E*
109 Q8 **Bad Hofgastein** Salzburg, NW Austria *47°11′N 13°07′E*
Bad Homburg *see* Bad Homburg vor der Höhe
101 G18 **Bad Homburg vor der Höhe** *var.* Bad Homburg. Hessen, W Germany *50°14′N 08°37′E*
101 E17 **Bad Honnef** Nordrhein-Westfalen, W Germany *50°39′N 07°13′E*
149 Q17 **Badīn** Sind, SE Pakistan *24°38′N 68°53′E*
21 S10 **Badin Lake** ⊠ North Carolina, SE USA
40 I8 **Badiraguato** Sinaloa, C Mexico *25°21′N 107°31′W*
109 R6 **Bad Ischl** Oberösterreich, N Austria *47°43′N 13°36′E*
Badjawa *see* Bajawa
Badje-Sohppar *see* Övre Soppero
101 J18 **Bad Kissingen** Bayern, SE Germany *50°12′N 10°04′E*
Bad Königswart *see* Lázně Kynžvart
101 F19 **Bad Kreuznach** Rheinland-Pfalz, SW Germany *49°51′N 07°52′E*
101 F24 **Bad Krozingen** Baden-Württemberg, SW Germany *47°55′N 07°43′E*
101 G16 **Bad Laasphe** Nordrhein-Westfalen, W Germany *50°57′N 08°24′E*
28 J6 **Badlands** *physical region* North Dakota/South Dakota, N USA
101 K16 **Bad Langensalza** Thüringen, C Germany *51°05′N 10°40′E*
109 T3 **Bad Leonfelden** Oberösterreich, N Austria *48°31′N 14°17′E*
101 I20 **Bad Mergentheim** Baden-Württemberg, SW Germany *49°30′N 09°46′E*
101 H17 **Bad Nauheim** Hessen, W Germany *50°22′N 08°45′E*
101 E17 **Bad Neuenahr-Ahrweiler** Rheinland-Pfalz, W Germany *50°33′N 07°07′E*
Bad Neustadt *see* Bad Neustadt an der Saale
101 J18 **Bad Neustadt an der Saale** *var.* Bad Neustadt. Berlin, C Germany *50°21′N 10°13′E*
Badnur *see* Betūl
100 H13 **Bad Oeynhausen** Nordrhein-Westfalen, NW Germany *52°12′N 08°48′E*
100 J9 **Bad Oldesloe** Schleswig-Holstein, N Germany *53°49′N 10°22′E*
77 Q16 **Badou** C Togo *07°37′N 00°37′E*
Bad Polzin *see* Połczyn-Zdrój
100 H13 **Bad Pyrmont** Niedersachsen, C Germany *51°58′N 09°16′E*
109 X9 **Bad Radkersburg** Steiermark, SE Austria *46°40′N 16°02′E*
139 V8 **Badrah** Wāsiţ, E Iraq *33°06′N 45°58′E*
Badrah *see* Tarialan
101 N24 **Bad Reichenhall** Bayern, SE Germany *47°43′N 12°52′E*
140 K8 **Badr Ḥunayn** Al Madīnah, W Saudi Arabia *23°46′N 38°45′E*
152 K8 **Badrīnāth** ▲ N India *30°44′N 79°29′E*
28 M10 **Bad River** ≈ South Dakota, N USA
30 K4 **Bad River** ≈ Wisconsin, N USA
100 H13 **Bad Salzuflen** Nordrhein-Westfalen, NW Germany *52°06′N 08°45′E*
101 J16 **Bad Salzungen** Thüringen, C Germany *50°48′N 10°15′E*
109 V8 **Bad Sankt Leonhard im Lavanttal** Kärnten, S Austria *46°55′N 14°51′E*
100 K9 **Bad Schwartau** Schleswig-Holstein, N Germany *53°55′N 10°42′E*
101 L24 **Bad Tölz** Bayern, SE Germany *47°44′N 11°34′E*
181 U1 **Badu Island** *island* Queensland, NE Australia
155 K25 **Badulla** Uva Province, C Sri Lanka *06°59′N 81°03′E*
109 X5 **Bad Vöslau** Niederösterreich, NE Austria *47°58′N 16°13′E*
101 I24 **Bad Waldsee** Baden-Württemberg, S Germany *47°54′N 09°44′E*
35 U11 **Badwater Basin** *depression* California, W USA
101 J20 **Bad Windsheim** Bayern, C Germany *49°30′N 10°25′E*
101 J23 **Bad Wörishofen** Bayern, S Germany *48°00′N 10°36′E*
100 G10 **Bad Zwischenahn** Niedersachsen, NW Germany *53°10′N 08°01′E*
104 M13 **Baena** Andalucía, S Spain *36°37′N 04°20′W*
163 V15 **Baengnyong-do** *prev.* Paengnyŏng. *island* NW South Korea
Baeterrae/Baeterrae Septimanorum *see* Béziers
Baetic Cordillera/Baetic Mountains *see* Béticos, Sistemas
Baetulo *see* Badalona
57 K18 **Baeza** Napo, NE Ecuador *0°30′S 77°52′W*
105 N13 **Baeza** Andalucía, S Spain *38°00′N 03°28′W*
79 D15 **Bafang** Ouest, W Cameroon *05°10′N 10°11′E*
76 H12 **Bafatá** C Guinea-Bissau *12°09′N 14°38′W*
149 U5 **Baffa** Khyber Pakhtunkhwa, NW Pakistan *34°30′N 73°18′E*
197 O11 **Baffin Basin** *undersea feature* N Labrador Sea
197 N12 **Baffin Bay** *bay* Canada/Greenland
25 T15 **Baffin Bay** *inlet* Texas, SW USA
196 M12 **Baffin Island** *island* Nunavut, NE Canada
79 E15 **Bafia** Centre, C Cameroon *04°49′N 11°14′E*
77 R14 **Bafilo** NE Togo *09°23′N 01°20′E*
76 J12 **Bafing** ≈ W Africa
76 J12 **Bafoulabé** Kayes, W Mali *13°43′N 10°49′W*
79 D15 **Bafoussam** Ouest, W Cameroon *05°31′N 10°25′E*
143 R9 **Bāfq** Yazd, C Iran *31°35′N 55°21′E*
136 L10 **Bafra** Samsun, N Turkey *41°34′N 35°56′E*
136 L10 **Bafra Burnu** *headland* N Turkey *41°42′N 36°02′E*
143 S12 **Bāft** Kermān, S Iran *29°12′N 56°36′E*
79 N18 **Bafwabalinga** Orientale, NE Dem. Rep. Congo *0°52′N 26°55′E*
79 N18 **Bafwaboli** Orientale, NE Dem. Rep. Congo *0°36′N 26°08′E*
79 N17 **Bafwasende** Orientale, NE Dem. Rep. Congo *01°00′N 27°09′E*
42 K13 **Bagaces** Guanacaste, NW Costa Rica *10°31′N 85°18′W*
153 O12 **Bagaha** Bihār, N India *27°08′N 84°04′E*
155 F16 **Bāgalkot** Karnātaka, W India *16°11′N 75°42′E*
81 J22 **Bagamoyo** Pwani, E Tanzania *06°26′S 38°55′E*
Bagan Datok *see* Bagan Datuk
168 J8 **Bagan Datuk** *var.* Bagan Datok. Perak, Peninsular Malaysia *03°58′N 100°47′E*
171 R7 **Baganga** Mindanao, S Philippines *07°31′N 126°34′E*
168 J9 **Bagansiapiapi** *var.* Pasirpangarayan. Sumatera, W Indonesia *02°06′N 100°52′E*
162 M8 **Baganuur** *var.* Nüürst. Töv, C Mongolia *47°44′N 108°22′E*
Bagaria *see* Bagheria
77 T11 **Bagaroua** Tahoua, W Niger *14°34′N 04°24′E*
79 I20 **Bagata** Bandundu, W Dem. Rep. Congo *03°47′S 17°57′E*
Bagdad *see* Baghdad
123 O13 **Bagdarin** Respublika Buryatiya, S Russian Federation *54°27′N 113°34′E*
61 G17 **Bagé** Rio Grande do Sul, S Brazil *31°22′S 54°06′W*
Bagenalstown *see* Muine Bheag
153 T16 **Bagerhat** *var.* Bagherhat. Khulna, S Bangladesh *22°40′N 89°48′E*
103 P16 **Bages et de Sigean, Étang de** ⊗ S France
33 W17 **Baggs** Wyoming, C USA *41°02′N 107°39′W*
154 F11 **Bāgh** Madhya Pradesh, C India *22°22′N 74°49′E*
139 T8 **Baghdād** *var.* Bagdad, *Eng.* Baghdad. ● Baghdād, C Iraq *33°20′N 44°26′E*
139 T8 **Baghdād** *off.* Muḥāfa at Baghdād, *var.* Amānat al 'Āṣimah. ◈ *governorate* C Iraq
139 T8 **Baghdād** ✈ (Baghdād) Baghdād, C Iraq *33°20′N 44°26′E*
Baghdad *see* Baghdād
Baghdād, Muḥāfa at *see* Baghdād
Bagherhat *see* Bagerhat
107 J23 **Bagheria** *var.* Bagaria. Sicilia, Italy, C Mediterranean Sea *38°05′N 13°31′E*
143 S10 **Bāghīn** Kermān, C Iran *30°50′N 57°00′E*
149 Q3 **Baghlān** Baghlān, NE Afghanistan *36°11′N 68°44′E*
149 Q3 **Baghlān** *var.* Bāghlān. ◈ *province* NE Afghanistan
Baghlān *see* Baghlān
148 M7 **Bāghrān** Helmand, S Afghanistan *32°55′N 64°57′E*
29 T4 **Bagley** Minnesota, N USA *47°31′N 95°24′W*
106 H10 **Bagnacavallo** Emilia-Romagna, C Italy *44°00′N 12°59′E*
102 K16 **Bagnères-de-Bigorre** Hautes-Pyrénées, S France *43°04′N 00°09′E*
102 L17 **Bagnères-de-Luchon** Hautes-Pyrénées, S France *42°46′N 00°34′E*
106 F11 **Bagni di Lucca** Toscana, C Italy *44°01′N 10°38′E*
106 H11 **Bagno di Romagna** Emilia-Romagna, C Italy *43°51′N 11°57′E*
103 R14 **Bagnols-sur-Cèze** Gard, S France *44°10′N 04°37′E*
162 M14 **Bag Nur** ⊗ N China
166 L8 **Bago** *var.* Pegu. Bago, SW Myanmar (Burma) *17°18′N 96°31′E*
171 P6 **Bago** *off.* Bago City. Negros, C Philippines *10°30′N 122°49′E*
166 L7 **Bago** *var.* Pegu. ◈ *region* S Myanmar (Burma)
Bago City *see* Bago
76 M13 **Bagoé** ≈ Ivory Coast/Mali
Bagrāmē *see* Bagrāmī
149 R5 **Bagrāmī** *var.* Bagrāmē. Kābōl, E Afghanistan *34°29′N 69°16′E*
119 B14 **Bagrationovsk** *Ger.* Preussisch Eylau. Kaliningradskaya Oblast', W Russian Federation *54°24′N 20°39′E*
Bagrax *see* Bohu
Bagrax Hu *see* Bosten Hu
56 C10 **Bagua** Amazonas, NE Peru *05°37′S 78°36′W*
171 O2 **Baguio** *off.* Baguio City. Luzon, N Philippines *16°25′N 120°36′E*
Baguio City *see* Baguio
77 V9 **Bagzane, Monts** ▲ N Niger *17°48′N 08°43′E*
Bāḥah, Minṭaqat al *see* Al Bāḥah
Bahama Islands *see* Bahamas, The
Bahamas *see* Bahamas, The
Bahamas, Commonwealth of The *see* Bahamas, The
0 L13 **Bahamas, The** *var.* Bahama Islands. *island group* N West Indies
44 H3 **Bahamas, The** *off.* Commonwealth of The Bahamas. ◆ *commonwealth republic* N West Indies
153 S15 **Baharampur** *prev.* Berhampore. West Bengal, NE India *24°06′N 88°19′E*
146 E12 **Baharly** *var.* Bäherden, *Rus.* Bakharden; *prev.* Bakherden. Ahal Welaýaty, C Turkmenistan *38°30′N 57°18′E*
149 U10 **Bahāwalnagar** Punjab, E Pakistan *30°00′N 73°03′E*
149 T11 **Bahāwalpur** Punjab, E Pakistan *29°25′N 71°40′E*
136 L16 **Bahçe** Osmaniye, S Turkey *37°14′N 36°34′E*
160 J8 **Ba He** ≈ C China
Bäherden *see* Baharly
59 N16 **Bahia** *off.* Estado da Bahia. ◈ *state* E Brazil
61 B24 **Bahía Blanca** Buenos Aires, E Argentina *38°43′S 62°19′W*
40 L15 **Bahía Bufadero** Michoacán, SW Mexico
63 J19 **Bahía Bustamante** Chubut, SE Argentina *45°06′S 66°30′W*
40 D5 **Bahía de los Ángeles** Baja California Norte, NW Mexico
40 C6 **Bahía de Tortugas** Baja California Sur, NW Mexico *27°42′N 114°54′W*
Bahia, Estado da *see* Bahia
42 J4 **Bahía, Islas de la** *Eng.* Bay Islands. *island group* N Honduras
40 E5 **Bahía Kino** Sonora, NW Mexico *28°48′N 111°55′W*
40 E9 **Bahía Magdalena** *var.* Puerto Magdalena. Baja California Sur, NW Mexico *24°34′N 112°07′W*
54 C8 **Bahía Solano** *var.* Ciudad Mutis, Solano. Chocó, W Colombia *06°13′N 77°27′W*
80 I11 **Bahir Dar** *var.* Bahr Dar, Bahrdar Giyorgis. Āmara, N Ethiopia *11°34′N 37°23′E*
141 X8 **Bahlā'** *var.* Bahlah, Bahlat. NW Oman *22°58′N 57°16′E*
Bāhla *see* Bahlā'
Bahlah/Bahlat *see* Bahlā'
152 M11 **Bahraich** Uttar Pradesh, N India *27°35′N 81°36′E*
143 M14 **Bahrain** *off.* State of Bahrain, Dawlat al Bahrayn, *Ar.* Al Baḥrayn, *prev.* Bahrein; *anc.* Tylos, Tyros. ◆ *monarchy* SW Asia
142 M14 **Bahrain** ✈ C Bahrain *26°15′N 50°39′E*
142 M15 **Bahrain, Gulf of** *gulf* Persian Gulf, NW Arabian Sea
Bahrain, State of *see* Bahrain
138 I7 **Baḥrat Mallāḥah** ⊗ W Syria
Bahrayn, Dawlat al *see* Bahrain
Bahr Dar/Bahrdar Giyorgis *see* Bahir Dar
Bahrein *see* Bahrain
Bahr el, Azraq *see* Blue Nile
Bahr el Gebel *see* Central Equatoria
Bahr el Jebel *see* Central Equatoria
80 E13 **Bahr ez Zaref** ≈ Jonglei, E South Sudan
67 R8 **Bahr Kameur** ≈ N Central African Republic
Bahr Tabariya, Sea of *see* Tiberias, Lake
143 W15 **Bāhū Kalāt** Sīstān va Balūchestān, SE Iran *25°42′N 61°28′E*
118 N13 **Bahushewsk** *Rus.* Bogushëvsk. Vitsyebskaya Voblasts', NE Belarus *54°51′N 30°13′E*
Bai *see* Tagow Bāy

◆ Country ● Country Capital ◇ Dependent Territory ○ Dependent Territory Capital ◈ Administrative Regions ✈ International Airport ▲ Mountain ⛰ Mountain Range 🌋 Volcano ≈ River ⊗ Lake ⊠ Reservoir

◆ Country
● Country Capital
◇ Dependent Territory
○ Dependent Territory Capital
◆ Administrative Regions
✕ International Airport
▲ Mountain
▲ Mountain Range
✕ Volcano
≈ River
◎ Lake
◎ Reservoir

BANG HIENG – BASLE

Bang Hieng *see* Xé Banghiang
169 O13 **Bangka-Belitung** *off.* Propinsi Bangka-Belitung. ◈ *province* W Indonesia
169 P11 **Bangkai, Tanjung** *var.* Bankai. *headland* Borneo, N Indonesia 0°21′N 108°53′E
169 S16 **Bangkalan** Pulau Madura, C Indonesia 07°05′S 112°44′E
169 N12 **Bangka, Pulau** *island* W Indonesia
169 N13 **Bangka, Selat** *strait* Sumatera, W Indonesia
169 N13 **Bangka, Selat** *var.* Selat Likupang. *strait* Sulawesi, N Indonesia
168 J11 **Bangkinang** Sumatera, W Indonesia 0°21′N 100°52′E
168 K12 **Bangko** Sumatera, W Indonesia 02°05′S 102°20′E
Bangkok *see* Ao Krung Thep
Bangkok, Bight of *see* Krung Thep, Ao
153 T14 **Bangladesh** *off.* People's Republic of Bangladesh; *prev.* East Pakistan. ◆ *republic* S Asia
Bangladesh, People's Republic of *see* Bangladesh
167 V13 **Ba Ngoi** Khanh Hoa, S Vietnam 11°56′N 109°07′E
Ba Ngoi *see* Cam Ranh
Bangong Co *see* Pangong Tso
97 I18 **Bangor** NW Wales, United Kingdom 53°13′N 04°08′W
97 G15 **Bangor** *Ir.* Beannchar. E Northern Ireland, United Kingdom 54°40′N 05°40′W
19 R6 **Bangor** Maine, NE USA 44°48′N 68°47′W
18 I14 **Bangor** Pennsylvania, NE USA 40°52′N 75°12′W
67 R8 **Bangoran** ≈ S Central African Republic
Bang Phra *see* Trat
Bang Pla Soi *see* Chon Buri
25 Q8 **Bangs** Texas, SW USA 31°43′N 99°07′W
167 N13 **Bang Saphan** *var.* Bang Saphan Yai. Prachuap Khiri Khan, SW Thailand 11°10′N 99°33′E
Bang Saphan Yai *see* Bang Saphan
36 I8 **Bangs, Mount** ▲ Arizona, SW USA 36°47′N 113°51′W
93 E15 **Bangsund** Nord-Trøndelag, C Norway 64°22′N 11°22′E
171 O2 **Bangued** Luzon, N Philippines 17°36′N 120°40′E
79 I15 **Bangui** ● (Central African Republic) Ombella-Mpoko, SW Central African Republic 04°21′N 18°32′E
79 I15 **Bangui** ✈ Ombella-Mpoko, SW Central African Republic 04°19′N 18°34′E
83 N16 **Bangula** Southern, S Malawi 16°38′S 35°04′E
Bangwaketse *see* Southern
82 K12 **Bangweulu, Lake** *var.* Lake Bengweulu. ◎ N Zambia
121 V13 **Banhā** *var.* Benha. N Egypt 30°28′N 31°11′E
Ban Hat Yai *see* Hat Yai
167 Q7 **Ban Hin Heup** Viangchan, C Laos 18°37′N 102°19′E
Ban Houayxay/Ban Houei Sai *see* Houayxay
167 O12 **Ban Hua Hin** *var.* Hua Hin. Prachuap Khiri Khan, SW Thailand 12°34′N 99°58′E
79 L14 **Bani** Haute-Kotto, E Central African Republic 07°06′N 22°51′E
45 O9 **Baní** S Dominican Republic 18°19′N 70°21′W
77 N12 **Bani** ≈ S Mali
Bāmīān *see* Bāmyān
Banias *see* Bāniyās
77 S11 **Bani Bangou** Tillabéri, SW Niger 15°04′N 02°40′E
76 M12 **Banifing** *var.* Ngorolaka. ≈ Burkina Faso/Mali
Banijska Palanka *see* Glina
77 R13 **Banikoara** N Benin 11°18′N 02°26′E
75 W9 **Banī Mazār** *var.* Beni Mazâr. C Egypt 28°29′N 30°48′E
114 K8 **Baniski Lom** ≈ N Bulgaria
21 U7 **Banister River** ≈ Virginia, NE USA
121 V14 **Banī Suwayf** *var.* Beni Suef. N Egypt 29°09′N 31°04′E
75 O8 **Banī Walīd** NW Libya 31°46′N 13°59′E
138 H5 **Bāniyās** *var.* Banias, Baniyas, Paneas. Ṭarṭūs, W Syria 35°12′N 35°57′E
113 K14 **Banja** Serbia, W Serbia 43°33′N 19°35′E
Banjak, Kepulauan *see* Banyak, Kepulauan
112 J12 **Banja Koviljača** Serbia, W Serbia 44°31′N 19°11′E
112 G11 **Banja Luka** ◈ Republika Srpska, NW Bosnia and Herzegovina
169 T13 **Banjarmasin** *prev.* Bandjarmasin. Borneo, C Indonesia 03°22′S 114°33′E
76 F11 **Banjul** *prev.* Bathurst. ● (Gambia) W Gambia 13°26′N 16°43′W
76 F11 **Banjul** ✈ W Gambia 13°18′N 16°39′W
Bank *see* Bankä
137 Y13 **Bankä** *Rus.* Bank. SE Azerbaijan 39°25′N 49°13′E
167 S11 **Ban Kadian** *var.* Ban Kadiene. Champasak, S Laos 14°25′N 105°42′E
Ban Kadiene *see* Ban Kadian
Bankai *see* Bangkai, Tanjung
166 M14 **Ban Kam Phuam** Phangnga, SW Thailand 09°16′N 98°24′E
Ban Kantang *see* Kantang
77 O11 **Bankass** Mopti, S Mali 14°05′N 03°30′W
95 L19 **Bankeryd** Jönköping, S Sweden 57°51′N 14°07′E
83 K16 **Banket** Mashonaland West, N Zimbabwe 17°23′S 30°24′E
167 T11 **Ban Khamphô** Attapu, S Laos 14°35′N 106°18′E
23 O4 **Bankhead Lake** ▣ Alabama, S USA
77 Q11 **Bankilaré** Tillabéri, SW Niger 14°34′N 00°41′E
Banks, Îles *see* Banks Islands
10 I14 **Banks Island** *island* British Columbia, SW Canada
187 R12 **Banks Islands** *Fr.* Îles Banks. *island group* N Vanuatu
23 U8 **Banks Lake** ◎ Georgia, SE USA
32 K8 **Banks Lake** ▣ Washington, NW USA
185 I19 **Banks Peninsula** *peninsula* South Island, New Zealand
183 Q15 **Banks Strait** *strait* SW Tasman Sea
Ban Kui Nua *see* Kui Buri
153 R16 **Bānkura** West Bengal, NE India 23°14′N 87°05′E
167 S8 **Ban Lakxao** *var.* Lak Sao. Bolikhamxai, C Laos 18°10′N 104°58′E
167 O16 **Ban Lam Phai** Songkhla, SW Thailand 06°43′N 100°57′E
Ban Mae Sot *see* Mae Sot
Ban Mae Suai *see* Mae Suai
Ban Mak Khaeng *see* Udon Thani
166 M3 **Banmauk** Sagaing, N Myanmar (Burma) 24°26′N 95°54′E
Banmo *see* Bhamo
167 T10 **Ban Mun-Houamuang** S Laos 15°11′N 106°44′E
97 F14 **Bann** *var.* Lower Bann, Upper Bann. ≈ N Northern Ireland, United Kingdom
167 S10 **Ban Nadou** Salavan, S Laos 15°51′N 105°38′E
167 S9 **Ban Nakala** Savannakhét, S Laos 16°14′N 105°09′E
167 Q8 **Ban Nakha** Viangchan, C Laos 18°13′N 102°29′E
167 S9 **Ban Nakham** Khammouan, S Laos 17°10′N 105°25′E
167 P7 **Ban Namoun** Xaignabouli, N Laos 18°40′N 101°34′E
167 O17 **Ban Nang Sata** Yala, SW Thailand 06°15′N 101°13′E
167 N15 **Ban Na San** Surat Thani, SW Thailand 08°53′N 99°17′E
167 R7 **Ban Nasi** Xiangkhoang, N Laos 19°37′N 103°33′E
44 I3 **Bannerman Town** Eleuthera Island, C The Bahamas 24°38′N 76°09′W
35 V15 **Banning** California, W USA 33°55′N 116°52′W
Banningville *see* Bandundu
167 S11 **Ban Nongsim** Champasak, S Laos 14°45′N 106°00′E
149 S7 **Bannu** *prev.* Edwardesabad. Khyber Pakhtunkhwa, NW Pakistan 33°00′N 70°36′E
Bañolas *see* Banyoles
56 C7 **Baños** Tungurahua, C Ecuador 01°26′S 78°24′W
Bánovce *see* Bánovce nad Bebravou
111 I19 **Bánovce nad Bebravou** *var.* Bánovce, *Hung.* Bán. Trenčiansky Kraj, W Slovakia 48°43′N 18°15′E
112 I12 **Banovići** ◈ Federacija Bosne I Hercegovine, E Bosnia and Herzegovina
Banow *see* Andarāb
Ban Pak Phanang *see* Pak Phanang
167 O7 **Ban Pan Nua** Lampang, NW Thailand 18°51′N 99°57′E
167 Q9 **Ban Phai** Khon Kaen, E Thailand 16°00′N 102°42′E
167 Q8 **Ban Phônhông** *var.* Phônhông. C Laos 18°29′N 102°26′E
167 T9 **Ban Phou A Douk** Khammouan, C Laos 17°12′N 106°07′E
167 Q8 **Ban Phu** Uthai Thani, W Thailand
167 O11 **Ban Pong** Ratchaburi, W Thailand 13°49′N 99°53′E
190 I3 **Banraeaba** Tarawa, W Kiribati 01°20′N 173°02′E
167 N10 **Ban Sai Yok** Kanchanaburi, W Thailand 14°24′N 98°54′E
Ban Sattahip/Ban Sattahipp *see* Sattahip
Ban Sichon *see* Sichon
Ban Si Racha *see* Si Racha
111 J19 **Banská Bystrica** *Ger.* Neusohl, *Hung.* Besztercebánya. Banskobystricky Kraj, C Slovakia 48°46′N 19°08′E
111 K20 **Banskobystrický Kraj** ◈ *region* C Slovakia
167 R8 **Ban Sôppheung** Bolikhamxai, C Laos 18°33′N 104°18′E
Ban Sop Prap *see* Sop Prap
152 G15 **Bānswāra** Rājasthān, N India 23°32′N 74°28′E
167 N15 **Ban Ta Khun** Surat Thani, SW Thailand 08°53′N 98°52′E
Ban Takua Pa *see* Takua Pa
167 S8 **Ban Talak** Khammouan, C Laos 17°33′N 105°40′E
77 R15 **Bantè** W Benin 08°25′N 01°58′E
167 Q11 **Bânteay Méan Choăy** *var.* Sisŏphŏn. Bătdâmbâng, NW Cambodia 13°37′N 102°58′E
167 N16 **Banten** *off.* Propinsi Banten. ◈ *province* W Indonesia
Propinsi Banten *see* Banten
167 Q8 **Ban Thabôk** Bolikhamxai, C Laos 18°21′N 103°12′E
167 T9 **Ban Tôp** Savannakhét, S Laos 16°07′N 106°07′E
97 B21 **Bantry** *Ir.* Beanntraí. Cork, SW Ireland 51°41′N 09°27′W
97 A21 **Bantry Bay** *Ir.* Bá Bheanntraí. *bay* SW Ireland
155 F19 **Bantvāl** *var.* Bantwāl. Karnātaka, E India 12°57′N 75°04′E
Bantwāl *see* Bantvāl
114 N9 **Banya** Burgas, E Bulgaria 42°46′N 27°49′E
168 G10 **Banyak, Kepulauan** *prev.* Kepulauan Banjak. *island group* NW Indonesia
105 U8 **Banya, La** *headland* E Spain 40°34′N 00°37′E
79 E14 **Banyo** Adamaoua, NW Cameroon 06°47′N 11°50′E
105 X4 **Banyoles** *var.* Bañolas. Cataluña, NE Spain 42°07′N 02°46′E
167 N16 **Ban Yong Sata** Trang, SW Thailand 07°09′N 99°42′E
195 X14 **Banzare Coast** *physical region* Antarctica
173 Q14 **Banzare Seamounts** *undersea feature* S Indian Ocean
Banzart *see* Bizerte
163 Q12 **Baochang** *var.* Taibus Qi. Nei Mongol Zizhiqu, N China 41°55′N 115°22′E
161 O3 **Baoding** *var.* Pao-ting; *prev.* Tsingyuan. Hebei, E China 38°47′N 115°30′E
Baoebaoe *see* Baubau
Baoi, Oileán *see* Dursey Island
160 J6 **Baoji** *var.* Pao-chi, Paoki. Shaanxi, C China 34°23′N 107°16′E
163 U9 **Baokang** *var.* Hoqin Zuoyi Zhongji. Nei Mongol Zizhiqu, N China 44°08′N 123°18′E
186 L8 **Baolo** Santa Isabel, N Solomon Islands 07°41′S 158°47′E
167 U13 **Bao Lôc** Lâm Đông, S Vietnam 11°33′N 107°48′E
163 Z7 **Baoqing** Heilongjiang, NE China 46°15′N 132°12′E
Baoqing *see* Shaoyang
79 H15 **Baoro** Nana-Mambéré, W Central African Republic 05°40′N 16°00′E
160 E12 **Baoshan** *var.* Pao-shan. Yunnan, SW China 25°05′N 99°07′E
163 N13 **Baotou** *var.* Pao-t'ou, Paotow. Nei Mongol Zizhiqu, N China 40°38′N 109°59′E
76 L14 **Baoulé** ≈ S Mali
76 K12 **Baoulé** ≈ W Mali
Bao Yèn *see* Phô Rang
Ba-Pahalaborwa *see* Phalaborwa
103 O2 **Bapaume** Pas-de-Calais, N France 50°06′N 02°50′E
14 J13 **Baptiste Lake** ◎ Ontario, SE Canada
Bapu *see* Meigu
Baqanas *see* Bakanas
Baqbaqty *see* Bakbakty
159 P14 **Baqên** *var.* Dartang. Xizang Zizhiqu, W China 31°50′N 94°08′E
138 F14 **Bāqir, Jabal** ▲▲ S Jordan
139 T7 **Ba'qūbah** *var.* Qubba. Diyālá, C Iraq 33°45′N 44°40′E
Ba'qūbah *see* Diyālá
62 H5 **Baquedano** Antofagasta, N Chile 23°20′S 69°50′W
Baquerizo Moreno *see* Puerto Baquerizo Moreno
113 J18 **Bar** *It.* Antivari. S Montenegro 42°02′N 19°09′E
116 M6 **Bar** Vinnyts'ka Oblast', C Ukraine 49°05′N 27°40′E
80 E10 **Bara** Northern Kordofan, C Sudan 13°42′N 30°21′E
81 M18 **Baraawe** *It.* Brava. Shabeellaha Hoose, S Somalia 01°10′N 43°59′E
152 M12 **Bāra Banki** Uttar Pradesh, N India 26°56′N 81°11′E
30 L8 **Baraboo** Wisconsin, N USA 43°27′N 89°45′W
30 K8 **Baraboo Range** *hill range* Wisconsin, N USA
Baracaldo *see* San Vicente de Barakaldo
15 Y6 **Barachois** Québec, SE Canada 48°37′N 64°14′W
44 J7 **Baracoa** Guantánamo, E Cuba 20°23′N 74°31′W
61 C19 **Baradero** Buenos Aires, E Argentina 33°50′S 59°30′W
183 R6 **Baradine** New South Wales, SE Australia 30°55′S 149°03′E
Baraf Daja Islands *see* Damar, Kepulauan
Baragarh *see* Bargarh
81 I17 **Baragoi** Samburu, W Kenya 01°39′N 36°46′E
45 N9 **Barahona** SW Dominican Republic 18°13′N 71°07′W
153 W13 **Barail Range** ▲▲ NE India
Baraka *see* Barka
80 G10 **Barakat** Gezira, C Sudan 14°18′N 33°32′E
Barakī *see* Barakī Barak
149 Q6 **Barakī Barak** *var.* Barakī, Baraki Rajan. Lōgar, E Afghanistan 33°58′N 68°58′E
Baraki Rajan *see* Barakī Barak
154 N11 **Bārākot** Odisha, E India 21°35′N 85°00′E
55 S7 **Barama River** ≈ N Guyana
155 E14 **Bārāmati** Mahārāshtra, W India 18°12′N 74°39′E
152 H5 **Bārāmūla** Jammu and Kashmir, NW India 34°15′N 74°24′E
119 N14 **Baran'** Vitsyebskaya Voblasts', NE Belarus 54°29′N 30°18′E
152 I14 **Bārān** Rājasthān, N India 25°08′N 76°32′E
Barānān, Shākh-i *see* Beranan, Shax-î
119 I17 **Baranavichy** *Pol.* Baranowicze, *Rus.* Baranovichi. Brestskaya Voblasts', SW Belarus 53°08′N 26°02′E
123 T6 **Baranikha** Chukotskiy Avtonomnyy Okrug, NE Russian Federation 68°29′N 168°13′E
75 Y11 **Baranīs** *var.* Berenice, Mînâ Baranîs. SE Egypt 23°58′N 35°29′E
116 M4 **Baranivka** Zhytomyrs'ka Oblast', N Ukraine 50°16′N 27°40′E
39 W14 **Baranof Island** *island* Alexander Archipelago, Alaska, USA
Baranovichi/Baranowicze *see* Baranavichy
111 N15 **Baranów Sandomierski** Podkarpackie, SE Poland 50°28′N 21°31′E
111 I26 **Baranya** *off.* Baranya Megye. ◈ *county* S Hungary
Baranya Megye *see* Baranya
153 R13 **Barāri** Bihār, NE India 25°31′N 87°23′E
22 L10 **Barataria Bay** *bay* Louisiana, S USA
Barat Daya, Kepulauan *see* Damar, Kepulauan
118 L12 **Baravukha** *Rus.* Borovukha. Vitsyebskaya Voblasts', N Belarus 55°34′N 28°36′E
54 E11 **Baraya** Huila, C Colombia 03°11′N 75°04′W
59 M21 **Barbacena** Minas Gerais, SE Brazil 21°13′S 43°47′W
54 B13 **Barbacoas** Nariño, SW Colombia 01°38′N 78°08′W
54 L6 **Barbacoas** Aragua, N Venezuela 09°29′N 66°58′W
45 Z13 **Barbados** ◆ *commonwealth republic* SE West Indies
47 S3 **Barbados** *island* Barbados
105 U11 **Barbaria, Cap de** *var.* Cabo de Berberia. *headland* Formentera, E Spain 38°39′N 01°24′E
114 N13 **Barbaros** Tekirdağ, NW Turkey 40°55′N 27°28′E
74 A11 **Barbas, Cap** *headland* S Western Sahara 22°14′N 16°45′W
105 T5 **Barbastro** Aragón, NE Spain 42°02′N 00°07′E
104 K16 **Barbate** ≈ SW Spain
104 K16 **Barbate de Franco** Andalucía, S Spain 36°11′N 05°55′W
83 K21 **Barberton** Mpumalanga, NE South Africa 25°48′S 31°03′E
31 U12 **Barberton** Ohio, N USA 41°02′N 81°37′W
102 K12 **Barbezieux-St-Hilaire** Charente, W France 45°28′N 00°09′W
54 G9 **Barbosa** Boyaca, C Colombia 05°57′N 73°37′W
21 N7 **Barbourville** Kentucky, S USA 36°52′N 83°54′W
45 W9 **Barbuda** *island* N Antigua and Barbuda
181 W8 **Barcaldine** Queensland, E Australia 23°33′S 145°21′E
Barcarozsnyó *see* Râşnov
104 I11 **Barcarrota** Extremadura, W Spain 38°31′N 06°51′W
Barcău *see* Berettyó
Barce *see* Al Marj
107 L23 **Barcellona** *var.* Barcellona Pozzo di Gotto. Sicilia, Italy, C Mediterranean Sea 38°10′N 15°15′E
Barcellona Pozzo di Gotto *see* Barcellona
105 W6 **Barcelona** *anc.* Barcino, Barcinona. Cataluña, E Spain 41°25′N 02°10′E
55 N5 **Barcelona** Anzoátegui, NE Venezuela 10°08′N 64°43′W
105 S5 **Barcelona** ◈ *province* Cataluña, NE Spain
105 W6 **Barcelona** ✈ Cataluña, E Spain 41°25′N 02°10′E
103 U14 **Barcelonnette** Alpes-de-Haute-Provence, SE France 44°24′N 06°37′E
58 E12 **Barcelos** Amazonas, N Brazil 0°59′S 62°58′W
104 G5 **Barcelos** Braga, N Portugal 41°32′N 08°37′W
110 I10 **Barcin** *Ger.* Bartschin. Kujawski-pomorskie, C Poland 52°51′N 17°55′E
Barcino/Barcinona *see* Barcelona
Barcoo *see* Cooper Creek
111 H26 **Barcs** Somogy, SW Hungary 45°58′N 17°26′E
137 W11 **Bärdä** *Rus.* Barda. C Azerbaijan 40°25′N 47°07′E
Barda *see* Bärdä
78 H5 **Bardaï** Tibesti, N Chad 21°21′N 16°56′E
139 R2 **Bardaṛash** Dahūk, N Iraq 36°32′N 43°36′E
92 K3 **Bárðarbunga** ▲ C Iceland 64°39′N 17°30′W
92 K2 **Bárðárdalur** *valley* C Iceland
139 Q7 **Bardasah** Al Anbār, SW Iraq 34°02′N 42°28′E
153 S16 **Barddhamān** West Bengal, NE India 23°10′N 88°03′E
111 N18 **Bardejov** *Ger.* Bartfeld, *Hung.* Bártfa. Presovský Kraj, E Slovakia 49°17′N 21°18′E
105 R4 **Bárdenas Reales** *physical region* N Spain
Bardera/Bardere *see* Baardheere
Bardesīr *see* Bardsīr
106 E9 **Bardi** Emilia-Romagna, C Italy 44°39′N 09°44′E
106 A8 **Bardonecchia** Piemonte, W Italy 45°04′N 06°40′E
97 H19 **Bardsey Island** *island* NW Wales, United Kingdom
143 S11 **Bardsīr** *var.* Bardesīr, Mashīz. Kermān, C Iran 29°58′N 56°29′E
20 L6 **Bardstown** Kentucky, S USA 37°49′N 85°29′W
Barduli *see* Barletta
20 G7 **Bardwell** Kentucky, S USA 36°52′N 89°01′W
152 K11 **Bareilly** *var.* Bareli. Uttar Pradesh, N India 28°20′N 79°24′E
Bareli *see* Bareilly
Baren *see* Jiashi
98 H13 **Barendrecht** Zuid-Holland, SW Netherlands 51°52′N 04°31′E
102 M3 **Barentin** Seine-Maritime, N France 49°33′N 00°57′E
92 N3 **Barentsburg** Spitsbergen, W Svalbard 78°01′N 14°19′E
Barentsevo More/Barents Havet *see* Barents Sea
92 O3 **Barentsøya** *island* E Svalbard
197 T11 **Barents Plain** *undersea feature* N Barents Sea
125 P3 **Barents Sea** *Nor.* Barents Havet, *Rus.* Barentsevo More. *sea* Arctic Ocean
197 U14 **Barents Trough** *undersea feature* SW Barents Sea 75°00′N 29°00′E
80 I9 **Barentu** W Eritrea 15°08′N 37°35′E
102 J3 **Barfleur** Manche, N France 49°41′N 01°18′W
102 J3 **Barfleur, Pointe de** *headland* N France 49°46′N 01°09′W
Barfrush/Barfurush *see* Bābol
158 H14 **Barga** Xizang Zizhiqu, W China 30°51′N 81°20′E
80 Q12 **Bargaal** *prev.* Baargaal. Bari, NE Somalia 11°12′N 51°04′E
154 M12 **Bargarh** *var.* Baragarh. Odisha, E India 21°25′N 83°35′E
105 N9 **Bargas** Castilla-La Mancha, C Spain 39°56′N 04°00′W
81 I15 **Bargē** Southern Nationalities, S Ethiopia 06°11′N 37°04′E
106 A9 **Barge** Piemonte, NE Italy 44°49′N 07°21′E
153 U16 **Barguna** Barisal, S Bangladesh 22°09′N 90°07′E
Bärguşad *see* Vorotan
123 N13 **Barguzin** Respublika Buryatiya, S Russian Federation 53°37′N 109°47′E
153 O13 **Barhaj** Uttar Pradesh, N India 26°16′N 83°44′E
183 N10 **Barham** New South Wales, SE Australia 35°39′S 144°09′E
152 J12 **Barhan** Uttar Pradesh, N India 27°21′N 78°11′E
19 S7 **Bar Harbor** Mount Desert Island, Maine, NE USA 44°23′N 68°14′W
153 R14 **Barharwa** Jhārkhand, NE India 24°52′N 87°47′E
153 P15 **Barhi** Jhārkhand, N India 24°19′N 85°25′E
107 O17 **Bari** *var.* Bari delle Puglie; *anc.* Barium. Puglia, SE Italy 41°06′N 16°52′E
80 P12 **Bari** *Off.* Gobolka Bari. ◈ *region* NE Somalia
167 T14 **Ba Ria** *var.* Châu Thanh. Ba Ria-Vung Tau, S Vietnam 10°30′N 107°10′E
81 G21 **Bariadi** Simiyu, NE Tanzania 02°48′S 33°59′E
Bāridah *see* Al Bāridah
Bari delle Puglie *see* Bari
Bari, Gobolka *see* Bari
Barikot *see* Barīkowṯ
149 T4 **Barīkowṯ** *var.* Barikot. Kunar, NE Afghanistan 35°18′N 71°36′E
42 C4 **Barillas** *var.* Santa Cruz Barillas. Huehuetenango, NW Guatemala 15°50′N 91°20′W
54 J6 **Barinas** Barinas, W Venezuela 08°36′N 70°15′W
54 I7 **Barinas** *off.* Estado Barinas; *prev.* Zamora. ◈ *state* C Venezuela
Barinas, Estado *see* Barinas
81 H18 **Baringo** ◈ *county* C Kenya
54 I6 **Barinitas** Barinas, NW Venezuela 08°45′N 70°26′W
154 P11 **Bāripada** Odisha, E India 21°56′N 86°43′E
60 K9 **Bariri** São Paulo, S Brazil 22°04′S 48°46′W
75 W11 **Bārīs** *var.* Bârîs. S Egypt 24°28′N 30°39′E
152 G14 **Bari Sādri** Rājasthān, N India 24°25′N 74°28′E
153 U16 **Barisal** Barisal, S Bangladesh 22°41′N 90°20′E
153 U16 **Barisal** ◈ *division* S Bangladesh
168 I10 **Barisan, Pegunungan** ▲▲ Sumatera, W Indonesia
169 T12 **Barito, Sungai** ≈ Borneo, C Indonesia
Barium *see* Bari
Bårjås *see* Porjus
80 I9 **Barka** *var.* Baraka, *Ar.* Khawr Barakah. *seasonal river* Eritrea/Sudan
Barka *see* Al Marj
160 H8 **Barkam** Sichuan, C China 31°56′N 102°22′E
118 J9 **Barkava** C Latvia 56°43′N 26°34′E
10 M15 **Barkerville** British Columbia, SW Canada 53°06′N 121°35′W
14 J12 **Bark Lake** ◎ Ontario, SE Canada
20 H7 **Barkley, Lake** ▣ Kentucky/Tennessee, S USA
10 K17 **Barkley Sound** *inlet* British Columbia, W Canada
83 J24 **Barkly East** *Afr.* Barkly-Oos. Eastern Cape, SE South Africa 30°58′S 27°35′E
Barkly-Oos *see* Barkly East
181 S4 **Barkly Tableland** *plateau* Northern Territory/Queensland, N Australia
Barkly-Wes *see* Barkly West
83 H22 **Barkly West** *Afr.* Barkly-Wes. Northern Cape, N South Africa 28°32′S 24°32′E
159 O5 **Barkol** *var.* Barkol Kazak Zizhixian. Xinjiang Uygur Zizhiqu, NW China 43°37′N 93°01′E
159 O5 **Barkol Hu** ◎ NW China
Barkol Kazak Zizhixian *see* Barkol
30 J3 **Bark Point** *headland* Wisconsin, N USA 46°53′N 91°11′W
25 P11 **Barksdale** Texas, SW USA 29°43′N 100°03′W
Bar Kunar *see* Asmār
116 L11 **Bârlad** *prev.* Bîrlad. Vaslui, E Romania 46°12′N 27°39′E
116 M11 **Bârlad** *prev.* Bîrlad. ≈ E Romania
76 D9 **Barlavento, Ilhas de** *var.* Windward Islands. *island group* N Cape Verde
103 R5 **Bar-le-Duc** *var.* Bar-sur-Ornain. Meuse, NE France 48°46′N 05°10′E
180 K11 **Barlee, Lake** ◎ Western Australia
180 H8 **Barlee Range** ▲▲ Western Australia
107 N16 **Barletta** *anc.* Barduli. Puglia, SE Italy 41°20′N 16°17′E
110 E10 **Barlinek** *Ger.* Berlinchen. Zachodnio-pomorskie, NW Poland 53°N 15°11′E
27 S11 **Barling** Arkansas, C USA 35°19′N 94°18′W
171 U12 **Barma** Papua Barat, E Indonesia 01°55′S 132°57′E
183 Q9 **Barmedman** New South Wales, SE Australia 34°09′S 147°21′E
Barmen-Elberfeld *see* Wuppertal
152 D12 **Bārmer** Rājasthān, NW India 25°43′N 71°25′E
182 K9 **Barmera** South Australia 34°14′S 140°26′E
97 I19 **Barmouth** NW Wales, United Kingdom 52°44′N 04°04′W
154 F10 **Barnagar** Madhya Pradesh, C India 23°01′N 75°28′E
152 H9 **Barnāla** Punjab, NW India 30°26′N 75°33′E
97 L15 **Barnard Castle** N England, United Kingdom 54°35′N 01°55′W
183 O6 **Barnato** New South Wales, SE Australia 31°39′S 145°01′E
112 I13 **Barnaul** Altayskiy Kray, C Russian Federation 53°21′N 83°45′E
109 V8 **Bärnbach** Steiermark, SE Austria 47°05′N 15°07′E
18 K16 **Barnegat** New Jersey, NE USA 39°43′N 74°12′W
23 S4 **Barnesville** Georgia, SE USA 33°03′N 84°09′W
29 R6 **Barnesville** Minnesota, N USA 46°39′N 96°25′W
31 U13 **Barnesville** Ohio, N USA 39°59′N 81°10′W
98 K11 **Barneveld** *var.* Barnveld. Gelderland, C Netherlands 52°08′N 05°34′E
25 O9 **Barnhart** Texas, SW USA 31°07′N 101°09′W
27 P8 **Barnsdall** Oklahoma, C USA 36°33′N 96°09′W
97 M17 **Barnsley** N England, United Kingdom 53°34′N 01°28′W
19 Q12 **Barnstable** Massachusetts, NE USA 41°42′N 70°17′W
97 I23 **Barnstaple** SW England, United Kingdom 51°05′N 04°04′W
Barnveld *see* Barneveld
21 Q14 **Barnwell** South Carolina, SE USA 33°14′N 81°21′W
77 U15 **Baro** Niger, C Nigeria 08°35′N 06°28′E
67 U8 **Baro** *var.* Baro Wenz. ≈ Ethiopia/Sudan
Baro *see* Baro Wenz
Baroda *see* Vadodara
149 U2 **Baroghil Pass** *var.* Kowtal-e Barowghīl. *pass* Afghanistan/Pakistan
119 Q17 **Baron'ki** *Rus.* Boron'ki. Mahilyowskaya Voblasts', E Belarus 53°09′N 32°08′E
182 J9 **Barossa Valley** *valley* South Australia
Baroui *see* Salisbury
81 H14 **Baro Wenz** *var.* Baro, Nahr Barū. ≈ Ethiopia/Sudan
Baro Wenz *see* Baro
Barowghīl, Kowtal-e *see* Baroghil Pass
153 U12 **Barpeta** Assam, NE India 26°19′N 91°05′E
31 S7 **Barques, Pointe Aux** *headland* Michigan, N USA 44°04′N 82°57′W
54 I5 **Barquisimeto** Lara, NW Venezuela 10°03′N 69°18′W
59 N16 **Barra** Bahia, E Brazil 11°06′S 43°15′W
96 E9 **Barra** *island* NW Scotland, United Kingdom
183 T5 **Barraba** New South Wales, SE Australia 30°24′S 150°37′E
60 L9 **Barra Bonita** São Paulo, S Brazil 22°30′S 48°35′W
64 J12 **Barracuda Fracture Zone** *var.* Fifteen Twenty Fracture Zone. *tectonic feature* SW Atlantic Ocean
64 G11 **Barracuda Ridge** *undersea feature* N Atlantic Ocean
43 N12 **Barra del Colorado** Limón, NE Costa Rica 10°44′N 83°35′W
43 N9 **Barra de Río Grande** Región Autónoma Atlántico Sur, E Nicaragua 12°56′N 83°30′W
82 A11 **Barra do Cuanza** Luanda, NW Angola 09°12′S 13°08′E
60 O9 **Barra do Piraí** Rio de Janeiro, SE Brazil 22°30′S 43°47′W
61 D16 **Barra do Quaraí** Rio Grande do Sul, SE Brazil 31°03′S 58°10′W
59 G14 **Barra do São Manuel** Pará, N Brazil 07°12′S 58°03′W
83 N19 **Barra Falsa, Ponta da** *headland* S Mozambique 22°57′S 35°36′E
96 E10 **Barra Head** *headland* NW Scotland, United Kingdom 56°46′N 07°37′W
60 O9 **Barra Mansa** Rio de Janeiro, SE Brazil 22°25′S 44°03′W
57 D14 **Barranca** Lima, W Peru 10°46′S 77°46′W
54 F8 **Barrancabermeja** Santander, N Colombia 07°01′N 73°51′W
54 H4 **Barrancas** La Guajira, N Colombia 10°59′N 72°46′W
54 J6 **Barrancas** Barinas, NW Venezuela 08°47′N 70°07′W
55 Q6 **Barrancas** Monagas, NE Venezuela 08°45′N 62°12′W
54 F6 **Barranco de Loba** Bolívar, N Colombia 08°56′N 74°07′W
104 I12 **Barrancos** Beja, S Portugal 38°08′N 06°59′W
62 N7 **Barranqueras** Chaco, N Argentina 27°29′S 58°54′W
54 E4 **Barranquilla** Atlántico, N Colombia 10°59′N 74°48′W
83 N20 **Barra, Ponta da** *headland* S Mozambique 23°46′S 35°33′E
105 P11 **Barrax** Castilla-La Mancha, C Spain 39°04′N 02°12′W
19 N11 **Barre** Massachusetts, NE USA 42°24′N 72°06′W
18 M7 **Barre** Vermont, NE USA 44°09′N 72°25′W
59 M17 **Barreiras** Bahia, E Brazil 12°09′S 44°58′W
104 F11 **Barreiro** Setúbal, W Portugal 38°40′N 09°05′W
65 C26 **Barren Island** *island* S Falkland Islands
20 K7 **Barren River Lake** ▣ Kentucky, S USA
60 L7 **Barretos** São Paulo, S Brazil 20°33′S 48°33′W
11 P14 **Barrhead** Alberta, SW Canada 54°10′N 114°22′W
14 G14 **Barrie** Ontario, S Canada 44°22′N 79°42′W
11 N16 **Barrière** British Columbia, SW Canada 51°10′N 120°06′W
14 H8 **Barrière, Lac** ◎ Québec, SE Canada
182 L6 **Barrier Range** *hill range* New South Wales, SE Australia
42 G3 **Barrier Reef** *reef* E Belize
188 C16 **Barrigada** C Guam 13°27′N 144°48′E
Barrington Island *see* Santa Fe, Isla
183 T7 **Barrington Tops** ▲ New South Wales, SE Australia 32°06′S 151°18′E
183 O4 **Barringun** New South Wales, SE Australia 29°02′S 145°45′E
59 K18 **Barro Alto** Goiás, S Brazil 15°07′S 48°56′W
59 N14 **Barro Duro** Piauí, NE Brazil 05°49′S 42°30′W
30 I5 **Barron** Wisconsin, N USA 45°24′N 91°50′W
14 J12 **Barron** ≈ Ontario, SE Canada
61 H15 **Barros Cassal** Rio Grande do Sul, S Brazil 29°12′S 52°33′W
45 P14 **Barrouallie** Saint Vincent , W Saint Vincent and the Grenadines 13°14′N 61°17′W
39 O4 **Barrow** Alaska, USA 71°17′N 156°47′W
97 E20 **Barrow** *Ir.* An Bhearú. ≈ SE Ireland
181 Q6 **Barrow Creek Roadhouse** Northern Territory, N Australia 21°30′S 133°52′E
97 J16 **Barrow-in-Furness** NW England, United Kingdom 54°07′N 03°14′W
180 G7 **Barrow Island** *island* Western Australia
39 O4 **Barrow, Point** *headland* Alaska, USA 71°23′N 156°28′W
11 V14 **Barrows** Manitoba, S Canada 52°49′N 101°36′W
97 J22 **Barry** S Wales, United Kingdom 51°24′N 03°18′W
14 J12 **Barry's Bay** Ontario, SE Canada 45°30′N 77°41′W
144 K14 **Barsakel'mes, Ostrov** *island* SW Kazakhstan
Baršć Łužyca *see* Forst
147 S14 **Barsem** S Tajikistan 37°36′N 71°43′E
145 V11 **Barshatas** Vostochnyy Kazakhstan, E Kazakhstan 48°13′N 78°33′E
155 F14 **Bārsi** Mahārāshtra, W India 18°14′N 75°42′E
100 I13 **Barsinghausen** Niedersachsen, C Germany 53°19′N 09°30′E
147 X8 **Barskoon** Issyk-Kul'skaya Oblast', E Kyrgyzstan 42°07′N 77°34′E
100 F10 **Barssel** Niedersachsen, NW Germany 53°10′N 07°46′E
35 U14 **Barstow** California, W USA 34°52′N 117°00′W
24 L8 **Barstow** Texas, SW USA 31°27′N 103°23′W
103 R6 **Bar-sur-Aube** Aube, NE France 48°13′N 04°43′E
Bar-sur-Ornain *see* Bar-le-Duc
103 Q6 **Bar-sur-Seine** Aube, N France 48°06′N 04°22′E
147 S13 **Bartang** S Tajikistan 38°06′N 71°48′E
147 T13 **Bartang** ≈ SE Tajikistan
Bartenstein *see* Bartoszyce
Bártfa/Bartfeld *see* Bardejov
100 N7 **Barth** Mecklenburg-Vorpommern, NE Germany 54°21′N 12°43′E
27 W13 **Bartholomew, Bayou** ≈ Arkansas/Louisiana, S USA
55 T8 **Bartica** N Guyana 06°24′N 58°36′W
136 H10 **Bartın** Bartın, NW Turkey 41°37′N 32°20′E
136 H10 **Bartın** ◈ *province* NW Turkey
181 W4 **Bartle Frere** ▲ Queensland, E Australia 17°15′S 145°43′E
27 P8 **Bartlesville** Oklahoma, C USA 36°44′N 95°59′W
29 P14 **Bartlett** Nebraska, C USA 41°51′N 98°32′W
20 E10 **Bartlett** Tennessee, S USA 35°12′N 89°52′W
25 T9 **Bartlett** Texas, SW USA 30°47′N 97°25′W
36 L13 **Bartlett Reservoir** ▣ Arizona, SW USA
19 N6 **Barton** Vermont, NE USA 44°44′N 72°09′W
110 L7 **Bartoszyce** *Ger.* Bartenstein. Warmińsko-mazurskie, NE Poland 54°16′N 20°49′E
23 W12 **Bartow** Florida, SE USA 27°54′N 81°50′W
Bartschin *see* Barcin
168 J10 **Barumun, Sungai** ≈ Sumatera, W Indonesia
Barū, Nahr *see* Baro Wenz
169 S17 **Barung, Nusa** *island* S Indonesia
168 H9 **Barus** Sumatera, NW Indonesia 02°02′N 98°20′E
162 I9 **Baruunbayan-Ulaan** *var.* Höövör. Övörhangay, C Mongolia 45°10′N 101°19′E
Baruunsuu *see* Tsogttsetsiy
163 P8 **Baruun-Urt** Sühbaatar, E Mongolia 46°40′N 113°17′E
43 P15 **Barú, Volcán** *var.* Volcán de Chiriquí. 🌋 W Panama 08°49′N 82°32′W
99 K21 **Barvaux** Luxembourg, SE Belgium 50°21′N 05°30′E
42 M13 **Barva, Volcán** 🌋 NW Costa Rica 10°07′N 84°08′W
117 W6 **Barvinkove** Kharkivs'ka Oblast', E Ukraine 48°54′N 37°03′E
154 G11 **Barwāh** Madhya Pradesh, C India 22°17′N 76°01′E
Bärwalde Neumark *see* Mieszkowice
154 F11 **Barwāni** Madhya Pradesh, C India 22°02′N 74°56′E
183 P5 **Barwon River** ≈ New South Wales, SE Australia
119 L15 **Barysaw** *Rus.* Borisov. Minskaya Voblasts', NE Belarus 54°14′N 28°30′E
127 Q6 **Barysh** Ul'yanovskaya Oblast', W Russian Federation 53°32′N 47°06′E
117 Q4 **Baryshivka** Kyyivs'ka Oblast', N Ukraine 50°21′N 31°21′E
114 G8 **Barzia** *var.* Bŭrziya. NW Bulgaria
79 J17 **Basankusu** Equateur, NW Dem. Rep. Congo 01°12′N 19°50′E
117 N11 **Basarabeasca** *Rus.* Bessarabka. SE Moldova 46°22′N 28°56′E
116 M14 **Basarabi** Constanţa, SW Romania 44°10′N 28°26′E
40 H6 **Basaseachic** Chihuahua, NW Mexico 28°18′N 108°13′W
Basauri *see* Arizgoiti
61 D18 **Basavilbaso** Entre Ríos, E Argentina 32°23′S 58°55′W
79 F21 **Bas-Congo** *off.* Région du Bas-Congo; *prev.* Bas-Zaïre. ◈ *region* SW Dem. Rep. Congo
108 E6 **Basel** *Eng.* Basle, *Fr.* Bâle. Basel Stadt, NW Switzerland 47°33′N 07°36′E
Baselland *see* Basel Landschaft
108 E7 **Basel Landschaft** *prev.* Baselland. ◈ *canton* NW Switzerland
108 E6 **Basel Stadt** *former canton* Basel. ◈ *canton* NW Switzerland
143 T14 **Bashākerd, Kūhhā-ye** ▲▲ SE Iran
11 Q15 **Bashaw** Alberta, SW Canada 52°40′N 112°53′W
146 K16 **Bashbedeng** Mary Welaýaty, S Turkmenistan 35°44′N 63°07′E
161 T15 **Bashi Channel** *Chin.* Pa-shih Hai-hsia. *channel* Philippines/Taiwan
Bashkiria *see* Bashkortostan, Respublika
122 F11 **Bashkortostan, Respublika** *prev.* Bashkiria. ◈ *autonomous republic* W Russian Federation
127 N6 **Bashmakovo** Penzenskaya Oblast', W Russian Federation 53°13′N 43°00′E
146 J10 **Bashsakarba** Lebap Welaýaty, NE Turkmenistan 40°25′N 62°16′E
117 R9 **Bashtanka** Mykolayivs'ka Oblast', S Ukraine 47°24′N 32°27′E
22 H8 **Basile** Louisiana, S USA 30°29′N 92°36′W
107 M18 **Basilicata** ◈ *region* S Italy
33 V13 **Basin** Wyoming, C USA 44°22′N 108°02′W
97 N22 **Basingstoke** S England, United Kingdom 51°16′N 01°08′W
143 U8 **Başīrān** Khorāsān-e Jonūbī, E Iran 31°57′N 59°07′E
112 B10 **Baška** *It.* Bescanuova. Primorje-Gorski Kotar, NW Croatia 44°58′N 14°46′E
137 T15 **Başkale** Van, SE Turkey 38°03′N 43°59′E
14 L10 **Baskatong, Réservoir** ▣ Québec, SE Canada
137 O14 **Baskil** Elazığ, E Turkey 38°38′N 38°47′E
Basle *see* Basel

◆ Country
● Country Capital
◇ Dependent Territory
○ Dependent Territory Capital
◈ Administrative Regions
✈ International Airport
▲ Mountain
▲▲ Mountain Range
🌋 Volcano
≈ River
◎ Lake
▣ Reservoir

Bāsoda – Beijing

154 H9 **Bāsoda** Madhya Pradesh, C India 23°54′N 77°58′E
79 L17 **Basoko** Orientale, N Dem. Rep. Congo 01°14′N 23°26′E
Basque Country, The see País Vasco
Basra see Al Başrah
Basra see Al Başrah
Başrah, Muḥāfa at al see Al Başrah
103 U5 **Bas-Rhin** ◈ department NE France
Bassam see Grand-Bassam
11 Q16 **Bassano** Alberta, SW Canada 50°48′N 112°28′W
106 H7 **Bassano del Grappa** Veneto, NE Italy 45°45′N 11°45′E
77 Q15 **Bassar** var. Bassari. NW Togo 09°15′N 00°47′E
Bassari see Bassar
172 L9 **Bassas da India** island group W Madagascar
108 D7 **Bassecourt** Jura, W Switzerland 47°20′N 07°16′E
Bassein see Pathein
79 J15 **Basse-Kotto** ◈ prefecture S Central African Republic
102 J5 **Basse-Normandie** Eng. Lower Normandy. ◈ region N France
45 Q11 **Basse-Pointe** N Martinique 14°52′N 61°07′W
76 H12 **Basse Santa Su** E Gambia 13°18′N 14°10′W
Basse-Saxe see Niedersachsen
45 X6 **Basse-Terre** ○ (Guadeloupe) Basse Terre, SW Guadeloupe 16°08′N 61°40′W
45 V10 **Basseterre** ● (Saint Kitts and Nevis) Saint Kitts, Saint Kitts and Nevis 17°16′N 62°45′W
45 X6 **Basse Terre** island W Guadeloupe
29 O13 **Bassett** Nebraska, C USA 42°34′N 99°32′W
21 S7 **Bassett** Virginia, NE USA 36°45′N 79°59′W
37 N15 **Bassett Peak** ▲ Arizona, SW USA 32°30′N 110°16′W
76 M10 **Bassikounou** Hodh ech Chargui, SE Mauritania 15°55′N 05°59′W
77 R15 **Bassila** W Benin 08°25′N 01°58′E
Bass, Îlots de see Marotiri
31 O11 **Bass Lake** Indiana, N USA 41°12′N 86°35′W
183 O14 **Bass Strait** strait SE Australia
100 H11 **Bassum** Niedersachsen, NW Germany 52°52′N 08°44′E
29 X3 **Basswood Lake** ◎ Canada/USA
95 J21 **Båstad** Skåne, S Sweden 56°25′N 12°50′E
Basṭah see Beste
153 N12 **Basti** Uttar Pradesh, N India 26°48′N 82°44′E
103 X14 **Bastia** Corse, France, C Mediterranean Sea 42°42′N 09°27′E
99 L23 **Bastogne** Luxembourg, SE Belgium 50°N 05°43′E
22 I5 **Bastrop** Louisiana, S USA 32°46′N 91°54′W
25 T11 **Bastrop** Texas, SW USA 30°07′N 97°21′W
93 J15 **Bastuträsk** Västerbotten, N Sweden 64°47′N 20°05′E
119 J19 **Bastyn'** Rus. Bostyn'. Brestskaya Voblasts', SW Belarus 52°23′N 26°45′E
Basuo see Dongfang
Basutoland see Lesotho
119 O15 **Basya** ∼ E Belarus
Bas-Zaïre see Bas-Congo
79 D17 **Bata** NW Equatorial Guinea 01°51′N 09°48′E
79 D17 **Bata** ✕ S Equatorial Guinea 01°55′N 09°48′E
Batae Coritanorum see Leicester
123 Q8 **Batagay** Respublika Sakha (Yakutiya), NE Russian Federation 67°36′N 134°44′E
123 P8 **Batagay-Alyta** Respublika Sakha (Yakutiya), NE Russian Federation 67°48′N 130°15′E
112 L10 **Batajnica** Vojvodina, N Serbia 44°55′N 20°17′E
136 H15 **Bataklık Gölü** ◎ S Turkey
114 H11 **Batak, Yazovir** ⊠ SW Bulgaria
152 H7 **Batāla** Punjab, N India 31°48′N 75°12′E
104 F9 **Batalha** Leiria, C Portugal 39°40′N 08°50′W
79 N17 **Batama** Orientale, NE Dem. Rep. Congo 0°54′N 26°25′E
123 Q10 **Batamay** Respublika Sakha (Yakutiya), NE Russian Federation 63°28′N 129°33′E
160 F9 **Batang** var. Bazhong. Sichuan, C China 30°04′N 99°10′E
79 I14 **Batangafo** Ouham, NW Central African Republic 07°19′N 18°22′E
171 P8 **Batangas** off. Batangas City. Luzon, N Philippines 13°47′N 121°03′E
Batangas City see Batangas
Bătania see Battonya
171 Q10 **Batan Islands** island group N Philippines
60 L8 **Batatais** São Paulo, S Brazil 20°54′S 47°37′W
18 E10 **Batavia** New York, NE USA 43°00′N 78°11′W
Batavia see Jakarta
173 T9 **Batavia Seamount** undersea feature E Indian Ocean 27°42′S 100°36′E
126 L12 **Bataysk** Rostovskaya Oblast', SW Russian Federation 47°10′N 39°46′E
14 B9 **Batchawana** ∼ Ontario, S Canada
14 B9 **Batchawana Bay** Ontario, S Canada 46°55′N 84°36′W
167 Q12 **Bătdâmbâng** prev. Battambang. Bătdâmbâng, NW Cambodia 13°06′N 103°13′E
79 G20 **Batéké, Plateaux** plateau S Congo
183 S11 **Batemans Bay** New South Wales, SE Australia 35°45′S 150°09′E
21 Q13 **Batesburg** South Carolina, SE USA 33°54′N 81°33′W
28 K12 **Batesland** South Dakota, N USA 43°05′N 102°07′W
27 V10 **Batesville** Arkansas, C USA 35°45′N 91°39′W
31 Q14 **Batesville** Indiana, N USA 39°18′N 85°13′W
22 L2 **Batesville** Mississippi, S USA 34°18′N 89°56′W
25 Q13 **Batesville** Texas, SW USA 28°56′N 99°38′W
44 L13 **Bath** E Jamaica 17°57′N 76°22′W
97 L22 **Bath** hist. Akermanceaster; anc. Aquae Calidae, Aquae Solis. SW England, United Kingdom 51°23′N 02°22′W
19 Q8 **Bath** Maine, NE USA 43°54′N 69°49′W
18 F11 **Bath** New York, NE USA 42°20′N 77°16′W
Bath see Berkeley Springs
78 I10 **Batha** off. Région du Batha. ◈ region C Chad
78 I10 **Batha** seasonal river C Chad
Batha, Région du see Batha
141 Y8 **Baṭḥā', Wādī al** dry watercourse NE Oman
152 H9 **Bathinda** Punjab, NW India 30°14′N 74°54′E
98 M11 **Bathmen** Overijssel, E Netherlands 52°15′N 06°16′E
45 Z14 **Bathsheba** E Barbados 13°13′N 59°31′W
183 R8 **Bathurst** New South Wales, SE Australia 33°32′S 149°35′E
13 O13 **Bathurst** New Brunswick, SE Canada 47°37′N 65°40′W
Bathurst see Banjul
8 H6 **Bathurst, Cape** headland Northwest Territories, NW Canada 70°33′N 128°00′W
196 L8 **Bathurst Inlet** Nunavut, N Canada 66°23′N 107°00′W
196 L8 **Bathurst Inlet** inlet Nunavut, N Canada
181 N1 **Bathurst Island** island Northern Territory, N Australia
197 O9 **Bathurst Island** island Parry Islands, Nunavut, N Canada
77 O14 **Batié** SW Burkina Faso 09°53′N 02°53′W
Batinah see Al Bāṭinah
141 Y9 **Bāṭin, Wādī al** dry watercourse SW Asia
15 P9 **Batiscan** ∼ Québec, SE Canada
136 F16 **Batı Toroslar** ▲▲ SW Turkey
Batjan see Bacan, Pulau
147 R11 **Batken** Batenskaya Oblast', SW Kyrgyzstan 40°03′N 70°50′E
Batken Oblasty see Batkenskaya Oblast'
147 Q11 **Batkenskaya Oblast'** Kir. Batken Oblasty. ◈ province SW Kyrgyzstan
Batlle y Ordóñez see José Batlle y Ordóñez
183 Q10 **Batlow** New South Wales, SE Australia 35°32′S 148°09′E
137 Q15 **Batman** var. İluh. Batman, SE Turkey 37°52′N 41°06′E
137 Q15 **Batman** ◈ province SE Turkey
74 L6 **Batna** NE Algeria 35°34′N 06°10′E
163 O7 **Batnorov** var. Dundbürd. Hentiy, E Mongolia 47°55′N 111°37′E
Batoe see Batu, Kepulauan
162 K7 **Bat-Öldziy** var. Övt. Övörhangay, C Mongolia 46°50′N 102°15′E
Bat-Öldziyt see Dzaamar
22 J8 **Baton Rouge** state capital Louisiana, S USA 30°28′N 91°09′W
79 G15 **Batouri** Est, E Cameroon 04°26′N 14°27′E
138 G14 **Batrā', Jibāl al** ▲▲ S Jordan
138 G6 **Batroûn** var. Al Batrūn. N Lebanon 34°15′N 35°42′E
Batsch see Bač
Batsevichi see Batsevichy
119 M17 **Batsevichy** Rus. Batsevichi. Mahilyowskaya Voblasts', E Belarus 53°24′N 29°14′E
92 M7 **Båtsfjord** Finnmark, N Norway 70°37′N 29°42′E
Batshireet see Hentiy
162 L7 **Batsümber** var. Mandal. Töv, C Mongolia 48°24′N 106°47′E
Battambang see Bătdâmbâng
195 X3 **Batterbee, Cape** headland Antarctica
155 L24 **Batticaloa** Eastern Province, E Sri Lanka 07°44′N 81°43′E
99 L19 **Battice** Liège, E Belgium 50°39′N 05°50′E
107 L18 **Battipaglia** Campania, S Italy 40°36′N 14°59′E
11 R15 **Battle** ∼ Alberta/Saskatchewan, SW Canada
Battle Born State see Nevada
31 Q10 **Battle Creek** Michigan, N USA 42°20′N 85°10′W
27 T7 **Battlefield** Missouri, C USA 37°07′N 93°22′W
11 S15 **Battleford** Saskatchewan, S Canada 52°45′N 108°20′W
29 S6 **Battle Lake** Minnesota, N USA 46°16′N 95°42′W
35 U3 **Battle Mountain** Nevada, W USA 40°37′N 116°55′W
111 M25 **Battonya** Rom. Bătania. Békés, SE Hungary 46°16′N 21°00′E
162 J7 **Battsengel** var. Jargalant. Arhangay, C Mongolia 47°46′N 101°56′E
168 D11 **Batu, Kepulauan** prev. Batoe. island group W Indonesia
137 Q10 **Batumi** W Georgia 41°39′N 41°38′E
168 K10 **Batu Pahat** prev. Bandar Penggaram. Johor, Peninsular Malaysia 01°51′N 102°56′E
171 O12 **Baturebe** Sulawesi, N Indonesia 01°43′S 121°43′E
122 J12 **Baturino** Tomskaya Oblast', C Russian Federation 57°46′N 85°08′E
117 R3 **Baturyn** Chernihivs'ka Oblast', N Ukraine 51°20′N 32°54′E
138 F10 **Bat Yam** Tel Aviv, C Israel 32°01′N 34°45′E
127 Q4 **Batyrevo** Chuvashskaya Respublika, W Russian Federation 55°04′N 47°34′E
Batys Qazaqstan Oblysy see Zapadnyy Kazakhstan
102 F5 **Batz, Île de** island NW France
169 Q10 **Bau** Sarawak, East Malaysia 01°25′N 110°10′E
171 N2 **Bauang** Luzon, N Philippines 16°33′N 120°19′E
171 P14 **Baubau** var. Baoebaoe. Pulau Buton, C Indonesia 05°30′S 122°37′E
77 W14 **Bauchi** Bauchi, NE Nigeria 10°18′N 09°46′E
77 W14 **Bauchi** ◈ state C Nigeria
102 H7 **Baud** Morbihan, NW France 47°52′N 02°59′W
29 T2 **Baudette** Minnesota, N USA 48°42′N 94°36′W
193 S9 **Bauer Basin** undersea feature E Pacific Ocean 10°00′S 101°45′W
187 R14 **Bauer Field** var. Port Vila. ✕ (Port-Vila) Éfaté, C Vanuatu 17°42′S 168°21′E
13 T9 **Bauld, Cape** headland Newfoundland and Labrador, E Canada 51°35′N 55°22′W
103 T8 **Baume-les-Dames** Doubs, E France 47°22′N 06°20′E
101 I15 **Baunatal** Hessen, C Germany 51°15′N 09°25′E
107 D18 **Baunei** Sardegna, Italy, C Mediterranean Sea 40°04′N 09°36′E
57 M15 **Baures, Río** ∼ N Bolivia
60 K9 **Bauru** São Paulo, S Brazil 22°19′S 49°07′W
Baushar see Bawshar
118 G10 **Bauska** Ger. Bauske. S Latvia 56°25′N 24°11′E
Bauske see Bauska
101 Q15 **Bautzen** Lus. Budyšin. Sachsen, E Germany 51°11′N 14°29′E
145 Q16 **Bauyrzhan Momyshuly** Kaz. Baūyrzhan Momyshuly; prev. Burnoye. Zhambyl, S Kazakhstan 42°36′N 70°46′E
Bauzanum see Bolzano
Bavaria see Bayern
109 N7 **Bavarian Alps** Ger. Bayrische Alpen. ▲▲ Austria/Germany
Bavière see Bayern
40 H4 **Bavispe, Río** ∼ NW Mexico
127 T5 **Bavly** Respublika Tatarstan, W Russian Federation 54°20′N 53°21′E
169 P13 **Bawal, Pulau** island N Indonesia
169 T12 **Bawan** Borneo, C Indonesia 01°36′S 113°55′E
183 O12 **Baw Baw, Mount** ▲ Victoria, SE Australia 37°49′S 146°16′E
169 S15 **Bawean, Pulau** island S Indonesia
75 V9 **Bawīṭī** var. Bawîṭi. N Egypt 28°19′N 28°53′E
Bawîṭi see Bawīṭī
77 Q13 **Bawku** N Ghana 11°00′N 00°12′W
Bawlake see Bawlakhe
167 N7 **Bawlakhe** var. Bawlake. Kayah State, C Myanmar (Burma) 19°10′N 97°19′E
169 H11 **Bawo Ofuloa** Pulau Tanahmasa, W Indonesia 0°10′S 98°24′E
141 Y8 **Bawshar** var. Baushar. NE Oman 23°32′N 58°24′E
Baxian see Bazhou
Ba Xian see Bazhou
158 M8 **Baxkorgan** Xinjiang Uygur Zizhiqu, W China 39°05′N 90°00′E
23 V6 **Baxley** Georgia, SE USA 31°46′N 82°21′W
159 R15 **Baxoi** var. Baima. Xizang Zizhiqu, W China 30°01′N 96°53′E
29 W14 **Baxter** Iowa, C USA 41°49′N 93°09′W
29 U6 **Baxter** Minnesota, N USA 46°21′N 94°18′W
27 R8 **Baxter Springs** Kansas, C USA 37°01′N 94°45′W
81 M17 **Bay** off. Gobolka Bay. ◈ region SW Somalia
Bay see Baicheng
44 H7 **Bayamo** Granma, E Cuba 20°21′N 76°38′W
45 U5 **Bayamón** E Puerto Rico 18°24′N 66°09′W
163 W8 **Bayan** Heilongjiang, NE China 46°05′N 127°24′E
170 L16 **Bayan** prev. Bajan. Pulau Lombok, C Indonesia 08°16′S 116°28′E
162 M8 **Bayan** var. Maanĭt. Töv, C Mongolia 47°14′N 107°34′E
Bayan see Hölönbuyr, Dornod, Mongolia
Bayan see Ihhet, Dornogovĭ, Mongolia
Bayan see Bayan-Uul, Govĭ-Altay, Mongolia
Bayan see Bayanhutag, Hentiy, Mongolia
Bayan see Bürentogtoh, Hövsgol, Mongolia
152 I12 **Bayāna** Rājasthān, N India 26°55′N 77°18′E
149 N5 **Bāyān, Band-e** ▲▲ C Afghanistan
162 H8 **Bayanbulag** Bayanhongor, C Mongolia 46°46′N 98°07′E
Bayanbulag see Ömnödelger
158 J5 **Bayanbulak** Xinjiang Uygur Zizhiqu, W China 43°05′N 84°05′E
162 L7 **Bayanchandmanĭ** var. Ihsüüj. Töv, C Mongolia 48°12′N 106°23′E
162 J11 **Bayandalay** var. Dalay. Ömnögovĭ, S Mongolia 43°27′N 103°30′E
163 O9 **Bayandelger** var. Shireet. Sühbaatar, SE Mongolia 45°33′N 112°19′E
162 I5 **Bayandzürh** var. Altraga. Hövsgöl, N Mongolia 50°08′N 98°54′E
Bayan Gol see Dengkou, China
Bayangol see Bugat, Mongolia
162 I9 **Bayangovĭ** var. Örgön. Bayanhongor, C Mongolia 44°43′N 100°23′E
159 R12 **Bayan Har Shan** var. Bayan Khar. ▲▲ C China
162 G6 **Bayanhayrhan** var. Altanbulag. Dzavhan, N Mongolia 49°16′N 96°22′E
162 I8 **Bayanhongor** Bayanhongor, C Mongolia 46°08′N 100°42′E
162 H9 **Bayanhongor** ◈ province C Mongolia
162 K14 **Bayan Hot** var. Alxa Zuoqi. Nei Mongol Zizhiqu, N China 38°49′N 105°40′E
163 O8 **Bayanhutag** var. Bayan. Hentiy, C Mongolia 47°13′N 110°57′E
163 T9 **Bayan Huxu** var. Horqin Zuoyi Zhongqi. Nei Mongol Zizhiqu, N China 45°02′N 121°28′E
Bayan Khar see Bayan Har Shan
163 N12 **Bayan Kuang** prev. Bayan Obo. Nei Mongol Zizhiqu, N China 41°45′N 109°58′E
168 J7 **Bayan Lepas** ✕ (George Town) Pinang, Peninsular Malaysia 05°18′N 100°15′E
162 I10 **Bayanlig** var. Hatansuudal. Bayanhongor, C Mongolia 44°34′N 100°41′E
162 K13 **Bayan Mod** Nei Mongol Zizhiqu, N China 40°45′N 104°29′E
163 N8 **Bayanmönh** var. Ulaan-Ereg. Hentiy, E Mongolia 46°50′N 109°39′E
162 L12 **Bayannur** var. Linhe. Nei Mongol Zizhiqu, N China 40°46′N 107°27′E
Bayan Nuru see Xar Burd
162 E5 **Bayannuur** var. Tsul-Ulaan. Bayan-Ölgiy, W Mongolia 48°51′N 91°13′E
Bayan Obo see Bayan Kuang
43 V15 **Bayano, Lago** ◎ E Panama
162 C5 **Bayan-Ölgiy** ◈ province NW Mongolia
162 H9 **Bayan-Öndör** var. Bulgan. Bayanhongor, C Mongolia 44°48′N 98°39′E
162 K8 **Bayan-Öndör** var. Bumbat. Övörhangay, C Mongolia 46°30′N 104°08′E
162 L8 **Bayan-Önjüül** var. Ihhayrhan. Töv, C Mongolia 46°57′N 105°51′E
163 O7 **Bayan-Ovoo** var. Javhlant. Hentiy, E Mongolia 47°46′N 112°06′E
162 L11 **Bayan-Ovoo** var. Erdenetsogt. Ömnögovĭ, S Mongolia 42°54′N 106°16′E
Bayansayr see Baatsagaan
159 Q9 **Bayan Shan** ▲ C China 37°36′N 96°23′E
162 J9 **Bayanteeg** Övörhangay, C Mongolia 45°39′N 101°30′E
162 G5 **Bayantes** var. Altay. Dzavhan, N Mongolia 49°40′N 96°21′E
Bayantöhöm see Büren
162 M8 **Bayantsagaan** var. Dzogsool. Töv, C Mongolia 46°46′N 107°18′E
163 P7 **Bayantümen** var. Tsagaanders. Dornod, NE Mongolia 48°03′N 114°16′E
Bayan-Uhaa see Ih-Uul
163 R10 **Bayan Ul** var. Xi Ujimqin Qi. Nei Mongol Zizhiqu, N China 44°31′N 117°36′E
Bayan-Ulaan see Dzüünbayan-Ulaan
163 O7 **Bayan-Uul** var. Javarthushuu. Dornod, NE Mongolia 49°05′N 112°40′E
162 F7 **Bayan-Uul** var. Bayan. Govĭ-Altay, W Mongolia 47°05′N 95°13′E
162 M8 **Bayanuur** var. Tsul-Ulaan. Töv, C Mongolia 47°44′N 108°22′E
28 J14 **Bayard** Nebraska, C USA 41°45′N 103°19′W
37 P15 **Bayard** New Mexico, SW USA 32°45′N 108°07′W
103 T13 **Bayard, Col** pass SE France
Bayasgalant see Mönhhaan
136 J12 **Bayat** Çorum, N Turkey 40°34′N 34°07′E
171 P6 **Bayawan** Negros, C Philippines 09°22′N 122°50′E
143 R10 **Bayāẕ** Kermān, C Iran 30°41′N 55°29′E
171 Q6 **Baybay** Leyte, C Philippines 10°41′N 124°49′E
21 X10 **Bayboro** North Carolina, SE USA 35°08′N 76°49′W
137 P12 **Bayburt** Bayburt, NE Turkey 40°15′N 40°16′E
137 P12 **Bayburt** ◈ province NE Turkey
31 R8 **Bay City** Michigan, N USA 43°35′N 83°52′W
25 V12 **Bay City** Texas, SW USA 28°59′N 96°00′W
Baydarata Bay see Baydaratskaya Guba
122 J7 **Baydaratskaya Guba** var. Baydarata Bay. bay N Russian Federation
81 M16 **Baydhabo** var. Baydhowa, Isha Baydhabo, It. Baidoa. Bay, SW Somalia 03°08′N 43°39′E
Baydhowa see Baydhabo
101 N21 **Bayerischer Wald** ▲▲ SE Germany
101 K21 **Bayern** Eng. Bavaria, Fr. Bavière. ◈ state SE Germany
147 V9 **Bayetovo** Narynskaya Oblast', C Kyrgyzstan 41°14′N 74°55′E
102 K4 **Bayeux** anc. Augustodurum. Calvados, N France 49°16′N 00°42′W
14 E15 **Bayfield** ∼ Ontario, S Canada
145 O15 **Baygekum** Kaz. Bäygequm. Kzylorda, S Kazakhstan 44°15′N 66°34′E
Bäygequm see Baygekum
136 C14 **Bayındır** Izmır, SW Turkey 38°12′N 27°40′E
138 H12 **Bāyir** var. Bā'ir. Ma'ān, S Jordan 30°46′N 36°40′E
Bay Islands see Bahía, Islas de la
Bayizhen see Nyingchi
139 R5 **Bayjī** var. Baiji. Şalāḩ ad Dīn, N Iraq 34°56′N 43°29′E
Baykadam see Saudakent
123 N13 **Baykal, Ozero** Eng. Lake Baikal. ◎ S Russian Federation
123 M14 **Baykal'sk** Irkutskaya Oblast', S Russian Federation 51°30′N 104°03′E
137 R15 **Baykan** Siirt, SE Turkey 38°08′N 41°43′E
139 S2 **Baykhmah** var. Bēkma. Arbīl, E Iraq 36°40′N 44°15′E
123 L11 **Baykit** Krasnoyarskiy Kray, C Russian Federation 61°37′N 96°23′E
145 N12 **Baykonur** var. Baykonyr. Karaganda, C Kazakhstan 47°50′N 75°33′E
Baykonur see Baykonyr
144 M14 **Baykonyr** var. Baykonur, Kaz. Bayqongyr; prev. Leninsk. Kyzylorda, S Kazakhstan 45°38′N 63°20′E
Baykonyr see Baykonur
158 E7 **Baykurt** Xinjiang Uygur Zizhiqu, W China 39°56′N 75°33′E
14 I9 **Bay, Lac** ◎ Québec, SE Canada
127 W6 **Baymak** Respublika Bashkortostan, W Russian Federation 52°34′N 58°20′E
23 O8 **Bay Minette** Alabama, S USA 30°52′N 87°46′W
143 O17 **Baynūnah** desert W United Arab Emirates
184 O8 **Bay of Plenty** off. Bay of Plenty Region. ◈ region North Island, New Zealand
Bay of Plenty Region see Bay of Plenty
191 Z3 **Bay of Wrecks** bay Kiritimati, E Kiribati
Bayonnaise Rocks see Beyonēsu-retsugan
102 I15 **Bayonne** anc. Lapurdum. Pyrénées-Atlantiques, SW France 43°30′N 01°28′W
22 H4 **Bayou D'Arbonne Lake** ⊠ Louisiana, S USA
23 N9 **Bayou La Batre** Alabama, S USA 30°24′N 88°15′W
Bayou State see Mississippi
Bayqadam see Saudakent
Bayqongyr see Baykonyr
Bayram-Ali see Baýramaly
146 J14 **Baýramaly** var. Bayramaly; prev. Bayram-Ali. Mary Welaýaty, S Turkmenistan 37°33′N 62°08′E
101 L19 **Bayreuth** var. Baireuth. Bayern, SE Germany 49°57′N 11°34′E
Bayrische Alpen see Bavarian Alps
Bayrūt see Beyrouth
22 L9 **Bay Saint Louis** Mississippi, S USA 30°18′N 89°19′W
Baysān see Beit She'an
Bayshint see Öndörshireet
14 H13 **Bays, Lake of** ◎ Ontario, S Canada
22 M6 **Bay Springs** Mississippi, S USA 31°58′N 89°17′W
Bay State see Massachusetts
Baysun see Boysun
14 H13 **Baysville** Ontario, S Canada 45°10′N 79°03′W
141 N15 **Bayt al Faqīh** W Yemen 14°30′N 43°20′E
158 M4 **Baytik Shan** ▲▲ China/Mongolia
Bayt Laḩm see Bethlehem
25 W11 **Baytown** Texas, SW USA 29°43′N 94°59′W
169 V11 **Bayur, Tanjung** headland Borneo, N Indonesia 0°43′S 117°32′E
121 N14 **Bayy al Kabīr, Wādī** dry watercourse NW Libya
145 P17 **Bayyrkum** Kaz. Bayyrqum; prev. Bairkum. Yuzhnyy Kazakhstan, S Kazakhstan 41°57′N 68°05′E
Bayyrqum see Bayyrkum
105 P14 **Baza** Andalucía, S Spain 37°30′N 02°45′W
137 X10 **Bazardüzü Dağı** Rus. Gora Bazardyuzyu. ▲ N Azerbaijan 41°13′N 47°50′E
Bazardyuzyu, Gora see Bazardüzü Dağı
Bazargic see Dobrich
83 N18 **Bazaruto, Ilha do** island SE Mozambique
102 K14 **Bazas** Gironde, SW France 44°27′N 00°11′W
105 O14 **Baza, Sierra de** ▲▲ S Spain
160 J8 **Bazhong** var. Bazhou. Sichuan, C China 31°55′N 106°44′E
Bazhong see Batang
161 P3 **Bazhou** prev. Baxian, Ba Xian. Hebei, E China 39°05′N 116°24′E
Bazhou see Bazhong
14 M9 **Bazin** ∼ Québec, SE Canada
Bazin see Pezinok
139 Q7 **Bāzīyah** Al Anbār, C Iraq 33°50′N 42°41′E
138 H6 **Bcharré** var. Bcharreh, Bsharrī, Bsherri. NE Lebanon 34°16′N 36°01′E
Bcharreh see Bcharré
28 J5 **Beach** North Dakota, N USA 46°55′N 104°00′W
182 K12 **Beachport** South Australia 37°29′S 140°03′E
97 O23 **Beachy Head** headland SE England, United Kingdom 50°44′N 00°16′E
18 K13 **Beacon** New York, NE USA 41°30′N 73°54′W
63 J25 **Beagle Channel** channel Argentina/Chile
181 O1 **Beagle Gulf** gulf Northern Territory, N Australia
Bealach an Doirín see Ballaghaderreen
Bealach Cláir see Ballyclare
Bealach Féich see Ballybofey
172 J3 **Bealanana** Mahajanga, NE Madagascar 14°33′S 48°44′E
Béal an Átha see Ballina
Béal an Átha Móir see Ballinamore
Béal an Mhuirhead see Belmullet
Béal Átha Beithe see Ballybay
Béal Átha Conaill see Ballyconnell
Beál Átha hAmhnais see Ballyhaunis
Béal Átha na Sluaighe see Ballinasloe
Béal Átha Seanaidh see Ballyshannon
Bealdovuopmi see Peltovuoma
Béal Feirste see Belfast
Béal Tairbirt see Belturbet
Beanna Boirche see Mourne Mountains
Beannchar see Banagher, Ireland
Beannchar see Bangor, Northern Ireland, UK
Beanntraí see Bantry
Bearalváhki see Berlevåg
23 N2 **Bear Creek** ∼ Alabama/Mississippi, S USA
30 J13 **Bear Creek** ∼ Illinois, N USA
27 U13 **Bearden** Arkansas, C USA 33°43′N 92°37′W
195 Q10 **Beardmore Glacier** glacier Antarctica
30 K13 **Beardstown** Illinois, N USA 40°01′N 90°25′W
28 L14 **Bear Hill** ▲ Nebraska, C USA 41°24′N 101°49′W
Bear Island see Bjørnøya
14 H12 **Bear Lake** Ontario, S Canada 45°28′N 79°31′W
36 M1 **Bear Lake** ◎ Idaho/Utah, NW USA
39 U11 **Bear, Mount** ▲ Alaska, USA 61°16′N 141°09′W
102 J16 **Béarn** cultural region SW France
194 J11 **Bear Peninsula** peninsula Antarctica
152 I7 **Beās** ∼ India/Pakistan
105 P3 **Beasain** País Vasco, N Spain 43°03′N 02°11′W
105 O12 **Beas de Segura** Andalucía, S Spain 38°16′N 02°54′W
45 N10 **Beata, Cabo** headland SW Dominican Republic 17°34′N 71°25′W
45 N10 **Beata, Isla** island SW Dominican Republic
64 F11 **Beata Ridge** undersea feature N Caribbean Sea 16°00′N 72°30′W
29 R17 **Beatrice** Nebraska, C USA 40°14′N 96°43′W
83 L16 **Beatrice** Mashonaland East, NE Zimbabwe 18°15′S 30°55′E
11 N11 **Beatton** ∼ British Columbia, W Canada
11 N11 **Beatton River** British Columbia, W Canada 57°35′N 121°45′W
35 V10 **Beatty** Nevada, W USA 36°53′N 116°44′W
21 N6 **Beattyville** Kentucky, S USA 37°33′N 83°44′W
173 X16 **Beau Bassin** W Mauritius 20°13′S 57°27′E
103 R15 **Beaucaire** Gard, S France 43°49′N 04°37′E
14 I8 **Beauchastel, Lac** ◎ Québec, SE Canada
14 I10 **Beauchêne, Lac** ◎ Québec, SE Canada
183 V3 **Beaudesert** Queensland, E Australia 28°00′S 152°27′E
182 M12 **Beaufort** Victoria, SE Australia 37°27′S 143°24′E
21 X11 **Beaufort** North Carolina, SE USA 34°44′N 76°41′W
21 R15 **Beaufort** South Carolina, SE USA 32°23′N 80°40′W
38 M11 **Beaufort Sea** sea Arctic Ocean
Beaufort-Wes see Beaufort West
83 G25 **Beaufort West** Afr. Beaufort-Wes. Western Cape, SW South Africa 32°21′S 22°35′E
103 N7 **Beaugency** Loiret, C France 47°46′N 01°38′E
19 R1 **Beau Lake** ⊠ Maine, NE USA
96 I8 **Beauly** N Scotland, United Kingdom 57°29′N 04°29′W
99 G21 **Beaumont** Hainaut, S Belgium 50°12′N 04°13′E
185 E23 **Beaumont** Otago, South Island, New Zealand 45°48′S 169°32′E
22 M7 **Beaumont** Mississippi, S USA 31°10′N 88°55′W
25 X10 **Beaumont** Texas, SW USA 30°05′N 94°06′W
102 M15 **Beaumont-de-Lomagne** Tarn-et-Garonne, S France 43°54′N 01°00′E
102 L6 **Beaumont-sur-Sarthe** Sarthe, NW France 48°15′N 00°07′E
103 R8 **Beaune** Côte d'Or, C France 47°02′N 04°50′E
15 R9 **Beaupré** Québec, SE Canada 47°03′N 70°52′W
102 J8 **Beaupréau** Maine-et-Loire, NW France 47°13′N 00°57′W
99 I22 **Beauraing** Namur, SE Belgium 50°07′N 04°57′E
103 R12 **Beaurepaire** Isère, E France 45°20′N 05°03′E
11 Y16 **Beausejour** Manitoba, S Canada 50°04′N 96°30′W
103 N4 **Beauvais** anc. Bellovacum, Caesaromagus. Oise, N France 49°27′N 02°04′E
11 S13 **Beauval** Saskatchewan, C Canada 55°10′N 107°37′W
102 I9 **Beauvoir-sur-Mer** Vendée, NW France 46°54′N 02°03′W
39 R8 **Beaver** Alaska, USA 66°22′N 147°31′W
26 J8 **Beaver** Oklahoma, C USA 36°48′N 100°32′W
18 B14 **Beaver** Pennsylvania, NE USA 40°39′N 80°19′W
36 K6 **Beaver** Utah, W USA 38°16′N 112°38′W
10 L9 **Beaver** ∼ British Columbia/Yukon, W Canada
11 S13 **Beaver** ∼ Saskatchewan, C Canada
29 N17 **Beaver City** Nebraska, C USA 40°08′N 99°49′W
10 G6 **Beaver Creek** Yukon, W Canada 62°20′N 140°45′W
31 R14 **Beavercreek** Ohio, N USA 39°42′N 83°58′W
39 S8 **Beaver Creek** ∼ Alaska, USA
26 H3 **Beaver Creek** ∼ Kansas/Nebraska, C USA
28 J5 **Beaver Creek** ∼ Montana/North Dakota, N USA
29 Q14 **Beaver Creek** ∼ Nebraska, C USA
25 Q4 **Beaver Creek** ∼ Texas, SW USA
30 M8 **Beaver Dam** Wisconsin, N USA 43°28′N 88°49′W
30 M8 **Beaver Dam Lake** ◎ Wisconsin, N USA
18 B14 **Beaver Falls** Pennsylvania, NE USA 40°45′N 80°20′W
33 P12 **Beaverhead Mountains** ▲▲ Idaho/Montana, NW USA
33 Q12 **Beaverhead River** ∼ Montana, NW USA
65 A25 **Beaver Island** island W Falkland Islands
31 P5 **Beaver Island** island Michigan, N USA
27 S9 **Beaver Lake** ⊠ Arkansas, C USA
11 N13 **Beaverlodge** Alberta, W Canada 55°11′N 119°29′W
18 I8 **Beaver River** ∼ New York, NE USA
26 J8 **Beaver River** ∼ Oklahoma, C USA
18 B13 **Beaver River** ∼ Pennsylvania, NE USA
65 A25 **Beaver Settlement** Beaver Island, W Falkland Islands 51°30′S 61°15′W
Beaver State see Oregon
14 H14 **Beaverton** Ontario, S Canada 44°24′N 79°07′W
32 G11 **Beaverton** Oregon, NW USA 45°29′N 122°49′W
152 G12 **Beāwar** Rājasthān, N India 26°08′N 74°22′E
Bebas, Dasht-i see Bābūs, Dasht-e
60 L8 **Bebedouro** São Paulo, S Brazil 20°58′S 48°28′W
101 I16 **Bebra** Hessen, C Germany 50°59′N 09°46′E
41 W12 **Becal** Campeche, SE Mexico 19°49′N 90°28′W
15 Q11 **Bécancour** ∼ Québec, SE Canada
97 Q19 **Beccles** E England, United Kingdom 52°27′N 01°32′E
112 L9 **Bečej** Ger. Altbetsche, Hung. Óbecse, Rácz-Becse; prev. Magyar-Becse, Stari Bečej. Vojvodina, N Serbia 45°36′N 20°02′E
104 I3 **Becerréa** Galicia, NW Spain 42°51′N 07°10′W
74 H7 **Béchar** prev. Colomb-Béchar. W Algeria 31°38′N 02°11′W
39 O14 **Becharof Lake** ◎ Alaska, USA
116 H15 **Bechet** var. Bechetu. Dolj, SW Romania 43°45′N 23°57′E
Bechetu see Bechet
21 R6 **Beckley** West Virginia, NE USA 37°46′N 81°12′W
101 G14 **Beckum** Nordrhein-Westfalen, W Germany 51°45′N 08°03′E
25 X7 **Beckville** Texas, SW USA 32°14′N 94°27′W
35 X4 **Becky Peak** ▲ Nevada, W USA 39°59′N 114°33′W
116 I9 **Beclean** Hung. Bethlen; prev. Betlen. Bistriţa-Năsăud, N Romania 47°10′N 24°11′E
Bécs see Wien
111 H18 **Bečva** Ger. Betschau, Pol. Beczwa. ∼ E Czech Republic
Beczwa see Bečva
103 P15 **Bédarieux** Hérault, S France 43°37′N 03°10′E
120 B10 **Beddouza, Cap** headland W Morocco 32°35′N 09°16′W
80 I13 **Bedelē** Oromīya, C Ethiopia 08°25′N 36°21′E
147 Y8 **Bedel Pass** Rus. Pereval Bedel. pass China/Kyrgyzstan
Bedel, Pereval see Bedel Pass
95 H22 **Beder** Midtjylland, C Denmark 56°03′N 10°13′E
97 N20 **Bedford** E England, United Kingdom 52°08′N 00°29′W
31 O15 **Bedford** Indiana, N USA 38°51′N 86°29′W
29 U16 **Bedford** Iowa, C USA 40°40′N 94°43′W
20 L4 **Bedford** Kentucky, S USA 38°36′N 85°18′W
18 D15 **Bedford** Pennsylvania, NE USA 40°00′N 78°29′W
21 T6 **Bedford** Virginia, NE USA 37°20′N 79°31′W
97 N20 **Bedfordshire** cultural region E England, United Kingdom
127 N5 **Bednodem'yanovsk** Penzenskaya Oblast', W Russian Federation 53°55′N 43°14′E
98 N5 **Bedum** Groningen, NE Netherlands 53°18′N 06°36′E
27 V11 **Beebe** Arkansas, C USA 35°04′N 91°52′W
Beechy Group see Chichijima-rettō
45 T9 **Beef Island** ✕ (Road Town) Tortola, E British Virgin Islands 18°25′N 64°31′W
Beehive State see Utah
99 L18 **Beek** Limburg, SE Netherlands 50°56′N 05°47′E
99 L18 **Beek** ✕ (Maastricht) Limburg, SE Netherlands 50°55′N 05°47′E
99 K14 **Beek-en-Donk** Noord-Brabant, S Netherlands 51°31′N 05°37′E
138 F13 **Be'er Menuha** prev. Be'ér Menuẖa. Southern, S Israel 30°22′N 35°09′E
Be'ér Menuẖa see Be'er Menuha
99 D16 **Beernem** West-Vlaanderen, NW Belgium 51°09′N 03°18′E
99 I16 **Beerse** Antwerpen, N Belgium 51°20′N 04°52′E
Beersheba see Be'er Sheva
138 E11 **Be'er Sheva** var. Beersheba, Ar. Bir es Saba; prev. Be'ér Sheva'. Southern, S Israel 31°15′N 34°47′E
Be'ér Sheva' see Be'er Sheva
98 J13 **Beesd** Gelderland, C Netherlands 51°52′N 05°12′E
99 M16 **Beesel** Limburg, SE Netherlands 51°16′N 06°02′E
83 J21 **Beestekraal** North-West, N South Africa 25°21′S 27°40′E
194 J7 **Beethoven Peninsula** peninsula Alexander Island, Antarctica
Beetstersweach see Beetsterzwaag
98 M6 **Beetsterzwaag** Fris. Beetstersweach. Fryslân, N Netherlands 53°03′N 06°04′E
25 S13 **Beeville** Texas, SW USA 28°25′N 97°47′W
79 J18 **Befale** Equateur, NW Dem. Rep. Congo 0°25′N 20°48′E
Befandriana see Befandriana Avaratra
172 J3 **Befandriana Avaratra** var. Befandriana, Befandriana Nord. Mahajanga, NW Madagascar 15°14′S 48°33′E
Befandriana Nord see Befandriana Avaratra
79 K18 **Befori** Equateur, N Dem. Rep. Congo 0°09′N 22°18′E
172 I7 **Befotaka** Fianarantsoa, S Madagascar 23°49′S 47°00′E
183 R11 **Bega** New South Wales, SE Australia 36°43′S 149°50′E
102 G5 **Bégard** Côtes d'Armor, NW France 48°37′N 03°18′W
112 M9 **Begejski Kanal** canal Vojvodina, NE Serbia
94 G13 **Begna** ∼ S Norway
Begoml' see Byahoml'
Begovat see Bekobod
153 Q13 **Begusarai** Bihār, NE India 25°25′N 86°08′E
143 R9 **Behābād** Yazd, C Iran 32°23′N 59°50′E
Behagle see Laï
55 Z10 **Béhague, Pointe** headland E French Guiana 04°38′N 51°52′W
Behar see Bihār
142 M10 **Behbahān** var. Behbehān. Khūzestān, SW Iran 30°38′N 50°07′E
Behbehān see Behbahān
44 G3 **Behring Point** Andros Island, W The Bahamas 24°28′N 77°44′W
143 P4 **Behshahr** prev. Ashraf. Māzandarān, N Iran 36°42′N 53°36′E
163 V6 **Bei'an** Heilongjiang, NE China 48°16′N 126°29′E
Beibunar see Sredishte
Beibu Wan see Tonkin, Gulf of
Beida see Al Bayḑā'
80 H13 **Beigi** Oromīya, C Ethiopia 09°13′N 34°48′E
160 L16 **Beihai** Guangxi Zhuangzu Zizhiqu, S China 21°29′N 109°10′E
159 Q10 **Bei Hulsan Hu** ◎ C China
161 N13 **Bei Jiang** ∼ S China
161 O2 **Beijing** var. Pei-ching, Eng. Peking; prev. Pei-p'ing. ● (China) Beijing Shi, E China 39°58′N 116°23′E
161 P2 **Beijing** ✕ Beijing Shi, N China 39°54′N 116°22′E
Beijing see Beijing Shi, China

◆ Country
● Country Capital
◇ Dependent Territory
○ Dependent Territory Capital
◈ Administrative Regions
✕ International Airport
▲ Mountain
▲▲ Mountain Range
Volcano
∼ River
◎ Lake
⊠ Reservoir

BEIJING SHI – BERKÅK

161 O2 **Beijing Shi** *var.* Beijing, Jing, Pei-ching, *Eng.* Peking; *prev.* Pei-p'ing. ⬖ *municipality* E China
76 G8 **Beïla** Trarza, W Mauritania 18°07′N 15°56′W
98 N7 **Beilen** Drenthe, NE Netherlands 52°52′N 06°27′E
160 L15 **Beiliu** *var.* Lingcheng. Guangxi Zhuangzu Zizhiqu, S China 22°50′N 110°22′E
159 O12 **Beilu He** ≈ W China
Beilul *see* Beylul
163 U12 **Beining** *prev.* Beizhen. Liaoning, NE China 41°34′N 121°51′E
96 H8 **Beinn Dearg** ▲ N Scotland, United Kingdom 57°47′N 04°52′W
Beinn MacDuibh *see* Ben Macdui
160 I12 **Beipan Jiang** ≈ S China
163 T12 **Beipiao** Liaoning, NE China 41°49′N 120°45′E
83 N17 **Beira** Sofala, C Mozambique 19°45′S 34°56′E
83 N17 **Beira** ✈ Sofala, C Mozambique 19°39′S 35°05′E
104 I7 **Beira Alta** *former province* N Portugal
104 H9 **Beira Baixa** *former province* C Portugal
104 G8 **Beira Litoral** *former province* N Portugal
Beirut *see* Beyrouth
Beisān *see* Beit She'an
11 Q16 **Beiseker** Alberta, SW Canada 51°20′N 113°34′W
Beitai Ding *see* Wutai Shan
83 K19 **Beitbridge** Matabeleland South, S Zimbabwe 22°10′S 30°02′E
Beit Lekhem *see* Bethlehem
138 G9 **Beit She'an** *Ar.* Baysān, Beisān; *anc.* Scythopolis, *prev.* Bet She'an. Northern, N Israel 32°30′N 35°30′E
116 G10 **Beiuş** *Hung.* Belényes. Bihor, NW Romania 46°40′N 22°21′E
Beizhen *see* Beining
104 H12 **Beja** *anc.* Pax Julia. Beja, SE Portugal 38°01′N 07°52′W
74 M5 **Béja** *var.* Bājah. N Tunisia 36°45′N 09°04′E
104 G13 **Beja** ⬖ *district* S Portugal
120 I9 **Béjaïa** *var.* Bejaïa, *Fr.* Bougie; *anc.* Saldae. NE Algeria 36°49′N 05°03′E
Bejaïa *see* Béjaïa
104 K8 **Béjar** Castilla y León, N Spain 40°24′N 05°45′W
Bejraburi *see* Phetchaburi
Bekaa Valley *see* El Beqaa
Bekabad *see* Bekobod
Békás *see* Bicaz
169 O15 **Bekasi** Jawa, C Indonesia 06°14′S 106°595′E
Bek-Budi *see* Qarshi
Bekdaş/Bekdash *see* Garabogaz
147 T10 **Bek-Dzhar** Oshskaya Oblast', SW Kyrgyzstan 40°22′N 73°08′E
111 N24 **Békés** *Rom.* Bichiş. Békés, SE Hungary 46°45′N 21°09′E
111 M24 **Békés** *off.* Békés Megye. ⬖ *county* SE Hungary
111 M24 **Békéscsaba** *Rom.* Bichiş-Ciaba. Békés, SE Hungary 46°40′N 21°05′E
Békés Megye *see* Békés
172 H7 **Bekily** Toliara, S Madagascar 24°12′S 45°20′E
165 W4 **Bekkai** *var.* Betsukai. Hokkaidō, NE Japan 43°23′N 145°07′E
Bēkma *see* Baykhmah
147 Q11 **Bekobod** *Rus.* Bekabad; *prev.* Begovat. Toshkent Viloyati, E Uzbekistan 40°17′N 69°11′E
127 O7 **Bekovo** Penzenskaya Oblast', W Russian Federation 52°27′N 43°41′E
Bel *see* Beliu
152 M13 **Bela** Uttar Pradesh, N India 25°55′N 82°00′E
149 N15 **Bela** Baluchistān, SW Pakistan 26°12′N 66°20′E
79 F15 **Bélabo** Est, C Cameroon 04°54′N 13°10′E
112 N10 **Bela Crkva** *Ger.* Weisskirchen, *Hung.* Fehértemplom. Vojvodina, W Serbia 44°55′N 21°28′E
173 Y16 **Bel Air** *var.* Rivière Sèche. E Mauritius
104 L12 **Belalcázar** Andalucía, S Spain 38°33′N 05°07′W
113 P15 **Bela Palanka** Serbia, SE Serbia 43°13′N 22°19′E
119 H16 **Belarus** *off.* Republic of Belarus, *var.* Belorussia, *Latv.* Baltkrievija; *prev.* Belorussian SSR, *Rus.* Belorusskaya SSR. ◆ *republic* E Europe
Belarus, Republic of *see* Belarus
Belau *see* Palau
59 H21 **Bela Vista** Mato Grosso do Sul, SW Brazil 22°04′S 56°25′W
83 L21 **Bela Vista** Maputo, S Mozambique 26°20′S 32°40′E
168 I8 **Belawan** Sumatera, W Indonesia 03°46′N 98°44′E
Běla Woda *see* Weisswasser
127 U4 **Belaya** ≈ W Russian Federation
123 R7 **Belaya Gora** Respublika Sakha (Yakutiya), NE Russian Federation 68°25′N 146°12′E
126 M11 **Belaya Kalitiva** Rostovskaya Oblast', SW Russian Federation 48°09′N 40°43′E
125 R14 **Belaya Kholunitsa** Kirovskaya Oblast', NW Russian Federation 58°54′N 50°52′E
Belaya Tserkov' *see* Bila Tserkva
77 V11 **Belbédji** Zinder, S Niger 14°35′N 08°00′E
111 K14 **Bełchatów** *var.* Belchatow. Łódzski, C Poland 51°23′N 19°20′E
Belchatow *see* Bełchatów
Belcher, Îles *see* Belcher Islands
12 H7 **Belcher Islands** *Fr.* Îles Belcher. *island group* Nunavut, SE Canada
105 S6 **Belchite** Aragón, NE Spain 41°18′N 00°45′W
29 O2 **Belcourt** North Dakota, N USA 48°50′N 99°44′W
31 P9 **Belding** Michigan, N USA 43°06′N 85°13′W
127 U5 **Belebey** Respublika Bashkortostan, W Russian Federation 54°04′N 54°13′E
81 N16 **Beledweyne** *var.* Belet Huen, *It.* Belet Uen. Hiiraan, C Somalia 04°39′N 45°12′E
146 B10 **Belek** Balkan Welaýaty, W Turkmenistan 39°57′N 53°51′E
58 L12 **Belém** *var.* Pará. *state capital* Pará, N Brazil 01°27′S 48°29′W
65 I14 **Belém Ridge** *undersea feature* C Atlantic Ocean
62 I7 **Belén** Catamarca, NW Argentina 27°36′N 67°00′W
54 G9 **Belén** Boyacá, C Colombia 06°01′N 72°55′W
42 J11 **Belén** Rivas, SW Nicaragua 11°30′N 85°55′W
62 O5 **Belén** Concepción, C Paraguay 23°25′S 57°14′W
61 D16 **Belén** Salto, N Uruguay 30°47′S 57°47′W
37 R12 **Belen** New Mexico, SW USA 34°37′N 106°46′W
61 D20 **Belén de Escobar** Buenos Aires, E Argentina 34°21′S 58°47′W
114 J7 **Belene** Pleven, N Bulgaria 43°39′N 25°09′E
114 J7 **Belene, Ostrov** *island* N Bulgaria
43 R15 **Belén, Río** ≈ C Panama
Belényes *see* Beiuş
Embalse de Belesar *see* Belesar, Encoro de
104 H3 **Belesar, Encoro de** *Sp.* Embalse de Belesar. ⊠ NW Spain
Belet Huen/Belet Uen *see* Beledweyne
126 J5 **Belëv** Tul'skaya Oblast', W Russian Federation 53°48′N 36°07′E
19 R7 **Belfast** Maine, NE USA 44°25′N 69°02′W
97 G15 **Belfast** *Ir.* Béal Feirste. ◐ E Northern Ireland, United Kingdom 54°35′N 05°55′W
97 G15 **Belfast Aldergrove** ✈ E Northern Ireland, United Kingdom 54°37′N 06°11′W
97 G15 **Belfast Lough** *Ir.* Loch Lao. *inlet* E Northern Ireland, United Kingdom
28 K5 **Belfield** North Dakota, N USA 46°53′N 103°12′W
103 U7 **Belfort** Territoire-de-Belfort, E France 47°38′N 06°52′E
Belgard *see* Białogard
155 E17 **Belgaum** Karnātaka, W India 15°52′N 74°30′E
Belgian Congo *see* Congo (Democratic Republic of)
België/Belgique *see* Belgium
99 F20 **Belgium** *off.* Kingdom of Belgium, *Dut.* België, *Fr.* Belgique. ◆ *monarchy* NW Europe
Belgium, Kingdom of *see* Belgium
126 J8 **Belgorod** Belgorodskaya Oblast', W Russian Federation 50°38′N 36°37′E
Belgorod-Dnestrovskiy *see* Bilhorod-Dnistrovs'kyy
126 J8 **Belgorodskaya Oblast'** ⬖ *province* W Russian Federation
Belgrad *see* Beograd
29 T8 **Belgrade** Minnesota, N USA 45°27′N 94°59′W
33 S11 **Belgrade** Montana, NW USA 45°46′N 111°10′W
Belgrade *see* Beograd
Belgrano, Cabo *see* Meredith, Cape
195 N5 **Belgrano II** *Argentinian research station* Antarctica 77°50′S 35°25′W
21 X9 **Belhaven** North Carolina, SE USA 35°36′N 76°50′W
107 I23 **Belice** *anc.* Hypsas. ≈ Sicilia, Italy, C Mediterranean Sea
Belice *see* Belize/Belize City
Beli Drim *see* Drini i Bardhë
Beligrad *see* Berat
188 C8 **Beliliou** *prev.* Peleliu. *island* S Palau
114 L8 **Beli Lom, Yazovir** ⊠ NE Bulgaria
112 I8 **Beli Manastir** *Hung.* Pélmonostor; *prev.* Monostor. Osijek-Baranja, NE Croatia 45°46′N 16°38′E
102 J13 **Bélin-Béliet** Gironde, SW France 44°30′N 00°48′W
79 F17 **Bélinga** Ogooué-Ivindo, NE Gabon 01°05′N 13°12′E
21 S4 **Belington** West Virginia, NE USA 39°01′N 79°57′W
127 O6 **Belinskiy** Penzenskaya Oblast', W Russian Federation 52°58′N 43°25′E
169 N12 **Belinyu** Pulau Bangka, W Indonesia 01°37′S 105°45′E
169 O13 **Belitung, Pulau** *island* W Indonesia
116 F10 **Beliu** *Hung.* Bel. Arad, W Romania 46°31′N 21°57′E
114 I9 **Beli Vit** ≈ NW Bulgaria
42 G2 **Belize** *Sp.* Belice; *prev.* British Honduras, Colony of Belize. ◆ *commonwealth republic* Central America
42 F2 **Belize** *Sp.* Belice. ⬖ *district* NE Belize
42 G2 **Belize** ≈ Belize/Guatemala
Belize *see* Belize City
42 G2 **Belize City** *var.* Belize, *Sp.* Belice. Belize, NE Belize 17°29′N 88°10′W
42 G2 **Belize City** ✈ Belize, NE Belize 17°31′N 88°15′W
Belize, Colony of *see* Belize
Beljak *see* Villach
39 N16 **Belkofski** Alaska, USA 55°07′N 162°04′W
123 O6 **Bel'kovskiy, Ostrov** *island* Novosibirskiye Ostrova, NE Russian Federation
14 J8 **Bell** ≈ Québec, SE Canada
10 J15 **Bella Bella** British Columbia, SW Canada 52°04′N 128°07′W
102 M10 **Bellac** Haute-Vienne, C France 46°07′N 01°04′E
10 K15 **Bella Coola** British Columbia, SW Canada 52°23′N 126°46′W
106 D6 **Bellagio** Lombardia, N Italy 45°58′N 09°15′E
31 P6 **Bellaire** Michigan, N USA 44°59′N 85°12′W
106 D6 **Bellano** Lombardia, N Italy 46°06′N 09°21′E
155 G17 **Bellary** *var.* Ballari. Karnātaka, S India 15°11′N 76°54′E
183 S5 **Bellata** New South Wales, SE Australia 29°58′S 149°49′E
61 D16 **Bella Unión** Artigas, N Uruguay 30°18′S 57°35′W
61 C14 **Bella Vista** Corrientes, NE Argentina 28°30′S 59°03′W
62 J7 **Bella Vista** Tucumán, N Argentina 27°05′S 65°19′W
62 P4 **Bella Vista** Amambay, C Paraguay 22°08′S 56°20′W
56 B10 **Bellavista** Cajamarca, N Peru 05°43′S 78°48′W
56 D11 **Bellavista** San Martín, N Peru 07°04′S 76°35′W
183 U6 **Bellbrook** New South Wales, SE Australia 30°48′S 152°32′E
27 V5 **Belle** Missouri, C USA 38°17′N 91°43′W
21 Q5 **Belle** West Virginia, NE USA 38°13′N 81°32′W
31 R13 **Bellefontaine** Ohio, N USA 40°22′N 83°45′W
18 F14 **Bellefonte** Pennsylvania, NE USA 40°54′N 77°43′W
28 J9 **Belle Fourche** South Dakota, N USA 44°40′N 103°50′W
28 J9 **Belle Fourche Reservoir** ⊠ South Dakota, N USA
28 K9 **Belle Fourche River** ≈ South Dakota/Wyoming, N USA
103 S10 **Bellegarde-sur-Valserine** Ain, E France 46°06′N 05°49′E
23 Y14 **Belle Glade** Florida, SE USA 26°40′N 80°40′W
102 G8 **Belle Île** *island* NW France
13 T9 **Belle Isle** *island* Belle Isle, Newfoundland and Labrador, E Canada
13 S10 **Belle Isle, Strait of** *strait* Newfoundland and Labrador, E Canada
Bellenz *see* Bellinzona
29 W14 **Belle Plaine** Iowa, C USA 41°54′N 92°16′W
29 V9 **Belle Plaine** Minnesota, N USA 44°39′N 93°47′W
14 I9 **Belleterre** Québec, SE Canada 47°24′N 78°40′W
14 J15 **Belleville** Ontario, SE Canada 44°10′N 77°22′W
103 R10 **Belleville** Rhône, E France 46°09′N 04°42′E
30 K15 **Belleville** Illinois, N USA 38°31′N 89°58′W
27 N3 **Belleville** Kansas, C USA 39°51′N 97°38′W
29 Z13 **Bellevue** Iowa, C USA 42°15′N 90°25′W
29 S15 **Bellevue** Nebraska, C USA 41°08′N 95°53′W
31 S11 **Bellevue** Ohio, N USA 41°16′N 82°50′W
25 S5 **Bellevue** Texas, SW USA 33°38′N 98°00′W
32 H8 **Bellevue** Washington, NW USA 47°36′N 122°12′W
55 Y11 **Bellevue de l'Inini, Montagnes** ▲▲ S French Guiana
103 S11 **Belley** Ain, E France 45°46′N 05°41′E
Bellin *see* Kangirsuk
183 V6 **Bellingen** New South Wales, SE Australia 30°27′S 152°53′E
97 L14 **Bellingham** N England, United Kingdom 55°09′N 02°16′W
32 H6 **Bellingham** Washington, NW USA 48°46′N 122°29′W
Belling Hausen Mulde *see* Southeast Pacific Basin
194 H2 **Bellingshausen** *Russian research station* South Shetland Islands, Antarctica 61°57′S 58°23′W
Bellingshausen *see* Motu One
Bellingshausen Abyssal Plain *see* Bellingshausen Plain
196 R14 **Bellingshausen Plain** *var.* Bellingshausen Abyssal Plain. *undersea feature* SE Pacific Ocean 64°00′S 90°00′W
194 I8 **Bellingshausen Sea** *sea* Antarctica
98 P6 **Bellingwolde** Groningen, NE Netherlands 53°07′N 07°10′E
108 H11 **Bellinzona** *Ger.* Bellenz. Ticino, S Switzerland 46°12′N 09°02′E
25 T8 **Bellmead** Texas, SW USA 31°36′N 97°02′W
54 E8 **Bello** Antioquia, W Colombia 06°19′N 75°34′W
61 B21 **Bellocq** Buenos Aires, E Argentina 35°55′S 61°32′W
Bello Horizonte *see* Belo Horizonte
186 L10 **Bellona** *var.* Mungiki. *island* S Solomon Islands
Bellovacum *see* Beauvais
182 D7 **Bell, Point** *headland* South Australia 32°13′S 133°08′E
20 F9 **Bells** Tennessee, S USA 35°42′N 89°05′W
25 U5 **Bells** Texas, SW USA 33°36′N 96°24′W
92 N3 **Bellsund** *inlet* SW Svalbard
106 H6 **Belluno** Veneto, NE Italy 46°08′N 12°13′E
62 L11 **Bell Ville** Córdoba, C Argentina 32°35′S 62°41′W
83 E26 **Bellville** Western Cape, SW South Africa 33°50′S 18°43′E
25 U11 **Bellville** Texas, SW USA 29°57′N 96°15′W
104 L12 **Belméz** Andalucía, S Spain 38°16′N 05°12′W
29 V12 **Belmond** Iowa, C USA 42°51′N 93°36′W
18 E11 **Belmont** New York, NE USA 42°14′N 78°02′W
21 R10 **Belmont** North Carolina, SE USA 35°13′N 81°01′W
59 O18 **Belmonte** Bahia, E Brazil 15°53′S 38°54′W
104 I8 **Belmonte** Castelo Branco, C Portugal 40°21′N 07°20′W
105 P10 **Belmonte** Castilla-La Mancha, C Spain 39°34′N 02°43′W
42 G2 **Belmopan** ● (Belize) Cayo, C Belize 17°13′N 88°48′W
97 B16 **Belmullet** *Ir.* Béal an Mhuirhead. Mayo, W Ireland 54°14′N 09°59′W
99 E20 **Belœil** Hainaut, SW Belgium 50°33′N 03°45′E
123 R13 **Belogorsk** Amurskaya Oblast', SE Russian Federation 50°53′N 128°24′E
Belogorsk *see* Bilohirs'k
114 F7 **Belogradchik** Vidin, NW Bulgaria 43°37′N 22°42′E
172 H8 **Beloha** Toliara, S Madagascar 25°09′S 45°04′E
59 M20 **Belo Horizonte** *prev.* Bello Horizonte. *state capital* Minas Gerais, SE Brazil 19°54′S 43°54′W
26 M3 **Beloit** Kansas, C USA 39°27′N 98°06′W
30 L9 **Beloit** Wisconsin, N USA 42°31′N 89°01′W
Belokorovichi *see* Novi Bilokorovychi
124 J8 **Belomorsk** Respublika Kareliya, NW Russian Federation 64°30′N 34°43′E
124 J8 **Belomorsko-Baltiyskiy Kanal** *Eng.* White Sea-Baltic Canal, White Sea Canal. *canal* NW Russian Federation
153 V15 **Belonia** Tripura, NE India 23°15′N 91°25′E
Beloozersk *see* Byelaazyorsk
Belopol'ye *see* Bilopillya
105 O4 **Belorado** Castilla y León, N Spain 42°25′N 03°11′W
126 L14 **Belorechensk** Krasnodarskiy Kray, SW Russian Federation 44°46′N 39°53′E
127 W5 **Beloretsk** Respublika Bashkortostan, W Russian Federation 53°56′N 58°26′E
Belorussia/Belorussian SSR *see* Belarus
Belorusskaya Gryada *see* Byelaruskaya Hrada
Belorusskaya SSR *see* Belarus
Beloshchel'ye *see* Nar'yan-Mar
114 N8 **Beloslav** Varna, E Bulgaria 43°13′N 27°42′E
Belostok *see* Białystok
Belo-sur-Tsiribihina *see* Belo Tsiribihina
172 H5 **Belo Tsiribihina** *var.* Belo-sur-Tsiribihina. Toliara, W Madagascar 19°40′S 44°30′E
Belovár *see* Bjelovar
Belovezhskaya, Pushcha *see* Białowieska, Puszcza/Byelavyezhskaya, Pushcha
114 H10 **Belovo** Pazardzhik, C Bulgaria 42°10′N 24°01′E
Belovodsk *see* Bilovods'k
122 H9 **Beloyarskiy** Khanty-Mansiyskiy Avtonomnyy Okrug-Yugra, N Russian Federation 63°40′N 66°31′E
124 K7 **Beloye More** *Eng.* White Sea. *sea* NW Russian Federation
124 K13 **Beloye, Ozero** ◎ NW Russian Federation
114 J10 **Belozem** Plovdiv, C Bulgaria 42°11′N 25°00′E
124 K13 **Belozërsk** Vologodskaya Oblast', NW Russian Federation 59°59′N 37°49′E
108 D8 **Belp** Bern, W Switzerland 46°54′N 07°31′E
108 D8 **Belp** ✈ (Bern) Bern, C Switzerland 46°55′N 07°29′E
107 L24 **Belpasso** Sicilia, Italy, C Mediterranean Sea 37°35′N 14°59′E
31 U14 **Belpre** Ohio, N USA 39°14′N 81°34′W
98 M8 **Belterwijde** ◎ N Netherlands
27 R4 **Belton** Missouri, C USA 38°48′N 94°31′W
21 P11 **Belton** South Carolina, SE USA 34°31′N 82°29′W
25 T9 **Belton** Texas, SW USA 31°04′N 97°30′W
25 S9 **Belton Lake** ⊠ Texas, SW USA
Bel'tsy *see* Bălţi
97 E16 **Belturbet** *Ir.* Béal Tairbirt. Cavan, N Ireland 54°06′N 07°26′W
Beluchistan *see* Balochistan
145 Z9 **Belukha, Gora** ▲ Kazakhstan/Russian Federation 49°50′N 86°44′E
107 M20 **Belvedere Marittimo** Calabria, SW Italy 39°37′N 15°52′E
30 L10 **Belvidere** Illinois, N USA 42°15′N 88°50′W
18 J14 **Belvidere** New Jersey, NE USA 40°50′N 75°05′W
Bely *see* Belyy
127 V8 **Belyayevka** Orenburgskaya Oblast', W Russian Federation 51°25′N 56°26′E
Belynichi *see* Byalynichy
124 H17 **Belyy** *var.* Bely, Beyj. Tverskaya Oblast', W Russian Federation 55°51′N 32°57′E
126 I6 **Belyye Berega** Bryanskaya Oblast', W Russian Federation 53°11′N 34°42′E
122 J6 **Belyy, Ostrov** *island* N Russian Federation
122 J11 **Belyy Yar** Tomskaya Oblast', C Russian Federation 58°26′N 84°57′E
100 N13 **Belzig** Brandenburg, NE Germany 52°09′N 12°37′E
22 K4 **Belzoni** Mississippi, S USA 33°10′N 90°29′W
172 H4 **Bemaraha** *var.* Plateau du Bemaraha. ▲▲ W Madagascar
Bemaraha, Plateau du *see* Bemaraha
82 B10 **Bembe** Uíge, NW Angola 07°03′S 14°25′E
77 S14 **Bembèrèkè** *var.* Bimbéréké. N Benin 10°10′N 02°41′E
104 K12 **Bembézar** ≈ SW Spain
104 J3 **Bembibre** Castilla y León, N Spain 42°37′N 06°25′W
29 T4 **Bemidji** Minnesota, N USA 47°27′N 94°53′W
98 L12 **Bemmel** Gelderland, SE Netherlands 51°53′N 05°54′E
171 T13 **Bemu** Pulau Seram, E Indonesia 03°21′S 129°58′E
Benāb *see* Bonāb
105 T5 **Benabarre** *var.* Benavarn. Aragón, NE Spain 42°06′N 00°28′E
Benaco *see* Garda, Lago di
79 L20 **Bena-Dibele** Kasai-Oriental, C Dem. Rep. Congo 04°01′S 22°50′E
105 R9 **Benagéber, Embalse de** E Spain
183 O11 **Benalla** Victoria, SE Australia 36°33′S 146°00′E
104 M14 **Benamejí** Andalucía, S Spain 37°16′N 04°33′W
Benares *see* Vārānasi
Benavarn *see* Benabarre
104 F10 **Benavente** Santarém, C Portugal 38°59′N 08°49′W
104 K5 **Benavente** Castilla y León, N Spain 42°N 05°40′W
25 S15 **Benavides** Texas, SW USA 27°36′N 98°24′W
96 F8 **Benbecula** *island* NW Scotland, United Kingdom
Bencovazzo *see* Benkovac
32 H13 **Bend** Oregon, NW USA 44°04′N 121°19′W
182 K7 **Benda Range** ▲▲ South Australia
183 T6 **Bendemeer** New South Wales, SE Australia 30°54′S 151°12′E
Bender *see* Tighina
Bender Beila/Bender Beyla *see* Bandarbeyla
Bender Cassim/Bender Qaasim *see* Boosaaso
Bendery *see* Tighina
183 N11 **Bendigo** Victoria, SE Australia 36°46′S 144°19′E
118 E10 **Bēne** SW Latvia 56°30′N 23°04′E
98 K13 **Beneden-Leeuwen** Gelderland, C Netherlands 51°52′N 05°32′E
101 L24 **Benediktenwand** ▲ S Germany 47°39′N 11°28′E
Benemérita de San Cristóbal *see* San Cristóbal
77 N12 **Bénéna** Ségou, S Mali 13°04′N 04°20′W
172 I7 **Benenitra** Toliara, S Madagascar 23°25′S 45°06′E
Beneschau *see* Benešov
Beneški Zaliv *see* Venice, Gulf of
111 D17 **Benešov** *Ger.* Beneschau. Středočeský Kraj, W Czech Republic 49°48′N 14°41′E
107 L17 **Benevento** *anc.* Beneventum, Malventum. Campania, S Italy 41°07′N 14°45′E
Beneventum *see* Benevento
173 S3 **Bengal, Bay of** *bay* N Indian Ocean
Bengalooru *see* Bangalore
Bengaluru *see* Bangalore
79 M17 **Bengamisa** Orientale, N Dem. Rep. Congo 0°58′N 25°11′E
Bengasi *see* Banghāzī
Bengazi *see* Banghāzī
161 P7 **Bengbu** *var.* Peng-pu. Anhui, E China 32°57′N 117°17′E
32 L9 **Benge** Washington, NW USA 46°55′N 118°01′W
Benghazi *see* Banghāzī
168 K10 **Bengkalis** Pulau Bengkalis, W Indonesia 01°27′N 102°10′E
168 K10 **Bengkalis, Pulau** *island* W Indonesia
169 Q10 **Bengkayang** Borneo, C Indonesia 0°45′N 109°28′E
168 K14 **Bengkulu** *prev.* Bengkoeloe, Benkoelen, Benkulen. Sumatera, W Indonesia 03°46′S 102°16′E
168 K13 **Bengkulu** *off.* Propinsi Bengkulu; *prev.* Bengkoeloe, Benkoelen, Benkulen. ⬖ *province* W Indonesia
Bengkulu, Propinsi *see* Bengkulu
82 A11 **Bengo** ⬖ *province* W Angola
95 J16 **Bengtsfors** Västra Götaland, S Sweden 59°03′N 12°14′E
82 B13 **Benguela** *var.* Benguella. Benguela, W Angola 12°35′S 13°30′E
83 A14 **Benguela** ⬖ *province* W Angola
Benguella *see* Benguela
138 F10 **Ben Gurion** ✈ Tel Aviv, C Israel 32°04′N 34°45′E
Bengweulu, Lake *see* Bangweulu, Lake
Benha *see* Banhā
192 F6 **Benham Seamount** *undersea feature* W Philippine Sea 15°48′N 124°15′E
96 H6 **Ben Hope** ▲ N Scotland, United Kingdom 58°25′N 04°36′W
79 P18 **Beni** Nord-Kivu, NE Dem. Rep. Congo 0°31′N 29°30′E
74 H8 **Beni Abbès** W Algeria 30°07′N 02°09′W
105 T8 **Benicarló** Valenciana, E Spain 40°25′N 00°25′E
105 T9 **Benicasim** *Cat.* Benicàssim. Valenciana, E Spain 40°03′N 00°03′E
Benicàssim *see* Benicasim
105 T12 **Benidorm** Valenciana, SE Spain 38°33′N 00°09′W
Beni Mazâr *see* Banī Mazār
121 C11 **Beni-Mellal** C Morocco 32°20′N 06°21′W
77 R14 **Benin** *off.* Republic of Benin; *prev.* Dahomey. ◆ *republic* W Africa
77 S17 **Benin, Bight of** *gulf* W Africa
77 U16 **Benin City** Edo, SW Nigeria 06°23′N 05°40′E
Benin, Republic of *see* Benin
57 K16 **Beni, Río** ≈ N Bolivia
120 F10 **Beni Saf** *var.* Beni-Saf. NW Algeria 35°19′N 01°23′W
Beni-Saf *see* Beni Saf
Benishangul *see* Bīnshangul Gumuz
105 T11 **Benissa** Valenciana, E Spain 38°43′N 00°03′E
Beni Suef *see* Banī Suwayf
11 V15 **Benito** Manitoba, S Canada 51°57′N 101°24′W
Benito *see* Uolo, Río
61 C23 **Benito Juárez** Buenos Aires, E Argentina 37°43′S 59°48′W
41 P14 **Benito Juárez Internacional** ✈ (México) México, S Mexico 19°24′N 99°02′W
25 P5 **Benjamin** Texas, SW USA 33°35′N 99°49′W
58 B13 **Benjamin Constant** Amazonas, N Brazil 04°22′S 70°02′W
40 F4 **Benjamín Hill** Sonora, NW Mexico 30°13′N 111°08′W
63 F19 **Benjamín, Isla** *island* Archipiélago de los Chonos, S Chile
164 Q4 **Benkei-misaki** *headland* Hokkaidō, NE Japan 42°49′N 140°10′E
28 L17 **Benkelman** Nebraska, C USA 40°04′N 101°30′W
96 I7 **Ben Klibreck** ▲ N Scotland, United Kingdom 58°15′N 04°23′W
Benkoelen/Bengkoeloe *see* Bengkulu
112 D13 **Benkovac** *It.* Bencovazzo. Zadar, SW Croatia 44°02′N 15°36′E
Benkulen *see* Bengkulu
96 I11 **Ben Lawers** ▲ C Scotland, United Kingdom 56°33′N 04°13′W
96 J9 **Ben Macdui** *var.* Beinn MacDuibh. ▲ C Scotland, United Kingdom 57°02′N 03°42′W
96 G11 **Ben More** ▲ W Scotland, United Kingdom 56°26′N 06°00′W
96 I11 **Ben More** ▲ C Scotland, United Kingdom 56°22′N 04°31′W
96 H7 **Ben More Assynt** ▲ N Scotland, United Kingdom 58°09′N 04°51′W
185 E20 **Benmore, Lake** ◎ South Island, New Zealand
98 L12 **Bennekom** Gelderland, SE Netherlands 52°00′N 05°40′E
123 Q5 **Bennetta, Ostrov** *island* Novosibirskiye Ostrova, NE Russian Federation
21 T11 **Bennettsville** South Carolina, SE USA 34°36′N 79°40′W
96 H10 **Ben Nevis** ▲ N Scotland, United Kingdom 56°80′N 05°00′W
184 M9 **Benneydale** Waikato, North Island, New Zealand 38°31′S 175°22′E
Bennichab *see* Bennichchâb
76 H8 **Bennichchâb** *var.* Bennichab. Inchiri, W Mauritania 19°26′N 15°21′W
18 L10 **Bennington** Vermont, NE USA 42°51′N 73°09′W
185 E20 **Ben Ohau Range** ▲▲ South Island, New Zealand
83 J21 **Benoni** Gauteng, NE South Africa 26°04′S 28°18′E
172 J2 **Be, Nosy** *var.* Nossi-Bé. *island* NW Madagascar
Bénoué *see* Benue
42 F2 **Benque Viejo del Carmen** Cayo, W Belize 17°04′N 89°08′W
101 G19 **Bensheim** Hessen, W Germany 49°41′N 08°38′E
37 N16 **Benson** Arizona, SW USA 31°55′N 110°16′W
29 S8 **Benson** Minnesota, N USA 45°19′N 95°36′W
21 U10 **Benson** North Carolina, SE USA 35°22′N 78°33′W
171 N15 **Benteng** Pulau Selayar, C Indonesia 06°07′S 120°28′E
83 A14 **Bentiaba** Namibe, SW Angola 14°18′S 12°27′E
181 T4 **Bentinck Island** *island* Wellesley Islands, Queensland, N Australia
80 E13 **Bentiu** Unity, N South Sudan 09°14′N 29°49′E
138 G8 **Bent Jbaïl** *var.* Bint Jubayl. S Lebanon 33°07′N 35°26′E
11 Q15 **Bentley** Alberta, SW Canada 52°27′N 114°02′W
61 I15 **Bento Gonçalves** Rio Grande do Sul, S Brazil 29°12′S 51°34′W
27 U12 **Benton** Arkansas, C USA 34°34′N 92°35′W
30 L16 **Benton** Illinois, N USA 38°00′N 88°55′W
20 H7 **Benton** Kentucky, S USA 36°51′N 88°21′W
22 G5 **Benton** Louisiana, S USA 32°41′N 93°44′W
27 Y7 **Benton** Missouri, C USA 37°05′N 89°34′W
20 M10 **Benton** Tennessee, S USA 35°10′N 84°39′W
31 O10 **Benton Harbor** Michigan, N USA 42°07′N 86°27′W
27 S9 **Bentonville** Arkansas, C USA 36°23′N 94°13′W
77 V16 **Benue** ⬖ *state* SE Nigeria
78 F13 **Benue** *Fr.* Bénoué. ≈ Cameroon/Nigeria
163 V12 **Benxi** *prev.* Pen-ch'i, Penhsihu, Penki. Liaoning, NE China 41°20′N 123°45′E
Benyakoni *see* Byenyakoni
112 K10 **Beočin** Vojvodina, N Serbia 45°13′N 19°43′E
Beodericsworth *see* Bury St Edmunds
112 M11 **Beograd** *Eng.* Belgrade, *Ger.* Belgrad; *anc.* Singidunum. ● (Serbia) Serbia, N Serbia 44°48′N 20°27′E
112 L11 **Beograd** *Eng.* Belgrade. ✈ Serbia, N Serbia 44°45′N 20°21′E
76 M16 **Béoumi** C Ivory Coast 07°40′N 05°34′W
35 V3 **Beowawe** Nevada, W USA 40°33′N 116°31′W
164 E14 **Beppu** Ōita, Kyūshū, SW Japan 33°18′N 131°30′E
187 X15 **Beqa** *prev.* Mbengga. *island* W Fiji
45 Y14 **Bequia** *island* C Saint Vincent and the Grenadines
139 U4 **Beranan, Shax-î** *var.* Shākh-i Barānān. ▲▲ E Iraq
113 L16 **Berane** *prev.* Ivangrad. E Montenegro 42°51′N 19°51′E
113 L21 **Berat** *var.* Berati, *SCr.* Beligrad. Berat, C Albania 40°43′N 19°58′E
113 L21 **Berat** ⬖ *district* C Albania
Berătău *see* Berettyó
Berati *see* Berat
Beraun *see* Berounka, Czech Republic
Beraun *see* Beroun, Czech Republic
171 U13 **Berau, Teluk** *var.* MacCluer Gulf. *bay* Papua, E Indonesia
80 G8 **Berber** River Nile, NE Sudan 18°01′N 34°00′E
80 N12 **Berbera** Woqooyi Galbeed, NW Somalia 10°24′N 45°02′E
79 H16 **Berbérati** Mambéré-Kadéï, SW Central African Republic 04°14′N 15°50′E
Berberia, Cabo de *see* Barbaria, Cap de
55 T9 **Berbice River** ≈ NE Guyana
Berchid *see* Berrechid
103 N2 **Berck-Plage** Pas-de-Calais, N France 50°24′N 01°35′E
25 T13 **Berclair** Texas, SW USA 28°33′N 97°32′W
117 W10 **Berda** ≈ SE Ukraine
Berdichev *see* Berdychiv
123 P10 **Berdigestyakh** Respublika Sakha (Yakutiya), NE Russian Federation 62°02′N 127°03′E
122 J12 **Berdsk** Novosibirskaya Oblast', C Russian Federation 54°42′N 82°56′E
117 W10 **Berdyans'k** *Rus.* Berdyansk; *prev.* Osipenko. Zaporiz'ka Oblast', SE Ukraine 46°46′N 36°49′E
117 W10 **Berdyans'ka Kosa** *spit* SE Ukraine
117 V10 **Berdyans'ka Zatoka** *gulf* S Ukraine
117 N5 **Berdychiv** *Rus.* Berdichev. Zhytomyrs'ka Oblast', N Ukraine 49°54′N 28°39′E
20 M6 **Berea** Kentucky, S USA 37°34′N 84°18′W
Beregovo/Beregszász *see* Berehove
116 G8 **Berehove** *Cz.* Berehovo, *Hung.* Beregszász, *Rus.* Beregovo. Zakarpats'ka Oblast', W Ukraine 48°13′N 22°39′E
Berehovo *see* Berehove
186 D9 **Bereina** Central, S Papua New Guinea 08°29′S 146°30′E
146 C11 **Bereket** *prev. Rus.* Gazandzhyk, Kazandzhik, *Turkm.* Gazanjyk. Balkan Welaýaty, W Turkmenistan 39°17′N 55°27′E
45 O12 **Berekua** S Dominica 15°14′N 61°19′W
77 O16 **Berekum** W Ghana 07°27′N 02°35′W
Berenice *see* Baranīs
11 O14 **Berens** ≈ Manitoba/Ontario, C Canada
11 X14 **Berens River** Manitoba, C Canada 52°22′N 97°00′W
29 R12 **Beresford** South Dakota, N USA 43°02′N 96°45′W
116 J4 **Berestechko** Volyns'ka Oblast', NW Ukraine 50°21′N 25°06′E
116 M11 **Bereşti** Galaţi, E Romania 46°04′N 27°54′E
117 U6 **Berestova** ≈ E Ukraine
Beretău *see* Berettyó
111 N23 **Berettyó** *Rom.* Barcău; *prev.* Berătău, Beretău. ≈ Hungary/Romania
111 N23 **Berettyóújfalu** Hajdú-Bihar, E Hungary 47°15′N 21°33′E
Berëza/Bereza Kartuska *see* Byaroza
117 Q4 **Berezan'** Kyyivs'ka Oblast', N Ukraine 50°18′N 31°30′E
117 Q10 **Berezanka** Mykolayivs'ka Oblast', S Ukraine 46°51′N 31°24′E
116 J6 **Berezhany** *Pol.* Brzeżany. Ternopil's'ka Oblast', W Ukraine 49°29′N 25°00′E
Berezina *see* Byarezina
Berezino *see* Byerazino
117 P10 **Berezivka** *Rus.* Berezovka. Odes'ka Oblast', SW Ukraine 47°12′N 30°56′E
117 Q2 **Berezna** Chernihivs'ka Oblast', NE Ukraine 51°35′N 31°50′E
116 L3 **Berezne** Rivnens'ka Oblast', NW Ukraine 51°00′N 26°46′E
117 R9 **Bereznehuvate** Mykolayivs'ka Oblast', S Ukraine 47°18′N 32°52′E
125 N10 **Bereznik** Arkhangel'skaya Oblast', NW Russian Federation 62°50′N 42°40′E
125 U13 **Berezniki** Permskiy Kray, NW Russian Federation 59°26′N 56°49′E
Berëzovka *see* Byarozawka, Belarus
Berezovka *see* Berezivka, Ukraine
122 H9 **Berezovo** Khanty-Mansiyskiy Avtonomnyy Okrug-Yugra, N Russian Federation 63°48′N 64°38′E
127 O9 **Berezovskaya** Volgogradskaya Oblast', SW Russian Federation 50°17′N 43°58′E
123 S13 **Berezovyy** Khabarovskiy Kray, E Russian Federation 51°42′N 135°39′E
83 E25 **Berg** ≈ W South Africa
Berg *see* Berg bei Rohrbach
105 V4 **Berga** Cataluña, NE Spain 42°06′N 01°41′E
95 N20 **Berga** Kalmar, S Sweden 57°13′N 16°03′E
136 B13 **Bergama** İzmir, W Turkey 39°08′N 27°10′E
106 E7 **Bergamo** *anc.* Bergomum. Lombardia, N Italy 45°42′N 09°40′E
105 P3 **Bergara** País Vasco, N Spain 43°05′N 02°25′W
109 S3 **Berg bei Rohrbach** *var.* Berg. Oberösterreich, N Austria 48°34′N 14°02′E
100 O6 **Bergen** Mecklenburg-Vorpommern, NE Germany 54°25′N 13°25′E
101 I11 **Bergen** Niedersachsen, NW Germany 52°49′N 09°57′E
98 H8 **Bergen** Noord-Holland, NW Netherlands 52°40′N 04°42′E
94 C13 **Bergen** Hordaland, S Norway 60°24′N 05°19′E
Bergen *see* Mons
55 W9 **Berg en Dal** Brokopondo, C Suriname 05°15′N 55°20′W
99 G15 **Bergen op Zoom** Noord-Brabant, S Netherlands 51°30′N 04°17′E
102 L13 **Bergerac** Dordogne, SW France 44°51′N 00°30′E
99 J16 **Bergeyk** Noord-Brabant, S Netherlands 51°19′N 05°21′E
101 D16 **Bergheim** Nordrhein-Westfalen, W Germany 50°57′N 06°39′E
55 X10 **Bergi** Sipaliwini, E Suriname 04°36′N 54°24′W
101 E16 **Bergisch Gladbach** Nordrhein-Westfalen, W Germany 50°59′N 07°09′E
101 F14 **Bergkamen** Nordrhein-Westfalen, W Germany 51°32′N 07°41′E
95 N21 **Bergkvara** Kalmar, S Sweden 56°22′N 16°04′E
Bergomum *see* Bergamo
98 K13 **Bergse Maas** ≈ S Netherlands
95 P15 **Bergshamra** Stockholm, C Sweden 59°37′N 18°40′E
94 N10 **Bergsjö** Gävleborg, C Sweden 62°00′N 17°10′E
93 J14 **Bergsviken** Norrbotten, N Sweden 65°16′N 21°24′E
Bergum *see* Burgum
98 M6 **Bergumer Meer** ◎ N Netherlands
94 N12 **Bergviken** ◎ C Sweden
168 M11 **Berhala, Selat** *strait* Sumatera, W Indonesia
Berhampore *see* Baharampur
99 J17 **Beringen** Limburg, NE Belgium 51°04′N 05°14′E
39 T12 **Bering Glacier** *glacier* Alaska, USA
Beringov Proliv *see* Bering Strait
192 L2 **Bering Sea** *sea* N Pacific Ocean
38 L9 **Bering Strait** *Rus.* Beringov Proliv. *strait* Bering Sea/Chukchi Sea
Berislav *see* Beryslav
105 O15 **Berja** Andalucía, S Spain 36°51′N 02°56′W
94 H9 **Berkåk** Sør-Trøndelag, S Norway 62°50′N 10°01′E

◆ Country
● Country Capital
◇ Dependent Territory
○ Dependent Territory Capital
⬖ Administrative Regions
✈ International Airport
▲ Mountain
▲▲ Mountain Range
⛰ Volcano
≈ River
◎ Lake
⊠ Reservoir

BERKEL – BINEFAR

98 N11 **Berkel** ≈ Germany/Netherlands
35 N8 **Berkeley** California, W USA *37°52′N 122°16′W*
65 E24 **Berkeley Sound** *sound* NE Falkland Islands
21 V2 **Berkeley Springs** ***var.*** Bath. West Virginia, NE USA *39°38′N 78°14′W*
195 N6 **Berkner Island** *island* Antarctica
114 G8 **Berkovitsa** Montana, NW Bulgaria *43°15′N 23°05′E*
97 M22 **Berkshire** *former county* S England, United Kingdom
99 H17 **Berlaar** Antwerpen, N Belgium *51°08′N 04°39′E*
Berlanga ***see*** Berlanga de Duero
105 P6 **Berlanga de Duero** ***var.*** Berlanga. Castilla y León, N Spain *41°28′N 02°51′W*
0 I16 **Berlanga Rise** *undersea feature* E Pacific Ocean *08°30′N 93°30′W*
99 F17 **Berlare** Oost-Vlaanderen, NW Belgium *51°02′N 04°01′E*
104 E9 **Berlenga, Ilha da** *island* C Portugal
92 M7 **Berlevåg** ***Lapp.*** Bearalváhki. Finnmark, N Norway *70°51′N 29°04′E*
100 O12 **Berlin** ● (Germany) Berlin, NE Germany *52°31′N 13°26′E*
21 Z4 **Berlin** Maryland, NE USA *38°19′N 75°13′W*
19 O7 **Berlin** New Hampshire, NE USA *44°27′N 71°13′W*
18 D16 **Berlin** Pennsylvania, NE USA *39°54′N 78°57′W*
30 L7 **Berlin** Wisconsin, N USA *43°57′N 88°59′W*
100 O12 **Berlin** ◐ *state* NE Germany
Berlinchen ***see*** Barlinek
31 U12 **Berlin Lake** ⊠ Ohio, N USA
183 R11 **Bermagui** New South Wales, SE Australia *36°26′S 150°01′E*
40 L8 **Bermejillo** Durango, C Mexico *25°55′N 103°39′W*
62 L5 **Bermejo, Río** ≈ N Argentina
62 I10 **Bermejo, Río** ≈ W Argentina
62 M6 **Bermejo viejo, Río** ≈ N Argentina
105 P2 **Bermeo** País Vasco, N Spain *43°25′N 02°44′W*
104 K6 **Bermillo de Sayago** Castilla y León, N Spain *41°22′N 06°08′W*
106 E6 **Bermina, Pizzo** ***Rmsch.*** Piz Bernina. ▲ Italy/Switzerland *46°22′N 09°52′E* ***see also*** Bermina, Piz
64 A12 **Bermuda** ***var.*** Bermuda Islands, Bermudas; ***prev.*** Somers Islands. ◇ *UK crown colony* NW Atlantic Ocean
1 N11 **Bermuda** ***var.*** Great Bermuda, Long Island, Main Island. *island* Bermuda
Bermuda Islands ***see*** Bermuda
Bermuda-New England Seamount Arc ***see*** New England Seamounts
1 N11 **Bermuda Rise** *undersea feature* C Sargasso Sea *32°30′N 65°00′W*
Bermudas ***see*** Bermuda
108 D8 **Bern** ***Fr.*** Berne. ● (Switzerland) Bern, W Switzerland *46°57′N 07°26′E*
108 D9 **Bern** ***Fr.*** Berne. ◐ *canton* W Switzerland
37 R11 **Bernalillo** New Mexico, SW USA *35°18′N 106°33′W*
14 H12 **Bernard Lake** ◎ Ontario, S Canada
61 B18 **Bernardo de Irigoyen** Santa Fe, NE Argentina *32°09′S 61°06′W*
18 J14 **Bernardsville** New Jersey, NE USA *40°43′N 74°34′W*
63 K14 **Bernasconi** La Pampa, C Argentina *37°55′S 63°44′W*
100 O12 **Bernau** Brandenburg, NE Germany *52°41′N 13°36′E*
102 L4 **Bernay** Eure, N France *49°05′N 00°36′E*
101 L14 **Bernburg** Sachsen-Anhalt, C Germany *51°47′N 11°45′E*
109 X5 **Berndorf** Niederösterreich, NE Austria *47°58′N 16°08′E*
31 Q12 **Berne** Indiana, N USA *40°39′N 84°57′W*
Berne ***see*** Bern
108 D10 **Berner Alpen** ***var.*** Berner Oberland, ***Eng.*** Bernese Oberland. ▲▲ SW Switzerland
Berner Oberland/Bernese Oberland ***see*** Berner Alpen
109 Y2 **Bernhardsthal** Niederösterreich, N Austria *48°40′N 16°51′E*
22 H4 **Bernice** Louisiana, S USA *32°49′N 92°39′W*
27 Y8 **Bernie** Missouri, C USA *36°40′N 89°58′W*
180 G9 **Bernier Island** *island* Western Australia
Bernina Pass ***see*** Bernina, Passo del
108 J10 **Bernina, Passo del** ***Eng.*** Bernina Pass. *pass* SE Switzerland
108 J10 **Bernina, Piz** ***It.*** Pizzo Bernina. ▲ Italy/Switzerland *46°22′N 09°55′E* ***see also*** Bermina, Pizzo
Bernina, Piz ***see*** Bermina, Pizzo
99 E20 **Bérnissart** Hainaut, SW Belgium *50°29′N 03°37′E*
101 E18 **Bernkastel-Kues** Rheinland-Pfalz, W Germany *49°55′N 07°04′E*
Beroea ***see*** Ḥalab
172 H6 **Beroroha** Toliara, SW Madagascar *21°40′S 45°10′E*
Béroubouay ***see*** Gbéroubouè
111 C17 **Beroun** ***Ger.*** Beraun. Středočeský Kraj, W Czech Republic *49°58′N 14°05′E*
111 C16 **Berounka** ***Ger.*** Beraun. ≈ W Czech Republic
113 Q18 **Berovo** E FYR Macedonia *41°45′N 22°50′E*
74 F6 **Berrechid** ***var.*** Berchid. W Morocco *33°16′N 07°32′W*
103 R15 **Berre, Étang de** ◎ SE France
103 S15 **Berre-l'Étang** Bouches-du-Rhône, SE France *43°28′N 05°10′E*
182 K9 **Berri** South Australia *34°16′S 140°35′E*
31 O10 **Berrien Springs** Michigan, N USA *41°57′N 86°20′W*
183 O10 **Berrigan** New South Wales, SE Australia *35°41′S 145°50′E*
103 N9 **Berry** *cultural region* C France
35 N7 **Berryessa, Lake** ◎ California, W USA
44 G2 **Berry Islands** *island group* N The Bahamas
27 T9 **Berryville** Arkansas, C USA *36°22′N 93°35′W*
21 V3 **Berryville** Virginia, NE USA *39°08′N 77°59′W*
83 D21 **Berseba** Karas, S Namibia *26°00′S 17°46′E*
117 O8 **Bershad'** Vinnyts'ka Oblast', C Ukraine *48°20′N 29°30′E*
28 L3 **Berthold** North Dakota, N USA *48°16′N 101°48′W*
37 T3 **Berthoud** Colorado, C USA *40°18′N 105°04′W*
37 S4 **Berthoud Pass** *pass* Colorado, C USA
79 F15 **Bertoua** Est, E Cameroon *04°34′N 13°42′E*
25 S10 **Bertram** Texas, SW USA *30°44′N 98°03′W*
63 G22 **Bertrand, Cerro** ▲ S Argentina *50°00′S 73°27′W*
99 J23 **Bertrix** Luxembourg, SE Belgium *49°52′N 05°15′E*
191 P3 **Beru** ***var.*** Peru. *atoll* Tungaru, W Kiribati
Beruni ***see*** Beruniy
146 I9 **Beruniy** ***var.*** Biruni, ***Rus.*** Beruni. Qoraqalpog'iston Respublikasi, W Uzbekistan *41°48′N 60°39′E*
58 F13 **Beruri** Amazonas, NW Brazil *03°44′S 61°13′W*
18 H14 **Berwick** Pennsylvania, NE USA *41°03′N 76°13′W*
96 K12 **Berwick** *cultural region* SE Scotland, United Kingdom
96 L12 **Berwick-upon-Tweed** N England, United Kingdom *55°46′N 02°W*
117 S10 **Beryslav** ***Rus.*** Berislav. Khersons'ka Oblast', S Ukraine *46°51′N 33°26′E*
Berytus ***see*** Beyrouth
172 H4 **Besalampy** Mahajanga, W Madagascar *16°43′S 44°29′E*
103 T8 **Besançon** ***anc.*** Besontium, Vesontio. Doubs, E France *47°14′N 06°01′E*
103 P10 **Besbre** ≈ C France
Bescanuova ***see*** Baška
Besdan ***see*** Bezdan
Besed' ***see*** Byesyedz'
147 R10 **Beshariq** ***Rus.*** Besharyk; ***prev.*** Kirovo. Farg'ona Viloyati, E Uzbekistan *40°26′N 70°33′E*
Besharyk ***see*** Beshariq
146 L9 **Beshbuloq** ***Rus.*** Beshulak. Navoiy Viloyati, N Uzbekistan *41°55′N 64°13′E*
Beshenkovichi ***see*** Byeshankovichy
146 M13 **Beshkent** Qashqadaryo Viloyati, S Uzbekistan *38°47′N 65°42′E*
Beshulak ***see*** Beshbuloq
112 L10 **Beška** Vojvodina, N Serbia *45°09′N 20°04′E*
Beskra ***see*** Biskra
127 O16 **Beslan** Respublika Severnaya Osetiya, SW Russian Federation *43°12′N 44°33′E*
113 P16 **Besna Kobila** ▲ SE Serbia *42°30′N 22°16′E*
137 N16 **Besni** Adıyaman, S Turkey *37°42′N 37°53′E*
Besontium ***see*** Besançon
121 Q2 **Beşparmak Dağları** ***Eng.*** Kyrenia Mountains. ▲▲ N Cyprus
Bessarabka ***see*** Basarabeasca
92 O2 **Bessels, Kapp** *headland* C Svalbard *78°36′N 21°43′E*
23 P4 **Bessemer** Alabama, S USA *33°24′N 86°57′W*
30 K3 **Bessemer** Michigan, N USA *46°28′N 90°03′W*
21 Q10 **Bessemer City** North Carolina, SE USA *35°16′N 81°16′W*
102 M10 **Bessines-sur-Gartempe** Haute-Vienne, C France *46°06′N 01°22′E*
99 K15 **Best** Noord-Brabant, S Netherlands *51°31′N 05°24′E*
25 N9 **Best** Texas, SW USA *31°13′N 101°34′W*
139 U2 **Beste** ***Ar.*** Basṭah. As Sulaymānīyah, E Iraq *36°20′N 45°14′E*
125 O11 **Bestuzhevo** Arkhangel'skaya Oblast', NW Russian Federation *61°36′N 43°54′E*
123 M11 **Bestyakh** Respublika Sakha (Yakutiya), NE Russian Federation *61°25′N 129°05′E*
Beszterce ***see*** Bistriţa
Besztercebánya ***see*** Banská Bystrica
172 I5 **Betafo** Antananarivo, C Madagascar *19°50′S 46°50′E*
104 H2 **Betanzos** Galicia, NW Spain *43°17′N 08°17′W*
104 G2 **Betanzos, Ría de** *estuary* NW Spain
79 G15 **Bétaré Oya** Est, E Cameroon *05°34′N 14°09′E*
105 S9 **Bétera** Valenciana, E Spain *39°35′N 00°28′W*
77 R15 **Bétérou** C Benin *09°13′N 02°18′E*
83 K21 **Bethal** Mpumalanga, NE South Africa *26°27′S 29°28′E*
30 K15 **Bethalto** Illinois, N USA *38°54′N 90°02′W*
83 D21 **Bethanie** ***var.*** Bethanien, Bethany. Karas, S Namibia *26°32′S 17°11′E*
Bethanien ***see*** Bethanie
27 S2 **Bethany** Missouri, C USA *40°15′N 94°03′W*
27 N10 **Bethany** Oklahoma, C USA *35°31′N 97°37′W*
Bethany ***see*** Bethanie
39 N12 **Bethel** Alaska, USA *60°47′N 161°45′W*
19 P7 **Bethel** Maine, NE USA *44°24′N 70°47′W*
21 W9 **Bethel** North Carolina, SE USA *35°46′N 77°21′W*
18 B15 **Bethel Park** Pennsylvania, NE USA *40°21′N 80°03′W*
21 W3 **Bethesda** Maryland, NE USA *39°00′N 77°05′W*
83 J22 **Bethlehem** Free State, C South Africa *28°12′S 28°16′E*
18 I14 **Bethlehem** Pennsylvania, NE USA *40°36′N 75°22′W*
138 F10 **Bethlehem** ***var.*** Beit Lekhem, ***Ar.*** Bayt Laḥm, ***Heb.*** Bet Leḥem. C West Bank *31°43′N 35°12′E*
Bethlen ***see*** Beclean
83 I24 **Bethulie** Free State, C South Africa *30°30′S 25°59′E*
103 O1 **Béthune** Pas-de-Calais, N France *50°32′N 02°38′E*
102 M3 **Béthune** ≈ N France
104 M14 **Béticos, Sistemas** ***var.*** Sistema Penibético, ***Eng.*** Baetic Cordillera, Baetic Mountains. ▲▲ S Spain
54 I6 **Betijoque** Trujillo, NW Venezuela *09°25′N 70°45′W*
59 M20 **Betim** Minas Gerais, SE Brazil *19°56′S 44°10′W*
190 H3 **Betio** Tarawa, W Kiribati *01°21′N 172°56′E*
172 H7 **Betioky** Toliara, S Madagascar *23°42′S 44°22′E*
Bet Leḥem ***see*** Bethlehem
Betlen ***see*** Beclean
167 O17 **Betong** Yala, SW Thailand *05°45′N 101°05′E*
79 I16 **Bétou** Likouala, N Congo *03°08′N 18°31′E*
145 P14 **Betpakdala** ***Kaz.*** Betpaqdala ***prev.*** Betpak-Dala. *plateau* S Kazakhstan
Betpak-Dala ***see*** Betpakdala
Betpaqdala ***see*** Betpakdala
172 H7 **Betroka** Toliara, S Madagascar *23°15′S 46°07′E*
Betschau ***see*** Bečva
Bet She'an ***see*** Beit She'an
15 T6 **Betsiamites** Québec, SE Canada *48°56′N 68°40′W*
15 T6 **Betsiamites** ≈ Québec, SE Canada
172 I4 **Betsiboka** ≈ N Madagascar
Betsukai ***see*** Bekkai
99 M25 **Bettembourg** Luxembourg, S Luxembourg *49°31′N 06°06′E*
99 M23 **Bettendorf** Diekirch, NE Luxembourg *49°53′N 06°13′E*
29 Z14 **Bettendorf** Iowa, C USA *41°31′N 90°31′W*
Bette, Pic ***see*** Bette, Picco
75 R13 **Bette, Picco** ***var.*** Bīkkū Bīttī, Pic Bette. ▲ S Libya *22°02′N 19°07′E*
153 P12 **Bettiah** Bihār, N India *26°49′N 84°30′E*
39 Q7 **Bettles** Alaska, USA *66°54′N 151°40′W*
95 N17 **Bettna** Södermanland, C Sweden *58°52′N 16°40′E*
154 H11 **Betūl** ***prev.*** Badnur. Madhya Pradesh, C India *21°55′N 77°54′E*
154 H9 **Betwa** ≈ C India
101 F16 **Betzdorf** Rheinland-Pfalz, W Germany *50°47′N 07°50′E*
82 C9 **Béu** Uíge, NW Angola *06°15′S 15°32′E*
31 P6 **Beulah** Michigan, N USA *43°35′N 83°52′W*
28 L5 **Beulah** North Dakota, N USA *47°16′N 101°48′W*
98 M8 **Beulakerwijde** ◎ N Netherlands
98 L13 **Beuningen** Gelderland, SE Netherlands *51°52′N 05°47′E*
Beuthen ***see*** Bytom
103 N7 **Beuvron** ≈ C France
99 F16 **Beveren** Oost-Vlaanderen, N Belgium *51°13′N 04°15′E*
21 T9 **B. Everett Jordan Reservoir** ***var.*** Jordan Lake. ⊠ North Carolina, SE USA
97 N17 **Beverley** E England, United Kingdom *53°51′N 00°26′W*
Beverley ***see*** Beverly
99 J17 **Beverlo** Limburg, NE Belgium *51°06′N 05°14′E*
19 P11 **Beverly** Massachusetts, NE USA *42°33′N 70°49′W*
32 J9 **Beverly** ***var.*** Beverley. Washington, NW USA *46°50′N 119°57′W*
35 S15 **Beverly Hills** California, W USA *34°02′N 118°25′W*
101 I14 **Beverungen** Nordrhein-Westfalen, C Germany *51°39′N 09°22′E*
98 H9 **Beverwijk** Noord-Holland, W Netherlands *52°29′N 04°40′E*
108 C10 **Bex** Vaud, W Switzerland *46°15′N 07°00′E*
97 P23 **Bexhill** ***var.*** Bexhill-on-Sea. SE England, United Kingdom *50°50′N 00°28′E*
Bexhill-on-Sea ***see*** Bexhill
136 E17 **Bey Dağları** ▲▲ SW Turkey
Beyj ***see*** Belyy
136 E10 **Beykoz** İstanbul, NW Turkey *41°09′N 29°06′E*
76 K15 **Beyla** SE Guinea *08°43′N 08°41′W*
137 X12 **Beyläqan** ***prev.*** Zhdanov. SW Azerbaijan *39°43′N 47°38′E*
80 L10 **Beylul** ***var.*** Beilul. SE Eritrea *13°10′N 42°27′E*
144 H14 **Beyneu** ***Kaz.*** Beyneū. Mangistau, SW Kazakhstan *45°20′N 55°11′E*
Beyneū ***see*** Beyneu
165 X14 **Beyonēsu-retsugan** ***Eng.*** Bayonnaise Rocks. *island group* SE Japan
136 G12 **Beypazarı** Ankara, NW Turkey *40°10′N 31°56′E*
155 F21 **Beypore** Kerala, SW India *11°10′N 75°49′E*
138 G7 **Beyrouth** ***var.*** Bayrūt, ***Eng.*** Beirut; ***anc.*** Berytus. ● (Lebanon) W Lebanon *33°55′N 35°31′E*
138 G7 **Beyrouth** ✈ W Lebanon *33°52′N 35°30′E*
136 G15 **Beyşehir** Konya, SW Turkey *37°40′N 31°43′E*
136 G15 **Beyşehir Gölü** ◎ C Turkey
108 J7 **Bezau** Vorarlberg, NW Austria *47°24′N 09°55′E*
112 J8 **Bezdan** ***Ger.*** Besdan, ***Hung.*** Bezdán. Vojvodina, NW Serbia *45°51′N 19°00′E*
Bezdezh ***see*** Byezdzyezh
124 G15 **Bezhanitsy** Pskovskaya Oblast', W Russian Federation *56°57′N 29°53′E*
124 K15 **Bezhetsk** Tverskaya Oblast', W Russian Federation *57°47′N 36°42′E*
103 P16 **Béziers** ***anc.*** Baeterrae, Baeterrae Septimanorum, Julia Beterrae. Hérault, S France *43°21′N 03°13′E*
Bezmein ***see*** Abadan
Bezwada ***see*** Vijayawāda
154 P12 **Bhadrak** ***var.*** Bhadrakh. Odisha, E India *21°04′N 86°30′N*
Bhadrakh ***see*** Bhadrak
155 F19 **Bhadra Reservoir** ⊠ SW India
155 F18 **Bhadrāvati** Karnātaka, SW India *13°52′N 75°43′E*
153 R14 **Bhāgalpur** Bihār, NE India *25°14′N 86°59′E*
Bhairab ***see*** Bhairab Bazar
153 U14 **Bhairab Bazar** ***var.*** Bhairab. Dhaka, C Bangladesh *24°04′N 91°00′E*
153 O11 **Bhairahawā** Western, S Nepal *27°31′N 83°27′E*
149 S8 **Bhakkar** Punjab, E Pakistan *31°40′N 71°08′E*
153 P11 **Bhaktapur** Central, C Nepal *27°47′N 85°21′E*
167 N3 **Bhamo** ***var.*** Banmo. Kachin State, N Myanmar (Burma) *24°15′N 97°15′E*
Bhāmragad ***see*** Bhāmragarh
154 K13 **Bhāmragarh** ***var.*** Bhāmragad. Mahārāshtra, C India *19°28′N 80°39′E*
154 J12 **Bhandāra** Mahārāshtra, C India *21°10′N 79°41′E*
Bhārat ***see*** India
152 J12 **Bharatpur** ***prev.*** Bhurtpore. Rājasthān, N India *27°14′N 77°29′E*
154 D11 **Bharūch** Gujarāt, W India *21°48′N 72°55′E*
155 E18 **Bhatkal** Karnātaka, W India *13°59′N 74°34′E*
153 O13 **Bhatni** ***var.*** Bhatni Junction. Uttar Pradesh, N India *26°23′N 83°56′E*
Bhatni Junction ***see*** Bhatni
153 S16 **Bhātpāra** West Bengal, NE India *22°55′N 88°30′E*
149 U7 **Bhaun** Punjab, E Pakistan *32°53′N 72°48′E*
Bhaunagar ***see*** Bhāvnagar
Bhavānīpātna ***see*** Bhawānipatna
155 H21 **Bhavānīsāgar** ⊠ S India
154 D11 **Bhāvnagar** ***prev.*** Bhaunagar. Gujarāt, W India *21°46′N 72°14′E*
154 M13 **Bhawānipatna** ***var.*** Bhavānīpātna. Odisha, E India *19°56′N 83°09′E*
Bheanntraí, Bá ***see*** Bantry Bay
Bheara, Béal an ***see*** Gweebarra Bay
154 K12 **Bhilai** Chhattīsgarh, C India *21°12′N 81°26′E*
152 G13 **Bhīlwāra** Rājasthān, N India *25°23′N 74°39′E*
155 E14 **Bhīma** ≈ S India
155 K16 **Bhīmavaram** Andhra Pradesh, E India *16°34′N 81°35′E*
154 I7 **Bhind** Madhya Pradesh, C India *26°33′N 78°47′E*
152 E13 **Bhīnmāl** Rājasthān, N India *25°01′N 72°22′E*
Bhir ***see*** Bīd
83 J25 **Bhisho** ***prev.*** Bisho. Eastern Cape, S South Africa *32°46′S 27°21′E* ***see also*** Bhisho
154 D13 **Bhiwandi** Mahārāshtra, W India *19°21′N 73°08′E*
152 H10 **Bhiwāni** Haryāna, N India *28°50′N 76°10′E*
152 L13 **Bhognīpur** Uttar Pradesh, N India *26°12′N 79°48′E*
153 U16 **Bhola** Barisal, S Bangladesh *22°40′N 91°36′E*
154 H10 **Bhopāl** *state capital* Madhya Pradesh, C India *23°17′N 77°25′E*
155 J14 **Bhopālpatnam** Chhattīsgarh, C India *18°51′N 80°22′E*
155 E14 **Bhor** Mahārāshtra, W India *18°10′N 73°55′E*
154 O12 **Bhubaneshwar** ***prev.*** Bhubaneswar, Bhuvaneshwar. *state capital* Odisha, E India *20°16′N 85°51′E*
Bhubaneswar ***see*** Bhubaneshwar
154 B9 **Bhuj** Gujarāt, W India *23°16′N 69°40′E*
Bhuket ***see*** Phuket
Bhurtpore ***see*** Bharatpur
Bhusaval ***see*** Bhusāwal
154 G12 **Bhusāwal** ***prev.*** Bhusaval. Mahārāshtra, C India *21°01′N 75°50′E*
153 T12 **Bhutan** ***off.*** Kingdom of Bhutan, ***var.*** Druk-yul. ◆ *monarchy* S Asia
Bhutan, Kingdom of ***see*** Bhutan
Bhuvaneshwar ***see*** Bhubaneshwar
143 T15 **Bīābān, Kūh-e** ▲▲ S Iran
77 V18 **Biafra, Bight of** ***var.*** Bight of Bonny. *bay* W Africa
171 W12 **Biak** Papua, E Indonesia *01°10′S 136°05′E*
171 W12 **Biak, Pulau** *island* E Indonesia
110 P12 **Biała Podlaska** Lubelskie, E Poland *52°03′N 23°08′E*
110 F7 **Białogard** ***Ger.*** Belgard. Zachodnio-pomorskie, NW Poland *54°01′N 15°59′E*
110 P10 **Białowieska, Puszcza** ***Bel.*** Byelavyezhskaya Pushcha, ***Rus.*** Belovezhskaya Pushcha. *physical region* Belarus/Poland ***see also*** Byelavyezhskaya, Pushcha
Białowieska, Puszcza ***see*** Byelavyezhskaya Pushcha
110 G8 **Biały Bór** ***Ger.*** Baldenburg. Zachodnio-pomorskie, NW Poland *53°53′N 16°49′E*
110 P9 **Białystok** ***Rus.*** Belostok, Bielostok. Podlaskie, NE Poland *53°08′N 23°09′E*
107 L24 **Biancavilla** ***prev.*** Inessa. Sicilia, Italy, C Mediterranean Sea *37°38′N 14°52′E*
Bianco, Monte ***see*** Blanc, Mont
Bianjiang ***see*** Xunke
76 L15 **Biankouma** W Ivory Coast *07°44′N 07°37′W*
167 R7 **Bia, Phou** ***var.*** Pou Bia. ▲ C Laos *18°59′N 103°09′E*
Bia, Pou ***see*** Bia, Phou
143 R5 **Bīārjmand** Semnān, N Iran *36°05′N 55°50′E*
105 P4 **Biarra** ≈ NE Spain
102 I15 **Biarritz** Pyrénées-Atlantiques, SW France *43°25′N 01°40′W*
108 H10 **Biasca** Ticino, S Switzerland *46°22′N 08°59′E*
61 E17 **Biassini** Salto, N Uruguay *31°18′S 57°05′W*
Biasteri ***see*** Laguardia
165 S3 **Bibai** Hokkaidō, NE Japan *43°21′N 141°53′E*
83 B15 **Bibala** ***Port.*** Vila Arriaga. Namibe, SW Angola *14°46′S 13°21′E*
104 I4 **Bibéi** ≈ NW Spain
Biberach ***see*** Biberach an der Riss
101 I23 **Biberach an der Riss** ***var.*** Biberach, ***Ger.*** Biberach an der Riß. Baden-Württemberg, S Germany *48°06′N 09°48′E*
108 E7 **Biberist** Solothurn, NW Switzerland *47°11′N 07°34′E*
77 O16 **Bibiani** SW Ghana *06°28′N 02°20′W*
112 C13 **Bibinje** Zadar, SW Croatia *44°04′N 15°17′E*
Biblical Gebal ***see*** Jbaïl
116 I5 **Bibrka** ***Pol.*** Bóbrka, ***Rus.*** Bobrka. L'vivs'ka Oblast', NW Ukraine *49°39′N 24°16′E*
117 N10 **Bîc** ≈ S Moldova
113 M18 **Bicaj** Kukës, NE Albania *42°00′N 20°24′E*
116 K10 **Bicaz** ***Hung.*** Békás. Neamţ, NE Romania *46°53′N 26°05′E*
183 Q16 **Bicheno** Tasmania, SE Australia *41°56′S 148°15′E*
Bichiș ***see*** Békés
Bichiș-Ciaba ***see*** Békéscsaba
Bichitra ***see*** Phichit
137 P8 **Bich'vinta** ***Prev.*** Bichvint'a, ***Rus.*** Pitsunda. NW Georgia *43°12′N 40°21′E*
Bichvint'a ***see*** Bich'vinta
15 T7 **Bic, Île du** *island* Québec, SE Canada
32 J10 **Bickleton** Washington, NW USA *46°00′N 120°16′W*
36 L6 **Bicknell** Utah, W USA *38°20′N 111°32′W*
171 S11 **Bicoli** Pulau Halmahera, E Indonesia *0°34′N 128°33′E*
111 J22 **Bicske** Fejér, C Hungary *47°30′N 18°36′E*
155 F14 **Bīd** ***prev.*** Bhir. Mahārāshtra, W India *19°17′N 75°22′E*
77 U15 **Bida** Niger, C Nigeria *09°06′N 06°02′E*
155 H15 **Bīdar** Karnātaka, C India *17°56′N 77°35′E*
141 Y8 **Bidbid** NE Oman *23°25′N 58°08′E*
19 P9 **Biddeford** Maine, NE USA *43°29′N 70°27′W*
98 L9 **Biddinghuizen** Flevoland, C Netherlands *52°28′N 05°41′E*
33 X11 **Biddle** Montana, NW USA *45°04′N 105°21′W*
97 J23 **Bideford** SW England, United Kingdom *51°01′N 04°12′W*
82 D13 **Bié** ◐ *province* C Angola
35 O2 **Bieber** California, W USA *41°07′N 121°09′W*
110 O9 **Biebrza** ≈ NE Poland
165 T3 **Biei** Hokkaidō, NE Japan *43°33′N 142°28′E*
108 D8 **Biel** ***Fr.*** Bienne. Bern, W Switzerland *47°09′N 07°16′E*
100 G13 **Bielefeld** Nordrhein-Westfalen, NW Germany *52°01′N 08°32′E*
108 D8 **Bieler See** ***Fr.*** Lac de Bienne. ◎ W Switzerland
Bielitz/Bielitz-Biala ***see*** Bielsko-Biała
106 C7 **Biella** Piemonte, N Italy *45°34′N 08°04′E*
Bielostok ***see*** Białystok
111 J17 **Bielsko-Biała** ***Ger.*** Bielitz, Bielitz-Biala. Śląskie, S Poland *49°49′N 19°01′E*
110 P10 **Bielsk Podlaski** Białystok, E Poland *52°45′N 23°11′E*
Bien Bien ***see*** Điên Biên Phu
Biên Đông ***see*** South China Sea
11 V17 **Bienfait** Saskatchewan, S Canada *49°06′N 102°47′W*
167 T14 **Biên Hoa** Đông Nai, S Vietnam *10°58′N 106°50′E*
Bienne ***see*** Biel
Bienne, Lac de ***see*** Bieler See
12 K8 **Bienville, Lac** ◎ Québec, C Canada
82 D13 **Bié, Planalto do** ***var.*** Bié Plateau. *plateau* C Angola
Bié Plateau ***see*** Bié, Planalto do
108 B9 **Bière** Vaud, W Switzerland *46°32′N 06°19′E*
98 O4 **Bierum** Groningen, NE Netherlands *53°25′N 06°51′E*
98 I13 **Biesbos** ***var.*** Biesbosch. *wetland* S Netherlands
Biesbosch ***see*** Biesbos
99 H21 **Biesme** Namur, S Belgium *50°19′N 04°43′E*
101 H21 **Bietigheim-Bissingen** Baden-Württemberg, SW Germany *48°57′N 09°07′E*
99 I23 **Bièvre** Namur, SE Belgium *49°56′N 05°01′E*
79 D18 **Bifoun** Moyen-Ogooué, NW Gabon *0°15′S 10°24′E*
165 T2 **Bifuka** Hokkaidō, NE Japan *44°28′N 142°20′E*
136 C11 **Biga** Çanakkale, NW Turkey *40°13′N 27°14′E*
136 C13 **Bigadiç** Balıkesir, W Turkey *39°24′N 28°08′E*
26 J7 **Big Basin** *basin* Kansas, C USA
185 B20 **Big Bay** *bay* South Island, New Zealand
31 O5 **Big Bay de Noc** ◎ Michigan, N USA
31 N3 **Big Bay Point** *headland* Michigan, N USA *46°51′N 87°40′W*
33 R10 **Big Belt Mountains** ▲▲ Montana, NW USA
29 N10 **Big Bend Dam** *dam* South Dakota, N USA
24 K12 **Big Bend National Park** *national park* Texas, S USA
22 K5 **Big Black River** ≈ Mississippi, S USA
27 O3 **Big Blue River** ≈ Kansas/Nebraska, C USA
24 M10 **Big Canyon** ≈ Texas, SW USA
33 N12 **Big Creek** Idaho, NW USA *45°05′N 115°20′W*
23 N8 **Big Creek Lake** ⊠ Alabama, S USA
23 X15 **Big Cypress Swamp** *wetland* Florida, SE USA
39 S9 **Big Delta** Alaska, USA *64°09′N 145°50′W*
30 K6 **Big Eau Pleine Reservoir** ⊠ Wisconsin, N USA
19 P5 **Bigelow Mountain** ▲ Maine, NE USA *45°09′N 70°17′W*
162 G9 **Biger** ***var.*** Jargalant. Govĭ-Altay, W Mongolia *45°39′N 97°10′E*
29 U3 **Big Falls** Minnesota, N USA *48°13′N 93°48′W*
33 P8 **Bigfork** Montana, NW USA *48°03′N 114°04′W*
29 U3 **Big Fork River** ≈ Minnesota, N USA
11 S15 **Biggar** Saskatchewan, S Canada *52°03′N 107°59′W*
180 L3 **Bigge Island** *island* Western Australia
35 O5 **Biggs** California, W USA *39°24′N 121°44′W*
32 I11 **Biggs** Oregon, NW USA *45°39′N 120°49′W*
14 K13 **Big Gull Lake** ◎ Ontario, SE Canada
37 P16 **Big Hachet Peak** ▲ New Mexico, SW USA *31°38′N 108°24′W*
33 P11 **Big Hole River** ≈ Montana, NW USA
33 V13 **Bighorn Basin** *basin* Wyoming, C USA
33 U11 **Bighorn Lake** ◎ Montana/Wyoming, N USA
33 W13 **Bighorn Mountains** ▲▲ Wyoming, C USA
36 J13 **Big Horn Peak** ▲ Arizona, SW USA *33°40′N 113°01′W*
33 V11 **Bighorn River** ≈ Montana/Wyoming, NW USA
9 S7 **Big Island** *island* Nunavut, NE Canada
39 O16 **Big Koniuji Island** *island* Shumagin Islands, Alaska, USA
25 N9 **Big Lake** Texas, SW USA *31°12′N 101°29′W*
19 T5 **Big Lake** ◎ Maine, NE USA
30 I3 **Big Manitou Falls** *waterfall* Wisconsin, N USA
35 R2 **Big Mountain** ▲ Nevada, W USA *41°18′N 119°03′W*
108 G10 **Bignasco** Ticino, S Switzerland *46°21′N 08°37′E*
29 R16 **Big Nemaha River** ≈ Nebraska, C USA
76 G12 **Bignona** SW Senegal *12°49′N 16°16′W*
Bigorra ***see*** Tarbes
Bigosovo ***see*** Bihosava
35 S10 **Big Pine** California, W USA *37°09′N 118°18′W*
35 Q14 **Big Pine Mountain** ▲ California, W USA *34°41′N 119°37′W*
27 V6 **Big Piney Creek** ≈ Missouri, C USA
65 M24 **Big Point** *headland* N Tristan da Cunha *37°10′S 12°18′W*
31 P8 **Big Rapids** Michigan, N USA *43°42′N 85°28′W*
30 K6 **Big Rib River** ≈ Wisconsin, N USA
14 L14 **Big Rideau Lake** ◎ Ontario, SE Canada
11 T14 **Big River** Saskatchewan, C Canada *53°48′N 106°55′W*
27 X5 **Big River** ≈ Missouri, C USA
31 N7 **Big Sable Point** *headland* Michigan, N USA *44°03′N 86°30′W*
33 S7 **Big Sandy** Montana, NW USA *48°08′N 110°09′W*
25 W6 **Big Sandy** Texas, SW USA *32°34′N 95°06′W*
37 V5 **Big Sandy Creek** ≈ Colorado, C USA
29 Q16 **Big Sandy Creek** ≈ Nebraska, C USA
29 V5 **Big Sandy Lake** ◎ Minnesota, N USA
36 J11 **Big Sandy River** ≈ Arizona, SW USA
21 P5 **Big Sandy River** ≈ S USA
23 V6 **Big Satilla Creek** ≈ Georgia, SE USA
29 R12 **Big Sioux River** ≈ Iowa/South Dakota, N USA
35 U7 **Big Smoky Valley** *valley* Nevada, W USA
25 N7 **Big Spring** Texas, SW USA *32°15′N 101°30′W*
19 Q5 **Big Squaw Mountain** ▲ Maine, NE USA *45°28′N 69°42′W*
21 O7 **Big Stone Gap** Virginia, NE USA *36°52′N 82°45′W*
29 Q8 **Big Stone Lake** ◎ Minnesota/South Dakota, N USA
22 K4 **Big Sunflower River** ≈ Mississippi, S USA
33 T11 **Big Timber** Montana, NW USA *45°50′N 109°57′W*
12 D8 **Big Trout Lake** Ontario, C Canada *53°40′N 90°00′W*
14 I12 **Big Trout Lake** ◎ Ontario, SE Canada
35 O2 **Big Valley Mountains** ▲▲ California, W USA
25 Q13 **Big Wells** Texas, SW USA *28°34′N 99°34′W*
14 F11 **Bigwood** Ontario, S Canada *46°03′N 80°37′W*
112 D11 **Bihać** ◐ Federacija Bosne I Hercegovine, NW Bosnia and Herzegovina
153 P14 **Bihār** ***prev.*** Behar. ◐ *state* N India
Bihār ***see*** Bihār Sharīf
81 F20 **Biharamulo** Kagera, NW Tanzania *02°37′S 31°20′E*
153 R13 **Bihārīganj** Bihār, NE India *25°44′N 86°59′E*
153 P14 **Bihār Sharīf** ***var.*** Bihār. Bihār, N India *25°13′N 85°31′E*
116 F10 **Bihor** ◐ *county* NW Romania
165 V3 **Bihoro** Hokkaidō, NE Japan *43°50′N 144°05′E*
118 K11 **Bihosava** ***Rus.*** Bigosovo. Vitsyebskaya Voblasts', NW Belarus *55°50′N 27°46′E*
Bijagos Archipelago ***see*** Bijagós, Arquipélago dos
76 G13 **Bijagós, Arquipélago dos** ***var.*** Bijagos Archipelago. *island group* W Guinea-Bissau
155 F16 **Bijāpur** Karnātaka, C India *16°50′N 75°42′E*
142 K5 **Bījār** Kordestān, W Iran *35°52′N 47°39′E*
112 J11 **Bijeljina** Republika Srpska, NE Bosnia and Herzegovina *44°46′N 19°13′E*
113 K15 **Bijelo Polje** E Montenegro *43°03′N 19°44′E*
160 I11 **Bijie** Guizhou, S China *27°15′N 105°16′E*
152 J10 **Bijnor** Uttar Pradesh, N India *29°22′N 78°09′E*
152 F11 **Bīkāner** Rājasthān, NW India *28°01′N 73°22′E*
189 V3 **Bikar Atoll** ***var.*** Pikaar. *atoll* Ratak Chain, N Marshall Islands
190 H3 **Bikeman** *atoll* Tungaru, W Kiribati
190 I3 **Bikenebu** Tarawa, W Kiribati
123 S14 **Bikin** Khabarovskiy Kray, SE Russian Federation *46°45′N 134°06′E*
123 S14 **Bikin** ≈ SE Russian Federation
189 R3 **Bikini Atoll** ***var.*** Pikinni. *atoll* Ralik Chain, NW Marshall Islands
83 L17 **Bikita** Masvingo, E Zimbabwe *20°06′S 31°41′E*
Bīkkū Bīttī ***see*** Bette, Picco
79 I19 **Bikoro** Equateur, W Dem. Rep. Congo *0°45′S 18°09′E*
141 Z9 **Bilād Banī Bū 'Alī** NE Oman *22°02′N 59°18′E*
141 Z9 **Bilād Banī Bū Ḥasan** NE Oman *22°09′N 59°14′E*
141 X9 **Bilād Manaḥ** ***var.*** Manaḥ. NE Oman *22°44′N 57°36′E*
77 Q12 **Bilanga** C Burkina Faso *12°35′N 00°08′W*
152 F12 **Bilāra** Rājasthān, N India *26°10′N x73°48′E*
152 K10 **Bilāri** Uttar Pradesh, N India *28°37′N 78°48′E*
138 J5 **Bil'ās, Jabal al** ▲▲ C Syria
154 L11 **Bilāspur** Chhattīsgarh, C India *22°06′N 82°08′E*
152 I8 **Bilāspur** Himāchal Pradesh, N India *31°18′N 76°48′E*
168 J9 **Bila, Sungai** ≈ Sumatera, W Indonesia
137 Y13 **Biläsuvar** ***Rus.*** Bilyasuvar; ***prev.*** Pushkino. SE Azerbaijan *39°26′N 48°34′E*
117 O5 **Bila Tserkva** ***Rus.*** Belaya Tserkov'. Kyyivs'ka Oblast', N Ukraine *49°49′N 30°08′E*
167 N11 **Bilauktaung Range** ***var.*** Thanintari Taungdan. ▲▲ Myanmar (Burma)/Thailand
105 O2 **Bilbao** ***Basq.*** Bilbo. País Vasco, N Spain *43°15′N 02°56′W*
Bilbo ***see*** Bilbao
92 H2 **Bíldudalur** Vestfirðir, NW Iceland *65°40′N 23°35′W*
113 I16 **Bileća** ◐ Republika Srpska, S Bosnia and Herzegovina
136 E12 **Bilecik** Bilecik, NW Turkey *39°59′N 29°54′E*
136 F12 **Bilecik** ◐ *province* NW Turkey
116 E11 **Biled** ***Ger.*** Billed, ***Hung.*** Billéd. Timiș, W Romania *45°55′N 20°55′E*
111 O15 **Biłgoraj** Lubelskie, E Poland *50°31′N 22°41′E*
117 P11 **Bilhorod-Dnistrovs'kyy** ***Rus.*** Belgorod-Dnestrovskiy, ***Rom.*** Cetatea Albă, ***prev.*** Akkerman; ***anc.*** Tyras. Odes'ka Oblast', SW Ukraine *46°10′N 30°19′E*
79 M16 **Bili** Orientale, N Dem. Rep. Congo *04°07′N 25°09′E*
123 T6 **Bilibino** Chukotskiy Avtonomnyy Okrug, NE Russian Federation *67°56′N 166°45′E*
166 M8 **Bilin** Mon State, S Myanmar (Burma) *17°14′N 97°12′E*
113 N21 **Bilisht** ***var.*** Bilishti. Korçë, SE Albania *40°36′N 21°00′E*
Bilishti ***see*** Bilisht
183 N10 **Billabong Creek** ***var.*** Moulamein Creek. *seasonal river* New South Wales, SE Australia
182 G4 **Billa Kalina** South Australia *29°57′S 136°13′E*
197 Q17 **Bill Baileys Bank** *undersea feature* N Atlantic Ocean *60°35′N 10°15′W*
Billed/Billéd ***see*** Biled
153 N14 **Billi** Uttar Pradesh, N India *24°30′N 82°59′E*
97 M15 **Billingham** N England, United Kingdom *54°36′N 01°17′W*
33 U11 **Billings** Montana, NW USA *45°47′N 108°32′W*
95 J16 **Billingsfors** Västra Götaland, S Sweden *58°57′N 12°14′E*
Bill of Cape Clear, The ***see*** Clear, Cape
28 L9 **Billsburg** South Dakota, N USA *44°22′N 101°40′W*
95 F23 **Billund** Syddtjylland, W Denmark *55°44′N 09°07′E*
36 L11 **Bill Williams Mountain** ▲ Arizona, SW USA *35°12′N 112°12′W*
36 I12 **Bill Williams River** ≈ Arizona, SW USA
77 Y8 **Bilma** Agadez, NE Niger *18°22′N 13°01′E*
77 Y8 **Bilma, Grand Erg de** *desert* NE Niger
181 Y9 **Biloela** Queensland, E Australia *24°27′S 150°31′E*
112 G8 **Bilo Gora** ▲▲ N Croatia
117 U13 **Bilohirs'k** ***Rus.*** Belogorsk; ***prev.*** Karasubazar. Avtonomna Respublika Krym, S Ukraine *45°04′N 34°35′E*
Bilokorovychi ***see*** Novi Bilokorovychi
Bilokurakine ***see*** Bilokurakyne
117 X5 **Bilokurakyne** ***var.*** Bilokurakine. Luhans'ka Oblast', E Ukraine *49°32′N 38°44′E*
117 T3 **Bilopillya** ***Rus.*** Belopol'ye. Sums'ka Oblast', NE Ukraine *51°09′N 34°17′E*
117 Y6 **Bilovods'k** ***Rus.*** Belovodsk. Luhans'ka Oblast', E Ukraine *49°11′N 39°34′E*
22 M9 **Biloxi** Mississippi, S USA *30°24′N 88°53′W*
117 R10 **Bilozerka** Khersons'ka Oblast', S Ukraine *46°36′N 32°23′E*
117 W7 **Bilozers'ke** Donets'ka Oblast', E Ukraine *48°29′N 37°03′E*
98 J11 **Bilthoven** Utrecht, C Netherlands *52°07′N 05°12′E*
78 K9 **Biltine** Wadi Fira, E Chad *14°30′N 20°53′E*
Biltine, Préfecture de ***see*** Wadi Fira
Bilüü ***see*** Ulaanhus
Bilwi ***see*** Puerto Cabezas
Bilyasuvar ***see*** Biläsuvar
117 O11 **Bilyayivka** Odes'ka Oblast', SW Ukraine *46°28′N 30°11′E*
99 K18 **Bilzen** Limburg, NE Belgium *50°52′N 05°31′E*
Bimbéréké ***see*** Bembèrèkè
183 R10 **Bimberi Peak** ▲ New South Wales, SE Australia *35°42′S 148°46′E*
77 Q15 **Bimbila** E Ghana *08°54′N 00°05′E*
79 I15 **Bimbo** Ombella-Mpoko, SW Central African Republic *04°19′N 18°27′E*
44 F2 **Bimini Islands** *island group* W The Bahamas
154 I9 **Bīna** Madhya Pradesh, C India *24°09′N 78°10′E*
143 T4 **Bīnālūd, Kūh-e** ▲▲ NE Iran
99 F20 **Binche** Hainaut, S Belgium *50°25′N 04°10′E*
Bindloe Island ***see*** Marchena, Isla
83 L16 **Bindura** Mashonaland Central, NE Zimbabwe *17°20′S 31°21′E*
105 T5 **Binefar** Aragón, NE Spain *41°51′N 00°17′E*

◆ Country
● Country Capital
◇ Dependent Territory
○ Dependent Territory Capital
◐ Administrative Regions
✈ International Airport
▲ Mountain
▲▲ Mountain Range
Volcano
≈ River
◎ Lake
⊠ Reservoir

BINGA – BOCONÓ

83 J16 **Binga** Matabeleland North, W Zimbabwe *17°40′S 27°22′E*
183 T5 **Bingara** New South Wales, SE Australia *29°54′S 150°36′E*
101 F18 **Bingen am Rhein** Rheinland-Pfalz, SW Germany *49°58′N 07°54′E*
26 M11 **Binger** Oklahoma, C USA *35°19′N 98°19′W*
Bingerau *see* Węgrów
Bin Ghalfān, Jazā'ir *see* Ḩalāniyāt, Juzur al
19 Q6 **Bingham** Maine, NE USA *45°01′N 69°51′W*
18 H11 **Binghamton** New York, NE USA *42°06′N 75°55′W*
Bin Ghanīmah, Jabal *see* Bin Ghunaymah, Jabal
75 P11 **Bin Ghunaymah, Jabal** *var.* Jabal Bin Ghanīmah. ▲▲ C Libya
139 U3 **Bingird** As Sulaymānīyah, NE Iraq *36°03′N 45°03′E*
Bingmei *see* Congjiang
137 P14 **Bingöl** Bingöl, E Turkey *38°54′N 40°29′E*
137 P14 **Bingöl** ◈ *province* E Turkey
161 R6 **Binhai** *var.* Dongkan. Jiangsu, E China *34°00′N 119°51′E*
167 V11 **Binh Đinh** *var.* An Nhon. Binh Đinh, C Vietnam *13°53′N 109°07′E*
Binh Sơn *see* Châu Ô
Binimani *see* Bintimani
168 I8 **Binjai** Sumatera, W Indonesia *03°37′N 98°30′E*
183 R6 **Binnaway** New South Wales, SE Australia *31°34′S 149°24′E*
108 E6 **Binningen** Basel Landschaft, NW Switzerland *47°32′N 07°35′E*
80 H12 **Bīnshangul Gumuz** *var.* Benishangul. ◈ W Ethiopia
168 J8 **Bintang, Banjaran** ▲▲ Peninsular Malaysia
168 M10 **Bintan, Pulau** *island* Kepulauan Riau, W Indonesia
76 J14 **Bintimani** *var.* Binimani. ▲ NE Sierra Leone *09°21′N 11°09′W*
Bint Jubayl *see* Bent Jbaïl
169 S9 **Bintulu** Sarawak, East Malaysia *03°12′N 113°01′E*
169 S9 **Bintuni** *prev.* Steenkool. Papua Barat, E Indonesia *02°03′S 133°45′E*
163 W8 **Binxian** *prev.* Binzhou. Heilongjiang, NE China *45°44′N 127°27′E*
160 K14 **Binyang** *var.* Binzhou. Guangxi Zhuangzu Zizhiqu, S China *23°15′N 108°40′E*
161 Q4 **Binzhou** Shandong, E China *37°23′N 118°03′E*
Binzhou *see* Binyang
Binzhou *see* Binxian
63 G14 **Bío Bío** *var.* Región del Bío Bío. ◈ *region* C Chile
Bío Bío, Región del *see* Bío Bío
63 G14 **Bío Bío, Río** ≈ C Chile
79 C16 **Bioco, Isla de** *var.* Bioko, *Eng.* Fernando Po, *Sp.* Fernando Póo; *prev.* Macías Nguema Biyogo. *island* NW Equatorial Guinea
112 D13 **Biograd na Moru** *It.* Zaravecchia. Zadar, SW Croatia *43°57′N 15°27′E*
Bioko *see* Bioco, Isla de
113 F14 **Biokovo** ▲▲ S Croatia
Biorra *see* Birr
Bipontium *see* Zweibrücken
143 W13 **Bīrag, Kūh-e** ▲▲ SE Iran
75 O10 **Birāk** *var.* Brak. C Libya *27°32′N 14°17′E*
139 S10 **Bi'r al Islām** Karbalā', C Iraq *32°15′N 43°40′E*
154 N11 **Biramitrapur** *var.* Birmitrapur. Odisha, E India *22°24′N 84°42′E*
139 T11 **Bi'r an Niṣf** An Najaf, S Iraq *31°22′N 44°07′E*
78 L12 **Birao** Vakaga, NE Central African Republic *10°14′N 22°49′E*
146 J10 **Birata** *Rus.* Darganata, Dargan-Ata. Lebap Welaýaty, NE Turkmenistan *40°30′N 62°09′E*
158 M6 **Biratar Bulak** *well* NW China
153 R12 **Birātnagar** Eastern, SE Nepal *26°28′N 87°16′E*
165 R5 **Biratori** Hokkaidō, NE Japan *42°35′N 142°07′E*
39 S8 **Birch Creek** Alaska, USA *66°17′N 145°54′W*
38 M11 **Birch Creek** ≈ Alaska, USA
11 T14 **Birch Hills** Saskatchewan, S Canada *52°58′N 105°22′W*
182 M10 **Birchip** Victoria, SE Australia *36°01′S 142°55′E*
29 X4 **Birch Lake** ◎ Minnesota, N USA
11 Q11 **Birch Mountains** ▲▲ Alberta, W Canada
11 V15 **Birch River** Manitoba, S Canada *52°22′N 101°03′W*
44 H12 **Birchs Hill** *hill* W Jamaica
39 R11 **Birchwood** Alaska, USA *61°24′N 149°28′W*
188 I5 **Bird Island** *island* S Northern Mariana Islands
137 N16 **Birecik** Şanlıurfa, S Turkey *37°03′N 37°59′E*
152 M10 **Birendranagar** *var.* Surkhet. Mid Western, W Nepal *28°35′N 81°36′E*
Bir es Saba *see* Be'er Sheva
74 A12 **Bir-Gandouz** SW Western Sahara *21°35′N 16°27′W*
153 P12 **Birganj** Central, C Nepal *27°03′N 84°53′E*
81 B14 **Biri** ≈ W South Sudan
Bi'r Ibn Hirmās *see* Al Bi'r
143 U8 **Bīrjand** Khorāsān-e Jonūbī, E Iran *32°54′N 59°14′E*
Birkaland *see* Pirkanmaa
139 T11 **Birkat Ḩāmid** *well* S Iraq
95 F18 **Birkeland** Aust-Agder, S Norway *58°18′N 08°13′E*
101 E19 **Birkenfeld** Rheinland-Pfalz, SW Germany *49°39′N 07°10′E*
97 K18 **Birkenhead** NW England, United Kingdom *53°24′N 03°02′W*
109 W7 **Birkfeld** Steiermark, SE Austria *47°21′N 15°40′E*
182 A2 **Birksgate Range** ▲▲ South Australia
Bîrlad *see* Bârlad
145 S15 **Birlik** *var.* Novotroickoje, Novotroitskoye; *prev.* Brlik. Zhambyl, SE Kazakhstan *43°39′N 73°45′E*
97 K20 **Birmingham** C England, United Kingdom *52°30′N 01°50′W*
23 P4 **Birmingham** Alabama, S USA *33°30′N 86°47′W*
97 M20 **Birmingham** ✈ C England, United Kingdom *52°27′N 01°46′W*
Birmitrapur *see* Biramitrapur
Bir Moghrein *see* Bîr Mogreïn
76 J4 **Bîr Mogreïn** *var.* Bir Moghrein; *prev.* Fort-Trinquet. Tiris Zemmour, N Mauritania *25°10′N 11°35′W*
191 S4 **Birnie Island** *atoll* Phoenix Islands, C Kiribati
77 S12 **Birnin Gaouré** *var.* Birni-Ngaouré. Dosso, SW Niger *12°59′N 03°02′E*
Birni-Ngaouré *see* Birnin Gaouré
77 S12 **Birnin Kebbi** Kebbi, NW Nigeria *12°28′N 04°08′E*
77 T12 **Birnin Konni** *var.* Birni-Nkonni. Tahoua, SW Niger *13°51′N 05°15′E*
Birni-Nkonni *see* Birnin Konni
77 W13 **Birnin Kudu** Jigawa, N Nigeria *11°28′N 09°29′E*
123 S16 **Birobidzhan** Yevreyskaya Avtonomnaya Oblast', SE Russian Federation *48°42′N 132°55′E*
97 D18 **Birr** *var.* Parsonstown, *Ir.* Biorra. C Ireland *53°06′N 07°55′W*
183 P4 **Birrie River** ≈ New South Wales/Queensland, SE Australia
108 D7 **Birse** ≈ NW Switzerland
Birsen *see* Biržai
108 E6 **Birsfelden** Basel Landschaft, NW Switzerland *47°33′N 07°37′E*
127 U4 **Birsk** Respublika Bashkortostan, W Russian Federation *55°24′N 55°33′E*
119 F14 **Birštonas** Kaunas, C Lithuania *54°37′N 24°00′E*
159 P14 **Biru** Xinjiang Uygur Zizhiqu, W China *31°30′N 93°56′E*
Biruni *see* Beruniy
122 L12 **Biryusa** ≈ C Russian Federation
122 L12 **Biryusinsk** Irkutskaya Oblast', C Russian Federation *55°52′N 97°48′E*
118 G10 **Biržai** *Ger.* Birsen. Panevėžys, NE Lithuania *56°12′N 24°47′E*
121 P16 **Birżebbuġa** SE Malta *35°50′N 14°32′E*
Bisanthe *see* Tekirdağ
171 R12 **Bisa, Pulau** *island* Maluku, E Indonesia
37 N17 **Bisbee** Arizona, SW USA *31°27′N 109°55′W*
29 O2 **Bisbee** North Dakota, N USA *48°36′N 99°21′W*
Biscaia, Baía de *see* Biscay, Bay of
102 I13 **Biscarrosse et de Parentis, Étang de** ◎ SW France
104 M1 **Biscay, Bay of** *Sp.* Golfo de Vizcaya, *Port.* Baía de Biscaia. *bay* France/Spain
23 Z16 **Biscayne Bay** *bay* Florida, SE USA
64 M7 **Biscay Plain** *undersea feature* SE Bay of Biscay *07°15′W 45°00′N*
107 N17 **Bisceglie** Puglia, SE Italy *41°14′N 16°31′E*
Bischoflack *see* Škofja Loka
Bischofsburg *see* Biskupiec
109 Q7 **Bischofshofen** Salzburg, NW Austria *47°25′N 13°13′E*
101 P15 **Bischofswerda** Sachsen, E Germany *51°07′N 14°13′E*
103 V5 **Bischwiller** Bas-Rhin, NE France *48°46′N 07°52′E*
21 T10 **Biscoe** North Carolina, SE USA *35°20′N 79°46′W*
194 G5 **Biscoe Islands** *island group* Antarctica
14 E9 **Biscotasi Lake** ◎ Ontario, S Canada
14 E9 **Biscotasing** Ontario, S Canada *47°16′N 82°04′W*
54 J6 **Biscucuy** Portuguesa, NW Venezuela *09°22′N 69°59′W*
99 M24 **Bissen** Luxembourg, C Luxembourg *49°47′N 06°04′E*
114 K11 **Biser** Haskovo, S Bulgaria *41°52′N 25°59′E*
113 D15 **Biševo** *It.* Busi. *island* SW Croatia
141 N12 **Bīshah, Wādī** *dry watercourse* C Saudi Arabia
147 U7 **Bishkek** *var.* Pishpek; *prev.* Frunze. ● (Kyrgyzstan) Chuyskaya Oblast', N Kyrgyzstan *42°54′N 74°27′E*
147 U7 **Bishkek** ✈ Chuyskaya Oblast', N Kyrgyzstan *42°55′N 74°37′E*
153 R16 **Bishnupur** West Bengal, NE India *23°05′N 87°20′E*
Bisho *see* Bhisho
35 S9 **Bishop** California, W USA *37°22′N 118°24′W*
25 S15 **Bishop** Texas, SW USA *27°36′N 97°49′W*
97 L15 **Bishop Auckland** N England, United Kingdom *54°41′N 01°41′W*
Bishop's Lynn *see* King's Lynn
97 O21 **Bishop's Stortford** E England, United Kingdom *51°45′N 00°11′E*
21 S12 **Bishopville** South Carolina, SE USA *34°18′N 80°15′W*
138 M5 **Bishrī, Jabal** ▲▲ E Syria
163 U4 **Bishui** Heilongjiang, NE China *52°06′N 123°42′E*
81 G17 **Bisina, Lake** *prev.* Lake Salisbury. ◎ E Uganda
Biskara *see* Biskra
74 L6 **Biskra** *var.* Beskra, Biskara. NE Algeria *34°51′N 05°44′E*
110 M8 **Biskupiec** *Ger.* Bischofsburg. Warmińsko-Mazurskie, NE Poland *53°52′N 20°57′E*
171 R7 **Bislig** Mindanao, S Philippines *08°10′N 126°19′E*
27 X6 **Bismarck** Missouri, C USA *37°46′N 90°37′W*
28 M5 **Bismarck** *state capital* North Dakota, N USA *46°49′N 100°47′W*
186 D5 **Bismarck Archipelago** *island group* NE Papua New Guinea
129 Z16 **Bismarck Plate** *tectonic feature* W Pacific Ocean
186 D7 **Bismarck Range** ▲▲ N Papua New Guinea
186 E6 **Bismarck Sea** *sea* W Pacific Ocean
137 P15 **Bismil** Dıyarbakır, SE Turkey *37°53′N 40°38′E*
43 N6 **Bismuna, Laguna** *lagoon* NE Nicaragua
Bisnulok *see* Phitsanulok
171 R10 **Bisoa, Tanjung** *headland* Pulau Halmahera, N Indonesia *02°15′N 127°57′E*
28 K7 **Bison** South Dakota, N USA *45°31′N 102°27′W*
93 H17 **Bispgården** Jämtland, C Sweden *63°00′N 16°40′E*
76 G13 **Bissau** ● (Guinea-Bissau) W Guinea-Bissau *11°52′N 15°39′W*
76 G13 **Bissau** ✈ W Guinea-Bissau *11°53′N 15°41′W*
Bissojohka *see* Børselv
76 G12 **Bissorã** W Guinea-Bissau *12°16′N 15°35′W*
11 O10 **Bistcho Lake** ◎ Alberta, W Canada
22 G5 **Bistineau, Lake** ⊠ Louisiana, S USA
Bistrica *see* Ilirska Bistrica
116 I9 **Bistrița** *Ger.* Bistritz, *Hung.* Beszterce; *prev.* Nösen. Bistrița-Năsăud, N Romania *47°10′N 24°31′E*
116 K10 **Bistrița** *Ger.* Bistritz. ≈ NE Romania
116 I9 **Bistrița-Năsăud** ◈ *county* N Romania
Bistritz *see* Bistrița
Bistritz ober Pernstein *see* Bystřice nad Pernštejnem
152 L11 **Biswān** Uttar Pradesh, N India *27°30′N 81°00′E*
110 M7 **Bisztynek** Warmińsko-Mazurskie, NE Poland *54°05′N 20°53′E*
79 E17 **Bitam** Woleu-Ntem, N Gabon *02°05′N 11°30′E*
101 D18 **Bitburg** Rheinland-Pfalz, SW Germany *49°58′N 06°31′E*
103 U4 **Bitche** Moselle, NE France *49°01′N 07°27′E*
78 I11 **Bitkine** Guéra, C Chad *11°59′N 18°13′E*
137 R15 **Bitlis** Bitlis, SE Turkey *38°23′N 42°04′E*
137 R14 **Bitlis** ◈ *province* E Turkey
Bitoeng *see* Bitung
113 N20 **Bitola** *Turk.* Monastir; *prev.* Bitolj. S FYR Macedonia *41°01′N 21°22′E*
Bitolj *see* Bitola
107 O17 **Bitonto** *anc.* Butuntum. Puglia, SE Italy *41°07′N 16°41′E*
77 Q13 **Bitou** *var.* Bittou. SE Burkina Faso *11°19′N 00°17′W*
155 C20 **Bitra Island** *island* Lakshadweep, India, N Indian Ocean
101 M14 **Bitterfeld** Sachsen-Anhalt, E Germany *51°37′N 12°18′E*
32 O9 **Bitterroot Range** ▲▲ Idaho/Montana, NW USA
33 P10 **Bitterroot River** ≈ Montana, NW USA
107 D18 **Bitti** Sardegna, Italy, C Mediterranean Sea *40°30′N 09°31′E*
Bittou *see* Bitou
171 Q11 **Bitung** *prev.* Bitoeng. Sulawesi, C Indonesia *01°28′N 125°13′E*
60 I12 **Bituruna** Paraná, S Brazil *26°11′S 51°34′W*
77 Y13 **Biu** Borno, E Nigeria *10°35′N 12°13′E*
Biumba *see* Byumba
164 J13 **Biwa-ko** ◎ Honshū, SW Japan
171 X14 **Biwarlaut** Papua, E Indonesia *05°44′S 138°14′E*
27 P10 **Bixby** Oklahoma, C USA *35°56′N 95°52′W*
122 J13 **Biya** ≈ S Russian Federation
122 J13 **Biysk** Altayskiy Kray, S Russian Federation *52°34′N 85°09′E*
164 H13 **Bizen** Okayama, Honshū, SW Japan *34°45′N 134°10′E*
Bizerta *see* Bizerte
120 K10 **Bizerte** *Ar.* Banzart, *Eng.* Bizerta. N Tunisia *37°18′N 09°48′E*
105 O2 **Bizkaia** *Cast.* Vizcaya. ◈ *province* País Vasco, N Spain
92 G2 **Bjargtangar** *headland* W Iceland *63°30′N 24°29′W*
Bjärnå *see* Perniö
95 K22 **Bjärnum** Skåne, S Sweden *56°15′N 13°45′E*
93 I16 **Bjästa** Västernorrland, C Sweden *63°12′N 18°30′E*
113 I14 **Bjelašnica** ▲ SE Bosnia and Herzegovina *43°13′N 18°18′E*
112 C10 **Bjelolasica** ▲ NW Croatia *45°13′N 14°56′E*
112 F8 **Bjelovar** *Hung.* Belovár. Bjelovar-Bilogora, N Croatia *45°54′N 16°49′E*
112 F8 **Bjelovar-Bilogora** *off.* Bjelovarsko-Bilogorska Županija. ◈ *province* NE Croatia
Bjelovarsko-Bilogorska Županija *see* Bjelovar-Bilogora
92 H10 **Bjerkvik** Nordland, C Norway *68°31′N 16°08′E*
95 G21 **Bjerringbro** Midtjylland, NW Denmark *56°23′N 09°40′E*
Bjeshkët e Namuna *see* North Albanian Alps
95 L14 **Björbo** Dalarna, C Sweden *60°28′N 14°44′E*
95 I15 **Bjørkelangen** Akershus, S Norway *59°54′N 11°33′E*
95 O14 **Björklinge** Uppsala, C Sweden *60°03′N 17°33′E*
95 P14 **Bjørkø-Arholma** Stockholm, C Sweden *59°51′N 19°01′E*
93 I14 **Björksele** Västerbotten, N Sweden *64°58′N 18°30′E*
93 I16 **Björna** Västernorrland, C Sweden *63°34′N 18°38′E*
95 C14 **Bjørnafjorden** *fjord* S Norway
95 L16 **Björneborg** Värmland, C Sweden *59°13′N 14°15′E*
Björneborg *see* Pori
95 E14 **Bjornesfjorden** ◎ S Norway
92 M9 **Bjørnevatn** Finnmark, N Norway *69°40′N 29°57′E*
197 T13 **Bjørnøya** *Eng.* Bear Island. *island* N Norway
93 I15 **Bjurholm** Västerbotten, N Sweden *63°57′N 19°16′E*
95 J22 **Bjuv** Skåne, S Sweden *56°05′N 12°57′E*
76 M12 **Bla** Ségou, W Mali *12°58′N 05°45′W*
181 W8 **Blackall** Queensland, E Australia *24°26′S 145°32′E*
29 V2 **Black Bay** *lake bay* Minnesota, N USA
27 N9 **Black Bear Creek** ≈ Oklahoma, C USA
97 K17 **Blackburn** NW England, United Kingdom *53°45′N 02°29′W*
39 T11 **Blackburn, Mount** ▲ Alaska, USA *61°43′N 143°25′W*
35 N5 **Black Butte Lake** ⊠ California, W USA
194 J5 **Black Coast** *physical region* Antarctica
11 Q16 **Black Diamond** Alberta, SW Canada *50°42′N 114°09′W*
18 K11 **Black Dome** ▲ New York, NE USA *42°16′N 74°07′W*
113 L18 **Black Drin** *Alb.* Lumi i Drinit të Zi, *SCr.* Crni Drim. ≈ Albania/FYR Macedonia
29 U4 **Blackduck** Minnesota, N USA *47°45′N 94°33′W*
12 D6 **Black Duck** ≈ Ontario, C Canada
33 R14 **Blackfoot** Idaho, NW USA *43°11′N 112°20′W*
33 P9 **Blackfoot River** ≈ Montana, NW USA
Black Forest *see* Schwarzwald
28 J10 **Blackhawk** South Dakota, N USA *44°09′N 103°18′W*
28 I10 **Black Hills** ▲▲ South Dakota/Wyoming, N USA
11 T10 **Black Lake** ◎ Saskatchewan, C Canada
22 G6 **Black Lake** ⊠ Louisiana, S USA
31 Q5 **Black Lake** ◎ Michigan, N USA
18 I7 **Black Lake** ◎ New York, NE USA
26 F7 **Black Mesa** ▲ Oklahoma, C USA *37°00′N 103°07′W*
21 P10 **Black Mountain** North Carolina, SE USA *35°37′N 82°19′W*
35 P13 **Black Mountain** ▲ California, W USA *35°22′N 120°21′W*
37 Q2 **Black Mountain** ▲ Colorado, C USA *40°47′N 107°23′W*
96 K1 **Black Mountains** ▲▲ SE Wales, United Kingdom
36 H10 **Black Mountains** ▲▲ Arizona, SW USA
21 O7 **Black Mountains** ▲ Kentucky, E USA *36°54′N 82°53′W*
33 Q16 **Black Pine Peak** ▲ Idaho, NW USA *42°07′N 113°07′W*
97 K17 **Blackpool** NW England, United Kingdom *53°50′N 03°03′W*
37 Q14 **Black Range** ▲▲ New Mexico, SW USA
44 I12 **Black River** W Jamaica *18°02′N 77°52′W*
14 J14 **Black River** ≈ Ontario, SE Canada
129 U12 **Black River** *Chin.* Babian Jiang, Lixian Jiang, *Fr.* Rivière Noire, *Vtn.* Sông Đa. ≈ China/Vietnam
44 I12 **Black River** ≈ W Jamaica
39 T7 **Black River** ≈ Alaska, USA
37 N13 **Black River** ≈ Arizona, SW USA
27 X7 **Black River** ≈ Arkansas/Missouri, C USA
22 I7 **Black River** ≈ Louisiana, S USA
31 S8 **Black River** ≈ Michigan, N USA
31 Q5 **Black River** ≈ Michigan, N USA
18 I8 **Black River** ≈ New York, NE USA
21 T13 **Black River** ≈ South Carolina, SE USA
30 J7 **Black River** ≈ Wisconsin, N USA
30 J7 **Black River Falls** Wisconsin, N USA *44°18′N 90°51′W*
35 R3 **Black Rock Desert** *desert* Nevada, W USA
Black Sand Desert *see* Garagum
21 S7 **Blacksburg** Virginia, NE USA *37°15′N 80°25′W*
136 H10 **Black Sea** *var.* Euxine Sea, *Bul.* Cherno More, *Rom.* Marea Neagră, *Rus.* Chernoye More, *Turk.* Karadeniz, *Ukr.* Chorne More. *sea* Asia/Europe
117 Q10 **Black Sea Lowland** *Ukr.* Prychornomor'ska Nyzovyna. *depression* SE Europe
33 S17 **Blacks Fork** ≈ Wyoming, C USA
23 V7 **Blackshear** Georgia, SE USA *31°18′N 82°14′W*
23 S6 **Blackshear, Lake** ⊠ Georgia, SE USA
97 A16 **Blacksod Bay** *Ir.* Cuan an Fhóid Duibh. *inlet* W Ireland
21 V7 **Blackstone** Virginia, NE USA *37°04′N 78°00′W*
77 O14 **Black Volta** *var.* Borongo, Mouhoun, Moun Hou, *Fr.* Volta Noire. ≈ W Africa
23 O5 **Black Warrior River** ≈ Alabama, S USA
181 X8 **Blackwater** Queensland, E Australia *23°34′S 148°51′E*
97 D20 **Blackwater** *Ir.* An Abhainn Mhór. ≈ S Ireland
27 T4 **Blackwater River** ≈ Missouri, C USA
21 W7 **Blackwater River** ≈ Virginia, NE USA
Blackwater State *see* Nebraska
27 N8 **Blackwell** Oklahoma, C USA *36°48′N 97°16′W*
25 P7 **Blackwell** Texas, SW USA *32°05′N 100°19′W*
99 J15 **Bladel** Noord-Brabant, S Netherlands *51°22′N 05°13′E*
Bladenmarkt *see* Bălăuşeri
114 G11 **Blagoevgrad** *prev.* Gorna Dzhumaya. Blagoevgrad, W Bulgaria *42°01′N 23°05′E*
114 G11 **Blagoevgrad** ◈ *province* SW Bulgaria
123 Q14 **Blagoveshchensk** Amurskaya Oblast', SE Russian Federation *50°19′N 127°30′E*
127 V4 **Blagoveshchensk** Respublika Bashkortostan, W Russian Federation *55°03′N 56°01′E*
102 I7 **Blain** Loire-Atlantique, NW France *47°26′N 01°47′W*
29 V8 **Blaine** Minnesota, N USA *45°09′N 93°13′W*
32 H6 **Blaine** Washington, NW USA *48°59′N 122°45′W*
11 T15 **Blaine Lake** Saskatchewan, S Canada *52°49′N 106°48′W*
29 S14 **Blair** Nebraska, C USA *41°32′N 96°07′W*
96 J10 **Blairgowrie** C Scotland, United Kingdom *56°19′N 03°25′W*
18 C15 **Blairsville** Pennsylvania, NE USA *40°25′N 79°12′W*
116 H11 **Blaj** *Ger.* Blasendorf, *Hung.* Balázsfalva. Alba, SW Romania *46°10′N 23°57′E*
64 F9 **Blake-Bahama Ridge** *undersea feature* W Atlantic Ocean *29°00′N 73°30′W*
23 S7 **Blakely** Georgia, SE USA *31°22′N 84°55′W*
64 E10 **Blake Plateau** *var.* Blake Terrace. *undersea feature* W Atlantic Ocean *31°00′N 79°00′W*
30 M1 **Blake Point** *headland* Michigan, N USA *48°11′N 88°25′W*
Blake Terrace *see* Blake Plateau
61 B24 **Blanca, Bahía** *bay* E Argentina
56 C12 **Blanca, Cordillera** ▲▲ W Peru
105 T12 **Blanca, Costa** *physical region* SE Spain
37 S7 **Blanca Peak** ▲ Colorado, C USA *37°34′N 105°29′W*
24 I9 **Blanca, Sierra** ▲ Texas, SW USA *31°15′N 105°26′W*
120 K9 **Blanc, Cap** *headland* N Tunisia *37°20′N 09°41′E*
Blanc, Cap *see* Nouâdhibou, Râs
31 R12 **Blanchard River** ≈ Ohio, N USA
182 E8 **Blanche, Cape** *headland* South Australia *33°03′S 134°10′E*
182 J4 **Blanche, Lake** ◎ South Australia
31 R14 **Blanchester** Ohio, N USA *39°17′N 83°59′W*
182 J9 **Blanchetown** South Australia *34°21′S 139°36′E*
45 U13 **Blanchisseuse** Trinidad, Trinidad and Tobago *10°47′N 61°18′W*
103 T11 **Blanc, Mont** *It.* Monte Bianco. ▲ France/Italy *45°45′N 06°51′E*
25 R11 **Blanco** Texas, SW USA *30°06′N 98°25′W*
42 K14 **Blanco, Cabo** *headland* NW Costa Rica *09°34′N 85°06′W*
32 D14 **Blanco, Cape** *headland* Oregon, NW USA *42°49′N 124°33′W*
62 H10 **Blanco, Río** ≈ W Argentina
56 F10 **Blanco, Río** ≈ NE Peru
15 O9 **Blanc, Réservoir** ⊠ Québec, SE Canada
21 R7 **Bland** Virginia, NE USA *37°06′N 81°08′W*
92 I2 **Blanda** ≈ N Iceland
37 O7 **Blanding** Utah, W USA *37°37′N 109°28′W*
105 X5 **Blanes** Cataluña, NE Spain *41°41′N 02°48′E*
103 N3 **Blangy-sur-Bresle** Seine-Maritime, N France *49°55′N 01°37′E*
111 C18 **Blanice** *Ger.* Blanitz. ≈ SE Czech Republic
Blanitz *see* Blanice
99 C16 **Blankenberge** West-Vlaanderen, NW Belgium *51°19′N 03°08′E*
101 D17 **Blankenheim** Nordrhein-Westfalen, W Germany *50°26′N 06°41′E*
25 R8 **Blanket** Texas, SW USA *31°49′N 98°47′W*
55 O3 **Blanquilla, Isla** *var.* La Blanquilla. *island* N Venezuela
Blanquilla, La *see* Blanquilla, Isla
61 F18 **Blanquillo** Durazno, C Uruguay *32°53′S 55°37′W*
111 G18 **Blansko** *Ger.* Blanz. Jihomoravský Kraj, SE Czech Republic *49°22′N 16°39′E*
83 N15 **Blantyre** *var.* Blantyre-Limbe. Southern, S Malawi *15°45′S 35°04′E*
83 N15 **Blantyre** ✈ Southern, S Malawi *15°34′S 35°03′E*
Blantyre-Limbe *see* Blantyre
Blanz *see* Blansko
98 J10 **Blaricum** Noord-Holland, C Netherlands *52°16′N 05°15′E*
Blasendorf *see* Blaj
Blatnitsa *see* Durankulak
113 F15 **Blato** *It.* Blatta. Dubrovnik-Neretva, S Croatia *42°57′N 16°47′E*
Blatta *see* Blato
108 E10 **Blatten** Valais, SW Switzerland *46°22′N 08°00′E*
101 J20 **Blaufelden** Baden-Württemberg, SW Germany *49°21′N 10°01′E*
95 E23 **Blåvands Huk** *headland* W Denmark *55°33′N 08°04′E*
102 G6 **Blavet** ≈ NW France
102 J12 **Blaye** Gironde, SW France
183 R8 **Blayney** New South Wales, SE Australia *33°33′S 149°13′E*
65 D25 **Bleaker Island** *island* SE Falkland Islands
109 T10 **Bled** *Ger.* Veldes. NW Slovenia *46°23′N 14°06′E*
99 D20 **Bléharies** Hainaut, SW Belgium *50°31′N 03°25′E*
109 U9 **Bleiburg** *Slvn.* Pliberk. Kärnten, S Austria *46°36′N 14°49′E*
101 L17 **Bleiloch-stausee** ⊠ C Germany
98 H12 **Bleiswijk** Zuid-Holland, W Netherlands *52°01′N 04°32′E*
95 L22 **Blekinge** ◈ *county* S Sweden
14 D17 **Blenheim** Ontario, S Canada *42°20′N 81°59′W*
185 K15 **Blenheim** Marlborough, South Island, New Zealand *41°32′S 174°E*
99 M15 **Blerick** Limburg, SE Netherlands *51°22′N 06°10′E*
Blesae *see* Blois
25 V13 **Blessing** Texas, SW USA *28°52′N 96°12′W*
14 I10 **Bleu, Lac** ◎ Québec, SE Canada
Blibba *see* Blitta
120 H10 **Blida** *var.* El Boulaida, El Boulaïda. N Algeria *36°30′N 02°50′E*
95 P15 **Blidö** Stockholm, C Sweden *59°37′N 18°55′E*
95 K18 **Blidsberg** Västra Götaland, S Sweden *57°55′N 13°30′E*
185 A21 **Bligh Sound** *sound* South Island, New Zealand
187 X14 **Bligh Water** *strait* NW Fiji
14 D11 **Blind River** Ontario, S Canada *46°12′N 82°59′W*
31 R11 **Blissfield** Michigan, N USA *41°49′N 83°51′W*
77 R15 **Blitta** *prev.* Blibba. C Togo *08°19′N 00°59′E*
19 O13 **Block Island** *island* Rhode Island, NE USA
19 O13 **Block Island Sound** *sound* Rhode Island, NE USA
98 H10 **Bloemendaal** Noord-Holland, W Netherlands *52°23′N 04°39′E*
83 H23 **Bloemfontein** *var.* Mangaung. ● (South Africa-judicial capital) Free State, C South Africa *29°07′S 26°14′E*
83 I22 **Bloemhof** North-West, NW South Africa *27°39′S 25°37′E*
102 M7 **Blois** *anc.* Blesae. Loir-et-Cher, C France *47°36′N 01°20′E*
98 L8 **Blokzijl** Overijssel, N Netherlands *52°46′N 05°58′E*
95 N20 **Blomstermåla** Kalmar, S Sweden *56°58′N 16°19′E*
92 I2 **Blönduós** Norðurland Vestra, N Iceland *65°39′N 20°15′W*
110 L11 **Błonie** Mazowieckie, C Poland *52°13′N 20°36′E*
97 C14 **Bloody Foreland** *Ir.* Cnoc Fola. *headland* NW Ireland *55°09′N 08°18′W*
31 N15 **Bloomfield** Indiana, N USA *39°01′N 86°58′W*
29 X16 **Bloomfield** Iowa, C USA *40°45′N 92°24′W*
27 Y8 **Bloomfield** Missouri, C USA *36°54′N 89°58′W*
37 P9 **Bloomfield** New Mexico, SW USA *36°42′N 108°00′W*
25 U7 **Blooming Grove** Texas, SW USA *32°05′N 96°43′W*
29 W10 **Blooming Prairie** Minnesota, N USA *43°52′N 93°03′W*
30 L13 **Bloomington** Illinois, N USA *40°28′N 88°59′W*
31 O15 **Bloomington** Indiana, N USA *39°10′N 86°31′W*
29 V9 **Bloomington** Minnesota, N USA *44°50′N 93°18′W*
25 U13 **Bloomington** Texas, SW USA *28°39′N 96°53′W*
18 H14 **Bloomsburg** Pennsylvania, NE USA *40°59′N 76°27′W*
181 X7 **Bloomsbury** Queensland, NE Australia *20°47′S 148°35′E*
169 R16 **Blora** Jawa, C Indonesia *06°55′S 111°29′E*
18 G12 **Blossburg** Pennsylvania, NE USA *41°38′N 77°00′W*
25 V5 **Blossom** Texas, SW USA *33°39′N 95°23′W*
123 T5 **Blossom, Mys** *headland* Ostrov Vrangelya, NE Russian Federation *70°49′N 178°49′E*
23 R8 **Blountstown** Florida, SE USA *30°26′N 85°03′W*
21 P8 **Blountville** Tennessee, S USA *36°31′N 82°19′W*
21 Q9 **Blowing Rock** North Carolina, SE USA *36°15′N 81°53′W*
108 J8 **Bludenz** Vorarlberg, W Austria *47°10′N 09°50′E*
36 L6 **Blue Bell Knoll** ▲ Utah, W USA *38°11′N 111°31′W*
23 Y12 **Blue Cypress Lake** ◎ Florida, SE USA
29 U11 **Blue Earth** Minnesota, N USA *43°38′N 94°06′W*
21 Q7 **Bluefield** Virginia, NE USA *37°15′N 81°16′W*
21 R7 **Bluefield** West Virginia, NE USA *37°16′N 81°13′W*
43 N10 **Bluefields** Región Autónoma Atlántico Sur, SE Nicaragua *12°01′N 83°47′W*
43 N10 **Bluefields, Bahía de** *bay* W Caribbean Sea
29 Z14 **Blue Grass** Iowa, C USA *41°30′N 90°46′W*
Bluegrass State *see* Kentucky
Blue Hen State *see* Delaware
19 S7 **Blue Hill** Maine, NE USA *44°25′N 68°36′W*
29 P16 **Blue Hill** Nebraska, C USA *40°19′N 98°27′W*
30 J5 **Blue Hills** *hill range* Wisconsin, N USA
34 L3 **Blue Lake** California, W USA *40°52′N 124°00′W*
Blue Law State *see* Connecticut
37 Q6 **Blue Mesa Reservoir** ⊠ Colorado, C USA
27 S12 **Blue Mountain** ▲ Arkansas, C USA *34°42′N 94°04′W*
19 O6 **Blue Mountain** ▲ New Hampshire, NE USA *44°48′N 71°26′W*
18 K8 **Blue Mountain** ▲ New York, NE USA *43°52′N 74°24′W*
18 H15 **Blue Mountain** *ridge* Pennsylvania, NE USA
44 H10 **Blue Mountain Peak** ▲ E Jamaica *18°02′N 76°34′W*
183 S8 **Blue Mountains** ▲▲ New South Wales, SE Australia
32 L11 **Blue Mountains** ▲▲ Oregon/Washington, NW USA
80 G12 **Blue Nile** ◈ *state* E Sudan
80 H12 **Blue Nile** *var.* Abai, Bahr el, Azraq, *Amh.* Ābay Wenz, *Ar.* An Nīl al Azraq. ≈ Ethiopia/Sudan
8 J7 **Bluenose Lake** ◎ Nunavut, NW Canada
27 O3 **Blue Rapids** Kansas, C USA *39°39′N 96°38′W*
23 S1 **Blue Ridge** Georgia, SE USA *34°51′N 84°19′W*
17 S11 **Blue Ridge** *var.* Blue Ridge Mountains. ▲▲ North Carolina/Virginia, USA
23 S1 **Blue Ridge Lake** ⊠ Georgia, SE USA
Blue Ridge Mountains *see* Blue Ridge
11 N15 **Blue River** British Columbia, SW Canada *52°03′N 119°21′W*
27 O12 **Blue River** ≈ Oklahoma, C USA
27 R4 **Blue Springs** Missouri, C USA *39°01′N 94°16′W*
21 R6 **Bluestone Lake** ⊠ West Virginia, NE USA
185 C25 **Bluff** Southland, South Island, New Zealand *46°36′S 168°22′E*
37 O8 **Bluff** Utah, W USA *37°15′N 109°36′W*
21 P8 **Bluff City** Tennessee, S USA *36°28′N 82°15′W*
65 E24 **Bluff Cove** East Falkland, Falkland Islands *51°45′S 58°11′W*
25 S7 **Bluff Dale** Texas, SW USA *32°18′N 98°01′W*
183 N15 **Bluff Hill Point** *headland* Tasmania, SE Australia *41°03′S 144°35′E*
31 Q12 **Bluffton** Indiana, N USA *40°44′N 85°10′W*
31 R12 **Bluffton** Ohio, N USA *40°54′N 83°53′W*
25 T7 **Blum** Texas, SW USA *32°08′N 97°24′W*
101 G24 **Blumberg** Baden-Württemberg, SW Germany *47°48′N 08°31′E*
60 K13 **Blumenau** Santa Catarina, S Brazil *26°55′S 49°07′W*
29 N9 **Blunt** South Dakota, N USA *44°30′N 99°58′E*
32 H15 **Bly** Oregon, NW USA *42°22′N 121°04′W*
39 R13 **Blying Sound** *sound* Alaska, USA
97 M14 **Blyth** N England, United Kingdom *55°07′N 01°30′W*
35 Y16 **Blythe** California, W USA *33°35′N 114°36′W*
27 Y9 **Blytheville** Arkansas, C USA *35°56′N 89°55′W*
117 V7 **Blyznyuky** Kharkivs'ka Oblast', E Ukraine *48°51′N 36°32′E*
95 G16 **Bø** Telemark, S Norway *59°24′N 09°04′E*
76 I15 **Bo** S Sierra Leone *07°58′N 11°45′W*
171 O4 **Boac** Marinduque, N Philippines *13°26′N 121°50′E*
42 K10 **Boaco** Boaco, S Nicaragua *12°28′N 85°45′W*
42 J10 **Boaco** ◈ *department* C Nicaragua
79 I15 **Boali** Ombella-Mpoko, SW Central African Republic *04°52′N 18°00′E*
Boalsert *see* Bolsward
31 V12 **Boardman** Ohio, N USA *41°01′N 80°39′W*
32 J11 **Boardman** Oregon, NW USA *45°50′N 119°42′W*
14 F13 **Boat Lake** ◎ Ontario, S Canada
58 F10 **Boa Vista** *state capital* Roraima, NW Brazil *02°51′N 60°43′W*
76 D9 **Boa Vista** *island* Ilhas de Barlavento, E Cape Verde
23 Q2 **Boaz** Alabama, S USA *34°12′N 86°10′W*
160 L15 **Bobai** Guangxi Zhuangzu Zizhiqu, S China *22°09′N 109°57′E*
172 J1 **Bobaomby, Tanjona** *Fr.* Cap d'Ambre. *headland* N Madagascar *11°58′S 49°13′E*
155 M14 **Bobbili** Andhra Pradesh, E India *18°32′N 83°29′E*
106 D9 **Bobbio** Emilia-Romagna, C Italy *44°48′N 09°27′E*
14 I14 **Bobcaygeon** Ontario, SE Canada *44°32′N 78°33′W*
Bober *see* Bóbr
103 O5 **Bobigny** Seine-St-Denis, N France *48°55′N 02°27′E*
77 N13 **Bobo-Dioulasso** SW Burkina Faso *11°12′N 04°21′W*
110 G8 **Bobolice** *Ger.* Bublitz. Zachodnio-pomorskie, NW Poland *53°56′N 16°37′E*
83 J19 **Bobonong** Central, E Botswana *21°58′S 28°26′E*
171 R11 **Bobopayo** Pulau Halmahera, E Indonesia *01°07′N 127°26′E*
113 J15 **Bobotov Kuk** ▲ N Montenegro *43°06′N 19°00′E*
114 G10 **Bobov Dol** *var.* Bobovdol. Kyustendil, W Bulgaria *42°21′N 22°59′E*
Bobovdol *see* Bobov Dol
119 M15 **Bobr** Minskaya Voblasts', NW Belarus *54°20′N 29°16′E*
119 M15 **Bobr** ≈ C Belarus
111 E14 **Bóbr** *Eng.* Bobrawa, *Ger.* Bober. ≈ SW Poland
Bobrawa *see* Bóbr
Bobrik *see* Bobryk
Bobrinets *see* Bobrynets'
Bobrka/Bóbrka *see* Bibrka
126 L8 **Bobrov** Voronezhskaya Oblast', W Russian Federation *51°10′N 40°03′E*
117 Q4 **Bobrovytsya** Chernihivs'ka Oblast', N Ukraine *50°43′N 31°24′E*
Bobruysk *see* Babruysk
119 J19 **Bobryk** *Rus.* Bobrik. ≈ SW Belarus
117 Q8 **Bobrynets'** *Rus.* Bobrinets. Kirovohrads'ka Oblast', C Ukraine *48°02′N 32°10′E*
14 K14 **Bobs Lake** ◎ Ontario, SE Canada
54 I6 **Bobures** Zulia, NW Venezuela *09°15′N 71°10′W*
42 H1 **Boca Bacalar Chico** *headland* N Belize *15°05′N 82°12′W*
112 G11 **Bočac** ◈ Republika Srpska, NW Bosnia and Herzegovina
41 R14 **Boca del Río** Veracruz-Llave, S Mexico *19°08′N 96°08′W*
55 O4 **Boca de Pozo** Nueva Esparta, NE Venezuela *11°00′N 64°23′W*
59 C15 **Boca do Acre** Amazonas, N Brazil *08°45′S 67°23′W*
55 N12 **Boca Mavaca** Amazonas, S Venezuela *02°30′N 65°11′W*
79 G14 **Bocaranga** Ouham-Pendé, W Central African Republic *07°07′N 15°40′E*
23 Z15 **Boca Raton** Florida, SE USA *26°22′N 80°05′W*
43 P14 **Bocas del Toro** Bocas del Toro, NW Panama *09°20′N 82°15′W*
43 P15 **Bocas del Toro** *off.* Provincia de Bocas del Toro. ◈ *province* NW Panama
43 P15 **Bocas del Toro, Archipiélago de** *island group* NW Panama
Bocas del Toro, Provincia de *see* Bocas del Toro
42 L7 **Bocay** Jinotega, N Nicaragua *14°19′N 85°08′W*
105 N6 **Boceguillas** Castilla y León, N Spain *41°20′N 03°39′W*
Bocheykovo *see* Bacheykava
111 L17 **Bochnia** Małopolskie, SE Poland *49°58′N 20°27′E*
99 K16 **Bocholt** Limburg, NE Belgium *51°10′N 05°37′E*
101 D14 **Bocholt** Nordrhein-Westfalen, W Germany *51°50′N 06°37′E*
101 E15 **Bochum** Nordrhein-Westfalen, W Germany *51°29′N 07°13′E*
103 Y15 **Bocognano** Corse, France, C Mediterranean Sea *42°04′N 09°03′E*
54 I6 **Boconó** Trujillo, NW Venezuela *09°17′N 70°17′W*

◆ Country
● Country Capital
◇ Dependent Territory
○ Dependent Territory Capital
◈ Administrative Regions
✈ International Airport
▲ Mountain
▲▲ Mountain Range
Volcano
≈ River
◎ Lake
⊠ Reservoir

BOCŞA – BORSOD-ABAÚJ-ZEMPLÉN MEGYE

116 F12 **Bocşa** *Ger.* Bokschen, *Hung.* Boksánbánya. Caraş-Severin, SW Romania 45°23′N 21°47′E
79 H15 **Boda** Lobaye, SW Central African Republic 04°17′N 17°25′E
94 L12 **Boda** Dalarna, C Sweden 61°00′N 15°15′E
95 O20 **Böda** Kalmar, S Sweden 57°16′N 17°04′E
95 L19 **Bodafors** Jönköping, S Sweden 57°30′N 14°40′E
123 O12 **Bodaybo** Irkutskaya Oblast', E Russian Federation 57°52′N 114°05′E
22 G5 **Bodcau, Bayou** *var.* Bodcau Creek. Louisiana, S USA
Bodcau Creek *see* Bodcau, Bayou
44 D8 **Bodden Town** *var.* Boddentown. Grand Cayman, SW Cayman Islands 19°20′N 81°14′W
Boddentown *see* Bodden Town
101 K14 **Bode** C Germany
34 L7 **Bodega Head** *headland* California, W USA 38°16′N 123°04′W
Bodegas *see* Babahoyo
98 H11 **Bodegraven** Zuid-Holland, C Netherlands 52°05′N 04°45′E
78 H8 **Bodélé** *depression* W Chad
92 J13 **Boden** Norrbotten, N Sweden 65°50′N 21°44′E
Bodensee *see* Constance, Lake, C Europe
65 M15 **Bode Verde Fracture Zone** *tectonic feature* E Atlantic Ocean
155 H14 **Bodhan** Telangana, C India 18°40′N 77°51′E
Bodī *see* Jinst
155 H22 **Bodināyakkanūr** Tamil Nādu, SE India 10°02′N 77°18′E
108 H10 **Bodio** Ticino, S Switzerland 46°23′N 08°55′E
Bodjonegoro *see* Bojonegoro
97 I24 **Bodmin** SW England, United Kingdom 50°29′N 04°43′W
97 I24 **Bodmin Moor** *moorland* SW England, United Kingdom
92 G12 **Bodø** Nordland, C Norway 67°17′N 14°22′E
59 H20 **Bodoquena, Serra da** ▲ SW Brazil
136 B16 **Bodrum** Muğla, SW Turkey 37°01′N 27°28′E
Bodzafordulő *see* Întorsura Buzăului
99 L14 **Boekel** Noord-Brabant, SE Netherlands 51°35′N 05°42′E
Boeloekoemba *see* Bulukumba
103 Q11 **Boën** Loire, E France 45°45′N 04°01′E
79 K18 **Boende** Equateur, C Dem. Rep. Congo 0°12′S 20°54′E
25 R11 **Boerne** Texas, SW USA 29°47′N 98°44′W
Boeroe *see* Buru, Pulau
Boetoeng *see* Buton, Pulau
22 I5 **Boeuf River** Arkansas/Louisiana, S USA
76 H14 **Boffa** W Guinea 10°12′N 14°02′W
Bó Finne, Inis *see* Inishbofin
Boga *see* Bogë
166 L9 **Bogale** Ayeyawady, SW Myanmar (Burma) 16°16′N 95°21′E
22 L8 **Bogalusa** Louisiana, S USA 30°47′N 89°51′W
77 Q12 **Bogandé** C Burkina Faso 13°02′N 00°08′W
79 I15 **Bogangolo** Ombella-Mpoko, C Central African Republic 05°36′N 18°17′E
183 Q7 **Bogan River** New South Wales, SE Australia
25 W5 **Bogata** Texas, SW USA 33°28′N 95°12′W
111 D14 **Bogatynia** *Ger.* Reichenau. Dolnośląskie, SW Poland 50°53′N 14°55′E
136 K13 **Boğazlıyan** Yozgat, C Turkey 39°13′N 35°17′E
79 J17 **Bogbonga** Equateur, NW Dem. Rep. Congo 01°36′N 19°24′E
158 J14 **Bogcang Zangbo** W China
162 I9 **Bogd** *var.* Horiult. Bayanhongor, C Mongolia 45°09′N 100°50′E
162 J10 **Bogd** *var.* Hovd. Övörhangay, C Mongolia 44°43′N 102°08′E
158 L5 **Bogda Feng** ▲ NW China 43°51′N 88°14′E
114 I9 **Bogdan** ▲ C Bulgaria 42°37′N 24°28′E
113 Q20 **Bogdanci** SE FYR Macedonia 41°12′N 22°34′E
158 M5 **Bogda Shan** *var.* Po-ko-to Shan. ▲▲ NW China
113 K17 **Bogë** *var.* Boga. Shkodër, N Albania 42°25′N 19°38′E
Bogeda'er *see* Wenquan
Bogendorf *see* Łuków
95 G23 **Bogense** Syddtjylland, C Denmark 55°34′N 10°06′E
183 T3 **Boggabilla** New South Wales, SE Australia 28°37′S 150°21′E
183 S6 **Boggabri** New South Wales, SE Australia 30°44′S 150°00′E
186 D6 **Bogia** Madang, N Papua New Guinea 04°16′S 144°56′E
97 N23 **Bognor Regis** SE England, United Kingdom 50°47′N 00°41′W
Bogodukhov *see* Bohodukhiv
181 V15 **Bogong, Mount** ▲ Victoria, SE Australia 36°43′S 147°19′E
169 O16 **Bogor** *Dut.* Buitenzorg. Jawa, C Indonesia 06°34′S 106°45′E
126 L5 **Bogoroditsk** Tul'skaya Oblast', W Russian Federation 53°46′N 38°09′E
127 O3 **Bogorodsk** Nizhegorodskaya Oblast', W Russian Federation 56°06′N 43°29′E
Bogorodskoje *see* Bogorodskoye
123 S12 **Bogorodskoye** Khabarovskiy Kray, SE Russian Federation 52°22′N 140°33′E
125 R15 **Bogorodskoye** *var.* Bogorodskoje. Kirovskaya Oblast', NW Russian Federation 57°50′N 50°41′E
54 F10 **Bogotá** *prev.* Santa Fe, Santa Fe de Bogotá. ● (Colombia) Cundinamarca, C Colombia 04°38′N 74°05′W
153 T14 **Bogra** Rajshahi, N Bangladesh 24°52′N 89°28′E
Bogschan *see* Boldu
122 L12 **Boguchany** Krasnoyarskiy Kray, C Russian Federation 58°20′N 97°20′E
126 M9 **Boguchar** Voronezhskaya Oblast', W Russian Federation 49°57′N 40°34′E
76 H10 **Bogué** Brakna, SW Mauritania 16°36′N 14°15′W
22 K8 **Bogue Chitto** Louisiana/Mississippi, S USA
Bogushëvsk *see* Bahushewsk
Boguslav *see* Bohuslav
44 K12 **Bog Walk** C Jamaica 18°06′N 77°01′W
161 Q3 **Bo Hai** *var.* Gulf of Chihli. *gulf* NE China
161 R3 **Bohai Haixia** *strait* NE China
161 Q3 **Bohai Wan** *bay* NE China
111 C17 **Bohemia** *Cz.* Čechy, *Ger.* Böhmen. W Czech Republic
111 B18 **Bohemian Forest** *Cz.* Ceský Les, Šumava, *Ger.* Böhmerwald. ▲▲ C Europe
Bohemian-Moravian Highlands *see* Českomoravská Vrchovina
77 R16 **Bohicon** S Benin 07°14′N 02°04′E
109 S11 **Bohinjska Bistrica** *Ger.* Wocheiner Feistritz. NW Slovenia 46°16′N 13°55′E
Bohkká *see* Pokka
Böhmen *see* Bohemia
Böhmerwald *see* Bohemian Forest
Böhmisch-Krumau *see* Český Krumlov
Böhmisch-Leipa *see* Česká Lípa
Böhmisch-Mährische Höhe *see* Českomoravská Vrchovina
Böhmisch-Trübau *see* Česká Třebová
117 U5 **Bohodukhiv** *Rus.* Bogodukhov. Kharkivs'ka Oblast', E Ukraine 50°10′N 35°32′E
171 Q6 **Bohol** *island* C Philippines
171 Q7 **Bohol Sea** *var.* Mindanao Sea. *sea* S Philippines
116 I7 **Bohorodchany** Ivano-Frankivs'ka Oblast', W Ukraine 48°46′N 24°31′E
Böhöt *see* Öndörshil
158 K6 **Bohu** *var.* Bagrax. Xinjiang Uygur Zizhiqu, NW China 42°00′N 86°28′E
111 I17 **Bohumín** *Ger.* Oderberg; *prev.* Neuoderberg, Nový Bohumín. Moravskoslezský Kraj, E Czech Republic 49°55′N 18°20′E
117 P6 **Bohuslav** *Rus.* Boguslav. Kyyivs'ka Oblast', N Ukraine 49°33′N 30°53′E
58 F11 **Boiaçu** Roraima, N Brazil 0°27′S 61°46′W
107 K16 **Boiano** Molise, C Italy 41°28′N 14°28′E
15 R8 **Boilleau** Québec, SE Canada 48°06′N 70°49′W
59 O17 **Boipeba, Ilha de** *island* SE Brazil
104 G3 **Boiro** Galicia, NW Spain 42°39′N 08°53′W
31 Q5 **Bois Blanc Island** *island* Michigan, N USA
29 R7 **Bois de Sioux River** Minnesota, N USA
33 N14 **Boise** *var.* Boise City. *state capital* Idaho, NW USA 43°39′N 116°14′W
26 G8 **Boise City** Oklahoma, C USA 36°44′N 102°31′W
Boise City *see* Boise
33 N14 **Boise River, Middle Fork** Idaho, NW USA
Bois, Lac des *see* Woods, Lake of the
Bois-le-Duc *see* 's-Hertogenbosch
11 W17 **Boissevain** Manitoba, S Canada 49°14′N 100°02′W
15 T7 **Boisvert, Pointe au** *headland* Québec, SE Canada 48°34′N 69°07′W
100 K10 **Boizenburg** Mecklenburg-Vorpommern, N Germany 53°23′N 10°43′E
Bojador *see* Boujdour
113 K18 **Bojana** *Alb.* Bunë. Albania/Montenegro *see also* Bunë
Bojana *see* Bunë, Lumi i
143 S3 **Bojnūrd** *var.* Bujnurd. Khorāsān-e Shemālī, N Iran 37°31′N 57°24′E
169 R16 **Bojonegoro** *prev.* Bodjonegoro. Jawa, C Indonesia 07°06′S 111°50′E
189 T1 **Bokaak Atoll** *var.* Bokak, Taongi. *atoll* Ratak Chain, NE Marshall Islands
Bokak *see* Bokaak Atoll
146 K8 **Bo'kantov Tog'lari** *Rus.* Gory Bukantau. ▲▲ N Uzbekistan
153 Q15 **Bokāro** Jhārkhand, N India 23°46′N 85°55′E
79 I18 **Bokatola** Equateur, NW Dem. Rep. Congo 0°37′S 18°45′E
76 H13 **Boké** W Guinea 10°56′N 14°18′W
Bokhara *see* Buxoro
183 Q4 **Bokharra River** New South Wales/Queensland, SE Australia
147 X8 **Bokonbayevo** *Kir.* Kajisay; *prev.* Kadzhi-Say. Issyk-Kul'skaya Oblast', NE Kyrgyzstan 42°07′N 76°59′E
78 H11 **Bokoro** Hadjer-Lamis, W Chad 12°23′N 17°03′E
79 K19 **Bokota** Equateur, NW Dem. Rep. Congo 0°56′S 22°24′E
167 N13 **Bokpyin** Taninthayi, S Myanmar (Burma) 11°16′N 98°47′E
Boksánbánya/Bokschen *see* Bocşa
83 F21 **Bokspits** Kgalagadi, SW Botswana 26°50′S 20°41′E
79 K18 **Bokungu** Equateur, C Dem. Rep. Congo 0°41′S 22°19′E
146 F12 **Bokurdak** *Rus.* Bakhardok. Ahal Welaýaty, C Turkmenistan 38°51′N 58°34′E
78 G10 **Bol** Lac, W Chad 13°27′N 14°40′E
76 G13 **Bolama** SW Guinea-Bissau 11°35′N 15°30′W
Bolangir *see* Balāngīr
Bolanos *see* Bolanos, Mount, Guam
Bolaños *see* Bolaños de Calatrava, Spain
105 N11 **Bolaños de Calatrava** *var.* Bolaños. Castilla-La Mancha, C Spain 38°55′N 03°39′W
188 B17 **Bolanos, Mount** *var.* Bolanos. ▲ S Guam 13°18′N 144°41′E
40 L12 **Bolaños, Río** C Mexico
115 M14 **Bolayır** Çanakkale, NW Turkey 40°31′N 26°46′E
102 L3 **Bolbec** Seine-Maritime, N France 49°34′N 00°31′E
116 L13 **Boldu** *var.* Bogschan. Buzău, SE Romania 45°18′N 27°15′E
146 H8 **Boldumsaz** *prev.* Kalinin, Kalininsk, Porsy. Daşoguz Welaýaty, N Turkmenistan 42°12′N 59°33′E
158 I4 **Bole** *var.* Bortala. Xinjiang Uygur Zizhiqu, NW China 44°52′N 82°06′E
77 O15 **Bole** NW Ghana 09°02′N 02°29′W
79 J19 **Boleko** Equateur, W Dem. Rep. Congo 01°27′S 19°52′E
111 E14 **Bolesławiec** *Ger.* Bunzlau. Dolnośląskie, SW Poland 51°16′N 15°34′E
127 R4 **Bolgar** *prev.* Kuybyshev. Respublika Tatarstan, W Russian Federation 54°58′N 49°03′E
77 P14 **Bolgatanga** N Ghana 10°45′N 00°52′W
Bolgrad *see* Bolhrad
117 N12 **Bolhrad** *Rus.* Bolgrad. Odes'ka Oblast', SW Ukraine 45°42′N 28°35′E
163 Y8 **Boli** Heilongjiang, NE China 45°45′N 130°32′E
79 I19 **Bolia** Bandundu, W Dem. Rep. Congo 01°34′S 18°24′E
93 J14 **Boliden** Västerbotten, N Sweden 64°52′N 20°20′E
171 T13 **Bolifar** Pulau Seram, E Indonesia 03°08′S 130°34′E
171 N2 **Bolinao** Luzon, N Philippines 16°22′N 119°52′E
54 C12 **Bolívar** Cauca, SW Colombia 01°52′N 76°56′W
27 T6 **Bolivar** Missouri, C USA 37°37′N 93°25′W
20 F10 **Bolivar** Tennessee, S USA 35°17′N 88°59′W
54 F7 **Bolívar** *off.* Departmento de Bolívar. ◈ *province* N Colombia
56 A13 **Bolívar** ◈ *province* C Ecuador
55 N9 **Bolívar** *off.* Estado Bolívar. ◈ *state* SE Venezuela
Bolívar, Departmento de *see* Bolívar
Bolívar, Estado *see* Bolívar
25 X12 **Bolivar Peninsula** *headland* Texas, SW USA 29°26′N 94°41′W
54 I6 **Bolívar, Pico** ▲ W Venezuela 08°33′N 71°05′W
57 K17 **Bolivia** *off.* Plurinational State of Bolivia; *prev.* Republic of Bolivia. ◆ *republic* W South America
Bolivia, Plurinatinoal State of *see* Bolivia
Bolivia, Republic of *see* Bolivia
112 O13 **Boljevac** Serbia, E Serbia 43°50′N 21°57′E
Bolkenhain *see* Bolków
126 J5 **Bolkhov** Orlovskaya Oblast', W Russian Federation 53°28′N 36°00′E
111 F14 **Bolków** *Ger.* Bolkenhain. Dolnośląskie, SW Poland 50°55′N 15°49′E
182 K3 **Bollards Lagoon** South Australia 28°58′S 140°52′E
103 R14 **Bollène** Vaucluse, SE France 44°16′N 04°45′E
94 N12 **Bollnäs** Gävleborg, C Sweden 61°18′N 16°27′E
181 W10 **Bollon** Queensland, C Australia 28°07′S 147°28′E
192 L12 **Bollons Tablemount** *undersea feature* S Pacific Ocean 49°40′S 176°10′W
93 H17 **Bollstabruk** Västernorrland, C Sweden 63°00′N 17°41′E
Bolluilos de Par del Condado *see* Bollulos Par del Condado
104 J14 **Bollulos Par del Condado** *var.* Bolluilos de Par del Condado. Andalucía, S Spain 37°20′N 06°32′W
95 K21 **Bolmen** ◎ S Sweden
137 T10 **Bolnisi** S Georgia 41°28′N 44°34′E
79 H19 **Bolobo** Bandundu, W Dem. Rep. Congo 02°10′S 16°17′E
106 G10 **Bologna** Emilia-Romagna, N Italy 44°30′N 11°20′E
124 I15 **Bologoye** Tverskaya Oblast', W Russian Federation 57°54′N 34°04′E
79 J18 **Bolomba** Equateur, NW Dem. Rep. Congo 0°27′N 19°13′E
41 X13 **Bolónchén de Rejón** *var.* Bolonchén de Rejón. Campeche, SE Mexico 20°00′N 89°34′W
114 J13 **Boloústra, Akrotírio** *headland* NE Greece 40°56′N 24°58′E
167 L8 **Bolovén, Phouphiang** *Fr.* Plateau des Bolovens. *plateau* S Laos
Bolovens, Plateau des *see* Bolovén, Phouphiang
106 H13 **Bolsena** Lazio, C Italy 42°39′N 11°59′E
107 G14 **Bolsena, Lago di** ◎ C Italy
126 B3 **Bol'shakovo** *Ger.* Kreuzingen; *prev.* Gross-Skaisgirren. Kaliningradskaya Oblast', W Russian Federation 54°53′N 21°38′E
Bol'shaya Berëstovitsa *see* Vyalikaya Byerastavitsa
127 S7 **Bol'shaya Chernigovka** Samarskaya Oblast', W Russian Federation 52°07′N 50°49′E
127 S7 **Bol'shaya Glushitsa** Samarskaya Oblast', W Russian Federation 52°22′N 50°29′E
124 J4 **Bol'shaya Imandra, Ozero** ◎ NW Russian Federation
Bol'shaya Khobda *see* Kobda
126 M12 **Bol'shaya Martynovka** Rostovskaya Oblast', SW Russian Federation 47°19′N 41°40′E
122 K12 **Bol'shaya Murta** Krasnoyarskiy Kray, C Russian Federation 56°51′N 93°10′E
125 V4 **Bol'shaya Rogovaya** NW Russian Federation
125 U7 **Bol'shaya Synya** NW Russian Federation
145 V9 **Bol'shaya Vladimirovka** Vostochnyy Kazakhstan, E Kazakhstan 50°53′N 79°29′E
123 V11 **Bol'sheretsk** Kamchatskiy Kray, E Russian Federation 52°20′N 156°24′E
127 W3 **Bol'sheust'ikinskoye** Respublika Bashkortostan, W Russian Federation 56°00′N 58°13′E
Bol'shevik *see* Bal'shavik
122 L5 **Bol'shevik, Ostrov** *island* Severnaya Zemlya, N Russian Federation
125 U4 **Bol'shezemel'skaya Tundra** *physical region* NW Russian Federation
144 J13 **Bol'shiye Barsuki, Peski** *desert* SW Kazakhstan
123 T7 **Bol'shoy Anyuy** NE Russian Federation
123 N7 **Bol'shoy Begichev, Ostrov** *island* NE Russian Federation
123 S15 **Bol'shoye Kamen'** Primorskiy Kray, SE Russian Federation 43°06′N 132°21′E
127 O4 **Bol'shoye Murashkino** Nizhegorodskaya Oblast', W Russian Federation 55°46′N 44°48′E
127 W4 **Bol'shoy Iremel'** ▲ W Russian Federation 54°31′N 58°47′E
127 R7 **Bol'shoy Irgiz** W Russian Federation
123 Q6 **Bol'shoy Lyakhovskiy, Ostrov** *island* NE Russian Federation
123 Q11 **Bol'shoy Nimnyr** Respublika Sakha (Yakutiya), NE Russian Federation 57°55′N 125°34′E
Bol'shoy Rozhan *see* Vyaliki Rozhan
Bol'shoy Uzen' *see* Karaozen
40 K6 **Bolsón de Mapimí** ▲▲ NW Mexico
98 K6 **Bolsward** *Fris.* Boalsert. Fryslân, N Netherlands 53°04′N 05°31′E
105 T4 **Boltaña** Aragón, NE Spain 42°28′N 00°02′E
14 G15 **Bolton** Ontario, S Canada 43°52′N 79°45′W
97 K17 **Bolton** *prev.* Bolton-le-Moors. NW England, United Kingdom 53°35′N 02°26′W
21 V12 **Bolton** North Carolina, SE USA 34°22′N 78°26′W
Bolton-le-Moors *see* Bolton
136 G11 **Bolu** Bolu, NW Turkey 40°45′N 31°38′E
136 G11 **Bolu** ◈ *province* NW Turkey
186 G9 **Bolubolu** Goodenough Island, S Papua New Guinea 09°22′S 150°22′E
92 H1 **Bolungarvík** Vestfirðir, NW Iceland 66°09′N 23°17′W
159 O10 **Boluntay** Qinghai, W China 36°30′N 92°11′E
159 P8 **Boluozhuanjing**, Aksay Kazakzu Zizhixian. Gansu, N China 39°25′N 94°09′E
136 F14 **Bolvadin** Afyon, W Turkey 38°43′N 31°02′E
114 M10 **Bolyarovo** *prev.* Pashkeni. Yambol, E Bulgaria 42°09′N 26°49′E
106 G6 **Bolzano** *Ger.* Bozen; *anc.* Bauzanum. Trentino-Alto Adige, N Italy 46°30′N 11°22′E
79 F22 **Boma** Bas-Congo, W Dem. Rep. Congo 05°42′S 13°05′E
183 R12 **Bombala** New South Wales, SE Australia 36°54′S 149°15′E
104 F10 **Bombarral** Leiria, C Portugal 39°15′N 09°09′W
Bombay *see* Mumbai
171 U13 **Bomberai, Semenanjung** *cape* Papua Barat, E Indonesia
81 F18 **Bombo** S Uganda 0°36′N 32°33′E
162 I8 **Bömbögör** *var.* Dzadgay. Bayanhongor, C Mongolia 46°12′N 99°29′E
79 I17 **Bomboma** Equateur, NW Dem. Rep. Congo 02°23′N 19°03′E
81 I19 **Bomet** ◈ *county* W Kenya
59 I14 **Bom Futuro** Pará, N Brazil 06°27′S 54°44′W
159 Q15 **Bomi** *var.* Bowo, Zhamo. Xizang Zizhiqu, W China 29°43′N 96°12′E
79 N17 **Bomili** Orientale, NE Dem. Rep. Congo 01°45′N 27°01′E
59 N17 **Bom Jesus da Lapa** Bahia, E Brazil 13°16′S 43°23′W
60 Q8 **Bom Jesus do Itabapoana** Rio de Janeiro, SE Brazil 21°07′S 41°43′W
95 C15 **Bømlafjorden** *fjord* S Norway
95 B15 **Bømlo** *island* S Norway
12 Q12 **Bomnak** Amurskaya Oblast', SE Russian Federation 54°43′N 128°50′E
79 I17 **Bomongo** Equateur, NW Dem. Rep. Congo 01°22′N 18°21′E
61 K14 **Bom Retiro** Santa Catarina, S Brazil 27°45′S 49°31′W
79 L15 **Bomu** *var.* Mbomou, Mbomu, M'Bomu. Central African Republic/Dem. Rep. Congo
142 J3 **Bonāb** *var.* Benāb, Bunab. Āžarbāyjān-e Sharqī, N Iran 37°30′N 46°03′E
45 S9 **Bonaire** ◈ *Dutch special municipality* S Caribbean Sea
45 Q16 **Bonaire** *island* Lesser Antilles
39 U11 **Bona, Mount** ▲ Alaska, USA 61°22′N 141°45′W
183 Q12 **Bonang** Victoria, SE Australia 37°13′S 148°43′E
42 L7 **Bonanza** Región Autónoma Atlántico Norte, NE Nicaragua 13°59′N 84°30′W
37 O4 **Bonanza** Utah, W USA 40°01′N 109°12′W
45 O9 **Bonao** C Dominican Republic 18°55′N 70°25′W
180 L3 **Bonaparte Archipelago** *island group* Western Australia
32 K6 **Bonaparte, Mount** ▲ Washington, NW USA 48°47′N 119°07′W
39 N11 **Bonasila Dome** ▲ Alaska, USA 62°24′N 160°28′W
45 T15 **Bonasse** Trinidad, Trinidad and Tobago 10°02′N 61°48′W
15 X7 **Bonaventure** Québec, SE Canada 48°03′N 65°30′W
15 X7 **Bonaventure** Québec, SE Canada
13 V11 **Bonavista** Newfoundland, Newfoundland and Labrador, SE Canada 48°38′N 53°08′W
13 U11 **Bonavista Bay** *inlet* NW Atlantic Ocean
79 E19 **Bonda** Ogooué-Lolo, C Gabon 0°50′S 12°28′E
127 N6 **Bondari** Tambovskaya Oblast', W Russian Federation 52°58′N 42°02′E
106 G9 **Bondeno** Emilia-Romagna, C Italy 44°53′N 11°24′E
30 L4 **Bond Falls Flowage** ◎ Michigan, N USA
79 L16 **Bondo** Orientale, N Dem. Rep. Congo 03°52′N 23°41′E
171 N17 **Bondokodi** Pulau Sumba, S Indonesia 09°36′S 119°01′E
77 O15 **Bondoukou** E Ivory Coast 08°03′N 02°45′W
Bondoukui/Bondoukuy *see* Boundoukui
169 T17 **Bondowoso** Jawa, C Indonesia 07°54′S 113°50′E
33 S14 **Bondurant** Wyoming, C USA 43°14′N 110°26′W
Bône *see* Annaba, Algeria
Bone *see* Watampone, Indonesia
30 I5 **Bone Lake** ◎ Wisconsin, N USA
171 P14 **Bonelipu** Pulau Buton, C Indonesia 04°42′S 123°09′E
171 O15 **Bonerate, Kepulauan** *var.* Macan. *island group* C Indonesia
29 O12 **Bonesteel** South Dakota, N USA 43°01′N 98°55′W
62 I8 **Bonete, Cerro** ▲ N Argentina 27°58′S 68°22′W
171 O14 **Bone, Teluk** *bay* Sulawesi, C Indonesia
108 D6 **Bonfol** Jura, NW Switzerland 47°28′N 07°08′E
153 U12 **Bongaigaon** Assam, NE India 26°30′N 90°31′E
79 K17 **Bongandanga** Equateur, NW Dem. Rep. Congo 01°28′N 21°03′E
78 L13 **Bongo, Massif des** *var.* Chaîne des Mongos. ▲▲ NE Central African Republic
78 G12 **Bongor** Mayo-Kébbi Est, SW Chad 10°18′N 15°20′E
77 N16 **Bongouanou** E Ivory Coast 06°39′N 04°12′W
167 V11 **Bông Sơn** *var.* Hoai Nhơn. Bình Định, C Vietnam 14°28′N 109°00′E
25 U5 **Bonham** Texas, SW USA 33°36′N 96°12′W
Bonhard *see* Bonyhád
103 U6 **Bonhomme, Col du** *pass* NE France
103 Y16 **Bonifacio** Corse, France, C Mediterranean Sea 41°24′N 09°09′E
Bonifacio, Bocche de/Bonifacio, Bouches de *see* Bonifacio, Strait of
103 Y16 **Bonifacio, Strait of** *Fr.* Bouches de Bonifacio, *It.* Bocche di Bonifacio. *strait* C Mediterranean Sea
23 Q8 **Bonifay** Florida, SE USA 30°49′N 85°42′W
Bonin Islands *see* Ogasawara-shotō
192 H5 **Bonin Trench** *undersea feature* NW Pacific Ocean
23 W15 **Bonita Springs** Florida, SE USA 26°19′N 81°48′W
42 I5 **Bonito, Pico** ▲ N Honduras 15°33′N 86°55′W
101 E17 **Bonn** Nordrhein-Westfalen, W Germany 50°44′N 07°06′E
92 H11 **Bonnåsjøen** Nordland, C Norway 67°35′N 15°39′E
14 J12 **Bonnechere** Ontario, SE Canada 45°39′N 77°36′W
14 J12 **Bonnechere** Ontario, SE Canada
33 N7 **Bonners Ferry** Idaho, NW USA 48°41′N 116°19′W
27 R4 **Bonner Springs** Kansas, C USA 39°03′N 94°52′W
102 L6 **Bonnétable** Sarthe, NW France 48°09′N 00°24′E
27 X6 **Bonne Terre** Missouri, C USA 37°55′N 90°33′W
10 J5 **Bonnet Plume** Yukon, NW Canada
102 M6 **Bonneval** Eure-et-Loir, C France 48°12′N 01°23′E
103 T10 **Bonneville** Haute-Savoie, E France 46°05′N 06°25′E
36 J3 **Bonneville Salt Flats** *salt flat* Utah, W USA
77 U18 **Bonny** Rivers, S Nigeria 04°25′N 07°13′E
Bonny, Bight of *see* Biafra, Bight of
37 W4 **Bonny Reservoir** ▣ Colorado, C USA
11 R14 **Bonnyville** Alberta, SW Canada 54°16′N 110°46′W
107 C18 **Bono** Sardegna, Italy, C Mediterranean Sea 40°24′N 09°01′E
Bononia *see* Vidin, Bulgaria
Bononia *see* Boulogne-sur-Mer, France
107 B18 **Bonorva** Sardegna, Italy, C Mediterranean Sea 40°27′N 08°46′E
30 M15 **Bonpas Creek** Illinois, N USA
190 I3 **Bonriki** Tarawa, W Kiribati 01°23′N 173°09′E
183 T4 **Bonshaw** New South Wales, SE Australia 29°06′S 151°15′E
76 I16 **Bonthe** SW Sierra Leone 07°32′N 12°30′W
171 N2 **Bontoc** Luzon, N Philippines 17°04′N 120°58′E
25 Y9 **Bon Wier** Texas, SW USA 30°43′N 93°40′W
111 J25 **Bonyhád** *Ger.* Bonhard. Tolna, S Hungary 46°20′N 18°31′E
Bonzabaai *see* Bonza Bay
83 J25 **Bonza Bay** *Afr.* Bonzabaai. Eastern Cape, S South Africa 32°58′S 27°58′E
182 D7 **Bookabie** South Australia 31°49′S 132°41′E
182 H6 **Bookaloo** South Australia 31°56′S 137°21′E
37 P5 **Book Cliffs** *cliff* Colorado/Utah, W USA
25 P1 **Booker** Texas, SW USA 36°27′N 100°32′W
76 K15 **Boola** SE Guinea 08°22′N 08°41′W
183 O8 **Booligal** New South Wales, SE Australia 33°56′S 144°54′E
99 G17 **Boom** Antwerpen, N Belgium 51°05′N 04°24′E
43 N6 **Boom** *var.* Boon. Región Autónoma Atlántico Norte, NE Nicaragua 14°52′N 83°36′W
183 S3 **Boomi** New South Wales, SE Australia 28°43′S 149°35′E
Boon *see* Boom
162 H9 **Bööncagaan Nuur** ◎ S Mongolia
29 V13 **Boone** Iowa, C USA 42°04′N 93°52′W
21 Q8 **Boone** North Carolina, SE USA 36°13′N 81°41′W
27 S11 **Booneville** Arkansas, C USA 35°09′N 93°57′W
21 N6 **Booneville** Kentucky, S USA 37°26′N 83°45′W
23 N2 **Booneville** Mississippi, S USA 34°39′N 88°34′W
21 V3 **Boonsboro** Maryland, NE USA 39°30′N 77°39′W
34 L6 **Boonville** California, W USA 38°58′N 123°21′W
31 N16 **Boonville** Indiana, N USA 38°03′N 87°16′W
27 U4 **Boonville** Missouri, C USA 38°58′N 92°43′W
18 I9 **Boonville** New York, NE USA 43°28′N 75°17′W
80 M12 **Boorama** Awdal, NW Somalia 09°58′N 43°15′E
183 O6 **Booroondarra, Mount** *hill* New South Wales, SE Australia
183 N9 **Booroorban** New South Wales, SE Australia 34°55′S 144°45′E
183 R9 **Boorowa** New South Wales, SE Australia 34°26′S 148°42′E
99 H17 **Boortmeerbeek** Vlaams Brabant, C Belgium 50°58′N 04°27′E
80 P11 **Boosaaso** *var.* Bandar Kassim, Bender Qaasim, Bosaso, *It.* Bender Cassim. Bari, N Somalia 11°26′N 49°37′E
19 Q8 **Boothbay Harbor** Maine, NE USA 43°52′N 69°35′W
Boothia Felix *see* Boothia Peninsula
9 N6 **Boothia, Gulf of** *gulf* Nunavut, NE Canada
9 N6 **Boothia Peninsula** *prev.* Boothia Felix. *peninsula* Nunavut, NE Canada
79 E18 **Booué** Ogooué-Ivindo, NE Gabon 0°03′S 11°58′E
101 J21 **Bopfingen** Baden-Württemberg, S Germany 48°51′N 10°21′E
101 F18 **Boppard** Rheinland-Pfalz, W Germany 50°13′N 07°36′E
62 M4 **Boquerón** *off.* Departamento de Boquerón. ◈ *department* W Paraguay
Boquerón, Departamento de *see* Boquerón
43 P15 **Boquete** *var.* Bajo Boquete. Chiriquí, W Panama 08°45′N 82°26′W
40 J6 **Boquilla, Presa de la** ▣ N Mexico
40 L5 **Boquillas** *var.* Boquillas del Carmen. Coahuila, NE Mexico 29°10′N 102°55′W
Boquillas del Carmen *see* Boquillas
112 P12 **Bor** Serbia, E Serbia 44°05′N 22°07′E
81 F15 **Bor** Jonglei, E South Sudan 06°12′N 31°33′E
95 L20 **Bor** Jönköping, S Sweden 57°04′N 14°10′E
136 J15 **Bor** Niğde, S Turkey 37°49′N 35°00′E
191 S10 **Bora-Bora** *island* Îles Sous le Vent, W French Polynesia
167 Q9 **Borabu** Maha Sarakham, E Thailand 16°01′N 103°06′E
172 K4 **Boraha, Nosy** *island* E Madagascar
33 P13 **Borah Peak** ▲ Idaho, NW USA 44°21′N 113°53′W
145 U16 **Boralday** *prev.* Burunday. Almaty, SE Kazakhstan 43°21′N 76°48′E
145 Y11 **Boran** *prev.* Buran. Vostochnyy Kazakhstan, E Kazakhstan 48°00′N 85°09′E
144 G13 **Borankul** *prev.* Opornyy. Mangistau, SW Kazakhstan 46°09′N 54°32′E
95 J19 **Borås** Västra Götaland, S Sweden 57°44′N 12°55′E
143 N11 **Borāzjān** *var.* Borazjān. Būshehr, S Iran 29°19′N 51°12′E
Borazjān *see* Borāzjān
58 G13 **Borba** Amazonas, N Brazil 04°39′S 59°35′W
104 H11 **Borba** Évora, S Portugal 38°48′N 07°28′W
Borbetomagus *see* Worms
55 O7 **Borbón** Bolívar, E Venezuela 07°55′N 64°03′W
59 Q15 **Borborema, Planalto da** *plateau* NE Brazil
116 M14 **Borcea, Braţul** S Romania
Borchalo *see* Marneuli
195 R15 **Borchgrevink Coast** *physical region* Antarctica
137 Q11 **Borçka** Artvin, NE Turkey 41°24′N 41°38′E
98 N11 **Borculo** Gelderland, E Netherlands 52°07′N 06°31′E
182 G10 **Borda, Cape** *headland* South Australia 35°45′S 136°34′E
102 K13 **Bordeaux** *anc.* Burdigala. Gironde, SW France 44°49′N 00°33′W
11 T15 **Borden** Saskatchewan, S Canada 52°23′N 107°10′W
14 D8 **Borden Lake** ◎ Ontario, S Canada
9 N4 **Borden Peninsula** *peninsula* Baffin Island, Nunavut, NE Canada
182 K11 **Bordertown** South Australia 36°21′S 140°48′E
92 H2 **Borðeyri** Norðurland Vestra, NW Iceland 65°12′N 21°09′W
95 B18 **Bordhoy** *Dan.* Bordø. *island* NE Faroe Islands
106 B11 **Bordighera** Liguria, NW Italy 43°48′N 07°40′E
74 K5 **Bordj-Bou-Arreridj** *var.* Bordj Bou Arreridj, Bordj Bou Arréridj. N Algeria 36°02′N 04°49′E
74 L10 **Bordj Omar Driss** E Algeria 28°09′N 06°52′E
143 N13 **Bord Khūn** Hormozgān, S Iran
Bordø *see* Bordhoy
147 V7 **Bordunskiy** Chuyskaya Oblast', N Kyrgyzstan 42°37′N 75°31′E
95 M17 **Borensberg** Östergötland, S Sweden 58°33′N 15°15′E
Borgå *see* Porvoo
92 H3 **Borgarnes** Vesturland, W Iceland 64°33′N 21°55′W
93 G14 **Børgefjell** ▲▲ C Norway
98 O7 **Borger** Drenthe, NE Netherlands 52°54′N 06°48′E
25 N2 **Borger** Texas, SW USA 35°40′N 101°24′W
95 N20 **Borgholm** Kalmar, S Sweden 56°50′N 16°41′E
107 N22 **Borgia** Calabria, SW Italy 38°48′N 16°28′E
99 J18 **Borgloon** Limburg, NE Belgium 50°48′N 05°21′E
Borg Massif *see* Borgmassivet
195 P2 **Borgmassivet** *Eng.* Borg Massif. ▲▲ Antarctica
22 L9 **Borgne, Lake** ◎ Louisiana, S USA
106 C7 **Borgomanero** Piemonte, NE Italy 45°42′N 08°33′E
106 G10 **Borgo Panigale** ✈ (Bologna) Emilia-Romagna, N Italy 44°33′N 11°16′E
107 J15 **Borgorose** Lazio, C Italy 42°10′N 13°15′E
106 A9 **Borgo San Dalmazzo** Piemonte, N Italy 44°19′N 07°29′E
106 G11 **Borgo San Lorenzo** Toscana, C Italy 43°58′N 11°22′E
106 C7 **Borgosesia** Piemonte, NE Italy 45°41′N 08°21′E
106 E9 **Borgo Val di Taro** Emilia-Romagna, C Italy 44°29′N 09°48′E
106 G6 **Borgo Valsugana** Trentino-Alto Adige, N Italy 46°04′N 11°31′E
Borhoyn Tal *see* Dzamïn-Üüd
167 R8 **Borikhan** *var.* Borikhane. Bolikhamxai, C Laos 18°36′N 103°43′E
Borikhane *see* Borikhan
144 G8 **Borili** *prev.* Burlin. Zapadnyy Kazakhstan, NW Kazakhstan 51°25′N 52°42′E
Borislav *see* Boryslav
127 N8 **Borisoglebsk** Voronezhskaya Oblast', W Russian Federation 51°23′N 42°00′E
Borisov *see* Barysaw
Borisovgrad *see* Parvomay
Borispol' *see* Boryspil'
172 I3 **Boriziny** *prev./Fr.* Port-Bergé. Mahajanga, NW Madagascar 15°31′S 47°40′E
105 Q5 **Borja** Aragón, NE Spain 41°50′N 01°32′W
Borjas Blancas *see* Les Borges Blanques
137 S10 **Borjomi** *Rus.* Borzhomi. C Georgia 41°50′N 43°24′E
118 L12 **Borkavichy** *Rus.* Borkovichi. Vitsyebskaya Voblasts', N Belarus 55°40′N 28°20′E
101 H16 **Borken** Hessen, C Germany 51°01′N 09°16′E
101 E14 **Borken** Nordrhein-Westfalen, W Germany 51°51′N 06°51′E
92 H10 **Borkenes** Troms, N Norway 68°46′N 16°10′E
78 H7 **Borkou** *off.* Région du Borkou. ◈ *region* N Chad
Borkou, Région du *see* Borkou
Borkovichi *see* Borkavichy
100 E9 **Borkum** *island* NW Germany
81 K17 **Bor, Lagh** *var.* Lak Bor. *dry watercourse* NE Kenya
Bor, Lak *see* Bor, Lagh
95 M14 **Borlänge** Dalarna, C Sweden 60°29′N 15°25′E
106 C9 **Bormida** NW Italy
106 F6 **Bormio** Lombardia, N Italy 46°27′N 10°24′E
101 M16 **Borna** Sachsen, E Germany 51°07′N 12°30′E
98 O10 **Borne** Overijssel, E Netherlands 52°18′N 06°45′E
99 F17 **Bornem** Antwerpen, N Belgium 51°06′N 04°14′E
169 S10 **Borneo** *island* Brunei/Indonesia/Malaysia
101 E16 **Bornheim** Nordrhein-Westfalen, W Germany 50°46′N 06°58′E
95 L24 **Bornholm** ◈ *county* E Denmark
95 L24 **Bornholm** *island* E Denmark
77 Y13 **Borno** ◈ *state* NE Nigeria
104 K15 **Bornos** Andalucía, S Spain 36°50′N 05°42′W
162 L7 **Bornuur** Töv, C Mongolia 48°28′N 106°15′E
117 O4 **Borodyanka** Kyyivs'ka Oblast', N Ukraine 50°40′N 29°54′E
158 I5 **Borohoro Shan** ▲▲ NW China
77 O13 **Boromo** SW Burkina Faso 11°47′N 02°54′W
35 T13 **Boron** California, W USA 35°00′N 117°42′W
Borongo *see* Black Volta
Boron'ki *see* Baron'ki
Borosjenő *see* Ineu
Borossebes *see* Sebiş
76 L15 **Borotou** NW Ivory Coast 08°46′N 07°30′W
117 W6 **Borova** Kharkivs'ka Oblast', E Ukraine 49°22′N 37°39′E
114 H8 **Borovan** Vratsa, NW Bulgaria 43°25′N 23°45′E
124 I14 **Borovichi** Novgorodskaya Oblast', W Russian Federation 58°24′N 33°56′E
Borovlje *see* Ferlach
114 K8 **Borovo** Ruse, N Bulgaria 43°28′N 25°46′E
112 J9 **Borovo** Vukovar-Srijem, NE Croatia 45°22′N 18°57′E
Borovoye *see* Burabay
126 K4 **Borovsk** Kaluzhskaya Oblast', W Russian Federation 55°12′N 36°22′E
145 N7 **Borovskoye** Kostanay, N Kazakhstan 53°48′N 64°17′E
Borovukha *see* Baravukha
95 L23 **Borrby** Skåne, S Sweden 55°27′N 14°10′E
105 T9 **Borriana** *var.* Burriana. Valenciana, E Spain 39°54′N 00°05′W
181 R3 **Borroloola** Northern Territory, N Australia 16°09′S 136°18′E
116 F9 **Borş** Bihor, NW Romania 47°07′N 21°49′E
116 I9 **Borşa** *Hung.* Borsa. Maramureş, N Romania 47°40′N 24°37′E
116 J10 **Borsec** *Ger.* Bad Borseck, *Hung.* Borszék. Harghita, C Romania 46°58′N 25°32′E
92 K8 **Børselv** *Lapp.* Bissojohka. Finnmark, N Norway 70°18′N 25°35′E
113 L23 **Borsh** *var.* Borshi. Vlorë, S Albania 40°04′N 19°51′E
Borshchev *see* Borshchiv
116 K7 **Borshchiv** *Pol.* Borszczów, *Rus.* Borshchev. Ternopil's'ka Oblast', W Ukraine 48°48′N 26°00′E
Borshi *see* Borsh
111 L20 **Borsod-Abaúj-Zemplén** *off.* Borsod-Abaúj-Zemplén Megye. ◈ *county* NE Hungary
Borsod-Abaúj-Zemplén Megye *see* Borsod-Abaúj-Zemplén

◆ Country
● Country Capital
◇ Dependent Territory
○ Dependent Territory Capital
◈ Administrative Regions
✈ International Airport
▲ Mountain
▲▲ Mountain Range
Volcano
River
◎ Lake
▣ Reservoir

BORSSELE – BRIDEL

99 E15 **Borssele** Zeeland, SW Netherlands 51°26′N 03°45′E
Borszczów see Borshchiv
Borszék see Borsec
Bortala see Bole
103 O12 **Bort-les-Orgues** Corrèze, C France 45°28′N 02°31′E
Bor u České Lípy see Nový Bor
Bor-Üdzüür see Altay
143 N9 **Borūjen** Chahār MaḤall va Bakhtīārī, C Iran 32°N 51°09′E
142 L7 **Borūjerd** var. Burujird. Lorestān, W Iran 33°55′N 48°46′E
116 H6 **Boryslav** Pol. Borysław, Rus. Borislav. L'vivs'ka Oblast', NW Ukraine 49°18′N 23°28′E
Borysław see Boryslav
117 P4 **Boryspil'** Rus. Borispol'. Kyyivs'ka Oblast', N Ukraine 50°21′N 30°59′E
117 P4 **Boryspil'** Rus. Borispol'. ✈ (Kyyiv) Kyyivs'ka Oblast', N Ukraine 50°21′N 30°46′E
Borzhomi see Borjomi
117 R3 **Borzna** Chernihivs'ka Oblast', NE Ukraine 51°15′N 32°25′E
123 O14 **Borzya** Zabaykal'skiy Kray, S Russian Federation 50°18′N 116°24′E
107 B18 **Bosa** Sardegna, Italy, C Mediterranean Sea 40°18′N 08°28′E
112 F10 **Bosanska Dubica** var. Kozarska Dubica. ⬖ Republika Srpska, NW Bosnia and Herzegovina
112 G10 **Bosanska Gradiška** var. Gradiška. ⬖ Republika Srpska, N Bosnia and Herzegovina
112 F10 **Bosanska Kostajnica** var. Srpska Kostajnica. ⬖ Republika Srpska, NW Bosnia and Herzegovina
112 E11 **Bosanska Krupa** var. Krupa, Krupa na Uni. ⬖ Federacija Bosne I Hercegovine, NW Bosnia and Herzegovina
112 H10 **Bosanski Brod** var. Srpski Brod. ⬖ Republika Srpska, N Bosnia and Herzegovina
112 E10 **Bosanski Novi** var. Novi Grad. Republika Srpska, NW Bosnia and Herzegovina 45°03′N 16°23′E
112 E11 **Bosanski Petrovac** var. Petrovac. Federacija Bosne I Hercegovine, NW Bosnia and Herzegovina 44°34′N 16°21′E
112 I10 **Bosanski Šamac** var. Šamac. Republika Srpska, N Bosnia and Herzegovina 45°03′N 18°27′E
112 E12 **Bosansko Grahovo** var. Grahovo, Hrvatsko Grahovi. Federacija Bosne I Hercegovine, W Bosnia and Herzegovina 44°10′N 16°22′E
Bosaso see Boosaaso
186 B7 **Bosavi, Mount** 🌋 W Papua New Guinea 06°33′S 142°50′E
160 J14 **Bose** Guangxi Zhuangzu Zizhiqu, S China 23°55′N 106°32′E
161 Q5 **Boshan** Shandong, E China 36°32′N 117°47′E
113 P16 **Bosilegrad** prev. Bosiljgrad. Serbia, SE Serbia 42°30′N 22°30′E
Bosiljgrad see Bosilegrad
Bösing see Pezinok
98 H12 **Boskoop** Zuid-Holland, C Netherlands 52°04′N 04°40′E
111 G18 **Boskovice** Ger. Boskowitz. Jihomoravský Kraj, SE Czech Republic 49°30′N 16°39′E
Boskowitz see Boskovice
112 I10 **Bosna** ≈ N Bosnia and Herzegovina
113 G14 **Bosne I Hercegovine, Federacija** ⬖ republic Bosnia and Herzegovina
112 H12 **Bosnia and Herzegovina** off. Republic of Bosnia and Herzegovina. ◆ republic SE Europe
Bosnia and Herzegovina, Republic of see Bosnia and Herzegovina
79 J16 **Bosobolo** Equateur, NW Dem. Rep. Congo 04°11′N 19°55′E
165 O14 **Bōsō-hantō** peninsula Honshū, S Japan
Bosora see Buşrá ash Shām
Bosphorus/Bosporus see İstanbul Boğazı
Bosporus Cimmerius see Kerch Strait
Bosporus Thracius see İstanbul Boğazı
Bosra see Buşrá ash Shām
79 H14 **Bossangoa** Ouham, C Central African Republic 06°32′N 17°25′E
Bossé Bangou see Bossey Bangou
79 I15 **Bossembélé** Ombella-Mpoko, C Central African Republic 05°13′N 17°39′E
79 H15 **Bossentélé** Ouham-Pendé, W Central African Republic 05°36′N 16°37′E
77 R12 **Bossey Bangou** var. Bossé Bangou. Tillabéri, SW Niger 13°22′N 01°18′E
22 G5 **Bossier City** Louisiana, S USA 32°31′N 93°43′W
83 D20 **Bossiesvlei** Hardap, S Namibia 25°02′S 16°48′E
77 Y11 **Bosso** Diffa, SE Niger 13°42′N 13°18′E
61 F15 **Bossoroca** Rio Grande do Sul, S Brazil 28°45′S 54°54′W
158 J10 **Bostan** Xinjiang Uygur Zizhiqu, W China 41°20′N 83°15′E
142 K3 **Bostānābād** Āžarbāyjān-e Sharqī, N Iran 37°52′N 46°51′E
158 K6 **Bosten Hu** var. Bagrax Hu. ◎ NW China
97 O18 **Boston** prev. St.Botolph's Town. E England, United Kingdom 52°59′N 00°01′W
19 O11 **Boston** state capital Massachusetts, NE USA 42°22′N 71°04′W
146 I9 **Bo'ston** Rus. Bustan. Qoraqalpog'iston Respublikasi, W Uzbekistan 41°49′N 60°51′E
10 M17 **Boston Bar** British Columbia, SW Canada 49°54′N 121°22′W
27 T10 **Boston Mountains** ▲▲ Arkansas, C USA
15 P8 **Bostonnais** ≈ Québec, SE Canada
Bostyn' see Bastyn'
112 J10 **Bosut** ≈ E Croatia
154 C11 **Botād** Gujarāt, W India 22°12′N 71°44′E
145 S10 **Botakara** Kas. Botaqara; prev. Ul'yanovskiy. Karaganda, C Kazakhstan 50°05′N 73°45′E
183 T9 **Botany Bay** inlet New South Wales, SE Australia
Botaqara see Botakara
83 G18 **Boteti** var. Botletle. ≈ N Botswana
114 J9 **Botev** ▲ C Bulgaria 42°45′N 24°57′E
114 H9 **Botevgrad** prev. Orkhaniye. Sofia, W Bulgaria 42°55′N 23°47′E
93 J16 **Bothnia, Gulf of** Fin. Pohjanlahti, Swe. Bottniska Viken. gulf N Baltic Sea
183 P17 **Bothwell** Tasmania, SE Australia 42°24′S 147°01′E
104 H5 **Boticas** Vila Real, N Portugal 41°41′N 07°40′W
55 W10 **Boti-Pasi** Sipaliwini, C Suriname 04°15′N 55°27′W
Botletle see Boteti
127 P16 **Botlikh** Chechenskaya Respublika, SW Russian Federation 42°39′N 46°12′E
117 N10 **Botna** ≈ E Moldova
116 I9 **Botoşani** Hung. Botosány. Botoşani, NE Romania 47°44′N 26°41′E
116 K8 **Botoşani** ⬖ county NE Romania
Botosány see Botoşani
147 P12 **Bototog', Tizmasi** Rus. Khrebet Babatag. ▲▲ Tajikistan/Uzbekistan
161 P4 **Botou** prev. Bozhen. Hebei, E China 38°09′N 116°37′E
99 M20 **Botrange** ▲ E Belgium 50°30′N 06°03′E
107 O21 **Botricello** Calabria, SW Italy 38°56′N 16°51′E
83 I23 **Botshabelo** Free State, C South Africa 29°15′S 26°51′E
93 J15 **Botsmark** Västerbotten, N Sweden 64°15′N 20°15′E
83 G19 **Botswana** off. Republic of Botswana. ◆ republic S Africa
Botswana, Republic of see Botswana
29 N2 **Bottineau** North Dakota, N USA 48°50′N 100°28′W
Bottniska Viken see Bothnia, Gulf of
60 L9 **Botucatu** São Paulo, S Brazil 22°52′S 48°30′W
76 M16 **Bouaflé** C Ivory Coast 06°59′N 05°45′W
77 N16 **Bouaké** var. Bwake. C Ivory Coast 07°42′N 05°00′W
79 G14 **Bouar** Nana-Mambéré, W Central African Republic 05°58′N 15°38′E
74 H7 **Bouarfa** NE Morocco 32°33′N 01°54′W
111 B19 **Boubín** ▲ SW Czech Republic 49°00′N 13°51′E
79 I14 **Bouca** Ouham, W Central African Republic 06°57′N 18°18′E
15 T5 **Boucher** ≈ Québec, SE Canada
103 R15 **Bouches-du-Rhône** ⬖ department SE France
74 C9 **Bou Craa** var. Bu Craa. NW Western Sahara 26°32′N 12°52′W
77 O9 **Boû Djébéha** oasis C Mali
108 C8 **Boudry** Neuchâtel, W Switzerland 46°57′N 06°46′E
79 F21 **Bouenza** ⬖ province S Congo
186 J7 **Bougainville** off. Autonomous Region of Bougainville; prev. North Solomons. ⬖ autonomous region Bougainville, NE Papua New Guinea Oceania
Bougainville, Autonomous Region of see Bougainville
180 L2 **Bougainville, Cape** cape Western Australia
65 E24 **Bougainville, Cape** headland East Falkland, Falkland Islands 51°18′S 58°28′W
Bougainville, Détroit de see Bougainville Strait
186 J7 **Bougainville Island** island NE Papua New Guinea
186 I8 **Bougainville Strait** strait N Solomon Islands
187 Q13 **Bougainville Strait** Fr. Détroit de Bougainville. strait C Vanuatu
120 I9 **Bougaroun, Cap** headland NE Algeria 37°07′N 06°18′E
77 R8 **Boughessa** Kidal, NE Mali 20°05′N 02°13′E
Bougie see Béjaïa
76 L13 **Bougouni** Sikasso, SW Mali 11°25′N 07°28′W
99 J24 **Bouillon** Luxembourg, SE Belgium 49°47′N 05°04′E
74 K5 **Bouira** var. Bouïra. N Algeria 36°22′N 03°55′E
74 D8 **Bou-Izakarn** SW Morocco 29°12′N 09°43′W
74 B9 **Boujdour** var. Bojador. W Western Sahara 26°06′N 14°29′W
74 G5 **Boukhalef** ✈ (Tanger) N Morocco 35°45′N 05°53′W
Boukombé see Boukoumbé
77 R14 **Boukoumbé** var. Boukombé. C Benin 10°13′N 01°09′E
76 G6 **Boû Lanouâr** Dakhlet Nouâdhibou, W Mauritania 21°17′N 16°29′W
37 T4 **Boulder** Colorado, C USA 40°02′N 105°18′W
33 R10 **Boulder** Montana, NW USA 46°14′N 112°07′W
35 X12 **Boulder City** Nevada, W USA 35°58′N 114°49′W
181 T7 **Boulia** Queensland, C Australia 23°02′S 139°58′E
15 N10 **Boullé** ≈ Québec, SE Canada
102 J9 **Boulogne** ≈ NW France
Boulogne see Boulogne-sur-Mer
102 L16 **Boulogne-sur-Gesse** Haute-Garonne, S France 43°18′N 00°38′E
103 N1 **Boulogne-sur-Mer** var. Boulogne; anc. Bononia, Gesoriacum, Gessoriacum. Pas-de-Calais, N France 50°43′N 01°36′E
77 Q12 **Boulsa** C Burkina Faso 12°41′N 00°29′W
77 W11 **Boultoum** Zinder, C Niger 14°43′N 10°22′E
187 Y14 **Bouma** Taveuni, N Fiji 16°49′S 179°50′W
79 G16 **Boumba** ≈ SE Cameroon
76 J9 **Boûmdeïd** var. Boumdeït. Assaba, S Mauritania 17°26′N 11°21′W
Boumdeït see Boûmdeïd
115 C17 **Boumistós** ▲ W Greece 38°48′N 20°59′E
77 O15 **Bouna** NE Ivory Coast 09°16′N 03°00′W
19 P4 **Boundary Bald Mountain** ▲ Maine, NE USA 45°45′N 70°10′W
35 S8 **Boundary Peak** ▲ Nevada, W USA 37°50′N 118°21′W
76 M14 **Boundiali** N Ivory Coast 09°30′N 06°31′W
79 G19 **Boundji** Cuvette, C Congo 01°05′S 15°18′E
77 O13 **Boundoukui** var. Bondoukui, Bondoukuy. W Burkina Faso 11°51′N 03°47′W
36 L2 **Bountiful** Utah, W USA 40°53′N 111°52′W
Bounty Basin see Bounty Trough
191 Q16 **Bounty Bay** bay Pitcairn Island, C Pacific Ocean
192 L12 **Bounty Islands** island group S New Zealand
175 Q13 **Bounty Trough** var. Bounty Basin. undersea feature S Pacific Ocean
187 P17 **Bourail** Province Sud, C New Caledonia 21°35′S 165°29′E
27 V5 **Bourbeuse River** ≈ Missouri, C USA
103 Q9 **Bourbon-Lancy** Saône-et-Loire, C France 46°39′N 03°48′E
31 N11 **Bourbonnais** Illinois, N USA 41°08′N 87°52′W
103 O10 **Bourbonnais** cultural region C France
103 S7 **Bourbonne-les-Bains** Haute-Marne, N France 48°00′N 05°43′E
Bourbon Vendée see la Roche-sur-Yon
74 M8 **Bourdj Messaouda** E Algeria 30°18′N 09°19′E
77 Q10 **Bourem** Gao, C Mali 16°56′N 00°21′W
Bourg see Bourg-en-Bresse
103 N11 **Bourganeuf** Creuse, C France 45°57′N 01°47′E
Bourgas see Burgas
Bourge-en-Bresse see Bourg-en-Bresse
103 S10 **Bourg-en-Bresse** var. Bourg, Bourge-en-Bresse. Ain, E France 46°12′N 05°13′E
103 O8 **Bourges** anc. Avaricum. Cher, C France 47°06′N 02°24′E
103 T11 **Bourget, Lac du** ◎ E France
103 P8 **Bourgogne** Eng. Burgundy. ⬖ region E France
103 S11 **Bourgoin-Jallieu** Isère, E France 45°34′N 05°16′E
103 R14 **Bourg-St-Andéol** Ardèche, E France 44°24′N 04°36′E
103 U11 **Bourg-St-Maurice** Savoie, E France 45°37′N 06°49′E
108 C11 **Bourg St. Pierre** Valais, SW Switzerland 45°54′N 07°10′E
76 H8 **Boû Rjeimât** well W Mauritania
183 P5 **Bourke** New South Wales, SE Australia 30°08′S 145°57′E
97 M24 **Bournemouth** S England, United Kingdom 50°43′N 01°54′W
99 M23 **Bourscheid** Diekirch, NE Luxembourg 49°55′N 06°04′E
74 K6 **Bou Saâda** var. Bou Saada. N Algeria 35°10′N 04°09′E
36 I13 **Bouse Wash** ≈ Arizona, SW USA
103 N10 **Boussac** Creuse, C France 46°20′N 02°12′E
102 M16 **Boussens** Haute-Garonne, S France 43°11′N 00°58′E
78 H12 **Bousso** prev. Fort-Bretonnet. Chari-Baguirmi, S Chad 10°32′N 16°45′E
76 H9 **Boutilimit** Trarza, SW Mauritania 17°33′N 14°42′W
65 D21 **Bouvet Island** ◇ Norwegian dependency S Atlantic Ocean
77 U11 **Bouza** Tahoua, SW Niger 14°25′N 06°09′E
109 R10 **Bovec** Ger. Flitsch, It. Plezzo. NW Slovenia 46°21′N 13°33′E
98 J8 **Bovenkarspel** Noord-Holland, NW Netherlands 52°33′N 05°03′E
29 V5 **Bovey** Minnesota, N USA 47°18′N 93°25′W
32 M9 **Bovill** Idaho, NW USA 46°50′N 116°24′W
24 L4 **Bovina** Texas, SW USA 34°30′N 102°52′W
107 M17 **Bovino** Puglia, SE Italy 41°14′N 15°19′E
61 C17 **Bovril** Entre Ríos, E Argentina 31°24′S 59°25′W
28 L2 **Bowbells** North Dakota, N USA 48°48′N 102°15′W
11 Q16 **Bow City** Alberta, SW Canada 50°27′N 112°16′W
29 O8 **Bowdle** South Dakota, N USA 45°27′N 99°39′W
181 X6 **Bowen** Queensland, NE Australia 20°S 148°10′E
192 L2 **Bowers Ridge** undersea feature S Bering Sea 50°00′N 180°00′W
25 S5 **Bowie** Texas, SW USA 33°33′N 97°51′W
11 R17 **Bow Island** Alberta, SW Canada 49°53′N 111°24′W
Bowkān see Būkān
20 J7 **Bowling Green** Kentucky, S USA 37°00′N 86°29′W
27 V3 **Bowling Green** Missouri, C USA 39°21′N 91°11′W
31 R11 **Bowling Green** Ohio, N USA 41°22′N 83°40′W
21 W5 **Bowling Green** Virginia, NE USA 38°02′N 77°22′W
28 J6 **Bowman** North Dakota, N USA 46°11′N 103°26′W
9 Q7 **Bowman Bay** bay NW Atlantic Ocean
194 I5 **Bowman Coast** physical region Antarctica
28 J7 **Bowman-Haley Lake** ⊠ North Dakota, N USA
195 Z11 **Bowman Island** island Antarctica
Bowo see Bomi
183 S9 **Bowral** New South Wales, SE Australia 34°28′S 150°52′E
186 E8 **Bowutu Mountains** ▲▲ C Papua New Guinea
83 I16 **Bowwood** Southern, S Zambia 17°09′S 26°16′E
28 I12 **Box Butte Reservoir** ⊠ Nebraska, C USA
28 J10 **Box Elder** South Dakota, N USA 44°06′N 103°04′W
95 M18 **Boxholm** Östergötland, S Sweden 58°12′N 15°05′E
Bo Xian/Boxian see Bozhou
161 Q4 **Boxing** Shandong, E China 37°06′N 118°05′E
99 L14 **Boxmeer** Noord-Brabant, SE Netherlands 51°39′N 05°57′E
99 J14 **Boxtel** Noord-Brabant, S Netherlands 51°36′N 05°20′E
136 J10 **Boyabat** Sinop, N Turkey 41°27′N 34°45′E
54 F9 **Boyacá** off. Departamento de Boyacá. ⬖ province C Colombia
Boyacá, Departamento de see Boyacá
117 O4 **Boyarka** Kyyivs'ka Oblast', N Ukraine 50°19′N 30°20′E
22 H7 **Boyce** Louisiana, S USA 31°23′N 92°40′W
114 H8 **Boychinovtsi** Montana, NW Bulgaria 43°28′N 23°20′E
33 U11 **Boyd** Montana, NW USA 45°27′N 109°03′W
25 S6 **Boyd** Texas, SW USA 33°01′N 97°33′W
21 V8 **Boydton** Virginia, NE USA 36°40′N 78°26′W
Boyer Ahmadī va Kohkīlūyeh see Kohgīlūyeh va Bowyer Aḩmad
29 T13 **Boyer River** ≈ Iowa, C USA
21 W8 **Boykins** Virginia, NE USA 36°35′N 77°11′W
11 Q13 **Boyle** Alberta, SW Canada 54°38′N 112°45′W
97 D16 **Boyle** Ir. Mainistirna Búille. C Ireland 53°58′N 08°18′W
97 F17 **Boyne** Ir. An Bhóinn. ≈ E Ireland
31 Q5 **Boyne City** Michigan, N USA 45°13′N 85°00′W
23 Z14 **Boynton Beach** Florida, SE USA 26°31′N 80°04′W
147 O13 **Boysun** Rus. Baysun. Surkhondaryo Viloyati, S Uzbekistan 38°14′N 67°08′E
Bozau see Întorsura Buzăului
136 B12 **Bozcaada** Island Çanakkale, NW Turkey
136 C14 **Boz Dağları** ▲▲ W Turkey
33 S11 **Bozeman** Montana, NW USA 45°40′N 111°02′W
Bozen see Bolzano
79 J16 **Bozene** Equateur, NW Dem. Rep. Congo 02°56′N 19°15′E
Bozhen see Botou
161 P7 **Bozhou** var. Boxian, Bo Xian. Anhui, E China 33°46′N 115°44′E
136 H16 **Bozkır** Konya, S Turkey 37°10′N 32°15′E
136 K13 **Bozok Yaylasī** plateau C Turkey
79 H14 **Bozoum** Ouham-Pendé, W Central African Republic 06°17′N 16°26′E
137 N16 **Bozova** Sanlıurfa, S Turkey 37°23′N 38°33′E
Bozrah see Buşrá ash Shām
136 E12 **Bozüyük** Bilecik, NW Turkey 39°55′N 30°02′E
106 B9 **Bra** Piemonte, NW Italy 44°42′N 07°51′E
194 G4 **Brabant Island** island Antarctica
99 I20 **Brabant Walloon** ⬖ province C Belgium
113 F15 **Brač** var. Brach, It. Brazza; anc. Brattia. island S Croatia
Bracara Augusta see Braga
107 H15 **Bracciano** Lazio, C Italy 42°04′N 12°12′E
107 H14 **Bracciano, Lago di** ◎ C Italy
14 H13 **Bracebridge** Ontario, S Canada 45°02′N 79°19′W
Brach see Brač
93 G17 **Bräcke** Jämtland, C Sweden 62°43′N 15°30′E
25 P12 **Brackettville** Texas, SW USA 29°19′N 100°27′W
97 N22 **Bracknell** S England, United Kingdom 51°26′N 00°46′W
61 K14 **Braço do Norte** Santa Catarina, S Brazil 28°16′S 49°11′W
116 G11 **Brad** Hung. Brád. Hunedoara, SW Romania 45°52′N 23°00′E
107 N18 **Bradano** ≈ S Italy
23 V13 **Bradenton** Florida, SE USA 27°30′N 82°34′W
45 V10 **Brades** de facto capital, de jure capital. Previous capital destroyed by volcano in 1995 Plymouth. ● (Montserrat) SW Montserrat 16°45′N 62°12′W
14 H14 **Bradford** Ontario, S Canada 44°09′N 79°34′W
97 L17 **Bradford** N England, United Kingdom 53°48′N 01°45′W
27 W10 **Bradford** Arkansas, C USA 35°25′N 91°27′W
18 D12 **Bradford** Pennsylvania, NE USA 41°57′N 78°38′W
27 T15 **Bradley** Arkansas, C USA 33°06′N 93°39′W
25 P7 **Bradshaw** Texas, SW USA 32°06′N 99°52′W
25 Q9 **Brady** Texas, SW USA 31°08′N 99°22′W
25 Q9 **Brady Creek** ≈ Texas, SW USA
95 G22 **Brædstrup** Syddanmark, C Denmark 55°58′N 09°38′E
96 J10 **Braemar** NE Scotland, United Kingdom 57°12′N 02°52′W
116 K8 **Brăeşti** Botoşani, NW Romania 47°50′N 26°26′E
104 G5 **Braga** anc. Bracara Augusta. Braga, NW Portugal 41°32′N 08°26′W
104 G5 **Braga** ⬖ district N Portugal
116 J15 **Bragadiru** Teleorman, S Romania 43°43′N 25°32′E
61 C20 **Bragado** Buenos Aires, E Argentina 35°10′S 60°29′W
104 J5 **Bragança** Eng. Braganza; anc. Julio Briga. Bragança, NE Portugal 41°47′N 06°46′W
104 I5 **Bragança** ⬖ district N Portugal
60 N9 **Bragança Paulista** São Paulo, S Brazil 22°55′S 46°30′W
Braganza see Bragança
Bragin see Brahin
29 V7 **Braham** Minnesota, N USA 45°43′N 93°10′W
Brahe see Brda
Brahestad see Raahe
119 O20 **Brahin** Rus. Bragin. Homyel'skaya Voblasts', SE Belarus 51°47′N 30°16′E
153 U15 **Brahmanbaria** Chittagong, E Bangladesh 23°58′N 91°04′E
154 O12 **Brāhmani** ≈ E India
154 N13 **Brahmapur** Odisha, E India 19°21′N 84°51′E
129 S10 **Brahmaputra** var. Padma, Tsangpo, Ben. Jamuna, Chin. Yarlung Zangbo Jiang, Ind. Bramaputra, Dihang, Siang. ≈ S Asia
97 H19 **Braich y Pwll** headland NW Wales, United Kingdom 52°47′N 04°46′W
183 R10 **Braidwood** New South Wales, SE Australia 35°26′S 149°48′E
30 M11 **Braidwood** Illinois, N USA 41°16′N 88°12′W
116 M13 **Brăila** Brăila, E Romania 45°17′N 27°57′E
116 L13 **Brăila** ⬖ county SE Romania
99 G19 **Braine-l'Alleud** Brabant Walloon, C Belgium 50°41′N 04°22′E
99 F19 **Braine-le-Comte** Hainaut, SW Belgium 50°37′N 04°08′E
29 U6 **Brainerd** Minnesota, N USA 46°22′N 94°10′W
99 J19 **Braives** Liège, E Belgium 50°37′N 05°09′E
83 H23 **Brak** ≈ C South Africa
Brak see Birāk
99 E18 **Brakel** Oost-Vlaanderen, SW Belgium 50°50′N 03°48′E
98 J13 **Brakel** Gelderland, C Netherlands 51°49′N 05°05′E
76 H9 **Brakna** ⬖ region S Mauritania
95 J17 **Brålanda** Västra Götaland, S Sweden 58°32′N 12°18′E
Bramaputra see Brahmaputra
95 F23 **Bramming** Syddtjylland, W Denmark 55°28′N 08°42′E
14 G15 **Brampton** Ontario, S Canada 43°42′N 79°46′W
100 F12 **Bramsche** Niedersachsen, NW Germany 52°25′N 07°58′E
116 J12 **Bran** Ger. Törzburg, Hung. Törcsvár. Braşov, S Romania 45°31′N 25°23′E
29 W8 **Branch** Minnesota, N USA 45°29′N 92°57′W
21 R14 **Branchville** South Carolina, SE USA 33°15′N 80°49′W
47 Y6 **Branco, Cabo** headland E Brazil 07°08′S 34°45′W
58 F11 **Branco, Rio** ≈ N Brazil
108 J8 **Brand** Vorarlberg, W Austria 47°07′N 09°45′E
83 B18 **Brandberg** ▲ NW Namibia 21°20′S 14°22′E
95 H14 **Brandbu** Oppland, S Norway 60°24′N 10°30′E
95 F22 **Brande** Midtjylland, W Denmark 55°57′N 09°08′E
Brandebourg see Brandenburg
100 M12 **Brandenburg** var. Brandenburg an der Havel. Brandenburg, NE Germany 52°25′N 12°34′E
20 K5 **Brandenburg** Kentucky, S USA 38°00′N 86°11′W
100 N12 **Brandenburg** off. Freie und Hansestadt Hamburg, Fr. Brandebourg. ⬖ state NE Germany
Brandenburg an der Havel see Brandenburg
83 I23 **Brandfort** Free State, C South Africa 28°42′S 26°28′E
11 W16 **Brandon** Manitoba, S Canada 49°50′N 99°57′W
23 V12 **Brandon** Florida, SE USA 27°56′N 82°17′W
22 L6 **Brandon** Mississippi, S USA 32°16′N 90°01′W
97 A20 **Brandon Mountain** Ir. Cnoc Bréanainn. ▲ SW Ireland 52°13′N 10°16′W
Brandsen see Coronel Brandsen
95 I14 **Brandval** Hedmark, S Norway 60°18′N 12°01′E
83 F24 **Brandvlei** Northern Cape, W South Africa 30°27′S 20°29′E
23 U9 **Branford** Florida, SE USA 29°57′N 82°54′W
110 K7 **Braniewo** Ger. Braunsberg. Warmińsko-mazurskie, N Poland 54°24′N 19°50′E
194 H3 **Bransfield Strait** strait Antarctica
37 U8 **Branson** Colorado, C USA 37°01′N 103°52′W
27 T8 **Branson** Missouri, C USA 36°38′N 93°13′W
14 G16 **Brantford** Ontario, S Canada 43°09′N 80°17′W
102 L12 **Brantôme** Dordogne, SW France 45°21′N 00°37′E
182 L12 **Branxholme** Victoria, SE Australia 37°51′S 141°48′E
Brasil see Brazil
59 C16 **Brasiléia** Acre, W Brazil 10°59′S 48°45′W
59 K18 **Brasília** ● (Brazil) Distrito Federal, C Brazil 15°45′S 47°57′W
Brasil, Repùblica Federativa do see Brazil
Braslav see Braslaw
118 J12 **Braslaw** Pol. Brasław, Rus. Braslav. Vitsyebskaya Voblasts', N Belarus 55°38′N 27°02′E
116 J12 **Braşov** Ger. Kronstadt, Hung. Brassó; prev. Oraşul Stalin. Braşov, C Romania 45°40′N 25°35′E
116 I12 **Braşov** ⬖ county C Romania
77 U18 **Brass** Bayelsa, S Nigeria 04°19′N 06°21′E
99 H16 **Brasschaat** var. Brasschaet. Antwerpen, N Belgium 51°17′N 04°30′E
Brasschaet see Brasschaat
169 V8 **Brassey, Banjaran** var. Brassey Range. ▲▲ East Malaysia
Brassey Range see Brassey, Banjaran
Brassó see Braşov
23 T1 **Brasstown Bald** ▲ Georgia, SE USA 34°52′N 83°48′W
113 K22 **Brataj** Vlorë, SW Albania 40°18′N 19°37′E
114 J10 **Bratan** var. Morozov. ▲ C Bulgaria 42°31′N 25°08′E
114 H10 **Bratia** var. Bratiya. C Bulgaria 42°36′N 24°08′E
111 F21 **Bratislava** Ger. Pressburg, Hung. Pozsony. ● (Slovakia) Bratislavský Kraj, W Slovakia 48°10′N 17°10′E
111 H21 **Bratislavský Kraj** ⬖ region W Slovakia
Bratiya see Bratia
122 M12 **Bratsk** Irkutskaya Oblast', C Russian Federation 56°20′N 101°50′E
117 Q8 **Brats'ke** Mykolayivs'ka Oblast', S Ukraine 47°52′N 31°34′E
122 M13 **Bratskoye Vodokhranilishche** Eng. Bratsk Reservoir. ⊠ S Russian Federation
Bratsk Reservoir see Bratskoye Vodokhranilishche
Brattia see Brač
94 D9 **Brattvåg** Møre og Romsdal, S Norway 62°36′N 06°21′E
112 K12 **Bratunac** ⬖ Republika Srpska, E Bosnia and Herzegovina
114 J10 **Bratya Daskalovi** prev. Grozdovo. Stara Zagora, C Bulgaria 42°13′N 25°21′E
109 U2 **Braunau** ≈ N Austria
Braunau see Braunau am Inn
109 Q4 **Braunau am Inn** var. Braunau. Oberösterreich, N Austria 48°16′N 13°03′E
Braunsberg see Braniewo
100 J13 **Braunschweig** Eng./Fr. Brunswick. Niedersachsen, N Germany 52°16′N 10°32′E
Brava see Baraawe
105 Y6 **Brava, Costa** coastal region NE Spain
43 V16 **Brava, Punta** headland E Panama 08°21′N 78°22′W
95 N17 **Bråviken** inlet S Sweden
56 B10 **Bravo, Cerro** ▲ N Peru 05°33′S 79°10′W
Bravo del Norte, Río/ Bravo, Río see Grande, Rio
35 X17 **Brawley** California, W USA 32°58′N 115°31′W
97 G18 **Bray** Ir. Bré. E Ireland 53°12′N 06°06′W
59 G16 **Brazil** off. Federative Republic of Brazil, Port. Repùblica Federativa do Brasil, Sp. Brasil; prev. United States of Brazil. ◆ federal republic South America
65 K15 **Brazil Basin** var. Brazilian Basin, Brazil'skaya Kotlovina. undersea feature W Atlantic Ocean 15°00′S 25°00′W
Brazil, Federative Republic of see Brazil
Brazilian Basin see Brazil Basin
Brazilian Highlands see Central, Planalto
Brazil'skaya Kotlovina see Brazil Basin
Brazil, United States of see Brazil
25 U10 **Brazos River** ≈ Texas, SW USA
Brazza see Brač
79 G21 **Brazzaville** ● (Congo) Capital District, S Congo 04°14′S 15°14′E
79 G21 **Brazzaville** ✈ Pool, S Congo 04°15′S 15°15′E
112 J11 **Brčko** Brčko Distrikt, NE Bosnia and Herzegovina 44°52′N 18°49′E
112 J11 **Brčko Distrikt** ⬖ self-governing district Bosnia and Herzegovina
110 H8 **Brda** Ger. Brahe. ≈ N Poland
Bré see Bray
185 A23 **Breaksea Sound** sound South Island, New Zealand
184 L4 **Bream Bay** bay North Island, New Zealand
184 L4 **Bream Head** headland North Island, New Zealand 35°51′S 174°35′E
Bréanainn, Cnoc see Brandon Mountain
45 S6 **Brea, Punta** headland W Puerto Rico 17°56′N 66°55′W
22 I9 **Breaux Bridge** Louisiana, S USA 30°16′N 91°54′W
116 J13 **Breaza** Prahova, SE Romania 45°13′N 25°38′E
169 P16 **Brebes** Jawa, C Indonesia 06°54′S 109°00′E
96 K10 **Brechin** E Scotland, United Kingdom 56°45′N 02°38′W
99 H15 **Brecht** Antwerpen, N Belgium 51°21′N 04°33′E
37 S4 **Breckenridge** Colorado, C USA 39°28′N 106°02′W
29 R6 **Breckenridge** Minnesota, N USA 46°15′N 96°35′W
25 R6 **Breckenridge** Texas, SW USA 32°45′N 98°56′W
97 J21 **Brecknock** cultural region SE Wales, United Kingdom
63 G25 **Brecknock, Península** headland S Chile 54°39′S 71°48′W
111 G19 **Břeclav** Ger. Lundenburg. Jihomoravský Kraj, SE Czech Republic 49°05′N 16°51′E
97 J21 **Brecon** E Wales, United Kingdom 51°58′N 03°26′W
97 J21 **Brecon Beacons** ▲▲ S Wales, United Kingdom
99 I14 **Breda** Noord-Brabant, S Netherlands 51°35′N 04°46′E
95 K20 **Bredaryd** Jönköping, S Sweden 57°10′N 13°45′E
83 F26 **Bredasdorp** Western Cape, SW South Africa 34°32′S 20°02′E
93 H16 **Bredbyn** Västernorrland, N Sweden 63°28′N 18°04′E
122 F11 **Bredy** Chelyabinskaya Oblast', C Russian Federation 52°23′N 60°24′E
99 K17 **Bree** Limburg, NE Belgium 51°08′N 05°36′E
67 T15 **Breede** ≈ S South Africa
98 I7 **Breezand** Noord-Holland, NW Netherlands 52°52′N 04°47′E
113 P18 **Bregalnica** ≈ E FYR Macedonia
108 I6 **Bregenz** anc. Brigantium. Vorarlberg, W Austria 47°31′N 09°46′E
108 J7 **Bregenzer Wald** ▲▲ W Austria
114 F6 **Bregovo** Vidin, NW Bulgaria 44°07′N 22°40′E
102 H5 **Bréhat, Île de** island NW France
92 H2 **Breiðafjörður** bay W Iceland
92 L3 **Breiðdalsvík** Austurland, E Iceland 64°48′N 14°02′W
108 H9 **Breil** Ger. Brigels. Graubünden, S Switzerland 46°46′N 09°04′E
92 J8 **Breivikbotn** Finnmark, N Norway 70°36′N 22°19′E
94 I9 **Brekken** Sør-Trøndelag, S Norway 62°39′N 11°49′E
94 G7 **Brekstad** Sør-Trøndelag, S Norway 63°42′N 09°40′E
94 B10 **Bremangerlandet** island S Norway
Brême see Bremen
100 H11 **Bremen** Fr. Brême. Bremen, NW Germany 53°06′N 08°48′E
23 R3 **Bremen** Georgia, SE USA 33°43′N 85°09′W
31 O11 **Bremen** Indiana, N USA 41°24′N 86°07′W
100 H10 **Bremen** off. Freie Hansestadt Bremen, Fr. Brême. ⬖ state N Germany
100 G9 **Bremerhaven** Bremen, NW Germany 53°33′N 08°35′E
Bremersdorp see Manzini
32 G8 **Bremerton** Washington, NW USA 47°34′N 122°37′W
100 H10 **Bremervörde** Niedersachsen, NW Germany 53°29′N 09°06′E
25 U9 **Bremond** Texas, SW USA 31°10′N 96°40′W
25 U10 **Brenham** Texas, SW USA 30°09′N 96°24′W
108 M8 **Brenner** Tirol, W Austria 47°10′N 11°51′E
Brenner, Col du/Brennero, Passo del see Brenner Pass
108 M8 **Brenner Pass** var. Brenner Sattel, Fr. Col du Brenner, Ger. Brennerpass, It. Passo del Brennero. pass Austria/Italy
Brennerpass see Brenner Pass
Brenner Sattel see Brenner Pass
108 G10 **Brenno** ≈ SW Switzerland
106 F7 **Breno** Lombardia, N Italy 45°58′N 10°18′E
23 O5 **Brent** Alabama, S USA 32°54′N 87°10′W
106 H7 **Brenta** ≈ NE Italy
97 P21 **Brentwood** E England, United Kingdom 51°38′N 00°21′E
18 L14 **Brentwood** Long Island, New York, NE USA 40°46′N 73°12′W
106 F7 **Brescia** anc. Brixia. Lombardia, N Italy 45°33′N 10°13′E
99 D15 **Breskens** Zeeland, SW Netherlands 51°24′N 03°33′E
Breslau see Wrocław
106 H5 **Bressanone** Ger. Brixen. Trentino-Alto Adige, N Italy 46°44′N 11°41′E
96 M2 **Bressay** island NE Scotland, United Kingdom
102 K9 **Bressuire** Deux-Sèvres, W France 46°50′N 00°29′W
119 F20 **Brest** Pol. Brześć nad Bugiem, Rus. Brest-Litovsk; prev. Brześć Litewski. Brestskaya Voblasts', SW Belarus 52°06′N 23°42′E
102 F5 **Brest** Finistère, NW France 48°24′N 04°31′W
Brest-Litovsk see Brest
112 A10 **Brestova** Istra, NW Croatia 45°09′N 14°13′E
Brestskaya Oblast' see Brestskaya Voblasts'
119 G19 **Brestskaya Voblasts'** prev. Rus. Brestskaya Oblast'. ⬖ province SW Belarus
102 G6 **Bretagne** Eng. Brittany, Lat. Britannia Minor. ⬖ region NW France
116 G12 **Bretea-Română** Hung. Oláhbrettye; prev. Bretea-Romînă. Hunedoara, W Romania 45°39′N 23°00′E
Bretea-Romînă see Bretea-Română
103 O3 **Breteuil** Oise, N France 49°37′N 02°18′E
102 I10 **Breton, Pertuis** inlet W France
22 L10 **Breton Sound** sound Louisiana, S USA
184 K2 **Brett, Cape** headland North Island, New Zealand 35°11′S 174°21′E
101 G21 **Bretten** Baden-Württemberg, SW Germany 49°01′N 08°42′E
99 K15 **Breugel** Noord-Brabant, S Netherlands 51°30′N 05°30′E
106 B6 **Breuil-Cervinia** It. Cervinia. Valle d'Aosta, NW Italy 45°57′N 07°37′E
98 I11 **Breukelen** Utrecht, C Netherlands 52°11′N 05°01′E
21 P10 **Brevard** North Carolina, SE USA 35°13′N 82°46′W
38 L9 **Brevig Mission** Alaska, USA 65°19′N 166°29′W
95 G16 **Brevik** Telemark, S Norway 59°05′N 09°42′E
183 P5 **Brewarrina** New South Wales, SE Australia 30°01′S 146°50′E
19 R6 **Brewer** Maine, NE USA 44°46′N 68°44′W
29 T11 **Brewster** Minnesota, N USA 43°43′N 95°28′W
29 N14 **Brewster** Nebraska, C USA 41°57′N 99°52′W
31 U12 **Brewster** Ohio, N USA 40°42′N 81°36′W
Brewster, Kap see Kangikajik
183 O8 **Brewster, Lake** ◎ New South Wales, SE Australia
23 P7 **Brewton** Alabama, S USA 31°06′N 87°04′W
Brezhnev see Naberezhnyye Chelny
109 W12 **Brežice** Ger. Rann. E Slovenia 45°54′N 15°35′E
114 G9 **Breznik** Pernik, W Bulgaria 42°45′N 22°54′E
111 K19 **Brezno** Ger. Bries, Briesen, Hung. Breznóbánya; prev. Brezno nad Hronom. Banskobystrický Kraj, C Slovakia 48°49′N 19°40′E
Breznóbánya/Brezno nad Hronom see Brezno
116 I12 **Brezoi** Vâlcea, SW Romania 45°18′N 24°15′E
114 J10 **Brezovo** prev. Abrashlare. Plovdiv, C Bulgaria 42°19′N 25°05′E
79 K14 **Bria** Haute-Kotto, C Central African Republic 06°30′N 22°00′E
103 U13 **Briançon** anc. Brigantio. Hautes-Alpes, SE France 44°55′N 06°37′E
36 K7 **Brian Head** ▲ Utah, W USA 37°40′N 112°49′W
103 O7 **Briare** Loiret, C France 47°35′N 02°46′E
183 V2 **Bribie Island** island Queensland, E Australia
43 O14 **Bribrí** Limón, E Costa Rica 09°37′N 82°51′W
116 L8 **Briceni** var. Brinceni, Rus. Brichany. N Moldova 48°21′N 27°02′E
Bricgstow see Bristol
Brichany see Briceni
99 M24 **Bridel** Luxembourg, C Luxembourg 49°40′N 06°03′E

◆ Country
● Country Capital
◇ Dependent Territory
○ Dependent Territory Capital
⬖ Administrative Regions
✈ International Airport
▲ Mountain
▲▲ Mountain Range
🌋 Volcano
≈ River
◎ Lake
⊠ Reservoir

Bridgend – Bull Island

97 J22 Bridgend S Wales, United Kingdom 51°30′N 03°37′W
14 I14 Bridgenorth Ontario, SE Canada 44°21′N 78°22′W
23 Q1 Bridgeport Alabama, S USA 34°57′N 85°42′W
35 R8 Bridgeport California, W USA 38°14′N 119°15′W
18 L13 Bridgeport Connecticut, NE USA 41°10′N 73°12′W
31 N15 Bridgeport Illinois, N USA 38°42′N 87°45′W
28 J14 Bridgeport Nebraska, C USA 41°37′N 103°07′W
25 S6 Bridgeport Texas, SW USA 33°12′N 97°45′W
21 S3 Bridgeport West Virginia, NE USA 39°17′N 80°15′W
25 S5 Bridgeport, Lake ⊠ Texas, SW USA
33 U11 Bridger Montana, NW USA 45°16′N 108°55′W
18 J17 Bridgeton New Jersey, NE USA 39°24′N 75°10′W
180 J14 Bridgetown Western Australia 34°01′S 116°07′E
45 Y14 Bridgetown ● (Barbados) SW Barbados 13°05′N 59°36′W
183 P17 Bridgewater Tasmania, SE Australia 42°47′S 147°15′E
13 P16 Bridgewater Nova Scotia, SE Canada 44°19′N 64°30′W
19 P12 Bridgewater Massachusetts, NE USA 41°59′N 70°58′W
29 Q11 Bridgewater South Dakota, N USA 43°33′N 97°30′W
21 U5 Bridgewater Virginia, NE USA 38°22′N 78°58′W
19 P8 Bridgton Maine, NE USA 44°04′N 70°43′W
97 K23 Bridgwater SW England, United Kingdom 51°08′N 03°W
97 K22 Bridgwater Bay bay SW England, United Kingdom
97 O16 Bridlington E England, United Kingdom 54°05′N 00°12′W
97 O16 Bridlington Bay bay E England, United Kingdom
183 P15 Bridport Tasmania, SE Australia 41°03′S 147°26′E
97 K24 Bridport S England, United Kingdom 50°44′N 02°43′W
103 O5 Brie cultural region N France
Brieg see Brzeg
Briel see Brielle
98 G12 Brielle var. Briel, Bril, Eng. The Brill. Zuid-Holland, SW Netherlands 51°54′N 04°10′E
108 E9 Brienz Bern, C Switzerland 46°45′N 08°00′E
108 E9 Brienzer See ◎ SW Switzerland
Bries/Briesen see Brezno
Brietzig see Brzesko
103 S4 Briey Meurthe-et-Moselle, NE France 49°15′N 05°57′E
108 E10 Brig Fr. Brigue, It. Briga. Valais, SW Switzerland 46°19′N 08°E
Briga see Brig
101 G24 Brigach ≈ S Germany
18 K17 Brigantine New Jersey, NE USA 39°23′N 74°21′W
Brigantio see Briançon
Brigantium see Bregenz
Brigels see Breil
25 S9 Briggs Texas, SW USA 30°52′N 97°55′W
36 L1 Brigham City Utah, W USA 41°30′N 112°00′W
14 J15 Brighton Ontario, SE Canada 44°01′N 77°44′W
97 O23 Brighton SE England, United Kingdom 50°50′N 00°10′W
37 T4 Brighton Colorado, C USA 39°58′N 104°46′W
30 K15 Brighton Illinois, N USA 39°01′N 90°09′W
103 T16 Brignoles Var, SE France 43°25′N 06°03′E
Brigue see Brig
105 O7 Brihuega Castilla-La Mancha, C Spain 40°45′N 02°52′W
112 A10 Brijuni It. Brioni. island group NW Croatia
76 G12 Brikama W Gambia 13°13′N 16°37′W
Bril see Brielle
Brill, The see Brielle
101 G15 Brilon Nordrhein-Westfalen, W Germany 51°24′N 08°34′E
Brinceni see Briceni
107 Q18 Brindisi anc. Brundisium, Brundusium. Puglia, SE Italy 40°39′N 17°55′E
27 W11 Brinkley Arkansas, C USA 34°53′N 91°11′W
Brioni see Brijuni
103 P12 Brioude anc. Brivas. Haute-Loire, C France 45°18′N 03°23′E
Briovera see St-Lô
183 U2 Brisbane state capital Queensland, E Australia 27°30′S 153°E
183 V2 Brisbane ✈ Queensland, E Australia 27°30′S 153°00′E
25 P2 Briscoe Texas, SW USA 35°34′N 100°17′W
106 H10 Brisighella Emilia-Romagna, C Italy 44°12′N 11°45′E
108 G11 Brissago Ticino, S Switzerland 46°07′N 08°40′E
97 K22 Bristol anc. Bricgstow. SW England, United Kingdom 51°27′N 02°35′W
18 M12 Bristol Connecticut, NE USA 41°40′N 72°56′W
23 R9 Bristol Florida, SE USA 30°25′N 84°58′W
19 N9 Bristol New Hampshire, NE USA 43°34′N 71°42′W
29 Q8 Bristol South Dakota, N USA 45°18′N 97°45′W
21 P8 Bristol Tennessee, S USA 36°36′N 82°11′W
18 M8 Bristol Vermont, NE USA 44°07′N 73°00′W
39 N14 Bristol Bay bay Alaska, USA
97 I22 Bristol Channel inlet England/Wales, United Kingdom
35 W14 Bristol Lake ◎ California, W USA
27 P10 Bristow Oklahoma, C USA 35°49′N 96°23′W
86 C10 Britain var. Great Britain. island United Kingdom
Britannia Minor see Bretagne
10 L12 British Columbia Fr. Colombie-Britannique. ◆ province SW Canada
British Guiana see Guyana
British Honduras see Belize
173 Q7 British Indian Ocean Territory ◇ UK dependent territory C Indian Ocean
86 B9 British Isles island group NW Europe
10 I1 British Mountains ▲▲ Yukon, NW Canada
British North Borneo see Sabah
British Solomon Islands Protectorate see Solomon Islands
45 S8 British Virgin Islands var. Virgin Islands. ◇ UK dependent territory E West Indies
83 J21 Brits North-West, N South Africa 25°39′S 27°47′E
83 H24 Britstown Northern Cape, W South Africa 30°36′S 23°30′E
14 F12 Britt Ontario, S Canada 45°46′N 80°34′W
29 V12 Britt Iowa, C USA 43°06′N 93°48′W
Brittany see Bretagne
29 Q7 Britton South Dakota, N USA 45°47′N 97°45′W
Briva Curretia see Brive-la-Gaillarde
Briva Isarae see Pontoise
Brivas see Brioude
Brive see Brive-la-Gaillarde
102 M12 Brive-la-Gaillarde prev. Brive; anc. Briva Curretia. Corrèze, C France 45°09′N 01°31′E
105 O4 Briviesca Castilla y León, N Spain 42°33′N 03°19′W
Brixen see Bressanone
Brixia see Brescia
Brlik see Birlik
Brněnský Kraj see Jihomoravský Kraj
111 G18 Brno Ger. Brünn. Jihomoravský Kraj, SE Czech Republic 49°11′N 16°35′E
96 G7 Broad Bay bay NW Scotland, United Kingdom
25 X8 Broaddus Texas, SW USA 31°18′N 94°16′W
183 O12 Broadford Victoria, SE Australia 37°07′S 145°04′E
96 G9 Broadford N Scotland, United Kingdom 57°14′N 05°54′W
96 J13 Broad Law ▲ S Scotland, United Kingdom 55°30′N 03°22′W
23 U3 Broad River ≈ Georgia, SE USA
21 N8 Broad River ≈ North Carolina/South Carolina, SE USA
181 Y8 Broadsound Range ▲▲ Queensland, E Australia
33 X11 Broadus Montana, NW USA 45°28′N 105°22′W
21 U4 Broadway Virginia, NE USA 38°36′N 78°48′W
118 E9 Broceni SW Latvia 56°41′N 22°31′E
11 U11 Brochet Manitoba, C Canada 57°55′N 101°40′W
11 U10 Brochet, Lac ◎ Manitoba, C Canada
15 S5 Brochet, Lac au ◎ Québec, SE Canada
101 K14 Brocken ▲ C Germany 51°48′N 10°38′E
19 O12 Brockton Massachusetts, NE USA 42°04′N 71°01′W
14 L14 Brockville Ontario, SE Canada 44°35′N 75°44′W
18 D13 Brockway Pennsylvania, NE USA 41°14′N 78°45′W
Brod/Bród see Slavonski Brod
9 N5 Brodeur Peninsula peninsula Baffin Island, Nunavut, NE Canada
96 H13 Brodick W Scotland, United Kingdom 55°34′N 05°10′W
Brod na Savi see Slavonski Brod
110 K9 Brodnica Ger. Buddenbrock. Kujawski-pomorskie, C Poland 53°15′N 19°23′E
Brod-Posavina see Slavonski Brod-Posavina
Brodsko-Posavska Županija see Slavonski Brod-Posavina
116 J5 Brody L'vivs'ka Oblast', NW Ukraine 50°05′N 25°08′E
98 I10 Broek-in-Waterland Noord-Holland, C Netherlands 52°27′N 04°59′E
32 L13 Brogan Oregon, NW USA 44°15′N 117°34′W
110 N10 Brok Mazowieckie, C Poland 52°42′N 21°53′E
27 P9 Broken Arrow Oklahoma, C USA 36°03′N 95°47′W
183 T9 Broken Bay bay New South Wales, SE Australia
29 N15 Broken Bow Nebraska, C USA 41°24′N 99°38′W
27 R13 Broken Bow Oklahoma, C USA 34°01′N 94°44′W
27 R12 Broken Bow Lake ⊠ Oklahoma, C USA
182 L6 Broken Hill New South Wales, SE Australia 31°58′S 141°27′E
173 S10 Broken Ridge undersea feature S Indian Ocean 31°30′S 95°00′E
186 C6 Broken Water Bay bay W Bismarck Sea
55 W10 Brokopondo Brokopondo, NE Suriname 05°04′N 55°00′W
55 W10 Brokopondo ◆ district C Suriname
Bromberg see Bydgoszcz
95 L22 Bromölla Skåne, S Sweden 56°04′N 14°28′E
97 L20 Bromsgrove W England, United Kingdom 52°20′N 02°03′W
95 G20 Brønderslev Nordjylland, N Denmark 57°16′N 09°58′E
106 D8 Broni Lombardia, N Italy 45°04′N 09°18′E
10 K11 Bronlund Peak ▲ British Columbia, W Canada 57°27′N 126°43′W
93 F14 Brønnøysund Nordland, C Norway 65°38′N 12°15′E
23 V10 Bronson Florida, SE USA 29°25′N 82°38′W
31 Q11 Bronson Michigan, N USA 41°52′N 85°11′W
25 X8 Bronson Texas, SW USA 31°20′N 94°00′W
107 L24 Bronte Sicilia, Italy, C Mediterranean Sea 37°47′N 14°50′E
25 P8 Bronte Texas, SW USA 31°53′N 100°17′W
25 Y9 Brookeland Texas, SW USA 31°05′N 93°57′W
170 M7 Brooke's Point Palawan, W Philippines 08°54′N 117°54′E
27 T3 Brookfield Missouri, C USA 39°46′N 93°04′W
22 K7 Brookhaven Mississippi, S USA 31°34′N 90°26′W
32 E16 Brookings Oregon, NW USA 42°03′N 124°16′W
29 R10 Brookings South Dakota, N USA 44°19′N 96°46′W
29 W14 Brooklyn Iowa, C USA 41°43′N 92°27′W
29 U8 Brooklyn Park Minnesota, N USA 45°06′N 93°18′W
21 U7 Brookneal Virginia, NE USA 37°03′N 78°56′W
11 R16 Brooks Alberta, SW Canada 50°35′N 111°54′W
25 V11 Brookshire Texas, SW USA 29°47′N 95°57′W
38 L8 Brooks Mountain ▲ Alaska, USA 65°31′N 167°24′W
38 M11 Brooks Range ▲▲ Alaska, USA
31 O12 Brookston Indiana, N USA 40°34′N 86°53′W
23 V11 Brooksville Florida, SE USA 28°33′N 82°23′W
23 N4 Brooksville Mississippi, S USA 33°13′N 88°34′W
180 J13 Brookton Western Australia 32°24′S 117°04′E
31 Q14 Brookville Indiana, N USA 39°25′N 85°00′W
18 D13 Brookville Pennsylvania, NE USA 41°07′N 79°05′W
31 Q14 Brookville Lake ⊠ Indiana, N USA
180 K5 Broome Western Australia 17°58′S 122°15′E
37 S4 Broomfield Colorado, C USA 39°55′N 105°05′W
Broos see Orăștie
96 J7 Brora N Scotland, United Kingdom 57°59′N 04°00′W
96 I7 Brora ≈ N Scotland, United Kingdom
95 F23 Brørup Syddtjylland, W Denmark 55°29′N 09°01′E
95 L23 Brösarp Skåne, S Sweden 55°43′N 14°10′E
116 J9 Broșteni Suceava, NE Romania 47°14′N 25°43′E
102 M6 Brou Eure-et-Loir, C France 48°12′N 01°10′E
Broucsella see Brussel/Bruxelles
Broughton Bay see Tongjosŏn-man
Broughton Island see Qikiqtarjuaq
138 G7 Broummâna C Lebanon 33°53′N 35°39′E
22 I9 Broussard Louisiana, S USA 30°09′N 91°57′W
98 E13 Brouwersdam dam SW Netherlands
98 E13 Brouwershaven Zeeland, SW Netherlands 51°44′N 03°50′E
117 P4 Brovary Kyyivs'ka Oblast', N Ukraine 50°30′N 30°45′E
95 G20 Brovst Nordjylland, N Denmark 57°06′N 09°32′E
31 S8 Brown City Michigan, N USA 43°12′N 82°50′W
24 M6 Brownfield Texas, SW USA 33°11′N 102°16′W
33 Q7 Browning Montana, NW USA 48°33′N 113°00′W
33 R6 Brown, Mount ▲ Montana, NW USA 48°52′N 111°08′W
0 M9 Browns Bank undersea feature NW Atlantic Ocean 42°40′N 66°05′W
31 O14 Brownsburg Indiana, N USA 39°50′N 86°24′W
18 J16 Browns Mills New Jersey, NE USA 39°58′N 74°33′W
44 J12 Browns Town C Jamaica 18°28′N 77°22′W
31 P15 Brownstown Indiana, N USA 38°52′N 86°02′W
29 R8 Browns Valley Minnesota, N USA 45°36′N 96°49′W
20 K7 Brownsville Kentucky, S USA 37°10′N 86°18′W
20 F9 Brownsville Tennessee, S USA 35°35′N 89°15′W
25 T17 Brownsville Texas, SW USA 25°56′N 97°28′W
55 W10 Brownsweg Brokopondo, C Suriname
29 U9 Brownton Minnesota, N USA 44°43′N 94°21′W
19 R5 Brownville Junction Maine, NE USA 45°20′N 69°04′W
25 R8 Brownwood Texas, SW USA 31°42′N 98°59′W
25 R8 Brownwood Lake ⊠ Texas, SW USA
104 I9 Brozas Extremadura, W Spain 39°37′N 06°48′W
119 M18 Brozha Mahilyowskaya Voblasts', E Belarus 52°57′N 29°07′E
103 O2 Bruay-en-Artois Pas-de-Calais, N France 50°31′N 02°30′E
103 P2 Bruay-sur-l'Escaut Nord, N France 50°24′N 03°33′E
14 F13 Bruce Peninsula peninsula Ontario, S Canada
20 H9 Bruceton Tennessee, S USA 36°02′N 88°14′W
25 T9 Bruceville Texas, SW USA 31°17′N 97°15′W
101 G21 Bruchsal Baden-Württemberg, SW Germany 49°07′N 08°35′E
109 Q7 Bruck Salzburg, NW Austria 47°18′N 12°51′E
Bruck see Bruck an der Mur
109 Y4 Bruck an der Leitha Niederösterreich, NE Austria 48°02′N 16°47′E
109 V7 Bruck an der Mur var. Bruck. Steiermark, C Austria 47°25′N 15°17′E
101 M24 Bruckmühl Bayern, SE Germany 47°52′N 11°54′E
168 E7 Brueuh, Pulau island NW Indonesia
Bruges see Brugge
108 F6 Brugg Aargau, NW Switzerland 47°29′N 08°13′E
99 C16 Brugge Fr. Bruges. West-Vlaanderen, NW Belgium 51°13′N 03°14′E
109 R9 Bruggen Kärnten, S Austria 46°46′N 13°13′E
101 E16 Brühl Nordrhein-Westfalen, W Germany 50°50′N 06°55′E
99 F14 Bruinisse Zeeland, SW Netherlands 51°40′N 04°04′E
169 R9 Bruit, Pulau island East Malaysia
14 K10 Brûlé, Lac ◎ Québec, SE Canada
30 M4 Brule River ≈ Michigan/Wisconsin, N USA
99 H23 Brûly Namur, S Belgium 49°55′N 04°31′E
59 N17 Brumado Bahia, E Brazil 14°14′S 41°38′W
98 M11 Brummen Gelderland, E Netherlands 52°05′N 06°10′E
94 H13 Brumunddal Hedmark, S Norway 60°54′N 11°00′E
23 Q6 Brundidge Alabama, S USA 31°43′N 85°49′W
Brundisium/Brundusium see Brindisi
33 N15 Bruneau River ≈ Idaho, NW USA
Bruneck see Brunico
169 T8 Brunei off. Brunei Darussalam, Mal. Negara Brunei Darussalam. ◆ monarchy SE Asia
169 T7 Brunei Bay var. Teluk Brunei. bay N Brunei
Brunei Darussalam see Brunei
Brunei, Teluk see Brunei Bay
Brunei Town see Bandar Seri Begawan
106 H5 Brunico Ger. Bruneck. Trentino-Alto Adige, N Italy 46°49′N 11°57′E
Brünn see Brno
185 G17 Brunner, Lake ◎ South Island, New Zealand
99 M18 Brunssum Limburg, SE Netherlands 50°57′N 05°59′E
23 W7 Brunswick Georgia, SE USA 31°09′N 81°30′W
19 Q8 Brunswick Maine, NE USA 43°54′N 69°58′W
21 V3 Brunswick Maryland, NE USA 39°18′N 77°37′W
27 T3 Brunswick Missouri, C USA 39°25′N 93°07′W
31 T11 Brunswick Ohio, N USA 41°14′N 81°50′W
Brunswick see Braunschweig
63 H24 Brunswick, Península headland S Chile 53°30′S 71°27′W
111 H17 Bruntál Ger. Freudenthal. Moravskoslezský Kraj, E Czech Republic 50°00′N 17°27′E
195 N3 Brunt Ice Shelf ice shelf Antarctica
Brusa see Bursa
114 G7 Brusartsi Montana, NW Bulgaria 43°39′N 23°04′E
37 U3 Brush Colorado, C USA 40°15′N 103°37′W
42 M5 Brus Laguna Gracias a Dios, E Honduras 15°46′N 84°32′W
60 K13 Brusque Santa Catarina, S Brazil 27°07′S 48°54′W
Brussa see Bursa
99 E18 Brussel var. Brussels, Fr. Bruxelles, Ger. Brüssel; anc. Broucsella. ● (Belgium) Brussels, C Belgium 50°52′N 04°21′E see also Bruxelles
Brussel see Bruxelles
Brüssel/Brussels see Brussel/Bruxelles
117 O5 Brusyliv Zhytomyrs'ka Oblast', N Ukraine 50°16′N 29°31′E
183 Q12 Bruthen Victoria, SE Australia 37°43′S 147°49′E
Bruttium see Calabria
Brüx see Most
99 E18 Bruxelles var. Brussels, Dut. Brussel, Ger. Brüssel; anc. Broucsella. ● Brussels, C Belgium 50°52′N 04°21′E see also Brussel
Bruxelles see Brussel
54 J7 Bruzual Apure, W Venezuela 07°59′N 69°18′W
31 Q11 Bryan Ohio, N USA 41°30′N 84°34′W
25 U10 Bryan Texas, SW USA 30°40′N 96°23′W
194 J4 Bryan Coast physical region Antarctica
122 L11 Bryanka Krasnoyarskiy Kray, C Russian Federation 59°01′N 93°13′E
117 Y7 Bryanka Luhans'ka Oblast', E Ukraine 48°30′N 38°45′E
182 J8 Bryan, Mount ▲ South Australia 33°25′S 138°59′E
126 I6 Bryansk Bryanskaya Oblast', W Russian Federation 53°16′N 34°07′E
126 H6 Bryanskaya Oblast' ◆ province W Russian Federation
194 J5 Bryant, Cape headland Antarctica
27 U8 Bryant Creek ≈ Missouri, C USA
36 K8 Bryce Canyon canyon Utah, W USA
119 O15 Bryli Mahilyowskaya Voblasts', E Belarus 53°54′N 30°33′E
95 C17 Bryne Rogaland, S Norway 58°43′N 05°40′E
25 R6 Bryson Texas, SW USA 33°09′N 98°23′W
21 N10 Bryson City North Carolina, SE USA 35°26′N 83°27′W
14 K11 Bryson, Lac ◎ Québec, SE Canada
126 K13 Bryukhovetskaya Krasnodarskiy Kray, SW Russian Federation 45°49′N 38°01′E
111 H15 Brzeg Ger. Brieg; anc. Civitas Altae Ripae. Opolskie, S Poland 50°52′N 17°27′E
111 G14 Brzeg Dolny Ger. Dyhernfurth. Dolnośląskie, SW Poland 51°15′N 16°40′E
Brześć Litewski/Brześć nad Bugiem see Brest
111 L17 Brzesko Ger. Brietzig. Małopolskie, SE Poland 49°59′N 20°34′E
Brzeżany see Berezhany
110 K12 Brzeziny Łódzkie, C Poland 51°49′N 19°41′E
Brzostowica Wielka see Vyalikaya Byerastavitsa
111 O17 Brzozów Podkarpackie, SE Poland 49°38′N 22°00′E
Bsharrī/Bsherri see Bcharré
187 X14 Bua Vanua Levu, N Fiji 16°48′S 178°36′E
95 J20 Bua Halland, S Sweden 57°14′N 12°07′E
82 M13 Bua ≈ C Malawi
Bua see Čiovo
81 L18 Bu'aale It. Buale. Jubbada Dhexe, SW Somalia 0°52′N 42°37′E
Buache, Mount see Mutunte, Mount
189 Q8 Buada Lagoon lagoon Nauru, C Pacific Ocean
186 M8 Buala Santa Isabel, E Solomon Islands 08°06′S 159°31′E
Buale see Bu'aale
190 H1 Buariki atoll Tungaru, W Kiribati
167 Q10 Bua Yai var. Ban Bua Yai. Nakhon Ratchasima, E Thailand 15°35′N 102°25′E
75 P8 Bu'ayrāt al Ḩasūn var. Buwayrāt al Ḩasūn. C Libya 31°22′N 15°41′E
76 H13 Buba S Guinea-Bissau 11°36′N 14°55′W
171 P11 Bubaa Sulawesi, N Indonesia 0°32′N 122°27′E
81 D20 Bubanza NW Burundi 03°04′S 29°22′E
83 K18 Bubi prev. Bubye. ≈ S Zimbabwe
142 L11 Būbīyan, Jazīrat island E Kuwait
Bublitz see Bobolice
Bubye see Bubi
187 Y13 Buca prev. Mbutha. Vanua Levu, N Fiji 16°39′S 179°51′E
136 F16 Bucak Burdur, SW Turkey 37°28′N 30°37′E
54 G8 Bucaramanga Santander, N Colombia 07°08′N 73°10′W
107 M18 Buccino Campania, S Italy 40°37′N 15°25′E
116 K9 Bucecea Botoșani, NE Romania 47°45′N 26°30′E
116 J6 Buchach Pol. Buczacz. Ternopil's'ka Oblast', W Ukraine 49°04′N 25°23′E
183 Q12 Buchan Victoria, SE Australia 37°26′S 148°11′E
76 J17 Buchanan prev. Grand Bassa. SW Liberia 05°53′N 10°03′W
23 R3 Buchanan Georgia, SE USA 33°48′N 85°11′W
31 O11 Buchanan Michigan, N USA 41°49′N 86°21′W
21 T6 Buchanan Virginia, NE USA 37°31′N 79°40′W
25 R10 Buchanan Dam Texas, SW USA 30°42′N 98°24′W
25 R10 Buchanan, Lake ⊠ Texas, SW USA
96 L8 Buchan Ness headland NE Scotland, United Kingdom 57°28′N 01°46′W
13 T12 Buchans Newfoundland and Labrador, SE Canada 48°49′N 56°53′W
Bucharest see București
101 H20 Buchen Baden-Württemberg, SW Germany 49°31′N 09°18′E
100 I10 Buchholz in der Nordheide Niedersachsen, NW Germany 53°19′N 09°52′E
108 F7 Buchs Aargau, N Switzerland 47°24′N 08°04′E
108 I8 Buchs Sankt Gallen, NE Switzerland 47°10′N 09°28′E
100 H13 Bückeburg Niedersachsen, NW Germany 52°16′N 09°03′E
36 K14 Buckeye Arizona, SW USA 33°22′N 112°34′W
Buckeye State see Ohio
21 S4 Buckhannon West Virginia, NE USA 38°59′N 80°14′W
25 T9 Buckholts Texas, SW USA 30°52′N 97°07′W
96 K8 Buckie NE Scotland, United Kingdom 57°39′N 02°56′W
14 M12 Buckingham Québec, SE Canada 45°35′N 75°25′W
21 U6 Buckingham Virginia, NE USA 37°33′N 78°34′W
97 N21 Buckinghamshire cultural region SE England, United Kingdom
39 N8 Buckland Alaska, USA 65°58′N 161°07′W
182 G7 Buckleboo South Australia 32°55′S 136°11′E
26 K7 Bucklin Kansas, C USA 37°33′N 99°37′W
27 T3 Bucklin Missouri, C USA 39°46′N 92°53′W
36 I12 Buckskin Mountains ▲▲ Arizona, SW USA
19 R7 Bucksport Maine, NE USA 44°34′N 68°46′W
82 A9 Buco Zau Cabinda, NW Angola 04°45′S 12°34′E
Bu Craa see Bou Craa
116 K14 București Eng. Bucharest, Ger. Bukarest, prev. Altenburg; anc. Cetatea Damboviței. ● (Romania) București, S Romania 44°27′N 26°06′E
31 S12 Bucyrus Ohio, N USA 40°47′N 82°57′W
Buczacz see Buchach
94 E9 Bud Møre og Romsdal, S Norway 62°55′N 06°55′E
25 S11 Buda Texas, SW USA 30°05′N 97°50′W
119 O18 Buda-Kashalyova Rus. Buda-Koshelëvo. Homyel'skaya Voblasts', SE Belarus 52°43′N 30°34′E
Buda-Koshelëvo see Buda-Kashalyova
166 L4 Budalin Sagaing, C Myanmar (Burma) 22°24′N 95°08′E
111 J22 Budapest off. Budapest Főváros, SCr. Budimpešta. ● (Hungary) Pest, N Hungary 47°30′N 19°03′E
Budapest Főváros see Budapest
92 I2 Búðardalur Vesturland, W Iceland 65°07′N 21°45′W
152 K11 Budaun Uttar Pradesh, N India 28°02′N 79°07′E
141 O9 Budayyi'ah oasis C Saudi Arabia
195 Y12 Budd Coast physical region Antarctica
Buddenbrock see Brodnica
107 C17 Budduso Sardegna, Italy, C Mediterranean Sea 40°37′N 09°19′E
97 I23 Bude SW England, United Kingdom 50°50′N 04°33′W
22 J7 Bude Mississippi, S USA 31°27′N 90°51′W
Budějovický Kraj see Jihočeský Kraj
99 K16 Budel Noord-Brabant, SE Netherlands 51°17′N 05°35′E
100 I8 Büdelsdorf Schleswig-Holstein, N Germany 54°20′N 09°40′E
127 O14 Budënnovsk Stavropol'skiy Kray, SW Russian Federation 44°46′N 44°07′E
116 K14 Budești Călărași, SE Romania 44°13′N 26°30′E
Budge-Budge see Baj Baj
Budgewoi see Budgewoi Lake
183 T8 Budgewoi Lake var. Budgewoi. New South Wales, SE Australia 33°14′S 151°34′E
Budimpešta see Budapest
79 J16 Budjala Equateur, NW Dem. Rep. Congo 02°39′N 19°42′E
106 G10 Budrio Emilia-Romagna, C Italy 44°33′N 11°34′E
Budslav see Budslaw
119 K14 Budslaw Rus. Budslav. Minskaya Voblasts', N Belarus 54°47′N 27°27′E
Budua see Budva
169 R9 Budu, Tanjung headland East Malaysia 02°51′N 111°42′E
113 J17 Budva It. Budua. W Montenegro 42°17′N 18°49′E
Budweis see České Budějovice
Budyšin see Bautzen
79 D16 Buea Sud-Ouest, SW Cameroon 04°09′N 09°13′E
103 S13 Buëch ≈ SE France
18 J17 Buena New Jersey, NE USA 39°30′N 74°55′W
62 K12 Buena Esperanza San Luis, C Argentina 34°45′S 65°15′W
54 C11 Buenaventura Valle del Cauca, W Colombia 03°54′N 77°02′W
40 I4 Buenaventura Chihuahua, N Mexico 29°50′N 107°30′W
57 M18 Buena Vista Santa Cruz, C Bolivia 17°28′S 63°37′W
40 G10 Buenavista Baja California Sur, NW Mexico 23°39′N 109°41′W
37 S5 Buena Vista Colorado, C USA 38°50′N 106°07′W
23 S5 Buena Vista Georgia, SE USA 32°19′N 84°31′W
21 T6 Buena Vista Virginia, NE USA 37°44′N 79°22′W
44 F5 Buena Vista, Bahia de bay N Cuba
35 R13 Buena Vista Lake Bed ◎ California, W USA
105 P8 Buendía, Embalse de ⊠ C Spain
63 F16 Bueno, Río ≈ S Chile
62 N12 Buenos Aires hist. Santa Maria del Buen Aire. ● (Argentina) Buenos Aires, E Argentina 34°40′S 58°30′W
43 O15 Buenos Aires Puntarenas, SE Costa Rica 09°10′N 83°23′W
61 C20 Buenos Aires off. Provincia de Buenos Aires. ◆ province E Argentina
63 H19 Buenos Aires, Lago var. Lago General Carrera. ◎ Argentina/Chile
Buenos Aires, Provincia de see Buenos Aires
54 C13 Buesaco Nariño, SW Colombia 01°22′N 77°07′W
29 U8 Buffalo Minnesota, N USA 45°11′N 93°50′W
26 T6 Buffalo Missouri, C USA 37°38′N 93°05′W
18 D10 Buffalo New York, NE USA 42°53′N 78°53′W
27 K8 Buffalo Oklahoma, C USA 36°51′N 99°38′W
28 J7 Buffalo South Dakota, N USA 45°25′N 103°35′W
25 V8 Buffalo Texas, SW USA 31°25′N 96°04′W
33 W12 Buffalo Wyoming, C USA 44°21′N 106°40′W
29 U11 Buffalo Center Iowa, C USA 43°23′N 93°57′W
24 M3 Buffalo Lake ⊠ Texas, SW USA
30 K7 Buffalo Lake ⊠ Wisconsin, N USA
11 S12 Buffalo Narrows Saskatchewan, C Canada 55°52′N 108°28′W
27 U9 Buffalo River ≈ Arkansas, C USA
29 R5 Buffalo River ≈ Minnesota, N USA
20 I10 Buffalo River ≈ Tennessee, S USA
30 J6 Buffalo River ≈ Wisconsin, N USA
44 L12 Buff Bay E Jamaica 18°18′N 76°40′W
23 T3 Buford Georgia, SE USA 34°07′N 84°00′W
28 J3 Buford North Dakota, N USA 48°00′N 103°58′W
33 Y17 Buford Wyoming, C USA 41°05′N 105°17′W
116 J14 Buftea Ilfov, S Romania 44°34′N 25°58′E
84 I9 Bug Bel. Zakhodni Buh, Eng. Western Bug, Rus. Zapadnyy Bug, Ukr. Zakhidnyy Buh. ≈ E Europe
54 D11 Buga Valle del Cauca, W Colombia 03°53′N 76°17′W
Buga see Dörvöljin
103 O17 Bugarach, Pic du ▲ S France 42°52′N 02°23′E
162 F8 Bugat var. Bayangol. Govĭ-Altay, SW Mongolia 45°33′N 94°22′E
146 B12 Bugdaýly Rus. Bugdayly. Balkan Welaýaty, W Turkmenistan 38°42′N 54°14′E
Bugdayly see Bugdaýly
Buggs Island Lake see John H. Kerr Reservoir
Bughotu see Santa Isabel
171 O14 Bugingkalo Sulawesi, C Indonesia 04°49′S 121°42′E
64 P6 Bugio island Madeira, Portugal, NE Atlantic Ocean
92 M8 Bugøynes Finnmark, N Norway 69°57′N 29°34′E
125 Q3 Bugrino Nenetskiy Avtonomnyy Okrug, NW Russian Federation 68°48′N 49°12′E
127 T5 Bugul'ma Respublika Tatarstan, W Russian Federation 54°31′N 52°45′E
Büğür see Luntai
127 T6 Buguruslan Orenburgskaya Oblast', W Russian Federation 53°38′N 52°30′E
159 R9 Buh He ≈ C China
101 F22 Bühl Baden-Württemberg, SW Germany 48°42′N 08°07′E
33 O15 Buhl Idaho, NW USA 42°36′N 114°45′W
116 K10 Buhuși Bacău, E Romania 46°41′N 26°45′E
Buie d'Istria see Buje
97 J20 Builth Wells E Wales, United Kingdom 52°07′N 03°28′W
186 J8 Buin Bougainville Island, NE Papua New Guinea 06°52′S 155°42′E
108 J9 Buin, Piz ▲ Austria/Switzerland 46°51′N 10°07′E
127 Q4 Buinsk Chuvashskaya Respublika, W Russian Federation 55°09′N 47°00′E
127 Q4 Buinsk Respublika Tatarstan, W Russian Federation 54°58′N 48°16′E
163 R8 Buir Nuur Mong. Buyr Nuur. ◎ China/Mongolia see also Buyr Nuur
Buir Nuur see Buyr Nuur
98 M5 Buitenpost Fris. Bûtenpost. Fryslân, N Netherlands 53°15′N 06°09′E
Buitenzorg see Bogor
83 F19 Buitepos Omaheke, E Namibia 22°17′S 19°59′E
105 N7 Buitrago del Lozoya Madrid, C Spain 41°00′N 03°38′W
Buj see Buy
104 M13 Bujalance Andalucía, S Spain 37°54′N 04°23′W
113 O17 Bujanovac Serbia, SE Serbia 42°29′N 21°44′E
105 S6 Bujaraloz Aragón, NE Spain 41°29′N 00°10′W
112 A9 Buje It. Buie d'Istria. Istra, NW Croatia 45°23′N 13°40′E
Bujnurd see Bojnūrd
81 D21 Bujumbura prev. Usumbura. ● (Burundi) W Burundi 03°25′S 29°24′E
81 D20 Bujumbura ✈ W Burundi 03°21′S 29°19′E
186 J6 Buka Bougainville, Papua New Guinea 05°24′S 154°40′E
159 N11 Buka Daban var. Bukadaban Feng. ▲ C China 36°09′N 90°52′E
Bukadaban Feng see Buka Daban
186 J6 Buka Island island NE Papua New Guinea
81 F18 Bukakata S Uganda 0°18′S 31°57′E
79 N24 Bukama Katanga, SE Dem. Rep. Congo 09°13′S 25°52′E
142 J4 Būkān var. Bowkān. Āzarbāyjān-e Gharbī, NW Iran 36°31′N 46°10′E
Bukantau, Gory see Bo'kantov Tog'lari
Bukarest see București
79 O19 Bukavu prev. Costermansville. Sud-Kivu, E Dem. Rep. Congo 02°19′S 28°49′E
81 F21 Bukene Tabora, NW Tanzania 04°15′S 32°51′E
141 W8 Bū Khābī var. Bakhābī. NW Oman 23°29′N 56°06′E
Bukhara see Buxoro
Bukharskaya Oblast' see Buxoro Viloyati
168 M14 Bukitkemuning Sumatera, W Indonesia 04°43′S 104°27′E
168 I11 Bukittinggi prev. Fort de Kock. Sumatera, W Indonesia 0°18′S 100°20′E
111 L21 Bükk ▲▲ NE Hungary
81 F19 Bukoba Kagera, NW Tanzania 01°19′S 31°49′E
113 N20 Bukovo S FYR Macedonia 40°59′N 21°20′E
108 G6 Bülach Zürich, NW Switzerland 47°31′N 08°30′E
Būlaevo see Bulayevo
Bulag see Tünel, Hövsgöl, Mongolia
Bulag see Möngönmorĭt, Töv, Mongolia
Bulagiyn Denj see Bulgan
183 U7 Bulahdelah New South Wales, SE Australia 32°24′S 152°13′E
171 P4 Bulan Luzon, N Philippines 12°40′N 123°55′E
137 N11 Bulancak Giresun, N Turkey 40°57′N 38°14′E
152 J10 Bulandshahr Uttar Pradesh, N India 28°30′N 77°49′E
137 R14 Bulanık Muş, E Turkey 39°04′N 42°16′E
127 V7 Bulanovo Orenburgskaya Oblast', W Russian Federation 52°27′N 55°08′E
83 J17 Bulawayo var. Buluwayo. Bulawayo, SW Zimbabwe 20°08′S 28°37′E
83 J17 Bulawayo ✈ Matabeleland North, SW Zimbabwe 20°00′S 28°36′E
145 Q6 Bulayevo Kaz. Būlaevo. Severnyy Kazakhstan, N Kazakhstan 54°55′N 70°29′E
136 D15 Buldan Denizli, SW Turkey 38°03′N 28°50′E
154 G12 Buldāna Mahārāshtra, C India 20°31′N 76°18′E
38 E16 Buldir Island island Aleutian Islands, Alaska, USA
Buldur see Burdur
162 I8 Bulgan var. Bulagiyn Denj. Arhangay, C Mongolia 47°14′N 100°56′E
162 D7 Bulgan var. Jargalant. Bayan-Ölgiy, W Mongolia 46°56′N 91°07′E
162 K6 Bulgan Bulgan, N Mongolia 50°31′N 101°30′E
162 F7 Bulgan var. Bürenhayrhan. Hovd, W Mongolia 46°04′N 91°34′E
162 J10 Bulgan Ömnögovĭ, S Mongolia 44°07′N 103°28′E
162 J7 Bulgan ◆ province N Mongolia
Bulgan see Bayan-Öndör, Bayanhongor, C Mongolia
Bulgan see Darvi, Hovd, Mongolia
Bulgan see Tsagaan-Üür, Hövsgöl, Mongolia
114 H10 Bulgaria off. Republic of Bulgaria, Bul. Bŭlgariya; prev. People's Republic of Bulgaria. ◆ republic SE Europe
Bulgaria, People's Republic of see Bulgaria
Bulgaria, Republic of see Bulgaria
Bŭlgariya see Bulgaria
Bŭlgarka see Balgarka
171 S11 Buli Pulau Halmahera, E Indonesia 0°56′N 128°17′E
171 S11 Buli, Teluk bay Pulau Halmahera, E Indonesia
160 J13 Buliu He ≈ S China
Bullange see Büllingen
Bulla, Ostrov see Xərə Zirə Adasi
104 M11 Bullaque ≈ C Spain
105 Q13 Bullas Murcia, SE Spain 38°02′N 01°40′W
80 M12 Bullaxaar Woqooyi Galbeed, NW Somalia 10°28′N 44°15′E
108 C9 Bulle Fribourg, SW Switzerland 46°37′N 07°04′E
185 G15 Buller ≈ South Island, New Zealand
183 P12 Buller, Mount ▲ Victoria, SE Australia 37°10′S 146°31′E
36 H11 Bullhead City Arizona, SW USA 35°07′N 114°32′W
99 N21 Büllingen Fr. Bullange. Liège, E Belgium 50°23′N 06°15′E
Bullion State see Missouri
21 T14 Bull Island island South Carolina, SE USA

230
◆ Country
● Country Capital
◇ Dependent Territory
○ Dependent Territory Capital
◆ Administrative Regions
✈ International Airport
▲ Mountain
▲▲ Mountain Range
Volcano
≈ River
◎ Lake
⊠ Reservoir

BULLOO RIVER OVERFLOW – CADIZ LAKE

182 M4 **Bulloo River Overflow** *wetland* New South Wales, SE Australia
184 M12 **Bulls** Manawatu-Wanganui, North Island, New Zealand 40°10′S 175°22′E
21 T14 **Bulls Bay** *bay* South Carolina, SE USA
27 U9 **Bull Shoals Lake** ⊠ Arkansas/Missouri, C USA
181 Q2 **Bulman** Northern Territory, N Australia 13°39′S 134°21′E
162 I6 **Bulnayn Nuruu** ▲ N Mongolia
171 O11 **Bulowa, Gunung** ▲ Sulawesi, N Indonesia 0°33′N 123°39′E
Bulqiza *see* Bulqizë
113 L19 **Bulqizë** *var.* Bulqiza. Dibër, C Albania 41°30′N 20°16′E
Bulsar *see* Valsād
171 N14 **Bulukumba** *prev.* Boeloekoemba. Sulawesi, C Indonesia 05°35′S 120°13′E
147 O11 **Bulungh'ur** *Rus.* Bulungur; *prev.* Krasnogvardeysk. Samarqand Viloyati, C Uzbekistan 39°46′N 67°18′E
79 I21 **Bulungu** Bandundu, SW Dem. Rep. Congo 04°36′S 18°34′E
Bulungur *see* Bulungh'ur
Buluwayo *see* Bulawayo
79 K17 **Bumba** Equateur, N Dem. Rep. Congo 02°14′N 22°25′E
121 R12 **Bumbah, Khalīj al** *gulf* N Libya
Bumbat *see* Bayan-Öndör
81 F19 **Bumbire Island** *island* N Tanzania
169 V8 **Bum Bun, Pulau** *island* East Malaysia
81 J17 **Buna** Wajir, NE Kenya 02°40′N 39°34′E
25 Y10 **Buna** Texas, SW USA 30°25′N 94°00′W
Bunab *see* Bonāb
Bunai *see* M'bunai
147 S13 **Bunay** S Tajikistan 38°29′N 71°41′E
180 I13 **Bunbury** Western Australia 33°24′S 115°44′E
97 E14 **Buncrana** *Ir.* Bun Cranncha. NW Ireland 55°08′N 07°27′W
Bun Cranncha *see* Buncrana
181 Z9 **Bundaberg** Queensland, E Australia 24°50′S 152°16′E
183 T5 **Bundarra** New South Wales, SE Australia 30°12′S 151°06′E
100 G13 **Bünde** Nordrhein-Westfalen, NW Germany 52°12′N 08°34′E
152 H13 **Būndi** Rājasthān, N India 25°28′N 75°42′E
Bun Dobhráin *see* Bundoran
97 D15 **Bundoran** *Ir.* Bun Dobhráin. NW Ireland 54°30′N 08°11′W
Bunë *see* Bojana
113 K18 **Bunë, Lumi i** *SCr.* Bojana. ≈ Albania/Montenegro *see also* Bojana
171 Q8 **Bunga** ≈ Mindanao, S Philippines
168 I12 **Bungalaut, Selat** *strait* W Indonesia
167 R8 **Bung Kan** Nong Khai, E Thailand 18°19′N 103°39′E
181 N4 **Bungle Bungle Range** ▲ Western Australia
82 C10 **Bungo** Uíge, NW Angola 07°30′S 15°24′E
81 G18 **Bungoma** Bungoma, W Kenya 0°34′N 34°34′E
81 G18 **Bungoma** ◆ *county* W Kenya
164 F15 **Bungo-suidō** *strait* SW Japan
164 E14 **Bungo-Takada** Ōita, Kyūshū, SW Japan 33°34′N 131°28′E
100 K8 **Bungsberg** *hill* N Germany
Bungur *see* Bunyu
79 P17 **Bunia** Orientale, NE Dem. Rep. Congo 01°33′N 30°16′E
35 U6 **Bunker Hill** ▲ Nevada, W USA 39°16′N 117°06′W
22 I7 **Bunkie** Louisiana, S USA 30°58′N 92°12′W
23 X10 **Bunnell** Florida, SE USA 29°28′N 81°15′W
105 S10 **Buñol** Valenciana, E Spain 39°25′N 00°47′W
98 K11 **Bunschoten** Utrecht, C Netherlands 52°15′N 05°23′E
136 K14 **Bünyan** Kayseri, C Turkey 38°51′N 35°50′E
169 W8 **Bunyu** *var.* Bungur. Borneo, N Indonesia 03°33′N 117°50′E
169 W8 **Bunyu, Pulau** *island* N Indonesia
Bunzlau *see* Bolesławiec
Buoddobohki *see* Patoniva
123 P7 **Buor-Khaya, Guba** *bay* N Russian Federation
123 P7 **Buor-Khaya, Guba** *bay* N Russian Federation
171 Z15 **Bupul** Papua, E Indonesia 07°24′S 140°57′E
80 P12 **Buraan** Bari, N Somalia 10°03′N 49°08′E
145 Q7 **Burabay** *prev.* Borovoye. Akmola, N Kazakhstan 53°07′N 70°20′E
Buraida *see* Buraydah
Buraimi *see* Al Buraymī
Buran *see* Boran
158 G15 **Burang** Xizang Zizhiqu, W China 30°28′N 81°13′E
Burao *see* Burco
138 H8 **Burāq** Dar'ā, S Syria 33°11′N 36°28′E
141 O6 **Buraydah** *var.* Buraida. Al Qaşīm, N Saudi Arabia 26°50′N 44°E
35 S15 **Burbank** California, W USA 34°10′N 118°25′W
31 N11 **Burbank** Illinois, N USA 41°45′N 87°48′W
183 Q8 **Burcher** New South Wales, SE Australia 33°29′S 147°16′E
80 N13 **Burco** *var.* Burao, Bur'o. Togdheer, NW Somalia 09°29′N 45°31′E
162 K8 **Bürd** *var.* Ongon. Övörhangay, C Mongolia 46°58′N 103°45′E
146 L13 **Burdalyk** Lebap Welayaty, E Turkmenistan 38°31′N 64°21′E
181 W6 **Burdekin River** ≈ Queensland, NE Australia
27 O7 **Burden** Kansas, C USA 37°18′N 96°45′W
Burdigala *see* Bordeaux
136 E15 **Burdur** *var.* Buldur. Burdur, SW Turkey 37°44′N 30°17′E
136 E15 **Burdur** *var.* Buldur. ◆ *province* SW Turkey
136 E15 **Burdur Gölü** *salt lake* SW Turkey
65 H21 **Burdwood Bank** *undersea feature* SW Atlantic Ocean
80 I12 **Burē** Āmara, N Ethiopia 10°43′N 37°09′E
80 H13 **Burē** Oromīya, C Ethiopia 08°13′N 35°09′E
93 J15 **Bureå** Västerbotten, N Sweden 64°36′N 21°15′E
162 K7 **Büreghangay** *var.* Darhan. Bulgan, C Mongolia 48°07′N 103°54′E
101 G14 **Büren** Nordrhein-Westfalen, W Germany 51°34′N 08°34′E
162 L8 **Büren** *var.* Bayantöhöm. Töv, C Mongolia 46°57′N 105°09′E
162 K6 **Bürengiyn Nuruu** ▲ N Mongolia
Bürenhayrhan *see* Bulgan
162 I6 **Bürentogtoh** *var.* Bayan. Hövsgol, C Mongolia 49°36′N 99°36′E
149 U10 **Būrewāla** *var.* Mandi Būrewāla. Punjab, E Pakistan 30°05′N 72°47′E
92 J9 **Burfjord** Troms, N Norway 69°55′N 21°54′E
100 L13 **Burg** *var.* Burg an der Ihle, Burg bei Magdeburg. Sachsen-Anhalt, C Germany 52°17′N 11°51′E
Burg an der Ihle *see* Burg
114 N10 **Burgas** *var.* Bourgas. Burgas, E Bulgaria 42°31′N 27°30′E
114 M10 **Burgas** ◆ *province* E Bulgaria
114 N9 **Burgas** ✈ Burgas, E Bulgaria 42°35′N 27°33′E
114 N10 **Burgaski Zaliv** *gulf* E Bulgaria
114 M10 **Burgasko Ezero** *lagoon* E Bulgaria
21 V11 **Burgaw** North Carolina, SE USA 34°33′N 77°56′W
Burg bei Magdeburg *see* Burg
108 E8 **Burgdorf** Bern, NW Switzerland 47°03′N 07°38′E
109 Y7 **Burgenland** *off.* Land Burgenland. ◆ *state* SE Austria
13 S13 **Burgeo** Newfoundland, Newfoundland and Labrador, SE Canada 47°37′N 57°38′W
83 I24 **Burgersdorp** Eastern Cape, SE South Africa 31°00′S 26°20′E
83 K20 **Burgersfort** Mpumalanga, NE South Africa 24°39′S 30°18′E
101 N23 **Burghausen** Bayern, SE Germany 48°10′N 12°48′E
139 O5 **Burghūth, Sabkhat al** ⊚ E Syria
101 M20 **Burglengenfeld** Bayern, SE Germany 49°11′N 12°01′E
41 P9 **Burgos** Tamaulipas, C Mexico 24°55′N 98°46′W
105 N4 **Burgos** Castilla y León, N Spain 42°21′N 03°41′W
105 N4 **Burgos** ◆ *province* Castilla y León, N Spain
Burgstadlberg *see* Hradiště
95 P20 **Burgsvik** Gotland, SE Sweden 57°01′N 18°18′E
98 L6 **Burgum** *Dutch.* Bergum. Fryslân, N Netherlands 53°12′N 05°59′E
Burgundy *see* Bourgogne
159 Q11 **Burhan Budai Shan** ▲ C China
136 B12 **Burhaniye** Balıkesir, W Turkey 39°29′N 26°59′E
154 G12 **Burhānpur** Madhya Pradesh, C India 21°18′N 76°14′E
127 W7 **Buribay** Respublika Bashkortostan, W Russian Federation 51°57′N 58°11′E
43 O17 **Burica, Punta** *headland* Costa Rica/Panama 08°02′N 82°53′W
167 Q10 **Buriram** *var.* Buri Ram, Puriramya. Buri Ram, E Thailand 15°01′N 103°06′E
Buri Ram *see* Buriram
105 S10 **Burjassot** Valenciana, E Spain 39°33′N 00°26′W
81 N16 **Burka Giibi** Hiiraan, C Somalia 03°52′N 45°07′E
147 X8 **Burkan** ≈ E Kyrgyzstan
25 R4 **Burkburnett** Texas, SW USA 34°06′N 98°34′W
29 O12 **Burke** South Dakota, N USA 43°09′N 99°18′W
10 K15 **Burke Channel** *channel* British Columbia, W Canada
194 J10 **Burke Island** *island* Antarctica
20 L7 **Burkesville** Kentucky, S USA 36°48′N 85°21′W
181 T4 **Burketown** Queensland, NE Australia 17°49′S 139°28′E
25 Q8 **Burkett** Texas, SW USA 32°01′N 99°17′W
25 Y9 **Burkeville** Texas, SW USA 30°58′N 93°41′W
21 V7 **Burkeville** Virginia, NE USA 37°11′N 78°12′W
Burkina *see* Burkina Faso
77 O12 **Burkina Faso** *off.* Burkina Faso; *prev.* Upper Volta. ◆ *republic* W Africa
Burkina Faso *see* Burkina Faso
194 L13 **Burks, Cape** *headland* Antarctica
14 H12 **Burk's Falls** Ontario, S Canada 45°38′N 79°25′W
101 H23 **Burladingen** Baden-Württemberg, S Germany 48°18′N 09°05′E
25 T7 **Burleson** Texas, SW USA 32°32′N 97°19′W
33 P15 **Burley** Idaho, NW USA 42°32′N 113°47′W
Burlin *see* Borili
14 G16 **Burlington** Ontario, S Canada 43°19′N 79°48′W
37 W4 **Burlington** Colorado, C USA 39°17′N 102°17′W
29 Y15 **Burlington** Iowa, C USA 40°48′N 91°05′W
27 P5 **Burlington** Kansas, C USA 38°11′N 95°46′W
21 T9 **Burlington** North Carolina, SE USA 36°05′N 79°27′W
28 M3 **Burlington** North Dakota, N USA 48°16′N 101°25′W
18 L7 **Burlington** Vermont, NE USA 44°28′N 73°14′W
30 M9 **Burlington** Wisconsin, N USA 42°38′N 88°12′W
27 Q1 **Burlington Junction** Missouri, C USA 40°27′N 95°04′W
Burma *see* Myanmar (Burma)
10 L17 **Burnaby** British Columbia, SW Canada 49°16′N 122°58′W
117 O12 **Burnas, Ozero** ⊚ SW Ukraine
25 S10 **Burnet** Texas, SW USA 30°46′N 98°14′W
35 O3 **Burney** California, W USA 40°52′N 121°42′W
183 O16 **Burnie** Tasmania, SE Australia 41°07′S 145°52′E
97 L17 **Burnley** NW England, United Kingdom 53°48′N 02°14′W
Burnoye *see* Bauyrzhan Momyshuly
153 R15 **Burnpur** West Bengal, NE India 23°39′N 86°55′E
32 K14 **Burns** Oregon, NW USA 43°35′N 119°03′W
26 K11 **Burns Flat** Oklahoma, C USA 35°21′N 99°10′W
20 M7 **Burnside** Kentucky, S USA 36°55′N 84°34′W
8 K8 **Burnside** ≈ Nunavut, NW Canada
32 L15 **Burns Junction** Oregon, NW USA 42°46′N 117°51′W
10 L13 **Burns Lake** British Columbia, SW Canada 54°14′N 125°45′W
29 V9 **Burnsville** Minnesota, N USA 44°49′N 93°14′W
21 P9 **Burnsville** North Carolina, SE USA 35°56′N 82°18′W
21 R4 **Burnsville** West Virginia, NE USA 38°50′N 80°39′W
14 I13 **Burnt River** ≈ Ontario, SE Canada
14 I11 **Burntroot Lake** ⊚ Ontario, SE Canada
11 W12 **Burntwood** ≈ Manitoba, C Canada
Bur'o *see* Burco
158 L2 **Burqin** Xinjiang Uygur Zizhiqu, NW China 47°42′N 86°50′E
182 J8 **Burra** South Australia 33°41′S 138°54′E
183 S9 **Burragorang, Lake** ⊠ New South Wales, SE Australia
96 K5 **Burray** *island* NE Scotland, United Kingdom
113 L19 **Burrel** *var.* Burreli. Dibër, C Albania 41°36′N 20°00′E
Burreli *see* Burrel
183 R8 **Burrendong Reservoir** ⊠ New South Wales, SE Australia
183 R5 **Burren Junction** New South Wales, SE Australia 30°06′S 149°01′E
Burriana *see* Borriana
183 R10 **Burrinjuck Reservoir** ⊠ New South Wales, SE Australia
36 J12 **Burro Creek** ≈ Arizona, SW USA
40 M5 **Burro, Serranías del** ▲ NW Mexico
62 K7 **Burruyacú** Tucumán, N Argentina 26°30′S 64°45′W
136 E12 **Bursa** *var.* Brussa, *prev.* Brusa; *anc.* Prusa. Bursa, NW Turkey 40°12′N 29°04′E
136 D12 **Bursa** *var.* Brusa, Brussa. ◆ *province* NW Turkey
75 Y9 **Būr Safājah** *var.* Būr Safājah. E Egypt 26°43′N 33°55′E
Būr Safājah *see* Būr Safājah
75 W7 **Būr Sa'īd** *var.* Port Said. N Egypt 31°17′N 32°18′E
81 O14 **Bur Tinle** Nugaal, C Somalia 07°50′N 48°01′E
31 Q5 **Burt Lake** ⊚ Michigan, N USA
118 H7 **Burtnieks** *var.* Burtnieks Ezers. ⊚ N Latvia
Burtnieks Ezers *see* Burtnieks
31 Q9 **Burton** Michigan, N USA 43°00′N 84°16′W
Burton on Trent *see* Burton upon Trent
97 M19 **Burton upon Trent** *var.* Burton on Trent, Burton-upon-Trent. C England, United Kingdom 52°48′N 01°36′W
93 J15 **Burträsk** Västerbotten, N Sweden 64°31′N 20°40′E
Burubaytal *see* Burybaytal
Burujird *see* Borūjerd
Burultokay *see* Fuhai
141 R15 **Burūm** SE Yemen 14°22′N 48°53′E
Burunday *see* Boralday
81 D21 **Burundi** *off.* Republic of Burundi; *prev.* Kingdom of Burundi, Urundi. ◆ *republic* C Africa
Burundi, Kingdom of *see* Burundi
Burundi, Republic of *see* Burundi
171 R13 **Buru, Pulau** *prev.* Boeroe. *island* E Indonesia
77 T17 **Burutu** Delta, S Nigeria 05°18′N 05°32′E
10 G7 **Burwash Landing** Yukon, W Canada 61°26′N 139°12′W
29 O14 **Burwell** Nebraska, C USA 41°46′N 99°04′W
97 L17 **Bury** NW England, United Kingdom 53°36′N 02°17′W
123 N13 **Buryatiya, Respublika** *prev.* Buryatskaya ASSR. ◆ *autonomous republic* S Russian Federation
Buryatskaya ASSR *see* Buryatiya, Respublika
145 S14 **Burybaytal** *prev.* Burubaytal. Zhambyl, SE Kazakhstan 44°56′N 73°59′E
117 S3 **Buryn'** Sums'ka Oblast', NE Ukraine 51°13′N 33°50′E
97 P20 **Bury St Edmunds** *hist.* Beodericsworth. E England, United Kingdom 52°15′N 00°43′E
Būrziya *see* Barzia
106 D9 **Busalla** Liguria, NW Italy 44°35′N 08°55′E
163 Z16 **Busan** *off.* Pusan-gwangyŏksi, *var.* Vusan; *prev.* Pusan, *Jap.* Fusan. SE South Korea 35°11′N 129°04′E
139 N5 **Buşayrah** Dayr az Zawr, E Syria 35°10′N 40°25′E
Buševa *see* Baba
143 N12 **Būshehr** *off.* Ostān-e Būshehr. ◆ *province* SW Iran
Būshehr/Bushire *see* Bandar-e Būshehr
Būshehr, Ostān-e *see* Būshehr
25 N2 **Bushland** Texas, SW USA 35°11′N 102°04′W
30 J12 **Bushnell** Illinois, N USA 40°33′N 90°30′W
Busi *see* Biševo
81 H18 **Busia** Western, W Kenya 0°28′N 34°07′E
81 G18 **Busia** SE Uganda 01°20′N 34°48′E
81 G18 **Busia** ◆ *county* W Kenya
Busiasch *see* Buziaş
79 K16 **Businga** Equateur, NW Dem. Rep. Congo 03°20′N 20°53′E
79 J18 **Busira** ≈ NW Dem. Rep. Congo
116 I5 **Bus'k** *Rus.* Busk. L'vivs'ka Oblast', W Ukraine 49°59′N 24°34′E
95 E14 **Buskerud** ◆ *county* S Norway
113 F14 **Buško Jezero** ⊠ SW Bosnia and Herzegovina
111 M15 **Busko-Zdrój** Świętokrzyskie, C Poland 50°28′N 20°44′E
Busra *see* Al Başrah, Iraq
Buşrá *see* Buşrá ash Shām, Syria
138 H9 **Buşrá ash Shām** *var.* Bosora, Bosra, Bozrah, Buşrá. Dar'ā, S Syria 32°31′N 36°29′E
180 I13 **Busselton** Western Australia 33°43′S 115°15′E
81 C14 **Busseri** ≈ W South Sudan
106 E9 **Busseto** Emilia-Romagna, C Italy 45°00′N 10°06′E
106 A8 **Bussoleno** Piemonte, NE Italy 45°11′N 07°07′E
Bussora *see* Al Başrah
41 N7 **Bustamante** Nuevo León, NE Mexico 26°29′N 100°30′W
63 I23 **Bustamante, Punta** *headland* S Argentina 51°35′S 68°58′W
Bustan *see* Bo'ston
116 J12 **Buşteni** Prahova, SE Romania 45°23′N 25°32′E
106 D7 **Busto Arsizio** Lombardia, N Italy 45°37′N 08°51′E
147 Q10 **Büston** *Rus.* Buston. NW Tajikistan 40°31′N 69°21′E
100 H8 **Büsum** Schleswig-Holstein, N Germany 54°08′N 08°52′E
98 J10 **Bussum** Noord-Holland, C Netherlands 52°17′N 05°10′E
79 M16 **Buta** Orientale, N Dem. Rep. Congo 02°50′N 24°41′E
81 E20 **Butare** *prev.* Astrida. S Rwanda 02°39′S 29°44′E
191 O2 **Butaritari** *atoll* Tungaru, W Kiribati
Butawal *see* Butwal
96 H13 **Bute** *cultural region* SW Scotland, United Kingdom
162 K6 **Büteeliyn Nuruu** ▲ N Mongolia
10 L16 **Bute Inlet** *fjord* British Columbia, W Canada
96 H12 **Bute, Island of** *island* SW Scotland, United Kingdom
79 P18 **Butembo** Nord-Kivu, NE Dem. Rep. Congo 0°09′N 29°17′E
Bütenpost *see* Buitenpost
107 K25 **Butera** Sicilia, Italy, C Mediterranean Sea 37°12′N 14°12′E
99 M20 **Bütgenbach** Liège, E Belgium 50°25′N 06°12′E
Butha Qi *see* Zalantun
166 J5 **Buthidaung** Rakhine State, W Myanmar (Burma) 20°50′N 92°25′E
61 I16 **Butiá** Rio Grande do Sul, S Brazil 30°09′S 51°55′W
81 F17 **Butiaba** NW Uganda 01°48′N 31°21′E
23 N6 **Butler** Alabama, S USA 32°05′N 88°13′W
23 S5 **Butler** Georgia, SE USA 32°33′N 84°14′W
31 Q11 **Butler** Indiana, N USA 41°25′N 84°52′W
27 R5 **Butler** Missouri, C USA 38°17′N 94°21′W
18 B14 **Butler** Pennsylvania, NE USA 40°51′N 79°52′W
194 K5 **Butler Island** *island* Antarctica
21 U8 **Butner** North Carolina, SE USA 36°07′N 78°45′W
171 P14 **Buton, Pulau** *var.* Pulau Butung; *prev.* Boetoeng. *island* C Indonesia
Bütow *see* Bytów
113 L23 **Butrintit, Liqeni i** ⊚ S Albania
23 N3 **Buttahatchee River** ≈ Alabama/Mississippi, S USA
33 Q10 **Butte** Montana, NW USA 46°01′N 112°33′W
29 O12 **Butte** Nebraska, C USA 42°54′N 98°51′W
168 J7 **Butterworth** Pinang, Peninsular Malaysia 05°24′N 100°22′E
83 J25 **Butterworth** *var.* Gcuwa. Eastern Cape, SE South Africa 32°20′S 28°09′E
13 O3 **Button Islands** *island group* Nunavut, NE Canada
35 R13 **Buttonwillow** California, W USA 35°24′N 119°26′W
171 Q7 **Butuan** *off.* Butuan City. Mindanao, S Philippines 08°57′N 125°33′E
Butuan City *see* Butuan
Butung, Pulau *see* Buton, Pulau
Butuntum *see* Bitonto
126 M8 **Buturlinovka** Voronezhskaya Oblast', W Russian Federation 50°48′N 40°33′E
153 O11 **Butwal** *var.* Butawal. Western, C Nepal 27°41′N 83°28′E
101 G17 **Butzbach** Hessen, W Germany 50°26′N 08°40′E
100 L9 **Bützow** Mecklenburg-Vorpommern, N Germany 53°49′N 11°58′E
80 N13 **Buuhoodle** Togdheer, N Somalia 08°18′N 46°15′E
81 N16 **Buulobarde** *var.* Buulo Berde. Hiiraan, C Somalia 03°52′N 45°37′E
Buulo Berde *see* Buulobarde
80 P12 **Buuraha Cal Miskaat** ▲ NE Somalia
81 L19 **Buur Gaabo** Jubbada Hoose, S Somalia 01°14′S 41°48′E
99 M22 **Buurgplaatz** ▲ N Luxembourg 50°09′N 06°02′E
162 H8 **Buutsagaan** *var.* Buyant. Bayanhongor, C Mongolia 46°07′N 98°45′E
Buwayrāt al Hasūn *see* Bu'ayrāt al Ḩasūn
146 L11 **Buxoro** *var.* Bokhara, *Rus.* Bukhara. Buxoro Viloyati, C Uzbekistan 39°51′N 64°23′E
146 J11 **Buxoro Viloyati** *Rus.* Bukharskaya Oblast'. ◆ *province* C Uzbekistan
100 I10 **Buxtehude** Niedersachsen, NW Germany 53°29′N 09°42′E
97 L18 **Buxton** C England, United Kingdom 53°18′N 01°52′W
124 M14 **Buy** *var.* Buj. Kostromskaya Oblast', NW Russian Federation 58°27′N 41°31′E
Buyanbat *see* Jargalant
162 D6 **Buyant** Bayan-Ölgiy, W Mongolia 48°31′N 89°36′E
Buyant *see* Buutsagaan, Bayanhongor, Mongolia
Buyant *see* Otgon, Dzavhan, Mongolia
Buyant *see* Galshar, Hentiy, Mongolia
163 N10 **Buyant-Uhaa** Dornogovĭ, SE Mongolia 44°52′N 110°12′E
162 M7 **Buyant Ukha** ✈ (Ulaanbaatar) Töv, N Mongolia
127 Q16 **Buynaksk** Respublika Dagestan, SW Russian Federation 42°53′N 47°03′E
119 L20 **Buynavichy** *Rus.* Buynovichi. Homyel'skaya Voblasts', SE Belarus 51°52′N 28°33′E
Buynovichi *see* Buynavichy
76 L16 **Buyo** SW Ivory Coast 06°16′N 07°03′W
76 L16 **Buyo, Lac de** ⊠ W Ivory Coast
163 R7 **Buyr Nuur** *var.* Buir Nur. ⊚ China/Mongolia *see also* Buir Nur
Buyr Nuur *see* Buir Nur
137 T13 **Büyükağrı Dağı** *var.* Aghri Dagh, Agri Dagi, Koh I Noh, Masis, *Eng.* Great Ararat, Mount Ararat. ▲ E Turkey 39°43′N 44°19′E
137 R15 **Büyük Çayı** ≈ NE Turkey
114 O13 **Büyük Çekmece** İstanbul, NW Turkey 41°02′N 28°35′E
114 N12 **Büyükkarıştıran** Kırklareli, NW Turkey 41°17′N 27°33′E
115 L14 **Büyükkemikli Burnu** *cape* NW Turkey
136 E15 **Büyükmenderes Nehri** ≈ SW Turkey
Büyükzap Suyu *see* Great Zab
102 M9 **Buzançais** Indre, C France 46°53′N 01°25′E
116 K13 **Buzău** Buzău, SE Romania 45°08′N 26°51′E
116 K13 **Buzău** ◆ *county* SE Romania
116 L12 **Buzău** ≈ E Romania
75 S11 **Buzaymah** *var.* Bzīmah. SE Libya 24°53′N 22°01′E
164 E13 **Buzen** Fukuoka, Kyūshū, SW Japan 33°37′N 131°06′E
116 F12 **Buziaş** *Ger.* Busiasch, *Hung.* Buziásfürdő; *prev.* Buziás. Timiş, W Romania 45°38′N 21°36′E
Buziás *see* Buziaş
Buziásfürdő *see* Buziaş
83 M18 **Búzi, Rio** ≈ C Mozambique
117 Q10 **Buz'kyy Lyman** *bay* S Ukraine
Büzmeyin *see* Abadan
127 T6 **Buzuluk** Orenburgskaya Oblast', W Russian Federation 52°47′N 52°16′E
127 N8 **Buzuluk** ≈ SW Russian Federation
Buzuluk *see* Buzylyk
145 O8 **Buzylyk** *prev.* Buzuluk. Akmola, C Kazakhstan 51°53′N 66°09′E
19 P12 **Buzzards Bay** Massachusetts, NE USA 41°45′N 70°37′W
19 P13 **Buzzards Bay** *bay* Massachusetts, NE USA
83 G16 **Bwabata** Caprivi, NE Namibia 17°52′S 22°39′E
186 H10 **Bwagaoia** Misima Island, SE Papua New Guinea 10°39′S 152°48′E
Bwake *see* Bouaké
187 R13 **Bwatnapne** Pentecost, C Vanuatu 15°42′S 168°07′E
119 K14 **Byahoml'** *Rus.* Begoml'. Vitsyebskaya Voblasts', N Belarus 54°44′N 28°04′E
114 K8 **Byala** Ruse, N Bulgaria 43°27′N 25°44′E
114 N9 **Byala** *prev.* Ak-Dere. Varna, E Bulgaria 42°52′N 27°53′E
Byala Reka *see* Erythropótamos
114 H8 **Byala Slatina** Vratsa, NW Bulgaria 43°28′N 23°56′E
119 N15 **Byalynichy** *Rus.* Belynichi. Mahilyowskaya Voblasts', E Belarus 54°00′N 29°42′E
Byan Tumen *see* Choybalsan
119 L14 **Byarezina** *prev.* Byerezino, *Rus.* Berezina. ≈ C Belarus
119 G19 **Byaroza** *Pol.* Bereza Kartuska, *Rus.* Berëza. Brestskaya Voblasts', SW Belarus 52°32′N 24°59′E
119 H16 **Byarozawka** *Rus.* Berëzovka. Hrodzyenskaya Voblasts', W Belarus 53°45′N 25°30′E
Bybles *see* Jbail
111 O14 **Bychawa** Lubelskie, SE Poland 51°06′N 22°34′E
Bychikha *see* Bychykha
118 N11 **Bychykha** *Rus.* Bychikha. Vitsyebskaya Voblasts', NE Belarus 55°41′N 29°59′E
111 I14 **Byczyna** *Ger.* Pitschen. Opolskie, S Poland 51°06′N 18°13′E
110 I10 **Bydgoszcz** *Ger.* Bromberg. Kujawski-pomorskie, C Poland 53°06′N 18°00′E
119 H19 **Byelaazyorsk** *Rus.* Beloozersk. Brestskaya Voblasts', SW Belarus 52°28′N 25°10′E
119 I17 **Byelaruskaya Hrada** *Rus.* Belorusskaya Gryada. *ridge* N Belarus
119 G18 **Byelavyezhskaya Pushcha** *Pol.* Puszcza Białowieska, *Rus.* Belovezhskaya Pushcha. *forest* Belarus/Poland *see also* Białowieska, Puszcza
Byelavyezhskaya, Pushcha *see* Białowieska, Puszcza
119 H15 **Byenyakoni** *Rus.* Benyakoni. Hrodzyenskaya Voblasts', W Belarus 54°15′N 25°22′E
119 M16 **Byerazino** *Rus.* Berezino. Minskaya Voblasts', C Belarus 53°50′N 29°00′E
118 L13 **Byerazino** *Rus.* Berezino. Vitsyebskaya Voblasts', N Belarus 54°54′N 28°12′E
Byerezino *see* Byarezina
118 M13 **Byeshankovichy** *Rus.* Beshenkovichi. Vitsyebskaya Voblasts', N Belarus 55°03′N 29°27′E
31 U13 **Byesville** Ohio, N USA 39°58′N 81°32′W
119 P18 **Byesyedz'** *Rus.* Besed'. ≈ SE Belarus
119 H19 **Byezdzyezh** *Rus.* Bezdezh. Brestskaya Voblasts', SW Belarus 52°19′N 25°18′E
93 J15 **Bygdeå** Västerbotten, N Sweden 64°03′N 20°49′E
94 F12 **Bygdin** ⊚ S Norway
93 J15 **Bygdsiljum** Västerbotten, N Sweden 64°20′N 20°31′E
95 E17 **Bygland** Aust-Agder, S Norway 58°46′N 07°50′E
95 E17 **Byglandsfjord** Aust-Agder, S Norway 58°42′N 07°51′E
119 N16 **Bykhaw** *Rus.* Bykhov. Mahilyowskaya Voblasts', E Belarus 53°31′N 30°15′E
Bykhov *see* Bykhaw
127 P9 **Bykovo** Volgogradskaya Oblast', SW Russian Federation 49°52′N 45°24′E
123 P7 **Bykovskiy** Respublika Sakha (Yakutiya), NE Russian Federation 71°57′N 129°07′E
195 R12 **Byrd Glacier** *glacier* Antarctica
14 K10 **Byrd, Lac** ⊚ Québec, SE Canada
183 P5 **Byrock** New South Wales, SE Australia 30°40′S 146°24′E
30 L10 **Byron** Illinois, N USA 42°06′N 89°15′W
183 V4 **Byron Bay** New South Wales, SE Australia 28°39′S 153°34′E
183 V4 **Byron, Cape** *headland* New South Wales, E Australia 28°37′S 153°40′E
63 F21 **Byron, Isla** *island* S Chile
Byron Island *see* Nikunau
65 B24 **Byron Sound** *sound* NW Falkland Islands
122 M6 **Byrranga, Gory** ▲ N Russian Federation
93 J14 **Byske** Västerbotten, N Sweden 64°58′N 21°10′E
111 K18 **Bystrá** ▲ N Slovakia 49°10′N 19°49′E
111 F18 **Bystřice nad Pernštejnem** *Ger.* Bistritz ober Pernstein. Vysočina, C Czech Republic 49°32′N 16°16′E
Bystrovka *see* Kemin
111 G16 **Bystrzyca Kłodzka** *Ger.* Habelschwerdt. Wałbrzych, SW Poland 50°19′N 16°39′E
111 I18 **Bytča** Žilinský Kraj, N Slovakia 49°15′N 18°32′E
119 L15 **Bytcha** Minskaya Voblasts', NE Belarus 54°19′N 28°24′E
Byteń/Byten' *see* Bytsyen'
111 J16 **Bytom** *Ger.* Beuthen. Śląskie, S Poland 50°21′N 18°51′E
110 H7 **Bytów** *Ger.* Bütow. Pomorskie, N Poland 54°10′N 17°30′E
119 H18 **Bytsyen'** *Pol.* Byteń, *Rus.* Byten'. Brestskaya Voblasts', SW Belarus 52°53′N 25°30′E
81 E19 **Byumba** *var.* Biumba. N Rwanda 01°37′S 30°05′E
Byuzmeyin *see* Abadan
119 O20 **Byval'ki** Homyel'skaya Voblasts', SE Belarus 51°51′N 30°38′E
95 O20 **Byxelkrok** Kalmar, S Sweden 57°18′N 17°01′E
Byzantium *see* İstanbul
Bzīmah *see* Buzaymah

C

62 O6 **Caacupé** Cordillera, S Paraguay 25°23′S 57°05′W
62 P6 **Caaguazú** *off.* Departamento de Caaguazú. ◆ *department* C Paraguay
Caaguazú, Departamento de *see* Caaguazú
82 C13 **Caála** *var.* Kaala, Robert Williams, *Port.* Vila Robert Williams. Huambo, C Angola 12°51′S 15°33′E
62 P7 **Caazapá** Caazapá, S Paraguay 26°09′S 56°21′W
62 P7 **Caazapá** *off.* Departamento de Caazapá. ◆ *department* SE Paraguay
Caazapá, Departamento de *see* Caazapá
81 P15 **Cabaad, Raas** *headland* C Somalia 06°13′N 49°01′E
55 N10 **Cabadisocaña** Amazonas, S Venezuela 04°28′N 64°45′W
44 F5 **Cabaiguán** Sancti Spíritus, C Cuba 22°04′N 79°32′W
Caballeria, Cabo *see* Cavalleria, Cap de
37 Q14 **Caballo Reservoir** ⊠ New Mexico, SW USA
40 L6 **Caballos Mesteños, Llano de los** *plain* N Mexico
104 L2 **Cabanaquinta** Asturias, N Spain 43°10′N 05°37′W
42 B9 **Cabañas** ◆ *department* E El Salvador
171 O3 **Cabanatuan** *off.* Cabanatuan City. Luzon, N Philippines 15°27′N 120°57′E
Cabanatuan City *see* Cabanatuan
15 T8 **Cabano** Québec, SE Canada 47°40′N 68°56′W
104 L11 **Cabeza del Buey** Extremadura, W Spain 38°44′N 05°13′W
45 V5 **Cabezas de San Juan** *headland* E Puerto Rico 18°23′N 65°37′W
105 N2 **Cabezón de la Sal** Cantabria, N Spain 43°19′N 04°14′W
Cabhán *see* Cavan
61 B23 **Cabildo** Buenos Aires, E Argentina 38°28′S 61°50′W
Cabillonum *see* Chalon-sur-Saône
54 H5 **Cabimas** Zulia, NW Venezuela 10°26′N 71°27′W
82 A9 **Cabinda** *var.* Kabinda. Cabinda, NW Angola 05°34′S 12°12′E
82 A9 **Cabinda** *var.* Kabinda. ◆ *province* NW Angola
33 N7 **Cabinet Mountains** ▲ Idaho/Montana, NW USA
82 B11 **Cabiri** Bengo, NW Angola 08°50′S 13°42′E
63 J20 **Cabo Blanco** Santa Cruz, SE Argentina 47°13′S 65°43′W
82 P13 **Cabo Delgado** *off.* Província de Cabo Delgado. ◆ *province* NE Mozambique
14 L9 **Cabonga, Réservoir** ⊠ Québec, SE Canada
27 V7 **Cabool** Missouri, C USA 37°07′N 92°06′W
183 V2 **Caboolture** Queensland, E Australia 27°05′S 152°50′E
Cabora Bassa, Lake *see* Cahora Bassa, Albufeira de
40 F3 **Caborca** Sonora, NW Mexico 30°44′N 112°06′W
Cabo San Lucas *see* San Lucas
27 V11 **Cabot** Arkansas, C USA 34°58′N 92°01′W
14 F12 **Cabot Head** *headland* Ontario, S Canada 45°13′N 81°17′W
13 R13 **Cabot Strait** *strait* E Canada
Cabo Verde, Ilhas do *see* Cape Verde
104 M14 **Cabra** Andalucía, S Spain 37°28′N 04°28′W
107 B19 **Cabras** Sardegna, Italy, C Mediterranean Sea 39°55′N 08°30′E
188 A15 **Cabras Island** *island* W Guam
45 O8 **Cabrera** N Dominican Republic 19°40′N 69°54′W
104 J4 **Cabrera** ≈ NW Spain
105 X10 **Cabrera, Illa de** *anc.* Capraria. *island* Islas Baleares, Spain, W Mediterranean Sea
105 Q15 **Cabrera, Sierra** ▲ S Spain
11 S16 **Cabri** Saskatchewan, S Canada 50°38′N 108°28′W
105 R10 **Cabriel** ≈ E Spain
54 M7 **Cabruta** Guárico, C Venezuela 07°39′N 66°19′W
171 N2 **Cabugao** Luzon, N Philippines 17°55′N 120°29′E
54 G10 **Cabuyaro** Meta, C Colombia 04°21′N 72°47′W
60 I13 **Caçador** Santa Catarina, S Brazil 26°47′S 51°00′W
42 G8 **Cacaguatique, Cordillera** *var.* Cordillera. ▲ NE El Salvador
112 L13 **Čačak** Serbia, C Serbia 43°52′N 20°23′E
55 Y10 **Cacao** NE French Guiana 04°37′N 52°29′W
61 H16 **Caçapava do Sul** Rio Grande do Sul, S Brazil 30°28′S 53°29′W
21 U3 **Cacapon River** ≈ West Virginia, NE USA
107 J23 **Caccamo** Sicilia, Italy, C Mediterranean Sea 37°56′N 13°40′E
107 A17 **Caccia, Capo** *headland* Sardegna, Italy, C Mediterranean Sea 40°34′N 08°09′E
146 H15 **Çäçe** *var.* Chäche, *Rus.* Chaacha. Ahal Welaýaty, S Turkmenistan 36°49′N 60°33′E
59 G18 **Cáceres** Mato Grosso, W Brazil 16°05′S 57°40′W
104 J10 **Cáceres** *Ar.* Qazris. Extremadura, W Spain 39°29′N 06°23′W
104 J9 **Cáceres** ◆ *province* Extremadura, W Spain
Cachacrou *see* Scotts Head Village
61 C21 **Cacharí** Buenos Aires, E Argentina 36°24′S 59°32′W
26 L12 **Cache** Oklahoma, C USA 34°37′N 98°37′W
10 M16 **Cache Creek** British Columbia, SW Canada 50°49′N 121°20′W
35 N6 **Cache Creek** ≈ California, W USA
37 S3 **Cache La Poudre River** ≈ Colorado, C USA
Cacheo *see* Cacheu
27 W11 **Cache River** ≈ Arkansas, C USA
30 L17 **Cache River** ≈ Illinois, N USA
76 G12 **Cacheu** *var.* Cacheo. W Guinea-Bissau 12°12′N 16°10′W
59 I15 **Cachimbo** Pará, NE Brazil 09°21′S 54°58′W
59 H15 **Cachimbo, Serra do** ▲ C Brazil
82 D13 **Cachingues** Bié, C Angola 13°05′S 16°48′E
54 G7 **Cáchira** Norte de Santander, N Colombia 07°44′N 73°07′W
61 H16 **Cachoeira do Sul** Rio Grande do Sul, S Brazil 29°58′S 52°54′W
59 O20 **Cachoeiro de Itapemirim** Espírito Santo, SE Brazil 20°51′S 41°07′W
82 E12 **Cacolo** Lunda Sul, NE Angola 10°09′S 19°21′E
83 C14 **Caconda** Huíla, C Angola 13°43′S 15°03′E
82 A9 **Cacongo** Cabinda, NW Angola 05°13′S 12°08′E
35 U9 **Cactus Peak** ▲ Nevada, W USA 37°42′N 116°51′W
82 A11 **Cacuaco** Luanda, NW Angola 08°47′S 13°21′E
82 B14 **Cacula** Huíla, SW Angola 14°33′S 14°04′E
82 R12 **Caculuvar** ≈ SW Angola
59 O19 **Caçumba, Ilha** *island* SE Brazil
55 N10 **Cacuri** Amazonas, S Venezuela
81 N17 **Cadale** Shabeellaha Dhexe, E Somalia 02°48′N 46°19′E
105 X4 **Cadaqués** Cataluña, NE Spain 42°17′N 03°16′E
111 J18 **Čadca** *Hung.* Csaca. Žilinský Kraj, N Slovakia 49°27′N 18°46′E
27 P13 **Caddo** Oklahoma, C USA 34°07′N 96°15′W
25 R6 **Caddo** Texas, SW USA 32°42′N 98°40′W
25 X6 **Caddo Lake** ⊠ Louisiana/Texas, SW USA
27 S12 **Caddo Mountains** ▲ Arkansas, C USA
41 O8 **Cadereyta** Nuevo León, NE Mexico 25°35′N 99°54′W
97 J19 **Cader Idris** ▲ NW Wales, United Kingdom 52°43′N 03°57′W
182 F3 **Cadibarrawirracanna, Lake** *salt lake* South Australia
14 I7 **Cadillac** Québec, SE Canada 48°12′N 78°23′W
11 T17 **Cadillac** Saskatchewan, S Canada 49°43′N 107°41′W
102 K13 **Cadillac** Gironde, SW France 44°37′N 00°16′W
31 P7 **Cadillac** Michigan, N USA 44°15′N 85°23′W
105 V4 **Cadí, Toretta de** *prev.* Torre de Cadí. ▲ NE Spain 42°16′N 01°38′E
Torre de Cadí *see* Cadí, Toretta de
171 P5 **Cadiz** *off.* Cadiz City. Negros, C Philippines 10°58′N 123°18′E
104 J15 **Cádiz** *anc.* Gades, Gadier, Gadir, Gadire. Andalucía, SW Spain 36°32′N 06°18′W
20 H7 **Cadiz** Kentucky, S USA 36°52′N 87°50′W
31 U13 **Cadiz** Ohio, N USA 40°16′N 81°00′W
104 K15 **Cádiz** ◆ *province* Andalucía, SW Spain
104 I15 **Cádiz, Bahía de** *bay* SW Spain
Cadiz City *see* Cadiz
104 H15 **Cádiz, Golfo de** *Eng.* Gulf of Cadiz. *gulf* Portugal/Spain
Cadiz, Gulf of *see* Cádiz, Golfo de
35 X14 **Cadiz Lake** ⊚ California, W USA

◆ Country ● Country Capital ◇ Dependent Territory ○ Dependent Territory Capital ◆ Administrative Regions ✈ International Airport ▲ Mountain ▲ Mountain Range Volcano ≈ River ⊚ Lake ⊠ Reservoir

182 E2 Cadney Homestead South Australia 27°52´S 134°03´E
Cadurcum see Cahors
Caecae see Xaixai
102 K4 Caen Calvados, N France 49°10´N 00°20´W
Caene/Caenepolis see Qinā
Caerdydd see Cardiff
Caer Glou see Gloucester
Caer Gybi see Holyhead
Caer Luel see Carlisle
97 I18 Caernarfon var. Carnarvon, Caernarvon. NW Wales, United Kingdom 53°08´N 04°16´W
97 H18 Caernarfon Bay bay NW Wales, United Kingdom
97 I19 Caernarvon var. Carnarvon NW Wales, United Kingdom
Caernarvon see Caernarfon
Caesaraugusta see Zaragoza
Caesarea Mazaca see Kayseri
Caesaroriga see Talavera de la Reina
Caesarodunum see Tours
Caesaromagus see Beauvais
Caesena see Cesena
59 N17 Caetité Bahia, E Brazil 14°04´S 42°29´W
62 J6 Cafayate Salta, N Argentina 26°02´S 66°00´W
171 O2 Cagayan ≈ Luzon, N Philippines
171 Q7 Cagayan de Oro off. Cagayan de Oro City. Mindanao, S Philippines 08°29´N 124°38´E
Cagayan de Oro City see Cagayan de Oro
170 M8 Cagayan de Tawi Tawi island S Philippines
171 N6 Cagayan Islands island group C Philippines
31 O14 Cagles Mill Lake ⊠ Indiana, N USA
106 I12 Cagli Marche, C Italy 43°33´N 12°39´E
107 C20 Cagliari anc. Caralis. Sardegna, Italy, C Mediterranean Sea 39°15´N 09°06´E
107 C20 Cagliari, Golfo di gulf Sardegna, Italy, C Mediterranean Sea
103 U15 Cagnes-sur-Mer Alpes-Maritimes, SE France 43°40´N 07°09´E
54 L5 Cagua Aragua, N Venezuela 10°09´N 67°27´W
171 O1 Cagua, Mount ▲ Luzon, N Philippines 18°10´N 122°03´E
54 F13 Caguán, Río ≈ SW Colombia
45 U6 Caguas E Puerto Rico 18°14´N 66°02´W
146 C9 Çagyl Rus. Chagyl. Balkan Welaýaty, NW Turkmenistan 40°48´N 55°21´E
23 P5 Cahaba River ≈ Alabama, S USA
42 E5 Cahabón, Río ≈ C Guatemala
83 B15 Cahama Cunene, SW Angola 16°16´S 14°23´E
97 B21 Caha Mountains Ir. An Cheacha. ▲ SW Ireland
97 D20 Caher Ir. An Cathair. S Ireland 52°21´N 07°58´W
97 A21 Caherciveen Ir. Cathair Saidhbhín. SW Ireland 51°56´N 10°12´W
30 K15 Cahokia Illinois, N USA 38°34´N 90°11´W
83 L15 Cahora Bassa, Albufeira de var. Lake Cabora Bassa. ⊠ NW Mozambique
97 G20 Cahore Point Ir. Rinn Chathóir. headland SE Ireland 52°33´N 06°11´W
102 M14 Cahors anc. Cadurcum. Lot, S France 44°26´N 01°27´E
56 D9 Cahuapanas, Río ≈ N Peru
116 M12 Cahul Rus. Kagul. S Moldova 45°53´N 28°13´E
Cahul, Lacul see Kahul, Ozero
83 N16 Caia Sofala, C Mozambique 17°50´S 35°21´E
59 J19 Caiapó, Serra do ▲ C Brazil
44 F5 Caibarién Villa Clara, C Cuba 22°31´N 79°29´W
55 O5 Caicara Monagas, NE Venezuela 09°52´N 63°38´W
54 L5 Caicara del Orinoco Bolívar, C Venezuela 07°38´N 66°10´W
59 P14 Caicó Rio Grande do Norte, E Brazil 06°25´S 37°04´W
44 M6 Caicos Islands island group W Turks and Caicos Islands
44 L5 Caicos Passage strait The Bahamas/Turks and Caicos Islands
161 O9 Caidian prev. Hanyang. Hubei, C China 30°37´N 114°02´E
Caiffa see Hefa
180 M12 Caiguna Western Australia 32°14´S 125°33´E
40 J11 Caimanero, Laguna del var. Laguna del Camaronero. lagoon E Pacific Ocean
117 N10 Căinari Rus. Kaynary. C Moldova 46°43´N 29°00´E
57 L19 Caine, Río ≈ C Bolivia
195 N4 Caiphas see Hefa
Caird Coast physical region Antarctica
96 J9 Cairn Gorm ▲ C Scotland, United Kingdom 57°07´N 03°38´W
Cairngorm Mountains ▲ C Scotland, United Kingdom
39 P12 Cairn Mountain ▲ Alaska, USA 61°07´N 155°23´W
181 W4 Cairns Queensland, NE Australia 16°51´S 145°43´E
121 V13 Cairo var. El Qāhira, Ar. Al Qāhirah. ● (Egypt) N Egypt 30°01´N 31°18´E
23 T8 Cairo Georgia, SE USA 30°52´N 84°12´W
30 L16 Cairo Illinois, N USA 37°00´N 89°11´W
75 V8 Cairo ✈ C Egypt 30°06´N 31°30´E
Caiseal see Cashel
Caisleán an Bharraigh see Castlebar
Caisleán na Finne see ...
96 J6 Caithness cultural region N Scotland, United Kingdom
83 D15 Caiundo Kuando Kubango, S Angola 15°43´S 17°28´E
56 C7 Cajamarca prev. Caxamarca. Cajamarca, NW Peru 07°09´S 78°32´W

56 B11 Cajamarca off. Departamento de Cajamarca. ◆ department N Peru
Cajamarca, Departamento de see Cajamarca
103 N14 Cajarc Lot, S France 44°28´N 01°51´E
42 G6 Cajón, Represa El ⊠ NW Honduras
58 N12 Caju, Ilha do island NE Brazil
159 R10 Caka Yanhu ⊙ C China
112 E7 Čakovec Ger. Csakathurn, Hung. Csáktornya; prev. Ger. Tschakathurn. Medimurje, N Croatia 46°24´N 16°29´E
77 V17 Calabar Cross River, S Nigeria 04°56´N 08°25´E
14 K13 Calabogie Ontario, SE Canada 45°18´N 76°46´W
54 L6 Calabozo Guárico, C Venezuela 08°58´N 67°28´W
107 N20 Calabria ◆ region SW Italy
104 M16 Calaborra, Punta de headland S Spain 36°30´N 04°38´W
116 G14 Calafat Dolj, SW Romania 43°59´N 22°57´E
Calafate see El Calafate
105 Q4 Calahorra La Rioja, N Spain 42°19´N 01°58´W
103 N1 Calais Pas-de-Calais, N France 51°N 01°54´E
19 T5 Calais Maine, NE USA 45°09´N 67°15´W
Calais, Pas de see Dover, Strait of
Calalen see Kallalen
62 H4 Calama Antofagasta, N Chile 22°26´S 68°54´W
Calamianes see Calamian Group
170 M5 Calamian Group var. Calamianes. island group W Philippines
105 R7 Calamocha Aragón, NE Spain 40°54´N 01°18´W
29 N14 Calamus River ≈ Nebraska, C USA
116 G12 Cālan Ger. Kalan, Hung. Pusztakalán. Hunedoara, SW Romania 45°45´N 22°59´E
105 S7 Calanda Aragón, NE Spain 40°56´N 00°15´W
168 F9 Calang Sumatera, W Indonesia 04°37´N 95°37´E
171 N4 Calapan Mindoro, N Philippines 13°24´N 121°08´E
116 M9 Calaras var. Călăras, Rus. Kalarash. C Moldova 47°19´N 28°13´E
116 L13 Călăraşi Călăraşi, SE Romania 44°11´N 26°52´E
116 K14 Călăraşi ◆ county SE Romania
54 E10 Calarca Quindío, W Colombia 04°31´N 75°38´W
105 Q12 Calasparra Murcia, SE Spain 38°14´N 01°41´W
107 I23 Calatafimi Sicilia, Italy, C Mediterranean Sea 37°54´N 12°52´E
105 Q6 Calatayud Aragón, NE Spain 41°21´N 01°39´W
171 O4 Calauag Luzon, N Philippines 13°57´N 122°18´E
35 N8 Calaveras River ≈ California, W USA
171 N4 Calavite, Cape headland Mindoro, N Philippines 13°25´N 120°16´E
171 Q8 Calbayog off. Calbayog City. Samar, C Philippines 12°08´N 124°36´E
Calbayog City see Calbayog
22 Q9 Calcasieu Lake ⊙ Louisiana, S USA
22 H8 Calcasieu River ≈ Louisiana, S USA
56 B6 Calceta Manabí, W Ecuador 00°51´S 80°07´W
61 B16 Calchaquí Santa Fe, C Argentina 29°56´S 63°00´W
62 J6 Calchaquí, Río ≈ NW Argentina
58 J10 Calçoene Amapá, N Brazil 02°28´N 50°55´W
153 S16 Calcutta var. Kolkata. West Bengal, NE India 22°30´N 88°20´E see also Kolkata
Calcutta see Kolkata
54 E6 Caldas off. Departamento de Caldas. ◆ province W Colombia
104 F10 Caldas da Rainha Leiria, W Portugal 39°24´N 09°08´W
Caldas, Departamento de see Caldas
104 G3 Caldas de Reis var. Caldas de Reyes. Galicia, NW Spain 42°36´N 08°38´W
Caldas de Reyes see Caldas de Reis
58 F13 Caldeirão Amazonas, NW Brazil 03°18´S 60°22´W
62 G7 Caldera Atacama, N Chile 27°03´S 70°51´W
42 L14 Caldera Puntarenas, W Costa Rica 09°55´N 84°51´W
105 N10 Calderina ▲ C Spain
137 T13 Çaldıran Van, E Turkey 39°09´N 43°54´E
32 M14 Caldwell Idaho, NW USA 43°40´N 116°41´W
27 N8 Caldwell Kansas, C USA 37°02´N 97°36´W
14 G15 Caledon Ontario, S Canada 43°51´N 79°58´W
83 I23 Caledon var. Mohokare. ≈ Lesotho/South Africa
42 G1 Caledonia Corozal, N Belize 18°14´N 88°29´W
14 F16 Caledonia Ontario, S Canada 43°04´N 79°57´W
29 X11 Caledonia Minnesota, N USA 43°37´N 91°30´W
105 X5 Calella var. Calella de la Costa. Cataluña, NE Spain 41°37´N 02°40´E
Calella de la Costa see Calella
23 P4 Calera Alabama, S USA 33°06´N 86°45´W
63 I19 Caleta Olivia Santa Cruz, SE Argentina 46°21´S 67°37´W
35 X17 Calexico California, USA 32°40´N 115°28´W
97 H16 Calf of Man island SW Isle of Man
11 Q16 Calgary Alberta, SW Canada 51°05´N 114°05´W
11 Q16 Calgary ✈ Alberta, SW Canada 51°07´N 114°03´W
37 W8 Calhan Colorado, C USA 39°00´N 104°18´W
24 M4 Calhoun Georgia, SE USA 34°30´N 84°56´W

20 I6 Calhoun Kentucky, S USA 37°32´N 87°15´W
22 J2 Calhoun City Mississippi, S USA 33°51´N 89°18´W
21 P12 Calhoun Falls South Carolina, SE USA 34°05´N 82°36´W
54 D11 Cali Valle del Cauca, W Colombia 03°24´N 76°30´W
27 V9 Calico Rock Arkansas, C USA 36°07´N 92°08´W
155 F21 Calicut var. Kozhikode. Kerala, SW India 11°17´N 75°49´E see also Kozhikode
35 R9 Caliente Nevada, W USA 37°37´N 114°30´W
27 U5 California Missouri, C USA 38°39´N 92°35´W
18 B15 California Pennsylvania, NE USA 40°02´N 79°52´W
35 Q12 California off. State of California, also known as El Dorado, The Golden State. ◆ state W USA
35 P11 California Aqueduct aqueduct California, W USA
35 T13 California City California, W USA 35°06´N 117°55´W
40 F6 California, Golfo de Eng. Gulf of California; prev. Sea of Cortez. gulf W Mexico
California, Gulf of see California, Golfo de
137 Y13 Cälilabad var. Dzhalilabad; prev. Astrakhan-Bazar. S Azerbaijan 39°15´N 48°30´E
116 L12 Cälimänesti Vâlcea, SW Romania 45°14´N 24°20´E
116 J9 Cälimani, Muntii ▲ N Romania
Calinisc see Cupcina
35 X17 Calipatria California, USA 33°07´N 115°30´W
34 M7 Calistoga California, W USA 38°34´N 122°37´W
83 G25 Calitzdorp Western Cape, SW South Africa 33°32´S 21°41´E
41 W12 Calkini Campeche, E Mexico 20°21´N 90°03´W
182 K4 Callabonna Creek var. Tilcha Creek. seasonal river New South Wales/South Australia
182 J4 Callabonna, Lake ⊙ South Australia
102 I6 Callac Côtes d'Armor, NW France 48°28´N 03°22´W
35 U5 Callaghan, Mount ▲ Nevada, W USA 39°38´N 116°52´W
Callain see Callan
97 E19 Callan Ir. Callainn. S Ireland 52°33´N 07°23´W
14 H11 Callander Ontario, S Canada 46°14´N 79°21´W
96 I11 Callander C Scotland, United Kingdom 56°15´N 04°16´W
98 H7 Callantsoog Noord-Holland, NW Netherlands 52°51´N 04°41´E
57 D14 Callao Callao, W Peru 12°03´S 77°10´W
57 D15 Callao dep. Departamento del Callao. ◆ constitutional province W Peru
Callao, Departamento del see Callao
58 F11 Callaria, Río ≈ E Peru
Callatis see Mangalia
11 Q13 Calling Lake Alberta, W Canada 55°12´N 113°07´W
32 M12 Callison Idaho, NW USA 44°34´N 116°42´W
105 T11 Callosa d'En Sarrià var. Callosa de Ensarriá. Valenciana, E Spain 38°40´N 00°08´W
105 S12 Callosa de Segura Valenciana, E Spain 38°07´N 00°53´W
29 X11 Calmar Iowa, C USA 43°10´N 91°51´W
Calmar see Kalmar
43 S14 Calobre Veraguas, C Panama 08°18´N 80°49´W
23 X14 Caloosahatchee River ≈ Florida, SE USA
183 V1 Caloundra Queensland, E Australia 26°48´S 153°08´E
Calp see Calpe
105 O20 Calpe Cat. Calp. Valenciana, E Spain 38°39´N 00°03´E
41 P14 Calpulalpan Tlaxcala, S Mexico 19°36´N 98°28´W
107 K25 Caltagirone Sicilia, Italy, C Mediterranean Sea 37°14´N 14°31´E
107 K25 Caltanissetta Sicilia, Italy, C Mediterranean Sea 37°30´N 14°01´E
83 I16 Caluango Lunda Norte, NE Angola 08°16´S 19°36´E
82 C12 Calucinga Bié, C Angola 11°18´S 16°12´E
82 B14 Calulo Kwanza Sul, NW Angola 09°58´S 14°56´E
80 Q11 Caluula Bari, NE Somalia 11°55´N 50°52´E
42 J6 Calvados ◆ department N France
186 I10 Calvados Chain, The island group SE Papua New Guinea
27 N8 Calvert Texas, SW USA 30°58´N 96°40´W
21 R12 Calvert City Kentucky, S USA 37°01´N 88°21´W
103 X14 Calvi Corse, France, C Mediterranean Sea 42°34´N 08°44´E
40 L12 Calvillo Aguascalientes, C Mexico 21°51´N 102°18´W
83 F24 Calvinia Northern Cape, W South Africa 31°25´S 19°47´E
104 K8 Calvitero ▲ W Spain
101 G22 Calw Baden-Württemberg, SW Germany 48°43´N 08°43´E
Calydon see Kalýdon
105 N11 Calzada de Calatrava Castilla-La Mancha, C Spain 38°42´N 03°46´W
Cama see Kama
186 I10 Cama ...
82 C11 Camabatela Kwanza Norte, NW Angola 08°13´S 15°23´E
64 Q5 Camacha Porto Santo, Madeira, Portugal, NE Atlantic Ocean 33°04´N 16°17´W
10 M12 Camachigama, Lac ⊙ ...
40 M9 Camacho Zacatecas, C Mexico 24°23´N 102°20´W
82 D13 Camacupa var. General Machado, Port. Vila General Machado. Bié, C Angola 12°01´S 17°31´E
54 L7 Camaguán Guárico, C Venezuela 08°09´N 67°37´W

44 G6 Camagüey prev. Puerto Príncipe. Camagüey, C Cuba 21°24´N 77°55´W
44 G5 Camagüey, Archipiélago de island group C Cuba
40 D5 Camalli, Sierra de ▲ NW Mexico 28°21´N 113°26´W
57 G18 Camana var. Camaná. Arequipa, SW Peru 16°35´S 72°42´W
25 Z14 Camanche Iowa, C USA 41°47´N 90°15´W
35 P8 Camanche Reservoir ⊠ California, W USA
61 I16 Camaquã Rio Grande do Sul, S Brazil 30°51´S 51°47´W
61 H16 Camaquã, Río ≈ S Brazil
64 P6 Câmara de Lobos Madeira, Portugal, NE Atlantic Ocean 32°39´N 16°58´W
58 O13 Camamu Bahia, E Brazil 13°55´S 40°50´W
41 O8 Camargo Tamaulipas, C Mexico 26°16´N 98°49´W
103 R15 Camargue physical region SE France
104 F2 Camariñas Galicia, NW Spain 43°07´N 09°10´W
63 J18 Camarones Chubut, S Argentina 44°48´S 65°42´W
63 J18 Camarones, Bahía bay S Argentina
104 J14 Camas Andalucía, S Spain 37°24´N 06°01´W
167 S15 Ca Mau var. Quan Long. Minh Hai, S Vietnam 09°11´N 105°09´E
137 N11 Çam Burnu headland N Turkey 41°07´N 37°48´E
183 S9 Camden New South Wales, SE Australia 34°04´S 150°40´E
23 N4 Camden Alabama, S USA 31°59´N 87°17´W
27 U14 Camden Arkansas, C USA 33°32´N 92°49´W
21 Y3 Camden Delaware, NE USA 39°06´N 75°30´W
19 R7 Camden Maine, NE USA 44°12´N 69°04´W
18 I16 Camden New Jersey, NE USA 39°55´N 75°07´W
18 J9 Camden New York, NE USA 43°21´N 75°45´W
21 R12 Camden South Carolina, SE USA 34°14´N 80°37´W
20 H8 Camden Tennessee, S USA 36°03´N 88°07´W
25 X9 Camden Texas, SW USA 30°55´N 94°43´W
23 S5 Camden Bay bay S Beaufort Sea
27 U6 Camdenton Missouri, C USA 38°01´N 92°44´W
Camellia State see Alabama
18 M7 Camels Hump ▲ Vermont, NE USA 44°18´N 72°50´W
62 L7 Camenca Rus. Kamenka. N Moldova 48°01´N 28°42´E
Camerinum see Camerino
79 I20 Cameron Louisiana, C USA 29°48´N 93°19´W
25 T9 Cameron Texas, SW USA 30°51´N 96°58´W
30 J5 Cameron Wisconsin, N USA 45°24´N 91°44´W
10 M12 Cameron ◆ British Columbia, SW Canada
185 A24 Cameron Mountains ▲ South Island, New Zealand
79 D15 Cameroon off. Republic of Cameroon, Fr. Cameroun. ◆ republic C Africa
79 D15 Cameroon Mountain ▲ SW Cameroon 04°11´N 09°10´E
Cameroon, Republic of see Cameroon

Cameroon Ridge see Camerounaise, Dorsale
Cameroun see Cameroon
79 E14 Camerounaise, Dorsale Eng. Cameroon Ridge. ridge NW Cameroon
136 B15 Çamiçi Gölü ⊙ SW Turkey
171 N3 Camiling Luzon, N Philippines 15°41´N 120°22´E
23 T7 Camilla Georgia, SE USA 31°13´N 84°12´W
104 G3 Caminha Viana do Castelo, N Portugal 41°52´N 08°50´W
35 P7 Camino California, W USA 38°43´N 120°39´W
107 J24 Cammarata Sicilia, Italy, C Mediterranean Sea 37°36´N 13°39´E
42 K10 Camoapa Boaco, S Nicaragua 12°25´N 85°30´W
58 O13 Camocim Ceará, E Brazil 02°55´S 40°50´W
106 D10 Camogli Liguria, NW Italy 44°21´N 09°09´E
181 S5 Camooweal Queensland, C Australia 19°57´S 138°14´E
151 Q22 Camorta island Nicobar Islands, India, NE Indian Ocean
42 I6 Campamento Olancho, C Honduras 14°36´N 86°38´W
61 D19 Campana Buenos Aires, E Argentina 34°10´S 58°57´W
63 F21 Campana, Isla island S Chile
104 K11 Campanario Extremadura, W Spain 38°52´N 05°36´W
107 L17 Campania Eng. Champagne. ◆ region S Italy
27 Y8 Campbell Missouri, C USA 36°29´N 90°04´W
185 K15 Campbell, Cape headland South Island, New Zealand 41°45´S 174°16´E
14 J14 Campbellford Ontario, SE Canada 44°18´N 77°47´W
14 J14 Campbell Hill hill Ohio, N USA
192 K13 Campbell Island island S New Zealand
175 P13 Campbell Plateau undersea feature SW Pacific Ocean 51°00´S 170°00´E
10 K17 Campbell River Vancouver Island, British Columbia, SW Canada 49°59´N 125°18´W
20 L6 Campbellsville Kentucky, S USA 37°20´N 85°21´W
13 O13 Campbellton New Brunswick, SE Canada 48°00´N 66°41´W
183 S9 Campbelltown New South Wales, SE Australia 34°04´S 150°46´E
183 P16 Campbell Town Tasmania, SE Australia 41°57´S 147°30´E
96 G13 Campbeltown W Scotland, United Kingdom 55°26´N 05°38´W
41 W13 Campeche Campeche, SE Mexico 19°47´N 90°29´W
41 W14 Campeche ◆ state SE Mexico
41 T14 Campeche, Bahía de Eng. Bay of Campeche. bay E Mexico
Campeche, Banco de see Campeche Bank
64 C11 Campeche Bank Sp. Banco de Campeche, Sonda de Campeche. undersea feature S Gulf of Mexico 22°00´N 90°00´W
Campeche, Bay of see Campeche, Bahía de
Campeche, Sonda de see Campeche Bank
44 H7 Campechuela Granma, E Cuba 20°15´N 77°17´W
182 M13 Camperdown Victoria, SE Australia 38°15´S 143°10´E
167 U6 Câm Pha Quang Ninh, N Vietnam 21°04´N 107°20´E
116 H10 Câmpia Turzii Ger. Jerischmarkt, Hung. Aranyosgyéres; prev. Cimpia Turzii, Ghiris, Gyéres. Cluj, NW Romania 46°33´N 23°53´E
104 K12 Campillo de Llerena Extremadura, W Spain 38°30´N 05°48´W
104 L15 Campillos Andalucía, S Spain 37°04´N 04°51´W
116 I13 Câmpina prev. Cimpina. Prahova, SE Romania 45°08´N 25°44´E
59 Q15 Campina Grande Paraíba, E Brazil 07°15´S 35°50´W
60 L9 Campinas São Paulo, S Brazil 22°54´S 47°06´W
59 I16 Campina Verde Minas Gerais, SE Brazil 19°34´S 49°28´W
38 L10 Camp Kulowiye Saint Lawrence Island, Alaska, USA 63°15´N 168°45´W
79 D17 Campo var. Kampo. Sud, SW Cameroon 02°22´N 09°50´E
Campo see Kampo
116 J13 Campo Alegre de Lourdes Bahia, E Brazil 09°28´S 43°01´W
59 L16 Campobasso Molise, C Italy 41°34´N 14°40´E
107 H24 Campobello di Mazara Sicilia, Italy, C Mediterranean Sea 37°38´N 12°45´E
Campo Criptana see Campo de Criptana
59 O10 Campo de Criptana var. Campo Criptana. Castilla-La Mancha, C Spain 39°25´N 03°07´W
54 E5 Campo de la Cruz Atlántico, N Colombia 10°23´N 74°52´W
105 P11 Campo de Montiel physical region C Spain
61 H14 Campo Erê Santa Catarina, S Brazil 26°24´S 53°04´W
62 L7 Campo Gallo Santiago del Estero, N Argentina 26°32´S 62°51´W
11 T14 Campo Grande state capital Mato Grosso do Sul, SW Brazil 20°24´S 54°35´W
11 L13 Campo Largo Paraná, S Brazil 25°27´S 49°31´W
58 N13 Campo Maior Piauí, E Brazil 04°50´S 42°12´W
104 I10 Campo Maior Portalegre, C Portugal 39°01´N 07°04´W
61 F20 Campo Mourão Paraná, S Brazil 24°01´S 52°24´W
60 L9 Campos Belos Goiás, S Brazil 13°11´S 46°47´W
60 N9 Campos do Jordão São Paulo, S Brazil 22°45´S 45°36´W
63 F14 Campos dos Goytacazes Rio de Janeiro, SE Brazil 21°46´S 41°21´W

60 I13 Campos Novos Santa Catarina, S Brazil 27°23´S 51°11´W
59 O14 Campos Sales Ceará, E Brazil 07°01´S 40°21´W
25 Q9 Camp San Saba Texas, SW USA 30°59´N 99°16´W
21 N6 Campton Kentucky, S USA 37°44´N 83°33´W
116 I13 Câmpulung prev. Câmpulung-Muscel, Cimpulung. Arges, S Romania 45°16´N 25°03´E
116 J9 Câmpulung Moldovenesc Ger. Kimpolung, Hung. Hosszúmezejő. Suceava, NE Romania 47°32´N 25°34´E
Câmpulung-Muscel see Câmpulung
Camulodunum see Colchester
Camus ...
Cana see Chaniá
167 S14 Cân Thô Cân Tho, S Vietnam 10°03´N 105°46´E
Canada ◆ commonwealth republic N America
197 P5 Canada Basin undersea feature Arctic Ocean 80°00´N 145°00´W
197 P6 Canada Plain undersea feature Arctic Ocean
Canada Rosquin Santa Fe, C Argentina 32°04´S 61°35´W
8 L12 Canadian Shield physical region Canada
63 I18 Cañadón Grande, Sierra ▲ S Argentina
136 B11 Çanakkale var. Dardanelli; prev. Chanak, Kale Sultanie. Çanakkale, W Turkey 40°09´N 26°25´E
136 B11 Çanakkale ◆ province NW Turkey
136 B11 Çanakkale Boǧazı Eng. Dardanelles. strait NW Turkey
187 O17 Canala Province Nord, C New Caledonia 21°31´S 165°57´E
59 A15 Canamari Amazonas, W Brazil 07°37´S 72°33´W
58 G10 Canandaigua New York, NE USA 42°52´N 77°18´W
18 F10 Canandaigua Lake ⊙ New York, NE USA
40 G3 Cananea Sonora, NW Mexico 30°59´N 110°20´W
60 N9 Cananéia São Paulo, S Brazil 25°00´S 47°56´W
14 I12 Canaan Lake ⊙ Ontario, SE Canada
Canaan see Kenaan
12 M7 Caniapiscau ≈ Québec, E Canada
12 M8 Caniapiscau, Réservoir de Québec, C Canada
107 J24 Canicatti Sicilia, Italy, C Mediterranean Sea 37°22´N 13°51´E
136 L11 Çanik Daǧları ▲ N Turkey
105 P14 Caniles Andalucía, S Spain 37°24´N 02°41´W
59 B16 Canindé Acre, W Brazil 10°55´S 69°45´W
62 P6 Canindeyú var. Canendiyú, Canindiyú. ◆ department E Paraguay
Canindiyú see Canindeyú
194 J10 Canisteo Peninsula peninsula Antarctica
18 F11 Canisteo River ≈ New York, NE USA
40 M10 Cañitas var. Cañitas de Felipe Pescador. Zacatecas, C Mexico 23°35´N 102°39´W
Cañitas de Felipe Pescador see Cañitas
105 P15 Canjáyar Andalucía, S Spain 37°00´N 02°45´W
136 I12 Çankırı var. Chankiri; anc. Gangra, Germanicopolis. Çankın, N Turkey 40°36´N 33°35´E
136 I11 Çankırı var. Chankiri ◆ province N Turkey
171 P6 Canlaon Volcano ▲ Negros, C Philippines
11 P16 Canmore Alberta, SW Canada 51°07´N 115°18´W
96 F9 Canna island NW Scotland, United Kingdom
155 F20 Cannanore var. Kannur, Jagatsinghapur. Kerala, SW India 11°53´N 75°23´E see also Kannur
31 O17 Cannelton Indiana, N USA 37°54´N 86°44´W
103 U15 Cannes Alpes-Maritimes, SE France 43°33´N 06°59´E
106 C6 Cannobio Piemonte, NE Italy 46°04´N 08°39´E
97 L19 Cannock C England, United Kingdom 52°41´N 02°03´W
28 M6 Cannonball River ≈ North Dakota, N USA
29 W9 Cannon Falls Minnesota, N USA 44°30´N 92°54´W
18 I11 Cannonsville Reservoir ⊠ New York, NE USA
183 R12 Cann River Victoria, SE Australia 37°34´S 149°11´E
61 I16 Canoas Rio Grande do Sul, S Brazil 29°55´S 51°10´W
14 I12 Canoe Lake ⊙ Ontario, SE Canada
60 J12 Canoinhas Santa Catarina, S Brazil 26°10´S 50°27´W
37 T6 Canon City Colorado, C USA 38°25´N 105°14´W
55 P8 Cano Negro Bolívar, SE Venezuela
173 X15 Canonniers Point headland N Mauritius
23 W6 Canoochee River ≈ Georgia, SE USA
11 V15 Canora Saskatchewan, S Canada 51°38´N 102°28´W
45 Y14 Canouan island Saint Vincent and the Grenadines
13 R8 Canso Nova Scotia, SE Canada 45°20´N 61°00´W
104 K3 Cantabria ◆ autonomous community N Spain
104 K3 Cantábrica, Cordillera ▲ N Spain
Cantabrigia see Cambridge
103 O12 Cantal ◆ department C France
105 N6 Cantalejo Castilla y León, N Spain 41°15´N 03°57´W
103 O12 Cantal, Monts du ▲ C France
104 G8 Cantanhede Coimbra, C Portugal 40°18´N 08°37´W
55 O6 Cantaura Anzoátegui, NE Venezuela 09°22´N 64°24´W
116 M11 Cantemir Rus. Kantemir. S Moldova 46°17´N 28°12´E
97 Q22 Canterbury anc. Durovernum, Lat. Cantuaria. SE England, United Kingdom 51°17´N 01°05´E
185 F19 Canterbury off. Canterbury Region. ◆ region South Island, New Zealand
185 H20 Canterbury Bight bight South Island, New Zealand
185 H19 Canterbury Plains plain South Island, New Zealand
Canterbury Region see Canterbury
104 K13 Cantillana Andalucía, S Spain 37°36´N 05°49´W
59 N15 Canto do Buriti Piauí, E Brazil 08°07´S 42°58´W
23 S2 Canton Georgia, SE USA 34°14´N 84°29´W
30 K12 Canton Illinois, N USA 40°33´N 90°02´W
22 L5 Canton Mississippi, S USA 32°36´N 90°02´W

Cañete see San Vicente de Cañete
27 P8 Caney Kansas, C USA 37°00´N 95°56´W
27 P8 Caney River ≈ Kansas/Oklahoma, C USA
105 S3 Canfranc-Estación Aragón, NE Spain 42°42´N 00°31´W
83 E14 Cangandala Port. Vila de Aljustrel. Malanje, N Angola 13°40´S 19°47´E
82 G12 Cangandala Malanje, NW Angola 09°47´S 16°27´E
104 G4 Cangas Galicia, NW Spain 42°16´N 08°46´W
104 J2 Cangas del Narcea Asturias, N Spain 43°10´N 06°33´W
Cangas de Onís see Cangues d'Onís
161 S11 Cangnan var. Lingxi. Zhejiang, SE China 27°29´N 120°23´E
82 C10 Cangola Uíge, NW Angola 07°54´S 15°57´E
83 E14 Cangombe Moxico, E Angola 14°27´S 20°05´E
63 H21 Cangrejo, Cerro ▲ Argentina 49°19´S 72°18´W
61 H17 Canguçu Rio Grande do Sul, S Brazil 31°24´S 52°40´W
104 L2 Cangues d'Onís var. Cangas de Onís. Asturias, N Spain 43°21´N 05°08´W
161 P3 Cangzhou Hebei, E China 38°19´N 116°54´E
12 M7 Caniapiscau ≈ Québec, E Canada
12 M8 Caniapiscau, Réservoir de Québec, C Canada
107 J24 Canicatti Sicilia, Italy, C Mediterranean Sea 37°22´N 13°51´E
136 L11 Çanik Daǧları ▲ N Turkey
105 P14 Caniles Andalucía, S Spain 37°24´N 02°41´W
59 B16 Canindé Acre, W Brazil 10°55´S 69°45´W
62 P6 Canindeyú var. Canendiyú, Canindiyú. ◆ department E Paraguay
Canindiyú see Canindeyú
194 J10 Canisteo Peninsula peninsula Antarctica
18 F11 Canisteo River ≈ New York, NE USA
40 M10 Cañitas var. Cañitas de Felipe Pescador. Zacatecas, C Mexico 23°35´N 102°39´W
Cañitas de Felipe Pescador see Cañitas
105 P15 Canjáyar Andalucía, S Spain 37°00´N 02°45´W
136 I12 Çankırı var. Chankiri; anc. Gangra, Germanicopolis. Çankın, N Turkey 40°36´N 33°35´E
136 I11 Çankırı var. Chankiri ◆ province N Turkey
171 P6 Canlaon Volcano ▲ Negros, C Philippines
11 P16 Canmore Alberta, SW Canada 51°07´N 115°18´W
96 F9 Canna island NW Scotland, United Kingdom
155 F20 Cannanore var. Kannur, Jagatsinghapur. Kerala, SW India 11°53´N 75°23´E see also Kannur
31 O17 Cannelton Indiana, N USA 37°54´N 86°44´W
103 U15 Cannes Alpes-Maritimes, SE France 43°33´N 06°59´E
106 C6 Cannobio Piemonte, NE Italy 46°04´N 08°39´E
97 L19 Cannock C England, United Kingdom 52°41´N 02°03´W
28 M6 Cannonball River ≈ North Dakota, N USA
29 W9 Cannon Falls Minnesota, N USA 44°30´N 92°54´W
18 I11 Cannonsville Reservoir ⊠ New York, NE USA
183 R12 Cann River Victoria, SE Australia 37°34´S 149°11´E
61 I16 Canoas Rio Grande do Sul, S Brazil 29°55´S 51°10´W
14 I12 Canoe Lake ⊙ Ontario, SE Canada
60 J12 Canoinhas Santa Catarina, S Brazil 26°10´S 50°27´W
37 T6 Canon City Colorado, C USA 38°25´N 105°14´W
55 P8 Cano Negro Bolívar, SE Venezuela
173 X15 Canonniers Point headland N Mauritius
23 W6 Canoochee River ≈ Georgia, SE USA
11 V15 Canora Saskatchewan, S Canada 51°38´N 102°28´W
45 Y14 Canouan island Saint Vincent and the Grenadines
13 R8 Canso Nova Scotia, SE Canada 45°20´N 61°00´W
104 K3 Cantabria ◆ autonomous community N Spain
104 K3 Cantábrica, Cordillera ▲ N Spain
Cantabrigia see Cambridge
103 O12 Cantal ◆ department C France
105 N6 Cantalejo Castilla y León, N Spain 41°15´N 03°57´W
103 O12 Cantal, Monts du ▲ C France
104 G8 Cantanhede Coimbra, C Portugal 40°18´N 08°37´W
55 O6 Cantaura Anzoátegui, NE Venezuela 09°22´N 64°24´W
116 M11 Cantemir Rus. Kantemir. S Moldova 46°17´N 28°12´E
97 Q22 Canterbury anc. Durovernum, Lat. Cantuaria. SE England, United Kingdom 51°17´N 01°05´E
185 F19 Canterbury off. Canterbury Region. ◆ region South Island, New Zealand
185 H20 Canterbury Bight bight South Island, New Zealand
185 H19 Canterbury Plains plain South Island, New Zealand
Canterbury Region see Canterbury
104 K13 Cantillana Andalucía, S Spain 37°36´N 05°49´W
59 N15 Canto do Buriti Piauí, E Brazil 08°07´S 42°58´W
23 S2 Canton Georgia, SE USA 34°14´N 84°29´W
30 K12 Canton Illinois, N USA 40°33´N 90°02´W
22 L5 Canton Mississippi, S USA 32°36´N 90°02´W

Canton – Castelo do Bode, Barragem do

27 V2 **Canton** Missouri, C USA 40°07′N 91°31′W
18 J7 **Canton** New York, NE USA 44°36′N 75°10′W
21 O10 **Canton** North Carolina, SE USA 35°31′N 82°50′W
31 U12 **Canton** Ohio, N USA 40°48′N 81°23′W
26 L9 **Canton** Oklahoma, C USA 36°03′N 98°35′W
18 G12 **Canton** Pennsylvania, NE USA 41°38′N 76°49′W
29 R11 **Canton** South Dakota, N USA 43°19′N 96°33′W
25 V7 **Canton** Texas, SW USA 32°33′N 95°51′W
Canton *see* Guangzhou
Canton Island *see* Kanton
26 L9 **Canton Lake** ⊠ Oklahoma, C USA
106 D7 **Cantù** Lombardia, N Italy 45°44′N 09°98′E
Cantuaria/Cantwaraburh *see* Canterbury
39 R10 **Cantwell** Alaska, USA 63°23′N 148°57′W
59 O16 **Canudos** Bahia, E Brazil 09°51′S 39°08′W
47 T7 **Canumã, Rio** ≈ N Brazil
Canusium *see* Puglia, Canosa di
24 G7 **Canutillo** Texas, SW USA 31°53′N 106°34′W
25 N3 **Canyon** Texas, SW USA 34°58′N 101°56′W
33 S12 **Canyon** Wyoming, C USA 44°44′N 110°30′W
32 K13 **Canyon City** Oregon, NW USA 44°24′N 118°58′W
33 R10 **Canyon Ferry Lake** ◎ Montana, NW USA
25 S11 **Canyon Lake** ⊠ Texas, SW USA
167 T5 **Cao Băng** *var.* Caobang. Cao Băng, N Vietnam 22°40′N 106°16′E
Caobang *see* Cao Băng
160 J12 **Caodu He** ≈ S China
Caohai *see* Weining
167 S14 **Cao Lanh** Đông Thap, S Vietnam 10°35′N 105°25′E
82 C11 **Caombo** Malanje, NW Angola 08°42′S 16°33′E
Caorach, Cuan na g *see* Sheep Haven
Caozhou *see* Heze
171 Q12 **Capalulu** Pulau Mangole, E Indonesia 01°51′S 125°53′E
54 K8 **Capanaparo, Río** ≈ Colomiba/Venezuela
58 L12 **Capanema** Pará, NE Brazil 01°08′S 47°07′W
60 L10 **Capão Bonito do Sul** São Paulo, S Brazil 24°01′S 48°23′W
60 I13 **Capão Doce, Morro do** ▲ S Brazil 26°37′S 51°28′W
54 I4 **Capatárida** Falcón, N Venezuela 11°11′N 70°37′W
102 I15 **Capbreton** Landes, SW France 43°40′N 01°25′W
Cap-Breton, Île du *see* Cape Breton Island
15 W6 **Cap-Chat** Québec, SE Canada 49°04′N 66°43′W
15 P11 **Cap-de-la-Madeleine** Québec, SE Canada 46°22′N 72°31′W
103 N13 **Capdenac** Aveyron, S France 44°35′N 02°06′E
Cap des Palmès *see* Palmas, Cape
183 Q15 **Cape Barren Island** *island* Furneaux Group, Tasmania, SE Australia
65 O18 **Cape Basin** *undersea feature* S Atlantic Ocean 37°00′S 07°00′E
13 R14 **Cape Breton Island** *Fr.* Île du Cap-Breton. *island* Nova Scotia, SE Canada
23 Y11 **Cape Canaveral** Florida, SE USA 28°24′N 80°36′W
21 Y6 **Cape Charles** Virginia, NE USA 37°16′N 76°01′W
77 P17 **Cape Coast** *prev.* Cape Coast Castle. S Ghana 05°10′N 01°13′W
Cape Coast Castle *see* Cape Coast
19 Q12 **Cape Cod Bay** *bay* Massachusetts, NE USA
23 W15 **Cape Coral** Florida, SE USA 26°33′N 81°57′W
181 R4 **Cape Crawford Roadhouse** Northern Territory, N Australia 16°39′S 135°44′E
9 Q7 **Cape Dorset** *var.* Kingait. Baffin Island, Nunavut, NE Canada 76°14′N 76°32′W
21 N8 **Cape Fear River** ≈ North Carolina, SE USA
27 Y7 **Cape Girardeau** Missouri, C USA 37°19′N 89°31′W
21 T14 **Cape Island** *island* South Carolina, SE USA
186 A6 **Capella** ▲ NW Papua New Guinea 05°00′S 141°09′E
98 H12 **Capelle aan den IJssel** Zuid-Holland, SW Netherlands 51°56′N 04°36′E
83 C15 **Capelongo** Huíla, C Angola 14°45′S 15°02′E
18 J17 **Cape May** New Jersey, NE USA 38°54′N 74°54′W
18 J17 **Cape May Court House** New Jersey, NE USA 39°03′N 74°46′W
Cape Palmas *see* Harper
8 I16 **Cape Parry** Northwest Territories, N Canada 70°10′N 124°33′W
65 P19 **Cape Rise** *undersea feature* SW Indian Ocean 42°00′S 15°00′E
Cape Saint Jacques *see* Vung Tau
Capesterre *see* Capesterre-Belle-Eau
45 Y6 **Capesterre-Belle-Eau** *var.* Capesterre. Basse Terre, S Guadeloupe 16°03′N 61°34′W
83 D26 **Cape Town** *var.* Ekapa, *Afr.* Kaapstad, Kapstad. ● (South Africa-legislative capital) Western Cape, SW South Africa 33°56′S 18°28′E
83 E26 **Cape Town** ✈ Western Cape, SW South Africa 31°51′S 21°06′E
76 D9 **Cape Verde** *off.* Republic of Cape Verde, *Port.* Cabo Verde, Ilhas do Cabo Verde. ◆ *republic* E Atlantic Ocean
64 L11 **Cape Verde Basin** *undersea feature* E Atlantic Ocean 15°00′N 30°00′W
66 K5 **Cape Verde Islands** *island group* E Atlantic Ocean
64 L10 **Cape Verde Plain** *undersea feature* E Atlantic Ocean 23°00′N 26°00′W
Cape Verde Plateau/Cape Verde Rise *see* Cape Verde Terrace
Cape Verde, Republic of *see* Cape Verde
64 L11 **Cape Verde Terrace** *var.* Cape Verde Plateau, Cape Verde Rise. *undersea feature* E Atlantic Ocean 18°00′N 20°00′W
181 V2 **Cape York Peninsula** *peninsula* Queensland, N Australia
44 M8 **Cap-Haïtien** *var.* Le Cap. N Haiti 19°44′N 72°12′W
43 T15 **Capira** West Panamá, C Panama 08°48′N 79°51′W
14 K8 **Capitachouane** ≈ Québec, SE Canada
14 L8 **Capitachouane, Lac** ◎ Québec, SE Canada
37 T13 **Capitan** New Mexico, SW USA 33°33′N 105°34′W
194 G3 **Capitán Arturo Prat** *Chilean research station* South Shetland Islands, Antarctica 62°24′S 59°42′W
37 S13 **Capitan Mountains** ▲▲ New Mexico, SW USA
62 M3 **Capitán Pablo Lagerenza** *var.* Mayor Pablo Lagerenza. Chaco, N Paraguay 19°55′S 60°46′W
37 T13 **Capitan Peak** ▲ New Mexico, SW USA 33°35′N 105°15′W
188 H5 **Capitol Hill** ● (Northern Mariana Islands-legislative capital) Saipan, S Northern Mariana Islands
60 I9 **Capivara, Represa** ⊠ S Brazil
61 J16 **Capivari** Rio Grande do Sul, S Brazil 30°08′S 50°32′W
113 H15 **Čapljina** Federicija Bosna I Hercegovina, S Bosnia and Herzegovina 43°07′N 17°42′E
83 M15 **Capoche** *var.* Kapoche. ≈ Mozambique/Zambia
Capo Delgado, Província de *see* Cabo Delgado
107 K17 **Capodichino** ✈ (Napoli) Campania, S Italy 40°53′N 14°15′E
Capodistria *see* Koper
106 E12 **Capraia, Isola di** *island* Arcipelago Toscano, C Italy
107 B16 **Caprara, Punta** *var.* Punta dello Scorno. *headland* Isola Asinara, W Italy 41°07′N 08°19′E
Capraria *see* Cabrera, Illa de
14 F10 **Capreol** Ontario, S Canada 46°43′N 80°56′W
107 K18 **Capri** Campania, S Italy 40°33′N 14°14′E
175 S9 **Capricorn Tablemount** *undersea feature* W Pacific Ocean 18°34′S 172°12′W
107 J18 **Capri, Isola di** *island* S Italy
83 G16 **Caprivi** ◐ *district* NE Namibia
Caprivi Concession *see* Caprivi Strip
83 F16 **Caprivi Strip** *Ger.* Caprivizipfel; *prev.* Caprivi Concession. *cultural region* NE Namibia
Caprivizipfel *see* Caprivi Strip
25 O5 **Cap Rock Escarpment** *cliffs* Texas, SW USA
15 R10 **Cap-Rouge** Québec, SE Canada 46°45′N 71°18′W
Cap Saint-Jacques *see* Vung Tau
38 F12 **Captain Cook** Hawaii, USA, C Pacific Ocean 19°30′N 155°55′W
183 R10 **Captains Flat** New South Wales, SE Australia 35°37′S 149°28′E
102 K14 **Captieux** Gironde, SW France 44°16′N 00°15′W
107 K17 **Capua** Campania, S Italy 41°06′N 14°13′E
54 F14 **Caquetá** *off.* Departamanto del Caquetá. ◐ *province* S Colombia
Caquetá, Departamanto del *see* Caquetá
54 E13 **Caquetá, Río** *var.* Rio Japurá, Yapurá. ≈ Brazil/Colombia *see also* Japurá, Rio
Caquetá, Río *see* Japurá, Rio
CAR *see* Central African Republic
Cara *see* Kara
57 I16 **Carabaya, Cordillera** ▲▲ E Peru
54 K5 **Carabobo** *off.* Estado Carabobo. ◐ *state* N Venezuela
Carabobo, Estado *see* Carabobo
116 I14 **Caracal** Olt, S Romania 44°07′N 24°18′E
58 F10 **Caracaraí** Rondônia, W Brazil 01°47′N 61°11′W
54 L5 **Caracas** ● (Venezuela) Distrito Federal, N Venezuela 10°29′N 66°54′W
54 I5 **Carache** Trujillo, N Venezuela 09°40′N 70°15′W
60 N10 **Caraguatatuba** São Paulo, S Brazil 23°37′S 45°24′W
48 I7 **Carajás, Serra dos** ▲▲ N Brazil
Caralis *see* Cagliari
54 E9 **Caramanta** Antioquia, W Colombia 05°36′N 75°38′W
171 P4 **Caramoan** Catanduanes Island, N Philippines 13°47′N 123°49′E
Caramurat *see* Mihail Kogălniceanu
116 F12 **Caransebeş** *Ger.* Karansebesch, *Hung.* Karánsebes. Caraş-Severin, SW Romania 45°23′N 22°13′E
Carapella *see* Carapelle
107 M16 **Carapelle** *var.* Carapella. ≈ SE Italy
55 O9 **Carapo** Bolívar, SE Venezuela
13 P13 **Caraquet** New Brunswick, SE Canada 47°48′N 64°59′W
Caras *see* Caraz
116 F12 **Caraşova** *Hung.* Krassóvár. Caraş-Severin, SW Romania 45°11′N 21°51′E
116 F12 **Caraş-Severin** ◐ *county* SW Romania
42 M5 **Caratasca, Laguna de** *lagoon* NE Honduras
58 C13 **Carauari** Amazonas, NW Brazil 04°55′S 66°57′W
Caravaca *see* Caravaca de la Cruz
105 Q12 **Caravaca de la Cruz** *var.* Caravaca. Murcia, SE Spain 38°06′N 01°51′W
106 E7 **Caravaggio** Lombardia, N Italy 45°31′N 09°39′E
107 C18 **Caravai, Passo di** *pass* Sardegna, Italy, C Mediterranean Sea
59 O19 **Caravelas** Bahia, E Brazil 17°45′S 39°15′W
56 C12 **Caraz** *var.* Caras. Ancash, W Peru 09°03′S 77°47′W
61 H14 **Carazinho** Rio Grande do Sul, S Brazil 28°16′S 52°46′W
42 J11 **Carazo** ◐ *department* SW Nicaragua
Carballiño *see* O Carballiño
104 G2 **Carballo** Galicia, NW Spain 43°13′N 08°41′W
11 W16 **Carberry** Manitoba, S Canada 49°52′N 99°20′W
40 F4 **Carbó** Sonora, NW Mexico 29°41′N 111°00′W
107 C20 **Carbonara, Capo** *headland* Sardegna, Italy, C Mediterranean Sea 39°06′N 09°31′E
37 Q5 **Carbondale** Colorado, C USA 39°24′N 107°12′W
30 L17 **Carbondale** Illinois, N USA 37°43′N 89°13′W
27 Q4 **Carbondale** Kansas, C USA 38°49′N 95°41′W
18 I13 **Carbondale** Pennsylvania, NE USA 41°34′N 75°30′W
13 V12 **Carbonear** Newfoundland, Newfoundland and Labrador, SE Canada 47°45′N 53°16′W
105 Q9 **Carboneras de Guadazaón** *var.* Carboneras de Guadazón. Castilla-La Mancha, C Spain 39°54′N 01°50′W
Carboneras de Guadazón *see* Carboneras de Guadazaón
23 O3 **Carbon Hill** Alabama, S USA 33°53′N 87°31′W
107 B20 **Carbonia** *var.* Carbonia Centro. Sardegna, Italy, C Mediterranean Sea 39°11′N 08°31′E
Carbonia Centro *see* Carbonia
63 I22 **Carbon, Laguna del** *depression* S Argentina
105 S10 **Carcaixent** Valenciana, E Spain 39°08′N 00°28′W
Carcaso *see* Carcassonne
65 B24 **Carcass Island** *island* NW Falkland Islands
103 O16 **Carcassonne** *anc.* Carcaso. Aude, S France 43°13′N 02°21′E
105 R12 **Carche** ▲ S Spain 38°24′N 01°11′W
56 A13 **Carchi** ◐ *province* N Ecuador
10 I8 **Carcross** Yukon, W Canada 60°11′N 134°41′W
Cardamomes, Chaîne des *see* Krâvanh, Chuŏr Phnum
155 G22 **Cardamom Hills** ▲▲ SW India
Cardamom Mountains *see* Krâvanh, Chuŏr Phnum
104 M12 **Cardeña** Andalucía, S Spain 38°16′N 04°20′W
44 D4 **Cárdenas** Matanzas, W Cuba 23°02′N 81°12′W
41 O11 **Cárdenas** San Luis Potosí, C Mexico 22°03′N 99°30′W
41 U15 **Cárdenas** Tabasco, SE Mexico 18°00′N 93°21′W
63 H21 **Cardiel, Lago** ◎ S Argentina
97 K22 **Cardiff** *Wel.* Caerdydd. ◐ S Wales, United Kingdom 51°30′N 03°13′W
97 J22 **Cardiff-Wales** ✈ S Wales, United Kingdom 51°24′N 03°22′W
97 I21 **Cardigan** *Wel.* Aberteifi. SW Wales, United Kingdom 52°06′N 04°40′W
97 I20 **Cardigan** *cultural region* W Wales, United Kingdom
97 I20 **Cardigan Bay** *bay* W Wales, United Kingdom
19 N8 **Cardigan, Mount** ▲ New Hampshire, NE USA 43°39′N 71°52′W
14 M13 **Cardinal** Ontario, SE Canada 44°48′N 75°22′W
105 V5 **Cardona** Cataluña, NE Spain 41°55′N 01°41′E
61 E19 **Cardona** Soriano, SW Uruguay 33°53′S 57°18′W
105 V4 **Cardoner** ≈ NE Spain
11 Q17 **Cardston** Alberta, SW Canada 49°14′N 113°19′W
181 W5 **Cardwell** Queensland, NE Australia 18°24′S 146°06′E
116 G8 **Carei** *Ger.* Gross-Karol, Karol, *Hung.* Nagykároly; *prev.* Careii-Mari. Satu Mare, NW Romania 47°40′N 22°28′E
Careii-Mari *see* Carei
58 F13 **Careiro** Amazonas, NW Brazil 03°40′S 60°23′W
102 J4 **Carentan** Manche, N France 49°18′N 01°15′W
104 M2 **Cares** ≈ N Spain
33 P14 **Carey** Idaho, NW USA 43°17′N 113°58′W
31 S12 **Carey** Ohio, N USA 40°57′N 83°22′W
25 P4 **Carey** Texas, SW USA 34°28′N 100°18′W
180 L11 **Carey, Lake** ◎ Western Australia
173 O8 **Cargados Carajos Bank** *undersea feature* C Indian Ocean
102 G6 **Carhaix-Plouguer** Finistère, NW France 48°16′N 03°35′W
61 A22 **Carhué** Buenos Aires, E Argentina 37°10′S 62°45′W
55 O5 **Cariaco** Sucre, NE Venezuela 10°33′N 63°37′W
107 O20 **Cariati** Calabria, SW Italy 39°30′N 16°57′E
2 H17 **Caribbean Plate** *tectonic feature*
44 I11 **Caribbean Sea** *sea* W Atlantic Ocean
11 N15 **Cariboo Mountains** ▲▲ British Columbia, SW Canada
11 W9 **Caribou** Manitoba, C Canada 59°27′N 97°43′W
19 S2 **Caribou** Maine, NE USA 46°51′N 68°00′W
11 P10 **Caribou Mountains** ▲▲ Alberta, SW Canada
Caribrod *see* Dimitrovgrad
40 I6 **Carichic** Chihuahua, N Mexico 27°57′N 107°01′W
103 R3 **Carignan** Ardennes, N France 49°38′N 05°08′E
183 Q5 **Carinda** New South Wales, SE Australia 30°26′S 147°45′E
105 R6 **Cariñena** Aragón, NE Spain 41°20′N 01°13′W
107 I23 **Carini** Sicilia, Italy, C Mediterranean Sea 38°06′N 13°09′E
107 K17 **Carinola** Campania, S Italy 41°14′N 14°03′E
Carinthi *see* Kärnten
55 O5 **Caripe** Monagas, NE Venezuela 10°13′N 63°30′W
55 P5 **Caripito** Monagas, NE Venezuela 10°03′N 63°05′W
15 W7 **Carleton** Québec, SE Canada 48°07′N 66°07′W
31 S10 **Carleton** Michigan, N USA 42°03′N 83°23′W
13 O14 **Carleton, Mount** ▲ New Brunswick, SE Canada 47°10′N 66°54′W
14 L13 **Carleton Place** Ontario, SE Canada 45°08′N 76°09′W
35 V3 **Carlin** Nevada, W USA 40°40′N 116°09′W
30 K14 **Carlinville** Illinois, N USA 39°16′N 89°52′W
97 K14 **Carlisle** *anc.* Caer Luel, Luguvallium, Luguvallum. NW England, United Kingdom 54°54′N 02°55′W
27 V11 **Carlisle** Arkansas, C USA 34°46′N 91°45′W
31 N15 **Carlisle** Indiana, N USA 38°57′N 87°23′W
29 V14 **Carlisle** Iowa, C USA 41°30′N 93°29′W
21 N5 **Carlisle** Kentucky, S USA 38°19′N 84°02′W
18 F15 **Carlisle** Pennsylvania, NE USA 40°10′N 77°10′W
21 Q11 **Carlisle** South Carolina, SE USA 34°35′N 81°27′W
38 J17 **Carlisle Island** *island* Aleutian Islands, Alaska, USA
27 R7 **Carl Junction** Missouri, C USA 37°10′N 94°34′W
107 A20 **Carloforte** Sardegna, Italy, C Mediterranean Sea 39°10′N 08°17′E
Carlopago *see* Karlobag
61 B21 **Carlos Casares** Buenos Aires, E Argentina 35°39′S 61°28′W
61 E18 **Carlos Reyles** Durazno, C Uruguay 33°00′S 56°30′W
61 A21 **Carlos Tejedor** Buenos Aires, E Argentina 35°25′S 62°25′W
97 F19 **Carlow** *Ir.* Ceatharlach. SE Ireland 52°50′N 06°55′W
97 F19 **Carlow** *Ir.* Cheatharlach. *cultural region* SE Ireland
96 F7 **Carloway** NW Scotland, United Kingdom 58°17′N 06°48′W
35 U17 **Carlsbad** California, W USA 33°09′N 117°21′W
37 U15 **Carlsbad** New Mexico, SW USA 32°24′N 104°15′W
Carlsbad *see* Karlovy Vary
129 N13 **Carlsberg Ridge** *undersea feature* S Arabian Sea 06°00′N 61°00′E
Carlsruhe *see* Karlsruhe
29 W6 **Carlton** Minnesota, N USA 46°39′N 92°25′W
11 V17 **Carlyle** Saskatchewan, S Canada 49°39′N 102°18′W
30 L15 **Carlyle** Illinois, N USA 38°36′N 89°22′W
30 L15 **Carlyle Lake** ⊠ Illinois, N USA
10 H7 **Carmacks** Yukon, W Canada 62°04′N 136°21′W
106 B9 **Carmagnola** Piemonte, NW Italy 44°50′N 07°43′E
11 X16 **Carman** Manitoba, S Canada 49°32′N 97°59′W
Carmana/Carmania *see* Kermān
97 I21 **Carmarthen** SW Wales, United Kingdom 51°52′N 04°19′W
97 I21 **Carmarthen** *cultural region* SW Wales, United Kingdom
97 I22 **Carmarthen Bay** *inlet* SW Wales, United Kingdom
103 N14 **Carmaux** Tarn, S France 44°03′N 02°09′E
35 N11 **Carmel** California, W USA 36°32′N 121°54′W
31 O13 **Carmel** Indiana, N USA 39°58′N 86°07′W
18 L13 **Carmel** New York, NE USA 41°25′N 73°40′W
97 H18 **Carmel Head** *headland* NW Wales, United Kingdom 53°24′N 04°35′W
42 E2 **Carmelita** Petén, N Guatemala 17°33′N 90°11′W
61 D19 **Carmelo** Colonia, SW Uruguay 34°00′S 58°20′W
41 V14 **Carmen** *var.* Ciudad del Carmen. Campeche, SE Mexico 18°38′N 91°50′W
61 A25 **Carmen de Patagones** Buenos Aires, E Argentina 40°45′S 63°00′W
40 F8 **Carmen, Isla** *island* NW Mexico
40 M5 **Carmen, Sierra del** ▲▲ NW Mexico
30 M16 **Carmi** Illinois, N USA 38°05′N 88°09′W
35 O7 **Carmichael** California, W USA 38°36′N 121°21′W
Carmiel *see* Karmi’el
25 U11 **Carmine** Texas, SW USA 30°07′N 96°40′W
104 K14 **Carmona** Andalucía, S Spain 37°28′N 05°38′W
Carmona *see* Uíge
Carnaro *see* Kvarner
180 G9 **Carnarvon** Western Australia 24°57′S 113°38′E
14 I13 **Carnarvon** Ontario, SE Canada 45°03′N 78°41′W
83 G24 **Carnarvon** Northern Cape, W South Africa 30°59′S 22°08′E
Carnarvon *see* Caernarfon
180 K9 **Carnarvon Range** ▲▲ Western Australia
Carn Domhnach *see* Carndonagh
96 E13 **Carndonagh** *Ir.* Carn Domhnach. NW Ireland 55°15′N 07°15′W
11 V17 **Carnduff** Saskatchewan, S Canada 49°11′N 101°50′W
26 L11 **Carnegie** Oklahoma, C USA 35°06′N 98°36′W
180 L9 **Carnegie, Lake** *salt lake* Western Australia
193 U8 **Carnegie Ridge** *undersea feature* E Pacific Ocean 01°00′S 85°00′W
96 H9 **Carn Eige** ▲ N Scotland, United Kingdom 57°18′N 05°04′W
182 F5 **Carnes** South Australia 30°12′S 134°31′E
194 J12 **Carney Island** *island* Antarctica
18 H16 **Carneys Point** New Jersey, NE USA 39°38′N 75°29′W
Carniche, Alpi *see* Karnische Alpen
151 Q21 **Car Nicobar** *island* Nicobar Islands, India, NE Indian Ocean
79 H15 **Carnot** Mambéré-Kadéï, W Central African Republic 04°58′N 15°55′E
182 F10 **Carnot, Cape** *headland* South Australia 34°57′S 135°39′E
96 K11 **Carnoustie** E Scotland, United Kingdom 56°30′N 02°42′W
97 F20 **Carnsore Point** *Ir.* Ceann an Chairn. *headland* SE Ireland 52°10′N 06°22′W
8 H7 **Carnwath** ≈ Northwest Territories, NW Canada
31 R8 **Caro** Michigan, N USA 43°29′N 83°24′W
23 Z15 **Carol City** Florida, SE USA 25°56′N 80°15′W
59 L14 **Carolina** Maranhão, E Brazil 07°20′S 47°25′W
45 U5 **Carolina** E Puerto Rico 18°22′N 65°57′W
21 V12 **Carolina Beach** North Carolina, SE USA 34°02′N 77°53′W
Caroline Island *see* Millennium Island
189 N15 **Caroline Islands** *island group* C Micronesia
129 Z14 **Caroline Plate** *tectonic feature*
192 H7 **Caroline Ridge** *undersea feature* E Philippine Sea 08°00′N 150°00′E
Carolopois *see* Châlons-en-Champagne
45 V14 **Caroni Arena Dam** ⊠ Trinidad, Trinidad and Tobago
Caronie, Monti *see* Nebrodi, Monti
55 P7 **Caroní, Río** ≈ E Venezuela
45 U14 **Caroni River** ≈ Trinidad, Trinidad and Tobago
Caronium *see* A Coruña
54 J5 **Carora** Lara, N Venezuela 10°12′N 70°07′W
86 F12 **Carpathian Mountains** *var.* Carpathians, *Cz./Pol.* Karpaty, *Ger.* Karpaten. ▲▲ E Europe
Carpathians *see* Carpathian Mountains
Carpathos/Carpathus *see* Kárpathos
116 H12 **Carpaţii Meridionalii** *var.* Alpi Transilvaniei, Carpaţii Sudici, *Eng.* South Carpathians, Transylvanian Alps, *Ger.* Südkarpaten, Transsylvanische Alpen, *Hung.* Déli-Kárpátok, Erdélyi-Havasok. ▲▲ C Romania
Carpaţii Sudici *see* Carpaţii Meridionalii
174 L7 **Carpentaria, Gulf of** *gulf* N Australia
Carpentoracte *see* Carpentras
103 R14 **Carpentras** *anc.* Carpentoracte. Vaucluse, SE France 44°03′N 05°03′E
106 F9 **Carpi** Emilia-Romagna, N Italy 44°47′N 10°53′E
116 E11 **Cărpiniş** *Hung.* Gyertyámos. Timiş, W Romania 45°46′N 20°53′E
35 R14 **Carpinteria** California, W USA 34°24′N 119°30′W
23 S9 **Carrabelle** Florida, SE USA 29°51′N 84°39′W
Carraig Aonair *see* Fastnet Rock
Carraig Fhearghais *see* Carrickfergus
Carraig Mhachaire Rois *see* Carrickmacross
Carraig na Siúire *see* Carrick-on-Suir
Carrantual *see* Carrauntoohil
106 E10 **Carrara** Toscana, C Italy 44°05′N 10°07′E
61 F20 **Carrasco** ✈ (Montevideo) Canelones, S Uruguay 34°51′S 56°00′W
105 P9 **Carrascosa del Campo** Castilla-La Mancha, C Spain 40°02′N 02°35′W
54 H4 **Carrasquero** Zulia, NW Venezuela 11°00′N 72°01′W
183 O9 **Carrathool** New South Wales, SE Australia 34°25′S 145°30′E
Carrauntohil *see* Carrauntoohil
97 B21 **Carrauntoohil** *Ir.* Carrantual, Carrauntohil, Corrán Tuathail. ▲ SW Ireland 51°98′N 09°53′W
45 Y15 **Carriacou** *island* N Grenada
97 G15 **Carrickfergus** *Ir.* Carraig Fhearghais. NE Northern Ireland, United Kingdom 54°43′N 05°49′W
97 F16 **Carrickmacross** *Ir.* Carraig Mhachaire Rois. N Ireland 53°58′N 06°43′W
97 D16 **Carrick-on-Shannon** *Ir.* Cora Droma Rúisc. NW Ireland 53°37′N 08°05′W
97 E20 **Carrick-on-Suir** *Ir.* Carraig na Siúire. S Ireland 52°21′N 07°25′W
182 I7 **Carrieton** South Australia 32°27′S 138°33′E
40 L7 **Carrillo** Chihuahua, N Mexico 25°53′N 103°54′W
29 O4 **Carrington** North Dakota, N USA 47°27′N 99°07′W
104 M4 **Carrión** ≈ N Spain
104 M4 **Carrión de los Condes** Castilla y León, N Spain 42°20′N 04°37′W
25 P13 **Carrizo Springs** Texas, SW USA 28°33′N 99°54′W
37 S13 **Carrizozo** New Mexico, SW USA 33°38′N 105°52′W
29 T13 **Carroll** Iowa, C USA 42°04′N 94°52′W
23 N4 **Carrollton** Alabama, S USA 33°13′N 88°05′W
23 R3 **Carrollton** Georgia, SE USA 33°33′N 85°04′W
30 K14 **Carrollton** Illinois, N USA 39°18′N 90°24′W
20 L4 **Carrollton** Kentucky, S USA 38°41′N 85°09′W
31 R8 **Carrollton** Michigan, N USA 43°27′N 83°55′W
27 T3 **Carrollton** Missouri, C USA 39°20′N 93°30′W
31 U12 **Carrollton** Ohio, N USA 40°34′N 81°05′W
25 T6 **Carrollton** Texas, SW USA 32°57′N 96°53′W
11 U14 **Carrot** ≈ Saskatchewan, S Canada
11 U14 **Carrot River** Saskatchewan, C Canada 53°18′N 103°32′W
18 J7 **Carry Falls Reservoir** ⊠ New York, NE USA
136 L11 **Çarşamba** Samsun, N Turkey 41°13′N 36°43′E
28 L6 **Carson** North Dakota, N USA 46°26′N 101°34′W
35 Q6 **Carson City** *state capital* Nevada, W USA 39°10′N 119°46′W
35 R6 **Carson River** ≈ Nevada, W USA
35 S5 **Carson Sink** *salt flat* Nevada, W USA
11 Q16 **Carstairs** Alberta, SW Canada 51°35′N 114°02′W
Carstensz, Puntjak *see* Jaya, Puncak
54 E5 **Cartagena** *var.* Cartagena de los Indes. Bolívar, NW Colombia 10°24′N 75°33′W
105 R13 **Cartagena** *anc.* Carthago Nova. Murcia, SE Spain 37°36′N 00°59′W
54 E13 **Cartagena de Chaira** Caquetá, S Colombia 01°19′N 74°52′W
Cartagena de los Indes *see* Cartagena
54 D10 **Cartago** Valle del Cauca, W Colombia 04°45′N 75°55′W
43 N14 **Cartago** Cartago, C Costa Rica 09°50′N 83°54′W
42 M14 **Cartago** *off.* Provincia de Cartago. ◐ *province* C Costa Rica
Cartago, Provincia de *see* Cartago
25 O11 **Carta Valley** Texas, SW USA 29°46′N 100°37′W
104 F10 **Cartaxo** Santarém, C Portugal 39°10′N 08°47′W
104 I14 **Cartaya** Andalucía, S Spain 37°16′N 07°09′W
Carteret Islands *see* Tulun Islands
29 S15 **Carter Lake** Iowa, C USA 41°17′N 95°55′W
23 S3 **Cartersville** Georgia, SE USA 34°10′N 84°48′W
185 M14 **Carterton** Wellington, North Island, New Zealand 41°01′S 175°30′E
30 J13 **Carthage** Illinois, N USA 40°25′N 91°09′W
22 L5 **Carthage** Mississippi, S USA 32°43′N 89°31′W
27 R7 **Carthage** Missouri, C USA 37°10′N 94°20′W
18 I8 **Carthage** New York, NE USA 43°58′N 75°36′W
21 T10 **Carthage** North Carolina, SE USA 35°21′N 79°27′W
20 K8 **Carthage** Tennessee, S USA 36°14′N 85°59′W
25 X7 **Carthage** Texas, SW USA 32°10′N 94°21′W
74 M5 **Carthage** ✈ (Tunis) N Tunisia 36°51′N 10°12′E
Carthago Nova *see* Cartagena
14 E10 **Cartier** Ontario, S Canada 46°40′N 81°31′W
13 S8 **Cartwright** Newfoundland and Labrador, E Canada 53°40′N 57°W
55 P9 **Caruana de Montaña** Bolívar, SE Venezuela 05°16′N 63°12′W
59 Q15 **Caruaru** Pernambuco, E Brazil 08°15′S 35°55′W
55 P5 **Carúpano** Sucre, NE Venezuela 10°39′N 63°14′W
Carusbur *see* Cherbourg
58 M12 **Carutapera** Maranhão, E Brazil 01°12′S 45°57′W
27 Y9 **Caruthersville** Missouri, C USA 36°11′N 89°40′W
103 O1 **Carvin** Pas-de-Calais, N France 50°31′N 03°00′E
58 E12 **Carvoeiro** Amazonas, NW Brazil 01°24′S 61°59′W
104 E10 **Carvoeiro, Cabo** *headland* C Portugal 39°19′N 09°27′W
21 U9 **Cary** North Carolina, SE USA 35°47′N 78°46′W
182 M3 **Caryapundy Swamp** *wetland* New South Wales/Queensland, SE Australia
65 E24 **Carysfort, Cape** *headland* East Falkland, Falkland Islands 51°26′S 57°50′W
74 F6 **Casablanca** *Ar.* Dar-el-Beida. NW Morocco 33°39′N 07°31′W
60 M8 **Casa Branca** São Paulo, S Brazil 21°47′S 47°05′W
36 L14 **Casa Grande** Arizona, SW USA 32°52′N 111°45′W
106 C8 **Casale Monferrato** Piemonte, NW Italy 45°08′N 08°27′E
106 E8 **Casalpusterlengo** Lombardia, N Italy 45°10′N 09°37′E
54 H10 **Casanare** *off.* Intendencia de Casanare. ◐ *province* C Colombia
Casanare, Intendencia de *see* Casanare
55 P5 **Casanay** Sucre, NE Venezuela 10°30′N 63°25′W
24 K11 **Casa Piedra** Texas, SW USA 29°43′N 104°03′W
107 Q19 **Casarano** Puglia, SE Italy 40°01′N 18°10′E
42 J11 **Casares** Carazo, W Nicaragua 11°37′N 86°19′W
105 R10 **Casas Ibáñez** Castilla-La Mancha, C Spain 39°17′N 01°28′W
61 I14 **Casca** Rio Grande do Sul, S Brazil 28°39′S 51°55′W
172 I17 **Cascade** Mahé, NE Seychelles 04°39′S 55°29′E
33 N13 **Cascade** Idaho, NW USA 44°31′N 116°02′W
29 Y13 **Cascade** Iowa, C USA 42°18′N 91°00′W
33 R9 **Cascade** Montana, NW USA 47°15′N 111°46′W
185 B20 **Cascade Point** *headland* South Island, New Zealand 44°00′S 168°23′E
32 G13 **Cascade Range** ▲▲ Oregon/Washington, NW USA
33 N12 **Cascade Reservoir** ⊠ Idaho, NW USA
0 E8 **Cascadia Basin** *undersea feature* NE Pacific Ocean 47°00′N 127°30′W
104 E11 **Cascais** Lisboa, C Portugal 38°41′N 09°25′W
15 W7 **Cascapédia** ≈ Québec, SE Canada
59 I22 **Cascavel** Ceará, E Brazil 04°10′S 38°15′W
60 G11 **Cascavel** Paraná, S Brazil 24°56′S 53°28′W
106 I13 **Cascia** Umbria, C Italy 42°45′N 13°01′E
106 F11 **Cascina** Toscana, C Italy 43°40′N 10°33′E
19 Q8 **Casco Bay** *bay* Maine, NE USA
194 J7 **Case Island** *island* Antarctica
106 B8 **Caselle** ✈ (Torino) Piemonte, NW Italy 45°06′N 07°41′E
107 K17 **Caserta** Campania, S Italy 41°05′N 14°20′E
15 N8 **Casey** Québec, SE Canada 47°50′N 74°09′W
30 M14 **Casey** Illinois, N USA 39°18′N 87°59′W
195 Y12 **Casey** *Australian research station* Antarctica 65°58′S 111°04′E
195 W3 **Casey Bay** *bay* Antarctica
80 Q11 **Caseyr, Raas** *headland* NE Somalia 11°51′S 51°16′E
97 D20 **Cashel** *Ir.* Caiseal. S Ireland 52°31′N 07°53′W
54 G6 **Casigua** Zulia, W Venezuela 08°46′N 72°30′W
61 B19 **Casilda** Santa Fe, C Argentina 33°05′S 61°10′W
Casim *see* General Toshevo
183 V4 **Casino** New South Wales, SE Australia 28°50′S 153°02′E
107 J16 **Cassino** *prev.* San Germano; *anc.* Casinum. Lazio, C Italy 41°29′N 13°50′E
Casinum *see* Cassino
111 E17 **Čáslav** *Ger.* Tschaslau. Střední Čechy, C Czech Republic 49°54′N 15°23′E
56 C13 **Casma** Ancash, C Peru 09°30′S 78°18′W
167 S7 **Ca, Sông** ≈ N Vietnam
107 K17 **Casoria** Campania, S Italy 40°54′N 14°28′E
105 T6 **Caspe** Aragón, NE Spain 41°14′N 00°03′W
33 X15 **Casper** Wyoming, C USA 42°48′N 106°22′W
84 M10 **Caspian Depression** *Kaz.* Kaspīy Mangy Oypaty, *Rus.* Prikaspiyskaya Nizmennost’. *depression* Kazakhstan/Russian Federation
130 D10 **Caspian Sea** *Az.* Xäzär Dänizi, *Kaz.* Kaspiy Tengizi, *Per.* Baḩr-e Khazar, Daryā-ye Khazar, *Rus.* Kaspiyskoye More. *inland sea* Asia/Europe
83 L14 **Cassacatiza** Tete, NW Mozambique 14°20′S 32°24′E
Cassai *see* Kasai
82 F13 **Cassamba** Moxico, E Angola 13°07′S 20°22′E
107 N20 **Cassano allo Ionio** Calabria, SE Italy 39°46′N 16°16′E
31 S8 **Cass City** Michigan, N USA 43°36′N 83°10′W
Cassel *see* Kassel
14 M13 **Casselman** Ontario, SE Canada 45°18′N 75°05′W
29 R5 **Casselton** North Dakota, N USA 46°53′N 97°10′W
Cássia *see* Santa Rita de Cassia
10 J9 **Cassiar** British Columbia, W Canada 59°16′N 129°40′W
10 K10 **Cassiar Mountains** ▲▲ British Columbia, W Canada
83 C15 **Cassinga** Huíla, SW Angola 15°08′S 16°05′E
29 T4 **Cass Lake** Minnesota, N USA 47°22′N 94°36′W
29 T4 **Cass Lake** ◎ Minnesota, N USA
31 P10 **Cassopolis** Michigan, N USA 41°56′N 86°00′W
31 S8 **Cass River** ≈ Michigan, N USA
27 S8 **Cassville** Missouri, C USA 36°42′N 93°52′W
Castamoni *see* Kastamonu
58 L12 **Castanhal** Pará, NE Brazil 01°16′S 47°55′W
104 G8 **Castanheira de Pêra** Leiria, C Portugal 40°01′N 08°12′W
41 N7 **Castaños** Coahuila, NE Mexico 26°48′N 101°26′W
108 I10 **Castasegna** Graubünden, SE Switzerland 46°21′N 09°30′E
106 D8 **Casteggio** Lombardia, N Italy 45°02′N 09°10′E
107 K23 **Castelbuono** Sicilia, Italy, C Mediterranean Sea 37°56′N 14°05′E
107 K15 **Castel di Sangro** Abruzzo, C Italy 41°46′N 14°03′E
106 H7 **Castelfranco Veneto** Veneto, NE Italy 45°40′N 11°55′E
102 K14 **Casteljaloux** Lot-et-Garonne, SW France 44°19′N 00°03′E
107 L18 **Castellabate** *var.* Santa Maria di Castellabate. Campania, S Italy 40°16′N 14°57′E
107 I23 **Castellammare del Golfo** Sicilia, Italy, C Mediterranean Sea 38°02′N 12°53′E
107 H22 **Castellammare, Golfo di** *gulf* Sicilia, Italy, C Mediterranean Sea
103 U15 **Castellane** Alpes-de-Haute-Provence, SE France 43°49′N 06°34′E
107 O18 **Castellaneta** Puglia, SE Italy 40°38′N 16°57′E
106 E9 **Castel l’Arquato** Emilia-Romagna, C Italy 44°52′N 09°51′E
61 E21 **Castelli** Buenos Aires, E Argentina 36°07′S 57°47′W
105 S8 **Castelló de la Plana** *var.* Castellón de la Plana. ◐ *province* Valenciana, E Spain
Castelló de la Plana *see* Castellón de la Plana
Castellón *see* Castellón de la Plana
105 T9 **Castellón de la Plana** *var.* Castellón, *Cat.* Castelló de la Plana. Valenciana, E Spain 39°59′N 00°03′W
Castellón de la Plana *see* Castelló de la Plana
105 S7 **Castellote** Aragón, NE Spain 40°46′N 00°18′W
103 N16 **Castelnaudary** Aude, S France 43°18′N 01°57′E
102 L16 **Castelnau-Magnoac** Hautes-Pyrénées, S France 43°18′N 00°30′E
106 F10 **Castelnovo ne’ Monti** Emilia-Romagna, C Italy 44°26′N 10°24′E
Castelnuovo *see* Herceg-Novi
104 H9 **Castelo Branco** Castelo Branco, C Portugal 39°50′N 07°30′W
104 H8 **Castelo Branco** ◐ *district* C Portugal
104 I10 **Castelo de Vide** Portalegre, C Portugal 39°25′N 07°27′W
104 G9 **Castelo do Bode, Barragem do** ⊠ C Portugal

◆ Country
● Country Capital
◇ Dependent Territory
○ Dependent Territory Capital
◐ Administrative Regions
✈ International Airport
▲ Mountain
▲▲ Mountain Range
⛰ Volcano
≈ River
◎ Lake
⊠ Reservoir

◆ Country ◇ Dependent Territory ■ Administrative Regions ▲ Mountain ☆ Volcano ◙ Lake
● Country Capital ○ Dependent Territory Capital ✈ International Airport ▲ Mountain Range ✍ River □ Reservoir

Chāībāsa – Chenstokhov

153 Q16 **Chāībāsa** Jhārkhand, N India 22°31′N 85°50′E
79 E19 **Chaillu, Massif du** ⛰ C Gabon
167 O10 **Chai Nat** *var.* Chainat, Jainat, Jayanath. Chai Nat, C Thailand 15°10′N 100°10′E
Chainat *see* Chai Nat
65 M14 **Chain Fracture Zone** *tectonic feature* E Atlantic Ocean
173 N5 **Chain Ridge** *undersea feature* W Indian Ocean 06°00′N 54°00′E
Chairn, Ceann an *see* Carnsore Point
158 L5 **Chaiwopu** Xinjiang Uygur Zizhiqu, W China 43°32′N 87°55′E
167 Q10 **Chaiyaphum** *var.* Jayabum. Chaiyaphum, C Thailand 15°46′N 101°55′E
62 N10 **Chajarí** Entre Ríos, E Argentina 30°45′S 57°57′W
42 C5 **Chajul** Quiché, W Guatemala 15°28′N 91°02′W
83 K16 **Chakari** Mashonaland West, N Zimbabwe 18°05′S 29°51′E
148 J9 **Chakhānsūr** Nīmrōz, SW Afghanistan 31°11′N 62°06′E
Chakhānsūr *see* Nīmrōz
Chakhcharan *see* Chaghcharān
149 V8 **Chak Jhumra** *var.* Jhumra. Punjab, E Pakistan 31°33′N 73°14′E
146 I16 **Chaknakdysonga** Ahal Welaýaty, S Turkmenistan 35°39′N 61°24′E
153 P16 **Chakradharpur** Jhāarkhand, N India 22°42′N 85°38′E
152 J8 **Chakrāta** Uttarakhand, N India 30°42′N 77°52′E
149 U7 **Chakwāl** Punjab, NE Pakistan 32°56′N 72°53′E
57 F17 **Chala** Arequipa, SW Peru 15°52′S 74°13′W
102 K12 **Chalais** Charente, W France 45°16′N 00°02′E
108 D10 **Chalais** Valais, SW Switzerland 46°18′N 07°37′E
115 J20 **Chalándri** *var.* Halandri; *prev.* Khalándrion. *prehistoric site* Sýros, Kykládes, Greece, Aegean Sea
188 H6 **Chalan Kanoa** Saipan, S Northern Mariana Islands 15°08′S 145°43′E
188 C16 **Chalan Pago** C Guam
Chalap Dalam/Chalap Dalan *see* Chehel Abdālān, Kūh-e
42 F7 **Chalatenango** Chalatenango, N El Salvador 14°04′N 88°53′W
42 A9 **Chalatenango** ◈ *department* NW El Salvador
83 P15 **Chalaua** Nampula, NE Mozambique 16°04′S 39°08′E
81 I16 **Chalbi Desert** *desert* N Kenya
42 D7 **Chalchuapa** Santa Ana, W El Salvador 13°59′N 89°41′W
Chalcidice *see* Chalkidikí
Chalcis *see* Chalkída
Chālderān *see* Siāh Chashmeh
103 N6 **Châlette-sur-Loing** Loiret, C France 48°01′N 02°45′E
15 X8 **Chaleur Bay** *Fr.* Baie des Chaleurs. *bay* New Brunswick/Québec, E Canada
Chaleurs, Baie des *see* Chaleur Bay
57 G16 **Chalhuanca** Apurímac, S Peru 14°17′S 73°15′W
154 F12 **Chālisgaon** Mahārāshtra, C India 20°29′N 75°10′E
115 N23 **Chálki** *island* Dodekánisa, Greece, Aegean Sea
115 F16 **Chalkiádes** Thessalía, C Greece 39°24′N 22°25′E
115 H18 **Chalkída** *var.* Halkida, *prev.* Khalkís; *anc.* Chalcis. Evvoia, E Greece 38°27′N 23°38′E
115 G14 **Chalkidikí** *var.* Khalkidhikí; *anc.* Chalcidice. *peninsula* NE Greece
185 A24 **Chalky Inlet** *inlet* South Island, New Zealand
39 S7 **Chalkyitsik** Alaska, USA 66°39′N 143°43′W
102 I9 **Challans** Vendée, NW France 46°51′N 01°52′W
57 K19 **Challapata** Oruro, SW Bolivia 18°50′S 66°45′W
192 H6 **Challenger Deep** *undersea feature* W Pacific Ocean 11°20′N 142°12′E
Challenger Deep *see* Mariana Trench
193 S11 **Challenger Fracture Zone** *tectonic feature* SE Pacific Ocean
192 K11 **Challenger Plateau** *undersea feature* E Tasman Sea
33 P13 **Challis** Idaho, NW USA 44°31′N 114°14′W
22 L9 **Chalmette** Louisiana, S USA 29°56′N 89°57′W
124 J11 **Chalna** Respublika Kareliya, NW Russian Federation 61°53′N 33°59′E
103 Q5 **Châlons-en-Champagne** *prev.* Châlons-sur-Marne, *hist.* Arcae Remorum; *anc.* Carolopois. Marne, NE France 48°58′N 04°22′E
Châlons-sur-Marne *see* Châlons-en-Champagne
103 R9 **Chalon-sur-Saône** *anc.* Cabillonum. Saône-et-Loire, C France 46°47′N 04°51′E
Chaltel, Cerro *see* Fitzroy, Monte
102 M11 **Châlus** Haute-Vienne, C France 45°38′N 01°00′E
143 N4 **Chālūs** Māzandarān, N Iran 36°40′N 51°25′E
101 N20 **Cham** Bayern, SE Germany 49°13′N 12°40′E
108 F7 **Cham** Zug, N Switzerland 47°11′N 08°28′E
37 R8 **Chama** New Mexico, SW USA 36°54′N 106°34′W
Cha Mai *see* Thung Song
83 E22 **Chamaites** Karas, S Namibia 27°15′S 17°52′E
149 O9 **Chaman** Baluchistān, SW Pakistan 30°55′N 66°27′E
37 R9 **Chama, Rio** ≈ New Mexico, SW USA
152 I6 **Chamba** Himāchal Pradesh, N India 32°33′N 76°10′E
81 I25 **Chamba** Ruvuma, S Tanzania 11°33′S 37°01′E
150 H12 **Chambal** ≈ C India
11 U16 **Chamberlain** Saskatchewan, S Canada 50°49′N 105°29′W
29 O11 **Chamberlain** South Dakota, N USA 43°48′N 99°19′W
19 R3 **Chamberlain Lake** ◎ Maine, NE USA
39 S5 **Chamberlin, Mount** ▲ Alaska, USA 69°16′N 144°54′W
37 O11 **Chambers** Arizona, SW USA 35°11′N 109°25′W
18 F16 **Chambersburg** Pennsylvania, NE USA 39°54′N 77°39′W
31 N5 **Chambers Island** *island* Wisconsin, N USA
103 T11 **Chambéry** *anc.* Camberia. Savoie, E France 45°34′N 05°56′E
82 L12 **Chambeshi** Muchinga, NE Zambia 10°55′S 31°07′E
82 L12 **Chambeshi** ≈ NE Zambia
74 M6 **Chambi, Jebel** *var.* Jabal ash Sha'nabī. ▲ W Tunisia 35°16′N 08°39′E
15 Q7 **Chambord** Québec, SE Canada 48°25′N 72°02′W
139 U11 **Chamcham** Al Muthanná, S Iraq 31°17′N 45°05′E
Chamchamāļ *see* Chemchemal
Chamdmanī *see* Altanshiree
40 J14 **Chamela** Jalisco, SW Mexico 19°31′N 105°02′W
42 G5 **Chamelecón, Río** ≈ NW Honduras
62 J9 **Chamical** La Rioja, C Argentina 30°21′S 66°19′W
115 L23 **Chamilí** *island* Kykládes, Greece, Aegean Sea
167 Q13 **Châmnar** Kaôh Kông, SW Cambodia 11°45′N 103°32′E
152 K9 **Chamoli** Uttarakhand, N India 30°22′N 79°19′E
103 U11 **Chamonix-Mont-Blanc** Haute-Savoie, E France 45°55′N 06°52′E
154 L11 **Chāmpa** Chhattīsgarh, C India 22°02′N 82°42′E
10 H8 **Champagne** Yukon, W Canada 60°48′N 136°22′W
103 Q5 **Champagne** *cultural region* N France
Champagne *see* Campania
103 Q5 **Champagne-Ardenne** ◈ *region* N France
103 S9 **Champagnole** Jura, E France 46°44′N 05°55′E
30 M13 **Champaign** Illinois, N USA 40°07′N 88°15′W
167 S10 **Champasak** Champasak, S Laos 14°50′N 105°51′E
103 U6 **Champ de Feu** ▲ NE France 48°24′N 07°15′E
13 O7 **Champdoré, Lac** ◎ Québec, NE Canada
42 B6 **Champerico** Retalhuleu, SW Guatemala 14°18′N 91°54′W
108 C11 **Champéry** Valais, SW Switzerland 46°12′N 06°52′E
18 L6 **Champlain** New York, NE USA 44°58′N 73°25′W
18 L9 **Champlain Canal** *canal* New York, NE USA
15 P13 **Champlain, Lac** ◎ Québec, Canada/USA *see also* Champlain, Lake
18 L7 **Champlain, Lake** ◎ Canada/USA *see also* Champlain, Lac
103 S7 **Champlitte** Haute-Saône, E France 47°36′N 05°31′E
41 W13 **Champotón** Campeche, SE Mexico 19°18′N 90°43′W
104 G10 **Chamusca** Santarém, C Portugal 39°21′N 08°29′W
119 O20 **Chamyarysy** *Rus.* Chemerisy. Homyel'skaya Voblasts', SE Belarus 51°42′N 30°27′E
127 P5 **Chamzinka** Respublika Mordoviya, W Russian Federation 54°22′N 45°22′E
Chanáil Mhór, An *see* Grand Canal
Chanak *see* Çanakkale
62 G7 **Chañaral** Atacama, N Chile 26°19′S 70°34′W
104 H13 **Chança, Rio** *var.* Chanza. ≈ Portugal/Spain
57 D14 **Chancay** Lima, W Peru 11°36′S 77°14′W
Chan-chiang/Chanchiang *see* Zhanjiang
62 G13 **Chanco** Maule, C Chile 35°43′S 72°35′W
39 R7 **Chandalar** Alaska, USA 67°30′N 148°29′W
39 R6 **Chandalar River** ≈ Alaska, USA
152 L10 **Chandan Chauki** Uttar Pradesh, N India 28°32′N 80°43′E
153 S16 **Chandannagar** *prev.* Chandernagore. West Bengal, E India 22°52′N 88°21′E
152 K10 **Chandausi** Uttar Pradesh, N India 28°27′N 78°43′E
22 M10 **Chandeleur Islands** *island group* Louisiana, S USA
22 M9 **Chandeleur Sound** *sound* N Gulf of Mexico
Chandernagore *see* Chandannagar
152 I8 **Chandīgarh** *state capital* Punjab, N India 30°41′N 76°51′E
153 Q16 **Chāndil** Jhārkhand, NE India 22°58′N 86°04′E
182 D2 **Chandler** South Australia 26°59′S 133°22′E
15 Y7 **Chandler** Québec, SE Canada 48°21′N 64°41′W
36 L14 **Chandler** Arizona, SW USA 33°18′N 111°50′W
27 O10 **Chandler** Oklahoma, C USA 35°43′N 96°54′W
25 V7 **Chandler** Texas, SW USA 32°18′N 95°28′W
39 Q6 **Chandler River** ≈ Alaska, USA
56 H13 **Chandles, Río** ≈ E Peru
162 H9 **Chandmanī** *var.* Talshand. Govī-Altayī, C Mongolia 45°21′N 98°00′E
162 E7 **Chandmanī** *var.* Urdgol. Hovd, W Mongolia 47°39′N 92°46′E
14 J13 **Chandos Lake** ◎ Ontario, SE Canada
153 U15 **Chandpur** Chittagong, C Bangladesh 23°13′N 90°43′E
154 I13 **Chandrapur** Mahārāshtra, C India 19°58′N 79°21′E
83 J15 **Changa** Southern, S Zambia 16°24′S 28°27′E
Chang'an *see* Rong'an, Guangxi Zhuangzu Zizhiqu, S China
Changan *see* Xi'an, Shaanxi, C China
155 G23 **Changanācheri** *var.* Changannassery. Kerala, SW India 09°26′N 76°31′E *see also* Changannassery
83 M19 **Changane** ≈ S Mozambique
83 M16 **Changara** Tete, NW Mozambique 16°54′S 33°15′E
163 X11 **Changbai** *var.* Changbai Chaoxianzu Zizhixian. Jilin, NE China 41°25′N 128°08′E
Changbai Chaoxianzu Zizhixian *see* Changbai
163 X11 **Changbai Shan** ⛰ NE China
163 V10 **Changchun** *var.* Ch'angch'un, Ch'ang-ch'un; *prev.* Hsinking. *province capital* Jilin, NE China 43°53′N 125°18′E
Ch'angch'un/Ch'ang-ch'un *see* Changchun
160 M10 **Changde** Hunan, S China 29°04′N 111°42′E
Changhua *see* Zhanghua
168 L10 **Changi** ✈ (Singapore) E Singapore 01°22′N 103°58′E
158 L5 **Changji** Xinjiang Uygur Zizhiqu, NW China 44°02′N 87°12′E
160 L17 **Changjiang** *var.* Changjiang Lizu Zizhixian, Shiliu. Hainan, S China 19°16′N 109°09′E
157 R11 **Chang Jiang** *var.* Yangtze Kiang, *Eng.* Yangtze. ≈ C China
157 N12 **Chang Jiang** *Eng.* Yangtze. ≈ SW China
161 S8 **Changjiang Kou** *delta* E China
Changjiang Lizu Zizhixian *see* Changjiang
Changkiakow *see* Zhangjiakou
167 F12 **Chang, Ko** *island* S Thailand
161 Q2 **Changli** Hebei, E China 39°44′N 119°13′E
163 V10 **Changling** Jilin, NE China 44°15′N 124°03′E
Changning *see* Xunwu
161 N11 **Changsha** *var.* Ch'angsha, Ch'ang-sha. *province capital* Hunan, S China 28°10′N 113°E
Ch'angsha/Ch'ang-sha *see* Changsha
161 Q10 **Changshan** Zhejiang, SE China 28°54′N 118°30′E
163 V14 **Changshan Qundao** *island group* NE China
161 S8 **Changshu** *var.* Ch'ang-shu. Jiangsu, E China 31°39′N 120°45′E
Ch'ang-shu *see* Changshu
163 V11 **Changtu** Liaoning, NE China 42°50′N 123°59′E
43 P14 **Changuinola** Bocas del Toro, NW Panama 09°28′N 82°31′W
159 N9 **Changweiliang** Qinghai, W China 38°24′N 92°08′E
160 K6 **Changwu** *var.* Zhaoren. Shaanxi, C China 35°12′N 107°46′E
163 U13 **Changxing Dao** *island* N China
160 M9 **Changyang** *var.* Longzhouping. Hubei, C China 30°45′N 111°13′E
163 W14 **Changyŏn** SW North Korea 38°19′N 125°15′E
161 N5 **Changzhi** Shanxi, C China 36°10′N 113°02′E
161 R8 **Changzhou** Jiangsu, E China 31°45′N 119°58′E
115 H24 **Chaniá** *var.* Hania, Khaniá, *Eng.* Canea; *anc.* Cydonia. Kríti, Greece, E Mediterranean Sea 35°31′N 24°00′E
62 J5 **Chañi, Nevado de** ▲ NW Argentina 24°09′S 65°44′W
115 H24 **Chanión, Kólpos** *gulf* Kríti, Greece, E Mediterranean Sea
Chankiri *see* Çankırı
30 M11 **Channahon** Illinois, N USA 41°25′N 88°13′W
155 H20 **Channapatna** Karnātaka, E India 12°43′N 77°14′E
97 K26 **Channel Islands** *Fr.* Iles Normandes. *island group* S English Channel
35 R16 **Channel Islands** *island group* California, W USA
13 S13 **Channel-Port aux Basques** Newfoundland and Labrador, SE Canada 47°35′N 59°02′W
Channel, The *see* English Channel
97 Q23 **Channel Tunnel** *tunnel* France/United Kingdom
24 M2 **Channing** Texas, SW USA 35°41′N 102°21′W
Chantabun/Chantaburi *see* Chanthaburi
104 H3 **Chantada** Galicia, NW Spain 42°36′N 07°46′W
167 P12 **Chanthaburi** *var.* Chantabun, Chantaburi. Chantaburi, S Thailand 12°35′N 102°08′E
103 O4 **Chantilly** Oise, N France 49°12′N 02°28′E
139 V12 **Chanūn as Sa'ūdī** Dhī Qār, S Iraq 31°04′N 46°00′E
27 Q6 **Chanute** Kansas, C USA 37°40′N 95°27′W
Chanza *see* Chança, Rio
Ch'ao-an/Chaochow *see* Chaozhou
161 P8 **Chao Hu** ◎ E China
167 P11 **Chao Phraya, Mae Nam** ≈ C Thailand
163 T8 **Chaor He** *prev.* Qulin Gol. ≈ NE China
Chaouèn *see* Chefchaouen
161 P14 **Chaoyang** Guangdong, S China 23°17′N 116°33′E
163 T12 **Chaoyang** Liaoning, NE China 41°34′N 120°29′E
Chaoyang *see* Jiayin, Heilongjiang, China
Chaoyang *see* Huinan, Jilin China
161 Q14 **Chaozhou** *var.* Chaoan, Chao'an, Ch'ao-an; *prev.* Chaochow. Guangdong, SE China 23°42′N 116°36′E
58 N13 **Chapadinha** Maranhão, E Brazil 03°45′S 43°23′W
12 K12 **Chapais** Québec, SE Canada 49°47′N 74°54′W
40 L13 **Chapala** Jalisco, SW Mexico 20°20′N 103°10′W
40 L13 **Chapala, Lago de** ◎ C Mexico
146 F13 **Chapan, Gora** ▲ C Turkmenistan 37°48′N 58°03′E
57 M18 **Chapare, Río** ≈ C Bolivia
54 E11 **Chaparral** Tolima, C Colombia 03°45′N 75°30′W
144 F9 **Chapayev** Zapadnyy Kazakhstan, NW Kazakhstan 50°12′N 51°09′E
123 O11 **Chapayevo** Respublika Sakha (Yakutiya), NE Russian Federation 60°03′N 117°19′E
127 R6 **Chapayevsk** Samarskaya Oblast', W Russian Federation 52°57′N 49°42′E
60 H13 **Chapecó** Santa Catarina, S Brazil 27°14′S 52°41′W
60 I13 **Chapecó, Rio** ≈ S Brazil
20 J9 **Chapel Hill** Tennessee, S USA 35°38′N 86°40′W
44 J12 **Chapelton** C Jamaica 18°05′N 77°16′W
14 C8 **Chapleau** Ontario, SE Canada 47°50′N 83°24′W
14 D7 **Chapleau** ≈ Ontario, S Canada
11 T16 **Chaplin** Saskatchewan, S Canada 50°27′N 106°37′W
126 M6 **Chaplygin** Lipetskaya Oblast', W Russian Federation 53°13′N 39°58′E
117 S11 **Chaplynka** Khersons'ka Oblast', S Ukraine 46°20′N 33°34′E
9 O6 **Chapman, Cape** *headland* Nunavut, NE Canada 69°15′N 89°09′W
25 T15 **Chapman Ranch** Texas, SW USA 27°32′N 97°25′W
Chapman's *see* Okwa
21 P5 **Chapmanville** West Virginia, NE USA 37°58′N 82°01′W
28 K15 **Chappell** Nebraska, C USA 41°05′N 102°28′W
Chapra *see* Chhapra
56 D9 **Chapuli, Río** ≈ N Peru
76 I6 **Châr** *well* N Mauritania
123 P12 **Chara** Zabaykal'skiy Kray, S Russian Federation 56°57′N 118°05′E
123 O11 **Chara** ≈ C Russian Federation
54 G8 **Charala** Santander, C Colombia 06°17′N 73°09′W
41 N10 **Charcas** San Luis Potosí, C Mexico 23°09′N 101°10′W
25 T13 **Charco** Texas, SW USA 28°42′N 97°35′E
194 H7 **Charcot Island** *island* Antarctica
64 M8 **Charcot Seamounts** *undersea feature* E Atlantic Ocean 11°30′W 45°00′N
Chardara *see* Shardara
Chardarinskoye Vodokhranilishche *see* Shardarinskoye Vodokhranilishche
31 U11 **Chardon** Ohio, N USA 41°34′N 81°12′W
44 K9 **Chardonnières** SW Haiti 18°16′N 74°10′W
Chardzhev *see* Türkmenabat
Chardzhevskaya Oblast *see* Lebap Welaýaty
Chardzhou/Chardzhui *see* Türkmenabat
102 L11 **Charente** ◈ *department* W France
102 J11 **Charente** ≈ W France
102 J10 **Charente-Maritime** ◈ *department* W France
137 U12 **Ch'arents'avan** C Armenia 40°23′N 44°41′E
78 I12 **Chari** *var.* Shari. ≈ Central African Republic/Chad
78 G11 **Chari-Baguirmi** *off.* Région du Chari-Baguirmi. ◈ *region* SW Chad
Chari-Baguirmi, Région du *see* Chari-Baguirmi
149 Q4 **Chārīkār** Parwān, NE Afghanistan 35°01′N 69°11′E
29 V15 **Chariton** Iowa, C USA 41°00′N 93°18′W
27 U3 **Chariton River** ≈ Missouri, C USA
55 T7 **Charity** NW Guyana 07°22′N 58°34′W
31 R7 **Charity Island** *island* Michigan, N USA
Chärjew *see* Türkmenabat
Chärjew Oblasty *see* Lebap Welaýaty
Charkhlik/Charkhliq *see* Ruoqiang
99 G20 **Charleroi** Hainaut, S Belgium 50°25′N 04°27′E
11 V12 **Charles** Manitoba, C Canada 55°27′N 100°58′W
15 R10 **Charlesbourg** Québec, SE Canada 46°50′N 71°15′W
21 Y7 **Charles, Cape** *headland* Virginia, NE USA 37°09′N 75°57′W
29 W12 **Charles City** Iowa, C USA 43°04′N 92°40′W
21 W6 **Charles City** Virginia, NE USA 37°21′N 77°05′W
103 O5 **Charles de Gaulle** ✈ (Paris) Seine-et-Marne, N France 49°04′N 02°36′E
12 K1 **Charles Island** *island* Nunavut, NE Canada
Charles Island *see* Santa María, Isla
30 K9 **Charles Mound** *hill* Illinois, N USA
185 A22 **Charles Sound** *sound* South Island, New Zealand
185 G15 **Charleston** West Coast, South Island, New Zealand 41°54′S 171°25′E
27 S11 **Charleston** Arkansas, C USA 35°19′N 94°02′W
30 M14 **Charleston** Illinois, N USA 39°30′N 88°10′W
22 L3 **Charleston** Mississippi, S USA 34°00′N 90°03′W
27 Z7 **Charleston** Missouri, C USA 36°54′N 89°22′W
21 T15 **Charleston** South Carolina, SE USA 32°48′N 79°57′W
21 Q5 **Charleston** *state capital* West Virginia, NE USA 38°21′N 81°38′W
14 L14 **Charleston Lake** ◎ Ontario, SE Canada
35 W11 **Charleston Peak** ▲ Nevada, W USA 36°16′N 115°40′W
45 W10 **Charlestown** Nevis, Saint Kitts and Nevis 17°08′N 62°37′W
31 P16 **Charlestown** Indiana, N USA 38°27′N 85°40′W
18 M9 **Charlestown** New Hampshire, NE USA 43°14′N 72°23′W
21 V3 **Charles Town** West Virginia, NE USA 39°18′N 77°54′W
181 W9 **Charleville** Queensland, E Australia 26°25′S 146°18′E
103 R3 **Charleville-Mézières** Ardennes, N France 49°45′N 04°43′E
31 P5 **Charlevoix** Michigan, N USA 45°19′N 85°15′W
31 Q6 **Charlevoix, Lake** ◎ Michigan, N USA
39 T9 **Charley River** ≈ Alaska, USA
64 J6 **Charlie-Gibbs Fracture Zone** *tectonic feature* N Atlantic Ocean
103 Q10 **Charlieu** Loire, E France 46°11′N 04°10′E
31 Q9 **Charlotte** Michigan, N USA 42°33′N 84°50′W
21 R10 **Charlotte** North Carolina, SE USA 35°14′N 80°51′W
20 I8 **Charlotte** Tennessee, S USA 36°11′N 87°18′W
25 R13 **Charlotte** Texas, SW USA 28°51′N 98°42′W
21 R10 **Charlotte** ✈ North Carolina, SE USA 35°12′N 80°54′W
45 T9 **Charlotte Amalie** *prev.* Saint Thomas. ○ (Virgin Islands (US)) Saint Thomas, N Virgin Islands (US) 18°22′N 64°56′W
21 U7 **Charlotte Court House** Virginia, NE USA 37°04′N 78°37′W
23 W14 **Charlotte Harbor** *inlet* Florida, SE USA
Charlotte Island *see* Abaiang
95 J15 **Charlottenberg** Värmland, C Sweden 59°53′N 12°17′E
Charlottenhof *see* Aegviidu
21 U5 **Charlottesville** Virginia, NE USA 38°02′N 78°29′W
13 Q14 **Charlottetown** *province capital* Prince Edward Island, Prince Edward Island, SE Canada 46°14′N 63°09′W
Charlotte Town *see* Roseau, Dominica
Charlotte Town *see* Gouyave, Grenada
45 Z16 **Charlotteville** Tobago, Trinidad and Tobago 11°16′N 60°33′W
182 M11 **Charlton** Victoria, SE Australia 36°18′S 143°19′E
12 H10 **Charlton Island** *island* Northwest Territories, C Canada
103 T6 **Charmes** Vosges, NE France 48°19′N 06°19′E
119 F19 **Charnawchytsy** *Rus.* Chernavchitsy. Brestskaya Voblasts', SW Belarus 52°13′N 23°44′E
15 R10 **Charny** Québec, SE Canada 46°43′N 71°15′W
149 T5 **Chārsadda** Khyber Pakhtunkhwa, NW Pakistan 34°12′N 71°46′E
Charshanga/Charshangngy/Charshangy *see* Köýtendag
Charsk *see* Shar
181 W6 **Charters Towers** Queensland, NE Australia 20°02′S 146°20′E
15 R12 **Chartierville** Québec, SE Canada 45°19′N 71°13′W
102 M6 **Chartres** *anc.* Autricum, Civitas Carnutum. Eure-et-Loir, C France 48°27′N 01°27′E
Charyn *see* Sharyn
Charyn *see* Sharyn
61 D21 **Chascomús** Buenos Aires, E Argentina 35°34′S 58°01′W
11 N16 **Chase** British Columbia, SW Canada 50°49′N 119°41′W
21 U7 **Chase City** Virginia, NE USA 36°48′N 78°27′W
19 S4 **Chase, Mount** ▲ Maine, NE USA 46°06′N 68°30′W
118 M13 **Chashniki** Vitsyebskaya Voblasts', N Belarus 54°52′N 29°10′E
115 D15 **Chásia** ⛰ C Greece
29 V9 **Chaska** Minnesota, N USA 44°47′N 93°36′W
185 D25 **Chaslands Mistake** *headland* South Island, New Zealand 46°37′S 169°21′E
125 R11 **Chasovo** Respublika Komi, NW Russian Federation 61°58′N 50°34′E
Chasovo *see* Vazhgort
124 H14 **Chastova** Novgorodskaya Oblast', NW Russian Federation 58°37′N 32°05′E
143 R3 **Chāt** Golestān, N Iran 37°52′N 55°27′E
Chatak *see* Chhatak
Chatang *see* Zhanang
39 R9 **Chatanika** Alaska, USA 65°06′N 147°28′W
39 R9 **Chatanika River** ≈ Alaska, USA
147 T8 **Chat-Bazar** Talasskaya Oblast', NW Kyrgyzstan 42°29′N 72°37′E
45 Y14 **Chateaubelair** Saint Vincent, W Saint Vincent and the Grenadines 13°15′N 61°05′W
102 J7 **Châteaubriant** Loire-Atlantique, NW France 47°43′N 01°22′W
103 Q8 **Château-Chinon** Nièvre, C France 47°04′N 03°50′E
108 C10 **Château d'Oex** Vaud, W Switzerland 46°28′N 07°09′E
102 L7 **Château-du-Loir** Sarthe, NW France 47°40′N 00°25′E
102 M6 **Châteaudun** Eure-et-Loir, C France 48°04′N 01°20′E
102 K7 **Château-Gontier** Mayenne, NW France 47°49′N 00°42′W
15 O13 **Châteauguay** Québec, SE Canada 45°22′N 73°44′W
102 F6 **Châteaulin** Finistère, NW France 48°12′N 04°07′W
103 N9 **Châteaumeillant** Cher, C France 46°33′N 02°10′E
102 K11 **Châteauneuf-sur-Charente** Charente, W France 45°34′N 00°33′W
102 M7 **Château-Renault** Indre-et-Loire, C France 47°34′N 00°52′E
103 N9 **Châteauroux** *prev.* Indreville. Indre, C France 46°50′N 01°43′E
103 T5 **Château-Salins** Moselle, NE France 48°50′N 06°29′E
103 P4 **Château-Thierry** Aisne, N France 49°03′N 03°24′E
99 H21 **Châtelet** Hainaut, S Belgium 50°24′N 04°32′E
Châtelherault *see* Châtellerault
102 L9 **Châtellerault** *var.* Châtelherault. Vienne, W France 46°49′N 00°33′E
29 X10 **Chatfield** Minnesota, N USA 43°51′N 92°11′W
13 O14 **Chatham** New Brunswick, SE Canada 47°02′N 65°30′W
14 D17 **Chatham** Ontario, S Canada 42°24′N 82°11′W
97 P22 **Chatham** SE England, United Kingdom 51°23′N 00°31′E
30 K14 **Chatham** Illinois, N USA 39°40′N 89°42′W
21 T7 **Chatham** Virginia, NE USA 36°49′N 79°26′W
63 F22 **Chatham, Isla** *island* S Chile
175 R12 **Chatham Island** *island* Chatham Islands, New Zealand
Chatham Island *see* San Cristóbal, Isla
Chatham Island Rise *see* Chatham Rise
175 R12 **Chatham Islands** *island group* New Zealand, SW Pacific Ocean
175 Q12 **Chatham Rise** *var.* Chatham Island Rise. *undersea feature* S Pacific Ocean
39 X13 **Chatham Strait** *strait* Alaska, USA
Chathóir, Rinn *see* Cahore Point
102 M9 **Châtillon-sur-Indre** Indre, C France 46°58′N 01°10′E
103 Q7 **Châtillon-sur-Seine** Côte d'Or, C France 47°51′N 04°30′E
147 S8 **Chatkal** *Uzb.* Chotqol. ≈ Kyrgyzstan/Uzbekistan
147 R9 **Chatkal Range** *Rus.* Chatkal'skiy Khrebet. ⛰ Kyrgyzstan/Uzbekistan
Chatkal'skiy Khrebet *see* Chatkal Range
23 N7 **Chatom** Alabama, S USA 31°28′N 88°15′W
Chatrapur *see* Chhatrapur
143 S10 **Chatrūd** Kermān, C Iran 30°39′N 56°57′E
23 S2 **Chatsworth** Georgia, SE USA 34°46′N 84°46′W
Chāttagām *see* Chittagong
23 S8 **Chattahoochee** Florida, SE USA 30°40′N 84°51′W
23 R8 **Chattahoochee River** ≈ SE USA
20 L10 **Chattanooga** Tennessee, S USA 35°05′N 85°16′W
147 V10 **Chatyr-Kël', Ozero** ◎ C Kyrgyzstan
147 W9 **Chatyr-Tash** Narynskaya Oblast', C Kyrgyzstan 40°54′N 76°22′E
15 R12 **Chaudière** ≈ Québec, SE Canada
167 S14 **Châu Đôc** *var.* Chauphu, Chau Phu. An Giang, S Vietnam 10°53′N 105°07′E
152 D13 **Chauhtan** *prev.* Chohtan. Rājasthān, NW India 25°27′N 71°08′E
166 L5 **Chauk** Magway, W Myanmar (Burma) 20°52′N 94°50′E
103 R6 **Chaumont** *prev.* Chaumont-en-Bassigny. Haute-Marne, N France 48°07′N 05°08′E
Chaumont-en-Bassigny *see* Chaumont
123 T5 **Chaunskaya Guba** *bay* NE Russian Federation
103 P3 **Chauny** Aisne, N France 49°37′N 03°13′E
167 U10 **Châu Ô** *var.* Binh Sơn. Quang Ngai, C Vietnam 15°18′N 108°45′E
Chau Phu *see* Châu Đôc
102 I5 **Chausey, Îles** *island group* N France
Chausy *see* Chavusy
18 C11 **Chautauqua Lake** ◎ New York, NE USA
ChâuThanh *see* Ba Ria
102 L9 **Chauvigny** Vienne, W France 46°35′N 00°37′E
124 L6 **Chavan'ga** Murmanskaya Oblast', NW Russian Federation 66°07′N 37°44′E
14 K10 **Chavannes, Lac** ◎ Québec, SE Canada
Chavantes, Represa de *see* Xavantes, Represa de
61 D15 **Chavarría** Corrientes, NE Argentina 28°57′S 58°35′W
Chavash Respubliki *see* Chuvashskaya Respublika
104 I5 **Chaves** *anc.* Aquae Flaviae. Vila Real, N Portugal 41°44′N 07°28′W
Chávez, Isla *see* Santa Cruz, Isla
82 G13 **Chavuma** North Western, NW Zambia 13°04′S 22°43′E
119 O16 **Chavusy** *Rus.* Chausy. Mahilyowskaya Voblasts', E Belarus 53°48′N 30°58′E
Chayan *see* Shayan
147 U8 **Chayek** Narynskaya Oblast', C Kyrgyzstan 41°54′N 74°28′E
139 T6 **Chāy Khānah** Diyālá, E Iraq 34°19′N 44°33′E
125 T16 **Chaykovskiy** Permskiy Kray, NW Russian Federation 56°45′N 54°09′E
167 T12 **Chbar** Môndól Kiri, E Cambodia 12°46′N 107°10′E
23 Q4 **Cheaha Mountain** ▲ Alabama, S USA 33°29′N 85°48′W
Cheatharlach *see* Carlow
21 S2 **Cheat River** ≈ NE USA
111 A16 **Cheb** *Ger.* Eger. Karlovarský Kraj, W Czech Republic 50°05′N 12°23′E
127 Q3 **Cheboksary** Chuvashskaya Respublika, W Russian Federation 56°06′N 47°15′E
31 Q5 **Cheboygan** Michigan, N USA 45°40′N 84°28′W
Chechaouèn *see* Chefchaouen
Chechenia *see* Chechenskaya Respublika
127 O15 **Chechenskaya Respublika** *Eng.* Chechenia, Chechnia, *Rus.* Chechnya. ◈ *autonomous republic* SW Russian Federation
67 N4 **Chech, Erg** *desert* Algeria/Mali
Chechevichi *see* Chachevichy
Che-chiang *see* Zhejiang
Chechnia/Chechnya *see* Chechenskaya Respublika
Chech'ŏn *see* Jecheon
111 L15 **Chęciny** Świętokrzyskie, S Poland 50°51′N 20°31′E
27 Q10 **Checotah** Oklahoma, C USA 35°28′N 95°31′W
13 R15 **Chedabucto Bay** *inlet* Nova Scotia, E Canada
166 J7 **Cheduba Island** *island* W Myanmar (Burma)
37 T5 **Cheesman Lake** ⊠ Colorado, C USA
195 S16 **Cheetham, Cape** *headland* Antarctica 70°26′S 162°40′E
74 G5 **Chefchaouen** *var.* Chaouèn, Chechaouèn, *Sp.* Xauen. N Morocco 35°10′N 05°16′W
Chefoo *see* Yantai
38 M12 **Chefornak** Alaska, USA 60°09′N 164°09′W
123 R13 **Chegdomyn** Khabarovskiy Kray, SE Russian Federation 51°09′N 132°58′E
76 M4 **Chegga** Tiris Zemmour, NE Mauritania 25°27′N 05°49′W
Cheghcheran *see* Chaghcharān
32 G9 **Chehalis** Washington, NW USA 46°39′N 122°57′W
32 G9 **Chehalis River** ≈ Washington, NW USA
148 M6 **Chehel Abdālān, Kūh-e** *var.* Chalap Dalam, *Pash.* Chalap Dalan. ⛰ C Afghanistan
115 D14 **Cheimadítis, Límni** *var.* Límni Cheimadítis. ◎ N Greece
Cheimadítis, Límni *see* Cheimadítis, Límni
103 U15 **Cheiron, Mont** ▲ SE France 43°49′N 07°00′E
163 Y17 **Cheju** ✈ S South Korea 33°31′N 126°29′E
Cheju *see* Jeju
Cheju-do *see* Jeju-do
Cheju-haehyeop *see* Jeju-haehyeop
Cheju Strait *see* Jeju-haehyeop
Chekiang *see* Zhejiang
Chekichler/Chekishlyar *see* Çekiçler
188 F8 **Chelab** Babeldaob, N Palau
147 N11 **Chelak** *Rus.* Chelek. Samarqand Viloyati, C Uzbekistan 39°55′N 66°45′E
32 J7 **Chelan, Lake** ⊠ Washington, NW USA
Chelek *see* Chelak
Cheleken *see* Hazar
Chélif/Chéliff *see* Chelif, Oued
74 J5 **Chelif, Oued** *var.* Chélif, Chéliff, Chellif, Shellif. ≈ N Algeria
Chelkar *see* Shalkar
Chelkar Ozero *see* Shalkar, Ozero
Chellif *see* Chelif, Oued
111 P14 **Chełm** *Rus.* Kholm. Lubelskie, SE Poland 51°08′N 23°29′E
110 I9 **Chełmno** *Ger.* Culm, Kulm. Kujawski-pomorskie, C Poland 53°21′N 18°27′E
115 E19 **Chelmós** *var.* Ároania. ⛰ S Greece
14 F10 **Chelmsford** Ontario, S Canada 46°33′N 81°16′W
97 P21 **Chelmsford** E England, United Kingdom 51°44′N 00°28′E
110 J9 **Chełmża** *Ger.* Culmsee, Kulmsee. Kujawski-pomorskie, C Poland 53°11′N 18°34′E
27 Q8 **Chelsea** Oklahoma, C USA 36°32′N 95°25′W
18 M8 **Chelsea** Vermont, NE USA 43°58′N 72°29′W
97 L21 **Cheltenham** C England, United Kingdom 51°54′N 02°04′W
105 R9 **Chelva** Valenciana, E Spain 39°45′N 01°00′W
122 G11 **Chelyabinsk** Chelyabinskaya Oblast', C Russian Federation 55°12′N 61°25′E
122 F1 **Chelyabinskaya Oblast'** ◈ *province* C Russian Federation
123 N5 **Chelyuskin, Mys** *headland* N Russian Federation 77°42′N 104°13′E
41 Y12 **Chemax** Yucatán, SE Mexico 20°41′N 87°54′W
83 N16 **Chemba** Sofala, C Mozambique 17°11′S 34°53′E
82 J13 **Chembe** Luapula, NE Zambia 11°58′S 28°45′E
139 T4 **Chemchemal** *Ar.* Juwārtā, *var.* Chamchamāļ. At Ta'mím, N Iraq 35°32′N 44°50′E
Chemenibit *see* Çemenibit
Chemerisy *see* Chamyarysy
116 K7 **Chemerivtsi** Khmel'nyts'ka Oblast', W Ukraine 49°00′N 26°21′E
102 J8 **Chemillé** Maine-et-Loire, NW France 47°15′N 00°42′W
173 X17 **Chemin Grenier** S Mauritius 20°29′S 57°28′E
101 N16 **Chemnitz** *prev.* Karl-Marx-Stadt. Sachsen, E Germany 50°50′N 12°55′E
Chemulpo *see* Incheon
32 H14 **Chemult** Oregon, NW USA 43°14′N 121°48′W
18 G12 **Chemung River** ≈ New York/Pennsylvania, NE USA
149 U8 **Chenāb** ≈ India/Pakistan
39 S9 **Chena Hot Springs** Alaska, USA 65°06′N 146°02′W
18 I11 **Chenango River** ≈ New York, NE USA
168 J7 **Chenderoh, Tasik** ◎ Peninsular Malaysia
15 Q11 **Chêne, Rivière du** ≈ Québec, SE Canada
32 L8 **Cheney** Washington, NW USA 47°29′N 117°34′W
26 M6 **Cheney Reservoir** ⊠ Kansas, C USA
Chengchiatun *see* Liaoyuan
Ch'eng-chou/Chengchow *see* Zhengzhou
161 P1 **Chengde** *var.* Jehol. Hebei, E China 41°N 117°57′E
160 I9 **Chengdu** *var.* Chengtu, Ch'eng-tu. *province capital* Sichuan, C China 30°41′N 104°03′E
161 Q14 **Chenghai** Guangdong, S China 23°30′N 116°42′E
Chenghsien *see* Zhengzhou
160 H13 **Chengjiang** Yunnan, SW China 24°40′N 102°55′E
Chengjiang *see* Taihe
160 L17 **Chengmai** *var.* Jinjiang. Hainan, S China 19°45′N 109°56′E
Chengtu/Ch'eng-tu *see* Chengdu
Chengwen *see* Chindu
159 W12 **Chengxian** *var.* Cheng Xiang. Gansu, C China 33°42′N 105°45′E
Cheng Xiang *see* Chengxian
Chengyang *see* Juxian
Chengzhong *see* Ningming
Chenkiang *see* Zhenjiang
155 J19 **Chennai** *prev.* Madras. *state capital* Tamil Nādu, S India 13°05′N 80°18′E
155 J19 **Chennai** ✈ Tamil Nādu, S India 13°07′N 80°13′E
103 R8 **Chenôve** Côte d'Or, C France 47°16′N 05°00′E
Chenstokhov *see* Częstochowa

◆ Country
● Country Capital
◇ Dependent Territory
○ Dependent Territory Capital
◈ Administrative Regions
✈ International Airport
▲ Mountain
⛰ Mountain Range
🌋 Volcano
≈ River
◎ Lake
⊠ Reservoir

CHENXI – CHMIELNIK

160 L11 **Chenxi** var. Chenyang. Hunan, S China 28°02′N 110°15′E
Chen Xian/Chenxian/Chen Xiang see Chenzhou
Chenyang see Chenxi
161 N12 **Chenzhou** var. Chenxian, Chen Xian, Chen Xiang. Hunan, S China 25°51′N 113°01′E
163 X15 **Cheonan** Jap. Tenan; prev. Ch'ŏnan. W South Korea 36°51′N 127°11′E
163 W13 **Cheongju** prev. Chŏngju. W North Korea 39°44′N 125°13′E
Cheo Reo see A Yun Pa
114 I11 **Chepelare** Smolyan, S Bulgaria 41°44′N 24°41′E
114 I11 **Chepelarska Reka** ≈ S Bulgaria
56 B11 **Chepén** La Libertad, C Peru 07°15′S 79°23′W
62 J10 **Chepes** La Rioja, C Argentina 31°19′S 66°40′W
161 O15 **Chep Lap Kok** ✈ S China 22°23′N 114°11′E
43 U14 **Chepo** Panamá, C Panama 09°09′N 79°03′W
Chepping Wycombe see High Wycombe
125 R14 **Cheptsa** ≈ NW Russian Federation
30 K3 **Chequamegon Point** headland Wisconsin, N USA 46°42′N 90°45′W
103 O8 **Cher** ◈ department C France
102 M8 **Cher** ≈ C France
Cherangani Hills see Cherangany Hills
81 H17 **Cherangany Hills** var. Cherangani Hills. ⛰ W Kenya
21 S11 **Cheraw** South Carolina, SE USA 34°42′N 79°52′W
102 I3 **Cherbourg** anc. Carusbur. Manche, N France 49°40′N 01°36′W
127 R5 **Cherdakly** Ul'yanovskaya Oblast', W Russian Federation 54°21′N 48°54′E
125 U12 **Cherdyn'** Permskiy Kray, NW Russian Federation 60°21′N 56°39′E
124 J14 **Cherekha** ≈ W Russian Federation
122 M13 **Cheremkhovo** Irkutskaya Oblast', S Russian Federation 53°16′N 102°44′E
Cheren see Keren
124 K14 **Cherepovets** Vologodskaya Oblast', NW Russian Federation 59°09′N 37°50′E
125 O11 **Cherevkovo** Arkhangel'skaya Oblast', NW Russian Federation 61°45′N 45°16′E
74 I6 **Chergui, Chott ech** salt lake NW Algeria
Cherikov see Cherykaw
117 P6 **Cherkas'ka Oblast'** var. Cherkasy, Rus. Cherkasskaya Oblast'. ◈ province C Ukraine
Cherkasskaya Oblast' see Cherkas'ka Oblast'
Cherkassy see Cherkasy
117 Q6 **Cherkasy** Rus. Cherkassy. Cherkas'ka Oblast', C Ukraine 49°26′N 32°05′E
Cherkasy see Cherkas'ka Oblast'
126 M15 **Cherkessk** Karachayevo-Cherkesskaya Respublika, SW Russian Federation 44°12′N 42°06′E
122 H12 **Cherlak** Omskaya Oblast', C Russian Federation 54°06′N 74°59′E
122 H12 **Cherlakskoye** Omskaya Oblast', C Russian Federation 53°42′N 74°23′E
125 U13 **Chermoz** Permskiy Kray, NW Russian Federation 58°49′N 56°07′E
Chernavchitsy see Charnawchytsy
125 T3 **Chernaya** Nenetskiy Avtonomnyy Okrug, NW Russian Federation 68°36′N 56°34′E
125 T4 **Chernaya** ≈ NW Russian Federation
Chernigov see Chernihiv
Chernigovskaya Oblast' see Chernihivs'ka Oblast'
117 Q2 **Chernihiv** Rus. Chernigov. Chernihivs'ka Oblast', NE Ukraine 51°28′N 31°19′E
Chernihiv see Chernihivs'ka Oblast'
117 V9 **Chernihivka** Zaporiz'ka Oblast', SE Ukraine 47°11′N 36°10′E
117 P2 **Chernihivs'ka Oblast'** var. Chernihiv, Rus. Chernigovskaya Oblast'. ◈ province NE Ukraine
114 I9 **Cherni Osŭm** ≈ N Bulgaria
116 J8 **Chernivets'ka Oblast'** var. Chernivtsi, Rus. Chernovitskaya Oblast'. ◈ province W Ukraine
114 I9 **Cherni Vit** ≈ NW Bulgaria
114 G10 **Cherni Vrah**, Cherni Vrŭkh. ▲ W Bulgaria 42°33′N 23°18′E
Cherni Vrŭkh see Cherni Vrah
116 K8 **Chernivtsi** Ger. Czernowitz, Rom. Cernăuţi, Rus. Chernovtsy. Chernivets'ka Oblast', W Ukraine 48°18′N 25°55′E
116 M7 **Chernivtsi** Vinnyts'ka Oblast', C Ukraine 48°33′N 28°06′E
Chernivtsi see Chernivets'ka Oblast'
Chernobyl' see Chornobyl'
Cherno More see Black Sea
Chernomorskoye see Chornomors'ke
145 T7 **Chernoretsk** prev. Chernoretskoye. Pavlodar, NE Kazakhstan 52°51′N 76°37′E
Chernoretskoye see Chernoretsk
Chernovitskaya Oblast' see Chernivets'ka Oblast'
Chernovtsy see Chernivtsi
145 U8 **Chernoye** Pavlodar, NE Kazakhstan 51°40′N 77°33′E
Chernoye More see Black Sea
125 U16 **Chernushka** Permskiy Kray, NW Russian Federation 56°30′N 56°07′E
117 N4 **Chernyakhiv** Rus. Chernyakhov. Zhytomyrs'ka Oblast', N Ukraine 50°30′N 28°38′E
Chernyakhov see Chernyakhiv
119 C14 **Chernyakhovsk** Ger. Insterburg. Kaliningradskaya Oblast', W Russian Federation 54°36′N 21°49′E
126 K8 **Chernyanka** Belgorodskaya Oblast', W Russian Federation 50°59′N 37°54′E
125 V5 **Chernyshëva, Gryada** ⛰ NW Russian Federation
144 J14 **Chernyshëva, Zaliv** gulf SW Kazakhstan
123 O10 **Chernyshevskiy** Respublika Sakha (Yakutiya), NE Russian Federation 62°57′N 112°29′E
127 P13 **Chernyye Zemli** plain SW Russian Federation
Chërnyy Irtysh see Ertix He, China/Kazakhstan
Chërnyy Irtysh see Kara Irtysh, Kazakhstan
127 V7 **Chernyy Otrog** Orenburgskaya Oblast', W Russian Federation 52°03′N 56°09′E
29 T12 **Cherokee** Iowa, C USA 42°45′N 95°33′W
26 M8 **Cherokee** Oklahoma, C USA 36°45′N 98°22′W
25 R9 **Cherokee** Texas, SW USA 30°56′N 98°42′W
21 O8 **Cherokee Lake** ⊠ Tennessee, S USA
Cherokees, Lake O' The see Grand Lake O' The Cherokees
44 H1 **Cherokee Sound** Great Abaco, N The Bahamas 26°16′N 77°03′W
153 V13 **Cherrapunji** Meghālaya, NE India 25°16′N 91°42′E
28 L9 **Cherry Creek** ≈ South Dakota, N USA
18 J16 **Cherry Hill** New Jersey, NE USA 39°55′N 75°01′W
27 Q7 **Cherryvale** Kansas, C USA 37°16′N 95°33′W
21 Q10 **Cherryville** North Carolina, SE USA 35°22′N 81°22′W
Cherski Range see Cherskogo, Khrebet
123 T6 **Cherskiy** Respublika Sakha (Yakutiya), NE Russian Federation 68°45′N 161°15′E
123 R8 **Cherskogo, Khrebet** var. Cherski Range. ⛰ NE Russian Federation
Cherso see Cres
126 L10 **Chertkovo** Rostovskaya Oblast', SW Russian Federation 49°22′N 40°10′E
Cherven' see Chervyen'
114 H8 **Cherven Bryag** Pleven, N Bulgaria 43°16′N 24°06′E
116 M4 **Chervonoarmiys'k** Zhytomyrs'ka Oblast', N Ukraine 50°27′N 28°15′E
Chervonograd see Chervonohrad
116 I4 **Chervonohrad** Rus. Chervonograd. L'vivs'ka Oblast', NW Ukraine 50°25′N 24°10′E
117 W6 **Chervonooskil's'ke Vodoskhovyshche** Rus. Krasnoosol'skoye Vodokhranilishche. ⊠ NE Ukraine
Chervonoye, Ozero see Chyrvonaye, Vozyera
117 S4 **Chervonozavods'ke** Poltavs'ka Oblast', C Ukraine 50°24′N 33°22′E
119 L16 **Chervyen'** Rus. Cherven'. Minskaya Voblasts', C Belarus 53°42′N 28°26′E
119 P16 **Cherykaw** Rus. Cherikov. Mahilyowskaya Voblasts', E Belarus 53°34′N 31°23′E
31 R9 **Chesaning** Michigan, N USA 43°10′N 84°07′W
21 X5 **Chesapeake Bay** inlet NE USA
Chesha Bay see Chëshskaya Guba
Cheshevlya see Tsyeshawlya
97 K18 **Cheshire** cultural region C England, United Kingdom
125 P5 **Chëshskaya Guba** var. Archangel Bay, Chesha Bay, Dvina Bay. bay NW Russian Federation
14 F14 **Chesley** Ontario, S Canada 44°17′N 81°06′W
21 Q10 **Chesnee** South Carolina, SE USA 35°09′N 81°51′W
97 K18 **Chester** Wel. Caerleon, hist. Legaceaster, Lat. Deva, Devana Castra. C England, United Kingdom 53°12′N 02°54′W
35 O4 **Chester** California, W USA 40°18′N 121°14′W
30 K16 **Chester** Illinois, N USA 37°54′N 89°49′W
33 S7 **Chester** Montana, NW USA 48°30′N 110°59′W
18 I16 **Chester** Pennsylvania, NE USA 39°51′N 75°21′W
21 R1 **Chester** South Carolina, SE USA 34°43′N 81°14′W
25 X9 **Chester** Texas, SW USA 30°55′N 94°36′W
21 W6 **Chester** Virginia, NE USA 37°22′N 77°27′W
21 R11 **Chester** West Virginia, NE USA 40°34′N 80°33′W
97 M18 **Chesterfield** C England, United Kingdom 53°15′N 01°25′W
21 S11 **Chesterfield** South Carolina, SE USA 34°44′N 80°04′W
21 W6 **Chesterfield** Virginia, NE USA 37°22′N 77°31′W
192 J9 **Chesterfield, Îles** island group NW New Caledonia
9 O9 **Chesterfield Inlet** Nunavut, NW Canada 63°19′N 90°57′W
9 O9 **Chesterfield Inlet** inlet Nunavut, N Canada
21 Y3 **Chester River** ≈ Delaware/Maryland, NE USA
21 X3 **Chestertown** Maryland, NE USA 39°13′N 76°04′W
19 R4 **Chesuncook Lake** ◎ Maine, NE USA
30 J5 **Chetek** Wisconsin, N USA 45°19′N 91°37′W
13 R14 **Chéticamp** Nova Scotia, SE Canada 46°14′N 61°19′W
27 Q8 **Chetopa** Kansas, C USA 37°02′N 95°05′W
41 Y14 **Chetumal** var. Payo Obispo. Quintana Roo, SE Mexico 18°32′N 88°16′W
Chetumal, Bahia/Chetumal, Bahía de see Chetumal Bay
42 G1 **Chetumal Bay** var. Bahia Chetumal, Bahía de Chetumal. bay Belize/Mexico
10 M13 **Chetwynd** British Columbia, W Canada 55°42′N 121°36′W
38 M11 **Chevak** Alaska, USA 61°31′N 165°35′W
36 M12 **Chevelon Creek** ≈ Arizona, SW USA
185 J17 **Cheviot** Canterbury, South Island, New Zealand 42°48′S 173°17′E
96 L13 **Cheviot Hills** hill range England/Scotland, United Kingdom
96 L13 **Cheviot, The** ▲ NE England, United Kingdom 55°28′N 02°10′W
14 M11 **Chevreuil, Lac du** ◎ Québec, SE Canada
81 I16 **Ch'ew Bahir** var. Lake Stefanie. ◎ Ethiopia/Kenya
32 L7 **Chewelah** Washington, NW USA 48°16′N 117°42′W
26 K10 **Cheyenne** Oklahoma, C USA 35°37′N 99°43′W
33 Z17 **Cheyenne** state capital Wyoming, C USA 41°08′N 104°46′W
26 L5 **Cheyenne Bottoms** ◎ Kansas, C USA
16 J8 **Cheyenne River** ≈ South Dakota/Wyoming, N USA
37 W5 **Cheyenne Wells** Colorado, C USA 38°49′N 102°21′W
108 C9 **Cheyres** Vaud, W Switzerland 46°48′N 06°48′E
Chezdi-Oşorheiu see Târgu Secuiesc
153 P13 **Chhapra** prev. Chapra. Bihār, N India 25°50′N 84°42′E
153 V13 **Chhatak** var. Chatak. Sylhet, NE Bangladesh 25°02′N 91°33′E
154 J9 **Chhatarpur** Madhya Pradesh, C India 24°54′N 79°35′E
154 N13 **Chhatrapur** prev. Chatrapur. Odisha, E India 19°26′N 85°02′E
154 K2 **Chhattīsgarh** ◈ state E India
154 L12 **Chhattīsgarh** plain C India
154 I11 **Chhindwāra** Madhya Pradesh, C India 22°04′N 78°58′E
153 T12 **Chhukha** SW Bhutan 27°02′N 89°36′E
Chiai see Jiayi
Chia-i see Jiayi
Chia-mu-ssu see Jiamusi
83 B15 **Chiange** Port. Vila de Almoster. Huíla, SW Angola 15°44′S 13°54′E
Chiang-hsi see Jiangxi
Chiang Kai-shek see Taiwan Taoyuan
167 P8 **Chiang Khan** Loei, E Thailand 17°51′N 101°43′E
167 O7 **Chiang Mai** var. Chiangmai, Chiengmai, Kiangmai. Chiang Mai, NW Thailand 18°48′N 98°59′E
167 O7 **Chiang Mai** ✈ Chiang Mai, NW Thailand 18°44′N 98°53′E
Chiangmai see Chiang Mai
167 O6 **Chiang Rai** var. Chianpai, Chienrai, Muang Chiang Rai. Chiang Rai, NW Thailand 19°56′N 99°51′E
Chiang-su see Jiangsu
Chianning/Chian-ning see Nanjing
Chianpai see Chiang Rai
106 G12 **Chianti** cultural region C Italy
Chiapa see Chiapa de Corzo
41 U16 **Chiapa de Corzo** var. Chiapa. Chiapas, SE Mexico 16°42′N 92°59′W
41 V16 **Chiapas** ◈ state SE Mexico
106 J12 **Chiaravalle** Marche, C Italy 43°36′N 13°19′E
107 N22 **Chiaravalle Centrale** Calabria, SW Italy 38°40′N 16°25′E
106 E7 **Chiari** Lombardia, N Italy 45°33′N 10°00′E
108 H12 **Chiasso** Ticino, S Switzerland 45°51′N 09°02′E
137 S9 **Chiatura** prev. Chiat'ura. C Georgia 42°13′N 43°11′E
Chiat'ura see Chiatura
41 P15 **Chiautla** var. Chiautla de Tapia. Puebla, S Mexico 18°16′N 98°31′W
Chiautla de Tapia see Chiautla
106 D10 **Chiavari** Liguria, NW Italy 44°19′N 09°19′E
106 E6 **Chiavenna** Lombardia, N Italy 46°19′N 09°22′E
Chiayi see Jiayi
Chiazza see Piazza Armerina
165 O14 **Chiba** var. Tiba. Chiba, Honshū, S Japan 35°37′N 140°06′E
165 O13 **Chiba** off. Chiba-ken, var. Tiba. ◈ prefecture Honshū, S Japan
83 M18 **Chibabava** Sofala, C Mozambique 20°17′S 33°39′E
Chiba-ken see Chiba
161 O10 **Chibi** prev. Puqi. Hubei, C China 29°45′N 113°55′E
83 B15 **Chibia** Port. João de Almeida, Vila João de Almeida. Huíla, SW Angola 15°11′S 13°41′E
83 M18 **Chiboma** Sofala, C Mozambique 20°06′S 33°54′E
82 J12 **Chibondo** Luapula, N Zambia 10°42′S 28°42′E
82 K11 **Chibote** Luapula, NE Zambia 09°52′S 29°33′E
12 K12 **Chibougamau** Québec, SE Canada 49°56′N 74°24′W
164 H11 **Chiburi-jima** island Oki-shotō, SW Japan
83 M20 **Chibuto** Gaza, S Mozambique 24°40′S 33°33′E
31 N11 **Chicago** Illinois, N USA 41°51′N 87°39′W
31 N11 **Chicago Heights** Illinois, N USA 41°30′N 87°38′W
15 W6 **Chic-Chocs, Monts** Eng. Shickshock Mountains. ⛰ Québec, SE Canada
39 W13 **Chichagof Island** island Alexander Archipelago, Alaska, USA
57 K20 **Chichas, Cordillera de** ⛰ SW Bolivia
41 X12 **Chichén-Itzá, Ruinas** ruins Yucatán, SE Mexico
97 N23 **Chichester** SE England, United Kingdom 50°50′N 00°48′W
42 C5 **Chichicastenango** Quiché, W Guatemala 14°55′N 91°06′W
42 I9 **Chichigalpa** Chinandega, NW Nicaragua 12°35′N 87°04′W
Ch'i-ch'i-ha-erh see Qiqihar
165 X16 **Chichijima-rettō** Eng. Beechy Group. island group SE Japan
54 K4 **Chichiriviche** Falcón, N Venezuela 10°58′N 68°17′W
39 R11 **Chickaloon** Alaska, USA 61°48′N 148°27′W
20 L10 **Chickamauga Lake** ⊠ Tennessee, S USA
23 N7 **Chickasawhay River** ≈ Mississippi, S USA
26 M11 **Chickasha** Oklahoma, C USA 35°03′N 97°57′W
39 T9 **Chicken** Alaska, USA 64°04′N 141°56′W
104 J16 **Chiclana de la Frontera** Andalucía, S Spain 36°26′N 06°09′W
56 B11 **Chiclayo** Lambayeque, NW Peru 06°47′S 79°47′W
35 N5 **Chico** California, W USA 39°42′N 121°51′W
83 L15 **Chicoa** Tete, NW Mozambique 15°45′S 32°25′E
83 M20 **Chicomo** Gaza, S Mozambique 24°29′S 34°15′E
18 M11 **Chicopee** Massachusetts, NE USA 42°08′N 72°34′W
63 I19 **Chico, Río** ≈ SE Argentina
63 I21 **Chico, Río** ≈ S Argentina
27 W14 **Chicot, Lake** ◎ Arkansas, C USA
15 R7 **Chicoutimi** Québec, SE Canada 48°24′N 71°04′W
15 Q8 **Chicoutimi** ≈ Québec, SE Canada
83 L19 **Chicualacuala** Gaza, SW Mozambique 22°06′S 31°42′E
83 B14 **Chicuma** Benguela, C Angola 13°33′S 14°41′E
155 J21 **Chidambaram** Tamil Nādu, SE India 11°25′N 79°42′E
196 K13 **Chidley, Cape** headland Newfoundland and Labrador, E Canada 60°25′N 64°39′W
101 N24 **Chiemsee** ◎ SE Germany
Chiengmai see Chiang Mai
Chienrai see Chiang Rai
106 B8 **Chieri** Piemonte, NW Italy 45°01′N 07°49′E
106 F8 **Chiese** ≈ N Italy
107 K14 **Chieti** var. Teate. Abruzzo, C Italy 42°22′N 14°10′E
99 E19 **Chièvres** Hainaut, SW Belgium 50°34′N 03°49′E
163 S12 **Chifeng** var. Ulanhad. Nei Mongol Zizhiqu, N China 42°17′N 118°56′E
82 F13 **Chifumage** ≈ E Angola
82 M13 **Chifunda** Muchinga, NE Zambia 11°57′S 32°36′E
Chiganak see Shyganak
39 P15 **Chiginagak, Mount** ▲ Alaska, USA 57°10′N 157°00′W
Chigirin see Chyhyryn
Chigirinskoye Vodokhranilishche see Chyhyrynske Vodoskhovishcha
41 P13 **Chignahuapan** Puebla, S Mexico 19°52′N 98°03′W
39 O15 **Chignik** Alaska, USA 56°18′N 158°24′W
83 M19 **Chigombe** ≈ S Mozambique
54 D7 **Chigorodó** Antioquia, NW Colombia 07°42′N 76°45′W
83 M19 **Chigubo** Gaza, S Mozambique 22°50′S 33°30′E
Chihertey see Altay
Chih-fu see Yantai
Chihli see Hebei
Chihli, Gulf of see Bo Hai
40 J6 **Chihuahua** Chihuahua, NW Mexico 28°40′N 106°06′W
40 I6 **Chihuahua** ◈ state N Mexico
Chiili see Shiyeli
26 M7 **Chikaskia River** ≈ Kansas/Oklahoma, C USA
155 H19 **Chik Ballāpur** Karnātaka, W India 13°28′N 77°42′E
124 G15 **Chikhacheva** Pskovskaya Oblast', W Russian Federation 57°17′N 29°51′E
155 F19 **Chikmagalūr** Karnātaka, W India 13°20′N 75°46′E
129 V7 **Chikoy** ≈ C Russian Federation
83 J15 **Chikumbi** Lusaka, C Zambia 15°11′S 28°20′E
82 M13 **Chikwa** Muchinga, NE Zambia 11°39′S 32°45′E
Chikwana see Chikwawa
83 N15 **Chikwawa** var. Chikwana. Southern, S Malawi 16°03′S 34°48′E
155 J16 **Chilakalūrupet** Andhra Pradesh, E India 16°09′N 80°13′E
146 L14 **Chilan** Lebap Welaýaty, E Turkmenistan 37°57′N 64°58′E
Chilapa see Chilapa de Alvarez
41 P16 **Chilapa de Alvarez** var. Chilapa. Guerrero, S Mexico 17°38′N 99°11′W
155 J25 **Chilaw** North Western Province, W Sri Lanka 07°34′N 79°48′E
57 D15 **Chilca** Lima, W Peru 12°35′S 76°41′W
23 Q4 **Childersburg** Alabama, S USA 33°16′N 86°21′W
25 P4 **Childress** Texas, SW USA 34°25′N 100°14′W
63 G14 **Chile** off. Republic of Chile. ◆ republic SW South America
47 R10 **Chile Basin** undersea feature E Pacific Ocean 33°00′S 80°00′W
63 H20 **Chile Chico** Aisén, W Chile 46°34′S 71°44′W
62 I9 **Chilecito** La Rioja, NW Argentina 29°10′S 67°30′W
62 H12 **Chilecito** Mendoza, W Argentina 33°53′S 69°03′W
83 L14 **Chilembwe** Eastern, E Zambia 13°54′S 31°38′E
Chile, Republic of see Chile
193 S11 **Chile Rise** undersea feature SE Pacific Ocean 40°00′S 90°00′W
117 N13 **Chilia, Braţul** ≈ SE Romania
Chilia-Nouă see Kiliya
Chilik see Shelek
Chilik see Shelek
154 O13 **Chilika Lake** var. Chilka Lake. ◎ E India
82 J13 **Chililabombwe** Copperbelt, C Zambia 12°20′S 27°52′E
Chi-lin see Jilin
Chilka Lake see Chilika Lake
10 H9 **Chilkoot Pass** pass British Columbia, W Canada
Chill Ala, Cuan see Killala Bay
62 G13 **Chillán** Bío Bío, C Chile 36°37′S 72°10′W
61 C22 **Chillar** Buenos Aires, E Argentina 37°16′S 59°58′W
Chill Chiaráin, Cuan see Kilkieran Bay
30 K12 **Chillicothe** Illinois, N USA 40°55′N 89°29′W
27 S3 **Chillicothe** Missouri, C USA 39°47′N 93°33′W
31 S14 **Chillicothe** Ohio, N USA 39°20′N 83°00′W
25 Q4 **Chillicothe** Texas, SW USA 34°15′N 99°31′W
10 M17 **Chilliwack** British Columbia, SW Canada 49°09′N 121°54′W
Chill Mhantáin, Ceann see Wicklow Head
Chill Mhantáin, Sléibhte see Wicklow Mountains
108 C10 **Chillon** Vaud, W Switzerland 46°24′N 06°56′E
Chil'mamedkum, Peski/Chilmämetgum see Çilmämmetgum
63 F17 **Chiloé, Isla de** var. Isla Grande de Chiloé. island W Chile
32 H15 **Chiloquin** Oregon, NW USA 42°33′N 121°33′W
41 O16 **Chilpancingo** var. Chilpancingo de los Bravos. Guerrero, S Mexico 17°33′N 99°30′W
Chilpancingo de los Bravos see Chilpancingo
97 N21 **Chiltern Hills** hill range S England, United Kingdom
30 M7 **Chilton** Wisconsin, N USA 44°04′N 88°10′W
82 F11 **Chiluage** Lunda Sul, NE Angola 09°32′S 21°48′E
82 N12 **Chilumba** prev. Deep Bay. Northern, N Malawi 10°27′S 34°12′E
Chilung see Jilong
83 N15 **Chilwa, Lake** var. Lago Chirua, Lake Shirwa. ◎ SE Malawi
167 R10 **Chi, Mae Nam** ≈ E Thailand
42 C6 **Chimaltenango** Chimaltenango, C Guatemala 14°40′N 90°48′W
42 A2 **Chimaltenango** off. Departamento de Chimaltenango. ◈ department S Guatemala
Chimaltenango, Departamento de see Chimaltenango
43 V15 **Chimán** Panamá, E Panama 08°42′N 78°35′W
83 M17 **Chimanimani** prev. Mandidzudzure, Melsetter. Manicaland, E Zimbabwe 19°48′S 32°52′E
99 G22 **Chimay** Hainaut, S Belgium 50°03′N 04°20′E
37 S10 **Chimayo** New Mexico, SW USA 36°00′N 105°55′W
Chimbay see Chimboy
56 A13 **Chimborazo** ◈ province C Ecuador
56 C7 **Chimborazo** ※ C Ecuador 01°29′S 78°50′W
56 C12 **Chimbote** Ancash, W Peru 09°04′S 78°34′W
146 H7 **Chimboy** Rus. Chimbay. Qoraqalpog'iston Respublikasi, NW Uzbekistan 43°03′N 59°52′E
186 D7 **Chimbu** var. Simbu. ◈ province C Papua New Guinea
54 F6 **Chimichagua** Cesar, N Colombia 09°19′N 73°51′W
Chimishliya see Cimişlia
Chimkent see Shymkent
Chimkentskaya Oblast' see Yuzhnyy Kazakhstan
28 I14 **Chimney Rock** rock Nebraska, C USA
83 M17 **Chimoio** Manica, C Mozambique 19°08′S 33°29′E
82 K11 **Chimpembe** Northern, NE Zambia 09°31′S 29°33′E
41 O8 **China** Nuevo León, NE Mexico 25°42′N 99°15′W
156 M9 **China** off. People's Republic of China, Chin. Chung-hua Jen-min Kung-ho-kuo, Zhonghua Renmin Gongheguo; prev. Chinese Empire. ◆ republic E Asia
19 Q7 **China Lake** ◎ Maine, NE USA
42 F8 **Chinameca** San Miguel, E El Salvador 13°30′N 88°20′W*
Chi-nan/Chinan see Jinan
42 H9 **Chinandega** Chinandega, NW Nicaragua 12°37′N 87°08′W
42 H9 **Chinandega** ◈ department NW Nicaragua
China, People's Republic of see China
China, Republic of see Taiwan
24 J11 **Chinati Mountains** ⛰ Texas, SW USA
Chinaz see Chinoz
57 E15 **Chincha Alta** Ica, SW Peru 13°25′S 76°07′W
11 N11 **Chinchaga** ≈ Alberta, SW Canada
Chin-chiang see Quanzhou
Chinchilla see Chinchilla de Monte Aragón
105 Q11 **Chinchilla de Monte Aragón** var. Chinchilla. Castilla-La Mancha, C Spain 38°56′N 01°44′W
54 D10 **Chinchiná** Caldas, W Colombia 04°59′N 75°37′W
105 O8 **Chinchón** Madrid, C Spain 40°08′N 03°26′W
41 Z14 **Chinchorro, Banco** island SE Mexico
Chin-chou/Chinchow see Jinzhou
21 Z5 **Chincoteague** Assateague Island, Virginia, NE USA 37°55′N 75°22′W
83 O17 **Chinde** Zambézia, NE Mozambique 18°35′S 36°28′E
Chin-do see Jin-do
159 R13 **Chindu** var. Chengwen; prev. Chuqung. Qinghai, C China 33°19′N 97°08′E
Chindwin see Chindwinn
166 M2 **Chindwinn** var. Chindwin. ≈ N Myanmar (Burma)
Chinese Empire see China
Chinghai see Qinghai
Ch'ing Hai see Qinghai Hu, China
Chingildi see Shengeldi
Chingirlau see Shyngghyrlau
82 J13 **Chingola** Copperbelt, C Zambia 12°31′S 27°53′E
Ching-Tao/Ch'ing-tao see Qingdao
82 C13 **Chinguar** Huambo, C Angola 12°33′S 16°25′E
76 I7 **Chinguețți** var. Chinguetti. Adrar, C Mauritania 20°25′N 12°24′W
Chinhae see Jinhae
166 K4 **Chin Hills** ⛰ W Myanmar (Burma)
83 K16 **Chinhoyi** prev. Sinoia. Mashonaland West, N Zimbabwe 17°22′S 30°12′E
Chinhsien see Jinzhou
39 Q14 **Chiniak, Cape** headland Kodiak Island, Alaska, USA 57°37′N 152°10′W
14 G10 **Chiniguchi Lake** ◎ Ontario, S Canada
149 U8 **Chiniot** Punjab, NE Pakistan 31°40′N 73°00′E
Chinju see Jinju
Chinkai see Jinhae
78 M13 **Chinko** ≈ E Central African Republic
37 O9 **Chinle** Arizona, SW USA 36°09′N 109°33′W
Chinmen Tao see Jinmen Dao
Chinnchâr see Shinshār
Chinnereth see Tiberias, Lake
164 C12 **Chino** var. Tino. Nagano, Honshū, S Japan 36°00′N 138°10′E
102 L8 **Chinon** Indre-et-Loire, C France 47°10′N 00°15′E
33 T7 **Chinook** Montana, NW USA 48°35′N 109°13′W
Chinook State see Washington
192 L4 **Chinook Trough** undersea feature N Pacific Ocean
36 K11 **Chino Valley** Arizona, SW USA 34°45′N 112°27′W
147 P10 **Chinoz** Rus. Chinaz. Toshkent Viloyati, E Uzbekistan 40°58′N 68°46′E
82 L12 **Chinsali** Muchinga, NE Zambia 10°33′S 32°05′E
166 K5 **Chin State** ◈ state W Myanmar (Burma)
Chinsura see Chunchura
Chin-tō see Jin-do
54 E6 **Chinú** Córdoba, NW Colombia 09°07′N 75°25′W
99 K24 **Chiny, Forêt de** forest SE Belgium
83 M15 **Chioco** Tete, NW Mozambique 16°22′S 32°50′E
106 H8 **Chioggia** anc. Fossa Claudia. Veneto, NE Italy 45°14′N 12°17′E
114 H12 **Chionótrypa** ▲ NE Greece 41°16′N 24°06′E
115 L18 **Chíos** var. Hios, Khíos, It. Scio, Turk. Sakiz-Adasi. Chíos, E Greece 38°23′N 26°07′E
115 K18 **Chíos** var. Khíos. island E Greece
83 M14 **Chipata** prev. Fort Jameson. Eastern, E Zambia 13°40′S 32°42′E
83 C14 **Chipindo** Huíla, C Angola 13°53′S 15°47′E
23 R8 **Chipley** Florida, SE USA 30°46′N 85°32′W
155 D15 **Chiplūn** Mahārāshtra, W India 17°32′N 73°32′E
81 H22 **Chipogolo** Dodoma, C Tanzania 06°52′S 36°03′E
23 R8 **Chipola River** ≈ Florida, SE USA
97 L22 **Chippenham** S England, United Kingdom 51°28′N 02°07′W
30 J6 **Chippewa Falls** Wisconsin, N USA 44°56′N 91°25′W
30 J4 **Chippewa, Lake** ⊠ Wisconsin, N USA
31 Q8 **Chippewa River** ≈ Michigan, N USA
30 I6 **Chippewa River** ≈ Wisconsin, N USA
Chipping Wycombe see High Wycombe
114 G8 **Chiprovtsi** Montana, NW Bulgaria 43°23′N 22°53′E
19 T4 **Chiputneticook Lakes** lakes Canada/USA
56 D13 **Chiquián** Ancash, W Peru 10°09′S 78°08′W
41 Y11 **Chiquilá** Quintana Roo, SE Mexico 21°25′N 87°20′W
42 E6 **Chiquimula** Chiquimula, SE Guatemala 14°46′N 89°32′W
42 A3 **Chiquimula** off. Departamento de Chiquimula. ◈ department SE Guatemala
Chiquimula, Departamento de see Chiquimula
42 D7 **Chiquimulilla** Santa Rosa, S Guatemala 14°06′N 90°23′W
54 F9 **Chiquinquirá** Boyacá, C Colombia 05°37′N 73°51′W
155 J17 **Chīrāla** Andhra Pradesh, E India 15°49′N 80°21′E
149 N4 **Chīras** Gōwr, N Afghanistan 35°15′N 65°39′E
152 H11 **Chirāwa** Rājasthān, N India 28°12′N 75°42′E
Chirchik see Chirchiq
147 Q9 **Chirchiq** Rus. Chirchik. Toshkent Viloyati, E Uzbekistan 41°30′N 69°32′E
147 P10 **Chirchiq** ≈ E Uzbekistan
Chire see Shire
83 L18 **Chiredzi** Masvingo, SE Zimbabwe 21°00′S 31°38′E
25 X8 **Chireno** Texas, SW USA 31°30′N 94°21′W
77 X7 **Chirfa** Agadez, NE Niger 21°01′N 12°41′E
37 O16 **Chiricahua Mountains** ⛰ Arizona, SW USA
37 O16 **Chiricahua Peak** ▲ Arizona, SW USA 31°51′N 109°17′W
54 F6 **Chiriguaná** Cesar, N Colombia 09°24′N 73°38′W
39 P15 **Chirikof Island** island Alaska, USA
43 P16 **Chiriquí** off. Provincia de Chiriquí. ◈ province SW Panama
43 P17 **Chiriquí, Golfo de** Eng. Chiriqui Gulf. gulf SW Panama
43 P15 **Chiriquí Grande** Bocas del Toro, W Panama 08°58′N 82°08′W
Chiriqui Gulf see Chiriquí, Golfo de
43 P15 **Chiriquí, Laguna de** lagoon NW Panama
Chiriquí, Provincia de see Chiriquí
43 O16 **Chiriquí Viejo, Río** ≈ W Panama
Chiriquí, Volcán de see Barú, Volcán
83 N15 **Chiromo** Southern, S Malawi 16°32′S 35°07′E
114 J10 **Chirpan** Stara Zagora, C Bulgaria 42°12′N 25°20′E
43 N14 **Chirripó Atlántico, Río** ≈ E Costa Rica
Chirripó, Cerro see Chirripó Grande, Cerro
Chirripó del Pacífico, Río see Chirripó, Río
43 N14 **Chirripó Grande, Cerro** var. Cerro Chirripó. ▲ SE Costa Rica 09°31′N 83°28′W
43 N13 **Chirripó, Río** var. Río Chirripó del Pacífico. ≈ NE Costa Rica
Chirua, Lago see Chilwa, Lake
83 J15 **Chirundu** Southern, S Zambia 16°03′S 28°50′E
29 W8 **Chisago City** Minnesota, N USA 45°22′N 92°53′W
83 J14 **Chisamba** Central, C Zambia 15°00′S 28°22′E
39 T10 **Chisana** Alaska, USA 62°09′N 142°07′W
82 I13 **Chisasa** North Western, NW Zambia 12°09′S 25°30′E
12 I9 **Chisasibi** prev. Fort George. Québec, C Canada 53°50′N 79°01′W
42 D4 **Chisec** Alta Verapaz, C Guatemala 15°50′N 90°18′W
127 U5 **Chishmy** Respublika Bashkortostan, W Russian Federation 54°33′N 55°21′E
29 V4 **Chisholm** Minnesota, N USA 47°29′N 92°52′W
149 U10 **Chishtiān** var. Chishtiān Mandi. Punjab, E Pakistan 29°44′N 72°54′E
Chishtiān Mandi see Chishtiān
160 I11 **Chishui He** ≈ C China
Chisimaio/Chisimayu see Kismaayo
117 N10 **Chişinău** Rus. Kishinev. ● (Moldova) C Moldova 47°N 28°51′E
117 N10 **Chişinău** ✈ S Moldova 46°54′N 28°56′E
Chişinău-Criş see Chişineu-Criş
116 F10 **Chişineu-Criş** Hung. Kisjenő; prev. Chişinău-Criş. Arad, W Romania 46°33′N 21°30′E
83 K14 **Chisomo** Central, C Zambia 13°30′S 30°37′E
106 A8 **Chisone** ≈ NW Italy
24 K12 **Chisos Mountains** ⛰ Texas, SW USA
39 T10 **Chistochina** Alaska, USA 62°34′N 144°39′W
127 R4 **Chistopol'** Respublika Tatarstan, W Russian Federation 55°20′N 50°39′E
145 O8 **Chistopol'ye** Severnyy Kazakhstan, N Kazakhstan 52°37′N 67°14′E
123 O13 **Chita** Zabaykal'skiy Kray, S Russian Federation 52°03′N 113°35′E
83 B16 **Chitado** Cunene, SW Angola 17°16′S 13°54′E
Chitaldroog/Chitaldrug see Chitradurga
83 C15 **Chitanda** ≈ S Angola
Chitangwiza see Chitungwiza
82 F10 **Chitato** Lunda Norte, NE Angola 07°23′S 20°46′E
83 C14 **Chitembo** Bié, C Angola 13°33′S 16°47′E
39 T11 **Chitina** Alaska, USA 61°31′N 144°26′W
39 T11 **Chitina River** ≈ Alaska, USA
82 M11 **Chitipa** Northern, NW Malawi 09°41′S 33°19′E
165 S4 **Chitose** var. Titose. Hokkaidō, NE Japan 42°51′N 141°40′E
155 G18 **Chitradurga** prev. Chitaldroog, Chitaldrug. Karnātaka, W India 14°16′N 76°23′E
149 T3 **Chitrāl** Khyber Pakhtunkhwa, NW Pakistan 35°51′N 71°47′E
43 S16 **Chitré** Herrera, S Panama 07°57′N 80°26′W
153 V16 **Chittagong** Ben. Chāttagām. Chittagong, SE Bangladesh 22°20′N 91°48′E
153 U16 **Chittagong** ◈ division E Bangladesh
153 Q15 **Chittaranjan** West Bengal, NE India 23°52′N 86°40′E
152 G14 **Chittaurgarh** var. Chittorgarh. Rājasthān, N India 24°54′N 74°42′E
155 I19 **Chittoor** Andhra Pradesh, E India 13°13′N 79°06′E
Chittorgarh see Chittaurgarh
155 G21 **Chittūr** Kerala, SW India 10°42′N 76°46′E
83 K16 **Chitungwiza** prev. Chitangwiza. Mashonaland East, NE Zimbabwe 18°S 31°06′E
62 H4 **Chiuchiu** Antofagasta, N Chile 22°13′S 68°34′W
82 F12 **Chiumbe** var. Tshiumbe. ≈ Angola/Dem. Rep. Congo
83 F15 **Chiume** Moxico, E Angola 15°08′S 21°09′E
82 K13 **Chiundaponde** Muchinga, NE Zambia 12°14′S 30°40′E
106 H13 **Chiusi** Toscana, C Italy 43°00′N 11°56′E
54 J5 **Chivacoa** Yaracuy, N Venezuela 10°10′N 68°54′W
106 B8 **Chivasso** Piemonte, NW Italy 45°13′N 07°54′E
83 L17 **Chivhu** prev. Enkeldoorn. Midlands, C Zimbabwe 19°01′S 30°54′E
61 C20 **Chivilcoy** Buenos Aires, E Argentina 34°55′S 60°00′W
Chiwei Yu see Sekibi-sho
82 N12 **Chiweta** Northern, N Malawi 10°36′S 34°09′E
42 D4 **Chixoy, Río** var. Río Negro, Río Salinas. ≈ Guatemala/Mexico
82 H13 **Chizela** North Western, NW Zambia 13°11′S 24°59′E
125 O5 **Chizha** Nenetskiy Avtonomnyy Okrug, NW Russian Federation 67°04′N 44°19′E
161 Q9 **Chizhou** var. Guichi. Anhui, E China 30°39′N 117°29′E
164 I12 **Chizu** Tottori, Honshū, SW Japan 35°15′N 134°14′E
Chkalov see Orenburg
74 J5 **Chlef** var. Ech Cheliff, Ech Chleff; prev. Al-Asnam, El Asnam, Orléansville. NW Algeria 36°11′N 01°21′E
115 G18 **Chlómo** ▲ C Greece 38°36′N 22°57′E
111 M15 **Chmielnik** Świętokrzyskie, C Poland 50°37′N 20°43′E

◆ Country ● Country Capital ◇ Dependent Territory ○ Dependent Territory Capital ◈ Administrative Regions ✈ International Airport ▲ Mountain ⛰ Mountain Range ※ Volcano ≈ River ◎ Lake ⊠ Reservoir

Chŏâm Khsant – Clavier

167 S11 **Chŏâm Khsant** Preăh Vihéar, N Cambodia 14°13′N 104°56′E
62 G10 **Choapa, Río** *var.* Choapo. 〰 C Chile
Choapas *see* Las Choapas
Choapo *see* Choapa, Río
Choarta *see* Chwarta
67 T13 **Chobe** 〰 N Botswana
14 K8 **Chochocouane** 〰 Québec, SE Canada
110 E13 **Chocianów** *Ger.* Kotzenan. Dolnośląskie, SW Poland 51°23′N 15°55′E
54 C9 **Chocó** *off.* Departamento del Chocó. ◈ *province* W Colombia
Chocó, Departamento del *see* Chocó
35 X16 **Chocolate Mountains** ⛰ California, W USA
21 W9 **Chocowinity** North Carolina, SE USA 35°33′N 77°03′W
27 N10 **Choctaw** Oklahoma, C USA 35°30′N 97°16′W
23 Q8 **Choctawhatchee Bay** *bay* Florida, SE USA
23 Q8 **Choctawhatchee River** 〰 Florida, SE USA
Chodau *see* Chodov
163 V14 **Ch'o-do** *island* SW North Korea
Chodorów *see* Khodoriv
111 A16 **Chodov** *Ger.* Chodau. Karlovarský Kraj, W Czech Republic 50°15′N 12°45′E
110 G10 **Chodzież** Wielkopolskie, C Poland 53°N 16°55′E
63 J15 **Choele Choel** Río Negro, C Argentina 39°19′S 65°42′W
83 L14 **Chofombo** Tete, NW Mozambique 14°43′S 31°48′E
Chohtan *see* Chauhtan
11 U14 **Choiceland** Saskatchewan, C Canada 53°28′N 104°26′W
186 K8 **Choiseul** ◈ *province* NW Solomon Islands
186 K8 **Choiseul** *var.* Lauru. *island* NW Solomon Islands
63 M23 **Choiseul Sound** *sound* East Falkland, Falkland Islands
40 H7 **Choix** Sinaloa, C Mexico 26°43′N 108°20′W
110 D10 **Chojna** Zachodnio-pomorskie, W Poland 52°56′N 14°25′E
110 H8 **Chojnice** *Ger.* Konitz. Pomorskie, N Poland 53°41′N 17°34′E
111 F14 **Chojnów** *Ger.* Hainau, Haynau. Dolnośląskie, SW Poland 51°16′N 15°55′E
167 Q10 **Chok Chai** Nakhon Ratchasima, C Thailand 14°45′N 102°10′E
80 I12 **Ch'ok'ē** *var.* Choke Mountains. ⛰ NW Ethiopia
25 R13 **Choke Canyon Lake** ▣ Texas, SW USA
Choke Mountains *see* Ch'ok'ē
Chokpar *see* Shokpar
147 W7 **Chok-Tal** *var.* Choktal. Issyk-Kul'skaya Oblast', E Kyrgyzstan 42°37′N 76°45′E
Choktal *see* Chok-Tal
Chókué *see* Chokwè
123 R7 **Chokurdakh** Respublika Sakha (Yakutiya), NE Russian Federation 70°38′N 148°18′E
83 L20 **Chokwè** *var.* Chókué. Gaza, S Mozambique 24°27′S 32°55′E
188 F8 **Chol** Babeldaob, N Palau
160 E8 **Chola Shan** ⛰ C China
102 J8 **Cholet** Maine-et-Loire, NW France 47°03′N 00°52′W
63 H17 **Cholila** Chubut, W Argentina 42°33′S 71°28′W
Cholo *see* Thyolo
147 V8 **Cholpon** Narynskaya Oblast', C Kyrgyzstan 42°07′N 75°25′E
147 X7 **Cholpon-Ata** Issyk-Kul'skaya Oblast', E Kyrgyzstan 42°39′N 77°05′E
41 P14 **Cholula** Puebla, S Mexico 19°03′N 98°19′W
42 I8 **Choluteca** Choluteca, S Honduras 13°15′N 87°10′W
42 H8 **Choluteca** ◈ *department* S Honduras
42 H8 **Choluteca, Río** 〰 SW Honduras
83 I15 **Choma** Southern, S Zambia 16°48′S 26°58′E
153 T11 **Chomo Lhari** ▲ NW Bhutan 27°59′N 89°24′E
167 N7 **Chom Thong** Chiang Mai, NW Thailand 18°25′N 98°44′E
111 B15 **Chomutov** *Ger.* Komotau. Ústecký Kraj, NW Czech Republic 50°28′N 13°24′E
123 N11 **Chona** 〰 C Russian Federation
Ch'ŏnan *see* Cheonan
167 P11 **Chon Buri** *prev.* Bang Pla Soi. Chon Buri, S Thailand 13°24′N 100°59′E
56 B6 **Chone** Manabí, W Ecuador 0°44′S 80°04′W
Chong'an *see* Wuyishan
163 W13 **Ch'ŏngch'ŏn-gang** 〰 W North Korea
163 Y11 **Ch'ŏngjin** NE North Korea 41°48′N 129°44′E
Chŏngju *see* Cheongju
161 S8 **Chongming Dao** *island* E China
160 J10 **Chongqing** *var.* Ch'ung-ching, Ch'ung-ch'ing, Chungking, Pahsien, Tchongking, Yuzhou. Chongqing Shi, C China 29°34′N 106°27′E
Chŏngup *see* Chŏnju
161 O10 **Chongyang** *var.* Tiancheng. Hubei, C China 29°35′N 114°03′E
160 J15 **Chongzuo** *prev.* Taiping. Guangxi Zhuangzu Zizhiqu, S China 22°18′N 107°23′E
163 Y16 **Chŏnju** *prev.* Chŏnju, Chŏngup, *Jap.* Seiyu. SW South Korea 35°51′N 127°08′E
Chŏnju *see* Chŏnju
Chŏnju *see* Jeonju
Chonnacht *see* Connaught
Chonogol *see* Erdenetsagaan
63 F19 **Chonos, Archipiélago de los** *island group* S Chile
42 K10 **Chontales** ◈ *department* S Nicaragua
167 T13 **Chơn Thanh** Sông Be, S Vietnam 11°25′N 106°38′E
158 K17 **Cho Oyu** *var.* Qowowuyag. ▲ China/Nepal 28°07′N 86°37′E
116 G7 **Chop** *Cz.* Čop, *Hung.* Csap. Zakarpats'ka Oblast', W Ukraine 48°26′N 22°13′E
21 Y3 **Choptank River** 〰 Maryland, NE USA
115 J22 **Chóra** *prev.* Íos. Íos, Kykládes, Greece, Aegean Sea 36°42′N 25°16′E
115 H25 **Chóra Sfakíon** *var.* Sfákia. Kríti, Greece, E Mediterranean Sea 35°12′N 24°05′E
Chorcaí, Cuan *see* Cork Harbour
43 P15 **Chorcha, Cerro** ▲ W Panama 08°39′N 82°07′W
Chorku *see* Chorkŭh
147 R11 **Chorkŭh** *Rus.* Chorku. N Tajikistan 40°04′N 70°30′E
97 K17 **Chorley** NW England, United Kingdom 53°40′N 02°38′W
Chorne More *see* Black Sea
117 R5 **Chornobay** Cherkas'ka Oblast', C Ukraine 49°40′N 32°20′E
117 O3 **Chornobyl'** *Rus.* Chernobyl'. Kyyivs'ka Oblast', N Ukraine 51°17′N 30°15′E
117 R12 **Chornomors'ke** *Rus.* Chernomorskoye. Avtonomna Respublika Krym, S Ukraine 45°29′N 32°45′E
117 R4 **Chornukhy** Poltavs'ka Oblast', C Ukraine 50°15′N 32°57′E
Chorokh/Chorokhi *see* Çoruh Nehri
110 O9 **Choroszcz** Podlaskie, NE Poland 53°10′N 23°E
116 K6 **Chortkiv** *Rus.* Chortkov. Ternopil's'ka Oblast', W Ukraine 49°01′N 25°46′E
Chortkov *see* Chortkiv
Chorum *see* Çorum
110 M9 **Chorzele** Mazowieckie, C Poland 53°16′N 20°53′E
111 J16 **Chorzów** *Ger.* Königshütte; *prev.* Królewska Huta. Śląskie, S Poland 50°17′N 18°58′E
163 W12 **Ch'osan** N North Korea 40°45′N 125°52′E
Chośebuz *see* Cottbus
Chōsen-kaikyō *see* Korea Strait
164 P14 **Chōshi** *var.* Tyôsi. Chiba, Honshū, S Japan 35°44′N 140°48′E
63 H14 **Chos Malal** Neuquén, W Argentina 37°23′S 70°16′W
Chosŏn-minjujuŭi-inmin-kanghwaguk *see* North Korea
110 E9 **Choszczno** *Ger.* Arnswalde. Zachodnio-pomorskie, NW Poland 53°10′N 15°24′E
153 O15 **Chota Nāgpur** *plateau* N India
33 R8 **Choteau** Montana, NW USA 47°48′N 112°40′W
Chotqol *see* Chatkal
14 M8 **Chouart** 〰 Québec, SE Canada
76 I7 **Choûm** Adrar, C Mauritania 21°19′N 12°59′W
27 Q9 **Chouteau** Oklahoma, C USA 36°11′N 95°20′W
21 X8 **Chowan River** 〰 North Carolina, SE USA
35 Q10 **Chowchilla** California, W USA 37°06′N 120°15′W
163 Q7 **Choybalsan** *var.* Hulstay. Dornod, NE Mongolia 48°25′N 114°56′E
163 P7 **Choybalsan** *prev.* Byan Tumen. Dornod, E Mongolia 48°03′N 114°32′E
162 M9 **Choyr** Govĭ Sumber, C Mongolia 46°20′N 108°21′E
185 I19 **Christchurch** Canterbury, South Island, New Zealand 43°31′S 172°39′E
97 M24 **Christchurch** S England, United Kingdom 50°44′N 01°45′W
185 I18 **Christchurch** ✈ Canterbury, South Island, New Zealand 43°28′S 172°33′E
44 J12 **Christiana** C Jamaica 18°13′N 77°29′W
83 H22 **Christiana** Free State, C South Africa 27°55′S 25°10′E
115 J23 **Christiána** *var.* Christiani. *island* Kykládes, Greece, Aegean Sea
Christiani *see* Christiána
Christiania *see* Oslo
14 G13 **Christian Island** *island* Ontario, S Canada
191 P16 **Christian, Point** *headland* Pitcairn Island, Pitcairn Islands 25°04′S 130°08′E
38 M11 **Christian River** 〰 Alaska, USA
Christiansand *see* Kristiansand
21 S7 **Christiansburg** Virginia, NE USA 37°07′N 80°26′W
95 G23 **Christiansfeld** Syddanmark, SW Denmark 55°21′N 09°30′E
Christianshåb *see* Qasigiannguit
39 X14 **Christian Sound** *inlet* Alaska, USA
45 T9 **Christiansted** Saint Croix, S Virgin Islands (US) 17°43′N 64°42′W
Christiansund *see* Kristiansund
25 R13 **Christine** Texas, SW USA 28°47′N 98°30′W
173 U7 **Christmas Island** ◇ *Australian external territory* E Indian Ocean
129 T17 **Christmas Island** *island* E Indian Ocean
Christmas Island *see* Kiritimati
192 M7 **Christmas Ridge** *undersea feature* C Pacific Ocean
30 L16 **Christopher** Illinois, N USA 37°58′N 89°03′W
25 P9 **Christoval** Texas, SW USA 31°09′N 100°30′W
111 F17 **Chrudim** Pardubický Kraj, C Czech Republic 49°58′N 15°49′E
115 K25 **Chrýsi** *island* SE Greece
121 N2 **Chrysochoú, Kólpos** *var.* Khrysokhou Bay. *bay* E Mediterranean Sea
114 I13 **Chrysoúpoli** *var.* Hrisoupoli; *prev.* Khrisoúpolis. Anatolikí Makedonía kai Thráki, NE Greece 40°59′N 24°42′E
111 K16 **Chrzanów** *var.* Chrzanow, *Ger.* Zaumgarten. Śląskie, S Poland 50°10′N 19°21′E
Chu *see* Shu
42 C5 **Chuacús, Sierra de** ⛰ W Guatemala
153 S15 **Chuadanga** Khulna, W Bangladesh 23°38′N 88°52′E
Chuan *see* Sichuan
Ch'uan-chou *see* Quanzhou
Chuankou *see* Minhe
39 O11 **Chuathbaluk** Alaska, USA 61°36′N 159°14′W
Chubek *see* Moskva
63 I17 **Chubut** *off.* Provincia de Chubut. ◈ *province* S Argentina
Chubut, Provincia de *see* Chubut
63 I17 **Chubut, Río** 〰 SE Argentina
43 V15 **Chucanti, Cerro** ▲ E Panama 08°48′N 78°27′W
Ch'u-chiang *see* Shaoguan
43 W15 **Chucunaque, Río** 〰 E Panama
Chudin *see* Chudzin
116 M5 **Chudniv** Zhytomyrs'ka Oblast', N Ukraine 50°02′N 28°06′E
124 H13 **Chudovo** Novgorodskaya Oblast', W Russian Federation 59°07′N 31°42′E
Chudskoye Ozero *see* Peipus, Lake
119 J18 **Chudzin** *Rus.* Chudin. Brestskaya Voblasts', SW Belarus 52°44′N 26°59′E
39 Q13 **Chugach Islands** *island group* Alaska, USA
39 S11 **Chugach Mountains** ⛰ Alaska, USA
164 G12 **Chūgoku-sanchi** ⛰ Honshū, SW Japan
Chugqênsumdo *see* Jigzhi
Chuguyev *see* Chuhuyiv
117 V5 **Chuhuyiv** *var.* Chuguyev. Kharkivs'ka Oblast', E Ukraine 49°51′N 36°44′E
61 H19 **Chuí** Rio Grande do Sul, S Brazil 33°45′S 53°23′W
Chuí *see* Chuy
Chu-Iliyskiye Gory *see* Gory Shu-Ile
Chukai *see* Cukai
Chukchi Avtonomnyy Okrug *see* Chukotskiy Avtonomnyy Okrug
Chukchi Peninsula *see* Chukotskiy Poluostrov
197 R6 **Chukchi Plain** *undersea feature* Arctic Ocean
197 R6 **Chukchi Plateau** *undersea feature* Arctic Ocean
197 R4 **Chukchi Sea** *Rus.* Chukotskoye More. *sea* Arctic Ocean
125 N14 **Chukhloma** Kostromskaya Oblast', NW Russian Federation 58°42′N 42°39′E
Chukotka *see* Chukotskiy Avtonomnyy Okrug
Chukot Range *see* Anadyrskiy Khrebet
123 V6 **Chukotskiy Avtonomnyy Okrug** *var.* Chukchi Avtonomnyy Okrug, Chukotka. ◈ *autonomous district* NE Russian Federation
123 W5 **Chukotskiy, Mys** *headland* NE Russian Federation 64°15′N 173°03′W
123 V5 **Chukotskiy Poluostrov** *Eng.* Chukchi Peninsula. *peninsula* NE Russian Federation
Chukotskoye More *see* Chukchi Sea
Chukurkak *see* Chuqurqoq
Chulakkurgan *see* Sholakkorgan
35 U17 **Chula Vista** California, W USA 32°38′N 117°04′W
123 Q12 **Chul'man** Respublika Sakha (Yakutiya), NE Russian Federation 56°50′N 124°47′E
56 B9 **Chulucanas** Piura, NW Peru 05°08′S 80°10′W
122 J12 **Chulym** 〰 C Russian Federation
152 K6 **Chumar** Jammu and Kashmir, N India 32°38′N 78°36′E
114 K9 **Chumerna** ▲ C Bulgaria 42°45′N 25°58′E
123 R12 **Chumikan** Khabarovskiy Kray, E Russian Federation 54°41′N 135°12′E
167 Q9 **Chum Phae** Khon Kaen, C Thailand 16°31′N 102°09′E
167 N13 **Chumphon** *var.* Jumporn. Chumphon, SW Thailand 10°30′N 99°11′E
167 O9 **Chumsaeng** *var.* Chum Saeng. Nakhon Sawan, C Thailand 15°52′N 100°18′E
Chum Saeng *see* Chumsaeng
122 L12 **Chuna** 〰 C Russian Federation
161 R9 **Chun'an** *var.* Qiandaohu; *prev.* Pailing. Zhejiang, SE China 29°37′N 118°59′E
Chunan *see* Zhunan
Chuncheng *see* Yangchun
163 Y14 **Chuncheon** *Jap.* Shunsen; *prev.* Ch'unch'ŏn. N South Korea 37°52′N 127°48′E
Ch'unch'ŏn *see* Chuncheon
153 S16 **Chunchura** *prev.* Chinsura. West Bengal, NE India 22°54′N 88°20′E
Chundzha *see* Shonzhy
Ch'ung-ch'ing/Ch'ung-ching *see* Chongqing
Chung-hua Jen-min Kung-ho-kuo *see* China
163 Y15 **Chungju** *Jap.* Chūshū; *prev.* Ch'ungju. C South Korea 36°57′N 127°50′E
Ch'ungju *see* Chungju
Chungking *see* Chongqing
161 T14 **Chungyang Shanmo** *Chin.* Taiwan Shan. ⛰ C Taiwan
149 V9 **Chūniān** Punjab, E Pakistan 30°57′N 74°01′E
122 L12 **Chunskiy** Irkutskaya Oblast', S Russian Federation 56°10′N 99°15′E
122 M11 **Chunya** 〰 C Russian Federation
124 J6 **Chupa** Respublika Kareliya, NW Russian Federation 66°15′N 33°02′E
125 P8 **Chuprovo** Respublika Komi, NW Russian Federation 64°16′N 46°27′E
57 G17 **Chuquibamba** Arequipa, SW Peru 15°47′S 72°44′W
62 H4 **Chuquicamata** Antofagasta, N Chile 22°20′S 68°56′W
57 L21 **Chuquisaca** ◈ *department* S Bolivia
Chuquisaca *see* Sucre
Chuqung *see* Chindu
146 I8 **Chuqurqoq** *Rus.* Chukurkak. Qoraqalpog'iston Respublikasi, NW Uzbekistan 42°44′N 61°33′E
127 T2 **Chur** Udmurtskaya Respublika, NW Russian Federation 57°06′N 52°57′E
108 I9 **Chur** *Fr.* Coire, *It.* Coira, *Rmsch.* Cuera, Quera; *anc.* Curia Rhaetorum. Graubünden, E Switzerland 46°52′N 09°32′E
123 Q10 **Churapcha** Respublika Sakha (Yakutiya), NE Russian Federation 61°59′N 132°06′E
11 V16 **Churchbridge** Saskatchewan, S Canada 50°55′N 101°53′W
21 O8 **Church Hill** Tennessee, S USA 36°31′N 82°42′W
11 X9 **Churchill** Manitoba, C Canada 58°46′N 94°10′W
11 X10 **Churchill** 〰 Manitoba/Saskatchewan, C Canada
13 P9 **Churchill** 〰 Newfoundland and Labrador, E Canada
11 Y9 **Churchill, Cape** *headland* Manitoba, C Canada 58°42′N 93°12′W
13 P9 **Churchill Falls** Newfoundland and Labrador, E Canada 53°38′N 64°00′W
11 S12 **Churchill Lake** ◎ Saskatchewan, C Canada
19 Q3 **Churchill Lake** ▣ Maine, NE USA
194 I5 **Churchill Peninsula** *peninsula* Antarctica
22 H8 **Church Point** Louisiana, S USA 30°24′N 92°13′W
29 O3 **Churchs Ferry** North Dakota, N USA 48°15′N 99°12′W
146 G12 **Churchuri** Ahal Welaýaty, C Turkmenistan 38°55′N 59°13′E
21 T5 **Churchville** Virginia, NE USA 38°13′N 79°10′W
152 G10 **Chūru** Rājasthān, NW India 28°18′N 75°00′E
54 J4 **Churuguara** Falcón, N Venezuela 10°52′N 69°35′W
167 U11 **Chư Sê** Gia Lai, C Vietnam 13°38′N 108°06′E
144 J12 **Chushkakul, Gory** ⛰ SW Kazakhstan
Chūshū *see* Chungju
37 O9 **Chuska Mountains** ⛰ Arizona/New Mexico, SW USA
Chu, Sông *see* Sam, Nam
125 V14 **Chusovoy** Permskiy Kray, NW Russian Federation 58°17′N 57°54′E
147 R10 **Chust** Namangan Viloyati, E Uzbekistan 40°58′N 71°12′E
Chust *see* Khust
15 U6 **Chute-aux-Outardes** Québec, SE Canada 49°07′N 68°25′W
117 U5 **Chutove** Poltavs'ka Oblast', C Ukraine 49°45′N 35°11′E
167 U11 **Chư Ty** *var.* Đức Cơ. Gia Lai, C Vietnam 13°48′N 107°41′E
189 O15 **Chuuk** *var.* Truk. ◈ *state* C Micronesia
189 P15 **Chuuk Islands** *var.* Hogoley Islands; *prev.* Truk Islands. *island group* Caroline Islands, C Micronesia
Chuvashia *see* Chuvashskaya Respublika
Chuvashiya *see* Chuvashskaya Respublika
127 P4 **Chuvashskaya Respublika** *var.* Chuvashiya, *Eng.* Chuvashia. ◈ *autonomous republic* W Russian Federation
Chuwārtah *see* Chwarta
Chu Xian/Chuxian *see* Chuzhou
160 G13 **Chuxiong** Yunnan, SW China 25°02′N 101°32′E
147 V7 **Chuy** Chuyskaya Oblast', N Kyrgyzstan 42°45′N 75°11′E
61 H19 **Chuy** *var.* Chuí. Rocha, E Uruguay 33°42′S 53°27′W
123 O11 **Chuya** Respublika Sakha (Yakutiya), NE Russian Federation 59°30′N 112°26′E
Chüy Oblasty *see* Chuyskaya Oblast'
147 U8 **Chuyskaya Oblast'** *Kir.* Chüy Oblasty. ◈ *province* N Kyrgyzstan
161 Q7 **Chuzhou** *var.* Chuxian, Chu Xian. Anhui, E China 32°20′N 118°18′E
139 U3 **Chwarta** *Ar.* Juwārtā, *var.* Chwārtā, Choarta, Chuwārtah. As Sulaymānīyah, NE Iraq 35°11′N 45°59′E
Chwārtā *see* Chwarta
119 N16 **Chyhirynskaye Vodaskhovishcha** *Rus.* Chigirinskoye Vodokhranilishche. ▣ E Belarus
117 R6 **Chyhyryn** *Rus.* Chigirin. Cherkas'ka Oblast', N Ukraine 49°03′N 32°40′E
119 J18 **Chyrvonaya Slabada** *Rus.* Krasnaya Slabada, Krasnaya Sloboda. Minskaya Voblasts', S Belarus 52°51′N 27°10′E
119 L19 **Chyrvonaye, Vozyera** *Rus.* Ozero Chervonoye. ◎ SE Belarus
117 N11 **Ciadîr-Lunga** *var.* Ceadâr-Lunga, *Rus.* Chadyr-Lunga. S Moldova 46°03′N 28°50′E
169 P16 **Ciamis** *prev.* Tjiamis. Jawa, C Indonesia 07°20′S 108°21′E
107 I16 **Ciampino** ✈ Lazio, C Italy 41°48′N 12°36′E
169 N16 **Cianjur** *prev.* Tjiandjoer. Jawa, C Indonesia 06°50′S 107°09′E
60 H10 **Cianorte** Paraná, S Brazil 23°42′S 52°31′W
Ciarraí *see* Kerry
112 N13 **Ćićevac** Serbia, E Serbia 43°44′N 21°25′E
187 Z14 **Cicia** *prev.* Thithia. *island* Lau Group, E Fiji
105 P4 **Cidacos** 〰 N Spain
136 I10 **Cide** Kastamonu, N Turkey 41°53′N 33°01′E
110 L10 **Ciechanów** *prev.* Zichenau. Mazowieckie, C Poland 52°53′N 20°37′E
110 O10 **Ciechanowiec** *Ger.* Rudelstadt. Podlaskie, E Poland 52°43′N 22°30′E
110 J10 **Ciechocinek** Kujawskio-pomorskie, C Poland 52°53′N 18°49′E
44 F6 **Ciego de Ávila** Ciego de Ávila, C Cuba 21°50′N 78°44′W
54 F4 **Ciénaga** Magdalena, N Colombia 11°01′N 74°15′W
54 E6 **Ciénaga de Oro** Córdoba, NW Colombia 08°54′N 75°39′W
44 E5 **Cienfuegos** Cienfuegos, C Cuba 22°10′N 80°27′W
104 F4 **Cíes, Illas** *island group* NW Spain
111 P16 **Cieszanów** Podkarpackie, SE Poland 50°15′N 23°09′E
111 J17 **Cieszyn** *Cz.* Těšín, *Ger.* Teschen. Śląskie, S Poland 49°45′N 18°35′E
105 R12 **Cieza** Murcia, SE Spain 38°14′N 01°25′W
136 F13 **Çifteler** Eskişehir, W Turkey 39°25′N 31°00′E
105 P7 **Cifuentes** Castilla-La Mancha, C Spain 40°47′N 02°37′W
Čiganak *see* Shyganak
105 P9 **Cigüela** 〰 C Spain
136 H14 **Cihanbeyli** Konya, C Turkey 38°40′N 32°55′E
136 H14 **Cihanbeyli Yaylası** *plateau* C Turkey
104 L10 **Cíjara, Embalse de** ▣ C Spain
169 P16 **Cikalong** Jawa, S Indonesia 07°46′S 108°13′E
169 N16 **Cikawung** Jawa, S Indonesia 06°49′S 105°29′E
187 Y13 **Cikobia** *prev.* Thikombia. *island* N Fiji
169 P17 **Cilacap** *prev.* Tjilatjap. Jawa, C Indonesia 07°44′S 109°E
173 O16 **Cilaos** C Réunion 21°08′S 55°28′E
137 S11 **Çıldır** Ardahan, NE Turkey 41°08′N 43°08′E
137 S11 **Çıldır Gölü** ◎ NE Turkey
160 M10 **Cili** Hunan, S China 29°24′N 110°59′E
Cilician Gates *see* Gülek Boğazı
121 V10 **Cilicia Trough** *undersea feature* E Mediterranean Sea
Cill Airne *see* Killarney
Cill Chainnigh *see* Kilkenny
Cill Chaoi *see* Kilkee
Cill Choca *see* Kilcock
Cill Dara *see* Kildare
105 N3 **Cilleruelo de Bezana** Castilla y León, N Spain 42°58′N 03°50′W
Cilli *see* Celje
Cill Mhantáin *see* Wicklow
Cill Rois *see* Kilrush
146 C11 **Çilmämmetgum** *Rus.* Peski Chil'mamedkum,, *Turkm.* Chilmämetgum. *desert* Balkan Welaýaty, W Turkmenistan
137 Z11 **Çilov Adası** *Rus.* Ostrov Zhiloy. *island* E Azerbaijan
26 J6 **Cimarron** Kansas, C USA 37°49′N 100°20′W
37 T9 **Cimarron** New Mexico, SW USA 36°30′N 104°55′W
26 M9 **Cimarron River** 〰 Kansas/Oklahoma, C USA
117 N11 **Cimişlia** *Rus.* Chimishliya. S Moldova 46°31′N 28°45′E
Cîmpia Turzii *see* Câmpia Turzii
Cîmpina *see* Câmpina
Cîmpulung *see* Câmpulung
Cîmpulung Moldovenesc *see* Câmpulung Moldovenesc
137 P15 **Çınar** Dıyarbakır, SE Turkey 37°45′N 40°22′E
54 J8 **Cinaruco, Río** 〰 Colombia/Venezuela
Cina Selatan, Laut *see* South China Sea
105 T5 **Cinca** 〰 NE Spain
112 G13 **Cincar** ▲ SW Bosnia and Herzegovina 43°55′N 17°05′E
31 Q15 **Cincinnati** Ohio, N USA 39°04′N 84°34′W
21 M4 **Cincinnati** ✈ Kentucky, S USA 39°03′N 84°39′W
Cinco de Outubro *see* Xá-Muteba
136 C15 **Çine** Aydın, SW Turkey 37°37′N 28°03′E
99 J21 **Ciney** Namur, SE Belgium 50°17′N 05°06′E
104 H6 **Cinfães** Viseu, N Portugal 41°04′N 08°06′W
106 J12 **Cingoli** Marche, C Italy 43°25′N 13°09′E
41 U16 **Cintalapa** *var.* Cintalapa de Figueroa. Chiapas, SE Mexico 16°42′N 93°40′W
Cintalapa de Figueroa *see* Cintalapa
103 X14 **Cinto, Monte** ▲ Corse, France, C Mediterranean Sea 42°22′N 08°57′E
Cintra *see* Sintra
105 Q5 **Cintruénigo** Navarra, N Spain 42°05′N 01°50′W
Cionn tSáile *see* Kinsale
116 K13 **Ciorani** Prahova, SE Romania 44°49′N 26°25′E
113 E14 **Čiovo** *It.* Bua. *island* S Croatia
Cipiúr *see* Kippure
63 I15 **Cipolletti** Río Negro, C Argentina 38°55′S 68°W
120 L7 **Circeo, Capo** *headland* C Italy 41°15′N 13°03′E
39 S8 **Circle** *var.* Circle City. Alaska, USA 65°51′N 144°04′W
33 X8 **Circle** Montana, NW USA 47°25′N 105°32′W
Circle City *see* Circle
31 S14 **Circleville** Ohio, N USA 39°36′N 82°57′W
36 K6 **Circleville** Utah, W USA 38°10′N 112°16′W
169 P16 **Cirebon** *prev.* Tjirebon. Jawa, S Indonesia 06°46′S 108°33′E
97 L21 **Cirencester** *anc.* Corinium, Corinium Dobunorum. C England, United Kingdom 51°44′N 01°59′W
Cirkvenica *see* Crikvenica
107 O20 **Cirò** Calabria, SW Italy 39°22′N 17°02′E
107 O20 **Cirò Marina** Calabria, S Italy 39°21′N 17°07′E
102 K14 **Ciron** 〰 SW France
Cirquenizza *see* Crikvenica
25 R7 **Cisco** Texas, SW USA 32°23′N 98°58′W
116 I12 **Cisnădie** *Ger.* Heltau, *Hung.* Nagydisznód. Sibiu, SW Romania 45°42′N 24°09′E
63 G18 **Cisnes, Río** 〰 S Chile
25 T11 **Cistern** Texas, SW USA 29°46′N 97°12′W
104 L3 **Cistierna** Castilla y León, N Spain 42°47′N 05°08′W
Citharista *see* la Ciotat
Citlaltépetl *see* Orizaba, Volcán Pico de
55 X10 **Citron** NW French Guiana 04°49′N 53°55′W
23 N7 **Citronelle** Alabama, S USA 31°05′N 88°13′W
35 O7 **Citrus Heights** California, W USA 38°42′N 121°18′W
106 H7 **Cittadella** Veneto, NE Italy 45°37′N 11°46′E
106 H13 **Città della Pieve** Umbria, C Italy 42°57′N 12°01′E
106 H12 **Città di Castello** Umbria, C Italy 43°27′N 12°13′E
107 I14 **Cittaducale** Lazio, C Italy 42°24′N 12°55′E
107 N22 **Cittanova** Calabria, SW Italy 38°21′N 16°05′E
Cittavecchia *see* Stari Grad
116 G10 **Ciucea** *Hung.* Csucsa. Cluj, NW Romania 46°58′N 22°50′E
116 M13 **Ciucurova** Tulcea, SE Romania 44°57′N 28°24′E
Ciudad Acuña *see* Villa Acuña
41 N15 **Ciudad Altamirano** Guerrero, S Mexico 18°20′N 100°40′W
42 G7 **Ciudad Barrios** San Miguel, NE El Salvador 13°46′N 88°13′W
54 I7 **Ciudad Bolívar** Barinas, NW Venezuela 08°22′N 70°37′W
55 N7 **Ciudad Bolívar** *prev.* Angostura. Bolívar, E Venezuela 08°08′N 63°31′W
40 K6 **Ciudad Camargo** Chihuahua, N Mexico 27°42′N 105°10′W
40 E8 **Ciudad Constitución** Baja California Sur, NW Mexico 25°09′N 111°43′W
Ciudad Cortés *see* Cortés
41 V17 **Ciudad Cuauhtémoc** Chiapas, SE Mexico 15°38′N 91°59′W
42 J9 **Ciudad Darío** *var.* Dario. Matagalpa, W Nicaragua 12°42′N 86°10′W
Ciudad de Dolores Hidalgo *see* Dolores Hidalgo
42 C6 **Ciudad de Guatemala** *Eng.* Guatemala City; *prev.* Santiago de los Caballeros. ● (Guatemala) Guatemala, C Guatemala 14°38′N 90°29′W
Ciudad del Carmen *see* Carmen
62 Q6 **Ciudad del Este** *prev.* Ciudad Presidente Stroessner, Presidente Stroessner, Puerto Presidente Stroessner. Alto Paraná, SE Paraguay 25°34′S 54°40′W
62 K5 **Ciudad de Libertador General San Martín** *var.* Libertador General San Martín. Jujuy, C Argentina 23°50′S 64°45′W
Ciudad Delicias *see* Delicias
41 O11 **Ciudad del Maíz** San Luis Potosí, C Mexico 22°26′N 99°36′W
Ciudad de México *see* México
54 J7 **Ciudad de Nutrias** Barinas, NW Venezuela 08°03′N 69°17′W
Ciudad de Panama *see* Panamá
55 P7 **Ciudad Guayana** *prev.* San Tomé de Guayana, Santo Tomé de Guayana. Bolívar, NE Venezuela 08°22′N 62°37′W
40 K14 **Ciudad Guzmán** Jalisco, SW Mexico 19°40′N 103°30′W
41 V17 **Ciudad Hidalgo** Chiapas, SE Mexico 14°40′N 92°11′W
41 N14 **Ciudad Hidalgo** Michoacán, SW Mexico 19°40′N 100°34′W
40 J3 **Ciudad Juárez** Chihuahua, N Mexico 31°39′N 106°26′W
40 L8 **Ciudad Lerdo** Durango, C Mexico 25°34′N 103°30′W
41 Q11 **Ciudad Madero** *var.* Villa Cecilia. Tamaulipas, C Mexico 22°18′N 97°56′W
41 P11 **Ciudad Mante** Tamaulipas, C Mexico 22°43′N 99°02′W
42 F2 **Ciudad Melchor de Mencos** *var.* Melchor de Mencos. Petén, NE Guatemala 17°03′N 89°12′W
41 P8 **Ciudad Miguel Alemán** Tamaulipas, C Mexico 26°20′N 98°56′W
Ciudad Mutis *see* Bahía Solano
40 G6 **Ciudad Obregón** Sonora, NW Mexico 27°32′N 109°53′W
54 I5 **Ciudad Ojeda** Zulia, NW Venezuela 10°12′N 71°17′W
55 P7 **Ciudad Piar** Bolívar, E Venezuela 07°25′N 63°19′W
Ciudad Porfirio Díaz *see* Piedras Negras
Ciudad Presidente Stroessner *see* Ciudad del Este
Ciudad Quesada *see* Quesada
105 N11 **Ciudad Real** Castilla-La Mancha, C Spain 38°59′N 03°55′W
105 N11 **Ciudad Real** ◈ *province* Castilla-La Mancha, C Spain
104 J7 **Ciudad-Rodrigo** Castilla y León, N Spain 40°36′N 06°33′W
42 A6 **Ciudad Tecún Umán** San Marcos, SW Guatemala 14°40′N 92°06′W
Ciudad Trujillo *see* Santo Domingo
41 P12 **Ciudad Valles** San Luis Potosí, C Mexico 21°59′N 99°01′W
41 O10 **Ciudad Victoria** Tamaulipas, C Mexico 23°44′N 99°07′W
42 C6 **Ciudad Vieja** Suchitepéquez, S Guatemala 14°30′N 90°46′W
116 L8 **Ciuhuru** *var.* Reuţel. 〰 N Moldova
105 Z8 **Ciutadella** *var.* Ciutadella de Menorca. Menorca, Spain, W Mediterranean Sea 40°N 03°50′E
Ciutadella Ciutadella de Menorca *see* Ciutadella
136 L11 **Civa Burnu** *headland* N Turkey 41°22′N 36°39′E
106 J7 **Cividale del Friuli** Friuli-Venezia Giulia, NE Italy 46°06′N 13°25′E
107 H14 **Civita Castellana** Lazio, C Italy 42°16′N 12°24′E
106 J12 **Civitanova Marche** Marche, C Italy 43°18′N 13°41′E
Civitas Altae Ripae *see* Brzeg
Civitas Carnutum *see* Chartres
Civitas Eburovicum *see* Évreux
Civitas Nemetum *see* Speyer
107 G15 **Civitavecchia** *anc.* Centum Cellae, Trajani Portus. Lazio, C Italy 42°05′N 11°47′E
102 L10 **Civray** Vienne, W France 46°10′N 00°18′E
136 E14 **Çivril** Denizli, W Turkey 38°18′N 29°43′E
161 O5 **Cixian** Hebei, E China 36°19′N 114°22′E
137 R16 **Cizre** Şırnak, SE Turkey 37°21′N 42°11′E
Clacton *see* Clacton-on-Sea
97 Q21 **Clacton-on-Sea** *var.* Clacton. E England, United Kingdom 51°48′N 01°09′E
22 H5 **Claiborne, Lake** ▣ Louisiana, S USA
102 L10 **Clain** 〰 W France
11 Q11 **Claire, Lake** ◎ Alberta, C Canada
25 O6 **Clairemont** Texas, SW USA 33°09′N 100°45′W
34 M3 **Clair Engle Lake** ▣ California, W USA
18 B15 **Clairton** Pennsylvania, NE USA 40°17′N 79°52′W
32 F7 **Clallam Bay** Washington, NW USA 48°13′N 124°16′W
103 P8 **Clamecy** Nièvre, C France 47°28′N 03°30′E
23 P5 **Clanton** Alabama, S USA 32°50′N 86°37′W
61 D17 **Clara** Entre Ríos, E Argentina 31°50′S 58°48′W
97 E18 **Clara** *Ir.* Clóirtheach. C Ireland 53°20′N 07°36′W
29 T9 **Clara City** Minnesota, N USA 44°57′N 95°22′W
61 D23 **Claraz** Buenos Aires, E Argentina 37°56′S 59°18′W
Clár Chlainne Mhuiris *see* Claremorris
182 I8 **Clare** South Australia 33°49′S 138°35′E
97 C19 **Clare** *Ir.* An Clár. *cultural region* W Ireland
97 C18 **Clare** 〰 W Ireland
97 A16 **Clare Island** *Ir.* Cliara. *island* W Ireland
44 J12 **Claremont** C Jamaica 18°23′N 77°11′W
29 W10 **Claremont** Minnesota, N USA 44°01′N 93°00′W
19 N9 **Claremont** New Hampshire, NE USA 43°21′N 72°18′W
27 Q9 **Claremore** Oklahoma, C USA 36°20′N 95°37′W
97 C17 **Claremorris** *Ir.* Clár Chlainne Mhuiris. W Ireland 53°47′N 09°W
185 J16 **Clarence** Canterbury, South Island, New Zealand 42°08′S 173°54′E
185 J16 **Clarence** 〰 South Island, New Zealand
65 F15 **Clarence Bay** *bay* Ascension Island, C Atlantic Ocean
63 H25 **Clarence, Isla** *island* S Chile
194 H2 **Clarence Island** *island* South Shetland Islands, Antarctica
183 V5 **Clarence River** 〰 New South Wales, SE Australia
44 J5 **Clarence Town** Long Island, C The Bahamas 23°03′N 74°57′W
27 W12 **Clarendon** Arkansas, C USA 34°41′N 91°19′W
25 O3 **Clarendon** Texas, SW USA 34°57′N 100°54′W
13 U12 **Clarenville** Newfoundland, Newfoundland and Labrador, SE Canada 48°10′N 54°00′W
11 Q17 **Claresholm** Alberta, SW Canada 50°02′N 113°33′W
29 T16 **Clarinda** Iowa, C USA 40°44′N 95°02′W
55 N5 **Clarines** Anzoátegui, NE Venezuela 09°56′N 65°11′W
29 V12 **Clarion** Iowa, C USA 42°43′N 93°43′W
18 C13 **Clarion** Pennsylvania, NE USA 41°11′N 79°21′W
193 O6 **Clarion Fracture Zone** *tectonic feature* NE Pacific Ocean
18 D13 **Clarion River** 〰 Pennsylvania, NE USA
29 Q9 **Clark** South Dakota, N USA 44°50′N 97°44′W
36 K11 **Clarkdale** Arizona, SW USA 34°46′N 112°03′W
15 W4 **Clarke City** Québec, SE Canada 50°09′N 66°36′W
183 Q15 **Clarke Island** *island* Furneaux Group, Tasmania, SE Australia
181 X6 **Clarke Range** ⛰ Queensland, E Australia
23 T2 **Clarkesville** Georgia, SE USA 34°36′N 83°31′W
29 S9 **Clarkfield** Minnesota, N USA 44°48′N 95°49′W
33 N7 **Clark Fork** Idaho, NW USA 48°06′N 116°10′W
33 N8 **Clark Fork** 〰 Idaho/Montana, NW USA
21 P13 **Clark Hill Lake** *var.* J.Storm Thurmond Reservoir. ▣ Georgia/South Carolina, SE USA
39 Q12 **Clark, Lake** ◎ Alaska, USA
35 W12 **Clark Mountain** ▲ California, W USA 35°30′N 115°34′W
37 S3 **Clark Peak** ▲ Colorado, C USA 40°36′N 105°57′W
14 D14 **Clark, Point** *headland* Ontario, S Canada 44°04′N 81°45′W
21 S3 **Clarksburg** West Virginia, NE USA 39°16′N 80°22′W
22 K2 **Clarksdale** Mississippi, S USA 34°12′N 90°34′W
33 U12 **Clarks Fork Yellowstone River** 〰 Montana/Wyoming, NW USA
29 R14 **Clarkson** Nebraska, C USA 41°42′N 97°07′W
39 O13 **Clarks Point** Alaska, USA 58°50′N 158°33′W
18 I13 **Clarks Summit** Pennsylvania, NE USA 41°29′N 75°42′W
32 M10 **Clarkston** Washington, NW USA 46°25′N 117°02′W
44 J12 **Clark's Town** C Jamaica 18°25′N 77°32′W
27 T10 **Clarksville** Arkansas, C USA 35°29′N 93°29′W
31 P13 **Clarksville** Indiana, N USA 40°01′N 85°54′W
20 I8 **Clarksville** Tennessee, S USA 36°32′N 87°22′W
25 W5 **Clarksville** Texas, SW USA 33°37′N 95°04′W
21 U8 **Clarksville** Virginia, NE USA 36°36′N 78°36′W
21 U11 **Clarkton** North Carolina, SE USA 34°28′N 78°39′W
61 C24 **Claromecó** *var.* Balneario Claromecó. Buenos Aires, E Argentina 38°51′S 60°01′W
25 N3 **Claude** Texas, SW USA 35°06′N 101°22′W
Clausentum *see* Southampton
171 O1 **Claveria** Luzon, N Philippines 18°36′N 121°04′E
99 J20 **Clavier** Liège, E Belgium 50°27′N 05°21′E

◆ Country
● Country Capital
◇ Dependent Territory
○ Dependent Territory Capital
◈ Administrative Regions
✈ International Airport
▲ Mountain
⛰ Mountain Range
🌋 Volcano
〰 River
◎ Lake
▣ Reservoir

◆ Country
● Country Capital
◇ Dependent Territory
○ Dependent Territory Capital
◆ Administrative Regions
✕ International Airport
▲ Mountain
▲ Mountain Range
🌋 Volcano
☒ River
☒ Lake
☒ Reservoir

Comoro Islands – Covington

172 K14 **Comoro Islands** *island group* W Indian Ocean
172 H13 **Comoros** *off.* Federal Islamic Republic of the Comoros, *Fr.* République Fédérale Islamique des Comores. ◆ *republic* W Indian Ocean
Comoros, Federal Islamic Republic of the *see* Comoros
10 L17 **Comox** Vancouver Island, British Columbia, SW Canada 49°40′N 124°55′W
103 O4 **Compiègne** Oise, N France 49°25′N 02°50′E
Complutum *see* Alcalá de Henares
Compniacum *see* Cognac
40 K12 **Compostela** Nayarit, C Mexico 21°12′N 104°52′W
Compostella *see* Santiago de Compostela
60 L11 **Comprida, Ilha** *island* S Brazil
117 N11 **Comrat** *Rus.* Komrat. S Moldova 46°18′N 28°40′E
25 O11 **Comstock** Texas, SW USA 29°39′N 101°10′W
31 P9 **Comstock Park** Michigan, N USA 43°00′N 85°40′W
193 N3 **Comstock Seamount** *undersea feature* N Pacific Ocean 48°15′N 156°55′W
Comum *see* Como
159 N17 **Cona** Xizang Zizhiqu, W China 27°59′N 91°54′E
76 H14 **Conakry** ● (Guinea) SW Guinea 09°31′N 13°43′W
76 H14 **Conakry** ✈ SW Guinea 09°37′N 13°32′W
Conamara *see* Connemara
Conca *see* Cuenca
25 Q12 **Concan** Texas, SW USA 29°27′N 99°43′W
102 F6 **Concarneau** Finistère, NW France 47°53′N 03°55′W
83 O17 **Conceição** Sofala, C Mozambique 18°47′S 36°18′E
59 K15 **Conceição do Araguaia** Pará, NE Brazil 08°15′S 49°15′W
58 F10 **Conceição do Maú** Roraima, W Brazil 03°35′N 59°52′W
61 D14 **Concepción** *var.* Concepcion. Corrientes, NE Argentina 28°25′S 57°54′W
62 J8 **Concepción** Tucumán, N Argentina 27°20′S 65°35′W
57 O17 **Concepción** Santa Cruz, E Bolivia 16°15′S 62°08′W
62 G13 **Concepción** Bío Bío, C Chile 36°47′S 73°01′W
54 E14 **Concepción** Putumayo, S Colombia 0°03′N 75°35′W
62 O5 **Concepción** *var.* Villa Concepción. Concepción, C Paraguay 23°26′S 57°24′W
62 O5 **Concepción** *off.* Departamento de Concepción. ◈ *department* E Paraguay
Concepción *see* La Concepción
Concepción de la Vega *see* La Vega
41 N9 **Concepción del Oro** Zacatecas, C Mexico 24°38′N 101°25′W
61 D18 **Concepción del Uruguay** Entre Ríos, E Argentina 32°30′S 58°15′W
Concepción, Departamento de *see* Concepción
42 K11 **Concepción, Volcán** ⊼ SW Nicaragua 11°31′N 85°37′W
44 J4 **Conception Island** *island* C The Bahamas
35 P14 **Conception, Point** *headland* California, W USA 34°27′N 120°28′W
54 H6 **Concha** Zulia, W Venezuela 09°02′N 71°45′W
60 L9 **Conchas** São Paulo, S Brazil 23°00′S 47°58′W
37 U11 **Conchas Dam** New Mexico, SW USA 35°21′N 104°11′W
37 U10 **Conchas Lake** ⊠ New Mexico, SW USA
102 M5 **Conches-en-Ouche** Eure, N France 49°00′N 01°00′E
37 N12 **Concho** Arizona, SW USA 34°28′N 109°33′W
40 J5 **Conchos, Río** ≋ NW Mexico
41 O8 **Conchos, Río** ≋ C Mexico
108 C8 **Concise** Vaud, W Switzerland 46°52′N 06°40′E
35 N8 **Concord** California, W USA 37°58′N 122°01′W
19 O9 **Concord** *state capital* New Hampshire, NE USA 43°10′N 71°32′W
21 R10 **Concord** North Carolina, SE USA 35°25′N 80°34′W
61 D17 **Concordia** Entre Ríos, E Argentina 31°25′S 58°W
60 I13 **Concórdia** Santa Catarina, S Brazil 27°14′S 52°01′W
54 D9 **Concordia** Antioquia, W Colombia 06°03′N 75°57′W
40 J10 **Concordia** Sinaloa, C Mexico 23°18′N 106°02′W
57 I19 **Concordia** Tacna, SW Peru 18°12′S 70°19′W
27 N3 **Concordia** Kansas, C USA 39°35′N 97°39′W
27 S4 **Concordia** Missouri, C USA 38°58′N 93°34′W
167 S7 **Con Cuông** Nghệ An, N Vietnam 19°02′N 104°54′E
167 T15 **Côn Đao Sơn** *var.* Con Son. *island* S Vietnam
Condate *see* Rennes, Ille-et-Vilaine, France
Condate *see* St-Claude, Jura, France
Condate *see* Montereau-Faut-Yonne, Seine-St-Denis, France
29 P8 **Conde** South Dakota, N USA 45°08′N 98°07′W
42 J8 **Condega** Estelí, NW Nicaragua 13°19′N 86°26′W
103 P2 **Condé-sur-l'Escaut** Nord, N France 50°27′N 03°36′E
102 K5 **Condé-sur-Noireau** Calvados, N France 48°52′N 00°31′W
Condivincum *see* Nantes
183 P8 **Condobolin** New South Wales, SE Australia 33°04′S 147°08′E
102 L15 **Condom** Gers, S France 43°56′N 00°23′E
32 J11 **Condon** Oregon, NW USA 45°15′N 120°10′W
54 D9 **Condoto** Chocó, W Colombia 05°06′N 76°37′W
23 P7 **Conecuh River** ≋ Alabama/Florida, SE USA
106 H7 **Conegliano** Veneto, NE Italy 45°53′N 12°18′E
61 C19 **Conesa** Buenos Aires, E Argentina 33°36′S 60°21′W
14 F15 **Conestogo** ≋ Ontario, S Canada
Confluentes *see* Koblenz
102 L10 **Confolens** Charente, W France 46°00′N 00°40′E
36 J4 **Confusion Range** ▲▲ Utah, W USA
62 N6 **Confuso, Río** ≋ C Paraguay
21 R12 **Congaree River** ≋ South Carolina, SE USA
Công Hoa Xa Hôi Chu Nghia Viêt Nam *see* Vietnam
160 K12 **Congjiang** *var.* Bingmei. Guizhou, S China 25°48′N 108°55′E
79 G18 **Congo** *off.* Republic of the Congo, *Fr.* Moyen-Congo; *prev.* Middle Congo. ◆ *republic* C Africa
79 K19 **Congo** *off.* Democratic Republic of Congo; *prev.* Zaire, Belgian Congo, Congo (Kinshasa). ◆ *republic* C Africa
67 T11 **Congo** *var.* Kongo, *Fr.* Zaire. ≋ C Africa
Congo *see* Zaire (province) Angola
68 G12 **Congo Basin** *drainage basin* W Dem. Rep. Congo
67 Q11 **Congo Canyon** *var.* Congo Seavalley, Congo Submarine Canyon. *undersea feature* E Atlantic Ocean 06°00′S 11°50′E
Congo Cone *see* Congo Fan
Congo/Congo (Kinshasa) *see* Congo (Democratic Republic of)
65 P15 **Congo Fan** *var.* Congo Cone. *undersea feature* E Atlantic Ocean 06°00′S 09°00′E
Congo Seavalley *see* Congo Canyon
Congo Submarine Canyon *see* Congo Canyon
Coni *see* Cuneo
63 H18 **Cónico, Cerro** ▲ SW Argentina 43°12′S 71°42′W
Conimbria/Conimbriga *see* Coimbra
Conjeeveram *see* Kānchipuram
11 R13 **Conklin** Alberta, C Canada 55°36′N 111°06′W
24 M1 **Conlen** Texas, SW USA 36°16′N 102°10′W
Con, Loch *see* Conn, Lough
Connacht *see* Connaught
97 B17 **Connaught** *var.* Connacht, *Ir.* Chonnacht, Cúige. *cultural region* W Ireland
31 V10 **Conneaut** Ohio, N USA 41°56′N 80°32′W
18 L13 **Connecticut** *off.* State of Connecticut, *also known as* Blue Law State, Constitution State, Land of Steady Habits, Nutmeg State. ◈ *state* NE USA
19 N8 **Connecticut** ≋ Canada/USA
19 O6 **Connecticut Lakes** *lakes* New Hampshire, NE USA
32 K9 **Connell** Washington, NW USA 46°39′N 118°51′W
97 B17 **Connemara** *Ir.* Conamara. *physical region* W Ireland
31 Q14 **Connersville** Indiana, N USA 39°38′N 85°15′W
97 B16 **Conn, Lough** *Ir.* Loch Con. ◎ W Ireland
35 X6 **Connors Pass** *pass* Nevada, W USA
181 X7 **Connors Range** ▲▲ Queensland, E Australia
56 E7 **Cononaco, Río** ≋ E Ecuador
29 W13 **Conrad** Iowa, C USA 42°13′N 92°52′W
33 R7 **Conrad** Montana, NW USA 48°10′N 111°58′W
25 W10 **Conroe** Texas, SW USA 30°19′N 95°28′W
25 V10 **Conroe, Lake** ⊠ Texas, SW USA
61 C17 **Conscripto Bernardi** Entre Ríos, E Argentina 31°03′S 59°05′W
59 M20 **Conselheiro Lafaiete** Minas Gerais, SE Brazil 20°40′S 43°48′W
Consentia *see* Cosenza
97 L14 **Consett** N England, United Kingdom 54°50′N 01°53′W
44 B5 **Consolación del Sur** Pinar del Río, W Cuba 22°32′N 83°32′W
Con Son *see* Côn Đao Sơn
11 R15 **Consort** Alberta, SW Canada 51°58′N 110°44′W
Constance *see* Konstanz
108 I6 **Constance, Lake** *Ger.* Bodensee. ◎ C Europe
104 G9 **Constância** Santarém, C Portugal 39°29′N 08°22′W
117 N14 **Constanța** *var.* Küstendje, *Eng.* Constanza, *Ger.* Konstanza, *Turk.* Küstence. Constanța, SE Romania 44°09′N 28°37′E
116 L14 **Constanța** ◈ *county* SE Romania
Constantia *see* Coutances
Constantia *see* Konstanz
104 K13 **Constantina** Andalucía, S Spain 37°54′N 05°36′W
74 L5 **Constantine** *var.* Qacentina, *Ar.* Qoussantîna. NE Algeria 36°23′N 06°44′E
39 O14 **Constantine, Cape** *headland* Alaska, USA 58°23′N 158°53′W
Constantinople *see* İstanbul
Constantiola *see* Oltenița
Constanz *see* Konstanz
Constanza *see* Constanța
62 G13 **Constitución** Maule, C Chile 35°20′S 72°28′W
61 D17 **Constitución** Salto, N Uruguay 31°05′S 57°51′W
Constitution State *see* Connecticut
105 N10 **Consuegra** Castilla-La Mancha, C Spain 39°28′N 03°36′W
181 X9 **Consuelo Peak** ▲ Queensland, E Australia 24°45′S 148°01′E
56 E11 **Contamana** Loreto, N Peru 07°19′S 75°04′W
Contrasto, Colle del *see* Contrasto, Portella del
107 K23 **Contrasto, Portella del** *var.* Colle del Contrasto. *pass* Sicilia, Italy, C Mediterranean Sea
54 G8 **Contratación** Santander, C Colombia 06°18′N 73°27′W
102 M8 **Contres** Loir-et-Cher, C France 47°24′N 01°30′E
107 O17 **Conversano** Puglia, SE Italy 40°58′N 17°07′E
27 U11 **Conway** Arkansas, C USA 35°05′N 92°27′W
19 O8 **Conway** New Hampshire, NE USA 43°58′N 71°05′W
21 U13 **Conway** South Carolina, SE USA 33°51′N 79°04′W
25 N2 **Conway** Texas, SW USA 35°10′N 101°23′W
27 U11 **Conway, Lake** ⊠ Arkansas, C USA
27 N7 **Conway Springs** Kansas, C USA 37°23′N 97°38′W
97 J18 **Conwy** N Wales, United Kingdom 53°17′N 03°51′W
23 T3 **Conyers** Georgia, SE USA 33°40′N 84°01′W
Coo *see* Kos
182 F4 **Coober Pedy** South Australia 29°01′S 134°47′E
181 P2 **Cooinda** Northern Territory, N Australia 12°54′S 132°31′E
182 B6 **Cook** South Australia 30°37′S 130°26′E
29 W4 **Cook** Minnesota, N USA 47°51′N 92°41′W
191 N6 **Cook, Baie de** *bay* Moorea, W French Polynesia
10 J16 **Cook, Cape** *headland* Vancouver Island, British Columbia, SW Canada 50°04′N 127°52′W
37 Q15 **Cookes Peak** ▲ New Mexico, SW USA 32°32′N 107°43′W
20 L8 **Cookeville** Tennessee, S USA 36°10′N 85°30′W
175 P9 **Cook Fracture Zone** *tectonic feature* S Pacific Ocean
Cook, Grand Récif de *see* Cook, Récif de
39 Q12 **Cook Inlet** *inlet* Alaska, USA
191 X2 **Cook Island** *island* Line Islands, E Kiribati
190 J14 **Cook Islands** ◇ *self-governing entity in free association with New Zealand* S Pacific Ocean
Cook, Mount *see* Aoraki
187 O15 **Cook, Récif de** *var.* Grand Récif de Cook. *reef* S New Caledonia
14 G14 **Cookstown** Ontario, S Canada 44°12′N 79°39′W
97 F15 **Cookstown** *Ir.* An Chorr Chríochach. C Northern Ireland, United Kingdom 54°39′N 06°45′W
185 K14 **Cook Strait** *var.* Raukawa. *strait* New Zealand
181 W3 **Cooktown** Queensland, NE Australia 15°28′S 145°15′E
183 P6 **Coolabah** New South Wales, SE Australia 31°03′S 146°42′E
182 J11 **Coola Coola Swamp** *wetland* South Australia
183 S7 **Coolah** New South Wales, SE Australia 31°49′S 149°43′E
183 P9 **Coolamon** New South Wales, SE Australia 34°49′S 147°13′E
183 T4 **Coolatai** New South Wales, SE Australia 29°16′S 150°45′E
180 K12 **Coolgardie** Western Australia 31°01′S 121°12′E
36 L14 **Coolidge** Arizona, SW USA 32°58′N 111°29′W
25 U8 **Coolidge** Texas, SW USA 31°45′N 96°39′W
183 Q11 **Cooma** New South Wales, SE Australia 36°16′S 149°09′E
Coomassie *see* Kumasi
183 R6 **Coonabarabran** New South Wales, SE Australia 31°19′S 149°18′E
182 J10 **Coonalpyn** South Australia 35°43′S 139°50′E
183 R6 **Coonamble** New South Wales, SE Australia 30°56′S 148°22′E
Coondapoor *see* Kundāpura
155 G21 **Coonoor** Tamil Nādu, SE India 11°21′N 76°46′E
29 U14 **Coon Rapids** Iowa, C USA 41°52′N 94°40′W
29 V8 **Coon Rapids** Minnesota, N USA 45°12′N 93°18′W
25 V5 **Cooper** Texas, SW USA 33°23′N 95°42′W
181 U9 **Cooper Creek** *var.* Barcoo, Cooper's Creek. *seasonal river* Queensland/South Australia
39 R12 **Cooper Landing** Alaska, USA 60°27′N 149°59′W
21 T14 **Cooper River** ≋ South Carolina, SE USA
Cooper's Creek *see* Cooper Creek
44 H1 **Coopers Town** Great Abaco, N The Bahamas 26°46′N 77°27′W
18 J10 **Cooperstown** New York, NE USA 42°43′N 74°56′W
29 P4 **Cooperstown** North Dakota, N USA 47°26′N 98°07′W
31 P9 **Coopersville** Michigan, N USA 43°03′N 85°55′W
182 D7 **Coorabie** South Australia 31°57′S 132°18′E
23 Q3 **Coosa River** ≋ Alabama/Georgia, S USA
32 E14 **Coos Bay** Oregon, NW USA 43°22′N 124°13′W
183 Q9 **Cootamundra** New South Wales, SE Australia 34°41′S 148°03′E
97 E16 **Cootehill** *Ir.* Muinchille. N Ireland 54°04′N 07°05′W
Čop *see* Chop
57 J17 **Copacabana** La Paz, W Bolivia 16°11′S 69°02′W
63 H14 **Copahué, Volcán** ⊼ C Chile 37°56′S 71°04′W
41 U16 **Copainalá** Chiapas, SE Mexico 17°04′N 93°13′W
32 F8 **Copalis Beach** Washington, NW USA 47°05′N 124°11′W
42 F6 **Copán** ◈ *department* W Honduras
Copán *see* Copán Ruinas
25 T14 **Copano Bay** *bay* NW Gulf of Mexico
42 F6 **Copán Ruinas** *var.* Copán. Copán, W Honduras 14°52′N 89°10′W
Copenhagen *see* København
107 Q19 **Copertino** Puglia, SE Italy 40°16′N 18°03′E
62 H7 **Copiapó** Atacama, N Chile 27°17′S 70°25′W
62 G8 **Copiapó, Bahía** *bay* N Chile
62 G7 **Copiapó, Río** ≋ N Chile
114 M12 **Çöpköy** Edirne, NW Turkey 41°14′N 26°51′E
182 I5 **Copley** South Australia 30°36′S 138°26′E
106 H9 **Copparo** Emilia-Romagna, C Italy 44°53′N 11°53′E
55 V10 **Coppename Rivier** *var.* Koppename. ≋ C Suriname
25 S9 **Copperas Cove** Texas, SW USA 31°07′N 97°54′W
82 J13 **Copperbelt** ◈ *province* C Zambia
39 S11 **Copper Center** Alaska, USA 61°57′N 145°21′W
8 K8 **Coppermine** ≋ Northwest Territories/Nunavut, N Canada
Coppermine *see* Kugluktuk
39 T11 **Copper River** ≋ Alaska, USA
Copper State *see* Arizona
116 I11 **Copșa Mică** *Ger.* Kleinkopisch, *Hung.* Kiskapus. Sibiu, C Romania 46°06′N 24°15′E
158 J14 **Coqên** Xizang Zizhiqu, W China 31°13′N 85°12′E
Coquilhatville *see* Mbandaka
32 E14 **Coquille** Oregon, NW USA 43°11′N 124°12′W
62 G9 **Coquimbo** Coquimbo, N Chile 30°S 71°18′W
62 G9 **Coquimbo** *off.* Región de Coquimbo. ◈ *region* C Chile
Coquimbo, Región de *see* Coquimbo
116 I15 **Corabia** Olt, S Romania 43°46′N 24°31′E
57 F17 **Coracora** Ayacucho, SW Peru 15°03′S 73°45′W
Cora Droma Rúisc *see* Carrick-on-Shannon
44 K9 **Corail** SW Haiti 18°34′N 73°53′W
183 V4 **Coraki** New South Wales, SE Australia 29°01′S 153°15′E
180 G8 **Coral Bay** Western Australia 23°02′S 113°51′E
23 Y16 **Coral Gables** Florida, SE USA 25°43′N 80°16′W
9 P8 **Coral Harbour** *var.* Salliq. Southampton Island, Nunavut, NE Canada 64°10′N 83°15′W
192 I9 **Coral Sea** *sea* SW Pacific Ocean
174 M7 **Coral Sea Basin** *undersea feature* N Coral Sea
192 H9 **Coral Sea Islands** ◇ *Australian external territory* SW Pacific Ocean
182 M12 **Corangamite, Lake** ◎ Victoria, SE Australia
Corantijn Rivier *see* Courantyne River
18 B14 **Coraopolis** Pennsylvania, NE USA 40°28′N 80°08′W
107 N17 **Corato** Puglia, SE Italy 41°09′N 16°25′E
103 O17 **Corbières** ▲▲ S France
103 P8 **Corbigny** Nièvre, C France 47°15′N 03°42′E
21 N7 **Corbin** Kentucky, S USA 36°57′N 84°06′W
104 L14 **Corbones** ≋ SW Spain
Corcaigh *see* Cork
35 R11 **Corcoran** California, W USA 36°06′N 119°33′W
47 T14 **Corcovado, Golfo** *gulf* S Chile
63 G18 **Corcovado, Volcán** ⊼ S Chile 43°13′S 72°45′W
104 F3 **Corcubión** Galicia, NW Spain 42°56′N 09°12′W
Corcyra Nigra *see* Korčula
60 Q9 **Cordeiro** Rio de Janeiro, SE Brazil 22°01′S 42°20′W
23 T6 **Cordele** Georgia, SE USA 31°59′N 83°49′W
26 L11 **Cordell** Oklahoma, C USA 35°17′N 98°59′W
103 N14 **Cordes** Tarn, S France 44°03′N 01°57′E
62 O6 **Cordillera** *off.* Departamento de la Cordillera. ◈ *department* C Paraguay
Cordillera *see* Cacaguatique, Cordillera
Cordillera, Departamento de la *see* Cordillera
182 K1 **Cordillo Downs** South Australia 26°44′S 140°37′E
62 K10 **Córdoba** Córdoba, C Argentina 31°25′S 64°11′W
41 R14 **Córdoba** Veracruz-Llave, E Mexico 18°55′N 96°55′W
104 M13 **Córdoba** *var.* Cordoba, *Eng.* Cordova; *anc.* Corduba. Andalucía, SW Spain 37°53′N 04°46′W
62 K11 **Córdoba** *off.* Provincia de Córdoba. ◈ *province* C Argentina
54 D7 **Córdoba** *off.* Departamento de Córdoba. ◈ *province* NW Colombia
104 L13 **Córdoba** ◈ *province* Andalucía, S Spain
Córdoba, Departamento de *see* Córdoba
Córdoba, Provincia de *see* Córdoba
62 K10 **Córdoba, Sierras de** ▲▲ C Argentina
23 O3 **Cordova** Alabama, S USA 33°45′N 87°10′W
39 S12 **Cordova** Alaska, USA 60°32′N 145°45′W
Cordova/Cordoba *see* Córdoba
Corduba *see* Córdoba
Corentyne River *see* Courantyne River
Corfu *see* Kérkyra
104 J9 **Coria** Extremadura, W Spain 39°59′N 06°32′W
104 J14 **Coria del Río** Andalucía, S Spain 37°18′N 06°04′W
183 S8 **Coricudgy, Mount** ▲ New South Wales, SE Australia 32°49′S 150°28′E
107 N20 **Corigliano Calabro** Calabria, SW Italy 39°36′N 16°32′E
Corinium/Corinium Dobunorum *see* Cirencester
23 N1 **Corinth** Mississippi, S USA 34°56′N 88°29′W
Corinth *see* Kórinthos
Corinth Canal *see* Dióryga Korinthou
Corinth, Gulf of/Corinthiacus Sinus *see* Korinthiakós Kólpos
Corinthus *see* Kórinthos
42 I9 **Corinto** Chinandega, NW Nicaragua 12°29′N 87°14′W
97 C21 **Cork** *Ir.* Corcaigh. S Ireland 51°54′N 08°28′W
97 C21 **Cork** *Ir.* Corcaigh. *cultural region* SW Ireland
97 C21 **Cork** ✈ Cork, SW Ireland 51°52′N 08°25′W
97 D21 **Cork Harbour** *Ir.* Cuan Chorcaí. *inlet* SW Ireland
107 I23 **Corleone** Sicilia, Italy, C Mediterranean Sea 37°49′N 13°18′E
114 N13 **Çorlu** Tekirdağ, NW Turkey 41°11′N 27°48′E
114 N12 **Çorlu Çayı** ≋ NW Turkey
Cormaiore *see* Courmayeur
11 V13 **Cormorant** Manitoba, C Canada 54°12′N 100°33′W
23 T2 **Cornelia** Georgia, SE USA 34°30′N 83°31′W
60 J10 **Cornélio Procópio** Paraná, S Brazil 23°07′S 50°40′W
55 V9 **Corneliskondre** Sipaliwini, N Suriname 05°21′N 56°10′W
30 J5 **Cornell** Wisconsin, N USA 45°09′N 91°10′W
13 S12 **Corner Brook** Newfoundland, Newfoundland and Labrador, E Canada 48°58′N 57°58′W
Corner Rise Seamounts *see* Corner Seamounts
64 I9 **Corner Seamounts** *var.* Corner Rise Seamounts. *undersea feature* NW Atlantic Ocean 35°30′N 51°30′W
116 M9 **Cornești** *Rus.* Korneshty. C Moldova 47°23′N 28°00′E
Corneto *see* Tarquinia
Cornhusker State *see* Nebraska
27 X8 **Corning** Arkansas, C USA 36°26′N 90°35′W
35 N5 **Corning** California, W USA 39°54′N 122°12′W
29 U15 **Corning** Iowa, C USA 40°58′N 94°46′W
18 G11 **Corning** New York, NE USA 42°08′N 77°03′W
Corn Islands *see* Maíz, Islas del
107 J14 **Corno Grande** ▲ C Italy 42°26′N 13°29′E
15 N13 **Cornwall** Ontario, SE Canada 45°02′N 74°45′W
97 H25 **Cornwall** *cultural region* SW England, United Kingdom
97 G25 **Cornwall, Cape** *headland* SW England, United Kingdom 50°11′N 05°39′W
54 J4 **Coro** *prev.* Santa Ana de Coro. Falcón, NW Venezuela 11°27′N 69°41′W
57 J18 **Corocoro** La Paz, W Bolivia 17°10′S 68°28′W
57 K17 **Coroico** La Paz, W Bolivia 16°09′S 67°45′W
184 M5 **Coromandel** Waikato, North Island, New Zealand 36°47′S 175°30′E
155 K20 **Coromandel Coast** *coast* E India
184 M5 **Coromandel Peninsula** *peninsula* North Island, New Zealand
184 M6 **Coromandel Range** ▲▲ North Island, New Zealand
171 N5 **Coron** Busuanga Island, W Philippines 12°02′N 120°10′E
35 T15 **Corona** California, W USA 33°52′N 117°34′W
37 T12 **Corona** New Mexico, SW USA 34°15′N 105°36′W
11 U17 **Coronach** Saskatchewan, S Canada 49°07′N 105°33′W
35 U17 **Coronado** California, W USA 32°41′N 117°10′W
43 N15 **Coronado, Bahía de** *bay* S Costa Rica
11 R15 **Coronation** Alberta, SW Canada 52°06′N 111°25′W
8 K7 **Coronation Gulf** *gulf* Nunavut, N Canada
194 I1 **Coronation Island** *island* Antarctica
39 X14 **Coronation Island** *island* Alexander Archipelago, Alaska, USA
61 B18 **Coronda** Santa Fe, C Argentina 31°58′S 60°56′W
63 F14 **Coronel** Bío Bío, C Chile 37°01′S 73°08′W
61 D20 **Coronel Brandsen** *var.* Brandsen. Buenos Aires, E Argentina 35°08′S 58°15′W
62 K4 **Coronel Cornejo** Salta, N Argentina 22°46′S 63°49′W
61 B24 **Coronel Dorrego** Buenos Aires, E Argentina 38°38′S 61°15′W
62 P6 **Coronel Oviedo** Caaguazú, SE Paraguay 25°24′S 56°30′W
61 B23 **Coronel Pringles** Buenos Aires, E Argentina 37°56′S 61°25′W
61 B23 **Coronel Suárez** Buenos Aires, E Argentina 37°30′S 61°52′W
61 E22 **Coronel Vidal** Buenos Aires, E Argentina 37°28′S 57°45′W
55 V9 **Coronie** ◈ *district* NW Suriname
57 G17 **Coropuna, Nevado** ▲ S Peru 15°31′S 72°31′W
Çorovoda *see* Çorovodë
113 L22 **Çorovodë** *var.* Çorovoda. Berat, S Albania 40°29′N 20°15′E
183 P11 **Corowa** New South Wales, SE Australia 36°01′S 146°22′E
42 G1 **Corozal** Corozal, N Belize 18°23′N 88°23′W
54 E6 **Corozal** Sucre, NW Colombia 09°18′N 75°19′W
42 G1 **Corozal** ◈ *district* N Belize
25 T14 **Corpus Christi** Texas, SW USA 27°48′N 97°24′W
25 T14 **Corpus Christi Bay** *inlet* Texas, SW USA
25 R14 **Corpus Christi, Lake** ⊠ Texas, SW USA
63 F16 **Corral** Los Ríos, C Chile 39°55′S 73°30′W
105 O9 **Corral de Almaguer** Castilla-La Mancha, C Spain 39°45′N 03°10′W
104 K6 **Corrales** Castilla y León, N Spain 41°22′N 05°44′W
37 R11 **Corrales** New Mexico, SW USA 35°11′N 106°37′W
Corrain Tuathail *see* Carrauntoohil
106 F9 **Correggio** Emilia-Romagna, C Italy 44°47′N 10°46′E
59 M16 **Corrente** Piauí, E Brazil 10°29′S 45°11′W
59 I19 **Correntes, Rio** ≋ SW Brazil
103 N12 **Corrèze** ◈ *department* C France
97 C17 **Corrib, Lough** *Ir.* Loch Coirib. ◎ W Ireland
61 C14 **Corrientes** Corrientes, NE Argentina 27°29′S 58°42′W
61 D15 **Corrientes** *off.* Provincia de Corrientes. ◈ *province* NE Argentina
44 A5 **Corrientes, Cabo** *headland* W Cuba 21°48′N 84°30′W
40 I13 **Corrientes, Cabo** *headland* SW Mexico 20°25′N 105°42′W
Corrientes, Provincia de *see* Corrientes
61 C16 **Corrientes, Río** ≋ NE Argentina
56 E8 **Corrientes, Río** ≋ Ecuador/Peru
25 W9 **Corrigan** Texas, SW USA 31°00′N 94°49′W
55 U9 **Corriverton** E Guyana 05°55′N 57°09′W
Corriza *see* Korçë
183 Q11 **Corryong** Victoria, SE Australia 36°14′S 147°54′E
103 F2 **Corse** *Eng.* Corsica. ◈ *region* France, C Mediterranean Sea
101 X13 **Corse** *Eng.* Corsica. *island* France, C Mediterranean Sea
103 Y12 **Corse, Cap** *headland* Corse, France, C Mediterranean Sea 43°01′N 09°25′E
103 X15 **Corse-du-Sud** ◈ *department* Corse, France, C Mediterranean Sea
29 P11 **Corsica** South Dakota, N USA 43°25′N 98°24′W
Corsica *see* Corse
25 U7 **Corsicana** Texas, SW USA 32°05′N 96°27′W
103 Y15 **Corte** Corse, France, C Mediterranean Sea 42°18′N 09°08′E
63 G16 **Corte Alto** Los Lagos, S Chile 40°58′S 73°04′W
104 I13 **Cortegana** Andalucía, S Spain 37°55′N 06°49′W
43 N15 **Cortés** *var.* Ciudad Cortés. Puntarenas, SE Costa Rica 08°59′N 83°32′W
42 G5 **Cortés** ◈ *department* NW Honduras
37 P8 **Cortez** Colorado, C USA 37°22′N 108°36′W
Cortez, Sea of *see* California, Golfo de
106 H6 **Cortina d'Ampezzo** Veneto, NE Italy 46°33′N 12°09′E
18 H11 **Cortland** New York, NE USA 42°34′N 76°09′W
31 V11 **Cortland** Ohio, N USA 41°19′N 80°43′W
106 H12 **Cortona** Toscana, C Italy 43°15′N 12°01′E
76 H13 **Corubal, Rio** ≋ E Guinea-Bissau
104 G10 **Coruche** Santarém, C Portugal 38°58′N 08°31′W
Çoruh *see* Rize
137 R11 **Çoruh Nehri** *Geo.* Chorokh, *Rus.* Chorokhi. ≋ Georgia/Turkey
136 K12 **Çorum** *var.* Chorum. Çorum, N Turkey 40°31′N 34°57′E
136 J12 **Çorum** *var.* Chorum. ◈ *province* N Turkey
59 H19 **Corumbá** Mato Grosso do Sul, S Brazil 19°S 57°35′W
14 D16 **Corunna** Ontario, S Canada 42°49′N 82°25′W
Corunna *see* A Coruña
32 F12 **Corvallis** Oregon, NW USA 44°35′N 122°36′W
64 M1 **Corvo** *var.* Ilha do Corvo. *island* Azores, Portugal, NE Atlantic Ocean
Corvo, Ilha do *see* Corvo
31 O16 **Corydon** Indiana, N USA 38°12′N 86°07′W
29 V16 **Corydon** Iowa, C USA 40°45′N 93°19′W
Cos *see* Kos
40 I9 **Cosalá** Sinaloa, C Mexico 24°25′N 106°39′W
41 R15 **Cosamaloapan** *var.* Cosamaloapan de Carpio. Veracruz-Llave, E Mexico 18°23′N 95°50′W
Cosamaloapan de Carpio *see* Cosamaloapan
107 N21 **Cosenza** *anc.* Consentia. Calabria, SW Italy 39°17′N 16°15′E
31 T13 **Coshocton** Ohio, N USA 40°16′N 81°53′W
42 H9 **Cosigüina, Punta** *headland* NW Nicaragua 12°53′N 87°42′W
29 T9 **Cosmos** Minnesota, N USA 44°56′N 94°42′W
103 O8 **Cosne-Cours-sur-Loire** Nièvre, C France 47°25′N 02°56′E
108 B9 **Cossonay** Vaud, W Switzerland 46°37′N 06°28′E
Cossyra *see* Pantelleria
47 R4 **Costa, Cordillera de la** *var.* Cordillera de Venezuela. ▲▲ N Venezuela
42 K13 **Costa Rica** *off.* Republic of Costa Rica. ◆ *republic* Central America
Costa Rica, Republic of *see* Costa Rica
43 N15 **Costeña, Fila** ▲▲ S Costa Rica
Costermansville *see* Bukavu
116 I14 **Costești** Argeș, SW Romania 44°40′N 24°53′E
37 S8 **Costilla** New Mexico, SW USA 36°58′N 105°31′W
35 O7 **Cosumnes River** ≋ California, W USA
101 O16 **Coswig** Sachsen, E Germany 51°07′N 13°36′E
101 M14 **Coswig** Sachsen-Anhalt, E Germany 51°53′N 12°26′E
Cosyra *see* Pantelleria
171 Q7 **Cotabato** Mindanao, S Philippines 07°13′N 124°12′E
56 C5 **Cotacachi** ▲ N Ecuador 0°29′N 78°17′W
57 L21 **Cotagaita** Potosí, S Bolivia 20°47′S 65°40′W
103 V15 **Côte d'Azur** *prev.* Nice. ✈ (Nice) Alpes-Maritimes, SE France 43°40′N 07°12′E
Côte d'Ivoire *see* Ivory Coast
Côte d'Ivoire, République de la *see* Ivory Coast
103 R7 **Côte d'Or** ◈ *department* E France
103 R8 **Côte d'Or** *cultural region* C France
Côte Française des Somalis *see* Djibouti
102 J4 **Cotentin** *peninsula* N France
102 G6 **Côtes d'Armor** *prev.* Côtes-du-Nord. ◈ *department* NW France
Côtes-du-Nord *see* Côtes d'Armor
Cöthen *see* Köthen
Côtière, Chaîne *see* Coast Mountains
40 M13 **Cotija** *var.* Cotija de la Paz. Michoacán, SW Mexico 19°49′N 102°39′W
Cotija de la Paz *see* Cotija
77 R16 **Cotonou** *var.* Kotonu. S Benin 06°24′N 02°26′E
77 R16 **Cotonou** ✈ S Benin 06°31′N 02°18′E
56 B6 **Cotopaxi** *prev.* León. ◈ *province* C Ecuador
56 C6 **Cotopaxi** ⊼ N Ecuador 0°42′S 78°24′W
Cotrone *see* Crotone
97 L21 **Cotswold Hills** *var.* Cotswolds. *hill range* S England, United Kingdom
Cotswolds *see* Cotswold Hills
32 F13 **Cottage Grove** Oregon, NW USA 43°48′N 123°03′W
21 S14 **Cottageville** South Carolina, SE USA 32°55′N 80°28′W
101 P14 **Cottbus** *Lus.* Chośebuz; *prev.* Kottbus. Brandenburg, E Germany 51°42′N 14°22′E
27 U9 **Cotter** Arkansas, C USA 36°16′N 92°30′W
106 A9 **Cottian Alps** *Fr.* Alpes Cottiennes, *It.* Alpi Cozie. ▲▲ France/Italy
Cottiennes, Alpes *see* Cottian Alps
Cotton State, The *see* Alabama
22 G4 **Cotton Valley** Louisiana, S USA 32°49′N 93°25′W
36 L12 **Cottonwood** Arizona, SW USA 34°43′N 112°00′W
32 M10 **Cottonwood** Idaho, NW USA 46°01′N 116°20′W
29 S9 **Cottonwood** Minnesota, N USA 44°37′N 95°41′W
25 Q7 **Cottonwood** Texas, SW USA 32°12′N 99°14′W
27 O5 **Cottonwood Falls** Kansas, C USA 38°21′N 96°33′W
36 L3 **Cottonwood Heights** Utah, W USA 40°37′N 111°48′W
29 S10 **Cottonwood River** ≋ Minnesota, N USA
45 O9 **Cotuí** C Dominican Republic 19°04′N 70°10′W
25 Q13 **Cotulla** Texas, SW USA 28°27′N 99°15′W
Cotyora *see* Ordu
102 I11 **Coubre, Pointe de la** *headland* W France 45°39′N 01°23′W
18 E12 **Coudersport** Pennsylvania, NE USA 41°45′N 78°00′W
15 S9 **Coudres, Île aux** *island* Québec, SE Canada
182 G11 **Couedic, Cape de** *headland* South Australia 36°04′S 136°43′E
Couentrey *see* Coventry
102 I6 **Couesnon** ≋ NW France
32 H10 **Cougar** Washington, NW USA 46°03′N 122°18′W
102 L10 **Couhé** Vienne, W France 46°18′N 00°10′E
32 K8 **Coulee City** Washington, NW USA 47°36′N 119°18′W
195 Q15 **Coulman Island** *island* Antarctica
103 P5 **Coulommiers** Seine-et-Marne, N France 48°49′N 03°04′E
14 K11 **Coulonge** ≋ Québec, SE Canada
14 K11 **Coulonge Est** ≋ Québec, SE Canada
35 Q9 **Coulterville** California, W USA 37°41′N 120°10′W
38 M9 **Council** Alaska, USA 64°54′N 163°40′W
32 M12 **Council** Idaho, NW USA 44°45′N 116°26′W
29 S15 **Council Bluffs** Iowa, C USA 41°16′N 95°52′W
27 O5 **Council Grove** Kansas, C USA 38°41′N 96°29′W
27 O5 **Council Grove Lake** ⊠ Kansas, C USA
32 G7 **Coupeville** Washington, NW USA 48°13′N 122°41′W
55 U12 **Courantyne River** *var.* Corantijn Rivier, Corentyne River. ≋ Guyana/Suriname
99 G21 **Courcelles** Hainaut, S Belgium 50°28′N 04°23′E
108 C7 **Courgenay** Jura, NW Switzerland 47°24′N 07°09′E
126 B2 **Courland Lagoon** *Ger.* Kurisches Haff, *Rus.* Kurskiy Zaliv. *lagoon* Lithuania/Russian Federation
118 B12 **Courland Spit** *Lith.* Kuršių Nerija, *Rus.* Kurshskaya Kosa. *spit* Lithuania/Russian Federation
106 A6 **Courmayeur** *prev.* Cormaiore. Valle d'Aosta, NW Italy 45°48′N 07°00′E
108 D7 **Courroux** Jura, NW Switzerland 47°22′N 07°23′E
10 K17 **Courtenay** Vancouver Island, British Columbia, SW Canada 49°40′N 124°58′W
21 W7 **Courtland** Virginia, NE USA 36°44′N 77°06′W
25 V10 **Courtney** Texas, SW USA 30°16′N 96°04′W
30 J4 **Court Oreilles, Lac** ◎ Wisconsin, N USA
Courtrai *see* Kortrijk
99 H19 **Court-Saint-Étienne** Walloon Brabant, C Belgium 50°38′N 04°34′E
22 G6 **Coushatta** Louisiana, S USA 32°00′N 93°20′W
172 I16 **Cousin** *island* Inner Islands, NE Seychelles
172 I16 **Cousine** *island* Inner Islands, NE Seychelles
102 J4 **Coutances** *anc.* Constantia. Manche, N France 49°04′N 01°27′W
102 K12 **Coutras** Gironde, SW France 45°01′N 00°07′W
45 U14 **Couva** Trinidad, Trinidad and Tobago 10°25′N 61°27′W
108 B8 **Couvet** Neuchâtel, W Switzerland 46°57′N 06°41′E
99 H22 **Couvin** Namur, S Belgium 50°03′N 04°30′E
116 K12 **Covasna** *Ger.* Kowasna, *Hung.* Kovászna. Covasna, E Romania 45°51′N 26°11′E
116 J11 **Covasna** ◈ *county* E Romania
14 E12 **Cove Island** *island* Ontario, S Canada
34 M5 **Covelo** California, W USA 39°46′N 123°16′W
97 M20 **Coventry** *anc.* Couentrey. C England, United Kingdom 52°25′N 01°30′W
Cove of Cork *see* Cobh
21 U5 **Covesville** Virginia, NE USA 37°52′N 78°41′W
104 I8 **Covilhã** Castelo Branco, E Portugal 40°17′N 07°30′W
23 T3 **Covington** Georgia, SE USA 33°34′N 83°52′W
31 N13 **Covington** Indiana, N USA 40°08′N 87°23′W
20 M3 **Covington** Kentucky, S USA 39°04′N 84°30′W

◆ Country ● Country Capital ◇ Dependent Territory ○ Dependent Territory Capital ◈ Administrative Regions ✈ International Airport ▲ Mountain ▲▲ Mountain Range ⊼ Volcano ≋ River ◎ Lake ⊠ Reservoir

Covington – Dacia Bank

22 K8 **Covington** Louisiana, S USA 30°28′N 90°06′W
31 Q13 **Covington** Ohio, N USA 40°07′N 84°21′W
20 F9 **Covington** Tennessee, S USA 35°32′N 89°40′W
21 S6 **Covington** Virginia, NE USA 37°48′N 80°01′W
183 Q8 **Cowal, Lake** *seasonal lake* New South Wales, SE Australia
11 W15 **Cowan** Manitoba, S Canada 51°59′N 100°36′W
18 F12 **Cowanesque River** New York/Pennsylvania, NE USA
180 L12 **Cowan, Lake** ◎ Western Australia
15 P13 **Cowansville** Québec, SE Canada 45°13′N 72°44′W
182 H8 **Cowell** South Australia 33°43′S 136°53′E
97 M23 **Cowes** S England, United Kingdom 50°45′N 01°19′W
27 Q10 **Coweta** Oklahoma, C USA 35°57′N 95°39′W
0 D6 **Cowie Seamount** *undersea feature* NE Pacific Ocean 54°15′N 149°30′W
32 G10 **Cowlitz River** Washington, NW USA
21 Q11 **Cowpens** South Carolina, SE USA 35°01′N 81°48′W
183 R8 **Cowra** New South Wales, SE Australia 33°50′S 148°45′E
Coxen Hole *see* Roatán
59 I19 **Coxim** Mato Grosso do Sul, S Brazil 18°28′S 54°45′W
59 I19 **Coxim, Rio** SW Brazil
Coxin Hole *see* Roatán
153 V17 **Cox's Bazar** Chittagong, S Bangladesh 21°25′N 91°59′E
76 H14 **Coyah** Conakry, W Guinea 09°45′N 13°26′W
40 K5 **Coyame** Chihuahua, N Mexico 29°29′N 105°07′W
24 L9 **Coyanosa Draw** Texas, SW USA
Coyhaique *see* Coihaique
42 C7 **Coyolate, Río** S Guatemala
Coyote State, The *see* South Dakota
40 I10 **Coyotitán** Sinaloa, C Mexico 23°48′N 106°37′W
41 N15 **Coyuca** *var.* Coyuca de Catalán. Guerrero, S Mexico 18°21′N 100°39′W
41 O16 **Coyuca** *var.* Coyuca de Benítez. Guerrero, S Mexico 17°01′N 100°08′W
Coyuca de Benítez/Coyuca de Catalán *see* Coyuca
29 N15 **Cozad** Nebraska, C USA 40°52′N 99°58′W
158 L14 **Cozhê** Xizang Zizhiqu, W China 31°53′N 87°51′E
Cozie, Alpi *see* Cottian Alps
Cozmeni *see* Kitsman'
40 E3 **Cozón, Cerro** ▲ NW Mexico 31°16′N 112°29′W
41 Z12 **Cozumel** Quintana Roo, E Mexico 20°29′N 86°54′W
41 Z12 **Cozumel, Isla** *island* SE Mexico
32 K8 **Crab Creek** Washington, NW USA
44 H12 **Crab Pond Point** *headland* W Jamaica 18°07′N 78°01′W
Cracovia/Cracow *see* Kraków
83 I25 **Cradock** Eastern Cape, S South Africa 32°07′S 25°38′E
39 Y14 **Craig** Prince of Wales Island, Alaska, USA 55°29′N 133°04′W
37 Q3 **Craig** Colorado, C USA 40°31′N 107°33′W
97 F15 **Craigavon** C Northern Ireland, United Kingdom 54°28′N 06°25′W
21 T5 **Craigsville** Virginia, NE USA 38°07′N 79°21′W
101 J21 **Crailsheim** Baden-Württemberg, S Germany 49°07′N 10°04′E
116 H14 **Craiova** Dolj, SW Romania 44°19′N 23°49′E
10 K12 **Cranberry Junction** British Columbia, SW Canada 55°35′N 128°21′W
18 J8 **Cranberry Lake** ◎ New York, NE USA
11 V13 **Cranberry Portage** Manitoba, C Canada 54°34′N 101°22′W
11 P17 **Cranbrook** British Columbia, SW Canada 49°29′N 115°48′W
30 M5 **Crandon** Wisconsin, N USA 45°34′N 88°54′W
32 K14 **Crane** Oregon, NW USA 43°24′N 118°35′W
24 M9 **Crane** Texas, SW USA 31°23′N 102°22′W
Crane *see* The Crane
25 S8 **Cranfills Gap** Texas, SW USA 31°46′N 97°49′W
19 O12 **Cranston** Rhode Island, NE USA 41°46′N 71°26′W
Cranz *see* Zelenogradsk
59 L15 **Craolândia** Tocantins, E Brazil 07°17′S 47°23′W
102 J7 **Craon** Mayenne, NW France 47°52′N 00°57′W
195 V16 **Crary, Cape** *headland* Antarctica
Crasna *see* Kraszna
32 G14 **Crater Lake** ◎ Oregon, NW USA
33 P14 **Craters of the Moon National Monument** *national park* Idaho, NW USA
59 O14 **Crateús** Ceará, E Brazil 05°10′S 40°39′W
Crathis *see* Crati
107 N20 **Crati** *anc.* Crathis. S Italy
11 U16 **Craven** Saskatchewan, S Canada 50°44′N 104°50′W
54 I8 **Cravo Norte** Arauca, E Colombia 06°17′N 70°15′W
28 J12 **Crawford** Nebraska, C USA 42°40′N 103°24′W
25 T8 **Crawford** Texas, SW USA 31°31′N 97°26′W
11 O17 **Crawford Bay** British Columbia, SW Canada 49°39′N 116°44′W
65 M19 **Crawford Seamount** *undersea feature* S Atlantic Ocean 40°30′S 10°00′W
31 O13 **Crawfordsville** Indiana, N USA 40°02′N 86°52′W
23 S9 **Crawfordville** Florida, SE USA 30°10′N 84°22′W
97 O23 **Crawley** SE England, United Kingdom 51°07′N 00°12′W
33 S10 **Crazy Mountains** ▲▲ Montana, NW USA
11 T11 **Cree** Saskatchewan, C Canada
37 R7 **Creede** Colorado, C USA 37°51′N 106°55′W
40 I6 **Creel** Chihuahua, N Mexico 27°45′N 107°36′W
11 S11 **Cree Lake** ◎ Saskatchewan, C Canada
11 V13 **Creighton** Saskatchewan, C Canada 54°46′N 101°54′W
29 Q13 **Creighton** Nebraska, C USA 42°28′N 97°54′W
103 O4 **Creil** Oise, N France 49°16′N 02°29′E
106 E8 **Crema** Lombardia, N Italy 45°22′N 09°40′E
106 E8 **Cremona** Lombardia, N Italy 45°08′N 10°02′E
Creole State *see* Louisiana
112 M10 **Crepaja** *Hung.* Cserépalja. Vojvodina, N Serbia 45°02′N 20°36′E
103 O4 **Crépy-en-Valois** Oise, N France 49°13′N 02°54′E
112 B10 **Cres** *It.* Cherso. Primorje-Gorski Kotar, NW Croatia 44°57′N 14°24′E
112 A11 **Cres** *It.* Cherso; *anc.* Crexa. *island* W Croatia
32 H14 **Crescent** Oregon, NW USA 43°27′N 121°40′W
34 K1 **Crescent City** California, W USA 41°45′N 124°14′W
23 W10 **Crescent City** Florida, SE USA 29°25′N 81°30′W
167 X10 **Crescent Group** *Chin.* Yongle Qundao, *Viet.* Nhom L i Liêm. *island group* C Paracel Islands
23 W10 **Crescent Lake** ◎ Florida, SE USA
29 X11 **Cresco** Iowa, C USA 43°22′N 92°06′W
61 B18 **Crespo** Entre Ríos, E Argentina 32°05′S 60°20′W
103 R13 **Crest** Drôme, E France 44°45′N 05°00′E
37 R5 **Crested Butte** Colorado, C USA 38°52′N 106°59′W
31 S12 **Crestline** Ohio, N USA 40°47′N 82°44′W
11 O17 **Creston** British Columbia, SW Canada 49°05′N 116°32′W
29 U15 **Creston** Iowa, C USA 41°03′N 94°21′W
33 V16 **Creston** Wyoming, C USA 41°40′N 107°43′W
37 S7 **Crestone Peak** ▲ Colorado, C USA 37°58′N 105°34′W
23 P8 **Crestview** Florida, SE USA 30°44′N 86°34′W
121 R10 **Cretan Trough** *undersea feature* Aegean Sea, C Mediterranean Sea
29 R16 **Crete** Nebraska, C USA 40°36′N 96°58′W
Crete *see* Kríti
103 O5 **Créteil** Val-de-Marne, N France 48°47′N 02°28′E
Crete, Sea of/Creticum, Mare *see* Kritikó Pélagos
105 X4 **Creus, Cap de** *headland* NE Spain 42°18′N 03°18′E
103 N10 **Creuse** ⬖ *department* C France
102 L9 **Creuse** C France
103 T4 **Creutzwald** Moselle, NE France 49°13′N 06°41′E
105 S12 **Crevillent** *prev.* Crevillente. Valenciana, E Spain 38°15′N 00°48′W
Crevillente *see* Crevillent
97 L18 **Crewe** C England, United Kingdom 53°05′N 02°27′W
21 V7 **Crewe** Virginia, NE USA 37°10′N 78°07′W
Crexa *see* Cres
43 Q15 **Cricamola, Río** NW Panama
61 K14 **Criciúma** Santa Catarina, S Brazil 28°39′S 49°23′W
96 J11 **Crieff** C Scotland, United Kingdom 56°22′N 03°49′W
112 B10 **Crikvenica** *It.* Cirquenizza; *prev.* Cirkvenica, Crjkvenica. Primorje-Gorski Kotar, NW Croatia 45°12′N 14°40′E
Crimea/Crimean Oblast *see* Krym, Avtonomna Respublika
101 M16 **Crimmitschau** *var.* Krimmitschau. Sachsen, E Germany 50°48′N 12°23′E
116 G11 **Crişcior** *Hung.* Kristyor. Hunedoara, W Romania 46°09′N 22°54′E
21 Y5 **Crisfield** Maryland, NE USA 37°58′N 75°51′W
31 P3 **Crisp Point** *headland* Michigan, N USA 46°45′N 85°15′W
59 L19 **Cristalina** Goiás, C Brazil 16°43′S 47°37′W
44 J7 **Cristal, Sierra del** ▲▲ E Cuba
43 T14 **Cristóbal** Colón, C Panama 09°18′N 79°52′W
54 F4 **Cristóbal Colón, Pico** ▲ N Colombia 10°52′N 73°46′W
Cristur/Cristuru Săcuiesc *see* Cristuru Secuiesc
116 I11 **Cristuru Secuiesc** *prev.* Cristur, Cristuru Săcuiesc, Sitaş Cristuru, *Ger.* Kreutz, *Hung.* Székelykeresztúr, Szitás-Keresztúr. Harghita, C Romania 46°17′N 25°02′E
116 F10 **Crişul Alb** *var.* Weisse Kreisch, *Ger.* Weisse Körös, *Hung.* Fehér-Körös. Hungary/Romania
116 F10 **Crişul Negru** *Ger.* Schwarze Körös, *Hung.* Fekete-Körös. Hungary/Romania
116 G10 **Crişul Repede** *var.* Schnelle Kreisch, *Ger.* Schnelle Körös, *Hung.* Sebes-Körös. Hungary/Romania
117 N10 **Criuleni** *Rus.* Kriulyany. C Moldova 47°12′N 29°09′E
Crivadia Vulcanului *see* Vulcan
Crjkvenica *see* Crikvenica
113 J17 **Crkvice** SW Montenegro 42°34′N 18°38′E
113 O17 **Crna Gora** *Alb.* Mali i Zi. ▲▲ FYR Macedonia/Serbia
Crna Gora *see* Montenegro
113 O20 **Crna Reka** S FYR Macedonia
Crni Drim *see* Black Drin
109 V10 **Črni vrh** ▲ NE Slovenia 46°28′N 15°14′E
109 V13 **Črnomelj** *Ger.* Tschernembl. SE Slovenia 45°32′N 15°12′E
97 A17 **Croagh Patrick** *Ir.* Cruach Phádraig. ▲ W Ireland 53°45′N 09°39′W
112 D9 **Croatia** *off.* Republic of Croatia, *Ger.* Kroatien, *SCr.* Hrvatska. ◆ *republic* SE Europe
Croatia, Republic of *see* Croatia
Croce, Picco di *see* Wilde Kreuzspitze
15 P8 **Croche** Québec, SE Canada
169 V7 **Crocker, Banjaran** *var.* Crocker Range. ▲▲ East Malaysia
Crocker Range *see* Crocker, Banjaran
25 V9 **Crockett** Texas, SW USA 31°21′N 95°30′W
67 V14 **Crocodile** *var.* Krokodil. N South Africa
Crocodile *see* Limpopo
20 I7 **Crofton** Kentucky, S USA 37°01′N 87°25′W
29 Q12 **Crofton** Nebraska, C USA 42°43′N 97°30′W
Croia *see* Krujë
103 R16 **Croisette, Cap** *headland* SE France 43°12′N 05°21′E
102 G8 **Croisic, Pointe du** *headland* NW France 47°16′N 02°42′W
103 S13 **Croix Haute, Col de la** *pass* E France
15 U5 **Croix, Pointe à la** *headland* Québec, SE Canada 49°16′N 67°46′W
14 F13 **Croker, Cape** *headland* Ontario, S Canada 44°56′N 80°57′W
181 P1 **Croker Island** *island* Northern Territory, N Australia
96 I8 **Cromarty** N Scotland, United Kingdom 57°40′N 04°02′W
99 M21 **Crombach** Liège, E Belgium 50°14′N 06°07′E
97 Q18 **Cromer** E England, United Kingdom 52°56′N 01°06′E
185 D22 **Cromwell** Otago, South Island, New Zealand 45°03′S 169°14′E
185 H16 **Cronadun** West Coast, South Island, New Zealand 42°03′S 171°52′E
39 O11 **Crooked Creek** Alaska, USA 61°52′N 158°06′W
44 K5 **Crooked Island** *island* SE The Bahamas
44 J5 **Crooked Island Passage** *channel* SE The Bahamas
32 I13 **Crooked River** Oregon, NW USA
29 R4 **Crookston** Minnesota, N USA 47°47′N 96°36′W
28 I10 **Crooks Tower** ▲ South Dakota, N USA 44°09′N 103°55′W
31 T14 **Crooksville** Ohio, N USA 39°46′N 82°05′W
183 R9 **Crookwell** New South Wales, SE Australia 34°28′S 149°27′E
14 L14 **Crosby** Ontario, SE Canada 44°39′N 76°13′W
97 K17 **Crosby** *var.* Great Crosby. NW England, United Kingdom 53°30′N 03°02′W
29 U6 **Crosby** Minnesota, N USA 46°30′N 93°58′W
28 K2 **Crosby** North Dakota, N USA 48°54′N 103°17′W
25 O5 **Crosbyton** Texas, SW USA 33°40′N 101°16′W
77 V16 **Cross** Cameroon/Nigeria
23 U10 **Cross City** Florida, SE USA 29°37′N 83°08′W
Crossen *see* Krosno Odrzańskie
27 V14 **Crossett** Arkansas, C USA 33°08′N 91°58′W
97 K15 **Cross Fell** ▲ N England, United Kingdom 54°42′N 02°30′W
11 P16 **Crossfield** Alberta, SW Canada 51°24′N 114°03′W
21 Q12 **Cross Hill** South Carolina, SE USA 34°18′N 81°58′W
19 U6 **Cross Island** *island* Maine, NE USA
11 X13 **Cross Lake** Manitoba, C Canada 54°38′N 97°35′W
22 F5 **Cross Lake** ⊠ Louisiana, S USA
36 I12 **Crossman Peak** ▲ Arizona, SW USA 34°33′N 114°09′W
25 Q7 **Cross Plains** Texas, SW USA 32°07′N 99°10′W
77 V17 **Cross River** ⬖ *state* SE Nigeria
20 L9 **Crossville** Tennessee, S USA 35°57′N 85°02′W
31 S8 **Croswell** Michigan, N USA 43°16′N 82°37′W
14 K13 **Crotch Lake** ◎ Ontario, SE Canada
Croton/Crotona *see* Crotone
107 O21 **Crotone** *var.* Cotrone; *anc.* Croton, Crotona. Calabria, SW Italy 39°05′N 17°07′E
33 V11 **Crow Agency** Montana, NW USA 45°35′N 107°28′W
183 U7 **Crowdy Head** *headland* New South Wales, SE Australia 31°52′S 152°45′E
25 Q4 **Crowell** Texas, SW USA 33°59′N 99°45′W
183 O6 **Crowl Creek** *seasonal river* New South Wales, SE Australia
22 H9 **Crowley** Louisiana, S USA 30°11′N 92°21′W
35 S9 **Crowley, Lake** ⊠ California, W USA
27 X10 **Crowleys Ridge** *hill range* Arkansas, C USA
31 N11 **Crown Point** Indiana, N USA 41°25′N 87°22′W
37 P10 **Crownpoint** New Mexico, SW USA 35°40′N 108°09′W
33 R10 **Crow Peak** ▲ Montana, NW USA 46°17′N 111°54′W
11 P17 **Crowsnest Pass** *pass* Alberta/British Columbia, SW Canada
29 T6 **Crow Wing River** Minnesota, N USA
97 O22 **Croydon** SE England, United Kingdom 51°21′N 00°06′W
Crozer, Mount *see* Finkol, Mount
173 P11 **Crozet Basin** *undersea feature* S Indian Ocean 39°00′S 60°00′E
173 O12 **Crozet Islands** *island group* French Southern and Antarctic Territories
173 N12 **Crozet Plateau** *var.* Crozet Plateaus. *undersea feature* SW Indian Ocean 46°00′S 51°00′E
Crozet Plateaus *see* Crozet Plateau
102 E6 **Crozon** Finistère, NW France 48°14′N 04°31′W
Cruacha Dubha, Na *see* Macgillycuddy's Reeks
Cruach Phádraig *see* Croagh Patrick
116 M14 **Crucea** Constanţa, SE Romania 44°30′N 28°18′E
44 E5 **Cruces** Cienfuegos, C Cuba 22°20′N 80°17′W
107 O20 **Crucoli Torretta** Calabria, SW Italy 39°26′N 17°03′E
41 P9 **Cruillas** Tamaulipas, C Mexico 24°43′N 98°26′W
64 K9 **Cruiser Tablemount** *undersea feature* E Atlantic Ocean 32°00′N 28°00′W
61 G14 **Cruz Alta** Rio Grande do Sul, S Brazil 28°38′S 53°38′W
44 G8 **Cruz, Cabo** *headland* S Cuba 19°50′N 77°43′W
60 N9 **Cruzeiro** São Paulo, S Brazil 22°33′S 44°59′W
60 H10 **Cruzeiro do Oeste** Paraná, S Brazil 23°45′S 53°03′W
59 A15 **Cruzeiro do Sul** Acre, W Brazil 07°40′S 72°39′W
23 U11 **Crystal Bay** *bay* Florida, SE USA NE Gulf of Mexico Atlantic Ocean
182 I8 **Crystal Brook** South Australia 33°24′S 138°10′E
11 X17 **Crystal City** Manitoba, S Canada 49°07′N 98°54′W
27 X5 **Crystal City** Missouri, C USA 38°13′N 90°22′W
25 P13 **Crystal City** Texas, SW USA 28°43′N 99°51′W
30 M4 **Crystal Falls** Michigan, N USA 46°06′N 88°20′W
23 Q8 **Crystal Lake** Florida, SE USA 30°26′N 85°41′W
31 O6 **Crystal Lake** ◎ Michigan, N USA
23 V11 **Crystal River** Florida, SE USA 28°54′N 82°35′W
37 Q5 **Crystal River** Colorado, C USA
22 K6 **Crystal Springs** Mississippi, S USA 31°59′N 90°21′W
Csaca *see* Čadca
Csakathurn/Csáktornya *see* Čakovec
Csap *see* Chop
Csepén *see* Čepin
Cserépalja *see* Crepaja
Csermő *see* Cermei
Csíkszereda *see* Miercurea-Ciuc
111 L24 **Csongrád** Csongrád, SE Hungary 46°42′N 20°09′E
111 L24 **Csongrád** *off.* Csongrád Megye. ⬖ *county* SE Hungary
Csongrád Megye *see* Csongrád
111 H22 **Csorna** Győr-Moson-Sopron, NW Hungary 47°37′N 17°14′E
Csucsa *see* Ciucea
111 G25 **Csurgó** Somogy, SW Hungary 46°16′N 17°09′E
Csurog *see* Čurug
54 L5 **Cúa** Miranda, N Venezuela 10°14′N 66°58′W
82 C11 **Cuale** Malanje, NW Angola 08°22′S 16°10′E
67 T12 **Cuando** *var.* Kwando. S Africa
Cuando Cubango *see* Kuando Kubango
83 E16 **Cuangar** Kuando Kubango, S Angola 17°34′S 18°39′E
82 D11 **Cuango** Lunda Norte, NE Angola 09°10′S 17°59′E
82 C10 **Cuango** Uíge, NW Angola 06°20′S 16°42′E
82 C10 **Cuango** *var.* Kwango. Angola/Dem. Rep. Congo *see also* Kwango
Cuango *see* Kwango
Cuan, Loch *see* Strangford Lough
82 C12 **Cuanza** *var.* Kwanza. C Angola
Cuanza Norte *see* Kwanza Norte
Cuanza Sul *see* Kwanza Sul
61 E16 **Cuareim, Río** *var.* Río Quaraí. Brazil/Uruguay *see also* Quaraí, Rio
Cuareim, Río *see* Quaraí, Rio
83 D15 **Cuatir** S Angola
40 M7 **Cuatro Ciénegas** *var.* Cuatro Ciénegas de Carranza. Coahuila, NE Mexico 27°00′N 102°03′W
Cuatro Ciénegas de Carranza *see* Cuatro Ciénegas
40 I6 **Cuauhtémoc** Chihuahua, N Mexico 28°22′N 106°52′W
41 P14 **Cuautla** Morelos, S Mexico 18°48′N 98°56′W
104 H12 **Cuba** Beja, S Portugal 38°10′N 07°54′W
27 W6 **Cuba** Missouri, C USA 38°03′N 91°24′W
37 R10 **Cuba** New Mexico, SW USA 36°01′N 106°57′W
44 E6 **Cuba** *off.* Republic of Cuba. ◆ *republic* W West Indies
47 O2 **Cuba** *island* W West Indies
82 B13 **Cubal** Benguela, W Angola 12°58′S 14°16′E
83 C15 **Cubango** *var.* Kuvango, *Port.* Vila Artur de Paiva, Vila da Ponte. Huíla, SW Angola 14°27′S 16°18′E
83 D16 **Cubango** *var.* Kavango, Kavengo, Kubango, Okavango, Okavanggo. S Africa *see also* Okavango
Cubango *see* Okavango
54 H8 **Cubará** Boyacá, N Colombia 07°01′N 72°07′W
Cuba, Republic of *see* Cuba
136 I12 **Çubuk** Ankara, N Turkey 40°13′N 33°02′E
83 D14 **Cuchi** Kuando Kubango, C Angola 14°40′S 16°58′E
42 C5 **Cuchumatanes, Sierra de los** ▲▲ W Guatemala
Cuculaya, Rio *see* Kukalaya, Rio
82 E12 **Cucumbi** *prev.* Trás-os-Montes. Lunda Sul, NE Angola 10°13′S 19°04′E
54 G7 **Cúcuta** *var.* San José de Cúcuta. Norte de Santander, N Colombia 07°55′N 72°31′W
31 N9 **Cudahy** Wisconsin, N USA 42°54′N 87°51′W
155 J21 **Cuddalore** Tamil Nādu, SE India 11°43′N 79°46′E
155 I18 **Cuddapah** Andhra Pradesh, S India 14°30′N 78°50′E
104 M6 **Cuéllar** Castilla y León, N Spain 41°24′N 04°19′W
82 D13 **Cuemba** *var.* Coemba. Bié, C Angola 12°09′S 18°07′E
56 B8 **Cuenca** Azuay, S Ecuador 02°54′S 79°W
105 Q9 **Cuenca** *anc.* Conca. Castilla-La Mancha, C Spain 40°04′N 02°07′W
105 P9 **Cuenca** ⬖ *province* Castilla-La Mancha, C Spain
40 L9 **Cuencamé** *var.* Cuencamé de Ceniceros. Durango, C Mexico 24°53′N 103°41′W
Cuencamé de Ceniceros *see* Cuencamé
105 Q8 **Cuenca, Serrania de** ▲▲ C Spain
Cuera *see* Chur
105 P5 **Cuerda del Pozo, Embalse de la** ⊠ N Spain
41 O14 **Cuernavaca** Morelos, S Mexico 18°57′N 99°15′W
25 T12 **Cuero** Texas, SW USA 29°06′N 97°19′W
44 I7 **Cueto** Holguín, E Cuba 20°43′N 75°54′W
41 Q13 **Cuetzalán** *var.* Cuetzalán del Progreso. Puebla, S Mexico 20°00′N 97°27′W
Cuetzalán del Progreso *see* Cuetzalán
105 Q14 **Cuevas de Almanzora** Andalucía, S Spain 37°19′N 01°52′W
Cuevas de Vinromá *see* Les Coves de Vinromá
116 H12 **Cugir** *Hung.* Kudzsir. Alba, SW Romania 45°48′N 23°25′E
59 H18 **Cuiabá** *prev.* Cuyabá. *state capital* Mato Grosso, SW Brazil 15°32′S 56°05′W
59 H19 **Cuiabá, Rio** SW Brazil
41 R15 **Cuicatlán** *var.* San Juan Bautista Cuicatlán. Oaxaca, SE Mexico 17°49′N 96°59′W
191 W16 **Cuidado, Punta** *headland* Easter Island, Chile, E Pacific Ocean 27°08′S 109°18′W
Cúige *see* Connaught
Cúige Laighean *see* Leinster
Cúige Mumhan *see* Munster
Cuihua *see* Daguan
98 L13 **Cuijck** Noord-Brabant, SE Netherlands 51°41′N 05°56′E
Cúil an tSúdaire *see* Portarlington
42 D7 **Cuilapa** Santa Rosa, S Guatemala 14°16′N 90°18′W
42 B5 **Cuilco, Río** W Guatemala
Cúil Mhuine *see* Collooney
Cúil Raithin *see* Coleraine
83 C14 **Cuima** Huambo, C Angola 13°16′S 15°39′E
83 E16 **Cuito** *var.* Kwito. SE Angola
83 E15 **Cuíto Cuanavale** Kuando Kubango, E Angola 15°01′S 19°07′E
41 N14 **Cuitzeo, Lago de** ◎ C Mexico
27 W4 **Cuivre River** Missouri, C USA
Çuka *see* Çukë
168 L8 **Cukai** *var.* Chukai, Kemaman. Terengganu, Peninsular Malaysia 04°15′N 103°25′E
113 L23 **Çukë** *var.* Çuka. Vlorë, S Albania 39°50′N 20°01′E
Cularo *see* Grenoble
33 Y7 **Culbertson** Montana, NW USA 48°09′N 104°30′W
28 M16 **Culbertson** Nebraska, C USA 40°13′N 100°50′W
183 P10 **Culcairn** New South Wales, SE Australia 35°41′S 147°01′E
45 W5 **Culebra** *var.* Dewey. E Puerto Rico 18°19′N 65°17′W
45 W6 **Culebra, Isla de** *island* E Puerto Rico
37 T8 **Culebra Peak** ▲ Colorado, C USA 37°07′N 105°11′W
104 J5 **Culebra, Sierra de la** ▲▲ NW Spain
98 J12 **Culemborg** Gelderland, C Netherlands 51°57′N 05°17′E
137 V14 **Culfa** *Rus.* Dzhul'fa. SW Azerbaijan 38°58′N 45°37′E
183 P4 **Culgoa River** New South Wales/Queensland, SE Australia
40 I9 **Culiacán** *var.* Culiacán Rosales, Culiacán-Rosales. Sinaloa, C Mexico 24°48′N 107°25′W
Culiacán-Rosales/Culiacán Rosales *see* Culiacán
105 P14 **Cúllar-Baza** Andalucía, S Spain 37°35′N 02°34′W
105 S10 **Cullera** Valenciana, E Spain 39°10′N 00°15′W
23 P3 **Cullman** Alabama, S USA 34°10′N 86°50′W
108 B10 **Cully** Vaud, W Switzerland 46°58′N 06°46′E
Culm *see* Chełmno
Culmsee *see* Chełmża
21 V4 **Culpeper** Virginia, NE USA 38°28′N 78°00′W
185 I17 **Culverden** Canterbury, South Island, New Zealand 42°46′S 172°51′E
83 H18 **Cum** *var.* Xhumo. Central, C Botswana 21°13′S 24°38′E
55 N5 **Cumaná** Sucre, NE Venezuela 10°29′N 64°12′W
55 O5 **Cumanacoa** Sucre, NE Venezuela 10°17′N 63°58′W
54 C13 **Cumbal, Nevado de** *elevation* S Colombia
21 O7 **Cumberland** Kentucky, S USA 36°55′N 83°00′W
21 U2 **Cumberland** Maryland, NE USA 39°40′N 78°47′W
21 V6 **Cumberland** Virginia, NE USA 37°31′N 78°16′W
187 P12 **Cumberland, Cape** *var.* Cape Nahoi. *headland* Espiritu Santo, N Vanuatu 14°39′S 166°35′E
11 V14 **Cumberland House** Saskatchewan, C Canada 53°57′N 102°21′W
23 W8 **Cumberland Island** *island* Georgia, SE USA
20 L7 **Cumberland, Lake** ⊠ Kentucky, S USA
9 R5 **Cumberland Peninsula** *peninsula* Baffin Island, Nunavut, NE Canada
2 N9 **Cumberland Plateau** *plateau* E USA
30 L1 **Cumberland Point** *headland* Michigan, N USA 47°51′N 89°14′W
21 O7 **Cumberland River** Kentucky/Tennessee, S USA
9 S6 **Cumberland Sound** *inlet* Baffin Island, Nunavut, NE Canada
96 I12 **Cumbernauld** S Scotland, United Kingdom 55°57′N 04°W
97 K15 **Cumbria** *cultural region* NW England, United Kingdom
97 K15 **Cumbrian Mountains** ▲▲ NW England, United Kingdom
23 S2 **Cumming** Georgia, SE USA 34°12′N 84°08′W
Cummin in Pommern *see* Kamień Pomorski
182 G9 **Cummins** South Australia 34°17′S 135°43′E
96 I13 **Cumnock** W Scotland, United Kingdom 55°32′N 04°28′W
40 G4 **Cumpas** Sonora, NW Mexico 30°N 109°48′W
136 H16 **Çumra** Konya, C Turkey 37°34′N 32°38′E
63 G15 **Cunco** Araucanía, C Chile 38°55′S 72°02′W
54 E9 **Cundinamarca** *off.* Departamento de Cundinamarca. ⬖ *province* C Colombia
Cundinamarca, Departamento de *see* Cundinamarca
41 U15 **Cunduacán** Tabasco, SE Mexico 18°00′N 93°07′W
83 C16 **Cunene** ⬖ *province* S Angola
83 A16 **Cunene** *var.* Kunene. Angola/Namibia *see also* Kunene
Cunene *see* Kunene
106 A9 **Cuneo** *Fr.* Coni. Piemonte, NW Italy 44°23′N 07°32′E
83 E15 **Cunjamba** Kuando Kubango, E Angola 15°22′S 20°07′E
181 V10 **Cunnamulla** Queensland, E Australia 28°09′S 145°44′E
Čunusavvon *see* Junosuando
Cuokkarašša *see* Čohkarášša
106 B7 **Cuorgne** Piemonte, NE Italy 45°23′N 07°34′E
96 K11 **Cupar** E Scotland, United Kingdom 56°19′N 03°01′W
116 L8 **Cupcina** *Rus.* Kupchino; *prev.* Calinisc, Kalinisk. N Moldova 48°07′N 27°22′E
54 C8 **Cupica** Chocó, W Colombia 06°43′N 77°31′W
54 C8 **Cupica, Golfo de** *gulf* W Colombia
112 N13 **Ćuprija** Serbia, E Serbia 43°57′N 21°21′E
Cura *see* Villa de Cura
45 S9 **Curaçao** *prev.* Dutch West Indies. ◇ *Dutch self-governing territory* S Caribbean Sea
45 P16 **Curaçao** *island* Lesser Antilles
56 H13 **Curanja, Río** E Peru
56 F7 **Curaray, Rio** Ecuador/Peru
116 K14 **Curcani** Călăraşi, SE Romania 44°11′N 26°39′E
182 H4 **Curdimurka** South Australia 29°27′S 136°56′E
103 P7 **Cure** C France
173 Y16 **Curepipe** C Mauritius 20°19′S 57°31′E
55 R6 **Curiapo** Delta Amacuro, NE Venezuela 10°03′N 63°05′W
Curia Rhaetorum *see* Chur
62 G12 **Curicó** Maule, C Chile 35°00′S 71°15′W
Curieta *see* Krk
172 I15 **Curieuse** *island* Inner Islands, NE Seychelles
59 C16 **Curitiba** Acre, W Brazil 10°08′S 69°00′W
60 K12 **Curitiba** *prev.* Curytiba. *state capital* Paraná, S Brazil 25°25′S 49°25′W
60 J13 **Curitibanos** Santa Catarina, S Brazil 27°18′S 50°35′W
183 S6 **Curlewis** New South Wales, SE Australia 31°09′S 150°18′E
182 J6 **Curnamona** South Australia 31°39′S 139°35′E
83 A15 **Curoca** SW Angola
183 T6 **Currabubula** New South Wales, SE Australia 31°17′S 150°43′E
59 Q14 **Currais Novos** Rio Grande do Norte, E Brazil 06°12′S 36°30′W
35 W7 **Currant** Nevada, W USA 38°43′N 115°27′W
35 W6 **Currant Mountain** ▲ Nevada, W USA 38°56′N 115°19′W
44 H2 **Current** Eleuthera Island, C The Bahamas 25°24′N 76°44′W
27 W8 **Current River** Arkansas/Missouri, C USA
182 M14 **Currie** Tasmania, SE Australia 39°59′S 143°51′E
21 Y8 **Currituck** North Carolina, SE USA 36°29′N 76°02′W
21 Y8 **Currituck Sound** *sound* North Carolina, SE USA
39 R11 **Curry** Alaska, USA 62°36′N 150°00′W
Curtbunar *see* Tervel
116 I13 **Curtea de Argeş** *var.* Curtea-de-Arges. Argeş, S Romania 45°06′N 24°40′E
Curtea-de-Arges *see* Curtea de Argeş
116 E10 **Curtici** *Ger.* Kurtitsch, *Hung.* Kürtös. Arad, W Romania 46°21′N 21°17′E
104 H2 **Curtis** Galicia, NW Spain 43°09′N 08°10′W
28 M16 **Curtis** Nebraska, C USA 40°36′N 100°27′W
183 O14 **Curtis Group** *island group* Tasmania, SE Australia
181 Y8 **Curtis Island** *island* Queensland, SE Australia
58 K11 **Curuá, Ilha do** *island* NE Brazil
47 U7 **Curuá, Rio** N Brazil
59 A14 **Curuçá, Rio** NW Brazil
112 L9 **Čurug** *Hung.* Csurog. Vojvodina, N Serbia 45°30′N 20°02′E
61 D16 **Curuzú Cuatiá** Corrientes, NE Argentina 29°50′S 58°05′W
59 M19 **Curvelo** Minas Gerais, SE Brazil 18°45′S 44°27′W
18 E14 **Curwensville** Pennsylvania, NE USA 40°57′N 78°29′W
30 M3 **Curwood, Mount** ▲ Michigan, N USA 46°42′N 88°14′W
Curytiba *see* Curitiba
Curzola *see* Korčula
42 A10 **Cuscatlán** ⬖ *department* C El Salvador
57 H15 **Cusco** *var.* Cuzco. Cusco, C Peru 13°35′S 72°02′W
57 H15 **Cusco** *off.* Departamento de Cusco, *var.* Cuzco. ⬖ *department* C Peru
Cusco, Departamento de *see* Cusco
27 O9 **Cushing** Oklahoma, C USA 36°01′N 96°46′W
25 W8 **Cushing** Texas, SW USA 31°48′N 94°50′W
40 I6 **Cusihuiriachic** Chihuahua, N Mexico 28°16′N 106°46′W
103 P10 **Cusset** Allier, C France 46°08′N 03°27′E
23 S6 **Cusseta** Georgia, SE USA 32°18′N 84°46′W
28 J10 **Custer** South Dakota, N USA 43°46′N 103°36′W
Cüstrin *see* Kostrzyn
33 Q7 **Cut Bank** Montana, NW USA 48°38′N 112°20′W
Cutch, Gulf of *see* Kachchh, Gulf of
23 S6 **Cuthbert** Georgia, SE USA 31°46′N 84°47′W
11 S15 **Cut Knife** Saskatchewan, S Canada 52°40′N 108°54′W
23 Y16 **Cutler Ridge** Florida, SE USA 25°34′N 80°21′W
22 K10 **Cut Off** Louisiana, S USA 29°32′N 90°20′W
63 I15 **Cutral-Có** Neuquén, C Argentina 38°56′S 69°13′W
107 O21 **Cutro** Calabria, SW Italy 39°01′N 16°59′E
183 O4 **Cuttaburra Channels** *seasonal river* New South Wales, SE Australia
154 O12 **Cuttack** Odisha, E India 20°28′N 85°53′E
83 C15 **Cuvelai** Cunene, SW Angola 15°40′S 15°48′E
79 G18 **Cuvette** *var.* Région de la Cuvette. ⬖ *province* C Congo
Cuvette, Région de la *see* Cuvette
173 V9 **Cuvier Basin** *undersea feature* E Indian Ocean
173 U9 **Cuvier Plateau** *undersea feature* E Indian Ocean
82 B12 **Cuvo** W Angola
100 H9 **Cuxhaven** Niedersachsen, NW Germany 53°51′N 08°43′E
Cuyabá *see* Cuiabá
Cuyuni, Río *see* Cuyuni River
55 S8 **Cuyuni River** *var.* Río Cuyuni. Guyana/Venezuela
Cuzco *see* Cusco
97 K22 **Cwmbran** *Wel.* Cwmbrân. SW Wales, United Kingdom 51°39′N 03°W
Cwmbrân *see* Cwmbran
28 K15 **C. W. McConaughy, Lake** ⊠ Nebraska, C USA
81 D20 **Cyangugu** SW Rwanda 02°27′S 29°00′E
110 D11 **Cybinka** *Ger.* Ziebingen. Lubuskie, W Poland 52°11′N 14°46′E
Cyclades *see* Kykládes
Cydonia *see* Chaniá
Cymru *see* Wales
20 M5 **Cynthiana** Kentucky, S USA 38°22′N 84°18′W
11 S17 **Cypress Hills** ▲▲ Alberta/Saskatchewan, SW Canada
Cypro-Syrian Basin *see* Cyprus Basin
121 U11 **Cyprus** *off.* Republic of Cyprus, *Gk.* Kypros, *Turk.* Kıbrıs, Kıbrıs Cumhuriyeti. ◆ *republic* E Mediterranean Sea
84 L14 **Cyprus** *Gk.* Kypros, *Turk.* Kıbrıs. *island* E Mediterranean Sea
121 W11 **Cyprus Basin** *var.* Cypro-Syrian Basin. *undersea feature* E Mediterranean Sea 34°00′N 34°00′E
Cyprus, Republic of *see* Cyprus
Cythera *see* Kýthira
Cythnos *see* Kýthnos
110 F9 **Czaplinek** *Ger.* Tempelburg. Zachodnio-pomorskie, NW Poland 53°33′N 16°14′E
Czarna Woda *see* Wda
110 G8 **Czarne** Pomorskie, N Poland 53°40′N 17°00′E
110 G10 **Czarnków** Wielkopolskie, C Poland 52°53′N 16°32′E
111 E17 **Czech Republic** *Cz.* Česká Republika. ◆ *republic* C Europe
Czegléd *see* Cegléd
110 G12 **Czempiń** Wielkopolskie, C Poland 52°10′N 16°46′E
Czenstochau *see* Częstochowa
Czerkow *see* Čerchov
Czernowitz *see* Chernivtsi
110 I8 **Czersk** Pomorskie, N Poland 53°48′N 17°58′E
111 J15 **Częstochowa** *Ger.* Czenstochau, Tschenstochau, *Rus.* Chenstokhov. Śląskie, S Poland 50°49′N 19°07′E
110 F10 **Człopa** *Ger.* Schloppe. Zachodnio-pomorskie, NW Poland 53°05′N 16°05′E
110 H8 **Człuchów** *Ger.* Schlochau. Pomorskie, NW Poland 53°41′N 17°20′E

D

163 V9 **Da'an** *var.* Dalai. Jilin, NE China 45°28′N 124°18′E
15 S10 **Daaquam** Québec, SE Canada 46°36′N 70°03′W
Daawo, Webi *see* Dawa Wenz
54 I4 **Dabajuro** Falcón, NW Venezuela 11°00′N 70°41′W
77 N15 **Dabakala** NE Ivory Coast 08°19′N 04°24′W
163 S11 **Daban** *var.* Bairin Youqi. Nei Mongol Zizhiqu, N China 43°33′N 118°40′E
111 K23 **Dabas** Pest, C Hungary 47°36′N 18°22′E
160 L8 **Daba Shan** ▲▲ C China
Dabba *see* Daocheng
140 J5 **Dabbāgh, Jabal** ▲ NW Saudi Arabia 27°52′N 35°48′E
54 D8 **Dabeiba** Antioquia, NW Colombia 07°01′N 76°18′W
154 E11 **Dabhoi** Gujarāt, W India 22°08′N 73°28′E
161 P8 **Dabie Shan** ▲▲ C China
76 J13 **Dabola** C Guinea 10°48′N 11°02′W
77 N17 **Dabou** S Ivory Coast 05°20′N 04°23′W
162 M15 **Dabqig** *prev.* Uxin Qi. Nei Mongol Zizhiqu, N China 38°29′N 108°48′E
110 P8 **Dąbrowa Białostocka** Podlaskie, NE Poland 53°38′N 23°18′E
111 M16 **Dąbrowa Tarnowska** Małopolskie, S Poland 50°10′N 21°E
119 M20 **Dabryn'** *Rus.* Dobryn'. Homyel'skaya Voblasts', SE Belarus 51°46′N 29°12′E
159 P10 **Dabsan Hu** ◎ C China
161 Q13 **Dabu** *var.* Huliao. Guangdong, S China 24°19′N 116°07′E
116 H15 **Dăbuleni** Dolj, SW Romania 43°48′N 24°05′E
152 G9 **Dabwāli** Haryāna, NW India 29°56′N 74°40′E
Dacca *see* Dhaka
101 L23 **Dachau** Bayern, SE Germany 48°15′N 11°26′E
Dachuan *see* Dazhou
Dacia Bank *see* Dacia Seamount

◆ Country
● Country Capital
◇ Dependent Territory
○ Dependent Territory Capital
⬖ Administrative Regions
✈ International Airport
▲ Mountain
▲▲ Mountain Range
Volcano
River
◎ Lake
⊠ Reservoir

64 M10 **Dacia Seamount** *var.* Dacia Bank. *undersea feature* E Atlantic Ocean 31°10´N 13°42´W
37 T3 **Dacono** Colorado, C USA 40°04´N 104°56´W
Đặc Tô *see* Đắk Tô
Dacura *see* Đákura
23 W12 **Dade City** Florida, SE USA 28°21´N 82°12´W
152 L10 **Dadeldhurā** *var.* Dandeldhura. Far Western, W Nepal 29°12´N 80°31´E
23 Q5 **Dadeville** Alabama, S USA 32°49´N 85°45´W
103 N15 **Dadon** ♦ S France
154 D12 **Dādra and Nagar Haveli** ♦ *union territory* W India
149 P14 **Dadu** Sind, SE Pakistan 26°42´N 67°48´E
167 U11 **Da Du Bôloc** Kon Tum, C Vietnam 14°06´N 107°40´E
160 G9 **Dadu He** ♒ C China
163 V15 **Daecheong-do** *prev.* Taechŏng-do. *island* NW South Korea
163 Y16 **Daegu** *Jap.* Taikyū; *prev.* Taegu. SE South Korea 35°55´N 128°33´E
163 Y15 **Daejeon** *Jap.* Taiden; *prev.* Taejŏn. C South Korea 36°20´N 127°28´E
Daerah Istimewa Aceh *see* Aceh
171 P4 **Daet** Luzon, N Philippines 14°06´N 122°57´E
160 I11 **Dafang** Guizhou, S China 27°07´N 105°40´E
Dafeng *see* Shanglin
153 W11 **Dafla Hills** ▲ NE India
11 U15 **Dafoe** Saskatchewan, S Canada 51°46´N 104°11´W
76 G10 **Dagana** N Senegal 16°28´N 15°35´W
Dagana *see* Massakory, Chad
Dagana *see* Dahana, Tajikistan
Dagcagoin *see* Zoigê
118 K11 **Dagda** SE Latvia 56°06´N 27°36´E
Dagden *see* Hiiumaa
Dagden-Sund *see* Soela Väin
127 P16 **Dagestan, Respublika** *prev.* Dagestanskaya ASSR, *Eng.* Daghestan. ♦ *autonomous republic* SW Russian Federation
Dagestanskaya ASSR *see* Dagestan, Respublika
127 R17 **Dagestanskiye Ogni** Respublika Dagestan, SW Russian Federation 42°09´N 48°08´E
Dagezhen *see* Fengning
185 A23 **Dagg Sound** *sound* South Island, New Zealand
141 Y8 **Daghmar** NE Oman 23°09´N 59°01´E
Dağliq Quarabağ *see* Nagorno-Karabakh
Dagö *see* Hiiumaa
54 D11 **Dagua** Valle del Cauca, W Colombia 03°39´N 76°40´W
160 H11 **Daguan** Yunnan, SW China 27°42´N 103°51´E
171 N3 **Dagupan** *off.* Dagupan City. Luzon, N Philippines 16°05´N 120°21´E
Dagupan City *see* Dagupan
159 N16 **Dagzê** *var.* Dêqên. Xizang Zizhiqu, W China 29°38´N 91°15´E
147 Q13 **Dahana** *Rus.* Dagana, Dakhana. SW Tajikistan 38°03´N 69°51´E
163 V10 **Dahei Shan** ▲ N China
163 T7 **Da Hinggan Ling** *Eng.* Great Khingan Range. ▲ NE China
Dahlac Archipelago *see* Dahlak Archipelago
80 K9 **Dahlak Archipelago** *var.* Dahlac Archipelago. *island group* E Eritrea
23 T2 **Dahlonega** Georgia, SE USA 34°31´N 83°59´W
101 O14 **Dahme** Brandenburg, E Germany 52°10´N 13°47´E
100 O13 **Dahme** ♒ E Germany
141 O14 **Dahm, Ramlat** *desert* NW Yemen
154 E14 **Dāhod** *prev.* Dohad. Gujarāt, W India 22°49´N 74°18´E
Dahomey *see* Benin
158 G10 **Dahongliutan** Xinjiang Uygur Zizhiqu, NW China 35°59´N 79°12´E
Dahra *see* Dara
Dahuaishu *see* Hongtong
139 R2 **Dahūk** *var.* Dohuk, Dahuk, *Kurd.* Dihok. Dahūk, ▲ N Iraq 36°52´N 43°01´E
139 R2 **Dahūk** *off.* Muḥāfaẓat Dahūk, *var.* Dohuk, *Kurd.* Dihok. *off. Kurd.* Parêzga-i Dihok. ♦ *governorate* N Iraq
Dahūk, Muḥāfaẓat at *see* Dahūk
116 J15 **Daia** Giurgiu, S Romania 44°00´N 25°59´E
165 P12 **Daigo** Ibaraki, Honshū, S Japan 36°43´N 140°22´E
O13 **Dai Hai** ◉ N China
Daihoku *see* Taibei
186 M8 **Dai Island** *island* N Solomon Islands
166 M8 **Daik-u** Bago, SW Myanmar (Burma) 17°46´N 96°40´E
138 H9 **Dā'il** Dar'ā, S Syria 32°45´N 36°08´E
167 U12 **Dai Lanh** Khanh Hoa, S Vietnam 12°49´N 109°20´E
161 N12 **Daishan** N China ...
105 N11 **Daimiel** Castilla-La Mancha, C Spain 39°04´N 03°37´W
115 F22 **Daimoniá** Pelopónnisos, S Greece 36°38´N 22°54´E
25 W6 **Daingerfield** Texas, SW USA 33°03´N 94°42´W
Daingin, Bá an *see* Dingle Bay
159 R13 **Dainkognubma** Xizang Zizhiqu, W China 32°50´N 92°58´E
164 K14 **Daiō-zaki** *headland* Honshū, SW Japan 34°15´N 136°54´E
Dairbhre *see* Valencia Island
61 B22 **Daireaux** Buenos Aires, E Argentina 36°34´S 61°40´W
Dairen *see* Dalian
Dairût *see* Dayrūt
25 X10 **Daisetta** Texas, SW USA 30°06´N 94°38´W
192 G5 **Daitō-jima** *island group* SW Japan

192 G5 **Daitō Ridge** *undersea feature* N Philippine Sea 25°30´N 133°00´E
161 N3 **Daixian** *var.* Dai Xian, Shangguan. Shanxi, C China 39°10´N 112°57´E
Dai Xian *see* Daixian
44 M8 **Dajabón** NW Dominican Republic 19°35´N 71°41´W
160 G8 **Dajin Chuan** ♒ C China
148 J6 **Dak** ♦ N Afghanistan
76 F11 **Dakar** ● (Senegal) W Senegal 14°44´N 17°27´W
76 F11 **Dakar** ✈ W Senegal 14°42´N 17°27´W
167 U10 **Đak Glei** *prev.* Đak Glây. Kon Tum, C Vietnam 15°05´N 107°42´E
153 U16 **Dakhin Shahbazpur Island** *island* S Bangladesh
76 F7 **Dakhlet Nouâdhibou** ♦ *region* NW Mauritania
Dakhla *see* Ad Dakhla
76 U11 **Đak Nông** *var.* Gia Nghia. S Vietnam 14°29´N 06°45´E
160 J11 **Dakoro** Maradi, S Niger 14°29´N 06°45´E
29 U12 **Dakota City** Iowa, C USA 42°42´N 94°13´W
29 R13 **Dakota City** Nebraska, C USA 42°25´N 96°25´W
112 I10 **Đakovica** *var.* Djakovo, *Hung.* Diakovár. Osijek-Baranja, E Croatia 45°18´N 18°24´E
Dakshin *see* Deccan
167 U11 **Đắk Tô** *var.* Đắc Tô. Kon Tum, C Vietnam 14°35´N 107°55´E
43 N7 **Đákura** *var.* Dacura. Región Autónoma Atlántico Norte, E Nicaragua 14°22´N 83°13´W
95 I14 **Dal** Akershus, S Norway 60°19´N 11°16´E
82 E12 **Đala** Lunda Sul, E Angola 11°04´S 20°15´E
108 J8 **Đalaas** Vorarlberg, W Austria 47°08´N 10°03´E
76 I13 **Đalaba** W Guinea 10°47´N 12°12´W
162 I12 **Dalain Hob** *var.* Ejin Qi. Nei Mongol Zizhiqu, N China 41°59´N 101°04´E
Dalai Nor *see* Hulun Nur
163 Q11 **Dalai Nur** *salt lake* N China
Dala-Jarna *see* Järna
95 M14 **Dalälven** ♒ C Sweden
136 C16 **Dalaman** Muğla, SW Turkey 36°47´N 28°47´E
136 D16 **Dalaman Çayı** ♒ SW Turkey
162 K11 **Dalandzadgad** Ömnögovĭ, S Mongolia 43°35´N 104°23´E
95 D17 **Dalane** *physical region* S Norway
189 Z2 **Dalap-Uliga-Djarrit** *var.* Delap-Uliga-Darrit, D-U-D. *island group* Ratak Chain, SE Marshall Islands
94 J12 **Dalarna** *prev.* Kopparberg. ♦ *county* C Sweden
94 L13 **Dalarna** *Eng.* Dalecarlia. *cultural region* C Sweden
95 P16 **Dalarö** Stockholm, C Sweden 59°07´N 18°25´E
167 U13 **Đà Lạt** Lâm Đồng, S Vietnam 11°56´N 108°25´E
Dalay *see* Bayandalay
148 L12 **Dalbandin** *var.* Dāl Bandin. Baluchistān, SW Pakistan 28°48´N 64°08´E
95 J17 **Dalbosjön** *lake bay* S Sweden
181 W7 **Dalby** Queensland, E Australia 27°11´S 151°12´E
95 D13 **Dale** Hordaland, S Norway 60°35´N 05°48´E
94 C12 **Dale** Sogn Og Fjordane, S Norway 61°22´N 05°24´E
32 K12 **Dale** Oregon, NW USA 44°58´N 118°56´W
25 T11 **Dale** Texas, SW USA 29°56´N 97°34´W
21 W4 **Dale City** Virginia, NE USA 38°38´N 77°18´W
Dalecarlia *see* Dalarna
21 P6 **Dale Hollow Lake** ◉ Kentucky/Tennessee, S USA
98 O8 **Dalen** Drenthe, NE Netherlands 52°42´N 06°45´E
95 E15 **Dalen** Telemark, S Norway 59°27´N 07°59´E
166 K14 **Daletme** Chin State, W Myanmar (Burma) 21°44´N 92°48´E
23 Q7 **Daleville** Alabama, S USA 31°18´N 85°42´W
94 D13 **Dalfsen** Overijssel, E Netherlands 52°31´N 06°16´E
114 M8 **Dalgopol** *var.* Dŭlgopol. Varna, E Bulgaria 43°05´N 27°24´E
24 M1 **Dalhart** Texas, SW USA 36°05´N 102°31´W
13 O13 **Dalhousie** New Brunswick, SE Canada 48°03´N 66°22´W
152 I6 **Dalhousie** Himāchal Pradesh, N India 32°32´N 76°01´E
160 F12 **Dali** *var.* Xiaguan. Yunnan, SW China 25°34´N 100°11´E
Dali *see* Idálion
163 U14 **Dalian** *var.* Dairen, Dalien, Jay Dalren, Lüda, Ta-lien, *Rus.* Dalny. Liaoning, NE China 38°53´N 121°37´E
Dalien *see* Dalian
105 O15 **Dalías** Andalucía, S Spain 36°49´N 02°50´W
112 J9 **Dalj** *Hung.* Dalja. Osijek-Baranja, E Croatia 45°29´N 19°00´E
Dalja *see* Dalj
32 F12 **Dallas** Oregon, NW USA 44°56´N 123°20´W
25 U6 **Dallas** Texas, SW USA 32°47´N 96°48´W
25 T6 **Dallas-Fort Worth** ✈ Texas, SW USA 32°54´N 97°02´W
154 K12 **Dalli Rājhara** *var.* Dhali Rājhara. Chhattīsgarh, C India 20°32´N 81°10´E
39 X15 **Dall Island** *island* Alexander Archipelago, Alaska, USA 54°45´N 132°48´W
38 M12 **Dall Lake** ◉ Alaska, USA
Dállogilli *see* Korpilombolo
S12 **Dallol Bosso** *seasonal river* W Niger
141 U7 **Dalmā** *island* W United Arab Emirates

113 E14 **Dalmacija** *Eng.* Dalmatia, *Ger.* Dalmatien, *It.* Dalmazia. *cultural region* S Croatia
Dalmatia/Dalmatien/Dalmazia *see* Dalmacija
123 S15 **Dal'negorsk** Primorskiy Kray, SE Russian Federation 44°27´N 135°30´E
Dalny *see* Dalian
76 M16 **Daloa** C Ivory Coast 06°56´N 06°28´W
160 J11 **Dalou Shan** ▲ S China
181 X7 **Dalrymple Lake** ◉ Queensland, E Australia
14 H14 **Dalrymple Lake** ◉ Ontario, S Canada
181 X7 **Dalrymple, Mount** ▲ Queensland, E Australia 21°01´S 148°34´E
93 K20 **Dalsbruk** *Fin.* Taalintehdas. Varsinais-Suomi, SW Finland 60°02´N 22°31´E
95 K19 **Dalsjöfors** Västra Götaland, S Sweden 57°43´N 13°05´E
95 J17 **Dals Långed** *var.* Långed. Västra Götaland, S Sweden 58°54´N 12°20´E
153 O15 **Dāltenganj** *prev.* Daltonganj. Jhārkhand, N India 24°02´N 84°07´E
23 R2 **Dalton** Georgia, SE USA 34°46´N 84°58´W
Daltonganj *see* Dāltenganj
195 X14 **Dalton Iceberg Tongue** *ice feature* Antarctica
92 J1 **Dalvík** Norðurland Eystra, N Iceland 65°58´N 18°31´W
Dálvvadis *see* Jokkmokk
35 N8 **Daly City** California, W USA 37°34´N 122°27´W
181 P2 **Daly River** ♦ Northern Territory, N Australia
181 Q3 **Daly Waters** Northern Territory, N Australia 16°21´S 133°22´E
119 F20 **Damachava** *var.* Damachëvo, *Pol.* Domaczewo, *Rus.* Domachëvo. Brestskaya Voblasts´, SW Belarus 51°45´N 23°36´E
Damachëvo *see* Damachava
Damachova *see* Damachava
77 W11 **Damagaram Takaya** Zinder, S Niger 14°02´N 09°28´E
154 D12 **Damān and Diu**, Damān and Diu. W India 20°25´N 72°58´E
154 B12 **Damān and Diu** ♦ *union territory* W India
75 V7 **Damanhûr** *anc.* Hermopolis Parva. N Egypt 31°03´N 30°28´E
153 Y11 **Damaqun Shan** ▲ E China
79 I15 **Damara** Ombella-Mpoko, S Central African Republic 04°58´N 18°45´E
83 D18 **Damaraland** *physical region* C Namibia
171 S15 **Damar, Kepulauan** *var.* Baraf Daja Islands, Kepulauan Barat Daya. *island group* C Indonesia
171 P6 **Damar, Pulau** *island* Maluku, E Indonesia 04°13´N 100°36´E
Damas *see* Dimashq
77 Y12 **Damasak** Borno, NE Nigeria 13°10´N 12°40´E
Damasco *see* Dimashq
21 Q8 **Damascus** Virginia, NE USA 36°37´N 81°46´W
Damascus *see* Dimashq
77 X13 **Damaturu** Yobe, NE Nigeria 11°44´N 11°58´E
171 R9 **Damau** Pulau Kaburuang, N Indonesia 03°46´N 126°49´E
143 Q5 **Dāmāvand, Qolleh-ye** ▲ N Iran 35°56´N 52°08´E
82 B10 **Damba** Uíge, NW Angola 06°44´S 15°20´E
114 M12 **Dambaslar** Tekirdağ, NW Turkey 41°13´N 27°13´E
116 J13 **Dâmboviţa** *prev.* Dîmboviţa. ♦ *county* SE Romania
116 J13 **Dâmboviţa** *prev.* Dîmboviţa. ♒ S Romania
173 Y15 **D'Ambre, Île** *island* NE Mauritius
155 K24 **Dambulla** Central Province, C Sri Lanka 07°51´N 80°40´E
44 K9 **Dame-Marie** SW Haiti 18°36´N 74°26´W
44 J9 **Dame Marie, Cap** *headland* SW Haiti 18°36´N 74°24´W
143 Q4 **Dāmghān** Semnān, N Iran 36°13´N 54°22´E
18 L6 **Damietta** *see* Dumyât
138 G10 **Damīyā** Al Balqā', NW Jordan 32°07´N 35°33´E
146 G11 **Damla** Daşoguz Welayaty, N Turkmenistan 40°05´N 59°15´E
100 G12 **Damme** Niedersachsen, NW Germany 52°31´N 08°12´E
153 R15 **Dāmodar** ♒ NE India
154 J9 **Damoh** Madhya Pradesh, C India 23°50´N 79°30´E
77 P15 **Damongo** NW Ghana 09°05´N 01°49´W
138 G7 **Damoûr** *var.* Ad Dāmûr. W Lebanon 33°36´N 35°22´E
171 N11 **Dampal, Teluk** *bay* Sulawesi, C Indonesia
180 H7 **Dampier** Western Australia 20°40´S 116°40´E
180 H6 **Dampier Archipelago** *island group* Western Australia
141 U14 **Damqawt** *var.* Damqut. E Yemen 16°35´N 52°39´E
159 O13 **Dam Qu** ♒ C China
Damqut *see* Damqawt
167 R13 **Dâmrei, Chuŏr Phnum** *Fr.* Chaîne de l'Éléphant. ▲ SW Cambodia
108 C7 **Damvant** Jura, NW Switzerland
21 Y5 **Damwoude** *Fris.* Damwâld. Fryslân, N Netherlands 53°18´N 05°59´E
98 L5 **Damwoude** *Fris.* Damwâld. Fryslân, N Netherlands 53°18´N 05°59´E
159 N15 **Damxung** *var.* Gongtang. Xizang Zizhiqu, W China 30°29´N 91°02´E
80 K11 **Danakil Desert** *var.* Afar Depression, Danakil Plain. *desert* E Africa
Danakil Plain *see* Danakil Desert
35 R8 **Dana, Mount** ▲ California, W USA 37°54´N 119°13´W
76 L16 **Danané** W Ivory Coast 07°16´N 08°09´W
167 U10 **Đa Nẵng** *prev.* Tourane. Quang Nam-Đa Nẵng, C Vietnam 16°04´N 108°14´E
160 F10 **Danba** *var.* Zhanggu. Sichuan, C China 30°54´N 101°49´E
166 L8 **Danba, Teluk** *see* Daneborg

18 L13 **Danbury** Connecticut, NE USA 41°21´N 73°27´W
25 W12 **Danbury** Texas, SW USA 29°13´N 95°20´W
35 X15 **Danby Lake** ◉ California, W USA
194 H4 **Danco Coast** *physical region* Antarctica
82 B11 **Dande** ♦ NW Angola
Dandeldhura *see* Dadeldhurā
155 E17 **Dandeli** Karnātaka, W India 15°18´N 74°42´E
183 O12 **Dandenong** Victoria, SE Australia 38°00´S 145°13´E
163 V13 **Dandong** *var.* Tan-tung; *prev.* An-tung. Liaoning, NE China 40°09´N 124°23´E
197 Q14 **Daneborg** ♦ N Greenland
25 V12 **Danevang** Texas, SW USA 29°03´N 96°11´W
14 L12 **Danford Lake** Québec, SE Canada 45°55´N 76°12´W
19 T4 **Danforth** Maine, NE USA 45°39´N 67°54´W
37 P3 **Danforth Hills** ▲ Colorado, C USA
Dangara *see* Danghara
159 P8 **Dangchang** Gansu, N China 34°01´N 104°19´E
159 P8 **Dangchengwan** *var.* Subei, Subei Mongolzu Zizhixian. Gansu, N China 39°33´N 94°50´E
82 B10 **Dange** Uíge, NW Angola 07°55´S 15°01´E
83 E26 **Danger Point** *headland* SW South Africa 34°33´S 19°17´E
147 Q12 **Danghara** *Rus.* Dangara. SW Tajikistan 38°05´N 69°14´E
159 P8 **Danghe Nanshan** ▲ W China
80 I12 **Dangila** *var.* Dānglā. Āmara, NW Ethiopia 11°16´N 36°51´E
159 P8 **Dangjin Shankou** *pass* N China
Dangla *see* Tanggula Shan
Dang La *see* Tanggula Shankou, China
Dānglā *see* Dangila, Ethiopia
153 Y11 **Dāngori** Assam, NE India 27°40´N 95°35´E
38 M9 **Dang Raek, Phanom/Dangrek, Chaîne des** *see* Dângrêk, Chuŏr Phnum
167 S11 **Dângrêk, Chuŏr Phnum** *var.* Phanom Dong Rak, *Fr.* Chaîne des Dangrek. ▲ Cambodia/Thailand
42 G3 **Dangriga** *prev.* Stann Creek. Stann Creek, E Belize 16°59´N 88°13´W
161 P6 **Dangshan** Anhui, E China 34°22´N 116°21´E
33 T15 **Daniel** Wyoming, C USA 42°49´N 110°04´W
83 H22 **Daniëlskuil** Northern Cape, N South Africa 28°11´S 23°33´E
19 N12 **Danielson** Connecticut, NE USA 41°48´N 71°53´W
124 M13 **Danilov** Yaroslavskaya Oblast´, W Russian Federation 58°11´N 40°11´E
127 O9 **Danilovka** Volgogradskaya Oblast´, SW Russian Federation 50°21´N 44°03´E
Danish West Indies *see* Virgin Islands (US)
160 L7 **Dan Jiang** ♒ C China
160 M7 **Danjiangkou Shuiku** ◉ C China
141 W8 **Dank** *var.* Dhank. NW Oman 23°34´N 56°19´E
152 J7 **Dankhar** Himāchal Pradesh, N India 32°08´N 78°12´E
126 L6 **Dankov** Lipetskaya Oblast´, W Russian Federation 53°17´N 39°07´E
42 H6 **Danlí** El Paraíso, S Honduras 14°02´N 86°34´W
18 L6 **Dannemora** New York, NE USA 44°42´N 73°42´W
95 O14 **Dannemora** Uppsala, C Sweden 60°13´N 17°49´E
100 K11 **Dannenberg** Niedersachsen, N Germany 53°05´N 11°06´E
184 N12 **Dannevirke** Manawatu-Wanganui, North Island, New Zealand 40°12´S 176°05´E
21 U8 **Dan River** ♦ Virginia, NE USA
167 P8 **Dan Sai** Loei, C Thailand 17°15´N 101°10´E
18 F10 **Dansville** New York, NE USA 42°34´N 77°40´W
154 J13 **Dantewāra** Chhattīsgarh, C India 18°54´N 81°21´E
Dantzig *see* Gdańsk
86 E12 **Danube** *Bul.* Dunav, *Cz.* Dunaj, *Ger.* Donau, *Hung.* Duna, *Rom.* Dunărea. ♒ C Europe
Danubian Plain *see* Dunavska Ravnina
166 L8 **Danubyu** Ayeyawady, SW Myanmar (Burma) 17°15´N 95°35´E
19 P11 **Danvers** Massachusetts, NE USA 42°34´N 70°54´W
27 T11 **Danville** Arkansas, C USA 35°03´N 93°22´W
22 N13 **Danville** Illinois, N USA 40°10´N 87°37´W
31 O14 **Danville** Indiana, N USA 39°45´N 86°31´W
29 Y15 **Danville** Iowa, C USA 40°52´N 91°18´W
20 M6 **Danville** Kentucky, S USA 37°38´N 84°45´W
18 G14 **Danville** Pennsylvania, NE USA 40°57´N 76°36´W
21 T8 **Danville** Virginia, NE USA 36°35´N 79°23´W
Danxian/Dan Xian *see* Danzhou
160 L17 **Danzhou** *prev.* Danxian, Dan Xian, Nada. S China 19°31´N 109°31´E
Danzhou *see* Yichuan
Danzig *see* Gdańsk
110 I6 **Danzig, Gulf of** *var.* Gulf of Gdańsk, *Pol.* Zatoka Gdańska, *Ger.* Danziger Bucht, *Rus.* Gdan'skaya Bukhta. *gulf* N Poland
183 T3 **Dao, Rio** ♒ N Portugal ...
181 O1 **Darling Downs** *hill range* Queensland, E Australia
2 M2 **Darling, Lake** ◉ North Dakota, N USA
180 I11 **Darling Range** ▲ Western Australia
182 L8 **Darling River** ♒ New South Wales, SE Australia

104 H7 **Daokou** *see* Huaxian
33 R6 **Dão, Rio** ♒ N Portugal
77 Y7 **Dao Timmi** Agadez, NE Niger 20°31´N 13°34´E
77 Q14 **Dapaong** N Togo 10°52´N 00°12´E
23 N8 **Daphne** Alabama, S USA 30°36´N 87°54´W
171 P7 **Dapitan** Mindanao, S Philippines 08°39´N 123°26´E
159 P9 **Da Qaidam** Qinghai, C China 37°50´N 95°18´E
163 V8 **Daqing** Heilongjiang, NE China 46°35´N 125°03´E
163 O13 **Daqing Shan** ▲ N China
163 T11 **Daqin Tal** *var.* Naiman Qi. Nei Mongol Zizhiqu, N China 42°51´N 120°41´E
Daqm *see* Duqm
160 G8 **Da Qu** *var.* Do Qu. ♒ C China
139 T5 **Dāqūq** *var.* Tāwūq. Kirkūk, N Iraq 35°08´N 44°27´E
76 G10 **Dara** *var.* Dahra. N Senegal 15°20´N 15°28´W
138 H9 **Dar'ā** *var.* Der'a, *Fr.* Déraa. Dar'ā, SW Syria 32°37´N 36°06´E
138 H8 **Dar'ā** *off.* Muḥāfaẓat Dar'ā, *var.* Dará, Der'a, Derra. ♦ *governorate* S Syria
Darabani *see* Darreh Shahr
143 Q12 **Dārāb** Fārs, S Iran 28°52´N 54°25´E
116 K8 **Darabani** Botoşani, NW Romania 48°10´N 26°39´E
Daraj *see* Dārah
142 M8 **Dārān** Eşfahān, W Iran 33°00´N 50°27´E
100 O15 **Darßlanden**, Jhārkhand ...
153 S15 **Darsana** *var.* Darshana. Khulna, N Bangladesh 23°32´N 88°49´E
Darshana *see* Darsana
100 N7 **Darss** *peninsula* NE Germany
100 N7 **Darsser Ort** *headland* NE Germany
97 J24 **Dart** ♒ SW England, United Kingdom
Dartang *see* Bagên
97 P22 **Dartford** SE England, United Kingdom 51°27´N 00°13´E
182 L12 **Dartmoor** SE Australia 37°56´S 141°18´E
97 J24 **Dartmoor** *moorland* SW England, United Kingdom
13 Q15 **Dartmouth** Nova Scotia, SE Canada 44°40´N 63°35´W
97 J24 **Dartmouth** SW England, United Kingdom 50°21´N 03°34´W
15 Y6 **Dartmouth** ♦ SE Canada
183 P17 **Dartmouth Reservoir** ◉ Victoria, SE Australia
186 C6 **Daru** Western, SW Papua New Guinea 09°05´S 143°10´E
112 C10 **Daruvar** *Hung.* Daruvár. Bjelovar-Bilogora, NE Croatia 45°34´N 17°12´E
Daruvár *see* Daruvar
Darvaza *see* Derweze, Turkmenistan
Darvaza *see* Darvoza, Turkmenistan
Darvazskiy Khrebet *see* Darvoz, Qatorkŭhi
Darvel Bay *see* Lahad Datu, Teluk
Darvel, Teluk *see* Lahad Datu, Teluk
147 O10 **Darvoza** *Rus.* Darvaza. Lebap Welayaty, NE Turkmenistan 40°59´N 57°19´E
147 R13 **Darvoz, Qatorkŭhi** *Rus.* Darvazskiy Khrebet. ▲ C Tajikistan
148 L9 **Darweshan** *var.* Garmser; *prev.* Darvishan. Helmand, S Afghanistan 31°13´N 64°12´E
63 J15 **Darwin** Río Negro, S Argentina 39°13´S 65°41´W
181 O1 **Darwin** *prev.* Palmerston, Port Darwin. *territory capital* Northern Territory, N Australia 12°28´S 130°52´E
65 D24 **Darwin** Falkland Islands 51°51´S 58°55´W
62 H8 **Darwin, Cordillera** ▲ S Chile
57 B17 **Darwin, Volcán** ⌃ Galápagos Islands, Ecuador, E Pacific Ocean
181 O1 **Darwin Settlement** *see* Darwin
149 S8 **Darya Khan** Punjab, E Pakistan 31°47´N 71°10´E
144 G15 **Dar'yalyktakyr, Ravnina** *plain* S Kazakhstan
143 T11 **Dārzīn** Kermān, S Iran 29°11´N 58°09´E
146 L8 **Dashhowuz Welayaty** *see* Daşoguz Welayaty
162 K7 **Dashinchilen** Bulgan, C Mongolia 47°49´N 104°09´E
119 O16 **Dashkawka** *Rus.* Dashkovka. Mahilyowskaya Voblasts´, E Belarus 53°44´N 30°16´E
Dashkhovuz *see* Daşoguz
Dashkhovuzskiy Velayat *see* Daşoguz Welayaty
Dashkovka *see* Dashkawka
148 J15 **Dasht** ♒ SW Pakistan
Dasht-e *see* Bābūs, Dasht-e
147 R13 **Dashtijum** *Rus.* Dashtidzhum. SW Tajikistan 38°05´N 70°11´E
149 W7 **Daska** Punjab, NE Pakistan
146 H8 **Daşoguz** *Rus.* Dashkhovuz, *Turkm.* Tashauz; *prev.* Tashauz. Daşoguz Welayaty, N Turkmenistan 41°51´N 59°58´E
146 H8 **Daşoguz** *Rus.* Dashkhovuz, *prev.* Dashkhovuzskiy Velayat. ♦ *province* N Turkmenistan
146 E9 **Daşoguz Welayaty** *var.* Dashhowuz Welayaty, *Rus.* Dashkhovuzskiy Velayat; *prev.* Tashauz, Daşoguz. ♦ *province* N Turkmenistan
24 J4 **Dawn** Texas, SW USA 34°55´N 102°12´W
140 M11 **Dawḥ, al Bāḥah**, SW Saudi Arabia 20°19´N 41°12´E
167 N9 **Dassa** Benin ...
Dassa-Zoumé *see* Dassa

97 M15 **Darlington** N England, United Kingdom 54°31´N 01°34´W
21 T12 **Darlington** South Carolina, SE USA 34°19´N 79°53´W
30 K9 **Darlington** Wisconsin, N USA 42°41´N 90°07´W
110 G7 **Darłowo** Zachodnio-pomorskie, NW Poland 54°24´N 16°21´E
101 G19 **Darmstadt** Hessen, SW Germany 49°52´N 08°39´E
75 S7 **Darnah** *var.* Dérna. NE Libya 32°46´N 22°39´E
182 M7 **Darnick** New South Wales, SE Australia 32°52´S 143°38´E
195 M7 **Darnley, Cape** *cape* Antarctica
105 S11 **Daroca** Aragón, NE Spain
147 S11 **Daroot-Korgon** *var.* Daraut-Kurgan. Oshskaya Oblast´, SW Kyrgyzstan
61 A23 **Darregueira** Buenos Aires, E Argentina 37°40´S 63°12´W
Darregueira *see* Darregueira
Darreh Gaz *see* Dargaz
142 K7 **Darreh Shahr** *var.* Darreh-ye Shahr. Īlām, W Iran 33°10´N 47°18´E
Darreh-ye Shahr *see* Darreh Shahr
32 I7 **Darrington** Washington, NW USA 48°15´N 121°36´W
25 P1 **Darrouzett** Texas, SW USA 36°27´N 100°19´W
153 S15 **Darsana** *var.* Darshana. Khulna, N Bangladesh ...
100 M3 **Darss** *peninsula* NE Germany
155 E14 **Dārwad** *prev.* Dhond. Mahārāshtra, W India 18°31´N 74°38´E
166 M12 **Daung Kyun** *island* S Myanmar (Burma)
1 W15 **Dauphin** Manitoba, S Canada 51°09´N 100°05´W
103 S13 **Dauphiné** *cultural region* E France
23 N9 **Dauphin Island** *island* Alabama, S USA
1 X15 **Dauphin River** Manitoba, S Canada 51°55´N 98°03´W
77 V12 **Daura** Katsina, N Nigeria 13°03´N 08°18´E
152 H12 **Dausa** *prev.* Daosa. Rājasthān, N India 26°41´N 76°21´E
Dauwa *see* Dawwah
Dāvāçi *see* Şabran
155 F18 **Dāvangere** Karnātaka, W India 14°28´N 75°59´E ...
171 Q8 **Davao** *off.* Davao City. Mindanao, S Philippines 07°06´N 125°36´E
Davao City *see* Davao
171 Q8 **Davao Gulf** *gulf* Mindanao, S Philippines
15 Q11 **Daveluyville** Québec, SE Canada 46°12´N 72°07´W
29 X14 **Davenport** Iowa, C USA 41°31´N 90°35´W
32 L8 **Davenport** Washington, NW USA 47°39´N 118°09´W
43 P16 **David** Chiriquí, W Panama 08°26´N 82°26´W
29 R15 **David City** Nebraska, C USA 41°15´N 97°07´W
David-Gorodok *see* Davyd-Haradok
11 T16 **Davidson** Saskatchewan, S Canada 51°15´N 105°59´W
21 R10 **Davidson** North Carolina, SE USA 35°29´N 80°49´W
39 S6 **Davidson Mountains** ▲ Alaska, USA
172 M8 **Davie Ridge** *undersea feature* W Indian Ocean 17°10´S 41°45´E
181 Q13 **Davies, Mount** ▲ South Australia 26°13´S 129°14´E
35 O7 **Davis** California, W USA 38°33´N 121°46´W
24 M12 **Davis** Oklahoma, C USA 34°30´N 97°06´W
195 M13 **Davis** *Australian research station* Antarctica 68°30´S 78°15´E
194 H3 **Davis Coast** *physical region* Antarctica
18 C16 **Davis, Mount** ▲ Pennsylvania, NE USA 39°48´N 79°11´W
24 K9 **Davis Mountains** ▲ Texas, SW USA
195 X3 **Davis Sea** *sea* Antarctica
65 O20 **Davis Seamounts** *undersea feature* E Atlantic Ocean
196 M13 **Davis Strait** *strait* Baffin Bay/Labrador Sea
127 V7 **Davlekanovo** Respublika Bashkortostan, W Russian Federation 54°13´N 55°06´E
108 I8 **Davos** *Rmsch.* Tavau. Graubünden, E Switzerland 46°48´N 09°50´E
Davvesiidda *see* Lebesby
119 J20 **Davyd-Haradok** *Pol.* Dawidgródek, *Rus.* David-Gorodok. Brestskaya Voblasts´, SW Belarus 52°03´N 27°13´E
163 U12 **Dawa** Liaoning, NE China 40°55´N 122°02´E
141 O11 **Dawāsir, Wādī ad** *dry watercourse* S Saudi Arabia
81 K15 **Dawa Wenz** *var.* Daua, Webi Daawo. ♒ E Africa
141 O11 **Dawāsir, Wādī ad** *dry watercourse*
Dawayimah, Birkat ad *see* Umm al Baqar, Hawr
167 N7 **Dawei** *var.* Tavoy, Htawei. (Tenasserim) S Myanmar (Burma) 14°02´N 98°12´E
119 K14 **Dawhinava** *Rus.* Dolginovo. Minskaya Voblasts´, N Belarus
Dawidgródek *see* Davyd-Haradok
141 Y9 **Dawkah** *var.* Dauka. SW Oman 18°33´N 54°02´E
Dawlat Qatar *see* Qatar
11 N15 **Dawson** Yukon, NW Canada 64°04´N 139°24´W
23 S6 **Dawson** Georgia, SE USA 31°46´N 84°27´W
29 S9 **Dawson** Minnesota, N USA 44°55´N 96°03´W
Dawson City *see* Dawson

29 U8 **Dassel** Minnesota, N USA 45°06´N 94°18´W
152 H3 **Dastegil Sar** ▲ N India
136 C16 **Datça** Muğla, SW Turkey 36°46´N 27°43´E
165 R4 **Date** Hokkaidō, NE Japan 42°28´N 140°51´E
154 I8 **Datia** *prev.* Duttia. Madhya Pradesh, C India 25°41´N 78°28´E
159 T10 **Datong** *var.* Datong Huizu Tuzu Zizhixian, Qiaotou. Qinghai, C China 37°01´N 101°33´E
161 N2 **Datong** *var.* Tatung, Ta-t'ung. Shanxi, C China 40°09´N 113°17´E
159 S8 **Datong He** ♒ C China
Datong Huizu Tuzu Zizhixian *see* Datong
159 S9 **Datong Shan** ▲ C China
169 O10 **Datu, Tanjung** *headland* Indonesia/Malaysia 02°01´N 109°37´E
Daua *see* Dawa Wenz
172 H16 **Dauban, Mount** ▲ Silhouette, NE Seychelles
149 T7 **Dāūd Khel** Punjab, E Pakistan 32°52´N 71°35´E
119 G15 **Daugai** Alytus, S Lithuania 54°22´N 24°20´E
Daugava *see* Western Dvina
118 J11 **Daugavpils** *Ger.* Dünaburg; *prev. Rus.* Dvinsk. SE Latvia 55°53´N 26°34´E
101 D18 **Daun** Rheinland-Pfalz, W Germany 50°13´N 06°50´E
Daulatabad *see* Malāyer
155 E14 **Daund** *prev.* Dhond. Mahārāshtra, W India 18°31´N 74°38´E
166 M12 **Daung Kyun** *island* S Myanmar (Burma)
1 W15 **Dauphin** Manitoba, S Canada
103 S13 **Dauphiné** *cultural region* E France
23 N9 **Dauphin Island** *island* Alabama, S USA
1 X15 **Dauphin River** Manitoba, S Canada
77 V12 **Daura** Katsina, N Nigeria
152 H12 **Dausa** *prev.* Daosa. Rājasthān, N India
Dauwa *see* Dawwah
Dāvāçi *see* Şabran
155 F18 **Dāvangere** Karnātaka, W India 14°28´N 75°59´E
171 Q8 **Davao** *off.* Davao City. Mindanao, S Philippines 07°06´N 125°36´E
Davao City *see* Davao
171 Q8 **Davao Gulf** *gulf* Mindanao, S Philippines
15 Q11 **Daveluyville** Québec, SE Canada
29 X14 **Davenport** Iowa, C USA
32 L8 **Davenport** Washington, NW USA
43 P16 **David** Chiriquí, W Panama
29 R15 **David City** Nebraska, C USA
David-Gorodok *see* Davyd-Haradok
11 T16 **Davidson** Saskatchewan, S Canada
21 R10 **Davidson** North Carolina, SE USA
39 S6 **Davidson Mountains** ▲ Alaska, USA
172 M8 **Davie Ridge** *undersea feature* W Indian Ocean
181 Q13 **Davies, Mount** ▲ South Australia
35 O7 **Davis** California, W USA
24 M12 **Davis** Oklahoma, C USA
195 M13 **Davis** *Australian research station* Antarctica
194 H3 **Davis Coast** *physical region* Antarctica
18 C16 **Davis, Mount** ▲ Pennsylvania, NE USA
24 K9 **Davis Mountains** ▲ Texas, SW USA
195 X3 **Davis Sea** *sea* Antarctica
65 O20 **Davis Seamounts** *undersea feature* E Atlantic Ocean
196 M13 **Davis Strait** *strait* Baffin Bay/Labrador Sea
127 V7 **Davlekanovo** Respublika Bashkortostan, W Russian Federation
108 I8 **Davos** *Rmsch.* Tavau. Graubünden, E Switzerland
Davvesiidda *see* Lebesby
119 J20 **Davyd-Haradok** *Pol.* Dawidgródek, *Rus.* David-Gorodok. Brestskaya Voblasts´, SW Belarus
163 U12 **Dawa** Liaoning, NE China
141 O11 **Dawāsir, Wādī ad** *dry watercourse* S Saudi Arabia
81 K15 **Dawa Wenz** *var.* Daua, Webi Daawo. ♒ E Africa
Dawayimah, Birkat ad *see* Umm al Baqar, Hawr
167 N7 **Dawei** *var.* Tavoy, Htawei. (Tenasserim) S Myanmar (Burma) 14°02´N 98°12´E
119 K14 **Dawhinava** *Rus.* Dolginovo. Minskaya Voblasts´, N Belarus
Dawidgródek *see* Davyd-Haradok
141 Y9 **Dawkah** *var.* Dauka. SW Oman 18°33´N 54°02´E
Dawlat Qatar *see* Qatar
11 N15 **Dawson** Yukon, NW Canada 64°04´N 139°24´W
23 S6 **Dawson** Georgia, SE USA 31°46´N 84°27´W
29 S9 **Dawson** Minnesota, N USA 44°55´N 96°03´W
Dawson City *see* Dawson

◆ Country   ◇ Dependent Territory   ◈ Administrative Regions   ▲ Mountain   ⌃ Volcano   ◉ Lake
● Country Capital   ○ Dependent Territory Capital   ✈ International Airport   ▲ Mountain Range   ♒ River   ▣ Reservoir

DAWSON CREEK – DEVOLL

11 N13 **Dawson Creek** British Columbia, W Canada 55°45′N 120°07′W
10 H7 **Dawson Range** Yukon, W Canada
181 Y9 **Dawson River** Queensland, E Australia
10 J15 **Dawsons Landing** British Columbia, SW Canada 51°33′N 127°38′W
20 I7 **Dawson Springs** Kentucky, S USA 37°10′N 87°41′W
23 S2 **Dawsonville** Georgia, SE USA 34°28′N 84°07′W
160 G8 **Dawu** *var.* Xianshui. Sichuan, C China 30°55′N 101°08′E
Dawu *see* Maqên
Dawukou *see* Huinong
141 Y10 **Dawwah** *var.* Dauwa. W Oman 20°36′N 58°52′E
102 J15 **Dax** *var.* Ax; *anc.* Aquae Augustae, Aquae Tarbelicae. Landes, SW France 43°43′N 01°03′W
Daxian *see* Dazhou
Daxiangshan *see* Gangu
Daxue *see* Wencheng
160 G9 **Daxue Shan** C China
Dayan *see* Lijiang
160 G12 **Dayao** *var.* Jinbi. Yunnan, SW China 25°41′N 101°23′E
Dayishan *see* Gaoyou
Dāykundī *see* Dāykundī
149 O6 **Dāykundī** *prev.* Dāykondī. province C Afghanistan
183 N12 **Daylesford** Victoria, SE Australia 37°24′S 144°07′E
35 U10 **Daylight Pass** *pass* California, W USA
61 D17 **Daymán, Río** N Uruguay
Dayong *see* Zhangjiajie
Dayr *see* Ad Dayr
138 G10 **Dayr 'Allā** *var.* Deir 'Alla. Al Balqā', N Jordan 32°39′N 36°06′E
139 N4 **Dayr az Zawr** *var.* Deir ez Zor. Dayr az Zawr, E Syria 35°12′N 40°12′E
138 M5 **Dayr az Zawr** *off.* Muḥāfaẓat Dayr az Zawr, *var.* Dayr Az-Zor. governorate E Syria
Dayr az Zawr, Muḥāfaẓat *see* Dayr az Zawr
Dayr Az-Zor *see* Dayr az Zawr
75 W9 **Dayrūṭ** *var.* Dairûṭ. C Egypt 27°34′N 30°48′E
11 Q15 **Daysland** Alberta, SW Canada 52°53′N 112°19′W
31 R14 **Dayton** Ohio, N USA 39°46′N 84°12′W
20 L10 **Dayton** Tennessee, S USA 35°30′N 85°01′W
25 W11 **Dayton** Texas, SW USA 30°03′N 94°53′W
32 L10 **Dayton** Washington, NW USA 46°19′N 117°58′W
23 X10 **Daytona Beach** Florida, SE USA 29°12′N 81°03′W
169 U12 **Dayu** Borneo, C Indonesia 01°59′S 115°04′E
161 O13 **Dayu Ling** S China
161 R7 **Da Yunhe** *Eng.* Grand Canal. *canal* E China
161 S11 **Dayu Shan** *island* SE China
160 K8 **Dazhou** *prev.* Dachuan, Daxian. Sichuan, C China 31°16′N 107°31′E
160 J9 **Dazhu** *var.* Zhuyang. Sichuan, C China 30°45′N 107°11′E
161 T13 **Dazhuoshui** *prev.* Tachoshui. N Taiwan 24°26′N 121°43′E
160 J9 **Dazu** *var.* Longgang. Chongqing Shi, C China 29°47′N 106°30′E
83 H24 **De Aar** Northern Cape, C South Africa 30°40′S 24°01′E
194 K5 **Deacon, Cape** *headland* Antarctica
39 R5 **Deadhorse** Alaska, USA 70°15′N 148°28′W
33 T12 **Dead Indian Peak** Wyoming, C USA 44°36′N 109°45′W
23 R9 **Dead Lake** Florida, SE USA
44 J4 **Deadman's Cay** Long Island, C The Bahamas 23°09′N 75°06′W
138 G11 **Dead Sea** *var.* Bahret Lut, Lacus Asphaltites, *Ar.* Al Baḥr al Mayyit, Baḥrat Lūt, *Heb.* Yam HaMelaḥ. *salt lake* Israel/Jordan
28 J9 **Deadwood** South Dakota, N USA 44°22′N 103°43′W
97 Q22 **Deal** SE England, United Kingdom 51°14′N 01°23′E
83 I22 **Dealesville** Free State, C South Africa 28°40′S 25°46′E
161 P10 **De'an** *var.* Puting. Jiangxi, S China 29°24′N 115°46′E
62 K9 **Deán Funes** Córdoba, C Argentina 30°25′S 64°22′W
194 L12 **Dean Island** *island* Antarctica
Deanuvuotna *see* Tanafjorden
31 S10 **Dearborn** Michigan, N USA 42°16′N 83°13′W
27 R3 **Dearborn** Missouri, C USA 39°31′N 94°46′W
Deargget *see* Tārendō
32 K9 **Deary** Idaho, NW USA 46°46′N 116°33′W
32 M9 **Deary** Washington, NW USA 46°42′N 116°36′W
10 J10 **Dease** British Columbia, W Canada
10 J10 **Dease Lake** British Columbia, W Canada 58°28′N 130°04′W
35 U11 **Death Valley** California, W USA 36°25′N 116°50′W
35 U11 **Death Valley** *valley* California, W USA
92 M8 **Deatnu** *Fin.* Tenojoki, *Nor.* Tana. Finland/Norway *see also* Tana, Tenojoki
Deatnu *see* Tana
102 L4 **Deauville** Calvados, N France 49°21′N 00°06′E
117 X7 **Debal'tseve** *Rus.* Debal'tsevo. Donets'ka Oblast', SE Ukraine 48°21′N 38°26′E
Debal'tsevo *see* Debal'tseve
113 M19 **Debar** *Ger.* Dibra, *Turk.* Debre. W FYR Macedonia 41°32′N 20°33′E
39 O9 **Debauch Mountain** Alaska, USA 64°31′N 159°52′W
De Behagle *see* Laï
25 X7 **De Berry** Texas, SW USA 32°18′N 94°09′W
Debessy *see* Debesy
127 T2 **Debesy** *prev.* Debessy. Udmurtskaya Respublika, NW Russian Federation 57°41′N 53°56′E
111 N16 **Dębica** Podkarpackie, SE Poland 50°04′N 21°24′E
De Bildt *see* De Bilt
98 J11 **De Bilt** *var.* De Bildt. Utrecht, C Netherlands 52°06′N 05°11′E
123 T9 **Debin** Magadanskaya Oblast', E Russian Federation 62°18′N 150°42′E
110 N13 **Dęblin** *Rus.* Ivangorod. Lubelskie, E Poland 51°34′N 21°50′E
110 D10 **Dębno** Zachodnio-pomorskie, NW Poland 52°43′N 14°42′E
39 S10 **Deborah, Mount** Alaska, USA 63°38′N 147°13′W
33 N8 **De Borgia** Montana, NW USA 47°23′N 115°24′W
Debra Birhan *see* Debre Birhan
Debra Marcos *see* Debre Mark'os
Debra Tabor *see* Debre Tabor
Debre *see* Debar
80 J13 **Debre Birhan** *var.* Debra Birhan. Āmara, N Ethiopia 09°45′N 39°40′E
111 N22 **Debrecen** *Ger.* Debreczin, *Rom.* Debreţin; *prev.* Debreczen. Hajdú-Bihar, E Hungary 47°32′N 21°38′E
Debreczen/Debreczin *see* Debrecen
80 I12 **Debre Mark'os** *var.* Debra Marcos. Āmara, N Ethiopia 10°18′N 37°48′E
113 N19 **Debrešte** SW FYR Macedonia 41°29′N 21°20′E
80 J11 **Debre Tabor** *var.* Debra Tabor. Āmara, N Ethiopia 11°46′N 38°06′E
Debreţin *see* Debrecen
80 J13 **Debre Zeyt** Oromiya, C Ethiopia 08°41′N 39°00′E
113 L16 **Deçan** *Serb.* Dečane; *prev.* Dečani. W Kosovo 42°33′N 20°18′E
Dečane *see* Deçan
Dečani *see* Deçan
23 P2 **Decatur** Alabama, S USA 34°36′N 86°58′W
23 S3 **Decatur** Georgia, SE USA 33°46′N 84°18′W
30 L13 **Decatur** Illinois, N USA 39°50′N 88°57′W
31 Q12 **Decatur** Indiana, N USA 40°50′N 84°57′W
22 M5 **Decatur** Mississippi, S USA 32°26′N 89°06′W
29 S14 **Decatur** Nebraska, C USA 42°00′N 96°19′W
25 S6 **Decatur** Texas, SW USA 33°14′N 97°35′W
20 H9 **Decaturville** Tennessee, S USA 35°35′N 88°08′W
103 O13 **Decazeville** Aveyron, S France 44°34′N 02°18′E
155 H17 **Deccan** *Hind.* Dakshin. *plateau* C India
14 J8 **Decelles, Réservoir** Québec, SE Canada
12 K2 **Déception** Québec, NE Canada 62°06′N 74°36′W
160 G11 **Dechang** *var.* Dezhou. Sichuan, C China 27°24′N 102°09′E
111 C15 **Děčín** *Ger.* Tetschen. Ústecký Kraj, NW Czech Republic 50°48′N 14°15′E
103 P9 **Decize** Nièvre, C France 46°51′N 03°25′E
98 I6 **De Cocksdorp** Noord-Holland, NW Netherlands 53°09′N 04°52′E
29 X11 **Decorah** Iowa, C USA 43°18′N 91°47′W
Dedeagaç/Dedeagach *see* Alexandroúpoli
188 C15 **Dededo** N Guam 13°30′N 144°51′E
98 N9 **Dedemsvaart** Overijssel, E Netherlands 52°36′N 06°28′E
19 O11 **Dedham** Massachusetts, NE USA 42°14′N 71°10′W
63 H19 **Dedo, Cerro** SW Argentina 44°46′S 71°48′W
77 O13 **Dédougou** W Burkina Faso 12°29′N 03°25′W
124 G15 **Dedovichi** Pskovskaya Oblast', W Russian Federation 57°31′N 29°53′E
Dedu *see* Wudalianchi
155 J24 **Deduru Oya** W Sri Lanka
83 N14 **Dedza** Central, S Malawi 14°20′S 34°24′E
83 N14 **Dedza Mountain** C Malawi 14°22′S 34°16′E
96 K9 **Dee** NE Scotland, United Kingdom
97 J19 **Dee** *Wel.* Afon Dyfrdwy. England/Wales, United Kingdom
Deep Bay *see* Chilumba
21 T3 **Deep Creek Lake** Maryland, NE USA
36 J4 **Deep Creek Range** Utah, W USA
27 P10 **Deep Fork River** Oklahoma, C USA
14 J11 **Deep River** Ontario, SE Canada 46°04′N 77°29′W
21 T10 **Deep River** North Carolina, SE USA
183 U4 **Deepwater** New South Wales, SE Australia 29°27′S 151°52′E
31 S14 **Deer Creek Lake** Ohio, N USA
23 Z15 **Deerfield Beach** Florida, SE USA 26°19′N 80°06′W
39 N8 **Deering** Alaska, USA 66°04′N 162°43′W
38 M16 **Deer Island** *island* Alaska, USA
19 S7 **Deer Isle** *island* Maine, NE USA
13 S11 **Deer Lake** Newfoundland and Labrador, SE Canada 49°11′N 57°27′W
99 D18 **Deerlijk** West-Vlaanderen, W Belgium 50°52′N 03°21′E
33 Q10 **Deer Lodge** Montana, NW USA 46°24′N 112°43′W
32 L8 **Deer Park** Washington, NW USA 47°55′N 117°28′W
29 U5 **Deer River** Minnesota, N USA 47°19′N 93°47′W
Deés *see* Dej
Defeng *see* Liping
31 R11 **Defiance** Ohio, N USA 41°17′N 84°21′W
23 Q8 **De Funiak Springs** Florida, SE USA 30°43′N 86°07′W
95 L23 **Degeberga** Skåne, S Sweden 55°48′N 14°06′E
104 H12 **Degebe, Ribeira** S Portugal
80 M13 **Degeh Bur** Sumalē, E Ethiopia 08°08′N 43°35′E
15 U9 **Dégelis** Québec, SE Canada 47°30′N 68°38′W
77 U17 **Degema** Rivers, S Nigeria 04°46′N 06°47′E
95 L16 **Degerfors** Örebro, C Sweden 59°14′N 14°26′E
193 R14 **De Gerlache Seamounts** *undersea feature* SE Pacific Ocean
101 N21 **Deggendorf** Bayern, SE Germany 48°50′N 12°58′E
80 I11 **Degoma** Āmara, N Ethiopia 12°22′N 37°36′E
De Gordyk *see* Gorredijk
27 T12 **De Gray Lake** Arkansas, C USA
180 J6 **De Grey River** Western Australia
126 M10 **Degtevo** Rostovskaya Oblast', SW Russian Federation 49°12′N 40°39′E
Dehbārez *see* Rūdān
Deh Bīd *see* Şafāshahr
142 M10 **Deh Dasht** Kohkīlūyeh va Būyer AḤmad, SW Iran 30°49′N 50°36′E
75 N8 **Dehibat** SE Tunisia 31°58′N 10°43′E
Dehli *see* Delhi
142 K8 **Dehlorān** Īlām, W Iran 32°41′N 47°18′E
147 N13 **Dehqonobod** *Rus.* Dekhkanabad. Qashqadaryo Viloyati, S Uzbekistan 38°24′N 66°31′E
152 J9 **Dehra Dūn** Uttaranchal, N India 30°19′N 78°04′E
153 O14 **Dehri** Bihār, N India 24°55′N 84°11′E
Deh Shū *see* Dīshū
163 W9 **Dehui** Jilin, NE China 44°23′N 125°42′E
99 D17 **Deinze** Oost-Vlaanderen, NW Belgium 50°59′N 03°32′E
Deir 'Alla *see* Dayr 'Allā
Deir ez Zor *see* Dayr az Zawr
Deirgeirt, Loch *see* Derg, Lough
116 H9 **Dej** *Hung.* Dés; *prev.* Deés. Cluj, NW Romania 47°08′N 23°55′E
95 K15 **Deje** Värmland, C Sweden 59°35′N 13°29′E
171 Y15 **De Jongs, Tanjung** *headland* Papua, SE Indonesia 06°56′S 138°32′E
De Jouwer *see* Joure
30 M10 **De Kalb** Illinois, N USA 41°55′N 88°45′W
22 M5 **De Kalb** Mississippi, S USA 32°46′N 88°39′W
25 W5 **De Kalb** Texas, SW USA 33°30′N 94°34′W
83 G18 **Dekar** *var.* D'Kar. Ghanzi, NW Botswana 21°31′S 21°55′E
Dekéleia *see* Dhekéleia
79 K20 **Dekese** Kasai-Occidental, C Dem. Rep. Congo 03°28′S 21°24′E
Dekhkanabad *see* Dehqonobod
79 I14 **Dékoa** Kémo, C Central African Republic 06°17′N 19°07′E
98 H6 **De Koog** Noord-Holland, NW Netherlands 53°06′N 04°43′E
30 M9 **Delafield** Wisconsin, N USA 43°03′N 88°22′W
61 C23 **De La Garma** Buenos Aires, E Argentina 37°58′S 60°25′W
14 K10 **Delahey, Lac** Québec, SE Canada
80 E11 **Delami** Southern Kordofan, C Sudan 11°51′N 30°30′E
23 X11 **De Land** Florida, SE USA 29°01′N 81°18′W
35 R12 **Delano** California, W USA 35°46′N 119°15′W
29 V8 **Delano** Minnesota, N USA 45°03′N 93°46′W
36 K6 **Delano Peak** Utah, W USA 38°22′N 112°21′W
Delap-Uliga-Darrit *see* Dalap-Uliga-Djarrit
Delārām *see* Dilārām
38 F17 **Delarof Islands** *island group* Aleutian Islands, Alaska, USA
30 M9 **Delavan** Wisconsin, N USA 42°37′N 88°37′W
31 S13 **Delaware** Ohio, N USA 40°18′N 83°06′W
18 I17 **Delaware** *off.* State of Delaware, *also known as* Blue Hen State, Diamond State, First State. *state* NE USA
18 I17 **Delaware Bay** *bay* NE USA
24 J8 **Delaware Mountains** Texas, SW USA
18 I12 **Delaware River** NE USA
27 Q3 **Delaware River** Kansas, C USA
18 J14 **Delaware Water Gap** *valley* New Jersey/Pennsylvania, NE USA
101 G14 **Delbrück** Nordrhein-Westfalen, W Germany 51°46′N 08°34′E
11 Q15 **Delburne** Alberta, SW Canada 52°09′N 113°11′W
172 M12 **Del Cano Rise** *undersea feature* SW Indian Ocean 45°15′S 44°15′E
113 Q18 **Delčevo** NE FYR Macedonia 41°57′N 22°45′E
Delcommune, Lac *see* Nzilo, Lac
98 O10 **Delden** Overijssel, E Netherlands 52°16′N 06°41′E
183 R12 **Delegate** New South Wales, SE Australia 37°04′S 148°57′E
De Lemmer *see* Lemmer
108 D7 **Delémont** *Ger.* Delsberg. Jura, NW Switzerland 47°22′N 07°21′E
25 R7 **De Leon** Texas, SW USA 32°06′N 98°33′W
115 F18 **Delfoí** Stereá Elláda, C Greece 38°28′N 22°31′E
98 G12 **Delft** Zuid-Holland, W Netherlands 52°01′N 04°22′E
155 J23 **Delft** *island* NW Sri Lanka
98 O5 **Delfzijl** Groningen, NE Netherlands 53°20′N 06°55′E
0 E9 **Delgada Fan** *undersea feature* NE Pacific Ocean 39°15′N 126°00′W
42 F7 **Delgado** San Salvador, SW El Salvador 13°44′N 89°07′W
82 Q12 **Delgado, Cabo** *headland* N Mozambique 10°41′S 40°40′E
162 G8 **Delger** *var.* Taygan. Govĭ-Altay, C Mongolia 46°20′N 97°22′E
163 O9 **Delgereh** *var.* Hongor. Dornogovĭ, SE Mongolia 45°49′N 111°20′E
162 J8 **Delgerhaan** *var.* Hujirt. Töv, C Mongolia 46°41′N 104°40′E
162 K9 **Delgerhangay** *var.* Hashaat. Dundgovĭ, C Mongolia 45°09′N 104°51′E
162 L9 **Delgertsogt** *var.* Amardalay. Dundgovĭ, C Mongolia 46°09′N 106°24′E
80 E6 **Delgo** Northern, N Sudan 20°08′N 30°35′E
159 R10 **Delhi** *var.* Delingha. Qinghai, C China 37°19′N 97°22′E
152 I10 **Delhi** *var.* Delhi, *Hind.* Dilli, *hist.* Shahjahanabad. *union territory capital* Delhi, N India 28°40′N 77°11′E
22 J5 **Delhi** Louisiana, S USA 32°28′N 91°29′W
18 J11 **Delhi** New York, NE USA 42°16′N 74°55′W
152 I10 **Delhi** *union territory* NW India
136 J17 **Deli Burnu** *headland* S Turkey 36°43′N 34°55′E
136 J12 **Delice Çayı** C Turkey
55 X10 **Délices** C French Guiana 04°45′N 53°45′W
40 J6 **Delicias** *var.* Ciudad Delicias. Chihuahua, N Mexico 28°09′N 105°22′W
143 N7 **Delījān** *var.* Dalijan, Dilijan. Markazī, W Iran 34°02′N 50°39′E
112 P12 **Deli Jovan** E Serbia
Déli-Kárpátok *see* Carpaţii Meridionali
8 I8 **Déline** *prev.* Fort Franklin. Northwest Territories, NW Canada 65°10′N 123°30′W
Delingha *see* Delhi
15 Q7 **Delisle** Québec, SE Canada 48°39′N 71°42′W
11 T15 **Delisle** Saskatchewan, S Canada 51°54′N 107°01′W
101 M15 **Delitzsch** Sachsen, E Germany 51°31′N 12°19′E
33 Q12 **Dell** Montana, NW USA 44°41′N 112°42′W
24 I7 **Dell City** Texas, SW USA 31°56′N 105°12′W
103 U7 **Delle** Territoire-de-Belfort, E France 47°30′N 07°00′E
36 J9 **Dellenbaugh, Mount** Arizona, SW USA 36°06′N 113°32′W
29 R11 **Dell Rapids** South Dakota, N USA 43°50′N 96°42′W
21 Y4 **Delmar** Maryland, NE USA 38°26′N 75°32′W
18 K11 **Delmar** New York, NE USA 42°37′N 73°49′W
100 G11 **Delmenhorst** Niedersachsen, NW Germany 53°03′N 08°38′E
112 C9 **Delnice** Primorje-Gorski Kotar, NW Croatia 45°24′N 14°49′E
37 R7 **Del Norte** Colorado, C USA 37°40′N 106°21′W
39 N6 **De Long Mountains** Alaska, USA
183 P16 **Deloraine** Tasmania, SE Australia 41°34′S 146°43′E
11 W17 **Deloraine** Manitoba, S Canada 49°12′N 100°28′W
31 O12 **Delphi** Indiana, N USA 40°35′N 86°40′W
31 Q12 **Delphos** Ohio, N USA 40°49′N 84°20′W
23 Z15 **Delray Beach** Florida, SE USA 26°27′N 80°04′W
25 O12 **Del Rio** Texas, SW USA 29°23′N 100°56′W
Delsberg *see* Delémont
94 N11 **Delsbo** Gävleborg, C Sweden 61°49′N 16°34′E
37 P6 **Delta** Colorado, C USA 38°44′N 108°04′W
36 K5 **Delta** Utah, W USA 39°21′N 112°34′W
77 T17 **Delta** *state* S Nigeria
55 Q6 **Delta Amacuro** *off.* Territorio Delta Amacuro. *federal district* NE Venezuela
Delta Amacuro, Territorio *see* Delta Amacuro
39 S9 **Delta Junction** Alaska, USA 64°02′N 145°43′W
23 X11 **Deltona** Florida, SE USA 28°54′N 81°15′W
183 T5 **Delungra** New South Wales, SE Australia 29°40′S 150°49′E
162 D6 **Delüün** *var.* Rashaant. Bayan-Ölgiy, W Mongolia 47°48′N 90°45′E
154 C12 **Delvāda** Gujarāt, W India 20°46′N 71°02′E
61 B21 **Del Valle** Buenos Aires, E Argentina 35°55′S 60°42′W
Delvina *see* Delvinë
115 C15 **Delvináki** *var.* Dhelvinákion; *prev.* Pogónion. Ípeiros, W Greece 39°57′N 20°28′E
113 L23 **Delvinë** *var.* Delvina, *It.* Delvino. Vlorë, S Albania 39°56′N 20°07′E
Delvino *see* Delvinë
116 I7 **Delyatyn** Ivano-Frankivs'ka Oblast', W Ukraine 48°32′N 24°38′E
127 U5 **Dëma** W Russian Federation
105 O5 **Demanda, Sierra de la** W Spain
39 T5 **Demarcation Point** *headland* Alaska, USA 69°40′N 141°19′W
79 K21 **Demba** Kasai-Occidental, C Dem. Rep. Congo 05°24′S 22°16′E
172 H13 **Dembéni** Grande Comore, NW Comoros 11°50′S 43°25′E
79 M15 **Dembia** Mbomou, SE Central African Republic 05°08′N 24°25′E
Dembidollo *see* Dembī Dolo
80 H13 **Dembī Dolo** *var.* Dembidollo. Oromīya, C Ethiopia 08°33′N 34°49′E
152 K6 **Demchok** *var.* Dêmqog. China/India 32°30′N 79°42′E *see also* Dêmqog
152 L6 **Demchok** *var.* Dêmqog. *disputed region* China/India *see also* Dêmqog
98 I12 **De Meern** Utrecht, C Netherlands 52°06′N 05°00′E
99 I17 **Demer** C Belgium
64 H12 **Demerara Plain** *undersea feature* W Atlantic Ocean 10°00′N 48°00′W
64 H12 **Demerara Plateau** *undersea feature* W Atlantic Ocean
55 T9 **Demerara River** NE Guyana
126 H3 **Demidov** Smolenskaya Oblast', W Russian Federation 55°15′N 31°30′E
37 Q15 **Deming** New Mexico, SW USA 32°17′N 107°46′W
32 H6 **Deming** Washington, NW USA 48°49′N 122°13′W
58 E10 **Demini, Rio** NW Brazil
136 D13 **Demirci** Manisa, W Turkey 39°03′N 28°40′E
113 P19 **Demir Kapija** *prev.* Železna Vrata. SE FYR Macedonia 41°25′N 22°15′E
114 N11 **Demirköy** Kırklareli, NW Turkey 41°48′N 27°49′E
100 N9 **Demmin** Mecklenburg-Vorpommern, NE Germany 53°53′N 13°03′E
23 O5 **Demopolis** Alabama, S USA 32°31′N 87°50′W
31 N11 **Demotte** Indiana, N USA 41°13′N 87°07′W
158 F13 **Dêmqog** *var.* Demchok. China/India 32°36′N 79°29′E *see also* Demchok
152 L6 **Dêmqog** *var.* Demchok. *disputed region* China/India *see also* Demchok
171 Y13 **Demta** Papua, E Indonesia 02°19′S 140°06′E
121 K11 **Dem'yanka** C Russian Federation
124 H15 **Demyansk** Novgorodskaya Oblast', W Russian Federation 57°39′N 32°31′E
122 H10 **Dem'yanskoye** Tyumenskaya Oblast', C Russian Federation 59°39′N 69°15′E
103 P2 **Denain** Nord, N France 50°19′N 03°24′E
39 S10 **Denali** Alaska, USA 63°08′N 147°33′W
Denali *see* McKinley, Mount
81 M14 **Denan** Sumalē, E Ethiopia 06°40′N 43°31′E
Denau *see* Denov
97 J18 **Denbigh** *Wel.* Dinbych. NE Wales, United Kingdom 53°11′N 03°25′W
97 J18 **Denbigh** *cultural region* N Wales, United Kingdom
98 I6 **Den Burg** Noord-Holland, NW Netherlands 53°03′N 04°47′E
99 F18 **Dender** *Fr.* Dendre. W Belgium
99 F18 **Denderleeuw** Oost-Vlaanderen, NW Belgium 50°53′N 04°05′E
99 F17 **Dendermonde** *Fr.* Termonde. Oost-Vlaanderen, NW Belgium 51°02′N 04°08′E
Dendre *see* Dender
194 I9 **Dendtler Island** *island* Antarctica
98 P10 **Denekamp** Overijssel, E Netherlands 52°23′N 07°E
77 W12 **Dengas** Zinder, S Niger 13°15′N 09°43′E
Dêngka *see* Têwo
Dêngkagoin *see* Têwo
162 L13 **Dengkou** *var.* Bayan Gol. Nei Mongol Zizhiqu, N China 40°15′N 106°59′E
159 Q14 **Dêngqên** *var.* Gyamotang. Xizang Zizhiqu, W China 31°28′N 95°28′E
Deng Xian *see* Dengzhou
160 M7 **Dengzhou** *prev.* Deng Xian. Henan, C China 32°48′N 111°59′E
Dengzhou *see* Penglai
Den Haag *see* 's-Gravenhage
180 H10 **Denham** Western Australia 25°56′S 113°35′E
98 N9 **Den Ham** Overijssel, E Netherlands 52°30′N 06°31′E
44 J12 **Denham, Mount** C Jamaica 18°13′N 77°33′W
22 J8 **Denham Springs** Louisiana, S USA 30°29′N 90°57′W
98 I7 **Den Helder** Noord-Holland, NW Netherlands 52°54′N 04°45′E
105 T11 **Dénia** Valenciana, E Spain 38°51′N 00°07′E
189 Q8 **Denig** W Nauru
183 N10 **Deniliquin** New South Wales, SE Australia 35°33′S 144°58′E
29 T14 **Denison** Iowa, C USA 42°00′N 95°22′W
25 U5 **Denison** Texas, SW USA 33°45′N 96°32′W
144 L8 **Denisovka** *prev.* Ordzhonikidze. Kostanay, N Kazakhstan 52°27′N 61°42′E
136 D15 **Denizli** Denizli, SW Turkey 37°46′N 29°05′E
136 D15 **Denizli** province SW Turkey
Denjong *see* Sikkim
183 S7 **Denman** New South Wales, SE Australia 32°24′S 150°43′E
195 Y10 **Denman Glacier** *glacier* Antarctica
21 R14 **Denmark** South Carolina, SE USA 33°19′N 81°08′W
95 G23 **Denmark** *off.* Kingdom of Denmark, *Dan.* Danmark; *anc.* Hafnia. *monarchy* N Europe
Denmark, Kingdom of *see* Denmark
92 H1 **Denmark Strait** *var.* Danmarksstraedet. *strait* Greenland/Iceland
45 T11 **Dennery** E Saint Lucia 13°55′N 60°53′W
98 I7 **Den Oever** Noord-Holland, NW Netherlands 52°56′N 05°01′E
147 O13 **Denov** *Rus.* Denau. Surkhondaryo Viloyati, S Uzbekistan 38°20′N 67°48′E
169 U17 **Denpasar** *prev.* Paloe. Bali, C Indonesia 08°40′S 115°14′E
116 E12 **Denta** Timiş, W Romania 45°20′N 21°15′E
21 Y3 **Denton** Maryland, NE USA 38°53′N 75°50′W
25 T6 **Denton** Texas, SW USA 33°11′N 97°08′W
186 G9 **D'Entrecasteaux Islands** *island group* SE Papua New Guinea
37 T4 **Denver** *state capital* Colorado, C USA 39°45′N 105°00′W
37 T4 **Denver** Colorado, C USA 39°50′N 104°40′W
24 L6 **Denver City** Texas, SW USA 32°57′N 102°49′W
152 J9 **Deoband** Uttar Pradesh, N India 29°41′N 77°40′E
Deoghar *see* Devghar
163 X15 **Deokjeok-gundo** *prev.* Tŏkchŏk-kundo. *island group* NW South Korea
154 E13 **Deolāli** Mahārāshtra, W India 19°55′N 73°49′E
154 I10 **Deori** Madhya Pradesh, C India 23°09′N 78°39′E
153 O12 **Deoria** Uttar Pradesh, N India 26°31′N 83°48′E
99 A17 **De Panne** West-Vlaanderen, W Belgium 51°06′N 02°35′E
Departamento del Quindío *see* Quindío
Departamento de Narino, *see* Nariño
54 M5 **Dependencia Federal** *off.* Territorio Dependencia Federal. *federal dependency* N Venezuela
Dependencia Federal, Territorio *see* Dependencia Federal
30 M7 **De Pere** Wisconsin, N USA 44°26′N 88°03′W
18 D10 **Depew** New York, NE USA 42°54′N 78°41′W
99 E17 **De Pinte** Oost-Vlaanderen, NW Belgium 51°00′N 03°37′E
25 V5 **Deport** Texas, SW USA 33°31′N 95°19′W
123 Q8 **Deputatskiy** Respublika Sakha (Yakutiya), NE Russian Federation 69°18′N 139°48′E
Dêqên *see* Dagzê
27 S13 **De Queen** Arkansas, C USA 34°02′N 94°20′W
22 G8 **De Quincy** Louisiana, S USA 30°27′N 93°25′W
81 J20 **Dera** *spring/well* S Kenya 02°39′S 39°52′E
Der'a/Derá/Déraa *see* Dar'ā
149 S10 **Dera Ghāzi Khān** *var.* Dera Ghāzikhān. Punjab, C Pakistan 30°01′N 70°37′E
Dera Ghāzikhān *see* Dera Ghāzi Khān
149 S8 **Dera Ismāīl Khān** Khyber Pakhtunkhwa, C Pakistan 31°51′N 70°56′E
Đeravica *see* Gjeravicë
116 L6 **Derazhnya** Khmel'nyts'ka Oblast', W Ukraine 49°16′N 27°24′E
127 R17 **Derbent** Respublika Dagestan, SW Russian Federation 42°01′N 48°16′E
147 N13 **Derbent** Surkhondaryo Viloyati, S Uzbekistan 38°15′N 66°59′E
Derbisīye *see* Darbāsīyah
79 M15 **Derbissaka** Mbomou, SE Central African Republic 05°43′N 24°48′E
180 L4 **Derby** Western Australia 17°18′S 123°37′E
97 M19 **Derby** C England, United Kingdom 52°55′N 01°30′W
27 N7 **Derby** Kansas, C USA 37°33′N 97°16′W
97 L18 **Derbyshire** *cultural region* C England, United Kingdom
112 O11 **Đerdap** *physical region* E Serbia
Dereli *see* Gönoi
162 L9 **Deren** *var.* Tsant. Dundgovĭ, C Mongolia 46°16′N 106°55′E
171 W13 **Derew** Papua, E Indonesia
127 R8 **Dergachi** Saratovskaya Oblast', W Russian Federation 51°15′N 48°58′E
Dergachi *see* Derhachi
97 C19 **Derg, Lough** *Ir.* Loch Deirgeirt. W Ireland
117 V5 **Derhachi** *Rus.* Dergachi. Kharkivs'ka Oblast', E Ukraine 50°09′N 36°11′E
22 G8 **De Ridder** Louisiana, S USA 30°51′N 93°18′W
137 P16 **Derik** Mardin, SE Turkey 37°22′N 40°16′E
83 E20 **Derm** Hardap, C Namibia 23°38′S 18°12′E
Dermentobe *see* Diirmentobe
27 W14 **Dermott** Arkansas, C USA 33°31′N 91°26′W
Dérna *see* Darnah
Dernberg, Cape *see* Dolphin Head
22 J11 **Dernieres, Isles** *island group* Louisiana, S USA
102 I4 **Déroute, Passage de la** *strait* Channel Islands/France
Derrá *see* Dar'ā
81 O16 **Derri** *prev.* Dirri. Galguduud, C Somalia 04°15′N 46°31′E
Derry *see* Londonderry
Dertona *see* Tortona
Dertosa *see* Tortosa
80 H8 **Derudeb** Red Sea, NE Sudan 17°31′N 36°07′E
112 H10 **Derventa** Republika Srpska, N Bosnia and Herzegovina 44°57′N 17°55′E
183 O16 **Derwent Bridge** Tasmania, SE Australia 42°10′S 146°13′E
183 O17 **Derwent, River** Tasmania, SE Australia
146 F10 **Derweze** *Rus.* Darvaza. Ahal Welaýaty, C Turkmenistan 40°10′N 58°27′E
Deržavinsk *see* Derzhavinsk
145 O9 **Derzhavinsk** *var.* Deržavinsk. Akmola, C Kazakhstan 51°06′N 66°15′E
Dés *see* Dej
57 J18 **Desaguadero** Puno, S Peru 16°35′S 69°05′W
57 J18 **Desaguadero, Río** Bolivia/Peru
191 W9 **Désappointement, Îles du** *island group* Îles Tuamotu, C French Polynesia
27 W11 **Des Arc** Arkansas, C USA 34°58′N 91°30′W
14 C10 **Desbarats** Ontario, S Canada 46°20′N 83°52′W
62 H13 **Descabezado Grande, Volcán** C Chile 35°34′S 70°40′W
102 L9 **Descartes** Indre-et-Loire, C France 46°58′N 00°40′E
11 T13 **Deschambault Lake** Saskatchewan, C Canada
Deschnaer Koppe *see* Velká Deštná
32 I11 **Deschutes River** Oregon, NW USA
80 J12 **Desē** *var.* Desse, *It.* Dessie. Āmara, N Ethiopia 11°02′N 39°39′E
63 I20 **Deseado, Río** S Argentina
106 F8 **Desenzano del Garda** Lombardia, N Italy 45°28′N 10°31′E
36 K3 **Deseret Peak** Utah, W USA 40°27′N 112°37′W
64 P6 **Deserta Grande** *island* Madeira, Portugal, NE Atlantic Ocean
64 P6 **Desertas, Ilhas** *island group* Madeira, Portugal, NE Atlantic Ocean
35 X16 **Desert Center** California, W USA 33°42′N 115°22′W
35 V15 **Desert Hot Springs** California, W USA 33°57′N 116°33′W
14 K10 **Désert, Lac** Québec, SE Canada
36 J2 **Desert Peak** Utah, W USA 41°03′N 113°22′W
31 R11 **Deshler** Ohio, N USA 41°12′N 83°55′W
Deshu *see* Dīshū
Desiderii Fanum *see* St-Dizier
106 D7 **Desio** Lombardia, N Italy 45°37′N 09°12′E
115 E15 **Deskáti** *var.* Dheskáti. Dytikí Makedonía, N Greece 39°55′N 21°49′E
28 L2 **Des Lacs River** North Dakota, N USA
27 X6 **Desloge** Missouri, C USA 37°52′N 90°31′W
11 Q12 **Desmarais** Alberta, W Canada 55°58′N 113°56′W
29 Q10 **De Smet** South Dakota, N USA 44°23′N 97°33′W
29 V14 **Des Moines** *state capital* Iowa, C USA 41°36′N 93°37′W
17 N9 **Des Moines River** C USA
117 P4 **Desna** Russian Federation/Ukraine
116 G14 **Desnăţui** S Romania
63 F24 **Desolación, Isla** *island* S Chile
29 V14 **De Soto** Iowa, C USA 41°31′N 94°00′W
23 Q4 **De Soto Falls** *waterfall* Alabama, S USA
83 I25 **Despatch** Eastern Cape, S South Africa 33°48′S 25°28′E
105 N12 **Despeñaperros, Desfiladero de** *pass* S Spain
31 N10 **Des Plaines** Illinois, N USA 42°01′N 87°52′W
115 J21 **Despotikó** *island* Kykládes, Greece, Aegean Sea
112 N12 **Despotovac** Serbia, E Serbia 44°06′N 21°25′E
101 M14 **Dessau** Sachsen-Anhalt, E Germany 51°51′N 12°15′E
Desse *see* Desē
99 J16 **Dessel** Antwerpen, N Belgium 51°15′N 05°07′E
Dessie *see* Desē
Destêrro *see* Florianópolis
23 P9 **Destin** Florida, SE USA 30°23′N 86°30′W
Deštná *see* Velká Deštná
193 T10 **Desventurados, Islas de los** *island group* W Chile
103 N1 **Desvres** Pas-de-Calais, N France 50°41′N 01°48′E
116 E12 **Deta** *Ger.* Detta. Timiş, W Romania 45°24′N 21°14′E
101 H14 **Detmold** Nordrhein-Westfalen, W Germany 51°55′N 08°52′E
31 S10 **Detroit** Michigan, N USA 42°20′N 83°03′W
25 W5 **Detroit** Texas, SW USA 33°39′N 95°16′W
31 S10 **Detroit** Canada/USA
29 S6 **Detroit Lakes** Minnesota, N USA 46°49′N 95°49′W
31 S10 **Detroit Metropolitan** Michigan, N USA 42°12′N 83°16′W
Detta *see* Deta
167 S10 **Det Udom** Ubon Ratchathani, E Thailand 14°54′N 105°03′E
111 K20 **Detva** *Hung.* Gyeva. Bankobýstrický Kraj, C Slovakia 48°35′N 19°25′E
154 G13 **Deūlgaon Rāja** Mahārāshtra, C India 20°04′N 76°08′E
99 L15 **Deurne** Noord-Brabant, SE Netherlands 51°28′N 05°47′E
99 H16 **Deurne** (Antwerpen) Antwerpen, N Belgium 51°10′N 04°28′E
Deutsch-Brod *see* Havlíčkův Brod
Deutschendorf *see* Poprad
Deutsch-Eylau *see* Iława
109 Y6 **Deutschkreutz** Burgenland, E Austria 47°37′N 16°37′E
Deutsch Krone *see* Wałcz
Deutschland/Deutschland, Bundesrepublik *see* Germany
109 V9 **Deutschlandsberg** Steiermark, SE Austria 46°52′N 15°13′E
Deutsch-Südwestafrika *see* Namibia
109 Y3 **Deutsch-Wagram** Niederösterreich, E Austria 48°19′N 16°33′E
Deux-Ponts *see* Zweibrücken
14 I11 **Deux Rivieres** Ontario, SE Canada 46°13′N 78°16′W
102 K9 **Deux-Sèvres** *department* W France
116 G11 **Deva** *Ger.* Diemrich, *Hung.* Déva. Hunedoara, W Romania 45°55′N 22°55′E
Déva *see* Deva
Deva *see* Chester
Devana *see* Aberdeen
Devana Castra *see* Chester
Devdelija *see* Gevgelija
136 L12 **Deveci Dağları** N Turkey
137 P15 **Devegeçidi Barajı** SE Turkey
136 K15 **Develi** Kayseri, C Turkey 38°22′N 35°28′E
98 M11 **Deventer** Overijssel, E Netherlands 52°15′N 06°10′E
15 O10 **Devenyns, Lac** Québec, SE Canada
96 K8 **Deveron** NE Scotland, United Kingdom
153 R14 **Devghar** *prev.* Deoghar. Jhārkhand, NE India 24°31′N 86°42′E
27 R10 **Devil's Den** *plateau* Arkansas, C USA
35 R7 **Devils Gate** *pass* California, W USA
30 J2 **Devils Island** *island* Apostle Islands, Wisconsin, N USA
Devil's Island *see* Diable, Île du
29 P3 **Devils Lake** North Dakota, N USA 48°08′N 98°50′W
31 R10 **Devils Lake** Michigan, N USA
29 O3 **Devils Lake** North Dakota, N USA
35 W13 **Devils Playground** *desert* California, W USA
25 O11 **Devils River** Texas, SW USA
33 Y12 **Devils Tower** Wyoming, C USA 44°33′N 104°45′W
114 I11 **Devin** *prev.* Dovlen. Smolyan, S Bulgaria 41°45′N 24°24′E
25 R12 **Devine** Texas, SW USA 29°08′N 98°54′W
152 H13 **Devli** Rājasthān, N India 25°47′N 75°23′E
Devne *see* Devnya
114 N8 **Devnya** *prev.* Devne. Varna, E Bulgaria 43°15′N 27°35′E
31 U14 **Devola** Ohio, N USA 39°28′N 81°28′W
Devoll *see* Devollit, Lumi i

◆ Country ● Country Capital ◇ Dependent Territory ○ Dependent Territory Capital ◈ Administrative Regions ✈ International Airport ▲ Mountain ▲ Mountain Range Volcano River Lake Reservoir

113 M21 **Devollit, Lumi i** *var.* Devoll. ≈ SE Albania
11 Q14 **Devon** Alberta, SW Canada *53°21′N 113°47′W*
97 I23 **Devon** *cultural region* SW England, United Kingdom
197 N10 **Devon Island** *prev.* North Devon Island. *island* Parry Islands, Nunavut, NE Canada
183 O16 **Devonport** Tasmania, SE Australia *41°14′S 146°21′E*
136 H11 **Devrek** Zonguldak, N Turkey *41°14′N 31°57′E*
154 G10 **Dewās** Madhya Pradesh, C India *22°59′N 76°03′E*
De Westerein *see* Zwaagwesteinde
27 P8 **Dewey** Oklahoma, C USA *36°48′N 95°56′W*
Dewey *see* Culebra
98 M8 **De Wijk** Drenthe, NE Netherlands *52°41′N 06°13′E*
27 W12 **De Witt** Arkansas, C USA *34°17′N 91°21′W*
29 Z14 **De Witt** Iowa, C USA *41°49′N 90°32′W*
29 R16 **De Witt** Nebraska, C USA *40°23′N 96°55′W*
97 M17 **Dewsbury** N England, United Kingdom *53°42′N 01°37′W*
161 Q10 **Dexing** Jiangxi, S China *28°51′N 117°36′E*
27 Y8 **Dexter** Missouri, C USA *36°48′N 89°57′W*
37 U14 **Dexter** New Mexico, SW USA *33°12′N 104°25′W*
160 I8 **Deyang** Sichuan, C China *31°08′N 104°23′E*
182 C4 **Dey-Dey, Lake** *salt lake* South Australia
143 S7 **Deyhūk** Khorāsān-e Jonūbī, E Iran *33°18′N 57°30′E*
Deynau *see* Galkynyş
Deyyer *see* Bandar-e Dayyer
142 L8 **Dezfūl** *var.* Dizful. Khūzestān, SW Iran *32°23′N 48°28′E*
129 X4 **Dezhneva, Mys** *headland* NE Russian Federation *66°08′N 69°40′W*
161 P4 **Dezhou** Shandong, E China *37°28′N 116°18′E*
Dezhou *see* Dechang
Dezh Shāhpūr *see* Marīvān
Dhaalu Atoll *see* South Nilandhe Atoll
Dhahran *see* Aẓ Ẓahrān
Dhahran Al Khobar *see* Aẓ Ẓahrān al Khubar
153 U14 **Dhaka** *prev.* Dacca. ● (Bangladesh) Dhaka, C Bangladesh *23°42′N 90°22′E*
153 T15 **Dhaka** ◈ *division* C Bangladesh
Dhali *see* Idálion
Dhalli Rajhara *see* Dalli Rājhara
141 O15 **Dhamār** W Yemen *14°31′N 44°25′E*
154 K12 **Dhamtari** Chhattīsgarh, C India *20°43′N 81°36′E*
153 Q15 **Dhanbād** Jhārkhand, NE India *23°48′N 86°27′E*
152 L10 **Dhangaḍhi** *var.* Dhangarhi. Far Western, W Nepal *28°45′N 80°38′E*
Dhangarhi *see* Dhangaḍhi
Dhank *see* Ḍank
153 R12 **Dhankuṭā** Eastern, E Nepal *27°06′N 87°21′E*
152 I6 **Dhaola Dhār** ▲▲ NE India
154 F10 **Dhār** Madhya Pradesh, C India *22°32′N 75°24′E*
153 R12 **Dharān** *var.* Dharan Bazar. Eastern, E Nepal *26°51′N 87°18′E*
Dharan Bazar *see* Dharān
155 H21 **Dhārāpuram** Tamil Nādu, SE India *10°45′N 77°33′E*
155 H20 **Dharmapuri** Tamil Nādu, SE India *12°11′N 78°07′E*
155 H18 **Dharmavaram** Andhra Pradesh, E India *14°27′N 77°44′E*
154 M11 **Dharmjaygarh** Chhattīsgarh, C India *22°27′N 83°16′E*
Dharmsāla *see* Dharmshāla
152 I7 **Dharmshāla** *prev.* Dharmsāla. Himāchal Pradesh, N India *32°14′N 76°24′E*
155 F17 **Dhārwād** *prev.* Dharwar. Karnātaka, SW India *15°30′N 75°04′E*
Dharwar *see* Dhārwād
Dhaulagiri *see* Dhawalāgiri
153 O10 **Dhawalāgiri** *var.* Dhaulagiri. ▲ C Nepal *28°45′N 83°27′E*
81 L18 **Dheere Laaq** *var.* Lak Dera, *It.* Lach Dera. *seasonal river* Kenya/Somalia
Dhekeleia Sovereign Base Area *see* Dhekelia Sovereign Base Area
121 Q3 **Dhekélia** *Eng.* Dhekelia, *Gk.* Dekéleia. *UK air base* SE Cyprus*35°00′N 33°45′E* *see* Dhekélia
121 Q3 **Dhekelia Sovereign Base Area** *var.* Dhekeleia Sovereign Base Area. *UK military installation* E Cyprus*34°59′N 33°45′E*
Dhelvinákion *see* Delvináki
113 M22 **Dhëmbelit, Maja e** ▲ S Albania *40°10′N 20°22′E*
154 O12 **Dhenkānāl** Odisha, E India *20°40′N 85°36′E*
Dheskáti *see* Deskáti
138 G11 **Dhībān** Mādabā, NW Jordan *31°30′N 35°47′E*
Dhidhimótikhon *see* Didymóteicho
Dhíkti Ori *see* Díkti
139 W11 **Dhī Qār** *off.* Muḥāfa at Dhī Qār, *var.* Al Muntafiq, An Nāṣirīyah. ◈ *governorate* SE Iraq
Dhī Qār, Muḥāfa at *see* Dhī Qār
138 I12 **Dhirwah, Wādī adh** *dry watercourse* C Jordan
Dhístomon *see* Dístomo
Dhodhekánisos *see* Dodekánisa
Dhodhóni *see* Dodóni
Dhofar *see* Ẓufār
Dhomokós *see* Domokós
Dhond *see* Daund
155 H17 **Dhone** Andhra Pradesh, C India *15°25′N 77°52′E*
154 B11 **Dhorāji** Gujarāt, W India *21°44′N 70°27′E*
Dhráma *see* Dráma
154 C10 **Dhrāngadhra** Gujarāt, W India *22°59′N 71°32′E*
Dhrepanon, Akrotírio *see* Drépano, Akrotírio
153 T13 **Dhuburi** Assam, NE India *26°06′N 89°55′E*
154 F12 **Dhule** *prev.* Dhulia. Mahārāshtra, C India *20°54′N 74°47′E*
Dhulia *see* Dhule
Dhún Dealgan, Cuan *see* Dundalk Bay
Dhún Droma, Cuan *see* Dundrum Bay
Dhún na nGall, Bá *see* Donegal Bay
Dhū Shaykh *see* Qazānīyah
80 Q13 **Dhuudo** Bari, NE Somalia *09°21′N 50°19′E*
81 N15 **Dhuusa Marreeb** *var.* Dusa Marreb, *It.* Dusa Mareb. Galguduud, C Somalia *05°33′N 46°24′E*
115 J24 **Día** *island* SE Greece
55 Y9 **Diable, Île du** *var.* Devil's Island. *island* N French Guiana
15 N12 **Diable, Rivière du** ≈ Québec, SE Canada
35 N8 **Diablo, Mount** ▲ California, W USA *37°52′N 121°57′W*
35 O9 **Diablo Range** ▲▲ California, W USA
24 I8 **Diablo, Sierra** ▲▲ Texas, SW USA
45 O11 **Diablotins, Morne** ▲ N Dominica *15°30′N 61°23′W*
77 N11 **Diafarabé** Mopti, C Mali *14°09′N 05°01′W*
77 N11 **Diaka** ≈ SW Mali
Diakovár *see* Đakovo
76 I12 **Dialakoto** S Senegal *13°21′N 13°19′W*
61 B18 **Diamante** Entre Ríos, E Argentina *32°05′S 60°40′W*
62 I12 **Diamante, Río** ≈ C Argentina
59 M19 **Diamantina** Minas Gerais, SE Brazil *18°17′S 43°37′W*
59 N17 **Diamantina, Chapada** ▲▲ E Brazil
173 U11 **Diamantina Fracture Zone** *tectonic feature* E Indian Ocean
181 T8 **Diamantina River** ≈ Queensland/South Australia
38 D9 **Diamond Head** *headland* O'ahu, Hawai'i, USA *21°15′N 157°48′W*
37 P2 **Diamond Peak** ▲ Colorado, C USA *40°56′N 108°56′W*
35 W5 **Diamond Peak** ▲ Nevada, W USA *39°34′N 115°46′W*
Diamond State *see* Delaware
76 J11 **Diamou** Kayes, SW Mali *14°04′N 11°16′W*
95 I23 **Dianalund** Sjælland, C Denmark *55°32′N 11°30′E*
65 G25 **Diana's Peak** ▲ C Saint Helena
160 M16 **Dianbai** *var.* Shuidong. Guangdong, S China *21°30′N 111°05′E*
160 G13 **Dian Chi** ▣ SW China
106 B10 **Diano Marina** Liguria, NW Italy *43°55′N 08°06′E*
163 V11 **Diaobingshan** *var.* Tiefa. Liaoning, NE China *42°25′N 123°39′E*
Diaoyu Dao *see* Uotsuri-shima
Diaoyutai *see* Senkaku-shotō
77 R13 **Diapaga** E Burkina Faso *12°09′N 01°48′E*
Diarbekr *see* Diyarbakır
107 J15 **Diavolo, Passo del** *pass* C Italy
61 B18 **Díaz** Santa Fe, C Argentina *32°22′S 61°05′W*
141 W6 **Dibā al Ḥiṣn** *var.* Dibāh, Dibba. Ash Shāriqah, NE United Arab Emirates *25°35′N 56°16′E*
dibaga *see* Dîbege
Dibāh *see* Dibā al Ḥiṣn
79 L22 **Dibaya** Kasai-Occidental, S Dem. Rep. Congo *06°31′S 22°57′E*
Dibba *see* Dibā al Ḥiṣn
195 W15 **Dibble Iceberg Tongue** *ice feature* Antarctica
139 S3 **Dîbege** *Ar.* Ad Dībakah, *var.* Dībaga. Arbīl, N Iraq *35°51′N 43°49′E*
113 L19 **Dibër** ◈ *district* E Albania
83 I20 **Dibete** Central, SE Botswana *23°45′S 26°26′E*
Dibio *see* Dijon
25 W9 **Diboll** Texas, SW USA *31°11′N 94°46′W*
Dibra *see* Debar
153 X11 **Dibrugarh** Assam, NE India *27°29′N 94°49′E*
54 G4 **Dibulla** La Guajira, N Colombia *11°14′N 73°22′W*
25 O5 **Dickens** Texas, SW USA *33°38′N 100°51′W*
19 R2 **Dickey** Maine, NE USA *47°04′N 69°05′W*
30 K9 **Dickeyville** Wisconsin, N USA *42°37′N 90°36′W*
28 K5 **Dickinson** North Dakota, N USA *46°54′N 102°48′W*
0 E6 **Dickins Seamount** *undersea feature* NE Pacific Ocean *54°30′N 137°00′W*
27 O13 **Dickson** Oklahoma, C USA *34°11′N 96°58′W*
20 I9 **Dickson** Tennessee, S USA *36°04′N 87°23′W*
Dicle *see* Tigris
Dicsőszentmárton *see* Târnăveni
98 M12 **Didam** Gelderland, E Netherlands *51°56′N 06°08′E*
163 Y8 **Didao** Heilongjiang, NE China *45°22′N 130°48′E*
76 L12 **Didiéni** Koulikoro, W Mali *13°48′N 08°01′W*
Didimo *see* Dídymo
Didimotiho *see* Didymóteicho
81 K17 **Didimtu** *spring/well* NE Kenya *02°58′N 40°07′E*
67 U9 **Didinga Hills** ▲▲ S Sudan
11 Q16 **Didsbury** Alberta, SW Canada *51°39′N 114°09′W*
152 G11 **Dīdwāna** Rājasthān, N India *27°23′N 74°36′E*
115 G20 **Dídymo** *var.* Didimo. ▲ S Greece *37°28′N 23°12′E*
114 L12 **Didymóteicho** *var.* Dhidhimótikhon, Didimotiho. Anatolikí Makedonía kai Thráki, NE Greece *41°22′N 26°29′E*
103 S13 **Die** Drôme, E France *44°46′N 05°21′E*
77 O13 **Diébougou** SW Burkina Faso *11°00′N 03°12′W*
Diedenhofen *see* Thionville
11 S16 **Diefenbaker, Lake** ▣ Saskatchewan, S Canada
62 H7 **Diego de Almagro** Atacama, N Chile *26°24′S 70°10′W*
63 F23 **Diego de Almagro, Isla** *island* S Chile
61 A20 **Diego de Alvear** Santa Fe, C Argentina *34°25′S 62°10′W*
173 Q7 **Diego Garcia** *island* S British Indian Ocean Territory
Diégo-Suarez *see* Antsirañana
99 M23 **Diekirch** Diekirch, C Luxembourg *49°52′N 06°10′E*
99 L23 **Diekirch** ◈ *district* N Luxembourg
76 K11 **Diéma** Kayes, W Mali *14°30′N 09°12′W*
101 H15 **Diemel** ≈ W Germany
98 I10 **Diemen** Noord-Holland, C Netherlands *52°21′N 04°58′E*
Diemrich *see* Deva
Điên Biên *see* Điên Biên Phu
167 R6 **Điên Biên Phu** *var.* Bien Bien, Điên Biên. Lai Châu, N Vietnam *21°23′N 103°02′E*
167 S7 **Diên Châu** Nghê An, N Vietnam *18°54′N 105°35′E*
99 K18 **Diepenbeek** Limburg, NE Belgium *50°54′N 05°25′E*
98 N11 **Diepenheim** Overijssel, E Netherlands *52°10′N 06°37′E*
98 M10 **Diepenveen** Overijssel, E Netherlands *52°18′N 06°09′E*
100 G12 **Diepholz** Niedersachsen, NW Germany *52°36′N 08°23′E*
102 M3 **Dieppe** Seine-Maritime, N France *49°55′N 01°05′E*
98 M12 **Dieren** Gelderland, E Netherlands *52°03′N 06°06′E*
27 S13 **Dierks** Arkansas, C USA *34°07′N 94°01′W*
99 J17 **Diest** Vlaams Brabant, C Belgium *50°58′N 05°03′E*
108 F7 **Dietikon** Zürich, NW Switzerland *47°24′N 08°25′E*
103 R13 **Dieulefit** Drôme, E France *44°30′N 05°01′E*
103 T5 **Dieuze** Moselle, NE France *48°49′N 06°41′E*
119 H15 **Dieveniškės** Vilnius, SE Lithuania *54°12′N 25°38′E*
98 N7 **Diever** Drenthe, NE Netherlands *52°49′N 06°19′E*
101 F17 **Diez** Rheinland-Pfalz, W Germany *50°22′N 08°01′E*
77 Y12 **Diffa** Diffa, SE Niger *13°19′N 12°37′E*
77 Y10 **Diffa** ◈ *department* SE Niger
99 L25 **Differdange** Luxembourg, SW Luxembourg *49°32′N 05°53′E*
13 O16 **Digby** Nova Scotia, SE Canada *44°37′N 65°47′W*
26 J5 **Dighton** Kansas, C USA *38°28′N 100°28′W*
Dignano d'Istria *see* Vodnjan
103 T14 **Digne** *var.* Digne-les-Bains. Alpes-de-Haute-Provence, SE France *44°05′N 06°14′E*
Digne-les-Bains *see* Digne
Digoel *see* Digul, Sungai
103 Q10 **Digoin** Saône-et-Loire, C France *46°30′N 04°E*
171 Q8 **Digos** Mindanao, S Philippines *06°46′N 125°21′E*
149 Q16 **Digri** Sind, SE Pakistan *25°11′N 69°10′E*
171 Y14 **Digul Barat, Sungai** ≈ Papua, E Indonesia
171 Y15 **Digul, Sungai** *prev.* Digoel. ≈ Papua, E Indonesia
171 Z14 **Digul Timur, Sungai** ≈ Papua, E Indonesia
153 X10 **Dihāng** ≈ NE India
Dihang *see* Brahmaputra
Dihok *see* Dahūk
Dihōk *see* Dahūk
Dihok *see* Dahūk
Dihok, Parêzga-î *see* Dahūk
81 L17 **Diinsoor** Bay, S Somalia *02°28′N 43°00′E*
144 M14 **Diirmentobe** *Kas.* Dīirmentöbe; *prev.* Dermentobe, Dyurment'yube. Kzyl-Orda, S Kazakhstan *45°46′N 63°42′E*
Dīirmentöbe *see* Diirmentobe
Dijlah *see* Tigris
99 H17 **Dijle** ≈ C Belgium
103 R8 **Dijon** *anc.* Dibio. Côte d'Or, C France *47°21′N 05°04′E*
93 H14 **Dikanäs** Västerbotten, N Sweden *65°15′N 16°00′E*
80 L12 **Dikhil** SW Djibouti *11°08′N 42°19′E*
136 B13 **Dikili** İzmir, W Turkey *39°05′N 26°52′E*
99 B17 **Diksmuide** *var.* Dixmuide, *Fr.* Dixmude. West-Vlaanderen, W Belgium *51°02′N 02°52′E*
122 K7 **Dikson** Krasnoyarskiy Kray, N Russian Federation *73°30′N 80°35′E*
115 K25 **Díkti** *var.* Dhíkti Ori. ▲▲ Kríti, Greece, E Mediterranean Sea
77 Z13 **Dikwa** Borno, NE Nigeria *12°00′N 13°57′E*
81 J15 **Dīla** Southern Nationalities, S Ethiopia *06°19′N 38°16′E*
148 L7 **Dilārām** *prev.* Delārām. Nīmrōz, SW Afghanistan *32°11′N 63°27′E*
99 G18 **Dilbeek** Vlaams Brabant, C Belgium *50°51′N 04°16′E*
171 Q16 **Dili** *var.* Dilli, Dilly. ● (East Timor) N East Timor *08°33′S 125°34′E*
77 Y11 **Dilia** *var.* Dillia. ≈ SE Niger
Dilijan *see* Delijān
167 U13 **Di Linh** Lâm Đông, S Vietnam *11°38′N 108°07′E*
101 G16 **Dillenburg** Hessen, W Germany *50°45′N 08°16′E*
25 Q13 **Dilley** Texas, SW USA *28°40′N 99°10′W*
Dilli *see* Dili, East Timor
Dilli *see* Delhi, India
Dillia *see* Dilia
80 E11 **Dilling** *var.* Ad Dalanj. Southern Kordofan, C Sudan *12°02′N 29°41′E*
101 D20 **Dillingen** Saarland, SW Germany *49°20′N 06°43′E*
Dillingen *see* Dillingen an der Donau
101 J22 **Dillingen an der Donau** *var.* Dillingen. Bayern, S Germany *48°34′N 10°29′E*
39 O13 **Dillingham** Alaska, USA *59°03′N 158°30′W*
33 Q12 **Dillon** Montana, NW USA *45°14′N 112°38′W*
21 T12 **Dillon** South Carolina, SE USA *34°25′N 79°22′W*
31 T13 **Dillon Lake** ▣ Ohio, N USA
Dilly *see* Dili
Dilman *see* Salmās
79 K24 **Dilolo** Katanga, S Dem. Rep. Congo *10°42′S 22°21′E*
115 J20 **Dílos** *island* Kykládes, Greece, Aegean Sea
141 Y11 **Ḍil', Ra's aḍ** *headland* E Oman *19°12′N 57°53′E*
29 R5 **Dilworth** Minnesota, N USA *46°53′N 96°38′W*
138 H7 **Dimashq** *var.* Ash Shām, Esh Sham, *Eng.* Damascus, *Fr.* Damas, *It.* Damasco. ● (Syria) Rīf Dimashq, SW Syria *33°30′N 36°19′E*
138 I7 **Dimashq** ✈ Rīf Dimashq, S Syria *33°30′N 36°19′E*
Dimashq, Muḥāfaẓat *see* Rīf Dimashq
79 L21 **Dimbelenge** Kasai-Occidental, C Dem. Rep. Congo *05°36′S 23°04′E*
77 N16 **Dimbokro** E Ivory Coast *06°43′N 04°46′W*
182 L11 **Dimboola** Victoria, SE Australia *36°29′S 142°03′E*
Dîmbovița *see* Dâmbovița
Dimitrov *see* Dymytrov
114 K11 **Dimitrovgrad** Haskovo, S Bulgaria *42°03′N 25°36′E*
127 R5 **Dimitrovgrad** Ul'yanovskaya Oblast', W Russian Federation *54°14′N 49°37′E*
113 Q15 **Dimitrovgrad** *prev.* Caribrod. Serbia, SE Serbia *43°01′N 22°46′E*
Dimitrovo *see* Pernik
Dimlang *see* Vogel Peak
24 M3 **Dimmitt** Texas, SW USA *34°32′N 102°20′W*
114 F7 **Dimovo** Vidin, NW Bulgaria *43°46′N 22°47′E*
59 A16 **Dimpolis** Acre, W Brazil *09°52′S 71°51′W*
115 O23 **Dimyliá** Ródos, Dodekánisa, Greece, Aegean Sea *36°17′N 27°59′E*
171 Q6 **Dinagat Island** *island* S Philippines
153 S13 **Dinajpur** Rajshahi, NW Bangladesh *25°38′N 88°40′E*
102 I6 **Dinan** Côtes d'Armor, NW France *48°27′N 02°02′W*
99 I21 **Dinant** Namur, S Belgium *50°16′N 04°55′E*
136 E15 **Dinar** Afyon, SW Turkey *38°05′N 30°09′E*
112 F13 **Dinara** ▲ W Croatia *43°49′N 16°42′E*
Dinara *see* Dinaric Alps
102 I5 **Dinard** Ille-et-Vilaine, NW France *48°38′N 02°04′W*
112 F13 **Dinaric Alps** *var.* Dinara. ▲▲ Bosnia and Herzegovina/ Croatia
143 N10 **Dīnār, Kūh-e** ▲ C Iran *30°51′N 51°36′E*
Dinbych *see* Denbigh
155 H22 **Dindigul** Tamil Nādu, SE India *10°23′N 78°E*
83 M19 **Dindiza** Gaza, S Mozambique *23°22′S 33°28′E*
79 H21 **Dinga** Bandundu, SW Dem. Rep. Congo *05°00′S 16°29′E*
149 V7 **Dinga** Punjab, E Pakistan *32°38′N 73°45′E*
Dingchang *see* Qinxian
158 L16 **Dinggyê** *var.* Gyangkar. Xizang Zizhiqu, W China *28°18′N 88°06′E*
97 A20 **Dingle** *Ir.* An Daingean. SW Ireland *52°09′N 10°16′W*
97 A20 **Dingle Bay** *Ir.* Bá an Daingin. *bay* SW Ireland
18 I13 **Dingmans Ferry** Pennsylvania, NE USA *41°12′N 74°51′W*
101 N22 **Dingolfing** Bayern, SE Germany *48°37′N 12°28′E*
171 O1 **Dingras** Luzon, N Philippines *18°06′N 120°43′E*
76 J13 **Dinguiraye** N Guinea *11°19′N 10°49′W*
96 I8 **Dingwall** N Scotland, United Kingdom *57°36′N 04°26′W*
159 V10 **Dingxi** Gansu, C China *35°36′N 104°33′E*
Ding Xian *see* Dingzhou
161 Q7 **Dingyuan** Anhui, E China *32°30′N 117°40′E*
161 O3 **Dingzhou** *prev.* Ding Xian. Hebei, E China *38°31′N 114°52′E*
167 U6 **Đinh Lâp** Lang Sơn, N Vietnam *21°33′N 107°03′E*
167 T13 **Đinh Quan** *var.* Tân Phu. Đông Nai, S Vietnam *11°11′N 107°20′E*
100 E13 **Dinkel** ≈ Germany/ Netherlands
101 J21 **Dinkelsbühl** Bayern, S Germany *49°06′N 10°18′E*
101 D14 **Dinslaken** Nordrhein-Westfalen, W Germany *51°34′N 06°43′E*
35 R11 **Dinuba** California, W USA *36°32′N 119°23′W*
21 W7 **Dinwiddie** Virginia, NE USA *37°02′N 77°40′W*
98 N13 **Dinxperlo** Gelderland, E Netherlands *51°51′N 06°30′E*
Dio *see* Díon
Diófás *see* Nucet
76 M12 **Dioïla** Koulikoro, W Mali *12°28′N 06°43′W*
115 F14 **Díon** *var.* Dio; *anc.* Dium. *site of ancient city* Kentrikí Makedonía, N Greece
115 G19 **Dióryga Korinthou** *Eng.* Corinth Canal. *canal* S Greece
76 G12 **Diouloulou** SW Senegal *13°00′N 16°34′W*
77 N11 **Dioura** Mopti, W Mali *14°48′N 05°20′W*
76 G11 **Diourbel** W Senegal *14°39′N 16°12′W*
152 L10 **Dipāyal** Far Western, W Nepal *29°09′N 80°46′E*
121 R1 **Dipkarpaz** *Gk.* Rizokarpaso, Rizokárpason. NE Cyprus *35°36′N 34°23′E*
149 R17 **Diplo** Sind, SE Pakistan *24°29′N 69°36′E*
171 P7 **Dipolog** *var.* Dipolog City. Mindanao, S Philippines *08°31′N 123°20′E*
Dipolog City *see* Dipolog
185 C23 **Dipton** Southland, South Island, New Zealand *45°55′S 168°21′E*
77 O10 **Diré** Tombouctou, C Mali *16°12′N 03°31′W*
80 L13 **Dirē Dawa** Dirē Dawa, E Ethiopia *09°35′N 41°53′E*
Dirfis *see* Dírfys
115 H18 **Dírfys** *var.* Dirfis. ▲▲ Évvioa, C Greece
75 N9 **Dirj** *var.* Daraj, Darj. W Libya *30°09′N 10°26′E*
180 G10 **Dirk Hartog Island** *island* Western Australia
77 Y8 **Dirkou** Agadez, NE Niger *18°45′N 13°00′E*
181 X11 **Dirranbandi** Queensland, E Australia *28°37′S 148°13′E*
Dirri *see* Derri
Dirschau *see* Tczew
37 N6 **Dirty Devil River** ≈ Utah, W USA
32 E10 **Disappointment, Cape** *headland* Washington, NW USA *46°16′N 124°06′W*
180 L8 **Disappointment, Lake** *salt lake* Western Australia
183 R12 **Disaster Bay** *bay* New South Wales, SE Australia
44 J11 **Discovery Bay** C Jamaica *18°27′N 77°24′W*
182 K13 **Discovery Bay** *inlet* SE Australia
67 Y15 **Discovery II Fracture Zone** *tectonic feature* SW Indian Ocean
Discovery Seamount/ Discovery Seamounts *see* Discovery Tablemounts
65 O19 **Discovery Tablemounts** *var.* Discovery Seamount, Discovery Seamounts. *undersea feature* SW Atlantic Ocean *42°00′S 00°10′E*
108 G9 **Disentis** *Rmsch.* Mustér. Graubünden, S Switzerland *46°43′N 08°52′E*
39 O10 **Dishna River** ≈ Alaska, USA
148 K10 **Dīshū** *var.* Deshu; *prev.* Deh Shū. Helmand, S Afghanistan *30°28′N 63°21′E*
Disko Bugt *see* Qeqertarsuup Tunua
195 X4 **Dismal Mountains** ▲▲ Antarctica
28 M14 **Dismal River** ≈ Nebraska, C USA
Disna *see* Dzisna
99 L19 **Dison** Liège, E Belgium *50°37′N 05°52′E*
153 V12 **Dispur** *state capital* Assam, NE India *26°03′N 91°52′E*
15 R11 **Disraeli** Québec, SE Canada *45°58′N 71°21′W*
115 F18 **Dístomo** *prev.* Dhístomon. Stereá Elláda, C Greece *38°25′N 22°40′E*
Dístos, Límni *see* Dýstos, Límni
59 L18 **Distrito Federal** *Eng.* Federal District. ◈ *federal district* C Brazil
41 P14 **Distrito Federal** ◈ *federal district* S Mexico
54 L4 **Distrito Federal** *off.* Territorio Distrito Federal. ◈ *federal district* N Venezuela
Distrito Federal, Territorio *see* Distrito Federal
116 J10 **Ditrău** *Hung.* Ditró. Harghita, C Romania *46°49′N 25°31′E*
Ditró *see* Ditrău
154 B12 **Diu** Damān and Diu, W India *20°42′N 70°59′E*
Dium *see* Díon
109 S13 **Divača** SW Slovenia *45°40′N 13°58′E*
102 K5 **Dives** ≈ N France
33 Q11 **Divide** Montana, NW USA *45°44′N 112°47′W*
Divin *see* Dzivin
83 N18 **Divinhe** Sofala, E Mozambique *20°41′S 34°46′E*
59 L20 **Divinópolis** Minas Gerais, SE Brazil *20°08′S 44°55′W*
127 N13 **Divnoye** Stavropol'skiy Kray, SW Russian Federation *45°54′N 43°18′E*
76 M17 **Divo** S Ivory Coast *05°50′N 05°22′W*
Divodurum Mediomatricum *see* Metz
137 N13 **Divriği** Sivas, C Turkey *39°23′N 38°06′E*
Diwaniyah *see* Ad Dīwānīyah
14 J10 **Dix Milles, Lac** ◎ Québec, SE Canada
14 M8 **Dix Milles, Lac des** ◎ Québec, SE Canada
Dixmude/Dixmuide *see* Diksmuide
35 N7 **Dixon** California, W USA *38°19′N 121°49′W*
30 L10 **Dixon** Illinois, N USA *41°51′N 89°26′W*
20 I6 **Dixon** Kentucky, S USA *37°30′N 87°39′W*
27 V6 **Dixon** Missouri, C USA *37°59′N 92°05′W*
37 S9 **Dixon** New Mexico, SW USA *36°10′N 105°49′W*
39 Y15 **Dixon Entrance** *strait* Canada/USA
18 D14 **Dixonville** Pennsylvania, NE USA *40°43′N 79°01′W*
137 T13 **Diyadin** Ağrı, E Turkey *39°33′N 43°41′E*
139 V7 **Diyālá** *off.* Muḥāfa at Diyālá, *var.* Ba'qūbah. ◈ *governorate* E Iraq
Diyālá, Muḥāfa at *see* Diyālá
Diyālá, Nahr *see* Sīrvān, Rūdkhāneh-ye
139 V5 **Diyālá, Sirwan Nahr** *var.* Rudkhaneh-ye Sīrvān, Sīrwān. ≈ Iran/Iraq *see also* Sīrvān, Rudkhaneh-ye
137 P15 **Diyarbakır** *var.* Diarbekr; *anc.* Amida. Diyarbakır, SE Turkey *37°55′N 40°14′E*
137 P15 **Diyarbakır** *var.* Diarbekr. ◈ *province* SE Turkey
Dizful *see* Dezfūl
79 F16 **Dja** ≈ SE Cameroon
Djadié *see* Zadié
77 X7 **Djado** Agadez, NE Niger *21°00′N 12°11′E*
77 X6 **Djado, Plateau du** ▲▲ NE Niger
Djailolo *see* Halmahera, Pulau
Djajapura *see* Jayapura
Djakarta *see* Jakarta
Djakovo *see* Đakovo
79 G20 **Djambala** Plateaux, C Congo *02°32′S 14°43′E*
Djambi *see* Hari, Batang
Djambi *see* Jambi
74 M9 **Djanet** E Algeria *28°43′N 08°57′E*
74 M11 **Djanet** *prev.* Fort Charlet. SE Algeria *24°34′N 09°33′E*
Djaul *see* Dyaul Island
Djawa *see* Jawa
Djéblé *see* Jablah
78 I10 **Djédaa** Batha, C Chad *13°31′N 18°34′E*
74 J6 **Djelfa** *var.* El Djelfa. N Algeria *34°43′N 03°14′E*
79 M14 **Djéma** Haut-Mbomou, E Central African Republic *06°04′N 25°20′E*
Djember *see* Jember
Djeneponto *see* Jeneponto
77 N12 **Djenné** *var.* Jenné. Mopti, C Mali *13°55′N 04°31′W*
Djérablous *see* Jarābulus
Djerba *see* Jerba, Île de
79 F15 **Djérem** ≈ C Cameroon
Djevdjelija *see* Gevgelija
77 P11 **Djibo** N Burkina Faso *14°09′N 01°38′W*
80 L12 **Djibouti** *var.* Jibuti. ● (Djibouti) E Djibouti *11°33′N 42°55′E*
80 L12 **Djibouti** *off.* Republic of Djibouti, *var.* Jibuti; *prev.* French Somaliland, French Territory of the Afars and Issas, *Fr.* Côte Française des Somalis, Territoire Français des Afars et des Issas. ◆ *republic* E Africa
80 L12 **Djibouti** ✈ C Djibouti *11°29′N 42°54′E*
Djibouti, Republic of *see* Djibouti
Djidjel/Djidjelli *see* Jijel
Djidji *see* Ivando
55 W10 **Djoemoe** Sipaliwini, C Suriname *04°00′N 55°27′W*
Djokjakarta *see* Yogyakarta
79 K21 **Djoku-Punda** Kasai-Occidental, S Dem. Rep. Congo *05°27′S 20°58′E*
79 K18 **Djolu** Equateur, N Dem. Rep. Congo *0°35′N 22°30′E*
Djombang *see* Jombang
Đjorče Petrov *see* Đorče Petrov
79 F17 **Djoua** ≈ Congo/Gabon
77 R14 **Djougou** W Benin *09°42′N 01°38′E*
79 F16 **Djoum** Sud, S Cameroon *02°38′N 12°51′E*
78 I8 **Djourab, Erg du** *desert* N Chad
79 P17 **Djugu** Orientale, NE Dem. Rep. Congo *01°55′N 30°31′E*
Djumbir *see* Ďumbier
92 L3 **Djúpivogur** Austurland, SE Iceland *64°40′N 14°18′W*
94 L13 **Djura** Dalarna, C Sweden *60°37′N 15°00′E*
Djurdjevac *see* Đurđevac
D'Kar *see* Dekar
197 U6 **Dmitriya Lapteva, Proliv** *strait* N Russian Federation
126 J7 **Dmitriyev-L'govskiy** Kurskaya Oblast', W Russian Federation *52°08′N 35°09′E*
Dmitriyevsk *see* Makiyivka
126 K3 **Dmitrov** Moskovskaya Oblast', W Russian Federation *56°23′N 37°30′E*
Dmitrovichi *see* Dzmitravichy
126 J6 **Dmitrovsk-Orlovskiy** Orlovskaya Oblast', W Russian Federation *52°28′N 35°01′E*
117 R3 **Dmytrivka** Chernihivs'ka Oblast', N Ukraine *50°56′N 32°57′E*
Dnepr *see* Dnieper
Dneprodzerzhinsk *see* Romaniv
Dneprodzerzhinskoye Vodokhranilishche *see* Dniprodzerzhyns'ke Vodoskhovyshche
Dnepropetrovsk *see* Dnipropetrovs'k
Dnepropetrovskaya Oblast' *see* Dnipropetrovs'ka Oblast'
Dneprorudnoye *see* Dniprorudne
Dneprovskiy Liman *see* Dniprovs'kyy Lyman
Dneprovsko-Bugskiy Kanal *see* Dnyaprowska-Buhski Kanal
Dnestr *see* Dniester
Dnestrovskiy Liman *see* Dnistrovs'kyy Lyman
86 H11 **Dnieper** *Bel.* Dnyapro, *Rus.* Dnepr, *Ukr.* Dnipro. ≈ E Europe
117 P3 **Dnieper Lowland** *Bel.* Prydnyaprowskaya Nizina, *Ukr.* Prydniprovs'ka Nyzovyna. *lowlands* Belarus/ Ukraine
116 M8 **Dniester** *Rom.* Nistru, *Rus.* Dnestr, *Ukr.* Dnister; *anc.* Tyras. ≈ Moldova/Ukraine
Dnipro *see* Dnieper
Dniprodzerzhyns'k *see* Romaniv
117 T7 **Dniprodzerzhyns'ke Vodoskhovyshche** *Rus.* Dneprodzerzhinskoye Vodokhranilishche. ▣ C Ukraine
117 U7 **Dnipropetrovs'k** *Rus.* Dnepropetrovsk; *prev.* Yekaterinoslav. Dnipropetrovs'ka Oblast', E Ukraine *48°28′N 35°E*
117 U8 **Dnipropetrovs'k** ✈ Dnipropetrovs'ka Oblast', S Ukraine *48°20′N 35°04′E*
Dnipropetrovs'k *see* Dnipropetrovs'ka Oblast'
117 T7 **Dnipropetrovs'ka Oblast'** *var.* Dnipropetrovs'k, *Rus.* Dnepropetrovskaya Oblast'. ◈ *province* E Ukraine
117 U9 **Dniprorudne** *Rus.* Dneprorudnoye. Zaporiz'ka Oblast', SE Ukraine *47°21′N 35°00′E*
117 Q11 **Dniprovs'kyy Lyman** *Rus.* Dneprovskiy Liman. *bay* S Ukraine
Dnister *see* Dniester
117 O11 **Dnistrovs'kyy Lyman** *Rus.* Dnestrovskiy Liman. *inlet* S Ukraine
124 G14 **Dno** Pskovskaya Oblast', W Russian Federation *57°48′N 29°58′E*
Dnyapro *see* Dnieper
119 H20 **Dnyaprowska-Buhski Kanal** *Rus.* Dneprovsko-Bugskiy Kanal. *canal* SW Belarus
13 O14 **Doaktown** New Brunswick, SE Canada *46°34′N 66°06′W*
78 H13 **Doba** Logone-Oriental, S Chad *08°40′N 16°50′E*
118 E9 **Dobele** *Ger.* Doblen. W Latvia *56°36′N 23°14′E*
101 N16 **Döbeln** Sachsen, E Germany *51°07′N 13°07′E*
171 U12 **Doberai, Jazirah** *Dut.* Vogelkop. *peninsula* Papua Barat, E Indonesia
110 F10 **Dobiegniew** *Ger.* Lubuskie, Woldenberg Neumark. Lubuskie, W Poland *52°58′N 15°43′E*
Doblen *see* Dobele
81 K18 **Dobli** *spring/well* SW Somalia *0°24′N 41°18′E*
112 H11 **Doboj** Republiks Srpska, N Bosnia and Herzegovina *44°45′N 18°03′E*
143 R12 **Doborjī** *var.* Fūrg. Fārs, S Iran *28°16′N 55°13′E*
110 L8 **Dobre Miasto** *Ger.* Guttstadt. Warmińsko-mazurskie, NE Poland *53°59′N 20°25′E*
114 N7 **Dobrich** *Rom.* Bazargic; *prev.* Tolbukhin. Dobrich, NE Bulgaria *43°35′N 27°49′E*
114 N7 **Dobrich** ◈ *province* NE Bulgaria
126 M8 **Dobrinka** Lipetskaya Oblast', W Russian Federation *52°10′N 40°30′E*
126 M7 **Dobrinka** Volgogradskaya Oblast', SW Russian Federation *50°52′N 41°48′E*
Dobrla Vas *see* Eberndorf
111 I15 **Dobrodzień** *Ger.* Guttentag. Opolskie, S Poland *50°43′N 18°24′E*
Dobrogea *see* Dobruja
117 W7 **Dobropillya** *Rus.* Dobropol'ye. Donets'ka Oblast', SE Ukraine *48°25′N 37°02′E*
Dobropol'ye *see* Dobropillya
117 P8 **Dobrovelychkivka** Kirovohrads'ka Oblast', C Ukraine *48°22′N 31°12′E*
Dobrudja/Dobrudzha *see* Dobruja
114 O7 **Dobruja** *var.* Dobrudja, *Bul.* Dobrudzha, *Rom.* Dobrogea. *physical region* Bulgaria/ Romania
119 P19 **Dobrush** Homyel'skaya Voblasts', SE Belarus *52°25′N 31°19′E*
125 U14 **Dobryanka** Permskiy Kray, NW Russian Federation *58°28′N 56°27′E*
117 P2 **Dobryanka** Chernihivs'ka Oblast', N Ukraine *52°03′N 31°09′E*
Dobryn' *see* Dabryn'
21 R8 **Dobson** North Carolina, SE USA *36°25′N 80°45′W*
59 N20 **Doce, Rio** ≈ SE Brazil
93 I16 **Docksta** Västernorrland, C Sweden *63°06′N 18°22′E*
41 N10 **Doctor Arroyo** Nuevo León, NE Mexico *23°40′N 100°09′W*
62 L4 **Doctor Pedro P. Peña** Boquerón, W Paraguay *22°22′S 62°23′W*
171 S11 **Dodaga** Pulau Halmahera, E Indonesia *01°06′N 128°10′E*
155 G21 **Dodda Betta** ▲ S India *11°28′N 76°44′E*
Dodecanese *see* Dodekánisa
115 M22 **Dodekánisa** *var.* Nóties Sporádes, *Eng.* Dodecanese; *prev.* Dhodhekánisos, Dodekanisos. *island group* SE Greece
Dodekanisos *see* Dodekánisa
26 J6 **Dodge City** Kansas, C USA *37°45′N 100°01′W*
30 K9 **Dodgeville** Wisconsin, N USA *42°57′N 90°08′W*
97 H25 **Dodman Point** *headland* SW England, United Kingdom *50°13′N 04°47′W*
81 J14 **Dodola** Oromīya, C Ethiopia *07°00′N 39°15′E*
81 H22 **Dodoma** ● (Tanzania) Dodoma, C Tanzania *06°11′S 35°45′E*
81 H22 **Dodoma** ◈ *region* C Tanzania
115 C16 **Dodóni** *var.* Dhodhóni. *site of ancient city* Ípeiros, W Greece
33 U7 **Dodson** Montana, NW USA *48°25′N 108°18′W*
25 P3 **Dodson** Texas, SW USA *34°46′N 100°01′W*
98 M12 **Doesburg** Gelderland, E Netherlands *52°01′N 06°08′E*
98 N12 **Doetinchem** Gelderland, E Netherlands *51°58′N 06°17′E*
158 L12 **Dogai Coring** *var.* Lake Montcalm. ◎ W China
137 N15 **Doğanşehir** Malatya, C Turkey *38°07′N 37°54′E*
84 E9 **Dogger Bank** *undersea feature* C North Sea *55°00′N 03°00′E*
23 S10 **Dog Island** *island* Florida, SE USA
14 C7 **Dog Lake** ◎ Ontario, S Canada
106 B9 **Dogliani** Piemonte, NE Italy *44°33′N 07°55′E*
164 H11 **Dōgo** *island* Oki-shotō, SW Japan
143 N10 **Do Gonbadān** *var.* Dow Gonbadān, Gonbadān. Kohkīlūyeh va Būyer AḤmad, SW Iran *30°12′N 50°48′E*
77 S12 **Dogondoutchi** Dosso, SW Niger *13°36′N 04°03′E*
137 T13 **Doğubayazıt** Ağrı, E Turkey *39°33′N 44°07′E*
137 P12 **Doğu Karadeniz Dağları** *var.* Anadolu Dağları. ▲▲ NE Turkey
158 K16 **Dogxung Zangbo** ≈ W China
Doha *see* Ad Dawḥah
Doha *see* Ad Dawḥah
Dohad *see* Dāhod
Dohuk *see* Dahūk
Dohuk *see* Dahūk
159 N16 **Doilungdêqên** *var.* Namka. Xizang Zizhiqu, W China *29°41′N 90°58′E*
114 F12 **Doïráni, Límni** *var.* Limni Doïranis, *Bul.* Ezero Doyransko. ◎ N Greece
Doire *see* Londonderry
99 H22 **Doische** Namur, S Belgium *50°09′N 04°43′E*
59 P17 **Dois de Julho** ✈ (Salvador) Bahia, NE Brazil *12°04′S 38°58′W*
60 H12 **Dois Vizinhos** Paraná, S Brazil *25°47′S 53°03′W*
80 H10 **Doka** Gedaref, E Sudan *13°30′N 35°47′E*
Doka *see* Kéita, Bahr
Dokan *see* Dûkan
94 H13 **Dokka** Oppland, S Norway *60°49′N 10°04′E*
98 L5 **Dokkum** Fryslân, N Netherlands *53°20′N 06°E*
98 L5 **Dokkumer Ee** ≈ N Netherlands
76 K13 **Doko** NE Guinea *11°46′N 08°58′W*
Dokshitsy *see* Dokshytsy
118 K13 **Dokshytsy** *Rus.* Dokshitsy. Vitsyebskaya Voblasts', N Belarus *54°54′N 27°46′E*
117 X8 **Dokuchayevs'k** *var.* Dokuchayevsk. Donets'ka Oblast', SE Ukraine *47°43′N 37°41′E*

◆ Country
● Country Capital
◇ Dependent Territory
○ Dependent Territory Capital
◈ Administrative Regions
✈ International Airport
▲ Mountain
▲▲ Mountain Range
⩚ Volcano
≈ River
◎ Lake
▣ Reservoir

DOKUCHAYEVSK – DUCHCOV

Dokuchayevsk *see* Dokuchayevs'k
Dolak, Pulau *see* Yos Sudarso, Pulau
29 *P9* **Doland** South Dakota, N USA *44°51′N 98°06′W*
63 *J18* **Dolavón** Chaco, S Argentina *43°16′S 65°44′W*
15 *P6* **Dolbeau** Québec, SE Canada *48°52′N 72°15′W*
15 *P6* **Dolbeau-Mistassini** Québec, SE Canada *48°54′N 72°13′W*
102 *I5* **Dol-de-Bretagne** Ille-et-Vilaine, NW France *48°33′N 01°45′W*
64 *J13* **Doldrums Fracture Zone** *tectonic feature* W Atlantic Ocean
103 *S8* **Dôle** Jura, E France *47°05′N 05°30′E*
97 *J19* **Dolgellau** NW Wales, United Kingdom *52°45′N 03°54′W*
Dolginovo *see* Dawhinava
Dolgi, Ostrov *see* Dolgiy, Ostrov
125 *U2* **Dolgiy, Ostrov** ***var.*** Ostrov Dolgi. *island* NW Russian Federation
162 *J9* **Dölgöön** Övörhangay, C Mongolia *45°57′N 103°14′E*
107 *C20* **Dolianova** Sardegna, Italy, C Mediterranean Sea *39°23′N 09°08′E*
Dolina *see* Dolyna
123 *T13* **Dolinsk** Ostrov Sakhalin, Sakhalinskaya Oblast', SE Russian Federation *47°20′N 142°52′E*
Dolinskaya *see* Dolyns'ka
79 *F21* **Dolisie** ***prev.*** Loubomo. Niari, S Congo *04°12′S 12°41′E*
116 *G14* **Dolj** ◈ *county* SW Romania
98 *P5* **Dollard** *bay* NW Germany
194 *J5* **Dolleman Island** *island* Antarctica
114 *K8* **Dolna Oryahovitsa** ***var.*** Dolna Oryakhovits. Veliko Tarnovo, N Bulgaria *43°09′N 25°44′E*
Dolna Oryakhovits *see* Dolna Oryahovitsa
114 *N9* **Dolni Chiflik** Varna, E Bulgaria *42°59′N 27°43′E*
114 *I8* **Dolni Dabnik** ***var.*** Dolni Dŭbnik. Pleven, N Bulgaria *43°24′N 24°25′E*
Dolni Dŭbnik *see* Dolni Dabnik
114 *F8* **Dolni Lom** Vidin, NW Bulgaria *43°31′N 22°46′E*
Dolnja Lendava *see* Lendava
129 *F14* **Dolnośląskie** ◈ *province* SW Poland
111 *K18* **Dolný Kubín** ***Hung.*** Alsókubin. Zílinský Kraj, N Slovakia *49°12′N 19°17′E*
106 *H8* **Dolo** Veneto, NE Italy *45°25′N 12°06′E*
Dolomites/Dolomiti *see* Dolomitiche, Alpi
106 *H6* **Dolomitiche, Alpi** ***var.*** Dolomiti, ***Eng.*** Dolomites. ▲▲ NE Italy
Dolonnur *see* Duolun
Doloon *see* Tsogt-Ovoo
61 *E21* **Dolores** Buenos Aires, E Argentina *36°21′S 57°39′W*
42 *E3* **Dolores** Petén, N Guatemala *16°33′N 89°26′W*
171 *Q5* **Dolores** Samar, C Philippines *12°01′N 125°27′E*
105 *S12* **Dolores** Valenciana, E Spain *38°09′N 00°45′W*
61 *D19* **Dolores** Soriano, SW Uruguay *33°34′S 58°15′W*
41 *N12* **Dolores Hidalgo** ***var.*** Ciudad de Dolores Hidalgo. Guanajuato, C Mexico *21°10′N 100°55′W*
8 *J7* **Dolphin and Union Strait** *strait* Northwest Territories/Nunavut, N Canada
65 *D23* **Dolphin, Cape** *headland* East Falkland, Falkland Islands *51°15′S 58°57′W*
44 *H12* **Dolphin Head** *hill* W Jamaica
83 *B21* **Dolphin Head** ***var.*** Cape Dernberg. *headland* SW Namibia *25°33′S 14°36′E*
110 *G12* **Dolsk** ***Ger.*** Dolzig. Weilkopolskie, C Poland *51°59′N 17°03′E*
167 *S8* **Đô Lương** Nghê An, N Vietnam *18°51′N 105°19′E*
116 *I6* **Dolyna** ***Rus.*** Dolina. Ivano-Frankivs'ka Oblast', W Ukraine *48°58′N 24°01′E*
117 *R8* **Dolyns'ka** ***Rus.*** Dolinskaya. Kirovohrads'ka Oblast', S Ukraine *48°06′N 32°46′E*
Dolzig *see* Dolsk
Domachëvo/Domaczewo *see* Damachava
117 *P9* **Domanivka** Mykolayivs'ka Oblast', S Ukraine *47°40′N 30°56′E*
153 *S13* **Domar** Rajshahi, N Bangladesh *26°08′N 88°57′E*
108 *I9* **Domat/Ems** Graubünden, SE Switzerland *46°50′N 09°28′E*
111 *A18* **Domažlice** ***Ger.*** Taus. Plzeňský Kraj, W Czech Republic *49°26′N 12°54′E*
127 *X8* **Dombarovskiy** Orenburgskaya Oblast', W Russian Federation *50°53′N 59°18′E*
94 *G10* **Dombås** Oppland, S Norway *62°04′N 09°07′E*
83 *M17* **Dombe** Manica, C Mozambique *19°59′S 33°24′E*
82 *A13* **Dombe Grande** Benguela, C Angola *12°57′S 13°07′E*
103 *R10* **Dombes** *physical region* E France
111 *I25* **Dombóvár** Tolna, S Hungary *46°24′N 18°09′E*
99 *D14* **Domburg** Zeeland, SW Netherlands *51°34′N 03°30′E*
58 *L13* **Dom Eliseu** Pará, NE Brazil *04°02′S 47°31′W*
Domel Island *see* Letsôk-aw Kyun
103 *O11* **Dôme, Puy de** ▲ C France *45°46′N 03°00′E*
36 *H13* **Dome Rock Mountains** ▲▲ Arizona, SW USA
Domesnes, Cape *see* Kolkasrags
62 *G8* **Domeyko** Atacama, N Chile *28°58′S 70°54′W*
62 *H5* **Domeyko, Cordillera** ▲▲ N Chile
102 *K5* **Domfront** Orne, N France *48°35′N 00°39′W*
171 *X13* **Dom, Gunung** ▲ Papua, E Indonesia *02°41′S 137°00′E*
45 *X11* **Dominica** ***off.*** Commonwealth of Dominica. ◆ *republic* E West Indies
47 *S3* **Dominica** *island* Dominica
Dominica Channel *see* Martinique Passage
Dominica, Commonwealth of *see* Dominica
43 *N15* **Dominical** Puntarenas, SE Costa Rica *09°16′N 83°52′W*
45 *Q8* **Dominican Republic** ◆ *republic* C West Indies
45 *X11* **Dominica Passage** *passage* E Caribbean Sea
99 *K14* **Dommel** ≈ S Netherlands
81 *O14* **Domo** Sumalē, E Ethiopia *07°53′N 46°55′E*
126 *L4* **Domodedovo** ✈ (Moskva) Moskovskaya Oblast', W Russian Federation *55°19′N 37°55′E*
106 *C6* **Domodossola** Piemonte, NE Italy *46°07′N 08°20′E*
115 *F17* **Domokós** ***var.*** Dhomokós. Stereá Elláda, C Greece *39°07′N 22°18′E*
172 *I14* **Domoni** Anjouan, SE Comoros *12°15′S 44°39′E*
61 *G16* **Dom Pedrito** Rio Grande do Sul, S Brazil *31°00′S 54°40′W*
Dompoe *see* Dompu
170 *M16* **Dompu** ***prev.*** Dompoe. Sumbawa, C Indonesia *08°30′S 118°28′E*
Domschale *see* Domžale
62 *H13* **Domuyo, Volcán** 🌋 W Argentina *36°36′S 70°22′W*
109 *U11* **Domžale** ***Ger.*** Domschale. C Slovenia *46°09′N 14°33′E*
127 *O10* **Don** ***var.*** Duna, Tanais. ≈ SW Russian Federation
96 *K9* **Don** ≈ NE Scotland, United Kingdom
182 *M11* **Donald** Victoria, SE Australia *36°27′S 143°03′E*
22 *J9* **Donaldsonville** Louisiana, S USA *30°06′N 90°59′W*
23 *S8* **Donalsonville** Georgia, SE USA *31°02′N 84°52′W*
Donau *see* Danube
101 *G23* **Donaueschingen** Baden-Württemberg, SW Germany *47°57′N 08°30′E*
101 *K22* **Donaumoos** *wetland* S Germany
101 *K22* **Donauwörth** Bayern, S Germany *48°43′N 10°46′E*
109 *U7* **Donawitz** Steiermark, SE Austria *47°23′N 15°00′E*
117 *X7* **Donbass** *industrial region* Russian Federation/Ukraine
104 *K11* **Don Benito** Extremadura, W Spain *38°57′N 05°52′W*
97 *M17* **Doncaster** ***anc.*** Danum. N England, United Kingdom *53°32′N 01°07′W*
44 *K12* **Don Christophers Point** *headland* C Jamaica *18°19′N 76°48′W*
55 *V9* **Donderkamp** Sipaliwini, NW Suriname *05°18′N 56°22′W*
82 *B12* **Dondo** Kwanza Norte, NW Angola *09°40′S 14°24′E*
171 *O12* **Dondo** Sulawesi, N Indonesia *0°54′S 121°33′E*
83 *N17* **Dondo** Sofala, C Mozambique *19°41′S 34°45′E*
155 *K26* **Dondra Head** *headland* S Sri Lanka *05°57′N 80°33′E*
Dondușani *see* Dondușeni
116 *M8* **Dondușeni** ***var.*** Dondușani, ***Rus.*** Dondyushany. N Moldova *48°13′N 27°38′E*
Dondyushany *see* Dondușeni
97 *D15* **Donegal** ***Ir.*** Dún na nGall. Donegal, NW Ireland *54°39′N 08°06′W*
97 *D14* **Donegal** ***Ir.*** Dún na nGall. *cultural region* NW Ireland
97 *C15* **Donegal Bay** ***Ir.*** Bá Dhún na nGall. *bay* NW Ireland
84 *K10* **Donets** ≈ Russian Federation/Ukraine
117 *X8* **Donets'k** ***Rus.*** Donetsk; ***prev.*** Stalino. Donets'ka Oblast', E Ukraine *47°58′N 37°50′E*
117 *W8* **Donets'k** ✈ Donets'ka Oblast', E Ukraine *48°03′N 37°44′E*
Donets'k *see* Donets'ka Oblast'
117 *W8* **Donets'ka Oblast'** ***var.*** Donets'k, ***Rus.*** Donetskaya Oblast'; ***prev. Rus.*** Stalins'kaya Oblast'. ◈ *province* SE Ukraine
Donetskaya Oblast' *see* Donets'ka Oblast'
67 *P8* **Donga** ≈ Cameroon/Nigeria
157 *O13* **Dongchuan** Yunnan, SW China *26°09′N 103°10′E*
161 *Q14* **Dongchuan Dao** ***prev.*** Dongshan Dao. *island* SE China
Dong Dao *see* Lincoln Island
99 *I14* **Dongen** Noord-Brabant, S Netherlands *51°38′N 04°56′E*
160 *K17* **Dongfang** ***var.*** Basuo. Hainan, S China *19°05′N 108°40′E*
163 *Z7* **Dongfanghong** Heilongjiang, NE China *46°13′N 133°13′E*
163 *W11* **Dongfeng** Jilin, NE China *42°39′N 125°33′E*
171 *N12* **Donggala** Sulawesi, C Indonesia *0°40′S 119°44′E*
163 *V13* **Donggang** ***var.*** Dadong; ***prev.*** Donggou. Liaoning, NE China *39°52′N 124°08′E*
Donggou *see* Donggang
161 *O14* **Dongguan** Guangdong, S China *23°03′N 113°43′E*
167 *T9* **Đông Ha** Quang Tri, C Vietnam *16°45′N 107°10′E*
163 *Y14* **Donghae** ***prev.*** Tonghae. NE South Korea *37°26′N 129°09′E*
Dong Hai *see* East China Sea
160 *M16* **Donghai Dao** *island* S China
Dong He *see* Omon Gol
Donghe *see* Wangcang
167 *T9* **Đông Hơi** Quang Binh, C Vietnam *17°32′N 106°35′E*
Donghua *see* Huating
Donghuang *see* Xishui
108 *H10* **Dongio** Ticino, S Switzerland *46°27′N 08°58′E*
Dongkan *see* Binhai
160 *L11* **Dongkou** Hunan, S China *27°06′N 110°35′E*
167 *S8* **Đông Lê** Quang Binh, C Vietnam *17°52′N 105°49′E*
Dongliao *see* Liaoyuan
Dong-nai *see* Đông Nai, Sông
167 *U13* **Đông Nai, Sông** ***var.*** Dong-nai, Dong Noi, Donnai. ≈ S Vietnam
161 *N14* **Dongnan Qiuling** *plateau* SE China
163 *Y9* **Dongning** Heilongjiang, NE China *44°01′N 131°03′E*
Dong Noi *see* Đông Nai, Sông
83 *C14* **Dongo** Huíla, C Angola *14°35′S 15°51′E*
80 *E7* **Dongola** ***var.*** Donqola, Dunqulah. Northern, N Sudan *19°10′N 30°27′E*
79 *I17* **Dongou** Likouala, NE Congo *02°05′N 18°E*
Đông Phu *see* Đông Xoai
Dongping *see* Anhua
Dong Rak, Phanom *see* Dângrêk, Chuŏr Phnum
Dongshan Dao *see* Dongchuan Dao
Dongsha Qundao *see* Tungsha Tao
Dongsheng *see* Ordos
161 *S13* **Dongshi** ***Jap.*** Tōsei; ***prev.*** Tungshih. N Taiwan *24°13′N 120°54′E*
161 *R7* **Dongtai** Jiangsu, E China *32°50′N 120°22′E*
161 *N10* **Dongting Hu** ***var.*** Tung-t'ing Hu. ◎ S China
161 *P10* **Dongxiang** ***var.*** Xiaogang. Jiangxi, S China *28°16′N 116°32′E*
167 *T13* **Đông Xoai** ***var.*** Đông Phu. Sông Be, S Vietnam *11°31′N 106°55′E*
161 *Q4* **Dongying** Shandong, E China *37°27′N 118°01′E*
27 *X8* **Doniphan** Missouri, C USA *36°39′N 90°51′W*
Donja Łužica *see* Niederlausitz
10 *G7* **Donjek** ≈ Yukon, W Canada
112 *E11* **Donji Lapac** Lika-Senj, W Croatia *44°33′N 15°58′E*
112 *H8* **Donji Miholjac** Osijek-Baranja, NE Croatia *45°45′N 18°10′E*
112 *P12* **Donji Milanovac** Serbia, E Serbia *44°27′N 22°06′E*
112 *G12* **Donji Vakuf** ***var.*** Srbobran. ◈ Federacija Bosne I Hercegovine, C Bosnia and Herzegovina
98 *M6* **Donkerbroek** Fryslân, N Netherlands *52°58′N 05°15′E*
167 *P11* **Don Muang** ✈ (Krung Thep) Nonthaburi, C Thailand *13°51′N 100°40′E*
25 *S17* **Donna** Texas, SW USA *26°10′N 98°03′W*
15 *Q10* **Donnacona** Québec, SE Canada *46°41′N 71°46′W*
Donnai *see* Đông Nai, Sông
29 *Y16* **Donnellson** Iowa, C USA *40°38′N 91°33′W*
11 *O13* **Donnelly** Alberta, W Canada *55°42′N 117°06′W*
35 *P6* **Donner Pass** *Pass* California, W USA
101 *F19* **Donnersberg** ▲ W Germany *49°37′N 07°54′E*
Donoso *see* Miguel de la Borda
105 *P2* **Donostia-San Sebastián** País Vasco, N Spain *43°19′N 01°59′W*
115 *K21* **Donoúsa** ***var.*** Donoússa. *island* Kykládes, Greece, Aegean Sea
Donoússa *see* Donoúsa
35 *P8* **Don Pedro Reservoir** ⊠ California, W USA
Donqola *see* Dongola
126 *L5* **Donskoy** Tul'skaya Oblast', W Russian Federation *54°02′N 38°27′E*
81 *L16* **Doolow** Sumalē, E Ethiopia *04°10′N 42°05′E*
39 *Q7* **Doonerak, Mount** ▲ Alaska, USA *67°54′N 150°33′W*
98 *J12* **Doorn** Utrecht, C Netherlands *52°02′N 05°21′E*
Doornik *see* Tournai
31 *N6* **Door Peninsula** *peninsula* Wisconsin, N USA
80 *P13* **Dooxo Nugaaleed** ***var.*** Nogal Valley. *valley* E Somalia
Do Qu *see* Da Qu
106 *B7* **Dora Baltea** ***anc.*** Duria Major. ≈ NW Italy
180 *K7* **Dora, Lake** *salt lake* Western Australia
106 *A8* **Dora Riparia** ***anc.*** Duria Minor. ≈ NW Italy
Dorbiljin *see* Emin
Dorbod/Dorbod Mongolzu Zizhixian *see* Taikang
Dorbod Mongolzu Zizhixian *see* Taikang
113 *N18* **Đorče Petrov** ***var.*** Ďjorče Petrov, Gorče Petrov. N Macedonia *42°01′N 21°21′E*
14 *F16* **Dorchester** Ontario, S Canada *43°00′N 81°04′W*
97 *L24* **Dorchester** ***anc.*** Durnovaria. S England, United Kingdom *50°43′N 02°26′W*
9 *P7* **Dorchester, Cape** *headland* Baffin Island, Nunavut, NE Canada *65°25′N 77°25′W*
83 *D19* **Dordabis** Khomas, C Namibia *22°57′S 17°39′E*
102 *L12* **Dordogne** ◈ *department* SW France
103 *N12* **Dordogne** ≈ W France
98 *H13* **Dordrecht** ***var.*** Dordt, Dort. Zuid-Holland, SW Netherlands *51°48′N 04°40′E*
Dordt *see* Dordrecht
103 *P11* **Dore** ≈ C France
11 *S13* **Doré Lake** Saskatchewan, C Canada *54°37′N 107°36′W*
103 *O12* **Dore, Monts** ▲▲ C France
101 *M23* **Dorfen** Bayern, SE Germany *48°16′N 12°06′E*
107 *D18* **Dorgali** Sardegna, Italy, C Mediterranean Sea *40°18′N 09°34′E*
159 *N11* **Dorgê Co** ***var.*** Elsen Nur. ◎ C China
162 *E6* **Dörgön** ***var.*** Seer. Hovd, W Mongolia *48°18′N 92°37′E*
162 *F7* **Dörgön Nuur** ◎ NW Mongolia
77 *Q12* **Dori** N Burkina Faso *14°03′N 00°02′W*
83 *E24* **Doring** ≈ S South Africa
101 *E16* **Dormagen** Nordrhein-Westfalen, W Germany *51°06′N 06°49′E*
103 *P4* **Dormans** Marne, N France *49°03′N 03°44′E*
108 *E6* **Dornach** Solothurn, NW Switzerland *47°29′N 07°37′E*
Dorna Watra *see* Vatra Dornei
108 *J7* **Dornbirn** Vorarlberg, W Austria *47°25′N 09°46′E*
96 *I7* **Dornoch** N Scotland, United Kingdom *57°52′N 04°01′W*
96 *J7* **Dornoch Firth** *inlet* N Scotland, United Kingdom
163 *P7* **Dornod** ◈ *province* E Mongolia
163 *N10* **Dornogovĭ** ◈ *province* SE Mongolia
77 *P10* **Doro** Tombouctou, S Mali *16°07′N 00°57′W*
116 *L14* **Dorobanțu** Călărași, S Romania *44°15′N 26°55′E*
111 *J22* **Dorog** Komárom-Esztergom, N Hungary *47°43′N 18°42′E*
126 *I4* **Dorogobuzh** Smolenskaya Oblast', W Russian Federation *54°56′N 33°16′E*
116 *K8* **Dorohoi** Botoșani, NE Romania *47°57′N 26°24′E*
93 *H15* **Dorotea** Västerbotten, N Sweden *64°17′N 16°30′E*
Dorpat *see* Tartu
180 *G10* **Dorre Island** *island* Western Australia
183 *U5* **Dorrigo** New South Wales, SE Australia *30°22′S 152°43′E*
35 *N1* **Dorris** California, W USA *41°58′N 121°54′W*
14 *H13* **Dorset** Ontario, SE Canada *45°12′N 78°52′W*
97 *K23* **Dorset** *cultural region* S England, United Kingdom
101 *E14* **Dorsten** Nordrhein-Westfalen, W Germany *51°38′N 06°58′E*
Dort *see* Dordrecht
101 *F15* **Dortmund** Nordrhein-Westfalen, W Germany *51°31′N 07°28′E*
100 *F12* **Dortmund-Ems-Kanal** *canal* W Germany
136 *L17* **Dörtyol** Hatay, S Turkey *36°51′N 36°11′E*
Do Rūd *see* Dow Rūd
79 *O15* **Doruma** Orientale, N Dem. Rep. Congo *04°35′N 27°43′E*
15 *O12* **Dorval** ✈ (Montréal) Québec, SE Canada *45°27′N 73°46′W*
162 *F7* **Dörvöljin** ***var.*** Buga. Dzavhan, W Mongolia *47°42′N 94°53′E*
45 *T5* **Dos Bocas, Lago** ◎ C Puerto Rico
104 *K14* **Dos Hermanas** Andalucía, S Spain *37°16′N 05°55′W*
Dospad Dagh *see* Rhodope Mountains
35 *P10* **Dos Palos** California, W USA *37°00′N 120°39′W*
114 *I11* **Dospat** Smolyan, S Bulgaria *41°39′N 24°10′E*
114 *H11* **Dospat, Yazovir** ⊠ SW Bulgaria
100 *M11* **Dosse** ≈ NE Germany
77 *S12* **Dosso** Dosso, SW Niger *13°03′N 03°10′E*
77 *S12* **Dosso** ◈ *department* SW Niger
144 *G12* **Dossor** Atyrau, SW Kazakhstan *47°31′N 53°01′E*
147 *O10* **Do'stlik** Jizzax Viloyati, C Uzbekistan *40°37′N 67°59′E*
147 *V9* **Dostuk** Narynskaya Oblast', C Kyrgyzstan *41°19′N 75°40′E*
145 *X13* **Dostyk** ***prev.*** Druzhba. Almaty, SE Kazakhstan *45°15′N 82°29′E*
23 *R7* **Dothan** Alabama, S USA *31°13′N 85°23′W*
39 *T9* **Dot Lake** Alaska, USA *63°39′N 144°10′W*
118 *F12* **Dotnuva** Kaunas, C Lithuania *55°23′N 23°53′E*
99 *D19* **Dottignies** Hainaut, W Belgium *50°43′N 03°16′E*
103 *P2* **Douai** ***prev.*** Douay; ***anc.*** Duacum. Nord, N France *50°22′N 03°04′E*
14 *L9* **Douaire, Lac** ◎ Québec, SE Canada
79 *D16* **Douala** ***var.*** Duala. Littoral, W Cameroon *04°04′N 09°43′E*
79 *D16* **Douala** ✈ Littoral, W Cameroon *03°57′N 09°48′E*
102 *F6* **Douarnenez** Finistère, NW France *48°05′N 04°20′W*
102 *E6* **Douarnenez, Baie de** *bay* NW France
Douay *see* Douai
25 *O6* **Double Mountain Fork Brazos River** ≈ Texas, SW USA
23 *O3* **Double Springs** Alabama, S USA *34°09′N 87°24′W*
103 *T8* **Doubs** ◈ *department* E France
108 *C8* **Doubs** ≈ France/Switzerland
185 *A22* **Doubtful Sound** *sound* South Island, New Zealand
184 *J2* **Doubtless Bay** *bay* North Island, New Zealand
25 *X9* **Doucette** Texas, SW USA *30°48′N 94°25′W*
102 *K8* **Doué-la-Fontaine** Maine-et-Loire, NW France *47°12′N 00°16′W*
77 *O11* **Douentza** Mopti, S Mali *14°59′N 02°57′W*
65 *E24* **Douglas** East Falkland, Falkland Islands
97 *I16* **Douglas** ○ (Isle of Man) E Isle of Man *54°09′N 04°28′W*
83 *H23* **Douglas** Northern Cape, C South Africa *29°04′S 23°47′E*
39 *X13* **Douglas** Alexander Archipelago, Alaska, USA *58°12′N 134°18′W*
37 *O17* **Douglas** Arizona, SW USA *31°20′N 109°32′W*
23 *U7* **Douglas** Georgia, SE USA *31°30′N 82°51′W*
33 *Y15* **Douglas** Wyoming, C USA *42°48′N 105°23′W*
21 *O7* **Douglas Cape** *headland* Alaska, N USA *64°59′N 166°41′W*
10 *J14* **Douglas Channel** *channel* British Columbia, W Canada
182 *G3* **Douglas Creek** *seasonal river* South Australia
31 *P5* **Douglas Lake** ◎ Michigan, N USA
21 *O9* **Douglas Lake** ⊠ Tennessee, S USA
39 *Q13* **Douglas, Mount** ▲ Alaska, USA *58°51′N 153°31′W*
194 *I6* **Douglas Range** ▲▲ Alexander Island, Antarctica
161 *S14* **Douliu** ***prev.*** Touliu. C Taiwan *23°44′N 120°27′E*
103 *O2* **Doullens** Somme, N France *50°09′N 02°21′E*
Douma *see* Dūmā
79 *F15* **Doumé** Est, E Cameroon *04°14′N 13°27′E*
99 *E21* **Dour** Hainaut , S Belgium *50°24′N 03°47′E*
59 *K18* **Dourada, Serra** ▲▲ S Brazil
59 *I21* **Dourados** Mato Grosso do Sul, S Brazil *22°09′S 54°52′W*
103 *N5* **Dourdan** Essonne, N France *48°33′N 01°58′E*
104 *I6* **Douro** ***Sp.*** Duero. ≈ Portugal/Spain *see also* Duero
Douro *see* Duero
104 *G6* **Douro Litoral** *former province* N Portugal
Douvres *see* Dover
102 *K15* **Douze** ≈ SW France
183 *P17* **Dover** Tasmania, SE Australia *43°19′S 147°01′E*
97 *Q22* **Dover** ***Fr.*** Douvres, ***Lat.*** Dubris Portus. SE England, United Kingdom *51°08′N 03°E*
21 *Y3* **Dover** *state capital* Delaware, NE USA *39°09′N 75°31′W*
19 *P9* **Dover** New Hampshire, NE USA *43°10′N 70°50′W*
18 *J14* **Dover** New Jersey, NE USA *40°51′N 74°33′W*
31 *U12* **Dover** Ohio, N USA *40°31′N 81°28′W*
20 *H8* **Dover** Tennessee, S USA *36°30′N 87°50′W*
97 *Q23* **Dover, Strait of** ***var.*** Straits of Dover, ***Fr.*** Pas de Calais. *strait* England, United Kingdom/France
Dover, Straits of *see* Dover, Strait of
Dovlen *see* Devin
94 *G11* **Dovre** Oppland, S Norway *61°59′N 09°16′E*
94 *G10* **Dovrefjell** *plateau* S Norway
Dovsk *see* Dowsk
83 *M14* **Dowa** Central, C Malawi *13°40′S 33°55′E*
31 *O10* **Dowagiac** Michigan, N USA *41°58′N 86°06′W*
Dow Gonbadān *see* Do Gonbadān
148 *M2* **Dowlatābād** Fāryāb, N Afghanistan *36°30′N 64°51′E*
97 *G16* **Down** *cultural region* SE Northern Ireland, United Kingdom
33 *R16* **Downey** Idaho, NW USA *42°25′N 112°06′W*
35 *P5* **Downieville** California, W USA *39°34′N 120°49′W*
97 *G16* **Downpatrick** ***Ir.*** Dún Pádraig. SE Northern Ireland, United Kingdom *54°20′N 05°43′W*
26 *M3* **Downs** Kansas, C USA *39°30′N 98°33′W*
18 *J12* **Downsville** New York, NE USA *42°03′N 74°59′W*
142 *L7* **Dow Rūd** ***var.*** Do Rūd, Durud. Lorestān, W Iran *33°28′N 49°04′E*
29 *V12* **Dows** Iowa, C USA *42°39′N 93°30′W*
119 *O17* **Dowsk** ***Rus.*** Dovsk. Homyel'skaya Voblasts', SE Belarus *53°09′N 30°28′E*
35 *Q4* **Doyle** California, W USA *40°00′N 120°06′W*
18 *I15* **Doylestown** Pennsylvania, NE USA *40°18′N 75°08′W*
Doyransko, Ezero *see* Doïráni, Límni
114 *I8* **Doyrentsi** Lovech, N Bulgaria *43°13′N 24°46′E*
164 *G11* **Dōzen** *island* Oki-shotō, SW Japan
14 *K9* **Dozois, Réservoir** ⊠ Québec, SE Canada
74 *D9* **Drâa** *seasonal river* S Morocco
Drâa, Hammada du *see* Dra, Hamada du
Drabble *see* José Enrique Rodó
117 *Q5* **Drabiv** Cherkas'ka Oblast', C Ukraine *49°57′N 32°10′E*
Drable *see* José Enrique Rodó
103 *S13* **Drac** ≈ E France
Drač/Draç *see* Durrës
60 *I8* **Dracena** São Paulo, S Brazil *21°27′S 51°30′W*
98 *M6* **Drachten** Fryslân, N Netherlands *53°07′N 06°06′E*
92 *H11* **Drag** ***Lapp.*** Ájluokta. Nordland, C Norway *68°02′N 16°E*
116 *L14* **Dragalina** Călărași, SE Romania *44°26′N 27°19′E*
116 *I14* **Draganesti-Olt** Olt, S Romania *44°06′N 25°00′E*
116 *J14* **Drăgănești-Vlașca** Teleorman, S Romania *44°05′N 25°39′E*
116 *I13* **Drăgășani** Vâlcea, SW Romania *44°40′N 24°16′E*
114 *G9* **Dragoman** Sofia, W Bulgaria *42°32′N 22°56′E*
115 *L25* **Dragonáda** *island* SE Greece
Dragonera, Isla *see* Sa Dragonera
45 *T14* **Dragon's Mouths, The** *strait* Trinidad and Tobago/Venezuela
95 *J23* **Dragør** Sjælland, E Denmark *55°36′N 12°42′E*
114 *F10* **Dragovishtitsa** Kysutendil, W Bulgaria *42°22′N 22°39′E*
103 *U15* **Draguignan** Var, SE France *43°31′N 06°31′E*
74 *E9* **Dra, Hamada du** ***var.*** Hammada du Drâa, Haut Plateau du Dra. *plateau* W Algeria
Dra, Haut Plateau du *see* Dra, Hamada du
119 *H19* **Drahichyn** ***Pol.*** Drohiczyn Poleski, ***Rus.*** Drogichin. Brestskaya Voblasts', SW Belarus *52°11′N 25°10′E*
29 *N4* **Drake** North Dakota, N USA *47°54′N 100°23′W*
83 *K23* **Drakensberg** ▲▲ Lesotho/South Africa
194 *F3* **Drake Passage** *passage* Atlantic Ocean/Pacific Ocean
114 *L8* **Dralfa** Tŭrgovishte, N Bulgaria *43°17′N 26°25′E*
114 *I12* **Dráma** ***var.*** Dhráma. Anatolikí Makedonía kai Thráki, NE Greece *41°09′N 24°10′E*
Dramburg *see* Drawsko Pomorskie
95 *H15* **Drammen** Buskerud, S Norway *59°44′N 10°12′E*
95 *H15* **Drammensfjorden** *fjord* S Norway
92 *H1* **Drangajökull** ▲ NW Iceland *66°13′N 22°18′W*
95 *F16* **Drangedal** Telemark, S Norway *59°05′N 09°05′E*
92 *I2* **Drangsnes** Vestfirðir, NW Iceland *65°42′N 21°27′W*
Drann *see* Dravinja
109 *T10* **Drau** ***var.*** Drava, ***Eng.*** Drave, ***Hung.*** Dráva. ≈ C Europe *see also* Drava
Drau *see* Drava
84 *I11* **Drava** ***var.*** Drau, ***Eng.*** Drave, ***Hung.*** Dráva. ≈ C Europe *see also* Drau
Dráva/Drave *see* Drau/Drava
109 *W10* **Dravinja** ***Ger.*** Drann. ≈ NE Slovenia
109 *V9* **Dravograd** ***Ger.*** Unterdrauburg; ***prev.*** Spodnji Dravograd. N Slovenia *46°36′N 15°00′E*
110 *F10* **Drawa** ≈ NW Poland
110 *F9* **Drawno** Zachodnio-pomorskie, NW Poland *53°12′N 15°44′E*
110 *F9* **Drawsko Pomorskie** ***Ger.*** Dramburg. Zachodnio-pomorskie, NW Poland *53°32′N 15°48′E*
29 *R3* **Drayton** North Dakota, N USA *48°34′N 97°10′W*
11 *P14* **Drayton Valley** Alberta, SW Canada *53°15′N 115°00′W*
186 *B6* **Dreikikir** East Sepik, NW Papua New Guinea *03°42′S 142°46′E*
Dreikirchen *see* Teiuș
98 *N7* **Drenthe** ◈ *province* NE Netherlands
115 *H15* **Drépano, Akrotírio** ***var.*** Akrotírio Dhrepanon. *headland* N Greece *39°56′N 23°57′E*
Drepanum *see* Trapani
14 *D17* **Dresden** Ontario, S Canada *42°34′N 82°09′W*
101 *O16* **Dresden** Sachsen, E Germany *51°03′N 13°43′E*
20 *G8* **Dresden** Tennessee, S USA *36°17′N 88°42′W*
118 *M11* **Dretun'** Vitsyebskaya Voblasts', N Belarus *55°41′N 29°13′E*
102 *M5* **Dreux** ***anc.*** Drocae, Durocasses. Eure-et-Loir, C France *48°44′N 01°23′E*
94 *I11* **Drevsjø** Hedmark, S Norway *61°52′N 12°01′E*
22 *K3* **Drew** Mississippi, S USA *33°48′N 90°31′W*
110 *F10* **Drezdenko** ***Ger.*** Driesen. Lubuskie, W Poland *52°51′N 15°50′E*
98 *J12* **Driebergen** ***var.*** Driebergen-Rijsenburg. Utrecht, C Netherlands *52°03′N 05°17′E*
Driebergen-Rijsenburg *see* Driebergen
Driesen *see* Drezdenko
97 *N16* **Driffield** E England, United Kingdom *54°00′N 00°28′W*
65 *D25* **Driftwood Point** *headland* East Falkland, Falkland Islands *52°15′S 59°00′W*
33 *S14* **Driggs** Idaho, NW USA *43°44′N 111°06′W*
Drin *see* Drinit, Lumi i
112 *K12* **Drina** ≈ Bosnia and Herzegovina/Serbia
Drin, Gulf of *see* Drinit, Gjiri i
113 *M16* **Drini i Bardhë** ***Serb.*** Beli Drim. ≈ Albania/Serbia
113 *K18* **Drinit, Gjiri i** ***var.*** Pellg i Drinit, ***Eng.*** Gulf of Drin. *gulf* NW Albania
113 *L17* **Drinit, Lumi i** ***var.*** Drin. ≈ NW Albania
Drinit, Pellg i *see* Drinit, Gjiri i
Drinit të Zi, Lumi i *see* Black Drin
113 *L22* **Dríno** ***var.*** Drino, Drínos Pótamos, ***Alb.*** Lumi i Drinos. ≈ Albania/Greece
Drinos, Lumi i/Drínos Pótamos *see* Dríno
25 *S11* **Dripping Springs** Texas, SW USA *30°11′N 98°04′W*
25 *S15* **Driscoll** Texas, SW USA *27°40′N 97°45′W*
22 *H5* **Driskill Mountain** ▲ Louisiana, S USA *32°25′N 92°54′W*
Drissa *see* Drysa
94 *G10* **Driva** ≈ S Norway
112 *E13* **Drniš** ***It.*** Šibenik-Knin. Šibenik-Knin, S Croatia *43°51′N 16°10′E*
95 *H15* **Drøbak** Akershus, S Norway *59°40′N 10°40′E*
116 *G13* **Drobeta-Turnu Severin** ***prev.*** Turnu Severin. Mehedinți, SW Romania *44°39′N 22°40′E*
Drocae *see* Dreux
116 *M8* **Drochia** ***Rus.*** Drokiya. N Moldova *48°02′N 27°46′E*
97 *F17* **Drogheda** ***Ir.*** Droichead Átha. NE Ireland *53°43′N 06°21′W*
Drogichin *see* Drahichyn
Drogobych *see* Drohobych
Drohiczyn Poleski *see* Drahichyn
116 *H6* **Drohobych** ***Pol.*** Drohobycz, ***Rus.*** Drogobych. L'vivs'ka Oblast', NW Ukraine *49°22′N 23°33′E*
Drohobycz *see* Drohobych
Droichead Átha *see* Drogheda
Droicheadna Bandan *see* Bandon
Droichead na Banna *see* Banbridge
Droim Mór *see* Dromore
Drokiya *see* Drochia
103 *R13* **Drôme** ◈ *department* E France
103 *S13* **Drôme** ≈ E France
97 *G15* **Dromore** ***Ir.*** Droim Mór. SE Northern Ireland, United Kingdom *54°25′N 06°09′W*
106 *A9* **Dronero** Piemonte, NE Italy *44°28′N 07°25′E*
102 *L12* **Dronne** ≈ SW France
195 *T3* **Dronning Fabiolafjella** ***var.*** Mount Victor. ▲ Antarctica *72°49′S 33°01′E*
195 *Q3* **Dronning Maud Land** *physical region* Antarctica
98 *K6* **Dronrijp** ***Fris.*** Dronryp. Fryslân, N Netherlands *53°12′N 05°37′E*
Dronryp *see* Dronrijp
98 *L9* **Dronten** Flevoland, C Netherlands *52°31′N 05°41′E*
Drontheim *see* Trondheim
102 *L13* **Dropt** ≈ SW France
149 *T4* **Drosh** Khyber Pakhtunkhwa, NW Pakistan *35°33′N 71°48′E*
Drossen *see* Ośno Lubuskie
Drug *see* Durg
Drujba *see* Pitnak
118 *I12* **Drūkšiai** ◎ NE Lithuania
Druk-yul *see* Bhutan
11 *Q16* **Drumheller** Alberta, SW Canada *51°28′N 112°42′W*
33 *Q10* **Drummond** Montana, NW USA *46°39′N 113°12′W*
31 *R4* **Drummond Island** *island* Michigan, N USA
Drummond Island *see* Tabiteuea
21 *X7* **Drummond, Lake** ◎ Virginia, NE USA
15 *P12* **Drummondville** Québec, SE Canada *45°52′N 72°28′W*
39 *T11* **Drum, Mount** ▲ Alaska, USA *62°11′N 144°37′W*
27 *O9* **Drumright** Oklahoma, C USA *35°59′N 96°36′W*
99 *J14* **Drunen** Noord-Brabant, S Netherlands *51°41′N 05°08′E*
Druskienniki *see* Druskininkai
119 *F15* **Druskininkai** ***Pol.*** Druskienniki. Alytus, S Lithuania *54°00′N 24°00′E*
98 *K13* **Druten** Gelderland, SE Netherlands *51°53′N 05°37′E*
118 *K11* **Druya** Vitsyebskaya Voblasts', NW Belarus *55°47′N 27°27′E*
117 *S2* **Druzhba** Sums'ka Oblast', NE Ukraine *52°01′N 33°56′E*
Druzhba *see* Dostyk, Kazakhstan
Druzhba *see* Pitnak, Uzbekistan
123 *R7* **Druzhina** Respublika Sakha (Yakutiya), NE Russian Federation *68°01′N 144°58′E*
117 *X7* **Druzhkivka** Donets'ka Oblast', E Ukraine *48°38′N 37°31′E*
112 *E12* **Drvar** Federacija Bosne I Hercegovine, W Bosnia and Herzegovina *44°21′N 16°24′E*
113 *G15* **Drvenik** Split-Dalmacija, SE Croatia *43°10′N 17°13′E*
114 *K9* **Dryanovo** Gabrovo, N Bulgaria *42°58′N 25°28′E*
26 *G7* **Dry Cimarron River** ≈ Kansas/Oklahoma, C USA
12 *B11* **Dryden** Ontario, C Canada *49°48′N 92°48′W*
24 *M11* **Dryden** Texas, SW USA *30°01′N 102°06′W*
195 *Q14* **Drygalski Ice Tongue** *ice feature* Antarctica
118 *L11* **Drysa** ***Rus.*** Drissa. ≈ N Belarus
23 *V17* **Dry Tortugas** *island* Florida, SE USA
79 *D15* **Dschang** Ouest, W Cameroon *05°28′N 10°02′E*
54 *J5* **Duaca** Lara, N Venezuela *10°22′N 69°08′W*
Duacum *see* Douai
Duala *see* Douala
45 *N9* **Duarte, Pico** ▲ C Dominican Republic *19°02′N 70°57′W*
140 *J5* **Ḍubā** Tabūk, NW Saudi Arabia *27°26′N 35°42′E*
Dubai *see* Dubayy
117 *N9* **Dubăsari** ***Rus.*** Dubossary. NE Moldova *47°16′N 29°07′E*
117 *N9* **Dubăsari Reservoir** ⊠ NE Moldova
8 *M10* **Dubawnt** ≈ Nunavut, NW Canada
8 *L9* **Dubawnt Lake** ◎ Northwest Territories/Nunavut, N Canada
30 *L6* **Du Bay, Lake** ⊠ Wisconsin, N USA
141 *U7* **Dubayy** ***Eng.*** Dubai. Dubayy, NE United Arab Emirates *25°11′N 55°18′E*
141 *W7* **Dubayy** ***Eng.*** Dubai. ✈ Dubayy, NE United Arab Emirates *25°15′N 55°22′E*
183 *R7* **Dubbo** New South Wales, SE Australia *32°16′S 148°41′E*
108 *G7* **Dübendorf** Zürich, NW Switzerland *47°23′N 08°37′E*
97 *F18* **Dublin** ***Ir.*** Baile Átha Cliath; ***anc.*** Eblana. ● (Ireland) Dublin, E Ireland *53°20′N 06°15′W*
23 *U5* **Dublin** Georgia, SE USA *32°32′N 82°54′W*
25 *R7* **Dublin** Texas, SW USA *32°05′N 98°20′W*
97 *G18* **Dublin** ***Ir.*** Baile Átha Cliath; ***anc.*** Eblana. *cultural region* E Ireland
97 *G18* **Dublin Airport** ✈ Dublin, E Ireland *53°25′N 06°18′W*
189 *V12* **Dublon** ***var.*** Tonoas. *island* Chuuk Islands, C Micronesia
126 *K2* **Dubna** Moskovskaya Oblast', W Russian Federation *56°45′N 37°09′E*
111 *G19* **Dubňany** ***Ger.*** Dubnian. Jihomoravský Kraj, SE Czech Republic *48°54′N 17°00′E*
Dubnian *see* Dubňany
111 *I19* **Dubnica nad Váhom** ***Hung.*** Máriatölgyes; ***prev.*** Dubnicz. Trenčiansky Kraj, W Slovakia *48°58′N 18°10′E*
Dubnicz *see* Dubnica nad Váhom
116 *K4* **Dubno** Rivnens'ka Oblast', NW Ukraine *50°28′N 25°40′E*
33 *R13* **Dubois** Idaho, NW USA *44°10′N 112°13′W*
18 *D13* **Du Bois** Pennsylvania, NE USA *41°07′N 78°45′W*
33 *T14* **Dubois** Wyoming, C USA *43°31′N 109°37′W*
Dubossary *see* Dubăsari
127 *O10* **Dubovka** Volgogradskaya Oblast', SW Russian Federation *49°10′N 48°49′E*
76 *H14* **Dubréka** SW Guinea *09°48′N 13°31′W*
14 *B7* **Dubreuilville** Ontario, S Canada *48°21′N 84°31′W*
Dubris Portus *see* Dover
119 *L20* **Dubrova** Homyel'skaya Voblasts', SE Belarus *51°47′N 28°13′E*
Dubrovačko-Neretvanska Županija *see* Dubrovnik-Neretva
126 *I5* **Dubrovka** Bryanskaya Oblast', W Russian Federation *53°44′N 33°27′E*
113 *H16* **Dubrovnik** ***It.*** Ragusa. Dubrovnik-Neretva, SE Croatia *42°40′N 18°06′E*
113 *I16* **Dubrovnik** ✈ Dubrovnik-Neretva, SE Croatia *42°34′N 18°17′E*
113 *F16* **Dubrovnik-Neretva** ***off.*** Dubrovačko-Neretvanska Županija. ◈ *province* SE Croatia
Dubrovno *see* Dubrowna
116 *L2* **Dubrovytsya** Rivnens'ka Oblast', NW Ukraine *51°34′N 26°37′E*
119 *O14* **Dubrowna** ***Rus.*** Dubrovno. Vitsyebskaya Voblasts', N Belarus *54°35′N 30°41′E*
29 *Z13* **Dubuque** Iowa, C USA *42°30′N 90°40′W*
118 *E12* **Dubysa** ≈ C Lithuania
Đưc Cơ *see* Chư Ty
191 *V12* **Duc de Gloucester, Îles du** ***Eng.*** Duke of Gloucester Islands. *island group* C French Polynesia
111 *C15* **Duchcov** ***Ger.*** Dux. Ústecký Kraj, NW Czech Republic *50°37′N 13°45′E*

◆ Country
● Country Capital
◇ Dependent Territory
○ Dependent Territory Capital
◈ Administrative Regions
✈ International Airport
▲ Mountain
▲▲ Mountain Range
🌋 Volcano
≈ River
◎ Lake
⊠ Reservoir

DUCHESNE – EAST HAVEN

37 N3 **Duchesne** Utah, W USA *40°09′N 110°24′W*
191 P17 **Ducie Island** *atoll* E Pitcairn Group of Islands
11 W15 **Duck Bay** Manitoba, S Canada *52°11′N 100°08′W*
23 X17 **Duck Key** *island* Florida Keys, Florida, SE USA
11 T14 **Duck Lake** Saskatchewan, S Canada *52°52′N 106°12′W*
11 V15 **Duck Mountain** ▲▲ Manitoba, S Canada
20 I9 **Duck River** ≈ Tennessee, S USA
20 M10 **Ducktown** Tennessee, S USA *35°01′N 84°24′W*
167 U10 **Đức Phô** Quang Ngai, C Vietnam *14°56′N 108°55′E*
Đưc Tho *see* Lin Camh
Đưc Trong *see* Liên Nghia
D-U-D *see* Dalap-Uliga-Djarrit
153 N15 **Dūddhinagar** *var.* Dūdhi. Uttar Pradesh, N India *24°09′N 83°16′E*
99 M25 **Dudelange** *var.* Forge du Sud, ***Ger.*** Dudelingen. Luxembourg, S Luxembourg *49°28′N 06°05′E*
Dudelingen *see* Dudelange
101 J15 **Duderstadt** Niedersachsen, C Germany *51°31′N 10°16′E*
Dūdhi *see* Dūddhinagar
122 K8 **Dudinka** Krasnoyarskiy Kray, N Russian Federation *69°27′N 86°13′E*
97 L20 **Dudley** C England, United Kingdom *52°30′N 02°05′W*
154 G13 **Dudna** ≈ C India
76 L16 **Duékoué** W Ivory Coast *05°50′N 05°22′W*
104 M5 **Dueñas** Castilla y León, N Spain *41°52′N 04°33′W*
104 K4 **Duerna** ≈ NW Spain
105 O6 **Duero** ***Port.*** Douro. ≈ Portugal/Spain ***see also*** Douro
Duero *see* Douro
Duesseldorf *see* Düsseldorf
21 P12 **Due West** South Carolina, SE USA *34°19′N 82°23′W*
195 P11 **Dufek Coast** *physical region* Antarctica
99 H17 **Duffel** Antwerpen, C Belgium *51°06′N 04°30′E*
35 S2 **Duffer Peak** ▲ Nevada, W USA *41°40′N 118°45′W*
187 Q9 **Duff Islands** *island group* E Solomon Islands
Dufour, Pizzo/Dufour, Punta *see* Dufour Spitze
108 E12 **Dufour Spitze** ***It.*** Pizzo Dufour, Punta Dufour. ▲ Italy/Switzerland *45°54′N 07°50′E*
112 D9 **Duga Resa** Karlovac, C Croatia *45°25′N 15°30′E*
22 H5 **Dugdemona River** ≈ Louisiana, S USA
154 J12 **Duggīpar** Mahārāshtra, C India *21°06′N 80°10′E*
112 B13 **Dugi Otok** *var.* Isola Grossa, ***It.*** Isola Lunga. *island* W Croatia
113 F14 **Dugopolje** Split-Dalmacija, S Croatia *43°35′N 16°35′E*
160 L8 **Du He** ≈ C China
54 M11 **Duida, Cerro** ▲ S Venezuela *03°21′N 65°45′W*
Duinekerke *see* Dunkerque
101 E15 **Duisburg** ***prev.*** Duisburg-Hamborn. Nordrhein-Westfalen, W Germany *51°25′N 06°47′E*
Duisburg-Hamborn *see* Duisburg
99 F14 **Duiveland** *island* SW Netherlands
98 M12 **Duiven** Gelderland, E Netherlands *51°57′N 06°02′E*
139 W10 **Dujaylah, Hawr ad** ◎ S Iraq
160 H9 **Dujiangyan** *var.* Guanxian, Guan Xian. Sichuan, C China *31°01′N 103°40′E*
81 L18 **Dujuuma** Shabeellaha Hoose, S Somalia *01°04′N 42°37′E*
139 T3 **Dûkan** ***Ar.*** Dūkān, ***var.*** Dokan. As Sulaymānīyah, E Iraq *35°55′N 44°58′E*
Dūkān *see* Dûkan
39 Z14 **Duke Island** *island* Alexander Archipelago, Alaska, USA
Dukelský Priesmy/ Dukelský Průsmyk *see* Dukla Pass
Duke of Gloucester Islands *see* Duc de Gloucester, Îles du
81 F14 **Duk Faiwil** Jonglei, E South Sudan *07°30′N 31°27′E*
141 T7 **Dukhān** C Qatar *25°29′N 50°48′E*
Dukhan Heights *see* Dukhān, Jabal
143 N16 **Dukhān, Jabal** *var.* Dukhan Heights. *hill range* S Qatar
127 Q7 **Dukhovnitskoye** Saratovskaya Oblast', W Russian Federation *52°31′N 48°32′E*
126 H4 **Dukhovshchina** Smolenskaya Oblast', W Russian Federation *55°15′N 32°22′E*
Dukielska, Przełęcz *see* Dukla Pass
111 N17 **Dukla** Podkarpackie, SE Poland *49°33′N 21°40′E*
Duklai Hág *see* Dukla Pass
111 N18 **Dukla Pass** ***Cz.*** Dukelský Průsmyk, ***Ger.*** Dukla-Pass, ***Hung.*** Duklai Hág, ***Pol.*** Przełęcz Dukielska, ***Slvk.*** Dukelský Priesmy. *pass* Poland/Slovakia
Dukla-Pass *see* Dukla Pass
Dukou *see* Panzhihua
118 I12 **Dūkštas** Utena, E Lithuania *55°32′N 26°21′E*
Dulaan *see* Herlenbayan-Ulaan
159 R10 **Dulan** *var.* Qagan Us. Qinghai, C China *36°11′N 97°51′E*
37 R8 **Dulce** New Mexico, SW USA *36°55′N 107°00′W*
43 N16 **Dulce, Golfo** *gulf* S Costa Rica
Dulce, Golfo *see* Izabal, Lago de
42 K6 **Dulce Nombre de Culmí** Olancho, C Honduras *15°09′N 85°37′W*
62 L9 **Dulce, Río** ≈ C Argentina
123 Q9 **Dulgalakh** ≈ NE Russian Federation
Dŭlgopol *see* Dalgopol
153 V14 **Dullabchara** Assam, NE India *24°25′N 92°22′E*
20 D3 **Dulles** ✈ (Washington DC)Virginia, NE USA *39°00′N 77°27′W*
101 E14 **Dülmen** Nordrhein-Westfalen, W Germany *51°51′N 07°17′E*
114 M7 **Dulovo** Silistra, NE Bulgaria *43°51′N 27°10′E*
29 W5 **Duluth** Minnesota, N USA *46°47′N 92°06′W*
138 H7 **Dūmā** ***Fr.*** Douma. Rīf Dimashq, SW Syria *33°33′N 36°24′E*
171 O8 **Dumagasa Point** *headland* Mindanao, S Philippines *07°01′N 121°54′E*
171 P6 **Dumaguete** *var.* Dumaguete City. Negros, C Philippines *09°16′N 123°17′E*
Dumaguete City *see* Dumaguete
168 J10 **Dumai** Sumatera, W Indonesia *01°39′N 101°28′E*
183 T4 **Dumaresq River** ≈ New South Wales/Queensland, SE Australia
27 W13 **Dumas** Arkansas, C USA *33°53′N 91°29′W*
25 N1 **Dumas** Texas, SW USA *35°51′N 101°57′W*
138 I7 **Dumayr** Rīf Dimashq, W Syria *33°36′N 36°28′E*
96 I12 **Dumbarton** W Scotland, United Kingdom *55°57′N 04°35′W*
96 I12 **Dumbarton** *cultural region* C Scotland, United Kingdom
187 Q17 **Dumbéa** Province Sud, S New Caledonia *22°11′S 166°27′E*
111 K19 **Ďumbier** ***Ger.*** Djumbir, ***Hung.*** Gyömbér. ▲ C Slovakia *48°54′N 19°36′E*
116 I11 **Dumbrăveni** ***Ger.*** Elisabethstedt, ***Hung.*** Erzsébetváros; ***prev.*** Ebesfalva, Eppeschdorf, Ibaşfalău. Sibiu, C Romania *46°14′N 24°34′E*
116 L12 **Dumbrăveni** Vrancea, E Romania *45°31′N 27°09′E*
97 J14 **Dumfries** S Scotland, United Kingdom *55°04′N 03°37′W*
97 J14 **Dumfries** *cultural region* SW Scotland, United Kingdom
153 R15 **Dumka** Jhārkhand, NE India *24°17′N 87°15′E*
Dümmer *see* Dümmersee
100 G12 **Dümmersee** *var.* Dümmer. ◎ NW Germany
14 J11 **Dumoine** ≈ Québec, SE Canada
14 J10 **Dumoine, Lac** ◎ Québec, SE Canada
195 V16 **Dumont d'Urville** *French research station* Antarctica *66°24′S 139°38′E*
195 W15 **Dumont d'Urville Sea** *sea* S Pacific Ocean
14 K11 **Dumont, Lac** ◎ Québec, SE Canada
75 W7 **Dumyāt** *var.* Dumyât, ***Eng.*** Damietta. N Egypt *31°26′N 31°48′E*
Duna *see* Danube, C Europe
Düna *see* Western Dvina
Duna *see* Don, Russian Federation
Dünaburg *see* Daugavpils
111 J24 **Dunaföldvár** Tolna, C Hungary *46°48′N 18°55′E*
Dunaj *see* Wien, Austria
Dunaj *see* Danube, C Europe
111 L18 **Dunajec** ≈ S Poland
111 H21 **Dunajská Streda** ***Hung.*** Dunaszerdahely. Trnavský Kraj, W Slovakia *48°N 17°28′E*
Dunapentele *see* Dunaújváros
Dunărea *see* Danube
116 M13 **Dunărea Veche, Braţul** ≈ SE Romania
117 N13 **Dunării, Delta** *delta* SE Romania
Dunaszerdahely *see* Dunajská Streda
111 J23 **Dunaújváros** ***prev.*** Dunapentele, Sztálinváros. Fejér, C Hungary *47°N 18°55′E*
Dunav *see* Danube
114 J8 **Dunavska Ravnina** ***Eng.*** Danubian Plain. *lowlands* N Bulgaria
114 G7 **Dunavtsi** Vidin, NW Bulgaria *43°54′N 22°49′E*
123 S15 **Dunay** Primorskiy Kray, SE Russian Federation *42°53′N 132°20′E*
Dunayevtsy *see* Dunayivtsi
116 L7 **Dunayivtsi** ***Rus.*** Dunayevtsy. Khmel'nyts'ka Oblast', NW Ukraine *48°56′N 26°50′E*
185 F22 **Dunback** Otago, South Island, New Zealand *45°22′S 170°37′E*
10 L17 **Duncan** Vancouver Island, British Columbia, SW Canada *48°46′N 123°10′W*
37 O15 **Duncan** Arizona, SW USA *32°43′N 109°06′W*
26 M12 **Duncan** Oklahoma, C USA *34°30′N 97°57′W*
Duncan Island *see* Pinzón, Isla
151 Q20 **Duncan Passage** *strait* Andaman Sea/Bay of Bengal
96 K6 **Duncansby Head** *headland* N Scotland, United Kingdom *58°37′N 03°01′W*
14 G12 **Dunchurch** Ontario, S Canada *45°36′N 79°54′W*
118 D7 **Dundaga** NW Latvia *57°29′N 22°19′E*
14 G14 **Dundalk** Ontario, S Canada *44°11′N 80°22′W*
97 F16 **Dundalk** ***Ir.*** Dún Dealgan. Louth, NE Ireland *54°01′N 06°25′W*
21 X3 **Dundalk** Maryland, NE USA *39°15′N 76°31′W*
97 F16 **Dundalk Bay** ***Ir.*** Cuan Dhún Dealgan. *bay* NE Ireland
14 G16 **Dundas** Ontario, S Canada *43°16′N 79°55′W*
180 L12 **Dundas, Lake** *salt lake* Western Australia
Dundbürd *see* Batnorov
Dún Dealgan *see* Dundalk
15 N13 **Dundee** Québec, SE Canada *45°01′N 74°27′W*
83 K22 **Dundee** KwaZulu/Natal, E South Africa *28°09′S 30°12′E*
96 K11 **Dundee** E Scotland, United Kingdom *56°28′N 03°W*
31 R10 **Dundee** Michigan, N USA *41°57′N 83°39′W*
25 R5 **Dundee** Texas, SW USA *33°43′N 98°52′W*
194 H3 **Dundee Island** *island* Antarctica
162 L9 **Dundgovĭ** ◈ *province* C Mongolia
97 G16 **Dundrum Bay** ***Ir.*** Cuan Dhún Droma. *inlet* NW Irish Sea
11 T15 **Dundurn** Saskatchewan, S Canada *51°43′N 106°27′W*
Dund-Us *see* Hovd
Dund-Us *see* Hovd
185 F23 **Dunedin** Otago, South Island, New Zealand *45°52′S 170°31′E*
183 R7 **Dunedoo** New South Wales, SE Australia *32°04′S 149°23′E*
97 D14 **Dunfanaghy** ***Ir.*** Dún Fionnachaidh. NW Ireland *55°11′N 07°59′W*
96 J12 **Dunfermline** C Scotland, United Kingdom *56°04′N 03°29′W*
Dún Fionnachaidh *see* Dunfanaghy
149 V10 **Dunga Bunga** Punjab, E Pakistan *29°51′N 73°19′E*
97 F15 **Dungannon** ***Ir.*** Dún Geanainn. C Northern Ireland, United Kingdom *54°31′N 06°46′W*
Dun Garbháin *see* Dungarvan
152 F15 **Dūngarpur** Rājasthān, N India *23°50′N 73°43′E*
97 E21 **Dungarvan** ***Ir.*** Dun Garbháin. Waterford, S Ireland *52°05′N 07°37′W*
101 N21 **Dungau** *cultural region* SE Germany
Dún Geanainn *see* Dungannon
97 P23 **Dungeness** *headland* SE England, United Kingdom *50°55′N 00°58′E*
63 I23 **Dungeness, Punta** *headland* S Argentina *52°25′S 68°25′W*
Dungloe *see* Dunglow
97 D14 **Dunglow** *var.* Dungloe, ***Ir.*** An Clochán Liath. Donegal, NW Ireland *54°57′N 08°22′W*
183 T7 **Dungog** New South Wales, SE Australia *32°24′S 151°45′E*
79 O16 **Dungu** Orientale, NE Dem. Rep. Congo *03°40′N 28°32′E*
168 L8 **Dungun** *var.* Kuala Dungun. Terengganu, Peninsular Malaysia *04°47′N 103°26′E*
80 I6 **Dungûnab** Red Sea, NE Sudan *21°10′N 37°09′E*
15 P13 **Dunham** Québec, SE Canada *45°08′N 72°48′W*
Dunheved *see* Launceston
Dunholme *see* Durham
163 X10 **Dunhua** Jilin, NE China *43°22′N 128°12′E*
159 P8 **Dunhuang** Gansu, N China *40°10′N 94°40′E*
182 L12 **Dunkeld** Victoria, SE Australia *37°41′S 142°19′E*
103 O1 **Dunkerque** ***Eng.*** Dunkirk, ***Flem.*** Duinekerke; ***prev.*** Dunquerque. Nord, N France *51°06′N 02°34′E*
97 K23 **Dunkery Beacon** ▲ SW England, United Kingdom *51°10′N 03°36′W*
18 C11 **Dunkirk** New York, NE USA *42°28′N 79°19′W*
Dunkirk *see* Dunkerque
77 P17 **Dunkwa** SW Ghana *05°59′N 01°45′W*
97 G18 **Dún Laoghaire** ***Eng.*** Dunleary; ***prev.*** Kingstown. E Ireland *53°17′N 06°08′W*
29 S14 **Dunlap** Iowa, C USA *41°51′N 95°36′W*
20 L10 **Dunlap** Tennessee, S USA *35°22′N 85°23′W*
Dunleary *see* Dún Laoghaire
Dún Mánmhaí *see* Dunmanway
97 B21 **Dunmanway** ***Ir.*** Dún Mánmhaí. Cork, SW Ireland *51°43′N 09°07′W*
18 I13 **Dunmore** Pennsylvania, NE USA *41°25′N 75°37′W*
21 U10 **Dunn** North Carolina, SE USA *35°18′N 78°36′W*
Dún na nGall *see* Donegal
23 V11 **Dunnellon** Florida, SE USA *29°03′N 82°27′W*
96 J6 **Dunnet Head** *headland* N Scotland, United Kingdom *58°40′N 03°27′W*
29 N14 **Dunning** Nebraska, C USA *41°49′N 100°04′W*
65 B24 **Dunnose Head Settlement** West Falkland, Falkland Islands *51°24′S 60°29′W*
14 G17 **Dunnville** Ontario, S Canada *42°54′N 79°36′W*
Dún Pádraig *see* Downpatrick
Dunquerque *see* Dunkerque
Dunqulah *see* Dongola
96 L12 **Duns** SE Scotland, United Kingdom *55°46′N 02°13′W*
29 N2 **Dunseith** North Dakota, N USA *48°48′N 100°03′W*
35 N2 **Dunsmuir** California, W USA *41°12′N 122°19′W*
97 N21 **Dunstable** ***Lat.*** Durocobrivae. E England, United Kingdom *51°53′N 00°32′W*
185 D21 **Dunstan Mountains** ▲▲ South Island, New Zealand
103 O9 **Dun-sur-Auron** Cher, C France *46°52′N 02°40′E*
185 F21 **Duntroon** Canterbury, South Island, New Zealand *44°52′S 170°40′E*
149 T10 **Dunyāpur** Punjab, E Pakistan *29°48′N 71°48′E*
163 U5 **Duobukur He** ≈ NE China
163 R12 **Duolun** *var.* Dolonnur. Nei Mongol Zizhiqu, N China *42°11′N 116°30′E*
167 R14 **Dương Đông** Kiên Giang, S Vietnam *10°15′N 103°58′E*
114 G10 **Dupnitsa** ***prev.*** Marek, Stanke Dimitrov. Kyustendil, W Bulgaria *42°16′N 23°07′E*
28 L8 **Dupree** South Dakota, N USA *45°03′N 101°36′W*
33 Q7 **Dupuyer** Montana, NW USA *48°13′N 112°34′W*
141 Y11 **Duqm** *var.* Daqm. E Oman *19°42′N 57°40′E*
63 F23 **Duque de York, Isla** *island* S Chile
181 N4 **Durack Range** ▲▲ Western Australia
136 K10 **Durağan** Sinop, N Turkey *41°26′N 35°03′E*
103 S15 **Durance** ≈ SE France
31 R9 **Durand** Michigan, N USA *42°54′N 83°58′W*
30 I6 **Durand** Wisconsin, N USA *44°37′N 91°56′W*
40 K10 **Durango** *var.* Victoria de Durango. Durango, W Mexico *24°03′N 104°38′W*
105 P3 **Durango** País Vasco, N Spain *43°10′N 02°38′W*
37 Q8 **Durango** Colorado, C USA *37°13′N 107°51′W*
40 J9 **Durango** ◈ *state* C Mexico
114 O7 **Durankulak** ***Rom.*** Răcari; ***prev.*** Blatnitsa, Duranulac. Dobrich, NE Bulgaria *43°41′N 28°31′E*
22 L4 **Durant** Mississippi, S USA *33°04′N 89°51′W*
27 P13 **Durant** Oklahoma, C USA *33°59′N 96°24′W*
Duranulac *see* Durankulak
105 N6 **Duratón** ≈ N Spain
61 E19 **Durazno** *var.* San Pedro de Durazno. Durazno, C Uruguay *33°22′S 56°31′W*
61 E19 **Durazno** ◈ *department* C Uruguay
Durazzo *see* Durrës
83 K23 **Durban** *var.* Port Natal. KwaZulu/Natal, E South Africa *29°51′S 31°E*
83 K23 **Durban** ✈ KwaZulu/Natal, E South Africa *29°55′S 31°01′E*
118 C9 **Durbe** ***Ger.*** Durben. W Latvia *56°34′N 21°22′E*
Durben *see* Durbe
99 K21 **Durbuy** Luxembourg, SE Belgium *50°21′N 05°27′E*
105 N15 **Dúrcal** Andalucía, S Spain *37°00′N 03°24′W*
112 F8 **Đurđevac** ***Ger.*** Sankt Georgen, ***Hung.*** Szentgyörgy; ***prev.*** Djurdjevac, Gjurgjevac. Koprivnica-Kri»zevci, N Croatia *46°02′N 17°03′E*
113 K15 **Đurđevica Tara** N Montenegro *43°09′N 19°18′E*
97 L24 **Durdle Door** *natural arch* S England, United Kingdom
158 L3 **Düre** Xinjiang Uygur Zizhiqu, W China *46°30′N 88°26′E*
101 D16 **Düren** ***anc.*** Marcodurum. Nordrhein-Westfalen, W Germany *50°48′N 06°30′E*
154 K12 **Durg** ***prev.*** Drug. Chhattīsgarh, C India *21°12′N 81°20′E*
153 U13 **Durgapur** Dhaka, N Bangladesh *25°10′N 90°41′E*
153 R15 **Durgāpur** West Bengal, NE India *23°30′N 87°20′E*
14 F14 **Durham** Ontario, S Canada *44°10′N 80°48′W*
97 M14 **Durham** ***hist.*** Dunholme. N England, United Kingdom *54°47′N 01°34′W*
21 U9 **Durham** North Carolina, SE USA *36°N 78°54′W*
97 L15 **Durham** *cultural region* N England, United Kingdom
168 J10 **Duri** Sumatera, W Indonesia *01°13′N 101°13′E*
Duria Major *see* Dora Baltea
Duria Minor *see* Dora Riparia
Durlas *see* Thurles
141 P8 **Ḑurmā** Ar Riyāḑ, C Saudi Arabia *24°37′N 46°06′E*
113 J15 **Durmitor** ▲ N Montenegro
96 H6 **Durness** N Scotland, United Kingdom *58°34′N 04°46′W*
109 Y3 **Dürnkrut** Niederösterreich, E Austria *48°28′N 16°50′E*
Durnovaria *see* Dorchester
Durobrivae *see* Rochester
Durocasses *see* Dreux
Durocobrivae *see* Dunstable
Durocortorum *see* Reims
Durostorum *see* Silistra
Durovernum *see* Canterbury
113 K20 **Durrës** *var.* Durrësi, Dursi, ***It.*** Durazzo, ***SCr.*** Drač, ***Turk.*** Draç. Durrës, W Albania *41°20′N 19°26′E*
113 K19 **Durrës** ◈ *district* W Albania
Durrësi *see* Durrës
97 A21 **Dursey Island** ***Ir.*** Oileán Baoi. *island* SW Ireland
Dursi *see* Durrës
Duru *see* Wuchuan
Durud *see* Dow Rūd
114 P12 **Durusu** İstanbul, NW Turkey *41°18′N 28°41′E*
114 O12 **Durusu Gölü** ◎ NW Turkey
138 I9 **Durūz, Jabal ad** ▲ SW Syria *37°00′N 32°30′E*
184 K13 **D'Urville Island** *island* C New Zealand
171 X12 **D'Urville, Tanjung** *headland* Papua, E Indonesia *01°26′S 137°52′E*
146 H14 **Duşak** ***Rus.*** Dushak. Ahal Welaýaty, S Turkmenistan *37°15′N 59°57′E*
Dusa Mareb/Dusa Marreb *see* Dhuusa Marreeb
118 I11 **Dusetos** Utena, NE Lithuania *55°44′N 25°49′E*
Dushak *see* Duşak
160 K12 **Dushan** Guizhou, S China *25°50′N 107°36′E*
147 P13 **Dushanbe** *var.* Dyushambe; ***prev.*** Stalinabad, ***Taj.*** Stalinobod. ● (Tajikistan) W Tajikistan *38°35′N 68°44′E*
147 P13 **Dushanbe** ✈ W Tajikistan *38°31′N 68°49′E*
137 T9 **Dusheti** ***prev.*** Dushet'i. E Georgia *42°07′N 44°44′E*
Dushet'i *see* Dusheti
18 H13 **Dushore** Pennsylvania, NE USA *41°30′N 76°23′W*
185 A23 **Dusky Sound** *sound* South Island, New Zealand
101 E15 **Düsseldorf** *var.* Duesseldorf. Nordrhein-Westfalen, W Germany *51°14′N 06°49′E*
147 P14 **Dŭstí** ***Rus.*** Dusti. SW Tajikistan *37°22′N 68°41′E*
194 I9 **Dustin Island** *island* Antarctica
Dutch East Indies *see* Indonesia
Dutch Guiana *see* Suriname
38 L17 **Dutch Harbor** Unalaska Island, Alaska, USA *53°51′N 166°33′W*
36 J3 **Dutch Mount** ▲ Utah, W USA *40°16′N 113°56′W*
Dutch New Guinea *see* Papua
Dutch West Indies *see* Curaçao
83 H20 **Dutlwe** Kweneng, S Botswana *23°58′S 23°56′E*
67 V16 **Du Toit Fracture Zone** *tectonic feature* SW Indian Ocean
125 U8 **Dutovo** Respublika Komi, NW Russian Federation *63°45′N 56°38′E*
77 V13 **Dutsan Wai** *var.* Dutsen Wai. Kaduna, C Nigeria *10°49′N 08°15′E*
77 W13 **Dutse** Jigawa, N Nigeria *11°43′N 09°25′E*
Dutsen Wai *see* Dutsan Wai
Duttia *see* Datia
14 E17 **Dutton** Ontario, S Canada *42°40′N 81°28′W*
36 L7 **Dutton, Mount** ▲ Utah, W USA *38°00′N 112°10′W*
83 J25 **Dutywa** ***prev.*** Idutywa. Eastern Cape, SE South Africa *32°06′S 28°20′E* ***see also*** Dutywa
162 E7 **Duut** Hovd, W Mongolia *47°28′N 91°52′E*
14 K11 **Duval, Lac** ◎ Québec, SE Canada
127 W3 **Duvan** Respublika Bashkortostan, W Russian Federation *55°42′N 57°56′E*
138 L9 **Duwaykhilat Satiḩ ar Ruwayshid** *seasonal river* SE Jordan
Dux *see* Duchcov
160 J13 **Duyang Shan** ▲▲ S China
167 T14 **Duyên Hai** Tra Vinh, S Vietnam *09°39′N 106°28′E*
160 K12 **Duyun** Guizhou, S China *26°16′N 107°29′E*
136 G11 **Düzce** Düzce, NW Turkey *40°51′N 31°09′E*
136 K14 **Düzce** ◈ *province* NW Turkey
Duzdab *see* Zāhedān
Duzenkyr, Khrebet *see* Duzkyr, Khrebet
146 I16 **Duzkyr, Khrebet** ***prev.*** Khrebet Duzenkyr. ▲▲ S Turkmenistan
114 K8 **Dve Mogili** Ruse, N Bulgaria *43°35′N 25°51′E*
Dvina Bay *see* Chëshskaya Guba
Dvinsk *see* Daugavpils
124 L7 **Dvinskaya Guba** *bay* NW Russian Federation
112 E10 **Dvor** Sisak-Moslavina, C Croatia *45°05′N 16°22′E*
117 W5 **Dvorichna** Kharkivs'ka Oblast', E Ukraine *49°52′N 37°43′E*
111 F16 **Dvůr Králové nad Labem** ***Ger.*** Königinhof an der Elbe. Královéhradecký Kraj, N Czech Republic *50°27′N 15°50′E*
154 A10 **Dwārka** Gujarāt, W India *22°14′N 68°58′E*
30 M12 **Dwight** Illinois, N USA *41°05′N 88°25′W*
98 N8 **Dwingeloo** Drenthe, NE Netherlands *52°49′N 06°20′E*
33 N10 **Dworshak Reservoir** ⊠ Idaho, NW USA
Dyal *see* Dyaul Island
Dyanev *see* Galkynyş
Dyatlovo *see* Dzyatlava
186 G5 **Dyaul Island** *var.* Djaul, Dyal. *island* NE Papua New Guinea
20 G8 **Dyer** Tennessee, S USA *36°04′N 88°59′W*
9 S5 **Dyer, Cape** *headland* Baffin Island, Nunavut, NE Canada *66°37′N 61°13′W*
20 F8 **Dyersburg** Tennessee, S USA *36°02′N 89°21′W*
29 Y13 **Dyersville** Iowa, C USA *42°29′N 91°07′W*
97 I21 **Dyfed** *cultural region* SW Wales, United Kingdom
Dyfrdwy, Afon *see* Dee
Dyhernfurth *see* Brzeg Dolny
111 E19 **Dyje** *var.* Thaya. ≈ Austria/ Czech Republic ***see also*** Thaya
Dyje *see* Thaya
117 T5 **Dykan'ka** Poltavs'ka Oblast', C Ukraine *49°48′N 34°33′E*
127 N16 **Dykhtau** ▲ SW Russian Federation *43°01′N 42°56′E*
111 A16 **Dyleń** ***Ger.*** Tillenberg. ▲ NW Czech Republic *49°58′N 12°31′E*
110 K9 **Dylewska Góra** ▲ N Poland *53°33′N 19°57′E*
117 O4 **Dymer** Kyyivs'ka Oblast', N Ukraine *50°50′N 30°20′E*
117 W7 **Dymytrov** ***Rus.*** Dimitrov. Donets'ka Oblast', SE Ukraine *48°18′N 37°19′E*
111 O17 **Dynów** Podkarpackie, SE Poland *49°49′N 22°14′E*
29 X13 **Dysart** Iowa, C USA *42°10′N 92°18′W*
Dysna *see* Dzisna
115 H18 **Dýstos, Límni** *var.* Límni Dístos. ◎ Évvoia, C Greece
115 D18 **Dytikí Elláda** ***Eng.*** Greece West, ***var.*** Dytikí Ellás. ◈ *region* C Greece
Dytikí Ellás *see* Dytikí Elláda
115 C14 **Dytikí Makedonía** ***Eng.*** Macedonia West. ◈ *region* N Greece
Dyurment'yube *see* Diirmentobe
127 U4 **Dyurtyuli** Respublika Bashkortostan, W Russian Federation *55°31′N 54°49′E*
Dyushambe *see* Dushanbe
162 K7 **Dzaamar** *var.* Bat-Öldziyt. Töv, C Mongolia *48°10′N 104°49′E*
Dzaanhushuu *see* Ihtamir
Dza Chu *see* Mekong
Dzadgay *see* Bömbögör
162 H8 **Dzag** Bayanhongor, C Mongolia *46°54′N 99°11′E*
Dzalaa *see* Shinejinst
163 O11 **Dzamïn-Üüd** *var.* Borhoyn Tal. Dornogovĭ, SE Mongolia *43°43′N 111°53′E*
172 J14 **Dzaoudzi** E Mayotte *12°48′S 45°18′E*
Dzaudzhikau *see* Vladikavkaz
162 G7 **Dzavhan** ◈ *province* NW Mongolia
162 G7 **Dzavhan Gol** ≈ NW Mongolia
162 G6 **Dzavhanmandal** *var.* Nuga. Dzavhan, W Mongolia *48°17′N 95°07′E*
Dzegstey *see* Ögiynuur
162 E7 **Dzereg** *var.* Altanteel. Hovd, W Mongolia *47°05′N 92°57′E*
127 O3 **Dzerzhinsk** Nizhegorodskaya Oblast', W Russian Federation *56°20′N 43°22′E*
Dzerzhinsk *see* Dzyarzhynsk
Dzerzhinsk *see* Romaniv
Dzerzhinskiy *see* Nar'yan-Mar
Dzerzhinskoye *see* Tokzhaylau
Dzerzhinskoye *see* Tokzhaylau
Dzerzhyns'k *see* Romaniv
Dzerzhyns'k *see* Romaniv
Džetygara *see* Zhitikara
Dzhailgan *see* Jayilgan
Dzhalagash *see* Zhalagash
147 T10 **Dzhalal-Abad** ***Kir.*** Jalal-Abad. Dzhalal-Abadskaya Oblast', W Kyrgyzstan *40°56′N 73°00′E*
147 S9 **Dzhalal-Abadskaya Oblast'** ***Kir.*** Jalal-Abad Oblasty. ◈ *province* W Kyrgyzstan
Dzhalilabad *see* Cälilabad
Dzhambeyty *see* Zhympity
Dzhambul *see* Taraz
Dzhambulskaya Oblast' *see* Zhambyl
144 D9 **Dzhanibek** ***prev.*** Dzhanybek, ***Kaz.*** Zhänibek. Zapadnyy Kazakhstan, W Kazakhstan *49°27′N 46°51′E*
Dzhankel'dy *see* Jongeldi
117 T12 **Dzhankoy** Avtonomna Respublika Krym, S Ukraine *45°40′N 34°20′E*
Dzhansugurov *see* Zhansugirov
147 R9 **Dzhany-Bazar** *var.* Yangibazar. Dzhalal-Abadskaya Oblast', W Kyrgyzstan *41°40′N 70°49′E*
Dzhanybek *see* Dzhanibek
123 P8 **Dzhardzhan** Respublika Sakha (Yakutiya), NE Russian Federation *68°47′N 123°51′E*
Dzharkurgan *see* Jarqo'rg'on
117 S11 **Dzharylhats'ka Zatoka** *gulf* S Ukraine
Dzhayilgan *see* Jayilgan
Dzhebel *see* Jebel
147 T14 **Dzhelandy** SE Tajikistan *37°34′N 72°35′E*
147 Y7 **Dzhergalan** ***Kir.*** Jyrgalan. Issyk-Kul'skaya Oblast', NE Kyrgyzstan *42°37′N 78°56′E*
Dzhetysay *see* Zhetysay
Dzhezkazgan *see* Zhezkazgan
Dzhigirbent *see* Jigerbent
Dzhirgatal' *see* Jirgatol
Dzhizak *see* Jizzax
Dzhizakskaya Oblast' *see* Jizzax Viloyati
123 P8 **Dzhugdzhur, Khrebet** ▲▲ E Russian Federation
Dzhul'fa *see* Culfa
Dzhuma *see* Juma
Dzhungarskiy Alatau *see* Zhetysuskiy Alatau
Dzhusaly *see* Zhosaly
146 J12 **Dzhynlykum, Peski** *desert* E Turkmenistan
110 L9 **Działdowo** Warmińsko-Mazurskie, C Poland *53°13′N 20°12′E*
111 L16 **Działoszyce** Świętokrzyskie, C Poland *50°21′N 20°19′E*
41 X11 **Dzidzantún** Yucatán, E Mexico
111 G15 **Dzierżoniów** ***Ger.*** Reichenbach. Dolnośląskie, SW Poland *50°43′N 16°40′E*
41 X11 **Dzilam de Bravo** Yucatán, E Mexico *21°24′N 88°52′W*
118 L12 **Dzisna** ***Rus.*** Disna. Vitsyebskaya Voblasts', N Belarus *55°33′N 28°13′E*
118 K12 **Dzisna** ***Lith.*** Dysna, ***Rus.*** Disna. ≈ Belarus/Lithuania
119 G20 **Dzivin** ***Rus.*** Divin. Brestskaya Voblasts', SW Belarus *51°58′N 24°33′E*
119 M15 **Dzmitravichy** ***Rus.*** Dmitrovichi. Minskaya Voblasts', C Belarus *53°58′N 29°14′E*
Dzogsool *see* Bayantsagaan
162 I5 **Dzöölön** *var.* Rinchinlhumbe. Hövsgöl, N Mongolia *51°06′N 99°40′E*
129 S8 **Dzungaria** *var.* Sungaria, Zungaria. *physical region* W China
Dzungarian Basin *see* Junggar Pendi
Dzür *see* Tes
162 J8 **Dzüünbayan-Ulaan** *var.* Bayan-Ulaan. Övörhangay, C Mongolia *46°38′N 102°30′E*
Dzüünbulag *see* Matad, Dornod, Mongolia
Dzüünbulag *see* Uulbayan, Sühbaatar, Mongolia
162 L8 **Dzuunmod** Töv, C Mongolia *47°45′N 107°00′E*
Dzuunmod *see* Ider
Dzüün Soyonï Nuruu *see* Vostochnyy Sayan
Dzüyl *see* Tonhil
Dzvina *see* Western Dvina
119 J16 **Dzyarzhynsk** ***Rus.*** Dzerzhinsk; ***prev.*** Kaydanovo. Minskaya Voblasts', C Belarus *53°41′N 27°09′E*
119 H17 **Dzyatlava** ***Pol.*** Zdzięciół, ***Rus.*** Dyatlovo. Hrodzyenskaya Voblasts', W Belarus *53°27′N 25°23′E*

E

E *see* Hubei
Éadan Doire *see* Edenderry
167 U12 **Ea Drăng** *var.* Ea H'leo. Đăc Lăc, S Vietnam *13°09′N 108°14′E*
37 W6 **Eads** Colorado, C USA *38°28′N 102°46′W*
37 O13 **Eagar** Arizona, SW USA *34°05′N 109°17′W*
39 T8 **Eagle** Alaska, USA *64°47′N 141°12′W*
13 S8 **Eagle** ≈ Newfoundland and Labrador, E Canada
10 I3 **Eagle** ≈ Yukon, NW Canada
29 T7 **Eagle Bend** Minnesota, N USA *46°10′N 95°02′W*
28 M8 **Eagle Butte** South Dakota, N USA *44°58′N 101°13′W*
29 V12 **Eagle Grove** Iowa, C USA *42°39′N 93°54′W*
19 R2 **Eagle Lake** Maine, NE USA *47°01′N 68°35′W*
25 U11 **Eagle Lake** Texas, SW USA *29°35′N 96°19′W*
12 A11 **Eagle Lake** ◎ Ontario, S Canada
35 P3 **Eagle Lake** ◎ California, W USA
19 R3 **Eagle Lake** ◎ Maine, NE USA
29 Y3 **Eagle Mountain** ▲ Minnesota, N USA *47°54′N 90°33′W*
25 T6 **Eagle Mountain Lake** ⊠ Texas, SW USA
37 S9 **Eagle Nest Lake** ⊠ New Mexico, SW USA
25 P13 **Eagle Pass** Texas, SW USA *28°44′N 100°31′W*
65 C25 **Eagle Passage** *passage* SW Atlantic Ocean
35 R8 **Eagle Peak** ▲ California, W USA *38°11′N 119°22′W*
35 Q2 **Eagle Peak** ▲ California, W USA *41°16′N 120°12′W*
37 P13 **Eagle Peak** ▲ New Mexico, SW USA *33°39′N 109°36′W*
10 I4 **Eagle Plain** Yukon, NW Canada *66°23′N 136°42′W*
32 G15 **Eagle Point** Oregon, NW USA *42°28′N 122°48′W*
186 P10 **Eagle Point** *headland* SE Papua New Guinea *10°31′S 149°53′E*
39 R11 **Eagle River** Alaska, USA *61°18′N 149°38′W*
30 M2 **Eagle River** Michigan, N USA *47°24′N 88°18′W*
30 L4 **Eagle River** Wisconsin, N USA *45°56′N 89°15′W*
21 S6 **Eagle Rock** Virginia, NE USA *37°40′N 79°46′W*
36 J13 **Eagletail Mountains** ▲▲ Arizona, SW USA
Ea H'leo *see* Ea Drăng
167 U12 **Ea Kar** Đăc Lăc, S Vietnam *12°47′N 108°26′E*
Eanjum *see* Anjum
Eanodat *see* Enontekiö
12 B10 **Ear Falls** Ontario, C Canada *50°38′N 93°13′W*
27 X10 **Earle** Arkansas, C USA *35°16′N 90°28′W*
35 R12 **Earlimart** California, W USA *35°52′N 119°17′W*
20 I6 **Earlington** Kentucky, S USA *37°16′N 87°30′W*
14 H8 **Earlton** Ontario, S Canada *47°41′N 79°46′W*
29 T13 **Early** Iowa, C USA *42°27′N 95°09′W*
96 J11 **Earn** ≈ N Scotland, United Kingdom
185 C21 **Earnslaw, Mount** ▲ South Island, New Zealand *44°34′S 168°26′E*
24 M4 **Earth** Texas, SW USA *34°13′N 102°24′W*
21 P11 **Easley** South Carolina, SE USA *34°49′N 82°36′W*
East *see* Est
East Açores Fracture Zone *see* East Azores Fracture Zone
97 P19 **East Anglia** *physical region* E England, United Kingdom
15 Q12 **East Angus** Québec, SE Canada *45°29′N 71°39′W*
195 V8 **East Antarctica** ***prev.*** Greater Antarctica. *physical region* Antarctica
18 E10 **East Aurora** New York, NE USA *42°44′N 78°36′W*
East Australian Basin *see* Tasman Basin
East Azerbaijan *see* Āžarbāyjān-e Sharqī
64 L9 **East Azores Fracture Zone** *var.* East Açores Fracture Zone. *tectonic feature* E Atlantic Ocean
22 M11 **East Bay** *bay* Louisiana, S USA
25 V11 **East Bernard** Texas, SW USA *29°32′N 96°04′W*
29 V8 **East Bethel** Minnesota, N USA *45°24′N 93°14′W*
East Borneo *see* Kalimantan Timur
97 P23 **Eastbourne** SE England, United Kingdom *50°46′N 00°16′E*
15 R11 **East-Broughton** Québec, SE Canada *46°14′N 71°05′W*
44 M6 **East Caicos** *island* E Turks and Caicos Islands
184 R7 **East Cape** *headland* North Island, New Zealand *37°40′S 178°31′E*
174 M4 **East Caroline Basin** *undersea feature* SW Pacific Ocean *04°00′N 146°45′E*
192 P4 **East China Sea** ***Chin.*** Dong Hai. *sea* W Pacific Ocean
97 P19 **East Dereham** E England, United Kingdom *52°41′N 00°55′E*
30 J9 **East Dubuque** Illinois, N USA *42°29′N 90°38′W*
11 S17 **Eastend** Saskatchewan, S Canada *49°29′N 108°48′W*
193 S10 **Easter Fracture Zone** *tectonic feature* E Pacific Ocean
Easter Island *see* Pascua, Isla de
153 Q12 **Eastern** ◈ *zone* E Nepal
155 K25 **Eastern** ◈ *province* E Sri Lanka
82 L13 **Eastern** ◈ *province* E Zambia
83 H24 **Eastern Cape** ***off.*** Eastern Cape Province, ***Afr.*** Oos-Kaap. ◈ *province* SE South Africa
Eastern Cape Province *see* Eastern Cape
80 C12 **Eastern Darfur** ◈ *state* SW Sudan
Eastern Desert *see* Sahara el Sharqîya
81 F15 **Eastern Equatoria** ◈ *state* SE South Sudan
Eastern Euphrates *see* Murat Nehri
155 J17 **Eastern Ghats** ▲▲ SE India
186 E7 **Eastern Highlands** ◈ *province* C Papua New Guinea
Eastern Region *see* Ash Sharqīyah
Eastern Sayans *see* Vostochnyy Sayan
Eastern Scheldt *see* Oosterschelde
Eastern Sierra Madre *see* Madre Oriental, Sierra
Eastern Transvaal *see* Mpumalanga
11 W14 **Easterville** Manitoba, C Canada *53°06′N 99°53′W*
Easterwâlde *see* Oosterwolde
63 M23 **East Falkland** *var.* Isla Soledad. *island* E Falkland Islands
19 P12 **East Falmouth** Massachusetts, NE USA *41°34′N 70°31′W*
East Fayu *see* Fayu
East Flanders *see* Oost-Vlaanderen
39 S6 **East Fork Chandalar River** ≈ Alaska, USA
29 U12 **East Fork Des Moines River** ≈ Iowa/Minnesota, C USA
East Frisian Islands *see* Ostfriesische Inseln
18 K10 **East Glenville** New York, NE USA *42°53′N 73°55′W*
29 R4 **East Grand Forks** Minnesota, N USA *47°54′N 97°59′W*
97 O23 **East Grinstead** SE England, United Kingdom *51°08′N 00°00′W*
18 M12 **East Hartford** Connecticut, NE USA *41°45′N 72°36′W*
18 M13 **East Haven** Connecticut, NE USA *41°16′N 72°52′W*

◆ Country
● Country Capital
◇ Dependent Territory
○ Dependent Territory Capital
◈ Administrative Regions
✈ International Airport
▲ Mountain
▲▲ Mountain Range
⩙ Volcano
≈ River
◎ Lake
⊠ Reservoir

East Indiaman Ridge – El Hawata

173 *T9* **East Indiaman Ridge** *undersea feature* E Indian Ocean
129 *V16* **East Indies** *island group* SE Asia
East Java *see* Jawa Timur
31 *Q6* **East Jordan** Michigan, N USA 45°09′N 85°07′W
East Kalimantan *see* Kalimantan Timur
East Kazakhstan *see* Vostochnyy Kazakhstan
96 *I12* **East Kilbride** S Scotland, United Kingdom 55°46′N 04°10′W
25 *R7* **Eastland** Texas, SW USA 32°23′N 98°50′W
31 *Q9* **East Lansing** Michigan, N USA 42°44′N 84°28′W
35 *X11* **East Las Vegas** Nevada, W USA 36°05′N 115°02′W
97 *M23* **Eastleigh** S England, United Kingdom 50°58′N 01°22′W
31 *V12* **East Liverpool** Ohio, N USA 40°37′N 80°34′W
83 *J25* **East London** *Afr.* Oos-Londen; *prev.* Emonti, Port Rex. Eastern Cape, S South Africa 33°S 27°54′E
96 *K12* **East Lothian** *cultural region* SE Scotland, United Kingdom
12 *I10* **Eastmain** Québec, E Canada 52°11′N 78°27′W
12 *J10* **Eastmain** 🌊 Québec, C Canada
15 *P13* **Eastman** Québec, SE Canada 45°19′N 72°18′W
23 *U6* **Eastman** Georgia, SE USA 32°12′N 83°10′W
175 *O3* **East Mariana Basin** *undersea feature* W Pacific Ocean
30 *K11* **East Moline** Illinois, N USA 41°30′N 90°26′W
186 *H7* **East New Britain** ◈ *province* E Papua New Guinea
29 *T15* **East Nishnabotna River** 🌊 Iowa, C USA
197 *V12* **East Novaya Zemlya Trough** *var.* Novaya Zemlya Trough. *undersea feature* W Kara Sea
East Nusa Tenggara *see* Nusa Tenggara Timur
21 *X4* **Easton** Maryland, NE USA 38°46′N 76°04′W
18 *I14* **Easton** Pennsylvania, NE USA 40°41′N 75°13′W
193 *R16* **East Pacific Rise** *undersea feature* E Pacific Ocean 20°00′S 115°00′W
East Pakistan *see* Bangladesh
31 *V12* **East Palestine** Ohio, N USA 40°49′N 80°32′W
30 *L12* **East Peoria** Illinois, N USA 40°40′N 89°34′W
23 *S3* **East Point** Georgia, SE USA 33°40′N 84°26′W
19 *U6* **Eastport** Maine, NE USA 44°54′N 66°59′W
27 *Z8* **East Prairie** Missouri, C USA 36°46′N 89°23′W
19 *O12* **East Providence** Rhode Island, NE USA 41°48′N 71°20′W
20 *L11* **East Ridge** Tennessee, S USA 35°00′N 85°15′W
97 *N16* **East Riding** *cultural region* N England, United Kingdom
18 *F9* **East Rochester** New York, NE USA 43°06′N 77°29′W
30 *K15* **East Saint Louis** Illinois, N USA 38°35′N 90°07′W
65 *K21* **East Scotia Basin** *undersea feature* SE Scotia Sea
129 *Y8* **East Sea** *var.* Sea of Japan, *Rus.* Yapanskoye More. Sea NW Pacific Ocean **see also** Japan, Sea of
186 *B6* **East Sepik** ◈ *province* NW Papua New Guinea
173 *N4* **East Sheba Ridge** *undersea feature* W Arabian Sea 14°30′N 56°15′E
East Siberian Sea *see* Vostochno-Sibirskoye More
18 *I14* **East Stroudsburg** Pennsylvania, NE USA 41°00′N 75°10′W
East Tasmanian Rise/ East Tasmania Plateau/ East Tasmania Rise *see* East Tasman Plateau
192 *I12* **East Tasman Plateau** *var.* East Tasmanian Rise, East Tasmania Plateau, East Tasmania Rise. *undersea feature* SW Tasman Sea
64 *L7* **East Thulean Rise** *undersea feature* N Atlantic Ocean
171 *R16* **East Timor** *var.* Loro Sae; *prev.* Portuguese Timor, Timor Timur. ◆ *country* S Indonesia
21 *Y6* **Eastville** Virginia, NE USA 37°22′N 75°58′W
35 *R7* **East Walker River** 🌊 California/Nevada, W USA
182 *D1* **Eateringinna Creek** 🌊 South Australia
37 *T3* **Eaton** Colorado, C USA 40°31′N 104°42′W
15 *Q12* **Eaton** 🌊 Québec, SE Canada
11 *S16* **Eatonia** Saskatchewan, S Canada 51°13′N 109°22′W
31 *Q10* **Eaton Rapids** Michigan, N USA 42°30′N 84°39′W
23 *U4* **Eatonton** Georgia, SE USA 33°19′N 83°23′W
32 *H9* **Eatonville** Washington, NW USA 46°51′N 122°19′W
30 *J6* **Eau Claire** Wisconsin, N USA 44°50′N 91°30′W
12 *J7* **Eau Claire, Lac à l'** ◎ Québec, SE Canada **Eau Claire, Lac à L'** *see* St. Clair, Lake
30 *L6* **Eau Claire River** 🌊 Wisconsin, N USA
188 *J16* **Eauripik Atoll** *atoll* Caroline Islands, C Micronesia
192 *H7* **Eauripik Rise** *undersea feature* W Pacific Ocean 03°00′N 142°00′E
102 *K15* **Eauze** Gers, S France 43°52′N 00°06′E
41 *P11* **Ébano** San Luis Potosí, C Mexico 22°16′N 98°26′W
97 *K21* **Ebbw Vale** SE Wales, United Kingdom 51°48′N 03°13′W
79 *E17* **Ebebiyín** NE Equatorial Guinea 02°08′N 11°15′E
95 *H22* **Ebeltoft** Midtjylland, C Denmark 56°11′N 10°42′E
109 *X5* **Ebenfurth** Niederösterreich, E Austria 47°53′N 16°22′E
18 *D14* **Ebensburg** Pennsylvania, NE USA 40°28′N 78°44′W
109 *S5* **Ebensee** Oberösterreich, N Austria 47°48′N 13°46′E
101 *H20* **Eberbach** Baden-Württemberg, SW Germany 49°28′N 08°58′E
121 *U8* **Eber Gölü** *salt lake* C Turkey
109 *U9* **Eberndorf** *Slvn.* Dobrla Vas. Kärnten, S Austria 46°33′N 14°35′E
109 *R4* **Eberschwang** Oberösterreich, N Austria 48°09′N 13°37′E
100 *O11* **Eberswalde-Finow** Brandenburg, E Germany 52°50′N 13°48′E
Ebesfalva *see* Dumbrăveni
165 *T4* **Ebetsu** *var.* Ebetu. Hokkaidō, NE Japan 43°08′N 141°37′E
Ebetu *see* Ebetsu
Ebinayon *see* Evinayong
158 *I4* **Ebinur Hu** ◎ NW China
138 *I3* **Ebla** *Ar.* Tell Mardīkh. *site of ancient city* Idlib, NW Syria
Eblana *see* Dublin
108 *H7* **Ebnat** Sankt Gallen, NE Switzerland 47°16′N 09°07′E
107 *L18* **Eboli** Campania, S Italy 40°37′N 15°03′E
79 *E16* **Ebolowa** Sud, S Cameroon 02°56′N 11°11′E
79 *N21* **Ebombo** Kasai-Oriental, C Dem. Rep. Congo 05°42′S 26°07′E
189 *T9* **Ebon Atoll** *var.* Epoon. *atoll* Ralik Chain, S Marshall Islands
Ebora *see* Évora
Eboracum *see* York
Eborodunum *see* Yverdon
101 *J19* **Ebrach** Bayern, C Germany 49°49′N 10°30′E
109 *X5* **Ebreichsdorf** Niederösterreich, E Austria 47°58′N 16°24′E
105 *S6* **Ebro** 🌊 NE Spain
105 *N3* **Ebro, Embalse del** ⊠ N Spain
120 *G7* **Ebro Fan** *undersea feature* W Mediterranean Sea
Eburacum *see* York
Ebusus *see* Ibiza
Ebusus *see* Eivissa
99 *F20* **Écaussinnes-d'Enghien** Hainaut, SW Belgium 50°34′N 04°10′E
Ecbatana *see* Hamadān
21 *Q6* **Eccles** West Virginia, NE USA 37°46′N 81°16′W
115 *L14* **Eceabat** Çanakkale, NW Turkey 40°12′N 26°22′E
171 *O2* **Echague** Luzon, N Philippines 16°42′N 121°37′E
Ech Cheliff/Ech Chleff *see* Chlef
Echeng *see* Ezhou
115 *C18* **Echinádes** *island group* W Greece
114 *J12* **Echínos** *var.* Ehinos, Ekhínos. Anatolikí Makedonía kai Thráki, NE Greece 41°16′N 25°00′E
Echizen *see* Takefu
164 *J12* **Echizen-misaki** *headland* Honshū, SW Japan 35°59′N 135°57′E
Echmiadzin *see* Vagharshapat
8 *J8* **Echo Bay** Northwest Territories, NW Canada 66°04′N 118°W
35 *Y11* **Echo Bay** Nevada, W USA 36°19′N 114°27′W
36 *L9* **Echo Cliffs** *cliff* Arizona, SW USA
14 *C10* **Echo Lake** ◎ Ontario, S Canada
35 *Q7* **Echo Summit** ▲ California, W USA 38°47′N 120°06′W
14 *L8* **Échouani, Lac** ◎ Québec, SE Canada
99 *L17* **Echt** Limburg, SE Netherlands 51°07′N 05°52′E
101 *H22* **Echterdingen** ✈ (Stuttgart) Baden-Württemberg, SW Germany 48°40′N 09°13′E
99 *N24* **Echternach** Grevenmacher, E Luxembourg 49°49′N 06°25′E
183 *N11* **Echuca** Victoria, SE Australia 36°10′S 144°20′E
104 *L14* **Écija** *anc.* Astigi. Andalucía, SW Spain 37°33′N 05°04′W
Eckengraf *see* Viesīte
100 *I7* **Eckernförde** Schleswig-Holstein, N Germany 54°28′N 09°49′E
100 *J7* **Eckernförder Bucht** *inlet* N Germany
102 *L7* **Écommoy** Sarthe, NW France 47°51′N 00°15′E
14 *L10* **Écorce, Lac de l'** ◎ Québec, SE Canada
15 *Q8* **Écorces, Rivière aux** 🌊 Québec, SE Canada
56 *C7* **Ecuador** *off.* Republic of Ecuador. ◆ *republic* NW South America
Ecuador, Republic of *see* Ecuador
95 *I17* **Ed** Västra Götaland, S Sweden 58°55′N 11°55′E
Ed *see* 'Īdī
98 *I9* **Edam** Noord-Holland, C Netherlands 52°30′N 05°02′E
96 *K4* **Eday** *island* NE Scotland, United Kingdom
25 *S17* **Edcouch** Texas, SW USA 26°17′N 97°57′W
80 *B12* **Ed Da'ein** Eastern Darfur, W Sudan 11°25′N 26°08′E
80 *G11* **Ed Damazin** *var.* Ad Damazīn. Blue Nile, E Sudan 11°45′N 34°20′E
80 *G8* **Ed Damer** *var.* Ad Dāmir, Ad Damar. River Nile, NE Sudan 17°37′N 33°59′E
80 *E8* **Ed Debba** Northern, N Sudan 18°02′N 30°56′E
80 *F10* **Ed Dueim** *var.* Ad Duwaym, Ad Duwēm. White Nile, C Sudan 13°38′N 32°36′E
183 *Q16* **Eddystone Point** *headland* Tasmania, SE Australia 41°01′S 148°18′E
97 *I25* **Eddystone Rocks** *rocks* SW England, United Kingdom
29 *W15* **Eddyville** Iowa, C USA 41°09′N 92°37′W
20 *H7* **Eddyville** Kentucky, S USA 37°03′N 88°02′W
98 *L12* **Ede** Gelderland, C Netherlands 52°03′N 05°40′E
77 *T16* **Ede** Osun, SW Nigeria 07°40′N 04°21′E
79 *D16* **Edéa** Littoral, SW Cameroon 03°47′N 10°08′E
111 *M20* **Edelény** Borsod-Abaúj-Zemplén, NE Hungary 48°18′N 20°40′E
183 *R12* **Eden** New South Wales, SE Australia 37°04′S 149°51′E
21 *T8* **Eden** North Carolina, SE USA 36°29′N 79°46′W
25 *P9* **Eden** Texas, SW USA 31°13′N 99°51′W
97 *K14* **Eden** 🌊 NW England, United Kingdom
83 *I23* **Edenburg** Free State, C South Africa 29°45′S 25°57′E
185 *D24* **Edendale** Southland, South Island, New Zealand 46°18′S 168°48′E
97 *E18* **Edenderry** *Ir.* Éadan Doire. Offaly, C Ireland 53°21′N 07°03′W
182 *L11* **Edenhope** Victoria, SE Australia 37°04′S 141°15′E
21 *X8* **Edenton** North Carolina, SE USA 36°04′N 76°39′W
101 *G16* **Eder** 🌊 NW Germany
101 *H15* **Edersee** ◎ W Germany
114 *E13* **Édessa** *var.* Edhessa. Kentrikí Makedonía, N Greece 40°48′N 22°03′E
Edessa *see* Şanlıurfa
Edfu *see* Idfū
29 *P16* **Edgar** Nebraska, C USA 40°22′N 97°58′W
19 *P13* **Edgartown** Martha's Vineyard, Massachusetts, NE USA 41°23′N 70°30′W
39 *X13* **Edgecumbe, Mount** ▲ Baranof Island, Alaska, USA 57°03′N 135°45′W
21 *Q13* **Edgefield** South Carolina, SE USA 33°50′N 81°57′W
29 *P6* **Edgeley** North Dakota, N USA 46°19′N 98°42′W
28 *I11* **Edgemont** South Dakota, N USA 43°18′N 103°49′W
92 *O3* **Edgeøya** *island* S Svalbard
27 *Q4* **Edgerton** Kansas, C USA 38°45′N 95°00′W
29 *S10* **Edgerton** Minnesota, N USA 43°52′N 96°07′W
21 *X3* **Edgewood** Maryland, NE USA 39°20′N 76°21′W
25 *V6* **Edgewood** Texas, SW USA 32°42′N 95°53′W
Édhessa *see* Édessa
29 *V9* **Edina** Minnesota, N USA 44°53′N 93°21′W
27 *U2* **Edina** Missouri, C USA 40°10′N 92°10′W
25 *S17* **Edinburg** Texas, SW USA 26°18′N 98°10′W
65 *M24* **Edinburgh** *var.* Settlement of Edinburgh. ○ (Tristan da Cunha) NW Tristan da Cunha 37°03′S 12°18′W
96 *J12* **Edinburgh** ◐ S Scotland, United Kingdom 55°57′N 03°13′W
31 *P14* **Edinburgh** Indiana, N USA 39°19′N 86°00′W
96 *J12* **Edinburgh** ✈ S Scotland, United Kingdom 55°57′N 03°22′W
116 *L8* **Edinet** *var.* Edineți, *Rus.* Yedintsy. NW Moldova 48°10′N 27°18′E
Edineți *see* Edineț
Edingen *see* Enghien
136 *B9* **Edirne** *Eng.* Adrianople; *anc.* Adrianopolis, Hadrianopolis. Edirne, NW Turkey 41°40′N 26°34′E
136 *B11* **Edirne** ◈ *province* NW Turkey
18 *K15* **Edison** New Jersey, NE USA 40°31′N 74°24′W
21 *S15* **Edisto Island** South Carolina, SE USA 32°34′N 80°17′W
21 *R14* **Edisto River** 🌊 South Carolina, SE USA
33 *S10* **Edith, Mount** ▲ Montana, NW USA 46°25′N 111°10′W
27 *N10* **Edmond** Oklahoma, C USA 35°40′N 97°30′W
32 *H8* **Edmonds** Washington, NW USA 47°48′N 122°22′W
11 *Q14* **Edmonton** *province capital* Alberta, SW Canada 53°34′N 113°25′W
20 *K7* **Edmonton** Kentucky, S USA 36°59′N 85°39′W
11 *Q14* **Edmonton** ✈ Alberta, SW Canada 53°22′N 113°43′W
29 *P3* **Edmore** North Dakota, N USA 48°22′N 98°26′W
13 *N13* **Edmundston** New Brunswick, SE Canada 47°22′N 68°20′W
25 *U12* **Edna** Texas, SW USA 29°00′N 96°41′W
39 *X14* **Edna Bay** Kosciusko Island, Alaska, USA 55°54′N 133°40′W
77 *U16* **Edo** ◈ *state* S Nigeria
106 *F6* **Edolo** Lombardia, N Italy 46°13′N 10°20′E
64 *L6* **Edoras Bank** *undersea feature* C Atlantic Ocean
96 *G7* **Edrachillis Bay** *bay* NW Scotland, United Kingdom
136 *B12* **Edremit** Balıkesir, NW Turkey 39°34′N 27°01′E
136 *B12* **Edremit Körfezi** *gulf* NW Turkey
95 *P14* **Edsbro** Stockholm, C Sweden 59°54′N 18°30′E
95 *N18* **Edsbruk** Kalmar, S Sweden 58°01′N 16°30′E
94 *M12* **Edsbyn** Gävleborg, C Sweden 61°22′N 15°45′E
11 *O14* **Edson** Alberta, SW Canada 53°36′N 116°28′W
62 *K13* **Eduardo Castex** La Pampa, C Argentina 35°55′S 64°18′W
58 *F12* **Eduardo Gomes** ✈ (Manaus) Amazonas, NW Brazil 05°55′S 35°15′W
Edwardesabad *see* Bannu
67 *U9* **Edward, Lake** *var.* Albert Edward Nyanza, Edward Nyanza, Lac Idi Amin, Lake Rutanzige. ◎ Uganda/Dem. Rep. Congo
Edward Nyanza *see* Edward, Lake
22 *K5* **Edwards** Mississippi, S USA 32°19′N 90°36′W
25 *O10* **Edwards Plateau** *plain* Texas, SW USA
30 *J11* **Edwards River** 🌊 Illinois, N USA
30 *K15* **Edwardsville** Illinois, N USA 38°48′N 89°57′W
195 *X4* **Edward VIII Gulf** *bay* Antarctica
195 *O13* **Edward VII Peninsula** *peninsula* Antarctica
10 *J11* **Edziza, Mount** ▲ British Columbia, W Canada 57°43′N 130°39′W
23 *H16* **Edzo** *prev.* Rae-Edzo. Northwest Territories, NW Canada 62°44′N 115°55′W
39 *N12* **Eek** Alaska, USA 60°13′N 162°01′W
99 *D16* **Eeklo** *var.* Eekloo. Oost-Vlaanderen, NW Belgium 51°11′N 03°34′E
Eekloo *see* Eeklo
39 *N12* **Eek River** 🌊 Alaska, USA
98 *N6* **Eelde** Drenthe, NE Netherlands 53°07′N 06°30′E
34 *L5* **Eel River** 🌊 California, W USA
31 *P12* **Eel River** 🌊 Indiana, N USA
Eems *see* Ems
98 *O4* **Eemshaven** Groningen, NE Netherlands 53°28′N 06°50′E
98 *O5* **Eems Kanaal** *canal* NE Netherlands
98 *M11* **Eerbeek** Gelderland, E Netherlands 52°07′N 06°04′E
99 *C17* **Eernegem** West-Vlaanderen, W Belgium 51°08′N 03°03′E
99 *J15* **Eersel** Noord-Brabant, S Netherlands 51°22′N 05°19′E
Eesti Vabariik *see* Estonia
187 *R14* **Efate** *var.* Efate, *Fr.* Vaté; *prev.* Sandwich Island. *island* C Vanuatu
Efate *see* Efate
109 *S4* **Eferding** Oberösterreich, N Austria 48°18′N 14°00′E
30 *M15* **Effingham** Illinois, N USA 39°07′N 88°32′W
117 *N15* **Eforie-Nord** Constanța, SE Romania 44°04′N 28°37′E
117 *N15* **Eforie-Sud** Constanța, E Romania 44°00′N 28°38′E
Efyrnwy, Afon *see* Vyrnwy
Eg *see* Hentiy
107 *G23* **Egadi, Isole** *island group* S Italy
35 *X6* **Egan Range** ▲▲ Nevada, W USA
14 *K12* **Eganville** Ontario, SE Canada 45°33′N 77°03′W
Ege Denizi *see* Aegean Sea
39 *O14* **Egegik** Alaska, USA 58°13′N 157°22′W
Egentliga Finland *see* Varsinais-Suomi
Egentliga Tavastland *see* Kanta-Häme
111 *L21* **Eger** *Ger.* Erlau. Heves, NE Hungary 47°54′N 20°22′E
Eger *see* Cheb, Czech Republic
Eger *see* Ohre, Czech Republic/Germany
173 *P8* **Egeria Fracture Zone** *tectonic feature* W Indian Ocean
95 *C17* **Egersund** Rogaland, S Norway 58°27′N 06°01′E
108 *J7* **Egg** Vorarlberg, NW Austria 47°27′N 09°55′E
101 *H14* **Egge-gebirge** ▲ C Germany
109 *Q4* **Eggelsberg** Oberösterreich, N Austria 48°04′N 13°00′E
109 *W2* **Eggenburg** Niederösterreich, NE Austria 48°29′N 15°49′E
101 *N22* **Eggenfelden** Bayern, SE Germany 48°24′N 12°45′E
18 *J17* **Egg Harbor City** New Jersey, NE USA 39°31′N 74°39′W
65 *G25* **Egg Island** *island* W Saint Helena
183 *N14* **Egg Lagoon** Tasmania, SE Australia 39°42′S 143°57′E
99 *I20* **Éghezèe** Namur, C Belgium 50°36′N 04°55′E
92 *L2* **Egilsstaðir** Austurland, E Iceland 65°14′N 14°21′W
Egina *see* Aígina
Egindibulaq *see* Yegindybulak
Egio *see* Aígio
103 *N12* **Egletons** Corrèze, C France 45°24′N 02°01′E
98 *H9* **Egmond aan Zee** Noord-Holland, NW Netherlands 52°37′N 04°37′E
Egmont *see* Taranaki, Mount
184 *J10* **Egmont, Cape** *headland* North Island, New Zealand 39°18′S 173°44′E
Egoli *see* Johannesburg
136 *I15* **Eğridir Gölü** ◎ W Turkey
Egri Palanka *see* Kriva Palanka
95 *G23* **Egtved** Syddanmark, C Denmark 55°34′N 09°18′E
123 *U5* **Egvekinot** Chukotskiy Avtonomnyy Okrug, NE Russian Federation 66°13′N 178°55′W
75 *V9* **Egypt** *off.* Arab Republic of Egypt, *Ar.* Jumhūrīyah Miṣr al 'Arabīyah, *prev.* United Arab Republic; *anc.* Aegyptus. ◆ *republic* NE Africa
30 *L17* **Egypt, Lake Of** ⊠ Illinois, N USA
162 *I14* **Ehen Hudag** *var.* Alx Youqi. Nei Mongol Zizhiqu, N China 39°12′N 101°40′E
164 *F14* **Ehime** *off.* Ehime-ken. ◈ *prefecture* Shikoku, SW Japan
Ehime-ken *see* Ehime
101 *I23* **Ehingen** Baden-Württemberg, S Germany 48°16′N 09°43′E
Ehinos *see* Echínos
21 *R14* **Ehrhardt** South Carolina, SE USA 33°06′N 81°00′W
108 *L7* **Ehrwald** Tirol, W Austria 47°24′N 10°52′E
191 *W6* **Eiao** *island* Îles Marquises, NE French Polynesia
105 *P2* **Eibar** País Vasco, N Spain 43°11′N 02°28′W
98 *O11* **Eibergen** Gelderland, E Netherlands 52°06′N 06°39′E
109 *V9* **Eibiswald** Steiermark, SE Austria 46°40′N 15°15′E
109 *P8* **Eichham** ▲ SW Austria 47°04′N 12°24′E
101 *J15* **Eichsfeld** *hill range* C Germany
101 *K21* **Eichstätt** Bayern, SE Germany 48°53′N 11°11′E
100 *H8* **Eider** 🌊 N Germany
94 *E13* **Eidfjord** Hordaland, S Norway 60°26′N 07°05′E
94 *D13* **Eidfjorden** *fjord* S Norway
94 *F9* **Eidsvåg** Møre og Romsdal, S Norway 62°46′N 08°00′E
95 *I14* **Eidsvoll** Akershus, S Norway 60°19′N 11°14′E
92 *N2* **Eidsvollfjellet** ▲ NW Svalbard 79°13′N 13°23′E
Eier-Berg *see* Suur Munamägi
101 *D18* **Eifel** *plateau* W Germany
108 *E9* **Eiger** ▲ C Switzerland 46°33′N 08°02′E
96 *G10* **Eigg** *island* W Scotland, United Kingdom
155 *D24* **Eight Degree Channel** *channel* India/Maldives
44 *G1* **Eight Mile Rock** Grand Bahama Island, N The Bahamas 26°28′N 78°43′W
194 *J9* **Eights Coast** *physical region* Antarctica
180 *K6* **Eighty Mile Beach** *beach* Western Australia
99 *L18* **Eijsden** Limburg, SE Netherlands 50°47′N 05°41′E
95 *G15* **Eikeren** ◎ S Norway
Eil *see* Eyl
Eilat *see* Elat
183 *O12* **Eildon** Victoria, SE Australia 37°17′S 145°57′E
183 *O12* **Eildon, Lake** ◎ Victoria, SE Australia
80 *E8* **Eilei** Northern Kordofan, C Sudan 16°33′N 30°54′E
101 *N15* **Eilenburg** Sachsen, E Germany 51°28′N 12°37′E
Eil Malk *see* Mecherchar
94 *H13* **Eina** Oppland, S Norway 60°37′N 10°37′E
138 *E12* **Ein Avdat** *prev.* En 'Avedat. *well* S Israel
101 *I14* **Einbeck** Niedersachsen, C Germany 51°49′N 09°52′E
99 *K15* **Eindhoven** Noord-Brabant, S Netherlands 51°26′N 05°30′E
108 *G8* **Einsiedeln** Schwyz, NE Switzerland 47°07′N 08°45′E
Eipel *see* Ipel'
Éire *see* Ireland
Éireann, Muir *see* Irish Sea
Eirik Outer Ridge *see* Eirik Ridge
64 *I6* **Eirik Ridge** *var.* Eirik Outer Ridge. *undersea feature* E Labrador Sea
92 *I3* **Eiríksjökull** ▲ C Iceland 64°47′N 20°23′W
59 *B14* **Eirunepé** Amazonas, N Brazil 06°38′S 69°53′W
99 *L17* **Eisden** Limburg, NE Belgium 51°05′N 05°42′E
83 *F18* **Eiseb** 🌊 Botswana/Namibia
Eisen *see* Yeongcheon
101 *J16* **Eisenach** Thüringen, C Germany 50°59′N 10°19′E
Eisenburg *see* Vasvár
109 *U6* **Eisenerz** Steiermark, SE Austria 47°33′N 14°53′E
100 *Q13* **Eisenhüttenstadt** Brandenburg, E Germany 52°09′N 14°36′E
109 *U10* **Eisenkappel** *Slvn.* Železna Kapela. Kärnten, S Austria 46°28′N 14°37′E
Eisenmarkt *see* Hunedoara
109 *Y5* **Eisenstadt** Burgenland, E Austria 47°50′N 16°32′E
Eishū *see* Yeongju
119 *H15* **Eišiškes** Vilnius, SE Lithuania 54°10′N 24°59′E
101 *L15* **Eisleben** Sachsen-Anhalt, C Germany 51°32′N 11°33′E
190 *I3* **Eita** Tarawa, W Kiribati 01°21′N 173°05′E
Eitape *see* Aitape
105 *V11* **Eivissa** *var.* Iviza, *Cast.* Ibiza; *anc.* Ebusus. Ibiza, Spain, W Mediterranean Sea 38°54′N 01°26′E
Eivissa *see* Ibiza
105 *R4* **Ejea de los Caballeros** Aragón, NE Spain 42°07′N 01°09′W
40 *E8* **Ejido Insurgentes** Baja California Sur, NW Mexico 25°18′N 111°51′W
Ejin Qi *see* Dalain Hob
Ejmiadzin/Ejmiatsin *see* Vagharshapat
77 *P16* **Ejura** C Ghana 07°23′N 01°22′W
41 *R16* **Ejutla** *var.* Ejutla de Crespo. Oaxaca, SE Mexico 16°33′N 96°40′W
Ejutla de Crespo *see* Ejutla
33 *Y10* **Ekalaka** Montana, NW USA 45°52′N 104°32′W
Ekapa *see* Cape Town
Ekaterinodar *see* Krasnodar
93 *L20* **Ekenäs** *Fin.* Tammisaari. Uusimaa, SW Finland 60°00′N 23°30′E
146 *B13* **Ekerem** *Rus.* Okarem. Balkan Welaýaty, W Turkmenistan 38°06′N 53°52′E
184 *M13* **Eketahuna** Manawatu-Wanganui, North Island, New Zealand 40°41′S 175°40′E
Ekhínos *see* Echínos
Ekiatapskiy Khrebet *see* Ekvyvatapskiy Khrebet
Ekibastuz *see* Yekibastuz
123 *R13* **Ekimchan** Amurskaya Oblast', SE Russian Federation 53°04′N 132°56′E
95 *O15* **Ekoln** ◎ C Sweden
80 *I7* **Ekowit** Red Sea, NE Sudan 18°46′N 37°07′E
95 *L19* **Eksjö** Jönköping, S Sweden 57°40′N 15°00′E
93 *I15* **Ekträsk** Västerbotten, N Sweden 64°28′N 19°49′E
39 *O13* **Ekuk** Alaska, USA 58°48′N 158°25′W
123 *U5* **Ekvyvatapskiy Khrebet** *prev.* Ekiatapskiy Khrebet. ▲ NE Russian Federation
12 *F9* **Ekwan** 🌊 Ontario, C Canada
39 *O13* **Ekwok** Alaska, USA 59°21′N 157°28′W
166 *M6* **Ela** Mandalay, C Myanmar (Burma) 19°37′N 96°15′E
El Aaiun *see* El Ayoun
81 *N15* **El Ābrēd** Sumalē, E Ethiopia 05°33′N 45°12′E
115 *F22* **Elafónisos** *island* S Greece
115 *F22* **Elafonísou, Porthmós** *strait* S Greece
El-Aïoun *see* El Ayoun
El 'Alamein *see* Al 'Alamayn
41 *Q12* **El Alazán** Veracruz-Llave, C Mexico 21°06′N 97°43′W
57 *J18* **El Alto** *var.* La Paz. ✈ (La Paz) La Paz, W Bolivia 16°31′S 68°07′W
Elam *see* Īlām
El Amparo *see* El Amparo de Apure
54 *I8* **El Amparo de Apure** *var.* El Amparo. Apure, C Venezuela 07°07′N 70°47′W
171 *R13* **Elara** Pulau Ambelau, E Indonesia 03°49′S 127°10′E
El Araïch/El Araîche *see* Larache
40 *D6* **El Arco** Baja California Norte, NW Mexico 28°03′N 113°25′W
El 'Arîsh *see* Al 'Arīsh
115 *L25* **Elása** *island* SE Greece
Elassón *see* Elassóna
115 *E15* **Elassóna** *prev.* Elassón. Thessalía, C Greece 39°53′N 22°10′E
105 *N2* **El Astillero** Cantabria, N Spain 43°23′N 03°45′W
138 *F14* **Elat** *var.* Eilat, Elath. Southern, S Israel 29°33′N 34°57′E
Elat, Gulf of *see* Aqaba, Gulf of
Elath *see* Elat, Israel
Elath *see* Al 'Aqabah, Jordan
115 *C17* **Eláti** ▲ Lefkáda, Iónia Nisiá, Greece, C Mediterranean Sea 38°43′N 20°38′E
188 *L16* **Elato Atoll** *atoll* Caroline Islands, C Micronesia
80 *C7* **El'Atrun** Northern Darfur, NW Sudan 18°11′N 26°40′E
74 *H6* **El Ayoun** *var.* El Aaiun, El-Aïoun, La Youne. NE Morocco 34°39′N 02°29′W
137 *N14* **Elâzığ** *var.* Elâzig, Elâziz. Elâzığ, E Turkey 38°41′N 39°14′E
137 *O14* **Elâzığ** *var.* Elâzig, Elâziz. ◈ *province* C Turkey
Elâzig/Elâziz *see* Elâzığ
23 *Q7* **Elba** Alabama, S USA 31°24′N 86°04′W
106 *E13* **Elba, Isola d'** *island* Archipelago Toscano, C Italy
123 *S13* **El'ban** Khabarovskiy Kray, E Russian Federation 50°03′N 136°34′E
54 *F6* **El Banco** Magdalena, N Colombia 09°00′N 74°01′W
El Barco *see* O Barco
104 *L8* **El Barco de Ávila** Castilla y León, N Spain 40°21′N 05°31′W
El Barco de Valdeorras *see* O Barco
138 *H7* **El Barouk, Jabal** ▲ C Lebanon
113 *L20* **Elbasan** *var.* Elbasani. Elbasan, C Albania 41°07′N 20°04′E
113 *L20* **Elbasan** ◈ *district* C Albania
Elbasani *see* Elbasan
54 *K6* **El Baúl** Cojedes, C Venezuela 08°59′N 68°16′W
86 *D11* **Elbe** *Cz.* Labe. 🌊 Czech Republic/Germany
100 *L13* **Elbe-Havel-Kanal** *canal* E Germany
100 *K9* **Elbe-Lübeck-Kanal** *canal* N Germany
57 *L15* **El Beni** *var.* El Beni. ◈ *department* N Bolivia
El Beni *see* El Beni
138 *H7* **El Beqaa** *var.* Al Biqā', Bekaa Valley. *valley* E Lebanon
25 *R6* **Elbert** Texas, SW USA 33°15′N 98°58′W
37 *R5* **Elbert, Mount** ▲ Colorado, C USA 39°07′N 106°26′W
23 *U3* **Elberton** Georgia, SE USA 34°06′N 82°52′W
100 *K11* **Elbe-Seiten-Kanal** *canal* N Germany
102 *M4* **Elbeuf** Seine-Maritime, N France 49°16′N 01°01′E
136 *M15* **Elbistan** Kahramanmaraş, S Turkey 38°14′N 37°11′E
110 *K7* **Elbląg** *var.* Elbląg, *Ger.* Elbing. Warmińsko-Mazurskie, NE Poland 54°10′N 19°25′E
43 *N10* **El Bluff** Región Autónoma Atlántico Sur, SE Nicaragua 12°00′N 83°40′W
63 *H17* **El Bolsón** Río Negro, W Argentina 41°59′S 71°35′W
105 *P11* **El Bonillo** Castilla-La Mancha, C Spain 38°57′N 02°32′W
El Bordo *see* Patía
El Boulaida/El Boulaïda *see* Blida
11 *T16* **Elbow** Saskatchewan, S Canada 51°07′N 106°30′W
29 *S7* **Elbow Lake** Minnesota, N USA 45°59′N 95°58′W
127 *N16* **El'brus** *var.* Gora El'brus. ▲ SW Russian Federation 42°29′E 43°21′N
El'brus, Gora *see* El'brus
126 *M15* **El'brusskiy** Karachayevo-Cherkesskaya Respublika, SW Russian Federation 43°36′N 42°06′E
El Buhayrat *see* Lakes
El Bur *see* Ceel Buur
98 *L10* **Elburg** Gelderland, E Netherlands 52°27′N 05°46′E
105 *O6* **El Burgo de Osma** Castilla y León, C Spain 41°36′N 03°04′W
Elburz Mountains *see* Alborz, Reshteh-ye Kūhhā-ye
35 *V17* **El Cajon** California, W USA 32°46′N 116°52′W
63 *H22* **El Calafate** *var.* Calafate. Santa Cruz, S Argentina 50°20′S 72°13′W
55 *Q8* **El Callao** Bolívar, E Venezuela 07°18′N 61°48′W
25 *U12* **El Campo** Texas, SW USA 29°12′N 96°16′W
54 *I7* **El Cantón** Barinas, W Venezuela 07°23′N 71°10′W
35 *Q8* **El Capitan** ▲ California, W USA 37°46′N 119°39′W
54 *H5* **El Carmelo** Zulia, NW Venezuela 10°20′N 71°48′W
62 *J5* **El Carmen** Jujuy, NW Argentina 24°24′S 65°16′W
54 *E5* **El Carmen de Bolívar** Bolívar, NW Colombia 09°43′N 75°07′W
55 *O8* **El Casabe** Bolívar, SE Venezuela 06°26′N 63°35′W
42 *M12* **El Castillo de La Concepción** Río San Juan, SE Nicaragua 11°01′N 84°24′W
El Cayo *see* San Ignacio
35 *X17* **El Centro** California, W USA 32°47′N 115°33′W
55 *N6* **El Chaparro** Anzoátegui, NE Venezuela 09°12′N 65°03′W
105 *S12* **Elche** *Cat.* Elx; *anc.* Ilici, *Lat.* Illicis. Valenciana, E Spain 38°16′N 00°41′W
105 *Q12* **Elche de la Sierra** Castilla-La Mancha, C Spain 38°27′N 02°03′W
41 *U15* **El Chichónal, Volcán** ☒ SE Mexico 17°20′N 93°12′W
40 *C2* **El Chinero** Baja California Norte, NW Mexico
181 *R1* **Elcho Island** *island* Wessel Islands, Northern Territory, N Australia
63 *H18* **El Corcovado** Chubut, SW Argentina 43°31′S 71°30′W
105 *R12* **Elda** Valenciana, E Spain 38°29′N 00°47′W
100 *M10* **Elde** 🌊 NE Germany
98 *L12* **Elden** Gelderland, E Netherlands 51°57′N 05°53′E
81 *J16* **El Der** *spring/well* S Ethiopia 03°55′N 39°48′E
El Dere *see* Ceel Dheere
40 *B2* **El Descanso** Baja California Norte, NW Mexico 32°08′N 116°51′W
40 *E3* **El Desemboque** Sonora, NW Mexico 30°33′N 112°59′W
54 *F5* **El Difícil** *var.* Ariguaní. Magdalena, N Colombia 09°55′N 74°12′W
123 *R10* **El'dikan** Respublika Sakha (Yakutiya), NE Russian Federation 60°46′N 135°04′E
El Djazaïr *see* Alger
El Djelfa *see* Djelfa
29 *X15* **Eldon** Iowa, C USA 40°55′N 92°13′W
27 *U5* **Eldon** Missouri, C USA 38°21′N 92°34′W
54 *E13* **El Doncello** Caquetá, S Colombia 01°43′N 75°17′W
29 *W13* **Eldora** Iowa, C USA 42°21′N 93°06′W
60 *G12* **Eldorado** Misiones, NE Argentina 26°24′S 54°38′W
40 *I9* **El Dorado** Sinaloa, C Mexico 24°19′N 107°23′W
27 *U14* **El Dorado** Arkansas, C USA 33°12′N 92°40′W
30 *M17* **Eldorado** Illinois, N USA 37°48′N 88°26′W
27 *O6* **El Dorado** Kansas, C USA 37°51′N 96°52′W
26 *K12* **Eldorado** Oklahoma, C USA 34°28′N 99°39′W
25 *O9* **Eldorado** Texas, SW USA 30°53′N 100°37′W
55 *Q8* **El Dorado** Bolívar, E Venezuela 06°45′N 61°37′W
54 *F10* **El Dorado** ✈ (Bogotá) Cundinamarca, C Colombia 01°15′N 71°52′W
El Dorado *see* California
27 *O6* **El Dorado Lake** ⊠ Kansas, C USA
27 *S6* **El Dorado Springs** Missouri, C USA 37°53′N 94°01′W
81 *H18* **Eldoret** Uasin Gishu, W Kenya 0°31′N 35°17′E
29 *Z14* **Eldridge** Iowa, C USA 41°39′N 90°34′W
95 *J21* **Eldsberga** Halland, S Sweden 56°36′N 13°00′E
25 *R4* **Electra** Texas, SW USA 34°01′N 98°55′W
37 *Q7* **Electra Lake** ⊠ Colorado, C USA
38 *B8* **'Ele'ele** *var.* Eleele. Kaua'i, Hawaii, USA, C Pacific Ocean 21°54′N 159°35′W
Eleele *see* 'Ele'ele
Elefantes *see* Lepelle
115 *H19* **Elefsína** *prev.* Elevsís. Attikí, C Greece 38°02′N 23°33′E
115 *G19* **Eleftheres** *anc.* Eleutherae. *site of ancient city* Attikí/Stereá Elláda, C Greece
114 *I13* **Eleftheroúpoli** *prev.* Elevtheroúpolis. Anatolikí Makedonía kai Thráki, NE Greece 40°55′N 24°15′E
74 *F10* **El Eglab** ▲▲ SW Algeria
118 *F10* **Eleja** C Latvia 56°24′N 23°41′E
Elek *see* Yelek
119 *G14* **Elektrėnai** Vilnius, SE Lithuania 54°47′N 24°35′E
126 *L3* **Elektrostal'** Moskovskaya Oblast', W Russian Federation 55°47′N 38°24′E
81 *H15* **Elemi Triangle** *disputed region* Kenya/Sudan
114 *K9* **Elena** Veliko Tarnovo, N Bulgaria 42°55′N 25°53′E
54 *G16* **El Encanto** Amazonas, S Colombia 01°45′S 73°12′W
37 *R14* **Elephant Butte Reservoir** ⊠ New Mexico, SW USA
Éléphant, Chaîne de l' *see* Dâmrei, Chuŏr Phnum
194 *G2* **Elephant Island** *island* South Shetland Islands, Antarctica
Elephant River *see* Olifants
El Escorial *see* San Lorenzo de El Escorial
Élesd *see* Aleşd
114 *F11* **Eleshnitsa** 🌊 W Bulgaria
137 *S13* **Eleşkirt** Ağrı, E Turkey 39°22′N 42°48′E
42 *F5* **El Estor** Izabal, E Guatemala 15°37′N 89°22′W
Eleutherae *see* Eléftheres
44 *I2* **Eleuthera Island** *island* N The Bahamas
37 *S5* **Elevenmile Canyon Reservoir** ⊠ Colorado, C USA
27 *W8* **Eleven Point River** 🌊 Arkansas/Missouri, C USA
Elevsís *see* Elefsína
Elevtheroúpolis *see* Eleftheroúpoli
80 *B10* **El Faiyûm** *see* Al Fayyūm
El Fasher *var.* Al Fāshir. Northern Darfur, W Sudan 13°37′N 25°22′E
El Fashn *see* Al Fashn
El Ferrol/El Ferrol del Caudillo *see* Ferrol
39 *W13* **Elfin Cove** Chichagof Island, Alaska, USA 58°09′N 136°16′W
105 *W4* **El Fluvià** 🌊 NE Spain
40 *H7* **El Fuerte** Sinaloa, W Mexico 26°28′N 108°35′W
80 *D11* **El Fula** Western Kordofan, C Sudan 11°44′N 28°20′E
El Gedaref *see* Gedaref
80 *A10* **El Geneina** *var.* Ajjinena, Al-Genain, Al Junaynah. Western Darfur, W Sudan 13°27′N 22°30′E
81 *I18* **Elgeyo/Marakwet** ◈ *county* W Kenya
96 *J8* **Elgin** NE Scotland, United Kingdom 57°39′N 03°20′W
30 *M10* **Elgin** Illinois, N USA 42°02′N 88°16′W
29 *P14* **Elgin** Nebraska, C USA 41°58′N 98°04′W
35 *Y9* **Elgin** Nevada, W USA 37°19′N 114°30′W
28 *L6* **Elgin** North Dakota, N USA 46°24′N 101°51′W
26 *M12* **Elgin** Oklahoma, C USA 34°46′N 98°17′W
25 *T10* **Elgin** Texas, SW USA 30°21′N 97°22′W
123 *R9* **El'ginskiy** Respublika Sakha (Yakutiya), NE Russian Federation 64°27′N 141°57′E
El Gîza *see* Giza
74 *J8* **El Goléa** *var.* Al Golea. C Algeria 30°35′N 02°59′E
40 *D2* **El Golfo de Santa Clara** Sonora, NW Mexico 31°48′N 114°40′W
81 *G18* **Elgon, Mount** ▲ E Uganda 01°07′N 34°29′E
94 *I10* **Elgpiggen** ▲ S Norway 62°13′N 11°18′E
105 *T4* **El Grado** Aragón, NE Spain 42°09′N 00°13′E
40 *L6* **El Guaje, Laguna** ⊠ NE Mexico
54 *H6* **El Guayabo** Zulia, W Venezuela 08°37′N 72°20′W
77 *O6* **El Guețțâra** *oasis* N Mali
76 *J6* **El Ḥammâmi** *desert* N Mauritania
76 *M5* **El Ḥank** *cliff* N Mauritania
El Haseke *see* Al Ḥasakah
80 *H10* **El Hawata** Gedaref, E Sudan 13°25′N 34°42′E

◆ Country
● Country Capital
◇ Dependent Territory
○ Dependent Territory Capital
◈ Administrative Regions
✈ International Airport
▲ Mountain
▲▲ Mountain Range
☒ Volcano
🌊 River
◎ Lake
⊠ Reservoir

El Higo – Erdenetsagaan

El Higo see Higos
114 L10 **Elhovo** var. Elkhovo; prev. Kizilagach. Yambol, E Bulgaria 42°10′N 26°34′E
171 T16 **Eliase** Pulau Selaru, E Indonesia 08°16′S 130°49′E
Elías Piña see Comendador
25 R6 **Eliasville** Texas, SW USA 32°55′N 98°46′W
Elichpur see Achalpur
37 V13 **Elida** New Mexico, SW USA 33°57′N 103°39′W
115 F18 **Elikónas** ⛰ C Greece
67 T10 **Elila** ≈ W Dem. Rep. Congo
39 N9 **Elim** Alaska, USA 64°37′N 162°15′W
Elimberrum see Auch
Eliocroca see Lorca
61 B16 **Elisa** Santa Fe, C Argentina 30°42′S 61°04′W
Elisabethstedt see Dumbrăveni
Élisabethville see Lubumbashi
127 O13 **Elista** Respublika Kalmykiya, SW Russian Federation 46°18′N 44°09′E
182 I9 **Elizabeth** South Australia 34°44′S 138°39′E
21 Q3 **Elizabeth** West Virginia, NE USA 39°04′N 81°24′W
19 Q9 **Elizabeth, Cape** headland Maine, NE USA 43°34′N 70°12′W
21 Y8 **Elizabeth City** North Carolina, SE USA 36°18′N 76°16′W
21 P8 **Elizabethton** Tennessee, S USA 36°22′N 82°15′W
30 M17 **Elizabethtown** Illinois, N USA 37°24′N 88°21′W
20 K6 **Elizabethtown** Kentucky, S USA 37°41′N 85°51′W
18 L7 **Elizabethtown** New York, NE USA 44°13′N 73°38′W
21 U11 **Elizabethtown** North Carolina, SE USA 34°36′N 78°36′W
18 G15 **Elizabethtown** Pennsylvania, NE USA 40°08′N 76°36′W
74 E6 **El-Jadida** prev. Mazagan. W Morocco 33°15′N 08°27′W
El Jafr see Jafr, Qāʻ al
80 F11 **El Jebelein** White Nile, E Sudan 12°38′N 32°51′E
110 N8 **Ełk** Ger. Lyck. Warmińsko-mazurskie, NE Poland 53°51′N 22°20′E
110 O8 **Ełk** ≈ NE Poland
29 Y12 **Elkader** Iowa, C USA 42°51′N 91°24′W
80 G9 **El Kamlin** Gezira, C Sudan 15°03′N 33°11′E
33 N11 **Elk City** Idaho, NW USA 45°50′N 115°28′W
26 K10 **Elk City** Oklahoma, C USA 35°24′N 99°24′W
27 P7 **Elk City Lake** ▣ Kansas, C USA
34 M5 **Elk Creek** California, W USA 39°34′N 122°34′W
28 J10 **Elk Creek** ≈ South Dakota, N USA
74 M5 **El Kef** var. Al Kāf, Le Kef. NW Tunisia 36°13′N 08°44′E
74 F7 **El Kelâa Srarhna** var. Kal al Sraghna. C Morocco 32°05′N 07°20′W
El Kerak see Al Karak
11 P17 **Elkford** British Columbia, SW Canada 49°58′N 114°57′W
El Khalil see Hebron
80 E7 **El Khandaq** Northern, N Sudan 18°34′N 30°34′E
El Khârga see Al Khārijah
31 P11 **Elkhart** Indiana, N USA 41°40′N 85°58′W
26 H7 **Elkhart** Kansas, C USA 37°00′N 101°51′W
25 V8 **Elkhart** Texas, SW USA 31°37′N 95°34′W
30 M7 **Elkhart Lake** ◎ Wisconsin, N USA
El Khartûm see Khartoum
37 Q3 **Elkhead Mountains** ⛰ Colorado, C USA
18 I12 **Elk Hill** ▲ Pennsylvania, NE USA 41°42′N 75°33′W
138 G8 **El Khiyam** var. Al Khiyām, Khiam. S Lebanon 33°12′N 35°42′E
29 S15 **Elkhorn** Nebraska, C USA 41°17′N 96°13′W
30 M9 **Elkhorn** Wisconsin, N USA 42°40′N 88°34′W
29 R14 **Elkhorn River** ≈ Nebraska, C USA
127 O16 **El'khotovo** Respublika Severnaya Osetiya, SW Russian Federation 43°18′N 44°17′E
Elkhovo see Elhovo
21 R8 **Elkin** North Carolina, SE USA 36°14′N 80°51′W
21 S4 **Elkins** West Virginia, NE USA 38°56′N 79°53′W
195 X3 **Elkins, Mount** ▲ Antarctica 66°25′S 53°54′E
14 G8 **Elk Lake** Ontario, S Canada 47°44′N 80°19′W
31 P6 **Elk Lake** ◎ Michigan, N USA
18 F12 **Elkland** Pennsylvania, NE USA 41°59′N 77°16′W
35 W3 **Elko** Nevada, W USA 40°48′N 115°46′W
11 R14 **Elk Point** Alberta, SW Canada 53°52′N 110°49′W
29 R12 **Elk Point** South Dakota, N USA 42°42′N 96°37′W
29 V8 **Elk River** Minnesota, N USA 45°18′N 93°34′W
20 J10 **Elk River** ≈ Alabama/Tennessee, S USA
21 R4 **Elk River** ≈ West Virginia, NE USA
20 I7 **Elkton** Kentucky, S USA 36°49′N 87°11′W
21 Y2 **Elkton** Maryland, NE USA 39°37′N 75°50′W
29 R10 **Elkton** South Dakota, N USA 44°14′N 96°28′W
20 I10 **Elkton** Tennessee, S USA 35°01′N 86°51′W
21 U5 **Elkton** Virginia, NE USA 38°22′N 78°35′W
El Kuneitra see Al Qunayṭirah
81 L15 **El Kure** Somali, E Ethiopia 05°37′N 42°05′E
80 D12 **El Lagowa** Western Kordofan, C Sudan 11°23′N 29°10′E
39 S12 **Ellamar** Alaska, USA 60°54′N 146°37′W
Ellás see Greece
23 S6 **Ellaville** Georgia, SE USA 32°14′N 84°18′W
197 P9 **Ellef Ringnes Island** island Nunavut, N Canada
29 V10 **Ellendale** Minnesota, N USA 43°53′N 93°19′W
29 P7 **Ellendale** North Dakota, N USA 45°57′N 98°33′W
36 M6 **Ellen, Mount** ▲ Utah, W USA 38°06′N 110°48′W
32 I9 **Ellensburg** Washington, NW USA 47°00′N 124°34′W
18 K12 **Ellenville** New York, NE USA 41°43′N 74°24′W
Ellep see Lib
21 T10 **Ellerbe** North Carolina, SE USA 35°04′N 79°45′W
197 P10 **Ellesmere Island** island Queen Elizabeth Islands, Nunavut, N Canada
185 H19 **Ellesmere, Lake** ◎ South Island, New Zealand
97 K18 **Ellesmere Port** C England, United Kingdom 53°17′N 02°54′W
31 O14 **Ellettsville** Indiana, N USA 39°13′N 86°37′W
99 E19 **Ellezelles** Hainaut, SW Belgium 50°44′N 03°40′E
8 L7 **Ellice** ≈ Nunavut, NE Canada
Ellice Islands see Tuvalu
Ellichpur see Achalpur
21 W3 **Ellicott City** Maryland, NE USA 39°16′N 76°48′W
23 S2 **Ellijay** Georgia, SE USA 34°42′N 84°28′W
27 W7 **Ellington** Missouri, C USA 37°14′N 90°58′W
26 L5 **Ellinwood** Kansas, C USA 38°21′N 98°34′W
83 J24 **Elliot** Eastern Cape, SE South Africa 31°20′S 27°51′E
14 D10 **Elliot Lake** Ontario, S Canada 46°24′N 82°38′W
181 X6 **Elliot, Mount** ▲ Queensland, E Australia 19°36′S 147°02′E
21 T5 **Elliott Knob** ▲ Virginia, NE USA 38°10′N 79°18′W
26 K4 **Ellis** Kansas, C USA 38°56′N 99°33′W
182 F8 **Elliston** South Australia 33°40′S 134°56′E
22 M7 **Ellisville** Mississippi, S USA 31°36′N 89°12′W
105 V5 **El Llobregat** ≈ NE Spain
96 L9 **Ellon** NE Scotland, United Kingdom 57°22′N 02°06′W
Ellore see Elūru
21 S13 **Elloree** South Carolina, SE USA 33°34′N 80°37′W
26 M4 **Ellsworth** Kansas, C USA 38°45′N 98°15′W
19 S7 **Ellsworth** Maine, NE USA 44°32′N 68°25′W
30 I6 **Ellsworth** Wisconsin, N USA 44°44′N 92°29′W
26 M11 **Ellsworth, Lake** ▣ Oklahoma, C USA
194 K9 **Ellsworth Land** physical region Antarctica
194 L9 **Ellsworth Mountains** ⛰ Antarctica
101 J21 **Ellwangen** Baden-Württemberg, S Germany 48°58′N 10°07′E
18 B14 **Ellwood City** Pennsylvania, NE USA 40°49′N 80°15′W
108 H8 **Elm** Glarus, NE Switzerland 46°55′N 09°09′E
32 G9 **Elma** Washington, NW USA 47°00′N 123°24′W
121 V13 **El Maḥalla el Kubra** var. Al Maḥallah al Kubrá, Mahalla el Kubra. N Egypt 30°59′N 31°10′E
74 E9 **El Mahbas** var. Mahbés. SW Western Sahara 27°26′N 09°09′W
63 H17 **El Maitén** Chubut, W Argentina 42°03′S 71°10′W
136 E16 **Elmalı** Antalya, SW Turkey 36°43′N 29°19′E
80 G10 **El Manaqil** Gezira, C Sudan 14°12′N 33°01′E
54 M12 **El Mango** Amazonas, S Venezuela 01°55′N 66°35′W
El Manşûra see Al Manşūrah
55 P8 **El Manteco** Bolívar, E Venezuela 07°27′N 62°32′W
29 O16 **Elm Creek** Nebraska, C USA 40°43′N 99°22′W
El Mediyya see Médéa
77 V9 **Elméki** Agadez, C Niger 17°52′N 08°07′E
108 K7 **Elmen** Tirol, W Austria 47°22′N 10°34′E
18 I16 **Elmer** New Jersey, NE USA 39°34′N 75°09′W
138 G6 **El Mina** var. Al Mīnā'. N Lebanon 34°28′N 35°49′E
El Minya see Al Minyā
14 F15 **Elmira** Ontario, S Canada 43°35′N 80°34′W
18 G11 **Elmira** New York, NE USA 42°06′N 76°50′W
36 K13 **El Mirage** Arizona, SW USA 33°36′N 112°19′W
29 O7 **Elm Lake** ▣ South Dakota, N USA
El Moján see San Rafael
105 N7 **El Molar** Madrid, C Spain 40°43′N 03°35′W
76 L7 **El Mrâyer** well C Mauritania
76 L5 **El Mreiti** well N Mauritania
76 L8 **El Mreyyé** desert E Mauritania
29 P8 **Elm River** ≈ North Dakota/South Dakota, N USA
100 I9 **Elmshorn** Schleswig-Holstein, N Germany 53°45′N 09°39′E
80 D12 **El Muglad** Western Kordofan, C Sudan 11°02′N 27°44′E
El Muwaqqar see Al Muwaqqar
14 G14 **Elmvale** Ontario, S Canada 44°34′N 79°53′W
30 K12 **Elmwood** Illinois, N USA 40°46′N 89°58′W
26 J8 **Elmwood** Oklahoma, C USA 36°37′N 100°31′W
103 P17 **Elne** anc. Illiberis. Pyrénées-Orientales, S France 42°36′N 02°58′E
54 F11 **El Nevado, Cerro** elevation C Colombia
171 N5 **El Nido** Palawan, W Philippines 11°10′N 119°25′E
62 I12 **El Nihuil** Mendoza, W Argentina 34°58′S 68°40′W
75 W7 **El Nouzha** ✈ (Alexandria) N Egypt 31°06′N 29°58′E
80 E10 **El Obeid** var. Al Obayyid, Al Ubayyiḍ. Northern Kordofan, C Sudan 13°11′N 30°10′E
41 O13 **El Oro** México, S Mexico 19°51′N 100°07′W
56 B8 **El Oro** ◐ province SW Ecuador
61 B19 **Elortondo** Santa Fe, C Argentina 33°42′S 61°37′W
54 J8 **Elorza** Apure, C Venezuela 07°02′N 69°30′W
El Ouâdi see El Oued
74 L7 **El Oued** var. Al Oued, El Ouâdi, El Wad. NE Algeria 33°20′N 06°53′E
36 L15 **Eloy** Arizona, SW USA 32°47′N 111°33′W
55 Q7 **El Palmar** Bolívar, E Venezuela 08°01′N 61°53′W
40 K8 **El Palmito** Durango, W Mexico 25°40′N 104°59′W
55 P7 **El Pao** Bolívar, E Venezuela 08°03′N 62°40′W
54 K5 **El Pao** Cojedes, N Venezuela 09°40′N 68°08′W
42 J7 **El Paraíso** El Paraíso, S Honduras 13°51′N 86°31′W
42 I7 **El Paraíso** ◐ department SE Honduras
30 L12 **El Paso** Illinois, N USA 40°44′N 89°01′W
24 G8 **El Paso** Texas, SW USA 31°45′N 106°30′W
24 G8 **El Paso** ✈ Texas, SW USA 31°48′N 106°24′W
105 U7 **El Perelló** Cataluña, NE Spain 40°53′N 00°43′E
55 P5 **El Pilar** Sucre, NE Venezuela 10°31′N 63°12′W
42 F7 **El Pital, Cerro** ▲ El Salvador/Honduras 14°19′N 89°06′W
35 Q9 **El Portal** California, W USA 37°40′N 119°46′W
40 J3 **El Porvenir** Chihuahua, N Mexico 31°15′N 105°48′W
43 U14 **El Porvenir** Kuna Yala, N Panama 09°33′N 78°56′W
105 W6 **El Prat de Llobregat** Cataluña, NE Spain 41°20′N 02°05′E
42 H5 **El Progreso** Yoro, NW Honduras 15°25′N 87°49′W
42 A2 **El Progreso** off. Departamento de El Progreso. ◐ department C Guatemala
El Progreso see Guastatoya
El Progreso, Departamento de see El Progreso
104 L9 **El Puente del Arzobispo** Castilla-La Mancha, C Spain 39°48′N 05°10′W
104 J15 **El Puerto de Santa María** Andalucía, S Spain 36°36′N 06°13′W
62 I8 **El Puesto** Catamarca, NW Argentina 27°57′S 67°37′W
El Qâhira see Cairo
El Qaşr see Al Qaşr
El Qatrani see Al Qaṭrānah
40 I10 **El Quelite** Sinaloa, C Mexico 23°37′N 106°26′W
62 G9 **Elqui, Río** ≈ N Chile
El Q'unayṭirah see Al Qunayṭirah
El Quneitra see Al Qunayṭirah
El Quseir see Al Quşayr
El Quweira see Al Quwayrah
141 O15 **El-Rahaba** ✈ (Şan'ā') W Yemen 15°28′N 44°12′E
42 M10 **El Rama** Región Autónoma Atlántico Sur, SE Nicaragua 12°09′N 84°15′W
43 W16 **El Real** var. El Real de Santa María. Darién, SE Panama 08°06′N 77°42′W
El Real de Santa María see El Real
26 M10 **El Reno** Oklahoma, C USA 35°32′N 95°57′W
40 K9 **El Rodeo** Durango, C Mexico 25°12′N 104°35′W
104 J13 **El Ronquillo** Andalucía, S Spain 37°43′N 06°09′W
11 S16 **Elrose** Saskatchewan, S Canada 51°07′N 107°59′W
30 K8 **Elroy** Wisconsin, N USA 43°43′N 90°16′W
25 S17 **Elsa** Texas, SW USA 26°17′N 97°59′W
El Şaff see Aş Şaff
40 J10 **El Salto** Durango, C Mexico 23°47′N 105°22′W
42 D8 **El Salvador** off. Republica de El Salvador. ◆ republic Central America
El Salvador, Republica de see El Salvador
54 K7 **El Samán de Apure** Apure, C Venezuela 07°54′N 68°44′W
14 D7 **Elsas** Ontario, S Canada 48°31′N 82°53′W
40 F3 **El Sásabe** var. Aduana del Sásabe. Sonora, NW Mexico 31°27′N 111°31′W
Elsass see Alsace
40 J5 **El Sáuz** Chihuahua, N Mexico 29°03′N 106°15′W
27 W4 **Elsberry** Missouri, C USA 39°10′N 90°46′W
45 P9 **El Seibo** var. Santa Cruz de El Seibo, Santa Cruz del Seibo. E Dominican Republic 18°45′N 69°04′W
42 B7 **El Semillero Barra Nahualate** Escuintla, SW Guatemala 14°01′N 91°28′W
Elsene see Ixelles
Elsen Nur see Dorgê Co
36 L6 **Elsinore** Utah, W USA 38°40′N 112°09′W
Elsinore see Helsingør
99 L18 **Elsloo** Limburg, SE Netherlands 50°57′N 05°46′E
60 G13 **El Soberbio** Misiones, NE Argentina 27°15′S 54°05′W
55 N6 **El Socorro** Guárico, C Venezuela 09°00′N 65°42′W
54 L6 **El Sombrero** Guárico, N Venezuela 09°25′N 67°06′W
98 L10 **Elspeet** Gelderland, E Netherlands 52°19′N 05°47′E
98 L12 **Elst** Gelderland, E Netherlands 51°55′N 05°51′E
101 O15 **Elsterwerda** Brandenburg, E Germany 51°27′N 13°32′E
40 J4 **El Sueco** Chihuahua, N Mexico 29°53′N 106°24′W
El Suweida see As Suwaydā'
El Suweis see Suez
54 D12 **El Tambo** Cauca, SW Colombia 02°25′N 76°50′W
175 T13 **Eltanin Fracture Zone** tectonic feature SE Pacific Ocean
105 X5 **El Ter** ≈ NE Spain
184 K11 **Eltham** Taranaki, North Island, New Zealand 39°26′S 174°25′E
55 O6 **El Tigre** Anzoátegui, NE Venezuela 08°55′N 64°15′W
El Tigrito see San José de Guanipa
54 J5 **El Tocuyo** Lara, N Venezuela 09°48′N 69°51′W
127 Q10 **El'ton** Volgogradskaya Oblast', SW Russian Federation 49°07′N 46°50′E
32 K10 **Eltopia** Washington, NW USA 46°33′N 118°59′W
El Toro see Mare de Déu del Toro
61 A18 **El Trébol** Santa Fe, C Argentina 32°12′S 61°40′W
40 J13 **El Tuito** Jalisco, SW Mexico 20°19′N 105°22′W
El Ţûr see Aţ Ţūr
161 S15 **Eluan Bi** Eng. Cape Olwanpi; prev. Oluan Pi. headland S Taiwan 21°57′N 120°48′E
155 K16 **Elūru** prev. Ellore. Andhra Pradesh, E India 16°45′N 81°10′E
118 H13 **Elva** Ger. Elwa. Tartumaa, SE Estonia 58°13′N 26°25′E
37 R9 **El Vado Reservoir** ▣ New Mexico, SW USA
43 S15 **El Valle** Coclé, C Panama 08°39′N 80°08′W
104 I11 **Elvas** Portalegre, C Portugal 38°53′N 07°10′W
54 K7 **El Venado** Apure, C Venezuela 07°25′N 68°46′W
105 V6 **El Vendrell** Cataluña, NE Spain 41°13′N 01°32′E
94 I13 **Elverum** Hedmark, S Norway 60°54′N 11°33′E
42 I9 **El Viejo** Chinandega, NW Nicaragua 12°39′N 87°11′W
54 G7 **El Viejo, Cerro** ▲ C Colombia 07°31′N 72°56′W
54 H6 **El Vigía** Mérida, NW Venezuela 08°38′N 71°39′W
105 Q4 **El Villar de Arnedo** La Rioja, N Spain 42°19′N 02°05′W
59 A14 **Elvira** Amazonas, W Brazil 07°12′S 69°56′W
Elwa see Elva
El Wad see El Oued
81 K17 **El Wak** Mandera, NE Kenya 02°46′N 40°57′E
33 R7 **Elwell, Lake** ▣ Montana, NW USA
31 P13 **Elwood** Indiana, N USA 40°16′N 85°50′W
27 R3 **Elwood** Kansas, C USA 39°43′N 94°52′W
29 N16 **Elwood** Nebraska, C USA 40°35′N 99°51′W
Elx see Elche
97 O20 **Ely** E England, United Kingdom 52°24′N 00°15′E
29 X4 **Ely** Minnesota, N USA 47°54′N 91°52′W
35 X6 **Ely** Nevada, W USA 39°15′N 114°53′W
El Yopal see Yopal
31 T11 **Elyria** Ohio, N USA 41°22′N 82°06′W
45 S9 **El Yunque** ▲ E Puerto Rico 18°15′N 65°46′W
101 F23 **Elz** ≈ SW Germany
187 R14 **Emae** island Shepherd Islands, C Vanuatu
118 I5 **Emajõgi** Ger. Embach. ≈ SE Estonia
Emāmrūd see Shāhrūd
Emām Şāḩeb see Imām Şāḩib
Emam Saheb see Imām Şāḩib
Emāmshahr see Shāhrūd
95 M20 **Emån** ≈ S Sweden
144 J11 **Emba** Kaz. Embi. Aktyubinsk, W Kazakhstan 48°50′N 58°10′E
Emba see Zhem
Embach see Emajõgi
62 K5 **Embarcación** Salta, N Argentina 23°15′S 64°05′W
30 M15 **Embarras River** ≈ Illinois, N USA
Embi see Emba
81 I19 **Embu** Embu, C Kenya 0°32′N 37°28′E
81 I19 **Embu** ◐ county C Kenya
100 E10 **Emden** Niedersachsen, NW Germany 53°22′N 07°12′E
160 H9 **Emei Shan** ▲ Sichuan, C China 29°32′N 103°21′E
29 Q4 **Emerado** North Dakota, N USA 47°55′N 97°21′W
181 X8 **Emerald** Queensland, E Australia 23°33′S 148°11′E
Emerald Isle see Montserrat
57 J15 **Emero, Río** ≈ W Bolivia
11 Y17 **Emerson** Manitoba, S Canada 49°01′N 97°07′W
29 T15 **Emerson** Iowa, C USA 41°00′N 95°22′W
29 R13 **Emerson** Nebraska, C USA 42°16′N 96°43′W
36 M5 **Emery** Utah, W USA 38°54′N 111°16′W
Emesa see Ḩimş
136 E13 **Emet** Kütahya, W Turkey 39°22′N 29°15′E
186 B8 **Emeti** Western, SW Papua New Guinea 07°54′S 143°18′E
35 V3 **Emigrant Pass** pass Nevada, W USA
78 I6 **Emi Koussi** ▲ N Chad 19°52′N 18°34′E
Emilia see Emilia-Romagna
41 V15 **Emiliano Zapata** Chiapas, SE Mexico 17°42′N 91°46′W
106 E9 **Emilia-Romagna** prev. Emilia; anc. Æmilia. ◐ region N Italy
158 J3 **Emin** var. Dorbiljin. Xinjiang Uygur Zizhiqu, NW China 46°30′N 83°42′E
149 W8 **Eminābād** Punjab, E Pakistan 32°02′N 73°51′E
21 L5 **Eminence** Kentucky, S USA 38°22′N 85°10′W
27 W7 **Eminence** Missouri, C USA 37°09′N 91°22′W
114 N9 **Emine, Nos** headland E Bulgaria 42°43′N 27°53′E
158 I3 **Emin He** ≈ NW China
186 G4 **Emirau Island** island N Papua New Guinea
136 F13 **Emirdağ** Afyon, W Turkey 39°01′N 31°09′E
83 K22 **eMkhondo** prev. Piet Retief. Mpumalanga, E South Africa 27°00′S 30°49′E see also Piet Retief
95 M21 **Emmaboda** Kalmar, S Sweden 56°36′N 15°30′E
118 E5 **Emmaste** Hiiumaa, W Estonia 58°43′N 22°36′E
18 I15 **Emmaus** Pennsylvania, NE USA 40°32′N 75°28′W
183 U4 **Emmaville** New South Wales, SE Australia 29°26′S 151°38′E
108 E9 **Emme** ≈ W Switzerland
98 L8 **Emmeloord** Flevoland, N Netherlands 52°43′N 05°46′E
98 O8 **Emmen** Drenthe, NE Netherlands 52°48′N 06°57′E
108 F8 **Emmen** Luzern, C Switzerland 47°03′N 08°14′E
101 F23 **Emmendingen** Baden-Württemberg, SW Germany 48°07′N 07°51′E
98 P8 **Emmer-Compascuum** Drenthe, NE Netherlands 52°47′N 07°03′E
101 D14 **Emmerich** Nordrhein-Westfalen, W Germany 51°49′N 06°16′E
29 U12 **Emmetsburg** Iowa, C USA 43°06′N 94°40′W
32 M14 **Emmett** Idaho, NW USA 43°52′N 116°30′W
38 M10 **Emmonak** Alaska, USA 62°46′N 164°31′W
Emona see Ljubljana
Emonti see East London
24 L12 **Emory Peak** ▲ Texas, SW USA 29°15′N 103°18′W
40 F6 **Empalme** Sonora, NW Mexico 27°57′N 110°49′W
83 L23 **Empangeni** KwaZulu/Natal, E South Africa 28°45′S 31°54′E
61 C14 **Empedrado** Corrientes, NE Argentina 27°59′S 58°47′W
192 K3 **Emperor Seamounts** undersea feature NW Pacific Ocean 42°00′N 170°00′E
192 L3 **Emperor Trough** undersea feature N Pacific Ocean
35 R4 **Empire** Nevada, W USA 40°26′N 119°21′W
Empire State of the South see Georgia
Emplawas see Amplawas
106 F11 **Empoli** Toscana, C Italy 43°43′N 10°57′E
27 P5 **Emporia** Kansas, C USA 38°24′N 96°10′W
21 W7 **Emporia** Virginia, NE USA 36°42′N 77°33′W
18 E13 **Emporium** Pennsylvania, NE USA 41°31′N 78°14′W
Empty Quarter see Ar Rub' al Khālī
100 E10 **Ems** Dut. Eems. ≈ NW Germany
100 F13 **Emsdetten** Nordrhein-Westfalen, NW Germany 52°11′N 07°32′E
Ems-Hunte Canal see Küstenkanal
100 F10 **Ems-Jade-Kanal** canal NW Germany
100 F11 **Emsland** cultural region NW Germany
182 D3 **Emu Junction** South Australia 28°39′S 132°13′E
163 T3 **Emur He** ≈ NE China
55 R8 **Enachu Landing** NW Guyana 06°09′N 60°02′W
93 F16 **Enafors** Jämtland, C Sweden 63°17′N 12°24′E
94 N11 **Enånger** Gävleborg, C Sweden 61°30′N 17°10′E
96 G7 **Enard Bay** bay NW Scotland, United Kingdom
Enareträsk see Inarijärvi
171 X14 **Enarotali** Papua, E Indonesia 03°55′S 136°21′E
En 'Avedat see Ein Avdat
165 T2 **Enbetsu** Hokkaidō, NE Japan 44°44′N 141°47′E
61 H16 **Encantadas, Serra das** ⛰ S Brazil
40 E7 **Encantado, Cerro** ▲ NW Mexico 26°46′N 112°33′W
62 P7 **Encarnación** Itapúa, S Paraguay 27°20′S 55°50′W
40 M12 **Encarnación de Díaz** Jalisco, SW Mexico 21°33′N 102°13′W
77 O17 **Enchi** SW Ghana 05°53′N 02°48′W
25 Q14 **Encinal** Texas, SW USA 28°02′N 99°21′W
35 U17 **Encinitas** California, W USA 33°02′N 117°17′W
25 S16 **Encino** Texas, SW USA 26°58′N 98°06′W
54 H6 **Encontrados** Zulia, NW Venezuela 09°04′N 72°16′W
182 I10 **Encounter Bay** inlet South Australia
61 F15 **Encruzilhada** Rio Grande do Sul, S Brazil 28°58′S 55°31′W
61 H16 **Encruzilhada do Sul** Rio Grande do Sul, S Brazil 30°30′S 52°32′W
111 M20 **Encs** Borsod-Abaúj-Zemplén, NE Hungary 48°21′N 21°09′E
193 P3 **Endeavour Seamount** undersea feature N Pacific Ocean 48°15′N 129°04′W
181 V1 **Endeavour Strait** strait Queensland, NE Australia
171 O16 **Endeh** Flores, S Indonesia 08°48′S 121°37′E
95 G23 **Endelave** island C Denmark
191 T4 **Enderbury Island** atoll Phoenix Islands, C Kiribati
11 N16 **Enderby** British Columbia, SW Canada 50°34′N 119°09′W
195 W4 **Enderby Land** physical region Antarctica
173 N14 **Enderby Plain** undersea feature S Indian Ocean
29 Q6 **Enderlin** North Dakota, N USA 46°37′N 97°36′W
Endersdorf see Jędrzejów
28 K16 **Enders Reservoir** ▣ Nebraska, C USA
18 H11 **Endicott** New York, NE USA 42°06′N 76°03′W
39 P7 **Endicott Mountains** ⛰ Alaska, USA
118 I5 **Endla Raba** wetland C Estonia
117 T9 **Enerhodar** Zaporiz'ka Oblast', SE Ukraine 47°30′N 34°40′E
57 F14 **Ene, Río** ≈ C Peru
189 N4 **Enewetak Atoll** var. Ānewetak, Eniwetok. atoll Ralik Chain, W Marshall Islands
114 L13 **Enez** Edirne, NW Turkey 40°44′N 26°05′E
21 W8 **Enfield** North Carolina, SE USA 36°10′N 77°40′W
186 B7 **Enga** ◐ province W Papua New Guinea
45 Q9 **Engaño, Cabo** headland E Dominican Republic 18°36′N 68°19′W
164 U3 **Engaru** Hokkaidō, NE Japan 44°06′N 143°30′E
108 F9 **Engelberg** Unterwalden, C Switzerland 46°51′N 08°25′E
21 Y9 **Engelhard** North Carolina, SE USA 35°30′N 76°00′W
127 P8 **Engel's** Saratovskaya Oblast', W Russian Federation 51°27′N 46°09′E
101 G24 **Engen** Baden-Württemberg, SW Germany 47°52′N 08°46′E
Engeten see Aiud
168 K15 **Enggano, Pulau** island W Indonesia
80 J8 **Enghershatu** ▲ N Eritrea 16°41′N 38°21′E
99 F19 **Enghien** Dut. Edingen. Hainaut, SW Belgium 50°42′N 04°03′E
27 V12 **England** Arkansas, C USA 34°32′N 91°58′W
97 M20 **England** Lat. Anglia. ◐ national region England, United Kingdom
14 H8 **Englehart** Ontario, S Canada 47°50′N 79°52′W
37 T4 **Englewood** Colorado, C USA 39°39′N 104°59′W
31 O16 **English** Indiana, N USA 38°20′N 86°28′W
39 Q13 **English Bay** Alaska, USA 59°21′N 151°55′W
English Bazar see Ingrāj Bāzār
97 N25 **English Channel** var. The Channel, Fr. la Manche. channel NW Europe
194 J7 **English Coast** physical region Antarctica
105 S11 **Enguera** Valenciana, E Spain 38°58′N 00°42′W
118 E8 **Engure** W Latvia 57°09′N 23°13′E
118 E8 **Engures Ezers** ◎ NW Latvia
137 R9 **Enguri** Rus. Inguri. ≈ NW Georgia
Engyum see Gangi
26 M9 **Enid** Oklahoma, C USA 36°25′N 97°53′W
22 L3 **Enid Lake** ▣ Mississippi, S USA
189 Y2 **Enigu** island Ratak Chain, SE Marshall Islands
Enikale Strait see Kerch Strait
147 Z8 **Enil'chek** Issyk-Kul'skaya Oblast', E Kyrgyzstan 42°04′N 79°01′E
115 F17 **Enipéfs** ≈ C Greece
165 S4 **Eniwa** Hokkaidō, NE Japan 42°53′N 141°14′E
Eniwetok see Enewetak Atoll
Enjiang see Yongfeng
Enkeldoorn see Chivhu
123 S11 **Enken, Mys** prev. Mys Enkan. headland NE Russian Federation 58°29′N 141°27′E
Enken, Mys see Enken, Mys
98 J8 **Enkhuizen** Noord-Holland, NW Netherlands 52°34′N 05°03′E
109 Q4 **Enknach** ≈ N Austria
95 N15 **Enköping** Uppsala, C Sweden 59°38′N 17°05′E
107 K24 **Enna** var. Castrogiovanni, Henna. Sicilia, Italy, C Mediterranean Sea 37°34′N 14°16′E
80 D11 **En Nahud** Western Kordofan, C Sudan 12°41′N 28°28′E
138 F8 **En Nâqoûra** var. An Nāqūrah. SW Lebanon 33°06′N 33°30′E
En Nazira see Natzrat
78 K8 **Ennedi** plateau NE Chad
78 J7 **Ennedi-Ouest** ◐ region N Chad
Ennedi-Ouest, Région de l' see Ennedi-Ouest
101 E15 **Ennepetal** Nordrhein-Westfalen, W Germany 51°18′N 07°23′E
183 P4 **Enngonia** New South Wales, SE Australia 29°19′S 145°52′E
97 C19 **Ennis** Ir. Inis. Clare, W Ireland 52°50′N 08°59′W
33 R11 **Ennis** Montana, NW USA 45°21′N 111°45′W
25 U7 **Ennis** Texas, SW USA 32°19′N 96°37′W
97 F20 **Enniscorthy** Ir. Inis Córthaidh. SE Ireland 52°30′N 06°34′W
97 E15 **Enniskillen** var. Inniskilling, Ir. Inis Ceithleann. SW Northern Ireland, United Kingdom 54°21′N 07°38′W
97 B19 **Ennistimon** Ir. Inis Díomáin. Clare, W Ireland 52°57′N 09°17′W
109 T4 **Enns** Oberösterreich, N Austria 48°13′N 14°28′E
109 T4 **Enns** ≈ C Austria
93 O16 **Eno** Pohjois-Karjala, SE Finland 62°45′N 30°15′E
24 M5 **Enochs** Texas, SW USA 33°51′N 102°46′W
93 N17 **Enonkoski** Etelä-Savo, E Finland 62°04′N 28°53′E
92 K10 **Enontekiö** Lapp. Eanodat. Lappi, N Finland 68°25′N 23°40′E
21 Q11 **Enoree** South Carolina, SE USA 34°39′N 81°58′W
21 P11 **Enoree River** ≈ South Carolina, SE USA
18 M6 **Enosburg Falls** Vermont, NE USA 44°54′N 72°50′W
171 N13 **Enrekang** Sulawesi, C Indonesia 03°33′S 119°46′E
45 N10 **Enriquillo** SW Dominican Republic 17°57′N 71°13′W
45 N9 **Enriquillo, Lago** ◎ SW Dominican Republic
98 L9 **Ens** Flevoland, N Netherlands 52°39′N 05°49′E
98 P11 **Enschede** Overijssel, E Netherlands 52°13′N 06°55′E
40 B2 **Ensenada** Baja California Norte, NW Mexico 31°52′N 116°32′W
101 E20 **Ensheim** ✈ (Saarbrücken) Saarland, W Germany 49°13′N 07°09′E
160 L9 **Enshi** Hubei, C China 30°16′N 109°26′E
164 L14 **Enshū-nada** gulf SW Japan
23 O8 **Ensley** Florida, SE USA 30°31′N 87°16′W
Enso see Svetogorsk
81 F18 **Entebbe** S Uganda 0°07′N 32°30′E
81 F18 **Entebbe** ✈ C Uganda 0°04′N 32°29′E
101 M18 **Entenbühl** ▲ Czech Republic/Germany 50°09′N 12°10′E
98 N10 **Enter** Overijssel, E Netherlands 52°19′N 06°34′E
23 Q7 **Enterprise** Alabama, S USA 31°19′N 85°50′W
32 L11 **Enterprise** Oregon, NW USA 45°25′N 117°18′W
36 J7 **Enterprise** Utah, W USA 37°33′N 113°42′W
32 J8 **Entiat** Washington, NW USA 47°39′N 120°15′W
105 P15 **Entinas, Punta de las** headland S Spain 36°40′N 02°44′W
108 F8 **Entlebuch** Luzern, W Switzerland 47°02′N 08°04′E
108 F8 **Entlebuch** valley C Switzerland
63 I22 **Entrada, Punta** headland S Argentina
103 O13 **Entraygues-sur-Truyère** Aveyron, S France 44°39′N 02°36′E
187 O14 **Entrecasteaux, Récifs d'** reef N New Caledonia
61 C17 **Entre Ríos** off. Provincia de Entre Ríos. ◐ province NE Argentina
42 K7 **Entre Ríos, Cordillera** ⛰ Honduras/Nicaragua
Entre Ríos, Provincia de see Entre Ríos
104 G9 **Entroncamento** Santarém, C Portugal 39°28′N 08°28′W
77 V16 **Enugu** Enugu, S Nigeria 06°24′N 07°24′E
77 U16 **Enugu** ◐ state SE Nigeria
123 V5 **Enurmino** Chukotskiy Avtonomnyy Okrug, NE Russian Federation 66°46′N 171°40′W
54 E9 **Envigado** Antioquia, W Colombia 06°09′N 75°38′W
59 B15 **Envira** Amazonas, W Brazil 07°12′S 69°59′W
Enyélé see Enyellé
79 I16 **Enyellé** var. Enyélé. Likouala, NE Congo 02°49′N 18°02′E
101 H21 **Enz** ≈ SW Germany
165 N13 **Enzan** var. Kōshū. Yamanashi, Honshū, S Japan 35°44′N 138°43′E
104 I2 **Eo** ≈ NW Spain
Eochaill see Youghal
Eochaille, Cuan see Youghal Bay
107 K22 **Eolie, Isole** var. Isole Lipari, Eng. Aeolian Islands, Lipari Islands. island group S Italy
189 U12 **Eot** island Chuuk, C Micronesia
Epáno Archánes/Epáno Arkhánai see Archánes
115 G14 **Epanomí** Kentrikí Makedonía, N Greece 40°25′N 22°57′E
98 M10 **Epe** Gelderland, E Netherlands 52°21′N 05°59′E
77 S16 **Epe** Lagos, S Nigeria 06°37′N 04°01′E
79 I17 **Epéna** Likouala, NE Congo 01°28′N 17°29′E
Eperies/Eperjes see Prešov
103 Q4 **Épernay** anc. Sparnacum. Marne, N France 49°02′N 03°58′E
36 L5 **Ephraim** Utah, W USA 39°21′N 111°35′W
18 H15 **Ephrata** Pennsylvania, NE USA 40°09′N 76°08′W
32 J8 **Ephrata** Washington, NW USA 47°19′N 119°33′W
187 R14 **Epi** var. Épi. island C Vanuatu
Épi see Epi
105 R6 **Épila** Aragón, NE Spain 41°34′N 01°19′W
103 T6 **Épinal** Vosges, NE France 48°10′N 06°28′E
Epiphania see Ḩamāh
Epirus see Ípeiros
121 P3 **Episkopí** SW Cyprus 34°37′N 32°53′E
Episkopi Bay see Episkopí, Kólpos
121 P3 **Episkopí, Kólpos** var. Episkopi Bay. bay SE Cyprus
Epoon see Ebon Atoll
Eporedia see Ivrea
Eppeschdorf see Dumbrăveni
101 H21 **Eppingen** Baden-Württemberg, SW Germany 49°09′N 08°54′E
83 E18 **Epukiro** Omaheke, E Namibia 21°40′S 19°09′E
29 Y13 **Epworth** Iowa, C USA 42°27′N 90°55′W
143 O10 **Eqlīd** var. Iqlid. Fārs, C Iran 30°54′N 52°40′E
Equality State see Wyoming
79 J18 **Equateur** off. Région de l' Equateur. ◐ region N Dem. Rep. Congo
Equateur, Région de l' see Equateur
151 K22 **Equatorial Channel** channel S Maldives
79 B17 **Equatorial Guinea** off. Equatorial Guinea, Republic of. ◆ republic C Africa
Equatorial Guinea, Republic of see Equatorial Guinea
121 V11 **Eratosthenes Tablemount** undersea feature E Mediterranean Sea 33°48′N 32°53′E
Erautini see Johannesburg
136 L12 **Erbaa** Tokat, N Turkey 40°42′N 36°37′E
101 E19 **Erbeskopf** ▲ W Germany 49°43′N 07°04′E
139 U4 **Erbet** Ar. 'Arbat, var. Arbat. As Sulaymānīyah, NE Iraq 35°26′N 45°34′E
Erbil see Arbīl
121 P2 **Ercan** ✈ (Nicosia) N Cyprus 35°07′N 33°30′E
Ercegnovi see Herceg-Novi
137 T14 **Erçek Gölü** ◎ E Turkey
137 S14 **Erciş** Van, E Turkey 39°02′N 43°21′E
136 K14 **Erciyes Dağı** anc. Argaeus. ▲ C Turkey 38°32′N 35°28′E
111 J22 **Érd** Ger. Hanselbeck. Pest, C Hungary 47°22′N 18°56′E
163 X11 **Erdaobaihe** prev. Baihe. Jilin, NE China 42°24′N 128°09′E
159 O12 **Erdaogou** Qinghai, W China 34°30′N 92°50′E
163 X11 **Erdao Jiang** ⛰ NE China
Erdăt-Sângeorz see Sângeorgiu de Pădure
136 C11 **Erdek** Balıkesir, NW Turkey 40°24′N 27°47′E
Erdély see Transylvania
Erdélyi-Havasok see Carpaţii Meridionalii
136 J17 **Erdemli** İçel, S Turkey 36°35′N 34°19′E
163 O10 **Erdene** var. Ulaan-Uul. Dornogovĭ, SE Mongolia 44°21′N 111°06′E
162 H9 **Erdene** var. Sangiyn Dalay. Govĭ-Altay, C Mongolia 45°12′N 97°51′E
162 E6 **Erdenebüren** var. Har-Us. Hovd, W Mongolia 48°30′N 91°25′E
162 K9 **Erdenedalay** var. Sangiyn Dalay. Dundgovĭ, C Mongolia 45°59′N 104°58′E
162 G7 **Erdenehayrhan** var. Altan. Dzavhan, W Mongolia 48°05′N 95°48′E
162 J7 **Erdenemandal** var. Öldziyt. Arhangay, C Mongolia 48°30′N 101°25′E
162 K6 **Erdenet** Orhon, N Mongolia 49°01′N 104°07′E
163 Q9 **Erdenetsagaan** var. Chonogol. Sühbaatar, E Mongolia 45°55′N 115°19′E

◆ Country
● Country Capital
◇ Dependent Territory
○ Dependent Territory Capital
◐ Administrative Regions
✈ International Airport
▲ Mountain
⛰ Mountain Range
Volcano
≈ River
◎ Lake
▣ Reservoir

ERDENETSOGT – FADIFFOLU

162 I8 **Erdenetsogt** Bayanhongor, C Mongolia 46°27′N 100°53′E
Erdenetsogt see Bayan-Ovoo
78 K7 **Erdi** plateau NE Chad
78 L7 **Erdi Ma** desert NE Chad
101 M23 **Erding** Bayern, SE Germany 48°18′N 11°54′E
Erdőszáda see Ardusat
Erdőszentgyörgy see Sângeorgiu de Pădure
102 I7 **Erdre** ≈ NW France
195 R13 **Erebus, Mount** ✱ Ross Island, Antarctica 78°11′S 165°09′E
61 H14 **Erechim** Rio Grande do Sul, S Brazil 27°35′S 52°15′W
163 O7 **Ereen Davaanï Nuruu** ▲▲ NE Mongolia
163 Q6 **Ereentsav** Dornod, NE Mongolia 49°51′N 115°41′E
136 I16 **Ereğli** Konya, S Turkey 37°30′N 34°02′E
115 A15 **Ereikoussa** island Iónia Nisiá, Greece, C Mediterranean Sea
163 O11 **Erenhot** var. Erlian. Nei Mongol Zizhiqu, NE China 43°35′N 112°E
104 M6 **Eresma** ≈ N Spain
115 K17 **Eresós** var. Eressós. Lésvos, E Greece 39°11′N 25°57′E
Eressós see Eresós
Ereymentaū see Yereymentau
99 K21 **Érezée** Luxembourg, SE Belgium 50°16′N 05°34′E
74 G7 **Erfoud** SE Morocco 31°29′N 04°18′W
101 D16 **Erft** ≈ W Germany
101 K16 **Erfurt** Thüringen, C Germany 50°59′N 11°02′E
137 P15 **Ergani** Diyarbakır, SE Turkey 38°17′N 39°44′E
Ergel see Hatanbulag
Ergene Çayı see Ergene Irmağı
136 C10 **Ergene Irmağı** var. Ergene Çayı. ≈ NW Turkey
118 I9 **Ērgļi** C Latvia 56°55′N 25°38′E
78 H11 **Erguig, Bahr** ≈ SW Chad
163 S5 **Ergun** var. Labudalin; prev. Ergun Youqi. Nei Mongol Zizhiqu, N China 50°13′N 120°09′E
Ergun see Gegan Gol
Ergun He see Argun
Ergun Youqi see Ergun
Ergun Zuoqi see Gegen Gol
160 F12 **Er Hai** ⊠ SW China
104 K4 **Ería** ≈ NW Spain
80 H8 **Eriba** Kassala, NE Sudan 16°37′N 36°04′E
96 I6 **Eriboll, Loch** inlet NW Scotland, United Kingdom
65 Q18 **Erica Seamount** undersea feature SW Indian Ocean 38°15′S 14°30′E
107 H23 **Erice** Sicilia, Italy, C Mediterranean Sea 38°02′N 12°35′E
104 E10 **Ericeira** Lisboa, C Portugal 38°58′N 09°25′W
96 H10 **Ericht, Loch** ◎ C Scotland, United Kingdom
26 J11 **Erick** Oklahoma, C USA 35°13′N 99°52′W
18 B11 **Erie** Pennsylvania, NE USA 42°07′N 80°04′W
18 E9 **Erie Canal** canal New York, NE USA
Érié, Lac see Erie, Lake
31 T10 **Erie, Lake** Fr. Lac Érié. ◎ Canada/USA
Erigabo see Ceerigaabo
77 N8 **'Erîgât** desert N Mali
Erigavo see Ceerigaabo
92 P2 **Erik Eriksenstretet** strait E Svalbard
11 X15 **Eriksdale** Manitoba, S Canada 50°52′N 98°07′W
189 V6 **Erikub Atoll** var. Ādkup. atoll Ratak Chain, C Marshall Islands
102 G4 **Er, Îles d'** island group NW France
Erimanthos see Erýmanthos
165 T6 **Erimo** Hokkaidō, NE Japan 42°01′N 143°07′E
165 T6 **Erimo-misaki** headland Hokkaidō, NE Japan 41°57′N 143°12′E
20 H8 **Erin** Tennessee, S USA 36°19′N 87°42′W
96 E9 **Eriskay** island NW Scotland, United Kingdom
Erithraí see Erythrés
80 I9 **Eritrea** off. State of Eritrea, Ērtra. ◆ transitional government E Africa
Eritrea, State of see Eritrea
Erivan see Yerevan
101 D16 **Erkelenz** Nordrhein-Westfalen, W Germany 51°04′N 06°19′E
95 P15 **Erken** ◎ C Sweden
101 K19 **Erlangen** Bayern, S Germany 49°36′N 11°E
160 G9 **Erlang Shan** ▲ C China 29°56′N 102°24′E
Erlau see Eger
109 V5 **Erlauf** ≈ NE Austria
181 Q8 **Erldunda Roadhouse** Northern Territory, N Australia 25°13′S 133°13′E
Erlian see Erenhot
27 T15 **Erling, Lake** ⊠ Arkansas, USA
109 O8 **Erlsbach** Tirol, W Austria 46°54′N 12°15′E
Ermak see Aksu
98 K10 **Ermelo** Gelderland, C Netherlands 52°18′N 05°38′E
83 K21 **Ermelo** Mpumalanga, NE South Africa 26°32′S 29°59′E
136 H17 **Ermenek** Karaman, S Turkey 36°38′N 32°55′E
Érmihályfalva see Valea lui Mihai
115 G20 **Ermióni** Pelopónnisos, S Greece 37°24′N 23°15′E
115 J20 **Ermoúpoli** var. Hermoupolis; prev. Ermoúpolis. Sýyros, Kykládes, Greece, Aegean Sea 37°26′N 24°55′E
Ermoúpolis see Ermoúpoli
Ernabella see Pukatja
155 G22 **Ernākulam** Kerala, SW India 10°04′N 76°18′E
102 J6 **Ernée** Mayenne, NW France 48°18′N 00°54′W
61 H14 **Ernestina, Barragem** ⊠ S Brazil
54 E4 **Ernesto Cortissoz** ✕ (Barranquilla) Atlántico, N Colombia
155 H21 **Erode** Tamil Nādu, SE India 11°21′N 77°43′E
Eroj see Iroj
83 C19 **Erongo** ◐ district W Namibia
99 F21 **Erquelinnes** Hainaut, S Belgium 50°18′N 04°08′E
74 G7 **Er-Rachidia** var. Ksar al Soule. E Morocco 31°58′N 04°22′W
80 E11 **Er Rahad** var. Ar Rahad. Northern Kordofan, C Sudan 12°43′N 30°39′E
Er Ramle see Ramla
83 O15 **Errego** Zambézia, NE Mozambique 16°02′S 37°11′E
105 Q2 **Errenteria** Cast. Rentería. País Vasco, N Spain 43°17′N 01°54′W
Er Rif/Er Riff see Rif
97 D14 **Errigal Mountain** Ir. An Earagail. ▲ N Ireland 55°03′N 08°09′W
97 A15 **Erris Head** Ir. Ceann Iorrais. headland W Ireland 54°18′N 10°01′W
187 S15 **Erromango** island S Vanuatu
Error Guyot see Error Tablemount
173 O4 **Error Tablemount** var. Error Guyot. undersea feature W Indian Ocean 10°20′N 56°05′E
80 G11 **Er Roseires** Blue Nile, E Sudan 11°52′N 34°23′E
Erseka see Ersekë
113 M22 **Ersekë** var. Erseka, Kolonjë. Korçë, SE Albania 40°19′N 20°39′E
Érsekújvár see Nové Zámky
29 S4 **Erskine** Minnesota, N USA 47°42′N 96°00′W
103 V6 **Erstein** Bas-Rhin, NE France 48°24′N 07°39′E
108 G9 **Erstfeld** Uri, C Switzerland 46°49′N 08°41′E
158 M3 **Ertai** Xinjiang Uygur Zizhiqu, NW China 46°04′N 90°06′E
126 M7 **Ertil'** Voronezhskaya Oblast', W Russian Federation 51°51′N 40°46′E
Ertis see Irtysh, C Asia
Ertis see Irtyshsk, Kazakhstan
158 K2 **Ertix He** Rus. Chërnyy Irtysh. ≈ China/Kazakhstan
Ērtra see Eritrea
21 P9 **Erwin** North Carolina, SE USA 35°19′N 78°40′W
Erydropótamos see Erythropótamos
115 E19 **Erýmanthos** var. Erimanthos. ▲ S Greece 37°57′N 21°51′E
115 G19 **Erythrés** prev. Erithraí. Stereá Elláda, C Greece 38°18′N 23°20′E
114 L12 **Erythropótamos** Bul. Byala Reka, var. Erydropótamos. ≈ Bulgaria/Greece
160 F12 **Eryuan** var. Yuhu. Yunnan, SW China 26°09′N 100°01′E
109 U6 **Erzbach** ≈ W Austria
Erzerum see Erzurum
101 N17 **Erzgebirge** Cz. Krušné Hory, Eng. Ore Mountains. ▲ Czech Republic/Germany see also Krušné Hory
Erzgebirge see Krušné Hory
122 L14 **Erzin** Respublika Tyva, S Russian Federation 50°17′N 95°03′E
137 O13 **Erzincan** var. Erzinjan. Erzincan, E Turkey 39°44′N 39°30′E
137 N13 **Erzincan** var. Erzinjan. ◐ province NE Turkey
Erzinjan see Erzincan
Erzsébetváros see Dumbrăveni
137 Q13 **Erzurum** prev. Erzerum. Erzurum, NE Turkey 39°57′N 41°17′E
137 Q12 **Erzurum** prev. Erzerum. ◐ province NE Turkey
186 G9 **Esa'ala** Normanby Island, SE Papua New Guinea 09°45′S 150°47′E
165 T2 **Esashi** Hokkaidō, NE Japan 44°57′N 142°32′E
165 Q9 **Esashi** var. Esasi. Iwate, Honshū, C Japan 39°13′N 141°11′E
165 Q5 **Esashi** Hokkaidō, N Japan 41°52′N 140°07′E
Esasi see Esashi
95 F23 **Esbjerg** Syddtjylland, W Denmark 55°28′N 08°28′E
Esbo see Espoo
36 L7 **Escalante** Utah, W USA 37°46′N 111°36′W
36 M7 **Escalante River** ≈ Utah, W USA
14 L12 **Escalier, Réservoir l'** ⊠ Québec, SE Canada
40 K7 **Escalón** Chihuahua, N Mexico 26°43′N 104°20′W
104 M8 **Escalona** Castilla-La Mancha, C Spain 40°10′N 04°24′W
23 O8 **Escambia River** ≈ Florida, SE USA
31 N5 **Escanaba** Michigan, N USA 45°45′N 87°03′W
31 N4 **Escanaba River** ≈ Michigan, N USA
105 R8 **Escandón, Puerto de** pass E Spain
41 W14 **Escárcega** Campeche, SE Mexico 18°33′N 90°41′W
171 O1 **Escarpada Point** headland Luzon, N Philippines 18°28′N 122°10′E
23 N8 **Escatawpa River** ≈ Alabama/Mississippi, S USA
103 P2 **Escaut** ≈ N France
Escaut see Scheldt
99 M25 **Esch-sur-Alzette** Luxembourg, S Luxembourg 49°30′N 05°59′E
101 J15 **Eschwege** Hessen, C Germany 51°10′N 10°03′E
101 D16 **Eschweiler** Nordrhein-Westfalen, W Germany 50°49′N 06°16′E
Esclaves, Grand Lac des see Great Slave Lake
45 O8 **Escocesa, Bahía** bay N Dominican Republic
43 W15 **Escocés, Punta** headland NE Panama 08°50′N 77°37′W
35 U17 **Escondido** California, W USA 33°07′N 117°05′W
42 M10 **Escondido, Río** ≈ SE Nicaragua
15 S7 **Escoumins, Rivière des** ≈ Québec, SE Canada
37 O13 **Escudilla Mountain** ▲ Arizona, SW USA 33°57′N 109°07′W
40 J11 **Escuinapa** var. Escuinapa de Hidalgo. Sinaloa, C Mexico 22°50′N 105°46′W
Escuinapa de Hidalgo see Escuinapa
42 C6 **Escuintla** Escuintla, S Guatemala 14°17′N 90°46′W
41 V17 **Escuintla** Chiapas, SE Mexico 15°20′N 92°40′W
42 A2 **Escuintla** off. Departamento de Escuintla. ◐ department S Guatemala
Escuintla, Departamento de see Escuintla
15 W7 **Escuminac** ≈ Québec, SE Canada
79 D16 **Eséka** Centre, SW Cameroon 03°40′N 10°48′E
136 I12 **Esenboğa** ✕ (Ankara) Ankara, C Turkey 40°05′N 33°01′E
136 D17 **Eşen Çayı** ≈ SW Turkey
146 B13 **Esenguly** Rus. Gasan-Kuli. Balkan Welaýaty, W Turkmenistan 37°29′N 53°57′E
105 T4 **Ésera** ≈ NE Spain
143 N8 **Eşfahān** Eng. Isfahan; anc. Aspadana. Eşfahān, C Iran 32°41′N 51°41′E
143 O7 **Eşfahān** off. Ostān-e Eşfahān. ◐ province C Iran
Eşfahān, Ostān-e see Eşfahān
105 N5 **Esgueva** ≈ N Spain
Eshkamesh see Ishkamish
Eshkāshem see Ishkāshim
83 L23 **Eshowe** KwaZulu/Natal, E South Africa 28°53′S 31°28′E
143 T5 **'Eshqābād** Khorāsān-e Razavī, NE Iran 36°00′N 59°01′E
Esh Sham see Rif Dimashq
Esh Sharā see Ash Sharāh
Esik see Yesik
Esil see Yesil'
Esil see Ishim, Kazakhstan/Russian Federation
183 V2 **Esk** Queensland, E Australia 27°15′S 152°23′E
184 O11 **Eskdale** Hawke's Bay, North Island, New Zealand 39°24′S 176°51′E
Eski Dzhumaya see Targovishte
92 L2 **Eskifjördur** Austurland, E Iceland 65°04′N 14°01′W
139 S3 **Eski Kaļak** var. Askī Kalak, Kalak. Arbīl, N Iraq 36°16′N 43°40′E
95 N16 **Eskilstuna** Södermanland, C Sweden 59°22′N 16°31′E
8 H6 **Eskimo Lakes** ◎ Northwest Territories, NW Canada
9 O10 **Eskimo Point** headland Nunavut, E Canada 61°19′N 93°49′W
Eskimo Point see Arviat
139 Q2 **Eski Mosul** Ninawá, N Iraq 36°31′N 42°45′E
Eski-Nookat see Nookat
136 F12 **Eskişehir** var. Eskishehr. Eskişehir, W Turkey 39°46′N 30°30′E
136 F13 **Eskişehir** var. Eski shehr. ◐ province NW Turkey
Eskishehr see Eskişehir
104 K5 **Esla** ≈ NW Spain
Eslāmābād see Eslāmābād-e Gharb
142 J6 **Eslāmābād-e Gharb** var. Eslāmābād; prev. Harunabad, Shāhābād. Kermānshāhān, W Iran 34°08′N 46°35′E
148 J4 **Eslām Qal'eh** Pash. Islam Qala. Herāt, W Afghanistan 34°41′N 61°03′E
95 K23 **Eslöv** Skåne, S Sweden 55°50′N 13°20′E
143 S12 **Esmā'īlābād** Kermān, S Iran 28°48′N 56°39′E
143 U8 **Esmā'īlābād** Khorāsān-e Jonūbī, E Iran 35°20′N 60°30′E
136 D14 **Eşme** Uşak, W Turkey 38°26′N 28°59′E
44 G6 **Esmeralda** Camagüey, E Cuba 21°51′N 78°10′W
63 F21 **Esmeralda, Isla** island S Chile
56 B5 **Esmeraldas** Esmeraldas, N Ecuador 0°55′N 79°40′W
56 B5 **Esmeraldas** ◐ province NW Ecuador
Esna see Isnā
14 B6 **Esnagi Lake** ◎ Ontario, S Canada
143 V14 **Espakeh** Sīstān va Balūchestān, SE Iran 26°54′N 60°09′E
103 O13 **Espalion** Aveyron, S France 44°31′N 02°45′E
España see Spain
14 E11 **Espanola** Ontario, S Canada 46°15′N 81°46′W
37 S10 **Espanola** New Mexico, SW USA 35°59′N 106°04′W
57 C18 **Española, Isla** var. Hood Island. island Galapagos Islands, Ecuador, E Pacific Ocean
104 M13 **Espejo** Andalucía, S Spain 37°40′N 04°34′W
94 C13 **Espeland** Hordaland, S Norway 60°22′N 05°27′E
100 G12 **Espelkamp** Nordrhein-Westfalen, NW Germany 52°22′N 08°37′E
38 M8 **Espenberg, Cape** headland Alaska, USA 66°33′N 163°36′W
180 L13 **Esperance** Western Australia 33°49′S 121°52′E
186 L9 **Esperance, Cape** headland Guadacanal, C Solomon Islands 09°09′S 159°38′E
57 P18 **Esperancita** Santa Cruz, E Bolivia
61 B17 **Esperanza** Santa Fe, C Argentina 31°29′S 61°00′W
40 G6 **Esperanza** Sonora, NW Mexico 27°37′N 109°51′W
24 H9 **Esperanza** Texas, SW USA 31°09′N 105°40′W
194 H3 **Esperanza** Argentinian research station Antarctica 63°29′S 56°53′W
104 E12 **Espichel, Cabo** headland S Portugal 38°25′N 09°15′W
54 E10 **Espinal** Tolima, C Colombia 04°08′N 74°53′W
48 K10 **Espinhaço, Serra do** ▲▲ SE Brazil
104 G6 **Espinho** Aveiro, N Portugal 41°01′N 08°38′W
59 N18 **Espinosa** Minas Gerais, SE Brazil 14°58′S 42°49′W
103 O15 **Espinouse** ▲▲ S France
60 Q8 **Espírito Santo** off. Estado do Espírito Santo. ◐ state E Brazil
Espírito Santo, Estado do see Espírito Santo
187 P13 **Espiritu Santo** var. Santo. island W Vanuatu
41 Z13 **Espíritu Santo, Bahía del** bay SE Mexico
40 F9 **Espíritu Santo, Isla del** island NW Mexico
41 Y12 **Espita** Yucatán, SE Mexico 21°00′N 88°17′W
15 Y7 **Espoir, Cap d'** headland Québec, SE Canada 48°24′N 64°21′W
Esponsede/Esponsende see Esposende
93 L20 **Espoo** Swe. Esbo. Uusimaa, S Finland 60°10′N 24°42′E
104 G5 **Esposende** var. Esponsede, Esponsende. Braga, N Portugal 41°32′N 08°47′W
83 M18 **Espungabera** Manica, SW Mozambique 20°29′S 32°48′E
63 H17 **Esquel** Chubut, SW Argentina 42°55′S 71°20′W
10 L17 **Esquimalt** Vancouver Island, British Columbia, SW Canada 48°26′N 123°27′W
61 C16 **Esquina** Corrientes, NE Argentina 30°00′S 59°30′W
42 E6 **Esquipulas** Chiquimula, SE Guatemala 14°36′N 89°22′W
42 K9 **Esquipulas** Matagalpa, C Nicaragua 12°30′N 85°55′W
94 I8 **Essandsjøen** ◎ S Norway
74 E7 **Essaouira** prev. Mogador. W Morocco 31°33′N 09°40′W
Esseg see Osijek
Es Semara see Smara
99 G15 **Essen** Antwerpen, N Belgium 51°28′N 04°28′E
101 E15 **Essen** var. Essen an der Ruhr. Nordrhein-Westfalen, W Germany 51°28′N 07°01′E
Essen an der Ruhr see Essen
74 I5 **Es Senia** ✕ (Oran) NW Algeria 35°34′N 00°42′W
55 T8 **Essequibo Islands** island group N Guyana
55 T11 **Essequibo River** ≈ C Guyana
14 C18 **Essex** Ontario, S Canada 42°10′N 82°50′W
29 T16 **Essex** Iowa, C USA 40°49′N 95°18′W
97 P21 **Essex** cultural region E England, United Kingdom
31 R8 **Essexville** Michigan, N USA 43°37′N 83°50′W
101 H22 **Esslingen** var. Esslingen am Neckar. Baden-Württemberg, SW Germany 48°45′N 09°19′E
Esslingen am Neckar see Esslingen
103 N6 **Essonne** ◐ department N France
Es Suweida see As Suwaydā'
79 F16 **Est** Eng. East. ◐ province SE Cameroon
104 I1 **Estaca de Bares, Punta de** point NW Spain
24 M5 **Estacado, Llano** plain New Mexico/Texas, SW USA
41 P12 **Estación Tamuín** San Luis Potosí, C Mexico 22°00′N 98°44′W
63 K25 **Estados, Isla de los** prev. Eng. Staten Island. island S Argentina
Estado Vargas see Vargas
143 P12 **Eşţahbān** Fārs, S Iran
14 F11 **Estaire** Ontario, S Canada 46°19′N 80°47′W
59 P16 **Estância** Sergipe, E Brazil 11°15′S 37°28′W
37 S12 **Estancia** New Mexico, SW USA 34°45′N 106°03′W
104 G7 **Estarreja** Aveiro, N Portugal 40°45′N 08°35′W
102 M17 **Estats, Pica d'** Sp. Pico d'Estats. ▲ France/Spain 42°39′N 01°24′E
Estats, Pico d' see Estats, Pica d'
83 K23 **Estcourt** KwaZulu/Natal, E South Africa 29°00′S 29°53′E
106 H8 **Este** anc. Ateste. Veneto, NE Italy 45°14′N 11°40′E
42 J9 **Estelí** Estelí, NW Nicaragua 13°05′N 86°21′W
42 J9 **Estelí** ◐ department NW Nicaragua
105 Q4 **Estella** Bas. Lizarra. Navarra, N Spain 42°41′N 02°02′W
29 R9 **Estelline** South Dakota, N USA 44°34′N 96°54′W
25 P4 **Estelline** Texas, SW USA 34°33′N 100°26′W
104 L14 **Estepa** Andalucía, S Spain 37°17′N 04°52′W
104 L16 **Estepona** Andalucía, S Spain 36°26′N 05°09′W
39 R9 **Ester** Alaska, USA 64°49′N 148°03′W
11 V16 **Esterhazy** Saskatchewan, S Canada 50°40′N 102°02′W
37 S3 **Estes Park** Colorado, C USA 40°22′N 105°31′W
11 V17 **Estevan** Saskatchewan, S Canada 49°07′N 103°05′W
29 T11 **Estherville** Iowa, C USA 43°24′N 94°49′W
21 R15 **Estill** South Carolina, SE USA 32°45′N 81°14′W
103 Q6 **Estissac** Aube, N France 48°17′N 03°51′E
15 T9 **Est, Lac de l'** ◎ Québec, SE Canada
Estland see Estonia
11 S16 **Eston** Saskatchewan, S Canada 51°09′N 108°42′W
118 G5 **Estonia** off. Republic of Estonia, Est. Eesti Vabariik, Ger. Estland, Latv. Igaunija; prev. Estonian SSR, Rus. Estonskaya SSR. ◆ republic NE Europe
Estonian SSR see Estonia
Estonia, Republic of see Estonia
Estonskaya SSR see Estonia
104 E11 **Estoril** Lisboa, W Portugal 38°42′N 09°23′W
59 L14 **Estreito** Maranhão, E Brazil 06°34′S 47°22′W
104 I8 **Estrela, Serra da** ▲▲ C Portugal
40 D3 **Estrella, Punta** headland NW Mexico 30°53′N 114°45′W
104 F10 **Estremadura** cultural and historical region W Portugal
Estremadura see Extremadura
104 H11 **Estremoz** Évora, S Portugal 38°50′N 07°35′W
79 D18 **Estuaire** off. Province de l'Estuaire, var. L'Estuaire. ◐ province NW Gabon
Estuaire, Province de l' see Estuaire
Eszék see Osijek
111 I22 **Esztergom** Ger. Gran; anc. Strigonium. Komárom-Esztergom, N Hungary 47°47′N 18°44′E
152 K11 **Etah** Uttar Pradesh, N India 27°33′N 78°39′E
189 R17 **Etal Atoll** atoll Mortlock Islands, C Micronesia
99 K24 **Étalle** Luxembourg, SE Belgium 49°41′N 05°36′E
103 N6 **Étampes** Essonne, N France 48°26′N 02°10′E
182 J1 **Etamunbanie, Lake** salt lake South Australia
103 N1 **Étaples** Pas-de-Calais, N France 50°31′N 01°39′E
152 K12 **Etāwah** Uttar Pradesh, N India 26°46′N 79°01′E
15 R10 **Etchemin** ≈ Québec, SE Canada
Etchmiadzin see Vagharshapat
40 G7 **Etchojoa** Sonora, NW Mexico 26°54′N 109°37′W
93 L19 **Etelä-Karjala** Swe. Södra Karelen, Eng. South Karelia. ◐ region S Finland
93 K17 **Etelä-Pohjanmaa** Swe. Södra Österbotten, Eng. South Ostrobothnia. ◐ region W Finland
93 M18 **Etelä-Savo** Swe. Södra Savolax. ◐ region SE Finland
83 B16 **Etengua** Kunene, NW Namibia 17°24′S 13°05′E
99 K25 **Éthe** Luxembourg, SE Belgium 49°34′N 05°32′E
11 W15 **Ethelbert** Manitoba, S Canada 51°30′N 100°22′W
80 H12 **Ethiopia** off. Federal Democratic Republic of Ethiopia; prev. Abyssinia, People's Democratic Republic of Ethiopia. ◆ republic E Africa
Ethiopia, Federal Democratic Republic of see Ethiopia
80 I13 **Ethiopian Highlands** var. Ethiopian Plateau. plateau N Ethiopia
Ethiopian Plateau see Ethiopian Highlands
Ethiopia, People's Democratic Republic of see Ethiopia
34 M2 **Etna** California, W USA 41°25′N 122°53′W
18 B14 **Etna** Pennsylvania, NE USA 40°29′N 79°55′W
94 G12 **Etna** ≈ S Norway
107 L24 **Etna, Monte** Eng. Mount Etna. ✱ Sicilia, Italy, C Mediterranean Sea 37°46′N 15°00′E
Etna, Mount see Etna, Monte
95 C15 **Etne** Hordaland, S Norway 59°40′N 05°55′E
Etoliko see Aitolikó
39 Y14 **Etolin Island** island Alexander Archipelago, Alaska, USA
38 L12 **Etolin Strait** strait Alaska, USA
83 C17 **Etosha Pan** salt lake N Namibia
79 G18 **Etoumbi** Cuvette Ouest, NW Congo 0°01′N 14°57′E
20 M10 **Etowah** Tennessee, S USA 35°19′N 84°31′W
23 S2 **Etowah River** ≈ Georgia, SE USA
146 B13 **Etrek** var. Gyzyletrek, Rus. Kizyl-Atrek. Balkan Welaýaty, W Turkmenistan 37°40′N 54°44′E
146 C13 **Etrek** Per. Rūd-e Atrak, Rus. Atrak, Atrek. ≈ Iran/Turkmenistan
102 L3 **Étretat** Seine-Maritime, N France 49°46′N 00°23′E
114 H9 **Etropole** Sofia, W Bulgaria 42°50′N 24°00′E
Etsch see Adige
Et Tafila see Aţ Ţafilah
99 M23 **Ettelbrück** Diekirch, C Luxembourg 49°51′N 06°06′E
189 V12 **Etten** atoll Chuuk Islands, C Micronesia
99 H14 **Etten-Leur** Noord-Brabant, S Netherlands 51°34′N 04°37′E
76 G7 **Et Tîdra** var. Ile Tidra. island Dakhlet Nouâdhibou, NW Mauritania
101 G21 **Ettlingen** Baden-Württemberg, SW Germany 48°57′N 08°25′E
102 M2 **Eu** Seine-Maritime, N France 50°01′N 01°24′E
193 W16 **'Eua** prev. Middleburg Island. island Tongatapu Group, SE Tonga
193 W15 **Eua Iki** island Tongatapu Group, S Tonga
Euboea see Évvoia
181 O12 **Eucla** Western Australia 31°41′S 128°51′E
31 U11 **Euclid** Ohio, N USA 41°34′N 81°33′W
27 W14 **Eudora** Arkansas, C USA 33°06′N 91°15′W
27 Q4 **Eudora** Kansas, C USA 38°56′N 95°06′W
182 J9 **Eudunda** South Australia 34°11′S 139°03′E
23 R6 **Eufaula** Alabama, S USA 31°53′N 85°09′W
27 Q11 **Eufaula** Oklahoma, C USA 35°16′N 95°36′W
27 Q11 **Eufaula Lake** var. Eufaula Reservoir. ⊠ Oklahoma, C USA
Eufaula Reservoir see Eufaula Lake
32 F13 **Eugene** Oregon, NW USA 44°03′N 123°05′W
40 B6 **Eugenia, Punta** headland NW Mexico 27°48′N 115°03′W
183 Q8 **Eugowra** New South Wales, SE Australia 33°28′S 148°21′E
104 I2 **Eume** ≈ NW Spain
104 H2 **Eume, Encoro do** ⊠ NW Spain
Eumolpias see Plovdiv
59 O18 **Eunápolis** Bahia, SE Brazil 16°20′S 39°36′W
22 H8 **Eunice** Louisiana, S USA 30°29′N 92°25′W
37 W15 **Eunice** New Mexico, SW USA 32°26′N 103°09′W
99 M19 **Eupen** Liège, E Belgium 50°38′N 06°02′E
130 B10 **Euphrates** Ar. Al-Furāt, Turk. Firat Nehri. ≈ SW Asia
138 L3 **Euphrates Dam** dam N Syria
22 M4 **Eupora** Mississippi, S USA 33°32′N 89°16′W
93 K19 **Eura** Satakunta, SW Finland 61°07′N 22°12′E
93 K19 **Eurajoki** Satakunta, SW Finland 61°13′N 21°45′E
0-1 **Eurasian Plate** tectonic feature
102 L4 **Eure** ◐ department N France
102 M4 **Eure** ≈ N France
102 M6 **Eure-et-Loir** ◐ department C France
34 K3 **Eureka** California, W USA 40°47′N 124°12′W
27 P6 **Eureka** Kansas, C USA 37°51′N 96°17′W
33 O6 **Eureka** Montana, NW USA 48°52′N 115°03′W
35 V5 **Eureka** Nevada, W USA 39°31′N 115°58′W
29 O7 **Eureka** South Dakota, N USA 45°46′N 99°37′W
36 L4 **Eureka** Utah, W USA 39°57′N 112°07′W
32 K10 **Eureka** Washington, NW USA 46°21′N 118°41′W
27 S9 **Eureka Springs** Arkansas, C USA 36°25′N 93°45′W
182 K6 **Eurinilla Creek** seasonal river South Australia
183 O11 **Euroa** Victoria, SE Australia 36°46′S 145°35′E
172 M9 **Europa, Île** island W Madagascar
104 L3 **Europa, Picos de** ▲ N Spain
104 L16 **Europa Point** headland S Gibraltar 36°07′N 05°20′W
84-85 **Europe** continent
98 F12 **Europoort** Zuid-Holland, W Netherlands 51°59′N 04°08′E
Euskadi see País Vasco
101 D17 **Euskirchen** Nordrhein-Westfalen, W Germany 50°40′N 06°47′E
23 W11 **Eustis** Florida, SE USA 28°51′N 81°41′W
182 M9 **Euston** New South Wales, SE Australia 34°34′S 142°45′E
23 N5 **Eutaw** Alabama, S USA 32°50′N 87°53′W
100 K8 **Eutin** Schleswig-Holstein, N Germany 54°08′N 10°38′E
10 K14 **Eutsuk Lake** ⊠ British Columbia, SW Canada
Euxine Sea see Black Sea
83 C16 **Evale** Cunene, SW Angola 16°36′S 15°46′E
37 T3 **Evans** Colorado, C USA 40°22′N 104°41′W
11 P14 **Evansburg** Alberta, SW Canada 53°34′N 114°57′W
29 X13 **Evansdale** Iowa, C USA 42°28′N 92°16′W
183 V4 **Evans Head** New South Wales, SE Australia 29°07′S 153°27′E
12 J11 **Evans, Lac** ◎ Québec, SE Canada
37 S5 **Evans, Mount** ▲ Colorado, C USA 39°15′N 106°10′W
9 Q6 **Evans Strait** strait Nunavut, N Canada
31 N10 **Evanston** Illinois, N USA 42°02′N 87°41′W
33 S17 **Evanston** Wyoming, C USA 41°16′N 110°57′W
14 D11 **Evansville** Manitoulin Island, Ontario, S Canada 45°48′N 82°34′W
31 N16 **Evansville** Indiana, N USA 37°58′N 87°33′W
30 L9 **Evansville** Wisconsin, N USA 42°46′N 89°16′W
25 S8 **Evant** Texas, SW USA 31°28′N 98°09′W
143 P13 **Evaz** Fārs, S Iran 27°48′N 53°58′E
29 W4 **Eveleth** Minnesota, N USA 47°27′N 92°32′W
182 E3 **Evelyn Creek** seasonal river South Australia
181 Q2 **Evelyn, Mount** ▲ Northern Territory, N Australia 13°28′S 132°50′E
122 K10 **Evenkiyskiy Avtonomnyy Okrug** ◐ autonomous district Krasnoyarskiy Kray, N Russian Federation
183 R13 **Everard, Cape** headland Victoria, SE Australia 37°48′S 149°21′E
182 F6 **Everard, Lake** salt lake South Australia
182 C2 **Everard Ranges** ▲▲ South Australia
153 R11 **Everest, Mount** Chin. Qomolangma Feng, Nep. Sagarmāthā. ▲ China/Nepal 27°59′N 86°57′E
18 E15 **Everett** Pennsylvania, NE USA 40°00′N 78°22′W
32 H7 **Everett** Washington, NW USA 47°59′N 122°12′W
99 E17 **Evergem** Oost-Vlaanderen, NW Belgium 51°07′N 03°43′E
23 X16 **Everglades City** Florida, SE USA 25°51′N 81°22′W
23 Y16 **Everglades, The** wetland Florida, SE USA
23 P7 **Evergreen** Alabama, S USA 31°25′N 86°55′W
37 T4 **Evergreen** Colorado, C USA 39°37′N 105°19′W
Evergreen State see Washington
97 L21 **Evesham** C England, United Kingdom 52°06′N 01°57′W
103 T10 **Évian-les-Bains** Haute-Savoie, E France 46°22′N 06°34′E
93 K16 **Evijärvi** Etelä-Pohjanmaa, W Finland 63°22′N 23°30′E
79 D17 **Evinayong** var. Ebinayon, Evinayoung. C Equatorial Guinea 01°28′N 10°17′E
Evinayoung see Evinayong
115 E18 **Évinos** ≈ C Greece
95 E17 **Evje** Aust-Agder, S Norway 58°35′N 07°49′E
Evmolpia see Plovdiv
104 H11 **Évora** anc. Ebora, Lat. Liberalitas Julia. Évora, C Portugal 38°34′N 07°54′W
104 G11 **Évora** ◐ district S Portugal
102 M4 **Évreux** anc. Civitas Eburovicum. Eure, N France 49°02′N 01°10′E
102 K6 **Évron** Mayenne, NW France 48°10′N 00°24′W
114 L13 **Évros** Bul. Maritsa, Turk. Meriç; anc. Hebrus. ≈ SE Europe see also Maritsa/Meriç
Évros see Meriç
Évros see Maritsa
115 F21 **Evrótas** ≈ S Greece
103 O5 **Évry** Essonne, N France 48°38′N 02°34′E
25 O8 **E. V. Spence Reservoir** ⊠ Texas, SW USA
115 I18 **Évvoia** Lat. Euboea. island C Greece
38 D9 **'Ewa Beach** var. Ewa Beach. O'ahu, Hawaii, USA, C Pacific Ocean 21°19′N 158°00′W
Ewa Beach see 'Ewa Beach
32 L9 **Ewan** Washington, NW USA 47°06′N 117°46′W
44 K12 **Ewarton** C Jamaica 18°11′N 77°06′W
81 J18 **Ewaso Ng'iro** var. Nyiro. ≈ C Kenya
29 P13 **Ewing** Nebraska, C USA 42°13′N 98°20′W
194 J5 **Ewing Island** island Antarctica
65 P17 **Ewing Seamount** undersea feature E Atlantic Ocean 23°20′S 08°45′E
158 L6 **Ewirgol** Xinjiang Uygur Zizhiqu, W China 42°56′N 87°39′E
79 G19 **Ewo** Cuvette, W Congo 0°55′S 14°49′E
27 S3 **Excelsior Springs** Missouri, C USA 39°20′N 94°13′W
97 J23 **Exe** ≈ SW England, United Kingdom
194 L12 **Executive Committee Range** ▲▲ Antarctica
14 E16 **Exeter** Ontario, S Canada 43°19′N 81°26′W
97 J24 **Exeter** anc. Isca Damnoniorum. SW England, United Kingdom 50°43′N 03°31′W
35 R11 **Exeter** California, W USA 36°17′N 119°08′W
19 P10 **Exeter** New Hampshire, NE USA 42°57′N 70°55′W
Exin see Kcynia
29 T14 **Exira** Iowa, C USA 41°36′N 94°55′W
97 J23 **Exmoor** moorland SW England, United Kingdom
21 Y6 **Exmore** Virginia, NE USA 37°31′N 75°48′W
180 G8 **Exmouth** Western Australia 22°01′S 114°06′E
97 J24 **Exmouth** SW England, United Kingdom 50°36′N 03°25′W
180 G8 **Exmouth Gulf** gulf Western Australia
173 V8 **Exmouth Plateau** undersea feature E Indian Ocean
83 K23 **eXobho** prev. Ixopo. KwaZulu/Natal, E South Africa 30°10′S 30°05′E
115 J20 **Exompourgo** ancient monument Tínos, Kykládes, Greece, Aegean Sea
104 I10 **Extremadura** var. Estremadura. ◐ autonomous community W Spain
78 F12 **Extrême-Nord** Eng. Extreme North. ◐ province N Cameroon
Extreme North see Extrême-Nord
44 I3 **Exuma Cays** islets C The Bahamas
44 I3 **Exuma Sound** sound C The Bahamas
81 H20 **Eyasi, Lake** ◎ N Tanzania
95 F17 **Eydehavn** Aust-Agder, S Norway 58°31′N 08°53′E
96 L12 **Eyemouth** SE Scotland, United Kingdom 55°52′N 02°07′W
96 G7 **Eye Peninsula** peninsula NW Scotland, United Kingdom
92 J4 **Eyjafjallajökull** ✱ S Iceland 63°37′N 19°37′W
80 Q13 **Eyl** It. Eil. Nugaal, E Somalia 08°03′N 49°49′E
103 N11 **Eymoutiers** Haute-Vienne, C France 45°45′N 01°43′E
Eyo (lower course) see Uolo, Rio
29 X10 **Eyota** Minnesota, N USA 44°00′N 92°13′W
182 H2 **Eyre Basin, Lake** salt lake South Australia
182 I1 **Eyre Creek** seasonal river Northern Territory/South Australia
174 L9 **Eyre, Lake** salt lake South Australia
185 C22 **Eyre Mountains** ▲▲ South Island, New Zealand
182 H3 **Eyre North, Lake** salt lake South Australia
182 G7 **Eyre Peninsula** peninsula South Australia
182 H4 **Eyre South, Lake** salt lake South Australia
95 B18 **Eysturoy** Dan. Østerø. island N Faroe Islands
61 D20 **Ezeiza** ✕ (Buenos Aires) Buenos Aires, E Argentina 34°49′S 58°30′W
Ezeres see Ezeriş
116 F12 **Ezeriş** Hung. Ezeres. Caraş-Severin, W Romania 45°21′N 21°55′E
161 O9 **Ezhou** prev. Echeng. Hubei, C China 30°23′N 114°52′E
125 R11 **Ezhva** Respublika Komi, NW Russian Federation 61°45′N 50°43′E
136 B12 **Ezine** Çanakkale, NW Turkey 39°46′N 26°20′E
Ezo see Hokkaidō
Ezra/Ezraa see Izra'

F

191 P7 **Faaa** Tahiti, W French Polynesia 17°32′S 149°36′W
191 P7 **Faaa** ✕ (Papeete) Tahiti, W French Polynesia 17°31′S 149°36′W
95 H24 **Faaborg** var. Fåborg. Syddtjylland, C Denmark 55°06′N 10°10′E
151 K19 **Faadhippolhu Atoll** var. Fadiffolu, Lhaviyani Atoll. atoll N Maldives
191 U10 **Faaite** atoll Îles Tuamotu, C French Polynesia
191 Q8 **Faaone** Tahiti, W French Polynesia 17°39′S 149°18′W
24 H8 **Fabens** Texas, SW USA 31°30′N 106°09′W
94 H12 **Fåberg** Oppland, S Norway 61°15′N 10°21′E
Fåborg see Faaborg
106 I12 **Fabriano** Marche, C Italy 43°20′N 12°54′E
145 U16 **Fabrichnoye** prev. Fabrichnyy. Almaty, SE Kazakhstan 43°12′N 76°19′E
Fabrichnyy see Fabrichnoye
54 F10 **Facatativá** Cundinamarca, C Colombia 04°49′N 74°22′W
77 X9 **Fachi** Agadez, C Niger 18°01′N 11°36′E
188 B16 **Facpi Point** headland W Guam
18 I13 **Factoryville** Pennsylvania, NE USA 41°34′N 75°45′W
78 K8 **Fada** Ennedi-Ouest, E Chad 17°14′N 21°32′E
77 Q13 **Fada-Ngourma** E Burkina Faso 12°05′N 00°26′E
123 N6 **Faddeya, Zaliv** bay N Russian Federation
123 Q5 **Faddeyevskiy, Poluostrov** island Novosibirskiye Ostrova, NE Russian Federation
141 W12 **Fadhī** S Oman 17°54′N 55°30′E
Fadiffolu see Faadhippolhu Atoll

◆ Country
● Country Capital
◇ Dependent Territory
○ Dependent Territory Capital
◐ Administrative Regions
✕ International Airport
▲ Mountain
▲▲ Mountain Range
✱ Volcano
≈ River
◎ Lake
⊠ Reservoir

FAENZA – FIERZA

106 H10 **Faenza** *anc.* Faventia. Emilia-Romagna, N Italy 44°17′N 11°53′E
Faeroe-Iceland Ridge *see* Faroe-Iceland Ridge
Faeroe Islands *see* Faroe Islands
Færøerne *see* Faroe Islands
Faeroe-Shetland Trough *see* Faroe-Shetland Trough
104 H6 **Fafe** Braga, N Portugal 41°27′N 08°11′W
80 K13 **Fafen Shet'** ≈ E Ethiopia
193 V15 **Fafo** *island* Tongatapu Group, S Tonga
192 I16 **Fagaloa Bay** *bay* Upolu, E Samoa
192 H15 **Fagamālo** Savai'i, N Samoa 13°27′S 172°22′W
116 I12 **Făgăraş** *Ger.* Fogarasch, *Hung.* Fogaras. Braşov, C Romania 45°50′N 24°59′E
191 W10 **Fagatau** *prev.* Fangatau. *atoll* Îles Tuamotu, C French Polynesia
191 X12 **Fagataufa** *prev.* Fangataufa. *island* Îles Tuamotu, SE French Polynesia
95 M20 **Fagerhult** Kalmar, S Sweden 57°07′N 15°40′E
94 G13 **Fagernes** Oppland, S Norway 60°59′N 09°14′E
92 I9 **Fagernes** Troms, N Norway 69°31′N 19°16′E
95 M14 **Fagersta** Västmanland, C Sweden 59°59′N 15°49′E
77 W13 **Faggo** *var.* Foggo. Bauchi, N Nigeria 11°22′N 09°55′E
Faghman *see* Fughmah
Fagibina, Lake *see* Faguibine, Lac
63 J25 **Fagnano, Lago** ◎ S Argentina
99 G22 **Fagne** *hill range* S Belgium
77 N10 **Faguibine, Lac** *var.* Lake Fagibina. ◎ NW Mali
Fahaheel *see* Al Fuḩayḩīl
Fahlun *see* Falun
143 U12 **Fahraj** Kermān, SE Iran 29°00′N 59°00′E
64 P5 **Faial** Madeira, Portugal, NE Atlantic Ocean 32°47′N 16°53′W
64 N2 **Faial** *var.* Ilha do Faial. *island* Azores, Portugal, NE Atlantic Ocean
Faial, Ilha do *see* Faial
108 G10 **Faido** Ticino, S Switzerland 46°30′N 08°48′E
Faifo *see* Hôi An
Failaka Island *see* Faylakah
190 G12 **Faioa, Île** *island* N Wallis and Futuna
181 W8 **Fairbairn Reservoir** ⊠ Queensland, E Australia
39 R9 **Fairbanks** Alaska, USA 64°48′N 147°47′W
21 U12 **Fair Bluff** North Carolina, SE USA 34°18′N 79°02′W
31 R14 **Fairborn** Ohio, N USA 39°48′N 84°03′W
23 S3 **Fairburn** Georgia, SE USA 33°34′N 84°34′W
30 M12 **Fairbury** Illinois, N USA 40°45′N 88°30′W
29 Q17 **Fairbury** Nebraska, C USA 40°08′N 97°10′W
29 T9 **Fairfax** Minnesota, N USA 44°31′N 94°43′W
27 O8 **Fairfax** Oklahoma, C USA 36°34′N 96°42′W
21 R14 **Fairfax** South Carolina, SE USA 32°57′N 81°14′W
35 N8 **Fairfield** California, W USA 38°14′N 122°03′W
33 O14 **Fairfield** Idaho, NW USA 43°20′N 114°45′W
30 M16 **Fairfield** Illinois, N USA 38°22′N 88°23′W
29 X15 **Fairfield** Iowa, C USA 41°00′N 91°57′W
33 R8 **Fairfield** Montana, NW USA 47°36′N 111°59′W
31 Q14 **Fairfield** Ohio, N USA 39°21′N 84°34′W
25 U8 **Fairfield** Texas, SW USA 31°43′N 96°10′W
27 T7 **Fair Grove** Missouri, C USA 37°22′N 93°09′W
19 P12 **Fairhaven** Massachusetts, NE USA 41°38′N 70°51′W
23 N8 **Fairhope** Alabama, S USA 30°31′N 87°54′W
96 L4 **Fair Isle** *island* NE Scotland, United Kingdom
185 F20 **Fairlie** Canterbury, South Island, New Zealand 44°06′S 170°50′E
29 U11 **Fairmont** Minnesota, N USA 43°40′N 94°27′W
29 Q16 **Fairmont** Nebraska, C USA 40°37′N 97°36′W
21 S3 **Fairmont** West Virginia, NE USA 39°28′N 80°08′W
31 P13 **Fairmount** Indiana, N USA 40°25′N 85°39′W
18 H10 **Fairmount** New York, NE USA 43°03′N 76°14′W
29 R7 **Fairmount** North Dakota, N USA 46°02′N 96°36′W
37 S5 **Fairplay** Colorado, C USA 39°13′N 106°00′W
18 F9 **Fairport** New York, NE USA 43°06′N 77°26′W
11 O12 **Fairview** Alberta, W Canada 56°03′N 118°28′W
26 L9 **Fairview** Oklahoma, C USA 36°16′N 98°29′W
36 L4 **Fairview** Utah, W USA 39°37′N 111°26′W
35 T6 **Fairview Peak** ▲ Nevada, W USA 39°13′N 118°09′W
188 H14 **Fais** *atoll* Caroline Islands, W Micronesia
149 U8 **Faisalābād** *prev.* Lyallpur. Punjab, NE Pakistan 31°26′N 73°06′E
Faisaliya *see* Fayşalīyah
28 L8 **Faith** South Dakota, N USA 45°01′N 102°02′W
153 N12 **Faizābād** Uttar Pradesh, N India 26°46′N 82°08′E
Faizabad/Faizābād *see* Feiẕābād
45 S9 **Fajardo** E Puerto Rico 18°20′N 65°39′W
139 R9 **Fajj, Wādī al** *dry watercourse* S Iraq
140 K4 **Fajr, Bi'r** *well* NW Saudi Arabia
191 W10 **Fakahina** *atoll* Îles Tuamotu, C French Polynesia
190 L10 **Fakaofo Atoll** *island* SE Tokelau
191 U10 **Fakarava** *atoll* Îles Tuamotu, C French Polynesia
127 T2 **Fakel** Udmurtskaya Respublika, NW Russian Federation 57°35′N 53°00′E
97 P19 **Fakenham** E England, United Kingdom 52°50′N 00°51′E
171 U13 **Fakfak** Papua Barat, E Indonesia 02°55′S 132°17′E
153 T12 **Fakīragrām** Assam, NE India 26°22′N 90°15′E
114 M10 **Fakiyska Reka** ≈ SE Bulgaria
95 J24 **Fakse** Sjælland, SE Denmark 55°16′N 12°08′E
95 J24 **Fakse Bugt** *bay* SE Denmark
95 J24 **Fakse Ladeplads** Sjælland, SE Denmark 55°14′N 12°11′E
163 V11 **Faku** Liaoning, NE China 42°30′N 123°27′E
76 J14 **Falaba** N Sierra Leone 09°54′N 11°22′W
102 K5 **Falaise** Calvados, N France 48°52′N 00°12′W
114 H12 **Falakró** ▲▲ NE Greece
189 T12 **Falalu** *island* Chuuk, C Micronesia
166 L4 **Falam** Chin State, W Myanmar (Burma) 22°58′N 93°45′E
143 N8 **Falāvarjān** Eşfahān, C Iran 32°33′N 51°28′E
116 M11 **Fălciu** Vaslui, E Romania 46°19′N 28°10′E
54 I4 **Falcón** *off.* Estado Falcón. ⬖ *state* NW Venezuela
106 J12 **Falconara Marittima** Marche, C Italy 43°37′N 13°23′E
Falcone, Capo del *see* Falcone, Punta del
107 A16 **Falcone, Punta del** *var.* Capo del Falcone. *headland* Sardegna, Italy, C Mediterranean Sea 40°57′N 8°12′E
Falcón, Estado *see* Falcón
11 Y16 **Falcon Lake** Manitoba, S Canada 49°44′N 95°18′W
Falcon Lake *see* Falcón, Presa/Falcon Reservoir
41 O7 **Falcón, Presa** *var.* Falcon Lake, Falcon Reservoir. ⊠ Mexico/USA *see also* Falcon Reservoir
Falcón, Presa *see* Falcon Reservoir
25 Q16 **Falcon Reservoir** *var.* Falcon Lake, Presa Falcón. ⊠ Mexico/USA *see also* Falcón, Presa
Falcon Reservoir *see* Falcón, Presa
190 L10 **Fale** *island* Fakaofo Atoll, SE Tokelau
192 F15 **Faleālupo** Savai'i, NW Samoa 13°30′S 172°46′W
190 B10 **Falefatu** *island* Funafuti Atoll, C Tuvalu
192 G15 **Fālelima** Savai'i, NW Samoa 13°30′S 172°41′W
95 N18 **Falerum** Östergötland, S Sweden 58°07′N 16°15′E
Faleshty *see* Făleşti
116 M9 **Făleşti** *Rus.* Faleshty. NW Moldova 47°33′N 27°43′E
25 S15 **Falfurrias** Texas, SW USA 27°17′N 98°10′W
11 O13 **Falher** Alberta, W Canada 55°45′N 117°18′W
Falkenau an der Eger *see* Sokolov
95 J21 **Falkenberg** Halland, S Sweden 56°55′N 12°30′E
Falkenberg *see* Niemodlin
Falkenburg in Pommern *see* Złocieniec
100 N12 **Falkensee** Brandenburg, NE Germany 52°34′N 13°04′E
96 J12 **Falkirk** C Scotland, United Kingdom 56°N 03°48′W
65 I20 **Falkland Escarpment** *undersea feature* SW Atlantic Ocean 50°00′S 45°00′W
63 K24 **Falkland Islands** *var.* Falklands, Islas Malvinas. ◇ *UK dependent territory* SW Atlantic Ocean
47 W14 **Falkland Islands** *island group* SW Atlantic Ocean
65 I20 **Falkland Plateau** *var.* Argentine Rise. *undersea feature* SW Atlantic Ocean 51°00′S 50°00′W
Falklands *see* Falkland Islands
63 M23 **Falkland Sound** *var.* Estrecho de San Carlos. *strait* C Falkland Islands
Falknov nad Ohří *see* Sokolov
115 H21 **Falkonéra** *island* S Greece
95 K18 **Falköping** Västra Götaland, S Sweden 58°10′N 13°31′E
139 U8 **Fallāḩ** Wāsiţ, E Iraq 32°58′N 45°09′E
35 U16 **Fallbrook** California, W USA 33°22′N 117°15′W
189 U12 **Falleallej Pass** *passage* Chuuk Islands, C Micronesia
93 J14 **Fällfors** Västerbotten, N Sweden 65°07′N 20°46′E
194 I6 **Fallières Coast** *physical region* Antarctica
100 I11 **Fallingbostel** Niedersachsen, NW Germany 52°52′N 09°42′E
33 X9 **Fallon** Montana, NW USA 46°49′N 105°07′W
35 S5 **Fallon** Nevada, W USA 39°29′N 118°47′W
19 O12 **Fall River** Massachusetts, NE USA 41°42′N 71°09′W
27 P6 **Fall River Lake** ⊠ Kansas, C USA
35 O3 **Fall River Mills** California, W USA 41°00′N 121°28′W
21 W4 **Falls Church** Virginia, NE USA 38°53′N 77°11′W
29 S17 **Falls City** Nebraska, C USA 40°03′N 95°36′W
25 S12 **Falls City** Texas, SW USA 28°58′N 98°01′W
Falluja *see* Al Fallūjah
77 S12 **Falmey** Dosso, SW Niger 12°29′N 02°58′E
45 W10 **Falmouth** Antigua, Antigua and Barbuda 17°02′N 61°47′W
44 J11 **Falmouth** W Jamaica 18°28′N 77°39′W
97 H25 **Falmouth** SW England, United Kingdom 50°08′N 05°04′W
20 M4 **Falmouth** Kentucky, S USA 38°40′N 84°20′W
19 P13 **Falmouth** Massachusetts, NE USA 41°31′N 70°36′W
21 W5 **Falmouth** Virginia, NE USA 38°19′N 77°28′W
189 U12 **Falos** *island* Chuuk, C Micronesia
83 E26 **False Bay** *Afr.* Valsbaai. *bay* SW South Africa
155 K17 **False Divi Point** *headland* E India 15°46′N 80°43′E
38 M16 **False Pass** Unimak Island, Alaska, USA 54°52′N 163°15′W
154 P12 **False Point** *headland* E India 20°23′N 86°52′E
105 U6 **Falset** Cataluña, NE Spain 41°08′N 00°49′E
95 I25 **Falster** *island* SE Denmark
116 K9 **Fălticeni** *Hung.* Falticsén. Suceava, NE Romania 47°27′N 26°20′E
Falticsén *see* Fălticeni
94 M13 **Falun** *var.* Fahlun. Kopparberg, C Sweden 60°36′N 15°36′E
Famagusta *see* Gazimağusa
Famagusta Bay *see* Gazimağusa Körfezi
62 I8 **Famatina** La Rioja, NW Argentina 28°58′S 67°46′W
99 J21 **Famenne** *physical region* SE Belgium
77 X15 **Fan** ≈ E Nigeria
76 M12 **Fana** Koulikoro, SW Mali 12°45′N 06°55′W
115 K19 **Fána** *ancient harbor* Chíos, SE Greece
189 V13 **Fanan** *island* Chuuk, C Micronesia
189 U12 **Fanapanges** *island* Chuuk, C Micronesia
115 L20 **Fanári, Akrotírio** *headland* Ikaría, Dodekánisa, Greece, Aegean Sea 37°40′N 26°21′E
45 Q13 **Fancy** Saint Vincent, Saint Vincent and the Grenadines 13°22′N 61°10′W
172 I5 **Fandriana** Fianarantsoa, SE Madagascar 20°14′S 47°21′E
167 O6 **Fang** Chiang Mai, NW Thailand 19°56′N 99°14′E
80 E13 **Fangak** Jonglei, E South Sudan 09°05′N 30°52′E
Fangatau *see* Fagatau
Fangataufa *see* Fagataufa
193 V15 **Fanga Uta** *bay* S Tonga
161 N7 **Fangcheng** Henan, C China 33°18′N 113°03′E
Fangcheng *see* Fangchenggang
160 K15 **Fangchenggang** *var.* Fangcheng Gezu Zizhixian; *prev.* Fangcheng. Guangxi Zhuangzu Zizhiqu, S China 21°49′N 108°21′E
Fangcheng Gezu Zizhixian *see* Fangchenggang
161 S15 **Fangshan** S Taiwan 22°19′N 120°41′E
163 X8 **Fangzheng** Heilongjiang, NE China 45°50′N 128°50′E
Fani *see* Fanit, Lumi i
119 K16 **Fanipal'** *Rus.* Fanipol'. Minskaya Voblasts', C Belarus 53°45′N 27°20′E
Fanipol' *see* Fanipal'
113 D22 **Fanit, Lumi i** *var.* Fani. ≈ N Albania
25 T13 **Fannin** Texas, SW USA 28°41′N 97°13′W
Fanning Island *see* Tabuaeran
94 G8 **Fannrem** Sør-Trøndelag, S Norway 63°16′N 09°48′E
106 I11 **Fano** *anc.* Colonia Julia Fanestris, Fanum Fortunae. Marche, C Italy 43°50′N 13°E
95 E23 **Fanø** *island* W Denmark
167 R5 **Fan Si Pan** ▲ N Vietnam 22°18′N 103°36′E
Fanum Fortunae *see* Fano
Fao *see* Al Fāw
141 W7 **Faq'** *var.* Al Faqa. Dubayy, E United Arab Emirates 24°42′N 55°37′E
Farab *see* Farap
185 G16 **Faraday, Mount** ▲ South Island, New Zealand 42°01′S 171°37′E
79 P16 **Faradje** Orientale, NE Dem. Rep. Congo 03°45′N 29°43′E
Faradofay *see* Tôlañaro
172 I7 **Farafangana** Fianarantsoa, SE Madagascar 22°50′S 47°50′E
148 J7 **Farāh** *var.* Farah, Fararud. Farāh, W Afghanistan 32°22′N 62°07′E
148 K7 **Farāh** ⬖ *province* W Afghanistan
148 J7 **Farāh Rūd** ≈ W Afghanistan
188 K7 **Farallon de Medinilla** *island* C Northern Mariana Islands
188 J2 **Farallon de Pajaros** *var.* Uracas. *island* N Northern Mariana Islands
76 J14 **Faranah** Haute-Guinée, S Guinea 10°02′N 10°44′W
146 K12 **Farap** *Rus.* Farab. Lebap Welaýaty, NE Turkmenistan 39°15′N 63°32′E
Fararud *see* Farāh
140 M13 **Farasān, Jazā'ir** *island group* SW Saudi Arabia
172 I5 **Faratsiho** Antananarivo, C Madagascar 19°24′S 46°57′E
188 K15 **Faraulep Atoll** *atoll* Caroline Islands, C Micronesia
99 H20 **Farciennes** Hainaut, S Belgium 50°26′N 04°33′E
105 O14 **Fardes** ≈ S Spain
191 S10 **Fare** Huahine, W French Polynesia 16°42′S 151°01′W
97 M23 **Fareham** S England, United Kingdom 50°51′N 01°10′W
39 P11 **Farewell** Alaska, USA 62°35′N 153°59′W
184 H13 **Farewell, Cape** *headland* South Island, New Zealand 40°30′S 172°39′E
Farewell, Cape *see* Nunap Isua
184 I13 **Farewell Spit** *spit* South Island, New Zealand
95 I17 **Färgelanda** Västra Götaland, S Sweden 58°34′N 11°59′E
Farghona, Wodii/ Farghona Valley *see* Fergana Valley
Farghona Wodiysi *see* Fergana Valley
23 V8 **Fargo** Georgia, SE USA 30°42′N 82°33′W
29 R5 **Fargo** North Dakota, N USA 46°53′N 96°47′W
147 S10 **Farg'ona** *Rus.* Fergana; *prev.* Novyy Margilan. Farg'ona Viloyati, E Uzbekistan 40°28′N 71°44′E
147 R10 **Farg'ona Viloyati** *Rus.* Ferganskaya Oblast'. ⬖ *province* E Uzbekistan
29 V10 **Faribault** Minnesota, N USA 44°18′N 93°16′W
152 J11 **Farīdābād** Haryāna, N India 28°26′N 77°190′E
152 H8 **Farīdkot** Punjab, NW India 30°42′N 74°47′E
153 T15 **Faridpur** Dhaka, C Bangladesh 23°29′N 89°50′E
121 P14 **Fārigh, Wādī al** ≈ N Libya
172 I4 **Farihy Alaotra** ◎ C Madagascar
94 M11 **Färila** Gävleborg, C Sweden 61°48′N 15°55′E
104 E9 **Farilhões** *island* C Portugal
76 G12 **Farim** NW Guinea-Bissau 12°30′N 15°09′W
Farish *see* Forish
141 T11 **Fāris, Qalamat** *well* SE Saudi Arabia
95 N21 **Färjestaden** Kalmar, S Sweden 56°38′N 16°30′E
149 R2 **Farkhār** Takhār, NE Afghanistan 36°39′N 69°43′E
147 Q14 **Farkhor** *Rus.* Parkhar. SW Tajikistan 37°32′N 69°22′E
116 F12 **Fârliug** *prev.* Fîrliug, *Hung.* Furluk. Caraş-Severin, SW Romania 45°21′N 21°55′E
115 M21 **Farmakonísi** *island* Dodekánisa, Greece, Aegean Sea
30 M13 **Farmer City** Illinois, N USA 40°14′N 88°38′W
31 N14 **Farmersburg** Indiana, N USA 39°14′N 87°21′W
25 U6 **Farmersville** Texas, SW USA 33°09′N 96°21′W
22 H5 **Farmerville** Louisiana, S USA 32°46′N 92°24′W
29 X16 **Farmington** Iowa, C USA 40°37′N 91°43′W
19 Q6 **Farmington** Maine, NE USA 44°40′N 70°09′W
29 V9 **Farmington** Minnesota, N USA 44°39′N 93°09′W
27 X6 **Farmington** Missouri, C USA 37°46′N 90°26′W
19 O9 **Farmington** New Hampshire, NE USA 43°23′N 71°04′W
37 P9 **Farmington** New Mexico, SW USA 36°44′N 108°13′W
36 L2 **Farmington** Utah, W USA 40°58′N 111°53′W
21 W9 **Farmville** North Carolina, SE USA 35°37′N 77°36′W
21 U6 **Farmville** Virginia, NE USA 37°17′N 78°25′W
97 N22 **Farnborough** S England, United Kingdom 51°17′N 00°46′W
97 N22 **Farnham** S England, United Kingdom 51°13′N 00°49′W
10 J7 **Faro** Yukon, W Canada 62°15′N 133°30′W
104 G14 **Faro** Faro, S Portugal 37°01′N 07°56′W
95 Q18 **Fårö** Gotland, SE Sweden 57°55′N 19°10′E
104 G14 **Faro** ⬖ *district* S Portugal
78 F13 **Faro** ≈ Cameroon/Nigeria
104 G14 **Faro** ✈ Faro, S Portugal 37°02′N 08°01′W
64 M5 **Faroe-Iceland Ridge** *var.* Faeroe-Iceland Ridge. *undersea ridge* NW Norwegian Sea
86 C8 **Faroe Islands** *var.* Faeroe Islands. *island group* N Atlantic Ocean
64 M5 **Faroe Islands** *var.* Faeroe Islands, *Dan.* Færøerne, *Faer.* Føroyar. *Self-governing territory of Denmark* N Atlantic Ocean
64 N6 **Faroe-Shetland Trough** *var.* Faeroe-Shetland Trough. *trough* NE Atlantic Ocean
Faro, Punta del *see* Peloro, Capo
95 Q18 **Fårösund** Gotland, SE Sweden 57°51′N 19°02′E
173 N7 **Farquhar Group** *island group* S Seychelles
18 B13 **Farrell** Pennsylvania, NE USA 41°12′N 80°28′W
152 K11 **Farrukhābād** Uttar Pradesh, N India 27°24′N 79°34′E
143 P11 **Fārs** *off.* Ostān-e Fārs; *anc.* Persis. ⬖ *province* S Iran
115 F16 **Fársala** Thessalía, C Greece 39°17′N 22°23′E
143 R4 **Fārsīān** Golestán, N Iran
Fars, Khalīj-e *see* Persian Gulf
95 G21 **Farsø** Nordjylland, N Denmark 56°47′N 09°21′E
Fārs, Ostān-e *see* Fārs
95 D18 **Farsund** Vest-Agder, S Norway 58°05′N 06°49′E
141 U14 **Fartak, Ra's** *headland* E Yemen 15°34′N 52°19′E
60 H13 **Fartura, Serra da** ▲▲ S Brazil
Farvel, Kap *see* Nunap Isua
24 L4 **Farwell** Texas, SW USA 34°23′N 103°03′W
194 I9 **Farwell Island** *island* Antarctica
152 L9 **Far Western** ⬖ *zone* W Nepal
148 M3 **Fāryāb** ⬖ *province* N Afghanistan
143 P12 **Fasā** Fārs, S Iran 28°55′N 53°39′E
141 U12 **Fasad, Ramlat** *desert* SW Oman
107 P17 **Fasano** Puglia, SE Italy 40°50′N 17°21′E
92 L3 **Fáskrúðsfjörður** Austurland, E Iceland 64°55′N 14°01′W
117 O5 **Fastiv** *Rus.* Fastov. Kyyivs'ka Oblast', NW Ukraine 50°08′N 29°59′E
97 B22 **Fastnet Rock** *Ir.* Carraig Aonair. *island* SW Ireland
Fastov *see* Fastiv
190 C9 **Fatato** *island* Funafuti Atoll, C Tuvalu
152 K12 **Fatehgarh** Uttar Pradesh, N India 27°22′N 79°38′E
149 U6 **Fatehjang** Punjab, E Pakistan 33°33′N 72°43′E
152 G11 **Fatehpur** Rājasthān, N India 27°59′N 74°58′E
152 L13 **Fatehpur** Uttar Pradesh, N India 25°56′N 80°55′E
126 J7 **Fatezh** Kurskaya Oblast', W Russian Federation 52°01′N 35°51′E
76 G11 **Fatick** W Senegal 14°19′N 16°27′W
104 G9 **Fátima** Santarém, W Portugal 39°37′N 08°39′W
136 M11 **Fatsa** Ordu, N Turkey 41°02′N 37°31′E
Fatshan *see* Foshan
190 D12 **Fatua, Pointe** *var.* Pointe Nord. *headland* Île Futuna, S Wallis and Futuna
191 X7 **Fatu Hiva** *island* Îles Marquises, NE French Polynesia
Fatunda *see* Fatundu
79 H21 **Fatundu** *var.* Fatunda. Bandundu, W Dem. Rep. Congo 04°08′S 17°13′E
29 O8 **Faulkton** South Dakota, N USA 45°02′N 99°07′W
116 L13 **Făurei** *prev.* Filimon Sîrbu. Brăila, SE Romania 45°05′N 27°15′E
92 G12 **Fauske** Nordland, C Norway 67°15′N 15°27′E
11 P13 **Faust** Alberta, W Canada 55°19′N 115°33′W
99 L23 **Fauvillers** Luxembourg, SE Belgium 49°52′N 05°40′E
107 J24 **Favara** Sicilia, Italy, C Mediterranean Sea 37°19′N 13°40′E
Faventia *see* Faenza
107 G23 **Favignana, Isola** *island* Isole Egadi, S Italy
12 D8 **Fawn** ≈ Ontario, SE Canada
Faxa Bay *see* Faxaflói
92 H3 **Faxaflói** *Eng.* Faxa Bay. *bay* W Iceland
78 I7 **Faya** *prev.* Faya-Largeau, Largeau. Borkou, N Chad 17°58′N 19°06′E
Faya-Largeau *see* Faya
187 Q16 **Fayaoué** Province des Îles Loyauté, C New Caledonia 20°41′S 166°31′E
138 M5 **Fayḑāt** *hill range* E Syria
23 O3 **Fayette** Alabama, S USA 33°40′N 87°49′W
29 X12 **Fayette** Iowa, C USA 42°50′N 91°48′W
22 J6 **Fayette** Mississippi, S USA 31°42′N 91°03′W
27 U4 **Fayette** Missouri, C USA 39°09′N 92°40′W
27 S9 **Fayetteville** Arkansas, C USA 36°04′N 94°10′W
21 U10 **Fayetteville** North Carolina, SE USA 35°03′N 78°53′W
20 J10 **Fayetteville** Tennessee, S USA 35°08′N 86°33′W
25 U11 **Fayetteville** Texas, SW USA 29°52′N 96°40′W
21 R5 **Fayetteville** West Virginia, NE USA 38°03′N 81°09′W
141 R4 **Faylakah** *var.* Failaka Island. *island* E Kuwait
139 T10 **Fayşalīyah** *var.* Faisaliya. Al Qādisīyah, S Iraq 31°48′N 44°36′E
189 P15 **Fayu** *var.* East Fayu. *island* Hall Islands, C Micronesia
152 G8 **Fāzilka** Punjab, NW India 30°26′N 74°04′E
Fdérick *see* Fdérik
76 I6 **Fdérik** *var.* Fdérick, *Fr.* Fort Gouraud. Tiris Zemmour, NW Mauritania 22°40′N 12°41′W
Feabhail, Loch *see* Foyle, Lough
97 B20 **Feale** ≈ SW Ireland
21 V12 **Fear, Cape** *headland* Bald Head Island, North Carolina, SE USA 33°50′N 77°57′W
35 O6 **Feather River** ≈ California, W USA
185 M14 **Featherston** Wellington, North Island, New Zealand 41°07′S 175°28′E
102 L3 **Fécamp** Seine-Maritime, N France 49°45′N 00°22′E
Fédala *see* Mohammedia
61 D17 **Federación** Entre Ríos, E Argentina 31°00′S 57°55′W
61 D17 **Federal** Entre Ríos, E Argentina 30°55′S 58°45′W
77 T15 **Federal Capital District** ⬖ *capital territory* C Nigeria
Federal Capital Territory *see* Australian Capital Territory
Federal District *see* Distrito Federal
21 Y4 **Federalsburg** Maryland, NE USA 38°41′N 75°46′W
74 M6 **Fedjaj, Chott el** *var.* Chott el Fejaj, Shaţţ al Fijāj. *salt lake* C Tunisia
94 B13 **Fedje** *island* S Norway
144 M7 **Fedorovka** Kostanay, N Kazakhstan 51°12′N 52°00′E
127 U6 **Fedorovka** Respublika Bashkortostan, W Russian Federation 53°09′N 55°07′E
Fëdory *see* Fyadory
117 U11 **Fedotova Kosa** *spit* SE Ukraine
189 V13 **Fefan** *atoll* Chuuk Islands, C Micronesia
111 O21 **Fehérgyarmat** Szabolcs-Szatmár-Bereg, E Hungary 47°59′N 22°29′E
Fehér-Körös *see* Crişul Alb
Fehértemplom *see* Bela Crkva
Fehérvölgy *see* Albac
100 L7 **Fehmarn** *island* N Germany
95 H25 **Fehmarn Belt** *Dan.* Femern Bælt, *Ger.* Fehmarnbelt. *strait* Denmark /Germany *see also* Femern Bælt
Fehmarnbelt *see* Fehmarn Belt/Femer Bælt
109 X8 **Fehring** Steiermark, SE Austria 46°56′N 16°00′E
59 B15 **Feijó** Acre, W Brazil 08°07′S 70°27′W
184 M12 **Feilding** Manawatu-Wanganui, North Island, New Zealand 40°15′S 175°34′E
Feira *see* Feira de Santana
59 O17 **Feira de Santana** *var.* Feira. Bahia, E Brazil 12°17′S 38°53′W
109 X7 **Feistritz** ≈ SE Austria
Feistritz *see* Ilirska Bistrica
161 P8 **Feixi** *var.* Shangpai; *prev.* Shangpaihe. Anhui, E China 31°40′N 117°08′E
149 S2 **Feiẕābād** *var.* Faizabad, Faizābād, Feyzābād, Fyzabad; *prev.* Feyẕābād. Badakhshān, NE Afghanistan 37°06′N 70°34′E
Fejaj, Chott el *see* Fedjaj, Chott el
111 I23 **Fejér** *off.* Fejér Megye. ⬖ *county* W Hungary
Fejér Megye *see* Fejér
95 I24 **Fejø** *island* SE Denmark
136 K15 **Feke** Adana, S Turkey 37°49′N 35°55′E
Feketehalom *see* Codlea
Fekete-Körös *see* Crişul Negru
105 Y9 **Felanitx** Mallorca, Spain, W Mediterranean Sea 39°28′N 03°08′E
109 T3 **Feldaist** ≈ N Austria
109 W8 **Feldbach** Steiermark, SE Austria 46°58′N 15°53′E
101 F24 **Feldberg** ▲ SW Germany 47°52′N 08°01′E
116 J12 **Feldioara** *Ger.* Marienburg, *Hung.* Földvár. Braşov, C Romania 45°49′N 25°36′E
108 I7 **Feldkirch** *anc.* Clunia. Vorarlberg, W Austria 47°15′N 09°38′E
109 S9 **Feldkirchen in Kärnten** *Slvn.* Trg. Kärnten, S Austria 46°42′N 14°01′E
Félegyháza *see* Kiskunfélegyháza
192 H16 **Feleolo** ✈ (Āpia) Upolu, C Samoa 13°49′S 171°59′W
104 H6 **Felgueiras** Porto, N Portugal 41°22′N 08°12′W
Felicitas Julia *see* Lisboa
172 J16 **Félicité** *island* Inner Islands, NE Seychelles
151 K20 **Felidhu Atoll** *atoll* C Maldives
41 Y13 **Felipe Carrillo Puerto** Quintana Roo, SE Mexico 19°34′N 88°02′W
97 Q21 **Felixstowe** E England, United Kingdom 51°58′N 01°20′E
103 N11 **Felletin** Creuse, C France 45°53′N 02°12′E
Fellin *see* Viljandi
Felsőbánya *see* Baia Sprie
Felsőmuzslya *see* Mužlja
Felsővisó *see* Vişeu de Sus
35 N10 **Felton** California, W USA 37°03′N 122°04′W
106 H7 **Feltre** Veneto, NE Italy 46°01′N 11°55′E
95 H25 **Femer Bælt** *Dan.* Fehmarn Belt, *Ger.* Fehmarnbelt. *strait* Denmark/Germany *see also* Fehmarn Belt
95 I24 **Femø** *island* SE Denmark
94 I10 **Femunden** ◎ S Norway
104 H2 **Fene** Galicia, NW Spain 43°28′N 08°10′W
14 I14 **Fenelon Falls** Ontario, SE Canada 44°34′N 78°43′W
189 U13 **Feneppi** *atoll* Chuuk Islands, C Micronesia
137 O11 **Fener Burnu** *headland* N Turkey 41°07′N 39°26′E
Fénérive *see* Fenoarivo Atsinanana
115 J14 **Fengári** ▲ Samothráki, E Greece 40°27′N 25°37′E
163 V13 **Fengcheng** *var.* Feng-cheng, Fenghwangcheng. Liaoning, NE China 40°28′N 124°01′E
Fengcheng *see* Lianjiang
Feng-cheng *see* Fengcheng
160 K11 **Fenggang** *var.* Longquan. Guizhou, S China 27°57′N 107°42′E
161 O7 **Feng He** ≈ C China
161 S9 **Fenghua** Zhejiang, SE China 29°40′N 121°25′E
Fenghwangcheng *see* Fengcheng
Fengjiaba *see* Wangcang
160 L9 **Fengjie** *var.* Yong'an. Sichuan, C China 31°03′N 109°31′E
160 M14 **Fengkai** *var.* Jiangkou. Guangdong, S China 23°26′N 111°28′E
161 T13 **Fenglin** *Jap.* Hōrin. C Taiwan 23°52′N 121°30′E
161 P1 **Fengning** *prev.* Dagezhen. Hebei, E China 41°12′N 116°37′E
160 E13 **Fengqing** *var.* Fengshan. Yunnan, SW China 24°38′N 99°54′E
161 O6 **Fengqiu** Henan, C China 35°02′N 114°24′E
161 Q2 **Fengrun** Hebei, E China 39°50′N 118°10′E
Fengshan *see* Luoyuan, Fujian, China
Fengshan *see* Fengqing, Yunnan, China
163 T4 **Fengshui Shan** ▲ NE China 52°20′N 123°22′E
161 P14 **Fengshun** Guangdong, S China 23°51′N 116°11′E
Fengtien *see* Liaoning, China
Fengtien *see* Shenyang, China
160 J7 **Fengxian** *var.* Feng Xian; *prev.* Shuangshipu. Shaanxi, C China 33°50′N 106°33′E
Feng Xian *see* Fengxian
Fengxiang *see* Luobei
Fengyizhen *see* Maoxian
163 P13 **Fengzhen** Nei Mongol Zizhiqu, N China 40°25′N 113°09′E
160 M6 **Fen He** ≈ C China
153 V15 **Feni** Chittagong, E Bangladesh 23°00′N 91°24′E
186 I6 **Feni Islands** *island group* NE Papua New Guinea
38 H17 **Fenimore Pass** *strait* Aleutian Islands, Alaska, USA
84 B9 **Feni Ridge** *undersea feature* N Atlantic Ocean 53°45′N 18°00′W
Fennern *see* Vändra
30 J9 **Fennimore** Wisconsin, N USA 42°58′N 90°39′W
172 J4 **Fenoarivo Atsinanana** *prev./Fr.* Fénérive. Toamasina, E Madagascar 20°52′S 46°52′E
95 I24 **Fensmark** Sjælland, SE Denmark 55°17′N 11°48′E
97 O19 **Fens, The** *wetland* E England, United Kingdom
31 R9 **Fenton** Michigan, N USA 42°48′N 83°42′W
190 K10 **Fenua Fala** *island* SE Tokelau
190 F12 **Fenuafo'ou, Île** *island* E Wallis and Futuna
190 L10 **Fenua Loa** *island* Fakaofo Atoll, E Tokelau
160 M4 **Fenyang** Shanxi, C China 37°14′N 111°40′E
117 U13 **Feodosiya** *var.* Kefe, *It.* Kaffa; *anc.* Theodosia. Avtonomna Respublika Krym, S Ukraine 45°03′N 35°24′E
94 I10 **Feragen** ◎ S Norway
74 L5 **Fer, Cap de** *headland* NE Algeria 37°05′N 07°10′E
31 O16 **Ferdinand** Indiana, N USA 38°13′N 86°51′W
Ferdinand *see* Montana, Bulgaria
Ferdinand *see* Mihail Kogălniceanu, Romania
Ferdinandsberg *see* Oţelu Roşu
143 T7 **Ferdows** *var.* Firdaus; *prev.* Tūn. Khorāsān-e Jonūbī, E Iran 34°00′N 58°09′E
103 Q5 **Fère-Champenoise** Marne, N France 48°45′N 03°59′E
Ferencz-József Csúcs *see* Gerlachovský štít
107 J16 **Ferentino** Lazio, C Italy 41°40′N 13°16′E
114 L13 **Féres** Anatolikí Makedonía kai Thráki, NE Greece 40°54′N 26°12′E
Fergana *see* Farg'ona
147 S10 **Fergana Valley** *var.* Farghona Valley, *Rus.* Ferganskaya Dolina, *Taj.* Wodii Farghona, *Uzb.* Farghona Wodiysi. *basin* Tajikistan/Uzbekistan
Ferganskaya Dolina *see* Fergana Valley
Ferganskaya Oblast' *see* Farg'ona Viloyati
147 U9 **Ferganskiy Khrebet** ▲▲ C Kyrgyzstan
14 F15 **Fergus** Ontario, S Canada 43°42′N 80°22′W
29 S6 **Fergus Falls** Minnesota, N USA 46°15′N 96°02′W
186 G9 **Fergusson Island** *var.* Kaluwawa. *island* SE Papua New Guinea
111 K22 **Ferihegy** ✈ (Budapest) Budapest, C Hungary 47°25′N 19°13′E
113 N17 **Ferizaj** *Serb.* Uroševac. C Kosovo 42°23′N 21°09′E
77 N14 **Ferkessédougou** N Ivory Coast 09°36′N 05°12′W
109 T10 **Ferlach** *Slvn.* Borovlje. Kärnten, S Austria 46°31′N 14°18′E
97 E16 **Fermanagh** *cultural region* SW Northern Ireland, United Kingdom
106 J13 **Fermo** *anc.* Firmum Picenum. Marche, C Italy 43°09′N 13°44′E
104 J6 **Fermoselle** Castilla y León, N Spain 41°19′N 06°24′W
97 D20 **Fermoy** *Ir.* Mainistir Fhear Maí. SW Ireland 52°08′N 08°16′W
23 W8 **Fernandina Beach** Amelia Island, Florida, SE USA 30°40′N 81°27′W
57 A17 **Fernandina, Isla** *var.* Narborough Island. *island* Galapagos Islands, Ecuador, E Pacific Ocean
47 Y5 **Fernando de Noronha** *island* E Brazil
Fernando Po/Fernando Póo *see* Bioco, Isla de
60 J7 **Fernandópolis** São Paulo, S Brazil 20°18′S 50°13′W
104 M13 **Fernán Núñez** Andalucía, S Spain 37°40′N 04°44′W
83 Q14 **Fernão Veloso, Baia de** *bay* NE Mozambique
34 K3 **Ferndale** California, W USA 40°34′N 124°16′W
32 H6 **Ferndale** Washington, NW USA 48°51′N 122°35′W
11 P17 **Fernie** British Columbia, SW Canada 49°30′N 115°00′W
35 R5 **Fernley** Nevada, W USA 39°35′N 119°15′W
Ferozepore *see* Fīrozpur
107 N18 **Ferrandina** Basilicata, S Italy 40°30′N 16°25′E
106 G9 **Ferrara** *anc.* Forum Alieni. Emilia-Romagna, N Italy 44°50′N 11°36′E
120 F9 **Ferrat, Cap** *headland* NW Algeria 35°52′N 00°24′W
107 D20 **Ferrato, Capo** *headland* Sardegna, Italy, C Mediterranean Sea 39°18′N 09°37′E
104 G12 **Ferreira do Alentejo** Beja, S Portugal 38°04′N 08°06′W
56 B11 **Ferreñafe** Lambayeque, W Peru 06°42′S 79°45′W
108 C12 **Ferret** Valais, SW Switzerland 45°57′N 07°04′E
102 I13 **Ferret, Cap** *headland* W France 44°37′N 01°15′W
22 I6 **Ferriday** Louisiana, S USA 31°37′N 91°33′W
Ferro *see* Hierro
107 D16 **Ferro, Capo** *headland* Sardegna, Italy, C Mediterranean Sea 41°09′N 09°31′E
104 H2 **Ferrol** *var.* El Ferrol; *prev.* El Ferrol del Caudillo. Galicia, NW Spain 43°29′N 08°14′W
56 B12 **Ferrol, Península de** *peninsula* W Peru
36 M5 **Ferron** Utah, W USA 39°05′N 111°07′W
21 S7 **Ferrum** Virginia, NE USA 36°54′N 80°01′W
23 O8 **Ferry Pass** Florida, SE USA 30°30′N 87°12′W
Ferryville *see* Menzel Bourguiba
29 S4 **Fertile** Minnesota, N USA 47°32′N 96°16′W
Fertő *see* Neusiedler See
Ferwerd *see* Ferwert
98 L5 **Ferwert** *Dutch.* Ferwerd. Fryslân, N Netherlands 53°21′N 05°47′E
74 G6 **Fès** *Eng.* Fez. N Morocco 34°06′N 04°57′W
79 I22 **Feshi** Bandundu, SW Dem. Rep. Congo 06°08′S 18°12′E
29 O4 **Fessenden** North Dakota, N USA 47°36′N 99°37′W
Festenberg *see* Twardogóra
27 X5 **Festus** Missouri, C USA 38°13′N 90°24′W
116 M14 **Feteşti** Ialomiţa, SE Romania 44°22′N 27°51′E
136 D17 **Fethiye** Muğla, SW Turkey 36°37′N 29°08′E
96 M1 **Fetlar** *island* NE Scotland, United Kingdom
95 I15 **Fetsund** Akershus, S Norway 59°55′N 11°03′E
12 L5 **Feuilles, Lac aux** ◎ Québec, E Canada
12 L5 **Feuilles, Rivière aux** ≈ Québec, E Canada
99 M23 **Feulen** Diekirch, C Luxembourg 49°52′N 06°03′E
103 Q11 **Feurs** Loire, E France 45°44′N 04°14′E
95 F18 **Fevik** Aust-Agder, S Norway 58°22′N 08°40′E
123 R13 **Fevral'sk** Amurskaya Oblast', SE Russian Federation 52°25′N 131°06′E
Feyzābād *see* Feiẕābād
Feyẕābād *see* Feiẕābād
Fez *see* Fès
97 J19 **Ffestiniog** NW Wales, United Kingdom 52°55′N 03°54′W
Fhóid Duibh, Cuan an *see* Blacksod Bay
62 I8 **Fiambalá** Catamarca, NW Argentina 27°45′S 67°37′W
172 I6 **Fianarantsoa** Fianarantsoa, C Madagascar 21°27′S 47°05′E
172 H6 **Fianarantsoa** ⬖ *province* SE Madagascar
78 G12 **Fianga** Mayo-Kébbi Est, SW Chad 09°57′N 15°09′E
Ficce *see* Fichē
80 J12 **Fichē** *It.* Ficce. Oromīya, C Ethiopia 09°48′N 38°43′E
101 N17 **Fichtelberg** ▲ Czech Republic/Germany 50°26′N 12°57′E
101 M18 **Fichtelgebirge** ▲▲ SE Germany
101 M19 **Fichtelnaab** ≈ SE Germany
106 E9 **Fidenza** Emilia-Romagna, N Italy 44°52′N 10°04′E
113 K21 **Fier** *var.* Fieri. Fier, SW Albania 40°44′N 19°34′E
113 K21 **Fier** ⬖ *district* W Albania
Fieri *see* Fier
Fierza *see* Fierzë

◆ Country
● Country Capital
◇ Dependent Territory
○ Dependent Territory Capital
⬖ Administrative Regions
✈ International Airport
▲ Mountain
▲▲ Mountain Range
Volcano
≈ River
◎ Lake
⊠ Reservoir

FIERZË – FOUR CORNERS

113 L17 **Fierzë** var. Fierza. Shkodër, N Albania 42°15′N 20°02′E
113 L17 **Fierzës, Liqeni i** ◎ N Albania
108 F10 **Fiesch** Valais, SW Switzerland 46°25′N 08°09′E
106 G11 **Fiesole** Toscana, C Italy 43°50′N 11°18′E
138 G12 **Fīfah** Aṭ Ṭafīlah, W Jordan 30°55′N 35°25′E
96 K11 **Fife** var. Kingdom of Fife. cultural region E Scotland, United Kingdom
Fife, Kingdom of see Fife
96 K11 **Fife Ness** headland E Scotland, United Kingdom 56°16′N 02°35′W
Fifteen Twenty Fracture Zone see Barracuda Fracture Zone
103 N13 **Figeac** Lot, S France 44°37′N 02°01′E
95 N19 **Figeholm** Kalmar, SE Sweden 57°12′N 16°34′E
Figig see Figuig
83 J18 **Figtree** Matabeleland South, SW Zimbabwe 20°24′S 28°21′E
104 F8 **Figueira da Foz** Coimbra, W Portugal 40°09′N 08°51′W
105 X4 **Figueres** Cataluña, E Spain 42°16′N 02°57′E
74 H7 **Figuig** var. Figig. E Morocco 32°09′N 01°13′W
Fijājj, Shaṭṭ al see Fedjaj, Chott el
187 Y15 **Fiji** off. Republic of Fiji, prev. Sovereign Democratic Republic of Fiji, prev. Republic of the Fiji Islands, Fij. Viti. ◆ republic SW Pacific Ocean
192 K9 **Fiji** island group SW Pacific Ocean
Fiji Islands, Republic of the see Fiji
175 Q8 **Fiji Plate** tectonic feature
Fiji, Republic of see Fiji
Fiji, Sovereign Democratic Republic of see Fiji
105 P14 **Filabres, Sierra de los** ▲▲ SE Spain
83 K18 **Filabusi** Matabeleland South, S Zimbabwe 20°34′S 29°20′E
42 K13 **Filadelfia** Guanacaste, W Costa Rica 10°28′N 85°33′W
111 K20 **Fil'akovo** Hung. Fülek. Banskobýstricky Kraj, C Slovakia 48°15′N 19°53′E
195 N5 **Filchner Ice Shelf** ice shelf Antarctica
14 J11 **Fildegrand** ≈ Québec, SE Canada
33 O15 **Filer** Idaho, NW USA 42°34′N 114°36′W
Filevo see Varbitsa
116 H14 **Filiași** Dolj, SW Romania 44°32′N 23°31′E
115 B16 **Filiátes** Ípeiros, W Greece 39°38′N 20°16′E
115 D21 **Filiatrá** Pelopónnisos, S Greece 37°09′N 21°35′E
107 K22 **Filicudi, Isola** island Isole Eolie, S Italy
141 Y10 **Filim** E Oman 20°37′N 58°11′E
Filimon Sîrbu see Făurei
77 S11 **Filingué** Tillabéri, W Niger 14°21′N 03°22′E
Filiouri see Líssos
114 I13 **Fílippoi** anc. Philippi. site of ancient city Anatolikí Makedonía kai Thráki, NE Greece
95 L15 **Filipstad** Värmland, C Sweden 59°44′N 14°10′E
108 I9 **Filisur** Graubünden, S Switzerland 46°40′N 09°43′E
94 E12 **Fillefjell** ▲▲ S Norway
35 R14 **Fillmore** California, W USA 34°23′N 118°56′W
36 K5 **Fillmore** Utah, W USA 38°57′N 112°19′W
14 J10 **Fils, Lac du** ◎ Québec, SE Canada
Filyos Çayı see Yenice Çayı
Fimbul Ice Shelf see Fimbulisen
195 Q2 **Fimbutheimen** physical region Antarctica
106 G9 **Finale Emilia** Emilia-Romagna, C Italy 44°50′N 11°17′E
106 C10 **Finale Ligure** Liguria, NW Italy 44°11′N 08°22′E
105 P14 **Fiñana** Andalucía, S Spain 37°09′N 02°47′W
21 S6 **Fincastle** Virginia, NE USA 37°30′N 79°54′W
99 M25 **Findel** ✈ (Luxembourg) Luxembourg, C Luxembourg 49°39′N 06°16′E
96 J9 **Findhorn** ≈ N Scotland, United Kingdom
31 R12 **Findlay** Ohio, N USA 41°02′N 83°40′W
18 G11 **Finger Lakes** ◎ New York, NE USA
83 L14 **Fingoè** Tete, NW Mozambique 15°10′S 31°51′E
136 E17 **Finike** Antalya, SW Turkey 36°18′N 30°08′E
102 F6 **Finistère** ⬖ department NW France
186 D7 **Finisterre Range** ▲▲ N Papua New Guinea
181 Q8 **Finke** Northern Territory, N Australia 25°37′S 134°35′E
109 S10 **Finkenstein** Kärnten, S Austria 46°34′N 13°53′E
189 Y15 **Finkol, Mount** var. Mount Crozer. ▲ Kosrae, E Micronesia 05°18′N 163°00′E
93 L17 **Finland** off. Republic of Finland, Fin. Suomen Tasavalta, Suomi. ◆ republic N Europe
124 F12 **Finland, Gulf of** Est. Soome Laht, Fin. Suomenlahti, Ger. Finnischer Meerbusen, Rus. Finskiy Zaliv, Swe. Finska Viken. gulf E Baltic Sea
Finland, Republic of see Finland
10 L11 **Finlay** ≈ British Columbia, W Canada
183 O10 **Finley** New South Wales, SE Australia 35°41′S 145°33′E
29 Q4 **Finley** North Dakota, N USA 47°30′N 97°50′W
Finnischer Meerbusen see Finland, Gulf of
92 K9 **Finnmark** ⬖ county N Norway
92 K9 **Finnmarksvidda** physical region N Norway
92 I9 **Finnsnes** Troms, N Norway 69°16′N 18°00′E
186 E7 **Finschhafen** Morobe, C Papua New Guinea 06°35′S 147°51′E
94 E13 **Finse** Hordaland, S Norway 60°35′N 07°33′E
Finska Viken/Finskiy Zaliv see Finland, Gulf of
95 M17 **Finspång** Östergötland, S Sweden 58°42′N 15°45′E
108 F10 **Finsteraarhorn** ▲ Switzerland 46°33′N 08°07′E
101 O14 **Finsterwalde** Brandenburg, E Germany 51°38′N 13°43′E
185 A23 **Fiordland** physical region South Island, New Zealand
106 E9 **Fiorenzuola d'Arda** Emilia-Romagna, C Italy 44°57′N 09°53′E
Firat Nehri see Euphrates
Firdaus see Ferdows
18 M14 **Fire Island** island New York, NE USA
106 G11 **Firenze** Eng. Florence; anc. Florentia. Toscana, C Italy 43°47′N 11°15′E
106 G10 **Firenzuola** Toscana, C Italy 44°07′N 11°22′E
14 C6 **Fire River** Ontario, S Canada 48°46′N 83°34′W
Fîrliug see Fârliug
61 B19 **Firmat** Santa Fe, C Argentina 33°29′S 61°29′W
103 Q12 **Firminy** Loire, E France 45°22′N 04°18′E
Firmum Picenum see Fermo
152 J12 **Fīrozābād** Uttar Pradesh, N India 27°09′N 78°24′E
152 G8 **Fīrozpur** var. Ferozepore. Punjab, NW India 30°55′N 74°38′E
First State see Delaware
143 O12 **Fīrūzābād** Fārs, S Iran 28°51′N 52°35′E
Fischamend see Fischamend Markt
109 Y4 **Fischamend Markt** var. Fischamend. Niederösterreich, NE Austria 48°08′N 16°37′E
109 W6 **Fischbacher Alpen** ▲▲ E Austria
Fischhausen see Primorsk
83 D21 **Fish** var. Vis. ≈ S Namibia
83 F24 **Fish** Afr. Vis. ≈ SW South Africa
11 X15 **Fisher Branch** Manitoba, S Canada 51°09′N 97°34′W
11 X15 **Fisher River** Manitoba, S Canada 51°25′N 97°23′W
19 N13 **Fishers Island** island New York, NE USA
37 U8 **Fishers Peak** ▲ Colorado, C USA 37°06′N 104°27′W
9 P9 **Fisher Strait** strait Nunavut, N Canada
97 H21 **Fishguard** Wel. Abergwaun. SW Wales, United Kingdom 51°59′N 04°49′W
19 R2 **Fish River Lake** ◎ Maine, NE USA
194 K6 **Fiske, Cape** headland Antarctica 74°27′S 60°28′W
103 P4 **Fismes** Marne, N France 49°19′N 03°41′E
104 F3 **Fisterra, Cabo** headland NW Spain 42°53′N 09°16′W
19 N11 **Fitchburg** Massachusetts, NE USA 42°34′N 71°48′W
96 L3 **Fitful Head** headland NE Scotland, United Kingdom 59°57′N 01°24′W
95 C14 **Fitjar** Hordaland, S Norway 59°55′N 05°19′E
192 H16 **Fito, Mauga** ▲ Upolu, C Samoa 13°57′S 171°42′W
23 U6 **Fitzgerald** Georgia, SE USA 31°42′N 83°15′W
180 M5 **Fitzroy Crossing** Western Australia 18°10′S 125°40′E
63 G21 **Fitzroy, Monte** var. Cerro Chaltel. ▲ S Argentina 49°18′S 73°06′W
181 Y8 **Fitzroy River** ≈ Queensland, E Australia
180 L5 **Fitzroy River** ≈ Western Australia
14 E12 **Fitzwilliam Island** island Ontario, S Canada
107 J15 **Fiuggi** Lazio, C Italy 41°47′N 13°16′E
Fiume see Rijeka
107 H15 **Fiumicino** Lazio, C Italy 41°46′N 12°13′E
Fiumicino see Leonardo da Vinci
106 E10 **Fivizzano** Toscana, C Italy 44°13′N 10°06′E
79 O21 **Fizi** Sud-Kivu, E Dem. Rep. Congo 04°15′S 28°57′E
Fizuli see Füzuli
92 I11 **Fjällåsen** Norrbotten, N Sweden 67°31′N 20°08′E
95 G20 **Fjerritslev** Nordjylland, N Denmark 57°06′N 09°17′E
F.J.S. see Franz Josef Strauss
95 L16 **Fjugesta** Örebro, C Sweden 59°10′N 14°50′E
Fladstrand see Frederikshavn
37 V5 **Flagler** Colorado, C USA 39°17′N 103°04′W
23 X10 **Flagler Beach** Florida, SE USA 29°28′N 81°07′W
36 L11 **Flagstaff** Arizona, SW USA 35°12′N 111°39′W
65 H24 **Flagstaff Bay** bay N Saint Helena, C Atlantic Ocean
19 P5 **Flagstaff Lake** ◎ Maine, NE USA
94 E13 **Flåm** Sogn Og Fjordane, S Norway 60°51′N 07°06′E
15 O8 **Flamand** ≈ Québec, SE Canada
30 J5 **Flambeau River** ≈ Wisconsin, N USA
97 O16 **Flamborough Head** headland E England, United Kingdom 54°06′N 00°03′W
100 N13 **Fläming** hill range NE Germany
16 H8 **Flaming Gorge Reservoir** ▣ Utah/Wyoming, NW USA
Flanders see Vlaanderen
Flandre see Vlaanderen
29 R10 **Flandreau** South Dakota, N USA 44°03′N 96°36′W
96 D6 **Flannan Isles** island group NW Scotland, United Kingdom
28 M6 **Flasher** North Dakota, N USA 46°25′N 101°12′W
93 G15 **Flåsjön** ◎ N Sweden
39 O11 **Flat** Alaska, USA 62°27′N 158°00′W
92 H1 **Flateyri** Vestfirðir, NW Iceland 66°03′N 23°28′W
33 P8 **Flathead Lake** ◎ Montana, NW USA
173 Y15 **Flat Island** Fr. Île Plate. island N Mauritius
25 T11 **Flatonia** Texas, SW USA 29°41′N 97°06′W
185 M14 **Flat Point** headland North Island, New Zealand 41°12′S 176°03′E
27 X6 **Flat River** Missouri, C USA 37°51′N 90°31′W
31 P8 **Flat River** ≈ Michigan, N USA
31 P14 **Flatrock River** ≈ Indiana, N USA
32 E6 **Flattery, Cape** headland Washington, NW USA 48°22′N 124°43′W
64 B12 **Flatts Village** var. The Flatts Village. C Bermuda 32°19′N 64°44′W
108 H7 **Flawil** Sankt Gallen, NE Switzerland 47°25′N 09°12′E
97 N22 **Fleet** S England, United Kingdom 51°16′N 00°50′W
97 K16 **Fleetwood** NW England, United Kingdom 53°55′N 03°02′W
18 H15 **Fleetwood** Pennsylvania, NE USA 40°27′N 75°49′W
95 D18 **Flekkefjord** Vest-Agder, S Norway 58°17′N 06°40′E
21 N5 **Flemingsburg** Kentucky, S USA 38°26′N 83°43′W
18 J15 **Flemington** New Jersey, NE USA 40°30′N 74°51′W
64 I7 **Flemish Cap** undersea feature NW Atlantic Ocean 47°00′N 45°00′W
95 N16 **Flen** Södermanland, C Sweden 59°04′N 16°39′E
100 I6 **Flensburg** Schleswig-Holstein, N Germany 54°47′N 09°26′E
100 J6 **Flensburger Förde** inlet Denmark/Germany
102 K5 **Flers** Orne, N France 48°45′N 00°34′W
95 C14 **Flesland** ✈ (Bergen) Hordaland, S Norway 60°18′N 05°15′E
Flessingue see Vlissingen
21 P10 **Fletcher** North Carolina, SE USA 35°24′N 82°29′W
31 R6 **Fletcher Pond** ◎ Michigan, N USA
102 L15 **Fleurance** Gers, S France 43°50′N 00°39′E
108 B8 **Fleurier** Neuchâtel, W Switzerland 46°55′N 06°37′E
99 H20 **Fleurus** Hainaut, S Belgium 50°28′N 04°33′E
103 N7 **Fleury-les-Aubrais** Loiret, C France 47°55′N 01°55′E
98 K10 **Flevoland** ⬖ province C Netherlands
Flickertail State see North Dakota
108 H9 **Flims** Glarus, NE Switzerland 46°50′N 09°16′E
182 F8 **Flinders Island** island Investigator Group, South Australia
183 P14 **Flinders Island** island Furneaux Group, Tasmania, SE Australia
182 I6 **Flinders Ranges** ▲▲ South Australia
181 U5 **Flinders River** ≈ Queensland, NE Australia
11 V13 **Flin Flon** Manitoba, C Canada 54°47′N 101°51′W
97 K18 **Flint** NE Wales, United Kingdom 53°15′N 03°10′W
31 R9 **Flint** Michigan, N USA 43°01′N 83°41′W
97 J18 **Flint** cultural region NE Wales, United Kingdom
27 O7 **Flint Hills** hill range Kansas, C USA
191 Y6 **Flint Island** island Line Islands, E Kiribati
23 S4 **Flint River** ≈ Georgia, SE USA
31 R9 **Flint River** ≈ Michigan, N USA
189 X12 **Flipper Point** headland C Wake Island 19°18′N 166°37′E
94 I13 **Flisa** Hedmark, S Norway 60°36′N 12°02′E
94 J13 **Flisa** ≈ S Norway
122 J5 **Flissingskiy, Mys** headland Novaya Zemlya, NW Russian Federation 76°43′N 69°01′E
Flitsch see Bovec
105 U6 **Flix** Cataluña, NE Spain 41°13′N 00°32′E
95 J19 **Floda** Västra Götaland, S Sweden 57°47′N 12°20′E
101 O16 **Flöha** ≈ E Germany
25 O4 **Flomot** Texas, SW USA 34°13′N 100°58′W
29 V5 **Floodwood** Minnesota, N USA 46°55′N 92°55′W
30 M15 **Flora** Illinois, N USA 38°40′N 88°29′W
103 P14 **Florac** Lozère, S France 44°18′N 03°35′E
23 Q8 **Florala** Alabama, S USA 31°00′N 86°19′W
103 S4 **Florange** Moselle, NE France 49°21′N 06°06′E
Floreana, Isla see Santa María, Isla
23 O2 **Florence** Alabama, S USA 34°48′N 87°40′W
36 L14 **Florence** Arizona, SW USA 33°01′N 111°23′W
37 T6 **Florence** Colorado, C USA 38°20′N 105°06′W
27 O5 **Florence** Kansas, C USA 38°13′N 96°56′W
20 M4 **Florence** Kentucky, S USA 39°00′N 84°37′W
32 E13 **Florence** Oregon, NW USA 43°58′N 124°06′W
21 T12 **Florence** South Carolina, SE USA 34°12′N 79°44′W
25 S9 **Florence** Texas, SW USA 30°50′N 97°47′W
Florence see Firenze
54 E13 **Florencia** Caquetá, S Colombia 01°37′N 75°37′W
99 H21 **Florennes** Namur, S Belgium 50°15′N 04°36′E
Florentia see Firenze
63 J18 **Florentino Ameghino, Embalse** ◎ S Argentina
99 J24 **Florenville** Luxembourg, SE Belgium 49°42′N 05°19′E
42 E3 **Flores** Petén, N Guatemala 16°55′N 89°56′W
61 E19 **Flores** ⬖ department S Uruguay
171 O16 **Flores** island Nusa Tenggara, C Indonesia
64 M1 **Flores** island Azores, Portugal, NE Atlantic Ocean
Floreshty see Florești
Flores, Lago de see Petén Itzá, Lago
Flores, Laut see Flores Sea
171 N15 **Flores Sea** Ind. Laut Flores. sea C Indonesia
116 M8 **Florești** Rus. Floreshty. N Moldova 47°52′N 28°19′E
25 S12 **Floresville** Texas, SW USA 29°09′N 98°10′W
59 N14 **Floriano** Piauí, E Brazil 06°45′S 43°W
61 K14 **Florianópolis** prev. Destêrro. state capital Santa Catarina, S Brazil 27°35′S 48°32′W
44 G6 **Florida** Camagüey, C Cuba 21°32′N 78°14′W
61 F19 **Florida** Florida, S Uruguay 34°04′S 56°14′W
61 F19 **Florida** ⬖ department S Uruguay
23 U9 **Florida** off. State of Florida, also known as Peninsular State, Sunshine State. ⬖ state SE USA
23 Y17 **Florida Bay** bay Florida, SE USA
54 G8 **Floridablanca** Santander, N Colombia 07°04′N 73°06′W
23 Y17 **Florida Keys** island group Florida, SE USA
37 Q16 **Florida Mountains** ▲▲ New Mexico, SW USA
64 D10 **Florida, Straits of** strait Atlantic Ocean/Gulf of Mexico
114 D13 **Flórina** var. Phlórina. Dytikí Makedonía, N Greece 40°48′N 21°26′E
27 X4 **Florissant** Missouri, C USA 38°48′N 90°20′W
94 C11 **Florø** Sogn Og Fjordane, S Norway 61°36′N 05°04′E
115 L22 **Floúda, Akrotírio** headland Astypálaia, Kykládes, Greece, Aegean Sea 36°38′N 26°23′E
21 S7 **Floyd** Virginia, NE USA 36°55′N 80°22′W
25 N4 **Floydada** Texas, SW USA 33°58′N 101°20′W
Flüela Wisshorn see Weisshorn
98 K7 **Fluessen** ◎ N Netherlands
105 S5 **Flúmen** ≈ NE Spain
107 C20 **Flumendosa** ≈ Sardegna, Italy, C Mediterranean Sea
31 R9 **Flushing** Michigan, N USA 43°03′N 83°51′W
Flushing see Vlissingen
25 O6 **Fluvanna** Texas, SW USA 32°54′N 101°06′W
186 B8 **Fly** ≈ Indonesia/Papua New Guinea
194 I10 **Flying Fish, Cape** headland Thurston Island, Antarctica 72°00′S 102°25′W
Flylân see Vlieland
Fly River see Western
193 Y15 **Foa** island Ha'apai Group, C Tonga
11 U15 **Foam Lake** Saskatchewan, S Canada 51°38′N 103°31′W
113 J14 **Foča** var. Srbinje. SE Bosnia and Herzegovina 43°32′N 18°46′E
116 L12 **Focșani** Vrancea, E Romania 45°45′N 27°13′E
Fogaras/Fogarasch see Făgăraș
107 M16 **Foggia** Puglia, SE Italy 41°28′N 15°31′E
Foggo see Faggo
76 D10 **Fogo** island Ilhas de Sotavento, SW Cape Verde
13 U11 **Fogo Island** island Newfoundland and Labrador, E Canada
109 U7 **Fohnsdorf** Steiermark, SE Austria 47°13′N 14°40′E
100 G7 **Föhr** island NW Germany
104 F14 **Fóia** ▲ S Portugal 37°19′N 08°39′W
14 I10 **Foins, Lac aux** ◎ Québec, SE Canada
103 N17 **Foix** Ariège, S France 42°58′N 01°39′E
126 I5 **Fokino** Bryanskaya Oblast', W Russian Federation 53°22′N 34°22′E
123 S15 **Fokino** Primorskiy Kray, W Russian Federation 42°58′N 132°25′E
Fola, Cnoc see Bloody Foreland
94 E13 **Folarskardnuten** ▲ S Norway 60°34′N 07°18′E
92 G11 **Folda** prev. Foldafjorden. fjord C Norway
Foldafjorden see Folda
93 F14 **Foldereid** Nord-Trøndelag, C Norway 64°58′N 12°09′E
Földvár see Feldioara
115 J22 **Folégandros** island Kykládes, Greece, Aegean Sea
23 O9 **Foley** Alabama, S USA 30°24′N 87°40′W
29 U7 **Foley** Minnesota, N USA 45°39′N 93°54′W
14 E7 **Foleyet** Ontario, S Canada 48°15′N 82°26′W
95 D14 **Folgefonna** glacier S Norway
106 I13 **Foligno** Umbria, C Italy 42°58′N 12°40′E
97 Q23 **Folkestone** SE England, United Kingdom 51°05′N 01°11′E
23 W8 **Folkston** Georgia, SE USA 30°49′N 82°00′W
94 H10 **Folldal** Hedmark, S Norway 62°08′N 10°00′E
25 P1 **Follett** Texas, SW USA 36°25′N 100°08′W
106 F13 **Follonica** Toscana, C Italy 42°55′N 10°45′E
21 T15 **Folly Beach** South Carolina, SE USA 32°39′N 79°56′W
35 O7 **Folsom** California, W USA 38°40′N 121°11′W
116 M12 **Foltești** Galați, E Romania 45°45′N 28°00′E
172 H14 **Fomboni** Mohéli, S Comoros 12°18′S 43°46′E
18 K10 **Fonda** New York, NE USA 42°57′N 74°24′W
11 S10 **Fond-du-Lac** Saskatchewan, C Canada 59°20′N 107°09′W
30 M8 **Fond du Lac** Wisconsin, N USA 43°48′N 88°27′W
11 T10 **Fond-du-Lac** ≈ Saskatchewan, C Canada
190 G8 **Fongafale** atoll C Tuvalu
107 C18 **Fonni** Sardegna, Italy, C Mediterranean Sea 40°07′N 09°17′E
189 V12 **Fono** island Chuuk, C Micronesia
54 G4 **Fonseca** La Guajira, N Colombia 10°53′N 72°51′W
Fonseca, Golfo de see Fonseca, Gulf of
42 H8 **Fonseca, Gulf of** Sp. Golfo de Fonseca. gulf C Central America
103 O6 **Fontainebleau** Seine-et-Marne, N France 48°24′N 02°42′E
63 G19 **Fontana, Lago** ◎ W Argentina
21 N10 **Fontana Lake** ▣ North Carolina, SE USA
107 L24 **Fontanarossa** ✈ (Catania) Sicilia, Italy, C Mediterranean Sea 37°28′N 15°04′E
11 N11 **Fontas** ≈ British Columbia, W Canada
58 D12 **Fonte Boa** Amazonas, N Brazil 02°32′S 66°01′W
102 J10 **Fontenay-le-Comte** Vendée, NW France 46°28′N 00°48′W
33 T16 **Fontenelle Reservoir** ▣ Wyoming, C USA
193 Y14 **Fonualei** island Vava'u Group, N Tonga
111 H24 **Fonyód** Somogy, W Hungary 46°43′N 17°32′E
Foochow see Fuzhou
39 Q10 **Foraker, Mount** ▲ Alaska, USA 62°57′N 151°24′W
187 R14 **Forari** Éfaté, C Vanuatu 17°42′S 168°33′E
103 U4 **Forbach** Moselle, NE France 49°11′N 06°54′E
183 Q8 **Forbes** New South Wales, SE Australia 33°24′S 148°00′E
77 T17 **Forcados** Delta, S Nigeria 05°16′N 05°25′E
103 S14 **Forcalquier** Alpes-de-Haute-Provence, SE France 43°57′N 05°46′E
101 K19 **Forchheim** Bayern, SE Germany 49°43′N 11°07′E
35 R13 **Ford City** California, W USA 35°09′N 119°27′W
94 D11 **Førde** Sogn Og Fjordane, S Norway 61°27′N 05°51′E
31 N4 **Ford River** ≈ Michigan, N USA
183 O4 **Fords Bridge** New South Wales, SE Australia 29°44′S 145°25′E
20 J6 **Fordsville** Kentucky, S USA 37°36′N 86°39′W
27 U13 **Fordyce** Arkansas, C USA 33°49′N 92°23′W
76 I14 **Forécariah** SW Guinea 09°28′N 13°06′W
197 O14 **Forel, Mont** ▲ SE Greenland 66°55′N 36°45′W
11 R17 **Foremost** Alberta, SW Canada 49°30′N 111°34′W
14 D16 **Forest** Ontario, S Canada 43°05′N 82°00′W
22 L5 **Forest** Mississippi, S USA 32°22′N 89°30′W
31 S12 **Forest** Ohio, N USA 40°47′N 83°26′W
29 V11 **Forest City** Iowa, C USA 43°15′N 93°38′W
21 Q10 **Forest City** North Carolina, SE USA 35°19′N 81°52′W
32 G11 **Forest Grove** Oregon, NW USA 45°31′N 123°06′W
183 P17 **Forestier Peninsula** peninsula Tasmania, SE Australia
29 V8 **Forest Lake** Minnesota, N USA 45°16′N 92°59′W
23 S3 **Forest Park** Georgia, SE USA 33°37′N 84°22′W
29 Q3 **Forest River** ≈ North Dakota, N USA
15 T6 **Forestville** Québec, SE Canada 48°45′N 69°04′W
103 Q11 **Forez, Monts du** ▲▲ C France
96 K10 **Forfar** E Scotland, United Kingdom 56°38′N 02°54′W
26 J8 **Forgan** Oklahoma, C USA 36°54′N 100°32′W
Forge du Sud see Dudelange
101 J24 **Forggensee** ◎ S Germany
147 N10 **Forish** Rus. Farish. Jizzax Viloyati, C Uzbekistan 40°33′N 66°52′E
20 F9 **Forked Deer River** ≈ Tennessee, S USA
32 F7 **Forks** Washington, NW USA 47°57′N 124°22′W
92 N2 **Forlandsundet** sound W Svalbard
106 H10 **Forlì** anc. Forum Livii. Emilia-Romagna, N Italy 44°14′N 12°02′E
29 Q7 **Forman** North Dakota, N USA 46°07′N 97°39′W
97 K17 **Formby** NW England, United Kingdom 53°34′N 03°05′W
105 V11 **Formentera** anc. Ophiusa, Lat. Frumentum. island Islas Baleares, Spain, W Mediterranean Sea
Formentor, Cabo de see Formentor, Cap de
105 Y9 **Formentor, Cap de** var. Cabo de Formentor, Cape Formentor. headland Mallorca, Spain, W Mediterranean Sea 39°57′N 03°12′E
Formentor, Cape see Formentor, Cap de
107 J16 **Formia** Lazio, C Italy 41°16′N 13°37′E
62 O7 **Formosa** Formosa, NE Argentina 26°07′S 58°14′W
62 M6 **Formosa** off. Provincia de Formosa. ⬖ province NE Argentina
Formosa/Formo'sa see Taiwan
Formosa, Provincia de see Formosa
59 I17 **Formosa, Serra** ▲▲ C Brazil
Formosa Strait see Taiwan Strait
95 H21 **Fornæs** headland C Denmark 56°26′N 10°57′E
25 U6 **Forney** Texas, SW USA 32°45′N 96°28′W
106 E9 **Fornovo di Taro** Emilia-Romagna, C Italy 44°42′N 10°07′E
117 T14 **Foros** Avtonomna Respublika Krym, S Ukraine 44°24′N 33°47′E
Føroyar see Faroe Islands
96 J8 **Forres** NE Scotland, United Kingdom 57°32′N 03°38′W
27 X11 **Forrest City** Arkansas, C USA 35°01′N 90°48′W
39 Y15 **Forrester Island** island Alexander Archipelago, Alaska, USA
25 N7 **Forsan** Texas, SW USA 32°06′N 101°22′W
181 V5 **Forsayth** Queensland, NE Australia 18°31′S 143°37′E
95 L19 **Forserum** Jönköping, S Sweden 57°42′N 14°28′E
95 K15 **Forshaga** Värmland, C Sweden 59°33′N 13°29′E
93 L19 **Forssa** Kanta-Häme, S Finland 60°49′N 23°40′E
101 Q14 **Forst** Lus. Baršć Łužyca. Brandenburg, E Germany 51°43′N 14°38′E
183 U7 **Forster-Tuncurry** New South Wales, SE Australia 32°11′S 152°30′E
23 T4 **Forsyth** Georgia, SE USA 33°00′N 83°57′W
27 T8 **Forsyth** Missouri, C USA 36°41′N 93°07′W
33 W10 **Forsyth** Montana, NW USA 46°16′N 106°40′W
149 U11 **Fort Abbās** Punjab, E Pakistan 29°12′N 73°00′E
12 G10 **Fort Albany** Ontario, C Canada 52°15′N 81°35′W
56 L13 **Fortaleza** Pando, N Bolivia 09°48′S 65°29′W
58 P13 **Fortaleza** prev. Ceará. state capital Ceará, NE Brazil 03°45′S 38°35′W
59 D16 **Fortaleza** Rondônia, W Brazil 08°45′S 64°06′W
56 C13 **Fortaleza, Río** ≈ W Peru
Fort-Archambault see Sarh
21 U3 **Fort Ashby** West Virginia, NE USA 39°30′N 78°46′W
96 I9 **Fort Augustus** N Scotland, United Kingdom 57°14′N 04°38′W
Fort-Bayard see Zhanjiang
33 S8 **Fort Benton** Montana, NW USA 47°49′N 110°40′W
35 Q1 **Fort Bidwell** California, W USA 41°50′N 120°07′W
34 L5 **Fort Bragg** California, W USA 39°25′N 123°48′W
31 N16 **Fort Branch** Indiana, N USA 38°15′N 87°34′W
Fort-Bretonnet see Bousso
33 T17 **Fort Bridger** Wyoming, C USA 41°18′N 110°19′W
Fort-Cappolani see Tidjikja
Fort-Carnot see Ikongo
Fort Charlet see Djanet
Fort-Chimo see Kuujjuaq
11 R10 **Fort Chipewyan** Alberta, C Canada 58°42′N 111°08′W
Fort Cobb Lake see Fort Cobb Reservoir
26 L11 **Fort Cobb Reservoir** var. Fort Cobb Lake. ▣ Oklahoma, C USA
37 T3 **Fort Collins** Colorado, C USA 40°35′N 105°05′W
14 K12 **Fort-Coulonge** Québec, SE Canada 45°50′N 76°45′W
Fort-Crampel see Kaga Bandoro
Fort-Dauphin see Tôlañaro
24 K10 **Fort Davis** Texas, SW USA 30°35′N 103°54′W
37 O10 **Fort Defiance** Arizona, SW USA 35°44′N 109°04′W
45 Q12 **Fort-de-France** prev. Fort-Royal. ○ (Martinique) W Martinique 14°36′N 61°05′W
45 P12 **Fort-de-France, Baie de** bay W Martinique
Fort de Kock see Bukittinggi
23 P6 **Fort Deposit** Alabama, S USA 31°58′N 86°34′W
29 U13 **Fort Dodge** Iowa, C USA 42°30′N 94°10′W
13 S10 **Forteau** Québec, E Canada 51°30′N 56°55′W
106 E11 **Forte dei Marmi** Toscana, C Italy 43°59′N 10°10′E
14 H17 **Fort Erie** Ontario, S Canada 42°55′N 78°56′W
180 H7 **Fortescue River** ≈ Western Australia
19 S2 **Fort Fairfield** Maine, NE USA 46°45′N 67°51′W
Fort-Foureau see Kousséri
12 A11 **Fort Frances** Ontario, S Canada 48°37′N 93°23′W
Fort Franklin see Déljne
23 R7 **Fort Gaines** Georgia, SE USA 31°36′N 85°03′W
37 T8 **Fort Garland** Colorado, C USA 37°22′N 105°24′W
21 P5 **Fort Gay** West Virginia, NE USA 38°06′N 82°35′W
Fort George see Chisasibi
Fort George see La Grande Rivière
27 Q10 **Fort Gibson** Oklahoma, C USA 35°48′N 95°15′W
27 Q9 **Fort Gibson Lake** ▣ Oklahoma, C USA
8 H7 **Fort Good Hope** var. Rádeyilikóé. Northwest Territories, NW Canada 66°16′N 128°37′W
23 V4 **Fort Gordon** Georgia, SE USA 33°25′N 82°09′W
Fort Gouraud see Fdérik
96 I11 **Forth** ≈ C Scotland, United Kingdom
Fort Hall see Murang'a
24 H8 **Fort Hancock** Texas, SW USA 31°18′N 105°49′W
Fort Hertz see Putao
96 K12 **Forth, Firth of** estuary E Scotland, United Kingdom
14 L14 **Forthton** Ontario, SE Canada 44°43′N 75°31′W
14 M8 **Fortier** ≈ Québec, SE Canada
Fortín General Eugenio Garay see General Eugenio A. Garay
Fort Jameson see Chipata
Fort Johnston see Mangochi
19 R1 **Fort Kent** Maine, NE USA 47°15′N 68°33′W
Fort-Lamy see N'Djaména
23 Z15 **Fort Lauderdale** Florida, SE USA 26°07′N 80°09′W
21 R11 **Fort Lawn** South Carolina, SE USA 34°43′N 80°46′W
8 H10 **Fort Liard** var. Liard. Northwest Territories, W Canada 60°14′N 123°28′W
44 M8 **Fort-Liberté** NE Haiti 19°42′N 71°51′W
21 N9 **Fort Loudoun Lake** ▣ Tennessee, S USA
37 T3 **Fort Lupton** Colorado, C USA 40°04′N 104°48′W
11 R12 **Fort MacKay** Alberta, C Canada 57°12′N 111°41′W
11 Q17 **Fort Macleod** var. MacLeod. Alberta, SW Canada 49°44′N 113°24′W
29 Y16 **Fort Madison** Iowa, C USA 40°37′N 91°15′W
Fort Manning see Mchinji
25 P9 **Fort McKavett** Texas, SW USA 30°50′N 100°07′W
11 R12 **Fort McMurray** Alberta, C Canada 56°44′N 111°23′W
8 G7 **Fort McPherson** var. McPherson. Northwest Territories, NW Canada 67°29′N 134°50′W
21 R11 **Fort Mill** South Carolina, SE USA 35°00′N 80°57′W
Fort-Millot see Ngouri
37 U3 **Fort Morgan** Colorado, C USA 40°14′N 103°48′W
23 W14 **Fort Myers** Florida, SE USA 26°39′N 81°52′W
23 W15 **Fort Myers Beach** Florida, SE USA 26°27′N 81°57′W
10 M10 **Fort Nelson** British Columbia, W Canada 58°48′N 122°44′W
10 M10 **Fort Nelson** ≈ British Columbia, W Canada
Fort Norman see Tulita
23 Q2 **Fort Payne** Alabama, S USA 34°23′N 85°43′W
33 W7 **Fort Peck** Montana, NW USA 48°00′N 106°28′W
33 V8 **Fort Peck Lake** ▣ Montana, NW USA
23 Y13 **Fort Pierce** Florida, SE USA 27°28′N 80°20′W
29 N10 **Fort Pierre** South Dakota, N USA 44°21′N 100°22′W
81 E18 **Fort Portal** SW Uganda 0°39′N 30°17′E
8 J10 **Fort Providence** var. Providence. Northwest Territories, W Canada 61°21′N 117°39′W
11 U16 **Fort Qu'Appelle** Saskatchewan, S Canada 50°50′N 103°52′W
Fort-Repoux see Akjoujt
8 K10 **Fort Resolution** var. Resolution. Northwest Territories, W Canada 61°10′N 113°39′W
33 T13 **Fortress Mountain** ▲ Wyoming, C USA 44°20′N 109°51′W
Fort Rosebery see Mansa
Fort Rousset see Owando
Fort-Royal see Fort-de-France
Fort Rupert see Waskaganish
8 H13 **Fort St. James** British Columbia, SW Canada 54°26′N 124°15′W
11 N12 **Fort St. John** British Columbia, W Canada 56°16′N 120°52′W
Fort Sandeman see Zhob
11 Q14 **Fort Saskatchewan** Alberta, SW Canada 53°42′N 113°12′W
27 R6 **Fort Scott** Kansas, C USA 37°52′N 94°43′W
12 E6 **Fort Severn** Ontario, C Canada 56°N 87°40′W
31 R12 **Fort Shawnee** Ohio, N USA 40°41′N 84°08′W
144 E14 **Fort-Shevchenko** Mangistau, W Kazakhstan 44°29′N 50°16′E
Fort-Sibut see Sibut
8 I10 **Fort Simpson** var. Simpson. Northwest Territories, W Canada 61°52′N 121°23′W
8 I10 **Fort Smith** Northwest Territories, W Canada 60°01′N 111°55′W
27 R10 **Fort Smith** Arkansas, C USA 35°23′N 94°24′W
37 T13 **Fort Stanton** New Mexico, SW USA 33°28′N 105°31′W
24 L9 **Fort Stockton** Texas, SW USA 30°54′N 102°54′W
37 U12 **Fort Sumner** New Mexico, SW USA 34°28′N 104°15′W
26 K8 **Fort Supply** Oklahoma, C USA 36°34′N 99°34′W
26 K8 **Fort Supply Lake** ▣ Oklahoma, C USA
29 O10 **Fort Thompson** South Dakota, N USA 44°01′N 99°22′W
Fort-Trinquet see Bîr Mogreïn
105 R12 **Fortuna** Murcia, SE Spain 38°11′N 01°07′W
34 K3 **Fortuna** California, W USA 40°35′N 124°07′W
28 J2 **Fortuna** North Dakota, N USA 48°53′N 103°46′W
23 T5 **Fort Valley** Georgia, SE USA 32°33′N 83°53′W
11 P11 **Fort Vermilion** Alberta, W Canada 58°22′N 115°59′W
Fort Victoria see Masvingo
31 P13 **Fortville** Indiana, N USA 39°55′N 85°51′W
23 P9 **Fort Walton Beach** Florida, SE USA 30°24′N 86°37′W
31 P12 **Fort Wayne** Indiana, N USA 41°08′N 85°08′W
96 H10 **Fort William** N Scotland, United Kingdom 56°49′N 05°07′W
25 T6 **Fort Worth** Texas, SW USA 32°44′N 97°19′W
28 M7 **Fort Yates** North Dakota, N USA 46°05′N 100°37′W
39 S7 **Fort Yukon** Alaska, USA 66°35′N 145°05′W
Forum Alieni see Ferrara
Forum Julii see Fréjus
Forum Livii see Forlì
143 Q15 **Forūr-e Bozorg, Jazīreh-ye** island S Iran
94 H7 **Fosen** physical region S Norway
161 N14 **Foshan** var. Fatshan, Fo-shan, Namhoi. Guangdong, S China 23°03′N 113°08′E
Fo-shan see Foshan
194 J6 **Fossil Bluff** UK research station Antarctica 71°30′S 68°30′W
Fossa Claudia see Chioggia
106 B9 **Fossano** Piemonte, NW Italy 44°33′N 07°43′E
99 H21 **Fosses-la-Ville** Namur, S Belgium 50°24′N 04°42′E
32 J12 **Fossil** Oregon, NW USA 45°01′N 120°14′W
Foss Lake see Foss Reservoir
106 I11 **Fossombrone** Marche, C Italy 43°42′N 12°48′E
26 K10 **Foss Reservoir** var. Foss Lake. ▣ Oklahoma, C USA
29 S4 **Fosston** Minnesota, N USA 47°34′N 95°45′W
183 O13 **Foster** Victoria, SE Australia 38°40′S 146°15′E
11 T12 **Foster Lakes** ◎ Saskatchewan, C Canada
31 S12 **Fostoria** Ohio, N USA 41°09′N 83°25′W
79 D19 **Fougamou** Ngounié, C Gabon 01°16′S 10°30′E
102 J6 **Fougères** Ille-et-Vilaine, NW France 48°21′N 01°12′W
Fou-hsin see Fuxin
27 S14 **Fouke** Arkansas, C USA 33°15′N 93°53′W
96 K2 **Foula** island NE Scotland, United Kingdom
65 D24 **Foul Bay** bay East Falkland, Falkland Islands
97 P21 **Foulness Island** island SE England, United Kingdom
185 F15 **Foulwind, Cape** headland South Island, New Zealand 41°45′S 171°28′E
79 E15 **Foumban** Ouest, NW Cameroon 05°43′N 10°50′E
172 H13 **Foumbouni** Grande Comore, NW Comoros 11°49′S 43°30′E
195 N8 **Foundation Ice Stream** glacier Antarctica
37 T6 **Fountain** Colorado, C USA 38°40′N 104°42′W
36 L4 **Fountain Green** Utah, W USA 39°37′N 111°37′W
21 P11 **Fountain Inn** South Carolina, SE USA 34°41′N 82°12′W
27 S11 **Fourche LaFave River** ≈ Arkansas, C USA
33 Z13 **Four Corners** Wyoming, C USA 44°04′N 104°08′W

◆ Country
● Country Capital
◇ Dependent Territory
○ Dependent Territory Capital
⬖ Administrative Regions
✈ International Airport
▲ Mountain
▲▲ Mountain Range
Volcano
≈ River
◎ Lake
▣ Reservoir

Fourmies – Gäddede

103 Q2 **Fourmies** Nord, N France 50°01′N 04°03′E
38 J17 **Four Mountains, Islands of** *island group* Aleutian Islands, Alaska, USA
173 P17 **Fournaise, Piton de la** ⛰ SE Réunion 21°14′S 55°43′E
14 J8 **Fournière, Lac** ◎ Québec, SE Canada
115 L20 **Foúrnoi** *island* Dodekánisa, Greece, Aegean Sea
64 K13 **Four North Fracture Zone** *tectonic feature* W Atlantic Ocean
Fouron-Saint-Martin *see* Sint-Martens-Voeren
30 L3 **Fourteen Mile Point** *headland* Michigan, N USA 46°59′N 89°07′W
Fou-shan *see* Fushun
76 I13 **Fouta Djallon** *var.* Futa Jallon. ▲▲ W Guinea
185 C25 **Foveaux Strait** *strait* S New Zealand
35 Q11 **Fowler** California, W USA 36°35′N 119°40′W
37 U6 **Fowler** Colorado, C USA 38°07′N 104°01′W
31 N12 **Fowler** Indiana, N USA 40°36′N 87°20′W
182 D7 **Fowlers Bay** *bay* South Australia
25 R13 **Fowlerton** Texas, SW USA 28°27′N 98°48′W
142 M3 **Fowman** *var.* Fuman, Fumen. Gīlān, NW Iran 37°15′N 49°19′E
65 C25 **Fox Bay East** West Falkland, Falkland Islands
65 C25 **Fox Bay West** West Falkland, Falkland Islands
14 J14 **Foxboro** Ontario, SE Canada 44°16′N 77°23′W
11 O14 **Fox Creek** Alberta, W Canada 54°25′N 116°57′W
64 G5 **Foxe Basin** *sea* Nunavut, N Canada
64 G5 **Foxe Channel** *channel* Nunavut, N Canada
95 I16 **Foxen** ◎ C Sweden
9 Q7 **Foxe Peninsula** *peninsula* Baffin Island, Nunavut, NE Canada
185 E19 **Fox Glacier** West Coast, South Island, New Zealand 43°28′S 170°00′E
38 L17 **Fox Islands** *island* Aleutian Islands, Alaska, USA
30 M10 **Fox Lake** Illinois, N USA 42°24′N 88°10′W
9 V12 **Fox Mine** Manitoba, C Canada 56°36′N 101°48′W
35 R3 **Fox Mountain** ▲ Nevada, W USA 41°01′N 119°30′W
65 E25 **Fox Point** *headland* East Falkland, Falkland Islands 51°55′S 58°24′W
30 M11 **Fox River** ≈ Illinois/Wisconsin, N USA
30 L7 **Fox River** ≈ Wisconsin, N USA
184 L13 **Foxton** Manawatu-Wanganui, North Island, New Zealand 40°27′S 175°18′E
11 S16 **Fox Valley** Saskatchewan, S Canada 50°29′N 109°29′W
11 W16 **Foxwarren** Manitoba, S Canada 50°30′N 101°09′W
97 E14 **Foyle, Lough** *Ir.* Loch Feabhail. *inlet* N Ireland
194 H5 **Foyn Coast** *physical region* Antarctica
104 I2 **Foz** Galicia, NW Spain 43°33′N 07°16′W
60 I12 **Foz do Areia, Represa de** ⊠ S Brazil
59 A16 **Foz do Breu** Acre, W Brazil 09°21′S 72°41′W
83 A16 **Foz do Cunene** Namibe, SW Angola 17°11′S 11°52′E
60 G12 **Foz do Iguaçu** Paraná, S Brazil 25°33′S 54°31′W
58 C12 **Foz do Mamoriá** Amazonas, NW Brazil 02°28′S 66°06′W
105 T6 **Fraga** Aragón, NE Spain 41°32′N 00°21′E
44 F5 **Fragoso, Cayo** *island* C Cuba
61 G18 **Fraile Muerto** Cerro Largo, NE Uruguay 32°30′S 54°30′W
99 H21 **Fraire** Namur, S Belgium 50°16′N 04°30′E
99 L21 **Fraiture, Baraque de** *hill* SE Belgium
Frakštát *see* Hlohovec
Fram Basin *see* Amundsen Basin
99 F20 **Frameries** Hainaut, S Belgium 50°25′N 03°41′E
19 O11 **Framingham** Massachusetts, NE USA 42°15′N 71°24′W
60 L7 **Franca** São Paulo, S Brazil 20°33′S 47°27′W
187 O15 **Français, Récif des** *reef* W New Caledonia
107 K14 **Francavilla al Mare** Abruzzo, C Italy 42°25′N 14°16′E
107 P18 **Francavilla Fontana** Puglia, SE Italy 40°32′N 17°35′E
102 M8 **France** *off.* French Republic, *It./Sp.* Francia; *prev.* Gaul, Gaule, *Lat.* Gallia. ◆ *republic* W Europe
45 O8 **Francés Viejo, Cabo** *headland* NE Dominican Republic 19°39′N 69°57′W
79 F19 **Franceville** *var.* Massoukou, Masuku. Haut-Ogooué, E Gabon 01°40′S 13°31′E
79 F19 **Franceville** ✈ Haut-Ogooué, E Gabon 01°38′S 13°24′E
Francfort *see* Frankfurt am Main
103 T8 **Franche-Comté** ◈ *region* E France
Francia *see* France
29 O11 **Francis Case, Lake** ⊠ South Dakota, N USA
60 H12 **Francisco Beltrão** Paraná, S Brazil 26°05′S 53°04′W
Francisco l. Madero *see* Villa Madero
61 A21 **Francisco Madero** Buenos Aires, E Argentina 35°52′S 62°03′W
42 H6 **Francisco Morazán** *prev.* Tegucigalpa. ◈ *department* C Honduras
83 J18 **Francistown** North East, NE Botswana 21°08′S 27°31′E
Franconian Forest *see* Frankenwald
Franconian Jura *see* Fränkische Alb
98 K6 **Franeker** *Fris.* Frentsjer. Fryslân, N Netherlands 53°11′N 05°33′E
Frankenalb *see* Fränkische Alb
101 H16 **Frankenberg** Hessen, C Germany 51°04′N 08°49′E
101 J20 **Frankenhöhe** *hill range* C Germany
31 R8 **Frankenmuth** Michigan, N USA 43°19′N 83°44′W
101 F20 **Frankenstein** *hill* W Germany
Frankenstein/Frankenstein in Schlesien *see* Ząbkowice Śląskie
101 G20 **Frankenthal** Rheinland-Pfalz, W Germany 49°32′N 08°22′E
101 L18 **Frankenwald** *Eng.* Franconian Forest. ▲▲ C Germany
44 J12 **Frankfield** C Jamaica 18°08′N 77°22′W
14 J14 **Frankford** Ontario, SE Canada 44°12′N 77°36′W
31 O13 **Frankfort** Indiana, N USA 40°16′N 86°30′W
27 O3 **Frankfort** Kansas, C USA 39°42′N 96°25′W
20 L5 **Frankfort** *state capital* Kentucky, S USA 38°12′N 84°52′W
Frankfort on the Main *see* Frankfurt am Main
Frankfurt *see* Frankfurt am Main, Germany
Frankfurt *see* Słubice, Poland
101 G18 **Frankfurt am Main** *var.* Frankfurt, *Fr.* Francfort; *prev. Eng.* Frankfort on the Main. Hessen, SW Germany 50°07′N 08°41′E
100 Q12 **Frankfurt an der Oder** Brandenburg, E Germany 52°20′N 14°32′E
101 L21 **Fränkische Alb** *var.* Frankenalb, *Eng.* Franconian Jura. ▲▲ S Germany
101 I18 **Fränkische Saale** ≈ C Germany
101 L19 **Fränkische Schweiz** *hill range* C Germany
23 R4 **Franklin** Georgia, SE USA 33°15′N 85°06′W
31 P14 **Franklin** Indiana, N USA 39°29′N 86°02′W
20 J7 **Franklin** Kentucky, S USA 36°42′N 86°35′W
22 I9 **Franklin** Louisiana, S USA 29°48′N 91°30′W
29 O17 **Franklin** Nebraska, C USA 40°06′N 98°57′W
21 N10 **Franklin** North Carolina, SE USA 35°12′N 83°23′W
18 C13 **Franklin** Pennsylvania, NE USA 41°24′N 79°49′W
20 J9 **Franklin** Tennessee, S USA 35°55′N 86°52′W
25 U9 **Franklin** Texas, SW USA 31°02′N 96°30′W
21 X7 **Franklin** Virginia, NE USA 36°41′N 76°58′W
21 T4 **Franklin** West Virginia, NE USA 38°39′N 79°21′W
30 M9 **Franklin** Wisconsin, N USA 42°53′N 88°00′W
8 I6 **Franklin Bay** *inlet* Northwest Territories, N Canada
32 K7 **Franklin D. Roosevelt Lake** ⊠ Washington, NW USA
35 W4 **Franklin Lake** ◎ Nevada, W USA
185 B22 **Franklin Mountains** ▲▲ South Island, New Zealand
39 R5 **Franklin Mountains** ▲▲ Alaska, USA
39 N4 **Franklin, Point** *headland* Alaska, USA 70°54′N 158°48′W
183 O17 **Franklin River** ≈ Tasmania, SE Australia
22 K8 **Franklinton** Louisiana, S USA 30°51′N 90°09′W
21 U9 **Franklinton** North Carolina, SE USA 36°06′N 78°27′W
Frankstadt *see* Frenštát pod Radhoštěm
25 V7 **Frankston** Texas, SW USA 32°03′N 95°30′W
33 U12 **Frannie** Wyoming, C USA 44°57′N 108°37′W
15 U5 **Franquelin** Québec, SE Canada 49°17′N 67°52′W
15 U5 **Franquelin** ≈ Québec, SE Canada
83 C18 **Fransfontein** Kunene, NW Namibia 20°12′S 15°01′E
93 H17 **Frånsta** Västernorrland, C Sweden 62°30′N 16°06′E
122 J3 **Frantsa-Iosifa, Zemlya** *Eng.* Franz Josef Land. *island group* N Russian Federation
185 E18 **Franz Josef Glacier** West Coast, South Island, New Zealand 43°22′S 170°11′E
Franz Josef Land *see* Frantsa-Iosifa, Zemlya
Franz-Josef Spitze *see* Gerlachovský štít
101 L23 **Franz Josef Strauss** *abbrev.* F.J.S.. ✈ (München) Bayern, SE Germany 48°07′N 11°43′E
107 A19 **Frasca, Capo della** *headland* Sardegna, Italy, C Mediterranean Sea 39°46′N 08°27′E
107 I15 **Frascati** Lazio, C Italy 41°48′N 12°41′E
11 N14 **Fraser** ≈ British Columbia, SW Canada
83 G24 **Fraserburg** Western Cape, SW South Africa 31°55′S 21°31′E
96 L8 **Fraserburgh** NE Scotland, United Kingdom 57°42′N 02°20′W
181 Z9 **Fraser Island** *var.* Great Sandy Island. *island* Queensland, E Australia
10 L14 **Fraser Lake** British Columbia, SW Canada 54°00′N 124°45′W
10 L15 **Fraser Plateau** *plateau* British Columbia, SW Canada
184 P10 **Frasertown** Hawke's Bay, North Island, New Zealand 38°58′S 177°25′E
99 E19 **Frasnes-lez-Buissenal** Hainaut, SW Belgium 50°40′N 03°37′E
108 I7 **Frastanz** Vorarlberg, NW Austria 47°13′N 09°38′E
14 B8 **Frater** Ontario, S Canada 47°19′N 84°28′W
Frauenbach *see* Baia Mare
Frauenburg *see* Saldus, Latvia
Frauenburg *see* Frombork, Poland
108 H6 **Frauenfeld** Thurgau, NE Switzerland 47°34′N 08°54′E
109 Z5 **Frauenkirchen** Burgenland, E Austria 47°50′N 16°57′E
61 D19 **Fray Bentos** Río Negro, W Uruguay 33°09′S 58°14′W
61 F19 **Fray Marcos** Florida, S Uruguay 34°13′S 55°43′W
29 S6 **Frazee** Minnesota, N USA 46°42′N 95°40′W
104 M5 **Frechilla** Castilla y León, N Spain 42°08′N 04°50′W
30 I4 **Frederic** Wisconsin, N USA 45°42′N 92°30′W
95 G23 **Fredericia** Syddanmark, C Denmark 55°34′N 09°47′E
21 W3 **Frederick** Maryland, NE USA 39°25′N 77°25′W
26 L12 **Frederick** Oklahoma, C USA 34°24′N 99°03′W
29 P7 **Frederick** South Dakota, N USA 45°49′N 98°31′W
29 X12 **Fredericksburg** Iowa, C USA 42°58′N 92°12′W
25 R10 **Fredericksburg** Texas, SW USA 30°17′N 98°52′W
21 W5 **Fredericksburg** Virginia, NE USA 38°16′N 77°27′W
39 X13 **Frederick Sound** *sound* Alaska, USA
27 X6 **Fredericktown** Missouri, C USA 37°33′N 90°19′W
60 H13 **Frederico Westphalen** Rio Grande do Sul, S Brazil 27°22′S 53°20′W
13 O15 **Fredericton** *province capital* New Brunswick, SE Canada 45°57′N 66°40′W
Frederiksborgs Amt *see* Hovedstaden
Frederikshåb *see* Paamiut
95 H19 **Frederikshavn** *prev.* Fladstrand. Nordjylland, N Denmark 57°28′N 10°33′E
95 J22 **Frederikssund** Hovedstaden, E Denmark 55°51′N 12°05′E
45 T9 **Frederiksted** Saint Croix, S Virgin Islands (US) 17°41′N 64°51′W
95 I22 **Frederiksværk** *var.* Frederiksværk og Hanehoved. Hovedstaden, E Denmark 55°58′N 12°02′E
Frederiksværk og Hanehoved *see* Frederiksværk
54 E9 **Fredonia** Antioquia, W Colombia 05°57′N 75°42′W
36 K8 **Fredonia** Arizona, SW USA 36°57′N 112°31′W
27 P7 **Fredonia** Kansas, C USA 37°32′N 95°50′W
18 C11 **Fredonia** New York, NE USA 42°26′N 79°19′W
35 P4 **Fredonyer Pass** *pass* California, W USA
93 I15 **Fredrika** Västerbotten, N Sweden 64°03′N 18°25′E
95 L14 **Fredriksberg** Dalarna, C Sweden 60°07′N 14°23′E
Fredrikshald *see* Halden
Fredrikshamn *see* Hamina
95 H16 **Fredrikstad** Østfold, S Norway 59°12′N 10°57′E
30 K16 **Freeburg** Illinois, N USA 38°25′N 89°54′W
18 K15 **Freehold** New Jersey, NE USA 40°14′N 74°14′W
18 H14 **Freeland** Pennsylvania, NE USA 41°01′N 75°54′W
182 J5 **Freeling Heights** ▲ South Australia 30°09′S 139°24′E
35 Q7 **Freel Peak** ▲ California, W USA 38°52′N 119°52′W
11 Z9 **Freels, Cape** *headland* Newfoundland and Labrador, E Canada 49°16′N 53°30′W
29 Q11 **Freeman** South Dakota, N USA 43°21′N 97°26′W
44 G1 **Freeport** Grand Bahama Island, N The Bahamas 26°28′N 78°43′W
30 L10 **Freeport** Illinois, N USA 42°18′N 89°37′W
25 W12 **Freeport** Texas, SW USA 28°57′N 95°21′W
44 G1 **Freeport** ✈ Grand Bahama Island, N The Bahamas 26°31′N 78°48′W
25 R14 **Freer** Texas, SW USA 27°52′N 98°37′W
83 I22 **Free State** *off.* Free State Province; *prev.* Orange Free State, *Afr.* Oranje Vrystaat. ◈ *province* C South Africa
Free State *see* Maryland
Free State Province *see* Free State
76 G15 **Freetown** ● (Sierra Leone) W Sierra Leone 08°27′N 13°16′W
172 J16 **Frégate** *island* Inner Islands, NE Seychelles
104 J12 **Fregenal de la Sierra** Extremadura, W Spain 38°10′N 06°39′W
182 C2 **Fregon** South Australia 26°44′S 132°03′E
102 H5 **Fréhel, Cap** *headland* NW France 48°41′N 02°21′W
94 F8 **Frei** Møre og Romsdal, S Norway 63°02′N 07°47′E
101 O16 **Freiberg** Sachsen, E Germany 50°55′N 13°21′E
101 O16 **Freiberger Mulde** ≈ E Germany
Freiburg *see* Freiburg im Breisgau, Germany
Freiburg *see* Fribourg, Switzerland
101 F23 **Freiburg im Breisgau** *var.* Freiburg, *Fr.* Fribourg-en-Brisgau. Baden-Württemberg, SW Germany 48°N 07°52′E
Freiburg in Schlesien *see* Świebodzice
Freie Hansestadt Bremen *see* Bremen
Freie und Hansestadt Hamburg *see* Brandenburg
101 L22 **Freising** Bayern, SE Germany 48°24′N 11°45′E
109 T3 **Freistadt** Oberösterreich, N Austria 48°31′N 14°31′E
Freistadtl *see* Hlohovec
101 O16 **Freital** Sachsen, E Germany 51°00′N 13°40′E
Freiwaldau *see* Jeseník
104 J6 **Freixo de Espada à Cinta** Bragança, N Portugal 41°05′N 06°49′W
103 U15 **Fréjus** *anc.* Forum Julii. Var, SE France 43°26′N 06°44′E
180 I13 **Fremantle** Western Australia 32°07′S 115°44′E
35 N9 **Fremont** California, W USA 37°34′N 122°01′W
31 Q11 **Fremont** Indiana, N USA 41°43′N 84°54′W
29 W15 **Fremont** Iowa, C USA 41°12′N 92°26′W
31 P8 **Fremont** Michigan, N USA 43°28′N 85°56′W
29 R15 **Fremont** Nebraska, C USA 41°25′N 96°30′W
31 S11 **Fremont** Ohio, N USA 41°21′N 83°08′W
33 T14 **Fremont Peak** ▲ Wyoming, C USA 43°07′N 109°37′W
36 M6 **Fremont River** ≈ Utah, W USA
21 O9 **French Broad River** ≈ Tennessee, S USA
21 N5 **Frenchburg** Kentucky, S USA 37°58′N 83°37′W
18 C12 **French Creek** ≈ Pennsylvania, NE USA
32 K15 **Frenchglen** Oregon, NW USA 42°49′N 118°55′W
55 Y10 **French Guiana** *var.* Guiana, Guyane. ◇ *French overseas department* N South America
French Guinea *see* Guinea
31 O15 **French Lick** Indiana, N USA 38°33′N 86°37′W
185 J14 **French Pass** Marlborough, South Island, New Zealand 40°57′S 173°49′E
191 T11 **French Polynesia** ◇ *French overseas territory* S Pacific Ocean
French Republic *see* France
14 F11 **French River** ≈ Ontario, S Canada
French Somaliland *see* Djibouti
173 P12 **French Southern and Antarctic Lands** *prev.* French Southern and Antarctic Territories, *Fr.* Terres Australes et Antarctiques Françaises. ◈ *French overseas territory* S Indian Ocean
French Southern and Antarctic Territories *see* French Southern and Antarctic Lands
French Sudan *see* Mali
French Territory of the Afars and Issas *see* Djibouti
French Togoland *see* Togo
74 J6 **Frenda** NW Algeria 35°04′N 01°03′E
111 I18 **Frenštát pod Radhoštěm** *Ger.* Frankstadt. Moravskoslezský Kraj, E Czech Republic 49°33′N 18°10′E
Frentsjer *see* Franeker
76 M17 **Fresco** S Ivory Coast 05°03′N 05°31′W
195 U16 **Freshfield, Cape** *headland* Antarctica
40 L10 **Fresnillo** *var.* Fresnillo de González Echeverría. Zacatecas, C Mexico 23°11′N 102°53′W
Fresnillo de González Echeverría *see* Fresnillo
35 Q10 **Fresno** California, W USA 36°45′N 119°48′W
Freu, Cabo del *see* Freu, Cap des
105 Y9 **Freu, Cap des** *var.* Cabo del Freu. *cape* Mallorca, Spain, W Mediterranean Sea
101 G22 **Freudenstadt** Baden-Württemberg, SW Germany 48°28′N 08°25′E
Freudenthal *see* Bruntál
183 Q17 **Freycinet Peninsula** *peninsula* Tasmania, SE Australia
76 H14 **Fria** W Guinea 10°27′N 13°38′W
83 A17 **Fria, Cape** *headland* NW Namibia 18°32′S 12°00′E
35 Q10 **Friant** California, W USA 36°56′N 119°44′W
62 K8 **Frías** Catamarca, N Argentina 28°41′S 65°00′W
108 D9 **Fribourg** *Ger.* Freiburg. Fribourg, W Switzerland 46°50′N 07°10′E
108 C9 **Fribourg** *Ger.* Freiburg. ◈ *canton* W Switzerland
Fribourg-en-Brisgau *see* Freiburg im Breisgau
32 G7 **Friday Harbor** San Juan Islands, Washington, NW USA 48°31′N 123°01′W
Friedau *see* Ormož
101 K23 **Friedberg** Bayern, S Germany 48°21′N 10°58′E
101 H18 **Friedberg** Hessen, W Germany 50°19′N 08°46′E
Friedeberg Neumark *see* Strzelce Krajeńskie
Friedek-Mistek *see* Frýdek-Místek
Friedland *see* Pravdinsk
101 I24 **Friedrichshafen** Baden-Württemberg, S Germany 47°39′N 09°29′E
Friedrichstadt *see* Jaunjelgava
29 Q16 **Friend** Nebraska, C USA 40°37′N 97°16′W
Friendly Islands *see* Tonga
55 W9 **Friendship** Coronie, N Suriname 05°50′N 56°16′W
30 L7 **Friendship** Wisconsin, N USA 43°58′N 89°48′W
109 T8 **Friesach** Kärnten, S Austria 46°58′N 14°24′E
Friesche Eilanden *see* Frisian Islands
101 F22 **Friesenheim** Baden-Württemberg, SW Germany 48°27′N 07°56′E
Friesische Inseln *see* Frisian Islands
Friesland *see* Fryslân
60 Q10 **Frio, Cabo** *headland* SE Brazil 23°01′S 41°59′W
24 M3 **Friona** Texas, SW USA 34°38′N 102°43′W
42 L12 **Frío, Río** ≈ N Costa Rica
25 R13 **Frio River** ≈ Texas, SW USA
99 M25 **Frisange** Luxembourg, S Luxembourg 49°31′N 06°12′E
Frisches Haff *see* Vistula Lagoon
36 J6 **Frisco Peak** ▲ Utah, W USA 38°31′N 113°17′W
84 F9 **Frisian Islands** *Dut.* Friesche Eilanden, *Ger.* Friesische Inseln. *island group* N Europe
18 L12 **Frissell, Mount** ▲ Connecticut, NE USA 42°01′N 73°25′W
95 J19 **Fristad** Västra Götaland, S Sweden 57°50′N 13°01′E
25 N2 **Fritch** Texas, SW USA 35°38′N 101°36′W
95 J19 **Fritsla** Västra Götaland, S Sweden 57°33′N 12°47′E
101 H16 **Fritzlar** Hessen, C Germany 51°09′N 09°16′E
106 H6 **Friuli-Venezia Giulia** ◈ *region* NE Italy
196 L13 **Frobisher Bay** *inlet* Baffin Island, Nunavut, NE Canada
Frobisher Bay *see* Iqaluit
11 S12 **Frobisher Lake** ◎ Saskatchewan, C Canada
94 G7 **Frohavet** *sound* C Norway
Frohenbruck *see* Veselí nad Lužnicí
109 V7 **Frohnleiten** Steiermark, SE Austria 47°17′N 15°20′E
99 G22 **Froidchapelle** Hainaut, S Belgium 50°10′N 04°18′E
127 O9 **Frolovo** Volgogradskaya Oblast', SW Russian Federation 49°46′N 43°38′E
110 K7 **Frombork** *Ger.* Frauenburg. Warmińsko-Mazurskie, NE Poland 54°21′N 19°40′E
97 L22 **Frome** SW England, United Kingdom 51°15′N 02°22′W
182 I4 **Frome Creek** *seasonal river* South Australia
182 J6 **Frome Downs** South Australia 31°17′S 139°48′E
182 J5 **Frome, Lake** *salt lake* South Australia
Fronicken *see* Wronki
104 H10 **Fronteira** Portalegre, C Portugal 39°03′N 07°39′W
40 M7 **Frontera** Coahuila, NE Mexico 26°55′N 101°27′W
41 U14 **Frontera** Tabasco, SE Mexico 18°32′N 92°39′W
40 G3 **Fronteras** Sonora, NW Mexico 30°51′N 109°33′W
103 Q16 **Frontignan** Hérault, S France 43°27′N 03°45′E
54 D8 **Frontino** Antioquia, NW Colombia 06°46′N 76°10′W
21 V4 **Front Royal** Virginia, NE USA 38°56′N 78°13′W
107 J16 **Frosinone** *anc.* Frusino. Lazio, C Italy 41°38′N 13°22′E
107 K16 **Frosolone** Molise, C Italy 41°34′N 14°25′E
25 U7 **Frost** Texas, SW USA 32°04′N 96°48′W
21 U2 **Frostburg** Maryland, NE USA 39°39′N 78°55′W
23 X13 **Frostproof** Florida, SE USA 27°45′N 81°31′W
Frostviken *see* Kvarnbergsvattnet
95 M15 **Frövi** Örebro, C Sweden 59°28′N 15°24′E
94 F7 **Frøya** *island* W Norway
37 P5 **Fruita** Colorado, C USA 39°10′N 108°42′W
28 J9 **Fruitdale** South Dakota, N USA 44°39′N 103°38′W
23 W11 **Fruitland Park** Florida, SE USA 28°51′N 81°54′W
Frumentum *see* Formentera
147 S11 **Frunze** Batkenskaya Oblast', SW Kyrgyzstan 40°07′N 71°40′E
Frunze *see* Bishkek
117 O9 **Frunzivka** Odes'ka Oblast', SW Ukraine 47°19′N 29°46′E
Frusino *see* Frosinone
108 E9 **Frutigen** Bern, W Switzerland 46°35′N 07°38′E
111 I17 **Frýdek-Místek** *Ger.* Friedek-Mistek. Moravskoslezský Kraj, E Czech Republic 49°40′N 18°22′E
98 K6 **Fryslân** *prev.* Friesland. ◈ *province* N Netherlands
193 V16 **Fua'amotu** Tongatapu, S Tonga 21°15′S 175°08′W
190 A9 **Fuafatu** *island* Funafuti Atoll, C Tuvalu
190 A9 **Fuagea** *island* Funafuti Atoll, C Tuvalu
190 B8 **Fualifeke** *atoll* C Tuvalu
190 A8 **Fualopa** *island* Funafuti Atoll, C Tuvalu
151 K22 **Fuammulah** *var.* Fuammulah, Gnaviyani. *atoll* S Maldives
Fuammulah *see* Fuammulah
161 R11 **Fu'an** Fujian, SE China 27°11′N 119°42′E
Fu-chien *see* Fujian
Fu-chou *see* Fuzhou
164 G13 **Fuchū** *var.* Hutyû. Hiroshima, Honshū, SW Japan 34°35′N 133°12′E
160 M13 **Fuchuan** *var.* Fuyang. Guangxi Zhuangzu Zizhiqu, S China 24°56′N 111°15′E
161 R9 **Fuchun Jiang** *var.* Tsien Tang. ≈ SE China
165 R8 **Fudai** Iwate, Honshū, C Japan 39°59′N 141°50′E
161 S11 **Fuding** *var.* Tongshan. Fujian, SE China 27°21′N 120°10′E
81 J20 **Fudua** *spring/well* S Kenya 02°13′S 39°43′E
104 M16 **Fuengirola** Andalucía, S Spain 36°32′N 04°38′W
104 J12 **Fuente de Cantos** Extremadura, W Spain 38°15′N 06°18′W
104 J11 **Fuente del Maestre** Extremadura, W Spain 38°31′N 06°26′W
104 L12 **Fuente Obejuna** Andalucía, S Spain 38°15′N 05°25′W
104 L6 **Fuentesaúco** Castilla y León, N Spain 41°14′N 05°30′W
62 O3 **Fuerte Olimpo** *var.* Olimpo. Alto Paraguay, NE Paraguay 21°02′S 57°51′W
40 H8 **Fuerte, Río** ≈ C Mexico
64 Q11 **Fuerteventura** *island* Islas Canarias, Spain, NE Atlantic Ocean
141 S14 **Fughmah** *var.* Faghman, Fugma. C Yemen 16°08′N 49°23′E
92 M2 **Fuglehuken** *headland* W Svalbard 78°54′N 10°30′E
Fuglø *see* Fugloy
95 B18 **Fugloy** *Dan.* Fuglø. *island* NE Faeroe Islands
197 T15 **Fugløya Bank** *undersea feature* E Norwegian Sea 71°00′N 19°20′E
Fugma *see* Fughmah
160 E11 **Fugong** Yunnan, SW China 27°00′N 98°48′E
81 K16 **Fugugo** *spring/well* NE Kenya 03°19′N 39°39′E
158 L2 **Fuhai** *var.* Burultokay. Xinjiang Uygur Zizhiqu, NW China 47°15′N 87°39′E
161 P10 **Fu He** ≈ S China
Fuhkien *see* Fujian
100 J9 **Fuhlsbüttel** ✈ (Hamburg) Hamburg, N Germany 53°37′N 09°57′E
101 L14 **Fuhne** ≈ C Germany
Fu-hsin *see* Fuxin
Fujairah *see* Al Fujayrah
164 M14 **Fuji** *var.* Huzi. Shizuoka, Honshū, S Japan 35°08′N 138°39′E
161 Q12 **Fujian** *var.* Fu-chien, Fuhkien, Fukien, Min, Fujian Sheng. ◈ *province* SE China
160 I9 **Fu Jiang** ≈ C China
Fujian Sheng *see* Fujian
164 M14 **Fujieda** *var.* Huzieda. Shizuoka, Honshū, S Japan 34°54′N 138°15′E
Fuji, Mount/Fujiyama *see* Fuji-san
163 Y7 **Fujin** Heilongjiang, NE China 47°12′N 132°01′E
164 M13 **Fujinomiya** *var.* Huzinomiya. Shizuoka, Honshū, S Japan 35°16′N 138°33′E
164 N13 **Fuji-san** *var.* Fujiyama, *Eng.* Mount Fuji. ▲ Honshū, SE Japan 35°23′N 138°44′E
165 N14 **Fujisawa** *var.* Huzisawa. Kanagawa, Honshū, S Japan 35°22′N 139°29′E
165 T3 **Fukagawa** *var.* Hukagawa. Hokkaidō, NE Japan 43°44′N 142°03′E
158 L5 **Fukang** Xinjiang Uygur Zizhiqu, W China 44°07′N 87°55′E
165 P7 **Fukaura** Aomori, Honshū, C Japan 40°38′N 139°55′E
193 W15 **Fukave** *island* Tongatapu Group, S Tonga
Fukien *see* Fujian
164 J13 **Fukuchiyama** *var.* Hukutiyama. Kyōto, Honshū, SW Japan 35°19′N 135°08′E
Fukue *see* Gotō
164 A13 **Fukue-jima** *island* Gotō-rettō, SW Japan
164 K12 **Fukui** *var.* Hukui. Fukui, Honshū, SW Japan 36°03′N 136°12′E
164 K12 **Fukui** *off.* Fukui-ken, *var.* Hukui. ◈ *prefecture* Honshū, SW Japan
Fukui-ken *see* Fukui
164 D13 **Fukuoka** *var.* Hukuoka, *hist.* Najima. Fukuoka, Kyūshū, SW Japan 33°36′N 130°24′E
164 D13 **Fukuoka** *off.* Fukuoka-ken, *var.* Hukuoka. ◈ *prefecture* Kyūshū, SW Japan
Fukuoka-ken *see* Fukuoka
165 Q6 **Fukushima** Hokkaidō, NE Japan 41°27′N 140°14′E
165 Q12 **Fukushima** *off.* Fukushima-ken, *var.* Hukusima. ◈ *prefecture* Honshū, C Japan
Fukushima-ken *see* Fukushima
164 G13 **Fukuyama** *var.* Hukuyama. Hiroshima, Honshū, SW Japan 34°29′N 133°21′E
76 G13 **Fulacunda** C Guinea-Bissau 11°44′N 15°03′W
129 P8 **Fūlādī, Kūh-e** ▲ E Afghanistan 34°38′N 67°32′E
187 Z15 **Fulaga** *island* Lau Group, E Fiji
101 I17 **Fulda** Hessen, C Germany 50°33′N 09°41′E
29 S10 **Fulda** Minnesota, N USA 43°52′N 95°36′W
101 I16 **Fulda** ≈ C Germany
Fülek *see* Fil'akovo
Fuli *see* Jixian
Fulin *see* Hanyuan
160 K10 **Fuling** Chongqing Shi, C China 29°45′N 107°23′E
35 T15 **Fullerton** California, W USA 33°53′N 117°55′W
29 P15 **Fullerton** Nebraska, C USA 41°21′N 97°58′W
108 M8 **Fulpmes** Tirol, W Austria 47°11′N 11°22′E
20 G8 **Fulton** Kentucky, S USA 36°31′N 88°52′W
23 N2 **Fulton** Mississippi, S USA 34°16′N 88°24′W
27 V4 **Fulton** Missouri, C USA 38°50′N 91°57′W
18 H9 **Fulton** New York, NE USA 43°18′N 76°22′W
Fuman/Fumen *see* Fowman
103 R3 **Fumay** Ardennes, N France 49°58′N 04°42′E
102 M13 **Fumel** Lot-et-Garonne, SW France 44°31′N 00°58′E
190 B10 **Funafara** *atoll* C Tuvalu
190 C9 **Funafuti** ✈ Funafuti Atoll, C Tuvalu 08°30′S 179°12′E
Funafuti *see* Vaiaku
190 C9 **Funafuti Atoll** ● (Tuvalu) C Tuvalu 08°31′N 179°11′E
190 C9 **Funafuti Atoll** *atoll* C Tuvalu
Funan *see* Fusui
190 B9 **Funangongo** *atoll* C Tuvalu
93 F17 **Funäsdalen** Jämtland, C Sweden 62°33′N 12°33′E
64 O6 **Funchal** Madeira, Portugal, NE Atlantic Ocean 32°40′N 16°55′W
64 P5 **Funchal** ✈ Madeira, Portugal, NE Atlantic Ocean 32°38′N 16°53′W
54 F5 **Fundación** Magdalena, N Colombia 10°31′N 74°09′W
104 I8 **Fundão** *var.* Fundáo. Castelo Branco, C Portugal 40°08′N 07°30′W
Fundáo *see* Fundão
13 O16 **Fundy, Bay of** *bay* Canada/USA
Fünen *see* Fyn
54 C13 **Fúnes** Nariño, SW Colombia 0°59′N 77°27′W
Fünfkirchen *see* Pécs
83 M19 **Funhalouro** Inhambane, S Mozambique 23°04′S 34°24′E
161 R6 **Funing** Jiangsu, E China 33°43′N 119°47′E
160 I14 **Funing** *var.* Xinhua. Yunnan, SW China 23°39′N 105°41′E
160 M7 **Funiu Shan** ▲▲ C China
77 U13 **Funtua** Katsina, N Nigeria 11°31′N 07°19′E
161 R12 **Fuqing** Fujian, SE China 25°40′N 119°23′E
83 M14 **Furancungo** Tete, NW Mozambique 14°51′S 33°39′E
116 I15 **Furculeşti** Teleorman, S Romania 43°51′N 25°07′E
Füred *see* Balatonfüred
165 W4 **Füren-ko** ◎ Hokkaidō, NE Japan
Fürg *see* Doborji
Furluk *see* Fârliug
Fürmanov/Furmanovka *see* Moyynkum
Furmanovo *see* Zhalpaktal
59 L20 **Furnas, Represa de** ⊠ SE Brazil
183 Q14 **Furneaux Group** *island group* Tasmania, SE Australia
Furnes *see* Veurne
160 J10 **Furong Jiang** ≈ S China
138 I5 **Furqlus** Ḥimṣ, W Syria 34°40′N 37°02′E
100 F12 **Fürstenau** Niedersachsen, NW Germany 52°30′N 07°40′E
109 X8 **Fürstenfeld** Steiermark, SE Austria 47°03′N 16°05′E
101 L23 **Fürstenfeldbruck** Bayern, S Germany 48°10′N 11°16′E
100 P12 **Fürstenwalde** Brandenburg, NE Germany 52°22′N 14°04′E
101 K20 **Fürth** Bayern, S Germany 49°29′N 10°59′E
109 W3 **Furth bei Göttweig** Niederösterreich, NW Austria 48°22′N 15°33′E
165 R3 **Furubira** Hokkaidō, NE Japan 43°14′N 140°38′E
94 L12 **Furudal** Dalarna, C Sweden 61°10′N 15°07′E
164 L12 **Furukawa** *var.* Hida. Gifu, Honshū, SW Japan 36°13′N 137°11′E
165 Q10 **Furukawa** *var.* Hurukawa, Ōsaki. Miyagi, Honshū, C Japan 38°36′N 140°57′E
54 F10 **Fusagasugá** Cundinamarca, C Colombia 04°22′N 74°21′W
Fusan *see* Busan
Fushë-Arëzi/Fushë-Arrësi *see* Fushë-Arrëz
113 L18 **Fushë-Arrëz** *var.* Fushë-Arëzi, Fushë-Arrësi. Shkodër, N Albania 42°05′N 20°01′E
113 N16 **Fushë Kosovë** *Serb.* Kosovo Polje. C Kosovo 42°40′N 21°07′E
Fushë-Kruja *see* Fushë-Krujë
113 K19 **Fushë-Krujë** *var.* Fushë-Kruja. Durrës, C Albania 41°30′N 19°43′E
163 V12 **Fushun** *var.* Fou-shan, Fu-shun. Liaoning, NE China 41°50′N 123°54′E
Fu-shun *see* Fushun
Fusin *see* Fuxin
108 G10 **Fusio** Ticino, S Switzerland 46°27′N 08°39′E
163 X11 **Fusong** Jilin, NE China 42°20′N 127°17′E
101 K24 **Füssen** Bayern, S Germany 47°34′N 10°43′E
160 K15 **Fusui** *var.* Xinning; *prev.* Funan. Guangxi Zhuangzu Zizhiqu, S China 22°39′N 107°49′E
Futa Jallon *see* Fouta Djallon
63 G18 **Futaleufú** Los Lagos, S Chile 43°14′S 71°50′W
112 K10 **Futog** Vojvodina, NW Serbia 45°15′N 19°43′E
165 O14 **Futtsu** *var.* Huttu. Chiba, Honshū, S Japan 35°11′N 139°52′E
187 S15 **Futuna** *island* S Vanuatu
190 D12 **Futuna, Île** *island* S Wallis and Futuna
161 Q11 **Futun Xi** ≈ SE China
160 L5 **Fuxian** *var.* Fu Xian. Shaanxi, C China 36°03′N 109°19′E
Fuxian *see* Wafangdian
Fu Xian *see* Fuxian
160 G13 **Fuxian Hu** ◎ SW China
163 U12 **Fuxin** *var.* Fou-hsin, Fu-hsin, Fusin. Liaoning, NE China 41°59′N 121°40′E
Fuxing *see* Wangmo
161 P7 **Fuyang** Anhui, E China 32°52′N 115°51′E
Fuyang *see* Fuchuan
161 O4 **Fuyang He** ≈ E China
163 U7 **Fuyu** Heilongjiang, NE China 47°48′N 124°26′E
163 Z6 **Fuyuan** Heilongjiang, NE China 48°20′N 134°22′E
Fuyu/Fu-yü *see* Songyuan
158 M3 **Fuyun** *var.* Koktokay. Xinjiang Uygur Zizhiqu, NW China 46°58′N 89°30′E
111 L22 **Füzesabony** Heves, E Hungary 47°46′N 20°25′E
161 R12 **Fuzhou** *var.* Foochow, Fu-chou. *province capital* Fujian, SE China 26°09′N 119°17′E
161 P11 **Fuzhou** *prev.* Linchuan. Jiangxi, S China 27°58′N 116°20′E
137 W13 **Füzuli** *Rus.* Fizuli. SW Azerbaijan 39°33′N 47°09′E
119 I20 **Fyadory** *Rus.* Fëdory. Brestskaya Voblasts', SW Belarus 51°57′N 26°24′E
95 G23 **Fyn** *Ger.* Fünen. *island* C Denmark
96 H12 **Fyne, Loch** *inlet* W Scotland, United Kingdom
95 E16 **Fyresvatnet** ◎ S Norway
FYR Macedonia/FYROM *see* Macedonia, FYR
Fyzabad *see* Feizābād

G

Gaafu Alifu Atoll *see* North Huvadhu Atoll
81 O14 **Gaalkacyo** *var.* Galka'yo, *It.* Galcaio. Mudug, C Somalia 06°42′N 47°24′E
146 J11 **Gabakly** *Rus.* Kabakly. Lebap Welaýaty, NE Turkmenistan 39°45′N 62°30′E
114 H8 **Gabare** Vratsa, NW Bulgaria 43°20′N 23°57′E
102 K15 **Gabas** ≈ SW France
Gabasumdo *see* Tongde
35 T7 **Gabbs** Nevada, W USA 38°51′N 117°55′W
82 B12 **Gabela** Kwanza Sul, W Angola 10°50′S 14°21′E
Gaberones *see* Gaborone
189 X14 **Gabert** *island* Caroline Islands, E Micronesia
74 M7 **Gabès** *var.* Qābis. E Tunisia 33°53′N 10°03′E
74 M6 **Gabès, Golfe de** *Ar.* Khalīj Qābis. *gulf* E Tunisia
Gablonz an der Neisse *see* Jablonec nad Nisou
Gablös *see* Cavalese
79 E18 **Gabon** *off.* Gabonese Republic. ◆ *republic* C Africa
Gabonese Republic *see* Gabon
83 I20 **Gaborone** *prev.* Gaberones. ● (Botswana) South East, SE Botswana 24°42′S 25°50′E
83 I20 **Gaborone** ✈ South East, SE Botswana 24°45′S 25°49′E
104 K8 **Gabriel y Galán, Embalse de** ⊠ W Spain
143 U15 **Gābrīk, Rūd-e** ≈ SE Iran
114 J9 **Gabrovo** Gabrovo, N Bulgaria 42°54′N 25°19′E
114 J9 **Gabrovo** ◈ *province* N Bulgaria
76 H12 **Gabú** *prev.* Nova Lamego. E Guinea-Bissau 12°16′N 14°09′W
29 O6 **Gackle** North Dakota, N USA 46°34′N 99°07′W
113 I15 **Gacko** Republika Srpska, S Bosnia and Herzegovina 43°08′N 18°29′E
155 F17 **Gadag** Karnātaka, W India 15°25′N 75°37′E
93 G15 **Gäddede** Jämtland, C Sweden 64°30′N 14°15′E

◆ Country ● Country Capital ◇ Dependent Territory ○ Dependent Territory Capital ◈ Administrative Regions ✈ International Airport ▲ Mountain ▲▲ Mountain Range ⛰ Volcano ≈ River ◎ Lake ⊠ Reservoir

251

GADÊ – GEDAREF

159 S12 **Gadê** *var.* Kequ; *prev.* Pagqên. Qinghai, C China *33°56′N 99°49′E*
Gades/Gadier/Gadir/Gadire *see* Cádiz
105 P15 **Gádor, Sierra de** ▲▲ S Spain
149 S15 **Gadra** Sind, SE Pakistan *25°39′N 70°28′E*
23 Q3 **Gadsden** Alabama, S USA *34°00′N 86°00′W*
36 H15 **Gadsden** Arizona, SW USA *32°33′N 114°45′W*
Gadyach *see* Hadyach
124 J3 **Gadzhiyevo** Murmanskaya Oblast', NW Russian Federation *69°16′N 33°20′E*
79 H15 **Gadzi** Mambéré-Kadéï, SW Central African Republic *04°46′N 16°42′E*
116 J13 **Găești** Dâmbovița, S Romania *44°42′N 25°19′E*
107 J17 **Gaeta** Lazio, C Italy *41°12′N 13°35′E*
107 J17 **Gaeta, Golfo di** *var.* Gulf of Gaeta. *gulf* C Italy
Gaeta, Gulf of *see* Gaeta, Golfo di
188 L14 **Gaferut** *atoll* Caroline Islands, W Micronesia
21 Q10 **Gaffney** South Carolina, SE USA *35°03′N 81°40′W*
Gäfle *see* Gävle
Gäfleborg *see* Gävleborg
74 M6 **Gafsa** *var.* Qafşah. W Tunisia *34°25′N 08°52′E*
Gafurov *see* Ghafurov
126 J3 **Gagarin** *prev.* Gzhatsk. Smolenskaya Oblast', W Russian Federation *53°33′N 35°00′E*
147 O10 **Gagarin** Jizzax Viloyati, C Uzbekistan *40°40′N 68°04′E*
116 M12 **Găgăuzia** ◈ *cultural region* S Moldavia
101 G21 **Gaggenau** Baden-Württemberg, SW Germany *48°48′N 08°19′E*
188 F16 **Gagil Tamil** *var.* Gagil-Tomil. *island* Caroline Islands, W Micronesia
Gagil-Tomil *see* Gagil Tamil
127 O4 **Gagino** Nizhegorodskaya Oblast', W Russian Federation *55°18′N 45°01′E*
107 Q19 **Gagliano del Capo** Puglia, SE Italy *39°49′N 18°22′E*
94 L13 **Gagnef** Dalarna, C Sweden *60°34′N 15°04′E*
76 M17 **Gagnoa** C Ivory Coast *06°11′N 05°56′W*
13 N10 **Gagnon** Québec, E Canada *51°56′N 68°16′W*
Gago Coutinho *see* Lumbala N'Guimbo
137 P8 **Gagra** NW Georgia *43°17′N 40°18′E*
31 S13 **Gahanna** Ohio, N USA *40°01′N 82°52′W*
143 R13 **Gahkom** Hormozgān, S Iran *28°14′N 55°48′E*
Gahnpa *see* Ganta
57 Q19 **Gaíba, Laguna** ◎ E Bolivia
153 T13 **Gaibandha** *var.* Gaibanda. Rajshahi, NW Bangladesh *25°21′N 89°36′E*
Gaibhlte, Cnoc Mór na n *see* Galtymore Mountain
109 R9 **Gail** ≈ S Austria
101 I21 **Gaildorf** Baden-Württemberg, S Germany *48°41′N 10°08′E*
103 N15 **Gaillac** *var.* Gaillac-sur-Tarn. Tarn, S France *43°54′N 01°54′E*
Gaillac-sur-Tarn *see* Gaillac
Gaillimh *see* Galway
Gaillimhe, Cuan na *see* Galway Bay
109 Q9 **Gailtaler Alpen** ▲▲ S Austria
63 J17 **Gaimán** Chaco, S Argentina *43°15′S 65°30′W*
20 K8 **Gainesboro** Tennessee, S USA *36°20′N 85°41′W*
23 V10 **Gainesville** Florida, SE USA *29°39′N 82°19′W*
23 T2 **Gainesville** Georgia, SE USA *34°18′N 83°49′W*
27 U8 **Gainesville** Missouri, C USA *36°37′N 92°28′W*
25 T5 **Gainesville** Texas, SW USA *33°39′N 97°08′W*
109 X5 **Gainfarn** Niederösterreich, NE Austria *47°59′N 16°11′E*
97 N18 **Gainsborough** E England, United Kingdom *53°24′N 00°48′W*
182 G6 **Gairdner, Lake** *salt lake* South Australia
Gaissane *see* Gáissát
92 L8 **Gáissát** *var.* Gaissane. ▲▲ N Norway
43 T15 **Gaital, Cerro** ▲ C Panama *08°37′N 80°04′W*
21 W3 **Gaithersburg** Maryland, NE USA *39°08′N 77°13′W*
163 U13 **Gaizhou** Liaoning, NE China *40°24′N 122°17′E*
Gaizina Kalns *see* Gaiziņkalns
118 H7 **Gaiziņkalns** *var.* Gaizina Kalns. ▲ E Latvia *56°51′N 25°58′E*
Gajac *see* Villeneuve-sur-Lot
197 T10 **Gakkel Ridge** *var.* Arctic Mid Oceanic Ridge; *prev.* Nansen Cordillera. *seamount range* Arctic Ocean
39 S10 **Gakona** Alaska, USA *62°21′N 145°16′W*
158 M16 **Gala** Xizang Zizhiqu, China *28°17′N 89°21′E*
Galaassiya *see* Galaosiyo
114 K10 **Galabovo** , Gŭlŭbovo. Stara Zagora, C Bulgaria *42°08′N 25°51′E*
Ğalāğıl *see* Jalājil
Galam, Pulau *see* Gelam, Pulau
62 J6 **Galán, Cerro** ▲ NW Argentina *25°54′S 66°45′W*
111 H21 **Galanta** *Hung.* Galánta. Trnavský Kraj, W Slovakia *48°12′N 17°45′E*
146 L11 **Galaosiyo** *Rus.* Galaassiya. Buxoro Viloyati, C Uzbekistan *39°53′N 64°25′E*
57 B17 **Galápagos** *off.* Provincia de Galápagos. ◈ *province* W Ecuador, E Pacific Ocean
193 P8 **Galapagos Fracture Zone** *tectonic feature* E Pacific Ocean
Galapagos Islands *see* Colón, Archipiélago de
Galápagos, Islas de los *see* Colón, Archipiélago de
Galápagos, Provincia de *see* Galápagos
193 S9 **Galapagos Rise** *undersea feature* E Pacific Ocean *15°00′S 97°00′W*
96 K13 **Galashiels** SE Scotland, United Kingdom *55°37′N 02°49′W*
116 M12 **Galați** *Ger.* Galatz. Galați, E Romania *45°27′N 28°00′E*
116 L12 **Galați** ◈ *county* E Romania
107 Q19 **Galatina** Puglia, SE Italy *40°10′N 18°10′E*
107 Q19 **Galatone** Puglia, SE Italy *40°09′N 18°05′E*
Galatz *see* Galați
21 R8 **Galax** Virginia, NE USA *36°40′N 88°56′W*
146 J16 **Galaýmor** *Rus.* Kala-i-Mor. Mary Welaýaty, S Turkmenistan *35°40′N 62°28′E*
Galcaio *see* Gaalkacyo
64 P11 **Gáldar** Gran Canaria, Islas Canarias, NE Atlantic Ocean *28°09′N 15°40′W*
94 F11 **Galdhøpiggen** ▲ S Norway *61°30′N 08°08′E*
40 I4 **Galeana** Chihuahua, N Mexico *30°08′N 107°38′W*
41 O9 **Galeana** Nuevo León, NE Mexico *24°45′N 99°59′W*
60 P9 **Galeão** ✈ (Rio de Janeiro) Rio de Janeiro, SE Brazil *22°48′S 43°16′W*
171 R10 **Galela** Pulau Halmahera, E Indonesia *01°52′N 127°48′E*
39 O9 **Galena** Alaska, USA *64°43′N 156°55′W*
30 K10 **Galena** Illinois, N USA *42°25′N 90°25′W*
27 R7 **Galena** Kansas, C USA *37°04′N 94°38′W*
27 T8 **Galena** Missouri, C USA *36°45′N 93°30′W*
45 V15 **Galeota Point** *headland* Trinidad, Trinidad and Tobago *10°07′N 60°59′W*
105 P13 **Galera** Andalucía, S Spain *37°45′N 02°33′W*
45 Y16 **Galera Point** *headland* Trinidad, Trinidad and Tobago *10°49′N 60°54′W*
56 A5 **Galera, Punta** *headland* NW Ecuador *0°49′N 80°03′W*
30 K12 **Galesburg** Illinois, N USA *40°57′N 90°22′W*
30 J7 **Galesville** Wisconsin, N USA *44°04′N 91°21′W*
18 F12 **Galeton** Pennsylvania, NE USA *41°43′N 77°38′W*
116 H9 **Gâlgău** *Hung.* Galgó; *prev.* Gîlgău. Sălaj, NW Romania *47°17′N 23°43′E*
Galgó *see* Gâlgău
Galgóc *see* Hlohovec
81 N15 **Galguduud** *off.* Gobolka Galguduud. ◈ *region* E Somalia
Galguduud, Gobolka *see* Galguduud
137 Q9 **Gali** W Georgia *42°40′N 41°39′E*
125 N14 **Galich** Kostromskaya Oblast', NW Russian Federation *58°21′N 42°21′E*
114 H7 **Galiche** Vratsa, NW Bulgaria *43°36′N 23°53′E*
104 H3 **Galicia** *anc.* Gallaecia. ◈ *autonomous community* NW Spain
64 M8 **Galicia Bank** *undersea feature* E Atlantic Ocean *11°45′W 42°40′N*
Galilee *see* HaGalil
181 W7 **Galilee, Lake** ◎ Queensland, NE Australia
Galilee, Sea of *see* Tiberias, Lake
106 E11 **Galileo Galilei** ✈ (Pisa) Toscana, C Italy *43°40′N 10°22′E*
31 S12 **Galion** Ohio, N USA *40°43′N 82°47′W*
Galka'yo *see* Gaalkacyo
146 K12 **Galkynyș** *prev. Rus.* Deynau, Dyanev, *Turkm.* Dänew. Lebap Welaýaty, NE Turkmenistan *39°16′N 63°10′E*
80 H11 **Gallabat** Gedaref, E Sudan *12°57′N 36°10′E*
Gallaecia *see* Galicia
147 O11 **G'allaorol** Jizzax Viloyati, C Uzbekistan *40°01′N 67°30′E*
106 C7 **Gallarate** Lombardia, NW Italy *45°39′N 08°47′E*
27 S2 **Gallatin** Missouri, C USA *39°54′N 93°57′W*
20 J8 **Gallatin** Tennessee, S USA *36°22′N 86°28′W*
33 R11 **Gallatin Peak** ▲ Montana, NW USA *45°22′N 111°21′W*
33 R12 **Gallatin River** ≈ Montana/Wyoming, NW USA
155 J26 **Galle** *prev.* Point de Galle. Southern Province, SW Sri Lanka *06°04′N 80°12′E*
105 S5 **Gállego** ≈ NE Spain
193 Q8 **Gallego Rise** *undersea feature* E Pacific Ocean *02°00′S 115°00′W*
Gallegos *see* Río Gallegos
63 H23 **Gallegos, Río** ≈ Argentina/Chile
Gallia *see* France
22 K10 **Galliano** Louisiana, S USA *29°26′N 90°18′W*
114 G13 **Gallikós** ≈ N Greece
37 S12 **Gallinas Peak** ▲ New Mexico, SW USA *34°14′N 105°47′W*
54 H3 **Gallinas, Punta** *headland* NE Colombia *12°27′N 71°44′W*
37 T11 **Gallinas River** ≈ New Mexico, SW USA
107 Q19 **Gallipoli** Puglia, SE Italy *40°08′N 18°E*
Gallipoli *see* Gelibolu
Gallipoli Peninsula *see* Gelibolu Yarımadası
31 T15 **Gallipolis** Ohio, N USA *38°49′N 82°14′W*
92 J12 **Gällivare** *Lapp.* Váhtjer. Norrbotten, N Sweden *67°08′N 20°39′E*
109 T4 **Gallneukirchen** Oberösterreich, N Austria *48°22′N 14°25′E*
93 G17 **Gällö** Jämtland, C Sweden *62°57′N 15°15′E*
105 Q7 **Gallo** ≈ C Spain
107 I23 **Gallo, Capo** *headland* Sicilia, Italy, C Mediterranean Sea *38°13′N 13°18′E*
37 P13 **Gallo Mountains** ▲▲ New Mexico, SW USA
18 G8 **Galloo Island** *island* New York, NE USA
97 H15 **Galloway, Mull of** *headland* S Scotland, United Kingdom *54°37′N 04°54′W*
37 P10 **Gallup** New Mexico, SW USA *35°32′N 108°45′W*
105 R5 **Gallur** Aragón, NE Spain *41°51′N 01°21′W*
Gâlma *see* Guelma
163 N9 **Galshar** *var.* Buyant. Hentiy, C Mongolia *46°15′N 110°50′E*
162 I6 **Galt** *var.* Ider. Hövsgöl, C Mongolia *48°45′N 99°52′E*
35 O8 **Galt** California, W USA *38°13′N 121°19′W*
74 C10 **Galtat-Zemmour** C Western Sahara *25°07′N 12°21′W*
95 G22 **Galten** Midtjylland, C Denmark *56°09′N 09°54′E*
Gålto *see* Kultsjön
97 D20 **Galtymore Mountain** *Ir.* Cnoc Mór na nGaibhlte. ▲ S Ireland *52°21′N 08°09′W*
97 D20 **Galty Mountains** *Ir.* Na Gaibhlte. ▲▲ S Ireland
30 K11 **Galva** Illinois, N USA *41°10′N 90°02′W*
25 X12 **Galveston** Texas, SW USA *29°17′N 94°48′W*
25 W11 **Galveston Bay** *inlet* Texas, SW USA
25 W12 **Galveston Island** *island* Texas, SW USA
61 B18 **Gálvez** Santa Fe, C Argentina *32°03′S 61°14′W*
97 C18 **Galway** *Ir.* Gaillimh. W Ireland *53°16′N 09°03′W*
97 B18 **Galway** *Ir.* Gaillimh. *cultural region* W Ireland
97 B18 **Galway Bay** *Ir.* Cuan na Gaillimhe. *bay* W Ireland
83 F18 **Gam** Otjozondjupa, NE Namibia *20°10′S 20°51′E*
164 L14 **Gamagōri** Aichi, Honshū, SW Japan *34°49′N 137°15′E*
54 F7 **Gamarra** Cesar, N Colombia *08°21′N 73°46′W*
Gámas *see* Kaamanen
158 L17 **Gamba** Xizang Zizhiqu, W China *28°13′N 88°32′E*
Gamba *see* Zamtang
77 P14 **Gambaga** NE Ghana *10°32′N 00°28′W*
80 G13 **Gambēla** Gambēla Hizboch, W Ethiopia *08°09′N 34°15′E*
81 H14 **Gambēla Hizboch** ◈ *federal region* W Ethiopia
38 K10 **Gambell** Saint Lawrence Island, Alaska, USA *63°44′N 171°41′W*
76 E12 **Gambia** *off.* Republic of The Gambia, The Gambia. ◆ *republic* W Africa
76 I12 **Gambia** *Fr.* Gambie. ≈ W Africa
64 K12 **Gambia Plain** *undersea feature* E Atlantic Ocean
Gambia, Republic of The *see* Gambia
Gambia, The *see* Gambia
Gambie *see* Gambia
31 T13 **Gambier** Ohio, N USA *40°22′N 82°24′W*
191 Y13 **Gambier, Îles** *island group* E French Polynesia
182 G10 **Gambier Islands** *island group* South Australia
79 H19 **Gamboma** Plateaux, E Congo *01°53′S 15°51′E*
79 G16 **Gamboula** Mambéré-Kadéï, SW Central African Republic *04°09′N 15°12′E*
37 P10 **Gamerco** New Mexico, SW USA *35°34′N 108°45′W*
137 V12 **Gamış Dağı** ▲ W Azerbaijan *40°18′N 46°15′E*
Gamlakarleby *see* Kokkola
95 N18 **Gamleby** Kalmar, S Sweden *57°54′N 16°25′E*
Gammelstad *see* Gammelstaden
93 J14 **Gammelstaden** *var.* Gammelstad. Norrbotten, N Sweden *65°38′N 22°05′E*
Gammouda *see* Sidi Bouzid
155 J25 **Gampaha** Western Province, W Sri Lanka *07°05′N 80°00′E*
155 K25 **Gampola** Central Province, C Sri Lanka *07°10′N 80°34′E*
167 S5 **Gâm, Sông** ≈ N Vietnam
92 L7 **Gamvik** Finnmark, N Norway *71°04′N 28°08′E*
150 H13 **Gan** Addu Atoll, C Maldives
Gan *see* Gansu, China
Gan *see* Jiangxi, China
Ganaane *see* Juba
37 O10 **Ganado** Arizona, SW USA *35°42′N 109°31′W*
25 U12 **Ganado** Texas, SW USA *29°02′N 96°30′W*
14 L14 **Gananoque** Ontario, SE Canada *44°21′N 76°11′W*
Ganāveh *see* Bandar-e Gonāveh
137 V11 **Gäncä** *Rus.* Gyandzha; *prev.* Kirovabad, Yelisavetpol. W Azerbaijan *40°42′N 46°23′E*
Ganchi *see* Ghonchí
Gand *see* Gent
82 B13 **Ganda** *var.* Mariano Machado, *Port.* Vila Mariano Machado. Benguela, W Angola *13°02′S 14°40′E*
79 L22 **Gandajika** Kasai-Oriental, S Dem. Rep. Congo *06°42′S 24°01′E*
153 O12 **Gandak** *Nep.* Nārāyāni. ≈ India/Nepal
13 U11 **Gander** Newfoundland and Labrador, SE Canada *48°56′N 54°33′W*
13 U11 **Gander** ✈ Newfoundland and Labrador, E Canada *49°03′N 54°49′W*
100 G11 **Ganderkesee** Niedersachsen, NW Germany *53°01′N 08°33′E*
105 T7 **Gandesa** Cataluña, NE Spain *41°03′N 00°26′E*
154 B10 **Gāndhīdhām** Gujarāt, W India *23°08′N 70°05′E*
154 D10 **Gāndhīnagar** *state capital* Gujarāt, W India *23°12′N 72°37′E*
154 F9 **Gāndhi Sāgar** ◎ C India
105 T11 **Gandia** *prev.* Gandía. Valenciana, E Spain *38°59′N 00°11′W*
Gandía *see* Gandia
159 O10 **Gang** Qinghai, W China
152 G9 **Gangānagar** Rājasthān, NW India *29°54′N 73°56′E*
152 I12 **Gangāpur** Rājasthān, N India *26°30′N 76°49′E*
153 S17 **Ganga Sāgar** West Bengal, NE India *21°39′N 88°05′E*
Gangavathi *see* Gangāwati
155 G17 **Gangāwati** *var.* Gangavathi. Karnātaka, C India *15°26′N 76°35′E*
159 S9 **Gangca** *var.* Shaliuhe. Qinghai, C China *37°21′N 100°09′E*
158 H14 **Gangdisê Shan** *Eng.* Kailas Range. ▲▲ W China
103 Q15 **Ganges** Hérault, S France *43°57′N 03°42′E*
153 P13 **Ganges** *Ben.* Padma. ≈ Bangladesh/India *see also* Padma
Ganges *see* Padma
Ganges Cone *see* Ganges Fan
173 S3 **Ganges Fan** *var.* Ganges Cone. *undersea feature* N Bay of Bengal *12°00′N 87°00′E*
153 U17 **Ganges, Mouths of the** *delta* Bangladesh/India
107 K23 **Gangi** *anc.* Engyum. Sicilia, Italy, C Mediterranean Sea *37°48′N 14°13′E*
163 Y14 **Gangneung** *Jap.* Kōryō; *prev.* Kangnŭng. NE South Korea *37°47′N 128°51′E*
152 K8 **Gangotri** Uttarakhand, N India *30°56′N 79°02′E*
Gangra *see* Çankırı
153 S11 **Gangtok** *state capital* Sikkim, N India *27°20′N 88°39′E*
159 W11 **Gangu** *var.* Daxiangshan. Gansu, C China *34°38′N 105°18′E*
163 U5 **Gan He** ≈ NE China
171 S12 **Gani** Pulau Halmahera, E Indonesia *0°45′S 128°13′E*
161 O12 **Gan Jiang** ≈ S China
163 U11 **Ganjig** *var.* Horqin Zuoyi Houqi. Nei Mongol Zizhiqu, N China *42°53′N 122°22′E*
146 H15 **Gannaly** Ahal Welaýaty, S Turkmenistan *37°02′N 60°43′E*
163 U7 **Gannan** Heilongjiang, NE China *47°58′N 123°36′E*
103 P10 **Gannat** Allier, C France *46°06′N 03°12′E*
33 T14 **Gannett Peak** ▲ Wyoming, C USA *43°10′N 109°39′W*
29 O10 **Gannvalley** South Dakota, N USA *44°01′N 98°59′W*
Ganqu *see* Lhünzhub
109 Y3 **Gänserndorf** Niederösterreich, NE Austria *48°21′N 16°43′E*
Gansos, Lago dos *see* Goose Lake
159 T9 **Gansu** *var.* Gan, Gansu Sheng, Kansu. ◈ *province* N China
Gansu Sheng *see* Gansu
76 K16 **Ganta** *var.* Gahnpa. NE Liberia *07°15′N 08°59′W*
182 H11 **Gantheaume, Cape** *headland* South Australia *36°04′S 137°28′E*
Gantsevichi *see* Hantsavichy
161 Q6 **Ganyu** *var.* Qingkou. Jiangsu, E China *34°52′N 119°11′E*
144 D12 **Ganyushkino** Atyrau, SW Kazakhstan *46°38′N 49°12′E*
161 O12 **Ganzhou** Jiangxi, S China *25°51′N 114°59′E*
Ganzhou *see* Zhangye
77 Q10 **Gao** Gao, E Mali *16°16′N 00°03′E*
77 R10 **Gao** ◈ *region* SE Mali
161 O10 **Gao'an** Jiangxi, S China *28°24′N 115°27′E*
Gaocheng *see* Litang
Gaoleshan *see* Xianfeng
161 R5 **Gaomi** Shandong, C China *36°23′N 119°45′E*
161 N5 **Gaoping** Shanxi, C China *35°51′N 112°55′E*
159 S8 **Gaotai** Gansu, N China *39°22′N 99°44′E*
Gaoth Dobhair *see* Gweedore
77 O14 **Gaoua** SW Burkina Faso *10°18′N 03°12′W*
76 I13 **Gaoual** N Guinea *11°44′N 13°14′W*
161 S14 **Gaoxiong** *var.* Kaohsiung, *Jap.* Takao, Takow. S Taiwan *22°36′N 120°17′E*
161 S14 **Gaoxiong** ✈ S Taiwan *22°26′N 120°32′E*
161 R7 **Gaoyou** *var.* Dayishan. Jiangsu, E China *32°48′N 119°26′E*
161 R7 **Gaoyou Hu** ◎ E China
160 M15 **Gaozhou** Guangdong, S China *21°56′N 110°49′E*
103 T13 **Gap** *anc.* Vapincum. Hautes-Alpes, SE France *44°33′N 06°05′E*
146 E9 **Gaplaňgyr Platosy** *Rus.* Plato Kaplangky. *ridge* Turkmenistan/Uzbekistan
156 G13 **Gar** *var.* Shiquanhe. Xizang Zizhiqu, W China *32°31′N 80°04′E*
Gar *see* Gar Xincun
Garabekevyul *see* Garabekewül
146 L13 **Garabekewül** *Rus.* Garabekevyul, Karabekaul. Lebap Welaýaty, E Turkmenistan *38°31′N 64°04′E*
146 K15 **Garabil Belentligi** *Rus.* Vozvyshennost' Karabil'. ▲▲ S Turkmenistan
146 A8 **Garabogaz** *Rus.* Bekdash. Balkan Welaýaty, NW Turkmenistan *41°33′N 52°33′E*
146 B9 **Garabogaz Aylagy** *Rus.* Zaliv Kara-Bogaz-Gol. *bay* NW Turkmenistan
146 A9 **Garabogazköl** *Rus.* Kara-Bogaz-Gol. Balkan Welaýaty, NW Turkmenistan *41°03′N 52°52′E*
43 V16 **Garachiné** Darién, SE Panama *08°03′N 78°22′W*
43 V16 **Garachiné, Punta** *headland* SE Panama *08°05′N 78°23′W*
146 K12 **Garagan** *Rus.* Karagan. Ahal Welaýaty, C Turkmenistan *38°16′N 57°34′E*
54 G10 **Garagoa** Boyacá, C Colombia *05°05′N 73°20′W*
146 A11 **Garagöl'** *Rus.* Karagel'. Balkan Welaýaty, W Turkmenistan *39°24′N 53°13′E*
146 F12 **Garagum** *var.* Garagumy, Qara Qum, *Eng.* Black Sand Desert, Kara Kum; *prev.* Peski Karakumy. *desert* C Turkmenistan
146 E12 **Garagum Kanaly** *var.* Kara Kum Canal, *Rus.* Karagumskiy Kanal, Karakumskiy Kanal. *canal* C Turkmenistan
Garagumy *see* Garagum
183 S4 **Garah** New South Wales, SE Australia *29°07′S 149°37′E*
64 O11 **Garajonay** ▲ Gomera, Islas Canarias, NE Atlantic Ocean *28°07′N 17°14′W*
76 L13 **Garalo** Sikasso, SW Mali *10°58′N 07°26′W*
Garam *see* Hron, Slovakia
146 L14 **Garamätnyýaz** *Rus.* Karamet-Niyaz. Lebap Welaýaty, E Turkmenistan *37°45′N 64°28′E*
Garamszentkereszt *see* Žiar nad Hronom
77 Q13 **Garango** S Burkina Faso *11°45′N 00°30′W*
59 Q15 **Garanhuns** Pernambuco, E Brazil *08°53′S 36°28′W*
188 H5 **Garapan** Saipan, S Northern Mariana Islands *15°12′S 145°43′E*
Gárasavvon *see* Karesuando
Gárassavon *see* Kaaresuvanto
78 J13 **Garba** Bamingui-Bangoran, N Central African Republic *09°09′N 20°24′E*
Garba *see* Jiulong
81 L16 **Garbahaarrey** *It.* Garba Harre. Gedo, SW Somalia *03°14′N 42°18′E*
Garba Harre *see* Garbahaarrey
81 J18 **Garba Tula** Isiolo, C Kenya *0°31′N 38°35′E*
27 N9 **Garber** Oklahoma, C USA *36°26′N 97°35′W*
34 L4 **Garberville** California, W USA *40°07′N 123°48′W*
Garbo *see* Lhozhag
100 I12 **Garbsen** Niedersachsen, N Germany *52°25′N 09°36′E*
60 K9 **Garça** São Paulo, S Brazil *22°14′S 49°36′W*
104 L10 **García de Solá, Embalse de** ⊠ C Spain
103 Q14 **Gard** ◈ *department* S France
103 Q14 **Gard** ≈ S France
106 F7 **Garda, Lago di** *var.* Benaco, *Eng.* Lake Garda, *Ger.* Gardasee. ◎ NE Italy
Garda, Lake *see* Garda, Lago di
Gardan Divāl *see* Gardan Dīwāl
149 Q5 **Gardan Dīwāl** *var.* Gardan Divāl. Wardak, C Afghanistan *34°30′N 68°15′E*
103 S15 **Gardanne** Bouches-du-Rhône, SE France *43°27′N 05°28′E*
Gardasee *see* Garda, Lago di
100 L12 **Gardelegen** Sachsen-Anhalt, C Germany *52°31′N 11°25′E*
14 B10 **Garden** ≈ Ontario, S Canada
23 X6 **Garden City** Georgia, SE USA *32°06′N 81°09′W*
26 I6 **Garden City** Kansas, C USA *37°57′N 100°54′W*
27 S5 **Garden City** Missouri, C USA *38°34′N 94°12′W*
25 N8 **Garden City** Texas, SW USA *31°51′N 101°30′W*
23 P3 **Gardendale** Alabama, S USA *33°39′N 86°48′W*
31 P5 **Garden Island** *island* Michigan, N USA
22 M11 **Garden Island Bay** *bay* Louisiana, S USA
31 O5 **Garden Peninsula** *peninsula* Michigan, N USA
Garden State, The *see* New Jersey
95 I14 **Gardermoen** Akershus, S Norway *60°10′N 11°04′E*
95 I14 **Gardermoen** ✈ (Oslo) Akershus, S Norway *60°12′N 11°05′E*
Gardeyz/Gardez *see* Gardēz
149 Q6 **Gardēz** *var.* Gardeyz, Gordiaz; *prev.* Gardīz. Paktiyā, E Afghanistan *33°35′N 69°14′E*
93 G14 **Gardiken** ◎ N Sweden
19 Q7 **Gardiner** Maine, NE USA *44°13′N 69°46′W*
33 S12 **Gardiner** Montana, NW USA *45°02′N 110°42′W*
19 N13 **Gardiners Island** *island* New York, NE USA
Gardner Island *see* Nikumaroro
19 T6 **Gardner Lake** ◎ Maine, NE USA
35 Q6 **Gardnerville** Nevada, W USA *38°55′N 119°44′W*
Gardo *see* Qardho
106 F7 **Gardone Val Trompia** Lombardia, N Italy *45°40′N 10°11′E*
Garegegasnjárga *see* Karigasniemi
38 F17 **Gareloi Island** *island* Aleutian Islands, Alaska, USA
Gares *see* Puente la Reina
106 B10 **Garessio** Piemonte, NE Italy *44°14′N 08°01′E*
32 M9 **Garfield** Washington, NW USA *47°00′N 117°07′W*
31 U11 **Garfield Heights** Ohio, N USA *41°25′N 81°36′W*
Gargaliani *see* Gargaliánoi
115 D21 **Gargaliánoi** *var.* Gargaliani. Pelopónnisos, S Greece *37°04′N 21°38′E*
107 N15 **Gargàno, Promontorio del** *headland* SE Italy *41°51′N 16°11′E*
108 J8 **Gargellen** Graubünden, W Switzerland *46°57′N 09°55′E*
93 I14 **Gargnäs** Västerbotten, N Sweden *65°19′N 18°00′E*
118 C11 **Gargždai** Klaipėda, W Lithuania *55°42′N 21°24′E*
154 J13 **Garhchiroli** Mahārāshtra, C India *20°14′N 79°58′E*
153 O15 **Garhwa** Jhārkhand, N India *24°07′N 83°52′E*
171 V13 **Gariau** Papua Barat, E Indonesia *03°43′S 134°54′E*
83 E24 **Garies** Northern Cape, W South Africa *30°30′S 18°00′E*
107 K17 **Garigliano** ≈ C Italy
81 K19 **Garissa** Garissa, E Kenya *0°27′S 39°39′E*
81 K19 **Garissa** ◈ *county* SE Kenya
21 V11 **Garland** North Carolina, SE USA *34°45′N 78°25′W*
25 T6 **Garland** Texas, SW USA *32°50′N 96°37′W*
36 L1 **Garland** Utah, W USA *41°43′N 112°07′W*
106 D8 **Garlasco** Lombardia, N Italy *45°12′N 08°59′E*
119 F14 **Garliava** Kaunas, S Lithuania *54°49′N 23°52′E*
Garm *see* Gharm
142 M9 **Garm, Āb-e** *var.* Rūd-e Khersān. ≈ SW Iran
Garmik *see* Germik
101 K25 **Garmisch-Partenkirchen** Bayern, S Germany *47°30′N 11°05′E*
143 O5 **Garmsār** *prev.* Qishlaq. Semnān, N Iran *35°18′N 52°22′E*
Garmser *see* Darwēshān
29 V12 **Garner** Iowa, C USA *43°06′N 93°36′W*
21 U9 **Garner** North Carolina, SE USA *35°42′N 78°36′W*
27 Q5 **Garnett** Kansas, C USA *38°16′N 95°15′W*
99 M25 **Garnich** Luxembourg, SW Luxembourg *49°37′N 05°57′E*
182 M8 **Garnpung, Lake** *salt lake* New South Wales, SE Australia
Garoe *see* Garoowe
Garoet *see* Garut
153 U13 **Gāro Hills** *hill range* NE India
102 K13 **Garonne** *anc.* Garumna. ≈ S France
80 P13 **Garoowe** *var.* Garoe. Nugaal, N Somalia *08°24′N 48°29′E*
78 F12 **Garoua** *var.* Garua. Nord, N Cameroon *09°17′N 13°22′E*
79 G14 **Garoua Boulaï** Est, E Cameroon *05°54′N 14°33′E*
77 O10 **Garou, Lac** ◎ C Mali
95 L16 **Garphyttan** Örebro, C Sweden *59°18′N 14°54′E*
29 R11 **Garretson** South Dakota, N USA *43°43′N 96°30′W*
31 Q11 **Garrett** Indiana, N USA *41°21′N 85°08′W*
33 Q10 **Garrison** Montana, NW USA *46°32′N 112°46′W*
28 M4 **Garrison** North Dakota, N USA *47°36′N 101°25′W*
25 X8 **Garrison** Texas, SW USA *31°49′N 94°29′W*
28 L4 **Garrison Dam** *dam* North Dakota, N USA
104 J9 **Garrovillas** Extremadura, W Spain *39°43′N 06°33′W*
Garrygala *see* Magtymguly
8 L8 **Garry Lake** ◎ Nunavut, N Canada
Gars *see* Gars am Kamp
109 W3 **Gars am Kamp** *var.* Gars. Niederösterreich, NE Austria *48°35′N 15°40′E*
81 K20 **Garsen** Tana River, S Kenya *02°16′S 40°07′E*
Garshy *see* Garșy
14 F10 **Garson** Ontario, S Canada *46°33′N 80°51′W*
109 T5 **Garsten** Oberösterreich, N Austria *48°00′N 14°24′E*
146 A9 **Garșy** *var.* Garshy, *Rus.* Karshi. Balkan Welaýaty, NW Turkmenistan *40°45′N 52°50′E*
Gartar *see* Qianning
102 M10 **Gartempe** ≈ C France
Gartog *see* Markam
Garua *see* Garoua
83 D21 **Garub** Karas, SW Namibia *26°33′S 16°00′E*
Garumna *see* Garonne
169 P16 **Garut** *prev.* Garoet. Jawa, C Indonesia *07°15′S 107°55′E*
185 C20 **Garvie Mountains** ▲▲ South Island, New Zealand
110 N12 **Garwolin** Mazowieckie, E Poland *51°54′N 21°36′E*
25 U12 **Garwood** Texas, SW USA *29°25′N 96°26′W*
158 G13 **Gar Xincun** *prev.* Gar. Xizang Zizhiqu, W China *32°04′N 80°01′E*
31 N11 **Gary** Indiana, N USA *41°36′N 87°21′W*
25 X7 **Gary** Texas, SW USA *32°00′N 94°21′W*
158 G13 **Garyarsa** Xizang Zizhiqu, SW China *31°44′N 80°20′E*
158 G13 **Gar Zangbo** ≈ W China
160 F8 **Garzê** Sichuan, C China *31°40′N 99°58′E*
54 E12 **Garzón** Huila, S Colombia *02°14′N 75°37′W*
Gasan-Kuli *see* Esenguly
31 P13 **Gas City** Indiana, N USA *40°29′N 85°36′W*
102 K15 **Gascogne** *Eng.* Gascony. *cultural region* S France
26 V5 **Gasconade River** ≈ Missouri, C USA
Gascony *see* Gascogne
180 H9 **Gascoyne Junction** Western Australia *25°06′S 115°10′E*
173 V8 **Gascoyne Plain** *undersea feature* E Indian Ocean
180 H9 **Gascoyne River** ≈ Western Australia
192 J11 **Gascoyne Tablemount** *undersea feature* N Tasman Sea *36°30′S 156°30′E*
67 U6 **Gash** *var.* Nahr al Qāsh. ≈ W Sudan
149 X3 **Gasherbrum** ▲ NE Pakistan *35°39′N 76°34′E*
Gas Hu *see* Gas Hure Hu
77 X12 **Gashua** Yobe, NE Nigeria *12°55′N 11°10′E*
159 N9 **Gas Hure Hu** *var.* Gas Hu. ◎ C China
Gásluokta *see* Kjøpsvik
186 G7 **Gasmata** New Britain, E Papua New Guinea *06°12′S 150°25′E*
23 V14 **Gasparilla Island** *island* Florida, SE USA
169 O13 **Gaspar, Selat** *strait* W Indonesia
15 Y6 **Gaspé** Québec, SE Canada *48°50′N 64°33′W*
15 Z6 **Gaspé, Cap de** *headland* Québec, SE Canada *48°45′N 64°10′W*
15 X6 **Gaspé, Péninsule de** *var.* Péninsule de la Gaspésie. *peninsula* Québec, SE Canada
Gaspésie, Péninsule de la *see* Gaspé, Péninsule de
77 W15 **Gassol** Taraba, E Nigeria *08°28′N 10°24′E*
Gastein *see* Badgastein
21 R10 **Gastonia** North Carolina, SE USA *35°14′N 81°12′W*
21 V8 **Gaston, Lake** ⊠ North Carolina/Virginia, SE USA
115 D19 **Gastoúni** Dytikí Elláda, S Greece *37°51′N 21°15′E*
63 I17 **Gastre** Chubut, S Argentina *42°20′S 69°10′W*
Gat *see* Ghāt
105 P15 **Gata, Cabo de** *cape* S Spain
Gata, Cape *see* Gátas, Akrotíri
105 T11 **Gata de Gorgos** Valenciana, E Spain *38°45′N 00°06′E*
116 E12 **Gătaia** *Ger.* Gataja, *Hung.* Gátalja; *prev.* Gáttája. Timiș, W Romania *45°24′N 21°26′E*
Gataja/Gátalja *see* Gătaia
121 P3 **Gátas, Akrotíri** *var.* Cape Gata. *headland* S Cyprus *34°34′N 33°03′E*
104 J8 **Gata, Sierra de** ▲▲ W Spain
124 G13 **Gatchina** Leningradskaya Oblast', NW Russian Federation *59°34′N 30°06′E*
21 P8 **Gate City** Virginia, NE USA *36°38′N 82°37′W*
97 M14 **Gateshead** NE England, United Kingdom *54°57′N 01°37′W*
21 X8 **Gatesville** North Carolina, SE USA *36°24′N 76°46′W*
25 S8 **Gatesville** Texas, SW USA *31°26′N 97°46′W*
14 L12 **Gatineau** Québec, SE Canada *45°29′N 75°40′W*
14 L11 **Gatineau** ≈ Ontario/Québec, SE Canada
21 N9 **Gatlinburg** Tennessee, S USA *35°42′N 83°30′W*
Gatooma *see* Kadoma
Gáttája *see* Gătaia
43 T14 **Gatún, Lago** ⊠ C Panama
59 N14 **Gaturiano** Piauí, NE Brazil *06°53′S 41°45′W*
97 O22 **Gatwick** ✈ (London) SE England, United Kingdom *51°10′N 00°12′W*
187 Y14 **Gau** *prev.* Ngau. *island* C Fiji
187 R12 **Gaua** *var.* Santa Maria. *Island* Banks Island, N Vanuatu
104 L16 **Gaucín** Andalucía, S Spain *36°31′N 05°19′W*
Gauhāti *see* Guwāhāti
118 I8 **Gauja** *Ger.* Aa. ≈ Estonia/Latvia
118 I7 **Gaujiena** NE Latvia *57°31′N 26°24′E*
94 H9 **Gauldalen** *valley* S Norway
21 R5 **Gauley River** ≈ West Virginia, NE USA
Gaul/Gaule *see* France
99 D19 **Gaurain-Ramecroix** Hainaut, SW Belgium *50°35′N 03°31′E*
95 F15 **Gaustatoppen** ▲ S Norway *59°50′N 08°39′E*
83 J21 **Gauteng** *off.* Gauteng Province; *prev.* Pretoria-Witwatersrand-Vereeniging. ◈ *province* NE South Africa
Gauteng *see* Johannesburg, South Africa
Gauteng *see* Germiston, South Africa
Gauteng Province *see* Gauteng
137 U11 **Gavarr** *prev.* Kamo. C Armenia *40°21′N 45°07′E*
143 P14 **Gāvbandī** Hormozgān, S Iran *27°07′N 53°21′E*
115 H25 **Gavdopoúla** *island* SE Greece
115 H26 **Gávdos** *island* SE Greece
102 K16 **Gave de Pau** *var.* Gave-de-Pay. ≈ SW France
Gave-de-Pay *see* Gave de Pau
102 J16 **Gave d'Oloron** ≈ SW France
99 E18 **Gavere** Oost-Vlaanderen, NW Belgium *50°56′N 03°41′E*
94 N13 **Gävle** *var.* Gäfle; *prev.* Gefle. Gävleborg, C Sweden *60°41′N 17°09′E*
94 M11 **Gävleborg** *var.* Gäfleborg, Gefleborg. ◈ *county* C Sweden
94 O13 **Gävlebukten** *bay* C Sweden
124 L16 **Gavrilov-Yam** Yaroslavskaya Oblast', W Russian Federation *57°19′N 39°52′E*
182 I9 **Gawler** South Australia *34°38′S 138°44′E*
182 G7 **Gawler Ranges** *hill range* South Australia
Gawso *see* Goaso
162 H11 **Gaxun Nur** ◎ N China
153 P14 **Gaya** Bihār, N India *24°48′N 85°E*
77 S13 **Gaya** Dosso, SW Niger *11°52′N 03°28′E*
Gaya *see* Kyjov
31 Q6 **Gaylord** Michigan, N USA *45°01′N 84°40′W*
29 U9 **Gaylord** Minnesota, N USA *44°33′N 94°13′W*
181 Y9 **Gayndah** Queensland, E Australia *25°37′S 151°31′E*
125 T12 **Gayny** Komi-Permyatskiy Okrug, NW Russian Federation *60°19′N 54°15′E*
Gaysin *see* Haysyn
Gayvorno *see* Hayvoron
138 E11 **Gaza** *Ar.* Ghazzah, *Heb.* 'Azza. NE Gaza Strip *31°30′N 34°E*
83 L20 **Gaza** *off.* Província de Gaza. ◈ *province* SW Mozambique
Gaz-Achak *see* Gazojak
147 Q9 **G'azalkent** *Rus.* Gazalkent. Toshkent Viloyati, E Uzbekistan *41°30′N 69°46′E*
Gazalkent *see* G'azalkent
Gazandzhyk/Gazanjyk *see* Bereket
77 V12 **Gazaoua** Maradi, S Niger *13°28′N 07°54′E*
Gaza, Província de *see* Gaza
138 E11 **Gaza Strip** *Ar.* Qita Ghazzah. *disputed region* SW Asia
136 M16 **Gaziantep** *var.* Gazi Antep; *prev.* Aintab, Antep. Gaziantep, S Turkey *37°04′N 37°21′E*
136 M17 **Gaziantep** *var.* Gazi Antep. ◈ *province* S Turkey
Gazi Antep *see* Gaziantep
114 M13 **Gaziköy** Tekirdağ, NW Turkey *40°45′N 27°18′E*
121 Q2 **Gazimağusa** *var.* Famagusta, *Gk.* Ammóchostos. E Cyprus *35°07′N 33°57′E*
121 Q2 **Gazimağusa Körfezi** *var.* Famagusta Bay, *Gk.* Kólpos Ammóchostos. *bay* E Cyprus
146 K11 **Gazli** Buxoro Viloyati, C Uzbekistan *40°09′N 63°28′E*
146 I9 **Gazojak** *Rus.* Gaz-Achak. Lebap Welaýaty, NE Turkmenistan *41°12′N 61°24′E*
79 K15 **Gbadolite** Equateur, NW Dem. Rep. Congo *04°14′N 20°59′E*
76 K16 **Gbanga** *var.* Gbarnga. N Liberia *07°02′N 09°30′W*
Gbarnga *see* Gbanga
77 S14 **Gbéroubouè** *var.* Béroubouay. N Benin *10°35′N 02°47′E*
77 W16 **Gboko** Benue, S Nigeria *07°21′N 08°57′E*
Gcuwa *see* Butterworth
110 J7 **Gdańsk** *Fr.* Dantzig, *Ger.* Danzig. Pomorskie, N Poland *54°22′N 18°35′E*
Gdan'skaya Bukhta/Gdańsk, Gulf of *see* Danzig, Gulf of
Gdańska, Zakota *see* Danzig, Gulf of
Gdingen *see* Gdynia
124 F13 **Gdov** Pskovskaya Oblast', W Russian Federation *58°43′N 27°51′E*
110 I6 **Gdynia** *Ger.* Gdingen. Pomorskie, N Poland *54°31′N 18°30′E*
26 M10 **Geary** Oklahoma, C USA *35°37′N 98°19′W*
Geavvú *see* Kevo
76 H12 **Gêba, Rio** ≈ C Guinea-Bissau
136 E11 **Gebze** Kocaeli, NW Turkey *40°48′N 29°26′E*
80 H10 **Gedaref** *var.* Al Qaḑārif, El Gedaref. Gedaref, E Sudan *14°03′N 35°24′E*
80 H10 **Gedaref** ◈ *state* E Sudan

◆ Country
● Country Capital
◇ Dependent Territory
○ Dependent Territory Capital
◈ Administrative Regions
✈ International Airport
▲ Mountain
▲▲ Mountain Range
⛰ Volcano
≈ River
◎ Lake
⊠ Reservoir

GEDID RAS EL FIL – GLÁMA

80 B11 **Gedid Ras el Fil** Southern Darfur, W Sudan 12°45′N 25°45′E
99 I23 **Gedinne** Namur, SE Belgium 49°57′N 04°55′E
136 E13 **Gediz** Kütahya, W Turkey 39°04′N 29°25′E
136 C14 **Gediz Nehri** ~ W Turkey
81 N14 **Gedlegubē** Sumalē, E Ethiopia 06°53′N 45°08′E
81 L17 **Gedo** *off.* Gobolka Gedo. ◈ *region* SW Somalia
Gedo, Gobolka *see* Gedo
95 I25 **Gedser** Sjælland, SE Denmark 54°34′N 11°57′E
99 I16 **Geel** *var.* Gheel. Antwerpen, N Belgium 51°10′N 04°59′E
183 N13 **Geelong** Victoria, SE Australia 38°10′S 144°21′E
Ge'e'mu *see* Golmud
99 I14 **Geertruidenberg** Noord-Brabant, S Netherlands 51°43′N 04°52′E
100 H10 **Geeste** ~ NW Germany
100 J10 **Geesthacht** Schleswig-Holstein, N Germany 53°25′N 10°22′E
183 P17 **Geeveston** Tasmania, SE Australia 43°12′S 146°54′E
Gefle *see* Gävle
Gefleborg *see* Gävleborg
163 S5 **Gegan Gol** *prev.* Ergun, Gen He, Zuoqi. NE China
163 T5 **Gegen Gol** *prev.* Ergun Zuoqi, Genhe. Nei Mongol Zizhiqu, N China 50°48′N 121°30′E
158 G13 **Gê'gyai** Xizang Zizhiqu, W China 32°29′N 81°04′E
77 X12 **Geidam** Yobe, NE Nigeria 12°52′N 11°55′E
11 T11 **Geikie** ~ Saskatchewan, C Canada
94 F13 **Geilo** Buskerud, S Norway 60°32′N 08°13′E
94 E10 **Geiranger** Møre og Romsdal, S Norway 62°07′N 07°12′E
101 I22 **Geislingen** *var.* Geislingen an der Steige. Baden-Württemberg, SW Germany 48°37′N 09°50′E
Geislingen an der Steige *see* Geislingen
81 F20 **Geita** Geita, NW Tanzania 02°52′S 32°12′E
81 F21 **Geita** *off.* Mkoa wa Geita. ◈ *region* N Tanzania
Geita, Mkoa wa *see* Geita
95 G15 **Geithus** Buskerud, S Norway 59°56′N 09°58′E
160 H14 **Gejiu** *var.* Kochiu. Yunnan, S China 23°22′N 103°07′E
Gëkdepe *see* Gökdepe
146 E9 **Geklengkui, Solonchak** *var.* Solonchak Goklenkuy. *salt marsh* NW Turkmenistan
81 D14 **Gel** ~ C South Sudan
107 K25 **Gela** *prev.* Terranova di Sicilia. Sicilia, Italy, C Mediterranean Sea 37°05′N 14°15′E
81 N14 **Geladī** SE Ethiopia 06°58′N 46°24′E
169 P13 **Gelam, Pulau** *var.* Pulau Galam. *island* N Indonesia
Gelaozu Miaozu Zhizhixian *see* Wuchuan
98 L11 **Gelderland** *prev. Eng.* Guelders. ◈ *province* E Netherlands
98 J13 **Geldermalsen** Gelderland, C Netherlands 51°53′N 05°17′E
101 D14 **Geldern** Nordrhein-Westfalen, W Germany 51°31′N 06°19′E
99 K15 **Geldrop** Noord-Brabant, SE Netherlands 51°25′N 05°34′E
99 L17 **Geleen** Limburg, SE Netherlands 50°57′N 05°49′E
126 K14 **Gelendzhik** Krasnodarskiy Kray, SW Russian Federation 44°34′N 38°06′E
Gelib *see* Jilib
136 B11 **Gelibolu** *Eng.* Gallipoli. Çanakkale, NW Turkey 40°25′N 26°41′E
115 L14 **Gelibolu Yarımadası** *Eng.* Gallipoli Peninsula. *peninsula* NW Turkey
81 O14 **Gellinsor** Galguduud, C Somalia 06°25′N 46°44′E
101 H18 **Gelnhausen** Hessen, C Germany 50°12′N 09°12′E
101 E14 **Gelsenkirchen** Nordrhein-Westfalen, W Germany 51°30′N 07°05′E
83 C20 **Geluk** Hardap, SW Namibia 24°35′S 15°48′E
99 H20 **Gembloux** Namur, Belgium 50°34′N 04°42′E
79 J16 **Gemena** Equateur, NW Dem. Rep. Congo 03°13′N 19°49′E
99 L14 **Gemert** Noord-Brabant, SE Netherlands 51°33′N 05°41′E
136 E11 **Gemlik** Bursa, NW Turkey 40°26′N 29°10′E
Gem of the Mountains *see* Idaho
106 J6 **Gemona del Friuli** Friuli-Venezia Giulia, NE Italy 46°18′N 13°12′E
Gem State *see* Idaho
Genalē Wenz *see* Juba
169 R10 **Genali, Danau** ◎ Borneo, N Indonesia
99 G19 **Genappe** Walloon Brabant, C Belgium 50°39′N 04°27′E
137 P14 **Genç** Bingöl, E Turkey 38°44′N 40°35′E
Genck *see* Genk
98 M9 **Genemuiden** Overijssel, E Netherlands 52°38′N 06°03′E
63 K14 **General Acha** La Pampa, C Argentina 37°25′S 64°38′W
61 C21 **General Alvear** Buenos Aires, E Argentina 36°03′S 60°01′W
62 I12 **General Alvear** Mendoza, W Argentina 34°59′S 67°40′W
61 B20 **General Arenales** Buenos Aires, E Argentina 34°21′S 61°20′W
61 D21 **General Belgrano** Buenos Aires, E Argentina 35°47′S 58°30′W
194 H3 **General Bernardo O'Higgins** *Chilean research station* Antarctica 63°09′S 57°13′W
41 O8 **General Bravo** Nuevo León, NE Mexico 25°47′N 99°04′W
62 M7 **General Capdevila** Chaco, N Argentina 27°25′S 61°30′W
General Carrera, Lago *see* Buenos Aires, Lago
41 N9 **General Cepeda** Coahuila, NE Mexico 25°18′N 101°24′W
63 K15 **General Conesa** Río Negro, E Argentina 40°06′S 64°26′W
61 G18 **General Enrique Martínez** Treinta y Tres, E Uruguay 33°13′S 53°47′W
62 L3 **General Eugenio A. Garay** *var.* Fortín General Eugenio Garay; *prev.* Yrendagüé. Nueva Asunción, NW Paraguay 20°30′S 61°56′W
61 C18 **General Galarza** Entre Ríos, E Argentina 32°43′S 59°24′W
61 E22 **General Guido** Buenos Aires, E Argentina 36°36′S 57°45′W
General José F.Uriburu *see* Zárate
61 E22 **General Juan Madariaga** Buenos Aires, E Argentina 37°00′S 57°09′W
41 O16 **General Juan N Alvarez** ✈ (Acapulco) Guerrero, S Mexico 16°47′N 99°47′W
61 B22 **General La Madrid** Buenos Aires, E Argentina 37°17′S 61°20′W
61 E21 **General Lavalle** Buenos Aires, E Argentina 36°25′S 56°56′W
General Machado *see* Camacupa
62 I8 **General Manuel Belgrano, Cerro** ▲ W Argentina 29°05′S 67°05′W
41 O8 **General Mariano Escobero** ✈ (Monterrey) Nuevo León, NE Mexico 25°47′N 100°00′W
61 B20 **General O'Brien** Buenos Aires, E Argentina 34°54′S 60°45′W
62 K13 **General Pico** La Pampa, C Argentina 35°43′S 63°45′W
62 M7 **General Pinedo** Chaco, N Argentina 27°17′S 61°20′W
61 B20 **General Pinto** Buenos Aires, E Argentina 34°45′S 61°50′W
61 E22 **General Pirán** Buenos Aires, E Argentina 37°16′S 57°46′W
43 N15 **General, Río** ~ S Costa Rica
63 I15 **General Roca** Río Negro, C Argentina 39°00′S 67°35′W
171 Q8 **General Santos** *off.* General Santos City. Mindanao, S Philippines 06°10′N 125°10′E
General Santos City *see* General Santos
41 O9 **General Terán** Nuevo León, NE Mexico 25°18′N 99°40′W
114 N7 **General Toshevo** *Rom.* I.G.Duca; *prev.* Casim, Kasimköj. Dobrich, NE Bulgaria 43°43′N 28°04′E
61 B20 **General Viamonte** Buenos Aires, E Argentina 35°01′S 61°00′W
61 A20 **General Villegas** Buenos Aires, E Argentina 35°02′S 63°01′W
Gênes *see* Genova
18 E11 **Genesee River** ~ New York/Pennsylvania, NE USA
30 K11 **Geneseo** Illinois, N USA 41°27′N 90°08′W
18 F10 **Geneseo** New York, NE USA 42°48′N 77°46′W
57 L14 **Geneshuaya, Río** ~ N Bolivia
23 Q8 **Geneva** Alabama, S USA 31°01′N 85°51′W
30 M10 **Geneva** Illinois, N USA 41°53′N 88°18′W
29 Q16 **Geneva** Nebraska, C USA 40°31′N 97°36′W
18 G10 **Geneva** New York, NE USA 42°52′N 76°58′W
31 U10 **Geneva** Ohio, NE USA 41°48′N 80°53′W
Geneva *see* Genève
108 B10 **Geneva, Lake** *Fr.* Lac de Genève, Lac Léman, le Léman, *Ger.* Genfer See. ◎ France/Switzerland
108 A10 **Genève** *Eng.* Geneva, *Ger.* Genf, *It.* Ginevra. Genève, SW Switzerland 46°13′N 06°09′E
108 A11 **Genève** *Eng.* Geneva, *Ger.* Genf, *It.* Ginevra. ◈ *canton* SW Switzerland
108 A10 **Genève** *var.* Geneva. ✈ Vaud, SW Switzerland 46°13′N 06°06′E
Genève, Lac de *see* Geneva, Lake
Genf *see* Genève
Genfer See *see* Geneva, Lake
Gen He *see* Gegan Gol
Genhe *see* Gegen Gol
Genichesk *see* Heniches'k
104 L14 **Genil** ~ S Spain
99 K18 **Genk** *var.* Genck. Limburg, NE Belgium 50°58′N 05°30′E
164 C13 **Genkai-nada** *gulf* Kyūshū, SW Japan
107 C19 **Gennargentu, Monti del** ▲▲ Sardegna, Italy, C Mediterranean Sea
99 M14 **Gennep** Limburg, SE Netherlands 51°43′N 05°58′E
30 M10 **Genoa** Illinois, N USA 42°06′N 88°41′W
29 Q15 **Genoa** Nebraska, C USA 41°27′N 97°43′W
Genoa *see* Genova
Genoa, Gulf of *see* Genova, Golfo di
106 D10 **Genova** *Eng.* Genoa, *Fr.* Gênes; *anc.* Genua. Liguria, NW Italy 44°28′N 09°E
106 D10 **Genova, Golfo di** *Eng.* Gulf of Genoa. *gulf* NW Italy
57 C17 **Genovesa, Isla** *var.* Tower Island. *island* Galapagos Islands, Ecuador, E Pacific Ocean
Genshū *see* Wonju
99 E17 **Gent** *Eng.* Ghent, *Fr.* Gand. Oost-Vlaanderen, NW Belgium 51°02′N 03°42′E
169 N16 **Genteng** Jawa, C Indonesia 07°21′S 106°20′E
100 M12 **Genthin** Sachsen-Anhalt, E Germany 52°24′N 12°10′E
27 R9 **Gentry** Arkansas, C USA 36°16′N 94°28′W
Genua *see* Genova
107 I15 **Genzano di Roma** Lazio, C Italy 41°42′N 12°42′E
163 Y17 **Geogeum-do** *prev.* Kŏgŭm-do. *island* S South Korea
163 Z16 **Geogeum-do** *Jap.* Kyōsai-tō; *prev.* Kŏje-do. *island* S South Korea
Geokchay *see* Göyçay
Geok-Tepe *see* Gökdepe
122 I3 **Georga, Zemlya** *Eng.* George Land. *island* Zemlya Frantsa-Iosifa, N Russian Federation
83 G26 **George** Western Cape, S South Africa 33°57′S 22°28′E
29 S11 **George** Iowa, C USA 43°20′N 96°00′W
13 O5 **George** ~ Newfoundland and Labrador/Québec, E Canada
George F L Charles *see* Vigie
65 C25 **George Island** *island* S Falkland Islands
183 R10 **George, Lake** ◎ New South Wales, SE Australia
81 E18 **George, Lake** ◎ SW Uganda
23 W10 **George, Lake** ◎ Florida, SE USA
18 L8 **George, Lake** ⊠ New York, NE USA
George Land *see* Georga, Zemlya
Georgenburg *see* Jurbarkas
George River *see* Kangiqsualujjuaq
64 G8 **Georges Bank** *undersea feature* W Atlantic Ocean 41°15′N 67°30′W
185 A21 **George Sound** *sound* South Island, New Zealand
65 F15 **Georgetown** ○ (Ascension Island) NW Ascension Island 17°56′S 14°25′W
181 V5 **Georgetown** Queensland, NE Australia 18°17′S 143°37′E
183 P15 **George Town** Tasmania, SE Australia 41°04′S 146°48′E
44 D8 **George Town** *var.* Georgetown. ○ (Cayman Islands) Grand Cayman, SW Cayman Islands 19°16′N 81°23′W
76 H12 **Georgetown** E Gambia 13°33′N 14°49′W
55 T8 **Georgetown** ● (Guyana) N Guyana 06°46′N 58°10′W
168 I7 **George Town** *var.* Penang, Pinang. Pinang, Peninsular Malaysia 05°28′N 100°20′E
45 Y14 **Georgetown** Saint Vincent, Saint Vincent and the Grenadines 13°19′N 61°09′W
44 I4 **George Town** Great Exuma Island, C The Bahamas 23°28′N 75°47′W
21 Y4 **Georgetown** Delaware, NE USA 38°39′N 75°22′W
23 R6 **Georgetown** Georgia, SE USA 31°52′N 85°04′W
20 M5 **Georgetown** Kentucky, S USA 38°13′N 84°30′W
21 T13 **Georgetown** South Carolina, SE USA 33°23′N 79°18′W
25 S10 **Georgetown** Texas, SW USA 30°39′N 97°42′W
55 T8 **Georgetown** ✈ N Guyana 06°46′N 58°10′W
Georgetown *see* George Town
195 U16 **George V Coast** *physical region* Antarctica
194 J7 **George VI Ice Shelf** *ice shelf* Antarctica
194 J6 **George VI Sound** *sound* Antarctica
195 T15 **George V Land** *physical region* Antarctica
25 S14 **George West** Texas, SW USA 28°21′N 98°08′W
137 R9 **Georgia** *off.* Republic of Georgia, *Geor.* Sak'art'velo, *Rus.* Gruzinskaya SSR, Gruziya. ◆ *republic* SW Asia
23 S5 **Georgia** *off.* State of Georgia, *also known as* Empire State of the South, Peach State. ◈ *state* SE USA
14 F12 **Georgian Bay** *lake bay* Ontario, S Canada
Georgia, Republic of *see* Georgia
10 L17 **Georgia, Strait of** *strait* British Columbia, W Canada
Georgi Dimitrov *see* Kostenets
Georgi Dimitrov, Yazovir *see* Koprinka, Yazovir
Georgiu-Dezh *see* Liski
145 W10 **Georgiyevka** Vostochnyy Kazakhstan, E Kazakhstan 49°19′N 81°35′E
Georgiyevka *see* Korday
127 N15 **Georgiyevsk** Stavropol'skiy Kray, SW Russian Federation 44°07′N 43°22′E
100 G13 **Georgsmarienhütte** Niedersachsen, NW Germany 52°13′N 08°02′E
195 O1 **Georg von Neumayer** *German research station* Antarctica 70°41′S 08°18′W
101 M16 **Gera** Thüringen, E Germany 50°51′N 12°13′E
101 K16 **Gera** ~ C Germany
99 E19 **Geraardsbergen** Oost-Vlaanderen, SW Belgium 50°47′N 03°53′E
115 F21 **Geráki** Pelopónnisos, S Greece 36°56′N 22°46′E
27 W5 **Gerald** Missouri, C USA 38°24′N 91°20′W
47 V8 **Geral de Goiás, Serra** ▲▲ E Brazil
185 G20 **Geraldine** Canterbury, South Island, New Zealand 44°06′S 171°14′E
180 H11 **Geraldton** Western Australia 28°48′S 114°40′E
12 E11 **Geraldton** Ontario, S Canada 49°44′N 86°59′W
60 J12 **Geral, Serra** ▲▲ S Brazil
103 U6 **Gérardmer** Vosges, NE France 48°05′N 06°54′E
Gerasa *see* Jarash
Gerdauen *see* Zheleznodorozhnyy
39 Q11 **Gerdine, Mount** ▲ Alaska, USA 61°40′N 152°21′W
136 H11 **Gerede** Bolu, N Turkey 40°48′N 32°13′E
136 H11 **Gerede Çayı** ~ N Turkey
148 M8 **Gereshk** Helmand, SW Afghanistan 31°50′N 64°32′E
101 L24 **Geretsried** Bayern, S Germany 47°51′N 11°28′E
105 P14 **Gérgal** Andalucía, S Spain 37°07′N 02°34′W
28 I14 **Gering** Nebraska, C USA 41°49′N 103°39′W
35 R3 **Gerlach** Nevada, W USA 40°38′N 119°21′W
Gerlachfalvi Csúcs/ Gerlachovka *see* Gerlachovský štít
111 L18 **Gerlachovský štít** *var.* Gerlachovka, *Ger.* Gerlsdorfer Spitze, *Hung.* Gerlachfalvi Csúcs; *prev.* Stalinov Štít, *Ger.* Franz-Josef Spitze, *Hung.* Ferencz-Jósef Csúcs. ▲ N Slovakia 49°12′N 20°09′E
108 E8 **Gerlafingen** Solothurn, NW Switzerland 47°10′N 07°35′E
Gerlsdorfer Spitze *see* Gerlachovský štít
Germak *see* Germik
German East Africa *see* Tanzania
Germanicopolis *see* Çankırı
Germanicum, Mare/ German Ocean *see* North Sea
Germanovichi *see* Hyermanavichy
German Southwest Africa *see* Namibia
20 E10 **Germantown** Tennessee, S USA 35°06′N 89°51′W
101 I15 **Germany** *off.* Federal Republic of Germany, Bundesrepublik Deutschland, *Ger.* Deutschland. ◆ *federal republic* N Europe
Germany, Federal Republic of *see* Germany
101 L23 **Germering** Bayern, SE Germany 48°07′N 11°22′E
139 V3 **Germik** *Ar.* Garmik, *var.* Germak. As Sulaymānīyah, E Iraq 35°49′N 46°09′E
83 J21 **Germiston** *var.* Gauteng. Gauteng, NE South Africa 26°15′S 28°10′E
Gernika *see* Gernika-Lumo
105 P2 **Gernika-Lumo** *var.* Gernika, Guernica, Guernica y Lumo. País Vasco, N Spain 43°19′N 02°40′W
164 L12 **Gero** Gifu, Honshū, SW Japan 35°48′N 137°15′E
115 F22 **Geroliménas** Pelopónnisos, S Greece 36°28′N 22°25′E
Gerona *see* Girona
99 H21 **Gerpinnes** Hainaut, S Belgium 50°20′N 04°38′E
102 L15 **Gers** ◈ *department* S France
102 L14 **Gers** ~ S France
Gerunda *see* Girona
158 I13 **Gêrzê** *var.* Luring. Xizang Zizhiqu, W China 32°19′N 84°05′E
136 K10 **Gerze** Sinop, N Turkey 41°48′N 35°13′E
Gesoriacum *see* Boulogne-sur-Mer
Gessoriacum *see* Boulogne-sur-Mer
99 J21 **Gesves** Namur, SE Belgium 50°24′N 05°04′E
93 J20 **Geta** Åland, SW Finland 60°22′N 19°49′E
105 N8 **Getafe** Madrid, C Spain 40°18′N 03°44′W
95 J21 **Getinge** Halland, S Sweden 56°46′N 12°42′E
18 F16 **Gettysburg** Pennsylvania, NE USA 39°49′N 77°13′W
29 N8 **Gettysburg** South Dakota, N USA 45°00′N 99°57′W
194 K12 **Getz Ice Shelf** *ice shelf* Antarctica
137 S15 **Gevaş** Van, SE Turkey 38°16′N 43°05′E
Gevgeli *see* Gevgelija
113 Q20 **Gevgelija** *var.* Đevđelija, Djevdjelija, *Turk.* Gevgeli. SE Macedonia 41°09′N 22°30′E
103 T10 **Gex** Ain, E France 46°21′N 06°02′E
92 I3 **Geysir** *physical region* SW Iceland
136 F11 **Geyve** Sakarya, NW Turkey 40°32′N 30°18′E
80 G10 **Gezira** ◈ *state* E Sudan
109 V3 **Gföhl** Niederösterreich, N Austria 48°30′N 15°27′E
83 H22 **Ghaap Plateau** *Afr.* Ghaapplato. *plateau* C South Africa
Ghaapplato *see* Ghaap Plateau
Ghaba *see* Al Ghābah
138 J8 **Ghāb, Tall** ▲ SE Syria 33°09′N 37°48′E
139 Q9 **Ghadaf, Wādī al** *dry watercourse* C Iraq
Ghadamès *see* Ghadāmis
74 M9 **Ghadāmis** *var.* Ghadamès, Rhadames. W Libya 30°08′N 09°30′E
141 Y10 **Ghadan** E Oman 20°20′N 57°58′E
75 O10 **Ghaddūwah** C Libya 26°36′N 14°26′E
147 Q11 **Ghafurov** *Rus.* Gafurov; *prev.* Sovetabad. NW Tajikistan 40°13′N 69°42′E
153 N12 **Ghāghara** ~ S Asia
149 P13 **Ghaibi Dero** Sind, SE Pakistan 27°35′N 67°42′E
141 Y10 **Ghalat** E Oman 21°06′N 58°51′E
139 W11 **Ghamūkah, Hawr** ◎ S Iraq
77 P15 **Ghana** *off.* Republic of Ghana. ◆ *republic* W Africa
141 X12 **Ghānah** *spring/well* S Oman 18°35′N 56°34′E
Ghanongga *see* Ranongga
Ghansi/Ghansiland *see* Ghanzi
83 F18 **Ghanzi** *var.* Khanzi. Ghanzi, W Botswana 21°39′S 21°38′E
83 G19 **Ghanzi** *var.* Ghansi, Ghansiland, Khanzi. ◈ *district* C Botswana
67 T14 **Ghanzi** *var.* Knanzi. ~ Botswana/South Africa
Ghap'an *see* Kapan
138 F13 **Gharandal** Al'Aqabah, SW Jordan 30°12′N 35°18′E
139 U14 **Gharbīyah, Sha'īb al** ~ S Iraq
Gharbt, Jabal al *see* Liban, Jebel
74 K7 **Ghardaïa** N Algeria 32°30′N 03°44′E
147 R12 **Gharm** *Rus.* Garm. C Tajikistan 39°03′N 70°25′E
149 P17 **Gharo** Sind, SE Pakistan 24°44′N 67°35′E
139 W10 **Gharrāf, Shaṭṭ al** ~ S Iraq
Gharvān *see* Gharyān
75 O7 **Gharyān** *var.* Gharvān. NW Libya 32°10′N 13°01′E
74 M11 **Ghāt** *var.* Gat. SW Libya 24°58′N 10°11′E
Ghawdex *see* Gozo
141 U8 **Ghayathi** Abū Ẓaby, W United Arab Emirates 23°51′N 53°01′E
Ghazāl, Baḥr al *see* Ghazal, Bahr el
78 H9 **Ghazal, Bahr el** *var.* Soro. *seasonal river* C Chad
80 E13 **Ghazal, Bahr el** *var.* Baḥr al Ghazāl. ~ N South Sudan
74 H6 **Ghazaouet** NW Algeria 35°08′N 01°50′W
152 J10 **Ghāziābād** Uttar Pradesh, N India 28°42′N 77°28′E
153 O13 **Ghāzīpur** Uttar Pradesh, N India 25°36′N 83°36′E
149 Q6 **Ghaznī** *var.* Ghazni. Ghaznī, E Afghanistan 33°31′N 68°24′E
149 P7 **Ghaznī** ◈ *province* SE Afghanistan
Ghazzah *see* Gaza
Gheel *see* Geel
Ghelîzâne *see* Relizane
Ghent *see* Gent
Gheorghe Braţul *see* Sfântu Gheorghe, Braţul
Gheorghe Gheorghiu-Dej *see* Oneşti
116 J10 **Gheorgheni** *prev.* Gheorghieni, Sîn-Miclăuş, *Ger.* Niklasmarkt, *Hung.* Gyergyószentmiklós. Harghita, C Romania 46°43′N 25°36′E
Gheorghieni *see* Gheorgheni
116 H10 **Gherla** *Ger.* Neuschliss, *Hung.* Szamosújvár; *prev.* Armenierstadt. Cluj, NW Romania 47°02′N 23°55′E
Gheweifat *see* Ghuwayfāt
Ghilan *see* Gīlān
107 C18 **Ghilarza** Sardegna, Italy, C Mediterranean Sea 40°09′N 08°50′E
Ghilizane *see* Relizane
Ghimbi *see* Gīmbī
Ghiriş *see* Câmpia Turzii
103 Y15 **Ghisonaccia** Corse, France, C Mediterranean Sea 42°00′N 09°25′E
147 Q11 **Ghonchí** *Rus.* Ganchi. NW Tajikistan 39°57′N 69°10′E
153 T13 **Ghoraghat** Rajshahi, NW Bangladesh 25°18′N 89°20′E
148 J5 **Ghōrīān** *prev.* Ghūrīān. Herāt, W Afghanistan 34°20′N 61°26′E
149 R13 **Ghotki** Sind, SE Pakistan 28°00′N 69°21′E
Ghowr *see* Gōwr
147 T13 **Ghŭdara** *var.* Gudara, *Rus.* Kudara. SE Tajikistan 38°28′N 72°39′E
153 R13 **Ghugri** ~ N India
147 S14 **Ghund** *Rus.* Gunt. ~ SE Tajikistan
Ghurābīyah, Sha'īb al *see* Gharbīyah, Sha'īb al
Ghurdaqah *see* Al Ghurdaqah
Ghūrīān *see* Ghōrīān
141 T8 **Ghuwayfāt** *var.* Gheweifat. Abū Ẓaby, W United Arab Emirates 24°06′N 51°40′E
121 O14 **Ghuzayyil, Sabkhat** *salt lake* N Libya
126 J3 **Ghzatsk** Smolenskaya Oblast', W Russian Federation 55°33′N 35°00′E
115 G17 **Giáltra** Évvoia, C Greece 38°21′N 22°58′E
Giamame *see* Jamaame
167 U13 **Gia Nghia** *var.* Đak Nông. Đăc Lăc, S Vietnam 11°58′N 107°42′E
114 F13 **Giannitsá** *var.* Yiannitsá. Kentrikí Makedonía, N Greece 40°49′N 22°24′E
107 F14 **Giannutri, Isola di** *island* Archipelago Toscano, C Italy
96 F13 **Giant's Causeway** *Ir.* Clochán an Aifir. *lava flow* N Northern Ireland, United Kingdom
167 S15 **Gia Rai** Minh Hai, S Vietnam 09°14′N 105°28′E
107 L24 **Giarre** Sicilia, Italy, C Mediterranean Sea 37°44′N 15°12′E
44 I7 **Gibara** Holguín, E Cuba 21°09′N 76°11′W
29 O16 **Gibbon** Nebraska, C USA 40°45′N 98°50′W
32 K11 **Gibbon** Oregon, NW USA 45°40′N 118°22′W
33 P11 **Gibbonsville** Idaho, NW USA 45°33′N 113°55′W
64 A13 **Gibbs Hill** *hill* S Bermuda
92 I9 **Gibostad** Troms, N Norway 69°21′N 18°01′E
104 I14 **Gibraleón** Andalucía, S Spain 37°23′N 06°58′W
104 L16 **Gibraltar** ○ (Gibraltar) S Gibraltar 36°08′N 05°21′W
104 L16 **Gibraltar** ◇ *UK dependent territory* SW Europe
Gibraltar, Détroit de/ Gibraltar, Estrecho de *see* Gibraltar, Strait of
104 J17 **Gibraltar, Strait of** *Fr.* Détroit de Gibraltar, *Sp.* Estrecho de Gibraltar. *strait* Atlantic Ocean/ Mediterranean Sea
31 S11 **Gibsonburg** Ohio, N USA 41°22′N 83°19′W
30 M13 **Gibson City** Illinois, N USA 40°27′N 88°24′W
180 L8 **Gibson Desert** *desert* Western Australia
10 L17 **Gibsons** British Columbia, SW Canada 49°24′N 123°32′W
149 N12 **Gīdār** Baluchistān, SW Pakistan 28°16′N 66°00′E
155 I17 **Giddalūr** Andhra Pradesh, E India 15°24′N 78°54′E
25 U10 **Giddings** Texas, SW USA 30°12′N 96°59′W
27 Y8 **Gideon** Missouri, C USA 36°27′N 89°55′W
81 I15 **Gīdolē** Southern Nationalities, S Ethiopia 05°31′N 37°26′E
Giebnegáisi *see* Kebnekaise
118 H13 **Giedraičiai** Utena, E Lithuania 55°05′N 25°16′E
103 O7 **Gien** Loiret, C France 47°40′N 02°37′E
101 G17 **Giessen** Hessen, W Germany 50°35′N 08°41′E
98 O6 **Gieten** Drenthe, NE Netherlands 53°00′N 06°43′E
23 Y13 **Gifford** Florida, SE USA 27°40′N 80°24′W
9 O5 **Gifford** ~ Baffin Island, Nunavut, NE Canada
100 J12 **Gifhorn** Niedersachsen, N Germany 52°28′N 10°33′E
11 P13 **Gift Lake** Alberta, W Canada 55°49′N 115°57′W
Giftokastro *see* Tsigansko Gradishte
164 L13 **Gifu** *var.* Gihu. Gifu, Honshū, SW Japan 35°24′N 136°46′E
164 K13 **Gifu** *off.* Gifu-ken, *var.* Gihu. ◈ *prefecture* Honshū, SW Japan
Gifu-ken *see* Gifu
126 M13 **Gigant** Rostovskaya Oblast', SW Russian Federation 46°29′N 41°18′E
40 E8 **Giganta, Sierra de la** ▲▲ NW Mexico
54 E12 **Gigante** Huila, S Colombia 02°24′N 75°34′W
114 I7 **Gigen** Pleven, N Bulgaria 43°40′N 24°31′E
Giggiga *see* Jijiga
96 G12 **Gigha Island** *island* SW Scotland, United Kingdom
107 E14 **Giglio, Isola del** *island* Archipelago Toscano, C Italy
Gihu *see* Gifu
146 L11 **G'ijduvon** *Rus.* Gizhduvon. Buxoro Viloyati, C Uzbekistan 40°06′N 64°38′E
104 L2 **Gijón** *var.* Xixón. Asturias, NW Spain 43°32′N 05°40′W
81 D20 **Gikongoro** SW Rwanda 02°30′S 29°32′E
36 K14 **Gila Bend** Arizona, SW USA 32°57′N 112°43′W
36 J14 **Gila Bend Mountains** ▲▲ Arizona, SW USA
36 I15 **Gila Mountains** ▲▲ Arizona, SW USA
37 N14 **Gila Mountains** ▲▲ Arizona, SW USA
142 M4 **Gīlān** *off.* Ostān-e Gīlān, *var.* Ghilan, Guilan. ◈ *province* NW Iran
Gīlān, Ostān-e *see* Gīlān
36 L14 **Gila River** ~ Arizona, SW USA
29 W4 **Gilbert** Minnesota, N USA 47°29′N 92°27′W
Gilbert Islands *see* Tungaru
10 L16 **Gilbert, Mount** ▲ British Columbia, SW Canada 50°49′N 124°03′W
181 U4 **Gilbert River** ~ Queensland, NE Australia
0 C6 **Gilbert Seamounts** *undersea feature* NE Pacific Ocean 52°50′N 150°10′W
33 S7 **Gildford** Montana, NW USA 48°34′N 110°21′W
83 P15 **Gilé** Zambézia, NE Mozambique 16°10′S 28°17′E
30 K4 **Gile Flowage** ◎ Wisconsin, N USA
182 G7 **Giles, Lake** *salt lake* South Australia
Gilf Kebir Plateau *see* Hadabat al Jilf al Kabīr
183 R6 **Gilgandra** New South Wales, SE Australia 31°43′S 148°39′E
Gilgău *see* Gâlgău
81 I19 **Gilgil** Nakuru, SW Kenya 0°29′S 36°19′E
183 S4 **Gil Gil Creek** ~ New South Wales, SE Australia
149 V3 **Gilgit** Jammu and Kashmir, NE Pakistan 35°54′N 74°20′E
149 V3 **Gilgit** ~ N Pakistan
11 X11 **Gillam** Manitoba, C Canada 56°25′N 94°45′W
95 J22 **Gilleleje** Hovedstaden, E Denmark 56°05′N 12°17′E
30 K14 **Gillespie** Illinois, N USA 39°07′N 89°49′W
27 W13 **Gillett** Arkansas, C USA 34°07′N 91°22′W
33 X12 **Gillette** Wyoming, C USA 44°17′N 105°30′W
97 P22 **Gillingham** SE England, United Kingdom 51°24′N 00°33′E
195 X6 **Gillock Island** *island* Antarctica
173 O16 **Gillot** ✈ (St-Denis) N Réunion 20°53′S 55°31′E
65 H25 **Gill Point** *headland* E Saint Helena 15°59′S 05°38′W
30 M12 **Gilman** Illinois, N USA 40°44′N 87°58′W
25 W6 **Gilmer** Texas, SW USA 32°44′N 94°58′W
Gilolo *see* Halmahera, Pulau
81 G14 **Gīlo Wenz** ~ SW Ethiopia
35 O10 **Gilroy** California, W USA 37°00′N 121°34′W
123 Q12 **Gilyuy** ~ SE Russian Federation
99 I14 **Gilze** Noord-Brabant, S Netherlands 51°33′N 04°56′E
165 R16 **Gima** Okinawa, Kume-jima, SW Japan
80 H13 **Gīmbī** *It.* Ghimbi. Oromīya, C Ethiopia 09°13′N 35°39′E
163 Y15 **Gimcheon** *prev.* Kimch'ŏn. C South Korea 36°08′N 128°06′E
163 Z16 **Gimhae** *prev.* Kim Hae. ✈ (Busan) SE South Korea 35°10′N 128°57′E
45 T12 **Gimie, Mount** ▲ C Saint Lucia 13°51′N 61°00′W
11 X16 **Gimli** Manitoba, S Canada 50°39′N 97°00′W
Gimma *see* Jīma
95 O14 **Gimo** Uppsala, C Sweden 60°11′N 18°12′E
102 L15 **Gimone** ~ S France
Gimpoe *see* Gimpu
171 N12 **Gimpu** *prev.* Gimpoe. Sulawesi, C Indonesia 01°38′S 120°00′E
182 F5 **Gina** South Australia 29°56′S 134°33′E
Ginevra *see* Genève
99 J19 **Gingelom** Limburg, NE Belgium 50°46′N 05°09′E
180 I12 **Gingin** Western Australia 31°22′S 115°51′E
171 Q7 **Gingoog** Mindanao, S Philippines 08°47′N 125°05′E
81 K14 **Gīnīr** Oromīya, C Ethiopia 07°12′N 40°43′E
Giohar *see* Jawhar
107 O17 **Gióia del Colle** Puglia, SE Italy 40°47′N 16°56′E
107 M22 **Gioia, Golfo di** *gulf* S Italy
Giona *see* Gkióna
115 I16 **Gioúra** *island* Vóreies Sporádes, Greece, Aegean Sea
107 O17 **Giovinazzo** Puglia, SE Italy 41°11′N 16°40′E
Gipeswic *see* Ipswich
105 P3 **Gipuzkoa** *Cast.* guipuzcoa. ◈ *province* País Vasco, N Spain
Giran *see* Ilan
30 K14 **Girard** Illinois, N USA 39°27′N 89°46′W
27 R7 **Girard** Kansas, C USA 37°30′N 94°50′W
25 O6 **Girard** Texas, SW USA 33°18′N 100°38′W
54 E10 **Girardot** Cundinamarca, C Colombia 04°19′N 74°47′W
172 M7 **Giraud Seamount** *undersea feature* SW Indian Ocean 09°57′S 46°55′E
83 A15 **Giraul** ~ SW Angola
96 L9 **Girdle Ness** *headland* NE Scotland, United Kingdom 57°09′N 02°04′W
137 N11 **Giresun** *var.* Kerasunt; *anc.* Cerasus, Pharnacia. Giresun, NE Turkey 40°55′N 38°35′E
137 N12 **Giresun** *var.* Kerasunt. ◈ *province* NE Turkey
137 N12 **Giresun Dağları** ▲▲ N Turkey
Girga *see* Jirjā
Girgeh *see* Jirjā
Girgenti *see* Agrigento
153 Q15 **Girīdīh** Jhārkhand, NE India 24°10′N 86°20′E
183 P6 **Girilambone** New South Wales, SE Australia 31°19′S 146°57′E
Girin *see* Jilin
121 W10 **Girne** *Gk.* Kerýneia, Kyrenia. N Cyprus 35°20′N 33°20′E
Giron *see* Kiruna
105 X5 **Girona** *var.* Gerona; *anc.* Gerunda. Cataluña, NE Spain 41°59′N 02°49′E
105 W5 **Girona** *var.* Gerona. ◈ *province* Cataluña, NE Spain
102 J12 **Gironde** ◈ *department* SW France
102 J11 **Gironde** *estuary* SW France
105 V5 **Gironella** Cataluña, NE Spain 42°02′N 01°53′E
103 N15 **Girou** ~ S France
97 H14 **Girvan** W Scotland, United Kingdom 55°14′N 04°53′W
24 M9 **Girvin** Texas, SW USA 31°05′N 102°24′W
184 Q9 **Gisborne** Gisborne, North Island, New Zealand 38°41′S 178°01′E
184 P9 **Gisborne** *off.* Gisborne District. ◈ *unitary authority* North Island, New Zealand
Gisborne District *see* Gisborne
Giseifu *see* Uijeongbu
Gisenye *see* Gisenyi
81 D19 **Gisenyi** *var.* Gisenye. NW Rwanda 01°42′S 29°18′E
95 K20 **Gislaved** Jönköping, S Sweden 57°19′N 13°30′E
103 N4 **Gisors** Eure, N France 49°18′N 01°46′E
Gissar *see* Hisor
147 P12 **Gissar Range** *Rus.* Gissarskiy Khrebet. ▲▲ Tajikistan/Uzbekistan
Gissarskiy Khrebet *see* Gissar Range
99 B16 **Gistel** West-Vlaanderen, W Belgium 51°09′N 02°58′E
108 F9 **Giswil** Obwalden, C Switzerland 46°49′N 08°11′E
115 B16 **Gitánes** *ancient monument* Ípeiros, W Greece
81 E20 **Gitarama** C Rwanda 02°05′S 29°45′E
81 E20 **Gitega** C Burundi 03°20′S 29°56′E
Githio *see* Gýtheio
108 H11 **Giubiasco** Ticino, S Switzerland 46°11′N 09°01′E
106 K13 **Giulianova** Abruzzi, C Italy 42°45′N 13°58′E
Giulie, Alpi *see* Julian Alps
Giumri *see* Gyumri
116 M13 **Giurgeni** Ialomiţa, SE Romania 44°45′N 27°48′E
116 J15 **Giurgiu** Giurgiu, S Romania 43°54′N 25°58′E
116 J14 **Giurgiu** ◈ *county* SE Romania
95 F22 **Give** Syddanmark, C Denmark 55°51′N 09°15′E
103 R2 **Givet** Ardennes, N France 50°08′N 04°50′E
103 R11 **Givors** Rhône, E France 45°35′N 04°47′E
83 K19 **Giyani** Limpopo, NE South Africa 23°20′S 30°37′E
80 I13 **Giyon** Oromīya, C Ethiopia 08°31′N 37°56′E
75 W8 **Giza** *var.* Al Jīzah, El Gîza, Gizeh. N Egypt 30°01′N 31°13′E
75 V8 **Giza, Pyramids of** *ancient monument* N Egypt
Gizhduvon *see* G'ijduvon
123 U8 **Gizhiga** Magadanskaya Oblast', E Russian Federation 61°58′N 160°16′E
123 T9 **Gizhiginskaya Guba** *bay* E Russian Federation
186 K8 **Gizo** Gizo, NW Solomon Islands 08°03′S 156°49′E
110 N7 **Giżycko** *Ger.* Lötzen. Warmińsko-Mazurskie, NE Poland 54°03′N 21°48′E
Gízymałów *see* Hrymayliv
113 M17 **Gjakovë** *Serb.* Đakovica. W Kosovo 42°22′N 20°30′E
94 F12 **Gjende** ◎ S Norway
113 L16 **Gjeravicë** *Serb.* Đeravica. ▲ S Serbia 42°33′N 20°08′E
95 F17 **Gjerstad** Aust-Agder, S Norway 58°54′N 09°03′E
113 O17 **Gjilan** *Serb.* Gnjilane. E Kosovo 42°27′N 21°28′E
Gjinokastër *see* Gjirokastër
113 L23 **Gjirokastër** *var.* Gjirokastra; *prev.* Gjinokastër, *Gk.* Argyrokastron, *It.* Argirocastro. Gjirokastër, S Albania 40°04′N 20°09′E
113 L22 **Gjirokastër** ◈ *district* S Albania
Gjirokastra *see* Gjirokastër
9 N7 **Gjoa Haven** *var.* Uqsuqtuuq. King William Island, Nunavut, NW Canada 68°38′N 95°57′W
94 H13 **Gjøvik** Oppland, S Norway 60°47′N 10°40′E
113 J22 **Gjuhëzës, Kepi i** *headland* SW Albania 40°25′N 19°19′E
Gjurgjevac *see* Đurđevac
115 E18 **Gkióna** *var.* Giona. ▲▲ C Greece
121 R3 **Gkréko, Akrotíri** *var.* Cape Greco, Pidálion. *cape* E Cyprus
99 I18 **Glabbeek-Zuurbemde** Vlaams Brabant, C Belgium 50°54′N 04°58′E
13 R14 **Glace Bay** Cape Breton Island, Nova Scotia, SE Canada 46°12′N 59°57′W
11 O16 **Glacier** British Columbia, SW Canada 51°12′N 117°33′W
39 W12 **Glacier Bay** *inlet* Alaska, USA
32 I7 **Glacier Peak** ▲ Washington, NW USA 48°06′N 121°06′W
159 N13 **Gladaindong Feng** ▲ C China 33°24′N 91°00′E
21 Q7 **Glade Spring** Virginia, NE USA 36°47′N 81°46′W
43 W7 **Gladewater** Texas, SW USA 32°32′N 94°57′W
181 Y8 **Gladstone** Queensland, E Australia 23°52′S 151°16′E
182 I8 **Gladstone** South Australia 33°16′S 138°21′E
11 X16 **Gladstone** Manitoba, S Canada 50°12′N 98°56′W
31 O5 **Gladstone** Michigan, N USA 45°51′N 87°01′W
27 R4 **Gladstone** Missouri, C USA 39°12′N 94°33′W
31 Q7 **Gladwin** Michigan, N USA 43°58′N 84°29′W
95 J15 **Glafsfjorden** ◎ C Sweden
92 H2 **Gláma** *physical region* NW Iceland
94 I12 **Gláma** *var.* Glommen. ~ S Norway

◆ Country
● Country Capital
◇ Dependent Territory
○ Dependent Territory Capital
◈ Administrative Regions
✈ International Airport
▲ Mountain
▲▲ Mountain Range
Volcano
~ River
◎ Lake
⊠ Reservoir

GLAMOČ – GOURIN

112 F13 Glamoč Federacija Bosne I Hercegovine, NE Boznia and Hercegovina 44°01′N 16°51′E
97 J22 Glamorgan cultural region S Wales, United Kingdom
95 G24 Glamsbjerg Syddtjylland, C Denmark 55°17′N 10°07′E
171 Q8 Glan Mindanao, S Philippines 05°49′N 125°11′E
109 T9 Glan ≈ SE Austria
101 F19 Glan ≈ W Germany
95 M17 Glan ◎ S Sweden
Glaris see Glarus
108 H9 Glarner Alpen Eng. Glarus Alps. ▲▲ E Switzerland
108 H8 Glarus Glarus, E Switzerland 47°03′N 09°04′E
108 H9 Glarus Fr. Glaris. ◐ canton C Switzerland
Glarus Alps see Glarner Alpen
27 N3 Glasco Kansas, C USA 39°21′N 97°50′W
96 I12 Glasgow S Scotland, United Kingdom 55°53′N 04°15′W
20 K7 Glasgow Kentucky, S USA 37°00′N 85°54′W
27 T4 Glasgow Missouri, C USA 39°13′N 92°51′W
33 W7 Glasgow Montana, NW USA 48°12′N 106°37′W
21 T6 Glasgow Virginia, NE USA 37°37′N 79°27′W
96 I12 Glasgow ✈ W Scotland, United Kingdom 55°52′N 04°27′W
11 S14 Glaslyn Saskatchewan, S Canada 53°20′N 108°18′W
18 I16 Glassboro New Jersey, NE USA 39°40′N 75°05′W
24 L10 Glass Mountains ▲▲ Texas, SW USA
97 K23 Glastonbury SW England, United Kingdom 51°09′N 02°43′W
Glatz see Kłodzko
101 N16 Glauchau Sachsen, E Germany 50°48′N 12°32′E
Glavn'a Morava see Velika Morava
Glavnik see Gllamnik
127 T1 Glazov Udmurtskaya Respublika, NW Russian Federation 58°06′N 52°38′E
Głda see Gwda
109 U8 Gleinalpe ▲▲ SE Austria
109 W8 Gleisdorf Steiermark, SE Austria 47°07′N 15°43′E
Gleiwitz see Gliwice
39 S11 Glenallen Alaska, USA 62°06′N 145°33′W
102 F7 Glénan, Îles island group NW France
185 G21 Glenavy Canterbury, South Island, New Zealand 44°53′S 171°04′E
10 H5 Glenboyle Yukon, NW Canada 63°55′N 138°43′W
21 X3 Glen Burnie Maryland, NE USA 39°09′N 76°37′W
36 L8 Glen Canyon canyon Utah, W USA
36 L8 Glen Canyon Dam dam Arizona, SW USA
30 K15 Glen Carbon Illinois, N USA 38°45′N 89°58′W
14 E17 Glencoe Ontario, S Canada 42°44′N 81°42′W
83 K22 Glencoe KwaZulu/Natal, E South Africa 28°10′S 30°15′E
29 U9 Glencoe Minnesota, N USA 44°46′N 94°09′W
96 H10 Glen Coe valley N Scotland, United Kingdom
36 K13 Glendale Arizona, SW USA 33°32′N 112°11′W
35 S15 Glendale California, W USA 34°09′N 118°20′W
182 G5 Glendambo South Australia 30°59′S 135°45′E
33 Y8 Glendive Montana, NW USA 47°08′N 104°42′W
33 Y15 Glendo Wyoming, C USA 42°27′N 105°01′W
55 S10 Glendor Mountains C Guyana
182 K12 Glenelg River ≈ South Australia/Victoria, SE Australia
29 P4 Glenfield North Dakota, N USA 47°25′N 98°33′W
25 V12 Glen Flora Texas, SW USA 29°22′N 96°12′W
181 P7 Glen Helen Northern Territory, N Australia 23°45′S 132°46′E
183 U5 Glen Innes New South Wales, SE Australia 29°42′S 151°45′E
31 P6 Glen Lake ◎ Michigan, N USA
10 I7 Glenlyon Peak ▲ Yukon, W Canada 62°32′N 134°51′W
37 N16 Glenn, Mount ▲ Arizona, SW USA 31°55′N 110°00′W
33 N15 Glenns Ferry Idaho, NW USA 42°57′N 115°18′W
23 W6 Glennville Georgia, SE USA 31°56′N 81°55′W
10 J10 Glenora British Columbia, W Canada 57°52′N 131°16′W
182 M11 Glenorchy Victoria, SE Australia 36°56′S 142°39′E
183 V5 Glenreagh New South Wales, SE Australia 30°04′S 153°00′E
33 X15 Glenrock Wyoming, C USA 42°51′N 105°52′W
96 K11 Glenrothes E Scotland, United Kingdom 56°11′N 03°09′W
18 L9 Glens Falls New York, NE USA 43°18′N 73°38′W
97 D14 Glenties Ir. Na Gleannta. Donegal, NW Ireland 54°47′N 08°17′W
28 L5 Glen Ullin North Dakota, N USA 46°49′N 101°49′W
21 R4 Glenville West Virginia, NE USA 38°57′N 80°51′W
27 T12 Glenwood Arkansas, C USA 34°19′N 93°33′W
29 S15 Glenwood Iowa, C USA 41°03′N 95°44′W
29 T7 Glenwood Minnesota, N USA 45°39′N 95°23′W
36 L5 Glenwood Utah, W USA 38°45′N 111°59′W
30 I5 Glenwood City Wisconsin, N USA 45°04′N 92°11′W
37 Q4 Glenwood Springs Colorado, C USA 39°33′N 107°21′W
108 F10 Gletsch Valais, S Switzerland 46°34′N 08°21′E
Glevum see Gloucester
29 U14 Glidden Iowa, C USA 42°03′N 94°43′W
112 E9 Glina var. Banijska Palanka. Sisak-Moslavina, NE Croatia 45°19′N 16°07′E
94 F11 Glittertind ▲ S Norway 61°24′N 08°19′E
111 J16 Gliwice Ger. Gleiwitz. Śląskie, S Poland 50°19′N 18°49′E
113 N16 Gllamnik Serb. Glavnik. N Kosovo 42°53′N 21°10′E
36 M14 Globe Arizona, SW USA 33°24′N 110°47′W
Globino see Hlobyne
108 L9 Glockturm ▲ SW Austria 46°54′N 10°42′E
116 L9 Glodeni Rus. Glodyany. N Moldova 47°47′N 27°33′E
109 S9 Glödnitz Kärnten, S Austria 46°57′N 14°03′E
Glodyany see Glodeni
Glogau see Głogów
109 W6 Gloggnitz Niederösterreich, E Austria 47°41′N 15°57′E
110 F13 Głogów Ger. Glogau, Glogow. Dolnośląskie, SW Poland 51°40′N 16°04′E
Glogow see Głogów
111 I16 Głogówek Ger. Oberglogau. Opolskie, S Poland 50°21′N 17°51′E
92 G12 Glomfjord Nordland, C Norway 66°49′N 14°00′E
Glomma see Gláma
Glommen see Gláma
93 I14 Glommersträsk Norrbotten, N Sweden 65°17′N 19°40′E
172 I1 Glorieuses, Îles Eng. Glorioso Islands. island (to France) N Madagascar
Glorioso Islands see Glorieuses, Îles
65 C25 Glorious Hill hill East Falkland, Falkland Islands
38 J12 Glory of Russia Cape headland Saint Matthew Island, Alaska, USA 60°36′N 172°57′W
22 J7 Gloster Mississippi, S USA 31°12′N 91°01′W
183 U7 Gloucester New South Wales, SE Australia 32°01′S 152°00′E
186 F7 Gloucester New Britain, E Papua New Guinea 05°30′S 148°30′E
97 L21 Gloucester hist. Caer Glou, Lat. Glevum. C England, United Kingdom 51°53′N 02°14′W
19 P10 Gloucester Massachusetts, NE USA 42°36′N 70°36′W
21 X6 Gloucester Virginia, NE USA 37°26′N 76°33′W
97 K21 Gloucestershire cultural region C England, United Kingdom
31 T14 Glouster Ohio, N USA 39°30′N 82°04′W
42 H3 Glovers Reef reef E Belize
18 K10 Gloversville New York, NE USA 43°03′N 74°20′W
110 K12 Głowno Łódź, C Poland 51°58′N 19°43′E
111 H16 Głubczyce Ger. Leobschütz. Opolskie, S Poland 50°13′N 17°50′E
126 L11 Glubokiy Rostovskaya Oblast', SW Russian Federation 48°34′N 40°16′E
145 W9 Glubokoye Vostochnyy Kazakhstan, E Kazakhstan 50°08′N 82°16′E
Glubokoye see Hlybokaye
111 H16 Głuchołazy Ger. Ziegenhais. Opolskie, S Poland 50°20′N 17°23′E
100 I9 Glückstadt Schleswig-Holstein, N Germany 53°47′N 09°26′E
Glukhov see Hlukhiv
Glushkevichi see Hlushkavichy
Glusk/Glussk see Hlusk
Glybokaya see Hlyboka
95 F21 Glyngøre Midtjylland, NW Denmark 56°45′N 08°55′E
127 Q9 Gmelinka Volgogradskaya Oblast', SW Russian Federation 50°50′N 46°51′E
109 R8 Gmünd Kärnten, S Austria 46°56′N 13°32′E
109 U2 Gmünd Niederösterreich, N Austria 48°47′N 14°59′E
Gmünd see Schwäbisch Gmünd
109 S5 Gmunden Oberösterreich, N Austria 47°56′N 13°48′E
Gmundner See see Traunsee
94 N10 Gnarp Gävleborg, C Sweden 62°03′N 17°19′E
109 W8 Gnas Steiermark, SE Austria 46°53′N 15°48′E
Gnaviyani see Fuammulah
Gnesen see Gniezno
95 O16 Gnesta Södermanland, C Sweden 59°05′N 17°20′E
110 H11 Gniezno Ger. Gnesen. Weilkopolskie, C Poland 52°33′N 17°35′E
Gnjilane see Gjilan
95 K20 Gnosjö Jönköping, S Sweden 57°22′N 13°44′E
155 E17 Goa prev. Old Goa, Vela Goa, Velha Goa. Goa, W India 15°31′N 73°56′E
155 E17 Goa var. Old Goa. ◐ state W India
Goabddális see Kåbdalis
42 H7 Goascorán, Río ≈ El Salvador/Honduras
77 O16 Goaso var. Gawso. W Ghana 06°49′N 02°27′W
81 K14 Goba Oromīya, C Ethiopia 07°02′N 39°58′E
83 C20 Gobabed Erongo, W Namibia 23°36′S 15°03′E
83 E19 Gobabis Omaheke, E Namibia 22°25′S 18°58′E
Gobannium see Abergavenny
64 M7 Goban Spur undersea feature NW Atlantic Ocean
63 H21 Gobernador Gregores Santa Cruz, S Argentina 48°43′S 70°21′W
61 F14 Gobernador Ingeniero Virasoro Corrientes, NE Argentina 28°06′S 56°00′W
162 L12 Gobi desert China/Mongolia
164 I14 Gobō Wakayama, Honshū, SW Japan 33°52′N 135°09′E
Gobolka Awdal see Awdal
Gobolka Sool see Sool
101 D14 Goch Nordrhein-Westfalen, W Germany 51°41′N 06°10′E
83 E20 Gochas Hardap, S Namibia 24°54′S 18°43′E
155 I14 Godāvari var. Godavari. ≈ C India
Godavari see Godāvari
155 L16 Godāvari, Mouths of the delta E India
15 V5 Godbout Québec, SE Canada 49°19′N 67°37′W
15 U5 Godbout ≈ Québec, SE Canada
15 U5 Godbout Est ≈ Québec, SE Canada
27 N6 Goddard Kansas, C USA 37°39′N 97°34′W
14 E15 Goderich Ontario, S Canada 43°43′N 81°43′W
Godhavn see Qeqertarsuaq
154 E10 Godhra Gujarāt, W India 22°49′N 73°40′E
Göding see Hodonín
111 K22 Gödöllő Pest, N Hungary 47°36′N 19°20′E
62 H11 Godoy Cruz Mendoza, W Argentina 32°59′S 68°49′W
11 Y11 Gods ≈ Manitoba, C Canada
11 X13 Gods Lake ◎ Manitoba, C Canada
11 Y13 Gods Lake Narrows Manitoba, C Canada 54°29′N 94°21′W
Godthaab/Godthåb see Nuuk
Godwin Austen, Mount see K2
Goede Hoop, Kaap de see Good Hope, Cape of
Goedgegun see Nhlangano
Goeie Hoop, Kaap die see Good Hope, Cape of
13 O7 Goélands, Lac aux ◎ Québec, SE Canada
98 E13 Goeree island SW Netherlands
99 F15 Goes Zeeland, SW Netherlands 51°30′N 03°55′E
Goettingen see Göttingen
19 O10 Goffstown New Hampshire, NE USA 43°01′N 71°34′W
14 E8 Gogama Ontario, S Canada 47°42′N 81°44′W
30 L3 Gogebic, Lake ◎ Michigan, N USA
30 K3 Gogebic Range hill range Michigan/Wisconsin, N USA
Gogi Lerr see Gogi, Mount
137 V13 Gogi, Mount Arm. Gogi Lerr, Az. Kükūdağ. ▲ Armenia/Azerbaijan 39°33′N 45°35′E
124 F12 Gogland, Ostrov island NW Russian Federation
111 I15 Gogolin Opolskie, S Poland 50°28′N 18°04′E
Gogonou see Gogounou
77 S14 Gogounou var. Gogonou. N Benin 10°50′N 02°50′E
152 I10 Gohāna Haryāna, N India 29°06′N 76°43′E
59 K18 Goianésia Goiás, C Brazil 15°21′S 49°02′W
59 K18 Goiânia prev. Goyania. state capital Goiás, C Brazil 16°43′S 49°18′W
59 K18 Goiás Goiás, C Brazil 15°57′S 50°07′W
59 J18 Goiás off. Estado de Goiás; prev. Goiaz, Goyaz. ◐ state C Brazil
Goiás, Estado de see Goiás
Goiaz see Goiás
Goidhoo Atoll see Horsburgh Atoll
159 R14 Goinsargoin Xizang Zizhiqu, W China 31°56′N 98°04′E
60 H10 Goio-Erê Paraná, SW Brazil 24°08′S 53°07′W
99 I15 Goirle Noord-Brabant, S Netherlands 51°31′N 05°04′E
104 H8 Góis Coimbra, N Portugal 40°10′N 08°06′W
165 Q8 Gojōme Akita, Honshū, NW Japan 39°55′N 140°07′E
149 U9 Gojra Punjab, E Pakistan 31°10′N 72°43′E
136 A11 Gökçeada var. Imroz Adasī, Gk. Imbros. island NW Turkey
Gökçeada see İmroz
146 F13 Gökdepe Rus. Gëkdepe, Geok-Tepe. Ahal Welaýaty, C Turkmenistan 38°05′N 58°08′E
136 I10 Gökırmak ≈ N Turkey
Gokleńkuy, Solonchak see Geklengkui, Solonchak
136 C16 Gökova Körfezi gulf SW Turkey
136 K15 Göksu ≈ S Turkey
136 L15 Göksun Kahramanmaraş, C Turkey 38°03′N 36°30′E
136 I17 Göksu Nehri ≈ S Turkey
83 J16 Gokwe Midlands, NW Zimbabwe 18°13′S 28°55′E
94 F13 Gol Buskerud, S Norway 60°42′N 08°57′E
153 X12 Golāghāt Assam, NE India 26°31′N 93°54′E
110 H10 Gołańcz Wielkopolskie, C Poland 52°57′N 17°17′E
138 G8 Golan Heights Ar. Al Jawlān, Heb. HaGolan. ▲▲ SW Syria
Golārā see Ārān-va-Bīdgol
Golaya Pristan see Hola Prystan'
143 T11 Golbāf Kermān, C Iran 29°51′N 57°44′E
136 M15 Gölbaşı Adıyaman, S Turkey 37°46′N 37°40′E
109 P9 Gölbner ▲ SW Austria 46°51′N 12°31′E
30 M17 Golconda Illinois, N USA 37°20′N 88°29′W
35 T3 Golconda Nevada, W USA 40°56′N 117°29′W
136 E11 Gölcük Kocaeli, NW Turkey 40°44′N 29°50′E
108 I7 Goldach Sankt Gallen, NE Switzerland 47°28′N 09°28′E
110 N7 Gołdap Ger. Goldap. Warmińsko-Mazurskie, NE Poland 54°19′N 22°23′E
32 E15 Gold Beach Oregon, NW USA 42°25′N 124°27′W
Goldberg see Złotoryja
183 V3 Gold Coast cultural region Queensland, E Australia
68 D12 Gold Coast coastal region S Ghana
39 R10 Gold Creek Alaska, USA 62°48′N 149°40′W
11 O16 Golden British Columbia, SW Canada 51°19′N 116°58′W
37 T4 Golden Colorado, C USA 39°40′N 105°12′W
184 I13 Golden Bay bay South Island, New Zealand
27 R7 Golden City Missouri, C USA 37°23′N 94°05′W
32 I11 Goldendale Washington, NW USA 45°49′N 120°49′W
Goldener Tisch see Zlatý Stôl
44 L13 Golden Grove E Jamaica 17°56′N 76°17′W
14 J12 Golden Lake ◎ Ontario, SE Canada
22 K10 Golden Meadow Louisiana, S USA 29°22′N 90°15′W
Golden Sands see Zlatni Pyasatsi
Golden State, The see California
83 K16 Golden Valley Mashonaland West, N Zimbabwe 18°11′S 29°50′E
35 U9 Goldfield Nevada, W USA 37°42′N 117°15′W
Goldingen see Kuldīga
Goldmarkt see Zlatna
10 K17 Gold River Vancouver Island, British Columbia, SW Canada 49°41′N 126°05′W
21 V10 Goldsboro North Carolina, SE USA 35°23′N 78°00′W
24 M8 Goldsmith Texas, SW USA 31°58′N 102°36′W
25 R8 Goldthwaite Texas, SW USA 31°28′N 98°35′W
137 R11 Göle Ardahan, NE Turkey 40°47′N 42°36′E
Golema Ada see Ostrovo
114 H9 Golema Planina ▲▲ W Bulgaria
114 F9 Golemi Vrah var. Golemi Vrŭkh. ▲ W Bulgaria 42°41′N 22°38′E
Golemi Vrŭkh see Golemi Vrah
110 D8 Goleniów Ger. Gollnow. Zachodnio-pomorskie, NW Poland 53°34′N 14°48′E
143 R3 Golestān off. Ostān-e Golestān. ◐ province N Iran
Golestān, Ostān-e see Golestān
35 Q14 Goleta California, W USA 34°26′N 119°50′W
43 O16 Golfito Puntarenas, SE Costa Rica 08°42′N 83°10′W
25 T13 Goliad Texas, SW USA 28°40′N 97°26′W
113 L14 Golija ▲▲ SW Serbia
Golinka see Gongbo'gyamda
113 O16 Goljak ▲▲ SE Serbia
136 M12 Gölköy Ordu, N Turkey 40°42′N 37°37′E
Gollel see Lavumisa
109 X3 Göllersbach ≈ NE Austria
Gollnow see Goleniów
Golmo see Golmud
159 P10 Golmud var. Ge'e'mu, Golmo, Chin. Ko-erh-mu. Qinghai, C China 36°23′N 94°56′E
103 Y14 Golo ≈ Corse, France, C Mediterranean Sea
Golovanevsk see Holovanivs'k
Golovchin see Halowchyn
39 N9 Golovin Alaska, USA 64°33′N 162°54′W
142 M7 Golpāyegān var. Gulpaigan. Eşfahān, W Iran 33°23′N 50°18′E
Golshan see Ţabas
Gol'shany see Hal'shany
96 J7 Golspie N Scotland, United Kingdom 57°59′N 03°56′W
112 O11 Golubac Serbia, NE Serbia 44°38′N 21°36′E
110 J9 Golub-Dobrzyń Kujawski-pomorskie, C Poland 53°07′N 19°03′E
145 S7 Golubovka Pavlodar, N Kazakhstan 53°07′N 74°11′E
82 B11 Golungo Alto Kwanza Norte, NW Angola 09°10′S 14°45′E
114 M8 Golyama Kamchia var. Golyama Kamchiya. ≈ E Bulgaria
Golyama Kamchiya see Golyama Kamchia
114 L8 Golyama Reka ≈ N Bulgaria
Golyama Syutka see Golyama Syutkya
114 H11 Golyama Syutkya var. Golyama Syutka. ▲ SW Bulgaria 41°55′N 24°03′E
114 I12 Golyam Perelik ▲ S Bulgaria 41°37′N 24°34′E
114 I11 Golyam Persenk ▲ S Bulgaria 41°50′N 24°33′E
79 P19 Goma Nord-Kivu, NE Dem. Rep. Congo 01°36′S 29°08′E
153 N13 Gomati var. Gumti. ≈ N India
77 X14 Gombe Gombe, E Nigeria 10°19′N 11°02′E
67 U10 Gombe var. Igombe. ≈ E Tanzania
77 Y14 Gombi Adamawa, E Nigeria 10°07′N 12°45′E
Gombroon see Bandar-e 'Abbās
Gomel' see Homyel'
Gomel'skaya Oblast' see Homyel'skaya Voblasts'
64 N11 Gomera island Islas Canarias, Spain, NE Atlantic Ocean
40 I5 Gómez Farías Chihuahua, N Mexico 29°25′N 107°46′W
40 L8 Gómez Palacio Durango, C Mexico 25°39′N 103°30′W
158 J13 Gomo Xizang Zizhiqu, W China 33°37′N 86°40′E
143 T6 Gonābād var. Gunabad. Khorāsān-e Razavī, NE Iran 34°30′N 59°18′E
44 L8 Gonaïves var. Les Gonaïves. N Haiti 19°26′N 72°41′W
123 Q12 Gonam ≈ NE Russian Federation
44 L9 Gonâve, Canal de la var. Canal de Sud. channel N Caribbean Sea
44 K9 Gonâve, Golfe de la gulf N Caribbean Sea
Gonāveh see Bandar-e Gonāveh
44 K9 Gonâve, Île de la island C Haiti
Gonbadān see Do Gonbadān
143 Q3 Gonbad-e Kāvūs var. Gunbad-i-Qawus. Golestān, N Iran 37°15′N 55°11′E
152 M12 Gonda Uttar Pradesh, N India 27°08′N 81°58′E
Gondar see Gonder
80 I11 Gonder var. Gondar. Āmara, NW Ethiopia 12°36′N 37°27′E
78 J13 Gondey Moyen-Chari, S Chad 09°07′N 19°10′E
154 J12 Gondia Mahārāshtra, C India 21°27′N 80°12′E
104 G6 Gondomar Porto, NW Portugal 41°10′N 08°35′W
136 C12 Gönen Balıkesir, W Turkey 40°06′N 27°39′E
136 C12 Gönen Çayı ≈ NW Turkey
159 O15 Gongbo'gyamda var. Golinka. Xizang Zizhiqu, W China 30°03′N 93°10′E
Gongchang see Longxi
159 N16 Gonggar var. Gyixong. Xizang Zizhiqu, W China 29°18′N 90°56′E
160 G9 Gongga Shan ▲ C China 29°50′N 101°55′E
159 T10 Gonghe var. Qabqa. Qinghai, C China 36°20′N 100°46′E
Gongjiang see Yudu
158 I5 Gongliu var. Tokkuztara. Xinjiang Uygur Zizhiqu, NW China 43°29′N 82°16′E
77 W14 Gongola ≈ E Nigeria
Gongoleh State see Jonglei
183 P5 Gongolgon New South Wales, SE Australia 30°19′S 146°57′E
159 Q6 Gongpoquan Gansu, N China 41°45′N 100°27′E
Gongquan see Gongxian
Gongtang see Damxung
160 I10 Gongxian var. Gongquan, Gong Xian. Sichuan, C China 28°25′N 104°51′E
Gong Xian see Gongxian
157 V10 Gongzhuling prev. Huaide. Jilin, NE China 43°30′N 124°48′E
159 S14 Gonjo Xizang Zizhiqu, W China 30°51′N 98°16′E
107 B20 Gonnesa Sardegna, Italy, C Mediterranean Sea 39°15′N 08°27′E
Gonni/Gónnos see Gónnoi
115 F15 Gónnoi var. Gonni, Gónnos; prev. Dereli. Thessalía, C Greece 39°52′N 22°27′E
Gōnoura see Iki
35 O11 Gonzales California, W USA 36°30′N 121°26′W
22 J9 Gonzales Louisiana, S USA 30°14′N 90°55′W
25 T12 Gonzales Texas, SW USA 29°31′N 97°29′W
41 P11 González Tamaulipas, C Mexico 22°50′N 98°25′W
21 V6 Goochland Virginia, NE USA 37°42′N 77°54′W
195 X14 Goodenough, Cape headland Antarctica 66°15′S 126°35′E
186 F9 Goodenough Island var. Morata. island SE Papua New Guinea
39 N8 Goodhope Bay bay Alaska, USA
83 D26 Good Hope, Cape of Afr. Kaap de Goede Hoop, Kaap die Goeie Hoop. headland SW South Africa 34°19′S 18°25′E
10 K10 Good Hope Lake British Columbia, W Canada 59°15′N 129°18′W
83 E23 Goodhouse Northern Cape, W South Africa 28°54′S 18°13′E
33 O15 Gooding Idaho, NW USA 42°56′N 114°42′W
26 H3 Goodland Kansas, C USA 39°20′N 101°43′W
173 Y15 Goodlands NW Mauritius 20°02′S 57°39′E
20 J8 Goodlettsville Tennessee, S USA 36°19′N 86°42′W
39 N13 Goodnews Alaska, USA 59°07′N 161°35′W
25 O3 Goodnight Texas, SW USA 35°00′N 101°07′W
183 Q4 Goodooga New South Wales, SE Australia 29°09′S 147°30′E
29 N4 Goodrich North Dakota, N USA 47°24′N 100°07′W
25 W10 Goodrich Texas, SW USA 30°36′N 94°57′W
29 X10 Goodview Minnesota, N USA 44°04′N 91°42′W
26 H8 Goodwell Oklahoma, C USA 36°36′N 101°38′W
97 N17 Goole E England, United Kingdom 53°43′N 00°46′W
183 O8 Goolgowi New South Wales, SE Australia 34°00′S 145°43′E
182 I10 Goolwa South Australia 35°31′S 138°43′E
181 Y11 Goondiwindi Queensland, E Australia 28°33′S 150°22′E
98 O11 Goor Overijssel, E Netherlands 52°13′N 06°33′E
Goose Bay see Happy Valley-Goose Bay
33 V13 Gooseberry Creek ≈ Wyoming, C USA
21 S14 Goose Creek South Carolina, SE USA 32°58′N 80°01′W
63 M23 Goose Green var. Prado del Ganso. East Falkland, Falkland Islands 51°52′S 59°W
16 D8 Goose Lake var. Lago dos Gansos. ◎ California/Oregon, W USA
29 Q4 Goose River ≈ North Dakota, N USA
153 T16 Gopalganj Dhaka, S Bangladesh 23°00′N 89°48′E
153 O12 Gopālganj Bihār, N India 26°28′N 84°26′E
Gopher State see Minnesota
101 I22 Göppingen Baden-Württemberg, SW Germany 48°42′N 09°39′E
110 G13 Góra Ger. Guhrau. Dolnośląskie, SW Poland 51°40′N 16°03′E
110 M12 Góra Kalwaria Mazowieckie, C Poland 52°00′N 21°14′E
153 O12 Gorakhpur Uttar Pradesh, N India 26°45′N 83°23′E
Gora Kyuren see Kürendag
Gorany see Harany
113 J14 Goražde Federacija Bosne I Hercegovine, SE Bosnia and Herzegovina 43°39′N 18°58′E
Gorbovichi see Harbavichy
Gorče Petrov see Đorče Petrov
0 E9 Gorda Ridges undersea feature NE Pacific Ocean 41°30′N 128°00′W
Gordiaz see Gardēz
78 K12 Gordil Vakaga, N Central African Republic 09°37′N 21°42′E
23 U5 Gordon Georgia, SE USA 32°52′N 83°19′W
28 K12 Gordon Nebraska, C USA 42°48′N 102°12′W
25 R7 Gordon Texas, SW USA 32°32′N 98°21′W
28 L13 Gordon Creek ≈ Nebraska, C USA
63 I25 Gordon, Isla island S Chile
183 O17 Gordon, Lake ◎ Tasmania, SE Australia
183 O17 Gordon River ≈ Tasmania, SE Australia
21 V5 Gordonsville Virginia, NE USA 38°08′N 78°11′W
78 H13 Goré Logone-Oriental, S Chad 07°55′N 16°38′E
80 H13 Gorē Oromīya, C Ethiopia 08°08′N 35°33′E
185 D24 Gore Southland, South Island, New Zealand 46°06′S 168°58′E
14 D11 Gore Bay Manitoulin Island, Ontario, S Canada 45°54′N 82°28′W
25 Q5 Goree Texas, SW USA 33°28′N 99°31′W
137 O11 Görele Giresun, NE Turkey 41°00′N 39°00′E
19 N6 Gore Mountain ▲ Vermont, NE USA 44°55′N 71°47′W
39 R13 Gore Point headland Alaska, USA 59°12′N 150°57′W
37 R4 Gore Range ▲▲ Colorado, C USA
97 F19 Gorey Ir. Guaire. Wexford, SE Ireland 52°40′N 06°18′W
143 R12 Gorgāb Kermān, S Iran
143 Q4 Gorgān var. Astarabad, Astrabad, Gurgan, prev. Asterābād; anc. Hyrcania. Golestān, N Iran 36°53′N 54°28′E
143 Q4 Gorgān, Rūd-e ≈ N Iran
76 I10 Gorgol ◐ region S Mauritania
106 D12 Gorgona, Isola di island Archipelago Toscano, C Italy
19 P8 Gorham Maine, NE USA 43°41′N 70°27′W
137 T10 Gori C Georgia 42°00′N 44°07′E
98 I13 Gorinchem var. Gorkum. Zuid-Holland, C Netherlands 51°50′N 04°59′E
137 V13 Goris SE Armenia 39°31′N 46°20′E
124 K16 Goritsy Tverskaya Oblast', W Russian Federation 57°09′N 36°44′E
106 J7 Gorizia Ger. Görz. Friuli-Venezia Giulia, NE Italy 45°57′N 13°37′E
116 G13 Gorj ◐ county SW Romania
109 W12 Gorjanci var. Uskočke Planine, Žumberak, Žumberačko Gorje, Ger. Uskokengebirge; prev. Sichelburger Gerbirge. ▲ Croatia/Slovenia Europe see also Žumberačko Gorje
Görkau see Jirkov
Gorki see Horki
Gor'kiy see Nizhniy Novgorod
Gorkum see Gorinchem
95 I23 Gørlev Sjælland, E Denmark 55°33′N 11°14′E
111 M17 Gorlice Małopolskie, S Poland 49°40′N 21°09′E
101 Q15 Görlitz Sachsen, E Germany 51°09′N 14°58′E
Görlitz see Zgorzelec
Gorlovka see Horlivka
25 R7 Gorman Texas, SW USA 32°12′N 98°40′W
21 T3 Gormania West Virginia, NE USA 39°16′N 79°18′W
Gorna Dzhumaya see Blagoevgrad
114 K8 Gorna Oryahovitsa , Gorna Oryakhovitsa. Veliko Tŭrnovo, N Bulgaria 43°07′N 25°40′E
Gorna Oryakhovitsa see Gorna Oryahovitsa
114 J8 Gorna Studena Veliko Tŭrnovo, N Bulgaria 43°26′N 25°21′E
Gornja Mužlja see Mužlja
109 X9 Gornja Radgona Ger. Oberradkersburg. NE Slovenia 46°39′N 16°00′E
112 M13 Gornji Milanovac Serbia, C Serbia 44°01′N 20°26′E
112 G13 Gornji Vakuf var. Uskoplje. Federacija Bosne I Hercegovine, SW Bosnia and Herzegovina 43°55′N 17°34′E
122 J13 Gorno-Altaysk Respublika Altay, S Russian Federation 51°59′N 85°56′E
Gorno-Altayskaya Respublika see Altay, Respublika
123 N12 Gorno-Chuyskiy Irkutskaya Oblast', C Russian Federation 57°33′N 111°38′E
125 V14 Gornozavodsk Permskiy Kray, NW Russian Federation 58°21′N 58°24′E
125 V14 Gornozavodsk Ostrov Sakhalin, Sakhalinskaya Oblast', SE Russian Federation 46°34′N 141°52′E
122 I13 Gornyak Altayskiy Kray, S Russian Federation 50°58′N 81°24′E
123 O14 Gornyy Chitinskaya Oblast', S Russian Federation 51°42′N 114°16′E
127 R8 Gornyy Saratovskaya Oblast', W Russian Federation 51°42′N 48°26′E
Gornyy Altay see Altay, Respublika
127 O10 Gornyy Balykley Volgogradskaya Oblast', SW Russian Federation 49°37′N 45°03′E
80 I13 Goroch'an ▲ W Ethiopia 09°09′N 37°16′E
116 J7 Gorodenka var. Horodenka. Ivano-Frankivs'ka Oblast', W Ukraine 48°41′N 25°28′E
127 O3 Gorodets Nizhegorodskaya Oblast', W Russian Federation 56°36′N 43°27′E
Gorodets see Haradzyets
Gorodeya see Haradzeya
127 P6 Gorodishche Penzenskaya Oblast', W Russian Federation 53°17′N 45°39′E
Gorodishche see Horodyshche
Gorodnya see Horodnya
Gorodok see Haradok
Gorodok/Gorodok Yagellonski see Horodok
126 M13 Gorodovikovsk Respublika Kalmykiya, SW Russian Federation 46°07′N 41°56′E
186 D7 Goroka Eastern Highlands, C Papua New Guinea 06°02′S 145°22′E
Gorokhov see Horokhiv
127 N3 Gorokhovets Vladimirskaya Oblast', W Russian Federation 56°12′N 42°40′E
77 Q11 Gorom-Gorom NE Burkina Faso 14°27′N 00°14′W
171 U13 Gorong, Kepulauan island group E Indonesia
83 M17 Gorongosa Sofala, C Mozambique 18°40′S 34°03′E
171 P11 Gorontalo Sulawesi, C Indonesia 00°33′N 123°05′E
171 O11 Gorontalo off. Propinsi Gorontalo. ◐ province N Indonesia
Propinsi Gorontalo see Gorontalo
Gorontalo, Teluk see Tomini, Teluk
110 L7 Górowo Iławeckie Ger. Landsberg. Warmińsko-Mazurskie, NE Poland 54°18′N 20°30′E
98 M7 Gorredijk Fris. De Gordyk. Fryslân, N Netherlands 53°00′N 06°04′E
84 C14 Gorringe Ridge undersea feature E Atlantic Ocean 36°40′N 11°35′W
98 M11 Gorssel Gelderland, E Netherlands 52°12′N 06°13′E
109 T8 Görtschitz ≈ S Austria
Goryn see Horyn'
145 S15 Gory Shu-Ile Kaz. Shū-Ile Taūlary; prev. Chu-Iliyskiye Gory. ▲▲ S Kazakhstan
Görz see Gorizia
110 E10 Gorzów Wielkopolski Ger. Landsberg, Landsberg an der Warthe. Lubuskie, W Poland 52°44′N 15°12′E
146 B10 Goşabo var. Goshoba, Rus. Koshoba. Balkan Welaýaty, NW Turkmenistan 40°28′N 54°11′E
108 G9 Göschenen Uri, C Switzerland 46°40′N 08°36′E
165 O11 Gosen Niigata, Honshū, C Japan 37°45′N 139°11′E
163 Y13 Goseong prev. Kosŏng. SE North Korea 38°41′N 128°14′E
183 T8 Gosford New South Wales, SE Australia 33°25′S 151°18′E
31 P11 Goshen Indiana, N USA 41°34′N 85°49′W
18 K13 Goshen New York, NE USA 41°24′N 74°17′W
Goshoba see Goşabo
Goshoba see Goşoba
165 Q7 Goshogawara var. Gosyogawara. Aomori, Honshū, C Japan 40°47′N 140°24′E
Goshquduq Qum see Tosquduq Qumlari
101 J14 Goslar Niedersachsen, C Germany 51°55′N 10°25′E
27 Y9 Gosnell Arkansas, C USA 35°57′N 89°58′W
146 B10 Goşoba var. Goshoba, Rus. Koshoba. Balkanskiy Velayat, NW Turkmenistan 40°28′N 54°11′E
112 C11 Gospić Lika-Senj, C Croatia 44°32′N 15°21′E
97 N23 Gosport S England, United Kingdom 50°48′N 01°08′W
94 D9 Gossa island S Norway
108 H7 Gossau Sankt Gallen, NE Switzerland 47°25′N 09°16′E
99 G20 Gosselies var. Goss'lies. Hainaut, S Belgium 50°28′N 04°26′E
77 P10 Gossi Tombouctou, C Mali 15°44′N 01°19′W
Goss'lies see Gosselies
113 N18 Gostivar W FYR Macedonia 41°48′N 20°55′E
Gostomel' see Hostomel'
110 G12 Gostyń var. Gostyn. Wielkopolskie, C Poland 51°52′N 17°00′E
Gostyn see Gostyń
110 K11 Gostynin Mazowiesckie, C Poland 52°25′N 19°27′E
Gosyogawara see Goshogawara
95 J18 Göta Älv ≈ S Sweden
95 N17 Göta kanal canal S Sweden
95 K18 Götaland cultural region S Sweden
95 H17 Göteborg Eng. Gothenburg. Västra Götaland, S Sweden 57°43′N 11°58′E
77 X16 Gotel Mountains ▲▲ E Nigeria
95 K17 Götene Västra Götaland, S Sweden 58°32′N 13°29′E
Gotera see San Francisco
101 K16 Gotha Thüringen, C Germany 50°57′N 10°43′E
29 N15 Gothenburg Nebraska, C USA 40°57′N 100°09′W
Gothenburg see Göteborg
77 R12 Gothèye Tillabéri, SW Niger 13°52′N 01°27′E
Gothland see Gotland
95 P19 Gotland var. Gothland, Gottland. ◐ county SE Sweden
95 O18 Gotland island SE Sweden
164 A14 Gotō var. Hukue; prev. Fukue. Nagasaki, Fukue-jima, SW Japan 32°41′N 128°52′E
164 B13 Gotō-rettō island group SW Japan
114 H12 Gotse Delchev prev. Nevrokop. Blagoevgrad, SW Bulgaria 41°33′N 23°42′E
95 P17 Gotska Sandön island SE Sweden
101 I15 Göttingen var. Goettingen. Niedersachsen, C Germany 51°33′N 09°55′E
Gottland see Gotland
93 I16 Gottne Västernorrland, C Sweden 63°27′N 18°25′E
Gottschee see Kočevje
Gottwaldov see Zlín
146 B11 Goturdepe Rus. Koturdepe. Balkan Welaýaty, W Turkmenistan 39°32′N 53°39′E
108 I7 Götzis Vorarlberg, NW Austria 47°21′N 09°40′E
98 H12 Gouda Zuid-Holland, C Netherlands 52°01′N 04°42′E
76 I11 Goudiri var. Goudiry. E Senegal 14°12′N 12°41′W
Goudiry see Goudiri
77 X12 Goudoumaria Diffa, S Niger 13°28′N 11°15′E
15 R9 Gouffre, Rivière du ≈ Québec, SE Canada
65 M19 Gough Fracture Zone tectonic feature S Atlantic Ocean
65 M19 Gough Island island Tristan da Cunha, S Atlantic Ocean
15 N8 Gouin, Réservoir ⊡ Québec, SE Canada
14 B10 Goulais River Ontario, S Canada 46°41′N 84°22′W
183 R9 Goulburn New South Wales, SE Australia 34°45′S 149°44′E
183 O11 Goulburn River ≈ Victoria, SE Australia
195 O10 Gould Coast physical region Antarctica
Goulimime see Guelmime
114 F13 Gouménissa Kentrikí Makedonía, N Greece 40°56′N 22°27′E
77 O10 Goundam Tombouctou, NW Mali 16°27′N 03°39′W
78 H12 Goundi Mandoul, S Chad 09°22′N 17°21′E
78 G12 Gounou-Gaya Mayo-Kébbi Est, SW Chad 09°37′N 15°30′E
Gourci see Goursi
Gourcy see Goursi
102 M13 Gourdon Lot, S France 44°45′N 01°22′E
77 W11 Gouré Zinder, SE Niger 13°59′N 10°16′E
102 G6 Gourin Morbihan, NW France 48°07′N 03°37′W

◆ Country ● Country Capital ◇ Dependent Territory ○ Dependent Territory Capital ◐ Administrative Regions ✈ International Airport ▲ Mountain ▲▲ Mountain Range ✕ Volcano ≈ River ◎ Lake ⊡ Reservoir

Gourma-Rharous – Grenaa

77 P10 **Gourma-Rharous** Tombouctou, C Mali 16°54′N 01°55′W
103 N4 **Gournay-en-Bray** Seine-Maritime, N France 49°29′N 01°42′E
78 J6 **Gouro** Ennedi-Ouest, N Chad 19°26′N 19°36′E
77 O12 **Goursi** *var.* Gourci, Gourcy. NW Burkina Faso 13°13′N 02°20′W
104 H8 **Gouveia** Guarda, N Portugal 40°29′N 07°35′W
18 I7 **Gouverneur** New York, NE USA 44°20′N 75°27′W
99 L21 **Gouvy** Luxembourg, E Belgium 50°10′N 05°55′E
45 R14 **Gouyave** *var.* Charlotte Town. NW Grenada 12°10′N 61°44′W
Goverla, Gora *see* Hoverla, Hora
59 N20 **Governador Valadares** Minas Gerais, SE Brazil 18°51′S 41°57′W
171 R8 **Governor Generoso** Mindanao, S Philippines 06°36′N 126°06′E
44 I2 **Governor's Harbour** Eleuthera Island, C The Bahamas 25°11′N 76°15′W
162 F9 **Govĭ-Altay** ◈ *province* SW Mongolia
162 I10 **Govĭ Altayn Nuruu** ▲▲ S Mongolia
154 L9 **Govind Ballabh Pant Sāgar** ⊠ C India
152 I7 **Govind Sāgar** ⊠ NE India
162 M8 **Govĭ-Sumber** ◈ *province* C Mongolia
Govurdak *see* Magdanly
18 D11 **Gowanda** New York, NE USA 42°25′N 78°55′W
148 J10 **Gowd-e Zereh, Dasht-e** *var.* Guad-i-Zirreh. *marsh* SW Afghanistan
14 F8 **Gowganda** Ontario, S Canada 47°41′N 80°46′W
14 G8 **Gowganda Lake** ◎ Ontario, S Canada
148 M5 **Gōwr** *prev.* Ghowr. ◈ *province* C Afghanistan
29 U13 **Gowrie** Iowa, C USA 42°16′N 94°17′W
Gowurdak *see* Magdanly
61 C15 **Goya** Corrientes, NE Argentina 29°10′S 59°15′W
Goyania *see* Goiânia
Goyaz *see* Goiás
137 X11 **Göyçay** *Rus.* Geokchay. C Azerbaijan 40°38′N 47°44′E
137 V11 **Göygöl** ; *prev.* Xanlar. NW Azerbaijan 40°37′N 46°18′E
146 D10 **Goymat** *Rus.* Koymat. Balkan Welaýaty, NW Turkmenistan 40°23′N 55°45′E
146 D10 **Goymatdag, Gory** *Rus.* Gory Koymatdag. *hill range* Balkan Welaýaty, NW Turkmenistan
136 F12 **Göynük** Bolu, NW Turkey 40°24′N 30°45′E
165 R9 **Goyō-san** ▲ Honshū, C Japan 39°12′N 141°40′E
78 K11 **Goz Beïda** Sila, SE Chad 12°06′N 21°22′E
146 M10 **G'ozg'on** *Rus.* Gazgan. Navoiy Viloyati, C Uzbekistan 40°36′N 65°29′E
158 H11 **Gozha Co** ◎ W China
121 O15 **Gozo** *var.* Ghawdex. *island* N Malta
80 H9 **Gōz Regeb** Kassala, NE Sudan 16°03′N 35°33′E
83 H25 **Graaff-Reinet** Eastern Cape, S South Africa 32°15′S 24°32′E
Graasten *see* Gråsten
76 L17 **Grabo** SW Ivory Coast 04°57′N 07°30′W
112 P11 **Grabovica** Serbia, E Serbia 44°30′N 22°29′E
110 I13 **Grabów nad Prosną** Wielkopolskie, C Poland 51°30′N 18°06′E
108 I8 **Grabs** Sankt Gallen, NE Switzerland 47°10′N 09°27′E
112 D12 **Gračac** Zadar, SW Croatia 44°18′N 15°52′E
112 I11 **Gračanica** Federacija Bosne I Hercegovine, NE Bosnia and Herzegovina 44°41′N 18°20′E
14 L11 **Gracefield** Québec, SE Canada 46°06′N 76°03′W
99 K19 **Grâce-Hollogne** Liège, E Belgium 50°38′N 05°30′E
23 R8 **Graceville** Florida, SE USA 30°57′N 85°31′W
29 R8 **Graceville** Minnesota, N USA 45°34′N 96°25′W
42 G6 **Gracias** Lempira, W Honduras 14°35′N 88°35′W
Gracias *see* Lempira
42 L5 **Gracias a Dios** ◈ *department* E Honduras
43 O6 **Gracias a Dios, Cabo de** *headland* Honduras/Nicaragua 15°00′N 83°10′W
64 O2 **Graciosa** *var.* Ilha Graciosa. *island* Azores, Portugal, NE Atlantic Ocean
64 Q11 **Graciosa** *island* Islas Canarias, Spain, NE Atlantic Ocean
Graciosa, Ilha *see* Graciosa
112 I11 **Gradačac** Federacija Bosne I Hercegovine, N Bosnia and Herzegovina 44°51′N 18°24′E
59 J15 **Gradaús, Serra dos** ▲▲ C Brazil
104 L3 **Gradefes** Castilla y León, N Spain 42°37′N 05°14′W
Gradiška *see* Bosanska Gradiška
Gradizhsk *see* Hradyz'k
106 J7 **Grado** Friuli-Venezia Giulia, NE Italy 45°41′N 13°24′E
Grado *see* Grau
113 P19 **Gradsko** C FYR Macedonia 41°34′N 21°56′E
37 V11 **Grady** New Mexico, SW USA 34°49′N 103°19′W
Grad Zagreb *see* Zagreb
29 T12 **Graettinger** Iowa, C USA 43°14′N 94°45′W
101 M23 **Grafing** Bayern, SE Germany 48°01′N 11°57′E
25 S6 **Graford** Texas, SW USA 32°56′N 98°15′W
183 V5 **Grafton** New South Wales, SE Australia 29°41′S 152°55′E
29 Q3 **Grafton** North Dakota, N USA 48°24′N 97°24′W
21 S3 **Grafton** West Virginia, NE USA 39°21′N 80°03′W
21 T9 **Graham** North Carolina, SE USA 36°05′N 79°25′W
25 R6 **Graham** Texas, SW USA 33°07′N 98°36′W
Graham Bell Island *see* Greem-Bell, Ostrov
10 I13 **Graham Island** *island* Queen Charlotte Islands, British Columbia, SW Canada
19 S6 **Graham Lake** ◎ Maine, NE USA
194 H4 **Graham Land** *physical region* Antarctica
37 N15 **Graham, Mount** ▲ Arizona, SW USA 32°42′N 109°52′W
Grahamstad *see* Grahamstown
83 I25 **Grahamstown** *Afr.* Grahamstad. Eastern Cape, S South Africa 33°18′S 26°32′E
Grahovo *see* Bosansko Grahovo
68 C12 **Grain Coast** *coastal region* S Liberia
169 S17 **Grajagan, Teluk** *bay* Jawa, S Indonesia
59 L14 **Grajaú** Maranhão, E Brazil 05°50′S 46°12′W
58 M13 **Grajaú, Rio** ∼ NE Brazil
110 O8 **Grajewo** Podlaskie, NE Poland 53°38′N 22°26′E
95 F24 **Gram** Syddanmark, SW Denmark 55°18′N 09°03′E
103 N13 **Gramat** Lot, S France 44°45′N 01°45′E
22 H5 **Grambling** Louisiana, S USA 32°31′N 92°43′W
115 C14 **Grámmos** ▲▲ Albania/Greece
96 I9 **Grampian Mountains** ▲▲ C Scotland, United Kingdom
182 L12 **Grampians, The** ▲▲ Victoria, SE Australia
98 O9 **Gramsbergen** Overijssel, E Netherlands 52°37′N 06°39′E
113 L21 **Gramsh** *var.* Gramshi. Elbasan, C Albania 40°52′N 20°12′E
Gramshi *see* Gramsh
Gran *see* Esztergom, Hungary
Gran *see* Hron
54 F11 **Granada** Meta, C Colombia 03°33′N 73°44′W
42 J10 **Granada** Granada, SW Nicaragua 11°55′N 85°58′W
105 N14 **Granada** Andalucía, S Spain 37°13′N 03°41′W
37 W6 **Granada** Colorado, C USA 38°00′N 102°18′W
42 J11 **Granada** ◈ *department* SW Nicaragua
105 N14 **Granada** ◈ *province* Andalucía, S Spain
63 I21 **Gran Antiplanicie Central** *plain* S Argentina
97 E17 **Granard** *Ir.* Gránard. C Ireland 53°47′N 07°30′W
Gránard *see* Granard
63 J20 **Gran Bajo** *basin* S Argentina
63 J15 **Gran Bajo del Gualicho** *basin* E Argentina
63 I21 **Gran Bajo de San Julián** *basin* SE Argentina
25 S7 **Granbury** Texas, SW USA 32°27′N 97°47′W
15 P12 **Granby** Québec, SE Canada 45°23′N 72°44′W
27 S8 **Granby** Missouri, C USA 36°55′N 94°14′W
37 S3 **Granby, Lake** ⊠ Colorado, C USA
64 O12 **Gran Canaria** *var.* Grand Canary. *island* Islas Canarias, Spain, NE Atlantic Ocean
47 T11 **Gran Chaco** *var.* Chaco. *lowland plain* South America
45 R14 **Grand Anse** SW Grenada 12°01′N 61°45′W
Grand-Anse *see* Portsmouth
44 G1 **Grand Bahama Island** *island* N The Bahamas
Grand Balé *see* Tui
103 U7 **Grand Ballon** *Ger.* Ballon de Guebwiller. ▲ NE France 47°53′N 07°06′E
13 T13 **Grand Bank** Newfoundland, Newfoundland and Labrador, SE Canada 47°06′N 55°48′W
64 I7 **Grand Banks of Newfoundland** *undersea feature* NW Atlantic Ocean 45°00′N 40°00′W
Grand Bassa *see* Buchanan
77 N17 **Grand-Bassam** *var.* Bassam. SE Ivory Coast 05°14′N 03°45′W
14 E16 **Grand Bend** Ontario, S Canada 43°17′N 81°46′W
76 L17 **Grand-Bérébi** *var.* Grand-Béréby. SW Ivory Coast 04°38′N 06°55′W
Grand-Béréby *see* Grand-Bérébi
45 X11 **Grand-Bourg** Marie-Galante, SE Guadeloupe 15°53′N 61°19′W
44 M6 **Grand Caicos** *var.* Middle Caicos. *island* C Turks and Caicos Islands
14 K12 **Grand Calumet, Île du** *island* Québec, SE Canada
97 E18 **Grand Canal** *Ir.* An Chanáil Mhór. *canal* C Ireland
Grand Canal *see* Da Yunhe
Grand Canary *see* Gran Canaria
36 K10 **Grand Canyon** Arizona, SW USA 36°01′N 112°10′W
36 J9 **Grand Canyon** *canyon* Arizona, SW USA
Grand Canyon State *see* Arizona
44 D8 **Grand Cayman** *island* SW Cayman Islands
11 R14 **Grand Centre** Alberta, SW Canada 54°25′N 110°13′W
76 L17 **Grand Cess** SE Liberia 04°36′N 08°12′W
108 D12 **Grand Combin** ▲ S Switzerland 45°58′N 07°27′E
32 K8 **Grand Coulee** Washington, NW USA 47°56′N 119°00′W
32 J8 **Grand Coulee** *valley* Washington, NW USA
45 X5 **Grand Cul-de-Sac Marin** *bay* N Guadeloupe
Grand Duchy of Luxembourg *see* Luxembourg
63 I22 **Grande, Bahía** *bay* S Argentina
11 N14 **Grande Cache** Alberta, W Canada 53°53′N 119°07′W
103 U12 **Grande Casse** ▲ E France 45°22′N 06°50′E
Grande Comore *see* Ngazidja
61 G18 **Grande, Cuchilla** *hill range* E Uruguay
45 S5 **Grande de Añasco, Río** ∼ W Puerto Rico
Grande de Chiloé, Isla *see* Chiloé, Isla de
58 J12 **Grande de Gurupá, Ilha** *river island* NE Brazil
57 K21 **Grande de Lipez, Río** ∼ SW Bolivia
45 U6 **Grande de Loíza, Río** ∼ E Puerto Rico
45 T5 **Grande de Manatí, Río** ∼ C Puerto Rico
42 L9 **Grande de Matagalpa, Río** ∼ C Nicaragua
40 K12 **Grande de Santiago, Río** *var.* Santiago. ∼ C Mexico
43 O15 **Grande de Térraba, Río** *var.* Río Térraba. ∼ SE Costa Rica
12 J9 **Grande Deux, Réservoir la** ⊠ Québec, E Canada
60 O10 **Grande, Ilha** *island* SE Brazil
11 O13 **Grande Prairie** Alberta, W Canada 55°10′N 118°52′W
74 I8 **Grand Erg Occidental** *desert* W Algeria
74 L9 **Grand Erg Oriental** *desert* Algeria/Tunisia
57 M18 **Grande, Río** ∼ C Bolivia
59 J20 **Grande, Rio** ∼ S Brazil
2 F15 **Grande, Rio** *var.* Río Bravo, *Sp.* Río Bravo del Norte, Bravo del Norte. ∼ Mexico/USA
15 Y7 **Grande-Rivière** Québec, SE Canada 48°27′N 64°37′W
15 Y6 **Grande Rivière** ∼ Québec, SE Canada
44 M8 **Grande-Rivière-du-Nord** N Haiti 19°36′N 72°10′W
62 K9 **Grande, Salina** *var.* Gran Salitral. *salt lake* C Argentina
15 S7 **Grandes-Bergeronnes** Québec, SE Canada 48°16′N 69°32′W
47 W6 **Grande, Serra** ▲▲ W Brazil
40 K4 **Grande, Sierra** ▲▲ N Mexico
103 S12 **Grandes Rousses** ▲▲ E France
63 K17 **Grandes, Salinas** *salt lake* E Argentina
45 Y5 **Grande Terre** *island* E West Indies
15 X5 **Grande-Vallée** Québec, SE Canada 49°14′N 65°08′W
45 Y5 **Grande Vigie, Pointe de la** *headland* Grande Terre, N Guadeloupe 16°31′N 61°27′W
13 N14 **Grand Falls** New Brunswick, SE Canada 47°02′N 67°46′W
13 T11 **Grand Falls** Newfoundland, Newfoundland and Labrador, SE Canada 48°57′N 55°48′W
24 L9 **Grandfalls** Texas, SW USA 31°20′N 102°51′W
21 P9 **Grandfather Mountain** ▲ North Carolina, SE USA 36°06′N 81°48′W
26 L13 **Grandfield** Oklahoma, C USA 34°15′N 98°40′W
11 N17 **Grand Forks** British Columbia, SW Canada 49°02′N 118°30′W
29 R4 **Grand Forks** North Dakota, N USA 47°54′N 97°03′W
31 O9 **Grand Haven** Michigan, N USA 43°03′N 86°13′W
Grandichi *see* Hrandzichy
29 P15 **Grand Island** Nebraska, C USA 40°55′N 98°20′W
31 O3 **Grand Island** *island* Michigan, N USA
22 K10 **Grand Isle** Louisiana, S USA 29°12′N 90°00′W
65 A23 **Grand Jason** *island* Jason Islands, NW Falkland Islands
37 P5 **Grand Junction** Colorado, C USA 39°03′N 108°33′W
20 F10 **Grand Junction** Tennessee, S USA 35°03′N 89°11′W
14 J9 **Grand-Lac-Victoria** Québec, SE Canada 47°33′N 77°28′W
14 J9 **Grand lac Victoria** ◎ Québec, SE Canada
77 N17 **Grand-Lahou** *var.* Grand Lahu. S Ivory Coast 05°09′N 05°01′W
Grand Lahu *see* Grand-Lahou
37 S3 **Grand Lake** Colorado, C USA 40°15′N 105°49′W
13 S11 **Grand Lake** ◎ Newfoundland and Labrador, E Canada
22 G9 **Grand Lake** ◎ Louisiana, S USA
31 R5 **Grand Lake** ◎ Michigan, N USA
31 Q13 **Grand Lake** ⊠ Ohio, N USA
27 R9 **Grand Lake O' The Cherokees** *var.* Lake O' The Cherokees. ⊠ Oklahoma, C USA
31 Q9 **Grand Ledge** Michigan, N USA 42°45′N 84°45′W
102 I8 **Grand-Lieu, Lac de** ◎ NW France
13 O15 **Grand Manan Channel** *channel* Canada/USA
13 O15 **Grand Manan Island** *island* New Brunswick, SE Canada
29 Y4 **Grand Marais** Minnesota, N USA 47°45′N 90°19′W
15 P10 **Grand-Mère** Québec, SE Canada 46°36′N 72°41′W
37 P5 **Grand Mesa** ▲▲ Colorado, C USA
108 C10 **Grand Muveran** ▲ W Switzerland 46°16′N 07°12′E
104 G12 **Grândola** Setúbal, S Portugal 38°10′N 08°34′W
Grand Paradis *see* Gran Paradiso
187 O15 **Grand Passage** *passage* N New Caledonia
77 R16 **Grand-Popo** S Benin 06°19′N 01°50′E
29 Z3 **Grand Portage** Minnesota, N USA 48°00′N 89°36′W
25 T6 **Grand Prairie** Texas, SW USA 32°45′N 97°00′W
11 W14 **Grand Rapids** Manitoba, C Canada 53°12′N 99°19′W
31 P9 **Grand Rapids** Michigan, N USA 42°57′N 86°40′W
29 V5 **Grand Rapids** Minnesota, N USA 47°14′N 93°31′W
14 L10 **Grand-Remous** Québec, SE Canada 46°36′N 75°53′W
14 F15 **Grand River** ∼ Ontario, S Canada
31 P9 **Grand River** ∼ Michigan, N USA
27 T3 **Grand River** ∼ Missouri, C USA
28 M7 **Grand River** ∼ South Dakota, N USA
45 Q11 **Grand' Rivière** N Martinique 14°52′N 61°11′W
32 F11 **Grand Ronde** Oregon, NW USA 45°03′N 123°43′W
32 L11 **Grand Ronde River** ∼ Oregon/Washington, NW USA
Grand-Saint-Bernard, Col du *see* Great Saint Bernard Pass
25 V6 **Grand Saline** Texas, SW USA 32°40′N 95°42′W
55 X10 **Grand-Santi** W French Guiana 04°19′N 54°24′W
Grandsee *see* Grandson
172 J16 **Grand Sœur** *island* Les Sœurs, NE Seychelles
108 B9 **Grandson** *prev.* Grandsee. Vaud, W Switzerland 46°49′N 06°39′E
33 S14 **Grand Teton** ▲ Wyoming, C USA 43°44′N 110°48′W
31 P5 **Grand Traverse Bay** *lake bay* Michigan, N USA
45 N6 **Grand Turk** ○ (Turks and Caicos Islands) Grand Turk Island, S Turks and Caicos Islands 21°24′N 71°08′W
45 N6 **Grand Turk Island** *island* SE Turks and Caicos Islands
103 S13 **Grand Veymont** ▲ E France 44°51′N 05°32′E
11 W15 **Grandview** Manitoba, S Canada 51°11′N 100°41′W
27 R4 **Grandview** Missouri, C USA 38°53′N 94°31′W
36 I10 **Grand Wash Cliffs** *cliff* Arizona, SW USA
14 J8 **Granet, Lac** ◎ Québec, SE Canada
95 L14 **Grangärde** Dalarna, C Sweden 60°15′N 15°00′E
44 H12 **Grange Hill** W Jamaica 18°19′N 78°11′W
96 J12 **Grangemouth** C Scotland, United Kingdom 56°01′N 03°44′W
25 T10 **Granger** Texas, SW USA 30°43′N 97°26′W
32 J10 **Granger** Washington, NW USA 46°20′N 120°11′W
33 T17 **Granger** Wyoming, C USA 41°37′N 109°58′W
Granges *see* Grenchen
95 L14 **Grängesberg** Dalarna, C Sweden 60°06′N 15°00′E
33 N11 **Grangeville** Idaho, NW USA 45°55′N 116°07′W
10 K13 **Granisle** British Columbia, SW Canada 54°55′N 126°14′W
30 K15 **Granite City** Illinois, N USA 38°42′N 90°09′W
29 S9 **Granite Falls** Minnesota, N USA 44°48′N 95°33′W
21 Q9 **Granite Falls** North Carolina, SE USA 35°48′N 81°25′W
36 K12 **Granite Mountain** ▲ Arizona, SW USA 34°38′N 112°34′W
33 T12 **Granite Peak** ▲ Montana, NW USA 45°09′N 109°48′W
35 T2 **Granite Peak** ▲ Nevada, W USA 41°40′N 117°35′W
36 J3 **Granite Peak** ▲ Utah, W USA 40°09′N 113°18′W
Granite State *see* New Hampshire
107 H24 **Granitola, Capo** *headland* Sicilia, Italy, C Mediterranean Sea 37°33′N 12°39′E
185 H15 **Granity** West Coast, South Island, New Zealand 41°37′S 171°53′E
Gran Lago *see* Nicaragua, Lago de
63 J18 **Gran Laguna Salada** ◎ S Argentina
Gran Malvina *see* West Falkland
95 L18 **Gränna** Jönköping, S Sweden 58°02′N 14°30′E
105 W5 **Granollers** *var.* Granollérs. Cataluña, NE Spain 41°37′N 02°18′E
Granollérs *see* Granollers
106 A7 **Gran Paradiso** *Fr.* Grand Paradis. ▲ NW Italy 45°31′N 07°13′E
Gran Pilastro *see* Hochfeiler
Gran Salitral *see* Grande, Salina
Gran San Bernardo, Passo di *see* Great Saint Bernard Pass
Gran Santiago *see* Santiago
107 J14 **Gran Sasso d'Italia** ▲▲ C Italy
100 N11 **Gransee** Brandenburg, NE Germany 53°00′N 13°10′E
28 L15 **Grant** Nebraska, C USA 40°50′N 101°43′W
27 R1 **Grant City** Missouri, C USA 40°29′N 94°25′W
97 N19 **Grantham** E England, United Kingdom 52°55′N 00°39′W
65 D24 **Grantham Sound** *sound* East Falkland, Falkland Islands
194 K13 **Grant Island** *island* Antarctica
45 Z14 **Grantley Adams** ✈ (Bridgetown) SE Barbados 13°04′N 59°29′W
35 S7 **Grant, Mount** ▲ Nevada, W USA 38°34′N 118°47′W
96 J9 **Grantown-on-Spey** N Scotland, United Kingdom 57°11′N 03°53′W
35 W8 **Grant Range** ▲▲ Nevada, W USA
37 Q11 **Grants** New Mexico, SW USA 35°09′N 107°50′W
30 I4 **Grantsburg** Wisconsin, N USA 45°47′N 92°40′W
32 F15 **Grants Pass** Oregon, NW USA 42°26′N 123°20′W
36 K3 **Grantsville** Utah, W USA 40°36′N 112°27′W
21 R4 **Grantsville** West Virginia, NE USA 38°55′N 81°07′W
102 I5 **Granville** Manche, N France 48°50′N 01°35′W
11 V12 **Granville Lake** ◎ Manitoba, C Canada
25 V8 **Grapeland** Texas, SW USA 31°29′N 95°28′W
25 T6 **Grapevine** Texas, SW USA 32°55′N 97°04′W
83 K20 **Graskop** Mpumalanga, NE South Africa 24°58′S 30°49′E
95 P14 **Gräsö** Uppsala, C Sweden 60°22′N 18°30′E
93 I19 **Gräsö** *island* C Sweden
103 U15 **Grasse** Alpes-Maritimes, SE France 43°42′N 06°52′E
18 E14 **Grassflat** Pennsylvania, NE USA 41°00′N 78°04′W
33 U9 **Grassrange** Montana, NW USA 47°02′N 108°48′W
18 J6 **Grass River** ∼ New York, NE USA
35 P6 **Grass Valley** California, W USA 39°12′N 121°04′W
183 N14 **Grassy** Tasmania, SE Australia 40°03′S 144°04′E
28 K4 **Grassy Butte** North Dakota, N USA 47°20′N 103°13′W
21 R5 **Grassy Knob** ▲ West Virginia, NE USA 38°04′N 80°31′W
95 G24 **Gråsten** *var.* Graasten. Syddanmark, SW Denmark 54°55′N 09°37′E
95 J18 **Grästorp** Västra Götaland, S Sweden 58°20′N 12°40′E
Gratianopolis *see* Grenoble
109 V8 **Gratwein** Steiermark, SE Austria 47°08′N 15°20′E
Gratz *see* Graz
104 K2 **Grau** *var.* Grado. Asturias, N Spain 43°23′N 06°04′W
108 I9 **Graubünden** *Fr.* Grisons, *It.* Grigioni. ◈ *canton* SE Switzerland
Graudenz *see* Grudziądz
103 N15 **Graulhet** Tarn, S France 43°45′N 01°58′E
105 T4 **Graus** Aragón, NE Spain 42°11′N 00°21′E
61 I16 **Gravataí** Rio Grande do Sul, S Brazil 29°55′S 51°00′W
98 L13 **Grave** Noord-Brabant, SE Netherlands 51°45′N 05°45′E
11 T17 **Gravelbourg** Saskatchewan, S Canada 49°53′N 106°33′W
103 N1 **Gravelines** Nord, N France 51°00′N 02°07′E
Graven *see* Grez-Doiceau
14 H13 **Gravenhurst** Ontario, S Canada 44°55′N 79°22′W
33 O10 **Grave Peak** ▲ Idaho, NW USA 46°24′N 114°43′W
102 I11 **Grave, Pointe de** *headland* W France 45°33′N 01°24′W
183 S4 **Gravesend** New South Wales, SE Australia 29°37′S 150°15′E
97 P22 **Gravesend** SE England, United Kingdom 51°27′N 00°24′E
107 N17 **Gravina in Puglia** Puglia, SE Italy 40°48′N 16°25′E
103 S8 **Gray** Haute-Saône, E France 47°28′N 05°34′E
23 T4 **Gray** Georgia, SE USA 33°00′N 83°31′W
195 V16 **Gray, Cape** *headland* Antarctica 67°30′N 143°30′E
32 F9 **Grayland** Washington, NW USA 46°46′N 124°07′W
39 N10 **Grayling** Alaska, USA 62°55′N 160°07′W
31 Q6 **Grayling** Michigan, N USA 44°40′N 84°43′W
32 F9 **Grays Harbor** *inlet* Washington, NW USA
21 O5 **Grayson** Kentucky, S USA 38°21′N 82°59′W
37 S4 **Grays Peak** ▲ Colorado, C USA 39°37′N 105°49′W
30 M16 **Grayville** Illinois, N USA 38°15′N 87°59′W
109 V8 **Graz** *prev.* Gratz. Steiermark, SE Austria 47°05′N 15°23′E
104 L15 **Grazalema** Andalucía, S Spain 36°46′N 05°23′W
113 P15 **Grdelica** Serbia, SE Serbia 42°54′N 22°05′E
44 H1 **Great Abaco** *var.* Abaco Island. *island* N The Bahamas
Great Admiralty Island *see* Manus Island
Great Alfold *see* Great Hungarian Plain
Great Ararat *see* Büyükağrı Dağı
181 U8 **Great Artesian Basin** *lowlands* Queensland, C Australia
181 O12 **Great Australian Bight** *bight* S Australia
64 E11 **Great Bahama Bank** *undersea feature* E Gulf of Mexico 23°15′N 78°00′W
184 M4 **Great Barrier Island** *island* N New Zealand
181 X4 **Great Barrier Reef** *reef* Queensland, NE Australia
18 L11 **Great Barrington** Massachusetts, NE USA 42°11′N 73°20′W
0 F10 **Great Basin** *basin* W USA
8 I8 **Great Bear Lake** *Fr.* Grand Lac de l'Ours. ◎ Northwest Territories, NW Canada
Great Belt *see* Storebælt
26 L5 **Great Bend** Kansas, C USA 38°22′N 98°47′W
Great Bermuda *see* Bermuda
97 A20 **Great Blasket Island** *Ir.* An Blascaod Mór. *island* SW Ireland
Great Britain *see* Britain
151 Q23 **Great Channel** *channel* Andaman Sea/Indian Ocean
166 J10 **Great Coco Island** *island* SW Myanmar (Burma)
Great Crosby *see* Crosby
21 X7 **Great Dismal Swamp** *wetland* North Carolina/Virginia, SE USA
33 V16 **Great Divide Basin** *basin* Wyoming, C USA
181 W7 **Great Dividing Range** ▲▲ NE Australia
14 D12 **Great Duck Island** *island* Ontario, S Canada
Great Elder Reservoir *see* Waconda Lake
44 G8 **Greater Antilles** *island group* West Indies
129 V16 **Greater Sunda Islands** *var.* Sunda Islands. *island group* Indonesia
184 I1 **Great Exhibition Bay** *inlet* North Island, New Zealand
44 H4 **Great Exuma Island** *island* C The Bahamas
33 R8 **Great Falls** Montana, NW USA 47°30′N 111°18′W
21 R11 **Great Falls** South Carolina, SE USA 34°34′N 80°54′W
84 F9 **Great Fisher Bank** *undersea feature* C North Sea 57°00′N 04°00′E
Great Glen *see* Mor, Glen
Great Grimsby *see* Grimsby
44 I4 **Great Guana Cay** *island* C The Bahamas
64 I5 **Great Hellefiske Bank** *undersea feature* N Atlantic Ocean
111 L24 **Great Hungarian Plain** *var.* Great Alfold, Plain of Hungary, *Hung.* Alföld. *plain* SE Europe
44 L7 **Great Inagua** *var.* Inagua Islands. *island* S The Bahamas
Great Indian Desert *see* Thar Desert
83 G25 **Great Karoo** *var.* Great Karoo, High Veld, *Afr.* Groot Karoo, Hoë Karoo. *plateau region* S South Africa
Great Karroo *see* Great Karoo
Great Kei *see* Nciba
Great Khingan Range *see* Da Hinggan Ling
14 E11 **Great La Cloche Island** *island* Ontario, S Canada
183 P16 **Great Lake** ◎ Tasmania, SE Australia
Great Lake *see* Tônlé Sap
11 R15 **Great Lakes** *lakes* Ontario, Canada/USA
Great Lakes State *see* Michigan
97 L20 **Great Malvern** W England, United Kingdom 52°07′N 02°19′W
184 M5 **Great Mercury Island** *island* N New Zealand
Great Meteor Seamount *see* Great Meteor Tablemount
64 K10 **Great Meteor Tablemount** *var.* Great Meteor Seamount. *undersea feature* E Atlantic Ocean 30°00′N 28°30′W
31 Q14 **Great Miami River** ∼ Ohio, N USA
151 Q24 **Great Nicobar** *island* Nicobar Islands, India, NE Indian Ocean
97 O19 **Great Ouse** *var.* Ouse. ∼ E England, United Kingdom
183 Q17 **Great Oyster Bay** *bay* Tasmania, SE Australia
44 I13 **Great Pedro Bluff** *headland* W Jamaica 17°51′N 77°44′W
21 T12 **Great Pee Dee River** ∼ North Carolina/South Carolina, SE USA
129 W9 **Great Plain of China** *plain* E China
0 F12 **Great Plains** *var.* High Plains. *plains* Canada/USA
37 W6 **Great Plains Reservoirs** ⊠ Colorado, C USA
19 Q13 **Great Point** *headland* Nantucket Island, Massachusetts, NE USA 41°23′N 70°03′W
68 I13 **Great Rift Valley** *var.* Rift Valley. *depression* Asia/Africa
81 I23 **Great Ruaha** ∼ S Tanzania
18 K10 **Great Sacandaga Lake** ⊠ New York, NE USA
108 C12 **Great Saint Bernard Pass** *Fr.* Col du Grand-Saint-Bernard, *It.* Passo del Gran San Bernardo. *pass* Italy/Switzerland
44 F1 **Great Sale Cay** *island* N The Bahamas
Great Salt Desert *see* Kavīr, Dasht-e
36 K1 **Great Salt Lake** *salt lake* Utah, W USA
36 J3 **Great Salt Lake Desert** *plain* Utah, W USA
26 M8 **Great Salt Plains Lake** ⊠ Oklahoma, C USA
75 T9 **Great Sand Sea** *desert* Egypt/Libya
180 L6 **Great Sandy Desert** *desert* Western Australia
Great Sandy Desert *see* Ar Rub' al Khālī
Great Sandy Island *see* Fraser Island
187 Y13 **Great Sea Reef** *reef* Vanua Levu, N Fiji
38 H17 **Great Sitkin Island** *island* Aleutian Islands, Alaska, USA
8 J10 **Great Slave Lake** *Fr.* Grand Lac des Esclaves. ◎ Northwest Territories, NW Canada
21 O10 **Great Smoky Mountains** ▲▲ North Carolina/Tennessee, SE USA
10 L11 **Great Snow Mountain** ▲ British Columbia, W Canada 57°22′N 124°08′W
Great Socialist People's Libyan Arab Jamahiriya *see* Libya
64 A12 **Great Sound** *sound* Bermuda, NW Atlantic Ocean
180 M10 **Great Victoria Desert** *desert* South Australia/Western Australia
194 H2 **Great Wall** *Chinese research station* South Shetland Islands, Antarctica 61°57′S 58°23′W
19 T7 **Great Wass Island** *island* Maine, NE USA
97 Q19 **Great Yarmouth** *var.* Yarmouth. E England, United Kingdom 52°37′N 01°44′E
139 S1 **Great Zab** *Ar.* Az Zāb al Kabīr, *Kurd.* Zē-i Bādīnān, *Turk.* Büyükzap Suyu. ∼ Iraq/Turkey
95 I17 **Grebbestad** Västra Götaland, S Sweden 58°42′N 11°15′E
Grebenka *see* Hrebinka
42 M13 **Grecia** Alajuela, C Costa Rica 10°04′N 84°19′W
61 E18 **Greco** Río Negro, W Uruguay 32°49′S 57°03′W
Greco, Cape *see* Gkréko, Akrotíri
104 L8 **Gredos, Sierra de** ▲▲ W Spain
18 F9 **Greece** New York, NE USA 43°12′N 77°41′W
115 E17 **Greece** *off.* Hellenic Republic, *Gk.* Ellás; *anc.* Hellas. ◆ *republic* SE Europe
Greece Central *see* Stereá Elláda
Greece West *see* Dytikí Elláda
37 T3 **Greeley** Colorado, C USA 40°21′N 104°41′W
29 P14 **Greeley** Nebraska, C USA 41°33′N 98°31′W
122 K3 **Greem-Bell, Ostrov** *Eng.* Graham Bell Island. *island* Zemlya Frantsa-Iosifa, N Russian Federation
30 M6 **Green Bay** Wisconsin, N USA 44°32′N 88°W
31 N6 **Green Bay** *lake bay* Michigan/Wisconsin, N USA
21 S5 **Greenbrier River** ∼ West Virginia, NE USA
29 S2 **Greenbush** Minnesota, N USA 48°42′N 96°10′W
183 R12 **Green Cape** *headland* New South Wales, SE Australia 37°15′S 150°03′E
31 O14 **Greencastle** Indiana, N USA 39°38′N 86°51′W
18 F16 **Greencastle** Pennsylvania, NE USA 39°47′N 77°43′W
27 T2 **Green City** Missouri, C USA 40°16′N 92°57′W
21 O9 **Greeneville** Tennessee, S USA 36°10′N 82°50′W
35 O11 **Greenfield** California, W USA 36°19′N 121°15′W
31 P14 **Greenfield** Indiana, N USA 39°47′N 85°46′W
29 U15 **Greenfield** Iowa, C USA 41°18′N 94°27′W
18 M11 **Greenfield** Massachusetts, NE USA 42°33′N 72°34′W
27 S7 **Greenfield** Missouri, C USA 37°25′N 93°50′W
31 S14 **Greenfield** Ohio, N USA 39°21′N 83°22′W
20 G8 **Greenfield** Tennessee, S USA 36°09′N 88°48′W
30 M9 **Greenfield** Wisconsin, N USA 42°55′N 87°59′W
27 T9 **Green Forest** Arkansas, C USA 36°19′N 93°24′W
37 T7 **Greenhorn Mountain** ▲ Colorado, C USA 37°50′N 104°59′W
Green Island *see* Lü Dao
186 J6 **Green Islands** *var.* Nissan Islands. *island group* NE Papua New Guinea
11 S14 **Green Lake** Saskatchewan, C Canada 54°15′N 107°51′W
30 L8 **Green Lake** ◎ Wisconsin, N USA
197 O14 **Greenland** *Dan.* Grønland, *Inuit* Kalaallit Nunaat. ◇ *Danish self-governing territory* NE North America
84 D4 **Greenland** *island* NE North America
197 R13 **Greenland Plain** *undersea feature* N Greenland Sea
197 R14 **Greenland Sea** *sea* Arctic Ocean
37 R4 **Green Mountain Reservoir** ⊠ Colorado, C USA
18 M8 **Green Mountains** ▲▲ Vermont, NE USA
Green Mountain State *see* Vermont
96 H12 **Greenock** W Scotland, United Kingdom 55°57′N 04°45′W
39 T5 **Greenough, Mount** ▲ Alaska, USA 69°15′N 141°37′W
186 A6 **Green River** West Sepik, NW Papua New Guinea 03°54′S 141°08′E
37 N5 **Green River** Utah, W USA 39°00′N 110°07′W
33 U17 **Green River** Wyoming, C USA 41°33′N 109°27′W
16 H9 **Green River** ∼ W USA
30 K11 **Green River** ∼ Illinois, N USA
20 J7 **Green River** ∼ Kentucky, C USA
28 K5 **Green River** ∼ North Dakota, N USA
37 N6 **Green River** ∼ Utah, W USA
33 T16 **Green River** ∼ Wyoming, C USA
20 L7 **Green River Lake** ⊠ Kentucky, S USA
23 O5 **Greensboro** Alabama, S USA 32°42′N 87°36′W
23 U3 **Greensboro** Georgia, SE USA 33°34′N 83°10′W
21 T9 **Greensboro** North Carolina, SE USA 36°04′N 79°48′W
31 P14 **Greensburg** Indiana, N USA 39°20′N 85°28′W
26 K6 **Greensburg** Kansas, C USA 37°36′N 99°17′W
20 L7 **Greensburg** Kentucky, S USA 37°14′N 85°30′W
18 C15 **Greensburg** Pennsylvania, NE USA 40°18′N 79°32′W
37 O13 **Greens Peak** ▲ Arizona, SW USA 34°06′N 109°34′W
21 V12 **Green Swamp** *wetland* North Carolina, SE USA
21 O4 **Greenup** Kentucky, S USA 38°34′N 82°49′W
36 M16 **Green Valley** Arizona, SW USA 31°49′N 111°00′W
76 K17 **Greenville** *var.* Sino, Sinoe. SE Liberia 05°01′N 09°03′W
23 P6 **Greenville** Alabama, S USA 31°49′N 86°37′W
23 T8 **Greenville** Florida, SE USA 30°28′N 83°37′W
23 S4 **Greenville** Georgia, SE USA 33°03′N 84°42′W
30 L15 **Greenville** Illinois, N USA 38°53′N 89°24′W
20 I7 **Greenville** Kentucky, S USA 37°11′N 87°11′W
19 Q5 **Greenville** Maine, NE USA 45°26′N 69°36′W
31 P9 **Greenville** Michigan, N USA 43°10′N 85°15′W
22 J4 **Greenville** Mississippi, S USA 33°24′N 91°03′W
21 W9 **Greenville** North Carolina, SE USA 35°36′N 77°23′W
31 Q13 **Greenville** Ohio, N USA 40°06′N 84°37′W
19 O12 **Greenville** Rhode Island, NE USA 41°52′N 71°33′W
21 P11 **Greenville** South Carolina, SE USA 34°51′N 82°24′W
25 U6 **Greenville** Texas, SW USA 33°09′N 96°07′W
31 T12 **Greenwich** Ohio, N USA 41°01′N 82°31′W
27 S11 **Greenwood** Arkansas, C USA 35°13′N 94°15′W
31 O14 **Greenwood** Indiana, N USA 39°38′N 86°06′W
22 K4 **Greenwood** Mississippi, S USA 33°30′N 90°11′W
21 P12 **Greenwood** South Carolina, SE USA 34°11′N 82°10′W
21 Q12 **Greenwood, Lake** ⊠ South Carolina, SE USA
21 P11 **Greer** South Carolina, SE USA 34°56′N 82°13′W
27 V10 **Greers Ferry Lake** ⊠ Arkansas, C USA
27 S13 **Greeson, Lake** ⊠ Arkansas, C USA
45 N8 **Gregorio Luperón** ✈ N Dominican Republic 19°43′N 70°43′W
29 O12 **Gregory** South Dakota, N USA 43°11′N 99°26′W
182 J3 **Gregory, Lake** *salt lake* South Australia
180 J9 **Gregory Lake** ◎ Western Australia
181 V5 **Gregory Range** ▲▲ Queensland, E Australia
Greifenberg/Greifenberg in Pommern *see* Gryfice
Greifenhagen *see* Gryfino
100 O8 **Greifswald** Mecklenburg-Vorpommern, NE Germany 54°04′N 13°24′E
100 O8 **Greifswalder Bodden** *bay* NE Germany
109 U4 **Grein** Oberösterreich, N Austria 48°14′N 14°50′E
101 M17 **Greiz** Thüringen, C Germany 50°40′N 12°11′E
125 V14 **Gremyachinsk** Permskiy Kray, NW Russian Federation 58°33′N 57°52′E
Grenå *see* Grenaa
95 H21 **Grenaa** *var.* Grenå. Midtjylland, C Denmark 56°25′N 10°53′E

◆ Country ● Country Capital ◇ Dependent Territory ○ Dependent Territory Capital ◈ Administrative Regions ✈ International Airport ▲ Mountain ▲▲ Mountain Range Volcano ∼ River ◎ Lake ⊠ Reservoir

Grenada – Gulin

22 L3 **Grenada** Mississippi, S USA 33°46′N 89°48′W
45 W15 **Grenada** ◆ *commonwealth republic* SE West Indies
47 S4 **Grenada** *island* Grenada
47 R4 **Grenada Basin** *undersea feature* W Atlantic Ocean 13°30′N 62°00′W
22 L3 **Grenada Lake** ⊠ Mississippi, S USA
45 Y14 **Grenadines, The** *island group* Grenada/St Vincent and the Grenadines
108 D7 **Grenchen** *Fr.* Granges. Solothurn, NW Switzerland 47°13′N 07°24′E
183 Q9 **Grenfell** New South Wales, SE Australia 33°54′S 148°09′E
11 V16 **Grenfell** Saskatchewan, S Canada 50°24′N 102°56′W
92 J1 **Grenivík** Norðurland Eystra, N Iceland 65°57′N 18°10′W
103 S12 **Grenoble** *anc.* Cularo, Gratianopolis. Isère, E France 45°11′N 05°42′E
28 J2 **Grenora** North Dakota, N USA 48°36′N 103°57′W
92 N8 **Grense-Jakobselv** Finnmark, N Norway 69°46′N 30°39′E
45 S14 **Grenville** E Grenada 12°07′N 61°37′W
32 G11 **Gresham** Oregon, NW USA 45°30′N 122°25′W
Gresk *see* Hresk
106 B7 **Gressoney-St-Jean** Valle d'Aosta, NW Italy 45°48′N 07°49′E
22 K9 **Gretna** Louisiana, S USA 29°54′N 90°03′W
21 T7 **Gretna** Virginia, NE USA 36°57′N 79°21′W
98 F13 **Grevelingen** *inlet* S North Sea
100 F13 **Greven** Nordrhein-Westfalen, NW Germany 52°07′N 07°38′E
115 D15 **Grevená** Dytikí Makedonía, N Greece 40°05′N 21°26′E
101 D16 **Grevenbroich** Nordrhein-Westfalen, W Germany 51°06′N 06°34′E
99 N24 **Grevenmacher** Grevenmacher, E Luxembourg 49°41′N 06°27′E
99 M24 **Grevenmacher** ◈ *district* E Luxembourg
100 K9 **Grevesmülhen** Mecklenburg-Vorpommern, N Germany 53°52′N 11°12′E
185 H16 **Grey** ≈ South Island, New Zealand
33 V12 **Greybull** Wyoming, C USA 44°29′N 108°03′W
33 U13 **Greybull River** ≈ Wyoming, C USA
65 A24 **Grey Channel** *sound* Falkland Islands
Greyerzer See *see* Gruyère, Lac de la
13 T10 **Grey Islands** *island group* Newfoundland and Labrador, E Canada
18 L10 **Greylock, Mount** ▲ Massachusetts, NE USA 42°38′N 73°09′W
185 G17 **Greymouth** West Coast, South Island, New Zealand 42°29′S 171°14′E
181 U10 **Grey Range** ▲▲ New South Wales/Queensland, E Australia
97 G18 **Greystones** *Ir.* Na Clocha Liatha. E Ireland 53°08′N 06°05′W
185 M14 **Greytown** Wellington, North Island, New Zealand 41°04′S 175°29′E
83 K23 **Greytown** KwaZulu/Natal, E South Africa 29°04′S 30°35′E
Greytown *see* San Juan del Norte
99 H19 **Grez-Doiceau** *Dut.* Graven. Walloon Brabant, C Belgium 50°43′N 04°41′E
115 J19 **Griá, Akrotírio** *headland* Ándros, Kykládes, Greece, Aegean Sea 37°54′N 24°57′E
127 N8 **Gribanovskiy** Voronezhskaya Oblast', W Russian Federation 51°27′N 41°53′E
78 I13 **Gribingui** ≈ N Central African Republic
35 O6 **Gridley** California, W USA 39°21′N 121°41′W
83 G23 **Griekwastad** *var.* Griquatown. Northern Cape, C South Africa 28°50′S 23°16′E
23 S4 **Griffin** Georgia, SE USA 33°15′N 84°17′W
183 O9 **Griffith** New South Wales, SE Australia 34°18′S 146°04′E
14 F13 **Griffith Island** *island* Ontario, S Canada
21 W10 **Grifton** North Carolina, SE USA 35°22′N 77°26′W
Grigioni *see* Graubünden
119 H14 **Grigiškes** Vilnius, SE Lithuania 54°42′N 25°00′E
117 N10 **Grigoriopol** C Moldova 47°09′N 29°18′E
147 X7 **Grigor'yevka** Issyk-Kul'skaya Oblast', E Kyrgyzstan 42°43′N 77°27′E
193 U8 **Grijalva Ridge** *undersea feature* E Pacific Ocean
41 U15 **Grijalva, Río** *var.* Tabasco. ≈ Guatemala/Mexico
98 N5 **Grijpskerk** Groningen, NE Netherlands 53°15′N 06°18′E
83 C22 **Grillenthal** Karas, SW Namibia 26°55′S 15°24′E
79 J15 **Grimari** Ouaka, C Central African Republic 05°44′N 20°02′E
Grimaylov *see* Hrymayliv
99 G18 **Grimbergen** Vlaams Brabant, C Belgium 50°56′N 04°22′E
183 N15 **Grim, Cape** *headland* Tasmania, SE Australia 40°42′S 144°42′E
100 N8 **Grimmen** Mecklenburg-Vorpommern, NE Germany 54°06′N 13°03′E
14 G16 **Grimsby** Ontario, S Canada 43°12′N 79°35′W
97 O17 **Grimsby** *prev.* Great Grimsby. E England, United Kingdom 53°35′N 00°05′W
92 J1 **Grímsey** *var.* Grimsey. *island* N Iceland
Grimsey *see* Grímsey
11 O12 **Grimshaw** Alberta, W Canada 56°11′N 117°37′W
95 F18 **Grimstad** Aust-Agder, S Norway 58°20′N 08°35′E
92 H4 **Grindavík** Suðurnes, W Iceland 65°57′N 18°10′W
108 F9 **Grindelwald** Bern, S Switzerland 46°37′N 08°04′E
95 F23 **Grindsted** Syddtjylland, W Denmark 55°46′N 08°56′E
29 W14 **Grinnell** Iowa, C USA 41°44′N 92°43′W
109 U10 **Grintovec** ▲ N Slovenia 46°21′N 14°31′E
Griquatown *see* Griekwastad
9 N4 **Grise Fiord** *var* Aujuittuq. Northwest Territories, Ellesmere Island, N Canada 76°10′N 83°15′W
182 H1 **Griselda, Lake** *salt lake* South Australia
Grisons *see* Graubünden
95 P14 **Grisslehamn** Stockholm, C Sweden 60°04′N 18°50′E
29 T15 **Griswold** Iowa, C USA 41°14′N 95°08′W
102 M1 **Griz Nez, Cap** *headland* N France 50°51′N 01°34′E
112 P13 **Grljan** Serbia, E Serbia 43°52′N 22°18′E
112 E11 **Grmeč** ▲▲ NW Bosnia and Herzegovina
99 H16 **Grobbendonk** Antwerpen, N Belgium 51°12′N 04°41′E
Grobin *see* Grobiņa
118 C10 **Grobiņa** *Ger.* Grobin. W Latvia 56°32′N 21°12′E
83 K20 **Groblersdal** Mpumalanga, NE South Africa 25°15′S 29°25′E
83 G23 **Groblershoop** Northern Cape, W South Africa 28°51′S 22°01′E
Gródek Jagielloński *see* Horodok
109 Q6 **Grödig** Salzburg, W Austria 47°42′N 13°06′E
111 H15 **Grodków** Opolskie, S Poland 50°42′N 17°23′E
Grodnenskaya Oblast' *see* Hrodzyenskaya Voblasts'
Grodno *see* Hrodna
110 L12 **Grodzisk Mazowiecki** Mazowieckie, C Poland 52°09′N 20°38′E
110 F12 **Grodzisk Wielkopolski** Wielkopolskie, C Poland 52°13′N 16°21′E
Grodzyanka *see* Hradzyanka
98 O12 **Groenlo** Gelderland, E Netherlands 52°02′N 06°36′E
83 E22 **Groenrivier** Karas, SE Namibia 27°27′S 18°52′E
25 U8 **Groesbeck** Texas, SW USA 31°31′N 96°35′W
98 L13 **Groesbeek** Gelderland, SE Netherlands 51°47′N 05°56′E
102 G7 **Groix, Îles de** *island group* NW France
110 M12 **Grójec** Mazowieckie, C Poland 51°51′N 20°52′E
65 K15 **Gröll Seamount** *undersea feature* C Atlantic Ocean 12°54′S 33°24′W
100 E13 **Gronau** *var.* Gronau in Westfalen. Nordrhein-Westfalen, NW Germany 52°13′N 07°02′E
Gronau in Westfalen *see* Gronau
93 F15 **Grong** Nord-Trøndelag, C Norway 64°29′N 12°19′E
95 N22 **Grönhögen** Kalmar, S Sweden 56°16′N 16°09′E
98 N5 **Groningen** Groningen, NE Netherlands 53°13′N 06°35′E
55 W9 **Groningen** Saramacca, N Suriname 05°45′N 55°31′W
98 N5 **Groningen** ◈ *province* NE Netherlands
Grønland *see* Greenland
108 H11 **Grono** Graubünden, S Switzerland 46°15′N 09°07′E
95 M20 **Grönskåra** Kalmar, S Sweden 57°04′N 15°45′E
25 O2 **Groom** Texas, SW USA 35°12′N 101°06′W
35 W9 **Groom Lake** ◎ Nevada, W USA
83 H25 **Groot** ≈ S South Africa
181 S2 **Groote Eylandt** *island* Northern Territory, N Australia
98 M6 **Grootegast** Groningen, NE Netherlands 53°11′N 06°12′E
83 D17 **Grootfontein** Otjozondjupa, N Namibia 19°32′S 18°05′E
83 E22 **Groot Karasberge** ▲▲ S Namibia
Groot Karoo *see* Great Karoo
Groot-Kei *see* Nciba
15 V6 **Grosses-Roches** Québec, SE Canada 48°55′N 67°06′W
109 V2 **Gross-Siegharts** Niederösterreich, N Austria 48°48′N 15°25′E
45 T10 **Gros Islet** N Saint Lucia 14°04′N 60°57′W
44 L8 **Gros-Morne** NW Haiti 19°45′N 72°46′W
13 S11 **Gros Morne** ▲ Newfoundland, Newfoundland and Labrador, E Canada 49°38′N 57°45′W
103 R9 **Grosne** ≈ C France
45 S12 **Gros Piton** ▲ SW Saint Lucia 13°48′N 61°04′W
Grossa, Isola *see* Dugi Otok
Grossbetschkerek *see* Zrenjanin
Grosse Isper *see* Grosse Ysper
Grosse Kokel *see* Târnava Mare
101 M21 **Grosse Laaber** *var.* Grosse Laber. ≈ SE Germany
Grosse Laber *see* Grosse Laaber
Grosse Morava *see* Velika Morava
101 O15 **Grossenhain** Sachsen, E Germany 51°18′N 13°31′E
109 Y4 **Grossenzersdorf** Niederösterreich, NE Austria 48°12′N 16°33′E
101 O21 **Grosser Arber** ▲ SE Germany 49°07′N 13°10′E
101 K17 **Grosser Beerberg** ▲ C Germany 50°39′N 10°45′E
101 G18 **Grosser Feldberg** ▲ W Germany 50°13′N 08°28′E
109 O8 **Grosser Löffler** *It.* Monte Lovello. ▲ Austria/Italy 47°02′N 11°56′E
109 N8 **Grosser Möseler** *var.* Mesule. ▲ Austria/Italy 47°01′N 11°52′E
100 J8 **Grosser Plöner See** ◎ N Germany
101 O21 **Grosser Rachel** ▲ SE Germany 48°59′N 13°23′E
Grosser Sund *see* Suur Väin
109 P8 **Grosses Wiesbachhorn** *var.* Wiesbachhorn. ▲ W Austria 47°09′N 12°44′E
106 F13 **Grosseto** Toscana, C Italy 42°45′N 11°07′E
101 M22 **Grosse Vils** ≈ SE Germany
109 U4 **Grosse Ysper** *var.* Grosse Isper. ≈ N Austria
101 G19 **Gross-Gerau** Hessen, W Germany 49°55′N 08°28′E
109 U3 **Gross Gerungs** Niederösterreich, N Austria 48°33′N 14°58′E
109 P8 **Grossglockner** ▲ W Austria 47°05′N 12°39′E
Grosskanizsa *see* Nagykanizsa
Gross-Karol *see* Carei
Grosskikinda *see* Kikinda
109 W9 **Grossklein** Steiermark, SE Austria 46°43′N 15°24′E
Grosskoppe *see* Velká Deštná
Grossmeseritsch *see* Velké Meziříčí
Grossmichel *see* Michalovce
101 H19 **Grossostheim** Bayern, C Germany 49°54′N 09°03′E
109 X7 **Grosspetersdorf** Burgenland, SE Austria 47°15′N 16°19′E
109 T5 **Grossraming** Oberösterreich, C Austria 47°54′N 14°34′E
101 P14 **Grossräschen** Brandenburg, E Germany 51°34′N 14°00′E
Grossrauschenbach *see* Revúca
Gross-Sankt-Johannis *see* Suure-Jaani
Gross-Schlatten *see* Abrud
Gross-Skaisgirren *see* Bol'shakovo
Gross-Steffelsdorf *see* Rimavská Sobota
Gross Strehlitz *see* Strzelce Opolskie
109 O8 **Grossvenediger** ▲ W Austria 47°07′N 12°19′E
Grosswardein *see* Oradea
Gross Wartenberg *see* Syców
109 U11 **Grosuplje** C Slovenia 46°00′N 14°36′E
99 H17 **Grote Nete** ≈ N Belgium
94 E10 **Grotli** Oppland, S Norway 62°02′N 07°36′E
19 N13 **Groton** Connecticut, NE USA 41°20′N 72°03′W
29 P8 **Groton** South Dakota, N USA 45°27′N 98°06′W
107 P18 **Grottaglie** Puglia, SE Italy 40°32′N 17°26′E
107 L17 **Grottaminarda** Campania, S Italy 41°04′N 15°02′E
106 K13 **Grottammare** Marche, C Italy 43°00′N 13°52′E
21 U5 **Grottoes** Virginia, NE USA 38°16′N 78°49′W
98 L6 **Grou** *Dutch.* Grouw. Fryslân, N Netherlands 53°07′N 05°51′E
13 N10 **Groulx, Monts** ▲▲ Québec, E Canada
14 E7 **Groundhog** ≈ Ontario, S Canada
36 J1 **Grouse Creek** Utah, W USA 41°41′N 113°52′W
36 J1 **Grouse Creek Mountains** ▲▲ Utah, W USA
Grouw *see* Grou
27 R8 **Grove** Oklahoma, C USA 36°35′N 94°46′W
31 S13 **Grove City** Ohio, N USA 39°52′N 83°05′W
18 B13 **Grove City** Pennsylvania, NE USA 41°09′N 80°02′W
23 O6 **Grove Hill** Alabama, S USA 31°42′N 87°46′W
33 S15 **Grover** Wyoming, C USA 42°48′N 110°57′W
35 P13 **Grover City** California, W USA 35°08′N 120°37′W
25 Y11 **Groves** Texas, SW USA 29°57′N 93°55′W
19 O7 **Groveton** New Hampshire, NE USA 44°35′N 71°28′W
25 W9 **Groveton** Texas, SW USA 31°04′N 95°08′W
36 J15 **Growler Mountains** ▲▲ Arizona, SW USA
Grozdovo *see* Bratya Daskalovi
127 P16 **Groznyy** Chechenskaya Respublika, SW Russian Federation 43°20′N 45°43′E
Grubeshov *see* Hrubieszów
112 G9 **Grubišno Polje** Bjelovar-Bilogora, NE Croatia 45°42′N 17°09′E
Grudovo *see* Sredets
110 J9 **Grudziądz** *Ger.* Graudenz. Kujawsko-pomorskie, C Poland 53°29′N 18°45′E
25 R17 **Grulla** *var.* La Grulla. Texas, SW USA 26°15′N 98°37′W
40 K14 **Grullo** Jalisco, SW Mexico 19°45′N 104°15′W
67 V10 **Grumeti** ≈ N Tanzania
95 K16 **Grums** Värmland, C Sweden 59°22′N 13°11′E
109 S5 **Grünau im Almtal** Oberösterreich, N Austria 47°51′N 13°56′E
101 H17 **Grünberg** Hessen, W Germany 50°36′N 08°57′E
Grünberg/Grünberg in Schlesien *see* Zielona Góra
92 H3 **Grundarfjörður** Vestfirðir, W Iceland 64°55′N 23°15′W
21 P7 **Grundy** Virginia, NE USA 37°17′N 82°06′W
29 W13 **Grundy Center** Iowa, C USA 42°21′N 92°46′W
Grüneberg *see* Zielona Góra
25 N1 **Gruver** Texas, SW USA 36°16′N 101°24′W
108 C9 **Gruyère, Lac de la** *Ger.* Greyerzer See. ◎ SW Switzerland
108 C9 **Gruyères** Fribourg, W Switzerland 46°34′N 07°04′E
118 E11 **Gruzdĭai** Šiauliai, N Lithuania 56°06′N 23°15′E
Gruzinskaya SSR/Gruziya *see* Georgia
Gryada Akkyr *see* Akgyr Erezi
126 L7 **Gryazi** Lipetskaya Oblast', W Russian Federation 52°27′N 39°56′E
124 M14 **Gryazovets** Vologodskaya Oblast', NW Russian Federation 58°52′N 40°12′E
111 M17 **Grybów** Małopolskie, SE Poland 49°35′N 20°54′E
94 M13 **Grycksbo** Dalarna, C Sweden 60°40′N 15°30′E
110 E8 **Gryfice** *Ger.* Greifenberg, Greifenberg in Pommern. Zachodnio-pomorskie, NW Poland 53°55′N 15°11′E
110 D9 **Gryfino** *Ger.* Greifenhagen. Zachodnio-pomorskie, NW Poland 53°15′N 14°30′E
92 H9 **Gryllefjord** Troms, N Norway 69°21′N 17°07′E
95 L15 **Grythyttan** Örebro, C Sweden 59°52′N 14°31′E
108 D10 **Gstaad** Bern, W Switzerland 46°30′N 07°16′E
43 P14 **Guabito** Bocas del Toro, NW Panama 09°30′N 82°35′W
44 G7 **Guacanayabo, Golfo de** *gulf* S Cuba
40 I7 **Guachochi** Chihuahua, N Mexico
104 J11 **Guadajira** ≈ SW Spain
104 M13 **Guadajoz** ≈ S Spain
40 L13 **Guadalajara** Jalisco, C Mexico 20°43′N 103°24′W
105 O8 **Guadalajara** *Ar.* Wad Al-Hajarah; *anc.* Arriaca. Castilla-La Mancha, C Spain 40°37′N 03°10′W
105 O7 **Guadalajara** ◈ *province* Castilla-La Mancha, C Spain
104 K12 **Guadalcanal** Andalucía, S Spain 38°06′N 05°49′W
186 L10 **Guadalcanal** *off.* Guadalcanal Province. ◈ *province* C Solomon Islands
186 M9 **Guadalcanal** *island* C Solomon Islands
Guadalcanal Province *see* Guadalcanal
105 O12 **Guadalén** ≈ S Spain
105 R13 **Guadalentín** ≈ SE Spain
104 K15 **Guadalete** ≈ SW Spain
105 O13 **Guadalimar** ≈ S Spain
105 P12 **Guadalmena** ≈ S Spain
104 L11 **Guadalmez** ≈ W Spain
105 S7 **Guadalope** ≈ E Spain
104 K13 **Guadalquivir** ≈ W Spain
104 J14 **Guadalquivir, Marismas del** *var.* Las Marismas. *wetland* SW Spain
40 M11 **Guadalupe** Zacatecas, C Mexico 22°47′N 102°30′W
57 E16 **Guadalupe** Ica, W Peru 13°59′S 75°49′W
104 L10 **Guadalupe** Extremadura, W Spain 39°26′N 05°18′W
36 L14 **Guadalupe** Arizona, SW USA 33°20′N 111°57′W
35 P13 **Guadalupe** California, W USA 34°55′N 120°34′W
Guadalupe *see* Canelones
40 J3 **Guadalupe Bravos** Chihuahua, N Mexico 31°22′N 106°04′W
40 A4 **Guadalupe, Isla** *island* NW Mexico
37 U15 **Guadalupe Mountains** ▲▲ New Mexico/Texas, SW USA
24 J8 **Guadalupe Peak** ▲ Texas, SW USA 31°53′N 104°51′W
25 R11 **Guadalupe River** ≈ SW USA
104 K10 **Guadalupe, Sierra de** ▲▲ W Spain
40 K9 **Guadalupe Victoria** Durango, C Mexico 24°30′N 104°08′W
40 I8 **Guadalupe y Calvo** Chihuahua, N Mexico 26°04′N 106°58′W
105 N7 **Guadarrama** Madrid, C Spain 40°40′N 04°06′W
105 N7 **Guadarrama** ≈ C Spain
104 M7 **Guadarrama, Puerto de** *pass* C Spain
105 N9 **Guadarrama, Sierra de** ▲▲ C Spain
105 Q9 **Guadazaón** ≈ C Spain
45 X10 **Guadeloupe** ◇ *French overseas department* E West Indies
47 S3 **Guadeloupe** *island group* E West Indies
45 W10 **Guadeloupe Passage** *passage* E Caribbean Sea
104 H13 **Guadiana** ≈ Portugal/Spain
105 O13 **Guadiana Menor** ≈ S Spain
105 Q8 **Guadiela** ≈ C Spain
105 O14 **Guadix** Andalucía, S Spain 37°19′N 03°08′W
Guad-i-Zirreh *see* Gowd-e Zereh, Dasht-e
193 T12 **Guafo Fracture Zone** *tectonic feature* SE Pacific Ocean
63 F18 **Guafo, Isla** *island* S Chile
42 I6 **Guaimaca** Francisco Morazán, C Honduras 14°34′N 86°49′W
54 J12 **Guainía** *off.* Comisaría del Guainía. ◈ *province* E Colombia
Guainía, Comisaría del *see* Guainía
54 K12 **Guainía, Río** ≈ Colombia/Venezuela
55 O9 **Guaiquinima, Cerro** *elevation* SE Venezuela
60 G10 **Guaíra** Paraná, S Brazil 24°05′S 54°15′W
60 L7 **Guaíra** São Paulo, S Brazil 20°17′S 48°21′W
62 O7 **Guairá** *off.* Departamento del Guairá. ◈ *department* S Paraguay
Guairá, Departamento del *see* Guairá
Guaire *see* Gorey
63 F18 **Guaiteca, Isla** *island* S Chile
44 G6 **Guajaba, Cayo** *headland* C Cuba 21°50′N 77°33′W
59 D16 **Guajará-Mirim** Rondônia, W Brazil 10°50′S 65°21′W
Guajira *see* La Guajira
Guajira, Departamento de La *see* La Guajira
54 H3 **Guajira, Península de la** *peninsula* N Colombia
42 J6 **Gualaco** Olancho, C Honduras 15°00′N 86°03′W
34 L7 **Gualala** California, W USA 38°45′N 123°33′W
42 E5 **Gualán** Zacapa, C Guatemala 15°06′N 89°22′W
61 C19 **Gualeguay** Entre Ríos, E Argentina 33°09′S 59°20′W
61 D18 **Gualeguaychú** Entre Ríos, E Argentina 33°03′S 58°31′W
61 C18 **Gualeguay, Río** ≈ E Argentina
63 K16 **Gualicho, Salina del** *salt lake* E Argentina
188 B15 **Guam** ◇ *US unincorporated territory* W Pacific Ocean
63 F19 **Guamblin, Isla** *island* Archipiélago de los Chonos, S Chile
61 A22 **Guaminí** Buenos Aires, E Argentina 37°01′S 62°28′W
40 H8 **Guamúchil** Sinaloa, C Mexico 25°23′N 108°01′W
54 H4 **Guana** *var.* Misión de Guana. Zulia, NW Venezuela 11°07′N 72°17′W
44 C4 **Guanabacoa** La Habana, W Cuba 23°02′N 82°12′W
42 K13 **Guanacaste** *off.* Provincia de Guanacaste. ◈ *province* NW Costa Rica
42 K12 **Guanacaste, Cordillera de** ▲▲ NW Costa Rica
Guanacaste, Provincia de *see* Guanacaste
40 J8 **Guanaceví** Durango, C Mexico 25°55′N 105°51′W
44 A5 **Guanahacabibes, Golfo de** *gulf* W Cuba
42 K4 **Guanaja, Isla de** *island* Islas de la Bahía, N Honduras
44 C4 **Guanajay** La Habana, W Cuba 22°56′N 82°42′W
41 N12 **Guanajuato** Guanajuato, C Mexico 21°N 101°19′W
40 M12 **Guanajuato** ◈ *state* C Mexico
54 J6 **Guanare** Portuguesa, N Venezuela 09°04′N 69°45′W
54 K7 **Guanare, Río** ≈ W Venezuela
54 J6 **Guanarito** Portuguesa, NW Venezuela 08°43′N 69°12′W
160 M3 **Guancen Shan** ▲▲ C China
62 I9 **Guandacol** La Rioja, W Argentina 29°32′S 68°37′W
44 A5 **Guane** Pinar del Río, W Cuba 22°12′N 84°05′W
161 N14 **Guangdong** *var.* Guangdong Sheng, Kuang-tung, Kwangtung, Yue. ◈ *province* S China
Guangdong Sheng *see* Guangdong
Guanghua *see* Laohekou
Guangju *see* Gwangju
161 Q9 **Guangming Ding** ▲ Anhui, China 30°06′N 118°04′E
160 I13 **Guangnan** *var.* Liancheng. Yunnan, SW China 24°07′N 104°54′E
161 N8 **Guangshui** *prev.* Yinshan. Hubei, C China 31°41′N 113°53′E
Guangxi *see* Guangxi Zhuangzu Zizhiqu
160 K14 **Guangxi Zhuangzu Zizhiqu** *var.* Guangxi, Gui, Kuang-hsi, Kwangsi, *Eng.* Kwangsi Chuang Autonomous Region. ◈ *autonomous region* S China
160 J8 **Guangyuan** *var.* Kuang-yuan, Kwangyuan. Sichuan, C China 32°27′N 105°49′E
161 N14 **Guangzhou** *var.* Kuang-chou, Kwangchow, *Eng.* Canton. *province capital* Guangdong, S China 23°11′N 113°19′E
59 N19 **Guanhães** Minas Gerais, SE Brazil 18°46′S 42°58′W
160 I12 **Guanling** *var.* Guanling Bouyeizu Miaozu Zizhixian. Guizhou, S China 26°00′N 105°40′E
Guanling Bouyeizu Miaozu Zizhixian *see* Guanling
55 N5 **Guanta** Anzoátegui, NE Venezuela 10°15′N 64°38′W
44 J8 **Guantánamo** Guantánamo, SE Cuba 20°06′N 75°16′W
44 J8 **Guantánamo, Bahia de** *Eng.* Guantanamo Bay. *US military base* SE Cuba 20°06′N 75°16′W
Guantanamo Bay *see* Guantánamo, Bahia de
Guanxian/Guan Xian *see* Dujiangyan
161 Q6 **Guanyun** *var.* Yishan. Jiangsu, E China 34°18′N 119°14′E
54 C12 **Guapí** Cauca, SW Colombia 02°36′N 77°54′W
43 N13 **Guápiles** Limón, NE Costa Rica 10°15′N 83°46′W
61 I15 **Guaporé** Rio Grande do Sul, S Brazil 28°55′S 51°53′W
47 S8 **Guaporé, Rio** *var.* Río Iténez. ≈ Bolivia/Brazil *see also* Río Iténez
Guaporé, Rio *see* Iténez, Río
56 B7 **Guaranda** Bolívar, C Ecuador 01°35′S 78°59′W
60 H11 **Guaraniaçu** Paraná, S Brazil 25°05′S 52°52′W
59 O20 **Guarapari** Espírito Santo, SE Brazil 20°39′S 40°31′W
60 I12 **Guarapuava** Paraná, S Brazil 25°22′S 51°28′W
60 J8 **Guararapes** São Paulo, S Brazil 21°16′S 50°37′W
105 S4 **Guara, Sierra de** ▲▲ NE Spain
60 N10 **Guaratinguetá** São Paulo, S Brazil 22°44′S 45°16′W
104 I7 **Guarda** Guarda, N Portugal 40°32′N 07°17′W
104 I7 **Guarda** ◈ *district* N Portugal
Guardak *see* Magdanly
104 M3 **Guardo** Castilla y León, N Spain 42°48′N 04°50′W
104 K11 **Guareña** Extremadura, W Spain 38°51′N 06°06′W
60 J11 **Guaricana, Pico** ▲ S Brazil 25°13′S 48°50′W
54 L6 **Guárico** *off.* Estado Guárico. ◈ *state* N Venezuela
Guárico, Estado *see* Guárico
44 J7 **Guarico, Punta** *headland* E Cuba 20°36′N 74°43′W
54 L7 **Guárico, Río** ≈ C Venezuela
60 M10 **Guarujá** São Paulo, SE Brazil 23°50′S 46°27′W
61 L22 **Guarulhos** ✈ (São Paulo) São Paulo, S Brazil 23°23′S 46°32′W
43 R17 **Guarumal** Veraguas, S Panama 07°48′N 81°15′W
Guasapa *see* Guasopa
40 H8 **Guasave** Sinaloa, C Mexico 25°33′N 108°29′W
54 I8 **Guasdualito** Apure, C Venezuela 07°15′N 70°40′W
55 Q7 **Guasipati** Bolívar, E Venezuela 07°28′N 61°58′W
186 I9 **Guasopa** *var.* Guasapa. Woodlark Island, SE Papua New Guinea 09°12′S 152°58′E
106 F9 **Guastalla** Emilia-Romagna, C Italy 44°54′N 10°38′E
42 D6 **Guastatoya** *var.* El Progreso. El Progreso, C Guatemala 14°51′N 90°01′W
42 D5 **Guatemala** *off.* Republic of Guatemala. ◆ *republic* Central America
42 A2 **Guatemala** *off.* Departamento de Guatemala. ◈ *department* S Guatemala
193 S7 **Guatemala Basin** *undersea feature* E Pacific Ocean 11°00′N 95°00′W
Guatemala City *see* Ciudad de Guatemala
Guatemala, Departamento de *see* Guatemala
Guatemala, Republic of *see* Guatemala
45 V14 **Guatuaro Point** *headland* Trinidad, Trinidad and Tobago 10°19′N 60°58′W
186 B8 **Guavi** ≈ SW Papua New Guinea
54 G13 **Guaviare** *off.* Comisaría Guaviare. ◈ *province* S Colombia
Guaviare, Comisaría *see* Guaviare
54 J11 **Guaviare, Río** ≈ E Colombia
61 E15 **Guaviravi** Corrientes, NE Argentina 29°20′S 56°50′W
54 G12 **Guayabero, Río** ≈ SW Colombia
45 U6 **Guayama** E Puerto Rico 17°59′N 66°07′W
42 J7 **Guayambre, Río** ≈ S Honduras
Guayanas, Macizo de las *see* Guiana Highlands
45 V6 **Guayanés, Punta** *headland* E Puerto Rico 18°03′N 65°48′W
42 J6 **Guayape, Río** ≈ C Honduras
56 B7 **Guayaquil** *var.* Santiago de Guayaquil. Guayas, SW Ecuador 02°13′S 79°54′W
Guayaquil *see* Simon Bolivar
56 A8 **Guayaquil, Golfo de** *var.* Gulf of Guayaquil. *gulf* SW Ecuador
Guayaquil, Gulf of *see* Guayaquil, Golfo de
56 A7 **Guayas** ◈ *province* W Ecuador
62 N7 **Guaycurú, Río** ≈ NE Argentina
40 F6 **Guaymas** Sonora, NW Mexico 27°56′N 110°54′W
45 U5 **Guaynabo** E Puerto Rico 18°19′N 66°05′W
80 H12 **Guba** Bīnishangul Gumuz, W Ethiopia 11°11′N 35°21′E
146 H8 **Gubadag** *Turkm.* Tel'man; *prev.* Tel'mansk. Daşoguz Welaýaty, N Turkmenistan 42°07′N 59°55′E
125 V13 **Gubakha** Permskiy Kray, NW Russian Federation 58°52′N 57°35′E
106 I12 **Gubbio** Umbria, C Italy 43°27′N 12°34′E
100 Q13 **Guben** *var.* Wilhelm-Pieck-Stadt. Brandenburg, E Germany 51°59′N 14°42′E
Guben *see* Gubin
110 D12 **Gubin** *Ger.* Guben. Lubuskie, W Poland 51°59′N 14°43′E
126 K8 **Gubkin** Belgorodskaya Oblast', W Russian Federation 51°16′N 37°32′E
162 J9 **Guchin-Us** *var.* Arguut. Övörhangay, C Mongolia 45°27′N 102°25′E
Gudara *see* Ghŭdara
105 S8 **Gúdar, Sierra de** ▲▲ E Spain
137 P8 **Gudauta** *prev.* Gudaut'a. NW Georgia 43°07′N 40°35′E
Gudaut'a *see* Gudauta
94 G12 **Gudbrandsdalen** *valley* S Norway
95 G21 **Gudenå** *var.* Gudenaa. ≈ C Denmark
Gudenaa *see* Gudenå
127 P16 **Gudermes** Chechenskaya Respublika, SW Russian Federation 43°23′N 46°06′E
155 J18 **Gūdūr** Andhra Pradesh, E India 14°10′N 79°51′E
146 B13 **Gudurolum** Balkan Welaýaty, W Turkmenistan 37°28′N 54°30′E
94 D13 **Gudvangen** Sogn Og Fjordane, S Norway 60°54′N 06°49′E
103 U7 **Guebwiller** Haut-Rhin, NE France 47°55′N 07°13′E
Guéckédou *see* Guékédou
14 K8 **Guéguen, Lac** ◎ Québec, SE Canada
76 J15 **Guékédou** *var.* Guéckédou. Guinée-Forestière, S Guinea 08°33′N 10°08′W
41 R16 **Guelatao** Oaxaca, SE Mexico
Guelders *see* Gelderland
78 G11 **Guélengdeng** Mayo-Kébbi Est, W Chad 10°55′N 15°31′E
74 L5 **Guelma** *var.* Gâlma. NE Algeria 36°29′N 07°25′E
74 D8 **Guelmime** *var.* Goulimime. SW Morocco 28°59′N 10°10′W
14 G15 **Guelph** Ontario, S Canada 43°34′N 80°16′W
102 I7 **Gueméné-Penfao** Loire-Atlantique, NW France 47°37′N 01°49′W
102 I7 **Guer** Morbihan, NW France 47°54′N 02°07′W
78 I11 **Guéra** *off.* Région du Guéra. ◈ *region* S Chad
102 H8 **Guérande** Loire-Atlantique, NW France 47°20′N 02°25′W
Guéra, Région du *see* Guéra
78 K9 **Guéréda** Wadi Fira, E Chad 14°30′N 22°05′E
103 N10 **Guéret** Creuse, C France 46°10′N 01°52′E
Guernica/Guernica y Lumo *see* Gernika-Lumo
33 Z15 **Guernsey** Wyoming, C USA 42°16′N 104°44′W
97 K25 **Guernsey** ◇ *British Crown Dependency* Channel Islands, NW Europe
97 K25 **Guernsey** *island* Channel Islands, NW Europe
76 J10 **Guérou** Assaba, S Mauritania 16°48′N 11°40′W
25 R16 **Guerra** Texas, SW USA 26°54′N 98°53′W
41 O15 **Guerrero** ◈ *state* S Mexico
40 D6 **Guerrero Negro** Baja California Sur, NW Mexico 27°56′N 114°04′W
103 P9 **Gueugnon** Saône-et-Loire, C France 46°36′N 04°03′E
76 M17 **Guéyo** S Ivory Coast 05°25′N 06°04′W
107 L15 **Guglionesi** Molise, C Italy 41°54′N 14°54′E
188 K5 **Guguan** *island* C Northern Mariana Islands
Guhrau *see* Góra
Gui *see* Guangxi Zhuangzu Zizhiqu
Guiana *see* French Guiana
47 V4 **Guiana Basin** *undersea feature* W Atlantic Ocean 11°00′N 52°00′W
48 G6 **Guiana Highlands** *var.* Macizo de las Guayanas. ▲▲ N South America
Guiba *see* Juba
102 I7 **Guichen** Ille-et-Vilaine, NW France 47°57′N 01°47′W
Guichi *see* Chizhou
61 E18 **Guichón** Paysandú, W Uruguay 32°30′S 57°13′W
77 U12 **Guidan-Roumji** Maradi, S Niger 13°40′N 06°41′E
Guidder *see* Guider
159 T10 **Guide** *var.* Heyin. Qinghai, C China 36°06′N 101°25′E
78 F12 **Guider** *var.* Guidder. Nord, N Cameroon 09°55′N 13°59′E
76 I11 **Guidimaka** ◈ *region* S Mauritania
77 W12 **Guidimouni** Zinder, S Niger 13°40′N 09°31′E
76 G10 **Guier, Lac de** *var.* Lac de Guiers. ◎ N Senegal
Guiers, Lac de *see* Guier, Lac de
160 L14 **Guigang** *var.* Guixian, Gui Xian. Guangxi Zhuangzu Zizhiqu, S China 23°06′N 109°36′E
76 L16 **Guiglo** W Ivory Coast 06°33′N 07°29′W
54 L5 **Güigüe** Carabobo, N Venezuela 10°05′N 67°48′W
83 M20 **Guijá** Gaza, S Mozambique 24°31′S 33°02′E
42 E7 **Güija, Lago de** ◎ El Salvador/Guatemala
160 L14 **Gui Jiang** *var.* Gui Shui. ≈ S China
104 K8 **Guijuelo** Castilla y León, N Spain 40°34′N 05°40′W
Guilan *see* Gīlān
97 N22 **Guildford** SE England, United Kingdom 51°14′N 00°35′W
19 R5 **Guildford** Maine, NE USA 45°10′N 69°22′W
19 O7 **Guildhall** Vermont, NE USA 44°34′N 71°36′W
103 R13 **Guilherand** Ardèche, E France 44°57′N 04°49′E
160 L13 **Guilin** *var.* Kuei-lin, Kweilin. Guangxi Zhuangzu Zizhiqu, S China 25°15′N 110°16′E
12 J6 **Guillaume-Delisle, Lac** ◎ Québec, NE Canada
103 U13 **Guillestre** Hautes-Alpes, SE France 44°41′N 06°39′E
104 H6 **Guimarães** *var.* Guimaráes. Braga, N Portugal 41°26′N 08°19′W
Guimaráes *see* Guimarães
58 D11 **Guimarães Rosas, Pico** ▲ NW Brazil
23 N3 **Guin** Alabama, S USA 33°58′N 87°54′W
Güina *see* Wina
76 I14 **Guinea** *off.* Republic of Guinea, *var.* Guinée; *prev.* French Guinea, People's Revolutionary Republic of Guinea. ◆ *republic* W Africa
64 N13 **Guinea Basin** *undersea feature* E Atlantic Ocean 0°00′N 05°00′W
76 E12 **Guinea-Bissau** *off.* Republic of Guinea-Bissau, *Fr.* Guinée-Bissau, *Port.* Guiné-Bissau; *prev.* Portuguese Guinea. ◆ *republic* W Africa
Guinea-Bissau, Republic of *see* Guinea-Bissau
66 K7 **Guinea Fracture Zone** *tectonic feature* E Atlantic Ocean
64 O13 **Guinea, Gulf of** *Fr.* Golfe de Guinée. *gulf* E Atlantic Ocean
Guinea, People's Revolutionary Republic of *see* Guinea
Guinea, Republic of *see* Guinea
Guiné-Bissau *see* Guinea-Bissau
Guinée *see* Guinea
Guinée-Bissau *see* Guinea-Bissau
Guinée, Golfe de *see* Guinea, Gulf of
44 C4 **Güines** La Habana, W Cuba 22°50′N 82°02′W
102 G5 **Guingamp** Côtes d'Armor, NW France 48°34′N 03°09′W
Guipúzcoa *see* Gipuzkoa
44 C5 **Güira de Melena** La Habana, W Cuba 22°47′N 82°33′W
74 G8 **Guir, Hamada du** *desert* Algeria/Morocco
55 P5 **Güiria** Sucre, NE Venezuela 10°37′N 62°21′W
Gui Shui *see* Gui Jiang
104 H2 **Guitiriz** Galicia, NW Spain 43°10′N 07°52′W
77 N17 **Guitri** S Ivory Coast 05°31′N 05°14′W
171 Q5 **Guiuan** Samar, C Philippines 11°02′N 125°45′E
Gui Xian/Guixian *see* Guigang
160 J12 **Guiyang** *var.* Kuei-Yang, Kuei-yang, Kueyang, Kweiyang; *prev.* Kweichu. *province capital* Guizhou, S China 26°33′N 106°45′E
160 J12 **Guizhou** *var.* Guizhou Sheng, Kuei-chou, Kweichow, Qian. ◈ *province* S China
Guizhou Sheng *see* Guizhou
102 J13 **Gujan-Mestras** Gironde, SW France 44°39′N 01°04′W
154 B10 **Gujarāt** *var.* Gujerat. ◈ *state* W India
149 V6 **Gūjar Khān** Punjab, E Pakistan 33°19′N 73°23′E
Gujerat *see* Gujarāt
149 V7 **Gujrānwāla** Punjab, NE Pakistan 32°11′N 74°09′E
149 V7 **Gujrāt** Punjab, E Pakistan 32°34′N 74°04′E
146 B8 **Gulandag** *Rus.* Gory Kulandag. ▲▲ Balkan Welaýaty, W Turkmenistan
159 U9 **Gulang** Gansu, C China 37°31′N 102°55′E
183 R6 **Gulargambone** New South Wales, SE Australia 31°19′S 148°31′E
155 G15 **Gulbarga** Karnātaka, C India 17°22′N 76°47′E
118 J8 **Gulbene** *Ger.* Alt-Schwanenburg. NE Latvia 57°10′N 26°44′E
147 U10 **Gul'cha** *Kir.* Gülchö. Oshskaya Oblast', SW Kyrgyzstan 40°16′N 73°27′E
Gülchö *see* Gul'cha
173 T10 **Gulden Draak Seamount** *undersea feature* E Indian Ocean 33°45′S 101°00′E
136 J16 **Gülek Boğazı** *var.* Cilician Gates. *pass* S Turkey
186 D8 **Gulf** ◈ *province* S Papua New Guinea
23 O9 **Gulf Breeze** Florida, SE USA 30°21′N 87°09′W
Gulf of Liaotung *see* Liaodong Wan
23 V13 **Gulfport** Florida, SE USA 27°45′N 82°42′W
22 M9 **Gulfport** Mississippi, S USA 30°22′N 89°06′W
23 O9 **Gulf Shores** Alabama, S USA 30°15′N 87°40′W
183 R7 **Gulgong** New South Wales, SE Australia 32°22′S 149°31′E
160 I11 **Gulin** Sichuan, C China 28°06′N 105°47′E

◆ Country
● Country Capital
◇ Dependent Territory
○ Dependent Territory Capital
◈ Administrative Regions
✈ International Airport
▲ Mountain
▲▲ Mountain Range
⩕ Volcano
≈ River
◎ Lake
⊠ Reservoir

GULIR – HALMAHERA, PULAU

171 U14 **Gulir** Pulau Kasiui, E Indonesia 04°27′S 131°41′E
Gulistan see Guliston
147 P10 **Guliston** Rus. Gulistan. Sirdaryo Viloyati, E Uzbekistan 40°29′N 68°46′E
163 T6 **Guliya Shan** ▲ NE China 49°42′N 122°22′E
Gulja see Yining
39 S11 **Gulkana** Alaska, USA 62°17′N 145°25′W
11 S17 **Gull Lake** Saskatchewan, S Canada 50°05′N 108°30′W
31 P10 **Gull Lake** ◎ Michigan, N USA
29 T6 **Gull Lake** ◎ Minnesota, N USA
95 L16 **Gullspång** Västra Götaland, S Sweden 58°58′N 14°04′E
136 B15 **Güllük Körfezi** prev. Akbük Limanı. bay W Turkey
152 H5 **Gulmarg** Jammu and Kashmir, NW India 34°04′N 74°25′E
Gulpaigan see Golpāyegān
99 L18 **Gulpen** Limburg, SE Netherlands 50°48′N 05°53′E
Gul'shad see Gul'shat
145 S13 **Gul'shat** var. Gul'shad. Karaganda, E Kazakhstan 46°37′N 74°22′E
81 F17 **Gulu** N Uganda 02°46′N 32°21′E
Gŭlŭbovo see Galabovo
114 I7 **Gulyantsi** Pleven, N Bulgaria 43°37′N 24°40′E
Gulyaypole see Hulyaypole
Guma see Pishan
Gümai see Darlag
79 K16 **Gumba** Equateur, NW Dem. Rep. Congo 02°58′N 21°23′E
Gumbinnen see Gusev
81 H24 **Gumbiro** Ruvuma, S Tanzania 10°19′S 35°40′E
146 B11 **Gumdag** prev. Kum-Dag. Balkan Welaýaty, W Turkmenistan 39°13′N 54°35′E
77 W12 **Gumel** Jigawa, N Nigeria 12°37′N 09°23′E
105 N5 **Gumiel de Hizán** Castilla y León, N Spain 41°46′N 03°42′W
153 P16 **Gumla** Jhārkhand, N India 23°03′N 84°36′E
Gumma see Gunma
101 F16 **Gummersbach** Nordrhein-Westfalen, W Germany 51°01′N 07°34′E
77 T13 **Gummi** Zamfara, NW Nigeria 12°07′N 05°07′E
Gumpolds see Humpolec
Gumti see Gomati
Gümülcine/Gümüljina see Komotiní
Gümüşane see Gümüşhane
137 O12 **Gümüşhane** var. Gümüşane, Gumushkhane. Gümüşhane, NE Turkey 40°31′N 39°27′E
137 O12 **Gümüşhane** var. Gümüşane, Gumushkhane. ◐ province NE Turkey
Gumushkhane see Gümüşhane
171 V14 **Gumzai** Pulau Kola, E Indonesia 05°27′S 134°38′E
154 H9 **Guna** Madhya Pradesh, C India 24°39′N 77°18′E
Gunabad see Gonābād
Gunan see Qijiang
Gunbad-i-Qawus see Gonbad-e Kāvūs
183 O9 **Gunbar** New South Wales, SE Australia 34°03′S 145°32′E
183 O9 **Gun Creek** seasonal river New South Wales, SE Australia
183 Q10 **Gundagai** New South Wales, SE Australia 35°06′S 148°03′E
79 K17 **Gundji** Equateur, N Dem. Rep. Congo 02°03′N 21°31′E
155 G20 **Gundlupet** Karnātaka, W India 11°48′N 76°42′E
136 G16 **Gündoğmuş** Antalya, S Turkey 36°50′N 32°07′E
137 O14 **Güney Doğu Toroslar** ▲▲ SE Turkey
79 J21 **Gungu** Bandundu, SW Dem. Rep. Congo 05°43′S 19°20′E
127 P17 **Gunib** Respublika Dagestan, SW Russian Federation 42°24′N 46°55′E
112 J11 **Gunja** Vukovar-Srijem, E Croatia 44°53′N 18°51′E
31 P9 **Gun Lake** ◎ Michigan, N USA
165 N12 **Gunma** off. Gunma-ken, var. Gumma. ◐ prefecture Honshū, S Japan
Gunma-ken see Gunma
197 P15 **Gunnbjørn Fjeld** var. Gunnbjørns Bjerge. ▲ C Greenland 69°03′N 29°36′W
Gunnbjørns Bjerge see Gunnbjørn Fjeld
183 S6 **Gunnedah** New South Wales, SE Australia 30°59′S 150°15′E
173 Y15 **Gunner's Quoin** var. Coin de Mire. island N Mauritius
37 R6 **Gunnison** Colorado, C USA 38°33′N 106°55′W
36 L5 **Gunnison** Utah, W USA 39°09′N 111°49′W
37 P5 **Gunnison River** ~ Colorado, C USA
21 X2 **Gunpowder River** ~ Maryland, NE USA
Güns see Kőszeg
163 X16 **Gunsan** var. Gunsan, Jap. Gunzan; prev. Kunsan. W South Korea 35°58′N 126°42′E
Gunsan see Gunsan
109 S4 **Gunskirchen** Oberösterreich, N Austria 48°07′N 13°54′E
Gunt see Ghund
155 H17 **Guntakal** Andhra Pradesh, C India 15°11′N 77°24′E
23 Q2 **Guntersville** Alabama, S USA 34°21′N 86°17′W
23 Q2 **Guntersville Lake** ⊠ Alabama, S USA
109 X4 **Guntramsdorf** Niederösterreich, E Austria 48°03′N 16°19′E
155 J16 **Guntūr** var. Guntur. Andhra Pradesh, SE India 16°20′N 80°27′E
168 H10 **Gunungsitoli** Pulau Nias, W Indonesia 01°11′N 97°35′E
155 M14 **Gunupur** Odisha, E India 19°04′N 83°52′E
101 J23 **Günz** ~ S Germany
Gunzan see Gunsan
101 J22 **Günzburg** Bayern, S Germany 48°26′N 10°18′E
101 K21 **Gunzenhausen** Bayern, S Germany 49°07′N 10°45′E
Guolüezhen see Lingbao
Guovdageaidnu see Kautokeino
161 P7 **Guoyang** Anhui, E China 33°30′N 116°12′E
116 G11 **Gurahonț** Hung. Honctő. Arad, W Romania 46°16′N 22°21′E
Gurahumora see Gura Humorului
116 K9 **Gura Humorului** Ger. Gurahumora. Suceava, NE Romania 47°31′N 26°00′E
146 H8 **Gurbansoltan Eje** prev. Ýylanly, Rus. Il'yaly. Daşoguz Welaýaty, N Turkmenistan 41°57′N 59°42′E
158 K4 **Gurbantünggüt Shamo** desert W China
152 H7 **Gurdāspur** Punjab, N India 32°04′N 75°28′E
27 T13 **Gurdon** Arkansas, C USA 33°55′N 93°09′W
Gurdzhaani see Gurjaani
Gurgan see Gorgān
152 I10 **Gurgaon** Haryāna, N India 28°27′N 77°01′E
59 M15 **Gurguéia, Rio** ~ NE Brazil
55 Q7 **Guri, Embalse de** ⊠ E Venezuela
137 V10 **Gurjaani** Rus. Gurdzhaani. E Georgia 41°42′N 45°47′E
109 T8 **Gurk** Kärnten, S Austria 46°52′N 14°17′E
109 T9 **Gurk** Slvn. Krka. ~ S Austria
Gurkfeld see Krško
114 K9 **Gurkovo** prev. Kolupchii. Stara Zagora, C Bulgaria 42°42′N 25°46′E
109 S9 **Gurktaler Alpen** ▲▲ S Austria
146 H8 **Gurlan** Rus. Gurlen. Xorazm Viloyati, W Uzbekistan 41°54′N 60°18′E
Gurlen see Gurlan
83 M16 **Guro** Manica, C Mozambique 17°28′S 33°18′E
136 M14 **Gürün** Sivas, C Turkey 38°44′N 37°15′E
59 K16 **Gurupi** Tocantins, C Brazil 11°44′S 49°01′W
58 L12 **Gurupi, Rio** ~ NE Brazil
152 E14 **Guru Sikhar** ▲ NW India 24°45′N 72°51′E
162 H8 **Gurvanbulag** var. Höviyn Am. Bayanhongor, C Mongolia 47°08′N 98°41′E
162 K7 **Gurvanbulag** var. Avdzaga. Bulgan, C Mongolia 47°43′N 103°30′E
162 I11 **Gurvantes** var. Urt. Ömnögovĭ, S Mongolia 43°16′N 101°00′E
Gur'yev/Gur'yevskaya Oblast' see Atyrau
77 U13 **Gusau** Zamfara, NW Nigeria 12°18′N 06°27′E
126 C3 **Gusev** Ger. Gumbinnen. Kaliningradskaya Oblast', W Russian Federation 54°36′N 22°14′E
146 J17 **Gushgy** Rus. Kushka. ~ Mary Welaýaty, S Turkmenistan
Gushiago see Gushiegu
77 Q14 **Gushiegu** var. Gushiago. NE Ghana 09°54′N 00°12′W
165 S17 **Gushikawa** Okinawa, Okinawa, SW Japan 26°21′N 127°50′E
113 L16 **Gusinje** E Montenegro 42°34′N 19°51′E
126 M4 **Gus'-Khrustal'nyy** Vladimirskaya Oblast', W Russian Federation 55°39′N 40°42′E
107 B19 **Guspini** Sardegna, Italy, C Mediterranean Sea 39°33′N 08°39′E
109 X8 **Güssing** Burgenland, SE Austria 47°03′N 16°19′E
109 V6 **Gusswerk** Steiermark, E Austria 47°43′N 15°18′E
92 O2 **Gustav Adolf Land** physical region NE Svalbard
195 X5 **Gustav Bull Mountains** ▲▲ Antarctica
39 W13 **Gustavus** Alaska, USA 58°24′N 135°44′W
92 O1 **Gustav V Land** physical region NE Svalbard
35 P9 **Gustine** California, W USA 37°14′N 121°00′W
25 R8 **Gustine** Texas, SW USA 31°51′N 98°24′W
100 M9 **Güstrow** Mecklenburg-Vorpommern, NE Germany 53°48′N 12°12′E
95 N18 **Gusum** Östergötland, S Sweden 58°15′N 16°30′E
Guta/Gúta see Kolárovo
Gutenstein see Ravne na Koroškem
101 G14 **Gütersloh** Nordrhein-Westfalen, W Germany 51°54′N 08°23′E
27 N10 **Guthrie** Oklahoma, C USA 35°53′N 97°26′W
25 P5 **Guthrie** Texas, SW USA 33°38′N 100°21′W
29 U14 **Guthrie Center** Iowa, C USA 41°40′N 94°30′W
41 Q13 **Gutiérrez Zamora** Veracruz-Llave, E Mexico 20°29′N 97°07′W
Gutta see Kolárovo
29 Y12 **Guttenberg** Iowa, C USA 42°47′N 91°06′W
Guttentag see Dobrodzień
Guttstadt see Dobre Miasto
162 G8 **Guulin** Govĭ-Altay, C Mongolia 46°33′N 97°21′E
153 V12 **Guwāhāti** prev. Gauhāti. Assam, NE India 26°09′N 91°42′E
139 R3 **Guwēr** var. Al Kuwayr, Al Quwayr, Quwair. Arbīl, N Iraq 36°03′N 43°30′E
146 A10 **Guwlumaýak** Rus. Kuuli-Mayak. Balkan Welaýaty, NW Turkmenistan 40°14′N 52°43′E
55 R9 **Guyana** off. Co-operative Republic of Guyana; prev. British Guiana. ◆ republic N South America
Guyana, Co-operative Republic of see Guyana
21 P5 **Guyandotte River** ~ West Virginia, NE USA
Guyane see French Guiana
Guyi see Sanjiang
26 H8 **Guymon** Oklahoma, C USA 36°42′N 101°30′W
146 K12 **Guynuk** Lebap Welaýaty, NE Turkmenistan 39°18′N 63°00′E
Guyong see Jiangle
21 O9 **Guyot, Mount** ▲ North Carolina/Tennessee, SE USA 35°42′N 83°15′W
183 U5 **Guyra** New South Wales, SE Australia 30°13′S 151°42′E
159 W10 **Guyuan** Ningxia, N China 35°57′N 106°13′E
Guzar see G'uzor
121 P2 **Güzelyurt** Gk. Kólpos Mórfu, Morphou. W Cyprus 35°12′N 33°E
121 N2 **Güzelyurt Körfezi** var. Morfou Bay, Morphou Bay, Gk. Kólpos Mórfou. bay W Cyprus
Guzhou see Rongjiang
40 I3 **Guzmán** Chihuahua, N Mexico 31°13′N 107°27′W
147 N13 **G'uzor** Rus. Guzar. Qashqadaryo Viloyati, S Uzbekistan 38°41′N 66°12′E
119 B14 **Gvardeysk** Ger. Tapaiu. Kaliningradskaya Oblast', W Russian Federation 54°39′N 21°02′E
Gvardeyskoye see Hvardiys'ke
183 R5 **Gwabegar** New South Wales, SE Australia 30°34′S 148°58′E
148 J16 **Gwādar** var. Gwadur. Baluchistān, SW Pakistan 25°09′N 62°21′E
148 J16 **Gwādar East Bay** bay SW Pakistan
148 J16 **Gwādar West Bay** bay SW Pakistan
Gwadur see Gwādar
83 J17 **Gwai** Matabeleland North, W Zimbabwe 19°17′S 27°37′E
154 I7 **Gwalior** Madhya Pradesh, C India 26°16′N 78°12′E
83 J18 **Gwanda** Matabeleland South, SW Zimbabwe 20°56′S 29°E
79 N15 **Gwane** Orientale, N Dem. Rep. Congo 04°40′N 25°51′E
163 X16 **Gwangju** off. Kwangju-gwangyŏksi, var. Guangju, Kwangchu, Jap. Kōshū; prev. Kwangju. SW South Korea 35°09′N 126°53′E
83 J17 **Gwayi** ~ W Zimbabwe
110 G8 **Gwda** var. Głda, Ger. Küddow. ~ NW Poland
97 C14 **Gweebarra Bay** Ir. Béal an Bheara. inlet W Ireland
97 D14 **Gweedore** Ir. Gaoth Dobhair. Donegal, NW Ireland 55°03′N 08°14′W
Gwelo see Gweru
97 K21 **Gwent** cultural region S Wales, United Kingdom
83 K17 **Gweru** prev. Gwelo. Midlands, C Zimbabwe 19°27′S 29°49′E
29 Q7 **Gwinner** North Dakota, N USA 46°10′N 97°42′W
77 Y13 **Gwoza** Borno, NE Nigeria 11°07′N 13°40′E
Gwy see Wye
183 R4 **Gwydir River** ~ New South Wales, SE Australia
97 I19 **Gwynedd** var. Gwyneth. cultural region NW Wales, United Kingdom
Gwyneth see Gwynedd
159 O16 **Gyaca** var. Ngarrab. Xizang Zizhiqu, W China 29°06′N 92°37′E
Gya'gya see Saga
Gyaijêpozhanggê see Zhidoi
Gyaisi see Jiulong
115 M22 **Gyalí** var. Yialí. island Dodekánisa, Greece, Aegean Sea
Gyamotang see Dêngqên
Gyandzha see Gäncä
Gyangkar see Dinggyê
158 M16 **Gyangzê** Xizang Zizhiqu, W China 28°50′N 89°38′E
158 L14 **Gyaring Co** ◎ W China
159 Q12 **Gyaring Hu** ◎ C China
115 I20 **Gyáros** var. Yioúra. island Kykládes, Greece, Aegean Sea
122 J7 **Gyda** Yamalo-Nenetskiy Avtonomnyy Okrug, N Russian Federation 70°55′N 78°34′E
122 J7 **Gydanskiy Poluostrov** Eng. Gyda Peninsula. peninsula N Russian Federation
Gyda Peninsula see Gydanskiy Poluostrov
Gyêgu see Yushu
163 W15 **Gyeonggi-man** prev. Kyŏnggi-man. bay NW South Korea
163 Z16 **Gyeongju** Jap. Keishū; prev. Kyŏngju. SE South Korea 35°49′N 129°09′E
Gyéres see Câmpia Turzii
Gyergyószentmiklós see Gheorgheni
Gyergyótölgyes see Tulgheş
Gyertyámos see Cărpiniş
Gyeva see Detva
Gyigang see Zayü
Gyixong see Gonggar
95 I23 **Gyldenløveshøy** hill range C Denmark
181 Z10 **Gympie** Queensland, E Australia 26°05′S 152°40′E
166 L7 **Gyobingauk** Bago, SW Myanmar (Burma) 18°14′N 95°39′E
111 M23 **Gyomaendrőd** Békés, SE Hungary 46°56′N 20°50′E
Gyömbér see Ďumbier
111 L22 **Gyöngyös** Heves, NE Hungary 47°44′N 19°49′E
111 H22 **Győr** Ger. Raab, Lat. Arrabona. Győr-Moson-Sopron, NW Hungary 47°41′N 17°40′E
111 G22 **Győr-Moson-Sopron** off. Győr-Moson-Sopron Megye. ◐ county NW Hungary
Győr-Moson-Sopron Megye see Győr-Moson-Sopron
11 X15 **Gypsumville** Manitoba, S Canada 51°47′N 98°38′W
12 M4 **Gyrfalcon Islands** island group Northwest Territories, NE Canada
95 N14 **Gysinge** Gävleborg, C Sweden 60°16′N 16°55′E
115 F22 **Gýtheio** var. Githio; prev. Yíthion. Pelopónnisos, S Greece 36°46′N 22°34′E
146 L13 **Gyuichbirleshik** Lebap Welaýaty, E Turkmenistan 38°10′N 64°33′E
111 N24 **Gyula** Rom. Jula. Békés, SE Hungary 46°39′N 21°17′E
Gyulafehérvár see Alba Iulia
Gyulovo see Roza
137 T11 **Gyumri** var. Giumri, Rus. Kumayri; prev. Aleksandropol', Leninakan. W Armenia 40°48′N 43°51′E
146 D13 **Gyunuzyndag, Gora** ▲ Balkan Welaýaty, W Turkmenistan 38°15′N 56°25′E
146 J15 **Gyzylbaydak** Rus. Krasnoye Znamya. Mary Welaýaty, S Turkmenistan 36°51′N 62°24′E
Gyzyletrek see Etrek
146 D10 **Gyzylgaýa** Rus. Kizyl-Kaya. Balkan Welaýaty, NW Turkmenistan 40°37′N 55°15′E
146 A10 **Gyzylsuw** Rus. Kizyl-Su. Balkan Welaýaty, W Turkmenistan 39°49′N 53°00′E
Gyzyrlabat see Serdar
Gzhatsk see Gagarin

H

153 T12 **Ha** W Bhutan 27°17′N 89°22′E
Haabai see Ha'apai Group
99 H17 **Haacht** Vlaams Brabant, C Belgium 50°58′N 04°38′E
109 T4 **Haag** Niederösterreich, NE Austria 48°07′N 14°32′E
194 L8 **Haag Nunataks** ▲ Antarctica
92 N2 **Haakon VII Land** physical region NW Svalbard
98 O11 **Haaksbergen** Overijssel, E Netherlands 52°09′N 06°45′E
99 E14 **Haamstede** Zeeland, SW Netherlands 51°43′N 03°45′E
193 Y15 **Ha'ano** island Ha'apai Group, C Tonga
193 Y15 **Ha'apai Group** var. Haabai. island group C Tonga
93 L15 **Haapajärvi** Pohjois-Pohjanmaa, C Finland 63°45′N 25°20′E
93 L17 **Haapamäki** Pirkanmaa, C Finland 62°11′N 24°32′E
93 L15 **Haapavesi** Pohjois-Pohjanmaa, C Finland 64°09′N 25°25′E
191 N7 **Haapiti** Moorea, W French Polynesia 17°33′S 149°52′W
118 F4 **Haapsalu** Ger. Hapsal. Läänemaa, W Estonia 58°58′N 23°32′E
Ha'Arava see 'Arabah, Wādī al
95 G24 **Haarby** var. Hårby. Syddtjylland, C Denmark 55°13′N 10°07′E
98 H10 **Haarlem** prev. Harlem. Noord-Holland, W Netherlands 52°23′N 04°39′E
185 D19 **Haast** West Coast, South Island, New Zealand 43°53′S 169°02′E
185 C20 **Haast** ~ South Island, New Zealand
185 D20 **Haast Pass** pass South Island, New Zealand
193 W16 **Ha'atua** 'Eau, E Tonga 21°23′S 174°57′W
149 P15 **Hab** ~ SW Pakistan
141 W7 **Habā** var. Al Haba. Dubayy, NE United Arab Emirates 25°01′N 55°37′E
158 K2 **Habahe** var. Kaba. Xinjiang Uygur Zizhiqu, NW China 48°04′N 86°20′E
141 U13 **Ḩabarūt** var. Habrut. SW Oman 17°19′N 52°45′E
81 J18 **Habaswein** Isiolo, NE Kenya 01°01′N 39°27′E
99 L24 **Habay-la-Neuve** Luxembourg, SE Belgium 49°43′N 05°38′E
139 S8 **Ḩabbānīyah, Buḩayrat** ◎ C Iraq
Habelschwerdt see Bystrzyca Kłodzka
153 V14 **Habiganj** Sylhet, NE Bangladesh 24°23′N 91°25′E
163 Q12 **Habirag** Nei Mongol Zizhiqu, N China 42°18′N 115°40′E
95 L19 **Habo** Västra Götaland, S Sweden 57°55′N 14°05′E
123 V14 **Habomai Islands** island group Kuril'skiye Ostrova, SE Russian Federation
165 S2 **Haboro** Hokkaidō, NE Japan 44°19′N 141°42′E
153 S16 **Habra** West Bengal, NE India 22°39′N 88°17′E
Habrut see Ḩabarūt
143 P17 **Ḩabshān** Abū Z̧aby, C United Arab Emirates 23°51′N 53°34′E
54 E14 **Hacha** Putumayo, S Colombia 0°02′S 75°30′W
165 X13 **Hachijō-jima** island Izu-shotō, SE Japan
164 L12 **Hachiman** Gifu, Honshū, SW Japan 35°46′N 136°57′E
165 P7 **Hachimori** Akita, Honshū, C Japan 40°22′N 139°59′E
165 R7 **Hachinohe** Aomori, Honshū, C Japan 40°30′N 141°29′E
165 X13 **Hachiōji** Tōkyō, Hachijō-jima, SE Japan 35°40′N 139°20′E
137 Y12 **Hacıqabal** prev. Qazımämmäd. SE Azerbaijan 40°03′N 48°56′E
93 G17 **Hackås** Jämtland, C Sweden 62°55′N 14°31′E
18 K14 **Hackensack** New Jersey, NE USA 40°51′N 73°57′W
75 U12 **Ḩadabat al Jilf al Kabīr** var. Gilf Kebir Plateau. plateau SW Egypt
Hadama see Nazrēt
141 W13 **Ḩadbaram** S Oman 17°27′N 55°13′E
139 U13 **Ḩaddānīyah** well S Iraq
96 K12 **Haddington** SE Scotland, United Kingdom 55°59′N 02°46′W
141 Z8 **Ḩadd, Ra's al** headland NE Oman 22°28′N 59°58′E
Haded see Xadeed
77 W12 **Hadejia** Jigawa, N Nigeria 12°22′N 10°02′E
77 W12 **Hadejia** ~ N Nigeria
138 F9 **Hadera** var. Khadera; prev. Ḩadera. Haifa, N Israel 32°26′N 34°55′E
Ḩadera see Hadera
95 G24 **Haderslev** Ger. Hadersleben. Syddanmark, SW Denmark 55°15′N 09°30′E
Hadersleben see Haderslev
151 J21 **Hadhdhunmathi Atoll** atoll S Maldives
141 W17 **Ḩadībūh** Suquṭrā, SE Yemen 12°38′N 54°05′E
158 K9 **Hadilik** Xinjiang Uygur Zizhiqu, W China 37°51′N 86°10′E
136 H16 **Hadım** Konya, S Turkey 36°58′N 32°27′E
140 K7 **Hadīyah** Al Madīnah, W Saudi Arabia 25°36′N 38°31′E
78 H11 **Hadjer-Lamis** off. Région du Hadjer-Lamis. ◐ region SW Chad
Hadjer-Lamis, Région du see Hadjer-Lamis
8 L5 **Hadley Bay** bay Victoria Island, Nunavut, N Canada
167 S6 **Ha Đông** var. Hadong. Ha Tây, N Vietnam 20°58′N 105°46′E
Hadong see Ha Đông
Hadramawt see Ḩaḑramawt
141 R15 **Ḩaḑramawt** Eng. Hadramaut. ▲▲ S Yemen
Hadria see Adria
Hadrianopolis see Edirne
Hadria Picena see Apricena
95 G22 **Hadsten** Midtjylland, C Denmark 56°19′N 10°03′E
95 G21 **Hadsund** Nordjylland, N Denmark 56°43′N 10°08′E
117 S4 **Hadyach** Rus. Gadyach. Poltavs'ka Oblast', NE Ukraine 50°21′N 34°00′E
114 N9 **Hadzhiyska Reka** var. Khadzhiyska Reka. ~ E Bulgaria
112 I13 **Hadžići** Federacija Bosne I Hercegovine, SE Bosnia and Herzegovina 43°49′N 18°12′E
163 W14 **Haeju** S North Korea 38°04′N 125°40′E
Haerbin/Haerhpin/Ha-erh-pin see Harbin
141 P5 **Ḩafar al Bāṭin** Ash Sharqīyah, N Saudi Arabia 28°25′N 45°59′E
11 T15 **Hafford** Saskatchewan, S Canada 52°43′N 107°19′W
136 M13 **Hafik** Sivas, N Turkey 39°53′N 37°24′E
149 V8 **Hāfizābād** Punjab, E Pakistan 32°03′N 73°42′E
92 H4 **Hafnarfjörður** Höfuðborgarsvæðið, W Iceland 64°03′N 21°57′W
Hafnia see Denmark
Hafnia see København
Hafren see Severn
Hafun see Xaafuun
Hafun, Ras see Xaafuun, Raas
80 G10 **Hag 'Abdullah** Sinnar, E Sudan 13°59′N 33°35′E
81 K18 **Hagadera** Garissa, E Kenya 0°06′N 40°23′E
138 G8 **HaGalil** Eng. Galilee. ▲▲ N Israel
14 G10 **Hagar** Ontario, S Canada 46°27′N 80°22′W
155 G18 **Hagari** var. Vedāvati. ~ W India
188 B16 **Hagåtña** , var. Agaña. ○ (Guam) NW Guam 13°27′N 144°45′E
100 M13 **Hagelberg** hill NE Germany
39 N14 **Hagemeister Island** island Alaska, USA
101 F15 **Hagen** Nordrhein-Westfalen, W Germany 51°22′N 07°27′E
100 K10 **Hagenow** Mecklenburg-Vorpommern, N Germany 53°27′N 11°10′E
10 K15 **Hagensborg** British Columbia, SW Canada 52°24′N 126°24′W
80 I13 **Hāgere Hiywet** var. Agere Hiywet, Ambo. Oromīya, C Ethiopia 09°00′N 37°55′E
33 O15 **Hagerman** Idaho, NW USA 42°48′N 114°53′W
37 U14 **Hagerman** New Mexico, SW USA 33°07′N 104°19′W
21 V2 **Hagerstown** Maryland, NE USA 39°39′N 77°44′W
14 G16 **Hagersville** Ontario, S Canada 42°58′N 80°03′W
102 J15 **Hagetmau** Landes, SW France 43°40′N 00°36′W
95 K14 **Hagfors** Värmland, C Sweden 60°03′N 13°45′E
93 G16 **Häggenås** Jämtland, C Sweden 63°24′N 14°53′E
164 E12 **Hagi** Yamaguchi, Honshū, SW Japan 34°25′N 131°22′E
167 S5 **Ha Giang** Ha Giang, N Vietnam 22°50′N 104°58′E
Hagios Evstrátios see Ágios Efstrátios
HaGolan see Golan Heights
103 T4 **Hagondange** Moselle, NE France 49°16′N 06°06′E
97 B18 **Hag's Head** Ir. Ceann Caillí. headland W Ireland 52°56′N 09°29′W
102 I3 **Hague, Cap de la** headland N France 49°43′N 01°56′W
103 V5 **Haguenau** Bas-Rhin, NE France 48°49′N 07°47′E
165 X16 **Hahajima-rettō** island group SE Japan
15 R8 **Há Há , Lac** ◎ Québec, SE Canada
172 H13 **Hahaya** ✕ (Moroni) Grande Comore, NW Comoros
22 K9 **Hahnville** Louisiana, S USA 29°58′N 90°24′W
83 E22 **Haib** Karas, S Namibia 28°12′S 18°19′E
Haibak see Aibak
149 N15 **Haibo** ~ SW Pakistan
Haibowan see Wuhai
163 U12 **Haicheng** Liaoning, NE China 40°53′N 122°45′E
Haicheng see Haifeng
Haicheng see Haiyuan
Haida see Nový Bor
Haidarabad see Hyderābād
Haidenschaft see Ajdovščina
167 T6 **Hai Dương** Hai Hưng, N Vietnam 20°56′N 106°21′E
138 F9 **Haifa** ◐ district NW Israel
Haifa see Hefa
Haifa, Bay of see Mifrats Hefa
161 P14 **Haifeng** var. Haicheng. Guangdong, S China 22°56′N 115°19′E
Haifong see Hai Phong
161 P3 **Hai He** ~ E China
Haikang see Leizhou
160 L17 **Haikou** var. Hai-k'ou, Hoihow, Fr. Hoï-Hao. province capital Hainan, S China 20°N 110°17′E
Hai-k'ou see Haikou
140 M6 **Ḩā'il** Ḩā'il, NW Saudi Arabia 27°N 42°50′E
141 N5 **Ḩā'il** off. Minţaqah Ḩā'il. ◐ province N Saudi Arabia
163 S6 **Hailar He** ~ NE China
33 P14 **Hailey** Idaho, NW USA 43°31′N 114°18′W
14 H9 **Haileybury** Ontario, S Canada 47°27′N 79°39′W
163 X9 **Hailin** Heilongjiang, NE China 44°37′N 129°24′E
Ḩā'il, Minţaqah see Ḩā'il
93 K14 **Hailuoto** Swe. Karlö. island W Finland
Haima see Haymā'
Haimen see Taizhou
160 M17 **Hainan** var. Hainan Sheng, Qiong. ◐ province S China
160 K17 **Hainan Dao** island S China
Hainan Sheng see Hainan
Hainan Strait see Qiongzhou Haixia
Hainasch see Ainaži
Hainau see Chojnów
99 E20 **Hainaut** ◐ province SW Belgium
Hainburg see Hainburg an der Donau
109 Z4 **Hainburg an der Donau** var. Hainburg. Niederösterreich, NE Austria 48°09′N 16°57′E
39 W12 **Haines** Alaska, USA 59°13′N 135°27′W
32 L12 **Haines** Oregon, NW USA 44°53′N 117°56′W
23 W12 **Haines City** Florida, SE USA 28°06′N 81°37′W
10 H8 **Haines Junction** Yukon, W Canada 60°45′N 137°30′W
109 W4 **Hainfeld** Niederösterreich, NE Austria 48°03′N 15°47′E
101 N16 **Hainichen** Sachsen, E Germany 50°58′N 13°08′E
Hai Ninh see Mong Cai
167 T6 **Hai Phong** var. Haifong, Haiphong. N Vietnam 20°50′N 106°41′E
Haiphong see Hai Phong
161 S12 **Haitan Dao** island SE China
44 K8 **Haiti** off. Republic of Haiti. ◆ republic C West Indies
Haiti, Republic of see Haiti
35 T11 **Haiwee Reservoir** ⊠ California, W USA
80 I7 **Haiya** Red Sea, NE Sudan 18°17′N 36°21′E
159 T10 **Haiyan** var. Sanjiaocheng. Qinghai, W China 36°55′N 100°54′E
160 M13 **Haiyang Shan** ▲▲ S China
159 V10 **Haiyuan** Ningxia, N China 36°32′N 105°31′E
Hajda see Nový Bor
111 M22 **Hajdú-Bihar** off. Hajdú-Bihar Megye. ◐ county E Hungary
Hajdú-Bihar Megye see Hajdú-Bihar
111 N22 **Hajdúböszörmény** Hajdú-Bihar, E Hungary 47°39′N 21°32′E
111 N22 **Hajdúhadház** Hajdú-Bihar, E Hungary 47°40′N 21°40′E
111 N21 **Hajdúnánás** Hajdú-Bihar, E Hungary 47°50′N 21°26′E
111 N22 **Hajdúszoboszló** Hajdú-Bihar, E Hungary 47°27′N 21°24′E
142 I3 **Ḩājī Ebrāhīm, Kūh-e** ▲ Iran/Iraq 36°53′N 44°56′E
165 O9 **Hajiki-zaki** headland Sado, C Japan 38°19′N 138°28′E
Hajine see Abū Ḩardān
153 P13 **Hājīpur** Bihār, N India 25°41′N 85°13′E
141 N14 **Ḩajjah** W Yemen 15°43′N 43°33′E
139 U11 **Ḩajjama** Al Muthannā, S Iraq 31°24′N 45°20′E
143 R12 **Ḩājjīābād** Hormozgān, C Iran
139 U14 **Ḩājj, Thaqb al** well S Iraq
113 L16 **Hajla** ▲ E Montenegro
110 P10 **Hajnówka** Ger. Hermhausen. Podlaskie, NE Poland 52°45′N 23°32′E
Haka see Hakha
Hakapehi see Punaauia
Hakâri see Hakkâri
138 F12 **HaKatan, HaMakhtesh** prev. HaMakhtesh HaQatan. ▲ S Israel
166 K4 **Hakha** var. Haka. Chin State, W Myanmar (Burma) 22°42′N 93°41′E
137 T16 **Hakkâri** var. Çölemerik, Hakâri. Hakkâri, SE Turkey 37°36′N 43°45′E
137 T16 **Hakkâri** var. Hakkari. ◐ province SE Turkey
Hakkari see Hakkâri
92 J12 **Hakkas** Norrbotten, N Sweden 66°53′N 21°36′E
164 J14 **Hakken-zan** ▲ Honshū, SW Japan 34°11′N 135°57′E
165 R7 **Hakkōda-san** ▲ Honshū, C Japan 40°40′N 140°49′E
165 T2 **Hako-dake** ▲ Hokkaidō, NE Japan 44°40′N 142°22′E
165 R5 **Hakodate** Hokkaidō, NE Japan 41°46′N 140°43′E
164 L11 **Hakui** Ishikawa, Honshū, SW Japan 36°55′N 136°46′E
190 B16 **Hakupu** SE Niue 19°06′S 169°50′E
164 L12 **Haku-san** ▲ Honshū, SW Japan 36°07′N 136°45′E
Hakusan see Mattō
Hal see Halle
149 Q15 **Hāla** Sind, SE Pakistan 25°47′N 68°28′E
138 J3 **Ḩalab** Eng. Aleppo, Fr. Alep; anc. Beroea. Ḩalab, NW Syria 36°14′N 37°10′E
138 J3 **Ḩalab** off. Muḩāfaz̧at Ḩalab, var. Aleppo, Halab. ◐ governorate NW Syria
138 J3 **Ḩalab** ✕ Ḩalab, NW Syria 36°12′N 37°10′E
Halab see Ḩalab
141 O8 **Ḩalabān** var. Halibān. Ar Riyāḑ, C Saudi Arabia 23°29′N 44°20′E
Ḩalabja see Helebce
Ḩalabjah see Helebce
Ḩalab, Muḩāfaz̧at see Ḩalab
146 L13 **Halaç** Rus. Khalach. Lebap Welaýaty, E Turkmenistan 38°05′N 64°46′E
190 A16 **Halagigie Point** headland W Niue
75 Z11 **Halaib** SE Egypt 22°10′N 36°33′E
76 Z11 **Halā'ib Triangle** ◐ disputed region S Egypt / N Sudan
190 G12 **Halalo** Île Uvea, N Wallis and Futuna 13°21′S 176°11′W
167 U10 **Ha Lam** Quang Nam-Đa Nẵng, C Vietnam 15°42′N 108°24′E
Halandri see Chalándri
141 X13 **Ḩalānīyāt, Juzur al** var. Jazā'ir Bin Ghalfān, Eng. Kuria Muria Islands. island group S Oman
141 W13 **Ḩalānīyāt, Khalīj al** Eng. Kuria Muria Bay. bay S Oman
Halas see Kiskunhalas
38 G11 **Hālawa** var. Halawa. Hawaii, USA, C Pacific Ocean 20°13′N 155°46′W
Halawa see Hālawa
38 F9 **Hālawa, Cape** var. Cape Halawa. headland Moloka'i, Hawai'i, USA 21°09′N 156°43′W
Cape Halawa see Hālawa, Cape
Halban see Tsetserleg
101 K14 **Halberstadt** Sachsen-Anhalt, C Germany 51°54′N 11°04′E
184 M12 **Halcombe** Manawatu-Wanganui, North Island, New Zealand 40°09′S 175°30′E
95 I16 **Halden** prev. Fredrikshald. Østfold, S Norway 59°08′N 11°20′E
100 L13 **Haldensleben** Sachsen-Anhalt, C Germany 52°18′N 11°25′E
Háldi see Halti
153 S17 **Haldia** West Bengal, NE India 22°04′N 88°02′E
152 K10 **Haldwāni** Uttarakhand, N India 29°13′N 79°31′E
163 P9 **Haldzan** var. Hatavch. Sühbaatar, E Mongolia 46°10′N 112°57′E
163 P9 **Haldzan** Sühbaatar, E Mongolia
38 F10 **Haleakalā** var. Haleakala. crater Maui, Hawai'i, USA
Haleakala see Haleakalā
25 N4 **Hale Center** Texas, SW USA 34°03′N 101°50′W
99 J18 **Halen** Limburg, NE Belgium 50°35′N 05°08′E
23 O2 **Haleyville** Alabama, S USA 34°13′N 87°37′W
77 O17 **Half Assini** SW Ghana 05°03′N 02°57′W
35 R8 **Half Dome** ▲ California, W USA 37°46′N 119°27′W
185 C25 **Halfmoon Bay** var. Oban. Stewart Island, Southland, New Zealand 46°53′S 168°08′E
182 E5 **Half Moon Lake** salt lake South Australia
163 R7 **Halhgol** Dornod, E Mongolia 47°57′N 118°07′E
163 S8 **Halhgol** var. Tsagaannuur. Dornod, E Mongolia 47°30′N 118°45′E
Haliacmon see Aliákmonas
Halibān see Ḩalabān
14 I13 **Haliburton** Ontario, SE Canada 45°03′N 78°20′W
14 I12 **Haliburton Highlands** var. Madawaska Highlands. hill range Ontario, SE Canada
13 Q15 **Halifax** province capital Nova Scotia, SE Canada 44°38′N 63°35′W
97 L17 **Halifax** N England, United Kingdom 53°44′N 01°52′W
21 W8 **Halifax** North Carolina, SE USA 36°19′N 77°37′W
21 U7 **Halifax** Virginia, NE USA 36°46′N 78°55′W
13 Q15 **Halifax** ✕ Nova Scotia, SE Canada 44°33′N 63°48′W
143 T13 **Halīl Rūd** seasonal river SE Iran
138 I6 **Ḩalīmah** ▲ Lebanon/Syria 34°12′N 36°37′E
162 G8 **Haliun** Govĭ-Altay, W Mongolia 45°55′N 96°06′E
118 I3 **Haljala** Ger. Halljal. Lääne-Virumaa, N Estonia 59°25′N 26°18′E
39 Q4 **Halkett, Cape** headland Alaska, USA 70°48′N 152°11′W
Halkida see Chalkída
96 J6 **Halkirk** N Scotland, United Kingdom 58°30′N 03°29′W
15 X7 **Hall** ~ Québec, SE Canada
Hall see Schwäbisch Hall
93 H15 **Hälla** Västerbotten, N Sweden 63°56′N 17°20′E
96 J6 **Halladale** ~ N Scotland, United Kingdom
95 J21 **Halland** ◐ county S Sweden
23 Z15 **Hallandale** Florida, SE USA 25°58′N 80°09′W
95 K22 **Hallandsås** physical region S Sweden
9 P6 **Hall Beach** var Sanirajak. Nunavut, N Canada 68°10′N 81°56′W
99 G19 **Halle** Fr. Hal. Vlaams Brabant, C Belgium 50°44′N 04°14′E
101 M15 **Halle** var. Halle an der Saale. Sachsen-Anhalt, C Germany 51°28′N 11°58′E
Halle an der Saale see Halle
35 W3 **Halleck** Nevada, W USA 40°57′N 115°27′W
95 L15 **Hällefors** Örebro, C Sweden 59°46′N 14°30′E
95 N16 **Hälleforsnäs** Södermanland, C Sweden 59°10′N 16°30′E
109 Q6 **Hallein** Salzburg, N Austria 47°41′N 13°06′E
101 L15 **Halle-Neustadt** Sachsen-Anhalt, C Germany 51°29′N 11°54′E
25 U12 **Hallettsville** Texas, SW USA 29°27′N 96°57′W
195 N4 **Halley** UK research station Antarctica 75°42′S 26°30′W
28 L4 **Halliday** North Dakota, N USA 47°19′N 102°19′W
37 S2 **Halligan Reservoir** ⊠ Colorado, C USA
100 G7 **Halligen** island group N Germany
94 G13 **Hallingdal** valley S Norway
38 J12 **Hall Island** island Alaska, USA
Hall Island see Maiana
189 P15 **Hall Islands** island group C Micronesia
118 H6 **Halliste** ~ S Estonia
Halljal see Haljala
93 I15 **Hällnäs** Västerbotten, N Sweden 64°20′N 19°41′E
29 R2 **Hallock** Minnesota, N USA 48°47′N 96°56′W
9 S6 **Hall Peninsula** peninsula Baffin Island, Nunavut, NE Canada
20 F9 **Halls** Tennessee, S USA 35°52′N 89°24′W
95 M16 **Hallsberg** Örebro, C Sweden 59°05′N 15°07′E
181 N5 **Halls Creek** Western Australia 18°17′S 127°39′E
182 L12 **Halls Gap** Victoria, SE Australia 37°09′S 142°30′E
95 N15 **Hallstahammar** Västmanland, C Sweden 59°37′N 16°13′E
109 R6 **Hallstatt** Salzburg, W Austria 47°32′N 13°39′E
109 R6 **Hallstätter See** ◎ C Austria
95 P14 **Hallstavik** Stockholm, C Sweden 60°12′N 18°45′E
25 X7 **Hallsville** Texas, SW USA 32°31′N 94°30′W
103 P1 **Halluin** Nord, N France 50°46′N 03°07′E
171 S12 **Halmahera, Laut** Eng. Halmahera Sea. sea E Indonesia
171 R11 **Halmahera, Pulau** prev. Djailolo, Gilolo, Jailolo. island E Indonesia

◆ Country ● Country Capital ◇ Dependent Territory ○ Dependent Territory Capital ◐ Administrative Regions ✕ International Airport ▲ Mountain ▲▲ Mountain Range ☈ Volcano ~ River ◎ Lake ⊠ Reservoir

Halmahera Sea – Haut, Isle au

Halmahera Sea see Halmahera, Laut
95 J21 **Halmstad** Halland, S Sweden 56°41′N 12°49′E
167 T6 **Ha Long** prev. Hòng Gai, var. Hon Gai, Hongay. Quang Ninh, N Vietnam 20°57′N 107°06′E
119 N15 **Halowchyn** Rus. Golovchin. Mahilyowskaya Voblasts', E Belarus 54°04′N 29°55′E
95 H20 **Hals** Nordjylland, N Denmark 57°00′N 10°19′E
94 F8 **Halsa** Møre og Romsdal, S Norway 63°04′N 08°13′E
119 I15 **Hal'shany** Rus. Gol'shany. Hrodzyenskaya Voblasts', W Belarus 54°15′N 26°01′E
Hälsingborg see Helsingborg
29 R5 **Halstad** Minnesota, N USA 47°21′N 96°49′W
27 N6 **Halstead** Kansas, C USA 38°00′N 97°30′W
99 G15 **Halsteren** Noord-Brabant, S Netherlands 51°32′N 04°16′E
93 L16 **Halsua** Keski-Pohjanmaa, W Finland 63°28′N 24°10′E
101 E14 **Haltern** Nordrhein-Westfalen, W Germany 51°45′N 07°10′E
92 J9 **Halti** var. Haltiatunturi, Lapp. Háldi. ▲ Finland/Norway 69°18′N 21°19′E
Haltiatunturi see Halti
116 J6 **Halych** Ivano-Frankivs'ka Oblast', W Ukraine 49°08′N 24°44′E
Halycus see Platani
103 P3 **Ham** Somme, N France 49°46′N 03°03′E
Hama see Ḩamāh
164 F12 **Hamada** Shimane, Honshū, SW Japan 34°54′N 132°07′E
142 L6 **Hamadān** anc. Ecbatana. Hamadān, W Iran 34°51′N 48°31′E
142 L6 **Hamadān** off. Ostān-e Hamadān. ◆ province W Iran
Hamadān, Ostān-e see Hamadān
138 I5 **Ḩamāh** var. Hama; anc. Epiphania, Bibl. Hamath. Ḩamāh, W Syria 35°09′N 36°44′E
138 I5 **Ḩamāh** off. Muḩāfaẓat Ḩamāh, var. Hama. ◆ governorate C Syria
Ḩamāh, Muḩāfaẓat see Ḩamāh
1665 S3 **Hamamasu** Hokkaidō, NE Japan 43°37′N 141°24′E
164 L14 **Hamamatsu** var. Hamamatu. Shizuoka, Honshū, S Japan 34°43′N 137°46′E
Hamamatu see Hamamatsu
165 W14 **Hamanaka** Hokkaidō, NE Japan 43°05′N 145°05′E
164 L14 **Hamana-ko** ◎ Honshū, S Japan
94 I13 **Hamar** prev. Storhammer. Hedmark, S Norway 60°57′N 10°55′E
141 U10 **Ḩamārīr al Kidan, Qalamat** well E Saudi Arabia
164 I12 **Hamasaka** Hyōgo, Honshū, SW Japan 35°37′N 134°27′E
Hamath see Ḩamāh
165 T1 **Hamatonbetsu** Hokkaidō, NE Japan 45°07′N 142°21′E
155 K26 **Hambantota** Southern Province, SE Sri Lanka 06°07′N 81°07′E
Hambourg see Hamburg
100 J9 **Hamburg** Hamburg, N Germany 53°33′N 10°03′E
27 V14 **Hamburg** Arkansas, C USA 33°13′N 91°50′W
29 S16 **Hamburg** Iowa, C USA 40°36′N 95°39′W
18 D10 **Hamburg** New York, NE USA 42°40′N 78°49′W
100 I10 **Hamburg** Fr. Hambourg. ◆ state N Germany
148 K5 **Hamdam Āb, Dasht-e** Pash. Dasht-i Hamdamab. ▲ W Afghanistan
Hamdamab, Dasht-i see Hamdam Āb, Dasht-e
18 M13 **Hamden** Connecticut, NE USA 41°23′N 72°55′W
140 K6 **Ḩamḑ, Wādī al** dry watercourse W Saudi Arabia
93 K18 **Hämeenkyrö** Pirkanmaa, W Finland 61°39′N 23°10′E
93 L19 **Hämeenlinna** Swe. Tavastehus. Kanta-Häme, S Finland 61°N 24°25′E
HaMela h, Yam see Dead Sea
Hamelin see Hameln
100 I13 **Hameln** Eng. Hamelin. Niedersachsen, N Germany 52°07′N 09°22′E
180 I8 **Hamersley Range** ▲ Western Australia
163 Y12 **Hamgyong-sanmaek** ▲ N North Korea
163 X13 **Hamhŭng** C North Korea 39°53′N 127°31′E
159 O6 **Hami** var. Ha-mi, Uigh. Kumul, Qomul. Xinjiang Uygur Zizhiqu, NW China 42°48′N 93°27′E
Ha-mi see Hami
139 X10 **Ḩāmid Amīn** Maysān, E Iraq 32°06′N 46°53′E
141 W11 **Ḩamīdān, Khawr** oasis SE Saudi Arabia
114 L12 **Hamidiye** Edirne, NW Turkey 41°09′N 26°40′E
182 L12 **Hamilton** Victoria, SE Australia 37°45′S 142°04′E
64 B12 **Hamilton** O (Bermuda) C Bermuda 32°18′N 64°48′W
14 G16 **Hamilton** Ontario, S Canada 43°15′N 79°50′W
184 M7 **Hamilton** Waikato, North Island, New Zealand 37°49′S 175°16′E
96 I12 **Hamilton** S Scotland, United Kingdom 55°47′N 04°03′W
23 N3 **Hamilton** Alabama, S USA 34°08′N 87°59′W
38 M10 **Hamilton** Alaska, USA 62°54′N 163°53′W
30 J13 **Hamilton** Illinois, N USA 40°24′N 91°20′W
27 S3 **Hamilton** Missouri, C USA 39°44′N 94°00′W
33 P10 **Hamilton** Montana, NW USA 46°15′N 114°09′W
25 S8 **Hamilton** Texas, SW USA 31°42′N 98°08′W
14 G16 **Hamilton** ✈ Ontario, SE Canada 43°12′N 79°54′W
64 I6 **Hamilton Bank** undersea feature SE Labrador Sea
182 E1 **Hamilton Creek** seasonal river South Australia
13 R8 **Hamilton Inlet** inlet Newfoundland and Labrador, E Canada
27 T12 **Hamilton, Lake** ▣ Arkansas, C USA
35 W6 **Hamilton, Mount** ▲ Nevada, W USA 39°15′N 115°30′W
75 S8 **Ḩamīm, Wādī al** ⟿ NE Libya
93 N19 **Hamina** Swe. Fredrikshamn. Kymenlaakso, S Finland 60°33′N 27°15′E
11 W16 **Hamiota** Manitoba, S Canada 50°13′N 100°37′W
152 L13 **Hamīrpur** Uttar Pradesh, N India 25°57′N 80°08′E
Hamīs Musait see Khamīs Mushayṭ
21 T11 **Hamlet** North Carolina, SE USA 34°52′N 79°41′W
25 P6 **Hamlin** Texas, SW USA 32°52′N 100°07′W
21 P5 **Hamlin** West Virginia, NE USA 38°16′N 82°07′W
31 O7 **Hamlin Lake** ◎ Michigan, N USA
101 F14 **Hamm** var. Hamm in Westfalen. Nordrhein-Westfalen, W Germany 51°39′N 07°49′E
Ḩammāmāt, Khalīj al see Hammamet, Golfe de
75 N5 **Hammamet, Golfe de** Ar. Khalīj al Ḩammāmāt. gulf NE Tunisia
139 R3 **Ḩammān al 'Alīl** Nīnawé, N Iraq 36°07′N 43°15′E
139 X12 **Ḩammār, Hawr al** ◎ SE Iraq
93 J20 **Hammarland** Åland, SW Finland 60°13′N 19°45′E
93 H16 **Hammarstrand** Jämtland, C Sweden 63°07′N 16°27′E
93 O17 **Hammaslahti** Pohjois-Karjala, SE Finland 62°26′N 29°58′E
99 F17 **Hamme** Oost-Vlaanderen, NW Belgium 51°06′N 04°08′E
100 H10 **Hamme** ⟿ NW Germany
95 G22 **Hammel** Midtjylland, C Denmark 56°15′N 09°53′E
101 I18 **Hammelburg** Bayern, C Germany 50°06′N 09°50′E
99 H18 **Hamme-Mille** Walloon Brabant, C Belgium 50°48′N 04°42′E
100 H10 **Hamme-Oste-Kanal** canal NW Germany
93 G16 **Hammerdal** Jämtland, C Sweden 63°34′N 15°19′E
92 K8 **Hammerfest** Finnmark, N Norway 70°40′N 23°44′E
101 D14 **Hamminkeln** Nordrhein-Westfalen, W Germany 51°43′N 06°36′E
Hamm in Westfalen see Hamm
26 K10 **Hammon** Oklahoma, C USA 35°37′N 99°22′W
31 N11 **Hammond** Indiana, N USA 41°35′N 87°30′W
22 K8 **Hammond** Louisiana, S USA 30°30′N 90°27′W
99 K20 **Hamoir** Liège, E Belgium 50°28′N 05°35′E
99 J21 **Hamois** Namur, SE Belgium 50°21′N 05°09′E
99 K16 **Hamont** Limburg, NE Belgium 51°15′N 05°33′E
185 F22 **Hampden** Otago, South Island, New Zealand 45°18′S 170°49′E
19 R6 **Hampden** Maine, NE USA 44°44′N 68°51′W
97 M23 **Hampshire** cultural region S England, United Kingdom
13 O15 **Hampton** New Brunswick, SE Canada 45°30′N 65°50′W
27 U14 **Hampton** Arkansas, C USA 33°33′N 92°28′W
29 V12 **Hampton** Iowa, C USA 42°44′N 93°12′W
19 P10 **Hampton** New Hampshire, NE USA 42°55′N 70°48′W
21 R14 **Hampton** South Carolina, SE USA 32°52′N 81°06′W
21 P8 **Hampton** Tennessee, S USA 36°16′N 82°10′W
21 X7 **Hampton** Virginia, NE USA 37°02′N 76°23′W
94 L11 **Hamra** Gävleborg, C Sweden 61°40′N 15°00′E
80 D10 **Hamrat esh Sheikh** Northern Kordofan, C Sudan 14°38′N 27°56′E
139 S5 **Ḩamrīn, Jabal** ▲ N Iraq
121 P16 **Ħamrun** C Malta 35°53′N 14°28′E
Ham Thuân Nam see Thuân Nam
Hāmūn, Daryācheh-ye see Şāberī, Hāmūn-e/Sīstān, Daryācheh-ye
Hamwih see Southampton
38 G10 **Hāna** var. Hana. Maui, Hawaii, USA, C Pacific Ocean 20°45′N 155°59′W
Hana see Hāna
21 S14 **Hanahan** South Carolina, SE USA 32°55′N 80°01′W
38 B8 **Hanalei** Kaua'i, Hawaii, USA, C Pacific Ocean 22°12′N 159°30′W
165 Q9 **Hanamaki** Iwate, Honshū, C Japan 39°25′N 141°04′E
38 F10 **Hanamanioa, Cape** headland Maui, Hawai'i, USA
190 B16 **Hanan** ✈ (Alofi) SW Niue
101 H18 **Hanau** Hessen, W Germany 50°06′N 08°56′E
162 M11 **Hanbogd** var. Ih Bulag. Ömnögovĭ, S Mongolia 43°04′N 107°42′E
8 L9 **Hanbury** ⟿ Northwest Territories, NW Canada
Hânceşti see Hînceşti
10 M15 **Hanceville** British Columbia, SW Canada 51°54′N 122°56′W
23 P3 **Hanceville** Alabama, S USA 34°03′N 86°46′W
Hancewicze see Hantsavichy
160 L6 **Hancheng** Shaanxi, C China 35°22′N 110°27′E
21 V2 **Hancock** Maryland, NE USA 39°42′N 78°10′W
30 M3 **Hancock** Michigan, N USA 47°07′N 88°34′W
29 S8 **Hancock** Minnesota, N USA 45°30′N 95°47′W
18 I12 **Hancock** New York, NE USA 41°57′N 75°15′W
80 Q12 **Handa** Bari, NE Somalia 10°35′N 51°09′E
161 O5 **Handan** var. Han-tan. Hebei, E China 36°35′N 114°28′E
95 P16 **Handen** Stockholm, C Sweden 59°12′N 18°09′E
81 J22 **Handeni** Tanga, E Tanzania 05°25′S 38°04′E
37 Q7 **Handies Peak** ▲ Colorado, C USA 37°54′N 107°30′W
111 J19 **Handlová** Ger. Krickerhäu, Hung. Nyitrabánya; prev. Kriegerhaj. Trenčiansky Kraj, C Slovakia 48°45′N 18°45′E
165 O13 **Haneda** ✈ (Tōkyō) Tōkyō, Honshū, S Japan 35°33′N 139°45′E
138 F13 **HaNegev** Eng. Negev. desert S Israel
Hanfeng see Kaixian
35 Q11 **Hanford** California, W USA 36°19′N 119°39′W
191 V16 **Hanga Roa** Easter Island, Chile, E Pacific Ocean 27°09′S 109°26′W
162 I7 **Hangay** var. Hunt. Arhangay, C Mongolia 47°49′N 99°24′E
162 H7 **Hangayn Nuruu** ▲ C Mongolia
Hang-chou/Hangchow see Hangzhou
95 K20 **Hänger** Jönköping, S Sweden 57°06′N 13°58′E
Hangö see Hanko
161 R9 **Hangzhou** var. Hang-chou, Hangchow. province capital Zhejiang, SE China 30°18′N 120°07′E
162 J4 **Hanh** var. Turt. Hövsgöl, N Mongolia 51°30′N 100°40′E
162 F5 **Hanhöhiy Uul** ▲ NW Mongolia
162 K10 **Hanhongor** var. Ögöömör. Ömnögovĭ, S Mongolia 43°47′N 104°31′E
146 I14 **Hanhowuz** Rus. Khauz-Khan. Ahal Welaýaty, S Turkmenistan 37°15′N 61°12′E
146 I14 **Hanhowuz Suw Howdany** Rus. Khauzkhanskoye Vodoranilishche. ▣ S Turkmenistan
137 P15 **Hani** Dıyarbakır, SE Turkey 38°26′N 40°23′E
Hania see Chaniá
141 R11 **Ḩanīsh al Kabīr, Jazīrat al** island SW Yemen
Hanka, Lake see Khanka, Lake
93 M17 **Hankasalmi** Keski-Suomi, C Finland 62°25′N 26°27′E
29 R7 **Hankinson** North Dakota, N USA 46°04′N 96°54′W
93 K20 **Hanko** Swe. Hangö. Uusimaa, SW Finland 59°50′N 23°E
Han-kou/Han-k'ou/Hankow see Wuhan
36 M6 **Hanksville** Utah, W USA 38°21′N 110°43′W
152 K6 **Hanle** Jammu and Kashmir, NW India 32°46′N 79°01′E
185 I17 **Hanmer Springs** Canterbury, South Island, New Zealand 42°31′S 172°49′E
11 R16 **Hanna** Alberta, SW Canada 51°38′N 111°56′W
27 V3 **Hannibal** Missouri, C USA 39°42′N 91°23′W
180 M3 **Hann, Mount** ▲ Western Australia 15°53′S 125°46′E
100 I12 **Hannover** Eng. Hanover. Niedersachsen, NW Germany 52°23′N 09°43′E
99 J19 **Hannut** Liège, C Belgium 50°40′N 05°05′E
95 L22 **Hanöbukten** bay S Sweden
167 T6 **Ha Nôi** Eng. Hanoi, Fr. Hanoï. ● (Vietnam) N Vietnam 21°01′N 105°52′E
14 F14 **Hanover** Ontario, S Canada 44°10′N 81°03′W
31 P15 **Hanover** Indiana, N USA 38°42′N 85°28′W
18 G16 **Hanover** Pennsylvania, NE USA 39°46′N 76°57′W
21 W6 **Hanover** Virginia, NE USA 37°44′N 77°21′W
Hanover see Hannover
63 G23 **Hanover, Isla** island S Chile
Hanselbeck see Érd
195 X5 **Hansen Mountains** ▲ Antarctica
160 M8 **Han Shui** ⟿ C China
152 H10 **Hānsi** Haryāna, NW India 29°06′N 76°01′E
95 F20 **Hanstholm** Midtjylland, NW Denmark 57°05′N 08°39′E
Han-tan see Handan
158 H6 **Hantengri Feng** var. Pik Khan-Tengri. ▲ China/Kazakhstan 42°17′N 80°11′E
see also Khan-Tengri, Pik
119 I18 **Hantsavichy** Pol. Hancewicze, Rus. Gantsevichi. Brestskaya Voblasts', SW Belarus 52°45′N 26°27′E
9 Q6 **Hantzsch** ⟿ Baffin Island, Nunavut, NE Canada
152 G9 **Hanumāngarh** Rājasthān, NW India 29°33′N 74°21′E
183 O9 **Hanwood** New South Wales, SE Australia 34°19′S 146°03′E
Hanyang see Wuhan
Hanyang see Caidian
160 H10 **Hanyuan** var. Fulin. Sichuan, C China 29°29′N 102°45′E
Hanyuan see Xihe
160 J7 **Hanzhong** Shaanxi, C China 33°12′N 107°06′E
191 W11 **Hao** atoll Îles Tuamotu, C French Polynesia
153 S16 **Hāora** prev. Howrah. West Bengal, NE India 22°35′N 88°20′E
78 K8 **Haouach, Ouadi** dry watercourse E Chad
92 K13 **Haparanda** Norrbotten, N Sweden 65°49′N 24°05′E
25 N3 **Happy** Texas, SW USA 34°44′N 101°51′W
34 M1 **Happy Camp** California, W USA 41°48′N 123°24′W
13 Q9 **Happy Valley-Goose Bay** prev. Goose Bay. Newfoundland and Labrador, E Canada 53°19′N 60°24′W
Hapsal see Haapsalu
152 J10 **Hāpur** Uttar Pradesh, N India 28°43′N 77°47′E
HaQatan, HaMakhtesh see HaKatan, HaMakhtesh
140 I4 **Ḩaql** Tabūk, NW Saudi Arabia 29°18′N 34°58′E
171 U14 **Har** Pulau Kai Besar, E Indonesia 05°21′S 133°09′E
Haraat see Tsagaandelger
141 R8 **Ḩaraḑ** var. Haradh. Ash Sharqīyah, E Saudi Arabia 24°08′N 49°02′E
Haradh see Ḩaraḑ
118 N12 **Haradok** Rus. Gorodok. Vitsyebskaya Voblasts', N Belarus 55°28′N 30°00′E
92 J13 **Harads** Norrbotten, N Sweden 66°04′N 21°04′E
119 G19 **Haradzyets** Rus. Gorodets. Brestskaya Voblasts', SW Belarus 52°12′N 24°40′E
119 J17 **Haradzyeya** Rus. Gorodeya. Minskaya Voblasts', C Belarus 53°19′N 26°32′E
191 V10 **Haraiki** atoll Îles Tuamotu, C French Polynesia
165 Q11 **Haramachi** Fukushima, Honshū, E Japan 37°40′N 140°55′E
118 M12 **Harany** Rus. Gorany. Vitsyebskaya Voblasts', N Belarus 55°25′N 29°03′E
83 L16 **Harare** prev. Salisbury. ● (Zimbabwe) Mashonaland East, NE Zimbabwe 17°47′S 31°04′E
83 L16 **Harare** ✈ Mashonaland East, NE Zimbabwe 17°51′S 31°06′E
78 J10 **Haraz-Djombo** Batha, C Chad 14°10′N 19°35′E
119 O16 **Harbavichy** Rus. Gorbovichi. Mahilyowskaya Voblasts', E Belarus 53°49′N 30°42′E
76 J16 **Harbel** W Liberia 06°19′N 10°20′W
163 W8 **Harbin** var. Haerbin, Ha-erh-pin, Kharbin; prev. Haerhpin, Pingkiang, Pinkiang. province capital Heilongjiang, NE China 45°45′N 126°41′E
31 S7 **Harbor Beach** Michigan, N USA 43°50′N 82°39′W
13 T13 **Harbour Breton** Newfoundland, Newfoundland and Labrador, E Canada 47°29′N 55°50′W
65 D25 **Harbours, Bay of** bay East Falkland, Falkland Islands
Hårby see Haarby
36 I13 **Harcuvar Mountains** ▲ Arizona, SW USA
108 J7 **Hard** Vorarlberg, NW Austria 47°29′N 09°42′E
154 H11 **Harda Khas** Madhya Pradesh, C India 22°22′N 77°06′E
95 D14 **Hardanger** physical region S Norway
95 D14 **Hardangerfjorden** fjord S Norway
94 E13 **Hardangerjøkulen** glacier S Norway
95 E14 **Hardangervidda** plateau S Norway
83 D20 **Hardap** ◆ district S Namibia
21 R15 **Hardeeville** South Carolina, SE USA 32°18′N 81°04′W
Hardegarijp see Hurdegaryp
98 O9 **Hardenberg** Overijssel, E Netherlands 52°34′N 06°38′E
183 Q9 **Harden-Murrumburrah** New South Wales, SE Australia 34°33′S 148°22′E
98 K10 **Harderwijk** Gelderland, C Netherlands 52°21′N 05°37′E
30 J14 **Hardin** Illinois, N USA 39°10′N 90°38′W
33 V11 **Hardin** Montana, NW USA 45°44′N 107°35′W
23 R5 **Harding, Lake** ▣ Alabama/Georgia, SE USA
20 J6 **Hardinsburg** Kentucky, S USA 37°46′N 86°29′W
98 I13 **Hardinxveld-Giessendam** Zuid-Holland, C Netherlands 51°52′N 04°49′E
11 R15 **Hardisty** Alberta, SW Canada 52°42′N 111°22′W
152 L12 **Hardoi** Uttar Pradesh, N India 27°23′N 80°06′E
Hardwar see Haridwār
23 U4 **Hardwick** Georgia, SE USA 33°03′N 83°13′W
27 W9 **Hardy** Arkansas, C USA 36°19′N 91°29′W
94 D10 **Hareid** Møre og Romsdal, S Norway 62°22′N 06°02′E
8 H7 **Hare Indian** ⟿ Northwest Territories, NW Canada
99 D18 **Harelbeke** var. Harlebeke. West-Vlaanderen, W Belgium 50°51′N 03°19′E
Harem see Ḩārim
100 E11 **Haren** Niedersachsen, NW Germany 52°47′N 07°16′E
98 N6 **Haren** Groningen, NE Netherlands 53°10′N 06°37′E
80 L13 **Hārer** E Ethiopia 09°17′N 42°19′E
95 P14 **Harg** Uppsala, C Sweden 60°13′N 18°25′E
Hargeisa see Hargeysa
80 M13 **Hargeysa** var. Hargeisa. Woqooyi Galbeed, NW Somalia 09°32′N 44°07′E
116 J10 **Harghita** ◆ county NE Romania
25 S17 **Hargill** Texas, SW USA 26°26′N 98°00′W
162 J8 **Harhorin** Övörhangay, C Mongolia 47°13′N 102°48′E
159 Q9 **Har Hu** ◎ C China
Hariana see Haryāna
141 P15 **Ḩarīb** W Yemen 15°08′N 45°35′E
168 M12 **Hari, Batang** prev. Djambi. ⟿ Sumatera, W Indonesia
152 J9 **Haridwār** prev. Hardwar. Uttarakhand, N India 29°58′N 78°09′E
155 F18 **Harihar** Karnātaka, W India 14°33′N 75°44′E
185 F18 **Harihari** West Coast, South Island, New Zealand 43°09′S 170°35′E
138 I3 **Ḩārim** var. Harem. Idlib, W Syria 36°30′N 36°30′E
98 F13 **Haringvliet** channel SW Netherlands
98 F13 **Haringvlietdam** dam SW Netherlands
149 U5 **Harīpur** Khyber Pakhtunkhwa, NW Pakistan 34°00′N 73°01′E
148 J4 **Harīrūd** var. Tedzhen, Turkm. Tejen. ⟿ Afghanistan/Iran see also Tejen
Harīrūd see Tejen
94 J11 **Härjåhågnen** Swe. Härjahågnen, var. Härjehågna. ▲ Norway/Sweden 61°43′N 12°07′E
Härjahågnen see Østrehogna
Härjåhågnen see Härjåhågnen
93 K18 **Harjavalta** Satakunta, W Finland 61°19′N 22°10′E
Härjehågna see Østrehogna
Härjehågna see Härjåhågnen
118 G4 **Harjumaa** var. Harju Maakond. ◆ province NW Estonia
Harju Maakond see Harjumaa
21 X11 **Harkers Island** North Carolina, SE USA 34°42′N 76°33′W
Harki see Hurkê
29 T14 **Harlan** Iowa, C USA 41°40′N 95°19′W
21 O7 **Harlan** Kentucky, S USA 36°50′N 83°19′W
29 N17 **Harlan County Lake** ▣ Nebraska, C USA
116 L9 **Hârlău** var. Hîrlău. Iaşi, NE Romania 47°26′N 26°54′E
Harlebeke see Harelbeke
33 U7 **Harlem** Montana, NW USA 48°31′N 108°46′W
Harlem see Haarlem
95 G22 **Harlev** Midtjylland, C Denmark 56°08′N 10°00′E
98 K6 **Harlingen** Fris. Harns. Fryslân, N Netherlands 53°10′N 05°25′E
25 T17 **Harlingen** Texas, SW USA 26°12′N 97°43′W
97 O21 **Harlow** E England, United Kingdom 51°47′N 00°07′E
33 T10 **Harlowton** Montana, NW USA 46°26′N 109°49′W
94 N11 **Harmånger** Gävleborg, C Sweden 61°55′N 17°19′E
114 K11 **Harmanli** var. Kharmanli. Haskovo, S Bulgaria 41°56′N 25°54′E
114 K11 **Harmanliyska Reka** var. Kharmanliyska Reka. ⟿ S Bulgaria
98 I11 **Harmelen** Utrecht, C Netherlands 52°06′N 04°58′E
29 X11 **Harmony** Minnesota, N USA 43°33′N 92°00′W
32 J14 **Harney Basin** basin Oregon, NW USA
32 J14 **Harney Lake** ◎ Oregon, NW USA
28 J10 **Harney Peak** ▲ South Dakota, N USA 43°52′N 103°31′W
93 H17 **Härnösand** var. Hernösand. Västernorrland, C Sweden 62°37′N 17°55′E
Harns see Harlingen
105 P4 **Haro** La Rioja, N Spain 42°34′N 02°52′W
40 F6 **Haro, Cabo** headland NW Mexico 27°50′N 110°55′W
94 D9 **Harøy** island S Norway
97 N21 **Harpenden** E England, United Kingdom 51°49′N 00°22′W
76 L18 **Harper** var. Cape Palmas. NE Liberia 04°25′N 07°43′W
26 M7 **Harper** Kansas, C USA 37°17′N 98°01′W
32 L13 **Harper** Oregon, NW USA 43°51′N 117°37′W
25 Q10 **Harper** Texas, SW USA 30°18′N 99°18′W
35 U13 **Harper Lake** salt flat California, W USA
39 T9 **Harper, Mount** ▲ Alaska, USA 64°18′N 143°54′W
95 J21 **Harplinge** Halland, S Sweden 56°45′N 12°45′E
36 J13 **Harquahala Mountains** ▲ Arizona, SW USA
141 T15 **Ḩarrah** SE Yemen 15°02′N 50°23′E
12 H11 **Harricana** ⟿ Québec, SE Canada
20 M9 **Harriman** Tennessee, S USA 35°57′N 84°33′W
13 R11 **Harrington Harbour** Québec, E Canada 50°34′N 59°29′W
64 B12 **Harrington Sound** bay Bermuda, NW Atlantic Ocean
96 F8 **Harris** physical region NW Scotland, United Kingdom
27 X10 **Harrisburg** Arkansas, C USA 35°33′N 90°43′W
30 M17 **Harrisburg** Illinois, N USA 37°44′N 88°32′W
28 I14 **Harrisburg** Nebraska, C USA 41°33′N 103°46′W
32 F12 **Harrisburg** Oregon, NW USA 44°16′N 123°10′W
18 G15 **Harrisburg** state capital Pennsylvania, NE USA 40°16′N 76°53′W
182 F6 **Harris, Lake** ◎ South Australia
23 W11 **Harris, Lake** ◎ Florida, SE USA
83 J22 **Harrismith** Free State, E South Africa 28°16′S 29°08′E
27 T9 **Harrison** Arkansas, C USA 36°13′N 93°07′W
31 Q7 **Harrison** Michigan, N USA 44°02′N 84°46′W
28 I12 **Harrison** Nebraska, C USA 42°42′N 103°53′W
39 Q5 **Harrison Bay** inlet Alaska, USA
22 I6 **Harrisonburg** Louisiana, S USA 31°44′N 91°51′W
21 U4 **Harrisonburg** Virginia, NE USA 38°27′N 78°54′W
13 R7 **Harrison, Cape** headland Newfoundland and Labrador, E Canada 54°55′N 57°48′W
27 R5 **Harrisonville** Missouri, C USA 38°40′N 94°21′W
Harris Ridge see Lomonosov Ridge
192 M3 **Harris Seamount** undersea feature N Pacific Ocean 46°09′N 161°25′W
96 F8 **Harris, Sound of** strait NW Scotland, United Kingdom
31 R6 **Harrisville** Michigan, N USA 44°41′N 83°19′W
21 R3 **Harrisville** West Virginia, NE USA 39°13′N 81°04′W
20 M6 **Harrodsburg** Kentucky, S USA 37°45′N 84°51′W
97 M16 **Harrogate** N England, United Kingdom 54°N 01°33′W
25 Q4 **Harrold** Texas, SW USA 34°05′N 99°02′W
27 S5 **Harry S. Truman Reservoir** ▣ Missouri, C USA
100 G13 **Harsewinkel** Nordrhein-Westfalen, W Germany 51°58′N 08°13′E
116 M14 **Hârşova** prev. Hirşova. Constanţa, SE Romania 44°41′N 27°56′E
92 H10 **Harstad** Troms, N Norway 68°48′N 16°31′E
31 O8 **Hart** Michigan, N USA 43°43′N 86°22′W
24 M4 **Hart** Texas, SW USA 34°23′N 102°07′W
10 I5 **Hart** ⟿ Yukon, NW Canada
83 F23 **Hartbees** ⟿ C South Africa
109 X7 **Hartberg** Steiermark, SE Austria 47°18′N 15°58′E
182 I10 **Hart, Cape** headland South Australia 35°54′S 138°01′E
95 E14 **Hårteigen** ▲ S Norway 60°11′N 07°01′E
23 Q7 **Hartford** Alabama, S USA 31°06′N 85°42′W
27 R11 **Hartford** Arkansas, C USA 35°01′N 94°22′W
18 M12 **Hartford** state capital Connecticut, NE USA 41°46′N 72°41′W
20 J6 **Hartford** Kentucky, S USA 37°26′N 86°57′W
31 P10 **Hartford** Michigan, N USA 42°12′N 85°54′W
29 R11 **Hartford** South Dakota, N USA 43°37′N 96°56′W
30 M8 **Hartford** Wisconsin, N USA 43°19′N 88°25′W
31 P13 **Hartford City** Indiana, N USA 40°27′N 85°22′W
29 Q13 **Hartington** Nebraska, C USA 42°37′N 97°15′W
13 N14 **Hartland** New Brunswick, SE Canada 46°18′N 67°31′W
97 H23 **Hartland Point** headland SW England, United Kingdom 51°01′N 04°33′W
97 M15 **Hartlepool** N England, United Kingdom 54°41′N 01°13′W
29 T12 **Hartley** Iowa, C USA 43°10′N 95°28′W
24 M1 **Hartley** Texas, SW USA 35°52′N 102°24′W
32 J15 **Hart Mountain** ▲ Oregon, NW USA 42°24′N 119°46′W
173 U10 **Hartog Ridge** undersea feature W Indian Ocean
93 M18 **Hartola** Päijät-Häme, S Finland 61°34′N 26°04′E
67 U14 **Harts** var. Hartz. ⟿ N South Africa
23 P2 **Hartselle** Alabama, S USA 34°26′N 86°56′W
23 S3 **Hartsfield Atlanta** ✈ Georgia, SE USA 33°38′N 84°24′W
27 Q11 **Hartshorne** Oklahoma, C USA 34°51′N 95°33′W
21 S12 **Hartsville** South Carolina, SE USA 34°22′N 80°04′W
20 K8 **Hartsville** Tennessee, S USA 36°23′N 86°11′W
27 U7 **Hartville** Missouri, C USA 37°15′N 92°30′W
23 U2 **Hartwell** Georgia, SE USA 34°21′N 82°55′W
21 O11 **Hartwell Lake** ▣ Georgia/South Carolina, SE USA
Hartz see Harts
Harunabad see Eslāmābād-e Gharb
Har-Us see Erdenebüren
162 F6 **Har Us Gol** ◎ Hovd, W Mongolia
162 E6 **Har Us Nuur** ◎ NW Mongolia
30 M10 **Harvard** Illinois, N USA 42°25′N 88°36′W
29 P16 **Harvard** Nebraska, C USA 40°37′N 98°06′W
37 R5 **Harvard, Mount** ▲ Colorado, C USA 38°55′N 106°19′W
31 N11 **Harvey** Illinois, N USA 41°36′N 87°39′W
29 N4 **Harvey** North Dakota, N USA 47°43′N 99°55′W
97 Q21 **Harwich** E England, United Kingdom 51°56′N 01°16′E
152 H10 **Haryāna** var. Hariana. ◆ state N India
141 Y9 **Ḩaryān, Ṭawī al** spring/well NE Oman 21°56′N 58°33′E
101 J14 **Harz** ▲ C Germany
Ḩasakah, Muḩāfaẓat al see Al Ḩasakah
165 Q9 **Hasama** Miyagi, Honshū, C Japan 38°42′N 141°09′E
137 Y13 **Häsänabad** prev. 26 Bakinskikh Komissarov. SE Azerbaijan 39°18′N 49°13′E
136 J15 **Hasan Dağı** ▲ C Turkey 38°09′N 34°15′E
139 T9 **Ḩasan Ibn Ḩassūn** An Najaf, C Iraq 32°24′N 44°13′E
149 R6 **Ḩasan Khēl** var. Ahmad Khel. Paktīyā, SE Afghanistan 33°46′N 69°37′E
100 F12 **Hase** ⟿ NW Germany
Haselberg see Krasnoznamensk
100 F12 **Haselünne** Niedersachsen, NW Germany 52°40′N 07°28′E
Hashaat see Delgerhangay
Hashemite Kingdom of Jordan see Jordan
139 V8 **Hāshimah** Wāsiţ, E Iraq 33°22′N 45°56′E
142 K3 **Hashtrūd** var. Azaran. Āẕarbāyjān-e Khāvarī, N Iran 37°34′N 47°10′E
141 W13 **Ḩāsik** S Oman 17°22′N 55°18′E
149 U10 **Hāsilpur** Punjab, E Pakistan 29°42′N 72°40′E
27 Q10 **Haskell** Oklahoma, C USA 35°49′N 95°40′W
25 Q6 **Haskell** Texas, SW USA 33°10′N 99°45′W
114 K11 **Haskovo** var. Khaskovo. Haskovo, S Bulgaria 41°56′N 25°35′E
114 K11 **Haskovo** var. Khaskovo. ◆ province S Bulgaria
114 M11 **Hasköy** Edirne, NW Turkey 41°37′N 26°51′E
95 L24 **Hasle** Bornholm, E Denmark 55°12′N 14°43′E
97 N23 **Haslemere** SE England, United Kingdom 51°06′N 00°45′W
102 I16 **Hasparren** Pyrénées-Atlantiques, SW France 43°23′N 01°18′W
Hassakeh see Al Ḩasakah
155 G19 **Hassan** Karnātaka, W India 13°01′N 76°03′E
36 J13 **Hassayampa River** ⟿ Arizona, SW USA
101 J18 **Hassberge** hill range C Germany
94 N10 **Hassela** Gävleborg, C Sweden 62°06′N 16°45′E
99 J18 **Hasselt** Limburg, NE Belgium 50°56′N 05°20′E
98 M9 **Hasselt** Overijssel, E Netherlands 52°36′N 06°06′E
Hassetché see Al Ḩasakah
101 J18 **Hassfurt** Bayern, C Germany 50°02′N 10°32′E
74 L9 **Hassi Bel Guebbour** E Algeria 28°41′N 06°29′E
74 L8 **Hassi Messaoud** E Algeria 31°41′N 06°10′E
95 K22 **Hässleholm** Skåne, S Sweden 56°09′N 13°45′E
Hasta Colonia/Hasta Pompeia see Asti
183 O13 **Hastings** Victoria, SE Australia 38°18′S 145°12′E
184 O11 **Hastings** Hawke's Bay, North Island, New Zealand 39°39′S 176°51′E
97 P23 **Hastings** SE England, United Kingdom 50°51′N 00°36′E
31 P9 **Hastings** Michigan, N USA 42°38′N 85°17′W
29 W9 **Hastings** Minnesota, N USA 44°44′N 92°51′W
29 P16 **Hastings** Nebraska, C USA 40°35′N 98°23′W
95 K22 **Hästveda** Skåne, S Sweden 56°16′N 13°55′E
92 J8 **Hasvik** Finnmark, N Norway 70°29′N 22°08′E
37 V6 **Haswell** Colorado, C USA 38°27′N 103°09′W
163 N11 **Hatanbulag** var. Ergel. Dornogovĭ, SE Mongolia 43°10′N 109°13′E
Hatansuudal see Bayanlig
Hatavch see Haldzan
136 K17 **Hatay** ◆ province S Turkey
37 R15 **Hatch** New Mexico, SW USA 32°40′N 107°10′W
36 K7 **Hatch** Utah, W USA 37°39′N 112°25′W
20 F9 **Hatchie River** ⟿ Tennessee, S USA
116 G12 **Haţeg** Ger. Wallenthal, Hung. Hátszeg; prev. Hatzeg, Hötzing. Hunedoara, SW Romania 45°35′N 22°57′E
165 O17 **Hateruma-jima** island Yaeyama-shotō, SW Japan
183 N8 **Hatfield** New South Wales, SE Australia 33°54′S 143°43′E
162 I5 **Hatgal** Hövsgöl, N Mongolia 50°24′N 100°12′E
153 V16 **Hathazari** Chittagong, SE Bangladesh 22°30′N 91°46′E
141 T13 **Hathūt, Hiṣā'** oasis NE Yemen
167 R14 **Ha Tiên** Kiên Giang, S Vietnam 10°24′N 104°30′E
167 T8 **Ha Tinh** Ha Tinh, N Vietnam 18°21′N 105°55′E
Ḩatira, Haré see Hatira, Harei
138 F12 **Hatira, Harei** prev. Haré Ḩatira. hill range S Israel
167 R6 **Hat Lot** var. Mai Son. Son La, N Vietnam 21°07′N 104°10′E
45 P16 **Hato Airport** ✈ (Willemstad) Curaçao 12°10′N 68°56′W
54 H9 **Hato Corozal** Casanare, C Colombia 06°08′N 71°45′W
Hato del Volcán see Volcán
45 P9 **Hato Mayor** E Dominican Republic 18°49′N 69°16′W
Hatra see Al Ḩaḑr
Hatria see Adria
Hátszeg see Haţeg
143 R16 **Haṭṭā** Dubayy, NE United Arab Emirates 24°50′N 56°06′E
182 L9 **Hattah** Victoria, SE Australia 34°49′S 142°18′E
98 M9 **Hattem** Gelderland, E Netherlands 52°29′N 06°04′E
21 Z10 **Hatteras** Hatteras Island, North Carolina, SE USA 35°13′N 75°39′W
21 Rr10 **Hatteras, Cape** headland North Carolina, SE USA 35°29′N 75°33′W
21 Z9 **Hatteras Island** island North Carolina, SE USA
64 F10 **Hatteras Plain** undersea feature W Atlantic Ocean 31°00′N 71°00′W
93 G14 **Hattfjelldal** Troms, N Norway 65°37′N 13°58′E
22 M7 **Hattiesburg** Mississippi, S USA 31°20′N 89°17′W
29 Q4 **Hatton** North Dakota, N USA 47°38′N 97°27′W
Hatton Bank see Hatton Ridge
64 L6 **Hatton Ridge** var. Hatton Bank. undersea feature N Atlantic Ocean 59°00′N 17°30′W
191 W6 **Hatutu** island Îles Marquises, NE French Polynesia
111 K22 **Hatvan** Heves, NE Hungary 47°40′N 19°39′E
167 O16 **Hat Yai** var. Ban Hat Yai. Songkhla, SW Thailand 07°01′N 100°27′E
Hatzeg see Haţeg
Hatzfeld see Jimbolia
80 N13 **Haud** plateau Ethiopia/Somalia
95 D18 **Hauge** Rogaland, S Norway 58°20′N 06°17′E
95 C15 **Haugesund** Rogaland, S Norway 59°24′N 05°17′E
109 X2 **Haugsdorf** Niederösterreich, NE Austria 48°41′N 16°04′E
184 M9 **Hauhungaroa Range** ▲ North Island, New Zealand
95 E15 **Haukeligrend** Telemark, S Norway 59°45′N 07°33′E
93 L14 **Haukipudas** Pohjois-Pohjanmaa, C Finland 65°11′N 25°21′E
93 M17 **Haukivesi** ◎ SE Finland
93 M17 **Haukivuori** Etelä-Savo, E Finland 62°02′N 27°11′E
Hauptkanal see Havelländ Grosse
187 N10 **Hauraha** Makira-Ulawa, SE Solomon Islands 10°47′S 162°00′E
184 L5 **Hauraki Gulf** gulf North Island, N New Zealand
185 B24 **Hauroko, Lake** ◎ South Island, New Zealand
167 S14 **Hâu, Sông** ⟿ S Vietnam
92 N12 **Hautajärvi** Lappi, NE Finland 66°30′N 29°01′E
74 F7 **Haut Atlas** Eng. High Atlas. ▲ C Morocco
79 M17 **Haut-Congo** off. Région du Haut-Congo; prev. Haut-Zaïre. ◆ region NE Dem. Rep. Congo
103 Y14 **Haute-Corse** ◆ department Corse, France, C Mediterranean Sea
102 L16 **Haute-Garonne** ◆ department S France
79 K14 **Haute-Kotto** ◆ prefecture E Central African Republic
103 P12 **Haute-Loire** ◆ department C France
103 R6 **Haute-Marne** ◆ department N France
102 M3 **Haute-Normandie** ◆ region N France
15 U6 **Hauterive** Québec, SE Canada 49°11′N 68°16′W
103 T13 **Hautes-Alpes** ◆ department SE France
103 S7 **Haute-Saône** ◆ department E France
103 T10 **Haute-Savoie** ◆ department E France
99 M20 **Hautes Fagnes** Ger. Hohes Venn. ▲ E Belgium
102 K16 **Hautes-Pyrénées** ◆ department S France
99 L23 **Haute Sûre, Lac de la** ▣ NW Luxembourg
102 M11 **Haute-Vienne** ◆ department C France
19 S8 **Haut, Isle au** island Maine, NE USA

◆ Country ● Country Capital ◇ Dependent Territory ○ Dependent Territory Capital ◆ Administrative Regions ✈ International Airport ▲ Mountain ▲ Mountain Range ⛰ Volcano ⟿ River ◎ Lake ▣ Reservoir

HAUT-MBOMOU – HEZE

79 M14 **Haut-Mbomou** ◈ *prefecture* SE Central African Republic
103 Q2 **Hautmont** Nord, N France 50°15′N 03°55′E
79 F19 **Haut-Ogooué** *off.* Province du Haut-Ogooué, *var.* Le Haut-Ogooué. ◈ *province* SE Gabon
Haut-Ogooué, Le *see* Haut-Ogooué
Haut-Ogooué, Province du *see* Haut-Ogooué
103 U7 **Haut-Rhin** ◈ *department* NE France
74 I6 **Hauts Plateaux** *plateau* Algeria/Morocco
Haut-Zaïre *see* Haut-Congo
38 D9 **Hau'ula** *var.* Hauula. O'ahu, Hawaii, USA, C Pacific Ocean 21°36′N 157°54′W
Hauula *see* Hau'ula
101 O22 **Hauzenberg** Bayern, SE Germany 48°39′N 13°37′E
30 K13 **Havana** Illinois, N USA 40°18′N 90°03′W
Havana *see* La Habana
97 N23 **Havant** S England, United Kingdom 50°51′N 00°59′W
35 Y14 **Havasu, Lake** ⊠ Arizona/ California, W USA
95 J23 **Havdrup** Sjælland, E Denmark 55°33′N 12°08′E
100 N10 **Havel** ≈ NE Germany
99 J21 **Havelange** Namur, SE Belgium 50°23′N 05°14′E
100 M11 **Havelberg** Sachsen-Anhalt, NE Germany 52°49′N 12°05′E
149 U5 **Havelian** Khyber Pakhtunkhwa, NW Pakistan 34°05′N 73°14′E
100 N12 **Havelländ Grosse** *var.* Hauptkanal. *canal* NE Germany
14 J14 **Havelock** Ontario, SE Canada 44°22′N 77°57′W
185 J14 **Havelock** Marlborough, South Island, New Zealand 41°17′S 173°46′E
21 X11 **Havelock** North Carolina, SE USA 34°52′N 76°54′W
184 O11 **Havelock North** Hawke's Bay, North Island, New Zealand 39°40′S 176°53′E
98 M8 **Havelte** Drenthe, NE Netherlands 52°46′N 06°14′E
27 N6 **Haven** Kansas, C USA 37°54′N 97°46′W
97 H21 **Haverfordwest** SW Wales, United Kingdom 51°50′N 04°57′W
97 P20 **Haverhill** E England, United Kingdom 52°05′N 00°26′E
19 O10 **Haverhill** Massachusetts, NE USA 42°46′N 71°02′W
93 G17 **Haverö** Västernorrland, C Sweden 62°25′N 15°04′E
111 I17 **Havířov** Moravskoslezský Kraj, E Czech Republic 49°47′N 18°30′E
111 E17 **Havlíčkův Brod** *Ger.* Deutsch-Brod; *prev.* Německý Brod. Vysočina, C Czech Republic 49°38′N 15°46′E
92 K7 **Havøysund** Finnmark, N Norway 70°59′N 24°39′E
99 F20 **Havré** Hainaut, S Belgium 50°29′N 04°03′E
33 T7 **Havre** Montana, NW USA 48°33′N 109°41′W
Havre *see* le Havre
13 P11 **Havre-St-Pierre** Québec, E Canada 50°16′N 63°36′W
136 B10 **Havsa** Edirne, NW Turkey 41°32′N 26°49′E
38 D8 **Hawai'i** *off.* State of Hawai'i, *also known as* Aloha State, Paradise of the Pacific, *var.* Hawaii. ◈ *state* USA, C Pacific Ocean
38 G12 **Hawai'i** *var.* Hawaii. *island* Hawaiian Islands, USA, C Pacific Ocean
192 M5 **Hawai'ian Islands** *prev.* Sandwich Islands. *island group* Hawaii, USA
192 L5 **Hawaiian Ridge** *undersea feature* N Pacific Ocean 24°00′N 165°00′W
193 N6 **Hawaiian Trough** *undersea feature* N Pacific Ocean
29 R12 **Hawarden** Iowa, C USA 43°00′N 96°29′W
Hawash *see* Āwash
139 P6 **Hawbayn al Gharbīyah** Al Anbār, C Iraq 34°24′N 42°06′E
185 D21 **Hawea, Lake** ◎ South Island, New Zealand
184 K11 **Hawera** Taranaki, North Island, New Zealand 39°36′S 174°16′E
20 J5 **Hawesville** Kentucky, S USA 37°53′N 86°47′W
38 G11 **Hawi** Hawaii, USA, C Pacific Ocean 20°14′N 155°50′W
38 G11 **Hāwī** *var.* Hawi. Hawaii, USA, C Pacific Ocean 20°13′N 155°49′E
Hawi *see* Hāwī
96 K13 **Hawick** SE Scotland, United Kingdom 55°24′N 02°49′W
139 S4 **Ḩawījah** Kirkūk, C Iraq 35°15′N 43°54′E
139 Y10 **Ḩawīzah, Hawr al** ◎ S Iraq
185 E21 **Hawkdun Range** ▲▲ South Island, New Zealand
184 P10 **Hawke Bay** *bay* North Island, New Zealand
182 I6 **Hawker** South Australia 31°54′S 138°25′E
184 N11 **Hawke's Bay** *off.* Hawkes Bay Region. ◈ *region* North Island, New Zealand
149 O16 **Hawkes Bay** *bay* SE Pakistan
Hawkes Bay Region *see* Hawke's Bay
15 N12 **Hawkesbury** Ontario, SE Canada 45°36′N 74°38′W
Hawkeye State *see* Iowa
23 T5 **Hawkinsville** Georgia, SE USA 32°16′N 83°28′W
14 B7 **Hawk Junction** Ontario, S Canada 48°05′N 84°34′W
21 N10 **Haw Knob** ▲ North Carolina/Tennessee, SE USA 35°18′N 84°01′W
21 Q9 **Hawksbill Mountain** ▲ North Carolina, SE USA 35°54′N 81°53′W
33 Z16 **Hawk Springs** Wyoming, C USA 41°48′N 104°17′W
Hawlēr *see* Arbīl
29 S5 **Hawley** Minnesota, N USA 46°53′N 96°18′W
25 P7 **Hawley** Texas, SW USA 32°36′N 99°47′W
141 R14 **Ḩawrā'** C Yemen 15°39′N 48°21′E
139 P7 **Ḩawrān, Wadi** *dry watercourse* W Iraq
21 T9 **Haw River** ≈ North Carolina, SE USA
139 U5 **Hawshqūrah** Diyālá, E Iraq 34°34′N 45°33′E
35 S7 **Hawthorne** Nevada, W USA 38°30′N 118°38′W
37 W3 **Haxtun** Colorado, C USA 40°36′N 102°38′W
183 N9 **Hay** New South Wales, SE Australia 34°31′S 144°51′E
11 O10 **Hay** ≈ W Canada
171 S13 **Haya** Pulau Seram, E Indonesia 03°22′S 129°31′E
165 R9 **Hayachine-san** ▲ Honshū, C Japan 39°31′N 141°28′E
103 S4 **Hayange** Moselle, NE France 49°19′N 06°04′E
HaYarden *see* Jordan
Hayastani Hanrapetut'yun *see* Armenia
Hayasui-seto *see* Hōyo-kaikyō
39 N9 **Haycock** Alaska, USA 65°12′N 161°10′W
36 M14 **Hayden** Arizona, SW USA 33°00′N 110°46′W
37 Q3 **Hayden** Colorado, C USA 40°29′N 107°15′W
28 M10 **Hayes** South Dakota, N USA 44°20′N 101°01′W
9 X13 **Hayes** ≈ Manitoba, C Canada
11 P12 **Hayes** ≈ Nunavut, NE Canada
28 M16 **Hayes Center** Nebraska, C USA 40°30′N 101°02′W
39 S10 **Hayes, Mount** ▲ Alaska, USA 63°37′N 146°43′W
21 N11 **Hayesville** North Carolina, SE USA 35°03′N 83°49′W
35 X10 **Hayford Peak** ▲ Nevada, W USA 36°40′N 115°10′W
34 M3 **Hayfork** California, W USA 40°33′N 123°10′W
Hayir, Qasr al *see* Ḩayr al Gharbī, Qaşr al
Haylaastay *see* Sühbaatar
14 I12 **Hay Lake** ◎ Ontario, SE Canada
141 X11 **Haymā'** *var.* Haima. C Oman 19°59′N 56°20′E
136 H13 **Haymana** Ankara, C Turkey 39°26′N 32°30′E
138 J7 **Ḩaymūr, Jabal** ▲ W Syria
Haynau *see* Chojnów
22 G4 **Haynesville** Louisiana, S USA 32°57′N 93°08′W
23 P6 **Hayneville** Alabama, S USA 32°13′N 86°34′W
114 M12 **Hayrabolu** Tekirdağ, NW Turkey 41°14′N 27°04′E
136 C10 **Hayrabolu Deresi** ≈ NW Turkey
138 J6 **Ḩayr al Gharbī, Qaşr al** *var.* Qasr al Hayir, Qasr al Hir al Gharbi. *ruins* Ḩimş, C Syria
138 L5 **Ḩayr ash Sharqī, Qaşr al** *var.* Qasr al Hir Ash Sharqī. *ruins* Ḩimş, C Syria
162 J7 **Hayrhan** *var.* Uubulan. Arhangay, C Mongolia 48°37′N 101°58′E
162 J9 **Hayrhandulaan** *var.* Mardzad. Övörhangay, C Mongolia 45°58′N 102°06′E
8 J10 **Hay River** Northwest Territories, W Canada 60°51′N 115°42′W
26 K4 **Hays** Kansas, C USA 38°53′N 99°20′W
28 K12 **Hay Springs** Nebraska, C USA 42°40′N 102°41′W
65 H25 **Haystack, The** ▲ NE Saint Helena 15°55′S 05°40′W
27 N7 **Haysville** Kansas, C USA 37°34′N 97°21′W
117 O7 **Haysyn** *Rus.* Gaysin. Vinnyts'ka Oblast', C Ukraine 48°50′N 29°29′E
27 Y9 **Hayti** Missouri, C USA 36°13′N 89°45′W
29 Q9 **Hayti** South Dakota, N USA 44°40′N 97°22′W
117 O8 **Hayvoron** *Rus.* Gayvorno. Kirovohrads'ka Oblast', C Ukraine 48°20′S 29°52′E
35 N9 **Hayward** California, W USA 37°40′N 122°07′W
30 J4 **Hayward** Wisconsin, N USA 46°02′N 91°26′W
97 O23 **Haywards Heath** SE England, United Kingdom 51°00′N 00°06′W
146 A11 **Hazar** *prev. Rus.* Cheleken. Balkan Welaýaty, W Turkmenistan 39°26′N 53°07′E
143 S11 **Hazārān, Kūh-e** *var.* Kūh-e ā Hazr. ▲ SE Iran 29°26′N 57°15′E
Hazarat Imam *see* Imām Şāḩib
21 O7 **Hazard** Kentucky, S USA 37°14′N 83°11′W
137 O15 **Hazar Gölü** ◎ C Turkey
153 P15 **Hazārībāg** *var.* Hazāribāgh. Jhārkhand, N India 24°00′N 85°23′E
Hazāribāgh *see* Hazārībāg
103 O1 **Hazebrouck** Nord, N France 50°43′N 02°33′E
30 K9 **Hazel Green** Wisconsin, N USA 42°33′N 90°26′W
192 K9 **Hazel Holme Bank** *undersea feature* S Pacific Ocean 12°49′S 174°30′E
10 K13 **Hazelton** British Columbia, SW Canada 55°15′N 127°38′W
29 N6 **Hazelton** North Dakota, N USA 46°27′N 100°17′W
35 R5 **Hazen** Nevada, W USA 39°33′N 119°02′W
28 L5 **Hazen** North Dakota, N USA 47°18′N 101°37′W
38 L12 **Hazen Bay** *bay* E Bering Sea
9 N1 **Hazen, Lake** ◎ Nunavut, N Canada
139 S5 **Hazīm, Bi'r** *well* C Iraq
23 V6 **Hazlehurst** Georgia, SE USA 31°51′N 82°35′W
22 K6 **Hazlehurst** Mississippi, S USA 31°51′N 90°24′W
18 K15 **Hazlet** New Jersey, NE USA 40°24′N 74°10′W
146 I9 **Hazorasp** *Rus.* Khazarosp. Xorazm Viloyati, W Uzbekistan 41°21′N 61°01′E
147 R13 **Hazratishoh, Qatorkŭhi** *var.* Khrebet Khazretishi, *Rus.* Khrebet Khozretishi. ▲▲ S Tajikistan
Hazr, Kūh-e ā *see* Hazārān, Kūh-e
149 U6 **Hazro** Punjab, E Pakistan 33°55′N 72°33′E
23 R7 **Headland** Alabama, S USA 31°21′N 85°20′W
182 C6 **Head of Bight** *headland* South Australia 31°33′S 131°05′E
33 N10 **Headquarters** Idaho, NW USA 46°38′N 115°52′W
34 M7 **Healdsburg** California, W USA 38°36′N 122°52′W
27 N13 **Healdton** Oklahoma, C USA 34°13′N 97°29′W
183 O12 **Healesville** Victoria, SE Australia 37°41′S 145°31′E
39 R10 **Healy** Alaska, USA 63°51′N 148°58′W
173 R13 **Heard and McDonald Islands** ◇ *Australian external territory* S Indian Ocean
173 R13 **Heard Island** *island* Heard and McDonald Islands, S Indian Ocean
25 U9 **Hearne** Texas, SW USA 30°52′N 96°35′W
12 F12 **Hearst** Ontario, S Canada 49°42′N 83°40′W
194 J5 **Hearst Island** *island* Antarctica
Heart of Dixie *see* Alabama
28 L5 **Heart River** ≈ North Dakota, N USA
31 T13 **Heath** Ohio, N USA 40°01′N 82°26′W
183 N11 **Heathcote** Victoria, SE Australia 36°57′S 144°43′E
97 N22 **Heathrow** ✈ (London) SE England, United Kingdom 51°28′N 00°27′W
21 X5 **Heathsville** Virginia, NE USA 37°55′N 76°29′W
27 R11 **Heavener** Oklahoma, C USA 34°53′N 94°36′W
25 R15 **Hebbronville** Texas, SW USA 27°19′N 98°41′W
163 Q13 **Hebei** *var.* Hebei Sheng, Hopeh, Hopei, Ji; *prev.* Chihli. ◈ *province* E China
Hebei Sheng *see* Hebei
36 M3 **Heber City** Utah, W USA 40°31′N 111°25′W
27 V10 **Heber Springs** Arkansas, C USA 35°30′N 92°01′W
161 N5 **Hebi** Henan, C China 35°57′N 114°08′E
32 F11 **Hebo** Oregon, NW USA 45°10′N 123°55′W
96 F9 **Hebrides, Sea of the** *sea* NW Scotland, United Kingdom
13 P5 **Hebron** Newfoundland and Labrador, E Canada 58°15′N 62°45′W
31 N11 **Hebron** Indiana, N USA 41°19′N 87°12′W
29 Q17 **Hebron** Nebraska, C USA 40°10′N 97°35′W
28 L5 **Hebron** North Dakota, N USA 46°54′N 102°03′W
138 F11 **Hebron** *var.* Al Khalīl, El Khalil, *Heb.* Hevron; *anc.* Kiriath-Arba. S West Bank 31°30′N 35°E
Hebrus *see* Évros/Maritsa/ Meriç
95 N14 **Heby** Västmanland, C Sweden 59°56′N 16°53′E
10 I14 **Hecate Strait** *strait* British Columbia, W Canada
41 W12 **Hecelchakán** Campeche, SE Mexico 20°09′N 90°04′W
160 K13 **Hechi** *var.* Jinchengjiang. Guangxi Zhuangzu Zizhiqu, S China 24°39′N 108°02′E
101 H23 **Hechingen** Baden-Württemberg, S Germany 48°20′N 08°58′E
99 K17 **Hechtel** Limburg, NE Belgium 51°07′N 05°24′E
160 J9 **Hechuan** *var.* Heyang. Chongqing Shi, C China 30°02′N 106°15′E
29 P7 **Hecla** South Dakota, N USA 45°52′N 98°09′W
9 N1 **Hecla, Cape** *headland* Nunavut, N Canada 82°00′N 64°00′W
29 T9 **Hector** Minnesota, N USA 44°44′N 94°43′W
93 F17 **Hede** Jämtland, C Sweden 62°25′N 13°33′E
Hede *see* Sheyang
95 M14 **Hedemora** Dalarna, C Sweden 60°18′N 15°58′E
92 K13 **Hedenäset** *Finn.* Hietaniemi. Norrbotten, N Sweden 66°12′N 23°40′E
95 G23 **Hedensted** Syddanmark, C Denmark 55°47′N 09°43′E
95 N14 **Hedesunda** Gävleborg, C Sweden 60°25′N 17°00′E
95 N14 **Hedesundafjärden** ◎ C Sweden
25 O3 **Hedley** Texas, SW USA 34°52′N 100°39′W
94 I12 **Hedmark** ◈ *county* S Norway
165 T16 **Hedo-misaki** *headland* Okinawa, SW Japan 26°55′N 128°15′E
29 X15 **Hedrick** Iowa, C USA 41°10′N 92°18′W
99 L16 **Heel** Limburg, SE Netherlands 51°12′N 06°01′E
189 Y12 **Heel Point** *point* Wake Island
98 H9 **Heemskerk** Noord-Holland, W Netherlands 52°31′N 04°40′E
98 M10 **Heerde** Gelderland, E Netherlands 52°24′N 06°02′E
98 L7 **Heerenveen** *Fris.* It Hearrenfean. Fryslân, N Netherlands 52°57′N 05°55′E
98 I8 **Heerhugowaard** Noord-Holland, NW Netherlands 52°40′N 04°50′E
92 O3 **Heer Land** *physical region* C Svalbard
99 M18 **Heerlen** Limburg, SE Netherlands 50°55′N 06°E
99 J19 **Heers** Limburg, NE Belgium 50°46′N 05°17′E
Heerwegen *see* Polkowice
98 K13 **Heesch** Noord-Brabant, S Netherlands 51°44′N 05°32′E
99 K15 **Heeze** Noord-Brabant, SE Netherlands 51°23′N 05°35′E
138 F8 **Hefa** *var.* Haifa, *hist.* Caiffa, Caiphas; *anc.* Sycaminum. Haifa, N Israel 32°49′N 34°59′E
Ḩefa, Mifraz *see* Mifrats Hefa
161 Q8 **Hefei** *var.* Hofei, *hist.* Luchow. *province capital* Anhui, E China 31°51′N 117°20′E
23 R3 **Heflin** Alabama, S USA 33°39′N 85°35′W
163 X7 **Hegang** Heilongjiang, NE China 47°18′N 130°16′E
164 L10 **Hegura-jima** *island* SW Japan
Heguri-jima *see* Heigun-tō
Hei *see* Heilongjiang
100 H8 **Heide** Schleswig-Holstein, N Germany 54°12′N 09°06′E
101 G20 **Heidelberg** Baden-Württemberg, SW Germany 49°24′N 08°41′E
83 J21 **Heidelberg** Gauteng, NE South Africa 26°31′S 28°21′E
22 M6 **Heidelberg** Mississippi, S USA 31°53′N 88°58′W
Heidenheim *see* Heidenheim an der Brenz
101 J22 **Heidenheim an der Brenz** *var.* Heidenheim. Baden-Württemberg, S Germany 48°41′N 10°09′E
109 U2 **Heidenreichstein** Niederösterreich, N Austria 48°53′N 15°07′E
164 F14 **Heigun-tō** *var.* Heguri-jima. *island* SW Japan
163 W5 **Heihe** *prev.* Ai-hun. Heilongjiang, NE China 50°13′N 127°29′E
162 S8 **Hei He** ≈ C China
Hei-ho *see* Nagqu
83 J22 **Heilbron** Free State, N South Africa 27°17′S 27°58′E
101 H21 **Heilbronn** Baden-Württemberg, SW Germany 49°09′N 09°13′E
Heiligenbeil *see* Mamonovo
109 Q8 **Heiligenblut** Tirol, W Austria 47°04′N 12°50′E
100 K7 **Heiligenhafen** Schleswig-Holstein, N Germany 54°22′N 10°57′E
Heiligenkreuz *see* Žiar nad Hronom
101 J15 **Heiligenstadt** Thüringen, C Germany 51°22′N 10°09′E
163 W8 **Heilongjiang** *var.* Hei, Heilongjiang Sheng, Hei-lung-chiang, Heilungkiang. ◈ *province* NE China
Heilong Jiang *see* Amur
Heilongjiang Sheng *see* Heilongjiang
98 H9 **Heiloo** Noord-Holland, NW Netherlands 52°36′N 04°43′E
Heilsberg *see* Lidzbark Warmiński
Hei-lung-chiang/ Heilungkiang *see* Heilongjiang
Heimaæy *see* Heimaey
92 I4 **Heimaey** *var.* Heimaæy. *island* S Iceland
94 H8 **Heimdal** Sør-Trøndelag, S Norway 63°21′N 10°23′E
Heinaste *see* Ainaži
93 N17 **Heinävesi** Etelä-Savo, E Finland 62°22′N 28°42′E
99 M22 **Heinerscheid** Diekirch, N Luxembourg 50°06′N 06°05′E
98 M10 **Heino** Overijssel, E Netherlands 52°26′N 06°13′E
93 M18 **Heinola** Päijät-Häme, S Finland 61°13′N 26°05′E
101 C16 **Heinsberg** Nordrhein-Westfalen, W Germany 51°02′N 06°01′E
163 U12 **Heishan** Liaoning, NE China 41°43′N 122°12′E
160 H8 **Heishui** *var.* Luhua. Sichuan, C China 32°08′N 102°54′E
99 H17 **Heist-op-den-Berg** Antwerpen, C Belgium 51°04′N 04°43′E
Heitō *see* Pingdong
171 X15 **Heitske** Papua, E Indonesia 07°02′S 138°45′E
Hejanah *see* Al Hījānah
Hejaz *see* Al Ḩijāz
160 M14 **He Jiang** ≈ S China
Hejiayan *see* Lüeyang
158 K6 **Hejing** Xinjiang Uygur Zizhiqu, NW China 42°21′N 86°19′E
Héjjasfalva *see* Vânători
Heka *see* Hoika
137 N14 **Hekimhan** Malatya, C Turkey 38°50′N 37°56′E
92 J4 **Hekla** ▲ S Iceland 63°56′N 19°42′W
Hekou *see* Yanshan, Jiangxi, China
Hekou *see* Yajiang, Sichuan, China
110 J6 **Hel** *Ger.* Hela. Pomorskie, N Poland 54°35′N 18°48′E
186 B6 **Hela** ◈ *province* W Papua New Guinea
Hela *see* Hel
93 F17 **Helagsfjället** ▲ C Sweden 62°57′N 12°31′E
159 W8 **Helan** *var.* Xigang. Ningxia, N China 38°33′N 106°21′E
162 K14 **Helan Shan** ▲▲ N China
99 M16 **Helden** Limburg, SE Netherlands 51°20′N 06°00′E
139 V4 **Helebce** *Ar.* Ḩalabjah, *var.* Halabja. As Sulaymānīyah, NE Iraq 35°11′N 45°59′E
27 X12 **Helena** Arkansas, C USA 34°32′N 90°34′W
33 R10 **Helena** *state capital* Montana, NW USA 46°36′N 112°02′W
96 H12 **Helensburgh** W Scotland, United Kingdom 56°00′N 04°45′W
184 K5 **Helensville** Auckland, North Island, New Zealand 36°42′S 174°26′E
95 L20 **Helgasjön** ◎ S Sweden
100 G8 **Helgoland** *Eng.* Heligoland. *island* NW Germany
Helgoland Bay *see* Helgoländer Bucht
100 G8 **Helgoländer Bucht** *var.* Helgoland Bay, Heligoland Bight. *bay* NW Germany
Heligoland *see* Helgoland
Heligoland Bight *see* Helgoländer Bucht
Heliopolis *see* Baalbek
92 I4 **Hella** Suðurland, SW Iceland 63°51′N 20°24′W
Hellas *see* Greece
143 N11 **Helleh, Rūd-e** ≈ S Iran
98 N10 **Hellendoorn** Overijssel, E Netherlands 52°22′N 06°27′E
Hellenic Republic *see* Greece
121 Q10 **Hellenic Trough** *undersea feature* Aegean Sea, C Mediterranean Sea 22°00′E 35°30′N
94 E10 **Hellesylt** Møre og Romsdal, S Norway 62°06′N 06°51′E
98 F13 **Hellevoetsluis** Zuid-Holland, SW Netherlands 51°49′N 04°08′E
105 Q12 **Hellín** Castilla-La Mancha, C Spain 38°31′N 01°43′W
115 H19 **Hellinikon** ✈ (Athína) Attikí, C Greece 37°53′N 23°43′E
32 M12 **Hells Canyon** *valley* Idaho/ Oregon, NW USA
148 L9 **Helmand** ◈ *province* S Afghanistan
148 K10 **Helmand, Daryā-ye** *var.* Rūd-e Hīrmand. ≈ Afghanistan/Iran *see also* Hīrmand, Rūd-e
Helmand, Daryā-ye *see* Hīrmand, Rūd-e
Helmantica *see* Salamanca
101 K15 **Helme** ≈ C Germany
99 L15 **Helmond** Noord-Brabant, S Netherlands 51°29′N 05°41′E
96 J7 **Helmsdale** N Scotland, United Kingdom 58°06′N 03°36′W
100 K13 **Helmstedt** Niedersachsen, N Germany 52°14′N 11°01′E
163 Y10 **Helong** Jilin, NE China 42°38′N 129°01′E
36 M4 **Helper** Utah, W USA 39°40′N 110°52′W
100 O10 **Helpter Berge** *hill* NE Germany
95 J22 **Helsingborg** *prev.* Hälsingborg. Skåne, S Sweden 56°N 12°48′E
Helsingfors *see* Helsinki
95 J22 **Helsingør** *Eng.* Elsinore. Hovedstaden, E Denmark 56°03′N 12°38′E
93 M20 **Helsinki** *Swe.* Helsingfors. ● (Finland) Uusimaa, S Finland 60°18′N 24°58′E
97 H25 **Helston** SW England, United Kingdom 50°04′N 05°17′W
Heltau *see* Cisnădie
61 C17 **Helvecia** Santa Fe, C Argentina 31°09′S 60°09′W
97 K15 **Helvellyn** ▲ NW England, United Kingdom 54°31′N 03°00′W
Helvetia *see* Switzerland
Helwān *see* Hilwān
97 N21 **Hemel Hempstead** E England, United Kingdom 51°46′N 00°28′W
35 U16 **Hemet** California, W USA 33°45′N 116°58′W
28 J13 **Hemingford** Nebraska, C USA 42°18′N 103°02′W
21 T13 **Hemingway** South Carolina, SE USA 33°45′N 79°25′W
92 G13 **Hemnesberget** Nordland, C Norway 66°14′N 13°40′E
25 Y8 **Hemphill** Texas, SW USA 31°21′N 93°50′W
25 V11 **Hempstead** Texas, SW USA 30°06′N 96°06′W
95 P20 **Hemse** Gotland, SE Sweden 57°12′N 18°22′E
94 F13 **Hemsedal** *valley* S Norway
161 N6 **Henan** *var.* Henan Sheng, Honan, Yu. ◈ *province* C China
184 L4 **Hen and Chickens** *island group* N New Zealand
Henan Mongolzu Zizhixian/Henan Sheng *see* Yêgainnyin
105 O7 **Henares** ≈ C Spain
165 P7 **Henashi-zaki** *headland* Honshū, C Japan 40°37′N 139°51′E
102 I16 **Hendaye** Pyrénées-Atlantiques, SW France 43°22′N 01°46′W
136 F11 **Hendek** Sakarya, NW Turkey 40°47′N 30°45′E
61 B21 **Henderson** Buenos Aires, E Argentina 36°18′S 61°43′W
20 I5 **Henderson** Kentucky, S USA 37°50′N 87°35′W
35 X11 **Henderson** Nevada, W USA 36°02′N 114°58′W
21 V8 **Henderson** North Carolina, SE USA 36°20′N 78°26′W
20 G10 **Henderson** Tennessee, S USA 35°27′N 88°40′W
25 W7 **Henderson** Texas, SW USA 32°11′N 94°48′W
30 J12 **Henderson Creek** ≈ Illinois, N USA
186 M9 **Henderson Field** ✈ (Honiara) Guadalcanal, C Solomon Islands 09°28′S 160°02′E
191 O17 **Henderson Island** *atoll* N Pitcairn Group of Islands
21 O10 **Hendersonville** North Carolina, SE USA 35°19′N 82°28′W
20 J8 **Hendersonville** Tennessee, S USA 36°18′N 86°37′W
143 O14 **Hendorābī, Jazīreh-ye** *island* S Iran
55 V10 **Hendrik Top** *var.* Hendriktop. *elevation* C Suriname
Hendriktop *see* Hendrik Top
Hendū Kosh *see* Hindu Kush
14 L12 **Heney, Lac** ◎ Québec, SE Canada
Hengchow *see* Hengyang
161 S15 **Hengch'un** S Taiwan 22°09′N 120°43′E
159 R16 **Hengduan Shan** ▲▲ SW China
98 N12 **Hengelo** Gelderland, E Netherlands 52°03′N 06°19′E
98 O10 **Hengelo** Overijssel, E Netherlands 52°16′N 06°46′E
Hengnan *see* Hengyang
161 N11 **Hengshan** Hunan, S China 27°17′N 112°51′E
160 L4 **Hengshan** Shaanxi, C China 37°57′N 109°17′E
161 O4 **Hengshui** Hebei, E China 37°42′N 115°39′E
161 N12 **Hengyang** *var.* Hengnan, Heng-yang; *prev.* Hengchow. Hunan, S China 26°55′N 112°34′E
Heng-yang *see* Hengyang
117 U11 **Heniches'k** *Rus.* Genichesk. Khersons'ka Oblast', S Ukraine 46°10′N 34°49′E
21 Z4 **Henlopen, Cape** *headland* Delaware, NE USA 38°48′N 75°06′W
Henna *see* Enna
94 M10 **Hennan** Gävleborg, C Sweden 62°01′N 15°55′E
102 G7 **Hennebont** Morbihan, NW France 47°48′N 03°17′W
30 L11 **Hennepin** Illinois, N USA 41°14′N 89°21′W
26 M9 **Hennessey** Oklahoma, C USA 36°06′N 97°54′W
100 N12 **Hennigsdorf** *var.* Hennigsdorf bei Berlin. Brandenburg, NE Germany 52°37′N 13°13′E
Hennigsdorf bei Berlin *see* Hennigsdorf
19 N9 **Henniker** New Hampshire, NE USA 43°10′N 71°47′W
25 S5 **Henrietta** Texas, SW USA 33°49′N 98°13′W
Henrique de Carvalho *see* Saurimo
30 L12 **Henry** Illinois, N USA 41°06′N 89°21′W
21 Y7 **Henry, Cape** *headland* Virginia, NE USA 36°55′N 76°01′W
27 P10 **Henryetta** Oklahoma, C USA 35°26′N 95°58′W
194 M7 **Henry Ice Rise** *ice cap* Antarctica
9 R5 **Henry Kater, Cape** *headland* Baffin Island, Nunavut, NE Canada 69°09′N 66°45′W
33 R13 **Henrys Fork** ≈ Idaho, NW USA
14 E15 **Hensall** Ontario, S Canada 43°25′N 81°28′W
100 J9 **Henstedt-Ulzburg** Schleswig-Holstein, N Germany 53°45′N 09°59′E
163 N7 **Hentiy** *var.* Batshireet, Eg. ◈ *province* N Mongolia
162 M7 **Hentiyn Nuruu** ▲▲ N Mongolia
183 P10 **Henty** New South Wales, SE Australia 35°33′S 147°03′E
Henzada *see* Hinthada
Heping *see* Huishui
101 G19 **Heppenheim** Hessen, W Germany 49°39′N 08°38′E
32 J11 **Heppner** Oregon, NW USA 45°21′N 119°32′W
160 L15 **Hepu** *var.* Lianzhou. Guangxi Zhuangzu Zizhiqu, S China 21°40′N 109°12′E
Heracleum *see* Irákleio
92 J2 **Heraðsvötn** ≈ C Iceland
Herakleion *see* Irákleio
148 K5 **Herāt** *var.* Herat; *anc.* Aria. Herāt, W Afghanistan 34°23′N 62°11′E
148 J5 **Herāt** ◈ *province* W Afghanistan
103 P14 **Hérault** ◈ *department* S France
103 P15 **Hérault** ≈ S France
11 T16 **Herbert** Saskatchewan, S Canada 50°27′N 107°09′W
185 F22 **Herbert** Otago, South Island, New Zealand 45°14′S 170°48′E
38 J17 **Herbert Island** *island* Aleutian Islands, Alaska, USA
Herbertshöhe *see* Kokopo
15 Q7 **Hérbertville** Québec, SE Canada 48°23′N 71°42′W
101 G17 **Herborn** Hessen, W Germany 50°40′N 08°18′E
113 I17 **Herceg-Novi** *It.* Castelnuovo; *prev.* Ercegnovi. SW Montenegro 42°28′N 18°35′E
11 X10 **Hercher** Manitoba, C Canada 57°25′N 94°12′W
186 E8 **Hercules Bay** *bay* E Papua New Guinea
92 K2 **Herðubreið** ▲ C Iceland 65°12′N 16°26′W
42 M13 **Heredia** Heredia, C Costa Rica 10°N 84°06′W
42 M12 **Heredia** *off.* Provincia de Heredia. ◈ *province* N Costa Rica
Heredia, Provincia de *see* Heredia
97 K21 **Hereford** W England, United Kingdom 52°04′N 02°43′W
24 M3 **Hereford** Texas, SW USA 34°49′N 102°25′W
15 Q13 **Hereford, Mont** ▲ Québec, SE Canada 45°04′N 71°38′W
97 K21 **Herefordshire** *cultural region* W England, United Kingdom
191 U11 **Hereheretue** *atoll* Îles Tuamotu, C French Polynesia
105 N10 **Herencia** Castilla-La Mancha, C Spain 39°22′N 03°12′W
99 H18 **Herent** Vlaams Brabant, C Belgium 50°54′N 04°40′E
99 I16 **Herentals** *var.* Herenthals. Antwerpen, N Belgium 51°11′N 04°50′E
Herenthals *see* Herentals
99 H17 **Herenthout** Antwerpen, N Belgium 51°09′N 04°45′E
95 J23 **Herfølge** Sjælland, E Denmark 55°25′N 12°09′E
100 G13 **Herford** Nordrhein-Westfalen, NW Germany 52°07′N 08°41′E
27 O5 **Herington** Kansas, C USA 38°37′N 96°55′W
108 H7 **Herisau** *Fr.* Hérisau. Ausser Rhoden, NE Switzerland 47°23′N 09°17′E
Hérisau *see* Herisau
Héristal *see* Herstal
99 J18 **Herk-de-Stad** Limburg, NE Belgium 50°57′N 05°12′E
Herkulesbad/ Herkulesfürdő *see* Băile Herculane
162 M8 **Herlenbayan-Ulaan** *var.* Dulaan. Hentiy, C Mongolia 47°09′N 108°48′E
Herlen Gol/Herlen He *see* Kerulen
35 Q4 **Herlong** California, W USA 40°07′N 120°06′W
97 L26 **Herm** *island* Channel Islands
109 R9 **Hermagor** *Slvn.* Šmohor. Kärnten, S Austria 46°37′N 13°24′E
29 S7 **Herman** Minnesota, N USA 45°49′N 96°08′W
96 L1 **Herma Ness** *headland* NE Scotland, United Kingdom 60°51′N 00°55′W
27 V4 **Hermann** Missouri, C USA 38°43′N 91°26′W
181 Q8 **Hermannsburg** Northern Territory, N Australia 23°59′S 132°55′E
Hermannstadt *see* Sibiu
94 E12 **Hermansverk** Sogn Og Fjordane, S Norway 61°11′N 06°52′E
138 H6 **Hermel** *var.* Hirmil. NE Lebanon 34°23′N 36°19′E
Hermhausen *see* Hajnówka
183 P6 **Hermidale** New South Wales, SE Australia 31°36′S 146°42′E
55 X9 **Herminadorp** Sipaliwini, NE Suriname 05°05′N 54°22′W
32 K11 **Hermiston** Oregon, NW USA 45°50′N 119°17′W
27 T6 **Hermitage** Missouri, C USA 37°57′N 93°21′W
186 D4 **Hermit Islands** *island group* N Papua New Guinea
25 O7 **Hermleigh** Texas, SW USA 32°37′N 100°44′W
138 G7 **Hermon, Mount** *Ar.* Jabal ash Shaykh. ▲ S Syria 33°30′N 33°30′E
Hermopolis Parva *see* Damanhūr
28 J10 **Hermosa** South Dakota, N USA 43°49′N 103°12′W
40 F5 **Hermosillo** Sonora, NW Mexico 28°59′N 110°53′W
Hermoupolis *see* Ermoúpoli
111 N20 **Hernád** *var.* Hornád, *Ger.* Kundert. ≈ Hungary/ Slovakia
61 C18 **Hernández** Entre Ríos, E Argentina 32°21′S 60°02′W
23 V11 **Hernando** Florida, SE USA 28°54′N 82°22′W
22 L1 **Hernando** Mississippi, S USA 34°49′N 89°59′W
105 Q2 **Hernani** País Vasco, N Spain 43°16′N 01°59′W
99 F19 **Herne** Vlaams Brabant, C Belgium 50°45′N 04°03′E
101 E14 **Herne** Nordrhein-Westfalen, W Germany 51°32′N 07°12′E
95 F22 **Herning** Midtjylland, W Denmark 56°08′N 08°59′E
Hernösand *see* Härnösand
121 U11 **Herodotus Basin** *undersea feature* E Mediterranean Sea
121 Q12 **Herodotus Trough** *undersea feature* C Mediterranean Sea
29 T11 **Heron Lake** Minnesota, N USA 43°48′N 95°18′W
Herowābād *see* Khalkhāl
95 G16 **Herre** Telemark, S Norway 59°06′N 09°34′E
29 N7 **Herreid** South Dakota, N USA 45°49′N 100°04′W
101 H22 **Herrenberg** Baden-Württemberg, S Germany 48°36′N 08°52′E
104 L14 **Herrera** Andalucía, S Spain 37°22′N 04°50′W
43 R17 **Herrera** *off.* Provincia de Herrera. ◈ *province* S Panama
104 L10 **Herrera del Duque** Extremadura, W Spain 39°10′N 05°03′W
104 M4 **Herrera de Pisuerga** Castilla y León, N Spain 42°35′N 04°20′W
Herrera, Provincia de *see* Herrera
41 Z13 **Herrero, Punta** *headland* SE Mexico 19°15′N 87°28′W
183 P16 **Herrick** Tasmania, SE Australia 41°07′S 147°53′E
30 L17 **Herrin** Illinois, N USA 37°48′N 89°01′W
20 M6 **Herrington Lake** ⊠ Kentucky, S USA
95 K18 **Herrljunga** Västra Götaland, S Sweden 58°05′N 13°02′E
103 N16 **Hers** ≈ S France
10 I1 **Herschel Island** *island* Yukon, NW Canada
99 I17 **Herselt** Antwerpen, C Belgium 51°04′N 04°53′E
18 G15 **Hershey** Pennsylvania, NE USA 40°17′N 76°39′W
99 K19 **Herstal** *Fr.* Héristal. Liège, E Belgium 50°40′N 05°38′E
97 O21 **Hertford** E England, United Kingdom 51°48′N 00°05′W
21 X8 **Hertford** North Carolina, SE USA 36°11′N 76°30′W
97 O21 **Hertfordshire** *cultural region* E England, United Kingdom
181 Z9 **Hervey Bay** Queensland, E Australia 25°17′S 152°48′E
101 O14 **Herzberg** Brandenburg, E Germany 51°42′N 13°15′E
99 E18 **Herzele** Oost-Vlaanderen, NW Belgium 50°52′N 03°52′E
101 K20 **Herzogenaurach** Bayern, SE Germany 49°34′N 10°52′E
109 W4 **Herzogenburg** Niederösterreich, NE Austria 48°18′N 15°43′E
Herzogenbusch *see* 's-Hertogenbosch
103 N2 **Hesdin** Pas-de-Calais, N France 50°21′N 02°00′E
160 K14 **Heshan** Guangxi Zhuangzu Zizhiqu, S China 23°45′N 108°58′E
159 X10 **Heshui** *var.* Xihuachi. Gansu, C China 35°42′N 108°06′E
99 M25 **Hespérange** Luxembourg, SE Luxembourg 49°34′N 06°10′E
35 U14 **Hesperia** California, W USA 34°25′N 117°17′W
37 P7 **Hesperus Mountain** ▲ Colorado, C USA 37°27′N 108°05′W
10 J6 **Hess** ≈ Yukon, NW Canada
Hesse *see* Hessen
101 J21 **Hesselberg** ▲ S Germany 49°04′N 10°32′E
95 I22 **Hesselo** *island* E Denmark
101 H17 **Hessen** *Eng./Fr.* Hesse. ◈ *state* C Germany
192 L6 **Hess Tablemount** *undersea feature* C Pacific Ocean 17°49′N 174°15′W
27 N6 **Hesston** Kansas, C USA 38°08′N 97°25′W
93 G15 **Hestkjøltoppen** ▲ C Norway 64°21′N 13°57′E
97 K18 **Heswall** NW England, United Kingdom 53°20′N 03°06′W
153 P12 **Hetaudā** Central, C Nepal
Hétfalu *see* Săcele
28 K7 **Hettinger** North Dakota, N USA 46°00′N 102°38′W
101 L14 **Hettstedt** Sachsen-Anhalt, C Germany 51°39′N 11°31′E
92 P3 **Heuglin, Kapp** *headland* SE Svalbard 78°15′N 22°49′E
163 W17 **Heuksan-jedo** *var.* Hŭksan-gundo. *island group* SW South Korea
187 N10 **Heuru** Makira-Ulawa, SE Solomon Islands 10°13′S 161°25′E
99 J17 **Heusden** Limburg, NE Belgium 51°02′N 05°17′E
98 J13 **Heusden** Noord-Brabant, S Netherlands 51°43′N 05°05′E
102 K3 **Hève, Cap de la** *headland* N France 49°28′N 00°13′W
99 H18 **Heverlee** Vlaams Brabant, C Belgium 50°52′N 04°41′E
111 L22 **Heves** Heves, NE Hungary 47°36′N 20°17′E
111 L22 **Heves** *off.* Heves Megye. ◈ *county* NE Hungary
Heves Megye *see* Heves
Hevron *see* Hebron
45 Y13 **Hewanorra** ✈ (Saint Lucia) S Saint Lucia 13°44′N 60°57′W
Hewlêr *see* Arbīl
Hewlêr *see* Arbīl
Hewlêr, Parêzga-î *see* Arbīl
Hexian *see* Hezhou
160 L6 **Heyang** Shaanxi, C China 35°14′N 110°02′E
Heyang *see* Hechuan
Heydebrech *see* Kędzierzyn-Kozle
Heydekrug *see* Šilutė
Heyin *see* Guide
97 K16 **Heysham** NW England, United Kingdom 54°02′N 02°54′W
161 O14 **Heyuan** *var.* Yuancheng. Guangdong, S China 23°41′N 114°45′E
182 L12 **Heywood** Victoria, SE Australia 38°09′S 141°38′E
180 K3 **Heywood Islands** *island group* Western Australia
161 O6 **Heze** *var.* Caozhou. Shandong, E China 35°16′N 115°27′E

◆ Country ● Country Capital ◇ Dependent Territory ○ Dependent Territory Capital ◈ Administrative Regions ✈ International Airport ▲ Mountain ▲▲ Mountain Range ⛰ Volcano ≈ River ◎ Lake ⊠ Reservoir

HEZHENG – HONGWANSI

159 U11 **Hezheng** Gansu, C China 35°29′N 103°36′E
160 M13 **Hezhou** var. Babu; prev. Hexian. Guangxi Zhuangzu Zizhiqu, S China 24°33′N 11°30′E
159 U11 **Hezuo** Gansu, C China 34°55′N 102°49′E
23 Z16 **Hialeah** Florida, SE USA 25°51′N 80°16′W
27 Q3 **Hiawatha** Kansas, C USA 39°51′N 95°34′W
36 M4 **Hiawatha** Utah, W USA 39°28′N 111°00′W
29 V4 **Hibbing** Minnesota, N USA 47°24′N 92°55′W
183 N17 **Hibbs, Point** headland Tasmania, SE Australia 42°37′S 145°15′E
Hibernia see Ireland
20 F8 **Hickman** Kentucky, S USA 36°33′N 89°11′W
21 Q9 **Hickory** North Carolina, SE USA 35°44′N 81°20′W
21 Q9 **Hickory, Lake** ◎ North Carolina, SE USA
184 Q7 **Hicks Bay** Gisborne, North Island, New Zealand 37°36′S 178°18′E
25 S8 **Hico** Texas, SW USA 31°58′N 98°01′W
Hida see Furukawa
165 T4 **Hidaka** Hokkaidō, NE Japan 42°53′N 142°24′E
164 I12 **Hidaka** Hyōgo, Honshū, SW Japan 35°27′N 134°43′E
165 T5 **Hidaka-sanmyaku** ▲▲ Hokkaidō, NE Japan
41 O6 **Hidalgo** var. Villa Hidalgo. Coahuila, NE Mexico 27°46′N 99°54′W
41 N8 **Hidalgo** Nuevo León, NE Mexico 29°59′N 100°27′W
41 O10 **Hidalgo** Tamaulipas, C Mexico 24°16′N 99°28′W
41 O13 **Hidalgo** ◈ state C Mexico
40 J7 **Hidalgo del Parral** var. Parral. Chihuahua, N Mexico 26°58′N 105°40′W
100 N7 **Hiddensee** island NE Germany
80 G6 **Hidiglib, Wadi** ≈ NE Sudan
109 U6 **Hieflau** Salzburg, E Austria 47°36′N 14°34′E
187 P16 **Hienghène** Province Nord, C New Caledonia 20°43′S 164°54′E
Hierosolyma see Jerusalem
64 N12 **Hierro** var. Ferro. island Islas Canarias, Spain, NE Atlantic Ocean
Hietaniemi see Hedenäset
164 G13 **Higashi-Hiroshima** var. Higasihirosima. Hiroshima, Honshū, SW Japan 34°27′N 132°43′E
164 C12 **Higashi-suidō** strait SW Japan
Higasihirosima see Higashi-Hiroshima
25 P1 **Higgins** Texas, SW USA 36°06′N 100°01′W
31 P7 **Higgins Lake** ◎ Michigan, N USA
27 S4 **Higginsville** Missouri, C USA 39°04′N 93°43′W
High Atlas see Haut Atlas
30 M5 **High Falls Reservoir** ⊠ Wisconsin, N USA
44 K12 **Highgate** C Jamaica 18°16′N 76°53′W
25 X11 **High Island** Texas, SW USA 29°35′N 94°24′W
31 O5 **High Island** island Michigan, N USA
30 K15 **Highland** Illinois, N USA 38°44′N 89°40′W
31 N10 **Highland Park** Illinois, N USA 42°10′N 87°48′W
21 O10 **Highlands** North Carolina, SE USA 35°04′N 83°10′W
11 O11 **High Level** Alberta, W Canada 58°31′N 117°08′W
29 O9 **Highmore** South Dakota, N USA 44°29′N 99°26′W
171 N3 **High Peak** ▲ Luzon, N Philippines 15°28′N 120°07′E
High Plains see Great Plains
21 S9 **High Point** North Carolina, SE USA 35°58′N 80°00′W
18 J13 **High Point** hill New Jersey, NE USA
11 P13 **High Prairie** Alberta, W Canada 55°27′N 116°28′W
11 Q16 **High River** Alberta, SW Canada 50°35′N 113°50′W
21 S9 **High Rock Lake** ⊠ North Carolina, SE USA
23 V9 **High Springs** Florida, SE USA 29°49′N 82°36′W
High Veld see Great Karoo
97 J24 **High Willhays** ▲ SW England, United Kingdom 50°39′N 03°58′W
97 N22 **High Wycombe** prev. Chepping Wycombe, Chipping Wycombe. SE England, United Kingdom 51°38′N 00°46′W
41 P12 **Higos** var. El Higo. Veracruz-Llave, E Mexico 21°48′N 98°25′W
102 I16 **Higuer, Cap** headland NE Spain 43°23′N 01°46′W
45 R5 **Higüero, Punta** headland W Puerto Rico 18°21′N 67°15′W
45 P9 **Higüey** var. Salvaleón de Higüey. E Dominican Republic 18°40′N 68°43′W
190 G11 **Hihifo** ✈ (Matā'utu) Île Uvea, N Wallis and Futuna
81 N16 **Hiiraan** off. Gobolka Hiiraan. ◈ region C Somalia
Hiiraan, Gobolka see Hiiraan
118 E4 **Hiiumaa** var. Hiiumaa Maakond. ◈ province W Estonia
118 D4 **Hiiumaa** Ger. Dagden, Swe. Dagö. island W Estonia
Hiiumaa Maakond see Hiiumaa
Hijanah see Al Hījānah
105 S6 **Híjar** Aragón, NE Spain 41°10′N 00°27′W
191 V10 **Hikueru** atoll Îles Tuamotu, C French Polynesia
184 K3 **Hikurangi** Northland, North Island, New Zealand 35°37′S 174°16′E
184 Q8 **Hikurangi** ▲ North Island, New Zealand 37°55′S 177°59′E
192 L11 **Hikurangi Trench** var. Hikurangi Trough. undersea feature SW Pacific Ocean
Hikurangi Trough see Hikurangi Trench
190 B15 **Hikutavake** NW Niue
121 Q12 **Hilāl, Ra's al** headland N Libya 32°55′N 22°09′E
61 A24 **Hilario Ascasubi** Buenos Aires, E Argentina 39°22′S 62°39′W
101 K17 **Hildburghausen** Thüringen, C Germany 50°26′N 10°44′E
101 E15 **Hilden** Nordrhein-Westfalen, W Germany 51°12′N 06°56′E
100 I13 **Hildesheim** Niedersachsen, N Germany 52°09′N 09°57′E
33 T9 **Hilger** Montana, NW USA 47°15′N 109°18′W
153 S13 **Hili** var. Hilli. Rajshahi, NW Bangladesh 25°16′N 89°04′E
Hilla see Al Ḩillah
45 O14 **Hillaby, Mount** ▲ N Barbados 13°12′N 59°34′W
Ḩillah, Al see Bābil
95 K19 **Hillared** Västra Götaland, S Sweden 57°37′N 13°10′E
195 R12 **Hillary Coast** physical region Antarctica
42 G2 **Hill Bank** Orange Walk, N Belize 17°36′N 88°43′W
33 O14 **Hill City** Idaho, NW USA 43°18′N 115°03′W
26 K3 **Hill City** Kansas, C USA 39°23′N 99°51′W
29 V5 **Hill City** Minnesota, N USA 46°59′N 93°36′W
28 J10 **Hill City** South Dakota, N USA 43°54′N 103°33′W
65 C24 **Hill Cove Settlement** West Falkland, Falkland Islands
98 H10 **Hillegom** Zuid-Holland, W Netherlands 52°18′N 04°35′E
95 J22 **Hillerød** Hovedstaden, E Denmark 55°56′N 12°19′E
36 M7 **Hillers, Mount** ▲ Utah, W USA 37°53′N 110°42′W
Hilli see Hili
29 R11 **Hills** Minnesota, N USA 43°31′N 96°21′W
30 L14 **Hillsboro** Illinois, N USA 39°09′N 89°29′W
27 N5 **Hillsboro** Kansas, C USA 38°21′N 97°12′W
27 X5 **Hillsboro** Missouri, C USA 38°13′N 90°33′W
19 N10 **Hillsboro** New Hampshire, NE USA 43°06′N 71°52′W
37 Q14 **Hillsboro** New Mexico, SW USA 32°55′N 107°33′W
29 R4 **Hillsboro** North Dakota, N USA 47°25′N 97°03′W
31 R14 **Hillsboro** Ohio, N USA 39°12′N 83°36′W
32 G11 **Hillsboro** Oregon, NW USA 45°32′N 122°59′W
25 T8 **Hillsboro** Texas, SW USA 32°01′N 97°08′W
30 K8 **Hillsboro** Wisconsin, N USA 43°40′N 90°21′W
23 Y14 **Hillsboro Canal** canal Florida, SE USA
45 Y15 **Hillsborough** Carriacou, N Grenada 12°28′N 61°28′W
97 G15 **Hillsborough** E Northern Ireland, United Kingdom 54°27′N 06°06′W
21 U9 **Hillsborough** North Carolina, SE USA 36°04′N 79°06′W
31 Q10 **Hillsdale** Michigan, N USA 41°55′N 84°37′W
183 O8 **Hillston** New South Wales, SE Australia 33°30′S 145°33′E
21 R7 **Hillsville** Virginia, NE USA 36°46′N 80°44′W
96 L2 **Hillswick** NE Scotland, United Kingdom 60°28′N 01°37′W
Hill Tippera see Tripura
38 H11 **Hilo** Hawaii, USA, C Pacific Ocean 19°42′N 155°04′W
18 F9 **Hilton** New York, NE USA 43°17′N 77°47′W
14 C10 **Hilton Beach** Ontario, S Canada 46°14′N 83°51′W
21 R16 **Hilton Head Island** South Carolina, SE USA 32°13′N 80°45′W
21 R16 **Hilton Head Island** island South Carolina, SE USA
99 J15 **Hilvarenbeek** Noord-Brabant, S Netherlands 51°29′N 05°08′E
98 J11 **Hilversum** Noord-Holland, C Netherlands 52°14′N 05°10′E
75 W8 **Hilwān** var. Helwân, Hilwân, Hulwan, Hulwân. N Egypt 29°51′N 31°20′E
Hilwân see Hilwān
152 J7 **Himāchal Pradesh** ◈ state NW India
Himalaya/Himalaya Shan see Himalayas
152 M9 **Himalayas** var. Himalaya, Chin. Himalaya Shan. ▲▲ S Asia
171 P6 **Himamaylan** Negros, C Philippines 10°04′N 122°52′E
93 K15 **Himanka** Pohjois-Pohjanmaa, W Finland 64°04′N 23°40′E
Himara see Himarë
113 L23 **Himarë** var. Himara. Vlorë, S Albania 40°06′N 19°45′E
138 M2 **Ḩimār, Wādī al** dry watercourse N Syria
154 D9 **Himatnagar** Gujarāt, W India 23°38′N 73°02′E
109 Y4 **Himberg** Niederösterreich, E Austria 48°05′N 16°27′E
164 I13 **Himeji** var. Himezi. Hyōgo, Honshū, SW Japan 34°47′N 134°32′E
164 E14 **Hime-jima** island SW Japan
Himezi see Himeji
164 L13 **Himi** Toyama, Honshū, SW Japan 36°54′N 136°59′E
109 S9 **Himmelberg** Kärnten, S Austria 46°45′N 14°01′E
138 I5 **Ḩimş** var. Homs; anc. Emesa. Ḩimş, C Syria 34°44′N 36°43′E
138 K6 **Ḩimş** off. Muḩāfazat Hims, var. Homs. ◈ governorate C Syria
138 I5 **Ḩimş, Buḩayrat** var. Buhayrat Qattinah. ◎ W Syria
171 R7 **Hinatuan** Mindanao, S Philippines 08°21′N 126°19′E
117 N10 **Hînceşti** var. Hânceşti; prev. Kotovsk. C Moldova 46°48′N 28°33′E
44 M9 **Hinche** C Haiti 19°07′N 72°00′W
181 X5 **Hinchinbrook Island** island Queensland, NE Australia
39 S12 **Hinchinbrook Island** island Alaska, USA
97 M19 **Hinckley** C England, United Kingdom 52°33′N 01°21′W
29 V7 **Hinckley** Minnesota, N USA 46°01′N 92°57′W
36 K5 **Hinckley** Utah, W USA 39°21′N 112°39′W
18 J9 **Hinckley Reservoir** ⊠ New York, NE USA
152 I12 **Hindaun** Rājasthān, N India 26°44′N 77°02′E
Hindenburg/Hindenburg in Oberschlesien see Zabrze
Hindiya see Al Hindīyah
21 O6 **Hindman** Kentucky, S USA 37°20′N 82°58′W
182 L10 **Hindmarsh, Lake** ◎ Victoria, SE Australia
185 G19 **Hinds** Canterbury, South Island, New Zealand 44°01′S 171°33′E
185 G19 **Hinds** ≈ South Island, New Zealand
95 H23 **Hindsholm** island C Denmark
149 S4 **Hindu Kush** Per. Hendū Kosh. ▲▲ Afghanistan/Pakistan
155 H19 **Hindupur** Andhra Pradesh, E India 13°46′N 77°33′E
11 O12 **Hines Creek** Alberta, W Canada 56°14′N 118°36′W
23 W6 **Hinesville** Georgia, SE USA 31°51′N 81°36′W
154 I12 **Hinganghāt** Mahārāshtra, C India 20°32′N 78°52′E
149 N15 **Hingol** ≈ SW Pakistan
154 H13 **Hingoli** Mahārāshtra, C India 19°45′N 77°08′E
137 R13 **Hınıs** Erzurum, E Turkey 39°22′N 41°44′E
92 O2 **Hinlopenstretet** strait N Svalbard
92 G10 **Hinnøya** Lapp. Iinnasuolu. island C Norway
108 H10 **Hinterrhein** ≈ SW Switzerland
166 L8 **Hinthada** var. Henzada. Ayeyawady, SW Myanmar (Burma) 17°36′N 95°26′E
11 O14 **Hinton** Alberta, SW Canada 53°24′N 117°35′W
26 M10 **Hinton** Oklahoma, C USA 35°28′N 98°21′W
21 R6 **Hinton** West Virginia, NE USA 37°42′N 80°54′W
Hios see Chíos
41 N8 **Hipólito** Coahuila, NE Mexico 25°42′N 101°22′W
Hipponium see Vibo Valentia
164 B13 **Hirado** Nagasaki, Hirado-shima, SW Japan 33°22′N 129°31′E
164 B13 **Hirado-shima** island SW Japan
165 P16 **Hirakubo-saki** headland Ishigaki-jima, SW Japan 24°36′N 124°19′E
154 M11 **Hīrākud Reservoir** ⊠ E India
Hir al Gharbi, Qasr al see Ḩayr al Gharbī, Qaşr al
165 Q16 **Hirara** Okinawa, Miyako-jima, SW Japan 24°48′N 125°17′E
Hir Ash Sharqī, Qasr al see Ḩayr ash Sharqī, Qaşr al
164 G12 **Hirata** Shimane, Honshū, SW Japan 35°25′N 132°45′E
136 I13 **Hirfanlı Barajı** ⊠ C Turkey
155 G18 **Hiriyūr** Karnātaka, W India 13°58′N 76°33′E
Hîrlău see Hârlău
148 K10 **Hīrmand, Rūd-e** var. Daryā-ye Helmand. ≈ Afghanistan/Iran see also Helmand, Daryā-ye
Hīrmand, Rūd-e see Helmand, Daryā-ye
Hirmil see Hermel
165 T5 **Hiroo** Hokkaidō, NE Japan 42°16′N 143°16′E
165 Q7 **Hirosaki** Aomori, Honshū, C Japan 40°34′N 140°28′E
164 G13 **Hiroshima** var. Hirosima. Hiroshima, Honshū, SW Japan 34°23′N 132°26′E
164 F13 **Hiroshima** off. Hiroshima-ken, var. Hirosima. ◈ prefecture Honshū, SW Japan
Hiroshima-ken see Hiroshima
Hirosima see Hiroshima
Hirschberg/Hirschberg im Riesengebirge/Hirschberg in Schlesien see Jelenia Góra
103 Q3 **Hirson** Aisne, N France 49°56′N 04°05′E
Hîrşova see Hârşova
95 G19 **Hirtshals** Nordjylland, N Denmark 57°34′N 09°58′E
152 H10 **Hisār** Haryāna, NW India 29°10′N 75°45′E
114 I10 **Hisarya** var. Khisarya. Plovdiv, C Bulgaria 42°33′N 24°43′E
162 K7 **Hishig Öndör** var. Maanĭt. Bulgan, C Mongolia 48°17′N 103°29′E
186 E9 **Hisiu** Central, SW Papua New Guinea 09°20′S 146°48′E
147 P13 **Hisor** Rus. Gissar. W Tajikistan 38°34′N 68°29′E
Hispalis see Sevilla
Hispana/Hispania see Spain
44 M7 **Hispaniola** island Dominion Republic/Haiti
64 F11 **Hispaniola Basin** var. Hispaniola Trough. undersea feature SW Atlantic Ocean
Hispaniola Trough see Hispaniola Basin
Histonium see Vasto
139 R7 **Hīt** Al Anbār, SW Iraq 33°38′N 42°50′E
165 P14 **Hita** Ōita, Kyūshū, SW Japan 33°19′N 130°55′E
165 P12 **Hitachi** var. Hitati. Ibaraki, Honshū, S Japan 36°40′N 140°42′E
165 P12 **Hitachiōta** Ibaraki, Honshū, S Japan 36°32′N 140°31′E
Hitati see Hitachi
97 O21 **Hitchin** E England, United Kingdom 51°57′N 00°17′W
191 Q7 **Hitiaa** Tahiti, W French Polynesia 17°35′S 149°17′W
164 D15 **Hitoyoshi** var. Hitoyosi. Kumamoto, Kyūshū, SW Japan 32°12′N 130°48′E
Hitoyosi see Hitoyoshi
94 F7 **Hitra** prev. Hitteren. island S Norway
Hitteren see Hitra
187 Q11 **Hiu** island Torres Islands, N Vanuatu
165 O11 **Hiuchiga-take** ▲ Honshū, C Japan 36°57′N 139°18′E
191 X7 **Hiva Oa** island Îles Marquises, N French Polynesia
20 M10 **Hiwassee Lake** ⊠ North Carolina, SE USA
20 M10 **Hiwassee River** ≈ SE USA
95 H20 **Hjallerup** Nordjylland, N Denmark 57°10′N 10°10′E
95 M16 **Hjälmaren** Eng. Lake Hjalmar. ◎ C Sweden
Hjalmar, Lake see Hjälmaren
95 C14 **Hjellestad** Hordaland, S Norway 60°15′N 05°13′E
95 D16 **Hjelmeland** Rogaland, S Norway 59°12′N 06°07′E
94 G10 **Hjerkinn** Oppland, S Norway 62°13′N 09°37′E
95 L18 **Hjo** Västra Götaland, S Sweden 58°18′N 14°17′E
95 G19 **Hjørring** Nordjylland, N Denmark 57°28′N 09°59′E
167 O1 **Hkakabo Razi** ▲ Myanmar (Burma)/China 28°17′N 97°28′E
166 M2 **Hkamti** var. Singkaling Hkamti. Sagaing, N Myanmar (Burma) 26°00′N 95°43′E
167 N1 **Hkring Bum** ▲ N Myanmar (Burma) 27°05′N 97°16′E
83 L21 **Hlathikulu** var. Hlatikulu. S Swaziland 26°58′S 31°19′E
Hlatikulu see Hlathikulu
Hliboka see Hlyboka
111 F17 **Hlinsko** var. Hlinsko v Čechách. Pardubický Kraj, C Czech Republic 49°46′N 15°54′E
Hlinsko v Čechách see Hlinsko
117 S6 **Hlobyne** Rus. Globino. Poltavs'ka Oblast', NE Ukraine 49°24′N 33°16′E
111 H20 **Hlohovec** Ger. Freistadtl, Hung. Galgóc; prev. Frakštát. Trnavský Kraj, W Slovakia 48°26′N 17°49′E
83 J23 **Hlotse** var. Leribe. NW Lesotho 28°55′S 28°01′E
111 I17 **Hlučín** Ger. Hultschin, Pol. Hulczyn. Moravskoslezský Kraj, E Czech Republic 49°54′N 18°11′E
117 S2 **Hlukhiv** Rus. Glukhov. Sums'ka Oblast', NE Ukraine 51°40′N 33°53′E
119 K21 **Hlushkavichy** Rus. Glushkevichi. Homyel'skaya Voblasts', SE Belarus 51°34′N 27°47′E
119 L18 **Hlusk** Rus. Glusk, Glussk. Mahilyowskaya Voblasts', E Belarus 52°54′N 28°41′E
116 K8 **Hlyboka** Ger. Hliboka, Rus. Glybokaya. Chernivets'ka Oblast', W Ukraine 48°04′N 25°56′E
118 K13 **Hlybokaye** Rus. Glubokoye. Vitsyebskaya Voblasts', N Belarus 55°08′N 27°41′E
77 Q16 **Ho** SE Ghana 06°36′N 00°28′E
167 S6 **Hoa Binh** Hoa Binh, N Vietnam 20°49′N 105°20′E
83 E20 **Hoachanas** Hardap, C Namibia 23°55′S 18°04′E
Hoai Nhơn see Bông Sơn
167 T8 **Hoa Lac** Quang Binh, C Vietnam 17°54′N 106°24′E
167 S5 **Hoang Liên Sơn** ▲▲ N Vietnam
Hoang Sa, Quân Đao see Paracel Islands
83 B17 **Hoanib** ≈ NW Namibia
33 S15 **Hoback Peak** ▲ Wyoming, C USA 43°04′N 110°34′W
183 P17 **Hobart** prev. Hobarton, Hobart Town. state capital Tasmania, SE Australia 42°54′S 147°18′E
26 L11 **Hobart** Oklahoma, C USA 35°03′N 99°04′W
183 P17 **Hobart** ✈ Tasmania, SE Australia 42°52′S 147°28′E
Hobarton/Hobart Town see Hobart
37 W14 **Hobbs** New Mexico, SW USA 32°42′N 103°08′W
194 L12 **Hobbs Coast** physical region Antarctica
23 Z14 **Hobe Sound** Florida, SE USA 27°03′N 80°08′W
Hobicaurikány see Uricani
54 E12 **Hobo** Huila, S Colombia 02°34′N 75°28′W
99 G16 **Hoboken** Antwerpen, N Belgium 51°12′N 04°22′E
158 K3 **Hoboksar** var. Hoboksar Mongol Zizhixian. Xinjiang Uygur Zizhiqu, NW China 46°48′N 85°42′E
Hoboksar Mongol Zizhixian see Hoboksar
95 G21 **Hobro** Nordjylland, N Denmark 56°39′N 09°51′E
21 X10 **Hobucken** North Carolina, SE USA 35°15′N 76°31′W
95 O20 **Hoburgen** headland SE Sweden 56°54′N 18°07′E
81 P15 **Hobyo** It. Obbia. Mudug, E Somalia 05°16′N 48°24′E
109 R8 **Hochalmspitze** ▲ SW Austria 47°00′N 13°19′E
109 Q4 **Hochburg** Oberösterreich, N Austria 48°10′N 12°57′E
108 F8 **Hochdorf** Luzern, N Switzerland 47°10′N 08°16′E
109 N8 **Hochfeiler** It. Gran Pilastro. ▲ Austria/Italy 46°59′N 11°42′E
167 T14 **Hô Chi Minh** var. Ho Chi Minh City; prev. Saigon. S Vietnam 10°46′N 106°43′E
Ho Chi Minh City see Hô Chi Minh
108 I7 **Höchst** Vorarlberg, NW Austria 47°28′N 09°40′E
Höchstadt see Höchstadt an der Aisch
101 K19 **Höchstadt an der Aisch** var. Höchstadt. Bayern, C Germany 49°43′N 10°48′E
108 L9 **Hochwilde** It. L'Altissima. ▲ Austria/Italy 46°45′N 11°00′E
109 S7 **Hochwildstelle** ▲ C Austria 47°21′N 13°53′E
31 T14 **Hocking River** ≈ Ohio, N USA
Hoctúm see Hoctún
41 X12 **Hoctún** var. Hoctúm. Yucatán, E Mexico 20°48′N 89°14′W
Hodeida see Al Ḩudaydah
20 K6 **Hodgenville** Kentucky, S USA 37°34′N 85°45′W
11 T17 **Hodgeville** Saskatchewan, S Canada 50°06′N 106°55′W
76 L9 **Hodh ech Chargui** ◈ region E Mauritania
Hodh el Garbi see Hodh el Gharbi
76 J10 **Hodh el Gharbi** var. Hodh el Garbi. ◈ region S Mauritania
111 L25 **Hódmezővásárhely** Csongrád, SE Hungary 46°27′N 20°18′E
74 J6 **Hodna, Chott El** var. Chott el-Hodna, Ar. Shatt al-Hodna. salt lake N Algeria
Hodna, Chott el-/Hodna, Shatt al- see Hodna, Chott El
111 G19 **Hodonín** Ger. Göding. Jihomoravský Kraj, SE Czech Republic 48°52′N 17°07′E
Hödrögö see Nömrög
Hodság/Hodschag see Odžaci
39 R7 **Hodzana River** ≈ Alaska, USA
Hoei see Huy
99 H19 **Hoeilaart** Vlaams Brabant, C Belgium 50°46′N 04°28′E
Hoë Karoo see Great Karoo
98 F12 **Hoek van Holland** Eng. Hook of Holland. Zuid-Holland, W Netherlands 52°00′N 04°07′E
98 L11 **Hoenderloo** Gelderland, E Netherlands 52°08′N 05°46′E
99 L18 **Hoensbroek** Limburg, SE Netherlands 50°55′N 05°55′E
163 Y11 **Hoeryŏng** NE North Korea 42°23′N 129°46′E
99 K18 **Hoeselt** Limburg, NE Belgium 50°50′N 05°30′E
98 K11 **Hoevelaken** Gelderland, C Netherlands 52°10′N 05°27′E
Hoey see Huy
101 M18 **Hof** Bayern, SE Germany 50°19′N 11°55′E
Höfdhakaupstadhur see Skagaströnd
Hofei see Hefei
101 G18 **Hofheim am Taunus** Hessen, W Germany 50°04′N 08°27′E
Hofmarkt see Odorheiu Secuiesc
92 L3 **Höfn** Austurland, SE Iceland 64°14′N 15°17′W
94 N13 **Hofors** Gävleborg, C Sweden 60°33′N 16°21′E
92 J6 **Hofsjökull** glacier C Iceland
92 J1 **Hofsós** Norðurland Vestra, N Iceland 65°54′N 19°25′W
164 E13 **Hōfu** Yamaguchi, Honshū, SW Japan 34°01′N 131°34′E
92 I4 **Höfuðborgarsvæðið** SW Iceland
Hofuf see Al Hufūf
95 J22 **Höganäs** Skåne, S Sweden 56°11′N 12°39′E
183 P14 **Hogan Group** island group Tasmania, SE Australia
23 R4 **Hogansville** Georgia, SE USA 33°10′N 84°55′W
39 P8 **Hogatza River** ≈ Alaska, USA
28 I14 **Hogback Mountain** ▲ Nebraska, C USA 41°40′N 103°44′W
95 G14 **Høgevarde** ▲ S Norway 60°19′N 09°27′E
Högfors see Karkkila
31 P5 **Hog Island** island Michigan, N USA
21 Y6 **Hog Island** island Virginia, NE USA
Hogoley Islands see Chuuk Islands
95 N20 **Högsby** Kalmar, S Sweden 57°10′N 16°03′E
36 K1 **Hogup Mountains** ▲▲ Utah, W USA
101 E17 **Hohe Acht** ▲ W Germany 50°23′N 07°00′E
Hohenelbe see Vrchlabí
108 I7 **Hohenems** Vorarlberg, W Austria 47°23′N 09°43′E
Hohenmauth see Vysoké Mýto
Hohensalza see Inowrocław
Hohenstadt see Zábřeh
Hohenstein in Ostpreussen see Olsztynek
20 I9 **Hohenwald** Tennessee, S USA 35°33′N 87°31′W
101 L17 **Hohenwarte-Stausee** ⊠ C Germany
Hohes Venn see Hautes Fagnes
109 Q8 **Hohe Tauern** ▲▲ W Austria
163 O13 **Hohhot** var. Huhehot, Huhohaote, Mong. Kukukhoto; prev. Kweisui, Kwesui. Nei Mongol Zizhiqu, N China 40°49′N 111°37′E
162 F7 **Hohmorĭt** var. Sayn-Ust. Govĭ-Altay, W Mongolia 47°23′N 94°19′E
103 U6 **Hohneck** ▲ NE France 48°04′N 07°01′E
77 Q16 **Hohoe** E Ghana 07°08′N 00°32′E
164 E12 **Hōhoku** Yamaguchi, Honshū, SW Japan 34°15′N 130°56′E
159 O11 **Hoh Sai Hu** ◎ C China
159 N11 **Hoh Xil Hu** ◎ C China
158 L11 **Hoh Xil Shan** ▲▲ W China
167 U10 **Hôi An** prev. Faifo. Quang Nam-Đa Năng, C Vietnam 15°54′N 108°19′E
Hoï-Hao/Hoihow see Haikou
159 S11 **Hoika** prev. Heka. Qinghai, W China 35°49′N 99°50′E
81 F17 **Hoima** W Uganda 01°25′N 31°22′E
26 L5 **Hoisington** Kansas, C USA 38°31′N 98°46′W
146 D12 **Hojagala** Rus. Khodzhakala. Balkan Welaýaty, W Turkmenistan 38°46′N 56°14′E
146 M13 **Hojambaz** Rus. Khodzhambas. Lebap Welaýaty, E Turkmenistan 38°11′N 64°33′E
95 H23 **Højby** Syddtjylland, C Denmark 55°20′N 10°27′E
95 F24 **Højer** Syddanmark, SW Denmark 54°57′N 08°43′E
164 E14 **Hōjō** var. Hôzyô. Ehime, Shikoku, SW Japan 33°58′N 132°47′E
184 J3 **Hokianga Harbour** inlet SE Tasman Sea
185 F17 **Hokitika** West Coast, South Island, New Zealand 42°44′S 170°58′E
165 U4 **Hokkai-dō** ◈ territory Hokkaidō, NE Japan
165 T3 **Hokkaidō** prev. Ezo, Yeso, Yezo. island NE Japan
95 G15 **Hokksund** Buskerud, S Norway 59°45′N 09°55′E
143 S4 **Hokmābād** Khorāsān-e Razavī, N Iran 36°37′N 57°34′E
Hokō see Pohang
Hoko-guntō/Hoko-shotō see Penghu Liedao
Hoktemberyan see Armavir
94 F13 **Hol** Buskerud, S Norway 60°36′N 08°18′E
81 K19 **Hola** Tana River, SE Kenya 01°06′S 40°01′E
117 R11 **Hola Prystan'** Rus. Golaya Pristan. Khersons'ka Oblast', S Ukraine 46°31′N 32°31′E
95 I23 **Holbæk** Sjælland, E Denmark 55°42′N 11°42′E
Holboo see Santmargats
183 P10 **Holbrook** New South Wales, SE Australia 35°45′S 147°18′E
37 N11 **Holbrook** Arizona, SW USA 34°54′N 110°09′W
27 S5 **Holden** Missouri, C USA 38°42′N 93°59′W
36 K5 **Holden** Utah, W USA 39°06′N 112°16′W
27 O11 **Holdenville** Oklahoma, C USA 35°05′N 96°25′W
29 O16 **Holdrege** Nebraska, C USA 40°28′N 99°28′W
35 X3 **Hole in the Mountain Peak** ▲ Nevada, W USA 40°54′N 115°06′W
155 G20 **Hole Narsipur** Karnātaka, W India 12°46′N 76°14′E
111 H18 **Holešov** Ger. Holleschau. Zlínský Kraj, E Czech Republic 49°20′N 17°35′E
45 N14 **Holetown** prev. Jamestown. W Barbados 13°11′N 59°38′W
31 Q12 **Holgate** Ohio, N USA 41°12′N 84°06′W
44 I7 **Holguín** Holguín, SE Cuba 20°51′N 76°16′W
23 V12 **Holiday** Florida, SE USA 28°11′N 82°44′W
39 O12 **Holitna River** ≈ Alaska, USA
94 J13 **Höljes** Värmland, C Sweden 60°54′N 12°34′E
109 X3 **Hollabrunn** Niederösterreich, NE Austria 48°33′N 16°06′E
36 L3 **Holladay** Utah, W USA 40°39′N 111°49′W
11 X16 **Holland** Manitoba, S Canada 49°36′N 98°52′W
31 O9 **Holland** Michigan, N USA 42°47′N 86°06′W
25 T9 **Holland** Texas, SW USA 30°52′N 97°24′W
Holland see Netherlands
22 K4 **Hollandale** Mississippi, S USA 33°10′N 90°51′W
Hollandia see Jayapura
Hollandsch Diep see Hollands Diep
99 H14 **Hollands Diep** var. Hollandsch Diep. channel SW Netherlands
Holleschau see Holešov
25 R5 **Holliday** Texas, SW USA 33°49′N 98°41′W
18 E15 **Hollidaysburg** Pennsylvania, NE USA 40°24′N 78°22′W
21 S6 **Hollins** Virginia, NE USA 37°20′N 79°56′W
26 J12 **Hollis** Oklahoma, C USA 34°42′N 99°56′W
35 O10 **Hollister** California, W USA 36°51′N 121°25′W
27 T8 **Hollister** Missouri, C USA 36°37′N 93°13′W
93 M19 **Hollola** Päijät-Häme, S Finland 61°N 25°32′E
98 K4 **Hollum** Fryslân, N Netherlands 53°27′N 05°38′E
95 J23 **Höllviken** prev. Höllviksnäs. Skåne, S Sweden 55°25′N 12°57′E
Höllviksnäs see Höllviken
37 W6 **Holly** Colorado, C USA 38°03′N 102°07′W
31 R9 **Holly** Michigan, N USA 42°47′N 83°37′W
21 S14 **Holly Hill** South Carolina, SE USA 33°19′N 80°24′W
21 W11 **Holly Ridge** North Carolina, SE USA 34°31′N 77°31′W
22 L1 **Holly Springs** Mississippi, S USA 34°47′N 89°25′W
23 Z15 **Hollywood** Florida, SE USA 26°00′N 80°09′W
8 J6 **Holman** Victoria Island, Northwest Territories, N Canada 70°42′N 117°45′W
92 I2 **Hólmavík** Vestfirðir, NW Iceland 65°42′N 21°43′W
30 J7 **Holmen** Wisconsin, N USA 43°57′N 91°14′W
23 R8 **Holmes Creek** ≈ Alabama/Florida, SE USA
95 H16 **Holmestrand** Vestfold, S Norway 59°29′N 10°20′E
93 J16 **Holmön** island N Sweden
95 E22 **Holmsland Klit** beach W Denmark
93 J16 **Holmsund** Västerbotten, N Sweden 63°42′N 20°26′E
95 Q18 **Holmudden** headland SE Sweden 57°59′N 19°14′E
138 F10 **Holon** var. Kholon; prev. Ḥolon. Tel Aviv, C Israel 32°01′N 34°46′E
Ḥolon see Holon
163 P7 **Hölönbuyr** var. Bayan. Dornod, E Mongolia 47°56′N 112°58′E
117 P8 **Holovanivs'k** Rus. Golovanevsk. Kirovohrads'ka Oblast', C Ukraine 48°21′N 30°26′E
95 F21 **Holstebro** Midtjylland, W Denmark 56°22′N 08°38′E
95 F23 **Holsted** Syddtjylland, W Denmark 55°30′N 08°54′E
29 T13 **Holstein** Iowa, C USA 42°29′N 95°32′W
Holsteinborg/Holsteinsborg/Holstenborg/Holstensborg see Sisimiut
21 O8 **Holston River** ≈ Tennessee, S USA
31 Q9 **Holt** Michigan, N USA 42°38′N 84°31′W
98 N10 **Holten** Overijssel, E Netherlands 52°16′N 06°25′E
27 P3 **Holton** Kansas, C USA 39°27′N 95°44′W
27 U5 **Holts Summit** Missouri, C USA 38°38′N 92°07′W
35 X17 **Holtville** California, W USA 32°48′N 115°22′W
98 L5 **Holwerd** Fris. Holwert. Fryslân, N Netherlands 53°22′N 05°51′E
Holwert see Holwerd
39 O11 **Holy Cross** Alaska, USA 62°12′N 159°46′W
37 R4 **Holy Cross, Mount Of The** ▲ Colorado, C USA 39°28′N 106°28′W
97 I18 **Holyhead** Wel. Caer Gybi. NW Wales, United Kingdom 53°19′N 04°38′W
97 H18 **Holy Island** island NW Wales, United Kingdom
96 L12 **Holy Island** island NE England, United Kingdom
37 W3 **Holyoke** Colorado, C USA 40°31′N 102°18′W
18 M11 **Holyoke** Massachusetts, NE USA 42°12′N 72°37′W
101 I14 **Holzminden** Niedersachsen, C Germany 51°49′N 09°27′E
81 G19 **Homa Bay** Homa Bay, W Kenya 0°31′S 34°30′E
81 G19 **Homa Bay** ◈ county W Kenya
Homāyūnshahr see Khomeynīshahr
77 P11 **Hombori** Mopti, S Mali 15°13′N 01°39′W
101 E20 **Homburg** Saarland, SW Germany 49°20′N 07°20′E
9 R5 **Home Bay** bay Baffin Bay, Nunavut, NE Canada
Homenau see Humenné
39 Q13 **Homer** Alaska, USA 59°38′N 151°33′W
22 H4 **Homer** Louisiana, S USA 32°47′N 93°03′W
18 H10 **Homer** New York, NE USA 42°38′N 76°10′W
23 V7 **Homerville** Georgia, SE USA 31°02′N 82°45′W
23 Y16 **Homestead** Florida, SE USA 25°28′N 80°28′W
27 O9 **Hominy** Oklahoma, C USA 36°24′N 96°24′W
94 H8 **Hommelvik** Sør-Trøndelag, S Norway 63°24′N 10°48′E
95 C16 **Hommersåk** Rogaland, S Norway 58°55′N 05°51′E
155 H15 **Homnābād** Karnātaka, C India 17°46′N 77°08′E
22 J7 **Homochitto River** ≈ Mississippi, S USA
83 N20 **Homoíne** Inhambane, SE Mozambique 23°51′S 35°04′E
112 O12 **Homoljske Planine** ▲▲ E Serbia
Homonna see Humenné
Homs see Al Khums, Libya
Homs see Ḩimş
119 P19 **Homyel'** Rus. Gomel'. Homyel'skaya Voblasts', SE Belarus 52°25′N 31°E
118 L12 **Homyel'** Vitsyebskaya Voblasts', N Belarus 55°20′N 28°52′E
119 L19 **Homyel'skaya Voblasts'** Rus. Gomel'skaya Oblast'. ◈ province SE Belarus
Honan see Henan, China
Honan see Luoyang, China
164 U4 **Honbetsu** Hokkaidō, NE Japan 43°09′N 143°46′E
Honctő see Gurahonţ
54 E9 **Honda** Tolima, C Colombia 05°12′N 74°45′W
83 D24 **Hondeklip** Afr. Hondeklipbaai. Northern Cape, W South Africa 30°15′S 17°17′E
Hondeklipbaai see Hondeklip
11 Q13 **Hondo** Alberta, W Canada 54°43′N 113°14′W
25 Q12 **Hondo** Texas, SW USA 29°21′N 99°09′W
42 G1 **Hondo** ≈ Central America
Hondo see Honshū
Hondo see Amakusa
42 G6 **Honduras** off. Republic of Honduras. ◆ republic Central America
Honduras, Golfo de see Honduras, Gulf of
42 H4 **Honduras, Gulf of** Sp. Golfo de Honduras. gulf W Caribbean Sea
Honduras, Republic of see Honduras
11 V12 **Hone** Manitoba, C Canada 56°13′N 101°12′W
21 P12 **Honea Path** South Carolina, SE USA 34°27′N 82°23′W
95 H14 **Hønefoss** Buskerud, S Norway 60°10′N 10°15′E
31 S12 **Honey Creek** ≈ Ohio, N USA
25 V5 **Honey Grove** Texas, SW USA 33°34′N 95°54′W
35 Q4 **Honey Lake** ◎ California, W USA
102 L4 **Honfleur** Calvados, N France 49°25′N 00°14′E
Hon Gai see Ha Long
161 O8 **Hong'an** prev. Huang'an. Hubei, C China 31°20′N 114°43′E
Hongay see Ha Long
Hông Gai see Ha Long
161 O15 **Honghai Wan** bay N South China Sea
Hông Hà, Sông see Red River
161 O7 **Hong He** ≈ C China
161 N9 **Hong Hu** ◎ C China
160 L11 **Hongjiang** Hunan, S China 27°09′N 109°58′E
Hongjiang see Wangcang
161 O15 **Hong Kong** off. Hong Kong Special Administrative Region, var. Hong Kong S.A.R., Chin. Xianggang Tebie Xingzhengqu. S China 22°17′N 114°09′E
Hong Kong S.A.R. see Hong Kong
Hong Kong Special Administrative Region see Hong Kong
160 L4 **Hongliu He** ≈ C China
160 L4 **Hongliu He** ≈ C China
159 P8 **Hongliuwan** var. Aksay, Aksay Kazakzu Zizhixian. Gansu, N China 39°25′N 94°09′E
159 P7 **Hongliuyuan** Gansu, N China 41°02′N 95°24′E
Hongor see Delgereh
161 S8 **Hongqiao** ✈ (Shanghai) Shanghai Shi, E China 31°28′N 121°08′E
160 K14 **Hongshui He** ≈ S China
160 M5 **Hongtong** var. Dahuaishu. Shanxi, C China 36°30′N 111°42′E
164 J15 **Hongū** Wakayama, Honshū, SW Japan 33°50′N 135°42′E
Honguedo, Détroit d' see Honguedo Passage
15 Y5 **Honguedo Passage** var. Honguedo Strait, Fr. Détroit d'Honguedo. strait Québec, E Canada
Honguedo Strait see Honguedo Passage
Hongwan see Hongwansi
159 S8 **Hongwansi** var. Sunan, Sunan Yugurzu Zizhixian; prev. Hongwan. Gansu, N China 38°55′N 99°29′E

◆ Country
● Country Capital
◇ Dependent Territory
○ Dependent Territory Capital
◈ Administrative Regions
✈ International Airport
▲ Mountain
▲▲ Mountain Range
Volcano
≈ River
◎ Lake
⊠ Reservoir

HONGWŎN – HUIXTLA

163 X13 Hongwŏn E North Korea 40°03′N 127°54′E
160 H7 Hongyuan var. Qiongxi; prev. Hurama. Sichuan, C China 32°49′N 102°40′E
161 Q7 Hongze Hu var. Hung-tse Hu. ◎ E China
186 L9 Honiara ● (Solomon Islands) Guadalcanal, C Solomon Islands 09°27′S 159°56′E
165 P8 Honjō var. Honzyô, Yurihonjô. Akita, Honshû, C Japan 39°23′N 140°03′E
93 K18 Honkajoki Satakunta, SW Finland 62°00′N 22°15′E
92 K7 Honningsvåg Finnmark, N Norway 70°58′N 25°59′E
95 I19 Hönö Västra Götaland, S Sweden 57°42′N 11°39′E
38 G11 Honoka'a Hawaii, USA, C Pacific Ocean 20°04′N 155°27′W
38 G11 Honoka'a var. Honokaa. Hawaii, USA, C Pacific Ocean 20°04′N 155°27′W
Honokaa see Honoka'a
38 D9 Honolulu state capital O'ahu, Hawaii, USA, C Pacific Ocean 21°18′N 157°52′W
38 H11 Honomú var. Honomu. Hawaii, USA, C Pacific Ocean 19°51′N 155°06′W
105 P10 Honrubia Castilla-La Mancha, C Spain 39°36′N 02°17′W
164 M12 Honshū var. Hondo, Honsyû. island SW Japan
Honsyû see Honshū
Honte see Westerschelde
Honzyô see Honjō
8 K8 Hood ≈ Nunavut, NW Canada
Hood Island see Española, Isla
32 H11 Hood, Mount ▲ Oregon, NW USA 45°22′N 121°41′W
32 H11 Hood River Oregon, NW USA 45°44′N 121°31′W
98 H10 Hoofddorp Noord-Holland, W Netherlands 52°18′N 04°41′E
99 G15 Hoogerheide Noord-Brabant, S Netherlands 51°25′N 04°19′E
98 N8 Hoogeveen Drenthe, NE Netherlands 52°44′N 06°30′E
98 O6 Hoogezand-Sappemeer Groningen, NE Netherlands 53°10′N 06°47′E
98 J8 Hoogkarspel Noord-Holland, NW Netherlands 52°42′N 04°59′E
98 N5 Hoogkerk Groningen, NE Netherlands 53°13′N 06°30′E
98 G13 Hoogvliet Zuid-Holland, SW Netherlands 51°51′N 04°23′E
26 I8 Hooker Oklahoma, C USA 36°51′N 101°12′W
97 E21 Hook Head Ir. Rinn Duáin. headland SE Ireland 52°07′N 06°55′W
Hook of Holland see Hoek van Holland
Hoolt see Tögrög
39 W13 Hoonah Chichagof Island, Alaska, USA 58°05′N 135°21′W
38 L11 Hooper Bay Alaska, USA 61°31′N 166°06′W
31 N13 Hoopeston Illinois, N USA 40°28′N 87°40′W
95 K22 Höör Skåne, S Sweden 55°55′N 13°33′E
98 I9 Hoorn Noord-Holland, NW Netherlands 52°38′N 05°04′E
18 L10 Hoosic River ≈ New York, NE USA
Hoosier State see Indiana
35 Y11 Hoover Dam dam Arizona/Nevada, W USA
Höövör see Baruunbayan-Ulaan
137 Q11 Hopa Artvin, NE Turkey 41°23′N 41°28′E
18 J14 Hopatcong New Jersey, NE USA 40°55′N 74°39′W
10 M17 Hope British Columbia, SW Canada 49°21′N 121°28′W
39 R12 Hope Alaska, USA 60°55′N 149°38′W
27 T14 Hope Arkansas, C USA 33°40′N 93°36′W
31 P14 Hope Indiana, N USA 39°18′N 85°46′W
29 Q5 Hope North Dakota, N USA 47°18′N 97°42′W
13 Q7 Hopedale Newfoundland and Labrador, NE Canada 55°26′N 60°14′W
Hopeh/Hopei see Hebei
180 K13 Hope, Lake salt lake Western Australia
41 X13 Hopelchén Campeche, SE Mexico 19°46′N 89°50′W
21 U11 Hope Mills North Carolina, SE USA 34°58′N 78°57′W
183 O7 Hope, Mount New South Wales, SE Australia 32°49′S 145°55′E
92 P4 Hopen island SE Svalbard
197 Q4 Hope, Point headland Alaska, USA
12 M3 Hopes Advance, Cap cape Québec, NE Canada
182 L10 Hopetoun Victoria, SE Australia 35°46′S 142°23′E
83 H23 Hopetown Northern Cape, W South Africa 29°37′S 24°05′E
21 W6 Hopewell Virginia, NE USA 37°16′N 77°15′W
109 O7 Hopfgarten im Brixental Tirol, W Austria 47°28′N 12°14′E
181 N8 Hopkins Lake salt lake Western Australia
182 M12 Hopkins River ≈ Victoria, SE Australia
20 I7 Hopkinsville Kentucky, S USA 36°50′N 87°30′W
34 M6 Hopland California, W USA 38°58′N 123°09′W
95 G24 Hoptrup Syddanmark, SW Denmark 55°09′N 09°27′E
Hoqin Zuoyi Zhongji see Baokang
32 F9 Hoquiam Washington, NW USA 46°58′N 123°53′W
29 R6 Horace North Dakota, N USA 46°44′N 96°54′W
117 T14 Hora Roman-Kosh ▲ S Ukraine 44°37′N 34°13′E
137 R12 Horasan Erzurum, NE Turkey 40°03′N 42°10′E
101 G22 Horb am Neckar Baden-Württemberg, S Germany 48°27′N 08°42′E
95 K23 Hörby Skåne, S Sweden 55°51′N 13°42′E
43 P16 Horconcitos Chiriquí, W Panama 08°20′N 82°10′W
95 C14 Hordaland ❖ county S Norway
116 H13 Horezu Vâlcea, SW Romania 45°06′N 24°00′E
108 G7 Horgen Zürich, N Switzerland 47°16′N 08°36′E
Horgo see Tariat
Hörin see Fenglin
163 O13 Horinger Nei Mongol Zizhiqu, N China 40°23′N 111°48′E
Horiult see Bogd
11 U17 Horizon Saskatchewan, S Canada 49°33′N 105°05′W
192 K9 Horizon Bank undersea feature S Pacific Ocean
192 L10 Horizon Deep undersea feature W Pacific Ocean
95 L14 Hörken Örebro, S Sweden 60°03′N 14°55′E
119 O15 Horki Rus. Gorki. Mahilyowskaya Voblasts', E Belarus 54°18′N 31°E
195 O10 Horlick Mountains ▲▲ Antarctica
117 X7 Horlivka Rom. Adâncata, Rus. Gorlovka. Donets'ka Oblast', E Ukraine 48°19′N 38°04′E
143 V11 Ḩormak Sīstān va Balūchestān, SE Iran 30°00′N 60°50′E
143 R13 Hormozgān off. Ostān-e Hormozgān. ❖ province S Iran
Hormozgān, Ostān-e see Hormozgān
Hormoz, Tangeh-ye see Hormuz, Strait of
141 W6 Hormuz, Strait of var. Strait of Ormuz, Per. Tangeh-ye Hormoz. strait Iran/Oman
109 W2 Horn Niederösterreich, NE Austria 48°40′N 15°40′E
95 M18 Horn Östergötland, S Sweden 57°54′N 15°49′E
8 J9 Horn ≈ Northwest Territories, NW Canada
Hornád see Hernád
8 I6 Hornaday ≈ Northwest Territories, NW Canada
92 H13 Hornavan ◎ N Sweden
65 C24 Hornby Mountains hill range West Falkland, Falkland Islands
Horn, Cape see Hornos, Cabo de
97 O18 Horncastle E England, United Kingdom 53°12′N 00°07′W
95 N14 Horndal Dalarna, C Sweden 60°16′N 16°25′E
93 I16 Hörnefors Västerbotten, N Sweden 63°37′N 19°54′E
18 F11 Hornell New York, NE USA 42°19′N 77°38′W
Horné Nové Mesto see Kysucké Nové Mesto
12 F12 Hornepayne Ontario, S Canada 49°14′N 84°48′W
94 D10 Hornindalsvatnet ◎ S Norway
101 G22 Hornisgrinde ▲ SW Germany 48°37′N 08°13′E
22 M9 Horn Island island Mississippi, S USA
Hornja Łužica see Oberlausitz
63 J26 Hornos, Cabo de Eng. Cape Horn. headland S Chile 55°52′S 67°00′W
117 S10 Hornostayivka Khersons'ka Oblast', S Ukraine 47°00′N 33°42′E
183 T9 Hornsby New South Wales, SE Australia 33°44′S 151°08′E
97 O16 Hornsea E England, United Kingdom 53°54′N 00°10′W
94 O11 Hornslandet peninsula C Sweden
95 H22 Hornslet Midtjylland, C Denmark 56°19′N 10°20′E
92 O4 Hornsundtind ▲ S Svalbard 76°54′N 16°07′E
Horochów see Horokhiv
Horodenka see Gorodenka
117 Q2 Horodnya Rus. Gorodnya. Chernihivs'ka Oblast', NE Ukraine 51°54′N 31°30′E
116 K6 Horodok Khmel'nyts'ka Oblast', W Ukraine 49°10′N 26°34′E
116 H5 Horodok Pol. Gródek Jagielloński, Rus. Gorodok, Gorodok Yagellonski. L'vivs'ka Oblast', NW Ukraine 49°48′N 23°39′E
117 Q6 Horodyshche Rus. Gorodishche. Cherkas'ka Oblast', C Ukraine 49°19′N 31°27′E
165 T3 Horokanai Hokkaidō, NE Japan 44°02′N 142°08′E
116 J4 Horokhiv Pol. Horochów, Rus. Gorokhov. Volyns'ka Oblast', NW Ukraine 50°31′N 24°50′E
165 T4 Horoshiri-dake var. Horosiri Dake. ▲ Hokkaidō, N Japan 42°43′N 142°41′E
Horosiri Dake see Horoshiri-dake
111 C17 Hořovice Ger. Horowitz. Střední Čechy, W Czech Republic 49°49′N 13°53′E
Horowitz see Hořovice
Horqin Zuoyi Houqi see Ganjig
Horqin Zuoyi Zhongqi see Bayan Huxu
62 O5 Horqueta Concepción, C Paraguay 23°24′S 56°53′W
55 O12 Horqueta Minas Amazonas, S Venezuela 02°20′N 63°32′W
95 J20 Horred Västra Götaland, S Sweden 57°22′N 12°25′E
151 J19 Horsburgh Atoll var. Goidhoo Atoll. atoll N Maldives
20 K7 Horse Cave Kentucky, S USA 37°10′N 85°54′W
37 V6 Horse Creek ≈ Colorado, C USA
27 S6 Horse Creek ≈ Missouri, C USA
18 G11 Horseheads New York, NE USA 42°10′N 76°49′W
37 P13 Horse Mount ▲ New Mexico, SW USA 33°58′N 108°10′W
95 G22 Horsens Syddanmark, C Denmark 55°53′N 09°53′E
65 F25 Horse Pasture Point headland W Saint Helena 15°57′S 05°46′W
33 N13 Horseshoe Bend Idaho, NW USA 43°55′N 116°11′W
36 L13 Horseshoe Reservoir ⊠ Arizona, SW USA
64 M9 Horseshoe Seamounts undersea feature E Atlantic Ocean 36°30′N 15°00′W
182 L11 Horsham Victoria, SE Australia 36°44′S 142°13′E
97 O23 Horsham SE England, United Kingdom 51°01′N 00°21′W
99 M15 Horst Limburg, SE Netherlands 51°30′N 06°05′E
64 N2 Horta Faial, Azores, Portugal, NE Atlantic Ocean 38°32′N 28°39′W
105 S12 Horta, Cap de l' Cast. Cabo Huertas. headland SE Spain 38°21′N 00°25′W
95 H16 Horten Vestfold, S Norway 59°25′N 10°25′E
111 M23 Hortobágy-Berettyó ≈ E Hungary
27 Q3 Horton Kansas, C USA 39°39′N 95°31′W
8 I7 Horton ≈ Northwest Territories, NW Canada
95 I23 Hørve Sjælland, E Denmark 55°46′N 11°28′E
95 L22 Hörvik Blekinge, S Sweden 56°01′N 14°45′E
Ḩorvot Ḩaluza see Horvot Halutsa
14 E7 Horwood Lake ◎ Ontario, S Canada
116 K4 Horyn' Rus. Goryn. ≈ NW Ukraine
81 I14 Hosa'ina var. Hosseina, It. Hosanna. Southern Nationalities, S Ethiopia 07°38′N 37°58′E
Hosanna see Hosa'ina
101 H18 Hösbach Bayern, C Germany 50°00′N 09°12′E
Hose Mountains see Hose, Pegunungan
169 T9 Hose, Pegunungan var. Hose Mountains. ▲▲ East Malaysia
148 L15 Hoshāb Balūchistān, SW Pakistan 26°01′N 63°51′E
154 H10 Hoshangābād Madhya Pradesh, C India 22°44′N 77°45′E
116 L4 Hoshcha Rivnens'ka Oblast', NW Ukraine 50°37′N 26°38′E
152 I7 Hoshiārpur Punjab, NW India 31°30′N 75°59′E
Hōshööt see Öldziyt
99 M23 Hosingen Diekirch, NE Luxembourg 50°01′N 06°05′E
186 G7 Hoskins New Britain, E Papua New Guinea 05°28′S 150°25′E
155 G17 Hospet Karnātaka, C India 15°16′N 76°20′E
104 K4 Hospital de Orbigo Castilla y León, N Spain 42°27′N 05°53′W
Hospitalet see L'Hospitalet de Llobregat
92 N13 Hossa Kainuu, E Finland 65°28′N 29°36′E
Hosseina see Hosa'ina
Hosszúmezjő see Câmpulung Moldovenesc
63 I25 Hoste, Isla island S Chile
117 O4 Hostomel' Rus. Gostomel'. Kyyivs'ka Oblast', N Ukraine 50°41′N 30°15′E
155 H20 Hosūr Tamil Nādu, SE India 12°45′N 77°51′E
167 N8 Hot Chiang Mai, NW Thailand 18°07′N 98°34′E
158 G10 Hotan var. Khotan, Chin. Ho-t'ien. Xinjiang Uygur Zizhiqu, NW China 37°10′N 79°51′E
158 H9 Hotan He ≈ NW China
83 G22 Hotazel Northern Cape, N South Africa 27°12′S 22°58′E
37 Q5 Hotchkiss Colorado, C USA 38°47′N 107°43′W
35 V7 Hot Creek Range ▲▲ Nevada, W USA
Hote see Hoti
171 T13 Hoti var. Hote. Pulau Seram, E Indonesia 02°58′S 130°19′E
Ho-t'ien see Hotan
Hotin see Khotyn
93 H15 Hoting Jämtland, C Sweden 64°07′N 16°14′E
162 L14 Hotong Qagan Nur ◎ N China
162 J8 Hotont Arhangay, C Mongolia 47°21′N 102°27′E
27 T12 Hot Springs Arkansas, C USA 34°31′N 93°03′W
28 J11 Hot Springs South Dakota, N USA 43°26′N 103°29′W
21 S5 Hot Springs Virginia, NE USA 38°00′N 79°50′W
35 Q4 Hot Springs Peak ▲ California, W USA 40°23′N 120°06′W
27 T12 Hot Springs Village Arkansas, C USA 34°39′N 93°03′W
Hotspur Bank see Hotspur Seamount
65 J16 Hotspur Seamount var. Hotspur Bank. undersea feature C Atlantic Ocean 18°00′S 35°00′W
8 J8 Hottah Lake ◎ Northwest Territories, NW Canada
44 K9 Hotte, Massif de la ▲▲ SW Haiti
99 K21 Hotton Luxembourg, SE Belgium 50°18′N 05°25′E
Hötzing see Haţeg
187 P17 Houaïlou Province Nord, C New Caledonia 21°17′S 165°37′E
74 K5 Houari Boumediène ✈ (Alger) N Algeria 36°38′N 03°15′E
167 P6 Houayxay var. Ban Houayxay. Bokèo, N Laos 20°17′N 100°27′E
103 N5 Houdan Yvelines, N France 48°48′N 01°36′E
99 F20 Houdeng-Goegnies var. Houdeng-Gœgnies. Hainaut, S Belgium 50°29′N 04°10′E
102 K14 Houeillès Lot-et-Garonne, SW France 44°15′N 00°02′E
99 L22 Houffalize Luxembourg, SE Belgium 50°08′N 05°47′E
30 M3 Houghton Michigan, N USA 47°07′N 88°34′W
31 Q7 Houghton Lake Michigan, N USA 44°18′N 84°45′W
31 Q7 Houghton Lake ◎ Michigan, N USA
19 T3 Houlton Maine, NE USA 46°09′N 67°50′W
160 M5 Houma Shanxi, C China 35°36′N 111°23′E
193 U16 Houma Tongatapu, S Tonga 21°18′S 174°55′W
22 J10 Houma Louisiana, S USA 29°35′N 90°44′W
196 V16 Houma Taloa headland Tongatapu, S Tonga 21°16′S 175°08′W
77 O13 Houndé SW Burkina Faso 11°34′N 03°31′W
102 J12 Hourtin-Carcans, Lac d' ◎ SW France
36 J5 House Range ▲▲ Utah, W USA
10 K13 Houston British Columbia, SW Canada 54°24′N 126°39′W
39 R11 Houston Alaska, USA 61°37′N 149°50′W
29 X10 Houston Minnesota, N USA 43°45′N 91°34′W
22 M3 Houston Mississippi, S USA 33°54′N 89°00′W
27 V7 Houston Missouri, C USA 37°19′N 91°59′W
25 W11 Houston Texas, SW USA 29°46′N 95°22′W
25 W11 Houston ✈ Texas, SW USA 30°03′N 95°18′W
98 J12 Houten Utrecht, C Netherlands 52°02′N 05°10′E
99 K17 Houthalen Limburg, NE Belgium 51°02′N 05°22′E
99 I22 Houyet Namur, SE Belgium 50°10′N 05°00′E
95 H22 Hov Midtjylland, C Denmark 55°54′N 10°13′E
95 L17 Hova Västra Götaland, S Sweden 58°52′N 14°13′E
162 E6 Hovd var. Dund-Us. Hovd, W Mongolia 48°06′N 91°22′E
162 J10 Hovd var. Dund-Us. Hovd, W Mongolia 48°06′N 91°22′E
162 E6 Hovd var. Khovd, Kobdo; prev. Jirgalanta. Hovd, W Mongolia 47°59′N 91°41′E
162 E7 Hovd ❖ province W Mongolia
Hovd see Bogd
162 C5 Hovd Gol ≈ NW Mongolia
97 O23 Hove SE England, United Kingdom 50°49′N 00°11′W
95 I22 Hovedstaden off. Frederiksborgs Amt. ❖ county E Denmark
29 N8 Hoven South Dakota, N USA 45°12′N 99°47′W
116 I8 Hoverla, Hora Rus. Gora Goverla. ▲ W Ukraine 48°09′N 24°30′E
Höviyn Am see Gurvanbulag
95 M21 Hovmantorp Kronoberg, S Sweden 56°47′N 15°08′E
163 N11 Hövsgöl Dornogovĭ, SE Mongolia 43°35′N 109°40′E
162 I5 Hövsgöl ❖ province N Mongolia
Hövsgol, Lake see Hövsgöl Nuur
162 J5 Hövsgöl Nuur var. Lake Hovsgol. ◎ N Mongolia
78 L9 Howa, Ouadi var. Wādi Howar. ≈ Chad/Sudan see also Howar, Wādi
Howa, Ouadi see Howar, Wādi
27 P7 Howard Kansas, C USA 37°27′N 96°16′W
29 Q10 Howard South Dakota, N USA 43°58′N 97°31′W
25 N10 Howard Draw valley Texas, SW USA
29 U8 Howard Lake Minnesota, N USA 45°03′N 94°03′W
80 B8 Howar, Wādi var. Ouadi Howa. ≈ Chad/Sudan see also Howa, Ouadi
Howar, Wādi see Howa, Ouadi
25 U5 Howe Texas, SW USA 33°29′N 96°38′W
183 R12 Howe, Cape headland New South Wales/Victoria, SE Australia 37°30′S 149°58′E
31 R9 Howell Michigan, N USA 42°36′N 83°55′W
28 L9 Howes South Dakota, N USA 44°34′N 102°03′W
83 K23 Howick KwaZulu/Natal, E South Africa 29°29′S 30°13′E
Howrah see Hāora
167 T9 Hô Xa prev. Vinh Linh. Quang Tri, C Vietnam 17°02′N 107°03′E
27 W9 Hoxie Arkansas, C USA 36°03′N 90°58′W
26 J3 Hoxie Kansas, C USA 39°21′N 100°27′W
101 I14 Höxter Nordrhein-Westfalen, W Germany 51°46′N 09°22′E
158 K6 Hoxud var. Tewulike. Xinjiang Uygur Zizhiqu, NW China 42°18′N 86°51′E
96 J5 Hoy island N Scotland, United Kingdom
43 S17 Hoya, Cerro ▲ S Panama 07°22′N 80°38′W
94 D12 Høyanger Sogn Og Fjordane, S Norway 61°13′N 06°05′E
101 P15 Hoyerswerda Lus. Wojerecy. Sachsen, E Germany 51°27′N 14°18′E
164 R14 Hōyo-kaikyō var. Hayasui-seto. strait SW Japan
104 J8 Hoyos Extremadura, W Spain 40°10′N 06°43′W
29 W4 Hoyt Lakes Minnesota, N USA 47°31′N 92°08′W
87 V2 Hoyvík Streymoy, N Faroe Islands
137 O14 Hozat Tunceli, E Turkey 39°09′N 39°13′E
Hözyô see Hōjō
167 N8 Hpa-an var. Pa-an. Kayin State, S Myanmar (Burma) 16°51′N 97°37′E
167 N8 Hpapun var. Papun. Kayin State, S Myanmar (Burma) 18°05′N 97°26′E
167 N7 Hpasawng var. Pasawng. Kayah State, C Myanmar (Burma) 18°50′N 97°16′E
Hpyu see Phyu
111 F16 Hradec Králové Ger. Königgrätz. Královéhradecký Kraj, N Czech Republic 50°13′N 15°50′E
Hradecký Kraj see Královehradecký Kraj
111 B16 Hradiště Ger. Burgstadlberg. ▲ NW Czech Republic 50°12′N 13°04′E
117 R6 Hradyz'k Rus. Gradizhsk. Poltavs'ka Oblast', NE Ukraine 49°14′N 33°07′E
119 M16 Hradzyanka Rus. Grodzyanka. Mahilyowskaya Voblasts', E Belarus 53°33′N 28°45′E
119 F16 Hrandzichy Rus. Grandichi. Hrodzyenskaya Voblasts', W Belarus 53°43′N 23°49′E
111 H18 Hranice Ger. Mährisch-Weisskirchen. Olomoucký Kraj, E Czech Republic 49°34′N 17°45′E
112 I13 Hrasnica Federacija Bosna I Hercegovina, SE Bosnia and Herzegovina 43°48′N 18°19′E
109 V11 Hrastnik C Slovenia 46°09′N 15°08′E
137 U12 Hrazdan Rus. Razdan. C Armenia 40°30′N 44°50′E
137 T12 Hrazdan var. Zanga, Rus. Razdan. ≈ C Armenia
117 R5 Hrebinka Rus. Grebenka. Poltavs'ka Oblast', NE Ukraine 50°08′N 32°27′E
119 K17 Hresk Rus. Gresk. Minskaya Voblasts', C Belarus 53°10′N 27°29′E
Hrisoupoli see Chrysoúpoli
119 F16 Hrodna Rus., Pol. Grodno. Hrodzyenskaya Voblasts', W Belarus 53°40′N 23°50′E
119 F16 Hrodzyenskaya Voblasts' Rus. Grodnenskaya Oblast'. ❖ province W Belarus
111 J21 Hron Ger. Gran, Hung. Garam. ≈ C Slovakia
111 Q14 Hrubieszów Rus. Grubeshov. Lubelskie, E Poland 50°49′N 23°53′E
112 F13 Hrvace Split-Dalmacija, SE Croatia 43°46′N 16°35′E
Hrvatska see Croatia
112 F10 Hrvatska Kostanjnica var. Kostajnica. Sisak-Moslavina, C Croatia 45°14′N 16°35′E
Hrvatski Grahovi see Bosansko Grahovo
116 K6 Hrymayliv Pol. Gizymałów, Rus. Grimaylov. Ternopil's'ka Oblast', W Ukraine 49°18′N 26°02′E
167 N4 Hseni var. Hsenwi. Shan State, E Myanmar (Burma) 23°20′N 97°59′E
Hsenwi see Hseni
Hsia-men see Xiamen
Hsiang-t'an see Xiangtan
Hsi Chiang see Xi Jiang
167 N6 Hsihseng Shan State, C Myanmar (Burma) 20°07′N 97°17′E
Hsinchu see Xinzhu
Hsing-K'ai Hu see Khanka, Lake
Hsi-ning/Hsining see Xining
Hsinking see Changchun
Hsin-yang see Xinyang
Hsinying see Xinying
167 N4 Hsipaw Shan State, C Myanmar (Burma) 22°32′N 97°12′E
Hsu-chou see Xuzhou
Hsüeh Shan see Xue Shan
Hu see Shanghai Shi
83 B18 Huab ≈ W Namibia
57 M21 Huacaya Chuquisaca, S Bolivia 20°45′S 63°42′W
57 J19 Huachacalla Oruro, SW Bolivia 18°47′S 68°23′W
159 X9 Huachi var. Rouyuan, Rouyuanchengzi. Gansu, C China 36°24′N 107°58′E
57 N16 Huachi, Laguna ◎ N Bolivia
57 D14 Huacho Lima, W Peru 11°05′S 77°36′W
163 Y7 Huachuan Heilongjiang, NE China 47°00′N 130°21′E
163 P12 Huade Nei Mongol Zizhiqu, N China 41°52′N 113°58′E
163 W10 Huadian Jilin, NE China 42°59′N 126°38′E
56 E13 Huagaruncho, Cordillera ▲▲ C Peru
Hua Hin see Ban Hua Hin
191 S10 Huahine island Îles Sous le Vent, W French Polynesia
Huahua, Rio see Wawa, Río
167 R8 Huai ≈ E Thailand
161 Q7 Huai'an var. Qingjiang. Jiangsu, E China 33°33′N 119°03′E
161 P6 Huaibei Anhui, E China 34°00′N 116°48′E
Huaide see Gongzhuling
157 T10 Huai He ≈ C China
160 L11 Huaihua Hunan, S China 27°36′N 109°57′E
161 N14 Huaiji Guangdong, S China 23°54′N 112°12′E
161 O2 Huailai var. Shacheng. Hebei, E China 40°22′N 115°34′E
161 P7 Huainan var. Huai-nan, Hwainan. Anhui, E China 32°37′N 116°57′E
Huai-nan see Huainan
161 N2 Huairen var. Yunzhong. Shanxi, C China 39°48′N 113°06′E
161 O7 Huaiyang Henan, C China 33°44′N 114°55′E
161 Q7 Huaiyin Jiangsu, E China 33°31′N 119°03′E
167 N16 Huai Yot Trang, SW Thailand 07°45′N 99°36′E
41 Q15 Huajuapan var. Huajuapan de León. Oaxaca, SE Mexico 17°50′N 97°48′W
Huajuapan de León see Huajuapan
41 O9 Hualahuises Nuevo León, NE Mexico 24°56′N 99°42′W
36 I11 Hualapai Mountains ▲▲ Arizona, SW USA
36 I11 Hualapai Peak ▲ Arizona, SW USA 35°04′N 113°54′W
62 J7 Hualfín Catamarca, N Argentina 27°15′S 66°53′W
161 T13 Hualian var. Hualien, Hwalien, Jap. Karen. C Taiwan 23°58′N 121°35′E
Hualien see Hualian
56 E10 Huallaga, Río ≈ N Peru
56 C11 Huamachuco La Libertad, C Peru 07°50′S 78°01′W
41 Q14 Huamantla Tlaxcala, S Mexico 19°18′N 97°57′W
82 C13 Huambo Port. Nova Lisboa. Huambo, C Angola 12°48′S 15°45′E
82 B13 Huambo ❖ province C Angola
41 P15 Huamuxtitlán Guerrero, S Mexico 17°49′N 98°34′W
163 Y8 Huanan Heilongjiang, NE China 46°21′N 130°43′E
63 H17 Huancache, Sierra ▲▲ SW Argentina
57 I17 Huancané Puno, SE Peru 15°10′S 69°44′W
57 F16 Huancapi Ayacucho, C Peru 13°40′S 74°05′W
57 E15 Huancavelica Huancavelica, SW Peru 12°45′S 75°03′W
57 E15 Huancavelica off. Departamento de Huancavelica. ❖ department W Peru
Huancavelica, Departamento de see Huancavelica
57 E14 Huancayo Junín, C Peru 12°03′S 75°14′W
57 K20 Huanchaca, Cerro ▲ S Bolivia 20°12′S 66°35′W
Huancheng see Huanxian
56 C12 Huandoy, Nevado ▲ W Peru 08°48′S 77°33′W
Huang'an see Hong'an
161 O8 Huangchuan Henan, C China 32°00′N 115°02′E
161 O9 Huanggang Hubei, C China 30°27′N 114°48′E
Huang Hai see Yellow Sea
157 Q8 Huang He var. Yellow River. ≈ C China
Huanghe see Madoi
161 Q4 Huanghe Kou delta E China
Huangheyan see Madoi
160 L5 Huangling Shaanxi, C China 35°40′N 109°14′E
161 O9 Huangpi Hubei, C China 30°53′N 114°22′E
163 P13 Huangqi Hai ◎ N China
161 Q9 Huangshan var. Tunxi. Anhui, E China 29°43′N 118°20′E
161 O9 Huangshi var. Huang-shih, Hwangshih. Hubei, C China 30°14′N 115°E
Huang-shih see Huangshi
160 L5 Huangtu Gaoyuan plateau C China
61 B22 Huanguelén Buenos Aires, E Argentina 37°02′S 61°57′W
161 S10 Huangyan Zhejiang, SE China 28°39′N 121°19′E
159 T10 Huangyuan Qinghai, C China 36°40′N 101°12′E
159 T10 Huangzhong var. Lushar. Qinghai, C China 36°31′N 101°32′E
163 W12 Huanren var. Huanren Manzu Zizhixian. Liaoning, NE China 41°16′N 125°25′E
Huanren Manzu Zizhixian see Huanren
57 F15 Huanta Ayacucho, C Peru 12°54′S 74°13′W
56 E13 Huánuco Huánuco, C Peru 09°58′S 76°16′W
56 D13 Huánuco off. Departamento de Huánuco. ❖ department C Peru
Huánuco, Departamento de see Huánuco
57 K19 Huanuni Oruro, W Bolivia 18°15′S 66°48′W
159 X9 Huanxian var. Huancheng. Gansu, C China 36°30′N 107°20′E
161 S12 Huaping Yu prev. Huap'ing Yu. island N Taiwan
62 H3 Huara Tarapacá, N Chile 19°59′S 69°42′W
57 D14 Huaral Lima, W Peru 11°31′S 77°10′W
Huarás see Huaraz
56 D13 Huaraz var. Huarás. Ancash, W Peru 09°31′S 77°32′W
57 I16 Huari Huari, Río ≈ S Peru
56 C13 Huarmey Ancash, W Peru 10°03′S 78°08′W
40 H4 Huásabas Sonora, NW Mexico 29°47′N 109°18′W
56 D8 Huasaga, Río ≈ Ecuador/Peru
167 O15 Hua Sai Nakhon Si Thammarat, SW Thailand 08°02′N 100°18′E
56 D12 Huascarán, Nevado ▲ W Peru 09°01′S 77°27′W
62 G8 Huasco Atacama, N Chile 28°30′S 71°15′W
62 G8 Huasco, Río ≈ C Chile
159 S11 Huashixia Qinghai, W China 35°13′N 99°12′E
40 G7 Huatabampo Sonora, NW Mexico 26°49′N 109°40′W
159 W10 Huating var. Donghua. Gansu, C China 35°13′N 106°39′E
167 S7 Huatt, Phou ▲ N Vietnam 19°45′N 104°48′E
41 Q14 Huatusco var. Huatusco de Chicuellar. Veracruz-Llave, C Mexico 19°13′N 96°57′W
Huatusco de Chicuellar see Huatusco
41 P13 Huauchinango Puebla, S Mexico 20°11′N 98°04′W
Huaunta see Wounta
41 R15 Huautla var. Huautla de Jiménez. Oaxaca, SE Mexico 18°10′N 96°51′W
Huautla de Jiménez see Huautla
161 O5 Huaxian var. Daokou, Hua Xian. Henan, C China 35°33′N 114°30′E
Hua Xian see Huaxian
Huazangsi see Tianzhu
29 V13 Hubbard Iowa, C USA 42°18′N 93°18′W
25 U8 Hubbard Texas, SW USA 31°52′N 96°43′W
25 Q6 Hubbard Creek Lake ⊠ Texas, SW USA
31 R6 Hubbard Lake ◎ Michigan, N USA
160 M9 Hubei var. E, Hubei Sheng, Hupeh, Hupei. ❖ province C China
Hubei Sheng see Hubei
109 P8 Huben Tirol, W Austria 46°55′N 12°35′E
31 R13 Huber Heights Ohio, N USA 39°50′N 84°07′W
155 F17 Hubli Karnātaka, SW India 15°20′N 75°14′E
163 X12 Huch'ang N North Korea 41°25′N 127°04′E
97 M18 Hucknall C England, United Kingdom 53°02′N 01°11′W
97 L17 Huddersfield N England, United Kingdom 53°39′N 01°47′W
95 O16 Huddinge Stockholm, C Sweden 59°15′N 17°57′E
94 N11 Hudiksvall Gävleborg, C Sweden 61°45′N 17°12′E
29 W13 Hudson Iowa, C USA 42°24′N 92°27′W
19 O11 Hudson Massachusetts, NE USA 42°24′N 71°34′W
31 Q11 Hudson Michigan, N USA 41°51′N 84°21′W
30 H6 Hudson Wisconsin, N USA 44°59′N 92°43′W
11 V14 Hudson Bay Saskatchewan, S Canada 52°51′N 102°23′W
12 G6 Hudson Bay bay NE Canada
195 T16 Hudson, Cape headland Antarctica 68°15′S 154°00′E
Hudson, Détroit d' see Hudson Strait
27 Q9 Hudson, Lake ⊠ Oklahoma, C USA
18 K9 Hudson River ≈ New Jersey/New York, NE USA
10 M12 Hudson's Hope British Columbia, W Canada 56°03′N 121°59′W
12 L2 Hudson Strait Fr. Détroit d'Hudson. strait Northwest Territories/Québec, NE Canada
Ḩudūd ash Shamālīyah, Minţaqat al see Al Ḩudūd ash Shamālīyah
Hudur see Xuddur
167 U9 Huê Thừa Thiên-Huê, C Vietnam 16°28′N 107°35′E
104 J7 Huebra ≈ W Spain
24 H8 Hueco Mountains ▲▲ Texas, SW USA
116 G10 Huedin Hung. Bánffyhunyad. Cluj, NW Romania 46°52′N 23°02′E
40 J10 Huehueto, Cerro ▲ C Mexico 24°04′N 105°42′W
42 B5 Huehuetenango Huehuetenango, W Guatemala 15°19′N 91°26′W
42 B4 Huehuetenango off. Departamento de Huehuetenango. ❖ department W Guatemala
Huehuetenango, Departamento de see Huehuetenango
40 L11 Huejuquilla Jalisco, SW Mexico 22°40′N 103°52′W
41 P12 Huejutla var. Huejutla de Reyes. Hidalgo, C Mexico 21°10′N 98°25′W
Huejutla de Reyes see Huejutla
102 G6 Huelgoat Finistère, NW France 48°22′N 03°45′W
105 O13 Huelma Andalucía, S Spain 37°39′N 03°28′W
104 I14 Huelva anc. Onuba. Andalucía, SW Spain 37°15′N 06°56′W
104 I13 Huelva ❖ province Andalucía, SW Spain
104 J13 Huelva ≈ SW Spain
105 Q14 Huercal-Overa Andalucía, S Spain 37°23′N 01°56′W
37 Q9 Huerfano Mountain ▲ New Mexico, SW USA 36°25′N 107°50′W
37 T7 Huerfano River ≈ Colorado, C USA
Huertas, Cabo see Horta, Cap de l'
105 R6 Huerva ≈ N Spain
105 S4 Huesca anc. Osca. Aragón, NE Spain 42°08′N 00°25′W
105 T4 Huesca ❖ province Aragón, NE Spain
105 P13 Huéscar Andalucía, S Spain 37°39′N 02°32′W
41 N15 Huetamo var. Huetamo de Núñez. Michoacán, SW Mexico 18°36′N 100°54′W
Huetamo de Núñez see Huetamo
105 P8 Huete Castilla-La Mancha, C Spain 40°09′N 02°42′W
23 P4 Hueytown Alabama, S USA 33°27′N 87°00′W
28 L16 Hugh Butler Lake ⊠ Nebraska, C USA
181 V6 Hughenden Queensland, NE Australia 20°57′S 144°16′E
182 A6 Hughes South Australia 30°41′S 129°31′E
39 P8 Hughes Alaska, USA 66°03′N 154°15′W
27 X11 Hughes Arkansas, C USA 34°57′N 90°28′W
25 W6 Hughes Springs Texas, SW USA 33°00′N 94°37′W
37 V5 Hugo Colorado, C USA 39°08′N 103°28′W
27 Q13 Hugo Oklahoma, C USA 34°01′N 95°31′W
27 Q13 Hugo Lake ⊠ Oklahoma, C USA
26 H7 Hugoton Kansas, C USA 37°11′N 101°22′W
Huhehot/Huhohaote see Hohhot
Huhttán see Kvikkjokk
161 R13 Hui'an var. Luocheng. Fujian, SE China 25°06′N 118°45′E
184 O9 Huiarau Range ▲▲ North Island, New Zealand
83 D22 Huib-Hoch Plateau plateau S Namibia
41 O13 Huichapán Hidalgo, C Mexico 20°24′N 99°40′W
Huicheng see Shexian
163 W13 Hŭich'ŏn C North Korea 40°09′N 126°17′E
83 B15 Huíla ❖ province SW Angola
54 E12 Huila off. Departamento del Huila. ❖ province S Colombia
Huila, Departamento del see Huila
54 D11 Huila, Nevado del elevation C Colombia
83 B15 Huíla Plateau plateau S Angola
160 G12 Huili Sichuan, C China 26°39′N 102°13′E
161 P4 Huimin Shandong, E China 37°29′N 117°30′E
163 W11 Huinan var. Chaoyang. Jilin, NE China 42°40′N 126°03′E
62 K12 Huinca Renancó Córdoba, C Argentina 34°51′S 64°22′W
159 V10 Huining var. Huishi. Gansu, C China 35°42′N 105°01′E
159 W8 Huinong var. Dawukou. Ningxia, N China 39°04′N 106°22′E
Huishi see Huining
160 J12 Huishui var. Heping. Guizhou, S China 26°07′N 106°39′E
102 L6 Huisne ≈ NW France
98 L12 Huissen Gelderland, SE Netherlands 51°57′N 05°57′E
159 N11 Huiten Nur ◎ C China
93 K19 Huittinen Satakunta, SW Finland 61°11′N 22°40′E
41 O15 Huitzuco var. Huitzuco de los Figueroa. Guerrero, S Mexico 18°18′N 99°22′W
Huitzuco de los Figueroa see Huitzuco
159 W11 Huixian var. Hui Xian. Gansu, C China 33°48′N 106°02′E
Hui Xian see Huixian
41 V17 Huixtla Chiapas, SE Mexico 15°09′N 92°30′W

◆ Country
● Country Capital
◇ Dependent Territory
○ Dependent Territory Capital
❖ Administrative Regions
✈ International Airport
▲ Mountain
▲▲ Mountain Range
⊼ Volcano
≈ River
◎ Lake
⊠ Reservoir

160 H12 **Huize** *var.* Zhongping. Yunnan, SW China 26°28′N 103°18′E
98 J10 **Huizen** Noord-Holland, C Netherlands 52°17′N 05°15′E
161 O14 **Huizhou** Guangdong, S China 23°02′N 114°28′E
162 J6 **Hujirt** Arhangay, C Mongolia 48°49′N 101°20′E
Hujirt *see* Tsetserleg, Övörhangay, Mongolia
Hujirt *see* Delgerhaan, Töv, Mongolia
Hukagawa *see* Fukagawa
Hŭksan-gundo *see* Heuksan-jedo
Hukue *see* Gotō
Hukui *see* Fukui
83 G20 **Hukuntsi** Kgalagadi, SW Botswana 23°59′S 21°44′E
Hukuoka *see* Fukuoka
Hukusima *see* Fukushima
Hukutiyama *see* Fukuchiyama
Hukuyama *see* Fukuyama
163 W8 **Hulan** Heilongjiang, NE China 45°59′N 126°37′E
163 W8 **Hulan He** NE China
31 Q4 **Hulbert Lake** Michigan, N USA
Hulczyn *see* Hlučín
Huliao *see* Dabu
163 Z8 **Hulin** Heilongjiang, NE China 45°48′N 133°06′E
Hulingol *see* Holin Gol
14 L12 **Hull** Québec, SE Canada 45°26′N 75°45′W
29 S12 **Hull** Iowa, C USA 43°11′N 96°07′W
Hull *see* Kingston upon Hull
Hull Island *see* Orona
99 F16 **Hulst** Zeeland, SW Netherlands 51°17′N 04°03′E
Hulstay *see* Choybalsan
Hultschin *see* Hlučín
95 M19 **Hultsfred** Kalmar, S Sweden 57°30′N 15°50′E
163 T13 **Huludao** *prev.* Jinxi, Lianshan. Liaoning, NE China 40°46′N 120°47′E
Hulun *see* Hulun Buir
163 S6 **Hulun Buir** *var.* Hailar; *prev.* Hulun. Nei Mongol Zizhiqu, N China 49°15′N 119°41′E
Hu-lun Ch'ih *see* Hulun Nur
163 Q6 **Hulun Nur** *var.* Hu-lun Ch'ih; *prev.* Dalai Nor. NE China
Hulwan/Hulwân *see* Hilwân
117 V8 **Hulyaypole** *Rus.* Gulyaypole. Zaporiz'ka Oblast', SE Ukraine 47°41′N 36°10′E
163 V4 **Huma** Heilongjiang, NE China 51°40′N 126°38′E
45 V6 **Humacao** E Puerto Rico 18°09′N 65°50′W
163 U4 **Huma He** NE China
62 J5 **Humahuaca** Jujuy, N Argentina 23°13′S 65°20′W
59 E14 **Humaitá** Amazonas, N Brazil 07°33′S 63°01′W
62 N7 **Humaitá** Ñeembucú, S Paraguay 27°02′S 58°31′W
83 H26 **Humansdorp** Eastern Cape, S South Africa 34°01′S 24°45′E
27 S6 **Humansville** Missouri, C USA 37°47′N 93°34′W
40 J8 **Humaya, Río** C Mexico
83 C16 **Humbe** Cunene, SW Angola 16°37′S 14°52′E
97 N17 **Humber** *estuary* E England, United Kingdom
97 N17 **Humberside** *cultural region* E England, United Kingdom
Humberto *see* Umberto
25 W11 **Humble** Texas, SW USA 29°58′N 95°15′W
11 U15 **Humboldt** Saskatchewan, S Canada 52°13′N 105°09′W
29 U12 **Humboldt** Iowa, C USA 42°42′N 94°13′W
27 Q6 **Humboldt** Kansas, C USA 37°48′N 95°26′W
29 S17 **Humboldt** Nebraska, C USA 40°09′N 95°56′W
35 S3 **Humboldt** Nevada, W USA 40°36′N 118°15′W
20 G9 **Humboldt** Tennessee, S USA 35°49′N 88°55′W
34 K3 **Humboldt Bay** *bay* California, W USA
35 S4 **Humboldt Lake** Nevada, W USA
35 S4 **Humboldt River** Nevada, W USA
35 T5 **Humboldt Salt Marsh** *wetland* Nevada, W USA
183 P11 **Hume, Lake** New South Wales/Victoria, SE Australia
111 N19 **Humenné** *Ger.* Homenau, *Hung.* Homonna. Prešovský Kraj, E Slovakia 48°57′N 21°54′E
29 V15 **Humeston** Iowa, C USA 40°51′N 93°30′W
54 J5 **Humocaro Bajo** Lara, N Venezuela 09°41′N 70°00′W
29 Q14 **Humphrey** Nebraska, C USA 41°38′N 97°29′W
35 S9 **Humphreys, Mount** California, W USA 37°11′N 118°39′W
36 L11 **Humphreys Peak** Arizona, SW USA 35°18′N 111°40′W
111 E17 **Humpolec** *Ger.* Gumpolds, Humpoletz. Vysočina, C Czech Republic 49°33′N 15°23′E
Humpoletz *see* Humpolec
93 K19 **Humppila** Kanta-Häme, S Finland 60°54′N 23°21′E
32 F8 **Humptulips** Washington, NW USA 47°13′N 123°57′W
42 H7 **Humuya, Río** W Honduras
75 P9 **Hūn** N Libya 29°06′N 15°56′E
92 I1 **Húnaflói** *bay* NW Iceland
160 M11 **Hunan** *var.* Hunan Sheng, Xiang. *province* S China
Hunan Sheng *see* Hunan
163 Y10 **Hunchun** Jilin, NE China 42°51′N 130°21′E
95 I22 **Hundested** Hovedstaden, E Denmark 55°58′N 11°53′E
Hundred Mile House *see* 100 Mile House
116 G12 **Hunedoara** *Ger.* Eisenmarkt, *Hung.* Vajdahunyad. Hunedoara, SW Romania 45°45′N 22°54′E
116 G12 **Hunedoara** *county* W Romania
101 I17 **Hünfeld** Hessen, C Germany 50°41′N 09°46′E
Hungarian People's Republic *see* Hungary
111 H23 **Hungary** *off.* Republic of Hungary, *Ger.* Ungarn, *Hung.* Magyarország, *Rom.* Ungaria, *SCr.* Mađarska, *Ukr.* Uhorshchyna; *prev.* Hungarian People's Republic. ◆ *republic* C Europe
Hungary, Plain of *see* Great Hungarian Plain
Hungary, Republic of *see* Hungary
Hungiy *see* Urgamal
163 X13 **Hŭngnam** E North Korea 39°50′N 127°36′E
33 P8 **Hungry Horse Reservoir** Montana, NW USA
Hungt'ou *see* Lan Yu
Hung-tse Hu *see* Hongze Hu
167 T6 **Hưng Yên** Hai Hưng, N Vietnam 20°38′N 106°05′E
Hunjiang *see* Baishan
95 I18 **Hunnebostrand** Västra Götaland, S Sweden 58°26′N 11°19′E
101 E19 **Hunsrück** W Germany
97 P18 **Hunstanton** E England, United Kingdom 52°57′N 00°27′E
155 G20 **Hunsūr** Karnātaka, E India 12°18′N 76°15′E
Hunt *see* Hangay
100 G12 **Hunte** NW Germany
29 Q5 **Hunter** North Dakota, N USA 47°10′N 97°11′W
25 S11 **Hunter** Texas, SW USA 29°47′N 98°01′W
185 D20 **Hunter** South Island, New Zealand
183 N15 **Hunter Island** *island* Tasmania, SE Australia
18 K11 **Hunter Mountain** New York, NE USA 42°10′N 74°13′W
185 B23 **Hunter Mountains** South Island, New Zealand
183 S7 **Hunter River** New South Wales, SE Australia
32 L7 **Hunters** Washington, NW USA 48°07′N 118°13′W
185 F20 **Hunters Hills, The** *hill range* South Island, New Zealand
184 M12 **Hunterville** Manawatu-Wanganui, North Island, New Zealand 39°55′S 175°34′E
31 N16 **Huntingburg** Indiana, N USA 38°18′N 86°57′W
97 O20 **Huntingdon** E England, United Kingdom 52°20′N 00°12′W
18 E15 **Huntingdon** Pennsylvania, NE USA 40°28′N 78°00′W
20 G9 **Huntingdon** Tennessee, S USA 36°00′N 88°25′W
97 O20 **Huntingdonshire** *cultural region* C England, United Kingdom
31 P12 **Huntington** Indiana, N USA 40°52′N 85°30′W
32 L13 **Huntington** Oregon, NW USA 44°22′N 117°18′W
25 X9 **Huntington** Texas, SW USA 31°16′N 94°34′W
36 M5 **Huntington** Utah, W USA 39°19′N 110°57′W
21 P5 **Huntington** West Virginia, NE USA 38°25′N 82°27′W
35 T16 **Huntington Beach** California, W USA 33°39′N 118°00′W
35 W4 **Huntington Creek** Nevada, W USA
184 L7 **Huntly** Waikato, North Island, New Zealand 37°34′S 175°09′E
96 K8 **Huntly** NE Scotland, United Kingdom 57°25′N 02°48′W
10 K8 **Hunt, Mount** Yukon, NW Canada 61°29′N 129°10′W
14 H12 **Huntsville** Ontario, S Canada 45°20′N 79°14′W
23 P2 **Huntsville** Alabama, S USA 34°44′N 86°35′W
27 S9 **Huntsville** Arkansas, C USA 36°04′N 93°46′W
27 U3 **Huntsville** Missouri, C USA 39°27′N 92°31′W
20 M8 **Huntsville** Tennessee, S USA 36°25′N 84°30′W
25 V10 **Huntsville** Texas, SW USA 30°43′N 95°34′W
36 L2 **Huntsville** Utah, W USA 41°16′N 111°47′W
41 W12 **Hunucmá** Yucatán, SE Mexico 20°59′N 89°55′W
149 W3 **Hunza** NE Pakistan
Hunza *see* Karīmābād
Hunze *see* Oostermoers Vaart
158 H4 **Huocheng** *var.* Shuiding. Xinjiang Uygur Zizhiqu, NW China 44°03′N 80°49′E
161 N6 **Huojia** Henan, C China 35°14′N 113°38′E
163 S9 **Huolin Gol** *prev.* Hulingol. Nei Mongol Zizhiqu, N China 45°36′N 119°54′E186 N14
Huon *reef* N New Caledonia
186 E7 **Huon Peninsula** *headland* C Papua New Guinea 06°24′S 147°50′E
Huoshao Dao *see* Lü Dao
Huoshao Tao *see* Lan Yu
Hupeh/Hupei *see* Hubei
Hurama *see* Hongyuan
95 H14 **Hurdalsjøen** *prev.* Hurdalssjøen. S Norway
Hurdalssjøen *see* Hurdalsjøen
14 E13 **Hurd, Cape** *headland* Ontario, S Canada 45°12′N 81°43′W
98 L5 **Hurdegaryp** *Dutch.* Hardegarijp. Fryslân, N Netherlands 53°13′N 05°57′E
29 N4 **Hurdsfield** North Dakota, N USA 47°24′N 99°55′W
Hüremt *see* Sayhan, Bulgan, Mongolia
Hüremt *see* Taragt, Övörhangay, Mongolia
Hurghada *see* Al Ghurdaqah
67 V9 **Huri Hills** NW Kenya
Hürkay *see* Hurkê
139 S1 **Hurkê** *Ar.* Hürkay, *var.* Harki. Dahūk, N Iraq 37°03′N 43°39′E
37 P15 **Hurley** New Mexico, SW USA 32°42′N 108°07′W
30 K4 **Hurley** Wisconsin, N USA 46°25′N 90°15′W
21 Y4 **Hurlock** Maryland, NE USA 38°37′N 75°51′W
162 K11 **Hürmen** *var.* Tsoohor. Ömnögovĭ, S Mongolia 43°15′N 104°04′E
29 P10 **Huron** South Dakota, N USA 44°19′N 98°13′W
31 S6 **Huron, Lake** Canada/USA
31 N3 **Huron Mountains** *hill range* Michigan, N USA
36 J8 **Hurricane** Utah, W USA 37°10′N 113°18′W
21 P5 **Hurricane** West Virginia, NE USA 38°25′N 82°01′W
36 J8 **Hurricane Cliffs** *cliff* Arizona, SW USA
23 V6 **Hurricane Creek** Georgia, SE USA
94 E12 **Hurrungane** S Norway 61°25′N 07°48′E
101 E16 **Hürth** Nordrhein-Westfalen, W Germany 50°52′N 06°49′E
Hurukawa *see* Furukawa
185 I17 **Hurunui** South Island, New Zealand
95 F21 **Hurup** Midtjylland, NW Denmark 56°46′N 08°26′E
117 T14 **Hurzuf** Avtonomna Respublika Krym, S Ukraine 44°33′N 34°18′E
Huş *see* Huşi
95 B19 **Húsavík** *Dan.* Husevig. Sandoy, C Faroe Islands 65°24′N 06°38′W
92 K1 **Húsavík** Norðurland Eystra, NE Iceland 66°03′N 17°20′W
Husevig *see* Húsavík
116 M10 **Huşi** *var.* Huş. Vaslui, E Romania 46°40′N 28°05′E
95 L19 **Huskvarna** Jönköping, S Sweden 57°47′N 14°15′E
39 P8 **Huslia** Alaska, USA 65°42′N 156°24′W
Husn *see* Al Ḥuṣn
95 C15 **Husnes** Hordaland, S Norway 59°52′N 05°46′E
94 D8 **Hustadvika** *sea area* S Norway
Husté *see* Khust
100 H7 **Husum** Schleswig-Holstein, N Germany 54°29′N 09°04′E
93 I16 **Husum** Västernorrland, C Sweden 63°21′N 19°12′E
116 K6 **Husyatyn** Ternopil's'ka Oblast', W Ukraine 49°04′N 26°10′E
Huszt *see* Khust
Hutag *see* Hutag-Öndör
162 K6 **Hutag-Öndör** *var.* Hutag. Bulgan, N Mongolia 49°22′N 102°50′E
26 M6 **Hutchinson** Kansas, C USA 38°03′N 97°56′W
29 U9 **Hutchinson** Minnesota, N USA 44°53′N 94°22′W
23 Y13 **Hutchinson Island** *island* Florida, SE USA
36 L11 **Hutch Mountain** Arizona, SW USA 34°49′N 111°22′W
141 O14 **Ḥūth** NW Yemen 16°14′N 44°E
186 I7 **Hutjena** Buka Island, NE Papua New Guinea 05°19′S 154°40′E
109 T8 **Hüttenberg** Kärnten, S Austria 46°58′N 14°33′E
25 T10 **Hutto** Texas, SW USA 30°32′N 97°33′W
Huttu *see* Futtsu
108 E8 **Huttwil** Bern, W Switzerland 47°06′N 07°48′E
158 K5 **Hutubi** Xinjiang Uygur Zizhiqu, NW China 44°10′N 86°51′E
161 N4 **Hutuo He** C China
Hutyû *see* Fuchū
185 E20 **Huxley, Mount** South Island, New Zealand 44°02′S 169°42′E
99 J20 **Huy** *Dut.* Hoei, Hoey. Liège, E Belgium 50°32′N 05°14′E
161 R8 **Huzhou** *var.* Wuxing. Zhejiang, SE China 30°52′N 120°06′E
Huzi *see* Fuji
Huzieda *see* Fujieda
Huzinomiya *see* Fujinomiya
Huzisawa *see* Fujisawa
92 I2 **Hvammstangi** Norðurland Vestra, N Iceland 65°22′N 20°54′W
92 K4 **Hvannadalshnúkur** *var.* Hvannadalshnúkur. S Iceland 64°01′N 16°39′W
Hvannadalshnúkur *see* Hvannadalshnjúkur
113 E15 **Hvar** *It.* Lesina. Split-Dalmacija, S Croatia 43°10′N 16°27′E
113 F15 **Hvar** *It.* Lesina; *anc.* Pharus. *island* S Croatia
117 T13 **Hvardiys'ke** *Rus.* Gvardeyskoye. Avtonomna Respublika Krym, S Ukraine 45°08′N 34°01′E
92 I4 **Hveragerði** Suðurland, SW Iceland 64°00′N 21°13′W
95 E22 **Hvide Sande** Midtjylland, W Denmark 56°00′N 08°08′E
92 I3 **Hvítá** C Iceland
95 G13 **Hvittingfoss** Buskerud, S Norway 59°28′N 10°00′E
92 I4 **Hvolsvöllur** Suðurland, SW Iceland 63°44′N 20°12′W
Hwach'ŏn-chŏsuji *see* Paro-ho
Hwainan *see* Huainan
Hwalien *see* Hualien
83 I16 **Hwange** *prev.* Wankie. Matabeleland North, W Zimbabwe 18°18′S 26°31′E
Hwang-Hae *see* Yellow Sea
Hwangshih *see* Huangshi
83 L17 **Hwedza** Mashonaland East, E Zimbabwe 18°35′S 31°35′E
63 G20 **Hyades, Cerro** S Chile 46°57′S 73°09′W
162 K6 **Hyalganat** *var.* Selenge. Bulgan, N Mongolia 49°34′N 104°18′E
19 Q12 **Hyannis** Massachusetts, NE USA 41°38′N 70°15′W
28 L13 **Hyannis** Nebraska, C USA 42°00′N 101°45′W
162 F6 **Hyargas Nuur** NW Mongolia
Hybla/Hybla Major *see* Paternò
39 Y14 **Hydaburg** Prince of Wales Island, Alaska, USA 55°10′N 132°44′W
185 F22 **Hyde** Otago, South Island, New Zealand 45°17′S 170°17′E
21 O7 **Hyden** Kentucky, S USA 37°08′N 83°23′W
18 K12 **Hyde Park** New York, NE USA 41°46′N 73°52′W
39 Z14 **Hyder** Alaska, USA 55°55′N 130°01′W
155 I15 **Hyderābād** *var.* Haidarabad. *state capital* Telangana/Andhra Pradesh, C India 17°22′N 78°26′E
149 Q16 **Hyderābād** *var.* Haidarabad. Sind, SE Pakistan 25°26′N 68°22′E
103 T16 **Hyères** *Var,* SE France 43°07′N 06°08′E
103 T16 **Hyères, Îles d'** *island group* S France
118 K12 **Hyermanavichy** *Rus.* Germanovichi. Vitsyebskaya Voblasts', N Belarus 55°24′N 27°48′E
163 X12 **Hyesan** NE North Korea 41°18′N 128°13′E
10 K8 **Hyland** Yukon, NW Canada
95 K20 **Hyltebruk** Halland, S Sweden 57°N 13°14′E
18 D16 **Hyndman** Pennsylvania, NE USA 39°49′N 78°42′W
33 P14 **Hyndman Peak** Idaho, NW USA 43°45′N 114°07′W
164 I13 **Hyōgo** *off.* Hyōgo-ken. ◐ *prefecture* Honshū, SW Japan
Hyōgo-ken *see* Hyōgo
Hypsas *see* Belice
Hyrcania *see* Gorgān
36 L1 **Hyrum** Utah, W USA 41°37′N 111°51′W
93 N14 **Hyrynsalmi** Kainuu, C Finland 64°41′N 28°30′E
33 V10 **Hysham** Montana, NW USA 46°16′N 107°14′W
11 N13 **Hythe** Alberta, W Canada 55°18′N 119°44′W
97 Q23 **Hythe** SE England, United Kingdom 51°05′N 01°04′E
Hyvinge *see* Hyvinkää
93 L19 **Hyvinkää** *Swe.* Hyvinge. Uusimaa, S Finland 60°37′N 24°50′E

I

116 J9 **Iacobeni** *Ger.* Jakobeny. Suceava, NE Romania 47°24′N 25°20′E
Iader *see* Zadar
172 I7 **Iakora** Fianarantsoa, SE Madagascar 23°04′S 46°40′E
116 K14 **Ialomiţa** *var.* Jalomitsa. ◐ *county* SE Romania
116 K14 **Ialomiţa** SE Romania
117 N10 **Ialoveni** *Rus.* Yaloveny. C Moldova 46°57′N 28°47′E
117 N11 **Ialpug** *var.* Ialpugul Mare, *Rus.* Yalpug. Moldova/Ukraine
Ialpugul Mare *see* Ialpug
23 T8 **Iamonia, Lake** Florida, SE USA
116 L13 **Ianca** Brăila, SE Romania 45°06′N 27°29′E
116 M10 **Iaşi** *Ger.* Jassy. Iaşi, NE Romania 47°08′N 27°38′E
116 L9 **Iaşi** *Ger.* Jassy. Yassy. ◐ *county* NE Romania
114 J13 **Íasmos** Anatolikí Makedonía kai Thráki, NE Greece 41°07′N 25°12′E
22 H6 **Iatt, Lake** Louisiana, S USA
58 B11 **Iauretê** Amazonas, NW Brazil 0°37′N 69°10′W
171 N3 **Iba** Luzon, N Philippines 15°25′N 119°55′E
77 S16 **Ibadan** Oyo, SW Nigeria 07°22′N 04°01′E
54 E10 **Ibagué** Tolima, C Colombia 04°27′N 75°14′W
60 J10 **Ibaiti** Paraná, S Brazil 23°49′S 50°15′W
36 J4 **Ibapah Peak** Utah, W USA 39°51′N 113°55′W
Ibar *see* Ibër
165 P13 **Ibaraki** *off.* Ibaraki-ken. ◐ *prefecture* Honshū, S Japan
Ibaraki-ken *see* Ibaraki
56 C5 **Ibarra** *var.* San Miguel de Ibarra. Imbabura, N Ecuador 0°23′S 78°08′W
Ibasfalău *see* Dumbrăveni
141 O16 **Ibb** W Yemen 13°55′N 44°10′E
100 F13 **Ibbenbüren** Nordrhein-Westfalen, NW Germany 52°17′N 07°43′E
79 H16 **Ibenga** N Congo
113 M15 **Ibër** *Serb.* Ibar. C Serbia
57 I14 **Iberia** Madre de Dios, E Peru 11°21′S 69°36′W
Iberia *see* Spain
66 M1 **Iberian Basin** *undersea feature* E Atlantic Ocean 39°00′N 16°00′W
Iberian Mountains *see* Ibérico, Sistema
84 D12 **Iberian Peninsula** *physical region* Portugal/Spain
64 M8 **Iberian Plain** *undersea feature* E Atlantic Ocean 13°30′W 43°45′N
Ibérica, Cordillera *see* Ibérico, Sistema
105 P6 **Ibérico, Sistema** *var.* Cordillera Ibérica, *Eng.* Iberian Mountains. NE Spain
12 K7 **Iberville Lac d'** Québec, NE Canada
77 T14 **Ibeto** Niger, W Nigeria 10°30′N 05°07′E
77 W15 **Ibi** Taraba, C Nigeria 08°13′N 09°46′E
105 S11 **Ibi** Valenciana, E Spain 38°38′N 00°35′W
59 L20 **Ibiá** Minas Gerais, SE Brazil 19°30′S 46°31′W
61 F15 **Ibicuí, Rio** S Brazil
61 C19 **Ibicuy** Entre Ríos, E Argentina 33°44′S 59°10′W
61 F16 **Ibirapuitã** S Brazil
105 V10 **Ibiza** *var.* Iviza, *Cast.* Eivissa; *anc.* Ebusus. *island* Islas Baleares, Spain, W Mediterranean Sea
Ibiza *see* Eivissa
138 J4 **Ibn Wardān, Qaṣr** *ruins* Ḥamāh, C Syria
Ibo *see* Sassandra
188 E9 **Ibobang** Babeldaob, N Palau
171 V13 **Ibonma** Papua Barat, E Indonesia 03°27′S 133°30′E
59 N17 **Ibotirama** Bahia, E Brazil 12°13′S 43°12′W
141 Y8 **Ibra** NE Oman 22°45′N 58°30′E
127 Q4 **Ibresi** Chuvashskaya Respublika, W Russian Federation 55°22′N 47°04′E
141 X8 **'Ibrī** NW Oman 23°12′N 56°28′E
164 C16 **Ibusuki** Kagoshima, Kyūshū, SW Japan 31°15′N 130°40′E
57 E16 **Ica** Ica, SW Peru 14°02′S 75°48′W
57 E16 **Ica** *off.* Departamento de Ica. ◐ *department* SW Peru
Ica, Departamento de *see* Ica
58 C11 **Içana** Amazonas, NW Brazil 0°22′N 67°25′W
Icaria *see* Ikaria
58 B13 **Içá, Rio** *var.* Río Putumayo. NW South America
see also Putumayo, Río
Içá, Rio *see* Putumayo, Río
136 I17 **İçel** *prev.* Ichili; *prev.* Mersin. ◐ *province* S Turkey
İçel *see* Mersin
92 I3 **Iceland** *off.* Republic of Iceland, *Dan.* Island, *Icel.* Ísland. ◆ *republic* N Atlantic Ocean
86 B6 **Iceland** *island* N Atlantic Ocean
64 L5 **Iceland Basin** *undersea feature* N Atlantic Ocean 61°00′N 19°00′W
Icelandic Plateau *see* Iceland Plateau
197 Q15 **Iceland Plateau** *var.* Icelandic Plateau. *undersea feature* S Greenland Sea 12°00′W 69°30′N
Iceland, Republic of *see* Iceland
155 E16 **Ichalkaranji** Mahārāshtra, W India 16°42′N 74°28′E
164 D15 **Ichifusa-yama** Kyūshū, SW Japan 32°18′N 131°05′E
Ichili *see* İçel
164 K13 **Ichinomiya** *var.* Itinomiya. Aichi, Honshū, SW Japan 35°18′N 136°48′E
165 Q9 **Ichinoseki** *var.* Itinoseki. Iwate, Honshū, C Japan 38°56′N 141°08′E
117 R3 **Ichnya** Chernihivs'ka Oblast', NE Ukraine 50°52′N 32°24′E
57 L17 **Ichoa, Río** C Bolivia
Iconium *see* Konya
Iculisma *see* Angoulême
39 U12 **Icy Bay** *inlet* Alaska, USA
39 N5 **Icy Cape** *headland* Alaska, USA 70°19′N 161°52′W
39 W13 **Icy Strait** *strait* Alaska, USA
27 R13 **Idabel** Oklahoma, C USA 33°54′N 94°50′W
29 T13 **Ida Grove** Iowa, C USA 42°21′N 95°28′W
77 U16 **Idah** Kogi, S Nigeria 07°06′N 06°45′E
33 N13 **Idaho** *off.* State of Idaho, *also known as* Gem of the Mountains, Gem State. ◆ *state* NW USA
33 N14 **Idaho City** Idaho, NW USA 43°48′N 115°51′W
33 R14 **Idaho Falls** Idaho, NW USA 43°28′N 112°01′W
121 P2 **Idálion** *var.* Dali, Dhali. C Cyprus 35°00′N 33°25′E
25 N5 **Idalou** Texas, SW USA 33°40′N 101°40′W
104 I9 **Idanha-a-Nova** Castelo Branco, C Portugal 39°55′N 07°15′W
101 E19 **Idar-Oberstein** Rheinland-Pfalz, SW Germany 49°43′N 07°19′E
118 J3 **Ida-Virumaa** *var.* Ida-Viru Maakond. ◐ *province* NE Estonia
Ida-Viru Maakond *see* Ida-Virumaa
124 J8 **Idel'** Respublika Kareliya, NW Russian Federation 64°08′N 34°12′E
79 C15 **Idenao** Sud-Ouest, SW Cameroon 04°04′N 09°01′E
Idensalmi *see* Iisalmi
162 I6 **Ider** *var.* Dzuunmod. Hövsgöl, C Mongolia 48°09′N 97°22′E
Ider *see* Galt
75 X10 **Idfū** *var.* Edfu. SE Egypt 24°55′N 32°52′E
Ídhi Óros *see* Ídi
Ídhra *see* Ýdra
80 L10 **'Īdī** *var.* Ed. SE Eritrea 13°54′N 41°39′E
168 H7 **Idi** Sumatera, W Indonesia 05°00′N 98°00′E
115 I25 **Ídi** *var.* Ídhi Óros. Kríti, Greece, E Mediterranean Sea
Idi Amin, Lac *see* Edward, Lake
106 G10 **Idice** N Italy
76 G9 **Idîni** Trarza, W Mauritania 17°58′N 15°40′W
79 J21 **Idiofa** Bandundu, SW Dem. Rep. Congo 05°00′S 19°38′E
39 O10 **Iditarod River** Alaska, USA
Idjavuotna *see* Ifjord
95 M14 **Idkerberget** Dalarna, C Sweden 60°22′N 15°15′E
138 I3 **Idlib** Idlib, NW Syria 35°57′N 36°38′E
138 I4 **Idlib** *off.* Muḥāfaẓat Idlib. ◐ *governorate* NW Syria
Idlib, Muḥāfaẓat *see* Idlib
Idra *see* Ýdra
94 J11 **Idre** Dalarna, C Sweden 61°52′N 12°45′E
Idria *see* Idrija
109 S11 **Idrija** *It.* Idria. W Slovenia 46°00′N 14°59′E
101 G18 **Idstein** Hessen, W Germany 50°10′N 08°16′E
Idutywa *see* Dutywa
Idzhevan *see* Ijevan
118 G9 **Iecava** S Latvia 56°36′N 24°10′E
165 T16 **Ie-jima** *var.* Ii-shima. *island* Nansei-shotō, SW Japan
99 B18 **Ieper** *Fr.* Ypres. West-Vlaanderen, W Belgium 50°51′N 02°53′E
115 K25 **Ierápetra** Kríti, Greece, E Mediterranean Sea 35°00′N 25°45′E
115 G22 **Iérax, Akrotírio** *headland* S Greece 36°45′N 23°06′E
Ierisós *see* Ierissós
115 H14 **Ierissós** *var.* Ierisós. Kentrikí Makedonía, N Greece 40°24′N 23°53′E
116 I11 **Iernut** *Hung.* Radnót. Mureş, C Romania 46°27′N 24°15′E
106 J12 **Iesi** *var.* Jesi. Marche, C Italy 43°33′N 13°16′E
92 K9 **Iešjavri** N Norway
Iesolo *see* Jesolo
188 K16 **Ifalik Atoll** *atoll* Caroline Islands, C Micronesia
172 I6 **Ifanadiana** Fianarantsoa, SE Madagascar 21°19′S 47°39′E
77 T16 **Ife** Osun, SW Nigeria 07°25′N 04°31′E
77 V8 **Iferouâne** Agadez, N Niger 19°05′N 08°24′E
Iferten *see* Yverdon
92 L8 **Ifjord** *Lapp.* Idjavuotna. Finnmark, N Norway 70°27′N 27°06′E
77 R8 **Ifôghas, Adrar des** *var.* Adrar des Iforas. NE Mali
Iforas, Adrar des *see* Ifôghas, Adrar des
182 D6 **Ifould lake** *salt lake* South Australia
74 G6 **Ifrane** C Morocco 33°31′N 05°09′W
171 S11 **Iga** Pulau Halmahera, E Indonesia 01°23′N 128°17′E
81 G18 **Iganga** SE Uganda 0°34′N 33°27′E
60 L7 **Igarapava** São Paulo, S Brazil 20°01′S 47°46′W
122 K9 **Igarka** Krasnoyarskiy Kray, N Russian Federation 67°31′N 86°33′E
Igaunija *see* Estonia
137 T12 **Iğdır** ◐ *province* NE Turkey
I.G.Duca *see* General Toshevo
Igel *see* Jihlava
94 N11 **Iggesund** Gävleborg, C Sweden 61°38′N 17°04′E
39 P7 **Igikpak, Mount** Alaska, USA 67°28′N 154°55′W
39 P13 **Igiugig** Alaska, USA 59°19′N 155°53′W
Iglau/Iglawa/Igława *see* Jihlava
107 B20 **Iglesias** Sardegna, Italy, C Mediterranean Sea 39°20′N 08°34′E
127 V4 **Iglino** Respublika Bashkortostan, W Russian Federation 54°51′N 56°29′E
Igló *see* Spišská Nová Ves
9 O6 **Igloolik** Nunavut, N Canada 69°24′N 81°55′W
12 B11 **Ignace** Ontario, S Canada 49°26′N 91°40′W
118 I12 **Ignalina** Utena, E Lithuania 55°20′N 26°10′E
127 Q5 **Ignatovka** Ul'yanovskaya Oblast', W Russian Federation 53°56′N 47°40′E
124 K12 **Ignatovo** Vologodskaya Oblast', NW Russian Federation 60°47′N 37°51′E
114 N11 **İğneada** Kırklareli, NW Turkey 41°54′N 27°58′E
121 S7 **İğneada Burnu** *headland* NW Turkey 41°54′N 28°03′E
Igombe *see* Gombe
115 B16 **Igoumenítsa** Ípeiros, W Greece 39°30′N 20°16′E
127 T2 **Igra** Udmurtskaya Respublika, NW Russian Federation 57°30′N 53°01′E
122 H9 **Igrim** Khanty-Mansiyskiy Avtonomnyy Okrug-Yugra, N Russian Federation 63°09′N 64°33′E
60 G12 **Iguaçu, Rio** *Sp.* Río Iguazú. Argentina/Brazil *see also* Iguazú, Río
Iguaçu, Rio *see* Iguazú, Río
59 I22 **Iguaçu, Salto do** *Sp.* Cataratas del Iguazú; *prev.* Victoria Falls. *waterfall* Argentina/Brazil *see also* Iguazú, Cataratas del
Iguaçu, Salto do *see* Iguazú, Cataratas del
41 O15 **Iguala** *var.* Iguala de la Independencia. Guerrero, S Mexico 18°21′N 99°31′W
105 V5 **Igualada** Cataluña, NE Spain 41°35′N 01°37′E
Iguala de la Independencia *see* Iguala
60 G12 **Iguazú, Cataratas del** *Port.* Salto do Iguaçu; *prev.* Victoria Falls. *waterfall* Argentina/Brazil *see also* Iguaçu, Salto do
Iguazú, Cataratas del *see* Iguaçu, Salto do
62 Q6 **Iguazú, Río** *var.* Rio Iguaçu. Argentina/Brazil *see also* Iguaçu, Rio
Iguazú, Río *see* Iguaçu, Rio
79 D19 **Iguéla** *prev.* Iguéla. Ogooué-Maritime, SW Gabon 02°00′S 09°23′E
Iguéla *see* Iguéla
Iguid, Erg *see* Iguîdi, 'Erg
67 M5 **Iguîdi, 'Erg** *var.* Erg Iguid. *desert* Algeria/Mauritania
172 K2 **Iharaña** *prev.* Vohémar. Antsiranana, NE Madagascar 13°22′S 50°00′E
151 K18 **Ihavandhippolhu Atoll** *var.* Ihavandiffulu Atoll. *atoll* N Maldives
Ihavandiffulu Atoll *see* Ihavandhippolhu Atoll
Ih Bulag *see* Hanbogd
165 T16 **Iheya-jima** *island* Nansei-shotō, SW Japan
Ihhayrhan *see* Bayan-Önjüül
163 N9 **Ihhet** *var.* Bayan. Dornogovĭ, SE Mongolia 46°15′N 110°16′E
172 I6 **Ihosy** Fianarantsoa, S Madagascar 22°23′S 46°09′E
Ihsüüj *see* Bayanchandmanĭ
162 I7 **Ihtamir** *var.* Dzaanhushuu. Arhangay, C Mongolia 47°36′N 101°06′E
114 H10 **Ihtiman** Sofia, W Bulgaria 42°26′N 23°49′E
162 H6 **Ih-Uul** *var.* Bayan-Uhaa. Dzavhan, C Mongolia 48°41′N 98°46′E
162 J6 **Ih-Uul** *var.* Selenge. Hövsgöl, N Mongolia 49°25′N 101°30′E
93 L14 **Ii** Pohjois-Pohjanmaa, C Finland 65°18′N 25°23′E
164 M13 **Iida** Nagano, Honshū, S Japan 35°32′N 137°48′E
93 M14 **Iijoki** C Finland
Iinnasuolu *see* Hinnøya
118 J4 **Iisaku** *Ger.* Isaak. Ida-Virumaa, NE Estonia 59°06′N 27°19′E
93 M16 **Iisalmi** *var.* Idensalmi. Pohjois-Savo, C Finland 63°32′N 27°10′E
Ii-shima *see* Ie-jima
165 N11 **Iiyama** Nagano, Honshū, S Japan 36°52′N 138°22′E
77 S16 **Ijebu-Ode** Ogun, SW Nigeria 06°46′N 03°57′E
137 U11 **Ijevan** *Rus.* Idzhevan. N Armenia 40°53′N 45°07′E
98 H9 **IJmuiden** Noord-Holland, W Netherlands 52°28′N 04°38′E
98 M12 **IJssel** *var.* Yssel. Netherlands
98 J8 **IJsselmeer** *prev.* Zuider Zee. N Netherlands
98 L9 **IJsselmuiden** Overijssel, E Netherlands 52°34′N 05°55′E
98 I12 **IJsselstein** Utrecht, C Netherlands 52°01′N 05°02′E
61 G14 **Ijuí** Rio Grande do Sul, S Brazil 28°23′S 53°55′W
61 G14 **Ijuí, Rio** S Brazil
189 R8 **Ijuw** NE Nauru 0°30′S 166°57′E
99 E16 **IJzendijke** Zeeland, SW Netherlands 51°20′N 03°36′E
99 A18 **Ijzer** W Belgium
Ikaahuk *see* Sachs Harbour
93 K18 **Ikaalinen** Pirkanmaa, W Finland 61°46′N 23°05′E
172 I6 **Ikalamavony** Fianarantsoa, SE Madagascar 21°10′S 46°35′E
Ikaluktutiak *see* Cambridge Bay
185 G16 **Ikamatua** West Coast, South Island, New Zealand 42°16′S 171°42′E
145 P16 **Ikan** *prev.* Staroikan. Yuzhnyy Kazakhstan, S Kazakhstan 43°09′N 68°34′E
77 U16 **Ikare** Ondo, SW Nigeria 07°36′N 05°52′E
115 L20 **Ikaría** *var.* Kariot, Nicaria, Nikaria; *anc.* Icaria. *island* Dodekánisa, Greece, Aegean Sea
95 F22 **Ikast** Midtjylland, W Denmark 56°09′N 09°10′E
184 O9 **Ikawhenua Range** North Island, New Zealand
165 U4 **Ikeda** Hokkaidō, NE Japan 42°54′N 143°25′E
164 H14 **Ikeda** Tokushima, Shikoku, SW Japan 34°00′N 133°47′E
77 S16 **Ikeja** Lagos, SW Nigeria 06°36′N 03°16′E
79 L19 **Ikela** Equateur, C Dem. Rep. Congo 01°11′S 23°16′E
164 C13 **Iki** *prev.* Gōnoura. Nagasaki, Iki, SW Japan 33°44′N 129°41′E
164 C13 **Iki** *island* SW Japan
127 O13 **Iki Burul** Respublika Kalmykiya, SW Russian Federation 45°48′N 44°44′E
137 P11 **İkizdere** Rize, NE Turkey 40°47′N 40°34′E
39 P14 **Ikolik, Cape** *headland* Kodiak Island, Alaska, USA 57°12′N 154°46′W
77 V17 **Ikom** Cross River, SE Nigeria 05°57′N 08°43′E
172 I6 **Ikongo** *prev.* Fort-Carnot. Fianarantsoa, SE Madagascar 21°52′S 47°27′E
39 P5 **Ikpikpuk River** Alaska, USA
190 H1 **Iku** *prev.* Lone Tree Islet. *atoll* Tungaru, W Kiribati
164 I12 **Ikuno** Hyōgo, Honshū, SW Japan 35°13′N 134°48′E
190 H16 **Ikurangi** Rarotonga, S Cook Islands 21°12′S 159°45′W
171 X14 **Ilaga** Papua, E Indonesia 03°54′S 137°30′E
171 O2 **Ilagan** Luzon, N Philippines 17°08′N 121°54′E
142 J7 **Īlām** *var.* Elam. Īlām, W Iran 33°37′N 46°27′E
153 R12 **Ilām** Eastern, E Nepal 26°52′N 87°58′E
142 J8 **Īlām** *off.* Ostān-e Īlām. ◐ *province* W Iran
Īlām, Ostān-e *see* Īlām
161 T13 **Ilan** *Jap.* Giran. N Taiwan 24°45′N 121°44′E
146 G9 **Ilanly Obvodnitel'nyy Kanal** *canal* N Turkmenistan
122 L12 **Ilanskiy** Krasnoyarskiy Kray, S Russian Federation 56°16′N 95°59′E
108 H9 **Ilanz** Graubünden, S Switzerland 46°46′N 09°10′E
77 S16 **Ilaro** Ogun, SW Nigeria 06°52′N 03°01′E
57 I17 **Ilave** Puno, S Peru 16°07′S 69°40′W
110 K8 **Iława** *Ger.* Deutsch-Eylau. Warmińsko-Mazurskie, NE Poland 53°36′N 19°35′E
121 P16 **Il-Bajja ta' Marsaxlokk** *var.* Marsaxlokk Bay. *bay* SE Malta
123 P10 **Ilbenge** Respublika Sakha (Yakutiya), NE Russian Federation 62°52′N 124°13′E
Ile *see* Ili/Ili He
Ile *see* Ili He
11 S13 **Île-à-la-Crosse** Saskatchewan, C Canada 55°29′N 108°00′W
79 J21 **Ilebo** *prev.* Port-Francqui. Kasai-Occidental, W Dem. Rep. Congo 04°19′S 20°32′E
103 N5 **Île-de-France** ◐ *region* N France
Ilek *see* Yelek
Ilerda *see* Lleida
77 T16 **Ilesha** Osun, SW Nigeria 07°35′N 04°49′E
187 Q16 **Îles Loyauté, Province des** ◐ *province* E New Caledonia
11 X12 **Ilford** Manitoba, C Canada 56°02′N 95°48′W
116 K14 **Ilfov** ◐ *county* S Romania
97 I23 **Ilfracombe** SW England, United Kingdom 51°12′N 04°10′W
136 I11 **Ilgaz Dağları** N Turkey
136 G15 **Ilgın** Konya, W Turkey 38°16′N 31°57′E
60 I7 **Ilha Solteira** São Paulo, S Brazil 20°28′S 51°19′W
104 G7 **Ílhavo** Aveiro, N Portugal 40°36′N 08°40′W
59 O18 **Ilhéus** Bahia, E Brazil 14°50′S 39°06′W
129 R7 **Ili** *var.* Ile, *Chin.* Ili He, *Rus.* Reka Ili. China/Kazakhstan *see also* Ili He
116 G11 **Ilia** *Hung.* Marosillye. Hunedoara, SW Romania 45°57′N 22°40′E
39 P13 **Iliamna** Alaska, USA 59°42′N 154°49′W
39 P13 **Iliamna Lake** Alaska, USA
137 N13 **İliç** Erzincan, C Turkey 39°27′N 38°34′E
Il'ichevsk *see* Şärur, Azerbaijan
Il'ichevsk *see* Illichivs'k, Ukraine
Ilici *see* Elche
37 V2 **Iliff** Colorado, C USA 40°46′N 103°04′W
171 Q7 **Iligan** *off.* Iligan City. Mindanao, S Philippines 08°12′N 124°16′E
171 Q7 **Iligan Bay** *bay* S Philippines
Iligan City *see* Iligan
158 I5 **Ili He** *var.* Ili, *Kaz.* Ile, *Rus.* Reka Ili. China/Kazakhstan *see also* Ile
Ili He *see* Ili
56 C6 **Iliniza** N Ecuador 0°37′S 78°41′W
Ilinski *see* Il'inskiy
125 U14 **Il'inskiy** *var.* Ilinski. Permskiy Kray, NW Russian Federation 58°33′N 55°31′E
125 U14 **Il'inskiy** Ostrov Sakhalin, Sakhalinskaya Oblast', SE Russian Federation 47°59′N 142°14′E
18 I10 **Ilion** New York, NE USA 43°01′N 75°02′W
38 E9 **'Ilio Point** *var.* Ilio Point. *headland* Moloka'i, Hawai'i, USA 21°13′N 157°15′W
Ilio Point *see* 'Ilio Point

◆ Country
● Country Capital
◇ Dependent Territory
○ Dependent Territory Capital
◐ Administrative Regions
✈ International Airport
▲ Mountain
▲ Mountain Range
Volcano
River
Lake
Reservoir

109 T13 **Ilirska Bistrica** *prev.* Bistrica, *Ger.* Feistritz, Illyrisch-Feistritz, *It.* Villa del Nevoso. SW Slovenia 45°34´N 14°12´E
137 Q16 **Ilisu Baraji** ◙ SE Turkey
155 G17 **Ilkal** Karnātaka, C India 15°59´N 76°08´E
97 M19 **Ilkeston** C.England, United Kingdom 52°59´N 01°18´W
121 O16 **Il-Kullana** *headland* SW Malta 35°49´N 14°26´E
108 J8 **Ill** ☒ W Austria
103 U6 **Ill** ☒ NE France
62 G10 **Illapel** Coquimbo, C Chile 31°40´S 71°13´W
Illaue Fartak Trench *see* Alula-Fartak Trench
182 C2 **Illbillee, Mount** ▲ South Australia 27°01´S 132°13´E
102 I6 **Ille-et-Vilaine** ◇ *department* NW France
77 T11 **Illéla** Tahoua, SW Niger 14°25´N 05°10´E
101 J24 **Iller** ☒ S Germany
121 J23 **Illertissen** Bayern, S Germany 48°13´N 10°08´E
105 X9 **Illes Baleares** ◆ *autonomous community* E Spain
105 N8 **Illescas** Castilla-La Mancha, C Spain 40°08´N 03°51´W
Ille-sur-la-Têt *see* Ille-sur-la-Têt
103 O17 **Ille-sur-la-Têt** *var.* Ille-sur-la-Têt. Pyrénées-Orientales, S France 42°40´N 02°37´E
Illiberis *see* Elne
117 P11 **Illichivs'k** *Rus.* Il'ichevsk. Odes'ka Oblast', SW Ukraine 46°18´N 30°36´E
Illiers *see* Elche
102 M6 **Illiers-Combray** Eure-et-Loir, C France 48°18´N 01°15´E
30 K12 **Illinois** *off.* State of Illinois, *also known as* Prairie State, Sucker State. ◇ *state* C USA
30 J13 **Illinois River** ☒ Illinois, N USA
117 N6 **Illintsi** Vinnyts'ka Oblast', C Ukraine 49°07´N 29°13´E
74 M10 **Illizi** SE Algeria 26°30´N 08°28´E
27 Y7 **Illmo** Missouri, C USA 37°13´N 89°30´W
Illurco *see* Lorca
Illuro *see* Mataró
Illyrisch-Feistritz *see* Ilirska Bistrica
101 K16 **Ilm** ☒ C Germany
101 K17 **Ilmenau** Thüringen, C Germany 50°40´N 10°55´E
124 H14 **Il'men', Ozero** ◎ NW Russian Federation
57 H18 **Ilo** Moquegua, SW Peru 17°42´S 71°20´W
171 O6 **Iloilo** *off.* Iloilo City. Panay Island, C Philippines 10°42´N 122°34´E
Iloilo City *see* Iloilo
112 K10 **Ilok** *Hung.* Újlak. Vojvodina, NW Serbia 45°12´N 19°23´E
93 O16 **Ilomantsi** Pohjois-Karjala, SE Finland 62°40´N 30°55´E
42 F8 **Ilopango, Lago de** *volcanic lake* C El Salvador
77 T15 **Ilorin** Kwara, W Nigeria 08°32´N 04°35´E
117 X8 **Ilovays'k** *Rus.* Ilovaysk. Donets'ka Oblast', SE Ukraine 47°55´N 38°14´E
Ilovaysk *see* Ilovays'k
127 O10 **Ilovlya** Volgogradskaya Oblast', SW Russian Federation 49°45´N 44°19´E
127 O10 **Ilovlya** ☒ SW Russian Federation
121 N15 **Il-Ponta ta' San Dimitri** *var.* Ras San Dimitri, San Dimitri Point. *headland* Gozo, NW Malta 36°04´N 14°12´E
126 K14 **Il'skiy** Krasnodarskiy Kray, SW Russian Federation 44°52´N 38°26´E
182 B2 **Iltur** South Australia 27°33´S 130°31´E
171 V13 **Ilugwa** Papua, E Indonesia 03°42´S 139°09´E
Iluh *see* Batman
118 I11 **Ilūkste** SE Latvia 55°58´N 26°21´E
196 N13 **Ilulissat** Qaasuitsup, C Greenland 68°13´N 51°06´W
171 V13 **Ilur** Pulau Gorong, E Indonesia 04°00´S 131°25´E
32 F10 **Ilwaco** Washington, NW USA 46°19´N 124°03´W
Il'yaly *var* Gurbansoltan Eje
Ilyasbaba Burnu *see* Tekke Burnu
125 U9 **Ilych** ☒ NW Russian Federation
121 O21 **Ilz** ☒ SE Germany
111 M14 **Iłża** Radom, SE Poland 51°09´N 21°15´E
164 G13 **Imabari** *var.* Imaharu. Ehime, Shikoku, SW Japan 34°04´N 132°59´E
Imaharu *see* Imabari
165 O12 **Imaichi** *var.* Imaiti. Tochigi, Honshū, S Japan 36°43´N 139°41´E
Imaiti *see* Imaichi
164 K12 **Imajō** Fukui, Honshū, SW Japan 35°45´N 136°10´E
139 R9 **Imām Ibn Hāshim** Karbalā', C Iraq 32°46´N 43°21´E
149 Q2 **Imām Şāḩib** *var.* Emam Saheb, Hazarat Imam; *prev.* Emām Şāḩeb. Kunduz, NE Afghanistan 37°11´N 68°55´E
139 T11 **Imān 'Abd Allāh** Al Qādisīyah, S Iraq 31°36´N 44°34´E
164 C13 **Imari** Saga, Kyūshū, SW Japan 32°51´N 132°48´E
164 C13 **Imari** Saga, Kyūshū, SW Japan 33°18´N 129°51´E
Imarssuak Mid-Ocean Seachannel *see* Imarssuak Seachannel
64 J6 **Imarssuak Seachannel** *var.* Imarssuak Mid-Ocean Seachannel. *channel* N Atlantic Ocean
93 N18 **Imatra** Etelä-Karjala, SE Finland 61°14´N 28°50´E
164 K13 **Imazu** Shiga, Honshū, SW Japan 35°25´N 136°00´E
56 C6 **Imbabura** ◆ *province* N Ecuador
55 U9 **Imbaimadai** W Guyana 05°44´N 60°23´W
61 Q7 **Imbituba** Santa Catarina, S Brazil 28°15´S 48°44´W
27 W2 **Imboden** Arkansas, C USA 36°12´N 91°10´W
Imbros *see* Gökçeada

Imeni 26 Bakinskikh Komissarov *see* Uzboý
125 N13 **Imeni Babushkina** Vologodskaya Oblast', NW Russian Federation 59°40´N 43°54´E
126 J7 **Imeni Karla Libknekhta** Kurskaya Oblast', W Russian Federation 51°36´N 35°28´E
Imeni Mollanepesa *see* Mollanepes Adyndaky
Imeni S. A. Niyazova *see* S. A.Nyyazow Adyndaky
Imeni Sverdlova Rudnik *see* Sverdlovs'k
188 E9 **Imeong** Babeldaob, N Palau
81 L14 **Imī** Sumalē, E Ethiopia 06°27´N 42°10´E
115 M21 **Imia** *Turk.* Kardak. *island* Dodekánisa, Greece, Aegean Sea
Imishli *see* Imişli
137 X12 **Imişli** *Rus.* Imishli. C Azerbaijan 39°54´N 48°04´E
185 X14 **Imiŋ-gang** ☒ North Korea/South Korea
51 S3 **Imlay** Nevada, W USA 40°39´N 118°10´W
31 S9 **Imlay City** Michigan, N USA 43°01´N 83°04´W
23 X15 **Immokalee** Florida, SE USA 26°24´N 81°25´W
77 U17 **Imo** ◆ *state* SE Nigeria
106 G10 **Imola** Emilia-Romagna, N Italy 44°22´N 11°43´E
186 A5 **Imonda** West Sepik, NW Papua New Guinea 03°21´S 141°10´E
Imoschi *see* Imotski
113 G14 **Imotski** *It.* Imoschi. Split-Dalmacija, SE Croatia 43°28´N 17°13´E
59 L14 **Imperatriz** Maranhão, NE Brazil 05°32´S 47°28´W
106 B10 **Imperia** Liguria, NW Italy 43°53´N 08°03´E
57 E15 **Imperial** Lima, W Peru 13°04´S 76°21´W
35 X17 **Imperial** California, W USA 32°51´N 115°34´W
28 L16 **Imperial** Nebraska, C USA 40°30´N 101°37´W
24 M9 **Imperial** Texas, SW USA 31°15´N 102°40´W
35 Y **Imperial Dam** *dam* California, W USA
79 I17 **Impfondo** Likouala, NE Congo 01°37´N 18°04´E
153 X14 **Imphāl** *state capital* Manipur, NE India 24°47´N 93°55´E
103 P9 **Imphy** Nièvre, C France 46°55´N 03°16´E
106 G11 **Impruneta** Toscana, C Italy 43°42´N 11°16´E
115 K15 **Imroz** *var.* Gökçeada. Çanakkale, NW Turkey 40°06´N 25°50´E
Imroz Adası *see* Gökçeada
108 L7 **Imst** Tirol, W Austria 47°14´N 10°45´E
40 F3 **Imuris** Sonora, NW Mexico 30°36´N 110°52´W
164 M13 **Ina** Nagano, Honshū, S Japan 35°50´N 137°58´E
65 M18 **Inaccessible Island** *island* W Tristan da Cunha
115 F20 **Inachos** ☒ S Greece
188 H6 **I Naftan, Puntan** *headland* Saipan, S Northern Mariana Islands
Inagua Islands *see* Little Inagua
Inagua Islands *see* Great Inagua
185 H15 **Inangahua** West Coast, South Island, New Zealand 41°51´S 171°58´E
57 I18 **Iñapari** Madre de Dios, E Peru 11°00´S 69°34´W
188 B17 **Inarajan** SE Guam 13°16´N 144°45´E
92 L10 **Inari** *Lapp.* Anár, Aanaar., Lapp’n, N Finland 68°54´N 27°06´E
92 L10 **Inarijärvi** *Lapp.* Inarinjärvi, *Swe.* Enareträsk. ◎ N Finland
92 L9 **Inarijoki** *Lapp.* Anárjohka. ☒ Finland/Norway
Inarinjärvi *see* Inarijärvi
165 P11 **Inawashiro-ko** *var.* Inawasiro Ko. ◎ Honshū, C Japan
Inawasiro Ko *see* Inawashiro-ko
105 X9 **Inca** Mallorca, Spain, W Mediterranean Sea 39°43´N 02°54´E
62 H7 **Inca de Oro** Atacama, N Chile 26°45´S 69°54´W
115 I25 **İnce Burnu** *cape* NW Turkey
136 K9 **İnce Burnu** *headland* N Turkey 42°06´N 34°57´E
136 H17 **İncekum Burnu** *headland* S Turkey 36°13´N 33°57´E
183 X15 **Incheon** *Jap.* Jinsen; *prev.* Chemulpo, Inch'ŏn. NW South Korea 37°27´N 126°41´E
161 X15 **Incheon** ✈ (Seoul) NW South Korea 37°27´N 126°42´E
76 G7 **Inchiri** ◆ *region* NW Mauritania
Inch'ŏn *see* Incheon
83 I24 **Inchope** Manica, C Mozambique 19°09´S 33°54´E
Incoronata *see* Kornat
91 Y15 **Incudine, Monte** ▲ Corse, France, C Mediterranean Sea 41°52´N 09°13´E
60 M10 **Indaiatuba** São Paulo, S Brazil 23°03´S 47°11´W
93 H17 **Indal** Västernorrland, C Sweden 62°36´N 17°06´E
93 H16 **Indalsälven** ☒ C Sweden
40 K8 **Indé** Durango, C Mexico 25°55´N 105°10´W
Indefatigable Island *see* Santa Cruz, Isla
35 S10 **Independence** California, W USA 36°48´N 118°13´W
23 X3 **Independence** Iowa, C USA 42°28´N 91°42´W
25 P7 **Independence** Kansas, C USA 37°13´N 95°43´W
20 M4 **Independence** Kentucky, S USA 38°56´N 84°32´W
27 T3 **Independence** Missouri, C USA 39°04´N 94°27´W
21 R6 **Independence** Virginia, NE USA 36°38´N 81°11´W
30 L6 **Independence** Wisconsin, N USA 44°21´N 91°25´W
197 R12 **Independence Fjord** *fjord* N Greenland
Independence Island *see* Malden Island
35 W2 **Independence Mountains** ▲ Nevada, W USA

57 K18 **Independencia** Cochabamba, C Bolivia 17°08´S 66°52´W
57 E16 **Independencia, Bahía de la** *bay* W Peru
Independencia, Monte *see* Adam, Mount
118 M12 **Independenţa** Galaţi, SE Romania 45°29´N 27°45´E
Independencia *see* Indragiri, Sungai
144 F11 **Inderbor** *prev.* Inderborskiy. Atyrau, W Kazakhstan 48°35´N 51°45´E
Inderborskiy *see* Inderbor
151 I14 **India** *off.* Republic of India, *var.* Indian Union, Union of India, *Hind.* Bhārat. ◆ *republic* S Asia
India *see* India
18 D14 **Indiana** Pennsylvania, NE USA 40°37´N 79°09´W
31 N13 **Indiana** *off.* State of Indiana, *also known as* Hoosier State. ◇ *state* N USA
31 O14 **Indianapolis** *state capital* Indiana, N USA 39°46´N 86°09´W
11 O10 **Indian Cabins** Alberta, W Canada 59°51´N 117°06´W
23 X15 **Indian Church** Orange Walk, N Belize 17°47´N 88°39´W
Indian Desert *see* Thar Desert
11 U16 **Indian Head** Saskatchewan, S Canada 50°32´N 103°41´W
31 O4 **Indian Lake** ◎ Michigan, N USA
18 K9 **Indian Lake** ◎ New York, NE USA
31 R13 **Indian Lake** ◎ Ohio, N USA
172-173 **Indian Ocean** *ocean*
29 V2 **Indianola** Iowa, C USA 41°21´N 93°33´W
22 K4 **Indianola** Mississippi, S USA 33°27´N 90°39´W
36 J6 **Indian Peak** ▲ Utah, W USA 38°18´N 113°52´W
23 Y13 **Indian River** *lagoon* Florida, SE USA
35 W10 **Indian Springs** Nevada, W USA 36°33´N 115°40´W
23 Y14 **Indiantown** Florida, SE USA 27°01´N 80°29´W
Indian Union *see* India
59 K19 **Indiara** Goiás, S Brazil 17°12´S 50°09´W
125 Q4 **Indiga** Nenetskiy Avtonomnyy Okrug, NW Russian Federation 67°40´N 49°01´E
123 R9 **Indigirka** ☒ NE Russian Federation
112 L10 **Inđija** *Hung.* India; *prev.* Indjija. Vojvodina, N Serbia 45°03´N 20°04´E
35 V16 **Indio** California, W USA 33°42´N 116°13´W
42 M12 **Indio, Río** ☒ SE Nicaragua
152 I10 **Indira Gandhi** ✈ (Delhi) Delhi, N India
151 Q23 **Indira Point** *headland* Andaman and Nicobar Islands, India, NE Indian Ocean 6°54´N 93°54´E
Indjija *see* Inđija
129 Q13 **Indo-Australian Plate** *tectonic feature*
173 N11 **Indomed Fracture Zone** *tectonic feature* SW Indian Ocean
170 L12 **Indonesia** *off.* Republic of Indonesia, *prev.* Dutch East Indies, Netherlands East Indies, United States of Indonesia. ◆ *republic* SE Asia
Indonesian Borneo *see* Kalimantan
Indonesia, Republic of *see* Indonesia
Indonesia, Republik *see* Indonesia
Indonesia, United States of *see* Indonesia
154 G10 **Indore** Madhya Pradesh, C India 22°42´N 75°51´E
168 L11 **Indragiri, Sungai** *var.* Batang Kuantan, Inderagiri. ☒ Sumatera, W Indonesia
Indramaiu *see* Indramayu
169 P15 **Indramayu** *prev.* Indramajoe, Indramaju. Jawa, C Indonesia 06°22´S 108°20´E
155 G10 **Indrāvati** ☒ S India
103 N9 **Indre** ◇ *department* C France
102 M8 **Indre** ☒ C France
94 D13 **Indre Ålvik** Hordaland, S Norway 60°26´N 06°32´E
102 L8 **Indre-et-Loire** ◇ *department* C France
Indreville *see* Châteauroux
152 J3 **Indus** *Chin.* Yindu He; *prev.* Yin-tu Ho. ☒ S Asia
173 P3 **Indus Cone** *see* Indus Fan
173 P3 **Indus Fan** *var.* Indus Cone. *undersea feature* N Arabian Sea 16°00´N 66°30´E
149 P17 **Indus, Mouths of the** *delta* S Pakistan
83 I24 **Indwe** Eastern Cape, SE South Africa 31°28´S 27°51´E
31 I10 **Inebolu** Kastamonu, N Turkey 41°55´N 33°54´E
77 P8 **I-n-Échaï** *oasis* C Mali
114 M13 **Inecik** Tekirdağ, NW Turkey 40°55´N 27°16´E
136 F10 **İnegöl** Bursa, NW Turkey 40°06´N 29°31´E
60 M15 **Inês, Monte** ...
108 M7 **Ineu** *Hung.* Borosjenő; *prev.* Ináu. Arad, W Romania 46°25´N 21°51´E
Ineul/Ineu, Vîrful *see* Ineu, Vârful
116 J9 **Ineu, Vârful** *var.* Ineul; *prev.* Virful Ineu. ▲ N Romania 47°31´N 24°52´E
21 P6 **Inez** Kentucky, S USA 37°53´N 82°33´W
74 E8 **Inezgane** ✈ (Agadir) W Morocco 30°35´N 09°27´W
40 M15 **Infiernillo, Presa del** ◎ S Mexico
Infiesto *see* L'Infiestu
105 L20 **Ingá** *Fin.* Inkoo. Uusimaa, S Finland 60°03´N 24°00´E
77 U10 **Ingal** *var.* I-n-Gall. Agadez, C Niger 16°52´N 06°57´E
I-n-Gall *see* Ingal
90 C18 **Ingelmunster** West-Vlaanderen, W Belgium 50°55´N 03°15´E

79 I18 **Ingende** Equateur, W Dem. Rep. Congo 0°15´S 18°58´E
62 L5 **Ingeniero Guillermo Nueva Juárez** Formosa, N Argentina 23°55´S 61°50´W
63 H16 **Ingeniero Jacobacci** Río Negro, C Argentina 41°18´S 69°35´W
14 F16 **Ingersoll** Ontario, S Canada 43°03´N 80°53´W
Ingettolgoy *see* Selenge
181 W5 **Ingham** Queensland, NE Australia 18°35´S 146°12´E
146 M11 **Ingichka** Samarqand Viloyati, C Uzbekistan 39°46´N 65°56´E
97 L16 **Ingleborough** ▲ N England, United Kingdom 54°07´N 02°22´W
25 T14 **Inglewood** Taranaki, North Island, New Zealand 39°07´S 174°13´E
35 S15 **Inglewood** California, W USA 33°57´N 118°21´W
101 L21 **Ingolstadt** Bayern, S Germany 48°46´N 11°26´E
33 V9 **Ingomar** Montana, N USA 46°34´N 107°21´W
13 R14 **Ingonish Beach** Cape Breton Island, Nova Scotia, SE Canada 46°42´N 60°22´W
153 S14 **Ingrāj Bāzār** *prev.* English Bazar. West Bengal, NE India 25°00´N 88°10´E
24 Q11 **Ingram** Texas, SW USA 30°04´N 99°14´W
195 X7 **Ingrid Christensen Coast** *physical region* Antarctica
74 K14 **I-n-Guezzam** S Algeria 19°35´N 05°49´E
Ingulets *see* Inhulets'
Inguri *see* Enguri
Ingushetia/Ingushetiya, Respublika *see* Ingushetiya, Respublika
127 O15 **Ingushetiya, Respublika** *var.* Respublika Ingushetiya, *Eng.* Ingushetia. ◆ *autonomous republic* SW Russian Federation
83 N20 **Inhambane** Inhambane, SE Mozambique 23°52´S 35°31´E
83 M20 **Inhambane** *off.* Província de Inhambane. ◇ *province* S Mozambique
Inhambane, Província de *see* Inhambane
83 N17 **Inhaminga** Sofala, C Mozambique 18°24´S 35°00´E
83 N20 **Inharrime** Inhambane, SE Mozambique 24°29´S 35°01´E
83 M18 **Inhassoro** Inhambane, E Mozambique 21°32´S 35°13´E
117 S9 **Inhulets'** *Rus.* Ingulets. Dnipropetrovs'ka Oblast', E Ukraine 47°43´N 33°12´E
117 R10 **Inhulets'** *Rus.* Ingulets. ☒ S Ukraine
105 Q10 **Iniesta** Castilla-La Mancha, C Spain 39°26´N 01°43´W
I-ning *see* Yining
54 K11 **Inírida, Río** ☒ E Colombia
Inis *see* Ennis
Inis Ceithleann *see* Enniskillen
Inis Córthaidh *see* Enniscorthy
Inis Diomáin *see* Ennistimon
97 A17 **Inishbofin** *Ir.* Inis Bó Finne. *island* W Ireland
97 B18 **Inisheer** *var.* Inishere, *Ir.* Inis Oírr. *island* W Ireland
97 B18 **Inishmaan** *Ir.* Inis Meáin. *island* W Ireland
97 A18 **Inishmore** *Ir.* Árainn. *island* W Ireland
96 E13 **Inishtrahull** *Ir.* Inis Trá Tholl. *island* NW Ireland
97 A17 **Inishturk** *Ir.* Inis Toirc. *island* W Ireland
Inkoo *see* Ingå
185 J16 **Inland Kaikoura Range** ▲ South Island, New Zealand
Inland Sea *see* Seto-naikai
21 P11 **Inman** South Carolina, SE USA 35°02´N 82°05´W
108 L7 **Inn** ☒ C Europe
197 O11 **Innaanganeq** *var.* Kap York. *headland* NW Greenland 75°54´N 66°27´W
182 K2 **Innamincka** South Australia 27°47´S 140°45´E
92 G12 **Inndyr** Nordland, C Norway 67°01´N 14°00´E
127 W5 **Inner** Republika Bashkortostan, W Russian Federation 54°11´N 57°37´E
96 F11 **Inner Hebrides** *island group* W Scotland, United Kingdom
172 H15 **Inner Islands** *var.* Central Group. *island group* NE Seychelles
Inner Mongolia/Inner Mongolian Autonomous Region *see* Nei Mongol Zizhiqu
109 I7 **Inner Rhoden** *former canton* Appenzell. ◇ NW Switzerland
96 G8 **Inner Sound** *strait* NW Scotland, United Kingdom
100 J13 **Innerste** ☒ C Germany
181 W5 **Innisfail** Queensland, NE Australia 17°31´S 146°03´E
11 Q15 **Innisfail** Alberta, SW Canada 52°01´N 113°59´W
39 O11 **Innoko River** ☒ Alaska, USA
108 M7 **Innsbruck** *var.* Innsbruck. Tirol, W Austria 47°17´N 11°25´E
Innsbruck *see* Innsbruck
79 I19 **Inongo** Bandundu, W Dem. Rep. Congo 01°55´S 18°20´E
110 I10 **Inowrocław** *Ger.* Hohensalza; *prev.* Inowraclaw, Kujawski-pomorskie, C Poland 52°47´N 18°15´E
Inowraclaw *see* Inowrocław
115 B17 **Inoússes** *island group* E Greece
57 I17 **Inquisivi** La Paz, W Bolivia 16°55´S 67°10´W
Inrin *see* Yuanlin
77 O5 **I-n-Sâkâne, 'Erg** *var.* I-n-Sakane. *desert* C Mali
74 J10 **I-n-Salah** *var.* In Salah. C Algeria 27°11´N 02°31´E
127 O5 **Insar** Respublika Mordoviya, W Russian Federation 53°52´N 44°25´E
189 X15 **Insiaf** Kosrae, E Micronesia

94 L13 **Insjön** Dalarna, C Sweden 60°40´N 15°05´E
Insterburg *see* Chernyakhovsk
Insula *see* Lille
116 L13 **Insurāţei** Brăila, SE Romania 44°55´N 27°40´E
125 V6 **Inta** Respublika Komi, NW Russian Federation 66°00´N 60°10´E
77 R9 **In-Tebezas** Kidal, E Mali 17°58´N 01°51´E
Interamna *see* Teramo
Interamna Nahars *see* Terni
28 L11 **Interior** South Dakota, N USA 43°42´N 101°57´W
108 E9 **Interlaken** Bern, SW Switzerland 46°41´N 07°51´E
29 Y14 **International Falls** Minnesota, N USA 48°38´N 93°26´W
167 O7 **Inthanon, Doi** ▲ NW Thailand 18°33´N 98°29´E
42 G7 **Intibucá** ◇ *department* SW Honduras
42 G8 **Intibucá** *var.* La Unión, SE El Salvador 13°10´N 88°03´W
61 B15 **Intiyaco** Santa Fe, C Argentina 28°43´S 60°04´W
116 K12 **Întorsura Buzăului** *Ger.* Bozau, *Hung.* Bodzaforduló. Covasna, E Romania 45°40´N 26°02´E
22 H9 **Intracoastal Waterway** *inland waterway system* Louisiana, S USA
25 V13 **Intracoastal Waterway** *inland waterway system* Texas, SW USA
108 G11 **Intragna** Ticino, S Switzerland 46°12´N 08°42´E
165 P14 **Inubō-zaki** *headland* Honshū, S Japan 35°42´N 140°51´E
164 E14 **Inukai** Ōita, Kyūshū, SW Japan 33°05´N 131°37´E
12 I5 **Inukjuak** *var.* Inoucdjouac; *prev.* Port Harrison. Québec, NE Canada 58°28´N 78°15´W
63 I24 **Inútil, Bahía** *bay* S Chile
11 R8 **Inuvik** *var.* Inuuvik. Northwest Territories, NW Canada 68°25´N 133°35´W
164 L13 **Inuyama** Aichi, Honshū, SW Japan 35°23´N 136°56´E
56 G13 **Inuya, Río** ☒ E Peru
125 U13 **In'va** ☒ NW Russian Federation
96 H11 **Inveraray** W Scotland, United Kingdom 56°13´N 05°05´W
185 C24 **Invercargill** Southland, South Island, New Zealand 46°26´S 168°21´E
183 T5 **Inverell** New South Wales, SE Australia 29°46´S 151°10´E
96 I8 **Invergordon** N Scotland, United Kingdom 57°42´N 04°02´W
11 P14 **Invermere** British Columbia, SW Canada 50°30´N 116°00´W
13 P14 **Inverness** Cape Breton Island, Nova Scotia, SE Canada 46°14´N 61°19´W
96 I8 **Inverness** N Scotland, United Kingdom 57°27´N 04°15´W
23 V11 **Inverness** Florida, SE USA 28°50´N 82°19´W
96 I9 **Inverness** *cultural region* NW Scotland, United Kingdom
96 K9 **Inverurie** NE Scotland, United Kingdom 57°14´N 02°14´W
182 F8 **Investigator Group** *island group* South Australia
173 T7 **Investigator Ridge** *undersea feature* E Indian Ocean 11°30´S 98°10´E
182 H10 **Investigator Strait** *strait* South Australia
29 R11 **Inwood** Iowa, C USA 43°16´N 96°25´W
Inyanga *see* Nyanga
83 J17 **Inyathi** Matabeleland North, SW Zimbabwe 19°39´S 28°54´E
35 T12 **Inyokern** California, W USA 35°37´N 117°48´W
35 T10 **Inyo Mountains** ▲ California, W USA
127 P6 **Inza** Ul'yanovskaya Oblast', W Russian Federation 53°51´N 46°21´E
127 N7 **Inzhavino** Tambovskaya Oblast', W Russian Federation 52°20´N 42°28´E
115 C16 **Ioánnina** *var.* Janina, Yannina. Ípeiros, W Greece 39°39´N 20°52´E
164 B17 **Iō-jima** *var.* Iwojima. *island* Nansei-shotō, SW Japan
124 L4 **Iokan'ga** ☒ NW Russian Federation
27 Q6 **Iola** Kansas, C USA 37°55´N 95°24´W
115 G16 **Iolkós** *anc.* Iolcus. *site of ancient city* Thessalía, C Greece
Iólotan' *see* Ýolöten
83 A16 **Iona** Namibe, SW Angola 16°54´S 12°39´E
96 F11 **Iona** *island* W Scotland, United Kingdom
116 M13 **Ion Corvin** Constanţa, SE Romania 44°07´N 27°50´E
Iona *see* Iona
121 O10 **Ionian Sea** *Gk.* Iónio Pélagos, *It.* Mar Ionio. *sea* C Mediterranean Sea
138 G9 **Ionian Basin** *var.* Ionia Basin. *undersea feature* Ionian Sea
115 B17 **Iónioi Nísia** *var.* Iónioi Nísoi, *Eng.* Ionian Islands. *island group* W Greece
121 O10 **Iónioi Nísoi** *Eng.* Ionian Islands. ◆ *region* W Greece
Iónio Pélagos *see* Ionian Sea
Iordan *see* Yordon

137 U10 **Iori** *var.* Qabırrı. ☒ Azerbaijan/Georgia
Iorrais, Ceann *see* Erris Head
116 L13 **Íos** *var.* Nio. *island* Kykládes, Greece, Aegean Sea
165 V6 **Io-Tori-shima** *prev.* Tori-shima. *island* Izu-shotō, SE Japan
115 I20 **Ioulís** *prev.* Kéa. Tziá, Kykládes, Greece, Aegean Sea 37°40´N 24°19´E
22 G9 **Iowa** Louisiana, S USA 30°12´N 93°00´W
29 V13 **Iowa** *off.* State of Iowa, *also known as* Hawkeye State. ◇ *state* C USA
29 Y14 **Iowa City** Iowa, C USA 41°40´N 91°32´W
29 Y14 **Iowa Falls** Iowa, C USA 42°31´N 93°15´W
25 R4 **Iowa Park** Texas, SW USA 33°57´N 98°40´W
29 Y14 **Iowa River** ☒ Iowa, C USA
119 M19 **Ipa** ☒ SE Belarus
59 N20 **Ipatinga** Minas Gerais, SE Brazil 19°32´S 42°30´W
127 N13 **Ipatovo** Stavropol'skiy Kray, SW Russian Federation 45°40´N 42°51´E
115 C16 **Ipeiros** *Eng.* Epirus. ◆ *region* W Greece
111 J21 **Ipel'** *var.* Ipoly, *Ger.* Eipel. ☒ Hungary/Slovakia
54 C13 **Ipiales** Nariño, SW Colombia 0°52´N 77°38´W
189 V14 **Ipis** *atoll* Chuuk Islands, C Micronesia
59 Q11 **Ipixuna** Amazonas, W Brazil 06°57´S 71°42´W
168 L8 **Ipoh** Perak, Peninsular Malaysia 04°36´N 101°02´E
Ipoly *see* Ipel'
187 S15 **Ipota** Erromango, S Vanuatu 18°54´S 169°17´E
79 K19 **Ippy** Ouaka, C Central African Republic 06°11´N 21°13´E
114 L13 **Ipsala** Edirne, NW Turkey 40°56´N 26°23´E
Ipsario *see* Ypsário
183 T5 **Ipswich** Queensland, E Australia 27°38´S 152°40´E
97 Q20 **Ipswich** *hist.* Gipeswic. E England, United Kingdom 52°05´N 01°08´E
29 O8 **Ipswich** South Dakota, N USA 45°24´N 99°00´W
14 C10 **Iron Bridge** Ontario, S Canada 46°16´N 83°12´W
20 H10 **Iron City** Tennessee, S USA 35°01´N 87°34´W
14 I3 **Irondale** Ontario, SE Canada
182 H4 **Iron Knob** South Australia 32°43´S 137°03´E
30 M5 **Iron Mountain** Michigan, N USA 45°49´N 88°03´W
30 M4 **Iron River** Michigan, N USA 46°05´N 88°38´W
27 X7 **Ironton** Missouri, C USA 37°37´N 90°40´W
31 S15 **Ironton** Ohio, N USA 38°32´N 82°41´W
30 K4 **Ironwood** Michigan, N USA 46°27´N 90°10´W
12 H13 **Iroquois Falls** Ontario, S Canada 48°47´N 80°41´W
31 N12 **Iroquois River** ☒ Illinois/Indiana, N USA
164 O15 **Irō-zaki** *headland* Honshū, S Japan 34°36´N 138°49´E
Irpen' *see* Irpin'
117 O4 **Irpin'** *Rus.* Irpen'. Kyyivs'ka Oblast', N Ukraine 50°31´N 30°16´E
117 O4 **Irpin'** *Rus.* Irpen'. ☒ N Ukraine
141 Q16 **'Irqah** SW Yemen 13°42´N 47°21´E
166 L6 **Irrawaddy** *var.* Ayeyarwady. ☒ W Myanmar (Burma)
Irrawaddy *see* Ayeyarwady
166 K8 **Irrawaddy, Mouths of the** *delta* SW Myanmar (Burma)
117 N4 **Irsha** ☒ N Ukraine
116 H7 **Irshava** Zakarpats'ka Oblast', W Ukraine 48°19´N 23°03´E
107 J16 **Irsina** Basilicata, S Italy 40°42´N 16°18´E
Irtish *see* Yertis
Irtysh *see* Yertis
79 P17 **Irumu** Orientale, E Dem. Rep. Congo 01°27´N 29°52´E
105 Q2 **Irún** Spain 43°20´N 01°47´W
Irún *see* Irun
Iruña *see* Pamplona
105 Q3 **Irurtzun** Navarra, N Spain
99 I13 **Irvine** W Scotland, United Kingdom 55°37´N 04°40´W
21 N6 **Irvine** Kentucky, S USA 37°40´N 83°58´W
25 T6 **Irving** Texas, SW USA 32°47´N 96°57´W
20 J5 **Irvington** Kentucky, S USA 37°52´N 86°16´W
164 C15 **Isa** *prev.* Ōkuchi, Ōkuti. Kagoshima, Kyūshū, SW Japan 32°04´N 130°36´E
Isaak *see* Iisaku
186 L8 **Isabel** *off.* Isabel Province. ◇ *province* N Solomon Islands
171 O3 **Isabela** Basilan Island, SW Philippines 06°41´N 122°02´E
45 N8 **Isabela** W Puerto Rico 18°30´N 67°02´W
44 J7 **Isabela, Cabo** *headland* NW Dominican Republic 19°54´N 71°03´W
57 A18 **Isabela, Isla** *var.* Albemarle Island. *island* Galapagos Islands, Ecuador, E Pacific Ocean
40 K9 **Isabela, Isla** *island* C Mexico
42 K9 **Isabella, Cordillera** ▲ NW Nicaragua
35 R11 **Isabella Lake** ◎ California, W USA
31 N2 **Isabelle, Point** *headland* Michigan, N USA 47°20´N 87°56´W
Isabel Province *see* Isabel
Isabel Segunda *see* Vieques
116 M13 **Isaccea** Tulcea, E Romania 45°16´N 28°28´E
92 H1 **Ísafjarðardjúp** *inlet* NW Iceland

64 A12 **Ireland Island South** *island* W Bermuda
Ireland, Republic of *see* Ireland
125 V15 **Iren'** ☒ NW Russian Federation
185 A22 **Irene, Mount** ▲ South Island, New Zealand 45°04´S 167°24´E
Irgalem *see* Yirga 'Alem
Irgiz *see* Yrghyz
Irian *see* New Guinea
Irian Barat *see* Papua
Irian Jaya *see* Papua
Irian Jaya Barat *see* Barat
Irian, Teluk *see* Cenderawasih, Teluk
78 K9 **Iriba** Wadi Fira, NE Chad 15°10´N 22°11´E
127 X7 **Iriklinskoye Vodokhranilishche** ◎ W Russian Federation
81 H23 **Iringa** Iringa, C Tanzania 07°49´S 35°39´E
81 H23 **Iringa** ◆ *region* S Tanzania
165 O16 **Iriomote-jima** *island* Sakishima-shotō, SW Japan
42 L4 **Iriona** Colón, NE Honduras 15°55´N 85°10´W
47 U7 **Iriri** ☒ C Brazil
58 J13 **Iriri, Rio** ☒ C Brazil
Iris *see* Yeşilırmak
35 W9 **Irish, Mount** ▲ Nevada, W USA 37°38´N 115°22´W
97 H17 **Irish Sea** *Ir.* Muir Éireann. *sea* C British Isles
139 U12 **Irjā' ash Shaykhīyah** Al Muthanná, S Iraq 30°49´N 44°58´E
147 N12 **Irkeshtam** Oshskaya Oblast', SW Kyrgyzstan 39°39´N 73°49´E
122 M13 **Irkutsk** Irkutskaya Oblast', S Russian Federation 52°18´N 104°15´E
122 M12 **Irkutskaya Oblast'** ◆ *province* S Russian Federation
Irlir, Gora *see* Irlir Tog'i
147 P11 **Irlir Tog'i** *var.* Gora Irlir. ▲ N Uzbekistan 42°43´N 63°24´E
Irminger Basin *see* Reykjanes Basin
21 R12 **Irmo** South Carolina, SE USA 34°05´N 81°10´W
102 E6 **Iroise** *sea* NW France
189 X2 **Iroj** *var.* Eroj. *island* Ratak Chain, SE Marshall Islands
182 H7 **Iron Baron** South Australia 33°01´S 137°13´E

125 V15 **Iren'** (continued)
(The remainder of column 8 continues from Ireland entries:)
64 A12 **Ireland Island North** *island* W Bermuda
92 H1 **Ísafjarðardjúp** *inlet* NW Iceland

◆ Country   ◇ Dependent Territory   ◆ Administrative Regions   ▲ Mountain   🌋 Volcano   ◎ Lake
● Country Capital   ○ Dependent Territory Capital   ✈ International Airport   ▲▲ Mountain Range   ☒ River   Reservoir

263

ÍSAFJARÐARDJÚP – JAMAARE

92 H1 **Ísafjarðardjúp** *inlet* NW Iceland
92 H1 **Ísafjörður** Vestfirðir, NW Iceland 66°04′N 23°09′W
164 C14 **Isahaya** Nagasaki, Kyūshū, SW Japan 32°51′N 130°02′E
149 S7 **Īsa Khel** Punjab, E Pakistan 32°39′N 71°20′E
172 H7 **Isalo** *var.* Massif de L'Isalo. ▲▲ SW Madagascar
79 K20 **Isandja** Kasai-Occidental, C Dem. Rep. Congo 03°03′S 21°57′E
187 R15 **Isangel** Tanna, S Vanuatu 19°34′S 169°17′E
79 M18 **Isangi** Orientale, C Dem. Rep. Congo 0°46′N 24°15′E
101 L24 **Isar** ≈ Austria/Germany
101 M23 **Isar-Kanal** *canal* SE Germany
Isbarta *see* Isparta
Isca Damnoniorum *see* Exeter
107 K18 **Ischia** *var.* Isola d'Ischia; *anc.* Aenaria. Campania, S Italy 40°44′N 13°57′E
107 J18 **Ischia, Isola d'** *island* S Italy
54 B12 **Iscuandé** *var.* Santa Bárbara. Nariño, SW Colombia 02°32′N 78°00′W
164 K14 **Ise** Mie, Honshū, SW Japan 34°29′N 136°43′E
100 J12 **Ise** ≈ N Germany
95 I23 **Isefjord** *fjord* E Denmark
Iseghem *see* Izegem
192 M14 **Iselin Seamount** *undersea feature* S Pacific Ocean 72°30′S 179°00′W
Isenhof *see* Püssi
106 E7 **Iseo** Lombardia, N Italy 45°40′N 10°03′E
103 U12 **Iseran, Col de l'** *pass* E France
103 S11 **Isère** ⬖ *department* E France
103 S12 **Isère** ≈ E France
101 F15 **Iserlohn** Nordrhein-Westfalen, W Germany 51°23′N 07°42′E
107 K16 **Isernia** *var.* Æsernia. Molise, C Italy 41°35′N 14°14′E
165 N12 **Isesaki** Gunma, Honshū, S Japan 36°19′N 139°11′E
129 Q5 **Iset'** ≈ C Russian Federation
77 S15 **Iseyin** Oyo, W Nigeria 07°56′N 03°33′E
Isfahan *see* Eşfahān
147 Q11 **Isfana** Batkenskaya Oblast', SW Kyrgyzstan 39°51′N 69°31′E
147 R11 **Isfara** N Tajikistan 40°06′N 70°34′E
149 O4 **Isfi Maidān** Gōwr, N Afghanistan 35°09′N 66°16′E
92 O3 **Isfjorden** *fjord* W Svalbard
Isgender *see* Kul'mach
Isha Baydhabo *see* Baydhabo
125 V11 **Isherim, Gora** ▲ NW Russian Federation 61°06′N 59°09′E
127 Q5 **Isheyevka** Ul'yanovskaya Oblast', W Russian Federation 54°27′N 48°18′E
165 P16 **Ishigaki** Okinawa, Ishigaki-jima, SW Japan 24°20′N 124°09′E
165 P16 **Ishigaki-jima** *island* Sakishima-shotō, SW Japan
165 R3 **Ishikari-wan** *bay* Hokkaidō, NE Japan
165 S16 **Ishikawa** *var.* Isikawa. Okinawa, Okinawa, SW Japan 26°25′N 127°47′E
164 K11 **Ishikawa** *off.* Ishikawa-ken, *var.* Isikawa. ⬖ *prefecture* Honshū, SW Japan
Ishikawa-ken *see* Ishikawa
122 H11 **Ishim** Tyumenskaya Oblast', C Russian Federation 56°13′N 69°25′E
127 V6 **Ishimbay** Respublika Bashkortostan, W Russian Federation 53°21′N 56°03′E
145 O9 **Ishimskoye** Akmola, C Kazakhstan 51°23′N 67°07′E
165 Q10 **Ishinomaki** *var.* Isinomaki. Miyagi, Honshū, C Japan 38°26′N 141°17′E
165 P13 **Ishioka** *var.* Isioka. Ibaraki, Honshū, S Japan 36°11′N 140°16′E
149 Q3 **Ishkamish** *prev.* Eshkamesh. Takhār, NE Afghanistan 36°25′N 69°11′E
149 T2 **Ishkāshim** *prev.* Eshkāshem. Badakhshān, NE Afghanistan 36°43′N 71°34′E
Ishkashim *see* Ishkoshim
Ishkashimskiy Khrebet *see* Ishkoshim, Qatorkŭhi
147 S15 **Ishkoshim** *Rus.* Ishkashim. S Tajikistan 36°46′N 71°35′E
147 S15 **Ishkoshim, Qatorkŭhi** *Rus.* Ishkashimskiy Khrebet. ▲▲ SE Tajikistan
31 N4 **Ishpeming** Michigan, N USA 46°29′N 87°40′W
147 N11 **Ishtixon** *Rus.* Ishtykhan. Samarqand Viloyati, C Uzbekistan 39°59′N 66°28′E
Ishtykhan *see* Ishtixon
Ishurdi *see* Iswardi
61 G17 **Isidoro Noblia** Cerro Largo, NE Uruguay 31°58′S 54°09′W
102 J4 **Isigny-sur-Mer** Calvados, N France 49°20′N 01°06′W
Isikawa *see* Ishikawa
136 C11 **Işıklar Dağı** ▲▲ NW Turkey
107 C19 **Isili** Sardegna, Italy, C Mediterranean Sea 39°46′N 09°06′E
122 H12 **Isil'kul'** Omskaya Oblast', C Russian Federation 54°52′N 71°07′E
Isinomaki *see* Ishinomaki
Isioka *see* Ishioka
81 I18 **Isiolo** Isiolo, C Kenya 0°20′N 37°36′E
81 J18 **Isiolo** ⬖ *county* C Kenya
79 O16 **Isiro** Orientale, NE Dem. Rep. Congo 02°51′N 27°47′E
92 P2 **Isispynten** *headland* NE Svalbard 79°51′N 26°44′E
123 P11 **Isit** Respublika Sakha (Yakutiya), NE Russian Federation 60°53′N 125°32′E
149 O2 **Iskabad Canal** *canal* N Afghanistan
147 Q9 **Iskandar** *Rus.* Iskander. Toshkent Viloyati, E Uzbekistan 41°32′N 69°46′E
Iskander *see* Iskandar
114 G10 **Iskar** *var.* Iskăr, Iskŭr. ≈ NW Bulgaria
Iskăr *see* Iskar
114 H10 **Iskar, Yazovir** *var.* Yazovir Iskŭr; *prev.* Yazovir Stalin. ⊠ W Bulgaria
121 Q2 **İskele** *var.* Trikomo, *Gk.* Tríkomon. E Cyprus 35°16′N 33°54′E
136 K17 **İskenderun** *Eng.* Alexandretta. Hatay, S Turkey 36°34′N 36°10′E
138 H2 **İskenderun Körfezi** *Eng.* Gulf of Alexandretta. *gulf* S Turkey
136 J11 **İskilip** Çorum, N Turkey 40°45′N 34°28′E
Iski-Nauket *see* Nookat
114 J11 **Iskra** *prev.* Popovo. Haskovo, S Bulgaria 41°55′N 25°12′E
Iskŭr *see* Iskar
Iskŭr, Yazovir *see* Iskar, Yazovir
41 S15 **Isla** Veracruz-Llave, SE Mexico 18°01′N 95°30′W
119 J15 **Islach** *Rus.* Isloch'. ≈ C Belarus
104 H14 **Isla Cristina** Andalucía, S Spain 37°12′N 07°20′W
Isla de León *see* San Fernando
149 U6 **Islāmābād** ● (Pakistan) Federal Capital Territory Islāmābād, NE Pakistan 33°40′N 73°08′E
149 V6 **Islāmābād** ✈ Federal Capital Territory Islāmābād, NE Pakistan 33°40′N 73°08′E
Islamabad *see* Anantnāg
149 R17 **Islāmkot** Sind, SE Pakistan 24°37′N 70°04′E
23 Y17 **Islamorada** Florida Keys, Florida, SE USA 24°55′N 80°37′W
153 P14 **Islāmpur** Bihār, N India 25°09′N 85°13′E
Islam Qala *see* Eslām Qal'eh
18 K16 **Island Beach** *spit* New Jersey, NE USA
19 S4 **Island Falls** Maine, NE USA 45°59′N 68°16′W
Island/Ísland *see* Iceland
182 H6 **Island Lagoon** ◎ South Australia
11 Y13 **Island Lake** ◎ Manitoba, C Canada
29 W5 **Island Lake Reservoir** ⊠ Minnesota, N USA
33 R13 **Island Park** Idaho, NW USA 44°27′N 111°21′W
19 N6 **Island Pond** Vermont, NE USA 44°48′N 71°51′W
184 K2 **Islands, Bay of** *inlet* North Island, New Zealand
103 R7 **Is-sur-Tille** Côte d'Or, C France 47°34′N 05°03′E
42 J3 **Islas de la Bahía** ⬖ *department* N Honduras
65 L20 **Islas Orcadas Rise** *undersea feature* S Atlantic Ocean
96 F12 **Islay** *island* SW Scotland, United Kingdom
116 I15 **Islaz** Teleorman, S Romania 43°44′N 24°45′E
29 V7 **Isle** Minnesota, N USA 46°08′N 93°28′W
102 M12 **Isle** ≈ W France
97 I16 **Isle of Man** ◇ *British Crown Dependency* NW Europe
97 I16 **Isle of Man** *island* NW Europe
21 X7 **Isle of Wight** Virginia, NE USA 36°54′N 76°41′W
97 M24 **Isle of Wight** *cultural region* S England, United Kingdom
191 Y3 **Isles Lagoon** ◎ Kiritimati, E Kiribati
37 R11 **Isleta Pueblo** New Mexico, SW USA 34°54′N 106°40′W
Isloch' *see* Islach
61 E19 **Ismael Cortinas** Flores, S Uruguay 33°57′S 57°05′W
Ismailia *see* Al Ismā'īlīya
Ismâ'ilîya *see* Al Ismā'īlīya
Ismailly *see* İsmayıllı
137 X11 **İsmayıllı** *Rus.* Ismailly. N Azerbaijan 40°47′N 48°09′E
Ismid *see* İzmit
147 S12 **Ismoili Somoní, Qullai** *prev.* Qullai Kommunizm. ▲ E Tajikistan
75 X10 **Isnā** *var.* Esna. SE Egypt 25°16′N 32°30′E
93 K18 **Isojoki** Etelä-Pohjanmaa, W Finland 62°07′N 22°00′E
82 M12 **Isoka** Muchinga, NE Zambia 10°08′S 32°43′E
Isola d'Ischia *see* Ischia
Isola d'Istria *see* Izola
Isonzo *see* Soča
15 U4 **Isoukustouc** ≈ Québec, SE Canada
136 F15 **Isparta** *var.* Isbarta. Isparta, SW Turkey 37°46′N 30°32′E
136 F15 **Isparta** *var.* Isbarta. ⬖ *province* SW Turkey
114 M7 **Isperih** *var.* Isperikh; *prev.* Kemanlar. Razgrad, N Bulgaria 43°43′N 26°49′E
Isperikh *see* Isperih
107 L26 **Ispica** Sicilia, Italy, C Mediterranean Sea 36°47′N 14°55′E
148 J14 **Ispikān** Baluchistān, SW Pakistan 26°21′N 62°15′E
137 Q12 **İspir** Erzurum, NE Turkey 40°29′N 41°02′E
138 E12 **Israel** *off.* State of Israel, *var.* Medinat Israel, *Heb.* Yisrael, Yisra'el. ◆ *republic* SW Asia
Israel, State of *see* Israel
Issa *see* Vis
55 S9 **Issano** C Guyana 05°49′N 59°28′W
76 M16 **Issia** SW Ivory Coast 06°33′N 06°33′W
Issiq Köl *see* Issyk-Kul', Ozero
103 P11 **Issoire** Puy-de-Dôme, C France 45°33′N 03°15′E
103 N9 **Issoudun** *anc.* Uxellodunum. Indre, C France 46°57′N 01°59′E
81 H22 **Issuna** Singida, C Tanzania 05°24′S 34°48′E
Issyk *see* Yesik
Issyk-Kul' *see* Balykchy
147 X7 **Issyk-Kul', Ozero** *var.* Issiq Köl, *Kir.* Ysyk-Köl. ◎ E Kyrgyzstan
147 X7 **Issyk-Kul'skaya Oblast'** *Kir.* Ysyk-Köl Oblasty. ⬖ *province* E Kyrgyzstan
149 Q7 **Īstādeh-ye Moqor, Āb-e-** *var.* Āb-i-Istāda. ◎ SE Afghanistan
136 D11 **İstanbul** *Bul.* Tsarigrad, *Eng.* Istanbul, *prev.* Constantinople; *anc.* Byzantium. Istanbul, NW Turkey 41°02′N 28°57′E
114 P12 **İstanbul** ⬖ *province* NW Turkey
114 P12 **İstanbul Boğazı** *var.* Bosporus Thracius, *Eng.* Bosphorus, Bosporus, *Turk.* Karadeniz Boğazi. *strait* NW Turkey
Istarska Županija *see* Istra
115 G19 **Isthmía** Pelopónnisos, S Greece 37°55′N 23°02′E
115 G17 **Istiaía** Évvoia, C Greece 38°57′N 23°09′E
54 D9 **Istmina** Chocó, W Colombia 05°09′N 76°42′W
23 W13 **Istokpoga, Lake** ◎ Florida, SE USA
112 A9 **Istra** *off.* Istarska Županija. ⬖ *province* NW Croatia
112 I10 **Istra** *Eng.* Istria, *Ger.* Istrien. *cultural region* NW Croatia
103 R15 **Istres** Bouches-du-Rhône, SE France 43°30′N 04°59′E
Istria/Istrien *see* Istra
153 T15 **Iswardi** *var.* Ishurdi. Rajshahi, W Bangladesh 24°10′N 89°04′E
127 V7 **Isyangulovo** Respublika Bashkortostan, W Russian Federation 52°10′N 56°38′E
62 O6 **Itá** Central, S Paraguay 25°29′S 57°21′W
59 O17 **Itaberaba** Bahia, E Brazil 12°34′S 40°21′W
59 M20 **Itabira** *prev.* Presidente Vargas. Minas Gerais, SE Brazil 19°39′S 43°14′W
59 O18 **Itabuna** Bahia, E Brazil 14°48′S 39°18′W
59 J18 **Itacaiu** Mato Grosso, S Brazil 14°49′S 51°21′W
58 G12 **Itacoatiara** Amazonas, N Brazil 03°06′S 58°22′W
54 D9 **Itagüí** Antioquia, W Colombia 06°12′N 75°40′W
60 D13 **Itá Ibaté** Corrientes, NE Argentina 27°27′S 57°24′W
60 G11 **Itaipú, Represa de** ⊠ Brazil/Paraguay
58 H13 **Itaituba** Pará, NE Brazil 04°15′S 55°56′W
60 K13 **Itajaí** Santa Catarina, S Brazil 26°50′S 48°39′W
Italia/Italiana, Republica/Italian Republic, The *see* Italy
Italian Somaliland *see* Somalia
25 T7 **Italy** Texas, SW USA 32°10′N 96°52′W
106 G12 **Italy** *off.* The Italian Republic, *It.* Italia, Repubblica Italiana. ◆ *republic* S Europe
59 O19 **Itamaraju** Bahia, E Brazil 16°58′S 39°32′W
59 C14 **Itamarati** Amazonas, W Brazil 06°13′S 68°17′W
59 M19 **Itambé, Pico de** ▲ SE Brazil 18°23′S 43°21′W
164 J13 **Itami** ✈ (Ōsaka) Ōsaka, Honshū, SW Japan 34°47′N 135°24′E
115 H15 **Ítamos** ▲ N Greece 40°06′N 23°51′E
153 W11 **Itānagar** *state capital* Arunāchal Pradesh, NE India 27°02′N 93°38′E
Itany *see* Litani
59 N19 **Itaobim** Minas Gerais, SE Brazil 16°34′S 41°27′W
59 P15 **Itaparica, Represa de** ⊠ E Brazil
58 M13 **Itapecuru-Mirim** Maranhão, E Brazil 03°24′S 44°20′W
60 Q8 **Itaperuna** Rio de Janeiro, SE Brazil 21°14′S 41°51′W
59 O18 **Itapetinga** Bahia, E Brazil 15°17′S 40°16′W
60 L10 **Itapetininga** São Paulo, S Brazil 23°36′S 48°07′W
60 K10 **Itapeva** São Paulo, S Brazil 23°58′S 48°54′W
47 W6 **Itapicuru, Rio** ≈ NE Brazil
58 O13 **Itapipoca** Ceará, E Brazil 03°29′S 39°35′W
60 M9 **Itapira** São Paulo, S Brazil 22°25′S 46°46′W
60 K8 **Itápolis** São Paulo, S Brazil 21°36′S 48°43′W
60 K10 **Itaporanga** São Paulo, S Brazil 23°43′S 49°28′W
62 P7 **Itapúa** *off.* Departamento de Itapúa. ⬖ *department* SE Paraguay
Itapúa, Departamento de *see* Itapúa
59 E15 **Itapuã do Oeste** Rondônia, W Brazil 09°21′S 63°07′W
61 E15 **Itaqui** Rio Grande do Sul, S Brazil 29°10′S 56°28′W
60 K10 **Itararé** São Paulo, S Brazil 24°07′S 49°16′W
60 K10 **Itararé, Rio** ≈ S Brazil
154 H11 **Itārsi** Madhya Pradesh, C India 22°39′N 77°48′E
25 T7 **Itasca** Texas, SW USA 32°09′N 97°09′W
Itassi *see* Vieille Case
60 D13 **Itatí** Corrientes, NE Argentina 27°16′S 58°15′W
60 K10 **Itatinga** São Paulo, S Brazil 23°08′S 48°36′W
115 F18 **Itéas, Kólpos** *gulf* C Greece
57 N15 **Iténez, Río** *var.* Rio Guaporé. ≈ Bolivia/Brazil *see also* Rio Guaporé
Iténez, Río *see* Guaporé, Rio
54 H11 **Iteviate, Río** ≈ C Colombia
100 I13 **Ith** *hill range* C Germany
31 Q8 **Ithaca** Michigan, N USA 43°17′N 84°36′W
18 H11 **Ithaca** New York, NE USA 42°26′N 76°30′W
115 C18 **Itháki** *island* Iónia Nísiá, Greece, C Mediterranean Sea
Itháki *see* Vathy
It Hearrenfean *see* Heerenveen
79 L17 **Itimbiri** ≈ N Dem. Rep. Congo
Itinomiya *see* Ichinomiya
Itinoseki *see* Ichinoseki
39 Q5 **Itkilik River** ≈ Alaska, USA
164 M11 **Itoigawa** Niigata, Honshū, C Japan 37°02′N 137°53′E
15 R6 **Itomamo, Lac** ◎ Québec, SE Canada
165 S17 **Itoman** Okinawa, SW Japan 26°05′N 127°40′E
102 M5 **Iton** ≈ N France
57 M16 **Itonamas Río** ≈ NE Bolivia
Itoupé, Mont *see* Sommet Tabulaire
Itseqqortoormiit *see* Ittoqqortoormiit
22 K4 **Itta Bena** Mississippi, S USA 33°30′N 90°19′W
107 B17 **Ittiri** Sardegna, Italy, C Mediterranean Sea 40°36′N 08°34′E
197 Q14 **Ittoqqortoormiit** *var.* Itseqqortoormiit, *Dan.* Scoresbysund, *Eng.* Scoresby Sound. Sermersooq, C Greenland 70°33′N 21°52′W
60 M10 **Itu** São Paulo, S Brazil 23°17′S 47°16′W
54 D8 **Ituango** Antioquia, NW Colombia 07°07′N 75°46′W
59 A14 **Ituí, Rio** ≈ NW Brazil
79 O20 **Itula** Sud-Kivu, E Dem. Rep. Congo 03°30′S 27°50′E
59 K19 **Itumbiara** Goiás, C Brazil 18°25′S 49°15′W
55 T9 **Ituni** E Guyana 05°24′N 58°18′W
41 X13 **Iturbide** Campeche, SE Mexico 19°41′N 89°29′W
Ituri *see* Aruwimi
123 V13 **Iturup, Ostrov** *island* Kuril'skiye Ostrova, SE Russian Federation
60 L7 **Ituverava** São Paulo, S Brazil 20°22′S 47°48′W
59 C15 **Ituxi, Rio** ≈ W Brazil
61 E14 **Ituzaingó** Corrientes, NE Argentina 27°34′S 56°44′W
101 K18 **Itz** ≈ C Germany
100 I9 **Itzehoe** Schleswig-Holstein, N Germany 53°56′N 09°31′E
23 N2 **Iuka** Mississippi, S USA 34°48′N 88°11′W
60 I11 **Ivaiporã** Paraná, S Brazil 24°16′S 51°46′W
60 I11 **Ivaí, Rio** ≈ S Brazil
92 L10 **Ivalo** *Lapp.* Avveel, Avvil. Lappi, N Finland 68°34′N 27°29′E
92 L10 **Ivalojoki** *Lapp.* Avveel. ≈ N Finland
119 H20 **Ivanava** *Pol.* Janów, Janów Poleski, *Rus.* Ivanovo. Brestskaya Voblasts', SW Belarus 52°09′N 25°32′E
79 F18 **Ivando** *var.* Djidji. ≈ Congo/Gabon
Ivangorod *see* Dęblin
Ivangrad *see* Berane
183 N7 **Ivanhoe** New South Wales, SE Australia 32°55′S 144°21′E
29 S9 **Ivanhoe** Minnesota, N USA 44°27′N 96°15′W
14 D8 **Ivanhoe** ≈ Ontario, S Canada
112 E8 **Ivanić-Grad** Sisak-Moslavina, N Croatia 45°43′N 16°23′E
117 T10 **Ivanivka** Khersons'ka Oblast', S Ukraine 46°43′N 34°28′E
117 P10 **Ivanivka** Odes'ka Oblast', SW Ukraine 46°57′N 30°26′E
113 L14 **Ivanjica** Serbia, C Serbia 43°36′N 20°14′E
112 G11 **Ivanjska** *var.* Potkozarje. Republika Srpska, NW Bosnia and Herzegovina 44°54′N 17°04′E
111 H21 **Ivanka** ✈ (Bratislava) Bratislavský Kraj, W Slovakia 48°10′N 17°13′E
117 O3 **Ivankiv** *Rus.* Ivankov. Kyyivs'ka Oblast', N Ukraine 50°55′N 29°53′E
Ivankov *see* Ivankiv
39 O15 **Ivanof Bay** Alaska, USA 55°55′N 159°28′W
116 J7 **Ivano-Frankivs'k** *Ger.* Stanislau, *Pol.* Stanisławów, *Rus.* Ivano-Frankovsk; *prev.* Stanislav. Ivano-Frankivs'ka Oblast', W Ukraine 48°55′N 24°45′E
Ivano-Frankivs'k *see* Ivano-Frankivs'ka Oblast'
116 I7 **Ivano-Frankivs'ka Oblast'** *var.* Ivano-Frankivs'k, *Rus.* Ivano-Frankovskaya Oblast'; *prev.* Stanislavskaya Oblast'. ⬖ *province* W Ukraine
Ivano-Frankovsk *see* Ivano-Frankivs'k
Ivano-Frankovskaya Oblast' *see* Ivano-Frankivs'ka Oblast'
124 M16 **Ivanovo** Ivanovskaya Oblast', W Russian Federation 57°02′N 40°58′E
Ivanovo *see* Ivanava
124 M16 **Ivanovskaya Oblast'** ⬖ *province* W Russian Federation
35 X12 **Ivanpah Lake** ◎ California, W USA
112 E7 **Ivanščica** ▲▲ NE Croatia
127 R7 **Ivanteyevka** Saratovskaya Oblast', W Russian Federation 52°13′N 49°06′E
Ivantsevichi/Ivatsevichi *see* Ivatsevichy
116 I4 **Ivanychi** Volyns'ka Oblast', NW Ukraine 50°37′N 24°22′E
119 H18 **Ivatsevichy** *Pol.* Iwacewicze, *Rus.* Ivantsevichi, Ivatsevichi. Brestskaya Voblasts', SW Belarus 52°43′N 25°21′E
114 L12 **Ivaylovgrad** Haskovo, S Bulgaria 41°32′N 26°06′E
114 K11 **Ivaylovgrad, Yazovir** ⊠ S Bulgaria
122 G9 **Ivdel'** Sverdlovskaya Oblast', C Russian Federation 60°42′N 60°07′E
Ivenets *see* Ivyanyets
116 L12 **Iveşti** Galaţi, E Romania 45°27′N 28°00′E
Ivgovuotna *see* Lyngen
Ivigtut *see* Ivittuut
59 I21 **Ivinheima** Mato Grosso do Sul, SW Brazil 22°16′S 53°52′W
196 M15 **Ivittuut** *var.* Ivigtut. Sermersooq, S Greenland 61°12′N 48°10′W
Iviza *see* Eivissa\Ibiza
172 I6 **Ivohibe** Fianarantsoa, SE Madagascar 22°28′S 46°53′E
Ivoire, Côte d' *see* Ivory Coast
76 L15 **Ivory Coast** *off.* Republic of the Ivory Coast, *Fr.* Côte d'Ivoire, République de la Côte d'Ivoire. ◆ *republic* W Africa
68 C12 **Ivory Coast** *Fr.* Côte d'Ivoire. *coastal region* S Ivory Coast
Ivory Coast, Republic of the *see* Ivory Coast
95 L22 **Ivösjön** ◎ S Sweden
106 B7 **Ivrea** *anc.* Eporedia. Piemonte, NW Italy 45°28′N 07°52′E
12 J2 **Ivujivik** Québec, NE Canada 62°26′N 77°49′W
119 J16 **Ivyanyets** *Rus.* Ivenets. Minskaya Voblasts', C Belarus 53°53′N 26°45′E
Iv'ye *see* Iwye
Iwacewicze *see* Ivatsevichy
165 R8 **Iwaizumi** Iwate, Honshū, NE Japan 39°48′N 141°46′E
165 P12 **Iwaki** Fukushima, Honshū, N Japan 37°01′N 140°52′E
164 F13 **Iwakuni** Yamaguchi, Honshū, SW Japan 34°08′N 132°06′E
165 S4 **Iwamizawa** Hokkaidō, NE Japan 43°12′N 141°47′E
165 R4 **Iwanai** Hokkaidō, NE Japan 42°51′N 140°21′E
165 Q10 **Iwanuma** Miyagi, Honshū, C Japan 38°06′N 140°51′E
164 L14 **Iwata** Shizuoka, Honshū, S Japan 34°42′N 137°51′E
165 R8 **Iwate** Iwate, Honshū, N Japan 40°03′N 141°12′E
165 R8 **Iwate** *off.* Iwate-ken. ⬖ *prefecture* Honshū, C Japan
Iwate-ken *see* Iwate
77 S16 **Iwo** Oyo, SW Nigeria 07°21′N 03°58′E
Iwojima *see* Iō-jima
119 I16 **Iwye** *Pol.* Iwje, *Rus.* Iv'ye. Hrodzyenskaya Voblasts', W Belarus 53°56′N 25°46′E
42 C4 **Ixcán, Río** ≈ Guatemala/Mexico
99 G18 **Ixelles** *Dut.* Elsene. Brussels, C Belgium 50°49′N 04°21′E
57 J16 **Ixiamas** La Paz, NW Bolivia 13°45′S 68°10′W
41 O13 **Ixmiquilpan** *var.* Ixmiquilpán. Hidalgo, C Mexico 20°30′N 99°15′W
Ixmiquilpán *see* Ixmiquilpan
Ixopo *see* eXobho
Ixtaccíhuatal, Volcán *see* Iztaccíhuati, Volcán
40 M16 **Ixtapa** Guerrero, S Mexico 17°38′N 101°29′W
41 S16 **Ixtepec** Oaxaca, SE Mexico 16°32′N 95°03′W
40 K12 **Ixtlán** *var.* Ixtlán del Río. Nayarit, C Mexico 21°02′N 104°21′W
Ixtlán del Río *see* Ixtlán
122 H11 **Iyevlevo** Tyumenskaya Oblast', C Russian Federation 57°36′N 67°20′E
164 F14 **Iyo** Ehime, Shikoku, SW Japan 33°43′N 132°42′E
164 E14 **Iyo-nada** *sea* S Japan
42 E4 **Izabal** *off.* Departamento de Izabal. ⬖ *department* E Guatemala
Izabal, Departamento de *see* Izabal
42 F5 **Izabal, Lago de** *prev.* Golfo Dulce. ◎ E Guatemala
143 O9 **Īzad Khvāst** Fārs, C Iran 31°31′N 52°09′E
41 X12 **Izamal** Yucatán, SE Mexico 20°58′N 89°00′W
127 Q16 **Izberbash** Respublika Dagestan, SW Russian Federation 42°32′N 47°51′E
99 C18 **Izegem** *prev.* Iseghem. West-Vlaanderen, W Belgium 50°55′N 03°13′E
142 M9 **Īzeh** Khūzestān, SW Iran 31°48′N 49°49′E
165 T16 **Izena-jima** *island* Nansei-shotō, SW Japan
114 N10 **Izgrev** Burgas, E Bulgaria 42°09′N 27°49′E
127 T2 **Izhevsk** *prev.* Ustinov. Udmurtskaya Respublika, NW Russian Federation 56°48′N 53°12′E
125 S7 **Izhma** Respublika Komi, NW Russian Federation 64°56′N 53°52′E
125 S7 **Izhma** ≈ NW Russian Federation
141 X8 **Izkī** NE Oman 22°45′N 57°36′E
Izmail *see* Izmayil
117 N13 **Izmayil** *Rus.* Izmail. Odes'ka Oblast', SW Ukraine 45°19′N 28°49′E
136 B14 **İzmir** *prev.* Smyrna. İzmir, W Turkey 38°25′N 27°10′E
136 C14 **İzmir** *prev.* Smyrna. ⬖ *province* W Turkey
136 E11 **İzmit** *var.* Ismid; *anc.* Astacus. Kocaeli, NW Turkey 40°47′N 29°55′E
104 M14 **Iznájar** Andalucía, S Spain 37°17′N 04°16′W
104 M14 **Iznajar, Embalse de** ⊠ S Spain
105 N14 **Iznalloz** Andalucía, S Spain 37°23′N 03°31′W
136 E11 **İznik** Bursa, NW Turkey 40°27′N 29°43′E
136 E12 **İznik Gölü** ◎ NW Turkey
126 M14 **Izobil'nyy** Stavropol'skiy Kray, SW Russian Federation 45°22′N 41°40′E
109 S13 **Izola** *It.* Isola d'Istria. SW Slovenia 45°31′N 13°40′E
138 H9 **Izra'** *var.* Ezra, Ezraa. Dar'ā, S Syria 32°52′N 36°15′E
41 P14 **Iztaccíhuati, Volcán** *var.* Volcán Ixtaccíhuatal. △ S Mexico 19°07′N 98°37′W
42 C7 **Iztapa** Escuintla, SE Guatemala 13°58′N 90°42′W
Izúcar de Matamoros *see* Matamoros
165 N14 **Izu-hantō** *peninsula* Honshū, S Japan
Izuhara *see* Tsushima
164 J14 **Izumiōtsu** Ōsaka, Honshū, SW Japan 34°29′N 135°25′E
164 I14 **Izumisano** Ōsaka, Honshū, SW Japan 34°23′N 135°18′E
164 G12 **Izumo** Shimane, Honshū, SW Japan 35°22′N 132°46′E
192 H5 **Izu Trench** *undersea feature* NW Pacific Ocean
122 K6 **Izvestiy TsIK, Ostrova** *island* N Russian Federation
114 G10 **Izvor** Pernik, W Bulgaria 42°27′N 22°53′E
116 L5 **Izyaslav** Khmel'nyts'ka Oblast', W Ukraine 50°08′N 26°53′E
117 W6 **Izyum** Kharkivs'ka Oblast', E Ukraine 49°12′N 37°19′E

J

93 M18 **Jaala** Kymenlaakso, S Finland 61°04′N 26°30′E
140 J5 **Jabal ash Shifā** *desert* NW Saudi Arabia
141 U8 **Jabal aẓ Ẓannah** *var.* Jebel Dhanna. Abū Ẓaby, W United Arab Emirates 24°10′N 52°36′E
138 E11 **Jabalīya** *var.* Jabāliyah. NE Gaza Strip 31°32′N 34°29′E
Jabāliyah *see* Jabalīya
105 N11 **Jabalón** ≈ C Spain
154 J10 **Jabalpur** *prev.* Jubbulpore. Madhya Pradesh, C India 23°10′N 79°59′E
141 N15 **Jabal Zuqar, Jazīrat** *var.* Az Zuqur. *island* SW Yemen
Jabar *see* Jawa Barat
Jabat *see* Jabwot
138 J3 **Jabbūl, Sabkhat al** *sabkha* NW Syria
181 P1 **Jabiru** Northern Territory, N Australia 12°44′S 132°48′E
138 H4 **Jablah** *var.* Jeble, *Fr.* Djéblé. Al Lādhiqīyah, W Syria 35°00′N 36°00′E
112 C11 **Jablanac** Lika-Senj, W Croatia 44°43′N 14°54′E
113 H14 **Jablanica** Federacija Bosne I Hercegovine, SW Bosnia and Herzegovina 43°39′N 17°43′E
113 M20 **Jablanica** *Alb.* Mali i Jabllanicës, *var.* Malet e Jabllanicës. ▲ Albania/FYR Macedonia *see also* Jabllanicës, Mali i
Jabllanicës, Malet e *see* Jablanica/Jabllanicës, Mali i
113 M20 **Jabllanicës, Mali i**, *Mac.* Jablanica. ▲ Albania/FYR Macedonia *see also* Jablanica
111 E15 **Jablonec nad Nisou** *Ger.* Gablonz an der Neisse. Liberecký Kraj, N Czech Republic 50°44′N 15°10′E
Jabłonków/Jablunkau *see* Jablunkov
110 J9 **Jabłonowo Pomorskie** Kujawski-pomorskie, C Poland 53°24′N 19°08′E
111 J17 **Jablunkov** *Ger.* Jablunkau, *Pol.* Jabłonków. Moravskoslezský Kraj, E Czech Republic 49°35′N 18°46′E
59 Q15 **Jaboatão** Pernambuco, E Brazil 08°05′S 35°W
60 L8 **Jaboticabal** São Paulo, S Brazil 21°15′S 48°17′W
189 U7 **Jabwot** *var.* Jabat, Jebat, Jōwat. *island* Ralik Chain, S Marshall Islands
105 S4 **Jaca** Aragón, NE Spain 42°34′N 00°33′W
42 B4 **Jacaltenango** Huehuetenango, W Guatemala 15°39′N 91°46′W
59 G14 **Jacaré-a-Canga** Pará, NE Brazil 05°59′S 57°32′W
60 N10 **Jacareí** São Paulo, S Brazil 23°18′S 45°55′W
59 I18 **Jaciara** Mato Grosso, W Brazil 15°59′S 54°57′W
59 E15 **Jaciparaná** Rondônia, W Brazil 09°20′S 64°28′W
19 P5 **Jackman** Maine, NE USA 45°35′N 70°14′W
35 X1 **Jackpot** Nevada, W USA 41°57′N 114°41′W
20 M8 **Jacksboro** Tennessee, S USA 36°19′N 84°11′W
25 S6 **Jacksboro** Texas, SW USA 33°13′N 98°11′W
23 N7 **Jackson** Alabama, S USA 31°30′N 87°53′W
35 P7 **Jackson** California, W USA 38°19′N 120°46′W
23 T4 **Jackson** Georgia, SE USA 33°17′N 83°58′W
21 O6 **Jackson** Kentucky, S USA 37°32′N 83°24′W
22 J8 **Jackson** Louisiana, S USA 30°50′N 91°13′W
31 Q10 **Jackson** Michigan, N USA 42°15′N 84°24′W
29 T11 **Jackson** Minnesota, N USA 43°38′N 95°00′W
22 K5 **Jackson** *state capital* Mississippi, S USA 32°19′N 90°12′W
27 Y7 **Jackson** Missouri, C USA 37°23′N 89°40′W
21 W8 **Jackson** North Carolina, SE USA 36°24′N 77°25′W
31 T15 **Jackson** Ohio, NE USA 39°03′N 82°40′W
20 G9 **Jackson** Tennessee, S USA 35°37′N 88°50′W
33 S14 **Jackson** Wyoming, C USA 43°28′N 110°45′W
185 C19 **Jackson Bay** *bay* South Island, New Zealand
186 E9 **Jackson Field** ✈ (Port Moresby) Central/National Capital District, S Papua New Guinea 09°28′S 147°12′E
185 C20 **Jackson Head** *headland* South Island, New Zealand 43°57′S 168°38′E
23 S8 **Jackson, Lake** ◎ Florida, SE USA
33 S13 **Jackson Lake** ◎ Wyoming, C USA
194 J6 **Jackson, Mount** ▲ Antarctica 71°43′S 63°45′W
37 U3 **Jackson Reservoir** ◎ Colorado, C USA
23 Q3 **Jacksonville** Alabama, S USA 33°48′N 85°45′W
27 V11 **Jacksonville** Arkansas, C USA 34°52′N 92°08′W
23 W8 **Jacksonville** Florida, SE USA 30°20′N 81°39′W
30 K14 **Jacksonville** Illinois, N USA 39°43′N 90°13′W
21 W11 **Jacksonville** North Carolina, SE USA 34°45′N 77°26′W
25 W7 **Jacksonville** Texas, SW USA 31°57′N 95°16′W
23 X9 **Jacksonville Beach** Florida, SE USA 30°17′N 81°23′W
44 L9 **Jacmel** *var.* Jaquemel. S Haiti 18°13′N 72°33′W
Jacob *see* Nkayi
149 Q12 **Jacobābād** Sind, SE Pakistan 28°16′N 68°30′E
55 T11 **Jacobs Ladder Falls** *waterfall* S Guyana
45 O11 **Jaco, Pointe** *headland* N Dominica 15°38′N 61°25′W
15 Q9 **Jacques-Cartier** ≈ Québec, SE Canada
13 P11 **Jacques-Cartier, Détroit de** *var.* Jacques-Cartier Passage. *strait* Gulf of St. Lawrence/St. Lawrence River, Canada
15 W6 **Jacques-Cartier, Mont** ▲ Québec, SE Canada 48°58′N 66°00′W
Jacques-Cartier Passage *see* Jacques-Cartier, Détroit de
61 H16 **Jacuí, Rio** ≈ S Brazil
60 L11 **Jacupiranga** São Paulo, S Brazil 24°42′S 48°00′W
100 G10 **Jade** ≈ NW Germany
100 G10 **Jadebusen** *bay* NW Germany
Jadotville *see* Likasi
Jadransko More/Jadransko Morje *see* Adriatic Sea
105 O7 **Jadraque** Castilla-La Mancha, C Spain 40°55′N 02°55′W
95 I22 **Jægerspris** Hovedstaden, E Denmark 55°52′N 11°59′E
56 C10 **Jaén** Cajamarca, N Peru 05°45′S 78°51′W
105 N13 **Jaén** Andalucía, SW Spain 37°46′N 03°48′W
105 N13 **Jaén** ⬖ *province* Andalucía, S Spain
95 C17 **Jæren** *physical region* S Norway
155 J23 **Jaffna** Northern Province, N Sri Lanka 09°42′N 80°03′E
155 K23 **Jaffna Lagoon** *lagoon* N Sri Lanka
19 N10 **Jaffrey** New Hampshire, NE USA 42°46′N 72°00′W
138 H13 **Jafr, Qā' al** *var.* El Jafr. *salt pan* S Jordan
152 J9 **Jagādhri** Haryāna, N India 30°11′N 77°18′E
118 H4 **Jägala** *var.* Jägala Jōgi, *Ger.* Jaggowal. ≈ NW Estonia
Jägala Jōgi *see* Jägala
Jagannath *see* Puri
155 L14 **Jagdalpur** Chhattīsgarh, C India 19°07′N 82°04′E
163 U5 **Jagdaqi** Nei Mongol Zizhiqu, N China 50°26′N 124°03′E
Jägerndorf *see* Krnov
Jaggowal *see* Jägala
139 O2 **Jaghjaghah, Nahr** ≈ N Syria
112 N13 **Jagodina** *prev.* Svetozarevo. Serbia, C Serbia 43°59′N 21°15′E
112 K12 **Jagodnja** ▲▲ W Serbia
101 I20 **Jagst** ≈ SW Germany
155 I14 **Jagtiāl** Telangana, C India 18°49′N 78°53′E
61 H18 **Jaguarão** Rio Grande do Sul, S Brazil 32°30′S 53°25′W
61 H18 **Jaguarão, Rio** *var.* Río Yaguarón. ≈ Brazil/Uruguay
60 K11 **Jaguariaíva** Paraná, S Brazil 24°15′S 49°44′W
44 D5 **Jagüey Grande** Matanzas, W Cuba 22°31′N 81°07′W
153 P14 **Jahānābād** Bihār, N India 25°13′N 84°59′E
Jahra *see* Al Jahrā'
143 P12 **Jahrom** *var.* Jahrum. Fārs, S Iran 28°35′N 53°32′E
Jahrum *see* Jahrom
Jailolo *see* Halmahera, Pulau
Jainat *see* Chai Nat
Jainti *see* Jayanti
152 H12 **Jaipur** *prev.* Jeypore. *state capital* Rājasthān, N India 26°54′N 75°47′E
153 T14 **Jaipurhat** *var.* Joypurhat. Rajshahi, NW Bangladesh 25°04′N 89°06′E
152 D11 **Jaisalmer** Rājasthān, NW India 26°55′N 70°56′E
154 O12 **Jājapur** *var.* Jajpur, Panikoilli. Odisha, E India 18°54′N 82°36′E
143 R4 **Jājarm** Khorāsān-e Shemālī, NE Iran 36°58′N 56°26′E
112 G12 **Jajce** Federacija Bosne I Hercegovine, W Bosnia and Herzegovina 44°20′N 17°16′E
Jaji *see* 'Alī Khēl
Jajpur *see* Jājapur
83 D17 **Jakalsberg** Otjozondjupa, N Namibia 19°23′S 17°28′E
169 O15 **Jakarta** *prev.* Djakarta, *Dut.* Batavia. ● (Indonesia) Jawa, C Indonesia 06°08′S 106°45′E
10 I8 **Jakes Corner** Yukon, W Canada 60°18′N 134°00′W
152 H9 **Jākhal** Haryāna, NW India 29°46′N 75°51′E
Jakobeny *see* Iacobeni
93 K16 **Jakobstad** *Fin.* Pietarsaari. Österbotten, W Finland 63°41′N 22°40′E
Jakobstadt *see* Jēkabpils
113 O18 **Jakupica** ▲▲ C FYR Macedonia
37 W15 **Jal** New Mexico, SW USA 32°07′N 103°10′W
141 P7 **Jalājil** *var.* Ğalāğıl. Ar Riyāḍ, C Saudi Arabia 25°43′N 45°22′E
149 S5 **Jalālābād** *var.* Jalalabad, Jelalabad. Nangarhār, E Afghanistan 34°26′N 70°28′E
Jalal-Abad *see* Dzhalal-Abad, Dzhalal-Abadskaya Oblast', Kyrgyzstan
Jalal-Abad Oblasty *see* Dzhalal-Abadskaya Oblast'
149 V7 **Jalālpur** Punjab, E Pakistan 32°39′N 74°11′E
149 T11 **Jalālpur Pīrwāla** Punjab, E Pakistan 29°30′N 71°20′E
152 H8 **Jalandhar** *prev.* Jullundur. Punjab, N India 31°20′N 75°37′E
42 J7 **Jalán, Río** ≈ S Honduras
42 E6 **Jalapa** Jalapa, C Guatemala 14°39′N 89°59′W
42 J7 **Jalapa** Nueva Segovia, NW Nicaragua 13°56′N 86°11′W
42 A3 **Jalapa** *off.* Departamento de Jalapa. ⬖ *department* SE Guatemala
Jalapa, Departamento de *see* Jalapa
42 E6 **Jalapa, Río** ≈ SE Guatemala
143 X13 **Jālaq** Sīstān va Balūchestān, SE Iran
93 K17 **Jalasjärvi** Etelä-Pohjanmaa, W Finland 62°30′N 22°50′E
149 O8 **Jaldak** Zābul, SE Afghanistan 32°00′N 66°45′E
60 J7 **Jales** São Paulo, S Brazil 20°15′S 50°34′W
154 P11 **Jaleshwar** *var.* Jaleswar. Odisha, NE India 21°51′N 87°15′E
Jaleswar *see* Jaleshwar
154 F12 **Jalgaon** Mahārāshtra, C India 21°01′N 75°34′E
139 W12 **Jalībah** Dhī Qār, S Iraq 30°37′N 46°31′E
139 W13 **Jalīb Shahāb** Al Muthanná, S Iraq 30°26′N 46°09′E
77 X15 **Jalingo** Taraba, E Nigeria 08°54′N 11°22′E
40 K13 **Jalisco** ⬖ *state* SW Mexico
154 G13 **Jālna** Mahārāshtra, W India 19°50′N 75°53′N
Jalomitsa *see* Ialomiţa
105 R5 **Jalón** ≈ N Spain
152 E13 **Jālor** Rājasthān, N India 25°21′N 72°43′E
112 K11 **Jalovik** Serbia, W Serbia 44°37′N 19°48′E
40 L12 **Jalpa** Zacatecas, C Mexico 21°40′N 103°W
153 S12 **Jalpāiguri** West Bengal, NE India 26°43′N 88°24′E
41 O12 **Jalpán** *var.* Jalpan. Querétaro de Arteaga, C Mexico 21°13′N 99°28′W
Jalpan *see* Jalpán
67 P2 **Jalta** *island* N Tunisia
75 S9 **Jālū** *var.* Jūlā. NE Libya 29°02′N 21°33′E
189 U8 **Jaluit Atoll** *var.* Jālwōj. *atoll* Ralik Chain, S Marshall Islands
Jālwōj *see* Jaluit Atoll
81 L18 **Jamaame** *It.* Giamame; *prev.* Margherita. Jubbada Hoose, S Somalia 0°00′N 42°43′E
77 W13 **Jamaare** ≈ NE Nigeria

◆ Country
● Country Capital
◇ Dependent Territory
○ Dependent Territory Capital
⬖ Administrative Regions
✈ International Airport
▲ Mountain
▲▲ Mountain Range
△ Volcano
≈ River
◎ Lake
⊠ Reservoir

JAMAICA – JINZHOU

44 G9 **Jamaica** ◆ *commonwealth republic* W West Indies
47 P3 **Jamaica** *island* W West Indies
44 I9 **Jamaica Channel** *channel* Haiti/Jamaica
153 T14 **Jamalpur** Dhaka, N Bangladesh 24°54′N 89°57′E
153 Q14 **Jamālpur** Bihār, NE India 25°19′N 86°30′E
168 L9 **Jamaluang** *var.* Jemaluang. Johor, Peninsular Malaysia 02°15′N 103°50′E
59 I14 **Jamanxim, Rio** ⌁ C Brazil
56 B8 **Jambeli, Canal de** *channel* S Ecuador
99 I20 **Jambes** Namur, SE Belgium 50°26′N 04°51′E
168 L12 **Jambi** *var.* Telanaipura; *prev.* Djambi. Sumatera, W Indonesia 01°34′S 103°37′E
168 K12 **Jambi** *off.* Propinsi Jambi, *var.* Djambi. ◆ *province* W Indonesia
Jambi, Propinsi *see* Jambi
Jamdena *see* Yamdena, Pulau
12 H8 **James Bay** *bay* Ontario/Québec, E Canada
63 F19 **James, Isla** *island* Archipiélago de los Chonos, S Chile
181 Q8 **James Ranges** ▲▲ Northern Territory, C Australia
29 P8 **James River** ⌁ North Dakota/South Dakota, N USA
21 X7 **James River** ⌁ Virginia, NE USA
194 H4 **James Ross Island** *island* Antarctica
182 I8 **Jamestown** South Australia 33°13′S 138°36′E
65 G25 **Jamestown** ○ (Saint Helena) NW Saint Helena 15°56′S 05°44′W
35 P8 **Jamestown** California, W USA 37°57′N 120°25′W
20 L7 **Jamestown** Kentucky, S USA 36°58′N 85°03′W
18 D11 **Jamestown** New York, NE USA 42°05′N 79°15′W
29 P5 **Jamestown** North Dakota, N USA 46°54′N 98°42′W
20 L8 **Jamestown** Tennessee, S USA 36°24′N 84°58′W
Jamestown *see* Holetown
15 N10 **Jamet** ⌁ Québec, SE Canada
41 Q17 **Jamiltepec** *var.* Santiago Jamiltepec. Oaxaca, SE Mexico 16°18′N 97°51′W
95 F20 **Jammerbugten** *bay* Skagerrak, E North Sea
152 H6 **Jammu** *prev.* Jummoo. *state capital* Jammu and Kashmir, NW India 32°43′N 74°54′E
152 I5 **Jammu and Kashmir** *var.* Jammu-Kashmir, Kashmir. ◆ *state* NW India
149 V4 **Jammu and Kashmīr** *disputed region* India/Pakistan
Jammu-Kashmir *see* Jammu and Kashmir
154 B10 **Jāmnagar** *prev.* Navanagar. Gujarāt, W India 22°28′N 70°06′E
149 S11 **Jāmpur** Punjab, E Pakistan 29°38′N 70°40′E
93 L18 **Jämsä** Keski-Suomi, C Finland 61°51′N 25°10′E
93 L18 **Jämsänkoski** Keski-Suomi, C Finland 61°54′N 25°10′E
153 Q16 **Jamshedpur** Jhārkhand, NE India 22°47′N 86°12′E
94 K9 **Jämtland** ◆ *county* C Sweden
153 Q14 **Jamūi** Bihār, NE India 24°57′N 86°14′E
Jamuna *see* Brahmaputra
153 T14 **Jamuna Nadi** ⌁ N Bangladesh
Jamundá *see* Nhamundá, Rio
54 D11 **Jamundí** Valle del Cauca, SW Colombia 03°16′N 76°31′W
153 Q12 **Janakpur** Central, C Nepal 26°45′N 85°55′E
59 N18 **Janaúba** Minas Gerais, SE Brazil 15°47′S 43°16′W
58 K11 **Janaucu, Ilha** *island* NE Brazil
143 Q7 **Jandaq** Eşfahān, C Iran 34°04′N 54°26′E
64 Q11 **Jandia, Punta de** *headland* Fuerteventura, Islas Canarias, Spain, NE Atlantic Ocean 28°03′N 14°32′W
59 B14 **Jandiatuba, Rio** ⌁ NW Brazil
105 N12 **Jándula** ⌁ S Spain
29 V10 **Janesville** Minnesota, N USA 44°07′N 93°43′W
30 L9 **Janesville** Wisconsin, N USA 42°42′N 89°02′W
83 N20 **Jangamo** Inhambane, SE Mozambique 24°04′S 35°25′E
155 J14 **Jangaon** Telangana, C India 18°47′N 79°25′E
153 S14 **Jangipur** West Bengal, NE India 24°31′N 88°03′E
Janina *see* Ioánnina
Janischken *see* Joniškis
112 J11 **Janja** NE Bosnia and Herzegovina 44°40′N 19°15′E
Jankovac *see* Jánoshalma
197 Q15 **Jan Mayen** ◇ *constituent part of Norway* N Atlantic Ocean
84 D5 **Jan Mayen** *island* N Atlantic Ocean
197 R15 **Jan Mayen Fracture Zone** *tectonic feature* Greenland Sea/Norwegian Sea
197 R15 **Jan Mayen Ridge** *undersea feature* Greenland Sea/Norwegian Sea 69°00′N 08°00′W
40 H3 **Janos** Chihuahua, N Mexico 30°50′N 108°10′W
111 K25 **Jánoshalma** *SCr.* Jankovac. Bács-Kiskun, S Hungary 46°19′N 19°16′E
Janów *see* Ivanava, Belarus
110 H10 **Janowiec Wielkopolski** *Ger.* Janowitz. Kujawski-pomorskie, C Poland 52°47′N 17°30′E
Janowitz *see* Janowiec Wielkopolski
Janow/Janów *see* Jonava, Lithuania
111 O15 **Janów Lubelski** Lubelski, E Poland 50°42′N 22°24′E
Janów Poleski *see* Ivanava
83 H25 **Jansenville** Eastern Cape, S South Africa 32°56′S 24°40′E
59 M18 **Januária** Minas Gerais, SE Brazil 15°28′S 44°23′W
Janūbīyah, Al Bādiyah al *see* Ash Shāmīyah
102 I7 **Janzé** Ille-et-Vilaine, NW France 47°55′N 01°28′W
154 F10 **Jaora** Madhya Pradesh, C India 23°40′N 75°10′E
131 Y9 **Japan** *var.* Nippon, *Jap.* Nihon. ◆ *monarchy* E Asia
129 Y9 **Japan** *island group* E Asia
192 H4 **Japan Basin** *undersea feature* N Sea of Japan 40°00′N 135°00′E
129 Y8 **Japan, Sea of** *var.* East Sea, *Rus.* Yapanskoye More. *sea* NW Pacific Ocean *see also* East Sea
192 H4 **Japan Trench** *undersea feature* NW Pacific Ocean 37°00′N 143°00′E
Japen *see* Yapen, Pulau
59 A15 **Japiim** *var.* Máncio Lima. Acre, W Brazil 08°00′S 73°39′W
58 D12 **Japurá** Amazonas, N Brazil 01°43′S 66°14′W
58 C12 **Japurá, Rio** *var.* Río Caquetá, Yapurá. ⌁ Brazil/Colombia *see also* Caquetá, Río
Japurá, Rio *see* Caquetá, Río
43 W17 **Jaqué** Darién, SE Panama 07°31′N 78°09′W
Jaquemel *see* Jacmel
Jarablos *see* Jarābulus
138 K2 **Jarābulus** *var.* Jarablos, Jerablus, *Fr.* Djérablous. Ḩalab, N Syria 36°51′N 38°02′E
60 K13 **Jaraguá do Sul** Santa Catarina, S Brazil 26°29′S 49°07′W
104 K9 **Jaraicejo** Extremadura, W Spain 39°40′N 05°49′W
104 K9 **Jaraíz de la Vera** Extremadura, W Spain 40°04′N 05°45′W
105 O7 **Jarama** ⌁ C Spain
63 J20 **Jaramillo** Santa Cruz, SE Argentina 47°10′S 67°07′W
Jarandilla de la Vega *see* Jarandilla de la Vera
104 K8 **Jarandilla de la Vera** *var.* Jarandilla de la Vega. Extremadura, W Spain 40°08′N 05°39′W
149 V9 **Jarānwāla** Punjab, E Pakistan 31°20′N 73°26′E
138 G9 **Jarash** *var.* Jerash; *anc.* Gerasa. Jarash, NW Jordan 32°17′N 35°54′E
138 G8 **Jarash** *off.* Muḩāfa at Jarash. ◆ *governorate* N Jordan
Jarash, Muḩāfa at *see* Jarash
Jarbah, Jazīrat *see* Jerba, Île de
94 N13 **Järbo** Gävleborg, C Sweden 60°43′N 16°40′E
Jardan *see* Yordon
44 F7 **Jardines de la Reina, Archipiélago de los** *island group* C Cuba
162 I8 **Jargalant** Bayanhongor, C Mongolia 47°14′N 99°43′E
162 K6 **Jargalant** Bulgan, N Mongolia 49°09′N 104°19′E
162 G7 **Jargalant** *var.* Buyanbat. Govĭ-Altay, W Mongolia 47°00′N 95°57′E
162 I6 **Jargalant** *var.* Orgil. Hövsgöl, C Mongolia 48°31′N 99°19′E
Jargalant *see* Battsengel
Jargalant *see* Bulgan, Bayan-Ölgiy, Mongolia
Jargalant *see* Biger, Govĭ-Altay, Mongolia
Jarīd, Shaţţ al *see* Jerid, Chott el
58 I11 **Jari, Rio** *var.* Jary. ⌁ N Brazil
141 N7 **Jarīr, Wādī al** *dry watercourse* C Saudi Arabia
Jarja *see* Yur'ya
94 L13 **Järna** *var.* Dala-Jarna. Dalarna, C Sweden 60°31′N 14°22′E
95 O16 **Järna** Stockholm, C Sweden 59°05′N 17°35′E
102 K11 **Jarnac** Charente, W France 45°41′N 00°10′W
110 H12 **Jarocin** Wielkopolskie, C Poland 51°59′N 17°30′E
111 F16 **Jaroměř** *Ger.* Jermer. Královéhradecký Kraj, N Czech Republic 50°22′N 15°55′E
Jaroslau *see* Jarosław
111 O16 **Jarosław** *Ger.* Jaroslau, *Rus.* Yaroslav. Podkarpackie, SE Poland 50°01′N 22°41′E
93 F16 **Järpen** Jämtland, C Sweden 63°21′N 13°30′E
147 O14 **Jarqo'rg'on** *Rus.* Dzharkurgan. Surkhondaryo Viloyati, S Uzbekistan 37°31′N 67°20′E
139 P2 **Jarrāh, Wadi** *dry watercourse* NE Syria
Jars, Plain of *see* Xiangkhoang, Plateau de
162 K14 **Jartai Yanchi** ⊚ N China
59 E16 **Jaru** Rondônia, W Brazil 10°24′S 62°45′W
Jarud Qi *see* Lubei
118 I4 **Järva-Jaani** *Ger.* Sankt-Johannis. Järvamaa, N Estonia 59°03′N 25°54′E
118 G5 **Järvakandi** *Ger.* Jerwakant. Raplamaa, NW Estonia 58°45′N 24°49′E
118 H4 **Järvamaa** *var.* Järva Maakond. ◆ *province* N Estonia
Järva Maakond *see* Järvamaa
93 L19 **Järvenpää** Uusimaa, S Finland 60°29′N 25°06′E
14 G17 **Jarvis** Ontario, S Canada 42°53′N 80°06′W
177 R8 **Jarvis Island** ◇ *US unincorporated territory* C Pacific Ocean
94 M11 **Järvsö** Gävleborg, C Sweden 61°43′N 16°25′E
Jary *see* Jari, Rio
112 M9 **Jaša Tomić** Vojvodina, NE Serbia 45°27′N 20°51′E
112 D12 **Jasenice** Zadar, SW Croatia 44°15′N 15°33′E
138 I11 **Jashshat al 'Adlah, Wādī al** *dry watercourse* C Jordan
77 Q16 **Jasikan** E Ghana 07°24′N 00°28′E
Jāsk *see* Bandar-e Jāsk
146 F6 **Jasliq** *Rus.* Zhaslyk. Qoraqalpog'iston Respublikasi, NW Uzbekistan 43°57′N 57°30′E
111 N17 **Jasło** Podkarpackie, SE Poland 49°45′N 21°28′E
11 U16 **Jasmin** Saskatchewan, S Canada 51°11′N 103°34′W
65 A23 **Jason Islands** *island group* NW Falkland Islands
194 I4 **Jason Peninsula** *peninsula* Antarctica
31 N15 **Jasonville** Indiana, N USA 39°09′N 87°12′W
11 O15 **Jasper** Alberta, SW Canada 52°55′N 118°05′W
14 L13 **Jasper** Ontario, SE Canada 44°50′N 75°57′W
23 O3 **Jasper** Alabama, S USA 33°49′N 87°16′W
27 T9 **Jasper** Arkansas, C USA 36°00′N 93°11′W
23 U8 **Jasper** Florida, SE USA 30°31′N 82°57′W
31 N16 **Jasper** Indiana, N USA 38°22′N 86°57′W
29 R11 **Jasper** Minnesota, N USA 43°51′N 96°24′W
27 S7 **Jasper** Missouri, C USA 37°20′N 94°18′W
20 K10 **Jasper** Tennessee, S USA 35°04′N 85°36′W
25 Y9 **Jasper** Texas, SW USA 30°55′N 94°00′W
11 O15 **Jasper National Park** *national park* Alberta/British Columbia, SW Canada
Jassy *see* Iaşi
113 N14 **Jastrebac** ▲▲ SE Serbia
112 D9 **Jastrebarsko** Zagreb, N Croatia 45°40′N 15°40′E
Jastrow *see* Jastrowie
110 G9 **Jastrowie** *Ger.* Jastrow. Wielkopolskie, C Poland 53°25′N 16°48′E
111 J17 **Jastrzębie-Zdrój** Śląskie, S Poland 49°58′N 18°34′E
111 L22 **Jászapáti** Jász-Nagykun-Szolnok, E Hungary 47°30′N 20°10′E
111 L22 **Jászberény** Jász-Nagykun-Szolnok, E Hungary 47°30′N 19°56′E
111 L23 **Jász-Nagykun-Szolnok** *off.* Jász-Nagykun-Szolnok Megye. ◆ *county* E Hungary
Jász-Nagykun-Szolnok Megye *see* Jász-Nagykun-Szolnok
59 J19 **Jataí** Goiás, C Brazil 17°58′S 51°45′W
58 G12 **Jatapu, Serra do** ▲▲
41 W16 **Jatate, Río** ⌁ SE Mexico
Jateng *see* Jawa Tengah
149 P17 **Jāti** Sind, SE Pakistan 24°20′N 68°18′E
44 F6 **Jatibonico** Sancti Spíritus, C Cuba 21°56′N 79°11′W
169 O16 **Jatiluhur, Danau** ⊚ Jawa, S Indonesia
Jatim *see* Jawa Timur
Jativa *see* Xàtiva
149 S11 **Jatoi** *prev.* Jattoi. Punjab, E Pakistan 29°29′N 70°58′E
Jattoi *see* Jatoi
60 L9 **Jaú** São Paulo, S Brazil 22°11′S 48°35′W
58 F11 **Jauaperi, Rio** ⌁ N Brazil
99 I19 **Jauche** Walloon Brabant, C Belgium 50°42′N 04°55′E
Jauer *see* Jawor
Jauf *see* Al Jawf
149 U7 **Jauharābād** Punjab, E Pakistan 32°16′N 72°17′E
57 E14 **Jauja** Junín, C Peru 11°48′S 75°30′W
41 O10 **Jaumave** Tamaulipas, C Mexico 23°28′N 99°22′W
118 H10 **Jaunjelgava** *Ger.* Friedrichstadt. S Latvia 56°38′N 25°03′E
Jaunlatgale *see* Pytalovo
118 I8 **Jaunpiebalga** NE Latvia 57°10′N 26°02′E
118 E9 **Jaunpils** C Latvia 56°45′N 23°03′E
153 N13 **Jaunpur** Uttar Pradesh, N India 25°44′N 82°41′E
29 N8 **Java** South Dakota, N USA 45°29′N 99°54′W
Java *see* Jawa
105 R9 **Javalambre** ▲ E Spain 40°02′N 01°06′W
173 V7 **Java Ridge** *undersea feature* E Indian Ocean
59 A14 **Javari, Rio** *var.* Yavarí. ⌁ Brazil/Peru
Javarthushuu *see* Bayan-Uul
169 Q15 **Java Sea** *Ind.* Laut Jawa. *sea* W Indonesia
173 U7 **Java Trench** *var.* Sunda Trench. *undersea feature* E Indian Ocean
143 Q10 **Javazm** *var.* Jowzam. Kermān, C Iran 30°31′N 55°01′E
105 T11 **Jávea** *Cat.* Xàbia. Valenciana, E Spain 38°48′N 00°10′E
Javhlant *see* Bayan-Ovoo
63 G20 **Javier, Isla** *island* S Chile
113 L14 **Javor** ▲▲ Bosnia and Herzegovina/Serbia
111 K20 **Javorie** *Hung.* Jávoros. ▲ S Slovakia 48°26′N 19°16′E
Jávoros *see* Javorie
93 J14 **Jävre** Norrbotten, N Sweden 65°07′N 21°31′E
192 E8 **Jawa** *Eng.* Java; *prev.* Djawa. *island* C Indonesia
169 O16 **Jawa Barat** *off.* Propinsi Jawa Barat, *var.* Jabar, *Eng.* West Java. ◆ *province* S Indonesia
Jawa Barat, Propinsi *see* Jawa Barat
Jawa, Laut *see* Java Sea
139 R3 **Jawān** Nīnawýa, NW Iraq 35°57′N 43°03′E
169 P16 **Jawa Tengah** *off.* Propinsi Jawa Tengah, *var.* Jateng, *Eng.* Central Java. ◆ *province* S Indonesia
Jawa Tengah, Propinsi *see* Jawa Tengah
169 R16 **Jawa Timur** *off.* Propinsi Jawa Timur, *var.* Jatim, *Eng.* East Java. ◆ *province* S Indonesia
Jawa Timur, Propinsi *see* Jawa Timur
81 N17 **Jawhar** *var.* Jowhar, *It.* Giohar. Shabeellaha Dhexe, S Somalia 02°37′N 45°30′E
111 F14 **Jawor** *Ger.* Jauer. Dolnośląskie, SW Poland 51°01′N 16°11′E
Jaworów *see* Yavoriv
111 J16 **Jaworzno** Śląskie, S Poland 50°13′N 19°11′E
Jaxartes *see* Syr Darya
27 R9 **Jay** Oklahoma, C USA 36°25′N 94°49′W
Jayabum *see* Chaiyaphum
Jayanath *see* Chai Nat
153 T12 **Jayanti** *prev.* Jainti. West Bengal, NE India 26°45′N 89°40′E
171 X14 **Jaya, Puncak** *prev.* Puntjak Carstensz, Puntjak Sukarno. ▲ Papua, E Indonesia 04°00′S 137°10′E
171 Z13 **Jayapura** *var.* Djajapura, *Dut.* Hollandia; *prev.* Kotabaru, Sukarnapura. Papua, E Indonesia 02°37′S 140°39′E
Jay Dairen *see* Dalian
Jayhawker State *see* Kansas
147 S12 **Jayilgan** *Rus.* Dzhailgan, Dzhayilgan. C Tajikistan 39°17′N 71°32′E
155 L14 **Jaypur** *var.* Jeypore, Jeypur. Odisha, E India 18°54′N 82°36′E
25 O6 **Jayton** Texas, SW USA 33°16′N 100°35′W
143 U13 **Jaz Mūrīān, Hāmūn-e** ⊚ SE Iran
138 M4 **Jazrah** Ar Raqqah, C Syria 35°56′N 39°02′E
138 G6 **Jbaïl** *var.* Jebeil, Jubayl, Jubeil; *anc.* Biblical Gebal, Bybles. W Lebanon 34°00′N 35°45′E
25 O7 **J. B. Thomas, Lake** ⊠ Texas, SW USA
Jdaïdé *see* Judaydah
35 X12 **Jean** Nevada, W USA 35°45′N 115°20′W
22 I9 **Jeanerette** Louisiana, S USA 29°54′N 91°39′W
44 L8 **Jean-Rabel** NW Haiti 19°48′N 73°05′W
143 T12 **Jebāl Bārez, Kūh-e** ▲▲ SE Iran
Jebat *see* Jabwot
77 T15 **Jebba** Kwara, W Nigeria 09°04′N 04°50′E
Jebeil *see* Jbaïl
116 E12 **Jebel** *Hung.* Széphely; *prev. Hung.* Zsebely. Timiş, W Romania 45°33′N 21°14′E
146 B11 **Jebel** *Rus.* Dzhebel. Balkan Welaýaty, W Turkmenistan 39°42′N 54°10′E
Jebel, Bahr el *see* White Nile
Jebel Dhanna *see* Jabal az Zannah
Jeble *see* Jablah
163 Y15 **Jecheon** *Jap.* Teisen; *prev.* Chech'ŏn. N South Korea 37°06′N 128°15′E
96 K13 **Jedburgh** SE Scotland, United Kingdom 55°29′N 02°34′W
Jedda *see* Jiddah
111 L15 **Jędrzejów** *Ger.* Endersdorf. Świętokrzyskie, C Poland 50°39′N 20°18′E
100 K12 **Jeetze** *var.* Jeetzel. ⌁ C Germany
Jeetzel *see* Jeetze
29 U14 **Jefferson** Iowa, C USA 42°01′N 94°22′W
21 Q8 **Jefferson** North Carolina, SE USA 36°24′N 81°33′W
25 X6 **Jefferson** Texas, SW USA 32°45′N 94°21′W
30 M9 **Jefferson** Wisconsin, N USA 43°01′N 88°48′W
27 U5 **Jefferson City** *state capital* Missouri, C USA 38°33′N 92°13′W
33 R10 **Jefferson City** Montana, NW USA 46°24′N 112°01′W
21 N9 **Jefferson City** Tennessee, S USA 36°07′N 83°29′W
35 U7 **Jefferson, Mount** ▲ Nevada, W USA 38°49′N 116°54′W
32 H12 **Jefferson, Mount** ▲ Oregon, NW USA 44°40′N 121°48′W
20 L5 **Jeffersontown** Kentucky, S USA 38°11′N 85°33′W
31 P16 **Jeffersonville** Indiana, N USA 38°16′N 85°45′W
33 V15 **Jeffrey City** Wyoming, C USA 42°29′N 107°49′W
77 T13 **Jega** Kebbi, NW Nigeria 12°15′N 04°21′E
Jehol *see* Chengde
163 X17 **Jeju** *Jap.* Saishū; *prev.* Cheju. S South Korea 33°31′N 126°34′E
163 Y17 **Jeju-do** *Jap.* Saishū; *prev.* Cheju-do, Quelpart. *island* S South Korea
163 Y17 **Jeju-haehyeop** *Eng.* Cheju Strait; *prev.* Cheju-haehyeop. *strait* S South Korea
62 P5 **Jejui-Guazú, Río** ⌁ E Paraguay
118 I10 **Jēkabpils** *Ger.* Jakobstadt. S Latvia 56°30′N 25°56′E
23 W7 **Jekyll Island** *island* Georgia, SE USA
169 R13 **Jelai, Sungai** ⌁ Borneo, N Indonesia
Jelalabad *see* Jalālābād
111 H14 **Jelcz-Laskowice** Dolnośląskie, SW Poland 51°01′N 17°24′E
111 E14 **Jelenia Góra** *Ger.* Hirschberg, Hirschberg im Riesengebirge, Hirschberg in Riesengebirge, Hirschberg in Schlesien. Dolnośląskie, SW Poland 50°54′N 15°48′E
153 S11 **Jelep La** *pass* N India
118 F9 **Jelgava** *Ger.* Mitau. C Latvia 56°38′N 23°47′E
112 L13 **Jelica** ▲▲ C Serbia
20 M8 **Jellico** Tennessee, S USA 36°33′N 84°06′W
95 G23 **Jelling** Syddanmark, C Denmark 55°45′N 09°24′E
169 N9 **Jemaja, Pulau** *island* W Indonesia
Jemaluang *see* Jamaluang
99 E20 **Jemappes** Hainaut, S Belgium 50°27′N 03°53′E
169 S17 **Jember** *prev.* Djember. Jawa, C Indonesia 08°07′S 113°45′E
99 I20 **Jemeppe-sur-Sambre** Namur, S Belgium 50°27′N 04°41′E
37 R10 **Jemez Pueblo** New Mexico, SW USA 35°36′N 106°43′W
158 K2 **Jeminay** *var.* Tuotiereke. Xinjiang Uygur Zizhiqu, NW China 47°28′N 85°49′E
189 U5 **Jemo Island** *atoll* Ratak Chain, C Marshall Islands
169 U11 **Jempang, Danau** ⊚ Borneo, N Indonesia
101 L16 **Jena** Thüringen, C Germany 50°56′N 11°35′E
22 I6 **Jena** Louisiana, S USA 31°40′N 92°07′W
108 I8 **Jenaz** Graubünden, SE Switzerland 46°56′N 09°43′E
109 N7 **Jenbach** Tirol, W Austria 47°24′N 11°47′E
171 N15 **Jeneponto** *prev.* Djeneponto. Sulawesi, C Indonesia 05°41′S 119°42′E
138 F9 **Jenin** N West Bank 32°28′N 35°17′E
21 P7 **Jenkins** Kentucky, S USA 37°10′N 82°37′W
27 P9 **Jenks** Oklahoma, C USA 36°01′N 95°58′W
Jenné *see* Djenné
109 X8 **Jennersdorf** Burgenland, SE Austria 46°57′N 16°08′E
22 H9 **Jennings** Louisiana, S USA 30°13′N 92°39′W
11 N7 **Jenny Lind Island** *island* Nunavut, N Canada
23 Y13 **Jensen Beach** Florida, SE USA 27°15′N 80°13′W
9 P6 **Jens Munk Island** *island* Nunavut, NE Canada
163 Y15 **Jeonju** *Jap.* Zenshū; *prev.* Chŏnju. SW South Korea 35°51′N 127°08′E
59 O17 **Jequié** Bahia, E Brazil 13°52′S 40°06′W
59 O18 **Jequitinhonha, Rio** ⌁ E Brazil
Jerablus *see* Jarābulus
74 H6 **Jerada** NE Morocco 34°16′N 02°07′W
Jerash *see* Jarash
75 N7 **Jerba, Île de** *var.* Djerba, Jazīrat Jarbah. *island* E Tunisia
44 K9 **Jérémie** SW Haiti 18°39′N 74°11′W
Jeréz *see* Jerez de García Salinas, Mexico
Jerez *see* Jeréz de la Frontera, Spain
40 L11 **Jerez de García Salinas** *var.* Jeréz. Zacatecas, C Mexico 22°40′N 103°00′W
104 J15 **Jerez de la Frontera** *var.* Jerez; *prev.* Xeres. Andalucía, SW Spain 36°41′N 06°08′W
104 I12 **Jerez de los Caballeros** Extremadura, W Spain 38°20′N 06°45′W
Jergucati *see* Jorgucat
138 G10 **Jericho** *Ar.* Arīḩā, *Heb.* Yeriḩo. E West Bank 31°51′N 35°27′E
74 M7 **Jerid, Chott el** *var.* Shaţţ al Jarīd. *salt lake* SW Tunisia
183 O10 **Jerilderie** New South Wales, SE Australia 35°24′S 145°43′E
Jerischmarkt *see* Câmpia Turzii
92 K11 **Jerisjärvi** ⊚ NW Finland
Jermak *see* Aksu
Jermentau *see* Yereymentau
Jermer *see* Jaroměř
36 K11 **Jerome** Arizona, SW USA 34°45′N 112°06′W
33 O15 **Jerome** Idaho, NW USA 42°43′N 114°31′W
97 K26 **Jersey** ◇ *British Crown Dependency* Channel Islands, NW Europe
97 L26 **Jersey** *island* Channel Islands, NW Europe
18 K14 **Jersey City** New Jersey, NE USA 40°42′N 74°01′W
18 F13 **Jersey Shore** Pennsylvania, NE USA 41°12′N 77°13′W
30 K14 **Jerseyville** Illinois, N USA 39°07′N 90°19′W
104 K8 **Jerte** ⌁ W Spain
138 F10 **Jerusalem** *Ar.* Al Quds, Al Quds ash Sharīf, *Heb.* Yerushalayim; *anc.* Hierosolyma. ● (Israel) Jerusalem, NE Israel 31°47′N 35°13′E
138 G10 **Jerusalem** ◆ *district* E Israel
183 S10 **Jervis Bay** New South Wales, SE Australia 35°09′S 150°42′E
183 S10 **Jervis Bay Territory** ◆ *territory* SE Australia
Jerwakant *see* Järvakandi
109 S10 **Jesenice** *Ger.* Assling. NW Slovenia 46°26′N 14°01′E
111 H16 **Jeseník** *Ger.* Freiwaldau. Olomoucký Kraj, E Czech Republic 50°14′N 17°12′E
Jesi *see* Iesi
106 I8 **Jesolo** *var.* Iesolo. Veneto, NE Italy 45°32′N 12°37′E
Jesselton *see* Kota Kinabalu
95 I14 **Jessheim** Akershus, S Norway 60°07′N 11°10′E
153 T15 **Jessore** Khulna, W Bangladesh 23°10′N 89°12′E
23 W6 **Jesup** Georgia, SE USA 31°36′N 81°54′W
41 S15 **Jesús Carranza** Veracruz-Llave, SE Mexico 17°30′N 95°01′W
62 K10 **Jesús María** Córdoba, C Argentina 30°59′S 64°05′W
26 K6 **Jetmore** Kansas, C USA 38°05′N 99°55′W
103 Q2 **Jeumont** Nord, N France 50°18′N 04°06′E
95 H14 **Jevnaker** Oppland, S Norway 60°15′N 10°25′E
Jewe *see* Jõhvi
25 V9 **Jewett** Texas, SW USA 31°21′N 96°08′W
19 N12 **Jewett City** Connecticut, NE USA 41°36′N 71°58′W
Jeypore *see* Jaipur, Rājasthān, India
Jeypore/Jeypur *see* Jaypur, Orissa, India
113 L17 **Jezercës, Maja e** ▲ N Albania 42°27′N 19°49′E
111 B18 **Jezerní Hora** ▲ SW Czech Republic 49°10′N 13°11′E
154 F10 **Jhābua** Madhya Pradesh, C India 22°44′N 74°37′E
152 H14 **Jhālāwār** Rājasthān, N India 24°37′N 76°12′E
Jhang/Jhang Sadar *see* Jhang Sadr
149 U9 **Jhang Sadr** *var.* Jhang, Jhang Sadar. Punjab, NE Pakistan 31°16′N 72°19′E
152 J13 **Jhānsi** Uttar Pradesh, N India 25°27′N 78°34′E
153 O16 **Jhārkhand** ◆ *state* NE India
154 M11 **Jhārsuguda** Odisha, E India 21°56′N 84°04′E
149 V7 **Jhelum** Punjab, NE Pakistan 32°55′N 73°42′E
129 P9 **Jhelum** ⌁ E Pakistan
Jhenaida *see* Jhenaidaha
153 T15 **Jhenaidaha** *var.* Jhenaida, Jhenida. Khulna, Bangladesh 23°32′N 89°09′E
Jhenaidaha *see* Jhenida
153 T15 **Jhenida** *var.* Jhenaidaha. Dhaka, W Bangladesh 23°34′N 89°39′E
Jhenida *see* Jhenaidaha
149 P16 **Jhimpir** Sind, SE Pakistan 25°00′N 68°01′E
Jhind *see* Jind
149 R16 **Jhudo** Sind, SE Pakistan 24°58′N 69°18′E
Jhumra *see* Chak Jhumra
152 H11 **Jhunjhunūn** Rājasthān, N India 28°05′N 75°30′E
Ji *see* Hebei, China
Ji *see* Jilin, China
Jiading *see* Xinfeng
153 S14 **Jiāganj** West Bengal, NE India 24°18′N 88°07′E
Jiaji *see* Qionghai
160 J7 **Jialing Jiang** ⌁ C China
163 Y7 **Jiamusi** *var.* Chia-mu-ssu, Kiamusze. Heilongjiang, NE China 46°46′N 130°19′E
161 O11 **Ji'an** Jiangxi, S China 27°08′N 115°00′E
163 W12 **Ji'an** Jilin, NE China 41°04′N 126°07′E
163 T13 **Jianchang** Liaoning, NE China 40°48′N 119°51′E
Jianchang *see* Nancheng
Jiancheng *see* Jianyang
160 F11 **Jianchuan** *var.* Jinhuan. Yunnan, SW China 26°28′N 99°49′E
158 M4 **Jiangjunmiao** Xinjiang Uygur Zizhiqu, W China 44°42′N 90°06′E
160 K11 **Jiangkou** *var.* Shuangjiang. Guizhou, S China 27°46′N 108°53′E
Jiangkou *see* Fengkai
161 Q12 **Jiangle** *var.* Guyong. Fujian, SE China 26°44′N 117°26′E
161 N15 **Jiangmen** Guangdong, S China 22°35′N 113°02′E
Jiangna *see* Yanshan
161 Q10 **Jiangshan** Zhejiang, SE China 28°41′N 118°33′E
161 Q7 **Jiangsu** *var.* Chiang-su, Jiangsu Sheng, Kiangsu, Su. ◆ *province* E China
Jiangsu *see* Nanjing
Jiangsu Sheng *see* Jiangsu
161 O11 **Jiangxi** *var.* Chiang-hsi, Gan, Jiangxi Sheng, Kiangsi. ◆ *province* S China
Jiangxi Sheng *see* Jiangxi
160 I8 **Jiangyou** *prev.* Zhongba. Sichuan, C China 31°52′N 104°52′E
161 N9 **Jianli** *var.* Rongcheng. Hubei, C China 29°51′N 112°50′E
161 Q11 **Jian'ou** Fujian, SE China 27°04′N 118°20′E
163 S12 **Jianping** *var.* Yebaishou. Liaoning, NE China 41°13′N 119°37′E
Jianshe *see* Baiyü
160 L9 **Jianshi** *var.* Yezhou. Hubei, C China 30°37′N 109°42′E
Jiantang *see* Xamgyi'nyilha
129 V11 **Jian Xi** ⌁ SE China
161 Q11 **Jianyang** Fujian, SE China 27°20′N 118°01′E
160 I9 **Jianyang** *var.* Jiancheng. Sichuan, C China 30°22′N 104°31′E
163 X10 **Jiaohe** Jilin, NE China 43°41′N 127°20′E
Jiaojiang *see* Taizhou
Jiaoxian *see* Jiaozhou
161 R5 **Jiaozhou** *prev.* Jiaoxian. Shandong, E China 36°17′N 120°00′E
161 N6 **Jiaozuo** Henan, C China 35°14′N 113°13′E
Jiashan *see* Mingguang
158 F8 **Jiashi** *var.* Baren, Payzawat. Xinjiang Uygur Zizhiqu, NW China 39°27′N 76°45′E
154 L9 **Jiāwān** Madhya Pradesh, C India 24°20′N 82°17′E
161 S9 **Jiaxing** Zhejiang, SE China 30°44′N 120°46′E
161 S14 **Jiayi** *var.* Chia-i, Chiai, Chiayi, Kiayi, *Jap.* Kagi. C Taiwan 23°29′N 120°27′E
163 X6 **Jiayin** *var.* Chaoyang. Heilongjiang, NE China 48°51′N 130°24′E
159 R8 **Jiayuguan** Gansu, N China 39°47′N 98°14′E
Jibhalanta *see* Uliastay
138 M4 **Jibli** Ar Raqqah, C Syria 35°49′N 39°23′E
116 H9 **Jibou** *Hung.* Zsibó. Sălaj, NW Romania 47°15′N 23°17′E
141 Z9 **Jibsh, Ra's al** *headland* E Oman 21°20′N 59°23′E
Jibuti *see* Djibouti
111 E15 **Jičín** *Ger.* Jitschin. Královéhradecký Kraj, N Czech Republic 50°27′N 15°20′E
140 K10 **Jiddah** *Eng.* Jedda. (Saudi Arabia) Makkah, W Saudi Arabia 21°34′N 39°13′E
141 W11 **Jiddat al Ḩarāsīs** *desert* C Oman
Jiesjavrre *see* Iešjavri
160 M4 **Jiexiu** Shanxi, C China 37°00′N 111°50′E
161 P14 **Jieyang** Guangdong, S China 23°32′N 116°20′E
119 F14 **Jieznas** Kaunas, S Lithuania 54°37′N 24°10′E
Jifa', Bi'r *see* Jif'īyah, Bi'r
141 P15 **Jif'īyah, Bi'r** *var.* Bi'r Jifa'. *well* C Yemen
77 W13 **Jigawa** ◆ *state* N Nigeria
146 J10 **Jigerbent** *Rus.* Dzhigirbent. Lebap Welaýaty, NE Turkmenistan 40°44′N 61°56′E
44 I7 **Jiguaní** Granma, E Cuba 20°24′N 76°26′W
159 T12 **Jigzhi** *var.* Chugqênsumdo. Qinghai, C China 33°23′N 101°25′E
Jih-k'a-tse *see* Xigazê
111 E18 **Jihlava** *Ger.* Iglau, *Pol.* Igława. Vysocina, S Czech Republic 49°22′N 15°36′E
111 E18 **Jihlava** *var.* Igel, *Ger.* Iglawa. ⌁ Vysočina, C Czech Republic
Jihlavský Kraj *see* Vysočina
111 C18 **Jihočeský Kraj** *prev.* Budějovický Kraj. ◆ *region* Czech Republic
111 G19 **Jihomoravský Kraj** *prev.* Brněnský Kraj. ◆ *region* SE Czech Republic
74 L5 **Jijel** *var.* Djidjel; *prev.* Djidjelli. NE Algeria 36°50′N 05°43′E
116 L9 **Jijia** ⌁ N Romania
80 L13 **Jijiga** *It.* Giggiga. Sumalē, E Ethiopia 09°21′N 42°53′E
105 S12 **Jijona** *var.* Xixona. Valenciana, E Spain 38°34′N 00°29′W
81 L18 **Jilib** *It.* Gelib. Jubbada Dhexe, S Somalia 0°18′N 42°48′E
163 W10 **Jilin** *var.* Chi-lin, Girin, Kirin; *prev.* Yungki, Yunki. Jilin, NE China 43°46′N 126°32′E
163 W10 **Jilin** *var.* Chi-lin, Girin, Ji, Jilin Sheng, Kirin. ◆ *province* NE China
163 W11 **Jilin Hada Ling** ▲▲ NE China
Jilin Sheng *see* Jilin
163 S4 **Jiliu He** ⌁ NE China
105 Q6 **Jiloca** ⌁ N Spain
161 T12 **Jilong** *var.* Keelung, *Jap.* Kirun, Kirun'; *prev.* Chilung, *prev. Sp.* Santissima Trinidad. N Taiwan 25°10′N 121°43′E
81 I14 **Jīma** *var.* Jimma, *It.* Gimma. Oromīya, C Ethiopia 07°42′N 36°51′E
44 M9 **Jimaní** W Dominican Republic 18°29′N 71°49′W
116 E11 **Jimbolia** *Ger.* Hatzfeld, *Hung.* Zsombolya. Timiş, W Romania 45°47′N 20°43′E
104 K16 **Jimena de la Frontera** Andalucía, S Spain 36°27′N 05°28′W
40 K7 **Jiménez** Chihuahua, N Mexico 27°09′N 104°54′W
41 N5 **Jiménez** Coahuila, NE Mexico 29°05′N 100°40′W
41 P9 **Jiménez** *var.* Santander Jiménez. Tamaulipas, C Mexico 24°11′N 98°29′W
40 L10 **Jiménez del Teul** Zacatecas, C Mexico 23°13′N 103°46′W
77 Y14 **Jimeta** Adamawa, E Nigeria 09°16′N 12°25′E
Jimma *see* Jīma
158 M5 **Jimsar** Xinjiang Uygur Zizhiqu, NW China 44°05′N 88°48′E
18 I14 **Jim Thorpe** Pennsylvania, NE USA 40°51′N 75°43′W
Jin *see* Shanxi
Jin *see* Tianjin Shi
161 P5 **Jinan** *var.* Chinan, Chi-nan, Tsinan. *province capital* Shandong, E China 36°43′N 116°58′E
Jin'an *see* Songpan
Jinbi *see* Dayao
159 T8 **Jinchang** Gansu, N China 38°31′N 102°07′E
161 N5 **Jincheng** Shanxi, C China 35°30′N 112°52′E
Jincheng *see* Wuding
Jinchengjiang *see* Hechi
152 I9 **Jīnd** *prev.* Jhind. Haryāna, NW India 29°29′N 76°22′E
183 Q11 **Jindabyne** New South Wales, SE Australia 36°28′S 148°36′E
163 X17 **Jin-do** *Jap.* Chin-tō; *prev.* Chin-do. *island* SW South Korea
111 O18 **Jindřichův Hradec** *Ger.* Neuhaus. Jihočeský Kraj, S Czech Republic 49°09′N 15°01′E
Jing *see* Beijing Shi
Jing *see* Jinghe, China
159 X10 **Jingchuan** Gansu, C China 35°20′N 107°45′E
161 Q10 **Jingdezhen** Jiangxi, S China 29°18′N 117°18′E
161 O12 **Jinggangshan** Jiangxi, S China 26°36′N 114°11′E
161 P3 **Jinghai** Tianjin Shi, E China 38°53′N 116°45′E
158 I4 **Jinghe** *var.* Jing. Xinjiang Uygur Zizhiqu, NW China 44°35′N 82°55′E
160 K6 **Jing He** ⌁ C China
160 F15 **Jinghong** *var.* Yunjinghong. Yunnan, SW China 22°03′N 100°56′E
160 M9 **Jingmen** Hubei, C China 30°58′N 112°09′E
163 X10 **Jingpo Hu** ⊚ NE China
160 M8 **Jing Shan** ▲ C China
159 V9 **Jingtai** *var.* Yitiaoshan. Gansu, C China 37°12′N 104°06′E
160 J14 **Jingxi** *var.* Xinjing. Guangxi Zhuangzu Zizhiqu, S China 23°10′N 106°22′E
Jing Xian *see* Jingzhou, Hunan, China
163 W11 **Jingyu** Jilin, NE China 42°23′N 126°48′E
159 V10 **Jingyuan** *var.* Wulan. Gansu, C China 36°35′N 104°40′E
160 M9 **Jingzhou** *prev.* Shashi, Shasshih, Shasi. Hubei, C China 30°21′N 112°09′E
160 L12 **Jingzhou** *var.* Jing Xian, Jingzhou Miaouzu Dongzu Zizhixian, Quyang. Hunan, S China 26°35′N 109°40′E
Jingzhou Miaouzu Dongzu Zizhixian *see* Jingzhou
163 Z16 **Jinhae** *Jap.* Chinkai; *prev.* Chinhae. S South Korea 35°06′N 128°48′E
Jinhe *see* Jinping
161 R10 **Jinhua** Zhejiang, SE China 29°15′N 119°36′E
Jinhuan *see* Jianchuan
161 P5 **Jining** Shandong, E China 35°25′N 116°35′E
Jining *see* Ulan Qab
81 G18 **Jinja** S Uganda 0°27′N 33°14′E
161 R13 **Jinjiang** *var.* Qingyang. Fujian, SE China 24°53′N 118°36′E
161 O11 **Jin Jiang** ⌁ S China
Jinjiang *see* Chengmai
163 Y16 **Jinju** *prev.* Chinju, *Jap.* Shinshū. S South Korea 35°12′N 128°06′E
171 V15 **Jin, Kepulauan** *island group* E Indonesia
161 R13 **Jinmen Dao** *var.* Chinmen Tao, Quemoy. *island* W Taiwan
42 J9 **Jinotega** Jinotega, NW Nicaragua 13°03′N 85°59′W
42 K7 **Jinotega** ◆ *department* N Nicaragua
42 J11 **Jinotepe** Carazo, SW Nicaragua 11°50′N 86°10′W
160 L13 **Jinping** *var.* Sanjiang. Guizhou, S China 26°42′N 109°13′E
160 H14 **Jinping** *var.* Jinhe. Yunnan, SW China 22°47′N 103°12′E
Jinsen *see* Incheon
160 J11 **Jinsha** Guizhou, S China 27°24′N 106°16′E
160 M10 **Jinshi** Hunan, S China 29°42′N 111°46′E
Jinshi *see* Xinning
162 I9 **Jinst** *var.* Bodĭ. Bayanhongor, C Mongolia 45°25′N 100°33′E
159 R7 **Jinta** Gansu, N China 40°01′N 98°57′E
161 Q12 **Jin Xi** ⌁ SE China
Jinxi *see* Huludao
Jinxian *see* Jinzhou
161 P6 **Jinxiang** Shandong, E China 35°08′N 116°19′E
161 P8 **Jinzhai** *var.* Meishan. Anhui, E China 31°42′N 115°47′E
161 N4 **Jinzhong** *var.* Yuci. Shanxi, C China 37°34′N 112°45′E
163 T12 **Jinzhou** *var.* Chin-chou, Chinchow; *prev.* Chinhsien. Liaoning, NE China 41°07′N 121°06′E

◆ Country ● Country Capital ◇ Dependent Territory ○ Dependent Territory Capital ◆ Administrative Regions ✈ International Airport ▲ Mountain ▲▲ Mountain Range ⚠ Volcano ⌁ River ⊚ Lake ⊠ Reservoir

JINZHOU – KÂHTA

163 U14 **Jinzhou** *prev.* Jinxian. Liaoning, NE China 39°04′N 121°45′E
Jinzhu *see* Daocheng
138 H12 **Jinz, Qāʿ al** ◎ C Jordan
47 S8 **Jiparaná, Rio** ≈ W Brazil
56 A7 **Jipijapa** Manabí, W Ecuador 01°23′S 80°35′W
42 F8 **Jiquilisco** Usulután, S El Salvador 13°19′N 88°35′W
Jirgalanta *see* Hovd
147 S12 **Jirgatol** *Rus.* Dzhirgatal'. C Tajikistan 39°13′N 71°09′E
75 X10 **Jirjā** *var.* Girga, Girgeh, Jirjā. C Egypt 26°17′N 31°58′E
Jirjā *see* Jirjā
111 B15 **Jirkov** *Ger.* Görkau. Ústecký Kraj, NW Czech Republic 50°30′N 13°27′E
143 T12 **Jīroft** *var.* Sabzawaran, Sabzvārān. Kermān, SE Iran 28°40′N 57°40′E
81 P14 **Jirriiban** Mudug, E Somalia 07°15′N 48°55′E
160 L11 **Jishou** Hunan, S China 28°20′N 109°43′E
Jisr ash Shadadi *see* Ash Shadādah
116 I14 **Jitaru** Olt, S Romania 44°27′N 24°32′E
Jitschin *see* Jičín
116 H14 **Jiu** *Ger.* Schil, Schyl, *Hung.* Zsil, Zsily. ≈ S Romania
161 R11 **Jiufeng Shan** ▲▲ SE China
161 P9 **Jiujiang** Jiangxi, S China 29°45′N 115°59′E
161 O10 **Jiuling Shan** ▲▲ S China
160 G10 **Jiulong** *var.* Garba, *Tib.* Gyaisi. Sichuan, C China 29°00′N 101°30′E
161 Q13 **Jiulong Jiang** ≈ SE China
161 Q12 **Jiulong Xi** ≈ SE China
159 R8 **Jiuquan** *var.* Suzhou. Gansu, N China 39°47′N 98°30′E
160 K17 **Jiusuo** Hainan, S China 18°25′N 109°55′E
163 W10 **Jiutai** Jilin, NE China 44°01′N 125°51′E
160 K13 **Jiuwan Dashan** ▲▲ S China
160 I7 **Jiuzhaigou** *var.* Nongle; *prev.* Nanping. Sichuan, C China 33°25′N 104°05′E
186 C7 **Jiwaka** ⬖ *province* C Papua New Guinea
148 I16 **Jīwani** Baluchistān, SW Pakistan 25°05′N 61°46′E
163 Y8 **Jixi** Heilongjiang, NE China 45°17′N 131°01′E
163 Y7 **Jixian** *var.* Fuli. Heilongjiang, NE China 46°38′N 131°04′E
160 M5 **Jixian** *var.* Ji Xian. Shanxi, C China 36°15′N 110°41′E
Ji Xian *see* Jixian
Jiza *see* Al Jīzah
141 N13 **Jīzān** *var.* Qīzān. Jīzān, SW Saudi Arabia 17°50′N 42°50′E
141 N13 **Jīzān** *var.* Minṭaqat Jīzān. ⬖ *province* SW Saudi Arabia
Jīzān, Minṭaqat *see* Jīzān
140 K6 **Jizl, Wādī al** *dry watercourse* W Saudi Arabia
164 H12 **Jizō-zaki** *headland* Honshū, SW Japan 35°34′N 133°16′E
141 U14 **Jizʿ, Wādī al** *dry watercourse* E Yemen
147 O11 **Jizzax** *Rus.* Dzhizak. Jizzax Viloyati, C Uzbekistan 40°08′N 67°47′E
147 N10 **Jizzax Viloyati** *Rus.* Dzhizakskaya Oblast'. ⬖ *province* C Uzbekistan
60 I13 **Joaçaba** Santa Catarina, S Brazil 27°08′S 51°30′W
Joal *see* Joal-Fadiout
76 F11 **Joal-Fadiout** *prev.* Joal. W Senegal 14°09′N 16°50′W
76 E10 **João Barrosa** Boa Vista, E Cape Verde 16°01′N 22°44′W
João Belo *see* Xai-Xai
João de Almeida *see* Chibia
59 Q15 **João Pessoa** *prev.* Paraíba. *state capital* Paraíba, E Brazil 07°06′S 34°53′W
25 X7 **Joaquin** Texas, SW USA 31°58′N 94°03′W
62 K6 **Joaquín V. González** Salta, N Argentina 25°06′S 64°07′W
Joazeiro *see* Juazeiro
Job'urg *see* Johannesburg
109 O7 **Jochberger Ache** ≈ W Austria
Jo-ch'iang *see* Ruoqiang
92 K12 **Jock** Norrbotten, N Sweden 66°40′N 22°45′E
42 I5 **Jocón** Yoro, N Honduras 15°17′N 86°55′W
105 O13 **Jódar** Andalucía, S Spain 37°51′N 03°18′W
152 F12 **Jodhpur** Rājasthān, NW India 26°17′N 73°02′E
99 I19 **Jodoigne** Walloon Brabant, C Belgium 50°43′N 04°52′E
93 O16 **Joensuu** Pohjois-Karjala, SE Finland 62°36′N 29°45′E
37 W4 **Joes** Colorado, C USA 39°36′N 102°40′W
191 Z3 **Joe's Hill** *hill* Kiritimati, NE Kiribati
165 N11 **Jōetsu** *var.* Zyôetu. Niigata, Honshū, C Japan 37°09′N 138°13′E
83 M18 **Jofane** Inhambane, S Mozambique 21°16′S 34°21′E
153 R12 **Jogbani** Bihār, NE India 26°23′N 87°16′E
118 I5 **Jõgeva** *Ger.* Laisholm. Jõgevamaa, E Estonia 58°48′N 26°28′E
118 I4 **Jõgevamaa** *off.* Jõgeva Maakond. ⬖ *province* E Estonia
Jõgeva Maakond *see* Jõgevamaa
155 E18 **Jog Falls** *Waterfall* Karnātaka, W India
143 S4 **Joghatāy** Khorāsān-e Razavī, NE Iran 36°34′N 57°00′E
153 U12 **Jogīghopa** Assam, NE India 26°14′N 90°35′E
152 I7 **Jogindarnagar** Himāchal Pradesh, N India 31°51′N 76°47′E
Jogjakarta *see* Yogyakarta
164 L11 **Jōhana** Toyama, Honshū, SW Japan 36°30′N 136°53′E
83 J21 **Johannesburg** *var.* Egoli, Erautini, Gauteng, *abbrev.* Job'urg. Gauteng, NE South Africa 26°10′S 28°02′E
35 T13 **Johannesburg** California, W USA 35°20′N 117°37′W
Johannisburg *see* Pisz
149 P14 **Johi** Sind, SE Pakistan 26°46′N 67°28′E
55 T13 **Johi Village** S Guyana 01°48′N 58°33′W
45 W10 **John A. Osborne** ✈ (Plymouth) E Montserrat 16°45′N 62°09′W
32 K13 **John Day** Oregon, NW USA 44°25′N 118°57′W
32 I11 **John Day River** ≈ Oregon, NW USA
18 L14 **John F Kennedy** ✈ (New York) Long Island, New York, NE USA 40°39′N 73°45′W
21 V8 **John H. Kerr Reservoir** *var.* Buggs Island Lake, Kerr Lake. ⊠ North Carolina/Virginia, SE USA
37 V6 **John Martin Reservoir** ⊠ Colorado, C USA
96 K6 **John o'Groats** N Scotland, United Kingdom 58°38′N 03°03′W
27 P5 **John Redmond Reservoir** ⊠ Kansas, C USA
39 Q7 **John River** ≈ Alaska, USA
26 H6 **Johnson** Kansas, C USA 37°33′N 101°46′W
18 M7 **Johnson** Vermont, NE USA 44°39′N 72°40′W
18 D13 **Johnsonburg** Pennsylvania, NE USA 41°28′N 78°37′W
18 H11 **Johnson City** New York, NE USA 42°06′N 75°54′W
21 P8 **Johnson City** Tennessee, S USA 36°18′N 82°21′W
25 R10 **Johnson City** Texas, SW USA 30°17′N 98°27′W
35 S12 **Johnsondale** California, W USA 35°58′N 118°32′W
10 I8 **Johnsons Crossing** Yukon, W Canada 60°30′N 133°15′W
21 T13 **Johnsonville** South Carolina, SE USA 33°50′N 79°26′W
21 Q13 **Johnston** South Carolina, SE USA 33°49′N 81°48′W
192 M6 **Johnston Atoll** ◇ *US unincorporated territory* C Pacific Ocean
175 Q3 **Johnston Atoll** *atoll* C Pacific Ocean
30 L17 **Johnston City** Illinois, N USA 37°49′N 88°55′W
180 K12 **Johnston, Lake** *salt lake* Western Australia
31 S13 **Johnstown** Ohio, N USA 40°08′N 82°39′W
18 D15 **Johnstown** Pennsylvania, NE USA 40°20′N 78°56′W
168 L10 **Johor** *var.* Johore. ⬖ *state* Peninsular Malaysia
Johor Baharu *see* Johor Bahru
168 K10 **Johor Bahru** *var.* Johor Baharu, Johore Bahru. Johor, Peninsular Malaysia 01°29′N 103°44′E
Johore *see* Johor
Johore Bahru *see* Johor Bahru
118 K3 **Jõhvi** *Ger.* Jewe. Ida-Virumaa, NE Estonia 59°21′N 27°25′E
103 P7 **Joigny** Yonne, C France 47°58′N 03°24′E
Joinvile *see* Joinville
60 K12 **Joinville** *var.* Joinvile. Santa Catarina, S Brazil 26°20′S 48°55′W
103 R6 **Joinville** Haute-Marne, N France 48°26′N 05°07′E
194 H3 **Joinville Island** *island* Antarctica
41 O15 **Jojutla** *var.* Jojutla de Juárez. Morelos, S Mexico 18°38′N 99°10′W
Jojutla de Juárez *see* Jojutla
92 I12 **Jokkmokk** *Lapp.* Dálvvadis. Norrbotten, N Sweden 66°35′N 19°57′E
92 L2 **Jökuldalur** ≈ E Iceland
92 K2 **Jökulsá á Fjöllum** ≈ NE Iceland
Jokyakarta *see* Yogyakarta
30 M11 **Joliet** Illinois, N USA 41°33′N 88°05′W
15 O11 **Joliette** Québec, SE Canada 46°02′N 73°27′W
171 O8 **Jolo** Jolo Island, SW Philippines 06°02′N 121°00′E
94 D11 **Jølstervatnet** ◎ S Norway
169 S16 **Jombang** *prev.* Djombang. Jawa, S Indonesia 07°33′S 112°14′E
159 R14 **Jomda** Xizang Zizhiqu, W China 31°26′N 98°09′E
118 G13 **Jonava** *Ger.* Janow, *Pol.* Janów. Kaunas, C Lithuania 55°05′N 24°19′E
146 L11 **Jondor** *Rus.* Zhondor. Buxoro Viloyati, C Uzbekistan 39°46′N 64°11′E
159 V11 **Jonê** *var.* Liulin. Gansu, C China 34°36′N 103°39′E
27 X9 **Jonesboro** Arkansas, C USA 35°50′N 90°42′W
23 S4 **Jonesboro** Georgia, SE USA 33°31′N 84°21′W
30 L17 **Jonesboro** Illinois, N USA 37°25′N 89°19′W
22 H5 **Jonesboro** Louisiana, S USA 32°14′N 92°43′W
21 P8 **Jonesboro** Tennessee, S USA 36°17′N 82°28′W
19 T6 **Jonesport** Maine, NE USA 44°33′N 67°35′W
0 J4 **Jones Sound** *channel* Nunavut, N Canada
22 I6 **Jonesville** Louisiana, S USA 31°37′N 91°49′W
31 Q10 **Jonesville** Michigan, N USA 41°58′N 84°39′W
21 Q11 **Jonesville** South Carolina, SE USA 34°49′N 81°38′W
146 K10 **Jongeldi** *Rus.* Dzhankel'dy. Buxoro Viloyati, C Uzbekistan 40°50′N 63°16′E
81 F14 **Jonglei** Jonglei, E South Sudan 06°54′N 31°19′E
81 F14 **Jonglei** *var.* Gongoleh State. ⬖ *state* E South Sudan
81 F14 **Jonglei Canal** *canal* E South Sudan
118 F11 **Joniškėlis** Panevėžys, N Lithuania 56°02′N 24°10′E
118 F10 **Joniškis** *Ger.* Janischken. Šiaulai, N Lithuania 56°15′N 23°36′E
95 L19 **Jönköping** Jönköping, S Sweden 57°45′N 14°10′E
95 K20 **Jönköping** ⬖ *county* S Sweden
15 Q7 **Jonquière** Québec, SE Canada 48°25′N 71°16′W
41 V15 **Jonuta** Tabasco, SE Mexico 18°04′N 92°03′W
102 K12 **Jonzac** Charente-Maritime, W France 45°26′N 00°25′W
27 R7 **Joplin** Missouri, C USA 37°04′N 94°31′W
33 W8 **Jordan** Montana, NW USA 47°18′N 106°54′W
138 H12 **Jordan** *off.* Hashemite Kingdom of Jordan, *Ar.* Al Mamlaka al Urdunīya al Hashemīyah, Al Urdunn; *prev.* Transjordan. ◆ *monarchy* SW Asia
138 G9 **Jordan** *Ar.* Urdunn, *Heb.* HaYarden. ≈ SW Asia
Jordan Lake *see* B. Everett Jordan Reservoir
111 K17 **Jordanów** Małopolskie, S Poland 49°39′N 19°51′E
32 M15 **Jordan Valley** Oregon, NW USA 42°59′N 117°03′W
138 G9 **Jordan Valley** *valley* N Israel
57 D15 **Jorge Chávez Internacional** *var.* Lima. ✈ (Lima) Lima, W Peru 12°07′S 77°01′W
113 L23 **Jorgucat** *var.* Jergucati, Jorgucati. Gjirokastër, S Albania 39°57′N 20°14′E
Jorgucati *see* Jorgucat
153 X12 **Jorhāt** Assam, NE India 26°45′N 94°09′E
93 J14 **Jörn** Västerbotten, N Sweden 65°03′N 20°04′E
37 R14 **Jornada Del Muerto** *valley* New Mexico, SW USA
93 N17 **Joroinen** Etelä-Savo, E Finland 62°11′N 27°50′E
95 C16 **Jørpeland** Rogaland, S Norway 59°01′N 06°04′E
77 W14 **Jos** Plateau, C Nigeria 09°59′N 08°57′E
171 Q8 **Jose Abad Santos** *var.* Trinidad. Mindanao, S Philippines 05°51′N 125°35′E
61 F19 **José Batlle y Ordóñez** *var.* Batlle y Ordóñez. Florida, C Uruguay 33°28′S 55°08′W
63 H18 **José de San Martín** Chubut, S Argentina 44°04′S 70°29′W
61 E19 **José Enrique Rodó** *var.* Rodó, José E.Rodo; *prev.* Drabble, Drable. Soriano, SW Uruguay 33°43′S 57°33′W
José E.Rodo *see* José Enrique Rodó
Josefsdorf *see* Žabalj
44 C4 **José Martí** ✈ (La Habana) Cuidad de La Habana, N Cuba 23°03′N 82°22′W
61 F19 **José Pedro Varela** *var.* José P.Varela. Lavalleja, S Uruguay 33°30′S 54°28′W
181 N2 **Joseph Bonaparte Gulf** *gulf* N Australia
37 N11 **Joseph City** Arizona, SW USA 34°56′N 110°18′W
13 O9 **Joseph, Lake** ◎ Newfoundland and Labrador, E Canada
14 G13 **Joseph, Lake** ◎ Ontario, S Canada
186 C6 **Josephstaal** Madang, N Papua New Guinea 04°42′S 144°55′E
José P.Varela *see* José Pedro Varela
59 J14 **José Rodrigues** Pará, N Brazil 05°45′S 51°20′W
152 K9 **Joshīmath** Uttarakhand, N India 30°33′N 79°35′E
25 T7 **Joshua** Texas, SW USA 32°27′N 97°23′W
35 V15 **Joshua Tree** California, W USA 34°07′N 116°18′W
77 V14 **Jos Plateau** *plateau* C Nigeria
102 H6 **Josselin** Morbihan, NW France 47°57′N 02°35′W
Jos Sudarso *see* Yos Sudarso, Pulau
94 E11 **Jostedalsbreen** *glacier* S Norway
94 F12 **Jotunheimen** ▲▲ S Norway
138 G7 **Joûnié** *var.* Junīyah. W Lebanon 33°54′N 33°36′E
25 R13 **Jourdanton** Texas, SW USA 28°55′N 98°34′W
98 L7 **Joure** *Fris.* De Jouwer. Fryslân, N Netherlands 52°58′N 05°48′E
93 M18 **Joutsa** Keski-Suomi, C Finland 61°46′N 26°09′E
93 N18 **Joutseno** Etelä-Karjala, SE Finland 61°06′N 28°30′E
92 M12 **Joutsijärvi** Lappi, NE Finland 66°40′N 28°00′E
108 A9 **Joux, Lac de** ◎ W Switzerland
Jovakān *see* Jowkān
44 D5 **Jovellanos** Matanzas, W Cuba 22°49′N 81°11′W
153 V13 **Jowai** Meghālaya, NE India 25°25′N 92°21′E
Jōwat *see* Jabwot
Jowhar *see* Jawhar
143 O12 **Jowkān** *var.* Jovakān. Fārs, S Iran
Jowzam *see* Javazm
149 N2 **Jowzjān** ⬖ *province* N Afghanistan
Joypurhat *see* Jaipurhat
Józseffalva *see* Žabalj
J.Storm Thurmond Reservoir *see* Clark Hill Lake
45 T6 **Juana Díaz** C Puerto Rico 18°03′N 66°30′W
40 L9 **Juan Aldama** Zacatecas, C Mexico 24°20′N 103°23′W
0 E9 **Juan de Fuca Plate** *tectonic feature*
32 F7 **Juan de Fuca, Strait of** *strait* Canada/USA
Juan Fernandez Islands *see* Juan Fernández, Islas
193 S11 **Juan Fernández, Islas** *Eng.* Juan Fernandez Islands. *island group* W Chile
55 O4 **Juangriego** Nueva Esparta, NE Venezuela 11°06′N 63°59′W
56 D11 **Juanjuí** *var.* Juanjuy. San Martín, N Peru 07°10′S 76°44′W
Juanjuy *see* Juanjuí
93 N16 **Juankoski** Pohjois-Savo, C Finland 63°01′N 28°24′E
Juan Lacaze *see* Juan L. Lacaze
61 E20 **Juan L. Lacaze** *var.* Juan Lacaze, Puerto Sauce; *prev.* Sauce. Colonia, SW Uruguay 34°26′S 57°25′W
62 L5 **Juan Solá** Salta, N Argentina 23°30′S 62°42′W
63 F21 **Juan Stuven, Isla** *island* S Chile
59 H16 **Juará** Mato Grosso, W Brazil 11°10′S 57°28′W
41 N7 **Juárez** *var.* Villa Juárez. Coahuila, NE Mexico 27°39′N 100°43′W
40 C2 **Juárez, Sierra de** ▲▲ NW Mexico
59 O15 **Juazeiro** *prev.* Joazeiro. Bahia, E Brazil 09°25′S 40°30′W
59 P14 **Juazeiro do Norte** Ceará, E Brazil 07°10′S 39°18′W
81 F15 **Juba** *var.* Jūbā. ● Central Equatoria, S South Sudan 04°50′N 31°35′E
81 L17 **Juba** *Amh.* Genalē Wenz, *It.* Guiba, *Som.* Ganaane, Webi Jubba. ≈ Ethiopia/Somalia
194 H2 **Jubany** *Argentinian research station* Antarctica 61°57′S 58°23′W
Jubayl *see* Jbaïl
81 L18 **Jubbada Dhexe** *off.* Gobolka Jubbada Dhexe. ⬖ *region* SW Somalia
Jubbada Dhexe, Gobolka *see* Jubbada Dhexe
81 K18 **Jubbada Hoose** ⬖ *region* SW Somalia
Jubba, Webi *see* Juba
Jubbulpore *see* Jabalpur
Jubeil *see* Jbaïl
74 B9 **Juby, Cap** *headland* SW Morocco 27°58′N 12°56′W
105 R10 **Júcar** *var.* Jucar. ≈ C Spain
40 L12 **Juchipila** Zacatecas, C Mexico 21°25′N 103°06′W
41 S16 **Juchitán** *var.* Juchitán de Zaragosa. Oaxaca, SE Mexico 16°27′N 95°W
Juchitán de Zaragosa *see* Juchitán
138 G11 **Judaea** *cultural region* Israel/West Bank
138 F11 **Judaean Hills** *Heb.* Haré Yehuda. *hill range* E Israel
138 H8 **Judaydah** *Fr.* Jdaïdé. Rīf Dimashq, W Syria 33°17′N 36°15′E
139 P11 **Judayyidat Hāmir** Al Anbār, S Iraq 31°50′N 41°50′E
109 U8 **Judenburg** Steiermark, C Austria 47°09′N 14°43′E
33 T8 **Judith River** ≈ Montana, NW USA
27 V11 **Judsonia** Arkansas, C USA 35°16′N 91°38′W
141 P14 **Jufrah, Wādī al** *dry watercourse* NW Yemen
Jugar *see* Sêrxü
Jugoslavija *see* Serbia
42 K10 **Juigalpa** Chontales, S Nicaragua 12°04′N 85°21′W
100 E9 **Juist** *island* NW Germany
Juisui *see* Ruisui
59 M21 **Juiz de Fora** Minas Gerais, SE Brazil 21°47′S 43°23′W
62 J5 **Jujuy** *off.* Provincia de Jujuy. ⬖ *province* N Argentina
Jujuy *see* San Salvador de Jujuy
Jujuy, Provincia de *see* Jujuy
92 J11 **Jukkasjärvi** *Lapp.* Čohkkiras. Norrbotten, N Sweden 67°52′N 20°39′E
Jula *see* Gyula, Hungary
Jūlā *see* Jālū, Libya
37 W2 **Julesburg** Colorado, C USA 40°59′N 102°15′W
Julia Beterrae *see* Béziers
57 I17 **Juliaca** Puno, SE Peru 15°32′S 70°10′W
181 U6 **Julia Creek** Queensland, C Australia 20°40′S 141°49′E
35 V17 **Julian** California, W USA 33°04′N 116°36′W
98 H7 **Julianadorp** Noord-Holland, NW Netherlands 52°53′N 04°43′E
109 S11 **Julian Alps** *Ger.* Julische Alpen, *It.* Alpi Giulie, *Slvn.* Julijske Alpe. ▲▲ Italy/Slovenia
55 V11 **Juliana Top** ▲ C Suriname 03°39′N 56°36′W
Julianehåb *see* Qaqortoq
Julijske Alpe *see* Julian Alps
40 J6 **Julimes** Chihuahua, N Mexico 28°29′N 105°21′W
Julio Briga *see* Bragança
Juliobriga *see* Logroño
61 G15 **Júlio de Castilhos** Rio Grande do Sul, S Brazil 29°14′S 53°42′W
Juliomagus *see* Angers
Julische Alpen *see* Julian Alps
Jullundur *see* Jalandhar
147 N11 **Juma** *Rus.* Dzhuma. Samarqand Viloyati, C Uzbekistan 39°43′N 66°37′E
161 O3 **Juma He** ≈ E China
81 L18 **Jumba** *prev.* Jumboo. Jubbada Hoose, S Somalia 0°12′S 42°34′E
Jumboo *see* Jumba
35 Y11 **Jumbo Peak** ▲ Nevada, W USA 36°12′N 114°09′W
105 R12 **Jumilla** Murcia, SE Spain 38°28′N 01°19′W
153 N10 **Jumla** Mid Western, NW Nepal 29°22′N 82°13′E
Jummoo *see* Jammu
Jumna *see* Yamuna
Jumporn *see* Chumphon
30 K5 **Jump River** ≈ Wisconsin, N USA
154 B11 **Jūnāgadh** *var.* Junagarh. Gujarāt, W India 21°32′N 70°32′E
Junagarh *see* Jūnāgadh
161 Q6 **Junan** *var.* Shizilu. Shandong, E China 35°11′N 118°47′E
62 G11 **Juncal, Cerro** ▲ C Chile 33°03′S 70°02′W
25 Q10 **Junction** Texas, SW USA 30°31′N 99°48′W
36 K6 **Junction** Utah, W USA 38°14′N 112°13′W
27 O4 **Junction City** Kansas, C USA 39°02′N 96°51′W
32 F13 **Junction City** Oregon, NW USA 44°13′N 123°12′W
60 M10 **Jundiaí** São Paulo, S Brazil 23°10′S 46°54′W
39 X12 **Juneau** *state capital* Alaska, USA 58°13′N 134°11′W
30 M8 **Juneau** Wisconsin, N USA 43°23′N 88°42′W
105 U6 **Juneda** Cataluña, NE Spain 41°33′N 00°49′E
183 Q9 **Junee** New South Wales, SE Australia 34°51′S 147°33′E
35 R8 **June Lake** California, W USA 37°46′N 119°04′W
Jungbunzlau *see* Mladá Boleslav
158 L4 **Junggar Pendi** *Eng.* Dzungarian Basin. *basin* NW China
99 N24 **Junglinster** Grevenmacher, C Luxembourg 49°43′N 06°15′E
18 F14 **Juniata River** ≈ Pennsylvania, NE USA
61 B20 **Junín** Buenos Aires, E Argentina 34°36′S 61°02′W
57 E14 **Junín** Junín, C Peru 11°11′S 76°00′W
57 F14 **Junín** *off.* Departamento de Junín. ⬖ *department* C Peru
63 H15 **Junín de los Andes** Neuquén, W Argentina 39°57′S 71°05′W
Junín, Departamento de *see* Junín
57 D14 **Junín, Lago de** ◎ C Peru
Junīyah *see* Joûnié
Junkseylon *see* Phuket
160 I11 **Junlian** Sichuan, C China 28°11′N 104°31′E
25 O11 **Juno** Texas, SW USA 30°09′N 101°07′W
92 J11 **Junosuando** *Lapp.* Čunusavvon. Norrbotten, N Sweden 67°25′N 22°28′E
93 H16 **Junsele** Västernorrland, C Sweden 63°42′N 16°54′E
32 L14 **Juntura** Oregon, NW USA 43°43′N 118°05′W
93 N14 **Juntusranta** Kainuu, E Finland 65°12′N 29°30′E
118 H11 **Juodupė** Panevėžys, NE Lithuania 56°07′N 25°37′E
119 H14 **Juozapinės Kalnas** ▲ SE Lithuania 54°29′N 25°27′E
99 K19 **Juprelle** Liège, E Belgium 50°43′N 05°31′E
80 D13 **Jur** ≈ W South Sudan
103 S9 **Jura** ⬖ *department* E France
108 C7 **Jura** ⬖ *canton* NW Switzerland
108 B8 **Jura** *var.* Jura Mountains. ▲▲ France/Switzerland
96 G12 **Jura** *island* SW Scotland, United Kingdom
Juraciszki *see* Yuratsishki
54 C8 **Juradó** Chocó, NW Colombia 07°07′N 77°45′W
Jura Mountains *see* Jura
96 G12 **Jura, Sound of** *strait* W Scotland, United Kingdom
139 V15 **Juraybīyāt, Bi'r** *well* S Iraq
118 E13 **Jurbarkas** *Ger.* Georgenburg, Jurburg. Tauragė, W Lithuania 55°04′N 22°45′E
99 F20 **Jurbise** Hainaut, SW Belgium 50°33′N 03°54′E
Jurburg *see* Jurbarkas
118 F9 **Jūrmala** C Latvia 56°57′N 23°42′E
58 D13 **Juruá** Amazonas, NW Brazil 03°08′S 65°59′W
48 F7 **Juruá, Rio** *var.* Río Yuruá. ≈ Brazil/Peru
59 G16 **Juruena** Mato Grosso, W Brazil 10°32′S 58°38′W
59 G16 **Juruena, Rio** ≈ W Brazil
165 Q6 **Jūsan-ko** ◎ Honshū, C Japan
25 O6 **Justiceburg** Texas, SW USA 32°57′N 101°07′W
Justinianopolis *see* Kirşehir
62 K11 **Justo Daract** San Luis, C Argentina 33°52′S 65°12′W
59 C14 **Jutaí** Amazonas, W Brazil 05°10′S 68°45′W
58 C13 **Jutaí, Rio** ≈ NW Brazil
100 N13 **Jüterbog** Brandenburg, E Germany 51°58′N 13°06′E
42 E6 **Jutiapa** Jutiapa, S Guatemala 14°18′N 89°52′W
42 A3 **Jutiapa** *off.* Departamento de Jutiapa. ⬖ *department* SE Guatemala
Jutiapa, Departamento de *see* Jutiapa
42 J6 **Juticalpa** Olancho, C Honduras 14°39′N 86°12′W
82 I13 **Jutila** North Western, NW Zambia 12°33′S 26°09′E
Jutland *see* Jylland
84 F8 **Jutland Bank** *undersea feature* SE North Sea 56°50′N 07°20′E
93 N16 **Juuka** Pohjois-Karjala, E Finland 63°12′N 29°17′E
93 N17 **Juva** Etelä-Savo, E Finland 61°55′N 27°54′E
Juvavum *see* Salzburg
44 A6 **Juventud, Isla de la** *var.* Isla de Pinos, *Eng.* Isle of Youth; *prev.* The Isle of the Pines. *island* W Cuba
Juwārtā *see* Chwarta
Juwārtā *see* Chemchemal
161 Q5 **Juxian** *var.* Chengyang, Ju Xian. Shandong, E China 35°33′N 118°45′E
Ju Xian *see* Juxian
161 P6 **Juye** Shandong, E China 35°26′N 116°04′E
113 O15 **Južna Morava** *Ger.* Südliche Morava. ≈ SE Serbia
83 H20 **Jwaneng** Southern, S Botswana 24°35′N 24°45′E
95 I23 **Jyderup** Sjælland, E Denmark 55°40′N 11°25′E
95 F22 **Jylland** *Eng.* Jutland. *peninsula* W Denmark
Jyrgalan *see* Dzhergalan
93 M17 **Jyväskylä** Keski-Suomii, C Finland 62°08′N 25°47′E

K

38 D9 **Kaʿaʿawa** *var.* Kaaawa. O'ahu, Hawaii, USA, C Pacific Ocean 21°33′N 157°47′W
Kaaawa *see* Kaʿaʿawa
81 G16 **Kaabong** NE Uganda 03°30′N 34°08′E
Kaaden *see* Kadaň
Kaafu Atoll *see* Male' Atoll
55 V9 **Kaaimanston** Sipaliwini, N Suriname 05°06′N 56°04′W
Kaakhka *see* Kaka
Kaala *see* Caála
187 O16 **Kaala-Gomen** Province Nord, W New Caledonia 20°40′S 164°24′E
92 L9 **Kaamanen** *Lapp.* Gámas. Lappi, N Finland 69°05′N 27°16′E
Kaapstad *see* Cape Town
Kaarasjoki *see* Karasjok
Kaaresuanto *see* Karesuando
92 J10 **Kaaresuvanto** *Lapp.* Gárassavon. Lappi, N Finland 68°28′N 22°29′E
93 K19 **Kaarina** Varsinais-Suomi, SW Finland 60°24′N 22°25′E
99 I14 **Kaatsheuvel** Noord-Brabant, S Netherlands 51°39′N 05°02′E
93 N16 **Kaavi** Pohjois-Savo, C Finland 62°58′N 28°30′E
Kaba *see* Habahe
171 O14 **Kabaena, Pulau** *island* C Indonesia
Kabakly *see* Gabakly
76 J14 **Kabala** N Sierra Leone 09°40′N 11°36′W
81 E19 **Kabale** SW Uganda 01°15′S 29°58′E
55 U10 **Kabalebo Rivier** ≈ W Suriname
79 N22 **Kabalo** Katanga, SE Dem. Rep. Congo 06°02′S 26°55′E
79 O21 **Kabambare** Maniema, E Dem. Rep. Congo 04°40′S 27°41′E
145 W13 **Kabanbay** *Kaz.* Qabanbay; *prev.* Andreyevka, *Kaz.* Andreevka. Almaty, SE Kazakhstan 45°50′N 80°34′E
145 Q9 **Kabanbay Batyr** *prev.* Rozhdestvenka. Akmola, C Kazakhstan 50°51′N 71°25′E
187 Y15 **Kabara** *prev.* Kambara. *island* Lau Group, E Fiji
Kabardino-Balkaria *see* Kabardino-Balkarskaya Respublika
126 M15 **Kabardino-Balkarskaya Respublika** *Eng.* Kabardino-Balkaria. ⬖ *autonomous republic* SW Russian Federation
79 O19 **Kabare** Sud-Kivu, E Dem. Rep. Congo 02°13′S 28°40′E
171 T11 **Kabarei** Papua Barat, E Indonesia 0°01′S 130°58′E
171 P7 **Kabasalan** Mindanao, S Philippines 07°46′N 122°49′E
77 U15 **Kabba** Kogi, S Nigeria 07°48′N 06°07′E
92 I13 **Kåbdalis** *Lapp.* Goabddális. Norrbotten, N Sweden 66°08′N 20°03′E
138 M6 **Kabd aş Şārim** *hill range* E Syria
14 B7 **Kabenung Lake** ◎ Ontario, S Canada
29 W3 **Kabetogama Lake** ◎ Minnesota, N USA
Kabia, Pulau *see* Kabin, Pulau
79 M22 **Kabinda** Kasai-Oriental, SE Dem. Rep. Congo 06°09′S 24°29′E
Kabinda *see* Cabinda
171 O15 **Kabin, Pulau** *var.* Pulau Kabia. *island* W Indonesia
171 P16 **Kabir** Pulau Pantar, S Indonesia 08°15′S 124°12′E
149 T10 **Kabīrwāla** Punjab, E Pakistan 30°24′N 71°51′E
114 M9 **Kableshkovo** Burgas, E Bulgaria 42°65′N 27°34′E
78 I13 **Kabo** Ouham, NW Central African Republic 07°43′N 18°38′E
Kābol *see* Kābul
83 H14 **Kabompo** North Western, W Zambia 13°36′S 24°10′E
83 H14 **Kabompo** ≈ W Zambia
79 M22 **Kabongo** Katanga, SE Dem. Rep. Congo 07°20′S 25°34′E
120 K11 **Kaboudia, Rass** *headland* E Tunisia 35°13′N 11°09′E
124 J14 **Kabozha** Novgorodskaya Oblast', W Russian Federation 58°48′N 35°00′E
Kabūd Gonbad *see* Kalāt
142 L5 **Kabūd Rāhang** Hamadān, W Iran 35°12′N 48°44′E
82 L12 **Kabuko** Muchinga, NE Zambia 11°31′S 31°16′E
149 Q5 **Kābul** *prev.* Kābol. ● (Afghanistan) Kābul, E Afghanistan 34°34′N 69°08′E
149 Q5 **Kābul**; *prev.* Kābol. ⬖ *province* E Afghanistan
149 Q5 **Kābul** ✈ Kābul, E Afghanistan 34°31′N 69°11′E
149 R5 **Kabul** *var.* Daryā-ye Kābul. ≈ Afghanistan/Pakistan *see also* Kābul, Daryā-ye
Kabul *see* Kābul, Daryā-ye
149 S5 **Kābul, Daryā-ye** *var.* Kabul. ≈ Afghanistan/Pakistan *see also* Kabul
Kābul, Daryā-ye *see* Kabul
79 O25 **Kabunda** Katanga, SE Dem. Rep. Congo 12°21′S 29°14′E
171 R9 **Kaburuang, Pulau** *island* Kepulauan Talaud, N Indonesia
80 G8 **Kabushiya** River Nile, NE Sudan 16°54′N 33°41′E
83 J14 **Kabwe** Central, C Zambia 14°29′S 28°25′E
186 E7 **Kabwum** Morobe, C Papua New Guinea 06°04′S 147°09′E
113 N17 **Kaçanik** *Serb.* Kačanik. S Kosovo 42°13′N 21°16′E
Kačanik *see* Kaçanik
118 F13 **Kačerginė** Kaunas, C Lithuania 54°55′N 23°40′E
117 S13 **Kacha** Avtonomna Respublika Krym, S Ukraine 44°46′N 33°33′E
154 A10 **Kachchh, Gulf of** *var.* Gulf of Cutch, Gulf of Kutch. *gulf* W India
154 I11 **Kachchhīdhāna** Madhya Pradesh, C India 21°33′N 78°54′E
149 Q11 **Kachchh, Rann of** *var.* Rann of Kachh, Rann of Kutch. *salt marsh* India/Pakistan
39 Q13 **Kachemak Bay** *bay* Alaska, USA
Kachh, Rann of *see* Kachchh, Rann of
77 V14 **Kachia** Kaduna, C Nigeria 09°52′N 08°00′E
167 N2 **Kachin State** ⬖ *state* N Myanmar (Burma)
Kachiry *see* Kashyr
137 Q11 **Kaçkar Dağları** ▲▲ NE Turkey
155 C21 **Kadamatt Island** *island* Lakshadweep, India, N Indian Ocean
111 B15 **Kadaň** *Ger.* Kaaden. Ústecký Kraj, NW Czech Republic 50°24′N 13°16′E
1667 N11 **Kadan Kyun** *prev.* King Island. *island* Mergui Archipelago, S Myanmar (Burma)
187 X15 **Kadavu** *prev.* Kandavu. *island* S Fiji
187 X15 **Kadavu Passage** *channel* S Fiji
79 G16 **Kadéï** ≈ Cameroon/Central African Republic
Kadhimain *see* Al Kāẓimīyah
Kadijica *see* Kadiytsa
114 M13 **Kadıköy Barajı** ⊠ NW Turkey
182 I8 **Kadina** South Australia 33°59′S 137°43′E
136 H15 **Kadınhanı** Konya, C Turkey 38°15′N 32°14′E
76 M14 **Kadiolo** Sikasso, S Mali 10°30′N 05°43′W
136 L16 **Kadirli** Osmaniye, S Turkey 37°22′N 36°05′E
114 G11 **Kadiytsa** *Mac.* Kadijica. ▲ Bulgaria/FYR Macedonia 41°48′N 22°58′E
28 L10 **Kadoka** South Dakota, N USA 43°49′N 101°30′W
127 N5 **Kadom** Ryazanskaya Oblast', W Russian Federation 54°35′N 42°27′E
83 K16 **Kadoma** *prev.* Gatooma. Mashonaland West, C Zimbabwe !8°22′S 29°55′E
80 E12 **Kadugli** Southern Kordofan, S Sudan 11°N 29°44′E
77 V14 **Kaduna** Kaduna, C Nigeria 10°32′N 07°26′E
77 V14 **Kaduna** ⬖ *state* C Nigeria
77 V15 **Kaduna** ≈ N Nigeria
124 K14 **Kaduy** Vologodskaya Oblast', NW Russian Federation 59°10′N 37°11′E
154 E13 **Kadwa** ≈ W India
123 S9 **Kadykchan** Magadanskaya Oblast', E Russian Federation 62°54′N 146°53′E
Kadzharan *see* K'ajaran
125 T7 **Kadzherom** Respublika Komi, NW Russian Federation 64°42′N 55°51′E
Kadzhi-Say *see* Bokonbayevo
76 I10 **Kaédi** Gorgol, S Mauritania 16°12′N 13°32′W
78 G12 **Kaélé** Extrême-Nord, N Cameroon 10°05′N 14°28′E
38 C9 **Kaʿena Point** *var.* Kaena Point. *headland* O'ahu, Hawaiʿi, USA 21°34′N 158°16′W
184 J2 **Kaeo** Northland, North Island, New Zealand 35°03′S 173°40′E
163 X14 **Kaesŏng** *var.* Kaesŏng-si. S North Korea 37°58′N 126°31′E
Kaesŏng-si *see* Kaesŏng
Kaewieng *see* Kavieng
79 L24 **Kafakumba** Shaba, S Dem. Rep. Congo 09°39′S 23°43′E
Kafan *see* Kapan
77 V14 **Kafanchan** Kaduna, C Nigeria 09°32′N 08°18′E
Kaffa *see* Feodosiya
76 G11 **Kaffrine** C Senegal 14°07′N 15°27′W
Kafiréas, Akrotírio *see* Ntóro, Kávo
115 I19 **Kafiréos, Stenó** *strait* Évvoia/Kykládes, Greece, Aegean Sea
Kafirnigan *see* Kofarnihon
Kafo *see* Kafu
75 W7 **Kafr ash Shaykh** *var.* Kafrel Sheik, Kafr el Sheikh. N Egypt 31°07′N 30°56′E
Kafr el Sheikh *see* Kafr ash Shaykh
81 F17 **Kafu** *var.* Kafo. ≈ W Uganda
83 J15 **Kafue** Lusaka, SE Zambia 15°44′S 28°10′E
83 I14 **Kafue** ≈ C Zambia
67 T13 **Kafue Flats** *plain* C Zambia
164 K12 **Kaga** Ishikawa, Honshū, SW Japan 36°18′N 136°19′E
79 J14 **Kaga Bandoro** *prev.* Fort-Crampel. Nana-Grébizi, C Central African Republic 06°54′N 19°10′E
81 E18 **Kagadi** W Uganda 0°57′N 30°52′E
38 H17 **Kagalaska Island** *island* Aleutian Islands, Alaska, USA
Kagan *see* Kogon
Kaganovichabad *see* Kolkhozobod
Kagarlyk *see* Kaharlyk
164 H14 **Kagawa** *off.* Kagawa-ken. ⬖ *prefecture* Shikoku, SW Japan
Kagawa-ken *see* Kagawa
154 J13 **Kagaznagar** Telangana, C India 19°25′N 79°30′E
93 J14 **Kåge** Västerbotten, N Sweden 64°49′N 21°00′E
81 E19 **Kagera** *var.* Ziwa Magharibi, *Eng.* West Lake. ⬖ *region* NW Tanzania
81 E19 **Kagera** *var.* Akagera. ≈ Rwanda/Tanzania *see also* Akagera
76 L5 **Kâgheț** *var.* Karet. *physical region* N Mauritania
Kagi *see* Jiayi
137 S12 **Kağızman** Kars, NE Turkey 40°08′N 43°07′E
188 I6 **Kagman Point** *headland* Saipan, S Northern Mariana Islands
164 C16 **Kagoshima** *var.* Kagosima. Kagoshima, Kyūshū, SW Japan 31°37′N 130°33′E
164 C16 **Kagoshima** *off.* Kagoshima-ken, *var.* Kagosima. ⬖ *prefecture* Kyūshū, SW Japan
Kagoshima-ken *see* Kagoshima
Kagosima *see* Kagoshima
Kagul *see* Cahul
Kagul, Ozero *see* Kahul, Ozero
38 B8 **Kahala Point** *headland* Kauaʿi, Hawaiʿi, USA 22°08′N 159°17′W
81 F21 **Kahama** Shinyanga, NW Tanzania 03°48′S 32°36′E
117 P5 **Kaharlyk** *Rus.* Kagarlyk. Kyyivs'ka Oblast', N Ukraine 49°50′N 30°50′E
169 T13 **Kahayan, Sungai** ≈ Borneo, C Indonesia
79 I22 **Kahemba** Bandundu, SW Dem. Rep. Congo 07°20′S 19°00′E
185 A23 **Kaherekoau Mountains** ▲▲ South Island, New Zealand
143 W14 **Kahīrī** *var.* Kūhīrī. Sīstān va Balūchestān, SE Iran 26°55′N 61°04′E
101 L16 **Kahla** Thüringen, C Germany 50°49′N 11°33′E
101 G15 **Kahler Asten** ▲ W Germany 51°11′N 08°32′E
149 Q4 **Kahmard, Daryā-ye** *prev.* Darya-i-surkhab. ≈ NE Afghanistan
143 T13 **Kahnūj** Kermān, SE Iran 28°N 57°41′E
27 V1 **Kahoka** Missouri, C USA 40°24′N 91°44′W
38 E10 **Kahoʿolawe** *var.* Kahoolawe. *island* Hawaiʿi, USA, C Pacific Ocean
Kahoolawe *see* Kahoʿolawe
136 M16 **Kahramanmaraş** *var.* Kahraman Maraş, Maraş, Marash. Kahramanmaraş, S Turkey 37°34′N 36°54′E
136 L15 **Kahramanmaraş** *var.* Kahraman Maraş, Maraş, Marash. ⬖ *province* C Turkey
Kahraman Maraş *see* Kahramanmaraş
Kahror/Kahror Pakka *see* Kahror Pakka
149 T11 **Kahror Pakka** *var.* Kahror, Koror Pacca. Punjab, E Pakistan 29°38′N 71°59′E
137 N15 **Kâhta** Adıyaman, S Turkey 37°48′N 38°35′E

◆ Country
● Country Capital
◇ Dependent Territory
○ Dependent Territory Capital
⬖ Administrative Regions
✈ International Airport
▲ Mountain
▲▲ Mountain Range
Volcano
≈ River
◎ Lake
⊠ Reservoir

38 D8 **Kahuku** O'ahu, Hawaii, USA, C Pacific Ocean 21°40´N 157°57´W
38 D8 **Kahuku Point** *headland* O'ahu, Hawai'i, USA 21°42´N 157°59´W
116 M12 **Kahul, Ozero** *var.* Lacul Cahul, *Rus.* Ozero Kagul. ⊚ Moldova/Ukraine
143 V11 **Kahūrak** Sīstān va Balūchestān, SE Iran 29°25´N 59°38´E
184 G13 **Kahurangi Point** *headland* South Island, New Zealand 40°41´S 171°57´E
149 V6 **Kahūta** Punjab, E Pakistan 33°38´N 73°27´E
77 S14 **Kaiama** Kwara, W Nigeria 09°37´N 03°58´E
186 D7 **Kaiapit** Morobe, C Papua New Guinea 06°12´S 146°09´E
185 I18 **Kaiapoi** Canterbury, South Island, New Zealand 43°23´S 172°40´E
36 K9 **Kaibab Plateau** *plain* Arizona, SW USA
171 U13 **Kai Besar, Pulau** *island* Kepulauan Kai, E Indonesia
36 L9 **Kaibito Plateau** *plain* Arizona, SW USA
158 K6 **Kaidu He** *var.* Karaxahar. ⚹ NW China
55 S10 **Kaieteur Falls** *waterfall* C Guyana
161 O6 **Kaifeng** Henan, C China 34°47´N 114°20´E
184 J3 **Kaihu** Northland, North Island, New Zealand 35°47´S 173°39´E
Kaihua *see* Wenshan
171 U14 **Kai Kecil, Pulau** *island* Kepulauan Kai, E Indonesia
169 U16 **Kai, Kepulauan** *prev.* Kei Islands. *island group* Maluku, SE Indonesia
184 J3 **Kaikohe** Northland, North Island, New Zealand 35°25´S 173°48´E
185 J16 **Kaikoura** Canterbury, South Island, New Zealand 42°22´S 173°40´E
185 J16 **Kaikoura Peninsula** *peninsula* South Island, New Zealand
Kailas Range *see* Gangdisê Shan
160 K12 **Kaili** Guizhou, S China 26°34´N 107°58´E
38 F10 **Kailua** Maui, Hawaii, USA, C Pacific Ocean 20°53´N 156°13´W
Kailua *see* Kalaoa
38 G11 **Kailua-Kona** *var.* Kona. Hawaii, USA, C Pacific Ocean 19°43´N 155°58´W
186 B7 **Kaim** ⚹ W Papua New Guinea
171 X14 **Kaima** Papua, E Indonesia 05°36´S 138°39´E
184 M7 **Kaimai Range** ▲ North Island, New Zealand
114 E13 **Kaïmaktsalán** *var.* Kajmakčalan. ▲ Greece/FYR Macedonia 40°57´N 21°48´E *see also* Kajmakčalan
Kaïmaktsalán *see* Kajmakčalan
185 C20 **Kaimanawa Mountains** ▲ North Island, New Zealand
118 E4 **Käina** *Ger.* Keinis; *prev.* Keina. Hiiumaa, W Estonia 58°50´N 22°49´E
109 V7 **Kainach** ⚹ SE Austria
164 I14 **Kainan** Tokushima, Shikoku, SW Japan 33°34´N 134°20´E
164 H15 **Kainan** Wakayama, Honshū, SW Japan 34°09´N 135°12´E
147 U7 **Kaindy** *Kir.* Kayyngdy. Chuyskaya Oblast', N Kyrgyzstan 42°48´N 73°39´E
77 T14 **Kainji Dam** *dam* W Nigeria
77 T14 **Kainji Reservoir** *var.* Kainji Lake. ⊡ W Nigeria
186 D8 **Kaintiba** *var.* Kamina. Gulf, S Papua New Guinea 07°29´S 146°04´E
92 K12 **Kainulasjärvi** Norrbotten, N Sweden 67°00´N 22°37´E
93 M14 **Kainuu** *Swe.* Kajanaland. ◇ *region* N Finland
184 K5 **Kaipara Harbour** *harbor* North Island, New Zealand
152 I10 **Kairāna** Uttar Pradesh, N India 29°24´N 77°10´E
74 M6 **Kairouan** *var.* Al Qayrawān. E Tunisia 35°46´N 10°11´E
Kaisaria *see* Kayseri
101 F20 **Kaiserslautern** Rheinland-Pfalz, SW Germany 49°27´N 07°46´E
118 G13 **Kaišiadorys** Kaunas, S Lithuania 54°51´N 24°27´E
184 I2 **Kaitaia** Northland, North Island, New Zealand 35°07´S 173°13´E
185 E24 **Kaitangata** Otago, South Island, New Zealand 46°18´S 169°52´E
152 I9 **Kaithal** Haryana, NW India 29°47´N 76°26´E
Kaitong *see* Tongyu
169 X13 **Kait, Tanjung** *headland* Sumatera, W Indonesia 03°13´S 106°03´E
38 E9 **Kaiwi Channel** *channel* Hawai'i, USA, C Pacific Ocean
K9 **Kaixian** *var.* Hanfeng. Sichuan, C China 31°13´N 108°25´E
163 V11 **Kaiyuan** *var.* K'ai-yüan. Liaoning, NE China 42°33´N 124°04´E
160 H14 **Kaiyuan** *var.* K'ai-yüan Yunnan, SW China 23°42´N 103°14´E
K'ai-yüan *see* Kaiyuan
O9 **Kaiyuh Mountains** ▲ Alaska, USA
93 M15 **Kajaani** *Swe.* Kajana. Kainuu, C Finland 64°17´N 27°46´E
149 N7 **Kajakī, Band-e** ⊚ C Afghanistan
Kajan *see* Kayan, Sungai
Kajana *see* Kajaani
Kajanaland *see* Kainuu
137 V13 **K'ajaran** *Rus.* Kadzharan. SE Armenia 39°10´N 46°05´E
81 I19 **Kajiado** Rift Valley, S Kenya 01°51´S 36°48´E
81 I20 **Kajiado** ◇ *county* S Kenya
Kajisay *see* Bokonbayevo
113 O20 **Kajmakčalan** ▲ S FYR Macedonia 40°57´N 21°48´E *see also* Kaïmaktsalán
Kajmakčalan *see* Kaïmaktsalán

Kajnar *see* Kaynar
149 N6 **Kajrān** Dāykundī, C Afghanistan 33°12´N 65°28´E
149 N5 **Kaj Rūd** ⚹ C Afghanistan
146 G14 **Kaka** *Rus.* Kaakhka. Ahal Welaýaty, S Turkmenistan 37°20´N 59°37´E
12 C12 **Kakabeka Falls** Ontario, S Canada 48°24´N 89°40´W
83 F23 **Kakamas** Northern Cape, W South Africa 28°45´S 20°33´E
81 H18 **Kakamega** Kakamega, W Kenya 0°17´N 34°47´E
81 H18 **Kakamega** ◇ *county* W Kenya
112 H13 **Kakanj** Federacija Bosne I Hercegovine, C Bosnia and Herzegovina 44°06´N 18°07´E
185 F22 **Kakanui Mountains** ▲ South Island, New Zealand
184 K11 **Kakaramea** Taranaki, North Island, New Zealand 39°42´S 174°27´E
76 J16 **Kakata** C Liberia 06°35´N 10°19´W
184 M11 **Kakatahi** Manawatu-Wanganui, North Island, New Zealand 39°40´S 175°20´E
113 M23 **Kakavi** Gjirokastër, S Albania 39°55´N 20°19´E
147 O14 **Kakaydy** Surkhondaryo Viloyati, S Uzbekistan 37°37´N 67°30´E
164 F13 **Kake** Hiroshima, Honshū, SW Japan 34°37´N 132°17´E
39 X13 **Kake** Kupreanof Island, Alaska, USA 56°58´N 133°57´W
171 P14 **Kakea** Pulau Wowoni, C Indonesia 04°09´S 123°06´E
164 M14 **Kakegawa** Shizuoka, Honshū, S Japan 34°47´N 138°02´E
145 V16 **Kakeroma-jima** Kagoshima, SW Japan
143 T6 **Kākhak** Khorāsān-e Razavī, E Iran
118 L11 **Kakhanavichy** *Rus.* Kokhanovichi. Vitsyebskaya Voblasts', N Belarus 55°52´N 28°08´E
39 P13 **Kakhonak** Alaska, USA 59°26´N 154°48´W
117 S10 **Kakhovka** Khersons'ka Oblast', S Ukraine 46°40´N 33°30´E
117 U9 **Kakhovs'ke Vodoskhovyshche** *Rus.* Kakhovskoye Vodokhranilishche. ⊡ SE Ukraine
Kakhovskoye Vodokhranilishche *see* Kakhovs'ke Vodoskhovyshche
117 T11 **Kakhovs'kyy Kanal** *canal* S Ukraine
Kakia *see* Khakhea
155 L16 **Kākināda** *prev.* Cocanada. Andhra Pradesh, E India 16°56´N 82°13´E
Kākisalmi *see* Priozersk
164 I13 **Kakogawa** Hyōgo, Honshū, SW Japan 34°49´N 134°52´E
81 F18 **Kakoge** C Uganda 01°03´N 32°30´E
145 O7 **Kak, Ozero** ⊚ N Kazakhstan
Ka-Kem *see* Malyy Yenisey
Kakshaal-Too, Khrebet *see* Kokshaal-Tau
39 S5 **Kaktovik** Alaska, USA 70°08´N 143°37´W
165 Q11 **Kakuda** Miyagi, Honshū, C Japan 37°59´N 140°48´E
165 Q8 **Kakunodate** Akita, Honshū, C Japan 39°37´N 140°35´E
Kalaallit Nunaat *see* Greenland
149 T7 **Kālābāgh** Punjab, E Pakistan 33°00´N 71°35´E
171 Q16 **Kalabahi** Pulau Alor, S Indonesia 08°14´S 124°32´E
188 I5 **Kalabera** Saipan, S Northern Mariana Islands
83 G14 **Kalabo** Western, W Zambia 15°00´S 22°37´E
126 M9 **Kalach** Voronezhskaya Oblast', W Russian Federation 50°24´N 41°00´E
127 N10 **Kalach-na-Donu** Volgogradskaya Oblast', SW Russian Federation 48°45´N 43°29´E
166 K5 **Kaladan** ⚹ W Myanmar (Burma)
14 K14 **Kaladar** Ontario, SE Canada 44°38´N 77°06´W
38 G13 **Ka Lae** *var.* South Cape, South Point. *headland* Hawai'i, USA, C Pacific Ocean 18°54´N 155°40´W
83 G19 **Kalahari Desert** *desert* Southern Africa
38 B8 **Kalāheo** *var.* Kalaheo. Kaua'i, Hawaii, USA, C Pacific Ocean 21°55´N 159°31´W
Kalaheo *see* Kalāheo
Kalaïkhum *see* Qal'aikhum
Kala-i-Mor *see* Galaýmor
93 K15 **Kalajoki** Pohjois-Pohjanmaa, W Finland 64°15´N 23°57´E
Kalak *see* Eski Kalak
Kal al Sraghna *see* El Kelâa Srarhna
32 G10 **Kalama** Washington, NW USA 46°00´N 122°50´W
Kalámai *see* Kalámata
115 G14 **Kalamariá** Kentrikí Makedonía, N Greece 40°37´N 22°58´E
115 C15 **Kalamás** *var.* Thiamis; *prev.* Thyamis. ⚹ W Greece
115 E21 **Kalámata** *prev.* Kalámai. Pelopónnisos, S Greece 37°02´N 22°07´E
31 P10 **Kalamazoo** Michigan, N USA 42°17´N 85°35´W
31 P9 **Kalamazoo River** ⚹ Michigan, N USA
147 P14 **Kalininobod** *Rus.* Kalininabad. SW Tajikistan 37°49´N 68°55´E
181 S13 **Kalamits'ka Zatoka** *Rus.* Kalamitskiy Zaliv. *gulf* S Ukraine
Kalamitskiy Zaliv *see* Kalamits'ka Zatoka
115 H18 **Kalámos** *island* Attikí, C Greece 38°16´N 23°15´E
115 C18 **Kalámos** *island* Iónioi Nísia, Greece, C Mediterranean Sea
115 D15 **Kalampáka** *var.* Kalabaka. Thessalía, C Greece 39°43´N 21°36´E
81 S11 **Kalanchak** Khersons'ka Oblast', S Ukraine 46°14´N 33°19´E

38 G11 **Kalaoa** *var.* Kailua. Hawaii, USA, C Pacific Ocean 19°43´N 155°59´W
171 O15 **Kalaotoa, Pulau** *island* W Indonesia
155 J24 **Kala Oya** ⚹ NW Sri Lanka
Kalarash *see* Călăraşi
93 H17 **Kälarne** Jämtland, C Sweden 63°00´N 16°10´E
143 V15 **Kalār Rūd** ⚹ N Iran
169 R9 **Kalasin** *var.* Muang Kalasin. Kalasin, E Thailand 16°29´N 103°31´E
143 U4 **Kalāt** *var.* Kabūd Gonbad. Khorāsān-e Razavī, NE Iran 37°02´N 59°46´E
149 O11 **Kalāt** *var.* Kelat, Khelat. Baluchistān, SW Pakistan 29°01´N 66°58´E
Kalāt *see* Qalāt
115 J14 **Kalathriá, Ákrotírio** *headland* Samothráki, NE Greece 40°24´N 25°34´E
193 W147 **Kalau** *island* Tongatapu Group, SE Tonga
38 E9 **Kalaupapa** Moloka'i, Hawaii, USA, C Pacific Ocean 21°11´N 156°59´W
127 N13 **Kalaus** ⚹ SW Russian Federation
189 X2 **Kalalen** *var.* Calalen. *island* Ratak Chain, SE Marshall Islands
115 E19 **Kalávryta** *var.* Kalávrita. Dytikí Elláda, S Greece 38°02´N 22°06´E
141 Y10 **Kalbān** W Oman 20°19´N 58°40´E
180 H11 **Kalbarri** Western Australia 27°43´S 114°08´E
144 G10 **Kaldygayty** ⚹ W Kazakhstan
136 I12 **Kalecik** Ankara, N Turkey 40°08´N 33°27´E
79 O19 **Kalehe** Sud-Kivu, E Dem. Rep. Congo 02°05´S 28°52´E
79 P22 **Kalemie** *prev.* Albertville. Katanga, SE Dem. Rep. Congo 05°55´S 29°09´E
166 L4 **Kalemyo** Sagaing, W Myanmar (Burma) 23°11´N 94°03´E
82 H12 **Kalene Hill** North Western, NW Zambia 11°10´S 24°12´E
167 T11 **Kaléng** *prev.* Phumi Kaléng. Stœng Trêng, NE Cambodia 13°57´N 106°17´E
124 I7 **Kale Sultanie** *see* Çanakkale
166 L4 **Kalevala** Respublika Kareliya, NW Russian Federation 65°12´N 31°16´E
166 L4 **Kalewa** Sagaing, C Myanmar (Burma) 23°15´N 94°19´E
Kalgan *see* Zhangjiakou
39 Q12 **Kalgin Island** *island* Alaska, USA
180 L12 **Kalgoorlie** Western Australia 30°51´S 121°27´E
Kali *see* Sārda
115 E17 **Kaliakoúda** ▲ C Greece 38°47´N 21°42´E
114 O8 **Kaliakra, Nos** *headland* NE Bulgaria 43°22´N 28°28´E
115 F19 **Kalianói** Pelopónnisos, S Greece 37°55´N 22°28´E
115 N24 **Sali Limni** ▲ Kárpathos, SE Greece 35°34´N 27°08´E
79 N20 **Kalima** Maniema, E Dem. Rep. Congo 02°34´S 26°27´E
169 S11 **Kalimantan** *Eng.* Indonesian Borneo. ◇ *geopolitical region* Borneo, C Indonesia
169 Q11 **Kalimantan Barat** *off.* Propinsi Kalimantan Barat, *var.* Kalbar, *Eng.* West Borneo, West Kalimantan. ◇ *province* N Indonesia
Kalimantan Barat, Propinsi *see* Kalimantan Barat
169 T13 **Kalimantan Selatan** *off.* Propinsi Kalimantan Selatan, *var.* Kalsel, *Eng.* South Borneo, South Kalimantan. ◇ *province* N Indonesia
Kalimantan Selatan, Propinsi *see* Kalimantan Selatan
169 R12 **Kalimantan Tengah** *off.* Propinsi Kalimantan Tengah, *var.* Kalteng, *Eng.* Central Borneo, Central Kalimantan. ◇ *province* N Indonesia
Kalimantan Tengah, Propinsi *see* Kalimantan Tengah
169 U10 **Kalimantan Timur** *off.* Propinsi Kalimantan Timur, *var.* Kaltim, *Eng.* East Borneo, East Kalimantan. ◇ *province* N Indonesia
Kalimantan Timur, Propinsi *see* Kalimantan Timur
169 V9 **Kalimantan Utara** *off.* Propinsi Kalimantan Utara, *var.* Kaltara, *Eng.* North Kalimantan. ◇ *province* N Indonesia
Kalimantan Utara, Propinsi *see* Kalimantan Utara
Kálimnos *see* Kálymnos
153 S12 **Kālimpang** West Bengal, NE India 27°02´N 88°34´E
116 J6 **Kalinin** *see* Tver'
Kalinin *see* Boldumsaz
Kalininabad *see* Kalininobod
126 B3 **Kaliningrad** Kaliningradskaya Oblast', W Russian Federation 54°42´N 20°32´E
Kaliningrad *see* Kaliningradskaya Oblast'
126 A3 **Kaliningradskaya Oblast'** ◇ *province and enclave* W Russian Federation
Kalinino *see* Tashir
127 O8 **Kalininsk** Saratovskaya Oblast', W Russian Federation 51°N 44°25´E
Kalinink *see* Boldumsaz
Kalinisk *see* Cupcina
119 M19 **Kalinkavichy** *Rus.* Kalinkovichi. Homyel'skaya Voblasts', SE Belarus 52°08´N 29°19´E
Kalinkovichi *see* Kalinkavichy
81 G18 **Kaliro** S Uganda 00°54´N 33°30´E
Kalisch/Kalish *see* Kalisz
33 O7 **Kalispell** Montana, NW USA 48°12´N 114°18´W

110 I13 **Kalisz** *Ger.* Kalisch, *Rus.* Kalish; *anc.* Calisia. Wielkopolskie, C Poland 51°46´N 18°04´E
110 F9 **Kalisz Pomorski** *Ger.* Kallies. Zachodnio-pomorskie, NW Poland 53°55´N 15°55´E
126 M10 **Kalitva** ⚹ SW Russian Federation
81 F21 **Kaliua** Tabora, C Tanzania 05°03´S 31°48´E
92 K13 **Kalix** Norrbotten, N Sweden 65°51´N 23°14´E
92 K13 **Kalixälven** ⚹ N Sweden 65°47´N 20°20´E
145 T8 **Kalkaman** *var.* Kalqaman. Pavlodar, NE Kazakhstan 51°57´N 75°58´E
Kalkandelen *see* Tetovo
181 O4 **Kalkaringi** Northern Territory, N Australia 17°33´S 130°40´E
31 P6 **Kalkaska** Michigan, N USA 44°44´N 85°11´W
93 F16 **Kall** Jämtland, C Sweden 63°31´N 13°16´E
189 X2 **Kalalen** *var.* Calalen. *island* Ratak Chain, SE Marshall Islands
118 J5 **Kallaste** *Ger.* Krasnogor. Tartumaa, SE Estonia 58°40´N 27°12´E
93 N16 **Kallavesi** ⊚ SE Finland
115 F17 **Kallídromo** ▲ C Greece 56°14´N 15°17´E
95 M22 **Kallinge** Blekinge, S Sweden 56°14´N 15°17´E
115 L16 **Kalloní** Lésvos, E Greece 39°14´N 26°16´E
13 S13 **Kallsjön** ⊚ C Sweden
95 N21 **Kalmar** *var.* Calmar. Kalmar, S Sweden 56°40´N 16°22´E
95 M19 **Kalmar** *var.* Calmar. ◇ *county* S Sweden
95 N20 **Kalmarsund** *strait* S Sweden
99 H15 **Kalmthout** Antwerpen, N Belgium 51°24´N 04°27´E
Kalmykia/Kalmykiya-Khal'mg Tangch, Respublika *see* Kalmykiya, Respublika
127 O12 **Kalmykiya, Respublika** *var.* Respublika Kalmykiya-Khal'mg Tangch, *Eng.* Kalmykia; *prev.* Kalmytskaya ASSR. ◇ *autonomous republic* SW Russian Federation
Kalmytskaya ASSR *see* Kalmykiya, Respublika
118 F9 **Kalnciems** C Latvia 56°46´N 23°37´E
114 I10 **Kalnitsa** ⚹ SE Bulgaria
111 J24 **Kalocsa** Bács-Kiskun, S Hungary 46°31´N 19°00´E
114 J9 **Kalofer** Plovdiv, C Bulgaria 42°38´N 24°58´E
38 E10 **Kalohi Channel** *channel* C Pacific Ocean
83 I16 **Kalomo** Southern, S Zambia 17°02´S 26°29´E
29 X14 **Kalona** Iowa, C USA 41°28´N 91°42´W
115 K22 **Kalotási, Akrotírio** *cape* Amorgós, Kykládes, Greece, Aegean Sea
152 J8 **Kalpa** Himāchal Pradesh, N India 31°33´N 78°16´E
115 C15 **Kalpáki** Ípeiros, W Greece 39°53´N 20°38´E
155 C22 **Kalpeni Island** *island* Lakshadweep, India, N Indian Ocean
152 K13 **Kālpi** Uttar Pradesh, N India 26°07´N 79°44´E
158 G7 **Kalpin** Xinjiang Uygur Zizhiqu, NW China 40°35´N 78°52´E
149 P16 **Kalri Lake** ⊚ SE Pakistan
Kalsel *see* Kalimantan Selatan
143 R5 **Kāl Shūr** ⚹ N Iran
39 N11 **Kalskag** Alaska, USA 61°32´N 160°15´W
108 H7 **Kaltbrunn** Sankt Gallen, NE Switzerland 47°11´N 09°00´E
Kalteng *see* Kalimantan Tengah
Kaltim *see* Kalimantan Timur
Kaltdorf *see* Pruszków
77 X14 **Kaltungo** Gombe, E Nigeria 09°49´N 11°22´E
126 K4 **Kaluga** Kaluzhskaya Oblast', W Russian Federation 54°31´N 36°16´E
155 J26 **Kalu Ganga** ⚹ S Sri Lanka
82 J13 **Kalulushi** Copperbelt, C Zambia 12°50´S 28°03´E
180 M2 **Kalumburu** Western Australia 14°11´S 126°40´E
95 H23 **Kalundborg** Sjælland, E Denmark 55°42´N 11°06´E
116 K5 **Kalush** *Pol.* Kałusz. Ivano-Frankivs'ka Oblast', W Ukraine 49°02´N 24°20´E
Kałusz *see* Kalush
155 J26 **Kalutara** Western Province, SW Sri Lanka 06°35´N 79°59´E
126 I5 **Kaluzhskaya Oblast'** ◇ *province* W Russian Federation
119 E14 **Kalvarija** *Pol.* Kalwaria. Marijampolė, S Lithuania 54°25´N 23°13´E
93 K15 **Kälviä** Keski-Pohjanmaa, W Finland 63°53´N 23°30´E
110 D8 **Kalwang** Steiermark, SE Austria
154 D13 **Kalyān** Mahārāshtra, W India 19°17´N 73°11´E
116 K16 **Kalyazin** Tverskaya Oblast', W Russian Federation 57°15´N 37°55´E
115 J26 **Kalydón** *anc.* Calydon. *site of ancient city* Dytikí Elláda, C Greece
115 M21 **Kálymnos** *var.* Kálimnos. Kálymnos, Dodekánisa, Greece, Aegean Sea 36°57´N 26°59´E

115 M21 **Kálymnos** *var.* Kálimnos. *island* Dodekánisa, Greece, Aegean Sea
117 O5 **Kalynivka** Kyyivs'ka Oblast', N Ukraine 50°14´N 30°16´E
117 N6 **Kalynivka** Vinnyts'ka Oblast', C Ukraine 49°27´N 28°32´E
145 W15 **Kalzhat** *prev.* Kol'zhat. Almaty, SE Kazakhstan 43°29´N 80°37´E
42 M10 **Kama** Región Autónoma Atlántico Sur, SE Nicaragua 12°06´N 83°55´W
165 R9 **Kama** *var.* Kamaisi. Iwate, Honshū, C Japan 39°18´N 141°52´E
Kama Reservoir *see* Kamskoye Vodokhranilishche
118 H13 **Kamajai** Utena, E Lithuania 55°49´N 25°30´E
Kamajai *see* Toliejai
149 U9 **Kamālia** Punjab, NE Pakistan
83 I14 **Kamalondo** North Western, NW Zambia 13°42´S 25°38´E
136 I13 **Kaman** Kirşehir, C Turkey 39°22´N 33°43´E
79 O20 **Kamanyola** Sud-Kivu, E Dem. Rep. Congo 02°54´S 29°04´E
141 N14 **Kamarān** *island* W Yemen
55 R9 **Kamarang** W Guyana 05°49´N 60°38´W
Kāmāreddi/Kamareddy *see* Rāmāreddi
93 N16 **Kamas** ⚹ SE Finland
115 F17 **Kambos** *see* Kámpos
79 S13 **Kamba** Kebbi, NW Nigeria 11°50´N 03°44´E
76 I11 **Kambia** W Sierra Leone 09°09´N 12°53´W
79 N25 **Kambove** Katanga, SE Dem. Rep. Congo 10°50´S 26°39´E
Kambryk *see* Cambrai
123 V10 **Kamchatka** ⚹ E Russian Federation
Kamchatka *var.* Kamchatka, Poluostrov
123 U10 **Kamchatka, Poluostrov** *Eng.* Kamchatka. *peninsula* E Russian Federation
123 V10 **Kamchatskiy Kray** ◇ *province* E Russian Federation
123 V10 **Kamchatskiy Zaliv** *gulf* E Russian Federation
114 N9 **Kamchiya** *var.* Kamchiya. ⚹ E Bulgaria
114 L9 **Kamchiya, Yazovir** *var.* Yazovir Kamchiya. ⊡ E Bulgaria
Kamchiya, Yazovir *see* Kamchiya, Yazovir
149 T4 **Kāmdêsh** *var.* Kamdesh; *prev.* Kāmdeysh. Nūrestān, E Afghanistan 35°25´N 71°26´E
Kamdesh *see* Kāmdêsh
Kāmdeysh *see* Kāmdêsh
Kampo *see* Campo, Cameroon
Kamen' *see* Kamyen'
Kamenets *see* Kamyanets
Kamenets-Podol'skiy *see* Kamyanets'-Podil's'kyy
113 Q18 **Kamenica** NE Macedonia 42°03´N 22°34´E
113 O16 **Kamenica** *var.* Dardané, *Serb.* Kosovska Kamenica.
112 A11 **Kamenjak, Rt** *headland* NW Croatia
125 O6 **Kamenka** Arkhangel'skaya Oblast', NW Russian Federation 65°55´N 44°01´E
126 6 **Kamenka** Penzenskaya Oblast', W Russian Federation 53°12´N 44°49´E
127 L8 **Kamenka** Voronezhskaya Oblast', W Russian Federation 50°44´N 39°31´E
Kamenka *see* Taskala
Kamenka *see* Camenca
Kamenka *see* Kam"yanka
Kamenka-Bugskaya *see* Kam"yanka-Buz'ka
Kamenka Dneprovskaya *see* Kam"yanka-Dniprovs'ka
Kamen Kashirskiy *see* Kamin'-Kashyrs'kyy
Kamenka-Strumilov *see* Kam"yanka-Buz'ka
155 L15 **Kamennomostskiy** Respublika Adygeya, SW Russian Federation 44°13´N 40°12´E
126 L11 **Kamenolomni** Rostovskaya Oblast', SW Russian Federation 47°36´N 40°18´E
Kamenskoye *see* Romanov
11 V15 **Kamsack** Saskatchewan, S Canada 51°34´N 101°51´W
Kamskoye Vodokhranilishche *see* above
101 P14 **Kamenz** Sachsen, E Germany 51°15´N 14°06´E
127 R4 **Kamskoye Ust'ye** Respublika Tatarstan, W Russian Federation 55°13´N 49°11´E
125 U14 **Kamskoye Vodokhranilishche** *var.* Kama Reservoir. ⊡ NW Russian Federation
164 C11 **Kami-Agata** Nagasaki, Tsushima, SW Japan
114 H13 **Kamsar** Guinée-Maritime, W Guinea 10°33´N 14°34´W
33 N10 **Kamiah** Idaho, NW USA 46°13´N 116°01´W
43 V16 **Kámuk, Cerro** ▲ SE Costa Rica 09°15´N 83°01´W
116 K7 **Kam"yanets'-Podil's'kyy** *Rus.* Kamenets-Podol'skiy. Khmel'nyts'ka Oblast', W Ukraine 48°40´N 26°36´E
117 Q6 **Kam"yanka** *Rus.* Kamenka. Cherkas'ka Oblast', C Ukraine 49°03´N 32°06´E

116 I5 **Kam"yanka-Buz'ka** *prev.* Kamenka-Strumilov, *Pol.* Kaminka Strumiłowa, *Rus.* Kamenka-Bugskaya. L'vivs'ka Oblast', NW Ukraine
117 T9 **Kam"yanka-Dniprovs'ka** *Rus.* Kamenka Dneprovskaya. Zaporiz'ka Oblast', SE Ukraine 47°28´N 34°24´E
119 F19 **Kamyanyets** *Rus.* Kamenets. Brestskaya Voblasts', SW Belarus 52°27´N 23°51´E
118 M13 **Kamyen'** *Rus.* Kamen'. Vitsyebskaya Voblasts', N Belarus 55°01´N 28°53´E
127 P9 **Kamyshin** Volgogradskaya Oblast', SW Russian Federation 50°07´N 45°20´E
127 Q13 **Kamyzyak** Astrakhanskaya Oblast', SW Russian Federation 46°07´N 48°03´E
12 K8 **Kanaaupscow** ⚹ Québec, C Canada
36 K9 **Kanab** Utah, W USA 37°03´N 112°31´W
36 K9 **Kanab Creek** ⚹ Arizona/Utah, SW USA
187 Y14 **Kanacea** *var.* Kanathea. Taveuni, N Fiji 16°59´S 179°54´E
38 G17 **Kanaga Island** *island* Aleutian Islands, Alaska, USA
38 G17 **Kanaga Volcano** ▲ Kanaga Island, Alaska, USA 51°55´N 177°09´W
164 N14 **Kanagawa** *off.* Kanagawa-ken. ◇ *prefecture* Honshū, S Japan
Kanagawa-ken *see* Kanagawa
13 Q8 **Kanairiktok** ⚹ Newfoundland and Labrador, E Canada
Kanaky *see* New Caledonia
107 Q25 **Kanála** Sicilia, Italy, C Mediterranean Sea 36°46´N 12°03´E
192 K4 **Kammu Seamount** *undersea feature* N Pacific Ocean 32°09´N 173°00´E
109 U11 **Kanank** *Ger.* Stein. Slovenia 46°13´N 14°34´E
109 T10 **Kamniško-Savinjske Alpe** *var.* Steiner Alps, Sanntaler Alpen, *Ger.* Steiner Alpen. ▲ N Slovenia
Kamo *see* Gavarr
79 K22 **Kananga** *prev.* Luluabourg. Kasai-Occidental, S Dem. Rep. Congo 05°53´S 22°22´E
109 T10 **Kamniško-Savinjske Alpe** *see above*
127 Q4 **Kanash** Chuvashskaya Respublika, W Russian Federation 55°30´N 47°27´E
21 Q4 **Kanawha River** ⚹ West Virginia, NE USA
164 L13 **Kanayama** Gifu, Honshū, SW Japan 35°N 137°15´E
164 L13 **Kanazawa** Ishikawa, Honshū, SW Japan 36°33´N 136°40´E
166 L8 **Kanbalu** Sagaing, C Myanmar (Burma)
166 M5 **Kanbe** Yangon, SW Myanmar (Burma) 16°40´N 96°01´E
167 O11 **Kanchanaburi** Kanchanaburi, W Thailand 14°02´N 99°32´E
Kanchanjangha/Kanchenjunga *see* Kangchenjunga
155 J19 **Kānchipuram** *prev.* Conjeeveram. Tamil Nādu, SE India 12°50´N 79°44´E
149 N8 **Kandahār** *Per.* Qandahār. S Afghanistan 31°36´N 65°48´E
149 N9 **Kandahār** *Per.* Qandahār. ◇ *province* SE Afghanistan
167 S13 **Kândal** ⚹ Ta Khmau. 11°30´N 104°59´E
124 K5 **Kandalaksha** *var.* Kandalaksha, *Fin.* Kantalahti. Murmanskaya Oblast', NW Russian Federation 67°09´N 32°14´E
124 K6 **Kandalaksha Gulf/Kandalakshskaya Guba** *see* Kandalakshskiy Zaliv
124 K6 **Kandalakshskiy Zaliv** *var.* Kandalakshskaya Guba, *Eng.* Kandalaksha Gulf. *bay* NW Russian Federation
83 G17 **Kandalengoti** Ngamiland, NW Botswana 19°25´S 22°12´E
Kandalengoti *see* Kandalengoti
169 U13 **Kandangan** Borneo, C Indonesia 02°50´S 115°15´E
118 E8 **Kandau** *Ger.* Kandau. W Latvia 57°02´N 22°45´E
Kandava *Ger.* Kandau. *see* Kadavu
77 R14 **Kandé** *var.* Kanté. NE Togo 09°55´N 01°01´E
101 F23 **Kandel** ▲ SW Germany 48°03´N 08°00´E
186 C7 **Kandep** Enga, W Papua New Guinea 05°54´S 143°34´E
149 V12 **Kandh Kot** Sind, SE Pakistan 28°15´N 69°18´E
77 S13 **Kandi** N Benin 11°05´N 02°59´E
149 P14 **Kandiāro** Sind, SE Pakistan 27°02´N 68°16´E
136 F11 **Kandıra** Kocaeli, NW Turkey 41°05´N 30°08´E
183 S8 **Kandos** New South Wales, SE Australia 32°52´S 149°58´E
148 M16 **Kandrāch** *var.* Kanrach. Baluchistān, SW Pakistan 25°26´N 65°28´E
172 I4 **Kandreho** Mahajanga, C Madagascar 17°27´S 46°06´E
186 F7 **Kandrian** New Britain, E Papua New Guinea 06°14´S 149°32´E
155 K25 **Kandy** Central Province, C Sri Lanka 07°17´N 80°40´E
144 I10 **Kandyagash** *Kaz.* Qandyaghash; *prev.* Oktyab'rsk. Aktyubinsk, W Kazakhstan 49°25´N 57°24´E
18 D12 **Kane** Pennsylvania, NE USA 41°39´N 78°47´W
64 I11 **Kane Fracture Zone** *tectonic feature* NW Atlantic Ocean
78 G9 **Kaném** *off.* Région du Kanem. ◇ *region* W Chad
78 G9 **Kaném, Région du** *see* Kaném
38 D9 **Kāne'ohe** *var.* Kaneohe. O'ahu, Hawaii, USA, C Pacific Ocean 21°25´N 157°48´W
Kanestron, Akrótírio *see* Palioúri, Akrotírio
Kanëv *see* Kaniv
124 M5 **Kanevka** *var.* Kaneka. Murmanskaya Oblast', NW Russian Federation 67°07´N 39°57´E
126 K13 **Kanevskaya** Krasnodarskiy Kray, SW Russian Federation 46°07´N 38°57´E

◆ Country   ◇ Dependent Territory   ⬟ Administrative Regions   ▲ Mountain   🌋 Volcano   ⊚ Lake
● Country Capital   ◈ Dependent Territory Capital   ✕ International Airport   ▲▲ Mountain Range   ⚹ River   ⊡ Reservoir

KANEVSKOYE VODOKHRANILISHCHE – KĀSGANJ

Kanevskoye Vodokhranilishche *see* Kanivs'ke Vodoskhovyshche
165 P9 **Kaneyama** Yamagata, Honshū, C Japan 38°54′N 140°20′E
83 G20 **Kang** Kgalagadi, C Botswana 23°41′S 22°50′E
76 L13 **Kangaba** Koulikoro, SW Mali 11°57′N 08°24′W
136 M13 **Kangal** Sivas, C Turkey 39°15′N 37°23′E
Kangān *see* Bandar-e Kangān
168 J6 **Kangar** Perlis, Peninsular Malaysia 06°28′N 100°10′E
76 L13 **Kangaré** Sikasso, S Mali 11°39′N 08°10′W
182 F10 **Kangaroo Island** *island* South Australia
93 M17 **Kangasniemi** Etelä-Savo, E Finland 61°58′N 26°37′E
142 K6 **Kangāvar** *var.* Kangāwar. Kermānshāhān, W Iran 34°29′N 47°55′E
Kangāwar *see* Kangāvar
153 S11 **Kangchenjunga** *var.* Kānchenjunga, *Nep.* Kanchanjańghā. ▲ NE India 27°36′N 88°06′E
160 G9 **Kangding** *var.* Lucheng, *Tib.* Dardo. Sichuan, C China 30°03′N 101°56′E
169 U16 **Kangean, Kepulauan** *island group* S Indonesia
169 T16 **Kangean, Pulau** *island* Kepulauan Kangean, S Indonesia
67 U8 **Kangen** *var.* Kengen. ≈ E South Sudan
197 N14 **Kangerlussuaq** *Dan.* Søndre Strømfjord. ✈ Qeqqata, W Greenland 66°59′N 50°28′E
197 Q15 **Kangertittivaq** *Dan.* Scoresby Sund. *fjord* E Greenland
167 O2 **Kangfang** Kachin State, N Myanmar (Burma) 26°09′N 98°36′E
163 X12 **Kanggye** N North Korea 40°58′N 126°37′E
197 P15 **Kangikajik** *var.* Kap Brewster. *headland* E Greenland 70°10′N 22°00′W
13 N5 **Kangiqsualujjuaq** *prev.* George River, Port-Nouveau-Québec. Québec, E Canada 58°35′N 65°59′W
12 L2 **Kangiqsujuaq** *prev.* Maricourt, Wakeham Bay. Québec, NE Canada 61°35′N 72°00′W
12 M4 **Kangirsuk** *prev.* Bellin, Payne. Québec, E Canada 60°00′N 70°01′W
Kangle *see* Wanzai
158 M16 **Kangmar** Xizang Zizhiqu, W China 28°34′N 89°40′E
Kangnŭng *see* Gangneung
79 D18 **Kango** Estuaire, NW Gabon 0°17′N 10°00′E
152 I7 **Kāngra** Himāchal Pradesh, NW India 32°04′N 76°16′E
153 Q16 **Kangsabati Reservoir** ⊠ N India
159 O17 **Kangto** ▲ China/India 27°54′N 92°33′E
159 W12 **Kangxian** *var.* Kang Xian, Zuitai, Zuitaizi. Gansu, C China 33°21′N 105°40′E
Kang Xian *see* Kangxian
76 M15 **Kani** NW Ivory Coast 08°29′N 06°36′W
166 L4 **Kani** Sagaing, C Myanmar (Burma) 22°24′N 94°55′E
79 M23 **Kaniama** Katanga, S Dem. Rep. Congo 07°32′S 24°11′E
Kanibadam *see* Konibodom
169 V6 **Kanibongan** Sabah, East Malaysia 06°40′N 117°12′E
185 F17 **Kaniere** West Coast, South Island, New Zealand 42°45′S 171°00′E
185 G17 **Kaniere, Lake** ◎ South Island, New Zealand
188 E17 **Kanifaay** Yap, W Micronesia
125 O4 **Kanin Kamen'** ⛰ NW Russian Federation
125 N3 **Kanin Nos** Nenetskiy Avtonomnyy Okrug, NW Russian Federation 68°38′N 43°19′E
125 N3 **Kanin Nos, Mys** *cape* NW Russian Federation
125 O5 **Kanin, Poluostrov** *peninsula* NW Russian Federation
139 V8 **Kānī Sakht** Wāsiţ, E Iraq 33°19′N 46°04′E
139 T3 **Kānī Slēman** *Ar.* Kānī Sulaymān. Arbīl, N Iraq 35°54′N 44°35′E
Kānī Sulaymān *see* Kānī Slēman
165 Q6 **Kanita** Aomori, Honshū, C Japan 41°04′N 140°36′E
117 Q5 **Kaniv** *Rus.* Kanëv. Cherkas'ka Oblast', C Ukraine 49°46′N 31°28′E
182 K11 **Kaniva** Victoria, SE Australia 36°25′S 141°13′E
117 Q5 **Kanivs'ke Vodoskhovyshche** *Rus.* Kanevskoye Vodokhranilishche. ⊠ C Ukraine
112 L8 **Kanjiža** *Ger.* Altkanischa, *Hung.* Magyarkanizsa, Okanizsa; *prev.* Stara Kanjiža. Vojvodina, N Serbia 46°03′N 20°03′E
93 K18 **Kankaanpää** Satakunta, SW Finland 61°47′N 22°25′E
30 M12 **Kankakee** Illinois, N USA 41°07′N 87°51′W
31 O11 **Kankakee River** ≈ Illinois/Indiana, N USA
76 K14 **Kankan** E Guinea 10°25′N 09°19′W
154 K13 **Kānker** Chhattīsgarh, C India 20°19′N 81°29′E
76 J10 **Kankossa** Assaba, S Mauritania 15°54′N 11°31′W
169 N12 **Kanmaw Kyun** *var.* Kisseraing, Kithareng. *island* Mergui Archipelago, S Myanmar (Burma)
164 F12 **Kanmuri-yama** ▲ Kyūshū, SW Japan 34°28′N 132°03′E
21 R10 **Kannapolis** North Carolina, SE USA 35°30′N 80°36′W
93 L16 **Kannonkoski** Keski-Suomi, C Finland 62°59′N 25°20′E
93 K15 **Kannus** Keski-Pohjanmaa, W Finland 63°55′N 23°55′E
77 V13 **Kano** Kano, N Nigeria 11°56′N 08°31′E
77 V13 **Kano** ⬖ *state* N Nigeria
77 V13 **Kano** ✈ Kano, N Nigeria 11°56′N 08°26′E
164 G14 **Kan'onji** *var.* Kanonzi. Kagawa, Shikoku, SW Japan 34°08′N 133°38′E
Kanonzi *see* Kan'onji
26 M5 **Kanopolis Lake** ⊠ Kansas, C USA
36 K5 **Kanosh** Utah, W USA 38°48′N 112°26′W
169 R9 **Kanowit** Sarawak, East Malaysia 02°03′N 112°15′E
164 C16 **Kanoya** Kagoshima, Kyūshū, SW Japan 31°22′N 130°50′E
152 L13 **Kānpur** *Eng.* Cawnpore. Uttar Pradesh, N India 26°28′N 80°21′E
Kanrach *see* Kandrāch
164 I14 **Kansai** ✈ (Ōsaka) Ōsaka, Honshū, SW Japan 34°25′N 135°13′E
27 R9 **Kansas** Oklahoma, C USA 36°14′N 94°46′W
26 L5 **Kansas** *off.* State of Kansas, *also known as* Jayhawker State, Sunflower State. ⬖ *state* C USA
27 R4 **Kansas City** Kansas, C USA 39°07′N 94°38′W
27 R4 **Kansas City** Missouri, C USA 39°06′N 94°35′W
27 R3 **Kansas City** ✈ Missouri, C USA 39°18′N 94°45′W
27 P4 **Kansas River** ≈ Kansas, C USA
122 L14 **Kansk** Krasnoyarskiy Kray, S Russian Federation 56°11′N 95°32′E
Kansu *see* Gansu
147 V7 **Kant** Chuyskaya Oblast', N Kyrgyzstan 42°54′N 74°47′E
93 L19 **Kanta-Häme** *Swe.* Egentliga Tavastland. ⬖ *region* S Finland
Kantalahti *see* Kandalaksha
167 N16 **Kantang** *var.* Ban Kantang. Trang, SW Thailand 07°25′N 99°30′E
115 H25 **Kántanos** Kríti, Greece, E Mediterranean Sea 35°20′N 23°42′E
77 R12 **Kantchari** E Burkina Faso 12°47′N 01°37′E
Kanté *see* Kandé
Kantemir *see* Cantemir
126 L9 **Kantemirovka** Voronezhskaya Oblast', W Russian Federation 49°44′N 39°53′E
167 R11 **Kantharalak** Si Sa Ket, E Thailand 14°32′N 104°37′E
Kantipur *see* Kathmandu
39 Q9 **Kantishna River** ≈ Alaska, USA
191 S3 **Kanton** *var.* Abariringa, Canton Island; *prev.* Mary Island. *atoll* Phoenix Islands, C Kiribati
97 C20 **Kanturk** *Ir.* Ceann Toirc. Cork, SW Ireland 52°12′N 08°54′W
55 T11 **Kanuku Mountains** ⛰ S Guyana
165 O12 **Kanuma** Tochigi, Honshū, S Japan 36°34′N 139°44′E
83 H20 **Kanye** Southern, SE Botswana 24°55′S 25°14′E
83 H17 **Kanyu** North-West, C Botswana 20°04′S 24°36′E
166 M7 **Kanyutkwin** Bago, C Myanmar (Burma) 18°19′N 96°30′E
79 M24 **Kanzenze** Katanga, SE Dem. Rep. Congo 10°33′S 25°28′E
193 Y15 **Kao** *island* Kotu Group, W Tonga
167 Q13 **Kaôh Kŏng** *var.* Krŏng Kaôh Kŏng. Kaôh Kŏng, SW Cambodia 11°37′N 102°59′E
Kaohsiung *see* Gaoxiong
Kaokaona *see* Kirakira
83 B17 **Kaoko Veld** ⛰ N Namibia
76 G11 **Kaolack** *var.* Kaolak. W Senegal 14°09′N 16°08′W
Kaolak *see* Kaolack
Kaolan *see* Lanzhou
186 M8 **Kaolo** San Jorge, N Solomon Islands 08°24′S 159°35′E
83 H14 **Kaoma** Western, W Zambia 14°50′S 24°48′E
38 B8 **Kapa'a** *var.* Kapaa. Kaua'i, Hawaii, USA, C Pacific Ocean 22°04′N 159°19′W
Kapaa *see* Kapa'a
113 J16 **Kapa Moračka** ▲ C Montenegro 42°53′N 19°01′E
137 V13 **Kapan** *Rus.* Kafan; *prev.* Ghap'an. SE Armenia 39°13′N 46°25′E
82 L13 **Kapandashila** Muchinga, NE Zambia 12°43′S 31°00′E
79 L23 **Kapanga** Katanga, S Dem. Rep. Congo 08°22′S 22°37′E
Kapchagay *see* Kapshagay
Kapchagayskoye Vodokhranilishche *see* Vodokhranilishche Kapshagay
99 F15 **Kapelle** Zeeland, SW Netherlands 51°29′N 03°58′E
99 G16 **Kapellen** Antwerpen, N Belgium 51°19′N 04°25′E
95 P15 **Kapellskär** Stockholm, C Sweden 59°43′N 19°03′E
81 H18 **Kapenguria** West Pokot, W Kenya 01°14′N 35°08′E
109 V6 **Kapfenberg** Steiermark, C Austria 47°27′N 15°18′E
83 J14 **Kapiri Mposhi** Central, C Zambia 13°59′S 28°40′E
149 R4 **Kāpīsā** ⬖ *province* E Afghanistan
12 G10 **Kapiskau** ≈ Ontario, C Canada
184 K13 **Kapiti Island** *island* C New Zealand
78 K9 **Kapka, Massif du** ⛰ E Chad
Kaplamada *see* Kaubalatmada, Gunung
22 H9 **Kaplan** Louisiana, S USA 30°00′N 92°16′W
Kaplangky, Plato *see* Gaplaňgyr Platosy
111 D19 **Kaplice** *Ger.* Kaplitz. Jihočeský Kraj, S Czech Republic 48°42′N 14°27′E
Kaplitz *see* Kaplice
Kapoche *see* Capoche
171 T12 **Kapocol** Papua Barat, E Indonesia 01°59′S 130°11′E
167 N14 **Kapoe** Ranong, SW Thailand 09°33′N 98°37′E
Kapoeas *see* Kapuas, Sungai
81 G15 **Kapoeta** Eastern Equatoria, SE South Sudan 04°50′N 33°35′E
111 I25 **Kapos** ≈ S Hungary
111 H25 **Kaposvár** Somogy, SW Hungary 46°23′N 17°54′E
94 H13 **Kapp** Oppland, S Norway 60°42′N 10°49′E
100 I7 **Kappeln** Schleswig-Holstein, N Germany 54°41′N 09°56′E
Kaproncza *see* Koprivnica
109 P7 **Kaprun** Salzburg, C Austria 47°15′N 12°48′E
145 U15 **Kapshagay** *prev.* Kapchagay. Almaty, SE Kazakhstan 43°52′N 77°05′E
Kapstad *see* Cape Town
Kapsukas *see* Marijampolė
171 Y13 **Kaptiau** Papua, E Indonesia 02°23′S 139°51′E
119 L19 **Kaptsevichy** *Rus.* Koptsevichi. Homyel'skaya Voblasts', SE Belarus 52°14′N 28°19′E
Kapuas Hulu, Banjaran/Kapuas Hulu, Pegunungan *see* Kapuas Mountains
169 S10 **Kapuas Mountains** *Ind.* Banjaran Kapuas Hulu, Pegunungan Kapuas Hulu. ⛰ Indonesia/Malaysia
169 P11 **Kapuas, Sungai** ≈ Borneo, N Indonesia
169 T13 **Kapuas, Sungai** *prev.* Kapoeas. ≈ Borneo, C Indonesia
182 J9 **Kapunda** South Australia 34°23′S 138°51′E
152 H8 **Kapūrthala** Punjab, N India 31°20′N 75°26′E
12 G12 **Kapuskasing** Ontario, S Canada 49°25′N 82°26′W
14 D6 **Kapuskasing** ≈ Ontario, S Canada
127 P11 **Kapustin Yar** Astrakhanskaya Oblast', SW Russian Federation 48°36′N 45°49′E
82 K11 **Kaputa** Northern, NE Zambia 08°28′S 29°41′E
111 G22 **Kapuvár** Győr-Moson-Sopron, NW Hungary 47°35′N 17°01′E
Kapydzhik, Gora *see* Qazangödağ
119 J17 **Kapyl'** *Rus.* Kopyl'. Minskaya Voblasts', C Belarus 53°09′N 27°05′E
43 N9 **Kara** *var.* Cara. Región Autónoma Atlántico Sur, E Nicaragua 12°50′N 83°35′W
77 R14 **Kara** *var.* Lama-Kara. NE Togo 09°33′N 01°12′E
77 Q14 **Kara** ≈ N Togo
147 U7 **Kara-Balta** Chuyskaya Oblast', N Kyrgyzstan 42°51′N 73°51′E
144 L7 **Karabalyk** *var.* Komsomolets, *Kaz.* Komsomol. Kostanay, N Kazakhstan 53°47′N 61°58′E
144 G11 **Karabau** *Kaz.* Qarabaū. Atyrau, W Kazakhstan 48°29′N 53°05′E
146 E7 **Karabaur', Uval** *Kaz.* Korabavur Pastligi, *Uzb.* Qorabowur Kirlari. *physical region* Kazakhstan/Uzbekistan
Karabekaul *see* Garabekewül
Karabil', Vozvyshennost' *see* Garabil Belentligi
Kara-Bogaz-Gol *see* Garabogazköl
Kara-Bogaz-Gol, Zaliv *see* Garabogaz Aýlagy
145 R15 **Karaboget** *Kaz.* Qaraböget. Zhambyl, S Kazakhstan 44°36′N 72°03′E
136 H11 **Karabük** Karabük, NW Turkey 41°12′N 32°36′E
136 H11 **Karabük** ⬖ *province* NW Turkey
122 L12 **Karabula** Krasnoyarskiy Kray, C Russian Federation 58°01′N 97°17′E
145 V14 **Karabulak** *Kaz.* Qarabulaq. Taldykorgan, SE Kazakhstan 44°53′N 78°29′E
145 Y11 **Karabulak** *Kaz.* Qarabulaq. Vostochnyy Kazakhstan, E Kazakhstan 47°34′N 84°40′E
145 Q17 **Karabulak** *Kaz.* Qarabulaq. Yuzhnyy Kazakhstan, S Kazakhstan 42°31′N 69°47′E
Karabura *see* Yumin
136 C17 **Kara Burnu** *headland* SW Turkey 36°34′N 28°00′E
144 K10 **Karabutak** *Kaz.* Qarabutaq. Aktyubinsk, W Kazakhstan 49°55′N 60°05′E
136 D12 **Karacabey** Bursa, NW Turkey 40°14′N 28°22′E
114 O12 **Karacaköy** İstanbul, NW Turkey 41°24′N 28°21′E
114 M12 **Karacaoğlan** Kırklareli, NW Turkey 41°30′N 27°06′E
Karachay-Cherkessia *see* Karachayevo-Cherkesskaya Respublika
126 L15 **Karachayevo-Cherkesskaya Respublika** *Eng.* Karachay-Cherkessia. ⬖ *autonomous republic* SW Russian Federation
126 M15 **Karachayevsk** Karachayevo-Cherkesskaya Respublika, SW Russian Federation 43°43′N 41°53′E
126 J6 **Karachev** Bryanskaya Oblast', W Russian Federation 53°07′N 35°56′E
149 O16 **Karāchi** Sind, SE Pakistan 24°51′N 67°02′E
149 O16 **Karāchi** ✈ Sind, S Pakistan 24°51′N 67°02′E
Karácsonkő *see* Piatra-Neamț
155 E15 **Karād** Mahārāshtra, W India 17°19′N 74°15′E
136 H16 **Karadağ** ▲ S Turkey 37°00′N 33°00′E
147 T10 **Karadar'ya** *Uzb.* Qoradaryo. ≈ Kyrgyzstan/Uzbekistan
Karadeniz *see* Black Sea
Karadeniz Boğazi *see* İstanbul Boğazı
146 B13 **Karadepe** Balkan Welaýaty, W Turkmenistan 38°04′N 54°01′E
Karadzhar *see* Qorajar
Karaferiye *see* Véroia
Karagan *see* Garagan
Karaganda *see* Karagandy
Karagandinskaya Oblast' *see* Karagandy
145 R10 **Karagandy** *Kaz.* Qaraghandy; *prev.* Karaganda. Karaganda, C Kazakhstan 49°53′N 73°07′E
145 R10 **Karagandy** *off.* Karagandinskaya Oblast', *Kaz.* Qaraghandy Oblysy; *prev.* Karaganda. ⬖ *province* C Kazakhstan
145 T10 **Karagayly** *Kaz.* Qaraghayly. Karaganda, C Kazakhstan 49°25′N 75°31′E
Karagel' *see* Garagöl'
123 U9 **Karaginskiy, Ostrov** *island* E Russian Federation
197 T1 **Karaginskiy Zaliv** *bay* E Russian Federation
137 P13 **Karagöl Dağları** ⛰ NE Turkey
Karagumskiy Kanal *see* Garagum Kanaly
114 L13 **Karahisar** Edirne, NW Turkey 40°47′N 26°34′E
127 V3 **Karaidel'** Respublika Bashkortostan, W Russian Federation 55°50′N 56°55′E
114 L13 **Karaidemir Barajı** ⊠ NW Turkey
155 J21 **Kāraikāl** Puducherry, SE India 10°58′N 79°50′E
155 I22 **Kāraikkudi** Tamil Nādu, SE India 10°04′N 78°46′E
Kara Irtysh *see* Kara Yertis
143 N5 **Karaj** Alborz, N Iran 35°44′N 51°26′E
168 K8 **Karak** Pahang, Peninsular Malaysia 03°24′N 101°59′E
Karak *see* Al Karak
147 T11 **Kara-Kabak** Oshskaya Oblast', SW Kyrgyzstan 39°40′N 72°45′E
Kara-Kala *see* Magtymguly
Karakala *see* Oqqal'a
Karakalpakstan, Respublika *see* Qoraqalpog'iston Respublikasi
Karakalpakya *see* Qoraqalpog'iston
158 G10 **Karakax He** ≈ NW China
121 X8 **Karakaya Barajı** ⊠ C Turkey
171 Q9 **Karakelong, Pulau** *island* N Indonesia
Karakılısse *see* Ağrı
Karak, Muḩāfaẓat al *see* Al Karak
147 X8 **Karakol** *var.* Karakolka. Issyk-Kul'skaya Oblast', NE Kyrgyzstan 41°30′N 77°18′E
147 Y7 **Karakol** *prev.* Przheval'sk. Issyk-Kul'skaya Oblast', NE Kyrgyzstan 42°32′N 78°21′E
Kara-Köl *see* Kara-Kul'
Karakolka *see* Karakol
149 W2 **Karakoram Highway** *road* China/Pakistan
149 Z3 **Karakoram Pass** *Chin.* Karakoram Shankou. *pass* C Asia
152 I3 **Karakoram Range** ⛰ C Asia
Karakoram Shankou *see* Karakoram Pass
Karaköse *see* Ağrı
145 P14 **Karakoyyn, Ozero** *Kaz.* Qaraqoyyn. ◎ C Kazakhstan
83 F19 **Karakubis** Ghanzi, W Botswana 22°03′S 20°36′E
147 T9 **Kara-Kul'** *Kir.* Kara-Köl. Dzhalal-Abadskaya Oblast', W Kyrgyzstan 40°35′N 73°36′E
Karakul' *see* Qarokŭl, Tajikistan
Karakul' *see* Qorakoʻl, Uzbekistan
147 U10 **Kara-Kul'dzha** Oshskaya Oblast', SW Kyrgyzstan 40°32′N 73°50′E
127 T3 **Karakulino** Udmurtskaya Respublika, NW Russian Federation 56°02′N 53°45′E
Karakul', Ozero *see* Qarokŭl
Kara Kum *see* Garagum
Kara Kum Canal/Karakumskiy Kanal *see* Garagum Kanaly
Karakumy, Peski *see* Garagum
83 E17 **Karakuwisa** Okavango, NE Namibia 18°56′S 19°40′E
122 M13 **Karam** Irkutskaya Oblast', S Russian Federation 55°07′N 107°21′E
Karamai *see* Karamay
169 T14 **Karamain, Pulau** *island* N Indonesia
136 I16 **Karaman** Karaman, S Turkey 37°11′N 33°13′E
136 H16 **Karaman** ⬖ *province* S Turkey
114 M8 **Karamandere** ≈ NE Bulgaria
158 J4 **Karamay** *var.* Karamai, Kelamayi; *prev. Chin.* K'o-la-ma-i. Xinjiang Uygur Zizhiqu, NW China 45°33′N 84°45′E
169 U14 **Karambu** Borneo, N Indonesia 03°48′S 116°06′E
185 H14 **Karamea** West Coast, South Island, New Zealand 41°15′S 172°07′E
185 H14 **Karamea** ≈ South Island, New Zealand
185 G15 **Karamea Bight** *gulf* South Island, New Zealand
Karamet-Niyaz *see* Garamätnyýaz
158 K10 **Karamiran He** ≈ NW China
147 S11 **Karamyk** Oshskaya Oblast', SW Kyrgyzstan 39°28′N 71°45′E
169 U17 **Karangasem** Bali, S Indonesia 08°24′S 115°40′E
154 H12 **Kāranja** Mahārāshtra, C India 20°30′N 77°26′E
Karanpur *see* Karanpura
152 F9 **Karanpura** *var.* Karanpur. Rājasthān, NW India 29°46′N 73°30′E
Karánsebes/Karansebesch *see* Caransebeş
145 T14 **Karaoy** *Kaz.* Qaraoy. Almaty, SE Kazakhstan 45°52′N 74°44′E
144 E10 **Karaozen** *Kaz.* Ülkenözen; *prev.* Bol'shoy Uzen'. ≈ Kazakhstan/Russian Federation
114 N7 **Karapelit** *Rom.* Stejarul. Dobrich, NE Bulgaria 43°40′N 27°33′E
136 I15 **Karapınar** Konya, C Turkey 37°43′N 33°34′E
83 D22 **Karas** ⬖ *district* S Namibia
147 Y8 **Kara-Say** Issyk-Kul'skaya Oblast', NE Kyrgyzstan 41°34′N 77°56′E
83 E22 **Karasburg** Karas, S Namibia 27°59′S 18°46′E
Kara Sea *see* Karskoye More
92 K9 **Kárášjohka** *var.* Karašjokka. ≈ N Norway
Kárášjohka *see* Karasjok
92 L9 **Karasjok** *Fin.* Kaarasjoki, *Lapp.* Kárášjohka. Finnmark, N Norway 69°27′N 25°28′E
Karašjokka *see* Kárášjohka
Kara Strait *see* Karskiye Vorota, Proliv
145 N8 **Karasu** *Kaz.* Qarasū. Kostanay, N Kazakhstan 52°44′N 65°29′E
136 F11 **Karasu** Sakarya, NW Turkey 41°07′N 30°37′E
Kara Su *see* Mesta/Néstos
Karasubazar *see* Bilohirs'k
122 I12 **Karasuk** Novosibirskaya Oblast', C Russian Federation 53°41′N 78°04′E
145 U13 **Karatal** *Kaz.* Qaratal. ≈ SE Kazakhstan
136 K17 **Karataş** Adana, S Turkey 36°32′N 35°22′E
145 Q16 **Karatau** *Kaz.* Qarataū. Zhambyl, S Kazakhstan 43°09′N 70°28′E
Karatau *see* Karatau, Khrebet
145 P16 **Karatau, Khrebet** *var.* Karatau, *Kaz.* Qarataū. ⛰ S Kazakhstan
144 G13 **Karaton** *Kaz.* Qaraton. Atyrau, W Kazakhstan 46°33′N 53°31′E
164 C13 **Karatsu** *var.* Karatu. Saga, Kyūshū, SW Japan 33°28′N 129°48′E
Karatu *see* Karatsu
122 K8 **Karaul** Krasnoyarskiy Kray, N Russian Federation 70°07′N 83°12′E
Karaulbazar *see* Qorowulbozor
Karauzyak *see* Qoraoʻzak
115 D16 **Karáva** ▲ C Greece 39°19′N 21°33′E
Karavanke *see* Karawanken
115 F22 **Karavás** Kýthira, S Greece 36°21′N 22°57′E
113 J20 **Karavastasë, Laguna e** *var.* Kënet' e Karavastas, Kravasta Lagoon. *lagoon* W Albania
Karavastas, Kënet' e *see* Karavastasë, Laguna e
118 I5 **Käravere** Tartumaa, E Estonia 58°25′N 26°29′E
115 L23 **Karavonísia** *island* Kykládes, Greece, Aegean Sea
169 O15 **Karawang** *prev.* Krawang. Jawa, C Indonesia 06°13′S 107°16′E
109 T10 **Karawanken** *Slvn.* Karavanke. ⛰ Austria/Serbia
Karaxahar *see* Kaidu He
137 R13 **Karayazı** Erzurum, NE Turkey 39°40′N 42°09′E
145 Y11 **Kara Yertis** *prev.* Kara Irtysh *Rus.* Chërnyy Irtysh. ≈ NE Kazakhstan
145 Q12 **Karazhal** *Kaz.* Qarazhal. Karaganda, C Kazakhstan 48°002′N 70°52′E
139 S9 **Karbalā'** *var.* Kerbala, Kerbela. Karbalā', S Iraq 32°37′N 44°03′E
139 S9 **Karbalā'** *off.* Muḩāfa at Karbalā'. ⬖ *governorate* S Iraq
Karbalā', Muḩāfa at *see* Karbalā'
94 L11 **Kårböle** Gävleborg, C Sweden 61°59′N 15°16′E
111 M23 **Karcag** Jász-Nagykun-Szolnok, E Hungary 47°22′N 20°51′E
Kardak *see* Imia
114 N7 **Kardam** Dobrich, NE Bulgaria 43°45′N 28°06′E
Kardamila *see* Kardámyla
115 L18 **Kardámyla** *var.* Kardamila, Kardhámila. Chíos, E Greece 38°33′N 26°04′E
Kardeljevo *see* Ploče
Kardh *see* Qardho
Kardhámila *see* Kardámyla
Kardhítsa *see* Karditsa
115 E16 **Karditsa** *var.* Kardhítsa. Thessalía, C Greece 39°22′N 21°56′E
118 E4 **Kärdla** *Ger.* Kertel. Hiiumaa, W Estonia 59°00′N 22°42′E
Kărdzhali *see* Kardzhali
114 J11 **Kardzhali** *var.* Kŭrdzhali. Kardzhali, S Bulgaria 41°39′N 25°23′E
114 K11 **Kardzhali** *var* Kŭrdzhali. ⬖ *province* S Bulgaria
114 J11 **Kardzhali, Yazovir** *var.* Yazovir Kŭrdzhali. ⊠ S Bulgaria
Karelia *see* Kareliya, Respublika
119 I16 **Karelichy** *Pol.* Korelicze, *Rus.* Korelichi. Hrodzyenskaya Voblasts', W Belarus 53°34′N 26°08′E
124 I10 **Kareliya, Respublika** *prev.* Karel'skaya ASSR, *Eng.* Karelia. ⬖ *autonomous republic* NW Russian Federation
Karel'skaya ASSR *see* Kareliya, Respublika
81 E22 **Karema** Katavi, W Tanzania 06°50′S 30°25′E
Karen *see* Hualian
83 I14 **Karenda** Central, C Zambia 14°42′S 26°52′E
Karen State *see* Kayin State
92 J10 **Karesuando** *Fin.* Kaaresuanto, *Lapp.* Gárasavvon. Norrbotten, N Sweden 68°25′N 22°28′E
Karet *see* Kâgheț
Kareyz-e-Elyās/Kārez Iliās *see* Kārīz-e Elyās
122 J11 **Kargasok** Tomskaya Oblast', C Russian Federation 59°01′N 80°34′E
122 I12 **Kargat** Novosibirskaya Oblast', C Russian Federation 55°07′N 80°19′E
136 J11 **Kargı** Çorum, N Turkey 41°09′N 34°32′E
152 I5 **Kargil** Jammu and Kashmir, NW India 34°34′N 76°06′E
Kargilik *see* Yecheng
124 L11 **Kargopol'** Arkhangel'skaya Oblast', NW Russian Federation 61°30′N 38°53′E
110 F12 **Kargowa** *Ger.* Unruhstadt. Lubuskie, W Poland 52°05′N 15°50′E
77 X13 **Kari** Bauchi, E Nigeria 11°13′N 10°34′E
83 J15 **Kariba** Mashonaland West, N Zimbabwe 16°29′S 28°48′E
83 J16 **Kariba, Lake** ◎ Zambia/Zimbabwe
165 Q4 **Kariba-yama** ▲ Hokkaidō, NE Japan 42°36′N 139°55′E
83 C19 **Karibib** Erongo, C Namibia 21°56′S 15°51′E
Karies *see* Karyés
92 L9 **Karigasniemi** *Lapp.* Garegegasnjárga. Lappi, N Finland 69°24′N 25°52′E
184 J2 **Karikari, Cape** *headland* North Island, New Zealand 34°47′S 173°24′E
149 W3 **Karīmābād** *prev.* Hunza. Jammu and Kashmir, NE Pakistan 36°23′N 74°43′E
169 P12 **Karimata, Kepulauan** *island group* N Indonesia
169 P12 **Karimata, Pulau** *island* Kepulauan Karimata, N Indonesia
169 O11 **Karimata, Selat** *strait* W Indonesia
155 I14 **Karīmnagar** Telangana, C India 18°28′N 79°09′E
186 C7 **Karimui** Chimbu, C Papua New Guinea 06°19′S 144°48′E
169 Q15 **Karimunjawa, Pulau** *island* S Indonesia
80 N12 **Karin** Woqooyi Galbeed, N Somalia 10°48′N 45°46′E
Kariot *see* Ikaria
93 L20 **Karis** *Fin.* Karjaa. Uusimaa, SW Finland 60°05′N 23°39′E
Káristos *see* Kárystos
148 J4 **Kārīz-e Elyās** *var.* Kareyz-e-Elyās, Kārez Iliās. Herāt, NW Afghanistan 35°26′N 61°24′E
Karjaa *see* Karis
Karkaralinsk *see* Karkaraly
145 T10 **Karkaraly** *prev.* Karkaralinsk. Karaganda, E Kazakhstan 49°31′N 75°53′E
186 D6 **Karkar Island** *island* N Papua New Guinea
143 N7 **Karkas, Kūh-e** ⛰ C Iran
142 K8 **Karkheh, Rūd-e** ≈ SW Iran
115 L20 **Karkinágri** *var.* Karkinagrio. Ikaría, Dodekánisa, Greece, Aegean Sea 37°31′N 26°01′E
Karkinagrio *see* Karkinágri
117 R12 **Karkinits'ka Zatoka** *Rus.* Karkinitskiy Zaliv. *gulf* S Ukraine
Karkinitskiy Zaliv *see* Karkinits'ka Zatoka
93 L19 **Karkkila** *Swe.* Högfors. Uusimaa, S Finland 60°32′N 24°10′E
93 M19 **Kärkölä** Päijät-Häme, S Finland 60°52′N 25°17′E
182 G9 **Karkoo** South Australia 34°03′S 135°45′E
Karkūk *see* Āltūn Kūbrī
118 D5 **Kärla** *Ger.* Kergel. Saaremaa, W Estonia 58°20′N 22°15′E
Karleby *see* Kokkola
110 F7 **Karlino** *Ger.* Körlin an der Persante. Zachodnio-pomorskie, NW Poland 54°02′N 15°52′E
137 Q13 **Karlıova** Bingöl, E Turkey 39°16′N 41°01′E
117 U6 **Karlivka** Poltavs'ka Oblast', C Ukraine 49°27′N 35°08′E
Karl-Marx-Stadt *see* Chemnitz
Karlö *see* Hailuoto
112 C11 **Karlobag** *It.* Carlopago. Lika-Senj, W Croatia 44°31′N 15°06′E
112 D9 **Karlovac** *Ger.* Karlstadt, *Hung.* Károlyváros. Karlovac, C Croatia 45°29′N 15°31′E
112 C10 **Karlovac** *off.* Karlovačka Županija. ⬖ *province* C Croatia
Karlovačka Županija *see* Karlovac
111 A16 **Karlovarský Kraj** ⬖ W Czech Republic
115 M19 **Karlovási** *var.* Néon Karlovásion, Néon Karlovasi. Sámos, Dodekánisa, Greece, Aegean Sea 37°47′N 26°40′E
114 J9 **Karlovo** *prev.* Levskigrad. Plovdiv, C Bulgaria 42°38′N 24°49′E
111 A16 **Karlovy Vary** *Ger.* Karlsbad; *prev. Eng.* Carlsbad. Karlovarský Kraj, W Czech Republic 50°13′N 12°51′E
Karlsbad *see* Karlovy Vary
95 L17 **Karlsborg** Västra Götaland, S Sweden 58°32′N 14°32′E
Karlsburg *see* Alba Iulia
95 L22 **Karlshamn** Blekinge, S Sweden 56°10′N 14°50′E
95 L16 **Karlskoga** Örebro, C Sweden 59°19′N 14°33′E
95 M22 **Karlskrona** Blekinge, S Sweden 56°11′N 15°39′E
101 G21 **Karlsruhe** *var.* Carlsruhe. Baden-Württemberg, SW Germany 49°01′N 08°24′E
95 K16 **Karlstad** Värmland, C Sweden 59°22′N 13°36′E
29 R3 **Karlstad** Minnesota, N USA 48°34′N 96°31′W
101 I18 **Karlstadt** Bayern, C Germany 49°58′N 09°46′E
Karlstadt *see* Karlovac
39 Q14 **Karluk** Kodiak Island, Alaska, USA 57°34′N 154°27′W
Karluk *see* Qarluq
119 O17 **Karma** *Rus.* Korma. Homyel'skaya Voblasts', SE Belarus 53°07′N 30°48′E
155 F14 **Karmāla** Mahārāshtra, W India 18°26′N 75°08′E
146 M11 **Karmana** Navoiy Viloyati, C Uzbekistan 40°09′N 65°18′E
138 G8 **Karmi'el** *var.* Carmiel. Northern, N Israel 32°55′N 35°18′E
95 B16 **Karmøy** *island* S Norway
152 I9 **Karnāl** Haryāna, N India 29°41′N 76°58′E
153 W15 **Karnaphuli Reservoir** ⊠ NE India
155 F17 **Karnātaka** *var.* Kanara; *prev.* Maisur, Mysore. ⬖ *state* W India
25 S13 **Karnes City** Texas, SW USA 28°54′N 97°55′W
109 P9 **Karnische Alpen** *It.* Alpi Carniche. ⛰ Austria/Italy
114 M9 **Karnobat** Burgas, E Bulgaria 42°39′N 26°59′E
109 Q9 **Kärnten** *off.* Land Kärnten, *Eng.* Carinthia, *Slvn.* Koroška. ⬖ *state* S Austria
Karnul *see* Kurnool
83 K16 **Karoi** Mashonaland West, N Zimbabwe 16°50′S 29°40′E
Karol *see* Carei
Károly-Fehérvár *see* Alba Iulia
Károlyváros *see* Karlovac
82 M12 **Karonga** Northern, N Malawi 09°54′S 33°55′E
147 W10 **Karool-Dëbë** *Kas.* Karool-Döbö; *prev.* Karool-Tëbë. Narynskaya Oblast', C Kyrgyzstan 40°33′N 75°52′E
Karool-Döbö *see* Karool-Dëbë
Karool-Tëbë *see* Karool-Dëbë
182 J9 **Karoonda** South Australia 35°04′S 139°58′E
149 S9 **Karor** *var.* Koror Lāl Esan. Punjab, E Pakistan 31°15′N 70°58′E
Karosa *see* Karossa
171 N12 **Karossa** *var.* Karosa. Sulawesi, C Indonesia 01°38′S 119°21′E
Karpaten *see* Carpathian Mountains
115 L22 **Karpáthio Pélagos** *sea* Dodekánisa, Greece, Aegean Sea
115 N24 **Kárpathos** Kárpathos, SE Greece 35°30′N 27°13′E
115 N24 **Kárpathos** *It.* Scarpanto; *anc.* Carpathos, Carpathus. *island* SE Greece
Karpathos Strait *see* Karpathou, Stenó
115 N24 **Karpathou, Stenó** *var.* Karpathos Strait, Scarpanto Strait. *strait* Dodekánisa, Greece, Aegean Sea
Karpaty *see* Carpathian Mountains
115 E17 **Karpenísi** *prev.* Karpenísion. Stereá Elláda, C Greece 38°55′N 21°46′E
Karpenísion *see* Karpenísi
Karpilovka *see* Aktsyabrski
125 O8 **Karpogory** Arkhangel'skaya Oblast', NW Russian Federation 64°01′N 44°22′E
180 I7 **Karratha** Western Australia 20°44′S 116°52′E
137 S12 **Kars** *var.* Qars. Kars, NE Turkey 40°35′N 43°05′E
137 S12 **Kars** *var.* Qars. ⬖ *province* NE Turkey
145 O12 **Karsakpay** *Kaz.* Qarsaqbay. Karaganda, C Kazakhstan 47°51′N 66°42′E
93 L15 **Kärsämäki** Pohjois-Pohjanmaa, C Finland 63°58′N 25°49′E
Karsau *see* Kārsava
118 K9 **Kārsava** *Ger.* Karsau; *prev. Rus.* Korsovka. E Latvia 56°46′N 27°39′E
Karshi *see* Garşy, Turkmenistan
Karshi *see* Qarshi, Uzbekistan
Karshinskaya Step *see* Qarshi Choʻli
Karshinskiy Kanal *see* Qarshi Kanali
84 I5 **Karskiye Vorota, Proliv** *Eng.* Kara Strait. *strait* N Russian Federation
122 J6 **Karskoye More** *Eng.* Kara Sea. *sea* Arctic Ocean
93 L17 **Karstula** Keski-Suomi, C Finland 62°52′N 24°48′E
127 Q5 **Karsun** Ul'yanovskaya Oblast', W Russian Federation 54°12′N 47°00′E
122 F11 **Kartaly** Chelyabinskaya Oblast', C Russian Federation 53°02′N 60°42′E
18 E13 **Karthaus** Pennsylvania, NE USA 41°06′N 78°03′W
110 I7 **Kartuzy** Pomorskie, NW Poland 54°21′N 18°11′E
165 R8 **Karumai** Iwate, Honshū, C Japan 40°19′N 141°27′E
181 U4 **Karumba** Queensland, NE Australia 17°31′S 140°51′E
142 L10 **Kārūn** *var.* Rūd-e Kārūn. ≈ SW Iran
92 K13 **Karungi** Norrbotten, N Sweden 66°03′N 23°55′E
92 K13 **Karunki** Lappi, N Finland 66°01′N 24°06′E
Kārūn, Rūd-e *see* Kārūn
155 H21 **Karūr** Tamil Nādu, SE India 10°58′N 78°03′E
93 K17 **Karvia** Satakunta, SW Finland 62°07′N 22°34′E
111 J17 **Karviná** *Ger.* Karwin, *Pol.* Karwina; *prev.* Nová Karvinná. Moravskoslezský Kraj, E Czech Republic 49°50′N 18°30′E
155 E17 **Kārwār** Karnātaka, W India 14°50′N 74°09′E
108 M7 **Karwendelgebirge** ⛰ Austria/Germany
Karwin/Karwina *see* Karviná
115 I14 **Karyés** *var.* Karies. Ágion Óros, N Greece 40°15′N 24°15′E
115 I19 **Kárystos** *var.* Káristos. Évvoia, C Greece 38°01′N 24°25′E
136 E17 **Kaş** Antalya, SW Turkey 36°12′N 29°38′E
39 Y14 **Kasaan** Prince of Wales Island, Alaska, USA 55°32′N 132°24′W
164 I13 **Kasai** Hyōgo, Honshū, SW Japan 34°56′N 134°49′E
79 K21 **Kasai** *var.* Cassai, Kassai. ≈ Angola/Dem. Rep. Congo
79 K22 **Kasai-Occidental** *off.* Région Kasai Occidental. ⬖ *region* S Dem. Rep. Congo
Kasai Occidental, Région *see* Kasai-Occidental
79 L21 **Kasai-Oriental** *off.* Région Kasai Oriental. ⬖ *region* C Dem. Rep. Congo
Kasai Oriental, Région *see* Kasai-Oriental
79 L24 **Kasaji** Katanga, S Dem. Rep. Congo 10°22′S 23°29′E
82 L12 **Kasama** Northern, N Zambia 10°14′S 31°12′E
Kasan *see* Koson
83 H16 **Kasane** North-West, NE Botswana 17°48′S 20°06′E
81 E23 **Kasanga** Rukwa, W Tanzania 08°27′S 31°10′E
79 G21 **Kasangulu** Bas-Congo, W Dem. Rep. Congo 04°33′S 15°12′E
Kasansay *see* Kosonsoy
155 E20 **Kāsaragod** Kerala, SW India 12°30′N 74°59′E
Kasargen *see* Kasari
118 P13 **Kasari** *var.* Kasari Jõgi, *Ger.* Kasargen. ≈ W Estonia
Kasari Jõgi *see* Kasari
8 L11 **Kasba Lake** ◎ Northwest Territories, Nunavut N Canada
Kaschau *see* Košice
Kaseda *see* Minamisatsuma
83 I14 **Kasempa** North Western, NW Zambia 13°27′S 25°49′E
79 O24 **Kasenga** Katanga, SE Dem. Rep. Congo 10°22′S 28°37′E
79 P17 **Kasenye** *var.* Kasenyi. Orientale, NE Dem. Rep. Congo 01°23′N 30°25′E
Kasenyi *see* Kasenye
79 O19 **Kasese** Maniema, E Dem. Rep. Congo 01°36′S 27°31′E
81 E18 **Kasese** SW Uganda 0°10′N 30°06′E
152 J11 **Kāsganj** Uttar Pradesh, N India 27°48′N 78°38′E

◆ Country ● Country Capital ◇ Dependent Territory ○ Dependent Territory Capital ⬖ Administrative Regions ✈ International Airport ▲ Mountain ⛰ Mountain Range ✱ Volcano ≈ River ◎ Lake ⊠ Reservoir

◆ Country ◇ Dependent Territory ⋉ Administrative Regions ▲ Mountain ⊼ Volcano ⊛ Lake
● Country Capital ○ Dependent Territory Capital ✈ International Airport ▲▲ Mountain Range ∼ River ⊠ Reservoir

269

KENYA, REPUBLIC OF – KICKAPOO RIVER

Kenya, Republic of see Kenya
168 L7 **Kenyir, Tasik** var. Tasek Kenyir. ◎ Peninsular Malaysia
29 W10 **Kenyon** Minnesota, N USA 44°16′N 92°59′W
29 Y16 **Keokuk** Iowa, C USA 40°24′N 91°22′W
Keonjihargarh see Kendujhargarh
Kéos see Tziá
29 X16 **Keosauqua** Iowa, C USA 40°43′N 91°58′W
29 X15 **Keota** Iowa, C USA 41°21′N 91°57′W
21 O11 **Keowee, Lake** ⊠ South Carolina, SE USA
124 I7 **Kepa** var. Kepe. Respublika Kareliya, NW Russian Federation 65°09′N 32°15′E
Kepe see Kepa
189 O13 **Kepirohi Falls** waterfall Pohnpei, E Micronesia
185 B22 **Kepler Mountains** ▲▲ South Island, New Zealand
111 I14 **Kępno** Wielkopolskie, C Poland 51°17′N 17°57′E
65 C24 **Keppel Island** island N Falkland Islands
Keppel Island see Niuatoputapu
65 C23 **Keppel Sound** sound N Falkland Islands
Kepri see Kepulauan Riau
136 D12 **Kepsut** Balıkesir, NW Turkey 39°41′N 28°09′E
168 M11 **Kepulauan Riau** off. Propinsi Kepulauan Riau, var. Kepri. ⬖ province NW Indonesia
Kequ see Gadê
171 V13 **Kerai** Papua Barat, E Indonesia 03°53′S 134°30′E
Kerak see Al Karak
155 F22 **Kerala** ⬖ state S India
165 R16 **Kerama-rettō** island group SW Japan
183 N10 **Kerang** Victoria, SE Australia 35°46′S 144°01′E
Kerasunt see Giresun
115 H19 **Keratéa** var. Keratea. Attikí, C Greece 37°48′N 23°58′E
Keratea see Keratéa
93 M19 **Kerava** Swe. Kervo. Uusimaa, S Finland 60°25′N 25°10′E
Kerbala/Kerbela see Karbalā'
32 F15 **Kerby** Oregon, NW USA 42°10′N 123°39′W
117 W12 **Kerch** Rus. Kerch'. Avtonomna Respublika Krym, SE Ukraine 45°22′N 36°30′E
Kerch' see Kerch
Kerchens'ka Protska/Kerchenskiy Proliv see Kerch Strait
117 V13 **Kerchens'kyy Pivostriv** peninsula S Ukraine
121 V4 **Kerch Strait** var. Bosporus Cimmerius, Enikale Strait, Rus. Kerchenskiy Proliv, Ukr. Kerchens'ka Protska. strait Black Sea/Sea of Azov
Kerdilio see Kerdýlio
114 H13 **Kerdýlio** var. Kerdilio. ▲ N Greece 40°46′N 23°37′E
186 D8 **Kerema** Gulf, S Papua New Guinea 07°59′S 145°46′E
Keremitlik see Lyulyakovo
136 I9 **Kerempe Burnu** headland N Turkey 42°01′N 33°20′E
80 J9 **Keren** var. Cheren. C Eritrea 15°45′N 38°22′E
25 U7 **Kerens** Texas, SW USA 32°07′N 96°13′W
184 M6 **Kerepehi** Waikato, North Island, New Zealand 37°18′S 175°33′E
145 P10 **Kerey, Ozero** ◎ C Kazakhstan
Kergel see Kärla
173 Q12 **Kerguelen** island C French Southern and Antarctic Territories
173 Q13 **Kerguelen Plateau** undersea feature S Indian Ocean
115 C20 **Kerí** Zákynthos, Iónia Nisiá, Greece, C Mediterranean Sea 37°40′N 20°48′E
81 H19 **Kericho** Kericho, W Kenya 0°22′S 35°19′E
81 H19 **Kericho** ⬖ county W Kenya
184 K2 **Kerikeri** Northland, North Island, New Zealand 35°14′S 173°58′E
93 O17 **Kerimäki** Etelä-Savo, E Finland 61°56′N 29°18′E
168 K12 **Kerinci, Gunung** ▲ Sumatera, W Indonesia 02°00′S 101°40′E
Keriya see Yutian
158 H9 **Keriya He** ≈ NW China
98 J9 **Kerkbuurt** Noord-Holland, C Netherlands 52°29′N 05°08′E
98 J13 **Kerkdriel** Gelderland, C Netherlands 51°46′N 05°21′E
75 N6 **Kerkenah, Îles de** var. Kerkenna Islands, Ar. Juzur Qarqannah. island group E Tunisia
Kerkenna Islands see Kerkenah, Îles de
115 M20 **Kerketévs** ▲ Sámos, Dodekánisa, Greece, Aegean Sea 37°44′N 26°39′E
29 T8 **Kerkhoven** Minnesota, N USA 45°12′N 95°18′W
Kerki see Atamyrat
Kerkichi see Kerkiçi
146 M14 **Kerkiçi** Rus. Kerkichi. Lebap Welaýaty, E Turkmenistan 37°46′N 65°18′E
115 F16 **Kerkíneo** prehistoric site Thessalía, C Greece
114 G12 **Kerkíni, Límni** var. Límni Kerkinitis. ◎ N Greece
Kerkinitis Límni see Kerkíni, Límni
Kérkira see Kérkyra
99 M18 **Kerkrade** Limburg, SE Netherlands 50°53′N 06°04′E
Kerkuk see Āltūn Kūbrī
115 B16 **Kérkyra** var. Kérkira, Eng. Corfu. Kérkyra, Iónia Nisiá, Greece, C Mediterranean Sea 39°37′N 19°56′E
115 B16 **Kérkyra** ✈ Kérkyra, Iónia Nisiá, Greece, C Mediterranean Sea 39°36′N 19°55′E
115 A16 **Kérkyra** var. Kérkira, Eng. Corfu. island Iónia Nisiá, Greece, C Mediterranean Sea
192 K10 **Kermadec Islands** island group New Zealand, SW Pacific Ocean
175 R10 **Kermadec Ridge** undersea feature SW Pacific Ocean 30°30′S 178°30′W
175 R11 **Kermadec Trench** undersea feature SW Pacific Ocean
143 S10 **Kermān** var. Kirman; anc. Carmana. Kermān, C Iran 30°18′N 57°05′E
143 R11 **Kermān** off. Ostān-e Kermān, var. Kirman; anc. Carmania. ⬖ province SE Iran
143 U12 **Kermān, Biābān-e** desert SE Iran
Kermān, Ostān-e see Kermān
142 K6 **Kermānshāh** var. Qahremānshahr; prev. Bākhtarān. Kermānshāhān, W Iran 34°19′N 47°04′E
143 Q9 **Kermānshāh** Yazd, C Iran 34°19′N 47°04′E
142 J6 **Kermānshāh** off. Ostān-e Kermānshāh; prev. Bākhtarān. ⬖ province W Iran
Kermānshāh, Ostān-e see Kermānshāh
114 L10 **Kermen** Sliven, C Bulgaria 42°30′N 26°12′E
24 L8 **Kermit** Texas, SW USA 31°49′N 103°07′W
21 P6 **Kermit** West Virginia, NE USA 37°51′N 82°24′W
21 S9 **Kernersville** North Carolina, SE USA 36°12′N 80°13′W
35 S12 **Kern River** ≈ California, W USA
35 S12 **Kernville** California, W USA 35°44′N 118°25′W
115 K21 **Kéros** island Kykládes, Greece, Aegean Sea
76 K14 **Kérouané** SE Guinea 09°16′N 09°00′W
101 D16 **Kerpen** Nordrhein-Westfalen, W Germany 50°51′N 06°40′E
146 I11 **Kerpichli** Lebap Welaýaty, NE Turkmenistan 40°12′N 61°09′E
24 M1 **Kerrick** Texas, SW USA 36°29′N 102°14′W
Kerr Lake see John H. Kerr Reservoir
11 S15 **Kerrobert** Saskatchewan, S Canada 51°56′N 109°09′W
25 Q11 **Kerrville** Texas, SW USA 30°03′N 99°09′W
97 B20 **Kerry** Ir. Ciarraí. cultural region SW Ireland
21 S11 **Kershaw** South Carolina, SE USA 34°33′N 80°34′W
Kertel see Kärdla
95 H23 **Kerteminde** Syddtjylland, C Denmark 55°27′N 10°40′E
163 Q7 **Kerulen** Chin. Herlen He, Mong. Herlen Gol. ≈ China/Mongolia
Kervo see Kerava
Kerýneia see Girne
12 H11 **Kesagami Lake** ◎ Ontario, SE Canada
93 O17 **Kesälahti** Pohjois-Karjala, SE Finland 61°54′N 29°49′E
136 B11 **Keşan** Edirne, NW Turkey 40°52′N 26°37′E
165 R9 **Kesennuma** Miyagi, Honshū, C Japan 38°55′N 141°35′E
163 V7 **Keshan** Heilongjiang, NE China 48°00′N 125°46′E
30 M6 **Keshena** Wisconsin, N USA 44°54′N 88°37′W
136 I13 **Keskin** Kırıkkale, C Turkey 39°41′N 33°36′E
93 K16 **Keski-Pohjanmaa** Swe. Mellersta Österbotten, Eng. centralostrobothnia. ⬖ region W Finland
93 M17 **Keski-Suomi** Swe. Mellersta Finland, Eng. Central Finland. ⬖ region C Finland
Késmárk see Kežmarok
124 I6 **Kesten'ga** var. Kest Enga. Respublika Kareliya, NW Russian Federation 65°53′N 31°47′E
Kest Enga see Kesten'ga
98 K12 **Kesteren** Gelderland, C Netherlands 51°55′N 05°34′E
14 H14 **Keswick** Ontario, S Canada 44°15′N 79°26′W
97 K15 **Keswick** NW England, United Kingdom 54°30′N 03°04′W
111 H24 **Keszthely** Zala, SW Hungary 46°47′N 17°16′E
122 K11 **Ket'** ≈ C Russian Federation
77 R17 **Keta** SE Ghana 05°55′N 00°59′E
169 Q12 **Ketapang** Borneo, C Indonesia 01°50′S 109°59′E
127 O12 **Ketchenery** prev. Sovetskoye. Respublika Kalmykiya, SW Russian Federation 47°18′N 44°31′E
39 Y14 **Ketchikan** Revillagigedo Island, Alaska, USA 55°21′N 131°39′W
33 O14 **Ketchum** Idaho, NW USA 43°40′N 114°24′W
Kete/Kete Krakye see Kete-Krachi
77 Q15 **Kete-Krachi** var. Kete, Kete Krakye. E Ghana 07°50′N 00°03′W
98 L9 **Ketelmeer** channel E Netherlands
149 P17 **Keti Bandar** Sind, SE Pakistan 23°55′N 67°31′E
77 S16 **Kétou** SE Benin 07°25′N 02°36′E
110 M7 **Kętrzyn** Ger. Rastenburg. Warmińsko-Mazurskie, NE Poland 54°05′N 21°24′E
97 N20 **Kettering** C England, United Kingdom 52°24′N 00°44′W
31 R14 **Kettering** Ohio, N USA 39°41′N 84°10′W
18 F13 **Kettle Creek** ≈ Pennsylvania, NE USA
32 L7 **Kettle Falls** Washington, NW USA 48°36′N 118°03′W
14 D16 **Kettle Point** headland Ontario, S Canada 43°12′N 82°01′W
29 V6 **Kettle River** ≈ Minnesota, N USA
186 B7 **Ketu** ≈ W Papua New Guinea
18 G10 **Keuka Lake** ◎ New York, NE USA
Keupriya see Primorsko
93 L17 **Keuruu** Keski-Suomi, C Finland 62°15′N 24°34′E
Kevevára see Kovin
92 L9 **Kevo** Lapp. Geavvú. Lappi, N Finland 69°42′N 27°08′E
44 M6 **Kew** North Caicos, N Turks and Caicos Islands 21°52′N 71°57′W
30 K11 **Kewanee** Illinois, N USA 41°15′N 89°55′W
31 N7 **Kewaunee** Wisconsin, N USA 44°27′N 87°31′W
30 M3 **Keweenaw Bay** ◎ Michigan, N USA
31 N2 **Keweenaw Peninsula** peninsula Michigan, N USA
31 N2 **Keweenaw Point** peninsula Michigan, N USA
29 N12 **Keya Paha River** ≈ Nebraska/South Dakota, N USA
Keyaygyr see Këk-Aygyr
23 Z16 **Key Biscayne** Florida, SE USA 25°41′N 80°09′W
26 G8 **Keyes** Oklahoma, C USA 36°48′N 102°15′W
23 Y17 **Key Largo** Key Largo, Florida, SE USA 25°06′N 80°25′W
21 U3 **Keyser** West Virginia, NE USA 39°25′N 78°59′W
27 O9 **Keystone Lake** ⊠ Oklahoma, C USA
36 L16 **Keystone Peak** ▲ Arizona, SW USA 31°52′N 111°12′W
Keystone State see Pennsylvania
21 U7 **Keysville** Virginia, NE USA 37°02′N 78°28′W
27 T3 **Keytesville** Missouri, C USA 39°25′N 92°56′W
23 W17 **Key West** Florida Keys, Florida, SE USA 24°34′N 81°48′W
127 T1 **Kez** Udmurtskaya Respublika, NW Russian Federation 57°55′N 53°42′E
Kezdivásárhely see Târgu Secuiesc
122 M12 **Kezhma** Krasnoyarskiy Kray, C Russian Federation 58°57′N 101°00′E
111 L18 **Kežmarok** Ger. Käsmark, Hung. Késmárk. Prešovský Kraj, E Slovakia 49°09′N 20°25′E
Kfar Saba see Kfar Sava
138 F10 **Kfar Sava** var. Kfar Saba; prev. Kefar Sava. Central, C Israel 32°11′N 34°58′E
83 F20 **Kgalagadi** ⬖ district SW Botswana
83 I20 **Kgatleng** ⬖ district SE Botswana
188 F8 **Kgeklau** Babeldaob, N Palau
125 R6 **Khabarikha** var. Chabaricha. Respublika Komi, NW Russian Federation 65°52′N 52°19′E
123 S14 **Khabarovsk** Khabarovskiy Kray, SE Russian Federation 48°32′N 135°08′E
123 R11 **Khabarovskiy Kray** ⬖ territory E Russian Federation
141 W7 **Khabb** Abū Ẓaby, E United Arab Emirates 24°39′N 55°43′E
Khabour, Nahr al see Khābūr, Nahr al
Khabura see Al Khābūrah
139 N2 **Khābūr, Nahr al** var. Nahr al Khabour. ≈ Syria/Turkey
Khachmas see Xaçmaz
80 B12 **Khadari** ≈ W Sudan
Khadera see Hadera
141 X12 **Khādhil** var. Khudal. SE Oman 18°48′N 56°48′E
155 E14 **Khadki** prev. Kirkee. Mahārāshtra, W India 18°34′N 73°52′E
126 L14 **Khadyzhensk** Krasnodarskiy Kray, SW Russian Federation 44°26′N 39°31′E
Khadzhiyska Reka see Hadzhiyska Reka
117 P10 **Khadzhybeys'kyy Lyman** ◎ SW Ukraine
138 K3 **Khafsah** Ḥalab, N Syria 36°16′N 38°03′E
152 M13 **Khāga** Uttar Pradesh, N India 25°47′N 81°05′E
153 Q13 **Khagaria** Bihār, NE India 25°31′N 86°27′E
149 Q13 **Khairpur** Sind, SE Pakistan 27°30′N 68°50′E
122 K13 **Khakasiya, Respublika** prev. Khakasskaya Avtonomnaya Oblast', Eng. Khakassia. ⬖ autonomous republic C Russian Federation
Khakassia/Khakasskaya Avtonomnaya Oblast' see Khakasiya, Respublika
167 N9 **Kha Khaeng, Khao** ▲ W Thailand 16°13′N 99°03′E
83 G20 **Khakhea** var. Kakia. Southern, S Botswana 24°41′S 23°29′E
Khalach see Halaç
Khalándrion see Chalándri
75 T7 **Khalīj as Sallūm** Ar. Gulf of Salûm. gulf Egypt/Libya
75 X8 **Khalīj as Suways** var. Suez, Gulf of. gulf NE Egypt
127 W7 **Khalilovo** Orenburgskaya Oblast', W Russian Federation 51°25′N 58°13′E
Khalkabad see Xalqobod
142 L3 **Khalkhāl** prev. Herowābād. Ardabīl, NW Iran 37°36′N 48°36′E
Khalkidhikí see Chalkidikí
Khalkís see Chalkída
125 W3 **Khal'mer-Yu** Respublika Komi, NW Russian Federation 68°00′N 64°45′E
119 M14 **Khalopyenichy** Rus. Kholopenichi. Minskaya Voblasts', NE Belarus 54°31′N 28°58′E
Khalturin see Orlov
141 Y10 **Khalūf** var. Al Khaluf. E Oman 20°27′N 57°59′E
154 K10 **Khamaria** Madhya Pradesh, C India 23°07′N 80°54′E
154 D11 **Khambhāt** Gujarāt, W India 22°19′N 72°39′E
154 C12 **Khambhāt, Gulf of** Eng. Gulf of Cambay. gulf W India
167 U10 **Khâm Đưc** var. Phước Sơn. Quang Nam-Đa Năng, C Vietnam 15°28′N 107°49′E
154 G12 **Khāmgaon** Mahārāshtra, C India 20°41′N 76°34′E
141 O14 **Khamir** var. Khamr. W Yemen 16°N 43°56′E
141 N12 **Khamīs Mushayṭ** var. Hamīs Musait. 'Asīr, SW Saudi Arabia 18°19′N 42°41′E
155 J16 **Khammam** Telangana, India 17°16′N 80°13′E
123 P10 **Khampa** Respublika Sakha (Yakutiya), NE Russian Federation 63°43′N 123°02′E
Khamr see Khamir
83 C19 **Khan** ≈ W Namibia
149 Q2 **Khānābād** Kunduz, NE Afghanistan 36°42′N 69°08′E
Khân Abou Chamâte/Khan Abou Ech Cham see Khān Abū Shāmāt
138 I7 **Khān Abū Shāmāt** var. Khân Abou Chamâte, Khan Abou Ech Cham. Rīf Dimashq, W Syria 33°43′N 36°56′E
Khān al Baghdādī see Al Baghdādī
Khān al Maḥāwīl see Al Maḥāwīl
139 T7 **Khān al Mashāhidah** Baghdād, C Iraq 33°40′N 44°15′E
139 T10 **Khān al Muṣallá** An Najaf, S Iraq 32°09′N 44°20′E
139 U6 **Khānaqīn** Diyālá, E Iraq 34°22′N 45°22′E
139 T11 **Khān ar Ruḥbah** An Najaf, S Iraq 31°42′N 44°18′E
139 P2 **Khān as Sūr** Nīnawá, N Iraq 36°28′N 41°36′E
139 T8 **Khān Āzād** Baghdād, C Iraq 33°08′N 44°21′E
154 N13 **Khandaparha** prev. Khandpara. Odisha, E India 20°15′N 85°11′E
Khandpara see Khandaparha
149 T2 **Khandūd** var. Khandud, Wakhan. Badakhshān, NE Afghanistan 36°57′N 72°19′E
Khandud see Khandūd
154 G11 **Khandwa** Madhya Pradesh, C India 21°49′N 76°23′E
123 R10 **Khandyga** Respublika Sakha (Yakutiya), NE Russian Federation 62°39′N 135°30′E
149 T10 **Khānewāl** Punjab, NE Pakistan 30°18′N 71°56′E
149 S10 **Khāngarh** Punjab, E Pakistan 29°57′N 71°14′E
Khanh Hung see Soc Trăng
Khaniá see Chaniá
Khanka see Xonqa
163 Z8 **Khanka, Lake** var. Hsing-K'ai Hu, Lake Hanka, Chin. Xingkai Hu, Rus. Ozero Khanka. ◎ China/Russian Federation
Khanka, Ozero see Khanka, Lake
Khankendi see Xankändi
123 O9 **Khannya** ≈ NE Russian Federation
144 D10 **Khan Ordasy** prev. Urda. Zapadnyy Kazakhstan, W Kazakhstan 48°52′N 47°31′E
149 S12 **Khānpur** Punjab, E Pakistan 28°31′N 70°30′E
138 I4 **Khān Shaykhūn** var. Khan Sheikhun. Idlib, NW Syria 35°27′N 36°38′E
Khan Sheikhun see Khān Shaykhūn
Khanshynghys see Khrebet Khanshyngys
145 S15 **Khantau** Zhambyl, S Kazakhstan 44°13′N 73°47′E
145 W16 **Khan Tengri, Pik** ▲ SE Kazakhstan 42°17′N 80°11′E
Khan-Tengri, Pik see Hantengri Feng
Khanthabouli see Savannakhét
127 V8 **Khanty-Mansiysk** prev. Ostyako-Voguls'k. Khanty-Mansiyskiy Avtonomnyy Okrug-Yugra, C Russian Federation 61°01′N 69°E
125 V8 **Khanty-Mansiyskiy Avtonomnyy Okrug-Yugra** ⬖ autonomous district C Russian Federation
139 R4 **Khānūqah** Nīnawýé, C Iraq 35°25′N 43°15′E
138 E11 **Khān Yūnis** var. Khān Yūnus. S Gaza Strip 31°21′N 34°18′E
Khān Yūnus see Khān Yūnis
Khanzi see Ghanzi
Khān Zūr see Xan Sûr
167 N10 **Khao Laem Reservoir** ⊠ W Thailand
123 O14 **Khapcheranga** Zabaykal'skiy Kray, S Russian Federation 49°46′N 112°21′E
127 Q12 **Kharabali** Astrakhanskaya Oblast', SW Russian Federation 47°28′N 47°14′E
153 R16 **Kharagpur** West Bengal, NE India 22°30′N 87°19′E
139 V11 **Kharā'ib 'Abd al Karīm** Al Muthannā, S Iraq 31°07′N 45°33′E
143 Q8 **Kharānaq** Yazd, C Iran 31°54′N 54°21′E
Kharbin see Harbin
Kharchi see Mārwār
146 H13 **Khardzhagaz** Ahal Welaýaty, C Turkmenistan 37°54′N 60°10′E
154 F11 **Khargon** Madhya Pradesh, C India 21°49′N 75°39′E
149 V7 **Khāriān** Punjab, NE Pakistan 32°52′N 73°52′E
117 V5 **Kharkiv** Rus. Khar'kov. Kharkivs'ka Oblast', NE Ukraine 50°N 36°14′E
117 V5 **Kharkiv** ✈ Kharkivs'ka Oblast', E Ukraine 49°54′N 36°20′E
Kharkiv see Kharkivs'ka Oblast'
117 U5 **Kharkivs'ka Oblast'** var. Kharkiv, Rus. Khar'kovskaya Oblast'. ⬖ province E Ukraine
Khar'kov see Kharkiv
Khar'kovskaya Oblast' see Kharkivs'ka Oblast'
124 L3 **Kharlovka** Murmanskaya Oblast', NW Russian Federation 68°47′N 37°09′E
Kharmanli see Harmanli
Kharmanliyska Reka see Harmanliyska Reka
124 M13 **Kharovsk** Vologodskaya Oblast', NW Russian Federation 59°57′N 40°05′E
80 F9 **Khartoum** var. El Khartûm, Khartum. ● (Sudan) Khartoum, C Sudan 15°33′N 32°32′E
80 F9 **Khartoum** ⬖ state NE Sudan
80 F9 **Khartoum** ✈ Khartoum, C Sudan 15°36′N 32°37′E
80 F9 **Khartoum North** Khartoum, C Sudan 15°38′N 32°33′E
117 X8 **Khartsyz'k** Rus. Khartsyzsk. Donets'ka Oblast', SE Ukraine 48°01′N 38°10′E
117 X8 **Khartsyz'k** Donets'ka Oblast', E Ukraine 48°01′N 38°10′E
Khartsyzsk see Khartsyz'k
Khartum see Khartoum
Kharwazawk see Xêrzok
Khasab see Al Khaṣab
123 S15 **Khasan** Primorskiy Kray, SE Russian Federation 42°24′N 130°45′E
127 P16 **Khasavyurt** Respublika Dagestan, SW Russian Federation 43°16′N 46°33′E
143 W12 **Khāsh** prev. Vāsht. Sīstān va Balūchestān, SE Iran 28°15′N 61°11′E
148 K8 **Khāsh, Dasht-e** Eng. Khash Desert. desert SW Afghanistan
Khash Desert see Khāsh, Dasht-e
Khashim Al Qirba/Khashm al Qirbah see Khashm el Girba
80 H9 **Khashm el Girba** var. Khashim Al Qirba, Khashm al Qirbah. Kassala, E Sudan 15°00′N 35°59′E
138 G14 **Khashsh, Jabal al** ▲ S Jordan
137 S10 **Khashuri** C Georgia 41°59′N 43°36′E
153 V13 **Khāsi Hills** hill range NE India
Khaskovo see Haskovo
Khaskovo see Haskovo
122 M7 **Khatanga** ≈ N Russian Federation
Khatanga, Gulf of see Khatangskiy Zaliv
123 N7 **Khatangskiy Zaliv** var. Gulf of Khatanga. bay N Russian Federation
141 W7 **Khatmat al Malāḥah** N Oman 24°58′N 56°22′E
143 S16 **Khatmat al Malāḥah** Ash Shāriqah, E United Arab Emirates
123 V7 **Khatyrka** Chukotskiy Avtonomnyy Okrug, NE Russian Federation 62°03′N 175°09′E
Khauz-Khan see Hanhowuz
Khauzkhanskoye Vodoranilishche see Hanhowuz Suw Howdany
Khavaling see Khovaling
Khavast see Xovos
139 W10 **Khawrah, Nahr al** ≈ S Iraq
Khawr Barakah see Barka
141 W7 **Khawr Fakkān** var. Khor Fakkan. Ash Shāriqah, NE United Arab Emirates 25°22′N 56°19′E
140 L6 **Khaybar** Al Madīnah, NW Saudi Arabia 25°53′N 39°16′E
Khaybar, Kowtal-e see Khyber Pass
147 S11 **Khaydarkan** var. Khaydarken. Batkenskaya Oblast', SW Kyrgyzstan 39°56′N 71°17′E
Khaydarken see Khaydarkan
125 U2 **Khaypudyrskaya Guba** bay NW Russian Federation
Khayrūzuk see Xêrzok
Khazar, Baḥr-e/Khazar, Daryā-ye see Caspian Sea
Khazarosp see Hazorasp
Khazretishi, Khrebet see Hazratishoh, Qatorkŭhi
75 X11 **Khazzān Aswān** var. Aswan Dam. dam SE Egypt
Khelat see Kālat
74 F6 **Khemisset** NW Morocco 33°52′N 06°04′W
167 R10 **Khemmarat** var. Kemarat. Ubon Ratchathani, E Thailand 16°03′N 105°11′E
74 L6 **Khenchela** var. Khenchla. NE Algeria 35°22′N 07°09′E
Khenchla see Khenchela
74 G7 **Khénifra** C Morocco 32°59′N 05°37′W
Khersān, Rūd-e see Garm, Āb-e
117 R10 **Kherson** Khersons'ka Oblast', S Ukraine 46°39′N 32°38′E
Kherson see Khersons'ka Oblast'
117 S14 **Khersones, Mys** Rus. Mys Khersonesskiy. headland S Ukraine 44°34′N 33°24′E
Khersonesskiy, Mys see Khersones, Mys
117 R10 **Khersons'ka Oblast'** var. Kherson, Rus. Khersonskaya Oblast'. ⬖ province S Ukraine
Khersonskaya Oblast' see Khersons'ka Oblast'
122 L8 **Kheta** ≈ N Russian Federation
149 U7 **Khewra** Punjab, E Pakistan 32°41′N 73°04′E
Khiam see El Khiyam
124 J4 **Khibiny** ▲▲ NW Russian Federation
126 K3 **Khimki** Moskovskaya Oblast', W Russian Federation 55°57′N 37°48′E
147 S12 **Khingov** Rus. Obi-Khingou. ≈ C Tajikistan
Khíos see Chíos
149 R15 **Khipro** Sind, SE Pakistan 25°50′N 69°24′E
139 S10 **Khirr, Wādī al** dry watercourse S Iraq
Khisarya see Hisarya
Khiva/Khiwa see Xiva
167 N9 **Khlong Khlung** Kamphaeng Phet, W Thailand 16°15′N 99°41′E
167 N15 **Khlong Thom** Krabi, SW Thailand 07°55′N 99°09′E
167 P12 **Khlung** Chantaburi, S Thailand 12°25′N 102°12′E
Khmel'nik see Khmil'nyk
Khmel'nitskaya Oblast' see Khmel'nyts'ka Oblast'
Khmel'nitskiy see Khmel'nyts'kyy
116 K5 **Khmel'nyts'ka Oblast'** var. Khmel'nyts'kyy, Rus. Khmel'nitskaya Oblast'; prev. Kamenets-Podol'skaya Oblast'. ⬖ province NW Ukraine
116 L6 **Khmel'nyts'kyy** Rus. Khmel'nitskiy; prev. Proskurov. Khmel'nyts'ka Oblast', W Ukraine 49°24′N 26°59′E
Khmel'nyts'kyy see Khmel'nyts'ka Oblast'
116 M6 **Khmil'nyk** Rus. Khmel'nik. Vinnyts'ka Oblast', C Ukraine 49°36′N 27°59′E
Khobda see Kobda
137 R9 **Khobi** W Georgia 42°20′N 41°54′E
119 P15 **Khodasy** Rus. Khodosy. Mahilyowskaya Voblasts', E Belarus 53°56′N 31°29′E
116 I6 **Khodoriv** Pol. Chodorów, Rus. Khodorov. L'vivs'ka Oblast', NW Ukraine 49°20′N 24°19′E
Khodorov see Khodoriv
Khodosy see Khodasy
Khodzhakala see Hojagala
Khodzhambas see Hojambaz
Khodzhent see Khujand
Khodzheyli see Xo'jayli
Khoi see Khvoy
Khojend see Khujand
Khokand see Qo'qon
126 L8 **Khokhol'skiy** Voronezhskaya Oblast', W Russian Federation 51°33′N 38°43′E
167 P10 **Khok Samrong** Lop Buri, C Thailand 15°03′N 100°44′E
124 H15 **Kholm** Novgorodskaya Oblast', W Russian Federation 57°10′N 31°06′E
Kholm see Khulm
Kholm see Chełm
Kholmech' see Kholmyech
123 T13 **Kholmsk** Ostrov Sakhalin, Sakhalinskaya Oblast', SE Russian Federation 46°57′N 142°10′E
119 O19 **Kholmyech** Rus. Kholmech'. Homyel'skaya Voblasts', SE Belarus 52°09′N 30°37′E
Kholon see Holon
Kholopenichi see Khalopyenichy
83 D19 **Khomas** ⬖ district C Namibia
83 D19 **Khomas Hochland** var. Khomasplato. plateau C Namibia
Khomasplato see Khomas Hochland
Khomein see Khomeyn
142 M7 **Khomeyn** var. Khomein, Khumain. Markazī, W Iran 33°38′N 50°03′E
143 N8 **Khomeynīshahr** prev. Homāyūnshahr. Eşfahān, C Iran 32°42′N 51°28′E
Khoms see Al Khums
Khong Sedone see Muang Khôngxédôn
167 Q9 **Khon Kaen** var. Muang Khon Kaen. Khon Kaen, E Thailand 16°25′N 102°50′E
Khonqa see Xonqa
167 Q9 **Khon San** Khon Kaen, E Thailand 16°40′N 101°51′E
123 R8 **Khonuu** Respublika Sakha (Yakutiya), NE Russian Federation 66°24′N 143°15′E
127 N8 **Khopër** var. Khoper. ≈ SW Russian Federation
Khoper see Khopër
123 S14 **Khor** Khabarovskiy Kray, SE Russian Federation 47°44′N 134°48′E
143 U9 **Khorāsān-e Jonūbī** off. Ostan-e Khorāsān-e Jonūbī. ⬖ province E Iran
143 U6 **Khorāsān-e Raẕavī** off. Ostān-e Khorāsān-e Raẕavī, var. Khorassan, Khurasan. ⬖ province NE Iran
Khorāsān-e Raẕavī, Ostān-e see Khorāsān-e Raẕavī
143 S3 **Khorāsān-e Shomālī** off. Ostan-e Khorāsān-e Shomālī. ⬖ province NE Iran
Khorassan see Khorāsān-e Raẕavī
Khorat see Nakhon Ratchasima
154 O13 **Khordha** prev. Khurda. Odisha, E India 20°10′N 85°42′E
125 U4 **Khorey-Ver** Nenetskiy Avtonomnyy Okrug, NW Russian Federation 67°25′N 58°05′E
Khorezmskaya Oblast' see Xorazm Viloyati
Khor Fakkan see Khawr Fakkān
Khorgos see Korgas
123 N13 **Khorinsk** Respublika Buryatiya, S Russian Federation 52°13′N 109°52′E
83 C18 **Khorixas** Kunene, NW Namibia 20°23′S 14°55′E
141 O17 **Khormaksar** var. Aden. ✈ ('Adan) SW Yemen 12°56′N 45°00′E
Khormal see Xurmał
Khormuj see Khvormūj
Khorog see Khorugh
117 S5 **Khorol** Poltavs'ka Oblast', NE Ukraine 49°49′N 33°17′E
142 L7 **Khorramābād** var. Khurramabad. Lorestān, W Iran 33°29′N 48°21′E
143 R9 **Khorramdasht** Kermān, C Iran 31°41′N 56°10′E
142 K10 **Khorramshahr** var. Khurramshahr, Muhammerah; prev. Mohammerah. Khūzestān, SW Iran 30°30′N 48°09′E
147 S14 **Khorugh** Rus. Khorog. S Tajikistan 37°30′N 71°31′E
Khorvot Khalutsa see Horvot Halutsa
127 Q12 **Khosheutovo** Astrakhanskaya Oblast', SW Russian Federation 47°04′N 47°49′E
149 R6 **Khōst** prev. Khowst. Khōst, E Afghanistan 33°22′N 69°57′E
149 S6 **Khōst** prev. Khowst. ⬖ province E Afghanistan
Khotan see Hotan
Khotimsk see Khotsimsk
Khotin see Khotyn
119 R16 **Khotsimsk** Rus. Khotimsk. Mahilyowskaya Voblasts', E Belarus 53°24′N 32°35′E
116 K7 **Khotyn** Rom. Hotin, Rus. Khotin. Chernivets'ka Oblast', W Ukraine 48°29′N 26°30′E
74 F7 **Khouribga** C Morocco 32°55′N 06°51′W
147 Q13 **Khovaling** Rus. Khavaling. SW Tajikistan 38°22′N 69°54′E
Khovd see Hovd
Khowst see Khōst
Khoy see Khvoy
119 N20 **Khoyniki** Homyel'skaya Voblasts', SE Belarus 51°54′N 29°58′E
Khozretishi, Khrebet see Hazratishoh, Qatorkŭhi
145 V11 **Khrebet Khanshyngys** Kaz. Khanshyngghys; prev. Khrebet Kanchingiz. ▲▲ E Kazakhstan
145 X10 **Khrebet Kalba** Kaz. Qalba Zhotasy; prev. Kalbinskiy Khrebet. ▲▲ E Kazakhstan
Khrebet Kanchingiz see Khrebet Khanshyngys
Khrebet Ketmen see Khrebet Uzynkara
145 Y10 **Khrebet Naryn** Kaz. Naryn Zhotasy; prev. Narymskiy Khrebet. ▲▲ E Kazakhstan
145 W16 **Khrebet Uzynkara** prev. Khrebet Ketmen. ▲▲ SE Kazakhstan
Khrisoúpolis see Chrysoúpoli
144 J10 **Khromtau** Kaz. Khromtaū. Aktyubinsk, W Kazakhstan 50°14′N 58°22′E
Khromtaū see Khromtau
Khrysokhou Bay see Chrysochoú, Kólpos
117 O7 **Khrystynivka** Cherkas'ka Oblast', C Ukraine 48°49′N 29°55′E
167 R10 **Khuang Nai** Ubon Ratchathani, E Thailand 15°22′N 104°33′E
Khudal see Khādhil
Khudat see Xudat
149 W9 **Khudiān** Punjab, E Pakistan 30°59′N 74°19′E
Khudzhand see Khujand
83 G21 **Khuis** Kgalagadi, SW Botswana 26°37′S 21°50′E
147 Q11 **Khujand** var. Khodzhent, Khojend, Rus. Khudzhand; prev. Leninabad, Taj. Leninobod. N Tajikistan 40°17′N 69°37′E
167 R11 **Khukhan** Si Sa Ket, E Thailand 14°38′N 104°12′E
149 P2 **Khulm** var. Tashqurghan; prev. Kholm. Balkh, N Afghanistan 36°42′N 67°41′E
153 T16 **Khulna** Khulna, SW Bangladesh 22°48′N 89°32′E
153 T16 **Khulna** ⬖ division SW Bangladesh
Khumain see Khomeyn
Khums see Al Khums
149 W2 **Khunjerāb Pass** pass China/Pakistan
Khūnjerāb Pass see Kunjirap Daban
153 P16 **Khunti** Jhārkhand, N India 23°02′N 85°19′E
167 N7 **Khun Yuam** Mae Hong Son, NW Thailand 18°54′N 97°54′E
Khurais see Khurayş
Khurasan see Khorāsān-e Raẕavī
141 R7 **Khurayş** var. Khurais. Ash Sharqīyah, C Saudi Arabia 25°06′N 48°03′E
Khurda see Khordha
152 J11 **Khurja** Uttar Pradesh, N India 28°15′N 77°51′E
Khūrmāl see Xurmał
Khurramabad see Khorramābād
Khurramshahr see Khorramshahr
149 U7 **Khushāb** Punjab, NE Pakistan 32°16′N 72°18′E
116 H8 **Khust** var. Husté, Cz. Chust, Hung. Huszt. Zakarpats'ka Oblast', W Ukraine 48°11′N 23°19′E
80 D11 **Khuwei** Western Kordofan, C Sudan 13°02′N 29°13′E
149 O13 **Khuzdār** Baluchistān, SW Pakistan 27°49′N 66°39′E
142 L9 **Khūzestān** off. Ostān-e Khūzestān, var. Khuzistan, prev. Arabistan; anc. Susiana. ⬖ province SW Iran
Khūzestān, Ostān-e see Khūzestān
Khuzistan see Khūzestān
Khvājeh Ghār see Khwājeh Ghār
127 Q7 **Khvalynsk** Saratovskaya Oblast', W Russian Federation 52°30′N 48°06′E
143 N12 **Khvormūj** var. Khormuj. Būshehr, S Iran 28°32′N 51°22′E
142 I2 **Khvoy** var. Khoi, Khoy. Āzarbāyjān-e Bākhtarī, NW Iran 38°36′N 45°04′E
Khwajaghar/Khwaja-i-Ghar see Khwājeh Ghār
149 R2 **Khwājeh Ghār** var. Khwajaghar, Khwaja-i-Ghar; prev. Khvājeh Ghār. Takhār, NE Afghanistan 37°08′N 69°24′E
149 U4 **Khyber Pakhtunkhwa** prev. North-West Frontier Province. ⬖ province NW Pakistan
149 S5 **Khyber Pass** var. Kowtal-e Khaybar. pass Afghanistan/Pakistan
186 L8 **Kia** Santa Isabel, N Solomon Islands 07°34′S 158°31′E
183 S10 **Kiama** New South Wales, SE Australia 34°41′S 150°49′E
79 O22 **Kiambi** Katanga, SE Dem. Rep. Congo 07°15′S 28°01′E
81 I19 **Kiambu** ⬖ county C Kenya
27 Q12 **Kiamichi Mountains** ▲▲ Oklahoma, C USA
27 Q12 **Kiamichi River** ≈ Oklahoma, C USA
14 M10 **Kiamika, Réservoir** ⊠ Québec, SE Canada
Kiamusze see Jiamusi
39 N7 **Kiana** Alaska, USA 66°58′N 160°25′W
Kiangmai see Chiang Mai
Kiang-ning see Nanjing
Kiangsi see Jiangxi
Kiangsu see Jiangsu
93 M14 **Kiantajärvi** ◎ E Finland
115 F19 **Kiáto** prev. Kiáton. Pelopónnisos, S Greece 38°01′N 22°45′E
Kiáton see Kiáto
Kiayi see Jiayi
95 F22 **Kibæk** Midtjylland, W Denmark 56°03′N 08°52′E
67 T9 **Kibali** var. Uele (upper course). ≈ NE Dem. Rep. Congo
79 E20 **Kibangou** Niari, SW Congo 03°27′S 12°21′E
Kibarty see Kybartai
92 M8 **Kiberg** Finnmark, N Norway 70°17′N 30°47′E
79 N20 **Kibombo** Maniema, E Dem. Rep. Congo 03°52′S 25°59′E
81 E20 **Kibondo** Kigoma, NW Tanzania 03°34′S 30°41′E
81 J15 **Kibre Mengist** var. Adola. Oromīya, C Ethiopia 05°50′N 39°06′E
Kıbrıs see Cyprus
Kıbrıs/Kıbrıs Cumhuriyeti see Cyprus
81 E20 **Kibungo** var. Kibungu. SE Rwanda 02°09′S 30°30′E
Kibungu see Kibungo
113 N19 **Kičevo** SW FYR Macedonia 41°31′N 20°57′E
125 P13 **Kichmengskiy Gorodok** Vologodskaya Oblast', NW Russian Federation 60°00′N 45°52′E
30 J8 **Kickapoo River** ≈ Wisconsin, N USA

◆ Country ● Country Capital ◇ Dependent Territory ○ Dependent Territory Capital ⬖ Administrative Regions ✈ International Airport ▲ Mountain ▲▲ Mountain Range Volcano ≈ River ◎ Lake ⊠ Reservoir

Kicking Horse Pass – Kizlyar

11 P16 Kicking Horse Pass pass Alberta/British Columbia, SW Canada
77 R9 Kidal Kidal, C Mali 18°22′N 01°21′E
77 Q8 Kidal ◈ region NE Mali
171 Q7 Kidapawan Mindanao, S Philippines 07°02′N 125°04′E
97 L20 Kidderminster C England, United Kingdom 52°23′N 02°14′W
76 I11 Kidira E Senegal 14°28′N 12°13′W
184 O11 Kidnappers, Cape headland North Island, New Zealand 41°13′S 175°15′E
100 J8 Kiel Schleswig-Holstein, N Germany 54°21′N 10°05′E
111 L15 Kielce Rus. Keltsy. Świętokrzyskie, C Poland 50°53′N 20°39′E
100 K7 Kieler Bucht bay N Germany
100 J7 Kieler Förde inlet N Germany
167 U13 Kiên Đưc var. Đak Lap. Đăc Lăc, S Vietnam 11°59′N 107°30′E
79 N24 Kienge Katanga, SE Dem. Rep. Congo 10°33′S 27°33′E
100 Q12 Kietz Brandenburg, NE Germany 52°33′N 14°36′E
Kiev see Kyyiv
Kiev Reservoir see Kyyivs'ke Vodoskhovyshche
76 J10 Kiffa Assaba, S Mauritania 16°38′N 11°23′W
115 H19 Kifisiá Attikí, C Greece 38°04′N 23°49′E
115 F18 Kifisós ≈ C Greece
139 U5 Kifrī At Ta'mím, N Iraq 34°44′N 44°58′E
81 D20 Kigali ● (Rwanda) C Rwanda 01°59′S 30°02′E
81 E20 Kigali ✈ C Rwanda 01°43′S 30°01′E
137 P13 Kiğı Bingöl, E Turkey 39°19′N 40°20′E
81 E21 Kigoma Kigoma, W Tanzania 04°52′S 29°36′E
81 E21 Kigoma ◈ region W Tanzania
38 F10 Kihei var. Kihei. Maui, Hawaii, USA, C Pacific Ocean 20°47′N 156°28′W
93 K17 Kihniö Pirkanmaa, W Finland 62°11′N 23°10′E
118 F6 Kihnu var. Kihnu Saar, Ger. Kühnö. island SW Estonia
Kihnu Saar see Kihnu
38 A8 Kii Landing Ni'ihau, Hawaii, USA, C Pacific Ocean 21°58′N 160°03′W
93 L14 Kiiminki Pohjois-Pohjanmaa, C Finland 65°05′N 25°47′E
164 J14 Kii-Nagashima var. Nagashima. Mie, Honshū, SW Japan 34°10′N 136°18′E
164 J14 Kii-sanchi ▲▲ Honshū, SW Japan
92 L11 Kiistala Lappi, N Finland 67°52′N 25°19′E
164 I15 Kii-suidō strait S Japan
165 V16 Kikai-shima island Nansei-shotō, SW Japan
112 M8 Kikinda Ger. Grosskikinda, Hung. Nagykikinda; prev. Velika Kikinda. Vojvodina, N Serbia 45°48′N 20°29′E
Kikládhes see Kykládes
165 Q5 Kikonai Hokkaidō, NE Japan 41°40′N 140°25′E
186 C8 Kikori Gulf, S Papua New Guinea 07°25′S 144°13′E
186 C8 Kikori ≈ W Papua New Guinea
165 O14 Kikuchi var. Kikuti. Kumamoto, Kyūshū, SW Japan 33°00′N 130°49′E
Kikuti see Kikuchi
127 N8 Kikvidze Volgogradskaya Oblast', SW Russian Federation 50°47′N 42°58′E
14 I10 Kikwissi, Lac ◎ Québec, SE Canada
79 I21 Kikwit Bandundu, W Dem. Rep. Congo 05°S 18°53′E
95 K15 Kil Värmland, C Sweden 59°30′N 13°20′E
94 N12 Kilafors Gävleborg, C Sweden 61°13′N 16°34′E
38 B8 Kilauea Kaua'i, Hawaii, USA, C Pacific Ocean 22°12′N 159°24′W
38 H12 Kilauea Caldera var. Kilauea Caldera. crater Hawai'i, USA, C Pacific Ocean
Kilauea Caldera see Kilauea Caldera
109 V4 Kilb Niederösterreich, C Austria 48°06′N 15°21′E
39 O12 Kilbuck Mountains ▲▲ Alaska, USA
163 Y12 Kilchu NE North Korea 40°58′N 129°22′E
97 F18 Kilcock Ir. Cill Choca. Kildare, E Ireland 53°25′N 06°40′W
183 V2 Kilcoy Queensland, E Australia 26°58′S 152°30′E
97 F18 Kildare Ir. Cill Dara. E Ireland 53°10′N 06°55′W
97 F18 Kildare Ir. Cill Dara. cultural region E Ireland
124 K2 Kil'din, Ostrov island NW Russian Federation
25 W7 Kilgore Texas, SW USA 32°23′N 94°52′W
Kilien Mountains see Qilian Shan
114 K9 Kilifarevo Veliko Tŭrnovo, N Bulgaria 43°00′N 25°36′E
81 K20 Kilifi Kilifi, SE Kenya 03°37′S 39°50′E
81 J21 Kilifi ◈ county SE Kenya
189 U9 Kili Island var. Kōle. island Ralik Chain, S Marshall Islands
149 V2 Kilik Pass pass Afghanistan/China
Kilimane see Quelimane
81 I21 Kilimanjaro ◈ region E Tanzania
81 I20 Kilimanjaro var. Uhuru Peak. ⛰ NE Tanzania 03°01′S 37°14′E
Kilimbangara see Kolombangara
Kilinailau Islands see Tulun Islands
81 K23 Kilindoni Pwani, E Tanzania 07°56′S 39°40′E
118 H6 Kilingi-Nõmme Ger. Kurkund. Pärnumaa, SW Estonia 58°07′N 24°00′E
136 M17 Kilis Kilis, S Turkey 36°43′N 37°07′E
136 M16 Kilis ◈ province S Turkey
117 N12 Kiliya Rom. Chilia-Nouă. Odes'ka Oblast', SW Ukraine 45°30′N 29°16′E
97 B19 Kilkee Ir. Cill Chaoi. Clare, W Ireland 52°41′N 09°38′W
97 E19 Kilkenny Ir. Cill Chainnigh. Kilkenny, S Ireland 52°39′N 07°15′W
97 E19 Kilkenny Ir. Cill Chainnigh. cultural region S Ireland
97 B18 Kilkieran Bay Ir. Cuan Chill Chiaráin. bay W Ireland
114 G13 Kilkís Kentrikí Makedonía, N Greece 40°59′N 22°55′E
97 C15 Killala Bay Ir. Cuan Chill Ala. inlet NW Ireland
11 R15 Killam Alberta, SW Canada 52°45′N 111°46′W
183 U3 Killarney Queensland, E Australia 28°18′S 152°15′E
11 W17 Killarney Manitoba, S Canada 49°12′N 99°40′W
14 E11 Killarney Ontario, S Canada 45°58′N 81°27′W
97 B20 Killarney Ir. Cill Airne. Kerry, SW Ireland 52°03′N 09°30′W
28 K4 Killdeer North Dakota, N USA 47°21′N 102°45′W
28 J4 Killdeer Mountains ▲▲ North Dakota, N USA
45 V15 Killdeer River ≈ Trinidad, Trinidad and Tobago
25 S9 Killeen Texas, SW USA 31°07′N 97°44′W
39 P6 Killik River ≈ Alaska, USA
11 T7 Killinek Island island Nunavut, NE Canada
Killini see Kyllíni
115 C19 Killínis, Akrotírio headland S Greece 37°55′N 21°07′E
97 D15 Killybegs Ir. Na Cealla Beaga. NW Ireland 54°38′N 08°27′W
Kilmain see Quelimane
96 I13 Kilmarnock W Scotland, United Kingdom 55°37′N 04°30′W
21 X6 Kilmarnock Virginia, NE USA 37°42′N 76°22′W
125 S16 Kil'mez' Kirovskaya Oblast', NW Russian Federation 56°55′N 51°03′E
127 S2 Kil'mez' Udmurtskaya Respublika, NW Russian Federation 57°04′N 51°22′E
125 R16 Kil'mez' ≈ NW Russian Federation
67 V11 Kilombero ≈ S Tanzania
92 J10 Kilpisjärvi Lappi, N Finland 69°03′N 20°50′E
97 B19 Kilrush Ir. Cill Rois. Clare, W Ireland 52°39′N 09°29′W
79 O24 Kilwa Katanga, SE Dem. Rep. Congo 09°22′S 28°19′E
Kilwa see Kilwa Kivinje
81 J24 Kilwa Kivinje var. Kilwa. Lindi, SE Tanzania 08°45′S 39°21′E
81 J24 Kilwa Masoko Lindi, SE Tanzania 08°55′S 39°31′E
171 T13 Kilwo Pulau Seram, E Indonesia 03°36′S 130°48′E
114 P12 Kilyos Istanbul, NW Turkey 41°15′N 29°01′E
37 V8 Kim Colorado, C USA 37°12′N 103°22′W
145 O9 Kima prev. Kiyma. Akmola, C Kazakhstan 51°37′N 67°31′E
169 U7 Kimanis, Teluk bay Sabah, East Malaysia
182 H8 Kimba South Australia 33°09′S 136°26′E
28 I15 Kimball Nebraska, C USA 41°16′N 103°40′W
29 O11 Kimball South Dakota, N USA 43°45′N 98°57′W
79 I21 Kimbao Bandundu, SW Dem. Rep. Congo 05°27′S 17°40′E
186 F7 Kimbe New Britain, E Papua New Guinea 05°36′S 150°10′E
186 G7 Kimbe Bay inlet New Britain, E Papua New Guinea
11 P17 Kimberley British Columbia, SW Canada 49°40′N 115°58′W
83 H23 Kimberley Northern Cape, C South Africa 28°45′S 24°46′E
180 M4 Kimberley Plateau plateau Western Australia
33 P15 Kimberly Idaho, NW USA 42°31′N 114°21′W
163 Y12 Kimch'aek prev. Sŏngjin. E North Korea 40°42′N 129°13′E
Kimch'ŏn see Gimcheon
Kim Hae see Gimhae
Kimi see Kými
93 K20 Kimito Swe. Kemiö. Varsinais-Suomi, SW Finland 60°10′N 22°45′E
9 R7 Kimmirut prev. Lake Harbour. Baffin Island, Nunavut, NE Canada 62°48′N 69°49′W
165 R4 Kimobetsu Hokkaidō, NE Japan 42°47′N 140°55′E
115 I21 Kímolos island Kykládes, Greece, Aegean Sea
115 I21 Kímolou Sífnou, Stenó strait Kykládes, Greece, Aegean Sea
126 L5 Kimovsk Tul'skaya Oblast', W Russian Federation 53°59′N 38°34′E
Kimpolung see Câmpulung Moldovenesc
124 K16 Kimry Tverskaya Oblast', W Russian Federation 56°52′N 37°21′E
79 H21 Kimvula Bas-Congo, SW Dem. Rep. Congo 05°44′S 15°58′E
169 U6 Kinabalu, Gunung ▲ East Malaysia 05°52′N 116°08′E
Kinabatangan see Kinabatangan, Sungai
169 V7 Kinabatangan, Sungai var. Kinabatangan. ≈ East Malaysia
115 L21 Kínaros island Kykládes, Greece, Aegean Sea
11 O15 Kinbasket Lake ▣ British Columbia, SW Canada
96 I7 Kinbrace N Scotland, United Kingdom 58°16′N 02°59′W
14 E14 Kincardine Ontario, S Canada 44°11′N 81°38′W
96 K10 Kincardine cultural region E Scotland, United Kingdom
79 K21 Kinda Kasai-Occidental, SE Dem. Rep. Congo 04°48′S 21°50′E
79 M24 Kinda Katanga, SE Dem. Rep. Congo 09°20′S 25°06′E
166 L3 Kindat Sagaing, N Myanmar (Burma) 23°42′N 94°29′E
109 V6 Kindberg Steiermark, C Austria 47°31′N 15°27′E
22 H8 Kinder Louisiana, S USA 30°29′N 92°51′W
98 H13 Kinderdijk Zuid-Holland, SW Netherlands 51°52′N 04°37′E
97 M17 Kinder Scout ▲ C England, United Kingdom 53°25′N 01°52′W
11 S16 Kindersley Saskatchewan, S Canada 51°29′N 109°08′W
76 I14 Kindia Guinée-Maritime, SW Guinea 10°12′N 12°26′W
81 B11 Kindley Field air base E Bermuda
29 R6 Kindred North Dakota, N USA 46°39′N 97°01′W
79 N20 Kindu prev. Kindu-Port-Empain. Maniema, C Dem. Rep. Congo 02°57′S 25°54′E
Kindu-Port-Empain see Kindu
127 S6 Kinel' Samarskaya Oblast', W Russian Federation 53°14′N 50°40′E
125 N15 Kineshma Ivanovskaya Oblast', W Russian Federation 57°28′N 42°08′E
King see King William's Town
140 K10 King Abdul Aziz ✈ (Makkah) Makkah, W Saudi Arabia 21°44′N 39°08′E
140 L9 King Abdullah Economic City Makkah, W Saudi Arabia 22°24′N 39°05′E
Kingait see Cape Dorset
21 X6 King and Queen Court House Virginia, NE USA 37°40′N 76°49′W
King Charles Islands see Kong Karls Land
King Christian IX Land see Kong Christian IX Land
King Christian X Land see Kong Christian X Land
35 O11 King City California, W USA 36°12′N 121°09′W
27 R2 King City Missouri, C USA 40°03′N 94°31′W
38 M16 King Cove Alaska, USA 55°03′N 162°19′W
26 M10 Kingfisher Oklahoma, C USA 35°53′N 97°56′W
King Frederik VI Coast see Kong Frederik VI Kyst
King Frederik VIII Land see Kong Frederik VIII Land
65 B24 King George Bay bay West Falkland, Falkland Islands
194 G3 King George Island var. King George Land. island South Shetland Islands, Antarctica
12 I6 King George Islands island group Northwest Territories, C Canada
King George Land see King George Island
124 G13 Kingisepp Leningradskaya Oblast', NW Russian Federation 59°23′N 28°37′E
183 N14 King Island island Tasmania, SE Australia
10 J15 King Island island British Columbia, SW Canada
King Island see Kadan Kyun
Kingissepp see Kuressaare
141 Q7 King Khalid ✈ (Ar Riyāḍ) Ar Riyāḍ, C Saudi Arabia 25°00′N 46°40′E
35 S2 King Lear Peak ▲ Nevada, W USA 41°13′N 118°30′W
195 Y8 King Leopold and Queen Astrid Land physical region Antarctica
180 M4 King Leopold Ranges ▲▲ Western Australia
36 I11 Kingman Arizona, SW USA 35°12′N 114°02′W
26 M6 Kingman Kansas, C USA 37°39′N 98°07′W
192 L7 Kingman Reef ◇ US unincorporated territory C Pacific Ocean
79 N20 Kingombe Maniema, E Dem. Rep. Congo 02°37′S 26°39′E
182 F5 Kingoonya South Australia 30°56′S 135°20′E
194 J10 King Peninsula peninsula Antarctica
39 P13 King Salmon Alaska, USA 58°41′N 156°39′W
35 Q6 Kings Beach California, W USA 39°13′N 120°02′W
35 R11 Kingsburg California, W USA 36°30′N 119°33′W
182 I10 Kingscote South Australia 35°41′S 137°36′E
King's County see Offaly
194 H2 King Sejong South Korean research station Antarctica 61°57′S 58°23′W
183 T9 Kingsford Smith ✈ (Sydney) New South Wales, SE Australia 33°58′S 151°09′E
11 P17 Kingsgate British Columbia, SW Canada 48°58′N 116°09′W
23 W8 Kingsland Georgia, SE USA 30°48′N 81°41′W
29 S13 Kingsley Iowa, C USA 42°35′N 95°58′W
97 O19 King's Lynn var. Bishop's Lynn, Kings Lynn, Lynn, Lynn Regis. E England, United Kingdom 52°45′N 00°24′E
Kings Lynn see King's Lynn
21 Q10 Kings Mountain North Carolina, SE USA 35°15′N 81°20′W
180 K4 King Sound sound Western Australia
37 N2 Kings Peak ▲ Utah, W USA 40°43′N 110°27′W
21 O8 Kingsport Tennessee, S USA 36°32′N 82°33′W
35 R11 Kings River ≈ California, W USA
183 P17 Kingston Tasmania, SE Australia 42°57′S 147°18′E
14 K14 Kingston Ontario, SE Canada 44°14′N 76°30′W
44 K13 Kingston ● (Jamaica) E Jamaica 17°58′N 76°48′W
185 C22 Kingston Otago, South Island, New Zealand 45°20′S 168°45′E
19 P12 Kingston Massachusetts, NE USA 41°59′N 70°43′W
27 S3 Kingston Missouri, C USA 39°36′N 94°02′W
18 K12 Kingston New York, NE USA 41°55′N 74°00′W
31 S14 Kingston Ohio, N USA 39°28′N 82°54′W
19 O13 Kingston Rhode Island, NE USA 41°28′N 71°31′W
20 M9 Kingston Tennessee, S USA 35°52′N 84°30′W
35 W12 Kingston Peak ▲ California, W USA 35°43′N 115°54′W
182 J11 Kingston Southeast South Australia 36°51′S 139°53′E
97 N17 Kingston upon Hull var. Hull. E England, United Kingdom 53°45′N 00°20′W
97 N22 Kingston upon Thames SE England, United Kingdom 51°26′N 00°18′W
45 P14 Kingstown ● (Saint Vincent and the Grenadines) Saint Vincent, Saint Vincent and the Grenadines 13°09′N 61°14′W
Kingstown see Dún Laoghaire
21 T13 Kingstree South Carolina, SE USA 33°40′N 79°50′W
64 L8 Kings Trough undersea feature E Atlantic Ocean 22°00′W 43°48′N
14 C18 Kingsville Ontario, S Canada 42°03′N 82°43′W
25 S15 Kingsville Texas, SW USA 27°32′N 97°53′W
21 W6 King William Virginia, NE USA 37°42′N 77°03′W
9 N7 King William Island island Nunavut, N Canada
83 I25 King William's Town var. King, Kingwilliamstown. Eastern Cape, S South Africa 32°53′S 27°24′E
Kingwilliamstown see King William's Town
21 T3 Kingwood West Virginia, NE USA 39°27′N 79°43′W
136 C13 Kınık İzmir, W Turkey 39°05′N 27°25′E
79 G21 Kinkala Pool, S Congo 04°18′S 14°49′E
165 R10 Kinka-san headland Honshū, C Japan 38°17′N 141°34′E
184 M8 Kinleith Waikato, North Island, New Zealand 38°16′S 175°53′E
95 J19 Kinna Västra Götaland, S Sweden 57°32′N 12°42′E
96 L8 Kinnaird Head var. Kinnairds Head. headland NE Scotland, United Kingdom 58°39′N 03°22′W
Kinnairds Head see Kinnaird Head
95 K20 Kinnared Halland, S Sweden 57°01′N 13°04′E
92 L7 Kinnarodden headland N Norway 71°07′N 27°40′E
Kinneret, Yam see Tiberias, Lake
155 K24 Kinniyai Eastern Province, NE Sri Lanka 08°30′N 81°11′E
93 L16 Kinnula Keski-Suomi, C Finland 63°24′N 25°E
14 I8 Kinojévis ≈ Québec, SE Canada
164 I14 Kino-kawa ≈ Honshū, SW Japan
11 U11 Kinoosao Saskatchewan, C Canada 57°06′N 101°02′W
99 L17 Kinrooi Limburg, NE Belgium 51°09′N 05°48′E
96 J11 Kinross C Scotland, United Kingdom 56°14′N 03°27′W
96 J11 Kinross cultural region C Scotland, United Kingdom
97 C21 Kinsale Ir. Cionn tSáile. Cork, SW Ireland 51°42′N 08°32′W
95 D14 Kinsarvik Hordaland, S Norway 60°22′N 06°43′E
79 G21 Kinshasa prev. Léopoldville. ● Kinshasa, W Dem. Rep. Congo 04°21′S 15°16′E
79 G21 Kinshasa off. Ville de Kinshasa, var. Kinshasa City. ◈ region (Dem. Rep. Congo) SW Dem. Rep. Congo
79 G21 Kinshasa ✈ Kinshasa, SW Dem. Rep. Congo 04°23′S 15°30′E
Kinshasa City see Kinshasa
117 U9 Kins'ka ≈ SE Ukraine
26 K6 Kinsley Kansas, C USA 37°55′N 99°26′W
21 W10 Kinston North Carolina, SE USA 35°16′N 77°35′W
77 P15 Kintampo W Ghana 08°06′N 01°40′W
182 B1 Kintore, Mount ▲ South Australia 26°30′S 130°24′E
96 G13 Kintyre peninsula W Scotland, United Kingdom
96 G13 Kintyre, Mull of headland W Scotland, United Kingdom 55°16′N 05°46′W
166 M4 Kin-u Sagaing, C Myanmar (Burma) 22°47′N 95°36′E
12 G8 Kinushseo ≈ Ontario, C Canada
11 P13 Kinuso Alberta, W Canada 55°19′N 115°23′W
154 I13 Kinwat Mahārāshtra, C India 19°37′N 78°12′E
81 F16 Kinyeti ▲ S South Sudan 03°56′N 32°52′E
101 I17 Kinzig ≈ SW Germany
Kioga, Lake see Kyoga, Lake
26 M8 Kiowa Kansas, C USA 37°01′N 98°29′W
27 P12 Kiowa Oklahoma, C USA 34°43′N 95°54′W
Kiparissía see Kyparissía
14 H10 Kipawa, Lac ◎ Québec, SE Canada
81 G24 Kipengere Range ▲▲ SW Tanzania
81 E23 Kipili Rukwa, W Tanzania 07°30′S 30°39′E
81 K20 Kipini Tana River, SE Kenya 02°30′S 40°30′E
11 V16 Kipling Saskatchewan, S Canada 50°04′N 102°45′W
38 M13 Kipnuk Alaska, USA 59°56′N 164°02′W
97 F18 Kippure Ir. Cipiúr. ▲ E Ireland 53°10′N 06°22′W
79 N25 Kipushi Katanga, SE Dem. Rep. Congo 11°45′S 27°20′E
187 N10 Kirakira var. Kaokaona. Makira-Ulawa, SE Solomon Islands 10°28′S 161°54′E
155 K14 Kirandul var. Bailādila. Chhattīsgarh, C India 18°46′N 81°18′E
155 I21 Kīranūr Tamil Nādu, SE India 11°37′N 79°10′E
119 N21 Kiraw Rus. Kirovo. Homyel'skaya Voblasts', SE Belarus 51°30′N 29°25′E
119 M17 Kirawsk Rus. Kirovsk; prev. Startsy. Mahilyowskaya Voblasts', E Belarus 53°16′N 29°29′E
118 F5 Kirbla Läänemaa, W Estonia 58°45′N 23°57′E
25 Y9 Kirbyville Texas, SW USA 30°39′N 93°53′W
114 M12 Kircasalih Edirne, NW Turkey 41°24′N 26°48′E
109 W8 Kirchbach var. Kirchbach in Steiermark. Steiermark, SE Austria 46°55′N 15°40′E
Kirchbach in Steiermark see Kirchbach
108 H7 Kirchberg Sankt Gallen, NE Switzerland 47°24′N 09°03′E
109 S5 Kirchdorf an der Krems Oberösterreich, N Austria 47°55′N 14°08′E
Kirchheim see Kirchheim unter Teck
101 I22 Kirchheim unter Teck var. Kirchheim. Baden-Württemberg, SW Germany 48°39′N 09°28′E
139 T1 Kirdī Kawrāw, Qimmat var. Sar-i Kōrāwa. NE Iraq 37°08′N 44°39′E
Kirdzhali see Kardzhali
123 N13 Kirenga ≈ S Russian Federation
123 N12 Kirensk Irkutskaya Oblast', C Russian Federation 57°37′N 107°54′E
Kirghizia see Kyrgyzstan
145 S16 Kirghiz Range Rus. Kirgizskiy Khrebet; prev. Alexander Range. ▲▲ Kazakhstan/Kyrgyzstan
Kirghiz SSR see Kyrgyzstan
Kirghiz Steppe see Saryarka
Kirgizskaya SSR see Kyrgyzstan
Kirgizskiy Khrebet see Kirghiz Range
79 I19 Kiri Bandundu, W Dem. Rep. Congo 01°29′S 19°00′E
Kiriath-Arba see Hebron
191 R3 Kiribati off. Republic of Kiribati. ◆ republic C Pacific Ocean
Kiribati, Republic of see Kiribati
136 L17 Kırıkhan Hatay, S Turkey 36°30′N 36°20′E
136 I13 Kırıkkale Kırıkkale, C Turkey 39°50′N 33°31′E
136 C10 Kırıkkale ◈ province C Turkey
124 L13 Kirillov Vologodskaya Oblast', NW Russian Federation 59°52′N 38°24′E
Kirin see Jilin
81 I19 Kirinyaga ◈ county C Kenya
81 I18 Kirinyaga prev. Mount Kenya. ⛰ Kirinyaga, C Kenya 0°02′S 37°19′E
124 H13 Kirishi var. Kirisi. Leningradskaya Oblast', NW Russian Federation 59°28′N 32°02′E
164 C16 Kirishima-yama ▲ Kyūshū, SW Japan 31°58′N 130°51′E
Kirisi see Kirishi
191 Y2 Kiritimati ✈ Kiritimati, E Kiribati 02°00′N 157°30′W
191 Y2 Kiritimati prev. Christmas Island. atoll Line Islands, E Kiribati
186 G9 Kiriwina Island Eng. Trobriand Island. island SE Papua New Guinea
186 G9 Kiriwina Islands var. Trobriand Islands. island group S Papua New Guinea
96 K12 Kirkcaldy E Scotland, United Kingdom 56°07′N 03°10′W
97 I14 Kirkcudbright S Scotland, United Kingdom 54°50′N 04°03′W
97 I14 Kirkcudbright cultural region S Scotland, United Kingdom
Kirkee see Khadki
95 I14 Kirkenær Hedmark, S Norway 60°27′N 12°04′E
92 M8 Kirkenes Fin. Kirkkoniemi. Finnmark, N Norway 69°43′N 30°02′E
92 J4 Kirkjubæjarklaustur Suðurland, S Iceland 63°46′N 18°03′W
Kirk-Kilissa see Kırklareli
Kirkkoniemi see Kirkenes
93 L20 Kirkkonummi Swe. Kyrkslätt. Uusimaa, S Finland 60°06′N 24°20′E
14 G7 Kirkland Lake Ontario, S Canada 48°10′N 80°02′W
136 C9 Kırklareli prev. Kirk-Kilissa. Kırklareli, NW Turkey 41°45′N 27°12′E
136 I13 Kırklareli ◈ province NW Turkey
185 F20 Kirkliston Range ▲▲ South Island, New Zealand
14 D10 Kirkpatrick Lake ◎ Ontario, S Canada
195 Q11 Kirkpatrick, Mount ▲ Antarctica 84°37′S 164°36′E
27 U2 Kirksville Missouri, C USA 40°12′N 92°35′W
139 T4 Kirkūk var. Karkūk, Kerkuk. Kirkūk, NE Iraq 35°28′N 44°26′E
139 S4 Kirkūk off. Muḥāfa at at Kirkūk; prev. At Ta'mīm. ◈ governorate NE Iraq
Kirkūk, Muḥāfa at at see Kirkūk
139 U7 Kir Kush Diyālá, E Iraq 33°42′N 45°15′E
96 K5 Kirkwall NE Scotland, United Kingdom 58°59′N 02°58′W
83 H25 Kirkwood Eastern Cape, S South Africa 33°23′S 25°19′E
27 X5 Kirkwood Missouri, C USA 38°35′N 90°24′W
Kirman see Kermān
Kir Moab/Kir of Moab see Al Karak
126 I5 Kirov Kaluzhskaya Oblast', W Russian Federation 54°02′N 34°17′E
125 R14 Kirov prev. Vyatka. Kirovskaya Oblast', NW Russian Federation 58°35′N 49°39′E
Kirov see Balpyk Bi/ Ust'yevoye
Kirovabad see Gäncä
Kirovabad see Panj, Tajikistan
Kirovakan see Vanadzor
Kirov/Kirova see Kopbirlik, Kazakhstan
Kirovo see Kiraw, Belarus
Kirovo see Beshariq, Uzbekistan
125 R14 Kirovo-Chepetsk Kirovskaya Oblast', NW Russian Federation 58°33′N 50°06′E
Kirovogradskaya Oblast'/Kirovohrad see Kirovohrads'ka Oblast'
117 R7 Kirovohrad Rus. Kirovograd; prev. Kirovo, Yelizavetgrad, Zinov'yevsk. Kirovohrads'ka Oblast', C Ukraine 48°30′N 31°17′E
117 P7 Kirovohrads'ka Oblast' var. Kirovohrad, Rus. Kirovogradskaya Oblast'. ◈ province C Ukraine
Kirovo/Kirovograd see Kirovohrad
124 J4 Kirovsk Murmanskaya Oblast', NW Russian Federation 67°37′N 33°38′E
117 X7 Kirovs'k Luhans'ka Oblast', E Ukraine 48°40′N 38°39′E
Kirovsk see Kirawsk, Belarus
Kirovsk see Babadaýhan, Turkmenistan
122 E9 Kirovskaya Oblast' ◈ province NW Russian Federation
117 U13 Kirovs'ke Rus. Kirovskoye. Avtonomna Respublika Krym, S Ukraine 45°13′N 35°12′E
117 X8 Kirovs'ke Donets'ka Oblast', E Ukraine 48°12′N 38°20′E
Kirovskiy see Balpyk Bi
Kirovskiy see Ust'yevoye
Kirovskoye see Kyzyl-Adyr
Kirovskoye see Kirovs'ke
146 E11 Kirpili Ahal Welaýaty, C Turkmenistan 39°31′N 57°13′E
96 K10 Kirriemuir E Scotland, United Kingdom 56°38′N 03°01′W
125 S13 Kirs Kirovskaya Oblast', NW Russian Federation 59°22′N 52°20′E
127 N7 Kirsanov Tambovskaya Oblast', W Russian Federation 52°40′N 42°48′E
136 J14 Kırşehir anc. Justinianopolis. Kırşehir, C Turkey 39°09′N 34°08′E
136 J13 Kırşehir ◈ province C Turkey
149 P4 Kirthar Range ▲▲ S Pakistan
37 P9 Kirtland New Mexico, SW USA 36°43′N 108°21′W
92 J11 Kiruna Lapp. Giron. Norrbotten, N Sweden 67°50′N 20°16′E
79 M18 Kirundu Orientale, NE Dem. Rep. Congo 0°45′S 25°28′E
Kirun/Kirun' see Jilong
26 L3 Kirwin Reservoir ▣ Kansas, C USA
127 Q4 Kirya Chuvashskaya Respublika, W Russian Federation 55°04′N 46°50′E
138 G8 Kiryat Shmona prev. Qiryat Shemona. Northern, N Israel 33°13′N 35°35′E
95 M18 Kisa Östergötland, S Sweden 58°N 15°39′E
165 P9 Kisakata Akita, Honshū, C Japan 39°12′N 139°55′E
Kisalföld see Little Alföld
79 L18 Kisangani prev. Stanleyville. Orientale, NE Dem. Rep. Congo 0°30′N 25°14′E
39 N12 Kisaralik River ≈ Alaska, USA
165 O14 Kisarazu Chiba, Honshū, S Japan 35°23′N 139°57′E
111 I22 Kisbér Komárom-Esztergom, NW Hungary 47°30′N 18°00′E
11 V17 Kisbey Saskatchewan, S Canada 49°41′N 102°39′W
122 J13 Kiselevsk Kemerovskaya Oblast', S Russian Federation 54°00′N 86°38′E
153 R13 Kishanganj Bihār, NE India 26°06′N 87°57′E
152 G12 Kishangarh Rājasthān, N India 26°33′N 74°52′E
Kishegyes see Mali Iđoš
77 S15 Kishi Oyo, W Nigeria 09°01′N 03°53′E
Kishinev see Chişinău
Kishiözen see Saryozen
164 I14 Kishiwada var. Kisiwada. Ōsaka, Honshū, SW Japan 34°28′N 135°22′E
143 P14 Kīsh, Jazīreh-ye var. Qey. island S Iran
145 R7 Kishkenekol' prev. Kzyltu, Kaz. Qyzyltū. Kokshetau, N Kazakhstan 53°39′N 72°22′E
138 G9 Kishon, Nahal prev. Naḥal Qishon. ≈ N Israel
152 I6 Kishtwār Jammu and Kashmir, NW India 33°20′N 75°49′E
81 H19 Kisii Kisii, SW Kenya 0°40′S 34°47′E
81 H19 Kisii ◈ county W Kenya
81 J23 Kisiju Pwani, E Tanzania 07°25′S 39°20′E
Kisiwada see Kishiwada
Kisjenő see Chişineu-Criş
38 E17 Kiska Island island Aleutian Islands, Alaska, USA
Kiskapus see Copşa Mică
111 M22 Kiskőrei-víztároló ◎ E Hungary
Kis-Küküllo see Târnava Mică
111 L24 Kiskunfélegyháza var. Félegyháza. Bács-Kiskun, C Hungary 46°42′N 19°52′E
111 K25 Kiskunhalas var. Halas. Bács-Kiskun, S Hungary 46°26′N 19°29′E
111 K24 Kiskunmajsa Bács-Kiskun, S Hungary 46°30′N 19°46′E
127 N15 Kislovodsk Stavropol'skiy Kray, SW Russian Federation 43°55′N 42°45′E
81 L18 Kismaayo var. Chisimayu, Kismayu, It. Chisimaio. Jubbada Hoose, S Somalia 0°05′S 42°35′E
Kismayu see Kismaayo
164 M13 Kiso-sanmyaku ▲▲ Honshū, S Japan
115 H24 Kíssamos prev. Kastélli. Kríti, Greece, E Mediterranean Sea 35°30′N 23°39′E
Kisseraing see Kanmaw Kyun
76 K14 Kissidougou Guinée-Forestière, S Guinea 09°15′N 10°08′W
23 X12 Kissimmee Florida, SE USA 28°17′N 81°24′W
23 X12 Kissimmee, Lake ◎ Florida, SE USA
23 X13 Kissimmee River ≈ Florida, SE USA
11 V13 Kississing Lake ◎ Manitoba, C Canada
111 L24 Kistelek Csongrád, SE Hungary 46°27′N 19°58′E
Kistna see Krishna
111 M23 Kisújszállás Jász-Nagykun-Szolnok, E Hungary 47°14′N 20°45′E
164 G12 Kisuki var. Unnan. Shimane, Honshū, SW Japan 35°25′N 133°15′E
81 H18 Kisumu prev. Port Florence. Kisumu, W Kenya 0°02′N 34°42′E
81 H18 Kisumu ◈ county W Kenya
Kisutzaneustadtl see Kysucké Nové Mesto
111 O20 Kisvárda Ger. Kleinwardein. Szabolcs-Szatmár-Bereg, E Hungary 48°13′N 22°03′E
81 J24 Kiswere Lindi, SE Tanzania 09°24′S 39°37′E
Kiszucaújhely see Kysucké Nové Mesto
76 K12 Kita Kayes, W Mali 13°00′N 09°28′W
Kitaa ◈ province W Greenland
Kita-Akita see Takanosu
Kitab see Kitob
165 Q4 Kitahiyama Hokkaidō, NE Japan 42°25′N 139°55′E
165 P12 Kitaibaraki Ibaraki, Honshū, S Japan 36°46′N 140°45′E
165 X16 Kita-Iō-jima Eng. San Alessandro. island SE Japan
165 Q9 Kitakami Iwate, Honshū, C Japan 39°18′N 141°05′E
165 P11 Kitakata Fukushima, Honshū, C Japan 37°38′N 139°52′E
164 D13 Kitakyūshū var. Kitakyûsyû. Fukuoka, Kyūshū, SW Japan 33°51′N 130°49′E
Kitakyûsyû see Kitakyūshū
81 H18 Kitale Trans Nzoia, W Kenya 01°01′N 35°01′E
165 U3 Kitami Hokkaidō, NE Japan 43°52′N 143°51′E
165 T2 Kitami-sanchi ▲▲ Hokkaidō, NE Japan
37 W5 Kit Carson Colorado, C USA 38°45′N 102°47′W
180 M12 Kitchener Western Australia 31°03′S 124°00′E
14 F16 Kitchener Ontario, S Canada 43°28′N 80°27′W
93 O17 Kitee Pohjois-Karjala, SE Finland 62°06′N 30°09′E
81 G16 Kitgum N Uganda 03°17′N 32°54′E
Kithareng see Kanmaw Kyun
Kíthira see Kýthira
Kíthnos see Kýthnos
8 L8 Kitikmeot ◈ cultural region Nunavut, N Canada
10 J13 Kitimat British Columbia, SW Canada 54°05′N 128°38′W
92 L11 Kitinen ≈ N Finland
147 N12 Kitob Rus. Kitab. Qashqadaryo Viloyati, S Uzbekistan 39°06′N 66°47′E
116 K7 Kitsman' Ger. Kotzman, Rom. Cozmeni, Rus. Kitsman. Chernivets'ka Oblast', W Ukraine 48°30′N 25°50′E
164 E14 Kitsuki var. Kituki. Ōita, Kyūshū, SW Japan 33°24′N 131°36′E
18 C14 Kittanning Pennsylvania, NE USA 40°48′N 79°28′W
19 P10 Kittery Maine, NE USA 43°05′N 70°44′W
92 L11 Kittilä Lappi, N Finland 67°39′N 24°53′E
109 Z4 Kittsee Burgenland, E Austria 48°06′N 17°03′E
81 J19 Kitui Kitui, S Kenya 01°25′S 38°00′E
81 J20 Kitui ◈ county S Kenya
Kituki see Kitsuki
81 G22 Kitunda Tabora, C Tanzania 06°47′S 33°13′E
10 K13 Kitwanga British Columbia, SW Canada 55°07′N 128°03′W
82 J13 Kitwe var. Kitwe-Nkana. Copperbelt, C Zambia 12°48′S 28°13′E
Kitwe-Nkana see Kitwe
109 O7 Kitzbühel Tirol, W Austria 47°27′N 12°23′E
109 O7 Kitzbüheler Alpen ▲▲ W Austria
101 J19 Kitzingen Bayern, SE Germany 49°45′N 10°11′E
153 Q14 Kiul Bihār, NE India 25°10′N 86°06′E
186 A7 Kiunga Western, SW Papua New Guinea 06°10′S 141°15′E
93 M16 Kiuruvesi Pohjois-Savo, C Finland 63°38′N 26°40′E
38 M7 Kivalina Alaska, USA 67°44′N 164°32′W
9 O10 Kivalliq ◈ cultural region Nunavut, N Canada
92 L13 Kivalo ridge C Finland
116 J3 Kivertsi Pol. Kiwerce, Rus. Kivertsy. Volyns'ka Oblast', NW Ukraine 50°50′N 25°31′E
Kivertsy see Kivertsi
93 L16 Kivijärvi Keski-Suomi, C Finland 63°09′N 25°04′E
95 L23 Kivik Skåne, S Sweden 55°40′N 14°15′E
118 J3 Kiviõli Ida-Virumaa, NE Estonia 59°20′N 27°00′E
Kivu, Lac see Kivu, Lake
67 U10 Kivu, Lake Fr. Lac Kivu. ◎ Rwanda/Dem. Rep. Congo
186 C9 Kiwai Island island SW Papua New Guinea
39 N8 Kiwalik Alaska, USA 66°01′N 161°50′W
Kiwerce see Kivertsi
Kiyev see Kyyiv
145 R10 Kiyevka Karaganda, C Kazakhstan 50°15′N 71°33′E
Kiyevskaya Oblast' see Kyyivs'ka Oblast'
Kiyevskoye Vodokhranilishche see Kyyivs'ke Vodoskhovyshche
136 D10 Kıyıköy Kırklareli, NW Turkey 41°37′N 28°07′E
Kiyma see Kima
125 V13 Kizel Permskiy Kray, NW Russian Federation 58°59′N 57°37′E
125 O12 Kizema var. Kizëma. Arkhangel'skaya Oblast', NW Russian Federation 61°06′N 44°51′E
Kizëma see Kizema
Kizilagach see Elhovo
136 H12 Kızılcahamam Ankara, N Turkey 40°28′N 32°37′E
136 J10 Kızıl Irmak ≈ C Turkey
Kizil Kum see Kyzyl Kum
137 P16 Kızıltepe Mardin, SE Turkey 37°12′N 40°36′E
Ki Zil Uzen see Qezel Owzan, Rūd-e
127 Q16 Kizilyurt Respublika Dagestan, SW Russian Federation 43°13′N 46°54′E
127 Q15 Kizlyar Respublika Dagestan, SW Russian Federation 43°51′N 46°39′E

◆ Country ● Country Capital ◇ Dependent Territory ○ Dependent Territory Capital ◈ Administrative Regions ✈ International Airport ▲ Mountain ▲▲ Mountain Range ⛰ Volcano ≈ River ◎ Lake ▣ Reservoir

KIZNER – KONKOURÉ

127 S3 **Kizner** Udmurtskaya Respublika, NW Russian Federation *56°19′N 51°37′E*
Kizyl-Arvat *see* Serdar
Kizyl-Atrek *see* Etrek
Kizyl-Kaya *see* Gyzylgaýa
Kizyl-Su *see* Gyzylsuw
95 H16 **Kjerkøy** *island* S Norway
Kjølen *see* Kölen
92 L7 **Kjøllefjord** Finnmark, N Norway *70°55′N 27°19′E*
92 H11 **Kjøpsvik** ***Lapp.*** Gásluokta. Nordland, C Norway *68°06′N 16°21′E*
169 N12 **Klabat, Teluk** *bay* Pulau Bangka, W Indonesia
112 I12 **Kladanj** ⬖ Fedederacijia Bosna I Hercegovina, E Bosnia and Herzegovina
171 X16 **Kladar** Papua, E Indonesia *08°14′S 137°46′E*
111 C16 **Kladno** Středočeský, NW Czech Republic *50°10′N 14°05′E*
112 P11 **Kladovo** Serbia, E Serbia *44°37′N 22°36′E*
167 P12 **Klaeng** Rayong, S Thailand *12°48′N 101°41′E*
109 T9 **Klagenfurt** ***Slvn.*** Celovec. Kärnten, S Austria *46°38′N 14°20′E*
118 B11 **Klaipėda** ***Ger.*** Memel. Klaipeda, NW Lithuania *55°42′N 21°09′E*
118 C11 **Klaipėda** ⬖ *province* W Lithuania
Klaksvig *see* Klaksvík
95 B18 **Klaksvík** ***Dan.*** Klaksvig. Faroe Islands *62°13′N 06°34′W*
34 L2 **Klamath** California, W USA *41°31′N 124°02′W*
32 H16 **Klamath Falls** Oregon, NW USA *42°14′N 121°47′W*
34 M1 **Klamath Mountains** ▲▲ California/Oregon, W USA
34 L2 **Klamath River** ≈ California/Oregon, W USA
168 K9 **Klang** ***var.*** Kelang; ***prev.*** Port Swettenham. Selangor, Peninsular Malaysia *03°02′N 101°27′E*
94 J13 **Klarälven** ≈ Norway/Sweden
111 B15 **Klášterec nad Ohří** ***Ger.*** Klösterle an der Eger. Ústecky Kraj, NW Czech Republic *50°24′N 13°10′E*
111 B18 **Klatovy** ***Ger.*** Klattau. Plzeňský Kraj, W Czech Republic *49°24′N 13°16′E*
Klattau *see* Klatovy
Klausenburg *see* Cluj-Napoca
39 Y14 **Klawock** Prince of Wales Island, Alaska, USA *55°33′N 133°06′W*
98 P8 **Klazienaveen** Drenthe, NE Netherlands *52°43′N 07°E*
Kleck *see* Klyetsk
110 H11 **Kłecko** Weilkopolskie, C Poland *52°37′N 17°27′E*
110 I11 **Kleczew** Wielkopolskie, C Poland *52°22′N 18°12′E*
10 L15 **Kleena Kleene** British Columbia, SW Canada *51°55′N 124°54′W*
83 D20 **Klein Aub** Hardap, C Namibia *23°48′S 16°39′E*
Kleine Donau *see* Mosoni-Duna
101 O14 **Kleine Elster** ≈ E Germany
Kleine Kokel *see* Târnava Mică
99 I16 **Kleine Nete** ≈ N Belgium
Kleines Ungarisches Tiefland *see* Little Alföld
83 E22 **Klein Karas** Karas, S Namibia *27°36′S 18°05′E*
Kleinkopisch *see* Copşa Mică
Klein-Marien *see* Väike-Maarja
Kleinschlatten *see* Zlatna
83 D23 **Kleinsee** Northern Cape, W South Africa *29°43′S 17°03′E*
Kleinwardein *see* Kisvárda
115 C16 **Kleisoúra** Ípeiros, W Greece *39°21′N 20°52′E*
95 C17 **Klepp** Rogaland, S Norway *58°46′N 05°39′E*
83 I22 **Klerksdorp** North-West, N South Africa *26°52′S 26°39′E*
126 I5 **Kletnya** Bryanskaya Oblast', W Russian Federation *53°25′N 32°58′E*
Kletsk *see* Klyetsk
101 D14 **Kleve** ***Eng.*** Cleves, ***Fr.*** Clèves; ***prev.*** Cleve. Nordrhein-Westfalen, W Germany *51°47′N 06°11′E*
113 J16 **Kličevo** C Montenegro *42°45′N 18°58′E*
119 M16 **Klichaw** ***Rus.*** Klichev. Mahilyowskaya Voblasts', E Belarus *53°29′N 29°21′E*
Klichev *see* Klichaw
119 Q16 **Klimavichy** ***Rus.*** Klimovichi. Mahilyowskaya Voblasts', E Belarus *53°37′N 31°58′E*
114 M7 **Kliment** Shumen, NE Bulgaria *43°37′N 27°00′E*
Klimovichi *see* Klimavichy
93 G14 **Klimpfjäll** Västerbotten, N Sweden *65°05′N 14°50′E*
126 K3 **Klin** Moskovskaya Oblast', W Russian Federation *56°19′N 36°45′E*
Klina *see* Klinë
113 M16 **Klinë** ***Serb.*** Klina. W Kosovo *42°38′N 20°35′E*
111 B15 **Klínovec** ***Ger.*** Keilberg. ▲ NW Czech Republic *50°23′N 12°57′E*
95 P19 **Klintehamn** Gotland, SE Sweden *57°22′N 18°15′E*
127 R8 **Klintsovka** Saratovskaya Oblast', W Russian Federation *51°42′N 49°17′E*
126 H6 **Klintsy** Bryanskaya Oblast', W Russian Federation *52°46′N 32°21′E*
95 K22 **Klippan** Skåne, S Sweden *56°08′N 13°10′E*
92 G13 **Klippen** Västerbotten, N Sweden *65°50′N 15°07′E*
121 P2 **Klírou** W Cyprus *35°01′N 33°11′E*
114 I9 **Klisura** Plovdiv, C Bulgaria *42°40′N 24°28′E*
95 F20 **Klitmøller** Midtjylland, NW Denmark *57°01′N 08°29′E*
112 F11 **Ključ** Federacija Bosne I Hercegovine, NW Bosnia and Herzegovina *44°32′N 16°46′E*
111 J14 **Kłobuck** Śląskie, S Poland *50°56′N 18°55′E*
110 J11 **Kłodawa** Wielkopolskie, C Poland *52°14′N 18°55′E*
111 G16 **Kłodzko** ***Ger.*** Glatz. Dolnośląskie, SW Poland *50°27′N 16°37′E*
95 I14 **Kløfta** Akershus, S Norway *60°04′N 11°06′E*
112 P12 **Klokočevac** Serbia, E Serbia *44°19′N 22°11′E*
118 G3 **Klooga** ***Ger.*** Lodensee. Harjumaa, NW Estonia *59°18′N 24°10′E*
99 F15 **Kloosterzande** Zeeland, SW Netherlands *51°22′N 04°01′E*
113 L19 **Klos** ***var.*** Klosi. Dibër, C Albania *41°30′N 20°07′E*
Klosi *see* Klos
Klösterle an der Eger *see* Klášterec nad Ohří
109 X3 **Klosterneuburg** Niederösterreich, NE Austria *48°19′N 16°20′E*
108 J9 **Klosters** Graubünden, SE Switzerland *46°54′N 09°52′E*
108 G7 **Kloten** Zürich, N Switzerland *47°27′N 08°35′E*
108 G7 **Kloten** ✈ (Zürich) Zürich, N Switzerland *47°25′N 08°36′E*
100 K12 **Klötze** Sachsen-Anhalt, C Germany *52°37′N 11°09′E*
12 K3 **Klotz, Lac** ◎ Québec, NE Canada
101 O15 **Klotzsche** ✈ (Dresden) Sachsen, E Germany *51°06′N 13°44′E*
10 H7 **Kluane Lake** ◎ Yukon, W Canada
Kluang *see* Keluang
111 I14 **Kluczbork** ***Ger.*** Kreuzburg, Kreuzburg in Oberschlesien. Opolskie, S Poland *50°59′N 18°13′E*
39 W12 **Klukwan** Alaska, USA *59°24′N 135°49′W*
Klyastitsy *see* Klyastsitsy
118 L11 **Klyastsitsy** ***Rus.*** Klyastitsy. Vitsyebskaya Voblasts', N Belarus *55°53′N 28°36′E*
127 T5 **Klyavlino** Samarskaya Oblast', W Russian Federation *54°21′N 52°12′E*
84 K9 **Klyaz'in** ≈ W Russian Federation
127 N3 **Klyaz'ma** ≈ W Russian Federation
119 J17 **Klyetsk** ***Pol.*** Kleck, ***Rus.*** Kletsk. Minskaya Voblasts', SW Belarus *53°04′N 26°38′E*
147 S8 **Klyuchevka** Talasskaya Oblast', NW Kyrgyzstan *42°34′N 71°45′E*
123 V10 **Klyuchevskaya Sopka, Vulkan** ⛰ E Russian Federation *56°03′N 160°38′E*
95 D17 **Knaben** Vest-Agder, S Norway *58°46′N 07°04′E*
95 K21 **Knäred** Halland, S Sweden *56°30′N 13°21′E*
97 M16 **Knaresborough** N England, United Kingdom *54°01′N 01°35′W*
114 H8 **Knezha** Vratsa, NW Bulgaria *43°29′N 24°04′E*
25 O9 **Knickerbocker** Texas, SW USA *31°18′N 100°35′W*
28 K5 **Knife River** ≈ North Dakota, N USA
10 K16 **Knight Inlet** *inlet* British Columbia, W Canada
39 S12 **Knight Island** *island* Alaska, USA
97 K20 **Knighton** E Wales, United Kingdom *52°20′N 03°01′W*
35 O7 **Knights Landing** California, W USA *38°47′N 121°43′W*
112 E13 **Knin** Šibenik-Knin, S Croatia *44°03′N 16°12′E*
25 Q12 **Knippal** Texas, SW USA
109 U7 **Knittelfeld** Steiermark, C Austria *47°14′N 14°50′E*
95 O15 **Knivsta** Uppsala, C Sweden *59°43′N 17°49′E*
113 P14 **Knjaževac** Serbia, E Serbia *43°34′N 22°16′E*
27 S4 **Knob Noster** Missouri, C USA *38°47′N 93°33′W*
99 D15 **Knokke-Heist** West-Vlaanderen, NW Belgium *51°21′N 03°19′E*
95 H20 **Knøsen** *hill* N Denmark
Knosós *see* Knossos
115 J25 **Knossos** ***Gk.*** Knosós. *prehistoric site* Kríti, Greece, E Mediterranean Sea
25 N7 **Knott** Texas, SW USA *32°21′N 101°35′W*
194 K5 **Knowles, Cape** *headland* Antarctica *71°45′S 60°20′W*
31 O11 **Knox** Indiana, N USA *41°17′N 86°37′W*
29 O3 **Knox** North Dakota, N USA *48°19′N 99°43′W*
18 C13 **Knox** Pennsylvania, NE USA *41°13′N 79°33′W*
189 X8 **Knox Atoll** ***var.*** Ņadikdik, Narikrik. *atoll* Ratak Chain, SE Marshall Islands
10 H13 **Knox, Cape** *headland* Graham Island, British Columbia, SW Canada *54°05′N 133°02′W*
25 P5 **Knox City** Texas, SW USA *33°25′N 99°49′W*
195 Y11 **Knox Coast** *physical region* Antarctica
31 T12 **Knox Lake** ⊠ Ohio, N USA
23 T5 **Knoxville** Georgia, SE USA *32°44′N 83°58′W*
30 K12 **Knoxville** Illinois, N USA *40°54′N 90°16′W*
29 W15 **Knoxville** Iowa, C USA *41°19′N 93°06′W*
21 N9 **Knoxville** Tennessee, S USA *35°58′N 83°55′W*
197 P11 **Knud Rasmussen Land** *physical region* N Greenland
Knüll *see* Knüllgebirge
101 I16 **Knüllgebirge** ***var.*** Knüll. ▲▲ C Germany
124 I5 **Knyazhegubskoye Vodokhranilishche** ◎ NW Russian Federation
Knyazhevo *see* Sredishte
Knyazhitsy *see* Knyazhytsy
119 O15 **Knyazhytsy** ***Rus.*** Knyazhitsy. Mahilyowskaya Voblasts', E Belarus *54°10′N 30°28′E*
83 G26 **Knysna** Western Cape, SW South Africa *34°03′S 23°03′E*
Koartac *see* Quaqtaq
169 N13 **Koba** Pulau Bangka, W Indonesia *02°30′S 106°26′E*
Kobanî *see* 'Ayn al 'Arab
164 D16 **Kobayashi** ***var.*** Kobayasi. Miyazaki, Kyūshū, SW Japan *32°00′N 130°58′E*
Kobayasi *see* Kobayashi
144 I10 **Kobda** ***prev.*** Khobda, Novoalekseyevka. Aktyubinsk, W Kazakhstan *50°09′N 55°39′E*
144 H9 **Kobda** ***Kaz.*** Ülkenqobda; ***prev.*** Bol'shaya Khobda. ≈ Kazakhstan/Russian Federation
Kobdo *see* Hovd
164 I13 **Kōbe** Hyōgo, Honshū, SW Japan *34°40′N 135°10′E*
Kobelyaki *see* Kobelyaky
117 T6 **Kobelyaky** ***Rus.*** Kobelyaki. Poltavs'ka Oblast', NE Ukraine *49°10′N 34°13′E*
95 J22 **København** ***Eng.*** Copenhagen; ***anc.*** Hafnia. ● (Denmark) Sjælland, København, E Denmark *55°43′N 12°34′E*
76 K10 **Kobenni** Hodh el Gharbi, S Mauritania *15°58′N 09°24′W*
171 T13 **Kobi** Pulau Seram, E Indonesia *02°56′S 129°53′E*
101 F17 **Koblenz** ***prev.*** Coblenz, ***Fr.*** Coblence; ***anc.*** Confluentes. Rheinland-Pfalz, W Germany *50°21′N 07°36′E*
108 F6 **Koblenz** Aargau, N Switzerland *47°34′N 08°16′E*
Kobrin *see* Kobryn
171 V15 **Kobroor, Pulau** *island* Kepulauan Aru, E Indonesia
119 G19 **Kobryn** ***Rus.*** Kobrin. Brestskaya Voblasts', SW Belarus *52°13′N 24°21′E*
39 O7 **Kobuk** Alaska, USA *66°54′N 156°52′W*
39 O7 **Kobuk River** ≈ Alaska, USA
137 Q10 **Kobuleti** ***prev.*** K'obulet'i. W Georgia *41°47′N 41°47′E*
K'obulet'i *see* Kobuleti
123 P10 **Kobyay** Respublika Sakha (Yakutiya), NE Russian Federation *63°36′N 126°33′E*
136 E11 **Kocaeli** ⬖ *province* NW Turkey
113 P18 **Kočani** NE FYR Macedonia *41°55′N 22°25′E*
112 K12 **Koceljevo** Serbia, W Serbia *44°28′N 19°49′E*
109 U12 **Kočevje** ***Ger.*** Gottschee. S Slovenia *45°41′N 14°48′E*
153 T12 **Koch Bihār** West Bengal, NE India *26°19′N 89°26′E*
Kochchi *see* Cochin/Kochi
122 M9 **Kochechum** ≈ N Russian Federation
101 I20 **Kocher** ≈ SW Germany
125 T13 **Kochevo** Komi-Permyatskiy Okrug, NW Russian Federation *59°37′N 54°16′E*
164 G14 **Kōchi** ***var.*** Kôti. Kōchi, Shikoku, SW Japan *33°31′N 133°30′E*
164 G14 **Kōchi** ***off.*** Kōchi-ken, ***var.*** Kôti. ⬖ *prefecture* Shikoku, SW Japan
Kōchi-ken *see* Kōchi
Kochiu *see* Gejiu
Kochkor *see* Kochkorka
147 V8 **Kochkorka** ***Kir.*** Kochkor. Narynskaya Oblast', C Kyrgyzstan *42°09′N 75°42′E*
125 V5 **Kochmes** Respublika Komi, NW Russian Federation *66°10′N 60°46′E*
127 P15 **Kochubey** Respublika Dagestan, SW Russian Federation *44°25′N 46°33′E*
115 I17 **Kochýlas** ▲ Skńyros, Vóreies Sporádes, Greece, Aegean Sea *38°50′N 24°35′E*
110 O13 **Kock** Lubelskie, E Poland *51°39′N 22°26′E*
81 J19 **Kodacho** *spring/well* S Kenya *01°52′S 39°22′E*
155 K24 **Koddiyar Bay** *bay* NE Sri Lanka
39 Q14 **Kodiak** Kodiak Island, Alaska, USA *57°47′N 152°24′W*
39 Q14 **Kodiak Island** *island* Alaska, USA
154 B12 **Kodīnār** Gujarāt, W India *20°44′N 70°46′E*
124 M9 **Kodino** Arkhangel'skaya Oblast', NW Russian Federation *63°36′N 39°54′E*
122 M12 **Kodinsk** Krasnoyarskiy Kray, C Russian Federation *58°37′N 99°18′E*
80 F12 **Kodok** Upper Nile, NE South Sudan *09°51′N 32°07′E*
117 N8 **Kodyma** Odes'ka Oblast', SW Ukraine *48°05′N 29°09′E*
Koedoes *see* Kudus
99 B17 **Koekelare** West-Vlaanderen, W Belgium *51°07′N 02°58′E*
Koeln *see* Köln
Koepang *see* Kupang
Ko-erh-mu *see* Golmud
99 J17 **Koersel** Limburg, NE Belgium *51°04′N 05°17′E*
83 E21 **Koës** Karas, SE Namibia *25°59′S 19°08′E*
Koetai *see* Mahakam, Sungai
Koetaradja *see* Banda Aceh
36 I14 **Kofa Mountains** ▲▲ Arizona, SW USA
171 Y15 **Kofarau** Papua, E Indonesia *07°29′S 140°28′E*
147 P13 **Kofarnihon** ***Rus.*** Kofarnikhon; ***prev.*** Ordzhonikidzeabad, ***Taj.*** Orjonikidzeobod, Yangi-Bazar. W Tajikistan *38°32′N 68°56′E*
147 P14 **Kofarnihon** ***Rus.*** Kafirnigan. ≈ SW Tajikistan
Kofarnikhon *see* Kofarnihon
114 M11 **Kofçaz** Kırklareli, NW Turkey *41°58′N 27°12′E*
115 J25 **Kófinas** ▲ Kríti, Greece, E Mediterranean Sea *34°58′N 25°03′E*
121 P3 **Kofínou** ***var.*** Kophinou. S Cyprus *34°49′N 33°24′E*
109 V8 **Köflach** Steiermark, SE Austria *47°04′N 15°04′E*
77 Q17 **Koforidua** SE Ghana *06°01′N 00°12′W*
164 H12 **Kōfu** Tottori, Honshū, SW Japan *35°16′N 133°31′E*
164 M13 **Kōfu** ***var.*** Kôhu. Yamanashi, Honshū, S Japan *35°41′N 138°33′E*
81 F22 **Koga** Tabora, C Tanzania *06°08′S 32°20′E*
Kogălniceanu *see* Mihail Kogălniceanu
13 P6 **Kogaluk** ≈ Newfoundland and Labrador, E Canada
12 J4 **Kogaluk, Riviére** ≈ Québec, NE Canada
145 V14 **Kogaly** ***Kaz.*** Qoghaly; ***prev.*** Kugaly. Almaty, SE Kazakhstan *44°30′N 78°40′E*
122 I10 **Kogalym** Khanty-Mansiyskiy Avtonomnyy Okrug-Yugra, C Russian Federation *62°13′N 74°34′E*
95 J23 **Køge** Sjælland, E Denmark *55°28′N 12°12′E*
95 J23 **Køge Bugt** *bay* E Denmark
77 U16 **Kogi** ⬖ *state* C Nigeria
146 L11 **Kogon** ***Rus.*** Kagan. Buxoro Viloyati, C Uzbekistan *39°47′N 64°29′E*
Kŏgŭm-do *see* Geogeum-do
Kőhalom *see* Rupea
149 T6 **Kohāt** Khyber Pakhtunkhwa, NW Pakistan *33°37′N 71°30′E*
142 L10 **Kohgīlūyeh va Bowyer Aḩmad** ***off.*** Ostān-e Kohgīlūyeh va Bowyer Aḩmad, ***var.*** Boyer Ahmadī va Kohkīlūyeh. ⬖ *province* SW Iran
Kohgīlūyeh va Bowyer Aḩmad, Ostān-e *see* Kohgīlūyeh va Bowyer Aḩmad
118 G4 **Kohila** ***Ger.*** Koil. Raplamaa, NW Estonia *59°09′N 24°45′E*
153 X13 **Kohīma** *state capital* Nāgāland, E India *25°40′N 94°08′E*
Koh I Noh *see* Büyükağrı Dağı
Kohsān *see* Kūhestān
118 J3 **Kohtla-Järve** Ida-Virumaa, NE Estonia *59°22′N 27°21′E*
Kôhu *see* Kōfu
Kohyl'nyk *see* Cogîlnic
165 N11 **Koide** Niigata, Honshū, C Japan *37°13′N 138°58′E*
10 G7 **Koidern** Yukon, W Canada *61°55′N 140°22′W*
76 J15 **Koidu** E Sierra Leone *08°40′N 11°01′W*
118 I4 **Koigi** Järvamaa, C Estonia *58°51′N 25°45′E*
Koil *see* Kohila
172 H13 **Koimbani** Grande Comore, NW Comoros *11°37′S 43°23′E*
93 O16 **Koitere** ◎ E Finland
Koivisto *see* Primorsk
Kŏje-do *see* Geogeum-do
80 J13 **K'ok'a Hāyk'** ◎ C Ethiopia
Kokand *see* Qo'qon
182 F6 **Kokatha** South Australia *31°17′S 135°16′E*
146 M10 **Ko'kcha** ***Rus.*** Kokcha. Buxoro Viloyati, C Uzbekistan *40°30′N 64°58′E*
Kokcha *see* Ko'kcha
Kokchetav *see* Kokshetau
93 K18 **Kokemäenjoki** ≈ SW Finland
171 W14 **Kokenau** ***var.*** Kokonau. Papua, E Indonesia *04°38′S 136°24′E*
83 E22 **Kokerboom** Karas, SE Namibia *28°11′S 19°25′E*
119 N14 **Kokhanava** ***Rus.*** Kokhanovo. Vitsyebskaya Voblasts', NE Belarus *54°28′N 29°59′E*
Kokhanovichi *see* Kakhanavichy
Kokhanovo *see* Kokhanava
Kök-Janggak *see* Kok-Yangak
93 K16 **Kokkola** ***Swe.*** Karleby; ***prev. Swe.*** Gamlakarleby. Keski-Pohjanmaa, W Finland *63°50′N 23°10′E*
158 L3 **Kok Kuduk** *spring/well* N China *46°03′N 87°34′E*
118 H9 **Koknese** C Latvia *56°38′N 25°27′E*
77 T13 **Koko** Kebbi, W Nigeria *11°25′N 04°33′E*
186 E9 **Kokoda** Northern, S Papua New Guinea *08°52′S 147°44′E*
76 K12 **Kokofata** Kayes, W Mali *12°48′N 09°56′W*
39 N6 **Kokolik River** ≈ Alaska, USA
31 O13 **Kokomo** Indiana, N USA *40°29′N 86°07′W*
Kokonau *see* Kokenau
Koko Nor *see* Qinghai, China
Koko Nor *see* Qinghai Hu, China
186 H6 **Kokopo** ***var.*** Kopopo; ***prev.*** Herbertshöhe. New Britain, E Papua New Guinea *04°18′S 152°17′E*
Kokpekti *see* Kokpekty
Kokpekti *see* Kokpekty
145 X10 **Kokpekty** ***prev.*** Kokpekti. Vostochnyy Kazakhstan, E Kazakhstan *48°47′N 82°28′E*
145 X11 **Kokpekty** ***prev.*** Kokpekti. ≈ E Kazakhstan
39 P9 **Kokrines** Alaska, USA *64°58′N 154°42′W*
39 P9 **Kokrines Hills** ▲▲ Alaska, USA
145 P17 **Koksaray** Yuzhnyy Kazakhstan, S Kazakhstan *42°34′N 68°06′E*
147 X9 **Kokshaal-Tau** ***Rus.*** Khrebet Kakshaal-Too. ▲▲ China/Kyrgyzstan
145 P7 **Kokshetau** ***Kaz.*** Kökshetaū; ***prev.*** Kokchetav. Kokshetau, N Kazakhstan *53°18′N 69°25′E*
Kökshetaū *see* Kokshetau
99 A17 **Koksijde** West-Vlaanderen, W Belgium *51°07′N 02°40′E*
12 M5 **Koksoak** ≈ Québec, E Canada
83 K24 **Kokstad** KwaZulu/Natal, E South Africa *30°23′S 29°23′E*
145 V14 **Koksu** ***Kaz.*** Rūdnīchnyy. Almaty, SE Kazakhstan *44°39′N 78°57′E*
145 W15 **Koktal** ***Kaz.*** Köktal. Almaty, SE Kazakhstan *44°05′N 79°44′E*
145 Q12 **Koktas** ≈ C Kazakhstan
Kök-Tash *see* Këk-Tash
Koktokay *see* Fuyun
147 T9 **Kok-Yangak** ***Kir.*** Kök-Janggak. Dzhalal-Abadskaya Oblast', W Kyrgyzstan *41°02′N 73°11′E*
158 F9 **Kokyar** Xinjiang Uygur Zizhiqu, W China *37°41′N 77°15′E*
149 O13 **Kolāchi** ***var.*** Kulachi. ≈ SW Pakistan
76 J15 **Kolahun** N Liberia *08°24′N 10°02′W*
171 O14 **Kolaka** Sulawesi, C Indonesia *04°04′S 121°38′E*
K'o-la-ma-i *see* Karamay
Kola Peninsula *see* Kol'skiy Poluostrov
155 H19 **Kolār** Karnātaka, E India *13°10′N 78°10′E*
155 H19 **Kolār Gold Fields** Karnātaka, E India *12°56′N 78°16′E*
92 K11 **Kolari** Lappi, NW Finland *67°20′N 23°51′E*
111 I21 **Kolárovo** ***Ger.*** Gutta; ***prev.*** Guta, ***Hung.*** Gúta. Nitriansky Kraj, SW Slovakia *47°54′N 18°01′E*
113 K16 **Kolašin** E Montenegro *42°49′N 19°32′E*
Kolāyat *see* Shir Kolāyat
95 N15 **Kolbäck** Västmanland, C Sweden *59°33′N 16°15′E*
Kolbcha *see* Kowbcha
197 Q15 **Kolbeinsey Ridge** *undersea feature* Denmark Strait/Norwegian Sea *69°00′N 17°30′W*
Kolberg *see* Kołobrzeg
95 H15 **Kolbotn** Akershus, S Norway *62°15′N 10°24′E*
111 N16 **Kolbuszowa** Podkarpackie, SE Poland *50°12′N 22°07′E*
126 L3 **Kol'chugino** Vladimirskaya Oblast', W Russian Federation *56°19′N 39°24′E*
76 H12 **Kolda** S Senegal *12°58′N 14°58′W*
95 G23 **Kolding** Syddanmark, C Denmark *55°29′N 09°30′E*
79 K20 **Kole** Kasai-Oriental, SW Dem. Rep. Congo *03°30′S 22°28′E*
79 M17 **Kole** Orientale, N Dem. Rep. Congo *02°08′N 25°25′E*
Kōle *see* Kili Island
84 F6 **Kölen** ***Nor.*** Kjølen. ▲▲ Norway/Sweden
Kolepom, Pulau *see* Yos Sudarso, Pulau
118 H3 **Kolga Laht** ***Ger.*** Kolko-Wiek. *bay* N Estonia
125 Q3 **Kolguyev, Ostrov** *island* NW Russian Federation
155 E16 **Kolhāpur** Mahārāshtra, SW India *16°42′N 74°13′E*
151 K21 **Kolhumadulu** ***var.*** Thaa Atoll. *atoll* S Maldives
93 O16 **Koli** ***var.*** Kolinkylä. Pohjois-Karjala, E Finland *63°06′N 29°46′E*
39 O13 **Koliganek** Alaska, USA *59°43′N 157°16′W*
111 E16 **Kolín** ***Ger.*** Kolin. Střední Čechy, C Czech Republic *50°02′N 15°10′E*
Kolinkylä *see* Koli
190 E12 **Koliu** Île Futuna, W Wallis and Futuna
118 E7 **Kolka** NW Latvia *57°44′N 22°34′E*
118 E7 **Kolkasrags** ***prev. Eng.*** Cape Domesnes. *headland* NW Latvia *57°45′N 22°35′E*
153 S16 **Kolkata** ***prev.*** Calcutta. *state capital* West Bengal, NE India *22°30′N 88°20′E*
Kolkhozabad *see* Kolkhozobod
147 P14 **Kolkhozobod** ***Rus.*** Kolkhozabad; ***prev.*** Kaganovichabad, Tugalan. SW Tajikistan *37°33′N 68°34′E*
Kolki/Kołki *see* Kolky
Kolko-Wiek *see* Kolga Laht
116 K3 **Kolky** ***Pol.*** Kołki, ***Rus.*** Kolki. Volyns'ka Oblast', NW Ukraine *51°05′N 25°40′E*
155 G20 **Kollegāl** Karnātaka, W India *12°08′N 77°06′E*
98 M5 **Kollum** Fryslân, N Netherlands *53°17′N 06°09′E*
Kolmar *see* Colmar
101 E16 **Köln** ***var.*** Koeln, ***Eng./Fr.*** Cologne, ***prev.*** Cöln; ***anc.*** Colonia Agrippina, Oppidum Ubiorum. Nordrhein-Westfalen, W Germany *50°57′N 06°57′E*
110 N9 **Kolno** Podlaskie, NE Poland *53°24′N 21°57′E*
110 J12 **Koło** Wielkopolskie, C Poland *52°11′N 18°39′E*
38 B8 **Kōloa** ***var.*** Koloa. Kaua'i, Hawaii, USA, C Pacific Ocean *21°54′N 159°28′W*
Koloa *see* Kōloa
110 E7 **Kołobrzeg** ***Ger.*** Kolberg. Zachodnio-pomorskie, NW Poland *54°11′N 15°34′E*
126 H4 **Kolodnya** Smolenskaya Oblast', W Russian Federation *54°57′N 32°22′E*
190 E13 **Kolofau, Mont** ⛰ Île Alofi, S Wallis and Futuna *14°21′S 178°02′W*
125 O14 **Kologriv** Kostromskaya Oblast', NW Russian Federation *58°49′N 44°22′E*
76 L12 **Kolokani** Koulikoro, W Mali *13°35′N 08°01′W*
77 N13 **Koloko** W Burkina Faso *11°06′N 05°18′W*
186 K8 **Kolombangara** ***var.*** Kilimbangara, Nduke. *island* New Georgia Islands, NW Solomon Islands
Kolomea *see* Kolomyya
126 L4 **Kolomna** Moskovskaya Oblast', W Russian Federation *55°03′N 38°52′E*
116 J7 **Kolomyya** ***Ger.*** Kolomea. Ivano-Frankivs'ka Oblast', W Ukraine *48°31′N 25°00′E*
76 M13 **Kolondiéba** Sikasso, SW Mali *11°04′N 06°55′W*
193 V15 **Kolonga** Tongatapu, S Tonga *21°07′S 175°05′W*
189 U16 **Kolonia** ***var.*** Colonia. Pohnpei, E Micronesia *06°57′N 158°12′E*
Kolonja *see* Kolonjë
113 K21 **Kolonjë** ***var.*** Kolonja. Fier, C Albania *40°49′N 19°37′E*
Kolonjë *see* Ersekë
Kolosjoki *see* Nikel'
Kolotambu *see* Avuavu
193 U15 **Kolovai** Tongatapu, S Tonga *21°05′S 175°20′W*
Kolozsvár *see* Cluj-Napoca
112 C9 **Kolpa** ***Ger.*** Kulpa, ***SCr.*** Kupa. ≈ Croatia/Slovenia
122 J11 **Kolpashevo** Tomskaya Oblast', C Russian Federation *58°21′N 82°44′E*
124 H13 **Kolpino** Leningradskaya Oblast', NW Russian Federation *59°44′N 30°39′E*
100 M10 **Kölpinsee** ◎ NE Germany
Kólpos Mórfu *see* Güzelyurt
146 K8 **Ko'lquduq** ***Rus.*** Kulkuduk. Navoiy Viloyati, N Uzbekistan *42°36′N 63°24′E*
124 K5 **Kol'skiy Poluostrov** ***Eng.*** Kola Peninsula. *peninsula* NW Russian Federation
127 T6 **Koltubanovskiy** Orenburgskaya Oblast', W Russian Federation *53°00′N 52°00′E*
112 L11 **Kolubara** ≈ C Serbia
Kolupchii *see* Gurkovo
110 K13 **Koluszki** Łódźskie, C Poland *51°44′N 19°50′E*
125 T6 **Kolva** ≈ NW Russian Federation
93 E14 **Kolvereid** Nord-Trøndelag, W Norway *64°47′N 11°22′E*
79 M24 **Kolwezi** Katanga, S Dem. Rep. Congo *10°43′S 25°29′E*
123 S7 **Kolyma** ≈ NE Russian Federation
Kolyma Lowland *see* Kolymskaya Nizmennost'
Kolyma Range/Kolymskiy, Khrebet *see* Kolymskoye Nagor'ye
123 S7 **Kolymskaya Nizmennost'** ***Eng.*** Kolyma Lowland. *lowlands* NE Russian Federation
123 S7 **Kolymskoye** Respublika Sakha (Yakutiya), NE Russian Federation *68°42′N 158°46′E*
123 U8 **Kolymskoye Nagor'ye** ***var.*** Khrebet Kolymskiy, ***Eng.*** Kolyma Range. ▲▲ E Russian Federation
123 V5 **Kolyuchinskaya Guba** *bay* NE Russian Federation
Kol'zhat *see* Kalzhat
114 G8 **Kom** ▲ NW Bulgaria *43°10′N 23°02′E*
80 I13 **Koma** Oromīya, C Ethiopia *08°19′N 36°48′E*
77 X12 **Komadugu Gana** ≈ NE Nigeria
164 M13 **Komagane** Nagano, Honshū, S Japan *35°44′N 137°54′E*
79 P17 **Komanda** Orientale, NE Dem. Rep. Congo *01°25′N 29°43′E*
197 U1 **Komandorskaya Basin** ***var.*** Kamchatka Basin. *undersea feature* SW Bering Sea *57°00′N 168°00′E*
125 Pp9 **Komandorskiye Ostrova** ***Eng.*** Commander Islands. *island group* E Russian Federation
Kománfalva *see* Comăneşti
111 I22 **Komárno** ***Ger.*** Komorn, ***Hung.*** Komárom. Nitriansky Kraj, SW Slovakia *47°46′N 18°07′E*
111 I22 **Komárom** Komárom-Esztergom, NW Hungary *47°43′N 18°06′E*
Komárom *see* Komárno
111 I22 **Komárom-Esztergom** ***off.*** Komárom-Esztergom Megye. ⬖ *county* N Hungary
Komárom-Esztergom Megye *see* Komárom-Esztergom
164 K11 **Komatsu** ***var.*** Komatu. Ishikawa, Honshū, SW Japan *36°25′N 136°27′E*
Komatu *see* Komatsu
83 D17 **Kombat** Otjozondjupa, N Namibia *19°42′S 17°45′E*
Kombissiguiri *see* Kombissiri
77 P13 **Kombissiri** ***var.*** Kombissiguiri. C Burkina Faso *12°01′N 01°27′W*
188 E10 **Komebail Lagoon** *lagoon* N Palau
81 F20 **Kome Island** *island* N Tanzania
Komeyo *see* Wandai
117 P10 **Kominternivs'ke** Odes'ka Oblast', SW Ukraine *46°52′N 30°56′E*
125 R8 **Komi, Respublika** ⬖ *autonomous republic* NW Russian Federation
111 I25 **Komló** Baranya, SW Hungary *46°11′N 18°15′E*
Kommunarsk *see* Alchevs'k
Kommunizm, Qullai *see* Ismoili Somoní, Qullai
186 B7 **Komo** Hela, W Papua New Guinea *06°06′S 142°52′E*
170 M16 **Komodo, Pulau** *island* Nusa Tenggara, S Indonesia
77 N15 **Komoé** ***var.*** Komoé Fleuve. ≈ E Ivory Coast
Komoé Fleuve *see* Komoé
75 X11 **Kom Ombo** ***var.*** Kôm Ombo, Kawm Umbū. SE Egypt *24°26′N 32°57′E*
79 F20 **Komono** Lékoumou, SW Congo *03°15′S 13°14′E*
171 Y16 **Komoran** Papua, E Indonesia *08°14′S 138°51′E*
171 Y16 **Komoran, Pulau** *island* E Indonesia
Komorn *see* Komárno
Komornok *see* Comorâşte
Komosolabad *see* Komsomolobod
Komotau *see* Chomutov
114 K13 **Komotiní** ***var.*** Gümüljina, ***Turk.*** Gümülcine. Anatolikí Makedonía kai Thráki, NE Greece *41°07′N 25°27′E*
113 K16 **Komovi** ▲▲ E Montenegro
117 R8 **Kompaniyivka** Kirovohrads'ka Oblast', C Ukraine *48°16′N 32°12′E*
Kompong *see* Kâmpóng Chhnăng
Kompong Cham *see* Kâmpóng Cham
Kompong Kleang *see* Kâmpóng Khleăng
Kompong Som *see* Sihanoukville
Kompong Speu *see* Kâmpóng Spoe
Komrat *see* Comrat
144 G12 **Komsomol** ***prev.*** Komsomol'skiy. Atyrau, W Kazakhstan *47°18′N 53°37′E*
125 W4 **Komsomol** ***prev.*** Komsomol'skiy. Respublika Komi, NW Russian Federation *67°33′N 64°00′E*
122 K14 **Komsomolets, Ostrov** *island* Severnaya Zemlya, N Russian Federation
144 F13 **Komsomolets, Zaliv** *lake gulf* SW Kazakhstan
Komsomol/Komsomolets *see* Karabalyk, Kostanay, Kazakhstan
147 Q12 **Komsomolobod** ***Rus.*** Komosolabad. C Tajikistan *38°51′N 69°54′E*
124 M16 **Komsomol'sk** Ivanovskaya Oblast', W Russian Federation *56°58′N 40°15′E*
117 S6 **Komsomol's'k** Poltavs'ka Oblast', C Ukraine *49°01′N 33°37′E*
146 M11 **Komsomol'sk** Navoiy Viloyati, N Uzbekistan *40°14′N 65°10′E*
Komsomol'skiy *see* Komsomol
Komsomol'skiy *see* Komsomol
123 S13 **Komsomol'sk-na-Amure** Khabarovskiy Kray, SE Russian Federation *50°32′N 136°59′E*
Komsomol'sk-na-Ustyurte *see* Kubla-Ustyurt
144 K10 **Komsomol'skoye** Aktyubinsk, NW Kazakhstan
127 Q8 **Komsomol'skoye** Saratovskaya Oblast', W Russian Federation *50°45′N 47°00′E*
145 P10 **Kon** ≈ C Kazakhstan
Kona *see* Kailua-Kona
124 K16 **Konakovo** Tverskaya Oblast', W Russian Federation *56°42′N 36°44′E*
Konar *see* Kunar
143 V15 **Konārak** Sīstān va Balūchestān, SE Iran *25°26′N 60°23′E*
Konarhā *see* Kunar
27 O11 **Konawa** Oklahoma, C USA *34°57′N 96°45′W*
122 H10 **Konda** ≈ C Russian Federation
154 L13 **Kondagaon** Chhattīsgarh, C India *19°38′N 81°41′E*
14 K10 **Kondiaronk, Lac** ◎ Québec, SE Canada
180 J13 **Kondinin** Western Australia *32°31′S 118°15′E*
81 H21 **Kondoa** Dodoma, C Tanzania *04°54′S 35°46′E*
127 P6 **Kondol'** Penzenskaya Oblast', W Russian Federation *52°49′N 45°03′E*
114 N10 **Kondolovo** Burgas, E Bulgaria *42°07′N 27°43′E*
171 Z16 **Kondomirat** Papua, E Indonesia *08°57′S 140°55′E*
124 J10 **Kondopoga** Respublika Kareliya, NW Russian Federation *62°13′N 34°17′E*
Kondoz *see* Kunduz
155 J17 **Kondukūr** ***var.*** Kandukur. Andhra Pradesh, E India *15°17′N 79°49′E*
Kondūz *see* Kunduz
187 P16 **Koné** Province Nord, W New Caledonia *21°04′S 164°51′E*
146 E13 **Könekesir** Balkan Welaýaty, W Turkmenistan *38°16′N 56°51′E*
146 G8 **Köneürgenç** ***var.*** Köneürgench, ***Rus.*** Këneurgench; ***prev.*** Kunya-Urgench. Daşoguz Welaýaty, N Turkmenistan *42°21′N 59°09′E*
77 N15 **Kong** N Ivory Coast *09°10′N 04°33′W*
39 S5 **Kongakut River** ≈ Alaska, USA
197 O14 **Kong Christian IX Land** ***Eng.*** King Christian IX Land. *physical region* SE Greenland
197 P13 **Kong Christian X Land** ***Eng.*** King Christian X Land. *physical region* E Greenland
197 N13 **Kong Frederik IX Land** *physical region* SW Greenland
197 Q12 **Kong Frederik VIII Land** ***Eng.*** King Frederik VIII Land. *physical region* NE Greenland
197 N15 **Kong Frederik VI Kyst** ***Eng.*** King Frederik VI Coast. *physical region* SE Greenland
167 P13 **Kŏng, Kaôh** ***prev.*** Kas Kong. *island* SW Cambodia
92 P2 **Kong Karls Land** ***Eng.*** King Charles Islands. *island group* SE Svalbard
81 G14 **Kong Kong** ≈ E South Sudan
Kongo *see* Congo (river)
83 G16 **Kongola** Caprivi, NE Namibia *17°47′S 23°24′E*
79 N21 **Kongolo** Katanga, E Dem. Rep. Congo *05°20′S 26°58′E*
81 F14 **Kongor** Jonglei, E South Sudan *07°09′N 31°44′E*
197 Q14 **Kong Oscar Fjord** *fjord* E Greenland
77 P12 **Kongoussi** N Burkina Faso *13°19′N 01°31′W*
95 G15 **Kongsberg** Buskerud, S Norway *59°39′N 09°38′E*
92 Q2 **Kongsøya** *island* Kong Karls Land, E Svalbard
95 I14 **Kongsvinger** Hedmark, S Norway *60°10′N 12°00′E*
Kongtong *see* Pingliang
167 T11 **Kông, Tônlé** ***var.*** Xê Kong. ≈ Cambodia/Laos
158 E8 **Kongur Shan** ▲ NW China *38°39′N 75°21′E*
81 I22 **Kongwa** Dodoma, C Tanzania *06°13′S 36°28′E*
Kong, Xê *see* Kông, Tônlé
Konia *see* Konya
147 R11 **Konibodom** ***Rus.*** Kanibadam. N Tajikistan *40°16′N 70°20′E*
111 K15 **Koniecpol nad Pilicą** Śląskie, S Poland *50°47′N 19°45′E*
Konieh *see* Konya
Königgrätz *see* Hradec Králové
Königinhof an der Elbe *see* Dvůr Králové nad Labem
101 K23 **Königsbrunn** Bayern, S Germany *48°16′N 10°52′E*
101 O24 **Königssee** ◎ SE Germany
Königshütte *see* Chorzów
109 S8 **Königstuhl** ▲ S Austria *46°57′N 13°47′E*
109 U3 **Königswiesen** Oberösterreich, N Austria *48°25′N 14°48′E*
101 E17 **Königswinter** Nordrhein-Westfalen, W Germany *50°40′N 07°12′E*
146 M11 **Konimex** ***Rus.*** Kenimekh. Navoiy Viloyati, N Uzbekistan *40°14′N 65°10′E*
110 I12 **Konin** ***Ger.*** Kuhnau. Weilkopolskie, C Poland *52°13′N 18°17′E*
Koninkrijk der Nederlanden *see* Netherlands
113 L24 **Konispol** ***var.*** Konispoli. Vlorë, S Albania *39°40′N 20°10′E*
Konispoli *see* Konispol
115 C15 **Kónitsa** Ípeiros, W Greece *40°04′N 20°48′E*
Konitz *see* Chojnice
108 D8 **Köniz** Bern, W Switzerland *46°56′N 07°25′E*
113 H14 **Konjic** ⬖ Federacija Bosne I Hercegovine, S Bosnia and Herzegovina
92 J10 **Könkämäälven** ≈ Finland/Sweden
155 D14 **Konkan** *plain* W India
83 D22 **Konkiep** ≈ S Namibia
76 I14 **Konkouré** ≈ W Guinea

◆ Country
● Country Capital
◇ Dependent Territory
○ Dependent Territory Capital
⬖ Administrative Regions
✈ International Airport
▲ Mountain
▲▲ Mountain Range
⛰ Volcano
≈ River
◎ Lake
⊠ Reservoir

KONNA – KRASNOLESNYY

77 O11 **Konna** Mopti, S Mali 14°58′N 03°49′W
186 H6 **Konogaiang, Mount** ▲ New Ireland, NE Papua New Guinea 04°05′S 152°43′E
186 H5 **Konogogo** New Ireland, NE Papua New Guinea 03°25′S 152°09′E
108 E9 **Konolfingen** Bern, W Switzerland 46°53′N 07°36′E
77 P16 **Konongo** C Ghana 06°39′N 01°06′W
186 H5 **Konos** New Ireland, NE Papua New Guinea 03°09′S 151°47′E
124 M12 **Konosha** Arkhangel'skaya Oblast', NW Russian Federation 60°58′N 40°09′E
117 R3 **Konotop** Sums'ka Oblast', NE Ukraine 51°15′N 33°14′E
158 L7 **Konqi He** ≈ NW China
111 L14 **Końskie** Śiętokrzyskie, C Poland 51°12′N 20°23′E
Konstantinovka see Kostyantynivka
126 M11 **Konstantinovsk** Rostovskaya Oblast', SW Russian Federation 47°37′N 41°07′E
101 H24 **Konstanz** var. Constanz, Eng. Constance, hist. Kostnitz; anc. Constantia. Baden-Württemberg, S Germany 47°40′N 09°10′E
Konstanza see Constanța
77 T14 **Kontagora** Niger, W Nigeria 10°25′N 05°29′E
78 E13 **Kontcha** Nord, N Cameroon 08°00′N 12°13′E
99 G17 **Kontich** Antwerpen, N Belgium 51°08′N 04°27′E
93 O16 **Kontiolahti** Pohjois-Karjala, SE Finland 62°46′N 29°51′E
93 M15 **Kontiomäki** Kainuu, C Finland 64°20′N 28°09′E
167 U11 **Kon Tum** var. Kontum. Kon Tum, C Vietnam 14°23′N 108°00′E
Kontum see Kon Tum
Konur see Sulakyurt
136 H15 **Konya** var. Konieh, prev. Konia; anc. Iconium. Konya, C Turkey 37°51′N 32°30′E
136 H15 **Konya** var. Konia, Konieh. ◐ province C Turkey
151 E15 **Konya Reservoir** prev. Shivāji Sāgar. ⊠ W India
145 T13 **Konyrat** var. Kounradskiy, Kaz. Qongyrat. Karaganda, SE Kazakhstan 46°57′N 75°01′E
145 W15 **Konyrolen** Almaty, SE Kazakhstan 44°16′N 79°18′E
81 I19 **Konza** Kajiado, S Kenya 01°44′S 37°07′E
98 I9 **Koog aan den Zaan** Noord-Holland, C Netherlands 52°28′N 04°49′E
182 E7 **Koonibba** South Australia 31°55′S 133°23′E
31 O11 **Koontz Lake** Indiana, N USA 41°25′N 86°24′W
171 U12 **Koor** Papua Barat, E Indonesia 0°21′S 132°28′E
183 R9 **Koorawatha** New South Wales, SE Australia 34°03′S 148°33′E
118 J5 **Koosa** Tartumaa, E Estonia 58°31′N 27°06′E
33 N7 **Kootenai** var. Kootenay. ≈ Canada/USA see also Kootenay
Kootenai see Kootenay
11 P17 **Kootenay** var. Kootenai. ≈ Canada/USA see also Kootenai
Kootenay see Kootenai
83 F24 **Kootjieskolk** Northern Cape, W South Africa 31°16′S 20°21′E
113 M15 **Kopaonik** ▲▲ S Serbia
Kopar see Koper
92 K1 **Kópasker** Norðurland Eystra, N Iceland 66°15′N 16°23′W
92 H4 **Kópavogur** Höfuðborgarsvæðið, W Iceland 64°06′N 21°47′W
145 U13 **Kopbirlik** prev. Kīrov, Kirova. Almaty, SE Kazakhstan 46°24′N 77°16′E
109 S13 **Koper** It. Capodistria; prev. Kopar. SW Slovenia 45°32′N 13°43′E
95 C16 **Kopervik** Rogaland, S Norway 59°17′N 05°20′E
Köpetdag Gershi/ Kopetdag, Khrebet see Koppeh Dāgh
Kophinou see Kofínou
182 G8 **Kopi** South Australia 33°24′S 135°40′E
153 W12 **Kopili** ≈ NE India
95 M15 **Köping** Västmanland, C Sweden 59°31′N 16°00′E
113 K17 **Koplik** var. Kopliku. Shkodër, NW Albania 42°12′N 19°26′E
Kopliku see Koplik
Kopopo see Kokopo
94 I11 **Koppang** Hedmark, S Norway 61°34′N 11°04′E
Kopparberg see Dalarna
143 S3 **Koppeh Dāgh** Rus. Khrebet Kopetdag, Turkm. Köpetdag Gershi. ▲▲ Iran/Turkmenistan
Koppename see Coppename Rivier
95 J15 **Koppom** Värmland, C Sweden 59°42′N 12°07′E
Kopreinitz see Koprivnica
114 K9 **Koprinka, Yazovir** prev. Yazovir Georgi Dimitrov. ⊠ C Bulgaria
112 F7 **Koprivnica** Ger. Kopreinitz, Hung. Kaproncza. Koprivnica-Križevci, N Croatia 46°10′N 16°49′E
112 F8 **Koprivnica-Križevci** off. Koprivničko-Križevačka Županija. ◐ province N Croatia
111 I17 **Kopřivnice** Ger. Nesselsdorf. Moravskoslezský Kraj, E Czech Republic 49°36′N 18°09′E
Koprivničko-Križevačka Županija see Koprivnica-Križevci
Köprülü see Veles
Koptsevichi see Kaptsevichy
Kopyl' see Kapyl'
119 O14 **Kopys'** Vitsyebskaya Voblasts', NE Belarus 54°19′N 30°18′E
113 M18 **Korab** ▲ Albania/ FYR Macedonia 41°48′N 20°33′E
Korabavur Pastligi see Karabaur', Uval
124 M5 **Korabel'noye** Murmanskaya Oblast', NW Russian Federation 67°00′N 41°10′E
81 M14 **K'orahē** Sumalē, E Ethiopia 06°36′N 44°21′E
115 L16 **Kórakas, Akrotírio** cape Lésvos, E Greece
112 D9 **Korana** ≈ C Croatia
155 L14 **Korāput** Odisha, E India 18°48′N 82°41′E
Korat see Nakhon Ratchasima
167 Q9 **Korat Plateau** plateau E Thailand
Kōrāwa, Sar-I see Kirdī Kawrāw, Qimmat
154 L11 **Korba** Chhattīsgarh, C India 22°25′N 82°43′E
101 H15 **Korbach** Hessen, C Germany 51°16′N 08°52′E
Korça see Korçë
113 M21 **Korçë** var. Korça, Gk. Korytsa, It. Corriza; prev. Koritsa. Korçë, SE Albania 40°38′N 20°47′E
113 M21 **Korçë** ◐ district SE Albania
113 G15 **Korčula** It. Curzola. Dubrovnik-Neretva, S Croatia 42°57′N 17°08′E
113 F15 **Korčula** It. Curzola; anc. Corcyra Nigra. island S Croatia
113 F15 **Korčulanski Kanal** channel S Croatia
145 T6 **Korday** prev. Georgiyevka. Zhambyl, SE Kazakhstan 43°03′N 74°43′E
142 J5 **Kordestān** off. Ostān-e Kordestān, var. Kurdestan. ◐ province W Iran
Kordestān, Ostān-e see Kordestān
143 P4 **Kord Kūy** var. Kurd Kui. Golestán, N Iran 36°49′N 54°05′E
163 V13 **Korea Bay** bay China/North Korea
Korea, Democratic People's Republic of see North Korea
171 T15 **Koreare** Pulau Yamdena, E Indonesia 07°33′S 131°13′E
Korea, Republic of see South Korea
163 Z17 **Korea Strait** Jap. Chōsen-kaikyō, Kor. Taehan-haehyŏp. channel Japan/South Korea
Korelichi/Korelicze see Karelichy
80 J11 **Korem** Tigrai, N Ethiopia 12°32′N 39°29′E
77 U11 **Korén Adoua** ≈ C Niger
126 I7 **Korenevo** Kurskaya Oblast', W Russian Federation 51°21′N 34°53′E
126 L13 **Korenovsk** Krasnodarskiy Kray, SW Russian Federation 45°28′N 39°25′E
116 L4 **Korets'** Pol. Korzec, Rus. Korets. Rivnens'ka Oblast', NW Ukraine 50°38′N 27°12′E
Korets see Korets'
194 L7 **Korff Ice Rise** ice cap Antarctica
145 Q10 **Korgalzhyn** var. Kurgal'dzhino, Kurgal'dzhinsky, Kaz. Qorghalzhyn. Akmola, C Kazakhstan 50°33′N 69°58′E
145 W15 **Korgas** ; prev. Khorgos. Almaty, SE Kazakhstan 44°13′N 80°22′E
92 G13 **Korgen** Troms, N Norway 66°04′N 13°51′E
147 R9 **Korgon-Dēbē** Dzhalal-Abadskaya Oblast', W Kyrgyzstan 41°51′N 70°52′E
76 M14 **Korhogo** N Ivory Coast 09°29′N 05°39′W
115 F19 **Korinthiakós Kólpos** Eng. Gulf of Corinth; anc. Corinthiacus Sinus. gulf C Greece
115 F19 **Kórinthos** anc. Corinthus Eng. Corinth. Pelopónnisos, S Greece 37°56′N 22°55′E
113 M18 **Koritnik** ▲ S Serbia 42°06′N 20°34′E
Koritsa see Korçë
165 P11 **Kōriyama** Fukushima, Honshū, C Japan 37°25′N 140°20′E
136 E16 **Korkuteli** Antalya, SW Turkey 37°07′N 30°11′E
158 K6 **Korla** Chin. K'u-erh-lo. Xinjiang Uygur Zizhiqu, NW China 41°48′N 86°10′E
122 J10 **Korliki** Khanty-Mansiyskiy Avtonomnyy Okrug-Yugra, C Russian Federation 61°28′N 82°12′E
Körlin an der Persante see Karlino
Korma see Karma
14 D8 **Kormak** Ontario, S Canada 47°38′N 83°00′W
Kormakíti, Akrotíri/ Kormakiti, Cape/ Kormakítis see Koruçam Burnu
111 G23 **Körmend** Vas, W Hungary 47°02′N 16°35′E
Kōrmōr Şalāḩ ad Dīn, E Iraq 35°06′N 44°47′E
112 C13 **Kornat** It. Incoronata. island W Croatia
Korneshty see Cornești
109 X3 **Korneuburg** Niederösterreich, NE Austria 48°22′N 16°20′E
145 P7 **Korneyevka** Severnyy Kazakhstan, N Kazakhstan 54°01′N 68°30′E
95 I17 **Kornsjø** Østfold, S Norway 58°55′N 11°40′E
77 O11 **Koro** Mopti, S Mali 14°05′N 03°06′W
1873 Y14 **Koro** island C Fiji
186 B7 **Koroba** Hela, W Papua New Guinea 05°46′S 142°48′E
126 K8 **Korocha** Belgorodskaya Oblast', W Russian Federation 50°49′N 37°08′E
136 H12 **Köroğlu Dağları** ▲▲ C Turkey
183 V6 **Korogoro Point** headland New South Wales, SE Australia 31°03′S 153°04′E
81 J21 **Korogwe** Tanga, E Tanzania 05°10′S 38°30′E
182 L13 **Koroit** Victoria, SE Australia 38°17′S 142°22′E
187 X15 **Korolevu** Viti Levu, W Fiji 18°12′S 177°44′E
190 I17 **Koromiri** island S Cook Islands
171 Q8 **Koronadal** Mindanao, S Philippines 06°23′N 124°54′E
114 G13 **Koróneia, Límni** var. Límni Korónia. ◎ N Greece
115 E22 **Koróni** Pelopónnisos, S Greece 36°47′N 21°57′E
Korónia, Límni see Koróneia, Límni
110 I9 **Koronowo** Ger. Krone an der Brahe. Kujawski-pomorskie,C Poland 53°18′N 17°56′E
117 R2 **Korop** Chernihivs'ka Oblast', N Ukraine 51°35′N 32°57′E
115 H19 **Koropí** Attikí, C Greece 37°54′N 23°52′E
188 C8 **Koror** (Palau) Oreor, N Palau 07°21′N 134°28′E
Koror see Oreor
Koror Lāl Esan see Karor
Koror Pacca see Kahror Pakka
111 L23 **Körös** ≈ E Hungary
Kőrös see Križevci
Körösbánya see Baia de Criș
187 Y14 **Koro Sea** sea C Fiji
Koroška see Kärnten
117 N3 **Korosten'** Zhytomyrs'ka Oblast', NW Ukraine 50°56′N 28°39′E
Korostyshev see Korostyshiv
117 N4 **Korostyshiv** Rus. Korostyshev. Zhytomyrs'ka Oblast', N Ukraine 50°18′N 29°05′E
125 V3 **Korotaikha** ≈ NW Russian Federation
122 J9 **Korotchayevo** Yamalo-Nenetskiy Avtonomnyy Okrug, N Russian Federation 66°00′N 78°11′E
78 I8 **Koro Toro** Borkou, N Chad 16°01′N 18°27′E
39 N16 **Korovin Island** island Shumagin Islands, Alaska, USA
187 X14 **Korovou** Viti Levu, W Fiji 17°48′S 178°32′E
93 M17 **Korpilahti** Keski-Suomi, C Finland 62°02′N 25°34′E
92 K12 **Korpilombolo** Lapp. Dállogilli. Norrbotten, N Sweden 66°51′N 23°00′E
123 T13 **Korsakov** Ostrov Sakhalin, Sakhalinskaya Oblast', SE Russian Federation 46°41′N 142°45′E
93 J16 **Korsholm** Fin. Mustasaari. Österbotten, W Finland 63°05′N 21°43′E
95 I23 **Korsør** Sjælland, E Denmark 55°19′N 11°09′E
Korsovka see Kārsava
117 P6 **Korsun'-Shevchenkivs'kyy** Rus. Korsun'-Shevchenkovskiy. Cherkas'ka Oblast', C Ukraine 49°26′N 31°15′E
Korsun'-Shevchenkovskiy see Korsun'-Shevchenkivs'kyy
99 C17 **Kortemark** West-Vlaanderen, W Belgium 51°03′N 03°03′E
99 H18 **Kortenberg** Vlaams Brabant, C Belgium 50°53′N 04°33′E
99 K18 **Kortessem** Limburg, NE Belgium 50°52′N 05°22′E
99 E14 **Kortgene** Zeeland, SW Netherlands 51°34′N 03°48′E
80 F8 **Korti** Northern, N Sudan 18°06′N 31°33′E
99 C18 **Kortrijk** Fr. Courtrai. West-Vlaanderen, W Belgium 50°50′N 03°17′E
121 O2 **Koruçam Burnu** var. Cape Kormakiti, Kormakítis, Gk. Akrotíri Kormakíti. headland N Cyprus 35°24′N 32°55′E
183 O13 **Korumburra** Victoria, SE Australia 38°27′S 145°48′E
Koryak Range see Koryakskoye Nagor'ye
Koryakskiy Khrebet see Koryakskoye Nagor'ye
123 V8 **Koryakskiy Okrug** ◐ autonomous district E Russian Federation
123 V7 **Koryakskoye Nagor'ye** var. Koryakskiy Khrebet, Eng. Koryak Range. ▲▲ NE Russian Federation
125 P11 **Koryazhma** Arkhangel'skaya Oblast', NW Russian Federation 61°16′N 47°07′E
Kōryō see Gangneung
Korytsa see Korçë
117 Q2 **Koryukivka** Chernihivs'ka Oblast', N Ukraine 51°45′N 32°16′E
Korzec see Korets'
115 N21 **Kos** Kos, Dodekánisa, Greece, Aegean Sea 36°53′N 27°19′E
115 M21 **Kos** It. Coo; anc. Cos. island Dodekánisa, Greece, Aegean Sea
125 T12 **Kosa** Komi-Permyatskiy Okrug, NW Russian Federation 59°55′N 54°54′E
125 T13 **Kosa** ≈ NW Russian Federation
164 B12 **Kō-saki** headland Nagasaki, Tsushima, SW Japan 34°06′N 129°13′E
163 X13 **Kosan** SE North Korea 38°50′N 127°26′E
119 H18 **Kosava** Rus. Kosovo. Brestskaya Voblasts', SW Belarus 52°45′N 25°16′E
Kosch see Kose
Koschagyl see Kosshagyl
110 G12 **Kościan** Ger. Kosten. Wielkopolskie, C Poland 52°05′N 16°38′E
110 I7 **Kościerzyna** Pomorskie, NW Poland 54°07′N 17°55′E
22 L4 **Kosciusko** Mississippi, S USA 33°03′N 89°35′W
Kosciusko, Mount see Kosciuszko, Mount
183 R11 **Kosciuszko, Mount** prev. Mount Kosciusko. ▲ New South Wales, SE Australia 36°28′S 148°15′E
118 H4 **Kose** Ger. Kosch. Harjumaa, NW Estonia 59°11′N 25°10′E
25 U9 **Kosse** Texas, SW USA 31°16′N 96°38′W
114 G6 **Koshava** Vidin, NW Bulgaria 44°03′N 23°00′E
147 U9 **Kosh-Dēbē** var. Koshtebē. Narynskaya Oblast', C Kyrgyzstan 41°03′N 74°08′E
K'o-shih see Kashi
164 B12 **Koshikijima-rettō** var. Kosikizima Rettô. island group SW Japan
145 W13 **Koshkarkol', Ozero** ◎ SE Kazakhstan
30 L9 **Koshkonong, Lake** ◎ Wisconsin, N USA
Koshoba see Goșabo
Koshoba see Goșoba
164 M12 **Kōshoku** var. Kôsyoku. Nagano, Honshū, S Japan 36°33′N 138°09′E
Koshtebē see Kosh-Dēbē
Kōshū see Enzan
Kōshū see Gwangju
111 N19 **Košice** Ger. Kaschau, Hung. Kassa. Košický Kraj, E Slovakia 48°44′N 21°15′E
111 M20 **Košický Kraj** ◐ region E Slovakia
Kosikizima Rettô see Koshikijima-rettō
153 R12 **Kosi Reservoir** ⊠ E Nepal
116 J8 **Kosiv** Ivano-Frankivs'ka Oblast', W Ukraine 48°19′N 25°04′E
145 O11 **Koskol'** Kaz. Qosköl. Karaganda, C Kazakhstan 49°32′N 67°08′E
125 Q9 **Koslan** Respublika Komi, NW Russian Federation 63°27′N 48°52′E
Köslin see Koszalin
146 M12 **Koson** Rus. Kasan. Qashqadaryo Viloyati, S Uzbekistan 39°04′N 65°35′E
Kosŏng see Goseong
147 S9 **Kosonsoy** Rus. Kasansay. Namangan Viloyati, E Uzbekistan 41°15′N 71°28′E
113 M16 **Kosovo** prev. Autonomous Province of Kosovo and Metohija. ◆ republic SE Europe
Kosovo see Kosava
Kosovo and Metohija, Autonomous Province of see Kosovo
Kosovo Polje see Fushë Kosovë
Kosovska Kamenica see Kamenicë
Kosovska Mitrovica see Mitrovicë
189 X17 **Kosrae** ◐ state E Micronesia
189 Y14 **Kosrae** prev. Kusaie. island Caroline Islands, E Micronesia
109 P6 **Kössen** Tirol, W Austria 47°40′N 12°24′E
144 G12 **Kosshagyl** prev. Koschagyl, Kaz. Qosshaghyl. Atyrau, W Kazakhstan 46°52′N 53°46′E
76 M16 **Kossou, Lac de** ◎ C Ivory Coast
Kossukavak see Krumovgrad
Kostajnica see Hrvatska Kostanjnica
Kostamus see Kostomuksha
144 M7 **Kostanay** var. Kustanay, Kaz. Qostanay. Kostanay, N Kazakhstan 53°16′N 63°34′E
144 L8 **Kostanay** var. Kostanayskaya Oblast', Kaz. Qostanay Oblysy. ◐ province N Kazakhstan
Kostanayskaya Oblast' see Kostanay
Kosten see Kościan
114 H10 **Kostenets** prev. Georgi Dimitrov. Sofia, W Bulgaria 42°15′N 23°48′E
80 F10 **Kosti** White Nile, C Sudan 13°11′N 32°38′E
Kostnitz see Konstanz
124 H7 **Kostomuksha** Fin. Kostamus. Respublika Kareliya, NW Russian Federation 64°33′N 30°28′E
116 K3 **Kostopil'** Rus. Kostopol'. Rivnens'ka Oblast', NW Ukraine 50°20′N 26°29′E
Kostopol' see Kostopil'
124 M15 **Kostroma** Kostromskaya Oblast', NW Russian Federation 57°46′N 41°E
125 N14 **Kostroma** ≈ NW Russian Federation
125 N14 **Kostromskaya Oblast'** ◐ province NW Russian Federation
110 D11 **Kostrzyn** Ger. Cüstrin, Küstrin. Lubuskie, W Poland 52°35′N 14°40′E
110 H11 **Kostrzyn** Wielkopolskie, C Poland 52°23′N 17°13′E
117 X7 **Kostyantynivka** Rus. Konstantinovka. Donets'ka Oblast', SE Ukraine
Kostyukovichi see Kastsyukovichy
Kostyukovka see Kastsyukowka
Kôsyoku see Kōshoku
125 U6 **Kos'yu** Respublika Komi, NW Russian Federation 65°39′N 59°01′E
125 U6 **Kos'yu** ≈ NW Russian Federation
110 F7 **Koszalin** Ger. Köslin. Zachodnio-pomorskie, NW Poland 54°12′N 16°10′E
111 F22 **Kőszeg** Ger. Güns. Vas, W Hungary 47°23′N 16°33′E
152 H13 **Kota** prev. Kotah. Rājasthān, N India 25°14′N 75°52′E
Kota Baharu see Kota Bharu
Kota Bahru see Kota Bharu
169 U13 **Kotabaru** Pulau Laut, C Indonesia 03°15′S 116°15′E
168 K12 **Kota Baru** Sumatera, W Indonesia 01°07′S 101°43′E
Kotabaru see Jayapura
168 K6 **Kota Bharu** var. Kota Baharu, Kota Bahru. Kelantan, Peninsular Malaysia 06°07′N 102°15′E
Kotaboemi see Kotabumi
168 M14 **Kotabumi** prev. Kotaboemi. Sumatera, W Indonesia 04°50′S 104°54′E
149 S10 **Kot Addu** Punjab, E Pakistan 30°28′N 70°58′E
Kotah see Kota
169 U7 **Kota Kinabalu** prev. Jesselton. Sabah, East Malaysia 05°59′N 116°04′E
169 U7 **Kota Kinabalu** ✈ Sabah, East Malaysia 05°59′N 116°04′E
92 M12 **Kotala** Lappi, N Finland 67°01′N 29°00′E
Kotamobagoe see Kotamobagu
171 Q11 **Kotamobagu** prev. Kotamobagoe. Sulawesi, C Indonesia 0°46′N 124°21′E
155 L14 **Kotapad** var. Kotapārh. Odisha, E India 19°10′N 82°23′E
Kotapārh see Kotapad
166 N17 **Ko Ta Ru Tao** island SW Thailand
169 R13 **Kotawaringin, Teluk** bay Borneo, C Indonesia
149 Q13 **Kot Dīji** Sind, SE Pakistan 27°16′N 68°44′E
152 K9 **Kotdwāra** Uttarakhand, N India 29°44′N 78°33′E
125 Q14 **Kotel'nich** Kirovskaya Oblast', NW Russian Federation 58°19′N 48°12′E
127 N12 **Kotel'nikovo** Volgogradskaya Oblast', SW Russian Federation 47°37′N 43°07′E
123 Q6 **Kotel'nyy, Ostrov** island Novosibirskiye Ostrova, N Russian Federation
117 T5 **Kotel'va** Poltavs'ka Oblast', C Ukraine 50°04′N 34°46′E
101 M14 **Köthen** var. Cöthen. Sachsen-Anhalt, C Germany 51°46′N 11°59′E
Kôti see Kōchi
81 G17 **Kotido** NE Uganda 03°03′N 34°07′E
93 N19 **Kotka** Kymenlaakso, S Finland 60°28′N 26°55′E
125 P11 **Kotlas** Arkhangel'skaya Oblast', NW Russian Federation 61°14′N 46°43′E
38 M10 **Kotlik** Alaska, USA 63°01′N 163°33′W
77 Q17 **Kotoka** ✈ (Accra) S Ghana 05°41′N 00°10′W
Kotonu see Cotonou
113 J17 **Kotor** It. Cattaro. SW Montenegro 42°25′N 18°47′E
Kotor see Kotoriba
112 F7 **Kotoriba** Hung. Kotor. Medimurje, N Croatia 46°20′N 16°47′E
113 I17 **Kotorska, Boka** It. Bocche di Cattaro. bay SW Montenegro
112 H11 **Kotorsko** ◐ Republika Srpska, N Bosnia and Herzegovina
112 G11 **Kotor Varoš** ◐ Republika Srpska, N Bosnia and Herzegovina
Koto Sho/Kotosho see Lan Yu
126 M7 **Kotovsk** Tambovskaya Oblast', W Russian Federation 52°39′N 41°31′E
117 O9 **Kotovs'k** Rus. Kotovsk. Odes'ka Oblast', SW Ukraine 47°42′N 29°30′E
Kotovsk see Hîncești
119 G16 **Kotra** ≈ W Belarus
149 P16 **Kotri** Sind, SE Pakistan 25°22′N 68°18′E
109 Q9 **Kötschach** Kärnten, S Austria 46°41′N 12°57′E
155 K15 **Kottagūdem** Telangana, E India 17°36′N 80°40′E
155 F21 **Kottappadi** Kerala, SW India 11°38′N 76°03′E
155 G23 **Kottayam** Kerala, SW India 09°34′N 76°31′E
Kottbus see Cottbus
79 K15 **Kotto** ≈ Central African Republic/Dem. Rep. Congo
193 X15 **Kotu Group** island group W Tonga
Koturdepe see Goturdepe
122 M9 **Kotuy** ≈ N Russian Federation
83 M16 **Kotwa** Mashonaland East, NE Zimbabwe 16°58′S 32°46′E
39 N7 **Kotzebue** Alaska, USA 66°54′N 162°36′W
38 M7 **Kotzebue Sound** inlet Alaska, USA
Kotzenan see Chocianów
Kotzman see Kitsman'
77 R14 **Kouandé** NW Benin 10°20′N 01°42′E
79 J15 **Kouango** Ouaka, S Central African Republic 05°00′N 20°01′E
77 O13 **Koudougou** C Burkina Faso 12°15′N 02°23′W
98 K7 **Koudum** Fryslân, N Netherlands 52°55′N 05°26′E
115 L25 **Koufonísi** island SE Greece
115 K21 **Koufonísi** island Kykládes, Greece, Aegean Sea
38 M8 **Kougarok Mountain** ▲ Alaska, USA 65°41′N 165°29′W
79 E21 **Kouilou** ◐ province SW Congo
79 E20 **Kouilou** ≈ S Congo
167 Q11 **Koŭk Kduŏch** prev. Phumĭ Koŭk Kduŏch. Bătdâmbâng, NW Cambodia 13°16′N 103°08′E
121 O3 **Koúklia** SW Cyprus 34°42′N 32°35′E
79 E19 **Koulamoutou** Ogooué-Lolo, C Gabon 01°07′S 12°27′E
76 L12 **Koulikoro** Koulikoro, SW Mali 12°55′N 07°31′W
76 L11 **Koulikoro** ◐ region SW Mali
187 P16 **Koumac** Province Nord, W New Caledonia 20°34′S 164°18′E
165 N12 **Koumi** Nagano, Honshū, S Japan 36°06′N 138°27′E
78 I13 **Koumra** Mandoul, S Chad 08°56′N 17°32′E
Kounadougou see Koundougou
76 M15 **Kounahiri** C Ivory Coast 07°47′N 05°51′W
76 I12 **Koundâra** Moyenne-Guinée, NW Guinea 12°28′N 13°15′W
77 N13 **Koundougou** var. Kounadougou. C Burkina Faso 11°43′N 04°40′W
76 H11 **Koungheul** C Senegal 14°00′N 14°48′W
Kounradskiy see Konyrat
25 X10 **Kountze** Texas, SW USA 30°22′N 94°20′W
77 Q13 **Koupéla** C Burkina Faso 12°07′N 00°21′W
77 N13 **Kouri** Sikasso, SW Mali 12°09′N 04°46′W
55 Y9 **Kourou** N French Guiana 05°08′N 52°37′W
114 J12 **Kouroú** ≈ NE Greece
76 K14 **Kouroussa** C Guinea 10°40′N 09°50′W
Kousseir see Al Quşayr
78 G11 **Kousséri** prev. Fort-Foureau. Extrême-Nord, NE Cameroon 12°05′N 14°56′E
Kouteifé see Al Quţayfah
76 M13 **Koutiala** Sikasso, S Mali 12°20′N 05°23′W
76 M14 **Kouto** NW Ivory Coast 09°51′N 06°25′W
93 M19 **Kouvola** Kymenlaakso, S Finland 60°50′N 26°48′E
79 G18 **Kouyou** ≈ C Congo
112 M10 **Kovačica** Hung. Antafalva; prev. Kovacsicza. Vojvodina, N Serbia 45°08′N 20°36′E
Kovacsicza see Kovačica
Kővárhosszúfalu see Satulung
Kovászna see Covasna
124 I4 **Kovdor** Murmanskaya Oblast', NW Russian Federation 67°32′N 30°27′E
116 J3 **Kovel'** Pol. Kowel. Volyns'ka Oblast', NW Ukraine 51°14′N 24°43′E
112 M11 **Kovin** Hung. Kevevára; prev. Temes-Kubin. Vojvodina, NE Serbia 44°45′N 20°59′E
Kovno see Kaunas
127 N3 **Kovrov** Vladimirskaya Oblast', W Russian Federation 56°24′N 41°21′E
127 O5 **Kovylkino** Respublika Mordoviya, W Russian Federation 54°03′N 43°52′E
110 J11 **Kowal** Kujawsko-pomorskie, C Poland 52°31′N 19°09′E
110 J9 **Kowalewo Pomorskie** Ger. Schönsee. Kujawsko-pomorskie, N Poland 53°07′N 18°48′E
Kowasna see Covasna
119 M16 **Kowbcha** Rus. Kolbcha. Mahilyowskaya Voblasts', E Belarus 53°39′N 29°14′E
Koweit see Kuwait
Kowel see Kovel'
185 F17 **Kowhitirangi** West Coast, South Island, New Zealand 42°54′S 171°01′E
161 O15 **Kowloon** Hong Kong, S China
Kowno see Kaunas
159 N7 **Kox Kuduk** well NW China
136 D16 **Köyceğiz** Muğla, SW Turkey 36°57′N 28°40′E
125 N6 **Koyda** Arkhangel'skaya Oblast', NW Russian Federation 66°22′N 42°42′E
139 T3 **Koye** Ar. Kūysanjaq, var. Koi Sanjaq. Arbīl, N Iraq 36°05′N 44°38′E
Koymat see Goymat
Koymatdag, Gory see Goymatdag, Gory
151 E15 **Koyna Reservoir** ⊠ W India
165 P9 **Koyoshi-gawa** ≈ Honshū, C Japan
Koi Sanjaq see Koye
Koytash see Qo'ytosh
146 M14 **Köýtendag** prev. Rus. Charshanga, Charshangy, Turkm. Charshangngy. Lebap Welaýaty, E Turkmenistan 37°31′N 65°58′E
39 N9 **Koyuk** Alaska, USA 64°55′N 161°09′W
39 N9 **Koyuk River** ≈ Alaska, USA
39 O9 **Koyukuk** Alaska, USA 64°52′N 157°42′W
39 O9 **Koyukuk River** ≈ Alaska, USA
136 J13 **Kozaklı** Nevşehir, C Turkey 39°12′N 34°48′E
136 K16 **Kozan** Adana, S Turkey 37°27′N 35°47′E
115 E14 **Kozáni** Dytikí Makedonía, N Greece 40°19′N 21°48′E
112 F10 **Kozara** ▲▲ NW Bosnia and Herzegovina
Kozarska Dubica see Bosanska Dubica
117 P3 **Kozelets'** Rus. Kozelets. Chernihivs'ka Oblast', NE Ukraine 50°54′N 31°09′E
Kozelets see Kozelets'
117 S6 **Kozel'shchyna** Poltavs'ka Oblast', C Ukraine 49°13′N 33°49′E
126 J5 **Kozel'sk** Kaluzhskaya Oblast', W Russian Federation 54°04′N 35°51′E
151 F21 **Kozhikode** var. Calicut. Kerala, SW India 11°17′N 75°49′E see also Calicut
Kozhimiz, Gora see Kozhymiz, Gora
124 L9 **Kozhozero, Ozero** ◎ NW Russian Federation
125 T7 **Kozhva** Respublika Komi, NW Russian Federation 65°06′N 57°00′E
125 T7 **Kozhva** ≈ NW Russian Federation
125 U6 **Kozhym** Respublika Komi, NW Russian Federation 65°43′N 59°25′E
125 V9 **Kozhymiz, Gora** prev. Gora Kozhimiz. ▲ NW Russian Federation 63°13′N 58°54′E
110 N13 **Kozienice** Mazowieckie, C Poland 51°35′N 21°31′E
109 S13 **Kozina** SW Slovenia 45°36′N 13°56′E
114 H7 **Kozloduy** Vratsa, NW Bulgaria 43°48′N 23°42′E
127 Q3 **Kozlovka** Chuvashskaya Respublika, W Russian Federation 55°53′N 48°07′E
Kozlovshchina/ Kozlowszczyzna see Kazlowshchyna
127 P3 **Koz'modem'yansk** Respublika Mariy El, W Russian Federation 56°19′N 46°33′E
116 J6 **Kozova** Ternopil's'ka Oblast', W Ukraine 49°25′N 25°09′E
113 P20 **Kožuf** ▲▲ Greece/Macedonia
165 N15 **Kōzu-shima** island E Japan
Koz'yany see Kazyany
117 N5 **Kozyatyn** Rus. Kazatin. Vinnyts'ka Oblast', C Ukraine 49°41′N 28°49′E
77 Q16 **Kpalimé** var. Palimé. SW Togo 06°54′N 00°38′E
77 Q16 **Kpandu** E Ghana 07°00′N 00°18′E
99 F15 **Krabbendijke** Zeeland, SW Netherlands 51°25′N 04°07′E
167 N15 **Krabi** var. Muang Krabi. Krabi, SW Thailand 08°04′N 98°52′E
167 N13 **Kra Buri** Ranong, SW Thailand 10°25′N 98°48′E
167 S12 **Krâchéh** prev. Kratie. Krâchéh, E Cambodia 12°29′N 106°01′E
95 G17 **Kragerø** Telemark, S Norway 58°54′N 09°25′E
112 M13 **Kragujevac** Serbia, C Serbia 44°01′N 20°55′E
Krainburg see Kranj
166 N13 **Kra, Isthmus of** isthmus Malaysia/Thailand
112 D12 **Krajina** cultural region SW Croatia
Krakatau, Pulau see Rakata, Pulau
Krakau see Kraków
111 L16 **Kraków** Eng. Cracow, Ger. Krakau; anc. Cracovia. Małopolskie, S Poland 50°03′N 19°58′E
100 L9 **Krakower See** ◎ NE Germany
167 Q11 **Krâlănh** Siĕmréab, NW Cambodia 13°35′N 103°27′E
45 Q16 **Kralendijk** ○ Bonaire 12°07′N 68°13′W
112 B10 **Kraljevica** It. Porto Re. Primorje-Gorski Kotar, NW Croatia 45°15′N 14°36′E
112 M13 **Kraljevo** prev. Rankovićevo. Serbia, C Serbia 43°44′N 20°40′E
111 E16 **Královehradecký Kraj** prev. Hradecký Kraj. ◐ region N Czech Republic
Kralup an der Moldau see Kralupy nad Vltavou
111 C16 **Kralupy nad Vltavou** Ger. Kralup an der Moldau. Středočeský Kraj, NW Czech Republic 50°15′N 14°20′E
117 W7 **Kramators'k** Rus. Kramatorsk. Donets'ka Oblast', SE Ukraine 48°43′N 37°34′E
Kramatorsk see Kramators'k
93 H17 **Kramfors** Västernorrland, C Sweden 62°55′N 17°50′E
Kranéa see Kraniá
108 M7 **Kranebitten** ✈ (Innsbruck) Tirol, W Austria 47°18′N 11°21′E
115 D15 **Kraniá** var. Kranéa. Dytikí Makedonía, N Greece 39°54′N 21°21′E
115 G20 **Kranídi** Pelopónnisos, S Greece 37°21′N 23°09′E
109 T11 **Kranj** Ger. Krainburg. NW Slovenia 46°17′N 14°16′E
115 F16 **Krannón** battleground Thessalía, C Greece
Kranz see Zelenogradsk
112 D7 **Krapina** Krapina-Zagorje, N Croatia 46°12′N 15°52′E
112 E8 **Krapina** ≈ N Croatia
112 D8 **Krapina-Zagorje** off. Krapinsko-Zagorska Županija. ◐ province N Croatia
114 L7 **Krapinets** ≈ NE Bulgaria
Krapinsko-Zagorska Županija see Krapina-Zagorje
111 I15 **Krapkowice** Ger. Krappitz. Opolskie, SW Poland 50°29′N 17°56′E
Krappitz see Krapkowice
125 O12 **Krasavino** Vologodskaya Oblast', NW Russian Federation 60°56′N 46°27′E
122 H6 **Krasino** Novaya Zemlya, Arkhangel'skaya Oblast', N Russian Federation 70°45′N 54°16′E
123 S15 **Kraskino** Primorskiy Kray, SE Russian Federation 42°40′N 130°51′E
118 J11 **Krāslava** SE Latvia 55°56′N 27°08′E
119 M14 **Krasnaluki** Rus. Krasnoluki. Vitsyebskaya Voblasts', N Belarus 54°37′N 28°50′E
119 P17 **Krasnapollye** Rus. Krasnopol'ye. Mahilyowskaya Voblasts', E Belarus 53°20′N 31°24′E
126 L15 **Krasnaya Polyana** Krasnodarskiy Kray, SW Russian Federation 43°40′N 40°13′E
Krasnaya Slabada / Krasnaya Sloboda see Chyrvonaya Slabada
119 J15 **Krasnaye** Rus. Krasnoye. Minskaya Voblasts', C Belarus 54°14′N 27°05′E
111 O14 **Kraśnik** Ger. Kratznick. Lubelskie, E Poland 50°56′N 22°14′E
117 O9 **Krasni Okny** Odes'ka Oblast', SW Ukraine 47°33′N 29°28′E
127 P8 **Krasnoarmeysk** Saratovskaya Oblast', W Russian Federation 51°02′N 45°42′E
Krasnoarmeysk see Tayynsha
Krasnoarmeysk see Krasnoarmiys'k/Tayynsha
123 T6 **Krasnoarmeyskiy** Chukotskiy Avtonomnyy Okrug, NE Russian Federation 69°30′N 171°44′E
117 W7 **Krasnoarmiys'k** Rus. Krasnoarmeysk. Donets'ka Oblast', SE Ukraine 48°17′N 37°14′E
125 P11 **Krasnoborsk** Arkhangel'skaya Oblast', NW Russian Federation 61°31′N 45°57′E
126 K14 **Krasnodar** prev. Ekaterinodar, Yekaterinodar. Krasnodarskiy Kray, SW Russian Federation 45°06′N 39°01′E
126 K13 **Krasnodarskiy Kray** ◐ territory SW Russian Federation
117 Z7 **Krasnodon** Luhans'ka Oblast', E Ukraine 48°17′N 39°44′E
Krasnogor see Kallaste
127 T2 **Krasnogorskoye** Udmurtskaya Respublika, NW Russian Federation 57°42′N 52°29′E
Krasnograd see Krasnohrad
Krasnogvardeysk see Bulungh'ur
126 M13 **Krasnogvardeyskoye** Stavropol'skiy Kray, SW Russian Federation 45°49′N 41°31′E
Krasnogvardeyskoye see Krasnohvardiys'ke
117 U6 **Krasnohrad** Rus. Krasnograd. Kharkivs'ka Oblast', E Ukraine 49°22′N 35°28′E
117 S12 **Krasnohvardiys'ke** Rus. Krasnogvardeyskoye. Avtonomna Respublika Krym, S Ukraine 45°30′N 34°19′E
123 P14 **Krasnokamensk** Zabaykal'skiy Kray, S Russian Federation 50°03′N 118°01′E
125 U14 **Krasnokamsk** Permskiy Kray, W Russian Federation 58°08′N 55°48′E
127 U8 **Krasnokholm** Orenburgskaya Oblast', W Russian Federation 51°34′N 54°11′E
117 U5 **Krasnokuts'k** Rus. Krasnokutsk. Kharkivs'ka Oblast', E Ukraine 50°01′N 35°03′E
Krasnokutsk see Krasnokuts'k
126 L7 **Krasnolesnyy** Voronezhskaya Oblast', W Russian Federation 51°53′N 39°37′E

◆ Country
● Country Capital
◇ Dependent Territory
○ Dependent Territory Capital
◐ Administrative Regions
✈ International Airport
▲ Mountain
▲▲ Mountain Range
Volcano
≈ River
◎ Lake
⊠ Reservoir

KRASNOLUKI – KUPYANSK

Krasnoluki see Krasnaluki
Krasnoosol'skoye Vodokhranilishche see Chervonooskil's'ke Vodoskhovyshche
117 S11 **Krasnoperekops'k** Rus. Krasnoperekopsk. Avtonomna Respublika Krym, S Ukraine 45°56′N 33°47′E
Krasnoperekopsk see Krasnoperekops'k
117 U4 **Krasnopillya** Sums'ka Oblast', NE Ukraine 50°46′N 35°17′E
Krasnopol'ye see Krasnapollye
124 L5 **Krasnoshchel'ye** Murmanskaya Oblast', NW Russian Federation 67°22′N 37°03′E
127 O5 **Krasnoslobodsk** Respublika Mordoviya, W Russian Federation 54°24′N 43°51′E
127 T2 **Krasnoslobodsk** Volgogradskaya Oblast', SW Russian Federation 48°41′N 44°34′E
Krasnostav see Krasnystaw
127 V5 **Krasnousol'skiy** Respublika Bashkortostan, W Russian Federation 53°55′N 56°22′E
125 U12 **Krasnovishersk** Permskiy Kray, NW Russian Federation 60°22′N 57°04′E
Krasnovodsk see Türkmenbaşy
Krasnovodskiy Zaliv see Türkmenbasy Aylagy
146 B10 **Krasnovodskoye Plato** Turkm. Krasnowodsk Platosy. plateau NW Turkmenistan
Krasnowodsk Aylagy see Türkmenbasy Aylagy
Krasnowodsk Platosy see Krasnovodskoye Plato
122 K12 **Krasnoyarsk** Krasnoyarskiy Kray, S Russian Federation 56°05′N 92°46′E
127 X7 **Krasnoyarskiy** Orenburgskaya Oblast', W Russian Federation 51°56′N 59°54′E
122 K11 **Krasnoyarskiy Kray** ◆ territory C Russian Federation
Krasnoye see Krasnaye
Krasnoye Znamya see Gyzylbaydak
125 R11 **Krasnozatonskiy** Respublika Komi, NW Russian Federation 61°39′N 51°00′E
118 D13 **Krasnoznamensk** prev. Lasdehnen, Ger. Haselberg. Kaliningradskaya Oblast', W Russian Federation 54°57′N 22°28′E
126 K3 **Krasnoznamensk** Moskovskaya Oblast', W Russian Federation 55°40′N 37°05′E
117 R11 **Krasnoznam"yans'kyy Kanal** canal S Ukraine
111 P14 **Krasnystaw** Rus. Krasnostav. Lubelskie, SE Poland 51°N 23°10′E
126 H4 **Krasnyy** Smolenskaya Oblast', W Russian Federation 54°36′N 31°27′E
127 P2 **Krasnyye Baki** Nizhegorodskaya Oblast', W Russian Federation 57°07′N 45°12′E
127 Q13 **Krasnyye Barrikady** Astrakhanskaya Oblast', SW Russian Federation 46°14′N 47°48′E
124 K15 **Krasnyy Kholm** Tverskaya Oblast', W Russian Federation 58°04′N 37°05′E
127 Q8 **Krasnyy Kut** Saratovskaya Oblast', W Russian Federation 50°54′N 46°58′E
Krasnyy Liman see Krasnyy Lyman
117 Y7 **Krasnyy Luch** prev. Krindachevka. Luhans'ka Oblast', E Ukraine 48°09′N 38°52′E
117 X6 **Krasnyy Lyman** Rus. Krasnyy Liman. Donets'ka Oblast', SE Ukraine 49°00′N 37°50′E
127 R3 **Krasnyy Steklovar** Respublika Mariy El, W Russian Federation 56°14′N 48°49′E
127 P8 **Krasnyy Tekstil'shchik** Saratovskaya Oblast', W Russian Federation 51°35′N 45°49′E
127 R13 **Krasnyy Yar** Astrakhanskaya Oblast', SW Russian Federation 46°33′N 48°21′E
Krassóvár see Caraşova
116 L5 **Krasyliv** Khmel'nyts'ka Oblast', W Ukraine 49°38′N 26°59′E
111 O21 **Kraszna** Rom. Crasna. ~ Hungary/Romania
Kratie see Krâchéh
113 P17 **Kratovo** NE FYR Macedonia 42°06′N 22°08′E
Kratznick see Kraśnik
171 Y13 **Krau** Papua, E Indonesia 03°15′S 140°07′E
167 Q13 **Krâvanh, Chuŏr Phnum** Eng. Cardamom Mountains, Fr. Chaîne des Cardamomes. ▲▲ W Cambodia
Kravasta Lagoon see Karavastasë, Laguna e
Krawang see Karawang
127 Q15 **Kraynovka** Respublika Dagestan, SW Russian Federation 43°58′N 47°24′E
118 D12 **Kražiai** Šiauliai, C Lithuania 55°36′N 22°41′E
27 P11 **Krebs** Oklahoma, C USA 34°55′N 95°43′W
101 D15 **Krefeld** Nordrhein-Westfalen, W Germany 51°20′N 06°34′E
Kreisstadt see Krosno Odrzańskie
115 D17 **Kremastón, Technití Límni** ⊠ C Greece
Kremenchug see Kremenchuk
Kremenchugskoye Vodokhranilishche/Kremenchuk Reservoir see Kremenchuts'ke Vodoskhovyshche
117 S6 **Kremenchuk** Rus. Kremenchug. Poltavs'ka Oblast', NE Ukraine 49°04′N 33°27′E
117 R6 **Kremenchuts'ke Vodoskhovyshche** Eng. Kremenchuk Reservoir, Rus. Kremenchugskoye Vodokhranilishche. ⊠ C Ukraine
116 K5 **Kremenets'** Pol. Krzemieniec, Rus. Kremenets. Ternopil's'ka Oblast', W Ukraine 50°06′N 25°43′E
Kremennaya see Kreminna
117 X6 **Kreminna** Rus. Kremennaya. Luhans'ka Oblast', E Ukraine 49°03′N 38°15′E
37 R4 **Kremmling** Colorado, C USA 40°03′N 106°23′W
109 V3 **Krems** ~ NE Austria
Krems see Krems an der Donau
109 W3 **Krems an der Donau** var. Krems. Niederösterreich, N Austria 48°25′N 15°36′E
Kremsier see Kroměříž
109 S4 **Kremsmünster** Oberösterreich, N Austria 48°04′N 14°08′E
38 M17 **Krenitzin Islands** island Aleutian Islands, Alaska, USA
Kresena see Kresna
114 G11 **Kresna** var. Kresena. Blagoevgrad, SW Bulgaria 41°43′N 23°10′E
112 O12 **Kresnoljin** Serbia, E Serbia 44°22′N 21°36′E
25 N4 **Kress** Texas, SW USA 34°21′N 101°43′W
123 V6 **Kresta, Zaliv** bay E Russian Federation
115 D20 **Kréstena** prev. Selinoús. Dytikí Elláda, S Greece 37°36′N 21°36′E
124 H14 **Kresttsy** Novgorodskaya Oblast', W Russian Federation 58°15′N 32°28′E
Kretikon Delagos see Kritikó Pélagos
118 C11 **Kretinga** Ger. Krottingen. Klaipėda, NW Lithuania 55°53′N 21°13′E
Kreutz see Cristuru Secuiesc
Kreuz see Križevci, Croatia
Kreuz see Risti, Estonia
Kreuzburg/Kreuzburg in Oberschlesien see Kluczbork
Kreuzingen see Bol'shakovo
108 H6 **Kreuzlingen** Thurgau, NE Switzerland 47°38′N 09°12′E
101 K25 **Kreuzspitze** ▲ S Germany 47°30′N 10°55′E
101 F16 **Kreuztal** Nordrhein-Westfalen, W Germany 50°58′N 08°00′E
119 I15 **Kreva** Rus. Krevo. Hrodzyenskaya Voblasts', W Belarus 54°19′N 26°17′E
Krevo see Kreva
Kría Vrísi see Krýa Vrýsi
79 D16 **Kribi** Sud, SW Cameroon 02°53′N 09°57′E
Krichëv see Krychaw
Krickerhäu/Kriegerhaj see Handlová
109 W6 **Krieglach** Steiermark, E Austria 47°33′N 15°37′E
108 F8 **Kriens** Luzern, W Switzerland 47°03′N 08°17′E
Kriévija see Russian Federation
Krimmitschau see Crimmitschau
98 H12 **Krimpen aan den IJssel** Zuid-Holland, SW Netherlands 51°56′N 04°39′E
Krindachevka see Krasnyy Luch
115 G25 **Kríos, Akrotírio** headland Kríti, Greece, E Mediterranean Sea 35°17′N 23°31′E
155 J16 **Krishna** prev. Kistna. ~ C India
155 H20 **Krishnagiri** Tamil Nādu, SE India 12°33′N 78°11′E
155 K17 **Krishna, Mouths of the** delta SE India
153 S15 **Krishnanagar** West Bengal, N India 23°22′N 88°32′E
155 G20 **Krishnarājāsāgara** var. Paradip. ⊠ W India
95 N19 **Kristdala** Kalmar, S Sweden 57°24′N 16°12′E
Kristiania see Oslo
95 E18 **Kristiansand** var. Christiansand. Vest-Agder, S Norway 58°08′N 07°52′E
95 L22 **Kristianstad** Skåne, S Sweden 56°02′N 14°10′E
94 F8 **Kristiansund** var. Christiansund. Møre og Romsdal, S Norway 63°07′N 07°45′E
Kristiinankaupunki see Kristinestad
93 I14 **Kristineberg** Västerbotten, N Sweden 65°07′N 18°36′E
95 L16 **Kristinehamn** Värmland, C Sweden 59°17′N 14°09′E
93 J17 **Kristinestad** Fin. Kristiinankaupunki. Österbotten, W Finland 62°15′N 21°24′E
Kristyor see Crişcior
115 J25 **Kriti** Eng. Crete. ◆ region Greece, Aegean Sea
115 J24 **Kriti** Eng. Crete. island Greece, Aegean Sea
115 J23 **Kritikó Pélagos** var. Kretikon Delagos, Eng. Sea of Crete; anc. Mare Creticum. sea Greece, Aegean Sea
Kriulyany see Criuleni
112 I12 **Krivaja** ~ NE Bosnia and Herzegovina
Krivaja see Mali Iđoš
113 P17 **Kriva Palanka** Turk. Eğri Palanka. NE Macedonia 42°13′N 22°19′E
Krivichi see Kryvichy
114 H8 **Krivodol** Vratsa, NW Bulgaria 43°23′N 23°30′E
126 M10 **Krivorozh'ye** Rostovskaya Oblast', SW Russian Federation 48°51′N 40°49′E
Krivoshin see Kryvoshyn
Krivoy Rog see Kryvyy Rih
112 F7 **Križevci** Ger. Kreuz, Hung. Kőrös. Varaždin, NE Croatia 46°02′N 16°32′E
112 B10 **Krk** It. Veglia. Primorje-Gorski Kotar, NW Croatia 45°01′N 14°36′E
112 B10 **Krk** It. Veglia; anc. Curieta. island NW Croatia
109 V12 **Krka** ~ SE Slovenia
Krka see Gurk
109 R11 **Krn** ▲ NW Slovenia 46°15′N 13°37′E
111 H16 **Krnov** Ger. Jägerndorf. Moravskoslezský Kraj, E Czech Republic 50°05′N 17°40′E
Kroatien see Croatia
95 G14 **Krøderen** Buskerud, S Norway 60°06′N 09°48′E
95 G14 **Krøderen** ◎ S Norway
Kroi see Krui
95 N17 **Krokek** Östergötland, S Sweden 58°40′N 16°25′E
Krokodil see Crocodile
93 G16 **Krokom** Jämtland, C Sweden 63°20′N 14°30′E
117 S2 **Krolevets'** Rus. Krolevets. Sums'ka Oblast', NE Ukraine 51°34′N 33°24′E
Krolevets see Krolevets'
Królewska Huta see Chorzów
111 H18 **Kroměříž** Ger. Kremsier. Zlínský Kraj, E Czech Republic 49°18′N 17°24′E
98 I9 **Krommenie** Noord-Holland, C Netherlands 52°30′N 04°46′E
126 J6 **Kromy** Orlovskaya Oblast', W Russian Federation 52°41′N 35°45′E
101 L18 **Kronach** Bayern, E Germany 50°14′N 11°19′E
Krone an der Brahe see Koronowo
Krŏng Kaôh Kŏng see Kaôh Kŏng
95 K21 **Kronoberg** ◆ county S Sweden
123 V10 **Kronotskiy Zaliv** bay E Russian Federation
195 O2 **Kronprinsesse Märtha Kyst** physical region Antarctica
195 V3 **Kronprins Olav Kyst** physical region Antarctica
124 G12 **Kronshtadt** Leningradskaya Oblast', NW Russian Federation 60°01′N 29°42′E
Kronstadt see Braşov
83 I22 **Kroonstad** Free State, C South Africa 27°40′S 27°15′E
123 O12 **Kropotkin** Irkutskaya Oblast', C Russian Federation 58°30′N 115°21′E
126 L14 **Kropotkin** Krasnodarskiy Kray, SW Russian Federation 45°29′N 40°31′E
110 J11 **Krośniewice** Łódzskie, C Poland 52°14′N 19°10′E
111 N17 **Krosno** Ger. Krossen. Podkarpackie, SE Poland 49°40′N 21°46′E
110 E12 **Krosno Odrzańskie** Ger. Crossen, Kreisstadt. Lubuskie, W Poland 52°02′N 15°06′E
Krossen see Krosno
Krotoschin see Krotoszyn
110 H13 **Krotoszyn** Ger. Krotoschin. Wielkopolskie, C Poland 51°43′N 17°24′E
Krottingen see Kretinga
Krousón see Krousónas
115 J25 **Krousónas** prev. Krousón, Kroussón. Kríti, Greece, E Mediterranean Sea 35°14′N 24°59′E
Kroussón see Krousónas
Krraba see Krrabë
113 L20 **Krrabë** var. Krraba. Tiranë, C Albania 41°15′N 19°56′E
113 L17 **Krrabit, Mali i** ▲▲ N Albania
109 W12 **Krško** Ger. Gurkfeld; prev. Videm-Krško. E Slovenia 45°57′N 15°31′E
83 K19 **Kruger National Park** national park Northern, N South Africa
83 J21 **Krugersdorp** Gauteng, NE South Africa 26°06′S 27°46′E
38 D16 **Krugloi Point** headland Agattu Island, Alaska, USA 52°30′N 173°46′E
Krugloye see Kruhlaye
119 N15 **Kruhlaye** Rus. Krugloye. Mahilyowskaya Voblasts', E Belarus 54°15′N 29°48′E
168 L15 **Krui** var. Kroi. Sumatera, SW Indonesia 05°11′S 103°55′E
99 G16 **Kruibeke** Oost-Vlaanderen, N Belgium 51°10′N 04°18′E
83 G25 **Kruidfontein** Western Cape, SW South Africa 32°50′S 21°59′E
99 F15 **Kruiningen** Zeeland, SW Netherlands 51°28′N 04°01′E
Kruja see Krujë
113 L19 **Krujë** var. Kruja, It. Croia. Durrës, C Albania 41°30′N 19°48′E
Krulevshchina/Krulewshchyna see Krulyewshchyna
118 K13 **Krulyewshchyna** Rus. Krulevshchina, Krulewshchyna. Vitsyebskaya Voblasts', N Belarus 55°02′N 27°45′E
25 T6 **Krum** Texas, SW USA 33°15′N 97°14′W
101 J23 **Krumbach** Bayern, S Germany 48°12′N 10°21′E
113 M17 **Krumë** Kukës, NE Albania 42°11′N 20°25′E
Krummau see Český Krumlov
114 K12 **Krumovgrad** prev. Kossukavak. Yambol, E Bulgaria 41°27′N 25°40′E
114 K12 **Krumovitsa** ~ S Bulgaria
114 L10 **Krumovo** Yambol, E Bulgaria 42°16′N 26°25′E
167 O11 **Krung Thep, Ao** var. Bight of Bangkok. bay S Thailand
Krung Thep Mahanakhon see Ao Krung Thep
Krupa/Krupa na Uni see Bosanska Krupa
119 M15 **Krupki** Minskaya Voblasts', C Belarus 54°19′N 29°08′E
95 G24 **Kruså** var. Krusaa. Syddanmark, SW Denmark 54°50′N 09°25′E
Krusaa see Kruså
113 N14 **Kruševac** Serbia, C Serbia 43°37′N 21°20′E
113 N19 **Kruševo** SW FYR Macedonia 41°22′N 21°15′E
111 A15 **Krušné Hory** Eng. Ore Mountains, Ger. Erzgebirge. ▲▲ Czech Republic/Germany see also Erzgebirge
Krušné Hory see Erzgebirge
39 W13 **Kruzof Island** island Alexander Archipelago, Alaska, USA
114 F13 **Krýa Vrýsi** var. Kría Vrísi. Kentrikí Makedonía, N Greece 40°41′N 22°18′E
119 P16 **Krychaw** Rus. Krichëv. Mahilyowskaya Voblasts', E Belarus 53°42′N 31°43′E
64 K11 **Krylov Seamount** undersea feature E Atlantic Ocean 17°35′N 30°07′W
Krym see Krym, Avtonomna Respublika
117 S13 **Krym, Avtonomna Respublika** var. Krym, Eng. Crimea, Crimean Oblast; prev. Rus. Krymskaya ASSR, Krymskaya Oblast'. ◆ province SE Ukraine
126 K14 **Krymsk** Krasnodarskiy Kray, SW Russian Federation 44°56′N 38°02′E
Krymskaya ASSR/Krymskaya Oblast' see Krym, Avtonomna Respublika
117 T13 **Kryms'ki Hory** ▲▲ S Ukraine
117 T13 **Kryms'kyy Pivostriv** peninsula S Ukraine
111 M18 **Krynica** Ger. Tannenhof. Małopolskie, S Poland 49°25′N 20°56′E
117 P8 **Kryve Ozero** Odes'ka Oblast', SW Ukraine 47°54′N 30°19′E
119 K14 **Kryvichy** Rus. Krivichi. Minskaya Voblasts', C Belarus 54°43′N 27°17′E
119 I18 **Kryvoshyn** Rus. Krivoshin. Brestskaya Voblasts', SW Belarus 52°52′N 26°08′E
117 S8 **Kryvyy Rih** Rus. Krivoy Rog. Dnipropetrovs'ka Oblast', SE Ukraine 47°53′N 33°24′E
117 N8 **Kryzhopil'** Vinnyts'ka Oblast', C Ukraine 48°22′N 28°51′E
Krzemieniec see Kremenets'
111 J14 **Krzepice** Śląskie, S Poland 50°58′N 18°42′E
110 F10 **Krzyż Wielkopolskie** Wielkopolskie, W Poland 52°52′N 16°03′E
Ksar al Kabir see Ksar-el-Kebir
Ksar al Soule see Er-Rachidia
74 J5 **Ksar El Boukhari** N Algeria 35°55′N 02°47′E
74 G5 **Ksar-el-Kebir** var. Alcázar, Ksar al Kabir, Ksar-el-Kébir, Ar. Al-Kasr al-Kebir, Al-Qsar al-Kbir, Sp. Alcazarquivir. NW Morocco 35°04′N 05°56′W
Ksar-el-Kébir see Ksar-el-Kebir
110 H12 **Książ Wielkopolski** Ger. Xions. Weilkopolskie, W Poland 52°03′N 17°10′E
127 O3 **Kstovo** Nizhegorodskaya Oblast', W Russian Federation 56°07′N 44°12′E
169 T8 **Kuala Belait** W Brunei 04°48′N 114°12′E
Kuala Dungun see Dungun
169 S10 **Kualakeriau** Borneo, C Indonesia
169 S12 **Kualakuayan** Borneo, C Indonesia 02°01′S 112°35′E
168 K8 **Kuala Lipis** Pahang, Peninsular Malaysia 04°11′N 102°00′E
168 K9 **Kuala Lumpur** ● (Malaysia) Kuala Lumpur, Peninsular Malaysia 03°08′N 101°42′E
168 K9 **Kuala Lumpur International** ✈ Selangor, Peninsular Malaysia 02°51′N 101°45′E
Kuala Pelabohan Kelang see Pelabuhan Klang
169 U7 **Kuala Penyu** Sabah, East Malaysia 05°37′N 115°36′E
38 E9 **Kualapu'u** var. Kualapuu. Moloka'i, Hawaii, USA, C Pacific Ocean 21°09′N 157°02′W
Kualapuu see Kualapu'u
168 L7 **Kuala Terengganu** var. Kuala Trengganu. Terengganu, Peninsular Malaysia 05°20′N 103°07′E
168 L11 **Kualatungkal** Sumatera, W Indonesia 0°49′S 103°22′E
171 P11 **Kuandang** Sulawesi, N Indonesia 0°50′N 122°55′E
163 V12 **Kuandian** var. Kuandian Manzu Zizhixian. Liaoning, NE China 40°41′N 124°46′E
Kuandian Manzu Zizhixian see Kuandian
83 E15 **Kuando Kubango** var. Cuando Cubango. ◆ province SE Angola
Kuang-chou see Guangzhou
Kuang-hsi see Guangxi Zhuangzu Zizhiqu
Kuang-tung see Guangdong
Kuang-yuan see Guangyuan
Kuantan, Batang see Indragiri, Sungai
Kuanzhou see Qingjian
Kuba see Quba
Kubango see Cubango/Okavango
141 X8 **Kubārah** NW Oman 23°03′N 56°52′E
93 H16 **Kubbe** Västernorrland, C Sweden 63°31′N 18°04′E
80 A11 **Kubbum** Southern Darfur, W Sudan 11°47′N 23°47′E
124 L13 **Kubenskoye, Ozero** ◎ NW Russian Federation
146 G6 **Kubla-Ustyurt** Rus. Komsomol'sk-na-Ustyurte. Qoraqalpog'iston Respublikasi, NW Uzbekistan 44°06′N 58°14′E
164 G15 **Kubokawa** Kōchi, Shikoku, SW Japan 33°22′N 133°14′E
114 L7 **Kubrat** prev. Balbunar. Razgrad, N Bulgaria 43°48′N 26°31′E
112 O13 **Kučajske Planine** ▲▲ E Serbia
165 T1 **Kuccharo-ko** ◎ Hokkaidō, N Japan
112 O11 **Kučevo** Serbia, NE Serbia 44°29′N 21°42′E
Kuchan see Qūchān
169 Q10 **Kuching** prev. Sarawak. Sarawak, East Malaysia 01°32′N 110°20′E
169 Q10 **Kuching** ✈ Sarawak, East Malaysia 01°32′N 110°20′E
164 B17 **Kuchinoerabu-jima** island Nansei-shotō, SW Japan
Kuchinotsu see Minamishimabara
109 Q6 **Kuchl** Salzburg, NW Austria 47°37′N 13°12′E
148 L9 **Kūchnay Darwēshān** prev. Kūchnay Darweyshān. Helmand, S Afghanistan 31°02′N 64°10′E
Kūchnay Darweyshān see Kūchnay Darwēshān
Kuchurgan see Kuchurhan
117 O9 **Kuchurhan** Rus. Kuchurgan. ~ NE Ukraine
Kuçova see Kuçovë
113 L21 **Kuçovë** var. Kuçova; prev. Qyteti Stalin. Berat, C Albania 40°48′N 19°55′E
136 D11 **Küçük Çekmece** İstanbul, NW Turkey 41°01′N 28°47′E
164 F14 **Kudamatsu** var. Kudamatu. Yamaguchi, Honshū, SW Japan 34°00′N 131°53′E
Kudamatu see Kudamatsu
Kudara see Ghŭdara
169 V6 **Kudat** Sabah, East Malaysia 06°54′N 116°47′E
Küddow see Gwda
155 G17 **Kūdligi** Karnātaka, W India 14°58′N 76°24′E
Kudowa see Kudowa-Zdrój
111 F16 **Kudowa-Zdrój** Ger. Kudowa. Wałbrzych, SW Poland 50°27′N 16°20′E
117 P9 **Kudryavtsivka** Mykolayivs'ka Oblast', S Ukraine 47°18′N 31°02′E
169 R16 **Kudus** prev. Koedoes. Jawa, C Indonesia 06°46′S 110°48′E
125 T13 **Kudymkar** Permskiy Kray, NW Russian Federation 59°01′N 54°40′E
Kudzsir see Cugir
Kuei-chou see Guizhou
Kuei-lin see Guilin
Kuei-Yang/Kuei-yang see Guiyang
K'u-erh-lo see Korla
Kueyang see Guiyang
Kufa see Al Kūfah
136 E14 **Küfiçayı** ~ C Turkey
109 O6 **Kufstein** Tirol, W Austria 47°36′N 12°10′E
9 N7 **Kugaaruk** prev. Pelly Bay. Nunavut, N Canada 68°38′N 89°45′W
Kugaly see Kogaly
8 K8 **Kugluktuk** var. Qurlurtuuq; prev. Coppermine. Nunavut, NW Canada 67°49′N 115°12′W
143 Y13 **Kūhak** Sīstān va Balūchestān, SE Iran 27°10′N 63°15′E
143 R9 **Kūhbonān** Kermān, C Iran 31°23′N 56°16′E
148 J5 **Kūhestān** var. Kohsān. Herāt, W Afghanistan 34°40′N 61°11′E
Kūhīrī see Kahīrī
93 N15 **Kuhmo** Kainuu, E Finland 64°04′N 29°34′E
93 L18 **Kuhmoinen** Keski-Suomi, C Finland 61°32′N 25°09′E
Kuhnau see Konin
Kühnö see Kihnu
143 O8 **Kūhpāyeh** Eşfahān, C Iran 32°42′N 52°25′E
167 O12 **Kui Buri** var. Ban Kui Nua. Prachuap Khiri Khan, SW Thailand 12°10′N 99°49′E
Kuibyshev see Kuybyshevskoye Vodokhranilishche
82 D13 **Kuito** Port. Silva Porto. Bié, C Angola 12°21′S 16°55′E
39 X14 **Kuiu Island** island Alexander Archipelago, Alaska, USA
93 L13 **Kuivaniemi** Pohjois-Pohjanmaa, C Finland 65°34′N 25°13′E
196 M15 **Kujalleo** ◆ municipality S Greenland
Kujalleo, Kommune see Kujalleo
77 V14 **Kujama** Kaduna, C Nigeria 10°27′N 07°39′E
110 I10 **Kujawsko-pomorskie** ◆ province C Poland
165 R8 **Kuji** var. Kuzi. Iwate, Honshū, C Japan 40°12′N 141°47′E
Kujto, Ozero see Yushkozerskoye Vodokhranilishche
Kujū-renzan see Kujū-san
164 D14 **Kujū-san** var. Kujū-renzan. ▲ Kyūshū, SW Japan 33°07′N 131°13′E
43 N7 **Kukalaya, Rio** var. Rio Cuculaya, Rio Kukulaya. ~ NE Nicaragua
113 O16 **Kukavica** var. Vlajna. ▲ SE Serbia 42°46′N 21°58′E
113 M18 **Kukës** var. Kukësi. Kukës, NE Albania 42°03′N 20°25′E
113 L18 **Kukës** ◆ district NE Albania
Kukësi see Kukës
186 D8 **Kukipi** Gulf, S Papua New Guinea 08°11′S 146°09′E
127 S3 **Kukmor** Respublika Tatarstan, W Russian Federation 56°11′N 50°56′E
Kukong see Shaoguan
39 N6 **Kukpowruk River** ~ Alaska, USA
38 M6 **Kukpuk River** ~ Alaska, USA
Kükūdağ see Gogi, Mount
Kukukhoto see Hohhot
Kukulaya, Rio see Kukalaya, Rio
189 W12 **Kuku Point** headland NW Wake Island 19°19′N 166°36′E
146 G11 **Kukurtli** Ahal Welaýaty, C Turkmenistan 39°58′N 58°47′E
Kül see Kūl, Rūd-e
114 F7 **Kula** Vidin, NW Bulgaria 43°55′N 22°32′E
112 K9 **Kula** Vojvodina, NW Serbia 45°37′N 19°31′E
136 D14 **Kula** Manisa, W Turkey 38°33′N 28°38′E
149 S8 **Kulāchi** Khyber Pakhtunkhwa, NW Pakistan 31°58′N 70°30′E
Kulachi see Kolāchi
Kulagino see Yesbol
168 L10 **Kulai** Johor, Peninsular Malaysia 01°41′N 103°33′E
114 M7 **Kulak** ~ NE Bulgaria
153 T11 **Kula Kangri** var. Kulhakangri. ▲ Bhutan/China 28°06′N 90°19′E
144 E13 **Kulaly, Ostrov** island SW Kazakhstan
145 S16 **Kulan** Kaz. Qulan; prev. Lugovoy, Lugovoye. Zhambyl, S Kazakhstan 42°54′N 72°45′E
147 V9 **Kulanak** Narynskaya Oblast', C Kyrgyzstan 41°18′N 75°38′E
Gory Kulandag see Gulandag
153 V14 **Kulaura** Sylhet, NE Bangladesh 24°32′N 92°02′E
118 D9 **Kuldīga** Ger. Goldingen. W Latvia 56°57′N 21°59′E
Kuldja see Yining
Kul'dzhuktau, Gory see QuljuqtovTog'lari
127 N4 **Kulebaki** Nizhegorodskaya Oblast', W Russian Federation 55°25′N 42°31′E
112 E11 **Kulen Vakuf** var. Spasovo. ◆ Federacija Bosne I Hercegovine, NW Bosnia and Herzegovina
181 Q9 **Kulgera Roadhouse** Northern Territory, N Australia 25°49′S 133°30′E
Kulhakangri see Kula Kangri
127 T1 **Kuliga** Udmurtskaya Respublika, NW Russian Federation 58°14′N 53°49′E
Kulkuduk see Ko'lquduq
118 G4 **Kullamaa** Läänemaa, W Estonia 58°52′N 24°05′E
197 O12 **Kullorsuaq** var. Kuvdlorssuak. ◆ Qaasuitsup, C Greenland
29 O6 **Kulm** North Dakota, N USA 46°18′N 98°57′W
Kulm see Chełmno
146 D12 **Kul'mach** prev. Turkm. Isgender. Balkan Welaýaty, W Turkmenistan 39°04′N 55°49′E
101 L18 **Kulmbach** Bayern, SE Germany 50°07′N 11°27′E
Kulmsee see Chełmża
147 Q14 **Kŭlob** Rus. Kulyab. SW Tajikistan 37°55′N 68°46′E
92 M13 **Kuloharju** Lappi, N Finland 65°51′N 28°10′E
125 N7 **Kuloy** Arkhangel'skaya Oblast', NW Russian Federation 64°55′N 43°35′E
125 N7 **Kuloy** ~ NW Russian Federation
137 Q14 **Kulp** Diyarbakır, SE Turkey 38°32′N 41°01′E
Kulpa see Kolpa
77 P14 **Kulpawn** ~ N Ghana
143 R13 **Kūl, Rūd-e** var. Kūl. ~ S Iran
144 G12 **Kul'sary** Kaz. Qulsary. Atyrau, W Kazakhstan 46°59′N 54°02′E
153 R15 **Kulti** West Bengal, NE India 23°45′N 86°50′E
93 G14 **Kultsjön** Lapp. Gålto. ◎ N Sweden
136 I14 **Kulu** Konya, W Turkey 39°06′N 33°02′E
123 S9 **Kulu** ~ E Russian Federation
122 I13 **Kulunda** Altayskiy Kray, S Russian Federation 52°33′N 79°04′E
Kulunda Steppe see Ravnina Kulyndy
Kulundinskaya Ravnina see Ravnina Kulyndy
182 M9 **Kulwin** Victoria, SE Australia 35°04′S 142°37′E
Kulyab see Kŭlob
117 Q3 **Kulykivka** Chernihivs'ka Oblast', N Ukraine 51°23′N 31°39′E
Kum see Qom
164 F14 **Kuma** Ehime, Shikoku, SW Japan 33°36′N 132°53′E
127 P14 **Kuma** ~ SW Russian Federation
165 O12 **Kumagaya** Saitama, Honshū, S Japan 36°09′N 139°22′E
165 Q5 **Kumaishi** Hokkaidō, NE Japan 42°08′N 139°57′E
169 R13 **Kumai, Teluk** bay Borneo, C Indonesia
127 Y7 **Kumak** Orenburgskaya Oblast', W Russian Federation 51°16′N 60°06′E
164 C14 **Kumamoto** Kumamoto, Kyūshū, SW Japan 32°49′N 130°41′E
164 D15 **Kumamoto** off. Kumamoto-ken. ◆ prefecture Kyūshū, SW Japan
Kumamoto-ken see Kumamoto
164 J15 **Kumano** Mie, Honshū, SW Japan 33°54′N 136°08′E
Kumanova see Kumanovo
113 O17 **Kumanovo** Turk. Kumanova. N Macedonia 42°08′N 21°43′E
185 G17 **Kumara** West Coast, South Island, New Zealand 42°39′S 171°12′E
180 J8 **Kumarina Roadhouse** Western Australia 24°46′S 119°39′E
153 T15 **Kumarkhali** Khulna, W Bangladesh 23°54′N 89°16′E
77 P16 **Kumasi** prev. Coomassie. C Ghana 06°41′N 01°40′W
Kumayri see Gyumri
79 D15 **Kumba** Sud-Ouest, W Cameroon 04°39′N 09°26′E
114 N13 **Kumbağ** Tekirdağ, NW Turkey 40°51′N 27°26′E
155 J21 **Kumbakonam** Tamil Nādu, SE India 10°59′N 79°24′E
Kum-Dag see Gumdag
165 R16 **Kume-jima** island Nansei-shotō, SW Japan
127 V6 **Kumertau** Respublika Bashkortostan, W Russian Federation 52°48′N 55°48′E
Kumillā see Comilla
35 R4 **Kumiva Peak** ▲ Nevada, W USA 40°24′N 119°16′W
159 N7 **Kum Kuduk** Xinjiang Uygur Zizhiqu, W China 40°15′N 91°55′E
159 N8 **Kum Kuduk** well NW China
Kumkurgan see Qumqo'rg'on
95 M16 **Kumla** Örebro, C Sweden 59°08′N 15°09′E
136 E17 **Kumluca** Antalya, SW Turkey 36°23′N 30°17′E
100 N9 **Kummerower See** ◎ NE Germany
77 X14 **Kumo** Gombe, E Nigeria 10°03′N 11°14′E
145 O13 **Kumola** ~ C Kazakhstan
167 N1 **Kumon Range** ▲▲ N Myanmar (Burma)
83 F22 **Kums** Karas, SE Namibia 28°07′S 19°40′E
155 E18 **Kumta** Karnātaka, W India 14°25′N 74°24′E
38 H12 **Kumukahi, Cape** headland Hawai'i, USA, C Pacific Ocean 19°31′N 154°48′W
127 Q17 **Kumukh** Respublika Dagestan, SW Russian Federation 42°10′N 47°07′E
Kumul see Hami
158 L6 **Kümüx** Xinjiang Uygur Zizhiqu, W China
127 N9 **Kumylzhenskaya** Volgogradskaya Oblast', SW Russian Federation 49°54′N 42°35′E
141 W6 **Kumzār** N Oman 26°19′N 56°26′E
43 W15 **Kuna de Wargandí** ◆ special territory NE Panama
149 S4 **Kunar** Per. Konarhā; prev. Konar. ◆ province E Afghanistan
Kunashiri see Kunashir, Ostrov
123 U14 **Kunashir, Ostrov** var. Kunashiri. island Kuril'skiye Ostrova, SE Russian Federation
43 V14 **Kuna Yala** prev. San Blas. ◆ special territory NE Panama
118 J3 **Kunda** Lääne-Virumaa, NE Estonia 59°31′N 26°33′E
152 M13 **Kunda** Uttar Pradesh, N India 25°43′N 81°31′E
155 E19 **Kundāpura** var. Coondapoor. Karnātaka, W India 13°39′N 74°41′E
79 O24 **Kundelungu, Monts** ▲▲ S Dem. Rep. Congo
Kundert see Hernád
186 D7 **Kundiawa** Chimbu, W Papua New Guinea 06°00′S 144°57′E
Kundla see Sāvarkundla
Kunduk, Ozero see Sasyk, Ozero
Kunduk, Ozero Sasyk see Sasyk, Ozero
168 L10 **Kundur, Pulau** island W Indonesia
149 Q2 **Kunduz** var. Kondūz, Qondūz; prev. Kondoz, Kundūz. Kunduz, NE Afghanistan 36°49′N 68°50′E
149 Q2 **Kunduz** prev. Kondoz. ◆ province NE Afghanistan
Kunduz/Kundūz see Kunduz
Kuneitra see Al Qunayţirah
83 B18 **Kunene** ◆ district NE Namibia
83 A16 **Kunene** var. Cunene. ~ Angola/Namibia see also Cunene
Kunene see Cunene
Künes see Xinyuan
158 J5 **Künes He** ~ NW China
95 J19 **Kungälv** Västra Götaland, S Sweden 57°54′N 12°00′E
147 W7 **Kungei Ala-Tau** Rus. Khrebet Kyungëy Ala-Too, Kir. Küngöy Ala-Too. ▲▲ Kazakhstan/Kyrgyzstan
Küngöy Ala-Too see Kungei Ala-Tau
Kungrad see Qo'ng'irot
95 J19 **Kungsbacka** Halland, S Sweden 57°30′N 12°05′E
95 I18 **Kungshamn** Västra Götaland, S Sweden 58°21′N 11°15′E
95 M16 **Kungsör** Västmanland, C Sweden 59°25′N 16°05′E
79 J16 **Kungu** Equateur, NW Dem. Rep. Congo 02°47′N 19°12′E
125 V15 **Kungur** Permskiy Kray, NW Russian Federation 57°24′N 56°56′E
166 L9 **Kungyangon** Yangon, SW Myanmar (Burma) 16°27′N 96°00′E
111 M22 **Kunhegyes** Jász-Nagykun-Szolnok, E Hungary 47°22′N 20°36′E
167 O5 **Kunhing** Shan State, E Myanmar (Burma) 21°17′N 98°26′E
158 D9 **Kunjirap Daban** var. Khūnjerāb Pass. pass China/Pakistan see also Khūnjerāb Pass
Kunjirap Daban see Khunjerāb Pass
Kunlun Mountains see Kunlun Shan
158 H10 **Kunlun Shan** Eng. Kunlun Mountains. ▲▲ NW China
159 P11 **Kunlun Shankou** pass C China
160 G13 **Kunming** var. K'un-ming; prev. Yunnan. province capital Yunnan, SW China 25°04′N 102°41′E
K'un-ming see Kunming
Kuno see Kunoy
95 B18 **Kunoy** Dan. Kunø. island N Faroe Islands
Kunsan see Gunsan
111 L24 **Kunszentmárton** Jász-Nagykun-Szolnok, E Hungary 46°50′N 20°19′E
111 J23 **Kunszentmiklós** Bács-Kiskun, C Hungary 47°00′N 19°07′E
181 N3 **Kununurra** Western Australia 15°50′S 128°44′E
Kunyang see Pingyang
Kunya-Urgench see Köneürgenç
Kunyé see Pins, Île des
169 T11 **Kunyi** Borneo, C Indonesia 03°23′S 119°20′E
101 I20 **Künzelsau** Baden-Württemberg, S Germany 49°22′N 09°43′E
161 S10 **Kuocang Shan** ▲▲ SE China
Kuolajärvi see Kuoloyarvi
124 H5 **Kuoloyarvi** Finn. Kuolajärvi, var. Luolajarvi. Murmanskaya Oblast', NW Russian Federation 66°58′N 29°13′E
93 N16 **Kuopio** Pohjois-Savo, C Finland 62°54′N 27°41′E
93 K17 **Kuortane** Etelä-Pohjanmaa, W Finland 62°48′N 23°30′E
93 M18 **Kuortti** Etelä-Savo, E Finland 61°25′N 26°25′E
Kupa see Kolpa
171 P17 **Kupang** prev. Koepang. Timor, C Indonesia 10°13′S 123°38′E
39 Q5 **Kuparuk River** ~ Alaska, USA
Kupchino see Cupcina
186 E9 **Kupiano** Central, S Papua New Guinea 10°06′S 148°12′E
180 M4 **Kupingarri** Western Australia 16°46′S 125°57′E
122 I12 **Kupino** Novosibirskaya Oblast', C Russian Federation 54°22′N 77°09′E
118 H11 **Kupiškis** Panevėžys, NE Lithuania 55°51′N 24°58′E
114 L13 **Küplü** Edirne, NW Turkey 41°06′N 26°23′E
39 X13 **Kupreanof Island** island Alexander Archipelago, Alaska, USA
39 O16 **Kupreanof Point** headland Alaska, USA 55°34′N 159°36′W
112 G13 **Kupres** ◆ Federacija Bosne I Hercegovine, SW Bosnia and Herzegovina
117 W5 **Kup"yans'k** Rus. Kupyansk. Kharkivs'ka Oblast', E Ukraine 49°42′N 37°36′E
Kupyansk see Kup"yans'k

◆ Country
● Country Capital
◇ Dependent Territory
○ Dependent Territory Capital
◆ Administrative Regions
✈ International Airport
▲ Mountain
▲▲ Mountain Range
⛰ Volcano
~ River
◎ Lake
⊠ Reservoir

Kup"yans'k-Vuzlovyy – La Cuesta

117 W5 **Kup"yans'k-Vuzlovyy** Kharkivs'ka Oblast', E Ukraine 49°40′N 37°41′E
158 I6 **Kuqa** Xinjiang Uygur Zizhiqu, NW China 41°43′N 82°58′E
Kür see Kura
137 W11 **Kura** Az. Kür, Geor. Mtkvari, Turk. Kura Nehri. ≈ SW Asia
55 R8 **Kuracki** NW Guyana 06°52′N 60°13′W
Kura Kurk see Irbe Strait
147 Q10 **Kurama Range** Rus. Kuraminskiy Khrebet. ▲▲ Tajikistan/Uzbekistan
Kuraminskiy Khrebet see Kurama Range
Kura Nehri see Kura
119 J14 **Kuranyets** Rus. Kurenets. Minskaya Voblasts', C Belarus 54°33′N 26°57′E
164 H13 **Kurashiki** var. Kurasiki. Okayama, Honshū, SW Japan 34°35′N 133°44′E
154 L10 **Kurasia** Chhattīsgarh, C India 23°11′N 82°16′E
Kurasiki see Kurashiki
164 H12 **Kurayoshi** var. Kurayosi. Tottori, Honshū, SW Japan 35°27′N 133°52′E
Kurayosi see Kurayoshi
163 X6 **Kurbin He** ≈ NE China
Kurchum see Kurshim
Kurchum see Kurshim
137 X11 **Kürdämir** Rus. Kyurdamir. C Azerbaijan 40°21′N 48°08′E
Kurdestan see Kordestān
139 S1 **Kurdistan** cultural region SW Asia
Kurd Kui see Kord Kūy
155 F15 **Kurduvādi** Mahārāshtra, W India 18°06′N 75°31′E
Kŭrdzhali see Kardzhali
Kŭrdzhali see Kardzhali
Kŭrdzhali, Yazovir see Kardzhali, Yazovir
164 F13 **Kure** Hiroshima, Honshū, SW Japan 34°15′N 132°33′E
192 K5 **Kure Atoll** var. Ocean Island. atoll Hawaiian Islands, Hawaii, USA
136 J10 **Küre Dağları** ▲▲ N Turkey
146 C11 **Kürendag** Rus. Gora Kyuren. ▲ W Turkmenistan 39°05′N 55°09′E
Kurenets see Kuranyets
118 E6 **Kuressaare** Ger. Arensburg; prev. Kingissepp. Saaremaa, W Estonia 58°17′N 22°29′E
122 K9 **Kureyka** Krasnoyarskiy Kray, N Russian Federation 66°22′N 87°21′E
122 K9 **Kureyka** ≈ N Russian Federation
Kurgal'dzhino/ Kurgal'dzhinsky see Korgalzhyn
122 G11 **Kurgan** Kurganskaya Oblast', C Russian Federation 55°30′N 65°20′E
126 L14 **Kurganinsk** Krasnodarskiy Kray, SW Russian Federation 44°55′N 40°45′E
122 G11 **Kurganskaya Oblast'** ◐ province C Russian Federation
Kurgan-Tyube see Qŭrghonteppa
191 O2 **Kuria** prev. Woodle Island. island Tungaru, W Kiribati
Kuria Muria Bay see Ḥalānīyāt, Khalīj al
Kuria Muria Islands see Ḥalāniyāt, Juzur al
153 T13 **Kurigram** Rajshahi, N Bangladesh 25°49′N 89°39′E
93 K17 **Kurikka** Etelä-Pohjanmaa, W Finland 62°36′N 22°25′E
192 I3 **Kuril Basin** var. Kurile Basin. undersea basin NW Pacific Ocean
Kurile Basin see Kuril Basin
Kurile Islands see Kuril'skiye Ostrova
Kurile-Kamchatka Depression see Kuril-Kamchatka Trench
Kurile Trench see Kuril-Kamchatka Trench
Kuril Islands see Kuril'skiye Ostrova
192 J3 **Kuril-Kamchatka Trench** var. Kurile-Kamchatka Depression, Kurile Trench. trench NW Pacific Ocean
127 Q9 **Kurilovka** Saratovskaya Oblast', W Russian Federation 50°39′N 48°02′E
123 U13 **Kuril'sk** Jap. Shana. Kuril'skiye Ostrova, Sakhalinskaya Oblats', SE Russian Federation 45°10′N 147°51′E
122 G11 **Kuril'skiye Ostrova** Eng. Kuril Islands, Kurile Islands. island group SE Russian Federation
42 M9 **Kurinwas, Río** ≈ E Nicaragua
Kurisches Haff see Courland Lagoon
Kurkund see Kilingi-Nõmme
126 M4 **Kurlovskiy** Vladimirskaya Oblast', W Russian Federation 55°25′N 40°39′E
80 G12 **Kurmuk** Blue Nile, SE Sudan 10°36′N 34°16′E
Kurna see Al Qurnah
155 H17 **Kurnool** var. Karnul. Andhra Pradesh, S India 15°51′N 78°01′E
164 M11 **Kurobe** Toyama, Honshū, SW Japan 36°55′N 137°24′E
165 Q7 **Kuroishi** var. Kuroisi. Aomori, Honshū, C Japan 40°37′N 140°34′E
Kuroisi see Kuroishi
165 O12 **Kuroiso** Tochigi, Honshū, S Japan 36°58′N 140°02′E
165 Q4 **Kuromatsunai** Hokkaidō, NE Japan 42°40′N 140°18′E
164 B17 **Kuro-shima** island SW Japan
185 F21 **Kurow** Canterbury, South Island, New Zealand 44°44′S 170°29′E
127 N15 **Kursavka** Stavropol'skiy Kray, SW Russian Federation 44°28′N 42°31′E
118 E11 **Kuršėnai** Šiauliai, N Lithuania 56°00′N 22°56′E
145 X10 **Kurshim** prev. Kurchum. Vostochnyy Kazakhstan, E Kazakhstan 48°35′N 83°37′E
145 Y10 **Kurshim** prev. Kurchum. ≈ E Kazakhstan
Kurshskaya Kosa/Kuršių Nerija see Courland Spit
126 J7 **Kursk** Kurskaya Oblast', W Russian Federation 51°44′N 36°47′E
126 I7 **Kurskaya Oblast'** ◐ province W Russian Federation
Kurskiy Zaliv see Courland Lagoon
113 N15 **Kuršumlija** Serbia, S Serbia 43°09′N 21°16′E
137 R15 **Kurtalan** Siirt, SE Turkey 37°58′N 41°36′E
Kurtbunar see Tervel
Kurt-Dere see Valchi Dol
Kurtitsch/Kürtös see Curtici
145 U15 **Kurtty** ≈ SE Kazakhstan
93 L18 **Kuru** Pirkanmaa, W Finland 61°51′N 23°46′E
80 C13 **Kuru** ≈ W South Sudan
114 M13 **Kuru Dağı** ▲▲ NW Turkey
158 L7 **Kuruktag** ▲▲ NW China
83 G22 **Kuruman** Northern Cape, N South Africa 27°28′S 23°27′E
67 T14 **Kuruman** ≈ W South Africa
164 D14 **Kurume** Fukuoka, Kyūshū, SW Japan 33°15′N 130°27′E
123 N13 **Kurumkan** Respublika Buryatiya, S Russian Federation 54°13′N 110°21′E
155 J25 **Kurunegala** North Western Province, C Sri Lanka 07°28′N 80°23′E
55 T10 **Kurupukari** C Guyana 04°39′N 58°39′W
125 U10 **Kur"ya** Respublika Komi, NW Russian Federation 61°38′N 57°12′E
144 E15 **Kuryk** var. Yeraliyev, Kaz. Quryq. Mangistau, SW Kazakhstan 43°12′N 51°43′E
136 B15 **Kuşadası** Aydın, SW Turkey 37°50′N 27°16′E
115 M19 **Kuşadası Körfezi** gulf SW Turkey
164 A17 **Kusagaki-guntō** island SW Japan
Kusaie see Kosrae
145 T12 **Kusak** ≈ C Kazakhstan
Kusary see Qusar
167 P7 **Ku Sathan, Doi** ▲ NW Thailand 18°22′N 100°31′E
164 J13 **Kusatsu** var. Kusatu. Shiga, Honshū, SW Japan 35°02′N 136°00′E
Kusatu see Kusatsu
138 F11 **Kuseifa** Southern, C Israel 31°15′N 35°01′E
136 C12 **Kuş Gölü** ◎ NW Turkey
126 L12 **Kushchevskaya** Krasnodarskiy Kray, SW Russian Federation 46°35′N 39°40′E
164 D16 **Kushima** var. Kusima. Miyazaki, Kyūshū, SW Japan 31°28′N 131°14′E
164 I15 **Kushimoto** Wakayama, Honshū, SW Japan 33°28′N 135°45′E
165 V4 **Kushiro** var. Kusiro. Hokkaidō, NE Japan 42°58′N 144°24′E
148 K4 **Kushk** prev. Kūshk. Herāt, W Afghanistan 34°55′N 62°20′E
Kushka see Serhetabat
Kushka see Gushgy/ Serhetabat
Kushmurun see Kusmuryn
Kushmurun, Ozero see Kusmuryn, Ozero
127 U4 **Kushnarenkovo** Respublika Bashkortostan, W Russian Federation 55°07′N 55°24′E
Kushrabat see Qo'shrabot
Kushtia see Kustia
Kusima see Kushima
Kusiro see Kushiro
38 M13 **Kuskokwim Bay** bay Alaska, USA
39 P11 **Kuskokwim Mountains** ▲▲ Alaska, USA
39 N12 **Kuskokwim River** ≈ Alaska, USA
145 N8 **Kusmuryn** Kaz. Qusmuryn; prev. Kushmurun. Kostanay, N Kazakhstan 52°27′N 64°31′E
145 N8 **Kusmuryn, Ozero** Kaz. Qusmuryn; prev. Ozero Kushmurun. ◎ N Kazakhstan
108 G7 **Küsnacht** Zürich, N Switzerland 47°19′N 08°34′E
165 V4 **Kussharo-ko** var. Kussyaro. ◎ Hokkaidō, NE Japan
Küssnacht see Küssnacht am Rigi
108 F8 **Küssnacht am Rigi** var. Küssnacht. Schwyz, C Switzerland 47°03′N 08°25′E
Kussyaro see Kussharo-ko
Kustanay see Kostanay
Küstence/Küstendje see Constanța
100 F11 **Küstenkanal** var. Ems-Hunte Canal. canal NW Germany
153 T15 **Kustia** var. Kushtia. Khulna, W Bangladesh 23°54′N 89°07′E
Küstrin see Kostrzyn
171 R11 **Kusu** Pulau Halmahera, E Indonesia 0°51′N 127°41′E
170 L16 **Kuta** Pulau Lombok, S Indonesia 08°53′S 116°15′E
139 T4 **Kutabān** Kirkūk, N Iraq 35°21′N 44°45′E
136 E13 **Kütahya** prev. Kutaia. Kütahya, W Turkey 39°25′N 29°56′E
136 E13 **Kütahya** var. Kutaia. ◐ province W Turkey
Kutai see Mahakam, Sungai
Kutaia see Kütahya
137 R9 **Kutaisi** W Georgia 42°16′N 42°42′E
Kūt, Al see Wāsiṭ
Kūt al 'Amārah see Al Kūt
Kut al Hai/Kūt al Ḥayy see Al Ḥayy
Kut al Imara see Al Kūt
123 Q11 **Kutana** Respublika Sakha (Yakutiya), NE Russian Federation 59°05′N 131°43′E
Kutaradja/Kutaraja see Banda Aceh
165 R4 **Kutchan** Hokkaidō, NE Japan 42°54′N 140°46′E
Kutch, Gulf of see Kachchh, Gulf of
Kutch, Rann of see Kachchh, Rann of
112 F9 **Kutina** Sisak-Moslavina, NE Croatia 45°29′N 16°45′E
112 H9 **Kutjevo** Požega-Slavonija, NE Croatia 45°26′N 17°54′E
111 E17 **Kutná Hora** Ger. Kuttenberg. Střední Čechy, C Czech Republic 49°58′N 15°18′E
110 K12 **Kutno** Łódzkie, C Poland 52°14′N 19°23′E
Kuttenberg see Kutná Hora
79 I20 **Kutu** Bandundu, W Dem. Rep. Congo 02°42′S 18°10′E
153 V17 **Kutubdia Island** island SE Bangladesh
80 B10 **Kutum** Northern Darfur, W Sudan 14°10′N 24°40′E
147 Y7 **Kuturgu** Issyk-Kul'skaya Oblast', E Kyrgyzstan 42°45′N 78°04′E
12 M5 **Kuujjuaq** prev. Fort-Chimo. Québec, E Canada 58°10′N 68°15′W
12 I7 **Kuujjuarapik** Québec, C Canada 55°07′N 78°09′W
12 I7 **Kuujjuarapik** prev. Poste-de-la-Baleine. Québec, NE Canada 55°13′N 77°54′W
Kuuli-Mayak see Guwlumaýak
118 I6 **Kuulsemägi** ▲ S Estonia
92 N13 **Kuusamo** Pohjois-Pohjanmaa, E Finland 65°57′N 29°15′E
93 M19 **Kuusankoski** Kymenlaakso, S Finland 60°51′N 26°40′E
127 W7 **Kuvandyk** Orenburgskaya Oblast', W Russian Federation 51°27′N 57°18′E
Kuvango see Cubango
Kuvasay see Quvasoy
Kuvdlorssuak see Kullorsuaq
124 I16 **Kuvshinovo** Tverskaya Oblast', W Russian Federation 57°03′N 34°09′E
141 Q4 **Kuwait** off. State of Kuwait, var. Dawlat al Kuwait, Koweit, Kuweit. ◆ monarchy SW Asia
Kuwait see Al Kuwayt
Kuwait Bay see Kuwayt, Jūn al
Kuwait City see Al Kuwayt
Kuwait, Dawlat al see Kuwait
Kuwait, State of see Kuwait
Kuwajleen see Kwajalein Atoll
164 K13 **Kuwana** Mie, Honshū, SW Japan 35°04′N 136°40′E
139 X9 **Kuwayt** Maysān, E Iraq 32°26′N 47°12′E
142 K11 **Kuwayt, Jūn al** var. Kuwait Bay. bay E Kuwait
Kuweit see Kuwait
117 P10 **Kuyal'nyts'kyy Lyman** ◎ SW Ukraine
122 I12 **Kuybyshev** Novosibirskaya Oblast', C Russian Federation 55°28′N 77°55′E
Kuybyshev see Bolgar, Respublika Tatarstan, Russian Federation
Kuybyshev see Samara
117 W9 **Kuybysheve** Rus. Kuybyshevo. Zaporiz'ka Oblast', SE Ukraine 47°20′N 36°41′E
Kuybyshevo see Kuybysheve
Kuybyshev Reservoir see Kuybyshevskoye Vodokhranilishche
Kuybyshevskaya Oblast' see Samarskaya Oblast'
Kuybyshevskiy see Novoishimskiy
127 R4 **Kuybyshevskoye Vodokhranilishche** var. Kuibyshev, Eng. Kuybyshev Reservoir. ▣ W Russian Federation
123 S9 **Kuydusun** Respublika Sakha (Yakutiya), NE Russian Federation 63°15′N 143°10′E
125 U16 **Kuyeda** Permskiy Kray, NW Russian Federation 56°23′N 55°19′E
Kūysanjaq see Koye
158 J4 **Kuytun** Xinjiang Uygur Zizhiqu, NW China 44°25′N 84°55′E
122 M13 **Kuytun** Irkutskaya Oblast', S Russian Federation 54°18′N 101°28′E
55 S12 **Kuyuwini Landing** S Guyana 02°06′N 59°14′W
Kuzi see Kuji
38 M9 **Kuzitrin River** ≈ Alaska, USA
127 P6 **Kuznetsk** Penzenskaya Oblast', W Russian Federation 53°06′N 46°27′E
116 K3 **Kuznetsovs'k** Rivnens'ka Oblast', NW Ukraine 51°21′N 25°51′E
165 R8 **Kuzumaki** Iwate, Honshū, C Japan 40°04′N 141°26′E
95 H24 **Kværndrup** Syddtjylland, C Denmark 55°10′N 10°31′E
92 H9 **Kvaløya** island N Norway
92 K8 **Kvalsund** Finnmark, N Norway 70°30′N 23°56′E
94 G11 **Kvam** Oppland, S Norway 61°42′N 09°43′E
127 X7 **Kvarkeno** Orenburgskaya Oblast', W Russian Federation 52°09′N 59°44′E
93 G15 **Kvarnbergsvattnet** var. Frostviken. ◎ N Sweden
112 A11 **Kvarner** var. Carnaro, It. Quarnero. gulf W Croatia
112 B11 **Kvarnerić** channel W Croatia
39 O14 **Kvichak Bay** bay Alaska, USA
92 H12 **Kvikkjokk** Lapp. Huhttán. Norrbotten, N Sweden 66°58′N 17°45′E
95 D17 **Kvina** ≈ S Norway
92 Q1 **Kvitøya** island NE Svalbard
95 F16 **Kvitseid** Telemark, S Norway 59°23′N 08°31′E
79 H20 **Kwa** ≈ W Dem. Rep. Congo
83 L23 **KwaDukuza** prev. Stanger. KwaZulu/Natal, E South Africa 29°20′S 31°18′E see also Stanger
77 Q15 **Kwadwokurom** C Ghana 07°49′N 00°15′W
186 M8 **Kwailibesi** Malaita, N Solomon Islands 08°25′S 160°48′E
189 S6 **Kwajalein Atoll** var. Kuwajleen. atoll Ralik Chain, C Marshall Islands
55 W9 **Kwakoegron** Brokopondo, N Suriname 05°14′N 55°20′W
81 J21 **Kwale** Kwale, S Kenya 04°10′S 39°27′E
77 U17 **Kwale** Delta, S Nigeria 05°51′N 06°29′E
81 J21 **Kwale** ◐ county SE Kenya
79 H20 **Kwamouth** Bandundu, W Dem. Rep. Congo 03°11′S 16°16′E
Kwando see Cuando
Kwangchow see Guangzhou
Kwangchu see Gwangju
Kwangju see Gwangju
Kwangju-gwangyŏksi see Gwangju
79 H20 **Kwango** Port. Cuango. ≈ Angola/Dem. Rep. Congo see also Cuango
Kwango see Cuango
Kwangsi/Kwangsi Chuang Autonomous Region see Guangxi Zhuangzu Zizhiqu
Kwangtung see Guangdong
Kwangyuan see Guangyuan
81 F17 **Kwania, Lake** ◎ C Uganda
Kwanza see Cuanza
82 B11 **Kwanza Norte** prev. Cuanza Norte. ◐ province NE Angola
82 B12 **Kwanza Sul** prev. Cuanza Sul. ◐ province NE Angola
77 S15 **Kwara** ◐ state SW Nigeria
83 K22 **KwaZulu/Natal** off. KwaZulu/Natal Province; prev. Natal. ◐ province E South Africa
KwaZulu/Natal Province see KwaZulu/Natal
Kweichow see Guizhou
Kweichu see Guiyang
Kweilin see Guilin
Kweisui see Hohhot
Kweiyang see Guiyang
83 K17 **Kwekwe** prev. Que Que. Midlands, C Zimbabwe 18°56′S 29°49′E
83 G20 **Kweneng** ◐ district S Botswana
Kwesui see Hohhot
39 N12 **Kwethluk** Alaska, USA 60°48′N 161°26′W
39 N12 **Kwethluk River** ≈ Alaska, USA
110 J8 **Kwidzyń** Ger. Marienwerder. Pomorskie, N Poland 53°44′N 18°55′E
38 M13 **Kwigillingok** Alaska, USA 59°52′N 163°08′W
186 E9 **Kwikila** Central, S Papua New Guinea 09°51′S 147°43′E
79 I20 **Kwilu** ≈ W Dem. Rep. Congo
Kwito see Cuito
171 U12 **Kwoka, Gunung** ▲ Papua Barat, E Indonesia 0°34′S 132°25′E
78 I12 **Kyabé** Moyen-Chari, S Chad 09°28′N 18°54′E
183 O11 **Kyabram** Victoria, SE Australia 36°21′S 145°05′E
166 M9 **Kyaikkami** prev. Amherst. Mon State, S Myanmar (Burma) 16°03′N 97°36′E
166 L9 **Kyaiklat** Ayeyawady, SW Myanmar (Burma) 16°25′N 95°42′E
166 M8 **Kyaikto** Mon State, S Myanmar (Burma) 17°16′N 97°01′E
123 N14 **Kyakhta** Respublika Buryatiya, S Russian Federation 50°25′N 106°13′E
182 G8 **Kyancutta** South Australia 33°10′S 135°33′E
167 T8 **Ky Anh** Ha Tinh, N Vietnam 18°05′N 106°16′E
166 L5 **Kyaukpadaung** Mandalay, C Myanmar (Burma) 20°50′N 95°08′E
Kyaukpyu see Kyaunkpyu
166 M5 **Kyaukse** Mandalay, C Myanmar (Burma) 21°33′N 96°06′E
166 L8 **Kyaunggon** Ayeyawady, SW Myanmar (Burma) 17°04′N 95°12′E
166 J6 **Kyaunkpyu** var. Kyaukpyu. Rakhine State, W Myanmar (Burma) 19°27′N 93°33′E
119 E14 **Kybartai** Pol. Kibarty. Marijampolė, S Lithuania 54°37′N 22°44′E
152 I7 **Kyelang** Himāchal Pradesh, NW India 32°33′N 77°03′E
111 G19 **Kyjov** Ger. Gaya. Jihomoravský Kraj, SE Czech Republic 49°00′N 17°07′E
115 J21 **Kykládes** var. Kikládhes, Eng. Cyclades. island group SE Greece
25 S11 **Kyle** Texas, SW USA 29°59′N 97°52′W
96 G9 **Kyle of Lochalsh** N Scotland, United Kingdom 57°18′N 05°39′W
101 D18 **Kyll** ≈ W Germany
115 F19 **Kyllíni** var. Killini. ▲▲ S Greece
93 N20 **Kymenlaakso** Swe. Kymmenedalen. ◐ region S Finland
115 H18 **Kými** prev. Kími. Évvoia, C Greece 38°38′N 24°06′E
93 M19 **Kymijoki** ≈ S Finland
115 H18 **Kýmis, Akrotírio** headland Évvoia, C Greece 38°39′N 24°08′E
Kymmenedalen see Kymenlaakso
125 W14 **Kyn** Permskiy Kray, NW Russian Federation 57°48′N 58°38′E
183 N12 **Kyneton** Victoria, SE Australia 37°14′S 144°28′E
81 G17 **Kyoga, Lake** var. Lake Kioga. ◎ C Uganda
164 J12 **Kyōga-misaki** headland Honshū, SW Japan 35°46′N 135°13′E
183 V4 **Kyogle** New South Wales, SE Australia 28°37′S 153°00′E
Kyŏnggi-man see Gyeonggi-man
Kyŏngju see Gyeongju
Kyŏngsŏng see Seoul
Kyōsai-tō see Geogeum-do
81 F19 **Kyotera** S Uganda 0°38′S 31°34′E
164 J13 **Kyōto** Kyōto, Honshū, SW Japan 35°01′N 135°46′E
164 J13 **Kyōto** off. Kyōto-fu, var. Kyōto Hu. ◐ urban prefecture Honshū, SW Japan
Kyōto-fu/Kyōto Hu see Kyōto
115 D21 **Kyparissía** var. Kiparissía. Pelopónnisos, S Greece 37°15′N 21°40′E
115 D20 **Kyparissiakós Kólpos** gulf S Greece
Kyperounda see Kyperoúnta
121 P3 **Kyperoúnta** var. Kyperounda. C Cyprus 34°57′N 33°02′E
Kypros see Cyprus
115 H16 **Kyrá Panagía** island Vóreies Sporádes, Greece, Aegean Sea
Kyrenia see Girne
Kyrenia Mountains see Beşparmak Dağları
Kyrgyz Republic see Kyrgyzstan
147 U9 **Kyrgyzstan** off. Kyrgyz Republic, var. Kirghizia; prev. Kirgizskaya SSR, Kirghiz SSR, Republic of Kyrgyzstan. ◆ republic C Asia
Kyrgyzstan, Republic of see Kyrgyzstan
138 F11 **Kyriat Gat** prev. Qiryat Gat. Southern, C Israel 31°37′N 34°47′E
100 M11 **Kyritz** Brandenburg, NE Germany 52°56′N 12°24′E
94 G8 **Kyrkseterøra** Sør-Trøndelag, S Norway 63°17′N 09°06′E
Kyrkslätt see Kirkkonummi
125 U8 **Kyrta** Respublika Komi, NW Russian Federation 64°03′N 57°41′E
111 J18 **Kysucké Nové Mesto** prev. Horné Nové Mesto, Ger. Kisutzaneustadtl, Oberneustadtl, Hung. Kiszucaújhely. Zílinský Kraj, N Slovakia 49°18′N 18°48′E
117 N12 **Kytay, Ozero** ◎ SW Ukraine
115 F23 **Kýthira** var. Kíthira, It. Cerigo, Lat. Cythera. Kńythira, S Greece 41°39′N 26°30′E
115 F23 **Kýthira** var. Kíthira, It. Cerigo, Lat. Cythera. island S Greece
115 I20 **Kýthnos** Kńythnos, Kykládes, Greece, Aegean Sea 37°24′N 24°28′E
115 I20 **Kýthnos** var. Kíthnos, Thermiá, It. Termia; anc. Cythnos. island Kykládes, Greece, Aegean Sea
115 I20 **Kýthnou, Stenó** strait Kykládes, Greece, Aegean Sea
Kyungëy Ala-Too, Khrebet see Kungei Ala-Tau
Kyurdamir see Kürdämir
164 D15 **Kyūshū** var. Kyûsyû. island SW Japan
192 H6 **Kyushu-Palau Ridge** var. Kyusyu-Palau Ridge. undersea feature W Pacific Ocean 20°00′N 136°00′E
114 F10 **Kyustendil** anc. Pautalia. Kyustendil, W Bulgaria 42°17′N 22°42′E
114 G11 **Kyustendil** ◐ province W Bulgaria
Kyûsyû see Kyūshū
Kyusyu-Palau Ridge see Kyushu-Palau Ridge
123 P8 **Kyusyur** Respublika Sakha (Yakutiya), NE Russian Federation 70°36′N 127°19′E
183 P10 **Kywong** New South Wales, SE Australia 34°59′S 146°42′E
117 P4 **Kyyiv** Eng. Kiev, Rus. Kiyev. ● (Ukraine) Kyyivs'ka Oblast', N Ukraine 50°26′N 30°32′E
Kyyiv see Kyyivs'ka Oblast'
117 O4 **Kyyivs'ka Oblast'** var. Kyyiv, Rus. Kiyevskaya Oblast'. ◐ province N Ukraine
117 P3 **Kyyivs'ke Vodoskhovyshche** Eng. Kiev Reservoir, Rus. Kiyevskoye Vodokhranilishche. ▣ N Ukraine
93 L16 **Kyyjärvi** Keski-Suomi, C Finland 63°02′N 24°34′E
122 K14 **Kyzyl** Respublika Tyva, C Russian Federation 51°45′N 94°28′E
147 S8 **Kyzyl-Adyr** var. Kirovskoye. Talasskaya Oblast', NW Kyrgyzstan 42°37′N 71°34′E
145 V14 **Kyzylagash** Kaz. Qyzylaghash. Almaty, SE Kazakhstan 45°20′N 78°45′E
146 C13 **Kyzylbair** Balkan Welaýaty, W Turkmenistan 38°13′N 55°38′E
Kyzyl-Dzhiik, Pereval see Uzbel Shankou
145 S7 **Kyzylkak, Ozero** ◎ NE Kazakhstan
145 X11 **Kyzylkesek** Vostochnyy Kazakhstan, E Kazakhstan 47°56′N 82°02′E
147 S10 **Kyzyl-Kiya** Kir. Kyzyl-Kyya. Batkenskaya Oblast', SW Kyrgyzstan 40°15′N 72°07′E
144 L11 **Kyzylkol', Ozero** ◎ C Kazakhstan
122 K14 **Kyzyl Kum** var. Kizil Kum, Qizil Qum, Uzb. Qizilqum. desert Kazakhstan/Uzbekistan
Kyzyl-Kyya see Kyzyl-Kiya
145 N15 **Kyzylorda** var. Kzyl-Orda, Qizil Orda, Qyzylorda; prev. Kzylorda, Perovsk. Kyzylorda, S Kazakhstan 44°54′N 65°31′E
144 L14 **Kyzylorda** off. Kyzylordinskaya Oblast', Kaz. Qyzylorda Oblysy. ◐ province S Kazakhstan
Kyzylordinskaya Oblast' see Kyzylorda
Kyzylrabat see Qizilravote
Kyzylrabot see Qizilrabot
Kyzylsu see Kyzyl-Suu
147 X7 **Kyzyl-Suu** prev. Pokrovka. Issyk-Kul'skaya Oblast', NE Kyrgyzstan 42°20′N 77°55′E
147 S12 **Kyzyl-Suu** var. Kyzylsu. ≈ Kyrgyzstan/Tajikistan
147 X8 **Kyzyl-Tuu** Issyk-Kul'skaya Oblast', E Kyrgyzstan 42°06′N 76°54′E
145 Q12 **Kyzylzhar** Kaz. Qyzylzhar. Karaganda, C Kazakhstan 48°22′N 70°00′E
Kzyl-Orda see Kyzylorda
Kzylorda see Kyzylorda
Kzyltu see Kishkenekol'

L

109 X2 **Laa an der Thaya** Niederösterreich, NE Austria 48°44′N 16°23′E
63 K15 **La Adela** La Pampa, SE Argentina 38°57′S 64°02′W
Laagen see Numedalslågen
109 S5 **Laakirchen** Oberösterreich, N Austria 47°59′N 13°49′E
Laaland see Lolland
104 I11 **La Albuera** Extremadura, W Spain 38°43′N 06°49′W
105 O7 **La Alcarria** physical region C Spain
104 K14 **La Algaba** Andalucía, S Spain 37°27′N 06°01′W
105 P9 **La Almarcha** Castilla-La Mancha, C Spain 39°41′N 02°23′W
105 R6 **La Almunia de Doña Godina** Aragón, NE Spain 41°28′N 01°23′W
41 N5 **La Amistad, Presa** ▣ NW Mexico
118 F4 **Läänemaa** var. Lääne Maakond. ◐ province NW Estonia
Lääne Maakond see Läänemaa
118 I3 **Lääne-Virumaa** off. Lääne-Viru Maakond. ◐ province NE Estonia
Lääne-Viru Maakond see Lääne-Virumaa
62 J9 **La Antigua, Salina** salt lake W Argentina
99 E17 **Laarne** Oost-Vlaanderen, NW Belgium 51°03′N 03°50′E
80 O13 **Laas Caanood** Sool, N Somalia 08°33′N 47°44′E
41 O9 **La Ascensión** Nuevo León, NE Mexico 24°15′N 99°53′W
80 N12 **Laas Dhaareed** Togdheer, N Somalia 10°12′N 46°09′E
55 O4 **La Asunción** Nueva Esparta, NE Venezuela 11°06′N 63°53′W
Laatokka see Ladozhskoye, Ozero
100 I13 **Laatzen** Niedersachsen, NW Germany 52°19′N 09°46′E
38 E9 **La'au Point** var. Laau Point. headland Moloka'i, Hawai'i, USA 21°06′N 157°18′W
Laau Point see La'au Point
42 D6 **La Aurora** ✈ (Ciudad de Guatemala) Guatemala, C Guatemala 14°33′N 90°30′W
74 C9 **Laâyoune** var. Aaiún. ● (Western Sahara) NW Western Sahara 27°10′N 13°11′W
126 L14 **Laba** ≈ SW Russian Federation
40 M6 **La Babia** Coahuila, NE Mexico 28°39′N 102°00′W
15 R7 **La Baie** Québec, SE Canada 48°20′N 70°54′W
171 P16 **Labala** Pulau Lomblen, S Indonesia 08°30′S 123°27′E
62 K8 **La Banda** Santiago del Estero, N Argentina 27°44′S 64°14′W
La Banda Oriental see Uruguay
104 K4 **La Bañeza** Castilla y León, N Spain 42°18′N 05°54′W
167 T11 **Labăng** prev. Phumĭ Labăng. Rôtânôkiri, NE Cambodia 13°51′N 107°01′E
40 M13 **La Barca** Jalisco, SW Mexico 20°20′N 102°33′W
40 K14 **La Barra de Navidad** Jalisco, C Mexico 19°12′N 104°38′W
187 Y13 **Labasa** prev. Lambasa. Vanua Levu, N Fiji 16°25′S 179°24′E
102 H8 **la Baule-Escoublac** Loire-Atlantique, NW France 47°17′N 02°24′W
76 I13 **Labé** NW Guinea 11°19′N 12°17′W
Labe see Elbe
15 N11 **Labelle** Québec, SE Canada 46°15′N 74°43′W
23 X14 **La Belle** Florida, SE USA 26°45′N 81°26′W
10 H7 **Laberge, Lake** ◎ Yukon, W Canada
Labes see Łobez
Labiau see Polessk
112 A10 **Labin** It. Albona. Istra, NW Croatia 45°05′N 14°10′E
126 L14 **Labinsk** Krasnodarskiy Kray, SW Russian Federation 44°39′N 40°43′E
105 X5 **La Bisbal d'Empordà** Cataluña, NE Spain 41°58′N 03°02′E
119 P16 **Labkovichy** Rus. Lobkovichi. Mahilyowskaya Voblasts', E Belarus 53°50′N 31°45′E
15 S4 **La Blache, Lac de** ◎ Québec, SE Canada
171 P4 **Labo** Luzon, N Philippines 14°10′N 122°47′E
Laboehanbadjo see Labuhanbajo
Laborca see Laborec
111 N18 **Laborec** Hung. Laborca. ≈ E Slovakia
108 D11 **La Borgne** ≈ S Switzerland
45 T12 **Laborie** SW Saint Lucia 13°45′N 61°00′W
102 J14 **Labouheyre** Landes, SW France 44°12′N 00°55′W
62 L12 **Laboulaye** Córdoba, C Argentina 34°05′S 63°20′W
13 Q7 **Labrador** cultural region Newfoundland and Labrador, SW Canada
64 I6 **Labrador Basin** var. Labrador Sea Basin. undersea feature Labrador Sea 53°00′N 48°00′W
13 N9 **Labrador City** Newfoundland and Labrador, E Canada 52°56′N 66°52′W
13 Q5 **Labrador Sea** sea NW Atlantic Ocean
Labrador Sea Basin see Labrador Basin
Labrang see Xiahe
54 G9 **Labranzagrande** Boyacá, C Colombia 05°34′N 72°34′W
59 D14 **Lábrea** Amazonas, N Brazil 07°20′S 64°46′W
45 U15 **La Brea** Trinidad, Trinidad and Tobago 10°14′N 61°37′W
15 S6 **Labrieville** Québec, SE Canada 49°15′N 69°31′W
102 K14 **Labrit** Landes, SW France 44°03′N 00°29′W
108 C9 **La Broye** ≈ SW Switzerland
103 N15 **Labruguière** Tarn, S France 43°32′N 02°15′E
168 M11 **Labu** Pulau Singkep, W Indonesia 0°34′S 104°24′E
169 T7 **Labuan** var. Victoria. Labuan, East Malaysia 05°20′N 115°14′E
169 T7 **Labuan** ◐ federal territory East Malaysia
Labuan see Labuan, Pulau
169 T7 **Labuan, Pulau** var. Labuan. island East Malaysia
Labudalin see Ergun
171 N16 **Labuhanbajo** prev. Laboehanbadjo. Flores, S Indonesia 08°33′S 119°55′E
168 J9 **Labuhanbilik** Sumatera, N Indonesia 02°30′N 100°10′E
168 G8 **Labuhanhaji** Sumatera, W Indonesia 03°31′N 97°00′E
Labuk see Labuk, Sungai
Labuk Bay see Labuk, Teluk
169 V7 **Labuk, Sungai** var. Labuk. ≈ East Malaysia
169 W6 **Labuk, Teluk** var. Labuk Bay, Telukan Labuk. bay S Sulu Sea
Labuk, Telukan see Labuk, Teluk
166 K9 **Labutta** Ayeyawady, SW Myanmar (Burma) 16°08′N 94°45′E
122 I8 **Labytnangi** Yamalo-Nenetskiy Avtonomnyy Okrug, N Russian Federation 66°39′N 66°26′E
113 K19 **Laç** var. Laci. Lezhë, C Albania 41°37′N 19°37′E
78 F10 **Lac** off. Région du Lac. ◐ region W Chad
57 K19 **Lacajahuira, Rio** ≈ W Bolivia
La Calamine see Kelmis
62 G11 **La Calera** Valparaíso, C Chile 32°47′S 71°16′W
13 P11 **Lac-Allard** Québec, E Canada 50°37′N 63°26′W
104 L13 **La Campana** Andalucía, S Spain 37°35′N 05°24′W
102 J12 **Lacanau** Gironde, SW France 44°59′N 01°04′W
42 C2 **Lacandón, Sierra del** ▲▲ Guatemala/Mexico
La Cañiza see A Cañiza
41 W16 **Lacantún, Río** ≈ SE Mexico
103 Q3 **la Capelle** Aisne, N France 49°58′N 03°55′E
112 K10 **Laćarak** Vojvodina, NW Serbia 45°00′N 19°34′E
62 L11 **La Carlota** Córdoba, C Argentina 33°30′S 63°15′W
104 L13 **La Carlota** Andalucía, S Spain 37°40′N 04°56′W
105 N12 **La Carolina** Andalucía, S Spain 38°15′N 03°37′W
103 O15 **Lacaune** Tarn, S France 43°42′N 02°42′E
15 P7 **Lac-Bouchette** Québec, SE Canada 48°14′N 72°11′W
Laccadive Islands/ Laccadive Minicoy and Amindivi Islands, the see Lakshadweep
11 Y16 **Lac du Bonnet** Manitoba, S Canada 50°13′N 96°04′W
30 L4 **Lac du Flambeau** Wisconsin, N USA 45°58′N 89°51′W
15 P8 **Lac-Édouard** Québec, SE Canada 47°39′N 72°16′W
42 I4 **La Ceiba** Atlántida, N Honduras 15°45′N 86°29′W
54 E9 **La Ceja** Antioquia, W Colombia 06°02′N 75°30′W
182 J11 **Lacepede Bay** bay South Australia
32 G9 **Lacey** Washington, NW USA 47°01′N 122°49′W
103 P12 **la Chaise-Dieu** Haute-Loire, C France 45°19′N 03°41′E
114 G13 **Lachanás** Kentrikí Makedonía, N Greece 40°57′N 23°15′E
124 L11 **Lacha, Ozero** ◎ NW Russian Federation
103 O8 **la Charité-sur-Loire** Nièvre, C France 47°10′N 02°59′E
103 N9 **la Châtre** Indre, C France 46°35′N 01°59′E
108 C8 **La Chaux-de-Fonds** Neuchâtel, W Switzerland 47°07′N 06°51′E
Lach Dera see Dheere Laaq
108 G8 **Lachen** Schwyz, C Switzerland 47°12′N 08°51′E
183 Q8 **Lachlan River** ≈ New South Wales, SE Australia
43 T15 **La Chorrera** West Panamá, C Panama 08°51′N 79°46′W
15 V7 **Lac-Humqui** Québec, SE Canada 48°21′N 67°32′W
15 N12 **Lachute** Québec, SE Canada 45°39′N 74°21′W
Lachyn see Laçın
Laci see Laç
137 W13 **Laçın** Rus. Lachyn. SW Azerbaijan 39°36′N 46°34′E
103 S16 **la Ciotat** anc. Citharista. Bouches-du-Rhône, SE France 43°10′N 05°36′E
18 D10 **Lackawanna** New York, NE USA 42°49′N 78°49′W
11 Q13 **Lac La Biche** Alberta, SW Canada 54°46′N 111°59′W
Lac la Martre see Wha Ti
15 R12 **Lac-Mégantic** var. Mégantic. Québec, SE Canada 45°35′N 70°53′W
Lacobriga see Lagos
40 G5 **La Colorada** Sonora, NW Mexico 28°49′N 110°32′W
11 Q15 **Lacombe** Alberta, SW Canada 52°30′N 113°42′W
30 L12 **Lacon** Illinois, N USA 41°01′N 89°24′W
43 P16 **La Concepción** var. Concepción. Chiriquí, W Panama 08°31′N 82°39′W
54 H5 **La Concepción** Zulia, NW Venezuela 10°48′N 71°46′W
107 C19 **Laconi** Sardegna, Italy, C Mediterranean Sea 39°52′N 09°02′E
19 O9 **Laconia** New Hampshire, NE USA 43°32′N 71°29′W
61 H19 **La Coronilla** Rocha, E Uruguay 33°44′S 53°31′W
La Coruña see A Coruña
103 O11 **la Courtine** Creuse, C France 45°42′N 02°18′E
102 J16 **Lacq** Pyrénées-Atlantiques, SW France 43°25′N 00°37′W
Lac, Région du see Lac
15 P9 **La Croche** Québec, SE Canada 47°38′N 72°42′W
29 X3 **la Croix, Lac** ◎ Canada/USA
26 K5 **La Crosse** Kansas, C USA 38°32′N 99°19′W
21 V7 **La Crosse** Virginia, NE USA 36°41′N 78°03′W
32 L9 **La Crosse** Washington, NW USA 46°48′N 117°51′W
30 J7 **La Crosse** Wisconsin, N USA 43°46′N 91°12′W
54 C13 **La Cruz** Nariño, SW Colombia 01°33′N 76°58′W
42 K12 **La Cruz** Guanacaste, NW Costa Rica 11°05′N 85°39′W
40 I10 **La Cruz** Sinaloa, W Mexico 23°53′N 106°53′W
61 F19 **La Cruz** Florida, S Uruguay 33°54′S 56°11′W
42 M9 **La Cruz de Río Grande** Región Autónoma Atlántico Sur, E Nicaragua 13°04′N 84°12′W
54 J4 **La Cruz de Taratara** Falcón, N Venezuela 11°03′N 69°44′W
15 Q10 **Lac-St-Charles** Québec, SE Canada 46°56′N 71°24′W
40 M6 **La Cuesta** Coahuila, NE Mexico 28°45′N 102°26′W

◆ Country
● Country Capital
◇ Dependent Territory
○ Dependent Territory Capital
◐ Administrative Regions
✈ International Airport
▲ Mountain
▲▲ Mountain Range
⛰ Volcano
≈ River
◎ Lake
▣ Reservoir

◆ Country   ◇ Dependent Territory   ◆ Administrative Regions   ▲ Mountain   ▲ Volcano   ◎ Lake
● Country Capital   ◎ Dependent Territory Capital   ✈ International Airport   ▲ Mountain Range   ◄ River   ◎ Reservoir

Lansdale – Lebanon

18 I15 **Lansdale** Pennsylvania, NE USA 40°14′N 75°13′W
14 L14 **Lansdowne** Ontario, SE Canada 44°25′N 76°00′W
152 K9 **Lansdowne** Uttarakhand, N India 29°50′N 78°42′E
30 M3 **L'Anse** Michigan, N USA 46°45′N 88°27′W
15 S7 **L'Anse-St-Jean** Québec, SE Canada 48°14′N 70°13′W
29 Y11 **Lansing** Iowa, C USA 43°22′N 91°11′W
27 R4 **Lansing** Kansas, C USA 39°15′N 94°54′W
31 Q9 **Lansing** *state capital* Michigan, N USA 42°44′N 84°33′W
Länsi-Suomi ⬖ *province* W Finland
92 J12 **Lansjärv** Norrbotten, N Sweden 66°39′N 22°10′E
111 G17 **Lanškroun** *Ger.* Landskron. Pardubický Kraj, E Czech Republic 49°55′N 16°38′E
167 N16 **Lanta, Ko** *island* S Thailand
161 O15 **Lantau Island** *Cant.* Tai Yue Shan, *Chin.* Landao. *island* Hong Kong, S China
Lantian *see* Lianyuan
Lan-ts'ang Chiang *see* Mekong
Lantung, Gulf of *see* Liaodong Wan
171 O11 **Lanu** Sulawesi, N Indonesia 01°00′N 121°33′E
107 D19 **Lanusei** Sardegna, Italy, C Mediterranean Sea 39°55′N 09°31′E
102 H7 **Lanvaux, Landes de** *physical region* NW France
163 W8 **Lanxi** Heilongjiang, NE China 46°18′N 126°15′E
161 R10 **Lanxi** Zhejiang, SE China 29°12′N 119°27′E
La Nyanga *see* Nyanga
161 T15 **Lan Yu** *var.* Huoshao Tao, Hungt'ou, Lan Hsü, Lanyü, *Eng.* Orchid Island; *prev.* Kotosho, Koto Sho, Lan Yü. *island* SE Taiwan
Lanyü *see* Lan Yu
64 P11 **Lanzarote** *island* Islas Canarias, Spain, NE Atlantic Ocean
159 V10 **Lanzhou** *var.* Lan-chou, Lanchow, Lan-chow; *prev.* Kaolan. *province capital* Gansu, C China 36°01′N 103°52′E
106 B8 **Lanzo Torinese** Piemonte, NE Italy 45°18′N 07°26′E
171 O11 **Laoag** Luzon, N Philippines 18°11′N 120°34′E
171 Q5 **Laoang** Samar, C Philippines 12°29′N 125°01′E
167 R5 **Lao Cai** Lao Cai, N Vietnam 22°29′N 104°00′E
Laodicea/Laodicea ad Mare *see* Al Lādhiqīyah
Laoet *see* Laut, Pulau
163 T11 **Laoha He** ≈ NE China
160 M8 **Laohekou** *var.* Guanghua. Hubei, C China 32°20′N 111°42′E
Laoi, An *see* Lee
97 E19 **Laois** *prev.* Leix, Queen's County. *cultural region* C Ireland
163 W12 **Lao Ling** ▲▲ N China
64 Q11 **La Oliva** *var.* Oliva. Fuerteventura, Islas Canarias, Spain, NE Atlantic Ocean 28°36′N 13°53′W
Lao, Loch *see* Belfast Lough
Laolong *see* Longchuan
Lao Mangnai *see* Mangnai
103 P3 **Laon** *var.* la Laon; *anc.* Laudunum. Aisne, N France 49°34′N 03°37′E
Lao People's Democratic Republic *see* Laos
54 M3 **La Orchila, Isla** *island* N Venezuela
64 O11 **La Orotava** Tenerife, Islas Canarias, Spain, NE Atlantic Ocean 28°23′N 16°32′W
57 E14 **La Oroya** Junín, C Peru 11°36′S 75°54′W
167 Q7 **Laos** *off.* Lao People's Democratic Republic. ◆ *republic* SE Asia
161 R5 **Laoshan Wan** *bay* E China
163 Y10 **Laoye Ling** ▲▲ NE China
60 J12 **Lapa** Paraná, S Brazil 25°46′S 49°44′W
103 P10 **Lapalisse** Allier, C France 46°13′N 03°39′E
54 F9 **La Palma** Cundinamarca, C Colombia 05°23′N 74°24′W
42 F7 **La Palma** Chalatenango, N El Salvador 14°19′N 89°10′W
43 W16 **La Palma** Darién, SE Panama 08°24′N 78°09′W
64 N11 **La Palma** *island* Islas Canarias, Spain, NE Atlantic Ocean
104 J14 **La Palma del Condado** Andalucía, S Spain 37°23′N 06°33′W
61 F18 **La Paloma** Durazno, C Uruguay 32°54′S 55°36′W
61 G20 **La Paloma** Rocha, E Uruguay 34°37′S 54°08′W
61 A21 **La Pampa** *off.* Provincia de La Pampa. ⬖ *province* C Argentina
La Pampa, Provincia de *see* La Pampa
55 P8 **La Paragua** Bolívar, E Venezuela 06°53′N 63°16′W
119 O16 **Lapatsichy** *Rus.* Lopatichi. Mahilyowskaya Voblasts', E Belarus 53°34′N 30°53′E
61 C16 **La Paz** Entre Ríos, E Argentina 30°45′S 59°36′W
62 I11 **La Paz** Mendoza, C Argentina 33°30′S 67°36′W
57 J18 **La Paz** *var.* La Paz de Ayacucho. ● (Bolivia-seat of government) La Paz, W Bolivia 16°30′S 68°13′W
42 H6 **La Paz** La Paz, SW Honduras 14°20′N 87°40′W
40 F9 **La Paz** Baja California Sur, NW Mexico 24°07′N 110°18′W
61 F20 **La Paz** Canelones, S Uruguay 34°46′S 56°13′W
57 J16 **La Paz** ⬖ *department* W Bolivia
42 B9 **La Paz** ⬖ *department* S El Salvador
42 G7 **La Paz** ⬖ *department* SW Honduras
La Paz *see* El Alto, Bolivia
La Paz *see* Robles, Colombia
La Paz *see* La Paz Centro
40 F9 **La Paz, Bahía de** *bay* NW Mexico
42 I10 **La Paz Centro** *var.* La Paz. León, W Nicaragua 12°20′N 86°41′W
La Paz de Ayacucho *see* La Paz
54 J15 **La Pedrera** Amazonas, SE Colombia 01°19′S 69°31′W
31 S9 **Lapeer** Michigan, N USA 43°03′N 83°19′W
40 K6 **La Perla** Chihuahua, N Mexico 28°18′N 104°34′W
165 T1 **La Pérouse Strait** *Jap.* Sōya-kaikyō, *Rus.* Proliv Laperuza. *strait* Japan/Russian Federation
63 I14 **La Perra, Salitral de** *salt lake* C Argentina
Laperuza, Proliv *see* La Pérouse Strait
41 Q10 **La Pesca** Tamaulipas, C Mexico 23°49′N 97°45′W
40 M13 **La Piedad Cavadas** Michoacán, C Mexico 20°20′N 102°01′W
Lapines *see* Lafnitz
93 M16 **Lapinlahti** Pohjois-Savo, C Finland 63°21′N 27°25′E
Lápithos *see* Lapta
22 K9 **Laplace** Louisiana, S USA 30°04′N 90°28′W
45 X12 **La Plaine** SE Dominica 15°20′N 61°15′W
173 P16 **La Plaine-des-Palmistes** C Réunion
92 K11 **Lapland** *Fin.* Lappi, *Swe.* Lappland. *cultural region* N Europe
Lapland *see* Lappi
28 M8 **La Plant** South Dakota, N USA 45°06′N 100°40′W
61 D20 **La Plata** Buenos Aires, E Argentina 34°56′S 57°55′W
54 D12 **La Plata** Huila, SW Colombia 02°33′N 75°55′W
21 W4 **La Plata** Maryland, NE USA 38°32′N 76°59′W
La Plata *see* Sucre
45 U6 **la Plata, Río de** ≈ C Puerto Rico
105 W4 **La Pobla de Lillet** Cataluña, NE Spain 42°15′N 01°57′E
105 U4 **La Pobla de Segur** Cataluña, NE Spain 42°15′N 00°58′E
15 S9 **La Pocatière** Québec, SE Canada 47°21′N 70°04′W
104 K2 **La Pola** *prev.* Pola de Lena. Asturias, N Spain 43°10′N 05°49′W
104 L3 **La Pola de Gordón** Castilla y León, N Spain 42°50′N 05°38′W
104 L2 **La Pola Siero** *prev.* Pola de Siero. Asturias, N Spain 43°24′N 05°39′W
31 O11 **La Porte** Indiana, N USA 41°36′N 86°43′W
18 H13 **Laporte** Pennsylvania, NE USA 41°25′N 76°29′W
29 X13 **La Porte City** Iowa, C USA 42°19′N 92°11′W
62 J8 **La Posta** Catamarca, C Argentina 27°59′S 65°32′W
40 E8 **La Poza Grande** Baja California Sur, NW Mexico 25°50′N 112°00′W
93 K16 **Lappajärvi** Etelä-Pohjanmaa, W Finland 63°13′N 23°40′E
93 L16 **Lappajärvi** ◎ W Finland
93 N18 **Lappeenranta** *Swe.* Villmanstrand. Etelä-Karjala, SE Finland 61°04′N 28°15′E
93 J17 **Lappfjärd** *Fin.* Lapväärtti. Österbotten, W Finland 62°14′N 21°30′E
92 L12 **Lappi** *Swe.* Lappland, *Eng.* Lapland. ⬖ *region* N Finland
Lappi/Lappland *see* Lapland
Lappo *see* Lapua
61 C23 **Laprida** Buenos Aires, E Argentina 37°34′S 60°45′W
25 P13 **La Pryor** Texas, SW USA 28°56′N 99°51′W
136 B11 **Lâpseki** Çanakkale, NW Turkey 40°22′N 26°42′E
121 P2 **Lapta** *Gk.* Lápithos. NW Cyprus 35°20′N 33°11′E
Laptev Sea *see* Laptevykh, More
122 N6 **Laptevykh, More** *Eng.* Laptev Sea. *sea* Arctic Ocean
93 K16 **Lapua** *Swe.* Lappo. Etelä-Pohjanmaa, W Finland 62°57′N 23°E
105 P3 **La Puebla de Arganzón** País Vasco, N Spain 42°45′N 02°49′W
104 L14 **La Puebla de Cazalla** Andalucía, S Spain 37°14′N 05°18′W
104 M9 **La Puebla de Montalbán** Castilla-La Mancha, C Spain 39°52′N 04°22′W
54 I6 **La Puerta** Trujillo, NW Venezuela 09°08′N 70°46′W
Lapurdum *see* Bayonne
40 E7 **La Purísima** Baja California Sur, NW Mexico 26°10′N 112°05′W
Lapväärtti *see* Lappfjärd
110 O10 **Łapy** Podlaskie, NE Poland 53°N 22°54′E
80 D6 **Laqiya Arba'in** Northern, NW Sudan 20°01′N 28°01′E
62 J4 **La Quiaca** Jujuy, N Argentina 22°12′S 65°36′W
107 J14 **L'Aquila** *var.* Aquila, Aquila degli Abruzzi. Abruzzo, C Italy 42°21′N 13°24′E
143 Q13 **Lār** Fārs, S Iran 27°42′N 54°19′E
54 J5 **Lara** *off.* Estado Lara. ⬖ *state* NW Venezuela
104 G2 **Laracha** Galicia, NW Spain 43°14′N 08°34′W
74 G5 **Larache** *var.* al Araïch, El Araïch, *prev.* El Araïche; *anc.* Lixus. NW Morocco 35°12′N 06°10′W
Lara, Estado *see* Lara
103 T14 **Laragne-Montéglin** Hautes-Alpes, SE France 44°21′N 05°46′E
104 M13 **La Rambla** Andalucía, S Spain 37°37′N 04°44′W
33 Y17 **Laramie** Wyoming, C USA 41°18′N 105°35′W
33 X15 **Laramie Mountains** ▲▲ Wyoming, C USA
33 Y16 **Laramie River** ≈ Wyoming, C USA
60 H12 **Laranjeiras do Sul** Paraná, S Brazil 25°23′S 52°23′W
Larantoeka *see* Larantuka
171 P16 **Larantuka** *prev.* Larantoeka. Flores, C Indonesia 08°20′S 123°00′E
171 U15 **Larat** Pulau Larat, E Indonesia 07°07′S 131°46′E
171 U15 **Larat, Pulau** *island* Kepulauan Tanimbar, E Indonesia
95 P19 **Lärbro** Gotland, SE Sweden 57°46′N 18°49′E
106 A9 **Larche, Col de** *pass* France/Italy
14 H8 **Larder Lake** Ontario, S Canada 48°06′N 79°44′W
105 O2 **Laredo** Cantabria, N Spain 43°23′N 03°22′W
25 Q15 **Laredo** Texas, SW USA 27°30′N 99°30′W
40 H9 **La Reforma** Sinaloa, W Mexico 25°05′N 108°03′W
98 N11 **Laren** Gelderland, E Netherlands 52°12′N 06°22′E
98 J11 **Laren** Noord-Holland, C Netherlands 52°15′N 05°13′E
102 K13 **la Réole** Gironde, SW France 44°34′N 00°00′W
La Réunion *see* Réunion
Largeau *see* Faya
103 U13 **l'Argentière-la-Bessée** Hautes-Alpes, SE France 44°49′N 06°34′E
149 O4 **Lar Gerd** *var.* Largird. Balkh, N Afghanistan 35°36′N 66°48′E
Largird *see* Lar Gerd
23 V12 **Largo** Florida, SE USA 27°55′N 82°47′W
37 Q9 **Largo, Canon** *valley* New Mexico, SW USA
44 D6 **Largo, Cayo** *island* W Cuba
23 Z17 **Largo, Key** *island* Florida Keys, Florida, SE USA
96 H12 **Largs** W Scotland, United Kingdom 55°48′N 04°50′W
102 I16 **la Rhune** *var.* Larrún. ▲ France/Spain 43°19′N 01°36′W *see also* Larrún
la Rhune *see* Larrún
la Riege *see* Ariège
29 Q4 **Larimore** North Dakota, N USA 47°54′N 97°37′W
107 L15 **Larino** Molise, C Italy 41°46′N 14°50′E
Lario *see* Como, Lago di
62 J9 **La Rioja** La Rioja, NW Argentina 29°26′S 66°50′W
105 O4 **La Rioja** ⬖ *autonomous community* N Spain
62 I9 **La Rioja** *off.* Provincia de La Rioja. ⬖ *province* NW Argentina
La Rioja, Provincia de *see* La Rioja
115 F16 **Lárisa** *var.* Larissa. Thessalía, C Greece 39°38′N 22°27′E
Larissa *see* Lárisa
149 Q13 **Lārkāna** *var.* Larkhana. Sind, SE Pakistan 27°32′N 68°18′E
Larkhana *see* Lārkāna
Larnaca *see* Lárnaka
121 Q3 **Lárnaka** *var.* Larnaca, Larnax. SE Cyprus 34°55′N 33°39′E
121 Q3 **Lárnaka** ✈ SE Cyprus 34°52′N 33°38′E
Larnax *see* Lárnaka
97 G14 **Larne** *Ir.* Latharna. E Northern Ireland, United Kingdom 54°51′N 05°49′W
26 L5 **Larned** Kansas, C USA 38°12′N 99°05′W
104 L3 **La Robla** Castilla y León, N Spain 42°48′N 05°37′W
104 J10 **La Roca de la Sierra** Extremadura, W Spain 39°06′N 06°41′W
99 K22 **La Roche-en-Ardenne** Luxembourg, SE Belgium 50°11′N 05°35′E
102 L11 **la Rochefoucauld** Charente, W France 45°43′N 00°23′E
102 J10 **la Rochelle** *anc.* Rupella. Charente-Maritime, W France 46°09′N 01°07′W
102 I9 **la Roche-sur-Yon** *prev.* Bourbon Vendée, Napoléon-Vendée. Vendée, NW France 46°40′N 01°26′W
105 Q10 **La Roda** Castilla-La Mancha, C Spain 39°13′N 02°10′W
104 L14 **La Roda de Andalucía** Andalucía, S Spain 37°12′N 04°45′W
45 P9 **La Romana** E Dominican Republic 18°25′N 69°00′W
11 T13 **La Ronge** Saskatchewan, C Canada 55°07′N 105°18′W
11 U13 **La Ronge, Lac** ◎ Saskatchewan, C Canada
22 K10 **Larose** Louisiana, S USA 29°34′N 90°22′W
42 M7 **La Rosita** Región Autónoma Atlántico Norte, NE Nicaragua 13°55′N 84°23′W
181 Q3 **Larrimah** Northern Territory, N Australia 15°30′S 133°12′E
62 N11 **Larroque** Entre Ríos, E Argentina 33°05′S 59°06′W
105 Q2 **Larrún** *Fr.* la Rhune. ▲ France/Spain 43°18′N 01°35′W *see also* la Rhune
Larrún *see* la Rhune
195 X6 **Lars Christensen Coast** *physical region* Antarctica
39 Q14 **Larsen Bay** Kodiak Island, Alaska, USA 57°32′N 153°58′W
194 I5 **Larsen Ice Shelf** *ice shelf* Antarctica
8 M6 **Larsen Sound** *sound* Nunavut, N Canada
La Rúa *see* A Rúa de Valdeorras
102 K16 **Laruns** Pyrénées-Atlantiques, SW France 43°00′N 00°25′W
95 G16 **Larvik** Vestfold, S Norway 59°04′N 10°02′E
La-sa *see* Lhasa
171 S13 **Lasahata** Pulau Seram, E Indonesia 02°52′S 128°27′E
Lasahau *see* Lasihao
37 O6 **La Sal** Utah, W USA 38°19′N 109°14′W
14 C17 **La Salle** Ontario, S Canada 42°13′N 83°05′W
30 L11 **La Salle** Illinois, N USA 41°19′N 89°06′W
45 O9 **Las Americas** ✈ (Santo Domingo) S Dominican Republic 18°24′N 69°38′W
37 V6 **Las Animas** Colorado, C USA 38°04′N 103°13′W
108 D10 **La Sarine** *var.* Sarine. ≈ SW Switzerland
108 B9 **La Sarraz** Vaud, W Switzerland 46°40′N 06°32′E
12 H12 **La Sarre** Québec, SE Canada 48°49′N 79°12′W
54 L3 **Las Aves, Islas** *var.* Islas de Aves. *island group* N Venezuela
55 N7 **Las Bonitas** Bolívar, C Venezuela 07°50′N 65°40′W
104 K15 **Las Cabezas de San Juan** Andalucía, S Spain 36°59′N 05°56′W
61 G19 **Lascano** Rocha, E Uruguay 33°40′S 54°12′W
62 I5 **Lascar, Volcán** 🌋 N Chile 23°22′S 67°33′W
41 T15 **Las Choapas** *var.* Choapas. Veracruz-Llave, SE Mexico 17°51′N 94°00′W
37 R15 **Las Cruces** New Mexico, SW USA 32°19′N 106°49′W
Lasdehnen *see* Krasnoznamensk
La Selle *see* Selle, Pic de la
62 G9 **La Serena** Coquimbo, C Chile 29°54′S 71°18′W
104 K11 **La Serena** *physical region* W Spain
105 V4 **La Seu d'Urgell** *var.* La Seu d'Urgell; *prev.* Seo de Urgel. Cataluña, NE Spain 42°22′N 01°27′E
La Seu d'Urgell *see* La Seu d'Urgell
103 T16 **la Seyne-sur-Mer** Var, SE France 43°07′N 05°53′E
61 D21 **Las Flores** Buenos Aires, E Argentina 36°03′S 59°08′W
62 H9 **Las Flores** San Juan, W Argentina 30°14′S 69°10′W
11 S14 **Lashburn** Saskatchewan, S Canada 53°09′N 109°37′W
62 I11 **Las Heras** Mendoza, W Argentina 32°48′S 68°50′W
148 M8 **Lashkar Gāh** *var.* Lash-Kar-Gar'. Helmand, S Afghanistan 31°35′N 64°21′E
Lash-Kar-Gar' *see* Lashkar Gāh
171 P14 **Lasihao** *var.* Lasahau. Pulau Muna, C Indonesia 05°01′S 122°23′E
107 N21 **La Sila** ▲▲ SW Italy
63 H23 **La Silueta, Cerro** ▲ S Chile 52°22′S 72°09′W
42 L9 **La Sirena** Región Autónoma Atlántico Sur, E Nicaragua 12°59′N 84°35′W
110 J13 **Łask** Łódzkie, C Poland 51°36′N 19°06′E
109 V11 **Laško** *Ger.* Tüffer. C Slovenia 46°08′N 15°13′E
63 H14 **Las Lajas** Neuquén, W Argentina 38°31′S 70°22′W
63 H15 **Las Lajas, Cerro** ▲ W Argentina 38°49′S 70°42′W
62 M6 **Las Lomitas** Formosa, N Argentina 24°45′S 60°35′W
41 V16 **Las Margaritas** Chiapas, SE Mexico 16°15′N 91°58′W
Las Marismas *see* Guadalquivir, Marismas del
54 M6 **Las Mercedes** Guárico, N Venezuela 09°08′N 66°27′W
42 F6 **Las Minas, Cerro** ▲ W Honduras 14°33′N 88°41′W
105 O11 **La Solana** Castilla-La Mancha, C Spain 38°56′N 03°14′W
45 Q14 **La Soufrière** 🌋 Saint Vincent, Saint Vincent and the Grenadines 13°20′N 61°11′W
102 M10 **la Souterraine** Creuse, C France 46°15′N 01°28′E
62 N7 **Las Palmas** Chaco, N Argentina 27°08′S 58°45′W
43 Q16 **Las Palmas** Veraguas, W Panama 08°09′N 81°28′W
64 P12 **Las Palmas** *var.* Las Palmas de Gran Canaria. Gran Canaria, Islas Canarias, Spain, NE Atlantic Ocean 28°08′N 15°27′W
64 P12 **Las Palmas** ⬖ *province* Islas Canarias, Spain, NE Atlantic Ocean
64 Q12 **Las Palmas** ✈ Gran Canaria, Islas Canarias, Spain, NE Atlantic Ocean
Las Palmas de Gran Canaria *see* Las Palmas
40 D6 **Las Palomas** Baja California Norte, W Mexico 31°44′N 107°37′W
105 P10 **Las Pedroñeras** Castilla-La Mancha, C Spain 39°27′N 02°41′W
106 E10 **La Spezia** Liguria, NW Italy 44°08′N 09°50′E
61 F20 **Las Piedras** Canelones, S Uruguay 34°42′S 56°14′W
63 J18 **Las Plumas** Chubut, S Argentina 43°46′S 67°15′W
61 B18 **Las Rosas** Santa Fe, C Argentina 32°27′S 61°30′W
Lassa *see* Lhasa
35 O4 **Lassen Peak** 🌋 California, W USA 40°27′N 121°28′W
194 K6 **Lassiter Coast** *physical region* Antarctica
109 V9 **Lassnitz** ≈ SE Austria
15 O12 **L'Assomption** Québec, SE Canada 45°48′N 73°27′W
15 N11 **L'Assomption** ≈ Québec, SE Canada
43 S17 **Las Tablas** Los Santos, S Panama 07°45′N 80°17′W
Lastarria, Volcán *see* Azufre, Volcán
37 V4 **Last Chance** Colorado, C USA 39°41′N 103°34′W
Last Frontier, The *see* Alaska
11 U16 **Last Mountain Lake** ◎ Saskatchewan, S Canada
62 H9 **Las Tórtolas, Cerro** ▲ W Argentina 29°57′S 69°49′W
61 C14 **Las Toscas** Santa Fe, C Argentina 28°22′S 59°20′W
79 F19 **Lastoursville** Ogooué-Lolo, E Gabon 0°50′S 12°43′E
113 F16 **Lastovo** *It.* Lagosta. *island* SW Croatia
113 F16 **Lastovski Kanal** *channel* SW Croatia
40 E6 **Las Tres Vírgenes, Volcán** 🌋 NW Mexico 27°27′N 112°34′W
40 F4 **Las Trincheras** Sonora, NW Mexico 30°21′N 111°27′W
55 N8 **Las Trincheras** Bolívar, E Venezuela 06°57′N 64°49′W
44 H7 **Las Tunas** *var.* Victoria de las Tunas. Las Tunas, E Cuba 20°58′N 76°59′W
La Suisse *see* Switzerland
40 J5 **Las Varas** Chihuahua, N Mexico 29°35′N 108°01′W
40 J12 **Las Varas** Nayarit, C Mexico 21°12′N 105°10′W
62 L10 **Las Varillas** Córdoba, E Argentina 31°54′S 62°45′W
35 X11 **Las Vegas** Nevada, W USA 36°09′N 115°10′W
37 T10 **Las Vegas** New Mexico, SW USA 35°35′N 105°15′W
187 P10 **Lata** Nendō, Solomon Islands 10°45′S 165°43′E
13 R10 **La Tabatière** Québec, E Canada 50°51′N 58°59′W
56 C6 **Latacunga** Cotopaxi, C Ecuador 0°58′S 78°36′W
194 I7 **Latady Island** *island* Antarctica
54 E14 **La Tagua** Putumayo, S Colombia 0°05′S 74°39′W
Latakia *see* Al Lādhiqīyah
92 J10 **Lätäseno** ≈ NW Finland
14 H9 **Latchford** Ontario, S Canada 47°20′N 79°45′W
14 J13 **Latchford Bridge** Ontario, SE Canada 45°16′N 77°29′W
193 Y14 **Late** *island* Vava'u Group, N Tonga
153 P15 **Lātehār** Jhārkhand, N India 23°48′N 84°28′E
15 R7 **Laterrière** Québec, SE Canada 48°17′N 71°10′W
102 J13 **la Teste** Gironde, SW France 44°38′N 01°04′W
25 V8 **Latexo** Texas, SW USA 31°24′N 95°28′W
18 L10 **Latham** New York, NE USA 42°45′N 73°45′W
Latharna *see* Larne
108 B9 **La Thielle** *var.* Thièle. ≈ W Switzerland
27 R3 **Lathrop** Missouri, C USA 39°33′N 94°19′W
107 I16 **Latina** *prev.* Littoria. Lazio, C Italy 41°28′N 12°53′E
41 R14 **La Tinaja** Veracruz-Llave, S Mexico
106 J7 **Latisana** Friuli-Venezia Giulia, NE Italy 45°47′N 13°01′E
Latium *see* Lazio
115 K25 **Lató** *site of ancient city* Kríti, Greece, E Mediterranean Sea
187 Q17 **La Tontouta** ✈ (Noumea) Province Sud, S New Caledonia 22°06′S 166°12′E
55 N4 **La Tortuga, Isla** *var.* Isla Tortuga. *island* N Venezuela
108 C10 **La Tour-de-Peilz** *var.* La Tour de Peilz. Vaud, SW Switzerland 46°28′N 06°52′E
La Tour de Peilz *see* La Tour-de-Peilz
103 S11 **La Tour-du-Pin** Isère, E France 45°34′N 05°25′E
102 J11 **la Tremblade** Charente-Maritime, W France 45°45′N 01°07′W
102 L10 **la Trimouille** Vienne, W France 46°27′N 01°02′E
42 J9 **La Trinidad** Estelí, NW Nicaragua 12°57′N 86°15′W
41 V16 **La Trinitaria** Chiapas, SE Mexico 16°02′N 92°00′W
45 Q11 **la Trinité** E Martinique 14°44′N 60°58′W
15 U7 **La Trinité-des-Monts** Québec, SE Canada 48°07′N 68°31′W
18 C15 **Latrobe** Pennsylvania, NE USA 40°18′N 79°19′W
183 P13 **La Trobe River** ≈ Victoria, SE Australia
Lattakia/Lattaquié *see* Al Lādhiqīyah
171 S13 **Latu** Pulau Seram, E Indonesia 03°24′S 128°37′E
15 P9 **La Tuque** Québec, SE Canada 47°26′N 72°47′W
155 G14 **Lātūr** Mahārāshtra, C India 18°24′N 76°34′E
118 G8 **Latvia** *off.* Republic of Latvia, *Ger.* Lettland, *Latv.* Latvija, Latvijas Republika; *prev.* Latvian SSR, *Rus.* Latviyskaya SSR. ◆ *republic* NE Europe
Latvian SSR/Latvija/Latvijas Republika/Latviyskaya SSR *see* Latvia
Latvia, Republic of *see* Latvia
186 H7 **Lau** New Britain, E Papua New Guinea 05°46′S 151°21′E
175 R9 **Lau Basin** *undersea feature* S Pacific Ocean
101 O15 **Lauchhammer** Brandenburg, E Germany 51°30′N 13°48′E
105 O3 **Laudio** *var.* Llodio. País Vasco, N Spain 43°08′N 02°59′W
Laudunum *see* Laon
Laudus *see* St-Lô
Lauenburg/Lauenburg in Pommern *see* Lębork
101 L20 **Lauf an der Pegnitz** Bayern, SE Germany 49°31′N 11°16′E
108 D7 **Laufen** Basel, NW Switzerland 47°26′N 07°31′E
109 P5 **Lauffen** Salzburg, NW Austria 47°54′N 12°57′E
92 I2 **Laugarbakki** Norðurland Vestra, N Iceland 65°18′N 20°51′W
92 I4 **Laugarvatn** Suðurland, SW Iceland 64°09′N 20°43′W
31 O3 **Laughing Fish Point** *headland* Michigan, N USA 46°31′N 87°01′W
187 Z14 **Lau Group** *island group* E Fiji
Lauis *see* Lugano
93 M17 **Laukaa** Keski-Suomi, C Finland 62°27′N 25°58′E
118 D12 **Laukuva** Tauragė, W Lithuania 55°37′N 22°12′E
Laun *see* Louny
183 P16 **Launceston** Tasmania, SE Australia 41°25′S 147°07′E
97 I24 **Launceston** *anc.* Dunheved. SW England, United Kingdom 50°38′N 04°21′W
54 C13 **La Unión** Nariño, SW Colombia 01°35′N 77°09′W
42 H8 **La Unión** La Unión, SE El Salvador 13°20′N 87°50′W
42 I6 **La Unión** Olancho, C Honduras 15°02′N 86°40′W
40 M15 **La Unión** Guerrero, S Mexico 17°58′N 101°49′W
41 Y14 **La Unión** Quintana Roo, E Mexico 18°00′N 101°48′W
105 S13 **La Unión** Murcia, SE Spain 37°37′N 00°54′W
54 L7 **La Unión** Barinas, C Venezuela 08°15′N 67°46′W
42 B10 **La Unión** ⬖ *department* E El Salvador
38 H11 **Laupāhoehoe** *var.* Laupahoehoe. Hawaii, USA, C Pacific Ocean 20°00′N 155°15′W
Laupahoehoe *see* Laupāhoehoe
101 I23 **Laupheim** Baden-Württemberg, S Germany 48°13′N 09°54′E
181 W3 **Laura** Queensland, NE Australia 15°37′S 144°34′E
189 X2 **Laura** *atoll* Majuro Atoll, SE Marshall Islands
Laurana *see* Lovran
54 L8 **La Urbana** Bolívar, C Venezuela 07°05′N 66°58′W
21 Y4 **Laurel** Delaware, NE USA 38°33′N 75°34′W
23 V14 **Laurel** Florida, SE USA 27°07′N 82°27′W
21 W3 **Laurel** Maryland, NE USA 39°06′N 76°51′W
22 M6 **Laurel** Mississippi, S USA 31°41′N 89°10′W
33 U11 **Laurel** Montana, NW USA 45°40′N 108°46′W
29 R13 **Laurel** Nebraska, C USA 42°25′N 97°04′W
18 H15 **Laureldale** Pennsylvania, NE USA 40°24′N 75°52′W
18 C16 **Laurel Hill** *ridge* Pennsylvania, NE USA
29 T12 **Laurens** Iowa, C USA 42°51′N 94°51′W
21 P11 **Laurens** South Carolina, SE USA 34°29′N 82°01′W
Laurentian Highlands *see* Laurentian Mountains
15 P10 **Laurentian Mountains** *var.* Laurentian Highlands, *Fr.* Les Laurentides. *plateau* Newfoundland and Labrador/Québec, Canada
15 O12 **Laurentides** Québec, SE Canada 45°51′N 73°49′W
Laurentides, Les *see* Laurentian Mountains
107 M19 **Lauria** Basilicata, S Italy 40°03′N 15°50′E
194 I1 **Laurie Island** *island* Antarctica
21 T11 **Laurinburg** North Carolina, SE USA 34°46′N 79°29′W
30 M2 **Laurium** Michigan, N USA 47°14′N 88°26′W
Lauru *see* Choiseul
108 B9 **Lausanne** *It.* Losanna. Vaud, SW Switzerland 46°32′N 06°39′E
101 Q16 **Lausche** *var.* Luže. ▲ Czech Republic/Germany 50°52′N 14°39′E *see also* Luže
Lausche *see* Luže
101 Q16 **Lausitzer Bergland** *var.* Lausitzer Gebirge, *Cz.* Gory Lužycke, Lužické Hory', *Eng.* Lusatian Mountains. ▲▲ E Germany
Lausitzer Gebirge *see* Lausitzer Bergland
Lausitzer Neisse *see* Neisse
103 T12 **Lautaret, Col du** *pass* SE France
63 G15 **Lautaro** Araucanía, C Chile 38°30′S 71°30′W
101 F21 **Lauter** ≈ W Germany
108 I7 **Lauterach** Vorarlberg, NW Austria 47°29′N 09°44′E
101 I17 **Lauterbach** Hessen, C Germany 50°37′N 09°24′E
108 E9 **Lauterbrunnen** Bern, C Switzerland 46°36′N 07°52′E
169 U14 **Laut Kecil, Kepulauan** *island group* N Indonesia
187 X14 **Lautoka** Viti Levu, W Fiji 17°36′S 177°28′E
169 O8 **Laut, Pulau** *prev.* Laoet. *island* Borneo, C Indonesia
169 V14 **Laut, Pulau** *island* Kepulauan Natuna, W Indonesia
169 U13 **Laut, Selat** *strait* Borneo, C Indonesia
168 H8 **Laut Tawar, Danau** ◎ Sumatera, NW Indonesia
189 V14 **Lauvergne Island** *island* Chuuk, C Micronesia
98 M5 **Lauwers Meer** ◎ N Netherlands
98 M4 **Lauwersoog** Groningen, NE Netherlands 53°25′N 06°14′E
102 M14 **Lauzerte** Tarn-et-Garonne, S France 44°15′N 01°08′E
25 U13 **Lavaca Bay** *bay* Texas, SW USA
25 U12 **Lavaca River** ≈ Texas, SW USA
15 O12 **Laval** Québec, SE Canada 45°32′N 73°44′W
102 J6 **Laval** Mayenne, NW France 48°04′N 00°46′W
15 T6 **Laval** ≈ Québec, SE Canada
105 S9 **La Vall d'Uixó** *var.* Vall D'Uxó. Valenciana, E Spain 39°49′N 00°15′W
61 F19 **Lavalleja** ⬖ *department* S Uruguay
15 O12 **Lavaltrie** Québec, SE Canada 45°56′N 73°14′W
186 M10 **Lavanggu** Rennell, S Solomon Islands 11°39′S 160°13′E
143 O14 **Lāvān, Jazīreh-ye** *island* S Iran
109 U8 **Lavant** ≈ S Austria
118 G5 **Lavassaare** *Ger.* Lawassaar. Pärnumaa, SW Estonia 58°29′N 24°22′E
104 L3 **La Vecilla de Curueño** Castilla y León, N Spain 42°51′N 05°24′W
45 N8 **La Vega** *var.* Concepción de la Vega. C Dominican Republic 19°15′N 70°33′W
La Vela *see* La Vela de Coro
54 J4 **La Vela de Coro** *var.* La Vela. Falcón, N Venezuela 11°30′N 69°33′W
103 N17 **Lavelanet** Ariège, S France 42°56′N 01°50′E
107 M17 **Lavello** Basilicata, S Italy 41°03′N 15°48′E
36 J8 **La Verkin** Utah, W USA 37°12′N 113°16′W
26 J8 **Laverne** Oklahoma, C USA 36°42′N 99°53′W
25 S12 **La Vernia** Texas, SW USA 29°19′N 98°07′W
93 K18 **Lavia** Satakunta, SW Finland 61°36′N 22°34′E
14 I12 **Lavieille, Lake** ◎ Ontario, SE Canada
94 C12 **Lavik** Sogn Og Fjordane, S Norway 61°06′N 05°25′E
La Vila Joíosa *see* Villajoyosa
33 U10 **Lavina** Montana, NW USA 46°18′N 108°55′W
194 H5 **Lavoisier Island** *island* Antarctica
23 U2 **Lavonia** Georgia, SE USA 34°26′N 83°06′W
103 R13 **la Voulte-sur-Rhône** Ardèche, E France 44°49′N 04°46′E
123 W5 **Lavrentiya** Chukotskiy Avtonomnyy Okrug, NE Russian Federation 65°33′N 171°12′W
115 H20 **Lávrio** *prev.* Lávrion. Attikí, C Greece 37°43′N 24°03′E
Lávrion *see* Lávrio
83 L22 **Lavumisa** *prev.* Gollel. SE Swaziland 27°18′S 31°55′E
149 T4 **Lawarai Pass** *pass* N Pakistan
Lawassaar *see* Lavassaare
141 P16 **Lawdar** SW Yemen 13°49′N 45°55′E
25 Q7 **Lawn** Texas, SW USA 32°07′N 99°45′W
195 Y4 **Law Promontory** *headland* Antarctica
77 O14 **Lawra** NW Ghana 10°40′N 02°49′W
185 E23 **Lawrence** Otago, South Island, New Zealand 45°53′S 169°43′E
31 P14 **Lawrence** Indiana, N USA 39°49′N 86°01′W
27 Q4 **Lawrence** Kansas, C USA 38°58′N 95°15′W
19 O10 **Lawrence** Massachusetts, NE USA 42°42′N 71°09′W
20 L5 **Lawrenceburg** Kentucky, S USA 38°02′N 84°53′W
20 I10 **Lawrenceburg** Tennessee, S USA 35°16′N 87°20′W
23 T3 **Lawrenceville** Georgia, SE USA 33°57′N 83°59′W
31 N15 **Lawrenceville** Illinois, N USA 38°43′N 87°40′W
21 V7 **Lawrenceville** Virginia, NE USA 36°45′N 77°50′W
27 S3 **Lawson** Missouri, C USA 39°26′N 94°12′W
26 L12 **Lawton** Oklahoma, C USA 34°35′N 98°20′W
140 I4 **Lawz, Jabal al** ▲ NW Saudi Arabia 28°45′N 35°20′E
95 L16 **Laxå** Örebro, C Sweden 59°00′N 14°37′E
125 T5 **Laya** ≈ NW Russian Federation
57 I19 **La Yarada** Tacna, SW Peru 18°14′S 70°30′W
141 S15 **Layjūn** C Yemen 15°27′N 49°16′E
141 Q9 **Laylā** *var.* Laila. Ar Riyāḍ, C Saudi Arabia 22°14′N 46°40′E
23 P4 **Lay Lake** ◎ Alabama, S USA
45 P14 **Layou** Saint Vincent, Saint Vincent and the Grenadines 13°11′N 61°16′W
La Youne *see* El Ayoun
192 L5 **Laysan Island** *island* Hawaiian Islands, Hawai'i, USA
36 L2 **Layton** Utah, W USA 41°03′N 112°00′W
34 L5 **Laytonville** California, N USA 39°39′N 123°30′W
172 H17 **Lazare, Pointe** *headland* Mahé, NE Seychelles 04°46′S 55°28′E
123 T12 **Lazarev** Khabarovskiy Kray, SE Russian Federation 52°11′N 141°18′E
112 L12 **Lazarevac** Serbia, C Serbia 44°25′N 20°17′E
65 N22 **Lazarev Sea** *sea* Antarctica
40 M15 **Lázaro Cárdenas** Michoacán, SW Mexico 17°56′N 102°11′W
119 F15 **Lazdijai** Alytus, S Lithuania 54°13′N 23°33′E
107 H15 **Lazio** *anc.* Latium. ⬖ *region* C Italy
111 A16 **Lázně Kynžvart** *Ger.* Bad Königswart. Karlovarský Kraj, W Czech Republic 50°00′N 12°40′E
Lazovsk *see* Singerei
167 R12 **Leach** Poŭthĭsăt, W Cambodia 12°19′N 103°45′E
27 X9 **Leachville** Arkansas, C USA 35°56′N 90°15′W
28 I9 **Lead** South Dakota, N USA 44°21′N 103°45′W
11 S16 **Leader** Saskatchewan, S Canada 50°55′N 109°31′W
19 S6 **Lead Mountain** ▲ Maine, NE USA 44°53′N 68°07′W
37 R5 **Leadville** Colorado, C USA 39°15′N 106°17′W
11 V12 **Leaf Rapids** Manitoba, C Canada 56°30′N 100°02′W
22 M7 **Leaf River** ≈ Mississippi, S USA
25 W11 **League City** Texas, SW USA 29°30′N 95°05′W
92 K8 **Leaibevuotna** *Nor.* Olderfjord. Finnmark, N Norway 70°29′N 24°58′E
23 N7 **Leakesville** Mississippi, S USA 31°09′N 88°33′W
25 Q11 **Leakey** Texas, SW USA 29°44′N 99°48′W
Leal *see* Lihula
83 G15 **Lealui** Western, W Zambia 15°12′S 22°59′E
Leamhcán *see* Lucan
14 C18 **Leamington** Ontario, S Canada 42°03′N 82°35′W
Leamington/Leamington Spa *see* Royal Leamington Spa
Leammi *see* Lemmenjoki
25 S10 **Leander** Texas, SW USA 30°34′N 97°51′W
60 F13 **Leandro N. Alem** Misiones, NE Argentina 27°34′S 55°15′W
97 A20 **Leane, Lough** *Ir.* Loch Léin. ◎ SW Ireland
180 G8 **Learmonth** Western Australia 22°17′S 114°03′E
Leau *see* Zoutleeuw
L'Eau d'Heure *see* Plate Taille, Lac de la
190 D12 **Leava** Île Futuna, S Wallis and Futuna
Leavdnja *see* Lakselv
27 R3 **Leavenworth** Kansas, C USA 39°19′N 94°55′W
32 I8 **Leavenworth** Washington, NW USA 47°36′N 120°39′W
92 L8 **Leavvajohka** *var.* Levajok. Finnmark, N Norway 69°57′N 26°18′E
27 R4 **Leawood** Kansas, C USA 38°57′N 94°37′W
110 H6 **Łeba** *Ger.* Leba. Pomorskie, N Poland 54°45′N 17°32′E
110 I6 **Łeba** *Ger.* Leba. ≈ N Poland
Leba *see* Łeba
101 D20 **Lebach** Saarland, SW Germany 49°25′N 06°54′E
Łeba, Jezioro *see* Łebsko, Jezioro
171 P8 **Lebak** Mindanao, S Philippines 06°28′N 124°03′E
Lebanese Republic *see* Lebanon
31 O13 **Lebanon** Indiana, N USA 40°03′N 86°28′W
20 L6 **Lebanon** Kentucky, S USA 37°33′N 85°15′W
27 U6 **Lebanon** Missouri, C USA 37°40′N 92°40′W
19 N9 **Lebanon** New Hampshire, NE USA 43°40′N 72°15′W
32 G12 **Lebanon** Oregon, NW USA 44°32′N 122°54′W
18 H15 **Lebanon** Pennsylvania, NE USA 40°20′N 76°24′W
20 J8 **Lebanon** Tennessee, S USA 36°11′N 86°19′W
21 P7 **Lebanon** Virginia, NE USA 36°52′N 82°07′W
138 G6 **Lebanon** *off.* Lebanese Republic, *Ar.* Al Lubnān, *Fr.* Liban. ◆ *republic* SW Asia

◆ Country
● Country Capital
◇ Dependent Territory
○ Dependent Territory Capital
⬖ Administrative Regions
✈ International Airport
▲ Mountain
▲▲ Mountain Range
🌋 Volcano
≈ River
◎ Lake
⊡ Reservoir

LEBANON JUNCTION – LEXINGTON

20 K6 **Lebanon Junction** Kentucky, S USA 37°49′N 85°43′W
Lebanon, Mount *see* Liban, Jebel
146 J10 **Lebap** Lebapskiy Velayat, NE Turkmenistan 41°04′N 61°49′E
Lebapskiy Velayat *see* Lebap Welaýaty
146 J11 **Lebap Welaýaty** *Rus.* Lebapskiy Velayat; *prev. Rus.* Chardzhevskaya Oblast, *Turkm.* Chärjew Oblasty. ◐ *province* E Turkmenistan
Lebasee *see* Łebsko, Jezioro
99 F17 **Lebbeke** Oost-Vlaanderen, NW Belgium 51°00′N 04°08′E
35 S14 **Lebec** California, W USA 34°51′N 118°52′W
Lebedin *see* Lebedyn
123 Q11 **Lebedinyy** Respublika Sakha (Yakutiya), NE Russian Federation 58°23′N 125°24′E
126 L6 **Lebedyan'** Lipetskaya Oblast', W Russian Federation 53°00′N 39°11′E
117 T4 **Lebedyn** *Rus.* Lebedin. Sums'ka Oblast', NE Ukraine 50°36′N 34°30′E
12 I12 **Lebel-sur-Quévillon** Québec, SE Canada 49°01′N 76°56′W
92 L8 **Lebesby** *Lapp.* Davvesiida. Finnmark, N Norway 70°31′N 27°00′E
102 M9 **le Blanc** Indre, C France 46°38′N 01°04′E
79 L15 **Lebo** Orientale, N Dem. Rep. Congo 04°30′N 23°58′E
27 P5 **Lebo** Kansas, C USA 38°22′N 95°50′W
110 H6 **Lębork** *var.* Lęборk, *Ger.* Lauenburg, Lauenburg in Pommern. Pomorskie, N Poland 54°32′N 17°43′E
103 O17 **le Boulou** Pyrénées-Orientales, S France 42°32′N 02°50′E
108 A9 **Le Brassus** Vaud, W Switzerland 46°35′N 06°14′E
104 J15 **Lebrija** Andalucía, S Spain 36°55′N 06°04′W
110 G6 **Łebsko, Jezioro** *Ger.* Lebasee; *prev.* Jezioro Łeba. ◎ N Poland
63 F14 **Lebu** Bío Bío, C Chile 37°38′S 73°43′W
Lebyazh'ye *see* Akku
104 F6 **Leça da Palmeira** Porto, N Portugal 41°12′N 08°43′W
103 U15 **le Cannet** Alpes-Maritimes, SE France 43°35′N 07°E
Le Cap *see* Cap-Haïtien
103 P2 **le Cateau-Cambrésis** Nord, N France 50°05′N 03°32′E
107 Q18 **Lecce** Puglia, SE Italy 40°23′N 18°11′E
106 D7 **Lecco** Lombardia, N Italy 45°51′N 09°23′E
29 V10 **Le Center** Minnesota, N USA 44°23′N 93°43′W
108 J7 **Lech** Vorarlberg, W Austria 47°14′N 10°10′E
101 K22 **Lech** ≈ Austria/Germany
115 D19 **Lechainá** *var.* Lehena, Lekhainá. Dytikí Elláda, S Greece 37°57′N 21°16′E
102 J11 **le Château d'Oléron** Charente-Maritime, W France 45°53′N 01°12′W
103 R3 **le Chesne** Ardennes, N France 49°33′N 04°42′E
103 R13 **le Cheylard** Ardèche, E France 44°55′N 04°27′E
108 K7 **Lechtaler Alpen** ▲▲ W Austria
100 H6 **Leck** Schleswig-Holstein, N Germany 54°45′N 09°00′E
14 L9 **Lecointre, Lac** ◎ Québec, SE Canada
22 H7 **Lecompte** Louisiana, S USA 31°05′N 92°24′W
103 Q9 **le Creusot** Saône-et-Loire, C France 46°48′N 04°27′E
Lecumberri *see* Lekunberri
110 P13 **Łęczna** Lubelskie, E Poland 51°20′N 22°52′E
110 J12 **Łęczyca** *Ger.* Lentschiza, *Rus.* Lenchitsa. Łódzkie, C Poland 52°04′N 19°10′E
100 F10 **Leda** ≈ NW Germany
109 Y9 **Ledava** ≈ NE Slovenia
99 F17 **Lede** Oost-Vlaanderen, NW Belgium 50°58′N 03°59′E
104 K6 **Ledesma** Castilla y León, N Spain 41°05′N 06°00′W
45 Q12 **le Diamant** SW Martinique 14°29′N 61°02′W
172 J16 **Le Digue** *island* Inner Islands, NE Seychelles
103 Q10 **le Donjon** Allier, C France 46°19′N 03°50′E
102 M10 **le Dorat** Haute-Vienne, C France 46°14′N 01°05′E
Ledo Salinarius *see* Lons-le-Saunier
11 Q14 **Leduc** Alberta, SW Canada 53°17′N 113°30′W
123 V7 **Ledyanaya, Gora** ▲ E Russian Federation 61°51′N 171°03′E
97 C21 **Lee** *Ir.* An Laoi. ≈ SW Ireland
29 U5 **Leech Lake** ◎ Minnesota, N USA
26 K10 **Leedey** Oklahoma, C USA 35°54′N 99°21′W
97 M17 **Leeds** N England, United Kingdom 53°50′N 01°35′W
23 P4 **Leeds** Alabama, S USA 33°33′N 86°32′W
29 O3 **Leeds** North Dakota, N USA 48°19′N 99°43′W
98 N6 **Leek** Groningen, NE Netherlands 53°10′N 06°24′E
99 K15 **Leende** Noord-Brabant, SE Netherlands 51°21′N 05°34′E
100 F10 **Leer** Niedersachsen, NW Germany 53°14′N 07°26′E
98 J13 **Leerdam** Zuid-Holland, C Netherlands 51°54′N 05°06′E
98 K12 **Leersum** Utrecht, C Netherlands 52°01′N 05°26′E
23 W11 **Leesburg** Florida, SE USA 28°48′N 81°52′W
21 V3 **Leesburg** Virginia, NE USA 39°09′N 77°34′W
27 R4 **Lees Summit** Missouri, C USA 38°55′N 94°21′W
22 G7 **Leesville** Louisiana, S USA 31°08′N 93°15′W
25 S12 **Leesville** Texas, SW USA 29°22′N 97°45′W
31 U13 **Leesville Lake** ⊠ Ohio, N USA
Leesville Lake *see* Smith Mountain Lake
183 P9 **Leeton** New South Wales, SE Australia 34°33′S 146°24′E
98 L6 **Leeuwarden** *Fris.* Ljouwert. Fryslân, N Netherlands 53°15′N 05°48′E
180 I14 **Leeuwin, Cape** *headland* Western Australia 34°18′S 115°03′E
35 R8 **Lee Vining** California, W USA 37°57′N 119°07′W
45 V8 **Leeward Islands** *island group* E West Indies
Leeward Islands *see* Sotavento, Ilhas de
Leeward Islands *see* Vent, Îles Sous le
79 G20 **Léfini** ≈ SE Congo
115 C17 **Lefkáda** *prev.* Levkás. Lefkáda, Iónia Nisiá, Greece, C Mediterranean Sea 38°50′N 20°42′E
115 B17 **Lefkáda** *It.* Santa Maura, *prev.* Levkás; *anc.* Leucas. *island* Iónia Nisiá, Greece, C Mediterranean Sea
115 H25 **Lefká Óri** ▲▲ Kríti, Greece, E Mediterranean Sea
115 B16 **Lefkímmi** *var.* Levkímmi. Kérkyra, Iónia Nisiá, Greece, C Mediterranean Sea 39°26′N 20°05′E
Lefkosía/Lefkoşa *see* Nicosia
25 O2 **Lefors** Texas, SW USA 35°26′N 100°48′W
45 R12 **le François** E Martinique 14°36′N 60°59′W
180 L12 **Lefroy, Lake** *salt lake* Western Australia
Legaceaster *see* Chester
105 N8 **Leganés** Madrid, C Spain 40°20′N 03°46′W
Legaspi *see* Legazpi City
Leghorn *see* Livorno
110 M11 **Legionowo** Mazowieckie, C Poland 52°25′N 20°56′E
99 K24 **Léglise** Luxembourg, SE Belgium 49°48′N 05°31′E
106 G8 **Legnago** Lombardia, NE Italy 45°13′N 11°18′E
106 D7 **Legnano** Veneto, NE Italy 45°36′N 08°54′E
111 F14 **Legnica** *Ger.* Liegnitz. Dolnośląskie, SW Poland 51°12′N 16°11′E
35 Q9 **Le Grand** California, W USA 37°12′N 120°15′W
103 Q15 **le Grau-du-Roi** Gard, S France 43°32′N 04°10′E
183 U3 **Legume** New South Wales, SE Australia 28°24′S 152°20′E
102 L4 **le Havre** *Eng.* Havre; *prev.* le Havre-de-Grâce. Seine-Maritime, N France 49°30′N 00°06′E
le Havre-de-Grâce *see* le Havre
Lehena *see* Lechainá
36 L3 **Lehi** Utah, W USA 40°23′N 111°51′W
18 I14 **Lehighton** Pennsylvania, NE USA 40°49′N 75°42′W
29 O6 **Lehr** North Dakota, N USA 46°15′N 99°21′W
38 A8 **Lehua Island** *island* Hawaiian Islands, Hawai'i, USA
149 S9 **Leiāh** Punjab, NE Pakistan 30°59′N 70°58′E
109 W9 **Leibnitz** Steiermark, SE Austria 46°48′N 15°33′E
97 M19 **Leicester** *Lat.* Batae Coritanorum. C England, United Kingdom 52°38′N 01°05′W
97 M19 **Leicestershire** *cultural region* C England, United Kingdom
Leicheng *see* Leizhou
98 H11 **Leiden** *prev.* Leyden; *anc.* Lugdunum Batavorum. Zuid-Holland, W Netherlands 52°09′N 04°30′E
98 H11 **Leiderdorp** Zuid-Holland, W Netherlands 52°08′N 04°32′E
98 G11 **Leidschendam** Zuid-Holland, W Netherlands 52°05′N 04°24′E
99 D18 **Leie** *Fr.* Lys. ≈ Belgium/France
Leifear *see* Lifford
184 L4 **Leigh** Auckland, North Island, New Zealand 36°17′S 174°48′E
97 K17 **Leigh** NW England, United Kingdom 53°30′N 02°33′W
182 I5 **Leigh Creek** South Australia 30°27′S 138°23′E
23 O2 **Leighton** Alabama, S USA 34°42′N 87°31′W
97 M21 **Leighton Buzzard** E England, United Kingdom 51°55′N 00°41′W
Léim an Bhradáin *see* Leixlip
Léim An Mhadaidh *see* Limavady
Léime, Ceann *see* Loop Head, Ireland
Léime, Ceann *see* Slyne Head, Ireland
101 G20 **Leimen** Baden-Württemberg, SW Germany 49°21′N 08°40′E
100 I13 **Leine** ≈ NW Germany
101 J15 **Leinefelde** Thüringen, C Germany 51°22′N 10°19′E
Léin, Loch *see* Leane, Lough
97 D19 **Leinster** *Ir.* Cúige Laighean. *cultural region* E Ireland
97 F19 **Leinster, Mount** *Ir.* Stua Laighean. ▲ SE Ireland 52°36′N 06°45′W
119 F15 **Leipalingis** Alytus, S Lithuania 54°05′N 23°52′E
92 J12 **Leipojärvi** Norrbotten, N Sweden 67°03′N 21°15′E
31 R12 **Leipsic** Ohio, N USA 41°06′N 83°58′W
Leipsic *see* Leipzig
115 M20 **Leipsoí** *island* Dodekánisa, Greece, Aegean Sea
101 M15 **Leipzig** *Pol.* Lipsk, *hist.* Leipsic; *anc.* Lipsia. Sachsen, E Germany 51°19′N 12°24′E
101 M15 **Leipzig Halle** ✈ Sachsen, E Germany 51°26′N 12°14′E
104 G9 **Leiria** *anc.* Collipo. Leiria, C Portugal 39°45′N 08°49′W
104 F9 **Leiria** ◐ *district* C Portugal
95 C15 **Leirvik** Hordaland, S Norway 59°49′N 05°27′E
118 E5 **Leisi** *Ger.* Laisberg. Saaremaa, W Estonia 58°33′N 22°41′E
104 J3 **Leitariegos, Puerto de** *pass* NW Spain
20 J6 **Leitchfield** Kentucky, S USA 37°28′N 86°19′W
109 Y5 **Leitha** *Hung.* Lajta. ≈ Austria/Hungary
Leitir Ceanainn *see* Letterkenny
Leitmeritz *see* Litoměřice
Leitomischl *see* Litomyšl
97 D16 **Leitrim** *Ir.* Liatroim. *cultural region* NW Ireland
Leix *see* Laois
97 F18 **Leixlip** *Eng.* Salmon Leap, *Ir.* Léim an Bhradáin. Kildare, E Ireland 53°23′N 06°32′W
64 N8 **Leixões** Porto, N Portugal 41°11′N 08°41′W
161 N12 **Leiyang** Hunan, S China 26°23′N 112°49′E
160 L16 **Leizhou** *var.* Haikang, Leicheng. Guangdong, S China 20°54′N 110°05′E
160 L16 **Leizhou Bandao** *var.* Luichow Peninsula. *peninsula* S China
98 H13 **Lek** ≈ SW Netherlands
114 I13 **Lekánis** ▲▲ NE Greece
172 H13 **Le Kartala** ▲ Grande Comore, NW Comoros
Le Kef *see* El Kef
79 G20 **Lékéti, Monts de la** ▲▲ S Congo
Lekhainá *see* Lechainá
92 G11 **Leknes** Nordland, C Norway 68°07′N 13°36′E
79 F20 **Lékoumou** ◐ *province* SW Congo
94 L13 **Leksand** Dalarna, C Sweden 60°44′N 15°E
124 H8 **Leksozero, Ozero** ◎ NW Russian Federation
105 Q3 **Lekunberri** *var.* Lecumberri. Navarra, N Spain 43°00′N 01°54′W
171 S11 **Lelai, Tanjung** *headland* Pulau Halmahera, N Indonesia 01°32′N 128°43′E
45 Q12 **le Lamentin** *var.* Lamentin. C Martinique 14°37′N 61°01′W
31 P6 **Leland** Michigan, N USA 45°01′N 85°44′W
22 J4 **Leland** Mississippi, S USA 33°24′N 90°54′W
95 J16 **Lelång** *var.* Lelången. ◎ S Sweden
Lelången *see* Lelång
Lel'chitsy *see* Lyel'chytsy
le Léman *see* Geneva, Lake
Leli *see* Tianlin
25 O3 **Lelia Lake** Texas, SW USA 34°52′N 100°42′W
113 I14 **Lelija** ▲ SE Bosnia and Herzegovina 43°25′N 18°31′E
108 C8 **Le Locle** Neuchâtel, W Switzerland 47°04′N 06°45′E
189 Y14 **Lelu** Kosrae, E Micronesia
Lelu *see* Lelu Island
189 Y14 **Lelu Island** *var.* Lelu. *island* Kosrae, E Micronesia
55 W9 **Lelydorp** Wanica, N Suriname 05°36′N 55°04′W
98 K9 **Lelystad** Flevoland, C Netherlands 52°30′N 05°26′E
63 K25 **Le Maire, Estrecho de** *strait* S Argentina
168 L10 **Lemang** Pulau Rangsang, W Indonesia 01°04′N 102°44′E
186 I7 **Lemankoa** Buka Island, NE Papua New Guinea 05°06′S 154°23′E
Léman, Lac *see* Geneva, Lake
102 L6 **le Mans** Sarthe, NW France 48°N 00°12′E
29 S12 **Le Mars** Iowa, C USA 42°47′N 96°10′W
109 S3 **Lembach im Mühlkreis** Oberösterreich, N Austria 48°28′N 13°53′E
101 G23 **Lemberg** ▲ SW Germany 48°09′N 08°47′E
Lemberg *see* L'viv
Lemdiyya *see* Médéa
121 P3 **Lemesós** *var.* Limassol. SW Cyprus 34°41′N 33°02′E
100 H13 **Lemgo** Nordrhein-Westfalen, W Germany 52°02′N 08°54′E
33 P13 **Lemhi Range** ▲▲ Idaho, NW USA
9 S6 **Lemieux Islands** *island group* Nunavut, NE Canada
171 O11 **Lemito** Sulawesi, N Indonesia 0°34′N 121°31′E
92 L10 **Lemmenjoki** *Lapp.* Leammi. ≈ NE Finland
98 L7 **Lemmer** *Fris.* De Lemmer. Fryslân, N Netherlands 52°50′N 05°43′E
28 L7 **Lemmon** South Dakota, N USA 45°54′N 102°08′W
36 M15 **Lemmon, Mount** ▲ Arizona, SW USA 32°26′N 110°47′W
Lemnos *see* Límnos
31 O14 **Lemon, Lake** ◎ Indiana, N USA
102 J5 **le Mont St-Michel** *castle* Manche, N France
35 Q11 **Lemoore** California, W USA 36°16′N 119°48′W
189 T13 **Lemotol Bay** *bay* Chuuk Islands, C Micronesia
45 Y5 **le Moule** *var.* Moule. Grande Terre, NE Guadeloupe 16°20′N 61°21′W
Lemovices *see* Limoges
Le Moyen-Ogooué *see* Moyen-Ogooué
12 M6 **le Moyne, Lac** ◎ Québec, E Canada
93 L18 **Lempäälä** Pirkanmaa, W Finland 61°14′N 23°47′E
42 E7 **Lempa, Río** ≈ Central America
42 F7 **Lempira** *prev.* Gracias. ◐ *department* SW Honduras
Lemsalu *see* Limbaži
107 N17 **Le Murge** ▲▲ SE Italy
125 V6 **Lemva** ≈ NW Russian Federation
95 F21 **Lemvig** Midtjylland, W Denmark 56°31′N 08°19′E
166 K8 **Lemyethna** Ayeyawady, SW Myanmar (Burma) 17°36′N 95°08′E
30 K10 **Lena** Illinois, N USA 42°22′N 89°49′W
129 V4 **Lena** ≈ NE Russian Federation
173 N13 **Lena Tablemount** *undersea feature* S Indian Ocean 51°06′S 56°54′E
Lenchitsa *see* Łęczyca
59 N17 **Lençóis** Bahia, E Brazil 12°36′S 41°24′W
60 K9 **Lençóis Paulista** São Paulo, S Brazil 22°35′S 48°51′W
109 Y9 **Lendava** *Hung.* Lendva, *Ger.* Unterlimbach; *prev.* Dolnja Lendava. NE Slovenia 46°33′N 16°27′E
83 F20 **Lendepas** Hardap, SE Namibia 24°41′S 19°58′E
124 H9 **Lendery** *Finn.* Lentiira. Respublika Kareliya, NW Russian Federation 63°20′N 31°18′E
Lendum *see* Lens
Lendva *see* Lendava
27 R4 **Lenexa** Kansas, C USA 38°57′N 94°43′W
109 Q5 **Lengau** Oberösterreich, N Austria 48°01′N 13°17′E
145 Q17 **Lenger** Yuzhnyy Kazakhstan, S Kazakhstan 42°10′N 69°54′E
Lenghu *see* Lenghuzhen
159 O9 **Lenghuzhen** *var.* Lenghu. Qinghai, C China 38°50′N 93°25′E
159 T9 **Lenglong Ling** ▲ N China 37°40′N 102°13′E
108 D7 **Lengnau** Bern, W Switzerland 47°12′N 07°23′E
95 M20 **Lenhovda** Kronoberg, S Sweden 57°00′N 15°16′E
Lenīn *see* Uzynkol', Kazakhstan
Lenin *see* Akdepe, Turkmenistan
Leninabad *see* Khujand
Leninakan *see* Gyumri
Lenina, Pik *see* Lenin Peak
117 V12 **Lenine** *Rus.* Lenino. Avtonomna Respublika Krym, S Ukraine 45°18′N 35°47′E
Lenīngor *see* Leninogorsk
Leningrad *see* Sankt-Peterburg
Leningrad *see* Mu'minobod
126 L13 **Leningradskaya** Krasnodarskiy Kray, SW Russian Federation 46°19′N 39°23′E
195 S16 **Leningradskaya** *Russian research station* Antarctica 69°30′S 159°51′E
124 H12 **Leningradskaya Oblast'** ◐ *province* NW Russian Federation
Leningradskiy *see* Mu'minobod
Lenino *see* Lyenina, Belarus
Lenino *see* Lenine, Ukraine
Leninobod *see* Khujand
145 X9 **Leninogorsk** *Kaz.* Lenīngor. Vostochnyy Kazakhstan, E Kazakhstan 50°20′N 83°34′E
147 T12 **Lenin Peak** *Rus.* Pik Lenina, *Taj.* Qullai Lenin. ▲ Kyrgyzstan/Tajikistan 39°20′N 72°50′E
147 S8 **Leninpol'** Talasskaya Oblast', NW Kyrgyzstan 42°29′N 71°54′E
Lenin, Qullai *see* Lenin Peak
127 P11 **Leninsk** Volgogradskaya Oblast', SW Russian Federation 48°41′N 45°18′E
Leninsk *see* Baykonyr, Kazakhstan
Leninsk *see* Akdepe, Turkmenistan
Leninsk *see* Asaka, Uzbekistan
145 T8 **Leninskiy** Pavlodar, E Kazakhstan 52°13′N 76°50′E
122 I13 **Leninsk-Kuznetskiy** Kemerovskaya Oblast', S Russian Federation 54°42′N 86°16′E
125 P15 **Leninskoje** Kirovskaya Oblast', NW Russian Federation 58°19′N 47°03′E
Leninskoye *see* Uzynkol'
Lenin-Turkmenski *see* Türkmenabat
Leninváros *see* Tiszaújváros
Lenkoran' *see* Länkäran
101 F15 **Lenne** ≈ W Germany
101 G16 **Lennestadt** Nordrhein-Westfalen, W Germany 51°07′N 08°04′E
29 R11 **Lennox** South Dakota, N USA 43°21′N 96°53′W
63 J25 **Lennox, Isla** *Eng.* Lennox Island. *island* S Chile
Lennox Island *see* Lennox, Isla
21 Q9 **Lenoir** North Carolina, SE USA 35°56′N 81°31′W
20 M9 **Lenoir City** Tennessee, S USA 35°48′N 84°15′W
108 C7 **Le Noirmont** Jura, NW Switzerland 47°14′N 06°57′E
14 L9 **Lenôtre, Lac** ◎ Québec, SE Canada
29 U15 **Lenox** Iowa, C USA 40°52′N 94°33′W
103 O2 **Lens** *anc.* Lendum, Lentium. Pas-de-Calais, N France 50°26′N 02°50′E
123 O11 **Lensk** Respublika Sakha (Yakutiya), NE Russian Federation 60°43′N 115°16′E
111 F24 **Lenti** Zala, SW Hungary 46°38′N 16°30′E
Lentia *see* Linz
93 N14 **Lentiira** Kainuu, E Finland 64°22′N 29°52′E
Lentiira *see* Lendery
107 L25 **Lentini** *anc.* Leontini. Sicilia, Italy, C Mediterranean Sea 37°17′N 15°00′E
Lentium *see* Lens
Lentschiza *see* Łęczyca
93 N15 **Lentua** ◎ E Finland
119 H14 **Lentvaris** *Pol.* Landwarów. Vilnius, SE Lithuania 24°39′N 24°58′E
108 F7 **Lenzburg** Aargau, N Switzerland 47°23′N 08°11′E
109 R5 **Lenzing** Oberösterreich, N Austria 47°58′N 13°34′E
77 P13 **Léo** SW Burkina Faso 11°07′N 02°08′W
109 V7 **Leoben** Steiermark, C Austria 47°23′N 15°06′E
Leobschütz *see* Głubczyce
44 L9 **Léogâne** S Haiti 18°32′N 72°37′W
171 O11 **Leok** Sulawesi, N Indonesia 01°10′N 121°20′E
29 O7 **Leola** South Dakota, N USA 45°41′N 98°58′W
97 K20 **Leominster** W England, United Kingdom 52°09′N 02°18′W
19 N11 **Leominster** Massachusetts, NE USA 42°29′N 71°43′W
102 I15 **Léon** Landes, SW France 43°54′N 01°17′W
40 M12 **León** *var.* León de los Aldamas. Guanajuato, C Mexico 21°05′N 101°43′W
42 I10 **León** León, NW Nicaragua 12°24′N 86°52′W
104 L4 **León** Castilla y León, NW Spain 42°34′N 05°34′W
29 V16 **Leon** Iowa, C USA 40°44′N 93°45′W
42 I9 **León** ◐ *department* W Nicaragua
104 K4 **León** ◐ *province* Castilla y León, NW Spain
León *see* Cotopaxi
25 V9 **Leona** Texas, SW USA 31°09′N 95°58′W
25 U5 **Leonard** Texas, SW USA 33°22′N 96°15′W
107 H15 **Leonardo da Vinci** *prev.* Fiumicino. ✈ (Roma) Lazio, C Italy 41°48′N 12°15′E
21 X5 **Leonardtown** Maryland, NE USA 38°17′N 76°35′W
25 Q13 **Leona River** ≈ Texas, SW USA
41 Z11 **Leona Vicario** Quintana Roo, SE Mexico 20°57′N 87°06′W
101 H21 **Leonberg** Baden-Württemberg, SW Germany 48°48′N 09°01′E
62 M3 **León, Cerro** ▲ NW Paraguay 20°21′S 60°16′W
León de los Aldamas *see* León
109 T4 **Leonding** Oberösterreich, N Austria 48°17′N 14°15′E
107 I14 **Leonessa** Lazio, C Italy 42°36′N 12°56′E
107 K24 **Leonforte** Sicilia, Italy, C Mediterranean Sea 37°38′N 14°23′E
183 O13 **Leongatha** Victoria, SE Australia 38°30′S 145°56′E
Leonídi *see* Leonídio
115 F21 **Leonídio** *var.* Leonídi. Pelopónnisos, S Greece 37°11′N 22°50′E
104 J4 **León, Montes de** ▲▲ NW Spain
180 K11 **Leonora** Western Australia 28°52′S 121°16′E
25 S8 **Leon River** ≈ Texas, SW USA
Leontini *see* Lentini
Léopold II, Lac *see* Mai-Ndombe, Lac
99 J17 **Leopoldsburg** Limburg, NE Belgium 51°07′N 05°16′E
Léopoldville *see* Kinshasa
26 I5 **Leoti** Kansas, C USA 38°28′N 101°22′W
116 M11 **Leova** *Rus.* Leovo. SW Moldova 46°31′N 28°16′E
Leovo *see* Leova
102 G8 **le Palais** Morbihan, NW France 47°20′N 03°08′W
27 X10 **Lepanto** Arkansas, C USA 35°34′N 90°21′W
169 N13 **Lepar, Pulau** *island* W Indonesia
104 I14 **Lepe** Andalucía, S Spain 37°15′N 07°12′W
Lepel' *see* Lyepyel'
83 E25 **Lepelle** *var.* Elefantes; *prev.* Olifants. ≈ SW South Africa
83 I20 **Lephepe** *var.* Lephephe. Kweneng, SE Botswana 23°20′S 25°50′E
Lephephe *see* Lephepe
161 Q10 **Leping** Jiangxi, S China 28°57′N 117°07′E
Lépontiennes, Alpes/Lepontine, Alpi *see* Lepontine Alps
108 G10 **Lepontine Alps** *Fr.* Alpes Lépontiennes, *It.* Alpi Lepontine. ▲▲ Italy/Switzerland
79 G20 **Le Pool** ◐ *province* S Congo
173 O16 **Le Port** NW Réunion
103 N1 **le Portel** Pas-de-Calais, N France 50°42′N 01°35′E
93 N17 **Leppävirta** Pohjois-Savo, C Finland 62°30′N 27°50′E
45 Q11 **le Prêcheur** NW Martinique 14°48′N 61°14′W
145 V13 **Lepsi** *prev.* Lepsy. Taldykorgan, SE Kazakhstan 46°14′N 78°56′E
145 V13 **Lepsi** *prev.* Lepsy. ≈ SE Kazakhstan
Lepsy *see* Lepsi
Le Puglie *see* Puglia
103 Q12 **le Puy** *prev.* le Puy-en-Velay, *hist.* Anicium, Podium Anicensis. Haute-Loire, C France 45°03′N 03°53′E
le Puy-en-Velay *see* le Puy
45 X11 **le Raizet** *var.* Le Raizet. ✈ (Pointe-à-Pitre) Grande Terre, C Guadeloupe 16°16′N 61°31′W
107 J24 **Lercara Friddi** Sicilia, Italy, C Mediterranean Sea 37°45′N 13°37′E
78 G12 **Léré** Mayo-Kébbi Ouest, SW Chad 09°41′N 14°17′E
Leribe *see* Hlotse
106 E10 **Lerici** Liguria, NW Italy 44°06′N 09°53′E
54 I14 **Lérida** Vaupés, SE Colombia 0°01′S 70°28′W
105 U5 **Lérida** ◐ *province* Cataluña, NE Spain
Lérida *see* Lleida
105 N5 **Lerma** Castilla y León, N Spain 42°02′N 03°46′W
40 M13 **Lerma, Río** ≈ C Mexico
Lerna *see* Lérni
115 F20 **Lérni** *var.* Lerna. *prehistoric site* Pelopónnisos, S Greece
45 R11 **le Robert** E Martinique 14°41′N 60°57′W
115 M21 **Léros** *island* Dodekánisa, Greece, Aegean Sea
30 L13 **Le Roy** Illinois, N USA 40°21′N 88°45′W
27 Q6 **Le Roy** Kansas, C USA 38°04′N 95°37′W
29 W11 **Le Roy** Minnesota, N USA 43°30′N 92°30′W
18 E10 **Le Roy** New York, NE USA 42°58′N 77°58′W
Lerrnayin Gharabakh *see* Nagorno-Karabakh
95 J19 **Lerum** Västra Götaland, S Sweden 57°46′N 12°12′E
96 M2 **Lerwick** NE Scotland, United Kingdom 60°09′N 01°09′W
45 Y6 **les Abymes** *var.* Abymes. Grande Terre, C Guadeloupe 16°16′N 61°31′W
les Albères *see* Albères, Chaîne des
102 M4 **les Andelys** Eure, N France 49°15′N 01°26′E
45 Q12 **les Anses-d'Arlets** SW Martinique 14°29′N 61°05′W
104 L2 **Les Arriondes** *prev.* Arriondas. Asturias, N Spain 43°23′N 05°11′W
105 U6 **Les Borges Blanques** *var.* Borjas Blancas. Cataluña, NE Spain 41°31′N 00°52′E
Lesbos *see* Lésvos
Les Cayes *see* Cayes
31 Q4 **Les Cheneaux Islands** *island* Michigan, N USA
105 T8 **Les Coves de Vinromá** *Cast.* Cuevas de Vinromá. Valenciana, E Spain 40°18′N 00°07′E
103 T12 **les Écrins** ▲ E France 44°54′N 06°25′E
108 C10 **Le Sépey** Vaud, W Switzerland 46°21′N 07°04′E
15 T7 **Les Escoumins** Québec, SE Canada 48°21′N 69°26′W
Les Gonaïves *see* Gonaïves
160 H9 **Leshan** Sichuan, C China 29°42′N 103°43′E
108 D11 **Les Haudères** Valais, SW Switzerland 46°02′N 07°27′E
102 J9 **les Herbiers** Vendée, NW France 46°52′N 01°01′W
Lesh/Leshi *see* Lezhë
125 O8 **Leshukonskoye** Arkhangel'skaya Oblast', NW Russian Federation 64°54′N 45°48′E
Lesina *see* Hvar
107 M15 **Lesina, Lago di** ◎ SE Italy
114 K13 **Lesítse** ▲▲ NE Greece
94 G10 **Lesja** Oppland, S Norway 62°07′N 08°56′E
95 L15 **Lesjöfors** Värmland, C Sweden 59°57′N 14°12′E
111 O18 **Lesko** Podkarpackie, SE Poland 49°28′N 22°19′E
113 O15 **Leskovac** Serbia, SE Serbia 43°00′N 21°58′E
113 M22 **Leskovik** *var.* Leskoviku. Korçë, S Albania 40°09′N 20°39′E
Leskoviku *see* Leskovik
33 P14 **Leslie** Idaho, NW USA 43°51′N 113°28′W
31 Q10 **Leslie** Michigan, N USA 42°27′N 84°25′W
Leśna/Lesnaya *see* Lyasnaya
Lesnaya *see* Lyasnaya
102 F5 **Lesneven** Finistère, NW France 48°35′N 04°19′W
112 J11 **Lešnica** Serbia, W Serbia 44°40′N 19°18′E
125 S13 **Lesnoy** Kirovskaya Oblast', NW Russian Federation 59°49′N 52°07′E
122 G10 **Lesnoy** Sverdlovskaya Oblast', C Russian Federation 58°40′N 59°48′E
122 K12 **Lesosibirsk** Krasnoyarskiy Kray, C Russian Federation 58°13′N 92°23′E
83 J23 **Lesotho** *off.* Kingdom of Lesotho; *prev.* Basutoland. ◆ *monarchy* S Africa
Lesotho, Kingdom of *see* Lesotho
102 J12 **Lesparre-Médoc** Gironde, SW France 45°18′N 00°57′W
108 C8 **Les Ponts-de-Martel** Neuchâtel, W Switzerland 47°00′N 06°45′E
103 P1 **Lesquin** ✈ Nord, N France 50°34′N 3°07′E
102 I9 **les Sables-d'Olonne** Vendée, NW France 46°30′N 01°47′W
109 S7 **Lessach** *var.* Lessachbach. ≈ E Austria
Lessachbach *see* Lessach
45 W11 **les Saintes** *var.* Îles des Saintes. *island group* S Guadeloupe
74 L5 **Les Salines** ✈ (Annaba) NE Algeria 36°45′N 07°57′E
99 J22 **Lesse** ≈ SE Belgium
95 M21 **Lessebo** Kronoberg, S Sweden 56°45′N 15°19′E
Lesser Antarctica *see* West Antarctica
45 P15 **Lesser Antilles** *island group* E West Indies
137 T10 **Lesser Caucasus** *Rus.* Malyy Kavkaz. ▲▲ SW Asia
Lesser Khingan Range *see* Xiao Hinggan Ling
11 P13 **Lesser Slave Lake** ◎ Alberta, W Canada
99 E19 **Lessines** Hainaut, SW Belgium 50°43′N 05°50′E
103 R16 **les Stes-Maries-de-la-Mer** Bouches-du-Rhône, SE France 43°27′N 04°26′E
14 G15 **Lester B. Pearson** *var.* Toronto. ✈ (Toronto) Ontario, S Canada 43°59′N 81°30′W
29 U9 **Lester Prairie** Minnesota, N USA 44°52′N 94°02′W
93 L16 **Lestijärvi** Keski-Pohjanmaa, W Finland 63°29′N 24°41′E
L'Estuaire *see* Estuaire
29 U9 **Le Sueur** Minnesota, N USA 44°27′N 93°53′W
108 B8 **Les Verrières** Neuchâtel, W Switzerland 46°54′N 06°29′E
115 L17 **Lésvos** *anc.* Lesbos. *island* E Greece
110 G12 **Leszno** *Ger.* Lissa. Wielkopolskie, C Poland 51°51′N 16°35′E
83 L20 **Letaba** Northern, NE South Africa 23°44′S 31°29′E
173 P17 **Le Tampon** SW Réunion
97 O21 **Letchworth** E England, United Kingdom 51°58′N 00°14′W
111 G25 **Letenye** Zala, SW Hungary 46°25′N 16°44′E
11 Q17 **Lethbridge** Alberta, SW Canada 49°43′N 112°48′W
55 S11 **Lethem** S Guyana 03°24′N 59°45′W
83 H18 **Letiahau** ≈ W Botswana
54 J18 **Leticia** Amazonas, S Colombia 04°09′N 69°57′W
171 S16 **Leti, Kepulauan** *island group* E Indonesia
83 I18 **Letlhakane** Central, C Botswana 21°26′S 25°36′E
83 H20 **Letlhakeng** Kweneng, SE Botswana 24°05′S 25°03′E
114 J8 **Letnitsa** Lovech, N Bulgaria 43°19′N 25°02′E
103 N1 **le Touquet-Paris-Plage** Pas-de-Calais, N France 50°31′N 01°36′E
166 L8 **Letpadan** Bago, SW Myanmar (Burma) 17°46′N 94°45′E
166 K6 **Letpan** Rakhine State, W Myanmar (Burma) 19°22′N 94°11′E
102 M2 **le Tréport** Seine-Maritime, N France 50°03′N 01°21′E
166 M12 **Letsôk-aw Kyun** *var.* Letsutan Island; *prev.* Domel Island. *island* Mergui Archipelago, S Myanmar (Burma)
Letsutan Island *see* Letsôk-aw Kyun
97 E14 **Letterkenny** *Ir.* Leitir Ceanainn. Donegal, NW Ireland 54°57′N 07°44′W
Lettland *see* Latvia
116 M6 **Letychiv** Khmel'nyts'ka Oblast', W Ukraine 49°24′N 27°39′E
Lëtzebuerg *see* Luxembourg
116 H14 **Leu** Dolj, SW Romania 44°10′N 24°01′E
Leucas *see* Lefkáda
103 P17 **Leucate** Aude, S France 42°55′N 03°03′E
103 P17 **Leucate, Étang de** ◎ S France
108 E10 **Leuk** Valais, SW Switzerland 46°18′N 07°46′E
108 E10 **Leukerbad** Valais, SW Switzerland 46°22′N 07°47′E
Leusden *see* Leusden-Centrum
98 K11 **Leusden-Centrum** *var.* Leusden. Utrecht, C Netherlands 52°08′N 05°25′E
Leutensdorf *see* Litvínov
Leutschau *see* Levoča
99 H18 **Leuven** *Fr.* Louvain, *Ger.* Löwen. Vlaams Brabant, C Belgium 50°53′N 04°42′E
99 I20 **Leuze** Namur, C Belgium 50°33′N 04°55′E
Leuze *see* Leuze-en-Hainaut
99 E19 **Leuze-en-Hainaut** *var.* Leuze. Hainaut, SW Belgium 50°36′N 03°37′E
Léva *see* Levice
Levádia *see* Livádeia
Levajok *see* Leavvajohka
36 L4 **Levan** Utah, W USA 39°33′N 111°51′W
93 E16 **Levanger** Nord-Trøndelag, C Norway 63°45′N 11°18′E
106 D10 **Levanto** Liguria, W Italy 44°12′N 09°33′E
107 H23 **Levanzo, Isola di** *island* Isole Egadi, S Italy
127 Q17 **Levashi** Respublika Dagestan, SW Russian Federation 42°27′N 47°19′E
24 M5 **Levelland** Texas, SW USA 33°35′N 102°23′W
39 P13 **Levelock** Alaska, USA 59°07′N 156°51′W
101 E16 **Leverkusen** Nordrhein-Westfalen, W Germany 51°02′N 07°E
111 J21 **Levice** *Ger.* Lewentz, *Hung.* Léva, Lewenz. Nitriansky Kraj, SW Slovakia 48°14′N 18°38′E
106 G6 **Levico Terme** Trentino-Alto Adige, N Italy 46°02′N 11°19′E
115 E20 **Levídi** Pelopónnisos, S Greece 37°39′N 22°13′E
103 P14 **le Vigan** Gard, S France 43°00′N 03°36′E
184 L13 **Levin** Manawatu-Wanganui, North Island, New Zealand 40°38′S 175°17′E
15 R10 **Lévis** *var.* Levis. Québec, SE Canada 46°47′N 71°12′W
Levis *see* Lévis
21 P6 **Levisa Fork** ≈ Kentucky/Virginia, S USA
115 L21 **Levítha** *island* Kykládes, Greece, Aegean Sea
18 L14 **Levittown** Long Island, New York, NE USA 40°42′N 73°29′W
18 J15 **Levittown** Pennsylvania, NE USA 40°09′N 74°50′W
Levkás *see* Lefkáda
Levkímmi *see* Lefkímmi
111 L19 **Levoča** *Ger.* Leutschau, *Hung.* Locse. Prešovský Kraj, E Slovakia 49°01′N 20°34′E
Lévrier, Baie du *see* Nouâdhibou, Dakhlet
103 N9 **Levroux** Indre, C France 47°00′N 01°37′E
114 J8 **Levski** Pleven, N Bulgaria 43°21′N 25°11′E
Levskigrad *see* Karlovo
126 L6 **Lev Tolstoy** Lipetskaya Oblast', W Russian Federation 53°12′N 39°28′E
187 X14 **Levuka** Ovalau, C Fiji 17°42′S 178°50′E
166 L6 **Lewe** Mandalay, C Myanmar (Burma) 19°40′N 96°04′E
Lewentz/Lewenz *see* Levice
97 O23 **Lewes** SE England, United Kingdom 50°52′N 00°01′E
21 Z4 **Lewes** Delaware, NE USA 38°46′N 75°08′W
29 Q12 **Lewis And Clark Lake** ⊠ Nebraska/South Dakota, N USA
18 G14 **Lewisburg** Pennsylvania, NE USA 40°57′N 76°52′W
20 J10 **Lewisburg** Tennessee, S USA 35°29′N 86°49′W
21 S6 **Lewisburg** West Virginia, NE USA 37°49′N 80°28′W
96 F6 **Lewis, Butt of** *headland* NW Scotland, United Kingdom 58°31′N 06°18′W
96 F7 **Lewis, Isle of** *island* NW Scotland, United Kingdom
35 U4 **Lewis, Mount** ▲ Nevada, W USA 40°22′N 116°50′W
185 H16 **Lewis Pass** *pass* South Island, New Zealand
33 P7 **Lewis Range** ▲▲ Montana, NW USA
23 O3 **Lewis Smith Lake** ⊠ Alabama, S USA
32 M10 **Lewiston** Idaho, NW USA 46°25′N 117°01′W
19 P7 **Lewiston** Maine, NE USA 44°08′N 70°14′W
29 X10 **Lewiston** Minnesota, N USA 43°58′N 91°52′W
18 D9 **Lewiston** New York, NE USA 43°10′N 79°02′W
36 L1 **Lewiston** Utah, W USA 41°58′N 111°52′W
30 K13 **Lewistown** Illinois, N USA 40°23′N 90°09′W
33 T9 **Lewistown** Montana, NW USA 47°04′N 109°26′W
27 T14 **Lewisville** Arkansas, C USA 33°21′N 93°38′W
25 T6 **Lewisville** Texas, SW USA 33°00′N 96°57′W
25 T6 **Lewisville, Lake** ⊠ Texas, SW USA
Le Woleu-Ntem *see* Woleu-Ntem
23 U3 **Lexington** Georgia, SE USA 33°51′N 83°04′W
20 M5 **Lexington** Kentucky, S USA 38°03′N 84°30′W
22 L4 **Lexington** Mississippi, S USA 33°06′N 90°03′W
27 S4 **Lexington** Missouri, C USA 39°11′N 93°52′W
29 N16 **Lexington** Nebraska, C USA 40°46′N 99°44′W
20 S9 **Lexington** North Carolina, SE USA 35°49′N 80°15′W
27 N11 **Lexington** Oklahoma, C USA 35°00′N 97°20′W
21 R12 **Lexington** South Carolina, SE USA 33°59′N 81°15′W
20 G9 **Lexington** Tennessee, S USA 35°39′N 88°24′W
25 T10 **Lexington** Texas, SW USA 30°25′N 97°00′W

◆ Country
● Country Capital
◇ Dependent Territory
○ Dependent Territory Capital
◐ Administrative Regions
✈ International Airport
▲ Mountain
▲▲ Mountain Range
Volcano
≈ River
◎ Lake
⊠ Reservoir

LEXINGTON – LITTLE CURRENT

21 T6 Lexington Virginia, NE USA 37°47′N 79°27′W
21 X5 Lexington Park Maryland, NE USA 38°16′N 76°27′W
Leyden see Leiden
102 J14 Leyre ≈ SW France
171 Q5 Leyte island C Philippines
171 Q6 Leyte Gulf gulf E Philippines
111 O16 Leżajsk Podkarpackie, SE Poland 50°15′N 22°25′E
Lezha see Lezhë
113 K18 Lezhë var. Lezha; prev. Lesh, Leshi. Lezhë, NW Albania 41°46′N 19°40′E
113 K18 Lezhë ◈ district NW Albania
103 O16 Lézignan-Corbières Aude, S France 43°12′N 02°46′E
126 J7 L'gov Kurskaya Oblast', W Russian Federation 51°38′N 35°17′E
159 P15 Lhari Xizang Zizhiqu, W China 30°34′N 93°40′E
159 N16 Lhasa var. La-sa, Lassa. Xizang Zizhiqu, W China 29°41′N 91°10′E
159 O15 Lhasa He ≈ W China
Lhaviyani Atoll see Faadhippolhu Atoll
158 K16 Lhazê var. Quxar. Xizang Zizhiqu, W China 29°07′N 87°32′E
158 K14 Lhazhong Xizang Zizhiqu, W China 31°58′N 86°43′E
168 H7 Lhoksukon Sumatera, W Indonesia 05°04′N 97°19′E
159 Q15 Lhorong var. Zito. Xizang Zizhiqu, W China 30°51′N 95°41′E
105 W6 L'Hospitalet de Llobregat var. Hospitalet. Cataluña, NE Spain 41°21′N 02°06′E
153 R11 Lhotse ▲ China/Nepal 28°00′N 86°55′E
159 N17 Lhozhag var. Garbo. Xizang Zizhiqu, W China 28°21′N 90°47′E
159 O16 Lhünzê var. Xingba. Xizang Zizhiqu, W China 28°25′N 92°30′E
159 N15 Lhünzhub var. Ganqu. Xizang Zizhiqu, W China 30°14′N 91°20′E
167 N8 Li Lamphun, NW Thailand 17°46′N 98°54′E
115 L21 Liádi var. Livádi. island Kykládes, Greece, Aegean Sea
161 P12 Liancheng var. Lianfeng. Fujian, SE China 25°47′N 116°42′E
Liancheng see Lianjiang, Guangdong, China
Liancheng see Qinglong, Guizhou, China
Liancheng see Guangnan, Yunnan, China
Lianfeng see Liancheng
160 K9 Liangping var. Liangshan. Sichuan, C China 30°40′N 107°46′E
Liangshan see Liangping
Liangzhou see Wuwei
161 O9 Liangzi Hu ◎ C China
161 R12 Lianjiang var. Fengcheng. Fujian, SE China 26°14′N 119°33′E
160 L15 Lianjiang var. Liancheng. Guangdong, S China 21°41′N 110°12′E
Lianjiang see Xingguo
161 O13 Lianping var. Yuanshan. Guangdong, S China 24°18′N 114°27′E
Lianshan see Huludao
Lian Xian see Lianzhou
160 M11 Lianyuan prev. Lantian. Hunan, S China 27°51′N 111°44′E
161 Q6 Lianyungang var. Xinpu. Jiangsu, E China 34°38′N 119°12′E
161 N13 Lianzhou var. Linxian; prev. Lian Xian. Guangdong, S China 24°48′N 112°26′E
Lianzhou see Hepu
Liao see Liaoning
161 P5 Liaocheng Shandong, E China 36°31′N 115°59′E
163 U13 Liaodong Bandao var. Liaotung Peninsula. peninsula NE China
163 T13 Liaodong Wan Eng. Gulf of Lantung, Gulf of Liaotung. gulf NE China
163 U11 Liao He ≈ NE China
163 U12 Liaoning var. Liao, Liaoning Sheng, Shengking, hist. Fengtien, Shenking. ◈ province NE China
Liaoning Sheng see Liaoning
Liaotung Peninsula see Liaodong Bandao
163 V12 Liaoyang var. Liao-yang. Liaoning, NE China 41°16′N 123°12′E
Liao-yang see Liaoyang
163 V11 Liaoyuan var. Dongliao, Shuang-liao, Jap. Chengchiatun. Jilin, NE China 42°52′N 125°09′E
163 U12 Liaozhong Liaoning, NE China 41°33′N 122°54′E
Liaqatabad see Piplān
10 M10 Liard ≈ W Canada
Liard see Fort Liard
10 L10 Liard River British Columbia, W Canada 59°23′N 126°05′W
149 O15 Liāri Baluchistān, SW Pakistan 25°43′N 66°28′E
Liatroim see Leitrim
189 S6 Lib var. Ellep. island Ralik Chain, C Marshall Islands
Liban see Lebanon
138 H6 Liban, Jebel Ar. Jabal al Gharbt, Jabal Lubnān, Eng. Mount Lebanon. ▲▲ C Lebanon
Libau see Liepāja
33 N7 Libby Montana, NW USA 48°25′N 115°33′W
79 I16 Libenge Equateur, NW Dem. Rep. Congo 03°39′N 18°39′E
26 I7 Liberal Kansas, C USA 37°03′N 100°56′W
27 R7 Liberal Missouri, C USA 37°33′N 94°31′W
Liberalitas Julia see Évora
111 D15 Liberec Ger. Reichenberg. Liberecký Kraj, N Czech Republic 50°45′N 15°05′E
111 D15 Liberecký Kraj ◈ region N Czech Republic
42 K12 Liberia Guanacaste, NW Costa Rica 10°36′N 85°26′W
76 K17 Liberia off. Republic of Liberia. ◆ republic W Africa
Liberia, Republic of see Liberia
61 D16 Libertad Corrientes, NE Argentina 30°01′S 57°51′W
61 E20 Libertad San José, S Uruguay 34°38′S 56°39′W
54 I7 Libertad Barinas, NW Venezuela 08°21′N 69°39′W
54 K6 Libertad Cojedes, N Venezuela 09°15′N 68°30′W
62 G12 Libertador off. Región del Libertador General Bernardo O'Higgins. ◈ region C Chile
Libertador General Bernardo O'Higgins, Región del see Libertador
Libertador General San Martín see Ciudad de Libertador General San Martín
20 L6 Liberty Kentucky, S USA 37°19′N 84°58′W
22 J7 Liberty Mississippi, S USA 31°09′N 90°49′W
27 R4 Liberty Missouri, C USA 39°15′N 94°22′W
18 J12 Liberty New York, NE USA 41°48′N 74°45′W
21 T9 Liberty North Carolina, SE USA 35°49′N 79°34′W
Libian Desert see Libyan Desert
99 J23 Libin Luxembourg, SE Belgium 50°01′N 05°13′E
Lībīyah, Aş Şaḥrā' al see Libyan Desert
160 K13 Libo var. Yuping. Guizhou, S China 25°28′N 107°52′E
Libohova see Libohovë
113 L23 Libohovë var. Libohova. Gjirokastër, S Albania 40°03′N 20°13′E
81 K18 Liboi Wajir, E Kenya 0°23′N 40°55′E
102 K13 Libourne Gironde, SW France 44°55′N 00°14′W
99 K23 Libramont Luxembourg, SE Belgium 49°55′N 05°21′E
113 M20 Librazhd var. Librazhdi. Elbasan, E Albania 41°10′N 20°22′E
Librazhdi see Librazhd
79 C18 Libreville ● (Gabon) Estuaire, NW Gabon 0°25′N 09°29′E
75 P10 Libya off. Great Socialist People's Libyan Arab Jamahiriya, Ar. Al Jamāhīrīyah al 'Arabīyah al Lībīyah ash Sha'bīyah al Ishtirākīy; prev. Libyan Arab Republic. ◆ Islamic state N Africa
Libyan Arab Republic see Libya
75 T11 Libyan Desert var. Libian Desert, Ar. Aş Şaḥrā' al Lībīyah. desert N Africa
75 T8 Libyan Plateau var. Aḍ Diffah. plateau Egypt/Libya
62 G12 Licantén Maule, C Chile 35°00′S 72°00′W
107 J25 Licata anc. Phintias. Sicilia, Italy, C Mediterranean Sea 37°07′N 13°57′E
137 P14 Lice Diyarbakır, SE Turkey 38°29′N 40°39′E
Licheng see Lipu
97 L19 Lichfield C England, United Kingdom 52°42′N 01°48′W
83 N14 Lichinga Niassa, N Mozambique 13°19′S 35°13′E
109 V3 Lichtenau Niederösterreich, N Austria 48°29′N 15°24′E
83 I21 Lichtenburg North-West, N South Africa 26°09′S 26°11′E
101 K18 Lichtenfels Bayern, SE Germany 50°09′N 11°04′E
98 O12 Lichtenvoorde Gelderland, E Netherlands 51°59′N 06°34′E
Lichtenwald see Sevnica
99 C17 Lichtervelde West-Vlaanderen, W Belgium 51°02′N 03°09′E
160 L9 Lichuan Hubei, C China 30°20′N 108°56′E
27 V7 Licking Missouri, C USA 37°30′N 91°51′W
20 M4 Licking River ≈ Kentucky, S USA
112 C11 Lički Osik Lika-Senj, C Croatia 44°36′N 15°24′E
Ličko-Senjska Županija see Lika-Senj
107 K19 Licosa, Punta headland S Italy 40°15′N 14°54′E
119 H16 Lida Hrodzyenskaya Voblasts', W Belarus 53°53′N 25°20′E
93 H17 Liden Västernorrland, C Sweden 62°43′N 16°49′E
29 R7 Lidgerwood North Dakota, N USA 46°04′N 97°09′W
Lidhorikion see Lidoríki
95 K21 Lidhult Kronoberg, S Sweden 56°49′N 13°25′E
95 P16 Lidingö Stockholm, C Sweden 59°22′N 18°10′E
95 K17 Lidköping Västra Götaland, S Sweden 58°30′N 13°10′E
Lido di Iesolo see Lido di Jesolo
106 I8 Lido di Jesolo var. Lido di Iesolo. Veneto, NE Italy 45°30′N 12°37′E
107 H15 Lido di Ostia Lazio, C Italy 41°42′N 12°19′E
Lidokhorikion see Lidoríki
115 E18 Lidoríki prev. Lidhorikion, Lidokhorikion. Stereá Elláda, C Greece 38°32′N 22°12′E
110 K9 Lidzbark Warmińsko-Mazurskie, NE Poland 53°15′N 19°49′E
110 L7 Lidzbark Warmiński Ger. Heilsberg. Olsztyn, N Poland 54°08′N 20°35′E
109 U3 Liebenau Oberösterreich, N Austria 48°33′N 14°48′E
181 P7 Liebig, Mount ▲ Northern Territory, C Australia 23°19′S 131°30′E
109 V8 Lieboch Steiermark, SE Austria 47°00′N 15°21′E
108 I8 Liechtenstein off. Principality of Liechtenstein. ◆ principality C Europe
Liechtenstein, Principality of see Liechtenstein
99 F18 Liedekerke Vlaams Brabant, C Belgium 50°51′N 04°05′E
99 K19 Liège Dut. Luik, Ger. Lüttich. Liège, E Belgium 50°38′N 05°35′E
99 K20 Liège Dut. Luik. ◈ province E Belgium
Liegnitz see Legnica
93 O16 Lieksa Pohjois-Karjala, E Finland 63°20′N 30°E
118 F10 Lielupe ≈ Latvia/Lithuania
118 G9 Lielvārde C Latvia 56°45′N 24°48′E
167 U13 Liên Hương var. Tuy Phong. Binh Thuân, S Vietnam 11°13′N 108°40′E
167 U13 Liên Nghia var. Liên Nghia var. Ðưc Trong. Lâm Ðông, S Vietnam 11°45′N 108°24′E
Liên Nghia see Liên Nghia
109 P9 Lienz Tirol, W Austria 46°50′N 12°45′E
118 B10 Liepāja Ger. Libau. W Latvia 56°32′N 21°02′E
99 H17 Lier Fr. Lierre. Antwerpen, N Belgium 51°08′N 04°35′E
95 H15 Lierbyen Buskerud, S Norway 59°57′N 10°14′E
99 L21 Lierneux Liège, E Belgium 50°12′N 05°51′E
Lierre see Lier
101 D18 Lieser ≈ W Germany
109 U7 Liesing ≈ E Austria
108 E6 Liestal Basel Landschaft, N Switzerland 47°29′N 07°43′E
Lietuva see Lithuania
Lievenhof see Līvāni
103 O2 Liévin Pas-de-Calais, N France 50°25′N 02°48′E
14 M9 Lièvre, Rivière du ≈ Québec, SE Canada
109 T6 Liezen Steiermark, C Austria 47°34′N 14°12′E
97 E14 Lifford Ir. Leifear. Donegal, NW Ireland 54°50′N 07°29′W
187 Q16 Lifou island Îles Loyauté, E New Caledonia
193 Y15 Lifuka island Ha'apai Group, C Tonga
171 P4 Ligao Luzon, N Philippines 13°16′N 123°30′E
Liger see Loire
42 H2 Lighthouse Reef reef E Belize
183 Q4 Lightning Ridge New South Wales, SE Australia 29°29′S 148°00′E
103 N9 Lignières Cher, C France 46°45′N 02°10′E
103 S5 Ligny-en-Barrois Meuse, NE France 48°42′N 05°22′E
83 P15 Ligonha ≈ NE Mozambique
31 P11 Ligonier Indiana, N USA 41°25′N 85°33′W
81 I25 Ligunga Ruvuma, S Tanzania 10°51′S 37°10′E
106 D9 Ligure, Appennino Eng. Ligurian Mountains. ▲▲ NW Italy
Ligure, Mar see Ligurian Sea
106 C9 Liguria ◈ region NW Italy
Ligurian Mountains see Ligure, Appennino
120 K6 Ligurian Sea Fr. Mer Ligurienne, It. Mar Ligure. sea N Mediterranean Sea
Ligurienne, Mer see Ligurian Sea
186 H5 Lihir Group island group NE Papua New Guinea
38 B8 Lihu'e var. Lihue. Kaua'i, Hawaii, USA 21°59′N 159°23′W
Lihue see Lihu'e
118 F5 Lihula Ger. Leal. Läänemaa, W Estonia 58°44′N 23°49′E
124 I2 Liinakhamari var. Linacmamari. Murmanskaya Oblast', NW Russian Federation 69°40′N 31°27′E
Liivi Laht see Riga, Gulf of
160 F11 Lijiang var. Dayan, Lijiang Naxizu Zizhixian. Yunnan, SW China 26°52′N 100°10′E
Lijiang Naxizu Zizhixian see Lijiang
112 C11 Lika-Senj off. Ličko-Senjska Županija. ◈ province W Croatia
79 N25 Likasi prev. Jadotville. Shaba, SE Dem. Rep. Congo 11°02′S 26°51′E
79 L16 Likati Orientale, N Dem. Rep. Congo 03°28′N 23°45′E
10 M15 Likely British Columbia, SW Canada 52°40′N 121°34′W
153 Y11 Likhapāni Assam, NE India 27°19′N 95°54′E
124 J16 Likhoslavl' Tverskaya Oblast', W Russian Federation 57°08′N 35°27′E
189 U5 Likiep Atoll atoll Ratak Chain, C Marshall Islands
95 D18 Liknes Vest-Agder, S Norway 58°20′N 06°58′E
79 H16 Likouala ◈ province NE Congo
79 H18 Likouala ≈ N Congo
79 H18 Likouala aux Herbes ≈ E Congo
190 B16 Liku E Niue 19°02′S 169°47′E
Likupang, Selat see Bangka, Selat
27 Y8 Lilbourn Missouri, C USA 36°35′N 89°37′W
103 X14 l'Île-Rousse Corse, France, C Mediterranean Sea 42°39′N 08°59′E
109 W5 Lilienfeld Niederösterreich, NE Austria 48°01′N 15°36′E
161 N11 Liling Hunan, S China 27°42′N 113°49′E
95 J18 Lilla Edet Västra Götaland, S Sweden 58°08′N 12°08′E
103 P1 Lille var. l'Isle, Dut. Rijssel, Flem. Ryssel, prev. Lisle; anc. Insula. Nord, N France 50°38′N 03°04′E
95 G24 Lillebælt var. Lille Bælt, Eng. Little Belt. strait S Denmark
Lille Bælt see Lillebælt
102 L3 Lillebonne Seine-Maritime, N France 49°30′N 00°34′E
94 H12 Lillehammer Oppland, S Norway 61°07′N 10°28′E
103 O1 Lillers Pas-de-Calais, N France 50°34′N 02°26′E
95 F18 Lillesand Aust-Agder, S Norway 58°15′N 08°23′E
95 I15 Lillestrøm Akershus, S Norway 59°58′N 11°05′E
93 F18 Lillhärdal Jämtland, C Sweden 61°51′N 14°04′E
21 U10 Lillington North Carolina, SE USA 35°25′N 78°50′W
105 O9 Lillo Castilla-La Mancha, C Spain 39°43′N 03°19′W
10 M16 Lillooet British Columbia, SW Canada 50°41′N 121°59′W
83 M14 Lilongwe ● (Malawi) Central, W Malawi 13°58′S 33°48′E
83 M14 Lilongwe ✈ Central, W Malawi 13°46′S 33°44′E
83 M14 Lilongwe ≈ W Malawi
171 P7 Liloy Mindanao, S Philippines 08°04′N 122°42′E
Lilybaeum see Marsala
182 J7 Lilydale South Australia 32°57′S 140°00′E
183 P16 Lilydale Tasmania, SE Australia 41°17′S 147°13′E
113 J14 Lim ≈ SE Europe
57 D15 Lima ● (Peru) Lima, W Peru 12°06′S 78°W
94 K13 Lima Dalarna, C Sweden 60°55′N 13°19′E
31 R12 Lima Ohio, NE USA 40°43′N 84°06′W
57 D14 Lima ◈ department W Peru
Lima see Jorge Chávez Internacional
137 Y13 Liman prev. Port-Il'ič. SE Azerbaijan 38°54′N 48°49′E
111 L17 Limanowa Małopolskie, S Poland 49°43′N 20°25′E
104 G5 Lima, Rio Sp. Limia. ≈ Portugal/Spain see also Limia
Lima, Rio see Limia
168 M11 Limas Pulau Sebangka, W Indonesia 0°09′N 104°31′E
Limassol see Lemesós
97 F14 Limavady Ir. Léim An Mhadaidh. NW Northern Ireland, United Kingdom 55°03′N 06°57′W
63 J14 Limay Mahuida La Pampa, C Argentina 37°09′S 66°40′W
63 H15 Limay, Río ≈ W Argentina
101 N16 Limbach-Oberfrohna Sachsen, E Germany 50°52′N 12°46′E
81 F22 Limba Limba ≈ C Tanzania
107 C17 Limbara, Monte ▲ Sardegna, Italy, C Mediterranean Sea 40°50′N 09°10′E
118 G7 Limbaži Est. Lemsalu. N Latvia 57°33′N 24°46′E
44 M8 Limbé N Haiti 19°44′N 72°25′W
99 L19 Limbourg Liège, E Belgium 50°37′N 05°56′E
99 K17 Limburg ◈ province NE Belgium
99 L16 Limburg ◈ province SE Netherlands
101 F17 Limburg an der Lahn Hessen, W Germany 50°22′N 08°04′E
94 K13 Limedsforsen Dalarna, C Sweden 60°52′N 13°25′E
60 L9 Limeira São Paulo, S Brazil 22°34′S 47°25′W
97 C19 Limerick Ir. Luimneach. Limerick, SW Ireland 52°40′N 08°38′W
97 C20 Limerick Ir. Luimneach. cultural region SW Ireland
19 S2 Limestone Maine, NE USA 46°52′N 67°49′W
25 U9 Limestone, Lake ◎ Texas, SW USA
39 P12 Lime Village Alaska, USA 61°21′N 155°26′W
95 F20 Limfjorden fjord N Denmark
95 J23 Limhamn Skåne, S Sweden 55°34′N 12°57′E
104 H5 Limia Port. Rio Lima. ≈ Portugal/Spain see also Lima, Rio
Limia see Lima, Rio
93 L14 Liminka Pohjois-Pohjanmaa, C Finland 64°48′N 25°19′E
Limín Vathéos see Sámos
115 G17 Límni Évvoia, C Greece 38°46′N 23°20′E
115 J15 Límnos anc. Lemnos. island E Greece
102 M11 Limoges anc. Augustoritum Lemovicensium, Lemovices. Haute-Vienne, C France 45°51′N 01°16′E
43 O13 Limón var. Puerto Limón. Limón, E Costa Rica 09°59′N 83°02′W
42 K4 Limón Colón, NE Honduras 15°50′N 85°31′W
37 U5 Limon Colorado, C USA 39°15′N 103°41′W
43 N13 Limón off. Provincia de Limón. ◈ province E Costa Rica
106 A10 Limone Piemonte Piemonte, NE Italy 44°12′N 07°37′E
Limones see Valdéz
Limón, Provincia de see Limón
Limonum see Poitiers
103 N11 Limousin ◈ region C France
103 N16 Limoux Aude, S France 43°03′N 02°13′E
83 J20 Limpopo off. Limpopo Province; prev. Northern, Northern Transvaal. ◈ province NE South Africa
83 L19 Limpopo var. Crocodile. ≈ S Africa
Limpopo Province see Limpopo
160 K17 Limu Ling ▲▲ S China
113 M20 Lin var. Lini. Elbasan, E Albania 41°03′N 20°37′E
Linacmamari see Liinakhamari
62 G13 Linares Maule, C Chile 35°50′S 71°37′W
54 C13 Linares Nariño, SW Colombia 01°24′N 77°30′W
41 O9 Linares Nuevo León, NE Mexico 24°54′N 99°38′W
105 N12 Linares Andalucía, S Spain 38°05′N 03°38′W
107 G15 Linaro, Capo headland C Italy 42°01′N 11°49′E
106 D8 Linate ✈ (Milano) Lombardia, N Italy 45°27′N 09°18′E
167 T8 Lin Camh prev. Ðưc Tho. Ha Tinh, N Vietnam 18°30′N 105°36′E
160 F13 Lincang Yunnan, SW China 23°55′N 100°03′E
Lincheng see Lingao
Linchuan see Fuzhou
61 B20 Lincoln Buenos Aires, E Argentina 34°54′S 61°30′W
185 H19 Lincoln Canterbury, South Island, New Zealand 43°37′S 172°30′E
97 N18 Lincoln anc. Lindum, Lindum Colonia. E England, United Kingdom 53°14′N 00°33′W
35 O6 Lincoln California, W USA 38°52′N 121°18′W
30 L13 Lincoln Illinois, N USA 40°09′N 89°21′W
26 M4 Lincoln Kansas, C USA 39°03′N 98°09′W
19 S5 Lincoln Maine, NE USA 45°22′N 68°30′W
27 T5 Lincoln Missouri, C USA 38°23′N 93°19′W
29 R16 Lincoln state capital Nebraska, C USA 40°46′N 96°43′W
32 F11 Lincoln City Oregon, NW USA 44°57′N 124°01′W
167 X10 Lincoln Island Chin. Dong Dao, Viet. Ðao Linh Côn. island E Paracel Islands
197 Q11 Lincoln Sea sea Arctic Ocean
97 N18 Lincolnshire cultural region E England, United Kingdom
21 R10 Lincolnton North Carolina, SE USA 35°27′N 81°16′W
25 V7 Lindale Texas, SW USA 32°31′N 95°24′W
101 I25 Lindau var. Lindau am Bodensee. Bayern, S Germany 47°33′N 09°41′E
Lindau am Bodensee see Lindau
123 P9 Linde ≈ NE Russian Federation
55 T9 Linden E Guyana 05°58′N 58°12′W
23 O6 Linden Alabama, S USA 32°18′N 87°48′W
20 H9 Linden Tennessee, S USA 35°38′N 87°50′W
25 X6 Linden Texas, SW USA 33°01′N 94°22′W
44 H2 Linden Pindling ✈ New Providence, C The Bahamas 25°00′N 77°26′W
18 J16 Lindenwold New Jersey, NE USA 39°47′N 74°58′W
95 M15 Lindesberg Örebro, C Sweden 59°36′N 15°15′E
95 D18 Lindesnes headland S Norway 57°58′N 07°03′E
Líndhos see Líndos
81 K24 Lindi Lindi, SE Tanzania 10°S 39°41′E
81 J24 Lindi ◈ region SE Tanzania
79 N17 Lindi ≈ NE Dem. Rep. Congo
163 V7 Lindian Heilongjiang, NE China 47°15′N 124°51′E
185 E21 Lindis Pass pass South Island, New Zealand
83 J22 Lindley Free State, C South Africa 27°52′S 27°55′E
95 J19 Lindome Västra Götaland, S Sweden 57°34′N 12°05′E
163 S10 Lindong var. Bairin Zuoqi. Nei Mongol Zizhiqu, N China 43°59′N 119°24′E
115 O23 Líndos var. Líndhos. Ródos, Dodekánisa, Greece, Aegean Sea 36°05′N 28°05′E
14 I14 Lindsay Ontario, SE Canada 44°21′N 78°44′W
35 R11 Lindsay California, W USA 36°11′N 119°06′W
33 X8 Lindsay Montana, NW USA 47°13′N 105°10′W
27 N11 Lindsay Oklahoma, C USA 34°50′N 97°37′W
27 N5 Lindsborg Kansas, C USA 38°34′N 97°39′W
95 N21 Lindsdal Kalmar, S Sweden 56°44′N 16°18′E
Lindum/Lindum Colonia see Lincoln
191 W3 Line Islands island group E Kiribati
Linëvo see Linova
160 M5 Linfen var. Lin-fen. Shanxi, C China 36°08′N 111°34′E
Lin-fen see Linfen
104 L2 L'Infiestu prev. Infiesto. Asturias, N Spain 43°21′N 05°21′W
155 F18 Linganamakki Reservoir ⊡ SW India
160 L17 Lingao var. Lincheng. Hainan, S China 19°44′N 109°23′E
171 N3 Lingayen Luzon, N Philippines 16°00′N 120°12′E
160 M6 Lingbao var. Guolüezhen. Henan, C China 34°34′N 110°50′E
94 N12 Lingbo Gävleborg, C Sweden 61°04′N 16°45′E
Lingcheng see Lingshan, Guangxi, China
Lingcheng see Beiliu, Guangxi, China
Lingeh see Bandar-e Lengeh
100 E12 Lingen var. Lingen an der Ems. Niedersachsen, NW Germany 52°31′N 07°19′E
Lingen an der Ems see Lingen
168 M11 Lingga, Kepulauan island group W Indonesia
168 L11 Lingga, Pulau island Kepulauan Lingga, W Indonesia
14 J14 Lingham Lake ◎ Ontario, SE Canada
94 M13 Linghed Dalarna, C Sweden 60°48′N 15°55′E
33 Z15 Lingle Wyoming, C USA 42°07′N 104°21′W
18 G15 Linglestown Pennsylvania, NE USA 40°20′N 76°46′W
160 M12 Lingling prev. Yongzhou, Zhishan. Hunan, S China 26°13′N 111°36′E
79 K18 Lingomo 11 Equateur, NW Dem. Rep. Congo 0°42′N 21°59′E
160 L15 Lingshan var. Lingcheng. Guangxi Zhuangzu Zizhiqu, S China 22°28′N 109°19′E
160 L17 Lingshui var. Lingshui Lizu Zizhixian. Hainan, S China 18°35′N 110°03′E
Lingshui Lizu Zizhixian see Lingshui
155 G16 Lingsugūr Karnātaka, C India 16°13′N 76°33′E
107 L23 Linguaglossa Sicilia, Italy, C Mediterranean Sea 37°51′N 15°06′E
76 H10 Linguère N Senegal 15°24′N 15°06′W
159 W8 Lingwu Ningxia, N China 38°04′N 106°21′E
Lingxi see Yongshun, Hunan, China
Lingxi see Cangnan, Zhejiang, China
Lingxian/Ling Xian see Yanling
163 S12 Lingyuan Liaoning, NE China 41°09′N 119°24′E
163 U4 Linhai Heilongjiang, NE China 51°30′N 124°18′E
161 S10 Linhai var. Taizhou. Zhejiang, SE China 28°54′N 121°08′E
59 O20 Linhares Espírito Santo, SE Brazil 19°22′S 40°04′W
Linh Côn, Ðao see Lincoln Island
Linhe see Bayannur
Lini see Lin
Linik, Chiyā-ē see Linkī, Chiya-i
Linjiang see Shanghang
139 S1 Linkī, Chiya-i Ar. Jabal Linkī, var. Chiyā-ē Linik, Chiyā-ī Linkī Kurezūr. ▲ N Iraq
Linkī, Jabal see Linkī, Chiya-i
95 M18 Linköping Östergötland, S Sweden 58°25′N 15°37′E
163 Y8 Linkou Heilongjiang, NE China 45°18′N 130°17′E
118 F11 Linkuva Šiaulai, N Lithuania 56°06′N 23°58′E
27 V5 Linn Missouri, C USA 38°29′N 91°51′W
25 S16 Linn Texas, SW USA 26°32′N 98°06′W
27 T2 Linneus Missouri, C USA 39°53′N 93°10′W
96 H10 Linnhe, Loch inlet W Scotland, United Kingdom
119 G19 Linova Rus. Linëvo. Brestskaya Voblasts', SW Belarus 52°29′N 24°30′E
161 O5 Linqing Shandong, E China 36°51′N 115°42′E
60 K8 Lins São Paulo, S Brazil 21°40′S 49°44′W
93 F17 Linsell Jämtland, C Sweden 62°10′N 14°E
160 J9 Linshui Sichuan, C China 30°24′N 106°54′E
44 K12 Linstead C Jamaica 18°08′N 77°02′W
159 U11 Lintan Gansu, N China 34°43′N 101°27′E
159 V11 Lintao var. Taoyang. Gansu, C China 35°23′N 103°54′E
15 S12 Lintère ≈ Québec, SE Canada
108 H8 Linth ≈ NW Switzerland
108 H8 Linthal Glarus, NE Switzerland 46°59′N 08°57′E
31 N15 Linton Indiana, N USA 39°01′N 87°10′W
29 N6 Linton North Dakota, N USA 46°16′N 100°13′W
163 R11 Linxi Nei Mongol Zizhiqu, N China 43°29′N 117°59′E
159 U11 Linxia var. Linxia Huizu Zizhizhou. Gansu, C China 35°34′N 103°08′E
Linxia Huizu Zizhizhou see Linxia
Linxian see Lianzhou
161 P4 Linyi var. Yishi. Shandong, E China 37°12′N 116°54′E
161 Q6 Linyi Shandong, S China 37°06′N 118°18′E
160 M6 Linyi Shanxi, C China 35°10′N 118°18′E
109 T4 Linz anc. Lentia. Oberösterreich, N Austria 48°19′N 14°18′E
159 S8 Linze var. Shahe; prev. Shahepu. Gansu, N China 39°06′N 100°03′E
44 J13 Lionel Town C Jamaica 17°49′N 77°14′W
103 Q16 Lion, Golfe du Eng. Gulf of Lion, Gulf of Lions; anc. Sinus Gallicus. gulf S France
Lion, Gulf of/Lions, Gulf of see Lion, Golfe du
83 K16 Lions Den Mashonaland West, N Zimbabwe 17°16′S 30°00′E
14 F13 Lion's Head Ontario, S Canada 44°59′N 81°16′W
Lios Ceannúir, Bá see Liscannor Bay
Lios Mór see Lismore
Lios na gCearrbhach see Lisburn
Lios Tuathail see Listowel
79 G17 Liouesso Sangha, N Congo 01°02′N 15°43′E
Liozno see Lyozna
171 O4 Lipa off. Lipa City. Luzon, N Philippines 13°57′N 121°10′E
Lipa City see Lipa
25 S7 Lipan Texas, SW USA 32°31′N 98°03′W
Lipari Islands/Lipari, Isole see Eolie, Isole
107 L22 Lipari, Isola island Isole Eolie, S Italy
116 L8 Lipcani Rus. Lipkany. N Moldova 48°16′N 26°47′E
93 N17 Liperi Pohjois-Karjala, SE Finland 62°33′N 29°29′E
126 L7 Lipetsk Lipetskaya Oblast', W Russian Federation 52°37′N 39°38′E
126 K6 Lipetskaya Oblast' ◈ province W Russian Federation
57 K22 Lípez, Cordillera de ▲▲ SW Bolivia
110 E10 Lipiany Ger. Lippehne. Zachodnio-pomorskie, W Poland 53°00′N 14°58′E
112 G9 Lipik Požega-Slavonija, NE Croatia 45°24′N 17°08′E
124 L12 Lipin Bor Vologodskaya Oblast', NW Russian Federation 60°12′N 38°04′E
160 L12 Liping var. Defeng. Guizhou, S China 26°16′N 109°08′E
Lipkany see Lipcani
119 H15 Lipnishki Hrodzyenskaya Voblasts', W Belarus 54°00′N 25°37′E
110 J10 Lipno Kujawsko-pomorskie, C Poland 52°52′N 19°11′E
116 F11 Lipova Hung. Lippa. Arad, W Romania 46°05′N 21°42′E
Lipovets see Lypovets'
Lippa see Lipova
101 E14 Lippe ≈ W Germany
Lippehne see Lipiany
101 G14 Lippstadt Nordrhein-Westfalen, W Germany 51°41′N 08°20′E
25 P1 Lipscomb Texas, SW USA 36°14′N 100°16′W
Lipsia/Lipsk see Leipzig
Liptau-Sankt-Nikolaus/Liptószentmiklós see Liptovský Mikuláš
111 K19 Liptovský Mikuláš Ger. Liptau-Sankt-Nikolaus, Hung. Liptószentmiklós. Zílinský Kraj, N Slovakia 49°06′N 19°36′E
183 O13 Liptrap, Cape headland Victoria, SE Australia 38°55′S 145°58′E
160 L13 Lipu var. Licheng. Guangxi Zhuangzu Zizhiqu, S China 24°25′N 110°15′E
81 G17 Lira N Uganda 02°15′N 32°55′E
57 F15 Lircay Huancavelica, C Peru 12°59′S 74°44′W
107 K17 Liri ≈ C Italy
144 M8 Lisakovsk Kostanay, NW Kazakhstan 52°32′N 62°32′E
79 K17 Lisala Equateur, N Dem. Rep. Congo 02°10′N 21°29′E
104 F11 Lisboa Eng. Lisbon; anc. Felicitas Julia, Olisipo. ● (Portugal) Lisboa, W Portugal 38°44′N 09°08′W
104 F10 Lisboa Eng. Lisbon. ◈ district C Portugal
19 N7 Lisbon New Hampshire, NE USA 44°11′N 71°52′W
29 Q6 Lisbon North Dakota, N USA 46°27′N 97°42′W
Lisbon see Lisboa
19 Q8 Lisbon Falls Maine, NE USA 44°00′N 70°03′W
97 G15 Lisburn Ir. Lios na gCearrbhach. E Northern Ireland, United Kingdom 54°31′N 06°03′W
38 L6 Lisburne, Cape headland Alaska, USA 68°52′N 166°13′W
97 B19 Liscannor Bay Ir. Bá Lios Ceannúir. inlet W Ireland
113 Q18 Lisec ▲ E FYR Macedonia 41°46′N 22°30′E
160 F13 Lishe Jiang ≈ SW China
Lishi see Lüliang
163 V10 Lishu Jilin, NE China 43°25′N 124°19′E
161 R10 Lishui Zhejiang, SE China 28°27′N 119°25′E
192 L5 Lisianski Island island Hawaiian Islands, Hawai'i, USA
Lisichansk see Lysychans'k
102 L4 Lisieux anc. Noviomagus. Calvados, N France 49°09′N 00°13′E
126 L8 Liski prev. Georgiu-Dezh. Voronezhskaya Oblast', W Russian Federation 51°00′N 39°36′E
103 N4 l'Isle-Adam Val-d'Oise, N France 49°07′N 02°13′E
Lisle/l'Isle see Lille
103 R15 l'Isle-sur-la-Sorgue Vaucluse, SE France 43°55′N 05°03′E
15 S9 L'Islet Québec, SE Canada 47°07′N 70°18′W
183 V4 Lismore New South Wales, SE Australia 28°48′S 153°12′E
182 M12 Lismore Victoria, SE Australia 37°59′S 143°18′E
97 D20 Lismore Ir. Lios Mór. S Ireland 52°10′N 07°10′W
Lissa see Vis, Croatia
Lissa see Leszno, Poland
98 H11 Lisse Zuid-Holland, W Netherlands 52°15′N 04°33′E
114 K13 Lissos var. Filiouri. ≈ NE Greece
95 D18 Lista peninsula S Norway
95 D18 Listafjorden fjord S Norway
195 R13 Lister, Mount ▲ Antarctica 78°12′S 161°46′E
126 M8 Listopadovka Voronezhskaya Oblast', W Russian Federation 51°54′N 41°08′E
14 F15 Listowel Ontario, S Canada 43°44′N 80°57′W
97 B20 Listowel Ir. Lios Tuathail. Kerry, SW Ireland 52°27′N 09°29′W
160 L14 Litang Guangxi Zhuangzu Zizhiqu, S China 23°09′N 109°08′E
160 F9 Litang var. Gaocheng. Sichuan, C China 30°03′N 100°12′E
160 F10 Litang Qu ≈ C China
55 X12 Litani var. Itany. ≈ French Guiana/Suriname
138 G8 Lītani, Nahr el var. Nahr al Litant. ≈ C Lebanon
Litant, Nahr al see Lītani, Nahr el
Litauen see Lithuania
30 K14 Litchfield Illinois, N USA 39°17′N 89°52′W
29 U8 Litchfield Minnesota, N USA 45°09′N 94°31′W
36 K13 Litchfield Park Arizona, SW USA 33°29′N 112°21′W
183 S8 Lithgow New South Wales, SE Australia 33°30′S 150°09′E
115 I26 Líthino, Akrotírio headland Kríti, Greece, E Mediterranean Sea 34°55′N 24°43′E
118 D12 Lithuania off. Republic of Lithuania, Ger. Litauen, Lith. Lietuva, Pol. Litwa, Rus. Litva; prev. Lithuanian SSR, Rus. Litovskaya SSR. ◆ republic NE Europe
Lithuanian SSR see Lithuania
Lithuania, Republic of see Lithuania
109 U11 Litija Ger. Littai. C Slovenia 46°03′N 14°50′E
18 H15 Lititz Pennsylvania, NE USA 40°09′N 76°18′W
115 F15 Litóchoro var. Litohoro, Litókhoron. Kentrikí Makedonía, N Greece 40°06′N 22°30′E
Litohoro/Litókhoron see Litóchoro
111 C15 Litoměřice Ger. Leitmeritz. Ústecký Kraj, NW Czech Republic 50°33′N 14°10′E
111 F17 Litomyšl Ger. Leitomischl. Pardubický Kraj, C Czech Republic 49°54′N 16°18′E
111 G17 Litovel Ger. Littau. Olomoucký Kraj, E Czech Republic 49°42′N 17°05′E
123 S13 Litovko Khabarovskiy Kray, SE Russian Federation 49°22′N 135°10′E
Litovskaya SSR see Lithuania
Littai see Litija
Littau see Litovel
44 G1 Little Abaco var. Abaco Island. island N The Bahamas
111 I21 Little Alföld Ger. Kleines Ungarisches Tiefland, Hung. Kisalföld, Slvk. Podunajská Rovina. plain Hungary/Slovakia
151 Q20 Little Andaman island Andaman Islands, India, NE Indian Ocean
26 M5 Little Arkansas River ≈ Kansas, C USA
184 L4 Little Barrier Island island N New Zealand
Little Belt see Lillebælt
38 M11 Little Black River ≈ Alaska, USA
27 O2 Little Blue River ≈ Kansas/Nebraska, C USA
44 D8 Little Cayman island E Cayman Islands
11 X11 Little Churchill ≈ Manitoba, C Canada
166 J10 Little Coco Island island SW Myanmar (Burma)
36 L10 Little Colorado River ≈ Arizona, SW USA
14 E11 Little Current Manitoulin Island, Ontario, S Canada 45°57′N 81°56′W

◆ Country
● Country Capital
◇ Dependent Territory
○ Dependent Territory Capital
◈ Administrative Regions
✈ International Airport
▲ Mountain
▲▲ Mountain Range
Volcano
≈ River
◎ Lake
⊡ Reservoir

LITTLE CURRENT – LÖRRACH

12 E11 **Little Current** ≈ Ontario, S Canada
38 L8 **Little Diomede Island** *island* Alaska, USA
44 I4 **Little Exuma** *island* C The Bahamas
29 U7 **Little Falls** Minnesota, N USA *45°59′N 94°21′W*
18 J10 **Little Falls** New York, NE USA *43°02′N 74°51′W*
24 M5 **Littlefield** Texas, SW USA *33°56′N 102°20′W*
29 V3 **Littlefork** Minnesota, N USA *48°24′N 93°33′W*
29 V3 **Little Fork River** ≈ Minnesota, N USA
11 N16 **Little Fort** British Columbia, SW Canada *51°27′N 120°15′W*
11 Y14 **Little Grand Rapids** Manitoba, C Canada *52°06′N 95°29′W*
97 N23 **Littlehampton** SE England, United Kingdom *50°48′N 00°33′W*
35 T2 **Little Humboldt River** ≈ Nevada, W USA
44 K6 **Little Inagua** *var.* Inagua Islands. *island* S The Bahamas
21 Q4 **Little Kanawha River** ≈ West Virginia, NE USA
83 F25 **Little Karoo** *plateau* S South Africa
39 O16 **Little Koniuji Island** *island* Shumagin Islands, Alaska, USA
44 H12 **Little London** W Jamaica *18°15′N 78°13′W*
13 R10 **Little Mecatina** *Fr.* Rivière du Petit Mécatina. ≈ Newfoundland and Labrador/Québec, E Canada
96 F8 **Little Minch, The** *strait* NW Scotland, United Kingdom
27 T13 **Little Missouri River** ≈ Arkansas, USA
28 J7 **Little Missouri River** ≈ NW USA
28 J3 **Little Muddy River** ≈ North Dakota, N USA
151 Q22 **Little Nicobar** *island* Nicobar Islands, India, NE Indian Ocean
27 R6 **Little Osage River** ≈ Missouri, C USA
97 P20 **Little Ouse** ≈ E England, United Kingdom
149 V2 **Little Pamir** *Pash.* Pāmīr-e Khord, *Rus.* Malyy Pamir. ▲▲ Afghanistan/Tajikistan
21 U12 **Little Pee Dee River** ≈ North Carolina/South Carolina, SE USA
27 V10 **Little Red River** ≈ Arkansas, C USA
Little Rhody *see* Rhode Island
185 I19 **Little River** Canterbury, South Island, New Zealand *43°45′S 172°49′E*
21 U12 **Little River** South Carolina, SE USA *33°52′N 78°36′W*
27 Y9 **Little River** ≈ Arkansas/Missouri, C USA
27 R13 **Little River** ≈ Arkansas/Oklahoma, USA
23 T7 **Little River** ≈ Georgia, SE USA
22 H6 **Little River** ≈ Louisiana, S USA
25 T10 **Little River** ≈ Texas, SW USA
27 V12 **Little Rock** *state capital* Arkansas, C USA *34°45′N 92°17′W*
31 N8 **Little Sable Point** *headland* Michigan, N USA *43°38′N 86°32′W*
103 U11 **Little Saint Bernard Pass** *Fr.* Col du Petit St-Bernard, *It.* Colle del Piccolo San Bernardo. *pass* France/Italy
36 K7 **Little Salt Lake** ◎ Utah, W USA
180 K8 **Little Sandy Desert** *desert* Western Australia
29 S13 **Little Sioux River** ≈ Iowa, C USA
38 E17 **Little Sitkin Island** *island* Aleutian Islands, Alaska, USA
11 O13 **Little Smoky** Alberta, W Canada *54°35′N 117°06′W*
11 O14 **Little Smoky** ≈ Alberta, W Canada
37 P3 **Little Snake River** ≈ Colorado, C USA
64 A12 **Little Sound** *bay* Bermuda, NW Atlantic Ocean
37 T4 **Littleton** Colorado, C USA *39°36′N 105°01′W*
19 N7 **Littleton** New Hampshire, NE USA *44°18′N 71°46′W*
18 D11 **Little Valley** New York, NE USA *42°15′N 78°47′W*
30 M15 **Little Wabash River** ≈ Illinois, N USA
14 D10 **Little White River** ≈ Ontario, S Canada
28 M12 **Little White River** ≈ South Dakota, N USA
25 R5 **Little Wichita River** ≈ Texas, SW USA
142 I4 **Little Zab** *Ar.* Nahraz Zāb aş Şaghīr, *Kurd.* Zē-i Kōya, *Per.* Rūdkhāneh-ye Zāb-e Kūchek. ≈ Iran/Iraq
79 D15 **Littoral** ◈ *province* W Cameroon
Littoria *see* Latina
Litva/Litwa *see* Lithuania
111 B15 **Litvínov** *Ger.* Leutensdorf. Ústecký Kraj, NW Czech Republic *50°38′N 13°30′E*
116 M6 **Lityn** Vinnyts'ka Oblast', C Ukraine *49°19′N 28°06′E*
Liu-chou/Liuchow *see* Liuzhou
163 W11 **Liuhe** Jilin, NE China *42°15′N 125°49′E*
Liujiaxia *see* Yongjing
Liulin *see* Jonê
Liupanshui *see* Lupanshui
83 Q15 **Liúpo** Nampula, NE Mozambique *15°36′S 39°57′E*
83 G14 **Liuwa Plain** *plain* W Zambia
160 L13 **Liuzhou** *var.* Liu-chou, Liuchow. Guangxi Zhuangzu Zizhiqu, S China *24°09′N 108°55′E*
116 H8 **Livada** *Hung.* Sárköz. Satu Mare, NW Romania *47°52′N 23°04′E*
115 J20 **Liváda, Akrotírio** *headland* Tínos, Kykládes, Greece, Aegean Sea *37°36′N 25°15′E*
115 F18 **Livádeia** *prev.* Levádia. Stereá Elláda, C Greece *38°24′N 22°51′E*
Livádi *see* Liádi
Livanátai *see* Livanátes
115 G18 **Livanátes** *prev.* Livanátai. Stereá Elláda, C Greece *38°43′N 23°03′E*
118 I10 **Līvāni** *Ger.* Lievenhof. SE Latvia *56°22′N 26°12′E*
65 E25 **Lively Island** *island* SE Falkland Islands
65 D25 **Lively Sound** *sound* SE Falkland Islands
39 R8 **Livengood** Alaska, USA *65°31′N 148°32′W*
106 I7 **Livenza** ≈ NE Italy
35 O6 **Live Oak** California, W USA *39°17′N 121°41′W*
23 U9 **Live Oak** Florida, SE USA *30°18′N 82°59′W*
35 O9 **Livermore** California, W USA *37°40′N 121°46′W*
20 I6 **Livermore** Kentucky, S USA *37°31′N 87°08′W*
19 Q7 **Livermore Falls** Maine, NE USA *44°30′N 70°09′W*
24 J10 **Livermore, Mount** ▲ Texas, SW USA *30°37′N 104°10′W*
13 P16 **Liverpool** Nova Scotia, SE Canada *44°03′N 64°43′W*
97 K17 **Liverpool** NW England, United Kingdom *53°25′N 02°55′W*
183 S7 **Liverpool Range** ▲▲ New South Wales, SE Australia
42 F4 **Livingston** Izabal, E Guatemala *15°50′N 88°44′W*
96 J12 **Livingston** C Scotland, United Kingdom *55°51′N 03°31′W*
23 N5 **Livingston** Alabama, S USA *32°35′N 88°12′W*
35 P9 **Livingston** California, W USA *37°22′N 120°45′W*
22 J8 **Livingston** Louisiana, S USA *30°30′N 90°45′W*
33 S11 **Livingston** Montana, NW USA *45°40′N 110°33′W*
20 L8 **Livingston** Tennessee, S USA *36°22′N 85°20′W*
25 W9 **Livingston** Texas, SW USA *30°42′N 94°58′W*
83 I16 **Livingstone** *var.* Maramba. Southern, S Zambia *17°51′S 25°48′E*
185 B22 **Livingstone Mountains** ▲▲ South Island, New Zealand
80 K13 **Livingstone Mountains** ▲▲ S Tanzania
82 N12 **Livingstonia** Northern, N Malawi *10°29′S 34°06′E*
194 G4 **Livingston Island** *island* Antarctica
25 W9 **Livingston, Lake** ⊠ Texas, SW USA
112 F13 **Livno** ◈ Federicija Bosna I Hercegovina, SW Bosnia and Herzegovina
126 K7 **Livny** Orlovskaya Oblast', W Russian Federation *52°25′N 37°42′E*
93 M14 **Livojoki** ≈ C Finland
31 R10 **Livonia** Michigan, N USA *42°22′N 83°22′W*
106 E11 **Livorno** *Eng.* Leghorn. Toscana, C Italy *43°32′N 10°18′E*
Livramento *see* Santana do Livramento
141 U8 **Liwā** *var.* Al Liwā'. *oasis region* S United Arab Emirates
81 I24 **Liwale** Lindi, SE Tanzania *09°46′S 37°56′E*
159 W9 **Liwang** Ningxia, N China *36°42′N 106°05′E*
83 N15 **Liwonde** Southern, S Malawi *15°01′S 35°15′E*
159 V11 **Lixian** *var.* Li Xian. Gansu, C China *34°15′N 105°07′E*
160 H8 **Lixian** *var.* Li Xian, Zagunao. Sichuan, C China *31°27′N 103°06′E*
Li Xian *see* Lixian
Lixian Jiang *see* Black River
115 B18 **Lixoúri** *prev.* Lixoúrion. Kefallinía, Iónia Nisiá, Greece, C Mediterranean Sea *38°14′N 20°24′E*
Lixoúrion *see* Lixoúri
Lixus *see* Larache
33 U15 **Lizard Head Peak** ▲ Wyoming, C USA *42°47′N 109°12′W*
97 H25 **Lizard Point** *headland* SW England, United Kingdom *49°57′N 05°12′W*
Lizarra *see* Estella
112 L12 **Ljig** Serbia, C Serbia *44°14′N 20°16′E*
Ljouwert *see* Leeuwarden
Ljubelj *see* Loibl Pass
109 U11 **Ljubljana** *Ger.* Laibach, *It.* Lubiana; *anc.* Aemona, Emona. ● (Slovenia) C Slovenia *46°03′N 14°29′E*
109 T11 **Ljubljana** ✈ C Slovenia *46°14′N 14°26′E*
113 N17 **Ljuboten** *Alb.* Luboten. ▲ S Serbia *42°12′N 21°06′E*
95 P19 **Ljugarn** Gotland, SE Sweden *57°23′N 18°45′E*
84 G7 **Ljungan** ≈ N Sweden
93 F17 **Ljungan** ≈ C Sweden
95 K21 **Ljungby** Kronoberg, S Sweden *56°49′N 13°55′E*
95 M17 **Ljungsbro** Östergötland, S Sweden *58°31′N 15°30′E*
95 I18 **Ljungskile** Västra Götaland, S Sweden *58°14′N 11°55′E*
94 M11 **Ljusdal** Gävleborg, C Sweden *61°50′N 16°10′E*
94 M11 **Ljusnan** ≈ C Sweden
94 N12 **Ljusne** Gävleborg, C Sweden *61°11′N 17°07′E*
95 P15 **Ljusterö** Stockholm, C Sweden *59°31′N 18°40′E*
109 X9 **Ljutomer** *Ger.* Luttenberg. NE Slovenia *46°31′N 16°12′E*
63 G15 **Llaima, Volcán** ※ S Chile *39°01′S 71°38′W*
105 X4 **Llançà** *var.* Llansá. Cataluña, NE Spain *42°23′N 03°08′E*
97 J21 **Llandovery** C Wales, United Kingdom *52°01′N 03°47′W*
97 J20 **Llandrindod Wells** E Wales, United Kingdom *52°15′N 03°23′W*
97 J18 **Llandudno** N Wales, United Kingdom *53°19′N 03°49′W*
97 I21 **Llanelli** *prev.* Llanelly. SW Wales, United Kingdom *51°41′N 04°12′W*
Llanelly *see* Llanelli
104 M2 **Llanes** Asturias, N Spain *43°25′N 04°46′W*
97 K19 **Llangollen** NE Wales, United Kingdom *52°58′N 03°10′W*
104 K2 **Llangréu** *var.* Langreo, Sama de Langreo. Asturias, N Spain *43°18′N 05°40′W*
25 R10 **Llano** Texas, SW USA *30°49′N 98°42′W*
25 Q10 **Llano River** ≈ Texas, SW USA
54 I9 **Llanos** *physical region* Colombia/Venezuela
63 G16 **Llanquihue, Lago** ◎ S Chile
Llansá *see* Llançà
105 U5 **Lleida** *Cast.* Lérida; *anc.* Ilerda. Cataluña, NE Spain *41°38′N 00°35′E*
104 K12 **Llerena** Extremadura, W Spain *38°13′N 06°00′W*
Linkî Kurezûr, Chiyā-î *see* Linkî, Chiya-î
105 S9 **Llíria** Valenciana, E Spain *39°38′N 00°36′W*
105 W4 **Llívia** Cataluña, NE Spain *42°27′N 02°00′E*
Llodio *see* Laudio
105 X5 **Lloret de Mar** Cataluña, NE Spain *41°42′N 02°51′E*
Llorri *see* Toretta de l'Orrí
10 L11 **Lloyd George, Mount** ▲ British Columbia, W Canada *57°46′N 124°57′W*
11 R14 **Lloydminster** Alberta/Saskatchewan, SW Canada *53°18′N 110°00′W*
104 K2 **Lluanco** *var.* Luanco. Asturias, N Spain *43°36′N 05°48′W*
105 X9 **Llucmajor** Mallorca, Spain, W Mediterranean Sea *39°29′N 02°53′E*
36 L6 **Loa** Utah, W USA *38°24′N 111°38′W*
169 S8 **Loagan Bunut** ◎ East Malaysia
38 G12 **Loa, Mauna** ▲ Hawai'i, USA *19°28′N 155°39′W*
Loanda *see* Luanda
79 J22 **Loange** ≈ S Dem. Rep. Congo
79 E21 **Loango** Kouilou, S Congo *04°38′S 11°50′E*
106 B10 **Loano** Liguria, NW Italy *44°07′N 08°15′E*
62 H4 **Loa, Río** ≈ N Chile
83 I20 **Lobatse** *var.* Lobatsi. Kgatleng, SE Botswana *25°11′S 25°40′E*
Lobatsi *see* Lobatse
101 Q15 **Löbau** Sachsen, E Germany *51°07′N 14°40′E*
79 H16 **Lobaye** ◈ *prefecture* SW Central African Republic
79 I16 **Lobaye** ≈ SW Central African Republic
99 G21 **Lobbes** Hainaut, S Belgium *50°21′N 04°16′E*
61 D23 **Lobería** Buenos Aires, E Argentina *38°08′S 58°48′W*
110 F8 **Łobez** *Ger.* Labes. Zacodnio-pomorskie, NW Poland *53°38′N 15°39′E*
82 A13 **Lobito** Benguela, W Angola *12°20′S 13°34′E*
Lobkovichi *see* Labkovichy
Lob Nor *see* Lop Nur
171 V13 **Lobo** Papua Barat, E Indonesia *03°41′S 134°06′E*
104 J11 **Lobón** Extremadura, W Spain *38°51′N 06°38′W*
61 D20 **Lobos** Buenos Aires, E Argentina *35°11′S 59°08′W*
40 E4 **Lobos, Cabo** *headland* NW Mexico *29°53′N 112°43′W*
40 F6 **Lobos, Isla** *island* NW Mexico
Lobositz *see* Lovosice
Lobsens *see* Łobżenica
Loburi *see* Lop Buri
110 H9 **Łobżenica** *Ger.* Lobsens. Wielkopolskie, C Poland *53°19′N 17°11′E*
108 G11 **Locarno** *Ger.* Luggarus. Ticino, S Switzerland *46°11′N 08°48′E*
96 E9 **Lochboisdale** NW Scotland, United Kingdom *57°08′N 07°17′W*
98 N11 **Lochem** Gelderland, E Netherlands *52°10′N 06°25′E*
102 M8 **Loches** Indre-et-Loire, C France *47°08′N 01°00′E*
Loch Garman *see* Wexford
96 H12 **Lochgilphead** W Scotland, United Kingdom *56°02′N 05°27′W*
96 H7 **Lochinver** N Scotland, United Kingdom *58°10′N 05°15′W*
96 F8 **Lochmaddy** NW Scotland, United Kingdom *57°35′N 07°10′W*
96 J10 **Lochnagar** ▲ C Scotland, United Kingdom *56°58′N 03°09′W*
99 E17 **Lochristi** Oost-Vlaanderen, NW Belgium *51°07′N 03°49′E*
96 H9 **Lochy, Loch** ◎ N Scotland, United Kingdom
182 G8 **Lock** South Australia *33°37′S 135°45′E*
97 J14 **Lockerbie** S Scotland, United Kingdom *55°11′N 03°27′W*
27 S13 **Lockesburg** Arkansas, C USA *33°58′N 94°10′W*
183 P10 **Lockhart** New South Wales, SE Australia *35°15′S 146°43′E*
25 S11 **Lockhart** Texas, SW USA *29°54′N 97°41′W*
18 F13 **Lock Haven** Pennsylvania, NE USA *41°08′N 77°27′W*
25 N4 **Lockney** Texas, SW USA *34°06′N 101°27′W*
100 O12 **Löcknitz** ≈ NE Germany
18 E9 **Lockport** New York, NE USA *43°09′N 78°40′W*
167 T13 **Lôc Ninh** Sông Be, S Vietnam *11°51′N 106°35′E*
107 N23 **Locri** Calabria, SW Italy *38°16′N 16°16′E*
Locse *see* Levoča
27 T2 **Locust Creek** ≈ Missouri, C USA
23 P3 **Locust Fork** ≈ Alabama, S USA
27 Q9 **Locust Grove** Oklahoma, C USA *36°12′N 95°10′W*
94 E11 **Lodalskåpa** ▲ S Norway *61°47′N 07°10′E*
183 N10 **Loddon River** ≈ Victoria, SE Australia
Lodensee *see* Klooga
103 P15 **Lodève** *anc.* Luteva. Hérault, S France *43°44′N 03°19′E*
124 I12 **Lodeynoye Pole** Leningradskaya Oblast', NW Russian Federation *60°41′N 33°29′E*
33 V11 **Lodge Grass** Montana, NW USA *45°19′N 107°20′W*
28 J15 **Lodgepole Creek** ≈ Nebraska/Wyoming, C USA
149 T11 **Lodhrān** Punjab, E Pakistan *29°32′N 71°40′E*
106 D8 **Lodi** Lombardia, NW Italy *45°19′N 09°30′E*
35 O8 **Lodi** California, W USA *38°07′N 121°17′W*
31 T12 **Lodi** Ohio, N USA *41°00′N 82°01′W*
92 H10 **Lødingen** *Lapp.* Lådik. Nordland, C Norway *68°25′N 16°00′E*
79 L20 **Lodja** Kasai-Oriental, C Dem. Rep. Congo *03°29′S 23°25′E*
37 O3 **Lodore, Canyon of** *canyon* Colorado, C USA
105 Q4 **Lodosa** Navarra, N Spain *42°26′N 02°05′W*
81 H16 **Lodwar** Turkana, NW Kenya *03°06′N 35°38′E*
110 K13 **Łódź** *Rus.* Lodz. Łódź, C Poland *51°51′N 19°26′E*
110 E12 **Łódzkie** ◈ *province* C Poland
167 P8 **Loei** *var.* Loey, Muang Loei. Loei, C Thailand *17°32′N 101°40′E*
98 I11 **Loenen** Utrecht, C Netherlands *52°13′N 05°01′E*
167 R9 **Loeng Nok Tha** Yasothon, E Thailand *16°12′N 104°31′E*
83 F24 **Loeriesfontein** Northern Cape, W South Africa *30°59′S 19°29′E*
Loewoek *see* Luwuk
Loey *see* Loei
76 J16 **Lofa** ≈ N Liberia
109 P6 **Lofer** Salzburg, C Austria *47°37′N 12°42′E*
92 F11 **Lofoten** *var.* Lofoten Islands. *island group* C Norway
Lofoten Islands *see* Lofoten
95 N18 **Loftahammar** Kalmar, S Sweden *57°55′N 16°45′E*
127 O10 **Log** Volgogradskaya Oblast', SW Russian Federation *49°32′N 43°52′E*
77 S12 **Loga** Dosso, SW Niger *13°40′N 03°15′E*
29 S14 **Logan** Iowa, C USA *41°38′N 95°47′W*
26 K3 **Logan** Kansas, C USA *39°39′N 99°34′W*
31 T14 **Logan** Ohio, N USA *39°32′N 82°24′W*
36 L1 **Logan** Utah, W USA *41°45′N 111°50′W*
21 P6 **Logan** West Virginia, NE USA *37°52′N 82°00′W*
35 Y10 **Logandale** Nevada, W USA *36°36′N 114°28′W*
19 O11 **Logan International** ✈ (Boston) Massachusetts, NE USA *42°22′N 71°00′W*
11 N16 **Logan Lake** British Columbia, SW Canada *50°28′N 120°42′W*
23 Q4 **Logan Martin Lake** ⊠ Alabama, S USA
10 G8 **Logan, Mount** ▲ Yukon, W Canada *60°32′N 140°34′W*
32 I7 **Logan, Mount** ▲ Washington, NW USA *48°32′N 120°57′W*
33 P7 **Logan Pass** *pass* Montana, NW USA
31 O12 **Logansport** Indiana, N USA *40°44′N 86°25′W*
22 F6 **Logansport** Louisiana, S USA *31°58′N 94°00′W*
149 Q5 **Lōgar** *prev.* Lowgar. ◈ *province* E Afghanistan
67 R11 **Loge** ≈ NW Angola
Logishin *see* Lahishyn
Log na Coille *see* Lugnaquillia Mountain
78 G11 **Logone** *var.* Logon. ≈ Cameroon/Chad
78 G13 **Logone-Occidental** *off.* Région du Logone-Occidental. ◈ *region* SW Chad
78 H13 **Logone Occidental** ≈ SW Chad
Logone-Occidental, Région du *see* Logone-Occidental
78 G13 **Logone-Oriental** *off.* Région du Logone-Oriental. ◈ *region* SW Chad
78 H13 **Logone Oriental** ≈ SW Chad
Logone Oriental *see* Pendé
Logone-Oriental, Région du *see* Logone-Oriental
L'Ogooué-Ivindo *see* Ogooué-Ivindo
L'Ogooué-Lolo *see* Ogooué-Lolo
L'Ogooué-Maritime *see* Ogooué-Maritime
Logoysk *see* Lahoysk
105 P4 **Logroño** *anc.* Vareia, *Lat.* Juliobriga. La Rioja, N Spain *42°28′N 02°26′W*
104 L10 **Logrosán** Extremadura, W Spain *39°21′N 05°29′W*
95 G20 **Løgstør** Nordjylland, N Denmark *56°57′N 09°19′E*
95 H22 **Løgten** Midtjylland, C Denmark *56°17′N 10°20′E*
95 F24 **Løgumkloster** Syddanmark, SW Denmark *55°04′N 08°58′E*
Lögurinn *see* Lagarfljót
153 P15 **Lohārdaga** Jhārkhand, N India *23°27′N 84°42′E*
152 H10 **Lohāru** Haryāna, N India *28°28′N 75°50′E*
101 D15 **Lohausen** ✈ (Düsseldorf) Nordrhein-Westfalen, W Germany *51°18′N 06°51′E*
189 O14 **Lohd** Pohnpei, E Micronesia
92 L12 **Lohiniva** Lappi, N Finland *67°09′N 25°04′E*
Lohiszyn *see* Lahishyn
93 L20 **Lohja** *var.* Lojo. Uusimaa, S Finland *60°14′N 24°07′E*
169 V11 **Lohjanan** Borneo, C Indonesia
25 Q9 **Lohn** Texas, SW USA *31°15′N 99°22′W*
100 G12 **Löhne** Niedersachsen, NW Germany *52°40′N 08°13′E*
Lohr *see* Lohr am Main
101 I18 **Lohr am Main** *var.* Lohr. Bayern, C Germany *50°00′N 09°30′E*
109 T10 **Loibl Pass** *Ger.* Loiblpass, *Slvn.* Ljubelj. *pass* Austria/Slovenia
Loiblpass *see* Loibl Pass
167 N6 **Loikaw** Kayah State, C Myanmar (Burma) *19°40′N 97°17′E*
93 K19 **Loimaa** Varsinais-Suomi, SW Finland *60°51′N 23°03′E*
103 O6 **Loing** ≈ C France
167 R6 **Loi, Phou** ▲ N Laos *20°18′N 103°14′E*
102 L7 **Loir** ≈ C France
103 Q11 **Loire** ◈ *department* E France
102 M7 **Loire** *var.* Liger. ≈ C France
102 I7 **Loire-Atlantique** ◈ *department* NW France
103 O7 **Loiret** ◈ *department* C France
102 M8 **Loir-et-Cher** ◈ *department* C France
101 L24 **Loisach** ≈ SE Germany
56 B9 **Loja** Loja, S Ecuador *03°59′S 79°16′W*
104 M14 **Loja** Andalucía, S Spain *37°10′N 04°09′W*
56 B9 **Loja** ◈ *province* S Ecuador
Lojo *see* Lohja
116 J4 **Lokachi** Volyns'ka Oblast', NW Ukraine *50°44′N 24°39′E*
79 M20 **Lokandu** Maniema, C Dem. Rep. Congo *02°34′S 25°44′E*
92 M11 **Lokan Tekojärvi** ◎ NE Finland
137 Z11 **Lökbatan** *Rus.* Lokbatan. E Azerbaijan *40°21′N 49°43′E*
Lokbatan *see* Lökbatan
99 F17 **Lokeren** Oost-Vlaanderen, NW Belgium *51°06′N 03°59′E*
Lokhvitsa *see* Lokhvytsya
117 S4 **Lokhvytsya** *Rus.* Lokhvitsa. Poltavs'ka Oblast', NE Ukraine *50°22′N 33°16′E*
81 H17 **Lokichar** Turkana, NW Kenya *02°23′N 35°40′E*
81 G16 **Lokichokio** Turkana, NW Kenya *04°16′N 34°22′E*
81 H16 **Lokitaung** Turkana, NW Kenya *04°15′N 35°45′E*
92 M11 **Lokka** Lappi, N Finland *67°48′N 27°41′E*
94 G8 **Løkken Verk** Sør-Trøndelag, S Norway *63°06′N 09°43′E*
124 G16 **Loknya** Pskovskaya Oblast', W Russian Federation *56°48′N 30°08′E*
77 V15 **Loko** Nassarawa, C Nigeria *08°00′N 07°48′E*
77 U15 **Lokoja** Kogi, C Nigeria *07°48′N 06°45′E*
81 H17 **Lokori** Turkana, W Kenya *01°56′N 36°03′E*
77 R16 **Lokossa** S Benin *06°38′N 01°43′E*
118 I3 **Loksa** *Ger.* Loxa. Harjumaa, NW Estonia *59°32′N 25°42′E*
9 T7 **Loks Land** *island* Nunavut, NE Canada
80 C13 **Lol** ≈ NW South Sudan
76 K15 **Lola** SE Guinea *07°52′N 08°29′W*
35 Q5 **Lola, Mount** ▲ California, W USA *39°27′N 120°20′W*
81 H20 **Loliondo** Arusha, NE Tanzania *02°03′S 35°46′E*
95 H25 **Lolland** *prev.* Laaland. *island* S Denmark
186 G6 **Lolobau Island** *island* E Papua New Guinea
79 E16 **Lolodorf** Sud, SW Cameroon *03°17′N 10°50′E*
114 G7 **Lom** *prev.* Lom-Palanka. Montana, NW Bulgaria *43°49′N 23°16′E*
114 G7 **Lom** ≈ NW Bulgaria
79 M19 **Lomami** ≈ C Dem. Rep. Congo
57 F17 **Lomas** Arequipa, SW Peru *15°29′S 74°54′W*
63 I23 **Lomas, Bahía** *bay* S Chile
61 D20 **Lomas de Zamora** Buenos Aires, E Argentina *34°53′S 58°26′W*
61 D20 **Loma Verde** Buenos Aires, E Argentina *35°16′S 58°24′W*
180 K4 **Lombadina** Western Australia *16°39′S 122°54′E*
106 E6 **Lombardia** *Eng.* Lombardy. ◈ *region* N Italy
Lombardy *see* Lombardia
102 M15 **Lombez** Gers, S France *43°29′N 00°54′E*
171 Q16 **Lomblen, Pulau** *island* Nusa Tenggara, S Indonesia
173 W7 **Lombok Basin** *undersea feature* E Indian Ocean *09°50′S 116°00′E*
170 L16 **Lombok, Pulau** *island* Nusa Tenggara, C Indonesia
77 Q16 **Lomé** ● (Togo) S Togo *06°08′N 01°13′E*
77 Q16 **Lomé** ✈ S Togo *06°08′N 01°13′E*
79 L19 **Lomela** Kasai-Oriental, C Dem. Rep. Congo *02°19′S 23°15′E*
79 L19 **Lomela** ≈ C Dem. Rep. Congo
25 R8 **Lometa** Texas, SW USA *31°13′N 98°23′W*
79 F16 **Lomié** Est, SE Cameroon *03°09′N 13°35′E*
30 M8 **Lomira** Wisconsin, N USA *43°36′N 88°26′W*
95 K23 **Lomma** Skåne, S Sweden *55°41′N 13°05′E*
99 J16 **Lommel** Limburg, N Belgium *51°14′N 05°19′E*
96 I11 **Lomond, Loch** ◎ C Scotland, United Kingdom
197 R9 **Lomonosov Ridge** *var.* Harris Ridge, *Rus.* Khrebet Homonsova. *undersea feature* Arctic Ocean *88°00′N 140°00′E*
Lomonsova, Khrebet *see* Lomonosov Ridge
Lom-Palanka *see* Lom
Lomphat *see* Lumphăt
35 P14 **Lompoc** California, W USA *34°39′N 120°29′W*
167 P9 **Lom Sak** *var.* Muang Lom Sak. Phetchabun, C Thailand *16°45′N 101°12′E*
110 N9 **Łomża** *Rus.* Lomzha. Podlaskie, NE Poland *53°11′N 22°04′E*
Lomzha *see* Łomża
Lonaula *see* Lonāvale
155 D14 **Lonāvale** *prev.* Lonaula. Mahārāshtra, W India *18°45′N 73°27′E*
63 G15 **Loncoche** Araucanía, C Chile *39°22′S 72°34′W*
63 H14 **Loncopue** Neuquén, W Argentina *38°04′S 70°43′W*
99 G17 **Londerzeel** Vlaams Brabant, C Belgium *51°00′N 04°19′E*
Londinium *see* London
14 E16 **London** Ontario, S Canada *42°59′N 81°13′W*
191 Y2 **London** Kiritimati, E Kiribati *02°00′N 157°28′W*
97 O22 **London** *anc.* Augusta, *Lat.* Londinium. ● (United Kingdom) SE England, United Kingdom *51°30′N 00°10′W*
21 N7 **London** Kentucky, S USA *37°07′N 84°05′W*
31 S13 **London** Ohio, NE USA *39°52′N 83°27′W*
25 Q10 **London** Texas, SW USA *30°40′N 99°33′W*
97 O22 **London City** ✈ SE England, United Kingdom *51°31′N 00°07′E*
97 E14 **Londonderry** *var.* Derry, *Ir.* Doire. NW Northern Ireland, United Kingdom *55°00′N 07°19′W*
97 F14 **Londonderry** *cultural region* NW Northern Ireland, United Kingdom
180 M2 **Londonderry, Cape** *cape* Western Australia
63 H25 **Londonderry, Isla** *island* S Chile
43 O7 **Londres, Cayos** *reef* NE Nicaragua
60 I10 **Londrina** Paraná, S Brazil *23°18′S 51°13′W*
27 N13 **Lone Grove** Oklahoma, C USA *34°11′N 97°15′W*
14 E12 **Lonely Island** *island* Ontario, S Canada
35 T8 **Lone Mountain** ▲ Nevada, W USA *38°01′N 117°28′W*
25 V6 **Lone Oak** Texas, SW USA *33°02′N 95°58′W*
35 T11 **Lone Pine** California, W USA *36°36′N 118°04′W*
Lone Star State *see* Texas
Lone Tree Islet *see* Iku
83 D14 **Longa** Kuando Kubango, C Angola *14°44′S 18°36′E*
82 B12 **Longa** ≈ W Angola
83 E15 **Longa** ≈ SE Angola
Long'an *see* Pingwu
163 W11 **Longang Shan** ▲ NE China
197 S4 **Longa, Proliv** *Eng.* Long Strait. *strait* NE Russian Federation
44 J13 **Long Bay** *bay* W Jamaica
21 V13 **Long Bay** *bay* North Carolina/South Carolina, E USA
35 T16 **Long Beach** California, W USA *33°46′N 118°11′W*
22 M9 **Long Beach** Mississippi, S USA *30°21′N 89°09′W*
18 L14 **Long Beach** Long Island, New York, NE USA *40°34′N 73°38′W*
32 F9 **Long Beach** Washington, NW USA *46°21′N 124°03′W*
18 K16 **Long Beach Island** *island* New Jersey, NE USA
65 M25 **Longbluff** *headland* SW Tristan da Cunha
23 U13 **Longboat Key** *island* Florida, SE USA
18 K15 **Long Branch** New Jersey, NE USA *40°18′N 73°59′W*
44 J5 **Long Cay** *island* SE The Bahamas
Longcheng *see* Xiaoxian
161 P14 **Longchuan** *var.* Laolong. Guangdong, S China *24°07′N 115°10′E*
Longchuan *see* Nanhua
Longchuan Jiang *see* Shweli
32 K12 **Long Creek** Oregon, NW USA *44°40′N 119°07′W*
159 W10 **Longde** Ningxia, N China *35°37′N 106°07′E*
183 P16 **Longford** Tasmania, SE Australia *41°41′S 147°03′E*
97 D17 **Longford** *Ir.* An Longfort. Longford, C Ireland *53°45′N 07°50′W*
97 E17 **Longford** *Ir.* An Longfort. *cultural region* C Ireland
Longgang *see* Dazu
161 P1 **Longhua** Hebei, E China *41°18′N 117°44′E*
169 U11 **Longiram** Borneo, C Indonesia *0°02′S 115°36′E*
12 H8 **Long Island** *island* Nunavut, C Canada
186 D7 **Long Island** *var.* Arop Island. *island* N Papua New Guinea
44 J4 **Long Island** *island* C The Bahamas
18 L14 **Long Island** *island* New York, NE USA
Long Island *see* Bermuda
18 M14 **Long Island Sound** *sound* NE USA
163 U7 **Longjiang** Heilongjiang, NE China *47°20′N 123°09′E*
160 K13 **Long Jiang** ≈ S China
163 Y10 **Longjing** *var.* Yanji. Jilin, NE China *42°48′N 129°26′E*
161 R4 **Longkou** Shandong, E China *37°40′N 120°21′E*
12 E11 **Longlac** Ontario, S Canada *49°47′N 86°34′W*
19 S1 **Long Lake** ◎ Maine, NE USA
31 O6 **Long Lake** ◎ Michigan, N USA
31 R5 **Long Lake** ◎ Michigan, N USA
29 N6 **Long Lake** ◎ North Dakota, N USA
30 J4 **Long Lake** ⊠ Wisconsin, N USA
99 K23 **Longlier** Luxembourg, SE Belgium *49°51′N 05°27′E*
160 I13 **Longlin** *var.* Longlin Gezu Zizhixian. Guangxi Zhuangzu Zizhiqu, S China *24°46′N 105°19′E*
Longlin Gezu Zizhixian *see* Longlin
37 T3 **Longmont** Colorado, C USA *40°09′N 105°07′W*
157 P10 **Longnan** *var.* Wudu. Gansu, C China *33°27′N 104°57′E*
29 N13 **Long Pine** Nebraska, C USA *42°32′N 99°42′W*
Longping *see* Luodian
14 F17 **Long Point** *headland* Ontario, S Canada *42°33′N 80°15′W*
14 K15 **Long Point** *headland* Ontario, SE Canada *43°56′N 76°53′W*
184 P10 **Long Point** *headland* North Island, New Zealand *39°07′S 177°41′E*
30 L2 **Long Point** *headland* Michigan, N USA *47°50′N 89°09′W*
14 G17 **Long Point Bay** *lake bay* Ontario, S Canada
29 T7 **Long Prairie** Minnesota, N USA *45°58′N 94°52′W*
Longquan *see* Fenggang
Longquàn *see* Yanggao
13 S11 **Long Range Mountains** *hill range* Newfoundland and Labrador, E Canada
65 H25 **Long Range Point** *headland* SE Saint Helena *16°00′S 05°41′W*
181 V8 **Longreach** Queensland, E Australia *23°31′S 144°18′E*
160 H7 **Longriba** Sichuan, C China *32°32′N 102°20′E*
160 L10 **Longshan** *var.* Min'an. Hunan, S China *29°25′N 109°28′E*
37 S3 **Longs Peak** ▲ Colorado, C USA *40°15′N 105°37′W*
Long Strait *see* Longa, Proliv
102 K8 **Longué** Maine-et-Loire, NW France *47°23′N 00°07′W*
13 P11 **Longue-Pointe** Québec, E Canada *50°20′N 64°13′W*
103 S4 **Longuyon** Meurthe-et-Moselle, NE France *49°25′N 05°37′E*
25 W7 **Longview** Texas, SW USA *32°30′N 94°45′W*
32 G10 **Longview** Washington, NW USA *46°08′N 122°56′W*
65 H25 **Longwood** C Saint Helena
25 P7 **Longworth** Texas, SW USA *32°37′N 100°20′W*
103 S3 **Longwy** Meurthe-et-Moselle, NE France *49°31′N 05°46′E*
159 V11 **Longxi** *var.* Gongchang. Gansu, C China *35°00′N 104°34′E*
Longxian *see* Wengyuan
167 S14 **Long Xuyên** *var.* Longxuyen. An Giang, S Vietnam *10°23′N 105°25′E*
Longxuyen *see* Long Xuyên
161 Q13 **Longyan** Fujian, SE China *25°06′N 117°02′E*
92 O3 **Longyearbyen** ○ (Svalbard) Spitsbergen, W Svalbard *78°12′N 15°39′E*
160 J15 **Longzhou** Guangxi Zhuangzu Zizhiqu, S China *22°22′N 106°46′E*
Longzhouping *see* Changyang
100 F12 **Löningen** Niedersachsen, NW Germany *52°43′N 07°42′E*
27 V11 **Lonoke** Arkansas, C USA *34°46′N 91°56′W*
95 L21 **Lönsboda** Skåne, S Sweden *56°24′N 14°19′E*
103 S9 **Lons-le-Saunier** *anc.* Ledo Salinarius. Jura, E France *46°41′N 05°32′E*
31 O15 **Loogootee** Indiana, N USA *38°40′N 86°54′W*
31 Q9 **Looking Glass River** ≈ Michigan, N USA
21 X11 **Lookout, Cape** *headland* North Carolina, SE USA *34°36′N 76°31′W*
39 O6 **Lookout Ridge** *ridge* Alaska, USA
Lookransar *see* Lūnkaransar
181 N11 **Loongana** Western Australia *30°53′S 127°15′E*
99 I14 **Loon op Zand** Noord-Brabant, S Netherlands *51°38′N 05°05′E*
97 A19 **Loop Head** *Ir.* Ceann Léime. *promontory* W Ireland
109 V4 **Loosdorf** Niederösterreich, NE Austria *48°13′N 15°25′E*
158 G10 **Lop** Xinjiang Uygur Zizhiqu, NW China *37°06′N 80°12′E*
112 J11 **Lopare** ◈ Republika Srpska, NE Bosnia and Herzegovina
Lopatichi *see* Lapatsichy
127 P7 **Lopatino** Penzenskaya Oblast', W Russian Federation *52°38′N 45°46′E*
167 P10 **Lop Buri** *var.* Loburi. Lop Buri, C Thailand *14°49′N 100°37′E*
25 R16 **Lopeno** Texas, SW USA *26°42′N 99°06′W*
79 C18 **Lopez, Cap** *headland* W Gabon *0°39′S 08°44′E*
98 I12 **Lopik** Utrecht, C Netherlands *51°58′N 04°57′E*
Lop Nor *see* Lop Nur
158 M7 **Lop Nur** *var.* Lob Nor, Lop Nor, Lo-pu Po. *seasonal lake* NW China
Lopnur *see* Yuli
79 K17 **Lopori** ≈ NW Dem. Rep. Congo
98 O5 **Loppersum** Groningen, NE Netherlands *53°20′N 06°45′E*
92 I8 **Lopphavet** *sound* N Norway
Lo-pu Po *see* Lop Nur
Lora *see* Lowrah
182 F3 **Lora Creek** *seasonal river* South Australia
104 K13 **Lora del Río** Andalucía, S Spain *37°39′N 05°32′W*
148 M11 **Lora, Hāmūn-i** *wetland* SW Pakistan
31 T11 **Lorain** Ohio, N USA *41°27′N 82°10′W*
25 O7 **Loraine** Texas, SW USA *32°24′N 100°42′W*
31 R13 **Loramie, Lake** ◎ Ohio, N USA
105 Q13 **Lorca** *Ar.* Lurka; *anc.* Eliocroca, *Lat.* Illurco. Murcia, S Spain *37°40′N 01°41′W*
192 I10 **Lord Howe Island** *island* E Australia
Lord Howe Island *see* Ontong Java Atoll
175 O10 **Lord Howe Rise** *undersea feature* SW Pacific Ocean
192 J10 **Lord Howe Seamounts** *undersea feature* W Pacific Ocean
37 P15 **Lordsburg** New Mexico, SW USA *32°19′N 108°42′W*
186 E5 **Lorengau** *var.* Lorungau. Manus Island, N Papua New Guinea *02°01′S 147°15′E*
25 N5 **Lorenzo** Texas, SW USA *33°40′N 101°31′W*
142 K7 **Lorestān** *off.* Ostān-e Lorestān, *var.* Luristan. ◈ *province* W Iran
Lorestān, Ostān-e *see* Lorestān
57 M17 **Loreto** El Beni, N Bolivia *15°13′S 64°44′W*
106 J12 **Loreto** Marche, C Italy *43°25′N 13°37′E*
40 F8 **Loreto** Baja California Sur, NW Mexico *25°59′N 111°22′W*
40 M11 **Loreto** Zacatecas, C Mexico *22°15′N 102°00′W*
56 E9 **Loreto** *off.* Departamento de Loreto. ◈ *department* NE Peru
Loreto, Departamento de *see* Loreto
81 K18 **Lorian Swamp** *swamp* E Kenya
54 E6 **Lorica** Córdoba, NW Colombia *09°14′N 75°50′W*
102 G7 **Lorient** *prev.* l'Orient. Morbihan, NW France *47°45′N 03°22′W*
l'Orient *see* Lorient
111 K22 **Lőrinci** Heves, NE Hungary *47°46′N 19°40′E*
14 G11 **Loring** Ontario, S Canada *45°55′N 79°59′W*
33 V6 **Loring** Montana, NW USA *48°49′N 107°48′W*
103 R13 **Loriol-sur-Drôme** Drôme, E France *44°46′N 04°51′E*
21 U12 **Loris** South Carolina, SE USA *34°03′N 78°53′W*
57 I18 **Loriscota, Laguna** ◎ S Peru
183 N13 **Lorne** Victoria, SE Australia *38°33′S 143°57′E*
96 G11 **Lorn, Firth of** *inlet* W Scotland, United Kingdom
Loro Sae *see* East Timor
101 F24 **Lörrach** Baden-Württemberg, S Germany *47°38′N 07°40′E*

◆ Country ● Country Capital ◇ Dependent Territory ○ Dependent Territory Capital ◈ Administrative Regions ✈ International Airport ▲ Mountain ▲▲ Mountain Range ※ Volcano ≈ River ◎ Lake ⊠ Reservoir

◆ Country   ◇ Dependent Territory   ◈ Administrative Regions   ▲ Mountain   ◬ Volcano   ◉ Lake
● Country Capital   ○ Dependent Territory Capital   ✈ International Airport   ▲ Mountain Range   ⬳ River   ▨ Reservoir

LUOHE – MADURA, SELAT

161 *N7* **Luohe** Henan, C China *33°37′N 114°00′E*
160 *M6* **Luo He** ≈ C China
160 *L5* **Luo He** ≈ C China
L i Liêm, Nhom *see* Crescent Group
Luolajarvi *see* Kuoloyarvi
Luong Nam Tha *see* Louangnamtha
160 *L13* **Luoqing Jiang** ≈ S China
161 *O8* **Luoshan** Henan, C China *32°12′N 114°30′E*
161 *O12* **Luoxiao Shan** ▲▲ S China
161 *N6* **Luoyang** *var.* Honan, Lo-yang. Henan, C China *34°41′N 112°25′E*
161 *R12* **Luoyuan** *var.* Fengshan. Fujian, SE China *26°29′N 119°32′E*
79 *F21* **Luozi** Bas-Congo, W Dem. Rep. Congo *04°57′S 14°08′E*
83 *J17* **Lupane** Matabeleland North, W Zimbabwe *18°54′S 27°44′E*
160 *I12* **Lupanshui** *var.* Liupanshui; *prev.* Shuicheng. Guizhou, S China *26°38′N 104°49′E*
169 *R10* **Lupar, Batang** ≈ East Malaysia
Lupatia *see* Altamura
116 *G12* **Lupeni** *Hung.* Lupény. Hunedoara, SW Romania *45°20′N 23°10′E*
Lupény *see* Lupeni
82 *N13* **Lupiliche** Niassa, N Mozambique *11°36′S 35°15′E*
83 *E14* **Lupire** Kuando Kubango, E Angola *14°39′S 19°39′E*
79 *L22* **Luputa** Kasai-Oriental, S Dem. Rep. Congo *07°07′S 23°43′E*
121 *P16* **Luqa** ✈ (Valletta) S Malta *35°53′N 14°27′E*
159 *U11* **Luqu** *var.* Ma'ai. Gansu, C China *34°34′N 102°27′E*
45 *U5* **Luquillo, Sierra de** ▲▲ E Puerto Rico
26 *L4* **Luray** Kansas, C USA *39°06′N 98°41′W*
21 *U4* **Luray** Virginia, NE USA *38°40′N 78°28′W*
103 *T7* **Lure** Haute-Saône, E France *47°42′N 06°30′E*
82 *D11* **Luremo** Lunda Norte, NE Angola *08°32′S 17°55′E*
97 *F15* **Lurgan** *Ir.* An Lorgain. S Northern Ireland, United Kingdom *54°28′N 06°20′W*
57 *K18* **Luribay** La Paz, W Bolivia *17°05′S 67°37′W*
Luring *see* Gêrzê
83 *Q14* **Lúrio** Nampula, NE Mozambique *13°32′S 40°34′E*
83 *P14* **Lúrio, Rio** ≈ NE Mozambique
Luristan *see* Lorestān
Lurka *see* Lorca
83 *J15* **Lusaka** ● (Zambia) Lusaka, SE Zambia *15°24′S 28°17′E*
83 *J15* **Lusaka** ⬖ *province* C Zambia
83 *J15* **Lusaka** ✈ Lusaka, C Zambia *15°10′S 28°22′E*
79 *L21* **Lusambo** Kasai-Oriental, C Dem. Rep. Congo *04°59′S 23°26′E*
186 *F8* **Lusancay Islands and Reefs** *island group* SE Papua New Guinea
79 *I21* **Lusanga** Bandundu, SW Dem. Rep. Congo *04°55′S 18°40′E*
79 *N21* **Lusangi** Maniema, E Dem. Rep. Congo *04°39′S 27°10′E*
Lusatian Mountains *see* Lausitzer Bergland
Lushar *see* Huangzhong
Lushnja *see* Lushnjë
113 *K21* **Lushnjë** *var.* Lushnja. Fier, C Albania *40°54′N 19°43′E*
81 *J21* **Lushoto** Tanga, E Tanzania *04°48′S 38°20′E*
102 *L10* **Lusignan** Vienne, W France *46°25′N 00°06′E*
33 *Z15* **Lusk** Wyoming, C USA *42°45′N 104°27′W*
Luso *see* Luena
102 *L10* **Lussac-les-Châteaux** Vienne, W France *46°23′N 00°44′E*
Lussin/Lussino *see* Lošinj
Lussinpiccolo *see* Mali Lošinj
108 *I7* **Lustenau** Vorarlberg, W Austria *47°26′N 09°42′E*
Lütao *see* Lü Dao
Lü Tao *see* Lü Dao
Lūt, Baḩrat/Lut, Bahret *see* Dead Sea
22 *K9* **Lutcher** Louisiana, S USA *30°02′N 90°42′W*
143 *T9* **Lūt, Dasht-e** *var.* Kavīr-e Lūt. *desert* E Iran
83 *F14* **Lutembo** Moxico, E Angola *13°30′S 21°21′E*
Lutetia/Lutetia Parisiorum *see* Paris
Luteva *see* Lodève
14 *G15* **Luther Lake** ◎ Ontario, S Canada
186 *K8* **Luti** Choiseul, NW Solomon Islands *07°13′S 157°01′E*
Lūt, Kavīr-e *see* Lūt, Dasht-e
97 *N21* **Luton** E England, United Kingdom *51°53′N 00°25′W*
97 *N21* **Luton** ✈ (London) SE England, United Kingdom *51°54′N 00°24′W*
108 *B10* **Lutry** Vaud, SW Switzerland *46°31′N 06°32′E*
8 *K10* **Łutselk'e** *prev.* Snowdrift. Northwest Territories, W Canada *62°24′N 110°42′W*
8 *K10* **Lutselk'e** *var.* Snowdrift. ≈ Northwest Territories, NW Canada
29 *Y4* **Lutsen** Minnesota, N USA *47°39′N 90°37′W*
116 *J4* **Luts'k** *Pol.* Łuck, *Rus.* Lutsk. Volyns'ka Oblast', NW Ukraine *50°45′N 25°23′E*
Lutsk *see* Luts'k
Luttenberg *see* Ljutomer
Lüttich *see* Liège
83 *G25* **Luttig** Western Cape, SW South Africa *32°33′S 22°13′E*
Lutto *see* Lotta
82 *E13* **Lutuai** Moxico, E Angola *12°38′S 20°06′E*
117 *Y7* **Lutuhyne** Luhans'ka Oblast', E Ukraine *48°24′N 39°12′E*
171 *V14* **Lutur, Pulau** *island* Kepulauan Aru, E Indonesia
23 *V12* **Lutz** Florida, SE USA *28°09′N 82°27′W*
Lutzow-Holm Bay *see* Lützow Holmbukta
195 *V2* **Lützow Holmbukta** *var.* Lutzow-Holm Bay. *bay* Antarctica
81 *L16* **Luuq** *It.* Lugh Ganana. Gedo, SW Somalia *03°42′N 42°34′E*
92 *M12* **Luusua** Lappi, NE Finland *66°28′N 27°16′E*
23 *Q6* **Luverne** Alabama, S USA *31°43′N 86°15′W*
29 *S11* **Luverne** Minnesota, N USA *43°39′N 96°12′W*
79 *O22* **Luvua** ≈ SE Dem. Rep. Congo
82 *F13* **Luvuei** Moxico, E Angola *13°08′S 21°09′E*
81 *H24* **Luwego** ≈ S Tanzania
82 *K12* **Luwingu** Northern, NE Zambia *10°13′S 29°58′E*
171 *P12* **Luwuk** *prev.* Loewoek. Sulawesi, C Indonesia *0°56′S 122°47′E*
23 *N3* **Luxapallila Creek** ≈ Alabama/Mississippi, S USA
99 *M25* **Luxembourg** ● (Luxembourg) Luxembourg, S Luxembourg *49°37′N 06°08′E*
99 *M25* **Luxembourg** *off.* Grand Duchy of Luxembourg, *var.* Lëtzebuerg, Luxemburg. ◆ *monarchy* NW Europe
99 *J23* **Luxembourg** ⬖ *province* SE Belgium
99 *L24* **Luxembourg** ⬖ *district* S Luxembourg
31 *N6* **Luxemburg** Wisconsin, N USA *44°32′N 87°42′W*
Luxemburg *see* Luxembourg
103 *U7* **Luxeuil-les-Bains** Haute-Saône, E France *47°49′N 06°22′E*
160 *E13* **Luxi** *prev.* Mangshi. Yunnan, SW China *24°27′N 98°31′E*
82 *E10* **Luxico** ≈ Angola/Dem. Rep. Congo
75 *X10* **Luxor** *Ar.* Al Uqşur. E Egypt *25°39′N 32°39′E*
75 *X10* **Luxor** ✈ C Egypt *25°39′N 32°48′E*
160 *M4* **Luya Shan** ▲▲ C China
102 *J15* **Luy de Béarn** ≈ SW France
102 *J15* **Luy de France** ≈ SW France
125 *P12* **Luza** Kirovskaya Oblast', NW Russian Federation *60°38′N 47°13′E*
125 *Q12* **Luza** ≈ NW Russian Federation
104 *I16* **Luz, Costa de la** *coastal region* SW Spain
111 *K20* **Luže** *var.* Lausche. ▲ Czech Republic/Germany *50°51′N 14°40′E* ***see also*** Lausche
Luže *see* Lausche
108 *F8* **Luzern** *Fr.* Lucerne, *It.* Lucerna. Luzern, C Switzerland *47°03′N 08°17′E*
108 *E8* **Luzern** *Fr.* Lucerne. ⬖ *canton* C Switzerland
160 *L13* **Luzhai** Guangxi Zhuangzu Zizhiqu, S China *24°31′N 109°46′E*
118 *K12* **Luzhki** Vitsyebskaya Voblasts', N Belarus *55°21′N 27°52′E*
160 *I10* **Luzhou** Sichuan, C China *28°55′N 105°25′E*
Lužická Nisa *see* Neisse
Lužické Hory' *see* Lausitzer Bergland
Lužnice *see* Lainsitz
171 *O2* **Luzon** *island* N Philippines
171 *N1* **Luzon Strait** *strait* Philippines/Taiwan
Lużyckie, Gory *see* Lausitzer Bergland
116 *I5* **L'viv** *Ger.* Lemberg, *Pol.* Lwów, *Rus.* L'vov. L'vivs'ka Oblast', W Ukraine *49°49′N 24°05′E*
L'viv *see* L'vivs'ka Oblast'
116 *I4* **L'vivs'ka Oblast'** *var.* L'viv, *Rus.* L'vovskaya Oblast'. ⬖ *province* NW Ukraine
L'vov *see* L'viv
L'vovskaya Oblast' *see* L'vivs'ka Oblast'
Lwena *see* Luena
Lwów *see* L'viv
110 *F11* **Lwówek** *Ger.* Neustadt bei Pinne. Wielkopolskie, C Poland *52°27′N 16°10′E*
111 *E14* **Lwówek Śląski** *Ger.* Löwenberg in Schlesien. Jelenia Góra, SW Poland *51°06′N 15°35′E*
119 *I18* **Lyakhavichy** *Rus.* Lyakhovichi. Brestskaya Voblasts', SW Belarus *53°02′N 26°16′E*
Lyakhovichi *see* Lyakhavichy
185 *B22* **Lyall, Mount** ▲ South Island, New Zealand *45°14′S 167°31′E*
Lyallpur *see* Faisalābād
Lyangar *see* Langar
124 *H11* **Lyaskelya** Respublika Kareliya, NW Russian Federation *61°42′N 31°06′E*
119 *I18* **Lyasnaya** *Rus.* Lesnaya. Brestskaya Voblasts', SW Belarus *52°59′N 25°46′E*
119 *F19* **Lyasnaya** *Pol.* Leśna, *Rus.* Lesnaya. ≈ SW Belarus
124 *H15* **Lychkovo** Novgorodskaya Oblast', W Russian Federation *57°55′N 32°24′E*
Lyck *see* Ełk
93 *I15* **Lycksele** Västerbotten, N Sweden *64°34′N 18°40′E*
18 *G13* **Lycoming Creek** ≈ Pennsylvania, NE USA
Lycopolis *see* Asyūţ
195 *N3* **Lyddan Island** *island* Antarctica
Lydenburg *see* Mashishing
119 *L20* **Lyel'chytsy** *Rus.* Lel'chitsy. Homyel'skaya Voblasts', SE Belarus *51°47′N 28°20′E*
119 *P14* **Lyenina** *Rus.* Lenino. Mahilyowskaya Voblasts', E Belarus *54°25′N 31°08′E*
118 *L13* **Lyepyel'** *Rus.* Lepel'. Vitsyebskaya Voblasts', N Belarus *54°54′N 28°44′E*
25 *S17* **Lyford** Texas, SW USA *26°24′N 97°47′W*
95 *E17* **Lygna** ≈ S Norway
18 *G14* **Lykens** Pennsylvania, NE USA *40°33′N 76°42′W*
115 *E21* **Lykódimo** ▲ S Greece *36°56′N 21°49′E*
97 *K24* **Lyme Bay** *bay* S England, United Kingdom
97 *K24* **Lyme Regis** S England, United Kingdom *50°44′N 02°56′W*
110 *L7* **Łyna** *Ger.* Alle. ≈ N Poland
29 *P12* **Lynch** Nebraska, C USA *42°49′N 98°27′W*
20 *J10* **Lynchburg** Tennessee, S USA *35°17′N 86°22′W*
21 *T6* **Lynchburg** Virginia, NE USA *37°24′N 79°09′W*
21 *T12* **Lynches River** ≈ South Carolina, SE USA
32 *H6* **Lynden** Washington, NW USA *48°57′N 122°27′W*
182 *I5* **Lyndhurst** South Australia *30°19′S 138°20′E*
27 *Q5* **Lyndon** Kansas, C USA *38°37′N 95°40′W*
19 *N7* **Lyndonville** Vermont, NE USA *44°31′N 71°58′W*
95 *D18* **Lyngdal** Vest-Agder, S Norway *58°10′N 07°08′E*
92 *I9* **Lyngen** *Lapp.* Ivgovuotna. *inlet* Arctic Ocean
95 *G17* **Lyngør** Aust-Agder, S Norway *58°38′N 09°05′E*
92 *I9* **Lyngseidet** Troms, N Norway *69°36′N 20°07′E*
19 *P11* **Lynn** Massachusetts, NE USA *42°28′N 70°57′W*
Lynn *see* King's Lynn
23 *R9* **Lynn Haven** Florida, SE USA *30°15′N 85°39′W*
11 *V11* **Lynn Lake** Manitoba, C Canada *56°51′N 101°01′W*
Lynn Regis *see* King's Lynn
118 *I13* **Lyntupy** Vitsyebskaya Voblasts', NW Belarus *55°03′N 26°19′E*
103 *R11* **Lyon** *Eng.* Lyons; *anc.* Lugdunum. Rhône, E France *45°46′N 04°50′E*
8 *I6* **Lyon, Cape** *headland* Northwest Territories, NW Canada *69°47′N 123°10′W*
18 *K6* **Lyon Mountain** ▲ New York, NE USA *44°42′N 73°52′W*
103 *Q11* **Lyonnais, Monts du** ▲▲ C France
65 *N25* **Lyon Point** *headland* SE Tristan da Cunha *37°06′S 12°13′W*
182 *E5* **Lyons** South Australia *30°40′S 133°50′E*
37 *T3* **Lyons** Colorado, C USA *40°13′N 105°16′W*
23 *V6* **Lyons** Georgia, SE USA *32°12′N 82°19′W*
26 *M5* **Lyons** Kansas, C USA *38°22′N 98°13′W*
29 *R14* **Lyons** Nebraska, C USA *41°56′N 96°28′W*
18 *G10* **Lyons** New York, NE USA *43°03′N 76°58′W*
Lyons *see* Lyon
118 *O13* **Lyozna** *Rus.* Liozno. Vitsyebskaya Voblasts', NE Belarus *55°02′N 30°48′E*
117 *S4* **Lypova Dolyna** Sums'ka Oblast', NE Ukraine *50°36′N 33°50′E*
117 *N6* **Lypovets'** *Rus.* Lipovets. Vinnyts'ka Oblast', C Ukraine *49°13′N 29°06′E*
Lys *see* Leie
111 *I18* **Lysá Hora** ▲ E Czech Republic *49°31′N 18°27′E*
95 *D16* **Lysefjorden** *fjord* S Norway
95 *I18* **Lysekil** Västra Götaland, S Sweden *58°16′N 11°26′E*
Lýsi *see* Akdoğan
33 *V14* **Lysite** Wyoming, C USA *43°16′N 107°42′W*
127 *P3* **Lyskovo** Nizhegorodskaya Oblast', W Russian Federation *56°04′N 45°01′E*
108 *D8* **Lyss** Bern, W Switzerland *47°04′N 07°19′E*
95 *H22* **Lystrup** Midtjylland, C Denmark *56°14′N 10°14′E*
125 *V14* **Lys'va** Permskiy Kray, NW Russian Federation *58°04′N 57°48′E*
117 *P6* **Lysyanka** Cherkas'ka Oblast', C Ukraine *49°15′N 30°50′E*
117 *X6* **Lysychans'k** *Rus.* Lisichansk. Luhans'ka Oblast', E Ukraine *48°52′N 38°27′E*
97 *K17* **Lytham St Anne's** NW England, United Kingdom *53°45′N 03°01′W*
185 *I19* **Lyttelton** South Island, New Zealand *43°35′S 172°44′E*
10 *M17* **Lytton** British Columbia, SW Canada *50°12′N 121°34′W*
119 *L18* **Lyuban'** Minskaya Voblasts', S Belarus *52°48′N 28°00′E*
119 *L18* **Lyubanskaye Vodaskhovishcha** *Rus.* Lyubanskoye Vodokhranilishche. ▣ C Belarus
Lyubanskoye Vodokhranilishche *see* Lyubanskaye Vodaskhovishcha
116 *M5* **Lyubar** Zhytomyrs'ka Oblast', N Ukraine *49°54′N 27°48′E*
Lyubashëvka *see* Lyubashivka
117 *O8* **Lyubashivka** *Rus.* Lyubashëvka. Odes'ka Oblast', SW Ukraine *47°49′N 30°18′E*
119 *I16* **Lyubcha** *Pol.* Lubcz. Hrodzyenskaya Voblasts', W Belarus *53°45′N 26°04′E*
126 *L4* **Lyubertsy** Moskovskaya Oblast', W Russian Federation *55°37′N 38°02′E*
116 *K2* **Lyubeshiv** Volyns'ka Oblast', NW Ukraine *51°46′N 25°33′E*
124 *M14* **Lyubim** Yaroslavskaya Oblast', NW Russian Federation *58°21′N 40°46′E*
114 *K11* **Lyubimets** Haskovo, S Bulgaria *41°51′N 26°03′E*
Lyublin *see* Lublin
116 *I3* **Lyuboml'** *Pol.* Luboml. Volyns'ka Oblast', NW Ukraine *51°12′N 24°01′E*
Lyubotin *see* Lyubotyn
117 *U5* **Lyubotyn** *Rus.* Lyubotin. Kharkivs'ka Oblast', E Ukraine *49°57′N 35°57′E*
126 *I5* **Lyudinovo** Kaluzhskaya Oblast', W Russian Federation *53°52′N 34°28′E*
127 *T2* **Lyuk** Udmurtskaya Respublika, NW Russian Federation *56°55′N 52°45′E*
114 *M9* **Lyulyakovo** *prev.* Keremitlik. Burgas, E Bulgaria *42°53′N 27°05′E*
119 *I18* **Lyusina** *Rus.* Lyusino. Brestskaya Voblasts', SW Belarus *52°38′N 26°31′E*
Lyusino *see* Lyusina

M

138 *G9* **Ma'ād** Irbid, N Jordan *32°37′N 35°36′E*
Ma'ai *see* Luqu
Maalahti *see* Malax
Maale *see* Male'
138 *G13* **Ma'ān** Ma'ān, SW Jordan *30°11′N 35°45′E*
138 *H13* **Ma'ān** *off.* Muḩāfaẓat Ma'ān, *var.* Ma'an, Ma'ān. ⬖ *governorate* S Jordan
93 *M16* **Maaninka** Pohjois-Savo, C Finland *63°10′N 27°19′E*
Maanĭt *see* Bayan, Töv, Mongolia
Maanĭt *see* Hishig Öndör, Bulgan, Mongolia
Ma'ān, Muḩāfaẓat *see* Ma'ān
93 *N15* **Maanselkä** Kainuu, C Finland *63°54′N 28°28′E*
161 *Q8* **Ma'anshan** Anhui, E China *31°45′N 118°32′E*
188 *F16* **Maap** *island* Caroline Islands, W Micronesia
118 *H3* **Maardu** *Ger.* Maart. Harjumaa, NW Estonia *59°28′N 24°56′E*
Ma'aret-en-Nu'man *see* Ma'arrat an Nu'mān
99 *K16* **Maarheeze** Noord-Brabant, SE Netherlands *51°19′N 05°37′E*
Maarianhamina *see* Mariehamn
138 *I4* **Ma'arrat an Nu'mān** *var.* Ma'aret-en-Nu'man, *Fr.* Maarret enn Naamâne. Idlib, NW Syria *35°40′N 36°40′E*
Maarret enn Naamâne *see* Ma'arrat an Nu'mān
98 *I11* **Maarssen** Utrecht, C Netherlands *52°08′N 05°03′E*
Maart *see* Maardu
99 *L17* **Maas** *Fr.* Meuse. ≈ W Europe ***see also*** Meuse
Maas *see* Meuse
99 *M15* **Maasbree** Limburg, SE Netherlands *51°22′N 06°03′E*
99 *L17* **Maaseik** *prev.* Maeseyck. Limburg, NE Belgium *51°05′N 05°48′E*
171 *Q6* **Maasin** Leyte, C Philippines *10°10′N 124°55′E*
99 *L17* **Maasmechelen** Limburg, NE Belgium *50°58′N 05°42′E*
98 *G12* **Maassluis** Zuid-Holland, SW Netherlands *51°55′N 04°15′E*
99 *L18* **Maastricht** *var.* Maestricht; *anc.* Traiectum ad Mosam, Traiectum Tungorum. Limburg, SE Netherlands *50°51′N 05°42′E*
183 *N18* **Maatsuyker Group** *island group* Tasmania, SE Australia
Maba *see* Qujiang
83 *L20* **Mabalane** Gaza, S Mozambique *23°43′S 32°37′E*
25 *V7* **Mabank** Texas, SW USA *32°22′N 96°06′W*
97 *O18* **Mablethorpe** E England, United Kingdom *53°21′N 00°14′E*
171 *V12* **Maboi** Papua Barat, E Indonesia *01°00′S 134°02′E*
83 *M19* **Mabote** Inhambane, S Mozambique *22°03′S 34°09′E*
32 *J10* **Mabton** Washington, NW USA *46°13′N 120°00′W*
83 *H20* **Mabutsane** Southern, S Botswana *24°24′S 23°34′E*
63 *G19* **Macá, Cerro** ▲ S Chile *45°07′S 73°11′W*
60 *Q9* **Macaé** Rio de Janeiro, SE Brazil *22°21′S 41°48′W*
82 *N13* **Macaloge** Niassa, N Mozambique *12°27′S 35°25′E*
Macan *see* Bonerate, Kepulauan
161 *N15* **Macao** *off.* Macao Special Administrative Region, *var.* Macao S.A.R., *Chin.* Aomen Tebie Xingzhengqu, *Port.* Região Administrativa Especial de Macau. Guangdong, SE China *22°06′N 113°30′E*
104 *H9* **Mação** Santarém, C Portugal *39°33′N 08°00′W*
Macao S.A.R. *see* Macao
Macao Special Administrative Region *see* Macao
58 *J11* **Macapá** *state capital* Amapá, N Brazil *0°04′N 51°04′W*
43 *S17* **Macaracas** Los Santos, S Panama *07°46′N 80°31′W*
55 *P6* **Macare, Caño** ≈ NE Venezuela
55 *Q6* **Macareo, Caño** ≈ NE Venezuela
Macarsca *see* Makarska
182 *L12* **Macarthur** Victoria, SE Australia *38°04′S 142°02′E*
MacArthur *see* Ormoc
56 *C7* **Macas** Morona Santiago, SE Ecuador *02°22′S 78°08′W*
Macassar *see* Makassar
59 *Q14* **Macau** Rio Grande do Norte, E Brazil *05°05′S 36°37′W*
Macău *see* Makó, Hungary
Macau, Região Administrativa Especial de *see* Macao
65 *E24* **Macbride Head** *headland* East Falkland, Falkland Islands *51°25′S 57°55′W*
23 *V9* **Macclenny** Florida, SE USA *30°16′N 82°07′W*
97 *L18* **Macclesfield** C England, United Kingdom *53°16′N 02°07′W*
192 *F6* **Macclesfield Bank** *undersea feature* N South China Sea *15°50′N 114°20′E*
MacCluer Gulf *see* Berau, Teluk
181 *N7* **Macdonald, Lake** *salt lake* Western Australia
181 *Q7* **Macdonnell Ranges** ▲▲ Northern Territory, C Australia
96 *K8* **Macduff** NE Scotland, United Kingdom *57°40′N 02°29′W*
104 *I6* **Macedo de Cavaleiros** Bragança, N Portugal *41°31′N 06°57′W*
Macedonia *see* Macedonia, FYR
Macedonia Central *see* Kentrikí Makedonía
Macedonia East and Thrace *see* Anatolikí Makedonía kai Thráki
113 *O19* **Macedonia, FYR** *off.* the Former Yugoslav Republic of Macedonia, *var.* Macedonia, *Mac.* Makedonija, *abbrev.* FYR Macedonia, FYROM. ◆ *republic* SE Europe
Macedonia, the Former Yugoslav Republic of *see* Macedonia, FYR
Macedonia West *see* Dytikí Makedonía
59 *Q16* **Maceió** *state capital* Alagoas, E Brazil *09°40′S 35°44′W*
76 *K15* **Macenta** SE Guinea *08°31′N 09°32′W*
106 *J12* **Macerata** Marche, C Italy *43°18′N 13°27′E*
11 *S11* **MacFarlane** ≈ Saskatchewan, C Canada
182 *H7* **Macfarlane, Lake** *var.* Lake Mcfarlane. ◎ South Australia
Macgillicuddy's Reeks Mountains *see* Macgillycuddy's Reeks
97 *B21* **Macgillycuddy's Reeks** *var.* Macgillicuddy's Reeks Mountains, *Ir.* Na Cruacha Dubha. ▲▲ SW Ireland
11 *X16* **MacGregor** Manitoba, S Canada *49°58′N 98°49′W*
149 *O10* **Mach** Baluchistān, SW Pakistan *29°52′N 67°20′E*
56 *C6* **Machachi** Pichincha, C Ecuador *0°33′S 78°34′W*
83 *M19* **Machaila** Gaza, S Mozambique *22°15′S 32°57′E*
Machaire Fíolta *see* Magherafelt
Machaire Rátha *see* Maghera
81 *I19* **Machakos** Machakos, S Kenya *01°31′S 37°16′E*
81 *I20* **Machakos** ⬖ *county* C Kenya
56 *B8* **Machala** El Oro, SW Ecuador *03°20′S 79°57′W*
83 *J19* **Machaneng** Central, SE Botswana *23°12′S 27°30′E*
83 *M18* **Machanga** Sofala, E Mozambique *20°56′S 35°04′E*
80 *G13* **Machar Marshes** *wetland* SE Sudan
102 *I8* **Machecoul** Loire-Atlantique, NW France *46°59′N 01°51′W*
161 *O8* **Macheng** Hubei, C China *31°10′N 115°00′E*
155 *J16* **Mācherla** Andhra Pradesh, C India *16°29′N 79°25′E*
153 *O11* **Māchhāpuchhre** ▲ C Nepal *28°30′N 83°57′E*
19 *T6* **Machias** Maine, NE USA *44°44′N 67°28′W*
19 *T6* **Machias River** ≈ Maine, NE USA
19 *R3* **Machias River** ≈ Maine, NE USA
64 *P5* **Machico** Madeira, Portugal, NE Atlantic Ocean *32°43′N 16°47′W*
155 *K16* **Machilīpatnam** *var.* Bandar Masulipatnam. Andhra Pradesh, E India *16°12′N 81°11′E*
54 *G5* **Machiques** Zulia, NW Venezuela *10°04′N 72°37′W*
57 *G15* **Machu Picchu** Cusco, C Peru *13°08′S 72°30′W*
83 *M20* **Macia** *var.* Vila de Macia. Gaza, S Mozambique *25°02′S 33°08′E*
Macías Nguema Biyogo *see* Bioco, Isla de
116 *M13* **Măcin** Tulcea, SE Romania *45°15′N 28°09′E*
183 *T4* **Macintyre River** ≈ New South Wales/Queensland, SE Australia
181 *Y7* **Mackay** Queensland, NE Australia *21°10′S 149°10′E*
181 *O7* **Mackay, Lake** *salt lake* Northern Territory/Western Australia
10 *M13* **Mackenzie** British Columbia, W Canada *55°18′N 123°09′W*
8 *I9* **Mackenzie** ≈ Northwest Territories, NW Canada
195 *Y6* **Mackenzie Bay** *bay* Antarctica
10 *J1* **Mackenzie Bay** *bay* NW Canada
2 *D9* **Mackenzie Delta** *delta* Northwest Territories, NW Canada
197 *P8* **Mackenzie King Island** *island* Queen Elizabeth Islands, Northwest Territories, N Canada
8 *H8* **Mackenzie Mountains** ▲▲ Northwest Territories, NW Canada
31 *Q5* **Mackinac, Straits of** ◎ Michigan, N USA
194 *K5* **Mackintosh, Cape** *headland* Antarctica *72°52′S 60°00′W*
11 *R15* **Macklin** Saskatchewan, S Canada *52°19′N 109°51′W*
183 *V6* **Macksville** New South Wales, SE Australia *30°39′S 152°54′E*
183 *V5* **Maclean** New South Wales, SE Australia *29°30′S 153°15′E*
83 *J24* **Maclear** Eastern Cape, SE South Africa *31°05′S 28°22′E*
183 *U6* **Macleay River** ≈ New South Wales, SE Australia
MacLeod *see* Fort Macleod
180 *G9* **Macleod, Lake** ◎ Western Australia
10 *I6* **Macmillan** ≈ Yukon, NW Canada
30 *J12* **Macomb** Illinois, N USA *40°27′N 90°40′W*
107 *B18* **Macomer** Sardegna, Italy, C Mediterranean Sea *40°15′N 08°47′E*
82 *Q13* **Macomia** Cabo Delgado, NE Mozambique *12°15′S 40°06′E*
103 *R10* **Mâcon** *anc.* Matisco, Matisco Ædourum. Saône-et-Loire, C France *46°19′N 04°49′E*
23 *T5* **Macon** Georgia, SE USA *32°49′N 83°41′W*
23 *N4* **Macon** Mississippi, S USA *33°06′N 88°33′W*
27 *U3* **Macon** Missouri, C USA *39°44′N 92°27′W*
22 *J6* **Macon, Bayou** ≈ Arkansas/Louisiana, S USA
82 *G13* **Macondo** Moxico, E Angola *12°31′S 23°45′E*
83 *M16* **Macossa** Manica, C Mozambique *17°51′S 33°54′E*
11 *T12* **Macoun Lake** ◎ Saskatchewan, C Canada
30 *K14* **Macoupin Creek** ≈ Illinois, N USA
Macouria *see* Tonate
83 *N18* **Macovane** Inhambane, SE Mozambique *21°30′S 35°07′E*
183 *N17* **Macquarie Harbour** *inlet* Tasmania, SE Australia
192 *J13* **Macquarie Island** *island* New Zealand, SW Pacific Ocean
183 *T8* **Macquarie, Lake** *lagoon* New South Wales, SE Australia
183 *Q6* **Macquarie Marshes** *wetland* New South Wales, SE Australia
175 *O13* **Macquarie Ridge** *undersea feature* SW Pacific Ocean *57°00′S 159°00′E*
183 *Q6* **Macquarie River** ≈ New South Wales, SE Australia
183 *P17* **Macquarie River** ≈ Tasmania, SE Australia
195 *V5* **Mac. Robertson Land** *physical region* Antarctica
97 *C21* **Macroom** *Ir.* Maigh Chromtha. Cork, SW Ireland *51°54′N 08°57′W*
42 *G5* **Macuelizo** Santa Bárbara, NW Honduras *15°21′N 88°31′W*
182 *G2* **Macumba River** ≈ South Australia
57 *I16* **Macusani** Puno, S Peru *14°05′S 70°24′W*
56 *E8* **Macusari, Río** ≈ N Peru
41 *U15* **Macuspana** Tabasco, SE Mexico *17°43′N 92°36′W*
138 *G10* **Mādabā** *var.* Ma'dabā, Madeba; *anc.* Medeba. Mādabā, NW Jordan *31°44′N 35°48′E*
138 *G11* **Mādabā** *off.* Muḩāfa at Mādabā, *var.* Ma'dabā. ⬖ *governorate* C Jordan
Ma'dabā *see* Mādabā
Ma'dabā *see* Mādabā
Mādabā, Muḩāfa at *see* Mādabā
172 *G2* **Madagascar** *off.* Democratic Republic of Madagascar, *Malg.* Madagasikara; *prev.* Malagasy Republic. ◆ *republic* W Indian Ocean
172 *I5* **Madagascar** *island* W Indian Ocean
128 *L17* **Madagascar Basin** *undersea feature* W Indian Ocean *27°00′S 53°00′E*
Madagascar, Democratic Republic of *see* Madagascar
128 *L16* **Madagascar Plain** *undersea feature* W Indian Ocean *19°00′S 52°00′E*
67 *Y14* **Madagascar Plateau** *var.* Madagascar Ridge, Madagascar Rise, *Rus.* Madagaskarskiy Khrebet. *undersea feature* W Indian Ocean *30°00′S 45°00′E*
Madagascar Rise/Madagascar Ridge *see* Madagascar Plateau
Madagasikara *see* Madagascar
Madagaskarskiy Khrebet *see* Madagascar Plateau
64 *N2* **Madalena** Pico, Azores, Portugal, NE Atlantic Ocean *38°32′N 28°15′W*
77 *Y6* **Madama** Agadez, NE Niger *21°54′N 13°43′E*
114 *J12* **Madan** Smolyan, S Bulgaria *41°29′N 24°56′E*
155 *I19* **Madanapalle** Andhra Pradesh, E India *13°33′N 78°31′E*
186 *D7* **Madang** Madang, N Papua New Guinea *05°14′S 145°45′E*
186 *C6* **Madang** ⬖ *province* N Papua New Guinea
146 *G7* **Madaniyat** *Rus.* Madeniyet. Qoraqalpog'iston Respublikasi, W Uzbekistan *42°48′N 59°00′E*
Madanīyīn *see* Médenine
77 *U11* **Madaoua** Tahoua, SW Niger *14°06′N 06°01′E*
Madaras *see* Vtáčnik
153 *U15* **Madaripur** Dhaka, C Bangladesh *23°09′N 90°11′E*
77 *U12* **Madarounfa** Maradi, S Niger *13°16′N 07°07′E*
Mađarska *see* Hungary
146 *B13* **Madau** Balkan Welaýaty, W Turkmenistan *38°11′N 54°46′E*
186 *H9* **Madau Island** *island* SE Papua New Guinea
19 *S1* **Madawaska** Maine, NE USA *47°19′N 68°19′W*
14 *J13* **Madawaska** ≈ Ontario, SE Canada
Madawaska Highlands *see* Haliburton Highlands
166 *M4* **Madaya** Mandalay, C Myanmar (Burma) *22°12′N 96°05′E*
107 *K17* **Maddaloni** Campania, S Italy *41°03′N 14°23′E*
29 *O3* **Maddock** North Dakota, N USA *47°57′N 99°31′W*
99 *I14* **Made** Noord-Brabant, S Netherlands *51°41′N 04°48′E*
Madeba *see* Mādabā
64 *L9* **Madeira** *var.* Ilha de Madeira. *island* Madeira, Portugal, NE Atlantic Ocean
Madeira, Ilha de *see* Madeira
64 *O5* **Madeira Islands** *Port.* Região Autoónoma da Madeira. ⬖ *autonomous region* Madeira, Portugal, NE Atlantic Ocean
64 *L9* **Madeira Plain** *undersea feature* E Atlantic Ocean
Madeira, Região Autoónoma da *see* Madeira Islands
64 *L9* **Madeira Ridge** *undersea feature* E Atlantic Ocean *35°30′N 15°45′W*
59 *F14* **Madeira, Rio** *var.* Río Madera. ≈ Bolivia/Brazil ***see also*** Madera, Río
Madeira, Rio *see* Madera, Río
101 *J25* **Mädelegabel** ▲ Austria/Germany *47°18′N 10°19′E*
15 *X6* **Madeleine** ≈ Québec, SE Canada
15 *X5* **Madeleine, Cap de la** *headland* Québec, SE Canada *49°13′N 65°20′W*
13 *Q13* **Madeleine, Îles de la** *Eng.* Magdalen Islands. *island group* Québec, E Canada
29 *U10* **Madelia** Minnesota, N USA *44°03′N 94°26′W*
35 *P3* **Madeline** California, W USA *41°02′N 120°28′W*
30 *K3* **Madeline Island** *island* Apostle Islands, Wisconsin, N USA
137 *O15* **Maden** Elazığ, SE Turkey *38°24′N 39°42′E*
145 *V12* **Madeniyet** Vostochnyy Kazakhstan, E Kazakhstan *47°51′N 78°37′E*
Madeniyet *see* Madaniyat
40 *H5* **Madera** Chihuahua, N Mexico *29°10′N 108°10′W*
35 *Q10* **Madera** California, W USA *36°57′N 120°02′W*
56 *L13* **Madera, Río** *Port.* Rio Madeira. ≈ Bolivia/Brazil ***see also*** Madeira, Rio
Madera, Río *see* Madeira, Rio
106 *D6* **Madesimo** Lombardia, N Italy *46°20′N 09°26′E*
141 *O14* **Madhāb, Wādī** *dry watercourse* NW Yemen
153 *R13* **Madhepura** *prev.* Madhipure. Bihār, NE India *25°56′N 86°48′E*
Madhipure *see* Madhepura
153 *Q13* **Madhubani** Bihār, N India *26°21′N 86°05′E*
153 *Q15* **Madhupur** Jhārkhand, NE India *24°17′N 86°38′E*
154 *I10* **Madhya Pradesh** *prev.* Central Provinces and Berar. ⬖ *state* C India
57 *K15* **Madidi, Río** ≈ W Bolivia
155 *F20* **Madikeri** *prev.* Mercara. Karnātaka, W India *12°29′N 75°40′E*
27 *O13* **Madill** Oklahoma, C USA *34°06′N 96°46′W*
79 *G21* **Madimba** Bas-Congo, SW Dem. Rep. Congo *04°58′S 15°08′E*
138 *M4* **Ma'din** Ar Raqqah, C Syria *35°45′N 39°36′E*
Madīnah, Minţaqat al *see* Al Madīnah
76 *M14* **Madinani** NW Ivory Coast *09°37′N 06°57′W*
141 *O17* **Madīnat ash Sha'b** *prev.* Al Ittiḩād. SW Yemen *12°52′N 44°55′E*
138 *K3* **Madīnat ath Thawrah** *var.* Ath Thawrah. Ar Raqqah, N Syria *35°36′N 39°00′E*
173 *O6* **Madingley Rise** *undersea feature* W Indian Ocean
79 *E21* **Madingo-Kayes** Kouilou, S Congo *04°27′S 11°43′E*
79 *F21* **Madingou** Bouenza, S Congo *04°10′S 13°33′E*
Madioen *see* Madiun
23 *U8* **Madison** Florida, SE USA *30°27′N 83°24′W*
23 *T3* **Madison** Georgia, SE USA *33°37′N 83°28′W*
31 *P15* **Madison** Indiana, N USA *38°44′N 85°22′W*
27 *P6* **Madison** Kansas, C USA *45′01′N 96′11′W*
19 *Q6* **Madison** Maine, NE USA *44°48′N 69°52′W*
29 *S9* **Madison** Minnesota, N USA *45°00′N 96°12′W*
22 *K5* **Madison** Mississippi, S USA *32°27′N 90°07′W*
29 *Q14* **Madison** Nebraska, C USA *41°49′N 97°27′W*
29 *R10* **Madison** South Dakota, N USA *44°00′N 97°06′W*
21 *V5* **Madison** Virginia, NE USA *38°23′N 78°16′W*
21 *Q5* **Madison** West Virginia, NE USA *38°03′N 81°50′W*
30 *L9* **Madison** *state capital* Wisconsin, N USA *43°04′N 89°22′W*
21 *T6* **Madison Heights** Virginia, NE USA *37°25′N 79°07′W*
20 *I6* **Madisonville** Kentucky, S USA *37°20′N 87°30′W*
20 *M10* **Madisonville** Tennessee, S USA *35°31′N 84°21′W*
25 *V9* **Madisonville** Texas, SW USA *30°58′N 95°56′W*
Madisonville *see* Taiohae
169 *R16* **Madiun** *prev.* Madioen. Jawa, C Indonesia *07°37′S 111°33′E*
Madjene *see* Majene
14 *J14* **Madoc** Ontario, SE Canada *44°31′N 77°27′W*
Madoera *see* Madura, Pulau
81 *J18* **Mado Gashi** Garissa, E Kenya *0°40′N 39°09′E*
159 *R11* **Madoi** *var.* Huanghe; *prev.* Huangheyan. Qinghai, C China *34°53′N 98°07′E*
189 *O13* **Madolenihmw** Pohnpei, E Micronesia
118 *I9* **Madona** *Ger.* Modohn. E Latvia *56°51′N 26°10′E*
107 *J23* **Madonie** ▲▲ Sicilia, Italy, C Mediterranean Sea
141 *Y11* **Madrakah, Ra's** *headland* E Oman *18°56′N 57°54′E*
32 *I12* **Madras** Oregon, NW USA *44°39′N 121°08′W*
Madras *see* Chennai
Madras *see* Tamil Nādu
57 *H14* **Madre de Dios** *off.* Departamento de Madre de Dios. ⬖ *department* E Peru
Madre de Dios, Departamento de *see* Madre de Dios
63 *F22* **Madre de Dios, Isla** *island* S Chile
57 *J14* **Madre de Dios, Río** ≈ Bolivia/Peru
0 *H15* **Madre del Sur, Sierra** ▲▲ S Mexico
41 *Q9* **Madre, Laguna** *lagoon* NE Mexico
25 *T16* **Madre, Laguna** *lagoon* Texas, SW USA
37 *Q12* **Madre Mount** ▲ New Mexico, SW USA *34°18′N 107°54′W*
0 *H13* **Madre Occidental, Sierra** *var.* Western Sierra Madre. ▲▲ C Mexico
0 *H13* **Madre Oriental, Sierra** *var.* Eastern Sierra Madre. ▲▲ C Mexico
41 *U17* **Madre, Sierra** *var.* Sierra de Soconusco. ▲▲ Guatemala/Mexico
37 *R2* **Madre, Sierra** ▲▲ Colorado/Wyoming, C USA
105 *N8* **Madrid** ● (Spain) Madrid, C Spain *40°25′N 03°43′W*
29 *V14* **Madrid** Iowa, C USA *41°52′N 93°49′W*
105 *N7* **Madrid** ⬖ *autonomous community* C Spain
105 *N10* **Madridejos** Castilla-La Mancha, C Spain *39°29′N 03°32′W*
104 *L7* **Madrigal de las Altas Torres** Castilla y León, N Spain *41°05′N 05°00′W*
104 *K10* **Madrigalejo** Extremadura, W Spain *39°08′N 05°36′W*
34 *L3* **Mad River** ≈ California, W USA
42 *J8* **Madriz** ⬖ *department* NW Nicaragua
104 *K10* **Madroñera** Extremadura, W Spain *39°25′N 05°46′W*
181 *N12* **Madura** Western Australia *31°52′S 127°01′E*
Madura *see* Madurai
155 *H22* **Madurai** *prev.* Madura, Mathurai. Tamil Nādu, S India *09°55′N 78°07′E*
169 *S16* **Madura, Pulau** *prev.* Madoera. *island* C Indonesia
169 *S16* **Madura, Selat** *strait* C Indonesia

◆ Country
● Country Capital
◇ Dependent Territory
○ Dependent Territory Capital
⬖ Administrative Regions
✈ International Airport
▲ Mountain
▲▲ Mountain Range
Volcano
≈ River
◎ Lake
▣ Reservoir

MADZHALIS – MALKO SHARKOVO, YAZOVIR

127 Q17 Madzhalis Respublika Dagestan, SW Russian Federation 42°12′N 47°46′E
114 K12 Madzharovo Haskovo, S Bulgaria 41°36′N 25°52′E
83 M14 Madzimoyo Eastern, E Zambia 13°42′S 32°34′E
165 O12 Maebashi var. Maebasi, Mayebashi. Gunma, Honshū, S Japan 36°24′N 139°02′E
Maebasi see Maebashi
167 O6 Mae Chan Chiang Rai, NW Thailand 20°13′N 99°52′E
167 N7 Mae Hong Son var. Maehongson, Muai To. Mae Hong Son, NW Thailand 19°16′N 97°56′E
Maehongson see Mae Hong Son
Mae Nam Khong see Mekong
167 Q7 Mae Nam Nan ≈ NW Thailand
167 O10 Mae Nam Tha Chin ≈ W Thailand
167 P7 Mae Nam Yom ≈ W Thailand
37 O3 Maeser Utah, W USA 40°28′N 109°35′W
Maeseyck see Maaseik
167 N9 Mae Sot var. Ban Mae Sot. Tak, W Thailand 16°44′N 98°32′E
44 H8 Maestra, Sierra ▲▲ E Cuba
Maestricht see Maastricht
167 O7 Mae Suai var. Ban Mae Suai. Chiang Rai, NW Thailand 19°43′N 99°30′E
167 O7 Mae Tho, Doi ▲ NW Thailand 18°56′N 99°20′E
172 I4 Maevatanana Mahajanga, C Madagascar 16°57′S 46°50′E
187 R13 Maéwo prev. Aurora. island C Vanuatu
171 S11 Mafa Pulau Halmahera, E Indonesia 0°01′N 127°50′E
83 I23 Mafeteng W Lesotho 29°48′S 27°15′E
99 J21 Maffe Namur, SE Belgium 50°21′N 05°19′E
183 P12 Maffra Victoria, SE Australia 37°59′S 147°03′E
81 K23 Mafia island E Tanzania
81 J23 Mafia Channel sea waterway E Tanzania
83 I21 Mafikeng off. Mahikeng. North-West, N South Africa 25°53′S 25°39′E
60 J12 Mafra Santa Catarina, S Brazil 26°08′S 49°47′W
104 F10 Mafra Lisboa, C Portugal 38°57′N 09°19′W
143 Q17 Mafraq Abū Ẓaby, C United Arab Emirates 24°21′N 54°33′E
Mafraq/Muḥāfaẓat al Mafraq see Al Mafraq
123 T10 Magadan Magadanskaya Oblast', E Russian Federation 59°38′N 150°50′E
123 T9 Magadanskaya Oblast' ◈ province E Russian Federation
108 G11 Magadino Ticino, S Switzerland 46°09′N 08°50′E
63 G23 Magallanes var. Región de Magallanes y de la Antártica Chilena. ◈ region S Chile
Magallanes see Punta Arenas
Magallanes, Estrecho de see Magellan, Strait of
Magallanes y de la Antártica Chilena, Región de see Magallanes
14 I10 Maganasipi, Lac ◎ Québec, SE Canada
54 F6 Magangué Bolívar, N Colombia 09°14′N 74°46′W
191 Y13 Magareva var. Mangareva. island Îles Tuamotu, SE French Polynesia
77 V12 Magaria Zinder, S Niger 13°00′N 08°55′E
186 F10 Magarida Central, SW Papua New Guinea 10°10′S 149°21′E
171 O2 Magat ≈ Luzon, N Philippines
27 T11 Magazine Mountain ▲ Arkansas, C USA 35°10′N 93°38′W
76 I15 Magburaka C Sierra Leone 08°44′N 11°57′W
123 Q13 Magdagachi Amurskaya Oblast', SE Russian Federation 53°25′N 125°41′E
62 O12 Magdalena Buenos Aires, E Argentina 35°05′S 57°30′W
57 M15 Magdalena El Beni, N Bolivia 13°22′S 64°07′W
40 F4 Magdalena Sonora, NW Mexico 30°38′N 110°59′W
37 Q13 Magdalena New Mexico, SW USA 34°07′N 107°14′W
54 F5 Magdalena off. Departamento del Magdalena. ◈ province N Colombia
40 E9 Magdalena, Bahía bay W Mexico
Magdalena, Departamento del see Magdalena
63 G19 Magdalena, Isla island Archipiélago de los Chonos, S Chile
40 D8 Magdalena, Isla island NW Mexico
47 P6 Magdalena, Río ≈ C Colombia
40 F4 Magdalena, Río ≈ NW Mexico
Magdalen Islands see Madeleine, Îles de la
147 N14 Magdanly Rus. Govurdak; prev. gowurdak, Guardak. Lebap Welaýaty, E Turkmenistan 37°50′N 66°06′E
100 L13 Magdeburg Sachsen-Anhalt, C Germany 52°08′N 11°39′E
22 L6 Magee Mississippi, S USA 31°52′N 89°43′W
169 Q16 Magelang Jawa, C Indonesia 07°28′S 110°11′E
192 K7 Magellan Rise undersea feature C Pacific Ocean
63 H24 Magellan, Strait of Sp. Estrecho de Magallanes. strait Argentina/Chile
106 D7 Magenta Lombardia, NW Italy 45°28′N 08°52′E
Mageroy see Magerøya
92 K7 Magerøya var. Mageroy, Lapp. Máhkarávju. island N Norway
164 C17 Mage-shima island Nansei-shotō, SW Japan
108 G11 Maggia Ticino, S Switzerland 46°15′N 08°42′E
108 G10 Maggia ≈ SW Switzerland
Maggiore, Lago see Maggiore, Lake
106 C6 Maggiore, Lake It. Lago Maggiore. ◎ Italy/Switzerland
44 I12 Maggotty W Jamaica 18°09′N 77°46′W
76 I10 Maghama Gorgol, S Mauritania 15°31′N 12°50′W
97 F14 Maghera Ir. Machaire Rátha. C Northern Ireland, United Kingdom 54°51′N 06°40′W
97 F15 Magherafelt Ir. Machaire Fíolta. C Northern Ireland, United Kingdom 54°45′N 06°36′W
188 H6 Magicienne Bay bay Saipan, S Northern Mariana Islands
105 O13 Magina ▲ S Spain 37°43′N 03°24′W
81 H24 Magingo Ruvuma, S Tanzania 09°57′S 35°23′E
112 H11 Maglaj ◈ Federacija Bosne I Hercegovine, N Bosnia and Herzegovina
107 Q19 Maglie Puglia, SE Italy 40°07′N 18°18′E
114 K10 Maglizh var. Müglizh. Stara Zagora, C Bulgaria 42°36′N 25°32′E
36 L2 Magna Utah, W USA 40°42′N 112°06′W
Magnesia see Manisa
14 G12 Magnetawan ≈ Ontario, S Canada
27 T14 Magnolia Arkansas, C USA 33°17′N 93°16′W
22 K7 Magnolia Mississippi, S USA 31°08′N 90°27′W
25 V10 Magnolia Texas, SW USA 30°12′N 95°46′W
Magnolia State see Mississippi
95 J15 Magnor Hedmark, S Norway 59°57′N 12°14′E
187 Y14 Mago prev. Mango. island Lau Group, E Fiji
83 L15 Màgoé Tete, NW Mozambique 15°50′S 31°42′E
15 Q13 Magog Québec, SE Canada 45°16′N 72°09′W
83 J15 Magoye Southern, S Zambia 16°00′S 27°34′E
41 Q12 Magozal Veracruz-Llave, C Mexico 21°33′N 97°57′W
14 B7 Magpie ≈ Ontario, S Canada
11 Q17 Magrath Alberta, SW Canada 49°27′N 112°52′W
105 R10 Magre ≈ Valenciana, E Spain
76 I9 Magta' Lahjar var. Magta Lahjar, Magta' Lahjar, Magtá Lahjar. Brakna, SW Mauritania 17°27′N 13°07′W
146 D12 Magtymguly prev. Garrygala, Rus. Kara-Kala. Balkan Welaýaty, W Turkmenistan 38°27′N 56°15′E
83 L20 Magude Maputo, S Mozambique 25°02′S 32°40′E
77 Y12 Magumeri Borno, NE Nigeria 12°07′N 12°48′E
189 O14 Magur Islands island group Caroline Islands, C Micronesia
166 L6 Magway var. Magwe. Magway, W Myanmar (Burma) 20°08′N 94°55′E
166 L6 Magway var. Magwe. ◈ region C Myanmar (Burma)
Magwe see Magway
Magyar-Becse see Bečej
Magyarkanizsa see Kanjiža
Magyarország see Hungary
Magyarzombor see Zimbor
142 J4 Mahābād var. Mehabad; prev. Sāūjbulāgh. Āžarbāyjān-e Gharbī, NW Iran 36°44′N 45°44′E
172 H5 Mahabo Toliara, W Madagascar 20°22′S 44°39′E
Maha Chai see Samut Sakhon
155 D14 Mahād Mahārāshtra, W India 18°04′N 73°21′E
81 N17 Mahadday Weyne Shabeellaha Dhexe, C Somalia 02°55′N 45°30′E
79 Q17 Mahagi Orientale, NE Dem. Rep. Congo 02°16′N 30°59′E
Mahāil see Muhāyil
172 I4 Mahajamba seasonal river NW Madagascar
152 G10 Mahājan Rājasthān, NW India 28°47′N 73°50′E
172 I3 Mahajanga var. Majunga. Mahajanga, NW Madagascar 15°40′S 46°20′E
172 I3 Mahajanga ◈ province W Madagascar
172 I3 Mahajanga ✈ Mahajanga, NW Madagascar 15°40′S 46°20′E
169 U10 Mahakam, Sungai var. Koetai, Kutai. ≈ Borneo, C Indonesia
83 I19 Mahalapye var. Mahalatswe. Central, SE Botswana 23°02′S 26°53′E
Mahalatswe see Mahalapye
Mahalla el Kubra see El Mahalla el Kubra
171 O13 Mahalona Sulawesi, C Indonesia 02°37′S 121°26′E
Mahameru see Semeru, Gunung
143 S11 Māhān Kermān, E Iran 30°00′N 57°00′E
154 N12 Mahānadi ≈ E India
172 J5 Mahanoro Toamasina, E Madagascar 19°53′S 48°48′E
153 P13 Mahārājganj Bihār, N India 26°07′N 84°31′E
154 G13 Mahārāshtra ◈ state W India
172 I4 Mahavavy seasonal river N Madagascar
155 K24 Mahaweli Ganga ≈ C Sri Lanka
Mahbés see El Mahbas
155 J15 Mahbūbābād Telangana, E India 17°35′N 80°00′E
155 H16 Mahbūbnagar Telangana, C India 16°46′N 78°01′E
140 M8 Mahd adh Dhahab Al Madīnah, W Saudi Arabia 23°33′N 40°56′E
55 S9 Mahdia C Guyana 05°16′N 59°08′W
75 N6 Mahdia var. Al Mahdīyah, Mehdia. NE Tunisia 35°14′N 11°06′E
155 F20 Mahe Fr. Mahé; prev. Mayyali. Puducherry, SW India 11°41′N 75°31′E
172 I16 Mahé ✈ Mahé, NE Seychelles 04°37′S 55°27′E
172 H16 Mahé island Inner Islands, NE Seychelles
Mahé see Mahe
173 Y17 Mahebourg SE Mauritius 20°24′S 57°42′E
152 I11 Mahendragarh prev. Mohendergarh. Haryāna, N India 28°17′N 76°14′E
152 L10 Mahendranagar Far Western, W Nepal 28°58′N 80°13′E
81 I23 Mahenge Morogoro, SE Tanzania 08°41′S 36°41′E
185 F22 Maheno Otago, South Island, New Zealand 45°10′S 170°51′E
154 D9 Mahesāna Gujarāt, W India 23°37′N 72°28′E
154 F11 Maheshwar Madhya Pradesh, C India 22°11′N 75°40′E
153 V17 Maheskhali Island var. Maiskhal Island. island SE Bangladesh
151 F14 Mahi ≈ N India
184 Q10 Mahia Peninsula peninsula North Island, New Zealand
Mahikeng see Mafikeng
119 O16 Mahilyow Rus. Mogilëv. Mahilyowskaya Voblasts', E Belarus 53°55′N 30°23′E
119 M16 Mahilyowskaya Voblasts' Rus. Mogilëvskaya Oblast'. ◈ province E Belarus
191 P7 Mahina Tahiti, W French Polynesia 17°29′S 149°27′W
185 E23 Mahinerangi, Lake ◎ South Island, New Zealand
Máhkarávju see Magerøya
83 L22 Mahlabatini KwaZulu/Natal, E South Africa 28°15′S 31°28′E
166 L5 Mahlaing Mandalay, C Myanmar (Burma) 21°03′N 95°44′E
109 X8 Mahldorf Steiermark, SE Austria 46°54′N 15°55′E
Maḥmūd-e 'Erāqī see Maḥmūd-e Rāqī
149 R4 Maḥmūd-e Rāqī var. Mahmūd-e 'Erāqī. Kāpīsā, NE Afghanistan 35°01′N 69°20′E
Mahmudiya see Al Maḥmūdīyah
29 S5 Mahnomen Minnesota, N USA 47°19′N 95°58′W
152 K14 Mahoba Uttar Pradesh, N India 25°18′N 79°53′E
Mahón see Maó
18 D14 Mahoning Creek Lake ◎ Pennsylvania, NE USA
105 Q10 Mahora Castilla-La Mancha, C Spain 39°13′N 01°44′W
Mähren see Moravia
Mährisch-Budwitz see Moravské Budějovice
Mährisch-Kromau see Moravský Krumlov
Mährisch-Neustadt see Uničov
Mährisch-Schönberg see Šumperk
Mährisch-Trübau see Moravská Třebová
Mährisch-Weisskirchen see Hranice
Māh-Shahr see Bandar-e Māhshahr
79 N19 Mahulu Maniema, E Dem. Rep. Congo 01°04′S 27°10′E
154 C12 Mahuva Gujarāt, W India 21°06′N 71°46′E
114 N11 Mahya Dağı ▲ NW Turkey 41°47′N 27°34′E
105 T6 Maials var. Mayals. Cataluña, NE Spain 41°22′N 00°30′E
191 O2 Maiana prev. Hall Island. atoll Tungaru, W Kiribati
191 S11 Maiao var. Tapuaemanu, Tubuai-Manu. island Îles du Vent, W French Polynesia
54 H4 Maicao La Guajira, N Colombia 11°23′N 72°16′W
Mai Ceu/Mai Chio see Maych'ew
103 U8 Maîche Doubs, E France 47°15′N 06°43′E
149 Q5 Maīdān Shahr var. Maydān Shahr; prev. Meydān Shahr. Wardak, E Afghanistan 34°27′N 68°48′E
97 N22 Maidenhead S England, United Kingdom 51°32′N 00°44′W
11 S15 Maidstone Saskatchewan, S Canada 53°06′N 109°21′W
97 P22 Maidstone SE England, United Kingdom 51°17′N 00°31′E
77 Y13 Maiduguri Borno, NE Nigeria 11°51′N 13°10′E
108 I8 Maienfeld Sankt Gallen, NE Switzerland 47°01′N 09°30′E
116 J12 Măieruş Hung. Szászmagyarós. Braşov, C Romania 45°55′N 25°30′E
Maigh Chromtha see Macroom
Maigh Eo see Mayo
55 N9 Maigualida, Sierra ▲▲ S Venezuela
154 K9 Maihar Madhya Pradesh, C India 24°18′N 80°46′E
154 K11 Maikala Range ▲▲ C India
67 T10 Maiko ≈ W Dem. Rep. Congo
Mailand see Milano
152 L11 Mailāni Uttar Pradesh, N India 28°17′N 80°20′E
149 U10 Māilsi Punjab, E Pakistan 29°46′N 72°15′E
147 R8 Maimak Talasskaya Oblast', NW Kyrgyzstan 42°40′N 71°12′E
Māimāna see Maīmanah
148 M3 Maīmanah var. Maimāna, Maymana; prev. Meymaneh. Fāryāb, NW Afghanistan 35°57′N 64°48′E
171 V13 Maimawa Papua Barat, E Indonesia 03°21′S 133°36′E
Maimuna see Al Maymūnah
101 G18 Main ≈ C Germany
115 F22 Maïna ancient monument Pelopónnisos, S Greece
115 E20 Maínalo ▲ S Greece
101 L22 Mainburg Bayern, SE Germany 48°40′N 11°48′E
Main Camp see Banana
14 E12 Main Channel lake channel Ontario, S Canada
79 I20 Mai-Ndombe, Lac prev. Lac Léopold II. ◎ W Dem. Rep. Congo
101 K20 Main-Donau-Kanal canal SE Germany
19 R6 Maine off. State of Maine, also known as Lumber State, Pine Tree State. ◈ state NE USA
102 K6 Maine cultural region NW France
102 J7 Maine-et-Loire ◈ department NW France
19 Q9 Maine, Gulf of gulf NE USA
77 X12 Maïné-Soroa Diffa, SE Niger 13°14′N 12°00′E
167 N2 Maingkwan var. Mungkawn. Kachin State, N Myanmar (Burma) 26°20′N 96°37′E
Main Island see Bermuda
Mainistir Fhear Maí see Fermoy
Mainistirna Búille see Boyle
Mainistir na Corann see Midleton
Mainistir na Féile see Abbeyfeale
96 J5 Mainland island N Scotland, United Kingdom
96 L2 Mainland island NE Scotland, United Kingdom
159 P16 Mainling var. Tungdor. Xizang Zizhiqu, W China 29°12′N 94°06′E
152 K12 Mainpuri Uttar Pradesh, N India 27°14′N 79°01′E
103 N5 Maintenon Eure-et-Loir, C France 48°35′N 01°34′E
172 H4 Maintirano Mahajanga, W Madagascar 18°01′S 44°03′E
93 M15 Mainua Kainuu, C Finland 64°05′N 27°28′E
101 G18 Mainz Fr. Mayence. Rheinland-Pfalz, SW Germany 50°00′N 08°16′E
76 I9 Maio var. Vila do Maio. Maio, S Cape Verde 15°07′N 23°12′W
76 E10 Maio var. Mayo. island Ilhas de Sotavento, SE Cape Verde
62 G12 Maipo, Río ≈ C Chile
62 H12 Maipo, Volcán W Argentina 34°09′S 69°51′W
61 E22 Maipú Buenos Aires, E Argentina 36°52′S 57°52′W
62 I11 Maipú Mendoza, E Argentina 33°00′S 68°46′W
62 H11 Maipú Santiago, C Chile 33°30′S 70°52′W
106 A9 Maira ≈ NW Italy
108 I10 Maira It. Mera. ≈ Italy/Switzerland
153 V12 Mairābāri Assam, NE India 26°28′N 92°22′E
44 K7 Maisí Guantánamo, E Cuba 20°13′N 74°08′W
118 H13 Maišiagala Vilnius, SE Lithuania 54°52′N 25°03′E
Maiskhal Island see Maheskhali Island
167 N13 Mai Sombun Chumphon, SW Thailand 10°49′N 99°13′E
Mai Son see Hat Lot
Maisur see Karnātaka, India
Maisur see Mysore, India
183 T8 Maitland New South Wales, SE Australia 32°33′S 151°33′E
182 I9 Maitland South Australia 34°21′S 137°42′E
14 F15 Maitland ≈ Ontario, S Canada
195 R1 Maitri Indian research station Antarctica 70°03′S 08°59′E
159 N15 Maizhokunggar Xizang Zizhiqu, W China 29°50′N 91°40′E
43 O10 Maíz, Islas del var. Corn Islands. island group SE Nicaragua
164 J12 Maizuru Kyōto, Honshū, SW Japan 35°30′N 135°20′E
54 F6 Majagual Sucre, N Colombia 08°36′N 74°39′W
41 Z13 Majahual Quintana Roo, E Mexico 18°43′N 87°43′W
Mäjeej see Mejit Island
171 N13 Majene prev. Madjene. Sulawesi, C Indonesia 03°33′S 118°59′E
43 V15 Majé, Serranía de ▲▲ E Panama
112 I11 Majevica ▲▲ NE Bosnia and Herzegovina
81 H15 Maji Southern Nationalities, S Ethiopia 06°11′N 35°32′E
141 X7 Majīs NW Oman 24°25′N 56°34′E
Majorca see Mallorca
105 X9 Major, Puig ▲ Mallorca, Spain, W Mediterranean Sea 39°50′N 2°50′E
Mājro see Majuro Atoll
Majunga see Mahajanga
189 Y3 Majuro ✈ Majuro Atoll, SE Marshall Islands 07°05′N 171°08′E
189 Y2 Majuro Atoll var. Mājro. atoll Ratak Chain, SE Marshall Islands
189 X2 Majuro Lagoon lagoon Majuro Atoll, SE Marshall Islands
76 H11 Maka C Senegal 13°40′N 14°12′W
79 F20 Makabana Niari, SW Congo 03°28′S 12°36′E
38 D9 Mākaha var. Makaha. O'ahu, Hawaii, USA, C Pacific Ocean 21°28′N 158°13′W
38 B8 Makahū'ena Point var. Makahuena Point. headland Kaua'i, Hawai'i, USA 21°52′N 159°28′W
38 D9 Makakilo City O'ahu, Hawaii, USA, C Pacific Ocean 21°21′N 158°05′W
83 H18 Makalamabedi Central, C Botswana 20°19′S 23°51′E
Makale see Mek'elē
158 K17 Makalu Chin. Makaru Shan. ▲ China/Nepal 27°53′N 87°09′E
81 G23 Makampi Mbeya, S Tanzania 08°00′S 33°17′E
Makanchi see Makanshy
145 X12 Makanshy prev. Makanchi. Vostochnyy Kazakhstan, E Kazakhstan 46°47′N 82°00′E
42 M8 Makantaka Región Autónoma Atlántico Norte, NE Nicaragua 13°13′N 84°04′W
190 B16 Makapu Point headland W Niue 18°59′S 169°56′E
185 C24 Makarewa Southland, South Island, New Zealand 46°17′S 168°16′E
117 O4 Makariv Kyyivs'ka Oblast', N Ukraine 50°28′N 29°49′E
185 D20 Makarora ≈ South Island, New Zealand
123 T13 Makarov Ostrov Sakhalin, Sakhalinskaya Oblast', SE Russian Federation 48°24′N 142°37′E
197 R9 Makarov Basin undersea feature Arctic Ocean
192 I5 Makarov Seamount undersea feature W Pacific Ocean 29°30′N 153°30′E
113 F15 Makarska It. Macarsca. Split-Dalmacija, SE Croatia 43°18′N 17°00′E
Makaru Shan see Makalu
125 O15 Makar'yev Kostromskaya Oblast', NW Russian Federation 57°52′N 43°46′E
82 L11 Makasa Northern, NE Zambia 09°42′S 31°54′E
Makasar see Makassar
Makasar, Selat see Makassar Straits
170 M14 Makassar var. Macassar, Makasar; prev. Ujungpandang. Sulawesi, C Indonesia 05°09′S 119°28′E
192 F7 Makassar Straits Ind. Makasar Selat. strait C Indonesia
144 G12 Makat Kaz. Maqat. Atyrau, SW Kazakhstan 47°40′N 53°28′E
191 T10 Makatea island Îles Tuamotu, C French Polynesia
139 U7 Makātū Diyālá, E Iraq 33°55′N 45°25′E
172 H6 Makay var. Massif du Makay. ▲▲ SW Madagascar
Makay, Massif du see Makay
114 J12 Makaza pass Bulgaria/Greece
Makedonija see Macedonia, FYR
190 B16 Makefu W Niue 18°59′S 169°55′W
191 V10 Makemo atoll Îles Tuamotu, C French Polynesia
76 I15 Makeni C Sierra Leone 08°57′N 12°02′W
Makenzen see Orlyak
Makeyevka see Makiyivka
127 Q16 Makhachkala prev. Petrovsk-Port. Respublika Dagestan, SW Russian Federation 42°58′N 47°30′E
Makhado see Louis Trichardt
144 F11 Makhambet Atyrau, W Kazakhstan 47°35′N 51°35′E
Makharadze see Ozurgeti
139 W13 Makhfar al Buşayyah Al Muthannā, S Iraq 30°09′N 46°09′E
Makhmūr see Mexmûr
138 I11 Makhrūq, Wadi al dry watercourse E Jordan
139 R4 Makḥūl, Jabal ▲ C Iraq
141 R13 Makhyah, Wādī dry watercourse N Yemen
171 V13 Maki Papua Barat, E Indonesia 03°00′S 134°10′E
185 G21 Makikihi Canterbury, South Island, New Zealand 44°36′S 171°09′E
191 O2 Makin prev. Pitt Island. atoll Tungaru, W Kiribati
81 I20 Makindu Makueni, S Kenya 02°15′S 37°49′E
145 Q8 Makinsk Akmola, N Kazakhstan 52°40′N 70°28′E
Makira see Makira-Ulawa
Makira see San Cristobal
187 N10 Makira-Ulawa prev. Makira. ◈ province SE Solomon Islands
117 X8 Makiyivka Rus. Makeyevka; prev. Dmitriyevsk. Donets'ka Oblast', E Ukraine 47°57′N 37°47′E
140 L10 Makkah Eng. Mecca. Makkah, W Saudi Arabia 21°28′N 39°50′E
140 M10 Makkah var. Minṭaqat Makkah. ◈ province W Saudi Arabia
Makkah, Minṭaqat see Makkah
13 R7 Makkovik Newfoundland and Labrador, NE Canada 55°06′N 59°07′W
98 K6 Makkum Fryslân, N Netherlands 53°03′N 05°25′E
111 M25 Makó Rom. Macău. Csongrád, SE Hungary 46°14′N 20°28′E
14 G9 Makobe Lake ◎ Ontario, S Canada
79 F18 Makokou Ogooué-Ivindo, NE Gabon 0°38′N 12°47′E
81 G23 Makongolosi Mbeya, S Tanzania 08°24′S 33°09′E
81 E19 Makota SW Uganda 0°37′S 30°12′E
79 G18 Makoua Cuvette, C Congo 0°01′S 15°40′E
110 M10 Maków Mazowiecki Mazowieckie, C Poland 52°51′N 21°06′E
111 K17 Maków Podhalański Małopolskie, S Poland 49°43′N 19°40′E
143 V14 Makran cultural region Iran/Pakistan
152 G12 Makrāna Rājasthān, N India 27°02′N 74°44′E
143 U15 Makran Coast coastal region SE Iran
119 F20 Makrany Rus. Mokrany. Brestskaya Voblasts', SW Belarus 51°50′N 24°15′E
Makrinoros see Makrynóros
115 H20 Makrónisosi island Kykládes, Greece, Aegean Sea
115 D17 Makrynóros var. Makrinoros. ▲▲ C Greece
115 G19 Makryplági ▲ C Greece 38°00′N 23°06′E
Maksamaa see Maxmo
Maksatha see Maksatikha
124 J15 Maksatikha var. Maksatha, Maksaticha. Tverskaya Oblast', W Russian Federation 57°49′N 35°46′E
154 G10 Maksi Madhya Pradesh, C India 23°20′N 76°36′E
142 I1 Mākū Āžarbāyjān-e Gharbī, NW Iran 39°20′N 44°50′E
81 I20 Makueni ◈ county S Kenya
153 Y11 Mākum Assam, NE India 27°28′N 95°28′E
Makun see Makung
161 R14 Makung prev. Mako, Makun. W Taiwan 23°35′N 119°35′E
164 B16 Makurazaki Kagoshima, Kyūshū, SW Japan 31°16′N 130°18′E
77 V15 Makurdi Benue, C Nigeria 07°42′N 08°36′E
38 L17 Makushin Volcano ▲ Unalaska Island, Alaska, USA 53°53′N 166°55′W
83 K16 Makwiro Mashonaland West, N Zimbabwe 17°58′S 30°25′E
57 D15 Mala Lima, W Peru 12°40′S 76°36′W
93 I14 Malå Västerbotten, N Sweden 65°12′N 18°45′E
Mala see Mallow, Ireland
Mala see Malaita, Solomon Islands
171 P8 Malabang Mindanao, S Philippines 07°35′N 124°03′E
155 E21 Malabār Coast coast SW India
79 C16 Malabo prev. Santa Isabel. ● (Equatorial Guinea) Isla de Bioco, NW Equatorial Guinea 03°43′N 08°52′E
79 C16 Malabo ✈ Isla de Bioco, N Equatorial Guinea 03°44′N 08°51′E
Malaca see Málaga
Malacca see Melaka
168 I7 Malacca, Strait of Ind. Selat Malaka. strait Indonesia/Malaysia
Malacka see Malacky
111 G20 Malacky Hung. Malacka. Bratislavský Kraj, W Slovakia 48°26′N 17°01′E
33 R16 Malad City Idaho, NW USA 42°10′N 112°16′W
117 Q4 Mala Divytsya Chernihivs'ka Oblast', N Ukraine 50°40′N 32°13′E
119 J15 Maladzyechna Pol. Molodeczno, Rus. Molodechno. Minskaya Voblasts', C Belarus 54°19′N 26°51′E
190 D12 Malaee Île Futuna, N Wallis and Futuna
190 G12 Mala'etoli Île Uvea, E Wallis and Futuna
54 G8 Málaga Santander, C Colombia 06°44′N 72°45′W
104 M15 Málaga anc. Malaca. Andalucía, S Spain 36°43′N 04°25′W
37 V15 Malaga New Mexico, SW USA 32°10′N 104°04′W
104 L15 Málaga ◈ province Andalucía, S Spain
104 M15 Málaga ✈ Andalucía, S Spain 36°38′N 04°36′W
Malagasy Republic see Madagascar
105 N10 Malagón Castilla-La Mancha, C Spain 39°10′N 03°51′W
97 G18 Malahide Ir. Mullach Íde. Dublin, E Ireland 53°27′N 06°09′W
187 N9 Malaita off. Malaita Province. ◈ province N Solomon Islands
187 N8 Malaita var. Mala. island N Solomon Islands
Malaita Province see Malaita
80 F13 Malakal Upper Nile, NE South Sudan 09°31′N 31°40′E
112 C10 Mala Kapela ▲▲ NW Croatia
25 V7 Malakoff Texas, SW USA 32°10′N 96°00′W
Malakula see Malekula
149 V7 Malakwāl var. Mālikwāla. Punjab, E Pakistan 32°32′N 73°18′E
186 E7 Malalamai Madang, W Papua New Guinea 05°49′S 146°44′E
171 O13 Malamala Sulawesi, C Indonesia 03°21′S 120°58′E
169 S17 Malang Jawa, C Indonesia 07°59′S 112°45′E
83 O14 Malanga Niassa, N Mozambique 13°27′S 36°05′E
Malange see Malanje
92 I9 Malangen sound N Norway
82 C11 Malanje var. Malange. Malanje, NW Angola 09°34′S 16°25′E
82 C11 Malanje var. Malange. ◈ province N Angola
148 M16 Malān, Rās cape SW Pakistan
77 S13 Malanville NE Benin 11°50′N 03°23′E
Malapane see Ozimek
155 F21 Malappuram Kerala, SW India 11°00′N 76°02′E
43 T17 Mala, Punta headland S Panama 07°28′N 79°58′W
148 L15 Malar Baluchistān, SW Pakistan 26°19′N 64°55′E
95 N16 Mälaren ◎ C Sweden
62 H13 Malargüe Mendoza, W Argentina 35°32′S 69°35′W
14 J8 Malartic Québec, SE Canada 48°09′N 78°09′W
119 F20 Malaryta Pol. Maloryta, Rus. Malorita. Brestskaya Voblasts', SW Belarus 51°47′N 24°05′E
63 J19 Malaspina Chubut, SE Argentina 44°56′S 66°52′W
10 G8 Malaspina Glacier glacier Yukon, W Canada
39 U12 Malaspina Glacier glacier Alaska, USA
137 N15 Malatya anc. Melitene. Malatya, SE Turkey 38°22′N 38°18′E
136 M14 Malatya ◈ province C Turkey
Malavate see Pédima
117 Q7 Mala Vyska Rus. Malaya Viska. Kirovohrads'ka Oblast', S Ukraine 48°37′N 31°36′E
83 M14 Malawi off. Republic of Malaŵi; prev. Nyasaland, Nyasaland Protectorate. ◆ republic S Africa
Malawi, Lake see Nyasa, Lake
Malaŵi, Republic of see Malawi
93 J17 Malax Fin. Maalahti. Österbotten, W Finland 62°55′N 21°30′E
124 H14 Malaya Vishera Novgorodskaya Oblast', W Russian Federation 58°52′N 32°12′E
Malaya Viska see Mala Vyska
171 Q7 Malaybalay Mindanao, S Philippines 08°10′N 125°08′E
142 L6 Malāyer prev. Daulatabad. Hamadān, W Iran 34°20′N 48°47′E
168 J7 Malay Peninsula peninsula Malaysia/Thailand
168 L7 Malaysia off. Malaysia, var. Federation of Malaysia; prev. the separate territories of Federation of Malaya, Sarawak and Sabah (North Borneo) and Singapore. ◆ monarchy SE Asia
Malaysia, Federation of see Malaysia
137 R14 Malazgirt Muş, E Turkey 39°09′N 42°30′E
15 R8 Malbaie ≈ Québec, SE Canada
77 T12 Malbaza Tahoua, S Niger 13°57′N 05°32′E
110 J7 Malbork Ger. Marienburg, Marienburg in Westpreussen. Pomorskie, N Poland 54°01′N 19°03′E
100 N9 Malchin Mecklenburg-Vorpommern, N Germany 53°43′N 12°46′E
100 M9 Malchiner See ◎ NE Germany
99 D16 Maldegem Oost-Vlaanderen, NW Belgium 51°12′N 03°27′E
98 L13 Malden Gelderland, SE Netherlands 51°47′N 05°51′E
19 O11 Malden Massachusetts, NE USA 42°25′N 71°04′W
27 Y8 Malden Missouri, C USA 36°33′N 89°58′W
191 X4 Malden Island prev. Independence Island. atoll E Kiribati
173 Q6 Maldives off. Maldivian Divehi, Republic of Maldives. ◆ republic N Indian Ocean
Maldives, Republic of see Maldives
Maldivian Divehi see Maldives
97 P21 Maldon E England, United Kingdom 51°44′N 00°40′E
61 G20 Maldonado Maldonado, S Uruguay 34°57′S 54°59′W
61 G20 Maldonado ◈ department S Uruguay
41 P17 Maldonado, Punta headland S Mexico 16°18′N 98°31′W
106 G6 Malè Trentino-Alto Adige, N Italy 46°21′N 10°51′E
151 K19 Male' Div. Maale. ● (Maldives) Male' Atoll, C Maldives 04°10′N 73°29′E
76 K13 Maléa var. Maléya. NE Guinea 11°46′N 09°43′W
Maléas, Ákra see Agriliá, Akrotírio
115 G22 Maléas, Akrotírio headland S Greece 36°25′N 23°11′E
151 K19 Male' Atoll var. Kaafu Atoll. atoll C Maldives
Malebo, Pool see Stanley Pool
154 E12 Mālegaon Mahārāshtra, W India 20°33′N 74°32′E
81 F15 Malek Jonglei, E South Sudan 06°04′N 31°36′E
187 Q13 Malekula var. Malakula; prev. Mallicolo. island W Vanuatu
189 Y15 Malem Kosrae, E Micronesia 05°16′N 163°01′E
83 O15 Malema Nampula, N Mozambique 14°57′S 37°28′E
79 N23 Malemba-Nkulu Katanga, SE Dem. Rep. Congo 08°01′S 26°48′E
139 T1 Małe Mela Ar. Marī Mīlā, var. Mārī Mīlah. Arbil, E Iraq 36°58′N 44°42′E
124 K9 Malen'ga Respublika Kareliya, NW Russian Federation 63°50′N 36°21′E
95 M20 Målerås Kalmar, S Sweden 56°55′N 15°34′E
103 O6 Malesherbes Loiret, C France 48°18′N 02°25′E
115 G18 Malesína Stereá Elláda, E Greece 38°37′N 23°15′E
Maléya see Maléa
127 O15 Malgobek Respublika Ingushetiya, SW Russian Federation 43°34′N 44°34′E
105 X5 Malgrat de Mar Cataluña, NE Spain 41°39′N 02°45′E
80 C9 Malha Northern Darfur, W Sudan 15°07′N 26°00′E
139 Q5 Malḥah var. Malḥāt. Şalāḥ ad Dīn, C Iraq 34°44′N 42°41′E
Malḥāt see Malḥah
32 K14 Malheur Lake ◎ Oregon, NW USA
32 L14 Malheur River ≈ Oregon, NW USA
76 I13 Mali NW Guinea 12°08′N 12°29′W
77 O9 Mali off. Republic of Mali, Fr. République du Mali; prev. French Sudan, Sudanese Republic. ◆ republic W Africa
171 Q16 Maliana W East Timor 08°57′S 125°25′E
167 O2 Mali Hka ≈ N Myanmar (Burma)
Mali Idjoš see Mali Iđoš
112 K8 Mali Iđoš var. Mali Idjoš, Hung. Kishegyes; prev. Krivaja. Vojvodina, N Serbia 45°43′N 19°40′E
113 M18 Mali i Sharrit Serb. Šar Planina. ▲▲ FYR Macedonia/Serbia
Mali i Zi see Crna Gora
112 K9 Mali Kanal canal N Serbia
171 P12 Maliku Sulawesi, N Indonesia 0°36′S 123°13′E
Malik, Wadi al see Milk, Wadi el
Mālikwāla see Malakwāl
167 N11 Mali Kyun var. Tavoy Island. island Mergui Archipelago, S Myanmar (Burma)
95 M19 Målilla Kalmar, S Sweden 57°24′N 15°49′E
112 B11 Mali Lošinj It. Lussinpiccolo. Primorje-Gorski Kotar, W Croatia 44°31′N 14°28′E
Malin see Malyn
171 P7 Malindang, Mount ▲ Mindanao, S Philippines 08°12′N 123°37′E
81 K20 Malindi Kilifi, SE Kenya 03°14′S 40°05′E
Malines see Mechelen
96 E13 Malin Head Ir. Cionn Mhálanna. headland NW Ireland 55°37′N 07°37′W
171 O11 Malino, Gunung ▲ Sulawesi, N Indonesia 0°44′N 120°45′E
113 M21 Maliq var. Maliqi. Korçë, SE Albania 40°45′N 20°45′E
Maliqi see Maliq
Mali, Republic of see Mali
Mali, République du see Mali
171 Q8 Malita Mindanao, S Philippines 06°13′N 125°39′E
Maljovica see Malyovitsa
154 G12 Malkāpur Mahārāshtra, C India 20°52′N 76°18′E
136 B10 Malkara Tekirdağ, NW Turkey 40°54′N 26°54′E
119 J19 Mal'kavichy Rus. Mal'kovichi. Brestskaya Voblasts', SW Belarus 52°31′N 26°36′E
Malkiye see Al Mālikīyah
114 L11 Malko Sharkovo, Yazovir ◎ SE Bulgaria

◆ Country
● Country Capital
◇ Dependent Territory
○ Dependent Territory Capital
◈ Administrative Regions
✈ International Airport
▲ Mountain
▲▲ Mountain Range
Volcano
≈ River
◎ Lake
Reservoir

MALKO TARNOVO – MAPLETON

114 N11 **Malko Tarnovo** *var.* Malko Tŭrnovo. Burgas, E Bulgaria *42°00′N 27°33′E*
Malko Tŭrnovo *see* Malko Tarnovo
Mal'kovichi *see* Mal'kavichy
183 R12 **Mallacoota** Victoria, SE Australia *37°34′S 149°45′E*
96 G10 **Mallaig** N Scotland, United Kingdom *57°04′N 05°48′W*
182 I9 **Mallala** South Australia *34°29′S 138°30′E*
75 W9 **Mallawī** *var.* Mallawi. C Egypt *27°44′N 30°50′E*
Mallawi *see* Mallawī
105 R5 **Mallén** Aragón, NE Spain *41°53′N 01°25′W*
106 F5 **Malles Venosta** *Ger.* Mals im Vinschgau. Trentino-Alto Adige, N Italy *46°40′N 10°37′E*
Mallicolo *see* Malekula
109 Q8 **Mallnitz** Salzburg, S Austria *46°58′N 13°09′E*
105 W9 **Mallorca** *Eng.* Majorca; *anc.* Baleares Major. *island* Islas Baleares, Spain, W Mediterranean Sea
97 C20 **Mallow** *Ir.* Mala. SW Ireland *52°08′N 08°39′W*
93 E15 **Malm** Nord-Trøndelag, C Norway *64°04′N 11°12′E*
95 L19 **Malmbäck** Jönköping, S Sweden *57°34′N 14°30′E*
92 J12 **Malmberget** *Lapp.* Malmivaara. Norrbotten, N Sweden *67°09′N 20°39′E*
99 M20 **Malmédy** Liège, E Belgium *50°26′N 06°02′E*
83 E25 **Malmesbury** Western Cape, SW South Africa *33°28′S 18°43′E*
Malmivaara *see* Malmberget
95 N16 **Malmköping** Södermanland, C Sweden *59°08′N 16°49′E*
95 K23 **Malmö** Skåne, S Sweden *55°36′N 13°E*
95 K23 **Malmo** ✈ Skåne, S Sweden *55°33′N 13°23′E*
45 Q16 **Malmok** *headland* N Bonaire *12°16′N 68°21′W*
95 M18 **Malmslätt** Östergötland, S Sweden *58°25′N 15°30′E*
125 R16 **Malmyzh** Kirovskaya Oblast', NW Russian Federation *56°30′N 50°37′E*
187 Q13 **Malo** *island* W Vanuatu
126 J7 **Maloarkhangel'sk** Orlovskaya Oblast', W Russian Federation *52°25′N 36°37′E*
Maļoeļap *see* Maloelap Atoll
189 V6 **Maloelap Atoll** *var.* Maļoeļap. *atoll* E Marshall Islands
Maloenda *see* Malunda
108 I10 **Maloja** Graubünden, S Switzerland *46°25′N 09°42′E*
82 L12 **Malole** Northern, NE Zambia *10°05′S 31°37′E*
171 O3 **Malolos** Luzon, N Philippines *14°51′N 120°49′E*
18 K6 **Malone** New York, NE USA *44°51′N 74°18′W*
79 K25 **Malonga** Katanga, S Dem. Rep. Congo *10°26′S 23°10′E*
111 L17 **Małopolskie** ⬖ *province* S Poland
Malorita/Maloryta *see* Malaryta
124 K9 **Maloshuyka** Arkhangel'skaya Oblast', NW Russian Federation *63°43′N 37°20′E*
Mal'ovitsa *see* Malyovitsa
145 V15 **Malovodnoye** Almaty, SE Kazakhstan *43°31′N 77°42′E*
94 C10 **Måløy** Sogn Og Fjordane, S Norway *61°57′N 05°06′E*
126 K4 **Maloyaroslavets** Kaluzhskaya Oblast', W Russian Federation *55°03′N 36°31′E*
122 G7 **Malozemel'skaya Tundra** *physical region* NW Russian Federation
104 J10 **Malpartida de Cáceres** Extremadura, W Spain *39°26′N 06°30′W*
104 K9 **Malpartida de Plasencia** Extremadura, W Spain *39°59′N 06°03′W*
106 C7 **Malpensa** ✈ (Milano) Lombardia, N Italy *45°41′N 08°40′E*
76 J6 **Malqteïr** *desert* N Mauritania
Mals im Vinschgau *see* Malles Venosta
118 J10 **Malta** SE Latvia *56°19′N 27°11′E*
33 V7 **Malta** Montana, NW USA *48°21′N 107°52′W*
120 M11 **Malta** *off.* Republic of Malta. ◆ *republic* C Mediterranean Sea
109 R8 **Malta** *var.* Maltabach. ≈ S Austria
120 M11 **Malta** *island* Malta, C Mediterranean Sea
Maltabach *see* Malta
Malta, Canale di *see* Malta Channel
120 M11 **Malta Channel** *It.* Canale di Malta. *strait* Italy/Malta
83 D20 **Maltahöhe** Hardap, SW Namibia *24°50′S 17°00′E*
Malta, Republic of *see* Malta
97 N16 **Malton** N England, United Kingdom *54°07′N 00°50′W*
171 R13 **Maluku** *off.* Propinsi Maluku, *Dut.* Molukken, *Eng.* Moluccas. ⬖ *province* E Indonesia
171 R13 **Maluku** *Dut.* Molukken, *Eng.* Moluccas; *prev.* Spice Islands. *island group* E Indonesia
Maluku, Laut *see* Molucca Sea
Maluku, Propinsi *see* Maluku
171 R11 **Maluku Utara** *off.* Propinsi Maluku Utara. ⬖ *province* E Indonesia
Maluku Utara, Propinsi *see* Maluku Utara
77 V13 **Malumfashi** Katsina, N Nigeria *11°51′N 07°39′E*
171 N13 **Malunda** *prev.* Maloenda. Sulawesi, C Indonesia *02°58′S 118°52′E*
94 K13 **Malung** Dalarna, C Sweden *60°40′N 13°45′E*
94 K13 **Malungsfors** Dalarna, C Sweden *60°43′N 13°34′E*
186 M8 **Maluu** *var.* Malu'u. Malaita, N Solomon Islands *08°22′S 160°39′E*
Malu'u *see* Maluu
155 D16 **Mālvan** Mahārāshtra, W India *16°05′N 73°28′E*
Malventum *see* Benevento
27 U12 **Malvern** Arkansas, C USA *34°21′N 92°50′W*
29 S15 **Malvern** Iowa, C USA *40°59′N 95°36′W*
44 I13 **Malvern** ▲ W Jamaica *17°59′N 77°42′W*
Malvina, Isla Gran *see* West Falkland
Malvinas, Islas *see* Falkland Islands
117 N4 **Malyn** *Rus.* Malin. Zhytomyrs'ka Oblast', N Ukraine *50°46′N 29°14′E*
114 G10 **Malyovitsa** *var.* Maljovica, Mal'ovitsa. ▲ W Bulgaria *42°12′N 23°19′E*
127 O11 **Malyye Derbety** Respublika Kalmykiya, SW Russian Federation *47°57′N 44°39′E*
Malyy Kavkaz *see* Lesser Caucasus
123 Q6 **Malyy Lyakhovskiy, Ostrov** *island* NE Russian Federation
Malyy Pamir *see* Little Pamir
122 N5 **Malyy Taymyr, Ostrov** *island* Severnaya Zemlya, N Russian Federation
Malyy Uzen' *see* Saryozen
122 L14 **Malyy Yenisey** *var.* Ka-Krem. ≈ S Russian Federation
127 S3 **Mamadysh** Respublika Tatarstan, W Russian Federation *55°46′N 51°22′E*
117 N14 **Mamaia** Constanța, E Romania *44°13′N 28°37′E*
187 W14 **Mamanuca Group** *island group* Yasawa Group, W Fiji
146 L13 **Mamash** Lebap Welaýaty, E Turkmenistan *38°24′N 64°12′E*
79 O17 **Mambasa** Orientale, NE Dem. Rep. Congo *01°20′N 29°05′E*
171 X13 **Mamberamo, Sungai** ≈ Papua, E Indonesia
79 G15 **Mambéré** ≈ SW Central African Republic
79 G15 **Mambéré-Kadéï** ⬖ *prefecture* SW Central African Republic
Mambij *see* Manbij
79 H18 **Mambili** ≈ W Congo
83 N18 **Mambone** *var.* Nova Mambone. Inhambane, E Mozambique *20°59′S 35°04′E*
171 O4 **Mamburao** Mindoro, N Philippines *13°16′N 120°36′E*
172 I16 **Mamelles** *island* Inner Islands, NE Seychelles
99 M25 **Mamer** Luxembourg, SW Luxembourg *49°37′N 06°01′E*
102 L6 **Mamers** Sarthe, NW France *48°21′N 00°22′E*
79 D15 **Mamfe** Sud-Ouest, W Cameroon *05°46′N 09°18′E*
145 P6 **Mamlyutka** Severnyy Kazakhstan, N Kazakhstan *54°54′N 68°36′E*
36 M15 **Mammoth** Arizona, SW USA *32°43′N 110°38′W*
33 S12 **Mammoth Hot Springs** Wyoming, C USA *44°57′N 110°40′W*
Mamoedjoe *see* Mamuju
119 A14 **Mamonovo** *Ger.* Heiligenbeil. Kaliningradskaya Oblast', W Russian Federation *54°28′N 19°57′E*
57 L14 **Mamoré, Rio** ≈ Bolivia/Brazil
76 I14 **Mamou** W Guinea *10°24′N 12°05′W*
22 H8 **Mamou** Louisiana, S USA *30°37′N 92°25′W*
172 I14 **Mamoudzou** ○ (Mayotte) C Mayotte *12°48′S 45°E*
172 I3 **Mampikony** Mahajanga, N Madagascar *16°03′S 47°39′E*
77 P16 **Mampong** C Ghana *07°06′N 01°20′W*
110 M7 **Mamry, Jezioro** *Ger.* Mauersee. ◎ NE Poland
171 N13 **Mamuju** *prev.* Mamoedjoe. Sulawesi, S Indonesia *02°41′S 118°55′E*
83 F19 **Mamuno** Ghanzi, W Botswana *22°15′S 20°02′E*
113 K19 **Mamuras** *var.* Mamurasi, Mamurras. Lezhë, C Albania *41°34′N 19°42′E*
Mamurasi/Mamurras *see* Mamuras
76 L16 **Man** W Ivory Coast *07°24′N 07°33′W*
55 X9 **Mana** NW French Guiana *05°40′N 53°49′W*
56 A6 **Manabí** ⬖ *province* W Ecuador
42 G4 **Manabique, Punta** *var.* Cabo Tres Puntas. *headland* E Guatemala *15°57′N 88°37′W*
54 G11 **Manacacías, Río** ≈ C Colombia
58 F13 **Manacapuru** Amazonas, N Brazil *03°16′S 60°37′W*
105 Y9 **Manacor** Mallorca, Spain, W Mediterranean Sea *39°35′N 03°12′E*
171 Q11 **Manado** *prev.* Menado. Sulawesi, C Indonesia *01°32′N 124°55′E*
188 H5 **Managaha** *island* S Northern Mariana Islands
99 G20 **Manage** Hainaut, S Belgium *50°30′N 04°14′E*
42 J10 **Managua** ● (Nicaragua) Managua, W Nicaragua *12°08′N 86°15′W*
42 J10 **Managua** ⬖ *department* W Nicaragua
42 J10 **Managua** ✈ Managua, W Nicaragua *12°07′N 86°11′W*
42 J10 **Managua, Lago de** *var.* Xolotlán. ◎ W Nicaragua
Manah *see* Bilād Manaḥ
18 K16 **Manahawkin** New Jersey, NE USA *39°39′N 74°12′W*
184 K11 **Manaia** Taranaki, North Island, New Zealand *39°33′S 174°07′E*
172 J6 **Manakara** Fianarantsoa, SE Madagascar *22°09′S 48°E*
152 J7 **Manāli** Himāchal Pradesh, NW India *32°12′N 77°06′E*
Ma, Nam *see* Sông Ma
Manama *see* Al Manāmah
186 D6 **Manam Island** *island* N Papua New Guinea
67 Y13 **Mananara Avaratra** ≈ SE Madagascar
182 M9 **Manangatang** Victoria, SE Australia *35°04′S 142°53′E*
172 I6 **Mananjary** Fianarantsoa, SE Madagascar *21°13′S 48°20′E*
76 L14 **Manankoro** Sikasso, SW Mali *10°33′N 07°25′W*
76 J12 **Manantali, Lac de** ▣ W Mali
Manáos *see* Manaus
185 B23 **Manapouri** Southland, South Island, New Zealand *45°33′S 167°38′E*
185 B23 **Manapouri, Lake** ◎ South Island, New Zealand
58 F13 **Manaquiri** Amazonas, NW Brazil *03°27′S 60°37′W*
Manar *see* Mannar
158 K5 **Manas** Xinjiang Uygur Zizhiqu, NW China *44°16′N 86°12′E*
153 U12 **Manās** *var.* Dangme Chu. ≈ Bhutan/India
153 P10 **Manāslu** *var.* Manaslu. ▲ C Nepal *28°33′N 84°33′E*
147 R8 **Manas, Gora** ▲ Kyrgyzstan/Uzbekistan *42°17′N 71°04′E*
158 K3 **Manas Hu** ◎ NW China
Manaslu *see* Manāslu
37 S8 **Manassa** Colorado, C USA *37°10′N 105°56′W*
21 W4 **Manassas** Virginia, NE USA *38°45′N 77°28′W*
45 T5 **Manatí** C Puerto Rico *18°26′N 66°29′W*
186 E8 **Manau** Northern, S Papua New Guinea *08°02′S 148°00′E*
54 H4 **Manaure** La Guajira, N Colombia *11°46′N 72°28′W*
58 F12 **Manaus** *prev.* Manáos. *state capital* Amazonas, NW Brazil *03°06′S 60°00′W*
136 G17 **Manavgat** Antalya, SW Turkey *36°47′N 31°28′E*
184 M13 **Manawatu** ≈ North Island, New Zealand
184 L11 **Manawatu-Wanganui** *off.* Manawatu-Wanganui Region. ⬖ *region* North Island, New Zealand
Manawatu-Wanganui Region *see* Manawatu-Wanganui
171 R7 **Manay** Mindanao, S Philippines *07°12′N 126°29′E*
138 K2 **Manbij** *var.* Mambij, *Fr.* Membidj. Ḥalab, N Syria *36°32′N 37°55′E*
105 N13 **Mancha Real** Andalucía, S Spain *37°47′N 03°37′W*
102 I4 **Manche** ⬖ *department* N France
97 L17 **Manchester** *Lat.* Mancunium. NW England, United Kingdom *53°30′N 02°15′W*
23 S5 **Manchester** Georgia, SE USA *32°51′N 84°37′W*
29 Y13 **Manchester** Iowa, C USA *42°28′N 91°27′W*
21 N7 **Manchester** Kentucky, S USA *37°09′N 83°46′W*
19 O10 **Manchester** New Hampshire, NE USA *42°59′N 71°26′W*
20 K10 **Manchester** Tennessee, S USA *35°28′N 86°05′W*
18 M9 **Manchester** Vermont, NE USA *43°09′N 73°03′W*
97 L18 **Manchester** ✈ NW England, United Kingdom *53°21′N 02°16′W*
149 P15 **Manchhar Lake** ◎ SE Pakistan
Man-chou-li *see* Manzhouli
129 X7 **Manchurian Plain** *plain* NE China
Máncio Lima *see* Japiim
Mancunium *see* Manchester
148 J15 **Mand** Baluchistān, SW Pakistan *26°06′N 61°58′E*
Mand *see* Mand, Rūd-e
81 H25 **Manda** Njombe, SW Tanzania *10°30′S 34°37′E*
172 H6 **Mandabe** Toliara, W Madagascar *21°02′S 44°56′E*
162 M10 **Mandah** *var.* Töhöm. Dornogovĭ, SE Mongolia *44°25′N 108°18′E*
95 E18 **Mandal** Vest-Agder, S Norway *58°02′N 07°30′E*
Mandal *see* Arbulag, Hövsgöl, Mongolia
Mandal *see* Batsümber, Töv, Mongolia
166 L5 **Mandalay** Mandalay, C Myanmar (Burma) *21°57′N 96°04′E*
166 M6 **Mandalay** ⬖ *region* C Myanmar (Burma)
162 L9 **Mandalgovĭ** Dundgovĭ, C Mongolia *45°47′N 106°18′E*
139 V7 **Mandalī** Diyālá, E Iraq *33°43′N 45°33′E*
162 K10 **Mandal-Ovoo** *var.* Sharhulsan. Ömnögovĭ, S Mongolia *44°43′N 104°06′E*
95 E18 **Mandalselva** ≈ S Norway
163 P11 **Mandalt** *var.* Sonid Zuoqi. Nei Mongol Zizhiqu, N China *43°49′N 113°36′E*
28 M5 **Mandan** North Dakota, N USA *46°49′N 100°53′W*
Mandargiri Hill *see* Mandār Hill
153 R14 **Mandār Hill** *prev.* Mandargiri Hill. Bihār, NE India *24°51′N 87°03′E*
170 M13 **Mandar, Teluk** *bay* Sulawesi, C Indonesia
107 C19 **Mandas** Sardegna, Italy, C Mediterranean Sea *39°40′N 09°07′E*
Mandasor *see* Mandsaur
81 L16 **Mandera** Mandera, NE Kenya *03°56′N 41°53′E*
81 K17 **Mandera** ⬖ *county* NE Kenya
33 V13 **Manderson** Wyoming, C USA *44°13′N 107°55′W*
44 J12 **Mandeville** C Jamaica *18°02′N 77°31′W*
22 K9 **Mandeville** Louisiana, S USA *30°21′N 90°04′W*
152 I7 **Mandi** Himāchal Pradesh, NW India *31°40′N 76°59′E*
76 K14 **Mandiana** E Guinea *10°37′N 08°39′W*
Mandi Būrewāla *see* Būrewāla
Mandidzudzure *see* Chimanimani
83 M15 **Mandié** Manica, NW Mozambique *16°27′S 33°28′E*
83 N14 **Mandimba** Niassa, N Mozambique *14°21′S 35°40′E*
57 Q19 **Mandioré, Laguna** ◎ E Bolivia
154 J10 **Mandla** Madhya Pradesh, C India *22°36′N 80°23′E*
83 M20 **Mandlakazi** *var.* Manjacaze. Gaza, S Mozambique *24°47′S 33°50′E*
95 E24 **Mandø** *var.* Manø. *island* W Denmark
Mandoúdhion/Mandoudi *see* Mantoúdi
78 I13 **Mandoul** *off.* Région du Mandoul. ⬖ *region* S Chad
Mandoul, Région du *see* Mandoul
115 G19 **Mándra** Attikí, C Greece *38°04′N 23°30′E*
172 I7 **Mandrare** ≈ S Madagascar
114 M10 **Mandra, Yazovir** *salt lake* SE Bulgaria
107 L23 **Mandrazzi, Portella** *pass* Sicilia, Italy, C Mediterranean Sea
172 J3 **Mandritsara** Mahajanga, N Madagascar *15°49′S 48°50′E*
143 O13 **Mand, Rūd-e** *var.* Mand. ≈ S Iran
154 F9 **Mandsaur** *prev.* Mandasor. Madhya Pradesh, C India *24°03′N 75°10′E*
154 F11 **Māndu** Madhya Pradesh, C India *22°22′N 75°24′E*
169 W8 **Mandul, Pulau** *island* N Indonesia
83 G15 **Mandundu** Western, W Zambia *16°34′S 22°18′E*
180 I13 **Mandurah** Western Australia *32°31′S 115°41′E*
107 P18 **Manduria** Puglia, SE Italy *40°24′N 17°38′E*
155 G20 **Mandya** Karnātaka, C India *12°34′N 76°55′E*
77 P12 **Mané** C Burkina Faso *12°59′N 01°21′W*
106 E8 **Manerbio** Lombardia, NW Italy *45°22′N 10°09′E*
Manevichi *see* Manevychi
116 K3 **Manevychi** *Pol.* Maniewicze, *Rus.* Manevichi. Volyns'ka Oblast', NW Ukraine *51°18′N 25°29′E*
107 N16 **Manfredonia** Puglia, SE Italy *41°38′N 15°54′E*
107 N16 **Manfredonia, Golfo di** *gulf* Adriatic Sea, N Mediterranean Sea
77 P13 **Manga** C Burkina Faso *11°41′N 01°04′W*
59 L16 **Mangabeiras, Chapada das** ▲▲ E Brazil
79 J20 **Mangai** Bandundu, W Dem. Rep. Congo *03°58′S 19°32′E*
190 L17 **Mangaia** *island group* S Cook Islands
184 M9 **Mangakino** Waikato, North Island, New Zealand *38°23′S 175°47′E*
116 M15 **Mangalia** *anc.* Callatis. Constanța, SE Romania *43°48′N 28°35′E*
78 J11 **Mangalmé** Guéra, SE Chad *12°26′N 19°37′E*
155 E19 **Mangalore** Karnātaka, W India *12°54′N 74°51′E*
Mangareva *see* Magareva
83 I23 **Mangaung** Free State, C South Africa *29°10′S 26°19′E*
Mangaung *see* Bloemfontein
154 K9 **Mangawān** Madhya Pradesh, C India *24°39′N 81°33′E*
184 M11 **Mangaweka** Manawatu-Wanganui, North Island, New Zealand *39°49′S 175°47′E*
184 N11 **Mangaweka** ▲ North Island, New Zealand *39°51′S 176°06′E*
79 P17 **Mangbwalu** Orientale, NE Dem. Rep. Congo *02°06′N 30°04′E*
139 R1 **Mangêsh** *Ar.* Māngīsh, *var.* Mangish. Dahūk, N Iraq *37°03′N 43°04′E*
101 L24 **Mangfall** ≈ SE Germany
169 P13 **Manggar** Pulau Belitung, W Indonesia *02°52′S 108°13′E*
Mangghystaū Ūstirti' *see* Mangystau, Plato
166 M2 **Mangin Range** ▲▲ N Myanmar (Burma)
Māngīsh *see* Mangêsh
Mangish *see* Mangêsh
Mangistau *see* Mangystau Zaliv
146 H8 **Mang'it** *Rus.* Mangit. Qoraqalpog'iston Respublikasi, W Uzbekistan *42°06′N 60°02′E*
Mangit *see* Mang'it
54 A13 **Manglares, Cabo** *headland* SW Colombia *01°36′N 79°02′W*
149 V6 **Mangla Reservoir** ▣ NE Pakistan
159 N9 **Mangnai** *var.* Lao Mangnai. Qinghai, C China *37°52′N 91°45′E*
Mango *see* Mago, Fiji
Mango *see* Sansanné-Mango, Togo
Mangoche *see* Mangochi
83 N14 **Mangochi** *var.* Mangoche; *prev.* Fort Johnston. Southern, SE Malawi *14°30′S 35°15′E*
77 N14 **Mangodara** SW Burkina Faso *09°49′N 04°22′W*
172 H6 **Mangoky** ≈ W Madagascar
171 Q12 **Mangole, Pulau** *island* Kepulauan Sula, E Indonesia
184 J2 **Mangonui** Northland, North Island, New Zealand *35°00′S 173°30′E*
Mangqystaū Oblysy *see* Mangystau
Mangqystaū Shyghanaghy *see* Mangystau Zaliv
Mangshi *see* Luxi
104 H7 **Mangualde** Viseu, N Portugal *40°36′N 07°46′W*
61 H18 **Mangueira, Lagoa** ◎ S Brazil
77 X6 **Manguéni, Plateau du** ▲ NE Niger
163 T4 **Mangui** Nei Mongol Zizhiqu, N China *52°02′N 122°13′E*
26 K11 **Mangum** Oklahoma, C USA *34°52′N 99°30′W*
79 O18 **Manguredjipa** Nord-Kivu, E Dem. Rep. Congo *0°28′N 28°33′E*
83 L16 **Mangwendi** Mashonaland East, E Zimbabwe *18°22′S 31°24′E*
Mangyshlak, Plato *see* Mangystau, Plato
Mangyshlakskiy Zaliv *see* Mangystau Zaliv
Mangyshlaskaya *see* Mangystau
144 F15 **Mangystau** *Kaz.* Mangqystaū Oblysy *prev.* Mangistau; *prev.* Mangyshlaskaya. ⬖ *province* SW Kazakhstan
144 F15 **Mangystau, Plato** *plateau* SW Kazakhstan
144 E14 **Mangystau Zaliv** *Kaz.* Mangqystaū Shyghanaghy; *prev.* Mangyshlakskiy Zaliv. *gulf* SW Kazakhstan
162 E7 **Manhan** *var.* Tögrög. Hovd, W Mongolia *47°24′N 92°06′E*
Manhan *see* Alag-Erdene
27 O4 **Manhattan** Kansas, C USA *39°11′N 96°35′W*
99 L21 **Manhay** Luxembourg, SE Belgium *50°13′N 05°43′E*
83 L21 **Manhiça** *prev.* Vila de Manhiça. Maputo, S Mozambique *25°25′S 32°49′E*
83 L21 **Manhoca** Maputo, S Mozambique *26°49′S 32°36′E*
59 N20 **Manhuaçu** Minas Gerais, SE Brazil *20°16′S 42°01′W*
117 W9 **Manhush** *prev.* Pershotravneve. Donets'ka Oblast', E Ukraine *47°03′N 37°20′E*
54 H10 **Maní** Casanare, C Colombia *04°50′N 72°15′W*
143 R11 **Mānī** Kermān, C Iran
83 M17 **Manica** *var.* Vila de Manica. Manica, W Mozambique *18°56′S 32°52′E*
83 M17 **Manica** *off.* Província de Manica. ⬖ *province* W Mozambique
83 L17 **Manicaland** ⬖ *province* E Zimbabwe
Manica, Província de *see* Manica
15 U5 **Manic Deux, Réservoir** ▣ Québec, SE Canada
Manich *see* Manych
59 F14 **Manicoré** Amazonas, N Brazil *05°48′S 61°16′W*
13 N11 **Manicouagan** Québec, SE Canada *50°40′N 68°46′W*
13 N11 **Manicouagan** ≈ Québec, SE Canada
15 U6 **Manicouagan, Péninsule de** *peninsula* Québec, SE Canada
13 N11 **Manicouagan, Réservoir** ◎ Québec, E Canada
15 T4 **Manic Trois, Réservoir** ▣ Québec, SE Canada
79 M20 **Maniema** *off.* Région du Maniema. ⬖ *region* E Dem. Rep. Congo
Maniema, Région du *see* Maniema
Maniewicze *see* Manevychi
160 F8 **Maniganggo** Sichuan, C China *32°01′N 99°04′E*
11 Y15 **Manigotagan** Manitoba, S Canada *51°06′N 96°18′W*
153 R13 **Manihāri** Bihār, N India *25°21′N 87°37′E*
191 U9 **Manihi** *island* Îles Tuamotu, C French Polynesia
190 L13 **Manihiki** *atoll* N Cook Islands
175 U8 **Manihiki Plateau** *undersea feature* C Pacific Ocean
196 M14 **Maniitsoq** *var.* Manîtsoq, *Dan.* Sukkertoppen. ⬖ Qeqqata, S Greenland
153 T15 **Manikganj** Dhaka, C Bangladesh *23°52′N 90°00′E*
152 M14 **Mānikpur** Uttar Pradesh, N India *25°04′N 81°06′E*
171 N4 **Manila** *off.* City of Manila. ● (Philippines) Luzon, N Philippines *14°34′N 120°59′E*
27 Y9 **Manila** Arkansas, C USA *35°52′N 90°10′W*
Manila, City of *see* Manila
189 N16 **Manila Reef** *reef* W Micronesia
183 T6 **Manilla** New South Wales, SE Australia *30°44′S 150°43′E*
192 P6 **Maniloa** *island* Tongatapu Group, S Tonga
123 U8 **Manily** Krasnoyarskiy Kray, E Russian Federation *62°33′N 165°03′E*
171 V12 **Manim, Pulau** *island* E Indonesia
168 I11 **Maninjau, Danau** ◎ Sumatera, W Indonesia
153 W13 **Manipur** ⬖ *state* NE India
153 X14 **Manipur Hills** *hill range* E India
136 C14 **Manisa** *var.* Manissa, *prev.* Saruhan; *anc.* Magnesia. Manisa, W Turkey *38°36′N 27°29′E*
136 C13 **Manisa** *var.* Manissa. ⬖ *province* W Turkey
Manissa *see* Manisa
31 O7 **Manistee** Michigan, N USA *44°14′N 86°19′W*
31 P7 **Manistee River** ≈ Michigan, N USA
31 O4 **Manistique** Michigan, N USA *45°57′N 86°15′W*
31 P4 **Manistique Lake** ◎ Michigan, N USA
11 W13 **Manitoba** ⬖ *province* S Canada
11 X16 **Manitoba, Lake** ◎ Manitoba, S Canada
11 X17 **Manitou** Manitoba, S Canada *49°12′N 98°28′W*
31 N2 **Manitou Island** *island* Michigan, N USA
14 H11 **Manitou Lake** ◎ Ontario, SE Canada
12 G15 **Manitoulin Island** *island* Ontario, S Canada
37 T5 **Manitou Springs** Colorado, C USA *38°51′N 104°56′W*
14 G12 **Manitouwabing Lake** ◎ Ontario, S Canada
12 E12 **Manitouwadge** Ontario, S Canada *49°14′N 85°51′W*
12 G15 **Manitowaning** Manitoulin Island, Ontario, S Canada *45°44′N 81°50′W*
14 B7 **Manitowik Lake** ◎ Ontario, S Canada
31 N7 **Manitowoc** Wisconsin, N USA *44°04′N 87°40′W*
Manîtsoq *see* Maniitsoq
139 O7 **Mānī', Wādī al** *dry watercourse* W Iraq
12 J14 **Maniwaki** Québec, SE Canada *46°22′N 75°58′W*
171 W13 **Maniwori** Papua, E Indonesia *02°49′S 136°00′E*
54 E10 **Manizales** Caldas, W Colombia *05°03′N 73°52′W*
112 F11 **Manjača** ▲ NW Bosnia and Herzegovina
Manjacaze *see* Mandlakazi
180 J14 **Manjimup** Western Australia *34°18′S 116°14′E*
109 V4 **Mank** Niederösterreich, C Austria *48°06′N 15°13′E*
79 I17 **Mankanza** Equateur, NW Dem. Rep. Congo *01°40′N 19°08′E*
153 N12 **Mankāpur** Uttar Pradesh, N India *27°03′N 82°12′E*
26 M3 **Mankato** Kansas, C USA *39°48′N 98°13′W*
29 U10 **Mankato** Minnesota, N USA *44°10′N 94°00′W*
117 O7 **Man'kivka** Cherkas'ka Oblast', C Ukraine *48°58′N 30°10′E*
76 M15 **Mankono** C Ivory Coast *08°01′N 06°09′W*
11 T17 **Mankota** Saskatchewan, S Canada *49°25′N 107°05′W*
155 K23 **Mankulam** Northern Province, N Sri Lanka *09°07′N 80°27′E*
162 L10 **Manlay** *var.* Üydzen. Ömnögovĭ, S Mongolia *44°08′N 106°48′E*
39 Q9 **Manley Hot Springs** Alaska, USA *65°00′N 150°37′W*
18 H10 **Manlius** New York, NE USA *43°00′N 75°58′W*
105 W5 **Manlleu** Cataluña, NE Spain *41°59′N 02°17′E*
29 V11 **Manly** Iowa, C USA *43°17′N 93°12′W*
154 E13 **Manmād** Mahārāshtra, W India *20°15′N 74°29′E*
182 J7 **Mannahill** South Australia *32°29′S 139°58′E*
155 J23 **Mannar** *var.* Manar. Northern Province, NW Sri Lanka *09°01′N 79°53′E*
155 I24 **Mannar, Gulf of** *gulf* India/Sri Lanka
155 J23 **Mannar Island** *island* N Sri Lanka
Mannersdorf *see* Mannersdorf am Leithagebirge
109 Y5 **Mannersdorf am Leithagebirge** *var.* Mannersdorf. Niederösterreich, E Austria *47°59′N 16°36′E*
109 Y6 **Mannersdorf an der Rabnitz** Burgenland, E Austria *47°25′N 16°32′E*
101 G20 **Mannheim** Baden-Württemberg, SW Germany *49°29′N 08°29′E*
11 O12 **Manning** Alberta, W Canada *56°53′N 117°39′W*
29 T14 **Manning** Iowa, C USA *41°54′N 95°03′W*
28 K5 **Manning** North Dakota, N USA *47°15′N 102°48′W*
21 S13 **Manning** South Carolina, SE USA *33°42′N 80°12′W*
191 Y2 **Manning, Cape** *headland* Kiritimati, NE Kiribati *02°02′N 157°26′W*
21 S3 **Mannington** West Virginia, NE USA *39°31′N 80°20′W*
182 A1 **Mann Ranges** ▲▲ South Australia
107 C19 **Mannu** ≈ Sardegna, Italy, C Mediterranean Sea
11 R14 **Mannville** Alberta, SW Canada *53°19′N 111°08′W*
76 J15 **Mano** ≈ Liberia/Sierra Leone
Manø *see* Mandø
61 F15 **Manoel Viana** Rio Grande do Sul, S Brazil *29°33′S 55°28′W*
39 O13 **Manokotak** Alaska, USA *59°00′N 158°58′W*
171 V12 **Manokwari** Papua Barat, E Indonesia *0°53′S 134°05′E*
79 N22 **Manono** Shaba, SE Dem. Rep. Congo *07°18′S 27°25′E*
25 T10 **Manor** Texas, SW USA *30°20′N 97°33′W*
97 D16 **Manorhamilton** *Ir.* Cluainín. Leitrim, NW Ireland *54°18′N 08°10′W*
103 S15 **Manosque** Alpes-de-Haute-Provence, SE France *43°50′N 05°47′E*
12 L11 **Manouane, Lac** ◎ Québec, SE Canada
163 W12 **Manp'o** *var.* Manp'ojin. NW North Korea *41°10′N 126°24′E*
Manp'ojin *see* Manp'o
191 T4 **Manra** *prev.* Sydney Island. *atoll* Phoenix Islands, C Kiribati
105 V5 **Manresa** Cataluña, NE Spain *41°43′N 01°50′E*
152 H9 **Mānsa** Punjab, NW India *30°00′N 75°25′E*
82 J12 **Mansa** *prev.* Fort Rosebery. Luapula, N Zambia *11°14′S 28°55′E*
76 G12 **Mansa Konko** C Gambia *13°26′N 15°29′W*
15 Q11 **Manseau** Québec, SE Canada *46°23′N 71°59′W*
149 U5 **Mānsehra** Khyber Pakhtunkhwa, NW Pakistan *34°23′N 73°18′E*
9 Q9 **Mansel Island** *island* Nunavut, NE Canada
183 O12 **Mansfield** Victoria, SE Australia *37°04′S 146°06′E*
97 M18 **Mansfield** C England, United Kingdom *53°09′N 01°11′W*
27 S11 **Mansfield** Arkansas, C USA *35°03′N 94°15′W*
22 G6 **Mansfield** Louisiana, S USA *32°02′N 93°42′W*
19 O12 **Mansfield** Massachusetts, NE USA *42°00′N 71°11′W*
31 T12 **Mansfield** Ohio, N USA *40°45′N 82°31′W*
18 G12 **Mansfield** Pennsylvania, NE USA *41°46′N 77°02′W*
18 M7 **Mansfield, Mount** ▲ Vermont, NE USA *44°31′N 72°49′W*
59 M16 **Mansidão** Bahia, E Brazil *10°46′S 44°04′W*
102 L11 **Mansle** Charente, W France *45°52′N 00°11′E*
76 G12 **Mansôa** C Guinea-Bissau *12°08′N 15°18′W*
47 V8 **Manso, Rio** ≈ C Brazil
Manṣūra *see* Al Manṣūrah
Mansurabad *see* Mehrān, Rūd-e
56 A6 **Manta** Manabí, W Ecuador *0°59′S 80°44′W*
56 A6 **Manta, Bahía de** *bay* W Ecuador
57 F14 **Mantaro, Río** ≈ C Peru
35 O8 **Manteca** California, W USA *37°48′N 121°13′W*
54 J7 **Mantecal** Apure, C Venezuela *07°34′N 69°07′W*
31 N11 **Manteno** Illinois, N USA *41°15′N 87°49′W*
21 Y9 **Manteo** Roanoke Island, North Carolina, SE USA *35°54′N 75°42′W*
Mantes-Gassicourt *see* Mantes-la-Jolie
103 N5 **Mantes-la-Jolie** *prev.* Mantes-Gassicourt, Mantes-sur-Seine; *anc.* Medunta. Yvelines, N France *48°59′N 01°43′E*
Mantes-sur-Seine *see* Mantes-la-Jolie
36 L5 **Manti** Utah, W USA *39°16′N 111°38′W*
Mantinea *see* Mantíneia
115 F20 **Mantíneia** *anc.* Mantinea. *site of ancient city* Pelopónnisos, S Greece
59 M21 **Mantiqueira, Serra da** ▲▲ S Brazil
29 W10 **Mantorville** Minnesota, N USA *44°04′N 92°45′W*
115 G17 **Mantoúdi** *var.* Mandoudi; *prev.* Mandoúdhion. Évvoia, C Greece *38°47′N 23°29′E*
Mantoue *see* Mantova
106 F8 **Mantova** *Eng.* Mantua, *Fr.* Mantoue. Lombardia, NW Italy *45°10′N 10°47′E*
93 M19 **Mäntsälä** Uusimaa, S Finland *60°38′N 25°21′E*
93 L17 **Mänttä** Pirkanmaa, W Finland *62°00′N 24°36′E*
Mantua *see* Mantova
125 O14 **Manturovo** Kostromskaya Oblast', NW Russian Federation *58°19′N 44°42′E*
93 M18 **Mäntyharju** Etelä-Savo, SE Finland *61°25′N 26°53′E*
92 M13 **Mäntyjärvi** Lappi, N Finland *66°00′N 27°35′E*
190 L16 **Manuae** *island* S Cook Islands
191 Q10 **Manuae** *atoll* Îles Sous le Vent, W French Polynesia
192 L16 **Manu'a Islands** *island group* E American Samoa
40 L5 **Manuel Benavides** Chihuahua, N Mexico *29°07′N 103°52′W*
61 D21 **Manuel J. Cobo** Buenos Aires, E Argentina *35°49′S 57°54′W*
58 M12 **Manuel Luís, Recife** *reef* E Brazil
59 I14 **Manuel Zinho** Pará, N Brazil *07°21′S 54°47′W*
191 V11 **Manuhagi** *prev.* Manuhangi. *atoll* Îles Tuamotu, C French Polynesia
Manuhangi *see* Manuhagi
185 E22 **Manuherikia** ≈ South Island, New Zealand
171 P13 **Manui, Pulau** *island* N Indonesia
Manukau *see* Manurewa
184 L6 **Manukau Harbour** *harbor* North Island, New Zealand
191 Z2 **Manulu Lagoon** ◎ Kiritimati, E Kiribati
182 J7 **Manunda Creek** *seasonal river* South Australia
57 K15 **Manupari, Río** ≈ N Bolivia
184 L6 **Manurewa** *var.* Manukau. Auckland, North Island, New Zealand *37°01′S 174°55′E*
57 K15 **Manurimi, Río** ≈ NW Bolivia
186 D5 **Manus** ⬖ *province* N Papua New Guinea
186 D5 **Manus Island** *var.* Great Admiralty Island. *island* N Papua New Guinea
171 T16 **Manuwui** Pulau Babar, E Indonesia *07°47′S 129°39′E*
29 Q3 **Manvel** North Dakota, N USA *48°07′N 97°15′W*
33 Z14 **Manville** Wyoming, C USA *42°45′N 104°38′W*
22 G6 **Many** Louisiana, S USA *31°34′N 93°28′W*
81 H21 **Manyara, Lake** ◎ NE Tanzania
126 L12 **Manych** *var.* Manich. ≈ SW Russian Federation
83 H14 **Manyinga** North Western, NW Zambia *13°28′S 24°18′E*
105 O11 **Manzanares** Castilla-La Mancha, C Spain *39°N 03°23′W*
44 H7 **Manzanillo** Granma, E Cuba *20°21′N 77°07′W*
40 K14 **Manzanillo** Colima, SW Mexico *19°00′N 104°19′W*
40 K14 **Manzanillo, Bahía** *bay* SW Mexico
37 S11 **Manzano Mountains** ▲▲ New Mexico, SW USA
37 R12 **Manzano Peak** ▲ New Mexico, SW USA *34°35′N 106°27′W*
163 R6 **Manzhouli** *var.* Man-chou-li. Nei Mongol Zizhiqu, N China *49°36′N 117°28′E*
Manzil Bū Ruqaybah *see* Menzel Bourguiba
139 X9 **Manzilīyah** Maysān, E Iraq *32°26′N 47°01′E*
83 L21 **Manzini** *prev.* Bremersdorp. C Swaziland *26°30′S 31°22′E*
83 L21 **Manzini** ✈ (Mbabane) C Swaziland *26°36′S 31°25′E*
78 G10 **Mao** Kanem, W Chad *14°06′N 15°17′E*
45 N8 **Mao** NW Dominican Republic *19°37′N 71°04′W*
105 Z9 **Maó** *Cast.* Mahón, *Eng.* Port Mahon; *anc.* Portus Magonis. Menorca, Spain, W Mediterranean Sea *39°54′N 04°15′E*
Maoemere *see* Maumere
159 W9 **Maojing** Gansu, N China *36°26′N 106°36′E*
171 Y14 **Maoke, Pegunungan** *Dut.* Sneeuw-gebergte, *Eng.* Snow Mountains. ▲▲ Papua, E Indonesia
Maol Réidh, Caoc *see* Mweelrea
160 M15 **Maoming** Guangdong, S China *21°46′N 110°51′E*
160 H8 **Maoxian** *var.* Mao Xian; *prev.* Fengyizhen. Sichuan, C China *31°42′N 103°48′E*
Mao Xian *see* Maoxian
83 L19 **Mapai** Gaza, SW Mozambique *22°52′S 32°00′E*
158 H15 **Mapam Yumco** ◎ W China
83 I15 **Mapanza** Southern, S Zambia *16°16′S 26°54′E*
54 J4 **Maparari** Falcón, N Venezuela *10°52′N 69°27′W*
41 U17 **Mapastepec** Chiapas, SE Mexico *15°28′N 93°00′W*
169 V9 **Mapat, Pulau** *island* N Indonesia
171 Y15 **Mapi** Papua, E Indonesia *07°02′S 139°24′E*
171 V11 **Mapia, Kepulauan** *island group* E Indonesia
40 L8 **Mapimí** Durango, C Mexico *25°50′N 103°50′W*
83 N19 **Mapinhane** Inhambane, SE Mozambique *22°14′S 35°07′E*
55 N7 **Mapire** Monagas, NE Venezuela *07°48′N 64°40′W*
11 S17 **Maple Creek** Saskatchewan, S Canada *49°55′N 109°28′W*
31 Q9 **Maple River** ≈ Michigan, N USA
29 P7 **Maple River** ≈ North Dakota/South Dakota, N USA
29 S13 **Mapleton** Iowa, C USA *42°10′N 95°47′W*
29 U10 **Mapleton** Minnesota, N USA *43°55′N 93°57′W*
29 R5 **Mapleton** North Dakota, N USA *46°51′N 97°04′W*
32 F13 **Mapleton** Oregon, NW USA *44°01′N 123°56′W*

◆ Country
● Country Capital
◇ Dependent Territory
○ Dependent Territory Capital
⬖ Administrative Regions
✈ International Airport
▲ Mountain
▲▲ Mountain Range
Volcano
≈ River
◎ Lake
▣ Reservoir

MAPLETON – MARYSVALE

36 L3 **Mapleton** Utah, W USA 40°07′N 111°37′W
192 K5 **Mapmaker Seamounts** *undersea feature* N Pacific Ocean 25°00′N 165°00′E
186 B6 **Maprik** East Sepik, NW Papua New Guinea 03°38′S 143°02′E
83 L21 **Maputo** *prev.* Lourenço Marques. ● (Mozambique) Maputo, S Mozambique 25°58′S 32°35′E
83 L21 **Maputo** ◈ *province* S Mozambique
67 V14 **Maputo** ≈ S Mozambique
83 L21 **Maputo** ✈ Maputo, S Mozambique 25°47′S 32°36′E
Maqat *see* Makat
113 M19 **Maqellarë** Dibër, C Albania 41°36′N 20°29′E
159 S12 **Maqên** *var.* Dawo; *prev.* Dawu. Qinghai, C China 34°32′N 100°17′E
159 S11 **Maqên Kangri** ▲ C China 34°44′N 99°25′E
141 X7 **Maqiz al Kurbā** N Oman 24°13′N 56°48′E
159 U12 **Maqu** *var.* Nyinma. Gansu, C China 34°02′N 102°00′E
104 M9 **Maqueda** Castilla-La Mancha, C Spain 40°04′N 04°22′W
82 B9 **Maquela do Zombo** Uíge, NW Angola 06°06′S 15°12′E
63 I16 **Maquinchao** Río Negro, C Argentina 41°19′S 68°47′W
29 Z13 **Maquoketa** Iowa, C USA 42°03′N 90°42′W
29 Y13 **Maquoketa River** ≈ Iowa, C USA
14 F13 **Mar** Ontario, S Canada 44°48′N 81°12′W
95 F14 **Mår** ◎ S Norway
81 G19 **Mara** ◈ *region* N Tanzania
58 D12 **Maraã** Amazonas, NW Brazil 01°48′S 65°21′W
191 P8 **Maraa** Tahiti, W French Polynesia 17°44′S 149°34′W
191 O8 **Maraa, Pointe** *headland* Tahiti, W French Polynesia 17°44′S 149°34′W
59 K14 **Marabá** Pará, NE Brazil 05°23′S 49°10′W
54 H5 **Maracaibo** Zulia, NW Venezuela 10°40′N 71°39′W
Maracaibo, Gulf of *see* Venezuela, Golfo de
54 H5 **Maracaibo, Lago de** *var.* Lake Maracaibo. *inlet* NW Venezuela
Maracaibo, Lake *see* Maracaibo, Lago de
58 K10 **Maracá, Ilha de** *island* NE Brazil
59 H20 **Maracaju, Serra de** ▲▲ S Brazil
58 I11 **Maracanaquará, Planalto** ▲▲ NE Brazil
54 L5 **Maracay** Aragua, N Venezuela 10°15′N 67°36′W
Marada *see* Marādah
75 R9 **Marādah** *var.* Marada. N Libya 29°16′N 19°29′E
77 U12 **Maradi** Maradi, S Niger 13°30′N 07°05′E
77 U11 **Maradi** ◈ *department* S Niger
81 E21 **Maragarazi** *var.* Muragarazi. ≈ Burundi/Tanzania
Maragha *see* Marāgheh
142 J3 **Marāgheh** *var.* Maragha. Āžarbāyjān-e Khāvarī, NW Iran 37°21′N 46°14′E
141 P7 **Marāh** *var.* Marrāt. Ar Riyāḑ, C Saudi Arabia 25°04′N 45°30′E
55 N11 **Marahuaca, Cerro** ▲ S Venezuela 03°37′N 65°25′W
27 R5 **Marais des Cygnes River** ≈ Kansas/Missouri, C USA
58 L11 **Marajó, Baía de** *bay* N Brazil
59 K12 **Marajó, Ilha de** *island* N Brazil
191 O2 **Marakei** *atoll* Tungaru, W Kiribati
Marakesh *see* Marrakech
81 I18 **Maralal** Samburu, C Kenya 01°05′N 36°42′E
83 G21 **Maralaleng** Kgalagadi, S Botswana 25°42′S 22°39′E
145 U8 **Maraldy, Ozero** ◎ NE Kazakhstan
182 C5 **Maralinga** South Australia 30°16′S 131°35′E
Máramarossziget *see* Sighetu Marmaţiei
187 N9 **Maramasike** *var.* Small Malaita. *island* N Solomon Islands
Maramba *see* Livingstone
194 H3 **Marambio** *Argentinian research station* Antarctica 64°22′S 57°18′W
116 H9 **Maramureş** ◈ *county* NW Romania
36 L15 **Marana** Arizona, SW USA 32°24′N 111°12′W
105 P7 **Maranchón** Castilla-La Mancha, C Spain 41°02′N 02°11′W
142 J2 **Marand** *var.* Merend. Āžarbāyjān-e Sharqī, NW Iran 38°25′N 45°40′E
Marandellas *see* Marondera
58 L13 **Maranhão** *off.* Estado do Maranhão. ◈ *state* E Brazil
104 H10 **Maranhão, Barragem do** ⊠ C Portugal
Maranhão, Estado do *see* Maranhão
149 O11 **Mārān, Koh-i** ▲ SW Pakistan 29°24′N 66°50′E
106 J7 **Marano, Laguna di** *lagoon* NE Italy
56 E9 **Marañón, Río** ≈ N Peru
102 J10 **Marans** Charente-Maritime, W France 46°19′N 00°58′W
83 M20 **Marão** Inhambane, S Mozambique 24°15′S 34°09′E
185 B23 **Mararoa** ≈ South Island, New Zealand
Maraş/Marash *see* Kahramanmaraş
107 M19 **Maratea** Basilicata, S Italy 39°57′N 15°44′E
104 G11 **Marateca** Setúbal, S Portugal 38°34′N 08°40′W
115 B20 **Marathiá, Akrotírio** *headland* Zákynthos, Iónia Nisiá, Greece, C Mediterranean Sea 37°39′N 20°49′E
12 E12 **Marathon** Ontario, S Canada 48°44′N 86°23′W
23 Y17 **Marathon** Florida Keys, Florida, SE USA 24°42′N 81°05′W
24 L10 **Marathon** Texas, SW USA 30°10′N 103°14′W
Marathón *see* Marathónas
115 H19 **Marathónas** *prev.* Marathón. Attikí, C Greece 38°09′N 23°57′E
169 W9 **Maratua, Pulau** *island* N Indonesia
59 O18 **Maraú** Bahia, SE Brazil 14°07′S 39°02′W
143 R3 **Marāveh Tappeh** Golestán, N Iran 37°53′N 55°57′E
24 L11 **Maravillas Creek** ≈ Texas, SW USA
186 D8 **Marawaka** Eastern Highlands, C Papua New Guinea 06°56′S 145°54′E
171 Q7 **Marawi** Mindanao, S Philippines 07°59′N 124°16′E
Mārāzā *see* Qobustan
Marbat *see* Mirbāţ
104 L16 **Marbella** Andalucía, S Spain 36°31′N 04°50′W
180 J7 **Marble Bar** Western Australia 21°13′S 119°48′E
36 L9 **Marble Canyon** *canyon* Arizona, SW USA
25 S10 **Marble Falls** Texas, SW USA 30°34′N 98°16′W
27 Y7 **Marble Hill** Missouri, C USA 37°18′N 89°58′W
33 T15 **Marbleton** Wyoming, C USA 42°31′N 110°06′W
Marburg *see* Marburg an der Lahn, Germany
Marburg *see* Maribor, Slovenia
101 H16 **Marburg an der Lahn** *hist.* Marburg. Hessen, W Germany 50°49′N 08°46′E
111 H23 **Marcal** ≈ W Hungary
42 G7 **Marcala** La Paz, SW Honduras 14°11′N 88°00′W
111 H24 **Marcali** Somogy, SW Hungary 46°33′N 17°29′E
83 A16 **Marca, Ponta da** *headland* SW Angola 16°31′S 11°42′E
59 I16 **Marcelândia** Mato Grosso, W Brazil 11°18′S 54°49′W
27 T3 **Marceline** Missouri, C USA 39°42′N 92°57′W
60 I13 **Marcelino Ramos** Rio Grande do Sul, S Brazil 27°31′S 51°57′W
55 Y12 **Marcel, Mont** ▲ S French Guiana 02°32′N 53°00′W
97 O19 **March** E England, United Kingdom 52°37′N 00°13′E
109 Z3 **March** *var.* Morava. ≈ C Europe *see also* Morava
March *see* Morava
106 I12 **Marche** *Eng.* Marches. ◈ *region* C Italy
103 N11 **Marche** *cultural region* C France
99 J21 **Marche-en-Famenne** Luxembourg, SE Belgium 50°13′N 05°21′E
104 K14 **Marchena** Andalucía, S Spain 37°20′N 05°24′W
57 B17 **Marchena, Isla** *var.* Bindloe Island. *island* Galapagos Islands, Ecuador, E Pacific Ocean
Marches *see* Marche
99 J20 **Marchin** Liège, E Belgium 50°30′N 05°17′E
181 S1 **Marchinbar Island** *island* Wessel Islands, Northern Territory, N Australia
62 L9 **Mar Chiquita, Laguna** ◎ C Argentina
103 Q10 **Marcigny** Saône-et-Loire, C France 46°16′N 04°04′E
23 W16 **Marco** Florida, SE USA 25°56′N 81°43′W
Marcodurum *see* Düren
59 O15 **Marcolândia** Pernambuco, E Brazil 07°21′S 40°40′W
106 I8 **Marco Polo** ✈ (Venezia) Veneto, NE Italy 45°30′N 12°21′E
Marcounda *see* Markounda
Marcq *see* Mark
116 M8 **Mărculeşti** *Rus.* Markuleshty. N Moldova 47°54′N 28°14′E
29 S12 **Marcus** Iowa, C USA 42°49′N 95°48′W
39 S11 **Marcus Baker, Mount** ▲ Alaska, USA 61°26′N 147°45′W
192 I5 **Marcus Island** *var.* Minami Tori Shima. *island* E Japan
18 K8 **Marcy, Mount** ▲ New York, NE USA 44°06′N 73°55′W
149 T5 **Mardān** Khyber Pakhtunkhwa, N Pakistan 34°14′N 71°59′E
63 N14 **Mar del Plata** Buenos Aires, E Argentina 38°S 57°32′W
137 Q16 **Mardin** Mardin, SE Turkey 37°19′N 40°43′E
137 Q16 **Mardin** ◈ *province* SE Turkey
137 Q16 **Mardin Dağları** ▲▲ SE Turkey
Mardzad *see* Hayrhandulaan
187 R17 **Maré** *island* Îles Loyauté, E New Caledonia
Marea Neagră *see* Black Sea
105 Z8 **Mare de Déu del Toro** *var.* El Toro. ▲ Menorca, Spain, W Mediterranean Sea 39°59′N 04°06′E
181 W4 **Mareeba** Queensland, NE Australia 17°03′S 145°30′E
96 G8 **Maree, Loch** ◎ N Scotland, United Kingdom
Mareeq *see* Mereeg
Marek *see* Dupnitsa
76 J11 **Maréna** Kayes, W Mali 14°36′N 10°57′W
190 I2 **Marenanuka** *atoll* Tungaru, W Kiribati
29 X14 **Marengo** Iowa, C USA 41°48′N 92°04′W
102 J11 **Marennes** Charente-Maritime, W France 45°47′N 01°04′W
107 G23 **Marettimo, Isola** *island* Isole Egadi, S Italy
24 K10 **Marfa** Texas, SW USA 30°19′N 104°03′W
57 P17 **Marfil, Laguna** ◎ E Bolivia
Margarets *see* Marharets'
25 Q4 **Margaret** Texas, SW USA 34°00′N 99°38′W
180 I14 **Margaret River** Western Australia 33°58′S 115°10′E
186 C7 **Margarima** Hela, W Papua New Guinea 06°00′S 143°23′E
55 N4 **Margarita, Isla de** *island* N Venezuela
115 I25 **Margarítes** Kríti, Greece, E Mediterranean Sea 35°19′N 24°40′E
97 Q22 **Margate** *prev.* Mergate. SE England, United Kingdom 51°24′N 01°24′E
23 Z15 **Margate** Florida, SE USA 26°14′N 80°12′W
Margelan *see* Marg'ilon
103 P13 **Margeride, Montagnes de la** ▲▲ C France
Margherita *see* Jamaame
107 N16 **Margherita di Savoia** Puglia, SE Italy 41°23′N 16°09′E
Margherita, Lake *see* Ābaya Hāyk'
81 E18 **Margherita Peak** *Fr.* Pic Marguerite. ▲ Uganda/Dem. Rep. Congo 0°28′N 29°58′E
149 O4 **Marghī** Bāmyān, N Afghanistan 35°10′N 66°26′E
116 G9 **Marghita** *Hung.* Margitta. Bihor, NW Romania 47°20′N 22°20′E
Margilan *see* Marg'ilon
147 S10 **Marg'ilon** *var.* Margelan, *Rus.* Margilan. Farg'ona Viloyati, E Uzbekistan 40°29′N 71°43′E
116 K8 **Marginea** Suceava, NE Romania 47°49′N 25°47′E
Margitta *see* Marghita
148 K9 **Mārgow, Dasht-e** *desert* SW Afghanistan
99 L18 **Margraten** Limburg, SE Netherlands 50°49′N 05°49′E
10 M15 **Marguerite** British Columbia, SW Canada 52°17′N 122°10′W
15 V3 **Marguerite** ≈ Quebec, SE Canada
194 I6 **Marguerite Bay** *bay* Antarctica
Marguerite, Pic *see* Margherita Peak
117 T9 **Marhanets'** *Rus.* Marganets. Dnipropetrovs'ka Oblast', E Ukraine 47°35′N 34°37′E
186 B9 **Mari** Western, SW Papua New Guinea 09°10′S 141°39′E
191 Y12 **Maria** *atoll* Groupe Actéon, SE French Polynesia
191 R12 **Maria** *island* Îles Australes, SW French Polynesia
40 I12 **María Cleofas, Isla** *island* C Mexico
62 H4 **María Elena** *var.* Oficina María Elena. Antofagasta, N Chile 22°18′S 69°40′W
95 G21 **Mariager** Midtjylland, C Denmark 56°39′N 09°59′E
61 C22 **María Ignacia** Buenos Aires, E Argentina 37°24′S 59°30′W
183 P17 **Maria Island** *island* Tasmania, SE Australia
40 H12 **María Madre, Isla** *island* C Mexico
40 I12 **María Magdalena, Isla** *island* C Mexico
192 H6 **Mariana Islands** *island group* Guam/Northern Mariana Islands
175 N3 **Mariana Trench** *var.* Challenger Deep. *undersea feature* W Pacific Ocean 15°00′N 147°30′E
153 X12 **Mariāni** Assam, NE India 26°39′N 94°18′E
27 X11 **Marianna** Arkansas, C USA 34°46′N 90°49′W
23 R8 **Marianna** Florida, SE USA 30°46′N 85°13′W
172 J16 **Marianne** *island* Inner Islands, NE Seychelles
95 M19 **Mariannelund** Jönköping, S Sweden 57°37′N 15°33′E
61 D15 **Mariano I. Loza** Corrientes, NE Argentina 29°22′S 58°12′W
Mariano Machado *see* Ganda
111 A16 **Mariánské Lázně** *Ger.* Marienbad. Karlovarský Kraj, W Czech Republic 49°57′N 12°43′E
Máriaradna *see* Radna
33 S7 **Marias River** ≈ Montana, NW USA
Maria-Theresiopel *see* Subotica
Máriatölgyes *see* Dubnica nad Váhom
184 H1 **Maria van Diemen, Cape** *headland* North Island, New Zealand 34°27′S 172°38′E
109 V5 **Mariazell** Steiermark, E Austria 47°47′N 15°20′E
141 P15 **Ma'rib** W Yemen 15°28′N 45°25′E
95 I25 **Maribo** Sjælland, S Denmark 54°47′N 11°30′E
109 W9 **Maribor** *Ger.* Marburg. NE Slovenia 46°34′N 15°40′E
Marica *see* Maritsa
35 R13 **Maricopa** California, W USA 35°03′N 119°24′W
Maricourt *see* Kangiqsujuaq
81 D15 **Maridi** Western Equatoria, SW South Sudan 04°55′N 29°30′E
194 M11 **Marie Byrd Land** *physical region* Antarctica
193 P14 **Marie Byrd Seamount** *undersea feature* N Amundsen Sea 70°00′S 118°00′W
45 X11 **Marie-Galante** *var.* Ceyre to the Caribs. *island* SE Guadeloupe
45 Y6 **Marie-Galante, Canal de** *channel* S Guadeloupe
93 J20 **Mariehamn** *Fin.* Maarianhamina. Åland, SW Finland 60°05′N 19°55′E
44 C4 **Mariel** La Habana, W Cuba 23°02′N 82°44′W
99 H22 **Mariembourg** Namur, S Belgium 50°07′N 04°30′E
Marienbad *see* Mariánské Lázně
Marienburg *see* Alūksne, Latvia
Marienburg *see* Malbork, Poland
Marienburg *see* Feldioara, Romania
Marienburg in Westpreussen *see* Malbork
Marienhausen *see* Viļaka
83 D20 **Mariental** Hardap, SW Namibia 24°35′S 17°56′E
18 D13 **Marienville** Pennsylvania, NE USA 41°27′N 79°07′W
Marienwerder *see* Kwidzyń
58 C12 **Marié, Rio** ≈ NW Brazil
95 K17 **Mariestad** Västra Götaland, S Sweden 58°42′N 13°50′E
23 S3 **Marietta** Georgia, SE USA 33°57′N 84°34′W
31 U14 **Marietta** Ohio, N USA 39°25′N 81°27′W
27 N13 **Marietta** Oklahoma, C USA 33°57′N 97°08′W
81 H18 **Marigat** Baringo, W Kenya 0°29′S 35°59′E
103 S16 **Marignane** Bouches-du-Rhône, SE France 43°25′N 05°12′E
Marignano *see* Melegnano
45 O11 **Marigot** NE Dominica 15°32′N 61°18′W
122 K12 **Mariinsk** Kemerovskaya Oblast', S Russian Federation 56°13′N 87°27′E
127 Q3 **Mariinskiy Posad** Respublika Mariy El, W Russian Federation 56°07′N 47°44′E
119 E14 **Marijampolė** *prev.* Kapsukas. Marijampolė, S Lithuania 54°33′N 23°21′E
Marikostenovo *see* Marikostinovo
114 G12 **Marikostinovo** *prev.* Marikostenovo. Blagoevgrad, SW Bulgaria 41°25′N 23°21′E
60 J9 **Marília** São Paulo, S Brazil 22°13′S 49°58′W
82 D11 **Marimba** Malanje, NW Angola 08°18′S 16°58′E
Mari Milā *see* Małe Mela
Mārī Milah *see* Małe Mela
104 G4 **Marín** Galicia, NW Spain 42°23′N 08°43′W
35 N10 **Marina** California, W USA 36°40′N 121°48′W
Mar"ina Gorka *see* Mar"ina Horka
119 L17 **Mar"ina Horka** *Rus.* Mar'ina Gorka. Minskaya Voblasts', C Belarus 53°31′N 28°09′E
171 O4 **Marinduque** *island* C Philippines
31 S9 **Marine City** Michigan, N USA 42°43′N 82°29′W
31 N6 **Marinette** Wisconsin, N USA 45°06′N 87°38′W
60 I10 **Maringá** Paraná, S Brazil 23°26′S 51°55′W
83 N16 **Maringuè** Sofala, C Mozambique 17°57′S 34°23′E
104 F9 **Marinha Grande** Leiria, C Portugal 39°45′N 08°55′W
107 I15 **Marino** Lazio, C Italy 41°46′N 12°40′E
59 A15 **Mário Lobão** Acre, W Brazil 08°21′S 72°58′W
23 O5 **Marion** Alabama, S USA 32°37′N 87°19′W
27 Y11 **Marion** Arkansas, C USA 35°12′N 90°12′W
30 L17 **Marion** Illinois, N USA 37°43′N 88°55′W
31 P13 **Marion** Indiana, N USA 40°32′N 85°40′W
29 X13 **Marion** Iowa, C USA 42°01′N 91°36′W
27 O5 **Marion** Kansas, C USA 38°22′N 97°02′W
20 H6 **Marion** Kentucky, S USA 37°19′N 88°06′W
21 P9 **Marion** North Carolina, SE USA 35°43′N 82°00′W
31 S12 **Marion** Ohio, N USA 40°35′N 83°08′W
21 T12 **Marion** South Carolina, SE USA 34°11′N 79°23′W
21 Q7 **Marion** Virginia, NE USA 36°51′N 81°30′W
27 O5 **Marion Lake** ⊠ Kansas, C USA
21 S13 **Marion, Lake** ⊠ South Carolina, SE USA
27 S8 **Marionville** Missouri, C USA 37°00′N 93°38′W
55 N7 **Maripa** Bolívar, E Venezuela 07°27′N 65°10′W
55 X11 **Maripasoula** W French Guiana 03°43′N 54°04′W
35 Q9 **Mariposa** California, W USA 37°28′N 119°59′W
61 G19 **Mariscala** Lavalleja, S Uruguay 34°03′S 54°47′W
62 M4 **Mariscal Estigarribia** Boquerón, NW Paraguay 22°03′S 60°39′W
56 C6 **Mariscal Sucre** *var.* Quito. ✈ (Quito) Pichincha, C Ecuador 0°21′S 78°37′W
30 K16 **Marissa** Illinois, N USA 38°15′N 89°45′W
103 U14 **Maritime Alps** *Fr.* Alpes Maritimes, *It.* Alpi Marittime. ▲▲ France/Italy
Maritimes, Alpes *see* Maritime Alps
Maritime Territory *see* Primorskiy Kray
114 K11 **Maritsa** *var.* Marica, *Gk.* Évros, *Turk.* Meriç; *anc.* Hebrus. ≈ SW Europe *see also* Évros/Meriç
Maritsa *see* Simeonovgrad, Bulgaria
Marittime, Alpi *see* Maritime Alps
Maritzburg *see* Pietermaritzburg
117 X9 **Mariupol'** *prev.* Zhdanov. Donets'ka Oblast', SE Ukraine 47°06′N 37°34′E
55 Q6 **Mariusa, Caño** ≈ NE Venezuela
142 J5 **Marīvān** *prev.* Dezh Shāhpūr. Kordestān, W Iran 35°30′N 46°09′E
127 R3 **Mariyets** Respublika Mariy El, W Russian Federation 56°31′N 49°48′E
118 G4 **Märjamaa** *Ger.* Merjama. Raplamaa, NW Estonia 58°54′N 24°21′E
99 I15 **Mark** *Fr.* Marcq. ≈ Belgium/Netherlands
81 N17 **Marka** *var.* Merca. Shabeellaha Hoose, S Somalia 01°43′N 44°45′E
145 Z10 **Markakol', Ozero** *Kaz.* Marqaköl. ◎ E Kazakhstan
76 M12 **Markala** Ségou, W Mali 13°38′N 06°07′W
159 S15 **Markam** *var.* Gartog. Xizang Zizhiqu, W China 29°40′N 98°33′E
95 K21 **Markaryd** Kronoberg, S Sweden 56°26′N 13°35′E
142 L7 **Markazī** *off.* Ostān-e Markazī. ◈ *province* W Iran
Markazī, Ostān-e *see* Markazī
14 F14 **Markdale** Ontario, S Canada 44°19′N 80°37′W
27 X10 **Marked Tree** Arkansas, C USA 35°31′N 90°25′W
98 N11 **Markelo** Overijssel, E Netherlands 52°15′N 06°30′E
98 J9 **Markermeer** ◎ C Netherlands
97 N20 **Market Harborough** C England, United Kingdom 52°30′N 00°57′W
97 N18 **Market Rasen** E England, United Kingdom 53°23′N 00°21′W
123 O10 **Markha** ≈ NE Russian Federation
12 H16 **Markham** Ontario, S Canada 43°54′N 79°16′W
25 V12 **Markham** Texas, SW USA 28°57′N 96°04′W
186 E7 **Markham** ≈ C Papua New Guinea
195 Q11 **Markham, Mount** ▲ Antarctica 82°58′S 163°30′E
110 M11 **Marki** Mazowieckie, C Poland 52°20′N 21°07′E
158 F8 **Markit** Xinjiang Uygur Zizhiqu, NW China 38°55′N 77°40′E
117 Y5 **Markivka** *Rus.* Markovka. Luhans'ka Oblast', E Ukraine 49°34′N 39°35′E
35 Q7 **Markleeville** California, W USA 38°41′N 119°46′W
98 L8 **Marknesse** Flevoland, N Netherlands 52°44′N 05°54′E
79 H14 **Markounda** *var.* Marcounda. Ouham, NW Central African Republic 07°38′N 17°00′E
Markovka *see* Markivka
123 U7 **Markovo** Chukotskiy Avtonomnyy Okrug, NE Russian Federation 64°43′N 170°13′E
127 P8 **Marks** Saratovskaya Oblast', W Russian Federation 51°40′N 46°44′E
22 K2 **Marks** Mississippi, S USA 34°15′N 90°16′W
22 I7 **Marksville** Louisiana, S USA 31°07′N 92°04′W
101 I19 **Marktheidenfeld** Bayern, C Germany 49°50′N 09°36′E
101 J24 **Marktoberdorf** Bayern, S Germany 47°45′N 10°36′E
101 M18 **Marktredwitz** Bayern, E Germany 50°N 12°04′E
Markt-Übelbach *see* Übelbach
27 V3 **Mark Twain Lake** ⊠ Missouri, C USA
Markuleshty *see* Mărculeşti
101 E14 **Marl** Nordrhein-Westfalen, W Germany 51°38′N 07°06′E
182 E2 **Marla** South Australia 27°19′S 133°35′E
181 Y8 **Marlborough** Queensland, E Australia 22°55′S 150°07′E
97 M22 **Marlborough** S England, United Kingdom 51°25′N 01°45′W
185 I15 **Marlborough** *off.* Marlborough District. ◈ *unitary authority* South Island, New Zealand
Marlborough District *see* Marlborough
103 P3 **Marle** Aisne, N France 49°44′N 03°47′E
31 S8 **Marlette** Michigan, N USA 43°20′N 83°05′W
25 T9 **Marlin** Texas, SW USA 31°20′N 96°55′W
21 S5 **Marlinton** West Virginia, NE USA 38°14′N 80°06′W
26 M12 **Marlow** Oklahoma, C USA 34°39′N 97°57′W
155 E17 **Marmagao** Goa, W India 15°26′N 73°50′E
Marmanda *see* Marmande
102 L13 **Marmande** *anc.* Marmanda. Lot-et-Garonne, SW France 44°30′N 00°10′E
136 C11 **Marmara** Balıkesir, NW Turkey 40°36′N 27°34′E
136 D11 **Marmara Denizi** *Eng.* Sea of Marmara. *sea* NW Turkey
114 N13 **Marmaraereğlisi** Tekirdağ, NW Turkey 40°59′N 27°57′E
Marmara, Sea of *see* Marmara Denizi
136 C16 **Marmaris** Muğla, SW Turkey 36°52′N 28°17′E
28 J6 **Marmarth** North Dakota, N USA 46°17′N 103°55′W
21 Q5 **Marmet** West Virginia, NE USA 38°12′N 81°31′W
106 H5 **Marmolada, Monte** ▲ N Italy 46°36′N 11°58′E
104 M13 **Marmolejo** Andalucía, S Spain 38°03′N 04°10′W
14 J14 **Marmora** Ontario, SE Canada 44°29′N 77°40′W
39 Q14 **Marmot Bay** *bay* Alaska, USA
103 Q4 **Marne** ◈ *department* N France
103 Q4 **Marne** ≈ N France
137 U10 **Marneuli** *prev.* Borchalo, Sarvani. S Georgia 41°28′N 44°45′E
78 I13 **Maro** Moyen-Chari, S Chad 08°25′N 18°46′E
54 L12 **Maroa** Amazonas, S Venezuela 02°40′N 67°33′W
172 J3 **Maroantsetra** Toamasina, NE Madagascar 15°23′S 49°44′E
191 W11 **Marokau** *atoll* Îles Tuamotu, C French Polynesia
172 J5 **Marolambo** Toamasina, E Madagascar 20°03′S 48°08′E
172 J2 **Maromokotro** ▲ N Madagascar
83 L16 **Marondera** *prev.* Marandellas. Mashonaland East, NE Zimbabwe 18°11′S 31°33′E
55 X9 **Maroni** *Dut.* Marowijne. ≈ French Guiana/Suriname
183 V2 **Maroochydore-Mooloolaba** Queensland, E Australia 26°36′S 153°04′E
171 N14 **Maros** Sulawesi, C Indonesia 04°59′S 119°35′E
116 H11 **Maros** *var.* Mureş, Mureşul, *Ger.* Marosch, Mieresch. ≈ Hungary/Romania *see also* Mureş
Marosch *see* Maros/Mureş
Marosheviz *see* Toplița
Marosillye *see* Ilia
Marosludas *see* Luduș
Marosújvár/Marosújvárakna *see* Ocna Mureș
Marosvásárhely *see* Târgu Mureș
191 V14 **Marotiri** *var.* Îlots de Bass, Morotiri. *island group* Îles Australes, SW French Polynesia
78 G12 **Maroua** Extrême-Nord, N Cameroon 10°35′N 14°20′E
55 X12 **Marouini Rivier** ≈ SE Suriname
172 I3 **Marovoay** Mahajanga, NW Madagascar 16°05′S 46°40′E
55 W9 **Marowijne** ◈ *district* NE Suriname
Marowijne *see* Maroni
Marqaköl *see* Markakol', Ozero
193 P8 **Marquesas Fracture Zone** *tectonic feature* E Pacific Ocean
Marquesas Islands *see* Marquises, Îles
23 W17 **Marquesas Keys** *island group* Florida, SE USA
29 Y12 **Marquette** Iowa, C USA 43°02′N 91°10′W
31 N3 **Marquette** Michigan, N USA 46°32′N 87°24′W
103 N1 **Marquise** Pas-de-Calais, N France 50°49′N 01°42′E
191 X7 **Marquises, Îles** *Eng.* Marquesas Islands. *island group* N French Polynesia
183 Q6 **Marra Creek** ≈ New South Wales, SE Australia
80 B10 **Marra Hills** *plateau* W Sudan
80 B11 **Marra, Jebel** ▲ W Sudan 12°59′N 24°16′E
74 E7 **Marrakech** *var.* Marakesh, *Eng.* Marrakesh; *prev.* Morocco. W Morocco 31°39′N 07°58′W
Marrakesh *see* Marrakech
Marrāt *see* Marāh
183 N15 **Marrawah** Tasmania, SE Australia 40°56′S 144°41′E
182 I4 **Marree** South Australia 29°40′S 138°06′E
81 L17 **Marrehan** ▲▲ SW Somalia
83 N17 **Marromeu** Sofala, C Mozambique 18°18′S 35°58′E
104 J17 **Marroquí, Punta** *headland* SW Spain 36°01′N 05°39′W
183 N8 **Marrowie Creek** *seasonal river* New South Wales, SE Australia
83 O14 **Marrupa** Niassa, N Mozambique 13°10′S 37°30′E
182 D1 **Marryat** South Australia 26°22′S 133°22′E
75 Y10 **Marsā al 'Alam** *var.* Marsa 'Alam. SE Egypt 25°03′N 33°44′E
Marsa 'Alam *see* Marsā al 'Alam
75 R8 **Marsá al Burayqah** *var.* Al Burayqah. N Libya 30°21′N 19°37′E
81 J17 **Marsabit** Marsabit, N Kenya 02°20′N 37°59′E
81 J17 **Marsabit** ◈ *county* N Kenya
107 H23 **Marsala** *anc.* Lilybaeum. Sicilia, Italy, C Mediterranean Sea 37°48′N 12°26′E
75 U7 **Marsā Maṭrūḩ** *var.* Matrûḩ; *anc.* Paraetonium. NW Egypt 31°21′N 27°15′E
Marsaxlokk Bay *see* Il-Bajja ta' Marsaxlokk
65 G15 **Mars Bay** *bay* Ascension Island, C Atlantic Ocean
101 H15 **Marsberg** Nordrhein-Westfalen, W Germany 51°28′N 08°51′E
11 R15 **Marsden** Saskatchewan, S Canada 52°50′N 109°45′W
98 H7 **Marsdiep** *strait* NW Netherlands
103 R16 **Marseille** *Eng.* Marseilles; *anc.* Massilia. Bouches-du-Rhône, SE France 43°19′N 05°22′E
Marseille-Marigrane *see* Provence
30 M11 **Marseilles** Illinois, N USA 41°19′N 88°42′W
Marseilles *see* Marseille
76 J16 **Marshall** W Liberia 06°10′N 10°23′W
39 N11 **Marshall** Alaska, USA 61°52′N 162°04′W
27 U9 **Marshall** Arkansas, C USA 35°54′N 92°40′W
31 N14 **Marshall** Illinois, N USA 39°23′N 87°41′W
31 Q10 **Marshall** Michigan, N USA 42°16′N 84°57′W
29 S9 **Marshall** Minnesota, N USA 44°26′N 95°48′W
27 T4 **Marshall** Missouri, C USA 39°07′N 93°12′W
21 O9 **Marshall** North Carolina, SE USA 35°48′N 82°43′W
25 X6 **Marshall** Texas, SW USA 32°33′N 94°22′W
189 S4 **Marshall Islands** *off.* Republic of the Marshall Islands. ◆ *republic* W Pacific Ocean
175 Q3 **Marshall Islands** *island group* W Pacific Ocean
Marshall Islands, Republic of the *see* Marshall Islands
192 K6 **Marshall Seamounts** *undersea feature* SW Pacific Ocean 10°00′N 165°00′E
29 W13 **Marshalltown** Iowa, C USA 42°01′N 92°54′W
19 P12 **Marshfield** Massachusetts, NE USA 42°04′N 70°40′W
27 T7 **Marshfield** Missouri, C USA 37°20′N 92°55′W
30 K6 **Marshfield** Wisconsin, N USA 44°41′N 90°10′W
44 H1 **Marsh Harbour** Great Abaco, W The Bahamas 26°31′N 77°03′W
19 S3 **Mars Hill** Maine, NE USA 46°31′N 67°51′W
21 P9 **Mars Hill** North Carolina, SE USA 35°49′N 82°33′W
22 H10 **Marsh Island** *island* Louisiana, S USA
21 S11 **Marshville** North Carolina, SE USA 34°59′N 80°22′W
15 W5 **Marsoui** Québec, SE Canada 49°12′N 65°58′W
15 R8 **Mars, Rivière à** ≈ Québec, SE Canada
95 O15 **Märsta** Stockholm, C Sweden 59°37′N 17°52′E
95 H24 **Marstal** Syddtjylland, C Denmark 54°52′N 10°32′E
95 I19 **Marstrand** Västra Götaland, S Sweden 57°54′N 11°31′E
25 U8 **Mart** Texas, SW USA 31°32′N 96°49′W
Martaban *see* Mottama
Martaban, Gulf of *see* Mottama, Gulf of
107 Q19 **Martano** Puglia, SE Italy 40°12′N 18°19′E
Martapoera *see* Martapura
169 T13 **Martapura** *prev.* Martapoera. Borneo, C Indonesia 03°25′S 114°51′E
99 L23 **Martelange** Luxembourg, SE Belgium 49°50′N 05°43′E
114 L7 **Marten** Ruse, N Bulgaria 43°57′N 26°08′E
14 H10 **Marten River** Ontario, S Canada 46°43′N 79°45′W
11 T15 **Martensville** Saskatchewan, S Canada 52°15′N 106°42′W
Marteskirch *see* Târnăveni
Martes Tolosane *see* Martres-Tolosane
115 K25 **Mártha** Kríti, Greece, E Mediterranean Sea 35°03′N 25°22′E
183 Q6 **Marthaguy Creek** ≈ New South Wales, SE Australia
19 P13 **Martha's Vineyard** *island* Massachusetts, NE USA
108 C11 **Martigny** Valais, SW Switzerland 46°06′N 07°04′E
103 R16 **Martigues** Bouches-du-Rhône, SE France 43°24′N 05°03′E
111 J19 **Martin** *Ger.* Sankt Martin, *Hung.* Turócszentmárton; *prev.* Turčiansky Svätý Martin. Žilinský Kraj, N Slovakia 49°03′N 18°54′E
28 L11 **Martin** South Dakota, N USA 43°10′N 101°43′W
20 G8 **Martin** Tennessee, S USA 36°20′N 88°51′W
105 S7 **Martín** ≈ E Spain
107 P18 **Martina Franca** Puglia, SE Italy 40°42′N 17°20′E
185 M14 **Martinborough** Wellington, North Island, New Zealand 41°12′S 175°28′E
25 S11 **Martindale** Texas, SW USA 29°49′N 97°49′W
35 N8 **Martinez** California, W USA 38°00′N 122°12′W
23 V3 **Martinez** Georgia, SE USA 33°31′N 82°04′W
41 Q13 **Martínez de La Torre** Veracruz-Llave, E Mexico 20°05′N 97°02′W
45 Y12 **Martinique** ◇ *French overseas department* E West Indies
1 O15 **Martinique** *island* E West Indies
Martinique Channel *see* Martinique Passage
45 X12 **Martinique Passage** *var.* Dominica Channel, Martinique Channel. *channel* Dominica/Martinique
23 Q5 **Martin Lake** ⊠ Alabama, S USA
115 G18 **Martíno** *prev.* Martínon. Stereá Elláda, C Greece 38°34′N 23°13′E
Martínon *see* Martíno
194 J11 **Martin Peninsula** *peninsula* Antarctica
39 S5 **Martin Point** *headland* Alaska, USA 70°06′N 143°04′W
109 V3 **Martinsberg** Niederösterreich, NE Austria 48°23′N 15°09′E
21 V3 **Martinsburg** West Virginia, NE USA 39°28′N 77°59′W
31 V13 **Martins Ferry** Ohio, N USA 40°06′N 80°43′W
Martinskirch *see* Târnăveni
31 O14 **Martinsville** Indiana, N USA 39°25′N 86°25′W
21 S8 **Martinsville** Virginia, NE USA 36°43′N 79°53′W
65 K16 **Martin Vaz, Ilhas** *island group* E Brazil
144 J9 **Martok** *prev.* Martuk. Aktyubinsk, NW Kazakhstan 50°45′N 56°30′E
184 M12 **Marton** Manawatu-Wanganui, North Island, New Zealand 40°05′S 175°22′E
105 N13 **Martos** Andalucía, S Spain 37°44′N 03°58′W
102 M16 **Martres-Tolosane** *var.* Martes Tolosane. Haute-Garonne, S France 43°13′N 01°00′E
92 M11 **Martti** Lappi, NE Finland 67°28′N 28°20′E
Martuk *see* Martok
137 U12 **Martuni** E Armenia 40°07′N 45°20′E
58 L11 **Marudá** Pará, E Brazil 05°25′S 49°04′W
169 V6 **Marudu, Teluk** *bay* East Malaysia
149 O8 **Ma'ruf** Kandahār, SE Afghanistan 31°34′N 67°06′E
164 H13 **Marugame** Kagawa, Shikoku, SW Japan 34°17′N 133°46′E
185 H16 **Maruia** ≈ South Island, New Zealand
98 M6 **Marum** Groningen, NE Netherlands 53°07′N 06°16′E
187 R13 **Marum, Mount** ▲ Ambrym, C Vanuatu 16°15′S 168°07′E
79 P23 **Marungu** ▲▲ SE Dem. Rep. Congo
191 Y12 **Marutea** *atoll* Groupe Actéon, C French Polynesia
143 O11 **Marv Dasht** *var.* Mervdasht. Fārs, S Iran 29°50′N 52°40′E
103 P13 **Marvejols** Lozère, S France 44°35′N 03°16′E
27 X12 **Marvell** Arkansas, C USA 34°33′N 90°52′W
36 L6 **Marvine, Mount** ▲ Utah, W USA 38°40′N 111°38′W
139 Q7 **Marwānīyah** Al Anbār, C Iraq 33°58′N 42°31′E
152 F13 **Mārwār** *var.* Kharchi, Marwar Junction. Rājasthān, N India 25°41′N 73°42′E
Marwar Junction *see* Mārwār
11 R14 **Marwayne** Alberta, SW Canada 53°30′N 110°25′W
146 I14 **Mary** *prev.* Merv. Mary Welaýaty, S Turkmenistan 37°25′N 61°48′E
Mary *see* Mary Welaýaty
181 Z9 **Maryborough** Queensland, E Australia 25°32′S 152°36′E
182 M11 **Maryborough** Victoria, SE Australia 37°05′S 143°47′E
Maryborough *see* Port Laoise
83 G23 **Marydale** Northern Cape, W South Africa 29°25′S 22°06′E
117 W8 **Mar"yinka** Donets'ka Oblast', E Ukraine 47°57′N 37°27′E
Mary Island *see* Kanton
21 W4 **Maryland** *off.* State of Maryland, *also known as* America in Miniature, Cockade State, Free State, Old Line State. ◈ *state* NE USA
Maryland, State of *see* Maryland
25 P7 **Maryneal** Texas, SW USA 32°12′N 100°25′W
97 J15 **Maryport** NW England, United Kingdom 54°45′N 03°28′W
13 U13 **Marystown** Newfoundland, Newfoundland and Labrador, SE Canada 47°10′N 55°10′W
36 K6 **Marysvale** Utah, W USA 38°26′N 112°14′W

◆ Country ● Country Capital ◇ Dependent Territory ○ Dependent Territory Capital ◈ Administrative Regions ✈ International Airport ▲ Mountain ▲▲ Mountain Range ⛰ Volcano ≈ River ◎ Lake ⊠ Reservoir

Marysville – Mbout

35 O6 **Marysville** California, W USA *39°07′N 121°35′W*
27 O3 **Marysville** Kansas, C USA *39°48′N 96°37′W*
31 S13 **Marysville** Michigan, N USA *42°54′N 82°29′W*
31 S9 **Marysville** Ohio, NE USA *40°13′N 83°22′W*
32 H7 **Marysville** Washington, NW USA *48°03′N 122°10′W*
27 R2 **Maryville** Missouri, C USA *40°20′N 94°53′W*
21 N9 **Maryville** Tennessee, S USA *35°45′N 83°59′W*
146 I15 **Mary Welaýaty** *var.* Mary, *Rus.* Maryyskiy Velayat. ◈ *province* S Turkmenistan
Maryyskiy Velayat *see* Mary Welaýaty
Marzūq *see* Murzuq
42 J11 **Masachapa** *var.* Puerto Masachapa. Managua, W Nicaragua *11°47′N 86°31′W*
81 G19 **Masai Mara National Reserve** *reserve* SW Kenya
81 I21 **Masai Steppe** *grassland* NW Tanzania
81 F19 **Masaka** SW Uganda *0°20′S 31°46′E*
169 T15 **Masalembo Besar, Pulau** *island* S Indonesia
137 Y13 **Masallı** *Rus.* Masally. S Azerbaijan *39°03′N 48°39′E*
Masally *see* Masallı
171 N13 **Masamba** Sulawesi, C Indonesia *02°33′S 120°20′E*
Masampo *see* Masan
163 Y16 **Masan** *prev.* Masampo. S South Korea *35°11′N 128°36′E*
Masandam Peninsula *see* Musandam Peninsula
81 J25 **Masasi** Mtwara, SE Tanzania *10°43′S 38°48′E*
Masawa/Massawa *see* Mits'iwa
42 J10 **Masaya** Masaya, W Nicaragua *11°59′N 86°06′W*
42 J10 **Masaya** ◈ *department* W Nicaragua
171 P5 **Masbate** Masbate, N Philippines *12°21′N 123°34′E*
171 P5 **Masbate** *island* C Philippines
74 I6 **Mascara** *var.* Mouaskar. NW Algeria *35°20′N 00°09′E*
173 O7 **Mascarene Basin** *undersea feature* W Indian Ocean *15°00′S 56°00′E*
173 O9 **Mascarene Islands** *island group* W Indian Ocean
173 N9 **Mascarene Plain** *undersea feature* W Indian Ocean *19°00′S 52°00′E*
173 O7 **Mascarene Plateau** *undersea feature* W Indian Ocean *10°00′S 60°00′E*
194 H5 **Mascart, Cape** *headland* Adelaide Island, Antarctica
62 J10 **Mascasín, Salinas de** *salt lake* C Argentina
40 K13 **Mascota** Jalisco, C Mexico *20°31′N 104°46′W*
15 O12 **Mascouche** Québec, SE Canada *45°46′N 73°37′W*
124 J9 **Masel'gskaya** Respublika Kareliya, NW Russian Federation *63°09′N 34°22′E*
83 J23 **Maseru** ● (Lesotho) W Lesotho *29°21′S 27°35′E*
83 J23 **Maseru** ✕ W Lesotho *29°27′S 27°37′E*
Mashaba *see* Mashava
160 K14 **Mashan** *var.* Baishan. Guangxi Zhuangzu Zizhiqu, S China *23°40′N 108°10′E*
83 K17 **Mashava** *prev.* Mashaba. Masvingo, SE Zimbabwe *20°03′S 30°29′E*
143 U4 **Mashhad** *var.* Meshed. Khorāsān-e Razavī, NE Iran *36°16′N 59°34′E*
165 S3 **Mashike** Hokkaidō, NE Japan *43°51′N 141°30′E*
83 K20 **Mashishing** *prev.* Lydenburg. Mpumalanga, NE South Africa *25°10′S 30°29′E*
Mashīz *see* Bardsīr
149 N14 **Mashkai** ⇝ SW Pakistan
143 X13 **Māshkel** *var.* Rūd-i Māshkel, Rūd-e Māshkīd. ⇝ Iran/Pakistan
148 K12 **Māshkel, Hāmūn-i** *salt marsh* SW Pakistan
Māshkel, Rūd-i/Māshkīd, Rūd-e *see* Māshkel
83 K15 **Mashonaland Central** ◈ *province* N Zimbabwe
83 K16 **Mashonaland East** ◈ *province* NE Zimbabwe
83 J16 **Mashonaland West** ◈ *province* NW Zimbabwe
Mashtagi *see* Maştağa
141 S14 **Masīlah, Wādī al** *dry watercourse* SE Yemen
79 I21 **Masi-Manimba** Bandundu, SW Dem. Rep. Congo *04°47′S 17°54′E*
81 F17 **Masindi** W Uganda *01°41′N 31°45′E*
81 I19 **Masinga Reservoir** ⊠ S Kenya
Masīra *see* Maşīrah, Jazīrat
Masira, Gulf of *see* Maşīrah, Khalīj
141 Y10 **Maşīrah, Jazīrat** *var.* Masīra. *island* E Oman
141 Y10 **Maşīrah, Khalīj** *var.* Gulf of Masira. *bay* E Oman
Masis *see* Büyükağrı Dağı
79 O19 **Masisi** Nord-Kivu, E Dem. Rep. Congo *01°25′S 28°50′E*
Masjed-e Soleymān *see* Masjed Soleymān
142 L9 **Masjed Soleymān** *var.* Masjed-e Soleymān, Masjid-i Sulaiman. Khūzestān, SW Iran *31°59′N 49°18′E*
Masjid-i Sulaiman *see* Masjed Soleymān
Maskat *see* Masqaţ
139 Q7 **Maskhān** Al Anbār, C Iraq *33°41′N 42°46′E*
141 X8 **Maskin** *var.* Miskin. NW Oman *23°28′N 56°46′E*
97 B17 **Mask, Lough** *Ir.* Loch Measca. ⊚ W Ireland
114 N10 **Maslen Nos** *headland* E Bulgaria *42°19′N 27°47′E*
172 K3 **Masoala, Tanjona** *headland* NE Madagascar *15°59′N 50°13′E*
Masohi *see* Amahai
31 Q9 **Mason** Michigan, N USA *42°33′N 84°25′W*
31 R14 **Mason** Ohio, N USA *39°21′N 84°18′W*
25 Q10 **Mason** Texas, SW USA *30°45′N 99°15′W*
21 P4 **Mason** West Virginia, NE USA *39°01′N 82°01′W*
185 B25 **Mason Bay** *bay* Stewart Island, New Zealand
30 K13 **Mason City** Illinois, N USA *40°12′N 89°42′W*
29 V12 **Mason City** Iowa, C USA *43°09′N 93°12′W*
18 B16 **Masontown** Pennsylvania, NE USA *39°49′N 79°53′W*
141 Y8 **Masqaţ** *var.* Maskat, *Eng.* Muscat. ● (Oman) NE Oman *23°35′N 58°36′E*
106 E10 **Massa** Toscana, C Italy *44°02′N 10°07′E*
18 M11 **Massachusetts** *off.* Commonwealth of Massachusetts, *also known as* Bay State, Old Bay State, Old Colony State. ◈ *state* NE USA
19 P11 **Massachusetts Bay** *bay* Massachusetts, NE USA
35 R2 **Massacre Lake** ⊚ Nevada, W USA
107 O18 **Massafra** Puglia, SE Italy *40°35′N 17°08′E*
108 G11 **Massagno** Ticino, S Switzerland *46°01′N 08°55′E*
78 G11 **Massaguet** Hadjer-Lamis, W Chad *12°28′N 15°26′E*
Massakori *see* Massakory
78 G10 **Massakory** *var.* Massakori; *prev.* Dagana. Hadjer-Lamis, W Chad *13°02′N 15°43′E*
78 H11 **Massalassef** Hadjer-Lamis, SW Chad *11°37′N 17°09′E*
106 F13 **Massa Marittima** Toscana, C Italy *43°03′N 10°55′E*
82 B11 **Massangano** Kwanza Norte, NW Angola *09°40′S 14°13′E*
83 M18 **Massangena** Gaza, S Mozambique *21°34′S 32°57′E*
80 K9 **Massawa Channel** *channel* E Eritrea
18 J6 **Massena** New York, NE USA *44°55′N 74°53′W*
78 H11 **Massenya** Chari-Baguirmi, SW Chad *11°21′N 16°09′E*
10 I13 **Masset** Graham Island, British Columbia, SW Canada *54°00′N 132°09′W*
102 L16 **Masseube** Gers, S France *43°26′N 00°33′E*
14 E11 **Massey** Ontario, S Canada *46°13′N 82°06′W*
103 P12 **Massiac** Cantal, C France *45°16′N 03°13′E*
103 P12 **Massif Central** *plateau* C France
Massif de L'Isalo *see* Isalo
Massilia *see* Marseille
31 U12 **Massillon** Ohio, N USA *40°48′N 81°31′W*
77 N12 **Massina** Ségou, W Mali *13°58′N 05°24′W*
83 N19 **Massinga** Inhambane, SE Mozambique *23°20′S 35°25′E*
83 L20 **Massingir** Gaza, SW Mozambique *23°51′S 31°58′E*
195 Z10 **Masson Island** *island* Antarctica
Massoukou *see* Franceville
137 Z11 **Maştağa** *Rus.* Mashtagi, Mastaga. E Azerbaijan *40°31′N 50°01′E*
Mastanli *see* Momchilgrad
184 M13 **Masterton** Wellington, North Island, New Zealand *40°56′S 175°40′E*
18 M14 **Mastic** Long Island, New York, NE USA *40°48′N 72°50′W*
149 O10 **Mastung** Baluchistān, SW Pakistan *29°44′N 66°56′E*
119 J20 **Mastva** *Rus.* Mostva. ⇝ SW Belarus
119 G17 **Masty** *Rus.* Mosty. Hrodzyenskaya Voblasts', W Belarus *53°25′N 24°32′E*
164 F12 **Masuda** Shimane, Honshū, SW Japan *34°40′N 131°50′E*
92 J11 **Masugnsbyn** Norrbotten, N Sweden *67°28′N 22°01′E*
Masuku *see* Franceville
83 K17 **Masvingo** *prev.* Fort Victoria, Nyanda, Victoria. Masvingo, SE Zimbabwe *20°05′S 30°50′E*
83 K18 **Masvingo** *prev.* Victoria. ◈ *province* SE Zimbabwe
138 H5 **Maşyāf** *Fr.* Misiaf. Ḩamāh, C Syria *35°04′N 36°21′E*
110 E9 **Maszewo** Zachodnio-pomorskie, NW Poland *53°29′N 15°01′E*
83 I17 **Matabeleland North** ◈ *province* W Zimbabwe
83 J18 **Matabeleland South** ◈ *province* S Zimbabwe
82 O13 **Mataca** Niassa, N Mozambique *12°27′S 36°13′E*
14 G8 **Matachewan** Ontario, S Canada *47°58′N 80°37′W*
163 Q8 **Matad** *var.* Dzüünbulag. Dornod, E Mongolia *46°48′N 115°21′E*
79 F22 **Matadi** Bas-Congo, W Dem. Rep. Congo *05°49′S 13°31′E*
25 O4 **Matador** Texas, SW USA *34°01′N 100°50′W*
42 J9 **Matagalpa** Matagalpa, C Nicaragua *12°53′N 85°56′W*
42 K9 **Matagalpa** ◈ *department* W Nicaragua
12 I12 **Matagami** Québec, S Canada *49°47′N 77°38′W*
25 U13 **Matagorda** Texas, SW USA *28°40′N 96°57′W*
25 U13 **Matagorda Bay** *inlet* Texas, SW USA
25 U14 **Matagorda Island** *island* Texas, SW USA
25 V13 **Matagorda Peninsula** *headland* Texas, SW USA *28°34′N 96°01′W*
191 Q8 **Mataiea** Tahiti, W French Polynesia *17°46′S 149°25′W*
191 T9 **Mataiva** *atoll* Îles Tuamotu, C French Polynesia
183 O7 **Matakana** New South Wales, SE Australia *32°59′S 145°53′E*
184 N7 **Matakana Island** *island* NE New Zealand
83 C15 **Matala** Huíla, SW Angola *14°45′S 15°02′E*
190 G12 **Matala'a Pointe** *headland* Île Uvea, N Wallis and Futuna *13°20′S 176°08′W*
155 K25 **Matale** Central Province, C Sri Lanka *07°29′N 80°38′E*
190 E12 **Matalesina, Pointe** *headland* Île Alofi, W Wallis and Futuna
76 I10 **Matam** NE Senegal *15°40′N 13°18′W*
184 M8 **Matamata** Waikato, North Island, New Zealand *37°49′S 175°45′E*
77 V12 **Matamey** Zinder, S Niger *13°27′N 08°27′E*
40 L8 **Matamoros** Coahuila, NE Mexico *25°34′N 103°13′W*
41 P15 **Matamoros** *var.* Izúcar de Matamoros. Puebla, S Mexico *18°38′N 98°30′W*
41 Q8 **Matamoros** Tamaulipas, C Mexico *25°50′N 97°31′W*
75 S13 **Ma'ṭan as Sārah** SE Libya *21°45′N 21°55′E*
82 J12 **Matanda** Luapula, N Zambia *11°24′S 28°25′E*
81 J24 **Matandu** ⇝ S Tanzania
15 V6 **Matane** Québec, SE Canada *48°50′N 67°31′W*
15 V6 **Matane** ⇝ Québec, SE Canada
77 S12 **Matankari** Dosso, SW Niger *13°39′N 04°03′E*
39 R11 **Matanuska River** ⇝ Alaska, USA
54 G7 **Matanza** Santander, N Colombia *07°22′N 73°02′W*
44 D4 **Matanzas** Matanzas, NW Cuba *23°N 81°32′W*
15 V7 **Matapédia** ⇝ Québec, SE Canada
15 V6 **Matapédia, Lac** ⊚ Québec, SE Canada
190 B17 **Mata Point** *headland* SE Niue *19°07′S 169°51′E*
190 D12 **Matapu, Pointe** *headland* Île Futuna, W Wallis and Futuna
62 G12 **Mataquito, Río** ⇝ C Chile
155 K26 **Matara** Southern Province, S Sri Lanka *05°58′N 80°33′E*
115 D18 **Mataránga** *var.* Matarángka. Dytikí Elláda, C Greece *38°32′N 21°28′E*
171 K16 **Mataram** Pulau Lombok, C Indonesia *08°36′S 116°07′E*
Matarángka *see* Mataránga
181 Q3 **Mataranka** Northern Territory, N Australia *14°55′S 133°03′E*
105 W6 **Mataró** *anc.* Illuro. Cataluña, E Spain *41°32′N 02°27′E*
184 O8 **Matata** Bay of Plenty, North Island, New Zealand *37°54′S 176°45′E*
192 K16 **Matātula, Cape** *headland* Tutuila, W American Samoa *14°15′S 170°35′W*
185 D24 **Mataura** Southland, South Island, New Zealand *46°12′S 168°53′E*
185 D24 **Mataura** ⇝ South Island, New Zealand
Mata Uta *see* Matā'utu
192 H16 **Matā'utu** Upolu, C Samoa *13°57′S 171°55′W*
190 G11 **Matā'utu** *var.* Mata Uta. ○ (Wallis and Futuna) Île Uvea, Wallis and Futuna *13°22′S 176°12′W*
190 G11 **Matā'utu, Baie de** *bay* Île Uvea, Wallis and Futuna
191 P7 **Matavai, Baie de** *bay* Tahiti, W French Polynesia
190 I16 **Matavera** Rarotonga, S Cook Islands *21°13′S 159°44′W*
191 V16 **Mataveri** Easter Island, Chile, E Pacific Ocean *27°10′S 109°27′W*
191 V17 **Mataveri** ✕ (Easter Island) Easter Island, Chile, E Pacific Ocean *27°10′S 109°27′W*
184 P9 **Matawai** Gisborne, North Island, New Zealand *38°23′S 177°31′E*
15 O10 **Matawin** ⇝ Québec, SE Canada
145 V13 **Matay** Almaty, SE Kazakhstan *45°53′N 78°45′E*
14 K8 **Matchi-Manitou, Lac** ⊚ Québec, SE Canada
41 O10 **Matehuala** San Luis Potosí, C Mexico *23°40′N 100°40′W*
45 V13 **Matelot** Trinidad, Trinidad and Tobago *10°48′N 61°06′W*
83 M15 **Matenge** Tete, NW Mozambique *15°22′S 33°47′E*
107 O18 **Matera** Basilicata, S Italy *40°39′N 16°35′E*
111 O21 **Mátészalka** Szabolcs-Szatmár-Bereg, E Hungary *47°58′N 22°17′E*
93 H17 **Matfors** Västernorrland, C Sweden *62°21′N 17°02′E*
102 K11 **Matha** Charente-Maritime, W France *45°50′N 00°13′W*
0 F15 **Mathematicians Seamounts** *undersea feature* E Pacific Ocean *15°00′N 111°00′W*
21 X6 **Mathews** Virginia, NE USA *37°26′N 76°20′W*
25 S14 **Mathis** Texas, SW USA *28°05′N 97°49′W*
152 J11 **Mathura** *prev.* Muttra. Uttar Pradesh, N India *27°30′N 77°42′E*
Mathurai *see* Madurai
171 R7 **Mati** Mindanao, S Philippines *06°58′N 126°11′E*
Matianus *see* Orūmīyeh, Daryācheh-ye
Matiara *see* Matiāri
149 Q15 **Matiāri** *var.* Matiara. Sind, SE Pakistan *25°38′N 68°29′E*
41 S16 **Matías Romero** Oaxaca, SE Mexico *16°53′N 95°02′W*
43 O13 **Matina** Limón, E Costa Rica *10°06′N 83°18′W*
14 D10 **Matinenda Lake** ⊚ Ontario, S Canada
19 R8 **Matinicus Island** *island* Maine, NE USA
Matisco/Matisco Ædourum *see* Mâcon
113 K19 **Matit, Lumi i** ⇝ NW Albania
149 Q16 **Mātli** Sind, SE Pakistan *25°06′N 68°37′E*
97 M18 **Matlock** C England, United Kingdom *53°08′N 01°32′W*
59 F18 **Mato Grosso** *prev.* Vila Bela da Santíssima Trindade. Mato Grosso, W Brazil *14°53′S 59°58′W*
59 G17 **Mato Grosso** *off.* Estado de Mato Grosso; *prev.* Matto Grosso. ◈ *state* W Brazil
60 H8 **Mato Grosso do Sul** *off.* Estado de Mato Grosso do Sul. ◈ *state* S Brazil
Mato Grosso do Sul, Estado de *see* Mato Grosso do Sul
Mato Grosso, Estado de *see* Mato Grosso
59 I18 **Mato Grosso, Planalto de** *plateau* C Brazil
83 L21 **Matola** Maputo, S Mozambique *25°57′S 32°27′E*
104 G6 **Matosinhos** *prev.* Matozinhos. Porto, NW Portugal *41°11′N 08°42′W*
Matou *see* Pingguo
55 Z10 **Matoury** NE French Guiana *04°49′N 52°17′W*
Matozinhos *see* Matosinhos
111 L21 **Mátra** ▲▲ N Hungary
141 Y8 **Maţraḩ** *var.* Mutrah. NE Oman *23°35′N 58°31′E*
116 L12 **Mătrăşeşti** Vrancea, E Romania *45°53′N 27°14′E*
108 M8 **Matrei am Brenner** Tirol, W Austria *47°09′N 11°28′E*
109 P8 **Matrei in Osttirol** Tirol, W Austria *47°04′N 12°31′E*
76 I15 **Matru** SW Sierra Leone *07°37′N 12°08′W*
Maṭrūḥ *see* Marsā Maṭrūḥ
165 U16 **Matsubara** *var.* Matubara. Kagoshima, Tokuno-shima, SW Japan *32°58′N 129°56′E*
161 S12 **Matsu Dao** *var.* Mazu Tao; *prev.* Matsu Tao. *island* NW Taiwan
164 G12 **Matsue** *var.* Matsuye, Matue. Shimane, Honshū, SW Japan *35°27′N 133°04′E*
165 Q6 **Matsumae** Hokkaidō, NE Japan *41°27′N 140°04′E*
164 M12 **Matsumoto** *var.* Matumoto. Nagano, Honshū, S Japan *36°18′N 137°58′E*
164 K14 **Matsusaka** *var.* Matsuzaka, Matusaka. Mie, Honshū, SW Japan *34°33′N 136°31′E*
Matsu Tao *see* Matsu Dao
Matsutō *see* Mattō
164 F14 **Matsuyama** *var.* Matuyama. Ehime, Shikoku, SW Japan *33°50′N 132°47′E*
Matsuye *see* Matsue
Matsuzaka *see* Matsusaka
164 M14 **Matsuzaki** Shizuoka, Honshū, S Japan *34°43′N 138°45′E*
14 F8 **Mattagami** ⇝ Ontario, S Canada
14 F8 **Mattagami Lake** ⊚ Ontario, S Canada
62 K12 **Mattaldi** Córdoba, C Argentina *34°26′S 64°14′W*
21 Y9 **Mattamuskeet, Lake** ⊚ North Carolina, SE USA
21 W6 **Mattaponi River** ⇝ Virginia, NE USA
14 I11 **Mattawa** Ontario, SE Canada *46°19′N 78°42′W*
14 I11 **Mattawa** ⇝ Ontario, SE Canada
19 S5 **Mattawamkeag** Maine, NE USA *45°30′N 68°20′W*
19 S4 **Mattawamkeag Lake** ⊚ Maine, NE USA
108 D11 **Matterhorn** *It.* Monte Cervino. ▲ Italy/Switzerland *45°58′N 07°36′E* *see also* Cervino, Monte
32 L12 **Matterhorn** *var.* Sacajawea Peak. ▲ Oregon, NW USA *45°12′N 117°18′W*
35 W1 **Matterhorn** ▲ Nevada, W USA *41°48′N 115°22′W*
Matterhorn *see* Cervino, Monte
35 R8 **Matterhorn Peak** ▲ California, W USA *38°06′N 119°19′W*
109 Y5 **Mattersburg** Burgenland, E Austria *47°45′N 16°24′E*
108 E11 **Matter Vispa** ⇝ S Switzerland
55 R7 **Matthews Ridge** N Guyana *07°30′N 60°07′W*
44 K7 **Matthew Town** Great Inagua, S The Bahamas *20°56′N 73°41′W*
109 Q4 **Mattighofen** Oberösterreich, NW Austria *48°07′N 13°09′E*
107 N16 **Mattinata** Puglia, SE Italy *41°41′N 16°01′E*
141 T9 **Maţţī, Sabkhat** *salt flat* Saudi Arabia/United Arab Emirates
18 M14 **Mattituck** Long Island, New York, NE USA *40°59′N 72°31′W*
164 L11 **Mattō** *var.* Hakusan, Matsutō. Ishikawa, Honshū, SW Japan *36°31′N 136°34′E*
Matto Grosso *see* Mato Grosso
30 M14 **Mattoon** Illinois, N USA *39°28′N 88°22′W*
57 L16 **Mattos, Río** ⇝ C Bolivia
Mattu *see* Metu
169 R9 **Matu** Sarawak, East Malaysia *02°39′N 111°31′E*
Matubara *see* Matsubara
57 E14 **Matucana** Lima, W Peru *11°54′S 76°25′W*
Matue *see* Matsue
187 Y15 **Matuku** *island* S Fiji
112 B9 **Matulji** Primorje-Gorski Kotar, NW Croatia *45°21′N 14°18′E*
Matumoto *see* Matsumoto
55 P5 **Maturín** Monagas, NE Venezuela *09°45′N 63°10′W*
Matusaka *see* Matsusaka
Matuyama *see* Matsuyama
126 K11 **Matveyev Kurgan** Rostovskaya Oblast', SW Russian Federation *47°31′N 38°55′E*
127 O8 **Matyshevo** Volgogradskaya Oblast', SW Russian Federation *50°53′N 44°09′E*
153 O13 **Mau** *var.* Maunāth Bhanjan. Uttar Pradesh, N India *25°57′N 83°33′E*
83 O14 **Maúa** Niassa, N Mozambique *13°53′S 37°10′E*
102 M17 **Maubermé, Pic de** *var.* Tuc de Moubermé, *Sp.* Pico Mauberme; *prev.* Tuc de Maubermé. ▲ France/Spain *42°48′N 00°54′E* *see also* Moubermé, Tuc de
Maubermé, Pic de *see* Moubermé, Tuc de
Mauberme, Pico *see* Maubermé, Pic de/Moubermé, Tuc de
Maubermé, Tuc de *see* Maubermé, Pic de/Moubermé, Tuc de
103 Q2 **Maubeuge** Nord, N France *50°16′N 04°00′E*
166 L8 **Maubin** Ayeyawady, SW Myanmar (Burma) *16°44′N 95°37′E*
152 L13 **Maudaha** Uttar Pradesh, N India *25°41′N 80°07′E*
183 N9 **Maude** New South Wales, SE Australia *34°30′S 144°20′E*
195 P3 **Maudheimvidda** *physical region* Antarctica
65 N22 **Maud Rise** *undersea feature* S Atlantic Ocean
109 Q4 **Mauerkirchen** Oberösterreich, NW Austria *48°11′N 13°08′E*
Mauersee *see* Mamry, Jezioro
188 K2 **Maug Islands** *island group* N Northern Mariana Islands
103 Q15 **Mauguio** Hérault, S France *43°37′N 04°01′E*
193 N5 **Maui** *island* Hawai'i, USA, C Pacific Ocean
190 M16 **Mauke** *atoll* S Cook Islands
62 G13 **Maule** *var.* Región del Maule. ◈ *region* C Chile
102 J9 **Mauléon** Deux-Sèvres, W France *46°55′N 00°45′W*
102 J16 **Mauléon-Licharre** Pyrénées-Atlantiques, SW France *43°14′N 00°51′W*
Maule, Región del *see* Maule
62 G13 **Maule, Río** ⇝ C Chile
63 G17 **Maullín** Los Lagos, S Chile *41°38′S 73°35′W*
31 R11 **Maumee** Ohio, N USA *41°34′N 83°40′W*
31 Q12 **Maumee River** ⇝ Indiana/Ohio, N USA
27 U11 **Maumelle** Arkansas, C USA *34°51′N 92°24′W*
27 T11 **Maumelle, Lake** ⊠ Arkansas, C USA
171 O16 **Maumere** *prev.* Maoemere. Flores, S Indonesia *08°35′S 122°13′E*
83 G17 **Maun** North-West, C Botswana *20°01′S 23°28′E*
Maunāth Bhanjan *see* Mau
Maunawai *see* Waimea
190 H16 **Maungaroa** ▲ Rarotonga, S Cook Islands *21°13′S 159°48′W*
184 K3 **Maungatapere** Northland, North Island, New Zealand *35°46′S 174°10′E*
184 K4 **Maungaturoto** Northland, North Island, New Zealand *36°06′S 174°21′E*
166 J5 **Maungdaw** *var.* Zullapara. Rakhine State, W Myanmar (Burma) *20°51′N 92°23′E*
191 R10 **Maupiti** *var.* Maurua. *island* Îles Sous le Vent, W French Polynesia
152 K14 **Mau Rānīpur** Uttar Pradesh, N India *25°14′N 79°07′E*
22 K9 **Maurepas, Lake** ⊚ Louisiana, S USA
103 T16 **Maures** ▲▲ SE France
103 O12 **Mauriac** Cantal, C France *45°13′N 02°21′E*
Maurice *see* Mauritius
65 J20 **Maurice Ewing Bank** *undersea feature* SW Atlantic Ocean *51°00′S 43°00′W*
182 C4 **Maurice, Lake** *salt lake* South Australia
18 I17 **Maurice River** ⇝ New Jersey, NE USA
25 Y10 **Mauriceville** Texas, SW USA *30°13′N 93°52′W*
98 K12 **Maurik** Gelderland, C Netherlands *51°57′N 05°25′E*
76 H8 **Mauritania** *off.* Islamic Republic of Mauritania, *Ar.* Mūrītānīyah. ◆ *republic* W Africa
Mauritania, Islamic Republic of *see* Mauritania
173 W15 **Mauritius** *off.* Republic of Mauritius, *Fr.* Maurice. ◆ *republic* W Indian Ocean
128 M17 **Mauritius** *island* W Indian Ocean
Mauritius, Republic of *see* Mauritius
173 N9 **Mauritius Trench** *undersea feature* W Indian Ocean
102 H6 **Mauron** Morbihan, NW France *48°06′N 02°16′W*
103 N13 **Maurs** Cantal, C France *44°45′N 02°12′E*
Maurua *see* Maupiti
Maury Mid-Ocean Channel *see* Maury Seachannel
64 L6 **Maury Seachannel** *var.* Maury Mid-Ocean Channel. *undersea feature* N Atlantic Ocean *56°33′N 24°00′W*
30 K8 **Mauston** Wisconsin, N USA *43°46′N 90°06′W*
109 R8 **Mauterndorf** Salzburg, NW Austria *47°09′N 13°39′E*
109 T4 **Mauthausen** Oberösterreich, N Austria *48°13′N 14°30′E*
109 Q9 **Mauthen** Kärnten, S Austria *46°39′N 12°58′E*
83 F15 **Mavinga** Kuando Kubango, SE Angola *15°44′S 20°21′E*
83 M17 **Mavita** Manica, W Mozambique *19°31′S 33°09′E*
115 K22 **Mavrópetra, Akrotírio** *headland* Santoríni, Kykládes, Greece, Aegean Sea *36°28′N 25°22′E*
115 F16 **Mavrovoúni** ▲ C Greece *39°37′N 22°45′E*
184 Q8 **Mawhai Point** *headland* North Island, New Zealand *38°08′S 178°24′E*
166 L3 **Mawlaik** Sagaing, C Myanmar (Burma) *23°40′N 94°26′E*
Mawlamyaing *see* Mawlamyine
166 M9 **Mawlamyine** *var.* Mawlamyaing, Moulmein. Mon State, S Myanmar (Burma) *16°30′N 97°39′E*
166 L8 **Mawlamyinegyunn** *var.* Moulmeingyun. Ayeyawady, SW Myanmar (Burma) *16°24′N 95°15′E*
141 N14 **Mawr, Wādī** *dry watercourse* NW Yemen
Mawşil, Al *see* Nīnawá
195 X5 **Mawson** *Australian research station* Antarctica *67°24′S 63°16′E*
195 X5 **Mawson Coast** *physical region* Antarctica
28 M4 **Max** North Dakota, N USA *47°48′N 101°18′W*
41 W12 **Maxcanú** Yucatán, SE Mexico *20°35′N 90°00′W*
Maxesibebi *see* Mount Ayliff
109 Q5 **Maxglan** ✕ (Salzburg) Salzburg, W Austria *47°46′N 13°00′E*
93 K16 **Maxmo** *Fin.* Maksamaa. Österbotten, W Finland *63°13′N 22°04′E*
21 T11 **Maxton** North Carolina, SE USA *34°47′N 79°34′W*
25 R8 **May** Texas, SW USA *31°58′N 98°54′W*
186 B6 **May** ⇝ NW Papua New Guinea
123 R10 **Maya** ⇝ E Russian Federation
151 Q19 **Māyābandar** Andaman and Nicobar Islands, India, E Indian Ocean *12°43′N 92°52′E*
Mayadin *see* Al Mayādīn
44 L5 **Mayaguana** *island* SE The Bahamas
44 L5 **Mayaguana Passage** *passage* SE The Bahamas
45 S6 **Mayagüez** W Puerto Rico *18°12′N 67°08′W*
45 R6 **Mayagüez, Bahía de** *bay* W Puerto Rico
Mayals *see* Maials
79 G20 **Mayama** Pool, SE Congo *03°50′S 14°52′E*
37 V8 **Maya, Mesa De** ▲ Colorado, C USA *37°06′N 103°30′W*
143 R4 **Mayamey** Semnān, N Iran *36°50′N 55°50′E*
42 F3 **Maya Mountains** *Sp.* Montañas Mayas. ▲▲ Belize/Guatemala
44 I7 **Mayarí** Holguín, E Cuba *20°41′N 75°42′W*
Mayas, Montañas *see* Maya Mountains
18 I17 **May, Cape** *headland* New Jersey, NE USA *38°55′N 74°57′W*
80 J11 **Maych'ew** *var.* Mai Chio, *It.* Mai Ceu. Tigray, N Ethiopia *12°55′N 39°30′E*
138 I2 **Maydān Ikbiz** Ḩalab, N Syria *36°51′N 36°40′E*
Maydān Shahr *see* Maīdān Shahr
80 O12 **Maydh** Sanaag, N Somalia *10°57′N 47°07′E*
Maydī *see* Mīdī
Mayebashi *see* Maebashi
Mayence *see* Mainz
102 K6 **Mayenne** Mayenne, NW France *48°18′N 00°37′W*
102 J6 **Mayenne** ◈ *department* NW France
102 J7 **Mayenne** ⇝ N France
36 K12 **Mayer** Arizona, SW USA *34°25′N 112°15′W*
22 J4 **Mayersville** Mississippi, S USA *32°54′N 91°04′W*
11 P14 **Mayerthorpe** Alberta, SW Canada *53°59′N 115°06′W*
21 S12 **Mayesville** South Carolina, SE USA *34°00′N 80°10′W*
185 G19 **Mayfield** Canterbury, South Island, New Zealand *43°50′S 171°24′E*
33 N14 **Mayfield** Idaho, NW USA *43°24′N 115°56′W*
20 G7 **Mayfield** Kentucky, S USA *36°45′N 88°40′W*
36 L5 **Mayfield** Utah, W USA *39°06′N 111°42′W*
Mayhan *see* Sant
37 T14 **Mayhill** New Mexico, SW USA *32°52′N 105°28′W*
Maykain *see* Maykayyn
145 T9 **Maykayyn** *prev.* Maykain, *Kaz.* Mayqayyng. Pavlodar, NE Kazakhstan *51°27′N 75°52′E*
126 L14 **Maykop** Respublika Adygeya, SW Russian Federation *44°36′N 40°07′E*
Maylibash *see* Maylybas
Mayli-Say *see* Mayluu-Suu
147 T9 **Mayluu-Suu** *prev.* Mayli-Say, *Kir.* Mayly-Say. Dzhalal-Abadskaya Oblast', W Kyrgyzstan *41°16′N 72°27′E*
144 L14 **Maylybas** *prev.* Maylibash. Kzylorda, S Kazakhstan *45°51′N 62°37′E*
Mayly-Say *see* Mayluu-Suu
Maymana *see* Maīmanah
Maymyo *see* Pyin-Oo-Lwin
123 V7 **Mayn** ⇝ NE Russian Federation
127 Q5 **Mayna** Ul'yanovskaya Oblast', W Russian Federation *54°04′N 47°20′E*
21 N8 **Maynardville** Tennessee, S USA *36°15′N 83°48′W*
14 J13 **Maynooth** Ontario, SE Canada *45°14′N 77°54′W*
10 I6 **Mayo** Yukon, NW Canada *63°37′N 135°48′W*
23 U9 **Mayo** Florida, SE USA *30°03′N 83°10′W*
97 B16 **Mayo** *Ir.* Maigh Eo. *cultural region* W Ireland
Mayo *see* Maio
78 G12 **Mayo-Kébbi Est** *off.* Région du Mayo-Kébbi Est. ◈ *region* SW Chad
Mayo-Kébbi Est, Région du *see* Mayo-Kébbi Est
78 G13 **Mayo-Kébbi Ouest** *off.* Région du mayo-Kébbi Ouest. ◈ *region* SW Chad
Mayo-Kébbi Ouest, Région du *see* Mayo-Kébbi Ouest
79 F19 **Mayoko** Niari, SW Congo *02°19′S 12°47′E*
171 P4 **Mayon Volcano** ⛰ Luzon, N Philippines *13°15′N 123°41′E*
61 A24 **Mayor Buratovich** Buenos Aires, E Argentina *39°15′S 62°35′W*
104 L4 **Mayorga** Castilla y León, N Spain *42°10′N 05°16′W*
184 N6 **Mayor Island** *island* NE New Zealand
Mayor Pablo Lagerenza *see* Capitán Pablo Lagerenza
173 I14 **Mayotte** ◇ *French overseas department* E Africa
Mayoumba *see* Mayumba
44 J13 **May Pen** C Jamaica *17°58′N 77°15′W*
Mayqayyng *see* Maykayyn
171 O1 **Mayraira Point** *headland* Luzon, N Philippines *18°36′N 120°47′E*
109 N8 **Mayrhofen** Tirol, W Austria *47°09′N 11°52′E*
186 A6 **May River** East Sepik, NW Papua New Guinea *04°24′S 141°52′E*
139 Y10 **Maysān** *off.* Muḩāfa at Maysān, *var.* Al 'Amārah, Mīsān. ◈ *governorate* SE Iraq
Maysān, Muḩāfa at *see* Maysān
123 R13 **Mayskiy** Amurskaya Oblast', SE Russian Federation *52°13′N 129°30′E*
127 O15 **Mayskiy** Kabardino-Balkarskaya Respublika, SW Russian Federation *43°37′N 44°04′E*
145 U9 **Mayskoye** Pavlodar, NE Kazakhstan *50°55′N 78°11′E*
18 J17 **Mays Landing** New Jersey, NE USA *39°27′N 74°44′W*
21 N4 **Maysville** Kentucky, S USA *38°38′N 83°46′W*
27 R2 **Maysville** Missouri, C USA *39°53′N 94°21′W*
79 D20 **Mayumba** *var.* Mayoumba. Nyanga, S Gabon *03°23′S 10°38′E*
31 S8 **Mayville** Michigan, N USA *43°18′N 83°16′W*
18 C11 **Mayville** New York, NE USA *42°15′N 79°32′W*
29 Q4 **Mayville** North Dakota, N USA *47°27′N 97°17′W*
Mayyali *see* Mahe
Mayyit, Al Baḩr al *see* Dead Sea
83 J15 **Mazabuka** Southern, S Zambia *15°52′S 27°46′E*
Mazaca *see* Kayseri
Mazagan *see* El-Jadida
32 J7 **Mazama** Washington, NW USA *48°34′N 120°26′W*
103 O15 **Mazamet** Tarn, S France *43°30′N 02°21′E*
143 O4 **Māzandarān** *off.* Ostān-e Māzandarān. ◈ *province* N Iran
Māzandarān, Ostān-e *see* Māzandarān
156 F7 **Mazar** Xinjiang Uygur Zizhiqu, NW China *36°28′N 77°00′E*
107 H24 **Mazara del Vallo** Sicilia, Italy, C Mediterranean Sea *37°39′N 12°36′E*
149 O2 **Mazār-e Sharīf** *var.* Mazār-i Sharif. Balkh, N Afghanistan *36°44′N 67°06′E*
Mazār-i Sharif *see* Mazār-e Sharīf
105 R13 **Mazarrón** Murcia, SE Spain *37°36′N 01°19′W*
105 R14 **Mazarrón, Golfo de** *gulf* SE Spain
55 S9 **Mazaruni River** ⇝ N Guyana
42 B6 **Mazatenango** Suchitepéquez, SW Guatemala *14°31′N 91°30′W*
40 I10 **Mazatlán** Sinaloa, C Mexico *23°15′N 106°24′W*
36 L12 **Mazatzal Mountains** ▲▲ Arizona, SW USA
118 D10 **Mažeikiai** Telšiai, NW Lithuania *56°19′N 22°22′E*
118 D7 **Mazirbe** NW Latvia *57°39′N 22°16′E*
40 G5 **Mazocahui** Sonora, NW Mexico *29°32′N 110°09′W*
57 I18 **Mazocruz** Puno, S Peru *16°41′S 69°42′W*
Mazoe, Rio *see* Mazowe
79 N21 **Mazomeno** Maniema, E Dem. Rep. Congo *04°54′S 27°13′E*
159 Q6 **Mazong Shan** ▲ N China *41°40′N 97°10′E*
83 L16 **Mazowe** *var.* Rio Mazoe. ⇝ Mozambique/Zimbabwe
110 M11 **Mazowieckie** ◈ *province* C Poland
Mazra'a *see* Al Mazra'ah
138 G6 **Mazraat Kfar Debiâne** C Lebanon *34°00′N 35°51′E*
118 H7 **Mazsalaca** *Est.* Väike-Salatsi, *Ger.* Salisburg. N Latvia *57°52′N 25°03′E*
110 L9 **Mazury** *physical region* NE Poland
Mazu Tao *see* Matsu Dao
119 M20 **Mazyr** *Rus.* Mozyr'. Homyel'skaya Voblasts', SE Belarus *52°04′N 29°15′E*
107 K25 **Mazzarino** Sicilia, Italy, C Mediterranean Sea *37°18′N 14°13′E*
Mba *see* Ba
83 L21 **Mbabane** ● (Swaziland) NW Swaziland *26°24′S 31°13′E*
Mbacké *see* Mbaké
77 N16 **Mbahiakro** E Ivory Coast *07°33′N 04°19′W*
79 I16 **Mbaïki** *var.* M'Baiki. Lobaye, SW Central African Republic *03°52′N 17°58′E*
M'Baiki *see* Mbaïki
79 F14 **Mbakaou, Lac de** ⊚ C Cameroon
76 G11 **Mbaké** *var.* Mbacké. W Senegal *14°47′N 15°54′W*
82 L11 **Mbala** *prev.* Abercorn. Northern, NE Zambia *08°50′S 31°23′E*
83 J18 **Mbalabala** *prev.* Balla Balla. Matabeleland South, SW Zimbabwe *20°27′S 29°03′E*
81 G18 **Mbale** E Uganda *01°04′N 34°12′E*
79 E16 **Mbalmayo** *var.* M'Balmayo. Centre, S Cameroon *03°30′N 11°31′E*
M'Balmayo *see* Mbalmayo
81 H25 **Mbamba Bay** Ruvuma, S Tanzania *11°15′S 34°44′E*
79 I18 **Mbandaka** *prev.* Coquilhatville. Equateur, NW Dem. Rep. Congo *0°07′N 18°12′E*
82 B9 **M'banza Kongo** Zaire Province, NW Angola *06°11′S 14°16′E*
79 G21 **Mbanza-Ngungu** Bas-Congo, W Dem. Rep. Congo *05°19′S 14°45′E*
67 V11 **Mbarangandu** ⇝ E Tanzania
81 E19 **Mbarara** SW Uganda *0°36′S 30°40′E*
79 L15 **Mbari** ⇝ SE Central African Republic
81 I24 **Mbarika Mountains** ▲▲ S Tanzania
78 F13 **Mbé** Nord, N Cameroon *07°51′N 13°36′E*
81 J24 **Mbemkuru** *var.* Mbwemkuru. ⇝ S Tanzania
Mbengga *see* Beqa
172 H13 **Mbéni** Grande Comore, NW Comoros
83 K18 **Mberengwa** Midlands, S Zimbabwe *20°29′S 29°55′E*
81 G24 **Mbeya** Mbeya, SW Tanzania *08°54′S 33°29′E*
81 G23 **Mbeya** ◈ *region* S Tanzania
83 J24 **Mbhashe** *prev.* Mbashe. ⇝ S South Africa
79 E19 **Mbigou** Ngounié, C Gabon *01°54′S 12°00′E*
Mbilua *see* Vella Lavella
79 F19 **Mbinda** Niari, SW Congo *02°07′S 12°52′E*
79 D17 **Mbini** W Equatorial Guinea *01°34′N 09°39′E*
Mbini *see* Uolo, Río
83 L18 **Mbizi** Masvingo, SE Zimbabwe *21°23′S 30°54′E*
81 G23 **Mbogo** Mbeya, W Tanzania *07°24′S 33°26′E*
79 N15 **Mboki** Haut-Mbomou, SE Central African Republic *05°18′N 25°52′E*
83 K21 **Mbombela** *prev.* Nelspruit. Mpumalanga, NE South Africa *25°28′S 30°58′E* *see also* Nelspruit
79 G18 **Mbomo** Cuvette, NW Congo *0°25′N 14°42′E*
79 L15 **Mbomou** ◈ *prefecture* SE Central African Republic
Mbomou/M'Bomu/Mbomu *see* Bomu
76 F11 **Mbour** W Senegal *14°22′N 16°54′W*
76 I10 **Mbout** Gorgol, S Mauritania *16°02′N 12°38′W*

◆ Country ● Country Capital ◇ Dependent Territory ○ Dependent Territory Capital ◈ Administrative Regions ✕ International Airport ▲ Mountain ▲▲ Mountain Range ⛰ Volcano ⇝ River ⊚ Lake ⊠ Reservoir

Mbrès – Mercara

79 J14 **Mbrès** var. Mbrés. Nana-Grébizi, C Central African Republic 06°40′N 19°46′E
Mbrés see Mbrès
79 L22 **Mbuji-Mayi** prev. Bakwanga. Kasai-Oriental, S Dem. Rep. Congo 06°05′S 23°30′E
81 H21 **Mbulu** Manyara, N Tanzania 03°45′S 35°33′E
186 E5 **M'bunai** var. Bunai. Manus Island, N Papua New Guinea 02°08′S 147°13′E
62 N8 **Mburucuyá** Corrientes, NE Argentina 28°03′S 58°15′W
Mbutha see Buca
Mbwemkuru see Mbemkuru
81 G21 **Mbwikwe** Singida, C Tanzania 05°19′S 34°09′E
13 O15 **McAdam** New Brunswick, SE Canada 45°34′N 67°20′W
25 O5 **McAdoo** Texas, SW USA 33°41′N 100°58′W
35 V2 **McAfee Peak** ▲ Nevada, W USA 41°31′N 115°57′W
27 P11 **McAlester** Oklahoma, C USA 34°56′N 95°46′W
25 S17 **McAllen** Texas, SW USA 26°12′N 98°14′W
21 S11 **McBee** South Carolina, SE USA 34°30′N 80°12′W
11 N14 **McBride** British Columbia, SW Canada 53°21′N 120°19′W
24 M9 **McCamey** Texas, SW USA 31°08′N 102°13′W
33 R15 **McCammon** Idaho, NW USA 42°38′N 112°10′W
35 X11 **McCarran** ✈ (Las Vegas) Nevada, W USA 36°04′N 115°07′W
39 T11 **McCarthy** Alaska, USA 61°25′N 142°55′W
30 M5 **McCaslin Mountain** hill Wisconsin, N USA
25 O2 **McClellan Creek** ~ Texas, SW USA
21 T14 **McClellanville** South Carolina, SE USA 33°07′N 79°27′W
195 R12 **McClintock, Mount** ▲ Antarctica 80°09′S 156°42′E
35 N2 **McCloud** California, W USA 41°15′N 122°09′W
35 N3 **McCloud River** ~ California, W USA
35 Q9 **McClure, Lake** ☒ California, W USA
197 O8 **McClure Strait** strait Northwest Territories, N Canada
29 N4 **McClusky** North Dakota, N USA 47°27′N 100°25′W
21 T11 **McColl** South Carolina, SE USA 34°40′N 79°33′W
22 K7 **McComb** Mississippi, S USA 31°14′N 90°27′W
18 E16 **McConnellsburg** Pennsylvania, NE USA 39°56′N 78°00′W
31 T14 **McConnelsville** Ohio, N USA 39°39′N 81°51′W
28 M17 **McCook** Nebraska, C USA 40°12′N 100°38′W
21 P13 **McCormick** South Carolina, SE USA 33°55′N 82°19′W
11 W16 **McCreary** Manitoba, S Canada 50°48′N 99°34′W
27 W11 **McCrory** Arkansas, C USA 35°15′N 91°12′W
25 T10 **McDade** Texas, SW USA 30°15′N 97°15′W
23 O8 **McDavid** Florida, SE USA 30°51′N 87°18′W
35 T1 **McDermitt** Nevada, W USA 41°57′N 117°43′W
23 S4 **McDonough** Georgia, SE USA 33°27′N 84°09′W
36 L12 **McDowell Mountains** ▲ Arizona, SW USA
20 H8 **McEwen** Tennessee, S USA 36°06′N 87°37′W
35 R12 **McFarland** California, W USA 35°41′N 119°14′W
McFarlane, Lake see Macfarlane, Lake
27 P12 **McGee Creek Lake** ☒ Oklahoma, C USA
27 W13 **McGehee** Arkansas, C USA 33°37′N 91°24′W
35 X5 **McGill** Nevada, W USA 39°24′N 114°46′W
14 K11 **McGillivray, Lac** ◎ Québec, SE Canada
39 P10 **McGrath** Alaska, USA 62°57′N 155°36′W
25 T8 **McGregor** Texas, SW USA 31°26′N 97°24′W
33 O12 **McGuire, Mount** ▲ Idaho, NW USA 45°10′N 114°36′W
83 M14 **Mchinji** prev. Fort Manning. Central, W Malawi 13°48′S 32°55′E
28 M7 **McIntosh** South Dakota, N USA 45°56′N 101°21′W
9 S7 **McKeand** ~ Baffin Island, Nunavut, NE Canada
191 R4 **McKean Island** island Phoenix Islands, C Kiribati
30 J13 **McKee Creek** ~ Illinois, N USA
18 C15 **McKeesport** Pennsylvania, NE USA 40°18′N 79°48′W
21 V7 **McKenney** Virginia, NE USA 36°57′N 77°42′W
20 G8 **McKenzie** Tennessee, S USA 36°07′N 88°31′W
185 B20 **McKerrow, Lake** ◎ South Island, New Zealand
39 Q10 **McKinley, Mount** var. Denali. ▲ Alaska, USA 63°04′N 151°00′W
39 R10 **McKinley Park** Alaska, USA 63°42′N 149°01′W
34 K3 **McKinleyville** California, W USA 40°56′N 124°06′W
25 U6 **McKinney** Texas, SW USA 33°14′N 96°37′W
26 I5 **McKinney, Lake** ◎ Kansas, C USA
28 M7 **McLaughlin** South Dakota, N USA 45°48′N 100°48′W
25 O2 **McLean** Texas, SW USA 35°13′N 100°36′W
30 M16 **Mcleansboro** Illinois, N USA 38°05′N 88°32′W
11 O13 **McLennan** Alberta, W Canada 55°42′N 116°50′W
14 L9 **McLennan, Lac** ◎ Québec, SE Canada
10 M13 **McLeod Lake** British Columbia, W Canada 55°03′N 123°02′W
8 L6 **M'Clintock Channel** channel Nunavut, N Canada
27 N10 **McLoud** Oklahoma, C USA 35°26′N 97°05′W
32 G15 **McLoughlin, Mount** ▲ Oregon, NW USA 42°27′N 122°18′W
37 U15 **McMillan, Lake** ☒ New Mexico, SW USA
32 G11 **McMinnville** Oregon, NW USA 45°14′N 123°12′W
20 K9 **McMinnville** Tennessee, S USA 35°40′N 85°49′W
195 R13 **McMurdo** US research station Antarctica 77°40′S 167°16′E
37 N13 **Mcnary** Arizona, SW USA 34°04′N 109°51′W
24 H9 **McNary** Texas, SW USA 31°15′N 105°46′W
27 N5 **McPherson** Kansas, C USA 38°22′N 97°41′W
McPherson see Fort McPherson
23 U6 **McRae** Georgia, SE USA 32°04′N 82°54′W
29 P4 **McVille** North Dakota, N USA 47°45′N 98°10′W
83 J25 **Mdantsane** Eastern Cape, SE South Africa 32°55′S 27°39′E
167 T6 **Me** Ninh Binh, N Vietnam 20°21′N 105°49′E
26 J7 **Meade** Kansas, C USA 37°17′N 100°21′W
39 O5 **Meade River** ~ Alaska, USA
35 Y11 **Mead, Lake** ☒ Arizona/Nevada, W USA
24 M5 **Meadow** Texas, SW USA 33°20′N 102°12′W
11 S14 **Meadow Lake** Saskatchewan, C Canada 54°90′N 108°30′W
35 Y10 **Meadow Valley Wash** ~ Nevada, W USA
22 J7 **Meadville** Mississippi, S USA 31°28′N 90°51′W
18 B12 **Meadville** Pennsylvania, NE USA 41°38′N 80°09′W
14 F14 **Meaford** Ontario, S Canada 44°35′N 80°35′W
Meáin, Inis see Inishmaan
104 G8 **Mealhada** Aveiro, N Portugal 40°22′N 08°27′W
13 R8 **Mealy Mountains** ▲ Newfoundland and Labrador, E Canada
11 O10 **Meander River** Alberta, W Canada 59°02′N 117°42′W
32 E11 **Meares, Cape** headland Oregon, NW USA 45°29′N 123°59′W
47 V6 **Mearim, Rio** ~ NE Brazil
Measca, Loch see Mask, Lough
97 F17 **Meath** Ir. An Mhí. cultural region E Ireland
11 T14 **Meath Park** Saskatchewan, S Canada 53°25′N 105°18′W
103 O5 **Meaux** Seine-et-Marne, N France 48°47′N 02°54′E
21 T9 **Mebane** North Carolina, SE USA 36°06′N 79°16′W
171 U12 **Mebo, Gunung** ▲ Papua Barat, E Indonesia 01°10′S 133°53′E
94 I8 **Mebonden** Sør-Trøndelag, S Norway 63°13′N 11°00′E
82 A10 **Mebridege** ~ NW Angola
35 W16 **Mecca** California, W USA 33°34′N 116°04′W
Mecca see Makkah
29 Y14 **Mechanicsville** Iowa, C USA 41°54′N 91°15′W
18 L10 **Mechanicville** New York, NE USA 42°54′N 73°41′W
99 H17 **Mechelen** Eng. Mechlin, Fr. Malines. Antwerpen, C Belgium 51°02′N 04°29′E
188 C8 **Mecherchar** var. Eil Malk. island Palau Islands, Palau
101 D17 **Mechernich** Nordrhein-Westfalen, W Germany 50°36′N 06°39′E
126 L12 **Mechetinskaya** Rostovskaya Oblast', SW Russian Federation 46°46′N 40°30′E
114 J11 **Mechka** ~ S Bulgaria
Mechlin see Mechelen
61 D23 **Mechongué** Buenos Aires, E Argentina 38°09′S 58°13′W
115 L14 **Meciidiye** Edirne, NW Turkey 40°39′N 26°33′E
101 I24 **Meckenbeuren** Baden-Württemberg, S Germany 47°42′N 09°34′E
100 L8 **Mecklenburger Bucht** bay N Germany
100 M10 **Mecklenburgische Seenplatte** wetland NE Germany
100 L9 **Mecklenburg-Vorpommern** ◆ state NE Germany
83 Q15 **Meconta** Nampula, NE Mozambique 15°01′S 39°52′E
111 I25 **Mecsek** ▲ SW Hungary
83 P14 **Mecubúri** ~ N Mozambique
83 Q14 **Mecúfi** Cabo Delgado, NE Mozambique 13°20′S 40°32′E
82 O13 **Mecula** Niassa, N Mozambique 12°03′S 37°37′E
168 I8 **Medan** Sumatera, E Indonesia 03°35′N 98°39′E
61 A24 **Médanos** var. Medanos. Buenos Aires, E Argentina 38°52′S 62°45′W
61 C19 **Médanos** Entre Ríos, E Argentina 33°28′S 59°07′W
155 K24 **Medawachchiya** North Central Province, N Sri Lanka 08°32′N 80°30′E
106 C8 **Mede** Lombardia, N Italy 45°06′N 08°43′E
74 J5 **Médéa** var. El Mediyya, Lemdiyya. N Algeria 36°15′N 02°48′E
Medeba see Mādabā
54 E8 **Medellín** Antioquia, NW Colombia 06°15′N 75°36′W
100 H9 **Medem** ~ NW Germany
98 J8 **Medemblik** Noord-Holland, NW Netherlands 52°47′N 05°06′E
75 N7 **Médenine** var. Madanīyīn. SE Tunisia 33°23′N 10°30′E
76 G9 **Mederdra** Trarza, SW Mauritania 16°56′N 15°40′W
Medeshamstede see Peterborough
42 F4 **Medesto Mendez** Izabal, NE Guatemala 15°54′N 89°13′W
19 O11 **Medford** Massachusetts, NE USA 42°25′N 71°08′W
27 N8 **Medford** Oklahoma, C USA 36°48′N 97°45′W
32 G15 **Medford** Oregon, NW USA 42°20′N 122°52′W
30 K6 **Medford** Wisconsin, N USA 45°08′N 90°22′W
39 P10 **Medfra** Alaska, USA 63°06′N 154°42′W
116 M14 **Medgidia** Constanţa, SE Romania 44°15′N 28°16′E
Medgyes see Mediaş
43 O5 **Media Luna, Arrecifes de la** reef E Honduras
60 G11 **Medianeira** Paraná, S Brazil 25°15′S 54°07′W
29 Y15 **Mediapolis** Iowa, C USA 41°00′N 91°09′W
116 I11 **Mediaş** Ger. Mediasch, Hung. Medgyes. Sibiu, C Romania 46°10′N 24°20′E
41 S15 **Medias Aguas** Veracruz-Llave, SE Mexico 17°40′N 95°02′W
Mediasch see Mediaş
106 G10 **Medicina** Emilia-Romagna, C Italy 44°29′N 11°41′E
33 X16 **Medicine Bow** Wyoming, C USA 41°52′N 106°11′W
37 S2 **Medicine Bow Mountains** ▲ Colorado/Wyoming, C USA
33 X16 **Medicine Bow River** ~ Wyoming, C USA
11 R17 **Medicine Hat** Alberta, SW Canada 50°03′N 110°41′W
26 L7 **Medicine Lodge** Kansas, C USA 37°18′N 98°35′W
26 L7 **Medicine Lodge River** ~ Kansas/Oklahoma, C USA
112 E7 **Medimurje** off. Medimurska Županija. ◆ province N Croatia
Medimurska Županija see Medimurje
54 G10 **Medina** Cundinamarca, C Colombia 04°31′N 73°21′W
18 E9 **Medina** New York, NE USA 43°13′N 78°23′W
29 O5 **Medina** North Dakota, N USA 46°53′N 99°18′W
31 T11 **Medina** Ohio, N USA 41°08′N 81°51′W
25 Q11 **Medina** Texas, SW USA 29°46′N 99°14′W
Medina see Al Madīnah
105 P6 **Medinaceli** Castilla y León, N Spain 41°10′N 02°26′W
104 L6 **Medina del Campo** Castilla y León, N Spain 41°18′N 04°55′W
104 L5 **Medina de Rioseco** Castilla y León, N Spain 41°53′N 05°03′W
Médina Gonassé see Médina Gounas
76 H12 **Médina Gounas** var. Médina Gonassé. S Senegal 13°06′N 13°49′W
25 S12 **Medina River** ~ Texas, SW USA
104 K16 **Medina Sidonia** Andalucía, S Spain 36°28′N 05°55′W
Medinat Israel see Israel
119 H14 **Medininkai** Vilnius, SE Lithuania 54°32′N 25°40′E
153 R16 **Medinīpur** West Bengal, NE India 22°25′N 87°24′E
Mediolanum see Saintes, France
Mediolanum see Milano, Italy
Mediomatrica see Metz
121 Q11 **Mediterranean Ridge** undersea feature C Mediterranean Sea 34°00′N 23°00′E
121 O16 **Mediterranean Sea** Fr. Mer Méditerranée. sea Africa/Asia/Europe
Méditerranée, Mer see Mediterranean Sea
79 N17 **Medje** Orientale, NE Dem. Rep. Congo 02°27′N 27°14′E
Medjerda, Oued see Mejerda
114 G7 **Medkovets** Montana, NW Bulgaria 43°39′N 23°22′E
93 J15 **Medle** Västerbotten, N Sweden 64°45′N 20°45′E
127 W7 **Mednogorsk** Orenburgskaya Oblast', W Russian Federation 51°24′N 57°37′E
123 W9 **Mednyy, Ostrov** island E Russian Federation
102 J12 **Médoc** cultural region SW France
159 Q16 **Mêdog** Xizang Zizhiqu, W China 29°26′N 95°26′E
28 J5 **Medora** North Dakota, N USA 46°56′N 103°40′W
79 E17 **Médouneu** Woleu-Ntem, N Gabon 0°58′N 10°50′E
106 I7 **Meduna** ~ NE Italy
Medunta see Mantes-la-Jolie
Medvedica see Medveditsa
124 J16 **Medveditsa** var. Medvedica. ~ W Russian Federation
127 O9 **Medveditsa** ~ SW Russian Federation
112 E8 **Medvednica** ▲ NE Croatia
125 R15 **Medvedok** Kirovskaya Oblast', NW Russian Federation 57°23′N 50°01′E
123 S6 **Medvezh'i, Ostrova** island group NE Russian Federation
124 J9 **Medvezh'yegorsk** Respublika Kareliya, NW Russian Federation 62°56′N 34°26′E
109 T11 **Medvode** Ger. Zwischenwässern. NW Slovenia 46°09′N 14°21′E
126 J4 **Medyn'** Kaluzhskaya Oblast', W Russian Federation 54°59′N 35°52′E
180 J10 **Meekatharra** Western Australia 26°37′S 118°35′E
37 Q4 **Meeker** Colorado, C USA 40°02′N 107°54′W
13 T12 **Meelpaeg Lake** ◎ Newfoundland, Newfoundland and Labrador, E Canada
Meemu Atoll see Mulakatholhu
Meenen see Menen
101 M16 **Meerane** Sachsen, E Germany 50°50′N 12°28′E
101 D15 **Meerbusch** Nordrhein-Westfalen, W Germany 51°19′N 06°43′E
98 I12 **Meerkerk** Zuid-Holland, C Netherlands 51°55′N 05°00′E
99 L18 **Meerssen** var. Mersen. Limburg, SE Netherlands 50°53′N 05°45′E
152 J10 **Meerut** Uttar Pradesh, N India 29°01′N 77°41′E
33 U13 **Meeteetse** Wyoming, C USA 44°10′N 108°53′W
99 K17 **Meeuwen** Limburg, NE Belgium 51°04′N 05°36′E
81 J16 **Mēga** Oromīya, C Ethiopia 04°03′N 38°15′E
81 J16 **Mēga Escarpment** escarpment S Ethiopia
Megála Kalívia see Megála Kalývia
115 E16 **Megála Kalývia** var. Megála Kalívia. Thessalía, C Greece 39°30′N 21°48′E
115 H14 **Megáli Panagiá** var. Megáli Panayía. Kentrikí Makedonía, N Greece 40°24′N 23°42′E
Megáli Panayía see Megáli Panagiá
Megáli Préspa, Límni see Prespa, Lake
114 K12 **Megálo Livádi** ▲ Bulgaria/Greece 41°18′N 25°51′E
115 E20 **Megalópoli** prev. Megalópolis. Pelopónnisos, S Greece 37°24′N 22°08′E
Megalópolis see Megalópoli
171 U12 **Megamo** Papua Barat, E Indonesia 0°55′S 131°46′E
115 C18 **Meganísi** island Iónia Nisiá, Greece, C Mediterranean Sea
Meganom, Mys see Mehanom, Mys
Mégantic see Lac-Mégantic
15 R12 **Mégantic, Mont** ▲ Québec, SE Canada 45°27′N 71°09′W
115 G19 **Mégara** Attikí, C Greece 38°00′N 23°20′E
25 R5 **Megargel** Texas, SW USA 33°27′N 98°55′W
98 K13 **Megen** Noord-Brabant, S Netherlands 51°49′N 05°34′E
153 U13 **Meghālaya** ◆ state NE India
153 U16 **Meghna Nadi** ~ S Bangladesh
137 V14 **Meghri** Rus. Megri. SE Armenia 38°57′N 46°15′E
115 Q23 **Megísti** var. Kastellórizon. island SE Greece
Megri see Meghri
Mehabad see Mahābād
116 F13 **Mehadia** Hung. Mehádia. Caraş-Severin, SW Romania 44°53′N 22°20′E
Mehádia see Mehadia
92 L7 **Mehamn** Finnmark, N Norway 71°01′N 27°46′E
117 U13 **Mehanom, Mys** Rus. Mys Meganom. headland S Ukraine 44°48′N 35°04′E
149 P14 **Mehar** Sind, SE Pakistan 27°12′N 67°51′E
180 J8 **Meharry, Mount** ▲ Western Australia 23°17′S 118°48′E
Mehdia see Mahdia
116 G14 **Mehedinţi** ◆ county SW Romania
153 S15 **Meherpur** Khulna, W Bangladesh 23°47′N 88°40′E
21 W8 **Meherrin River** ~ North Carolina/Virginia, SE USA
Meheso see Mī'ēso
191 T11 **Mehetia** island Îles du Vent, W French Polynesia
118 K6 **Mehikoorma** Tartumaa, E Estonia 58°14′N 27°29′E
Me Hka see Nmai Hka
143 N5 **Mehrabad** ✈ (Tehrān) Tehrān, N Iran 35°46′N 51°07′E
142 J7 **Mehrān** Īlām, W Iran 33°07′N 46°10′E
143 Q14 **Mehrān, Rūd-e** prev. Mansurabad. ~ W Iran
143 Q9 **Mehrīz** Yazd, C Iran 31°32′N 54°28′E
149 R5 **Mehtar Lām** var. Mehtarlām, Meterlam, Methariam, Metharlam. Laghmān, E Afghanistan 34°39′N 70°10′E
Mehtarlām see Mehtar Lām
103 N8 **Mehun-sur-Yèvre** Cher, C France 47°09′N 02°15′E
79 G14 **Meiganga** Adamaoua, NE Cameroon 06°31′N 14°07′E
160 H10 **Meigu** var. Bapu. Sichuan, C China 28°16′N 103°20′E
163 W11 **Meihekou** var. Hailong. Jilin, NE China 42°31′N 125°40′E
99 L15 **Meijel** Limburg, SE Netherlands 51°22′N 05°52′E
Meijiang see Ningdu
166 M5 **Meiktila** Mandalay, C Myanmar (Burma) 20°53′N 95°54′E
Meilbhe, Loch see Melvin, Lough
108 G7 **Meilen** Zürich, N Switzerland 47°17′N 08°39′E
Meilu see Wuchuan
101 J17 **Meiningen** Thüringen, C Germany 50°34′N 10°25′E
108 F9 **Meiringen** Bern, S Switzerland 46°42′N 08°13′E
Meishan see Jinzhai
101 O15 **Meissen** Ger. Meißen. Sachsen, E Germany 51°10′N 13°28′E
Meißen see Meissen
101 I15 **Meissner** ▲ C Germany 51°13′N 09°52′E
99 K25 **Meix-devant-Virton** Luxembourg, SE Belgium 49°36′N 05°27′E
Meixian see Meizhou
Meixing see Xinjin
161 P13 **Meizhou** var. Meixian, Mei Xian. Guangdong, S China 24°21′N 116°05′E
67 P2 **Mejerda** var. Oued Medjerda, Wādī Majardah. ~ Algeria/Tunisia see also Medjerda, Oued
42 F7 **Mejicanos** San Salvador, C El Salvador 13°50′N 89°13′W
Méjico see Mexico
62 G5 **Mejillones** Antofagasta, N Chile 23°03′S 70°25′W
189 V5 **Mejit Island** var. Mājeej. island Ratak Chain, NE Marshall Islands
79 F17 **Mékambo** Ogooué-Ivindo, NE Gabon 01°03′N 13°50′E
80 J10 **Mek'elē** var. Makale. Tigray, N Ethiopia 13°36′N 39°29′E
74 I10 **Mekerrhane, Sebkha** var. Sebkha Meqerghane, Sebkra Mekerrhane. salt flat C Algeria
Mekerrhane, Sebkra see Mekerrhane, Sebkha
76 G10 **Mékhé** NW Senegal 15°02′N 16°40′W
146 G14 **Mekhinli** Ahal Welaýaty, C Turkmenistan 37°28′N 59°20′E
15 P9 **Mékinac, Lac** ◎ Québec, SE Canada
Meklong see Samut Songkhram
74 G6 **Meknès** N Morocco 33°54′N 05°27′W
129 U12 **Mekong** var. Lan-ts'ang Chiang, Cam. Mékôngk, Chin. Lancang Jiang, Lao. Mènam Khong, Nam Khong, Th. Mae Nam Khong, Tib. Dza Chu, Vtn. Sông Tiên Giang. ~ SE Asia
Mékôngk see Mekong
167 T15 **Mekong, Mouths of the** delta S Vietnam
38 L12 **Mekoryuk** Nunivak Island, Alaska, USA 60°23′N 166°11′W
77 R14 **Mékrou** ~ N Benin
168 K9 **Melaka** var. Malacca. Melaka, Peninsular Malaysia 02°14′N 102°14′E
168 L9 **Melaka** var. Malacca. ◆ state Peninsular Malaysia
Melaka, Selat see Malacca, Strait of
175 O6 **Melanesia** island group W Pacific Ocean
175 P5 **Melanesian Basin** undersea feature W Pacific Ocean 0°05′N 160°35′E
171 R9 **Melanguane** Pulau Karakelang, N Indonesia 04°02′N 126°43′E
169 R11 **Melawi, Sungai** ~ Borneo, N Indonesia
183 N12 **Melbourne** state capital Victoria, SE Australia 37°51′S 144°56′E
27 V9 **Melbourne** Arkansas, C USA 36°04′N 91°54′W
23 Y12 **Melbourne** Florida, SE USA 28°04′N 80°36′W
29 W14 **Melbourne** Iowa, C USA 41°57′N 93°07′W
92 G10 **Melbu** Nordland, C Norway 68°31′N 14°50′E
Melchor de Mencos see Ciudad Melchor de Mencos
63 F19 **Melchor, Isla** island Archipiélago de los Chonos, S Chile
40 M9 **Melchor Ocampo** Zacatecas, C Mexico 24°45′N 101°38′W
14 C11 **Meldrum Bay** Manitoulin Island, Ontario, S Canada 45°55′N 83°06′W
Meleda see Mljet
106 D8 **Melegnano** prev. Marignano. Lombardia, N Italy 45°22′N 09°19′E
188 F9 **Melekeok** ● Babeldaob, N Palau 07°30′N 134°37′E
112 L9 **Melenci** Hung. Melencze. Vojvodina, N Serbia 45°32′N 20°18′E
Melencze see Melenci
127 N4 **Melenki** Vladimirskaya Oblast', W Russian Federation 55°21′N 41°37′E
127 V6 **Meleuz** Respublika Bashkortostan, W Russian Federation 52°55′N 55°54′E
12 L6 **Mélèzes, Rivière aux** ~ Québec, C Canada
78 I11 **Melfi** Guéra, S Chad 11°05′N 17°57′E
107 M17 **Melfi** Basilicata, S Italy 41°00′N 15°33′E
11 U14 **Melfort** Saskatchewan, S Canada 52°52′N 104°38′W
104 H4 **Melgaço** Viana do Castelo, N Portugal 42°07′N 08°15′W
105 N4 **Melgar de Fernamental** Castilla y León, N Spain 42°24′N 04°15′W
74 L6 **Melghir, Chott** var. Chott Melrhir. salt lake E Algeria
94 H8 **Melhus** Sør-Trøndelag, S Norway 63°17′N 10°18′E
104 H3 **Melide** Galicia, NW Spain 42°54′N 08°01′W
Meligalá see Meligalás
115 E21 **Meligalás** prev. Meligalá. Pelopónnisos, S Greece 37°13′N 21°58′E
60 L12 **Mel, Ilha do** island S Brazil
120 E10 **Melilla** anc. Rusaddir, Russadir. Melilla, Spain, N Africa 35°18′N 02°56′W
71 N1 **Melilla** enclave Spain, N Africa
63 G18 **Melimoyu, Monte** ▲ S Chile 44°06′S 72°49′W
169 V11 **Melintang, Danau** ◎ Borneo, N Indonesia
117 U7 **Melioratyvne** Dnipropetrovs'ka Oblast', E Ukraine 48°35′N 35°18′E
62 G11 **Melipilla** Santiago, C Chile 33°42′S 71°15′W
115 I25 **Mélissa, Akrotírio** headland Kríti, Greece, E Mediterranean Sea 35°06′N 24°33′E
11 W17 **Melita** Manitoba, S Canada 49°16′N 100°59′W
Melita see Mljet
Melitene see Malatya
107 M23 **Melito di Porto Salvo** Calabria, SW Italy 37°55′N 15°48′E
117 U10 **Melitopol'** Zaporiz'ka Oblast', SE Ukraine 46°49′N 35°23′E
109 V4 **Melk** Niederösterreich, NE Austria 48°14′N 15°21′E
95 K15 **Mellan-Fryken** ◎ C Sweden
99 E17 **Melle** Oost-Vlaanderen, NW Belgium 51°N 03°48′E
100 G13 **Melle** Niedersachsen, NW Germany 52°12′N 08°19′E
Mellersta Finland see Keski-Suomi
Mellersta Österbotten see Keski-Pohjanmaa
95 J17 **Mellerud** Västra Götaland, S Sweden 58°42′N 12°27′E
102 K10 **Melle-sur-Bretonne** Deux-Sèvres, W France 46°13′N 00°07′W
29 P8 **Mellette** South Dakota, N USA 45°07′N 98°29′W
121 O15 **Mellieħa** E Malta 35°58′N 14°21′E
80 B10 **Mellit** Northern Darfur, W Sudan 14°07′N 25°34′E
75 N7 **Mellita** ✈ SE Tunisia 33°47′N 10°51′E
63 G21 **Mellizo Sur, Cerro** ▲ S Chile 48°27′S 73°10′W
100 G9 **Mellum** island NW Germany
83 L22 **Melmoth** KwaZulu/Natal, E South Africa 28°35′S 31°25′E
111 D16 **Mělník** Ger. Melnik. Středočeský Kraj, NW Czech Republic 50°21′N 14°30′E
122 J12 **Mel'nikovo** Tomskaya Oblast', C Russian Federation 56°35′N 84°11′E
61 G18 **Melo** Cerro Largo, NE Uruguay 32°22′S 54°10′W
Melodunum see Melun
Melrhir, Chott see Melghir, Chott
183 P7 **Melrose** New South Wales, SE Australia 32°41′S 146°58′E
182 I7 **Melrose** South Australia 32°52′S 138°16′E
29 T7 **Melrose** Minnesota, N USA 45°40′N 94°46′W
33 Q11 **Melrose** Montana, NW USA 45°33′N 112°41′W
37 V12 **Melrose** New Mexico, SW USA 34°25′N 103°37′W
108 I8 **Mels** Sankt Gallen, NE Switzerland 47°03′N 09°26′E
Melsetter see Chimanimani
33 V9 **Melstone** Montana, NW USA 46°37′N 107°49′W
101 I16 **Melsungen** Hessen, C Germany 51°08′N 09°33′E
92 L12 **Meltaus** Lappi, NW Finland 66°54′N 25°18′E
97 N19 **Melton Mowbray** C England, United Kingdom 52°46′N 01°04′W
82 Q13 **Meluco** Cabo Delgado, NE Mozambique 12°31′S 39°39′E
103 O5 **Melun** anc. Melodunum. Seine-et-Marne, N France 48°32′N 02°40′E
80 F12 **Melut** Upper Nile, NE South Sudan 10°27′N 32°13′E
27 P5 **Melvern Lake** ☒ Kansas, C USA
11 V16 **Melville** Saskatchewan, S Canada 50°57′N 102°49′W
Melville Bay/Melville Bugt see Qimusseriarsuaq
45 O11 **Melville Hall** ✈ (Dominica) NE Dominica 15°33′N 61°19′W
181 O1 **Melville Island** island Northern Territory, N Australia
197 O8 **Melville Island** island Parry Islands, Northwest Territories, NW Canada
11 W9 **Melville, Lake** ◎ Newfoundland and Labrador, E Canada
9 O7 **Melville Peninsula** peninsula Nunavut, NE Canada
Melville Sound see Viscount Melville Sound
25 Q9 **Melvin** Texas, SW USA 31°12′N 99°34′W
97 D15 **Melvin, Lough** Ir. Loch Meilbhe. ◎ S Northern Ireland, United Kingdom/Ireland
169 S12 **Memala** Borneo, C Indonesia 01°44′S 112°36′E
113 L22 **Memaliaj** Gjirokastër, S Albania 40°21′N 19°56′E
83 Q14 **Memba** Nampula, NE Mozambique 14°07′S 40°33′E
83 Q14 **Memba, Baia de** inlet NE Mozambique
Membidj see Manbij
Memel see Neman, NE Europe
Memel see Klaipeda, Lithuania
101 J23 **Memmingen** Bayern, S Germany 47°59′N 10°11′E
27 U1 **Memphis** Missouri, C USA 40°28′N 92°11′W
20 E10 **Memphis** Tennessee, S USA 35°09′N 90°03′W
25 P3 **Memphis** Texas, SW USA 34°43′N 100°34′W
20 E10 **Memphis** ✈ Tennessee, S USA 35°02′N 89°57′W
15 Q13 **Memphrémagog, Lac** var. Lake Memphremagog. ◎ Canada/USA see also Lake Memphremagog
19 N6 **Memphremagog, Lake** var. Lac Memphrémagog. ◎ Canada/USA see also Memphrémagog, Lac
117 Q2 **Mena** Chernihivs'ka Oblast', NE Ukraine 51°30′N 32°15′E
27 S12 **Mena** Arkansas, C USA 34°40′N 94°15′W
Menaam see Menaldum
Menado see Manado
106 D6 **Menaggio** Lombardia, N Italy 46°03′N 09°14′E
29 T6 **Menahga** Minnesota, N USA 46°45′N 95°06′W
77 R10 **Ménaka** Goa, E Mali 15°55′N 02°25′E
98 K5 **Menaldum** Fris. Menaam. Fryslân, N Netherlands 53°14′N 05°38′E
Mènam Khong see Mekong
74 E7 **Menara** ✈ (Marrakech) C Morocco 31°36′N 08°00′W
25 Q9 **Menard** Texas, SW USA 30°56′N 99°48′W
193 Q12 **Menard Fracture Zone** tectonic feature E Pacific Ocean
30 M7 **Menasha** Wisconsin, N USA 44°13′N 88°25′W
Mencezi Garagum see Merkezi Garagumy
193 U9 **Mendaña Fracture Zone** tectonic feature E Pacific Ocean
169 S13 **Mendawai, Sungai** ~ Borneo, C Indonesia
103 P13 **Mende** anc. Mimatum. Lozère, S France 44°32′N 03°30′E
81 J14 **Mendebo** ▲ C Ethiopia
80 J9 **Mendefera** prev. Adi Ugri. S Eritrea 14°53′N 38°51′E
197 S7 **Mendeleyev Ridge** undersea feature Arctic Ocean
127 T3 **Mendeleyevsk** Respublika Tatarstan, W Russian Federation 55°54′N 52°19′E
101 F15 **Menden** Nordrhein-Westfalen, W Germany 51°26′N 07°47′E
22 L6 **Mendenhall** Mississippi, S USA 31°57′N 89°52′W
38 L13 **Mendenhall, Cape** headland Nunivak Island, Alaska, USA 59°45′N 166°10′W
41 P9 **Méndez** var. Villa de Méndez. Tamaulipas, C Mexico 25°06′N 98°32′W
80 H13 **Mendī** Oromīya C Ethiopia 09°43′N 35°07′E
186 C7 **Mendi** Southern Highlands, W Papua New Guinea 06°13′S 143°39′E
97 K22 **Mendip Hills** var. Mendips. hill range S England, United Kingdom
Mendips see Mendip Hills
34 L6 **Mendocino** California, W USA 39°18′N 123°48′W
34 J3 **Mendocino, Cape** headland California, W USA 40°26′N 124°24′W
0 B8 **Mendocino Fracture Zone** tectonic feature NE Pacific Ocean
35 P10 **Mendota** California, W USA 36°44′N 120°24′W
30 L11 **Mendota** Illinois, N USA 41°32′N 89°04′W
30 K8 **Mendota, Lake** ◎ Wisconsin, N USA
62 I11 **Mendoza** Mendoza, W Argentina 33°00′S 68°47′W
62 I12 **Mendoza** off. Provincia de Mendoza. ◆ province W Argentina
Mendoza, Provincia de see Mendoza
108 H12 **Mendrisio** Ticino, S Switzerland 45°53′N 08°59′E
168 L10 **Mendung** Pulau Mendol, W Indonesia 0°33′N 103°09′E
54 I5 **Mene de Mauroa** Falcón, NW Venezuela 10°39′N 71°04′W
54 I5 **Mene Grande** Zulia, NW Venezuela 09°51′N 70°57′W
136 B14 **Menemen** İzmir, W Turkey 38°34′N 27°03′E
99 C18 **Menen** var. Meenen, Fr. Menin. West-Vlaanderen, W Belgium 50°48′N 03°07′E
163 Q8 **Menengiyn Tal** plain E Mongolia
189 R9 **Meneng Point** headland SW Nauru 0°33′S 166°57′E
92 L10 **Menesjärvi** Lapp. Menešjávri. Lappi, N Finland 68°39′N 26°22′E
Menešjávri see Menesjärvi
107 I24 **Menfi** Sicilia, Italy, C Mediterranean Sea 37°36′N 12°59′E
161 P7 **Mengcheng** Anhui, E China 33°15′N 116°33′E
160 F15 **Menghai** Yunnan, SW China 22°02′N 100°18′E
160 F15 **Mengla** Yunnan, SW China 21°30′N 101°33′E
65 F24 **Menguera Point** headland East Falkland, Falkland Islands
160 M13 **Mengzhu Ling** ▲ S China
160 H14 **Mengzi** Yunnan, SW China 23°20′N 103°32′E
114 H13 **Meníkio** var. Menoíkio. ▲ NE Greece 40°50′N 12°40′E
Menin see Menen
182 L7 **Menindee** New South Wales, SE Australia 32°24′S 142°25′E
182 L7 **Menindee Lake** ◎ New South Wales, SE Australia
182 J10 **Meningie** South Australia 35°43′S 139°20′E
103 O5 **Mennecy** Essonne, N France 48°34′N 02°25′E
29 Q12 **Menno** South Dakota, N USA 43°14′N 97°34′W
114 H13 **Menoíkio** ▲ NE Greece
Menoíkio see Meníkio
31 N5 **Menominee** Michigan, N USA 45°06′N 87°36′W
30 M5 **Menominee River** ~ Michigan/Wisconsin, N USA
30 M8 **Menomonee Falls** Wisconsin, N USA 43°11′N 88°09′W
30 I6 **Menomonie** Wisconsin, N USA 44°52′N 91°55′W
83 D14 **Menongue** var. Vila Serpa Pinto, Port. Serpa Pinto. Kuando Kubango, C Angola 14°38′S 17°39′E
120 H8 **Menorca** Eng. Minorca; anc. Balearis Minor. island Islas Baleares, Spain, W Mediterranean Sea
105 S13 **Menor, Mar** lagoon SE Spain
39 S10 **Mentasta Lake** ◎ Alaska, USA
39 S10 **Mentasta Mountains** ▲ Alaska, USA
168 I13 **Mentawai, Kepulauan** island group W Indonesia
168 I12 **Mentawai, Selat** strait W Indonesia
168 M12 **Mentok** Pulau Bangka, W Indonesia 02°01′S 105°10′E
103 V15 **Menton** It. Mentone. Alpes-Maritimes, SE France 43°47′N 07°30′E
24 K8 **Mentone** Texas, SW USA 31°42′N 103°36′W
Mentone see Menton
31 U11 **Mentor** Ohio, N USA 41°40′N 81°20′W
169 U10 **Menyapa, Gunung** ▲ Borneo, N Indonesia 01°04′N 116°01′E
159 T9 **Menyuan** var. Menyuan Huizu Zizhixian. Qinghai, C China 37°27′N 101°33′E
Menyuan Huizu Zizhixian see Menyuan
74 M5 **Menzel Bourguiba** var. Manzil Bū Ruqaybah; prev. Ferryville. N Tunisia 37°09′N 09°51′E
136 M15 **Menzelet Barajı** ☒ C Turkey
127 T4 **Menzelinsk** Respublika Tatarstan, W Russian Federation 55°44′N 53°00′E
180 K11 **Menzies** Western Australia 29°42′S 121°04′E
195 V6 **Menzies, Mount** ▲ Antarctica 73°32′S 61°02′E
40 J6 **Meoqui** Chihuahua, N Mexico 28°18′N 105°30′W
83 N14 **Meponda** Niassa, NE Mozambique 13°20′S 34°53′E
98 M8 **Meppel** Drenthe, NE Netherlands 52°42′N 06°12′E
100 E12 **Meppen** Niedersachsen, NW Germany 52°42′N 07°18′E
Meqerghane, Sebkha see Mekerrhane, Sebkha
105 T6 **Mequinenza, Embalse de** ☒ NE Spain
30 M8 **Mequon** Wisconsin, N USA 43°13′N 87°57′W
Mera see Maira
182 D3 **Meramangye, Lake** salt lake South Australia
27 W5 **Meramec River** ~ Missouri, C USA
Meran see Merano
168 K13 **Merangin** ~ Sumatera, W Indonesia
106 G5 **Merano** Ger. Meran. Trentino-Alto Adige, N Italy 46°40′N 11°10′E
168 K8 **Merapuh Lama** Pahang, Peninsular Malaysia 04°37′N 101°58′E
106 D7 **Merate** Lombardia, N Italy 45°42′N 09°26′E
169 U13 **Meratus, Pegunungan** ▲ Borneo, N Indonesia
171 Y16 **Merauke, Sungai** ~ Papua, E Indonesia
182 L9 **Merbein** Victoria, SE Australia 34°11′S 142°03′E
99 F21 **Merbes-le-Château** Hainaut, S Belgium 50°19′N 04°09′E
Merca see Marka
54 C13 **Mercaderes** Cauca, SW Colombia 01°46′N 77°09′W
Mercara see Madikeri

◆ Country
● Country Capital
◇ Dependent Territory
○ Dependent Territory Capital
◆ Administrative Regions
✈ International Airport
▲ Mountain
▲ Mountain Range
Volcano
~ River
◎ Lake
☒ Reservoir

35 P9 **Merced** California, W USA 37°17′N 120°30′W
61 C20 **Mercedes** Buenos Aires, E Argentina 34°42′S 59°30′W
61 D15 **Mercedes** Corrientes, NE Argentina 29°09′S 58°05′W
61 D19 **Mercedes** Soriano, SW Uruguay 33°16′S 58°01′W
25 S17 **Mercedes** Texas, SW USA 26°09′N 97°54′W
Mercedes *see* Villa Mercedes
35 R9 **Merced Peak** ▲ California, W USA 37°34′N 119°30′W
35 P9 **Merced River** 🌊 California, W USA
18 B13 **Mercer** Pennsylvania, NE USA 41°14′N 80°14′W
99 G18 **Merchtem** Vlaams Brabant, C Belgium 50°57′N 04°14′E
15 O13 **Mercier** Québec, SE Canada 45°15′N 73°45′W
25 Q9 **Mercury** Texas, SW USA 31°23′N 99°09′W
184 M5 **Mercury Islands** *island group* N New Zealand
19 O9 **Meredith** New Hampshire, NE USA 43°36′N 71°28′W
65 B25 **Meredith, Cape** *var.* Cabo Belgrano. *headland* West Falkland, Falkland Islands 52°15′S 60°40′W
37 V6 **Meredith, Lake** ◎ Colorado, C USA
25 N2 **Meredith, Lake** ▣ Texas, SW USA
81 O16 **Mereeg** *var.* Mareeq, *It.* Meregh. Galguduud, E Somalia 03°47′N 47°19′E
117 V5 **Merefa** Kharkivs'ka Oblast', E Ukraine 49°49′N 36°05′E
Meregh *see* Mereeg
99 E17 **Merelbeke** Oost-Vlaanderen, NW Belgium 51°00′N 03°45′E
Merend *see* Marand
167 T12 **Méreuch** Môndól Kiri, E Cambodia 13°01′N 107°26′E
Mergate *see* Margate
Mergui *see* Myeik
Mergui Archipelago *see* Myeik Archipelago
114 L12 **Meriç** Edirne, NW Turkey 41°12′N 26°24′E
114 L12 **Meriç** *Bul.* Maritsa, *Gk.* Évros; *anc.* Hebrus. 🌊 SE Europe *see also* Évros/ Maritsa
41 X12 **Mérida** Yucatán, SW Mexico 20°58′N 89°35′W
104 J11 **Mérida** *anc.* Augusta Emerita. Extremadura, W Spain 38°55′N 06°20′W
54 I6 **Mérida** Mérida, W Venezuela 08°36′N 71°08′W
54 H7 **Mérida** *off.* Estado Mérida. ⬖ *state* W Venezuela
Mérida, Estado *see* Mérida
18 M13 **Meriden** Connecticut, NE USA 41°32′N 72°48′W
22 M5 **Meridian** Mississippi, S USA 32°24′N 88°43′W
25 S8 **Meridian** Texas, SW USA 31°56′N 97°40′W
102 J13 **Mérignac** Gironde, SW France 44°50′N 00°40′W
102 J13 **Mérignac** ✈ (Bordeaux) Gironde, SW France 44°51′N 00°44′W
93 J18 **Merikarvia** Satakunta, SW Finland 61°51′N 21°30′E
183 R12 **Merimbula** New South Wales, SE Australia 36°52′S 149°51′E
182 L9 **Meringur** Victoria, SE Australia 34°26′S 141°19′E
Merín, Laguna *see* Mirim Lagoon
97 I19 **Merioneth** *cultural region* W Wales, United Kingdom
188 A11 **Merir** *island* Palau Islands, N Palau
188 B17 **Merizo** SW Guam 13°15′N 144°40′E
Merjama *see* Märjamaa
Merke *see* Merki
25 P7 **Merkel** Texas, SW USA 32°28′N 100°00′W
146 E12 **Merkezi Garagumy** *var.* Mencezi Garagum, *Rus.* Tsentral'nyye Nizmennyye Garagumy. *desert* C Turkmenistan
145 S16 **Merki** *prev.* Merke. Zhambyl, S Kazakhstan 42°48′N 73°10′E
119 F15 **Merkinė** Alytus, S Lithuania 54°09′N 24°11′E
99 G16 **Merksem** Antwerpen, N Belgium 51°17′N 04°26′E
99 I15 **Merksplas** Antwerpen, N Belgium 51°22′N 04°54′E
Merkulovichi *see* Myerkulavichy
119 G15 **Merkys** 🌊 S Lithuania
32 F15 **Merlin** Oregon, NW USA 42°34′N 123°23′W
61 C20 **Merlo** Buenos Aires, E Argentina 34°39′S 58°45′W
138 G8 **Meron, Harei** *prev.* Haré Meron. ▲ N Israel 35°06′N 33°00′E
74 K6 **Merouane, Chott** *salt lake* NE Algeria
80 F7 **Merowe** Northern, N Sudan 18°29′N 31°49′E
180 J12 **Merredin** Western Australia 31°31′S 118°18′E
97 I14 **Merrick** ▲ S Scotland, United Kingdom 55°09′N 04°28′W
32 H16 **Merrill** Oregon, NW USA 42°00′N 121°37′W
30 L5 **Merrill** Wisconsin, N USA 45°12′N 89°43′W
31 N11 **Merrillville** Indiana, N USA 41°28′N 87°19′W
19 O10 **Merrimack River** 🌊 Massachusetts/New Hampshire, NE USA
28 L12 **Merriman** Nebraska, C USA 42°54′N 101°42′W
11 N17 **Merritt** British Columbia, SW Canada 50°09′N 120°49′W
23 Y12 **Merritt Island** Florida, SE USA 28°21′N 80°42′W
23 Y11 **Merritt Island** *island* Florida, SE USA
28 M12 **Merritt Reservoir** ▣ Nebraska, C USA
183 S7 **Merriwa** New South Wales, SE Australia 32°09′S 150°24′E
183 O8 **Merriwagga** New South Wales, SE Australia 33°51′S 145°38′E
22 G8 **Merryville** Louisiana, S USA 30°45′N 93°32′W
80 K9 **Mersa Fat'ma** E Eritrea 14°52′N 40°16′E
102 M7 **Mer St-Aubin** Loir-et-Cher, C France 47°42′N 01°31′E
99 M24 **Mersch** Luxembourg, C Luxembourg 49°45′N 06°06′E
101 M15 **Merseburg** Sachsen-Anhalt, C Germany 51°22′N 12°00′E
Mersen *see* Meerssen
97 K18 **Mersey** 🌊 NW England, United Kingdom
136 J17 **Mersin** *var.* İçel. İçel, S Turkey 36°50′N 34°39′E
Mersin *see* İçel
168 L9 **Mersing** Johor, Peninsular Malaysia 02°25′N 103°50′E
118 E8 **Mērsrags** NW Latvia 57°21′N 23°05′E
Merta *see* Merta City
152 G12 **Merta City** *var.* Merta. Rājasthān, N India 26°40′N 74°04′E
152 F12 **Merta Road** Rājasthān, N India 26°42′N 73°54′E
97 J21 **Merthyr Tydfil** S Wales, United Kingdom 51°46′N 03°23′W
104 H13 **Mértola** Beja, S Portugal 37°38′N 07°40′W
144 G14 **Mertvyy Kultuk, Sor** *salt flat* SW Kazakhstan
195 V16 **Mertz Glacier** *glacier* Antarctica
99 M24 **Mertzig** Diekirch, C Luxembourg 49°50′N 06°00′E
25 O9 **Mertzon** Texas, SW USA 31°16′N 100°50′W
103 N4 **Méru** Oise, N France 49°15′N 02°07′E
81 I18 **Meru** Meru, C Kenya 0°03′N 37°38′E
81 I19 **Meru** ⬖ *county* C Kenya
81 I20 **Meru, Mount** ▲ NE Tanzania 03°12′S 36°45′E
Merv *see* Mary
Mervdasht *see* Marv Dasht
136 K11 **Merzifon** Amasya, N Turkey 40°52′N 35°28′E
101 D20 **Merzig** Saarland, SW Germany 49°27′N 06°39′E
36 L14 **Mesa** Arizona, SW USA 33°25′N 111°49′W
29 V4 **Mesabi Range** ⛰ Minnesota, N USA
54 H6 **Mesa Bolívar** Mérida, NW Venezuela 08°30′N 71°38′W
107 Q18 **Mesagne** Puglia, SE Italy 40°33′N 17°49′E
39 P12 **Mesa Mountain** ▲ Alaska, USA 60°26′N 155°14′W
115 J25 **Mesará** *lowland* Kríti, Greece, E Mediterranean Sea
37 S14 **Mescalero** New Mexico, SW USA 33°09′N 105°46′W
101 G15 **Meschede** Nordrhein-Westfalen, W Germany 51°21′N 08°16′E
137 Q12 **Mescit Dağları** ⛰ NE Turkey
189 V13 **Mesegon** *island* Chuuk, C Micronesia
Meseritz *see* Międzyrzecz
54 F11 **Mesetas** Meta, C Colombia 03°14′N 74°09′W
Meshchera Lowland *see* Meshcherskaya Nizmennost'
Meshcherskaya Nizina *see* Meshcherskaya Nizmennost'
126 M4 **Meshcherskaya Nizmennost'** *var.* Meshcherskaya Nizina, *Eng.* Meshchera Lowland. *basin* W Russian Federation
126 J5 **Meshchovsk** Kaluzhskaya Oblast', W Russian Federation 54°21′N 35°23′E
125 R9 **Meshchura** Respublika Komi, NW Russian Federation 63°18′N 50°56′E
Meshed *see* Mashhad
Meshed-i-Sar *see* Bābolsar
80 E13 **Meshra'er Req** Warap, W South Sudan 08°30′N 29°27′E
37 R15 **Mesilla** New Mexico, SW USA 32°15′N 106°49′W
108 H10 **Mesocco** *Ger.* Misox. Ticino, S Switzerland 46°18′N 09°13′E
115 D18 **Mesolóngi** *prev.* Mesolóngion. Dytikí Elláda, W Greece 38°21′N 21°26′E
Mesolóngion *see* Mesolóngi
14 E8 **Mesomikenda Lake** ◎ Ontario, S Canada
61 D15 **Mesopotamia** *var.* Mesopotamia Argentina. *physical region* NE Argentina
Mesopotamia Argentina *see* Mesopotamia
35 Y10 **Mesquite** Nevada, W USA 36°47′N 114°04′W
82 Q13 **Messalo, Rio** *var.* Mualo. 🌊 NE Mozambique
Messana/Messene *see* Messina
99 L25 **Messancy** Luxembourg, SE Belgium 49°36′N 05°49′E
107 M23 **Messina** *var.* Messana, Messene; *anc.* Zancle. Sicilia, Italy, C Mediterranean Sea 38°12′N 15°33′E
Messina *see* Musina
Messina, Strait of *see* Messina, Stretto di
107 M23 **Messina, Stretto di** *Eng.* Strait of Messina. *strait* SW Italy
115 E21 **Messíni** Pelopónnisos, S Greece 37°03′N 22°00′E
115 E21 **Messinía** *peninsula* S Greece
115 E22 **Messiniakós Kólpos** *gulf* S Greece
122 J8 **Messoyakha** 🌊 N Russian Federation
114 H11 **Mesta** *Gk.* Néstos, *Turk.* Kara Su. 🌊 Bulgaria/Greece *see also* Néstos
Mesta *see* Néstos
Mestghanem *see* Mostaganem
137 R8 **Mest'ia** *prev.* Mestia, *var.* Mestiya. N Georgia 43°03′N 42°50′E
Mestia *see* Mest'ia
Mestiya *see* Mest'ia
115 K18 **Mestón, Akrotírio** *cape* Chíos, E Greece
106 H8 **Mestre** Veneto, NE Italy 45°30′N 12°14′E
59 M16 **Mestre, Espigão** ▲ E Brazil
169 N14 **Mesuji** 🌊 Sumatera, W Indonesia
Mesule *see* Grosser Möseler
10 J10 **Meszah Peak** ▲ British Columbia, W Canada 58°31′N 131°28′W
54 G11 **Meta** *off.* Departamento del Meta. ⬖ *province* C Colombia
15 Q8 **Metabetchouane** 🌊 Québec, SE Canada
Meta, Departamento del *see* Meta
9 S7 **Meta Incognita Peninsula** *peninsula* Baffin Island, Nunavut, NE Canada
22 K9 **Metairie** Louisiana, S USA 29°58′N 90°09′W
32 M6 **Metaline Falls** Washington, NW USA 48°51′N 117°21′W
62 K6 **Metán** Salta, N Argentina 25°29′S 64°57′W
82 N13 **Metangula** Niassa, N Mozambique 12°41′S 34°50′E
42 E7 **Metapán** Santa Ana, NW El Salvador 14°20′N 89°28′W
54 K9 **Meta, Río** 🌊 Colombia/ Venezuela
106 I11 **Metauro** 🌊 C Italy
80 H11 **Metema** Āmara, N Ethiopia 12°53′N 36°10′E
115 D15 **Metéora** *religious building* Thessalía, C Greece
65 O20 **Meteor Rise** *undersea feature* SW Indian Ocean 46°00′S 05°30′E
186 G5 **Meteran** New Hanover, NE Papua New Guinea 02°40′S 150°12′E
Meterlam *see* Mehtar Lām
115 G20 **Methanon** *peninsula* S Greece
Metharaw/Metharlam *see* Mehtar Lām
32 J6 **Methow River** 🌊 Washington, NW USA
19 O10 **Methuen** Massachusetts, NE USA 42°43′N 71°10′W
185 G19 **Methven** Canterbury, South Island, New Zealand 43°37′S 171°38′E
Metis *see* Metz
113 G15 **Metković** Dubrovnik-Neretva, SE Croatia 43°02′N 17°37′E
39 Y14 **Metlakatla** Annette Island, Alaska, USA 55°07′N 131°34′W
109 V13 **Metlika** *Ger.* Möttling. SE Slovenia 45°38′N 15°18′E
109 T8 **Metnitz** Kärnten, S Austria 46°58′N 14°09′E
27 W12 **Meto, Bayou** 🌊 Arkansas, C USA
168 M15 **Metro** Sumatera, W Indonesia 05°05′S 105°20′E
30 M17 **Metropolis** Illinois, N USA 37°09′N 88°43′W
Metropolitan *see* Santiago
35 N8 **Metropolitan Oakland** ✈ California, W USA 37°42′N 122°13′W
115 D15 **Métsovo** *prev.* Métsovon. Ípeiros, C Greece 39°47′N 21°12′E
Métsovon *see* Métsovo
23 V5 **Metter** Georgia, SE USA 32°24′N 82°03′W
99 H21 **Mettet** Namur, S Belgium 50°19′N 04°43′E
101 D20 **Mettlach** Saarland, SW Germany 49°28′N 06°37′E
80 H13 **Metu** *var.* Mattu, Mettu. Oromīya, C Ethiopia 08°18′N 35°39′E
138 G8 **Metula** *prev.* Metulla. Northern, N Israel 33°16′N 35°35′E
169 T10 **Metulang** Borneo, N Indonesia 01°28′N 114°40′E
Metulla *see* Metula
103 T4 **Metz** *anc.* Divodurum Mediomatricum, Mediomatrica, Metis. Moselle, NE France 49°07′N 06°09′E
101 H22 **Metzingen** Baden-Württemberg, S Germany 48°31′N 09°16′E
168 G8 **Meulaboh** Sumatera, W Indonesia 04°10′N 96°09′E
99 D18 **Meulebeke** West-Vlaanderen, W Belgium 50°57′N 03°18′E
103 U6 **Meurthe** 🌊 NE France
103 S5 **Meurthe-et-Moselle** ⬖ *department* NE France
103 S4 **Meuse** ⬖ *department* NE France
84 F10 **Meuse** *Dut.* Maas. 🌊 W Europe *see also* Maas
Meuse *see* Maas
Mexcala, Río *see* Balsas, Río
25 U8 **Mexia** Texas, SW USA 31°40′N 96°28′W
58 K11 **Mexiana, Ilha** *island* NE Brazil
40 C1 **Mexicali** Baja California Norte, NW Mexico 32°34′N 115°26′W
Mexicanos, Estados Unidos *see* Mexico
41 O14 **México** *var.* Ciudad de México, *Eng.* Mexico City. ● (Mexico) México, C Mexico 19°26′N 99°08′W
27 V4 **Mexico** Missouri, C USA 39°10′N 91°54′W
18 H9 **Mexico** New York, NE USA 43°27′N 76°14′W
40 L7 **Mexico** *off.* United Mexican States, *var.* Méjico, México, *Sp.* Estados Unidos Mexicanos. ◆ *federal republic* N Central America
41 O13 **México** ⬖ *state* S Mexico
México *see* Mexico
0 J13 **Mexico Basin** *var.* Sigsbee Deep. *undersea feature* C Gulf of Mexico 25°00′N 92°00′W
Mexico City *see* México
México, Golfo de *see* Mexico, Gulf of
44 B4 **Mexico, Gulf of** *Sp.* Golfo de México. *gulf* W Atlantic Ocean
139 R4 **Mexmûr** *Ar.* Makhmūr. Arbīl, N Iraq 35°47′N 43°32′E
Meyadine *see* Al Mayādīn
Meydān Shahr *see* Maīdān Shahr
39 Y14 **Meyers Chuck** Etolin Island, Alaska, USA 55°44′N 132°15′W
Meymaneh *see* Maīmanah
143 N7 **Meymeh** Eşfahān, C Iran 33°29′N 51°09′E
123 V7 **Meynypil'gyno** Chukotskiy Avtonomnyy Okrug, NE Russian Federation 62°33′N 177°00′E
108 A10 **Meyrin** Genève, SW Switzerland 46°14′N 06°05′E
166 L7 **Mezaligon** Ayeyawady, SW Myanmar (Burma) 17°53′N 95°12′E
41 O15 **Mezcala** Guerrero, S Mexico 17°55′N 99°34′W
114 H8 **Mezdra** Vratsa, NW Bulgaria 43°08′N 23°42′E
103 P16 **Mèze** Hérault, S France 43°26′N 03°37′E
125 O6 **Mezen'** Arkhangel'skaya Oblast', NW Russian Federation 65°54′N 44°10′E
125 P8 **Mezen'** 🌊 NW Russian Federation
Mezen, Bay of *see* Mezenskaya Guba
103 Q13 **Mézenc, Mont** ▲ C France 44°57′N 04°15′E
125 O8 **Mezenskaya Guba** *var.* Bay of Mezen. *bay* NW Russian Federation
Mezha *see* Myazha
122 H6 **Mezhdusharskiy, Ostrov** *island* Novaya Zemlya, N Russian Federation
Mezhëvo *see* Myezhava
Mezhgor'ye *see* Mizhhir"ya
117 V8 **Mezhova** Dnipropetrovs'ka Oblast', E Ukraine 48°15′N 36°44′E
10 J12 **Meziadin Junction** British Columbia, W Canada 56°06′N 129°15′W
111 G16 **Mezileské Sedlo** *var.* Przełęcz Międzyleska. *pass* Czech Republic/Poland
102 L14 **Mézin** Lot-et-Garonne, SW France 44°03′N 00°16′E
111 M24 **Mezőberény** Békés, SE Hungary 46°49′N 21°00′E
111 M25 **Mezőhegyes** Békés, SE Hungary 46°20′N 20°48′E
111 M25 **Mezőkovácsháza** Békés, SE Hungary 46°24′N 20°52′E
111 M21 **Mezőkövesd** Borsod-Abaúj-Zemplén, NE Hungary 47°49′N 20°32′E
Mezőtelegd *see* Tileagd
111 M23 **Mezőtúr** Jász-Nagykun-Szolnok, E Hungary 47°N 20°37′E
40 K10 **Mezquital** Durango, C Mexico 23°31′N 104°19′W
106 G6 **Mezzolombardo** Trentino-Alto Adige, N Italy 46°13′N 11°08′E
82 L13 **Mfuwe** Muchinga, N Zambia 13°00′S 31°45′E
121 O15 **Mġarr** Gozo, N Malta 36°01′N 14°18′E
126 H6 **Mglin** Bryanskaya Oblast', W Russian Federation 53°01′N 32°54′E
Mhálanna, Cionn *see* Malin Head
154 G10 **Mhow** Madhya Pradesh, C India 22°32′N 75°49′E
Miadzioł Nowy *see* Myadzyel
171 O6 **Miagao** Panay Island, C Philippines 10°40′N 122°15′E
41 R17 **Miahuatlán** *var.* Miahuatlán de Porfirio Díaz. Oaxaca, SE Mexico 16°21′N 96°36′W
Miahuatlán de Porfirio Díaz *see* Miahuatlán
104 K10 **Miajadas** Extremadura, W Spain 39°10′N 05°54′W
Miajlar *see* Myājlār
36 M14 **Miami** Arizona, SW USA 33°23′N 110°53′W
23 Z16 **Miami** Florida, SE USA 25°46′N 80°12′W
27 R8 **Miami** Oklahoma, C USA 36°53′N 94°54′W
25 O2 **Miami** Texas, SW USA 35°42′N 100°37′W
23 Z16 **Miami** ✈ Florida, SE USA 25°47′N 80°16′W
23 Z16 **Miami Beach** Florida, SE USA 25°47′N 80°08′W
23 Y15 **Miami Canal** *canal* Florida, SE USA
31 R14 **Miamisburg** Ohio, N USA 39°38′N 84°17′W
149 U10 **Miān Channūn** Punjab, E Pakistan 30°24′N 72°27′E
142 J4 **Miāndowāb** *var.* Mianduab, Miyāndoāb. Āzarbāyjān-e Gharbī, NW Iran 36°57′N 46°06′E
172 H5 **Miandrivazo** Toliara, C Madagascar 19°31′S 45°29′E
Mianduab *see* Miāndowāb
142 K3 **Mīāneh** *var.* Miyāneh. Āzarbāyjān-e Sharqī, NW Iran 37°23′N 47°45′E
161 T12 **Mianhua Yu** *prev.* Mienhua Yü. *island* N Taiwan
149 O16 **Miāni Hōr** *lagoon* S Pakistan
160 G10 **Mianning** Sichuan, C China 28°34′N 102°12′E
149 T7 **Miānwāli** Punjab, NE Pakistan 32°32′N 71°33′E
160 J7 **Mianxian** *var.* Mian Xian. Shaanxi, C China 33°12′N 106°36′E
Mian Xian *see* Mianxian
160 I8 **Mianyang** Sichuan, C China 31°29′N 104°43′E
Mianyang *see* Xiantao
161 R3 **Miaodao Qundao** *island group* E China
161 S13 **Miaoli** N Taiwan 24°33′N 120°48′E
122 F11 **Miass** Chelyabinskaya Oblast', C Russian Federation 55°00′N 59°55′E
110 G8 **Miastko** *Ger.* Rummelsburg in Pommern. Pomorskie, N Poland 54°N 16°58′E
Miava *see* Myjava
11 O15 **Mica Creek** British Columbia, SW Canada 51°58′N 118°29′W
160 J7 **Micang Shan** ⛰ C China
Mi Chai *see* Nong Khai
111 O19 **Michalovce** *Ger.* Grossmichel, *Hung.* Nagymihály. Košický Kraj, E Slovakia 48°46′N 21°55′E
99 M20 **Michel, Baraque** *hill* E Belgium
39 S5 **Michelson, Mount** ▲ Alaska, USA 69°19′N 144°16′W
45 P9 **Miches** E Dominican Republic 18°59′N 69°03′W
30 M4 **Michigamme, Lake** ◎ Michigan, N USA
30 M4 **Michigamme Reservoir** ▣ Michigan, N USA
31 N4 **Michigamme River** 🌊 Michigan, N USA
31 O7 **Michigan** *off.* State of Michigan, *also known as* Great Lakes State, Lake State, Wolverine State. ⬖ *state* N USA
31 O11 **Michigan City** Indiana, N USA 41°43′N 86°52′W
31 O8 **Michigan, Lake** ◎ N USA
31 P2 **Michipicoten Bay** *lake bay* Ontario, N Canada
14 A8 **Michipicoten Island** *island* Ontario, S Canada
14 B7 **Michipicoten River** Ontario, S Canada 47°56′N 84°48′W
Michurin *see* Tsarevo
126 M6 **Michurinsk** Tambovskaya Oblast', W Russian Federation 52°56′N 40°31′E
Mico, Punta/Mico, Punto *see* Monkey Point
42 L10 **Mico, Río** 🌊 SE Nicaragua
45 T12 **Micoud** SE Saint Lucia 13°49′N 60°54′W
189 N16 **Micronesia** *off.* Federated States of Micronesia. ◆ *federation* W Pacific Ocean
175 P4 **Micronesia** *island group* W Pacific Ocean
Micronesia, Federated States of *see* Micronesia
169 O9 **Midai, Pulau** *island* Kepulauan Natuna, W Indonesia
Mid-Atlantic Cordillera *see* Mid-Atlantic Ridge
65 M17 **Mid-Atlantic Ridge** *var.* Mid-Atlantic Cordillera, Mid-Atlantic Rise, Mid-Atlantic Swell. *undersea feature* Atlantic Ocean 0°00′N 20°00′W
Mid-Atlantic Rise/ Mid-Atlantic Swell *see* Mid-Atlantic Ridge
99 E15 **Middelburg** Zeeland, SW Netherlands 51°30′N 03°36′E
83 H24 **Middelburg** Eastern Cape, S South Africa 31°28′S 25°01′E
83 K21 **Middelburg** Mpumalanga, NE South Africa 25°47′S 29°28′E
95 G23 **Middelfart** Syddtjylland, C Denmark 55°30′N 09°44′E
98 G13 **Middelharnis** Zuid-Holland, SW Netherlands 51°45′N 04°10′E
99 B16 **Middelkerke** West-Vlaanderen, W Belgium 51°12′N 02°51′E
98 I9 **Middenbeemster** Noord-Holland, C Netherlands 52°33′N 04°55′E
98 I8 **Middenmeer** Noord-Holland, NW Netherlands 52°48′N 04°58′E
35 Q2 **Middle Alkali Lake** ◎ California, W USA
193 S6 **Middle America Trench** *undersea feature* E Pacific Ocean 15°00′N 95°00′W
151 P19 **Middle Andaman** *island* Andaman Islands, India, NE Indian Ocean
Middle Atlas *see* Moyen Atlas
21 R3 **Middlebourne** West Virginia, NE USA 39°30′N 80°53′W
23 W9 **Middleburg** Florida, SE USA 30°03′N 81°55′W
Middleburg Island *see* 'Eua
Middle Caicos *see* Grand Caicos
25 N8 **Middle Concho River** 🌊 Texas, SW USA
Middle Congo *see* Congo (Republic of)
39 R6 **Middle Fork Chandalar River** 🌊 Alaska, USA
39 Q7 **Middle Fork Koyukuk River** 🌊 Alaska, USA
33 O12 **Middle Fork Salmon River** 🌊 Idaho, NW USA
11 T15 **Middle Lake** Saskatchewan, S Canada 52°31′N 105°16′W
28 L13 **Middle Loup River** 🌊 Nebraska, C USA
185 E22 **Middlemarch** Otago, South Island, New Zealand 45°30′S 170°07′E
31 T15 **Middleport** Ohio, N USA 39°00′N 82°03′W
29 U14 **Middle Raccoon River** 🌊 Iowa, C USA
29 R3 **Middle River** 🌊 Minnesota, N USA
21 N8 **Middlesboro** Kentucky, S USA 36°37′N 83°42′W
97 M15 **Middlesbrough** N England, United Kingdom 54°35′N 01°14′W
42 G3 **Middlesex** Stann Creek, C Belize 17°00′N 88°31′W
97 N22 **Middlesex** *cultural region* SE England, United Kingdom
13 P15 **Middleton** Nova Scotia, SE Canada 44°56′N 65°04′W
20 F10 **Middleton** Tennessee, S USA 35°05′N 88°57′W
30 L9 **Middleton** Wisconsin, N USA 43°06′N 89°30′W
39 S13 **Middleton Island** *island* Alaska, USA
34 M7 **Middletown** California, W USA 38°44′N 122°39′W
21 Y2 **Middletown** Delaware, NE USA 39°25′N 75°39′W
18 K15 **Middletown** New Jersey, NE USA 40°23′N 74°08′W
18 K13 **Middletown** New York, NE USA 41°27′N 74°25′W
31 R14 **Middletown** Ohio, N USA 39°33′N 84°19′W
18 G15 **Middletown** Pennsylvania, NE USA 40°11′N 76°42′W
141 N14 **Midi** *var.* Maydī. NW Yemen 16°18′N 42°51′E
103 O16 **Midi, Canal du** *canal* S France
102 K17 **Midi de Bigorre, Pic du** ▲ S France 42°57′N 00°08′E
102 K17 **Midi d'Ossau, Pic du** ▲ SW France 42°51′N 00°27′W
173 R7 **Mid-Indian Basin** *undersea feature* N Indian Ocean 10°00′S 80°00′E
173 P7 **Mid-Indian Ridge** *var.* Central Indian Ridge. *undersea feature* C Indian Ocean 12°00′S 66°00′E
103 N14 **Midi-Pyrénées** ⬖ *region* S France
25 N8 **Midkiff** Texas, SW USA 31°35′N 101°51′W
14 G13 **Midland** Ontario, S Canada 44°45′N 79°53′W
31 R8 **Midland** Michigan, N USA 43°37′N 84°15′W
28 M10 **Midland** South Dakota, N USA 44°04′N 101°07′W
24 M8 **Midland** Texas, SW USA 32°N 102°05′W
83 K17 **Midlands** ⬖ *province* C Zimbabwe
97 D21 **Midleton** *Ir.* Mainistir na Corann. SW Ireland 51°55′N 08°10′W
25 T7 **Midlothian** Texas, SW USA 32°28′N 96°59′W
96 K12 **Midlothian** *cultural region* S Scotland, United Kingdom
172 I7 **Midongy Atsimo** Fianarantsoa, S Madagascar 21°58′S 47°47′E
102 K15 **Midou** 🌊 SW France
192 J6 **Mid-Pacific Mountains** *var.* Mid-Pacific Seamounts. *undersea feature* NW Pacific Ocean 20°00′N 178°00′W
Mid-Pacific Seamounts *see* Mid-Pacific Mountains
171 Q7 **Midsayap** Mindanao, S Philippines 07°12′N 124°31′E
95 F21 **Midtjylland** ⬖ *county* NW Denmark
36 L3 **Midway** Utah, W USA 40°30′N 111°28′W
192 L5 **Midway Islands** ◇ *US unincorporated territory* C Pacific Ocean
33 X14 **Midwest** Wyoming, C USA 43°24′N 106°15′W
27 N10 **Midwest City** Oklahoma, C USA 35°28′N 98°24′W
152 M10 **Mid Western** ⬖ *zone* W Nepal
98 P5 **Midwolda** Groningen, NE Netherlands 53°12′N 07°00′E
137 Q16 **Midyat** Mardin, SE Turkey 37°25′N 41°20′E
114 F8 **Midzhur** *SCr.* Midžor. ▲ Bulgaria/Serbia 43°24′N 22°41′E *see also* Midžor
Midzhur *see* Midžor
113 Q14 **Midžor** *Bul.* Midzhur. ▲ Bulgaria/Serbia 43°24′N 22°40′E *see also* Midzhur
Midžor *see* Midzhur
164 K14 **Mie** *off.* Mie-ken. ⬖ *prefecture* Honshū, SW Japan
111 L16 **Miechów** Małopolskie, S Poland 50°21′N 20°01′E
110 F11 **Międzychód** *Ger.* Mitteldorf. Wielkopolskie, C Poland 52°36′N 15°53′E
Międzyleska, Przełęcz *see* Mezileské Sedlo
110 O12 **Międzyrzec Podlaski** Lubelskie, E Poland 52°N 22°47′E
110 E11 **Międzyrzecz** *Ger.* Meseritz. Lubuskie, W Poland 52°26′N 15°33′E
Mie-ken *see* Mie
102 L16 **Miélan** Gers, S France 43°26′N 00°19′E
111 N16 **Mielec** Podkarpackie, SE Poland 50°18′N 21°27′E
95 L21 **Mien** ◎ S Sweden
Mienhua Yü *see* Mianhua Yu
41 O8 **Mier** Tamaulipas, C Mexico 26°28′N 99°10′W
116 J11 **Miercurea-Ciuc** *Ger.* Szeklerburg, *Hung.* Csíkszereda. Harghita, C Romania 46°24′N 25°48′E
Mieresch *see* Maros/Mureş
104 K2 **Mieres del Camín** *var.* Mieres del Camino. Asturias, NW Spain 43°15′N 05°46′W
Mieres del Camino *see* Mieres del Camín
99 K15 **Mierlo** Noord-Brabant, SE Netherlands 51°27′N 05°37′E
41 O10 **Mier y Noriega** Nuevo León, NE Mexico 23°24′N 100°06′W
Mies *see* Stříbro
80 K13 **Mī'ēso** *var.* Meheso, Miesso. Oromīya, C Ethiopia 09°13′N 40°47′E
Miesso *see* Mī'ēso
110 D10 **Mieszkowice** *Ger.* Bärwalde Neumark. Zachodnio-pomorskie, W Poland 52°45′N 14°24′E
18 G14 **Mifflinburg** Pennsylvania, NE USA 40°55′N 77°03′W
18 F14 **Mifflintown** Pennsylvania, NE USA 40°34′N 77°24′W
138 F8 **Mifrats Hefa** *Eng.* Bay of Haifa; *prev.* Mifraz Hefa. *bay* N Israel
81 G19 **Migori** ⬖ *county* W Kenya
41 R15 **Miguel Alemán, Presa** ▣ SE Mexico
40 L9 **Miguel Asua** *var.* Miguel Auza. Zacatecas, C Mexico 24°17′N 103°29′W
Miguel Auza *see* Miguel Asua
43 S15 **Miguel de la Borda** *var.* Donoso. Colón, C Panama 09°09′N 80°20′W
41 N13 **Miguel Hidalgo** ✈ (Guadalajara) Jalisco, SW Mexico 20°52′N 101°09′W
40 H7 **Miguel Hidalgo, Presa** ▣ W Mexico
116 J14 **Mihăileşti** Giurgiu, S Romania 44°20′N 25°54′E
116 M14 **Mihail Kogălniceanu** *var.* Kogălniceanu; *prev.* Caramurat, Ferdinand. Constanța, SE Romania 44°22′N 28°27′E
117 N14 **Mihai Viteazu** Constanța, SE Romania 44°37′N 28°41′E
136 G12 **Mihalıççık** Eskişehir, NW Turkey 39°52′N 31°30′E
164 G13 **Mihara** Hiroshima, Honshū, SW Japan 34°24′N 133°04′E
165 N14 **Mihara-yama** 🌋 Miyako-jima, SE Japan 34°43′N 139°23′E
105 S8 **Mijares** 🌊 E Spain
98 I11 **Mijdrecht** Utrecht, C Netherlands 52°12′N 04°52′E
165 S4 **Mikasa** Hokkaidō, NE Japan 43°15′N 141°57′E
Mikashevichi *see* Mikashevichy
119 K19 **Mikashevichy** *Pol.* Mikaszewicze, *Rus.* Mikashevichi. Brestskaya Voblasts', SW Belarus 52°13′N 27°28′E
Mikaszewicze *see* Mikashevichy
126 L5 **Mikhaylov** Ryazanskaya Oblast', W Russian Federation 54°12′N 39°03′E
Mikhaylovgrad *see* Montana
195 Z8 **Mikhaylov Island** *island* Antarctica
145 T6 **Mikhaylovka** Pavlodar, N Kazakhstan 53°49′N 76°31′E
127 N9 **Mikhaylovka** Volgogradskaya Oblast', SW Russian Federation 50°06′N 43°17′E
Mikhaylovka *see* Mykhaylivka
81 K24 **Mikindani** Mtwara, SE Tanzania 10°16′S 40°05′E
93 N18 **Mikkeli** *Swe.* Sankt Michel. Etelä-Savo, SE Finland 61°41′N 27°14′E
110 M8 **Mikołajki** *Ger.* Nikolaiken. Warmińsko-Mazurskie, NE Poland 53°49′N 21°31′E
Míkonos *see* Mýkonos
114 I9 **Mikre** Lovech, N Bulgaria 43°01′N 24°31′E
114 C13 **Mikrí Préspa, Límni** ◎ N Greece
125 P4 **Mikulkin, Mys** *headland* NW Russian Federation 67°50′N 46°36′E
81 I23 **Mikumi** Morogoro, SE Tanzania 07°22′S 37°00′E
125 R10 **Mikun'** Respublika Komi, NW Russian Federation 62°20′N 50°02′E
164 K13 **Mikuni** Fukui, Honshū, SW Japan 36°12′N 136°09′E
165 X13 **Mikura-jima** *island* E Japan
29 V7 **Milaca** Minnesota, N USA 45°45′N 93°40′W
62 J10 **Milagro** La Rioja, C Argentina 31°00′S 66°01′W
56 B7 **Milagro** Guayas, SW Ecuador 02°11′S 79°36′W
31 P4 **Milakokia Lake** ◎ Michigan, N USA
30 J1 **Milan** Illinois, N USA 41°27′N 90°33′W
31 R10 **Milan** Michigan, N USA 42°05′N 83°40′W
27 T2 **Milan** Missouri, C USA 40°12′N 93°08′W
37 Q11 **Milan** New Mexico, SW USA 35°10′N 107°53′W
20 G9 **Milan** Tennessee, S USA 35°55′N 88°45′W
Milan *see* Milano
95 F15 **Miland** Telemark, S Norway 59°57′N 08°48′E
83 N15 **Milange** Zambézia, NE Mozambique 16°09′S 35°44′E
106 D8 **Milano** *Eng.* Milan, *Ger.* Mailand; *anc.* Mediolanum. Lombardia, N Italy 45°28′N 09°10′E
25 U10 **Milano** Texas, SW USA 30°42′N 96°51′W
136 C15 **Milas** Muğla, SW Turkey 37°17′N 27°46′E
119 K21 **Milashavichy** *Rus.* Milashevichi. Homyel'skaya Voblasts', SE Belarus 51°39′N 27°56′E
Milashevichi *see* Milashavichy
119 I18 **Milavidy** *Rus.* Milovidy. Brestskaya Voblasts', SW Belarus 52°54′N 25°51′E
107 L23 **Milazzo** *anc.* Mylae. Sicilia, Italy, C Mediterranean Sea 38°13′N 15°15′E
29 R8 **Milbank** South Dakota, N USA 45°12′N 96°36′W
19 T7 **Milbridge** Maine, NE USA 44°31′N 67°55′W
100 L11 **Milde** 🌊 C Germany
14 F14 **Mildmay** Ontario, S Canada 44°03′N 81°07′W
182 L9 **Mildura** Victoria, SE Australia 34°13′S 142°09′E
137 X12 **Mil Düzü** *Rus.* Mil'skaya Ravnina, Mil'skaya Step'. *physical region* C Azerbaijan
160 H13 **Mile** *var.* Miyang. Yunnan, SW China 24°28′N 103°26′E
Mile *see* Mili Atoll
181 Y10 **Miles** Queensland, E Australia 26°41′S 150°15′E
25 P8 **Miles** Texas, SW USA 31°36′N 100°10′W
33 X9 **Miles City** Montana, NW USA 46°24′N 105°48′W
11 U17 **Milestone** Saskatchewan, S Canada 50°00′N 104°24′W
107 N22 **Mileto** Calabria, SW Italy 38°35′N 16°03′E
107 K16 **Miletto, Monte** ▲ C Italy 41°28′N 14°21′E
18 M13 **Milford** Connecticut, NE USA 41°12′N 73°01′W
21 Y3 **Milford** *var.* Milford City. Delaware, NE USA 38°54′N 75°25′W
29 T11 **Milford** Iowa, C USA 43°19′N 95°09′W
19 S6 **Milford** Maine, NE USA 44°57′N 68°37′W
29 R16 **Milford** Nebraska, C USA 40°46′N 97°03′W
19 O10 **Milford** New Hampshire, NE USA 42°49′N 71°38′W
18 J13 **Milford** Pennsylvania, NE USA 41°20′N 74°48′W
25 T7 **Milford** Texas, SW USA 32°07′N 96°57′W
36 K6 **Milford** Utah, W USA 38°22′N 112°57′W
Milford *see* Milford Haven
Milford City *see* Milford
97 H21 **Milford Haven** *prev.* Milford. SW Wales, United Kingdom 51°44′N 05°02′W
27 O4 **Milford Lake** ▣ Kansas, C USA
185 B21 **Milford Sound** Southland, South Island, New Zealand 44°41′S 167°57′E
185 B21 **Milford Sound** *inlet* South Island, New Zealand
Milhau *see* Millau
Milh, Bahr al *see* Razāzah, Buḩayrat ar
139 T10 **Milḩ, Wādī al** *dry watercourse* S Iraq
189 W8 **Mili Atoll** *var.* Mile. *atoll* Ratak Chain, SE Marshall Islands
110 H13 **Milicz** Dolnośląskie, SW Poland 51°32′N 17°15′E
107 L25 **Militello in Val di Catania** Sicilia, Italy, C Mediterranean Sea 37°17′N 14°47′E
11 R17 **Milk River** Alberta, SW Canada 49°10′N 112°06′W
44 J13 **Milk River** 🌊 C Jamaica
33 W7 **Milk River** 🌊 Montana, NW USA
80 D9 **Milk, Wadi el** *var.* Wadi al Malik. 🌊 C Sudan
99 L14 **Mill** Noord-Brabant, SE Netherlands 51°42′N 05°46′E
103 P14 **Millau** *var.* Milhau; *anc.* Æmilianum. Aveyron, S France 44°06′N 03°05′E
14 I14 **Millbrook** Ontario, SE Canada 44°09′N 78°26′W
23 U4 **Milledgeville** Georgia, SE USA 33°04′N 83°13′W
12 C12 **Mille Lacs, Lac des** ◎ Ontario, S Canada
29 V6 **Mille Lacs Lake** ◎ Minnesota, N USA
23 V4 **Millen** Georgia, SE USA 32°50′N 81°56′W
191 Y5 **Millennium Island** *prev.* Caroline Island, Thornton Island. *atoll* Line Islands, E Kiribati
29 O9 **Miller** South Dakota, N USA 44°31′N 98°59′W
30 K5 **Miller Dam Flowage** ◎ Wisconsin, N USA
39 U12 **Miller, Mount** ▲ Alaska, USA 60°29′N 142°16′W

◆ Country ● Country Capital ◇ Dependent Territory ○ Dependent Territory Capital ⬖ Administrative Regions ✈ International Airport ▲ Mountain ⛰ Mountain Range 🌋 Volcano 🌊 River ◎ Lake ▣ Reservoir

MILLEROVO – MOGILËVSKAYA OBLAST'

126 L10 **Millerovo** Rostovskaya Oblast', SW Russian Federation 48°57′N 40°26′E
37 N17 **Miller Peak** ▲ Arizona, SW USA 31°23′N 110°17′W
31 T12 **Millersburg** Ohio, N USA 40°33′N 81°55′W
18 G15 **Millersburg** Pennsylvania, NE USA 40°31′N 76°56′W
185 D23 **Millers Flat** Otago, South Island, New Zealand 45°42′S 169°25′E
25 Q8 **Millersview** Texas, SW USA 31°26′N 99°44′W
106 B10 **Millesimo** Piemonte, NE Italy 44°24′N 08°09′E
12 C12 **Milles Lacs, Lac des** ◎ Ontario, SW Canada
25 Q13 **Millett** Texas, SW USA 28°33′N 99°10′W
103 N11 **Millevaches, Plateau de** *plateau* C France
182 K12 **Millicent** South Australia 37°29′S 140°01′E
98 M13 **Millingen aan den Rijn** Gelderland, SE Netherlands 51°52′N 06°02′E
20 E10 **Millington** Tennessee, S USA 35°20′N 89°54′W
19 R4 **Millinocket** Maine, NE USA 45°38′N 68°45′W
19 R4 **Millinocket Lake** ◎ Maine, NE USA
195 Z11 **Mill Island** *island* Antarctica
183 T3 **Millmerran** Queensland, E Australia 27°53′S 151°15′E
109 R9 **Millstatt** Kärnten, S Austria 46°45′N 13°36′E
97 B19 **Milltown Malbay** *Ir.* Sráid na Cathrach. W Ireland 52°51′N 09°23′W
18 J17 **Millville** New Jersey, NE USA 39°24′N 75°01′W
27 S13 **Millwood Lake** ⊠ Arkansas, C USA
Milne Bank *see* Milne Seamounts
186 G10 **Milne Bay** ◈ *province* SE Papua New Guinea
64 J8 **Milne Seamounts** *var.* Milne Bank. *undersea feature* N Atlantic Ocean
29 Q6 **Milnor** North Dakota, N USA 46°15′N 97°27′W
19 R5 **Milo** Maine, NE USA 45°15′N 69°01′W
115 I22 **Mílos** *island* Kykládes, Greece, Aegean Sea
Mílos *see* Pláka
110 H11 **Miłosław** Wielkopolskie, C Poland 52°13′N 17°28′E
113 K19 **Milot** *var.* Miloti. Lezhë, C Albania 41°42′N 19°43′E
Miloti *see* Milot
117 Z5 **Milove** Luhans'ka Oblast', E Ukraine 49°22′N 40°09′E
Milovidy *see* Milavidy
182 L4 **Milparinka** New South Wales, SE Australia 29°48′S 141°57′E
35 N9 **Milpitas** California, W USA 37°25′N 121°54′W
Mil'skaya Ravnina/ Mil'skaya Step' *see* Mil Düzü
14 G15 **Milton** Ontario, S Canada 43°31′N 79°53′W
185 E24 **Milton** Otago, South Island, New Zealand 46°08′S 169°59′E
21 Y4 **Milton** Delaware, NE USA 38°48′N 75°21′W
23 P8 **Milton** Florida, SE USA 30°37′N 87°02′W
18 G14 **Milton** Pennsylvania, NE USA 41°01′N 76°49′W
18 L7 **Milton** Vermont, NE USA 44°37′N 73°04′W
32 K11 **Milton-Freewater** Oregon, NW USA 45°54′N 118°24′W
97 N21 **Milton Keynes** SE England, United Kingdom 52°N 00°43′W
27 N3 **Miltonvale** Kansas, C USA 39°21′N 97°27′W
161 N10 **Miluo** Hunan, S China 28°52′N 113°00′E
30 M9 **Milwaukee** Wisconsin, N USA 43°03′N 87°56′W
Milyang *see* Miryang
Mimatum *see* Mende
37 Q15 **Mimbres Mountains** ▲▲ New Mexico, SW USA
182 D2 **Mimili** South Australia 27°01′S 132°33′E
102 J14 **Mimizan** Landes, SW France 44°12′N 01°12′W
Mimmaya *see* Minmaya
79 E19 **Mimongo** Ngounié, C Gabon 01°36′S 11°44′E
Min *see* Fujian
35 T7 **Mina** Nevada, W USA 38°23′N 118°07′W
143 S14 **Mīnāb** Hormozgān, SE Iran 27°08′N 57°02′E
Mînâ Baranîs *see* Baranīs
149 R9 **Mīna Bāzār** Baluchistān, SW Pakistan 30°58′N 69°11′E
Minami-Awaji *see* Nandan
165 X17 **Minami-Iō-jima** *Eng.* San Augustine. *island* SE Japan
165 R5 **Minami-Kayabe** Hokkaidō, NE Japan 41°54′N 140°58′E
164 B16 **Minamisatsuma** *var.* Kaseda. Kagoshima, Kyūshū, SW Japan 31°25′N 130°17′E
164 C14 **Minamishimabara** *var.* Kuchinotsu. Nagasaki, Kyūshū, SW Japan 32°36′N 130°11′E
164 C17 **Minamitane** Kagoshima, Tanega-shima, SW Japan 30°23′N 130°54′E
Minami Tori Shima *see* Marcus Island
Min'an *see* Longshan
62 J4 **Mina Pirquitas** Jujuy, NW Argentina 22°48′S 66°24′W
173 O3 **Mīnā' Qābūs** NE Oman
61 F19 **Minas** Lavalleja, S Uruguay 34°20′S 55°15′W
13 P15 **Minas Basin** *bay* Nova Scotia, SE Canada
61 F17 **Minas de Corrales** Rivera, NE Uruguay 31°35′S 55°20′W
44 A5 **Minas de Matahambre** Pinar del Río, W Cuba 22°34′N 83°57′W
104 J13 **Minas de Ríotinto** Andalucía, S Spain 37°40′N 06°36′W
60 K7 **Minas Gerais** *off.* Estado de Minas Gerais. ◈ *state* E Brazil
Minas Gerais, Estado de *see* Minas Gerais
42 E5 **Minas, Sierra de las** ▲▲ E Guatemala
41 T15 **Minatitlán** Veracruz-Llave, E Mexico 17°59′N 94°32′W
166 L6 **Minbu** Magway, W Myanmar (Burma) 20°09′N 94°52′E
149 V10 **Minchinābād** Punjab, E Pakistan 30°10′N 73°40′E
63 G17 **Minchinmávida, Volcán** Ѫ S Chile 42°51′S 72°23′W
96 G7 **Minch, The** *var.* North Minch. *strait* NW Scotland, United Kingdom
106 F8 **Mincio** *anc.* Mincius. ≈ N Italy
Mincius *see* Mincio
26 M11 **Minco** Oklahoma, C USA 35°18′N 97°56′W
171 Q7 **Mindanao** *island* S Philippines
Mindanao Sea *see* Bohol Sea
101 J23 **Mindel** ≈ S Germany
101 J23 **Mindelheim** Bayern, S Germany 48°03′N 10°30′E
Mindello *see* Mindelo
76 C9 **Mindelo** *var.* Mindello; *prev.* Porto Grande. São Vicente, N Cape Verde 16°54′N 25°01′W
14 I13 **Minden** Ontario, SE Canada 44°54′N 78°41′W
100 H13 **Minden** *anc.* Minthun. Nordrhein-Westfalen, NW Germany 52°18′N 08°55′E
22 G5 **Minden** Louisiana, S USA 32°37′N 93°17′W
29 O16 **Minden** Nebraska, C USA 40°30′N 98°57′W
35 Q6 **Minden** Nevada, W USA 38°58′N 119°47′W
182 L8 **Mindona Lake** *seasonal lake* New South Wales, SE Australia
171 O4 **Mindoro** *island* N Philippines
171 N5 **Mindoro Strait** *strait* W Philippines
97 J23 **Minehead** SW England, United Kingdom 51°13′N 03°29′W
97 E21 **Mine Head** *Ir.* Mionn Ard. *headland* S Ireland 51°58′N 07°36′W
59 J19 **Mineiros** Goiás, C Brazil 17°34′S 52°33′W
25 V6 **Mineola** Texas, SW USA 32°39′N 95°29′W
25 S13 **Mineral** Texas, SW USA 28°32′N 97°54′W
127 N15 **Mineral'nyye Vody** Stavropol'skiy Kray, SW Russian Federation 44°13′N 43°06′E
30 K9 **Mineral Point** Wisconsin, N USA 42°54′N 90°09′W
25 S6 **Mineral Wells** Texas, SW USA 32°48′N 98°06′W
36 K6 **Minersville** Utah, W USA 38°12′N 112°56′W
31 U12 **Minerva** Ohio, N USA 40°43′N 81°06′W
107 N17 **Minervino Murge** Puglia, SE Italy 41°06′N 16°05′E
103 O16 **Minervois** *physical region* S France
158 I10 **Minfeng** *var.* Niya. Xinjiang Uygur Zizhiqu, NW China 37°07′N 82°43′E
79 O25 **Minga** Katanga, SE Dem. Rep. Congo 11°06′S 27°57′E
137 W11 **Mingäçevir** *Rus.* Mingechaur, Mingechevir. C Azerbaijan 40°46′N 47°02′E
137 W11 **Mingäçevir Su Anbarı** *Rus.* Mingechaurskoye Vodokhranilishche, Mingechevirskoye Vodokhranilishche. ⊠ NW Azerbaijan
166 L8 **Mingaladon** ✈ (Yangon) Yangon, SW Myanmar (Burma) 16°55′N 96°11′E
13 P11 **Mingan** Québec, E Canada 50°19′N 64°02′W
Mingāora *see* Saīdu
146 K8 **Mingbuloq** *Rus.* Mynbulak. Navoiy Viloyati, N Uzbekistan 42°18′N 62°53′E
146 K9 **Mingbuloq Botig'I** *Rus.* Vpadina Mynbulak. *depression* N Uzbekistan
Mingechaur/Mingechevir *see* Mingäçevir
Mingechaurskoye Vodokhranilishche/ Mingechevirskoye Vodokhranilishche *see* Mingäçevir Su Anbarı
161 Q7 **Mingguang** *prev.* Jiashan. Anhui, SE China 32°45′N 117°59′E
166 L4 **Mingin** Sagaing, C Myanmar (Burma) 22°51′N 94°30′E
105 Q10 **Minglanilla** Castilla-La Mancha, C Spain 39°32′N 01°36′W
31 V13 **Mingo Junction** Ohio, N USA 40°19′N 80°36′W
Mingora *see* Saīdu
163 V7 **Mingshui** Heilongjiang, NE China 47°10′N 125°53′E
Mingtekl Daban *see* Mintaka Pass
Mingu *see* Zhenfeng
83 Q14 **Minguri** Nampula, NE Mozambique 14°30′S 40°37′E
Mingzhou *see* Suide
159 U10 **Minhe** *var.* Chuankou; *prev.* Minhe Huizu Tuzu Zizhixian, Shangchuankou. Qinghai, C China 36°21′N 102°40′E
Minhe Huizu Tuzu Zizhixian *see* Minhe
166 L6 **Minhla** Magway, W Myanmar (Burma) 19°58′N 95°03′E
167 S14 **Minh Lương** Kiên Giang, S Vietnam 09°52′N 105°10′E
104 G5 **Minho** *former province* N Portugal
104 G5 **Minho, Rio** *Sp.* Miño. ≈ Portugal/Spain *see also* Miño
Minho, Rio *see* Miño
155 C24 **Minicoy Island** *island* SW India
33 P15 **Minidoka** Idaho, NW USA 42°45′N 113°29′W
118 C11 **Minija** ≈ W Lithuania
180 G9 **Minilya** Western Australia 23°45′S 114°03′E
14 E8 **Minisinakwa Lake** ◎ Ontario, S Canada
45 T12 **Ministre Point** *headland* S Saint Lucia 13°42′N 60°57′W
11 V15 **Minitonas** Manitoba, S Canada 52°07′N 101°02′W
Minius *see* Miño
186 D7 **Minj** Jiwaka, Papua New Guinea 05°55′S 144°37′E
161 R12 **Min Jiang** ≈ SE China
160 H10 **Min Jiang** ≈ C China
182 H9 **Minlaton** South Australia 34°52′S 137°33′E
159 S9 **Minle** Gansu, N China
165 Q6 **Minmaya** *var.* Mimmaya. Aomori, Honshū, C Japan 41°10′N 140°24′E
77 U14 **Minna** Niger, C Nigeria 09°33′N 06°33′E
165 P16 **Minna-jima** *island* Sakishima-shotō, SW Japan
27 N4 **Minneapolis** Kansas, C USA 39°08′N 97°43′W
29 U9 **Minneapolis** Minnesota, N USA 44°59′N 93°16′W
29 V8 **Minneapolis-Saint Paul** ✈ Minnesota, N USA 44°53′N 93°13′W
11 W16 **Minnedosa** Manitoba, S Canada 50°14′N 99°50′W
26 J7 **Minneola** Kansas, C USA 37°26′N 100°00′W
29 S7 **Minnesota** *off.* State of Minnesota, *also known as* Gopher State, New England of the West, North Star State. ◈ *state* N USA
29 S9 **Minnesota River** ≈ Minnesota/South Dakota, N USA
29 V9 **Minnetonka** Minnesota, N USA 44°55′N 93°28′W
29 O3 **Minnewaukan** North Dakota, N USA 48°04′N 99°14′W
182 F7 **Minnipa** South Australia 32°52′S 135°07′E
104 G5 **Miño** *var.* Mino, Minius, *Port.* Rio Minho. ≈ Portugal/Spain *see also* Minho, Rio
Miño *see* Minho, Rio
30 L4 **Minocqua** Wisconsin, N USA 45°53′N 89°42′W
30 L12 **Minonk** Illinois, N USA 40°54′N 89°01′W
Minorca *see* Menorca
28 M3 **Minot** North Dakota, N USA 48°15′N 101°19′W
159 U8 **Minqin** Gansu, N China 38°35′N 103°07′E
119 J16 **Minsk** ● (Belarus) Horad Minsk, C Belarus 53°52′N 27°34′E
Minsk-2 *see* Minsk National
Minskaya Oblast' *see* Minskaya Voblasts'
119 K16 **Minskaya Voblasts'** *prev. Rus.* Minskaya Oblast'. ◈ *province* C Belarus
Minskaya Vozvyshennost' *see* Minskaye Wzvyshsha
119 J16 **Minskaye Wzvyshsha** *Rus.* Minskaya Vozvyshennost'. ▲▲ C Belarus
Minsk, Gorod *see* Minsk, Horad
119 J16 **Minsk, Horad** *Russ.* Gorod Minsk. ◈ *province* C Belarus
110 N12 **Mińsk Mazowiecki** *var.* Nowo-Minsk. Mazowieckie, C Poland 52°10′N 21°31′E
119 L16 **Minsk National** *prev.* Minsk-2. ✈ Minskaya Voblasts', C Belarus 53°52′N 27°58′E
31 Q13 **Minster** Ohio, N USA 40°23′N 84°22′W
79 F15 **Minta** Centre, C Cameroon 04°34′N 12°54′E
149 W2 **Mintaka Pass** *Chin.* Mingtekl Daban. *pass* China/ Pakistan
115 D20 **Mínthi** ▲▲ S Greece
Minthun *see* Minden
13 O14 **Minto** New Brunswick, SE Canada 46°05′N 66°05′W
10 H6 **Minto** Yukon, W Canada 62°33′N 136°45′W
39 R9 **Minto** Alaska, USA 65°07′N 149°22′W
29 Q3 **Minto** North Dakota, N USA 48°17′N 97°22′W
12 K6 **Minto, Lac** ◎ Québec, C Canada
195 R16 **Minto, Mount** ▲ Antarctica 71°38′S 169°11′E
11 U17 **Minton** Saskatchewan, S Canada 49°12′N 104°33′W
189 R15 **Minto Reef** *atoll* Caroline Islands, C Micronesia
37 R4 **Minturn** Colorado, C USA 39°34′N 106°21′W
107 J16 **Minturno** Lazio, C Italy 41°15′N 13°47′E
122 K13 **Minusinsk** Krasnoyarskiy Kray, S Russian Federation 53°37′N 91°49′E
108 G11 **Minusio** Ticino, S Switzerland 46°11′N 08°47′E
79 E17 **Minvoul** Woleu-Ntem, N Gabon 02°08′N 12°12′E
141 R13 **Minwakh** N Yemen 16°55′N 48°04′E
159 V11 **Minxian** *var.* Min Xian, Minyang. Gansu, C China 34°20′N 104°09′E
Min Xian *see* Minxian
Minya *see* Al Minyā
Minyang *see* Minxian
31 R6 **Mio** Michigan, N USA 44°40′N 84°09′W
Mionn Ard *see* Mine Head
Miory *see* Myory
158 L5 **Miquan** Xinjiang Uygur Zizhiqu, NW China 44°04′N 87°40′E
119 I17 **Mir** Hrodzyenskaya Voblasts', W Belarus 53°25′N 26°28′E
106 H8 **Mira** Veneto, NE Italy 45°25′N 12°07′E
12 K15 **Mirabel** *var.* Montreal. ✈ (Montréal) Québec, SE Canada 45°27′N 73°47′W
60 Q8 **Miracema** Rio de Janeiro, SE Brazil 21°24′S 42°10′W
54 G9 **Miraflores** Boyacá, C Colombia 05°07′N 73°09′W
40 G10 **Miraflores** Baja California Sur, NW Mexico 23°24′N 109°45′W
44 L9 **Miragoâne** S Haiti 18°25′N 73°07′W
155 E16 **Miraj** Mahārāshtra, W India 16°51′N 74°42′E
61 E23 **Miramar** Buenos Aires, E Argentina 38°15′S 57°50′W
103 R15 **Miramas** Bouches-du-Rhône, SE France 43°33′N 05°02′E
102 K12 **Mirambeau** Charente-Maritime, W France 45°23′N 00°33′W
102 L13 **Miramont-de-Guyenne** Lot-et-Garonne, SW France 44°34′N 00°20′E
115 L25 **Mirampéllou Kólpos** *gulf* Kríti, Greece, E Mediterranean Sea
158 L8 **Miran** Xinjiang Uygur Zizhiqu, NW China 39°13′N 88°58′E
54 M5 **Miranda** *off.* Estado Miranda. ◈ *state* N Venezuela
Miranda de Corvo *see* Miranda do Corvo
105 O3 **Miranda de Ebro** La Rioja, N Spain 42°41′N 02°57′W
104 G8 **Miranda do Corvo** *var.* Miranda de Corvo. Coimbra, N Portugal 40°05′N 08°20′W
104 J6 **Miranda do Douro** Bragança, N Portugal 41°30′N 06°16′W
Miranda, Estado *see* Miranda
102 L15 **Mirande** Gers, S France 43°31′N 00°25′E
104 I6 **Mirandela** Bragança, N Portugal 41°28′N 07°10′W
25 R15 **Mirando City** Texas, SW USA 27°24′N 99°00′W
106 G9 **Mirandola** Emilia-Romagna, N Italy 44°52′N 11°04′E
60 I8 **Mirandópolis** São Paulo, S Brazil 21°10′S 51°03′W
104 G13 **Mira, Rio** ≈ S Portugal
60 K8 **Mirassol** São Paulo, S Brazil 20°50′S 49°30′W
104 J3 **Miravalles** ▲ NW Spain 42°52′N 06°45′W
42 L12 **Miravalles, Volcán** Ѫ NW Costa Rica 10°43′N 85°07′W
141 W13 **Mirbāṭ** *var.* Marbat. S Oman 17°03′N 54°44′E
44 M9 **Mirebalais** C Haiti 18°51′N 72°08′W
103 T6 **Mirecourt** Vosges, NE France 48°19′N 06°04′E
103 N16 **Mirepoix** Ariège, S France 43°05′N 01°51′E
Mirgorod *see* Myrhorod
139 W10 **Mīr Ḥājī Khalīl** Wāsiṭ, E Iraq 32°11′N 46°19′E
169 T8 **Miri** Sarawak, East Malaysia 04°23′N 113°59′E
77 W12 **Miria** Zinder, S Niger 13°39′N 09°15′E
182 F5 **Mirikata** South Australia 29°56′S 135°13′E
54 K4 **Mirimire** Falcón, N Venezuela 11°14′N 68°39′W
61 H18 **Mirim Lagoon** *var.* Lake Mirim, *Sp.* Laguna Merín. *lagoon* Brazil/Uruguay
Mirim, Lake *see* Mirim Lagoon
Mírina *see* Mýrina
172 H14 **Miringoni** Mohéli, S Comoros 12°17′S 93°39′E
143 W11 **Mīrjāveh** Sīstān va Balūchestān, SE Iran 29°04′N 61°24′E
195 Z9 **Mirny** *Russian research station* Antarctica 66°25′S 93°09′E
124 M10 **Mirnyy** Arkhangel'skaya Oblast', NW Russian Federation 62°50′N 40°20′E
123 O10 **Mirnyy** Respublika Sakha (Yakutiya), NE Russian Federation 62°30′N 113°58′E
Mironovka *see* Myronivka
110 F9 **Mirosławiec** Zachodnio-pomorskie, NW Poland 53°21′N 16°04′E
Mirovo *see* Vrattsa
100 N10 **Mirow** Mecklenburg-Vorpommern, N Germany 53°16′N 12°48′E
152 G6 **Mīrpur** Jammu and Kashmir, NW India 33°06′N 73°49′E
Mīrpur *see* New Mīrpur
149 P17 **Mīrpur Batoro** Sind, SE Pakistan 24°40′N 68°15′E
149 Q16 **Mīrpur Khās** Sind, SE Pakistan 25°31′N 69°01′E
149 P17 **Mīrpur Sakro** Sind, SE Pakistan 24°32′N 67°38′E
143 T14 **Mīr Shahdād** Hormozgān, S Iran 26°15′N 58°29′E
Mirtoan Sea *see* Mirtóo Pélagos
115 G21 **Mirtóo Pélagos** *Eng.* Mirtoan Sea; *anc.* Myrtoum Mare. *sea* S Greece
163 Z16 **Miryang** *var.* Milyang, *Jap.* Mitsuō. SE South Korea 35°30′N 128°46′E
Mirzachirla *see* Murzechirla
164 E14 **Misaki** Ehime, Shikoku, SW Japan 33°22′N 132°04′E
Mīsān *see* Maysān
41 Q13 **Misantla** Veracruz-Llave, E Mexico 19°54′N 96°51′W
165 R7 **Misawa** Aomori, Honshū, C Japan 40°42′N 141°26′E
57 G14 **Mishagua, Río** ≈ C Peru
163 Z8 **Mishan** Heilongjiang, NE China 45°30′N 131°53′E
31 O11 **Mishawaka** Indiana, N USA 41°40′N 86°10′W
39 N6 **Misheguk Mountain** ▲ Alaska, USA 68°13′N 161°11′W
165 N14 **Mishima** *var.* Misima. Shizuoka, Honshū, S Japan 35°08′N 138°54′E
164 E12 **Mi-shima** *island* SW Japan
127 V4 **Mishkino** Respublika Bashkortostan, W Russian Federation 55°31′N 55°57′E
153 Y10 **Mishmi Hills** *hill range* NE India
161 N11 **Mi Shui** ≈ S China
Misiaf *see* Maṣyāf
107 J23 **Misilmeri** Sicilia, Italy, C Mediterranean Sea 38°03′N 13°27′E
Misima *see* Mishima
14 C7 **Missinaibi Lake** ◎ Ontario, S Canada
Misión de Guana *see* Guana
60 F13 **Misiones** *off.* Provincia de Misiones. ◈ *province* NE Argentina
62 P8 **Misiones** *off.* Departamento de las Misiones. ◈ *department* S Paraguay
Misiones, Departamento de las *see* Misiones
Misiones, Provincia de *see* Misiones
Misión San Fernando *see* San Fernando
Miskin *see* Maskin
Miskito Coast *see* La Mosquitia
43 O7 **Miskitos, Cayos** *island group* NE Nicaragua
111 M21 **Miskolc** Borsod-Abaúj-Zemplén, NE Hungary 48°05′N 20°46′E
171 T12 **Misool, Pulau** *island* Papua Barat, E Indonesia
Misox *see* Mesocco
29 Y3 **Misquah Hills** *hill range* Minnesota, N USA
75 P7 **Miṣrātah** *var.* Misurata. NW Libya 32°23′N 15°06′E
75 P7 **Miṣrātah, Râs** *headland* N Libya 32°22′N 15°16′E
14 C7 **Missanabie** Ontario, S Canada 48°18′N 84°04′W
58 E10 **Missão Catrimani** Roraima, N Brazil 01°26′N 62°05′W
14 D6 **Missinaibi** ≈ Ontario, S Canada
11 T13 **Missinipe** Saskatchewan, C Canada 55°36′N 104°45′W
28 M11 **Mission** South Dakota, N USA 43°16′N 100°38′W
25 S17 **Mission** Texas, SW USA 26°13′N 98°19′W
12 F10 **Missisa Lake** ◎ Ontario, C Canada
18 M6 **Missisquoi Bay** *lake bay* Canada/USA
14 C10 **Mississagi** ≈ Ontario, S Canada
14 G15 **Mississauga** Ontario, S Canada 43°38′N 79°36′W
31 P12 **Mississinewa Lake** ⊠ Indiana, N USA
31 P12 **Mississinewa River** ≈ Indiana/Ohio, N USA
22 K4 **Mississippi** *off.* State of Mississippi, *also known as* Bayou State, Magnolia State. ◈ *state* SE USA
14 K13 **Mississippi** ≈ Ontario, SE Canada
47 N1 **Mississippi Fan** *undersea feature* N Gulf of Mexico 26°45′N 88°30′W
14 L13 **Mississippi Lake** ◎ Ontario, SE Canada
22 M10 **Mississippi Delta** *delta* Louisiana, S USA
0 J11 **Mississippi River** ≈ C USA
22 M9 **Mississippi Sound** *sound* Alabama/Mississippi, S USA
33 P9 **Missoula** Montana, NW USA 46°54′N 114°03′W
27 T5 **Missouri** *off.* State of Missouri, *also known as* Bullion State, Show Me State. ◈ *state* C USA
25 V11 **Missouri City** Texas, SW USA 29°37′N 95°32′W
0 J10 **Missouri River** ≈ C USA
15 Q6 **Mistassibi** ≈ Québec, SE Canada
15 P6 **Mistassini** ≈ Québec, SE Canada
12 J11 **Mistassini, Lac** ◎ Québec, SE Canada
109 Y3 **Mistelbach an der Zaya** Niederösterreich, NE Austria 48°34′N 16°33′E
107 L24 **Misterbianco** Sicilia, Italy, C Mediterranean Sea 37°31′N 15°01′E
95 N19 **Misterhult** Kalmar, S Sweden 57°28′N 16°34′E
12 K11 **Mistissini** *var.* Baie-du-Poste. Québec, SE Canada 50°20′N 73°50′W
57 H17 **Misti, Volcán** Ѫ S Peru 16°20′S 71°22′W
Mistras *see* Mystrás
107 K23 **Mistretta** *anc.* Amestratus. Sicilia, Italy, C Mediterranean Sea 37°56′N 14°22′E
164 F12 **Misumi** Shimane, Honshū, SW Japan 34°47′N 132°00′E
Misurata *see* Miṣrātah
83 O14 **Mitande** Niassa, N Mozambique 14°06′S 36°03′E
40 J13 **Mita, Punta de** *headland* C Mexico 20°46′N 105°31′W
55 W12 **Mitaraka, Massif du** ▲ NE South America 02°18′N 54°31′W
Mitau *see* Jelgava
181 X9 **Mitchell** Queensland, E Australia 26°29′S 148°00′E
14 E15 **Mitchell** Ontario, S Canada 43°28′N 81°11′W
28 I13 **Mitchell** Nebraska, C USA 41°56′N 103°48′W
32 J12 **Mitchell** Oregon, NW USA 44°34′N 120°09′W
29 P11 **Mitchell** South Dakota, N USA 43°42′N 98°01′W
23 P5 **Mitchell Lake** ⊠ Alabama, S USA
31 P7 **Mitchell, Lake** ◎ Michigan, N USA
21 P9 **Mitchell, Mount** ▲ North Carolina, SE USA 35°46′N 82°16′W
181 V3 **Mitchell River** ≈ Queensland, NE Australia
97 D20 **Mitchelstown** *Ir.* Baile Mhistéala. SW Ireland 52°20′N 08°16′W
14 M9 **Mitchinamécus, Lac** ◎ Québec, SE Canada
Mitèmboni *see* Mitemele, Río
79 D17 **Mitemele, Río** *var.* Mitèmboni, Temboni, Utamboni. ≈ S Equatorial Guinea
149 S12 **Mithān Kot** Punjab, E Pakistan 28°53′N 70°25′E
149 T7 **Mitha Tiwāna** Punjab, E Pakistan 32°18′N 72°08′E
149 R17 **Mithi** Sind, SE Pakistan 24°43′N 69°53′E
Míthimna *see* Míthymna
Mi Tho *see* My Tho
115 L16 **Míthymna** *var.* Míthimna. Lésvos, E Greece 39°20′N 26°12′E
190 L16 **Mitiaro** *island* S Cook Islands
Mitilíni *see* Mytilíni
15 U7 **Mitis** ≈ Québec, SE Canada
41 R16 **Mitla** Oaxaca, SE Mexico 16°56′N 96°19′W
165 P13 **Mito** Ibaraki, Honshū, S Japan 36°21′N 140°26′E
92 N2 **Mitra, Kapp** *headland* W Svalbard 79°07′N 11°11′E
184 M13 **Mitre** ▲ North Island, New Zealand 40°46′S 175°27′E
185 B21 **Mitre Peak** ▲ South Island, New Zealand 44°37′S 167°45′E
39 O15 **Mitrofania Island** *island* Alaska, USA
Mitrovica/Mitrovicë *see* Kosovska Mitrovica, Serbia
Mitrovica/Mitrowitz *see* Sremska Mitrovica, Serbia
113 M16 **Mitrovicë** *Serb.* Mitrovica, Kosovska Mitrovica, Titova Mitrovica. N Kosovo 42°54′N 20°52′E
172 H12 **Mitsamiouli** Grande Comore, NW Comoros 11°22′S 43°19′E
172 I3 **Mitsinjo** Mahajanga, NW Madagascar 16°00′S 45°52′E
80 J9 **Mits'iwa** *var.* Masawa, Massawa. E Eritrea 15°37′N 39°27′E
172 H13 **Mitsoudjé** Grande Comore, NW Comoros
138 F12 **Mitspe Ramon** *prev.* Mizpe Ramon. Southern, S Israel 30°38′N 34°47′E
165 T5 **Mitsuishi** Hokkaidō, NE Japan 42°12′N 142°40′E
165 O11 **Mitsuke** *var.* Mituke. Niigata, Honshū, C Japan 37°30′N 138°54′E
Mitsuō *see* Miryang
164 C12 **Mitsushima** Nagasaki, Tsushima, SW Japan 34°16′N 129°18′E
100 G12 **Mittelandkanal** *canal* NW Germany
108 J7 **Mittelberg** Vorarlberg, NW Austria 47°19′N 10°09′E
Mitteldorf *see* Międzychód
Mittelstadt *see* Baia Sprie
Mitterburg *see* Pazin
109 P7 **Mittersill** Salzburg, NW Austria 47°16′N 12°27′E
Mittimatalik *see* Pond Inlet
101 N16 **Mittweida** Sachsen, E Germany 50°59′N 12°57′E
54 J13 **Mitú** Vaupés, SE Colombia 01°07′N 70°05′W
Mituke *see* Mitsuke
Mitumba, Chaîne des/ Mitumba Range *see* Mitumba, Monts
79 O22 **Mitumba, Monts** *var.* Chaîne des Mitumba, Mitumba Range. ▲▲ E Dem. Rep. Congo
79 N23 **Mitwaba** Katanga, SE Dem. Rep. Congo 08°37′S 27°20′E
79 E18 **Mitzic** Woleu-Ntem, N Gabon 0°48′N 11°30′E
82 K11 **Miueru Wantipa, Lake** ◎ N Zambia
165 N14 **Miura** Kanagawa, Honshū, S Japan 35°08′N 139°37′E
165 Q10 **Miyagi** *off.* Miyagi-ken. ◈ *prefecture* Honshū, C Japan
Miyagi-ken *see* Miyagi
138 M7 **Miyāh, Wādī al** *dry watercourse* E Syria
165 X13 **Miyake** Tōkyō, Miyako-jima, SE Japan 34°35′N 135°33′E
165 Q16 **Miyake-jima** *island* Sakishima-shotō, SW Japan
165 R8 **Miyako** Iwate, Honshū, C Japan 39°39′N 141°57′E
164 D16 **Miyakonojō** *var.* Miyakonzyô. Miyazaki, Kyūshū, SW Japan 31°42′N 131°04′E
Miyakonzyô *see* Miyakonojō
165 Q16 **Miyako-shotō** *island group* SW Japan
144 G11 **Miyaly** Atyrau, W Kazakhstan 48°52′N 53°55′E
Miyāndoāb *see* Mīāndowāb
Miyāneh *see* Mīāneh
Miyang *see* Mile
164 D16 **Miyazaki** Miyazaki, Kyūshū, SW Japan 31°55′N 131°24′E
164 D16 **Miyazaki** *off.* Miyazaki-ken. ◈ *prefecture* Kyūshū, SW Japan
Miyazaki-ken *see* Miyazaki
164 J12 **Miyazu** Kyōto, Honshū, SW Japan 35°33′N 135°12′E
Miyory *see* Myory
164 G12 **Miyoshi** *var.* Miyosi. Hiroshima, Honshū, SW Japan 34°48′N 132°51′E
Miyosi *see* Miyoshi
Miza *see* Mizë
81 H14 **Mīzan Teferī** Southern Nationalities, S Ethiopia 06°57′N 35°30′E
Mizda *see* Mizdah
75 O8 **Mizdah** *var.* Mizda. NW Libya 31°26′N 12°59′E
113 K20 **Mizë** *var.* Miza. Fier, W Albania 40°58′N 19°32′E
97 A22 **Mizen Head** *Ir.* Carn Uí Néid. *headland* SW Ireland 51°26′N 09°50′W
116 H7 **Mizhhir"ya** *Rus.* Mezhgor'ye. Zakarpats'ka Oblast', W Ukraine 48°30′N 23°30′E
160 L4 **Mizhi** Shaanxi, C China 37°50′N 110°03′E
114 H7 **Mizia** *var.* Miziya. Vratsa, NW Bulgaria 43°42′N 23°52′E
116 K13 **Mizil** Prahova, SE Romania 45°00′N 26°29′E
Miziya *see* Mizia
153 W15 **Mizo Hills** *hill range* E India
153 W15 **Mizoram** ◈ *state* NE India
Mizpe Ramon *see* Mitspe Ramon
57 L19 **Mizque** Cochabamba, C Bolivia 17°57′S 65°10′W
57 M19 **Mizque, Río** ≈ C Bolivia
165 Q9 **Mizusawa** *var.* Ōshū. Iwate, Honshū, C Japan 39°10′N 141°07′E
95 M18 **Mjölby** Östergötland, S Sweden 58°19′N 15°10′E
95 G15 **Mjøndalen** Buskerud, S Norway 59°45′N 09°58′E
95 J19 **Mjörn** ◎ S Sweden
94 I13 **Mjøsa** *var.* Mjøsen. ◎ S Norway
Mjøsen *see* Mjøsa
81 G21 **Mkalama** Singida, C Tanzania 04°09′S 34°35′E
80 K13 **Mkata** ≈ C Tanzania
83 K14 **Mkushi** Central, C Zambia 13°40′S 29°26′E
83 L22 **Mkuze** KwaZulu/Natal, E South Africa 27°37′S 32°03′E
81 J22 **Mkwaja** Tanga, E Tanzania 05°42′S 38°48′E
111 D16 **Mladá Boleslav** *Ger.* Jungbunzlau. Středočeský Kraj, N Czech Republic 50°26′N 14°55′E
112 M12 **Mladenovac** Serbia, C Serbia 44°27′N 20°42′E
114 L11 **Mladinovo** Haskovo, S Bulgaria 41°57′N 26°13′E
113 O17 **Mlado Nagoričane** N FYR Macedonia 42°11′N 21°49′E
Mlanje *see* Mulanje
112 N12 **Mlava** ≈ E Serbia
110 L9 **Mława** Mazowieckie, C Poland 53°07′N 20°23′E
113 G16 **Mljet** *It.* Meleda; *anc.* Melita. *island* S Croatia
167 S11 **Mlu Prey** *prev.* Phumĭ Mlu Prey. Preăh Vihéar, N Cambodia 13°48′N 105°16′E
116 K4 **Mlyniv** Rivnens'ka Oblast', NW Ukraine 50°31′N 25°36′E
83 I21 **Mmabatho** North-West, N South Africa 25°51′S 25°37′E
83 I19 **Mmashoro** Central, E Botswana 21°56′S 26°39′E
44 J7 **Moa** Holguín, E Cuba 20°42′N 74°57′W
76 J15 **Moa** ≈ Guinea/Sierra Leone
37 O6 **Moab** Utah, W USA 38°35′N 109°34′W
181 V1 **Moa Island** *island* Queensland, NE Australia
187 Y15 **Moala** *island* S Fiji
83 L21 **Moamba** Maputo, SW Mozambique 25°35′S 32°13′E
79 F19 **Moanda** *var.* Mouanda. Haut-Ogooué, SE Gabon 01°31′S 13°07′E
83 M15 **Moatize** Tete, NW Mozambique 16°04′S 33°43′E
79 P22 **Moba** Katanga, E Dem. Rep. Congo 07°03′S 29°52′E
Mobay *see* Montego Bay
79 K15 **Mobaye** Basse-Kotto, S Central African Republic 04°19′N 21°17′E
79 K15 **Mobayi-Mbongo** Equateur, NW Dem. Rep. Congo 04°21′N 21°10′E
25 P2 **Mobeetie** Texas, SW USA 35°33′N 100°25′W
27 U3 **Moberly** Missouri, C USA 39°25′N 92°26′W
23 N8 **Mobile** Alabama, S USA 30°42′N 88°03′W
23 N9 **Mobile Bay** *bay* Alabama, S USA
23 N8 **Mobile River** ≈ Alabama, S USA
29 N8 **Mobridge** South Dakota, N USA 45°32′N 100°25′W
Mobutu Sese Seko, Lac *see* Albert, Lake
45 N8 **Moca** N Dominican Republic 19°26′N 70°33′W
83 Q15 **Moçambique** Nampula, NE Mozambique 15°00′S 40°44′E
Moçâmedes *see* Namibe
167 S6 **Môc Châu** Sơn La, N Vietnam 20°49′N 104°38′E
187 Z15 **Moce** *island* Lau Group, E Fiji
Mocha *see* Al Mukhā
193 T11 **Mocha Fracture Zone** *tectonic feature* SE Pacific Ocean
63 F14 **Mocha, Isla** *island* C Chile
56 C12 **Moche, Río** ≈ W Peru
167 S14 **Môc Hoa** Long An, S Vietnam 10°46′N 105°56′E
83 I20 **Mochudi** Kgatleng, SE Botswana 24°25′S 26°07′E
82 Q13 **Mocímboa da Praia** *var.* Vila de Mocímboa da Praia. Cabo Delgado, N Mozambique 11°17′S 40°21′E
94 L13 **Mockfjärd** Dalarna, C Sweden 60°30′N 14°57′E
21 R9 **Mocksville** North Carolina, SE USA 35°53′N 80°33′W
32 F8 **Moclips** Washington, NW USA 47°11′N 124°13′W
82 C13 **Môco** *var.* Morro de Môco. ▲ W Angola 12°36′S 15°09′E
54 D13 **Mocoa** Putumayo, SW Colombia 01°07′N 76°38′W
60 M8 **Mococa** São Paulo, S Brazil 21°30′S 47°00′W
Môco, Morro de *see* Môco
40 H8 **Mocorito** Sinaloa, C Mexico 25°24′N 107°55′W
40 J4 **Moctezuma** Chihuahua, N Mexico 30°10′N 106°28′W
41 N11 **Moctezuma** San Luis Potosí, C Mexico 22°46′N 101°06′W
40 G4 **Moctezuma** Sonora, NW Mexico 29°50′N 109°40′W
41 P12 **Moctezuma, Río** ≈ C Mexico
Mó, Cuan *see* Clew Bay
83 O16 **Mocuba** Zambézia, NE Mozambique 16°50′S 37°02′E
103 U12 **Modane** Savoie, E France 45°14′N 06°41′E
106 F9 **Modena** *anc.* Mutina. Emilia-Romagna, N Italy 44°39′N 10°55′E
36 I7 **Modena** Utah, W USA 37°46′N 113°54′W
35 O9 **Modesto** California, W USA 37°38′N 121°02′W
107 L25 **Modica** *anc.* Motyca. Sicilia, Italy, C Mediterranean Sea 36°52′N 14°45′E
83 J20 **Modimolle** *prev.* Nylstroom. Limpopo, NE South Africa 24°39′N 28°23′E
79 K17 **Modjamboli** Equateur, N Dem. Rep. Congo 02°27′N 22°03′E
109 X4 **Mödling** Niederösterreich, NE Austria 48°06′N 16°18′E
Modohn *see* Madona
Modot *see* Tsenhermandal
171 V14 **Modowi** Papua Barat, E Indonesia 04°05′S 134°39′E
112 I12 **Modračko Jezero** ⊠ NE Bosnia and Herzegovina
112 I10 **Modriča** Republika Srpska, N Bosnia and Herzegovina 44°57′N 18°17′E
183 O13 **Moe** Victoria, SE Australia 38°11′S 146°18′E
Moearatewe *see* Muaratewe
Moei, Mae Nam *see* Thaungyin
94 H13 **Moelv** Hedmark, S Norway 60°55′N 10°47′E
92 I10 **Moen** Troms, N Norway 69°08′N 18°35′E
Möen *see* Møn, Denmark
Moen *see* Weno, Micronesia
Moena *see* Muna, Pulau
36 M10 **Moenkopi Wash** ≈ Arizona, SW USA
185 F22 **Moeraki Point** *headland* South Island, New Zealand 45°23′S 170°52′E
99 F16 **Moerbeke** Oost-Vlaanderen, NW Belgium 51°11′N 03°57′E
99 H14 **Moerdijk** Noord-Brabant, S Netherlands 51°42′N 04°37′E
Moero, Lac *see* Mweru, Lake
101 D15 **Moers** *var.* Mörs. Nordrhein-Westfalen, W Germany 51°27′N 06°36′E
Moesi *see* Musi, Air
Moeskroen *see* Mouscron
96 J13 **Moffat** S Scotland, United Kingdom 55°29′N 03°36′W
185 C22 **Moffat Peak** ▲ South Island, New Zealand 44°57′S 168°10′E
79 N19 **Moga** Sud-Kivu, E Dem. Rep. Congo 02°16′S 26°54′E
152 H8 **Moga** Punjab, N India 30°49′N 75°13′E
Mogadiscio/Mogadishu *see* Muqdisho
Mogador *see* Essaouira
104 J6 **Mogadouro** Bragança, N Portugal 41°20′N 06°43′W
167 N2 **Mogaung** Kachin State, N Myanmar (Burma) 25°20′N 96°54′E
110 L13 **Mogielnica** Mazowieckie, C Poland 51°40′N 20°42′E
Mogilëv *see* Mahilyow
Mogilev-Podol'skiy *see* Mohyliv-Podil's'kyy
Mogilëvskaya Oblast' *see* Mahilyowskaya Voblasts'

◆ Country
● Country Capital
◇ Dependent Territory
○ Dependent Territory Capital
◈ Administrative Regions
✈ International Airport
▲ Mountain
▲▲ Mountain Range
Ѫ Volcano
≈ River
◎ Lake
⊠ Reservoir

MOGILNO – MONTROSE

110 I11 **Mogilno** Kujawsko-pomorskie, C Poland 52°39′N 17°58′E
83 Q15 **Mogincual** Nampula, NE Mozambique 15°33′S 40°28′E
114 E13 **Moglenítsas** ≈ N Greece
106 H8 **Mogliano Veneto** Veneto, NE Italy 45°34′N 12°14′E
113 M21 **Moglicë** Korçë, SE Albania 40°43′N 20°22′E
123 O13 **Mogocha** Zabaykal'skiy Kray, S Russian Federation 53°39′N 119°47′E
122 J11 **Mogochin** Tomskaya Oblast', C Russian Federation 57°42′N 83°24′E
80 F13 **Mogogh** Jonglei, E South Sudan 08°26′N 31°19′E
171 U12 **Mogoi** Papua Barat, E Indonesia 01°44′S 133°13′E
166 M4 **Mogok** Mandalay, C Myanmar (Burma) 22°55′N 96°29′E
37 P14 **Mogollon Mountains** ▲▲ New Mexico, SW USA
36 M12 **Mogollon Rim** cliff Arizona, SW USA
61 E23 **Mogotes, Punta** headland E Argentina 38°03′S 57°31′W
42 J8 **Mogotón** ▲ NW Nicaragua 13°45′N 86°22′W
104 I14 **Moguer** Andalucía, S Spain 37°15′N 06°52′W
111 J26 **Mohács** Baranya, SW Hungary 46°N 18°40′E
185 C20 **Mohaka** ≈ North Island, New Zealand
28 M2 **Mohall** North Dakota, N USA 48°45′N 101°30′W
Moḩammadābād see Dargaz
143 U12 **Moḩammadābād-e Rīgān** Kermān, SE Iran 28°39′N 59°01′E
74 F6 **Mohammedia** prev. Fédala. NW Morocco 33°46′N 07°16′W
74 F6 **Mohammed V** ✈ (Casablanca) W Morocco 33°07′N 08°28′W
Mohammerah see Khorramshahr
36 H10 **Mohave, Lake** ⊠ Arizona/Nevada, W USA
36 I12 **Mohave Mountains** ▲▲ Arizona, SW USA
36 I15 **Mohawk Mountains** ▲▲ Arizona, SW USA
18 J10 **Mohawk River** ≈ New York, NE USA
163 T3 **Mohe** var. Xilinji. Heilongjiang, NE China 53°01′N 122°26′E
95 L20 **Moheda** Kronoberg, S Sweden 57°00′N 14°34′E
Mohéli see Mwali
Mohendergarh see Mahendragarh
38 K12 **Mohican, Cape** headland Nunivak Island, Alaska, USA 60°12′N 167°25′W
Mohn see Muhu
101 G15 **Möhne** ≈ W Germany
101 G15 **Möhne-Stausee** ⊠ W Germany
92 P2 **Mohn, Kapp** headland NW Svalbard 79°26′N 25°44′E
197 S14 **Mohns Ridge** undersea feature Greenland Sea/Norwegian Sea 72°30′N 05°00′E
57 I17 **Moho** Puno, SE Peru 15°21′S 69°32′W
Mohokare see Caledon
95 L17 **Moholm** Västra Götaland, S Sweden 58°37′N 14°04′E
36 J11 **Mohon Peak** ▲ Arizona, SW USA 34°55′N 113°07′W
81 J23 **Mohoro** Pwani, E Tanzania 08°09′S 39°10′E
Mohra see Moravice
Mohrungen see Morąg
116 M7 **Mohyliv-Podil's'kyy** Rus. Mogilev-Podol'skiy. Vinnyts'ka Oblast', C Ukraine 48°29′N 27°49′E
95 D17 **Moi** Rogaland, S Norway 58°27′N 06°32′E
Moili see Mwali
116 K11 **Moinești** Hung. Mojnest. Bacău, E Romania 46°27′N 26°31′E
Móinteach Mílic see Mountmellick
14 J14 **Moira** ≈ Ontario, SE Canada
92 G13 **Mo i Rana** Nordland, C Norway 66°19′N 14°10′E
153 X14 **Moirāng** Manipur, NE India 24°29′N 93°45′E
115 J25 **Moíres** Kríti, Greece, E Mediterranean Sea 35°03′N 24°51′E
118 H6 **Mõisaküla** Ger. Moiseküll. Viljandimaa, S Estonia 58°05′N 25°12′E
Moiseküll see Mõisaküla
15 W4 **Moisie** Québec, E Canada 50°12′N 66°06′W
15 W3 **Moisie** ≈ Québec, SE Canada
102 M14 **Moissac** Tarn-et-Garonne, S France 44°07′N 01°05′E
78 I13 **Moïssala** Mandoul, S Chad 08°21′N 17°46′E
55 O7 **Moitaco** Bolívar, E Venezuela 08°00′N 64°22′W
95 P15 **Möja** Stockholm, C Sweden 59°25′N 18°55′E
105 Q14 **Mojácar** Andalucía, S Spain 37°09′N 01°50′W
35 T13 **Mojave** California, W USA 35°03′N 118°10′W
35 V13 **Mojave Desert** plain California, W USA
35 V13 **Mojave River** ≈ California, W USA
60 L9 **Moji-Mirim** var. Moji-Mirim. São Paulo, S Brazil 22°26′S 46°55′W
Moji-Mirim see Moji-Mirim
113 K15 **Mojkovac** E Montenegro 42°57′N 19°34′E
Mojnest see Moinești
Mōka see Mooka
153 Q13 **Mokāma** prev. Mokameh, Mukama. Bihār, N India 25°24′N 85°55′E
79 O25 **Mokambo** Katanga, SE Dem. Rep. Congo 12°23′S 28°21′E
Mokameh see Mokāma
38 D9 **Mōkapu Point** var. Mokapu Point. headland O'ahu, Hawai'i, USA 21°27′N 157°43′W
184 L9 **Mokau** Waikato, North Island, New Zealand 38°42′S 174°37′E
184 L9 **Mokau** ≈ North Island, New Zealand
35 P7 **Mokelumne River** ≈ California, W USA
83 J23 **Mokhotlong** NE Lesotho 29°19′S 29°06′E
Mokil Atoll see Mwokil Atoll
95 N14 **Möklinta** Västmanland, C Sweden 60°04′N 16°34′E
Mokna see Mokra Gora
184 L4 **Mokohinau Islands** island group N New Zealand
153 X12 **Mokokchūng** Nāgāland, NE India 26°20′N 94°30′E
78 F12 **Mokolo** Extrême-Nord, N Cameroon 10°49′N 13°54′E
83 J20 **Mokopane** prev. Potgietersrus. Limpopo, NE South Africa 24°09′S 28°58′E
185 D24 **Mokoreta** ≈ South Island, New Zealand
163 X17 **Mokpo** Jap. Moppo; prev. Mokp'o. SW South Korea 34°50′N 126°26′E
Mokp'o see Mokpo
113 L16 **Mokra Gora** Alb. Mokna. ▲▲ S Serbia
Mokrany see Makrany
127 O5 **Moksha** ≈ W Russian Federation
143 X12 **Mok Sukhteh-ye Pāyīn** Sīstān va Balūchestān, SE Iran
Moktama see Mottama
77 T14 **Mokwa** Niger, W Nigeria 09°19′N 05°01′E
99 J16 **Mol** prev. Moll. Antwerpen, N Belgium 51°11′N 05°07′E
107 O17 **Mola di Bari** Puglia, SE Italy 41°03′N 17°05′E
Molai see Moláoi
41 P13 **Molango** Hidalgo, C Mexico 20°48′N 98°44′W
115 F22 **Moláoi** var. Molai. Pelopónnisos, S Greece 36°48′N 22°51′E
41 Z12 **Molas del Norte, Punta** var. Punta Molas. headland SE Mexico 20°34′N 86°43′W
Molas, Punta see Molas del Norte, Punta
105 R11 **Molatón** ▲ C Spain 38°58′N 01°19′W
97 K18 **Mold** NE Wales, United Kingdom 53°10′N 03°08′W
Moldau see Vltava, Czech Republic
Moldau see Moldova
Moldavia see Moldova
Moldavian SSR/Moldavskaya SSR see Moldova
94 E9 **Molde** Møre og Romsdal, S Norway 62°44′N 07°08′E
Moldotau, Khrebet see Moldo-Too, Khrebet
147 V9 **Moldo-Too, Khrebet** prev. Khrebet Moldotau. ▲▲ C Kyrgyzstan
116 L9 **Moldova** off. Republic of Moldova, var. Moldavia; prev. Moldavian SSR, Rus. Moldavskaya SSR. ◆ republic SE Europe
116 K9 **Moldova** Eng. Moldavia, Ger. Moldau. former province NE Romania
116 K9 **Moldova** ≈ N Romania
116 F13 **Moldova Nouă** Ger. Neumoldowa, Hung. Újmoldova. Caraș-Severin, SW Romania 44°45′N 21°39′E
Moldova, Republic of see Moldova
116 F13 **Moldova Veche** Ger. Altmoldowa, Hung. Ómoldova. Caraș-Severin, SW Romania 44°45′N 21°13′E
Moldoveanul see Vârful Moldoveanu
83 I20 **Molepolole** Kweneng, SE Botswana 24°25′S 25°30′E
44 L8 **Môle-St-Nicolas** NW Haiti 19°46′N 73°19′W
118 H13 **Molėtai** Utena, E Lithuania 55°14′N 25°25′E
107 O17 **Molfetta** Puglia, SE Italy 41°12′N 16°35′E
171 P11 **Molibagu** Sulawesi, N Indonesia 0°25′N 123°57′E
62 G12 **Molina** Maule, C Chile 35°06′S 71°18′W
105 Q7 **Molina de Aragón** Castilla-La Mancha, C Spain 40°50′N 01°54′W
105 R13 **Molina de Segura** Murcia, SE Spain 38°03′N 01°11′W
30 J11 **Moline** Illinois, N USA 41°30′N 90°31′W
27 P7 **Moline** Kansas, C USA 37°21′N 96°18′W
79 P23 **Moliro** Katanga, SE Dem. Rep. Congo 08°11′S 30°31′E
107 K16 **Molise** ◈ region S Italy
95 K15 **Molkom** Värmland, C Sweden 59°36′N 13°43′E
109 Q9 **Möll** ≈ S Austria
Moll see Mol
146 I14 **Mollanepes Adyndaky** Rus. Imeni Mollanepesa. Mary Welaýaty, S Turkmenistan 37°36′N 61°54′E
95 J22 **Mölle** Skåne, S Sweden 56°15′N 12°19′E
57 H18 **Mollendo** Arequipa, SW Peru 17°02′S 72°01′W
105 U5 **Mollerussa** Cataluña, NE Spain 41°37′N 00°53′E
108 H8 **Mollis** Glarus, NE Switzerland 47°05′N 09°03′E
95 J19 **Mölndal** Västra Götaland, S Sweden 57°39′N 12°05′E
95 J19 **Mölnlycke** Västra Götaland, S Sweden 57°39′N 12°09′E
117 U9 **Molochans'k** Rus. Molochansk. Zaporiz'ka Oblast', SE Ukraine 47°10′N 35°38′E
117 U10 **Molochna** Rus. Molochnaya. ≈ S Ukraine
Molochnaya see Molochna
117 U10 **Molochnyy Lyman** bay N Black Sea
Molodechno/Molodeczno see Maladzyechna
195 V3 **Molodezhnaya** Russian research station Antarctica 67°33′S 46°12′E
124 J14 **Mologa** ≈ NW Russian Federation
38 E9 **Moloka'i** var. Molokai. island Hawaiian Islands, Hawai'i, USA
175 X3 **Molokai Fracture Zone** tectonic feature NE Pacific Ocean
124 K15 **Molokovo** Tverskaya Oblast', W Russian Federation 58°10′N 36°43′E
125 Q14 **Moloma** ≈ NW Russian Federation
183 R8 **Molong** New South Wales, SE Australia 33°07′S 148°52′E
83 H21 **Molopo** seasonal river Botswana/South Africa
115 F17 **Mólos** Stereá Elláda, C Greece 38°48′N 22°39′E
171 O11 **Molosipat** Sulawesi, N Indonesia 0°28′N 121°08′E
Molotov see Severodvinsk, Arkhangel'skaya Oblast', Russian Federation
Molotov see Perm', Permskaya Oblast', Russian Federation
79 G17 **Moloundou** Est, SE Cameroon 02°03′N 15°14′E
103 U5 **Molsheim** Bas-Rhin, NE France 48°33′N 07°30′E
11 X13 **Molson Lake** ◎ Manitoba, C Canada
Moluccas see Maluku
171 Q12 **Molucca Sea** Ind. Laut Maluku. sea E Indonesia
Molukken see Maluku
83 O15 **Molumbo** Zambézia, N Mozambique 15°33′S 36°19′E
171 T15 **Molu, Pulau** island Maluku, E Indonesia
83 P16 **Moma** Nampula, NE Mozambique 16°42′S 39°12′E
171 X14 **Momats** ≈ Papua, E Indonesia
42 J11 **Mombacho, Volcán** Ѫ SW Nicaragua 11°49′N 85°58′W
81 K21 **Mombasa** Mombasa, SE Kenya 04°04′N 39°40′E
81 K21 **Mombasa** ◈ county SE Kenya
81 J21 **Mombasa** ✈ Mombasa, SE Kenya 04°01′S 39°31′E
Mombetsu see Monbetsu
114 J12 **Momchilgrad** prev. Mastanli. Kardzhali, S Bulgaria 41°33′N 25°25′E
99 F23 **Momignies** Hainaut, S Belgium 50°02′N 04°10′E
54 E6 **Momil** Córdoba, NW Colombia 09°15′N 75°40′W
42 I10 **Momotombo, Volcán** Ѫ W Nicaragua 12°25′N 86°33′W
56 B5 **Mompiche, Ensenada de** bay NW Ecuador
79 K18 **Mompono** Equateur, NW Dem. Rep. Congo 0°11′N 21°31′E
54 F6 **Mompós** Bolívar, NW Colombia 09°15′N 74°29′W
95 J24 **Møn** prev. Möen. island SE Denmark
36 L4 **Mona** Utah, W USA 39°49′N 111°52′W
Mona, Canal de la see Mona Passage
96 E8 **Monach Islands** island group NW Scotland, United Kingdom
103 V14 **Monaco** var. Monaco-Ville; anc. Monoecus. ● (Monaco) S Monaco 43°46′N 07°23′E
103 V14 **Monaco** off. Principality of Monaco. ◆ monarchy W Europe
Monaco see München
Monaco Basin see Canary Basin
Monaco, Principality of see Monaco
Monaco-Ville see Monaco
96 I9 **Monadhliath Mountains** ▲▲ N Scotland, United Kingdom
55 O6 **Monagas** off. Estado Monagas. ◈ state NE Venezuela
Monagas, Estado see Monagas
97 F16 **Monaghan** Ir. Muineachán. Monaghan, N Ireland 54°15′N 06°58′W
97 E16 **Monaghan** Ir. Muineachán. cultural region N Ireland
43 S16 **Monagrillo** Herrera, S Panama 08°00′N 80°28′W
24 L8 **Monahans** Texas, SW USA 31°35′N 102°54′W
45 Q9 **Mona, Isla** island W Puerto Rico
45 Q9 **Mona Passage** Sp. Canal de la Mona. channel Dominican Republic/Puerto Rico
43 O14 **Mona, Punta** headland E Costa Rica 09°44′N 82°48′W
155 K25 **Monaragala** Uva Province, SE Sri Lanka 06°52′N 81°22′E
33 S9 **Monarch** Montana, NW USA 47°04′N 110°51′W
10 H14 **Monarch Mountain** ▲ British Columbia, SW Canada 51°59′N 125°56′W
Monasterio see Monesterio
Monasterzyska see Monastyrys'ka
Monastir see Bitola
Monastyriska see Monastyrys'ka
117 O7 **Monastyryshche** Cherkas'ka Oblast', C Ukraine 48°59′N 29°47′E
116 J6 **Monastyrys'ka** Pol. Monasterzyska, Rus. Monastyriska. Ternopil's'ka Oblast', W Ukraine 49°05′N 25°10′E
79 E15 **Monatélé** Centre, SW Cameroon 04°16′N 11°12′E
165 U2 **Monbetsu** var. Mombetsu, Monbetu. Hokkaidō, NE Japan 44°23′N 143°22′E
Monbetu see Monbetsu
106 B8 **Moncalieri** Piemonte, NW Italy 45°00′N 07°41′E
104 G4 **Monção** Viana do Castelo, N Portugal 42°03′N 08°29′W
105 Q5 **Moncayo** ▲ N Spain 41°43′N 01°51′W
105 Q5 **Moncayo, Sierra del** ▲▲ N Spain
124 J4 **Monchegorsk** Murmanskaya Oblast', NW Russian Federation 67°56′N 32°47′E
101 D15 **Mönchengladbach** prev. München-Gladbach. Nordrhein-Westfalen, W Germany 51°12′N 06°35′E
104 F14 **Monchique** Faro, S Portugal 37°19′N 08°33′W
104 G14 **Monchique, Serra de** ▲▲ S Portugal
21 S14 **Moncks Corner** South Carolina, SE USA 33°12′N 80°00′W
41 N7 **Monclova** Coahuila, NE Mexico 26°55′N 101°25′W
Moncorvo see Torre de Moncorvo
13 P14 **Moncton** New Brunswick, SE Canada 46°04′N 64°50′W
104 F8 **Mondego, Cabo** headland N Portugal 40°10′N 08°58′W
104 G8 **Mondego, Rio** ≈ N Portugal
104 I2 **Mondoñedo** Galicia, NW Spain 43°25′N 07°22′W
99 N25 **Mondorf-les-Bains** Grevenmacher, SE Luxembourg 49°30′N 06°16′E
102 M7 **Mondoubleau** Loir-et-Cher, C France 48°00′N 00°49′E
106 B9 **Mondovì** Piemonte, NW Italy 44°23′N 07°56′E
30 J6 **Mondovi** Wisconsin, N USA 44°34′N 91°40′W
Mondragón see Arrasate
107 J17 **Mondragone** Campania, S Italy 41°07′N 13°53′E
109 R5 **Mondsee** ◎ N Austria
115 G22 **Monemvasiá** var. Monemvasía. Pelopónnisos, S Greece 36°22′N 23°03′E
18 B15 **Monessen** Pennsylvania, NE USA 40°07′N 79°51′W
104 J12 **Monesterio** var. Monasterio. Extremadura, W Spain 38°05′N 06°16′W
14 L8 **Monet** Québec, SE Canada 48°09′N 75°37′W
27 S8 **Monett** Missouri, C USA 36°55′N 93°55′W
27 X9 **Monette** Arkansas, C USA 35°53′N 90°20′W
14 G11 **Monetville** Ontario, S Canada 46°08′N 80°24′W
106 J7 **Monfalcone** Friuli-Venezia Giulia, NE Italy 45°49′N 13°32′E
104 H10 **Monforte** Portalegre, C Portugal 39°03′N 07°26′W
104 I4 **Monforte de Lemos** Galicia, NW Spain 42°32′N 07°30′W
79 L16 **Monga** Orientale, N Dem. Rep. Congo 04°12′N 22°49′E
81 I24 **Monga** Lindi, SE Tanzania 09°05′S 37°51′E
81 F15 **Mongalla** Central Equatoria, S South Sudan 05°12′N 31°42′E
153 U11 **Mongar** E Bhutan 27°16′N 91°07′E
167 U6 **Mong Cai** var. Hai Ninh. Quang Ninh, N Vietnam 21°33′N 107°56′E
180 I11 **Mongers Lake** salt lake Western Australia
186 K8 **Mongga** Kolombangara, NW Solomon Islands 07°51′S 157°00′E
167 O6 **Möng Hpayak** Shan State, E Myanmar (Burma) 20°56′N 100°00′E
Monghyr see Munger
106 B10 **Mongioie** ▲ NW Italy 44°13′N 07°46′E
153 T16 **Mongla** var. Mungla. Khulna, S Bangladesh 22°18′N 89°34′E
188 C15 **Mongmong** C Guam
167 N6 **Möng Nai** Shan State, E Myanmar (Burma) 20°28′N 97°51′E
78 I11 **Mongo** Guéra, C Chad 12°12′N 18°40′E
76 I14 **Mongo** ≈ N Sierra Leone
163 I8 **Mongolia** Mong. Mongol Uls. ◆ republic E Asia
129 V8 **Mongolia, Plateau of** plateau E Mongolia
Mongolküre see Zhaosu
Mongol Uls see Mongolia
79 E17 **Mongomo** E Equatorial Guinea 01°39′N 11°18′E
162 M7 **Möngönmorit** var. Bulag. Töv, C Mongolia 48°09′N 108°33′E
77 Y12 **Mongonu** var. Monguno. Borno, NE Nigeria 12°42′N 13°37′E
Mongora see Saidu
78 K11 **Mongororo** Sila, SE Chad 12°03′N 22°26′E
Mongos, Chaîne des see Bongo, Massif des
79 I16 **Mongoumba** Lobaye, SW Central African Republic 03°39′N 18°30′E
Mongrove, Punta see Cayacal, Punta
83 G15 **Mongu** Western, W Zambia 15°13′S 23°09′E
76 I10 **Mônguel** Gorgol, SW Mauritania 16°25′N 13°08′W
Monguno see Mongonu
167 N4 **Mong Yai** Shan State, E Myanmar (Burma) 22°25′N 98°02′E
167 O5 **Möng Yang** Shan State, E Myanmar (Burma) 21°52′N 99°31′E
167 N3 **Möng Yu** Shan State, E Myanmar (Burma) 24°00′N 97°57′E
Mönhbulag see Yösöndzüyl
163 O8 **Mönhhaan** var. Bayasgalant. Sühbaatar, E Mongolia 46°55′N 112°11′E
162 E7 **Mönhhayrhan** var. Tsenher. Hovd, W Mongolia 47°07′N 92°04′E
Mönh Sarĭdag see Munku-Sardyk, Gora
186 P9 **Moni** ≈ S Papau New Guinea
115 I15 **Moní Megístis Lávras** monastery Kentrikí Makedonía, N Greece
115 F18 **Moní Osíou Loúkas** monastery Stereá Elláda, C Greece
54 F9 **Moniquirá** Boyacá, C Colombia 05°54′N 73°35′W
103 Q12 **Monistrol-sur-Loire** Haute-Loire, C France 45°19′N 04°12′E
35 V7 **Monitor Range** ▲▲ Nevada, W USA
115 I14 **Moní Vatopedíou** monastery Kentrikí Makedonía, N Greece
Monkchester see Newcastle upon Tyne
83 N14 **Monkey Bay** Southern, SE Malawi 14°09′S 34°53′E
43 N11 **Monkey Point** var. Punta Mico, Punte Mono, Punto Mico. headland SE Nicaragua 11°37′N 83°39′W
Monkey River see Monkey River Town
42 G3 **Monkey River Town** var. Monkey River. Toledo, SE Belize 16°22′N 88°29′W
14 M13 **Monkland** Ontario, SE Canada 45°11′N 74°51′W
79 J19 **Monkoto** Equateur, NW Dem. Rep. Congo 01°39′S 20°41′E
97 K21 **Monmouth** Wel. Trefynwy. SE Wales, United Kingdom 51°50′N 02°43′W
30 J12 **Monmouth** Illinois, N USA 40°54′N 90°39′W
32 F12 **Monmouth** Oregon, NW USA 44°51′N 123°13′W
97 K21 **Monmouth** cultural region SE Wales, United Kingdom
98 I10 **Monnickendam** Noord-Holland, C Netherlands 52°28′N 05°02′E
77 R15 **Mono** ≈ C Togo
Monoecus see Monaco
35 R8 **Mono Lake** ◎ California, W USA
115 O23 **Monólithos** Ródos, Dodekánisa, Greece, Aegean Sea 36°08′N 27°45′E
19 Q12 **Monomoy Island** island Massachusetts, NE USA
31 O12 **Monon** Indiana, N USA 40°52′N 86°54′W
29 Y12 **Monona** Iowa, C USA 43°03′N 91°23′W
30 L9 **Monona** Wisconsin, N USA 43°03′N 89°18′W
18 B15 **Monongahela** Pennsylvania, NE USA 40°10′N 79°54′W
18 B16 **Monongahela River** ≈ NE USA
107 P17 **Monopoli** Puglia, SE Italy 40°57′N 17°18′E
Mono, Punte see Monkey Point
111 K23 **Monor** Pest, C Hungary 47°21′N 18°27′E
Monostor see Beli Manastir
78 K8 **Monou** Ennedi-Ouest, NE Chad 16°22′N 22°15′E
105 S12 **Monóvar** Cat. Monòver. Valenciana, E Spain 38°26′N 00°50′W
Monòver see Monóvar
105 R7 **Monreal del Campo** Aragón, NE Spain 40°47′N 01°20′W
107 I23 **Monreale** Sicilia, Italy, C Mediterranean Sea 38°05′N 13°17′E
23 T3 **Monroe** Georgia, SE USA 33°47′N 83°42′W
29 W14 **Monroe** Iowa, C USA 41°31′N 93°06′W
22 I5 **Monroe** Louisiana, S USA 32°32′N 92°06′W
31 S10 **Monroe** Michigan, N USA 41°55′N 83°24′W
18 K13 **Monroe** New York, NE USA 41°18′N 74°09′W
21 S11 **Monroe** North Carolina, SE USA 35°00′N 80°35′W
36 L6 **Monroe** Utah, W USA 38°37′N 112°07′W
32 H7 **Monroe** Washington, NW USA 47°51′N 121°58′W
30 L9 **Monroe** Wisconsin, N USA 42°35′N 89°39′W
27 V3 **Monroe City** Missouri, C USA 39°39′N 91°43′W
31 O15 **Monroe Lake** ⊠ Indiana, N USA
23 O7 **Monroeville** Alabama, S USA 31°31′N 87°19′W
18 C15 **Monroeville** Pennsylvania, NE USA 40°24′N 79°44′W
76 J16 **Monrovia** ● (Liberia) W Liberia 06°18′N 10°48′W
76 J16 **Monrovia** ✈ W Liberia 06°22′N 10°50′W
105 T7 **Monroyo** Aragón, NE Spain 40°47′N 00°03′W
99 F20 **Mons** Dut. Bergen. Hainaut, S Belgium 50°28′N 03°58′E
104 I8 **Monsanto** Castelo Branco, C Portugal 40°02′N 07°07′W
106 H8 **Monselice** Veneto, NE Italy 45°15′N 11°47′E
166 M9 **Mon State** ◈ state S Myanmar (Burma)
98 G12 **Monster** Zuid-Holland, W Netherlands 52°01′N 04°10′E
95 N20 **Mönsterås** Kalmar, S Sweden 57°03′N 16°27′E
101 F17 **Montabaur** Rheinland-Pfalz, W Germany 50°25′N 07°48′E
106 G8 **Montagnana** Veneto, NE Italy 45°14′N 11°31′E
35 N1 **Montague** California, W USA 41°43′N 122°31′W
25 S5 **Montague** Texas, SW USA 33°40′N 97°44′W
183 S11 **Montague Island** island New South Wales, SE Australia
39 S12 **Montague Island** island Alaska, USA
39 S13 **Montague Strait** strait N Gulf of Alaska
102 J8 **Montaigu** Vendée, NW France 46°58′N 01°18′W
Montaigu see Scherpenheuvel
105 S7 **Montalbán** Aragón, NE Spain 40°49′N 00°48′W
106 G13 **Montalcino** Toscana, C Italy 43°01′N 11°34′E
104 H5 **Montalegre** Vila Real, N Portugal 41°49′N 07°48′W
114 G8 **Montana** prev. Ferdinand, Mikhaylovgrad. Montana, NW Bulgaria 43°25′N 23°14′E
108 D10 **Montana** Valais, SW Switzerland 46°23′N 07°29′E
39 R11 **Montana** Alaska, USA 62°06′N 150°03′W
114 G8 **Montana** ◈ province NW Bulgaria
33 T9 **Montana** off. State of Montana, also known as Mountain State, Treasure State. ◆ state NW USA
104 J10 **Montánchez** Extremadura, W Spain 39°15′N 06°07′W
Montañita see La Montañita
15 Q8 **Mont-Apica** Québec, SE Canada 47°57′N 71°24′W
104 G10 **Montargil** Portalegre, C Portugal 39°05′N 08°10′W
104 G10 **Montargil, Barragem de** ⊠ C Portugal
103 O7 **Montargis** Loiret, C France 48°N 02°44′E
103 O4 **Montataire** Oise, N France 49°16′N 02°24′E
102 M14 **Montauban** Tarn-et-Garonne, S France 44°01′N 01°20′E
19 N14 **Montauk** Long Island, New York, NE USA 41°01′N 71°58′W
19 N14 **Montauk Point** headland Long Island, New York, NE USA 41°04′N 71°51′W
103 Q7 **Montbard** Côte d'Or, C France 47°35′N 04°25′E
103 U7 **Montbéliard** Doubs, E France 47°31′N 06°49′E
25 W11 **Mont Belvieu** Texas, SW USA 29°51′N 94°53′W
105 U6 **Montblanc** prev. Montblanch. Cataluña, NE Spain 41°23′N 01°10′E
Montblanch see Montblanc
103 Q11 **Montbrison** Loire, E France 45°37′N 04°04′E
Montcalm, Lake see Dogai Coring
103 Q9 **Montceau-les-Mines** Saône-et-Loire, C France 46°40′N 04°19′E
103 U12 **Mont Cenis, Col du** pass E France
102 K15 **Mont-de-Marsan** Landes, SW France 43°54′N 00°30′W
103 O3 **Montdidier** Somme, N France 49°39′N 02°35′E
187 Q17 **Mont-Dore** Province Sud, S New Caledonia 22°18′S 166°34′E
20 K10 **Monteagle** Tennessee, S USA 35°15′N 85°47′W
57 M20 **Monteagudo** Chuquisaca, S Bolivia 19°48′S 63°57′W
41 R16 **Monte Albán** ruins Oaxaca, S Mexico
105 R11 **Montealegre del Castillo** Castilla-La Mancha, C Spain 38°48′N 01°18′W
59 N18 **Monte Azul** Minas Gerais, SE Brazil 15°13′S 42°53′W
14 M12 **Montebello** Québec, SE Canada 45°40′N 74°56′W
106 H7 **Montebelluna** Veneto, NE Italy 45°46′N 12°03′E
60 G13 **Montecarlo** Misiones, NE Argentina 26°38′S 54°45′W
61 D16 **Monte Caseros** Corrientes, NE Argentina 30°15′S 57°39′W
60 J13 **Monte Castelo** Santa Catarina, S Brazil 26°34′S 50°12′W
106 F11 **Montecatini Terme** Toscana, C Italy 43°53′N 10°46′E
42 H7 **Montecillos, Cordillera de** ▲▲ W Honduras
62 I12 **Monte Comén** Mendoza, W Argentina 34°35′S 67°53′W
44 M8 **Monte Cristi** var. San Fernando de Monte Cristi. NW Dominican Republic 19°52′N 71°39′W
58 C13 **Monte Cristo** Amazonas, W Brazil 03°14′S 68°00′W
107 E14 **Montecristo, Isola di** island Archipelago Toscano, C Italy
Monte Croce Carnico, Passo di see Plöcken Pass
58 J12 **Monte Dourado** Pará, NE Brazil 0°48′S 52°32′W
40 L11 **Monte Escobedo** Zacatecas, C Mexico 22°19′N 103°30′W
106 I13 **Montefalco** Umbria, C Italy 42°54′N 12°40′E
107 H14 **Montefiascone** Lazio, C Italy 42°33′N 12°01′E
105 N14 **Montefrío** Andalucía, S Spain 37°19′N 04°00′W
44 I11 **Montego Bay** var. Mobay. W Jamaica 18°28′N 77°55′W
Montego Bay see Sangster
104 J8 **Montehermoso** Extremadura, W Spain 40°05′N 06°20′W
104 F10 **Montejunto, Serra de** ▲ C Portugal 39°10′N 09°01′W
Monteleone di Calabria see Vibo Valentia
54 E7 **Montelíbano** Córdoba, NW Colombia 08°02′N 75°29′W
103 R13 **Montélimar** anc. Acunum Acusio, Montilium Adhemari. Drôme, E France 44°33′N 04°45′E
104 K15 **Montellano** Andalucía, S Spain 37°00′N 05°34′W
35 Y2 **Montello** Nevada, W USA 41°18′N 114°10′W
30 L8 **Montello** Wisconsin, N USA 43°47′N 89°20′W
63 J18 **Montemayor, Meseta de** plain SE Argentina
41 O9 **Montemorelos** Nuevo León, NE Mexico 25°10′N 99°52′W
104 G11 **Montemor-o-Novo** Évora, S Portugal 38°38′N 08°13′W
104 G8 **Montemor-o-Velho** var. Montemor-o-Vélho. Coimbra, N Portugal 40°11′N 08°41′W
Montemor-o-Vélho see Montemor-o-Velho
104 H7 **Montemuro, Serra de** ▲ N Portugal 40°59′N 07°59′W
102 K12 **Montendre** Charente-Maritime, W France 45°17′N 00°24′W
61 I15 **Montenegro** Rio Grande do Sul, S Brazil 29°40′S 51°32′W
113 J16 **Montenegro** Serb. Crna Gora. ◆ republic SW Europe
62 G10 **Monte Patria** Coquimbo, N Chile 30°40′S 71°00′W
45 O9 **Monte Plata** E Dominican Republic 18°50′N 69°47′W
83 P14 **Montepuez** Cabo Delgado, N Mozambique 13°09′S 39°00′E
83 P14 **Montepuez** ≈ N Mozambique
106 G13 **Montepulciano** Toscana, C Italy 43°02′N 11°51′E
62 L6 **Monte Quemado** Santiago del Estero, N Argentina 25°48′S 62°52′W
103 O6 **Montereau-Faut-Yonne** anc. Condate. Seine-St-Denis, N France 48°23′N 02°57′E
35 N11 **Monterey** California, W USA 36°36′N 121°53′W
20 L9 **Monterey** Tennessee, S USA 36°09′N 85°16′W
21 T5 **Monterey** Virginia, NE USA 38°24′N 79°36′W
Monterey see Monterrey
35 N10 **Monterey Bay** bay California, W USA
54 D6 **Montería** Córdoba, NW Colombia 08°45′N 75°54′W
57 N18 **Montero** Santa Cruz, C Bolivia 17°20′S 63°15′W
62 J7 **Monteros** Tucumán, C Argentina 27°12′S 65°30′W
104 I5 **Monterrei** Galicia, NW Spain 41°56′N 07°27′W
41 O8 **Monterrey** var. Monterey. Nuevo León, NE Mexico 25°41′N 100°16′W
32 F9 **Montesano** Washington, NW USA 46°58′N 123°37′W
107 M19 **Montesano sulla Marcellana** Campania, S Italy 40°15′N 15°41′E
107 N16 **Monte Sant' Angelo** Puglia, SE Italy 41°43′N 15°58′E
59 O16 **Monte Santo** Bahia, E Brazil 10°25′S 39°18′W
107 D18 **Monte Santu, Capo di** headland Sardegna, Italy, C Mediterranean Sea 40°05′N 09°43′E
59 M19 **Montes Claros** Minas Gerais, SE Brazil 16°45′S 43°52′W
107 K14 **Montesilvano Marina** Abruzzo, C Italy 42°28′N 14°07′E
23 P4 **Montevallo** Alabama, S USA 33°06′N 86°51′W
106 G12 **Montevarchi** Toscana, C Italy 43°32′N 11°34′E
61 F20 **Montevideo** ● (Uruguay) Montevideo, S Uruguay 34°55′S 56°10′W
29 S9 **Montevideo** Minnesota, N USA 44°56′N 95°43′W
37 S7 **Monte Vista** Colorado, C USA 37°33′N 106°08′W
23 T5 **Montezuma** Georgia, SE USA 32°18′N 84°01′W
29 W14 **Montezuma** Iowa, C USA 41°35′N 92°31′W
26 J6 **Montezuma** Kansas, C USA 37°33′N 100°25′W
103 U12 **Montgenèvre, Col de** pass France/Italy
97 K20 **Montgomery** E Wales, United Kingdom 52°38′N 03°05′W
23 Q5 **Montgomery** state capital Alabama, S USA 32°22′N 86°18′W
29 V9 **Montgomery** Minnesota, N USA 44°26′N 93°34′W
18 G13 **Montgomery** Pennsylvania, NE USA 41°08′N 76°52′W
21 Q5 **Montgomery** West Virginia, NE USA 38°07′N 81°19′W
97 K19 **Montgomery** cultural region E Wales, United Kingdom
Montgomery see Sāhīwāl
27 V4 **Montgomery City** Missouri, C USA 38°57′N 91°27′W
35 S8 **Montgomery Pass** pass Nevada, W USA
102 K12 **Montguyon** Charente-Maritime, W France 45°12′N 00°13′W
108 C10 **Monthey** Valais, SW Switzerland 46°15′N 06°56′E
27 V13 **Monticello** Arkansas, C USA 33°38′N 91°49′W
23 T4 **Monticello** Florida, SE USA 30°33′N 83°52′W
23 T8 **Monticello** Georgia, SE USA 33°18′N 83°40′W
30 M13 **Monticello** Illinois, N USA 40°01′N 88°34′W
31 O12 **Monticello** Indiana, N USA 40°45′N 86°46′W
29 Y13 **Monticello** Iowa, C USA 42°14′N 91°11′W
20 L7 **Monticello** Kentucky, S USA 36°50′N 84°50′W
29 V8 **Monticello** Minnesota, N USA 45°19′N 93°45′W
22 K7 **Monticello** Mississippi, S USA 31°33′N 90°06′W
27 V2 **Monticello** Missouri, C USA 40°07′N 91°42′W
18 J12 **Monticello** New York, NE USA 41°39′N 74°41′W
37 O7 **Monticello** Utah, W USA 37°52′N 109°20′W
106 F8 **Montichiari** Lombardia, N Italy 45°24′N 10°27′E
102 M12 **Montignac** Dordogne, SW France 45°24′N 01°54′E
99 G21 **Montignies-le-Tilleul** var. Montigny-le-Tilleul. Hainaut, S Belgium 50°22′N 04°23′E
14 J8 **Montigny, Lac de** ◎ Québec, SE Canada
103 S6 **Montigny-le-Roi** Haute-Marne, N France 48°02′N 05°28′E
Montigny-le-Tilleul see Montignies-le-Tilleul
43 R16 **Montijo** Veraguas, S Panama 07°59′N 80°58′W
104 F11 **Montijo** Setúbal, W Portugal 38°42′N 08°59′W
104 J11 **Montijo** Extremadura, W Spain 38°55′N 06°38′W
Montilium Adhemari see Montélimar
104 M13 **Montilla** Andalucía, S Spain 37°36′N 04°39′W
102 L3 **Montivilliers** Seine-Maritime, N France 49°31′N 00°10′E
15 U7 **Mont-Joli** Québec, SE Canada 48°36′N 68°14′W
14 M10 **Mont-Laurier** Québec, SE Canada 46°33′N 75°31′W
15 X5 **Mont-Louis** Québec, SE Canada 49°15′N 65°46′W
103 N17 **Mont-Louis** var. Mont Louis. Pyrénées-Orientales, S France 42°30′N 02°08′E
103 O10 **Montluçon** Allier, C France 46°21′N 02°37′E
15 R10 **Montmagny** Québec, SE Canada 46°59′N 70°31′W
103 S3 **Montmédy** Meuse, NE France 49°31′N 05°21′E
103 P5 **Montmirail** Marne, N France 48°53′N 03°31′E
15 R9 **Montmorency** ≈ Québec, SE Canada
102 M10 **Montmorillon** Vienne, W France 46°26′N 00°52′E
107 J14 **Montorio al Vomano** Abruzzo, C Italy 42°31′N 13°39′E
104 M13 **Montoro** Andalucía, S Spain 38°00′N 04°21′W
33 S16 **Montpelier** Idaho, NW USA 42°19′N 111°18′W
29 P6 **Montpelier** North Dakota, N USA 46°40′N 98°34′W
18 M7 **Montpelier** state capital Vermont, NE USA 44°16′N 72°32′W
103 Q15 **Montpellier** Hérault, S France 43°37′N 03°52′E
102 L12 **Montpon-Ménestérol** Dordogne, SW France 45°01′N 00°10′E
12 K15 **Montréal** Eng. Montreal. Québec, SE Canada 45°30′N 73°36′W
14 G8 **Montreal** ≈ Ontario, S Canada
Montreal see Mirabel
11 T14 **Montreal Lake** ◎ Saskatchewan, C Canada
14 B9 **Montreal River** Ontario, S Canada 47°13′N 84°36′W
103 N2 **Montreuil** Pas-de-Calais, N France 50°28′N 01°46′E
102 K8 **Montreuil-Bellay** Maine-et-Loire, NW France 47°07′N 00°10′W
108 C10 **Montreux** Vaud, SW Switzerland 46°27′N 06°55′E
108 B9 **Montricher** Vaud, W Switzerland 46°37′N 06°24′E
96 K10 **Montrose** E Scotland, United Kingdom 56°43′N 02°29′W
27 W14 **Montrose** Arkansas, C USA 33°18′N 91°29′W
37 Q6 **Montrose** Colorado, C USA 38°29′N 107°53′W

◆ Country
● Country Capital
◇ Dependent Territory
○ Dependent Territory Capital
◈ Administrative Regions
✈ International Airport
▲ Mountain
▲▲ Mountain Range
Ѫ Volcano
≈ River
◎ Lake
⊠ Reservoir

MONTROSE – MOYALĒ

29 Y16 **Montrose** Iowa, C USA 40°31′N 91°24′W
18 H12 **Montrose** Pennsylvania, NE USA 41°49′N 75°53′W
21 X5 **Montross** Virginia, NE USA 38°04′N 76°51′W
15 O12 **Mont-St-Hilaire** Québec, SE Canada 45°34′N 73°10′W
103 S3 **Mont-St-Martin** Meurthe-et-Moselle, NE France 49°31′N 05°51′E
45 V10 **Montserrat** *var.* Emerald Isle. ◇ *UK dependent territory* E West Indies
105 V5 **Montserrat** ▲ NE Spain 41°39′N 01°44′E
104 M7 **Montuenga** Castilla y León, N Spain 41°04′N 04°38′W
99 M19 **Montzen** Liège, E Belgium 50°42′N 05°59′E
37 N8 **Monument Valley** *valley* Arizona/Utah, SW USA
166 L4 **Monywa** Sagaing, C Myanmar (Burma) 22°05′N 95°12′E
106 D7 **Monza** Lombardia, N Italy 45°35′N 09°16′E
83 J15 **Monze** Southern, S Zambia 16°20′S 27°29′E
105 T5 **Monzón** Aragón, NE Spain 41°54′N 00°12′E
25 T9 **Moody** Texas, SW USA 31°18′N 97°21′W
98 L13 **Mook** Limburg, SE Netherlands 51°45′N 05°52′E
165 O12 **Mooka** *var.* Mōka. Tochigi, Honshū, S Japan 36°27′N 139°59′E
182 K3 **Moomba** South Australia 28°07′S 140°12′E
14 G13 **Moon** ⚡ Ontario, S Canada
Moon *see* Muhu
181 Y10 **Moonie** Queensland, E Australia 27°46′S 150°22′E
193 O5 **Moonless Mountains** *undersea feature* E Pacific Ocean 30°40′N 140°00′W
182 L13 **Moonlight Head** *headland* Victoria, SE Australia 38°47′S 143°12′E
Moon-Sund *see* Väinameri
182 H8 **Moonta** South Australia 34°03′S 137°36′E
Moor *see* Mór
180 I12 **Moora** Western Australia 30°23′S 116°05′E
98 H12 **Moordrecht** Zuid-Holland, C Netherlands 51°59′N 04°40′E
33 T9 **Moore** Montana, NW USA 47°00′N 109°40′W
27 N11 **Moore** Oklahoma, C USA 35°21′N 97°30′W
25 R12 **Moore** Texas, SW USA 29°03′N 99°01′W
191 S10 **Moorea** *island* Îles du Vent, W French Polynesia
21 U3 **Moorefield** West Virginia, NE USA 39°04′N 78°59′W
23 X14 **Moore Haven** Florida, SE USA 26°49′N 81°05′W
180 J11 **Moore, Lake** ◎ Western Australia
19 N7 **Moore Reservoir** ⊠ New Hampshire/Vermont, NE USA
44 G1 **Moores Island** *island* N The Bahamas
21 R10 **Mooresville** North Carolina, SE USA 35°34′N 80°48′W
29 R5 **Moorhead** Minnesota, N USA 46°51′N 96°44′W
22 K4 **Moorhead** Mississippi, S USA 33°27′N 90°30′W
99 F18 **Moorsel** Oost-Vlaanderen, C Belgium 50°58′N 04°06′E
99 C18 **Moorslede** West-Vlaanderen, W Belgium 50°53′N 03°03′E
18 L8 **Moosalamoo, Mount** ▲ Vermont, NE USA 43°55′N 73°03′W
101 M22 **Moosburg in der Isar** Bayern, SE Germany 48°28′N 11°55′E
33 S14 **Moose** Wyoming, C USA 43°38′N 110°42′W
12 H11 **Moose** ⚡ Ontario, S Canada
12 H10 **Moose Factory** Ontario, S Canada 51°16′N 80°32′W
19 Q4 **Moosehead Lake** ◎ Maine, NE USA
11 U16 **Moose Jaw** Saskatchewan, S Canada 50°23′N 105°35′W
11 V14 **Moose Lake** Manitoba, C Canada 53°42′N 100°22′W
29 W6 **Moose Lake** Minnesota, N USA 46°28′N 92°46′W
19 P6 **Mooselookmeguntic Lake** ◎ Maine, NE USA
39 R12 **Moose Pass** Alaska, USA 60°28′N 149°21′W
19 P5 **Moose River** ⚡ Maine, NE USA
18 J9 **Moose River** ⚡ New York, NE USA
11 V16 **Moosomin** Saskatchewan, S Canada 50°09′N 101°41′W
12 H10 **Moosonee** Ontario, SE Canada 51°18′N 80°40′W
19 N12 **Moosup** Connecticut, NE USA 41°42′N 71°51′W
83 N16 **Mopeia** Zambézia, NE Mozambique 17°59′S 35°43′E
83 H18 **Mopipi** Central, C Botswana 21°07′S 24°55′E
Moppo *see* Mokpo
77 N11 **Mopti** Mopti, C Mali 14°30′N 04°15′W
77 O11 **Mopti** ◆ *region* S Mali
57 H18 **Moquegua** Moquegua, SE Peru 17°07′S 70°55′W
57 H18 **Moquegua** *off.* Departamento de Moquegua. ◆ *department* S Peru
Moquegua, Departamento de *see* Moquegua
111 I23 **Mór** *Ger.* Moor. Fejér, C Hungary 47°21′N 18°12′E
78 G11 **Mora** Extrême-Nord, N Cameroon 11°02′N 14°07′E
104 G11 **Mora** Évora, S Portugal 38°56′N 08°10′W
105 N9 **Mora** Castilla-La Mancha, C Spain 39°40′N 03°46′W
94 L12 **Mora** Dalarna, C Sweden 61°N 14°30′E
29 V7 **Mora** Minnesota, N USA 45°52′N 93°18′W
37 T10 **Mora** New Mexico, SW USA 35°56′N 105°16′W
113 J17 **Morača** ⚡ S Montenegro
152 K10 **Morādābād** Uttar Pradesh, N India 28°50′N 78°45′E
105 U6 **Móra d'Ebre** *var.* Mora de Ebro. Cataluña, NE Spain 41°05′N 00°38′E
Mora de Ebro *see* Móra d'Ebre
105 S8 **Mora de Rubielos** Aragón, NE Spain 40°15′N 00°45′W
172 H4 **Morafenobe** Mahajanga, W Madagascar 17°49′S 44°54′E
110 K8 **Morąg** *Ger.* Mohrungen. Warmińsko-Mazurskie, N Poland 53°55′N 19°56′E
111 L25 **Mórahalom** Csongrád, S Hungary 46°14′N 19°52′E
105 N11 **Moral de Calatrava** Castilla-La Mancha, C Spain 38°50′N 03°34′W
63 G19 **Moraleda, Canal** *strait* SE Pacific Ocean
54 J3 **Morales** Bolívar, N Colombia 08°17′N 73°52′W
54 D12 **Morales** Cauca, SW Colombia 02°46′N 76°44′W
42 F5 **Morales** Izabal, E Guatemala 15°28′N 88°46′W
172 J5 **Moramanga** Toamasina, E Madagascar 18°57′S 48°13′E
27 Q6 **Moran** Kansas, C USA 37°55′N 95°10′W
25 Q7 **Moran** Texas, SW USA 32°33′N 99°10′W
181 X7 **Moranbah** Queensland, NE Australia 22°01′S 148°08′E
44 L13 **Morant Bay** E Jamaica 17°53′N 76°25′W
96 G10 **Morar, Loch** ◎ N Scotland, United Kingdom
Morata *see* Goodenough Island
105 Q12 **Moratalla** Murcia, SE Spain 38°11′N 01°53′W
108 C8 **Morat, Lac de** *Ger.* Murtensee. ◎ W Switzerland
84 I11 **Morava** *var.* March. ⚡ C Europe *see also* March
Morava *see* March
Morava *see* Moravia, Czech Republic
Morava *see* Velika Morava, Serbia
29 W15 **Moravia** Iowa, C USA 40°53′N 92°49′W
111 F18 **Moravia** *Cz.* Morava, *Ger.* Mähren. *cultural region* E Czech Republic
111 H17 **Moravice** *Ger.* Mohra. ⚡ NE Czech Republic
116 E12 **Moravița** Timiș, SW Romania 45°15′N 21°17′E
111 G17 **Moravská Třebová** *Ger.* Mährisch-Trübau. Pardubický Kraj, C Czech Republic 49°47′N 16°40′E
111 E19 **Moravské Budějovice** *Ger.* Mährisch-Budwitz. Vysočina, C Czech Republic 49°03′N 15°48′E
111 H17 **Moravskoslezský Kraj** *prev.* Ostravský Kraj. ◆ *region* E Czech Republic
111 F19 **Moravský Krumlov** *Ger.* Mährisch-Kromau. Jihomoravský Kraj, SE Czech Republic 48°58′N 16°30′E
96 J8 **Moray** *cultural region* N Scotland, United Kingdom
96 J8 **Moray Firth** *inlet* N Scotland, United Kingdom
42 B10 **Morazán** ◆ *department* NE El Salvador
154 C10 **Morbi** Gujarāt, W India 22°51′N 70°49′E
102 G7 **Morbihan** ◆ *department* NW France
Mörbisch *see* Mörbisch am See
109 Y5 **Mörbisch am See** *var.* Mörbisch. Burgenland, E Austria 47°43′N 16°40′E
95 N21 **Mörbylånga** Kalmar, S Sweden 56°31′N 16°25′E
102 J14 **Morcenx** Landes, SW France 44°04′N 00°55′W
Morcheh Khort *see* Mürcheh Khvort
163 T5 **Mordaga** Nei Mongol Zizhiqu, N China 51°15′N 120°47′E
11 X17 **Morden** Manitoba, S Canada 49°12′N 98°05′W
Mordovia *see* Mordoviya, Respublika
127 N5 **Mordoviya, Respublika** *prev.* Mordovskaya ASSR, *Eng.* Mordovia, Mordvinia. ◆ *autonomous republic* W Russian Federation
126 M7 **Mordovo** Tambovskaya Oblast', W Russian Federation 52°05′N 40°49′E
Mordovskaya ASSR/ Mordvinia *see* Mordoviya, Respublika
Morea *see* Pelopónnisos
28 K8 **Moreau River** ⚡ South Dakota, N USA
97 K16 **Morecambe** NW England, United Kingdom 54°04′N 02°53′W
97 K16 **Morecambe Bay** *inlet* NW England, United Kingdom
183 S4 **Moree** New South Wales, SE Australia 29°29′S 149°53′E
21 N5 **Morehead** Kentucky, S USA 38°11′N 83°27′W
21 X11 **Morehead City** North Carolina, SE USA 34°43′N 76°43′W
27 Y8 **Morehouse** Missouri, C USA 36°51′N 89°41′W
108 E10 **Mörel** Valais, SW Switzerland 46°22′N 08°03′E
54 D13 **Morelia** Caquetá, S Colombia 01°30′N 75°43′W
41 N14 **Morelia** Michoacán, S Mexico 19°40′N 101°11′W
105 T7 **Morella** Valenciana, E Spain 40°37′N 00°06′W
40 I7 **Morelos** Chihuahua, N Mexico 26°37′N 107°37′W
41 O15 **Morelos** ◆ *state* S Mexico
154 H7 **Morena** Madhya Pradesh, C India 26°30′N 78°04′E
104 L12 **Morena, Sierra** ▲ S Spain
37 O14 **Morenci** Arizona, SW USA 33°05′N 109°21′W
31 R11 **Morenci** Michigan, N USA 41°43′N 84°13′W
116 J13 **Moreni** Dâmbovița, S Romania 44°59′N 25°39′E
94 D9 **Møre og Romsdal** ◆ *county* S Norway
10 I14 **Moresby Island** *island* Queen Charlotte Islands, British Columbia, SW Canada
183 W2 **Moreton Island** *island* Queensland, E Australia
103 O3 **Moreuil** Somme, N France 49°47′N 02°28′E
35 V7 **Morey Peak** ▲ Nevada, W USA 38°40′N 116°16′W
125 U4 **More-Yu** ⚡ NW Russian Federation
103 T9 **Morez** Jura, E France 46°33′N 06°01′E
Morfou Bay/Mórfou, Kólpos *see* Güzelyurt Körfezi
182 J8 **Morgan** South Australia 34°02′S 139°39′E
23 S7 **Morgan** Georgia, SE USA 31°31′N 84°34′W
25 S8 **Morgan** Texas, SW USA 32°01′N 97°36′W
22 J10 **Morgan City** Louisiana, S USA 29°42′N 91°12′W
20 H6 **Morganfield** Kentucky, S USA 37°41′N 87°55′W
35 O10 **Morgan Hill** California, W USA 37°05′N 121°38′W
21 Q9 **Morganton** North Carolina, SE USA 35°44′N 81°43′W
20 J7 **Morgantown** Kentucky, S USA 37°12′N 86°42′W
21 S2 **Morgantown** West Virginia, NE USA 39°38′N 79°57′W
108 B10 **Morges** Vaud, SW Switzerland 46°31′N 06°30′E
Morghāb, Daryā-ye *see* Murgap
Morghāb, Daryā-ye *see* Murghāb, Daryā-ye
96 I9 **Mor, Glen** *var.* Glen Albyn, Great Glen. *valley* N Scotland, United Kingdom
103 T5 **Morhange** Moselle, NE France 48°56′N 06°37′E
158 M5 **Mori** *var.* Mori Kazak Zizhixian. Xinjiang Uygur Zizhiqu, NW China 43°48′N 90°21′E
165 R5 **Mori** Hokkaidō, NE Japan 42°04′N 140°36′E
35 Y6 **Moriah, Mount** ▲ Nevada, W USA 39°16′N 114°10′W
37 S11 **Moriarty** New Mexico, SW USA 34°59′N 106°03′W
54 J12 **Morichal** Guaviare, E Colombia 02°09′N 70°35′W
Mori Kazak Zizhixian *see* Mori
Morin Dawa Daurzu Zizhiqi *see* Nirji
11 Q14 **Morinville** Alberta, SW Canada 53°48′N 113°38′W
165 R8 **Morioka** Iwate, Honshū, C Japan 39°42′N 141°08′E
183 T8 **Morisset** New South Wales, SE Australia 33°07′S 151°32′E
165 Q8 **Moriyoshi-zan** ▲ Honshū, C Japan 39°58′N 140°32′E
92 K13 **Morjärv** Norrbotten, N Sweden 66°03′N 22°45′E
127 R3 **Morki** Respublika Mariy El, W Russian Federation 56°27′N 49°01′E
123 N10 **Morkoka** ⚡ NE Russian Federation
102 F5 **Morlaix** Finistère, NW France 48°35′N 03°50′W
95 M20 **Mörlunda** Kalmar, S Sweden 57°19′N 15°52′E
107 N19 **Mormanno** Calabria, SW Italy 39°54′N 15°58′E
36 L11 **Mormon Lake** ◎ Arizona, SW USA
35 Y10 **Mormon Peak** ▲ Nevada, W USA 36°59′N 114°25′W
Mormon State *see* Utah
45 Y5 **Morne-à-l'Eau** Grande Terre, N Guadeloupe 16°20′N 61°31′W
29 Y15 **Morning Sun** Iowa, C USA 41°06′N 91°15′W
193 S12 **Mornington Abyssal Plain** *undersea feature* SE Pacific Ocean 50°00′S 90°00′W
63 F22 **Mornington, Isla** *island* S Chile
181 T4 **Mornington Island** *island* Wellesley Islands, Queensland, N Australia
115 E18 **Mórnos** ⚡ C Greece
149 P14 **Moro** Sind, SE Pakistan 26°36′N 67°59′E
32 I11 **Moro** Oregon, NW USA 45°30′N 120°46′W
186 E8 **Morobe** Morobe, C Papua New Guinea 07°46′S 147°35′E
186 E8 **Morobe** ◆ *province* C Papua New Guinea
31 N12 **Morocco** Indiana, N USA 40°57′N 87°27′W
74 E8 **Morocco** *off.* Kingdom of Morocco, *Ar.* Al Mamlakah. ◆ *monarchy* N Africa
Morocco *see* Marrakech
Morocco, Kingdom of *see* Morocco
81 I22 **Morogoro** Morogoro, E Tanzania 06°49′S 37°40′E
81 H24 **Morogoro** ◆ *region* SE Tanzania
171 Q7 **Moro Gulf** *gulf* S Philippines
41 N13 **Moroleón** Guanajuato, C Mexico 20°00′N 101°13′W
172 H6 **Morombe** Toliara, W Madagascar 21°47′S 43°21′E
44 G5 **Morón** Ciego de Ávila, C Cuba 22°08′N 78°39′W
163 N8 **Mörön** Hentiy, C Mongolia 47°21′N 110°21′E
162 I6 **Mörön** Hövsgöl, N Mongolia 49°39′N 100°08′E
54 K5 **Morón** Carabobo, N Venezuela 10°29′N 68°11′W
Morón *see* Morón de la Frontera
56 D8 **Morona, Río** ⚡ N Peru
56 C8 **Morona Santiago** ◆ *province* E Ecuador
172 H5 **Morondava** Toliara, W Madagascar 20°19′S 44°17′E
104 K14 **Morón de la Frontera** *var.* Morón. Andalucía, S Spain 37°07′N 05°27′W
172 G13 **Moroni** ● (Comoros) Grande Comore, NW Comoros 11°41′S 43°16′E
171 S10 **Morotai, Pulau** *island* Maluku, E Indonesia
Morotiri *see* Marotiri
81 H17 **Moroto** NE Uganda 02°32′N 34°41′E
Morozov *see* Bratan
126 M11 **Morozovsk** Rostovskaya Oblast', SW Russian Federation 48°21′N 41°54′E
97 L14 **Morpeth** N England, United Kingdom 55°10′N 01°41′W
Morphou *see* Güzelyurt
Morphou Bay *see* Güzelyurt Körfezi
28 I13 **Morrill** Nebraska, C USA 41°57′N 103°55′W
27 U11 **Morrilton** Arkansas, C USA 35°09′N 92°45′W
11 Q16 **Morrin** Alberta, SW Canada 51°40′N 112°45′W
184 M7 **Morrinsville** Waikato, North Island, New Zealand 37°41′S 175°32′E
11 X16 **Morris** Manitoba, S Canada 49°22′N 97°21′W
30 M11 **Morris** Illinois, N USA 41°21′N 88°25′W
29 S8 **Morris** Minnesota, N USA 45°32′N 95°53′W
14 M13 **Morrisburg** Ontario, SE Canada 44°55′N 75°07′W
197 R11 **Morris Jesup, Kap** *headland* N Greenland 83°33′N 32°40′W
182 B1 **Morris, Mount** ▲ South Australia 26°04′S 131°03′E
30 K10 **Morrison** Illinois, N USA 41°48′N 89°58′W
36 K13 **Morristown** Arizona, SW USA 33°48′N 112°34′W
18 J14 **Morristown** New Jersey, NE USA 40°48′N 74°29′W
21 O8 **Morristown** Tennessee, S USA 36°13′N 83°18′W
42 L11 **Morrito** Río San Juan, SW Nicaragua 11°37′N 85°05′W
35 P13 **Morro Bay** California, W USA 35°21′N 120°51′W
95 L22 **Mörrum** Blekinge, S Sweden 56°11′N 14°45′E
83 N16 **Morrumbala** Zambézia, NE Mozambique 17°17′S 35°35′E
83 N20 **Morrumbene** Inhambane, SE Mozambique 23°41′S 35°25′E
95 F21 **Mors** *island* NW Denmark
Mörs *see* Moers
25 N1 **Morse** Texas, SW USA 36°03′N 101°28′W
127 N6 **Morshansk** Tambovskaya Oblast', W Russian Federation 53°27′N 41°46′E
102 L5 **Mortagne-au-Perche** Orne, N France 48°32′N 00°31′E
102 J8 **Mortagne-sur-Sèvre** Vendée, NW France 47°00′N 00°57′W
104 G8 **Mortágua** Viseu, N Portugal 40°24′N 08°14′W
102 J5 **Mortain** Manche, N France 48°39′N 00°51′W
106 C8 **Mortara** Lombardia, N Italy 45°15′N 08°44′E
59 J17 **Mortes, Rio das** ⚡ C Brazil
182 M12 **Mortlake** Victoria, SE Australia 38°06′S 142°48′E
Mortlock Group *see* Takuu Islands
189 Q17 **Mortlock Islands** *prev.* Nomoi Islands. *island group* C Micronesia
29 T9 **Morton** Minnesota, N USA 44°33′N 94°58′W
22 L5 **Morton** Mississippi, S USA 32°21′N 89°39′W
24 M5 **Morton** Texas, SW USA 33°40′N 102°45′W
32 H9 **Morton** Washington, NW USA 46°33′N 122°16′W
0 D7 **Morton Seamount** *undersea feature* NE Pacific Ocean 50°15′N 142°45′W
45 U15 **Moruga** Trinidad, Trinidad and Tobago 10°04′N 61°16′W
183 P9 **Morundah** New South Wales, SE Australia 34°57′S 146°18′E
191 X12 **Moruroa** *var.* Mururoa. *atoll* Îles Tuamotu, SE French Polynesia
183 S11 **Moruya** New South Wales, SE Australia 35°55′S 150°04′E
103 Q8 **Morvan** *physical region* C France
185 G21 **Morven** Canterbury, South Island, New Zealand 44°51′S 171°07′E
183 O13 **Morwell** Victoria, SE Australia 38°14′S 146°25′E
125 N6 **Morzhovets, Ostrov** *island* NW Russian Federation
126 J4 **Mosal'sk** Kaluzhskaya Oblast', W Russian Federation 54°30′N 34°55′E
101 H20 **Mosbach** Baden-Württemberg, SW Germany 49°21′N 09°06′E
95 E18 **Mosby** Vest-Agder, S Norway 58°12′N 07°55′E
33 V9 **Mosby** Montana, NW USA 46°58′N 107°53′W
32 M9 **Moscow** Idaho, NW USA 46°43′N 117°00′W
20 F10 **Moscow** Tennessee, S USA 35°04′N 89°27′W
Moscow *see* Moskva
101 D19 **Mosel** *Fr.* Moselle. ⚡ W Europe *see also* Moselle
Mosel *see* Moselle
103 T4 **Moselle** ◆ *department* NE France
103 T6 **Moselle** *Ger.* Mosel. ⚡ W Europe *see also* Mosel
Moselle *see* Mosel
32 K9 **Moses Lake** ◎ Washington, NW USA
83 I18 **Mosetse** Central, E Botswana 20°40′S 26°38′E
92 H4 **Mosfellsbær** Höfudhborgarsvædhid, SW Iceland 65°09′N 21°43′W
183 N8 **Mossgiel** New South Wales, SE Australia 33°16′S 144°34′E
185 F23 **Mossgiel** Otago, South Island, New Zealand 45°51′S 170°22′E
124 M11 **Mosha** ⚡ NW Russian Federation
81 I20 **Moshi** Kilimanjaro, NE Tanzania 03°21′S 37°19′E
110 G12 **Mosina** Wielkolpolskie, C Poland 52°15′N 16°50′E
30 L6 **Mosinee** Wisconsin, N USA 44°45′N 89°39′W
92 F13 **Mosjøen** Nordland, C Norway 65°49′N 13°12′E
123 S12 **Moskal'vo** Ostrov Sakhalin, Sakhalinskaya Oblast', SE Russian Federation 53°36′N 142°31′E
92 I13 **Moskosel** Norrbotten, N Sweden 65°52′N 19°30′E
126 K4 **Moskovskaya Oblast'** ◆ *province* W Russian Federation
Moskovskiy *see* Moskva
126 J3 **Moskva** *Eng.* Moscow. ● (Russian Federation) Gorod Moskva, W Russian Federation 55°45′N 37°42′E
147 Q14 **Moskva** *Rus.* Moskovskiy; *prev.* Chubek. SW Tajikistan 37°41′N 69°33′E
126 L4 **Moskva** ⚡ W Russian Federation
83 I20 **Mosomane** Kgatleng, SE Botswana 24°04′S 26°15′E
Moson and Magyaróvár *see* Mosonmagyaróvár
111 H21 **Mosoni-Duna** *Ger.* Kleine Donau. ⚡ NW Hungary
111 H21 **Mosonmagyaróvár** *Ger.* Wieselburg-Ungarisch-Altenburg; *prev.* Moson and Magyaróvár, *Ger.* Wieselburg and Ungarisch-Altenburg. Győr-Moson-Sopron, NW Hungary 47°52′N 17°15′E
Mospino *see* Mospyne
117 X8 **Mospyne** *Rus.* Mospino. Donets'ka Oblast', E Ukraine 47°53′N 38°03′E
54 B12 **Mosquera** Nariño, SW Colombia 02°32′N 78°24′W
37 U10 **Mosquero** New Mexico, SW USA 35°46′N 103°57′W
Mosquito Coast *see* La Mosquitia
31 U11 **Mosquito Creek Lake** ⊠ Ohio, N USA
Mosquito Gulf *see* Mosquitos, Golfo de los
23 X11 **Mosquito Lagoon** *wetland* Florida, SE USA
43 N10 **Mosquito, Punta** *headland* E Nicaragua 12°18′N 83°38′W
43 W14 **Mosquito, Punta** *headland* NE Panama 09°06′N 77°52′W
43 Q15 **Mosquitos, Golfo de los** *Eng.* Mosquito Gulf. *gulf* N Panama
95 H16 **Moss** Østfold, S Norway 59°25′N 10°40′E
Mossâmedes *see* Namibe
22 G8 **Moss Bluff** Louisiana, S USA 30°18′N 93°11′W
185 C23 **Mossburn** Southland, South Island, New Zealand 45°40′S 168°15′E
83 G26 **Mosselbaai** *var.* Mosselbai, *Eng.* Mossel Bay. Western Cape, SW South Africa 34°11′S 22°08′E
Mosselbai/Mossel Bay *see* Mosselbaai
79 F20 **Mossendjo** Niari, SW Congo 02°57′S 12°40′E
101 H22 **Mössingen** Baden-Württemberg, S Germany 48°22′N 09°01′E
181 W4 **Mossman** Queensland, NE Australia 16°34′S 145°27′E
59 P14 **Mossoró** Rio Grande do Norte, NE Brazil 05°11′S 37°20′W
23 N9 **Moss Point** Mississippi, S USA 30°24′N 88°31′W
183 S9 **Moss Vale** New South Wales, SE Australia 34°33′S 150°20′E
32 G9 **Mossyrock** Washington, NW USA 46°32′N 122°30′W
111 B15 **Most** *Ger.* Brüx. Ústecký Kraj, NW Czech Republic 50°30′N 13°37′E
162 E7 **Möst** *var.* Ulaantolgoy. Hovd, W Mongolia 46°39′N 92°50′E
121 P16 **Mosta** *var.* Musta. C Malta 35°54′N 14°25′E
74 I5 **Mostaganem** *var.* Mestghanem. NW Algeria 35°54′N 00°05′E
113 H14 **Mostar** Federacija Bosne I Hercegovine, S Bosnia and Herzegovina 43°21′N 17°47′E
61 J17 **Mostardas** Rio Grande do Sul, S Brazil 31°02′S 50°51′W
116 K14 **Mostiștea** ⚡ S Romania
Mostva *see* Mastva
Mosty *see* Masty
116 H5 **Mostys'ka** L'vivs'ka Oblast', W Ukraine 49°47′N 23°09′E
Mosul *see* Al Mawṣil
95 F15 **Møsvatnet** ◎ S Norway
80 J12 **Mot'a** Āmara, N Ethiopia 11°03′N 38°03′E
79 H16 **Motaba** ⚡ N Congo
105 O10 **Mota del Cuervo** Castilla-La Mancha, C Spain 39°30′N 02°52′W
104 L5 **Mota del Marqués** Castilla y León, N Spain 41°38′N 05°11′W
42 F5 **Motagua, Río** ⚡ Guatemala/Honduras
119 H19 **Motal'** Brestskaya Voblasts', SW Belarus 52°13′N 25°34′E
95 L17 **Motala** Östergötland, S Sweden 58°34′N 15°05′E
191 X7 **Motane** *island* Îles Marquises, NE French Polynesia
152 K13 **Moth** Uttar Pradesh, N India 25°44′N 78°56′E
Mother of Presidents/ Mother of States *see* Virginia
96 I12 **Motherwell** C Scotland, United Kingdom 55°48′N 04°W
153 P12 **Motīhāri** Bihār, N India 26°40′N 84°55′E
105 Q10 **Motilla del Palancar** Castilla-La Mancha, C Spain 39°34′N 01°55′W
184 N7 **Motiti Island** *island* NE New Zealand
65 E25 **Motley Island** *island* SE Falkland Islands
83 J19 **Motloutse** ⚡ E Botswana
41 V17 **Motozintla de Mendoza** Chiapas, SE Mexico 15°21′N 92°14′W
105 N15 **Motril** Andalucía, S Spain 36°45′N 03°30′W
116 G13 **Motru** Gorj, SW Romania 44°49′N 22°56′E
165 Q4 **Motsuta-misaki** *headland* Hokkaidō, NE Japan 42°36′N 139°48′E
28 L6 **Mott** North Dakota, N USA 46°21′N 102°17′W
166 M9 **Mottama** *var.* Martaban, Moktama. Mon State, S Myanmar (Burma) 16°32′N 97°35′E
166 L9 **Mottama, Gulf of** *var.* Gulf of Martaban. *gulf* S Myanmar (Burma)
Möttling *see* Metlika
107 O18 **Mottola** Puglia, SE Italy 40°38′N 17°02′E
184 P8 **Motu** ⚡ North Island, New Zealand
185 I14 **Motueka** Tasman, South Island, New Zealand 41°08′S 173°00′E
185 I14 **Motueka** ⚡ South Island, New Zealand
Motu Iti *see* Tupai
41 X12 **Motul** *var.* Motul de Felipe Carrillo Puerto. Yucatán, SE Mexico 21°06′N 89°17′W
Motul de Felipe Carrillo Puerto *see* Motul
191 U17 **Motu Nui** *island* Easter Island, Chile, E Pacific Ocean
191 Q10 **Motu One** *var.* Bellingshausen. *atoll* Îles Sous le Vent, W French Polynesia
190 I16 **Motutapu** *island* E Cook Islands
193 V15 **Motu Tapu** *island* Tongatapu Group, S Tonga
184 L5 **Motutapu Island** *island* N New Zealand
Motyca *see* Modica
Mouanda *see* Moanda
Mouaskar *see* Mascara
105 U3 **Moubermé, Tuc de** *Fr.* Pic de Maubermé, *Sp.* Pico Mauberme; *prev.* Tuc de Maubermé. ▲ France/Spain 42°48′N 00°57′E *see also* Maubermé, Pic de
Moubermé, Tuc de *see* Maubermé, Pic de
45 N7 **Mouchoir Passage** *passage* SE Turks and Caicos Islands
76 I9 **Moudjéria** Tagant, SW Mauritania 17°52′N 12°20′W
108 C9 **Moudon** Vaud, W Switzerland 46°41′N 06°49′E
Mouhoun *see* Black Volta
79 E19 **Mouila** Ngounié, C Gabon 01°50′S 11°02′E
79 K14 **Mouka** Haute-Kotto, C Central African Republic 07°12′N 21°52′E
Moukden *see* Shenyang
183 N10 **Moulamein** New South Wales, SE Australia 35°06′S 144°03′E
Moulamein Creek *see* Billabong Creek
74 F6 **Moulay-Bousselham** NW Morocco 35°00′N 06°22′W
Moulhoulé *see* le Moule
80 M11 **Moulhoulé** N Djibouti 12°34′N 43°06′E
103 P9 **Moulins** Allier, C France 46°34′N 03°20′E
Moulmein *see* Mawlamyine
Moulmeingyun *see* Mawlamyinegyunn
74 G6 **Moulouya** *var.* Mulucha, Muluya, Mulwiya. *seasonal river* NE Morocco
23 O2 **Moulton** Alabama, S USA 34°28′N 87°18′W
29 W16 **Moulton** Iowa, C USA 40°41′N 92°40′W
25 T11 **Moulton** Texas, SW USA 29°34′N 97°08′W
23 T7 **Moultrie** Georgia, SE USA 31°10′N 83°47′W
21 S14 **Moultrie, Lake** ⊠ South Carolina, SE USA
22 K3 **Mound Bayou** Mississippi, S USA 33°52′N 90°43′W
30 L17 **Mound City** Illinois, N USA 37°06′N 89°09′W
27 R6 **Mound City** Kansas, C USA 38°07′N 94°49′W
27 Q2 **Mound City** Missouri, C USA 40°07′N 95°13′W
29 N7 **Mound City** South Dakota, N USA 45°44′N 100°03′W
78 H13 **Moundou** Logone-Occidental, SW Chad 08°35′N 16°01′E
27 P10 **Mounds** Oklahoma, C USA 35°52′N 96°03′W
21 R2 **Moundsville** West Virginia, NE USA 39°54′N 80°44′W
167 R11 **Moŭng** *prev.* Phumĭ Moŭng. Siĕmréab, NW Cambodia 13°45′N 103°35′E
167 Q12 **Moŭng Roessei** Bătdâmbâng, W Cambodia 12°47′N 103°28′E
Moun Hou *see* Black Volta
8 H8 **Mountain** ⚡ Northwest Territories, NW Canada
37 S12 **Mountainair** New Mexico, SW USA 34°31′N 106°14′W
35 V1 **Mountain City** Nevada, W USA 41°48′N 115°58′W
21 Q8 **Mountain City** Tennessee, S USA 36°28′N 81°48′W
27 U7 **Mountain Grove** Missouri, C USA 37°07′N 92°15′W
27 U9 **Mountain Home** Arkansas, C USA 36°19′N 92°24′W
33 N15 **Mountain Home** Idaho, NW USA 43°07′N 115°42′W
25 Q11 **Mountain Home** Texas, SW USA 30°11′N 99°19′W
29 W4 **Mountain Iron** Minnesota, N USA 47°31′N 92°37′W
29 T10 **Mountain Lake** Minnesota, N USA 43°57′N 94°54′W
23 S3 **Mountain Park** Georgia, SE USA 34°04′N 84°24′W
35 W12 **Mountain Pass** *pass* California, W USA
27 T12 **Mountain Pine** Arkansas, C USA 34°34′N 93°10′W
39 Y14 **Mountain Point** Annette Island, Alaska, USA 55°17′N 131°31′W
Mountain State *see* Montana
Mountain State *see* West Virginia
27 V7 **Mountain View** Arkansas, C USA 35°52′N 92°07′W
38 H12 **Mountain View** Hawaii, USA, C Pacific Ocean 19°32′N 155°03′W
27 V10 **Mountain View** Missouri, C USA 37°00′N 91°42′W
38 M11 **Mountain Village** Alaska, USA 62°06′N 163°42′W
21 R8 **Mount Airy** North Carolina, SE USA 36°30′N 80°36′W
83 K24 **Mount Ayliff** *Xh.* Maxesibebi. Eastern Cape, SE South Africa 30°48′S 29°23′E
29 U16 **Mount Ayr** Iowa, C USA 40°42′N 94°14′W
182 J9 **Mount Barker** South Australia 35°06′S 138°52′E
180 J14 **Mount Barker** Western Australia 34°42′S 117°40′E
183 P11 **Mount Beauty** Victoria, SE Australia 36°47′S 147°12′E
14 E16 **Mount Brydges** Ontario, S Canada 42°54′N 81°29′W
31 N16 **Mount Carmel** Illinois, N USA 38°23′N 87°46′W
30 K10 **Mount Carroll** Illinois, N USA 42°05′N 89°59′W
31 S9 **Mount Clemens** Michigan, N USA 42°36′N 82°52′W
185 E19 **Mount Cook** Canterbury, South Island, New Zealand 43°47′S 170°06′E
83 L16 **Mount Darwin** Mashonaland Central, NE Zimbabwe 16°45′S 31°39′E
19 S7 **Mount Desert Island** *island* Maine, NE USA
23 W11 **Mount Dora** Florida, SE USA 28°48′N 81°38′W
182 G5 **Mount Eba** South Australia 30°11′S 135°40′E
25 W8 **Mount Enterprise** Texas, SW USA 31°53′N 94°40′W
182 J4 **Mount Fitton** South Australia 29°55′S 139°26′E
83 J24 **Mount Fletcher** Eastern Cape, SE South Africa 30°41′S 28°30′E
14 F15 **Mount Forest** Ontario, S Canada 43°58′N 80°44′W
182 K12 **Mount Gambier** South Australia 37°47′S 140°49′E
181 W5 **Mount Garnet** Queensland, NE Australia 17°41′S 145°07′E
21 P6 **Mount Gay** West Virginia, NE USA 37°49′N 82°00′W
31 S12 **Mount Gilead** Ohio, N USA 40°33′N 82°49′W
186 C7 **Mount Hagen** Western Highlands, C Papua New Guinea 05°54′S 144°13′E
18 J16 **Mount Holly** New Jersey, NE USA 39°59′N 74°46′W
21 R10 **Mount Holly** North Carolina, SE USA 35°18′N 81°01′W
27 T12 **Mount Ida** Arkansas, C USA 34°32′N 93°38′W
181 T6 **Mount Isa** Queensland, C Australia 20°48′S 139°32′E
21 U4 **Mount Jackson** Virginia, NE USA 38°45′N 78°38′W
18 D12 **Mount Jewett** Pennsylvania, NE USA 41°43′N 78°37′W
18 L13 **Mount Kisco** New York, NE USA 41°12′N 73°42′W
18 B15 **Mount Lebanon** Pennsylvania, NE USA 40°21′N 80°03′W
182 J8 **Mount Lofty Ranges** ▲ South Australia
180 J10 **Mount Magnet** Western Australia 28°09′S 117°52′E
184 N7 **Mount Maunganui** Bay of Plenty, North Island, New Zealand 37°39′S 176°11′E
97 E18 **Mountmellick** *Ir.* Móinteach Mílic. Laois, C Ireland 53°07′N 07°20′W
30 L10 **Mount Morris** Illinois, N USA 42°03′N 89°25′W
31 R9 **Mount Morris** Michigan, N USA 43°07′N 83°42′W
18 F10 **Mount Morris** New York, NE USA 42°43′N 77°51′W
18 B16 **Mount Morris** Pennsylvania, NE USA 39°43′N 80°06′W
30 K15 **Mount Olive** Illinois, N USA 39°04′N 89°43′W
21 V10 **Mount Olive** North Carolina, SE USA 35°12′N 78°03′W
21 N4 **Mount Olivet** Kentucky, S USA 38°32′N 84°01′W
29 Y15 **Mount Pleasant** Iowa, C USA 40°57′N 91°33′W
31 Q8 **Mount Pleasant** Michigan, N USA 43°36′N 84°46′W
18 C15 **Mount Pleasant** Pennsylvania, NE USA 40°07′N 79°33′W
21 T14 **Mount Pleasant** South Carolina, SE USA 32°47′N 79°51′W
20 I9 **Mount Pleasant** Tennessee, S USA 35°32′N 87°11′W
25 W6 **Mount Pleasant** Texas, SW USA 33°10′N 94°49′W
36 L4 **Mount Pleasant** Utah, W USA 39°33′N 111°27′W
63 N23 **Mount Pleasant** ✈ (Stanley) East Falkland, Falkland Islands
97 G25 **Mount's Bay** *inlet* SW England, United Kingdom
35 N2 **Mount Shasta** California, W USA 41°18′N 122°19′W
30 J13 **Mount Sterling** Illinois, N USA 39°59′N 90°44′W
21 N5 **Mount Sterling** Kentucky, S USA 38°03′N 83°56′W
18 E15 **Mount Union** Pennsylvania, NE USA 40°21′N 77°51′W
23 V6 **Mount Vernon** Georgia, SE USA 32°10′N 82°35′W
30 L16 **Mount Vernon** Illinois, N USA 38°19′N 88°54′W
20 M6 **Mount Vernon** Kentucky, S USA 37°20′N 84°20′W
27 S7 **Mount Vernon** Missouri, C USA 37°05′N 93°49′W
31 T13 **Mount Vernon** Ohio, N USA 40°23′N 82°29′W
32 K13 **Mount Vernon** Oregon, NW USA 44°22′N 119°07′W
25 W6 **Mount Vernon** Texas, SW USA 33°11′N 95°13′W
32 H7 **Mount Vernon** Washington, NW USA 48°25′N 122°19′W
20 L5 **Mount Washington** Kentucky, S USA 38°03′N 85°33′W
182 F8 **Mount Wedge** South Australia 33°29′S 135°08′E
30 L14 **Mount Zion** Illinois, N USA 39°46′N 88°52′W
181 Y9 **Moura** Queensland, NE Australia 24°34′S 149°57′E
58 F12 **Moura** Amazonas, NW Brazil 01°32′S 61°43′W
104 H12 **Moura** Beja, S Portugal 38°08′N 07°27′W
104 I12 **Mourão** Évora, S Portugal 38°22′N 07°20′W
76 L11 **Mourdiah** Koulikoro, W Mali 14°28′N 07°31′W
78 K7 **Mourdi, Dépression du** *desert lowland* Chad/Sudan
102 J16 **Mourenx** Pyrénées-Atlantiques, SW France 43°24′N 00°37′W
Mourgana *see* Mourgkána
115 C15 **Mourgkána** *var.* Mourgana. ▲ Albania/Greece 39°48′N 20°24′E
97 G16 **Mourne Mountains** *Ir.* Beanna Boirche. ▲ SE Northern Ireland, United Kingdom
115 I15 **Moúrtzeflos, Akrotírio** *headland* Límnos, E Greece 40°00′N 25°02′E
99 C19 **Mouscron** *Dut.* Moeskroen. Hainaut, W Belgium 50°44′N 03°14′E
Mouse River *see* Souris River
78 H10 **Moussoro** Bahr el Gazel, W Chad 13°41′N 16°31′E
103 T11 **Moûtiers** Savoie, E France 45°28′N 06°31′E
172 J14 **Moutsamoudou** *var.* Moutsamudou. Anjouan, SE Comoros 12°10′S 44°25′E
Moutsamoudou *see* Moutsamoudou
74 K11 **Mouydir, Monts de** ▲ S Algeria
79 F20 **Mouyondzi** Bouenza, S Congo 03°58′S 13°57′E
115 E16 **Mouzáki** *prev.* Mouzákion. Thessalía, C Greece 39°25′N 21°40′E
Mouzákion *see* Mouzáki
29 S15 **Moville** Iowa, C USA 42°30′N 96°04′W
82 E13 **Moxico** ◆ *province* E Angola
172 I14 **Moya** Anjouan, SE Comoros 12°18′S 44°27′E
40 L12 **Moyahua** Zacatecas, C Mexico 21°18′N 103°09′W
81 J16 **Moyalē** Oromīya, C Ethiopia 03°34′N 38°58′E

◆ Country ◇ Dependent Territory ◆ Administrative Regions ▲ Mountain ⛰ Volcano ◎ Lake
● Country Capital ○ Dependent Territory Capital ✈ International Airport ▲ Mountain Range ⚡ River ⊠ Reservoir

MOYAMBA – MUSALA, PULAU

76 *I15* **Moyamba** W Sierra Leone *08°04′N 12°30′W*
74 *G7* **Moyen Atlas** ***Eng.*** Middle Atlas. ▲▲ N Morocco
78 *H13* **Moyen-Chari** ***off.*** Région du Moyen-Chari. ◈ *region* S Chad
Moyen-Chari, Région du *see* Moyen-Chari
Moyen-Congo *see* Congo (Republic of)
83 *J24* **Moyeni** ***var.*** Quthing. SW Lesotho *30°25′S 27°43′E*
79 *D18* **Moyen-Ogooué** ***off.*** Province du Moyen-Ogooué, ***var.*** Le Moyen-Ogooué. ◈ *province* C Gabon
Moyen-Ogooué, Province du *see* Moyen-Ogooué
103 *S4* **Moyeuvre-Grande** Moselle, NE France *49°15′N 06°03′E*
33 *N7* **Moyie Springs** Idaho, NW USA *48°43′N 116°15′W*
146 *G6* **Mo'ynoq** ***Rus.*** Muynak. Qoraqalpog'iston Respublikasi, NW Uzbekistan *43°45′N 59°03′E*
81 *F16* **Moyo** NW Uganda *03°38′N 31°43′E*
56 *D10* **Moyobamba** San Martín, NW Peru *06°04′S 76°56′W*
78 *H10* **Moyto** Hadjer-Lamis, W Chad *12°35′N 16°33′E*
158 *G9* **Moyu** ***var.*** Karakax. Xinjiang Uygur Zizhiqu, NW China *37°16′N 79°39′E*
122 *M9* **Moyyero** ≈ N Russian Federation
145 *S15* **Moyynkum** ***var.*** Furmanovka, ***Kaz.*** Fūrmanov. Zhambyl, S Kazakhstan *44°15′N 72°55′E*
145 *Q15* **Moyynkum, Peski** ***Kaz.*** Moyynqum. *desert* S Kazakhstan
Moyynqum *see* Moyynkum, Peski
145 *S12* **Moyynty** Karaganda, C Kazakhstan *47°10′N 73°24′E*
145 *S12* **Moyynty** ≈ Karaganda, C Kazakhstan
Mozambika, Lakandranon' i *see* Mozambique Channel
83 *M18* **Mozambique** ***off.*** Republic of Mozambique; ***prev.*** People's Republic of Mozambique, Portuguese East Africa. ◆ *republic* S Africa
Mozambique Basin *see* Natal Basin
Mozambique, Canal de *see* Mozambique Channel
83 *P17* **Mozambique Channel** ***Fr.*** Canal de Mozambique, ***Mal.*** Lakandranon' i Mozambika. *strait* W Indian Ocean
172 *L11* **Mozambique Escarpment** ***var.*** Mozambique Scarp. *undersea feature* SW Indian Ocean *33°00′S 36°30′E*
Mozambique, People's Republic of *see* Mozambique
172 *L10* **Mozambique Plateau** ***var.*** Mozambique Rise. *undersea feature* SW Indian Ocean *32°00′S 35°00′E*
Mozambique, Republic of *see* Mozambique
Mozambique Rise *see* Mozambique Plateau
Mozambique Scarp *see* Mozambique Escarpment
127 *O15* **Mozdok** Respublika Severnaya Osetiya, SW Russian Federation *43°48′N 44°42′E*
57 *K17* **Mozetenes, Serranías de** ▲▲ C Bolivia
126 *J4* **Mozhaysk** Moskovskaya Oblast', W Russian Federation *55°31′N 36°01′E*
127 *T3* **Mozhga** Udmurtskaya Respublika, NW Russian Federation *56°24′N 52°13′E*
Mozyr' *see* Mazyr
79 *P22* **Mpala** Katanga, E Dem. Rep. Congo *06°43′S 29°28′E*
79 *G19* **Mpama** ≈ C Congo
81 *E22* **Mpanda** Katavi, W Tanzania *06°21′S 31°01′E*
82 *L11* **Mpande** Northern, NE Zambia *09°13′S 31°42′E*
83 *J18* **Mphoengs** Matabeleland South, SW Zimbabwe *21°04′S 27°56′E*
81 *F18* **Mpigi** S Uganda *0°14′N 32°19′E*
82 *L13* **Mpika** Muchinga, NE Zambia *11°50′S 31°30′E*
83 *J14* **Mpima** Central, C Zambia *14°25′S 28°34′E*
82 *J13* **Mpongwe** Copperbelt, C Zambia *13°25′S 28°13′E*
82 *K11* **Mporokoso** Northern, N Zambia *09°22′S 30°06′E*
79 *H20* **Mpouya** Plateaux, SE Congo *02°38′S 16°13′E*
77 *P16* **Mpraeso** C Ghana *06°36′N 00°43′W*
82 *L11* **Mpulungu** Northern, N Zambia *08°50′S 31°06′E*
83 *K21* **Mpumalanga** ***prev.*** Eastern Transvaal, ***Afr.*** Oos-Transvaal. ◈ *province* NE South Africa
83 *D16* **Mpungu** Okavango, N Namibia *17°36′S 18°16′E*
81 *I22* **Mpwapwa** Dodoma, C Tanzania *06°21′S 36°29′E*
Mqinvartsveri *see* Kazbek
110 *M8* **Mrągowo** ***Ger.*** Sensburg. Warmińsko-Mazurskie, NE Poland *53°53′N 21°19′E*
127 *V6* **Mrakovo** Respublika Bashkortostan, W Russian Federation *52°43′N 56°36′E*
172 *I13* **Mramani** Anjouan, E Comoros *12°18′N 44°39′E*
166 *K5* **Mrauk-oo** ***var.*** Mrauk U, Myohaung. Rakhine State, W Myanmar (Burma) *20°35′N 93°12′E*
Mrauk U *see* Mrauk-oo
112 *F12* **Mrkonjić Grad** ◈ Republika Srpska, W Bosnia and Herzegovina
110 *H9* **Mrocza** Kujawsko-pomorskie, C Poland *53°15′N 17°38′E*
124 *I14* **Msta** ≈ NW Russian Federation
Mstislavl' *see* Mstsislaw
119 *P15* **Mstsislaw** ***Rus.*** Mstislavl'. Mahilyowskaya Voblasts', E Belarus *54°01′N 31°43′E*
83 *J24* **Mthatha** ***prev.*** Umtata. Eastern Cape, SE South Africa *31°33′S 28°47′E* ***see also*** Umtata
Mtkvari *see* Kura
Mtoko *see* Mutoko
126 *K6* **Mtsensk** Orlovskaya Oblast', W Russian Federation *53°17′N 36°34′E*
81 *K24* **Mtwara** Mtwara, SE Tanzania *10°17′S 40°11′E*
81 *J25* **Mtwara** ◈ *region* SE Tanzania
104 *G14* **Mu** ▲ S Portugal *37°24′N 08°04′W*
193 *V15* **Mu'a** Tongatapu, S Tonga *21°11′S 175°07′W*
Muai To *see* Mae Hong Son
83 *P16* **Mualama** Zambézia, NE Mozambique *16°51′S 38°21′E*
Mualo *see* Messalo, Rio
79 *E22* **Muanda** Bas-Congo, SW Dem. Rep. Congo *05°53′S 12°17′E*
Muang Chiang Rai *see* Chiang Rai
167 *R6* **Muang Ham** Houaphan, N Laos *20°19′N 104°00′E*
167 *S8* **Muang Hinboun** Khammouan, C Laos *17°37′N 104°37′E*
Muang Kalasin *see* Kalasin
Muang Khammouan *see* Thakhèk
167 *S11* **Muang Không** Champasak, S Laos *14°08′N 105°48′E*
167 *S10* **Muang Khôngxédôn** ***var.*** Khong Sedone. Salavan, S Laos *15°34′N 105°46′E*
Muang Khon Kaen *see* Khon Kaen
167 *Q6* **Muang Khoua** Phôngsali, N Laos *21°07′N 102°31′E*
Muang Krabi *see* Krabi
Muang Lampang *see* Lampang
Muang Lamphun *see* Lamphun
Muang Loei *see* Loei
Muang Lom Sak *see* Lom Sak
Muang Nakhon Sawan *see* Nakhon Sawan
167 *Q6* **Muang Namo** Oudômxai, N Laos *20°58′N 101°46′E*
Muang Nan *see* Nan
167 *Q6* **Muang Ngoy** Louangphabang, N Laos *20°43′N 102°42′E*
167 *Q5* **Muang Ou Tai** Phôngsali, N Laos *22°06′N 101°59′E*
Muang Pak Lay *see* Pak Lay
Muang Pakxan *see* Pakxan
167 *T10* **Muang Pakxong** Champasak, S Laos *15°10′N 106°17′E*
167 *S9* **Muang Phalan** ***var.*** Muang Phalane. Savannakhét, S Laos *16°40′N 105°33′E*
Muang Phalane *see* Muang Phalan
Muang Phan *see* Phan
Muang Phayao *see* Phayao
Muang Phichit *see* Phichit
167 *T9* **Muang Phin** Savannakhét, S Laos *16°31′N 106°01′E*
Muang Phitsanulok *see* Phitsanulok
Muang Phrae *see* Phrae
Muang Roi Et *see* Roi Et
Muang Sakon Nakhon *see* Sakon Nakhon
Muang Samut Prakan *see* Samut Prakan
167 *P6* **Muang Sing** Louang Namtha, N Laos *21°12′N 101°09′E*
Muang Ubon *see* Ubon Ratchathani
Muang Uthai Thani *see* Uthai Thani
167 *P7* **Muang Vangviang** Viangchan, C Laos *18°53′N 102°27′E*
Muang Xaignabouri *see* Xaignabouli
Muang Xay *see* Oudômxai
167 *S9* **Muang Xépôn** ***var.*** Sepone. Savannakhét, S Laos *16°40′N 106°15′E*
168 *K10* **Muar** ***var.*** Bandar Maharani. Johor, Peninsular Malaysia *02°01′N 102°35′E*
168 *J9* **Muara** Sumatera, W Indonesia *02°18′N 98°54′E*
168 *L13* **Muarabeliti** Sumatera, W Indonesia *03°13′S 103°00′E*
168 *K12* **Muarabungo** Sumatera, W Indonesia *01°28′S 102°06′E*
168 *L13* **Muaraenim** Sumatera, W Indonesia *03°40′S 103°48′E*
169 *T11* **Muarajuloi** Borneo, C Indonesia *0°12′S 114°03′E*
169 *U12* **Muarakaman** Borneo, C Indonesia *0°09′S 116°43′E*
168 *H12* **Muarasigep** Pulau Siberut, W Indonesia *01°01′S 98°48′E*
168 *L12* **Muaratembesi** Sumatera, W Indonesia *01°40′S 103°08′E*
169 *T12* **Muaratewe** ***var.*** Muarateweh; ***prev.*** Moearatewe. Borneo, C Indonesia *0°58′S 114°52′E*
Muarateweh *see* Muaratewe
169 *U10* **Muarawahau** Borneo, N Indonesia *01°03′N 116°48′E*
138 *G13* **Mubārak, Jabal** ▲ S Jordan *29°19′N 35°13′E*
153 *N13* **Mubārakpur** Uttar Pradesh, N India *26°05′N 83°19′E*
Mubarek *see* Muborak
81 *F18* **Mubende** SW Uganda *0°35′N 31°24′E*
77 *Y14* **Mubi** Adamawa, NE Nigeria *10°15′N 13°18′E*
146 *M12* **Muborak** ***Rus.*** Mubarek. Qashqadaryo Viloyati, S Uzbekistan *39°17′N 65°10′E*
171 *U12* **Mubrani** Papua Barat, E Indonesia *0°42′S 133°25′E*
82 *L12* **Muchinga** ◈ *province* NE Zambia
67 *U12* **Muchinga Escarpment** *escarpment* NE Zambia
127 *N7* **Muchkapskiy** Tambovskaya Oblast', W Russian Federation *51°51′N 42°25′E*
96 *G10* **Muck** *island* W Scotland, United Kingdom
82 *Q13* **Mucojo** Cabo Delgado, N Mozambique *12°05′S 40°30′E*
82 *F12* **Muconda** Lunda Sul, NE Angola *10°37′S 21°19′E*
54 *I10* **Muco, Río** ≈ E Colombia
83 *O16* **Mucubela** Zambézia, NE Mozambique *16°51′S 37°48′E*
42 *J5* **Mucupina, Monte** ▲ N Honduras *15°07′N 86°36′W*
136 *J14* **Mucur** Kirşehir, C Turkey *39°05′N 34°25′E*
143 *U8* **Mūd** Khorāsān-e Janūbī, E Iran *32°41′N 59°30′E*
163 *Y9* **Mudanjiang** ***var.*** Mu-tan-chiang. Heilongjiang, NE China *44°33′N 129°40′E*
163 *Y9* **Mudan Jiang** ≈ NE China
136 *D11* **Mudanya** Bursa, NW Turkey *40°23′N 28°53′E*
28 *K8* **Mud Butte** South Dakota, N USA *45°00′N 102°51′W*
155 *G16* **Muddebihāl** Karnātaka, C India *16°26′N 76°07′E*
27 *P12* **Muddy Boggy Creek** ≈ Oklahoma, C USA
36 *M6* **Muddy Creek** ≈ Utah, W USA
37 *V7* **Muddy Creek Reservoir** ▣ Colorado, C USA
33 *W15* **Muddy Gap** Wyoming, C USA *42°21′N 107°27′W*
35 *Y11* **Muddy Peak** ▲ Nevada, W USA *36°17′N 114°40′W*
183 *R7* **Mudgee** New South Wales, SE Australia *32°37′S 149°36′E*
29 *S3* **Mud Lake** ◎ Minnesota, N USA
29 *P7* **Mud Lake Reservoir** ▣ South Dakota, N USA
167 *N9* **Mudon** Mon State, S Myanmar (Burma) *16°17′N 97°40′E*
81 *O14* **Mudug** ***off.*** Gobolka Mudug. ◈ *region* NE Somalia
81 *O14* **Mudug** ***var.*** Mudugh. *plain* N Somalia
Mudug, Gobolka *see* Mudug
Mudugh *see* Mudug
83 *Q15* **Muecate** Nampula, NE Mozambique *14°56′S 39°38′E*
82 *Q13* **Mueda** Cabo Delgado, NE Mozambique *11°40′S 39°31′E*
42 *L10* **Muelle de los Bueyes** Región Autónoma Atlántico Sur, SE Nicaragua *12°03′N 84°34′W*
Muenchen *see* München
83 *M14* **Muende** Tete, NW Mozambique *14°22′S 33°00′E*
25 *T5* **Muenster** Texas, SW USA *33°39′N 97°22′W*
Muenster *see* Münster
43 *O6* **Muerto, Cayo** *reef* NE Nicaragua
41 *T17* **Muerto, Mar** *lagoon* SE Mexico
64 *F11* **Muertos Trough** *undersea feature* N Caribbean Sea
83 *H14* **Mufaya Kuta** Western, NW Zambia *14°30′S 24°18′E*
82 *J13* **Mufulira** Copperbelt, C Zambia *12°33′S 28°16′E*
161 *O10* **Mufu Shan** ▲▲ C China
Mugalla *see* Yutian
Mugalzhar Taūlary *see* Mugodzhary, Gory
137 *Y12* **Muğan Düzü** ***Rus.*** Muganskaya Ravnina, Muganskaya Step'. *physical region* S Azerbaijan
Muganskaya Ravnina/Muganskaya Step' *see* Muğan Düzü
106 *K8* **Múggia** Friuli-Venezia Giulia, NE Italy *45°36′N 13°48′E*
153 *N14* **Mughal Sarāi** Uttar Pradesh, N India *25°18′N 83°07′E*
Mughla *see* Muğla
141 *W11* **Mughshin** ***var.*** Muqshin. S Oman *19°26′N 54°38′E*
147 *S12* **Mughsu** ***Rus.*** Muksu. ≈ C Tajikistan
164 *H14* **Mugi** Tokushima, Shikoku, SW Japan *33°39′N 134°24′E*
136 *C16* **Muğla** ***var.*** Mughla. Muğla, SW Turkey *37°13′N 28°22′E*
136 *C16* **Muğla** ***var.*** Mughla. ◈ *province* SW Turkey
Müglizh *see* Maglizh
144 *J11* **Mugodzhary, Gory** ***Kaz.*** Mugalzhar Taūlary. ▲▲ W Kazakhstan
83 *O15* **Mugulama** Zambézia, NE Mozambique *16°01′S 37°33′E*
Muḩāfazat Hims *see* Ḩimş
139 *U9* **Muḩammad** Wāsiţ, E Iraq *32°46′N 45°14′E*
139 *R8* **Muḩammadīyah** Al Anbār, C Iraq *33°22′N 42°48′E*
80 *I6* **Muhammad Qol** Red Sea, NE Sudan *20°53′N 37°09′E*
75 *Y9* **Muḩammad, Râs** *headland* E Egypt *27°45′N 34°18′E*
Muhammerah *see* Khorramshahr
140 *M12* **Muhāyil** ***var.*** Mahāil. 'Asīr, SW Saudi Arabia *18°34′N 42°01′E*
139 *O7* **Muḩaywīr** Al Anbār, W Iraq *33°35′N 41°06′E*
101 *H21* **Mühlacker** Baden-Württemberg, SW Germany *48°57′N 08°51′E*
Mühlbach *see* Sebeş
Mühldorf *see* Mühldorf am Inn
101 *N23* **Mühldorf am Inn** ***var.*** Mühldorf. Bayern, SE Germany *48°14′N 12°32′E*
101 *J15* **Mühlhausen** ***var.*** Mühlhausen in Thüringen. Thüringen, C Germany *51°13′N 10°28′E*
Mühlhausen in Thüringen *see* Mühlhausen
195 *Q2* **Mühlig-Hofmanfjella** ***Eng.*** Mülig-Hofmann Mountains. ▲▲ Antarctica
93 *L14* **Muhos** Pohjois-Pohjanmaa, C Finland *64°48′N 26°00′E*
138 *K6* **Mūḩ, Sabkhat al** ◎ C Syria
118 *E5* **Muhu** ***Ger.*** Mohn, Moon. *island* W Estonia
81 *F19* **Muhutwe** Kagera, NW Tanzania *01°31′S 31°41′E*
Muhu Väin *see* Väinameri
98 *J10* **Muiden** Noord-Holland, C Netherlands *52°19′N 05°04′E*
193 *W15* **Mui Hopohoponga** *headland* Tongatapu, S Tonga *21°09′S 175°02′W*
Muinchille *see* Cootehill
Muineachán *see* Monaghan
97 *F19* **Muine Bheag** ***Eng.*** Bagenalstown. Carlow, SE Ireland *52°42′N 06°57′W*
56 *B5* **Muisne** Esmeraldas, NW Ecuador *0°35′N 79°58′W*
83 *P14* **Muite** Nampula, NE Mozambique *14°02′S 39°06′E*
41 *Z11* **Mujeres, Isla** *island* E Mexico
116 *G7* **Mukacheve** ***Hung.*** Munkács, ***Rus.*** Mukachevo. Zakarpats'ka Oblast', W Ukraine *48°27′N 22°45′E*
Mukachevo *see* Mukacheve
169 *R9* **Mukah** Sarawak, East Malaysia *02°56′N 112°02′E*
Mukalla *see* Al Mukallā
Mukama *see* Mokāma
Mukāshafa/Mukashshafah *see* Mukayshīfah
139 *S6* **Mukayshīfah** ***var.*** Mukāshafa, Mukashshafah. Şalāḩ ad Dīn, N Iraq *34°24′N 43°44′E*
167 *R9* **Mukdahan** Mukdahan, E Thailand *16°31′N 104°43′E*
Mukden *see* Shenyang
165 *Y15* **Mukojima-rettō** ***Eng.*** Parry group. *island group* SE Japan
146 *M14* **Mukry** Lebap Welaýaty, E Turkmenistan *37°39′N 65°37′E*
Muksu *see* Mughsu
Muktagacha *see* Muktagachha
153 *U14* **Muktagachha** ***var.*** Muktagacha. N Bangladesh *24°46′N 90°16′E*
82 *K13* **Mukuku** Central, C Zambia *12°10′S 29°50′E*
82 *K11* **Mukupa Kaoma** Northern, NE Zambia *09°55′S 30°19′E*
81 *I18* **Mukutan** Baringo, W Kenya *01°06′N 36°16′E*
83 *F16* **Mukwe** Caprivi, NE Namibia *18°01′S 21°24′E*
105 *R13* **Mula** Murcia, SE Spain *38°02′N 01°29′W*
151 *K20* **Mulakatholhu** ***var.*** Meemu Atoll, Mulaku Atoll. *atoll* C Maldives
Mulaku Atoll *see* Mulakatholhu
83 *J15* **Mulalika** Lusaka, C Zambia *15°37′S 28°48′E*
163 *X8* **Mulan** Heilongjiang, NE China *45°57′N 128°00′E*
83 *N15* **Mulanje** ***var.*** Mlanje. Southern, S Malawi *16°05′S 35°29′E*
40 *H5* **Mulatos** Sonora, NW Mexico *28°42′N 108°44′W*
23 *P3* **Mulberry Fork** ≈ Alabama, S USA
39 *P12* **Mulchatna River** ≈ Alaska, USA
125 *W4* **Mul'da** Respublika Komi, NW Russian Federation *67°29′N 63°55′E*
101 *M14* **Mulde** ≈ E Germany
27 *R10* **Muldrow** Oklahoma, C USA *35°25′N 94°34′W*
40 *E7* **Mulegé** Baja California Sur, NW Mexico *26°54′N 112°00′W*
108 *I10* **Mulegns** Graubünden, S Switzerland *46°30′N 09°36′E*
79 *M21* **Mulenda** Kasai-Oriental, C Dem. Rep. Congo *04°19′S 24°55′E*
24 *M4* **Muleshoe** Texas, SW USA *34°13′N 102°43′W*
83 *O15* **Mulevala** Zambézia, NE Mozambique *16°25′S 37°35′E*
183 *P5* **Mulgoa Creek** *seasonal river* New South Wales, SE Australia
105 *O15* **Mulhacén** ***var.*** Cerro de Mulhacén. ▲ S Spain *37°07′N 03°11′W*
Mulhacén, Cerro de *see* Mulhacén
Mülhausen *see* Mulhouse
101 *E24* **Mülheim** Baden-Württemberg, SW Germany *47°50′N 07°37′E*
101 *E15* **Mülheim** ***var.*** Mulheim an der Ruhr. Nordrhein-Westfalen, W Germany *51°25′N 06°50′E*
Mulheim an der Ruhr *see* Mülheim
103 *U7* **Mulhouse** ***Ger.*** Mülhausen. Haut-Rhin, NE France *47°45′N 07°20′E*
160 *G11* **Muli** ***var.*** Qiaowa, Muli Zangzu Zizhixian. Sichuan, C China *27°49′N 101°10′E*
171 *X15* **Muli** *channel* Papua, E Indonesia
Mülig-Hofmann Mountains *see* Mühlig-Hofmanfjella
163 *Y9* **Muling** Heilongjiang, NE China *44°54′N 130°35′E*
Muli Zangzu Zizhixian *see* Muli
Mullach Íde *see* Malahide
Mullaitivu *see* Mullaittivu
155 *K23* **Mullaittivu** ***var.*** Mullaitivu. Northern Province, N Sri Lanka *09°15′N 80°48′E*
33 *N8* **Mullan** Idaho, NW USA *47°28′N 115°48′W*
28 *M13* **Mullen** Nebraska, C USA *42°02′N 101°01′W*
183 *Q6* **Mullengudgery** New South Wales, SE Australia *31°42′S 147°24′E*
21 *Q6* **Mullens** West Virginia, NE USA *37°34′N 81°22′W*
Müller-gerbergte *see* Muller, Pegunungan
169 *T10* **Muller, Pegunungan** ***Dut.*** Müller-gerbergte. ▲▲ Borneo, C Indonesia
31 *Q5* **Mullett Lake** ◎ Michigan, N USA
18 *J16* **Mullica River** ≈ New Jersey, NE USA
25 *R8* **Mullin** Texas, SW USA *31°33′N 98°40′W*
97 *E17* **Mullingar** ***Ir.*** An Muileann gCearr. C Ireland *53°32′N 07°20′W*
21 *T12* **Mullins** South Carolina, SE USA *34°12′N 79°15′W*
96 *G11* **Mull, Isle of** *island* W Scotland, United Kingdom
127 *R5* **Mullovka** Ul'yanovskaya Oblast', W Russian Federation *54°13′N 49°19′E*
95 *K19* **Mullsjö** Västra Götaland, S Sweden *57°56′N 13°55′E*
183 *V4* **Mullumbimby** New South Wales, SE Australia *28°34′S 153°28′E*
83 *H15* **Mulobezi** Western, SW Zambia *16°48′S 25°11′E*
83 *C15* **Mulondo** Huíla, SW Angola *15°41′S 15°09′E*
83 *G15* **Mulonga Plain** *plain* W Zambia
79 *N23* **Mulongo** Katanga, SE Dem. Rep. Congo *07°44′S 26°57′E*
149 *T10* **Multān** Punjab, E Pakistan *30°12′N 71°30′E*
93 *L17* **Multia** Keski-Suomi, C Finland *62°27′N 24°49′E*
Mulucha *see* Moulouya
83 *J14* **Mulungushi** Central, C Zambia *14°15′S 28°27′E*
83 *K14* **Mulungwe** Central, C Zambia *13°57′S 29°51′E*
Muluya *see* Moulouya
27 *N7* **Mulvane** Kansas, C USA *37°28′N 97°14′W*
183 *O10* **Mulwala** New South Wales, SE Australia *35°59′S 146°00′E*
Mulwiya *see* Moulouya
182 *K6* **Mulyungarie** South Australia *31°29′S 140°45′E*
154 *D13* **Mumbai** ***prev.*** Bombay. *state capital* Mahārāshtra, W India *18°56′N 72°51′E*
154 *D13* **Mumbai** ✈ Mahārāshtra, W India *19°10′N 72°51′E*
83 *D14* **Mumbué** Bié, C Angola *13°52′S 17°15′E*
186 *E8* **Mumeng** Morobe, C Papua New Guinea *06°57′S 146°34′E*
171 *V12* **Mumi** Papua Barat, E Indonesia *01°33′S 134°09′E*
Muminabad/Mŭ'minobod *see* Mu'minobod
147 *Q13* **Mu'minobod** ***Rus.*** Leningradskiy, Muminabad; ***prev.*** Leningrad. SW Tajikistan *38°03′N 69°50′E*
127 *Q13* **Mumra** Astrakhanskaya Oblast', SW Russian Federation *45°46′N 47°46′E*
41 *X12* **Muna** Yucatán, SE Mexico *20°29′N 89°41′W*
123 *O9* **Muna** ≈ NE Russian Federation
152 *C12* **Munābāo** Rājasthān, NW India *25°46′N 70°19′E*
Munamägi *see* Suur Munamägi
171 *O14* **Muna, Pulau** ***prev.*** Moena. *island* C Indonesia
101 *L18* **Münchberg** Bayern, E Germany *50°10′N 11°50′E*
101 *L23* **München** ***var.*** Muenchen, ***Eng.*** Munich, ***It.*** Monaco. Bayern, SE Germany *48°09′N 11°34′E*
München-Gladbach *see* Mönchengladbach
108 *E6* **Münchenstein** Basel Landschaft, NW Switzerland *47°31′N 07°38′E*
10 *L10* **Muncho Lake** British Columbia, W Canada *58°52′N 125°40′W*
31 *P13* **Muncie** Indiana, N USA *40°11′N 85°22′W*
18 *G13* **Muncy** Pennsylvania, NE USA *41°10′N 76°46′W*
11 *Q14* **Mundare** Alberta, SW Canada *53°34′N 112°20′W*
25 *Q5* **Munday** Texas, SW USA *33°27′N 99°37′W*
31 *N10* **Mundelein** Illinois, N USA *42°15′N 88°00′W*
101 *I15* **Münden** Niedersachsen, C Germany *52°16′N 08°54′E*
105 *Q12* **Mundo** ≈ S Spain
82 *B12* **Munenga** Kwanza Sul, NW Angola *10°03′S 14°40′E*
105 *P11* **Munera** Castilla-La Mancha, C Spain *39°03′N 02°29′W*
20 *E9* **Munford** Tennessee, S USA *35°27′N 89°49′W*
20 *K7* **Munfordville** Kentucky, S USA *37°17′N 85°55′W*
182 *D5* **Mungala** South Australia *30°36′S 132°57′E*
83 *M16* **Mungári** Manica, C Mozambique *17°09′S 33°33′E*
79 *O16* **Mungbere** Orientale, NE Dem. Rep. Congo *02°38′N 28°30′E*
153 *Q13* **Munger** ***prev.*** Monghyr. Bihār, NE India *25°23′N 86°28′E*
182 *I2* **Mungeranie** South Australia *28°02′S 138°42′E*
Mu Nggava *see* Rennell
169 *O10* **Mungguresak, Tanjung** *headland* Borneo, N Indonesia *01°57′N 109°19′E*
Mungiki *see* Bellona
183 *R4* **Mungindi** New South Wales, SE Australia *28°59′S 149°00′E*
Mungkawn *see* Maingkwan
Mungla *see* Mongla
82 *C13* **Mungo** Huambo, W Angola *11°49′S 16°16′E*
188 *F16* **Munguuy Bay** *bay* Yap, W Micronesia
82 *E13* **Munhango** Bié, C Angola *12°12′S 18°34′E*
Munich *see* München
105 *S7* **Muniesa** Aragón, NE Spain *41°02′N 00°49′W*
31 *O4* **Munising** Michigan, N USA *46°24′N 86°39′W*
Munkács *see* Mukacheve
95 *I17* **Munkedal** Västra Götaland, S Sweden *58°28′N 11°38′E*
95 *K15* **Munkfors** Värmland, C Sweden *59°50′N 13°35′E*
122 *M14* **Munku-Sardyk, Gora** ***var.*** Mönh Sarĭdag. ▲ Mongolia/Russian Federation *51°45′N 100°22′E*
99 *E18* **Munkzwalm** Oost-Vlaanderen, NW Belgium *50°53′N 03°44′E*
167 *R10* **Mun, Mae Nam** ≈ E Thailand
153 *U15* **Munshiganj** Dhaka, C Bangladesh *23°32′N 90°32′E*
108 *D8* **Münsingen** Bern, W Switzerland *46°53′N 07°34′E*
103 *U6* **Munster** Haut-Rhin, NE France *48°03′N 07°09′E*
100 *J11* **Münster** Niedersachsen, NW Germany *52°59′N 10°07′E*
100 *F13* **Münster** ***var.*** Muenster, Münster in Westfalen. Nordrhein-Westfalen, W Germany *51°58′N 07°38′E*
108 *F10* **Münster** Valais, S Switzerland *46°31′N 08°18′E*
97 *B20* **Munster** ***Ir.*** Cúige Mumhan. *cultural region* S Ireland
Münsterberg in Schlesien *see* Ziębice
Münster in Westfalen *see* Münster
100 *E13* **Münsterland** *cultural region* NW Germany
100 *F13* **Münster-Osnabruck** ✈ Nordrhein-Westfalen, NW Germany *52°08′N 07°41′E*
Muntafiq, Al *see* Dhī Qār
31 *R4* **Munuscong Lake** ◎ Michigan, N USA
83 *K17* **Munyati** ≈ C Zimbabwe
109 *R3* **Münzkirchen** Oberösterreich, N Austria *48°29′N 13°37′E*
92 *K11* **Muodoslompolo** Norrbotten, N Sweden *67°57′N 23°31′E*
92 *M13* **Muojärvi** ◎ NE Finland
167 *S6* **Mương Khên** Hoa Binh, N Vietnam *20°34′N 105°18′E*
Muong Sai *see* Oudômxai
167 *Q7* **Muong Xiang Ngeun** ***var.*** Xieng Ngeun. Louangphabang, N Laos *19°43′N 102°09′E*
92 *K11* **Muonio** Lappi, N Finland *67°58′N 23°40′E*
Muonioälv/Muoniojoki *see* Muonionjoki
92 *K11* **Muonionjoki** ***var.*** Muoniojoki, ***Swe.*** Muonioälv. ≈ Finland/Sweden
Muorjek *see* Murjek
83 *N17* **Mupa** ≈ C Mozambique
83 *E16* **Mupini** Okavango, NE Namibia *17°55′S 19°34′E*
80 *F8* **Muqaddam, Wadi** ≈ N Sudan
138 *K9* **Muqāţ** Al Mafraq, E Jordan *32°28′N 38°04′E*
81 *N17* **Muqdisho** ***Eng.*** Mogadishu, ***It.*** Mogadiscio. ● (Somalia) Banaadir, S Somalia *02°06′N 45°27′E*
81 *N17* **Muqdisho** ✈ Banaadir, E Somalia *01°58′N 45°18′E*
Muqshin *see* Mughshin
109 *T8* **Mur** ***SCr.*** Mura. ≈ C Europe
109 *X9* **Mura** ≈ NE Slovenia
Mura *see* Mur
137 *T14* **Muradiye** Van, E Turkey *39°N 43°44′E*
Muragarazi *see* Maragarazi
165 *O10* **Murakami** Niigata, Honshū, C Japan *38°13′N 139°28′E*
63 *G22* **Murallón, Cerro** ▲ S Argentina *49°49′S 73°25′W*
81 *E20* **Muramvya** C Burundi *03°18′S 29°41′E*
81 *I19* **Murang'a** ***prev.*** Fort Hall. Murang'a, SW Kenya *0°43′S 37°10′E*
81 *I19* **Murang'a** ◈ *county* C Kenya
81 *H16* **Murangering** Turkana, NW Kenya *03°48′N 35°29′E*
Murapara *see* Murupara
140 *M5* **Murār, Bi'r al** *well* NW Saudi Arabia
125 *Q13* **Murashi** Kirovskaya Oblast', NW Russian Federation *59°27′N 48°02′E*
103 *O12* **Murat** Cantal, C France *45°07′N 02°52′E*
114 *N12* **Muratlı** Tekirdağ, NW Turkey *41°12′N 27°30′E*
137 *R14* **Murat Nehri** ***var.*** Eastern Euphrates; ***anc.*** Arsanias. ≈ NE Turkey
107 *D20* **Muravera** Sardegna, Italy, C Mediterranean Sea *39°24′N 09°34′E*
165 *P10* **Murayama** Yamagata, Honshū, C Japan *38°29′N 140°21′E*
121 *R13* **Muraysah, Ra's al** *headland* N Libya *31°58′N 25°00′E*
104 *I6* **Murça** Vila Real, N Portugal *41°24′N 07°28′W*
80 *Q11* **Murcanyo** Bari, NE Somalia *11°39′N 50°27′E*
143 *N8* **Mūrcheh Khvort** ***var.*** Morcheh Khort. Eşfahān, C Iran *33°07′N 51°26′E*
185 *H15* **Murchison** Tasman, South Island, New Zealand *41°48′S 172°19′E*
185 *B22* **Murchison Mountains** ▲▲ South Island, New Zealand
180 *I10* **Murchison River** ≈ Western Australia
105 *R13* **Murcia** Murcia, SE Spain *37°59′N 01°08′W*
105 *Q13* **Murcia** ◈ *autonomous community* SE Spain
103 *O13* **Mur-de-Barrez** Aveyron, S France *44°48′N 02°39′E*
182 *G8* **Murdinga** South Australia *33°46′S 135°46′E*
28 *M10* **Murdo** South Dakota, N USA *43°53′N 100°42′W*
15 *X6* **Murdochville** Québec, SE Canada *48°57′N 65°30′W*
109 *W9* **Mureck** Steiermark, SE Austria *46°42′N 15°46′E*
114 *M13* **Mürefte** Tekirdağ, NW Turkey *40°40′N 27°15′E*
116 *I10* **Mureş** ◈ *county* N Romania
84 *J11* **Mureş** ≈ Hungary/Romania
Mureş *see* Maros
Mureşul *see* Maros/Mureş
102 *M16* **Muret** Haute-Garonne, S France *43°28′N 01°19′E*
27 *T13* **Murfreesboro** Arkansas, C USA *34°04′N 93°42′W*
21 *W8* **Murfreesboro** North Carolina, SE USA *36°26′N 77°06′W*
20 *J9* **Murfreesboro** Tennessee, S USA *35°50′N 86°25′W*
Murgab *see* Morghāb, Daryā-ye/Murgap
Murgab *see* Murghob
146 *I14* **Murgap** Mary Welaýaty, S Turkmenistan *37°19′N 61°48′E*
146 *J16* **Murgap** ***var.*** Deryasy Murgap, Murghab, ***Pash.*** Daryā-ye Morghāb, ***Rus.*** Murgab. ≈ Afghanistan/Turkmenistan ***see also*** Morghāb, Daryā-ye
Murgap *see* Murghāb, Daryā-ye
Murgap, Deryasy *see* Morghāb, Daryā-ye/Murgap
114 *H9* **Murgash** ▲ W Bulgaria *42°51′N 23°58′E*
Murghab *see* Morghāb, Daryā-ye/Murgap
148 *M4* **Murghāb, Daryā-ye** ***Rus.*** Murgab, Murghab, ***Turk.*** Murgap, Deryasy Murgap; ***prev.*** Morghāb, Daryā-ye. ≈ Afghanistan/Turkmenistan ***see also*** Murgap
147 *U13* **Murghob** ***Rus.*** Murgab. SE Tajikistan *38°11′N 74°E*
147 *U13* **Murghob** ***Rus.*** Murgab. ≈ SE Tajikistan
181 *Z10* **Murgon** Queensland, E Australia *26°08′S 152°04′E*
190 *I16* **Muri** Rarotonga, S Cook Islands *21°15′S 159°44′W*
108 *F7* **Muri** Aargau, W Switzerland *47°16′N 08°21′E*
108 *D8* **Muri** ***var.*** Muri bei Bern. Bern, W Switzerland *46°55′N 07°30′E*
104 *K3* **Murias de Paredes** Castilla y León, N Spain *42°51′N 06°11′W*
Muri bei Bern *see* Muri
82 *F11* **Muriege** Lunda Sul, NE Angola *09°55′S 21°12′E*
189 *P14* **Murilo Atoll** *atoll* Hall Islands, C Micronesia
Mūrītānīyah *see* Mauritania
100 *N10* **Müritz** ***var.*** Müritzee. ◎ NE Germany
Müritzee *see* Müritz
100 *L10* **Müritz-Elde-Wasserstrasse** *canal* N Germany
184 *K6* **Muriwai Beach** Auckland, North Island, New Zealand *36°56′S 174°28′E*
92 *J13* **Murjek** ***Lapp.*** Muorjek. Norrbotten, N Sweden *66°27′N 20°54′E*
124 *J3* **Murmansk** Murmanskaya Oblast', NW Russian Federation *68°59′N 33°08′E*
124 *I4* **Murmanskaya Oblast'** ◈ *province* NW Russian Federation
197 *V14* **Murmansk Rise** *undersea feature* SW Barents Sea *71°00′N 37°00′E*
124 *J3* **Murmashi** Murmanskaya Oblast', NW Russian Federation *68°49′N 32°43′E*
126 *M5* **Murmino** Ryazanskaya Oblast', W Russian Federation *54°31′N 40°01′E*
101 *K24* **Murnau** Bayern, SE Germany *47°41′N 11°12′E*
103 *X16* **Muro, Capo di** *headland* Corse, France, C Mediterranean Sea *41°45′N 08°40′E*
107 *M18* **Muro Lucano** Basilicata, S Italy *40°48′N 15°33′E*
127 *N4* **Murom** Vladimirskaya Oblast', W Russian Federation *55°33′N 42°03′E*
122 *I11* **Muromtsevo** Omskaya Oblast', C Russian Federation *56°18′N 75°15′E*
165 *R5* **Muroran** Hokkaidō, NE Japan *42°20′N 140°58′E*
104 *G3* **Muros** Galicia, NW Spain *42°47′N 09°04′W*
104 *F3* **Muros e Noia, Ría de** *estuary* NW Spain
164 *H15* **Muroto** Kōchi, Shikoku, SW Japan *33°16′N 134°10′E*
164 *H15* **Muroto-zaki** Shikoku, SW Japan
116 *L7* **Murovani Kurylivtsi** Vinnyts'ka Oblast', C Ukraine *48°43′N 27°31′E*
110 *G11* **Murowana Goślina** Wielkolpolskie, W Poland *52°33′N 16°59′E*
32 *M14* **Murphy** Idaho, NW USA *43°14′N 116°36′W*
21 *N10* **Murphy** North Carolina, SE USA *35°05′N 84°02′W*
35 *P8* **Murphys** California, W USA *38°07′N 120°27′W*
30 *L17* **Murphysboro** Illinois, N USA *37°45′N 89°20′W*
29 *V15* **Murray** Iowa, C USA *41°03′N 93°56′W*
20 *H8* **Murray** Kentucky, S USA *36°35′N 88°20′W*
182 *J10* **Murray Bridge** South Australia *35°10′S 139°17′E*
175 *X2* **Murray Fracture Zone** *tectonic feature* NE Pacific Ocean
192 *H11* **Murray, Lake** ◎ SW Papua New Guinea
21 *P12* **Murray, Lake** ▣ South Carolina, SE USA
10 *K8* **Murray, Mount** ▲ Yukon, NW Canada *60°49′N 128°57′W*
Murray Range *see* Murray Ridge
173 *O3* **Murray Ridge** ***var.*** Murray Range. *undersea feature* N Arabian Sea *21°45′N 61°50′E*
183 *N10* **Murray River** ≈ SE Australia
182 *K10* **Murrayville** Victoria, SE Australia *35°17′S 141°12′E*
149 *U5* **Murree** Punjab, E Pakistan *33°55′N 73°26′E*
101 *I21* **Murrhardt** Baden-Württemberg, S Germany *49°00′N 09°34′E*
183 *O9* **Murrumbidgee River** ≈ New South Wales, SE Australia
83 *P15* **Murrupula** Nampula, NE Mozambique *15°26′S 38°46′E*
183 *T7* **Murrurundi** New South Wales, SE Australia *31°47′S 150°51′E*
109 *X9* **Murska Sobota** ***Ger.*** Olsnitz. NE Slovenia *46°41′N 16°09′E*
154 *G12* **Murtajāpur** ***prev.*** Murtazapur. Mahārāshtra, C India *20°43′N 77°28′E*
77 *S16* **Murtala Muhammed** ✈ (Lagos) Ogun, SW Nigeria *06°31′N 03°12′E*
Murtazapur *see* Murtajāpur
108 *C8* **Murten** Neuchâtel, W Switzerland *46°55′N 07°06′E*
Murtensee *see* Morat, Lac de
182 *L11* **Murtoa** Victoria, SE Australia *36°39′S 142°27′E*
92 *N13* **Murtovaara** Pohjois-Pohjanmaa, E Finland *65°40′N 29°25′E*
Murua Island *see* Woodlark Island
155 *D14* **Murud** Mahārāshtra, W India *18°27′N 72°56′E*
184 *O9* **Murupara** ***var.*** Murapara. Bay of Plenty, North Island, New Zealand *38°27′S 176°41′E*
Mururoa *see* Moruroa
Murviedro *see* Sagunto
154 *J9* **Murwāra** Madhya Pradesh, N India *23°50′N 80°23′E*
183 *V4* **Murwillumbah** New South Wales, SE Australia *28°20′S 153°24′E*
146 *H11* **Murzechirla** ***prev.*** Mirzachirla. Ahal Welaýaty, C Turkmenistan *39°33′N 60°02′E*
Murzuk *see* Murzuq
75 *O11* **Murzuq** ***var.*** Marzūq, Murzuk. SW Libya *25°55′N 13°55′E*
Murzuq, Edeyin *see* Murzuq, Idhān
75 *N11* **Murzuq, Ḩammādat** *plateau* W Libya
75 *O11* **Murzuq, Idhān** ***var.*** Edeyin Murzuq. *desert* SW Libya
109 *W6* **Mürzzuschlag** Steiermark, E Austria *47°35′N 15°41′E*
137 *Q14* **Muş** ***var.*** Mush. Muş, E Turkey *38°45′N 41°30′E*
137 *Q14* **Muş** ***var.*** Mush. ◈ *province* E Turkey
118 *G11* **Mūša** ≈ Latvia/Lithuania
186 *F9* **Musa** ≈ S Papua New Guinea
Mûsa, Gebel *see* Mūsá, Jabal
Musaiyib *see* Al Musayyib
75 *X8* **Mūsá, Jabal** ***var.*** Gebel Mûsa. ▲ NE Egypt *28°33′N 33°51′E*
149 *R9* **Musā Khel** ***var.*** Mūsa Khel Bāzār. Baluchistān, SW Pakistan *30°53′N 69°52′E*
Mūsa Khel Bāzār *see* Musā Khel
139 *Z13* **Mūsá, Khowr-e** *bay* Iraq/Kuwait
114 *H10* **Musala** ▲ W Bulgaria *42°12′N 23°36′E*
168 *H10* **Musala, Pulau** *island* W Indonesia

◆ Country
● Country Capital
◇ Dependent Territory
○ Dependent Territory Capital
◈ Administrative Regions
✈ International Airport
▲ Mountain
▲▲ Mountain Range
Volcano
≈ River
◎ Lake
▣ Reservoir

MUSALE – NAMPULA

83 *I15* **Musale** Southern, S Zambia 15°27′S 26°50′E
141 *Y9* **Muşalla** NE Oman 22°20′N 58°03′E
141 *W6* **Musandam Peninsula** *Ar.* Masandam Peninsula. *peninsula* N Oman
Musay'īd *see* Umm Sa'īd
Muscat *see* Masqaṭ
Muscat and Oman *see* Oman
29 *Y14* **Muscatine** Iowa, C USA 41°25′N 91°03′W
Muscat Sīb Airport *see* Seeb
31 *O15* **Muscatuck River** Indiana, N USA
30 *K8* **Muscoda** Wisconsin, N USA 43°11′N 90°27′W
185 *F19* **Musgrave, Mount** ▲ South Island, New Zealand 43°48′S 170°43′E
181 *P9* **Musgrave Ranges** ▲ South Australia
Mush *see* Muş
138 *H12* **Mushayyish, Qaşr al** *castle* Ma'ān, C Jordan
79 *H20* **Mushie** Bandundu, W Dem. Rep. Congo 03°00′S 16°55′E
168 *M13* **Musi, Air** *prev.* Moesi. Sumatera, W Indonesia
192 *M4* **Musicians Seamounts** *undersea feature* N Pacific Ocean
83 *K19* **Musina** *prev.* Messina. Limpopo, NE South Africa 22°18′S 30°02′E
54 *D8* **Musinga, Alto** ▲ NW Colombia 06°49′N 76°24′W
29 *T2* **Muskeg Bay** *lake bay* Minnesota, N USA
31 *O8* **Muskegon** Michigan, N USA 43°13′N 86°15′W
31 *O8* **Muskegon Heights** Michigan, N USA 43°12′N 86°14′W
31 *P8* **Muskegon River** Michigan, N USA
31 *T14* **Muskingum River** Ohio, N USA
95 *P16* **Muskö** Stockholm, C Sweden 58°58′N 18°10′E
Muskogean *see* Tallahassee
27 *Q10* **Muskogee** Oklahoma, C USA 35°45′N 95°21′W
14 *H13* **Muskoka, Lake** ◎ Ontario, S Canada
80 *H8* **Musmar** Red Sea, NE Sudan 18°13′N 35°40′E
83 *K14* **Musofu** Central, C Zambia 13°31′S 29°02′E
81 *G19* **Musoma** Mara, N Tanzania 01°31′S 33°49′E
82 *L13* **Musoro** Central, C Zambia 13°21′S 31°04′E
186 *F4* **Mussau Island** *island* NE Papua New Guinea
98 *P7* **Musselkanaal** Groningen, NE Netherlands 52°55′N 07°01′E
33 *V9* **Musselshell River** Montana, NW USA
82 *C12* **Mussende** Kwanza Sul, NW Angola 10°33′S 16°02′E
102 *L12* **Mussidan** Dordogne, SW France 45°03′N 00°22′E
99 *L25* **Musson** Luxembourg, SE Belgium 49°33′N 05°42′E
152 *J9* **Mussoorie** Uttarakhand, N India 30°26′N 78°04′E
Musta *see* Mosta
152 *M13* **Mustafābād** Uttar Pradesh, N India 25°54′N 81°17′E
136 *D12* **Mustafakemalpaşa** Bursa, NW Turkey 40°03′N 28°25′E
Mustafa-Pasha *see* Svilengrad
81 *M15* **Mustahīl** Sumalē, E Ethiopia 05°18′N 44°34′E
24 *M7* **Mustang Draw** *valley* Texas, SW USA
25 *T14* **Mustang Island** *island* Texas, SW USA
Mustasaari *see* Korsholm
Mustér *see* Disentis
63 *I19* **Musters, Lago** ◎ S Argentina
45 *Y14* **Mustique** *island* C Saint Vincent and the Grenadines
118 *I6* **Mustla** Viljandimaa, S Estonia 58°12′N 25°50′E
118 *J4* **Mustvee** *Ger.* Tschorna. Jõgevamaa, E Estonia 58°51′N 26°59′E
42 *L9* **Musún, Cerro** ▲ NE Nicaragua 13°01′N 85°02′W
183 *T7* **Muswellbrook** New South Wales, SE Australia 32°17′S 150°55′E
111 *M18* **Muszyna** Małopolskie, SE Poland 49°21′N 20°54′E
75 *V10* **Mūţ** *var.* Mut. C Egypt 25°28′N 28°58′E
136 *I17* **Mut** İçel, S Turkey 36°38′N 33°27′E
109 *V9* **Muta** N Slovenia 46°37′N 15°09′E
190 *B15* **Mutalau** N Niue 18°56′S 169°50′E
Mu-tan-chiang *see* Mudanjiang
82 *I13* **Mutanda** North Western, NW Zambia 12°24′S 26°13′E
59 *O17* **Mutá, Ponta do** *headland* E Brazil 13°54′S 38°54′W
83 *L17* **Mutare** *var.* Mutari; *prev.* Umtali. Manicaland, E Zimbabwe 18°55′S 32°36′E
Mutari *see* Mutare
54 *D8* **Mutatá** Antioquia, NW Colombia 07°16′N 76°32′W
Muthannár, Muḩāfa at al *see* Al Muthannā
Mutina *see* Modena
83 *L16* **Mutoko** *prev.* Mtoko. Mashonaland East, NE Zimbabwe 17°24′S 32°13′E
81 *J20* **Mutomo** Kitui, S Kenya 01°50′S 38°13′E
Mutrah *see* Maţraḩ
79 *M24* **Mutshatsha** Katanga, S Dem. Rep. Congo 10°40′S 24°26′E
165 *R6* **Mutsu** *var.* Mutu. Aomori, Honshū, N Japan 41°18′N 141°11′E
165 *R6* **Mutsu-wan** *bay* N Japan
108 *E6* **Muttenz** Basel Landschaft, NW Switzerland 47°31′N 07°39′E
185 *A26* **Muttonbird Islands** *island group* SW New Zealand
Muttra *see* Mathura
Mutu *see* Mutsu
83 *O15* **Mutuáli** Nampula, N Mozambique 14°51′S 37°01′E
82 *D13* **Mutumbo** Bié, C Angola 13°10′S 17°22′E
189 *Y14* **Mutunte, Mount** *var.* Mount Buache. ▲ Kosrae, E Micronesia 05°21′N 163°00′E
155 *K24* **Mutur** Eastern Province, E Sri Lanka 08°27′N 81°15′E
92 *L13* **Muurola** Lappi, NW Finland 66°22′N 25°20′E
162 *M14* **Mu Us Shadi** *var.* Ordos Desert; *prev.* Mu Us Shamo. *desert* N China
Mu Us Shamo *see* Mu Us Shadi
82 *B11* **Muxima** Bengo, NW Angola 09°33′S 13°58′E
124 *I8* **Muyezerskiy** Respublika Kareliya, NW Russian Federation 63°54′N 32°00′E
81 *E20* **Muyinga** NE Burundi 02°54′S 30°19′E
42 *K9* **Muy Muy** Matagalpa, C Nicaragua 12°43′N 85°35′W
Muynak *see* Mo'ynoq
79 *N22* **Muyumba** Katanga, SE Dem. Rep. Congo 07°13′S 27°02′E
149 *V5* **Muzaffarābād** Jammu and Kashmir, NE Pakistan 34°23′N 73°34′E
149 *S10* **Muzaffargarh** Punjab, E Pakistan 30°04′N 71°15′E
152 *J9* **Muzaffarnagar** Uttar Pradesh, N India 29°28′N 77°42′E
153 *P13* **Muzaffarpur** Bihār, N India 26°07′N 85°23′E
158 *H6* **Muzat He** W China
83 *L15* **Muze** Tete, NW Mozambique 15°05′S 31°16′E
122 *H8* **Muzhi** Yamalo-Nenetskiy Avtonomnyy Okrug, N Russian Federation 65°25′N 64°28′E
102 *H7* **Muzillac** Morbihan, NW France 47°34′N 02°30′W
Muzkol, Khrebet *see* Muzqūl, Qatorkūhi
112 *L9* **Mužlja** *Hung.* Felsőmuzslya; *prev.* Gornja Mužlja. Vojvodina, N Serbia 45°21′N 20°25′E
54 *F9* **Muzo** Boyacá, C Colombia 05°34′N 74°07′W
83 *J15* **Muzoka** Southern, S Zambia 16°39′S 27°18′E
39 *Y15* **Muzon, Cape** *headland* Dall Island, Alaska, USA 54°39′N 132°41′W
40 *M6* **Múzquiz** Coahuila, NE Mexico 27°54′N 101°30′W
147 *U13* **Muzqūl, Qatorkūhi** *Rus.* Khrebet Muzkol. SE Tajikistan
Muztag *see* Muztag Feng
158 *D8* **Muztagata** ▲ NW China 38°16′N 75°03′E
158 *K10* **Muztag Feng** *var.* Muztag. ▲ W China 36°26′N 87°15′E
83 *K17* **Mvuma** *prev.* Umvuma. Midlands, C Zimbabwe 19°17′S 30°32′E
172 *H13* **Mwali** *var.* Moili, *Fr.* Mohéli. *island* S Comoros
82 *L13* **Mwanya** Eastern, E Zambia 12°40′S 32°15′E
79 *N23* **Mwanza** Katanga, SE Dem. Rep. Congo 07°49′S 26°49′E
81 *F20* **Mwanza** ◆ *region* N Tanzania
81 *G20* **Mwanza** ◆ *state capital* Mwanza, NW Tanzania
82 *M13* **Mwase Lundazi** Eastern, E Zambia 12°26′S 33°20′E
97 *B17* **Mweelrea** *Ir.* Caoc Maol Réidh. ▲ W Ireland 53°37′N 09°47′W
79 *K21* **Mweka** Kasai-Occidental, C Dem. Rep. Congo 04°52′S 21°38′E
82 *K12* **Mwenda** Luapula, N Zambia 10°30′S 30°21′E
79 *L22* **Mwene-Ditu** Kasai-Oriental, S Dem. Rep. Congo 07°06′S 23°34′E
83 *L18* **Mwenezi** S Zimbabwe
79 *O20* **Mwenga** Sud-Kivu, E Dem. Rep. Congo 03°00′S 28°28′E
82 *K11* **Mweru, Lake** *var.* Lac Moero. ◎ Dem. Rep. Congo/ Zambia
82 *H13* **Mwinilunga** North Western, NW Zambia 11°44′S 24°24′E
189 *V16* **Mwokil Atoll** *prev.* Mokil Atoll. *atoll* Caroline Islands, E Micronesia
Myadel' *see* Myadzyel
118 *J13* **Myadzyel** *Pol.* Miadzioł Nowy, *Rus.* Myadel'. Minskaya Voblasts', N Belarus 54°51′N 26°51′E
152 *C12* **Myājlār** *var.* Miajlar. Rājasthān, NW India 26°16′N 70°21′E
123 *T9* **Myakit** Magadanskaya Oblast', E Russian Federation 61°23′N 151°58′E
23 *W13* **Myakka River** Florida, SE USA
124 *L14* **Myaksa** Vologodskaya Oblast', NW Russian Federation 58°54′N 38°15′E
183 *U8* **Myall Lake** ◎ New South Wales, SE Australia
166 *L7* **Myanaung** Ayeyawady, SW Myanmar (Burma) 18°17′N 95°19′E
166 *M4* **Myanmar (Burma)** *off.* Republic of the Union of Myanmar; *prev.* Union of Myanmar, *var.* Burma. ◆ *transitional democracy* SE Asia
Myanmar, Republic of the Union of *see* Myanmar (Burma)
Myanmar, Union of *see* Myanmar (Burma)
166 *K8* **Myaungmya** Ayeyawady, SW Myanmar (Burma) 16°33′N 94°55′E
Myaydo *see* Aunglan
118 *N11* **Myazha** *Rus.* Mezha. Vitsyebskaya Voblasts', NE Belarus 55°41′N 30°25′E
167 *N12* **Myeik** *var.* Mergui. Tanintharyi, S Myanmar (Burma) 12°26′N 98°34′E
166 *M12* **Myeik Archipelago** *var.* Mergui Archipelago. *island group* S Myanmar (Burma)
119 *O18* **Myerkulavichy** *Rus.* Merkulovichi. Homyel'skaya Voblasts', SE Belarus 52°58′N 30°36′E
119 *N14* **Myezhava** *Rus.* Mezhëvo. Vitsyebskaya Voblasts', NE Belarus 54°38′N 30°20′E
Myggenaes *see* Mykines
166 *L5* **Myingyan** Mandalay, C Myanmar (Burma) 21°25′N 95°20′E
167 *N12* **Myitkyina** Kachin State, N Myanmar (Burma) 25°24′N 97°25′E
166 *M5* **Myittha** Mandalay, C Myanmar (Burma) 21°21′N 96°06′E
111 *H19* **Myjava** *Hung.* Miava. Trenčiansky Kraj, W Slovakia 48°45′N 17°35′E
Myjeldino *see* Myyëldino
117 *U9* **Mykhaylivka** *Rus.* Mikhaylovka. Zaporiz'ka Oblast', SE Ukraine 47°16′N 35°14′E
95 *A18* **Mykines** *Dan.* Myggenaes. *island* W Faroe Islands
116 *I5* **Mykolayiv** L'vivs'ka Oblast', W Ukraine 49°34′N 23°58′E
117 *Q10* **Mykolayiv** *Rus.* Nikolayev. Mykolayivs'ka Oblast', S Ukraine 46°58′N 31°59′E
117 *Q10* **Mykolayiv** ✕ Mykolayivs'ka Oblast', S Ukraine 47°02′N 31°54′E
Mykolayiv *see* Mykolayivs'ka Oblast'
117 *S13* **Mykolayivka** Avtonomna Respublika Krym, S Ukraine 44°58′N 33°37′E
117 *P9* **Mykolayivka** Odes'ka Oblast', SW Ukraine 47°34′N 30°48′E
117 *P9* **Mykolayivs'ka Oblast'** *var.* Mykolayiv, *Rus.* Nikolayevskaya Oblast'. ◆ *province* S Ukraine
115 *J20* **Mýkonos** Mýkonos, Kykládes, Greece, Aegean Sea 37°27′N 25°20′E
115 *K20* **Mýkonos** *var.* Míkonos. *island* Kykládes, Greece, Aegean Sea
125 *R7* **Myla** Respublika Komi, NW Russian Federation 65°24′N 50°51′E
Mylae *see* Milazzo
93 *M19* **Myllykoski** Kymenlaakso, S Finland 60°45′N 26°52′E
153 *U14* **Mymensing** *var.* Mymensingh. Dhaka, N Bangladesh 24°45′N 90°23′E
Mymensingh *see* Mymensing
93 *K19* **Mynämäki** Varsinais-Suomi, SW Finland 60°41′N 22°00′E
145 *S14* **Mynaral** *Kaz.* Myngaral. Zhambyl, S Kazakhstan 45°25′N 73°37′E
Mynbulak *see* Mingbuloq
Mynbulak, Vpadina *see* Mingbuloq Botig'I
Myngaral *see* Mynaral
Myohaung *see* Mrauk-oo
163 *W13* **Myohyang-sanmaek** ▲ C North Korea
164 *M11* **Myōkō-san** ▲ Honshū, S Japan 36°54′N 138°05′E
83 *J15* **Myooye** Central, C Zambia 15°11′S 27°10′E
118 *K12* **Myory** *prev.* Miyory, *Rus.* Miory. Vitsyebskaya Voblasts', N Belarus 55°39′N 27°39′E
92 *J4* **Mýrdalsjökull** *glacier* S Iceland
92 *G10* **Myre** Nordland, C Norway 68°54′N 15°04′E
117 *S5* **Myrhorod** *Rus.* Mirgorod. Poltavs'ka Oblast', NE Ukraine 49°58′N 33°37′E
115 *J15* **Mýrina** *var.* Mírina. Límnos, SE Greece 39°52′N 25°04′E
117 *P5* **Myronivka** *Rus.* Mironovka. Kyyivs'ka Oblast', N Ukraine 49°40′N 30°59′E
21 *U13* **Myrtle Beach** South Carolina, SE USA 33°41′N 78°53′W
32 *F14* **Myrtle Creek** Oregon, NW USA 43°01′N 123°19′W
183 *P11* **Myrtleford** Victoria, SE Australia 36°34′S 146°45′E
32 *E14* **Myrtle Point** Oregon, NW USA 43°04′N 124°08′W
115 *K25* **Mýrtos** Kríti, Greece, E Mediterranean Sea 35°00′N 25°34′E
Myrtoum Mare *see* Mirtóo Pélagos
93 *G17* **Myrviken** Jämtland, C Sweden 62°59′N 14°19′E
95 *I15* **Mysen** Østfold, S Norway 59°33′N 11°20′E
124 *L15* **Myshkin** Yaroslavskaya Oblast', NW Russian Federation 57°47′N 38°28′E
111 *K17* **Myślenice** Małopolskie, S Poland 49°50′N 19°55′E
110 *D10* **Myślibórz** Zachodnio-pomorskie, NW Poland 52°55′N 14°51′E
155 *G20* **Mysore** *var.* Maisur. Karnātaka, W India 12°18′N 76°37′E
Mysore *see* Karnātaka
115 *F21* **Mystrás** *var.* Mistras. Pelopónnisos, S Greece 37°03′N 22°22′E
111 *K15* **Myszków** Śląskie, S Poland 50°36′N 19°20′E
167 *T14* **My Tho** *var.* Mi Tho. Tiên Giang, S Vietnam 10°21′N 106°21′E
Mytilene *see* Mytilíni
115 *L17* **Mytilíni** *var.* Mitilíni; *anc.* Mytilene. Lésvos, E Greece 39°06′N 26°33′E
126 *K3* **Mytishchi** Moskovskaya Oblast', W Russian Federation 56°00′N 37°51′E
37 *N3* **Myton** Utah, W USA 40°11′N 110°03′W
92 *K2* **Mývatn** ◎ C Iceland
125 *T11* **Myyëldino** *var.* Myjeldino. Respublika Komi, NW Russian Federation 61°46′N 54°48′E
82 *M13* **Mzimba** Northern, NW Malawi 11°56′S 33°36′E
82 *M12* **Mzuzu** Northern, N Malawi 11°23′S 34°03′E

N

101 *M19* **Naab** SE Germany
98 *G12* **Naaldwijk** Zuid-Holland, W Netherlands 52°00′N 04°13′E
38 *G12* **Nā'ālehu** *var.* Naalehu. Hawaii, USA, C Pacific Ocean 19°04′N 155°36′W
93 *K19* **Naantali** *Swe.* Nådendal. Varsinais-Suomi, SW Finland 60°28′N 22°05′E
98 *J10* **Naarden** Noord-Holland, C Netherlands 52°18′N 05°10′E
109 *U4* **Naarn** Austria
97 *F18* **Naas** *Ir.* An Nás, Nás na Ríogh. Kildare, C Ireland 53°13′N 06°39′W
92 *M9* **Näätämöjoki** *Lapp.* Njávdám. NE Finland
83 *E23* **Nababeep** *var.* Nababiep. Northern Cape, W South Africa 29°36′S 17°46′E
Nababiep *see* Nababeep
Nabadwip *see* Navadwip
164 *J14* **Nabari** Mie, Honshū, SW Japan 34°37′N 136°05′E
Nabatié *see* Nabatīyé
138 *G8* **Nabatīyé** *var.* An Nabaṭīyah at Taḥtā, Nabatié, Nabatiyet et Tahta. SW Lebanon 33°18′N 35°36′E
Nabatiyet et Tahta *see* Nabatīyé
187 *X14* **Nabavatu** Vanua Levu, N Fiji 16°35′S 178°55′E
190 *I2* **Nabeina** *island* Tungaru, W Kiribati
127 *T4* **Naberezhnyye Chelny** *prev.* Brezhnev. Respublika Tatarstan, W Russian Federation 55°43′N 52°21′E
39 *T10* **Nabesna** Alaska, USA 62°22′N 143°00′W
39 *T10* **Nabesna River** Alaska, USA
75 *N5* **Nabeul** *var.* Nābul. NE Tunisia 36°32′N 10°45′E
152 *I9* **Nābha** Punjab, NW India 30°22′N 76°12′E
171 *W13* **Nabire** Papua, E Indonesia 03°23′S 135°31′E
141 *O15* **Nabī Shu'ayb, Jabal an** ▲ W Yemen 15°24′N 44°04′E
138 *F10* **Nablus** *var.* Nābulus, *Heb.* Shekhem; *anc.* Neapolis, *Bibl.* Shechem. N West Bank 32°13′N 35°16′E
187 *X14* **Nabouwalu** Vanua Levu, N Fiji 17°00′S 178°43′E
Nābul *see* Nabeul
Nābulus *see* Nablus
187 *Y13* **Nabuna** Vanua Levu, N Fiji 16°13′S 179°46′E
83 *Q14* **Nacala** Nampula, NE Mozambique 14°30′S 40°37′E
42 *H8* **Nacaome** Valle, S Honduras 13°30′N 87°31′W
Na Cealla Beaga *see* Killybegs
Na-Ch'ii *see* Nagqu
164 *J15* **Nachikatsuura** *var.* Nachi-Katsuura. Wakayama, Honshū, SE Japan 33°37′N 135°54′E
Nachi-Katsuura *see* Nachikatsuura
81 *J24* **Nachingwea** Lindi, SE Tanzania 10°21′S 38°46′E
111 *F16* **Náchod** Královéhradecký Kraj, N Czech Republic 50°26′N 16°10′E
Na Clocha Liatha *see* Greystones
40 *G3* **Naco** Sonora, NW Mexico 31°16′N 109°56′W
25 *X8* **Nacogdoches** Texas, SW USA 31°36′N 94°40′W
40 *G4* **Nacozari de García** Sonora, NW Mexico 30°27′N 109°43′W
Nada *see* Danzhou
77 *O14* **Nadawli** NW Ghana 10°30′N 02°40′W
104 *I3* **Nadela** Galicia, NW Spain 42°58′N 07°33′W
Nådendal *see* Naantali
144 *M7* **Nadezhdinka** *prev.* Nadezhdinskiy. Kostanay, N Kazakhstan 53°46′N 63°44′E
Nadezhdinskiy *see* Nadezhdinka
Nadgan *see* Nadqān, Qalamat
187 *W14* **Nadi** *prev.* Nandi. Viti Levu, W Fiji 17°47′S 177°32′E
187 *X14* **Nadi** *prev.* Nandi. ✕ Viti Levu, W Fiji 17°46′S 177°28′E
154 *D10* **Nadiād** Gujarāt, W India 22°42′N 72°55′E
Nadikdik *see* Knox Atoll
116 *E11* **Nădlac** *Ger.* Nadlak, *Hung.* Nagylak. Arad, W Romania 46°10′N 20°47′E
Nadlak *see* Nădlac
74 *H6* **Nador** *prev.* Villa Nador. NE Morocco 35°10′N 05°22′W
141 *S9* **Nadqān, Qalamat** *var.* Nadgan. *well* E Saudi Arabia
111 *N22* **Nádudvar** Hajdú-Bihar, E Hungary 47°36′N 21°09′E
121 *O15* **Nadur** Gozo, N Malta 36°03′N 14°18′E
187 *X13* **Naduri** *prev.* Nanduri. Vanua Levu, N Fiji 16°26′S 179°08′E
116 *I7* **Nadvirna** *Pol.* Nadwórna, *Rus.* Nadvornaya. Ivano-Frankivs'ka Oblast', W Ukraine 48°27′N 24°30′E
124 *J8* **Nadvoitsy** Respublika Kareliya, NW Russian Federation 63°53′N 34°17′E
Nadvornaya/Nadwórna *see* Nadvirna
122 *I9* **Nadym** Yamalo-Nenetskiy Avtonomnyy Okrug, N Russian Federation 65°25′N 72°40′E
122 *I9* **Nadym** C Russian Federation
186 *E7* **Nadzab** Morobe, C Papua New Guinea 06°36′S 146°46′E
95 *C17* **Nærbø** Rogaland, S Norway 58°40′N 05°39′E
95 *I24* **Næstved** Sjælland, SE Denmark 55°12′N 11°47′E
77 *X13* **Nafada** Gombe, E Nigeria 11°02′N 11°18′E
108 *H8* **Näfels** Glarus, NE Switzerland 47°06′N 09°04′E
115 *E18* **Náfpaktos** *var.* Návpaktos. Dytikí Elláda, C Greece 38°23′N 21°50′E
115 *F20* **Náfplio** *prev.* Návplion. Pelopónnisos, S Greece 37°34′N 22°50′E
139 *U6* **Naft Khāneh** Diyālá, E Iraq 34°01′N 45°26′E
149 *N13* **Nāg** Baluchistān, SW Pakistan 27°43′N 65°31′E
171 *P4* **Naga** *off.* Naga City; *prev.* Nueva Caceres. Luzon, N Philippines 13°36′N 123°10′E
Nagaarzê *see* Nagarzê
Naga City *see* Naga
12 *F11* **Nagagami** Ontario, S Canada
164 *F14* **Nagahama** Ehime, Shikoku, SW Japan 33°36′N 132°29′E
153 *X12* **Nāga Hills** ▲ NE India
165 *P10* **Nagai** Yamagata, Honshū, C Japan 38°08′N 140°00′E
Na Gaibhlte *see* Galty Mountains
39 *N16* **Nagai Island** *island* Shumagin Islands, Alaska, USA
153 *X12* **Nāgāland** ◆ *state* NE India
164 *M11* **Nagano** Nagano, Honshū, S Japan 36°39′N 138°11′E
164 *M12* **Nagano** *off.* Nagano-ken. ◆ *prefecture* Honshū, S Japan
Nagano-ken *see* Nagano
165 *N11* **Nagaoka** Niigata, Honshū, C Japan 37°26′N 138°48′E
153 *W12* **Nagaon** *prev.* Nowgong. Assam, NE India 26°21′N 92°41′E
155 *J21* **Nāgappattinam** *var.* Negapatam, Negapattinam. Tamil Nādu, SE India 10°45′N 79°50′E
Nagara Nayok *see* Nakhon Nayok
Nagara Panom *see* Nakhon Phanom
Nagara Pathom *see* Nakhon Pathom
Nagara Sridharmaraj *see* Nakhon Si Thammarat
Nagara Svarga *see* Nakhon Sawan
155 *H16* **Nāgārjuna Sāgar** ⊠ E India
42 *I10* **Nagarote** León, SW Nicaragua 12°15′N 86°35′W
158 *M16* **Nagarzê** *var.* Nagaarzê. Xizang Zizhiqu, W China 28°57′N 90°26′E
164 *C14* **Nagasaki** Nagasaki, Kyūshū, SW Japan 32°45′N 129°52′E
164 *C14* **Nagasaki** *off.* Nagasaki-ken. ◆ *prefecture* Kyūshū, SW Japan
Nagasaki-ken *see* Nagasaki
Nagashima *see* Kii-Nagashima
164 *E12* **Nagato** Yamaguchi, Honshū, SW Japan 34°22′N 131°10′E
152 *F11* **Nāgaur** Rājasthān, NW India 27°12′N 73°48′E
154 *F10* **Nāgda** Madhya Pradesh, C India 23°30′N 75°29′E
98 *L8* **Nagele** Flevoland, N Netherlands 52°39′N 05°43′E
155 *H24* **Nāgercoil** Tamil Nādu, SE India 08°11′N 77°30′E
153 *X12* **Nāginimāra** Nāgāland, NE India 26°44′N 94°51′E
Na Gleannta *see* Glenties
165 *T16* **Nago** Okinawa, Okinawa, SW Japan 26°36′N 127°59′E
154 *K9* **Nāgod** Madhya Pradesh, C India 24°34′N 80°34′E
155 *J26* **Nagoda** Southern Province, S Sri Lanka 06°13′N 80°13′E
101 *G22* **Nagold** Baden-Württemberg, SW Germany 48°33′N 08°43′E
137 *V12* **Nagorno-Karabakh** *var.* Nagorno- Karabakhskaya Avtonomnaya Oblast, *Arm.* Lerrnayin Gharabakh, *Az.* Dağliq Quarabağ, *Rus.* Nagornyy Karabakh. *former autonomous region* SW Azerbaijan
Nagorno- Karabakhskaya Avtonomnaya Oblast *see* Nagorno-Karabakh
123 *Q12* **Nagornyy** Respublika Sakha (Yakutiya), NE Russian Federation 55°53′N 124°58′E
Nagornyy Karabakh *see* Nagorno-Karabakh
125 *R13* **Nagorsk** Kirovskaya Oblast', NW Russian Federation 59°18′N 50°49′E
164 *K13* **Nagoya** Aichi, Honshū, SW Japan 35°10′N 136°53′E
154 *I12* **Nāgpur** Mahārāshtra, C India 21°09′N 79°06′E
156 *K10* **Nagqu** *Chin.* Na-Ch'ii; *prev.* Hei-ho. Xizang Zizhiqu, W China 31°30′N 91°57′E
152 *J8* **Nāg Tibba Range** ▲ N India
45 *O8* **Nagua** NE Dominican Republic 19°25′N 69°49′W
111 *H25* **Nagyatád** Somogy, SW Hungary 46°15′N 17°25′E
Nagybánya *see* Baia Mare
Nagybecskerek *see* Zrenjanin
Nagydisznód *see* Cisnădie
Nagyenyed *see* Aiud
111 *N21* **Nagykálló** Szabolcs-Szatmár-Bereg, E Hungary 47°50′N 21°47′E
111 *G25* **Nagykanizsa** *Ger.* Grosskanizsa. Zala, SW Hungary 46°27′N 17°E
Nagykároly *see* Carei
111 *K22* **Nagykáta** Pest, C Hungary 47°25′N 19°45′E
Nagykikinda *see* Kikinda
111 *K23* **Nagykőrös** Pest, C Hungary 47°01′N 19°46′E
Nagy-Küküllő *see* Târnava Mare
Nagylak *see* Nădlac
Nagymihály *see* Michalovce
Nagyrőce *see* Revúca
Nagysomkút *see* Şomcuta Mare
Nagysurány *see* Šurany
Nagyszalonta *see* Salonta
Nagyszeben *see* Sibiu
Nagyszentmiklós *see* Sânnicolau Mare
Nagyszöllös *see* Vynohradiv
Nagyszombat *see* Trnava
Nagytapolcsány *see* Topoľčany
Nagyvárad *see* Oradea
165 *S17* **Naha** Okinawa, Okinawa, SW Japan 26°10′N 127°40′E
152 *J8* **Nāhan** Himāchal Pradesh, NW India 30°33′N 77°18′E
Nahang, Rūd-e *see* Nihing
138 *F8* **Nahariya** *prev.* Nahariyya. Northern, N Israel 33°01′N 35°05′E
Nahariyya *see* Nahariya
142 *L6* **Nahāvand** *var.* Nehavend. Hamadān, W Iran 34°13′N 48°21′E
101 *F19* **Nahe** SW Germany
Na H-Iarmhidhe *see* Westmeath
189 *O13* **Nahnalaud** ▲ Pohnpei, E Micronesia
Nahoi, Cape *see* Cumberland, Cape
Nahtavárr *see* Nattavaara
63 *H16* **Nahuel Huapí, Lago** ◎ W Argentina
23 *W7* **Nahunta** Georgia, SE USA 31°11′N 81°58′W
40 *J6* **Naica** Chihuahua, N Mexico 27°53′N 105°30′W
11 *U15* **Naicam** Saskatchewan, S Canada 52°26′N 104°30′W
Naiman Qi *see* Daqin Tal
158 *M4* **Naimin Bulak** *spring* NW China
13 *P6* **Nain** Newfoundland and Labrador, NE Canada 56°33′N 61°46′W
143 *P8* **Nā'īn** Eşfahān, C Iran 32°52′N 53°05′E
152 *K10* **Naini Tāl** Uttarakhand, N India 29°22′N 79°26′E
154 *J11* **Nainpur** Madhya Pradesh, C India 22°26′N 80°10′E
96 *J8* **Nairn** N Scotland, United Kingdom 57°36′N 03°51′W
96 *I8* **Nairn** *cultural region* NE Scotland, United Kingdom
81 *I19* **Nairobi** ● (Kenya) Nairobi City, S Kenya 01°17′S 36°50′E
81 *I19* **Nairobi** ✕ Nairobi City, S Kenya 01°21′S 37°01′E
81 *I19* **Nairobi City** ◆ *county* Nairobi City, Kenya, Africa
82 *P13* **Nairoto** Cabo Delgado, NE Mozambique 12°22′S 39°05′E
118 *G3* **Naissaar** *island* N Estonia
Naissus *see* Niš
187 *Z14* **Naitaba** *var.* Naitauba; *prev.* Naitamba. *island* Lau Group, E Fiji
Naitamba/Naitauba *see* Naitaba
81 *I19* **Naivasha** Nakuru, SW Kenya 0°44′S 36°26′E
81 *H19* **Naivasha, Lake** ◎ SW Kenya
Najaf *see* An Najaf
143 *N8* **Najafābād** *var.* Nejafabad. Eşfahān, C Iran 32°38′N 51°23′E
Najaf, Muḩāfa at an *see* An Najaf
141 *N7* **Najd** *var.* Nejd. *cultural region* C Saudi Arabia
105 *O4* **Nájera** La Rioja, N Spain 42°25′N 02°45′W
105 *P4* **Najerilla** N Spain
163 *U7* **Naji** *var.* Arun Qi. Nei Mongol Zizhiqu, N China 48°05′N 123°28′E
152 *J9* **Najībābād** Uttar Pradesh, N India 29°37′N 78°19′E
Najima *see* Fukuoka
163 *Y11* **Najin** NE North Korea 42°13′N 130°16′E
139 *T9* **Najm al Ḩassūn** Bābil, C Iraq 32°24′N 44°13′E
141 *O13* **Najrān** *var.* Abā as Su'ūd. Najrān, S Saudi Arabia 17°31′N 44°09′E
141 *P12* **Najrān** *var.* Minţaqat al Najrān. ◆ *province* S Saudi Arabia
Najrān, Minţaqat al *see* Najrān
165 *T2* **Nakagawa** Hokkaidō, NE Japan 44°49′N 142°04′E
38 *F9* **Nākālele Point** *var.* Nakalele Point. *headland* Maui, Hawai'i, USA 21°01′N 156°35′W
164 *D13* **Nakama** Fukuoka, Kyūshū, SW Japan 33°53′N 130°48′E
Nakambé *see* White Volta
Nakamti *see* Nek'emtē
164 *F15* **Nakamura** *var.* Shimanto. Kōchi, Shikoku, SW Japan 33°00′N 132°55′E
186 *H7* **Nakanai Mountains** ▲ New Britain, E Papua New Guinea
164 *H11* **Nakano-shima** *island* Oki-shotō, SW Japan
165 *Q6* **Nakasato** Aomori, Honshū, C Japan 40°58′N 140°26′E
165 *T5* **Nakasatsunai** Hokkaidō, NE Japan 42°42′N 143°09′E
165 *W4* **Nakashibetsu** Hokkaidō, NE Japan 43°31′N 144°58′E
81 *F18* **Nakasongola** C Uganda 01°19′N 32°28′E
165 *T1* **Nakatonbetsu** Hokkaidō, NE Japan 44°58′N 142°18′E
164 *L13* **Nakatsugawa** *var.* Nakatugawa. Gifu, Honshū, SW Japan 35°30′N 137°29′E
Nakatugawa *see* Nakatsugawa
Nakdong *see* Nakdong-gang
163 *Y15* **Nakdong-gang** *var.* Nakdong, *Jap.* Rakutō-kō; *prev.* Naktong-gang. C South Korea
Nakel *see* Nakło nad Notecią
Nakfa *see* Nak'fa
80 *J8* **Nakh'fa** *var.* Nakfa. N Eritrea 16°38′N 38°26′E
Nakhichevan' *see* Naxçıvan
123 *S15* **Nakhodka** Primorskiy Kray, SE Russian Federation 42°46′N 132°48′E
122 *J8* **Nakhodka** Yamalo-Nenetskiy Avtonomnyy Okrug, N Russian Federation 67°48′N 77°48′E
Nakhon Navok *see* Nakhon Nayok
167 *P11* **Nakhon Nayok** *var.* Nagara Nayok, Nakhon Navok. Nakhon Nayok, C Thailand 14°15′N 101°12′E
167 *O11* **Nakhon Pathom** *var.* Nagara Pathom, Nakorn Pathom. Nakhon Pathom, W Thailand 13°49′N 100°06′E
167 *R8* **Nakhon Phanom** *var.* Nagara Panom. Nakhon Phanom, E Thailand 17°22′N 104°46′E
167 *Q10* **Nakhon Ratchasima** *var.* Khorat, Korat. Nakhon Ratchasima, E Thailand 15°N 102°06′E
167 *O9* **Nakhon Sawan** *var.* Muang Nakhon Sawan, Nagara Svarga. Nakhon Sawan, W Thailand 15°42′N 100°06′E
167 *N15* **Nakhon Si Thammarat** *var.* Nagara Sridharmaraj, Nakhon Sithamnaraj. Nakhon Si Thammarat, SW Thailand 08°24′N 99°58′E
Nakhon Sithamnaraj *see* Nakhon Si Thammarat
139 *Y11* **Nakhrash** Al Başrah, SE Iraq 31°13′N 47°24′E
10 *J9* **Nakina** British Columbia, W Canada 59°12′N 132°48′W
110 *H9* **Nakło nad Notecią** *Ger.* Nakel. Kujawsko-pomorskie, C Poland 53°08′N 17°35′E
39 *P13* **Naknek** Alaska, USA 58°45′N 157°01′W
152 *H8* **Nakodar** Punjab, NW India 31°06′N 75°31′E
82 *M11* **Nakonde** Muchinga, NE Zambia 09°22′S 32°47′E
Nakorn Pathom *see* Nakhon Pathom
95 *H24* **Nakskov** Sjælland, SE Denmark 54°50′N 11°10′E
Naktong-gang *see* Nakdong-gang
81 *H18* **Nakuru** Nakuru, SW Kenya 0°16′S 36°04′E
81 *I19* **Nakuru** ◆ *county* W Kenya
81 *H19* **Nakuru, Lake** ◎ Nakuru, C Kenya
11 *O17* **Nakusp** British Columbia, SW Canada 50°14′N 117°48′W
149 *N15* **Nāl** W Pakistan
162 *M7* **Nalayh** Töv, C Mongolia 47°48′N 107°17′E
153 *V12* **Nalbāri** Assam, NE India 26°36′N 91°49′E
63 *G19* **Nalcayec, Isla** *island* Archipiélago de los Chonos, S Chile
127 *N15* **Nal'chik** Kabardino-Balkarskaya Respublika, SW Russian Federation 43°30′N 43°39′E
155 *I16* **Nalgonda** Telangana, C India 17°04′N 79°15′E
153 *S14* **Nalhāti** West Bengal, NE India 24°19′N 87°53′E
153 *U14* **Nalitabari** Dhaka, N Bangladesh 25°06′N 90°11′E
155 *I17* **Nallamala Hills** ▲ E India
136 *G12* **Nallıhan** Ankara, NW Turkey 40°12′N 31°22′E
104 *K2* **Nalón** NW Spain
167 *N3* **Nalong** Kachin State, N Myanmar (Burma) 24°42′N 97°27′E
75 *N8* **Nālūt** NW Libya 31°52′N 10°59′E
171 *T14* **Nama** Pulau Manawoka, E Indonesia 04°07′S 131°22′E
189 *Q16* **Nama** *island* C Micronesia
83 *O16* **Namacurra** Zambézia, NE Mozambique 17°31′S 37°03′E
188 *F9* **Namai Bay** *bay* Babeldaob, N Palau
29 *W2* **Namakan Lake** ◎ Canada/ USA
143 *O6* **Namak, Daryācheh-ye** *marsh* N Iran
143 *T6* **Namak, Kavīr-e** *salt pan* NE Iran
167 *O6* **Namakkwe** Shan State, E Myanmar (Burma) 19°45′N 99°01′E
Namaksār, Kowl-e/ Namakzār, Daryācheh-ye *see* Namakzar
148 *J5* **Namakzar** *Pash.* Daryācheh-ye Namakzār, Kowl-e Namaksār. *marsh* Afghanistan/Iran
171 *V15* **Namalau** Pulau Jursian, E Indonesia 05°50′S 134°43′E
81 *I20* **Namanga** Kajiado, S Kenya 02°33′S 36°48′E
147 *S10* **Namangan** Namangan Viloyati, E Uzbekistan 40°59′N 71°34′E
Namanganskaya Oblast' *see* Namangan Viloyati
147 *R10* **Namangan Viloyati** *Rus.* Namanganskaya Oblast'. ◆ *province* E Uzbekistan
83 *Q14* **Namapa** Nampula, NE Mozambique 13°43′S 39°48′E
83 *C21* **Namaqualand** *physical region* S Namibia
81 *G18* **Namasagali** C Uganda 01°02′N 32°58′E
186 *H6* **Namatanai** New Ireland, NE Papua New Guinea 03°40′S 152°26′E
83 *I14* **Nambala** Central, C Zambia 15°04′S 26°56′E
81 *J23* **Nambanje** Lindi, SE Tanzania 08°37′S 38°21′E
183 *V2* **Nambour** Queensland, E Australia 26°40′S 152°52′E
183 *V6* **Nambucca Heads** New South Wales, SE Australia 30°37′S 153°00′E
159 *N15* **Nam Co** ◎ W China
167 *R5* **Nâm Cum** Lai Châu, N Vietnam 22°37′N 103°12′E
Namdik *see* Namorik Atoll
167 *T6* **Nam Định** Nam Ha, N Vietnam 20°25′N 106°12′E
99 *I20* **Namêche** Namur, SE Belgium 50°29′N 05°02′E
30 *J4* **Namekagon Lake** ◎ Wisconsin, N USA
188 *F10* **Namekakl Passage** *passage* Babeldaob, N Palau
Namen *see* Namur
83 *P15* **Nametil** Nampula, NE Mozambique 15°46′S 39°21′E
163 *X14* **Nam-gang** C North Korea
163 *Y16* **Nam-gang** S South Korea
163 *Y17* **Namhae-do** *Jap.* Nankai-tō. *island* S South Korea
Namhoi *see* Foshan
83 *C19* **Namib Desert** *desert* W Namibia
83 *A15* **Namibe** *Port.* Moçâmedes, Mossâmedes. Namibe, SW Angola 15°10′S 12°09′E
83 *A15* **Namibe** ◆ *province* SE Angola
83 *C18* **Namibia** *off.* Republic of Namibia, *var.* South West Africa, *Afr.* Suidwes-Afrika, *Ger.* Deutsch-Südwestafrika; *prev.* German Southwest Africa, South-West Africa. ◆ *republic* S Africa
65 *O17* **Namibia Plain** *undersea feature* S Atlantic Ocean
Namibia, Republic of *see* Namibia
165 *Q11* **Namie** Fukushima, Honshū, C Japan 37°29′N 140°58′E
165 *Q7* **Namioka** Aomori, Honshū, C Japan 40°43′N 140°34′E
40 *I5* **Namiquipa** Chihuahua, N Mexico 29°15′N 107°25′W
159 *P15* **Namjagbarwa Feng** ▲ W China 29°39′N 95°00′E
Namka *see* Doilungdêqên
Nam Khong *see* Mekong
171 *R13* **Namlea** Pulau Buru, E Indonesia 03°12′S 127°06′E
158 *L16* **Namling** Xizang Zizhiqu, W China 29°40′N 88°58′E
Namnetes *see* Nantes
167 *R8* **Nam Ngum** C Laos
Namo *see* Namu Atoll
183 *R5* **Namoi River** New South Wales, SE Australia
189 *Q17* **Namoluk Atoll** *atoll* Mortlock Islands, C Micronesia
189 *O15* **Namonuito Atoll** *atoll* Caroline Islands, C Micronesia
189 *T9* **Namorik Atoll** *var.* Namdik. *atoll* Ralik Chain, S Marshall Islands
167 *Q6* **Nam Ou** N Laos
32 *M14* **Nampa** Idaho, NW USA 43°32′N 116°33′W
76 *M11* **Nampala** Ségou, W Mali 15°21′N 05°32′W
163 *W14* **Namp'o** SW North Korea 38°46′N 125°25′E
83 *P15* **Nampula** Nampula, NE Mozambique 15°09′S 39°14′E

◆ Country
● Country Capital
◇ Dependent Territory
○ Dependent Territory Capital
◆ Administrative Regions
✕ International Airport
▲ Mountain
▲ Mountain Range
Volcano
River
◎ Lake
⊠ Reservoir

NAMPULA – NEDER RIJN

83 P15 **Nampula** *off.* Província de Nampula. ◆ *province* NE Mozambique
Nampula, Província de *see* Nampula
163 W13 **Namsan-ni** NW North Korea *40°25′N 125°01′E*
Namslau *see* Namysłów
93 E15 **Namsos** Nord-Trøndelag, C Norway *64°28′N 11°31′E*
93 F14 **Namsskogan** Nord-Trøndelag, C Norway *64°57′N 13°04′E*
167 O6 **Nam Teng** ⤳ E Myanmar (Burma)
167 P6 **Nam Tha** ⤳ N Laos
123 Q10 **Namtsy** Respublika Sakha (Yakutiya), NE Russian Federation *62°42′N 129°30′E*
167 N4 **Namtu** Shan State, E Myanmar (Burma) *23°04′N 97°26′E*
10 J15 **Namu** British Columbia, SW Canada *51°46′N 127°49′W*
189 T7 **Namu Atoll** *var.* Namo. *atoll* Ralik Chain, C Marshall Islands
187 Y15 **Namuka-i-lau** *island* Lau Group, E Fiji
83 O15 **Namuli, Mont** ▲ NE Mozambique *15°15′S 37°33′E*
83 P14 **Namuno** Cabo Delgado, N Mozambique *13°39′S 38°50′E*
99 I20 **Namur** *Dut.* Namen. Namur, SE Belgium *50°28′N 04°52′E*
99 H21 **Namur** *Dut.* Namen. ◆ *province* S Belgium
83 D17 **Namutoni** Kunene, N Namibia *18°49′S 16°55′E*
163 Y16 **Namwon** *Jap.* Nangen; *prev.* Namwŏn. S South Korea *35°24′N 127°20′E*
Namwŏn *see* Namwon
111 H14 **Namysłów** *Ger.* Namslau. Opole, SW Poland *51°03′N 17°41′E*
167 P7 **Nan** *var.* Muang Nan. Nan, NW Thailand *18°47′N 100°50′E*
79 G15 **Nana** ⤳ W Central African Republic
165 R5 **Nanae** Hokkaidō, NE Japan *41°55′N 140°40′E*
79 I14 **Nana-Grébizi** ◆ *prefecture* N Central African Republic
10 L17 **Nanaimo** Vancouver Island, British Columbia, SW Canada *49°08′N 123°58′W*
38 C9 **Nānākuli** *var.* Nanakuli. Oʻahu, Hawaii, USA, C Pacific Ocean *21°23′N 158°09′W*
79 G15 **Nana-Mambéré** ◆ *prefecture* W Central African Republic
161 R13 **Nan'an** Fujian, SE China *24°57′N 118°22′E*
183 U2 **Nanango** Queensland, E Australia *26°42′S 151°58′E*
164 L11 **Nanao** Ishikawa, Honshū, SW Japan *37°03′N 136°58′E*
161 Q14 **Nan'ao Dao** *island* S China
164 L10 **Nanatsu-shima** *island* SW Japan
56 F8 **Nanay, Río** ⤳ NE Peru
160 J8 **Nanbu** Sichuan, C China *31°19′N 106°02′E*
163 X7 **Nancha** Heilongjiang, NE China *47°09′N 129°17′E*
161 P10 **Nanchang** *var.* Nan-ch'ang, Nanch'ang-hsien. *province capital* Jiangxi, S China *28°38′N 115°58′E*
Nan-ch'ang *see* Nanchang
Nanch'ang-hsien *see* Nanchang
161 P11 **Nancheng** *var.* Jianchang. Jiangxi, S China *27°37′N 116°37′E*
Nan-ching *see* Nanjing
160 J9 **Nanchong** Sichuan, C China *30°47′N 106°03′E*
160 J10 **Nanchuan** Chongqing Shi, C China *29°06′N 107°13′E*
103 T5 **Nancy** Meurthe-et-Moselle, NE France *48°40′N 06°11′E*
185 A22 **Nancy Sound** *sound* South Island, New Zealand
152 L9 **Nanda Devi** ▲ NW India *30°27′N 80°00′E*
42 J11 **Nandaime** Granada, SW Nicaragua *11°45′N 86°02′W*
160 K13 **Nandan** *var.* Minami-Awaji. Guangxi Zhuangzu Zizhiqu, S China *25°03′N 107°31′E*
155 H14 **Nānded** Mahārāshtra, C India *19°11′N 77°21′E*
183 S5 **Nandewar Range** ▲ New South Wales, SE Australia
81 H18 **Nandi** ◆ *county* W Kenya
Nandi *see* Nadi
160 E13 **Nanding He** ⤳ China/Vietnam
Nándorhgy *see* Oţelu Roşu
154 E11 **Nandurbār** Mahārāshtra, W India *21°22′N 74°18′E*
Nanduri *see* Naduri
155 I17 **Nandyāl** Andhra Pradesh, E India *15°30′N 78°28′E*
161 P11 **Nanfeng** *var.* Qincheng. Jiangxi, S China *27°15′N 116°16′E*
Nang *see* Nangxian
79 E15 **Nanga Eboko** Centre, C Cameroon *04°38′N 12°21′E*
Nangah Serawai *see* Nangaserawai
149 W4 **Nanga Parbat** ▲ India/Pakistan *35°15′N 74°36′E*
169 R11 **Nangapinoh** Borneo, C Indonesia *0°21′S 111°44′E*
149 R5 **Nangarhār** ◆ *province* E Afghanistan
169 S11 **Nangaserawai** *var.* Nangah Serawai. Borneo, C Indonesia *0°20′S 112°26′E*
169 Q12 **Nangatayap** Borneo, C Indonesia *01°30′S 110°33′E*
Nangen *see* Namwon
103 P5 **Nangis** Seine-et-Marne, N France *48°36′N 03°02′E*
163 X13 **Nangnim-sanmaek** ▲ C North Korea
161 O4 **Nangong** Hebei, E China *37°22′N 115°20′E*
159 Q14 **Nangqên** *var.* Xangda. Qinghai, C China *32°05′N 96°28′E*
167 Q10 **Nang Rong** Buri Ram, E Thailand *14°37′N 102°48′E*
159 O16 **Nangxian** *var.* Nang. Xizang Zizhiqu, W China *29°04′N 93°03′E*
Nan Hai *see* South China Sea
160 L8 **Nan He** ⤳ C China
160 F12 **Nanhua** *var.* Longchuan. Yunnan, SW China *25°15′N 101°15′E*
Naniwa *see* Ōsaka
155 G20 **Nanjangūd** Karnātaka, W India *12°07′N 76°40′E*
161 Q8 **Nanjing** *var.* Nan-ching, Nanking; *prev.* Chianning, Chian-ning, Kiang-ning, Jiangsu. *province capital* Jiangsu, E China *32°03′N 118°47′E*
Nankai-tō *see* Namhae-do
161 O12 **Nankang** *var.* Rongjiang. Jiangxi, S China *25°42′N 114°45′E*
Nanking *see* Nanjing
161 N13 **Nan Ling** ▲ S China
160 L15 **Nanliu Jiang** ⤳ S China
189 P13 **Nan Madol** *ruins* Temwen Island, E Micronesia
160 K15 **Nanning** *var.* Nan-ning; *prev.* Yung-ning. Guangxi Zhuangzu Zizhiqu, S China *22°50′N 108°19′E*
Nan-ning *see* Nanning
196 M15 **Nanortalik** Kujalleo, S Greenland *60°08′N 45°14′W*
Nanouki *see* Aranuka
160 H13 **Nanpan Jiang** ⤳ S China
152 M11 **Nānpāra** Uttar Pradesh, N India *27°51′N 81°30′E*
161 Q12 **Nanping** *var.* Nan-p'ing; *prev.* Yenping. Fujian, SE China *26°40′N 118°07′E*
Nan-p'ing *see* Nanping
Nanping *see* Jiuzhaigou
Nanpu *see* Pucheng
161 R12 **Nanri Dao** *island* SE China
165 S16 **Nansei-shotō** *Eng.* Ryukyu Islands. *island group* SW Japan
Nansei Syotō Trench *see* Ryukyu Trench
197 T10 **Nansen Basin** *undersea feature* Arctic Ocean
Nansen Cordillera *see* Gakkel Ridge
129 T9 **Nan Shan** ▲ C China
Nansha Qundao *see* Spratly Islands
12 K3 **Nantais, Lac** ⊚ Québec, NE Canada
103 N5 **Nanterre** Hauts-de-Seine, N France *48°53′N 02°13′E*
102 I8 **Nantes** *Bret.* Naoned; *anc.* Condivincum, Namnetes. Loire-Atlantique, NW France *47°12′N 01°32′W*
14 G17 **Nanticoke** Ontario, S Canada *42°49′N 80°04′W*
18 H13 **Nanticoke** Pennsylvania, NE USA *41°12′N 76°00′W*
21 Y4 **Nanticoke River** ⤳ Delaware/Maryland, NE USA
11 Q17 **Nanton** Alberta, SW Canada *50°21′N 113°47′W*
161 S8 **Nantong** Jiangsu, E China *32°00′N 120°52′E*
161 S13 **Nantou** *prev.* Nant'ou. W Taiwan *23°54′N 120°51′E*
Nant'ou *see* Nantou
103 S10 **Nantua** Ain, E France *46°10′N 05°34′E*
19 Q13 **Nantucket** Nantucket Island, Massachusetts, NE USA *41°15′N 70°05′W*
19 Q13 **Nantucket Island** *island* Massachusetts, NE USA
19 Q13 **Nantucket Sound** *sound* Massachusetts, NE USA
82 P13 **Nantulo** Cabo Delgado, N Mozambique *12°30′S 39°03′E*
189 O12 **Nanuh** Pohnpei, E Micronesia
190 D6 **Nanumaga** *var.* Nanumanga. *atoll* NW Tuvalu
Nanumanga *see* Nanumaga
190 D5 **Nanumea Atoll** *atoll* NW Tuvalu
59 O19 **Nanuque** Minas Gerais, SE Brazil *17°49′S 40°21′W*
171 R10 **Nanusa, Kepulauan** *island group* N Indonesia
Nanwei Dao *see* Spratly Island
163 U4 **Nanweng He** ⤳ NE China
160 I10 **Nanxi** Sichuan, C China *28°54′N 104°59′E*
161 N10 **Nanxian** *var.* Nan Xian, Nanzhou. Hunan, S China *29°23′N 112°18′E*
Nan Xian *see* Nanxian
161 N7 **Nanyang** *var.* Nan-yang. Henan, C China *32°59′N 112°29′E*
Nan-yang *see* Nanyang
161 P6 **Nanyang Hu** ⊚ E China
165 P10 **Nan'yō** Yamagata, Honshū, C Japan *38°04′N 140°06′E*
81 I18 **Nanyuki** Laikipia, C Kenya *0°01′N 37°05′E*
160 M8 **Nanzhang** Hubei, C China *31°47′N 111°48′E*
Nanzhou *see* Nanxian
105 T11 **Nao, Cabo de la** *headland* E Spain *38°43′N 00°13′E*
12 M9 **Naococane, Lac** ⊚ Québec, E Canada
153 S14 **Naogaon** Rajshahi, NW Bangladesh *24°49′N 88°59′E*
Naokot *see* Naukot
187 R13 **Naone** Maewo, C Vanuatu *15°03′S 168°06′E*
Naoned *see* Nantes
115 E14 **Náousa** Kentrikí Makedonía, N Greece *40°38′N 22°04′E*
35 N8 **Napa** California, W USA *38°15′N 122°17′W*
39 O11 **Napaimiut** Alaska, USA *61°32′N 158°46′W*
39 N12 **Napakiak** Alaska, USA *60°42′N 161°57′W*
122 J7 **Napalkovo** Yamalo-Nenetskiy Avtonomnyy Okrug, N Russian Federation *70°06′N 73°43′E*
12 I16 **Napanee** Ontario, SE Canada *44°13′N 76°57′W*
39 N12 **Napaskiak** Alaska, USA *60°42′N 161°46′W*
167 S5 **Na Phác** Cao Bằng, N Vietnam *22°24′N 105°54′E*
184 O11 **Napier** Hawke's Bay, North Island, New Zealand *39°30′S 176°55′E*
195 X3 **Napier Mountains** ▲ Antarctica
15 O13 **Napierville** Québec, SE Canada *45°12′N 73°25′W*
23 W15 **Naples** Florida, SE USA *26°08′N 81°48′W*
25 W5 **Naples** Texas, SW USA *33°12′N 94°40′W*
Naples *see* Napoli
160 I14 **Napo** Guangxi Zhuangzu Zizhiqu, S China *23°21′N 105°47′E*
56 C6 **Napo** ◆ *province* NE Ecuador
29 O6 **Napoleon** North Dakota, N USA *46°30′N 99°46′W*
31 R11 **Napoleon** Ohio, N USA *41°23′N 84°07′W*
Napoléon-Vendée *see* la Roche-sur-Yon
22 J9 **Napoleonville** Louisiana, S USA *29°55′N 91°01′W*
107 K17 **Napoli** *Eng.* Naples, *Ger.* Neapel; *anc.* Neapolis. Campania, S Italy *40°52′N 14°15′E*
107 J18 **Napoli, Golfo di** *gulf* S Italy
57 F7 **Napo, Río** ⤳ Ecuador/Peru
191 W9 **Napuka** *island* Îles Tuamotu, C French Polynesia
142 J3 **Naqadeh** Āzarbāyjān-e Bākhtarī, NW Iran *36°57′N 45°24′E*
139 U6 **Naqnah** Diyālá, E Iraq *34°13′N 45°33′E*
Nar *see* Nera
164 J14 **Nara** Nara, Honshū, SW Japan *34°41′N 135°49′E*
76 L11 **Nara** Koulikoro, W Mali *15°04′N 07°19′W*
149 R14 **Nāra Canal** *irrigation canal* S Pakistan
182 K11 **Naracoorte** South Australia *37°02′S 140°45′E*
183 P8 **Naradhan** New South Wales, SE Australia *33°37′S 146°19′E*
Naradhivas *see* Narathiwat
56 B8 **Naranjal** Guayas, W Ecuador *02°43′S 79°38′W*
57 Q19 **Naranjos** Santa Cruz, E Bolivia
41 Q12 **Naranjos** Veracruz-Llave, E Mexico *21°21′N 97°41′W*
159 Q6 **Naran Sebstein Bulag** *spring* NW China
164 B14 **Narao** Nagasaki, Nakadōri-jima, SW Japan *32°40′N 129°03′E*
155 J16 **Narasaraopet** Andhra Pradesh, E India *16°16′N 80°06′E*
158 J5 **Narat** Xinjiang Uygur Zizhiqu, W China *43°20′N 84°02′E*
167 P17 **Narathiwat** *var.* Naradhivas. Narathiwat, SW Thailand *06°25′N 101°48′E*
37 V10 **Nara Visa** New Mexico, SW USA *35°35′N 103°06′W*
Nārāyāni *see* Gandak
Narbada *see* Narmada
Narbo Martius *see* Narbonne
103 P16 **Narbonne** *anc.* Narbo Martius. Aude, S France *43°11′N 03°E*
Narborough Island *see* Fernandina, Isla
104 J2 **Narcea** ⤳ NW Spain
152 J9 **Narendranagar** Uttarakhand, N India *30°10′N 78°21′E*
Nares Abyssal Plain *see* Nares Plain
64 G11 **Nares Plain** *var.* Nares Abyssal Plain. *undersea feature* NW Atlantic Ocean *23°30′N 63°00′W*
Nares Stræde *see* Nares Strait
197 P10 **Nares Strait** *Dan.* Nares Stræde. *strait* Canada/Greenland
110 O9 **Narew** ⤳ E Poland
155 F17 **Nargund** Karnātaka, W India *15°43′N 75°23′E*
83 D20 **Narib** Hardap, S Namibia *24°11′S 17°46′E*
Narikrik *see* Knox Atoll
Narin Gol *see* Omon Gol
54 B13 **Nariño** *off.* Departamento de Nariño. ◆ *province* SW Colombia
165 P13 **Narita** Chiba, Honshū, S Japan *35°46′N 140°20′E*
165 P13 **Narita** ✈ (Tōkyō) Chiba, Honshū, S Japan *35°45′N 140°23′E*
Nariya *see* An Nuʻayrīyah
162 F5 **Narijn Gol** ⤳ Mongolia/Russian Federation
162 J8 **Nariynteel** *var.* Tsagaan-Ovoo. Övörhangay, C Mongolia *45°57′N 101°25′E*
152 J8 **Narkanda** Himāchal Pradesh, NW India *31°14′N 77°27′E*
92 L13 **Narkaus** Lappi, NW Finland *66°13′N 26°09′E*
154 E11 **Narmada** *var.* Narbada. ⤳ C India
152 H11 **Narnaul** *var.* Nārnaul. Haryāna, N India *28°04′N 76°10′E*
107 I14 **Narni** Umbria, C Italy *42°31′N 12°31′E*
107 J24 **Naro** Sicilia, Italy, C Mediterranean Sea *37°18′N 13°48′E*
Narodichi *see* Narodychi
125 V7 **Narodnaya, Gora** ▲ NW Russian Federation *65°04′N 60°12′E*
117 N3 **Narodychi** *Rus.* Narodichi. Zhytomyrs'ka Oblast', N Ukraine *51°11′N 29°01′E*
126 J4 **Naro-Fominsk** Moskovskaya Oblast', W Russian Federation *55°25′N 36°41′E*
81 H19 **Narok** Narok, SW Kenya *01°04′S 35°54′E*
81 H20 **Narok** ◆ *county* SW Kenya
104 H2 **Narón** Galicia, NW Spain *43°31′N 08°08′W*
183 S11 **Narooma** New South Wales, SE Australia *36°16′S 150°08′E*
Narova *see* Narva
Narovlya *see* Narowlya
149 W8 **Nārowāl** Punjab, E Pakistan *32°04′N 74°54′E*
119 N20 **Narowlya** *Rus.* Narovlya. Homyel'skaya Voblasts', SE Belarus *51°48′N 29°30′E*
93 J17 **Närpes** *Fin.* Närpiö. Österbotten, W Finland *62°28′N 21°19′E*
Närpiö *see* Närpes
183 S5 **Narrabri** New South Wales, SE Australia *30°21′S 149°48′E*
183 P9 **Narrandera** New South Wales, SE Australia *34°46′S 146°32′E*
183 Q4 **Narran Lake** ⊚ New South Wales, SE Australia
183 Q4 **Narran River** ⤳ New South Wales/Queensland, SE Australia
180 J13 **Narrogin** Western Australia *32°53′S 117°17′E*
183 Q7 **Narromine** New South Wales, SE Australia *32°16′S 148°15′E*
21 R6 **Narrows** Virginia, NE USA *37°19′N 80°48′W*
196 M15 **Narsarsuaq** ✈ Kujalleo, S Greenland *61°07′N 45°03′W*
154 I10 **Narsimhapur** Madhya Pradesh, C India *22°58′N 79°15′E*
153 U15 **Narsingdi** *var.* Narsinghdi. Dhaka, C Bangladesh *23°56′N 90°40′E*
Narsinghdi *see* Narsingdi
154 H9 **Narsinghgarh** Madhya Pradesh, C India *23°42′N 77°08′E*
163 Q11 **Nart** Nei Mongol Zizhiqu, N China *42°54′N 115°55′E*
Nartës, Gjol i/Nartës, Laguna e *see* Nartës, Liqeni i
113 J22 **Nartës, Liqeni i** *var.* Gjol i Nartës, Laguna e Nartës. ⊚ SW Albania
115 F17 **Narthákí** ▲ C Greece *39°12′N 22°24′E*
127 O15 **Nartkala** Kabardino-Balkarskaya Respublika, SW Russian Federation *43°34′N 43°55′E*
118 K3 **Narva** Ida-Virumaa, NE Estonia *59°23′N 28°12′E*
118 K4 **Narva** *prev.* Narova. ⤳ Estonia/Russian Federation
118 J3 **Narva Bay** *Est.* Narva Laht, *Ger.* Narwa-Bucht, *Rus.* Narvskiy Zaliv. *bay* Estonia/Russian Federation
Narva Laht *see* Narva Bay
124 F13 **Narva Reservoir** *Est.* Narva Veehoidla, *Rus.* Narvskoye Vodokhranilishche. ⊠ Estonia/Russian Federation
Narva Veehoidla *see* Narva Reservoir
92 H10 **Narvik** Nordland, C Norway *68°26′N 17°24′E*
Narvskiy Zaliv *see* Narva Bay
Narvskoye Vodokhranilishche *see* Narva Reservoir
Narwa-Bucht *see* Narva Bay
152 I9 **Narwāna** Haryāna, NW India *29°36′N 76°11′E*
125 R4 **Nar'yan-Mar** *prev.* Beloshchel'ye, Dzerzhinskiy. Nenetskiy Avtonomnyy Okrug, NW Russian Federation *67°38′N 53°E*
122 J12 **Narym** Tomskaya Oblast', C Russian Federation *58°59′N 81°20′E*
Narymskiy Khrebet *see* Khrebet Naryn
147 W9 **Naryn** Narynskaya Oblast', C Kyrgyzstan *41°24′N 76°E*
147 U8 **Naryn** ⤳ Kyrgyzstan/Uzbekistan
145 W16 **Narynkol** *Kaz.* Narynqol. Almaty, SE Kazakhstan *42°45′N 80°12′E*
Naryn Oblasty *see* Narynskaya Oblast'
Narynqol *see* Narynkol
147 V9 **Narynskaya Oblast'** *Kir.* Naryn Oblasty. ◆ *province* C Kyrgyzstan
Naryn Zhotasy *see* Khrebet Naryn
126 J6 **Naryshkino** Orlovskaya Oblast', W Russian Federation *53°00′N 35°41′E*
95 L14 **Nås** Dalarna, C Sweden *60°28′N 14°30′E*
10 J11 **Nass** ⤳ British Columbia, SW Canada
92 G13 **Nasafjellet** *Lapp.* Násávárre. ▲ C Norway *66°29′N 15°23′E*
93 H16 **Näsåker** Västernorrland, C Sweden *63°27′N 16°55′E*
187 Y14 **Nasau** Koro, C Fiji *17°20′S 179°26′E*
190 J13 **Nassau** *island* N Cook Islands
116 I9 **Năsăud** *Ger.* Nussdorf, *Hung.* Naszód. Bistriţa-Năsăud, N Romania *47°16′N 24°24′E*
Násávárre *see* Nasafjellet
103 P13 **Nasbinals** Lozère, S France *44°40′N 03°03′E*
Na Sceirí *see* Skerries
Nase *see* Naze
185 E22 **Naseby** Otago, South Island, New Zealand *45°02′S 170°09′E*
143 R10 **Nāşerīyeh** Kermān, C Iran
25 X5 **Nash** Texas, SW USA *33°26′N 94°04′W*
154 E13 **Nāshik** *prev.* Nāsik. Mahārāshtra, W India *20°05′N 73°48′E*
56 E7 **Nashiño, Río** ⤳ Ecuador/Peru
29 W12 **Nashua** Iowa, C USA *42°57′N 92°32′W*
33 W7 **Nashua** Montana, NW USA *48°06′N 106°16′W*
19 O10 **Nashua** New Hampshire, NE USA *42°45′N 71°26′W*
27 S13 **Nashville** Arkansas, C USA *33°57′N 93°50′W*
23 U7 **Nashville** Georgia, SE USA *31°12′N 83°15′W*
30 L16 **Nashville** Illinois, N USA *38°20′N 89°22′W*
31 O14 **Nashville** Indiana, N USA *39°13′N 86°15′W*
21 V9 **Nashville** North Carolina, SE USA *35°58′N 78°00′W*
20 J8 **Nashville** *state capital* Tennessee, S USA *36°11′N 86°48′W*
20 J9 **Nashville** ✈ Tennessee, S USA *36°06′N 86°44′W*
64 H10 **Nashville Seamount** *undersea feature* NW Atlantic Ocean *30°00′N 57°20′W*
112 H9 **Našice** Osijek-Baranja, E Croatia *45°29′N 18°05′E*
110 M11 **Nasielsk** Mazowieckie, C Poland *52°33′N 20°46′E*
93 K18 **Näsijärvi** ⊚ SW Finland
Nāsik *see* Nāshik
80 G13 **Nasir** Upper Nile, NE South Sudan *08°37′N 33°06′E*
148 K15 **Nasīrābād** Baluchistān, SW Pakistan *26°15′N 62°32′E*
Nasir, Buhayrat/Nâşir,Buḥeiret *see* Nasser, Lake
Nāsiri *see* Ahvāz
Nasiriya *see* An Nāşirīyah
Nāşiriyah, An *see* Dhi Qār
Nás na Ríogh *see* Naas
107 L23 **Naso** Sicilia, Italy, C Mediterranean Sea *38°07′N 14°46′E*
Nasratabad *see* Zābol
77 V15 **Nassarawa** Nassarawa, C Nigeria *08°33′N 07°42′E*
44 H2 **Nassau** ● (The Bahamas) New Providence, N The Bahamas *25°03′N 77°21′W*
23 W8 **Nassau Sound** *sound* Florida, SE USA
108 L7 **Nassereith** Tirol, W Austria *47°19′N 10°51′E*
80 F5 **Nasser, Lake** *var.* Buhayrat Nasir, Buḥayrat Nāşir, Buḥeiret Nâşir. ⊠ Egypt/Sudan
95 L19 **Nässjö** Jönköping, S Sweden *57°39′N 14°40′E*
99 K22 **Nassogne** Luxembourg, SE Belgium *50°08′N 05°19′E*
12 J6 **Nastapoka Islands** *island group* Northwest Territories, C Canada
93 M19 **Nastola** Päijät-Häme, S Finland *60°57′N 25°56′E*
171 O4 **Nasugbu** Luzon, N Philippines *14°03′N 120°39′E*
94 N11 **Näsviken** Gävleborg, C Sweden *61°46′N 16°55′E*
Naszód *see* Năsăud
83 I17 **Nata** Central, NE Botswana *20°11′S 26°10′E*
54 E11 **Natagaima** Tolima, C Colombia *03°38′N 75°07′W*
59 Q14 **Natal** *state capital* Rio Grande do Norte, E Brazil *05°46′S 35°15′W*
168 I11 **Natal** Sumatera, N Indonesia *0°32′N 99°07′E*
Natal *see* KwaZulu/Natal
173 L10 **Natal Basin** *var.* Mozambique Basin. *undersea feature* W Indian Ocean *30°00′S 40°00′E*
25 R12 **Natalia** Texas, SW USA *29°11′N 98°51′W*
67 W15 **Natal Valley** *undersea feature* SW Indian Ocean *31°00′S 33°15′E*
Natanya *see* Netanya
143 O7 **Naţanz** Eşfahān, C Iran *33°31′N 51°55′E*
13 Q11 **Natashquan** Québec, E Canada *50°10′N 61°50′W*
13 Q10 **Natashquan** ⤳ Newfoundland and Labrador/Québec, E Canada
22 J7 **Natchez** Mississippi, S USA *31°34′N 91°24′W*
22 G6 **Natchitoches** Louisiana, S USA *31°45′N 93°05′W*
108 E10 **Naters** Valais, S Switzerland *46°22′N 08°00′E*
Nathanya *see* Netanya
92 O3 **Nathorst Land** *physical region* W Svalbard
186 E9 **National Capital District** ◆ *province* S Papua New Guinea
35 U17 **National City** California, W USA *32°40′N 117°06′W*
184 M10 **National Park** Manawatu-Wanganui, North Island, New Zealand *39°11′S 175°22′E*
77 R14 **Natitingou** NW Benin *10°21′N 01°26′E*
40 B5 **Natividad, Isla** *island* NW Mexico
165 Q10 **Natori** Miyagi, Honshū, C Japan *38°12′N 140°51′E*
18 C14 **Natrona Heights** Pennsylvania, NE USA *40°37′N 79°42′W*
81 H20 **Natron, Lake** ⊚ Kenya/Tanzania
Natsrat *see* Natzrat
166 L7 **Nattalin** Bago, C Myanmar (Burma) *18°25′N 95°34′E*
92 J12 **Nattavaara** *Lapp.* Nahtavárr. Norrbotten, N Sweden *66°45′N 20°58′E*
109 S3 **Natternbach** Oberösterreich, N Austria *48°26′N 13°44′E*
95 M22 **Nättraby** Blekinge, S Sweden *56°12′N 15°30′E*
169 P10 **Natuna Besar, Pulau** *island* Kepulauan Natuna, W Indonesia
Natuna Islands *see* Natuna, Kepulauan
169 O9 **Natuna, Kepulauan** *var.* Natuna Islands. *island group* W Indonesia
169 N9 **Natuna, Laut** *Eng.* Natuna Sea. *sea* W Indonesia
Natuna Sea *see* Natuna, Laut
21 N6 **Natural Bridge** *tourist site* Kentucky, C USA
173 V11 **Naturaliste Fracture Zone** *tectonic feature* E Indian Ocean
174 J10 **Naturaliste Plateau** *undersea feature* E Indian Ocean
138 G9 **Natzrat** *var.* Natsrat, *Ar.* En Nazira, *Eng.* Nazareth; *prev.* Nazerat. Northern, N Israel *32°42′N 35°18′E*
Nau *see* Nov
103 O14 **Naucelle** Aveyron, S France *44°10′N 02°19′E*
83 D20 **Nauchas** Hardap, C Namibia *23°40′S 16°19′E*
108 K9 **Nauders** Tirol, W Austria *46°52′N 10°31′E*
Naugard *see* Nowogard
118 F12 **Naujamiestis** Panevėžys, C Lithuania *55°42′N 24°10′E*
118 E10 **Naujoji Akmenė** Šiaulai, NW Lithuania *56°20′N 22°57′E*
149 R16 **Naukot** *var.* Naokot. Sind, SE Pakistan *24°52′N 69°27′E*
101 L16 **Naumburg** *var.* Naumburg an der Saale. Sachsen-Anhalt, C Germany *51°09′N 11°48′E*
Naumburg am Queis *see* Nowogrodziec
Naumburg an der Saale *see* Naumburg
191 W15 **Naunau** *ancient monument* Easter Island, Chile, E Pacific Ocean
138 G10 **Nāʻūr** ʻAmmān, W Jordan *31°52′N 35°50′E*
189 Q8 **Nauru** *off.* Republic of Nauru; *prev.* Pleasant Island. ◆ *republic* W Pacific Ocean
175 P5 **Nauru** *island* W Pacific Ocean
189 Q9 **Nauru International** ✈ S Nauru
Nauru, Republic of *see* Nauru
Nausari *see* Navsāri
19 Q12 **Nauset Beach** *beach* Massachusetts, NE USA
Naushahra *see* Nowshera
149 P14 **Naushahro Firoz** Sind, SE Pakistan *26°51′N 68°11′E*
Naushara *see* Nowshera
187 X14 **Nausori** Viti Levu, W Fiji *18°01′S 178°31′E*
56 F9 **Nauta** Loreto, N Peru *04°31′S 73°36′W*
153 O12 **Nautanwa** Uttar Pradesh, N India *27°26′N 83°25′E*
41 R13 **Nautla** Veracruz-Llave, E Mexico *20°15′N 96°45′W*
Nauzad *see* Now Zād
41 N6 **Nava** Coahuila, NE Mexico *28°28′N 100°45′W*
Navabad *see* Navobod
104 L6 **Nava del Rey** Castilla y León, N Spain *41°19′N 05°04′W*
153 S15 **Navadwīp** *prev.* Nabadwip. West Bengal, NE India *23°24′N 88°23′E*
104 M9 **Navahermosa** Castilla-La Mancha, C Spain *39°39′N 04°25′W*
119 I16 **Navahrudak** *Pol.* Nowogródek, *Rus.* Novogrudok. Hrodzyenskaya Voblasts', W Belarus *53°36′N 25°50′E*
119 I16 **Navahrudskaye Wzvyshsha** *Rus.* Novogrudskaya Vozvyshennost'. ▲ W Belarus
36 M8 **Navajo Mount** ▲ Utah, W USA *37°00′N 110°52′W*
37 Q9 **Navajo Reservoir** ⊠ New Mexico, SW USA
104 K9 **Navalmoral de la Mata** Extremadura, W Spain *39°54′N 05°33′W*
104 K10 **Navalvillar de Pelea** Extremadura, W Spain *39°05′N 05°27′W*
97 F17 **Navan** *Ir.* An Uaimh. E Ireland *53°39′N 06°41′W*
Navanagar *see* Jāmnagar
118 L12 **Navapolatsk** *Rus.* Novopolotsk. Vitsyebskaya Voblasts', N Belarus *55°34′N 28°35′E*
149 P6 **Nāvar, Dasht-e** *Pash.* Dasht-i-Nawar. *desert* C Afghanistan
123 W6 **Navarin, Mys** *headland* NE Russian Federation *62°18′N 179°06′E*
63 I25 **Navarino, Isla** *island* S Chile
105 Q4 **Navarra** *Eng./Fr.* Navarre. ◆ *autonomous community* N Spain
Navarre *see* Navarra
105 P4 **Navarrete** La Rioja, N Spain *42°26′N 02°34′W*
61 C20 **Navarro** Buenos Aires, E Argentina *35°00′S 59°15′W*
105 O12 **Navas de San Juan** Andalucía, S Spain *38°11′N 03°19′W*
25 V10 **Navasota** Texas, SW USA *30°23′N 96°05′W*
25 U9 **Navasota River** ⤳ Texas, SW USA
44 I9 **Navassa Island** ◇ US *unincorporated territory* C West Indies
119 L19 **Navasyolki** *Rus.* Novosëlki. Homyel'skaya Voblasts', SE Belarus *52°24′N 28°33′E*
119 H17 **Navayel'nya** *Pol.* Nowojelnia, *Rus.* Novoyel'nya. Hrodzyenskaya Voblasts', W Belarus *53°28′N 25°35′E*
171 Y13 **Naver** Papua, E Indonesia *03°27′S 139°45′E*
118 H5 **Navesti** ⤳ C Estonia
104 J2 **Navia** Asturias, N Spain *43°33′N 06°43′W*
104 J2 **Navia** ⤳ NW Spain
59 I21 **Naviraí** Mato Grosso do Sul, SW Brazil *23°01′S 54°09′W*
126 I6 **Navlya** Bryanskaya Oblast', W Russian Federation *52°47′N 34°28′E*
187 X13 **Navoalevu** Vanua Levu, N Fiji *16°22′S 179°28′E*
147 P13 **Navobod** *Rus.* Navabad. W Tajikistan *38°37′N 68°42′E*
147 R12 **Navobod** *Rus.* Navabad, Novabad. C Tajikistan *39°00′N 70°06′E*
Navoi *see* Navoiy
146 M11 **Navoiy** *Rus.* Navoi. Navoiy Viloyati, C Uzbekistan *40°05′N 65°23′E*
Navoiyskaya Oblast' *see* Navoiy Viloyati
146 K8 **Navoiy Viloyati** *Rus.* Navoiyskaya Oblast'. ◆ *province* N Uzbekistan
40 G7 **Navojoa** Sonora, NW Mexico *27°04′N 109°28′W*
Navolat *see* Navolato
40 H9 **Navolato** *var.* Navolat. Sinaloa, C Mexico *24°46′N 107°42′W*
187 Q13 **Navonda** Ambae, C Vanuatu *15°21′S 167°58′E*
Návpaktos *see* Náfpaktos
Návplion *see* Náfplio
77 P14 **Navrongo** N Ghana *10°51′N 01°03′W*
154 D12 **Navsāri** *var.* Nausari. Gujarāt, W India *20°55′N 72°55′E*
187 X15 **Navua** Viti Levu, W Fiji *18°15′S 178°10′E*
138 H8 **Nawá** Darʻā, S Syria *32°53′N 36°03′E*
Nawabashah *see* Nawābshāh
153 S14 **Nawabganj** Rajshahi, NW Bangladesh *24°35′N 88°21′E*
153 N12 **Nawābganj** Uttar Pradesh, N India *26°52′N 82°09′E*
149 Q15 **Nawābshāh** *var.* Nawabashah. Sind, S Pakistan *26°15′N 68°26′E*
153 P14 **Nawāda** Bihār, N India *24°54′N 85°33′E*
152 H11 **Nawalgarh** Rājasthān, N India *27°48′N 75°21′E*
Nawāl, Sabkhat an *see* Noual, Sebkhet en
Nawar, Dasht-i- *see* Nāvar, Dasht-e
167 N4 **Nawnghkio** *var.* Nawngkio. Shan State, C Myanmar (Burma) *22°17′N 96°50′E*
Nawngkio *see* Nawnghkio
137 U13 **Naxçıvan** *Rus.* Nakhichevan'. SW Azerbaijan *39°14′N 45°24′E*
160 I10 **Naxi** Sichuan, C China *28°50′N 105°20′E*
115 K21 **Náxos** *var.* Naxos. Náxos, Kykládes, Greece, Aegean Sea *37°07′N 25°24′E*
115 K21 **Náxos** *island* Kykládes, Greece, Aegean Sea
40 J11 **Nayarit** ◆ *state* C Mexico
187 Y14 **Nayau** *island* Lau Group, E Fiji
143 S8 **Nāy Band** Khorāsān-e Jonūbī, E Iran *32°26′N 57°30′E*
165 T2 **Nayoro** Hokkaidō, NE Japan *44°22′N 142°27′E*
166 M7 **Nay Pyi Taw** ◆ *union territory* C Myanmar (Burma)
104 F9 **Nazaré** *var.* Nazare. Leiria, C Portugal *39°36′N 09°04′W*
Nazare *see* Nazaré
24 M4 **Nazareth** Texas, SW USA *34°32′N 102°06′W*
Nazareth *see* Natzrat
173 O8 **Nazareth Bank** *undersea feature* W Indian Ocean
40 K9 **Nazas** Durango, C Mexico *25°15′N 104°06′W*
57 F16 **Nazca** Ica, S Peru *14°53′S 74°54′W*
0 L17 **Nazca Plate** *tectonic feature*
193 U9 **Nazca Ridge** *undersea feature* E Pacific Ocean *22°00′S 82°00′W*
165 V15 **Naze** *var.* Nase. Kagoshima, Amami-ōshima, SW Japan *28°21′N 129°30′E*
Nazerat *see* Natzrat
137 R14 **Nazik Gölü** ⊚ E Turkey
136 C15 **Nazilli** Aydın, SW Turkey *37°55′N 28°20′E*
137 P14 **Nazımiye** Tunceli, E Turkey *39°12′N 39°51′E*
Nazinon *see* Red Volta
10 L15 **Nazko** British Columbia, SW Canada *52°57′N 123°44′W*
127 O16 **Nazran'** Respublika Ingushetiya, SW Russian Federation *43°14′N 44°47′E*
80 J13 **Nazrēt** *var.* Adama, Hadama. Oromīya, C Ethiopia *08°31′N 39°20′E*
Nazwāh *see* Nizwa
82 J13 **Nchanga** Copperbelt, C Zambia *12°30′S 27°53′E*
82 J11 **Nchelenge** Luapula, N Zambia *09°20′S 28°50′E*
Ncheu *see* Ntcheu
83 J25 **Nciba** *Eng.* Great Kei; *prev.* Groot-Kei. ⤳ S South Africa
Ndaghamcha, Sebkra de *see* Te-n-Dghâmcha, Sebkhet
81 G21 **Ndala** Tabora, C Tanzania *04°45′S 33°15′E*
82 B11 **N'Dalatando** *Port.* Salazar, Vila Salazar. Kwanza Norte, NW Angola *09°19′S 14°48′E*
77 S14 **Ndali** C Benin *09°50′N 02°46′E*
81 E18 **Ndeke** SW Uganda *0°11′S 30°04′E*
78 J13 **Ndélé** Bamingui-Bangoran, N Central African Republic *08°24′N 20°41′E*
79 E19 **Ndendé** Ngounié, S Gabon *02°21′S 11°20′E*
79 E20 **Ndindi** Nyanga, S Gabon *03°47′S 11°06′E*
78 G11 **N'Djaména** *var.* Ndjamena; *prev.* Fort-Lamy. ● (Chad) Ville de N'Djaména, W Chad *12°08′N 15°02′E*
78 G11 **N'Djaména** ✈ Ville de N'Djaména, W Chad *12°09′N 15°00′E*
Ndjamena *see* N'Djaména
N'Djaména, Région de la Ville de *see* N'Djaména, Ville de
78 G11 **N'Djaména, Ville de** ◆ *region* SW Chad
79 D18 **Ndjolé** Moyen-Ogooué, W Gabon *0°07′S 10°45′E*
82 J13 **Ndola** Copperbelt, C Zambia *12°59′S 28°35′E*
Ndrhamcha, Sebkha de *see* Te-n-Dghâmcha, Sebkhet
79 L15 **Ndu** Orientale, N Dem. Rep. Congo *04°36′N 22°49′E*
81 H21 **Nduguti** Singida, C Tanzania *04°19′S 34°40′E*
186 M9 **Nduindui** Guadalcanal, C Solomon Islands *09°46′S 159°54′E*
Nduke *see* Kolombangara
Ndzouani *see* Nzwani
115 F16 **Néa Anchíalos** *var.* Nea Anhialos, Néa Ankhíalos. Thessalía, C Greece *39°16′N 22°49′E*
Nea Anhialos/Néa Ankhíalos *see* Néa Anchíalos
115 H18 **Néa Artáki** Évvoia, C Greece *38°31′N 23°39′E*
97 F15 **Neagh, Lough** ⊚ E Northern Ireland, United Kingdom
32 F7 **Neah Bay** Washington, NW USA *48°21′N 124°39′W*
115 J22 **Nea Kaméni** *island* Kykládes, Greece, Aegean Sea
181 O8 **Neale, Lake** ⊚ Northern Territory, C Australia
182 G2 **Neales River** *seasonal river* South Australia
115 G14 **Néa Moudanía** *var.* Néa Moudhaniá. Kentrikí Makedonía, N Greece *40°14′N 23°17′E*
Néa Moudhaniá *see* Néa Moudanía
116 K10 **Neamţ** ◆ *county* NE Romania
Neapel *see* Napoli
115 D14 **Neápoli** *prev.* Neápolis. Dytikí Makedonía, N Greece *40°19′N 21°23′E*
115 K25 **Neápoli** Kríti, Greece, E Mediterranean Sea *35°15′N 25°37′E*
115 G22 **Neápoli** Pelopónnisos, S Greece *36°29′N 23°05′E*
Neápolis *see* Neápoli, Greece
Neapolis *see* Napoli, Italy
Neapolis *see* Nablus, West Bank
38 D16 **Near Islands** *island group* Aleutian Islands, Alaska, USA
97 J21 **Neath** S Wales, United Kingdom *51°40′N 03°48′W*
114 H13 **Néa Zíchni** *var.* Néa Zíkhni; *prev.* Néa Zíkhna. Kentrikí Makedonía, NE Greece *41°02′N 23°50′E*
Néa Zíkhna/Néa Zíkhni *see* Néa Zíchni
42 C5 **Nebaj** Quiché, W Guatemala *15°25′N 91°05′W*
77 P13 **Nebbou** S Burkina Faso *11°22′N 01°49′W*
Nebitdag *see* Balkanabat
54 M13 **Neblina, Pico da** ▲ NW Brazil *0°49′N 66°31′W*
124 I13 **Nebolchi** Novgorodskaya Oblast', W Russian Federation *59°08′N 33°19′E*
36 L4 **Nebo, Mount** ▲ Utah, W USA *39°47′N 111°46′W*
28 L14 **Nebraska** *off.* State of Nebraska, *also known as* Blackwater State, Cornhusker State, Tree Planters State. ◆ *state* C USA
29 S16 **Nebraska City** Nebraska, C USA *40°38′N 95°52′W*
107 K23 **Nebrodi, Monti** *var.* Monti Caronie. ▲ Sicilia, Italy, C Mediterranean Sea
10 L14 **Nechako** ⤳ British Columbia, SW Canada
29 Q2 **Neche** North Dakota, N USA *48°57′N 97°33′W*
25 V8 **Neches** Texas, SW USA *31°51′N 95°28′W*
25 W8 **Neches River** ⤳ Texas, SW USA
101 H20 **Neckar** ⤳ SW Germany
101 H20 **Neckarsulm** Baden-Württemberg, SW Germany *49°12′N 09°13′E*
192 L5 **Necker Island** *island* C British Virgin Islands
175 U3 **Necker Ridge** *undersea feature* N Pacific Ocean
61 D23 **Necochea** Buenos Aires, E Argentina *38°34′S 58°42′W*
104 H2 **Neda** Galicia, NW Spain *43°29′N 08°09′W*
115 E20 **Néda** *var.* Nédas. ⤳ S Greece
Nédas *see* Néda
114 J12 **Nedelino** Smolyan, S Bulgaria *41°27′N 25°05′E*
25 Y11 **Nederland** Texas, SW USA *29°58′N 93°59′W*
Nederland *see* Netherlands
98 K12 **Neder Rijn** *Eng.* Lower Rhine. ⤳ C Netherlands

◆ Country ● Country Capital ◇ Dependent Territory ○ Dependent Territory Capital ◆ Administrative Regions ✈ International Airport ▲ Mountain ▲ Mountain Range Volcano ⤳ River ⊚ Lake ⊠ Reservoir

NEDERWEERT – NEW TOWN

99 L16 **Nederweert** Limburg, SE Netherlands 51°17′N 05°45′E

95 G16 **Nedre Tokke** ◎ S Norway

Nedrigaylov see Nedryhayliv

117 S3 **Nedryhayliv** Rus. Nedrigaylov. Sums'ka Oblast', NE Ukraine 50°51′N 33°54′E

98 O11 **Neede** Gelderland, E Netherlands 52°08′N 06°36′E

33 T13 **Needle Mountain** ▲ Wyoming, C USA 44°03′N 109°33′W

35 Y14 **Needles** California, W USA 34°50′N 114°37′W

97 M24 **Needles, The** rocks S England, United Kingdom

62 O7 **Ñeembucú** off. Departamento de Ñeembucú. ◐ department SW Paraguay

Ñeembucú, Departamento de see Ñeembucú

30 M7 **Neenah** Wisconsin, N USA 44°09′N 88°26′W

11 W16 **Neepawa** Manitoba, S Canada 50°14′N 99°29′W

99 K16 **Neerpelt** Limburg, NE Belgium 51°13′N 05°26′E

74 M6 **Nefta** ✈ W Tunisia 34°03′N 08°05′E

126 L15 **Neftegorsk** Krasnodarskiy Kray, SW Russian Federation 44°21′N 39°40′E

127 U3 **Neftekamsk** Respublika Bashkortostan, W Russian Federation 56°07′N 54°13′E

127 O14 **Neftekumsk** Stavropol'skiy Kray, SW Russian Federation 44°45′N 45°00′E

Neftezavodsk see Seýdi

82 C10 **Negage** var. N'Gage. Uíge, NW Angola 07°47′S 15°27′E

Negapatam/Negapattinam see Nāgappattinam

169 T17 **Negara** Bali, Indonesia 08°21′S 114°35′E

169 T13 **Negara** Borneo, C Indonesia 02°40′S 115°05′E

Negara Brunei Darussalam see Brunei

31 N4 **Negaunee** Michigan, N USA 46°30′N 87°36′W

81 J15 **Negēlē** var. Negelli, It. Neghelli. Oromīya, C Ethiopia 05°13′N 39°43′E

Negelli see Negēlē

Negeri Pahang Darul Makmur see Pahang

Negeri Selangor Darul Ehsan see Selangor

168 K9 **Negeri Sembilan** var. Negri Sembilan. ◐ state Peninsular Malaysia

92 P3 **Negerpynten** headland S Svalbard 77°15′N 22°40′E

Negev see HaNegev

Neghelli see Negēlē

116 I12 **Negoiu** var. Negoiul. ▲ S Romania 45°34′N 24°34′E

Negoiul see Negoiu

82 P13 **Negomane** var. Negomano. Cabo Delgado, N Mozambique 11°22′S 38°32′E

Negomano see Negomane

155 J25 **Negombo** Western Province, SW Sri Lanka 07°13′N 79°51′E

191 W11 **Negonego** prev. Nengonengo. atoll Îles Tuamotu, C French Polynesia

Negoreloye see Nyeharelaye

112 P12 **Negotin** Serbia, E Serbia 44°14′N 22°31′E

113 P19 **Negotino** C Macedonia 41°29′N 22°04′E

56 A10 **Negra, Punto** headland NW Peru 06°03′S 81°08′W

104 G3 **Negreira** Galicia, NW Spain 42°54′N 08°46′W

116 L10 **Negreşti** Vaslui, E Romania 46°50′N 27°28′E

Negreşti see Negreşti-Oaş

116 H8 **Negreşti-Oaş** Hung. Avasfelsőfalu; prev. Negreşti. Satu Mare, NE Romania 47°56′N 23°22′E

44 H12 **Negril** W Jamaica 18°16′N 78°21′W

Negri Sembilan see Negeri Sembilan

63 K15 **Negro, Río** ≈ E Argentina

62 N7 **Negro, Río** ≈ NE Argentina

57 N17 **Negro, Río** ≈ E Bolivia

48 F6 **Negro, Rio** ≈ N South America

61 E18 **Negro, Río** ≈ Brazil/Uruguay

62 O5 **Negro, Río** ≈ C Paraguay

Negro, Río see Chixoy, Río, Guatemala/Mexico

Negro, Río see Sico Tinto, Río, Honduras

171 P6 **Negros** island C Philippines

116 M15 **Negru Vodă** Constanţa, SE Romania 43°49′N 28°12′E

13 P13 **Neguac** New Brunswick, SE Canada 47°16′N 65°04′W

14 B7 **Negwazu, Lake** ◎ Ontario, S Canada

Négyfalu see Săcele

32 F10 **Nehalem** Oregon, NW USA 45°42′N 123°55′W

32 F10 **Nehalem River** ≈ Oregon, NW USA

Nehavend see Nahāvand

143 V9 **Nehbandān** Khorāsān-e Jonūbī, E Iran 31°00′N 60°00′E

163 V6 **Nehe** Heilongjiang, NE China 48°28′N 124°52′E

193 Y14 **Neiafu** 'Uta Vava'u, N Tonga 18°36′S 173°58′W

45 N9 **Neiba** var. Neyba. SW Dominican Republic 18°31′N 71°25′W

Néid, Carn Uí see Mizen Head

92 M9 **Neiden** Finnmark, N Norway 69°41′N 29°23′E

Neidín see Kenmare

Néifinn see Nephin

103 S10 **Neige, Crêt de la** ▲ E France 46°18′N 05°58′E

173 O16 **Neiges, Piton des** ▲ C Réunion 21°05′S 55°28′E

15 R9 **Neiges, Rivière des** ≈ Québec, SE Canada

160 I10 **Neijiang** Sichuan, C China 29°32′N 105°03′E

30 K6 **Neillsville** Wisconsin, N USA 44°34′N 90°36′W

Nei Monggol Zizhiqu/Nei Mongol see Nei Mongol Zizhiqu

163 Q10 **Nei Mongol Gaoyuan** plateau NE China

163 O12 **Nei Mongol Zizhiqu** var. Nei Mongol, Eng. Inner Mongolia, Inner Mongolian Autonomous Region; prev. Nei Monggol Zizhiqu. ◐ autonomous region N China

161 O4 **Neiqiu** Hebei, E China 37°22′N 114°34′E

Neiriz see Neyrīz

101 Q16 **Neisse** Pol. Nisa Cz. Lužická Nisa, Ger. Lausitzer Neisse, Nysa Łużycka. ≈ C Europe

Neisse see Nysa

54 E11 **Neiva** Huila, S Colombia 02°58′N 75°15′W

160 M7 **Neixiang** Henan, C China 33°08′N 111°50′E

Nejafabad see Najafābād

11 V9 **Nejanilini Lake** ◎ Manitoba, C Canada

Nejd see Najd

80 I13 **Nek'emtē** var. Lakemti, Nakamti. Oromīya, C Ethiopia 09°06′N 36°31′E

126 M9 **Nekhayevskaya** Volgogradskaya Oblast', SW Russian Federation 50°25′N 41°44′E

30 K7 **Nekoosa** Wisconsin, N USA 44°19′N 89°54′W

Neksø Bornholm see Nexø

104 H7 **Nelas** Viseu, N Portugal 40°32′N 07°52′W

124 H16 **Nelidovo** Tverskaya Oblast', W Russian Federation 56°13′N 32°45′E

29 P13 **Neligh** Nebraska, C USA 42°07′N 98°01′W

123 R11 **Nel'kan** Khabarovskiy Kray, E Russian Federation 57°44′N 136°09′E

92 M10 **Nellim** var. Nellimö, Lapp. Njellim. Lappi, N Finland 68°49′N 28°18′E

Nellimö see Nellim

155 J18 **Nellore** Andhra Pradesh, E India 14°29′N 80°E

61 B17 **Nelson** Santa Fe, C Argentina 31°16′S 60°45′W

11 O17 **Nelson** British Columbia, SW Canada 49°29′N 117°17′W

185 I14 **Nelson** Nelson, South Island, New Zealand 41°17′S 173°17′E

97 L17 **Nelson** NW England, United Kingdom 53°51′N 02°13′W

29 P17 **Nelson** Nebraska, C USA 40°12′N 98°04′W

185 J14 **Nelson** ◐ unitary authority South Island, New Zealand

11 X12 **Nelson** ≈ Manitoba, C Canada

183 U8 **Nelson Bay** New South Wales, SE Australia 32°48′S 152°10′E

182 K13 **Nelson, Cape** headland Victoria, SE Australia 38°25′S 141°33′E

63 G23 **Nelson, Estrecho** strait SE Pacific Ocean

11 W12 **Nelson House** Manitoba, C Canada 55°49′N 98°51′W

30 J4 **Nelson Lake** ◎ Wisconsin, N USA

31 T14 **Nelsonville** Ohio, N USA 39°27′N 82°13′W

27 S2 **Nelsoon River** ≈ Iowa/Missouri, C USA

83 K21 **Nelspruit** Mpumalanga, NE South Africa 25°28′S 30°58′E

Nelspruit see Mbombela

76 L10 **Néma** Hodh ech Chargui, SE Mauritania 16°32′N 07°12′W

118 D13 **Neman** Ger. Ragnit. Kaliningradskaya Oblast', W Russian Federation 55°01′N 22°00′E

84 I9 **Neman** Bel. Nyoman, Ger. Memel, Lith. Nemunas, Pol. Niemen. ≈ NE Europe

Nemausus see Nîmes

115 F19 **Neméa** Pelopónnisos, S Greece 37°49′N 22°40′E

Německý Brod see Havlíčkův Brod

14 D7 **Nemegosenda** ≈ Ontario, S Canada

14 D8 **Nemegosenda Lake** ◎ Ontario, S Canada

119 H14 **Nemencinė** Vilnius, SE Lithuania 54°50′N 25°29′E

Nemetocenna see Arras

Nemirov see Nemyriv

103 O6 **Nemours** Seine-et-Marne, N France 48°16′N 02°41′E

Nemunas see Neman

165 W4 **Nemuro** Hokkaidō, NE Japan 43°20′N 145°35′E

165 W4 **Nemuro-hantō** peninsula Hokkaidō, NE Japan

165 W3 **Nemuro-kaikyō** strait Japan/Russian Federation

165 W4 **Nemuro-wan** bay N Japan

116 H5 **Nemyriv** Rus. Nemirov. L'vivs'ka Oblast', NW Ukraine 50°08′N 23°28′E

117 N7 **Nemyriv** Rus. Nemirov. Vinnyts'ka Oblast', C Ukraine 48°58′N 28°50′E

97 D19 **Nenagh** Ir. An tAonach. Tipperary, C Ireland 52°52′N 08°12′W

39 R9 **Nenana** Alaska, USA 64°33′N 149°05′W

39 R9 **Nenana River** ≈ Alaska, USA

187 P10 **Nendö** var. Swallow Island. island Santa Cruz Islands, E Solomon Islands

97 O19 **Nene** ≈ E England, United Kingdom

125 R4 **Nenetskiy Avtonomnyy Okrug** ◐ autonomous district Arkhangel'skaya Oblast', NW Russian Federation

Nengonengo see Negonego

163 V6 **Nenjiang** Heilongjiang, NE China 49°11′N 125°18′E

163 U6 **Nen Jiang** var. Nonni. ≈ NE China

189 P16 **Neoch** atoll Caroline Islands, C Micronesia

115 D18 **Neochóri** Dytikí Elláda, C Greece 38°23′N 21°14′E

27 Q7 **Neodesha** Kansas, C USA 37°25′N 95°40′W

29 S14 **Neola** Iowa, C USA 41°27′N 95°40′W

115 E16 **Néo Monastíri** var. Néon Monastiri. Thessalía, C Greece 39°22′N 21°55′E

Néon Karlovasi/Néon Karlovásion see Karlovási

Néon Monastiri see Néo Monastíri

27 R8 **Neosho** Missouri, C USA 36°53′N 94°24′W

27 Q7 **Neosho River** ≈ Kansas/Oklahoma, C USA

123 N12 **Nepa** ≈ C Russian Federation

153 N10 **Nepal** off. Nepal. ◆ monarchy S Asia

Nepal see Nepal

152 M11 **Nepālganj** Mid Western, SW Nepal 28°04′N 81°37′E

14 L13 **Nepean** Ontario, SE Canada 45°19′N 75°54′W

36 L4 **Nephi** Utah, W USA 39°43′N 111°50′W

97 B16 **Nephin** Ir. Néifinn. ▲ W Ireland 54°00′N 09°21′W

67 T9 **Nepoko** ≈ NE Dem. Rep. Congo

18 K15 **Neptune** New Jersey, NE USA 40°10′N 74°03′W

182 G10 **Neptune Islands** island group South Australia

107 I14 **Nera** anc. Nar. ≈ C Italy

102 L14 **Nérac** Lot-et-Garonne, SW France 44°08′N 00°21′E

111 D16 **Neratovice** Ger. Neratowitz. Středocesky Kraj, C Czech Republic 50°16′N 14°31′E

Neratowitz see Neratovice

123 O13 **Nercha** ≈ S Russian Federation

123 O13 **Nerchinsk** Zabaykal'skiy Kray, S Russian Federation 52°01′N 116°25′E

123 P14 **Nerchinskiy Zavod** Zabaykal'skiy Kray, S Russian Federation 51°13′N 119°25′E

124 M15 **Nerekhta** Kostromskaya Oblast', NW Russian Federation 57°27′N 40°33′E

118 H10 **Nereta** S Latvia 56°12′N 25°18′E

106 K13 **Nereto** Abruzzo, C Italy 42°49′N 13°50′E

113 H15 **Neretva** ≈ Bosnia and Herzegovina/Croatia

115 C17 **Nerikós** ruins Lefkáda, Iónia Nísiá, Greece, C Mediterranean Sea

83 F15 **Neriquinha** Kuando Kubango, SE Angola 15°44′S 21°34′E

118 I13 **Neris** Bel. Viliya, Pol. Wilia; prev. Pol. Wilja. ≈ Belarus/Lithuania

Neris see Viliya

105 N15 **Nerja** Andalucía, S Spain 36°45′N 03°35′W

124 L16 **Nerl'** ≈ W Russian Federation

105 P12 **Nerpio** Castilla-La Mancha, C Spain 38°08′N 02°18′W

104 J13 **Nerva** Andalucía, S Spain 37°40′N 06°31′W

98 L4 **Nes** Fryslân, N Netherlands 53°28′N 05°46′E

94 G13 **Nesbyen** Buskerud, S Norway 60°36′N 09°35′E

114 M9 **Nesebar** var. Nesebŭr. Burgas, E Bulgaria 42°40′N 27°43′E

Nesebŭr see Nesebar

Neshcherdo, Ozero see Nyeshcharda, Vozyera

92 L2 **Neskaupstaður** Austurland, E Iceland 65°08′N 13°45′W

92 F13 **Nesna** Nordland, C Norway 66°11′N 12°54′E

26 K5 **Ness City** Kansas, C USA 38°27′N 99°54′W

Nesselsdorf see Kopřivnice

108 H7 **Nesslau** Sankt Gallen, NE Switzerland 47°13′N 09°12′E

96 I9 **Ness, Loch** ◎ N Scotland, United Kingdom

Nesterov see Zhovkva

114 I12 **Néstos** Bul. Mesta, Turk. Kara Su. ≈ Bulgaria/Greece see also Mesta

Néstos see Mesta

95 C14 **Nesttun** Hordaland, S Norway 60°19′N 05°16′E

Nesvizh see Nyasvizh

138 F9 **Netanya** var. Natanya, Nathanya. Central, C Israel 32°20′N 34°51′E

98 I9 **Netherlands** off. Kingdom of the Netherlands, var. Holland, Dut. Koninkrijk der Nederlanden, Nederland. ◆ monarchy NW Europe

Netherlands East Indies see Indonesia

Netherlands Guiana see Suriname

Netherlands, Kingdom of the see Netherlands

Netherlands New Guinea see Papua

116 L4 **Netishyn** Khmel'nyts'ka Oblast', W Ukraine 50°20′N 26°38′E

138 E11 **Netivot** Southern, S Israel 31°26′N 34°36′E

107 O21 **Neto** ≈ S Italy

9 Q6 **Nettilling Lake** ◎ Baffin Island, Nunavut, N Canada

29 V3 **Nett Lake** ◎ Minnesota, N USA

107 I16 **Nettuno** Lazio, C Italy 41°27′N 12°40′E

Netum see Noto

41 U16 **Netzahualcóyotl, Presa** ⊠ SE Mexico

Netze see Noteć

Neu Amerika see Puławy

Neubetsche see Novi Bečej

Neubidschow see Nový Bydžov

Neubistritz see Nová Bystřice

100 N9 **Neubrandenburg** Mecklenburg-Vorpommern, NE Germany 53°33′N 13°16′E

101 K22 **Neuburg an der Donau** Bayern, S Germany 48°43′N 11°10′E

108 C8 **Neuchâtel** Ger. Neuenburg. Neuchâtel, W Switzerland 46°59′N 06°55′E

108 C8 **Neuchâtel** Ger. Neuenburg. ◐ canton W Switzerland

108 C8 **Neuchâtel, Lac de** Ger. Neuenburger See. ◎ W Switzerland

Neudorf see Spišská Nová Ves

100 L10 **Neue Elde** canal N Germany

Neuenburg see Neuchâtel

Neuenburg an der Elbe see Nymburk

Neuenburger See see Neuchâtel, Lac de

108 F7 **Neuenhof** Aargau, N Switzerland 47°27′N 08°17′E

100 H11 **Neuenland** ✈ (Bremen) Bremen, NW Germany 53°03′N 08°46′E

Neuenstadt see La Neuveville

101 C18 **Neuerburg** Rheinland-Pfalz, W Germany 50°01′N 06°13′E

99 K24 **Neufchâteau** Luxembourg, SE Belgium 49°50′N 05°26′E

103 S6 **Neufchâteau** Vosges, NE France 48°21′N 05°42′E

102 M3 **Neufchâtel-en-Bray** Seine-Maritime, N France 49°44′N 01°26′E

109 S3 **Neufelden** Oberösterreich, N Austria 48°27′N 14°01′E

Neugradisk see Nova Gradiška

Neuhaus see Jindřichův Hradec

Neuhäusel see Nové Zámky

108 G6 **Neuhausen** var. Neuhausen am Rheinfall. Schaffhausen, N Switzerland 47°24′N 08°37′E

Neuhausen am Rheinfall see Neuhausen

101 I17 **Neuhof** Hessen, C Germany 50°26′N 09°34′E

Neuhof see Zgierz

Neukuhren see Pionerskiy

Neu-Langenburg see Tukuyu

109 W4 **Neulengbach** Niederösterreich, NE Austria 48°10′N 15°53′E

113 G15 **Neum** Federacija Bosne I Hercegovine, S Bosnia and Herzegovina 42°57′N 17°33′E

Neumark see Nowy Targ, Małopolskie, Poland

Neumark see Nowe Miasto Lubawskie, Warmińsko-Mazurskie, Poland

Neumarkt see Neumarkt im Hausruckkreis, Oberösterreich, Austria

Neumarkt see Neumarkt am Wallersee, Salzburg, Austria

Neumarkt see Sroda Śląska, Dolnośląskie, Poland

Neumarkt see Târgu Secuiesc, Covasna, Romania

Neumarkt see Târgu Mureş

109 Q5 **Neumarkt am Wallersee** var. Neumarkt. Salzburg, NW Austria 47°55′N 13°16′E

109 R4 **Neumarkt im Hausruckkreis** var. Neumarkt. Oberösterreich, N Austria 48°16′N 13°40′E

101 L20 **Neumarkt in der Oberpfalz** Bayern, SE Germany 49°16′N 11°28′E

Neumarktl see Tržič

Neumoldowa see Moldova Nouă

100 J8 **Neumünster** Schleswig-Holstein, N Germany 54°04′N 09°59′E

109 X5 **Neunkirchen** var. Neunkirchen am Steinfeld. Niederösterreich, E Austria 47°44′N 16°05′E

101 E20 **Neunkirchen** Saarland, SW Germany 49°21′N 07°11′E

Neunkirchen am Steinfeld see Neunkirchen

Neuoderberg see Bohumín

63 I15 **Neuquén** Neuquén, SE Argentina 39°03′S 68°36′W

63 H14 **Neuquén** off. Provincia de Neuquén. ◐ province W Argentina

Neuquén, Provincia de see Neuquén

63 H14 **Neuquén, Río** ≈ W Argentina

Neurode see Nowa Ruda

100 N11 **Neuruppin** Brandenburg, NE Germany 52°56′N 12°49′E

Neusalz an der Oder see Nowa Sól

Neu Sandec see Nowy Sącz

101 K22 **Neusäss** Bayern, S Germany 48°24′N 10°49′E

Neusatz see Novi Sad

Neuschliss see Gherla

21 N8 **Neuse River** ≈ North Carolina, SE USA

109 Z5 **Neusiedl am See** Burgenland, E Austria 47°58′N 16°51′E

111 G22 **Neusiedler See** Hung. Fertő. ◎ Austria/Hungary

Neusohl see Banská Bystrica

101 D15 **Neuss** anc. Novaesium, Novesium. Nordrhein-Westfalen, W Germany 51°12′N 06°42′E

Neuss see Nyon

Neustadt see Neustadt bei Coburg, Bayern, Germany

Neustadt see Neustadt an der Aisch, Bayern, Germany

Neustadt see Prudnik, Opole, Poland

Neustadt see Baia Mare, Maramureş, Romania

100 I12 **Neustadt am Rübenberge** Niedersachsen, N Germany 52°30′N 09°28′E

101 J19 **Neustadt an der Aisch** var. Neustadt. Bayern, C Germany 49°34′N 10°36′E

Neustadt an der Haardt see Neustadt an der Weinstrasse

101 F20 **Neustadt an der Weinstrasse** prev. Neustadt an der Haardt, hist. Niewenstat; anc. Nova Civitas. Rheinland-Pfalz, SW Germany 49°21′N 08°09′E

101 K18 **Neustadt bei Coburg** var. Neustadt. Bayern, C Germany 50°19′N 11°06′E

Neustadt bei Pinne see Lwówek

Neustadt in Oberschlesien see Prudnik

Neustadtl see Novo mesto

Neustadtl in Mähren see Nové Město na Moravě

Neustettin see Szczecinek

108 M8 **Neustift im Stubaital** var. Stubaital. Tirol, W Austria 47°07′N 11°26′E

100 N10 **Neustrelitz** Mecklenburg-Vorpommern, NE Germany 53°22′N 13°05′E

Neutitschein see Nový Jičín

Neutra see Nitra

101 J22 **Neu-Ulm** Bayern, S Germany 48°23′N 10°02′E

Neuveville see La Neuveville

103 N12 **Neuvic** Corrèze, C France 45°23′N 02°16′E

Neuwarp see Nowe Warpno

100 G9 **Neuwerk** island NW Germany

101 E17 **Neuwied** Rheinland-Pfalz, W Germany 50°26′N 07°28′E

Neuzen see Terneuzen

124 H12 **Neva** ≈ NW Russian Federation

29 V14 **Nevada** Iowa, C USA 42°01′N 93°27′W

27 R6 **Nevada** Missouri, C USA 37°51′N 94°22′W

35 R5 **Nevada** off. State of Nevada, also known as Battle Born State, Sagebrush State, Silver State. ◐ state W USA

35 P6 **Nevada City** California, W USA 39°15′N 121°02′W

105 O14 **NevadaSierra** ▲▲ S Spain

35 P6 **Nevada, Sierra** ▲▲ W USA

62 I13 **Nevado, Sierra del** ▲▲ W Argentina

124 G16 **Nevel'** Pskovskaya Oblast', W Russian Federation 56°01′N 29°54′E

123 T14 **Nevel'sk** Ostrov Sakhalin, Sakhalinskaya Oblast', SE Russian Federation 46°41′N 141°54′E

123 Q13 **Never** Amurskaya Oblast', SE Russian Federation 53°58′N 124°04′E

127 Q6 **Neverkino** Penzenskaya Oblast', W Russian Federation 52°53′N 46°46′E

103 P9 **Nevers** anc. Noviodunum. Nièvre, C France 47°N 03°09′E

18 J12 **Neversink River** ≈ New York, NE USA

183 Q6 **Nevertire** New South Wales, SE Australia 31°52′S 147°42′E

113 H15 **Nevesinje** ◐ Republika Srpska, S Bosnia and Herzegovina

118 G12 **Nevežis** ≈ C Lithuania

138 F11 **Neve Zohar** prev. Newé Zohar. Southern, E Israel 31°07′N 35°23′E

126 M14 **Nevinnomyssk** Stavropol'skiy Kray, SW Russian Federation 44°39′N 41°57′E

45 W10 **Nevis** island Saint Kitts and Nevis

Nevoso, Monte see Veliki Snežnik

Nevrokop see Gotse Delchev

136 J14 **Nevşehir** var. Nevshehr. Nevşehir, C Turkey 38°38′N 34°43′E

136 J14 **Nevşehir** var. Nevshehr. ◐ province C Turkey

Nevshehr see Nevşehir

122 G10 **Nev'yansk** Sverdlovskaya Oblast', C Russian Federation 57°26′N 60°15′E

81 J25 **Newala** Mtwara, SE Tanzania 10°59′S 39°18′E

31 P16 **New Albany** Indiana, N USA 38°17′N 85°50′W

22 M2 **New Albany** Mississippi, S USA 34°29′N 89°00′W

29 Y11 **New Albin** Iowa, C USA 43°30′N 91°17′W

55 U8 **New Amsterdam** E Guyana 06°17′N 57°31′W

183 Q4 **New Angledool** New South Wales, SE Australia 29°06′S 147°54′E

21 Y2 **Newark** Delaware, NE USA 39°42′N 75°45′W

18 K14 **Newark** New Jersey, NE USA 40°42′N 74°12′W

18 G10 **Newark** New York, NE USA 43°01′N 77°04′W

31 T13 **Newark** Ohio, N USA 40°03′N 82°24′W

Newark see Newark-on-Trent

35 W5 **Newark Lake** ◎ Nevada, W USA

97 N18 **Newark-on-Trent** var. Newark. C England, United Kingdom 53°05′N 00°49′W

22 M7 **New Augusta** Mississippi, S USA 31°12′N 89°03′W

19 P12 **New Bedford** Massachusetts, NE USA 41°38′N 70°55′W

32 G11 **Newberg** Oregon, NW USA 45°18′N 122°58′W

21 X10 **New Bern** North Carolina, SE USA 35°05′N 77°04′W

20 F8 **Newbern** Tennessee, S USA 36°06′N 89°15′W

31 P4 **Newberry** Michigan, N USA 46°21′N 85°30′W

21 Q12 **Newberry** South Carolina, SE USA 34°17′N 81°39′W

18 F15 **New Bloomfield** Pennsylvania, NE USA 40°24′N 77°08′W

25 X5 **New Boston** Texas, SW USA 33°27′N 94°25′W

25 S11 **New Braunfels** Texas, SW USA 29°43′N 98°09′W

31 Q13 **New Bremen** Ohio, N USA 40°26′N 84°22′W

97 F18 **Newbridge** Ir. An Droichead Nua. Kildare, C Ireland 53°11′N 06°48′W

18 B14 **New Brighton** Pennsylvania, NE USA 40°44′N 80°18′W

18 M12 **New Britain** Connecticut, NE USA 41°37′N 72°45′W

186 G7 **New Britain** island E Papua New Guinea

192 I8 **New Britain Trench** undersea feature W Pacific Ocean

18 J15 **New Brunswick** New Jersey, NE USA 40°29′N 74°27′W

15 V8 **New Brunswick** Fr. Nouveau-Brunswick. ◐ province SE Canada

18 K13 **Newburgh** New York, NE USA 41°30′N 74°00′W

97 M22 **Newbury** S England, United Kingdom 51°25′N 01°20′W

19 P10 **Newburyport** Massachusetts, NE USA 42°49′N 70°53′W

77 T14 **New Bussa** Niger, W Nigeria 09°50′N 04°32′E

187 O17 **New Caledonia** var. Kanaky, Fr. Nouvelle-Calédonie. ◇ French self-governing territory of special status SW Pacific Ocean

187 O15 **New Caledonia** island SW Pacific Ocean

175 O10 **New Caledonia Basin** undersea feature W Pacific Ocean

183 T8 **Newcastle** New South Wales, SE Australia 32°55′S 151°46′E

13 O14 **Newcastle** New Brunswick, SE Canada 47°01′N 65°36′W

14 I15 **Newcastle** Ontario, SE Canada 43°55′N 78°35′W

83 K22 **Newcastle** KwaZulu/Natal, E South Africa 27°45′S 29°55′E

97 G16 **Newcastle** Ir. An Caisleán Nua. SE Northern Ireland, United Kingdom 54°12′N 05°54′W

31 P13 **New Castle** Indiana, N USA 39°56′N 85°21′W

20 L5 **New Castle** Kentucky, S USA 38°28′N 85°10′W

27 N11 **Newcastle** Oklahoma, C USA 35°15′N 97°36′W

18 B13 **New Castle** Pennsylvania, NE USA 41°00′N 80°22′W

25 R6 **Newcastle** Texas, SW USA 33°11′N 98°44′W

36 J7 **Newcastle** Utah, W USA 37°40′N 113°31′W

21 S6 **New Castle** Virginia, NE USA 37°31′N 80°09′W

33 Z13 **Newcastle** Wyoming, C USA 43°52′N 104°14′W

97 L14 **Newcastle** ✈ NE England, United Kingdom 55°03′N 01°42′W

Newcastle see Newcastle upon Tyne

97 L18 **Newcastle-under-Lyme** C England, United Kingdom 53°N 02°14′W

97 M14 **Newcastle upon Tyne** var. Newcastle, hist. Monkchester, Lat. Pons Aelii. NE England, United Kingdom 54°59′N 01°35′W

181 Q4 **Newcastle Waters** Northern Territory, N Australia 17°20′S 133°26′E

Newchwang see Yingkou

18 K13 **New City** New York, NE USA 41°08′N 73°57′W

31 U13 **Newcomerstown** Ohio, N USA 40°16′N 81°36′W

18 G15 **New Cumberland** Pennsylvania, NE USA 40°13′N 76°52′W

21 R1 **New Cumberland** West Virginia, NE USA 40°30′N 80°35′W

152 I10 **New Delhi** ● (India) Delhi, N India 28°35′N 77°15′E

11 O17 **New Denver** British Columbia, SW Canada 49°58′N 117°21′W

28 J9 **Newell** South Dakota, N USA 44°42′N 103°25′W

21 Q13 **New Ellenton** South Carolina, SE USA 33°25′N 81°41′W

22 J6 **Newellton** Louisiana, S USA 32°04′N 91°14′W

28 K6 **New England** North Dakota, N USA 46°32′N 102°52′W

19 P8 **New England** cultural region NE USA

New England of the West see Minnesota

183 U5 **New England Range** ▲▲ New South Wales, SE Australia

64 G9 **New England Seamounts** var. Bermuda-New England Seamount Arc. undersea feature W Atlantic Ocean 38°00′N 61°00′W

38 M14 **Newenham, Cape** headland Alaska, USA 58°39′N 162°10′W

Newé Zohar see Neve Zohar

18 D9 **Newfane** New York, NE USA 43°16′N 78°40′W

97 M23 **New Forest** physical region S England, United Kingdom

13 T12 **Newfoundland** Fr. Terre-Neuve. island Newfoundland and Labrador, SE Canada

13 R9 **Newfoundland and Labrador** Fr. Terre Neuve. ◐ province E Canada

65 J8 **Newfoundland Basin** undersea feature NW Atlantic Ocean 45°00′N 40°00′W

64 I8 **Newfoundland Ridge** undersea feature NW Atlantic Ocean

64 J8 **Newfoundland Seamounts** undersea feature N Sargasso Sea

18 G16 **New Freedom** Pennsylvania, NE USA 39°43′N 76°41′W

186 K9 **New Georgia** island New Georgia Islands, NW Solomon Islands

186 K8 **New Georgia Islands** island group NW Solomon Islands

186 L8 **New Georgia Sound** var. The Slot. sound E Solomon Sea

30 L9 **New Glarus** Wisconsin, N USA 42°50′N 89°38′W

13 Q15 **New Glasgow** Nova Scotia, SE Canada 45°36′N 62°38′W

New Goa see Panaji

186 A6 **New Guinea** Dut. Nieuw Guinea, Ind. Irian. island Indonesia/Papua New Guinea

192 H8 **New Guinea Trench** undersea feature SW Pacific Ocean

32 I6 **Newhalem** Washington, NW USA 48°40′N 121°18′W

39 P13 **Newhalen** Alaska, USA 59°43′N 154°54′W

29 X13 **Newhall** Iowa, C USA 42°00′N 91°58′W

14 F16 **New Hamburg** Ontario, S Canada 43°24′N 80°37′W

19 N9 **New Hampshire** off. State of New Hampshire, also known as Granite State. ◐ state NE USA

29 W12 **New Hampton** Iowa, C USA 43°03′N 92°19′W

186 G5 **New Hanover** island NE Papua New Guinea

97 P23 **Newhaven** SE England, United Kingdom 50°48′N 00°00′E

18 M13 **New Haven** Connecticut, NE USA 41°18′N 72°55′W

31 Q12 **New Haven** Indiana, N USA 41°02′N 84°59′W

27 W5 **New Haven** Missouri, C USA 38°34′N 91°15′W

10 K13 **New Hazelton** British Columbia, SW Canada 55°15′N 127°30′W

New Hebrides see Vanuatu

175 P9 **New Hebrides Trench** undersea feature N Coral Sea

18 H15 **New Holland** Pennsylvania, NE USA 40°06′N 76°05′W

22 I9 **New Iberia** Louisiana, S USA 30°00′N 91°51′W

186 G5 **New Ireland** ◐ province NE Papua New Guinea

186 G5 **New Ireland** island NE Papua New Guinea

65 A24 **New Island** island W Falkland Islands

18 J15 **New Jersey** off. State of New Jersey, also known as The Garden State. ◐ state NE USA

18 C14 **New Kensington** Pennsylvania, NE USA 40°33′N 79°45′W

21 W6 **New Kent** Virginia, NE USA 37°32′N 76°59′W

27 O8 **Newkirk** Oklahoma, C USA 36°54′N 97°03′W

21 Q9 **Newland** North Carolina, SE USA 36°04′N 81°50′W

28 L6 **New Leipzig** North Dakota, N USA 46°21′N 101°54′W

14 H9 **New Liskeard** Ontario, S Canada 47°31′N 79°41′W

22 G7 **Newllano** Louisiana, S USA 31°06′N 93°16′W

19 N13 **New London** Connecticut, NE USA 41°21′N 72°04′W

29 Y15 **New London** Iowa, C USA 40°55′N 91°24′W

29 T8 **New London** Minnesota, N USA 45°18′N 94°56′W

27 V3 **New London** Missouri, C USA 39°34′N 91°24′W

30 M7 **New London** Wisconsin, N USA 44°25′N 88°44′W

27 Y8 **New Madrid** Missouri, C USA 36°34′N 89°32′W

180 J8 **Newman** Western Australia 23°18′S 119°45′E

194 M13 **Newman Island** island Antarctica

14 H15 **Newmarket** Ontario, S Canada 44°03′N 79°27′W

97 P20 **Newmarket** E England, United Kingdom 52°18′N 00°28′E

19 P10 **Newmarket** New Hampshire, NE USA 43°04′N 70°53′W

21 U4 **New Market** Virginia, NE USA 38°39′N 78°40′W

21 R2 **New Martinsville** West Virginia, NE USA 39°39′N 80°52′W

31 U14 **New Matamoras** Ohio, N USA 39°32′N 81°04′W

32 M12 **New Meadows** Idaho, NW USA 44°57′N 116°16′W

26 R12 **New Mexico** off. State of New Mexico, also known as Land of Enchantment, Sunshine State. ◐ state SW USA

149 V6 **New Mīrpur** var. Mīrpur. Sind, SE Pakistan 33°11′N 73°46′E

151 N15 **New Moore Island** Island E India

23 S4 **Newnan** Georgia, SE USA 33°22′N 84°48′W

183 P17 **New Norfolk** Tasmania, SE Australia 42°46′S 147°02′E

22 K9 **New Orleans** Louisiana, S USA 30°00′N 90°01′W

22 K9 **New Orleans** ✈ Louisiana, S USA 29°59′N 90°17′W

18 K12 **New Paltz** New York, NE USA 41°44′N 74°04′W

31 U12 **New Philadelphia** Ohio, N USA 40°29′N 81°27′W

184 K10 **New Plymouth** Taranaki, North Island, New Zealand 39°04′S 174°06′E

97 M24 **Newport** S England, United Kingdom 50°42′N 01°18′W

97 K22 **Newport** SE Wales, United Kingdom 51°35′N 03°W

27 W10 **Newport** Arkansas, C USA 35°36′N 91°16′W

31 N13 **Newport** Indiana, N USA 39°52′N 87°24′W

20 M3 **Newport** Kentucky, S USA 39°05′N 84°27′W

29 W9 **Newport** Minnesota, N USA 44°52′N 93°00′W

32 F12 **Newport** Oregon, NW USA 44°39′N 124°04′W

19 O13 **Newport** Rhode Island, NE USA 41°29′N 71°17′W

21 O9 **Newport** Tennessee, S USA 35°58′N 83°13′W

19 N6 **Newport** Vermont, NE USA 44°56′N 72°13′W

32 M7 **Newport** Washington, NW USA 48°08′N 117°05′W

21 X7 **Newport News** Virginia, NE USA 36°59′N 76°26′W

97 N20 **Newport Pagnell** SE England, United Kingdom 52°05′N 00°44′W

23 U12 **New Port Richey** Florida, SE USA 28°14′N 82°42′W

29 V9 **New Prague** Minnesota, N USA 44°32′N 93°34′W

44 H3 **New Providence** island N The Bahamas

97 I20 **New Quay** SW Wales, United Kingdom 52°13′N 04°22′W

97 H24 **Newquay** SW England, United Kingdom 50°27′N 05°03′W

29 V10 **New Richland** Minnesota, N USA 43°53′N 93°29′W

15 X7 **New-Richmond** Québec, SE Canada 48°12′N 65°52′W

31 R15 **New Richmond** Ohio, N USA 38°57′N 84°16′W

30 I5 **New Richmond** Wisconsin, N USA 45°09′N 92°31′W

42 G1 **New River** ≈ N Belize

55 T12 **New River** ≈ SE Guyana

21 R6 **New River** ≈ West Virginia, NE USA

42 G1 **New River Lagoon** ◎ N Belize

22 J8 **New Roads** Louisiana, S USA 30°42′N 91°26′W

18 L14 **New Rochelle** New York, NE USA 40°55′N 73°44′W

29 O4 **New Rockford** North Dakota, N USA 47°40′N 99°08′W

97 P23 **New Romney** SE England, United Kingdom 50°58′N 00°56′E

97 F20 **New Ross** Ir. Ros Mhic Thriúin. Wexford, SE Ireland 52°24′N 06°56′W

97 F16 **Newry** Ir. An tIúr. SE Northern Ireland, United Kingdom 54°11′N 06°20′W

28 M5 **New Salem** North Dakota, N USA 46°51′N 101°24′W

New Sarum see Salisbury

29 W14 **New Sharon** Iowa, C USA 41°28′N 92°39′W

New Siberian Islands see Novosibirskiye Ostrova

23 X11 **New Smyrna Beach** Florida, SE USA 29°01′N 80°55′W

183 O7 **New South Wales** ◐ state SE Australia

39 O13 **New Stuyahok** Alaska, USA 59°27′N 157°18′W

21 N8 **New Tazewell** Tennessee, S USA 36°26′N 83°36′W

152 K9 **New Tehri** prev. Tehri. Uttarakhand, N India 30°12′N 78°29′E

38 M12 **Newtok** Alaska, USA 60°56′N 164°37′W

23 S7 **Newton** Georgia, SE USA 31°18′N 84°20′W

29 W14 **Newton** Iowa, C USA 41°42′N 93°03′W

27 N6 **Newton** Kansas, C USA 38°02′N 97°22′W

19 O11 **Newton** Massachusetts, NE USA 42°19′N 71°10′W

22 M5 **Newton** Mississippi, S USA 32°19′N 89°09′W

18 J14 **Newton** New Jersey, NE USA 41°03′N 74°45′W

21 R9 **Newton** North Carolina, SE USA 35°42′N 81°14′W

25 Y9 **Newton** Texas, SW USA 30°51′N 93°45′W

97 J24 **Newton Abbot** SW England, United Kingdom 50°33′N 03°34′W

96 K13 **Newton St Boswells** SE Scotland, United Kingdom 55°34′N 02°40′W

97 I14 **Newton Stewart** S Scotland, United Kingdom 54°58′N 04°30′W

92 O2 **Newtontoppen** ▲ C Svalbard 78°57′N 17°34′E

97 J20 **Newtown** E Wales, United Kingdom 52°32′N 03°19′W

28 K3 **New Town** North Dakota, N USA 47°58′N 102°30′W

◆ Country
● Country Capital
◇ Dependent Territory
○ Dependent Territory Capital
◐ Administrative Regions
✈ International Airport
▲ Mountain
▲▲ Mountain Range
Volcano
≈ River
◎ Lake
⊠ Reservoir

NEWTOWNABBEY – NOMAD

97 G15 **Newtownabbey** Ir. Baile na Mainistreach. E Northern Ireland, United Kingdom 54°40′N 05°57′W
97 G15 **Newtownards** Ir. Baile Nua na hArda. SE Northern Ireland, United Kingdom 54°36′N 05°41′W
29 U10 **New Ulm** Minnesota, N USA 44°20′N 94°28′W
28 K10 **New Underwood** South Dakota, N USA 44°05′N 102°46′W
25 V10 **New Waverly** Texas, SW USA 30°32′N 95°28′W
18 K14 **New York** New York, NE USA 40°45′N 73°57′W
18 G10 **New York** ◈ state NE USA
35 X13 **New York Mountains** ▲▲ California, W USA
184 K12 **New Zealand** ◆ commonwealth republic SW Pacific Ocean
95 M24 **Nexø** var. Neksø Bornholm, E Denmark 55°04′N 15°09′E
125 O15 **Neya** Kostromskaya Oblast', NW Russian Federation 58°19′N 43°51′E
Neyba see Neiba
143 Q12 **Neyrīz** var. Neiriz, Niriz. Fārs, S Iran 29°14′N 54°18′E
143 T4 **Neyshābūr** var. Nishapur. Khorāsān-Razavī, NE Iran 36°15′N 58°47′E
155 J21 **Neyveli** Tamil Nādu, SE India 11°36′N 79°26′E
Nezhin see Nizhyn
33 N10 **Nezperce** Idaho, NW USA 46°14′N 116°15′W
22 H8 **Nezpique, Bayou** ≈ Louisiana, S USA
77 Y13 **Ngadda** ≈ NE Nigeria
N'Gage see Negage
185 G16 **Ngahere** West Coast, South Island, New Zealand 42°22′S 171°29′E
77 Z12 **Ngala** Borno, NE Nigeria 12°19′N 14°11′E
158 K16 **Ngamring** Xizang Zizhiqu, W China 29°16′N 87°10′E
81 K19 **Ngangerabeli Plain** plain SE Kenya
158 I14 **Ngangla Ringco** ◎ W China
158 H13 **Nganglong Kangri** ▲ W China 32°55′N 81°00′E
158 K15 **Ngangzê Co** ◎ W China
79 F14 **Ngaoundéré** var. N'Gaoundéré. Adamaoua, N Cameroon 07°20′N 13°35′E
N'Gaoundéré see Ngaoundéré
81 E20 **Ngara** Kagera, NW Tanzania 02°30′S 30°40′E
188 F8 **Ngardmau Bay** bay Babeldaob, N Palau
188 F7 **Ngaregur** island Palau Islands, N Palau
Ngarrab see Gyaca
184 L7 **Ngaruawahia** Waikato, North Island, New Zealand 37°41′S 175°10′E
184 N11 **Ngaruroro** ≈ North Island, New Zealand
190 I16 **Ngatangiia** Rarotonga, S Cook Islands 21°14′S 159°44′W
184 M6 **Ngatea** Waikato, North Island, New Zealand 37°16′S 175°29′E
166 L8 **Ngathainggyaung** Ayeyawady, SW Myanmar (Burma) 17°22′N 95°04′E
Ngatik see Ngetik Atoll
Ngau see Gau
Ngawa see Aba
172 G12 **Ngazidja** Fr. Grande Comore, var. Njazidja. island NW Comoros
188 C7 **Ngcheangel** var. Kayangel Islands. island Palau Islands, N Palau
188 E10 **Ngchemiangel** Babeldaob, N Palau
188 C8 **Ngeaur** var. Angaur. island Palau Islands, S Palau
188 E10 **Ngerkeai** Babeldaob, N Palau
188 F9 **Ngermechau** Babeldaob, N Palau 07°35′N 134°39′E
188 C8 **Ngeruktabel** prev. Urukthapel. island Palau Islands, S Palau
188 F8 **Ngetbong** Babeldaob, N Palau 07°37′N 134°35′E
189 T17 **Ngetik Atoll** var. Ngatik; prev. Los Jardines. atoll Caroline Islands, E Micronesia
188 E10 **Ngetkip** Babeldaob, N Palau
Nghia Dan see Thai Hoa
N'Giva see Ondjiva
79 G20 **Ngo** Plateaux, SE Congo 02°28′S 15°43′E
167 S7 **Ngoc Lắc** Thanh Hoa, N Vietnam 20°06′N 105°21′E
79 G17 **Ngoko** ≈ Cameroon/Congo
81 H19 **Ngorengore** Narok, SW Kenya 01°01′S 35°26′E
159 Q11 **Ngoring Hu** ◎ C China
Ngorolaka see Banifing
81 H20 **Ngorongoro Crater** crater N Tanzania
79 D19 **Ngounié** off. Province de la Ngounié, var. La Ngounié. ◈ province S Gabon
79 D19 **Ngounié** ≈ Congo/Gabon
Ngounié, Province de la see Ngounié
78 H10 **Ngoura** var. NGoura. Hadjer-Lamis, W Chad 12°52′N 16°27′E
NGoura see Ngoura
78 G10 **Ngouri** var. NGouri; prev. Fort-Millot. Lac, W Chad 13°42′N 15°19′E
NGouri see Ngouri
77 Y10 **Ngourti** Diffa, E Niger 15°22′N 13°13′E
77 Y11 **Nguigmi** var. N'Guigmi. Diffa, SE Niger 14°17′N 13°07′E
N'Guigmi see Nguigmi
Nguimbo see Lumbala N'Guimbo
188 F15 **Ngulu Atoll** atoll Caroline Islands, W Micronesia
187 R14 **Nguna** island C Vanuatu
N'Gunza see Sumbe
169 U17 **Ngurah Rai** ✈ (Bali) Bali, S Indonesia 08°40′S 115°14′E
77 W12 **Nguru** Yobe, NE Nigeria 12°55′N 10°31′E
Ngwaketze see Southern
83 I16 **Ngweze** ≈ S Zambia
83 M17 **Nhamatanda** Sofala, C Mozambique 19°16′S 34°10′E
58 G12 **Nhamundá, Rio** var. Jamundá, Yamundá. ≈ N Brazil
60 J7 **Nhandeara** São Paulo, S Brazil 20°40′S 50°03′W
82 D12 **Nharêa** var. N'Harea, Nhareia. Bié, W Angola 11°38′S 16°58′E
N'Harea see Nharêa
Nhareia see Nharêa
167 V12 **Nha Trang** Khanh Hoa, S Vietnam 12°15′N 109°10′E
182 L11 **Nhill** Victoria, SE Australia 36°21′S 141°38′E
83 L22 **Nhlangano** prev. Goedgegun. SW Swaziland 27°06′S 31°12′E
181 S1 **Nhulunbuy** Northern Territory, N Australia 12°16′S 136°46′E
77 N10 **Niafounké** Tombouctou, W Mali 15°54′N 03°58′W
31 N5 **Niagara** Wisconsin, N USA 45°45′N 87°57′W
14 H16 **Niagara** ≈ Ontario, S Canada
14 G15 **Niagara Escarpment** hill range Ontario, S Canada
14 H16 **Niagara Falls** Ontario, S Canada 43°05′N 79°06′W
18 D9 **Niagara Falls** New York, NE USA 43°06′N 79°04′W
14 H16 **Niagara Falls** waterfall Canada/USA
76 K12 **Niagassola** var. Nyagassola. Haute-Guinée, NE Guinea 12°24′N 09°03′W
77 R12 **Niamey** ● (Niger) Niamey, SW Niger 13°28′N 02°03′E
77 R12 **Niamey** ✈ Niamey, SW Niger 13°28′N 02°14′E
77 R14 **Niamtougou** N Togo 09°50′N 01°08′E
79 O16 **Niangara** Orientale, NE Dem. Rep. Congo 03°45′N 27°54′E
77 O10 **Niangay, Lac** ◎ E Mali
77 N14 **Niangoloko** SW Burkina Faso 10°15′N 04°53′W
27 U6 **Niangua River** ≈ Missouri, C USA
79 O17 **Nia-Nia** Orientale, NE Dem. Rep. Congo 01°26′N 27°38′E
19 N13 **Niantic** Connecticut, NE USA 41°19′N 72°11′W
163 U7 **Nianzishan** Heilongjiang, NE China 47°31′N 122°53′E
79 E20 **Niari** ◈ province SW Congo
168 H10 **Nias, Pulau** island W Indonesia
82 O13 **Niassa** off. Província do Niassa. ◈ province N Mozambique
Niassa, Província do see Niassa
191 U10 **Niau** island Îles Tuamotu, C French Polynesia
95 G20 **Nibe** Nordjylland, N Denmark 56°59′N 09°39′E
189 Q8 **Nibok** N Nauru 0°31′S 166°55′E
118 C10 **Nīca** W Latvia 56°21′N 21°03′E
Nicaea see Nice
42 J9 **Nicaragua** off. Republic of Nicaragua. ◆ republic Central America
42 K11 **Nicaragua, Lago de** var. Cocibolca, Gran Lago, Eng. Lake Nicaragua. ◎ S Nicaragua
Nicaragua, Lake see Nicaragua, Lago de
64 D11 **Nicaraguan Rise** undersea feature NW Caribbean Sea 16°00′N 80°00′W
Nicaragua, Republic of see Nicaragua
Nicaria see Ikaría
107 N21 **Nicastro** Calabria, SW Italy 38°59′N 16°20′E
103 V15 **Nice** It. Nizza; anc. Nicaea. Alpes-Maritimes, SE France 43°43′N 07°13′E
Nice see Côte d'Azur
Nicephorium see Ar Raqqah
12 M9 **Nichicun, Lac** ◎ Québec, E Canada
164 D16 **Nichinan** var. Nitinan. Miyazaki, Kyūshū, SW Japan 31°36′N 131°23′E
44 E4 **Nicholas Channel** channel N Cuba
Nicholas II Land see Severnaya Zemlya
149 U2 **Nicholas Range** Pash. Selselehye Kuhe Vākhān, Taj. Qatorkŭhi Vakhon. ▲▲ Afghanistan/Tajikistan
20 M6 **Nicholasville** Kentucky, S USA 37°52′N 84°34′W
44 G2 **Nicholls Town** Andros Island, NW The Bahamas 25°07′N 78°01′W
21 U12 **Nichols** South Carolina, SE USA 34°13′N 79°09′W
55 U9 **Nickerie** ◈ district NW Suriname
55 V9 **Nickerie Rivier** ≈ NW Suriname
151 P22 **Nicobar Islands** island group India, E Indian Ocean
116 L9 **Nicolae Bălcescu** Botoşani, NE Romania 47°33′N 26°52′E
15 P11 **Nicolet** Québec, SE Canada 46°13′N 72°37′W
15 Q12 **Nicolet** ≈ Québec, SE Canada
31 Q4 **Nicolet, Lake** ◎ Michigan, N USA
29 U10 **Nicollet** Minnesota, N USA 44°16′N 94°11′W
61 F19 **Nico Pérez** Florida, S Uruguay 33°30′S 55°10′W
Nicopolis see Nikopol, Bulgaria
Nicopolis see Nikópoli, Greece
121 P2 **Nicosia** Gk. Lefkosía, Turk. Lefkoşa. ● (Cyprus) C Cyprus 35°10′N 33°23′E
107 K24 **Nicosia** Sicilia, Italy, C Mediterranean Sea 37°45′N 14°24′E
107 N22 **Nicotera** Calabria, SW Italy 38°33′N 15°55′E
42 K13 **Nicoya** Guanacaste, W Costa Rica 10°09′N 85°26′W
42 L14 **Nicoya, Golfo de** gulf W Costa Rica
42 L14 **Nicoya, Península de** peninsula NW Costa Rica
Nictheroy see Niterói
118 B12 **Nida** Ger. Nidden. Klaipėda, SW Lithuania 55°20′N 21°00′E
111 L15 **Nida** ≈ S Poland
Nidaros see Trondheim
108 D8 **Nidau** Bern, W Switzerland 47°07′N 07°15′E
101 H17 **Nidda** ≈ W Germany
Nidden see Nida
95 F17 **Nidelva** ≈ S Norway
108 F9 **Nidwalden** ◈ canton C Switzerland
110 L9 **Nidzica** Ger. Niedenburg. Warmińsko-Mazurskie, NE Poland 53°22′N 20°27′E
100 H6 **Niebüll** Schleswig-Holstein, N Germany 54°47′N 08°51′E
Niedenburg see Nidzica
99 N25 **Niederanven** Luxembourg, C Luxembourg 49°39′N 06°15′E
103 V4 **Niederbronn-les-Bains** Bas-Rhin, NE France 48°57′N 07°37′E
Niederdonau see Niederösterreich
109 S7 **Niedere Tauern** ▲ C Austria
101 P14 **Niederlausitz** Eng. Lower Lusatia, Lus. Donja Łužica. physical region E Germany
109 U5 **Niederösterreich** off. Land Niederösterreich, Eng. Lower Austria, Ger. Niederdonau; prev. Lower Danube. ◈ state NE Austria
Niederösterreich, Land see Niederösterreich
100 G12 **Niedersachsen** Eng. Lower Saxony, Fr. Basse-Saxe. ◈ state NW Germany
79 D17 **Niefang** var. Sevilla de Niefang. NW Equatorial Guinea 01°52′N 10°12′E
83 G23 **Niekerkshoop** Northern Cape, W South Africa 29°21′S 22°49′E
99 G17 **Niel** Antwerpen, N Belgium 51°07′N 04°20′E
Niélé see Niellé
76 M14 **Niellé** var. Niélé. N Ivory Coast 10°12′N 05°38′W
79 O22 **Niemba** Katanga, E Dem. Rep. Congo 05°58′S 28°24′E
111 G15 **Niemcza** Ger. Nimptsch. Dolnośląskie, SW Poland 50°45′N 16°52′E
Niemen see Neman
92 J13 **Niemisel** Norrbotten, N Sweden 66°00′N 22°00′E
111 H15 **Niemodlin** Ger. Falkenberg. Opolskie, SW Poland 50°37′N 17°45′E
76 M13 **Niéna** Sikasso, SW Mali 11°24′N 06°20′W
100 H12 **Nienburg** Niedersachsen, N Germany 52°37′N 09°12′E
100 N13 **Nieplitz** ≈ NE Germany
111 L16 **Niepołomice** Małopolskie, S Poland 50°02′N 20°12′E
101 D14 **Niers** ≈ Germany/ Netherlands
101 Q15 **Niesky** Lus. Niska. Sachsen, E Germany 51°16′N 14°49′E
Nieśwież see Nyasvizh
Nieuport see Nieuwpoort
98 O8 **Nieuw-Amsterdam** Drenthe, NE Netherlands 52°43′N 06°52′E
55 W9 **Nieuw Amsterdam** Commewijne, NE Suriname 05°53′N 55°05′W
99 M14 **Nieuw-Bergen** Limburg, SE Netherlands 51°36′N 06°04′E
98 O7 **Nieuw-Buinen** Drenthe, NE Netherlands 52°57′N 06°55′E
98 J12 **Nieuwegein** Utrecht, C Netherlands 52°03′N 05°06′E
98 P6 **Nieuwe Pekela** Groningen, NE Netherlands 53°04′N 06°58′E
98 P5 **Nieuweschans** Groningen, NE Netherlands 53°10′N 07°10′E
Nieuw Guinea see New Guinea
98 I11 **Nieuwkoop** Zuid-Holland, C Netherlands 52°09′N 04°46′E
98 M9 **Nieuwleusen** Overijssel, S Netherlands 52°34′N 06°16′E
98 J11 **Nieuw-Loosdrecht** Noord-Holland, C Netherlands 52°12′N 05°08′E
55 U9 **Nieuw Nickerie** Nickerie, NW Suriname 05°56′N 57°W
98 P5 **Nieuwolda** Groningen, NE Netherlands 53°15′N 06°58′E
99 B17 **Nieuwpoort** var. Nieuport. West-Vlaanderen , W Belgium 51°08′N 02°45′E
99 G14 **Nieuw-Vossemeer** Noord-Brabant, S Netherlands 51°34′N 04°13′E
98 P7 **Nieuw-Weerdinge** Drenthe, NE Netherlands 52°51′N 07°00′E
40 L10 **Nieves** Zacatecas, C Mexico 24°00′N 102°57′W
64 O11 **Nieves, Pico de las** ▲ Gran Canaria, Islas Canarias, Spain, NE Atlantic Ocean 27°58′N 15°34′W
103 P8 **Nièvre** ◈ department C France
Niewenstat see Neustadt an der Weinstrasse
136 J15 **Niğde** Niğde, C Turkey 37°58′N 34°42′E
136 J15 **Niğde** ◈ province C Turkey
83 J21 **Nigel** Gauteng, NE South Africa 26°25′S 28°28′E
77 V10 **Niger** off. Republic of Niger. ◆ republic W Africa
77 T14 **Niger** ◈ state C Nigeria
67 P8 **Niger** ≈ W Africa
Niger Cone see Niger Fan
67 P9 **Niger Delta** delta S Nigeria
67 P9 **Niger Fan** var. Niger Cone. undersea feature E Atlantic Ocean 04°15′N 05°00′E
77 T13 **Nigeria** off. Federal Republic of Nigeria. ◆ federal republic W Africa
Nigeria, Federal Republic of see Nigeria
77 T17 **Niger, Mouths of the** delta S Nigeria
Niger, Republic of see Niger
185 C24 **Nightcaps** Southland, South Island, New Zealand 45°58′S 168°03′E
14 F7 **Night Hawk Lake** ◎ Ontario, S Canada
65 M19 **Nightingale Island** island S Tristan da Cunha, S Atlantic Ocean
38 M12 **Nightmute** Alaska, USA 60°28′N 164°43′W
114 G13 **Nigríta** Kentrikí Makedonía, NE Greece 40°54′N 23°29′E
148 J15 **Nihing** Per. Rūd-e Nahang. ≈ Iran/Pakistan
191 V10 **Nihiru** atoll Îles Tuamotu, C French Polynesia
Nihommatsu see Nihonmatsu
Nihon see Japan
165 P11 **Nihonmatsu** var. Nihommatsu, Nihonmatu. Fukushima, Honshū, C Japan 37°34′N 140°25′E
Nihonmatu see Nihonmatsu
62 I12 **Nihuil, Embalse del** ▣ W Argentina
165 O10 **Niigata** Niigata, Honshū, C Japan 37°55′N 139°01′E
165 O11 **Niigata** off. Niigata-ken. ◈ prefecture Honshū, C Japan
Niigata-ken see Niigata
165 G14 **Niihama** Ehime, Shikoku, SW Japan 33°57′N 133°15′E
38 A8 **Ni'ihau** var. Niihau. island Hawai'i, USA, C Pacific Ocean
165 X12 **Nii-jima** island E Japan
165 H12 **Niimi** Okayama, Honshū, SW Japan 35°00′N 133°27′E
165 O10 **Niitsu** var. Niitu. Niigata, Honshū, C Japan 37°48′N 139°09′E
Niitu see Niitsu
105 P15 **Níjar** Andalucía, S Spain 36°57′N 02°13′W
98 K11 **Nijkerk** Gelderland, C Netherlands 52°13′N 05°30′E
99 H16 **Nijlen** Antwerpen, N Belgium 51°10′N 04°40′E
98 L13 **Nijmegen** Ger. Nimwegen; anc. Noviomagus. Gelderland, SE Netherlands 51°50′N 05°52′E
98 N10 **Nijverdal** Overijssel, E Netherlands 52°22′N 06°28′E
190 G16 **Nikao** Rarotonga, S Cook Islands
Nikaria see Ikaría
124 I2 **Nikel'** Finn. Kolosjoki. Murmanskaya Oblast', NW Russian Federation 69°25′N 30°12′E
171 Q17 **Nikiniki** Timor, S Indonesia 10°N 124°30′E
129 Q15 **Nikitin Seamount** undersea feature E Indian Ocean 05°48′S 84°48′E
77 S14 **Nikki** E Benin 09°55′N 03°12′E
Niklasmarkt see Gheorgheni
39 P10 **Nikolai** Alaska, USA 63°00′N 154°22′W
Nikolaiken see Mikołajki
Nikolainkaupunki see Vaasa
Nikolayev see Mykolayiv
145 O6 **Nikolayevka** Severnyy Kazakhstan, N Kazakhstan
Nikolayevka see Zhetigen
127 P9 **Nikolayevsk** Volgogradskaya Oblast', SW Russian Federation 50°03′N 45°30′E
Nikolayevskaya Oblast' see Mykolayivs'ka Oblast'
123 S12 **Nikolayevsk-na-Amure** Khabarovskiy Kray, SE Russian Federation 53°04′N 140°39′E
127 P6 **Nikol'sk** Penzenskaya Oblast', W Russian Federation 53°46′N 46°03′E
125 O13 **Nikol'sk** Vologodskaya Oblast', NW Russian Federation 59°35′N 45°31′E
Nikol'sk see Ussuriysk
38 K17 **Nikolski** Umnak Island, Alaska, USA 52°56′N 168°52′W
Nikol'skiy see Satpayev
127 V7 **Nikol'skoye** Orenburgskaya Oblast', W Russian Federation 52°01′N 55°48′E
Nikol'sk-Ussuriyskiy see Ussuriysk
114 J7 **Nikopol** anc. Nicopolis. Pleven, N Bulgaria 43°43′N 24°55′E
117 S9 **Nikopol'** Dnipropetrovs'ka Oblast', SE Ukraine 47°34′N 34°23′E
115 C17 **Nikópoli** anc. Nicopolis. site of ancient city Ípeiros, W Greece
136 M12 **Niksar** Tokat, N Turkey 40°35′N 36°59′E
143 V14 **Nīkshahr** Sīstān va Balūchestān, SE Iran 26°15′N 60°10′E
113 J16 **Nikšić** C Montenegro 42°47′N 18°56′E
191 R4 **Nikumaroro** ; prev. Gardner Island. atoll Phoenix Islands, C Kiribati
191 P3 **Nikunau** var. Nukunau; prev. Byron Island. atoll Tungaru, W Kiribati
155 G21 **Nilambūr** Kerala, SW India 11°17′N 76°15′E
35 X16 **Niland** California, W USA 33°14′N 115°31′W
80 G8 **Nile** former province NW Uganda
67 T3 **Nile** Ar. Nahr an Nil. ≈ N Africa
75 W7 **Nile Delta** delta N Egypt
67 T3 **Nile Fan** undersea feature E Mediterranean Sea 33°00′N 31°00′E
31 O11 **Niles** Michigan, N USA 41°49′N 86°15′W
31 V11 **Niles** Ohio, N USA 41°10′N 80°46′W
155 F20 **Nīleswaram** Kerala, SW India 12°18′N 75°07′E
14 K10 **Nilgaut, Lac** ◎ Québec, SE Canada
149 O6 **Nīlī** Dāykundī, C Afghanistan 33°43′N 66°07′E
158 I5 **Nilka** Xinjiang Uygur Zizhiqu, NW China 43°46′N 82°33′E
Nīl, Nahr an see Nile
93 N16 **Nilsiä** Pohjois-Savo, C Finland 63°13′N 28°00′E
154 F9 **Nimach** Madhya Pradesh, C India 24°27′N 74°56′E
152 G14 **Nimbāhera** Rājasthān, N India 24°38′N 74°45′E
76 L15 **Nimba, Monts** var. Nimba Mountains. ▲▲ W Africa
Nimba Mountains see Nimba, Monts
Nimburg see Nymburk
103 Q15 **Nîmes** anc. Nemausus, Nismes. Gard, S France 43°49′N 04°20′E
152 H11 **Nīm ka Thāna** Rājasthān, N India 27°42′N 75°50′E
183 R11 **Nimmitabel** New South Wales, SE Australia 36°34′S 149°18′E
Nimptsch see Niemcza
195 R11 **Nimrod Glacier** glacier Antarctica
148 K8 **Nīmrōz** var. Nimroze; prev. Chakhānsūr, Nīmrūz. ◈ province SW Afghanistan
Nimroze see Nīmrōz
Nīmrūz see Nīmrōz
81 F16 **Nimule** Eastern Equatoria, S South Sudan 03°35′N 32°03′E
Nimwegen see Nijmegen
139 Q3 **Nīnawá** off. Muḩāfa at Ninawá, var. Al Mawşil, Nineveh. ◈ governorate NW Iraq
Nīnawá, Muḩāfa at see Nīnawá
155 C23 **Nine Degree Channel** channel India/Maldives
18 G9 **Ninemile Point** headland New York, NE USA 43°31′N 76°22′W
173 S8 **Ninetyeast Ridge** undersea feature E Indian Ocean 04°00′S 90°00′E
183 P13 **Ninety Mile Beach** beach Victoria, SE Australia
184 I2 **Ninety Mile Beach** beach North Island, New Zealand
21 P12 **Ninety Six** South Carolina, SE USA 34°10′N 82°01′W
Nineveh see Nīnawá
163 Y9 **Ning'an** Heilongjiang, NE China 44°20′N 129°28′E
161 S9 **Ningbo** var. Ning-po, Yin-hsien; prev. Ninghsien. Zhejiang, SE China 29°54′N 121°33′E
161 U12 **Ningde** Fujian, SE China 26°48′N 119°33′E
161 P12 **Ningdu** var. Meijiang. Jiangxi, S China 26°28′N 115°53′E
Ning'er see Pu'er
186 A7 **Ningerum** Western, SW Papua New Guinea 05°40′S 141°10′E
161 R9 **Ningguo** Anhui, E China 30°33′N 118°58′E
161 S9 **Ninghai** Zhejiang, SE China 29°18′N 121°26′E
Ning-hsia see Ningxia
Ninghsien see Ningbo
160 J15 **Ningming** var. Chengzhong. Guangxi Zhuangzu Zizhiqu, S China 22°07′N 106°43′E
160 H11 **Ningnan** var. Pisha. Sichuan, C China 26°59′N 102°49′E
Ning-po see Ningbo
Ningsia/Ningsia Hui/ Ningsia Hui Autonomous Region see Ningxia
160 J5 **Ningxia** off. Ningxia Huizu Zizhiqu, var. Ning-hsia, Ningsia, Eng. Ningsia Hui, Ningsia Hui Autonomous Region. ◈ autonomous region N China
Ningxia Huizu Zizhiqu see Ningxia
159 X10 **Ningxian** var. Xinning. Gansu, N China 35°30′N 108°05′E
167 T7 **Ninh Binh** Ninh Binh, N Vietnam 20°14′N 106°00′E
167 V12 **Ninh Hoa** Khanh Hoa, S Vietnam 12°28′N 109°07′E
186 C4 **Ninigo Group** island group N Papua New Guinea
39 Q12 **Ninilchik** Alaska, USA 60°03′N 151°40′W
27 N7 **Ninnescah River** ≈ Kansas, C USA
195 U16 **Ninnis Glacier** glacier Antarctica
165 R8 **Ninohe** Iwate, Honshū, C Japan 40°16′N 141°18′E
99 F18 **Ninove** Oost-Vlaanderen, C Belgium 50°50′N 04°02′E
171 O4 **Ninoy Aquino** ✈ (Manila) Luzon, N Philippines 14°26′N 121°00′E
Nio see Íos
29 P12 **Niobrara** Nebraska, C USA 42°43′N 97°59′W
28 M12 **Niobrara River** ≈ Nebraska/Wyoming, C USA
79 I20 **Nioki** Bandundu, W Dem. Rep. Congo 02°44′S 17°42′E
76 M11 **Niono** Ségou, C Mali 14°18′N 05°59′W
76 K11 **Nioro** var. Nioro du Sahel. Kayes, W Mali 15°13′N 09°39′W
76 G11 **Nioro du Rip** SW Senegal 13°44′N 15°48′W
Nioro du Sahel see Nioro
102 K10 **Niort** Deux-Sèvres, W France 46°21′N 00°25′W
172 H14 **Nioumachoua** Mohéli, S Comoros 12°21′S 43°43′E
186 C7 **Nipa** Southern Highlands, W Papua New Guinea 06°11′S 143°27′E
11 U14 **Nipawin** Saskatchewan, S Canada 52°23′N 104°01′W
12 D12 **Nipigon** Ontario, S Canada 49°02′N 88°15′W
12 D11 **Nipigon, Lake** ◎ Ontario, S Canada
11 S13 **Nipin** ≈ Saskatchewan, C Canada
14 G11 **Nipissing, Lake** ◎ Ontario, S Canada
35 P13 **Nipomo** California, W USA 35°02′N 120°28′W
Nippon see Japan
138 K6 **Niqniqīyah, Jabal an** ▲ C Syria
62 I9 **Niquivil** San Juan, W Argentina 30°25′S 68°42′W
171 Y13 **Nirabotong** Papua, E Indonesia 02°35′S 140°08′E
Niriz see Neyrīz
163 U7 **Nirji** var. Morin Dawa Daurzu Zizhiqi. Nei Mongol Zizhiqu, N China 48°21′N 124°32′E
155 I14 **Nirmal** Telangana, C India 19°04′N 78°21′E
153 Q13 **Nirmāli** Bihār, NE India 26°18′N 86°35′E
113 O14 **Niš** Eng. Nish, Ger. Nisch; anc. Naissus. Serbia, SE Serbia 43°21′N 21°53′E
104 H9 **Nisa** Portalegre, C Portugal 39°31′N 07°39′W
Nisa see Neisse
141 P4 **Nişāb** Al Ḩudūd ash Shamālīyah, N Saudi Arabia 29°11′N 44°43′E
141 Q15 **Nişāb** var. Anşāb. SW Yemen 14°24′N 46°47′E
113 P14 **Nišava** Bul. Nishava. ≈ Bulgaria/Serbia see also Nishava
Nišava see Nishava
107 K25 **Niscemi** Sicilia, Italy, C Mediterranean Sea 37°09′N 14°23′E
Nisch/Nish see Niš
165 R4 **Niseko** Hokkaidō, NE Japan 42°50′N 140°43′E
Nishapur see Neyshābūr
114 G9 **Nishava** var. Nišava. ≈ Bulgaria/Serbia see also Nišava
Nishava see Nišava
118 L11 **Nishcha** ≈ N Belarus
165 C17 **Nishinoomote** Kagoshima, Tanega-shima, SW Japan 30°42′N 130°59′E
165 X15 **Nishino-shima** Eng. Rosario. island Ogasawara-shotō, SE Japan
165 I13 **Nishiwaki** var. Nisiwaki. Hyōgo, Honshū, SW Japan 34°59′N 134°58′E
141 U14 **Nishtūn** SE Yemen 15°47′N 52°08′E
Nisibin see Nusaybin
Nisiros see Nísyros
Nisiwaki see Nishiwaki
Niska see Niesky
113 O14 **Niška Banja** Serbia, SE Serbia 43°18′N 22°01′E
12 D6 **Niskibi** ≈ Ontario, C Canada
111 O15 **Nisko** Podkrapackie, SE Poland 50°31′N 22°09′E
10 H7 **Nisling** ≈ Yukon, W Canada
99 H22 **Nismes** Namur, S Belgium 50°04′N 04°31′E
Nismes see Nîmes
116 M10 **Nisporeni** Rus. Nisporeny. W Moldova 47°04′N 28°10′E
Nisporeny see Nisporeni
95 K20 **Nissan** ≈ S Sweden
Nissan Islands see Green Islands
95 F16 **Nisser** ◎ S Norway
95 E21 **Nissum Bredning** inlet NW Denmark
29 U6 **Nisswa** Minnesota, N USA 46°31′N 94°17′W
Nistru see Dniester
115 M22 **Nísyros** var. Nisiros. island Dodekánisa, Greece, Aegean Sea
118 H8 **Nītaure** C Latvia 57°05′N 25°12′E
60 P10 **Niterói** prev. Nictheroy. Rio de Janeiro, SE Brazil 22°54′S 43°06′W
14 F16 **Nith** ≈ Ontario, S Canada
96 J13 **Nith** ≈ S Scotland, United Kingdom
Nitinan see Nichinan
111 I21 **Nitra** Ger. Neutra, Hung. Nyitra. Nitriansky Kraj, SW Slovakia 48°20′N 18°05′E
111 I20 **Nitra** Ger. Neutra, Hung. Nyitra. ≈ W Slovakia
111 I21 **Nitriansky Kraj** ◈ region SW Slovakia
21 Q5 **Nitro** West Virginia, NE USA 38°24′N 81°50′W
Niuatobutabu see Niuatoputapu
193 X13 **Niuatoputapu** var. Niuatobutabu; prev. Keppel Island. island N Tonga
193 U15 **Niu'Aunofa** headland Tongatapu, S Tonga 21°03′S 175°19′W
Niuchwang see Yingkou
190 B16 **Niue** ◇ self-governing territory in free association with New Zealand S Pacific Ocean
190 F10 **Niulakita** var. Nurakita. atoll S Tuvalu
190 E6 **Niutao** atoll NW Tuvalu
93 L15 **Nivala** Pohjois-Pohjanmaa, C Finland 63°56′N 25°00′E
102 I15 **Nive** ≈ SW France
99 G19 **Nivelles** Walloon Brabant, C Belgium 50°36′N 04°04′E
103 P8 **Nivernais** cultural region C France
15 N8 **Niverville, Lac** ◎ Québec, SE Canada
27 T7 **Nixa** Missouri, C USA 37°02′N 93°17′W
35 R5 **Nixon** Nevada, W USA 39°48′N 119°24′W
25 S12 **Nixon** Texas, SW USA 29°16′N 97°45′W
Niya see Minfeng
Niyazov see Nyýazow
155 H14 **Nizāmābād** Telangana, C India 18°40′N 78°05′E
155 H15 **Nizām Sāgar** ◎ C India
125 N16 **Nizhegorodskaya Oblast'** ◈ province W Russian Federation
Nizhnegorskiy see Nyzhn'ohirs'kyy
127 S4 **Nizhnekamsk** Respublika Tatarstan, W Russian Federation 55°36′N 51°45′E
127 U3 **Nizhnekamskoye Vodokhranilishche** ▣ W Russian Federation
123 S14 **Nizhneleninskoye** Yevreyskaya Avtonomnaya Oblast', SE Russian Federation 47°50′N 132°30′E
122 L13 **Nizhneudinsk** Irkutskaya Oblast', S Russian Federation 54°48′N 98°51′E
122 I10 **Nizhnevartovsk** Khanty-Mansiyskiy Avtonomnyy Okrug-Yugra, C Russian Federation 60°57′N 76°40′E
123 Q7 **Nizhneyansk** Respublika Sakha (Yakutiya), NE Russian Federation 71°25′N 135°59′E
127 Q11 **Nizhniy Baskunchak** Astrakhanskaya Oblast', SW Russian Federation 48°15′N 46°49′E
127 O6 **Nizhniy Lomov** Penzenskaya Oblast', W Russian Federation 53°32′N 43°39′E
127 P3 **Nizhniy Novgorod** prev. Gor'kiy. Nizhegorodskaya Oblast', W Russian Federation 56°17′N 44°E
125 T8 **Nizhniy Odes** Respublika Komi, NW Russian Federation 63°42′N 54°59′E
Nizhniy Pyandzh see Panji Poyon
122 G10 **Nizhniy Tagil** Sverdlovskaya Oblast', C Russian Federation 57°57′N 59°51′E
125 T9 **Nizhnyaya-Omra** Respublika Komi, NW Russian Federation 62°46′N 55°54′E
125 P5 **Nizhnyaya Pësha** Nenetskiy Avtonomnyy Okrug, NW Russian Federation 66°54′N 47°37′E
117 Q3 **Nizhyn** Rus. Nezhin. Chernihivs'ka Oblast', NE Ukraine 51°03′N 31°54′E
136 M17 **Nizip** Gaziantep, S Turkey 37°02′N 37°47′E
141 X8 **Nizwa** var. Nazwāh. NE Oman 23°00′N 57°50′E
Nizza see Nice
106 C9 **Nizza Monferrato** Piemonte, NE Italy 44°47′N 08°22′E
Njávdám see Näätämöjoki
Njazidja see Ngazidja
Njellim see Nellim
81 H24 **Njombe** Njombe, S Tanzania 09°20′S 34°47′E
81 H25 **Njombe** off. Mkoa wa Njombe. ◈ region S Tanzania
81 G23 **Njombe** ≈ C Tanzania
Njombe, Mkoa wa see Njombe
Njuk, Ozero see Nyuk, Ozero
Njuksenica see Nyuksenitsa
92 I10 **Njunis** ▲ N Norway 68°47′N 19°24′E
Njurunda see Njurundabommen
93 H17 **Njurundabommen** prev. Njurunda. Västernorrland, C Sweden 62°15′N 17°24′E
94 N11 **Njutånger** Gävleborg, C Sweden 61°37′N 17°04′E
79 D14 **Nkambe** Nord-Ouest, NW Cameroon 06°35′N 10°44′E
Nkata Bay see Nkhata Bay
79 F21 **Nkayi** prev. Jacob. Bouenza, S Congo 04°11′S 13°17′E
83 J17 **Nkayi** Matabeleland North, W Zimbabwe 19°00′S 28°54′E
82 N13 **Nkhata Bay** var. Nkata Bay. Northern, N Malawi 11°37′S 34°20′E
81 E22 **Nkonde** Kigoma, N Tanzania 06°16′S 30°17′E
79 D15 **Nkongsamba** var. N'Kongsamba. Littoral, W Cameroon 04°59′N 09°53′E
N'Kongsamba see Nkongsamba
83 E16 **Nkurenkuru** Okavango, N Namibia 17°38′S 18°39′E
77 Q15 **Nkwanta** E Ghana 08°18′N 00°27′E
167 O2 **Nmai Hka** var. Me Hka. ≈ N Myanmar (Burma)
Noardwâlde see Noordwolde
39 N7 **Noatak** Alaska, USA 67°34′N 162°58′W
39 N7 **Noatak River** ≈ Alaska, USA
164 E15 **Nobeoka** Miyazaki, Kyūshū, SW Japan 32°34′N 131°37′E
27 N11 **Noble** Oklahoma, C USA 35°08′N 97°23′W
31 P13 **Noblesville** Indiana, N USA 40°03′N 86°00′W
165 R5 **Noboribetsu** var. Noboribetu. Hokkaidō, NE Japan 42°27′N 141°08′E
Noboribetu see Noboribetsu
59 H18 **Nobres** Mato Grosso, W Brazil 14°44′S 56°15′W
107 N21 **Nocera Terinese** Calabria, S Italy 39°03′N 16°10′E
41 Q16 **Nochixtlán** var. Asunción Nochixtlán. Oaxaca, SE Mexico 17°29′N 97°17′W
25 S5 **Nocona** Texas, SW USA 33°47′N 97°43′W
63 K21 **Nodales, Bahía de los** bay S Argentina
27 Q2 **Nodaway River** ≈ Iowa/ Missouri, C USA
27 R8 **Noel** Missouri, C USA 36°33′N 94°29′W
40 H3 **Nogales** Chihuahua, NW Mexico 18°50′N 97°12′W
40 F3 **Nogales** Sonora, NW Mexico 31°17′N 110°53′W
36 M17 **Nogales** Arizona, SW USA 31°20′N 110°55′W
Nogal Valley see Dooxo Nugaaleed
102 K15 **Nogaro** Gers, S France 43°46′N 00°01′W
110 J7 **Nogat** ≈ N Poland
164 D12 **Nōgata** Fukuoka, Kyūshū, SW Japan 33°46′N 130°42′E
127 P15 **Nogayskaya Step'** steppe SW Russian Federation
102 M6 **Nogent-le-Rotrou** Eure-et-Loir, C France 48°19′N 00°50′E
103 O4 **Nogent-sur-Oise** Oise, N France 49°16′N 02°28′E
103 P6 **Nogent-sur-Seine** Aube, N France 48°30′N 03°31′E
122 L10 **Noginsk** Krasnoyarskiy Kray, N Russian Federation 64°28′N 91°09′E
126 L3 **Noginsk** Moskovskaya Oblast', W Russian Federation 55°51′N 38°23′E
123 T12 **Nogliki** Ostrov Sakhalin, Sakhalinskaya Oblast', SE Russian Federation 51°44′N 143°14′E
164 K12 **Nōgōhaku-san** ▲ Honshū, SW Japan 35°46′N 136°30′E
162 D5 **Nogoonnuur** Bayan-Ölgiy, NW Mongolia 49°31′N 89°48′E
61 C18 **Nogoyá** Entre Ríos, E Argentina 32°25′S 59°50′W
111 K21 **Nógrád** off. Nógrád Megye. ◈ county N Hungary
Nógrád Megye see Nógrád
105 U5 **Noguera Pallaresa** ≈ NE Spain
105 U4 **Noguera Ribagorçana** ≈ NE Spain
101 E19 **Nohfelden** Saarland, SW Germany 49°35′N 07°08′E
38 A8 **Nohili Point** headland Kaua'i, Hawai'i, USA 22°03′N 159°48′W
104 G3 **Noia** Galicia, NW Spain 42°48′N 08°52′W
103 N16 **Noire, Montagne** ▲▲ S France
14 J10 **Noire, Rivière** ≈ Québec, SE Canada
15 P12 **Noire, Rivière** ≈ Québec, SE Canada
Noire, Rivi`ere see Black River
102 G6 **Noires, Montagnes** ▲▲ NW France
102 H8 **Noirmoutier-en-l'Île** Vendée, NW France 47°00′N 02°15′W
102 H8 **Noirmoutier, Île de** island NW France
187 Q10 **Noka** Nendö, E Solomon Islands 10°42′S 165°57′E
83 G17 **Nokaneng** North West, NW Botswana 19°40′S 22°12′E
93 L18 **Nokia** Pirkanmaa, W Finland 61°29′N 23°30′E
148 K11 **Nok Kundi** Baluchistān, SW Pakistan 28°49′N 62°39′E
30 L14 **Nokomis** Illinois, N USA 39°18′N 89°17′W
30 K5 **Nokomis, Lake** ▣ Wisconsin, N USA
78 G9 **Nokou** Kanem, W Chad 14°36′N 14°45′E
187 Q12 **Nokuku** Espiritu Santo, W Vanuatu 14°56′S 166°34′E
95 J18 **Nol** Västra Götaland, S Sweden 57°55′N 12°03′E
79 H16 **Nola** Sangha-Mbaéré, SW Central African Republic 03°29′N 16°05′E
25 P7 **Nolan** Texas, SW USA 32°15′N 100°15′W
125 R15 **Nolinsk** Kirovskaya Oblast', NW Russian Federation 57°35′N 49°54′E
Nolsø see Nólsoy
95 B19 **Nólsoy** Dan. Nolsø. island E Faroe Islands
186 B7 **Nomad** Western, SW Papua New Guinea 06°11′S 142°13′E

◆ Country ● Country Capital ◇ Dependent Territory ○ Dependent Territory Capital ◈ Administrative Regions ✈ International Airport ▲ Mountain ▲▲ Mountain Range ☼ Volcano ≈ River ◎ Lake ▣ Reservoir

Noma-zaki – Noupoort

164 B16 **Noma-zaki** Kyūshū, SW Japan
40 K10 **Nombre de Dios** Durango, C Mexico 23°51′N 104°14′W
42 I5 **Nombre de Dios, Cordillera** ▲▲ N Honduras
38 M9 **Nome** Alaska, USA 64°30′N 165°24′W
29 Q6 **Nome** North Dakota, N USA 46°39′N 97°49′W
38 M9 **Nome, Cape** *headland* Alaska, USA 64°25′N 165°00′W
162 K11 **Nomgon** *var.* Sangiyn Dalay. Ömnögovĭ, S Mongolia 42°50′N 105°04′E
14 M11 **Nominingue, Lac** ◎ Québec, SE Canada
Nomoi Islands *see* Mortlock Islands
164 B16 **Nomo-zaki** *headland* Kyūshū, SW Japan 32°34′N 129°45′E
162 G6 **Nömrög** *var.* Hödrögö. Dzavhan, N Mongolia 48°51′N 96°48′E
193 X15 **Nomuka** *island* Nomuka Group, C Tonga
193 X15 **Nomuka Group** *island group* W Tonga
189 Q15 **Nomwin Atoll** *atoll* Hall Islands, C Micronesia
8 L10 **Nonacho Lake** ◎ Northwest Territories, NW Canada
Nondaburi *see* Nonthaburi
39 P12 **Nondalton** Alaska, USA 59°58′N 154°51′W
163 V10 **Nong'an** Jilin, NE China 44°25′N 125°10′E
169 P10 **Nong Bua Khok** Nakhon Ratchasima, C Thailand 15°23′N 101°51′E
167 Q9 **Nong Bua Lamphu** Udon Thani, E Thailand 17°11′N 102°27′E
167 R7 **Nông Hèt** Xiangkhoang, N Laos 19°27′N 104°02′E
Nongkaya *see* Nong Khai
167 Q8 **Nong Khai** *var.* Mi Chai, Nongkaya. Nong Khai, E Thailand 17°52′N 102°44′E
Nongle *see* Jiuzhaigou
167 N14 **Nong Met** Surat Thani, SW Thailand 09°27′N 99°09′E
83 L22 **Nongoma** KwaZulu/Natal, E South Africa 27°54′S 31°40′E
167 P9 **Nong Phai** Phetchabun, C Thailand 15°58′N 101°02′E
153 U13 **Nongstoin** Meghālaya, NE India 25°24′N 91°19′E
83 C19 **Nonidas** Erongo, N Namibia 22°36′S 14°40′E
Nonni *see* Nen Jiang
40 I7 **Nonoava** Chihuahua, N Mexico 27°24′N 106°18′W
191 O3 **Nonouti** *prev.* Sydenham Island. *atoll* Tungaru, W Kiribati
167 O11 **Nonthaburi** *var.* Nondaburi, Nontha Buri. Nonthaburi, C Thailand 13°48′N 100°11′E
Nontha Buri *see* Nonthaburi
102 L11 **Nontron** Dordogne, SW France 45°34′N 00°41′E
147 T10 **Nookat** *var.* Iski-Nauket; *prev.* Eski-Nookat. Oshskaya Oblast', SW Kyrgyzstan 40°18′N 72°29′E
181 P1 **Noonamah** Northern Territory, N Australia 12°46′S 131°08′E
28 K2 **Noonan** North Dakota, N USA 48°51′N 102°57′W
Noonu *see* South Miladhunmadulu Atoll
99 E14 **Noord-Beveland** *var.* North Beveland. *island* SW Netherlands
99 J14 **Noord-Brabant** *Eng.* North Brabant. ◆ *province* S Netherlands
98 H7 **Noorder Haaks** *spit* NW Netherlands
98 H9 **Noord-Holland** *Eng.* North Holland. ◆ *province* NW Netherlands
Noordhollandsch Kanaal *see* Noordhollands Kanaal
98 H8 **Noordhollands Kanaal** *var.* Noordhollandsch Kanaal. *canal* NW Netherlands
Noord-Kaap *see* Northern Cape
98 L8 **Noordoostpolder** *island* N Netherlands
45 P16 **Noordpunt** *headland* C Curaçao 12°21′N 69°08′W
98 I8 **Noord-Scharwoude** Noord-Holland, NW Netherlands 52°42′N 04°48′E
Noordwes *see* North-West
98 G11 **Noordwijk aan Zee** Zuid-Holland, W Netherlands 52°15′N 04°25′E
98 H11 **Noordwijkerhout** Zuid-Holland, W Netherlands 52°16′N 04°30′E
98 M7 **Noordwolde** *Fris.* Noardwâlde. Fryslân, N Netherlands 52°54′N 06°10′E
Noordzee *see* North Sea
98 H10 **Noordzee-Kanaal** *canal* NW Netherlands
93 K18 **Noormarkku** *Swe.* Normark. Satakunta, SW Finland 61°35′N 21°54′E
39 N8 **Noorvik** Alaska, USA 66°50′N 161°01′W
10 J17 **Nootka Sound** *inlet* British Columbia, W Canada
82 A9 **Nóqui** Zaire Province, NW Angola 05°54′S 13°30′E
95 L15 **Nora** Örebro, C Sweden 59°31′N 15°02′E
147 Q13 **Norak** *Rus.* Nurek. W Tajikistan 38°23′N 69°14′E
13 I13 **Noranda** Québec, SE Canada 48°16′N 79°03′W
29 W12 **Nora Springs** Iowa, C USA 43°08′N 93°00′W
95 M14 **Norberg** Västmanland, C Sweden 60°04′N 15°56′E
14 K13 **Norcan Lake** ◎ Ontario, SE Canada
197 R12 **Nord** N Greenland 81°38′N 12°51′W
78 F13 **Nord** *Eng.* North. ◆ *province* N Cameroon
103 P2 **Nord** ◆ *department* N France
92 P1 **Nordaustlandet** *island* NE Svalbard
95 G24 **Nordborg** *Ger.* Nordburg. Syddanmark, SW Denmark 55°04′N 09°41′E
Nordburg *see* Nordborg
95 F23 **Nordby** Syddtjylland, W Denmark 55°27′N 08°25′E
11 P15 **Nordegg** Alberta, SW Canada 52°27′N 116°06′W
100 E9 **Norden** Niedersachsen, NW Germany 53°36′N 07°12′E
100 G10 **Nordenham** Niedersachsen, NW Germany 53°30′N 08°29′E
122 M6 **Nordenshel'da, Arkhipelag** *island group* N Russian Federation
92 O3 **Nordenskiold Land** *physical region* W Svalbard
100 E9 **Norderney** *island* NW Germany
100 J9 **Norderstedt** Schleswig-Holstein, N Germany 53°42′N 09°59′E
94 D11 **Nordfjord** *fjord* S Norway
94 C11 **Nordfjord** *physical region* S Norway
94 D11 **Nordfjordeid** Sogn og Fjordane, S Norway 61°54′N 06°E
92 G11 **Nordfold** Nordland, C Norway 67°48′N 15°16′E
Nordfriesische Inseln *see* North Frisian Islands
100 H7 **Nordfriesland** *cultural region* N Germany
Nordgrønland *see* Avannaarsua
101 K15 **Nordhausen** Thüringen, C Germany 51°31′N 10°48′E
25 T13 **Nordheim** Texas, SW USA 28°55′N 97°36′W
94 C13 **Nordhordland** *physical region* S Norway
100 E12 **Nordhorn** Niedersachsen, NW Germany 52°26′N 07°04′E
172 H16 **Nord, Île du** *island* Inner Islands, NE Seychelles
95 F20 **Nordjylland** ◆ *county* N Denmark
92 K7 **Nordkapp** *Eng.* North Cape. *headland* N Norway 25°47′E 71°10′N
92 O1 **Nordkapp** *headland* N Svalbard 80°31′N 19°58′E
79 N19 **Nord Kivu** *off.* Région du Nord Kivu. ◆ *region* E Dem. Rep. Congo
Nord Kivu, Région du *see* Nord-Kivu
92 G12 **Nordland** ◆ *county* C Norway
101 J21 **Nördlingen** Bayern, S Germany 48°49′N 10°28′E
93 I16 **Nordmaling** Västerbotten, N Sweden 63°35′N 19°30′E
95 K15 **Nordmark** Värmland, C Sweden 59°52′N 14°04′E
Nord, Mer du *see* North Sea
94 F8 **Nordmøre** *physical region* S Norway
100 I8 **Nord-Ostsee-Kanal** *canal* N Germany
0 J3 **Nordostrundingen** *cape* NE Greenland
79 D14 **Nord-Ouest** *Eng.* North-West. ◆ *province* NW Cameroon
Nord-Ouest, Territoires du *see* Northwest Territories
103 N2 **Nord-Pas-de-Calais** ◆ *region* N France
101 F19 **Nordpfälzer Bergland** ▲▲ W Germany
Nord, Pointe *see* Fatua, Pointe
187 P16 **Nord, Province** ◆ *province* C New Caledonia
101 D14 **Nordrhein-Westfalen** *Eng.* North Rhine-Westphalia, *Fr.* Rhénanie du Nord-Westphalie. ◆ *state* W Germany
Nordsee/Nordsjøen/Nordsøen *see* North Sea
100 H7 **Nordstrand** *island* N Germany
93 E15 **Nord-Trøndelag** ◆ *county* C Norway
92 I1 **Norðurfjörður** Vestfirðir, NW Iceland 66°01′N 21°33′W
92 J1 **Norðurland Eystra** ◆ *region* N Iceland
92 I2 **Norðurland Vestra** ◆ *region* N Iceland
97 E19 **Nore** *Ir.* An Fheoir. ≈ S Ireland
29 Q14 **Norfolk** Nebraska, C USA 42°01′N 97°25′W
21 X7 **Norfolk** Virginia, NE USA 36°51′N 76°17′W
97 P19 **Norfolk** *cultural region* E England, United Kingdom
192 K10 **Norfolk Island** ◇ *Australian self-governing territory* SW Pacific Ocean
175 P9 **Norfolk Ridge** *undersea feature* W Pacific Ocean
27 U8 **Norfork Lake** ⊠ Arkansas/Missouri, C USA
98 N6 **Norg** Drenthe, NE Netherlands 53°04′N 06°28′E
Norge *see* Norway
95 D14 **Norheimsund** Hordaland, S Norway 60°22′N 06°09′E
25 S16 **Norias** Texas, SW USA 26°47′N 97°45′W
164 L12 **Norikura-dake** ▲ Honshū, S Japan 36°06′N 137°33′E
122 K8 **Noril'sk** Krasnoyarskiy Kray, N Russian Federation 69°21′N 88°02′E
14 I13 **Norland** Ontario, SE Canada 44°46′N 78°48′W
21 V8 **Norlina** North Carolina, SE USA 36°26′N 78°11′W
30 L13 **Normal** Illinois, N USA 40°30′N 88°59′W
27 N11 **Norman** Oklahoma, C USA 35°13′N 97°27′W
Norman *see* Tulita
186 G9 **Normanby Island** *island* SE Papua New Guinea
Normandes, Îles *see* Channel Islands
58 G9 **Normandia** Roraima, N Brazil 03°57′N 59°39′W
102 L5 **Normandie** *Eng.* Normandy. *cultural region* N France
102 J5 **Normandie, Collines de** *hill range* NW France
Normandy *see* Normandie
25 V9 **Normangee** Texas, SW USA 31°01′N 96°06′W
21 Q10 **Norman, Lake** ⊠ North Carolina, SE USA
44 K13 **Norman Manley** ✈ (Kingston) E Jamaica 17°55′N 76°46′W
181 U5 **Norman River** ≈ Queensland, NE Australia
181 U4 **Normanton** Queensland, NE Australia 17°49′S 141°08′E
8 I8 **Norman Wells** Northwest Territories, NW Canada 65°18′N 126°42′W
12 H12 **Normétal** Québec, S Canada 48°59′N 79°23′W
163 O7 **Norovlin** *var.* Uldz. Hentiy, NE Mongolia 48°47′N 112°01′E
11 V15 **Norquay** Saskatchewan, S Canada 51°51′N 102°04′W
Norra Karelen *see* Pohjois-Karjala
93 G15 **Norråker** Jämtland, C Sweden 64°25′N 15°40′E
94 N12 **Norrala** Gävleborg, C Sweden 61°22′N 17°04′E
Norra Ny *see* Stöllet
Norra Österbotten *see* Pohjois-Pohjanmaa
Norra Savolax *see* Pohjois-Savo
92 G13 **Norra Storfjället** ▲ N Sweden 65°57′N 15°15′E
92 I13 **Norrbotten** ◆ *county* N Sweden
94 N11 **Norrdellen** ◎ C Sweden
95 G23 **Nørre Aaby** *var.* Nørre Åby. Syddtjylland, C Denmark 55°28′N 09°53′E
Nørre Åby *see* Nørre Aaby
95 I24 **Nørre Alslev** Sjælland, SE Denmark 54°54′N 11°53′E
95 E23 **Nørre Nebel** Syddtjylland, W Denmark 55°45′N 08°16′E
95 G20 **Nørresundby** Nordjylland, N Denmark 57°05′N 09°55′E
21 N8 **Norris Lake** ⊠ Tennessee, S USA
18 I15 **Norristown** Pennsylvania, NE USA 40°07′N 75°20′W
95 N17 **Norrköping** Östergötland, S Sweden 58°35′N 16°10′E
Norrmark *see* Noormarkku
94 N13 **Norrsundet** Gävleborg, C Sweden 60°55′N 17°09′E
95 P15 **Norrtälje** Stockholm, C Sweden 59°46′N 18°42′E
180 L12 **Norseman** Western Australia 32°16′S 121°46′E
93 I14 **Norsjö** Västerbotten, N Sweden 64°55′N 19°30′E
95 G16 **Norsjø** ◎ S Norway
123 R13 **Norsk** Amurskaya Oblast', SE Russian Federation 52°20′N 129°57′E
Norske Havet *see* Norwegian Sea
187 Q13 **Norsup** Malekula, C Vanuatu 16°05′S 167°24′E
191 V15 **Norte, Cabo** *headland* Easter Island, Chile, E Pacific Ocean 27°03′S 109°24′W
54 F7 **Norte de Santander** *off.* Departamento de Norte de Santander. ◆ *province* N Colombia
Norte de Santander, Departamento de *see* Norte de Santander
61 E21 **Norte, Punta** *headland* E Argentina 36°17′S 56°46′W
21 R13 **North** South Carolina, SE USA 33°37′N 81°06′W
North *see* Nord
18 L10 **North Adams** Massachusetts, NE USA 42°40′N 73°06′W
113 L17 **North Albanian Alps** *Alb.* Bjeshkët e Namuna, *SCr.* Prokletije. ▲▲ SE Europe
97 M15 **Northallerton** N England, United Kingdom 54°20′N 01°26′W
180 J12 **Northam** Western Australia 31°40′S 116°40′E
83 J20 **Northam** Northern, N South Africa 24°56′S 27°18′E
1 **North America** *continent*
1 N12 **North American Basin** *undersea feature* W Sargasso Sea 30°00′N 60°00′W
0 C5 **North American Plate** *tectonic feature*
18 M11 **North Amherst** Massachusetts, NE USA 42°24′N 72°31′W
97 N20 **Northampton** C England, United Kingdom 52°14′N 00°54′W
97 M20 **Northamptonshire** *cultural region* C England, United Kingdom
151 P18 **North Andaman** *island* Andaman Islands, India, NE Indian Ocean
65 D25 **North Arm** East Falkland, Falkland Islands 52°06′S 59°21′W
21 Q13 **North Augusta** South Carolina, SE USA 33°30′N 81°58′W
173 W8 **North Australian Basin** *Fr.* Bassin Nord de l' Australie. *undersea feature* E Indian Ocean
31 R11 **North Baltimore** Ohio, N USA 41°10′N 83°40′W
11 T15 **North Battleford** Saskatchewan, S Canada 52°47′N 108°19′W
14 H11 **North Bay** Ontario, S Canada 46°20′N 79°28′W
12 H6 **North Belcher Islands** *island group* Belcher Islands, Nunavut, C Canada
29 R15 **North Bend** Nebraska, C USA 41°27′N 96°46′W
32 E14 **North Bend** Oregon, NW USA 43°24′N 124°13′W
96 K12 **North Berwick** SE Scotland, United Kingdom 56°04′N 02°44′W
North Beveland *see* Noord-Beveland
North Borneo *see* Sabah
183 P5 **North Bourke** New South Wales, SE Australia 30°03′S 145°56′E
North Brabant *see* Noord-Brabant
182 F2 **North Branch Neales** *seasonal river* South Australia
44 M6 **North Caicos** *island* NW Turks and Caicos Islands
26 L10 **North Canadian River** ≈ Oklahoma, C USA
31 U12 **North Canton** Ohio, N USA 40°52′N 81°24′W
13 R13 **North, Cape** *headland* Cape Breton Island, Nova Scotia, SE Canada 47°06′N 60°24′W
184 I1 **North Cape** *headland* North Island, New Zealand 34°23′S 173°02′E
186 G5 **North Cape** *headland* New Ireland, NE Papua New Guinea 02°33′S 150°48′E
North Cape *see* Nordkapp
18 J17 **North Cape May** New Jersey, NE USA 38°59′N 74°55′W
12 C9 **North Caribou Lake** ◎ Ontario, C Canada
21 U10 **North Carolina** *off.* State of North Carolina, *also known as* Old North State, Tar Heel State, Turpentine State. ◆ *state* SE USA
North Celebes *see* Sulawesi Utara
155 J24 **North Central** ◆ *province* N Sri Lanka
31 S4 **North Channel** *lake channel* Canada/USA
97 G14 **North Channel** *strait* Northern Ireland/Scotland, United Kingdom
21 S14 **North Charleston** South Carolina, SE USA 32°53′N 79°59′W
31 N10 **North Chicago** Illinois, N USA 42°19′N 87°50′W
195 Y10 **Northcliffe Glacier** *glacier* Antarctica
31 Q14 **North College Hill** Ohio, N USA 39°13′N 84°33′W
25 O8 **North Concho River** ≈ Texas, SW USA
19 O8 **North Conway** New Hampshire, NE USA 44°03′N 71°06′W
27 V14 **North Crossett** Arkansas, C USA 33°10′N 91°56′W
28 L4 **North Dakota** *off.* State of North Dakota, *also known as* Flickertail State, Peace Garden State, Sioux State. ◆ *state* N USA
North Devon Island *see* Devon Island
97 O22 **North Downs** *hill range* SE England, United Kingdom
18 C11 **North East** Pennsylvania, NE USA 42°13′N 79°49′W
83 I18 **North East** ◆ *district* NE Botswana
65 G15 **North East Bay** *bay* Ascension Island, C Atlantic Ocean
38 L10 **Northeast Cape** *headland* Saint Lawrence Island, Alaska, USA 63°16′N 168°50′W
North East Frontier Agency/North East Frontier Agency of Assam *see* Arunāchal Pradesh
65 E25 **North East Island** *island* E Falkland Islands
189 V11 **Northeast Island** *island* Chuuk, C Micronesia
44 L12 **North East Point** *headland* E Jamaica 18°09′N 76°19′W
191 Z2 **Northeast Point** *headland* Kiritimati, E Kiribati 10°23′S 105°45′E
44 L6 **Northeast Point** *headland* Great Inagua, S The Bahamas 21°18′N 73°01′W
44 K5 **Northeast Point** *headland* Acklins Island, SE The Bahamas 22°43′N 73°50′W
44 H2 **Northeast Providence Channel** *channel* N The Bahamas
101 J14 **Northeim** Niedersachsen, C Germany 51°42′N 10°E
29 X14 **North English** Iowa, C USA 41°30′N 92°04′W
138 G8 **Northern** ◆ *district* N Israel
82 M12 **Northern** ◆ *region* N Malawi
186 F8 **Northern** *var.* Oro. ◆ *province* S Papua New Guinea
155 J23 **Northern** ◆ *province* N Sri Lanka
80 D7 **Northern** ◆ *state* N Sudan
82 K12 **Northern** ◆ *province* NE Zambia
Northern *see* Limpopo
80 B13 **Northern Bahr el Ghazal** ◆ *state* NW South Sudan
Northern Border Region *see* Al Ḥudūd ash Shamālīyah
83 F24 **Northern Cape** *off.* Northern Cape Province, *Afr.* Noord-Kaap. ◆ *province* W South Africa
Northern Cape Province *see* Northern Cape
190 K14 **Northern Cook Islands** *island group* N Cook Islands
80 B8 **Northern Darfur** ◆ *state* NW Sudan
Northern Dvina *see* Severnaya Dvina
97 F14 **Northern Ireland** *var.* The Six Counties. *cultural region* Northern Ireland, United Kingdom
97 F14 **Northern Ireland** *var.* The Six Counties. ◆ *political division* Northern Ireland, United Kingdom
80 D9 **Northern Kordofan** ◆ *state* C Sudan
187 Z14 **Northern Lau Group** *island group* Lau Group, NE Fiji
188 K3 **Northern Mariana Islands** ◇ *US commonwealth territory* W Pacific Ocean
Northern Rhodesia *see* Zambia
Northern Sporades *see* Vóreies Sporádes
182 D1 **Northern Territory** ◆ *territory* N Australia
Northern Transvaal *see* Limpopo
Northern Ural Hills *see* Severnyye Uvaly
84 I9 **North European Plain** *plain* N Europe
27 V2 **North Fabius River** ≈ Missouri, C USA
65 D24 **North Falkland Sound** *sound* N Falkland Islands
29 V9 **Northfield** Minnesota, N USA 44°27′N 93°10′W
19 O9 **Northfield** New Hampshire, NE USA 43°26′N 71°34′W
175 Q8 **North Fiji Basin** *undersea feature* N Coral Sea
97 Q22 **North Foreland** *headland* SE England, United Kingdom 51°22′N 01°26′E
35 P6 **North Fork American River** ≈ California, W USA
39 R7 **North Fork Chandalar River** ≈ Alaska, USA
28 K7 **North Fork Grand River** ≈ North Dakota/South Dakota, N USA
21 O6 **North Fork Kentucky River** ≈ Kentucky, S USA
39 Q7 **North Fork Koyukuk River** ≈ Alaska, USA
39 Q10 **North Fork Kuskokwim River** ≈ Alaska, USA
26 K11 **North Fork Red River** ≈ Oklahoma/Texas, SW USA
26 K3 **North Fork Solomon River** ≈ Kansas, C USA
23 W14 **North Fort Myers** Florida, SE USA 26°40′N 81°52′W
31 P5 **North Fox Island** *island* Michigan, N USA
100 G6 **North Frisian Islands** *var.* Nordfriesische Inseln. *island group* N Germany
197 N9 **North Geomagnetic Pole** *pole* Arctic Ocean
18 M13 **North Haven** Connecticut, NE USA 41°25′N 72°51′W
184 J5 **North Head** *headland* North Island, New Zealand 36°23′S 174°01′E
18 L6 **North Hero** Vermont, NE USA 44°49′N 73°14′W
35 O7 **North Highlands** California, W USA 38°40′N 121°25′W
North Holland *see* Noord-Holland
81 I16 **North Horr** Marsabit, N Kenya 03°17′N 37°08′E
151 K21 **North Huvadhu Atoll** *var.* Gaafu Alifu Atoll. *atoll* S Maldives
65 A24 **North Island** *island* W Falkland Islands
184 N9 **North Island** *island* N New Zealand
21 U14 **North Island** *island* South Carolina, SE USA
31 O11 **North Judson** Indiana, N USA 41°12′N 86°44′W
North Kalimantan *see* Kalimantan Utara
North Karelia *see* Pohjois-Karjala
North Kazakhstan *see* Severnyy Kazakhstan
31 V10 **North Kingsville** Ohio, N USA 41°54′N 80°41′W
163 Y13 **North Korea** *off.* Democratic People's Republic of Korea, *Kor.* Chosŏn-minjujuŭi-inmin-kanghwaguk. ◆ *republic* E Asia
153 X11 **North Lakhimpur** Assam, NE India 27°10′N 94°00′E
184 J3 **Northland** *off.* Northland Region. ◆ *region* North Island, New Zealand
192 K11 **Northland Plateau** *undersea feature* S Pacific Ocean
Northland Region *see* Northland
35 X11 **North Las Vegas** Nevada, W USA 36°12′N 115°07′W
31 O11 **North Liberty** Indiana, N USA 41°36′N 86°22′W
29 X14 **North Liberty** Iowa, C USA 41°45′N 91°36′W
27 V12 **North Little Rock** Arkansas, C USA 34°46′N 92°15′W
28 M13 **North Loup River** ≈ Nebraska, C USA
151 K18 **North Maalhosmadulu Atoll** *var.* North Malosmadulu Atoll, Raa Atoll. *atoll* N Maldives
31 U10 **North Madison** Ohio, N USA 41°48′N 81°03′W
North Malosmadulu Atoll *see* North Maalhosmadulu Atoll
31 P12 **North Manchester** Indiana, N USA 41°00′N 85°45′W
31 P6 **North Manitou Island** *island* Michigan, N USA
29 U10 **North Mankato** Minnesota, N USA 44°11′N 94°03′W
23 Z15 **North Miami** Florida, SE USA 25°54′N 80°11′W
151 K18 **North Miladhunmadulu Atoll** *var.* Shaviyani Atoll. *atoll* N Maldives
North Minch *see* Minch, The
23 W15 **North Naples** Florida, SE USA 26°13′N 81°47′W
175 P8 **North New Hebrides Trench** *undersea feature* N Coral Sea
23 Y15 **North New River Canal** ≈ Florida, SE USA
151 K20 **North Nilandhe Atoll** *atoll* C Maldives
36 L2 **North Ogden** Utah, W USA 41°18′N 111°57′W
North Ossetia *see* Severnaya Osetiya-Alaniya, Respublika
North Ostrobothnia *see* Pohjois-Pohjanmaa
35 S10 **North Palisade** ▲ California, W USA 37°06′N 118°31′W
189 U11 **North Pass** *passage* Chuuk Islands, C Micronesia
28 M15 **North Platte** Nebraska, C USA 41°07′N 100°46′W
33 X17 **North Platte River** ≈ C USA
65 G14 **North Point** *headland* Ascension Island, C Atlantic Ocean
172 I16 **North Point** *headland* Mahé, NE Seychelles 04°23′S 55°28′E
31 R5 **North Point** *headland* Michigan, N USA 45°21′N 83°30′W
31 S6 **North Point** *headland* Michigan, N USA 45°01′N 83°16′W
39 S9 **North Pole** Alaska, USA 64°42′N 147°09′W
197 R9 **North Pole** *pole* Arctic Ocean
23 O4 **Northport** Alabama, S USA 33°13′N 87°34′W
23 W14 **North Port** Florida, SE USA 27°03′N 82°15′W
32 L6 **Northport** Washington, NW USA 48°54′N 117°48′W
32 L12 **North Powder** Oregon, NW USA 45°00′N 117°56′W
29 U13 **North Raccoon River** ≈ Iowa, C USA
North Rhine-Westphalia *see* Nordrhein-Westfalen
97 M16 **North Riding** *cultural region* N England, United Kingdom
96 G5 **North Rona** *island* NW Scotland, United Kingdom
96 K4 **North Ronaldsay** *island* NE Scotland, United Kingdom
36 L2 **North Salt Lake** Utah, W USA 40°51′N 111°54′W
11 P15 **North Saskatchewan** ≈ Alberta/Saskatchewan, S Canada
35 X5 **North Schell Peak** ▲ Nevada, W USA 39°25′N 114°34′W
North Scotia Ridge *see* South Georgia Ridge
86 D10 **North Sea** *Dan.* Nordsøen, *Dut.* Noordzee, *Fr.* Mer du Nord, *Ger.* Nordsee, *Nor.* Nordsjøen; *prev.* German Ocean, *Lat.* Mare Germanicum. *sea* NW Europe
35 T6 **North Shoshone Peak** ▲ Nevada, W USA 39°08′N 117°28′W
North Siberian Lowland/North Siberian Plain *see* Severo-Sibirskaya Nizmennost'
29 R13 **North Sioux City** South Dakota, N USA 42°31′N 96°28′W
North Solomons *see* Bougainville
96 K4 **North Sound, The** *sound* N Scotland, United Kingdom
183 T4 **North Star** New South Wales, SE Australia 28°55′S 150°25′E
North Star State *see* Minnesota
183 V3 **North Stradbroke Island** *island* Queensland, E Australia
North Sulawesi *see* Sulawesi Utara
North Sumatra *see* Sumatera Utara
14 D17 **North Sydenham** ≈ Ontario, S Canada
18 H9 **North Syracuse** New York, NE USA 43°07′N 76°07′W
184 K9 **North Taranaki Bight** *gulf* North Island, New Zealand
12 H9 **North Twin Island** *island* Nunavut, C Canada
96 E8 **North Uist** *island* NW Scotland, United Kingdom
97 L14 **Northumberland** *cultural region* N England, United Kingdom
181 Y7 **Northumberland Isles** *island group* Queensland, NE Australia
13 Q14 **Northumberland Strait** *strait* SE Canada
32 G14 **North Umpqua River** ≈ Oregon, NW USA
45 Q13 **North Union** Saint Vincent, Saint Vincent and the Grenadines 13°15′N 61°07′W
10 L17 **North Vancouver** British Columbia, SW Canada 49°21′N 123°05′W
18 K9 **Northville** New York, NE USA 43°13′N 74°08′W
97 Q19 **North Walsham** E England, United Kingdom 52°49′N 01°22′E
39 T10 **Northway** Alaska, USA 62°57′N 141°56′W
83 G17 **North-West** ◆ *district* NW Botswana
83 G21 **North-West** *off.* North-West Province, *Afr.* Noordwes. ◆ *province* N South Africa
North-West *see* Nord-Ouest
64 I6 **Northwest Atlantic Mid-Ocean Canyon** *undersea feature* N Atlantic Ocean
180 G8 **North West Cape** *cape* Western Australia
38 J9 **Northwest Cape** *headland* Saint Lawrence Island, Alaska, USA 63°46′N 171°45′W
155 J24 **North Western** ◆ *province* W Sri Lanka
82 H13 **North Western** ◆ *province* W Zambia
North-West Frontier Province *see* Khyber Pakhtunkhwa
96 H8 **North West Highlands** ▲▲ N Scotland, United Kingdom
192 J4 **Northwest Pacific Basin** *undersea feature* NW Pacific Ocean 40°00′N 150°00′E
191 Y2 **Northwest Point** *headland* Kiritimati, E Kiribati 10°25′S 105°35′E
44 G1 **Northwest Providence Channel** *channel* N The Bahamas
North-West Province *see* North-West
13 Q8 **North West River** Newfoundland and Labrador, E Canada 53°30′N 60°10′W
8 J9 **Northwest Territories** *Fr.* Territoires du Nord-Ouest. ◆ *territory* NW Canada
97 K18 **Northwich** C England, United Kingdom 53°16′N 02°32′W
25 Q5 **North Wichita River** ≈ Texas, SW USA
18 J17 **North Wildwood** New Jersey, NE USA 39°00′N 74°45′W
21 R9 **North Wilkesboro** North Carolina, SE USA 36°09′N 81°09′W
19 P8 **North Windham** Maine, NE USA 43°51′N 70°25′W
197 Q6 **Northwind Plain** *undersea feature* Arctic Ocean
29 V11 **Northwood** Iowa, C USA 43°26′N 93°13′W
29 Q4 **Northwood** North Dakota, N USA 47°43′N 97°34′W
97 M15 **North York Moors** *moorland* N England, United Kingdom
25 V9 **North Zulch** Texas, SW USA 30°54′N 96°06′W
26 K2 **Norton** Kansas, C USA 39°51′N 99°53′W
31 S13 **Norton** Ohio, N USA 40°25′N 83°04′W
21 P7 **Norton** Virginia, NE USA 36°56′N 82°37′W
39 N9 **Norton Bay** *bay* Alaska, USA
Norton de Matos *see* Balombo
31 O9 **Norton Shores** Michigan, N USA 43°10′N 86°15′W
38 M10 **Norton Sound** *inlet* Alaska, USA
27 Q3 **Nortonville** Kansas, C USA 39°25′N 95°19′W
102 I8 **Nort-sur-Erdre** Loire-Atlantique, NW France 47°27′N 01°30′W
195 N2 **Norvegia, Cape** *headland* Antarctica 71°16′S 12°25′W
18 L13 **Norwalk** Connecticut, NE USA 41°08′N 73°28′W
29 V14 **Norwalk** Iowa, C USA 41°28′N 93°40′W
31 S11 **Norwalk** Ohio, N USA 41°14′N 82°37′W
19 P7 **Norway** Maine, NE USA 44°13′N 70°30′W
31 N5 **Norway** Michigan, N USA 45°47′N 87°54′W
93 E17 **Norway** *off.* Kingdom of Norway, *Nor.* Norge. ◆ *monarchy* N Europe
11 X13 **Norway House** Manitoba, C Canada 53°59′N 97°50′W
Norway, Kingdom of *see* Norway
197 R16 **Norwegian Basin** *undersea feature* NW Norwegian Sea 68°00′N 02°00′W
84 D6 **Norwegian Sea** *var.* Norske Havet. *sea* NE Atlantic Ocean
197 S17 **Norwegian Trench** *undersea feature* NE North Sea 59°00′N 04°30′E
14 F16 **Norwich** Ontario, S Canada 42°57′N 80°37′W
97 Q19 **Norwich** E England, United Kingdom 52°38′N 01°18′E
19 N13 **Norwich** Connecticut, NE USA 41°30′N 72°02′W
18 I11 **Norwich** New York, NE USA 42°31′N 75°31′W
29 U9 **Norwood** Minnesota, N USA 44°46′N 93°55′W
31 Q15 **Norwood** Ohio, N USA 39°07′N 84°27′W
14 H11 **Nosbonsing, Lake** ◎ Ontario, S Canada
Nösen *see* Bistrița
165 T1 **Noshappu-misaki** *headland* Hokkaidō, NE Japan 45°26′N 141°38′E
165 P7 **Noshiro** *var.* Nosiro; *prev.* Noshirominato. Akita, Honshū, C Japan 40°11′N 140°02′E
Noshirominato/Nosiro *see* Noshiro
117 Q3 **Nosivka** *Rus.* Nosovka. Chernihivs'ka Oblast', NE Ukraine 50°55′N 31°37′E
67 T14 **Nosop** *var.* Nossob, Nossop. ≈ Botswana/Namibia
83 E20 **Nossop** ≈ E Namibia
125 S4 **Nosovaya** Nenetskiy Avtonomnyy Okrug, NW Russian Federation 68°12′N 54°33′E
Nosovka *see* Nosivka
143 V11 **Noṣratābād** Sīstān va Balūchestān, E Iran 29°53′N 59°57′E
95 J18 **Nossebro** Västra Götaland, S Sweden 58°12′N 12°42′E
96 K6 **Noss Head** *headland* N Scotland, United Kingdom 58°29′N 03°03′W
Nossi-Bé *see* Be, Nosy
Nossob/Nossop *see* Nosop
172 J2 **Nosy Be** ✈ Antsiranana, N Madagascar 23°36′S 47°36′E
172 J6 **Nosy Varika** Fianarantsoa, SE Madagascar 20°36′S 48°31′E
14 L10 **Notawassi** ≈ Québec, SE Canada
14 M9 **Notawassi, Lac** ◎ Québec, SE Canada
36 J5 **Notch Peak** ▲ Utah, W USA 39°08′N 113°24′W
110 G10 **Noteć** *Ger.* Netze. ≈ NW Poland
Nóties Sporádes *see* Dodekánisa
115 J22 **Nótion Aigaíon** *Eng.* Aegean South. ◆ *region* E Greece
115 H18 **Nótios Evvoïkós Kólpos** *gulf* E Greece
115 B16 **Nótio Stenó Kérkyras** *strait* W Greece
107 L25 **Noto** *anc.* Netum. Sicilia, Italy, C Mediterranean Sea 36°53′N 15°05′E
164 M10 **Noto** Ishikawa, Honshū, SW Japan 37°18′N 137°11′E
95 G15 **Notodden** Telemark, S Norway 59°35′N 09°18′E
107 L25 **Noto, Golfo di** *gulf* Sicilia, Italy, C Mediterranean Sea
164 L10 **Noto-hantō** *peninsula* Honshū, SW Japan
164 L11 **Noto-jima** *island* SW Japan
13 T11 **Notre Dame Bay** *bay* Newfoundland, Newfoundland and Labrador, E Canada
15 P6 **Notre-Dame-de-Lorette** Québec, SE Canada 49°05′N 72°24′W
14 L11 **Notre-Dame-de-Pontmain** Québec, SE Canada 46°18′N 75°37′W
15 T8 **Notre-Dame-du-Lac** Québec, SE Canada 47°36′N 68°48′W
15 Q6 **Notre-Dame-du-Rosaire** Québec, SE Canada 48°48′N 71°27′W
15 U8 **Notre-Dame, Monts** ▲▲ Québec, S Canada
77 R16 **Notsé** S Togo 06°59′N 01°12′E
14 G14 **Nottawasaga** ≈ Ontario, S Canada
14 G14 **Nottawasaga Bay** *lake bay* Ontario, S Canada
12 I11 **Nottaway** ≈ Québec, SE Canada
23 S1 **Nottely Lake** ⊠ Georgia, SE USA
95 H16 **Nøtterøy** *island* S Norway
97 M19 **Nottingham** C England, United Kingdom 52°58′N 01°10′W
9 E14 **Nottingham Island** *island* Nunavut, NE Canada
97 N18 **Nottinghamshire** *cultural region* C England, United Kingdom
21 V7 **Nottoway** Virginia, NE USA 37°07′N 78°03′W
21 V7 **Nottoway River** ≈ Virginia, NE USA
76 G7 **Nouâdhibou** *prev.* Port-Étienne. Dakhlet Nouâdhibou, W Mauritania 20°54′N 17°01′W
76 G7 **Nouâdhibou** ✈ Dakhlet Nouâdhibou, W Mauritania 20°59′N 17°02′W
76 F7 **Nouâdhibou, Dakhlet** *prev.* Baie du Lévrier. *bay* W Mauritania
76 F7 **Nouâdhibou, Râs** *prev.* Cap Blanc. *headland* NW Mauritania 20°48′N 17°03′W
76 G9 **Nouakchott** ● (Mauritania) Nouakchott District, SW Mauritania 18°09′N 15°58′W
76 G9 **Nouakchott** ✈ Trarza, SW Mauritania 18°18′N 15°54′W
120 J11 **Noual, Sebkhet en** *var.* Sabkhat an Nawāl. *salt flat* C Tunisia
76 G8 **Nouâmghâr** *var.* Nouamrhar. Dakhlet Nouâdhibou, W Mauritania 19°22′N 16°31′W
Nouamrhar *see* Nouâmghâr
Nouă Sulița *see* Novoselytsya
187 Q17 **Nouméa** ○ (New Caledonia) Province Sud, S New Caledonia 22°13′S 166°29′E
79 E15 **Noun** ≈ C Cameroon
77 N12 **Nouna** W Burkina Faso 12°44′N 03°54′W
83 H24 **Noupoort** Northern Cape, C South Africa 31°11′S 24°58′E

◆ Country ● Country Capital ◇ Dependent Territory ○ Dependent Territory Capital ◆ Administrative Regions ✈ International Airport ▲ Mountain ▲▲ Mountain Range Volcano ≈ River ◎ Lake ⊠ Reservoir
297

NOUVEAU-BRUNSWICK – NZÉRÉKORÉ

Nouveau-Brunswick see New Brunswick
Nouveau-Comptoir see Wemindji
15 T4 **Nouvel, Lacs** ◎ Québec, SE Canada
15 W7 **Nouvelle** Québec, SE Canada 48°07′N 66°16′W
15 W7 **Nouvelle** ≈ Québec, SE Canada
Nouvelle-Calédonie see New Caledonia
Nouvelle Écosse see Nova Scotia
103 R3 **Nouzonville** Ardennes, N France 49°49′N 04°45′E
147 Q11 **Nov** Rus. Nau. NW Tajikistan 40°10′N 69°16′E
59 I21 **Nova Alvorada** Mato Grosso do Sul, SW Brazil 21°25′S 54°19′W
Novabad see Navobod
111 D19 **Nová Bystřice** Ger. Neubistritz. Jihočeský Kraj, S Czech Republic 49°N 15°05′E
116 H13 **Novaci** Gorj, SW Romania 45°07′N 23°37′E
Nova Civitas see Neustadt an der Weinstrasse
Novaesium see Neuss
60 H10 **Nova Esperança** Paraná, S Brazil 23°09′S 52°13′W
106 H11 **Novafeltria** Marche, C Italy 43°54′N 12°18′E
60 Q9 **Nova Friburgo** Rio de Janeiro, SE Brazil 22°16′S 42°34′W
82 D12 **Nova Gaia** var. Cambundi-Catembo. Malanje, NE Angola 10°09′S 17°31′E
109 S12 **Nova Gorica** W Slovenia 45°57′N 13°40′E
112 G10 **Nova Gradiška** Ger. Neugradisk, Hung. Újgradiska. Brod-Posavina, NE Croatia 45°15′N 17°23′E
60 K7 **Nova Granada** São Paulo, S Brazil 20°33′S 49°19′W
60 O10 **Nova Iguaçu** Rio de Janeiro, SE Brazil 22°31′S 44°05′W
117 S10 **Nova Kakhovka** Rus. Novaya Kakhovka. Khersons'ka Oblast', SE Ukraine 46°45′N 33°20′E
Nová Karvinná see Karviná
Nova Lamego see Gabú
Nova Lisboa see Huambo
112 C11 **Novalja** Lika-Senj, W Croatia 44°33′N 14°53′E
119 M14 **Novalukoml'** Rus. Novolukoml'. Vitsyebskaya Voblasts', N Belarus 54°40′N 29°09′E
Nova Mambone see Mambone
83 P16 **Nova Nabúri** Zambézia, NE Mozambique 16°47′S 38°55′E
117 Q9 **Nova Odesa** var. Novaya Odessa. Mykolayivs'ka Oblast', S Ukraine 47°19′N 31°45′E
60 H10 **Nova Olímpia** Paraná, S Brazil 23°28′S 53°12′W
61 I15 **Nova Prata** Rio Grande do Sul, S Brazil 28°45′S 51°37′W
14 H12 **Novar** Ontario, S Canada 45°26′N 79°14′W
106 C7 **Novara** anc. Novaria. Piemonte, NW Italy 45°27′N 08°36′E
Novaria see Novara
13 P15 **Nova Scotia** Fr. Nouvelle Écosse. ◐ province SE Canada
0 M9 **Nova Scotia** physical region SE Canada
34 M8 **Novato** California, W USA 38°06′N 122°35′W
192 M7 **Nova Trough** undersea feature W Pacific Ocean
116 L7 **Nova Ushytsya** Khmel'nyts'ka Oblast', W Ukraine 48°50′N 27°16′E
83 M17 **Nova Vanduzi** Manica, C Mozambique 18°54′S 33°18′E
117 U5 **Nova Vodolaha** Rus. Novaya Vodolaga. Kharkivs'ka Oblast', E Ukraine 49°43′N 35°49′E
123 O12 **Novaya Chara** Zabaykal'skiy Kray, S Russian Federation 56°45′N 117°58′E
122 M12 **Novaya Igirma** Irkutskaya Oblast', C Russian Federation 57°08′N 103°52′E
Novaya Kakhovka see Nova Kakhovka
Novaya Kazanka see Zhanakazan
124 I12 **Novaya Ladoga** Leningradskaya Oblast', NW Russian Federation 60°03′N 32°15′E
127 R5 **Novaya Malykla** Ul'yanovskaya Oblast', W Russian Federation 54°13′N 49°55′E
Novaya Odessa see Nova Odesa
123 Q5 **Novaya Sibir', Ostrov** island Novosibirskiye Ostrova, NE Russian Federation
Novaya Vodolaga see Nova Vodolaha
122 I6 **Novaya Zemlya** island group N Russian Federation
Novaya Zemlya Trough see East Novaya Zemlya Trough
114 K10 **Nova Zagora** Sliven, C Bulgaria 42°29′N 26°00′E
105 S12 **Novelda** Valenciana, E Spain 38°24′N 00°45′W
111 H19 **Nové Mesto nad Váhom** Ger. Waagneustadtl, Hung. Vágújhely. Trenčiansky Kraj, W Slovakia 48°46′N 17°50′E
111 F17 **Nové Město na Moravě** Ger. Neustadtl in Mähren. Vysočina, C Czech Republic 49°34′N 16°05′E
Novesium see Neuss
111 I21 **Nové Zámky** Ger. Neuhäusel, Hung. Érsekújvár. Nitriansky Kraj, SW Slovakia 49°00′N 18°10′E
Novgorod see Velikiy Novgorod
Novgorod-Severskiy see Novhorod-Sivers'kyy
122 C7 **Novgorodskaya Oblast'** ◐ province W Russian Federation
117 R8 **Novhorodka** Kirovohrads'ka Oblast', C Ukraine 48°21′N 32°38′E
117 R2 **Novhorod-Sivers'kyy** Rus. Novgorod-Severskiy. Chernihivs'ka Oblast', NE Ukraine 52°00′N 33°15′E
31 R10 **Novi** Michigan, N USA 42°28′N 83°28′W
Novi see Novi Vinodolski
112 L9 **Novi Bečej** prev. Új-Becse, Vološinovo, Ger. Neubetsche, Hung. Törökbecse. Vojvodina, N Serbia 45°36′N 20°09′E
116 M3 **Novi Bilokorovychi** Rus. Belokorovichi; prev. Bilokorovychi. Zhytomyrs'ka Oblast', N Ukraine 51°07′N 28°02′E
25 Q8 **Novice** Texas, SW USA 32°00′N 99°38′W
112 A9 **Novigrad** Istra, NW Croatia 45°19′N 13°33′E
Novi Grad see Bosanski Novi
114 G9 **Novi Iskar** Sofia Grad, W Bulgaria 42°46′N 23°19′E
106 C9 **Novi Ligure** Piemonte, NW Italy 44°46′N 08°47′E
99 L22 **Noville** Luxembourg, SE Belgium 50°04′N 05°46′E
194 I10 **Noville Peninsula** peninsula Thurston Island, Antarctica
Noviodunum see Soissons, Aisne, France
Noviodunum see Nevers, Nièvre, France
Noviodunum see Nyon, Vaud, Switzerland
Noviomagus see Lisieux, Calvados, France
Noviomagus see Nijmegen, Netherlands
114 M8 **Novi Pazar** Shumen, NE Bulgaria 43°20′N 27°12′E
113 M15 **Novi Pazar** Turk. Yenipazar. Serbia, S Serbia 43°09′N 20°31′E
112 K10 **Novi Sad** Ger. Neusatz, Hung. Újvidék. Vojvodina, N Serbia 45°16′N 19°49′E
117 T6 **Novi Sanzhary** Poltavs'ka Oblast', C Ukraine 49°21′N 34°18′E
112 H12 **Novi Travnik** prev. Pučarevo. Federacija Bosne I Hercegovine, C Bosnia and Herzegovina 44°12′N 17°39′E
112 B10 **Novi Vinodolski** var. Novi. Primorje-Gorski Kotar, NW Croatia 45°08′N 14°46′E
58 F12 **Novo Airão** Amazonas, N Brazil 02°06′S 61°20′W
Novoalekseyevka see Kobda
127 N9 **Novoanninskiy** Volgogradskaya Oblast', SW Russian Federation 50°31′N 42°43′E
58 F13 **Novo Aripuanã** Amazonas, N Brazil 05°05′S 60°20′W
117 P7 **Novoarkhangel's'k** Kirovohrads'ka Oblast', C Ukraine 48°39′N 30°48′E
117 Y6 **Novoaydar** Luhans'ka Oblast', E Ukraine 49°00′N 39°00′E
117 X9 **Novoazovs'k** Rus. Novoazovsk. Donets'ka Oblast', E Ukraine 47°07′N 38°06′E
123 R14 **Novobureyskiy** Amurskaya Oblast', SE Russian Federation 49°42′N 129°46′E
127 Q3 **Novocheboksarsk** Chuvashskaya Respublika, W Russian Federation 56°07′N 47°33′E
127 R5 **Novocheremshansk** Ul'yanovskaya Oblast', W Russian Federation 54°23′N 50°08′E
126 L12 **Novocherkassk** Rostovskaya Oblast', SW Russian Federation 47°23′N 40°E
127 R6 **Novodevich'ye** Samarskaya Oblast', W Russian Federation 53°33′N 48°51′E
124 M8 **Novodvinsk** Arkhangel'skaya Oblast', NW Russian Federation 64°22′N 40°49′E
Novograd-Volynskiy see Novohrad-Volyns'kyy
Novogrudok see Navahrudak
Novogrudskaya Vozvyshennost' see Navahrudskaye Wzvyshsha
61 I15 **Novo Hamburgo** Rio Grande do Sul, S Brazil 29°42′S 51°07′W
59 H16 **Novo Horizonte** Mato Grosso, W Brazil 11°19′S 57°11′W
60 K8 **Novo Horizonte** São Paulo, S Brazil 21°27′S 49°14′W
116 M4 **Novohrad-Volyns'kyy** Rus. Novograd-Volynskiy. Zhytomyrs'ka Oblast', N Ukraine 50°34′N 27°32′E
145 O7 **Novoishimskiy** prev. Kuybyshevskiy. Severnyy Kazakhstan, N Kazakhstan 53°15′N 66°51′E
Novokazalinsk see Ayteke Bi
126 M8 **Novokhopersk** Voronezhskaya Oblast', W Russian Federation 51°09′N 41°34′E
127 R6 **Novokuybyshevsk** Samarskaya Oblast', W Russian Federation 53°06′N 49°56′E
122 J13 **Novokuznetsk** prev. Stalinsk. Kemerovskaya Oblast', S Russian Federation 53°45′N 87°12′E
195 R1 **Novolazarevskaya** Russian research station Antarctica 70°42′S 11°31′E
Novolukoml' see Novalukoml'
109 V12 **Novo mesto** Ger. Rudolfswert; prev. Ger. Neustadtl. SE Slovenia 45°49′N 15°09′E
126 K15 **Novomikhaylovskiy** Krasnodarskiy Kray, SW Russian Federation 44°18′N 38°49′E
112 L8 **Novo Miloševo** Vojvodina, N Serbia 45°43′N 20°20′E
Novomirgorod see Novomyrhorod
126 L5 **Novomoskovsk** Tul'skaya Oblast', W Russian Federation 54°05′N 38°23′E
117 U7 **Novomoskovs'k** Rus. Novomoskovsk. Dnipropetrovs'ka Oblast', E Ukraine 48°38′N 35°15′E
117 V8 **Novomykolayivka** Zaporiz'ka Oblast', SE Ukraine 47°58′N 35°54′E
117 Q7 **Novomyrhorod** Rus. Novomirgorod. Kirovohrads'ka Oblast', S Ukraine 48°46′N 31°39′E
127 N8 **Novonikolayevskiy** Volgogradskaya Oblast', SW Russian Federation 50°55′N 42°24′E
127 P10 **Novonikol'skoye** Volgogradskaya Oblast', SW Russian Federation 49°23′N 45°06′E
127 X7 **Novoorsk** Orenburgskaya Oblast', W Russian Federation 51°21′N 59°03′E
126 M13 **Novopokrovskaya** Krasnodarskiy Kray, SW Russian Federation 45°58′N 40°43′E
Novopolotsk see Navapolatsk
117 Y5 **Novopskov** Luhans'ka Oblast', E Ukraine 49°33′N 39°07′E
Novoradomsk see Radomsko
Novo Redondo see Sumbe
127 R8 **Novorepnoye** Saratovskaya Oblast', W Russian Federation 51°04′N 48°34′E
126 K14 **Novorossiysk** Krasnodarskiy Kray, SW Russian Federation 44°50′N 37°38′E
Novorossiyskiy/Novorossiyskoye see Akzhar
124 F15 **Novorzhev** Pskovskaya Oblast', W Russian Federation 57°01′N 29°19′E
Novoselitsa see Novoselytsya
117 S12 **Novoselivs'ke** Avtonomna Respublika Krym, S Ukraine 45°26′N 33°37′E
Novosëlki see Navasyolki
114 G6 **Novo Selo** Vidin, NW Bulgaria 44°08′N 22°48′E
113 M14 **Novo Selo** Serbia, C Serbia 43°39′N 20°54′E
116 K8 **Novoselytsya** Rom. Nouă Suliţa, Rus. Novoselitsa. Chernivets'ka Oblast', W Ukraine 48°14′N 26°18′E
127 U7 **Novosergiyevka** Orenburgskaya Oblast', W Russian Federation 52°04′N 53°40′E
126 L11 **Novoshakhtinsk** Rostovskaya Oblast', SW Russian Federation 47°48′N 39°51′E
122 J12 **Novosibirsk** Novosibirskaya Oblast', C Russian Federation 55°04′N 83°05′E
122 J12 **Novosibirskaya Oblast'** ◐ province C Russian Federation
122 M4 **Novosibirskiye Ostrova** Eng. New Siberian Islands. island group N Russian Federation
126 K6 **Novosil'** Orlovskaya Oblast', W Russian Federation 53°00′N 37°59′E
124 G16 **Novosokol'niki** Pskovskaya Oblast', W Russian Federation 56°21′N 30°07′E
127 Q6 **Novospasskoye** Ul'yanovskaya Oblast', W Russian Federation 53°08′N 47°48′E
Novotroickoje see Birlik
127 X8 **Novotroitsk** Orenburgskaya Oblast', W Russian Federation 51°10′N 58°18′E
Novotroitskoye see Brlik, Kazakhstan
Novotroitskoye see Novotroyits'ke, Ukraine
117 T11 **Novotroyits'ke** Rus. Novotroitskoye. Khersons'ka Oblast', S Ukraine 46°21′N 34°21′E
Novoukrainka see Novoukrayinka
117 Q8 **Novoukrayinka** Rus. Novoukrainka. Kirovohrads'ka Oblast', C Ukraine 48°19′N 31°33′E
127 Q5 **Novoul'yanovsk** Ul'yanovskaya Oblast', W Russian Federation 54°10′N 48°19′E
127 W8 **Novoural'sk** Orenburgskaya Oblast', W Russian Federation 51°19′N 56°57′E
Novo-Urgench see Urganch
116 I4 **Novovolyns'k** Rus. Novovolynsk. Volyns'ka Oblast', NW Ukraine 50°46′N 24°09′E
117 S9 **Novovorontsovka** Khersons'ka Oblast', S Ukraine 47°28′N 33°55′E
147 Y7 **Novovoznesenovka** Issyk-Kul'skaya Oblast', E Kyrgyzstan 42°36′N 78°44′E
125 R14 **Novovyatsk** Kirovskaya Oblast', NW Russian Federation 58°30′N 49°42′E
Novoyel'nya see Navayel'nya
117 O6 **Novozhyvotiv** Vinnyts'ka Oblast', C Ukraine 49°16′N 29°31′E
126 H6 **Novozybkov** Bryanskaya Oblast', W Russian Federation 52°36′N 31°58′E
112 F9 **Novska** Sisak-Moslavina, NE Croatia 45°20′N 16°58′E
Nový Bohumín see Bohumín
111 D15 **Nový Bor** Ger. Haida; prev. Bor u České Lípy, Hajda. Liberecký Kraj, N Czech Republic 50°46′N 14°32′E
111 E16 **Nový Bydžov** Ger. Neubidschow. Královéhradecký Kraj, N Czech Republic 50°15′N 15°27′E
119 G18 **Novy Dvor** Rus. Novyy Dvor. Hrodzyenskaya Voblasts', W Belarus 53°48′N 24°34′E
111 I17 **Nový Jičín** Ger. Neutitschein. Moravskoslezský Kraj, E Czech Republic 49°36′N 18°00′E
118 K12 **Novy Pahost** Rus. Novyy Pogost. Vitsyebskaya Voblasts', NW Belarus 55°30′N 27°29′E
Novyy Bug see Novyy Buh
117 R9 **Novyy Buh** Rus. Novyy Bug. Mykolayivs'ka Oblast', S Ukraine 47°40′N 32°30′E
117 Q4 **Novyy Bykiv** Chernihivs'ka Oblast', N Ukraine 50°36′N 31°39′E
Novyy Dvor see Novy Dvor
Novyye Aneny see Anenii Noi
127 P7 **Novyye Burasy** Saratovskaya Oblast', W Russian Federation 52°10′N 46°00′E
Novyy Margilan see Farg'ona
126 K8 **Novyy Oskol** Belgorodskaya Oblast', W Russian Federation 50°43′N 37°55′E
Novyy Pogost see Novy Pahost
127 R2 **Novyy Tor"yal** Respublika Mariy El, W Russian Federation 56°59′N 48°53′E
123 N12 **Novyy Uoyan** Respublika Buryatiya, S Russian Federation 56°06′N 111°27′E
122 J9 **Novyy Urengoy** Yamalo-Nenetskiy Avtonomnyy Okrug, N Russian Federation 66°06′N 76°25′E
Novyy Uzen' see Zhanaozen
111 N16 **Nowa Dęba** Podkarpackie, SE Poland 50°31′N 21°53′E
111 G15 **Nowa Ruda** Ger. Neurode. Dolnośląskie, SW Poland 50°34′N 16°30′E
110 F12 **Nowa Sól** var. Nowasól, Ger. Neusalz an der Oder. Lubuskie, W Poland 51°47′N 15°43′E
Nowasól see Nowa Sól
27 Q8 **Nowata** Oklahoma, C USA 36°42′N 95°36′W
142 M6 **Nowbarān** Markazī, W Iran 35°07′N 49°51′E
110 J8 **Nowe** Kujawski-pomorskie, N Poland 53°40′N 18°44′E
110 K9 **Nowe Miasto Lubawskie** Ger. Neumark. Warmińsko-Mazurskie, NE Poland 53°24′N 19°36′E
110 L13 **Nowe Miasto nad Pilicą** Mazowieckie, C Poland 51°37′N 20°34′E
110 D8 **Nowe Warpno** Ger. Neuwarp. Zachodnio-pomorskie, NW Poland 53°52′N 14°12′E
Nowgong see Nagaon
110 E8 **Nowogard** var. Nowógard, Ger. Naugard. Zachodnio-pomorskie, NW Poland 53°41′N 15°09′E
110 N9 **Nowogród** Podlaskie, NE Poland 53°14′N 21°52′E
Nowogródek see Navahrudak
111 E14 **Nowogrodziec** Ger. Naumburg am Queis. Dolnośląskie, SW Poland 51°12′N 15°24′E
Nowojelnia see Navayel'nya
Nowo-Minsk see Mińsk Mazowiecki
33 V13 **Nowood River** ≈ Wyoming, C USA
Nowo-Święciany see Švenčionėliai
183 S10 **Nowra-Bomaderry** New South Wales, SE Australia 34°51′S 150°41′E
149 T5 **Nowshera** var. Naushahra, Naushara. Khyber Pakhtunkhwa, NE Pakistan 34°00′N 72°00′E
110 J7 **Nowy Dwór Gdański** Ger. Tiegenhof. Pomorskie, N Poland 54°12′N 19°03′E
110 L11 **Nowy Dwór Mazowiecki** Mazowieckie, C Poland 52°26′N 20°43′E
111 M17 **Nowy Sącz** Ger. Neu Sandec. Małopolskie, S Poland 49°36′N 20°42′E
111 L18 **Nowy Targ** Ger. Neumark. Małopolskie, S Poland 49°28′N 20°00′E
110 F11 **Nowy Tomyśl** var. Nowy Tomysl. Wielkopolskie, C Poland 52°18′N 16°07′E
Nowy Tomysl see Nowy Tomyśl
148 M7 **Now Zād** var. Nauzad. Helmand, S Afghanistan 32°22′N 64°32′E
23 N4 **Noxubee River** ≈ Alabama/Mississippi, S USA
122 I10 **Noyabr'sk** Yamalo-Nenetskiy Avtonomnyy Okrug, N Russian Federation 63°08′N 75°19′E
102 I8 **Noyant** Maine-et-Loire, NW France 47°28′N 00°08′W
39 X14 **Noyes Island** island Alexander Archipelago, Alaska, USA
103 O3 **Noyon** Oise, N France 49°35′N 03°E
102 I7 **Nozay** Loire-Atlantique, NW France 47°34′N 01°36′W
82 L12 **Nsando** Northern, NE Zambia 10°22′S 31°14′E
83 N16 **Nsanje** Southern, S Malawi 16°57′S 35°10′E
77 Q17 **Nsawam** SE Ghana 05°47′N 00°19′W
79 E16 **Nsimalen** ✈ Centre, C Cameroon 19°15′N 81°22′E
82 K12 **Nsombo** Northern, NE Zambia 10°35′S 29°58′E
82 H13 **Ntambu** North Western, NW Zambia 12°21′S 25°03′E
83 N14 **Ntcheu** var. Ncheu. Central, S Malawi 14°49′S 34°37′E
79 D17 **Ntem** prev. Campo, Kampo. ≈ Cameroon/Equatorial Guinea
83 I14 **Ntemwa** North Western, NW Zambia 14°03′S 26°13′E
Ntlenyana, Mount see Thabana Ntlenyana
79 I19 **Ntomba, Lac** var. Lac Tumba. ◎ NW Dem. Rep. Congo
115 I19 **Ntóro, Kávo** prev. Akrotírio Kafiréas. cape Évvoia, C Greece
81 E19 **Ntungamo** SW Uganda 0°54′S 30°16′E
81 E18 **Ntusi** SW Uganda 0°05′N 31°13′E
83 H18 **Ntwetwe Pan** salt lake NE Botswana
93 M15 **Nuasjärvi** ◎ C Finland
80 F11 **Nuba Mountains** ▲▲ C Sudan
68 J9 **Nubian Desert** desert NE Sudan
116 G10 **Nucet** Hung. Diófás. Bihor, W Romania 46°28′N 22°35′E
Nu Chiang see Salween
145 U9 **Nuclear Testing Ground** nuclear site Pavlodar, E Kazakhstan
56 E9 **Nucuray, Río** ≈ N Peru
25 R15 **Nueces River** ≈ Texas, SW USA
9 V9 **Nueltin Lake** ◎ Manitoba/Northwest Territories, C Canada
99 K15 **Nuenen** Noord-Brabant, S Netherlands 51°29′N 05°36′E
62 G6 **Nuestra Señora, Bahía** bay N Chile
61 D14 **Nuestra Señora Rosario de Caa Catí** Corrientes, NE Argentina 27°48′S 57°42′W
54 J9 **Nueva Antioquia** Vichada, E Colombia 06°04′N 69°30′W
Nueva Caceres see Naga
41 O7 **Nueva Ciudad Guerrera** Tamaulipas, C Mexico 26°32′N 99°13′W
55 N4 **Nueva Esparta** off. Estado Nueva Esparta. ◐ state NE Venezuela
Nueva Esparta, Estado see Nueva Esparta
44 C5 **Nueva Gerona** Isla de la Juventud, S Cuba 21°49′N 82°49′W
42 H8 **Nueva Guadalupe** San Miguel, E El Salvador 13°30′N 88°21′W
42 M11 **Nueva Guinea** Región Autónoma Atlántico Sur, SE Nicaragua 11°40′N 84°22′W
61 D19 **Nueva Helvecia** Colonia, SW Uruguay 34°16′S 57°53′W
63 J25 **Nueva, Isla** island S Chile
40 M14 **Nueva Italia** Michoacán, SW Mexico 19°01′N 102°06′W
56 D6 **Nueva Loja** var. Lago Agrio. Sucumbíos, NE Ecuador
42 F6 **Nueva Ocotepeque** prev. Ocotepeque. Ocotepeque, W Honduras 14°25′N 89°10′W
61 D19 **Nueva Palmira** Colonia, SW Uruguay 33°53′S 58°25′W
41 N6 **Nueva Rosita** Coahuila, NE Mexico 27°58′N 101°11′W
42 E7 **Nueva San Salvador** prev. Santa Tecla. La Libertad, SW El Salvador 13°40′N 89°18′W
42 J8 **Nueva Segovia** ◐ department NW Nicaragua
Nueva Tabarca see Plana, Isla
Nueva Villa de Padilla see Nuevo Padilla
61 B21 **Nueve de Julio** Buenos Aires, E Argentina 35°29′S 60°52′W
44 H6 **Nuevitas** Camagüey, E Cuba 21°34′N 77°18′W
61 D18 **Nuevo Berlín** Río Negro, W Uruguay 32°59′S 58°03′W
40 I4 **Nuevo Casas Grandes** Chihuahua, N Mexico 30°23′N 107°54′W
43 T14 **Nuevo Chagres** Colón, C Panama 09°14′N 80°05′W
41 W15 **Nuevo Coahuila** Campeche, E Mexico 17°53′N 90°46′W
63 K17 **Nuevo, Golfo** gulf S Argentina
41 O7 **Nuevo Laredo** Tamaulipas, NE Mexico 27°28′N 99°32′W
41 N8 **Nuevo León** ◐ state NE Mexico
41 P10 **Nuevo Padilla** var. Nueva Villa de Padilla. Tamaulipas, C Mexico 24°01′N 98°48′W
56 E6 **Nuevo Rocafuerte** Orellana, E Ecuador 0°59′S 75°27′W
Nuga see Dzavhanmandal
80 O13 **Nugaal** off. Gobolka Nugaal. ◐ region N Somalia
Nugaal, Gobolka see Nugaal
185 E24 **Nugget Point** headland South Island, New Zealand 46°26′S 169°49′E
186 J5 **Nuguria Islands** island group E Papua New Guinea
184 P10 **Nuhaka** Hawke's Bay, North Island, New Zealand 39°03′S 177°43′E
138 M10 **Nuhaydayn, Wādī an** dry watercourse W Iraq
190 E7 **Nui Atoll** atoll W Tuvalu
Nu Jiang see Salween
Nûk see Nuuk
182 G7 **Nukey Bluff** hill South Australia
Nukha see Şäki
123 T9 **Nukh Yablonevyy, Gora** ▲ E Russian Federation 60°26′N 151°45′E
186 K7 **Nukiki** Choiseul, NW Solomon Islands 06°45′S 156°30′E
186 B6 **Nuku** West Sepik, NW Papua New Guinea 03°48′S 142°23′E
193 W15 **Nuku** island Tongatapu Group, NE Tonga
193 Y16 **Nuku'alofa** ● (Tonga) Tongatapu, S Tonga 21°08′S 175°13′W
193 U15 **Nuku'alofa** Tongatapu, S Tonga 21°09′S 175°14′W
190 G12 **Nukuatea** island N Wallis and Futuna
190 F7 **Nukufetau Atoll** atoll C Tuvalu
190 G12 **Nukuhifala** island E Wallis and Futuna
191 W7 **Nuku Hiva** island Îles Marquises, NE French Polynesia
191 W7 **Nuku Hiva Island** island Îles Marquises, N French Polynesia
190 F9 **Nukulaelae Atoll** var. Nukulailai. atoll E Tuvalu
Nukulailai see Nukulaelae Atoll
190 G11 **Nukuloa** island N Wallis and Futuna
186 L6 **Nukumanu Islands** prev. Tasman Group. island group NE Papua New Guinea
Nukunau see Nikunau
190 J9 **Nukunonu Atoll** island C Tokelau
190 J9 **Nukunonu Village** Nukunonu Atoll, C Tokelau
189 S18 **Nukuoro Atoll** atoll Caroline Islands, S Micronesia
146 H8 **Nukus** Qoraqalpog'iston Respublikasi, W Uzbekistan 42°29′N 59°32′E
190 G11 **Nukutapu** island N Wallis and Futuna
39 O9 **Nulato** Alaska, USA 64°43′N 158°06′W
39 O10 **Nulato Hills** ▲▲ Alaska, USA
105 T9 **Nules** Valenciana, E Spain 39°52′N 00°10′W
Nuling see Sultan Kudarat
182 C6 **Nullarbor** South Australia 31°28′S 130°57′E
180 M11 **Nullarbor Plain** plateau South Australia/Western Australia
163 S12 **Nulu'erhu Shan** ▲▲ N China
77 X14 **Numan** Adamawa, E Nigeria 09°26′N 11°58′E
165 S3 **Numata** Hokkaidō, NE Japan 43°48′N 141°55′E
81 C15 **Numatinna** ≈ W South Sudan
95 F14 **Numedalen** valley S Norway
95 G14 **Numedalslågen** var. Laagen. ≈ S Norway
93 L19 **Nummela** Uusimaa, S Finland 60°21′N 24°20′E
183 O11 **Numurkah** Victoria, SE Australia 36°04′S 145°28′E
196 L16 **Nunap Isua** var. Uummannarsuaq, Dan. Kap Farvel, Eng. Cape Farewell. cape S Greenland
9 N8 **Nunavut** ◐ territory N Canada
54 H9 **Nunchia** Casanare, C Colombia 05°37′N 72°13′W
97 M20 **Nuneaton** C England, United Kingdom 52°32′N 01°28′W
153 W14 **Nungba** Manipur, NE India 24°46′N 93°25′E
38 L12 **Nunivak Island** island Alaska, USA
152 I5 **Nun Kun** ▲ NW India 34°01′N 76°04′E
98 L10 **Nunspeet** Gelderland, E Netherlands 52°21′N 05°45′E
107 C18 **Nuoro** Sardegna, Italy, C Mediterranean Sea 40°20′N 09°20′E
75 R12 **Nuqayy, Jabal** hill range S Libya
54 C9 **Nuquí** Chocó, W Colombia 05°44′N 77°16′W
143 O4 **Nūr** Māzandarān, N Iran 36°32′N 52°00′E
145 Q9 **Nura** ≈ N Kazakhstan
143 N11 **Nūrābād** Fārs, C Iran 30°08′N 51°30′E
Nurakita see Niulakita
Nurata see Nurota
Nuratau, Khrebet see Nurota Tizmasi
136 L17 **Nur Dağları** ▲▲ S Turkey
Nurek see Norak
Nuremberg see Nürnberg
Nürestān see Nūristān
136 M15 **Nurhak** Kahramanmaraş, S Turkey 37°57′N 37°21′E
182 J9 **Nuriootpa** South Australia 34°28′S 139°00′E
149 S4 **Nūristān** prev. Nūrestān. ◐ province C Afghanistan
127 S5 **Nurlat** Respublika Tatarstan, W Russian Federation 54°26′N 50°48′E
93 N15 **Nurmes** Pohjois-Karjala, E Finland 63°31′N 29°10′E
101 K20 **Nürnberg** Eng. Nuremberg. Bayern, S Germany 49°27′N 11°05′E
101 K20 **Nürnberg** ✈ Bayern, SE Germany 49°29′N 11°04′E
146 M10 **Nurota** Rus. Nurata. Navoiy Viloyati, C Uzbekistan 40°41′N 65°43′E
147 N10 **Nurota Tizmasi** Rus. Khrebet Nuratau. ▲▲ C Uzbekistan
149 T8 **Nūrpur** Punjab, E Pakistan 31°54′N 71°55′E
183 P6 **Nurri, Mount** hill New South Wales, SE Australia
25 T13 **Nursery** Texas, SW USA 28°55′N 97°05′W
169 V17 **Nusa Tenggara Barat** off. Propinsi Nusa Tenggara Barat, Eng. West Nusa Tenggara. ◐ province S Indonesia
Nusa Tenggara Barat, Propinsi see Nusa Tenggara Barat
171 O16 **Nusa Tenggara Timur** off. Propinsi Nusa Tenggara Timur, Eng. East Nusa Tenggara. ◐ province S Indonesia
Nusa Tenggara Timur, Propinsi see Nusa Tenggara Timur
171 U14 **Nusawulan** Papua Barat, E Indonesia 04°03′S 132°56′E
137 Q16 **Nusaybin** var. Nisibin. Manisa, SE Turkey 37°08′N 41°11′E
39 O14 **Nushagak Bay** bay Alaska, USA
39 O13 **Nushagak Peninsula** headland Alaska, USA 58°39′N 159°03′W
39 O13 **Nushagak River** ≈ Alaska, USA
160 E11 **Nu Shan** ▲▲ SW China
149 N11 **Nushki** Baluchistān, SW Pakistan 29°33′N 66°01′E
Nussdorf see Năsăud
112 J9 **Nuštar** Vukovar-Srijem, E Croatia 45°20′N 18°48′E
99 L18 **Nuth** Limburg, SE Netherlands 50°55′N 05°52′E
100 N13 **Nuthe** ≈ NE Germany
Nutmeg State see Connecticut
39 T10 **Nutzotin Mountains** ▲▲ Alaska, USA
64 I5 **Nuuk** var. Nûk, Dan. Godthaab, Godthåb. ○ (Greenland) Sermersooq, SW Greenland 64°15′N 51°35′W
92 L13 **Nuupas** Lappi, NW Finland 66°01′N 26°19′E
191 O7 **Nuupere, Pointe** headland Moorea, W French Polynesia 17°35′S 149°47′W
191 O7 **Nuuroa, Pointe** headland Tahiti, W French Polynesia
Nüürst see Baganuur
Nuwara see Nuwara Eliya
155 K25 **Nuwara Eliya** var. Nuwara. Central Province, S Sri Lanka 06°58′N 80°46′E
182 E7 **Nuyts Archipelago** island group South Australia
83 F17 **Nxaunxau** North West, NW Botswana 18°57′S 21°18′E
39 N12 **Nyac** Alaska, USA 61°00′N 159°56′W
122 H9 **Nyagan'** Khanty-Mansiyskiy Avtonomnyy Okrug-Yugra, N Russian Federation 62°10′N 65°32′E
Nyagassola see Niagassola
Nyagquka see Yajiang
81 I18 **Nyahururu** Nyandarua, C Kenya 0°04′N 36°22′E
182 M10 **Nyah West** Victoria, SE Australia 35°14′S 143°18′E
158 M15 **Nyainqêntanglha Feng** ▲ W China 30°20′N 90°28′E
159 N15 **Nyainqêntanglha Shan** ▲▲ W China
80 B11 **Nyala** Southern Darfur, W Sudan 12°01′N 24°50′E
83 M16 **Nyamapanda** Mashonaland East, NE Zimbabwe 16°59′S 32°52′E
81 H19 **Nyamira** ◐ county W Kenya
81 H25 **Nyamtumbo** Ruvuma, S Tanzania 10°33′S 36°08′E
Nyanda see Masvingo
81 I19 **Nyandarua** ◐ county C Kenya
124 M11 **Nyandoma** Arkhangel'skaya Oblast', NW Russian Federation 61°39′N 40°10′E
83 M16 **Nyanga** prev. Inyanga. Manicaland, E Zimbabwe 18°13′S 32°46′E
79 D20 **Nyanga** off. Province de la Nyanga, var. La Nyanga. ◐ province SW Gabon
79 E20 **Nyanga** ≈ Congo/Gabon
Nyanga, Province de la see Nyanga
81 F20 **Nyantakara** Kagera, NW Tanzania 03°05′S 31°23′E
81 E21 **Nyanza-Lac** S Burundi 04°16′S 29°38′E
68 J14 **Nyasa, Lake** var. Lake Malawi; prev. Lago Nyassa. ◎ E Africa
Nyasaland/Nyasaland Protectorate see Malawi
Nyassa, Lago see Nyasa, Lake
119 J17 **Nyasvizh** Pol. Nieśwież, Rus. Nesvizh. Minskaya Voblasts', C Belarus 53°13′N 26°40′E
166 M8 **Nyaunglebin** Bago, SW Myanmar (Burma) 17°59′N 94°44′E
166 M5 **Nyaung-u** Magway, C Myanmar (Burma) 21°03′N 95°44′E
95 H24 **Nyborg** Syddtjylland, C Denmark 55°19′N 10°48′E
95 N21 **Nybro** Kalmar, S Sweden 56°45′N 15°54′E
119 J16 **Nyeharelaye** Rus. Negoreloye. Minskaya Voblasts', C Belarus 53°36′N 27°04′E
195 W3 **Nye Mountains** ▲▲ Antarctica
81 I19 **Nyeri** Nyeri, C Kenya 0°25′S 36°56′E
81 I19 **Nyeri** ◐ county C Kenya
118 M11 **Nyeshcharda, Vozyera** Rus. Ozero Neshcherdo. ◎ N Belarus
92 O2 **Ny-Friesland** physical region N Svalbard
95 L14 **Nyhammar** Dalarna, C Sweden 60°19′N 14°55′E
160 F7 **Nyikog Qu** ≈ C China
83 L14 **Nyimba** Eastern, E Zambia 14°33′S 30°49′E
159 P15 **Nyingchi** var. Bayizhen. Xizang Zizhiqu, W China 29°27′N 94°43′E
159 P15 **Nyingchi** var. Pula. Xizang Zizhiqu, W China 29°34′N 94°33′E
Nyinma see Maqu
111 O21 **Nyírbátor** Szabolcs-Szatmár-Bereg, E Hungary 47°50′N 22°09′E
111 N21 **Nyíregyháza** Szabolcs-Szatmár-Bereg, NE Hungary 47°57′N 21°43′E
Nyiro see Ewaso Ng'iro
Nyitra see Nitra
Nyitrabánya see Handlová
93 K16 **Nykarleby** Fin. Uusikaarlepyy. Österbotten, W Finland 63°22′N 22°30′E
95 F21 **Nykøbing** Midtjylland, NW Denmark 56°48′N 08°52′E
95 I25 **Nykøbing** Sjælland, SE Denmark 54°47′N 11°53′E
95 I22 **Nykøbing** Sjælland, C Denmark 55°56′N 11°41′E
95 N17 **Nyköping** Södermanland, S Sweden 58°45′N 17°03′E
95 L15 **Nykroppa** Värmland, C Sweden 59°37′N 14°18′E
Nyland see Uusimaa
Nylstroom see Modimolle
183 P7 **Nymagee** New South Wales, SE Australia 32°06′S 146°19′E
183 V5 **Nymboida** New South Wales, SE Australia 29°57′S 152°45′E
183 U5 **Nymboida River** ≈ New South Wales, SE Australia
111 D16 **Nymburk** var. Neuenburg an der Elbe, Ger. Nimburg. Středočeský Kraj, C Czech Republic 50°12′N 15°00′E
95 O16 **Nynäshamn** Stockholm, C Sweden 58°54′N 17°55′E
183 Q6 **Nyngan** New South Wales, SE Australia 31°36′S 147°07′E
Nyoman see Neman
108 A10 **Nyon** Ger. Neuss; anc. Noviodunum. Vaud, SW Switzerland 46°23′N 06°15′E
79 D16 **Nyong** ≈ SW Cameroon
103 S14 **Nyons** Drôme, E France 44°22′N 05°06′E
79 D14 **Nyos, Lac** Eng. Lake Nyos. ◎ NW Cameroon
Nyos, Lake see Nyos, Lac
125 U11 **Nyrob** var. Nyrov. Permskiy Kray, NW Russian Federation 60°41′N 56°42′E
Nyrov see Nyrob
111 H15 **Nysa** Ger. Neisse. Opolskie, S Poland 50°28′N 17°20′E
32 M13 **Nyssa** Oregon, NW USA 43°52′N 116°59′W
Nysa Łużycka see Neisse
Nyslott see Savonlinna
Nystad see Uusikaupunki
95 I25 **Nysted** Sjælland, SE Denmark 54°40′N 11°41′E
125 U14 **Nytva** Permskiy Kray, NW Russian Federation 57°56′N 55°22′E
165 P8 **Nyūdō-zaki** headland Honshū, C Japan 39°59′N 139°40′E
125 P9 **Nyukhcha** Arkhangel'skaya Oblast', NW Russian Federation 63°24′N 46°34′E
124 H8 **Nyuk, Ozero** var. Ozero Njuk. ◎ NW Russian Federation
125 O12 **Nyuksenitsa** var. Njuksenica. Vologodskaya Oblast', NW Russian Federation 60°25′N 44°12′E
79 O22 **Nyunzu** Katanga, SE Dem. Rep. Congo 05°55′S 28°00′E
123 O10 **Nyurba** Respublika Sakha (Yakutiya), NE Russian Federation 63°17′N 118°15′E
123 O11 **Nyuya** Respublika Sakha (Yakutiya), NE Russian Federation 60°33′N 116°10′E
146 K12 **Nyýazow** Rus. Niyazov. Lebap Welaýaty, NE Turkmenistan 39°13′N 63°16′E
117 T10 **Nyzhni Sirohozy** Khersons'ka Oblast', S Ukraine 46°49′N 34°21′E
117 U12 **Nyzhn'ohirs'kyy** Rus. Nizhnegorskiy. Avtonomna Respublika Krym, S Ukraine 45°26′N 34°42′E
NZ see New Zealand
81 G21 **Nzega** Tabora, C Tanzania 04°13′S 33°11′E
76 K15 **Nzérékoré** SE Guinea 07°45′N 08°49′W

◆ Country
● Country Capital
◇ Dependent Territory
○ Dependent Territory Capital
◐ Administrative Regions
✈ International Airport
▲ Mountain
▲▲ Mountain Range
Volcano
≈ River
◎ Lake
Reservoir

N'Zeto – Old Harbor

82 A10 **N'Zeto** *prev.* Ambrizete. Zaire Province, NW Angola *07°14′S 12°52′E*
79 M24 **Nzilo, Lac** *prev.* Lac Delcommune. ◎ SE Dem. Rep. Congo
172 I13 **Nzwani** *Fr.* Anjouan, *var.* Ndzouani. *island* SE Comoros

O

29 O11 **Oacoma** South Dakota, N USA *43°49′N 99°25′W*
29 N9 **Oahe Dam** *dam* South Dakota, N USA
28 M9 **Oahe, Lake** ⊠ North Dakota/South Dakota, N USA
38 C9 **Oa'hu** *var.* Oahu. *island* Hawai'ian Islands, Hawai'i, USA
165 V4 **O-Akan-dake** ▲ Hokkaidō, NE Japan *43°26′N 144°09′E*
182 K8 **Oakbank** South Australia *33°07′S 140°36′E*
19 P13 **Oak Bluffs** Martha's Vineyard, New York, NE USA *41°25′N 70°32′W*
36 K4 **Oak City** Utah, W USA *39°22′N 112°19′W*
37 R3 **Oak Creek** Colorado, C USA *40°16′N 106°57′W*
35 P8 **Oakdale** California, W USA *37°46′N 120°51′W*
22 H8 **Oakdale** Louisiana, S USA *30°49′N 92°39′W*
29 P7 **Oakes** North Dakota, N USA *46°08′N 98°05′W*
22 J4 **Oak Grove** Louisiana, S USA *32°51′N 91°25′W*
97 N19 **Oakham** C England, United Kingdom *52°41′N 00°45′W*
32 H7 **Oak Harbor** Washington, NW USA *48°17′N 122°38′W*
21 R5 **Oak Hill** West Virginia, NE USA *37°59′N 81°09′W*
35 N8 **Oakland** California, W USA *37°48′N 122°16′W*
29 T15 **Oakland** Iowa, C USA *41°18′N 95°22′W*
19 Q7 **Oakland** Maine, NE USA *44°32′N 69°43′W*
21 T3 **Oakland** Maryland, NE USA *39°24′N 79°25′W*
29 R14 **Oakland** Nebraska, C USA *41°50′N 96°28′W*
31 N11 **Oak Lawn** Illinois, N USA *41°43′N 87°45′W*
33 P16 **Oakley** Idaho, NW USA *42°13′N 113°54′W*
26 I4 **Oakley** Kansas, C USA *39°08′N 100°53′W*
31 N10 **Oak Park** Illinois, N USA *41°53′N 87°46′W*
11 X16 **Oak Point** Manitoba, S Canada *50°23′N 97°00′W*
32 G13 **Oakridge** Oregon, NW USA *43°45′N 122°27′W*
20 M9 **Oak Ridge** Tennessee, S USA *36°02′N 84°12′W*
184 K10 **Oakura** Taranaki, North Island, New Zealand *39°07′S 173°58′E*
22 L7 **Oak Vale** Mississippi, S USA *31°26′N 89°57′W*
14 G16 **Oakville** Ontario, S Canada *43°27′N 79°41′W*
25 V8 **Oakwood** Texas, SW USA *31°34′N 95°51′W*
185 F22 **Oamaru** Otago, South Island, New Zealand *45°10′S 170°51′E*
96 F13 **Oa, Mull of** *headland* W Scotland, United Kingdom *55°35′N 06°20′W*
171 O11 **Oan** Sulawesi, N Indonesia *01°16′N 121°25′E*
185 J17 **Oaro** Canterbury, South Island, New Zealand *42°29′S 173°30′E*
35 X2 **Oasis** Nevada, W USA *41°01′N 114°29′W*
195 S15 **Oates Land** *physical region* Antarctica
183 P17 **Oatlands** Tasmania, SE Australia *42°21′S 147°23′E*
36 I11 **Oatman** Arizona, SW USA *35°03′N 114°19′W*
41 R16 **Oaxaca** *var.* Oaxaca de Juárez; *prev.* Antequera. Oaxaca, SE Mexico *17°04′N 96°41′W*
41 Q16 **Oaxaca** ◐ *state* SE Mexico
Oaxaca de Juárez *see* Oaxaca
122 I19 **Ob'** ≈ C Russian Federation
145 X9 **Oba** *prev.* Uba. E Kazakhstan
14 G9 **Obabika Lake** ◎ Ontario, S Canada
Obagan *see* Ubagan
118 M12 **Obal'** *Rus.* Obol'. Vitsyebskaya Voblasts', N Belarus *55°22′N 29°17′E*
79 E16 **Obala** Centre, SW Cameroon *04°09′N 11°32′E*
14 C6 **Oba Lake** ◎ Ontario, S Canada
164 J12 **Obama** Fukui, Honshū, SW Japan *35°32′N 135°45′E*
96 H11 **Oban** W Scotland, United Kingdom *56°25′N 05°29′W*
Oban *see* Halfmoon Bay
Obando *see* Puerto Inírida
104 I4 **O Barco** *var.* El Barco, El Barco de Valdeorras, O Barco de Valdeorras. Galicia, NW Spain *42°24′N 07°00′W*
O Barco de Valdeorras *see* O Barco
Obbia *see* Hobyo
93 J16 **Obbola** Västerbotten, N Sweden *63°41′N 20°16′E*
Obbrovazzo *see* Obrovac
Obchuga *see* Abchuha
Obdorsk *see* Salekhard
Óbecse *see* Bečej
118 I11 **Obeliai** Panevėžys, NE Lithuania *55°57′N 25°47′E*
60 F13 **Oberá** Misiones, NE Argentina *27°29′S 55°08′W*
108 E8 **Oberburg** Bern, W Switzerland *47°00′N 07°37′E*
109 Q9 **Oberdrauburg** Salzburg, S Austria *46°45′N 12°59′E*
Oberglogau *see* Głogówek
109 W4 **Ober Grafendorf** Niederösterreich, NE Austria *48°09′N 15°33′E*
101 E15 **Oberhausen** Nordrhein-Westfalen, W Germany *51°27′N 06°50′E*
Oberhollabrunn *see* Tulln
Oberlaibach *see* Vrhnika
101 Q15 **Oberlausitz** *var.* Hornja Łužica. *physical region* E Germany
26 J2 **Oberlin** Kansas, C USA *39°49′N 100°33′W*
22 H8 **Oberlin** Louisiana, S USA *30°37′N 92°45′W*
31 T11 **Oberlin** Ohio, N USA *41°17′N 82°13′W*
103 U5 **Obernai** Bas-Rhin, NE France *48°28′N 07°30′E*
109 R4 **Obernberg am Inn** Oberösterreich, N Austria *48°19′N 13°20′E*
Oberndorf *see* Oberndorf am Neckar
101 G23 **Oberndorf am Neckar** *var.* Oberndorf. Baden-Württemberg, SW Germany *48°18′N 08°32′E*
109 Q5 **Oberndorf bei Salzburg** Salzburg, W Austria *47°57′N 12°57′E*
Oberneustadtl *see* Kysucké Nové Mesto
183 S8 **Oberon** New South Wales, SE Australia *33°42′S 149°50′E*
109 Q4 **Oberösterreich** *off.* Land Oberösterreich, *Eng.* Upper Austria. ◐ *state* NW Austria
Oberösterreich, Land *see* Oberösterreich
Oberpahlen *see* Põltsamaa
101 M19 **Oberpfälzer Wald** ▲ SE Germany
109 Y6 **Oberpullendorf** Burgenland, E Austria *47°32′N 16°30′E*
Oberradkersburg *see* Gornja Radgona
101 G18 **Oberursel** Hessen, W Germany *50°12′N 08°34′E*
109 Q8 **Obervellach** Salzburg, S Austria *46°56′N 13°10′E*
109 X7 **Oberwart** Burgenland, SE Austria *47°18′N 16°12′E*
Oberwischau *see* Vişeu de Sus
109 T7 **Oberwölz** *var.* Oberwölz-Stadt. Steiermark, SE Austria *47°12′N 14°20′E*
Oberwölz-Stadt *see* Oberwölz
31 S13 **Obetz** Ohio, N USA *39°52′N 82°57′W*
Ob', Gulf of *see* Obskaya Guba
58 H12 **Óbidos** Pará, NE Brazil *01°52′S 55°30′W*
104 F10 **Óbidos** Leiria, C Portugal *39°21′N 09°09′W*
Obidovichi *see* Abidavichy
147 Q13 **Obigarm** W Tajikistan *38°42′N 69°34′E*
165 T2 **Obihiro** Hokkaidō, NE Japan *42°56′N 143°10′E*
Obi-Khingou *see* Khingov
147 P13 **Obikiik** SW Tajikistan *38°07′N 68°36′E*
Obilić *see* Obiliq
113 N16 **Obiliq** *Serb.* Obilić. N Kosovo *42°50′N 20°57′E*
127 O12 **Obil'noye** Respublika Kalmykiya, SW Russian Federation *47°31′N 44°24′E*
20 F8 **Obion** Tennessee, S USA *36°15′N 89°11′W*
20 F8 **Obion River** ≈ Tennessee, S USA
171 S12 **Obi, Pulau** *island* Maluku, E Indonesia
165 S2 **Obira** Hokkaidō, NE Japan *44°01′N 141°39′E*
127 N11 **Oblivskaya** Rostovskaya Oblast', SW Russian Federation *48°34′N 42°31′E*
123 R14 **Obluch'ye** Yevreyskaya Avtonomnaya Oblast', SE Russian Federation *48°59′N 131°18′E*
126 K4 **Obninsk** Kaluzhskaya Oblast', W Russian Federation *55°06′N 36°40′E*
114 J8 **Obnova** Pleven, N Bulgaria *43°26′N 25°04′E*
79 N15 **Obo** Haut-Mbomou, E Central African Republic *05°20′N 26°29′E*
159 T9 **Obo** Qinghai, C China *37°57′N 101°03′E*
80 M11 **Obock** E Djibouti *11°57′N 43°09′E*
Obol' *see* Obal'
Obolyanka *see* Abalyanka
171 V13 **Obome** Papua Barat, E Indonesia *03°42′S 133°21′E*
110 G11 **Oborniki** Wielkolpolskie, W Poland *52°38′N 16°48′E*
79 G19 **Obouya** Cuvette, C Congo *0°56′S 15°41′E*
126 J8 **Oboyan'** Kurskaya Oblast', W Russian Federation *51°12′N 36°15′E*
124 M9 **Obozerskiy** Arkhangel'skaya Oblast', NW Russian Federation *63°26′N 40°20′E*
112 L11 **Obrenovac** Serbia, N Serbia *44°39′N 20°12′E*
112 D12 **Obrovac** *It.* Obbrovazzo. Zadar, SW Croatia *44°12′N 15°40′E*
Obrovo *see* Abrova
35 Q3 **Observation Peak** ▲ California, W USA *40°48′N 120°07′W*
122 J8 **Obskaya Guba** *Eng.* Gulf of Ob. *gulf* N Russian Federation
173 N13 **Ob' Tablemount** *undersea feature* S Indian Ocean *50°16′S 51°59′E*
173 T10 **Ob' Trench** *undersea feature* E Indian Ocean
77 P16 **Obuasi** S Ghana *06°15′N 01°36′W*
117 P5 **Obukhiv** *Rus.* Obukhov. Kyyivs'ka Oblast', N Ukraine *50°05′N 30°37′E*
Obukhov *see* Obukhiv
125 U14 **Obva** ≈ NW Russian Federation
108 F8 **Obwalden** ◐ *canton* C Switzerland
117 V10 **Obytichna Kosa** *spit* SE Ukraine
117 V10 **Obytichna Zatoka** *gulf* SE Ukraine
114 N9 **Obzor** Burgas, E Bulgaria *42°52′N 27°53′E*
105 O3 **Oca** ≈ N Spain
23 W10 **Ocala** Florida, SE USA *29°11′N 82°08′W*
40 M7 **Ocampo** Coahuila, NE Mexico *27°18′N 102°24′W*
54 G7 **Ocaña** Norte de Santander, N Colombia *08°16′N 73°21′W*
105 N9 **Ocaña** Castilla-La Mancha, C Spain *39°57′N 03°30′W*
104 H4 **O Carballiño** *Cast.* Carballiño. Galicia, NW Spain *42°26′N 08°05′W*
37 T9 **Ocate** New Mexico, SW USA *36°09′N 105°03′W*
Ocavango *see* Okavango
57 D14 **Occidental, Cordillera** ▲ W South America
21 Q6 **Oceana** West Virginia, NE USA *37°41′N 81°37′W*
21 Z4 **Ocean City** Maryland, NE USA *38°20′N 75°05′W*
18 J17 **Ocean City** New Jersey, NE USA *39°15′N 74°33′W*
10 K15 **Ocean Falls** British Columbia, SW Canada *52°24′N 127°42′W*
Ocean Island *see* Banaba
Ocean Island *see* Kure Atoll
64 J9 **Oceanographer Fracture Zone** *tectonic feature* NW Atlantic Ocean
35 U17 **Oceanside** California, W USA *33°12′N 117°23′W*
22 M9 **Ocean Springs** Mississippi, S USA *30°24′N 88°49′W*
Ocean State *see* Rhode Island
25 O9 **O C Fisher Lake** ⊠ Texas, SW USA
117 Q10 **Ochakiv** *Rus.* Ochakov. Mykolayivs'ka Oblast', S Ukraine *46°36′N 31°33′E*
Ochakov *see* Ochakiv
Ochamchira *see* Ochamchire
137 Q9 **Ochamchire** *Rus.* Ochamchira; *prev.* Och'amch'ire. W Georgia *42°45′N 41°30′E*
Och'amch'ire *see* Ochamchire
Ochansk *see* Okhansk
125 T15 **Ochër** Permskiy Kray, NW Russian Federation *57°54′N 54°40′E*
115 I19 **Óchi** ▲ Évvoia, C Greece *38°03′N 24°27′E*
165 W4 **Ochiishi-misaki** *headland* Hokkaidō, NE Japan *43°10′N 145°29′E*
23 S9 **Ochlockonee River** ≈ Florida/Georgia, SE USA
44 K12 **Ocho Rios** C Jamaica *18°24′N 77°06′W*
Ochrida *see* Ohrid
Ochrida, Lake *see* Ohrid, Lake
101 J19 **Ochsenfurt** Bayern, C Germany *49°39′N 10°03′E*
23 U7 **Ocilla** Georgia, SE USA *31°35′N 83°15′W*
94 N13 **Ockelbo** Gävleborg, C Sweden *60°51′N 16°46′E*
Ocker *see* Oker
95 I19 **Öckerö** Västra Götaland, S Sweden *57°43′N 11°39′E*
23 U6 **Ocmulgee River** ≈ Georgia, SE USA
116 H11 **Ocna Mureş** *Hung.* Marosújvár; *prev.* Ocna Mureşului, *prev. Hung.* Marosújvárakna. Alba, C Romania *46°25′N 23°53′E*
Ocna Mureşului *see* Ocna Mureş
116 H11 **Ocna Sibiului** *Ger.* Salzburg, *Hung.* Vizakna. Sibiu, C Romania *45°52′N 23°59′E*
116 H13 **Ocnele Mari** *prev.* Vioara. Vâlcea, S Romania *45°03′N 24°18′E*
116 L7 **Ocniţa** *Rus.* Oknitsa. N Moldova *48°25′N 27°30′E*
23 U4 **Oconee, Lake** ⊠ Georgia, SE USA
23 U5 **Oconee River** ≈ Georgia, SE USA
30 M9 **Oconomowoc** Wisconsin, N USA *43°06′N 88°29′W*
30 M6 **Oconto** Wisconsin, N USA *44°55′N 87°52′W*
30 M6 **Oconto Falls** Wisconsin, N USA *44°52′N 88°06′W*
30 M6 **Oconto River** ≈ Wisconsin, N USA
104 I3 **O Corgo** Galicia, NW Spain *42°56′N 07°25′W*
41 V16 **Ocosingo** Chiapas, SE Mexico *17°04′N 92°15′W*
42 J8 **Ocotal** Nueva Segovia, NW Nicaragua *13°38′N 86°28′W*
42 F6 **Ocotepeque** ◐ *department* W Honduras
Ocotepeque *see* Nueva Ocotepeque
40 L13 **Ocotlán** Jalisco, SW Mexico *20°21′N 102°42′W*
41 R16 **Ocotlán** *var.* Ocotlán de Morelos. Oaxaca, SE Mexico *16°49′N 96°49′W*
Ocotlán de Morelos *see* Ocotlán
41 U16 **Ocozocuautla** Chiapas, SE Mexico *16°46′N 93°22′W*
21 Y10 **Ocracoke Island** *island* North Carolina, SE USA
102 I3 **Octeville** Manche, N France *49°37′N 01°39′W*
October Revolution Island *see* Oktyabr'skoy Revolyutsii, Ostrov
43 R17 **Ocú** Herrera, S Panama *07°55′N 80°43′W*
83 Q14 **Ocua** Cabo Delgado, NE Mozambique *13°37′S 39°44′E*
Ocumare *see* Ocumare del Tuy
54 M5 **Ocumare del Tuy** *var.* Ocumare. Miranda, N Venezuela *10°07′N 66°47′W*
77 P17 **Oda** SE Ghana *05°55′N 00°56′W*
165 G12 **Ōda** *var.* Oda. Shimane, Honshū, SW Japan *35°10′N 132°29′E*
92 K3 **Ódáðahraun** *lava flow* C Iceland
165 Q7 **Ōdate** Akita, Honshū, C Japan *40°18′N 140°34′E*
165 N14 **Odawara** Kanagawa, Honshū, S Japan *35°15′N 139°08′E*
95 D14 **Odda** Hordaland, S Norway *60°03′N 06°34′E*
95 G22 **Odder** Midtjylland, C Denmark *55°59′N 10°10′E*
Oddur *see* Xuddur
29 T13 **Odebolt** Iowa, C USA *42°18′N 95°15′W*
104 H14 **Odeleite** Faro, S Portugal *37°02′N 07°29′W*
25 Q4 **Odell** Texas, SW USA *34°19′N 99°24′W*
25 T14 **Odem** Texas, SW USA *27°57′N 97°34′W*
104 F13 **Odemira** Beja, S Portugal *37°35′N 08°38′W*
136 C14 **Ödemiş** İzmir, SW Turkey *38°11′N 27°58′E*
Ödenburg *see* Sopron
83 I22 **Odendaalsrus** Free State, C South Africa *27°52′S 26°42′E*
Odenpäh *see* Otepää
95 H23 **Odense** Syddtjylland, C Denmark *55°24′N 10°23′E*
101 H19 **Odenwald** ▲ W Germany
84 H10 **Oder** *Cz./Pol.* Odra. ≈ C Europe
Oderberg *see* Bohumín
100 P11 **Oderbruch** *wetland* Germany/Poland
Oderhaff *see* Szczeciński, Zalew
100 O11 **Oder-Havel-Kanal** *canal* NE Germany
Oderhellen *see* Odorheiu Secuiesc
100 P13 **Oder-Spree-Kanal** *canal* NE Germany
Odertal *see* Zdzieszowice
106 I7 **Oderzo** Veneto, NE Italy *45°48′N 12°33′E*
117 P10 **Odesa** *Rus.* Odessa. Odes'ka Oblast', SW Ukraine *46°29′N 30°44′E*
24 M8 **Odessa** Texas, SW USA *31°51′N 102°22′W*
32 K8 **Odessa** Washington, NW USA *47°19′N 118°41′W*
Odesa *see* Odes'ka Oblast'
95 L18 **Ödeshög** Östergötland, S Sweden *58°13′N 14°40′E*
117 O9 **Odes'ka Oblast'** *var.* Odesa, *Rus.* Odesskaya Oblast'. ◐ *province* SW Ukraine
Odessa *see* Odesa
Odesskaya Oblast' *see* Odes'ka Oblast'
122 H12 **Odesskoye** Omskaya Oblast', C Russian Federation *54°15′N 72°45′E*
Odessus *see* Varna
102 F6 **Odet** ≈ NW France
104 I14 **Odiel** ≈ SW Spain
76 L14 **Odienné** NW Ivory Coast *09°32′N 07°35′W*
171 O4 **Odiongan** Tablas Island, C Philippines *12°23′N 122°01′E*
153 P17 **Odisha** *prev.* Orissa. ◐ *state* NE India
116 L12 **Odobeşti** Vrancea, E Romania *45°46′N 27°06′E*
110 H13 **Odolanów** *Ger.* Adelnau. Wielkopolskie, C Poland *51°35′N 17°42′E*
167 R13 **Ôdôngk** Kâmpóng Spoe, S Cambodia *11°48′N 104°45′E*
25 N6 **O'donnell** Texas, SW USA *32°57′N 101°49′W*
98 O7 **Odoorn** Drenthe, NE Netherlands *52°52′N 06°49′E*
Odorhei *see* Odorheiu Secuiesc
116 J11 **Odorheiu Secuiesc** *Ger.* Oderhellen, *Hung.* Vámosudvarhely; *prev.* Odorhei, *Ger.* Hofmarkt. Harghita, C Romania *46°18′N 25°19′E*
Odra *see* Oder
112 J9 **Odžaci** *Ger.* Hodschag, *Hung.* Hodság. Vojvodina, NW Serbia *45°31′N 19°15′E*
59 N14 **Oeiras** Piauí, E Brazil *07°00′S 42°07′W*
104 F11 **Oeiras** Lisboa, C Portugal *38°41′N 09°18′W*
101 G14 **Oelde** Nordrhein-Westfalen, W Germany *51°49′N 08°09′E*
28 J11 **Oelrichs** South Dakota, N USA *43°08′N 103°13′W*
101 M17 **Oelsnitz** Sachsen, E Germany *50°22′N 12°12′E*
Oels/Oels in Schlesien *see* Oleśnica
29 X12 **Oelwein** Iowa, C USA *42°40′N 91°54′W*
Oeniadae *see* Oiniádes
191 N17 **Oeno Island** *atoll* Pitcairn Group of Islands, C Pacific Ocean
Oesel *see* Saaremaa
108 L7 **Oetz** *var.* Ötz. Tirol, W Austria *47°15′N 10°56′E*
137 P11 **Of** Trabzon, NE Turkey *40°57′N 40°17′E*
30 K15 **O'Fallon** Illinois, N USA *38°35′N 89°54′W*
27 W4 **O'Fallon** Missouri, C USA *38°54′N 90°31′W*
107 N16 **Ofanto** ≈ S Italy
97 D18 **Offaly** *Ir.* Ua Uíbh Fhailí; *prev.* King's County. *cultural region* C Ireland
101 H18 **Offenbach** *var.* Offenbach am Main. Hessen, W Germany *50°06′N 08°46′E*
Offenbach am Main *see* Offenbach
101 F22 **Offenburg** Baden-Württemberg, SW Germany *48°28′N 07°57′E*
182 C2 **Officer Creek** *seasonal river* South Australia
Oficina María Elena *see* María Elena
Oficina Pedro de Valdivia *see* Pedro de Valdivia
115 K22 **Ofidoússa** *island* Kykládes, Greece, Aegean Sea
Ofiral *see* Sharm ash Shaykh
92 H10 **Ofotfjorden** *fjord* N Norway
192 L16 **Ofu** *island* Manua Islands, E American Samoa
165 R9 **Ōfunato** Iwate, Honshū, C Japan *39°04′N 141°43′E*
165 P8 **Oga** Akita, Honshū, C Japan *39°56′N 139°47′E*
Ogaadeen *see* Ogaden
165 Q9 **Ōgachi** Akita, Honshū, C Japan *39°03′N 140°26′E*
165 P9 **Ōgachi-tōge** *pass* Honshū, C Japan
81 N14 **Ogaden** *Som.* Ogaadeen. *plateau* Ethiopia/Somalia
165 P8 **Oga-hantō** *peninsula* Honshū, C Japan
165 K13 **Ōgaki** Gifu, Honshū, SW Japan *35°22′N 136°35′E*
28 L15 **Ogallala** Nebraska, C USA *41°09′N 101°44′W*
168 M14 **Ogan, Air** ≈ Sumatera, W Indonesia
165 Y15 **Ogasawara-shotō** *Eng.* Bonin Islands. *island group* SE Japan
14 J9 **Ogascanane, Lac** ◎ Québec, SE Canada
165 R7 **Ogawara-ko** ◎ Honshū, C Japan
77 T15 **Ogbomosho** *var.* Ogmoboso. Oyo, W Nigeria *08°10′N 04°16′E*
29 U13 **Ogden** Iowa, C USA *42°03′N 94°01′W*
36 L2 **Ogden** Utah, W USA *41°09′N 111°58′W*
18 I6 **Ogdensburg** New York, NE USA *44°42′N 75°25′W*
23 W5 **Ogeechee River** ≈ Georgia, SE USA
Oger *see* Ogre
165 N10 **Ogi** Niigata, Sado, C Japan *37°49′N 138°16′E*
10 H5 **Ogilvie** Yukon, NW Canada *63°34′N 139°44′W*
10 H4 **Ogilvie** ≈ Yukon, NW Canada
10 H5 **Ogilvie Mountains** ▲ Yukon, NW Canada
Oginskiy Kanal *see* Ahinski Kanal
162 J7 **Ögiynuur** *var.* Dzegstey. Arhangay, C Mongolia *47°38′N 102°31′E*
146 F6 **Og'iyon Sho'rxogi** *wetland* NW Uzbekistan
146 B10 **Oglanly** Balkan Welaýaty, W Turkmenistan *39°56′N 54°25′E*
23 T5 **Oglethorpe** Georgia, SE USA *32°17′N 84°03′W*
23 T2 **Oglethorpe, Mount** ▲ Georgia, SE USA *34°29′N 84°19′W*
106 F7 **Oglio** *anc.* Ollius. ≈ N Italy
Ogmoboso *see* Ogbomosho
103 T8 **Ognon** ≈ E France
123 R13 **Ogodzha** Amurskaya Oblast', S Russian Federation *52°51′N 132°49′E*
77 W16 **Ogoja** Cross River, S Nigeria *06°37′N 08°48′E*
12 C10 **Ogoki** ≈ Ontario, S Canada
12 D11 **Ogoki Lake** ◎ Ontario, C Canada
Ögöömör *see* Hanhongor
79 F19 **Ogooué** ≈ Congo/Gabon
79 E18 **Ogooué-Ivindo** *off.* Province de l'Ogooué-Ivindo, *var.* L'Ogooué-Ivindo. ◐ *province* N Gabon
Ogooué-Ivindo, Province de l' *see* Ogooué-Ivindo
79 E19 **Ogooué-Lolo** *off.* Province de l'Ogooué-Lolo, *var.* L'Ogooué-Lolo. ◐ *province* C Gabon
Ogooué-Lolo, Province de l' *see* Ogooué-Lolo
79 C19 **Ogooué-Maritime** *off.* Province de l'Ogooué-Maritime, *var.* L'Ogooué-Maritime. ◐ *province* W Gabon
Ogooué-Maritime, Province de l' *see* Ogooué-Maritime
165 D14 **Ogōri** Fukuoka, Kyūshū, SW Japan *33°24′N 130°34′E*
114 H7 **Ogosta** ≈ NW Bulgaria
112 Q9 **Ogražden** *Bul.* Ograzhden. ▲ Bulgaria/FYR Macedonia *see also* Ograzhden
114 G12 **Ograzhden** *Mac.* Orgražden. ▲ Bulgaria/FYR Macedonia *see also* Orgražden
Ograzhden *see* Ogražden
118 G9 **Ogre** *Ger.* Oger. C Latvia *56°49′N 24°36′E*
118 H9 **Ogre** ≈ C Latvia
112 C10 **Ogulin** Karlovac, NW Croatia *45°15′N 15°13′E*
77 S16 **Ogun** ◐ *state* SW Nigeria
Ogurdzhaly, Ostrov *see* Ogurjaly Adasy
146 A12 **Ogurjaly Adasy** *Rus.* Ogurdzhaly, Ostrov. *island* W Turkmenistan
77 U16 **Ogwashi-Uku** Delta, S Nigeria *06°08′N 06°38′E*
185 B23 **Ohai** Southland, South Island, New Zealand *45°56′S 167°59′E*
147 Q10 **Ohangaron** *Rus.* Akhangaran. Toshkent Viloyati, E Uzbekistan *40°56′N 69°37′E*
147 Q10 **Ohangaron** *Rus.* Akhangaran. ≈ E Uzbekistan
83 C16 **Ohangwena** ◐ *district* N Namibia
30 M10 **O'Hare** ✈ (Chicago) Illinois, N USA *41°59′N 87°56′W*
165 R6 **Ōhata** Aomori, Honshū, C Japan *41°23′N 141°09′E*
184 L13 **Ohau** Manawatu-Wanganui, North Island, New Zealand *40°40′S 175°15′E*
185 E20 **Ohau, Lake** ◎ South Island, New Zealand
Ohcejohka *see* Utsjoki
99 J20 **Ohey** Namur, SE Belgium *50°26′N 05°07′E*
191 X15 **O'Higgins, Cabo** *headland* Easter Island, Chile, E Pacific Ocean *27°05′S 109°15′W*
O'Higgins, Lago *see* San Martín, Lago
31 S12 **Ohio** *off.* State of Ohio, *also known as* Buckeye State. ◐ *state* N USA
0 L10 **Ohio River** ≈ N USA
101 H16 **Ohm** ≈ C Germany
193 W16 **Ohonua** 'Eua, E Tonga *21°20′S 174°57′W*
23 V5 **Ohoopee River** ≈ Georgia, SE USA
100 L12 **Ohre** *Ger.* Eger. ≈ Czech Republic/Germany
Ohri *see* Ohrid
113 M20 **Ohrid** *Turk.* Ochrida, Ohri. SW FYR Macedonia *41°07′N 20°48′E*
113 M20 **Ohrid, Lake** *var.* Lake Ochrida, *Alb.* Liqeni i Ohrit, *Mac.* Ohridsko Ezero. ◎ Albania/FYR Macedonia
Ohridsko Ezero/Ohrit, Liqeni i *see* Ohrid, Lake
184 L9 **Ohura** Manawatu-Wanganui, North Island, New Zealand *38°51′S 174°58′E*
58 J9 **Oiapoque** Amapá, E Brazil *03°54′N 51°46′W*
58 J10 **Oiapoque, Rio** *var.* Fleuve l'Oyapok, Oyapock. ≈ Brazil/French Guiana *see also* Oyapok, Fleuve l'
Oiapoque, Rio *see* Oyapok, Fleuve l'
54 G8 **Oiba** Santander, C Colombia *06°16′N 73°18′W*
15 O9 **Oies, Île aux** *island* Québec, SE Canada
92 L13 **Oijärvi** Pohjois-Pohjanmaa, C Finland *65°38′N 26°05′E*
92 L12 **Oikarainen** Lappi, N Finland *66°30′N 25°46′E*
188 F10 **Oikuul** Babeldaob, N Palau
18 C13 **Oil City** Pennsylvania, NE USA *41°25′N 79°42′W*
18 C12 **Oil Creek** ≈ Pennsylvania, NE USA
35 R13 **Oildale** California, W USA *35°25′N 119°01′W*
Oileán Ciarraí *see* Castleisland
Oil Islands *see* Chagos Archipelago
115 D18 **Oiniádes** *anc.* Oeniadae. *site of ancient city* Dytikí Elláda, W Greece
115 L18 **Oinoússes** *island* E Greece
Oírr, Inis *see* Inisheer
99 J15 **Oirschot** Noord-Brabant, S Netherlands *51°30′N 05°18′E*
103 N4 **Oise** ◐ *department* N France
103 P3 **Oise** ≈ N France
99 J14 **Oisterwijk** Noord-Brabant, S Netherlands *51°35′N 05°12′E*
45 O14 **Oistins** S Barbados *13°04′N 59°33′W*
165 E4 **Ōita** Ōita, Kyūshū, SW Japan *33°15′N 131°35′E*
165 D14 **Ōita** *off.* Ōita-ken. ◐ *prefecture* Kyūshū, SW Japan
Ōita-ken *see* Ōita
115 E17 **Oíti** ▲ C Greece *38°48′N 22°12′E*
165 S4 **Oiwake** Hokkaidō, NE Japan *42°54′N 141°49′E*
35 R14 **Ojai** California, W USA *34°25′N 119°15′W*
94 K13 **Öje** Dalarna, C Sweden *60°49′N 13°54′E*
93 J14 **Öjebyn** Norrbotten, N Sweden *65°20′N 21°26′E*
165 B13 **Ojika-jima** *island* SW Japan
40 K5 **Ojinaga** Chihuahua, N Mexico *29°31′N 104°26′W*
40 M11 **Ojo Caliente** *var.* Ojocaliente. Zacatecas, C Mexico *22°35′N 102°18′W*
Ojocaliente *see* Ojo Caliente
40 D6 **Ojo de Liebre, Laguna** *var.* Laguna Scammon, Scammon Lagoon. *lagoon* NW Mexico
62 I7 **Ojos del Salado, Cerro** ▲ W Argentina *27°04′S 68°34′W*
105 R7 **Ojos Negros** Aragón, NE Spain *40°43′N 01°30′W*
40 M12 **Ojuelos de Jalisco** Aguascalientes, C Mexico *21°52′N 101°40′W*
127 N4 **Oka** ≈ W Russian Federation
83 D19 **Okahandja** Otjozondjupa, C Namibia *21°58′S 16°55′E*
184 L9 **Okahukura** Manawatu-Wanganui, North Island, New Zealand *38°48′S 175°13′E*
184 J3 **Okaihau** Northland, North Island, New Zealand *35°19′S 173°45′E*
83 D18 **Okakarara** Otjozondjupa, N Namibia *20°33′S 17°20′E*
13 P5 **Okak Islands** *island group* Newfoundland and Labrador, E Canada
10 M17 **Okanagan** ≈ British Columbia, SW Canada
11 N17 **Okanagan Lake** ◎ British Columbia, SW Canada
Ókanizsa *see* Kanjiža
83 C16 **Okankolo** Oshikoto, N Namibia *17°57′S 16°28′E*
32 K6 **Okanogan River** ≈ Washington, NW USA
83 D18 **Okaputa** Otjozondjupa, N Namibia *20°09′S 16°56′E*
149 V9 **Okāra** Punjab, E Pakistan *30°49′N 73°31′E*
26 M10 **Okarche** Oklahoma, C USA *35°43′N 97°58′W*
Okarem *see* Ekerem
189 X14 **Okat Harbor** *harbor* Kosrae, E Micronesia
22 M5 **Okatibbee Creek** ≈ Mississippi, S USA
83 C17 **Okaukuejo** Kunene, N Namibia *19°10′S 15°23′E*
Okavanggo *see* Cubango/Okavango
83 E17 **Okavango** ◐ *district* NW Namibia
83 G17 **Okavango** *var.* Cubango, Kavango, Kavengo, Kubango, Okavanggo, *Port.* Ocavango. ≈ S Africa *see also* Cubango
Okavango *see* Cubango
83 G17 **Okavango Delta** *wetland* N Botswana
164 M12 **Okaya** Nagano, Honshū, S Japan *36°03′N 138°00′E*
164 H13 **Okayama** Okayama, Honshū, SW Japan *34°40′N 133°54′E*
164 H13 **Okayama** *off.* Okayama-ken. ◐ *prefecture* Honshū, SW Japan
Okayama-ken *see* Okayama
164 L14 **Okazaki** Aichi, Honshū, C Japan *34°58′N 137°10′E*
110 M12 **Okęcie** ✈ (Warszawa) Mazowieckie, C Poland *52°08′N 20°57′E*
23 Y13 **Okeechobee** Florida, SE USA *27°14′N 80°49′W*
23 Y14 **Okeechobee, Lake** ◎ Florida, SE USA
26 M9 **Okeene** Oklahoma, C USA *36°07′N 98°19′W*
23 V8 **Okefenokee Swamp** *wetland* Georgia, SE USA
97 J24 **Okehampton** SW England, United Kingdom *50°44′N 04°W*
27 P10 **Okemah** Oklahoma, C USA *35°25′N 96°20′W*
77 U16 **Okene** Kogi, S Nigeria *07°31′N 06°15′E*
100 K13 **Oker** *var.* Ocker. ≈ NW Germany
101 J14 **Oker-Stausee** ◎ C Germany
123 T12 **Okha** Ostrov Sakhalin, Sakhalinskaya Oblast', SE Russian Federation *53°33′N 142°55′E*
125 U15 **Okhansk** *var.* Ochansk. Permskiy Kray, NW Russian Federation *57°44′N 55°20′E*
123 S10 **Okhotsk** Khabarovskiy Kray, E Russian Federation *59°21′N 143°15′E*
192 J2 **Okhotsk, Sea of** *sea* NW Pacific Ocean
117 T4 **Okhtyrka** *Rus.* Akhtyrka. Sums'ka Oblast', NE Ukraine *50°19′N 34°54′E*
83 E23 **Okiep** Northern Cape, W South Africa *29°39′S 17°53′E*
Oki-guntō *see* Oki-shotō
164 H11 **Oki-kaikyō** *strait* SW Japan
165 P16 **Okinawa** Okinawa, SW Japan *26°20′N 127°47′E*
165 S16 **Okinawa** *off.* Okinawa-ken. ◐ *prefecture* Okinawa, SW Japan
165 S16 **Okinawa** *island* SW Japan
Okinawa-ken *see* Okinawa
165 U16 **Okinoerabu-jima** *island* Nansei-shotō, SW Japan
164 F15 **Okino-shima** *island* SW Japan
164 H11 **Oki-shotō** *var.* Oki-guntō. *island group* SW Japan
77 T16 **Okitipupa** Ondo, SW Nigeria *06°33′N 04°43′E*
166 L8 **Okkan** Bago, SW Myanmar (Burma) *17°30′N 95°52′E*
27 N10 **Oklahoma** *off.* State of Oklahoma, *also known as* The Sooner State. ◐ *state* C USA
27 N11 **Oklahoma City** *state capital* Oklahoma, C USA *35°28′N 97°31′W*
25 Q4 **Oklaunion** Texas, SW USA *34°07′N 99°07′W*
23 W10 **Oklawaha River** ≈ Florida, SE USA
27 P10 **Okmulgee** Oklahoma, C USA *35°38′N 95°59′W*
Oknitsa *see* Ocniţa
22 M3 **Okolona** Mississippi, S USA *34°00′N 88°45′W*
165 U2 **Okoppe** Hokkaidō, NE Japan *44°27′N 143°06′E*
11 Q16 **Okotoks** Alberta, SW Canada *50°46′N 113°57′W*
80 H6 **Oko, Wadi** ≈ NE Sudan
79 G19 **Okoyo** Cuvette, W Congo *01°28′S 15°04′E*
77 S15 **Okpara** ≈ Benin/Nigeria
92 J8 **Øksfjord** Finnmark, N Norway *70°13′N 22°22′E*
125 R4 **Oksino** Nenetskiy Avtonomnyy Okrug, NW Russian Federation *67°33′N 52°15′E*
92 G13 **Oksskolten** ▲ C Norway *66°00′N 14°18′E*
Oksu *see* Oqsu
144 M8 **Oktaybr'skiy** Kostanay, N Kazakhstan
186 B7 **Ok Tedi** Western, W Papua New Guinea
Oktemberyan *see* Armavir
166 M7 **Oktwin** Bago, C Myanmar (Burma) *18°47′N 96°21′E*
127 R6 **Oktyabr'sk** Samarskaya Oblast', W Russian Federation *53°13′N 48°36′E*
Oktyabr'sk *see* Kandyagash
125 N12 **Oktyabr'skiy** Arkhangel'skaya Oblast', NW Russian Federation *61°03′N 43°16′E*
122 E10 **Oktyabr'skiy** Kamchatskiy Kray, E Russian Federation *52°35′N 156°18′E*
127 T5 **Oktyabr'skiy** Respublika Bashkortostan, W Russian Federation *54°28′N 53°29′E*
127 O11 **Oktyabr'skiy** Volgogradskaya Oblast', SW Russian Federation *48°00′N 43°35′E*
Oktyabr'skiy *see* Aktsyabrski
127 V7 **Oktyabr'skoye** Orenburgskaya Oblast', W Russian Federation *52°22′N 55°39′E*
122 M5 **Oktyabr'skoy Revolyutsii, Ostrov** *Eng.* October Revolution Island. *island* Severnaya Zemlya, N Russian Federation
Ōkuchi *see* Isa
124 I14 **Okulovka** *var.* Okulovka. Novgorodskaya Oblast', W Russian Federation *58°24′N 33°16′E*
Okulovka *see* Okulovka
165 Q4 **Okushiri-tō** *var.* Okusiri Tô. *island* NE Japan
Okusiri Tô *see* Okushiri-tō
77 S15 **Okuta** Kwara, W Nigeria *09°18′N 03°09′E*
Ōkuti *see* Isa
83 F19 **Okwa** *var.* Chapman's. ≈ Botswana/Namibia
123 T10 **Ola** Magadanskaya Oblast', E Russian Federation *59°36′N 151°18′E*
27 T11 **Ola** Arkansas, C USA *35°01′N 93°13′W*
Ola *see* Ala
35 T11 **Olacha Peak** ▲ California, W USA *36°15′N 118°07′W*
92 J1 **Ólafsfjörður** Norðurland Eystra, N Iceland *66°04′N 18°36′W*
92 H3 **Ólafsvík** Vesturland, W Iceland *64°52′N 23°45′W*
Oláhbrettye *see* Bretea-Română
Oláhszentgyörgy *see* Sângeorz-Băi
Oláh-Toplicza *see* Toplița
118 F9 **Olaine** C Latvia *56°47′N 23°56′E*
35 T11 **Olancha** California, W USA *36°16′N 118°00′W*
42 J5 **Olanchito** Yoro, C Honduras *15°30′N 86°34′W*
42 J6 **Olancho** ◐ *department* E Honduras
95 O20 **Öland** *island* S Sweden
95 O19 **Ölands norra udde** *headland* S Sweden *57°21′N 17°06′E*
95 N22 **Ölands södra udde** *headland* S Sweden *56°12′N 16°26′E*
182 K7 **Olary** South Australia *32°18′S 140°16′E*
27 R4 **Olathe** Kansas, C USA *38°52′N 94°50′W*
61 C22 **Olavarría** Buenos Aires, E Argentina *36°57′S 60°20′W*
92 O2 **Olav V Land** *physical region* C Svalbard
111 H14 **Oława** *Ger.* Ohlau. Dolnośląskie, SW Poland *50°57′N 17°18′E*
107 D17 **Olbia** *prev.* Terranova Pausania. Sardegna, Italy, C Mediterranean Sea *40°55′N 09°30′E*
44 G5 **Old Bahama Channel** *channel* The Bahamas/Cuba
Old Bay State/Old Colony State *see* Massachusetts
10 H2 **Old Crow** Yukon, NW Canada *67°34′N 139°55′W*
Old Dominion *see* Virginia
Oldebekeap *see* Oldeberkoop
98 M7 **Oldeberkoop** *Fris.* Aldeberkeap. Fryslân, N Netherlands *52°55′N 06°07′E*
98 L10 **Oldebroek** Gelderland, E Netherlands *52°27′N 05°54′E*
98 L8 **Oldemarkt** Overijssel, N Netherlands *52°49′N 05°58′E*
94 E11 **Olden** Sogn Og Fjordane, C Norway *61°52′N 06°44′E*
100 G10 **Oldenburg** Niedersachsen, NW Germany *53°09′N 08°13′E*
100 K8 **Oldenburg** *var.* Oldenburg in Holstein. Schleswig-Holstein, N Germany *54°17′N 10°55′E*
Oldenburg in Holstein *see* Oldenburg
98 P10 **Oldenzaal** Overijssel, E Netherlands *52°19′N 06°53′E*
Olderfjord *see* Leaibevuotna
18 J8 **Old Forge** New York, NE USA *43°42′N 74°59′W*
Old Goa *see* Goa
97 L17 **Oldham** NW England, United Kingdom *53°36′N 02°W*
39 Q14 **Old Harbor** Kodiak Island, Alaska, USA *57°12′N 153°18′W*

◆ Country ● Country Capital ◇ Dependent Territory ○ Dependent Territory Capital ◐ Administrative Regions ✈ International Airport ▲ Mountain ▲ Mountain Range Volcano ≈ River ◎ Lake ⊠ Reservoir

Old Harbour – Oriente

44 J13 **Old Harbour** C Jamaica 17°56′N 77°06′W
97 C22 **Old Head of Kinsale** *Ir.* An Seancheann. *headland* SW Ireland 51°37′N 08°33′W
20 J8 **Old Hickory Lake** ⊞ Tennessee, S USA
Old Line State *see* Maryland
Old North State *see* North Carolina
81 I17 **Ol Doinyo Lengeyo** ▲▲ C Kenya
11 Q16 **Olds** Alberta, SW Canada 51°50′N 114°06′W
19 O7 **Old Speck Mountain** ▲ Maine, NE USA 44°34′N 70°55′W
19 S6 **Old Town** Maine, NE USA 44°55′N 68°39′W
11 T17 **Old Wives Lake** ◎ Saskatchewan, S Canada
162 J7 **Öldziyt** *var.* Höshööt. Arhangay, C Mongolia 48°06′N 102°34′E
162 I8 **Öldziyt** *var.* Ulaan-Uul. Bayanhongor, C Mongolia 46°03′N 100°52′E
162 L10 **Öldziyt** *var.* Rashaant. Dundgovĭ, C Mongolia 44°54′N 106°32′E
162 K8 **Öldziyt** *var.* Sangiyn Dalay. Övörhangay, C Mongolia 46°35′N 103°18′E
Öldziyt *see* Erdenemandal, Arhangay, Mongolia
Öldziyt *see* Sayhandulaan, Dornogovĭ, Mongolia
188 H6 **Oleai** *var.* San Jose. Saipan, S Northern Mariana Islands
18 E11 **Olean** New York, NE USA 42°04′N 78°24′W
110 O7 **Olecko** *Ger.* Treuburg. Warmińsko-Mazurskie, NE Poland 54°02′N 22°29′E
106 C7 **Oleggio** Piemonte, NE Italy 45°36′N 08°37′E
123 P11 **Olëkma** Amurskaya Oblast', SE Russian Federation 57°00′N 120°27′E
123 P12 **Olëkma** ≈ C Russian Federation
123 P11 **Olëkminsk** Respublika Sakha (Yakutiya), NE Russian Federation 60°25′N 120°25′E
117 W7 **Oleksandrivka** Donets'ka Oblast', E Ukraine 48°42′N 36°56′E
117 R7 **Oleksandrivka** *Rus.* Aleksandrovka. Kirovohrads'ka Oblast', C Ukraine 48°59′N 32°14′E
117 Q9 **Oleksandrivka** Mykolayivs'ka Oblast', S Ukraine 47°42′N 31°17′E
117 S7 **Oleksandriya** *Rus.* Aleksandriya. Kirovohrads'ka Oblast', C Ukraine 48°42′N 33°07′E
93 B20 **Ølen** Hordaland, S Norway 59°36′N 05°48′E
124 J4 **Olenegorsk** Murmanskaya Oblast', NW Russian Federation 68°06′N 33°15′E
123 N9 **Olenëk** Respublika Sakha (Yakutiya), NE Russian Federation 68°28′N 112°18′E
123 N9 **Olenëk** ≈ NE Russian Federation
123 O7 **Olenëkskiy Zaliv** *bay* N Russian Federation
124 K6 **Olenitsa** Murmanskaya Oblast', NW Russian Federation 66°27′N 35°21′E
102 I11 **Oléron, Île d'** *island* W France
111 H14 **Oleśnica** *Ger.* Oels, Oels in Schlesien. Dolnośląskie, SW Poland 51°13′N 17°20′E
111 I15 **Olesno** *Ger.* Rosenberg. Opolskie, S Poland 50°53′N 18°23′E
116 M3 **Olevs'k** *Rus.* Olevsk. Zhytomyrs'ka Oblast', N Ukraine 51°12′N 27°38′E
Olevsk *see* Olevs'k
123 S15 **Ol'ga** Primorskiy Kray, SE Russian Federation 43°41′N 135°06′E
Olga, Mount *see* Kata Tjuṯa
92 P2 **Olgastretet** *strait* E Svalbard
162 D5 **Ölgiy** Bayan-Ölgiy, W Mongolia 48°57′N 89°59′E
95 F23 **Ølgod** Syddtjylland, W Denmark 55°44′N 08°37′E
104 H14 **Olhão** Faro, S Portugal 37°01′N 07°50′W
93 L14 **Olhava** Pohjois-Pohjanmaa, C Finland 65°28′N 25°25′E
105 V5 **Oliana** Cataluña, NE Spain 42°04′N 01°19′E
112 B12 **Olib** *It.* Ulbo. *island* W Croatia
83 B16 **Olifa** Kunene, NW Namibia 17°25′S 14°27′E
83 E20 **Olifants** *var.* Elephant River. ≈ E Namibia
Olifants *see* Lepelle
83 I20 **Olifants Drift** *var.* Oliphants Drift. Kgatleng, SE Botswana 24°13′S 26°52′E
83 G22 **Olifantshoek** Northern Cape, N South Africa 27°56′S 22°45′E
188 L15 **Olimarao Atoll** *atoll* Caroline Islands, C Micronesia
Ólimbos *see* Ólympos
Olimpo *see* Fuerte Olimpo
59 Q15 **Olinda** Pernambuco, E Brazil 08°S 34°51′W
Olinthos *see* Ólynthos
Oliphants Drift *see* Olifants Drift
Olisipo *see* Lisboa
Olita *see* Alytus
105 Q4 **Olite** Navarra, N Spain 42°29′N 01°40′W
62 K10 **Oliva** Córdoba, C Argentina 32°03′S 63°34′W
105 T11 **Oliva** Valenciana, E Spain 38°55′N 00°09′W
Oliva *see* La Oliva
104 I12 **Oliva de la Frontera** Extremadura, W Spain 38°17′N 06°54′W
Olivares *see* Olivares de Júcar
62 H9 **Olivares, Cerro de** ▲ N Chile 30°25′S 69°52′W
105 P9 **Olivares de Júcar** *var.* Olivares. Castilla-La Mancha, C Spain 39°45′N 02°21′W
22 L1 **Olive Branch** Mississippi, S USA 34°58′N 89°49′W
21 O5 **Olive Hill** Kentucky, S USA 38°18′N 83°10′W
35 O6 **Olivehurst** California, W USA 39°05′N 121°33′W
104 G7 **Oliveira de Azeméis** Aveiro, N Portugal 40°49′N 08°29′W
104 I11 **Olivenza** Extremadura, W Spain 38°41′N 07°06′W
11 N17 **Oliver** British Columbia, SW Canada 49°10′N 119°37′W
103 N7 **Olivet** Loiret, C France 47°53′N 01°53′E
29 Q12 **Olivet** South Dakota, N USA 43°14′N 97°40′W
29 T9 **Olivia** Minnesota, N USA 44°46′N 94°59′W
185 C20 **Olivine Range** ▲▲ South Island, New Zealand
108 H10 **Olivone** Ticino, S Switzerland 46°32′N 08°55′E
Ölkeyek *see* Ol'keyyek
144 L11 **Ol'keyyek** *Kaz.* Ölkeyek; *prev.* Ul'kayak. ≈ C Kazakhstan
127 O9 **Ol'khovka** Volgogradskaya Oblast', SW Russian Federation 49°54′N 44°36′E
111 K16 **Olkusz** Małopolskie, S Poland 50°18′N 19°33′E
22 I6 **Olla** Louisiana, S USA 31°54′N 92°14′W
62 I4 **Ollagüe, Volcán** *var.* Oyahue, Volcán Oyahue. ⛰ N Chile 21°25′S 68°10′W
189 U13 **Ollan** *island* Chuuk, C Micronesia
188 F7 **Ollei** Babeldaob, N Palau 07°43′N 134°37′E
Ollius *see* Oglio
108 C10 **Ollon** Vaud, W Switzerland 46°19′N 07°00′E
147 Q10 **Olmaliq** *Rus.* Almalyk. Toshkent Viloyati, E Uzbekistan 40°51′N 69°39′E
104 M6 **Olmedo** Castilla y León, N Spain 41°17′N 04°41′W
56 B10 **Olmos** Lambayeque, W Peru 06°00′S 79°43′W
Olmütz *see* Olomouc
30 M15 **Olney** Illinois, N USA 38°43′N 88°05′W
25 R5 **Olney** Texas, SW USA 33°22′N 98°45′W
95 L22 **Olofström** Blekinge, S Sweden 56°16′N 14°33′E
187 N9 **Olomburi** Malaita, N Solomon Islands 09°00′S 161°09′E
111 H17 **Olomouc** *Ger.* Olmütz, *Pol.* Ołomuniec. Olomoucký Kraj, E Czech Republic 49°36′N 17°13′E
111 H18 **Olomoucký Kraj** ⬖ *region* E Czech Republic
Ołomuniec *see* Olomouc
122 D7 **Olonets** Respublika Kareliya, NW Russian Federation 60°58′N 33°01′E
171 N3 **Olongapo** *off.* Olongapo City. Luzon, N Philippines 14°52′N 120°16′E
Olongapo City *see* Olongapo
102 J16 **Oloron-Ste-Marie** Pyrénées-Atlantiques, SW France 43°12′N 00°35′W
192 L16 **Olosega** *island* Manua Islands, E American Samoa
105 W4 **Olot** Cataluña, NE Spain 42°11′N 02°30′E
146 K12 **Olot** *Rus.* Alat. Buxoro Viloyati, C Uzbekistan 39°22′N 63°42′E
112 I12 **Olovo** Federacija Bosne I Hercegovine, E Bosnia and Herzegovina 44°08′N 18°35′E
123 O14 **Olovyannaya** Zabaykal'skiy Kray, S Russian Federation 50°59′N 115°24′E
123 T7 **Oloy** ≈ NE Russian Federation
101 F16 **Olpe** Nordrhein-Westfalen, W Germany 51°02′N 07°51′E
109 N8 **Olperer** ▲ SW Austria 47°03′N 11°36′E
Olshanka *see* Vil'shanka
Ol'shany *see* Al'shany
Olsnitz *see* Murska Sobota
98 M10 **Olst** Overijssel, E Netherlands 52°19′N 06°06′E
110 L8 **Olsztyn** *Ger.* Allenstein. Warmińsko-Mazurskie, N Poland 53°46′N 20°28′E
110 L8 **Olsztynek** *Ger.* Hohenstein in Ostpreussen. Warmińsko-Mazurskie, N Poland 53°35′N 20°17′E
116 I14 **Olt** ⬖ *county* SW Romania
116 I14 **Olt** *var.* Oltul, *Ger.* Alt. ≈ S Romania
108 E7 **Olten** Solothurn, NW Switzerland 47°22′N 07°55′E
116 K14 **Oltenița** *prev. Eng.* Oltenitsa; *anc.* Constantiola. Călărași, SE Romania 44°05′N 26°40′E
Oltenitsa *see* Oltenița
116 H14 **Olteț** ≈ S Romania
24 M4 **Olton** Texas, SW USA 34°10′N 102°07′W
137 R12 **Oltu** Erzurum, NE Turkey 40°34′N 41°59′E
Oltul *see* Olt
146 G7 **Oltynko'l** Qoraqalpog'iston Respublikasi, NW Uzbekistan 43°04′N 58°51′E
Oluan Pi *see* Eluan Bi
Ólubló *see* Stará L'ubovňa
137 R11 **Olur** Erzurum, NE Turkey 40°49′N 42°08′E
104 L15 **Olvera** Andalucía, S Spain 36°56′N 05°15′W
Ol'viopol' *see* Pervomays'k
Olwanpi, Cape *see* Eluan Bi
115 D20 **Olympía** Dytikí Elláda, S Greece 37°39′N 21°36′E
32 G9 **Olympia** *state capital* Washington, NW USA 47°02′N 122°54′W
182 H5 **Olympic Dam** South Australia 30°25′S 136°56′E
32 F7 **Olympic Mountains** ▲▲ Washington, NW USA
121 O3 **Ólympos** *var.* Troodos, *Eng.* Mount Olympus. ▲ C Cyprus 34°55′N 32°49′E
115 F15 **Ólympos** *var.* Ólimbos, *Eng.* Mount Olympus. ▲ N Greece 40°04′N 22°24′E
115 L17 **Ólympos** ▲ Lésvos, E Greece 39°03′N 26°20′E
16 C5 **Olympus, Mount** ▲ Washington, NW USA 47°48′N 123°42′W
Olympus, Mount *see* Ólympos
115 G14 **Ólynthos** *var.* Olinthos; *anc.* Olynthus. *site of ancient city* Kentrikí Makedonía, N Greece
Olynthus *see* Ólynthos
117 Q3 **Olyshivka** Chernihivs'ka Oblast', N Ukraine 51°13′N 31°19′E
123 W8 **Olyutorskiy, Mys** *headland* E Russian Federation 59°56′N 170°22′E
123 V8 **Olyutorskiy Zaliv** *bay* E Russian Federation
186 M10 **Om** ≈ W Papua New Guinea
129 S6 **Om'** ≈ N Russian Federation
158 I13 **Oma** Xizang Zizhiqu, W China 32°30′N 83°14′E
165 R6 **Ōma** Aomori, Honshū, C Japan 41°31′N 140°54′E
125 P6 **Oma** ≈ NW Russian Federation
164 M12 **Ōmachi** *var.* Ômati. Nagano, Honshū, S Japan 36°30′N 137°51′E
165 Q8 **Ōmagari** Akita, Honshū, C Japan 39°29′N 140°29′E
97 E15 **Omagh** *Ir.* An Ómaigh. W Northern Ireland, United Kingdom 54°36′N 07°18′W
29 S15 **Omaha** Nebraska, C USA 41°14′N 95°57′W
83 E19 **Omaheke** ⬖ *district* W Namibia
141 W10 **Oman** *off.* Sultanate of Oman, *Ar.* Salṭanat 'Umān; *prev.* Muscat and Oman. ◆ *monarchy* SW Asia
129 O10 **Oman Basin** *var.* Bassin d'Oman. *undersea feature* N Indian Ocean 23°20′N 63°00′E
Oman, Bassin d' *see* Oman Basin
129 N10 **Oman, Gulf of** *Ar.* Khalīj 'Umān. *gulf* N Arabian Sea
Oman, Sultanate of *see* Oman
184 J3 **Omapere** Northland, North Island, New Zealand 35°32′S 173°24′E
185 E20 **Omarama** Canterbury, South Island, New Zealand 44°29′S 169°57′E
112 F11 **Omarska** ⬖ Republika Srpska, NW Bosnia and Herzegovina
83 C18 **Omaruru** Erongo, NW Namibia 21°28′S 15°56′E
83 C19 **Omaruru** ≈ W Namibia
83 E17 **Omatako** ≈ NE Namibia
Ômati *see* Ōmachi
83 E18 **Omawewozonyanda** Omaheke, E Namibia 21°30′S 19°34′E
165 R6 **Ōma-zaki** *headland* Honshū, C Japan 41°32′N 140°53′E
Omba *see* Ambae
Ombai *see* Alor, Pulau
83 C16 **Ombalantu** Omusati, N Namibia 17°33′S 14°58′E
79 H15 **Ombella-Mpoko** ⬖ *prefecture* S Central African Republic
Ombetsu *see* Onbetsu
83 B17 **Ombombo** Kunene, NW Namibia 18°43′S 13°53′E
79 D19 **Omboué** Ogooué-Maritime, W Gabon 01°38′S 09°20′E
106 G13 **Ombrone** ≈ C Italy
80 F9 **Omdurman** *var.* Umm Durmān. Khartoum, C Sudan 15°37′N 32°29′E
165 N13 **Ōme** Tōkyō, Honshū, S Japan 35°48′N 139°17′E
106 C6 **Omegna** Piemonte, NE Italy 45°54′N 08°25′E
183 P12 **Omeo** Victoria, SE Australia 37°09′S 147°36′E
138 F11 **Omer** Southern, C Israel 31°16′N 34°51′E
41 P16 **Ometepec** Guerrero, S Mexico 16°39′N 98°23′W
42 K11 **Ometepe, Isla de** *island* S Nicaragua
Om Hager *see* Om Ḥajer
80 I10 **Om Ḥajer** *var.* Om Hager. SW Eritrea 14°19′N 36°46′E
165 J13 **Ōmihachiman** Shiga, Honshū, SW Japan 35°08′N 136°04′E
10 L12 **Omineca Mountains** ▲▲ British Columbia, W Canada
113 F14 **Omiš** *It.* Almissa. Split-Dalmacija, S Croatia 43°27′N 16°41′E
112 B10 **Omišalj** Primorje-Gorski Kotar, NW Croatia 45°10′N 14°33′E
83 D19 **Omitara** Khomas, C Namibia 22°18′S 18°01′E
41 O16 **Omitlán, Río** ≈ S Mexico
39 X14 **Ommaney, Cape** *headland* Baranof Island, Alaska, USA 56°10′N 134°40′W
98 N9 **Ommen** Overijssel, E Netherlands 52°31′N 06°25′E
163 N7 **Ömnödelger** *var.* Bayanbulag. Hentiy, C Mongolia 47°54′N 109°51′E
162 K11 **Ömnögovĭ** ⬖ *province* S Mongolia
191 X7 **Omoa** Fatu Hira, NE French Polynesia 10°30′S 138°41′E
Omo Botego *see* Omo Wenz
Ómoldova *see* Moldova Veche
123 T7 **Omolon** Chukotskiy Avtonomnyy Okrug, NE Russian Federation 65°11′N 160°33′E
123 T7 **Omolon** ≈ NE Russian Federation
123 Q8 **Omoloy** ≈ NE Russian Federation
162 I12 **Omon Gol** *Chin.* Dong He, *var.* Narin Gol. ≈ N China
165 P8 **Omono-gawa** ≈ Honshū, C Japan
81 I14 **Omo Wenz** *var.* Omo Botego. ≈ Ethiopia/Kenya
122 H12 **Omsk** Omskaya Oblast', C Russian Federation 55°N 73°22′E
122 H11 **Omskaya Oblast'** ⬖ *province* C Russian Federation
165 U2 **Ōmu** Hokkaidō, NE Japan 44°36′N 142°55′E
110 M9 **Omulew** ≈ NE Poland
116 J12 **Omul, Vârful** *prev.* Vîrful Omu. ▲ C Romania 45°24′N 25°26′E
83 D16 **Omundaungilo** Ohangwena, N Namibia 17°28′S 16°39′E
164 C14 **Ōmura** Nagasaki, Kyūshū, SW Japan 32°56′N 129°58′E
83 B17 **Omusati** ⬖ *district* N Namibia
164 C14 **Ōmuta** Fukuoka, Kyūshū, SW Japan 33°01′N 130°27′E
125 S14 **Omutninsk** Kirovskaya Oblast', NW Russian Federation 58°37′N 52°08′E
Omu, Vîrful *see* Omul, Vârful
29 V7 **Onamia** Minnesota, N USA 46°04′N 93°40′W
21 Y5 **Onancock** Virginia, NE USA 37°42′N 75°45′W
14 E10 **Onaping Lake** ◎ Ontario, S Canada
30 M12 **Onarga** Illinois, N USA 40°39′N 88°00′W
15 R6 **Onatchiway, Lac** ◎ Québec, SE Canada
29 S14 **Onawa** Iowa, C USA 42°01′N 96°06′W
165 U5 **Onbetsu** *var.* Ombetsu. Hokkaidō, NE Japan 42°52′N 143°54′E
83 B16 **Oncócua** Cunene, SW Angola 16°37′S 13°23′E
105 S9 **Onda** Valenciana, E Spain 39°58′N 00°17′W
111 N18 **Ondava** ≈ NE Slovakia
83 C16 **Ondjiva** Cunene, S Angola 17°02′S 15°42′E
77 T16 **Ondo** Ondo, SW Nigeria 07°07′N 04°50′E
77 T16 **Ondo** ⬖ *state* SW Nigeria
163 N8 **Öndörhaan** *var.* Undur Khan; *prev.* Tsetsen Khan. Hentiy, E Mongolia 47°21′N 110°42′E
162 M9 **Öndörshil** *var.* Böhöt. Dundgovĭ, C Mongolia 45°13′N 108°12′E
162 L8 **Öndörshireet** *var.* Bayshint. Töv, C Mongolia 47°22′N 105°04′E
162 I7 **Öndör-Ulaan** *var.* Teel. Arhangay, C Mongolia 48°01′N 100°30′E
83 D18 **Ondundazongonda** Otjozondjupa, N Namibia 20°28′S 18°00′E
151 K21 **One and Half Degree Channel** *channel* S Maldives
187 Z15 **Oneata** *island* Lau Group, E Fiji
124 L9 **Onega** Arkhangel'skaya Oblast', NW Russian Federation 63°54′N 37°59′E
124 L9 **Onega** ≈ NW Russian Federation
Onega Bay *see* Onezhskaya Guba
Onega, Lake *see* Onezhskoye Ozero
18 I10 **Oneida** New York, NE USA 43°05′N 75°39′W
20 M8 **Oneida** Tennessee, S USA 36°30′N 84°30′W
18 H9 **Oneida Lake** ◎ New York, NE USA
29 P13 **O'Neill** Nebraska, C USA 42°28′N 98°38′W
123 V12 **Onekotan, Ostrov** *island* Kuril'skiye Ostrova, SE Russian Federation
23 P3 **Oneonta** Alabama, S USA 33°57′N 86°28′W
18 J11 **Oneonta** New York, NE USA 42°27′N 75°03′W
190 I16 **Oneroa** *island* S Cook Islands
116 K11 **Onești** *Hung.* Onyest; *prev.* Gheorghe Gheorghiu-Dej. Bacău, E Romania 46°14′N 26°46′E
193 V15 **Onevai** *island* Tongatapu Group, S Tonga
108 A11 **Onex** Genève, SW Switzerland 46°12′N 06°04′E
124 K8 **Onezhskaya Guba** *Eng.* Onega Bay. *bay* NW Russian Federation
122 D7 **Onezhskoye Ozero** *Eng.* Lake Onega. ◎ NW Russian Federation
83 C16 **Ongandjera** Omusati, N Namibia 17°49′S 15°06′E
184 N12 **Ongaonga** Hawke's Bay, North Island, New Zealand 39°57′S 176°21′E
Ongi *see* Sayhan-Ovoo, Dundgovĭ, Mongolia
Ongi *see* Uyanga
163 W14 **Ongjin** SW North Korea 37°56′N 125°22′E
155 J17 **Ongole** Andhra Pradesh, E India 15°33′N 80°03′E
Ongon *see* Bürd
Ongtüstik Qazaqstan Oblysy *see* Yuzhnyy Kazakhstan
99 I21 **Onhaye** Namur, S Belgium 50°15′N 04°51′E
166 M8 **Onhne** Bago, SW Myanmar (Burma) 17°02′N 96°28′E
137 S9 **Oni** N Georgia 42°36′N 43°13′E
29 N9 **Onida** South Dakota, N USA 44°42′N 100°03′W
164 F15 **Onigajō-yama** ▲ Shikoku, SW Japan 33°10′N 132°37′E
172 H7 **Onilahy** ≈ S Madagascar
77 U16 **Onitsha** Anambra, S Nigeria 06°09′N 06°48′E
164 K12 **Ōno** Fukui, Honshū, SW Japan 35°59′N 136°30′E
164 I13 **Ono** Hyōgo, Honshū, SW Japan 34°52′N 134°55′E
187 K15 **Ono** *island* SW Fiji
164 E13 **Onoda** Yamaguchi, Honshū, SW Japan 34°00′N 131°11′E
187 Z16 **Ono-i-lau** *island* SE Fiji
164 D13 **Ōnojō** *var.* Ônozyô. Fukuoka, Kyūshū, SW Japan 33°34′N 130°29′E
163 O7 **Onon Gol** ≈ N Mongolia
Ononte *see* Orontes
55 N6 **Onoto** Anzoátegui, NE Venezuela 09°36′N 65°12′W
191 O3 **Onotoa** *prev.* Clerk Island. *atoll* Tungaru, W Kiribati
Ônozyô *see* Ōnojō
83 E23 **Onseepkans** Northern Cape, W South Africa 28°44′S 19°18′E
104 F4 **Ons, Illa de** *island* NW Spain
180 H7 **Onslow** Western Australia 21°42′S 115°08′E
21 W11 **Onslow Bay** *bay* North Carolina, E USA
98 P6 **Onstwedde** Groningen, NE Netherlands 53°01′N 07°04′E
164 C16 **On-take** ▲ Kyūshū, SW Japan 31°35′N 130°39′E
35 T15 **Ontario** California, W USA 34°03′N 117°39′W
32 M13 **Ontario** Oregon, NW USA 44°01′N 116°57′W
12 D10 **Ontario** ⬖ *province* S Canada
11 P14 **Ontario, Lake** ◎ Canada/USA
0 L9 **Ontario Peninsula** *peninsula* Canada/USA
Onteniente *see* Ontinyent
105 S11 **Ontinyent** *var.* Onteniente. Valenciana, E Spain 38°49′N 00°37′W
93 N15 **Ontojärvi** ◎ E Finland
30 L3 **Ontonagon** Michigan, N USA 46°52′N 89°18′W
30 L3 **Ontonagon River** ≈ Michigan, N USA
186 M7 **Ontong Java Atoll** *prev.* Lord Howe Island. *atoll* N Solomon Islands
175 N5 **Ontong Java Rise** *undersea feature* W Pacific Ocean 01°00′N 157°00′E
Onuba *see* Huelva
55 W9 **Onverwacht** Para, N Suriname 05°36′N 55°12′W
Onyest *see* Onești
Oodeypore *see* Udaipur
182 J7 **Oodla Wirra** South Australia 32°52′S 139°05′E
182 F2 **Oodnadatta** South Australia 27°34′S 135°27′E
182 C5 **Ooldea** South Australia 30°29′S 131°50′E
27 Q8 **Oologah Lake** ⊞ Oklahoma, C USA
Oos-Kaap *see* Eastern Cape
Oos-Londen *see* East London
99 E17 **Oostakker** Oost-Vlaanderen, NW Belgium 51°06′N 03°46′E
99 D15 **Oostburg** Zeeland, SW Netherlands 51°20′N 03°30′E
98 K9 **Oostelijk-Flevoland** *polder* C Netherlands
99 B16 **Oostende** *Eng.* Ostend, *Fr.* Ostende. West-Vlaanderen, NW Belgium 51°13′N 02°55′E
99 B16 **Oostende** ✈ West-Vlaanderen, NW Belgium 51°12′N 02°55′E
98 L12 **Oosterbeek** Gelderland, SE Netherlands 51°59′N 05°51′E
99 I14 **Oosterhout** Noord-Brabant, S Netherlands 51°38′N 04°51′E
98 O6 **Oostermoers Vaart** *var.* Hunze. ≈ NE Netherlands
99 F14 **Oosterschelde** *Eng.* Eastern Scheldt. *inlet* SW Netherlands
99 E14 **Oosterscheldedam** *dam* SW Netherlands
98 M7 **Oosterwolde** *Fris.* Easterwâlde. Fryslân, N Netherlands 53°01′N 06°15′E
98 I9 **Oosthuizen** Noord-Holland, NW Netherlands 52°34′N 05°00′E
99 H16 **Oostmalle** Antwerpen, N Belgium 51°18′N 04°44′E
Oos-Transvaal *see* Mpumalanga
99 E15 **Oost-Souburg** Zeeland, SW Netherlands 51°28′N 03°36′E
99 E17 **Oost-Vlaanderen** *Eng.* East Flanders. ⬖ *province* NW Belgium
98 J5 **Oost-Vlieland** Fryslân, N Netherlands 53°19′N 05°02′E
98 F12 **Oostvoorne** Zuid-Holland, SW Netherlands 51°55′N 04°06′E
Ootacamund *see* Udagamandalam
98 O10 **Ootmarsum** Overijssel, E Netherlands 52°25′N 06°55′E
10 K14 **Ootsa Lake** ⊞ British Columbia, SW Canada
Ooty *see* Udagamandalam
114 L8 **Opaka** Targovishte, N Bulgaria 43°26′N 26°12′E
79 M18 **Opala** Orientale, C Dem. Rep. Congo 0°40′S 24°20′E
125 Q13 **Oparino** Kirovskaya Oblast', NW Russian Federation 59°52′N 48°14′E
14 H8 **Opasatica, Lac** ◎ Québec, SE Canada
112 B9 **Opatija** *It.* Abbazia. Primorje-Gorski Kotar, NW Croatia 45°18′N 14°15′E
111 N15 **Opatów** Šuiętokrzyskie, C Poland 50°45′N 21°27′E
111 I17 **Opava** *Ger.* Troppau. Moravskoslezský Kraj, E Czech Republic 49°56′N 17°53′E
111 H16 **Opava** *Ger.* Oppa. ≈ NE Czech Republic
Ópazova *see* Stara Pazova
Ópécska *see* Pecica
14 E8 **Opeepeesway Lake** ◎ Ontario, S Canada
23 R5 **Opelika** Alabama, S USA 32°39′N 85°22′W
22 I8 **Opelousas** Louisiana, S USA 30°31′N 92°04′W
186 G6 **Open Bay** *bay* New Britain, E Papua New Guinea
14 I12 **Opeongo Lake** ◎ Ontario, SE Canada
99 K17 **Opglabbeek** Limburg, NE Belgium 51°04′N 05°39′E
33 W6 **Opheim** Montana, NW USA 48°50′N 106°24′W
39 P10 **Ophir** Alaska, USA 63°08′N 156°31′W
Ophiusa *see* Formentera
79 N18 **Opienge** Orientale, E Dem. Rep. Congo 0°15′N 27°25′E
185 G20 **Opihi** ≈ South Island, New Zealand
12 J9 **Opinaca** ≈ Québec, C Canada
12 J10 **Opinaca, Réservoir** ⊞ Québec, E Canada
117 T5 **Opishnya** *Rus.* Oposhnya. Poltavs'ka Oblast', NE Ukraine 49°58′N 34°37′E
98 I8 **Opmeer** Noord-Holland, NW Netherlands 52°43′N 04°56′E
77 V17 **Opobo** Akwa Ibom, S Nigeria 04°36′N 07°37′E
124 F16 **Opochka** Pskovskaya Oblast', W Russian Federation 56°42′N 28°40′E
110 L13 **Opoczno** Łodzkie, C Poland 51°24′N 20°18′E
111 I15 **Opole** *Ger.* Oppeln. Opolskie, S Poland 50°40′N 17°56′E
111 H15 **Opolskie** ⬖ *province* S Poland
Opornyy *see* Borankul
104 G4 **O Porriño** *var.* Porriño. Galicia, NW Spain 42°10′N 08°38′W
Oporto *see* Porto
Oposhnya *see* Opishnya
184 P8 **Opotiki** Bay of Plenty, North Island, New Zealand 38°02′S 177°18′E
23 Q7 **Opp** Alabama, S USA 31°16′N 86°14′W
Oppa *see* Opava
94 G9 **Oppdal** Sør-Trøndelag, S Norway 62°36′N 09°41′E
Oppeln *see* Opole
107 N23 **Oppido Mamertina** Calabria, SW Italy 38°17′N 15°58′E
Oppidum Ubiorum *see* Köln
94 F12 **Oppland** ⬖ *county* S Norway
118 J12 **Opsa** Vitsyebskaya Voblasts', NW Belarus 55°32′N 26°50′E
26 I8 **Optima Lake** ⊞ Oklahoma, C USA
184 J11 **Opunake** Taranaki, North Island, New Zealand 39°27′S 173°52′E
191 N6 **Opunohu, Baie d'** *bay* Moorea, W French Polynesia
83 B17 **Opuwo** Kunene, NW Namibia 18°03′S 13°54′E
146 H6 **Oqqal'a** *var.* Akkala, *Rus.* Karakala. Qoraqalpog'iston Respublikasi, NW Uzbekistan 43°43′N 59°25′E
147 V13 **Oqsu** *Rus.* Oksu. ≈ SE Tajikistan
147 P14 **Oqtogh, Qatorkŭhi** *Rus.* Khrebet Aktau. ▲▲ SW Tajikistan
146 M11 **Oqtosh** *Rus.* Aktash. Samarqand Viloyati, C Uzbekistan 39°23′N 65°46′E
147 N11 **Oqtov Tizmasi** *var.* Khrebet Aktau. ▲▲ C Uzbekistan
30 J12 **Oquawka** Illinois, N USA 40°56′N 90°56′W
144 J10 **Or'** *Kaz.* Or. ≈ Kazakhstan/Russian Federation
36 M15 **Oracle** Arizona, SW USA 32°36′N 110°46′W
147 N13 **O'radaryo** *Rus.* Uradar'ya. ≈ S Uzbekistan
116 F9 **Oradea** *prev.* Oradea Mare, *Ger.* Grosswardein, *Hung.* Nagyvárad. Bihor, NW Romania 47°03′N 21°56′E
Oradea Mare *see* Oradea
Orahovac *see* Rahovec
112 H9 **Orahovica** Virovitica-Podravina, NE Croatia 45°33′N 17°54′E
152 K13 **Orai** Uttar Pradesh, N India 26°00′N 79°26′E
92 K12 **Orajärvi** Lappi, NW Finland 66°54′N 24°04′E
138 F9 **Or'Akiva** *prev.* Or 'Aqiva. Haifa, N Israel 32°40′N 34°58′E
Oral *see* Ural'sk
74 I5 **Oran** *var.* Ouahran, Wahran. NW Algeria 35°42′N 00°37′W
183 R8 **Orange** New South Wales, SE Australia 33°16′S 149°06′E
103 R14 **Orange** *anc.* Arausio. Vaucluse, SE France 44°06′N 04°52′E
25 Y10 **Orange** Texas, SW USA 30°05′N 93°43′W
21 V5 **Orange** Virginia, NE USA 38°14′N 78°07′W
21 R13 **Orangeburg** South Carolina, SE USA 33°28′N 80°53′W
58 J9 **Orange, Cabo** *headland* NE Brazil 04°24′N 51°33′W
29 S12 **Orange City** Iowa, C USA 43°00′N 96°03′W
Orange Cone *see* Orange Fan
172 J10 **Orange Fan** *var.* Orange Cone. *undersea feature* SW Indian Ocean 32°00′S 12°00′E
Orange Free State *see* Free State
25 S14 **Orange Grove** Texas, SW USA 27°57′N 97°56′W
18 K13 **Orange Lake** New York, NE USA 41°32′N 74°06′W
23 V10 **Orange Lake** ◎ Florida, SE USA
Orange Mouth/Orangemund *see* Oranjemund
23 W9 **Orange Park** Florida, SE USA 30°10′N 81°42′W
83 E23 **Orange River** *Afr.* Oranjerivier. ≈ S Africa
14 G15 **Orangeville** Ontario, S Canada 43°55′N 80°06′W
36 M5 **Orangeville** Utah, W USA 39°14′N 111°03′W
42 G1 **Orange Walk** Orange Walk, N Belize 18°06′N 88°30′W
42 F1 **Orange Walk** ⬖ *district* NW Belize
100 N11 **Oranienburg** Brandenburg, NE Germany 52°46′N 13°15′E
98 O7 **Oranjekanaal** *canal* NE Netherlands
83 D23 **Oranjemund** *var.* Orangemund; *prev.* Orange Mouth. Karas, SW Namibia 28°33′S 16°28′E
Oranjerivier *see* Orange River
45 N16 **Oranjestad** ○ (Aruba) W Aruba 12°31′N 70°W
Oranje Vrystaat *see* Free State
Orany *see* Varėna
83 H18 **Orapa** Central, C Botswana 21°16′S 25°22′E
Or 'Aqiva *see* Or'Akiva
112 I10 **Orašje** ⬖ Federacija Bosna I Herecegovina, N Bosnia and Herzegovina
116 G11 **Orăștie** *Ger.* Broos, *Hung.* Szászváros. Hunedoara, W Romania 45°50′N 23°11′E
Orașul Stalin *see* Brașov
111 K18 **Orava** *Hung.* Árva, *Pol.* Orawa. ≈ N Slovakia
Oravainen *see* Oravais
93 K16 **Oravais** *Fin.* Oravainen. Österbotten, W Finland 63°18′N 22°25′E
Oravicabánya *see* Oravița
116 F13 **Oravița** *Ger.* Orawitza, *Hung.* Oravicabánya. Caraș-Severin, SW Romania 45°02′N 21°43′E
Orawa *see* Orava
185 B24 **Orawia** Southland, South Island, New Zealand 46°03′S 167°49′E
Orawitza *see* Oravița
103 P16 **Orb** ≈ S France
106 C9 **Orba** ≈ NW Italy
158 H12 **Orba Co** ◎ W China
108 B9 **Orbe** Vaud, W Switzerland 46°42′N 06°28′E
107 G14 **Orbetello** Toscana, C Italy 42°28′N 11°15′E
104 K3 **Orbigo** ≈ NW Spain
183 Q12 **Orbost** Victoria, SE Australia 37°44′S 148°28′E
95 O14 **Örbyhus** Uppsala, C Sweden 60°15′N 17°43′E
194 I1 **Orcadas** *Argentinian research station* South Orkney Islands, Antarctica 60°37′S 44°48′W
105 P12 **Orcera** Andalucía, S Spain 38°20′N 02°36′W
33 P9 **Orchard Homes** Montana, NW USA 46°52′N 114°01′W
37 P5 **Orchard Mesa** Colorado, C USA 39°02′N 108°33′W
18 D10 **Orchard Park** New York, NE USA 42°46′N 78°44′W
Orchid Island *see* Lan Yu
115 G18 **Orchómenos** *var.* Orhomenos, Orkhómenos, *prev.* Skripón; *anc.* Orchomenus. Stereá Elláda, C Greece 38°29′N 22°58′E
Orchomenus *see* Orchómenos
106 B7 **Orco** ≈ NW Italy
103 R8 **Or, Côte d'** *physical region* C France
29 O14 **Ord** Nebraska, C USA 41°36′N 98°55′W
119 O15 **Ordats'** *Rus.* Ordat'. Mahilyowskaya Voblasts', E Belarus 54°10′N 30°42′E
36 K8 **Orderville** Utah, W USA 37°16′N 112°38′W
104 H2 **Ordes** Galicia, NW Spain 43°04′N 08°25′W
35 V14 **Ord Mountain** ▲ California, W USA 34°41′N 116°46′W
163 N14 **Ordos** *prev.* Dongsheng. Nei Mongol Zizhiqu, N China 39°51′N 110°00′E
Ordos Desert *see* Mu Us Shadi
188 B16 **Ordot** C Guam
137 N11 **Ordu** *anc.* Cotyora. Ordu, N Turkey 41°N 37°52′E
136 M11 **Ordu** ⬖ *province* N Turkey
137 V14 **Ordubad** SW Azerbaijan 38°55′N 46°00′E
Orduña *see* Urduña
37 U6 **Ordway** Colorado, C USA 38°13′N 103°45′W
117 T9 **Ordzhonikidze** Dnipropetrovs'ka Oblast', E Ukraine 47°39′N 34°08′E
Ordzhonikidze *see* Denisovka, Kazakhstan
Ordzhonikidze *see* Vladikavkaz, Russian Federation
Ordzhonikidze *see* Yenakiyeve, Ukraine
Ordzhonikidzeabad *see* Kofarnihon
55 U9 **Orealla** E Guyana 05°13′N 57°17′W
113 G15 **Orebić** *It.* Sabbioncello. Dubrovnik-Neretva, S Croatia 42°58′N 17°12′E
95 M16 **Örebro** Örebro, C Sweden 59°18′N 15°12′E
95 L16 **Örebro** ⬖ *county* C Sweden
25 W6 **Ore City** Texas, SW USA 32°48′N 94°43′W
30 L10 **Oregon** Illinois, N USA 42°00′N 89°19′W
27 Q2 **Oregon** Missouri, C USA 39°59′N 95°08′W
31 R11 **Oregon** Ohio, N USA 41°38′N 83°29′W
32 H13 **Oregon** *off.* State of Oregon, *also known as* Beaver State, Sunset State, Valentine State, Webfoot State. ⬖ *state* NW USA
32 G11 **Oregon City** Oregon, NW USA 45°21′N 122°36′W
Oregon, State of *see* Oregon
95 P14 **Öregrund** Uppsala, C Sweden 60°19′N 18°30′E
Orekhov *see* Orikhiv
126 L3 **Orekhovo-Zuyevo** Moskovskaya Oblast', W Russian Federation 55°46′N 39°01′E
Orekhovsk *see* Arekhawsk
126 J6 **Orël** Orlovskaya Oblast', W Russian Federation 52°57′N 36°06′E
Orel *see* Oril'
56 E11 **Orellana** Loreto, N Peru 06°53′S 75°10′W
56 E6 **Orellana** ⬖ *province* NE Ecuador
104 L11 **Orellana, Embalse de** ⊞ W Spain
36 L3 **Orem** Utah, W USA 40°18′N 111°42′W
Ore Mountains *see* Erzgebirge/Krušné Hory
127 V7 **Orenburg** *prev.* Chkalov. Orenburgskaya Oblast', W Russian Federation 51°46′N 55°12′E
127 V7 **Orenburg** ✈ Orenburgskaya Oblast', W Russian Federation 51°54′N 55°15′E
127 T7 **Orenburgskaya Oblast'** ⬖ *province* W Russian Federation
Orense *see* Ourense
188 C8 **Oreor** *var.* Koror. *island* N Palau
185 B24 **Orepuki** Southland, South Island, New Zealand 46°17′S 167°45′E
114 L12 **Orestiáda** *prev.* Orestiás. Anatolikí Makedonía kai Thráki, NE Greece 41°30′N 26°31′E
Orestiás *see* Orestiáda
Öresund/Øresund *see* Sound, The
185 C23 **Oreti** ≈ South Island, New Zealand
184 L5 **Orewa** Auckland, North Island, New Zealand 36°34′S 174°43′E
65 A25 **Orford, Cape** *headland* West Falkland, Falkland Islands 52°00′S 61°04′W
44 B5 **Órganos, Sierra de los** ▲▲ W Cuba
37 R15 **Organ Peak** ▲ New Mexico, SW USA 32°17′N 106°35′W
105 N9 **Orgaz** Castilla-La Mancha, C Spain 39°39′N 03°52′W
Orgeyev *see* Orhei
Orgil *see* Jargalant
105 O15 **Orgiva** *var.* Orjiva. Andalucía, S Spain 36°54′N 03°25′W
163 O10 **Örgön** *var.* Senj. Dornogovĭ, SE Mongolia 44°34′N 110°58′E
Örgön *see* Bayangovĭ
Orgražden *see* Ograzhden
117 N9 **Orhei** *var.* Orheiu, *Rus.* Orgeyev. N Moldova 47°25′N 28°48′E
Orheiu *see* Orhei
105 R3 **Orhi** *var.* Orhy, Pico de Orhy, Pic d'Orhy. ▲ France/Spain 42°55′N 01°01′W *see also* Orhy
Orhi *see* Orhy
Orhomenos *see* Orchómenos
162 K6 **Orhon** ⬖ *province* N Mongolia
162 L6 **Orhon Gol** ≈ N Mongolia
102 J16 **Orhy** *var.* Orhi, Pico de Orhy, Pic d'Orhy. ▲ France/Spain 43°00′N 01°00′W *see also* Orhi
Orhy *see* Orhi
Orhy, Pic d'/Orhy, Pico de *see* Orhi/Orhy
34 L2 **Orick** California, W USA 41°16′N 124°03′W
32 L6 **Orient** Washington, NW USA 48°51′N 118°14′W
48 D6 **Oriental, Cordillera** ▲▲ Bolivia/Peru
48 D6 **Oriental, Cordillera** ▲▲ C Colombia
57 H16 **Oriental, Cordillera** ▲▲ C Peru
63 M15 **Oriente** Buenos Aires, E Argentina 38°45′S 60°37′W

◆ Country
● Country Capital
◇ Dependent Territory
○ Dependent Territory Capital
⬖ Administrative Regions
✈ International Airport
▲ Mountain
▲▲ Mountain Range
⛰ Volcano
≈ River
◎ Lake
⊞ Reservoir

Orihuela – Ourinhos

105 R12 **Orihuela** Valenciana, E Spain 38°05´N 00°56´W
117 V9 **Orikhiv** *Rus.* Orekhov. Zaporiz'ka Oblast', SE Ukraine 47°32´N 35°48´E
113 K22 **Orikum** *var.* Orikumi. Vlorë, SW Albania 40°20´N 19°28´E
Orikumi *see* Orikum
117 V6 **Oril'** *Rus.* Orel. ≈ E Ukraine
14 H14 **Orillia** Ontario, S Canada 44°36´N 79°26´W
93 M19 **Orimattila** Päijät-Häme, S Finland 60°48´N 25°40´E
33 Y15 **Orin** Wyoming, C USA 42°39´N 105°10´W
47 R4 **Orinoco, Río** ≈ Colombia/ Venezuela
186 C9 **Oriomo** Western, SW Papua New Guinea 08°53´S 143°13´E
30 K11 **Orion** Illinois, N USA 41°21´N 90°22´W
29 Q5 **Oriska** North Dakota, N USA 46°54´N 97°46´W
Orissa *see* Odisha
Orissaar *see* Orissaare
118 E5 **Orissaare** *Ger.* Orissaar. Saaremaa, W Estonia 58°34´N 23°05´E
107 B19 **Oristano** Sardegna, Italy, C Mediterranean Sea 39°54´N 08°35´E
107 A19 **Oristano, Golfo di** *gulf* Sardegna, Italy, C Mediterranean Sea
54 D13 **Orito** Putumayo, SW Colombia 0°49´N 76°57´W
93 L18 **Orivesi** Häme, W Finland 61°39´N 24°21´E
93 N17 **Orivesi** ◎ Etelä-Savo, SE Finland
58 H12 **Oriximiná** Pará, NE Brazil 01°45´S 55°50´W
41 Q14 **Orizaba** Veracruz-Llave, E Mexico 18°51´N 97°08´W
41 Q14 **Orizaba, Volcán Pico de** *var.* Citlaltépetl. ▲ S Mexico 19°00´N 97°15´W
95 I16 **Ørje** Østfold, S Norway 59°28´N 11°40´E
113 I16 **Orjen** ▲▲ Bosnia and Herzegovina/Montenegro
Orjiva *see* Orgiva
Orjonikidzeobod *see* Kofarnihon
94 G8 **Orkanger** Sør-Trøndelag, S Norway 63°17´N 09°52´E
94 G8 **Orkdalen** *valley* S Norway
95 K22 **Örkelljunga** Skåne, S Sweden 56°17´N 13°20´E
Orkhaniye *see* Botevgrad
Orkhómenos *see* Orchómenos
94 H9 **Orkla** ≈ S Norway
Orkney *see* Orkney Islands
65 J22 **Orkney Deep** *undersea feature* Scotia Sea/Weddell Sea
96 J4 **Orkney Islands** *var.* Orkney, Orkneys. *island group* N Scotland, United Kingdom
Orkneys *see* Orkney Islands
24 K8 **Orla** Texas, SW USA 31°48´N 103°55´W
35 N5 **Orland** California, W USA 39°43´N 122°12´W
23 X11 **Orlando** Florida, SE USA 28°32´N 81°23´W
23 X12 **Orlando** ✈ Florida, SE USA 28°24´N 81°16´W
107 K23 **Orlando, Capo d'** *headland* Sicilia, Italy, C Mediterranean Sea 38°10´N 14°44´E
Orlau *see* Orlová
103 N6 **Orléanais** *cultural region* C France
103 N7 **Orléans** *anc.* Aurelianum. Loiret, C France 47°54´N 01°53´E
34 L2 **Orleans** California, W USA 41°16´N 123°36´W
19 Q12 **Orleans** Massachusetts, NE USA 41°48´N 69°57´W
15 R10 **Orléans, Île d'** *island* Québec, SE Canada
Orléansville *see* Chlef
111 F16 **Orlice** *Ger.* Adler. ≈ NE Czech Republic
122 L13 **Orlik** Respublika Buryatiya, S Russian Federation 52°32´N 99°36´E
125 Q14 **Orlov** *prev.* Khalturin. Kirovskaya Oblast', NW Russian Federation 58°34´N 48°57´E
111 I17 **Orlová** *Ger.* Orlau, *Pol.* Orłowa. Moravskoslezský Kraj, E Czech Republic 49°50´N 18°21´E
Orlov, Mys *see* Orlovskiy, Mys
126 I6 **Orlovskaya Oblast'** ◐ *province* W Russian Federation
124 M5 **Orlovskiy, Mys** *var.* Mys Orlov. *headland* NW Russian Federation 67°14´N 41°17´E
Orłowa *see* Orlová
103 O5 **Orly** ✈ (Paris) Essonne, N France 48°43´N 02°24´E
119 G16 **Orlya** Hrodzyenskaya Voblasts', W Belarus 53°30´N 24°59´E
114 M7 **Orlyak** *prev.* Makenzen, Trubchular, *Rom.* Trupcilar. Dobrich, NE Bulgaria 43°39´N 27°21´E
148 L16 **Ormāra** Baluchistān, SW Pakistan 25°14´N 64°36´E
171 P5 **Ormoc** *off.* Ormoc City, *var.* MacArthur. Leyte, C Philippines 11°02´N 124°35´E
Ormoc City *see* Ormoc
23 X10 **Ormond Beach** Florida, SE USA 29°16´N 81°04´W
109 X10 **Ormož** *Ger.* Friedau. NE Slovenia 46°24´N 16°09´E
14 J13 **Ormsby** Ontario, SE Canada 44°52´N 77°45´W
97 K17 **Ormskirk** NW England, United Kingdom 53°35´N 02°54´W
Ormsö *see* Vormsi
15 N13 **Ormstown** Québec, SE Canada 45°08´N 73°57´W
Ormuz, Strait of *see* Hormuz, Strait of
103 T8 **Ornans** Doubs, E France 47°06´N 06°06´E
102 K5 **Orne** ◐ *department* N France
102 K5 **Orne** ≈ N France
92 G12 **Ørnes** Nordland, C Norway 66°51´N 13°43´E
110 L7 **Orneta** Warmińsko-Mazurskie, NE Poland 54°07´N 20°10´E
95 P16 **Ornö** Stockholm, C Sweden 59°03´N 18°28´E
37 Q3 **Orno Peak** ▲ Colorado, C USA 40°06´N 107°06´W
93 I16 **Örnsköldsvik** Västernorrland, C Sweden 63°16´N 18°45´E
163 X13 **Oro** E North Korea 39°59´N 127°27´E
Oro *see* Northern
45 T6 **Orocovis** C Puerto Rico 18°13´N 66°22´W
54 H10 **Orocué** Casanare, E Colombia 04°51´N 71°21´W
77 N13 **Orodara** SW Burkina Faso 11°00´N 04°54´W
105 S4 **Oroel, Peña de** ▲ N Spain 42°30´N 00°31´W
33 N10 **Orofino** Idaho, NW USA 46°28´N 116°15´W
162 I9 **Orog Nuur** ◎ S Mongolia
35 U14 **Oro Grande** California, W USA 34°36´N 117°19´W
37 S15 **Orogrande** New Mexico, SW USA 32°24´N 106°04´W
191 Q7 **Orohena, Mont** ⛰ Tahiti, W French Polynesia 17°37´S 149°27´W
Orolaunum *see* Arlon
Orol Dengizi *see* Aral Sea
189 S15 **Oroluk Atoll** *atoll* Caroline Islands, C Micronesia
80 J13 **Oromīya** *var.* Oromo. ◐ C Ethiopia
Oromo *see* Oromīya
13 O15 **Oromocto** New Brunswick, SE Canada 45°50´N 66°28´W
191 S4 **Orona** *prev.* Hull Island. *atoll* Phoenix Islands, C Kiribati
191 V17 **Orongo** *ancient monument* Easter Island, Chile, E Pacific Ocean
138 I3 **Orontes** *var.* Ononte, Nahr el Aassi, *Ar.* Nahr al 'Aṣī. ≈ SW Asia
104 L9 **Oropesa** Castilla-La Mancha, C Spain 39°55´N 05°10´W
Oropesa *see* Oropesa del Mar
105 T8 **Oropesa del Mar** *var.* Oropesa, Orpesa, *Cat.* Orpes. Valenciana, E Spain 40°06´N 00°07´E
Oropeza *see* Cochabamba
Oroqen Zizhiqi *see* Alihe
171 P7 **Oroquieta** *var.* Oroquieta City. Mindanao, S Philippines 08°27´N 123°46´E
Oroquieta City *see* Oroquieta
40 J8 **Oro, Río del** ≈ C Mexico
59 O14 **Orós, Açude** ▣ E Brazil
107 D18 **Orosei, Golfo di** *gulf* Tyrrhenian Sea, C Mediterranean Sea
111 M24 **Orosháza** Békés, SE Hungary 46°33´N 20°40´E
Orosirá Rodhópis *see* Rhodope Mountains
111 I22 **Oroszlány** Komárom-Esztergom, W Hungary 47°28´N 18°16´E
188 B16 **Orote Peninsula** *peninsula* W Guam
123 T9 **Orotukan** Magadanskaya Oblast', E Russian Federation 62°18´N 150°46´E
35 O5 **Oroville** California, W USA 39°29´N 121°35´W
32 K6 **Oroville** Washington, NW USA 48°56´N 119°25´W
35 O5 **Oroville, Lake** ▣ California, W USA
0 G15 **Orozco Fracture Zone** *tectonic feature* E Pacific Ocean
Orpes *see* Oropesa del Mar
Orpesa *see* Oropesa del Mar
64 I7 **Orphan Knoll** *undersea feature* NW Atlantic Ocean 51°00´N 47°00´W
29 V3 **Orr** Minnesota, N USA 48°03´N 92°48´W
95 M21 **Orrefors** Kalmar, S Sweden 56°48´N 15°45´E
182 I7 **Orroroo** South Australia 32°46´S 138°38´E
31 T12 **Orrville** Ohio, N USA 40°50´N 81°45´W
94 L12 **Orsa** Dalarna, C Sweden 61°07´N 14°40´E
Orschowa *see* Orşova
Orschütz *see* Orzyc
119 O14 **Orsha** Vitsyebskaya Voblasts', NE Belarus 54°30´N 30°26´E
127 Q2 **Orshanka** Respublika Mariy El, W Russian Federation 56°54´N 47°54´E
108 C11 **Orsières** Valais, SW Switzerland 46°00´N 07°09´E
127 X8 **Orsk** Orenburgskaya Oblast', W Russian Federation 51°13´N 58°35´E
116 F13 **Orşova** *Ger.* Orschowa, *Hung.* Orsova. Mehedinţi, SW Romania 44°42´N 22°22´E
94 D10 **Ørsta** Møre og Romsdal, S Norway 62°12´N 06°09´E
95 O15 **Örsundsbro** Uppsala, C Sweden 59°45´N 17°19´E
136 D16 **Ortaca** Muğla, SW Turkey 36°49´N 28°43´E
83 I21 **O.R. Tambo** ✈ (Johannesburg) Gauteng, NE South Africa 26°08´S 28°01´E
107 M16 **Orta Nova** Puglia, SE Italy 41°20´N 15°43´E
136 I17 **Orta Toroslar** ▲▲ S Turkey
54 E11 **Ortega** Tolima, W Colombia 03°57´N 75°11´W
104 H1 **Ortegal, Cabo** *headland* NW Spain 43°46´N 07°54´W
Ortelsburg *see* Szczytno
102 J15 **Orthez** Pyrénées-Atlantiques, SW France 43°29´N 00°46´W
60 J10 **Ortigueira** Paraná, S Brazil 24°10´S 50°55´W
104 H1 **Ortigueira** Galicia, NW Spain 43°40´N 07°50´W
106 H5 **Ortisei** *Ger.* Sankt-Ulrich. Trentino-Alto Adige, N Italy 46°35´N 11°42´E
40 F6 **Ortiz** Sonora, NW Mexico 28°18´N 110°40´W
54 L5 **Ortiz** Guárico, N Venezuela 09°37´N 67°20´W
Ortler *see* Ortles
106 F5 **Ortles** *Ger.* Ortler. ▲ N Italy 46°29´N 10°33´E
107 K14 **Ortona** Abruzzo, C Italy 42°21´N 14°24´E
57 K14 **Ortón, Río** ≈ N Bolivia
29 R8 **Ortonville** Minnesota, N USA 45°18´N 96°26´W
147 W8 **Orto-Tokoy** Issyk-Kul'skaya Oblast', NE Kyrgyzstan 42°20´N 76°03´E
93 I15 **Örträsk** Västerbotten, N Sweden 64°10´N 19°00´E
100 J12 **Örtze** ≈ NW Germany
Oruba *see* Aruba
142 I3 **Orūmīyeh** *var.* Rizaiyeh, Urmia, Urmiyeh; *prev.* Regā'īyeh. Āẕarbāyjān-e Gharbī, NW Iran 37°33´N 45°06´E
142 J3 **Orūmīyeh, Daryācheh-ye** *var.* Matianus, Sha Hi, Urumi Yeh, *Eng.* Lake Urmia; *prev.* Daryācheh-ye Regā'īyeh. ◎ NW Iran
57 K19 **Oruro** Oruro, W Bolivia 17°58´S 67°06´W
57 J19 **Oruro** ◐ *department* W Bolivia
95 I18 **Orust** *island* S Sweden
Orūzgān *see* Urūzgān
106 H13 **Orvieto** *anc.* Velsuna. Umbria, C Italy 42°43´N 12°06´E
194 K7 **Orville Coast** *physical region* Antarctica
114 H7 **Oryahovo** *var.* Oryakhovo. Vratsa, NW Bulgaria 43°44´N 23°58´E
Oryakhovo *see* Oryahovo
Oryokko *see* Yalu
117 R5 **Orzhytsya** Poltavs'ka Oblast', C Ukraine 49°48´N 32°40´E
110 M9 **Orzyc** *Ger.* Orschütz. ≈ NE Poland
110 N8 **Orzysz** *Ger.* Arys. Warmińsko-Mazurskie, NE Poland 53°49´N 21°54´E
98 K13 **Oss** Noord-Brabant, S Netherlands 51°46´N 05°32´E
94 I10 **Os** Hedmark, S Norway 62°29´N 11°14´E
125 U15 **Osa** Permskiy Kray, NW Russian Federation 57°16´N 55°22´E
115 F15 **Óssa** ▲▲ C Greece
104 H11 **Ossa** ▲ S Portugal 38°43´N 07°33´W
29 W11 **Osage** Iowa, C USA 43°16´N 92°48´W
27 U5 **Osage Beach** Missouri, C USA 38°09´N 92°37´W
27 P5 **Osage City** Kansas, C USA 38°37´N 95°49´W
27 U7 **Osage Fork River** ≈ Missouri, C USA
27 U5 **Osage River** ≈ Missouri, C USA
164 J13 **Ōsaka** *hist.* Naniwa. Ōsaka, Honshū, SW Japan 34°38´N 135°28´E
164 I13 **Ōsaka** *off.* Ōsaka-fu, *var.* Ōsaka Hu. ◐ *urban prefecture* Honshū, SW Japan
Ōsaka-fu/Ōsaka Hu *see* Ōsaka
145 R10 **Osakarovka** Karaganda, C Kazakhstan 50°32´N 72°39´E
Ōsaki *see* Furukawa
29 T7 **Osakis** Minnesota, N USA 45°51´N 95°08´W
114 I9 **Osam** *var.* Osŭm. ≈ N Bulgaria
43 N16 **Osa, Peninsula de** *peninsula* S Costa Rica
60 M10 **Osasco** São Paulo, S Brazil 23°32´S 46°46´W
27 R5 **Osawatomie** Kansas, C USA 38°30´N 94°57´W
26 L3 **Osborne** Kansas, C USA 39°26´N 98°42´W
173 S8 **Osborn Plateau** *undersea feature* E Indian Ocean
95 L21 **Osby** Skåne, S Sweden 56°24´N 14°00´E
Osca *see* Huesca
92 N2 **Oscar II Land** *physical region* W Svalbard
27 Y10 **Osceola** Arkansas, C USA 35°43´N 89°58´W
29 V15 **Osceola** Iowa, C USA 41°01´N 93°45´W
27 S6 **Osceola** Missouri, C USA 38°01´N 93°41´W
29 Q15 **Osceola** Nebraska, C USA 41°09´N 97°28´W
101 N15 **Oschatz** Sachsen, E Germany 51°17´N 13°10´E
100 K13 **Oschersleben** Sachsen-Anhalt, C Germany 52°02´N 11°14´E
31 R7 **Oscoda** Michigan, N USA 44°25´N 83°19´W
Ösel *see* Saaremaa
94 H6 **Osen** Sør-Trøndelag, S Norway 64°17´N 10°29´E
94 I12 **Osensjøen** ◎ S Norway
164 A14 **Ōse-zaki** Fukue-jima, SW Japan
147 T10 **Osh** Oshskaya Oblast', SW Kyrgyzstan 40°34´N 72°46´E
83 C16 **Oshakati** Oshana, N Namibia 17°46´S 15°43´E
83 C16 **Oshana** ◐ *district* N Namibia
14 H15 **Oshawa** Ontario, SE Canada 43°54´N 78°50´W
165 R10 **Oshika-hantō** *peninsula* Honshū, C Japan
83 C16 **Oshikango** Ohangwena, N Namibia 17°29´S 15°54´E
83 C17 **Oshikoto** *var.* Otjikoto. ◐ *district* N Namibia
165 P5 **Ō-shima** *island* NE Japan
165 N14 **Ō-shima** *island* S Japan
165 Q5 **Oshima-hantō** ▲▲ Hokkaidō, NE Japan
83 D17 **Oshivelo** Oshikoto, N Namibia 18°37´S 17°10´E
28 K14 **Oshkosh** Nebraska, C USA 41°25´N 102°21´W
30 M7 **Oshkosh** Wisconsin, N USA 44°01´N 88°32´W
Oshmyany *see* Ashmyany
Osh Oblasty *see* Oshskaya Oblast'
77 T16 **Oshogbo** *var.* Osogbo. Osun, W Nigeria 07°42´N 04°31´E
147 S11 **Oshskaya Oblast'** *Kir.* Osh Oblasty. ◐ *province* SW Kyrgyzstan
Ōshū *see* Mizusawa
Oshun *see* Osun
79 J20 **Oshwe** Bandundu, C Dem. Rep. Congo 03°24´S 19°32´E
Osiek *see* Osijek
112 I9 **Osijek** *prev.* Osiek, Osjek, *Ger.* Esseg, *Hung.* Eszék. Osijek-Baranja, E Croatia 45°33´N 18°41´E
112 I9 **Osijek-Baranja** *off.* Osječko-Baranjska Županija. ◐ *province* E Croatia
106 J12 **Osimo** Marche, C Italy 43°28´N 13°29´E
122 M12 **Osinovka** Irkutskaya Oblast', C Russian Federation 56°19´N 101°55´E
Osintorf *see* Asintorf
112 N11 **Osipaonica** Serbia, NE Serbia 44°34´N 21°00´E
Osipenko *see* Berdyans'k
Osipovichi *see* Asipovichy
Osječko-Baranjska Županija *see* Osijek-Baranja
Osjek *see* Osijek
29 W15 **Oskaloosa** Iowa, C USA 41°17´N 92°38´W
27 Q4 **Oskaloosa** Kansas, C USA 39°14´N 95°21´W
95 N20 **Oskarshamn** Kalmar, S Sweden 57°16´N 16°25´E
95 J21 **Oskarström** Halland, S Sweden 56°48´N 13°00´E
14 M8 **Oskélanéo** Québec, SE Canada 48°06´N 75°12´W
Öskemen *see* Ust'-Kamenogorsk
117 W5 **Oskil** *Rus.* Oskil. ≈ Russian Federation/Ukraine
Oskil *see* Oskil
93 D20 **Oslo** *prev.* Christiania, Kristiania. ● (Norway) Oslo, S Norway 59°54´N 10°44´E
93 D20 **Oslo** ◐ *county* S Norway
93 D21 **Oslofjorden** *fjord* S Norway
155 G15 **Osmānābād** Mahārāshtra, C India 18°09´N 76°06´E
136 J11 **Osmancık** Çorum, N Turkey 40°58´N 34°50´E
136 L16 **Osmaniye** Osmaniye, S Turkey 37°04´N 36°15´E
136 L16 **Osmaniye** ◐ *province* S Turkey
95 O16 **Ösmo** Stockholm, C Sweden 58°58´N 17°55´E
118 E3 **Osmussaar** *island* W Estonia
100 G13 **Osnabrück** Niedersachsen, NW Germany 52°09´N 07°42´E
110 D11 **Ośno Lubuskie** *Ger.* Drossen. Lubuskie, W Poland 52°28´N 14°51´E
Osogbo *see* Oshogbo
113 P19 **Osogov Mountains** *var.* Osogovske Planine, Osogovski Planina, *Mac.* Osogovski Planini. ▲▲ Bulgaria/FYR Macedonia
Osogovske Planine/ Osogovski Planina/ Osogovski Planini *see* Osogov Mountains
165 R6 **Osore-zan** ▲ Honshū, C Japan 41°18´N 141°06´E
Oşorhei *see* Târgu Mureş
61 J16 **Osório** Rio Grande do Sul, S Brazil 29°53´S 50°17´W
63 G16 **Osorno** Los Lagos, C Chile 40°39´S 73°05´W
104 M4 **Osorno** Castilla y León, N Spain 42°24´N 04°22´W
11 N17 **Osoyoos** British Columbia, SW Canada 49°02´N 119°31´W
95 C14 **Osøyro** Hordaland, S Norway 60°11´N 05°30´E
54 J6 **Ospino** Portuguesa, N Venezuela 09°17´N 69°26´W
23 X6 **Ossabaw Island** *island* Georgia, SE USA
23 X6 **Ossabaw Sound** *sound* Georgia, SE USA
183 O16 **Ossa, Mount** ▲ Tasmania, SE Australia 41°55´S 146°03´E
104 H11 **Ossa, Serra d'** ▲▲ SE Portugal
77 U16 **Osse** ≈ S Nigeria
30 J6 **Osseo** Wisconsin, N USA 44°33´N 91°13´W
109 S9 **Ossiacher See** ◎ S Austria
18 K13 **Ossining** New York, NE USA 41°10´N 73°50´W
123 V9 **Ossora** Krasnoyarskiy Kray, E Russian Federation 59°16´N 163°02´E
124 I15 **Ostashkov** Tverskaya Oblast', W Russian Federation 57°08´N 33°10´E
100 H9 **Oste** ≈ NW Germany
Ostee *see* Baltic Sea
Ostend/Ostende *see* Oostende
117 P3 **Oster** Chernihivs'ka Oblast', N Ukraine 50°57´N 30°55´E
93 J16 **Österbotten** *Fin.* Pohjanmaa, *Eng.* Ostrobothnia. ◐ *region* W Finland
95 O14 **Österbybruk** Uppsala, C Sweden 60°13´N 17°55´E
95 M19 **Österbymo** Östergotland, S Sweden 57°49´N 15°15´E
94 K12 **Österdalälven** ◎ C Sweden
94 I12 **Østerdalen** *valley* S Norway
95 L18 **Östergötland** ◐ *county* S Sweden
100 H10 **Osterholz-Scharmbeck** Niedersachsen, NW Germany 53°13´N 08°46´E
Östermark *see* Teuva
Östermyra *see* Seinäjoki
Osterode *see* Eysturoy
101 J14 **Osterode am Harz** Niedersachsen, C Germany 51°43´N 10°15´E
Osterode/Osterode in Ostpreussen *see* Ostróda
Østerøy *see* Osterøyni
94 C13 **Osterøyni** *prev.* Osterøy. *island* S Norway
Österreich *see* Austria
93 G16 **Östersund** Jämtland, C Sweden 63°10´N 14°44´E
95 N14 **Östervåla** Västmanland, C Sweden 60°10´N 17°13´E
101 H22 **Ostfildern** Baden-Württemberg, SW Germany 48°43´N 09°16´E
95 H16 **Østfold** ◐ *county* S Norway
100 E9 **Ostfriesische Inseln** *Eng.* East Frisian Islands. *island group* NW Germany
100 F10 **Ostfriesland** *historical region* NW Germany
95 P14 **Östhammar** Uppsala, C Sweden 60°15´N 18°25´E
Ostia Aterni *see* Pescara
106 G8 **Ostiglia** Lombardia, N Italy 45°04´N 11°09´E
95 J14 **Östmark** Värmland, C Sweden 60°15´N 12°45´E
95 K22 **Östra Ringsjön** ◎ S Sweden
111 I17 **Ostrava** Moravskoslezský Kraj, E Czech Republic 49°50´N 18°15´E
Ostravský Kraj *see* Moravskoslezský Kraj
94 J11 **Østrehogna** *Swe.* Härjahågnen, Härjehågna. ▲ Norway/Sweden 61°43´N 12°07´E
Ostrobothnia *see* Österbotten
110 K8 **Ostróda** *Ger.* Osterode, Osterode in Ostpreussen. Warmińsko-Mazurskie, NE Poland 53°42´N 19°59´E
Ostrog/Ostróg *see* Ostroh
126 L8 **Ostrogozhsk** Voronezhskaya Oblast', W Russian Federation 50°52´N 39°00´E
116 L4 **Ostroh** *Pol.* Ostróg, *Rus.* Ostrog. Rivnens'ka Oblast', NW Ukraine 50°20´N 26°29´E
110 N9 **Ostrołęka** *Ger.* Wiesenhof, *Rus.* Ostrolenka. Mazowieckie, C Poland 53°06´N 21°34´E
Ostrolenka *see* Ostrołęka
111 A16 **Ostrov** *Ger.* Schlackenwerth. Karlovarský Kraj, W Czech Republic 50°18´N 12°56´E
124 F15 **Ostrov** *Latv.* Austrava. Pskovskaya Oblast', W Russian Federation 57°21´N 28°18´E
Ostrovets *see* Ostrowiec Świętokrzyski
113 M21 **Ostrovicës, Mali i** ▲ SE Albania 40°36´N 20°25´E
165 Z2 **Ostrov Iturup** *island* NE Russian Federation
124 M4 **Ostrovnoy** Murmanskaya Oblast', NW Russian Federation 68°00´N 39°40´E
114 L7 **Ostrovo** *prev.* Golema Ada. Razgrad, N Bulgaria 43°40´N 26°37´E
125 N15 **Ostrovskoye** Kostromskaya Oblast', NW Russian Federation 57°46´N 42°18´E
Ostrów *see* Ostrów Wielkopolski
Ostrowiec *see* Ostrowiec Świętokrzyski
111 M14 **Ostrowiec Świętokrzyski** *var.* Ostrowiec, *Rus.* Ostrovets. Świętokrzyskie, C Poland 50°55´N 21°23´E
110 P13 **Ostrów Lubelski** Lubelskie, E Poland 51°29´N 22°57´E
110 N10 **Ostrów Mazowiecka** *var.* Ostrów Mazowiecki. Mazowieckie, NE Poland 52°49´N 21°53´E
Ostrów Mazowiecki *see* Ostrów Mazowiecka
Ostrowo *see* Ostrów Wielkopolski
110 H13 **Ostrów Wielkopolski** *var.* Ostrów, *Ger.* Ostrowo. Wielkopolskie, C Poland 51°40´N 17°47´E
Ostryna *see* Astryna
110 I13 **Ostrzeszów** Wielkopolskie, C Poland 51°25´N 17°55´E
107 P18 **Ostuni** Puglia, SE Italy 40°44´N 17°35´E
Ostyako-Voguls'k *see* Khanty-Mansiysk
Osum *see* Osumit, Lumi i
Osŭm *see* Osam
164 C17 **Ōsumi-hantō** ▲▲ Kyūshū, SW Japan
164 C17 **Ōsumi-kaikyō** *strait* SW Japan
113 L22 **Osumit, Lumi i** *var.* Osum. ≈ SE Albania
77 T16 **Osun** *var.* Oshun. ◐ *state* SW Nigeria
104 L14 **Osuna** Andalucía, S Spain 37°14´N 05°06´W
60 J8 **Osvaldo Cruz** São Paulo, S Brazil 21°49´S 50°52´W
Osveya *see* Asvyeya
18 J7 **Oswegatchie River** ≈ New York, NE USA
27 Q7 **Oswego** Kansas, C USA 37°11´N 95°10´W
18 H9 **Oswego** New York, NE USA 43°27´N 76°13´W
97 K19 **Oswestry** W England, United Kingdom 52°51´N 03°06´W
111 J16 **Oświęcim** *Ger.* Auschwitz. Małopolskie, S Poland 50°02´N 19°11´E
185 E22 **Otago** *off.* Otago Region. ◐ *region* South Island, New Zealand
185 F23 **Otago Peninsula** *peninsula* South Island, New Zealand
Otago Region *see* Otago
165 F13 **Ōtake** Hiroshima, Honshū, SW Japan 34°13´N 132°12´E
184 L13 **Otaki** Wellington, North Island, New Zealand 40°46´S 175°08´E
93 M15 **Otanmäki** Kainuu, C Finland 64°07´N 27°04´E
145 T15 **Otar** Zhambyl, SE Kazakhstan 43°30´N 75°13´E
165 R4 **Otaru** Hokkaidō, NE Japan 43°14´N 140°59´E
185 C24 **Otatara** Southland, South Island, New Zealand 46°26´S 168°18´E
185 C24 **Otautau** Southland, South Island, New Zealand 46°10´S 168°01´E
93 M18 **Otava** Etelä-Savo, E Finland 61°37´N 27°07´E
111 B18 **Otava** *Ger.* Wottawa. ≈ SW Czech Republic
56 C6 **Otavalo** Imbabura, N Ecuador 0°13´N 78°15´W
83 D17 **Otavi** Otjozondjupa, N Namibia 19°35´S 17°25´E
165 P12 **Ōtawara** Tochigi, Honshū, S Japan 36°52´N 140°01´E
83 B16 **Otchinjau** Cunene, SW Angola 16°31´S 13°54´E
116 F12 **Oţelu Roşu** *Ger.* Ferdinandsberg, *Hung.* Nándorhgy. Caras-Severin, SW Romania 45°30´N 22°22´E
185 E21 **Otematata** Canterbury, South Island, New Zealand 44°37´S 170°12´E
118 I6 **Otepää** *Ger.* Odenpäh. Valgamaa, SE Estonia 58°01´N 26°30´E
144 G14 **Otes** *Kaz.* Say-Ötesh; *prev.* Say-Utës. Mangistau, SW Kazakhstan 44°20´N 53°32´E
162 H7 **Otgon** *var.* Buyant. Dzavhan, C Mongolia 47°14´N 97°14´E
32 K9 **Othello** Washington, NW USA 46°49´N 119°10´W
83 K23 **oThongathi** *prev.* Tongaat, *var.* uThongathi. KwaZulu/ Natal, E South Africa 29°35´S 31°07´E *see also* Tongaat
115 A15 **Othonoí** *island* Iónia Nisiá, Greece, C Mediterranean Sea
Othris *see* Óthrys
115 F17 **Óthrys** *var.* Othris. ▲▲ C Greece
77 Q14 **Oti** ≈ N Togo
40 K10 **Otinapa** Durango, C Mexico 24°01´N 104°58´W
185 G17 **Otira** West Coast, South Island, New Zealand 42°52´S 171°33´E
37 V3 **Otis** Colorado, C USA 40°09´N 102°57´W
12 L10 **Otish, Monts** ▲▲ Québec, E Canada
83 C17 **Otjikondo** Kunene, N Namibia 19°50´S 15°23´E
Otjikoto *see* Oshikoto
83 E18 **Otjinene** Omaheke, NE Namibia 21°10´S 18°43´E
83 D18 **Otjiwarongo** Otjozondjupa, N Namibia 20°29´S 16°36´E
83 D18 **Otjosondu** *var.* Otjosundu. Otjozondjupa, C Namibia 21°19´S 17°51´E
Otjosundu *see* Otjosondu
83 D18 **Otjozondjupa** ◐ *district* C Namibia
112 C11 **Otočac** Lika-Senj, W Croatia 44°52´N 15°13´E
Otog Qi *see* Ulan
112 J10 **Otok** Vukovar-Srijem, E Croatia 45°10´N 18°52´E
116 K14 **Otopeni** ✈ (Bucureşti) Ilfov, S Romania 44°34´N 26°09´E
184 L8 **Otorohanga** Waikato, North Island, New Zealand 38°10´S 175°14´E
12 D9 **Otoskwin** ≈ Ontario, C Canada
165 G14 **Ōtoyo** Kōchi, Shikoku, SW Japan 33°45´N 133°42´E
95 E16 **Otra** ≈ S Norway
107 R19 **Otranto** Puglia, SE Italy 40°08´N 18°28´E
Otranto, Canale d' *see* Otranto, Strait of
107 Q18 **Otranto, Strait of** *It.* Canale d'Otranto. *strait* Albania/Italy
111 H18 **Otrokovice** *Ger.* Otrokowitz. Zlínský Kraj, E Czech Republic 49°13´N 17°33´E
Otrokowitz *see* Otrokovice
31 P10 **Otsego** Michigan, N USA 42°27´N 85°42´W
31 Q6 **Otsego Lake** ◎ Michigan, N USA
18 I11 **Otselic River** ≈ New York, NE USA
164 J14 **Ōtsu** *var.* Ōtu. Shiga, Honshū, SW Japan 35°03´N 135°49´E
94 G11 **Otta** Oppland, S Norway 61°46´N 09°33´E
189 U13 **Otta** *island* Chuuk, C Micronesia
94 F11 **Otta** ≈ S Norway
189 U13 **Otta Pass** *passage* Chuuk Islands, C Micronesia
95 J22 **Ottarp** Skåne, S Sweden 55°55´N 12°55´E
14 L12 **Ottawa** ● (Canada) Ontario, SE Canada 45°24´N 75°41´W
30 L11 **Ottawa** Illinois, N USA 41°21´N 88°50´W
27 Q5 **Ottawa** Kansas, C USA 38°35´N 95°16´W
31 R12 **Ottawa** Ohio, N USA 41°01´N 84°03´W
14 L12 **Ottawa** *var.* Uplands. ✈ Ontario, SE Canada 45°19´N 75°39´W
14 M12 **Ottawa** *Fr.* Outaouais. ≈ Ontario/Québec, SE Canada
9 R10 **Ottawa Islands** *island group* Nunavut, C Canada
18 L8 **Otter Creek** ≈ Vermont, NE USA
36 L6 **Otter Creek Reservoir** ▣ Utah, W USA
98 L11 **Otterlo** Gelderland, E Netherlands 52°06´N 05°46´E
94 D9 **Otterøya** *island* S Norway
29 S6 **Otter Tail Lake** ◎ Minnesota, N USA
29 R7 **Otter Tail River** ≈ Minnesota, C USA
95 H23 **Otterup** Syddtjylland, C Denmark 55°31´N 10°25´E
99 H19 **Ottignies** Wallon Brabant, C Belgium 50°40´N 04°34´E
101 L23 **Ottobrunn** Bayern, SE Germany 48°02´N 11°40´E
29 X15 **Ottumwa** Iowa, C USA 41°00´N 92°24´W
Ōtu *see* Ōtsu
83 B16 **Otuazuma** Kunene, NW Namibia 17°52´S 13°16´E
77 V16 **Oturkpo** Benue, S Nigeria 07°12´N 08°06´E
193 Y15 **Otu Tolu Group** *island group* SE Tonga
182 M13 **Otway, Cape** *headland* Victoria, SE Australia 38°52´S 143°31´E
63 H24 **Otway, Seno** *inlet* S Chile
Ötz *see* Oetz
108 L8 **Ötztaler Ache** ≈ W Austria
108 L9 **Ötztaler Alpen** *It.* Alpi Venoste. ▲▲ SW Austria
27 T12 **Ouachita, Lake** ▣ Arkansas, C USA
27 R11 **Ouachita Mountains** ▲▲ Arkansas/Oklahoma, C USA
27 U13 **Ouachita River** ≈ Arkansas/Louisiana, C USA
Ouadaï *see* Ouaddaï
76 J7 **Ouadâne** *var.* Ouadane. Adrar, C Mauritania 20°57´N 11°35´W
78 K13 **Ouadda** Haute-Kotto, N Central African Republic 08°02´N 22°22´E
78 J10 **Ouaddaï** *off.* Région du Ouaddaï, *var.* Ouadaï, Wadai. ◐ *region* SE Chad
Ouaddaï, Région du *see* Ouaddaï
77 P13 **Ouagadougou** *var.* Wagadugu. ● (Burkina Faso) C Burkina 12°20´N 01°32´W
77 P13 **Ouagadougou** ✈ C Burkina Faso 12°21´N 01°27´W
77 O12 **Ouahigouya** NW Burkina Faso 13°31´N 02°20´W
Ouahran *see* Oran
79 J14 **Ouaka** ◐ *prefecture* C Central African Republic
79 J15 **Ouaka** ≈ S Central African Republic
Oualam *see* Ouallam
76 M9 **Oualâta** *var.* Oualata. Hodh ech Chargui, SE Mauritania 17°18´N 07°00´W
77 R11 **Ouallam** *var.* Oualam. Tillabéri, W Niger 14°23´N 02°09´E
172 H14 **Ouanani** Mohéli, S Comoros 12°19´S 93°98´E
55 Z10 **Ouanary** E French Guiana 04°11´N 51°40´W
78 L13 **Ouanda Djallé** Vakaga, NE Central African Republic 08°53´N 22°47´E
79 N14 **Ouando** Haut-Mbomou, SE Central African Republic 05°57´N 25°57´E
79 L15 **Ouango** Mbomou, S Central African Republic 04°19´N 22°30´E
77 N14 **Ouangolodougou** *var.* Wangolodougou. N Ivory Coast 09°58´N 05°09´W
172 I13 **Ouani** Anjouan, SE Comoros
79 M15 **Ouara** ≈ E Central African Republic
76 K7 **Ouarâne** *desert* C Mauritania
15 O11 **Ouareau** ≈ Québec, SE Canada
74 K7 **Ouargla** *var.* Wargla. NE Algeria 32°N 05°16´E
74 F8 **Ouarzazate** S Morocco 30°54´N 06°55´W
77 Q11 **Ouatagouna** Gao, E Mali 15°06´N 00°41´E
74 G6 **Ouazzane** *var.* Ouezzane, *Ar.* Wazan, Wazzan. N Morocco 34°52´N 05°35´W
Oubangui *see* Ubangi
Oubangui-Chari *see* Central African Republic
Oubangui-Chari, Territoire de l' *see* Central African Republic
Oubari, Edeyen d' *see* Awbāri, Idhān
98 G13 **Oud-Beijerland** Zuid-Holland, SW Netherlands 51°50´N 04°25´E
98 F13 **Ouddorp** Zuid-Holland, SW Netherlands 51°49´N 03°55´E
77 P9 **Oudeïka** *oasis* C Mali
98 G13 **Oude Maas** ≈ SW Netherlands
99 E18 **Oudenaarde** *Fr.* Audenarde. Oost-Vlaanderen, SW Belgium 50°50´N 03°37´E
99 H14 **Oudenbosch** Noord-Brabant, S Netherlands 51°35´N 04°32´E
98 P6 **Oude Pekela** Groningen, NE Netherlands 53°06´N 07°00´E
Ouderkerk *see* Ouderkerk aan den Amstel
98 I10 **Ouderkerk aan den Amstel** *var.* Ouderkerk. Noord-Holland, C Netherlands 52°18´N 04°54´E
98 I6 **Oudeschild** Noord-Holland, NW Netherlands 53°01´N 04°51´E
99 G14 **Oude-Tonge** Zuid-Holland, SW Netherlands 51°40´N 04°13´E
98 I12 **Oudewater** Utrecht, C Netherlands 52°02´N 04°54´E
Oudjda *see* Oujda
Oudkerk *see* Aldtsjerk
167 Q6 **Oudômxai** *var.* Muang Xay, Muong Sai, Xai. Oudômxai, N Laos 20°41´N 102°00´E
102 J7 **Oudon** ≈ NW France
98 I9 **Oudorp** Noord-Holland, NW Netherlands 52°39´N 04°47´E
83 G25 **Oudtshoorn** Western Cape, SW South Africa 33°35´S 22°14´E
99 I16 **Oud-Turnhout** Antwerpen, N Belgium 51°19´N 05°01´E
74 F7 **Oued-Zem** C Morocco 32°53´N 06°30´W
187 P16 **Ouégoa** Province Nord, C New Caledonia 20°22´S 164°25´E
76 L13 **Ouéléssébougou** *var.* Ouolossébougou. Koulikoro, SW Mali 11°58´N 07°51´W
77 N16 **Ouellé** E Ivory Coast 07°18´N 04°01´W
77 R16 **Ouémé** ≈ C Benin
77 O13 **Ouessa** S Burkina Faso 11°02´N 02°44´W
102 D5 **Ouessant, Île d'** *Eng.* Ushant. *island* NW France
79 H17 **Ouésso** Sangha, NW Congo 01°38´N 16°03´E
79 D15 **Ouest** *Eng.* West. ◐ *province* W Cameroon
190 G11 **Ouest, Baie del'** *bay* Îles Wallis, E Wallis and Futuna
15 Y7 **Ouest, Pointe de l'** *headland* Québec, SE Canada 48°08´N 64°57´W
Ouezzane *see* Ouazzane
99 K20 **Ouffet** Liège, E Belgium 50°30´N 05°31´E
79 H14 **Ouham** ◐ *prefecture* NW Central African Republic
78 I13 **Ouham** ≈ Central African Republic/Chad
79 G14 **Ouham-Pendé** ◐ *prefecture* W Central African Republic
77 R16 **Ouidah** *Eng.* Whydah, Wida. S Benin 06°23´N 02°08´E
74 H6 **Oujda** *Ar.* Oudjda, Ujda. NE Morocco 34°45´N 01°53´W
76 I7 **Oujeft** Adrar, C Mauritania 20°05´N 13°00´W
93 L15 **Oulainen** Pohjois-Pohjanmaa, C Finland 64°14´N 24°50´E
Ould Yanja *see* Ould Yenjé
76 J10 **Ould Yenjé** *var.* Ould Yanja. Guidimaka, S Mauritania 15°33´N 11°43´W
93 L14 **Oulu** *Swe.* Uleåborg. Pohjois-Pohjanmaa, C Finland 65°01´N 25°28´E
Oulu *Swe.* Uleåborg. ◐ *province* N Finland
93 L15 **Oulujärvi** *Swe.* Uleträsk. ◎ C Finland
93 M14 **Oulujoki** *Swe.* Uleälv. ≈ C Finland
93 L14 **Oulunsalo** Pohjois-Pohjanmaa, C Finland 64°55´N 25°19´E
106 A8 **Oulx** Piemonte, NE Italy 45°05´N 06°41´E
78 J9 **Oum-Chalouba** Ennedi-Ouest, NE Chad 15°48´N 20°46´E
76 M16 **Oumé** C Ivory Coast 06°25´N 05°23´W
74 F7 **Oum er Rbia** ≈ C Morocco
78 J10 **Oum-Hadjer** Batha, E Chad 13°18´N 19°41´E
92 K10 **Ounasjoki** ≈ N Finland
78 J7 **Ounianga Kébir** Ennedi-Ouest, N Chad 19°06´N 20°29´E
Ouolossébougou *see* Ouéléssébougou
Oup *see* Auob
99 K19 **Oupeye** Liège, E Belgium 50°42´N 05°38´E
99 N21 **Our** ≈ NW Europe
37 Q7 **Ouray** Colorado, C USA 38°01´N 107°40´W
103 R7 **Ource** ≈ C France
104 G9 **Ourém** Santarém, C Portugal 39°40´N 08°32´W
104 H4 **Ourense** *Cast.* Orense, *Lat.* Aurium. Galicia, NW Spain 42°20´N 07°52´W
104 I4 **Ourense** *Cast.* Orense. ◐ *province* Galicia, NW Spain
59 O15 **Ouricuri** Pernambuco, E Brazil 07°51´S 40°05´W
60 J9 **Ourinhos** São Paulo, S Brazil 22°59´S 49°52´W

◆ Country ● Country Capital ◇ Dependent Territory ○ Dependent Territory Capital ◐ Administrative Regions ✈ International Airport ▲ Mountain ▲▲ Mountain Range ⛰ Volcano ≈ River ◎ Lake ▣ Reservoir
301

OURIQUE – PAMPELUNA

104 G13 **Ourique** Beja, S Portugal 37°38′N 08°13′W
59 M20 **Ouro Preto** Minas Gerais, NE Brazil 20°25′S 43°30′W
Ours, Grand Lac de l' *see* Great Bear Lake
99 K20 **Ourthe** ≈ E Belgium
165 Q9 **Ōu-sanmyaku** ▲▲ Honshū, C Japan
97 M17 **Ouse** ≈ N England, United Kingdom
Ouse *see* Great Ouse
102 H7 **Oust** ≈ NW France
Outaouais *see* Ottawa
15 T4 **Outardes Quatre, Réservoir** ⊠ Québec, SE Canada
15 T5 **Outardes, Rivière aux** ≈ Québec, SE Canada
96 E8 **Outer Hebrides** *var.* Western Isles. *island group* NW Scotland, United Kingdom
30 K3 **Outer Island** *island* Apostle Islands, Wisconsin, N USA
35 S16 **Outer Santa Barbara Passage** *passage* California, SW USA
83 C18 **Outjo** Kunene, N Namibia 20°08′S 16°08′E
11 T16 **Outlook** Saskatchewan, S Canada 51°30′N 107°03′W
93 N16 **Outokumpu** Pohjois-Karjala, E Finland 62°43′N 29°05′E
96 M2 **Out Skerries** *island group* NE Scotland, United Kingdom
187 Q16 **Ouvéa** *island* Îles Loyauté, NE New Caledonia
103 S14 **Ouvèze** ≈ SE France
182 L9 **Ouyen** Victoria, SE Australia 35°07′S 142°19′E
39 Q14 **Ouzinkie** Kodiak Island, Alaska, USA 57°54′N 152°27′W
137 O13 **Ovacık** Tunceli, E Turkey 39°23′N 39°13′E
106 C9 **Ovada** Piemonte, NE Italy 44°41′N 08°39′E
187 X14 **Ovalau** *island* C Fiji
62 G9 **Ovalle** Coquimbo, N Chile 30°33′S 71°16′W
83 C17 **Ovamboland** *physical region* N Namibia
54 L10 **Ovana, Cerro** ▲ S Venezuela 04°41′N 66°54′W
104 G7 **Ovar** Aveiro, N Portugal 40°52′N 08°38′W
114 L10 **Ovcharitsa, Yazovir** ⊠ SE Bulgaria
54 E6 **Ovejas** Sucre, NW Colombia 09°32′N 75°14′W
101 E16 **Overath** Nordrhein-Westfalen, W Germany 50°55′N 07°16′E
98 F13 **Overflakkee** *island* SW Netherlands
99 H19 **Overijse** Vlaams Brabant, C Belgium 50°46′N 04°32′E
98 N10 **Overijssel** ◈ *province* E Netherlands
98 M9 **Overijssels Kanaal** *canal* E Netherlands
92 K13 **Överkalix** Norrbotten, N Sweden 66°19′N 22°49′E
27 R4 **Overland Park** Kansas, C USA 38°57′N 94°41′W
99 L14 **Overloon** Noord-Brabant, SE Netherlands 51°35′N 05°54′E
99 K16 **Overpelt** Limburg, NE Belgium 51°13′N 05°24′E
35 Y10 **Overton** Nevada, W USA 36°32′N 114°25′W
25 W7 **Overton** Texas, SW USA 32°16′N 94°58′W
92 K13 **Övertorneå** Norrbotten, N Sweden 66°22′N 23°40′E
95 N18 **Överum** Kalmar, S Sweden 57°58′N 16°20′E
92 G13 **Överuman** ◎ N Sweden
117 P11 **Ovidiopol'** Odes'ka Oblast', SW Ukraine 46°15′N 30°27′E
116 M14 **Ovidiu** Constanţa, SE Romania 44°16′N 28°34′E
45 N10 **Oviedo** SW Dominican Republic 17°47′N 71°22′W
104 K2 **Oviedo** *anc.* Asturias. Asturias, NW Spain 43°21′N 05°50′W
104 K2 **Oviedo** ✈ Asturias, N Spain 43°21′N 05°50′W
Ovilava *see* Wels
118 D7 **Ovišī** W Latvia 57°34′N 21°43′E
146 K10 **Ovminzatovo Tog'lari** *Rus.* Gory Auminzatau. ▲▲ N Uzbekistan
Övögdiy *see* Telmen
Ovoot *see* Darīganga
157 O4 **Övörhangay** ◈ *province* C Mongolia
94 E12 **Øvre Årdal** Sogn Og Fjordane, S Norway 61°18′N 07°48′E
95 J14 **Övre Fryken** ◎ C Sweden
92 J11 **Övre Soppero** *Lapp.* Badje-Sohppar. Norrbotten, N Sweden 68°07′N 21°40′E
117 N3 **Ovruch** Zhytomyrs'ka Oblast', N Ukraine 51°20′N 58°50′E
Övt *see* Bat-Öldziy
185 E24 **Owaka** Otago, South Island, New Zealand 46°27′S 169°42′E
79 H18 **Owando** *prev.* Fort Rousset. Cuvette, C Congo 0°29′S 15°55′E
164 J14 **Owase** Mie, Honshū, SW Japan 34°04′N 136°11′E
27 P9 **Owasso** Oklahoma, C USA 36°16′N 95°51′W
29 V10 **Owatonna** Minnesota, N USA 44°04′N 93°13′W
173 O4 **Owen Fracture Zone** *tectonic feature* W Arabian Sea
185 H15 **Owen, Mount** ▲ South Island, New Zealand 41°32′S 172°33′E
185 H15 **Owen River** Tasman, South Island, New Zealand 41°40′S 172°28′E
44 D8 **Owen Roberts** ✈ Grand Cayman, Cayman Islands 19°15′N 81°22′W
20 I6 **Owensboro** Kentucky, S USA 37°46′N 87°07′W
35 T11 **Owens Lake** *salt flat* California, W USA
14 F14 **Owen Sound** Ontario, S Canada 44°34′N 80°56′W
14 F13 **Owen Sound** ◎ Ontario, S Canada
35 T10 **Owens River** ≈ California, W USA
186 F9 **Owen Stanley Range** ▲▲ S Papua New Guinea
27 V5 **Owensville** Missouri, C USA 38°21′N 91°30′W
20 M4 **Owenton** Kentucky, S USA 38°33′N 84°50′W
77 U17 **Owerri** Imo, S Nigeria 05°19′N 07°07′E
184 M10 **Owhango** Manawatu-Wanganui, North Island, New Zealand 39°01′S 175°22′E
21 N5 **Owingsville** Kentucky, S USA 38°09′N 83°46′W
77 T16 **Owo** Ondo, SW Nigeria 07°10′N 05°31′E
31 R9 **Owosso** Michigan, N USA 43°00′N 84°10′W
35 V1 **Owyhee** Nevada, W USA 41°57′N 116°07′W
32 L14 **Owyhee, Lake** ◎ Oregon, NW USA
32 L15 **Owyhee River** ≈ Idaho/Oregon, NW USA
92 K1 **Öxarfjörður** *var.* Axarfjördhur. *fjord* N Iceland
94 K12 **Oxberg** Dalarna, C Sweden 61°07′N 14°10′E
11 V17 **Oxbow** Saskatchewan, S Canada 49°16′N 102°12′W
95 O17 **Oxelösund** Södermanland, S Sweden 58°40′N 17°10′E
185 H18 **Oxford** Canterbury, South Island, New Zealand 43°18′S 172°10′E
97 M21 **Oxford** *Lat.* Oxonia. S England, United Kingdom 51°46′N 01°15′W
23 Q3 **Oxford** Alabama, S USA 33°36′N 85°50′W
22 L2 **Oxford** Mississippi, S USA 34°23′N 89°30′W
29 N16 **Oxford** Nebraska, C USA 40°15′N 99°37′W
18 I11 **Oxford** New York, NE USA 42°21′N 75°39′W
21 U8 **Oxford** North Carolina, SE USA 36°22′N 78°37′W
31 Q14 **Oxford** Ohio, N USA 39°30′N 84°45′W
18 H16 **Oxford** Pennsylvania, NE USA 39°46′N 75°57′W
11 X12 **Oxford House** Manitoba, C Canada 54°55′N 95°13′W
29 Y13 **Oxford Junction** Iowa, C USA 41°58′N 90°57′W
11 X12 **Oxford Lake** ◎ Manitoba, C Canada
97 M21 **Oxfordshire** *cultural region* S England, United Kingdom
Oxia *see* Oxyá
41 X12 **Oxkutzcab** Yucatán, SE Mexico 20°18′N 89°26′W
35 R15 **Oxnard** California, W USA 34°12′N 119°10′W
Oxonia *see* Oxford
14 I12 **Oxtongue** ≈ Ontario, SE Canada
Oxus *see* Amu Darya
115 E15 **Oxyá** *var.* Oxia. ▲ C Greece 39°46′N 21°56′E
164 L11 **Oyabe** Toyama, Honshū, SW Japan 36°42′N 136°52′E
Oyahue/Oyahue, Volcán *see* Ollagüe, Volcán
165 O12 **Oyama** Tochigi, Honshū, S Japan 36°19′N 139°46′E
47 U5 **Oyapock** ≈ E French Guiana
Oyapock *see* Oiapoque, Rio/Oyapok, Fleuve l'
55 Z10 **Oyapok, Baie de L'** *bay* Brazil/French Guiana South America W Atlantic Ocean
55 Z11 **Oyapok, Fleuve l'** *var.* Rio Oiapoque, Oyapock. ≈ Brazil/French Guiana *see also* Oiapoque, Rio
Oyapok, Fleuve l' *see* Oiapoque, Rio
79 E17 **Oyem** Woleu-Ntem, N Gabon 01°34′N 11°31′E
11 R16 **Oyen** Alberta, SW Canada 51°20′N 110°28′W
95 I15 **Øyeren** ◎ S Norway
Oygon *see* Tüdevtey
96 I7 **Oykel** ≈ N Scotland, United Kingdom
123 R9 **Oymyakon** Respublika Sakha (Yakutiya), NE Russian Federation 63°28′N 142°22′E
79 H19 **Oyo** Cuvette, C Congo 01°17′S 16°00′E
77 S15 **Oyo** Oyo, W Nigeria 07°51′N 03°57′E
77 S15 **Oyo** ◈ *state* SW Nigeria
56 D13 **Oyón** Lima, C Peru 10°39′S 76°44′W
103 S10 **Oyonnax** Ain, E France 46°16′N 05°39′E
146 L10 **Oyoqog'itma** *Rus.* Ayakagytma. Buxoro Viloyati, C Uzbekistan 40°37′N 64°26′E
146 M9 **Oyoqquduq** *Rus.* Ayakkuduk. Navoiy Viloyati, N Uzbekistan 41°16′N 65°12′E
32 F9 **Oysterville** Washington, NW USA 46°33′N 124°03′W
95 D14 **Øystese** Hordaland, S Norway 60°23′N 06°13′E
145 S16 **Oytal** Zhambyl, S Kazakhstan 42°54′N 73°21′E
147 U10 **Oy-Tal** Oshskaya Oblast', SW Kyrgyzstan 40°23′N 74°04′E
147 T10 **Oy-Tal** ≈ SW Kyrgyzstan
145 Q15 **Oyyk** *prev.* Uyuk. Zhambyl, S Kazakhstan 43°46′N 70°55′E
144 H10 **Oyyl** *prev.* Uil. Aktyubinsk, W Kazakhstan 49°06′N 54°41′E
144 H10 **Oyyl** *prev.* Uil. ≈ W Kazakhstan
Ozarichi *see* Azarychy
23 R7 **Ozark** Alabama, S USA 31°27′N 85°38′W
27 S10 **Ozark** Arkansas, C USA 35°30′N 93°50′W
27 T8 **Ozark** Missouri, C USA 37°01′N 93°12′W
27 T8 **Ozark Plateau** *plain* Arkansas/Missouri, C USA
27 T6 **Ozarks, Lake of the** ⊠ Missouri, C USA
192 L10 **Ozbourn Seamount** *undersea feature* W Pacific Ocean 26°00′S 174°49′W
111 L20 **Ózd** Borsod-Abaúj-Zemplén, NE Hungary 48°15′N 20°18′E
112 D11 **Ozeblin** ▲ C Croatia 44°37′N 15°52′E
123 V11 **Ozernovskiy** Kamchatskiy Kray, E Russian Federation 51°28′N 156°32′E
144 M7 **Ozërnoye** *var.* Ozërnyy. Kostanay, N Kazakhstan 53°29′N 63°14′E
124 J15 **Ozërnyy** Tverskaya Oblast', W Russian Federation 57°55′N 33°45′E
Ozërnyy *see* Ozërnoye
Ozero Azhbulat *see* Ozero Ul'ken Azhibolat
Ozero Segozero *see* Segozerskoye Vodokhranilishche
115 D18 **Ozerós, Límni** ◎ W Greece
145 T7 **Ozero Ul'ken Azhibolat** *prev.* Ozero Azhbulat. ◎ NE Kazakhstan
122 G11 **Ozërsk** Chelyabinskaya Oblast', C Russian Federation 55°44′N 60°59′E
119 D14 **Ozersk** *prev.* Darkehnen, *Ger.* Angerapp. Kaliningradskaya Oblast', W Russian Federation 54°23′N 21°59′E
126 L4 **Ozery** Moskovskaya Oblast', W Russian Federation 54°51′N 38°37′E
Özgön *see* Uzgen
107 C17 **Ozieri** Sardegna, Italy, C Mediterranean Sea 40°35′N 09°01′E
111 I15 **Ozimek** *Ger.* Malapane. Opolskie, SW Poland 50°41′N 18°16′E
127 R8 **Ozinki** Saratovskaya Oblast', W Russian Federation 51°16′N 49°45′E
25 O10 **Ozona** Texas, SW USA 30°43′N 101°13′W
Ozorkov *see* Ozorków
110 J12 **Ozorków** *Rus.* Ozorkov. Łódźz, C Poland 52°00′N 19°17′E
164 F14 **Ōzu** Ehime, Shikoku, SW Japan 33°30′N 132°33′E
137 R10 **Ozurgeti** *prev.* Makharadze, Ozurget'i. W Georgia 41°57′N 42°01′E
Ozurget'i *see* Ozurgeti

P

99 J17 **Paal** Limburg, NE Belgium 51°03′N 05°08′E
196 M14 **Paamiut** *var.* Pâmiut, *Dan.* Frederikshåb. Sermersooq, S Greenland 61°59′N 49°40′W
Pa-an *see* Hpa-an
101 L22 **Paar** ≈ SE Germany
83 E26 **Paarl** Western Cape, SW South Africa 33°45′S 18°58′E
93 L15 **Paavola** Pohjois-Pohjanmaa, C Finland 64°34′N 25°15′E
96 E8 **Pabbay** *island* NW Scotland, United Kingdom
153 T15 **Pabna** Rajshahi, W Bangladesh 24°02′N 89°15′E
109 U4 **Pabneukirchen** Oberösterreich, N Austria 48°19′N 14°49′E
118 H13 **Pabradė** *Pol.* Podbrodzie. Vilnius, SE Lithuania 54°58′N 25°43′E
56 L13 **Pacahuaras, Río** ≈ N Bolivia
Pacaraima, Sierra/Pacaraim, Serra *see* Pakaraima Mountains
56 B11 **Pacasmayo** La Libertad, W Peru 07°27′S 79°33′W
42 D6 **Pacaya, Volcán de** ⛰ S Guatemala 14°19′N 90°36′W
115 K23 **Pacheía** *var.* Pachía. *island* Kykládes, Greece, Aegean Sea
Pachía *see* Pacheía
107 L26 **Pachino** Sicilia, Italy, C Mediterranean Sea 36°43′N 15°06′E
56 F12 **Pachitea, Río** ≈ C Peru
154 I11 **Pachmarhi** Madhya Pradesh, C India 22°36′N 78°18′E
121 P3 **Páchna** *var.* Pakhna. SW Cyprus 34°47′N 32°48′E
115 H25 **Páchnes** ▲ Kríti, Greece, E Mediterranean Sea 35°19′N 24°00′E
54 F9 **Pacho** Cundinamarca, C Colombia 05°09′N 74°08′W
154 F12 **Pāchora** Mahārāshtra, C India 20°52′N 75°28′E
41 P13 **Pachuca** *var.* Pachuca de Soto. Hidalgo, C Mexico 20°05′N 98°46′W
Pachuca de Soto *see* Pachuca
27 W5 **Pacific** Missouri, C USA 38°28′N 90°44′W
192 L14 **Pacific-Antarctic Ridge** *undersea feature* S Pacific Ocean 62°00′S 157°00′W
32 F8 **Pacific Beach** Washington, NW USA 47°09′N 124°12′W
35 N10 **Pacific Grove** California, W USA 36°35′N 121°54′W
29 S15 **Pacific Junction** Iowa, C USA 41°01′N 95°48′W
192-193 **Pacific Ocean** *ocean*
129 Z10 **Pacific Plate** *tectonic feature*
113 J15 **Pačir** ▲ N Montenegro 43°19′N 19°07′E
182 L5 **Packsaddle** New South Wales, SE Australia 30°42′S 141°55′E
32 H9 **Packwood** Washington, NW USA 46°37′N 121°38′W
Padalung *see* Phatthalung
168 J12 **Padang** Sumatera, W Indonesia 01°S 100°21′E
168 L9 **Padang Endau** Pahang, Peninsular Malaysia 02°38′N 103°37′E
Padangpandjang *see* Padangpanjang
168 I11 **Padangpanjang** *prev.* Padangpandjang. Sumatera, W Indonesia 0°30′S 100°26′E
168 I10 **Padangsidempuan** *prev.* Padangsidimpoean. Sumatera, W Indonesia 01°23′N 99°15′E
Padangsidimpoean *see* Padangsidempuan
124 I9 **Padany** Respublika Kareliya, NW Russian Federation 63°18′N 33°20′E
93 M18 **Padasjoki** Päijät-Häme, S Finland 61°20′N 25°21′E
57 M22 **Padcaya** Tarija, S Bolivia 21°52′S 64°46′W
101 H14 **Paderborn** Nordrhein-Westfalen, NW Germany 51°43′N 08°45′E
Padeşul/Padeş, Vîrful *see* Padeş, Vârful
116 F12 **Padeş, Vârful** *var.* Padeşul; *prev.* Vîrful Padeş. ▲ W Romania 45°39′N 22°19′E
112 L10 **Padinska Skela** Serbia, N Serbia 44°58′N 20°25′E
153 S14 **Padma** *var.* Ganges. ≈ Bangladesh/India *see also* Ganges
Padma *see* Brahmaputra
Padma *see* Ganges
106 H8 **Padova** *Eng.* Padua; *anc.* Patavium. Veneto, NE Italy 45°24′N 11°52′E
82 A10 **Padrão, Ponta do** *headland* NW Angola 06°06′S 12°18′E
25 T16 **Padre Island** *island* Texas, SW USA
104 G3 **Padrón** Galicia, NW Spain 42°44′N 08°40′W
118 K13 **Padsvillye** *Rus.* Podsvil'ye. Vitsyebskaya Voblasts', N Belarus 55°09′N 27°58′E
182 K11 **Padthaway** South Australia 36°39′S 140°30′E
Padua *see* Padova
20 G7 **Paducah** Kentucky, S USA 37°03′N 88°36′W
25 P4 **Paducah** Texas, SW USA 34°01′N 100°18′W
105 N15 **Padul** Andalucía, S Spain 37°02′N 03°37′W
191 P8 **Paea** Tahiti, W French Polynesia 17°41′S 149°35′W
185 L14 **Paekakariki** Wellington, North Island, New Zealand 41°00′S 174°58′E
163 X11 **Paektu-san** *var.* Baitou Shan. ▲ China/North Korea 42°00′N 128°03′W
Paengnyŏng *see* Baengnyong-do
184 M7 **Paeroa** Waikato, North Island, New Zealand 37°23′S 175°39′E
54 D12 **Páez** Cauca, SW Colombia 02°37′N 76°00′W
121 O3 **Páfos** *var.* Paphos. W Cyprus 34°46′N 32°26′E
121 O3 **Páfos** ✈ SW Cyprus 34°46′N 32°25′E
83 L19 **Pafúri** Gaza, SW Mozambique 22°27′S 31°21′E
112 C12 **Pag** *It.* Pago. Lika-Senj, SW Croatia 44°26′N 15°01′E
112 B11 **Pag** *It.* Pago. *island* Zadar, C Croatia
171 P7 **Pagadian** Mindanao, S Philippines 07°47′N 123°22′E
168 J13 **Pagai Selatan, Pulau** *island* Kepulauan Mentawai, W Indonesia
168 J13 **Pagai Utara, Pulau** *island* Kepulauan Mentawai, W Indonesia
188 K4 **Pagan** *island* C Northern Mariana Islands
115 G16 **Pagasitikós Kólpos** *gulf* E Greece
36 L8 **Page** Arizona, SW USA 36°54′N 111°28′W
29 Q5 **Page** North Dakota, N USA 47°09′N 97°33′W
118 D13 **Pagėgiai** *Ger.* Pogegen. Tauragė, SW Lithuania 55°08′N 21°54′E
21 S11 **Pageland** South Carolina, SE USA 34°46′N 80°23′W
81 G16 **Pager** ≈ NE Uganda
149 Q5 **Paghmān** Kābul, E Afghanistan 34°33′N 68°55′E
Pago *see* Pag
188 C16 **Pago Bay** *bay* E Guam, W Pacific Ocean
115 M20 **Pagóndas** *var.* Pagóndhas. Sámos, Dodekánisa, Greece, Aegean Sea 37°41′N 26°50′E
Pagóndhas *see* Pagóndas
192 J16 **Pago Pago** ○ (American Samoa) Tutuila, W American Samoa 14°16′S 170°43′W
37 R8 **Pagosa Springs** Colorado, C USA 37°13′N 107°01′W
Pagqên *see* Gadê
38 H12 **Pāhala** *var.* Pahala. Hawaii, USA, C Pacific Ocean 19°12′N 155°28′W
168 K8 **Pahang** *var.* Negeri Pahang Darul Makmur. ◈ *state* Peninsular Malaysia
Pahang *see* Pahang, Sungai
168 L8 **Pahang, Sungai** *var.* Pahang, Sungei Pahang. ≈ Peninsular Malaysia
149 S8 **Pahārpur** Khyber Pakhtunkhwa, NW Pakistan 32°06′N 71°00′E
185 B24 **Pahia Point** *headland* South Island, New Zealand 46°19′S 167°42′E
184 M13 **Pahiatua** Manawatu-Wanganui, North Island, New Zealand 40°30′S 175°49′E
38 H12 **Pāhoa** *var.* Pahoa. Hawaii, USA, C Pacific Ocean 19°29′N 154°56′W
23 Y14 **Pahokee** Florida, SE USA 26°49′N 80°40′W
35 X9 **Pahranagat Range** ▲▲ Nevada, W USA
35 W11 **Pahrump** Nevada, W USA 36°11′N 115°58′W
Pahsien *see* Chongqing
35 V9 **Pahute Mesa** ▲▲ Nevada, W USA
167 N7 **Pai** Mae Hong Son, NW Thailand 19°24′N 98°26′E
38 F10 **Pa'ia** *var.* Paia. Maui, Hawaii, USA, C Pacific Ocean 20°54′N 156°22′W
Paia *see* Pa'ia
Pai-ch'eng *see* Baicheng
118 H4 **Paide** *Ger.* Weissenstein. Järvamaa, N Estonia 58°55′N 25°36′E
97 J24 **Paignton** SW England, United Kingdom 50°26′N 03°34′W
184 K3 **Paihia** Northland, North Island, New Zealand 35°18′S 174°06′E
93 M18 **Päijänne** ◎ C Finland
Päijänne-Tavastland *see* Päijät-Häme
93 M19 **Päijät-Häme** *Swe.* Päijänne-Tavastland. ◈ *region* S Finland
114 F13 **Páïko** ▲▲ N Greece
57 M17 **Paila, Río** ≈ C Bolivia
167 Q12 **Pailĭn** Bătdâmbâng, W Cambodia 12°51′N 102°34′E
Pailing *see* Chun'an
54 F6 **Pailitas** Cesar, N Colombia 08°58′N 73°38′W
38 F9 **Pailolo Channel** *channel* Hawai'i, USA, C Pacific Ocean
93 K19 **Paimio** *Swe.* Pemar. Varsinais-Suomi, SW Finland 60°27′N 22°42′E
165 O16 **Paimi-saki** *var.* Yaeme-saki. *headland* Iriomote-jima, SW Japan 24°18′N 123°40′E
102 G5 **Paimpol** Côtes d'Armor, NW France 48°47′N 03°03′W
168 J12 **Painan** Sumatera, W Indonesia 01°22′S 100°33′E
63 G23 **Paine, Cerro** ▲ S Chile 51°01′S 72°57′W
31 U11 **Painesville** Ohio, N USA 41°43′N 81°15′W
31 S14 **Paint Creek** ≈ Ohio, N USA
36 L10 **Painted Desert** *desert* Arizona, SW USA
Paint Hills *see* Wemindji
30 M4 **Paint River** ≈ Michigan, N USA
25 P8 **Paint Rock** Texas, SW USA 31°32′N 99°56′W
21 O6 **Paintsville** Kentucky, S USA 37°48′N 82°48′W
Paisance *see* Piacenza
96 I12 **Paisley** W Scotland, United Kingdom 55°50′N 04°26′W
32 I15 **Paisley** Oregon, NW USA 42°40′N 120°31′W
105 O3 **País Vasco** *Basq.* Euskadi, *Eng.* The Basque Country, *Sp.* Provincias Vascongadas. ◈ *autonomous community* N Spain
56 A9 **Paita** Piura, NW Peru 05°11′S 81°09′W
169 V6 **Paitan, Teluk** *bay* Sabah, East Malaysia
104 H7 **Paiva, Rio** ≈ N Portugal
92 K12 **Pajala** Norrbotten, N Sweden 67°12′N 23°19′E
104 K3 **Pajares, Puerto de** *pass* NW Spain
54 G9 **Pajarito** Boyacá, C Colombia 05°18′N 72°43′W
54 G4 **Pajaro** La Guajira, S Colombia 11°41′N 72°37′W
Pakanbaru *see* Pekanbaru
55 Q10 **Pakaraima Mountains** *var.* Serra Pacaraim, Sierra Pacaraima. ▲▲ N South America
Pākaur *see* Pākur
167 P10 **Pak Chong** Nakhon Ratchasima, C Thailand 14°38′N 101°22′E
123 V8 **Pakhachi** Krasnoyarskiy Kray, E Russian Federation 60°36′N 168°59′E
Pakhna *see* Páchna
189 U16 **Pakin Atoll** *atoll* Caroline Islands, E Micronesia
149 Q12 **Pakistan** *off.* Islamic Republic of Pakistan, *var.* Islami Jamhuriya e Pakistan. ◆ *republic* S Asia
Pakistan, Islamic Republic of *see* Pakistan
Pakistan, Islami Jamhuriya e *see* Pakistan
167 P8 **Pak Lay** *var.* Muang Pak Lay. Xaignabouli, C Laos 18°06′N 101°21′E
Paknam *see* Samut Prakan
166 L5 **Pakokku** Magway, C Myanmar (Burma) 21°20′N 95°05′E
110 I10 **Pakość** *Ger.* Pakosch. Kujawski-pomorskie, C Poland 52°47′N 18°03′E
Pakosch *see* Pakość
149 V10 **Pākpattan** Punjab, E Pakistan 30°20′N 73°27′E
167 O15 **Pak Phanang** *var.* Ban Pak Phanang. Nakhon Si Thammarat, SW Thailand 08°20′N 100°10′E
112 G9 **Pakrac** *Hung.* Pakrácz. Požega-Slavonija, NE Croatia 45°26′N 17°09′E
Pakrácz *see* Pakrac
118 F11 **Pakruojis** Šiaulai, N Lithuania 56°N 23°51′E
111 J24 **Paks** Tolna, S Hungary 46°38′N 18°51′E
Pak Sane *see* Pakxan
Paksé *see* Pakxé
167 Q10 **Pak Thong Chai** Nakhon Ratchasima, C Thailand 14°43′N 102°01′E
Paktīā *see* Paktiyā
149 Q7 **Paktīkā** ◈ *province* SE Afghanistan
149 R6 **Paktiyā** *prev.* Paktīā. ◈ *province* SE Afghanistan
171 N12 **Pakuli** Sulawesi, C Indonesia 01°14′S 119°55′E
153 S14 **Pākur** *var.* Pākaur. Jharkhand, N India 24°48′N 87°14′E
81 F17 **Pakwach** NW Uganda 02°28′N 31°28′E
167 R8 **Pakxan** *var.* Muang Pakxan, Pak Sane. Bolikhamxai, C Laos 18°27′N 103°38′E
167 S10 **Pakxé** *var.* Paksé. Champasak, S Laos 15°09′N 105°49′E
78 G12 **Pala** Mayo-Kébbi Ouest, SW Chad 09°22′N 14°54′E
61 A17 **Palacios** Santa Fe, C Argentina 30°43′S 61°37′W
25 V13 **Palacios** Texas, SW USA 28°42′N 96°13′W
105 X5 **Palafrugell** Cataluña, NE Spain 41°55′N 03°10′E
107 L24 **Palagonia** Sicilia, Italy, C Mediterranean Sea 37°20′N 14°45′E
113 E17 **Palagruža** *It.* Pelagosa. *island* SW Croatia
115 G20 **Palaiá Epídavros** Pelopónnisos, S Greece 37°38′N 23°09′E
121 P3 **Palaichóri** *var.* Palekhori. C Cyprus 34°55′N 33°06′E
115 H25 **Palaiochóra** Kríti, Greece, E Mediterranean Sea 35°14′N 23°37′E
115 A15 **Palaiolastrítsa** *religious building* Kérkyra, Iónia Nisiá, Greece, C Mediterranean Sea
115 J19 **Palaiópoli** Ándros, Kykládes, Greece, Aegean Sea 37°49′N 24°49′E
103 N5 **Palaiseau** Essonne, N France 48°41′N 02°14′E
154 N11 **Pāla Laharha** Odisha, E India 21°25′N 85°18′E
83 G19 **Palamakoloi** Ghanzi, C Botswana 23°10′S 22°22′E
115 E16 **Palamás** Thessalía, C Greece 39°28′N 22°05′E
105 X5 **Palamós** Cataluña, NE Spain 41°51′N 03°06′E
118 J5 **Palamuse** *Ger.* Sankt-Bartholomäi. Jõgevamaa, E Estonia 58°41′N 26°35′E
183 Q14 **Palana** Tasmania, SE Australia 39°48′S 147°54′E
123 U9 **Palana** Krasnoyarskiy Kray, E Russian Federation 59°05′N 159°59′E
118 C11 **Palanga** *Ger.* Polangen. Klaipėda, NW Lithuania 05°54′N 21°05′E
143 V10 **Palangān, Kūh-e** ▲▲ E Iran
Palangkaraja *see* Palangkaraya
169 T12 **Palangkaraya** *prev.* Palangkaraja. Borneo, C Indonesia 02°16′S 113°55′E
155 H22 **Palani** Tamil Nādu, SE India 10°30′N 77°24′E
Palanka *see* Bačka Palanka
154 D9 **Pālanpur** Gujarāt, W India 24°12′N 72°29′E
Palantia *see* Palencia
83 I19 **Palapye** Central, SE Botswana 22°37′S 27°06′E
155 J19 **Pālār** ≈ SE India
104 H3 **Palas de Rei** Galicia, NW Spain 42°52′N 07°51′W
123 T9 **Palatka** Magadanskaya Oblast', E Russian Federation 60°09′N 150°33′E
23 W10 **Palatka** Florida, SE USA 29°39′N 81°38′W
188 B9 **Palau** *var.* Belau. ◆ *republic* W Pacific Ocean
Palau *see* Palau Islands
129 Y14 **Palau Islands** *var.* Palau. *island group* N Palau
192 G16 **Palauli Bay** *bay* Savai'i, C Samoa, C Pacific Ocean
167 N11 **Palaw** Taninthayi, S Myanmar (Burma) 12°57′N 98°39′E
170 M6 **Palawan** *island* W Philippines
171 N6 **Palawan Passage** *passage* W Philippines
192 E7 **Palawan Trough** *undersea feature* S South China Sea 07°00′N 115°00′E
155 H23 **Pālayankottai** Tamil Nādu, SE India 08°42′N 77°46′E
107 L25 **Palazzola Acreide** *anc.* Acrae. Sicilia, Italy, C Mediterranean Sea 37°04′N 14°54′E
118 G3 **Paldiski** *prev.* Baltiski, *Eng.* Baltic Port, *Ger.* Baltischport. Harjumaa, NW Estonia 59°22′N 24°08′E
112 I13 **Pale** Republika Srpska, SE Bosnia and Herzegovina 43°49′N 18°35′E
Palekhori *see* Palaichóri
168 L13 **Palembang** Sumatera, W Indonesia 02°59′S 104°45′E
63 G18 **Palena** Los Lagos, S Chile 43°40′S 71°50′W
63 G18 **Palena, Río** ≈ S Chile
104 M5 **Palencia** *anc.* Palantia, Pallantia. Castilla y León, NW Spain 42°23′N 04°32′W
104 M3 **Palencia** ◈ *province* Castilla y León, N Spain
35 X15 **Palen Dry Lake** ◎ California, W USA
41 V15 **Palenque** Chiapas, SE Mexico 17°32′N 91°59′W
41 V15 **Palenque** *var.* Ruinas de Palenque. *ruins* Chiapas, SE Mexico
45 O9 **Palenque, Punta** *headland* S Dominican Republic 18°13′N 70°08′W
Palenque, Ruinas de *see* Palenque
Palerme *see* Palermo
107 I23 **Palermo** *Fr.* Palerme; *anc.* Panhormus, Panormus. Sicilia, Italy, C Mediterranean Sea 38°08′N 13°23′E
25 V8 **Palestine** Texas, SW USA 31°45′N 93°39′W
25 V7 **Palestine, Lake** ⊠ Texas, SW USA
107 I15 **Palestrina** Lazio, C Italy 41°49′N 12°53′E
166 K5 **Paletwa** Chin State, W Myanmar (Burma) 21°25′N 92°49′E
155 G21 **Pālghāt** *var.* Palakkad. Kerala, SW India 10°46′N 76°42′E *see also* Palakkad
152 F13 **Pāli** Rājasthān, N India 25°48′N 73°21′E
167 N16 **Palian** Trang, SW Thailand
189 O12 **Palikir** ● (Micronesia) Pohnpei, E Micronesia 06°58′N 158°13′E
Palimé *see* Kpalimé
107 L19 **Palinuro, Capo** *headland* S Italy 40°02′N 15°16′E
115 H15 **Palioúri, Akrotírio** *var.* Akrotírio Kanestron. *headland* N Greece 39°55′N 23°45′E
33 R14 **Palisades Reservoir** ⊠ Idaho, NW USA
99 J23 **Paliseul** Luxembourg, SE Belgium 49°55′N 05°09′E
154 C11 **Pālitāna** Gujarāt, W India 21°30′N 71°50′E
118 F4 **Palivere** Läänemaa, W Estonia 58°59′N 23°58′E
41 V14 **Palizada** Campeche, SE Mexico 18°15′N 92°03′W
93 L18 **Pälkäne** Pirkanmaa, W Finland 61°22′N 24°16′E
155 J22 **Palk Strait** *strait* India/Sri Lanka
155 J23 **Pallai** Northern Province, NW Sri Lanka 09°35′N 80°20′E
Pallantia *see* Palencia
106 C6 **Pallanza** Piemonte, NE Italy 45°57′N 08°32′E
127 Q9 **Pallasovka** Volgogradskaya Oblast', SW Russian Federation 50°06′N 46°52′E
Pallene/Pallíni *see* Kassándra
185 L15 **Palliser Bay** *bay* North Island, New Zealand
185 L15 **Palliser, Cape** *headland* North Island, New Zealand 41°37′S 175°16′E
191 U9 **Palliser, Îles** *island group* Îles Tuamotu, C French Polynesia
82 Q12 **Palma** Cabo Delgado, N Mozambique 10°46′S 40°30′E
105 X9 **Palma** *var.* Palma de Mallorca. Mallorca, Spain, W Mediterranean Sea 39°35′N 02°39′E
105 X9 **Palma** ✈ Mallorca, Spain, W Mediterranean Sea
105 X10 **Palma, Badia de** *bay* Mallorca, Spain, W Mediterranean Sea
104 L13 **Palma del Río** Andalucía, S Spain 37°42′N 05°16′W
Palma de Mallorca *see* Palma
107 J25 **Palma di Montechiaro** Sicilia, Italy, C Mediterranean Sea 37°12′N 13°46′E
106 J7 **Palmanova** Friuli-Venezia Giulia, NE Italy 45°54′N 13°20′E
54 J7 **Palmarito** Apure, C Venezuela 07°36′N 70°08′W
43 N15 **Palmar Sur** Puntarenas, SE Costa Rica 08°54′N 83°27′W
60 I12 **Palmas** Paraná, S Brazil 26°29′S 52°00′W
59 K16 **Palmas** *var.* Palmas do Tocantins. Tocantins, C Brazil 10°24′S 48°19′W
76 L18 **Palmas, Cape** *Fr.* Cap des Palmès. *headland* SW Ivory Coast 04°18′N 07°31′W
Palmas do Tocantins *see* Palmas
54 D11 **Palmaseca** ✈ (Cali) Valle del Cauca, SW Colombia 03°31′N 76°27′W
107 B21 **Palmas, Golfo di** *gulf* Sardegna, Italy, C Mediterranean Sea
44 I7 **Palma Soriano** Santiago de Cuba, E Cuba 20°10′N 76°00′W
23 Y12 **Palm Bay** Florida, SE USA 28°01′N 80°35′W
35 T14 **Palmdale** California, W USA 34°34′N 118°07′W
61 H14 **Palmeira das Missões** Rio Grande do Sul, S Brazil 27°54′S 53°20′W
82 A11 **Palmeirinhas, Ponta das** *headland* NW Angola 09°04′S 13°02′E
39 R11 **Palmer** Alaska, USA 61°36′N 149°06′W
19 N11 **Palmer** Massachusetts, NE USA 42°09′N 72°19′W
25 U7 **Palmer** Texas, SW USA 32°25′N 96°40′W
194 H4 **Palmer** *US research station* Antarctica 64°37′S 64°01′W
15 R11 **Palmer** ≈ Québec, SE Canada
37 T5 **Palmer Lake** Colorado, C USA 39°07′N 104°55′W
194 J6 **Palmer Land** *physical region* Antarctica
14 F15 **Palmerston** Ontario, SE Canada 43°51′N 80°49′W
185 F22 **Palmerston** Otago, South Island, New Zealand 45°27′S 170°42′E
190 K15 **Palmerston** *island* S Cook Islands
Palmerston *see* Darwin
184 M12 **Palmerston North** Manawatu-Wanganui, North Island, New Zealand 40°20′S 175°52′E
23 V13 **Palmetto** Florida, SE USA 27°31′N 82°34′W
The Palmetto State *see* South Carolina
107 M22 **Palmi** Calabria, SW Italy 38°21′N 15°51′E
54 D11 **Palmira** Valle del Cauca, W Colombia 03°33′N 76°17′W
56 F8 **Palmira, Río** ≈ N Peru
61 D19 **Palmitas** Soriano, SW Uruguay 33°27′S 57°48′W
Palmnicken *see* Yantarnyy
35 V15 **Palm Springs** California, W USA 33°48′N 116°33′W
27 V2 **Palmyra** Missouri, C USA 39°48′N 91°31′W
18 G10 **Palmyra** New York, NE USA 43°02′N 77°13′W
18 G15 **Palmyra** Pennsylvania, NE USA 40°18′N 76°35′W
21 V5 **Palmyra** Virginia, NE USA 37°53′N 78°17′W
Palmyra *see* Tudmur
192 L7 **Palmyra Atoll** ◇ *US incorporated territory* C Pacific Ocean
154 P12 **Palmyras Point** *headland* E India 20°46′N 87°00′E
35 N9 **Palo Alto** California, W USA 37°26′N 122°08′W
25 O1 **Palo Duro Creek** ≈ Texas, SW USA
Paloe *see* Denpasar, Bali, C Indonesia
Paloe *see* Palu
168 L9 **Paloh** Johor, Peninsular Malaysia 02°10′N 103°11′E
80 F12 **Paloich** Upper Nile, NE South Sudan 10°29′N 32°31′E
40 I3 **Palomas** Chihuahua, N Mexico 31°45′N 107°38′W
107 I15 **Palombara Sabina** Lazio, C Italy 42°04′N 12°45′E
105 S13 **Palos, Cabo de** *headland* SE Spain 37°38′N 00°42′W
104 I14 **Palos de la Frontera** Andalucía, S Spain 37°14′N 06°53′W
60 G11 **Palotina** Paraná, S Brazil 24°16′S 53°49′W
32 M9 **Palouse** Washington, NW USA 46°54′N 117°04′W
32 L9 **Palouse River** ≈ Washington, NW USA
35 Y16 **Palo Verde** California, W USA 33°25′N 114°43′W
57 E16 **Palpa** Ica, W Peru 14°35′S 75°09′W
95 M16 **Pålsboda** Örebro, C Sweden 59°04′N 15°21′E
93 M15 **Paltamo** Kainuu, C Finland 64°25′N 27°50′E
171 N12 **Palu** *prev.* Paloe. Sulawesi, C Indonesia 0°54′S 119°52′E
137 P14 **Palu** Elazığ, E Turkey 38°43′N 39°56′E
152 I11 **Palwal** Haryāna, N India 28°15′N 77°18′E
123 U6 **Palyavaam** ≈ NE Russian Federation
77 Q13 **Pama** SE Burkina Faso 11°13′N 00°46′E
172 J14 **Pamandzi** ✈ (Mamoudzou) Petite-Terre, E Mayotte
Pamangkat *see* Pemangkat
143 R11 **Pā Mazār** Kermān, C Iran
83 N19 **Pambarra** Inhambane, SE Mozambique 21°57′S 35°06′E
171 X12 **Pamdai** Papua, E Indonesia 01°58′S 137°19′E
103 N16 **Pamiers** Ariège, S France 43°07′N 01°37′E
147 T14 **Pamir** *var.* Daryā-ye Pāmīr, *Taj.* Dar''yoi Pomir. ≈ Afghanistan/Tajikistan *see also* Pāmīr, Daryā-ye
Pamir *see* Pāmīr, Daryā-ye
149 U1 **Pāmīr, Daryā-ye** *var.* Pamir, *Taj.* Dar''yoi Pomir. ≈ Afghanistan/Tajikistan *see also* Pamir
Pāmīr, Daryā-ye *see* Pamir
Pāmīr-e Khord *see* Little Pamir
Pamir/Pāmīr, Daryā-ye *see* Pamirs
129 Q8 **Pamirs** *Pash.* Daryā-ye Pāmīr, *Rus.* Pamir. ▲▲ C Asia
Pâmiut *see* Paamiut
21 X10 **Pamlico River** ≈ North Carolina, SE USA
21 Y10 **Pamlico Sound** *sound* North Carolina, SE USA
25 O2 **Pampa** Texas, SW USA 35°32′N 100°58′W
Pampa Aullagas, Lago *see* Poopó, Lago
61 B21 **Pampa Húmeda** *grassland* E Argentina
56 A10 **Pampa las Salinas** *salt lake* NW Peru
57 F15 **Pampas** Huancavelica, C Peru 12°22′S 74°53′W
62 K13 **Pampas** *plain* C Argentina
55 O4 **Pampatar** Nueva Esparta, NE Venezuela 11°03′N 63°51′W
Pampeluna *see* Pamplona

◆ Country
● Country Capital
◇ Dependent Territory
○ Dependent Territory Capital
◈ Administrative Regions
✈ International Airport
▲ Mountain
▲▲ Mountain Range
⛰ Volcano
≈ River
◎ Lake
⊠ Reservoir

PAMPILHOSA DA SERRA – PATNONGON

104 H8 **Pampilhosa da Serra** var. Pampilhosa de Serra. Coimbra, N Portugal 40°03′N 07°58′W
173 Y15 **Pamplemousses** N Mauritius 20°06′S 57°34′E
54 G7 **Pamplona** Norte de Santander, N Colombia 07°24′N 72°38′W
105 Q3 **Pamplona** Basq. Iruña, prev. Pampeluna; anc. Pompaelo. Navarra, N Spain 42°49′N 01°39′W
114 I11 **Pamporovo** prev. Vasil Kolarov. Smolyan, S Bulgaria 41°39′N 24°45′E
136 D15 **Pamukkale** Denizli, W Turkey 37°51′N 29°13′E
21 W5 **Pamunkey River** ≈ Virginia, NE USA
152 K5 **Pamzal** Jammu and Kashmir, NW India 34°17′N 78°50′E
30 L14 **Pana** Illinois, N USA 39°23′N 89°04′W
41 Y11 **Panabá** Yucatán, SE Mexico 21°20′N 88°16′W
35 Y8 **Panaca** Nevada, W USA 37°47′N 114°24′W
115 E19 **Panachaikó** ▲▲ S Greece
14 F11 **Panache Lake** ◎ Ontario, S Canada
114 I10 **Panagyurishte** Pazardzhik, C Bulgaria 42°30′N 24°11′E
168 M16 **Panaitan, Pulau** island S Indonesia
115 D18 **Panaitolikó** ▲▲ C Greece
155 E17 **Panaji** var. Pangim, Panjim, New Goa. state capital Goa, W India 15°31′N 73°52′E
43 T15 **Panamá** var. Ciudad de Panama, Eng. Panama City. ● (Panama) Panamá, C Panama 08°57′N 79°33′W
43 T14 **Panama** off. Republic of Panama. ◆ republic Central America
43 U14 **Panamá** off. Provincia de Panamá. ◈ province E Panama
43 U15 **Panamá, Bahía de** bay N Gulf of Panama
193 T7 **Panama Basin** undersea feature E Pacific Ocean 05°00′N 83°30′W
43 T15 **Panama Canal** canal E Panama
23 R9 **Panama City** Florida, SE USA 30°09′N 85°39′W
43 T14 **Panama City** ✈ Panamá, C Panama 09°02′N 79°24′W
Panama City see Panamá
23 Q9 **Panama City Beach** Florida, SE USA 30°10′N 85°48′W
43 T17 **Panamá, Golfo de** var. Gulf of Panama. gulf S Panama
Panama, Gulf of see Panamá, Golfo de
Panama, Isthmus of see Panama, Istmo de
43 T15 **Panama, Istmo de** Eng. Isthmus of Panama; prev. Isthmus of Darien. isthmus E Panama
Panamá, Provincia de see Panamá
Panama, Republic of see Panama
35 U11 **Panamint Range** ▲▲ California, W USA
107 L22 **Panarea, Isola** island Isole Eolie, S Italy
106 G9 **Panaro** ≈ N Italy
171 P5 **Panay Island** island C Philippines
35 W7 **Pancake Range** ▲▲ Nevada, W USA
112 M11 **Pančevo** Ger. Pantschowa, Hung. Pancsova. Vojvodina, N Serbia 44°53′N 20°40′E
113 M15 **Pančićev Vrh** ▲ SW Serbia 43°16′N 20°49′E
116 L12 **Panciu** Vrancea, E Romania 45°54′N 27°08′E
116 F10 **Pâncota** Hung. Pankota; prev. Pîncota. Arad, W Romania 46°20′N 21°45′E
Pancsova see Pančevo
83 N20 **Panda** Inhambane, SE Mozambique 24°02′S 34°45′E
171 X12 **Pandaidori, Kepulauan** island group E Indonesia
25 N11 **Pandale** Texas, SW USA 30°09′N 101°34′W
169 P12 **Pandang Tikar, Pulau** island N Indonesia
61 F20 **Pan de Azúcar** Maldonado, S Uruguay 34°45′S 55°14′W
118 H11 **Pandėlys** Panevėžys, NE Lithuania 56°04′N 25°18′E
155 F15 **Pandharpur** Mahārāshtra, W India 17°42′N 75°24′E
182 J1 **Pandie Pandie** South Australia 26°06′S 139°26′E
171 O12 **Pandiri** Sulawesi, C Indonesia 01°32′S 120°47′E
61 F20 **Pando** Canelones, S Uruguay 34°44′S 55°58′W
57 J14 **Pando** ◈ department N Bolivia
192 K9 **Pandora Bank** undersea feature W Pacific Ocean
95 G20 **Pandrup** Nordjylland, N Denmark 57°14′N 09°42′E
79 J15 **Pandu** Equateur, NW Dem. Rep. Congo 05°03′N 19°14′E
153 V12 **Pandu** Assam, NE India 26°08′N 91°37′E
Paneas see Bāniyās
59 F15 **Panelas** Mato Grosso, W Brazil 09°06′S 60°41′W
118 G12 **Panevėžys** Panevėžys, C Lithuania 55°44′N 24°21′E
118 G11 **Panevėžys** ◈ province NW Lithuania
Panfilov see Zharkent
127 N9 **Panfilovo** Volgogradskaya Oblast', SW Russian Federation 50°25′N 42°55′E
79 N17 **Panga** Orientale, N Dem. Rep. Congo 01°52′N 26°18′E
193 Y15 **Pangai** Lifuka, C Tonga 19°50′S 174°23′W
114 H13 **Pangaío** ▲▲ N Greece
79 G20 **Pangala** Pool, S Congo 03°26′S 14°38′E
81 J22 **Pangani** Tanga, E Tanzania 05°27′S 39°00′E
81 I21 **Pangani** ≈ NE Tanzania
186 K8 **Panggoe** Choiseul, NW Solomon Islands 07°00′S 157°05′E
79 N20 **Pangi** Maniema, E Dem. Rep. Congo 03°12′S 26°39′E
Pangim see Panaji
168 H8 **Pangkalanbrandan** Sumatera, W Indonesia 04°00′N 98°15′E
Pangkalanbun see Pangkalanbuun
169 R13 **Pangkalanbuun** var. Pangkalanbun. Borneo, C Indonesia 02°43′S 111°38′E
169 N12 **Pangkalpinang** Pulau Bangka, W Indonesia 02°05′S 106°09′E
11 U17 **Pangman** Saskatchewan, S Canada 49°37′N 104°33′W
Pang-Nga see Phang-Nga
9 S6 **Pangnirtung** Baffin Island, Nunavut, NE Canada 66°05′N 65°45′W
152 K6 **Pangong Tso** var. Bangong Co. ◎ China/India see also Bangong Co
Pangong Tso see Banggong Co
36 K7 **Panguitch** Utah, W USA 37°49′N 112°26′W
186 J7 **Panguna** Bougainville Island, NE Papua New Guinea 06°22′S 155°20′E
171 N8 **Pangutaran Group** island group Sulu Archipelago, SW Philippines
25 N2 **Panhandle** Texas, SW USA 35°21′N 101°24′W
Panhormus see Palermo
171 W14 **Paniai, Danau** ◎ Papua, E Indonesia
79 L21 **Pania-Mutombo** Kasai-Oriental, C Dem. Rep. Congo 05°09′S 23°49′E
187 P16 **Panié, Mont** ▲ C New Caledonia 20°33′S 164°41′E
Panikoilli see Jājapur
152 I10 **Pānīpat** Haryāna, N India 29°18′N 77°00′E
147 Q14 **Panj** Rus. Pyandzh; prev. Kirovabad. SW Tajikistan 37°39′N 69°55′E
147 P15 **Panj** Rus. Pyandzh. ≈ Afghanistan/Tajikistan
149 O5 **Panjāb** Bāmyān, C Afghanistan 34°21′N 67°00′E
147 O12 **Panjakent** Rus. Pendzhikent. W Tajikistan 39°28′N 67°33′E
148 L14 **Panjgūr** Baluchistān, SW Pakistan 26°58′N 64°05′E
Panjim see Panaji
163 U12 **Panjin** Liaoning, NE China 41°11′N 122°05′E
147 P14 **Panji Poyon** Rus. Nizhniy Pyandzh. SW Tajikistan 37°14′N 68°32′E
149 Q4 **Panjshayr** prev. Panjshīr. ≈ E Afghanistan
149 S4 **Panjshīr** ◈ province NE Afghanistan
Panjshīr see Panjshayr
Pankota see Pâncota
77 W14 **Pankshin** Plateau, C Nigeria 09°21′N 09°27′E
163 Y10 **Pan Ling** ▲▲ N China
Panlong Jiang see Lô, Sông
154 J9 **Panna** Madhya Pradesh, C India 24°43′N 80°11′E
99 M16 **Panningen** Limburg, SE Netherlands 51°20′N 05°59′E
149 R13 **Pāno Āqil** Sind, SE Pakistan 27°55′N 69°18′E
121 P3 **Páno Léfkara** S Cyprus 34°52′N 33°18′E
121 O3 **Páno Panagiá** var. Pano Panayia. W Cyprus 34°55′N 32°38′E
Pano Panayia see Páno Panagiá
Panopolis see Akhmīm
29 U14 **Panora** Iowa, C USA 41°41′N 94°21′W
60 I8 **Panorama** São Paulo, S Brazil 21°22′S 51°51′W
115 I24 **Pánormos** Kríti, Greece, E Mediterranean Sea 35°24′N 24°42′E
Panormus see Palermo
163 W11 **Panshi** Jilin, NE China 42°56′N 126°02′E
Panshi Yu see Passu Keah
59 H19 **Pantanal** var. Pantanalmato-Grossense. swamp SW Brazil
Pantanalmato-Grossense see Pantanal
61 H16 **Pântano Grande** Rio Grande do Sul, S Brazil 30°12′S 52°24′W
171 Q16 **Pantar, Pulau** island Kepulauan Alor, S Indonesia
21 X9 **Pantego** North Carolina, SE USA 35°34′N 76°39′E
107 G25 **Pantelleria** anc. Cossyra, Cosyra. Sicilia, Italy, C Mediterranean Sea 36°47′N 12°00′E
107 G25 **Pantelleria, Isola di** island SW Italy
Pante Makasar/Pante Macassar/Pante Makassar see Ponte Macassar
152 K10 **Pantnagar** Uttarakhand, N India 29°00′N 79°28′E
115 A15 **Pantokrátoras** ▲ Kérkyra, Iónia Nisiá, Greece, C Mediterranean Sea 39°45′N 19°51′E
Pantschowa see Pančevo
41 P11 **Pánuco** Veracruz-Llave, E Mexico 22°01′N 98°13′W
41 P11 **Pánuco, Río** ≈ C Mexico
160 I12 **Panxian** Guizhou, S China 25°45′N 104°39′E
168 I10 **Panyabungan** Sumatera, N Indonesia 0°55′N 99°30′E
77 W14 **Panyam** Plateau, C Nigeria 09°28′N 09°13′E
157 N13 **Panzhihua** prev. Dukou, Tu-k'ou. Sichuan, C China 26°35′N 101°41′E
79 I22 **Panzi** Bandundu, SW Dem. Rep. Congo 07°10′S 17°55′E
42 E5 **Panzós** Alta Verapaz, E Guatemala 15°21′N 89°40′W
Pao-chi/Paoki see Baoji
Pao-king see Shaoyang
107 N20 **Paola** Calabria, SW Italy 39°21′N 16°03′E
121 P16 **Paola** E Malta 35°52′N 14°30′E
27 R5 **Paola** Kansas, C USA 38°34′N 94°54′W
31 O15 **Paoli** Indiana, N USA 38°35′N 86°25′W
187 R14 **Paonangisu** Éfaté, C Vanuatu 17°33′S 168°23′E
171 S13 **Paoni** var. Pauni. Pulau Seram, E Indonesia 02°48′S 129°03′E
37 Q5 **Paonia** Colorado, C USA 38°52′N 107°35′W
191 O7 **Paopao** Moorea, W French Polynesia 17°28′S 149°48′W
Pao-shan see Baoshan
Pao-ting see Baoding
Pao-t'ou/Paotow see Baotou
79 H14 **Paoua** Ouham-Pendé, W Central African Republic 07°22′N 16°25′E
Pap see Pop
111 H23 **Pápa** Veszprém, W Hungary 47°20′N 17°29′E
42 J12 **Papagayo, Golfo de** gulf NW Costa Rica
38 H11 **Pāpa'ikou** var. Papaikou. Hawaii, USA, C Pacific Ocean 19°45′N 155°06′W
41 R15 **Papaloapan, Río** ≈ S Mexico
184 L6 **Papakura** Auckland, North Island, New Zealand 37°03′S 174°57′E
41 Q13 **Papantla** var. Papantla de Olarte. Veracruz-Llave, E Mexico 20°30′N 97°21′W
Papantla de Olarte see Papantla
191 P8 **Papara** Tahiti, W French Polynesia 17°45′S 149°33′W
184 K4 **Paparoa** Northland, North Island, New Zealand 36°06′S 174°12′E
185 G16 **Paparoa Range** ▲▲ South Island, New Zealand
115 K20 **Pápas, Akrotírio** headland Ikaría, Dodekánisa, Greece, Aegean Sea 37°31′N 25°58′E
96 L2 **Papa Stour** island NE Scotland, United Kingdom
184 L6 **Papatoetoe** Auckland, North Island, New Zealand 36°58′S 174°52′E
185 E25 **Papatowai** Otago, South Island, New Zealand 46°33′S 169°33′E
96 K4 **Papa Westray** island NE Scotland, United Kingdom
191 T10 **Papeete** ○ (French Polynesia) Tahiti, W French Polynesia 17°32′S 149°34′W
100 F11 **Papenburg** Niedersachsen, NW Germany 53°04′N 07°24′E
98 H13 **Papendrecht** Zuid-Holland, SW Netherlands 51°50′N 04°42′E
191 Q7 **Papenoo** Tahiti, W French Polynesia 17°29′S 149°25′W
191 Q7 **Papenoo Rivière** ≈ Tahiti, W French Polynesia
191 N7 **Papetoai** Moorea, W French Polynesia 17°29′S 149°52′W
92 L3 **Papey** island E Iceland
Paphos see Páfos
40 H5 **Papigochic, Río** ≈ NW Mexico
118 E10 **Papilė** Šiauliai, NW Lithuania 56°08′N 22°51′E
29 S15 **Papillion** Nebraska, C USA 41°09′N 96°02′W
15 T5 **Papinachois** ≈ Québec, SE Canada
171 X13 **Papua** var. Irian Barat, West Irian, West New Guinea, West Papua; prev. Dutch New Guinea, Irian Jaya, Netherlands New Guinea. ◈ province E Indonesia
Papua and New Guinea, Territory of see Papua New Guinea
171 V10 **Papua Barat** off. Propinsi Papua Barat; prev. Irian Jaya Barat, Eng. West Papua. ◈ province E Indonesia
186 C9 **Papua, Gulf of** gulf S Papua New Guinea
186 C8 **Papua New Guinea** off. Independent State of Papua New Guinea; prev. Territory of Papua and New Guinea. ◆ commonwealth republic NW Melanesia
Papua New Guinea, Independent State of see Papua New Guinea
192 H8 **Papua Plateau** undersea feature N Coral Sea
112 G9 **Papuk** ▲▲ NE Croatia
Papun see Hpapun
42 L14 **Paquera** Puntarenas, W Costa Rica 09°52′N 84°56′W
58 I13 **Pará** off. Estado do Pará. ◈ state NE Brazil
55 V9 **Para** ◈ district N Suriname
Pará see Belém
180 I8 **Paraburdoo** Western Australia 23°07′S 117°40′E
57 E16 **Paracas, Península de** peninsula W Peru
59 L19 **Paracatu** Minas Gerais, NE Brazil 17°14′S 46°52′W
192 E6 **Paracel Islands** Chin. Xisha Qundao, Viet. Quân Đao Hoang Sa. ◇ disputed territory SE Asia
182 I6 **Parachilna** South Australia 31°09′S 138°23′E
149 R6 **Pārachinār** Khyber Pakhtunkhwa, NW Pakistan 33°56′N 70°04′E
112 N13 **Paraćin** Serbia, C Serbia 43°51′N 21°25′E
Paradip see Krishnarājāsāgara
14 K8 **Paradis** Québec, SE Canada 48°13′N 76°36′W
39 N11 **Paradise** var. Paradise Hill. Alaska, USA 62°28′N 160°09′W
35 O5 **Paradise** California, W USA 39°42′N 121°39′W
35 X11 **Paradise** Nevada, W USA 36°05′N 115°10′W
Paradise Hill see Paradise
37 R11 **Paradise Hills** New Mexico, SW USA 35°12′N 106°42′W
Paradise of the Pacific see Hawai'i
36 L13 **Paradise Valley** Arizona, SW USA 33°31′N 111°56′W
35 T2 **Paradise Valley** Nevada, W USA 41°30′N 117°30′W
115 O22 **Paradisi** ✈ (Ródos) Ródos, Dodekánisa, Greece, Aegean Sea 36°24′N 28°08′E
154 P12 **Pārādwīp** Odisha, E India 20°17′N 86°42′E
Pará, Estado do see Pará
Paraetonium see Marsā Maṭroūḥ
117 R4 **Parafiyivka** Chernihivs'ka Oblast', N Ukraine 50°53′N 32°40′E
36 K7 **Paragonah** Utah, W USA 37°53′N 112°46′W
27 X9 **Paragould** Arkansas, C USA 36°02′N 90°30′W
47 X8 **Paraguaçu** var. Paraguassú. ≈ E Brazil
60 J9 **Paraguaçu Paulista** São Paulo, S Brazil 22°22′S 50°35′W
54 H4 **Paraguaipoa** Zulia, NW Venezuela 11°21′N 71°58′W
62 O6 **Paraguarí** Paraguarí, S Paraguay 25°36′S 57°06′W
62 O7 **Paraguarí** off. Departamento de Paraguarí. ◈ department S Paraguay
Paraguarí, Departamento de see Paraguarí
57 O16 **Paraguá, Río** ≈ NE Bolivia
55 O8 **Paragua, Río** ≈ SE Venezuela
Paraguassú see Paraguaçu
62 N5 **Paraguay** ◆ republic C South America
47 U10 **Paraguay** var. Río Paraguay. ≈ C South America
Paraguay, Río see Paraguay
Parahiba/Parahyba see Paraíba
59 P15 **Paraíba** off. Estado da Paraíba; prev. Parahiba, Parahyba. ◈ state E Brazil
Paraíba see João Pessoa
60 P9 **Paraíba do Sul, Rio** ≈ SE Brazil
Paraíba, Estado da see Paraíba
Parainen see Pargas
43 N14 **Paraíso** Cartago, C Costa Rica 09°51′N 83°50′W
41 U14 **Paraíso** Tabasco, SE Mexico 18°26′N 93°10′W
57 O17 **Paraiso, Río** ≈ E Bolivia
Parajd see Praid
77 S14 **Parakou** C Benin 09°23′N 02°40′E
115 F20 **Paralía Tyrou** Pelopónnisos, S Greece 37°17′N 22°50′E
121 Q2 **Paralímni** E Cyprus 35°02′N 34°00′E
115 G18 **Paralímni, Límni** ◎ C Greece
55 W8 **Paramaribo** ● (Suriname) Paramaribo, N Suriname 05°52′N 55°14′W
55 W9 **Paramaribo** ◈ district N Suriname
55 W9 **Paramaribo** ✈ Paramaribo, N Suriname 05°52′N 55°14′W
Paramithiá see Paramythiá
56 C13 **Paramonga** Lima, W Peru 10°42′S 77°50′W
123 V12 **Paramushir, Ostrov** island SE Russian Federation
115 C16 **Paramythiá** var. Paramithiá. Ípeiros, W Greece 39°28′N 20°31′E
62 M10 **Paraná** Entre Ríos, E Argentina 31°48′S 60°29′W
60 H11 **Paraná** off. Estado do Paraná. ◈ state S Brazil
47 U11 **Paraná** var. Alto Paraná. ≈ C South America
Paraná, Estado do see Paraná
60 K12 **Paranaguá** Paraná, S Brazil 25°32′S 48°36′W
59 J20 **Paranaíba, Rio** ≈ E Brazil
61 C19 **Paraná Ibicuy, Río** ≈ E Argentina
59 H15 **Paranaíta** Mato Grosso, W Brazil 09°35′S 57°01′W
60 H9 **Paranapanema, Rio** ≈ S Brazil
60 K11 **Paranapiacaba, Serra do** ▲▲ S Brazil
60 H9 **Paranavaí** Paraná, S Brazil 23°02′S 52°36′W
143 N5 **Parandak** Markazī, W Iran 35°19′N 50°40′E
114 I12 **Paranésti** var. Paranestio. Anatolikí Makedonía kai Thráki, NE Greece 41°16′N 24°31′E
Paranestio see Paranésti
191 W11 **Paraoa** atoll Îles Tuamotu, C French Polynesia
184 L13 **Paraparaumu** Wellington, North Island, New Zealand 40°55′S 175°01′E
57 N20 **Parapeti, Río** ≈ SE Bolivia
54 L10 **Paraque, Cerro** ▲ W Venezuela 06°00′S 67°00′W
154 I11 **Parāsiya** Madhya Pradesh, C India 22°11′N 78°50′E
115 M23 **Paraspóri, Akrotírio** headland Kárpathos, SE Greece 35°54′N 27°15′E
60 O10 **Parati** Rio de Janeiro, SE Brazil 23°15′S 44°42′W
59 K14 **Parauapebas** Pará, N Brazil 06°03′S 49°48′W
103 Q10 **Paray-le-Monial** Saône-et-Loire, C France 46°27′N 04°07′E
Parbatsar see Parvatsar
154 G13 **Parbhani** Mahārāshtra, C India 19°16′N 76°51′E
100 L10 **Parchim** Mecklenburg-Vorpommern, N Germany 53°26′N 11°51′E
Parchwitz see Prochowice
110 P13 **Parczew** Lubelskie, E Poland 51°40′N 23°E
60 L8 **Pardo, Rio** ≈ S Brazil
111 E16 **Pardubice** Ger. Pardubitz. Pardubický Kraj, C Czech Republic 50°01′N 15°47′E
111 E17 **Pardubický Kraj** ◈ region N Czech Republic
Pardubitz see Pardubice
119 F16 **Parechcha** Pol. Porzecze, Rus. Porech'ye. Hrodzyenskaya Voblasts', W Belarus 53°53′N 24°08′E
59 F17 **Parecis, Chapada dos** var. Serra dos Parecis. ▲▲ W Brazil
Parecis, Serra dos see Parecis, Chapada dos
104 M4 **Paredes de Nava** Castilla y León, N Spain 42°09′N 04°42′W
189 U12 **Parem** island Chuuk, C Micronesia
189 O12 **Parem Island** island E Micronesia
184 I1 **Parengarenga Harbour** inlet North Island, New Zealand
15 N8 **Parent** Québec, SE Canada 47°55′N 74°36′W
102 J14 **Parentis-en-Born** Landes, SW France 44°22′N 01°04′W
Parenzo see Poreč
185 G20 **Pareora** Canterbury, South Island, New Zealand 44°28′S 171°12′E
171 N14 **Parepare** Sulawesi, C Indonesia 04°S 119°40′E
115 B16 **Párga** Ípeiros, W Greece 39°18′N 20°19′E
93 K20 **Pargas** Swe. Parainen. Varsinais-Suomi, SW Finland 60°18′N 22°20′E
64 O5 **Pargo, Ponta do** headland Madeira, Portugal, NE Atlantic Ocean 32°48′N 17°17′W
Paria, Golfo de see Paria, Gulf of
55 N6 **Pariaguán** Anzoátegui, NE Venezuela 08°51′N 64°43′W
45 X17 **Paria, Gulf of** var. Golfo de Paria. gulf Trinidad and Tobago/Venezuela
57 I15 **Pariamanu, Río** ≈ E Peru
36 L8 **Paria River** ≈ Utah, W USA
Parichi see Parychy
40 M14 **Paricutín, Volcán** ☗ C Mexico 19°25′N 102°20′W
43 P16 **Parida, Isla** island SW Panama
55 T8 **Parika** NE Guyana 06°51′N 58°25′W
93 O18 **Parikkala** Etelä-Karjala, SE Finland 61°33′N 29°34′E
58 E10 **Parima, Serra** var. Sierra Parima. ▲ Brazil/Venezuela see also Parima, Sierra
55 N11 **Parima, Sierra** var. Serra Parima. ▲ Brazil/Venezuela see also Parima, Serra
57 F17 **Parinacochas, Laguna** ◎ SW Peru
56 A9 **Pariñas, Punta** headland NW Peru 04°45′S 81°22′W
58 H12 **Parintins** Amazonas, N Brazil 02°38′S 56°45′W
103 O5 **Paris** anc. Lutetia, Lutetia Parisiorum, Parisii. ● (France) Paris, N France 48°52′N 02°19′E
191 Y2 **Paris** Kiritimati, E Kiribati 01°55′N 157°30′W
27 S11 **Paris** Arkansas, C USA 35°17′N 93°46′W
33 S16 **Paris** Idaho, NW USA 42°14′N 111°24′W
31 N14 **Paris** Illinois, N USA 39°36′N 87°42′W
20 M5 **Paris** Kentucky, S USA 38°13′N 84°15′W
27 V3 **Paris** Missouri, C USA 39°28′N 92°00′W
20 H8 **Paris** Tennessee, S USA 36°19′N 88°20′W
25 V5 **Paris** Texas, SW USA 33°41′N 95°33′W
Parisii see Paris
43 S16 **Parita** Herrera, S Panama 08°01′N 80°30′W
43 S16 **Parita, Bahía de** bay S Panama
93 K18 **Parkano** Pirkanmaa, W Finland 62°03′N 23°E
Parkan/Párkány see Štúrovo
27 N6 **Park City** Kansas, C USA 37°48′N 97°19′W
36 L3 **Park City** Utah, W USA 40°39′N 111°30′W
36 I12 **Parker** Arizona, SW USA 34°07′N 114°16′W
23 R9 **Parker** Florida, SE USA 30°07′N 85°36′W
29 R11 **Parker** South Dakota, N USA 43°24′N 97°08′W
35 Z14 **Parker Dam** California, W USA 34°17′N 114°08′W
29 W13 **Parkersburg** Iowa, C USA 42°34′N 92°47′W
21 Q3 **Parkersburg** West Virginia, NE USA 39°17′N 81°33′W
29 T7 **Parkers Prairie** Minnesota, N USA 46°09′N 95°19′W
171 P8 **Parker Volcano** ☗ Mindanao, S Philippines 06°09′N 124°52′E
181 W13 **Parkes** New South Wales, SE Australia 33°10′S 148°10′E
30 K4 **Park Falls** Wisconsin, N USA 45°57′N 90°25′W
Parkhar see Farkhor
14 E16 **Parkhill** Ontario, S Canada 43°11′N 81°39′W
29 T5 **Park Rapids** Minnesota, N USA 46°55′N 95°03′W
29 Q3 **Park River** North Dakota, N USA 48°24′N 97°44′W
29 Q11 **Parkston** South Dakota, N USA 43°24′N 97°58′W
10 L17 **Parksville** Vancouver Island, British Columbia, SW Canada 49°13′N 124°13′W
37 S3 **Parkview Mountain** ▲ Colorado, C USA 40°19′N 106°08′W
105 N8 **Parla** Madrid, C Spain 40°13′N 03°48′W
29 S8 **Parle, Lac qui** ⊠ Minnesota, N USA
155 G14 **Parli Vaijnāth** Mahārāshtra, C India 18°53′N 76°36′E
106 F9 **Parma** Emilia-Romagna, N Italy 44°50′N 10°20′E
31 T11 **Parma** Ohio, N USA 41°24′N 81°43′W
Parnahyba see Parnaíba
58 N13 **Parnaíba** var. Parnahyba. Piauí, E Brazil 02°58′S 41°46′W
65 J14 **Parnaíba Ridge** undersea feature C Atlantic Ocean
58 N13 **Parnaíba, Rio** ≈ NE Brazil
115 F18 **Parnassós** ▲▲ C Greece
185 J17 **Parnassus** Canterbury, South Island, New Zealand 42°41′S 173°18′E
182 H10 **Parndana** South Australia 35°48′S 137°13′E
115 H19 **Párnitha** ▲▲ C Greece
Parnon see Párnonas
115 F21 **Párnonas** var. Parnon. ▲▲ S Greece
118 G5 **Pärnu** Ger. Pernau, Latv. Pērnava; prev. Rus. Pernov. Pärnumaa, SW Estonia 58°24′N 24°32′E
118 G6 **Pärnu** var. Parnu Jõgi, Ger. Pernau. ≈ SW Estonia
118 G5 **Pärnu-Jaagupi** Ger. Sankt-Jakobi. Pärnumaa, SW Estonia 58°36′N 24°30′E
Parnu Jõgi see Pärnu
118 G5 **Pärnu Laht** Ger. Pernauer Bucht. bay SW Estonia
118 F5 **Pärnumaa** var. Pärnu Maakond. ◈ province SW Estonia
Pärnu Maakond see Pärnumaa
153 T11 **Paro** W Bhutan 27°23′N 89°31′E
153 T11 **Paro** ✈ (Thimphu) W Bhutan 27°23′N 89°31′E
185 G17 **Paroa** West Coast, South Island, New Zealand 42°31′S 171°10′E
163 X14 **Paro-ho** var. Hwach'ŏn-chŏsuji; prev. P'aro-ho. ⊠ N South Korea
P'aro-ho see Paro-ho
115 J21 **Pároikiá** prev. Páros. Páros, Kykládes, Greece, Aegean Sea 37°04′N 25°06′E
183 N6 **Paroo River** seasonal river New South Wales/Queensland, SE Australia
Paropamisus Range see Sefīd Kūh, Selseleh-ye
Paropamisus Range see Sefīd Kūh, Selseleh-ye
115 J21 **Páros** island Kykládes, Greece, Aegean Sea
Páros see Pároikiá
36 K7 **Parowan** Utah, W USA 37°50′N 112°49′W
103 U13 **Parpaillon** ▲▲ SE France
108 I9 **Parpan** Graubünden, S Switzerland 46°46′N 09°32′E
62 G13 **Parral** Maule, C Chile 36°08′S 71°52′W
Parral see Hidalgo del Parral
183 T9 **Parramatta** New South Wales, SE Australia 33°49′S 150°59′E
21 Y6 **Parramore Island** island Virginia, NE USA
40 M8 **Parras** var. Parras de la Fuente. Coahuila, NE Mexico 25°26′N 102°07′W
Parras de la Fuente see Parras
42 M14 **Parrita** Puntarenas, S Costa Rica 09°30′N 84°20′W
Parry group see Mukojima-rettō
14 G13 **Parry Island** island Ontario, S Canada
197 O9 **Parry Islands** island group Nunavut, NW Canada
14 G12 **Parry Sound** Ontario, S Canada 45°21′N 80°03′W
110 F7 **Parsęta** Ger. Persante. ≈ NW Poland
28 L3 **Parshall** North Dakota, N USA 47°57′N 102°07′W
27 Q7 **Parsons** Kansas, C USA 37°20′N 95°15′W
20 H9 **Parsons** Tennessee, S USA 35°39′N 88°07′W
21 T3 **Parsons** West Virginia, NE USA 39°06′N 79°43′W
Parsonstown see Birr
100 P11 **Parsteiner See** ◎ NE Germany
107 I24 **Partanna** Sicilia, Italy, C Mediterranean Sea 37°43′N 12°54′E
108 J8 **Partenen** Graubünden, E Switzerland 46°58′N 10°01′E
102 K9 **Parthenay** Deux-Sèvres, W France 46°39′N 00°13′W
95 J19 **Partille** Västra Götaland, S Sweden 57°43′N 12°12′E
107 I23 **Partinico** Sicilia, Italy, C Mediterranean Sea 38°03′N 13°07′E
111 I20 **Partizánske** prev. Šimonovany, Hung. Simony. Trenčiansky Kraj, W Slovakia 48°35′N 18°23′E
58 H11 **Paru de Oeste, Rio** ≈ N Brazil
182 K9 **Paruna** South Australia 34°45′S 140°43′E
58 I11 **Paru, Rio** ≈ N Brazil
Parvān see Parwān
155 M14 **Pārvatipuram** Andhra Pradesh, E India 17°01′N 81°47′E
152 G12 **Parvatsar** prev. Parbatsar. Rājasthān, N India 26°52′N 74°49′E
114 J11 **Parvomay**, Pŭrvomay; prev. Borisovgrad. Plovdiv, C Bulgaria 42°06′N 25°13′E
149 Q5 **Parwān** prev. Parvān. E Afghanistan
158 I15 **Paryang** Xizang Zizhiqu, W China 30°04′N 83°28′E
119 M18 **Parychy** Rus. Parichi. Homyel'skaya Voblasts', SE Belarus 52°48′N 29°25′E
83 J21 **Parys** Free State, C South Africa 26°55′S 27°28′E
35 T15 **Pasadena** California, W USA 34°09′N 118°09′W
25 W11 **Pasadena** Texas, SW USA 29°41′N 95°13′W
56 B8 **Pasaje** El Oro, SW Ecuador 03°23′S 79°50′W
137 T9 **Pasanauri** prev. P'asanauri. N Georgia 42°21′N 44°40′E
P'asanauri see Pasanauri
168 I13 **Pasapuat** Pulau Pagai Utara, W Indonesia 02°36′S 99°58′E
Pasawng see Hpasawng
114 L13 **Paşayiğit** Edirne, NW Turkey 40°58′N 26°38′E
23 N9 **Pascagoula** Mississippi, S USA 30°21′N 88°32′W
22 M8 **Pascagoula River** ≈ Mississippi, S USA
116 F12 **Paşcani** Hung. Páskán. Iaşi, NE Romania 47°14′N 26°46′E
109 T4 **Pasching** Oberösterreich, N Austria 48°16′N 14°10′E
32 K10 **Pasco** Washington, NW USA 46°13′N 119°06′W
56 E13 **Pasco** off. Departamento de Pasco. ◈ department C Peru
Pasco, Departamento de see Pasco
191 N11 **Pascua, Isla de** var. Rapa Nui, Easter Island. island E Pacific Ocean
63 G21 **Pascua, Río** ≈ S Chile
103 N1 **Pas-de-Calais** ◈ department N France
100 P10 **Pasewalk** Mecklenburg-Vorpommern, NE Germany 53°31′N 13°59′E
11 T10 **Pasfield Lake** ◎ Saskatchewan, C Canada
Pa-shih Hai-hsia see Bashi Channel
Pashkeni see Bolyarovo
Pashmakli see Smolyan
153 X10 **Pāsighāt** Arunāchel Pradesh, NE India 28°08′N 95°13′E
137 Q12 **Pasinler** Erzurum, NE Turkey 39°59′N 41°41′E
Pasi Oloy, Qatorkŭhi see Zaalayskiy Khrebet
42 E3 **Pasión, Río de la** ≈ N Guatemala
168 J12 **Pasirganting** Sumatera, W Indonesia 02°04′S 100°51′E
Pasirpangarayan see Bagansiapiapi
168 K6 **Pasir Puteh** var. Pasir Putih. Kelantan, Peninsular Malaysia 05°50′N 102°24′E
Pasir Putih see Pasir Puteh
169 R9 **Pasir, Tanjung** headland East Malaysia 02°24′N 111°12′E
95 N20 **Påskallavik** Kalmar, S Sweden 57°10′N 16°25′E
Páskán see Paşcani
Paskevicha, Zaliv see Tushchybas, Zaliv
110 K7 **Pasłęk** Ger. Preußisch Holland. Warmińsko-Mazurskie, NE Poland 54°03′N 19°40′E
110 K7 **Pasłęka** Ger. Passarge. ≈ N Poland
148 K16 **Pasni** Baluchistān, SW Pakistan 25°13′N 63°30′E
63 I18 **Paso de Indios** Chubut, S Argentina 43°52′S 69°06′W
54 L7 **Paso del Caballo** Guárico, N Venezuela 08°19′N 67°08′W
61 E15 **Paso de los Libres** Corrientes, NE Argentina 29°43′S 57°09′W
61 E18 **Paso de los Toros** Tacuarembó, C Uruguay 32°45′S 56°30′W
Pasoeroean see Pasuruan
35 P12 **Paso Robles** California, W USA 35°37′N 120°42′W
15 Y7 **Paspébiac** Québec, SE Canada 48°03′N 65°10′W
11 U14 **Pasquia Hills** ▲▲ Saskatchewan, S Canada
149 W7 **Pasrūr** Punjab, E Pakistan 32°12′N 74°42′E
30 M1 **Passage Island** island Michigan, N USA
65 B24 **Passage Islands** island group W Falkland Islands
8 K5 **Passage Point** headland Banks Island, Northwest Territories, NW Canada 73°31′N 115°12′W
Passarge see Pasłęka
115 C15 **Passarón** ancient monument Ípeiros, W Greece
Passarowitz see Požarevac
101 O22 **Passau** Bayern, SE Germany 48°34′N 13°28′E
22 M9 **Pass Christian** Mississippi, S USA 30°19′N 89°15′W
107 L26 **Passero, Capo** headland Sicilia, Italy, C Mediterranean Sea 36°40′N 15°09′E
171 P5 **Passi** Panay Island, C Philippines 11°05′N 122°37′E
61 H14 **Passo Fundo** Rio Grande do Sul, S Brazil 28°16′S 52°20′W
60 H13 **Passo Fundo, Barragem de** ⊠ S Brazil
61 H15 **Passo Real, Barragem de** ⊠ S Brazil
59 L20 **Passos** Minas Gerais, NE Brazil 20°45′S 46°38′W
167 X10 **Passu Keah** Chin. Panshi Yu, Viet. Đao Bach Quy. island S Paracel Islands
118 J13 **Pastavy** Pol. Postawy, Rus. Postavy. Vitsyebskaya Voblasts', NW Belarus 55°07′N 26°50′E
56 D7 **Pastaza** ◈ province E Ecuador
56 D9 **Pastaza, Río** ≈ Ecuador/Peru
61 A21 **Pasteur** Buenos Aires, E Argentina 35°10′S 62°14′W
15 V3 **Pasteur** ≈ Québec, SE Canada
147 Q12 **Pastigav** Rus. Pastigov. W Tajikistan 39°27′N 69°16′E
Pastigov see Pastigav
54 C13 **Pasto** Nariño, SW Colombia 01°12′N 77°17′W
38 M10 **Pastol Bay** bay Alaska, USA
37 O8 **Pastora Peak** ▲ Arizona, SW USA 36°48′N 109°10′W
105 O8 **Pastrana** Castilla-La Mancha, C Spain 40°24′N 02°55′W
169 S16 **Pasuruan** prev. Pasoeroean. Jawa, C Indonesia 07°38′S 112°44′E
118 F11 **Pasvalys** Panevėžys, N Lithuania 56°03′N 24°24′E
111 K21 **Pásztó** Nógrád, N Hungary 47°57′N 19°41′E
189 U12 **Pata** var. Patta. atoll Chuuk Islands, C Micronesia
36 M16 **Patagonia** Arizona, SW USA 31°32′N 110°45′W
63 H20 **Patagonia** physical region Argentina/Chile
Patalung see Phatthalung
154 D9 **Pātan** Gujarāt, W India 23°51′N 72°11′E
154 J10 **Pātan** Madhya Pradesh, C India 23°19′N 79°41′E
171 S11 **Patani** Pulau Halmahera, E Indonesia 0°19′N 128°46′E
Patani see Pattani
15 V7 **Patapédia Est** ≈ Québec, SE Canada
116 K13 **Pătâriagele** prev. Pătîriagele. Buzău, SE Romania 45°19′N 26°21′E
Patavium see Padova
182 I5 **Patawarta Hill** ▲ South Australia 30°57′S 138°42′E
182 L10 **Patchewollock** Victoria, SE Australia 35°24′S 142°11′E
184 K11 **Patea** Taranaki, North Island, New Zealand 39°48′S 174°35′E
184 K11 **Patea** ≈ North Island, New Zealand
77 U15 **Pategi** Kwara, C Nigeria 08°39′N 05°46′E
81 K20 **Pate Island** var. Patta Island. island SE Kenya
105 S10 **Paterna** Valenciana, E Spain 39°30′N 00°24′W
109 R9 **Paternion** Slvn. Špatrjan. Kärnten, S Austria 46°40′N 13°43′E
107 L24 **Paternò** anc. Hybla, Hybla Major. Sicilia, Italy, C Mediterranean Sea 37°34′N 14°55′E
32 J7 **Pateros** Washington, NW USA 48°01′N 119°55′W
18 J14 **Paterson** New Jersey, NE USA 40°55′N 74°12′W
32 J10 **Paterson** Washington, NW USA 45°56′N 119°37′W
185 C25 **Paterson Inlet** inlet Stewart Island, New Zealand
98 N6 **Paterswolde** Drenthe, NE Netherlands 53°07′N 06°32′E
152 H7 **Pathānkot** Himāchal Pradesh, N India 32°16′N 75°43′E
166 K8 **Pathein** var. Bassein. Ayeyawady, SW Myanmar (Burma) 16°46′N 94°45′E
33 W15 **Pathfinder Reservoir** ⊠ Wyoming, C USA
167 O11 **Pathum Thani** var. Patumdhani, Prathum Thani. Pathum Thani, C Thailand 14°03′N 100°29′E
54 C12 **Patía** var. El Bordo. Cauca, SW Colombia 02°07′N 76°57′W
152 I9 **Patiāla** var. Puttiala. Punjab, NW India 30°21′N 76°27′E
54 B12 **Patía, Río** ≈ SW Colombia
188 D15 **Pati Point** headland NE Guam 13°36′N 144°39′E
Pătîriagele see Pătâriagele
56 C13 **Pativilca** Lima, W Peru 10°44′S 77°45′W
166 M1 **Patkai Bum** var. Patkai Range. ▲▲ Myanmar (Burma)/India
Patkai Range see Patkai Bum
115 L20 **Pátmos** Pátmos, Dodekánisa, Greece, Aegean Sea 37°18′N 26°32′E
115 L20 **Pátmos** island Dodekánisa, Greece, Aegean Sea
153 P13 **Patna** var. Azimabad. state capital Bihār, N India 25°36′N 85°11′E
154 M12 **Patnāgarh** Odisha, E India 20°42′N 83°12′E
171 O5 **Patnongon** Panay Island, C Philippines 10°56′N 122°03′E

◆ Country
● Country Capital
◇ Dependent Territory
○ Dependent Territory Capital
◈ Administrative Regions
✈ International Airport
▲ Mountain
▲▲ Mountain Range
☗ Volcano
≈ River
◎ Lake
⊠ Reservoir

137 S13 **Patnos** Ağrı, E Turkey 39°14′N 42°52′E
60 H12 **Pato Branco** Paraná, S Brazil 26°20′S 52°40′W
31 O16 **Patoka Lake** ◎ Indiana, N USA
92 L9 **Patoniva** *Lapp.* Buoddobohki. Lappi, N Finland 69°44′N 27°01′E
113 K21 **Patos** *var.* Patosi. Fier, SW Albania 40°40′N 19°37′E
Patos *see* Patos de Minas
59 K19 **Patos de Minas** *var.* Patos. Minas Gerais, NE Brazil 18°35′S 46°32′W
Patosi *see* Patos
61 I17 **Patos, Lagoa dos** *lagoon* S Brazil
62 J9 **Patquía** La Rioja, C Argentina 30°02′S 66°54′W
115 E19 **Pátra** *Eng.* Patras; *prev.* Pátrai. Dytikí Elláda, S Greece 38°14′N 21°45′E
115 D18 **Patraïkós Kólpos** *gulf* S Greece
Pátrai/Patras *see* Pátra
92 G2 **Patreksfjörður** Vestfirðir, W Iceland 65°33′N 23°54′W
24 M7 **Patricia** Texas, SW USA 32°34′N 102°00′W
63 F21 **Patricio Lynch, Isla** *island* S Chile
Patta *see* Pata
Patta Island *see* Pate Island
167 O16 **Pattani** *var.* Patani. Pattani, SW Thailand 06°50′N 101°20′E
167 P12 **Pattaya** Chon Buri, S Thailand 12°57′N 100°53′E
19 S4 **Patten** Maine, NE USA 45°58′N 68°27′W
35 O9 **Patterson** California, W USA 37°27′N 121°07′W
22 J10 **Patterson** Louisiana, S USA 29°41′N 91°18′W
35 R7 **Patterson, Mount** ▲ California, W USA 38°27′N 119°16′W
31 P4 **Patterson, Point** *headland* Michigan, N USA 45°58′N 85°39′W
107 L23 **Patti** Sicilia, Italy, C Mediterranean Sea 38°08′N 14°58′E
107 L23 **Patti, Golfo di** *gulf* Sicilia, Italy
93 L14 **Pattijoki** Pohjois-Pohjanmaa, W Finland 64°41′N 24°40′E
193 Q4 **Patton Escarpment** *undersea feature* E Pacific Ocean
27 S2 **Pattonsburg** Missouri, C USA 40°03′N 94°08′W
0 D6 **Patton Seamount** *undersea feature* NE Pacific Ocean 54°40′N 150°30′W
10 J12 **Pattullo, Mount** ▲ British Columbia, W Canada 56°18′N 129°43′W
153 U16 **Patuakhali** *var.* Patukhali. Barisal, S Bangladesh 22°20′N 90°20′E
42 M5 **Patuca, Río** ≈ E Honduras
Patukhali *see* Patuakhali
Patumdhani *see* Pathum Thani
40 M14 **Pátzcuaro** Michoacán, SW Mexico 19°30′N 101°38′W
42 C6 **Patzicía** Chimaltenango, S Guatemala 14°38′N 90°52′W
102 K16 **Pau** Pyrénées-Atlantiques, SW France 43°18′N 00°22′W
102 J12 **Pauillac** Gironde, SW France 45°12′N 00°44′W
166 L5 **Pauk** Magway, W Myanmar (Burma) 21°25′N 94°30′E
8 I6 **Paulatuk** Northwest Territories, NW Canada 69°23′N 124°W
42 K5 **Paulayá, Río** ≈ NE Honduras
22 M6 **Paulding** Mississippi, S USA 32°01′N 89°01′W
31 Q12 **Paulding** Ohio, N USA 41°08′N 84°34′W
29 S12 **Paullina** Iowa, C USA 42°58′N 95°41′W
59 P15 **Paulo Afonso** Bahia, E Brazil 09°21′S 38°14′W
38 M16 **Pauloff Harbor** *var.* Pavlor Harbour. Sanak Island, Alaska, USA 54°26′N 162°43′W
27 N12 **Pauls Valley** Oklahoma, C USA 34°46′N 97°14′W
166 L7 **Paungde** Bago, C Myanmar (Burma) 18°30′N 95°30′E
Pauni *see* Paoni
152 K9 **Pauri** Uttaranchal, N India 30°08′N 78°48′E
Pautalia *see* Kyustendil
142 J5 **Pāveh** Kermānshāhān, NW Iran 35°02′N 46°15′E
114 I9 **Pavel Banya** Stara Zagora, C Bulgaria 42°35′N 25°19′E
126 L5 **Pavelets** Ryazanskaya Oblast', W Russian Federation 53°47′N 39°22′E
106 D8 **Pavia** *anc.* Ticinum. Lombardia, N Italy 45°10′N 09°10′E
118 C9 **Pāvilosta** W Latvia 56°52′N 21°12′E
125 P14 **Pavino** Kostromskaya Oblast', NW Russian Federation 59°10′N 46°09′E
114 J8 **Pavlikeni** Veliko Tarnovo, N Bulgaria 43°14′N 25°20′E
145 T8 **Pavlodar** Pavlodar, NE Kazakhstan 52°21′N 76°59′E
145 S9 **Pavlodar** *off.* Pavlodarskaya Oblast', *Kaz.* Pavlodar Oblysy. ◈ *province* NE Kazakhstan
Pavlodar Oblysy/Pavlodarskaya Oblast' *see* Pavlodar
Pavlograd *see* Pavlohrad
117 U7 **Pavlohrad** *Rus.* Pavlograd. Dnipropetrovs'ka Oblast', E Ukraine 48°32′N 35°50′E
Pavlor Harbour *see* Pauloff Harbor
145 R9 **Pavlovka** Akmola, C Kazakhstan 51°22′N 72°35′E
127 V4 **Pavlovka** Respublika Bashkortostan, W Russian Federation 55°28′N 56°36′E
127 Q7 **Pavlovka** Ul'yanovskaya Oblast', W Russian Federation 52°40′N 47°08′E
127 N3 **Pavlovo** Nizhegorodskaya Oblast', W Russian Federation 55°59′N 43°03′E
126 L9 **Pavlovsk** Voronezhskaya Oblast', W Russian Federation 50°26′N 40°08′E
126 L13 **Pavlovskaya** Krasnodarskiy Kray, SW Russian Federation 46°06′N 39°52′E
117 S7 **Pavlysh** Kirovohrads'ka Oblast', C Ukraine 48°54′N 33°20′E
106 F10 **Pavullo nel Frignano** Emilia-Romagna, C Italy 44°19′N 10°52′E
27 P8 **Pawhuska** Oklahoma, C USA 36°42′N 96°21′W
21 U13 **Pawleys Island** South Carolina, SE USA 33°27′N 79°07′W
30 K14 **Pawnee** Illinois, N USA 39°35′N 89°34′W
27 O9 **Pawnee** Oklahoma, C USA 36°21′N 96°50′W
37 U2 **Pawnee Buttes** ▲ Colorado, C USA 40°49′N 103°58′W
29 S17 **Pawnee City** Nebraska, C USA 40°06′N 96°09′W
26 K5 **Pawnee River** ≈ Kansas, C USA
167 N6 **Pawn, Nam** ≈ C Myanmar (Burma)
31 O10 **Paw Paw** Michigan, N USA 42°12′N 86°09′W
31 O10 **Paw Paw Lake** Michigan, N USA 42°12′N 86°16′W
19 O12 **Pawtucket** Rhode Island, NE USA 41°52′N 71°22′W
Pax Augusta *see* Badajoz
115 I25 **Paximádia** *island* SE Greece
Pax Julia *see* Beja
115 B16 **Paxoí** *island* Iónia Nisiá, Greece, C Mediterranean Sea
39 S10 **Paxson** Alaska, USA 62°58′N 145°27′W
147 O11 **Paxtakor** Jizzax Viloyati, C Uzbekistan 40°21′N 67°54′E
30 M13 **Paxton** Illinois, N USA 40°27′N 88°06′W
124 J11 **Pay** Respublika Kareliya, NW Russian Federation 61°10′N 34°24′E
166 M8 **Payagyi** Bago, SW Myanmar (Burma) 17°28′N 96°32′E
108 C9 **Payerne** *Ger.* Peterlingen. Vaud, W Switzerland 46°49′N 06°57′E
32 M13 **Payette** Idaho, NW USA 44°04′N 116°55′W
32 M13 **Payette River** ≈ Idaho, NW USA
125 V2 **Pay-Khoy, Khrebet** ▲▲ NW Russian Federation
Payne *see* Kangirsuk
12 K4 **Payne, Lac** ◎ Québec, NE Canada
29 T8 **Paynesville** Minnesota, N USA 45°22′N 94°42′W
169 S8 **Payong, Tanjung** *cape* East Malaysia
Payo Obispo *see* Chetumal
61 D18 **Paysandú** Paysandú, W Uruguay 32°21′S 58°05′W
61 D17 **Paysandú** ◈ *department* W Uruguay
102 I7 **Pays de la Loire** ◈ *region* NW France
36 L12 **Payson** Arizona, SW USA 34°13′N 111°19′W
36 L4 **Payson** Utah, W USA 40°02′N 111°43′W
125 W4 **Payyer, Gora** ▲ NW Russian Federation 66°49′N 64°33′E
Payzawat *see* Jiashi
137 Q11 **Pazar** Rize, NE Turkey 41°10′N 40°53′E
136 F10 **Pazarbaşı Burnu** *headland* NW Turkey 41°12′N 30°18′E
136 M16 **Pazarcık** Kahramanmaraş, S Turkey 37°31′N 37°19′E
114 I10 **Pazardzhik** *prev.* Tatar Pazardzhik. Pazardzhik, SW Bulgaria 42°11′N 24°21′E
64 H11 **Pazardzhik** ◈ *province* C Bulgaria
54 H9 **Paz de Ariporo** Casanare, E Colombia 05°54′N 71°52′W
112 A10 **Pazin** *Ger.* Mitterburg, *It.* Pisino. Istra, NW Croatia 45°14′N 13°56′E
42 D7 **Paz, Río** ≈ El Salvador/Guatemala
113 O18 **Pčinja** ≈ N Macedonia
193 V15 **Pea** Tongatapu, S Tonga 21°10′S 175°14′W
27 O6 **Peabody** Kansas, C USA 38°10′N 97°06′W
11 O12 **Peace** ≈ Alberta/British Columbia, W Canada
Peace Garden State *see* North Dakota
11 Q10 **Peace Point** Alberta, C Canada 59°11′N 112°12′W
11 O12 **Peace River** Alberta, W Canada 56°15′N 117°18′W
23 W13 **Peace River** ≈ Florida, SE USA
11 N17 **Peachland** British Columbia, SW Canada 49°49′N 119°48′W
36 J10 **Peach Springs** Arizona, SW USA 35°33′N 113°27′W
Peach State *see* Georgia
23 S4 **Peachtree City** Georgia, SE USA 33°24′N 84°36′W
189 Y13 **Peacock Point** *point* SE Wake Island
97 M18 **Peak District** *physical region* C England, United Kingdom
183 Q7 **Peak Hill** New South Wales, SE Australia 32°39′S 148°12′E
65 G15 **Peak, The** ▲ C Ascension Island
105 O13 **Peal de Becerro** Andalucía, S Spain 37°55′N 03°08′W
189 X11 **Peale Island** *island* N Wake Island
37 O6 **Peale, Mount** ▲ Utah, W USA 38°26′N 109°13′W
39 O4 **Peard Bay** *bay* Alaska, USA
23 Q7 **Pea River** ≈ Alabama/Florida, S USA
25 W11 **Pearland** Texas, SW USA 29°33′N 95°17′W
38 D9 **Pearl City** O'ahu, Hawaii, USA, C Pacific Ocean 21°24′N 157°58′W
38 D9 **Pearl Harbor** *inlet* O'ahu, Hawai'i, USA, C Pacific Ocean
Pearl Islands *see* Perlas, Archipiélago de las
Pearl Lagoon *see* Perlas, Laguna de
22 M5 **Pearl River** ≈ Louisiana/Mississippi, S USA
25 Q13 **Pearsall** Texas, SW USA 28°54′N 99°07′W
23 U7 **Pearson** Georgia, SE USA 31°18′N 82°51′W
25 P4 **Pease River** ≈ Texas, SW USA
12 F7 **Peawanuck** Ontario, C Canada 54°55′N 85°31′W
12 E8 **Peawanuk** ≈ Ontario, S Canada
83 P16 **Pebane** Zambézia, NE Mozambique 17°14′S 38°10′E
65 C23 **Pebble Island** *island* N Falkland Islands
65 C23 **Pebble Island Settlement** Pebble Island, N Falkland Islands 51°20′S 59°40′W
Peć *see* Pejë
25 R8 **Pecan Bayou** ≈ Texas, SW USA
22 H10 **Pecan Island** Louisiana, S USA 29°39′N 92°26′W
60 L12 **Peças, Ilha das** *island* S Brazil
30 L10 **Pecatonica River** ≈ Illinois/Wisconsin, N USA
108 G10 **Peccia** Ticino, S Switzerland 46°24′N 08°39′E
Pechenegi *see* Pechenihy
Pechenezhskoye Vodokhranilishche *see* Pecheniz'ke Vodoskhovyshche
124 I2 **Pechenga** *Fin.* Petsamo. Murmanskaya Oblast', NW Russian Federation 69°34′N 31°14′E
117 V5 **Pechenihy** *Rus.* Pechenegi. Kharkivs'ka Oblast', E Ukraine 49°49′N 36°57′E
117 V5 **Pecheniz'ke Vodoskhovyshche** *Rus.* Pechenezhskoye Vodokhranilishche. ⊠ E Ukraine
125 U7 **Pechora** Respublika Komi, NW Russian Federation 65°09′N 57°09′E
125 R6 **Pechora** ≈ NW Russian Federation
Pechora Bay *see* Pechorskaya Guba
Pechora Sea *see* Pechorskoye More
125 S3 **Pechorskaya Guba** *Eng.* Pechora Bay. *bay* NW Russian Federation
122 H7 **Pechorskoye More** *Eng.* Pechora Sea. *sea* NW Russian Federation
116 E11 **Pecica** *Ger.* Petschka, *Hung.* Ópécska. Arad, W Romania 46°09′N 21°06′E
24 K8 **Pecos** Texas, SW USA 31°25′N 103°30′W
25 N11 **Pecos River** ≈ New Mexico/Texas, SW USA
111 I25 **Pécs** *Ger.* Fünfkirchen, *Lat.* Sopianae. Baranya, SW Hungary 46°05′N 18°11′E
43 T17 **Pedasí** Los Santos, S Panama 07°36′N 80°04′W
Pedde *see* Pedja
183 O17 **Pedder, Lake** ◎ Tasmania, SE Australia
44 M10 **Pedernales** SW Dominican Republic 18°02′N 71°41′W
55 Q5 **Pedernales** Delta Amacuro, NE Venezuela 09°58′N 62°15′W
25 R10 **Pedernales River** ≈ Texas, SW USA
62 H6 **Pedernales, Salar de** *salt lake* N Chile
Pedhoulas *see* Pedoulás
55 X11 **Pédima** *var.* Malavate. SW French Guiana 03°15′N 54°08′W
182 F1 **Pedirka** South Australia 26°41′S 135°11′E
171 S11 **Pediwang** Pulau Halmahera, E Indonesia 01°29′N 127°57′E
118 I5 **Pedja** *var.* Pedja Jõgi, *Ger.* Pedde. ≈ E Estonia
Pedja Jõgi *see* Pedja
121 O3 **Pedoulás** *var.* Pedhoulas. W Cyprus 34°58′N 32°51′E
59 N18 **Pedra Azul** Minas Gerais, NE Brazil 16°02′S 41°17′W
104 I3 **Pedrafita, Porto de** *var.* Puerto de Piedrafita. *pass* NW Spain
76 E9 **Pedra Lume** Sal, NE Cape Verde 16°47′N 22°54′W
43 P16 **Pedregal** Chiriquí, W Panama 09°04′N 79°25′W
54 J4 **Pedregal** Falcón, N Venezuela 11°04′N 70°08′W
40 L9 **Pedriceña** Durango, C Mexico 25°08′N 103°46′W
60 L11 **Pedro Barros** São Paulo, S Brazil 24°12′S 47°22′W
39 Q13 **Pedro Bay** Alaska, USA 59°47′N 154°06′W
62 H4 **Pedro de Valdivia** *var.* Oficina Pedro de Valdivia. Antofagasta, N Chile 22°33′S 69°38′W
62 P4 **Pedro Juan Caballero** Amambay, E Paraguay 22°34′S 55°41′W
63 L15 **Pedro Luro** Buenos Aires, E Argentina 39°30′S 62°38′W
105 O10 **Pedro Muñoz** Castilla-La Mancha, C Spain 39°25′N 02°56′W
155 J22 **Pedro, Point** *headland* NW Sri Lanka 09°54′N 80°08′E
182 K9 **Peebinga** South Australia 34°56′S 140°56′E
96 J13 **Peebles** SE Scotland, United Kingdom 55°40′N 03°15′W
31 S15 **Peebles** Ohio, N USA 38°57′N 83°23′W
96 J12 **Peebles** *cultural region* SE Scotland, United Kingdom
18 K13 **Peekskill** New York, NE USA 41°17′N 73°54′W
97 I16 **Peel** W Isle of Man 54°13′N 04°04′W
8 G7 **Peel** ≈ Northwest Territories/Yukon, NW Canada
8 K5 **Peel Point** *headland* Victoria Island, Northwest Territories, NW Canada 73°22′N 114°33′W
8 M5 **Peel Sound** *passage* Nunavut, N Canada
100 N9 **Peene** ≈ NE Germany
99 K17 **Peer** Limburg, NE Belgium 51°08′N 05°29′E
14 H14 **Pefferlaw** Ontario, S Canada 44°18′N 79°11′W
185 I18 **Pegasus Bay** *bay* South Island, New Zealand
121 O3 **Pégeia** *var.* Peyia. SW Cyprus 34°52′N 32°24′E
109 V7 **Peggau** Steiermark, SE Austria 47°10′N 15°20′E
101 L19 **Pegnitz** Bayern, SE Germany 49°45′N 11°33′E
101 L19 **Pegnitz** ≈ SE Germany
105 T11 **Pego** Valenciana, E Spain 38°51′N 00°08′W
Pegu *see* Bago
Pegu *see* Bago
189 N13 **Pehleng** Pohnpei, E Micronesia
114 M12 **Pehlivanköy** Kırklareli, NW Turkey 41°21′N 26°55′E
77 R14 **Péhonko** C Benin 10°14′N 01°57′E
61 B21 **Pehuajó** Buenos Aires, E Argentina 35°48′S 61°53′W
Pei-ching *see* Beijing/Beijing Shi
100 J13 **Peine** Niedersachsen, C Germany 52°19′N 10°14′E
Pei-p'ing *see* Beijing/Beijing Shi
Peipsi Järv/Peipus-See *see* Peipus, Lake
118 J5 **Peipus, Lake** *Est.* Peipsi Järv, *Ger.* Peipus-See, *Rus.* Chudskoye Ozero. ◎ Estonia/Russian Federation
115 H19 **Peiraías** *prev.* Piraiévs, *Eng.* Piraeus. Attikí, C Greece 37°57′N 23°42′E
Peisern *see* Pyzdry
60 I8 **Peixe, Rio do** ≈ S Brazil
59 I16 **Peixoto de Azevedo** Mato Grosso, W Brazil 10°18′S 55°03′W
168 O11 **Pejantan, Pulau** *island* W Indonesia
113 L16 **Pejë** *Serb.* Peć. W Kosovo 42°40′N 20°19′E
112 N11 **Pek** ≈ E Serbia
Pèk *see* Phônsaven
169 Q16 **Pekalongan** Jawa, C Indonesia 06°54′S 109°37′E
168 K11 **Pekanbaru** *var.* Pakanbaru. Sumatera, W Indonesia 0°31′N 101°27′E
30 L12 **Pekin** Illinois, N USA 40°34′N 89°38′W
Peking *see* Beijing/Beijing Shi
Pelabohan Kelang/Pelabuhan Kelang *see* Pelabuhan Klang
168 J9 **Pelabuhan Klang** *var.* Kuala Pelabohan Kelang, Pelabohan Kelang, Pelabuhan Kelang, Port Klang, Port Swettenham. Selangor, Peninsular Malaysia 02°57′N 101°24′E
120 L11 **Pelagie, Isole** *island group* SW Italy
Pelagosa *see* Palagruža
22 L5 **Pelahatchie** Mississippi, S USA 32°19′N 89°48′W
169 T14 **Pelaihari** *var.* Pleihari. Borneo, C Indonesia 03°48′S 114°45′E
103 U14 **Pelat, Mont** ▲ SE France 44°16′N 06°46′E
116 F12 **Peleaga, Vârful** *prev.* Vîrful Peleaga. ▲ W Romania 45°23′N 22°52′E
Peleaga, Vîrful *see* Peleaga, Vârful
123 O11 **Peleduy** Respublika Sakha (Yakutiya), NE Russian Federation 59°39′N 112°36′E
14 C18 **Pelee Island** *island* Ontario, S Canada
45 Q11 **Pelée, Montagne** ▲ N Martinique 14°47′N 61°10′W
14 D18 **Pelee, Point** *headland* Ontario, S Canada 41°56′N 82°30′W
171 P12 **Pelei** Pulau Peleng, N Indonesia 01°26′S 123°27′E
Peleliu *see* Beliliou
171 P12 **Peleng, Pulau** *island* Kepulauan Banggai, N Indonesia
23 T7 **Pelham** Georgia, SE USA 31°07′N 84°09′W
111 E18 **Pelhřimov** *Ger.* Pilgram. Vysočina, C Czech Republic 49°26′N 15°14′E
39 W13 **Pelican** Chichagof Island, Alaska, USA 57°52′N 136°05′W
191 Z3 **Pelican Lagoon** ◎ Kiritimati, E Kiribati
29 V3 **Pelican Lake** ◎ Minnesota, N USA
29 U6 **Pelican Lake** ◎ Minnesota, N USA
30 L5 **Pelican Lake** ◎ Wisconsin, N USA
44 G1 **Pelican Point** Grand Bahama Island, N The Bahamas 26°39′N 78°09′W
83 B19 **Pelican Point** *headland* W Namibia 22°55′S 14°25′E
29 S6 **Pelican Rapids** Minnesota, N USA 46°34′N 96°04′W
Pelican State *see* Louisiana
11 U13 **Pelikan Narrows** Saskatchewan, C Canada 55°11′N 102°51′W
115 L18 **Pelinaío** ▲ Chíos, E Greece 38°31′N 26°01′E
Pelinnaeum *see* Pelinnaío
115 E16 **Pelinnaío** *anc.* Pelinnaeum. *ruins* Thessalía, C Greece
113 N20 **Pelister** ▲ SW FYR Macedonia 41°00′N 21°12′E
113 G15 **Pelješac** *peninsula* S Croatia
92 M12 **Pelkosenniemi** Lappi, NE Finland 67°06′N 27°30′E
29 W15 **Pella** Iowa, C USA 41°24′N 92°55′W
114 F13 **Pélla** *site of ancient city* Kentrikí Makedonía, N Greece
23 Q3 **Pell City** Alabama, S USA 33°35′N 86°17′W
61 A22 **Pellegrini** Buenos Aires, E Argentina 36°16′S 63°07′W
92 K12 **Pello** Lappi, NW Finland 66°47′N 24°E
100 G7 **Pellworm** *island* N Germany
10 H6 **Pelly** ≈ Yukon, NW Canada
Pelly Bay *see* Kugaaruk
10 I8 **Pelly Mountains** ▲▲ Yukon, W Canada
Pélmonostor *see* Beli Manastir
37 P13 **Pelona Mountain** ▲ New Mexico, SW USA 33°40′N 108°06′W
Peloponnese/Peloponnesus *see* Pelopónnisos
115 E20 **Pelopónnisos** *Eng.* Peloponnese. ◈ *region* S Greece
115 E20 **Pelopónnisos** *var.* Morea, *Eng.* Peloponnese; *anc.* Peloponnesus. *peninsula* S Greece
107 L23 **Peloritani, Monti** *anc.* Pelorus and Neptunius. ▲▲ Sicilia, Italy, C Mediterranean Sea
107 M22 **Peloro, Capo** *var.* Punta del Faro. *headland* S Italy 38°15′N 15°39′E
Pelorus and Neptunius *see* Peloritani, Monti
61 H17 **Pelotas** Rio Grande do Sul, S Brazil 31°45′S 52°20′W
61 I14 **Pelotas, Rio** ≈ S Brazil
92 K10 **Peltovuoma** *Lapp.* Bealdovuopmi. Lappi, N Finland 68°23′N 24°12′E
19 R4 **Pemadumcook Lake** ◎ Maine, NE USA
169 Q16 **Pemalang** Jawa, C Indonesia 06°53′S 109°07′E
169 P10 **Pemangkat** *var.* Pamangkat. Borneo, C Indonesia 01°11′N 109°00′E
Pemar *see* Paimio
168 I9 **Pematangsiantar** Sumatera, W Indonesia 02°59′N 99°01′E
83 Q14 **Pemba** *prev.* Port Amelia, Porto Amélia. Cabo Delgado, NE Mozambique 13°S 40°35′E
81 J22 **Pemba** ◈ *region* E Tanzania
81 K21 **Pemba** *island* E Tanzania
83 Q14 **Pemba, Baia de** *inlet* NE Mozambique
81 J21 **Pemba Channel** *channel* E Tanzania
180 J14 **Pemberton** Western Australia 34°27′S 116°09′E
10 M16 **Pemberton** British Columbia, SW Canada 50°19′N 122°49′W
29 Q2 **Pembina** North Dakota, N USA 48°58′N 97°14′W
11 P15 **Pembina** ≈ Alberta, SW Canada
29 Q2 **Pembina** ≈ Canada/USA
171 X16 **Pembre** Papua, E Indonesia 07°49′S 138°01′E
14 K12 **Pembroke** Ontario, SE Canada 45°49′N 77°08′W
97 H21 **Pembroke** SW Wales, United Kingdom 51°41′N 04°55′W
23 W6 **Pembroke** Georgia, SE USA 32°09′N 81°35′W
21 U11 **Pembroke** North Carolina, SE USA 34°40′N 79°12′W
21 R7 **Pembroke** Virginia, NE USA 37°19′N 80°38′W
97 H21 **Pembroke** *cultural region* SW Wales, United Kingdom
Pembuang, Sungai *see* Seruyan, Sungai
43 S15 **Peña Blanca, Cerro** ▲ C Panama 08°39′N 80°39′W
104 K8 **Peña de Francia, Sierra de la** ▲▲ W Spain
104 G6 **Penafiel** *var.* Peñafiel. Porto, N Portugal 41°12′N 08°17′W
105 N6 **Peñafiel** Castilla y León, N Spain 41°36′N 04°07′W
Peñafiel *see* Penafiel
Peñagolosa *see* Penyagolosa
105 N7 **Peñalara, Pico de** ▲ C Spain 40°52′N 03°55′W
171 X16 **Penambo, Banjaran** *var.* Banjaran Tama Abu, Penambo Range. ▲▲ Indonesia/Malaysia
Penambo Range *see* Penambo, Banjaran
41 O10 **Peña Nevada, Cerro** ▲ C Mexico 23°46′N 99°52′W
Penang *see* Pinang
Penang *see* George Town
Penang *see* Pinang, Pulau, Peninsular Malaysia
60 J8 **Penápolis** São Paulo, S Brazil 21°23′S 50°02′W
104 L7 **Peñaranda de Bracamonte** Castilla y León, N Spain 40°54′N 05°13′W
105 S8 **Peñarroya** ▲ E Spain 40°24′N 00°42′W
104 L12 **Peñarroya-Pueblonuevo** Andalucía, S Spain 38°21′N 05°18′W
97 K22 **Penarth** S Wales, United Kingdom 51°27′N 03°11′W
104 K1 **Peñas, Cabo de** *headland* N Spain 43°39′N 05°52′W
63 F20 **Penas, Golfo de** *gulf* S Chile
Pen-ch'i *see* Benxi
79 H14 **Pendé** *var.* Logone Oriental. ≈ Central African Republic/Chad
76 I14 **Pendembu** E Sierra Leone 09°06′N 12°12′W
29 R13 **Pender** Nebraska, C USA 42°06′N 96°42′W
Penderma *see* Bandırma
32 K11 **Pendleton** Oregon, NW USA 45°40′N 118°47′W
32 M7 **Pend Oreille, Lake** ◎ Idaho, NW USA
32 M7 **Pend Oreille River** ≈ Idaho/Washington, NW USA
Pendzhikent *see* Panjakent
Peneius *see* Pineiós
104 G8 **Penela** Coimbra, N Portugal 40°02′N 08°23′W
14 G13 **Penetanguishene** Ontario, S Canada 44°45′N 79°55′W
151 H15 **Penganga** ≈ C India
P'engchia Yu *see* Pengjia Yu
79 M21 **Penge** Kasai-Oriental, C Dem. Rep. Congo 05°29′S 24°38′E
Penghu Archipelago/P'enghu Ch'üntao/Penghu Islands *see* Penghu Liedao
161 R14 **Penghu Liedao** *var.* P'enghu Ch'üntao, Penghu Islands, P'enghu Liehtao, *Eng.* Penghu Archipelago, Pescadores, *Jap.* Hoko-guntō, Hoko-shotō. *island group* W Taiwan
P'enghu Liehtao *see* Penghu Liedao
161 S14 **Penghu Shuidao** *var.* Pescadores Channel, P'enghu Shuitao. *channel* W Taiwan
P'enghu Shuitao *see* Penghu Shuidao
161 T12 **Pengjia Yu** *prev.* P'engchia Yu. *island* N Taiwan
161 R4 **Penglai** *var.* Dengzhou. Shandong, E China 37°50′N 120°45′E
Peng-pu *see* Bengbu
Penhsihu *see* Benxi
Penibético, Sistema *see* Béticos, Sistemas
104 F10 **Peniche** Leiria, W Portugal 39°21′N 09°23′W
169 U17 **Penida, Nusa** *island* S Indonesia
Peninsular State *see* Florida
105 T8 **Peníscola** *var.* Peñíscola. Valenciana, E Spain 40°22′N 00°24′E
40 M13 **Pénjamo** Guanajuato, C Mexico 20°26′N 101°44′W
Penki *see* Benxi
102 F7 **Penmarch, Pointe de** *headland* NW France 47°46′N 04°34′W
107 L15 **Penna, Punta della** *headland* C Italy 42°10′N 14°43′E
107 K14 **Penne** Abruzzo, C Italy 42°28′N 13°57′E
Penner *see* Penneru
155 J18 **Penneru** *var.* Penner. ≈ C India
182 I10 **Penneshaw** South Australia 35°45′S 137°57′E
18 C14 **Penn Hills** Pennsylvania, NE USA 40°28′N 79°53′W
Penninae, Alpes/Pennine, Alpi *see* Pennine Alps
108 D11 **Pennine Alps** *Fr.* Alpes Pennines, *It.* Alpi Pennine, *Lat.* Alpes Penninae. ▲▲ Italy/Switzerland
Pennine Chain *see* Pennines
97 L15 **Pennines** *var.* Pennine Chain. ▲▲ N England, United Kingdom
Pennines, Alpes *see* Pennine Alps
21 O8 **Pennington Gap** Virginia, NE USA 36°45′N 83°01′W
18 I16 **Penns Grove** New Jersey, NE USA 39°42′N 75°27′W
18 I16 **Pennsville** New Jersey, NE USA 39°37′N 75°29′W
18 E14 **Pennsylvania** *off.* Commonwealth of Pennsylvania, *also known as* Keystone State. ◈ *state* NE USA
18 G10 **Penn Yan** New York, NE USA 42°39′N 77°02′W
124 H16 **Peno** Tverskaya Oblast', W Russian Federation 56°55′N 32°44′E
19 R7 **Penobscot Bay** *bay* Maine, NE USA
19 S5 **Penobscot River** ≈ Maine, NE USA
182 K12 **Penola** South Australia 37°24′S 140°50′E
40 K9 **Peñón Blanco** Durango, C Mexico 25°12′N 100°50′W
182 E7 **Penong** South Australia 31°57′S 133°01′E
43 S16 **Penonomé** Coclé, C Panama 08°29′N 80°22′W
190 L13 **Penrhyn** *atoll* N Cook Islands
192 M9 **Penrhyn Basin** *undersea feature* C Pacific Ocean
183 S9 **Penrith** New South Wales, SE Australia 33°45′S 150°48′E
97 K15 **Penrith** NW England, United Kingdom 54°40′N 02°44′W
23 O9 **Pensacola** Florida, SE USA 30°25′N 87°13′W
23 O9 **Pensacola Bay** *bay* Florida, SE USA
195 N7 **Pensacola Mountains** ▲▲ Antarctica
182 L12 **Penshurst** Victoria, SE Australia 37°54′S 142°19′E
187 R13 **Pentecost** *Fr.* Pentecôte. *island* C Vanuatu
15 V4 **Pentecôte** ≈ Québec, SE Canada
Pentecôte *see* Pentecost
15 V4 **Pentecôte, Lac** ◎ Québec, SE Canada
8 H15 **Penticton** British Columbia, SW Canada 49°29′N 119°38′W
96 J6 **Pentland Firth** *strait* N Scotland, United Kingdom
96 J12 **Pentland Hills** *hill range* S Scotland, United Kingdom
171 Q12 **Penu** Pulau Taliabu, E Indonesia 01°43′S 125°09′E
155 H18 **Penukonda** Andhra Pradesh, E India 14°04′N 77°38′E
166 L7 **Penwegon** Bago, C Myanmar (Burma) 18°14′N 96°34′E
24 M8 **Penwell** Texas, SW USA 31°45′N 102°32′W
105 S8 **Penyagolosa** *var.* Peñagolosa. ▲ E Spain 40°10′N 00°15′W
97 J21 **Pen y Fan** ▲ SE Wales, United Kingdom 51°52′N 03°25′W
97 L16 **Pen-y-ghent** ▲ N England, United Kingdom 54°11′N 02°15′W
127 O6 **Penza** Penzenskaya Oblast', W Russian Federation 53°11′N 45°E
97 G25 **Penzance** SW England, United Kingdom 50°08′N 05°33′W
127 N6 **Penzenskaya Oblast'** ◈ *province* W Russian Federation
123 U7 **Penzhina** ≈ E Russian Federation
123 U9 **Penzhinskaya Guba** *bay* E Russian Federation
Penzig *see* Pieńsk
36 K13 **Peoria** Arizona, SW USA 33°34′N 112°14′W
30 L12 **Peoria** Illinois, N USA 40°42′N 89°35′W
30 L12 **Peoria Heights** Illinois, N USA 40°45′N 89°34′W
31 N11 **Peotone** Illinois, N USA 41°19′N 87°47′W
18 J11 **Pepacton Reservoir** ⊠ New York, NE USA
76 I15 **Pepel** W Sierra Leone 08°39′N 13°04′W
30 I6 **Pepin, Lake** ⊠ Minnesota/Wisconsin, N USA
99 L20 **Pepinster** Liège, E Belgium 50°34′N 05°49′E
113 L20 **Peqin** *var.* Peqini. Elbasan, C Albania 41°03′N 19°46′E
Peqini *see* Peqin
40 D7 **Pequeña, Punta** *headland* NW Mexico 26°13′N 112°34′W
168 J8 **Perak** ◈ *state* Peninsular Malaysia
105 R7 **Perales del Alfambra** Aragón, NE Spain 40°38′N 01°00′W
115 C15 **Pérama** *var.* Perama. Ípeiros, W Greece 39°42′N 20°51′E
Perama *see* Pérama
92 M13 **Perä-Posio** Lappi, NE Finland 66°10′N 27°56′E
15 Z6 **Percé** Québec, SE Canada 48°32′N 64°14′W
15 Z6 **Percé, Rocher** *island* Québec, S Canada
102 L5 **Perche, Collines de** ▲▲ N France
109 X4 **Perchtoldsdorf** Niederösterreich, NE Austria 48°06′N 16°16′E
180 L6 **Percival Lakes** *lakes* Western Australia
105 T3 **Perdido, Monte** ▲ NE Spain 42°41′N 00°01′E
23 O8 **Perdido River** ≈ Alabama/Florida, S USA
Perece Vela Basin *see* West Mariana Basin
116 G7 **Perechyn** Zakarpats'ka Oblast', W Ukraine 48°45′N 22°28′E
54 E10 **Pereira** Risaralda, W Colombia 04°47′N 75°46′W
60 I7 **Pereira Barreto** São Paulo, S Brazil 20°37′S 51°07′W
59 G15 **Pereirinha** Pará, N Brazil 08°18′S 57°30′W
127 N10 **Perelazovskiy** Volgogradskaya Oblast', SW Russian Federation 49°10′N 42°30′E
127 S7 **Perelyub** Saratovskaya Oblast', W Russian Federation 51°52′N 50°19′E
31 P7 **Pere Marquette River** ≈ Michigan, N USA
Peremyshl *see* Przemyśl
116 I5 **Peremyshlyany** L'vivs'ka Oblast', W Ukraine 49°42′N 24°33′E
Pereshchepino *see* Pereshchepyne
116 L9 **Pereshchepyne** *Rus.* Pereshchepino. Dnipropetrovs'ka Oblast', E Ukraine 48°59′N 35°22′E
124 L16 **Pereslavl'-Zalesskiy** Yaroslavskaya Oblast', W Russian Federation 56°42′N 38°45′E
117 Y7 **Pereval's'k** Luhans'ka Oblast', E Ukraine 48°28′N 38°54′E
127 U7 **Perevolotskiy** Orenburgskaya Oblast', W Russian Federation 51°54′N 54°05′E
Pereyaslav-Khmel'nitskiy *see* Pereyaslav-Khmel'nyts'kyy
117 Q5 **Pereyaslav-Khmel'nyts'kyy** *Rus.* Pereyaslav-Khmel'nitskiy. Kyyivs'ka Oblast', N Ukraine 50°05′N 31°28′E
109 U4 **Perg** Oberösterreich, N Austria 48°15′N 14°38′E
61 B19 **Pergamino** Buenos Aires, E Argentina 33°56′S 60°38′W
106 G6 **Pergine Valsugana** *Ger.* Persen. Trentino-Alto Adige, N Italy 46°04′N 11°13′E
29 S6 **Perham** Minnesota, N USA 46°35′N 95°34′W
93 L16 **Perho** Keski-Pohjanmaa, W Finland 63°15′N 24°25′E
116 E11 **Periam** *Ger.* Perjamosch, *Hung.* Perjámos. Timiş, W Romania 46°02′N 20°54′E
15 Q6 **Péribonca** ≈ Québec, SE Canada
12 L11 **Péribonca, Lac** ◎ Québec, SE Canada
15 Q6 **Péribonca, Petite Rivière** ≈ Québec, SE Canada
15 Q7 **Péribonka** Québec, SE Canada 48°45′N 72°01′W
40 I9 **Pericos** Sinaloa, C Mexico 25°03′N 107°42′W
169 Q10 **Perigi** Borneo, C Indonesia
102 L12 **Périgueux** *anc.* Vesuna. Dordogne, SW France 45°12′N 00°41′E
54 G5 **Perijá, Serranía de** ▲▲ Columbia/Venezuela
115 H17 **Peristéra** *island* Vóreies Sporádes, Greece, Aegean Sea
63 H20 **Perito Moreno** Santa Cruz, S Argentina 46°35′S 71°W
155 G22 **Periyāl** *var.* Periyār. ≈ SW India
Periyār *see* Periyāl
155 G23 **Periyār Lake** ◎ S India
Perjámos/Perjamosch *see* Periam
27 O9 **Perkins** Oklahoma, C USA 35°58′N 97°01′W
116 L7 **Perkivtsi** Chernivets'ka Oblast', W Ukraine 48°28′N 26°48′E
43 U15 **Perlas, Archipiélago de las** *Eng.* Pearl Islands. *island group* SE Panama
43 O10 **Perlas, Cayos de** *reef* SE Nicaragua
43 N9 **Perlas, Laguna de** *Eng.* Pearl Lagoon. *lagoon* E Nicaragua
43 N10 **Perlas, Punta de** *headland* E Nicaragua 12°22′N 83°30′W
100 L11 **Perleberg** Brandenburg, N Germany 53°04′N 11°52′E
Perlepe *see* Prilep
168 I6 **Perlis** ◈ *state* Peninsular Malaysia
125 U14 **Perm'** *prev.* Molotov. Permskiy Kray, NW Russian Federation 58°01′N 56°10′E
113 M22 **Përmet** *var.* Përmeti, Prëmet. Gjirokastër, S Albania 40°12′N 20°24′E
Përmeti *see* Përmet
125 U15 **Permskiy Kray** ◈ *province* NW Russian Federation
59 P15 **Pernambuco** *off.* Estado de Pernambuco. ◈ *state* E Brazil
Pernambuco *see* Recife
Pernambuco Abyssal Plain *see* Pernambuco Plain
Pernambuco, Estado de *see* Pernambuco
47 Y6 **Pernambuco Plain** *var.* Pernambuco Abyssal Plain. *undersea feature* E Atlantic Ocean 07°30′S 27°00′W
65 K15 **Pernambuco Seamounts** *undersea feature* C Atlantic Ocean
182 H6 **Pernatty Lagoon** *salt lake* South Australia
Pernau *see* Pärnu
Pernauer Bucht *see* Pärnu Laht
Pērnava *see* Pärnu
114 G9 **Pernik** *prev.* Dimitrovo. Pernik, W Bulgaria 42°36′N 23°02′E
114 G10 **Pernik** ◈ *province* W Bulgaria
93 K20 **Perniö** *Swe.* Bjärnå. Varsinais-Suomi, SW Finland 60°13′N 23°10′E
109 X5 **Pernitz** Niederösterreich, E Austria 47°54′N 15°58′E
Pernov *see* Pärnu
103 O3 **Péronne** Somme, N France 49°56′N 02°57′E
14 L8 **Péronne, Lac** ◎ Québec, SE Canada
106 A8 **Perosa Argentina** Piemonte, NE Italy 45°02′N 07°10′E
41 Q14 **Perote** Veracruz-Llave, E Mexico 19°32′N 97°16′W
Pérouse *see* Perugia
191 W15 **Pérouse, Bahía de la** *bay* Easter Island, Chile, E Pacific Ocean
Perovsk *see* Kyzylorda
103 O17 **Perpignan** Pyrénées-Orientales, S France 42°41′N 02°53′E
113 M20 **Përrenjas** *var.* Përrenjasi, Prenjas, Prenjasi. Elbasan, E Albania 41°04′N 20°34′E
Përrenjasi *see* Përrenjas
92 O2 **Perriertoppen** ▲ C Svalbard 79°10′N 17°01′E

◆ Country
● Country Capital
◇ Dependent Territory
○ Dependent Territory Capital
◈ Administrative Regions
✈ International Airport
▲ Mountain
▲▲ Mountain Range
Volcano
≈ River
◎ Lake
⊠ Reservoir

PERRIN – PINE FALLS

25 S6 **Perrin** Texas, SW USA 32°59′N 98°03′W
23 Y16 **Perrine** Florida, SE USA 25°36′N 80°21′W
37 S12 **Perro, Laguna del** ◎ New Mexico, SW USA
102 G5 **Perros-Guirec** Côtes d'Armor, NW France 48°49′N 03°28′W
23 T9 **Perry** Florida, SE USA 30°07′N 83°34′W
23 T5 **Perry** Georgia, SE USA 32°27′N 83°43′W
29 U14 **Perry** Iowa, C USA 41°50′N 94°06′W
18 E10 **Perry** New York, NE USA 42°43′N 78°00′W
27 N9 **Perry** Oklahoma, C USA 36°17′N 97°18′W
27 Q3 **Perry Lake** ⊠ Kansas, C USA
31 R11 **Perrysburg** Ohio, N USA 41°33′N 83°37′W
25 O1 **Perryton** Texas, SW USA 36°23′N 100°48′W
39 O15 **Perryville** Alaska, USA 55°55′N 159°08′W
27 U11 **Perryville** Arkansas, C USA 35°00′N 92°48′W
27 Y6 **Perryville** Missouri, C USA 37°43′N 89°51′W
Persante *see* Parsęta
Persen *see* Pergine Valsugana
Pershay *see* Pyarshai
117 V7 **Pershotravens'k** Dnipropetrovs'ka Oblast', E Ukraine 48°19′N 36°22′E
Pershotravneve *see* Manhush
Persia *see* Iran
141 T5 **Persian Gulf** *var.* The Gulf, *Ar.* Khalīj al 'Arabī, *Per.* Khalīj-e Fars. *Gulf* SW Asia *see also* Gulf, The
141 T5 **Persian Gulf** *var.* Gulf, The, *Ar.* Khalīj al 'Arabī, *Per.* Khalīj-e Fars. *gulf* SW Asia *see also* Persian Gulf
Persis *see* Fārs
95 K22 **Perstorp** Skåne, S Sweden 56°08′N 13°23′E
137 O14 **Pertek** Tunceli, C Turkey 38°53′N 39°19′E
183 P16 **Perth** Tasmania, SE Australia 41°39′S 147°11′E
180 I13 **Perth** *state capital* Western Australia 31°58′S 115°49′E
14 L13 **Perth** Ontario, SE Canada 44°54′N 76°15′W
96 J11 **Perth** C Scotland, United Kingdom 56°24′N 03°28′W
96 J10 **Perth** *cultural region* C Scotland, United Kingdom
180 I12 **Perth** ✈ Western Australia 31°51′S 116°06′E
173 V10 **Perth Basin** *undersea feature* SE Indian Ocean 28°30′S 110°00′E
103 S15 **Pertuis** Vaucluse, SE France 43°42′N 05°30′E
103 Y16 **Pertusato, Capo** *headland* Corse, France, C Mediterranean Sea 41°22′N 09°10′E
30 L11 **Peru** Illinois, N USA 41°18′N 89°09′W
31 P12 **Peru** Indiana, N USA 40°45′N 86°04′W
57 E13 **Peru** *off.* Republic of Peru. ◆ *republic* W South America
Peru *see* Beru
193 T9 **Peru Basin** *undersea feature* E Pacific Ocean 15°00′S 85°00′W
193 U8 **Peru-Chile Trench** *undersea feature* E Pacific Ocean 20°00′S 73°00′W
112 F13 **Perućko Jezero** ◎ S Croatia
106 H13 **Perugia** *Fr.* Pérouse; *anc.* Perusia. Umbria, C Italy 43°06′N 12°24′E
Perugia, Lake of *see* Trasimeno, Lago
61 D15 **Perugorría** Corrientes, NE Argentina 29°21′S 58°35′W
60 M11 **Peruíbe** São Paulo, S Brazil 24°18′S 47°01′W
155 B21 **Perumalpār** *reef* India, N Indian Ocean
Peru, Republic of *see* Peru
Perusia *see* Perugia
99 D20 **Péruwelz** Hainaut, SW Belgium 50°30′N 03°35′E
137 R15 **Pervari** Siirt, SE Turkey 37°55′N 42°32′E
127 O4 **Pervomaysk** Nizhegorodskaya Oblast', W Russian Federation 54°52′N 43°49′E
117 X7 **Pervomays'k** Luhans'ka Oblast', E Ukraine 48°38′N 38°36′E
117 P8 **Pervomays'k** *prev.* Ol'viopol'. Mykolayivs'ka Oblast', S Ukraine 48°02′N 30°51′E
117 S12 **Pervomays'ke** Avtonomna Respublika Krym, S Ukraine 45°43′N 33°49′E
127 V7 **Pervomayskiy** Orenburgskaya Oblast', W Russian Federation 51°32′N 54°58′E
126 M6 **Pervomayskiy** Tambovskaya Oblast', W Russian Federation 53°15′N 40°20′E
117 V6 **Pervomays'kyy** Kharkivs'ka Oblast', E Ukraine 49°24′N 36°12′E
122 F10 **Pervoural'sk** Sverdlovskaya Oblast', C Russian Federation 56°58′N 59°50′E
123 V11 **Pervyy Kuril'skiy Proliv** *strait* E Russian Federation
99 I19 **Perwez** Walloon Brabant, C Belgium 50°39′N 04°49′E
106 I11 **Pesaro** *anc.* Pisaurum. Marche, C Italy 43°55′N 12°53′E
35 N9 **Pescadero** California, W USA 37°15′N 122°23′W
Pescadores *see* Penghu Liedao
Pescadores Channel *see* Penghu Shuidao
107 K14 **Pescara** *anc.* Aternum, Ostia Aterni. Abruzzo, C Italy 42°28′N 14°13′E
107 K15 **Pescara** ≈ C Italy
106 F11 **Pescia** Toscana, C Italy 43°54′N 10°41′E
108 C8 **Peseux** Neuchâtel, W Switzerland 46°59′N 06°53′E
125 P6 **Pësha** ≈ NW Russian Federation
149 T5 **Peshāwar** Khyber Pakhtunkhwa, N Pakistan 34°01′N 71°33′E
149 T6 **Peshāwar** ✈ Khyber Pakhtunkhwa, N Pakistan 34°01′N 71°40′E
113 M19 **Peshkopi** *var.* Peshkopia, Peshkopija. Dibër, NE Albania 41°40′N 20°25′E
Peshkopia/Peshkopija *see* Peshkopi
114 I11 **Peshtera** Pazardzhik, C Bulgaria 42°02′N 24°18′E
31 N6 **Peshtigo** Wisconsin, N USA 45°04′N 87°43′W
31 N6 **Peshtigo River** ≈ Wisconsin, N USA
Peski *see* Pyaski
125 S13 **Peskovka** Kirovskaya Oblast', NW Russian Federation 59°04′N 52°17′E
103 S8 **Pesmes** Haute-Saône, E France 47°17′N 05°33′E
104 H6 **Peso da Régua** *var.* Pêso da Regua. Vila Real, N Portugal 41°10′N 07°47′W
40 F5 **Pesqueira** Sonora, NW Mexico 29°22′N 110°58′W
102 J13 **Pessac** Gironde, SW France 44°46′N 00°42′W
111 J23 **Pest** *off.* Pest Megye. ◈ *county* C Hungary
Pest Megye *see* Pest
124 J14 **Pestovo** Novgorodskaya Oblast', W Russian Federation 58°37′N 35°48′E
40 M15 **Petacalco, Bahía** *bay* W Mexico
Petach-Tikva *see* Petah Tikva
138 F10 **Petah Tikva** *var.* Petach-Tikva, Petah Tiqva, Petakh Tikva; *prev.* Petah Tiqwa. Tel Aviv, C Israel 32°05′N 34°53′E
Petah Tiqwa *see* Petah Tikva
93 L17 **Petäjävesi** Keski-Suomi, C Finland 62°17′N 25°10′E
Petakh Tikva/Petah Tiqva *see* Petah Tikva
22 M7 **Petal** Mississippi, S USA 31°21′N 89°15′W
115 I19 **Petalioí** *island* C Greece
115 H19 **Petalión, Kólpos** *gulf* E Greece
115 J19 **Pétalo** ▲ Ándros, Kykládes, Greece, Aegean Sea 37°51′N 24°50′E
34 M8 **Petaluma** California, W USA 38°15′N 122°37′W
99 L25 **Pétange** Luxembourg, SW Luxembourg 49°33′N 05°53′E
54 M5 **Petare** Miranda, N Venezuela 10°31′N 66°50′W
41 N16 **Petatlán** Guerrero, S Mexico 17°31′N 101°16′W
83 L14 **Petauke** Eastern, E Zambia 14°12′S 31°16′E
14 J12 **Petawawa** Ontario, SE Canada 45°54′N 77°18′W
14 J11 **Petawawa** ≈ Ontario, SE Canada
Petchaburi *see* Phetchaburi
42 D2 **Petén** *off.* Departamento del Petén. ◈ *department* N Guatemala
Petén, Departamento del *see* Petén
42 D2 **Petén Itzá, Lago** *var.* Lago de Flores. ◎ N Guatemala
30 K7 **Petenwell Lake** ⊠ Wisconsin, N USA
14 D6 **Peterbell** Ontario, S Canada 48°34′N 83°19′W
182 I7 **Peterborough** South Australia 32°59′S 138°51′E
14 I14 **Peterborough** Ontario, SE Canada 44°19′N 78°20′W
97 N20 **Peterborough** *prev.* Medeshamstede. E England, United Kingdom 52°35′N 00°15′W
19 N10 **Peterborough** New Hampshire, NE USA 42°51′N 71°54′W
96 L8 **Peterhead** NE Scotland, United Kingdom 57°30′N 01°46′W
Peterhof *see* Luboń
Peter I øy *see* Peter I Øy
193 Q14 **Peter I Øy** ◇ *Norwegian dependency* Antarctica
194 H9 **Peter I Øy** *var.* Peter I øy. *island* Antarctica
97 M14 **Peterlee** N England, United Kingdom 54°45′N 01°18′W
Peterlingen *see* Payerne
197 P14 **Petermann Bjerg** ▲ C Greenland 73°16′N 27°59′W
11 S12 **Peter Pond Lake** ◎ Saskatchewan, C Canada
39 X13 **Petersburg** Mytkof Island, Alaska, USA 56°43′N 132°51′W
30 K13 **Petersburg** Illinois, N USA 40°01′N 89°52′W
31 N16 **Petersburg** Indiana, N USA 38°30′N 87°16′W
29 Q3 **Petersburg** North Dakota, N USA 47°59′N 97°59′W
25 N5 **Petersburg** Texas, SW USA 33°52′N 101°36′W
21 V7 **Petersburg** Virginia, NE USA 37°14′N 77°24′W
21 T4 **Petersburg** West Virginia, NE USA 39°01′N 79°09′W
100 H12 **Petershagen** Nordrhein-Westfalen, NW Germany 52°22′N 08°58′E
55 S9 **Peters Mine** *var.* Peter's Mine. N Guyana 06°13′N 59°18′W
107 O21 **Petilia Policastro** Calabria, SW Italy 39°07′N 16°48′E
44 M9 **Pétionville** S Haiti 18°29′N 72°16′W
45 X6 **Petit-Bourg** Basse Terre, C Guadeloupe 16°12′N 61°36′W
15 Y5 **Petit-Cap** Québec, SE Canada 49°00′N 64°26′W
45 Y6 **Petit Cul-de-Sac Marin** *bay* C Guadeloupe
44 M9 **Petite-Rivière-de-l'Artibonite** C Haiti 19°10′N 72°30′W
173 X16 **Petite Rivière Noire, Piton de la** ▲ C Mauritius
15 R9 **Petite-Rivière-St-François** Québec, SE Canada 47°18′N 70°34′W
44 L9 **Petit-Goâve** S Haiti 18°27′N 72°51′W
Petitjean *see* Sidi-Kacem
13 N10 **Petit Lac Manicouagan** ◎ Québec, E Canada
19 T7 **Petit Manan Point** *headland* Maine, NE USA 44°23′N 67°54′W
Petit Mécatina, Rivière du *see* Little Mecatina
11 N10 **Petitot** ≈ Alberta/British Columbia, W Canada
45 S12 **Petit Piton** ▲ SW Saint Lucia 13°49′N 61°03′W
Petit-Popo *see* Aného
Petit St-Bernard, Col du *see* Little Saint Bernard Pass
13 O8 **Petitsikapau Lake** ◎ Newfoundland and Labrador, E Canada
92 L11 **Petkula** Lappi, N Finland 67°41′N 26°44′E
41 X12 **Peto** Yucatán, SE Mexico 20°09′N 88°55′W
62 G10 **Petorca** Valparaíso, C Chile 32°18′S 70°49′W
31 Q5 **Petoskey** Michigan, N USA 45°51′N 88°03′W
138 G14 **Petra** *archaeological site* Ma'ān, W Jordan
Petra *see* Wādī Mūsā
115 F14 **Pétras, Sténa** *pass* N Greece
123 S16 **Petra Velikogo, Zaliv** *bay* SE Russian Federation
Petrel *see* Petrer
14 K15 **Petre, Point** *headland* Ontario, SE Canada 43°49′N 77°07′W
105 S12 **Petrer** *var.* Petrel. Valenciana, E Spain 38°28′N 00°46′W
125 U11 **Petretsovo** Permskiy Kray, NW Russian Federation 61°22′N 57°21′E
114 G12 **Petrich** Blagoevgrad, SW Bulgaria 41°25′N 23°12′E
187 P15 **Petrie, Récif** *reef* N New Caledonia
37 N11 **Petrified Forest** *prehistoric site* Arizona, SW USA
Petrikau *see* Piotrków Trybunalski
Petrikov *see* Pyetrykaw
116 H12 **Petrila** *Hung.* Petrilla. Hunedoara, W Romania 45°27′N 23°25′E
Petrilla *see* Petrila
112 E9 **Petrinja** Sisak-Moslavina, C Croatia 45°27′N 16°14′E
Petroaleksandrovsk *see* To'rtkok'l
Petrőcz *see* Bački Petrovac
124 G12 **Petrodvorets** *Fin.* Pietarhovi. Leningradskaya Oblast', NW Russian Federation 59°53′N 29°52′E
Petrograd *see* Sankt-Peterburg
Petrokov *see* Piotrków Trybunalski
54 G6 **Petrólea** Norte de Santander, NE Colombia 08°30′N 72°35′W
14 D16 **Petrolia** Ontario, S Canada 42°54′N 82°07′W
25 S4 **Petrolia** Texas, SW USA 34°00′N 98°13′W
59 O15 **Petrolina** Pernambuco, E Brazil 09°22′S 40°30′W
45 T6 **Petrona, Punta** *headland* C Puerto Rico 17°57′N 66°23′W
Petropavl *see* Petropavlovsk
117 V7 **Petropavlivka** Dnipropetrovs'ka Oblast', E Ukraine 48°28′N 36°28′E
145 P6 **Petropavlovsk** *Kaz.* Petropavl. Severnyy Kazakhstan, N Kazakhstan 54°47′N 69°06′E
123 V11 **Petropavlovsk-Kamchatskiy** Kamchatskiy Kray, E Russian Federation 53°03′N 158°43′E
60 P9 **Petrópolis** Rio de Janeiro, SE Brazil 22°30′S 43°28′W
116 H12 **Petroşani** *var.* Petroşeni, *Ger.* Petroschen, *Hung.* Petrozsény. Hunedoara, W Romania 45°25′N 23°22′E
Petroschen/Petroşeni *see* Petroşani
Petroskoi *see* Petrozavodsk
112 N12 **Petrovac** Serbia, E Serbia 42°22′N 21°25′E
Petrovac *see* Bosanski Petrovac
113 J17 **Petrovac na Moru** S Montenegro 42°11′N 19°00′E
Petrovac/Petrovácz *see* Bački Petrovac
117 S8 **Petrove** Kirovohrads'ka Oblast', C Ukraine 48°22′N 33°12′E
113 O18 **Petrovec** C FYR Macedonia 41°57′N 21°37′E
Petrovgrad *see* Zrenjanin
127 P7 **Petrovsk** Saratovskaya Oblast', W Russian Federation 52°20′N 45°23′E
Petrovsk-Port *see* Makhachkala
127 P9 **Petrov Val** Volgogradskaya Oblast', SW Russian Federation 50°10′N 45°16′E
124 J11 **Petrozavodsk** *Fin.* Petroskoi. Respublika Kareliya, NW Russian Federation 61°46′N 34°19′E
Petrozsény *see* Petroşani
83 D20 **Petrusdal** Hardap, C Namibia 23°42′S 17°23′E
117 T7 **Petrykivka** Dnipropetrovs'ka Oblast', E Ukraine 48°44′N 34°42′E
Petsamo *see* Pechenga
Petschka *see* Pecica
Pettau *see* Ptuj
109 S5 **Pettenbach** Oberösterreich, C Austria 47°58′N 14°03′E
25 S13 **Pettus** Texas, SW USA 28°34′N 97°49′W
122 G12 **Petukhovo** Kurganskaya Oblast', C Russian Federation 55°04′N 67°49′E
Petuna *see* Songyuan
109 R4 **Peuerbach** Oberösterreich, N Austria 48°19′N 13°45′E
62 G12 **Peumo** Libertador, C Chile 34°20′S 71°12′W
123 T6 **Pevek** Chukotskiy Avtonomnyy Okrug, NE Russian Federation 69°41′N 170°19′E
27 X5 **Pevely** Missouri, C USA 38°16′N 90°24′W
Peyia *see* Pégeia
102 J15 **Peyrehorade** Landes, SW France 43°33′N 01°05′W
124 J14 **Peza** ≈ NW Russian Federation
103 P16 **Pézenas** Hérault, S France 43°28′N 03°25′E
111 H20 **Pezinok** *Ger.* Bösing, *Hung.* Bazin. Bratislavský Kraj, W Slovakia 48°17′N 17°15′E
101 L22 **Pfaffenhofen an der Ilm** Bayern, SE Germany 48°31′N 11°30′E
108 G7 **Pfäffikon** Schwyz, C Switzerland 47°11′N 08°46′E
101 F20 **Pfälzer Wald** *hill range* W Germany
101 N22 **Pfarrkirchen** Bayern, SE Germany 48°25′N 12°56′E
101 G21 **Pforzheim** Baden-Württemberg, SW Germany 48°53′N 08°42′E
101 H24 **Pfullendorf** Baden-Württemberg, S Germany 47°55′N 09°16′E
108 K8 **Pfunds** Tirol, W Austria 46°56′N 10°30′E
101 G19 **Pfungstadt** Hessen, W Germany 49°48′N 08°36′E
83 L20 **Phalaborwa** *var.* Ba-Pahalaborwa. Limpopo, NE South Africa 23°59′S 31°04′E
152 E11 **Phalodi** Rājasthān, NW India 27°06′N 72°22′E
152 E12 **Phalsund** Rājasthān, NW India 26°22′N 71°56′E
155 E15 **Phaltan** Mahārāshtra, W India 18°01′N 74°31′E
167 O7 **Phan** *var.* Muang Phan. Chiang Rai, NW Thailand 19°34′N 99°44′E
167 O14 **Phangan, Ko** *island* SW Thailand
166 M15 **Phang-Nga** *var.* Pang-Nga, Phangnga. Phangnga, SW Thailand 08°29′N 98°31′E
Phangnga *see* Phang-Nga
Phan Rang/Phanrang *see* Phan Rang-Thap Cham
167 V13 **Phan Rang-Thap Cham** *var.* Phanrang, Phan Rang, Phan Rang Thap Cham. Ninh Thuân, S Vietnam 11°34′N 109°00′E
167 U13 **Phan Ri** Binh Thuân, S Vietnam 11°11′N 108°31′E
167 U13 **Phan Thiêt** Binh Thuân, S Vietnam 10°56′N 108°06′E
Pharnacia *see* Giresun
25 S17 **Pharr** Texas, SW USA 26°11′N 98°10′W
Pharus *see* Hvar
167 N16 **Phatthalung** *var.* Padalung, Patalung. Phatthalung, SW Thailand 07°38′N 100°04′E
167 O7 **Phayao** *var.* Muang Phayao. Phayao, NW Thailand 19°10′N 99°55′E
11 U10 **Phelps Lake** ◎ Saskatchewan, C Canada
21 X9 **Phelps Lake** ◎ North Carolina, SE USA
23 R5 **Phenix City** Alabama, S USA 32°28′N 85°00′W
Phet Buri *see* Phetchaburi
167 O11 **Phetchaburi** *var.* Bejraburi, Petchaburi, Phet Buri. Phetchaburi, SW Thailand 13°05′N 99°58′E
167 O9 **Phichit** *var.* Bichitra, Muang Phichit, Pichit. Phichit, C Thailand 16°29′N 100°21′E
22 M5 **Philadelphia** Mississippi, S USA 32°45′N 89°06′W
18 I7 **Philadelphia** New York, NE USA 44°10′N 75°40′W
18 I16 **Philadelphia** Pennsylvania, NE USA 40°N 75°10′W
18 I16 **Philadelphia** ✈ Pennsylvania, NE USA 39°51′N 75°13′W
Philadelphia *see* 'Amman
28 L10 **Philip** South Dakota, N USA 44°02′N 101°39′W
99 H22 **Philippeville** Namur, S Belgium 50°12′N 04°33′E
Philippeville *see* Skikda
21 S3 **Philippi** West Virginia, NE USA 39°08′N 80°03′W
Philippi *see* Fílippoi
195 Y9 **Philippi Glacier** *glacier* Antarctica
192 G6 **Philippine Basin** *undersea feature* W Pacific Ocean 17°00′N 132°00′E
129 X12 **Philippine Plate** *tectonic feature*
171 O5 **Philippines** *off.* Republic of the Philippines. ◆ *republic* SE Asia
129 X13 **Philippines** *island group* W Pacific Ocean
171 P3 **Philippine Sea** *sea* W Pacific Ocean
Philippines, Republic of the *see* Philippines
192 F6 **Philippine Trench** *undersea feature* W Philippine Sea
83 H23 **Philippolis** Free State, C South Africa 30°16′S 25°16′E
Philippopolis *see* Plovdiv
Philippopolis *see* Shahbā', Syria
45 V9 **Philipsburg** ○ Sint Maarten 17°58′N 63°02′W
33 P10 **Philipsburg** Montana, NW USA 46°19′N 113°17′W
39 R6 **Philip Smith Mountains** ▲▲ Alaska, USA
152 H8 **Phillaur** Punjab, N India 31°02′N 75°50′E
183 N13 **Phillip Island** *island* Victoria, SE Australia
25 N2 **Phillips** Texas, SW USA 35°39′N 101°21′W
30 K5 **Phillips** Wisconsin, N USA 45°42′N 90°23′W
26 K3 **Phillipsburg** Kansas, C USA 39°45′N 99°19′W
18 I14 **Phillipsburg** New Jersey, NE USA 40°39′N 75°09′W
21 S7 **Philpott Lake** ⊠ Virginia, NE USA
Phintias *see* Licata
167 P9 **Phitsanulok** *var.* Bisnulok, Muang Phitsanulok, Pitsanulok. Phitsanulok, C Thailand 16°49′N 100°15′E
Phlórina *see* Flórina
Phnom Penh *see* Phnum Penh
167 S13 **Phnum Penh** *var.* Phnom Penh. ● (Cambodia) Phnum Penh, S Cambodia 11°35′N 104°55′E
167 S11 **Phnum Tbêng Meanchey** Preăh Vihéar, N Cambodia 13°45′N 104°58′E
36 K13 **Phoenix** *state capital* Arizona, SW USA 33°27′N 112°04′W
Phoenix Island *see* Rawaki
191 R3 **Phoenix Islands** *island group* C Kiribati
18 I15 **Phoenixville** Pennsylvania, NE USA 40°07′N 75°31′W
83 K22 **Phofung** *var.* Mont-aux-Sources. ▲ N Lesotho 28°47′S 28°52′E
167 Q10 **Phon** Khon Kaen, E Thailand 15°47′N 102°35′E
167 Q5 **Phôngsali** *var.* Phong Saly. Phôngsali, N Laos 21°40′N 102°04′E
Phong Saly *see* Phôngsali
Phônhông *see* Ban Phônhông
167 R7 **Phônsaven** *var.* Pèk, Xieng Khouang; *prev.* Xiangkhoang. Xiangkhoang, N Laos 19°19′N 103°23′E
167 R5 **Phô Rang** *var.* Bao Yèn. Lao Cai, N Vietnam 22°12′N 104°27′E
Phort Láirge, Cuan *see* Waterford Harbour
Phou Louang *see* Annamite Mountains
167 N10 **Phra Chedi Sam Ong** Kanchanaburi, W Thailand 15°18′N 98°26′E
167 O8 **Phrae** *var.* Muang Phrae, Prae. Phrae, NW Thailand 18°07′N 100°09′E
Phra Nakhon Si Ayutthaya *see* Ayutthaya
167 M14 **Phra Thong, Ko** *island* SW Thailand
Phu Cường *see* Thu Dâu Môt
166 M15 **Phuket** *var.* Bhuket, Puket, *Mal.* Ujung Salang; *prev.* Junkseylon, Salang. Phuket, SW Thailand 07°52′N 98°22′E
166 M15 **Phuket** ✈ Phuket, SW Thailand 08°03′N 98°16′E
166 M15 **Phuket, Ko** *island* SW Thailand
154 N12 **Phulabāni** *prev.* Phulbani. Odisha, E India 20°30′N 84°18′E
Phulbani *see* Phulabāni
167 U9 **Phu Lôc** Th,a Thiên-Huê, C Vietnam 16°13′N 107°53′E
167 R13 **Phumĭ Chŏâm** Kâmpóng Spœ, SW Cambodia 11°42′N 103°58′E
Phumĭ Kalêng *see* Kalêng
Phumĭ Kâmpóng Trâbêk *see* Kâmpóng Trâbêk
Phumĭ Koŭk Kduŏch *see* Koŭk Kduŏch
Phumĭ Labăng *see* Labăng
Phumĭ Mlu Prey *see* Mlu Prey
Phumĭ Moŭng *see* Moŭng
Phumĭ Prâmaôy *see* Prâmaôy
Phumĭ Samĭt *see* Samĭt
Phumĭ Sâmraông *see* Sâmraông
Phumĭ Siĕmbok *see* Siĕmbok
Phumĭ Thalabârĭvăt *see* Thalabârĭvăt
Phumĭ Veal Renh *see* Veal Renh
Phumĭ Yeay Sên *see* Yeay Sên
Phum Kompong Trabek *see* Kâmpóng Trâbêk
Phum Samrong *see* Sâmraông
167 V11 **Phu My** Binh Đinh, C Vietnam 14°07′N 109°05′E
Phung Hiêp *see* Tân Hiêp
153 T12 **Phuntsholing** SW Bhutan 26°52′N 89°26′E
167 V13 **Phư ơ c Dân** Ninh Thuân, S Vietnam 11°28′N 108°53′E
167 R15 **Phươc Long** Minh Hai, S Vietnam 09°27′N 105°25′E
Phươc Sơn *see* Khâm Đưc
167 R14 **Phu Quôc, Đao** *var.* Phu Quoc Island. *island* S Vietnam
Phu Quoc Island *see* Phu Quôc, Đao
167 S6 **Phu Tho** Vinh Phu, N Vietnam 21°23′N 105°13′E
Phu Vinh *see* Tra Vinh
166 M7 **Phyu** *var.* Hpyu, Pyu. Bago, C Myanmar (Burma) 18°29′N 96°28′E
189 T13 **Piaanu Pass** *passage* Chuuk Islands, C Micronesia
106 E8 **Piacenza** *Fr.* Paisance; *anc.* Placentia. Emilia-Romagna, N Italy 45°02′N 09°42′E
107 K14 **Pianella** Abruzzo, C Italy 42°23′N 14°04′E
107 M15 **Pianosa, Isola** *island* Archipelago Toscano, C Italy
171 U13 **Piar** Papua Barat, E Indonesia 02°49′S 132°46′E
45 U14 **Piarco** *var.* Port of Spain. ✈ (Port-of-Spain) Trinidad, Trinidad and Tobago 10°36′N 61°21′W
110 M12 **Piaseczno** Mazowieckie, C Poland 52°03′N 21°00′E
116 I15 **Piatra** Teleorman, S Romania 43°49′N 25°10′E
116 L10 **Piatra-Neamţ** *Hung.* Karácsonkő. Neamţ, NE Romania 46°54′N 26°23′E
Piauhy *see* Piauí
59 N15 **Piauí** *off.* Estado do Piauí; *prev.* Piauhy. ◈ *state* E Brazil
Piauí, Estado do *see* Piauí
106 I7 **Piave** ≈ NE Italy
107 K24 **Piazza Armerina** *var.* Chiazza. Sicilia, Italy, C Mediterranean Sea 37°23′N 14°22′E
81 G14 **Pibor** *Amh.* Pībor Wenz. ≈ Ethiopia/South Sudan
81 G14 **Pibor Post** Jonglei, E South Sudan 06°50′N 33°06′E
Pībor Wenz *see* Pibor
Pibrans *see* Příbram
36 K11 **Picacho Butte** ▲ Arizona, SW USA 35°12′N 112°44′W
40 D4 **Picachos, Cerro** ▲ NW Mexico 29°15′N 114°04′W
103 O4 **Picardie** *Eng.* Picardy. ◈ *region* N France
Picardy *see* Picardie
22 L8 **Picayune** Mississippi, S USA 30°31′N 89°40′W
Piccolo San Bernardo, Colle di *see* Little Saint Bernard Pass
PicdeBälaitous *see* Balaïtous
62 K5 **Pichanal** Salta, N Argentina 23°18′S 64°10′W
147 P12 **Pichandar** W Tajikistan 38°44′N 68°51′E
27 R8 **Picher** Oklahoma, C USA 36°59′N 94°49′W
62 G12 **Pichilemu** Libertador, C Chile 34°25′S 72°00′W
40 F9 **Pichilingue** Baja California Sur, NW Mexico 24°20′N 110°17′W
56 B6 **Pichincha** ◈ *province* N Ecuador
56 C6 **Pichincha** ▲ N Ecuador 0°12′S 78°39′W
Pichit *see* Phichit
41 U15 **Pichucalco** Chiapas, SE Mexico 17°32′N 93°07′W
22 L5 **Pickens** Mississippi, S USA 32°52′N 89°58′W
21 O11 **Pickens** South Carolina, SE USA 34°53′N 82°42′W
14 G11 **Pickerel** ≈ Ontario, S Canada
14 H15 **Pickering** Ontario, S Canada 43°50′N 79°03′W
97 N16 **Pickering** N England, United Kingdom 54°14′N 00°47′W
31 S13 **Pickerington** Ohio, N USA 39°52′N 82°45′W
12 C10 **Pickle Lake** Ontario, C Canada 51°30′N 90°10′W
29 P12 **Pickstown** South Dakota, N USA 43°02′N 98°31′W
25 V6 **Pickton** Texas, SW USA 33°01′N 95°19′W
23 N1 **Pickwick Lake** ⊠ S USA
64 N2 **Pico** *var.* Ilha do Pico. *island* Azores, Portugal, NE Atlantic Ocean
63 J19 **Pico de Salamanca** Chubut, SE Argentina 45°26′S 67°26′W
1 P9 **Pico Fracture Zone** *tectonic feature* NW Atlantic Ocean
Pico, Ilha do *see* Pico
59 O14 **Picos** Piauí, E Brazil 07°05′S 41°24′W
63 I20 **Pico Truncado** Santa Cruz, SE Argentina 46°45′S 68°00′W
183 S9 **Picton** New South Wales, SE Australia 34°12′S 150°36′E
14 K15 **Picton** Ontario, SE Canada 43°59′N 77°09′W
185 K14 **Picton** Marlborough, South Island, New Zealand 41°18′S 174°00′E
63 H15 **Pićun Leufú, Arroyo** ≈ SW Argentina
Pidálion *see* Gkréko, Akrotíri
155 K25 **Pidurutalagala** ▲ S Sri Lanka 07°03′N 80°47′E
116 K6 **Pidvolochys'k** Ternopil's'ka Oblast', W Ukraine 49°31′N 26°09′E
107 K16 **Piedimonte Matese** Campania, S Italy 41°20′N 14°30′E
27 X7 **Piedmont** Missouri, C USA 37°09′N 90°42′W
21 P11 **Piedmont** South Carolina, SE USA 34°42′N 82°27′W
17 S12 **Piedmont** *escarpment* E USA
Piedmont *see* Piemonte
31 U13 **Piedmont Lake** ⊠ Ohio, N USA
104 M11 **Piedrabuena** Castilla-La Mancha, C Spain 39°02′N 04°10′W
Piedrafita, Puerto de *see* Pedrafita, Porto de
104 L8 **Piedrahita** Castilla y León, N Spain 40°27′N 05°20′W
41 N6 **Piedras Negras** *var.* Ciudad Porfirio Díaz. Coahuila, NE Mexico 28°40′N 100°32′W
61 E21 **Piedras, Punta** *headland* E Argentina 35°27′S 57°04′W
57 I14 **Piedras, Río de las** ≈ E Peru
111 J16 **Piekary Śląskie** Śląskie, S Poland 50°24′N 18°58′E
93 M17 **Pieksämäki** Etelä-Savo, E Finland 62°18′N 27°10′E
109 V5 **Pielach** ≈ NE Austria
93 M16 **Pielavesi** Pohjois-Savo, C Finland 63°14′N 26°45′E
93 M16 **Pielavesi** ◎ C Finland
93 N16 **Pielinen** *var.* Pielisjärvi. ◎ E Finland
Pielisjärvi *see* Pielinen
106 A8 **Piemonte** *Eng.* Piedmont. ◈ *region* NW Italy
111 L18 **Pieniny** ▲▲ S Poland
111 E14 **Pieńsk** *Ger.* Penzig. Dolnośląskie, SW Poland 51°14′N 15°03′E
29 Q13 **Pierce** Nebraska, C USA 42°12′N 97°31′W
11 R14 **Pierceland** Saskatchewan, C Canada
115 E14 **Piéria** ▲▲ N Greece
29 N10 **Pierre** *state capital* South Dakota, N USA 44°22′N 100°21′W
102 K16 **Pierrefitte-Nestalas** Hautes-Pyrénées, S France 42°57′N 00°04′W
103 R14 **Pierrelatte** Drôme, E France 44°22′N 04°40′E
15 P11 **Pierreville** Québec, SE Canada 46°05′N 72°48′W
15 O7 **Pierriche** ≈ Québec, SE Canada
111 H20 **Piešt'any** *Ger.* Pistyan, *Hung.* Pöstyén. Tranavský Kraj, W Slovakia 48°37′N 17°48′E
109 X5 **Piesting** ≈ E Austria
Pietarhovi *see* Petrodvorets
Pietari *see* Sankt-Peterburg
Pietarsaari *see* Jakobstad
83 K23 **Pietermaritzburg** *var.* Maritzburg. KwaZulu/Natal, E South Africa 29°35′S 30°23′E
Pietersburg *see* Polokwane
107 K24 **Pietraperzia** Sicilia, Italy, C Mediterranean Sea 37°25′N 14°08′E
107 N22 **Pietra Spada, Passo della** *pass* SW Italy
Piet Retief *see* eMkhondo
116 J10 **Pietrosul, Vârful** *prev.* Vîrful Pietrosu. ▲ N Romania 47°06′N 25°09′E
106 I6 **Pieve di Cadore** Veneto, NE Italy 46°27′N 12°22′E
14 C18 **Pigeon Bay** *lake bay* Ontario, S Canada
27 X8 **Piggott** Arkansas, C USA 36°22′N 90°11′W
83 L21 **Piggs Peak** NW Swaziland 25°58′S 31°17′E
Pigs, Bay of *see* Cochinos, Bahía de
61 A23 **Pigüé** Buenos Aires, E Argentina 37°38′S 62°27′W
41 O12 **Piguícas** ▲ C Mexico 21°08′N 99°37′W
193 W15 **Piha Passage** *passage* S Tonga
Pihkva Järv *see* Pskov, Lake
93 N18 **Pihlajavesi** ◎ SE Finland
93 J18 **Pihlava** Satakunta, SW Finland 61°33′N 21°36′E
93 L16 **Pihtipudas** Keski-Suomi, C Finland 63°20′N 25°37′E
40 L14 **Pihuamo** Jalisco, SW Mexico 19°20′N 103°21′W
189 U11 **Piis Moen** *var.* Pis. *atoll* Chuuk Islands, C Micronesia
41 U17 **Pijijiapán** Chiapas, SE Mexico 15°42′N 93°12′W
98 G12 **Pijnacker** Zuid-Holland, W Netherlands 52°01′N 04°26′E
42 H5 **Pijol, Pico** ▲ NW Honduras 15°07′N 87°39′W
Pikaar *see* Bikar Atoll
124 I13 **Pikalevo** Leningradskaya Oblast', NW Russian Federation 59°33′N 34°04′E
188 M15 **Pikelot** *island* Caroline Islands, C Micronesia
30 M5 **Pike River** ≈ Wisconsin, N USA
37 T5 **Pikes Peak** ▲ Colorado, C USA 38°51′N 105°06′W
21 P6 **Pikeville** Kentucky, S USA 37°29′N 82°33′W
20 L9 **Pikeville** Tennessee, S USA 35°35′N 85°11′W
Pikinni *see* Bikini Atoll
79 H18 **Pikounda** Sangha, C Congo 0°30′N 16°44′E
110 G9 **Piła** *Ger.* Schneidemühl. Wielkopolskie, C Poland 53°09′N 16°44′E
62 N6 **Pilagá, Riacho** ≈ NE Argentina
61 D20 **Pilar** Buenos Aires, E Argentina 34°28′S 58°55′W
62 N7 **Pilar** *var.* Villa del Pilar. Ñeembucú, S Paraguay 26°55′S 58°20′W
62 N6 **Pilcomayo, Rio** ≈ C South America
147 R12 **Pildon** *Rus.* Pil'don. C Tajikistan 39°10′N 71°00′E
Piles *see* Pylés
Pilgram *see* Pelhřimov
152 L10 **Pīlībhīt** Uttar Pradesh, N India 28°37′N 79°48′E
110 M13 **Pilica** ≈ C Poland
115 G16 **Pílio** ▲▲ C Greece
111 J22 **Pilisvörösvár** Pest, N Hungary 47°38′N 18°55′E
65 G15 **Pillar Bay** *bay* Ascension Island, C Atlantic Ocean
183 P17 **Pillar, Cape** *headland* Tasmania, SE Australia 43°13′S 147°58′E
Pillau *see* Baltiysk
183 R5 **Pilliga** New South Wales, SE Australia 30°22′S 148°53′E
44 H8 **Pilón** Granma, E Cuba 19°54′N 77°20′W
Pilos *see* Pýlos
11 W17 **Pilot Mound** Manitoba, S Canada 49°12′N 98°49′W
21 S8 **Pilot Mountain** North Carolina, SE USA 36°23′N 80°28′W
39 O14 **Pilot Point** Alaska, USA 57°33′N 157°34′W
25 T5 **Pilot Point** Texas, SW USA 33°24′N 96°57′W
32 K11 **Pilot Rock** Oregon, NW USA 45°28′N 118°49′W
38 M11 **Pilot Station** Alaska, USA 61°56′N 162°52′W
Pilsen *see* Plzeň
111 K18 **Pilsko** ▲ N Slovakia 49°31′N 19°21′E
Pilten *see* Piltene
118 D8 **Piltene** *Ger.* Pilten. W Latvia 57°14′N 21°41′E
111 M16 **Pilzno** Podkarpackie, SE Poland 49°58′N 21°18′E
Pilzno *see* Plzeň
37 N14 **Pima** Arizona, SW USA 32°49′N 109°50′W
58 H13 **Pimenta** Pará, N Brazil 04°32′S 56°17′W
59 F16 **Pimenta Bueno** Rondônia, W Brazil 11°40′S 61°14′W
56 B11 **Pimentel** Lambayeque, W Peru 06°51′S 79°53′W
105 S6 **Pina** Aragón, NE Spain 41°28′N 00°31′W
119 I20 **Pina** ≈ SW Belarus
40 E2 **Pinacate, Sierra del** ▲ NW Mexico 31°49′N 113°30′W
63 H22 **Pináculo, Cerro** ▲ S Argentina 50°46′S 72°07′W
191 X11 **Pinaki** *atoll* Îles Tuamotu, E French Polynesia
37 N15 **Pinaleno Mountains** ▲▲ Arizona, SW USA
171 P4 **Pinamalayan** Mindoro, N Philippines 13°00′N 121°30′E
169 Q10 **Pinang** Borneo, C Indonesia 0°36′N 109°11′W
168 J7 **Pinang** *var.* Penang. ◈ *state* Peninsular Malaysia
Pinang *see* George Town
Pinang *see* Pinang, Pulau, Peninsular Malaysia
168 J7 **Pinang, Pulau** *var.* Penang, Pinang; *prev.* Prince of Wales Island. *island* Peninsular Malaysia
44 B5 **Pinar del Río** Pinar del Río, W Cuba 22°24′N 83°42′W
114 N11 **Pınarhisar** Kırklareli, NW Turkey 41°37′N 27°32′E
171 O3 **Pinatubo, Mount** ⛰ Luzon, N Philippines 15°08′N 120°21′E
11 Y16 **Pinawa** Manitoba, S Canada 50°09′N 95°52′W
11 Q17 **Pincher Creek** Alberta, SW Canada 49°31′N 113°53′W
30 L16 **Pinckneyville** Illinois, N USA 38°04′N 89°22′W
Pîncota *see* Pâncota
111 L15 **Pińczów** Świętokrzyskie, C Poland 50°30′N 20°30′E
149 U7 **Pind Dādan Khān** Punjab, E Pakistan 32°36′N 73°07′E
Píndhos/Píndhos Óros *see* Píndos
149 V8 **Pindi Bhattiān** Punjab, E Pakistan 31°54′N 73°20′E
149 U6 **Pindi Gheb** Punjab, E Pakistan 33°16′N 72°21′E
115 D15 **Píndos** *var.* Píndhos Óros, *Eng.* Pindus Mountains; *prev.* Píndhos. ▲▲ C Greece
Pindus Mountains *see* Píndos
18 J16 **Pine Barrens** *physical region* New Jersey, NE USA
27 V12 **Pine Bluff** Arkansas, C USA 34°15′N 92°00′W
23 X11 **Pine Castle** Florida, SE USA 28°28′N 81°22′W
29 V7 **Pine City** Minnesota, N USA 45°49′N 92°55′W
181 P2 **Pine Creek** Northern Territory, N Australia 13°51′S 131°51′E
35 V4 **Pine Creek** ≈ Nevada, W USA
18 F13 **Pine Creek** ≈ Pennsylvania, NE USA
27 Q13 **Pine Creek Lake** ⊠ Oklahoma, C USA
33 T15 **Pinedale** Wyoming, C USA 42°52′N 109°51′W
11 X15 **Pine Dock** Manitoba, S Canada 51°34′N 96°47′W
11 Y16 **Pine Falls** Manitoba, S Canada 50°29′N 96°12′W

◆ Country
● Country Capital
◇ Dependent Territory
○ Dependent Territory Capital
◈ Administrative Regions
✈ International Airport
▲ Mountain
▲▲ Mountain Range
⛰ Volcano
≈ River
◎ Lake
⊠ Reservoir

PINE FLAT LAKE – POJOAQUE

35 *R10* **Pine Flat Lake** ⊠ California, W USA
125 *N8* **Pinega** Arkhangel'skaya Oblast', NW Russian Federation *64°40′N 43°24′E*
125 *N8* **Pinega** ≈ NW Russian Federation
15 *N12* **Pine Hill** Québec, SE Canada *45°44′N 74°30′W*
11 *T12* **Pinehouse Lake** ◎ Saskatchewan, C Canada
21 *T10* **Pinehurst** North Carolina, SE USA *35°12′N 79°28′W*
115 *D19* **Pineiós** ≈ S Greece
115 *E16* **Pineiós** *var.* Piniós; *anc.* Peneius. ≈ C Greece
29 *W10* **Pine Island** Minnesota, N USA *44°12′N 92°39′W*
23 *V15* **Pine Island** *island* Florida, SE USA
194 *K10* **Pine Island Glacier** *glacier* Antarctica
25 *X9* **Pineland** Texas, SW USA *31°15′N 93°58′W*
23 *V13* **Pinellas Park** Florida, SE USA *27°50′N 82°42′W*
10 *M13* **Pine Pass** *pass* British Columbia, W Canada
8 *J10* **Pine Point** Northwest Territories, W Canada *60°52′N 114°30′W*
28 *K12* **Pine Ridge** South Dakota, N USA *43°01′N 102°33′W*
29 *U6* **Pine River** Minnesota, N USA *46°43′N 94°24′W*
31 *Q8* **Pine River** ≈ Michigan, N USA
30 *M4* **Pine River** ≈ Wisconsin, N USA
106 *A8* **Pinerolo** Piemonte, NE Italy *44°56′N 07°21′E*
115 *I15* **Píneś, Akrotírio** *var.* Akrotírio Pínnes. *headland* N Greece *40°06′N 24°19′E*
25 *W6* **Pines, Lake O' the** ⊠ Texas, SW USA
Pines, The Isle of the *see* Juventud, Isla de la
Pine Tree State *see* Maine
21 *N7* **Pineville** Kentucky, S USA *36°47′N 83°43′W*
22 *H7* **Pineville** Louisiana, S USA *31°19′N 92°25′W*
27 *R8* **Pineville** Missouri, C USA *36°36′N 94°23′W*
21 *R10* **Pineville** North Carolina, SE USA *35°04′N 80°53′W*
21 *Q6* **Pineville** West Virginia, NE USA *37°35′N 81°34′W*
33 *V8* **Piney Buttes** *physical region* Montana, NW USA
163 *W9* **Ping'an** Jilin, NE China *44°36′N 127°13′E*
160 *H14* **Pingbian** *var.* Pingbian Miaozu Zizhixian, Yuping. Yunnan, SW China *22°51′N 103°28′E*
Pingbian Miaozu Zizhixian *see* Pingbian
157 *S9* **Pingdingshan** Henan, C China *33°52′N 113°20′E*
161 *S14* **Pingdong** *Jap.* Heitō; *prev.* P'ingtung. S Taiwan *22°40′N 120°30′E*
161 *R4* **Pingdu** Shandong, E China *36°50′N 119°55′E*
189 *W16* **Pingelap Atoll** *atoll* Caroline Islands, E Micronesia
160 *K14* **Pingguo** *var.* Matou. Guangxi Zhuangzu Zizhiqu, S China *23°24′N 107°30′E*
161 *Q13* **Pinghe** *var.* Xiaoxi. Fujian, SE China *24°30′N 117°19′E*
P'ing-hsiang *see* Pingxiang
161 *N10* **Pingjiang** Hunan, S China *28°44′N 113°33′E*
Pingkiang *see* Harbin
160 *L8* **Pingli** Shaanxi, C China *32°27′N 109°21′E*
159 *W10* **Pingliang** *var.* Kongtong, P'ing-liang. Gansu, C China *35°27′N 106°38′E*
P'ing-liang *see* Pingliang
159 *W8* **Pingluo** Ningxia, N China *38°55′N 106°31′E*
Pingma *see* Tiandong
167 *O7* **Ping, Mae Nam** ≈ W Thailand
161 *Q1* **Pingquan** Hebei, E China *41°02′N 118°35′E*
29 *P5* **Pingree** North Dakota, N USA *47°07′N 98°54′W*
Pingsiang *see* Pingxiang
P'ingtung *see* Pingdong
160 *I8* **Pingwu** *var.* Long'an. Sichuan, C China *32°33′N 104°32′E*
160 *J15* **Pingxiang** Guangxi Zhuangzu Zizhiqu, S China *22°03′N 106°44′E*
161 *O11* **Pingxiang** *var.* P'ing-hsiang; *prev.* Pingsiang. Jiangxi, S China *27°42′N 113°50′E*
Pingxiang *see* Tongwei
161 *S11* **Pingyang** *var.* Kunyang. Zhejiang, SE China *27°46′N 120°37′E*
161 *P5* **Pingyi** Shandong, E China *35°30′N 117°38′E*
161 *P5* **Pingyin** Shandong, E China *36°18′N 116°24′E*
60 *H13* **Pinhalzinho** Santa Catarina, S Brazil *26°53′S 52°57′W*
60 *I12* **Pinhão** Paraná, S Brazil *25°46′S 51°32′W*
61 *H17* **Pinheiro Machado** Rio Grande do Sul, S Brazil *31°34′S 53°22′W*
104 *I7* **Pinhel** Guarda, N Portugal *40°47′N 07°03′W*
Piniós *see* Pineiós
168 *I11* **Pini, Pulau** *island* Kepulauan Batu, W Indonesia
109 *Y7* **Pinka** ≈ SE Austria
109 *X7* **Pinkafeld** Burgenland, SE Austria *47°23′N 16°08′E*
Pinkiang *see* Harbin
10 *M12* **Pink Mountain** British Columbia, W Canada *57°10′N 122°36′W*
166 *M3* **Pinlebu** Sagaing, N Myanmar (Burma) *24°02′N 95°21′E*
38 *J12* **Pinnacle Island** *island* Alaska, USA
180 *I12* **Pinnacles, The** *tourist site* Western Australia
182 *K10* **Pinnaroo** South Australia *35°17′S 140°54′E*
Pinne *see* Pniewy
100 *I9* **Pinneberg** Schleswig-Holstein, N Germany *53°40′N 09°49′E*
Pínnes, Akrotírio *see* Píneś, Akrotírio
Pinos, Isla de *see* Juventud, Isla de la
35 *R14* **Pinos, Mount** ▲ California, W USA *34°48′N 119°09′W*
105 *R12* **Pinoso** Valenciana, E Spain *38°25′N 01°02′W*
105 *N14* **Pinos-Puente** Andalucía, S Spain *37°16′N 03°46′W*
41 *Q17* **Pinotepa Nacional** *var.* Santiago Pinotepa Nacional. Oaxaca, SE Mexico *16°20′N 98°02′W*
114 *F13* **Pinovo** ▲ N Greece *41°06′N 22°19′E*
187 *R17* **Pins, Île des** *var.* Kunyé. *island* E New Caledonia
119 *I20* **Pinsk** *Pol.* Pińsk. Brestskaya Voblasts', SW Belarus *52°07′N 26°07′E*
14 *D18* **Pins, Pointe aux** *headland* Ontario, S Canada *42°14′N 81°53′W*
57 *B16* **Pinta, Isla** *var.* Abingdon. *island* Galapagos Islands, Ecuador, E Pacific Ocean
125 *Q12* **Pinyug** Kirovskaya Oblast', NW Russian Federation *60°12′N 47°45′E*
57 *B17* **Pinzón, Isla** *var.* Duncan Island. *island* Galapagos Islands, Ecuador, E Pacific Ocean
35 *Y8* **Pioche** Nevada, W USA *37°57′N 114°30′W*
106 *F13* **Piombino** Toscana, C Italy *42°54′N 10°30′E*
0 *C9* **Pioneer Fracture Zone** *tectonic feature* NE Pacific Ocean
122 *L5* **Pioner, Ostrov** *island* Severnaya Zemlya, N Russian Federation
118 *A13* **Pionerskiy** *Ger.* Neukuhren. Kaliningradskaya Oblast', W Russian Federation *54°57′N 20°16′E*
110 *N13* **Pionki** Mazowieckie, C Poland *51°30′N 21°27′E*
184 *L9* **Piopio** Waikato, North Island, New Zealand *38°27′S 175°00′E*
110 *K13* **Piotrków Trybunalski** *Ger.* Petrikau, *Rus.* Petrokov. Łodzkie, C Poland *51°25′N 19°42′E*
152 *F12* **Pīpār Road** Rājasthān, N India *26°25′N 73°29′E*
115 *I16* **Pipéri** *island* Vóreies Sporádes, Greece, Aegean Sea
29 *S10* **Pipestone** Minnesota, N USA *44°00′N 96°19′W*
12 *C9* **Pipestone** ≈ Ontario, C Canada
61 *E21* **Pipinas** Buenos Aires, E Argentina *35°32′S 57°20′W*
149 *T7* **Piplān** *prev.* Liaqatabad. Punjab, E Pakistan *32°17′N 71°24′E*
15 *R5* **Pipmuacan, Réservoir** ⊠ Québec, SE Canada
Piqan *see* Shanshan
31 *R13* **Piqua** Ohio, N USA *40°08′N 84°14′W*
105 *P5* **Piqueras, Puerto de** *pass* N Spain
60 *H11* **Piquiri, Rio** ≈ S Brazil
60 *L9* **Piracicaba** São Paulo, S Brazil *22°45′S 47°40′W*
Piraeus/Piraiévs *see* Peiraías
60 *K10* **Piraju** São Paulo, S Brazil *23°12′S 49°24′W*
60 *K9* **Pirajuí** São Paulo, S Brazil *21°58′S 49°27′W*
63 *G21* **Pirámide, Cerro** ▲ S Chile *49°06′S 73°32′W*
Piramiva *see* Pyramíva
109 *R13* **Piran** *It.* Pirano. SW Slovenia *45°35′N 13°35′E*
62 *N6* **Pirané** Formosa, N Argentina *25°42′S 59°06′W*
59 *J18* **Piranhas** Goiás, S Brazil *16°24′S 51°51′W*
Pirano *see* Piran
142 *J4* **Pīrānshahr** Āžarbāyjān-e Gharbī, NW Iran *36°41′N 45°08′E*
59 *M19* **Pirapora** Minas Gerais, NE Brazil *17°20′S 44°54′W*
60 *I9* **Pirapòzinho** São Paulo, S Brazil *22°17′S 51°31′W*
61 *G19* **Pirarajá** Lavalleja, S Uruguay *33°44′S 54°45′W*
60 *L9* **Pirassununga** São Paulo, S Brazil *21°58′S 47°23′W*
45 *V6* **Pirata, Monte** ▲ E Puerto Rico *18°06′N 65°33′W*
60 *I13* **Piratuba** Santa Catarina, S Brazil *27°26′S 51°47′W*
114 *I9* **Pirdop** *prev.* Strednogorie. Sofia, W Bulgaria *42°44′N 24°09′E*
191 *P7* **Pirea** Tahiti, W French Polynesia
59 *K18* **Pirenópolis** Goiás, S Brazil *15°48′S 49°00′W*
153 *S13* **Pirganj** Rajshahi, NW Bangladesh *25°51′N 88°25′E*
Pirgi *see* Pyrgí
Pírgos *see* Pýrgos
61 *F20* **Piriápolis** Maldonado, S Uruguay *34°51′S 55°15′W*
114 *G11* **Pirin** ▲▲ SW Bulgaria
Pirineos *see* Pyrenees
58 *N13* **Piripiri** Piauí, E Brazil *04°15′S 41°46′W*
118 *H4* **Pirita** *var.* Pirita Jõgi. ≈ NW Estonia
Pirita Jõgi *see* Pirita
54 *J6* **Píritu** Portuguesa, N Venezuela *09°21′N 69°16′W*
93 *L18* **Pirkanmaa** *Swe.* Birkaland. ◈ *region* W Finland
93 *L18* **Pirkkala** Pirkanmaa, W Finland *61°27′N 23°47′E*
101 *F20* **Pirmasens** Rheinland-Pfalz, SW Germany *49°12′N 07°37′E*
101 *P16* **Pirna** Sachsen, E Germany *50°57′N 13°56′E*
Piroe *see* Piru
113 *Q15* **Pirot** Serbia, SE Serbia *43°12′N 22°34′E*
152 *H6* **Pīr Panjāl Range** ▲▲ NE India
43 *W16* **Pirre, Cerro** ▲ SE Panama *07°54′N 77°42′W*
137 *Y11* **Pirsaat** *Rus.* Pirsagat. ≈ E Azerbaijan
Pirsagat *see* Pirsaat
143 *V11* **Pīr Shūrān, Selseleh-ye** ▲▲ SE Iran
92 *M12* **Pirttikoski** Lappi, N Finland *66°20′N 27°08′E*
Pirttikylä *see* Pörtom
171 *R13* **Piru** *prev.* Piroe. Pulau Seram, E Indonesia *03°01′S 128°10′E*
Piryatin *see* Pyryatyn
Pis *see* Piis Moen
106 *F11* **Pisa** *var.* Pisae. Toscana, C Italy *43°43′N 10°23′E*
Pisae *see* Pisa
189 *V12* **Pisar** *atoll* Chuuk Islands, C Micronesia
Pisaurum *see* Pesaro
14 *M10* **Piscatosine, Lac** ◎ Québec, SE Canada
109 *W7* **Pischeldorf** Steiermark, SE Austria *47°11′N 15°48′E*
Pischk *see* Simeria
107 *L19* **Pisciotta** Campania, S Italy *40°07′N 15°13′E*
57 *E16* **Pisco** Ica, SW Peru *13°46′S 76°12′W*
116 *G9* **Pişcolt** *Hung.* Piskolt. Satu Mare, NW Romania *47°35′N 22°18′E*
57 *E16* **Pisco, Río** ≈ E Peru
111 *C18* **Písek** Budějovický Kraj, S Czech Republic *49°19′N 14°07′E*
31 *R14* **Pisgah** Ohio, N USA *39°19′N 84°22′W*
Pisha *see* Ningnan
158 *F9* **Pishan** *var.* Guma. Xinjiang Uygur Zizhiqu, NW China *37°36′N 78°45′E*
117 *N8* **Pishchanka** Vinnyts'ka Oblast', C Ukraine *48°12′N 28°52′E*
113 *K21* **Pishë** Fier, SW Albania *40°40′N 19°22′E*
143 *X14* **Pīshīn** Sīstān va Balūchestān, SE Iran *26°05′N 61°46′E*
149 *O9* **Pishin** Khyber Pakhtunkhwa, NW Pakistan *30°33′N 67°01′E*
149 *N11* **Pishin Lora** *var.* Psein Lora, *Pash.* Pseyn Bowr. ≈ SW Pakistan
Pishma *see* Pizhma
Pishpek *see* Bishkek
171 *O14* **Pising** Pulau Kabaena, C Indonesia *05°07′S 121°50′E*
Pisino *see* Pazin
Piski *see* Simeria
Piskolt *see* Pişcolt
147 *Q9* **Piskom** *Rus.* Pskem. ≈ E Uzbekistan
Piskom Tizmasi *see* Pskemskiy Khrebet
35 *P13* **Pismo Beach** California, W USA *35°08′N 120°38′W*
77 *P12* **Pissila** C Burkina Faso *13°07′N 00°51′W*
62 *H8* **Pissis, Monte** ▲ N Argentina *27°45′S 68°43′W*
41 *X12* **Piste** Yucatán, E Mexico *20°40′N 88°34′W*
107 *O18* **Pisticci** Basilicata, S Italy *40°23′N 16°33′E*
106 *F11* **Pistoia** *anc.* Pistoria, Pistoriæ. Toscana, C Italy *43°57′N 10°53′E*
32 *E15* **Pistol River** Oregon, NW USA *42°13′N 124°23′W*
Pistoria/Pistoriæ *see* Pistoia
15 *U5* **Pistuacanis** ≈ Québec, SE Canada
Pistyan *see* Piešťany
104 *M5* **Pisuerga** ≈ N Spain
110 *N8* **Pisz** *Ger.* Johannisburg. Warmińsko-Mazurskie, NE Poland *53°37′N 21°49′E*
76 *I13* **Pita** NW Guinea *11°05′N 12°15′W*
54 *D12* **Pitalito** Huila, S Colombia *01°51′N 76°01′W*
60 *I11* **Pitanga** Paraná, S Brazil *24°45′S 51°43′W*
182 *M9* **Pitarpunga Lake** *salt lake* New South Wales, SE Australia
Pitcairn Group of Islands *see* Pitcairn, Henderson, Ducie and Oeno Islands
193 *P10* **Pitcairn, Henderson, Ducie and Oeno Islands** *var.* Pitcairn Group of Islands. ◇ *UK overseas territory* C Pacific Ocean
191 *O14* **Pitcairn Island** *island* S Pitcairn Group of Islands
93 *J14* **Piteå** Norrbotten, N Sweden *65°19′N 21°30′E*
92 *I13* **Piteälven** ≈ N Sweden
116 *I13* **Pitești** Argeș, S Romania *44°53′N 24°49′E*
Pithagorio *see* Pythagóreio
180 *I12* **Pithara** Western Australia *30°31′S 116°38′E*
103 *N6* **Pithiviers** Loiret, C France *48°10′N 02°15′E*
152 *L9* **Pithorāgarh** Uttarakhand, N India *29°35′N 80°12′E*
188 *B16* **Piti** W Guam *13°28′N 144°42′E*
106 *G13* **Pitigliano** Toscana, C Italy *42°38′N 11°40′E*
40 *F3* **Pitiquito** Sonora, NW Mexico *30°39′N 112°00′W*
Pitkäranta *see* Pitkyaranta
38 *M11* **Pitkas Point** Alaska, USA *62°01′N 163°17′W*
124 *H11* **Pitkyaranta** *Fin.* Pitkäranta. Respublika Kareliya, NW Russian Federation *61°34′N 31°27′E*
96 *J10* **Pitlochry** C Scotland, United Kingdom *56°47′N 03°48′W*
18 *I16* **Pitman** New Jersey, NE USA *39°43′N 75°06′W*
146 *I9* **Pitnak** *var.* Drujba, *Rus.* Druzhba. Xorazm Viloyati, W Uzbekistan *41°14′N 61°13′E*
112 *G8* **Pitomača** Virovitica-Podravina, NE Croatia *45°57′N 17°14′E*
35 *O2* **Pit River** ≈ California, W USA
63 *G15* **Pitrufquén** Araucanía, S Chile *38°59′S 72°40′W*
Pitsanulok *see* Phitsanulok
Pitschen *see* Byczyna
Pitsunda *see* Bich'vinta
109 *X6* **Pitten** ≈ E Austria
10 *J14* **Pitt Island** *island* British Columbia, W Canada
Pitt Island *see* Makin
22 *M3* **Pittsboro** Mississippi, S USA *33°55′N 89°20′W*
21 *T9* **Pittsboro** North Carolina, SE USA *35°46′N 79°12′W*
27 *R7* **Pittsburg** Kansas, C USA *37°24′N 94°42′W*
25 *W6* **Pittsburg** Texas, SW USA *33°00′N 94°58′W*
18 *B14* **Pittsburgh** Pennsylvania, NE USA *40°26′N 80°W*
30 *J14* **Pittsfield** Illinois, N USA *39°36′N 90°48′W*
19 *R6* **Pittsfield** Maine, NE USA *44°46′N 69°22′W*
18 *L11* **Pittsfield** Massachusetts, NE USA *42°27′N 73°15′W*
183 *U3* **Pittsworth** Queensland, E Australia *27°43′S 151°36′E*
62 *I8* **Pituil** La Rioja, NW Argentina *28°33′S 67°24′W*
56 *A10* **Piura** Piura, NW Peru *05°11′S 80°41′W*
56 *A9* **Piura** *off.* Departamento de Piura. ◈ *department* NW Peru
Piura, Departamento de *see* Piura
35 *S13* **Piute Peak** ▲ California, W USA *35°27′N 118°24′W*
113 *J15* **Piva** ≈ NW Montenegro
117 *V5* **Pivdenne** Kharkivs'ka Oblast', E Ukraine *49°52′N 36°04′E*
117 *P8* **Pivdennyy Buh** *Rus.* Yuzhnyy Bug. ≈ S Ukraine
54 *F5* **Pivijay** Magdalena, N Colombia *10°31′N 74°36′W*
109 *T13* **Pivka** *prev.* Šent Peter, *Ger.* Sankt Peter, *It.* San Pietro del Carso. SW Slovenia *45°41′N 14°12′E*
117 *U13* **Pivnichno-Kryms'kyy Kanal** *canal* S Ukraine
113 *J15* **Pivsko Jezero** ◎ NW Montenegro
111 *M18* **Piwniczna** Małopolskie, S Poland *49°26′N 20°43′E*
35 *R12* **Pixley** California, W USA *35°58′N 119°18′W*
125 *Q15* **Pizhma** *var.* Pishma. ≈ NW Russian Federation
13 *U13* **Placentia** Newfoundland, Newfoundland and Labrador, SE Canada *47°12′N 53°58′W*
Placentia *see* Piacenza
13 *U13* **Placentia Bay** *inlet* Newfoundland, Newfoundland and Labrador, SE Canada
171 *P5* **Placer** Masbate, N Philippines *11°54′N 123°54′E*
35 *P7* **Placerville** California, W USA *38°42′N 120°48′W*
44 *F5* **Placetas** Villa Clara, C Cuba *22°18′N 79°40′W*
113 *Q18* **Plačkovica** ▲▲ E Macedonia
36 *L2* **Plain City** Utah, W USA *41°18′N 112°05′W*
22 *G4* **Plain Dealing** Louisiana, S USA *32°54′N 93°42′W*
31 *O14* **Plainfield** Indiana, N USA *39°42′N 86°18′W*
18 *K14* **Plainfield** New Jersey, NE USA *40°37′N 74°25′W*
33 *O8* **Plains** Montana, NW USA *47°27′N 114°52′W*
24 *L6* **Plains** Texas, SW USA *33°12′N 102°50′W*
29 *X10* **Plainview** Minnesota, N USA *44°10′N 92°10′W*
29 *Q13* **Plainview** Nebraska, C USA *42°21′N 97°47′W*
25 *N4* **Plainview** Texas, SW USA *34°12′N 101°43′W*
26 *K4* **Plainville** Kansas, C USA *39°13′N 99°18′W*
115 *I22* **Pláka** *var.* Mílos. Mílos, Kykládes, Greece, Aegean Sea *36°44′N 24°25′E*
115 *J15* **Pláka, Akrotírio** *headland* Límnos, E Greece *40°02′N 25°25′E*
113 *N19* **Plakenska Planina** ▲▲ SW Macedonia
44 *K5* **Plana Cays** *islets* SE The Bahamas
105 *S12* **Plana, Isla** *var.* Nueva Tabarca. *island* E Spain
59 *L18* **Planaltina** Goiás, S Brazil *15°35′S 47°28′W*
83 *O14* **Planalto Moçambicano** *plateau* N Mozambique
112 *N10* **Plandište** Vojvodina, NE Serbia *45°13′N 21°07′E*
100 *N13* **Plane** ≈ NE Germany
54 *E6* **Planeta Rica** Córdoba, NW Colombia *08°24′N 75°39′W*
29 *P11* **Plankinton** South Dakota, N USA *43°43′N 98°28′W*
30 *M11* **Plano** Illinois, N USA *41°39′N 88°32′W*
25 *U6* **Plano** Texas, SW USA *33°01′N 96°42′W*
23 *W12* **Plant City** Florida, SE USA *28°01′N 82°06′W*
22 *J9* **Plaquemine** Louisiana, S USA *30°17′N 91°13′W*
104 *K9* **Plasencia** Extremadura, W Spain *40°02′N 06°05′W*
110 *P7* **Płaska** Podlaskie, NE Poland *53°55′N 23°18′E*
112 *C10* **Plaški** Karlovac, C Croatia *45°04′N 15°21′E*
113 *N19* **Plasnica** SW FYR Macedonia *41°28′N 21°07′E*
13 *N14* **Plaster Rock** New Brunswick, SE Canada *46°55′N 67°24′W*
107 *J24* **Platani** *anc.* Halycus. ≈ Sicilia, Italy, C Mediterranean Sea
115 *G17* **Plataniá** Thessalía, C Greece *39°09′N 23°15′E*
115 *G24* **Plátanos** Kríti, Greece, E Mediterranean Sea *35°27′N 23°34′E*
65 *H18* **Plata, Río de la** *var.* River Plate. *estuary* Argentina/Uruguay
77 *V15* **Plateau** ◈ *state* C Nigeria
79 *G19* **Plateaux** *var.* Région des Plateaux. ◈ *province* C Congo
Plateaux, Région des *see* Plateaux
Plate, Île *see* Flat Island
92 *P1* **Platen, Kapp** *headland* NE Svalbard *80°30′N 22°46′E*
Plate, River *see* Plata, Río de la
99 *G22* **Plate Taille, Lac de la** *var.* L'Eau d'Heure. ◎ SE Belgium
Plathe *see* Płoty
39 *N13* **Platinum** Alaska, USA *59°01′N 161°49′W*
54 *F5* **Plato** Magdalena, N Colombia *09°47′N 74°47′W*
29 *O11* **Platte** South Dakota, N USA *43°20′N 98°51′W*
27 *R3* **Platte City** Missouri, C USA *39°22′N 94°47′W*
Plattensee *see* Balaton
27 *R3* **Platte River** ≈ Iowa/Missouri, USA
29 *Q15* **Platte River** ≈ Nebraska, C USA
37 *T3* **Platteville** Colorado, C USA *40°13′N 104°49′W*
30 *K9* **Platteville** Wisconsin, N USA *42°44′N 90°27′W*
101 *N21* **Plattling** Bayern, SE Germany *48°45′N 12°52′E*
27 *R3* **Plattsburg** Missouri, C USA *39°33′N 94°27′W*
18 *L6* **Plattsburgh** New York, NE USA *44°42′N 73°29′W*
29 *S15* **Plattsmouth** Nebraska, C USA *41°00′N 95°52′W*
101 *M17* **Plauen** *var.* Plauen im Vogtland. Sachsen, E Germany *50°31′N 12°08′E*
Plauen im Vogtland *see* Plauen
100 *M10* **Plauer See** ◎ NE Germany
113 *L16* **Plav** E Montenegro *42°36′N 19°57′E*
118 *I10* **Pļaviņas** *Ger.* Stockmannshof. S Latvia *56°37′N 25°40′E*
126 *K5* **Plavsk** Tul'skaya Oblast', W Russian Federation *53°42′N 37°20′E*
41 *Z12* **Playa del Carmen** Quintana Roo, E Mexico *20°37′N 87°04′W*
40 *J12* **Playa Los Corchos** Nayarit, SW Mexico *21°91′N 105°28′W*
37 *P16* **Playas Lake** ◎ New Mexico, SW USA
41 *S15* **Playa Vicenté** Veracruz-Llave, SE Mexico *17°42′N 95°01′W*
Plây Cu *see* Plei Ku
28 *L3* **Plaza** North Dakota, N USA *48°00′N 102°00′W*
63 *I15* **Plaza Huincul** Neuquén, C Argentina *38°55′S 69°14′W*
36 *L3* **Pleasant Grove** Utah, W USA *40°21′N 111°44′W*
29 *V14* **Pleasant Hill** Iowa, C USA *41°34′N 93°31′W*
27 *R4* **Pleasant Hill** Missouri, C USA *38°47′N 94°16′W*
Pleasant Island *see* Nauru
36 *K13* **Pleasant, Lake** ⊠ Arizona, SW USA
19 *P8* **Pleasant Mountain** ▲ Maine, NE USA *44°01′N 70°47′W*
27 *R5* **Pleasanton** Kansas, C USA *38°09′N 94°43′W*
25 *R12* **Pleasanton** Texas, SW USA *28°58′N 98°28′W*
185 *G20* **Pleasant Point** Canterbury, South Island, New Zealand *44°16′S 171°09′E*
19 *R5* **Pleasant River** ≈ Maine, NE USA
18 *J17* **Pleasantville** New Jersey, NE USA *39°22′N 74°31′W*
103 *N12* **Pléaux** Cantal, C France *45°08′N 02°10′E*
111 *B19* **Plechý** *var.* Plöckenstein. ▲ Austria/Czech Republic *48°45′N 13°50′E*
Pleebo *see* Plibo
Pleihari *see* Pelaihari
167 *U11* **Plei Ku** *prev.* Plây Cu. Gia Lai, C Vietnam *13°57′N 108°01′E*
101 *M16* **Pleisse** ≈ E Germany
Plencia *see* Plentzia
184 *O7* **Plenty, Bay of** *bay* North Island, New Zealand
33 *Y6* **Plentywood** Montana, NW USA *48°46′N 104°33′W*
105 *O2* **Plentzia** *var.* Plencia. País Vasco, N Spain *43°25′N 02°56′W*
102 *H5* **Plérin** Côtes d'Armor, NW France *48°33′N 02°46′W*
124 *M10* **Plesetsk** Arkhangel'skaya Oblast', NW Russian Federation *62°41′N 40°14′E*
Pleshchenitsy *see* Plyeshchanitsy
Pleskau *see* Pskov
Pleskauer See *see* Pskov, Lake
Pleskava *see* Pskov
112 *E8* **Pleso International** ✈ (Zagreb) Zagreb, NW Croatia *45°45′N 16°00′E*
Pless *see* Pszczyna
15 *Q11* **Plessisville** Québec, SE Canada *46°14′N 71°46′W*
110 *H12* **Pleszew** Wielkopolskie, C Poland *51°54′N 17°47′E*
12 *L10* **Plétipi, Lac** ◎ Québec, SE Canada
101 *F15* **Plettenberg** Nordrhein-Westfalen, W Germany *51°13′N 07°52′E*
114 *I8* **Pleven** *prev.* Plevna. Pleven, N Bulgaria *43°25′N 24°36′E*
114 *I8* **Pleven** ◈ *province* N Bulgaria
Plevlja/Plevlje *see* Pljevlja
Plevna *see* Pleven
Plezzo *see* Bovec
Pliberk *see* Bleiburg
76 *L17* **Plibo** *var.* Pleebo. SE Liberia *04°38′N 07°41′W*
121 *R11* **Pliny Trench** *undersea feature* C Mediterranean Sea
118 *K13* **Plisa** *Rus.* Plissa. Vitsyebskaya Voblasts', N Belarus *55°13′N 27°57′E*
Plissa *see* Plisa
112 *D11* **Plitvica Selo** Lika-Senj, W Croatia *44°53′N 15°36′E*
112 *D11* **Plješevica** ▲ C Croatia
113 *K14* **Pljevlja** *prev.* Plevlja, Plevlje. N Montenegro *43°21′N 19°21′E*
Ploça *see* Ploçë
Plocce *see* Ploče
113 *K22* **Ploçë** *var.* Ploça. Vlorë, SW Albania *40°24′N 19°41′E*
113 *G15* **Ploče** *It.* Plocce; *prev.* Kardeljevo. Dubrovnik-Neretva, SE Croatia *43°02′N 17°25′E*
110 *K11* **Płock** *Ger.* Plozk. Mazowieckie, C Poland *52°32′N 19°40′E*
109 *Q10* **Plöcken Pass** *Ger.* Plöckenpass, *It.* Passo di Monte Croce Carnico. *pass* SW Austria
Plöckenpass *see* Plöcken Pass
Plöckenstein *see* Plechý
99 *B19* **Ploegsteert** Hainaut, W Belgium *50°45′N 02°52′E*
102 *H6* **Ploërmel** Morbihan, NW France *47°57′N 02°24′W*
Ploești *see* Ploiești
116 *K13* **Ploiești** *prev.* Ploești. Prahova, SE Romania *44°56′N 26°03′E*
115 *L17* **Plomári** *prev.* Plomárion. Lésvos, E Greece *38°58′N 26°24′E*
Plomárion *see* Plomári
103 *O12* **Plomb du Cantal** ▲ C France *45°03′N 02°48′E*
183 *V6* **Plomer, Point** *headland* New South Wales, SE Australia *31°19′S 153°00′E*
100 *J8* **Plön** Schleswig-Holstein, N Germany *54°10′N 10°25′E*
110 *L11* **Płońsk** Mazowieckie, C Poland *52°38′N 20°23′E*
119 *J20* **Plotnitsa** Brestskaya Voblasts', SW Belarus *52°03′N 26°39′E*
110 *E8* **Płoty** *Ger.* Plathe. Zachodnio-pomorskie, NW Poland *53°48′N 15°16′E*
102 *G7* **Plouay** Morbihan, NW France *47°54′N 03°14′W*
111 *D15* **Ploučnice** *Ger.* Polzen. ≈ NE Czech Republic
114 *I10* **Plovdiv** *prev.* Eumolpias; *anc.* Evmolpia, Philippopolis, *Lat.* Trimontium. Plovdiv, C Bulgaria *42°09′N 24°47′E*
114 *I11* **Plovdiv** ◈ *province* C Bulgaria
30 *L6* **Plover** Wisconsin, N USA *44°30′N 89°33′W*
Plozk *see* Płock
27 *U11* **Plumerville** Arkansas, C USA *35°09′N 92°38′W*
19 *P10* **Plum Island** *island* Massachusetts, NE USA
32 *M9* **Plummer** Idaho, NW USA *47°19′N 116°54′W*
83 *J18* **Plumtree** Matabeleland South, SW Zimbabwe *20°30′S 27°50′E*
118 *D11* **Plungė** Telšiai, W Lithuania *55°55′N 21°53′E*
113 *J15* **Plužine** NW Montenegro *43°08′N 18°49′E*
119 *K14* **Plyeshchanitsy** *Rus.* Pleshchenitsy. Minskaya Voblasts', N Belarus *54°25′N 27°50′E*
97 *I24* **Plymouth** SW England, United Kingdom *50°23′N 04°10′W*
31 *O11* **Plymouth** Indiana, N USA *41°20′N 86°19′W*
19 *P12* **Plymouth** Massachusetts, NE USA *41°57′N 70°40′W*
19 *N8* **Plymouth** New Hampshire, NE USA *43°43′N 71°39′W*
21 *X9* **Plymouth** North Carolina, SE USA *35°53′N 76°46′W*
30 *M8* **Plymouth** Wisconsin, N USA *43°48′N 87°58′W*
Plymouth *see* Brades
97 *J20* **Plynlimon** ▲ C Wales, United Kingdom *52°27′N 03°48′W*
124 *G14* **Plyussa** Pskovskaya Oblast', W Russian Federation *58°27′N 29°21′E*
111 *B17* **Plzeň** *Ger.* Pilsen, *Pol.* Pilzno. Plzeňský Kraj, W Czech Republic *49°45′N 13°23′E*
111 *B17* **Plzeňský Kraj** ◈ *region* W Czech Republic
110 *F11* **Pniewy** *Ger.* Pinne. Wielkolpolskie, W Poland *52°31′N 16°14′E*
77 *P13* **Pô** S Burkina Faso *11°11′N 01°10′W*
106 *D8* **Po** ≈ N Italy
42 *M13* **Poás, Volcán** ☗ NW Costa Rica *10°12′N 84°12′W*
77 *S16* **Pobè** S Benin *07°00′N 02°41′E*
123 *S8* **Pobeda, Gora** ▲ NE Russian Federation *65°28′N 145°44′E*
Pobeda Peak *see* Pobedy, Pik/Tomür Feng
147 *Z7* **Pobedy, Pik** *Chin.* Tomür Feng. ▲ China/Kyrgyzstan *42°02′N 80°02′E* *see also* Tomür Feng
Pobedy, Pik *see* Tomür Feng
110 *H11* **Pobiedziska** *Ger.* Pudewitz. Wielkolpolskie, C Poland *52°30′N 17°19′E*
Po, Bocche del *see* Po, Foci del
27 *W9* **Pocahontas** Arkansas, C USA *36°15′N 91°00′W*
29 *U12* **Pocahontas** Iowa, C USA *42°44′N 94°40′W*
33 *Q15* **Pocatello** Idaho, NW USA *42°52′N 112°27′W*
167 *S13* **Pochentong** ✈ (Phnum Penh) Phnum Penh, S Cambodia *11°24′N 104°52′E*
126 *I6* **Pochep** Bryanskaya Oblast', W Russian Federation *52°56′N 33°20′E*
126 *H4* **Pochinok** Smolenskaya Oblast', W Russian Federation *54°21′N 32°29′E*
41 *R17* **Pochutla** *var.* San Pedro Pochutla. Oaxaca, SE Mexico *15°45′N 96°30′W*
62 *I6* **Pocitos, Salar** *var.* Salar Quiróm. *salt lake* NW Argentina
101 *O22* **Pocking** Bayern, SE Germany *48°22′N 13°17′E*
186 *I10* **Pocklington Reef** *reef* SE Papua New Guinea
59 *P15* **Poço da Cruz, Açude** ⊠ E Brazil
27 *R11* **Pocola** Oklahoma, C USA *35°13′N 94°28′W*
21 *Y5* **Pocomoke City** Maryland, NE USA *38°04′N 75°34′W*
59 *L21* **Poços de Caldas** Minas Gerais, NE Brazil *21°48′S 46°33′W*
124 *H14* **Podberez'ye** Novgorodskaya Oblast', NW Russian Federation *58°42′N 31°22′E*
Podbrodzie *see* Pabradė
125 *U8* **Podcher'ye** Respublika Komi, NW Russian Federation *63°55′N 57°34′E*
111 *E16* **Poděbrady** *Ger.* Podiebrad. Středočeský Kraj, C Czech Republic *50°10′N 15°06′E*
126 *L9* **Podgorenskiy** Voronezhskaya Oblast', W Russian Federation *50°22′N 39°43′E*
113 *J17* **Podgorica** *prev.* Titograd. ● S Montenegro *42°25′N 19°16′E*
113 *K17* **Podgorica** ✈ S Montenegro *42°22′N 19°16′E*
109 *T13* **Podgrad** SW Slovenia *45°31′N 14°09′E*
Podiebrad *see* Poděbrady
116 *M5* **Podil's'ka Vysochina** *plateau* W Ukraine
Podium Anicensis *see* le Puy
122 *L11* **Podkamennaya Tunguska** *Eng.* Stony Tunguska. ≈ C Russian Federation
111 *N17* **Podkarpackie** ◈ *province* SE Poland
Pod Klošter *see* Arnoldstein
110 *P9* **Podlaskie** ◈ *province* Mazowieckie, C Poland
127 *Q8* **Podlesnoye** Saratovskaya Oblast', W Russian Federation *51°51′N 47°03′E*
126 *K4* **Podol'sk** Moskovskaya Oblast', W Russian Federation *55°24′N 37°30′E*
76 *H10* **Podor** N Senegal *16°40′N 14°57′W*
125 *P12* **Podosinovets** Kirovskaya Oblast', NW Russian Federation *60°15′N 47°06′E*
124 *I12* **Podporozh'ye** Leningradskaya Oblast', NW Russian Federation *60°52′N 34°00′E*
Podravska Slatina *see* Slatina
112 *J13* **Podromanlja** Republika Srpska, SE Bosnia and Herzegovina *43°55′N 18°46′E*
Podsvil'ye *see* Padsvillye
116 *L9* **Podu Iloaiei** *prev.* Podul Iloaiei. Iași, NE Romania *47°13′N 27°16′E*
113 *N15* **Podujevë** *Serb.* Podujevo. N Kosovo *42°56′N 21°13′E*
Podujevo *see* Podujevë
Podul Iloaiei *see* Podu Iloaiei
Podunajská Rovina *see* Little Alföld
124 *M12* **Podyuga** Arkhangel'skaya Oblast', NW Russian Federation *61°04′N 40°46′E*
56 *A9* **Poechos, Embalse** ◎ NW Peru
55 *W10* **Poeketi** Sipaliwini, E Suriname
100 *L8* **Poel** *island* N Germany
83 *M20* **Poelela, Lagoa** ◎ S Mozambique
Poerwodadi *see* Purwodadi
Poerwokerto *see* Purwokerto
Poerworedjo *see* Purworejo
Poetovio *see* Ptuj
83 *E23* **Pofadder** Northern Cape, W South Africa *29°09′S 19°25′E*
106 *I9* **Po, Foci del** *var.* Bocche del Po. ≈ NE Italy
116 *E12* **Pogăniș** ≈ W Romania
Pogegen *see* Pagėgiai
106 *G12* **Poggibonsi** Toscana, C Italy *43°28′N 11°09′E*
107 *I14* **Poggio Mirteto** Lazio, C Italy *42°17′N 12°42′E*
109 *V4* **Pöggstall** Niederösterreich, N Austria *48°19′N 15°10′E*
116 *L13* **Pogoanele** Buzău, SE Romania *44°55′N 27°00′E*
Pogónion *see* Delvináki
113 *M21* **Pogradec** *var.* Pogradeci. Korçë, SE Albania *40°54′N 20°40′E*
Pogradeci *see* Pogradec
123 *S15* **Pogranichnyy** Primorskiy Kray, SE Russian Federation *44°18′N 131°33′E*
38 *M16* **Pogromni Volcano** ☗ Unimak Island, Alaska, USA *54°34′N 164°41′W*
163 *Z15* **Pohang** *Jap.* Hokō; *prev.* P'ohang. E South Korea *36°02′N 129°20′E*
P'ohang *see* Pohang
15 *T9* **Pohénégamook, Lac** ◎ Québec, SE Canada
93 *L20* **Pohja** *Swe.* Pojo. Uusimaa, SW Finland *60°07′N 23°30′E*
Pohjanlahti *see* Bothnia, Gulf of
Pohjanmaa *see* Österbotten
93 *O16* **Pohjois-Karjala** *Swe.* Norra Karelen, *Eng.* North Karelia. ◈ *region* E Finland
93 *L14* **Pohjois-Pohjanmaa** *Swe.* Norra Österbotten, *Eng.* North Ostrobothnia. ◈ *region* N Finland
93 *M17* **Pohjois-Savo** *Swe.* Norra Savolax. ◈ *region* C Finland
189 *U16* **Pohnpei** ◈ *state* E Micronesia
189 *O12* **Pohnpei** ✈ Pohnpei, E Micronesia
189 *O12* **Pohnpei** *prev.* Ponape Ascension Island. *island* E Micronesia
111 *F19* **Pohořelice** *Ger.* Pohrlitz. Jihomoravský Kraj, SE Czech Republic *48°58′N 16°30′E*
109 *V10* **Pohorje** *Ger.* Bacher. ▲▲ N Slovenia
117 *N6* **Pohrebyshche** Vinnyts'ka Oblast', C Ukraine *49°31′N 29°16′E*
Pohrlitz *see* Pohořelice
161 *P9* **Po Hu** ⊠ E China
116 *G15* **Poiana Mare** Dolj, S Romania *43°55′N 23°02′E*
Poictiers *see* Poitiers
127 *N6* **Poim** Penzenskaya Oblast', W Russian Federation *53°03′N 43°11′E*
159 *N15* **Poindo** Xigzang Zizhiqu, W China *29°58′N 91°20′E*
195 *Y13* **Poinsett, Cape** *headland* Antarctica *65°35′S 113°00′E*
29 *R9* **Poinsett, Lake** ◎ South Dakota, N USA
22 *I10* **Point Au Fer Island** *island* Louisiana, S USA
39 *X14* **Point Baker** Prince of Wales Island, Alaska, USA *56°19′N 133°31′W*
25 *U13* **Point Comfort** Texas, SW USA *28°40′N 96°33′W*
Point de Galle *see* Galle
44 *K10* **Pointe à Gravois** *headland* SW Haiti *18°00′N 73°53′W*
22 *L10* **Pointe a la Hache** Louisiana, S USA *29°34′N 89°48′W*
45 *Y6* **Pointe-à-Pitre** Grande Terre, C Guadeloupe *16°14′N 61°32′W*
15 *U7* **Pointe-au-Père** Québec, SE Canada *48°31′N 68°27′W*
15 *V5* **Pointe-aux-Anglais** Québec, SE Canada *49°40′N 67°09′W*
45 *T10* **Pointe Du Cap** *headland* N Saint Lucia *14°06′N 60°56′W*
79 *E21* **Pointe-Noire** Kouilou, S Congo *04°46′S 11°53′E*
45 *X6* **Pointe Noire** Basse Terre, W Guadeloupe *16°14′N 61°47′W*
79 *E21* **Pointe-Noire** ✈ Kouilou, S Congo *04°45′S 11°55′E*
45 *U15* **Point Fortin** Trinidad, Trinidad and Tobago *10°12′N 61°41′W*
38 *M6* **Point Hope** Alaska, USA *68°21′N 166°48′W*
39 *N5* **Point Lay** Alaska, USA *69°42′N 162°57′W*
18 *B16* **Point Marion** Pennsylvania, NE USA *39°44′N 79°53′W*
18 *K16* **Point Pleasant** New Jersey, NE USA *40°04′N 74°00′W*
21 *P4* **Point Pleasant** West Virginia, NE USA *38°53′N 82°07′W*
45 *R14* **Point Salines** ✈ (St. George's) SW Grenada *12°00′N 61°47′W*
102 *L9* **Poitiers** *prev.* Poictiers; *anc.* Limonum. Vienne, W France *46°35′N 00°19′E*
102 *K9* **Poitou** *cultural region* W France
102 *K10* **Poitou-Charentes** ◈ *region* W France
103 *N3* **Poix-de-Picardie** Somme, N France *49°47′N 01°58′E*
Pojo *see* Pohja
37 *S10* **Pojoaque** New Mexico, SW USA *35°52′N 106°01′W*

◆ Country ● Country Capital ◇ Dependent Territory ○ Dependent Territory Capital ◈ Administrative Regions ✈ International Airport ▲ Mountain ▲▲ Mountain Range ☗ Volcano ≈ River ◎ Lake ⊠ Reservoir

Map Symbols
◆ Country ◇ Dependent Territory ✕ Administrative Regions ▲ Mountain ☒ Volcano ⊚ Lake
● Country Capital ○ Dependent Territory Capital ✈ International Airport ▲ Mountain Range ☒ River ☒ Reservoir

PORTO VELHO – PRÜM

59 E15 Porto Velho var. Velho. state capital Rondônia, W Brazil 08°45′S 63°54′W
56 A6 Portoviejo var. Puertoviejo. Manabí, W Ecuador 01°03′S 80°31′W
185 B26 Port Pegasus bay Stewart Island, New Zealand
14 H15 Port Perry Ontario, SE Canada 44°08′N 78°57′W
183 N12 Port Phillip Bay harbor Victoria, SE Australia
182 I8 Port Pirie South Australia 33°11′S 138°01′E
96 G9 Portree N Scotland, United Kingdom 57°26′N 06°12′W
Port Rex see East London
Port Rois see Portrush
44 K13 Port Royal E Jamaica 17°55′N 76°52′W
21 R15 Port Royal South Carolina, SE USA 32°22′N 80°41′W
21 R15 Port Royal Sound inlet South Carolina, SE USA
97 F14 Portrush Ir. Port Rois. N Northern Ireland, United Kingdom 55°12′N 06°40′W
Port Said see Būr Sa'īd
23 R9 Port Saint Joe Florida, SE USA 29°49′N 85°18′W
23 Y11 Port Saint John Florida, SE USA 28°28′N 80°46′W
103 R16 Port-St-Louis-du-Rhône Bouches-du-Rhône, SE France 43°22′N 04°48′E
44 K10 Port Salut SW Haiti 18°04′N 73°55′W
65 E24 Port Salvador inlet East Falkland, Falkland Islands
65 D24 Port San Carlos East Falkland, Falkland Islands 51°30′S 58°59′W
13 S10 Port Saunders Newfoundland, Newfoundland and Labrador, SE Canada 50°40′N 57°17′W
83 K24 Port Shepstone KwaZulu/Natal, E South Africa 30°44′S 30°28′E
45 O11 Portsmouth var. Grand-Anse. NW Dominica 15°34′N 61°27′W
97 N24 Portsmouth S England, United Kingdom 50°48′N 01°05′W
19 P10 Portsmouth New Hampshire, NE USA 43°04′N 70°47′W
31 S15 Portsmouth Ohio, N USA 38°43′N 83°00′W
21 X7 Portsmouth Virginia, NE USA 36°50′N 76°18′W
14 E17 Port Stanley Ontario, S Canada 42°39′N 81°12′W
Port Stanley see Stanley
65 B25 Port Stephens inlet West Falkland, Falkland Islands
65 B25 Port Stephens Settlement West Falkland, Falkland Islands
97 F14 Portstewart Ir. Port Stíobhaird. N Northern Ireland, United Kingdom 55°11′N 06°43′W
Port Stíobhaird see Portstewart
83 K24 Port St. Johns Eastern Cape, SE South Africa 31°37′S 29°32′E
80 I7 Port Sudan Red Sea, NE Sudan 19°37′N 37°14′E
22 L10 Port Sulphur Louisiana, S USA 29°28′N 89°41′W
Port Swettenham see Klang/Pelabuhan Klang
97 J22 Port Talbot S Wales, United Kingdom 51°36′N 03°47′W
92 L11 Porttipahdan Tekojärvi ⊚ N Finland
32 G7 Port Townsend Washington, NW USA 48°07′N 122°45′W
104 H9 Portugal off. Portuguese Republic. ◆ republic SW Europe
105 O2 Portugalete País Vasco, N Spain 43°19′N 03°01′W
54 J6 Portuguesa off. Estado Portuguesa. ◈ state N Venezuela
Portuguesa, Estado see Portuguesa
Portuguese East Africa see Mozambique
Portuguese Guinea see Guinea-Bissau
Portuguese Republic see Portugal
Portuguese Timor see East Timor
Portuguese West Africa see Angola
97 D18 Portumna Ir. Port Omna. Galway, W Ireland 53°06′N 08°13′W
Portus Cale see Porto
Portus Magnus see Almería
Portus Magonis see Maó
103 P17 Port-Vendres var. Port Vendres. Pyrénées-Orientales, S France 42°31′N 03°06′E
182 H9 Port Victoria South Australia 34°34′S 137°31′E
187 Q14 Port-Vila var. Vila. ● (Vanuatu) Éfaté, C Vanuatu 17°45′S 168°21′E
Port Vila see Bauer Field
182 I9 Port Wakefield South Australia 34°13′S 138°10′E
31 N8 Port Washington Wisconsin, N USA 43°23′N 87°54′W
57 J14 Porvenir Pando, NW Bolivia 11°15′S 68°43′W
63 I24 Porvenir Magallanes, S Chile 53°18′S 70°22′W
61 D18 Porvenir Paysandú, W Uruguay 32°23′S 57°59′W
93 M19 Porvoo Swe. Borgå. Uusimaa, S Finland 60°25′N 25°40′E
Porzecze see Parechcha
104 M10 Porzuna Castilla-La Mancha, C Spain 39°10′N 04°10′W
61 E14 Posadas Misiones, NE Argentina 27°27′S 55°52′W
104 L13 Posadas Andalucía, S Spain 37°48′N 05°06′W
Poschega see Požega
108 J11 Poschiavino ≈ Italy/Switzerland
108 J10 Poschiavo Ger. Puschlav. Graubünden, S Switzerland 46°19′N 10°02′E
112 D12 Posedarje Zadar, SW Croatia 44°12′N 15°27′E
Posen see Poznań
124 L14 Poshekhon'ye Yaroslavskaya Oblast', W Russian Federation 58°31′N 39°07′E
92 M13 Posio Lappi, NE Finland 66°06′N 28°16′E
Poskam see Zepu
Posnania see Poznań
1713 O12 Poso Sulawesi, C Indonesia 01°23′S 120°45′E
171 O12 Poso, Danau ⊚ Sulawesi, C Indonesia
137 R10 Posof Ardahan, NE Turkey 41°30′N 42°33′E
25 R6 Possum Kingdom Lake ⊠ Texas, SW USA
25 N6 Post Texas, SW USA 33°14′N 101°24′W
Postavy/Postawy see Pastavy
Poste-de-la-Baleine see Kuujjuarapik
99 M17 Posterholt Limburg, SE Netherlands 51°07′N 06°02′E
83 G22 Postmasburg Northern Cape, N South Africa 28°20′S 23°05′E
Pôsto Diuarum see Campo de Diauarum
59 I16 Pôsto Jacaré Mato Grosso, W Brazil 12°S 53°27′W
109 T12 Postojna Ger. Adelsberg, It. Postumia. SW Slovenia 45°48′N 14°12′E
Postumia see Postojna
29 X12 Postville Iowa, C USA 43°04′N 91°34′W
Pöstyén see Piešt'any
113 G14 Posušje Federacija Bosne I Hercegovine, SW Bosnia and Herzegovina 43°28′N 17°20′E
171 O16 Pota Flores, C Indonesia 08°21′S 120°50′E
115 G23 Potamós Antikýthira, S Greece 35°53′N 23°17′E
55 S9 Potaru River ≈ C Guyana
83 I21 Potchefstroom North-West, N South Africa 26°42′S 27°06′E
27 R11 Poteau Oklahoma, C USA 35°03′N 94°36′W
25 R12 Poteet Texas, SW USA 29°02′N 98°34′W
115 G14 Poteídaia site of ancient city Kentrikí Makedonía, N Greece
Potentia see Potenza
107 M18 Potenza anc. Potentia. Basilicata, S Italy 40°40′N 15°50′E
185 A24 Poteriteri, Lake ⊚ South Island, New Zealand
104 M2 Potes Cantabria, N Spain 43°10′N 04°41′W
Potgietersrus see Mokopane
25 S12 Poth Texas, SW USA 29°04′N 98°04′W
32 J9 Potholes Reservoir ⊠ Washington, NW USA
137 Q9 Poti prev. P'ot'i. W Georgia 42°10′N 41°42′E
P'ot'i see Poti
77 X13 Potiskum Yobe, NE Nigeria 11°38′N 11°07′E
Potkozarje see Ivanjska
32 M9 Potlatch Idaho, NW USA 46°55′N 116°51′W
33 N9 Pot Mountain ▲ Idaho, NW USA 46°44′N 115°24′W
113 H14 Potoci Federacija Bosne I Hercegovine, S Bosnia and Herzegovina 43°24′N 17°52′E
21 V3 Potomac River ≈ NE USA
Potoprens see Port-au-Prince
57 L20 Potosí Potosí, S Bolivia 19°35′S 65°51′W
42 H9 Potosí Chinandega, NW Nicaragua 12°58′N 87°30′W
27 W6 Potosi Missouri, C USA 37°57′N 90°49′W
57 K21 Potosí ◈ department SW Bolivia
62 H7 Potrerillos Atacama, N Chile 26°30′S 69°25′W
42 H5 Potrerillos Cortés, NW Honduras 15°10′N 87°58′W
62 H8 Potro, Cerro del ▲ N Chile 28°22′S 69°34′W
100 N12 Potsdam Brandenburg, NE Germany 52°24′N 13°04′E
18 J7 Potsdam New York, NE USA 44°40′N 74°58′W
109 X5 Pottendorf Niederösterreich, E Austria 47°55′N 16°23′E
109 X5 Pottenstein Niederösterreich, E Austria 47°58′N 16°07′E
18 I15 Pottstown Pennsylvania, NE USA 40°15′N 75°39′W
18 H14 Pottsville Pennsylvania, NE USA 40°40′N 76°10′W
155 L25 Pottuvil Eastern Province, SE Sri Lanka 06°53′N 81°49′E
149 U6 Potwar Plateau plateau NE Pakistan
102 J7 Pouancé Maine-et-Loire, W France 47°46′N 01°11′W
15 R6 Poulin de Courval, Lac ⊚ Québec, SE Canada
18 L9 Poultney Vermont, NE USA 43°31′N 73°12′W
187 O16 Poum Province Nord, W New Caledonia 20°15′S 164°03′E
59 L21 Pouso Alegre Minas Gerais, NE Brazil 22°13′S 45°56′W
192 I16 Poutasi Upolu, SE Samoa 14°00′S 171°43′W
167 R12 Poŭthĭsăt prev. Pursat. Poŭthĭsăt, W Cambodia 12°32′N 103°55′E
167 R12 Poŭthĭsăt, Stœng prev. Pursat. ≈ W Cambodia
102 J9 Pouzauges Vendée, NW France 46°47′N 00°54′W
Po, Valle del see Po Valley
106 F8 Po Valley It. Valle del Po. valley N Italy
111 I19 Považská Bystrica Ger. Waagbistritz, Hung. Vágbeszterce. Trenčiansky Kraj, W Slovakia 49°07′N 18°26′E
124 J10 Povenets Respublika Kareliya, NW Russian Federation 62°50′N 34°47′E
184 Q9 Poverty Bay inlet North Island, New Zealand
112 K12 Povlen ▲ W Serbia
104 G6 Póvoa de Varzim Porto, NW Portugal 41°22′N 08°46′W
127 N8 Povorino Voronezhskaya Oblast', W Russian Federation 51°10′N 42°16′E
Povungnituk see Puvirnituq
Rivière de Povungnituk see Puvirnituq, Riviere de
14 H11 Powassan Ontario, S Canada 46°05′N 79°21′W
35 U17 Poway California, W USA 32°57′N 117°02′W
33 W14 Powder River Wyoming, C USA 43°01′N 106°57′W
33 Y10 Powder River ≈ Montana/Wyoming, NW USA
32 L12 Powder River ≈ Oregon, NW USA
33 W13 Powder River Pass pass Wyoming, C USA
33 U12 Powell Wyoming, C USA 44°45′N 108°45′W
65 I22 Powell Basin undersea feature NW Weddell Sea
36 M8 Powell, Lake ⊚ Utah, W USA
37 R4 Powell, Mount ▲ Colorado, C USA 39°25′N 106°20′W
10 L17 Powell River British Columbia, SW Canada 49°54′N 124°34′W
31 N5 Powers Michigan, N USA 45°40′N 87°29′W
28 K2 Powers Lake North Dakota, N USA 48°33′N 102°37′W
21 V6 Powhatan Virginia, NE USA 37°33′N 77°56′W
31 V13 Powhatan Point Ohio, N USA 39°49′N 80°49′W
97 J20 Powys cultural region E Wales, United Kingdom
187 P17 Poya Province Nord, C New Caledonia 21°19′S 165°07′E
161 P10 Poyang Hu ⊚ S China
30 L7 Poygan, Lake ⊚ Wisconsin, N USA
109 Y2 Poysdorf Niederösterreich, NE Austria 48°40′N 16°38′E
112 N11 Požarevac Ger. Passarowitz. Serbia, NE Serbia 44°37′N 21°11′E
41 Q13 Poza Rica var. Poza Rica de Hidalgo. Veracruz-Llave, E Mexico 20°34′N 97°26′W
Poza Rica de Hidalgo see Poza Rica
112 L13 Požega prev. Slovenska Požega, Ger. Poschega, Hung. Pozsega. Požega-Slavonija, NE Croatia 45°19′N 17°42′E
112 H9 Požega-Slavonija off. Požeško-Slavonska Županija. ◈ province NE Croatia
Požeško-Slavonska Županija see Požega-Slavonija
125 U13 Pozhva Komi-Permyatskiy Okrug, NW Russian Federation 59°07′N 56°04′E
110 G11 Poznań Ger. Posen, Posnania. Wielkolpolskie, C Poland 52°24′N 16°56′E
105 O13 Pozo Alcón Andalucía, S Spain 37°43′N 02°55′W
62 H3 Pozo Almonte Tarapacá, N Chile 20°16′S 69°50′W
104 L12 Pozoblanco Andalucía, S Spain 38°23′N 04°48′W
105 Q11 Pozo Cañada Castilla-La Mancha, C Spain 38°49′N 01°45′W
62 N5 Pozo Colorado Presidente Hayes, C Paraguay 23°26′S 58°51′W
63 J20 Pozos, Punta headland S Argentina 47°55′S 65°46′W
Pozsega see Požega
Pozsony see Bratislava
55 N5 Pozuelos Anzoátegui, NE Venezuela 10°11′N 64°39′W
107 L26 Pozzallo Sicilia, Italy, C Mediterranean Sea 36°44′N 14°51′E
107 K17 Pozzuoli anc. Puteoli. Campania, S Italy 40°49′N 14°07′E
77 P17 Pra ≈ S Ghana
111 C19 Prachatice Ger. Prachatitz. Jihočeský Kraj, S Czech Republic 49°01′N 14°02′E
Prachatitz see Prachatice
167 P11 Prachin Buri var. Prachinburi. Prachin Buri, C Thailand 14°05′N 101°23′E
Prachinburi see Prachin Buri
Prachuab Girikhand see Prachuap Khiri Khan
167 O12 Prachuap Khiri Khan var. Prachuab Girikhand. Prachuap Khiri Khan, SW Thailand 11°50′N 99°49′E
111 H16 Praděd Ger. Altvater. ▲ NE Czech Republic 50°06′N 17°14′E
54 D11 Pradera Valle del Cauca, SW Colombia 03°23′N 76°11′W
103 O17 Prades Pyrénées-Orientales, S France 42°36′N 02°22′E
59 O19 Prado Bahia, SE Brazil 17°13′S 39°15′W
54 E11 Prado Tolima, C Colombia 03°45′N 74°55′W
Prado del Ganso see Goose Green
Prae see Phrae
95 I24 Præstø Sjælland, SE Denmark 55°08′N 12°03′E
Prag/Praga/Prague see Praha
27 O10 Prague Oklahoma, C USA 35°29′N 96°40′W
111 D16 Praha Eng. Prague, Ger. Prag, Pol. Praga. ● (Czech Republic) Středočeský Kraj, NW Czech Republic 50°06′N 14°26′E
116 J13 Prahova ◈ county SE Romania
116 J13 Prahova ≈ S Romania
76 E10 Praia ● (Cape Verde) Santiago, S Cape Verde 14°55′N 23°31′W
83 M21 Praia do Bilene Gaza, S Mozambique 25°18′S 33°10′E
83 M20 Praia do Xai-Xai Gaza, S Mozambique 25°04′S 33°43′E
116 J10 Praid Hung. Parajd. Harghita, C Romania 46°33′N 25°06′E
26 J3 Prairie Dog Creek ≈ Kansas/Nebraska, C USA
30 J9 Prairie du Chien Wisconsin, N USA 43°02′N 91°08′W
27 S9 Prairie Grove Arkansas, C USA 35°58′N 94°19′W
31 P10 Prairie River ≈ Michigan, N USA
Prairie State see Illinois
25 V11 Prairie View Texas, SW USA 30°05′N 95°59′W
167 Q10 Prakhon Chai Buri Ram, E Thailand 14°36′N 103°04′E
109 R4 Pram ≈ N Austria
167 Q12 Prâmaôy prev. Phumĭ Prâmaôy. Poŭthĭsăt, W Cambodia 12°13′N 103°05′E
109 S4 Prambachkirchen Oberösterreich, N Austria 48°18′N 13°50′E
118 H2 Prangli island N Estonia
154 J13 Prānhita ≈ C India
172 I15 Praslin island Inner Islands, NE Seychelles
115 O23 Prasonísi, Akrotírio cape Ródos, Dodekánisa, Greece, Aegean Sea
111 I14 Praszka Opolskie, S Poland 51°05′N 18°29′E
Pratas Island see Tungsha Tao
119 M18 Pratasy Rus. Protasy. Homyel'skaya Voblasts', SE Belarus 52°47′N 29°05′E
167 Q10 Prathai Nakhon Ratchasima, E Thailand 15°31′N 102°42′E
Prathet Thai see Thailand
Prathum Thani see Pathum Thani
63 F21 Prat, Isla island S Chile
106 G11 Prato Toscana, C Italy 43°53′N 11°05′E
103 O17 Prats-de-Mollo-la-Preste Pyrénées-Orientales, S France 42°25′N 02°28′E
26 L6 Pratt Kansas, C USA 37°40′N 98°45′W
108 E6 Pratteln Basel Landschaft, NW Switzerland 47°32′N 07°42′E
193 O2 Pratt Seamount undersea feature N Pacific Ocean 56°09′N 142°30′W
23 P5 Prattville Alabama, S USA 32°27′N 86°27′W
Praust see Pruszcz Gdański
119 B14 Pravdinsk Ger. Friedland. Kaliningradskaya Oblast', W Russian Federation 54°26′N 21°01′E
104 K2 Pravia Asturias, N Spain 43°30′N 06°06′W
118 L12 Prazaroki Rus. Prozoroki. Vitsyebskaya Voblasts', N Belarus 55°18′N 28°13′E
Prázsmár see Prejmer
167 S11 Preăh Vihéar Preăh Vihéar, N Cambodia 13°57′N 104°48′E
116 J12 Predeal Hung. Predeál. Braşov, C Romania 45°30′N 25°31′E
109 S8 Predlitz Steiermark, SE Austria 47°04′N 13°54′E
11 V15 Preeceville Saskatchewan, S Canada 51°58′N 102°40′W
Preenkuln see Priekule
102 K6 Pré-en-Pail Mayenne, NW France 48°27′N 00°15′W
109 T4 Pregarten Oberösterreich, N Austria 48°21′N 14°31′E
54 H7 Pregonero Táchira, NW Venezuela 08°02′N 71°35′W
118 J10 Preiļi Ger. Preli. SE Latvia 56°17′N 26°52′E
116 J12 Prejmer Ger. Tartlau, Hung. Prázsmár. Braşov, S Romania 45°42′N 25°49′E
113 J16 Prekornica ⛰ C Montenegro
Preli see Preiļi
Prëmet see Përmet
100 M12 Premnitz Brandenburg, NE Germany 52°33′N 12°22′E
25 S15 Premont Texas, SW USA 27°21′N 98°07′W
113 H14 Prenj ⛰ S Bosnia and Herzegovina
Prenjas/Prenjasi see Përrenjas
22 L7 Prentiss Mississippi, S USA 31°36′N 89°52′W
Preny see Prienai
100 O10 Prenzlau Brandenburg, NE Germany 53°19′N 13°52′E
123 N11 Preobrazhenka Irkutskaya Oblast', C Russian Federation 60°01′N 108°00′E
166 J9 Preparis Island island SW Myanmar (Burma)
Prerau see Přerov
111 H18 Přerov Ger. Prerau. Olomoucký Kraj, E Czech Republic 49°27′N 17°27′E
Preschau see Prešov
14 M14 Prescott Ontario, SE Canada 44°43′N 75°33′W
36 K12 Prescott Arizona, SW USA 34°33′N 112°26′W
27 T13 Prescott Arkansas, C USA 33°49′N 93°25′W
32 L10 Prescott Washington, NW USA 46°17′N 118°21′W
30 H6 Prescott Wisconsin, N USA 44°46′N 92°45′W
185 A24 Preservation Inlet inlet South Island, New Zealand
112 O7 Preševo Serbia, SE Serbia 42°20′N 21°38′E
29 N10 Presho South Dakota, N USA 43°54′N 100°03′W
58 M13 Presidente Dutra Maranhão, E Brazil 05°17′S 44°30′W
60 I8 Presidente Epitácio São Paulo, S Brazil 21°45′S 52°07′W
62 N5 Presidente Hayes off. Departamento de Presidente Hayes. ◈ department C Paraguay
Presidente Hayes, Departamento de see Presidente Hayes
60 I9 Presidente Prudente São Paulo, S Brazil 22°09′S 51°24′W
Presidente Stroessner see Ciudad del Este
Presidente Vargas see Itabira
60 I8 Presidente Venceslau São Paulo, S Brazil 21°52′S 51°51′W
193 O10 President Thiers Seamount undersea feature C Pacific Ocean 24°39′S 145°50′W
24 J11 Presidio Texas, SW USA 29°33′N 104°22′W
Preslav see Veliki Preslav
111 M19 Prešov var. Preschau, Ger. Eperies, Hung. Eperjes. Prešovský Kraj, E Slovakia 49°N 21°14′E
111 M19 Prešovský Kraj ◈ region E Slovakia
113 N20 Prespa, Lake Alb. Liqen i Prespës, Gk. Límni Megáli Préspa, Límni Prespa, Mac. Prespansko Ezero, Serb. Prespansko Jezero. ⊚ SE Europe
Prespa, Limni/Prespansko Ezero/Prespansko Jezero/Prespës, Liqen i see Prespa, Lake
19 S2 Presque Isle Maine, NE USA 46°40′N 68°01′W
18 B11 Presque Isle headland Pennsylvania, NE USA 42°09′N 80°06′W
Pressburg see Bratislava
77 P17 Prestea SW Ghana 05°22′N 02°07′W
111 B17 Přeštice Ger. Pschestitz. Plzeňský Kraj, W Czech Republic 49°36′N 13°19′E
97 K17 Preston NW England, United Kingdom 53°46′N 02°42′W
23 S6 Preston Georgia, SE USA 32°03′N 84°32′W
33 R16 Preston Idaho, NW USA 42°06′N 111°52′W
29 Z13 Preston Iowa, C USA 42°03′N 90°24′W
29 X11 Preston Minnesota, N USA 43°41′N 92°06′W
21 O6 Prestonsburg Kentucky, S USA 37°40′N 82°46′W
96 I13 Prestwick W Scotland, United Kingdom 55°31′N 04°39′W
83 145 Pretoria var. Epitoli. ● Gauteng, NE South Africa 25°41′S 28°12′E
Pretoria-Witwatersrand-Vereeniging see Gauteng
Pretusha see Pretushë
113 M21 Pretushë var. Pretusha. Korçë, SE Albania 40°50′N 20°45′E
Preussisch Eylau see Bagrationovsk
Preußisch Holland see Pasłęk
Preussisch-Stargard see Starogard Gdański
115 C17 Préveza Ípeiros, W Greece 38°59′N 20°44′E
37 V3 Prewitt Reservoir ⊠ Colorado, C USA
167 S13 Prey Vêng Prey Vêng, S Cambodia 11°30′N 105°20′E
Priaral'skiye Karakumy, Peski see Priaral'skiy Karakum
144 M12 Priaral'skiy Karakum prev. Priaral'skiye Karakumy, Peski. desert SW Kazakhstan
123 P14 Priargunsk Zabaykal'skiy Kray, S Russian Federation 50°25′N 119°12′E
38 K14 Pribilof Islands island group Alaska, USA
113 K14 Priboj Serbia, W Serbia 43°34′N 19°33′E
111 C17 Příbram Ger. Pibrans. Středočeský Kraj, W Czech Republic 49°41′N 14°02′E
36 M4 Price Utah, W USA 39°35′N 110°49′W
37 N5 Price River ≈ Utah, W USA
23 N8 Prichard Alabama, S USA 30°44′N 88°04′W
25 R8 Priddy Texas, SW USA 31°39′N 98°30′W
105 P8 Priego Castilla-La Mancha, C Spain 40°26′N 02°19′W
104 M14 Priego de Córdoba Andalucía, S Spain 37°27′N 04°12′W
118 C10 Priekule Ger. Preenkuln. SW Latvia 56°26′N 21°36′E
118 C12 Priekulė Ger. Prökuls. Klaipėda, W Lithuania 55°36′N 21°16′E
119 F14 Prienai Pol. Preny. Kaunas, S Lithuania 54°37′N 23°56′E
83 G23 Prieska Northern Cape, C South Africa 29°40′S 22°45′E
32 M7 Priest Lake ⊠ Idaho, NW USA
32 M7 Priest River Idaho, NW USA 48°10′N 117°02′W
104 M3 Prieta, Peña ▲ N Spain 43°01′N 04°42′W
40 J10 Prieto, Cerro ▲ C Mexico 24°10′N 105°21′W
111 J19 Prievidza var. Priewitz, Ger. Priwitz, Hung. Privigye. Trenčiansky Kraj, W Slovakia 48°47′N 18°35′E
Priewitz see Prievidza
112 F10 Prijedor ◈ Republika Srpska, NW Bosnia and Herzegovina
113 K14 Prijepolje Serbia, W Serbia 43°24′N 19°39′E
Prikaspiyskaya Nizmennost' see Caspian Depression
113 O19 Prilep Turk. Perlepe. S FYR Macedonia 41°21′N 21°34′E
108 B9 Prilly Vaud, SW Switzerland 46°32′N 06°38′E
Priluki see Pryluky
62 L10 Primero, Río ≈ C Argentina
29 S12 Primghar Iowa, C USA 43°05′N 95°37′W
112 B9 Primorje-Gorski Kotar off. Primorsko-Goranska Županija. ◈ province NW Croatia
118 A13 Primorsk Ger. Fischhausen. Kaliningradskaya Oblast', W Russian Federation 54°45′N 20°00′E
124 G12 Primorsk Fin. Koivisto. Leningradskaya Oblast', NW Russian Federation 60°20′N 28°39′E
123 S14 Primorskiy Kray prev. Eng. Maritime Territory. ◈ territory SE Russian Federation
114 N10 Primorsko prev. Keupriya. Burgas, E Bulgaria 42°15′N 27°45′E
126 K13 Primorsko-Akhtarsk Krasnodarskiy Kray, SW Russian Federation 46°03′N 38°44′E
Primorsko-Goranska Županija see Primorje-Gorski Kotar
Primorsk/Primorskoye see Prymors'k
113 D14 Primošten Šibenik-Knin, S Croatia 43°34′N 15°57′E
11 R13 Primrose Lake ⊚ Saskatchewan, C Canada
11 T14 Prince Albert Saskatchewan, S Canada 53°09′N 105°43′W
83 G25 Prince Albert Western Cape, SW South Africa 33°13′S 22°03′E
8 J5 Prince Albert Peninsula peninsula Victoria Island, Northwest Territories, NW Canada
8 J6 Prince Albert Sound inlet Northwest Territories, N Canada
J5 8 Prince Alfred, Cape headland Northwest Territories, NW Canada
9 P6 Prince Charles Island island Nunavut, NE Canada
195 W6 Prince Charles Mountains ⛰ Antarctica
Prince-Édouard, Île-du see Prince Edward Island
172 M13 Prince Edward Fracture Zone tectonic feature SW Indian Ocean
13 P14 Prince Edward Island Fr. Île-du Prince-Édouard. ◈ province SE Canada
13 Q14 Prince Edward Island Fr. Île-du Prince-Édouard. island SE Canada
173 M12 Prince Edward Islands island group S South Africa
21 X4 Prince Frederick Maryland, NE USA 38°32′N 76°35′W
10 M14 Prince George British Columbia, SW Canada 53°55′N 122°49′W
21 W6 Prince George Virginia, NE USA 37°13′N 77°13′W
197 O8 Prince Gustaf Adolf Sea sea Nunavut, N Canada
197 Q3 Prince of Wales, Cape headland Alaska, USA 65°39′N 168°12′W
181 V1 Prince of Wales Island island Queensland, E Australia
8 L5 Prince of Wales Island island Queen Elizabeth Islands, Nunavut, NW Canada
39 Y14 Prince of Wales Island island Alexander Archipelago, Alaska, USA
Prince of Wales Island see Pinang, Pulau
8 J5 Prince of Wales Strait strait Northwest Territories, N Canada
197 O8 Prince Patrick Island island Parry Islands, Northwest Territories, NW Canada
9 N5 Prince Regent Inlet channel Nunavut, N Canada
10 J13 Prince Rupert British Columbia, SW Canada 54°18′N 130°17′W
Prince's Island see Príncipe
21 Y5 Princess Anne Maryland, NE USA 38°12′N 75°42′W
Princess Astrid Coast see Prinsesse Astrid Kyst
181 W2 Princess Charlotte Bay bay Queensland, NE Australia
195 W7 Princess Elizabeth Land physical region Antarctica
10 J14 Princess Royal Island island British Columbia, SW Canada
45 U15 Princes Town Trinidad, Trinidad and Tobago 10°16′N 61°23′W
11 N17 Princeton British Columbia, SW Canada 49°25′N 120°35′W
30 L11 Princeton Illinois, N USA 41°22′N 89°27′W
31 N16 Princeton Indiana, N USA 38°21′N 87°33′W
29 Z14 Princeton Iowa, C USA 41°40′N 90°21′W
20 H7 Princeton Kentucky, S USA 37°06′N 87°52′W
29 V8 Princeton Minnesota, N USA 45°34′N 93°34′W
27 S1 Princeton Missouri, C USA 40°22′N 93°37′W
18 J15 Princeton New Jersey, NE USA 40°21′N 74°39′W
21 R6 Princeton West Virginia, NE USA 37°23′N 81°06′W
39 S12 Prince William Sound inlet Alaska, USA
67 P9 Príncipe var. Príncipe Island, Eng. Prince's Island. island N Sao Tome and Principe
Príncipe Island see Príncipe
32 I13 Prineville Oregon, NW USA 44°19′N 120°50′W
28 J11 Pringle South Dakota, N USA 43°34′N 103°34′W
25 N1 Pringle Texas, SW USA 35°55′N 101°28′W
99 H14 Prinsenbeek Noord-Brabant, S Netherlands 51°36′N 04°42′E
98 L6 Prinses Margriet Kanaal canal N Netherlands
195 R1 Prinsesse Astrid Kyst Eng. Princess Astrid Coast. physical region Antarctica
195 T2 Prinsesse Ragnhild Kyst physical region Antarctica
195 U2 Prins Harald Kyst physical region Antarctica
92 N2 Prins Karls Forland island W Svalbard
43 N8 Prinzapolka Región Autónoma Atlántico Norte, NE Nicaragua 13°19′N 83°35′W
42 L8 Prinzapolka, Río ≈ NE Nicaragua
122 H9 Priob'ye Khanty-Mansiyskiy Avtonomnyy Okrug-Yugra, N Russian Federation 62°25′N 65°36′E
104 H1 Prior, Cabo headland NW Spain 43°33′N 08°21′W
29 V9 Prior Lake Minnesota, N USA 44°42′N 93°25′W
124 H11 Priozersk Fin. Käkisalmi. Leningradskaya Oblast', NW Russian Federation 61°02′N 30°07′E
119 J20 Pripet Bel. Prypyats', Ukr. Prypyat'. ≈ Belarus/Ukraine
119 J20 Pripet Marshes wetland Belarus/Ukraine
113 N16 Prishtinë, Serb. Priština. C Kosovo 42°40′N 21°10′E
126 J8 Pristen' Kurskaya Oblast', W Russian Federation 51°15′N 36°47′E
Priština see Prishtinë
100 M10 Pritzwalk Brandenburg, NE Germany 53°10′N 12°11′E
103 R13 Privas Ardèche, E France 44°45′N 04°35′E
107 I16 Priverno Lazio, C Italy 41°28′N 13°11′E
Privigye see Prievidza
112 C12 Privlaka Zadar, SW Croatia 44°15′N 15°07′E
124 M15 Privolzhsk Ivanovskaya Oblast', NW Russian Federation 57°24′N 41°16′E
127 P7 Privolzhskaya Vozvyshennost' var. Volga Uplands. ⛰ W Russian Federation
127 P8 Privolzhskoye Saratovskaya Oblast', W Russian Federation 51°08′N 45°57′E
Priwitz see Prievidza
127 N13 Priyutnoye Respublika Kalmykiya, SW Russian Federation 46°08′N 43°33′E
113 M17 Prizren S Kosovo 42°14′N 20°46′E
107 I24 Prizzi Sicilia, Italy, C Mediterranean Sea 37°44′N 13°26′E
113 P18 Probištip NE FYR Macedonia 42°00′N 22°06′E
169 S16 Probolinggo Jawa, C Indonesia 07°45′S 113°12′E
Probstberg see Wyszków
111 F14 Prochowice Ger. Parchwitz. Dolnośląskie, SW Poland 51°15′N 16°22′E
29 W5 Proctor Minnesota, N USA 46°46′N 92°13′W
25 R8 Proctor Texas, SW USA 31°57′N 98°25′W
25 R8 Proctor Lake ⊠ Texas, SW USA
155 I18 Proddatūr Andhra Pradesh, E India 14°45′N 78°34′E
104 H9 Proença-a-Nova var. Proença a Nova. Castelo Branco, C Portugal 39°45′N 07°56′W
Proença a Nova see Proença-a-Nova
99 I21 Profondeville Namur, SE Belgium 50°22′N 04°52′E
41 W11 Progreso Yucatán, SE Mexico 21°14′N 89°41′W
123 R14 Progress Amurskaya Oblast', SE Russian Federation 49°40′N 129°30′E
127 O15 Prokhladnyy Kabardino-Balkarskaya Respublika, SW Russian Federation 43°48′N 44°02′E
Prokletije see North Albanian Alps
Prökuls see Priekulė
113 O15 Prokuplje Serbia, SE Serbia 43°15′N 21°35′E
124 H14 Proletariy Novgorodskaya Oblast', W Russian Federation 58°24′N 31°40′E
126 M12 Proletarsk Rostovskaya Oblast', SW Russian Federation 46°42′N 41°48′E
127 N13 Proletarskoye Vodokhranilishche salt lake SW Russian Federation
Prome see Pyay
60 J8 Promissão São Paulo, S Brazil 21°33′S 49°51′W
60 J8 Promissão, Represa de ⊠ S Brazil
125 V4 Promyshlennyy Respublika Komi, NW Russian Federation 67°36′N 64°E
119 O16 Pronya ≈ E Belarus
10 M11 Prophet River British Columbia, W Canada 58°07′N 122°39′W
30 K11 Prophetstown Illinois, N USA 41°40′N 89°56′W
Propinsi Kepulauan Riau see Kepulauan Riau
Propinsi Papua Barat see Papua Barat
59 P16 Propriá Sergipe, E Brazil 10°15′S 36°51′W
103 X16 Propriano Corse, France, C Mediterranean Sea 41°41′N 08°54′E
Prościejów see Prostějov
Proskurov see Khmel 'nyts'kyy
114 H12 Prosotsáni Anatolikí Makedonía kai Thráki, NE Greece 41°11′N 23°59′E
171 Q7 Prosperidad Mindanao, S Philippines 08°36′N 125°54′E
32 J10 Prosser Washington, NW USA 46°12′N 119°46′W
Prossnitz see Prostějov
111 G18 Prostějov Ger. Prossnitz, Pol. Prościejów. Olomoucký Kraj, E Czech Republic 49°29′N 17°08′E
117 V8 Prosyana Dnipropetrovs'ka Oblast', E Ukraine 48°07′N 36°22′E
111 L16 Proszowice Małopolskie, S Poland 50°12′N 20°15′E
Protasy see Pratasy
172 J11 Protea Seamount undersea feature SW Indian Ocean 36°50′S 18°05′E
115 D21 Próti island S Greece
114 N8 Provadia , Provadiya. Varna, E Bulgaria 43°10′N 27°29′E
Provadiya see Provadia
103 T14 Provence cultural region SE France
103 S15 Provence prev. Marseille-Marigrane. ✈ (Marseille) Bouches-du-Rhône, SE France 43°25′N 05°15′E
103 T14 Provence-Alpes-Côte d'Azur ◈ region SE France
20 H6 Providence Kentucky, S USA 37°23′N 87°47′W
19 N12 Providence state capital Rhode Island, NE USA 41°50′N 71°26′W
36 L1 Providence Utah, W USA 41°42′N 111°49′W
Providence see Fort Providence
Providence see Providence Atoll
67 X10 Providence Atoll var. Providence. atoll S Seychelles
14 D12 Providence Bay Manitoulin Island, Ontario, S Canada 45°39′N 82°16′W
23 R6 Providence Canyon valley Alabama/Georgia, S USA
22 I5 Providence, Lake ⊚ Louisiana, S USA
35 X13 Providence Mountains ⛰ California, W USA
44 L6 Providenciales island W Turks and Caicos Islands
19 Q12 Provincetown Massachusetts, NE USA 42°01′N 70°10′W
103 P5 Provins Seine-et-Marne, N France 48°34′N 03°18′E
36 L3 Provo Utah, W USA 40°14′N 111°39′W
11 R15 Provost Alberta, SW Canada 52°24′N 110°16′W
112 G13 Prozor Federacija Bosne I Hercegovine, SW Bosnia and Herzegovina 43°46′N 17°38′E
Prozoroki see Prazaroki
60 I11 Prudentópolis Paraná, S Brazil 25°12′S 50°58′W
39 R5 Prudhoe Bay Alaska, USA 70°16′N 148°18′W
39 R4 Prudhoe Bay bay Alaska, USA
111 H16 Prudnik Ger. Neustadt, Neustadt in Oberschlesien. Opole, SW Poland 50°20′N 17°34′E
119 J16 Prudy Minskaya Voblasts', C Belarus 53°47′N 26°32′E
101 D18 Prüm Rheinland-Pfalz, W Germany 50°15′N 06°27′E
101 D18 Prüm ≈ W Germany

◆ Country
● Country Capital
◇ Dependent Territory
○ Dependent Territory Capital
◈ Administrative Regions
✈ International Airport
▲ Mountain
⛰ Mountain Range
⨳ Volcano
≈ River
⊚ Lake
⊠ Reservoir

Prusa – Qal'at Aḥmad

Prusa *see* Bursa
110 J7 **Pruszcz Gdański** *Ger.* Praust. Pomorskie, N Poland 54°16′N 18°36′E
110 M12 **Pruszków** *Ger.* Kaltdorf. Mazowieckie, C Poland 52°09′N 20°49′E
116 K8 **Prut** *Ger.* Pruth. E Europe
Pruth *see* Prut
108 L8 **Prutz** Tirol, W Austria 47°07′N 10°42′E
Prużana *see* Pruzhany
119 G19 **Pruzhany** *Pol.* Prużana. Brestskaya Voblasts', SW Belarus 52°33′N 24°28′E
124 I11 **Pryazha** Respublika Kareliya, NW Russian Federation 61°42′N 33°39′E
117 U10 **Pryazovs'ke** Zaporiz'ka Oblast', SE Ukraine 46°43′N 35°39′E
Prychornomor'ska Nyzovyna *see* Black Sea Lowland
Prydniprovs'ka Nyzovyna/ Prydnyaprowskaya Nizina *see* Dnieper Lowland
195 Y7 **Prydz Bay** *bay* Antarctica
117 R4 **Pryluky** *Rus.* Priluki. Chernihivs'ka Oblast', NE Ukraine 50°35′N 32°23′E
117 V10 **Prymors'k** *Rus.* Primorsk; *prev.* Primorskoye. Zaporiz'ka Oblast', SE Ukraine 46°44′N 36°19′E
117 U13 **Prymors'kyy** Avtonomna Respublika Krym, S Ukraine 45°09′N 35°33′E
27 Q9 **Pryor** Oklahoma, C USA 36°19′N 95°19′W
33 U11 **Pryor Creek** Montana, NW USA
Pryp"yat'/Prypyats' *see* Pripet
110 M10 **Przasnysz** Mazowieckie, C Poland 53°01′N 20°51′E
111 K14 **Przedbórz** Łodzkie, S Poland 51°04′N 19°51′E
111 P17 **Przemyśl** *Rus.* Peremyshl. Podkarpackie, C Poland 49°47′N 22°47′E
111 O16 **Przeworsk** Podkarpackie, SE Poland 50°04′N 22°30′E
Przheval'sk *see* Karakol
110 L13 **Przysucha** Mazowieckie, SE Poland 51°22′N 20°36′E
115 H18 **Psachná** *var.* Psahna, Psakhná. Évvoia, C Greece 38°35′N 23°39′E
Psahna/Psakhná *see* Psachná
115 K18 **Psará** *island* E Greece
115 I16 **Psathoúra** *island* Vóreies Sporádes, Greece, Aegean Sea
Pschestitz *see* Přeštice
Psein Lora *see* Pishin Lora
117 S5 **Psel** *Rus.* Psël. Russian Federation/Ukraine
Psël *see* Psel
115 M21 **Psérimos** *island* Dodekánisa, Greece, Aegean Sea
Pseyn Bowr *see* Pishin Lora
Pskem *see* Piskom
147 R8 **Pskemskiy Khrebet** *Uzb.* Piskom Tizmasi. ▲▲ Kyrgyzstan/Uzbekistan
124 F14 **Pskov** *Ger.* Pleskau, *Latv.* Pleskava. Pskovskaya Oblast', W Russian Federation 58°32′N 31°15′E
118 K6 **Pskov, Lake** *Est.* Pihkva Järv, *Ger.* Pleskauer See, *Rus.* Pskovskoye Ozero. ◎ Estonia/Russian Federation
124 F15 **Pskovskaya Oblast'** ◈ *province* W Russian Federation
Pskovskoye Ozero *see* Pskov, Lake
112 G9 **Psunj** ▲▲ NE Croatia
111 J17 **Pszczyna** *Ger.* Pless. Śląskie, S Poland 49°59′N 18°54′E
Ptačnik/Ptacsnik *see* Vtáčnik
115 D17 **Ptéri** ▲ C Greece 39°08′N 21°32′E
Ptich' *see* Ptsich
115 E14 **Ptolemaïda** *prev.* Ptolemaïs. Dytikí Makedonía, N Greece 40°34′N 21°42′E
Ptolemaïs *see* Ptolemaïda, Greece
Ptolemaïs *see* 'Akko, Israel
119 M19 **Ptsich** *Rus.* Ptich'. Homyel'skaya Voblasts', SE Belarus 52°11′N 28°49′E
119 M18 **Ptsich** *Rus.* Ptich'. SE Belarus
109 X10 **Ptuj** *Ger.* Pettau; *anc.* Poetovio. NE Slovenia 46°26′N 15°54′E
61 A23 **Puán** Buenos Aires, E Argentina 37°35′S 62°45′W
192 H15 **Pu'apu'a** Savai'i, C Samoa 13°32′S 172°09′W
192 G15 **Puava, Cape** *headland* Savai'i, NW Samoa
Pubao *see* Baingoin
56 F12 **Pucallpa** Ucayali, C Peru 08°21′S 74°33′W
57 J17 **Pucarani** La Paz, NW Bolivia 16°25′S 68°29′W
Pučarevo *see* Novi Travnik
157 U12 **Pucheng** Shaanxi, SE China 35°00′N 109°34′E
160 L6 **Pucheng** *var.* Nanpu. Fujian, C China 27°59′N 118°31′E
125 N16 **Puchezh** Ivanovskaya Oblast', W Russian Federation 56°58′N 41°08′E
111 I19 **Púchov** *Hung.* Puhó. Trenčiansky Kraj, W Slovakia 49°08′N 18°15′E
116 J13 **Pucioasa** Dâmbovița, S Romania 45°04′N 25°23′E
110 I6 **Puck** Pomorskie, N Poland 54°43′N 18°24′E
30 L8 **Puckaway Lake** ◎ Wisconsin, N USA
63 G15 **Pucón** Araucanía, S Chile 39°18′S 71°52′W
93 M14 **Pudasjärvi** Pohjois-Pohjanmaa, C Finland 65°20′N 27°02′E
148 L8 **Pūdeh Tal, Shelleh-ye** SW Afghanistan
127 S1 **Pudem** Udmurtskaya Respublika, NW Russian Federation 58°18′N 52°08′E
Pudewitz *see* Pobiedziska
124 K11 **Pudozh** Respublika Kareliya, NW Russian Federation 61°48′N 36°30′E
97 M17 **Pudsey** N England, United Kingdom 53°48′N 01°40′W
Puduchcheri *see* Puducherry
151 I20 **Puducherry** *prev.* Pondicherry, *var.* Puduchcheri, *Fr.* Pondichéry. ◈ *union territory* India
151 H21 **Pudukkottai** Tamil Nādu, SE India 10°23′N 78°47′E
171 Z13 **Pue** Papua, E Indonesia 02°42′S 140°36′E
41 P14 **Puebla** *var.* Puebla de Zaragoza. Puebla, S Mexico 19°02′N 98°13′W
41 P15 **Puebla** ◈ *state* S Mexico
104 L11 **Puebla de Alcocer** Extremadura, W Spain 38°59′N 05°14′W
Puebla de Don Fabrique *see* Puebla de Don Fadrique
105 P13 **Puebla de Don Fadrique** *var.* Puebla de Don Fabrique. Andalucía, S Spain 37°58′N 02°25′W
104 J11 **Puebla de la Calzada** Extremadura, W Spain 38°54′N 06°38′W
104 J5 **Puebla de Sanabria** Castilla y León, N Spain 42°04′N 06°38′W
Puebla de Trives *see* A Pobla de Trives
Puebla de Zaragoza *see* Puebla
37 T6 **Pueblo** Colorado, C USA 38°15′N 104°37′W
37 N10 **Pueblo Colorado Wash** *valley* Arizona, SW USA
61 C16 **Pueblo Libertador** Corrientes, NE Argentina 30°13′S 59°23′W
40 J10 **Pueblo Nuevo** Durango, C Mexico 23°24′N 105°21′W
42 J8 **Pueblo Nuevo** Estelí, NW Nicaragua 13°21′N 86°30′W
54 J3 **Pueblo Nuevo** Falcón, N Venezuela 11°59′N 69°57′W
42 B6 **Pueblo Nuevo Tiquisate** *var.* Tiquisate. Escuintla, SW Guatemala 14°16′N 91°21′W
41 Q11 **Pueblo Viejo, Laguna de** *lagoon* E Mexico
63 J14 **Puelches** La Pampa, C Argentina 38°08′S 65°56′W
104 L14 **Puente-Genil** Andalucía, S Spain 37°23′N 04°45′W
105 Q3 **Puente la Reina** *Bas.* Gares. Navarra, N Spain 42°40′N 01°49′W
104 L12 **Puente Nuevo, Embalse de** ⊠ S Spain
57 D14 **Puente Piedra** Lima, W Peru 11°49′S 77°01′W
160 F14 **Pu'er** *var.* Ning'er. Yunnan, SW China 23°09′N 100°58′E
45 V6 **Puerca, Punta** *headland* E Puerto Rico 18°13′N 65°36′W
37 R12 **Puerco, Rio** New Mexico, SW USA
57 J17 **Puerto Acosta** La Paz, W Bolivia 15°33′S 69°15′W
63 G19 **Puerto Aisén** Aisén, S Chile 45°24′S 72°42′W
41 R17 **Puerto Ángel** Oaxaca, SE Mexico 15°39′N 96°29′W
Puerto Argentino *see* Stanley
41 T17 **Puerto Arista** Chiapas, SE Mexico 15°55′N 93°47′W
43 O16 **Puerto Armuelles** Chiriquí, SW Panama 08°19′N 82°51′W
Puerto Arrecife *see* Arrecife
54 D14 **Puerto Asís** Putumayo, SW Colombia 0°31′N 76°31′W
54 L9 **Puerto Ayacucho** Amazonas, SW Venezuela 05°45′N 67°37′W
57 C18 **Puerto Ayora** Galapagos Islands, Ecuador, E Pacific Ocean 0°45′S 90°19′W
57 C18 **Puerto Baquerizo Moreno** *var.* Baquerizo Moreno. Galapagos Islands, Ecuador, E Pacific Ocean 0°54′S 89°37′W
42 G4 **Puerto Barrios** Izabal, E Guatemala 15°42′N 88°34′W
Puerto Bello *see* Portobelo
54 F8 **Puerto Berrío** Antioquia, C Colombia 06°28′N 74°28′W
54 F9 **Puerto Boyaca** Boyacá, C Colombia 05°58′N 74°36′W
54 K4 **Puerto Cabello** Carabobo, N Venezuela 10°29′N 68°02′W
43 N7 **Puerto Cabezas** *var.* Bilwi. Región Autónoma Atlántico Norte, NE Nicaragua 14°05′N 83°22′W
54 L9 **Puerto Carreño** Vichada, E Colombia 06°08′N 67°30′W
54 E4 **Puerto Colombia** Atlántico, N Colombia 10°59′N 74°57′W
42 H4 **Puerto Cortés** Cortés, NW Honduras 15°50′N 87°55′W
54 J4 **Puerto Cumarebo** Falcón, N Venezuela 11°29′N 69°21′W
Puerto de Cabras *see* Puerto del Rosario
55 Q5 **Puerto de Hierro** Sucre, NE Venezuela 10°40′N 62°03′W
64 O11 **Puerto de la Cruz** Tenerife, Islas Canarias, Spain, NE Atlantic Ocean 28°24′N 16°33′W
64 Q11 **Puerto del Rosario** *var.* Puerto de Cabras. Fuerteventura, Islas Canarias, Spain, NE Atlantic Ocean 28°29′N 13°52′W
63 J20 **Puerto Deseado** Santa Cruz, SE Argentina 47°46′S 65°53′W
40 F8 **Puerto Escondido** Baja California Sur, NW Mexico 25°48′N 111°20′W
41 R17 **Puerto Escondido** Oaxaca, SE Mexico 15°48′N 96°57′W
60 G12 **Puerto Esperanza** Misiones, NE Argentina 26°01′S 54°39′W
54 H10 **Puerto Gaitán** Meta, C Colombia 04°20′N 72°10′W
Puerto Gallegos *see* Río Gallegos
60 G12 **Puerto Iguazú** Misiones, NE Argentina 25°39′S 54°35′W
56 F12 **Puerto Inca** Huánuco, N Peru 09°22′S 74°54′W
54 L11 **Puerto Inírida** *var.* Obando. Guainía, E Colombia 03°48′N 67°54′W
42 K13 **Puerto Jesús** Guanacaste, NW Costa Rica 10°08′N 85°26′W
41 Z11 **Puerto Juárez** Quintana Roo, SE Mexico 21°06′N 86°46′W
55 N5 **Puerto La Cruz** Anzoátegui, NE Venezuela 10°14′N 64°40′W
54 E14 **Puerto Leguízamo** Putumayo, S Colombia 0°14′N′S 74°45′W
43 N5 **Puerto Lempira** Gracias a Dios, E Honduras 15°14′N 83°48′W
Puerto Libertad *see* La Libertad
54 I11 **Puerto Limón** Meta, E Colombia 04°00′N 71°09′W
54 D13 **Puerto Limón** Putumayo, SW Colombia 01°02′N 76°30′W
Puerto Limón *see* Limón
105 N11 **Puertollano** Castilla-La Mancha, C Spain 38°41′N 04°07′W
63 K17 **Puerto Lobos** Chubut, SE Argentina 42°00′S 64°58′W
54 I3 **Puerto López** La Guajira, N Colombia 11°54′N 71°21′W
105 Q14 **Puerto Lumbreras** Murcia, SE Spain 37°35′N 01°49′W
41 V17 **Puerto Madero** Chiapas, SE Mexico 14°44′N 92°25′W
63 K17 **Puerto Madryn** Chubut, S Argentina 42°45′S 65°02′W
Puerto Magdalena *see* Bahía Magdalena
57 J15 **Puerto Maldonado** Madre de Dios, E Peru 12°37′S 69°11′W
Puerto Masachapa *see* Masachapa
Puerto México *see* Coatzacoalcos
63 G17 **Puerto Montt** Los Lagos, C Chile 41°28′S 72°57′W
41 Z12 **Puerto Morelos** Quintana Roo, SE Mexico 20°47′N 86°54′W
54 L10 **Puerto Nariño** Vichada, E Colombia 04°57′N 67°51′W
63 H23 **Puerto Natales** Magallanes, S Chile 51°42′S 72°28′W
43 X15 **Puerto Obaldía** Kuna Yala, NE Panama 08°38′N 77°26′W
44 H6 **Puerto Padre** Las Tunas, E Cuba 21°13′N 76°35′W
54 L9 **Puerto Páez** Apure, C Venezuela 06°10′N 67°30′W
40 E3 **Puerto Peñasco** Sonora, NW Mexico 31°20′N 113°35′W
55 N5 **Puerto Píritu** Anzoátegui, NE Venezuela 10°04′N 65°00′W
45 N8 **Puerto Plata** *var.* San Felipe de Puerto Plata. N Dominican Republic 19°46′N 70°42′W
Puerto Presidente Stroessner *see* Ciudad del Este
171 N6 **Puerto Princesa** *off.* Puerto Princesa City. Palawan, W Philippines 09°48′N 118°43′E
Puerto Princesa City *see* Puerto Princesa
Puerto Príncipe *see* Camagüey
Puerto Quellón *see* Quellón
60 F13 **Puerto Rico** Misiones, NE Argentina 26°48′S 54°59′W
57 K14 **Puerto Rico** Pando, N Bolivia 11°07′S 67°32′W
54 E12 **Puerto Rico** Caquetá, S Colombia 01°54′N 75°13′W
45 U5 **Puerto Rico** *off.* Commonwealth of Puerto Rico; *prev.* Porto Rico. ◇ *US commonwealth territory* C West Indies
64 F11 **Puerto Rico** *island* C West Indies
Puerto Rico, Commonwealth of *see* Puerto Rico
64 G11 **Puerto Rico Trench** *undersea feature* NE Caribbean Sea
54 I8 **Puerto Rondón** Arauca, E Colombia 06°16′N 71°05′W
Puerto San José *see* San José
63 J21 **Puerto San Julián** *var.* San Julián. Santa Cruz, SE Argentina 49°14′S 67°41′W
63 I22 **Puerto Santa Cruz** *var.* Santa Cruz. Santa Cruz, SE Argentina 50°05′S 68°31′W
Puerto Sauce *see* Juan L. Lacaze
57 Q20 **Puerto Suárez** Santa Cruz, E Bolivia 18°59′S 57°47′W
54 D13 **Puerto Umbría** Putumayo, SW Colombia 0°52′N 76°36′W
40 J13 **Puerto Vallarta** Jalisco, SW Mexico 20°36′N 105°15′W
63 G16 **Puerto Varas** Los Lagos, C Chile 41°20′S 73°00′W
42 M13 **Puerto Viejo** Heredia, NE Costa Rica 10°27′N 84°00′W
Puertoviejo *see* Portoviejo
57 B18 **Puerto Villamil** *var.* Villamil. Galapagos Islands, Ecuador, E Pacific Ocean 0°57′S 91°00′W
54 F8 **Puerto Wilches** Santander, N Colombia 07°22′N 73°53′W
63 H20 **Pueyrredón, Lago** *var.* Lago Cochrane. ◎ S Argentina
127 R7 **Pugachëv** Saratovskaya Oblast', W Russian Federation 52°06′N 48°50′E
127 T3 **Pugachëvo** Udmurtskaya Respublika, NW Russian Federation 56°38′N 53°03′E
32 H8 **Puget Sound** *sound* Washington, NW USA
107 O17 **Puglia** *var.* Le Puglie, *Eng.* Apulia. ◈ *region* SE Italy
107 N17 **Puglia, Canosa di** *anc.* Canusium. Puglia, SE Italy 41°13′N 16°04′E
118 I6 **Puhja** *Ger.* Kawelecht. Tartumaa, SE Estonia 58°20′N 26°19′E
Puhó *see* Púchov
105 V4 **Puigcerdà** Cataluña, NE Spain 42°25′N 01°53′E
Puigmal *see* Puigmal d'Err
103 N17 **Puigmal d'Err** *var.* Puigmal. ▲ S France 42°24′N 02°07′E
76 I16 **Pujehun** S Sierra Leone 07°23′N 11°44′W
Puka *see* Pukë
185 E20 **Pukaki, Lake** ◎ South Island, New Zealand
38 F10 **Pukalani** Maui, Hawaii, USA, C Pacific Ocean 20°50′N 156°20′W
190 J13 **Pukapuka** *atoll* N Cook Islands
191 X9 **Pukapuka** *atoll* Îles Tuamotu, E French Polynesia
Pukari Neem *see* Purekkari Neem
191 X11 **Pukarua** *var.* Pukaruha. *atoll* Îles Tuamotu, E French Polynesia
Pukaruha *see* Pukarua
14 A7 **Pukaskwa** Ontario, S Canada
11 V12 **Pukatawagan** Manitoba, C Canada 55°46′N 101°14′W
191 X16 **Pukatikei, Maunga** Easter Island, Chile, E Pacific Ocean
182 C1 **Pukatja** *var.* Ernabella. South Australia 26°18′S 132°13′E
163 Y12 **Pukch'ŏng** E North Korea 40°13′N 128°20′E
113 L18 **Pukë** *var.* Puka. Shkodër, N Albania 42°03′N 19°53′E
184 L6 **Pukekohe** Auckland, North Island, New Zealand 37°12′S 174°54′E
184 L7 **Pukemiro** Waikato, North Island, New Zealand 37°37′S 175°02′E
190 D12 **Puke, Mont** ▲ Île Futuna, W Wallis and Futuna
Puket *see* Phuket
185 C20 **Puketeraki Range** ▲▲ South Island, New Zealand
184 N13 **Puketoi Range** ▲▲ North Island, New Zealand
185 F21 **Pukeuri Junction** Otago, South Island, New Zealand 45°01′S 171°01′E
119 L16 **Pukhavichy** *Rus.* Pukhovichi. Minskaya Voblasts', C Belarus 53°32′N 28°15′E
Pukhovichi *see* Pukhavichy
124 M10 **Puksoozero** Arkhangel'skaya Oblast', NW Russian Federation 62°37′N 40°29′E
112 A10 **Pula** *It.* Pola; *prev.* Pulj. Istra, NW Croatia 44°53′N 13°51′E
Pula *see* Nyingchi
163 U14 **Pulandian** *var.* Xinjin. Liaoning, NE China 39°25′N 121°58′E
163 T14 **Pulandian Wan** *bay* NE China
189 O15 **Pulap Atoll** *atoll* Caroline Islands, C Micronesia
18 H9 **Pulaski** New York, NE USA 43°34′N 76°06′W
20 I10 **Pulaski** Tennessee, S USA 35°11′N 87°00′W
21 R7 **Pulaski** Virginia, NE USA 37°03′N 80°47′W
171 Y14 **Pulau, Sungai** Papua, E Indonesia
110 N13 **Puławy** *Ger.* Neu Amerika. Lubelskie, E Poland 51°25′N 21°57′E
149 R5 **Pul-e-'Alam** *prev.* Pol-e-'Alam. Lōgar, E Afghanistan 33°59′N 69°02′E
149 Q3 **Pul-e Khumrī** *prev.* Pol-e Khomrī. Baghlān, NE Afghanistan 35°55′N 68°45′E
146 I16 **Pulhatyn** *Rus.* Polekhatum; *prev.* Pul'-I-Khatum. Ahal Welaýaty, S Turkmenistan 36°01′N 61°08′E
101 E16 **Pulheim** Nordrhein-Westfalen, W Germany 51°00′N 06°48′E
155 J19 **Pulicat Lake** *lagoon* SE India
Pul'-I-Khatum *see* Pulhatyn
Pul-i-Sefid *see* Pol-e Sefīd
Pulj *see* Pula
109 W2 **Pulkau** NE Austria
93 L15 **Pulkkila** Pohjois-Pohjanmaa, C Finland 64°15′N 25°53′E
122 C7 **Pulkovo** ✈ (Sankt-Peterburg) Leningradskaya Oblast', NW Russian Federation 60°06′N 30°23′E
32 M9 **Pullman** Washington, NW USA 46°43′N 117°10′W
108 B10 **Pully** Vaud, SW Switzerland 46°31′N 06°40′E
40 F7 **Púlpita, Punta** *headland* NW Mexico 26°30′N 111°28′W
110 M10 **Pułtusk** Mazowieckie, C Poland 52°41′N 21°04′E
158 H10 **Pulu** Xinjiang Uygur Zizhiqu, W China 36°10′N 81°29′E
137 P13 **Pülümür** Tunceli, E Turkey 39°30′N 39°54′E
189 N16 **Pulusuk** *island* Caroline Islands, C Micronesia
189 N16 **Puluwat Atoll** *atoll* Caroline Islands, C Micronesia
25 N11 **Pumpville** Texas, SW USA 39°55′N 101°43′W
191 P7 **Punaauia** *var.* Hakapehi. Tahiti, W French Polynesia 17°38′S 149°37′W
56 B8 **Puná, Isla** *island* SW Ecuador
185 G16 **Punakaiki** West Coast, South Island, New Zealand 42°07′S 171°21′E
153 T11 **Punakha** C Bhutan 27°38′N 89°50′E
57 L18 **Punata** Cochabamba, C Bolivia 17°32′S 65°50′W
155 E14 **Pune** *prev.* Poona. Mahārāshtra, W India 18°32′N 73°52′E
83 M17 **Pungoè, Rio** *var.* Púnguè, Pungwe. C Mozambique
21 X10 **Pungo River** North Carolina, SE USA
Púnguè/Pungwe *see* Pungoè, Rio
79 N19 **Punia** Maniema, E Dem. Rep. Congo 01°28′S 26°25′E
62 H8 **Punilla, Sierra de la** ▲▲ W Argentina
161 P14 **Puning** Guangdong, S China 23°24′N 116°14′E
62 G10 **Punitaqui** Coquimbo, C Chile 30°50′S 71°29′W
149 T9 **Punjab** *prev.* West Punjab, Western Punjab. ◈ *province* E Pakistan
152 H8 **Punjab** *state* NW India
129 Q9 **Punjab Plains** *plain* N India
93 O17 **Punkaharju** *var.* Punkasalmi. Etelä-Savo, E Finland 61°45′N 29°21′E
Punkasalmi *see* Punkaharju
57 I17 **Puno** Puno, SE Peru 15°53′S 70°03′W
57 H17 **Puno** *off.* Departamento de Puno. ◈ *department* S Peru
Puno, Departamento de *see* Puno
61 B24 **Punta Alta** Buenos Aires, E Argentina 38°54′S 62°01′W
63 H24 **Punta Arenas** *prev.* Magallanes. Magallanes, S Chile 53°10′S 70°56′W
45 T6 **Punta, Cerro de** ▲ C Puerto Rico 18°10′N 66°36′W
43 T15 **Punta Chame** West Panamá, C Panama 08°39′N 79°42′W
57 G17 **Punta Colorada** Arequipa, SW Peru 16°17′S 72°31′W
40 F9 **Punta Coyote** Baja California Sur, NW Mexico
62 G8 **Punta de Díaz** Atacama, N Chile 28°03′S 70°36′W
61 G20 **Punta del Este** Maldonado, S Uruguay 34°59′S 54°58′W
63 K17 **Punta Delgada** Chubut, SE Argentina 42°46′S 63°40′W
55 O5 **Punta de Mata** Monagas, NE Venezuela 09°48′N 63°38′W
55 O4 **Punta de Piedras** Nueva Esparta, NE Venezuela 10°57′N 64°06′W
42 F4 **Punta Gorda** Toledo, SE Belize 16°07′N 88°47′W
43 N11 **Punta Gorda** Región Autónoma Atlántico Sur, SE Nicaragua 11°31′N 83°46′W
23 W14 **Punta Gorda** Florida, SE USA 26°55′N 82°03′W
42 M11 **Punta Gorda, Río** SE Nicaragua
62 H6 **Punta Negra, Salar de** *salt lake* N Chile
40 D5 **Punta Prieta** Baja California Norte, NW Mexico 28°56′N 114°11′W
42 L13 **Puntarenas** Puntarenas, W Costa Rica 09°58′N 84°50′W
42 L13 **Puntarenas** *off.* Provincia de Puntarenas. ◈ *province* W Costa Rica
Puntarenas, Provincia de *see* Puntarenas
80 P13 **Puntland** *cultural region* NE Somalia
54 J4 **Punto Fijo** Falcón, N Venezuela 11°42′N 70°13′W
105 S4 **Puntón de Guara** ▲ N Spain 42°18′N 00°13′W
18 D14 **Punxsutawney** Pennsylvania, NE USA 40°55′N 78°57′W
93 M14 **Puolanka** Kainuu, C Finland 64°51′N 27°42′E
57 J17 **Pupuya, Nevado** ▲ W Bolivia 15°04′S 69°01′W
Puqi *see* Chibi
57 F16 **Puquio** Ayacucho, S Peru 14°44′S 74°07′W
122 J9 **Pur** N Russian Federation
186 D7 **Purari** S Papua New Guinea
27 N11 **Purcell** Oklahoma, C USA 35°00′N 97°21′W
11 O16 **Purcell Mountains** ▲▲ British Columbia, SW Canada
105 P14 **Purchena** Andalucía, S Spain 37°21′N 02°21′W
27 S8 **Purdy** Missouri, C USA 36°49′N 93°55′W
118 I2 **Purekkari Neem** *prev.* Pukari Neem. *headland* N Estonia 59°33′N 24°49′E
37 U7 **Purgatoire River** Colorado, C USA
Purgstall *see* Purgstall an der Erlauf
109 V5 **Purgstall an der Erlauf** *var.* Purgstall. Niederösterreich, NE Austria 48°01′N 15°08′E
154 O13 **Puri** *var.* Jagannath. Odisha, E India 19°52′N 85°49′E
Puriramya *see* Buriram
109 X4 **Purkersdorf** Niederösterreich, NE Austria 48°13′N 16°12′E
98 I9 **Purmerend** Noord-Holland, C Netherlands 52°30′N 04°56′E
151 G16 **Pūrna** C India
Purnea *see* Pūrnia
153 R13 **Pūrnia** *prev.* Purnea. Bihār, NE India 25°47′N 87°28′E
Pursat *see* Poŭthĭsăt, Poŭthĭsăt, W Cambodia
Pursat *see* Poŭthĭsăt, Stœng, W, Cambodia
Purulia *see* Puruliya
150 L13 **Puruliya** *prev.* Purulia. West Bengal, NE India 23°20′N 86°24′E
47 G7 **Purus, Rio** *var.* Río Purús. Brazil/Peru
186 C9 **Purutu Island** *island* SW Papua New Guinea
93 N17 **Puruvesi** ◎ SE Finland
22 L7 **Purvis** Mississippi, S USA 31°08′N 89°24′W
Pŭrvomay *see* Parvomay
169 R16 **Purwodadi** *prev.* Poerwodadi. Jawa, C Indonesia 07°05′S 110°53′E
169 P16 **Purwokerto** *prev.* Poerwokerto. Jawa, C Indonesia 07°25′S 109°14′E
169 P16 **Purworejo** *prev.* Poerworedjo. Jawa, C Indonesia 07°45′S 110°04′E
20 H8 **Puryear** Tennessee, S USA 36°25′N 88°21′W
154 H13 **Pusad** Mahārāshtra, C India 19°56′N 77°40′E
Pusan *see* Busan
Pusan-gwangyŏksi *see* Busan
168 H7 **Pusatgajo, Pegunungan** ▲▲ Sumatera, NW Indonesia
Puschlav *see* Poschiavo
124 G13 **Pushkin** *prev.* Tsarskoye Selo. Leningradskaya Oblast', NW Russian Federation 59°42′N 30°24′E
126 L3 **Pushkino** Moskovskaya Oblast', W Russian Federation 55°57′N 37°45′E
127 Q8 **Pushkino** Saratovskaya Oblast', W Russian Federation 51°09′N 47°00′E
Pushkino *see* Biläsuvar
111 M22 **Püspökladány** Hajdú-Bihar, E Hungary 47°20′N 21°05′E
118 J3 **Püssi** *Ger.* Isenhof. Ida-Virumaa, NE Estonia 59°22′N 27°04′E
116 I5 **Pustomyty** L'vivs'ka Oblast', W Ukraine 49°43′N 23°55′E
124 F16 **Pustoshka** Pskovskaya Oblast', W Russian Federation 56°21′N 29°16′E
Pusztakalán *see* Călan
167 N1 **Putao** *prev.* Fort Hertz. Kachin State, N Myanmar (Burma) 27°22′N 97°24′E
184 M8 **Putaruru** Waikato, North Island, New Zealand 38°03′S 175°48′E
Puteoli *see* Pozzuoli
161 R12 **Putian** Fujian, SE China 25°32′N 119°02′E
107 O17 **Putignano** Puglia, SE Italy 40°51′N 17°08′E
Puting *see* De'an
Putivl' *see* Putyvl'
41 Q16 **Putla** *var.* Putla de Guerrero. Oaxaca, SE Mexico 17°01′N 97°56′W
Putla de Guerrero *see* Putla
19 N12 **Putnam** Connecticut, NE USA 41°56′N 71°52′W
25 Q7 **Putnam** Texas, SW USA 32°22′N 99°11′W
18 M10 **Putney** Vermont, NE USA 42°59′N 72°30′W
111 L20 **Putnok** Borsod-Abaúj-Zemplén, NE Hungary 48°18′N 20°25′E
Putorana, Gory/Putorana Mountains *see* Putorana, Plato
122 L8 **Putorana, Plato** *var.* Gory Putorana, *Eng.* Putorana Mountains. ▲▲ N Russian Federation
168 K9 **Putrajaya** ● (Malaysia) Kuala Lumpur, Peninsular Malaysia 02°57′N 101°42′E
62 H2 **Putre** Arica y Parinacota, N Chile 18°11′S 69°30′W
155 J24 **Puttalam** North Western Province, W Sri Lanka 08°02′N 79°55′E
155 J24 **Puttalam Lagoon** *lagoon* W Sri Lanka
99 H17 **Putte** Antwerpen, C Belgium 51°04′N 04°39′E
94 E10 **Puttegga** ▲ S Norway 62°13′N 07°40′E
98 K11 **Putten** Gelderland, C Netherlands 52°15′N 05°36′E
100 K7 **Puttgarden** Schleswig-Holstein, N Germany 54°30′N 11°13′E
Puttiala *see* Patiāla
101 D20 **Püttlingen** Saarland, SW Germany 49°16′N 06°52′E
54 D14 **Putumayo** *off.* Intendencia del Putumayo. ◈ *province* S Colombia
Putumayo, Intendencia del *see* Putumayo
48 E7 **Putumayo, Río** *var.* Içá, Rio. NW South America *see also* Içá, Rio
Putumayo, Río *see* Içá, Rio
169 P11 **Putus, Tanjung** *headland* Borneo, N Indonesia 0°27′S 109°04′E
116 J8 **Putyla** Chernivets'ka Oblast', W Ukraine 47°59′N 25°04′E
117 S3 **Putyvl'** *Rus.* Putivl'. Sums'ka Oblast', NE Ukraine 51°21′N 33°53′E
93 M18 **Puula** ◎ SE Finland
93 N18 **Puumala** Etelä-Savo, E Finland 61°31′N 28°12′E
118 I5 **Puurmani** *Ger.* Talkhof. Jõgevamaa, E Estonia 58°36′N 26°17′E
99 G17 **Puurs** Antwerpen, N Belgium 51°05′N 04°17′E
38 F10 **Pu'u 'Ula'ula** *var.* Red Hill. ▲ Maui, Hawai'i, USA 20°42′N 156°16′W
38 A8 **Pu'uwai** *var.* Puuwai. Ni'ihau, Hawaii, USA, C Pacific Ocean 21°54′N 160°11′W
12 J4 **Puvirnituq** *prev.* Povungnituk. Québec, NE Canada 60°10′N 77°20′W
12 J3 **Puvirnituq, Riviere de** *prev.* Rivière de Povungnituk. Québec, NE Canada
32 H8 **Puyallup** Washington, NW USA 47°11′N 122°17′W
161 O5 **Puyang** Henan, C China 35°40′N 115°00′E
103 O11 **Puy-de-Dôme** ◈ *department* C France
103 N15 **Puylaurens** Tarn, S France 43°33′N 02°01′E
102 M13 **Puy-l'Evêque** Lot, S France 44°31′N 01°10′E
103 N17 **Puymorens, Col de** *pass* S France
56 C7 **Puyo** Pastaza, C Ecuador 01°30′S 77°58′W
185 A24 **Puysegur Point** *headland* South Island, New Zealand 46°09′S 166°38′E
148 J8 **Pūzak, Hāmūn-e** *Pash.* Hāmūn-i-Puzak. ◎ SW Afghanistan
Puzak, Hāmūn-i- *see* Pūzak, Hāmūn-e
81 J23 **Pwani** *Eng.* Coast. ◈ *region* E Tanzania
79 O23 **Pweto** Katanga, SE Dem. Rep. Congo 08°28′S 28°52′E
97 I19 **Pwllheli** NW Wales, United Kingdom 52°54′N 04°23′W
189 O14 **Pwok** Pohnpei, E Micronesia
122 I9 **Pyakupur** N Russian Federation
124 M6 **Pyalitsa** Murmanskaya Oblast', NW Russian Federation 66°17′N 39°56′E
124 K10 **Pyal'ma** Respublika Kareliya, NW Russian Federation 62°24′N 35°56′E
Pyandzh *see* Panj
166 L9 **Pyapon** Ayeyawady, SW Myanmar (Burma) 16°15′N 95°40′E
119 J15 **Pyarshai** *Rus.* Pershay. Minskaya Voblasts', C Belarus 54°02′N 26°41′E
114 I10 **Pyasachnik, Yazovir** *var.* Yazovir Pyasŭchnik. ⊠ C Bulgaria
122 K8 **Pyasina** N Russian Federation
119 G17 **Pyaski** *Rus.* Peski; *prev.* Pyeski. Hrodzyenskaya Voblasts', W Belarus 53°21′N 24°38′E
Pyasŭchnik, Yazovir *see* Pyasachnik, Yazovir
Pyatikhatki *see* P"yatykhatky
117 S7 **P"yatykhatky** *Rus.* Pyatikhatki. Dnipropetrovs'ka Oblast', E Ukraine 48°23′N 33°43′E
166 M6 **Pyawbwe** Mandalay, C Myanmar (Burma) 20°39′N 96°04′E
166 L7 **Pyay** *var.* Prome, Pye. Bago, C Myanmar (Burma) 18°50′N 95°14′E
127 T3 **Pychas** Udmurtskaya Respublika, NW Russian Federation 56°30′N 52°33′E
Pye *see* Pyay
166 K6 **Pyechin** Chin State, W Myanmar (Burma) 20°01′N 93°36′E
163 X15 **Pyeongtaek** *prev.* P'yŏngt'aek. NW South Korea 37°00′N 127°04′E
Pyeski *see* Pyaski
119 L19 **Pyetrykaw** *Rus.* Petrikov. Homyel'skaya Voblasts', SE Belarus 52°08′N 28°30′E
93 O17 **Pyhäjärvi** ◎ SE Finland
93 M16 **Pyhäjärvi** ◎ C Finland
93 L15 **Pyhäjoki** Pohjois-Pohjanmaa, W Finland 64°28′N 24°15′E
93 L15 **Pyhäjoki** W Finland
93 M15 **Pyhäntä** Pohjois-Pohjanmaa, C Finland 64°07′N 26°19′E
93 M16 **Pyhäsalmi** Pohjois-Pohjanmaa, C Finland 63°38′N 26°E
93 O17 **Pyhäselkä** ◎ SE Finland
93 M19 **Pyhtää** *Swe.* Pyttis. Kymenlaakso, S Finland 60°29′N 26°40′E
166 M5 **Pyin-Oo-Lwin** *var.* Maymyo. Mandalay, C Myanmar (Burma) 22°03′N 96°30′E
115 N24 **Pylés** *var.* Piles. Kárpathos, SE Greece 35°31′N 27°08′E
115 D21 **Pýlos** *var.* Pilos. Pelopónnisos, S Greece 36°55′N 21°42′E
18 B12 **Pymatuning Reservoir** ⊠ Ohio/Pennsylvania, NE USA
P'yŏngt'aek *see* Pyeongtaek
163 V14 **P'yŏngyang** *var.* P'yŏngyang-si, *Eng.* Pyongyang. ● (North Korea) SW North Korea 39°04′N 125°46′E
P'yŏngyang-si *see* P'yŏngyang
35 Q4 **Pyramid Lake** ◎ Nevada, W USA
37 P15 **Pyramid Mountains** ▲▲ New Mexico, SW USA
37 R5 **Pyramid Peak** ▲ Colorado, C USA 39°04′N 106°57′W
115 D17 **Pyramíva** *var.* Piramiva. ▲ C Greece 39°08′N 21°18′E
Pyrenaei Montes *see* Pyrenees
86 B12 **Pyrenees** *Fr.* Pyrénées, *Sp.* Pirineos; *anc.* Pyrenaei Montes. ▲▲ SW Europe
102 J16 **Pyrénées-Atlantiques** ◈ *department* SW France
103 N17 **Pyrénées-Orientales** ◈ *department* S France
115 L19 **Pyrgí** *var.* Pirgi. Chíos, E Greece 38°13′N 26°01′E
115 D20 **Pýrgos** *var.* Pírgos. Dytikí Elláda, S Greece 37°40′N 21°27′E
Pyritz *see* Pyrzyce
115 E19 **Pýrros** S Greece
117 R4 **Pyryatyn** *Rus.* Piryatin. Poltavs'ka Oblast', NE Ukraine 50°14′N 32°31′E
110 D9 **Pyrzyce** *Ger.* Pyritz. Zachodnio-pomorskie, NW Poland 53°09′N 14°53′E
124 F15 **Pytalovo** *Latv.* Abrene; *prev.* Jaunlatgale. Pskovskaya Oblast', W Russian Federation 57°06′N 27°55′E
115 M20 **Pythagóreio** *var.* Pithagorio. Sámos, Dodekánisa, Greece, Aegean Sea 37°42′N 26°57′E
14 L11 **Pythonga, Lac** ◎ Québec, SE Canada
Pyttis *see* Pyhtää
Pyu *see* Phyu
166 M8 **Pyuntaza** Bago, SW Myanmar (Burma) 17°51′N 96°44′E
153 N11 **Pyuthān** Mid Western, W Nepal 28°09′N 82°50′E
110 H12 **Pyzdry** *Ger.* Peisern. Wielkopolskie, C Poland 52°10′N 17°42′E

Q

138 H13 **Qā' al Jafr** ◎ S Jordan
197 O11 **Qaanaaq** *var.* Qânâq, *Dan.* Thule. ◈ Qaasuitsup, N Greenland
197 P12 **Qaasuitsup** *off.* Qaasuitsup Kommunia. ◈ *municipality* NW Greenland
Qaasuitsup Kommunia *see* Qaasuitsup
Qabanbay *see* Kabanbay
138 G7 **Qabb Eliâs** E Lebanon 33°46′N 35°49′E
Qabil *see* Al Qābil
Qabırrı *see* Iori
Qābis *see* Gabès
Qābis, Khalīj *see* Gabès, Golfe de
Qabqa *see* Gonghe
141 S14 **Qabr Hūd** C Yemen 16°02′N 49°36′E
Qacentina *see* Constantine
Qādes *see* Qādis
148 L4 **Qādis** *prev.* Qādes. Bādghīs, NW Afghanistan 34°48′N 63°26′E
139 T11 **Qādisīyah** Al Qādisīyah, S Iraq 31°43′N 44°28′E
Qādisīyah, Muḥāfa at al *see* Al Qādisīyah
143 O4 **Qā'emshahr** *prev.* 'Alīābad, Shāhī. Māzandarān, N Iran 36°31′N 52°49′E
143 U7 **Qā'en** *var.* Qain, Qāyen. Khorāsān-e Jonūbī, E Iran 33°43′N 59°07′E
141 U13 **Qafa** *spring/well* SW Oman 17°46′N 52°55′E
Qafṣah *see* Gafsa
163 Q12 **Qagan Nur** *var.* Xulun Hobot Qagan, Zhengxiangbai Qi. Nei Mongol Zizhiqu, N China 42°10′N 114°57′E
163 V9 **Qagan Nur** ◎ NE China
163 Q11 **Qagan Nur** ◎ N China
Qagan Us *see* Dulan
158 H13 **Qagcaka** Xizang Zizhiqu, W China 32°32′N 81°52′E
Qagchêng *see* Xiangcheng
Qahremānshahr *see* Kermānshāh
159 Q10 **Qaidam He** C China
156 L8 **Qaidam Pendi** *basin* C China
Qain *see* Qā'en
Qala Āhangarān *see* Chaghcharān
Qaļā Diza *see* Qeładize
Qal'ah Sālih *see* Qal'at Şāliḩ
149 N4 **Qal'ah Shahr** *Pash.* Qala Shāhar; *prev.* Qal'eh Shahr. Sar-e Pul, N Afghanistan 35°34′N 65°38′E
148 L4 **Qal'ah-ye Now** *var.* Qala Nau; *prev.* Qal'eh-ye Now. Bādghīs, NW Afghanistan 35°N 63°08′E
147 R13 **Qal'aikhum** *Rus.* Kalaikhum. S Tajikistan 38°28′N 70°49′E
Qala Nau *see* Qal'ah-ye Now
141 V17 **Qalansīyah** Suquṭrā, W Yemen 12°40′N 53°30′E
Qala Panja *see* Qal'eh-ye Panjeh
Qala Shāhar *see* Qal'ah Shahr
149 O8 **Qalāt** *Per.* Kalāt. Zābul, S Afghanistan 32°10′N 66°54′E
139 W9 **Qal'at Aḥmad** Maysān, E Iraq 32°24′N 46°46′E

◆ Country
● Country Capital
◇ Dependent Territory
○ Dependent Territory Capital
◈ Administrative Regions
✈ International Airport
▲ Mountain
▲▲ Mountain Range
Volcano
River
◎ Lake
⊠ Reservoir

QAL'AT BĪSHAH – RADĀ'

141 N11 **Qal'at Bīshah** 'Asīr, SW Saudi Arabia *19°59′N 42°38′E*
138 H4 **Qal'at Burzay** Ḩamāh, W Syria *35°37′N 36°16′E*
Qal 'at Dīzah *see* Qeładize
139 W9 **Qal'at Ḩuşayn** Maysān, E Iraq *32°19′N 46°46′E*
139 V10 **Qal'at Majnūnah** Al Qādisīyah, S Iraq *31°39′N 45°44′E*
139 X11 **Qal'at Şāliḩ** *var.* Qal'ah Sālih. Maysān, E Iraq *31°30′N 47°24′E*
139 V10 **Qal'at Sukkar** Dhī Qār, SE Iraq *31°52′N 46°05′E*
Qalba Zhotasy *see* Khrebet Kalba
143 Q12 **Qal'eh Bīābān** Fārs, S Iran
Qal'eh Shahr *see* Qal'ah Shahr
Qal'eh-ye Now *see* Qal'ah-ye Now
149 T2 **Qal'eh-ye Panjeh** *var.* Qala Panja. Badakhshān, NE Afghanistan *36°56′N 72°15′E*
Qalqaman *see* Kalkaman
Qamanittuaq *see* Baker Lake
Qamar Bay *see* Qamar, Ghubbat al
141 U14 **Qamar, Ghubbat al** *Eng.* Qamar Bay. *bay* Oman/Yemen
141 V13 **Qamar, Jabal al** ▲▲ SW Oman
147 N12 **Qamashi** Qashqadaryo Viloyati, S Uzbekistan *38°52′N 66°30′E*
Qambar *see* Kambar
159 R14 **Qamdo** Xizang Zizhiqu, W China *31°09′N 97°09′E*
75 R7 **Qamīnis** NE Libya *31°48′N 20°04′E*
Qamishly *see* Al Qāmishlī
Qânâq *see* Qaanaaq
Qandahār *see* Kandahār
80 Q11 **Qandala** Bari, NE Somalia *11°30′N 50°00′E*
Qandyaghash *see* Kandyagash
138 L2 **Qanţārī** Ar Raqqah, N Syria *36°24′N 39°16′E*
Qapiciğ Daği *see* Qazangödağ
158 H5 **Qapqal** *var.* Qapqal Xibe Zizhixian. Xinjiang Uygur Zizhiqu, NW China *43°46′N 81°09′E*
Qapqal Xibe Zizhixian *see* Qapqal
Qapshagay Böyeni *see* Vodokhranilishche Kapshagay
Qapugtang *see* Zadoi
196 M15 **Qaqortoq** *Dan.* Julianehåb. S Greenland
139 T4 **Qara Anjīr** Kirkūk, N Iraq *35°30′N 44°37′E*
Qarabāgh *see* Qarah Bāgh
Qarabaŭ *see* Karabau
Qaraböget *see* Karaboget
Qarabulaq *see* Karabulak
Qarabutaq *see* Karabutak
Qaraghandy/Qaraghandy Oblysy *see* Karagandy
Qaraghayly *see* Karagayly
Qara Gol *see* Qere Gol
75 U8 **Qārah** *var.* Qāra. NW Egypt *29°34′N 26°28′E*
Qārah *see* Qārah
148 J4 **Qarah Bāgh** *var.* Qarabāgh. Herāt, NW Afghanistan *35°06′N 61°33′E*
Qarah Gawl *see* Qere Gol
138 G7 **Qaraoun, Lac de** *var.* Buḩayrat al Qir'awn. ⊚ S Lebanon
Qaraoy *see* Karaoy
Qaraqoyyn *see* Karakoyyn, Ozero
Qara Qum *see* Garagum
Qarasū *see* Karasu
Qaratal *see* Karatal
Qarataŭ *see* Karatau, Khrebet, Kazakhstan
Qarataŭ *see* Karatau, Zhambyl, Kazakhstan
Qaraton *see* Karaton
Qarazhal *see* Karazhal
80 P13 **Qardho** *var.* Kardh, *It.* Gardo. Bari, N Somalia *09°34′N 49°30′E*
142 M6 **Qareh Chāy** ∿ N Iran
142 K2 **Qareh Sū** ∿ NW Iran
Qariateïne *see* Al Qaryatayn
Qarkilik *see* Ruoqiang
147 O13 **Qarluq** *Rus.* Karluk. Surkhondaryo Viloyati, S Uzbekistan *38°17′N 67°39′E*
147 U12 **Qarokŭl** *Rus.* Karakul'. E Tajikistan *39°07′N 73°33′E*
147 T12 **Qarokŭl** *Rus.* Ozero Karakul'. ⊚ E Tajikistan
Qarqan *see* Qiemo
158 K9 **Qarqan He** ∿ NW China
Qarqannah, Juzur *see* Kerkenah, Îles de
149 O1 **Qarqīn** Jowzjān, N Afghanistan *37°25′N 66°03′E*
Qars *see* Kars
Qarsaqbay *see* Karsakpay
146 M12 **Qarshi** *Rus.* Karshi; *prev.* Bek-Budi. Qashqadaryo Viloyati, S Uzbekistan *38°54′N 65°48′E*
146 L12 **Qarshi Cho'li** *Rus.* Karshinskaya Step. *grassland* S Uzbekistan
146 M13 **Qarshi Kanali** *Rus.* Karshinskiy Kanal. *canal* Turkmenistan/Uzbekistan
Qaryatayn *see* Al Qaryatayn
Qāsh, Nahr al *see* Gash
146 M12 **Qashqadaryo Viloyati** *Rus.* Kashkadar'inskaya Oblast'. ⬖ *province* S Uzbekistan
Qasigianguit *see* Qasigiannguit
197 N13 **Qasigiannguit** *var.* Qasigianguit, *Dan.* Christianshåb. ⬖ Qaasuitsup, C Greenland
Qaşim, Mintaqat *see* Al Qaşim
75 V10 **Qaşr al Farāfirah** *var.* Qasr Farâfra. W Egypt *27°00′N 27°59′E*
139 P8 **Qaşr 'Amīj** Al Anbār, C Iraq *33°30′N 41°52′E*
139 R9 **Qaşr Darwīshāh** Karbalā', C Iraq *32°36′N 43°27′E*
142 J6 **Qaşr-e Shīrīn** Kermānshāhān, W Iran *34°32′N 45°36′E*
Qasr Farâfra *see* Qaşr al Farāfirah
Qassim *see* Al Qaşim
141 O16 **Qa'ṭabah** SW Yemen *13°51′N 44°42′E*
138 H7 **Qaţanā** *var.* Katana. Rīf Dimashq, S Syria *33°27′N 36°04′E*
143 N15 **Qatar** *off.* State of Qatar, *Ar.* Dawlat Qaţar. ◆ *monarchy* SW Asia
Qatar, State of *see* Qatar
Qatrana *see* Al Qaţrānah
143 Q12 **Qaţrūyeh** Fārs, S Iran *29°08′N 54°42′E*
Qattara Depression/Qaţţārah, Munkhafaḑ al *see* Qaţţārah, Munkhafaḑ al
75 U8 **Qaţţārah, Munkhafaḑ al** *var.* Munkhafaḑ al Qaţţārah, *Eng.* Qattara Depression. *desert* NW Egypt
Qaţţâra, Monkhafad el *see* Qaţţārah, Munkhafaḑ al
Qattinah, Buhayrat *see* Ḩimş, Buḩayrat
Qausuittuq *see* Resolute
Qaydār *see* Qeydār
Qāyen *see* Qā'en
Qaynar *see* Kaynar
147 Q11 **Qayroqqum** *Rus.* Kayrakkum. NW Tajikistan *40°16′N 69°46′E*
147 Q10 **Qayroqqum, Obanbori** *Rus.* Kayrakkumskoye Vodokhranilishche. ⊠ NW Tajikistan
137 V13 **Qazangödağ** *Rus.* Gora Kapydzhik, *Turk.* Qapiciğ Daği. ▲ SW Azerbaijan *39°18′N 46°00′E*
139 U7 **Qazānīyah** *var.* Dhū Shaykh. Diyālá, E Iraq *33°39′N 45°33′E*
Qazaqstan/Qazaqstan Respublikasy *see* Kazakhstan
137 T9 **Q'azbegi** *Rus.* Kazbegi; *prev.* Qazbegi. NE Georgia *42°39′N 44°36′E*
Qazbegi *see* Q'azbegi
149 P15 **Qāzi Ahmad** *var.* Kazi Ahmad. Sind, SE Pakistan *26°19′N 68°08′E*
Qazimämmäd *see* Hacıqabal
Qazris *see* Cáceres
142 M4 **Qazvīn** *var.* Kazvin. Qazvīn, N Iran *36°16′N 50°E*
142 M5 **Qazvīn** *off.* Ostān-e Qazvīn. ⬖ *province* N Iran
Qazvīn, Ostān-e *see* Qazvīn
139 U3 **Qeładize** *Ar.* Qal 'at Dīzah, *var.* Qalā Diza. As Sulaymānīyah, NE Iraq *36°11′N 45°07′E*
187 Z13 **Qelelevu Lagoon** *lagoon* NE Fiji
Qena *see* Qinā
113 L23 **Qeparo** Vlorë, S Albania *40°04′N 19°49′E*
Qeqertarssuaq *see* Qeqertarsuaq
197 N13 **Qeqertarsuaq** *var.* Qeqertarssuaq, *Dan.* Godhavn. ⬖ Qaasuitsup, S Greenland
196 M13 **Qeqertarsuaq** *island* W Greenland
197 N13 **Qeqertarsuup Tunua** *Dan.* Disko Bugt. *inlet* W Greenland
197 N14 **Qeqqata** *off.* Qeqqata Kommunia. ⬖ *municipality* W Greenland
Qeqqata Kommunia *see* Qeqqata
139 U4 **Qere Gol** *Ar.* Qarah Gawl, *var.* Qara Gol. As Sulaymānīyah, NE Iraq *35°21′N 45°38′E*
Qerveh *see* Qorveh
143 S14 **Qeshm** Hormozgān, S Iran *26°58′N 56°17′E*
143 R14 **Qeshm** *var.* Jazīreh-ye Qeshm, Qeshm Island. *island* S Iran
Qeshm Island/Qeshm, Jazīreh-ye *see* Qeshm
Qey *see* Kīsh, Jazīreh-ye
142 L4 **Qeydār** *var.* Qaydār. Zanjān, NW Iran *36°50′N 47°40′E*
142 K5 **Qezel Owzan, Rūd-e** *var.* Ki Zil Uzen, Qi Zil Uzun. ∿ NW Iran
Qian *see* Guizhou
161 Q2 **Qian'an** Heilongjiang, E China *45°00′N 124°00′E*
161 R10 **Qiandao Hu** *prev.* Xin'anjiang Shuiku. ⊠ SE China
Qiandaohu *see* Chun'an
Qian Gorlo/Qian Gorlos/Qian Gorlos Mongolzu Zizhixian/Quianguozhen *see* Qianguo
163 V9 **Qianguo** *var.* Qian Gorlo, Qian Gorlos, Qian Gorlos Mongolzu Zizhixian, Quianguozhen. Jilin, NE China *45°08′N 124°48′E*
161 N9 **Qianjiang** Hubei, C China *30°23′N 112°58′E*
160 K10 **Qianjiang** Sichuan, C China *29°30′N 108°45′E*
160 L14 **Qian Jiang** ∿ S China
160 G9 **Qianning** *var.* Gartar. Sichuan, C China *30°27′N 101°24′E*
163 U13 **Qian Shan** ▲▲ NE China
160 H10 **Qianwei** *var.* Yujin. Sichuan, C China *29°15′N 103°52′E*
160 J11 **Qianxi** Guizhou, S China *27°00′N 106°01′E*
Qiaotou *see* Datong
Qiaowa *see* Muli
159 Q7 **Qiaowan** Gansu, N China *40°37′N 96°40′E*
Qibili *see* Kebili
158 K9 **Qiemo** *var.* Qarqan. Xinjiang Uygur Zizhiqu, NW China *38°09′N 85°30′E*
160 J10 **Qijiang** *var.* Gunan. Chongqing Shi, C China *29°01′N 106°40′E*
159 N5 **Qijiaojing** Xinjiang Uygur Zizhiqu, NW China *43°29′N 91°35′E*
Qike *see* Xunke
9 N5 **Qikiqtaaluk** ⬖ *cultural region* Nunavut, N Canada
9 R5 **Qikiqtarjuaq** *prev.* Broughton Island. Nunavut, NE Canada *67°35′N 63°55′W*
149 P9 **Qila Saifullāh** Baluchistān, SW Pakistan *30°45′N 68°08′E*
159 S9 **Qilian** *var.* Babao. Qinghai, C China *38°09′N 100°08′E*
159 N8 **Qilian Shan** *var.* Kilien Mountains. ▲▲ N China
197 O11 **Qimusseriarsuaq** *Dan.* Melville Bugt, *Eng.* Melville Bay. *bay* NW Greenland
75 X10 **Qinā** *var.* Qena; *anc.* Caene, Caenepolis. E Egypt *26°12′N 32°49′E*
159 W11 **Qin'an** Gansu, C China *34°49′N 105°50′E*
Qincheng *see* Nanfeng
Qing *see* Qinghai
163 W7 **Qing'an** Heilongjiang, NE China *46°53′N 127°29′E*
159 X10 **Qingcheng** *var.* Xifeng. Gansu, C China *35°46′N 107°35′E*
161 R5 **Qingdao** *var.* Ching-Tao, Ch'ing-tao, Tsingtao, Tsintao, *Ger.* Tsingtau. Shandong, E China *36°31′N 120°55′E*
163 V8 **Qinggang** Heilongjiang, NE China *46°41′N 126°05′E*
Qinggil *see* Qinghe
159 P11 **Qinghai** *var.* Chinghai, Koko Nor, Qing, Qinghai Sheng, Tsinghai. ⬖ *province* C China
159 S10 **Qinghai Hu** *var.* Ch'ing Hai, Tsing Hai, *Mong.* Koko Nor. ⊚ C China
Qinghai Sheng *see* Qinghai
158 M3 **Qinghe** *var.* Qinggil. Xinjiang Uygur Zizhiqu, NW China *46°42′N 90°19′E*
160 L4 **Qingjian** *var.* Kuanzhou; *prev.* Xiuyan. Shaanxi, C China *37°10′N 110°09′E*
160 L9 **Qing Jiang** ∿ C China
Qingjiang *see* Huai'an
Qingkou *see* Ganyu
160 I12 **Qinglong** *var.* Liancheng. Guizhou, S China *25°49′N 105°10′E*
161 Q2 **Qinglong** Hebei, E China *40°24′N 118°57′E*
Qingshan *see* Wudalianchi
159 R12 **Qingshuihe** Qinghai, C China *33°47′N 97°10′E*
Qingyang *see* Jinjiang
161 N14 **Qingyuan** Guangdong, S China *23°42′N 113°02′E*
163 V11 **Qingyuan** *var.* Qingyuan Manzu Zizhixian. Liaoning, NE China *42°08′N 124°55′E*
Qingyuan *see* Shandan
Qingyuan *see* Weiyuan
Qingyuan Manzu Zizhixian *see* Qingyuan
158 L13 **Qingzang Gaoyuan** *var.* Xizang Gaoyuan, *Eng.* Plateau of Tibet. *plateau* W China
161 Q4 **Qingzhou** *prev.* Yidu. Shandong, E China *36°41′N 118°29′E*
157 R9 **Qin He** ∿ C China
161 Q2 **Qinhuangdao** Hebei, E China *39°57′N 119°31′E*
160 K7 **Qin Ling** ▲▲ C China
161 N5 **Qinxian** *var.* Dingchang, Qin Xian. Shanxi, C China *36°46′N 112°42′E*
Qin Xian *see* Qinxian
161 N6 **Qinyang** Henan, C China *35°05′N 112°56′E*
160 K15 **Qinzhou** Guangxi Zhuangzu Zizhiqu, S China *22°09′N 108°36′E*
Qiong *see* Hainan
160 L17 **Qionghai** *prev.* Jiaji. Hainan, S China *19°12′N 110°26′E*
160 H9 **Qionglai** Sichuan, C China *30°24′N 103°28′E*
160 H8 **Qionglai Shan** ▲▲ C China
Qiongxi *see* Hongyuan
160 L17 **Qiongzhou Haixia** *var.* Hainan Strait. *strait* S China
163 U7 **Qiqihar** *var.* Ch'i-ch'i-ha-erh, Tsitsihar; *prev.* Lungkiang. Heilongjiang, NE China *47°23′N 124°E*
Qir *see* Qīr-va-Kārzīn
158 H10 **Qira** Xinjiang Uygur Zizhiqu, NW China *37°05′N 80°45′E*
Qir'awn, Buḩayrat al *see* Qaraoun, Lac de
143 P12 **Qīr-va-Kārzīn** *var.* Qir. Fārs, S Iran *28°27′N 53°04′E*
Qiryat Gat *see* Kiryat Gat
Qiryat Shemona *see* Kiryat Shmona
Qishlaq *see* Garmsār
141 U14 **Qishn** SE Yemen *15°29′N 51°44′E*
Qishon, Naḩal *see* Kishon, Nahal
Qita Ghazzah *see* Gaza Strip
156 K5 **Qitai** Xinjiang Uygur Zizhiqu, NW China *44°N 89°34′E*
163 Y8 **Qitaihe** Heilongjiang, NE China *45°45′N 130°53′E*
141 W12 **Qitbīt, Wādī** *dry watercourse* S Oman
161 O5 **Qixian** *var.* Qi Xian, Zhaoge. Henan, C China *35°35′N 114°10′E*
Qi Xian *see* Qixian
Qīzān *see* Jīzān
Qizil Orda *see* Kyzylorda
Qizil Qum/Qizilqum *see* Kyzyl Kum
147 V14 **Qizilrabot** *Rus.* Kyzylrabot. SE Tajikistan *37°29′N 74°44′E*
146 J10 **Qizilravote** *Rus.* Kyzylrabat. Buxoro Viloyati, C Uzbekistan *40°35′N 62°09′E*
Qi Zil Uzun *see* Qezel Owzan, Rūd-e
139 S4 **Qizil Yār** Kirkūk, N Iraq *35°26′N 44°12′E*
164 J12 **Qkutango-hantō** *peninsula* Honshū, SW Japan
137 Y11 **Qobustan** *prev.* Märäzä. E Azerbaijan *40°32′N 48°56′E*
Qoghaly *see* Kogaly
Qogir Feng *see* K2
143 N6 **Qom** *var.* Kum, Qum. Qom, N Iran *34°43′N 50°54′E*
143 N6 **Qom** *off.* Ostān-e Qom. ⬖ *province* N Iran
Qomisheh *see* Shahreẕā
Qomolangma Feng *see* Everest, Mount
Qom, Ostān-e *see* Qom
142 M7 **Qom, Rūd-e** ∿ C Iran
Qomsheh *see* Shahreẕā
Qomul *see* Hami
Qondūz *see* Kunduz
146 G7 **Qo'ng'irot** *Rus.* Kungrad. Qoraqalpog'iston Respublikasi, NW Uzbekistan *43°01′N 58°49′E*
Qongyrat *see* Konyrat
Qoqek *see* Tacheng
147 R10 **Qo'qon** *var.* Khokand, *Rus.* Kokand. Farg'ona Viloyati, E Uzbekistan *40°34′N 70°55′E*
Qorabowur Kirlari *see* Karabaur', Uval
Qoradaryo *see* Karadar'ya
146 G6 **Qorajar** *Rus.* Karadzhar. Qoraqalpog'iston Respublikasi, NW Uzbekistan *43°34′N 58°35′E*
146 K12 **Qorako'l** *Rus.* Karakul'. Buxoro Viloyati, C Uzbekistan *39°27′N 63°45′E*
146 H7 **Qorao'zak** *Rus.* Karauzyak. Qoraqalpog'iston Respublikasi, NW Uzbekistan *43°07′N 60°03′E*
146 E5 **Qoraqalpog'iston** *Rus.* Karakalpakya. Qoraqalpog'iston Respublikasi, NW Uzbekistan *44°45′N 56°06′E*
146 G7 **Qoraqalpog'iston Respublikasi** *Rus.* Respublika Karakalpakstan. ⬖ *autonomous republic* NW Uzbekistan
Qorghalzhyn *see* Korgalzhyn
138 H6 **Qornet es Saouda** ▲ NE Lebanon *36°06′N 34°06′E*
146 L12 **Qorowulbozor** *Rus.* Karaulbazar. Buxoro Viloyati, C Uzbekistan *39°28′N 64°49′E*
142 K5 **Qorveh** *var.* Qerveh, Qurveh. Kordestān, W Iran *35°09′N 47°48′E*
147 N11 **Qo'shrabot** *Rus.* Kushrabat. Samarqand Viloyati, C Uzbekistan *40°15′N 66°40′E*
Qosköl *see* Koskol'
Qosshaghyl *see* Kosshagyl
Qostanay/Qostanay Oblysy *see* Kostanay
143 P12 **Qoţbābād** Fārs, S Iran *28°52′N 53°40′E*
143 R13 **Qoţbābād** Hormozgān, S Iran *27°49′N 56°00′E*
138 H6 **Qoubaïyât** *var.* Al Qubayyāt. N Lebanon *37°00′N 34°30′E*
Qoussantîna *see* Constantine
Qowowuyag *see* Cho Oyu
147 O11 **Qo'ytosh** *Rus.* Koytash. Jizzax Viloyati, C Uzbekistan *40°13′N 67°19′E*
146 G7 **Qozonketkan** *Rus.* Kazanketken. Qoraqalpog'iston Respublikasi, W Uzbekistan *42°59′N 59°21′E*
146 H6 **Qozoqdaryo** *Rus.* Kazakdar'ya. Qoraqalpog'iston Respublikasi, NW Uzbekistan *43°26′N 59°47′E*
19 N11 **Quabbin Reservoir** ⊠ Massachusetts, NE USA
100 F12 **Quakenbrück** Niedersachsen, NW Germany *52°41′N 07°57′E*
18 I15 **Quakertown** Pennsylvania, NE USA *40°26′N 75°17′W*
182 M10 **Quambatook** Victoria, SE Australia *35°52′S 143°28′E*
25 Q4 **Quanah** Texas, SW USA *34°17′N 99°46′W*
167 V10 **Quang Ngai** *var.* Quangngai, Quang Nghia. Quang Ngai, C Vietnam *15°09′N 108°50′E*
Quangngai *see* Quang Ngai
Quang Nghia *see* Quang Ngai
167 T9 **Quang Tri** *var.* Trièu Hai. Quang Tri, C Vietnam *16°46′N 107°11′E*
Quanjiang *see* Suichuan
Quan Long *see* Ca Mau
152 L4 **Quanshuigou** China/India *35°40′N 79°28′E*
161 R13 **Quanzhou** *var.* Ch'uan-chou, Tsinkiang; *prev.* Chin-chiang. Fujian, SE China *24°56′N 118°31′E*
160 M12 **Quanzhou** Guangxi Zhuangzu Zizhiqu, S China *25°59′N 111°02′E*
11 V16 **Qu'Appelle** ∿ Saskatchewan, S Canada
12 M3 **Quaqtaq** *prev.* Koartac. Québec, NE Canada *60°50′N 69°30′W*
61 E16 **Quaraí** Rio Grande do Sul, S Brazil *30°38′S 56°25′W*
59 H24 **Quaraí, Rio** *Sp.* Río Cuareim. ∿ Brazil/Uruguay *see also* Cuareim, Río
Quaraí, Rio *see* Cuareim, Río
171 N13 **Quarles, Pegunungan** ▲▲ Sulawesi, C Indonesia
Quarnero *see* Kvarner
107 C20 **Quartu Sant' Elena** Sardegna, Italy, C Mediterranean Sea *39°15′N 09°12′E*
29 X13 **Quasqueton** Iowa, C USA *42°23′N 91°45′W*
173 X16 **Quatre Bornes** W Mauritius *20°15′S 57°28′E*
172 I17 **Quatre Bornes** Mahé, NE Seychelles
137 X10 **Quba** *Rus.* Kuba. N Azerbaijan *41°22′N 48°30′E*
Qubba *see* Ba'qūbah
143 T3 **Qūchān** *var.* Kuchan. Khorāsān-e Raẕavī, NE Iran *37°12′N 58°28′E*
183 R10 **Queanbeyan** New South Wales, SE Australia *35°24′S 149°17′E*
15 Q10 **Québec** *var.* Quebec. *province capital* Québec, SE Canada *46°50′N 71°15′W*
14 K10 **Québec** *var.* Quebec. ⬖ *province* SE Canada
61 D17 **Quebracho** Paysandú, W Uruguay *31°58′S 57°53′W*
101 K14 **Quedlinburg** Sachsen-Anhalt, C Germany *51°48′N 11°09′E*
138 H10 **Queen Alia** ✈ ('Ammān) 'Ammān, C Jordan
10 L16 **Queen Bess, Mount** ▲ British Columbia, SW Canada *51°15′N 124°29′W*
10 I14 **Queen Charlotte** British Columbia, SW Canada *53°18′N 132°04′W*
65 B24 **Queen Charlotte Bay** *bay* West Falkland, W Falkland Islands
10 H14 **Queen Charlotte Islands** *Fr.* Îles de la Reine-Charlotte. *island group* British Columbia, SW Canada
10 I15 **Queen Charlotte Sound** *sea area* British Columbia, W Canada
10 J16 **Queen Charlotte Strait** *strait* British Columbia, W Canada
27 U1 **Queen City** Missouri, C USA *40°24′N 92°34′W*
25 X5 **Queen City** Texas, SW USA *33°09′N 94°09′W*
197 O9 **Queen Elizabeth Islands** *Fr.* Îles de la Reine-Élisabeth. *island group* Nunavut, N Canada
195 Y10 **Queen Mary Coast** *physical region* Antarctica
65 N24 **Queen Mary's Peak** ▲ C Tristan da Cunha
196 M8 **Queen Maud Gulf** *gulf* Arctic Ocean
195 P11 **Queen Maud Mountains** ▲▲ Antarctica
Queen's County *see* Laois
181 U7 **Queensland** ⬖ *state* N Australia
192 I9 **Queensland Plateau** *undersea feature* N Coral Sea
183 O16 **Queenstown** Tasmania, SE Australia *42°06′S 145°33′E*
185 C22 **Queenstown** Otago, South Island, New Zealand *45°01′S 168°44′E*
83 I24 **Queenstown** Eastern Cape, S South Africa *31°52′S 26°50′E*
Queenstown *see* Cobh
32 F8 **Queets** Washington, NW USA *47°31′N 124°19′W*
61 D18 **Queguay Grande, Río** ∿ W Uruguay
59 O16 **Queimadas** Bahia, E Brazil *10°59′S 39°38′W*
82 D11 **Quela** Malanje, NW Angola *09°18′S 17°07′E*
83 O16 **Quelimane** *var.* Kilimane, Kilmain, Quilimane. Zambézia, NE Mozambique *17°53′S 36°51′E*
63 G18 **Quellón** *var.* Puerto Quellón. Los Lagos, S Chile *43°05′S 73°38′W*
Quelpart *see* Jeju-do
37 P12 **Quemado** New Mexico, SW USA *34°18′N 108°29′W*
25 O12 **Quemado** Texas, SW USA *28°58′N 100°36′W*
44 K7 **Quemado, Punta de** *headland* E Cuba *20°13′N 74°07′W*
Quemoy *see* Jinmen Dao
62 K13 **Quemú Quemú** La Pampa, E Argentina *36°03′S 63°36′W*
155 E17 **Quepem** Goa, W India *15°13′N 74°03′E*
42 M14 **Quepos** Puntarenas, S Costa Rica *09°28′N 84°10′W*
Que Que *see* Kwekwe
61 D23 **Quequén** Buenos Aires, E Argentina *38°30′S 58°44′W*
61 D23 **Quequén Grande, Río** ∿ E Argentina
61 C23 **Quequén Salado, Río** ∿ E Argentina
Quera *see* Chur
41 N13 **Querétaro** Querétaro de Arteaga, C Mexico *20°36′N 100°24′W*
40 F4 **Querobabi** Sonora, NW Mexico *30°02′N 111°02′W*
42 M13 **Quesada** *var.* Ciudad Quesada, San Carlos. Alajuela, N Costa Rica *10°19′N 84°26′W*
105 O13 **Quesada** Andalucía, S Spain *37°52′N 03°05′W*
161 O7 **Queshan** Henan, C China *32°48′N 114°03′E*
10 M15 **Quesnel** British Columbia, SW Canada *52°59′N 122°30′W*
37 S9 **Questa** New Mexico, SW USA *36°41′N 105°37′W*
102 H7 **Questembert** Morbihan, NW France *47°39′N 02°24′W*
57 K22 **Quetena, Río** ∿ SW Bolivia
149 O10 **Quetta** Baluchistān, SW Pakistan *30°15′N 67°E*
Quetzalcoalco *see* Coatzacoalcos
Quetzaltenango *see* Quezaltenango
56 B6 **Quevedo** Los Ríos, C Ecuador *01°02′S 79°27′W*
42 B6 **Quezaltenango** *var.* Quetzaltenango. Quezaltenango, W Guatemala *14°50′N 91°30′W*
42 A2 **Quezaltenango** *off.* Departamento de Quezaltenango, *var.* Quetzaltenango. ⬖ *department* SW Guatemala
Quezaltenango, Departamento de *see* Quezaltenango
42 E6 **Quezaltepeque** Chiquimula, SE Guatemala *14°38′N 89°25′W*
170 M6 **Quezon** Palawan, W Philippines *09°13′N 118°01′E*
161 P5 **Qufu** Shandong, E China *35°37′N 117°05′E*
82 B12 **Quibala** Kwanza Sul, NW Angola *10°44′S 14°58′E*
82 B11 **Quibaxe** *var.* Quibaxi. Kwanza Norte, NW Angola *08°30′S 14°36′E*
Quibaxi *see* Quibaxe
54 D9 **Quibdó** Chocó, W Colombia *05°40′N 76°38′W*
102 G7 **Quiberon** Morbihan, NW France *47°30′N 03°07′W*
102 G7 **Quiberon, Baie de** *bay* NW France
54 J5 **Quíbor** Lara, N Venezuela *09°55′N 69°35′W*
42 C4 **Quiché** *off.* Departamento del Quiché. ⬖ *department* W Guatemala
Quiché, Departamento del *see* Quiché
99 E21 **Quiévrain** Hainaut, S Belgium *50°25′N 03°41′E*
40 I9 **Quila** Sinaloa, C Mexico *24°24′N 107°11′W*
83 B14 **Quilengues** Huíla, SW Angola *14°09′S 14°04′E*
Quilimane *see* Quelimane
57 G15 **Quillabamba** Cusco, C Peru *12°49′S 72°41′W*
57 L18 **Quillacollo** Cochabamba, C Bolivia *17°26′S 66°16′W*
62 H4 **Quillagua** Antofagasta, N Chile *21°33′S 69°32′W*
103 N17 **Quillan** Aude, S France *42°52′N 02°11′E*
11 U15 **Quill Lakes** ⊚ Saskatchewan, S Canada
62 G11 **Quillota** Valparaíso, C Chile *32°54′S 71°16′W*
155 G23 **Quilon** *var.* Kollam. Kerala, SW India *08°57′N 76°37′E* *see also* Kollam
181 V9 **Quilpie** Queensland, C Australia *26°39′S 144°15′E*
149 O4 **Quil-Qala** Bāmyān, N Afghanistan *35°13′N 67°02′E*
62 L7 **Quimilí** Santiago del Estero, C Argentina *27°35′S 62°25′W*
57 O19 **Quimome** Santa Cruz, E Bolivia *17°45′S 61°15′W*
102 F6 **Quimper** *anc.* Quimper Corentin. Finistère, NW France *48°00′N 04°05′W*
Quimper Corentin *see* Quimper
102 G7 **Quimperlé** Finistère, NW France *47°52′N 03°33′W*
32 F8 **Quinault** Washington, NW USA *47°27′N 123°53′W*
32 F8 **Quinault River** ∿ Washington, NW USA
35 P5 **Quincy** California, W USA *39°56′N 120°56′W*
23 S8 **Quincy** Florida, SE USA *30°35′N 84°34′W*
30 I13 **Quincy** Illinois, N USA *39°56′N 91°24′W*
19 O11 **Quincy** Massachusetts, NE USA *42°15′N 71°00′W*
32 J9 **Quincy** Washington, NW USA *47°13′N 119°51′W*
54 E10 **Quindío** *off.* Departamento del Quindío. ⬖ *province* C Colombia
54 E10 **Quindío, Nevado del** ▲ C Colombia *04°42′N 75°25′W*
62 J10 **Quines** San Luis, C Argentina *32°15′S 65°46′W*
39 N13 **Quinhagak** Alaska, USA *59°45′N 161°55′W*
76 G13 **Quinhámel** W Guinea-Bissau *11°52′N 15°52′W*
Qui Nhon/Quinhon *see* Quy Nhơn
Quinindé *see* Rosa Zárate
25 U6 **Quinlan** Texas, SW USA *32°54′N 96°08′W*
61 H17 **Quinta** Rio Grande do Sul, S Brazil *32°05′S 52°18′W*
105 O10 **Quintanar de la Orden** Castilla-La Mancha, C Spain *39°36′N 03°03′W*
41 X13 **Quintana Roo** ⬖ *state* SE Mexico
105 S6 **Quinto** Aragón, NE Spain *41°25′N 00°31′W*
108 G10 **Quinto** Ticino, S Switzerland *46°32′N 08°44′E*
27 Q11 **Quinton** Oklahoma, C USA *35°07′N 95°22′W*
62 K12 **Quinto, Río** ∿ C Argentina
82 A10 **Quinzau** Zaire Province, NW Angola *06°50′S 12°48′E*
14 H8 **Quinze, Lac des** ⊚ Québec, SE Canada
83 B15 **Quipungo** Huíla, C Angola *14°49′S 14°29′E*
62 G13 **Quirihue** Bío Bío, C Chile *36°15′S 72°35′W*
82 D12 **Quirima** Malanje, NW Angola *10°51′S 18°06′E*
183 T6 **Quirindi** New South Wales, SE Australia *31°29′S 150°40′E*
55 P5 **Quiriquire** Monagas, NE Venezuela *09°59′N 63°14′W*
14 D10 **Quirke Lake** ⊚ Ontario, S Canada
61 B21 **Quiroga** Buenos Aires, E Argentina *35°18′S 61°22′W*
104 I4 **Quiroga** Galicia, NW Spain *42°28′N 07°15′W*
Quiróm, Salar *see* Pocitos, Salar
56 B9 **Quiroz, Río** ∿ NW Peru
82 Q13 **Quissanga** Cabo Delgado, NE Mozambique *12°24′S 40°33′E*
83 M20 **Quissico** Inhambane, S Mozambique *24°42′S 34°44′E*
25 O4 **Quitaque** Texas, SW USA *34°22′N 101°03′W*
82 Q13 **Quiterajo** Cabo Delgado, NE Mozambique *11°37′S 40°22′E*
23 T6 **Quitman** Georgia, SE USA *30°46′N 83°33′W*
22 M6 **Quitman** Mississippi, S USA *32°02′N 88°43′W*
25 V6 **Quitman** Texas, SW USA *32°37′N 95°26′W*
56 C6 **Quito** ● (Ecuador) Pichincha, N Ecuador *0°14′S 78°30′W*
Quito *see* Mariscal Sucre
58 P13 **Quixadá** Ceará, E Brazil *04°57′S 39°04′W*
83 Q15 **Quixaxe** Nampula, NE Mozambique *15°15′S 40°07′E*
161 N13 **Qujiang** *var.* Maba. Guangdong, S China *24°47′N 113°34′E*
160 J9 **Qu Jiang** ∿ C China
161 R10 **Qu Jiang** ∿ SE China
160 H12 **Qujing** Yunnan, SW China *25°39′N 103°52′E*
Qulan *see* Kulan
Qulin Gol *see* Chaor He
146 L10 **QuljuqtovTog'lari** *Rus.* Gory Kul'dzhuktau. ▲▲ C Uzbekistan
Qulsary *see* Kul'sary
Qulyndy Zhazyghy *see* Ravnina Kulyndy
Qum *see* Qom
Qumälisch *see* Lubartów
159 P11 **Qumar He** ∿ C China
159 Q12 **Qumarlêb** *var.* Yuegai; *prev.* Yuegaitan. Qinghai, C China *34°06′N 95°54′E*
Qumisheh *see* Shahreẕā
147 O14 **Qumqo'rg'on** *Rus.* Kumkurgan. Surkhondaryo Viloyati, S Uzbekistan *37°54′N 67°31′E*
Qunaytirah/Qunayṭirah, Muḩāfaẓat al *see* Al Qunayṭirah
189 V12 **Quoi** *island* Chuuk, C Micronesia
9 N8 **Quoich** ∿ Nunavut, NE Canada
83 E26 **Quoin Point** *headland* SW South Africa *34°48′S 19°39′E*
182 I7 **Quorn** South Australia *32°22′S 138°03′E*
Qurein *see* Al Kuwayt
147 P14 **Qŭrghonteppa** *Rus.* Kurgan-Tyube. SW Tajikistan *37°51′N 68°42′E*
Qurlurtuuq *see* Kugluktuk
Qurveh *see* Qorveh
Quryq *see* Kuryk
Qusair *see* Al Quşayr
137 X10 **Qusar** *Rus.* Kusary. NE Azerbaijan *41°26′N 48°27′E*
Quşayr *see* Al Quşayr
Quseir *see* Al Quşayr
142 I2 **Qūshchī** Āẕarbāyjān-e Gharbī, N Iran *37°59′N 45°05′E*
Qusmuryn *see* Kushmurun, Kostanay, Kazakhstan
Qusmuryn *see* Kusmuryn, Ozero
Quṭayfah/Qutayfe/Quteife *see* Al Quṭayfah
Quthing *see* Moyeni
147 S10 **Quvasoy** *Rus.* Kuvasay. Farg'ona Viloyati, E Uzbekistan *40°17′N 71°53′E*
Quwair *see* Guwēr
Quxar *see* Lhazê
Qu Xian *see* Quzhou
159 N16 **Qüxü** *var.* Xoi. Xizang Zizhiqu, W China *29°25′N 90°48′E*
Quyang *see* Jingzhou
167 V11 **Quy Nhơn** *var.* Quinhon, Qui Nhon. Binh Ðinh, C Vietnam *13°47′N 109°11′E*
161 R10 **Quzhou** *var.* Qu Xian. Zhejiang, SE China *28°55′N 118°54′E*
Qyteti Stalin *see* Kuçovë
Qyzylaghash *see* Kyzylagash
Qyzylorda *see* Kyzylorda
Qyzyltū *see* Kishkenekol'
Qyzylzhar *see* Kyzylzhar

R

Raa Atoll *see* North Maalhosmadulu Atoll
109 R4 **Raab** Oberösterreich, N Austria *48°19′N 13°40′E*
109 X8 **Raab** *Hung.* Rába. ∿ Austria/Hungary *see also* Rába
Raab *see* Rába
Raab *see* Győr
109 V2 **Raabs an der Thaya** Niederösterreich, E Austria *48°51′N 15°28′E*
93 L14 **Raahe** *Swe.* Brahestad. Pohjois-Pohjanmaa, W Finland *64°42′N 24°31′E*
98 M10 **Raalte** Overijssel, E Netherlands *52°23′N 06°16′E*
99 I14 **Raamsdonksveer** Noord-Brabant, S Netherlands *51°42′N 04°54′E*
92 L12 **Raanujärvi** Lappi, NW Finland *66°39′N 24°40′E*
96 G9 **Raasay** *island* NW Scotland, United Kingdom
118 H3 **Raasiku** *Ger.* Rasik. Harjumaa, NW Estonia *59°22′N 25°11′E*
112 B11 **Rab** *It.* Arbe. Primorje-Gorski Kotar, NW Croatia *44°46′N 14°46′E*
112 B11 **Rab** *It.* Arbe. *island* NW Croatia
171 N16 **Raba** Sumbawa, S Indonesia *08°27′S 118°45′E*
111 G22 **Rába** *Ger.* Raab. ∿ Austria/Hungary *see also* Raab
Rába *see* Raab
112 A10 **Rabac** Istra, NW Croatia *45°03′N 14°09′E*
104 I2 **Rábade** Galicia, NW Spain *43°07′N 07°37′W*
80 F10 **Rabak** White Nile, C Sudan *13°12′N 32°44′E*
186 G9 **Rabaraba** Milne Bay, SE Papua New Guinea *10°00′S 149°50′E*
102 K16 **Rabastens-de-Bigorre** Hautes-Pyrénées, S France *43°23′N 00°10′E*
121 O16 **Rabat** W Malta *35°51′N 14°25′E*
74 F6 **Rabat** *var.* al Dar al Baida. ● (Morocco) NW Morocco *34°02′N 06°51′W*
Rabat *see* Victoria
186 H6 **Rabaul** New Britain, E Papua New Guinea *04°13′S 152°11′E*
Rabbah Ammon/Rabbath Ammon *see* 'Ammān
28 K8 **Rabbit Creek** ∿ South Dakota, N USA
14 H10 **Rabbit Lake** ⊚ Ontario, S Canada
187 Y14 **Rabi** *prev.* Rambi. *island* N Fiji
140 K9 **Rābigh** Makkah, W Saudi Arabia *22°51′N 39°E*
42 D5 **Rabinal** Baja Verapaz, C Guatemala *15°05′N 90°26′W*
169 R9 **Rabi, Pulau** *island* NW Indonesia, East Indies
111 L17 **Rabka** Małopolskie, S Poland *49°38′N 20°E*
155 F16 **Rabkavi** Karnātaka, W India *16°40′N 75°03′E*
Râbniţa *see* Rîbniţa
109 Y6 **Rabnitz** ∿ E Austria
124 J7 **Rabocheostrovsk** Respublika Kareliya, NW Russian Federation *64°58′N 34°46′E*
23 U1 **Rabun Bald** ▲ Georgia, SE USA *34°58′N 83°18′W*
75 S11 **Rabyānah** SE Libya *24°07′N 21°58′E*
75 S11 **Rabyānah, Ramlat** *var.* Rebiana Sand Sea, Şahrā' Rabyānah. *desert* SE Libya
Rabyānah, Şahrā' *see* Rabyānah, Ramlat
116 L11 **Răcăciuni** Bacău, E Romania *46°20′N 27°00′E*
Racaka *see* Riwoqê
107 J24 **Racalmuto** Sicilia, Italy, C Mediterranean Sea *37°25′N 13°44′E*
116 J14 **Răcari** Dâmboviţa, SE Romania *44°37′N 25°43′E*
Răcari *see* Durankulak
116 F13 **Răcăşdia** *Hung.* Rakasd. Caraş-Severin, SW Romania *44°58′N 21°36′E*
106 B9 **Racconigi** Piemonte, NE Italy *44°45′N 07°41′E*
31 T15 **Raccoon Creek** ∿ Ohio, N USA
13 V13 **Race, Cape** *headland* Newfoundland, Newfoundland and Labrador, E Canada *46°40′N 53°05′W*
22 K10 **Raceland** Louisiana, S USA *29°43′N 90°36′W*
19 Q12 **Race Point** *headland* Massachusetts, NE USA *42°03′N 70°14′W*
167 S14 **Rach Gia** Kiên Giang, S Vietnam *10°01′N 105°05′E*
167 S14 **Rach Gia, Vinh** *bay* S Vietnam
76 J8 **Rachïd** Tagant, C Mauritania *18°48′N 11°41′W*
110 L10 **Raciąż** Mazowieckie, C Poland *52°46′N 20°04′E*
111 I16 **Racibórz** *Ger.* Ratibor. Śląskie, S Poland *50°05′N 18°10′E*
31 N9 **Racine** Wisconsin, N USA *42°42′N 87°50′W*
14 D7 **Racine Lake** ⊚ Ontario, S Canada
111 J23 **Ráckeve** Pest, C Hungary *47°10′N 18°58′E*
Rácz-Becse *see* Bečej
141 O15 **Radā'** *var.* Ridā'. W Yemen *14°24′N 44°49′E*

◆ Country
● Country Capital
◇ Dependent Territory
○ Dependent Territory Capital
⬖ Administrative Regions
✈ International Airport
▲ Mountain
▲▲ Mountain Range
⛰ Volcano
∿ River
⊚ Lake
⊠ Reservoir

Radan – Recz

113 O15 Radan ▲ SE Serbia 42°59′N 21°31′E
63 J19 Rada Tilly Chubut, SE Argentina 45°54′S 67°33′W
116 K8 Rădăuţi Ger. Radautz, Hung. Rádóc. Suceava, N Romania 47°49′N 25°58′E
116 L8 Rădăuţi-Prut Botoşani, NE Romania 48°14′N 26°47′E
Radautz see Rădăuţi
Radbusa see Radbuza
111 A17 Radbuza Ger. Radbusa. ≈ SE Czech Republic
20 K6 Radcliff Kentucky, S USA 37°50′N 85°57′W
139 O2 Radd, Wādī ar dry watercourse N Syria
95 H16 Råde Østfold, S Norway 59°21′N 10°53′E
109 V11 Radeče Ger. Ratschach. C Slovenia 46°01′N 15°10′E
Radein see Radenci
116 J4 Radekhiv Pol. Radziechów, Rus. Radekhov. L'vivs'ka Oblast', W Ukraine 50°17′N 24°39′E
Radekhov see Radekhiv
109 X9 Radenci Ger. Radein; prev. Radinci. NE Slovenia 46°36′N 16°02′E
109 S9 Radenthein Kärnten, S Austria 46°48′N 13°42′E
Rádeyilikóé see Fort Good Hope
21 R7 Radford Virginia, NE USA 37°07′N 80°34′W
154 C9 Rādhanpur Gujarāt, W India 23°52′N 71°49′E
Radinci see Radenci
127 Q6 Radishchevo Ul'yanovskaya Oblast', W Russian Federation 52°49′N 47°54′E
12 I9 Radisson Québec, E Canada 53°47′N 77°35′W
11 P16 Radium Hot Springs British Columbia, SW Canada 50°39′N 116°09′W
116 F11 Radna Hung. Máriaradna. Arad, W Romania 46°05′N 21°41′E
Rádnávrre see Randijaure
114 K10 Radnevo Stara Zagora, C Bulgaria 42°17′N 25°58′E
97 J20 Radnor cultural region E Wales, United Kingdom
Radnót see Iernut
Rádóc see Rădăuţi
101 H24 Radolfzell am Bodensee Baden-Württemberg, S Germany 47°43′N 08°58′E
110 M13 Radom Mazowieckie, C Poland 51°23′N 21°08′E
116 I14 Radomireşti Olt, S Romania 44°06′N 25°00′E
111 K14 Radomsko Rus. Novoradomsk. Łódzkie, C Poland 51°04′N 19°25′E
117 N4 Radomyshl' Zhytomyrs'ka Oblast', N Ukraine 50°30′N 29°16′E
113 P19 Radoviš prev. Radovište. E Macedonia 41°39′N 22°26′E
Radovište see Radoviš
Radøy see Radøyni
94 B13 Radøyni prev. Radøy. island S Norway
109 R7 Radstadt Salzburg, NW Austria 47°24′N 13°31′E
182 E8 Radstock, Cape headland South Australia 33°11′S 134°18′E
109 U10 Raduha ▲ N Slovenia 46°24′N 14°46′E
119 G15 Radun' Hrodzyenskaya Voblasts', W Belarus 54°03′N 25°00′E
126 M3 Raduzhnyy Vladimirskaya Oblast', W Russian Federation 55°59′N 40°15′E
118 F11 Radviliškis Šiaulai, N Lithuania 55°48′N 23°32′E
11 U17 Radville Saskatchewan, S Canada 49°28′N 104°19′W
140 K7 Raḑwá, Jabal ▲ W Saudi Arabia 24°31′N 38°21′E
111 P16 Radymno Podkarpackie, SE Poland 49°57′N 22°49′E
116 J5 Radyvyliv Rivnens'ka Oblast', NW Ukraine 50°07′N 25°12′E
Radziechów see Radekhiv
110 I11 Radziejów Kujawskio-pomorskie, C Poland 52°36′N 18°33′E
110 O12 Radzyń Podlaski Lubelskie, E Poland 51°48′N 22°37′E
8 J7 Rae ≈ Nunavut, NW Canada
152 M13 Rāe Bareli Uttar Pradesh, N India 26°14′N 81°14′E
Rae-Edzo see Edzo
21 T11 Raeford North Carolina, SE USA 34°59′N 79°15′W
99 M19 Raeren Liège, E Belgium 50°42′N 06°06′E
9 N7 Rae Strait strait Nunavut, N Canada
184 L11 Raetihi Manawatu-Wanganui, North Island, New Zealand 39°29′S 175°16′E
Raevavae see Raivavae
Rafa see Rafah
62 M10 Rafaela Santa Fe, E Argentina 31°16′S 61°25′W
54 E5 Rafael Núñez ✈ (Cartagena) Bolívar, NW Colombia 10°27′N 75°31′W
138 E11 Rafah var. Rafa, Rafaḩ, Heb. Rafiah, Raphiah. SW Gaza Strip 31°18′N 34°15′E
79 L15 Rafaï Mbomou, SE Central African Republic 05°01′N 23°51′E
141 O4 Rafḩah Al Ḩudūd ash Shamālīyah, N Saudi Arabia 29°41′N 43°29′E
Rafiah see Rafah
143 R10 Rafsanjān Kermān, C Iran 30°25′N 56°E
80 B13 Raga Western Bahr el Ghazal, W South Sudan 08°28′N 25°41′E
19 S8 Ragged Island island Maine, NE USA
44 I5 Ragged Island Range island group S The Bahamas
184 L7 Raglan Waikato, North Island, New Zealand 37°48′S 174°54′E
22 G8 Ragley Louisiana, S USA 30°31′N 93°13′W
Ragnit see Neman
107 K25 Ragusa Sicilia, Italy, C Mediterranean Sea 36°56′N 14°42′E
Ragusa see Dubrovnik
Ragusavecchia see Cavtat
171 P14 Raha Pulau Muna, C Indonesia 04°50′S 122°43′E
119 N17 Rahachow Rus. Rogachëv. Homyel'skaya Voblasts', SE Belarus 53°03′N 30°03′E
67 U6 Rahad var. Nahr ar Rahad. ≈ W Sudan
Rahad, Nahr ar see Rahad
Rahaeng see Tak
138 F11 Rahat Southern, C Israel 31°20′N 34°43′E
140 L8 Rahaţ, Ḩarrat lava flow W Saudi Arabia
149 S12 Rahīmyār Khān Punjab, SE Pakistan 28°27′N 70°21′E
95 I14 Råholt Akershus, S Norway 60°16′N 11°10′E
113 M17 Rahovec Serb. Orahovac. W Kosovo 42°24′N 20°40′E
191 S10 Raiatea island Îles Sous le Vent, W French Polynesia
155 H16 Rāichūr Karnātaka, C India 16°15′N 77°20′E
Raidestos see Tekirdağ
153 S13 Rāiganj West Bengal, NE India 25°38′N 88°11′E
154 M11 Raigarh Chhattīsgarh, C India 21°53′N 83°28′E
183 O16 Railton Tasmania, SE Australia 41°24′S 146°28′E
36 L8 Rainbow Bridge natural arch Utah, W USA
23 Q3 Rainbow City Alabama, S USA 33°57′N 86°02′W
11 N11 Rainbow Lake Alberta, W Canada 58°30′N 119°24′W
21 R5 Rainelle West Virginia, NE USA 37°57′N 80°46′W
32 G10 Rainier Oregon, NW USA 46°05′N 122°55′W
32 H9 Rainier, Mount ⛰ Washington, NW USA 46°51′N 121°45′W
23 Q2 Rainsville Alabama, S USA 34°29′N 85°51′W
12 B11 Rainy Lake ◎ Canada/USA
12 A11 Rainy River Ontario, C Canada 48°44′N 94°33′W
Raippaluoto see Replot
154 K12 Raipur Chhattīsgarh, C India 21°16′N 81°42′E
154 H10 Raisen Madhya Pradesh, C India 23°21′N 77°49′E
15 N13 Raisin ≈ Ontario, SE Canada
31 R11 Raisin, River ≈ Michigan, N USA
191 U13 Raivavae var. Raevavae. Îles Australes, SW French Polynesia
149 W9 Rāiwind Punjab, E Pakistan 31°14′N 74°10′E
171 T12 Raja Ampat, Kepulauan island group E Indonesia
155 L16 Rājahmundry Andhra Pradesh, E India 17°05′N 81°42′E
155 I18 Rājampet Andhra Pradesh, E India 14°09′N 79°10′E
Rajang see Rajang, Batang
169 S9 Rajang, Batang var. Rajang. ≈ East Malaysia
149 S11 Rājanpur Punjab, E Pakistan 29°05′N 70°25′E
155 H23 Rājapālaiyam Tamil Nādu, SE India 09°26′N 77°36′E
152 E12 Rājasthān ◐ state NW India
153 T15 Rajbari Dhaka, C Bangladesh 23°47′N 89°39′E
153 R12 Rajbiraj Eastern, E Nepal 26°34′N 86°52′E
154 G9 Rājgarh Madhya Pradesh, C India 24°01′N 76°42′E
152 H10 Rājgarh Rājasthān, NW India 28°38′N 75°21′E
153 P14 Rājgīr Bihār, N India 25°01′N 85°26′E
110 O8 Rajgród Podlaskie, NE Poland 53°43′N 22°40′E
154 L12 Rājim Chhattīsgarh, C India 20°57′N 81°58′E
112 C11 Rajinac, Mali ▲ W Croatia 44°47′N 15°04′E
154 B10 Rājkot Gujarāt, W India 22°18′N 70°47′E
153 R14 Rājmahal Jhārkhand, NE India 25°03′N 87°49′E
153 Q14 Rājmahāl Hills hill range N India
154 K12 Rāj Nāndgaon Chhattīsgarh, C India 21°06′N 81°02′E
152 I8 Rājpura Punjab, NW India 30°29′N 76°40′E
153 S14 Rajshahi prev. Rampur Boalia. Rajshahi, W Bangladesh 24°24′N 88°40′E
153 S13 Rajshahi ◐ division NW Bangladesh
190 K13 Rakahanga atoll N Cook Islands
185 H19 Rakaia Canterbury, South Island, New Zealand 43°45′S 172°02′E
185 G19 Rakaia ≈ South Island, New Zealand
152 H3 Rakaposhi ▲ N India 36°06′N 74°31′E
Rakasd see Răcăşdia
169 N15 Rakata, Pulau var. Pulau Krakatau. island S Indonesia
141 U10 Rakbah, Qalamat ar well SE Saudi Arabia
166 K6 Rakhine State var. Arakan State. ◐ state W Myanmar (Burma)
116 I8 Rakhiv Zakarpats'ka Oblast', W Ukraine 48°05′N 24°15′E
141 V13 Rakhyūt SW Oman 16°41′N 53°09′E
192 K9 Rakiraki Viti Levu, W Fiji 17°22′S 178°10′E
126 J8 Rakitnoye Belgorodskaya Oblast', W Russian Federation 50°50′N 35°51′E
Rakka see Ar Raqqah
118 I4 Rakke Lääne-Virumaa, NE Estonia 58°58′N 26°14′E
95 I16 Rakkestad Østfold, S Norway 59°25′N 11°17′E
110 F12 Rakoniewice Ger. Rakwitz. Wielkopolskie, C Poland 52°09′N 16°10′E
Rakonitz see Rakovník
83 H18 Rakops Central, C Botswana 21°01′S 24°20′E
111 C16 Rakovník Ger. Rakonitz. Středočeský Kraj, W Czech Republic 50°07′N 13°44′E
114 J10 Rakovski Plovdiv, C Bulgaria 42°16′N 24°58′E
Rakutō-kō see Nakdong-gang
118 I3 Rakvere Ger. Wesenberg. Lääne-Virumaa, N Estonia 59°21′N 26°20′E
Rakwitz see Rakoniewice
22 L6 Raleigh Mississippi, S USA 32°01′N 89°30′W
21 U9 Raleigh state capital North Carolina, SE USA 35°46′N 78°38′W
21 Y11 Raleigh Bay bay North Carolina, SE USA
21 U9 Raleigh-Durham ✈ North Carolina, SE USA 35°54′N 78°45′W
189 S6 Ralik Chain island group Ralik Chain, W Marshall Islands
25 N5 Ralls Texas, SW USA 33°40′N 101°23′W
18 G13 Ralston Pennsylvania, NE USA 41°29′N 76°57′W
141 O16 Ramādah W Yemen 13°35′N 43°50′E
Ramadi see Ar Ramādī
105 N2 Ramales de la Victoria Cantabria, N Spain 43°15′N 03°28′W
138 F10 Ramallah C West Bank 31°55′N 35°12′E
61 C19 Ramallo Buenos Aires, E Argentina 33°30′S 60°01′W
155 H20 Rāmanagaram Karnātaka, E India 12°45′N 77°16′E
155 I23 Rāmanāthapuram Tamil Nādu, SE India 09°23′N 78°53′E
154 N12 Rāmapur Odisha, E India 21°48′N 84°00′E
155 I14 Rāmāreddi var. Kāmāreddi, Kamareddy. Telangana, C India 18°19′N 78°23′E
138 F10 Ramat Gan Tel Aviv, W Israel 32°04′N 34°48′E
103 T6 Rambervillers Vosges, NE France 48°15′N 06°50′E
Rambi see Rabi
103 N5 Rambouillet Yvelines, N France 48°39′N 01°50′E
186 E5 Rambutyo Island island N Papua New Guinea
153 Q12 Ramechhāp Central, C Nepal 27°20′N 86°05′E
183 R12 Rame Head headland Victoria, SE Australia 37°48′S 149°30′E
126 L4 Ramenskoye Moskovskaya Oblast', W Russian Federation 55°31′N 38°24′E
124 J15 Rameshki Tverskaya Oblast', W Russian Federation 57°21′N 36°05′E
153 P14 Rāmgarh Jhārkhand, N India 23°37′N 85°32′E
152 D11 Rāmgarh Rājasthān, NW India 27°29′N 70°38′E
142 M9 Rāmhormoz var. Ram Hormuz, Ramuz. Khūzestān, SW Iran 31°15′N 49°38′E
Ram Hormuz see Rāmhormoz
Ram, Jebel see Ramm, Jabal
138 F10 Ramla var. Ramle, Ramleh, Ar. Er Ramle. Central, C Israel 31°56′N 34°52′E
Ramle/Ramleh see Ramla
138 F14 Ramm, Jabal var. Jebel Ram. ▲ SW Jordan 29°34′N 35°24′E
152 K10 Rāmnagar Uttarakhand, N India 29°23′N 79°07′E
95 N15 Ramnäs Västmanland, C Sweden 59°46′N 16°16′E
Râmnicul-Sărat see Râmnicu Sărat
116 L12 Râmnicu Sărat prev. Râmnicul-Sărat, Rîmnicu-Sărat. Buzău, E Romania 45°24′N 27°06′E
116 I13 Râmnicu Vâlcea prev. Rîmnicu Vîlcea. Vâlcea, C Romania 45°04′N 24°32′E
Ramokgwebana see Ramokgwebane
83 J18 Ramokgwebane var. Ramokgwebana. Central, NE Botswana 20°38′S 27°40′E
126 L7 Ramon' Voronezhskaya Oblast', W Russian Federation 51°51′N 39°18′E
35 V17 Ramona California, W USA 33°02′N 116°52′W
56 A10 Ramón, Laguna ◎ NW Peru
14 G7 Ramore Ontario, S Canada 48°26′N 80°19′W
40 M11 Ramos San Luis Potosí, C Mexico 22°48′N 101°55′W
41 N8 Ramos Arizpe Coahuila, NE Mexico 25°35′N 100°59′W
40 J9 Ramos, Río de ≈ C Mexico
83 J21 Ramotswa South East, S Botswana 24°56′S 25°50′E
39 R8 Rampart Alaska, USA 65°30′N 150°10′W
8 H8 Ramparts ≈ Northwest Territories, NW Canada
152 K10 Rāmpur Uttar Pradesh, N India 28°48′N 79°03′E
154 F9 Rāmpura Madhya Pradesh, C India 24°30′N 75°32′E
Rampur Boalia see Rajshahi
166 K6 Ramree Island island W Myanmar (Burma)
141 W6 Rams var. Ar Rams. Ra's al Khaymah, NE United Arab Emirates 25°52′N 56°01′E
143 N4 Rāmsar prev. Sakhtsar. Māzandarān, N Iran 36°55′N 50°39′E
93 H16 Ramsele Västernorrland, N Sweden 63°33′N 16°35′E
21 T9 Ramseur North Carolina, SE USA 35°43′N 79°39′W
97 I16 Ramsey NE Isle of Man 54°19′N 04°24′W
97 I16 Ramsey Bay bay NE Isle of Man
14 E9 Ramsey Lake ◎ Ontario, S Canada
97 Q22 Ramsgate SE England, United Kingdom 51°20′N 01°25′E
94 M10 Ramsjö Gävleborg, C Sweden 62°10′N 15°40′E
154 I12 Rāmtek Mahārāshtra, C India 21°28′N 79°28′E
Ramtha see Ar Ramthā
Ramuz see Rāmhormoz
118 G12 Ramygala Panevėžys, C Lithuania 55°30′N 24°18′E
152 H14 Rāna Pratāp Sāgar ◎ N India
169 V7 Ranau Sabah, East Malaysia 05°56′N 116°43′E
168 L14 Ranau, Danau ◎ Sumatera, W Indonesia
62 H12 Rancagua Libertador, C Chile 34°10′S 70°45′W
99 G22 Rance Hainaut, S Belgium 50°09′N 04°16′E
102 H6 Rance ≈ NW France
60 J9 Rancharia São Paulo, S Brazil 22°13′S 50°53′W
153 P15 Rānchi Jhārkhand, N India 23°22′N 85°20′E
61 D21 Ranchos Buenos Aires, E Argentina 35°32′S 58°22′W
37 S9 Ranchos De Taos New Mexico, SW USA 36°21′N 105°36′W
63 G16 Ranco, Lago ◎ C Chile
95 C16 Randaberg Rogaland, S Norway 59°00′N 05°38′E
29 U7 Randall Minnesota, N USA 46°05′N 94°30′W
107 L23 Randazzo Sicilia, Italy, C Mediterranean Sea 37°52′N 14°57′E
95 G21 Randers Midtjylland, C Denmark 56°28′N 10°03′E
92 I12 Randijaure Lapp. Rádnávrre. ◎ N Sweden
21 T9 Randleman North Carolina, SE USA 35°49′N 79°48′W
19 O11 Randolph Massachusetts, NE USA 42°09′N 71°02′W
29 Q13 Randolph Nebraska, C USA 42°25′N 97°05′W
36 M1 Randolph Utah, W USA 41°40′N 111°10′W
100 P9 Randow ≈ NE Germany
95 H14 Randsfjorden ◎ S Norway
92 K13 Råneå Norrbotten, N Sweden 65°52′N 22°17′E
92 G12 Ranelva ≈ C Norway
93 F15 Ranemsletta Nord-Trøndelag, C Norway 64°36′N 11°55′E
76 H10 Ranérou C Senegal 15°17′N 14°00′W
Ráneš see Ringvassøya
185 E22 Ranfurly Otago, South Island, New Zealand 45°07′S 170°06′E
167 P17 Rangae Narathiwat, SW Thailand 06°15′N 101°45′E
153 V16 Rangamati Chittagong, SE Bangladesh 22°40′N 92°10′E
184 I2 Rangaunu Bay bay North Island, New Zealand
19 P6 Rangeley Maine, NE USA 44°58′N 70°37′W
37 O4 Rangely Colorado, C USA 40°05′N 108°48′W
25 R7 Ranger Texas, SW USA 32°28′N 98°40′W
14 C9 Ranger Lake Ontario, S Canada 46°51′N 83°34′W
14 C9 Ranger Lake ◎ Ontario, S Canada
153 V12 Rangia Assam, NE India 26°26′N 91°38′E
185 I18 Rangiora Canterbury, South Island, New Zealand 43°19′S 172°34′E
191 T9 Rangiroa atoll Îles Tuamotu, W French Polynesia
184 N9 Rangitaiki ≈ North Island, New Zealand
185 F19 Rangitata ≈ South Island, New Zealand
184 M12 Rangitikei ≈ North Island, New Zealand
184 L6 Rangitoto Island island N New Zealand
Rangkasbitoeng see Rangkasbitung
169 N16 Rangkasbitung prev. Rangkasbitoeng. Jawa, SW Indonesia 06°21′S 106°12′E
167 P9 Rang, Khao ▲ C Thailand 16°13′N 99°03′E
147 V13 Rangkŭl Rus. Rangkul'. SE Tajikistan 38°30′N 74°24′E
Rangkul' see Rangkŭl
Rangoon see Yangon
153 T13 Rangpur Rajshahi, N Bangladesh 25°46′N 89°20′E
155 F18 Rānibennur Karnātaka, W India 14°36′N 75°39′E
153 R15 Rāniganj West Bengal, NE India 23°34′N 87°13′E
149 Q13 Rānīpur Sind, SE Pakistan 27°17′N 68°34′E
Rānīyah see Ranye
25 N9 Rankin Texas, SW USA 31°14′N 101°56′W
9 O9 Rankin Inlet Nunavut, C Canada 62°52′N 92°14′W
183 P8 Rankins Springs New South Wales, SE Australia 33°51′S 146°16′E
Rankovićevo see Kraljevo
108 I7 Rankweil Vorarlberg, W Austria 47°17′N 09°40′E
Rann see Brežice
127 T8 Ranneye Orenburgskaya Oblast', W Russian Federation 51°28′N 52°29′E
96 I10 Rannoch, Loch ◎ C Scotland, United Kingdom
191 U17 Rano Kau var. Rano Kao. crater Easter Island, Chile, E Pacific Ocean
167 N14 Ranong Ranong, SW Thailand 09°59′N 98°40′E
186 J8 Ranongga var. Ghanongga. island NW Solomon Islands
191 W16 Rano Raraku ancient monument Easter Island, Chile, E Pacific Ocean
171 V12 Ransiki Papua Barat, E Indonesia 01°27′S 134°12′E
92 K12 Rantajärvi Norrbotten, N Sweden 66°45′N 23°39′E
93 N17 Rantasalmi Etelä-Savo, SE Finland 62°02′N 28°22′E
169 U13 Rantau Borneo, C Indonesia 02°56′S 115°09′E
168 L10 Rantau, Pulau var. Pulau Tebingtinggi. island W Indonesia
171 N13 Rantepao Sulawesi, C Indonesia 02°58′S 119°58′E
30 M13 Rantoul Illinois, N USA 40°19′N 88°08′W
93 L15 Rantsila Pohjois-Pohjanmaa, C Finland 64°31′N 25°40′E
92 L13 Ranua Lappi, NW Finland 65°55′N 26°34′E
Rānya see Ranye
139 T3 Ranye Ar. Rānīyah, var. Rānya. As Sulaymānīyah, NE Iraq 36°15′N 44°53′E
157 X3 Raohe Heilongjiang, NE China 46°49′N 134°00′E
74 H9 Raoui, Erg er desert W Algeria
193 O10 Rapa island Îles Australes, S French Polynesia
191 V14 Rapa Iti island Îles Australes, SW French Polynesia
106 D10 Rapallo Liguria, NW Italy 44°21′N 09°13′E
Rapa Nui see Pascua, Isla de
Raphiah see Rafah
21 V5 Rapidan River ≈ Virginia, NE USA
28 J10 Rapid City South Dakota, N USA 44°05′N 103°14′W
15 P8 Rapide-Blanc Québec, SE Canada 47°48′N 72°57′W
14 I8 Rapide-Deux Québec, SE Canada 47°56′N 78°33′W
118 K6 Räpina Ger. Rappin. Põlvamaa, SE Estonia 58°06′N 27°27′E
118 G4 Rapla Ger. Rappel. Raplamaa, NW Estonia 59°00′N 24°46′E
118 G4 Raplamaa var. Rapla Maakond. ◐ province NW Estonia
Rapla Maakond see Raplamaa
21 X6 Rappahonnock River ≈ Virginia, NE USA
Rappel see Rapla
108 G7 Rapperswil Sankt Gallen, NE Switzerland 47°14′N 08°50′E
Rappin see Räpina
153 N12 Rāpti ≈ N India
57 K16 Rápulo, Río ≈ E Bolivia
Raqqah/Raqqah, Muḩāfaẓat al see Ar Raqqah
18 J8 Raquette Lake ◎ New York, NE USA
18 J6 Raquette River ≈ New York, NE USA
191 V10 Raraka atoll Îles Tuamotu, C French Polynesia
191 V10 Raroia atoll Îles Tuamotu, C French Polynesia
190 H15 Rarotonga ✈ Rarotonga, S Cook Islands, C Pacific Ocean 21°15′S 159°45′W
190 H16 Rarotonga island S Cook Islands, C Pacific Ocean
147 P12 Rarz W Tajikistan 39°23′N 68°43′E
Ras al'Ain see Ra's al 'Ayn
139 N2 Ra's al 'Ayn var. Ras al'Ain. Al Ḩasakah, N Syria 36°52′N 40°05′E
138 H3 Ra's al Basīţ Al Lādhiqīyah, W Syria 35°57′N 35°55′E
Ra's al-Hafğī see Ra's al Khafjī
141 R5 Ra's al Khafjī var. Ra's al-Hafğī. Ash Sharqīyah, NE Saudi Arabia 28°22′N 48°30′E
Ras al-Khaimah/Ras al Khaimah see Ra's al Khaymah
143 R15 Ra's al Khaymah var. Ras al Khaimah. Ra's al Khaymah, NE United Arab Emirates 25°44′N 55°55′E
143 R15 Ra's al Khaymah var. Ras al-Khaimah. ✈ Ra's al Khaymah, NE United Arab Emirates 25°37′N 55°51′E
138 G13 Ra's an Naqb Ma'ān, S Jordan 30°00′N 35°29′E
61 B26 Rasa, Punta headland E Argentina 40°50′S 62°15′W
171 V12 Rasawi Papua Barat, E Indonesia 02°04′S 134°02′E
Râşcani see Rîşcani
80 J10 Ras Dashen Terara ▲ N Ethiopia 13°12′N 38°09′E
Rasdhoo Atoll see Rasdu Atoll
151 K19 Rasdu Atoll var. Rasdhoo Atoll. atoll C Maldives
118 E12 Raseiniai Kaunas, C Lithuania 55°23′N 23°06′E
75 X8 Ra's Ghārib var. Râs Ghârib. E Egypt 28°16′N 33°01′E
Râs Ghârib see Ra's Ghārib
162 J6 Rashaant Hövsgöl, N Mongolia 49°08′N 101°27′E
Rashaant see Delüün, Bayan-Ölgiy, Mongolia
Rashaant see Öldziyt, Dundgovĭ, Mongolia
75 V7 Rashīd Eng. Rosetta. N Egypt 31°25′N 30°25′E
139 Y11 Rashīd Al Başrah, E Iraq 31°15′N 47°31′E
142 M3 Rasht var. Resht. Gīlān, NW Iran 37°18′N 49°38′E
Ŗashwan see Reshwan
Rashwān see Reshwan
Rasik see Raasiku
113 M15 Raška Serbia, C Serbia 43°18′N 20°37′E
119 P15 Rasna Rus. Ryasna. Mahilyowskaya Voblasts', E Belarus 54°01′N 31°12′E
116 J12 Râşnov prev. Rîşno, Rozsnyó, Hung. Barcarozsnyó. Braşov, C Romania 45°35′N 25°27′E
118 L11 Rasony Rus. Rossony. Vitsyebskaya Voblasts', N Belarus 55°53′N 28°50′E
Ra's Shamrah see Ugarit
127 N7 Rasskazovo Tambovskaya Oblast', W Russian Federation 52°42′N 41°45′E
119 O16 Rasta Rus. Resta. ≈ E Belarus
Rastadt see Rastatt
Rastâne see Ar Rastān
141 S6 Ra's Tannūrah Eng. Ras Tanura. Ash Sharqīyah, NE Saudi Arabia 26°44′N 50°04′E
Ras Tanura see Ra's Tannūrah
101 G21 Rastatt var. Rastadt. Baden-Württemberg, SW Germany 48°51′N 08°13′E
Rastenburg see Kętrzyn
149 V7 Rasūlnagar Punjab, E Pakistan 32°20′N 73°51′E
189 U6 Ratak Chain island group Ratak Chain, E Marshall Islands
119 K15 Ratamka Rus. Ratomka. Minskaya Voblasts', C Belarus 53°56′N 27°21′E
93 G17 Ratän Jämtland, C Sweden 62°28′N 14°35′E
152 G11 Ratangarh Rājasthān, NW India 28°02′N 74°39′E
Rat Buri see Ratchaburi
167 O11 Ratchaburi var. Rat Buri. Ratchaburi, W Thailand 13°30′N 99°50′E
29 W15 Rathbun Lake ⊠ Iowa, C USA
Ráth Caola see Rathkeale
166 K5 Rathedaung Rakhine State, W Myanmar (Burma) 20°30′N 92°48′E
100 M12 Rathenow Brandenburg, NE Germany 52°37′N 12°21′E
97 C19 Rathkeale Ir. Ráth Caola. Limerick, SW Ireland 52°32′N 08°56′W
96 F13 Rathlin Island Ir. Reachlainn. island N Northern Ireland, United Kingdom
97 C20 Ráth Luirc Ir. An Ráth. Cork, SW Ireland 52°22′N 08°44′W
Ratibor see Racibórz
Ratisbon/Ratisbona/Ratisbonne see Regensburg
Rätische Alpen see Rhaetian Alps
38 E17 Rat Island island Aleutian Islands, Alaska, USA
38 E17 Rat Islands island group Aleutian Islands, Alaska, USA
154 F10 Ratlām prev. Rutlam. Madhya Pradesh, C India 23°23′N 75°04′E
155 D15 Ratnāgiri Mahārāshtra, W India 17°00′N 73°20′E
155 K26 Ratnapura Sabaragamuwa Province, S Sri Lanka 06°41′N 80°25′E
Ratne see Ratno
116 J2 Ratno var. Ratne. Volyns'ka Oblast', NW Ukraine 51°40′N 24°33′E
Ratomka see Ratamka
37 U8 Raton New Mexico, SW USA 36°54′N 104°27′W
139 O7 Ratqah, Wādī ar dry watercourse W Iraq
Ratschach see Radeče
167 O16 Rattaphum Songkhla, SW Thailand 07°07′N 100°16′E
26 L6 Rattlesnake Creek ≈ Kansas, C USA
94 L13 Rättvik Dalarna, C Sweden 60°53′N 15°12′E
100 K9 Ratzeburg Mecklenburg-Vorpommern, N Germany 53°41′N 10°48′E
100 K9 Ratzeburger See ◎ N Germany
10 J10 Ratz, Mount ▲ British Columbia, SW Canada 57°22′N 132°17′W
61 D22 Rauch Buenos Aires, E Argentina 36°45′S 59°05′W
41 U16 Raudales Chiapas, SE Mexico
Raudhatain see Ar Rawḑatayn
Raudnitz an der Elbe see Roudnice nad Labem
92 K1 Raufarhöfn Norðurland Eystra, NE Iceland 66°27′N 15°58′W
94 H13 Raufoss Oppland, S Norway 60°44′N 10°39′E
Raukawa see Cook Strait
184 Q8 Raukumara ▲ North Island, New Zealand 37°46′S 178°07′E
192 K11 Raukumara Plain undersea feature N Coral Sea
184 P8 Raukumara Range ▲▲ North Island, New Zealand
Rāulakela see Rāurkela
95 F15 Rauland Telemark, S Norway 59°41′N 07°57′E
93 J19 Rauma Swe. Raumo. Satakunta, SW Finland 61°09′N 21°30′E
94 F10 Rauma ≈ S Norway
Raumo see Rauma
118 H8 Rauna C Latvia 57°19′N 25°34′E
169 T17 Raung, Gunung ▲ Jawa, S Indonesia 08°00′S 114°07′E
154 N11 Rāurkela var. Rāulakela, Rourkela. Odisha, E India 22°13′N 84°53′E
95 J22 Raus Skåne, S Sweden 56°01′N 12°48′E
165 W3 Rausu Hokkaidō, NE Japan 44°00′N 145°06′E
165 W3 Rausu-dake ▲ Hokkaidō, NE Japan 44°06′N 145°04′E
116 M9 Răut var. Răuţel. ≈ C Moldova
93 M17 Rautalampi Pohjois-Savo, C Finland 62°37′N 26°48′E
93 N16 Rautavaara Pohjois-Savo, C Finland 63°30′N 28°19′E
Răuţel see Răut
93 O18 Rautjärvi Etelä-Karjala, SE Finland 61°21′N 29°20′E
Rautu see Sosnovo
191 V11 Ravahere atoll Îles Tuamotu, C French Polynesia
107 J25 Ravanusa Sicilia, Italy, C Mediterranean Sea 37°16′N 13°59′E
143 S9 Rāvar Kermān, C Iran 31°15′N 56°51′E
147 Q11 Ravat Batkenskaya Oblast', SW Kyrgyzstan 39°54′N 70°06′E
18 K11 Ravena New York, NE USA 42°28′N 73°49′W
106 H10 Ravenna Emilia-Romagna, N Italy 44°28′N 12°15′E
29 O15 Ravenna Nebraska, C USA 41°01′N 98°54′W
31 U11 Ravenna Ohio, N USA 41°09′N 81°14′W
101 I24 Ravensburg Baden-Württemberg, S Germany 47°47′N 09°37′E
181 W4 Ravenshoe Queensland, NE Australia 17°29′S 145°28′E
180 K13 Ravensthorpe Western Australia 33°37′S 120°03′E
21 Q4 Ravenswood West Virginia, NE USA 38°57′N 81°45′W
149 U9 Rāvi ≈ India/Pakistan
112 C9 Ravna Gora Primorje-Gorski Kotar, NW Croatia 45°20′N 14°54′E
109 U10 Ravne na Koroškem Ger. Gutenstein. N Slovenia 46°33′N 14°57′E
145 T7 Ravnina Kulyndy prev. Kulunda Steppe, Kaz. Qulyndy Zhazyghy, Rus. Kulundinskaya Ravnina. grassland Kazakhstan/Russian Federation
139 P6 Rāwah Al Anbār, W Iraq 34°32′N 41°54′E
191 T4 Rawaki prev. Phoenix Island. atoll Phoenix Islands, C Kiribati
149 U6 Rāwalpindi Punjab, NE Pakistan 33°38′N 73°06′E
110 L13 Rawa Mazowiecka Łódzkie, C Poland 51°47′N 20°16′E
Ŗawāndiz see Rewanduz
171 U12 Rawas Papua Barat, E Indonesia 01°07′S 132°12′E
139 O4 Rawḑah ◎ E Syria
110 G13 Rawicz Ger. Rawitsch. Wielkopolskie, C Poland 51°37′N 16°51′E
Rawitsch see Rawicz
180 M11 Rawlinna Western Australia 31°01′S 125°36′E
33 W16 Rawlins Wyoming, C USA 41°47′N 107°14′W
63 K17 Rawson Chubut, SE Argentina 43°22′S 65°01′W
159 R16 Rawu Xizang Zizhiqu, W China 29°30′N 96°42′E
153 P12 Raxaul Bihār, N India 26°58′N 84°51′E
28 K3 Ray North Dakota, N USA 48°19′N 103°11′W
169 S11 Raya, Bukit ▲ Borneo, C Indonesia 0°40′S 112°40′E
155 I18 Rāyachoti Andhra Pradesh, E India 14°03′N 78°43′E
Rāyadrug see Rāyagarha
Rāyagada see Rāyagarha
155 M14 Rāyagarha prev. Rāyadrug, var. Rāyagada. Odisha, E India 19°10′N 83°28′E
138 H7 Rayak var. Rayaq, Riyāq. E Lebanon 33°51′N 36°03′E
Rayaq see Rayak
139 T2 Rayat Ar. Rāyāt, var. Ŗāyat. Arbīl, E Iraq 36°39′N 44°56′E
Rāyāt see Rayat
Ŗāyat see Rayat
169 N12 Raya, Tanjung cape Pulau Bangka, W Indonesia
13 R13 Ray, Cape headland Newfoundland, Newfoundland and Labrador, E Canada 47°38′N 59°15′W
123 Q13 Raychikhinsk Amurskaya Oblast', SE Russian Federation 49°47′N 129°19′E
127 U5 Rayevskiy Respublika Bashkortostan, W Russian Federation 54°04′N 54°58′E
11 Q17 Raymond Alberta, SW Canada 49°30′N 112°41′W
22 K6 Raymond Mississippi, S USA 32°15′N 90°25′W
32 F9 Raymond Washington, NW USA 46°41′N 123°43′W
183 T8 Raymond Terrace New South Wales, SE Australia 32°47′S 151°45′E
25 T17 Raymondville Texas, SW USA 26°30′N 97°48′W
11 U16 Raymore Saskatchewan, S Canada 51°24′N 104°34′W
39 Q8 Ray Mountains ▲▲ Alaska, USA
22 H9 Rayne Louisiana, S USA 30°13′N 92°15′W
41 O12 Rayón San Luis Potosí, C Mexico 21°54′N 99°33′W
40 G4 Rayón Sonora, NW Mexico 29°45′N 110°33′W
167 P12 Rayong Rayong, S Thailand 12°42′N 101°17′E
25 T5 Ray Roberts, Lake ◎ Texas, SW USA
18 E15 Raystown Lake ⊠ Pennsylvania, NE USA
141 V13 Raysūt SW Oman 16°58′N 54°02′E
27 R4 Raytown Missouri, C USA 39°00′N 94°27′W
22 I5 Rayville Louisiana, S USA 32°29′N 91°45′W
142 L5 Razan Hamadān, W Iran 35°22′N 48°58′E
139 S9 Razāzah, Buḩayrat ar var. Baḩr al Milḩ. ◎ C Iraq
114 L9 Razboyna ▲ E Bulgaria 42°54′N 26°31′E
Razdan see Hrazdan
Razdolnoye see Rozdol'ne
Razelm, Lacul see Razim, Lacul
Ŗazga see Rezge
114 L8 Razgrad Razgrad, N Bulgaria 43°33′N 26°31′E
114 L8 Razgrad ◐ province NE Bulgaria
114 I10 Razhevo Konare var. Rŭzhevo Konare. Plovdiv, C Bulgaria 42°16′N 24°58′E
117 N13 Razim, Lacul prev. Lacul Razelm. lagoon NW Black Sea
Razkah see Rezge
114 G11 Razlog Blagoevgrad, SW Bulgaria 41°53′N 23°28′E
118 K10 Rāznas Ezers ◎ SE Latvia
102 E6 Raz, Pointe du headland NW France 48°06′N 04°52′W
Reachlainn see Rathlin Island
Reachrainn see Lambay Island
97 N22 Reading S England, United Kingdom 51°28′N 00°59′W
18 H15 Reading Pennsylvania, NE USA 40°20′N 75°55′W
48 C7 Real, Cordillera ▲▲ C Ecuador
62 K12 Realicó La Pampa, C Argentina 35°02′S 64°14′W
25 R15 Realitos Texas, SW USA 27°26′N 98°31′W
108 G9 Realp Uri, C Switzerland 46°36′N 08°32′E
167 Q12 Reăng Kesei Bătdâmbâng, W Cambodia 12°57′N 103°15′E
191 Y11 Reao atoll Îles Tuamotu, E French Polynesia
Reate see Rieti
Greater Antarctica see East Antarctica
180 L11 Rebecca, Lake ◎ Western Australia
Rebiana Sand Sea see Rabyānah, Ramlat
124 H8 Reboly Finn. Repola. Respublika Kareliya, NW Russian Federation 63°51′N 30°49′E
165 S1 Rebun Rebun-tō, NE Japan 45°19′N 141°02′E
165 S1 Rebun-tō island NE Japan
106 J12 Recanati Marche, C Italy 43°23′N 13°34′E
Rechitsa see Rechytsa
109 Y7 Rechnitz Burgenland, SE Austria 47°19′N 16°26′E
119 J20 Rechytsa Rus. Rechitsa. Brestskaya Voblasts', SW Belarus 51°51′N 26°48′E
119 O19 Rechytsa Rus. Rechitsa. Homyel'skaya Voblasts', SE Belarus 52°22′N 30°23′E
59 Q15 Recife prev. Pernambuco. state capital Pernambuco, E Brazil 08°06′S 34°53′W
83 I26 Recife, Cape Afr. Kaap Recife. headland S South Africa 34°03′S 25°37′E
Recife, Kaap see Recife, Cape
172 I16 Récifs, Îles aux island Inner Islands, NE Seychelles
101 E14 Recklinghausen Nordrhein-Westfalen, W Germany 51°37′N 07°12′E
100 M8 Recknitz ≈ NE Germany
99 K23 Recogne Luxembourg, SE Belgium 49°56′N 05°20′E
61 C15 Reconquista Santa Fe, C Argentina 29°08′S 59°38′W
195 O6 Recovery Glacier glacier Antarctica
59 G15 Recreio Mato Grosso, W Brazil 08°13′S 58°15′W
27 X9 Rector Arkansas, C USA 36°15′N 90°17′W
110 E9 Recz Ger. Reetz Neumark. Zachodnio-pomorskie, NW Poland 53°16′N 15°32′E

◆ Country ● Country Capital ◇ Dependent Territory ○ Dependent Territory Capital ◐ Administrative Regions ✈ International Airport ▲ Mountain ▲▲ Mountain Range ⛰ Volcano ≈ River ◎ Lake ⊠ Reservoir

REDANGE – RINCÓN DE LA VICTORIA

99 L24 **Redange** *var.* Redange-sur-Attert. Diekirch, W Luxembourg *49°46′N 05°53′E*
Redange-sur-Attert *see* Redange
18 C13 **Redbank Creek** ≈ Pennsylvania, NE USA
13 S9 **Red Bay** Québec, E Canada *51°40′N 56°37′W*
23 N2 **Red Bay** Alabama, S USA *34°26′N 88°08′W*
35 N4 **Red Bluff** California, W USA *40°09′N 122°14′W*
24 J8 **Red Bluff Reservoir** ⊠ New Mexico/Texas, SW USA
30 K16 **Red Bud** Illinois, N USA *38°12′N 89°59′W*
30 J5 **Red Cedar River** ≈ Wisconsin, N USA
11 R17 **Redcliff** Alberta, SW Canada *50°06′N 110°48′W*
83 K17 **Redcliff** Midlands, C Zimbabwe *19°00′S 29°49′E*
182 L9 **Red Cliffs** Victoria, SE Australia *34°21′S 142°12′E*
29 P17 **Red Cloud** Nebraska, C USA *40°05′N 98°31′W*
22 L8 **Red Creek** ≈ Mississippi, S USA
11 P15 **Red Deer** Alberta, SW Canada *52°15′N 113°48′W*
11 Q16 **Red Deer** ≈ Alberta, SW Canada
39 O11 **Red Devil** Alaska, USA *61°45′N 157°18′W*
35 N3 **Redding** California, W USA *40°33′N 122°26′W*
97 L20 **Redditch** W England, United Kingdom *52°19′N 01°56′W*
29 P9 **Redfield** South Dakota, N USA *44°51′N 98°31′W*
24 J12 **Redford** Texas, SW USA *29°31′N 104°19′W*
45 V13 **Redhead** Trinidad, Trinidad and Tobago *10°44′N 60°58′W*
182 I8 **Red Hill** South Australia *33°34′S 138°13′E*
Red Hill *see* Pu'u 'Ula'ula
26 K7 **Red Hills** *hill range* Kansas, C USA
13 T12 **Red Indian Lake** ◎ Newfoundland, Newfoundland and Labrador, E Canada
124 J16 **Redkino** Tverskaya Oblast', W Russian Federation *56°41′N 36°07′E*
12 A10 **Red Lake** Ontario, C Canada *51°00′N 93°55′W*
36 I10 **Red Lake** *salt flat* Arizona, SW USA
29 S4 **Red Lake Falls** Minnesota, N USA *47°52′N 96°16′W*
29 R4 **Red Lake River** ≈ Minnesota, N USA
35 U15 **Redlands** California, W USA *34°03′N 117°10′W*
18 G16 **Red Lion** Pennsylvania, NE USA *39°53′N 76°36′W*
33 U11 **Red Lodge** Montana, NW USA *45°11′N 109°15′W*
32 H13 **Redmond** Oregon, NW USA *44°16′N 121°10′W*
36 L5 **Redmond** Utah, W USA *39°00′N 111°51′W*
32 H8 **Redmond** Washington, NW USA *47°40′N 122°07′W*
Rednitz *see* Regnitz
29 T15 **Red Oak** Iowa, C USA *41°00′N 95°10′W*
18 K12 **Red Oaks Mill** New York, NE USA *41°39′N 73°52′W*
102 I7 **Redon** Ille-et-Vilaine, NW France *47°39′N 02°05′W*
45 W10 **Redonda** *island* SW Antigua and Barbuda
104 G4 **Redondela** Galicia, NW Spain *42°17′N 08°36′W*
104 H11 **Redondo** Évora, S Portugal *38°38′N 07°32′W*
39 Q12 **Redoubt Volcano** ▲ Alaska, USA *60°29′N 152°44′W*
11 Y16 **Red River** ≈ Canada/USA
129 U12 **Red River** *var.* Yuan, *Chin.* Yuan Jiang, *Vtn.* Sông Hông Hà. ≈ China/Vietnam
25 W4 **Red River** ≈ S USA
22 H7 **Red River** ≈ Louisiana, S USA
30 M6 **Red River** ≈ Wisconsin, N USA
Red Rock, Lake *see* Red Rock Reservoir
29 W14 **Red Rock Reservoir** *var.* Lake Red Rock. ⊠ Iowa, C USA
80 H7 **Red Sea** ◈ *state* NE Sudan
75 Y9 **Red Sea** *var.* Sinus Arabicus. *sea* Africa/Asia
21 T11 **Red Springs** North Carolina, SE USA *34°49′N 79°10′W*
8 I9 **Redstone** ≈ Northwest Territories, NW Canada
11 V17 **Redvers** Saskatchewan, S Canada *49°31′N 101°33′W*
77 P13 **Red Volta** *var.* Nazinon, *Fr.* Volta Rouge. ≈ Burkina Faso/Ghana
11 Q14 **Redwater** Alberta, SW Canada *53°57′N 113°06′W*
28 M16 **Red Willow Creek** ≈ Nebraska, C USA
29 W9 **Red Wing** Minnesota, N USA *44°33′N 92°31′W*
35 N9 **Redwood City** California, W USA *37°29′N 122°13′W*
29 T9 **Redwood Falls** Minnesota, N USA *44°33′N 95°07′W*
31 P7 **Reed City** Michigan, N USA *43°52′N 85°30′W*
28 K6 **Reeder** North Dakota, N USA *46°03′N 102°55′W*
35 R11 **Reedley** California, W USA *36°35′N 119°27′W*
33 T11 **Reedpoint** Montana, NW USA *45°41′N 109°33′W*
30 K8 **Reedsburg** Wisconsin, N USA *43°33′N 90°03′W*
32 E13 **Reedsport** Oregon, NW USA *43°42′N 124°06′W*
187 Q9 **Reef Islands** *island group* Santa Cruz Islands, E Solomon Islands
185 H16 **Reefton** West Coast, South Island, New Zealand *42°07′S 171°53′E*
20 F8 **Reelfoot Lake** ◎ Tennessee, S USA
97 D17 **Ree, Lough** *Ir.* Loch Rí. ◎ C Ireland
Reengus *see* Rīngas
35 U4 **Reese River** ≈ Nevada, W USA
98 M8 **Reest** ≈ E Netherlands
Reetz Neumark *see* Recz
Reevhtse *see* Røssvatnet
137 N13 **Refahiye** Erzincan, C Turkey *39°54′N 38°45′E*
23 N4 **Reform** Alabama, S USA *33°22′N 88°01′W*
95 K20 **Reftele** Jönköping, S Sweden *57°10′N 13°34′E*
25 T14 **Refugio** Texas, SW USA *28°19′N 97°18′W*
110 E8 **Rega** ≈ NW Poland
Regar *see* Tursunzoda
101 O21 **Regen** Bayern, SE Germany *48°57′N 13°10′E*
101 M20 **Regen** ≈ SE Germany
101 M21 **Regensburg** *Eng.* Ratisbon, *Fr.* Ratisbonne, *hist.* Ratisbona; *anc.* Castra Regina, Reginum. Bayern, SE Germany *49°01′N 12°06′E*
101 M21 **Regenstauf** Bayern, SE Germany *49°06′N 12°07′E*
148 M10 **Rēgestān** *var.* Registan *prev.* Rīgestān. S Afghanistan
74 I10 **Reggane** C Algeria *26°46′N 00°09′E*
98 N9 **Regge** ≈ E Netherlands
Reggio *see* Reggio nell'Emilia
Reggio Calabria *see* Reggio di Calabria
107 M23 **Reggio di Calabria** *var.* Reggio Calabria, *Gk.* Rhegion; *anc.* Regium, Rhegium. Calabria, SW Italy *38°06′N 15°39′E*
Reggio Emilia *see* Reggio nell'Emilia
106 F9 **Reggio nell'Emilia** *var.* Reggio Emilia, *abbrev.* Reggio; *anc.* Regium Lepidum. Emilia-Romagna, N Italy *44°42′N 10°37′E*
116 I10 **Reghin** *Ger.* Sächsisch-Reen, *Hung.* Szászrégen; *prev.* Reghinul Săsesc, *Ger.* Sächsisch-Regen. Mureș, C Romania *46°46′N 24°41′E*
Reghinul Săsesc *see* Reghin
11 U16 **Regina** *province capital* Saskatchewan, S Canada *50°25′N 104°39′W*
55 Z10 **Régina** E French Guiana *04°20′N 52°07′W*
11 U16 **Regina** ✈ Saskatchewan, S Canada *50°21′N 104°43′W*
11 U16 **Regina Beach** Saskatchewan, S Canada *50°44′N 105°03′W*
Reginum *see* Regensburg
Région du Haut-Congo *see* Haut-Congo
Registan *see* Rēgestān
60 L11 **Registro** São Paulo, S Brazil *24°30′S 47°50′W*
Regium *see* Reggio di Calabria
Regium Lepidum *see* Reggio nell'Emilia
101 K19 **Regnitz** *var.* Rednitz. ≈ SE Germany
40 K10 **Regocijo** Durango, W Mexico *23°35′N 105°11′W*
104 H12 **Reguengos de Monsaraz** Évora, S Portugal *38°25′N 07°32′W*
101 M18 **Rehau** Bayern, E Germany *50°15′N 12°03′E*
83 D19 **Rehoboth** Hardap, C Namibia *23°18′S 17°03′E*
21 Z4 **Rehoboth Beach** Delaware, NE USA *38°42′N 75°03′W*
138 F10 **Rehovot** ; *prev.* Rehovot. Central, C Israel *31°54′N 34°49′E*
Rehovot *see* Rehovot
81 J20 **Rei** *spring/well* S Kenya *03°24′S 39°18′E*
Reichenau *see* Rychnov nad Kněžnou
Reichenau *see* Bogatynia, Poland
101 M17 **Reichenbach** *var.* Reichenbach im Vogtland. Sachsen, E Germany *50°36′N 12°18′E*
Reichenbach *see* Dzierżoniów
Reichenbach im Vogtland *see* Reichenbach
Reichenberg *see* Liberec
181 O11 **Reid** Western Australia *30°49′S 128°24′E*
23 V6 **Reidsville** Georgia, SE USA *32°05′N 82°07′W*
21 T8 **Reidsville** North Carolina, SE USA *36°21′N 79°39′W*
Reifnitz *see* Ribnica
97 O22 **Reigate** SE England, United Kingdom *51°14′N 00°13′W*
Reikjavik *see* Reykjavík
102 I10 **Ré, Île de** *island* W France
37 N15 **Reiley Peak** ▲ Arizona, SW USA *32°24′N 110°09′W*
103 Q4 **Reims** *Eng.* Rheims; *anc.* Durocortorum, Remi. Marne, N France *49°16′N 04°01′E*
63 G23 **Reina Adelaida, Archipiélago** *island group* S Chile
45 O16 **Reina Beatrix** ✈ (Oranjestad) C Aruba *12°30′N 69°57′W*
108 F7 **Reinach** Aargau, W Switzerland *47°16′N 08°12′E*
108 E6 **Reinach** Basel Landschaft, NW Switzerland *47°30′N 07°36′E*
64 O11 **Reina Sofía** ✈ (Tenerife) Tenerife, Islas Canarias, Spain, NE Atlantic Ocean
29 W13 **Reinbeck** Iowa, C USA *42°19′N 92°36′W*
100 J10 **Reinbek** Schleswig-Holstein, N Germany *53°31′N 10°15′E*
11 U12 **Reindeer** ≈ Saskatchewan, C Canada
11 U11 **Reindeer Lake** ◎ Manitoba/Saskatchewan, C Canada
Reine-Charlotte, Îles de la *see* Queen Charlotte Islands
Reine-Élisabeth, Îles de la *see* Queen Elizabeth Islands
94 F13 **Reineskarvet** ▲ S Norway *60°38′N 07°48′E*
184 H1 **Reinga, Cape** *headland* North Island, New Zealand *34°24′S 172°40′E*
105 N3 **Reinosa** Cantabria, N Spain *43°01′N 04°09′W*
109 R8 **Reisseck** ▲ S Austria *46°57′N 13°21′E*
21 W3 **Reisterstown** Maryland, NE USA *39°27′N 76°46′W*
Reisui *see* Yeosu
98 N5 **Reitdiep** ≈ NE Netherlands
191 V10 **Reitoru** *atoll* Îles Tuamotu, C French Polynesia
95 M17 **Rejmyre** Östergötland, S Sweden *58°49′N 15°55′E*
Reka *see* Rijeka
Reka Ili *see* Ile/Ili He
Rekarne *see* Tumbo
Rekhovot *see* Rehovot
8 K9 **Reliance** Northwest Territories, C Canada *62°45′N 109°08′W*
33 U16 **Reliance** Wyoming, C USA *41°42′N 109°13′W*
74 I5 **Relizane** *var.* Ghelîzâne, Ghilizane. NW Algeria *35°45′N 00°33′E*
182 I7 **Remarkable, Mount** ▲ South Australia *32°46′S 138°08′E*
54 E8 **Remedios** Antioquia, N Colombia *07°02′N 74°42′W*
43 Q16 **Remedios** Veraguas, W Panama *08°13′N 81°48′W*
42 D8 **Remedios, Punta** *headland* SW El Salvador *13°31′N 89°48′W*
Remi *see* Reims
99 N25 **Remich** Grevenmacher, SE Luxembourg *49°33′N 06°23′E*
99 J19 **Remicourt** Liège, E Belgium *50°40′N 05°19′E*
14 H8 **Rémigny, Lac** ◎ Québec, SE Canada
55 Z10 **Rémire** NE French Guiana *04°52′N 52°16′W*
127 N13 **Remontnoye** Rostovskaya Oblast', SW Russian Federation *46°35′N 43°38′E*
171 U14 **Remoon** Pulau Kur, E Indonesia *05°18′S 131°59′E*
99 L20 **Remouchamps** Liège, E Belgium *50°29′N 05°43′E*
103 R15 **Remoulins** Gard, S France *43°56′N 04°34′E*
173 X16 **Rempart, Mont du** *hill* W Mauritius
101 E15 **Remscheid** Nordrhein-Westfalen, W Germany *51°10′N 07°11′E*
29 S12 **Remsen** Iowa, C USA *42°48′N 95°58′W*
94 I12 **Rena** Hedmark, S Norway *61°08′N 11°21′E*
94 I11 **Renåa** ≈ S Norway
Renaix *see* Ronse
118 H7 **Rencēni** N Latvia *57°43′N 25°25′E*
118 D9 **Renda** W Latvia *57°04′N 22°18′E*
107 N20 **Rende** Calabria, SW Italy *39°19′N 16°10′E*
99 K21 **Rendeux** Luxembourg, SE Belgium *50°15′N 05°28′E*
Rendina *see* Rentína
30 L16 **Rend Lake** ⊠ Illinois, N USA
186 K9 **Rendova** *island* New Georgia Islands, NW Solomon Islands
100 I8 **Rendsburg** Schleswig-Holstein, N Germany *54°18′N 09°40′E*
108 B9 **Renens** Vaud, SW Switzerland *46°32′N 06°36′E*
14 K12 **Renfrew** Ontario, SE Canada *45°28′N 76°44′W*
96 I12 **Renfrew** *cultural region* SW Scotland, United Kingdom
168 L11 **Rengat** Sumatera, W Indonesia *0°26′S 102°38′E*
153 W12 **Rengma Hills** ▲▲ NE India
62 H12 **Rengo** Libertador, C Chile *34°24′S 70°50′W*
116 M12 **Reni** Odes'ka Oblast', SW Ukraine *45°30′N 28°18′E*
80 F11 **Renk** Upper Nile, NE South Sudan *11°48′N 32°49′E*
93 L19 **Renko** Kanta-Häme, S Finland *60°52′N 24°16′E*
98 L12 **Renkum** Gelderland, SE Netherlands *51°58′N 05°43′E*
182 K9 **Renmark** South Australia *34°12′S 140°43′E*
186 L10 **Rennell** *var.* Mu Nggava. *island* S Solomon Islands
186 M9 **Rennell and Bellona** *prev.* Central. ◈ *province* S Solomon Islands
181 Q4 **Renner Springs Roadhouse** Northern Territory, N Australia *18°12′S 133°48′E*
102 I6 **Rennes** *Bret.* Roazon; *anc.* Condate. Ille-et-Vilaine, NW France *48°08′N 01°40′W*
195 S16 **Rennick Glacier** *glacier* Antarctica
11 Y16 **Rennie** Manitoba, S Canada *49°51′N 95°28′W*
35 Q5 **Reno** Nevada, W USA *39°32′N 119°49′W*
106 H10 **Reno** ≈ N Italy
35 Q5 **Reno-Cannon** ✈ Nevada, W USA *39°26′N 119°42′W*
83 F24 **Renoster** ≈ SW South Africa
15 T5 **Renouard, Lac** ◎ Québec, SE Canada
18 F13 **Renovo** Pennsylvania, NE USA *41°19′N 77°42′W*
161 O3 **Renqiu** Hebei, E China *38°49′N 116°02′E*
160 I9 **Renshou** Sichuan, C China *30°02′N 104°09′E*
31 N12 **Rensselaer** Indiana, N USA *40°57′N 87°09′W*
18 L11 **Rensselaer** New York, NE USA *42°38′N 73°44′W*
Rentería *see* Errenteria
115 E17 **Rentína** *var.* Rendina. Thessalía, C Greece *39°04′N 21°58′E*
29 T9 **Renville** Minnesota, N USA *44°48′N 95°13′W*
77 O13 **Réo** W Burkina Faso *12°20′N 02°28′W*
15 O12 **Repentigny** Québec, SE Canada *45°42′N 73°28′W*
146 K13 **Repetek** Lebap Welaýaty, E Turkmenistan *38°40′N 63°12′E*
93 J16 **Replot** *Fin.* Raippaluoto. *island* W Finland
Repola *see* Reboly
Reppen *see* Rzepin
Reps *see* Rupea
27 T7 **Republic** Missouri, C USA *37°07′N 93°28′W*
32 K7 **Republic** Washington, NW USA *48°39′N 118°44′W*
27 N3 **Republican River** ≈ Kansas/Nebraska, C USA
9 O7 **Repulse Bay** Northwest Terretories, N Canada *66°35′N 86°20′W*
56 F9 **Requena** Loreto, NE Peru *05°05′S 73°52′W*
105 R10 **Requena** Valenciana, E Spain *39°29′N 01°08′W*
103 O14 **Réquista** Aveyron, S France *44°00′N 02°31′E*
136 M12 **Reşadiye** Tokat, N Turkey *40°24′N 37°19′E*
Reschenpass *see* Resia, Passo di
Reschitza *see* Reşiţa
113 N20 **Resen** *Turk.* Resne. SW FYR Macedonia *41°07′N 21°00′E*
60 J11 **Reserva** Paraná, S Brazil *24°40′S 50°52′W*
11 V15 **Reserve** Saskatchewan, S Canada *52°24′N 102°37′W*
37 P13 **Reserve** New Mexico, SW USA *33°42′N 108°45′W*
Reshetilovka *see* Reshetylivka
117 S6 **Reshetylivka** *Rus.* Reshetilovka. Poltavs'ka Oblast', NE Ukraine *49°34′N 34°05′E*
Resht *see* Rasht
139 S2 **Reshwan** *Ar.* Rashwān, *var.* Rashwan. Arbīl, N Iraq *36°28′N 43°54′E*
106 F5 **Resia, Passo di** *Ger.* Reschenpass. *pass* Austria/Italy
Resicabánya *see* Reşiţa
62 N7 **Resistencia** Chaco, NE Argentina *27°27′S 58°56′W*
116 F12 **Reşiţa** *Ger.* Reschitza, *Hung.* Resicabánya. Caraş-Severin, W Romania *45°14′N 21°58′E*
Resne *see* Resen
197 N9 **Resolute** *Inuit* Qausuittuq. Cornwallis Island, Nunavut, N Canada *74°41′N 94°54′W*
Resolution *see* Fort Resolution
9 T7 **Resolution Island** *island* Nunavut, NE Canada
185 A23 **Resolution Island** *island* SW New Zealand
Resta *see* Rasta
15 W7 **Restigouche** Québec, SE Canada *48°02′N 66°42′W*
11 W17 **Reston** Manitoba, S Canada *49°33′N 101°03′W*
14 H11 **Restoule Lake** ◎ Ontario, S Canada
54 F10 **Restrepo** Meta, C Colombia *04°20′N 73°29′W*
42 B6 **Retalhuleu** Retalhuleu, SW Guatemala *14°31′N 91°40′W*
42 A1 **Retalhuleu** *off.* Departamento de Retalhuleu. ◈ *department* SW Guatemala
Retalhuleu, Departamento de *see* Retalhuleu
97 N18 **Retford** C England, United Kingdom *53°18′N 00°52′W*
103 Q3 **Rethel** Ardennes, N France *49°31′N 04°22′E*
Rethimno/Réthimnon *see* Réthymno
115 I25 **Réthymno** *prev.* Rethimno, Réthimnon. Kríti, Greece, E Mediterranean Sea *35°21′N 24°29′E*
Retiche, Alpi *see* Rhaetian Alps
99 J16 **Retie** Antwerpen, N Belgium *51°18′N 05°05′E*
111 J21 **Rétság** Nógrád, N Hungary *47°57′N 19°08′E*
109 W2 **Retz** Niederösterreich, NE Austria *48°46′N 15°58′E*
173 N15 **Réunion** *off.* La Réunion. ◇ *French overseas department* W Indian Ocean
128 L17 **Réunion** *island* W Indian Ocean
105 U6 **Reus** Cataluña, E Spain *41°10′N 01°06′E*
108 F7 **Reuss** ≈ NW Switzerland
99 J15 **Reusel** Noord-Brabant, S Netherlands *51°21′N 05°10′E*
Reuţel *see* Ciuhuru
101 H22 **Reutlingen** Baden-Württemberg, S Germany *48°30′N 09°13′E*
108 L7 **Reutte** Tirol, W Austria *47°30′N 10°44′E*
99 M16 **Reuver** Limburg, SE Netherlands *51°17′N 06°05′E*
28 K7 **Reva** South Dakota, N USA *45°30′N 103°03′W*
Reval/Revel *see* Tallinn
124 J4 **Revda** Murmanskaya Oblast', NW Russian Federation *67°57′N 34°29′E*
122 F6 **Revda** Sverdlovskaya Oblast', C Russian Federation *56°48′N 59°42′E*
103 N16 **Revel** Haute-Garonne, S France *43°27′N 01°59′E*
11 O16 **Revelstoke** British Columbia, SW Canada *51°02′N 118°12′W*
43 N13 **Reventazón, Río** ≈ E Costa Rica
106 G9 **Revere** Lombardia, N Italy *45°03′N 11°07′E*
39 Y14 **Revillagigedo Island** *island* Alexander Archipelago, Alaska, USA
193 R7 **Revillagigedo Islands** Mexico
103 R3 **Revin** Ardennes, N France *49°57′N 04°39′E*
92 O3 **Revnosa** *headland* C Svalbard *78°03′N 18°52′E*
Revolyutsii, Pik *see* Revolyutsiya, Qullai
147 T13 **Revolyutsiya, Qullai** *Rus.* Pik Revolyutsii. ▲ SE Tajikistan *38°40′N 72°26′E*
111 L19 **Revúca** *Ger.* Grossrauschenbach, *Hung.* Nagyrőce. Banskobystrický Kraj, C Slovakia *48°40′N 20°10′E*
154 K9 **Rewa** Madhya Pradesh, C India *24°32′N 81°18′E*
139 T2 **Rewanduz** *Ar.* Rawāndūz, *var.* Rawāndiz. Arbīl, N Iraq *36°38′N 44°32′E*
152 I11 **Rewāri** Haryāna, N India *28°14′N 76°38′E*
33 R14 **Rexburg** Idaho, NW USA *43°49′N 111°47′W*
78 G13 **Rey Bouba** Nord, NE Cameroon *08°40′N 14°11′E*
92 L3 **Reyðarfjörður** Austurland, E Iceland *65°02′N 14°12′W*
57 K16 **Reyes** El Beni, NW Bolivia *14°17′S 67°18′W*
34 L8 **Reyes, Point** *headland* California, W USA *37°59′N 123°01′W*
54 B12 **Reyes, Punta** *headland* SW Colombia *02°44′N 78°08′W*
136 L17 **Reyhanlı** Hatay, S Turkey *36°16′N 36°35′E*
43 U16 **Rey, Isla del** *island* Archipiélago de las Perlas, SE Panama
92 H2 **Reykhólar** Vestfirðir, W Iceland *65°28′N 22°12′W*
92 K2 **Reykjahlíð** Norðurland Eystra, NE Iceland *65°37′N 16°54′W*
197 O16 **Reykjanes Basin** *var.* Irminger Basin. *undersea feature* N Atlantic Ocean *62°30′N 33°30′W*
197 N17 **Reykjanes Ridge** *undersea feature* N Atlantic Ocean *62°00′N 27°00′W*
92 H4 **Reykjavík** *var.* Reikjavik. ● (Iceland) Höfudhborgarsvaedhi, W Iceland *64°08′N 21°54′W*
18 D13 **Reynoldsville** Pennsylvania, NE USA *41°04′N 78°51′W*
41 P8 **Reynosa** Tamaulipas, C Mexico *26°03′N 98°19′W*
Reza'īyeh *see* Orūmīyeh
Rezā'īyeh, Daryācheh-ye *see* Orūmīyeh, Daryācheh-ye
102 I8 **Rezé** Loire-Atlantique, NW France *47°10′N 01°36′W*
118 K10 **Rēzekne** *Ger.* Rositten; *prev. Rus.* Rezhitsa. SE Latvia *56°31′N 27°22′E*
139 U2 **Rezge** *Ar.* Razkah, *var.* Razga. As Sulaymānīyah, E Iraq *36°25′N 45°06′E*
Rezhitsa *see* Rēzekne
117 N9 **Rezina** NE Moldova *47°44′N 28°58′E*
114 N11 **Rezovo** *Turk.* Rezve. Burgas, E Bulgaria *42°00′N 28°00′E*
114 N11 **Rezovska Reka** *Turk.* Rezve Deresi. ≈ Bulgaria/Turkey *see also* Rezve Deresi
Rezovska Reka *see* Rezve Deresi
Rezve *see* Rezovo
114 N11 **Rezve Deresi** *Bul.* Rezovska Reka. ≈ Bulgaria/Turkey *see also* Rezovska Reka
Rezve Deresi *see* Rezovska Reka
Rhadames *see* Ghadāmis
Rhaedestus *see* Tekirdağ
108 J10 **Rhaetian Alps** *Fr.* Alpes Rhétiques, *Ger.* Rätische Alpen, *It.* Alpi Retiche. ▲▲ C Europe
108 I8 **Rhätikon** ▲▲ C Europe
101 G14 **Rheda-Wiedenbrück** Nordrhein-Westfalen, W Germany *51°51′N 08°17′E*
98 M12 **Rheden** Gelderland, E Netherlands *52°01′N 06°03′E*
Rhegion/Rhegium *see* Reggio di Calabria
Rheims *see* Reims
Rhein *see* Rhine
101 E17 **Rheinbach** Nordrhein-Westfalen, W Germany *50°37′N 06°57′E*
100 F13 **Rheine** *var.* Rheine in Westfalen. Nordrhein-Westfalen, NW Germany *52°17′N 07°27′E*
Rheine in Westfalen *see* Rheine
Rheinfeld *see* Rheinfelden
101 F24 **Rheinfelden** Baden-Württemberg, S Germany *47°34′N 07°46′E*
108 E6 **Rheinfelden** *var.* Rheinfeld. Aargau, N Switzerland *47°33′N 07°47′E*
101 E17 **Rheinisches Schiefergebirge** *var.* Rhine State Uplands, *Eng.* Rhenish Slate Mountains. ▲▲ W Germany
101 D18 **Rheinland-Pfalz** *Eng.* Rhineland-Palatinate, *Fr.* Rhénanie-Palatinat. ◈ *state* W Germany
101 G18 **Rhein Main** ✈ (Frankfurt am Main) Hessen, W Germany *50°03′N 08°33′E*
Rhénanie du Nord-Westphalie *see* Nordrhein-Westfalen
Rhénanie-Palatinat *see* Rheinland-Pfalz
98 K12 **Rhenen** Utrecht, C Netherlands *52°01′N 06°02′E*
Rhenish Slate Mountains *see* Rheinisches Schiefergebirge
Rhétiques, Alpes *see* Rhaetian Alps
100 N10 **Rhin** ≈ NE Germany
Rhin *see* Rhine
84 F10 **Rhine** *Dut.* Rijn, *Fr.* Rhin, *Ger.* Rhein. ≈ W Europe
30 L5 **Rhinelander** Wisconsin, N USA *45°39′N 89°23′W*
Rhineland-Palatinate *see* Rheinland-Pfalz
Rhine State Uplands *see* Rheinisches Schiefergebirge
100 N11 **Rhinkanal** *canal* NE Germany
81 F17 **Rhino Camp** NW Uganda *02°58′N 31°24′E*
74 D7 **Rhir, Cap** *headland* W Morocco *30°40′N 09°54′W*
106 D7 **Rho** Lombardia, N Italy *45°32′N 09°02′E*
19 N12 **Rhode Island** *off.* State of Rhode Island and Providence Plantations, *also known as* Little Rhody, Ocean State. ◈ *state* NE USA
19 O13 **Rhode Island** *island* Rhode Island, NE USA
19 O13 **Rhode Island Sound** *sound* Maine/Rhode Island, NE USA
Rhodes *see* Ródos
Rhode-Saint-Genèse *see* Sint-Genesius-Rode
84 L14 **Rhodes Basin** *undersea feature* E Mediterranean Sea *35°55′N 28°30′E*
Rhodesia *see* Zimbabwe
114 I12 **Rhodope Mountains** *var.* Rodhópi Óri, *Bul.* Rhodope Planina, Rodopi, *Gk.* Orosirá Rodhópis, *Turk.* Dospad Dagh. ▲▲ Bulgaria/Greece
Rhodope Planina *see* Rhodope Mountains
101 I18 **Rhön** ▲▲ C Germany
103 Q10 **Rhône** ◈ *department* E France
86 C12 **Rhône** ≈ France/Switzerland
103 R12 **Rhône-Alpes** ◈ *region* E France
98 G13 **Rhoon** Zuid-Holland, SW Netherlands *51°52′N 04°25′E*
96 G9 **Rhum** *var.* Rum. *island* W Scotland, United Kingdom
Rhuthun *see* Ruthin
97 J18 **Rhyl** NE Wales, United Kingdom *53°19′N 03°28′W*
59 K18 **Rialma** Goiás, S Brazil *15°22′S 49°35′W*
104 L3 **Riaño** Castilla y León, N Spain *42°59′N 05°00′W*
105 O9 **Riansáres** ≈ C Spain
152 H6 **Riāsi** Jammu and Kashmir, NW India *33°03′N 74°51′E*
168 K10 **Riau** *off.* Propinsi Riau. ◈ *province* W Indonesia
Riau Archipelago *see* Riau, Kepulauan
168 M11 **Riau, Kepulauan** *var.* Riau Archipelago, *Dut.* Riouw-Archipel. *island group* W Indonesia
Riau, Propinsi *see* Riau
105 O6 **Riaza** Castilla y León, N Spain *41°17′N 03°29′W*
105 N6 **Riaza** ≈ N Spain
81 K17 **Riba** *spring/well* NE Kenya *01°56′N 40°38′E*
104 H4 **Ribadavia** Galicia, NW Spain *42°17′N 08°08′W*
104 J2 **Ribadeo** Galicia, NW Spain *43°32′N 07°04′W*
104 L2 **Ribadesella** *var.* Ribeseya. Asturias, N Spain *43°27′N 05°04′W*
104 G10 **Ribatejo** *former province* C Portugal
83 P15 **Ribáuè** Nampula, N Mozambique *14°56′S 38°19′E*
97 K17 **Ribble** ≈ NW England, United Kingdom
95 F23 **Ribe** Syddtjylland, W Denmark *55°20′N 08°47′E*
Ribeira *see* Sanata Uxía de Ribeira
64 O5 **Ribeira Brava** Madeira, Portugal, NE Atlantic Ocean *32°39′N 17°04′W*
64 P3 **Ribeira Grande** São Miguel, Azores, Portugal, NE Atlantic Ocean *37°34′N 25°32′W*
60 L8 **Ribeirão Preto** São Paulo, S Brazil *21°09′S 47°48′W*
60 L11 **Ribeira, Rio** ≈ S Brazil
107 I24 **Ribera** Sicilia, Italy, C Mediterranean Sea *37°31′N 13°16′E*
57 L14 **Riberalta** El Beni, N Bolivia *11°01′S 66°04′W*
105 W4 **Ribes de Freser** Cataluña, NE Spain *42°18′N 02°11′E*
Ribeseya *see* Ribadesella
30 L6 **Rib Mountain** ▲ Wisconsin, N USA *44°55′N 89°41′W*
109 U12 **Ribnica** *Ger.* Reifnitz. S Slovenia *45°46′N 14°40′E*
117 N9 **Rîbniţa** *var.* Râbniţa, *Rus.* Rybnitsa. NE Moldova *47°46′N 29°01′E*
100 M8 **Ribnitz-Damgarten** Mecklenburg-Vorpommern, NE Germany *54°14′N 12°25′E*
111 D16 **Říčany** *Ger.* Ritschan. Středočesky Kraj, W Czech Republic *49°59′N 14°40′E*
29 U7 **Rice** Minnesota, N USA *45°42′N 94°10′W*
30 J5 **Rice Lake** Wisconsin, N USA *45°33′N 91°43′W*
14 E8 **Rice Lake** ◎ Ontario, S Canada
14 I15 **Rice Lake** ◎ Ontario, SE Canada
23 V3 **Richard B. Russell Lake** ⊠ Georgia, SE USA
25 U6 **Richardson** Texas, SW USA *32°55′N 96°44′W*
11 R11 **Richardson** ≈ Alberta, C Canada
10 I3 **Richardson Mountains** ▲▲ Yukon, NW Canada
185 C21 **Richardson Mountains** ▲▲ South Island, New Zealand
42 F3 **Richardson Peak** ▲ SE Belize *16°34′N 88°46′W*
76 G10 **Richard Toll** N Senegal *16°28′N 15°44′W*
28 L5 **Richardton** North Dakota, N USA *46°52′N 102°19′W*
14 F13 **Rich, Cape** *headland* Ontario, S Canada *44°42′N 80°37′W*
102 L8 **Richelieu** Indre-et-Loire, C France *47°01′N 00°18′E*
33 P15 **Richfield** Idaho, NW USA *43°03′N 114°11′W*
36 K5 **Richfield** Utah, W USA *38°45′N 112°05′W*
18 J10 **Richfield Springs** New York, NE USA *42°52′N 74°57′W*
18 M6 **Richford** Vermont, NE USA *44°59′N 72°37′W*
27 R6 **Rich Hill** Missouri, C USA *38°06′N 94°22′W*
13 P14 **Richibucto** New Brunswick, SE Canada *46°42′N 64°54′W*
108 G8 **Richisau** Glarus, NE Switzerland *47°00′N 08°54′E*
23 S6 **Richland** Georgia, SE USA *32°05′N 84°40′W*
27 U6 **Richland** Missouri, C USA *37°51′N 92°24′W*
25 U8 **Richland** Texas, SW USA *31°55′N 96°25′W*
32 K10 **Richland** Washington, NW USA *46°17′N 119°16′W*
30 K8 **Richland Center** Wisconsin, N USA *43°22′N 90°24′W*
21 W11 **Richlands** North Carolina, SE USA *34°52′N 77°33′W*
21 Q7 **Richlands** Virginia, NE USA *37°05′N 81°47′W*
25 R9 **Richland Springs** Texas, SW USA *31°16′N 98°56′W*
183 S8 **Richmond** New South Wales, SE Australia *33°36′S 150°44′E*
10 L17 **Richmond** British Columbia, SW Canada *49°07′N 123°09′W*
14 L13 **Richmond** Ontario, SE Canada *45°12′N 75°49′W*
15 Q12 **Richmond** Québec, SE Canada *45°39′N 72°07′W*
185 I14 **Richmond** Tasman, South Island, New Zealand *41°25′S 173°04′E*
35 N8 **Richmond** California, W USA *37°57′N 122°22′W*
31 Q14 **Richmond** Indiana, N USA *39°50′N 84°51′W*
20 M6 **Richmond** Kentucky, S USA *37°45′N 84°19′W*
27 S4 **Richmond** Missouri, C USA *39°15′N 93°59′W*
25 V11 **Richmond** Texas, SW USA *29°36′N 95°48′W*
36 L1 **Richmond** Utah, W USA *41°55′N 111°51′W*
21 W6 **Richmond** *state capital* Virginia, NE USA *37°33′N 77°28′W*
14 H15 **Richmond Hill** Ontario, S Canada *43°51′N 79°24′W*
185 J15 **Richmond Range** ▲▲ South Island, New Zealand
27 S12 **Rich Mountain** ▲ Arkansas, C USA *34°37′N 94°17′W*
31 S13 **Richwood** Ohio, N USA *40°25′N 83°18′W*
21 R5 **Richwood** West Virginia, NE USA *38°13′N 80°31′W*
104 K5 **Ricobayo, Embalse de** ⊠ NW Spain
Ricomagus *see* Riom
Ridā' *see* Radā'
127 T5 **Ridder** Respublika Tatarstan, W Russian Federation *54°34′N 52°27′E*
98 H13 **Ridderkerk** Zuid-Holland, SW Netherlands *51°52′N 04°35′E*
33 N16 **Riddle** Idaho, NW USA *42°07′N 116°09′W*
32 F14 **Riddle** Oregon, NW USA *42°57′N 123°21′W*
14 L13 **Rideau** ≈ Ontario, SE Canada
35 T12 **Ridgecrest** California, W USA *35°37′N 117°40′W*
18 L13 **Ridgefield** Connecticut, NE USA *41°16′N 73°30′W*
22 K5 **Ridgeland** Mississippi, S USA *32°25′N 90°07′W*
21 R15 **Ridgeland** South Carolina, SE USA *32°30′N 80°59′W*
20 F8 **Ridgely** Tennessee, S USA *36°15′N 89°29′W*
14 D17 **Ridgetown** Ontario, S Canada *42°27′N 81°52′W*
21 R12 **Ridgeway** South Carolina, SE USA *34°17′N 80°56′W*
Ridgeway *see* Ridgway
18 D13 **Ridgway** *var.* Ridgeway. Pennsylvania, NE USA *41°24′N 78°40′W*
11 W16 **Riding Mountain** ▲▲ Manitoba, S Canada
Ried *see* Ried im Innkreis
109 R4 **Ried im Innkreis** *var.* Ried. Oberösterreich, NW Austria *48°13′N 13°29′E*
109 X8 **Riegersburg** Steiermark, SE Austria *47°03′N 15°52′E*
108 E6 **Riehen** Basel-Stadt, NW Switzerland *47°35′N 07°39′E*
92 J9 **Riehppegáisá** *var.* Rieppe. ▲ N Norway *69°38′N 21°31′E*
99 K18 **Riemst** Limburg, NE Belgium *50°49′N 05°36′E*
Rieppe *see* Riehppegáisá
101 O15 **Riesa** Sachsen, E Germany *51°18′N 13°18′E*
63 H24 **Riesco, Isla** *island* S Chile
107 K25 **Riesi** Sicilia, Italy, C Mediterranean Sea *37°17′N 14°05′E*
83 I23 **Riet** ≈ SW South Africa
83 F25 **Riet** ≈ SW South Africa
118 D11 **Rietavas** Telšiai, W Lithuania *55°43′N 21°56′E*
83 F19 **Rietfontein** Omaheke, E Namibia *21°58′S 20°58′E*
107 I14 **Rieti** *anc.* Reate. Lazio, C Italy *42°24′N 12°51′E*
84 D14 **Rif** *var.* Riff, Er Rif, Er Riff. ▲▲ N Morocco
138 I8 **Rīf Dimashq** *off.* Muḩāfaẓat Dimashq, *var.* Damascus, *Ar.* Ash Sham, Ash Shām, Damasco, Esh Sham, *Fr.* Damas. ◈ *governorate* S Syria
Riff *see* Rif
37 Q4 **Rifle** Colorado, C USA *39°30′N 107°46′W*
31 R7 **Rifle River** ≈ Michigan, N USA
Rift Valley *see* Great Rift Valley
118 F9 **Rīga** *Eng.* Riga. ● C Latvia *56°57′N 24°08′E*
Rigaer Bucht *see* Riga, Gulf of
118 F6 **Riga, Gulf of** *Est.* Liivi Laht, *Ger.* Rigaer Bucht, *Latv.* Rīgas Jūras Līcis, *Rus.* Rizhskiy Zaliv; *prev. Est.* Riia Laht. *gulf* Estonia/Latvia
Rīgas Jūras Līcis *see* Riga, Gulf of
15 N12 **Rigaud** ≈ Ontario/Québec, SE Canada
33 R14 **Rigby** Idaho, NW USA *43°40′N 111°54′W*
Rīgestān *see* Rēgestān
32 M11 **Riggins** Idaho, NW USA *45°24′N 116°18′W*
13 R8 **Rigolet** Newfoundland and Labrador, NE Canada *51°10′N 58°25′W*
78 G9 **Rig-Rig** Kanem, W Chad *14°16′N 14°21′E*
118 F4 **Riguldi** Läänemaa, W Estonia *59°07′N 23°34′E*
Riia Laht *see* Riga, Gulf of
93 L19 **Riihimäki** Kanta-Häme, S Finland *60°45′N 24°45′E*
195 U2 **Riiser-Larsen Peninsula** *peninsula* Antarctica
65 P22 **Riiser-Larsen Sea** *sea* Antarctica
Riisser Larsen Ice Shelf *see* Riiser-Larsenisen
40 D2 **Riíto** Sonora, NW Mexico *32°06′N 114°57′W*
112 B9 **Rijeka** *Ger.* Sankt Veit am Flaum, *It.* Fiume, *Slvn.* Reka; *anc.* Tarsatica. Primorje-Gorski Kotar, NW Croatia *45°20′N 14°26′E*
99 I14 **Rijen** Noord-Brabant, S Netherlands *51°35′N 04°55′E*
99 H15 **Rijkevorsel** Antwerpen, N Belgium *51°23′N 04°43′E*
Rijn *see* Rhine
98 G11 **Rijnsburg** Zuid-Holland, W Netherlands *52°12′N 04°27′E*
Rijssel *see* Lille
98 N10 **Rijssen** Overijssel, E Netherlands *52°19′N 06°30′E*
98 G12 **Rijswijk** *Eng.* Ryswick. Zuid-Holland, W Netherlands *52°03′N 04°20′E*
92 I10 **Riksgränsen** Norrbotten, N Sweden *68°24′N 18°15′E*
165 U4 **Rikubetsu** Hokkaidō, NE Japan *43°30′N 143°43′E*
165 R9 **Rikuzen-Takata** Iwate, Honshū, C Japan *39°03′N 141°38′E*
27 O4 **Riley** Kansas, C USA *39°18′N 96°49′W*
99 I17 **Rillaar** Vlaams Brabant, C Belgium *50°58′N 04°58′E*
Rí, Loch *see* Ree, Lough
114 G11 **Rilska Reka** ≈ W Bulgaria
77 T12 **Rima** ≈ N Nigeria
141 N7 **Rimah, Wādī ar** *var.* Wādī ar Rummah. *dry watercourse* C Saudi Arabia
Rimaszombat *see* Rimavská Sobota
191 R12 **Rimatara** *island* Îles Australes, SW French Polynesia
111 L20 **Rimavská Sobota** *Ger.* Gross-Steffelsdorf, *Hung.* Rimaszombat. Bankobystrický Kraj, C Slovakia *48°24′N 20°01′E*
11 Q15 **Rimbey** Alberta, SW Canada *52°39′N 114°10′W*
95 P15 **Rimbo** Stockholm, C Sweden *59°44′N 18°21′E*
95 M18 **Rimforsa** Östergötland, S Sweden *58°06′N 15°40′E*
106 I11 **Rimini** *anc.* Ariminum. Emilia-Romagna, N Italy *44°03′N 12°33′E*
Rîmnicu-Sărat *see* Râmnicu Sărat
Rîmnicu Vîlcea *see* Râmnicu Vâlcea
149 Y3 **Rimo Muztāgh** ▲▲ India/Pakistan
15 U7 **Rimouski** Québec, SE Canada *48°26′N 68°32′W*
158 M16 **Rinbung** Xizang Zizhiqu, W China *29°15′N 89°40′E*
Rinchinlhumbe *see* Dzöölön
62 I5 **Rincón, Cerro** ▲ N Chile *24°01′S 67°19′W*
104 M15 **Rincón de la Victoria** Andalucía, S Spain *36°43′N 04°18′W*

◆ Country
● Country Capital
◇ Dependent Territory
○ Dependent Territory Capital
◈ Administrative Regions
✈ International Airport
▲ Mountain
▲▲ Mountain Range
⛰ Volcano
≈ River
◎ Lake
⊠ Reservoir

Rincón del Bonete, Lago Artificial de – Ronne Entrance

Rincón del Bonete, Lago Artificial de *see* Río Negro, Embalse del
105 Q4 **Rincón de Soto** La Rioja, N Spain 42°15′N 01°50′W
94 G8 **Rindal** Møre og Romsdal, S Norway 63°02′N 09°09′E
115 J20 **Ríneia** *island* Kykládes, Greece, Aegean Sea
152 H11 **Rīngas** *prev.* Reengus, Rīngus. Rājasthān, N India 27°18′N 75°27′E
95 H24 **Ringe** Syddtjylland, C Denmark 55°14′N 10°30′E
94 H11 **Ringebu** Oppland, S Norway 61°31′N 10°09′E
Ringen *see* Rõngu
186 K8 **Ringgi** Kolombangara, NW Solomon Islands 08°03′S 157°08′E
23 R1 **Ringgold** Georgia, SE USA 34°55′N 85°06′W
22 G5 **Ringgold** Louisiana, S USA 32°19′N 93°16′W
25 S5 **Ringgold** Texas, SW USA 33°47′N 97°56′W
95 E22 **Ringkøbing** Midtjylland, W Denmark 56°04′N 08°22′E
95 E22 **Ringkøbing Fjord** *fjord* W Denmark
33 S10 **Ringling** Montana, NW USA 46°15′N 110°48′W
27 N13 **Ringling** Oklahoma, C USA 34°12′N 97°35′W
94 H13 **Ringsaker** Hedmark, S Norway 60°54′N 10°45′E
95 I23 **Ringsted** Sjælland, E Denmark 55°28′N 11°48′E
Rīngus *see* Rīngas
92 I9 **Ringvassøya** *Lapp.* Ráneš. *island* N Norway
18 K13 **Ringwood** New Jersey, NE USA 41°06′N 74°15′W
Rinn Duáin *see* Hook Head
100 H13 **Rinteln** Niedersachsen, NW Germany 52°10′N 09°04′E
115 E18 **Río** Dytikí Elláda, S Greece 38°18′N 21°48′E
Rio *see* Rio de Janeiro
56 C7 **Riobamba** Chimborazo, C Ecuador 01°44′S 78°40′W
60 P9 **Rio Bonito** Rio de Janeiro, SE Brazil 22°42′S 42°38′W
59 C16 **Rio Branco** *state capital* Acre, W Brazil 09°59′S 67°49′W
61 H18 **Río Branco** Cerro Largo, NE Uruguay 32°32′S 53°28′W
Rio Branco, Território de *see* Roraima
41 P8 **Río Bravo** Tamaulipas, C Mexico 25°57′N 98°03′W
63 G16 **Río Bueno** Los Ríos, C Chile 40°20′S 72°55′W
55 P5 **Río Caribe** Sucre, NE Venezuela 10°43′N 63°06′W
54 M5 **Río Chico** Miranda, N Venezuela 10°18′N 66°00′W
63 H18 **Río Cisnes** Aisén, S Chile 44°29′S 71°15′W
60 L9 **Rio Claro** São Paulo, S Brazil 22°19′S 47°35′W
45 V14 **Rio Claro** Trinidad, Trinidad and Tobago 10°18′N 61°11′W
54 J5 **Río Claro** Lara, N Venezuela 09°54′N 69°23′W
63 K15 **Rio Colorado** Río Negro, E Argentina 39°01′S 64°05′W
62 K11 **Río Cuarto** Córdoba, C Argentina 33°06′S 64°20′W
60 P10 **Rio de Janeiro** *var.* Rio. *state capital* Rio de Janeiro, SE Brazil 22°53′S 43°17′W
60 P9 **Rio de Janeiro** *off.* Estado do Rio de Janeiro. ◈ *state* SE Brazil
Rio de Janeiro, Estado do *see* Rio de Janeiro
43 R17 **Río de Jesús** Veraguas, S Panama 07°58′N 81°01′W
34 K3 **Rio Dell** California, W USA 40°30′N 124°07′W
60 K13 **Rio do Sul** Santa Catarina, S Brazil 27°15′S 49°37′W
63 I23 **Río Gallegos** *var.* Gallegos, Puerto Gallegos. Santa Cruz, S Argentina 51°40′S 69°21′W
63 J24 **Río Grande** Tierra del Fuego, S Argentina 53°45′S 67°46′W
61 I18 **Rio Grande** *var.* São Pedro do Rio Grande do Sul. Rio Grande do Sul, S Brazil 32°03′S 52°08′W
40 L10 **Río Grande** Zacatecas, C Mexico 23°50′N 103°20′W
42 J9 **Río Grande** León, NW Nicaragua 12°59′N 86°34′W
45 V5 **Río Grande** E Puerto Rico 18°23′N 65°51′W
24 I9 **Rio Grande** ~ Texas, SW USA
25 R17 **Rio Grande City** Texas, SW USA 26°24′N 98°50′W
59 P14 **Rio Grande do Norte** *off.* Estado do Rio Grande do Norte. ◈ *state* E Brazil
Rio Grande do Norte, Estado do *see* Rio Grande do Norte
61 G15 **Rio Grande do Sul** *off.* Estado do Rio Grande do Sul. ◈ *state* S Brazil
Rio Grande do Sul, Estado do *see* Rio Grande do Sul
65 M17 **Rio Grande Fracture Zone** *tectonic feature* C Atlantic Ocean
65 J18 **Rio Grande Gap** *undersea feature* S Atlantic Ocean
Rio Grande Plateau *see* Rio Grande Rise
65 J18 **Rio Grande Rise** *var.* Rio Grande Plateau. *undersea feature* SW Atlantic Ocean 31°00′S 35°00′W
54 G4 **Ríohacha** La Guajira, N Colombia 11°23′N 72°47′W
43 S16 **Río Hato** Coclé, C Panama 08°21′N 80°10′W
25 T17 **Rio Hondo** Texas, SW USA 26°14′N 97°34′W
56 D10 **Rioja** San Martín, N Peru 06°02′S 77°10′W
41 Y11 **Río Lagartos** Yucatán, SE Mexico 21°35′N 88°08′W
103 P11 **Riom** *anc.* Ricomagus. Puy-de-Dôme, C France 45°54′N 03°06′E
104 F10 **Rio Maior** Santarém, C Portugal 39°20′N 08°55′W
103 O12 **Riom-ès-Montagnes** Cantal, C France 45°15′N 02°39′E
60 J12 **Rio Negro** Paraná, S Brazil 26°06′S 49°46′W
63 I15 **Río Negro** *off.* Provincia de Río Negro. ◈ *province* C Argentina
61 D18 **Río Negro** ◈ *department* W Uruguay
47 V12 **Río Negro, Embalse del** *var.* Lago Artificial de Rincón del Bonete. ⊠ C Uruguay
Río Negro, Provincia de *see* Río Negro
107 M17 **Rionero in Vulture** Basilicata, S Italy 40°55′N 15°40′E
137 S9 **Rioni** ~ W Georgia
105 P12 **Riópar** Castilla-La Mancha, C Spain 38°31′N 02°27′W
61 H16 **Rio Pardo** Rio Grande do Sul, S Brazil 29°41′S 52°25′W
37 R11 **Rio Rancho Estates** New Mexico, SW USA 35°14′N 106°40′W
42 L11 **Río San Juan** ◈ *department* S Nicaragua
54 E9 **Ríosucio** Caldas, W Colombia 05°26′N 75°44′W
54 C7 **Ríosucio** Chocó, NW Colombia 07°25′N 77°05′W
62 K10 **Río Tercero** Córdoba, C Argentina 32°15′S 64°08′W
42 K5 **Río Tinto, Sierra** ▲▲ NE Honduras
54 J5 **Río Tocuyo** Lara, N Venezuela 10°18′N 70°00′W
Riouw-Archipel *see* Riau, Kepulauan
59 J19 **Rio Verde** Goiás, C Brazil 17°50′S 50°55′W
41 O12 **Río Verde** *var.* Rioverde. San Luis Potosí, C Mexico 21°58′N 100°W
Rioverde *see* Río Verde
35 O8 **Rio Vista** California, W USA 38°09′N 121°42′W
112 M11 **Ripanj** Serbia, N Serbia 44°37′N 20°30′E
106 J13 **Ripatransone** Marche, C Italy 43°00′N 13°45′E
22 M2 **Ripley** Mississippi, S USA 34°43′N 88°57′W
31 R15 **Ripley** Ohio, N USA 38°45′N 83°51′W
20 F9 **Ripley** Tennessee, S USA 35°43′N 89°30′W
21 Q4 **Ripley** West Virginia, NE USA 38°49′N 81°44′W
105 W4 **Ripoll** Cataluña, NE Spain 42°12′N 02°12′E
97 M16 **Ripon** N England, United Kingdom 54°07′N 01°31′W
30 M7 **Ripon** Wisconsin, N USA 43°52′N 88°48′W
107 L24 **Riposto** Sicilia, Italy, C Mediterranean Sea 37°44′N 15°13′E
99 L14 **Rips** Noord-Brabant, SE Netherlands 51°31′N 05°49′E
54 D9 **Risaralda** *off.* Departamento de Risaralda. ◈ *province* C Colombia
Risaralda, Departamento de *see* Risaralda
116 L8 **Rîşcani** *var.* Râşcani, *Rus.* Ryshkany. NW Moldova 47°55′N 27°31′E
152 J9 **Rishikesh** Uttarakhand, N India 30°06′N 78°16′E
165 S1 **Rishiri-tō** *var.* Risiri Tô. *island* NE Japan
165 S1 **Rishiri-yama** ⛰ Rishiri-tō, NE Japan 45°11′N 141°11′E
25 R7 **Rising Star** Texas, SW USA 32°06′N 98°57′W
31 Q15 **Rising Sun** Indiana, N USA 38°58′N 84°53′W
Risiri Tô *see* Rishiri-tō
102 L4 **Risle** ~ N France
Rîşno *see* Râşnov
27 V13 **Rison** Arkansas, C USA 33°58′N 92°11′W
95 G17 **Risør** Aust-Agder, S Norway 58°44′N 09°15′E
92 H10 **Risøyhamn** Nordland, C Norway 69°00′N 15°37′E
101 I23 **Riss** ~ S Germany
118 G4 **Risti** *Ger.* Kreuz. Läänemaa, W Estonia 59°01′N 24°01′E
15 V8 **Ristigouche** ~ Québec, SE Canada
93 N18 **Ristiina** Etelä-Savo, E Finland 61°32′N 27°15′E
93 N14 **Ristijärvi** Kainuu, C Finland 64°30′N 28°15′E
188 C14 **Ritidian Point** *headland* N Guam 13°39′N 144°51′E
Ritschan *see* Říčany
35 R9 **Ritter, Mount** ▲ California, W USA 37°40′N 119°10′W
31 T12 **Rittman** Ohio, N USA 40°58′N 81°46′W
32 L9 **Ritzville** Washington, NW USA 47°07′N 118°22′W
Riva *see* Riva del Garda
61 A21 **Rivadavia** Buenos Aires, E Argentina 35°29′S 62°59′W
106 F7 **Riva del Garda** *var.* Riva. Trentino-Alto Adige, N Italy 45°54′N 10°50′E
106 B8 **Rivarolo Canavese** Piemonte, W Italy 45°21′N 07°42′E
42 K11 **Rivas** Rivas, SW Nicaragua 11°26′N 85°50′W
42 J11 **Rivas** ◈ *department* SW Nicaragua
103 R11 **Rive-de-Gier** Loire, E France 45°31′N 04°36′E
61 A22 **Rivera** Buenos Aires, E Argentina 37°13′S 63°14′W
61 F16 **Rivera** Rivera, NE Uruguay 30°54′S 55°31′W
61 F17 **Rivera** ◈ *department* NE Uruguay
35 P9 **Riverbank** California, W USA 37°43′N 120°59′W
76 K17 **River Cess** SW Liberia 05°28′N 09°32′W
28 M4 **Riverdale** North Dakota, N USA 47°29′N 101°22′W
30 I6 **River Falls** Wisconsin, N USA 44°52′N 92°38′W
11 T16 **Riverhurst** Saskatchewan, S Canada 50°52′N 106°49′W
183 O10 **Riverina** *physical region* New South Wales, SE Australia
80 G8 **River Nile** ◈ *state* NE Sudan
63 F19 **Rivero, Isla** *island* Archipiélago de los Chonos, S Chile
11 W16 **Rivers** Manitoba, S Canada 50°02′N 100°14′W
77 U17 **Rivers** ◈ *state* S Nigeria
185 D23 **Riversdale** Southland, South Island, New Zealand 45°54′S 168°44′E
83 F26 **Riversdale** Western Cape, SW South Africa 34°05′S 21°15′E
35 U15 **Riverside** California, W USA 33°58′N 117°25′W
25 W9 **Riverside** Texas, SW USA 30°51′N 95°24′W
37 U3 **Riverside Reservoir** ⊠ Colorado, C USA
10 K15 **Rivers Inlet** British Columbia, SW Canada 51°43′N 127°19′W
10 K15 **Rivers Inlet** *inlet* British Columbia, SW Canada
11 X15 **Riverton** Manitoba, S Canada 51°00′N 97°00′W
185 C24 **Riverton** Southland, South Island, New Zealand 46°20′S 168°02′E
30 L13 **Riverton** Illinois, N USA 39°50′N 89°31′W
36 L3 **Riverton** Utah, W USA 40°32′N 111°57′W
33 V15 **Riverton** Wyoming, C USA 43°01′N 108°22′W
14 G10 **River Valley** Ontario, S Canada 46°36′N 80°09′W
13 P14 **Riverview** New Brunswick, SE Canada 46°03′N 64°47′W
103 O17 **Rivesaltes** Pyrénées-Orientales, S France 42°46′N 02°48′E
36 H11 **Riviera** Arizona, SW USA 35°06′N 114°36′W
25 S15 **Riviera** Texas, SW USA 27°15′N 97°48′W
23 Z14 **Riviera Beach** Florida, SE USA 26°46′N 80°03′W
15 Q10 **Rivière-à-Pierre** Québec, SE Canada 46°59′N 72°12′W
15 T9 **Rivière-Bleue** Québec, SE Canada 47°26′N 69°02′W
15 T8 **Rivière-du-Loup** Québec, SE Canada 47°49′N 69°32′W
173 Y15 **Rivière du Rempart** NE Mauritius 20°06′S 57°41′E
45 R12 **Rivière-Pilote** S Martinique 14°29′N 60°54′W
173 O17 **Rivière St-Etienne, Pointe de la** *headland* SW Réunion
13 S10 **Rivière-St-Paul** Québec, E Canada 51°26′N 57°52′W
Rivière Sèche *see* Bel Air
116 K4 **Rivne** *Pol.* Równe, *Rus.* Rovno. Rivnens'ka Oblast', NW Ukraine 50°37′N 26°16′E
Rivne *see* Rivnens'ka Oblast'
116 K3 **Rivnens'ka Oblast'** *var.* Rivne, *Rus.* Rovenskaya Oblast'. ◈ *province* NW Ukraine
106 B8 **Rivoli** Piemonte, NW Italy 45°04′N 07°31′E
159 Q14 **Riwoqê** *var.* Racaka. Xizang Zizhiqu, W China 31°10′N 96°25′E
99 H19 **Rixensart** Walloon Brabant, C Belgium 50°43′N 04°32′E
Riyadh/Riyāḍ, Minṭaqat ar *see* Ar Riyāḍ
Riyāq *see* Rayak
Rizaiyeh *see* Orūmīyeh
137 P11 **Rize** Rize, NE Turkey 41°03′N 40°33′E
137 P11 **Rize** *prev.* Çoruh. ◈ *province* NE Turkey
161 R5 **Rizhao** Shandong, E China 35°23′N 119°32′E
Rizhskiy Zaliv *see* Riga, Gulf of
Rizokarpaso/Rizokárpason *see* Dipkarpaz
107 O21 **Rizzuto, Capo** *headland* S Italy 38°54′N 17°05′E
95 F15 **Rjukan** Telemark, S Norway 59°54′N 08°33′E
95 D16 **Rjuven** ▲▲ S Norway
76 H9 **Rkîz** Trarza, W Mauritania 16°50′N 15°20′W
115 Q23 **Ro** *prev.* Ágios Geórgios. *island* SE Greece
95 H14 **Roa** Oppland, S Norway 60°16′N 10°38′E
105 N5 **Roa** Castilla y León, N Spain 41°42′N 03°55′W
45 T9 **Road Town** ○ (British Virgin Islands) Tortola, C British Virgin Islands 18°28′N 64°39′W
96 F6 **Roag, Loch** *inlet* NW Scotland, United Kingdom
37 O5 **Roan Cliffs** *cliff* Colorado/Utah, W USA
21 P9 **Roan High Knob** *var.* Roan Mountain. ▲ North Carolina/Tennessee, SE USA 36°09′N 82°07′W
Roan Mountain *see* Roan High Knob
103 Q10 **Roanne** *anc.* Rodunma. Loire, E France 46°03′N 04°04′E
23 R4 **Roanoke** Alabama, S USA 33°09′N 85°22′W
21 S7 **Roanoke** Virginia, NE USA 37°16′N 79°57′W
21 Z9 **Roanoke Island** *island* North Carolina, SE USA
21 W8 **Roanoke Rapids** North Carolina, SE USA 36°27′N 77°39′W
21 X9 **Roanoke River** ~ North Carolina/Virginia, SE USA
37 O4 **Roan Plateau** *plain* Utah, W USA
37 R5 **Roaring Fork River** ~ Colorado, C USA
25 O5 **Roaring Springs** Texas, SW USA 33°54′N 100°51′W
42 J4 **Roatán** *var.* Coxen Hole, Coxin Hole. Islas de la Bahía, N Honduras 16°19′N 86°33′W
42 I4 **Roatán, Isla de** *island* Islas de la Bahía, N Honduras
Roat Kampuchea *see* Cambodia
Roazon *see* Rennes
143 T7 **Robāṭ-e Chāh Gonbad** Khorāsān-e Jonūbī, E Iran 33°24′N 57°43′E
143 R7 **Robāṭ-e Khān** Khorāsān-e Jonūbī, C Iran 33°24′N 56°04′E
143 T7 **Robāṭ-e Khvosh Āb** Khorāsān-e Jonūbī, E Iran 33°24′N 57°43′E
143 R8 **Robāṭ-e Posht-e Bādām** Yazd, NE Iran 33°01′N 55°34′E
143 Q8 **Robaṭ-e Rīzāb** Yazd, C Iran
175 S8 **Robbie Ridge** *undersea feature* W Pacific Ocean
21 T10 **Robbins** North Carolina, SE USA 35°25′N 79°35′W
183 N15 **Robbins Island** *island* Tasmania, SE Australia
21 N10 **Robbinsville** North Carolina, SE USA 35°18′N 83°49′W
182 J12 **Robe** South Australia 37°11′S 139°48′E
21 W9 **Robersonville** North Carolina, SE USA 35°49′N 77°15′W
45 V10 **Robert L. Bradshaw** ✈ (Basseterre) Saint Kitts, Saint Kitts and Nevis 17°16′N 62°43′W
25 P8 **Robert Lee** Texas, SW USA 31°50′N 100°30′W
35 V5 **Roberts Creek Mountain** ▲ Nevada, W USA 39°52′N 116°16′W
93 J15 **Robertsfors** Västerbotten, N Sweden 64°12′N 20°50′E
27 R11 **Robert S. Kerr Reservoir** ⊠ Oklahoma, C USA
38 L12 **Roberts Mountain** ▲ Nunivak Island, Alaska, USA 60°01′N 166°15′W
83 F26 **Robertson** Western Cape, SW South Africa 33°48′S 19°53′E
194 H4 **Robertson Island** *island* Antarctica
76 J16 **Robertsport** W Liberia 06°45′N 11°15′W
182 J8 **Robertstown** South Australia 34°00′S 139°04′E
Robert Williams *see* Caála
15 P7 **Roberval** Québec, SE Canada 48°31′N 72°16′W
31 N15 **Robinson** Illinois, N USA 39°00′N 87°44′W
193 U11 **Róbinson Crusoe, Isla** *island* Islas Juan Fernández, Chile, E Pacific Ocean
180 J9 **Robinson Range** ▲▲ Western Australia
182 M9 **Robinvale** Victoria, SE Australia 34°37′S 142°45′E
105 P11 **Robledo** Castilla-La Mancha, C Spain 38°45′N 02°27′W
54 G5 **Robles** *var.* La Paz, Robles La Paz. Cesar, N Colombia 10°24′N 73°11′W
Robles La Paz *see* Robles
11 V15 **Roblin** Manitoba, S Canada 51°15′N 101°20′W
11 S17 **Robsart** Saskatchewan, S Canada 49°22′N 109°15′W
11 N15 **Robson, Mount** ▲ British Columbia, SW Canada 53°09′N 119°16′W
25 T14 **Robstown** Texas, SW USA 27°47′N 97°40′W
25 P6 **Roby** Texas, SW USA 32°42′N 100°23′W
104 E11 **Roca, Cabo da** *cape* C Portugal
Rocadas *see* Xangongo
41 S14 **Roca Partida, Punta** *headland* C Mexico 18°43′N 95°11′W
47 X6 **Rocas, Atol das** *island* E Brazil
107 L18 **Roccadaspide** *var.* Rocca d'Aspide. Campania, S Italy 40°25′N 15°12′E
Rocca d'Aspide *see* Roccadaspide
107 K15 **Roccaraso** Abruzzo, C Italy 41°49′N 14°01′E
106 H10 **Rocca San Casciano** Emilia-Romagna, C Italy 44°06′N 11°51′E
106 G13 **Roccastrada** Toscana, C Italy 43°00′N 11°09′E
61 G20 **Rocha** Rocha, E Uruguay 34°30′S 54°22′W
61 G19 **Rocha** ◈ *department* E Uruguay
97 L17 **Rochdale** NW England, United Kingdom 53°38′N 02°09′W
102 L11 **Rochechouart** Haute-Vienne, C France 45°49′N 00°49′E
99 J22 **Rochefort** Namur, SE Belgium 50°10′N 05°13′E
102 J11 **Rochefort** *var.* Rochefort sur Mer. Charente-Maritime, W France 45°57′N 00°58′W
Rochefort sur Mer *see* Rochefort
125 N10 **Rochegda** Arkhangel'skaya Oblast', NW Russian Federation 62°37′N 43°21′E
30 L10 **Rochelle** Illinois, N USA 41°54′N 89°03′W
25 Q9 **Rochelle** Texas, SW USA 31°13′N 99°10′W
15 V3 **Rochers Ouest, Rivière aux** ~ Québec, SE Canada
97 O22 **Rochester** *anc.* Durobrivae. SE England, United Kingdom 51°24′N 00°30′E
31 O12 **Rochester** Indiana, N USA 41°03′N 86°13′W
29 W10 **Rochester** Minnesota, N USA 44°01′N 92°28′W
19 O9 **Rochester** New Hampshire, NE USA 43°18′N 70°58′W
18 F9 **Rochester** New York, NE USA 43°09′N 77°37′W
25 P5 **Rochester** Texas, SW USA 33°19′N 99°51′W
31 S9 **Rochester Hills** Michigan, N USA 42°39′N 83°04′W
Rocheuses, Montagnes/Rockies *see* Rocky Mountains
64 M6 **Rockall** *island* N Atlantic Ocean, United Kingdom
64 L6 **Rockall Bank** *undersea feature* N Atlantic Ocean
84 B8 **Rockall Rise** *undersea feature* N Atlantic Ocean 59°00′N 14°00′W
84 C9 **Rockall Trough** *undersea feature* N Atlantic Ocean 57°00′N 12°00′W
35 U2 **Rock Creek** ~ Nevada, W USA
25 T10 **Rockdale** Texas, SW USA 30°39′N 96°58′W
195 N12 **Rockefeller Plateau** *plateau* Antarctica
30 K11 **Rock Falls** Illinois, N USA 41°46′N 89°41′W
23 Q5 **Rockford** Alabama, S USA 32°53′N 86°11′W
30 L10 **Rockford** Illinois, N USA 42°16′N 89°06′W
15 Q12 **Rock Forest** Québec, SE Canada 45°21′N 71°58′W
11 T17 **Rockglen** Saskatchewan, S Canada 49°11′N 105°57′W
181 Y8 **Rockhampton** Queensland, E Australia 23°31′S 150°31′E
21 R11 **Rock Hill** South Carolina, SE USA 34°55′N 81°01′W
180 I13 **Rockingham** Western Australia 32°16′S 115°21′E
21 T11 **Rockingham** North Carolina, SE USA 34°56′N 79°47′W
30 J11 **Rock Island** Illinois, N USA 41°30′N 90°34′W
25 U12 **Rock Island** Texas, SW USA 29°31′N 96°33′W
14 C10 **Rock Lake** Ontario, S Canada 46°25′N 83°49′W
29 O2 **Rock Lake** North Dakota, N USA 48°45′N 99°12′W
14 I12 **Rock Lake** ◎ Ontario, SE Canada
14 M12 **Rockland** Ontario, SE Canada 45°33′N 75°16′W
19 R7 **Rockland** Maine, NE USA 44°08′N 69°06′W
182 L11 **Rocklands Reservoir** ⊠ Victoria, SE Australia
35 O7 **Rocklin** California, W USA 38°48′N 121°13′W
23 R3 **Rockmart** Georgia, SE USA 34°00′N 85°02′W
31 N16 **Rockport** Indiana, N USA 37°53′N 87°04′W
27 Q1 **Rock Port** Missouri, C USA 40°26′N 95°30′W
25 T14 **Rockport** Texas, SW USA 28°02′N 99°04′W
32 I7 **Rockport** Washington, NW USA 48°28′N 121°36′W
29 S11 **Rock Rapids** Iowa, C USA 43°25′N 96°10′W
30 K11 **Rock River** ~ Illinois/Wisconsin, N USA
44 I3 **Rock Sound** Eleuthera Island, C The Bahamas 24°52′N 76°10′W
25 P11 **Rocksprings** Texas, SW USA 30°02′N 100°14′W
33 U17 **Rock Springs** Wyoming, C USA 41°35′N 109°12′W
55 T9 **Rockstone** C Guyana 05°58′N 58°33′W
29 S12 **Rock Valley** Iowa, C USA 43°12′N 96°17′W
31 N14 **Rockville** Indiana, N USA 39°45′N 87°15′W
21 W3 **Rockville** Maryland, NE USA 39°05′N 77°10′W
25 U6 **Rockwall** Texas, SW USA 32°56′N 96°27′W
29 U13 **Rockwell City** Iowa, C USA 42°24′N 94°37′W
31 S10 **Rockwood** Michigan, N USA 42°04′N 83°15′W
20 M9 **Rockwood** Tennessee, S USA 35°52′N 84°41′W
25 Q8 **Rockwood** Texas, SW USA 31°29′N 99°23′W
37 U6 **Rocky Ford** Colorado, C USA 38°03′N 103°45′W
14 D9 **Rocky Island Lake** ◎ Ontario, S Canada
21 V9 **Rocky Mount** North Carolina, SE USA 35°56′N 77°48′W
21 S7 **Rocky Mount** Virginia, NE USA 37°00′N 79°53′W
33 Q8 **Rocky Mountain** ▲ Montana, NW USA 47°45′N 112°46′W
11 P15 **Rocky Mountain House** Alberta, SW Canada 52°24′N 114°52′W
37 T3 **Rocky Mountain National Park** *national park* Colorado, C USA
2 E12 **Rocky Mountains** *var.* Rockies, *Fr.* Montagnes Rocheuses. ▲▲ Canada/USA
42 H1 **Rocky Point** *headland* NE Belize 18°21′N 88°04′W
83 A17 **Rocky Point** *headland* NW Namibia 19°01′S 12°27′E
95 F14 **Rødberg** Buskerud, S Norway 60°16′N 09°00′E
95 I25 **Rødby** Sjælland, SE Denmark 54°42′N 11°24′E
95 I25 **Rødbyhavn** Sjælland, SE Denmark 54°39′N 11°24′E
13 T10 **Roddickton** Newfoundland, Newfoundland and Labrador, SE Canada 50°51′N 56°03′W
95 F23 **Rødding** Syddanmark, SW Denmark 55°22′N 09°04′E
95 M22 **Rödeby** Blekinge, S Sweden 56°16′N 15°35′E
98 N6 **Roden** Drenthe, NE Netherlands 53°08′N 06°26′E
62 H9 **Rodeo** San Juan, W Argentina 30°12′S 69°06′W
103 O14 **Rodez** *anc.* Segodunum. Aveyron, S France 44°21′N 02°34′E
Rodholívos *see* Rodolívos
Rodhópi Óri *see* Rhodope Mountains
Ródhos/Rodi *see* Ródos
107 N15 **Rodi Gargancio** Puglia, SE Italy 41°54′N 15°51′E
101 N20 **Roding** Bayern, SE Germany 49°12′N 12°30′E
113 J19 **Rodinit, Kepi i** *headland* W Albania 41°35′N 19°27′E
116 I9 **Rodnei, Munţii** ▲▲ N Romania
184 L4 **Rodney, Cape** *headland* North Island, New Zealand 36°16′S 174°48′E
38 L9 **Rodney, Cape** *headland* Alaska, USA 64°39′N 166°24′W
124 M16 **Rodniki** Ivanovskaya Oblast', W Russian Federation 57°04′N 41°45′E
119 Q16 **Rodnya** Mahilyowskaya Voblasts', E Belarus 53°31′N 32°07′E
Rodó *see* José Enrique Rodó
114 H13 **Rodolívos** *var.* Rodholívos. Kentrikí Makedonía, NE Greece 40°55′N 24°00′E
Rodopi *see* Rhodope Mountains
115 O22 **Ródos** *var.* Ródhos, *Eng.* Rhodes, *It.* Rodi. Ródos, Dodekánisa, Greece, Aegean Sea 36°26′N 28°14′E
115 O22 **Ródos** *var.* Ródhos, *Eng.* Rhodes, *It.* Rodi; *anc.* Rhodos. *island* Dodekánisa, Greece, Aegean Sea
Rodosto *see* Tekirdağ
59 A14 **Rodrigues** Amazonas, W Brazil 06°50′S 73°45′W
173 P8 **Rodrigues** *var.* Rodriquez. *island* E Mauritius
Rodriquez *see* Rodrigues
Rodunma *see* Roanne
180 I7 **Roebourne** Western Australia 20°49′S 117°04′E
83 J20 **Roedtan** Limpopo, NE South Africa 24°37′S 29°05′E
98 H11 **Roelofarendsveen** Zuid-Holland, W Netherlands 52°12′N 04°37′E
Roepat *see* Rupat, Pulau
Roer *see* Rur
99 M16 **Roermond** Limburg, SE Netherlands 51°12′N 06°E
99 C18 **Roeselare** *Fr.* Roulers; *prev.* Rousselaere. West-Vlaanderen, W Belgium 50°57′N 03°08′E
9 P8 **Roes Welcome Sound** *strait* Nunavut, N Canada
Roeteng *see* Ruteng
Rofreit *see* Rovereto
Rogachëv *see* Rahachow
57 L15 **Rogagua, Laguna** ◎ NW Bolivia
95 C16 **Rogaland** ◈ *county* S Norway
25 Y9 **Roganville** Texas, SW USA 30°49′N 93°54′W
109 W11 **Rogaška Slatina** *Ger.* Rohitsch-Sauerbrunn; *prev.* Rogatec-Slatina. E Slovenia 46°13′N 15°38′E
Rogatec-Slatina *see* Rogaška Slatina
112 J13 **Rogatica** Republika Srpska, SE Bosnia and Herzegovina 43°50′N 18°55′E
Rogatin *see* Rohatyn
93 F17 **Rogen** ◎ C Sweden
27 S9 **Rogers** Arkansas, C USA 36°19′N 94°07′W
29 P5 **Rogers** North Dakota, N USA 47°03′N 98°12′W
25 T9 **Rogers** Texas, SW USA 30°53′N 97°10′W
31 R5 **Rogers City** Michigan, N USA 45°25′N 83°49′W
Roger Simpson Island *see* Abemama
35 T14 **Rogers Lake** *salt flat* California, W USA
21 Q8 **Rogers, Mount** ▲ Virginia, NE USA 36°39′N 81°32′W
33 O16 **Rogerson** Idaho, NW USA 42°11′N 114°36′W
11 O16 **Rogers Pass** *pass* British Columbia, SW Canada
21 O8 **Rogersville** Tennessee, S USA 36°26′N 83°01′W
99 L16 **Roggel** Limburg, SE Netherlands 51°16′N 05°55′E
Roggeveen *see* Roggewein, Cabo
193 R10 **Roggeveen Basin** *undersea feature* E Pacific Ocean 31°30′S 95°30′W
191 X16 **Roggewein, Cabo** *var.* Roggeveen. *headland* Easter Island, Chile, E Pacific Ocean 27°07′S 109°15′W
103 Y13 **Rogliano** Corse, France, C Mediterranean Sea 42°58′N 09°25′E
107 N21 **Rogliano** Calabria, SW Italy 39°09′N 16°18′E
92 G12 **Rognan** Nordland, C Norway 67°04′N 15°21′E
100 K10 **Rögnitz** ~ N Germany
Rogozhina/Rogozhinë *see* Rrogozhinë
110 G10 **Rogoźno** Wielkopolskie, C Poland 52°46′N 16°58′E
32 E15 **Rogue River** ~ Oregon, NW USA
116 I6 **Rohatyn** *Rus.* Rogatin. Ivano-Frankivs'ka Oblast', W Ukraine 49°25′N 24°35′E
189 O14 **Rohi** Pohnpei, E Micronesia
Rohitsch-Sauerbrunn *see* Rogaška Slatina
149 Q13 **Rohri** Sind, SE Pakistan 27°39′N 68°57′E
152 I10 **Rohtak** Haryāna, N India 28°57′N 76°38′E
Roi Ed *see* Roi Et
167 R9 **Roi Et** *var.* Muang Roi Et, Roi Ed. Roi Et, E Thailand 16°05′N 103°38′E
191 U9 **Roi Georges, Îles du** *island group* Îles Tuamotu, C French Polynesia
153 Y10 **Roing** Arunāchal Pradesh, NE India 28°06′N 95°46′E
118 E7 **Roja** NW Latvia 57°31′N 22°44′E
61 B20 **Rojas** Buenos Aires, E Argentina 34°10′S 60°45′W
149 R12 **Rojhān** Punjab, E Pakistan 28°39′N 70°00′E
41 Q12 **Rojo, Cabo** *headland* C Mexico 21°33′N 97°19′W
45 Q10 **Rojo, Cabo** *headland* W Puerto Rico 17°57′N 67°10′W
168 K10 **Rokan Kiri, Sungai** ~ Sumatera, W Indonesia
118 I11 **Rokiškis** Panevėžys, NE Lithuania 55°58′N 25°35′E
165 R7 **Rokkasho** Aomori, Honshū, C Japan 40°59′N 141°22′E
111 B17 **Rokycany** *Ger.* Rokytzan. Plzeňský Kraj, W Czech Republic 49°45′N 13°36′E
117 P6 **Rokytne** Kyyivs'ka Oblast', N Ukraine 49°40′N 30°29′E
116 L3 **Rokytne** Rivnens'ka Oblast', NW Ukraine 51°19′N 27°09′E
Rokytzan *see* Rokycany
158 L11 **Rola Co** ◎ W China
29 V13 **Roland** Iowa, C USA 42°10′N 93°30′W
95 D15 **Røldal** Hordaland, S Norway 59°52′N 06°49′E
98 O7 **Rolde** Drenthe, NE Netherlands 52°58′N 06°39′E
29 O2 **Rolette** North Dakota, N USA 48°39′N 99°50′W
27 V6 **Rolla** Missouri, C USA 37°56′N 91°47′W
29 O2 **Rolla** North Dakota, N USA 48°51′N 99°37′W
108 A10 **Rolle** Vaud, W Switzerland 46°27′N 06°19′E
181 X8 **Rolleston** Queensland, E Australia 24°30′S 148°36′E
185 H19 **Rolleston** Canterbury, South Island, New Zealand 43°34′S 172°24′E
185 G18 **Rolleston Range** ▲▲ South Island, New Zealand
14 H8 **Rollet** Québec, SE Canada 47°56′N 79°14′W
22 J4 **Rolling Fork** Mississippi, S USA 32°54′N 90°52′W
20 L6 **Rolling Fork** ~ Kentucky, S USA
14 J11 **Rolphton** Ontario, SE Canada 46°09′N 77°43′W
Röm *see* Rømø
181 X10 **Roma** Queensland, E Australia 26°37′S 148°54′E
107 I15 **Roma** *Eng.* Rome. ● (Italy) Lazio, C Italy 41°53′N 12°30′E
95 P19 **Roma** Gotland, SE Sweden 57°31′N 18°28′E
21 T14 **Romain, Cape** *headland* South Carolina, SE USA 33°00′N 79°21′W
13 P11 **Romaine** ~ Newfoundland and Labrador/Québec, E Canada
25 R17 **Roma Los Saenz** Texas, SW USA 26°24′N 99°01′W
114 H8 **Roman** Vratsa, NW Bulgaria 43°09′N 23°56′E
116 L10 **Roman** *Hung.* Románvásár. Neamţ, NE Romania 46°46′N 26°56′E
64 M13 **Romanche Fracture Zone** *tectonic feature* E Atlantic Ocean
61 C15 **Romang** Santa Fe, C Argentina 29°30′S 59°46′W
171 R15 **Romang, Pulau** *var.* Pulau Roma. *island* Kepulauan Damar, E Indonesia
171 R15 **Romang, Selat** *strait* Nusa Tenggara, S Indonesia
116 J11 **Romania** *Bul.* Rumŭniya, *Ger.* Rumänien, *Hung.* Románia, *Rom.* România, *SCr.* Rumunjska, *Ukr.* Rumuniya, *prev.* Republica Socialistă România, Roumania, Rumania, Socialist Republic of Romania, *prev.Rom.* Romînia. ◆ *republic* SE Europe
România, Republica Socialistă *see* Romania
Romania, Socialist Republic of *see* Romania
117 T7 **Romaniv** *Rus.* Dneprodzerzhinsk, *prev.* Dniprodzerzhyns'k, *prev.* Kamenskoye. Dnipropetrovs'ka Oblast', E Ukraine 48°30′N 34°35′E
117 X7 **Romaniv** *Rus.* Dzerzhinsk; *prev.* Dzerzhyns'k. Donets'ka Oblast', SE Ukraine 48°21′N 37°50′E
116 M5 **Romaniv** *prev.* Dzerzhyns'k. Zhytomyrs'ka Oblast', N Ukraine 50°07′N 27°56′E
23 W16 **Romano, Cape** *headland* Florida, SE USA 25°51′N 81°40′W
44 G5 **Romano, Cayo** *island* C Cuba
123 O13 **Romanovka** Respublika Buryatiya, S Russian Federation 53°10′N 112°34′E
127 N8 **Romanovka** Saratovskaya Oblast', W Russian Federation 51°45′N 42°45′E
108 I6 **Romanshorn** Thurgau, NE Switzerland 47°34′N 09°23′E
103 R12 **Romans-sur-Isère** Drôme, E France 45°03′N 05°03′E
189 U12 **Romanum** *island* Chuuk, C Micronesia
Románvásár *see* Roman
39 S5 **Romanzof Mountains** ▲▲ Alaska, USA
Roma, Pulau *see* Romang, Pulau
103 S4 **Rombas** Moselle, NE France 49°15′N 06°04′E
23 R2 **Rome** Georgia, SE USA 34°01′N 85°02′W
18 I9 **Rome** New York, NE USA 43°13′N 75°28′W
Rome *see* Roma
31 S9 **Romeo** Michigan, N USA 42°48′N 83°00′W
Römerstadt *see* Rýmařov
Rometan *see* Romiton
103 P5 **Romilly-sur-Seine** Aube, N France 48°31′N 03°44′E
Romînia *see* Romania
146 L11 **Romiton** *Rus.* Rometan. Buxoro Viloyati, C Uzbekistan 39°56′N 64°21′E
21 U3 **Romney** West Virginia, NE USA 39°21′N 78°44′W
117 S4 **Romny** Sums'ka Oblast', NE Ukraine 50°45′N 33°30′E
95 E24 **Rømø** *Ger.* Röm. *island* SW Denmark
117 S5 **Romodan** Poltavs'ka Oblast', NE Ukraine 50°00′N 33°20′E
127 P5 **Romodanovo** Respublika Mordoviya, W Russian Federation 54°25′N 45°24′E
Romorantin *see* Romorantin-Lanthenay
103 N8 **Romorantin-Lanthenay** *var.* Romorantin. Loir-et-Cher, C France 47°22′N 01°44′E
94 F9 **Romsdal** *physical region* S Norway
94 F10 **Romsdalen** *valley* S Norway
94 E9 **Romsdalsfjorden** *fjord* S Norway
33 P8 **Ronan** Montana, NW USA 47°31′N 114°06′W
59 M14 **Roncador** Maranhão, E Brazil 05°48′S 45°08′W
186 M7 **Roncador Reef** *reef* N Solomon Islands
59 J17 **Roncador, Serra do** ▲▲ C Brazil
21 S6 **Ronceverte** West Virginia, NE USA 37°45′N 80°27′W
107 H14 **Ronciglione** Lazio, C Italy 42°16′N 12°15′E
104 L15 **Ronda** Andalucía, S Spain 36°45′N 05°10′W
94 G11 **Rondane** ▲▲ S Norway
104 L15 **Ronda, Serranía de** ▲▲ S Spain
95 H22 **Rønde** Midtjylland, C Denmark 56°18′N 10°28′E
Ronde, Île *see* Round Island
Rōṇdik *see* Rongrik Atoll
59 E16 **Rondônia** *off.* Estado de Rondônia; *prev.* Território de Rondônia. ◈ *state* W Brazil
Rondônia, Estado de *see* Rondônia
Rondônia, Território de *see* Rondônia
59 I18 **Rondonópolis** Mato Grosso, W Brazil 16°29′S 54°37′W
94 G11 **Rondslottet** ▲ S Norway 61°54′N 09°48′E
95 P20 **Ronehamn** Gotland, SE Sweden 57°10′N 18°30′E
160 L13 **Rong'an** *var.* Chang'an, Rongan. Guangxi Zhuangzu Zizhiqu, S China 25°14′N 109°20′E
Rongan *see* Rong'an
Rongcheng *see* Rongxian, Guangxi, China
Rongcheng *see* Jianli, Hubei, China
189 R4 **Rongelap Atoll** *var.* Rōnlap. *atoll* Ralik Chain, NW Marshall Islands
Rongerik *see* Rongrik Atoll
160 K12 **Rongjiang** *var.* Guzhou. Guizhou, S China 25°59′N 108°27′E
160 L13 **Rong Jiang** ~ S China
Rongjiang *see* Nankang
Rong, Kas *see* Rŭng, Kaôh
167 P8 **Rong Kwang** Phrae, NW Thailand 18°19′N 100°18′E
189 T4 **Rongrik Atoll** *var.* Rōṇdik, Rongerik. *atoll* Ralik Chain, N Marshall Islands
189 X2 **Rongrong** *island* SE Marshall Islands
160 L13 **Rongshui** *var.* Rongshui Miaozu Zizhixian. Guangxi Zhuangzu Zizhiqu, S China 25°05′N 109°09′E
Rongshui Miaozu Zizhixian *see* Rongshui
118 I6 **Rõngu** *Ger.* Ringen. Tartumaa, SE Estonia 58°10′N 26°17′E
Rongwo *see* Tongren
160 L15 **Rongxian** *var.* Rongzhou; *prev.* Rongcheng. Guangxi Zhuangzu Zizhiqu, S China 22°52′N 110°33′E
Rongzhag *see* Danba
Rongzhou *see* Rongxian
Roniu, Mont *see* Rooniu, Mont
189 N13 **Ronkiti** Pohnpei, E Micronesia 06°48′N 158°10′E
Rōnlap *see* Rongelap Atoll
95 L24 **Rønne** Bornholm, E Denmark 55°07′N 14°43′E
95 M22 **Ronneby** Blekinge, S Sweden 56°12′N 15°18′E
194 J7 **Ronne Entrance** *inlet* Antarctica

◆ Country
● Country Capital
◇ Dependent Territory
○ Dependent Territory Capital
◈ Administrative Regions
✈ International Airport
▲ Mountain
▲▲ Mountain Range
⛰ Volcano
~ River
◎ Lake
⊠ Reservoir

RONNE ICE SHELF – RUVUMA

194 L6 **Ronne Ice Shelf** *ice shelf* Antarctica
99 E19 **Ronse** *Fr.* Renaix. Oost-Vlaanderen, SW Belgium 50°45′N 03°36′E
30 K14 **Roodhouse** Illinois, N USA 39°28′N 90°22′W
83 C19 **Rooibank** Erongo, W Namibia 23°04′S 14°34′E
Rooke Island *see* Umboi Island
65 N24 **Rookery Point** *headland* NE Tristan da Cunha 37°03′S 12°15′W
191 R8 **Rooniu, Mont** *prev.* Mont Roniu. 🌋 Tahiti, W French Polynesia 17°49′S 149°12′W
171 V13 **Roon, Pulau** *island* E Indonesia
173 V7 **Roo Rise** *undersea feature* E Indian Ocean
152 J9 **Roorkee** Uttarakhand, N India 29°51′N 77°54′E
99 H15 **Roosendaal** Noord-Brabant, S Netherlands 51°32′N 04°29′E
25 P10 **Roosevelt** Texas, SW USA 30°28′N 100°06′W
37 N3 **Roosevelt** Utah, W USA 40°18′N 109°59′W
47 T8 **Roosevelt** ≈ W Brazil
195 O13 **Roosevelt Island** *island* Antarctica
10 L10 **Roosevelt, Mount** ▲ British Columbia, W Canada 58°28′N 125°22′W
11 P17 **Roosville** British Columbia, SW Canada 48°59′N 115°03′W
29 X10 **Root River** ≈ Minnesota, N USA
Ropar *see* Rūpnagar
111 N16 **Ropczyce** Podkarpackie, SE Poland 50°04′N 21°31′E
181 Q3 **Roper Bar** Northern Territory, N Australia 14°45′S 134°30′E
24 M5 **Ropesville** Texas, SW USA 33°24′N 102°09′W
102 K14 **Roquefort** Landes, SW France 44°01′N 00°18′W
61 C21 **Roque Pérez** Buenos Aires, E Argentina 35°25′S 59°24′W
58 E10 **Roraima** *off.* Estado de Roraima; *prev.* Território de Rio Branco, Território de Roraima. ◈ *state* N Brazil
Roraima, Estado de *see* Roraima
58 F9 **Roraima, Mount** ▲ N South America 05°10′N 60°36′W
Roraima, Território de *see* Roraima
94 I9 **Røros** Sør-Trøndelag, S Norway 62°37′N 11°25′E
108 I7 **Rorschach** Sankt Gallen, NE Switzerland 47°28′N 09°30′E
93 E14 **Rørvik** Nord-Trøndelag, C Norway 64°54′N 11°15′E
119 G17 **Ros'** *Rus.* Ross'. Hrodzyenskaya Voblasts', W Belarus 53°25′N 24°24′E
185 F17 **Ross** West Coast, South Island, New Zealand 42°54′S 170°52′E
119 G17 **Ros'** *Rus.* Ross'. ≈ W Belarus
10 J7 **Ross** ≈ Yukon, W Canada
117 O6 **Ros'** ≈ N Ukraine
44 K7 **Rosa, Lake** ◎ Great Inagua, S The Bahamas
32 M9 **Rosalia** Washington, NW USA 47°14′N 117°22′W
191 W15 **Rosalia, Punta** *headland* Easter Island, Chile, E Pacific Ocean 27°04′S 109°19′W
45 P12 **Rosalie** E Dominica 15°22′N 61°15′W
35 T14 **Rosamond** California, W USA 34°51′N 118°09′W
35 S14 **Rosamond Lake** *salt flat* California, W USA
96 H8 **Ross and Cromarty** *cultural region* N Scotland, United Kingdom
61 B18 **Rosario** Santa Fe, C Argentina 32°56′S 60°39′W
40 J11 **Rosario** Sinaloa, C Mexico 23°00′N 105°51′W
40 G6 **Rosario** Sonora, NW Mexico 27°53′N 109°18′W
62 O6 **Rosario** San Pedro, C Paraguay 24°26′S 57°06′W
61 E20 **Rosario** Colonia, SW Uruguay 34°20′S 57°26′W
54 H5 **Rosario** Zulia, NW Venezuela 10°18′N 72°19′W
Rosario *see* Nishino-shima
Rosario *see* Rosarito
40 B4 **Rosario, Bahía del** *bay* NW Mexico
62 K6 **Rosario de la Frontera** Salta, N Argentina 25°50′S 65°00′W
61 C18 **Rosario del Tala** Entre Ríos, E Argentina 32°20′S 59°10′W
61 F16 **Rosário do Sul** Rio Grande do Sul, S Brazil 30°15′S 54°55′W
59 H18 **Rosário Oeste** Mato Grosso, W Brazil 14°50′S 56°25′W
40 B1 **Rosarito** *var.* Rosario. Baja California Norte, NW Mexico 32°25′N 117°04′W
40 D5 **Rosarito** Baja California Norte, NW Mexico 28°27′N 113°58′W
40 F7 **Rosarito** Baja California Sur, NW Mexico 26°28′N 111°41′W
104 L9 **Rosarito, Embalse del** ⊠ W Spain
107 N22 **Rosarno** Calabria, SW Italy 38°29′N 15°59′E
56 B5 **Rosa Zárate** *var.* Quinindé. Esmeraldas, SW Ecuador 0°14′N 79°28′W
Roscianum *see* Rossano
29 O8 **Roscoe** South Dakota, N USA 45°24′N 99°19′W
25 P7 **Roscoe** Texas, SW USA 32°27′N 100°32′W
102 F5 **Roscoff** Finistère, NW France 48°43′N 04°00′W
Ros Comáin *see* Roscommon
97 C17 **Roscommon** *Ir.* Ros Comáin. C Ireland 53°38′N 08°11′W
31 Q7 **Roscommon** Michigan, N USA 44°30′N 84°35′W
97 C17 **Roscommon** *Ir.* Ros Comáin. *cultural region* C Ireland
Ros. Cré *see* Roscrea
97 D19 **Roscrea** *Ir.* Ros. Cré. C Ireland 52°57′N 07°47′W
14 H13 **Rosseau** Ontario, S Canada 45°15′N 79°38′W
45 X12 **Roseau** *prev.* Charlotte Town. ● (Dominica) SW Dominica 15°17′N 61°23′W
29 S2 **Roseau** Minnesota, N USA 48°51′N 95°45′W
173 Y16 **Rose Belle** SE Mauritius 20°24′S 57°36′E
183 O16 **Rosebery** Tasmania, SE Australia 41°51′S 145°33′E
21 U11 **Roseboro** North Carolina, SE USA 34°58′N 78°31′W
25 T9 **Rosebud** Texas, SW USA 31°04′N 96°58′W
33 W10 **Rosebud Creek** ≈ Montana, NW USA
32 F14 **Roseburg** Oregon, NW USA 43°13′N 123°21′W
22 J3 **Rosedale** Mississippi, S USA 33°51′N 91°01′W
99 H21 **Rosée** Namur, S Belgium 50°15′N 04°43′E
55 U8 **Rose Hall** E Guyana 06°14′N 57°30′W
173 X16 **Rose Hill** W Mauritius 20°14′S 57°29′E
80 H12 **Roseires, Reservoir** *var.* Lake Rusayris. ⊠ E Sudan
Rosenau *see* Rožnov pod Radhoštěm
Rosenau *see* Rožňava
25 V11 **Rosenberg** Texas, SW USA 29°33′N 95°48′W
Rosenberg *see* Olesno, Poland
Rosenberg *see* Ružomberok, Slovakia
100 I10 **Rosengarten** Niedersachsen, N Germany 53°24′N 09°54′E
101 M24 **Rosenheim** Bayern, S Germany 47°51′N 12°08′E
Rosenhof *see* Zilupe
105 X4 **Roses** Cataluña, NE Spain 42°15′N 03°11′E
105 X4 **Roses, Golf de** *gulf* NE Spain
107 K14 **Roseto degli Abruzzi** Abruzzo, C Italy 42°39′N 14°01′E
11 S16 **Rosetown** Saskatchewan, S Canada 51°34′N 107°59′W
Rosetta *see* Rashīd
35 O7 **Roseville** California, W USA 38°44′N 121°16′W
30 J12 **Roseville** Illinois, N USA 40°42′N 90°40′W
29 V8 **Roseville** Minnesota, N USA 45°00′N 93°09′W
29 R7 **Rosholt** South Dakota, N USA 45°51′N 96°42′W
106 F12 **Rosignano Marittimo** Toscana, C Italy 43°24′N 10°28′E
116 I14 **Roşiori de Vede** Teleorman, S Romania 44°06′N 25°00′E
114 K8 **Rositsa** ≈ N Bulgaria
Rositten *see* Rēzekne
95 J23 **Roskilde** Sjælland, E Denmark 55°39′N 12°07′E
Ros Láir *see* Rosslare
126 H5 **Roslavl'** Smolenskaya Oblast', W Russian Federation 54°N 32°57′E
32 I8 **Roslyn** Washington, NW USA 47°13′N 120°52′W
99 K14 **Rosmalen** Noord-Brabant, S Netherlands 51°43′N 05°21′E
Ros Mhic Thriúin *see* New Ross
113 P19 **Rosoman** C FYR Macedonia 41°31′N 21°55′E
102 F6 **Rosporden** Finistère, NW France 47°58′N 03°54′W
Ross' *see* Ros'
107 O20 **Rossano** *anc.* Roscianum. Calabria, SW Italy 39°35′N 16°38′E
22 L5 **Ross Barnett Reservoir** ⊠ Mississippi, S USA
11 W16 **Rossburn** Manitoba, S Canada 50°42′N 100°49′W
14 H13 **Rosseau, Lake** ◎ Ontario, S Canada
186 I10 **Rossel Island** *prev.* Yela Island. *island* SE Papua New Guinea
195 P12 **Ross Ice Shelf** *ice shelf* Antarctica
13 P16 **Rossignol, Lake** ◎ Nova Scotia, SE Canada
83 C19 **Rössing** Erongo, W Namibia 22°31′S 14°52′E
195 Q14 **Ross Island** *island* Antarctica
Rossitten *see* Rybachiy
Rossiyskaya Federatsiya *see* Russian Federation
11 N17 **Rossland** British Columbia, SW Canada 49°03′N 117°49′W
97 F20 **Rosslare** *Ir.* Ros Láir. Wexford, SE Ireland 52°16′N 06°23′W
97 F20 **Rosslare Harbour** Wexford, SE Ireland 52°15′N 06°20′W
101 M14 **Rosslau** Sachsen-Anhalt, E Germany 51°52′N 12°15′E
76 G10 **Rosso** Trarza, SW Mauritania 16°36′N 15°50′W
103 X14 **Rosso, Cap** *headland* Corse, France, C Mediterranean Sea 42°25′N 08°22′E
93 H16 **Rossön** Jämtland, C Sweden 63°54′N 16°21′E
97 K21 **Ross-on-Wye** W England, United Kingdom 51°55′N 02°34′W
Rossony *see* Rasony
126 L9 **Rossosh'** Voronezhskaya Oblast', W Russian Federation 50°10′N 39°34′E
181 Q7 **Ross River** Northern Territory, N Australia 23°36′S 134°30′E
10 J7 **Ross River** Yukon, W Canada 61°57′N 132°26′W
195 O15 **Ross Sea** *sea* Antarctica
92 G13 **Røssvatnet** *Lapp.* Reevhtse. ◎ C Norway
23 R1 **Rossville** Georgia, SE USA 34°59′N 85°22′W
Rostak *see* Ar Rustāq
143 P14 **Rostāq** Hormozgān, S Iran 26°48′N 53°50′E
117 N5 **Rostavytsya** ≈ N Ukraine
11 T15 **Rosthern** Saskatchewan, S Canada 52°40′N 106°20′W
100 M8 **Rostock** Mecklenburg-Vorpommern, NE Germany 54°05′N 12°08′E
124 L16 **Rostov** Yaroslavskaya Oblast', W Russian Federation 57°11′N 39°19′E
Rostov *see* Rostov-na-Donu
126 L12 **Rostov-na-Donu** *var.* Rostov, *Eng.* Rostov-on-Don. Rostovskaya Oblast', SW Russian Federation 47°16′N 39°45′E
Rostov-on-Don *see* Rostov-na-Donu
126 L10 **Rostovskaya Oblast'** ◈ *province* SW Russian Federation
93 J14 **Rosvik** Norrbotten, N Sweden 65°26′N 21°48′E
23 S3 **Roswell** Georgia, SE USA 34°01′N 84°21′W
37 U14 **Roswell** New Mexico, SW USA 33°23′N 104°31′W
94 K12 **Rot** Dalarna, C Sweden 61°16′N 14°04′E
101 I23 **Rot** ≈ S Germany
104 J15 **Rota** Andalucía, S Spain 36°39′N 06°20′W
188 K9 **Rota** *island* S Northern Mariana Islands
25 P6 **Rotan** Texas, SW USA 32°51′N 100°28′W
Rotcher Island *see* Tamana
100 I11 **Rotenburg** Niedersachsen, NW Germany 53°06′N 09°25′E
Rotenburg *see* Rotenburg an der Fulda
101 I16 **Rotenburg an der Fulda** *var.* Rotenburg. Thüringen, C Germany 51°00′N 09°43′E
101 L18 **Roter Main** ≈ E Germany
101 K20 **Roth** Bayern, SE Germany 49°15′N 11°06′E
101 G16 **Rothaargebirge** ▲▲ W Germany
Rothenburg *see* Rothenburg ob der Tauber
101 J20 **Rothenburg ob der Tauber** *var.* Rothenburg. Bayern, S Germany 49°23′N 10°10′E
194 H6 **Rothera** *UK research station* Antarctica 67°28′S 68°31′W
185 I17 **Rotherham** Canterbury, South Island, New Zealand 42°42′S 172°56′E
97 M17 **Rotherham** N England, United Kingdom 53°26′N 01°20′W
96 H12 **Rothesay** W Scotland, United Kingdom 55°49′N 05°03′W
108 E7 **Rothrist** Aargau, N Switzerland 47°18′N 07°54′E
194 H6 **Rothschild Island** *island* Antarctica
171 P17 **Roti, Pulau** *island* S Indonesia
95 H14 **Rotnes** Akershus, S Norway 60°08′N 10°45′E
183 O8 **Roto** New South Wales, SE Australia 33°04′S 145°27′E
184 N8 **Rotoiti, Lake** ◎ North Island, New Zealand
Rotomagus *see* Rouen
107 N19 **Rotondella** Basilicata, S Italy 40°12′N 16°30′E
103 X15 **Rotondo, Monte** ▲ Corse, France, C Mediterranean Sea 42°15′N 09°03′E
185 I15 **Rotoroa, Lake** ◎ South Island, New Zealand
184 N8 **Rotorua** Bay of Plenty, North Island, New Zealand 38°10′S 176°14′E
184 N8 **Rotorua, Lake** ◎ North Island, New Zealand
101 N22 **Rott** ≈ SE Germany
108 F10 **Rotten** ≈ S Switzerland
109 T6 **Rottenmann** Steiermark, E Austria 47°31′N 14°18′E
98 H12 **Rotterdam** Zuid-Holland, SW Netherlands 51°55′N 04°30′E
18 K10 **Rotterdam** New York, NE USA 42°46′N 73°57′W
95 M21 **Rottnen** ◎ S Sweden
98 N4 **Rottumeroog** *island* Waddeneilanden, NE Netherlands
98 N4 **Rottumerplaat** *island* Waddeneilanden, NE Netherlands
101 G23 **Rottweil** Baden-Württemberg, S Germany 48°10′N 08°38′E
191 O7 **Rotui, Mont** 🌋 Moorea, W French Polynesia 17°30′S 149°50′W
103 P1 **Roubaix** Nord, N France 50°42′N 03°10′E
111 C15 **Roudnice nad Labem** *Ger.* Raudnitz an der Elbe. Ústecký Kraj, NW Czech Republic 50°25′N 14°14′E
102 M4 **Rouen** *anc.* Rotomagus. Seine-Maritime, N France 49°26′N 01°05′E
171 X13 **Rouffaer Reserves** *reserve* Papua, E Indonesia
15 N10 **Rouge, Rivière** ≈ Québec, SE Canada
20 J6 **Rough River** ≈ Kentucky, S USA
20 J6 **Rough River Lake** ⊠ Kentucky, S USA
Rouhaïbé *see* Ar Ruḩaybah
102 K11 **Rouillac** Charente, W France 45°46′N 00°04′W
Roulers *see* Roeselare
Roumania *see* Romania
173 Y15 **Round Island** *var.* Île Ronde. *island* NE Mauritius
14 J12 **Round Lake** ◎ Ontario, SE Canada
35 U7 **Round Mountain** Nevada, W USA 38°42′N 117°04′W
25 R10 **Round Mountain** Texas, SW USA 30°25′N 98°20′W
183 U5 **Round Mountain** ▲ New South Wales, SE Australia 30°22′S 152°13′E
25 S10 **Round Rock** Texas, SW USA 30°30′N 97°40′W
33 U10 **Roundup** Montana, NW USA 46°27′N 108°32′W
55 Y10 **Roura** NE French Guiana 04°44′N 52°16′W
Rourkela *see* Rāurkela
96 J4 **Rousay** *island* N Scotland, United Kingdom
Rousselaere *see* Roeselare
103 O17 **Roussillon** *cultural region* S France
15 V7 **Routhierville** Québec, SE Canada 48°09′N 67°07′W
99 K25 **Rouvroy** Luxembourg, SE Belgium 49°33′N 05°28′E
14 I7 **Rouyn-Noranda** Québec, SE Canada 48°16′N 79°03′W
Rouyuan *see* Huachi
Rouyuanchengzi *see* Huachi
92 L12 **Rovaniemi** Lappi, N Finland 66°29′N 25°40′E
106 E7 **Rovato** Lombardia, N Italy 45°34′N 10°03′E
125 N11 **Rovdino** Arkhangel'skaya Oblast', NW Russian Federation 61°36′N 42°28′E
Roven'ki *see* Roven'ky
117 Y8 **Roven'ky** *var.* Roven'ki. Luhans'ka Oblast', E Ukraine 48°05′N 39°20′E
Rovenskaya Oblast' *see* Rivnens'ka Oblast'
Rovenskaya Sloboda *see* Rovyenskaya Slabada
106 G7 **Rovereto** *Ger.* Rofreit. Trentino-Alto Adige, N Italy 45°53′N 11°03′E
167 S12 **Rôviěng Tbong** Preăh Vihéar, N Cambodia 13°18′N 105°06′E
Rovigno *see* Rovinj
106 H8 **Rovigo** Veneto, NE Italy 45°04′N 11°48′E
112 A10 **Rovinj** *It.* Rovigno. Istra, NW Croatia 45°06′N 13°39′E
54 E10 **Rovira** Tolima, C Colombia 04°15′N 75°15′W
Rovno *see* Rivne
127 P9 **Rovnoye** Saratovskaya Oblast', W Russian Federation 50°43′N 46°03′E
82 Q12 **Rovuma, Rio** *var.* Ruvuma. ≈ Mozambique/Tanzania *see also* Ruvuma
Rovuma, Rio *see* Ruvuma
119 O19 **Rovyenskaya Slabada** *Rus.* Rovenskaya Sloboda. Homyel'skaya Voblasts', SE Belarus 52°13′N 30°19′E
183 R5 **Rowena** New South Wales, SE Australia 29°51′S 148°55′E
21 T11 **Rowland** North Carolina, SE USA 34°32′N 79°17′W
9 P5 **Rowley** ≈ Baffin Island, Nunavut, NE Canada
9 P6 **Rowley Island** *island* Nunavut, NE Canada
173 W8 **Rowley Shoals** *reef* NW Australia
Równe *see* Rivne
171 O4 **Roxas** Mindoro, N Philippines 12°36′N 121°29′E
171 P5 **Roxas City** Panay Island, C Philippines 11°33′N 122°43′E
21 U8 **Roxboro** North Carolina, SE USA 36°24′N 79°00′W
185 D23 **Roxburgh** Otago, South Island, New Zealand 45°32′S 169°18′E
96 K13 **Roxburgh** *cultural region* SE Scotland, United Kingdom
182 H5 **Roxby Downs** South Australia 30°29′S 136°56′E
95 M17 **Roxen** ◎ S Sweden
25 V5 **Roxton** Texas, SW USA 33°33′N 95°43′W
15 P12 **Roxton-Sud** Québec, SE Canada 45°30′N 72°35′W
33 U8 **Roy** Montana, NW USA 47°19′N 108°55′W
37 U10 **Roy** New Mexico, SW USA 35°56′N 104°12′W
97 E17 **Royal Canal** *Ir.* An Chanáil Ríoga. *canal* C Ireland
30 L1 **Royale, Isle** *island* Michigan, N USA
37 S6 **Royal Gorge** *valley* Colorado, C USA
97 M20 **Royal Leamington Spa** *var.* Leamington, Leamington Spa. C England, United Kingdom 52°18′N 01°31′W
97 O23 **Royal Tunbridge Wells** *var.* Tunbridge Wells. SE England, United Kingdom 51°08′N 00°16′E
24 L9 **Royalty** Texas, SW USA 31°21′N 102°51′W
102 J11 **Royan** Charente-Maritime, W France 45°37′N 01°01′W
65 B24 **Roy Cove Settlement** West Falkland, Falkland Islands 51°32′S 60°23′W
103 O3 **Roye** Somme, N France 49°42′N 02°46′E
95 H15 **Røyken** Buskerud, S Norway 59°47′N 10°21′E
93 F14 **Røyrvik** Nord-Trøndelag, C Norway 64°53′N 13°30′E
25 U6 **Royse City** Texas, SW USA 32°58′N 96°19′W
97 O21 **Royston** E England, United Kingdom 52°05′N 00°01′W
23 U2 **Royston** Georgia, SE USA 34°17′N 83°06′W
114 L10 **Roza** *prev.* Gyulovo. Yambol, E Bulgaria 42°29′N 26°30′E
113 L16 **Rožaje** E Montenegro 42°51′N 20°11′E
110 M10 **Różan** Mazowieckie, C Poland 52°36′N 21°27′E
117 O10 **Rozdil'na** Odes'ka Oblast', SW Ukraine 46°51′N 30°03′E
117 S12 **Rozdol'ne** *Rus.* Razdolnoye. Avtonomna Respublika Krym, S Ukraine 45°45′N 33°27′E
Rozhdestvenka *see* Kabanbay Batyr
116 I6 **Rozhnyativ** Ivano-Frankivs'ka Oblast', W Ukraine 48°58′N 24°07′E
116 J3 **Rozhyshche** Volyns'ka Oblast', NW Ukraine 50°54′N 25°16′E
Roznau am Radhost *see* Rožnov pod Radhoštěm
111 L19 **Rožňava** *Ger.* Rosenau, *Hung.* Rozsnyó. Košický Kraj, E Slovakia 48°41′N 20°32′E
116 K10 **Roznov** Neamţ, NE Romania 46°47′N 26°33′E
111 I18 **Rožnov pod Radhoštěm** *Ger.* Rosenau, Roznau am Radhost. Zlínský Kraj, E Czech Republic 49°28′N 18°09′E
Rózsahegy *see* Ružomberok
Rozsnyó *see* Râşnov, Romania
Rozsnyó *see* Rožňava, Slovakia
113 K18 **Rranxë** Shkodër, NW Albania 41°58′N 19°27′E
113 L18 **Rrëshen** *var.* Rresheni, Rrshen. Lezhë, C Albania 41°46′N 19°54′E
Rresheni *see* Rrëshen
Rrogozhina *see* Rrogozhinë
113 K20 **Rrogozhinë** *var.* Rogozhina, Rogozhinë, Rrogozhina. Tiranë, W Albania 41°04′N 19°40′E
Rrshen *see* Rrëshen
112 O13 **Rtanj** ▲ E Serbia 43°45′N 21°54′E
127 O7 **Rtishchevo** Saratovskaya Oblast', W Russian Federation 52°16′N 43°46′E
184 N12 **Ruahine Range** *var.* Ruarine. ▲▲ North Island, New Zealand
185 L14 **Ruamahanga** ≈ North Island, New Zealand
Ruanda *see* Rwanda
184 M10 **Ruapehu, Mount** 🌋 North Island, New Zealand 39°15′S 175°33′E
185 C25 **Ruapuke Island** *island* SW New Zealand
Ruarine *see* Ruahine Range
184 O9 **Ruatahuna** Bay of Plenty, North Island, New Zealand 38°38′S 176°56′E
184 Q8 **Ruatoria** Gisborne, North Island, New Zealand 37°54′S 178°18′E
184 K4 **Ruawai** Northland, North Island, New Zealand 36°08′S 174°04′E
15 N8 **Ruban** ≈ Québec, SE Canada
81 I22 **Rubeho Mountains** ▲▲ C Tanzania
165 U3 **Rubeshibe** Hokkaidō, NE Japan 43°49′N 143°37′E
Rubezhnoye *see* Rubizhne
113 L18 **Rubik** Lezhë, C Albania 41°46′N 19°48′E
54 H7 **Rubio** Táchira, W Venezuela 07°42′N 72°23′W
117 X6 **Rubizhne** *Rus.* Rubezhnoye. Luhans'ka Oblast', E Ukraine 49°01′N 38°22′E
81 F20 **Rubondo Island** *island* N Tanzania
122 I13 **Rubtsovsk** Altayskiy Kray, S Russian Federation 51°34′N 81°11′E
39 P9 **Ruby** Alaska, USA 64°44′N 155°29′W
35 W3 **Ruby Dome** ▲ Nevada, W USA 40°35′N 115°25′W
35 W4 **Ruby Lake** ◎ Nevada, W USA
35 W4 **Ruby Mountains** ▲▲ Nevada, W USA
33 Q12 **Ruby Range** ▲▲ Montana, NW USA
118 C10 **Rucava** SW Latvia 56°09′N 21°10′E
143 S13 **Rūdān** *var.* Dehbārez. Hormozgān, S Iran 27°30′N 57°10′E
Rudelstadt *see* Ciechanowiec
Rudensk *see* Rudzyensk
119 G14 **Rūdiškės** Vilnius, S Lithuania 54°31′N 24°49′E
95 H24 **Rudkøbing** Syddtjylland, C Denmark 54°57′N 10°43′E
125 S13 **Rudnichnyy** Kirovskaya Oblast', NW Russian Federation 59°37′N 52°28′E
Rūdnīchnyy *see* Koksu
Rudny *see* Rudnyy
126 H4 **Rudnya** Smolenskaya Oblast', W Russian Federation 54°55′N 31°10′E
127 O8 **Rudnya** Volgogradskaya Oblast', SW Russian Federation 50°54′N 44°27′E
144 M7 **Rudnyy** *var.* Rudny. Kostanay, N Kazakhstan 53°N 63°05′E
122 K3 **Rudol'fa, Ostrov** *island* Zemlya Frantsa-Iosifa, NW Russian Federation
Rudolf, Lake *see* Turkana, Lake
Rudolfswert *see* Novo mesto
101 L17 **Rudolstadt** Thüringen, C Germany 50°44′N 11°20′E
31 Q4 **Rudyard** Michigan, N USA 46°15′N 84°36′W
33 S7 **Rudyard** Montana, NW USA 48°33′N 110°37′W
119 K16 **Rudzyensk** *Rus.* Rudensk. Minskaya Voblasts', C Belarus 53°36′N 27°52′E
104 L6 **Rueda** Castilla y León, N Spain 41°24′N 04°58′W
114 F10 **Ruen** ▲ Bulgaria/FYR Macedonia 42°10′N 22°31′E
80 G10 **Rufa'a** Gezira, C Sudan 14°49′N 33°21′E
102 L10 **Ruffec** Charente, W France 46°01′N 00°10′E
21 R14 **Ruffin** South Carolina, SE USA 33°00′N 80°48′W
81 J23 **Rufiji** ≈ E Tanzania
61 A20 **Rufino** Santa Fe, C Argentina 34°16′S 62°45′W
76 F11 **Rufisque** W Senegal 14°44′N 17°18′W
83 K14 **Rufunsa** Lusaka, C Zambia 15°02′S 29°35′E
118 J9 **Rugāji** E Latvia 57°01′N 27°07′E
161 R7 **Rugao** Jiangsu, E China 32°27′N 120°35′E
97 M20 **Rugby** C England, United Kingdom 52°22′N 01°18′W
29 N3 **Rugby** North Dakota, N USA 48°24′N 100°00′W
100 N7 **Rügen** *headland* NE Germany 54°25′N 13°21′E
Ruhaybeh *see* Ar Ruḩaybah
81 E19 **Ruhengeri** NW Rwanda 01°39′S 29°16′E
Ruhja *see* Rūjiena
100 M10 **Ruhner Berg** *hill* N Germany
118 F7 **Ruhnu** *var.* Ruhnu Saar, *Swe.* Runö. *island* SW Estonia
Ruhnu Saar *see* Ruhnu
101 G15 **Ruhr** ≈ W Germany
91 W6 **Ruhr Valley** *industrial region* W Germany
161 S11 **Rui'an** *var.* Rui an. Zhejiang, SE China 27°51′N 120°39′E
Rui an *see* Rui'an
161 P10 **Ruichang** Jiangxi, S China 29°46′N 115°37′E
24 J11 **Ruidosa** Texas, SW USA 30°00′N 104°40′W
37 S14 **Ruidoso** New Mexico, SW USA 33°19′N 105°40′W
161 P12 **Ruijin** Jiangxi, S China 25°52′N 116°01′E
160 D13 **Ruili** Yunnan, SW China 24°04′N 97°49′E
98 N8 **Ruinen** Drenthe, NE Netherlands 52°46′N 06°19′E
99 D17 **Ruiselede** West-Vlaanderen, W Belgium 51°03′N 03°21′E
161 T13 **Ruisui** *prev.* Juisui. C Taiwan 23°43′N 121°28′E
64 P5 **Ruivo de Santana, Pico** ▲ Madeira, Portugal, NE Atlantic Ocean 32°46′N 16°57′W
40 J12 **Ruiz** Nayarit, SW Mexico 22°00′N 105°09′W
54 E10 **Ruiz, Nevado del** 🌋 W Colombia 04°53′N 75°22′W
138 J9 **Rujaylah, Ḩarrat ar** *salt lake* N Jordan
Rujen *see* Rūjiena
118 H7 **Rūjiena** *Est.* Ruhja, *Ger.* Rujen. N Latvia 57°54′N 25°22′E
79 I18 **Ruki** ≈ W Dem. Rep. Congo
81 E22 **Rukwa** ◈ *region* SW Tanzania
81 F23 **Rukwa, Lake** ◎ SE Tanzania
25 P6 **Rule** Texas, SW USA 33°10′N 99°53′W
22 K3 **Ruleville** Mississippi, S USA 33°43′N 90°33′W
Rum *see* Rhum
112 K10 **Ruma** Vojvodina, N Serbia 45°02′N 19°51′E
Rumadiya *see* Ar Ramādī
141 Q7 **Rumāḩ** Ar Riyāḑ, C Saudi Arabia 25°35′N 47°09′E
Rumaitha *see* Ar Rumaythah
Rumania/Rumänien *see* Romania
Rumänisch-Sankt-Georgen *see* Sângeorz-Băi
139 Y13 **Rumaylah** Al Başrah, SE Iraq 30°16′N 47°22′E
139 P2 **Rumaylah, Wādī** *dry watercourse* NE Syria
171 U13 **Rumbati** Papua Barat, E Indonesia 02°44′S 132°04′E
81 E14 **Rumbek** Lakes, C South Sudan 06°50′N 29°42′E
Rumburg *see* Rumburk
111 D14 **Rumburk** *Ger.* Rumburg. Ústecký Kraj, NW Czech Republic 50°58′N 14°35′E
44 J4 **Rum Cay** *island* C The Bahamas
99 M26 **Rumelange** Luxembourg, S Luxembourg 49°28′N 06°02′E
99 D20 **Rumes** Hainaut, SW Belgium 50°33′N 03°19′E
19 P7 **Rumford** Maine, NE USA 44°31′N 70°31′W
110 I6 **Rumia** Pomorskie, N Poland 54°36′N 18°21′E
113 J17 **Rumija** ▲▲ S Montenegro
103 T11 **Rumilly** Haute-Savoie, E France 45°52′N 05°57′E
139 O6 **Rūmiyah** Al Anbār, W Iraq 34°28′N 41°17′E
Rummah, Wādī ar *see* Rimah, Wādī ar
Rummelsburg in Pommern *see* Miastko
165 S3 **Rumoi** Hokkaidō, NE Japan 43°57′N 141°40′E
82 M12 **Rumphi** *var.* Rumpi. Northern, N Malawi 11°00′S 33°51′E
Rumpi *see* Rumphi
29 V7 **Rum River** ≈ Minnesota, N USA
188 F16 **Rumung** *island* Caroline Islands, W Micronesia
Rumuniya/Rumŭniya/Rumunjska *see* Romania
185 G16 **Runanga** West Coast, South Island, New Zealand 42°25′S 171°15′E
184 P7 **Runaway, Cape** *headland* North Island, New Zealand 37°33′S 177°59′E
97 K18 **Runcorn** C England, United Kingdom 53°20′N 02°44′W
118 K10 **Rundāni** *var.* Rundāni. E Latvia 56°19′N 27°51′E
Rundāni *see* Rundāni
83 L18 **Runde** *var.* Lundi. ≈ SE Zimbabwe
83 E16 **Rundu** *var.* Runtu. Okavango, NE Namibia 17°55′S 19°45′E
93 I16 **Rundvik** Västerbotten, N Sweden 63°31′N 19°22′E
81 G20 **Runere** Mwanza, N Tanzania 03°06′S 33°18′E
25 S13 **Runge** Texas, SW USA 28°52′N 97°42′W
167 Q13 **Rŭng, Kaôh** *prev.* Kas Rong. *island* SW Cambodia
79 O16 **Rungu** Orientale, NE Dem. Rep. Congo 03°11′N 27°52′E
81 F23 **Rungwa** Katavi, W Tanzania 07°18′S 31°40′E
81 G22 **Rungwa** Singida, C Tanzania 06°54′S 33°33′E
94 M14 **Runn** ◎ C Sweden
24 M4 **Running Water Draw** *valley* New Mexico/Texas, SW USA
Runö *see* Ruhnu
Runtu *see* Rundu
189 V12 **Ruo** *island* Caroline Islands, C Micronesia
158 L9 **Ruoqiang** *var.* Jo-ch'iang, *Uigh.* Charkhlik, Charkhliq, Qarkilik. Xinjiang Uygur Zizhiqu, NW China 38°59′N 88°08′E
159 S7 **Ruo Shui** ≈ N China
92 L8 **Ruostekfielbmá** *var.* Rustefjelbma Finnmark. Finnmark, N Norway 70°25′N 28°10′E
93 L18 **Ruovesi** Pirkanmaa, W Finland 61°59′N 24°05′E
112 B9 **Rupa** Primorje-Gorski Kotar, NW Croatia 45°29′N 14°15′E
182 M11 **Rupanyup** Victoria, SE Australia 36°38′S 142°37′E
168 K9 **Rupat, Pulau** *prev.* Roepat. *island* W Indonesia
168 K10 **Rupat, Selat** *strait* Sumatera, W Indonesia
116 J11 **Rupea** *Ger.* Reps, *Hung.* Kőhalom; *prev.* Cohalm. Braşov, C Romania 46°02′N 25°13′E
99 G17 **Rupel** ≈ N Belgium
Rupella *see* la Rochelle
33 P15 **Rupert** Idaho, NW USA 42°37′N 113°40′W
21 R5 **Rupert** West Virginia, NE USA 37°57′N 80°40′W
Rupert House *see* Waskaganish
12 J10 **Rupert, Rivière de** ≈ Québec, C Canada
152 I8 **Rūpnagar** *var.* Ropar. Punjab, India
194 M13 **Ruppert Coast** *physical region* Antarctica
100 N11 **Ruppiner Kanal** *canal* NE Germany
55 S11 **Rupununi River** ≈ S Guyana
101 D16 **Rur** *Dut.* Roer. ≈ Germany/Netherlands
58 H13 **Rurópolis Presidente Medici** Pará, N Brazil 04°05′S 55°26′W
191 S12 **Rurutu** *island* Îles Australes, SW French Polynesia
Rusaddir *see* Melilla
83 L17 **Rusape** Manicaland, E Zimbabwe 18°32′S 32°07′E
Rusayris, Lake *see* Roseires, Reservoir
Ruschuk/Rusçuk *see* Ruse
114 K7 **Ruse** *var.* Ruschuk, Rustchuk, *Turk.* Rusçuk. Ruse, N Bulgaria 43°50′N 25°59′E
109 W10 **Ruše** NE Slovenia 46°31′N 15°30′E
114 L7 **Ruse** ◈ *province* N Bulgaria
114 K7 **Rusenski Lom** ≈ N Bulgaria
97 G17 **Rush** *Ir.* An Ros. Dublin, E Ireland 53°32′N 06°06′W
161 S4 **Rushan** *var.* Xiacun. Shandong, E China 36°55′N 121°26′E
Rushan *see* Rŭshon
Rushanskiy Khrebet *see* Rushon, Qatorkŭhi
29 V7 **Rush City** Minnesota, N USA 45°41′N 92°56′W
37 V5 **Rush Creek** ≈ Colorado, C USA
29 X10 **Rushford** Minnesota, N USA 43°48′N 91°45′W
154 N13 **Rushikulya** ≈ E India
14 D8 **Rush Lake** ◎ Ontario, S Canada
30 M7 **Rush Lake** ◎ Wisconsin, N USA
28 J10 **Rushmore, Mount** ▲ South Dakota, N USA 43°52′N 103°27′W
147 S13 **Rŭshon** *Rus.* Rushan. S Tajikistan 37°58′N 71°31′E
147 S14 **Rushon, Qatorkŭhi** *Rus.* Rushanskiy Khrebet. ▲▲ SE Tajikistan
26 M12 **Rush Springs** Oklahoma, C USA 34°46′N 97°57′W
45 V15 **Rushville** Trinidad, Trinidad and Tobago 10°07′N 61°03′W
30 J13 **Rushville** Illinois, N USA 40°07′N 90°33′W
28 K12 **Rushville** Nebraska, C USA 42°41′N 102°28′W
183 O11 **Rushworth** Victoria, SE Australia 36°36′S 145°03′E
25 W8 **Rusk** Texas, SW USA 31°49′N 95°11′W
93 I14 **Rusksele** Västerbotten, N Sweden 64°49′N 18°55′E
118 C12 **Rusnė** Klaipėda, W Lithuania 55°18′N 21°19′E
114 M10 **Rusokastrenska Reka** ≈ E Bulgaria
Russadir *see* Melilla
109 X3 **Russbach** ≈ NE Austria
11 V16 **Russell** Manitoba, S Canada 50°47′N 101°17′W
184 K2 **Russell** Northland, North Island, New Zealand 35°17′S 174°07′E
26 L4 **Russell** Kansas, C USA 38°54′N 98°51′W
21 O4 **Russell** Kentucky, S USA 38°30′N 82°43′W
20 L7 **Russell Springs** Kentucky, S USA 37°02′N 85°03′W
23 O2 **Russellville** Alabama, S USA 34°30′N 87°43′W
27 T11 **Russellville** Arkansas, C USA 35°17′N 93°06′W
20 J7 **Russellville** Kentucky, S USA 36°50′N 86°54′W
101 G18 **Rüsselsheim** Hessen, W Germany 50°00′N 08°25′E
Russia *see* Russian Federation
Russian America *see* Alaska
122 J11 **Russian Federation** *off.* Russian Federation, *var.* Russia, *Latv.* Krievija, *Rus.* Rossiyskaya Federatsiya. ◆ *republic* Asia/Europe
Russian Federation *see* Russian Federation
39 N11 **Russian Mission** Alaska, USA 61°48′N 161°23′W
34 M7 **Russian River** ≈ California, W USA
122 J5 **Russkaya Gavan'** Novaya Zemlya, Arkhangel'skaya Oblast', N Russian Federation 76°13′N 62°48′E
122 J5 **Russkiy, Ostrov** *island* N Russian Federation
109 Y5 **Rust** Burgenland, E Austria 47°48′N 16°42′E
Rustaq *see* Ar Rustāq
137 U10 **Rustavi** *prev.* Rust'avi. SE Georgia 41°36′N 45°00′E
Rust'avi *see* Rustavi
21 T7 **Rustburg** Virginia, NE USA 37°17′N 79°07′W
Rustchuk *see* Ruse
Rustefjelbma Finnmark *see* Ruostekfielbmá
83 I21 **Rustenburg** North-West, N South Africa 25°40′S 27°15′E
22 H5 **Ruston** Louisiana, S USA 32°31′N 92°38′W
81 E21 **Rutana** SE Burundi 04°01′S 30°01′E
62 I4 **Rutana, Volcán** 🌋 N Chile 22°43′S 67°52′W
Rutanzige, Lake *see* Edward, Lake
Rutba *see* Ar Ruţbah
104 M14 **Rute** Andalucía, S Spain 37°20′N 04°23′W
171 N16 **Ruteng** *prev.* Roeteng. Flores, C Indonesia 08°35′S 120°28′E
194 L8 **Rutford Ice Stream** *ice feature* Antarctica
35 X6 **Ruth** Nevada, W USA 39°15′N 115°00′W
101 G15 **Rüthen** Nordrhein-Westfalen, W Germany 51°30′N 08°28′E
14 D17 **Rutherford** Ontario, S Canada 42°39′N 82°06′W
21 Q10 **Rutherfordton** North Carolina, SE USA 35°23′N 81°57′W
97 J18 **Ruthin** *Wel.* Rhuthun. NE Wales, United Kingdom 53°05′N 03°18′W
108 G7 **Rüti** Zürich, N Switzerland 47°16′N 08°51′E
Rutlam *see* Ratlām
18 M9 **Rutland** Vermont, NE USA 43°37′N 72°59′W
97 N19 **Rutland** *cultural region* C England, United Kingdom
21 N8 **Rutledge** Tennessee, S USA 36°16′N 83°31′W
158 G12 **Rutog** *var.* Rutög, Rutok. Xizang Zizhiqu, W China 33°27′N 79°43′E
Rutok *see* Rutog
79 P19 **Rutshuru** Nord-Kivu, E Dem. Rep. Congo 01°11′S 29°28′E
98 L8 **Rutten** Flevoland, N Netherlands 52°49′N 05°44′E
127 Q17 **Rutul** Respublika Dagestan, SW Russian Federation 41°35′N 47°30′E
93 L14 **Ruukki** Pohjois-Pohjanmaa, C Finland 64°40′N 25°35′E
98 N11 **Ruurlo** Gelderland, E Netherlands 52°04′N 06°27′E
143 S15 **Ru'ūs al Jibāl** *cape* Oman/United Arab Emirates
138 I7 **Ru'ūs aţ Ţiwāl, Jabal** ▲▲ W Syria
81 H23 **Ruvuma** ◈ *region* SE Tanzania

◆ Country
● Country Capital
◇ Dependent Territory
○ Dependent Territory Capital
◈ Administrative Regions
✈ International Airport
▲ Mountain
▲▲ Mountain Range
🌋 Volcano
≈ River
◎ Lake
⊠ Reservoir

81 I25 **Ruvuma** *var.* Rio Rovuma. ≈ Mozambique/Tanzania *see also* Rovuma, Rio
Ruvuma *see* Rovuma, Rio
Ruwais *see* Ar Ruways
138 L9 **Ruwayshid, Wadi ar** *dry watercourse* NE Jordan
141 Z10 **Ruways, Ra's ar** *headland* E Oman 20°58′N 59°00′E
79 P18 **Ruwenzori** ⛰ Dem. Rep. Congo/Uganda
141 Y8 **Ruwī** NE Oman 23°33′N 58°31′E
114 F9 **Ruy** ▲ Bulgaria/Serbia 42°52′N 22°35′E
Ruya *see* Luia, Rio
81 E20 **Ruyigi** E Burundi 03°28′S 30°19′E
127 P5 **Ruzayevka** Respublika Mordoviya, W Russian Federation 54°04′N 44°56′E
119 G18 **Ruzhany** Brestskaya Voblasts', SW Belarus 52°52′N 24°53′E
Rŭzhevo Konare *see* Razhevo Konare
Ruzhin *see* Ruzhyn
114 G7 **Ruzhintsi** Vidin, NW Bulgaria 43°38′N 22°50′E
161 N6 **Ruzhou** Henan, C China 34°10′N 112°51′E
117 N5 **Ruzhyn** *Rus.* Ruzhin. Zhytomyrs'ka Oblast', N Ukraine 49°42′N 29°01′E
111 K19 **Ružomberok** *Ger.* Rosenberg, *Hung.* Rózsahegy. Zílinský Kraj, N Slovakia 49°04′N 19°19′E
111 C16 **Ruzyně** ✈ (Praha) Praha, C Czech Republic
81 D19 **Rwanda** *off.* Rwandese Republic; *prev.* Ruanda. ◆ *republic* C Africa
Rwandese Republic *see* Rwanda
95 G22 **Ry** Midtjylland, C Denmark 56°06′N 09°46′E
Ryasna *see* Rasna
126 L5 **Ryazan'** Ryazanskaya Oblast', W Russian Federation 54°37′N 39°37′E
126 L5 **Ryazanskaya Oblast'** ◐ *province* W Russian Federation
126 M6 **Ryazhsk** Ryazanskaya Oblast', W Russian Federation 53°42′N 40°09′E
118 B13 **Rybachiy** *Ger.* Rossitten. Kaliningradskaya Oblast', W Russian Federation 55°09′N 20°49′E
124 J2 **Rybachiy, Poluostrov** *peninsula* NW Russian Federation
Rybach'ye *see* Balykchy
124 L15 **Rybinsk** *prev.* Andropov. Yaroslavskaya Oblast', W Russian Federation 58°03′N 38°53′E
124 K14 **Rybinskoye Vodokhranilishche** *Eng.* Rybinsk Reservoir, Rybinsk Sea. ⊠ W Russian Federation
Rybinsk Reservoir/Rybinsk Sea *see* Rybinskoye Vodokhranilishche
111 I16 **Rybnik** Śląskie, S Poland 50°05′N 18°31′E
Rybnitsa *see* Rîbnița
111 F16 **Rychnov nad Kněžnou** *Ger.* Reichenau. Královéhradecký Kraj, N Czech Republic 50°10′N 16°17′E
110 I12 **Rychwał** Wielkopolskie, C Poland 52°04′N 18°10′E
11 O13 **Rycroft** Alberta, W Canada 55°45′N 118°42′W
95 L21 **Ryd** Kronoberg, S Sweden 56°27′N 14°44′E
95 L20 **Rydaholm** Jönköping, S Sweden 56°57′N 14°19′E
194 I8 **Rydberg Peninsula** *peninsula* Antarctica
97 P23 **Rye** SE England, United Kingdom 50°57′N 00°42′E
33 T10 **Ryegate** Montana, NW USA 46°21′N 109°12′W
35 S3 **Rye Patch Reservoir** ⊠ Nevada, W USA
95 D15 **Ryfylke** *physical region* S Norway
95 H16 **Rygge** Østfold, S Norway 59°22′N 10°45′E
110 N13 **Ryki** Lubelskie, E Poland 51°38′N 21°57′E
Rykovo *see* Yenakiyeve
126 I7 **Ryl'sk** Kurskaya Oblast', W Russian Federation 51°34′N 34°41′E
183 S8 **Rylstone** New South Wales, SE Australia 32°48′S 149°58′E
111 H17 **Rýmařov** *Ger.* Römerstadt. Moravskoslezský Kraj, E Czech Republic 49°56′N 17°15′E
144 E11 **Ryn-Peski** *desert* W Kazakhstan
165 N10 **Ryōtsu** *var.* Ryôtu. Niigata, Sado, C Japan 38°06′N 138°28′E
Ryôtu *see* Ryōtsu
110 K10 **Rypin** Kujawsko-pomorskie, C Poland 53°03′N 19°25′E
Ryshkany *see* Rîşcani
Ryssel *see* Lille
Ryswick *see* Rijswijk
95 M24 **Rytterknægten** *hill* E Denmark
Ryukyu Islands *see* Nansei-shotō
192 G5 **Ryukyu Trench** *var.* Nansei Syotō Trench. *undersea feature* S East China Sea 24°45′N 128°00′E
110 D11 **Rzepin** *Ger.* Reppen. Lubuskie, W Poland 52°20′N 14°48′E
111 N16 **Rzeszów** Podkarpackie, SE Poland 50°03′N 22°01′E
124 I16 **Rzhev** Tverskaya Oblast', W Russian Federation 56°17′N 34°22′E
Rzhishchev *see* Rzhyshchiv
117 P5 **Rzhyshchiv** *Rus.* Rzhishchev. Kyyivs'ka Oblast', N Ukraine 49°58′N 31°02′E

S

138 E11 **Sa'ad** Southern, W Israel 31°27′N 34°31′E
109 P7 **Saalach** ≈ W Austria
101 L14 **Saale** ≈ C Germany
101 L17 **Saalfeld** *var.* Saalfeld an der Saale. Thüringen, C Germany 50°39′N 11°22′E
Saalfeld *see* Zalewo
Saalfeld an der Saale *see* Saalfeld
108 C8 **Saane** ≈ W Switzerland
101 D19 **Saar** *Fr.* Sarre. ≈ France/Germany
101 E20 **Saarbrücken** *Fr.* Sarrebruck. Saarland, SW Germany 49°13′N 07°01′E
Saarburg *see* Sarrebourg
118 D6 **Sääre** *var.* Sjar. Saaremaa, W Estonia 57°57′N 21°53′E
Saare *see* Saaremaa
118 D5 **Saaremaa** *off.* Saare Maakond. ◐ *province* W Estonia
118 E6 **Saaremaa** *Ger.* Oesel, Ösel; *prev.* Saare. *island* W Estonia
Saare Maakond *see* Saaremaa
92 L12 **Saarenkylä** Lappi, N Finland 66°31′N 25°51′E
Saargemund *see* Sarreguemines
93 L17 **Saarijärvi** Keski-Suomi, C Finland 62°42′N 25°16′E
Saar in Mähren *see* Žd'ár nad Sázavou
92 M10 **Saariselkä** *Lapp.* Suoločielgi. Lappi, N Finland 68°27′N 27°29′E
92 L10 **Saariselkä** *hill range* NE Finland
101 D20 **Saarland** *Fr.* Sarre. ◐ *state* SW Germany
Saarlautern *see* Saarlouis
101 D20 **Saarlouis** *prev.* Saarlautern. Saarland, SW Germany 49°19′N 06°45′E
108 E11 **Saaser Vispa** ≈ S Switzerland
137 X12 **Saatlı** *Rus.* Saatly. C Azerbaijan 39°57′N 48°24′E
Saatly *see* Saatlı
Saaz *see* Žatec
45 V9 **Saba** ◇ *Dutch special municipality* Sint Maarten
138 J7 **Sab' Ābār** *var.* Sab'a Biyar, Sa'b Bi'ār. Ḥimṣ, C Syria 33°46′N 37°41′E
Sab'a Biyar *see* Sab' Ābār
112 K11 **Šabac** Serbia, W Serbia 44°45′N 19°42′E
105 W5 **Sabadell** Cataluña, E Spain 41°33′N 02°07′E
164 K12 **Sabae** Fukui, Honshū, SW Japan 36°00′N 136°12′E
169 V7 **Sabah** *prev.* British North Borneo, North Borneo. ◐ *state* East Malaysia
168 J8 **Sabak** *var.* Sabak Bernam. Selangor, Peninsular Malaysia 03°45′N 100°59′E
Sabak Bernam *see* Sabak
38 D16 **Sabak, Cape** *headland* Agattu Island, Alaska, USA 52°21′N 173°43′E
81 J20 **Sabaki** ≈ S Kenya
142 L2 **Sabalān, Kuhhā-ye** ▲ NW Iran 38°21′N 47°47′E
154 H7 **Sabalgarh** Madhya Pradesh, C India 26°18′N 77°28′E
44 E4 **Sabana, Archipiélago de** *island group* C Cuba
42 H7 **Sabanagrande** *var.* Sabana Grande. Francisco Morazán, S Honduras 13°48′N 87°15′W
Sabana Grande *see* Sabanagrande
54 E5 **Sabanalarga** Atlántico, N Colombia 10°38′N 74°55′W
41 W14 **Sabancuy** Campeche, SE Mexico 18°58′N 91°11′W
45 N8 **Sabaneta** NW Dominican Republic 19°30′N 71°21′W
54 J4 **Sabaneta** Falcón, N Venezuela 11°17′N 70°00′W
188 H4 **Sabaneta, Puntan** *prev.* Ushi Point. *headland* Saipan, S Northern Mariana Islands 15°17′N 145°49′E
171 X14 **Sabang** Papua, E Indonesia 04°33′S 138°42′E
116 L10 **Săbăoani** Neamţ, NE Romania 47°01′N 26°51′E
155 J26 **Sabaragamuwa** ◐ *province* C Sri Lanka
Sabaria *see* Szombathely
154 D10 **Sābarmati** ≈ NW India
171 S10 **Sabatai** Pulau Morotai, E Indonesia 02°04′N 128°23′E
141 Q15 **Sab'atayn, Ramlat as** *desert* C Yemen
107 I16 **Sabaudia** Lazio, C Italy 41°17′N 13°02′E
57 J19 **Sabaya** Oruro, S Bolivia 19°09′S 68°21′W
Sa'b Bi'ār *see* Sab' Ābār
Sabbioncello *see* Orebić
148 I8 **Şāberī, Hāmūn-e** *var.* Daryācheh-ye Hāmun, Daryācheh-ye Sīstān. ◎ Afghanistan/Iran *see also* Sīstān, Daryācheh-ye
Şāberī, Hāmūn-e *see* Sīstān, Daryācheh-ye
27 P2 **Sabetha** Kansas, C USA 39°54′N 95°48′W
75 P10 **Sabhā** C Libya 27°02′N 14°26′E
67 V13 **Sabi** *var.* Save. ≈ Mozambique/Zimbabwe *see also* Save
Sabi *see* Save
118 E8 **Sabile** *Ger.* Zabeln. NW Latvia 57°03′N 22°33′E
31 R14 **Sabina** Ohio, N USA 39°29′N 83°38′W
40 I3 **Sabinal** Chihuahua, N Mexico 30°59′N 107°29′W
25 Q12 **Sabinal** Texas, SW USA 29°19′N 99°28′W
25 Q11 **Sabinal River** ≈ Texas, SW USA
105 S4 **Sabiñánigo** Aragón, NE Spain 42°31′N 00°22′W
41 N6 **Sabinas** Coahuila, NE Mexico 27°52′N 101°04′W
41 O8 **Sabinas Hidalgo** Nuevo León, NE Mexico 26°29′N 100°09′W
41 N6 **Sabinas, Río** ≈ NE Mexico
22 F9 **Sabine Lake** ◎ Louisiana/Texas, S USA
92 O3 **Sabine Land** *physical region* C Svalbard
25 W7 **Sabine River** ≈ Louisiana/Texas, SW USA
137 X12 **Sabirabad** C Azerbaijan 40°00′N 48°27′E
Sabkha *see* As Sabkhah
171 O4 **Sablayan** Mindoro, N Philippines 12°48′N 120°48′E
13 P16 **Sable, Cape** *headland* Newfoundland and Labrador, SE Canada 43°21′N 65°40′W
23 X17 **Sable, Cape** *headland* Florida, SE USA 25°12′N 81°06′W
13 R16 **Sable Island** *island* Nova Scotia, SE Canada
14 L11 **Sables, Lac des** ◎ Québec, SE Canada
14 E10 **Sables, Rivière aux** ≈ Ontario, S Canada
102 K7 **Sable-sur-Sarthe** Sarthe, NW France 47°49′N 00°19′W
125 U7 **Sablya, Gora** ▲ NW Russian Federation 64°46′N 58°52′E
77 U14 **Sabon Birnin Gwari** Kaduna, C Nigeria 10°43′N 06°39′E
77 V11 **Sabon Kafi** Zinder, C Niger 14°37′N 08°46′E
104 I6 **Sabor, Rio** ≈ N Portugal
14 J8 **Sabourin, Lac** ◎ Québec, SE Canada
137 Y10 **Şabran** *prev.* Däväçi. NE Azerbaijan 41°15′N 48°58′E
102 J14 **Sabres** Landes, SW France 44°07′N 00°46′W
195 X13 **Sabrina Coast** *physical region* Antarctica
140 M11 **Sabt al Ulāyā** 'Asīr, SW Saudi Arabia 19°33′N 41°58′E
104 I8 **Sabugal** Guarda, N Portugal 40°20′N 07°05′W
29 Z13 **Sabula** Iowa, C USA 42°04′N 90°10′W
141 N13 **Şabyā** Jīzān, SW Saudi Arabia 17°50′N 42°50′E
Sabzawar *see* Sabzevār
Sabzawaran *see* Jīroft
143 S4 **Sabzevār** *var.* Sabzawar. Khorāsān-e Razavī, NE Iran 36°13′N 57°38′E
Sabzvārān *see* Jīroft
Sacajawea Peak *see* Matterhorn
82 C9 **Sacandica** Uíge, NW Angola 06°01′S 15°57′E
42 A2 **Sacatepéquez** *off.* Departamento de Sacatepéquez. ◐ *department* S Guatemala
Sacatepéquez, Departamento de *see* Sacatepéquez
104 F11 **Sacavém** Lisboa, W Portugal 38°47′N 09°06′W
29 T13 **Sac City** Iowa, C USA 42°25′N 94°59′W
105 P8 **Sacedón** Castilla-La Mancha, C Spain 40°29′N 02°44′W
116 J12 **Săcele** *Ger.* Vierdörfer, *Hung.* Négyfalu; *prev. Ger.* Sieben Dörfer, *Hung.* Hétfalu. Braşov, C Romania 45°36′N 25°40′E
163 Y16 **Sacheon** *Jap.* Sansenhō; *prev.* Sach'ŏn, Samch'ŏnpŏ. S South Korea 34°55′N 128°07′E
12 C7 **Sachigo** ≈ Ontario, C Canada
12 C8 **Sachigo Lake** Ontario, C Canada 53°52′N 92°16′W
12 C8 **Sachigo Lake** ◎ Ontario, C Canada
Sach'ŏn *see* Sacheon
101 O15 **Sachsen** *Eng.* Saxony, *Fr.* Saxe. ◐ *state* E Germany
101 K14 **Sachsen-Anhalt** *Eng.* Saxony-Anhalt. ◐ *state* C Germany
109 R9 **Sachsenburg** Salzburg, S Austria 46°49′N 13°23′E
Sachsenfeld *see* Žalec
8 I5 **Sachs Harbour** *var.* Ikaahuk. Banks Island, Northwest Territories, N Canada 72°N 125°14′W
Sächsisch-Reen/Sächsisch-Regen *see* Reghin
18 H8 **Sackets Harbor** New York, NE USA 43°57′N 76°06′W
13 P14 **Sackville** New Brunswick, SE Canada 45°54′N 64°23′W
19 P9 **Saco** Maine, NE USA 43°32′N 70°25′W
19 P8 **Saco River** ≈ Maine/New Hampshire, NE USA
35 O7 **Sacramento** *state capital* California, W USA 38°35′N 121°30′W
37 T14 **Sacramento Mountains** ⛰ New Mexico, SW USA
35 N6 **Sacramento River** ≈ California, W USA
35 N5 **Sacramento Valley** *valley* California, W USA
36 I10 **Sacramento Wash** *valley* Arizona, SW USA
105 N15 **Sacratif, Cabo** *headland* S Spain 36°41′N 03°30′W
116 F9 **Săcueni** *prev.* Săcuieni, *Hung.* Székelyhid. Bihor, W Romania 47°20′N 22°05′E
Săcuieni *see* Săcueni
105 R4 **Sádaba** Aragón, NE Spain 42°15′N 01°16′W
Sá da Bandeira *see* Lubango
138 I6 **Şadad** Ḥimş, W Syria 34°19′N 36°52′E
141 O13 **Şa'dah** NW Yemen 16°59′N 43°45′E
167 O16 **Sadao** Songkhla, SW Thailand 06°39′N 100°30′E
142 L8 **Sadd-e Dez, Daryācheh-ye** ◎ W Iran
19 S3 **Saddleback Mountain** *hill* Maine, NE USA
19 P6 **Saddleback Mountain** ▲ Maine, NE USA 44°57′N 70°27′W
141 W13 **Sadḥ** S Oman 17°11′N 55°08′E
76 J11 **Sadiola** Kayes, W Mali 13°48′N 11°47′W
149 R12 **Sādiqābād** Punjab, E Pakistan 28°16′N 70°10′E
153 Y10 **Sadiya** Assam, NE India 27°49′N 95°38′E
139 W9 **Sa'dīyah, Hawr as** ◎ E Iraq
Sado *see* Sadoga-shima
165 N9 **Sadoga-shima** *var.* Sado. *island* C Japan
104 F12 **Sado, Rio** ≈ S Portugal
114 I8 **Sadovets** Pleven, N Bulgaria 43°19′N 24°21′E
114 J11 **Sadovo** Plovdiv, C Bulgaria 42°07′N 24°56′E
127 O11 **Sadovoye** Respublika Kalmykiya, SW Russian Federation 47°51′N 44°34′E
105 W9 **Sa Dragonera** *var.* Isla Dragonera. *island* Islas Baleares, Spain, W Mediterranean Sea
114 I10 **Saedinenie** *var.* Săedinenie, Sŭedinenie. Plovdiv, C Bulgaria 42°14′N 24°36′E
Săedinenie *see* Saedinenie
105 P9 **Saelices** Castilla-La Mancha, C Spain 39°55′N 02°49′W
Saena Julia *see* Siena
Saetabicula *see* Alzira
114 O12 **Safaalan** Tekirdağ, NW Turkey 41°26′N 28°07′E
Safad *see* Tsefat
Şafāqis *see* Sfax
143 P10 **Şafāshahr** *var.* Deh Bīd. Fārs, C Iran 30°50′N 53°50′E
192 I16 **Sāfata Bay** *bay* Upolu, Samoa, C Pacific Ocean
Safed *see* Tsefat
139 X11 **Şaffāf, Ḥawr aş** *marshy lake* S Iraq
95 J16 **Säffle** Värmland, C Sweden 59°08′N 12°55′E
37 N15 **Safford** Arizona, SW USA 32°46′N 109°41′W
74 E7 **Safi** W Morocco 32°19′N 09°14′W
126 I4 **Safonovo** Smolenskaya Oblast', W Russian Federation 55°05′N 33°12′E
136 H11 **Safranbolu** Karabük, NW Turkey 41°16′N 32°41′E
139 Y13 **Safwān** Al Başrah, SE Iraq 30°06′N 47°44′E
158 J16 **Saga** *var.* Gya'gya. Xizang Zizhiqu, W China 29°22′N 85°19′E
164 C14 **Saga** Saga, Kyūshū, SW Japan 33°14′N 130°16′E
164 C13 **Saga** *off.* Saga-ken. ◐ *prefecture* Kyūshū, SW Japan
165 P10 **Sagae** Yamagata, Honshū, C Japan 38°22′N 140°12′E
166 L3 **Sagaing** Sagaing, C Myanmar (Burma) 21°55′N 95°56′E
166 L5 **Sagaing** ◐ *region* N Myanmar (Burma)
Saga-ken *see* Saga
165 N13 **Sagamihara** Kanagawa, Honshū, S Japan 35°34′N 139°22′E
165 N14 **Sagami-nada** *inlet* SW Japan
Sagan *see* Żagań
29 Y3 **Saganaga Lake** ◎ Minnesota, N USA
155 F18 **Sāgar** Karnātaka, W India 14°09′N 75°02′E
154 I9 **Sāgar** *prev.* Saugor. Madhya Pradesh, C India 23°53′N 78°46′E
15 S8 **Sagard** Québec, SE Canada 48°01′N 70°03′W
Sagarmāthā *see* Everest, Mount
Sagebrush State *see* Nevada
143 V11 **Sāghand** Yazd, C Iran 32°33′N 55°12′E
19 N14 **Sag Harbor** Long Island, New York, NE USA 40°59′N 72°15′W
Saghez *see* Saqqez
31 R8 **Saginaw** Michigan, N USA 43°25′N 83°57′W
31 R8 **Saginaw Bay** *lake bay* Michigan, N USA
Sagiz *see* Sagyz
64 H6 **Saglek Bank** *undersea feature* W Labrador Sea
13 P5 **Saglek Bay** *bay* SW Labrador Sea
Saglouc/Sagluk *see* Salluit
103 X15 **Sagonne, Golfe de** *gulf* Corse, France, C Mediterranean Sea
105 P13 **Sagra** ▲ S Spain 37°59′N 02°33′W
104 F14 **Sagres** Faro, S Portugal 37°01′N 08°56′W
37 S7 **Saguache** Colorado, C USA 38°05′N 106°05′W
44 J7 **Sagua de Tánamo** Holguín, E Cuba 20°38′N 75°14′W
44 E5 **Sagua la Grande** Villa Clara, C Cuba 22°48′N 80°06′W
15 R7 **Saguenay** ≈ Québec, SE Canada
74 C9 **Saguia al Hamra** *var.* As Saqia al Hamra. ≈ N Western Sahara
105 S9 **Sagunto** *Cat.* Sagunt, *Ar.* Murviedro; *anc.* Saguntum. Valenciana, E Spain 39°40′N 00°17′W
Sagunt/Saguntum *see* Sagunto
144 H11 **Sagyz** *prev.* Sagiz. Atyrau, W Kazakhstan 48°12′N 54°56′E
138 H10 **Saḥāb** 'Ammān, NW Jordan 31°52′N 36°00′E
54 E6 **Sahagún** Córdoba, NW Colombia 08°58′N 75°30′W
104 L4 **Sahagún** Castilla y León, N Spain 42°23′N 05°02′W
141 X8 **Saḥam** N Oman 24°06′N 56°52′E
68 F9 **Sahara** *desert* Libya/Algeria
Sahara el Gharbîya *see* Şaḥrā' al Gharbīyah
75 X9 **Sahara el Sharqîya** *var.* Aş Şaḥrā' ash Sharqīyah, *Eng.* Arabian Desert, Eastern Desert. *desert* E Egypt
Saharan Atlas *see* Atlas Saharien
152 J9 **Sahāranpur** Uttar Pradesh, N India 29°58′N 77°33′E
64 L10 **Saharan Seamounts** *var.* Saharian Seamounts. *undersea feature* E Atlantic Ocean 25°00′N 20°00′W
Saharian Seamounts *see* Saharan Seamounts
153 Q13 **Saharsa** Bihār, NE India 25°54′N 86°36′E
67 O7 **Sahel** *physical region* C Africa
153 R14 **Sāhibganj** Jhārkhand, NE India 25°15′N 87°40′E
139 Q7 **Sahīlīyah** Al Anbār, C Iraq 33°43′N 42°42′E
138 H4 **Sāḥilīyah, Jibāl as** ⛰ NW Syria
114 M13 **Şahin** Tekirdağ, NW Turkey 41°01′N 26°51′E
149 U9 **Sāhīwāl** *prev.* Montgomery. Punjab, E Pakistan 30°40′N 73°05′E
149 U8 **Sāhīwāl** Punjab, E Pakistan 31°57′N 72°22′E
141 W11 **Saḥmah, Ramlat as** *desert* C Oman
75 U9 **Şaḥrā' al Gharbīyah** *var.* Sahara el Gharbîya, *Eng.* Western Desert. *desert* C Egypt
139 T13 **Şaḥrā' al Ḥijārah** *desert* S Iraq
40 H5 **Sahuaripa** Sonora, NW Mexico 29°02′N 109°14′W
36 M16 **Sahuarita** Arizona, SW USA 31°24′N 110°55′W
40 L13 **Sahuayo** *var.* Sahuayo de José Mariá Morelos; *prev.* Sahuayo de Díaz, Sahuayo de Porfirio Díaz. Michoacán, SW Mexico 20°05′N 102°42′W
Sahuayo de Díaz/Sahuayo de José Mariá Morelos/Sahuayo de Porfirio Díaz *see* Sahuayo
173 W8 **Sahul Shelf** *undersea feature* N Timor Sea
167 P17 **Sai Buri** Pattani, SW Thailand 06°42′N 101°37′E
74 I6 **Saïda** NW Algeria 34°50′N 00°10′E
138 G7 **Saïda** *var.* Şaydā, Sayida; *anc.* Sidon. W Lebanon 33°20′N 35°24′E
Sa'īdābād *see* Sīrjān
80 B13 **Sa'id Bundas** Western Bahr el Ghazal, W South Sudan 08°24′N 24°53′E
186 E7 **Saidor** Madang, N Papua New Guinea 05°38′S 146°28′E
153 S13 **Saidpur** *var.* Syedpur. Rajshahi, NW Bangladesh 25°48′N 89°E
149 U5 **Saīdu** *var.* Mingora, Mongora; *prev.* Mingāora. Khyber Pakhtunkhwa, N Pakistan 34°45′N 72°21′E
108 C7 **Saignelégier** Jura, NW Switzerland 47°18′N 07°03′E
164 H11 **Saigō** Shimane, Dōgo, SW Japan 36°12′N 133°18′E
Saigon *see* Hô Chi Minh
163 P11 **SaihanTal** *var.* Sonid Youqi. Nei Mongol Zizhiqu, N China 42°45′N 112°36′E
162 I12 **Saihan Toroi** Nei Mongol Zizhiqu, N China 41°44′N 100°29′E
Sai Hun *see* Syr Darya
92 M11 **Saija** Lappi, NE Finland 67°07′N 28°46′E
164 G14 **Saijō** Ehime, Shikoku, SW Japan 33°55′N 133°10′E
164 E15 **Saiki** Ōita, Kyūshū, SW Japan 32°57′N 131°52′E
93 N18 **Saimaa** ◎ SE Finland
93 N18 **Saimaa Canal** *Fin.* Saimaan Kanava, *Rus.* Saymenskiy Kanal. *canal* Finland/Russian Federation
Saimaan Kanava *see* Saimaa Canal
40 L10 **Saín Alto** Zacatecas, C Mexico 23°36′N 103°14′W
96 L12 **St Abb's Head** *headland* SE Scotland, United Kingdom 55°54′N 02°07′W
11 Y16 **St. Adolphe** Manitoba, S Canada 49°39′N 96°55′W
103 O15 **St-Affrique** Aveyron, S France 43°57′N 02°52′E
15 Q10 **St-Agapit** Québec, SE Canada 46°22′N 71°37′W
97 O21 **St Albans** *anc.* Verulamium. E England, United Kingdom 51°46′N 00°21′W
18 L6 **Saint Albans** Vermont, NE USA 44°49′N 73°07′W
21 Q5 **Saint Albans** West Virginia, NE USA 38°21′N 81°47′W
St. Alban's Head *see* St Aldhelm's Head
11 Q14 **St. Albert** Alberta, SW Canada 53°38′N 113°38′W
97 M24 **St Aldhelm's Head** *var.* St. Alban's Head. *headland* S England, United Kingdom 50°34′N 02°04′W
15 O11 **St-Alexis-des-Monts** Québec, SE Canada 46°30′N 73°08′W
103 P2 **St-Amand-les-Eaux** Nord, N France 50°27′N 03°25′E
103 O9 **St-Amand-Montrond** *var.* St-Amand-Mont-Rond. Cher, C France 46°43′N 02°29′E
173 P16 **St-André** NE Réunion
14 M12 **St-André-Avellin** Québec, SE Canada 45°45′N 75°04′W
Saint-André, Cap *see* Vilanandro, Tanjona
102 K12 **St-André-de-Cubzac** Gironde, SW France 45°01′N 00°26′W
96 K11 **St Andrews** E Scotland, United Kingdom 56°20′N 02°49′W
23 Q9 **Saint Andrews Bay** *bay* Florida, SE USA
23 W7 **Saint Andrew Sound** *sound* Georgia, SE USA
Saint Anna Trough *see* Svyataya Anna Trough
44 J11 **St. Ann's Bay** C Jamaica 18°26′N 77°12′W
13 T10 **St. Anthony** Newfoundland and Labrador, SE Canada 51°22′N 55°34′W
33 R13 **Saint Anthony** Idaho, NW USA 43°56′N 111°38′W
182 M11 **Saint Arnaud** Victoria, SE Australia 36°39′S 143°15′E
185 I15 **St.Arnaud Range** ⛰ South Island, New Zealand
13 R10 **St-Augustin** Québec, E Canada 51°13′N 58°39′W
23 X9 **Saint Augustine** Florida, SE USA 29°54′N 81°19′W
97 H24 **St Austell** SW England, United Kingdom 50°21′N 04°47′W
103 T4 **St-Avold** Moselle, NE France 49°06′N 06°43′E
103 N17 **St-Barthélemy** ⛰ S France
102 L17 **St-Béat** Haute-Garonne, S France 42°55′N 00°39′E
97 I15 **St Bees Head** *headland* NW England, United Kingdom 54°30′N 03°39′W
173 P16 **St-Benoit** E Réunion
103 T13 **St-Bonnet** Hautes-Alpes, SE France 44°41′N 06°04′E
St.Botolph's Town *see* Boston
97 G21 **St Brides Bay** *inlet* SW Wales, United Kingdom
102 H5 **St-Brieuc** Côtes d'Armor, NW France 48°31′N 02°45′W
102 H5 **St-Brieuc, Baie de** *bay* NW France
102 L7 **St-Calais** Sarthe, NW France 47°55′N 00°48′E
15 Q10 **St-Casimir** Québec, SE Canada 46°40′N 72°05′W
14 H16 **St. Catharines** Ontario, S Canada 43°10′N 79°15′W
45 S14 **St. Catherine, Mount** ▲ N Grenada 12°10′N 61°41′W
64 C11 **St Catherine Point** *headland* E Bermuda
23 X6 **Saint Catherines Island** *island* Georgia, SE USA
97 M24 **St Catherine's Point** *headland* S England, United Kingdom 50°34′N 01°17′W
103 N13 **St-Céré** Lot, S France 44°52′N 01°53′E
108 A10 **St. Cergue** Vaud, W Switzerland 46°25′N 06°10′E
103 R11 **St-Chamond** Loire, E France 45°29′N 04°32′E
33 S16 **Saint Charles** Idaho, NW USA 42°05′N 111°23′W
27 X4 **Saint Charles** Missouri, C USA 38°48′N 90°29′W
103 P13 **St-Chély-d'Apcher** Lozère, S France 44°51′N 03°16′E
Saint Christopher and Nevis, Federation of *see* Saint Kitts and Nevis
Saint Christopher-Nevis *see* Saint Kitts and Nevis
31 S9 **Saint Clair** Michigan, N USA 42°49′N 82°29′W
183 O17 **St. Clair, Lake** ◎ Tasmania, SE Australia
14 C17 **St. Clair, Lake** *var.* Lac à L'Eau Claire. ◎ Canada/USA
31 S10 **Saint Clair Shores** Michigan, N USA 42°30′N 82°53′W
103 S10 **St-Claude** *anc.* Condate. Jura, E France 46°23′N 05°52′E
45 X6 **St-Claude** Basse Terre, SW Guadeloupe 16°02′N 61°42′W
23 X12 **Saint Cloud** Florida, SE USA 28°15′N 81°15′W
29 U8 **Saint Cloud** Minnesota, N USA 45°34′N 94°10′W
45 T9 **Saint Croix** *island* S Virgin Islands (US)
30 J4 **Saint Croix Flowage** ⊠ Wisconsin, N USA
19 T5 **Saint Croix River** ≈ Canada/USA
29 W7 **Saint Croix River** ≈ Minnesota/Wisconsin, N USA
45 S14 **St.David's** SE Grenada 12°01′N 61°40′W
97 H21 **St David's** SW Wales, United Kingdom 51°53′N 05°16′W
97 G21 **St David's Head** *headland* SW Wales, United Kingdom 51°54′N 05°19′W
64 C12 **St David's Island** *island* E Bermuda
173 O16 **St-Denis** ○ (Réunion) NW Réunion 20°55′S 14°34′E
103 U6 **St-Dié** Vosges, NE France 48°17′N 06°57′E
103 R5 **St-Dizier** *anc.* Desiderii Fanum. Haute-Marne, N France 48°39′N 05°00′E
15 N11 **St-Donat** Québec, SE Canada 46°19′N 74°15′W
15 N11 **Ste-Adèle** Québec, SE Canada 45°58′N 74°10′W
15 N11 **Ste-Agathe-des-Monts** Québec, SE Canada 46°03′N 74°19′W
11 Y16 **Ste. Anne** Manitoba, S Canada 49°40′N 96°40′W
45 R12 **Ste-Anne** Grande Terre, E Guadeloupe 16°13′N 61°23′W
45 Y6 **Ste-Anne** SE Martinique 14°26′N 60°53′W
15 Q10 **Ste-Anne** ≈ Québec, SE Canada
172 I16 **Sainte Anne** *island* Inner Islands, NE Seychelles
15 W6 **Ste-Anne-des-Monts** Québec, SE Canada 49°07′N 66°29′W
14 M10 **Ste-Anne-du-Lac** Québec, SE Canada 46°51′N 75°20′W
15 S10 **Ste-Apolline** Québec, SE Canada 46°47′N 70°15′W
15 R10 **Ste-Claire** Québec, SE Canada 46°30′N 70°40′W
15 Q10 **Ste-Croix** Québec, SE Canada 46°36′N 71°42′W
108 B8 **Ste. Croix** Vaud, SW Switzerland 46°50′N 06°31′E
103 P14 **Ste-Énimie** Lozère, S France 44°21′N 03°26′E
27 Y6 **Sainte Genevieve** Missouri, C USA 37°57′N 90°01′W
103 S12 **St-Egrève** Isère, E France 45°15′N 05°41′E
39 T12 **Saint Elias, Cape** *headland* Kayak Island, Alaska, USA 59°48′N 144°36′W
39 U11 **St Elias, Mount** ▲ Alaska, USA 60°18′N 140°57′W
10 G8 **Saint Elias Mountains** ⛰ Canada/USA
55 Y10 **St-Élie** N French Guiana 04°50′N 53°21′W
103 O10 **St-Eloy-les-Mines** Puy-de-Dôme, C France 46°07′N 02°50′E
15 R10 **Ste-Marie** Québec, SE Canada 46°28′N 71°00′W
45 Q11 **Ste-Marie** NE Martinique 14°47′N 61°00′W
173 P16 **Ste-Marie** NE Réunion
103 U6 **Ste-Marie-aux-Mines** Haut-Rhin, NE France 48°16′N 07°12′E
Sainte Marie, Cap *see* Vohimena, Tanjona
102 L8 **Ste-Maure-de-Touraine** Indre-et-Loire, C France 47°06′N 00°38′E
103 R4 **Ste-Menehould** Marne, NE France 49°06′N 04°54′E
Ste-Perpétue *see* Ste-Perpétue-de-l'Islet
15 S9 **Ste-Perpétue-de-l'Islet** *var.* Ste-Perpétue. Québec, SE Canada 47°02′N 69°58′W
45 X11 **Ste-Rose** Basse Terre, N Guadeloupe 16°20′N 61°42′W
173 P16 **Ste-Rose** E Réunion
11 W15 **Ste. Rose du Lac** Manitoba, S Canada 51°04′N 99°31′W
102 J11 **Saintes** *anc.* Mediolanum. Charente-Maritime, W France 45°45′N 00°37′W
45 X7 **Saintes, Canal des** *channel* SW Guadeloupe
Saintes, Îles des *see* les Saintes
173 P16 **Ste-Suzanne** N Réunion
15 P10 **Ste-Thècle** Québec, SE Canada 46°48′N 72°31′W
103 Q12 **St-Étienne** Loire, E France 45°26′N 04°23′E
102 M4 **St-Étienne-du-Rouvray** Seine-Maritime, N France 49°22′N 01°07′E
Saint Eustatius *see* Sint Eustatius
14 M11 **Ste-Véronique** Québec, SE Canada 46°30′N 74°58′W
15 P7 **St-Félicien** Québec, SE Canada 48°38′N 72°29′W
15 O11 **St-Félix-de-Valois** Québec, SE Canada 46°10′N 73°26′W
103 Y14 **St-Florent** Corse, France, C Mediterranean Sea 42°41′N 09°19′E
103 Y14 **St-Florent, Golfe de** *gulf* Corse, France, C Mediterranean Sea
103 P6 **St-Florentin** Yonne, C France 48°00′N 03°46′E
103 N9 **St-Florent-sur-Cher** Cher, C France 47°00′N 02°13′E
103 P12 **St-Flour** Cantal, C France 45°02′N 03°05′E
26 H2 **Saint Francis** Kansas, C USA 39°45′N 101°31′W
83 H26 **St. Francis, Cape** *headland* S South Africa 34°11′S 24°45′E
27 X10 **Saint Francis River** ≈ Arkansas/Missouri, C USA
22 J8 **Saint Francisville** Louisiana, S USA 30°46′N 91°22′W
45 Y6 **St-François** Grande Terre, E Guadeloupe 16°15′N 61°17′W
15 Q12 **St-François** ≈ Québec, SE Canada
27 X7 **Saint Francois Mountains** ⛰ Missouri, C USA
St-Gall/Saint Gall/St. Gallen *see* Sankt Gallen
St-Gall *see* St-Gall/Saint Gall/St.Gallen
102 L16 **St-Gaudens** Haute-Garonne, S France 43°07′N 00°43′E
15 R12 **St-Gédéon** Québec, SE Canada 45°51′N 70°36′W
181 X10 **Saint George** Queensland, E Australia 28°05′S 148°40′E
64 B12 **St George** N Bermuda 32°24′N 64°42′W
38 K15 **Saint George** Saint George Island, Alaska, USA 56°34′N 169°30′W
21 S14 **Saint George** South Carolina, SE USA 33°12′N 80°34′W
36 J8 **Saint George** Utah, W USA 37°06′N 113°35′W
13 R12 **St. George, Cape** *headland* Newfoundland and Labrador, E Canada 48°26′N 59°17′W
186 I6 **St. George, Cape** *headland* New Ireland, NE Papua New Guinea 04°49′S 152°52′E
38 J15 **Saint George Island** *island* Pribilof Islands, Alaska, USA
23 S10 **Saint George Island** *island* Florida, SE USA
99 J19 **Saint-Georges** Liège, E Belgium 50°36′N 05°20′E
15 R11 **St-Georges** Québec, SE Canada 46°08′N 70°40′W
55 Z11 **St-Georges** E French Guiana 03°55′N 51°49′W
45 R14 **St. George's** ● (Grenada) SW Grenada 12°04′N 61°45′W
13 R12 **St. George's Bay** *inlet* Newfoundland and Labrador, E Canada
97 G21 **Saint George's Channel** *channel* Ireland/Wales, United Kingdom
186 H6 **St. George's Channel** *channel* NE Papua New Guinea
64 B11 **St George's Island** *island* E Bermuda
99 I21 **Saint-Gérard** Namur, S Belgium 50°20′N 04°47′E
St-Germain *see* St-Germain-en-Laye
15 P12 **St-Germain-de-Grantham** Québec, SE Canada 45°49′N 72°32′W
103 N5 **St-Germain-en-Laye** *var.* St-Germain. Yvelines, N France 48°53′N 02°04′E
102 H8 **St-Gildas, Pointe du** *headland* NW France 47°09′N 02°25′W
103 R15 **St-Gilles** Gard, S France 43°41′N 04°24′E
102 I9 **St-Gilles-Croix-de-Vie** Vendée, NW France 46°41′N 01°55′W
173 O16 **St-Gilles-les-Bains** W Réunion 21°02′S 55°14′E
102 M16 **St-Girons** Ariège, S France 42°58′N 01°07′E
Saint Gotthard *see* Szentgotthárd
108 G9 **St. Gotthard Tunnel** *tunnel* Ticino, S Switzerland
97 H22 **St Govan's Head** *headland* SW Wales, United Kingdom 51°35′N 04°55′W
34 M7 **Saint Helena** California, W USA 38°29′N 122°30′W
67 O12 **Saint Helena** *island* C Atlantic Ocean
Saint Helena *see* Saint Helena, Ascension and Tristan da Cunha
65 F24 **Saint Helena, Ascension and Tristan da Cunha** *terr.* Saint Helena, Ascension, Tristan da Cunha. ◐ *UK overseas territory* C Atlantic Ocean
65 M16 **Saint Helena Fracture Zone** *tectonic feature* C Atlantic Ocean
34 M7 **Saint Helena, Mount** ▲ California, W USA 38°40′N 122°39′W
21 S15 **Saint Helena Sound** *inlet* South Carolina, SE USA
31 Q7 **Saint Helen, Lake** ◎ Michigan, N USA
183 Q16 **Saint Helens** Tasmania, SE Australia 41°21′S 148°15′E
97 K18 **St Helens** NW England, United Kingdom 53°28′N 02°44′W
32 G10 **Saint Helens** Oregon, NW USA 45°54′N 122°50′W
32 H10 **Saint Helens, Mount** ⋀ Washington, NW USA 46°24′N 121°49′W
97 L26 **St Helier** ○ (Jersey) S Jersey, Channel Islands 49°12′N 02°07′W
15 S9 **St-Hilarion** Québec, SE Canada 47°34′N 70°24′W
99 K22 **Saint-Hubert** Luxembourg, SE Belgium 50°02′N 05°23′E
15 P12 **St-Hyacinthe** Québec, SE Canada 45°38′N 72°57′W
St.Iago de la Vega *see* Spanish Town
31 Q4 **Saint Ignace** Michigan, N USA 45°53′N 84°44′W
15 O10 **St-Ignace-du-Lac** Québec, SE Canada 46°43′N 73°49′W
12 D12 **St. Ignace Island** *island* Ontario, S Canada
108 C7 **St. Imier** Bern, W Switzerland 47°09′N 06°55′E
97 G25 **St Ives** SW England, United Kingdom 50°12′N 05°29′W
29 U10 **Saint James** Minnesota, N USA 44°00′N 94°36′W
10 I15 **St. James, Cape** *headland* Graham Island, British Columbia, SW Canada 51°57′N 131°04′W

◆ Country
● Country Capital
◇ Dependent Territory
○ Dependent Territory Capital
◐ Administrative Regions
✈ International Airport
▲ Mountain
⛰ Mountain Range
⋀ Volcano
≈ River
◎ Lake
⊠ Reservoir

ST-JEAN – SALZBURG ALPS

15 O13 **St-Jean** *var.* St-Jean-sur-Richelieu. Québec, SE Canada 45°15′N 73°16′W
55 X9 **St-Jean** NW French Guiana 05°25′N 54°05′W
Saint-Jean-d'Acre *see* Akko
102 K11 **St-Jean-d'Angély** Charente-Maritime, W France 45°57′N 00°31′W
103 N7 **St-Jean-de-Braye** Loiret, C France 47°54′N 01°58′E
102 I16 **St-Jean-de-Luz** Pyrénées-Atlantiques, SW France 43°24′N 01°40′W
103 T12 **St-Jean-de-Maurienne** Savoie, E France 45°17′N 06°21′E
102 I9 **St-Jean-de-Monts** Vendée, NW France 46°45′N 02°00′W
103 Q14 **St-Jean-du-Gard** Gard, S France 44°06′N 03°49′E
15 Q7 **St-Jean, Lac** ◎ Québec, SE Canada
102 I16 **St-Jean-Pied-de-Port** Pyrénées-Atlantiques, SW France 43°10′N 01°14′W
15 S9 **St-Jean-Port-Joli** Québec, SE Canada 47°13′N 70°16′W
St-Jean-sur-Richelieu *see* St-Jean
15 N12 **St-Jérôme** Québec, SE Canada 45°47′N 74°01′W
25 T5 **Saint Jo** Texas, SW USA 33°42′N 97°33′W
13 O15 **St. John** New Brunswick, SE Canada 45°16′N 66°03′W
26 L6 **Saint John** Kansas, C USA 37°59′N 98°44′W
19 Q2 **Saint John** *Fr.* Saint-John. ⟿ Canada/USA
76 K16 **Saint John** ⟿ C Liberia
45 T9 **Saint John** *island* C Virgin Islands (US)
Saint-John *see* Saint John
22 I6 **Saint John, Lake** ◎ Louisiana, S USA
45 W10 **St John's** ● (Antigua and Barbuda) Antigua, Antigua and Barbuda 17°06′N 61°50′W
13 V12 **St. John's** *province capital* Newfoundland and Labrador, E Canada 47°34′N 52°41′W
37 O12 **Saint Johns** Arizona, SW USA 34°28′N 109°22′W
31 Q9 **Saint Johns** Michigan, N USA 43°01′N 84°31′W
13 V12 **St. John's** ✈ Newfoundland and Labrador, E Canada 47°22′N 52°45′W
23 X11 **Saint Johns River** ⟿ Florida, SE USA
103 Q11 **St-Jost-St-Rambert** Loire, E France 45°30′N 04°13′E
13 S8 **St. Joseph** W Dominica 15°24′N 61°26′W
173 P17 **St-Joseph** S Réunion
22 J6 **Saint Joseph** Louisiana, S USA 31°56′N 91°14′W
31 O10 **Saint Joseph** Michigan, N USA 42°05′N 86°30′W
27 R3 **Saint Joseph** Missouri, C USA 39°46′N 94°49′W
20 I10 **Saint Joseph** Tennessee, S USA 35°02′N 87°29′W
22 R9 **Saint Joseph Bay** *bay* Florida, SE USA
15 R11 **St-Joseph-de-Beauce** Québec, SE Canada 46°20′N 70°52′W
12 C10 **St. Joseph, Lake** ◎ Ontario, C Canada
31 Q11 **Saint Joseph River** ⟿ N USA
14 C11 **Saint Joseph's Island** *island* Ontario, S Canada
15 N11 **St-Jovite** Québec, SE Canada 46°07′N 74°35′W
St Julian's *see* San Giljan
St-Julien *see* St-Julien-en-Genevois
103 T10 **St-Julien-en-Genevois** *var.* St-Julien. Haute-Savoie, E France 46°07′N 06°06′E
102 M11 **St-Junien** Haute-Vienne, C France 45°52′N 00°54′E
96 D8 **St Kilda** *island* NW Scotland, United Kingdom
45 V10 **Saint Kitts** *island* Saint Kitts and Nevis
45 U10 **Saint Kitts and Nevis** *off.* Federation of Saint Christopher and Nevis, *var.* Saint Christopher-Nevis. ◆ *commonwealth republic* E West Indies
11 X16 **St. Laurent** Manitoba, S Canada 50°20′N 97°55′W
St-Laurent *see* St-Laurent-du-Maroni
55 X9 **St-Laurent-du-Maroni** *var.* St-Laurent. NW French Guiana 05°29′N 54°03′W
St-Laurent, Fleuve *see* St. Lawrence
102 J12 **St-Laurent-Médoc** Gironde, SW France 45°11′N 00°50′W
13 N12 **St. Lawrence** *Fr.* Fleuve St-Laurent. ⟿ Canada/USA
13 Q12 **St. Lawrence, Gulf of** *gulf* NW Atlantic Ocean
38 K10 **Saint Lawrence Island** *island* Alaska, USA
14 M14 **Saint Lawrence River** ⟿ Canada/USA
99 L25 **Saint-Léger** Luxembourg, SE Belgium 49°36′N 05°39′E
13 N14 **St. Léonard** New Brunswick, SE Canada 47°10′N 67°55′W
15 P11 **St-Léonard** Québec, SE Canada 46°06′N 72°18′W
173 O17 **St-Leu** W Réunion 21°09′S 55°17′E
102 J4 **St-Lô** *anc.* Briovera, Laudus. Manche, N France 49°07′N 01°08′W
11 T15 **St. Louis** Saskatchewan, S Canada 52°50′N 105°43′W
103 V7 **St-Louis** Haut-Rhin, NE France 47°35′N 07°34′E
173 O17 **St-Louis** S Réunion
76 G10 **Saint Louis** NW Senegal 15°59′N 16°30′W
27 X4 **Saint Louis** Missouri, C USA 38°38′N 90°15′W
29 W5 **Saint Louis River** ⟿ Minnesota, N USA
103 T7 **St-Loup-sur-Semouse** Haute-Saône, E France 47°53′N 06°15′E
15 O12 **St-Luc** Québec, SE Canada 45°19′N 73°18′W
45 X13 **Saint Lucia** ◆ *commonwealth republic* SE West Indies
47 S3 **Saint Lucia** *island* SE West Indies
83 L22 **St. Lucia, Cape** *headland* E South Africa 28°29′S 32°26′E
45 Y13 **Saint Lucia Channel** *channel* Martinique/Saint Lucia
23 Y14 **Saint Lucie Canal** *canal* Florida, SE USA
23 Z13 **Saint Lucie Inlet** *inlet* Florida, SE USA
96 L2 **St Magnus Bay** *bay* N Scotland, United Kingdom
102 K10 **St-Maixent-l'École** Deux-Sèvres, W France 46°24′N 00°13′W
11 Y16 **St. Malo** Manitoba, S Canada 49°16′N 96°58′W
102 I5 **St-Malo** Ille-et-Vilaine, NW France 48°39′N 02°W
102 H4 **St-Malo, Golfe de** *gulf* NW France
44 L9 **St-Marc** C Haiti 19°08′N 72°41′W
44 L9 **St-Marc, Canal de** *channel* W Haiti
103 S12 **St-Marcellin-le-Mollard** Isère, E France 45°12′N 05°18′E
55 Y12 **Saint-Marcel, Mont** ▲ S French Guiana 2°32′N 53°00′W
96 K5 **St Margaret's Hope** NE Scotland, United Kingdom 58°50′N 02°57′W
32 M9 **Saint Maries** Idaho, NW USA 47°19′N 116°37′W
23 T9 **Saint Marks** Florida, SE USA 30°09′N 84°12′W
108 D11 **St. Martin** Valais, SW Switzerland 46°09′N 07°27′E
Saint Martin *see* Sint Maarten
31 O5 **Saint Martin Island** *island* Michigan, N USA
22 I9 **Saint Martinville** Louisiana, S USA 30°09′N 91°51′W
185 E20 **St. Mary, Mount** ▲ South Island, New Zealand 44°16′S 169°42′E
186 E8 **St. Mary, Mount** ▲ S Papua New Guinea 08°06′S 147°00′E
182 I6 **Saint Mary Peak** ▲ South Australia 31°25′S 138°39′E
183 Q16 **Saint Marys** Tasmania, SE Australia 41°34′S 148°13′E
14 E16 **St. Marys** Ontario, S Canada 43°15′N 81°08′W
38 M11 **Saint Marys** Alaska, USA 62°03′N 163°10′W
23 W8 **Saint Marys** Georgia, SE USA 30°44′N 81°30′W
27 P4 **Saint Marys** Kansas, C USA 39°09′N 96°00′W
31 Q4 **Saint Marys** Ohio, N USA 40°31′N 84°22′W
21 R3 **Saint Marys** West Virginia, NE USA 39°24′N 81°13′W
23 W8 **Saint Marys River** ⟿ Florida/Georgia, SE USA
31 Q4 **Saint Marys River** ⟿ Michigan, N USA
102 D6 **St-Mathieu, Pointe** *headland* NW France 48°17′N 04°56′W
38 J12 **Saint Matthew Island** *island* Alaska, USA
21 R13 **Saint Matthews** South Carolina, SE USA 33°40′N 80°44′W
St.Matthew's Island *see* Zadetkyi Kyun
186 G4 **St.Matthias Group** *island group* NE Papua New Guinea
108 C11 **St. Maurice** Valais, SW Switzerland 46°09′N 07°28′E
15 P9 **St-Maurice** ⟿ Québec, SE Canada
102 J13 **St-Médard-en-Jalles** Gironde, SW France 44°54′N 00°43′W
39 N10 **Saint Michael** Alaska, USA 63°28′N 162°02′W
15 N10 **St-Michel-des-Saints** Québec, SE Canada 46°39′N 73°54′W
103 S5 **St-Mihiel** Meuse, NE France 48°57′N 05°33′E
108 J10 **St. Moritz** *Ger.* Sankt Moritz, *Rmsch.* San Murezzan. Graubünden, SE Switzerland 46°30′N 09°51′E
102 H8 **St-Nazaire** Loire-Atlantique, NW France 47°17′N 02°12′W
Saint Nicholas *see* São Nicolau
Saint-Nicolas *see* Sint-Niklaas
103 N1 **St-Omer** Pas-de-Calais, N France 50°45′N 02°15′E
102 J11 **Saintonge** *cultural region* W France
15 S9 **St-Pacôme** Québec, SE Canada 47°22′N 69°56′W
15 S10 **St-Pamphile** Québec, SE Canada 46°57′N 69°46′W
15 S9 **St-Pascal** Québec, SE Canada 47°32′N 69°48′W
14 J11 **St-Patrice, Lac** ◎ Québec, SE Canada
11 R14 **St. Paul** Alberta, SW Canada 54°00′N 111°18′W
173 O16 **St-Paul** NW Réunion
38 K14 **Saint Paul** Saint Paul Island, Alaska, USA 57°08′N 170°13′W
29 V8 **Saint Paul** *state capital* Minnesota, N USA 45°N 93°10′W
29 P15 **Saint Paul** Nebraska, C USA 41°13′N 98°26′W
21 P7 **Saint Paul** Virginia, NE USA 36°53′N 82°18′W
77 Q17 **Saint Paul, Cape** *headland* S Ghana 05°44′N 00°55′E
103 O17 **St-Paul-de-Fenouillet** Pyrénées-Orientales, S France 42°49′N 02°29′E
65 K14 **Saint Paul Fracture Zone** *tectonic feature* E Atlantic Ocean
38 J14 **Saint Paul Island** *island* Pribilof Islands, Alaska, USA
102 J15 **St-Paul-les-Dax** Landes, SW France 43°45′N 01°01′W
21 U11 **Saint Pauls** North Carolina, SE USA 34°45′N 78°56′W
Saint Paul's Bay *see* San Pawl il Bahar
191 R16 **St Paul's Point** *headland* Pitcairn Island, Pitcairn Islands
29 U10 **Saint Peter** Minnesota, N USA 44°21′N 93°58′W
97 L26 **St Peter Port** ○ (Guernsey) C Guernsey, Channel Islands 49°28′N 02°33′W
23 V13 **Saint Petersburg** Florida, SE USA 27°47′N 82°37′W
Saint Petersburg *see* Sankt-Peterburg
23 V13 **Saint Petersburg Beach** Florida, SE USA 27°43′N 82°43′W
173 P17 **St-Philippe** SE Réunion 21°21′S 55°46′E
45 Q11 **St-Pierre** NW Martinique 14°44′N 61°11′W
173 O17 **St-Pierre** SW Réunion
13 S13 **St-Pierre and Miquelon** *Fr.* Îles St-Pierre et Miquelon. ◇ *French overseas collectivity* NE North America
15 P11 **St-Pierre, Lac** ◎ Québec, SE Canada
102 F5 **St-Pol-de-Léon** Finistère, NW France 48°42′N 04°00′W
103 O2 **St-Pol-sur-Ternoise** Pas-de-Calais, N France 50°22′N 02°21′E
St. Pons *see* St-Pons-de-Thomières
103 O16 **St-Pons-de-Thomières** *var.* St. Pons. Hérault, S France 43°28′N 02°48′E
103 P10 **St-Pourçain-sur-Sioule** Allier, C France 46°19′N 03°16′E
15 S11 **St-Prosper** Québec, SE Canada 46°14′N 70°28′W
103 P3 **St-Quentin** Aisne, N France 49°51′N 03°17′E
15 R10 **St-Raphaël** Québec, SE Canada 46°47′N 70°46′W
103 U15 **St-Raphaël** Var, SE France 43°26′N 06°46′E
15 Q10 **St-Raymond** Québec, SE Canada 46°53′N 71°49′W
33 O9 **Saint Regis** Montana, NW USA 47°18′N 115°06′W
18 J7 **Saint Regis River** ⟿ New York, NE USA
103 R15 **St-Rémy-de-Provence** Bouches-du-Rhône, SE France 43°48′N 04°49′E
102 M9 **St-Savin** Vienne, W France 46°34′N 00°53′E
Saint-Sébastien,Cap *see* Anorontany, Tanjona
23 X7 **Saint Simons Island** *island* Georgia, SE USA
191 Y2 **Saint Stanislas Bay** *bay* Kiritimati, E Kiribati
13 O15 **St. Stephen** New Brunswick, SE Canada 45°12′N 67°18′W
45 V9 **St Barthélemy** ◇ *French overseas collectivity* E Caribbean Sea
39 X12 **Saint Terese** Alaska, USA 58°28′N 134°46′W
14 E17 **St. Thomas** Ontario, S Canada 42°46′N 81°12′W
29 Q2 **Saint Thomas** North Dakota, N USA 48°37′N 97°28′W
45 T9 **Saint Thomas** *island* W Virgin Islands (US)
Saint Thomas *see* São Tomé, Sao Tome and Principe
Saint Thomas *see* Charlotte Amalie, Virgin Islands (US)
15 P10 **St-Tite** Québec, SE Canada 46°42′N 72°32′W
45 V9 **St Martin** ◇ *French overseas collectivity* E Caribbean Sea
Saint-Trond *see* Sint-Truiden
103 U16 **St-Tropez** Var, SE France 43°16′N 06°39′E
Saint Ubes *see* Setúbal
102 L2 **St-Valéry-en-Caux** Seine-Maritime, N France 49°53′N 00°42′E
103 Q9 **St-Vallier** Saône-et-Loire, C France 46°39′N 04°19′E
106 B7 **St-Vincent** Valle d'Aosta, NW Italy 45°47′N 07°42′E
45 Q14 **Saint Vincent** *island* N Saint Vincent and the Grenadines
Saint Vincent *see* São Vicente
45 W14 **Saint Vincent and the Grenadines** ◆ *commonwealth republic* SE West Indies
Saint-Vincent, Cap *see* Ankaboa, Tanjona
Saint-Vincent, Cape *see* São Vicente, Cabo de
102 I15 **St-Vincent-de-Tyrosse** Landes, SW France 43°39′N 01°16′W
182 I9 **Saint Vincent, Gulf** *gulf* South Australia
23 R10 **Saint Vincent Island** *island* Florida, SE USA
45 T12 **Saint Vincent Passage** *passage* Saint Lucia/Saint Vincent and the Grenadines
183 N18 **Saint Vincent, Point** *headland* Tasmania, SE Australia 43°19′S 145°50′E
Saint-Vith *see* Sankt-Vith
11 S14 **St. Walburg** Saskatchewan, S Canada 53°38′N 109°12′W
St Wolfgangsee *see* Wolfgangsee
102 M11 **St-Yrieix-la-Perche** Haute-Vienne, C France 45°31′N 01°12′E
Saint Yves *see* Setúbal
188 H5 **Saipan** *island* ● (Northern Mariana Islands) S Northern Mariana Islands
188 H6 **Saipan Channel** *channel* S Northern Mariana Islands
188 H6 **Saipan International** ✈ Saipan, S Northern Mariana Islands
74 G6 **Sais** ✈ (Fès) C Morocco 33°58′N 04°48′W
Saishū *see* Jeju-do
Saishū *see* Jeju
102 J16 **Saison** ⟿ SW France
169 R10 **Sai, Sungai** ⟿ Borneo, N Indonesia
165 N13 **Saitama** *off.* Saitama-ken. ◐ *prefecture* Honshū, S Japan
Saitama *see* Urawa
Saitama-ken *see* Saitama
Saiyid Abid *see* Sayyid 'Abid
57 J19 **Sajama, Nevado** ▲ W Bolivia 17°57′S 68°51′W
141 V13 **Sājir, Ras** *headland* S Oman 16°42′N 53°40′E
111 M20 **Sajószentpéter** Borsod-Abaúj-Zemplén, NE Hungary 48°13′N 20°43′E
83 F24 **Sak** ⟿ SW South Africa
81 J18 **Saka** Tana River, E Kenya 0°11′S 39°27′E
167 P11 **Sa Kaeo** Prachin Buri, C Thailand 13°47′N 102°03′E
164 J14 **Sakai** Ōsaka, Honshū, SW Japan 34°35′N 135°28′E
164 H14 **Sakaide** Kagawa, Shikoku, SW Japan 34°19′N 133°50′E
164 H12 **Sakaiminato** Tottori, Honshū, SW Japan 35°34′N 133°12′E
140 M3 **Sakākah** Al Jawf, NW Saudi Arabia 29°56′N 40°10′E
28 L4 **Sakakawea, Lake** ⊡ North Dakota, N USA
12 J9 **Sakami, Lac** ◎ Québec, C Canada
79 O26 **Sakania** Katanga, SE Dem. Rep. Congo 12°44′S 28°34′E
146 K12 **Sakar** Lebap Welaýaty, E Turkmenistan 38°57′N 63°46′E
172 H7 **Sakaraha** Toliara, SW Madagascar 22°54′S 44°31′E
146 I14 **Sakarçäge** *var.* Sakarchäge, *Rus.* Sakar-Chaga. Mary Welaýaty, C Turkmenistan 37°40′N 61°33′E
Sakar-Chaga/Sakarchäge *see* Sakarçäge
Sak'art'velo *see* Georgia
136 F11 **Sakarya** ◐ *province* NW Turkey
136 F12 **Sakarya Nehri** ⟿ NW Turkey
165 P9 **Sakata** Yamagata, Honshū, C Japan 38°54′N 139°51′E
123 P9 **Sakha (Yakutiya), Respublika** *var.* Respublika Yakutiya, *Eng.* Yakutia. ◐ *autonomous republic* NE Russian Federation
Sakhalin *see* Sakhalin, Ostrov
192 I3 **Sakhalin, Ostrov** *var.* Sakhalin. *island* SE Russian Federation
123 U12 **Sakhalinskaya Oblast'** ◐ *province* SE Russian Federation
123 T12 **Sakhalinskiy Zaliv** *gulf* E Russian Federation
Sakhnovshchina *see* Sakhnovshchyna
117 U6 **Sakhnovshchyna** *Rus.* Sakhnovshchina. Kharkivs'ka Oblast', E Ukraine 49°08′N 35°52′E
Sakhon Nakhon *see* Sakon Nakhon
Sakhtsar *see* Rāmsar
137 W10 **Şäki** *Rus.* Sheki; *prev.* Nukha. NW Azerbaijan 41°09′N 47°10′E
Saki *see* Saky
118 E13 **Šakiai** *Ger.* Schaken. Marijampolė, S Lithuania 54°57′N 23°04′E
165 O16 **Sakishima-shotō** *var.* Sakisima Syotô. *island group* SW Japan
Sakisima Syotô *see* Sakishima-shotō
Sakiz *see* Saqqez
Sakiz-Adasi *see* Chíos
155 F19 **Sakleshpur** Karnātaka, E India 12°58′N 75°45′E
167 S9 **Sakon Nakhon** *var.* Muang Sakon Nakhon, Sakhon Nakhon. Sakon Nakhon, E Thailand 17°10′N 104°08′E
149 P15 **Sakrand** Sind, SE Pakistan 26°06′N 68°20′E
83 F24 **Sak River** *Afr.* Sakrivier. Northern Cape, W South Africa 30°49′S 20°24′E
Sakrivier *see* Sak River
Saksaul'skiy *see* Saksaul'skoye
144 K13 **Saksaul'skoye** *var.* Saksaul'skiy, *Kaz.* Sekseüil. Kzylorda, S Kazakhstan 47°07′N 61°06′E
95 I25 **Sakskøbing** Sjælland, SE Denmark 54°48′N 11°39′E
165 N12 **Saku** Nagano, Honshū, S Japan 36°17′N 138°29′E
117 S13 **Saky** *Rus.* Saki. Avtonomna Respublika Krym, S Ukraine 45°09′N 33°36′E
76 E9 **Sal** *island* Ilhas de Barlavento, NE Cape Verde
127 N12 **Sal** ⟿ SW Russian Federation
111 I21 **Sal'a** *Hung.* Sellye, Vágsellye. Nitriansky Kraj, SW Slovakia 48°09′N 17°51′E
95 N15 **Sala** Västmanland, C Sweden 59°55′N 16°38′E
15 N13 **Salaberry-de-Valleyfield** *var.* Valleyfield. Québec, SE Canada 45°15′N 74°07′W
118 G7 **Salacgrīva** *Est.* Salatsi. N Latvia 57°45′N 24°21′E
107 M18 **Sala Consilina** Campania, S Italy 40°23′N 15°35′E
40 C2 **Salada, Laguna** ◎ NW Mexico
61 D14 **Saladas** Corrientes, NE Argentina 28°15′S 58°40′W
61 C21 **Saladillo** Buenos Aires, E Argentina 35°40′S 59°50′W
61 B16 **Saladillo, Río** ⟿ C Argentina
25 T9 **Salado** Texas, SW USA 30°57′N 97°32′W
63 J16 **Salado, Arroyo** ⟿ SE Argentina
61 D21 **Salado, Río** ⟿ E Argentina
62 J12 **Salado, Río** ⟿ C Argentina
41 N7 **Salado, Río** ⟿ NE Mexico
37 Q12 **Salado, Río** ⟿ New Mexico, SW USA
143 N6 **Salafchegān** *var.* Sarafjagān. Quom, N Iran 34°28′N 50°28′E
77 Q15 **Salaga** C Ghana 08°31′N 00°37′W
139 S6 **Şalāḩ ad Dīn** *off.* Muḩāfa at Şalāḩ ad Dīn, *var.* Salāhuddīn. ◐ *governorate* C Iraq
Şalāḩ ad Dīn, Muḩāfa at *see* Şalāḩ ad Dīn
Salāhuddīn *see* Şalāḩ ad Dīn
192 G5 **Sala'ilua** Savai'i, W Samoa 13°39′S 172°33′W
116 G9 **Sălaj** ◐ *county* NW Romania
83 H20 **Salajwe** Kweneng, SE Botswana 23°40′S 24°46′E
78 H9 **Salal** Bahr el Gazel, W Chad 14°48′N 17°12′E
80 I6 **Salala** Red Sea, NE Sudan 21°17′N 36°16′E
141 V13 **Şalālah** SW Oman 17°01′N 54°04′E
42 D5 **Salamá** Baja Verapaz, C Guatemala 15°06′N 90°18′W
42 J6 **Salamá** Olancho, C Honduras 14°48′N 86°34′W
62 G10 **Salamanca** Coquimbo, C Chile 31°47′S 70°58′W
41 N13 **Salamanca** Guanajuato, C Mexico 20°34′N 101°12′W
104 K7 **Salamanca** *anc.* Helmantica, Salmantica. Castilla y León, NW Spain 40°58′N 05°40′W
18 D11 **Salamanca** New York, NE USA 42°09′N 78°43′W
104 J7 **Salamanca** ◐ *province* Castilla y León, W Spain
63 J19 **Salamanca, Pampa de** *plain* S Argentina
78 J12 **Salamat** *off.* Région du Salamat. ◐ *region* SE Chad
78 I12 **Salamat, Bahr** ⟿ S Chad
Salamat, Région du *see* Salamat
54 F5 **Salamina** Magdalena, N Colombia 10°30′N 74°48′W
115 G19 **Salamína** *var.* Salamís. Salamína, C Greece 37°58′N 23°29′E
115 G19 **Salamína** *island* C Greece
Salamís *see* Salamína
138 I5 **Salamīyah** *var.* As Salamīyah. Ḩamāh, W Syria 35°01′N 37°02′E
31 P12 **Salamonie Lake** ⊡ Indiana, N USA
31 P12 **Salamonie River** ⟿ Indiana, N USA
Salang *see* Phuket
192 I16 **Salani** Upolu, SE Samoa 14°00′S 171°35′W
118 C11 **Salantai** Klaipėda, NW Lithuania 56°05′N 21°36′E
104 K2 **Salas** Asturias, N Spain 43°25′N 06°15′W
105 O5 **Salas de los Infantes** Castilla y León, N Spain 42°02′N 03°17′W
102 M16 **Salat** ⟿ S France
189 V13 **Salat** *island* Chuuk, C Micronesia
169 Q16 **Salatiga** Jawa, C Indonesia 07°15′S 110°34′E
189 V13 **Salat Pass** *passage* W Pacific Ocean
Salatsi *see* Salacgrīva
167 T10 **Salavan** *var.* Saravan, Saravane. Salavan, S Laos 15°43′N 106°26′E
127 V6 **Salavat** Respublika Bashkortostan, W Russian Federation 53°20′N 55°54′E
56 C12 **Salaverry** La Libertad, N Peru 08°14′S 78°55′W
171 T12 **Salawati, Pulau** *island* E Indonesia
193 R10 **Sala y Gomez** *island* Chile, E Pacific Ocean
Sala y Gomez Fracture Zone *see* Sala y Gomez Ridge
193 S10 **Sala y Gomez Ridge** *var.* Sala y Gomez Fracture Zone. *tectonic feature* SE Pacific Ocean
61 A22 **Salazar** Buenos Aires, E Argentina 36°20′S 62°11′W
54 G7 **Salazar** Norte de Santander, N Colombia 07°46′N 72°46′W
Salazar *see* N'Dalatando
173 P16 **Salazie** C Réunion 21°02′S 55°33′E
103 N8 **Salbris** Loir-et-Cher, C France 47°25′N 02°02′E
57 G15 **Salcantay, Nevado** ▲ C Peru 13°21′S 72°31′W
45 O8 **Salcedo** N Dominican Republic 19°26′N 70°25′W
39 S9 **Salcha River** ⟿ Alaska, USA
119 H15 **Šalčininkai** Vilnius, SE Lithuania 54°20′N 25°26′E
Saldae *see* Béjaïa
54 E11 **Saldaña** Tolima, C Colombia 03°57′N 75°01′W
104 M4 **Saldaña** Castilla y León, N Spain 42°31′N 04°44′W
83 E25 **Saldanha** Western Cape, SW South Africa 33°00′S 17°56′E
Salduba *see* Zaragoza
61 B23 **Saldungaray** Buenos Aires, E Argentina 38°15′S 61°45′W
118 D9 **Saldus** *Ger.* Frauenburg. W Latvia 56°40′N 22°29′E
183 P13 **Sale** Victoria, SE Australia 38°06′S 147°06′E
74 F6 **Salé** NW Morocco 34°07′N 06°40′W
74 F6 **Salé** ✈ (Rabat) W Morocco 34°09′N 06°30′W
Salehābād *see* Andimeshk
170 M16 **Saleh, Teluk** *bay* Nusa Tenggara, S Indonesia
122 H8 **Salekhard** *prev.* Obdorsk. Yamalo-Nenetskiy Avtonomnyy Okrug, N Russian Federation 66°33′N 66°35′E
192 H16 **Sālelologa** Savai'i, C Samoa 13°42′S 172°10′W
155 H21 **Salem** Tamil Nādu, SE India 11°38′N 78°08′E
27 V9 **Salem** Arkansas, C USA 36°21′N 91°49′W
30 L15 **Salem** Illinois, N USA 38°37′N 88°57′W
31 P15 **Salem** Indiana, N USA 38°38′N 86°06′W
19 P11 **Salem** Massachusetts, NE USA 42°30′N 70°51′W
27 V6 **Salem** Missouri, C USA 37°39′N 91°32′W
18 I16 **Salem** New Jersey, NE USA 39°33′N 75°26′W
31 U12 **Salem** Ohio, N USA 40°52′N 80°51′W
32 G12 **Salem** *state capital* Oregon, NW USA 44°57′N 123°01′W
29 Q11 **Salem** South Dakota, N USA 43°43′N 97°23′W
36 L4 **Salem** Utah, W USA 40°03′N 111°40′W
21 S7 **Salem** Virginia, NE USA 37°16′N 80°00′W
21 R3 **Salem** West Virginia, NE USA 39°15′N 80°32′W
107 H23 **Salemi** Sicilia, Italy, C Mediterranean Sea 37°48′N 12°48′E
Salemy *see* As Sālimī
94 K12 **Sälen** Dalarna, C Sweden 61°11′N 13°14′E
107 Q18 **Salentina, Campi** Puglia, SE Italy 40°23′N 18°01′E
107 Q18 **Salentina, Penisola** *peninsula* SE Italy
107 L18 **Salerno** *anc.* Salernum. Campania, S Italy 40°40′N 14°44′E
107 L18 **Salerno, Golfo di** *Eng.* Gulf of Salerno. *gulf* S Italy
Salerno, Gulf of *see* Salerno, Golfo di
Salernum *see* Salerno
97 K17 **Salford** NW England, United Kingdom 53°30′N 02°16′W
Salgir *see* Salhyr
111 K21 **Salgótarján** Nógrád, N Hungary 48°07′N 19°47′E
59 O15 **Salgueiro** Pernambuco, E Brazil 08°04′S 39°05′W
94 C13 **Salhus** Hordaland, S Norway 60°30′N 05°15′E
117 T12 **Salhyr** *Rus.* Salgir. ⟿ S Ukraine
171 Q9 **Salibabu, Pulau** *island* N Indonesia
37 S6 **Salida** Colorado, C USA 38°29′N 105°57′W
102 J15 **Salies-de-Béarn** Pyrénées-Atlantiques, SW France 43°28′N 00°55′W
136 C14 **Salihli** Manisa, W Turkey 38°29′N 28°08′E
119 K18 **Salihorsk** *Rus.* Soligorsk. Minskaya Voblasts', S Belarus 52°48′N 27°32′E
119 K18 **Salihorskaye Vodaskhovishcha** *Rus.* Soligorskoye Vodokhranilishche. ⊡ C Belarus
83 N14 **Salima** Central, C Malawi 13°44′S 34°21′E
166 L5 **Salin** Magway, W Myanmar (Burma) 20°30′N 94°40′E
27 N4 **Salina** Kansas, C USA 38°53′N 97°36′W
36 L5 **Salina** Utah, W USA 38°57′N 111°54′W
41 S17 **Salina Cruz** Oaxaca, SE Mexico 16°11′N 95°12′W
107 L22 **Salina, Isola** *island* Isole Eolie, S Italy
44 J5 **Salina Point** *headland* Acklins Island, SE The Bahamas 22°10′N 74°16′W
56 A7 **Salinas** Guayas, W Ecuador 02°15′S 80°58′W
40 M11 **Salinas** *var.* Salinas de Hidalgo. San Luis Potosí, C Mexico 22°36′N 101°41′W
45 T6 **Salinas** C Puerto Rico 17°57′N 66°18′W
35 O10 **Salinas** California, W USA 36°41′N 121°40′W
Salinas, Cabo de *see* Salines, Cap de ses
Salinas de Hidalgo *see* Salinas
82 A13 **Salinas, Ponta das** *headland* W Angola 12°50′S 12°57′E
45 O10 **Salinas, Punta** *headland* S Dominican Republic 18°11′N 70°32′W
Salinas, Río *see* Chixoy, Río
35 O11 **Salinas River** ⟿ California, W USA
22 H6 **Saline Lake** ◎ Louisiana, S USA
25 R17 **Salineno** Texas, SW USA 26°29′N 99°06′W
27 V14 **Saline River** ⟿ Arkansas, C USA
30 M17 **Saline River** ⟿ Illinois, N USA
105 X10 **Salines, Cap de ses** *var.* Cabo de Salinas. *headland* Mallorca, Spain, W Mediterranean Sea 39°15′N 03°03′E
Salisburg *see* Mazsalaca
45 O12 **Salisbury** *var.* Baroui. W Dominica 15°26′N 61°27′W
97 M23 **Salisbury** *var.* New Sarum. S England, United Kingdom 51°05′N 01°48′W
21 Y4 **Salisbury** Maryland, NE USA 38°22′N 75°37′W
27 T3 **Salisbury** Missouri, C USA 39°25′N 92°48′W
21 S9 **Salisbury** North Carolina, SE USA 35°40′N 80°28′W
Salisbury *see* Harare
9 Q7 **Salisbury Island** *island* Nunavut, NE Canada
Salisbury, Lake *see* Bisina, Lake
97 L23 **Salisbury Plain** *plain* S England, United Kingdom
21 R14 **Salkehatchie River** ⟿ South Carolina, SE USA
138 I9 **Şalkhad** As Suwaydā', SW Syria 32°29′N 36°42′E
92 M12 **Salla** Lappi, NE Finland 66°50′N 28°40′E
Sallan *see* Sørøya
103 U11 **Sallanches** Haute-Savoie, E France 45°55′N 06°37′E
105 V5 **Sallent** Cataluña, NE Spain 41°48′N 01°52′E
Salliq *see* Coral Harbour
61 A22 **Salliqueló** Buenos Aires, E Argentina 36°45′S 62°55′W
27 R10 **Sallisaw** Oklahoma, C USA 35°27′N 94°49′W
80 I7 **Sallom** Red Sea, NE Sudan 19°17′N 37°02′E
12 J2 **Salluit** *prev.* Saglouc, Sagluk. Québec, NE Canada 62°10′N 75°40′W
Sallyana *see* Şalyān
13 S11 **Sally's Cove** Newfoundland and Labrador, E Canada 49°43′N 58°00′W
139 W9 **Salmān Bin 'Arāzah** Maysān, E Iraq 32°33′N 46°36′E
Salmantica *see* Salamanca
142 I2 **Salmās** *prev.* Dilman, Shāpūr. Āžarbāyjān-e Gharbī, NW Iran 38°13′N 44°50′E
124 I11 **Salmi** Respublika Kareliya, NW Russian Federation 61°21′N 31°55′E
33 P12 **Salmon** Idaho, NW USA 45°10′N 113°54′W
11 N16 **Salmon Arm** British Columbia, SW Canada 50°41′N 119°18′W
192 L5 **Salmon Bank** *undersea feature* N Pacific Ocean 26°55′N 176°28′W
Salmon Leap *see* Leixlip
34 L2 **Salmon Mountains** ▲▲ California, W USA
14 J15 **Salmon Point** *headland* Ontario, SE Canada 43°51′N 77°15′W
33 N11 **Salmon River** ⟿ Idaho, NW USA
18 K6 **Salmon River** ⟿ New York, NE USA
33 N12 **Salmon River Mountains** ▲▲ Idaho, NW USA
18 I9 **Salmon River Reservoir** ⊡ New York, NE USA
93 K19 **Salo** Varsinais-Suomi, SW Finland 60°23′N 23°10′E
106 F7 **Salò** Lombardia, N Italy 45°37′N 10°30′E
Salona/Salonae *see* Solin
103 S15 **Salon-de-Provence** Bouches-du-Rhône, SE France 43°39′N 05°05′E
Salonica/Salonika *see* Thessaloníki
115 I14 **Salonikiós, Akrotírio** *var.* Akrotírio Salonikós. *headland* Thásos, E Greece 40°34′N 24°39′E
Salonikós, Akrotírio *see* Salonikiós, Akrotírio
116 F10 **Salonta** *Hung.* Nagyszalonta. Bihor, W Romania 46°49′N 21°40′E
104 I9 **Salor** ⟿ W Spain
105 U6 **Salou** Cataluña, NE Spain 41°05′N 01°08′E
76 H11 **Saloum** ⟿ C Senegal
42 H4 **Sal, Punta** *headland* NW Honduras 15°55′N 87°36′W
92 N3 **Salpynten** *headland* W Svalbard 78°12′N 12°11′E
138 I3 **Salqīn** Idlib, W Syria 36°09′N 36°27′E
93 F14 **Salsbruket** Nord-Trøndelag, C Norway 64°49′N 11°48′E
126 M13 **Sal'sk** Rostovskaya Oblast', SW Russian Federation 46°30′N 41°31′E
107 J25 **Salso** ⟿ Sicilia, Italy, C Mediterranean Sea
107 K25 **Salso** ⟿ Sicilia, Italy, C Mediterranean Sea
106 E9 **Salsomaggiore Terme** Emilia-Romagna, N Italy 44°49′N 09°58′E
Salt *see* As Salt
62 J6 **Salta** Salta, NW Argentina 24°47′S 65°23′W
62 K6 **Salta** *off.* Provincia de Salta. ◐ *province* N Argentina
Salta, Provincia de *see* Salta
97 I24 **Saltash** SW England, United Kingdom 50°26′N 04°14′W
24 I8 **Salt Basin** *basin* Texas, SW USA
11 V16 **Saltcoats** Saskatchewan, S Canada 51°06′N 102°12′W
30 L13 **Salt Creek** ⟿ Illinois, N USA
24 J9 **Salt Draw** ⟿ Texas, SW USA
97 F21 **Saltee Islands** *island group* SE Ireland
92 G12 **Saltfjorden** *inlet* C Norway
24 I8 **Salt Flat** Texas, SW USA 31°43′N 105°05′W
27 N8 **Salt Fork Arkansas River** ⟿ Oklahoma, C USA
31 T13 **Salt Fork Lake** ⊡ Ohio, N USA
26 J11 **Salt Fork Red River** ⟿ Oklahoma/Texas, C USA
95 J23 **Saltholm** *island* E Denmark
41 N8 **Saltillo** Coahuila, NE Mexico 25°30′N 101°W
182 L5 **Salt Lake** *salt lake* New South Wales, SE Australia
37 V15 **Salt Lake** ◎ New Mexico, SW USA
36 K2 **Salt Lake City** *state capital* Utah, W USA 40°45′N 111°54′W
61 C20 **Salto** Buenos Aires, E Argentina 34°18′S 60°17′W
61 D17 **Salto** Salto, N Uruguay 31°23′S 57°58′W
61 E17 **Salto** ◐ *department* N Uruguay
107 I14 **Salto** ⟿ C Italy
62 Q6 **Salto del Guairá** Canindeyú, E Paraguay 24°06′S 54°22′W
61 D17 **Salto Grande, Embalse de** *var.* Lago de Salto Grande. ⊡ Argentina/Uruguay
Salto Grande, Lago de *see* Salto Grande, Embalse de
35 W16 **Salton Sea** ◎ California, W USA
60 I12 **Salto Santiago, Represa de** ⊡ S Brazil
149 U7 **Salt Range** ▲▲ E Pakistan
36 M13 **Salt River** ⟿ Arizona, SW USA
20 L5 **Salt River** ⟿ Kentucky, S USA
27 V3 **Salt River** ⟿ Missouri, C USA
95 F17 **Saltrød** Aust-Agder, S Norway 58°28′N 08°49′E
95 P16 **Saltsjöbaden** Stockholm, C Sweden 59°15′N 18°20′E
92 G12 **Saltstraumen** Nordland, C Norway 67°16′N 14°42′E
21 Q7 **Saltville** Virginia, NE USA 36°52′N 81°48′W
Saluces/Saluciae *see* Saluzzo
21 Q12 **Saluda** South Carolina, SE USA 34°00′N 81°47′W
21 X6 **Saluda** Virginia, NE USA 37°36′N 76°36′W
21 Q12 **Saluda River** ⟿ South Carolina, SE USA
Salûm *see* As Sallūm
152 F14 **Sālūmbar** Rājasthān, N India 24°16′N 74°04′E
Salûm, Gulf of *see* Khalīj as Sallūm
171 O11 **Salumpaga** Sulawesi, N Indonesia 01°18′N 120°58′E
155 M14 **Sālūr** Andhra Pradesh, E India 18°31′N 83°16′E
55 Y9 **Salut, Îles du** *island group* N French Guiana
106 A9 **Saluzzo** *Fr.* Saluces; *anc.* Saluciae. Piemonte, NW Italy 44°39′N 07°29′E
63 F23 **Salvación, Bahía** *bay* S Chile
59 P17 **Salvador** *prev.* São Salvador. *state capital* Bahia, E Brazil 12°58′S 38°29′W
65 E24 **Salvador** East Falkland, Falkland Islands 51°28′S 58°22′W
22 K10 **Salvador, Lake** ◎ Louisiana, S USA
Salvaleón de Higüey *see* Higüey
104 F10 **Salvaterra de Magos** Santarém, C Portugal 39°01′N 08°47′W
41 N13 **Salvatierra** Guanajuato, C Mexico 20°14′N 100°52′W
105 P3 **Salvatierra** *Basq.* Agurain. País Vasco, N Spain 42°52′N 02°23′W
Salwa/Salwah *see* As Salwā
166 M7 **Salween** *Bur.* Thanlwin, *Chin.* Nu Chiang, Nu Jiang. ⟿ SE Asia
137 Y12 **Salyan** *Rus.* Sal'yany. SE Azerbaijan 39°36′N 48°57′E
153 N11 **Şalyān** *var.* Sallyana. Mid Western, W Nepal 28°22′N 82°10′E
Sal'yany *see* Salyan
21 O6 **Salyersville** Kentucky, S USA 37°43′N 83°06′W
109 V6 **Salza** ⟿ E Austria
109 Q7 **Salzach** ⟿ Austria/Germany
109 Q6 **Salzburg** *anc.* Juvavum. Salzburg, N Austria 47°48′N 13°03′E
109 O8 **Salzburg** *off.* Land Salzburg. ◐ *state* C Austria
Salzburg *see* Ocna Sibiului
Salzburg Alps *see* Salzburger Kalkalpen

◆ Country
● Country Capital
◇ Dependent Territory
○ Dependent Territory Capital
◐ Administrative Regions
✈ International Airport
▲ Mountain
▲▲ Mountain Range
⊼ Volcano
⟿ River
◎ Lake
⊡ Reservoir

SALZBURGER KALKALPEN – SAN JOSÉ

109 Q7 **Salzburger Kalkalpen** *Eng.* Salzburg Alps. ⛰ C Austria
Salzburg, Land *see* Salzburg
100 J13 **Salzgitter** *prev.* Watenstedt-Salzgitter. Niedersachsen, C Germany 52°07′N 10°24′E
101 G14 **Salzkotten** Nordrhein-Westfalen, W Germany 51°40′N 08°36′E
100 K11 **Salzwedel** Sachsen-Anhalt, N Germany 52°51′N 11°10′E
152 D11 **Sām** Rājasthān, NW India 26°50′N 70°30′E
Šamac *see* Bosanski Šamac
54 G9 **Samacá** Boyacá, C Colombia 05°28′N 73°33′W
40 I7 **Samachique** Chihuahua, N Mexico 27°17′N 107°28′W
141 Y8 **Şamad** NE Oman 22°47′N 58°12′E
Sama de Langreo *see* Sama, Spain
Samaden *see* Samedan
57 M19 **Samaipata** Santa Cruz, C Bolivia 18°08′S 63°53′W
Samakhixai *see* Attapu
Samakov *see* Samokov
42 B6 **Samalá, Río** ≈ SW Guatemala
40 J3 **Samalayuca** Chihuahua, N Mexico 31°25′N 106°30′W
155 L16 **Sāmalkot** Andhra Pradesh, E India 17°03′N 82°15′E
45 P8 **Samaná** *var.* Santa Bárbara de Samaná. E Dominican Republic 19°14′N 69°20′W
45 P8 **Samaná, Bahía de** *bay* E Dominican Republic
44 K4 **Samana Cay** *island* SE The Bahamas
136 K17 **Samandağı** Hatay, S Turkey 36°07′N 35°55′E
149 P3 **Samangān** ◈ *province* N Afghanistan
Samangān *see* Aibak
165 T5 **Samani** Hokkaidō, NE Japan 42°07′N 142°57′E
54 C13 **Samaniego** Nariño, SW Colombia 01°22′N 77°35′W
171 Q5 **Samar** *island* C Philippines
127 S6 **Samara** *prev.* Kuybyshev. Samarskaya Oblast', W Russian Federation 53°15′N 50°15′E
127 T7 **Samara** ≈ W Russian Federation
127 S6 **Samara** ✈ Samarskaya Oblast', W Russian Federation 53°11′N 50°27′E
117 V7 **Samara** ≈ E Ukraine
186 G10 **Samarai** Milne Bay, SE Papua New Guinea 10°36′S 150°39′E
Samarang *see* Semarang
123 T14 **Samarga** Khabarovskiy Kray, SE Russian Federation 47°43′N 139°08′E
138 G9 **Samarian Hills** *hill range* N Israel
54 L9 **Samariapo** Amazonas, C Venezuela 05°16′N 67°43′W
169 V11 **Samarinda** Borneo, C Indonesia 0°30′S 117°09′E
Samarkand *see* Samarqand
Samarkandskaya Oblast' *see* Samarqand Viloyati
Samarkandski/Samarkandskoye *see* Temirtau
Samarobriva *see* Amiens
147 N11 **Samarqand** *Rus.* Samarkand. Samarqand Viloyati, C Uzbekistan 39°40′N 66°56′E
146 M11 **Samarqand Viloyati** *Rus.* Samarkandskaya Oblast'. ◈ *province* C Uzbekistan
139 S6 **Sāmarrā'** Şalāḩ ad Dīn, C Iraq 34°13′N 43°52′E
127 R7 **Samarskaya Oblast'** *prev.* Kuybyshevskaya Oblast'. ◈ *province* W Russian Federation
153 Q13 **Samastīpur** Bihār, N India 25°52′N 85°47′E
76 L14 **Samatiguila** NW Ivory Coast 09°51′N 07°36′W
Samawa *see* As Samāwah
137 Y11 **Şamaxı** *Rus.* Shemakha. E Azerbaijan 40°38′N 48°34′E
79 K18 **Samba** Equateur, NW Dem. Rep. Congo 0°13′N 21°17′E
79 N21 **Samba** Maniema, E Dem. Rep. Congo 04°41′S 26°23′E
152 H6 **Samba** Jammu and Kashmir, NW India 32°32′N 75°08′E
169 W10 **Sambaliung, Pegunungan** ⛰ Borneo, N Indonesia
154 M11 **Sambalpur** Odisha, E India 21°28′N 84°04′E
67 X12 **Sambao** ≈ W Madagascar
169 Q10 **Sambas, Sungai** ≈ Borneo, N Indonesia
172 K2 **Sambava** Antsiranana, NE Madagascar 14°16′S 50°10′E
152 J10 **Sambhal** Uttar Pradesh, N India 28°35′N 78°34′E
152 H12 **Sāmbhar Salt Lake** ◎ N India
107 N21 **Sambiase** Calabria, SW Italy 38°58′N 16°16′E
116 H5 **Sambir** *Rus.* Sambor. L'vivs'ka Oblast', NW Ukraine 49°31′N 23°10′E
82 C13 **Sambo** Huambo, C Angola 13°07′S 16°06′E
Sambor *see* Sambir
61 E21 **Samborombón, Bahía** *bay* NE Argentina
99 H20 **Sambre** ≈ Belgium/France
43 V16 **Sambú, Río** ≈ SE Panama
81 I18 **Samburu** ◈ *county* N Kenya
163 Z14 **Samcheok** *Jap.* Sanchoku; *prev.* Samch'ŏk. NE South Korea 37°21′N 129°12′E
Samch'ŏk *see* Samcheok
Samch'ŏnpŏ *see* Sacheon
81 I21 **Same** Kilimanjaro, NE Tanzania 04°04′S 37°41′E
108 J10 **Samedan** *Ger.* Samaden. Graubünden, S Switzerland 46°31′N 09°51′E
82 K12 **Samfya** Luapula, N Zambia 11°22′S 29°34′E
141 W13 **Samḥān, Jabal** ⛰ SW Oman
115 C18 **Sámi** Kefallonía, Iónia Nisiá, Greece, C Mediterranean Sea 38°15′N 20°39′E
56 F10 **Samiria, Río** ≈ N Peru
Samirum *see* Semīrom
167 Q13 **Samit** *prev.* Phumĭ Samĭt. Kaôh Kŏng, SW Cambodia 10°54′N 103°09′E
137 V11 **Şämkir** *Rus.* Shamkhor. NW Azerbaijan 40°51′N 46°03′E
167 S7 **Sam, Nam** *Vtn.* Sông Chu. ≈ Laos/Vietnam
Samnān *see* Semnān
Sam Neua *see* Xam Nua
75 P10 **Samnū** C Libya 27°19′N 15°01′E
192 H15 **Samoa** *off.* Independent State of Samoa, *var.* Sāmoa; *prev.* Western Samoa. ◆ *monarchy* W Polynesia
192 L9 **Sāmoa** *island group* C Pacific Ocean
Sāmoa *see* Samoa
175 T9 **Samoa Basin** *undersea feature* W Pacific Ocean
Samoa, Independent State of *see* Samoa
112 D8 **Samobor** Zagreb, N Croatia 45°48′N 15°38′E
114 H10 **Samokov** *var.* Samakov. Sofia, W Bulgaria 42°19′N 23°34′E
111 H21 **Šamorín** *Ger.* Sommerein, *Hung.* Somorja. Trnavský Kraj, W Slovakia 48°01′N 17°18′E
115 M19 **Sámos** *prev.* Limín Vathéos. Sámos, Dodekánisa, Greece, Aegean Sea 37°45′N 26°58′E
115 M20 **Sámos** *island* Dodekánisa, Greece, Aegean Sea
Samosch *see* Szamos
168 I9 **Samosir, Pulau** *island* W Indonesia
Samothrace *see* Samothráki
115 K14 **Samothráki** Samothráki, NE Greece 40°28′N 25°31′E
115 J14 **Samothráki** *anc.* Samothrace. *island* NE Greece
115 A15 **Samothráki** *island* Iónia Nisiá, Greece, C Mediterranean Sea
Samotschin *see* Szamocin
Sampê *see* Xiangcheng
169 S13 **Sampit** Borneo, C Indonesia 02°30′S 112°30′E
169 S12 **Sampit, Sungai** ≈ Borneo, N Indonesia
Sampoku *see* Sanpoku
186 H7 **Sampun** New Britain, E Papua New Guinea 05°19′S 152°06′E
79 N24 **Sampwe** Katanga, SE Dem. Rep. Congo 09°17′S 27°22′E
167 R11 **Samraông** *prev.* Phumĭ Sâmraông, Phum Samrong. Siĕmréab, NW Cambodia 14°11′N 103°31′E
25 X8 **Sam Rayburn Reservoir** ⊠ Texas, SW USA
167 Q6 **Sam Sao, Phou** ⛰ Laos/Thailand
95 H22 **Samsø** *island* E Denmark
95 H23 **Samsø Bælt** *channel* E Denmark
167 T7 **Sâm Sơn** Thanh Hoa, N Vietnam 19°44′N 105°53′E
136 L11 **Samsun** *anc.* Amisus. Samsun, N Turkey 41°17′N 36°22′E
136 K11 **Samsun** ◈ *province* N Turkey
137 R9 **Samt'redia** *prev.* Samtredia. W Georgia 42°09′N 42°20′E
Samtredia *see* Samt'redia
59 E15 **Samuel, Represa de** ⊠ W Brazil
167 O14 **Samui, Ko** *island* SW Thailand
Samundari *see* Samundri
149 U9 **Samundri** *var.* Samundari. Punjab, E Pakistan 31°04′N 72°58′E
137 X10 **Samur** ≈ Azerbaijan/Russian Federation
137 Y11 **Samur-Abşeron Kanalı** *Rus.* Samur-Apsheronskiy Kanal. *canal* E Azerbaijan
Samur-Apsheronskiy Kanal *see* Samur-Abşeron Kanalı
167 O11 **Samut Prakan** *var.* Muang Samut Prakan, Paknam. Samut Prakan, C Thailand 13°36′N 100°36′E
167 O11 **Samut Sakhon** *var.* Maha Chai, Samut Sakorn, Tha Chin. Samut Sakhon, C Thailand 13°31′N 100°15′E
Samut Sakorn *see* Samut Sakhon
167 O11 **Samut Songkhram** *prev.* Meklong. Samut Songkhram, SW Thailand 13°25′N 100°01′E
77 N12 **San** Ségou, C Mali 13°21′N 04°57′W
111 O15 **San** ≈ SE Poland
141 O15 **Şan'ā'** *Eng.* Sana. ● (Yemen) W Yemen 15°24′N 44°14′E
112 F11 **Sana** ≈ NW Bosnia and Herzegovina
80 O12 **Sanaag** *off.* Gobolka Sanaag. ◈ *region* N Somalia
Sanaag, Gobolka *see* Sanaag
114 J8 **Sanadinovo** Pleven, N Bulgaria 43°33′N 25°00′E
195 P1 **Sanae** *South African research station* Antarctica 70°19′S 01°31′W
139 Y10 **Sanāf, Hawr as** ◎ S Iraq
79 E15 **Sanaga** ≈ C Cameroon
54 D12 **San Agustín** Huila, SW Colombia 01°53′N 76°14′W
171 R8 **San Agustin, Cape** *headland* Mindanao, S Philippines 06°17′N 126°12′E
37 Q13 **San Agustin, Plains of** *plain* New Mexico, SW USA
38 M16 **Sanak Islands** *island* Aleutian Islands, Alaska, USA
San Alessandro *see* Kita-Iō-jima
193 U10 **San Ambrosio, Isla** *Eng.* San Ambrosio Island. *island* W Chile
San Ambrosio Island *see* San Ambrosio, Isla
171 Q12 **Sanana** Pulau Sanana, E Indonesia 02°04′S 125°58′E
171 Q12 **Sanana, Pulau** *island* Maluku, E Indonesia
142 K5 **Sanandaj** *prev.* Sinneh. Kordestān, W Iran 35°18′N 47°01′E
35 P8 **San Andreas** California, W USA 38°10′N 120°40′W
2 C13 **San Andreas Fault** *fault* W USA
54 G8 **San Andrés** Santander, C Colombia 06°52′N 72°53′W
61 C20 **San Andrés de Giles** Buenos Aires, E Argentina 34°27′S 59°27′W
37 R14 **San Andres Mountains** ⛰ New Mexico, SW USA
41 S15 **San Andrés Tuxtla** *var.* Tuxtla. Veracruz-Llave, E Mexico 18°28′N 95°15′W
25 P8 **San Angelo** Texas, SW USA 31°28′N 100°26′W
107 A20 **San Antioco, Isola di** *island* W Italy
42 F4 **San Antonio** Toledo, S Belize 16°13′N 89°02′W
62 G11 **San Antonio** Valparaíso, C Chile 33°35′S 71°38′W
188 H6 **San Antonio** Saipan, S Northern Mariana Islands
37 R13 **San Antonio** New Mexico, SW USA 33°53′N 106°52′W
25 R12 **San Antonio** Texas, SW USA 29°25′N 98°30′W
54 M11 **San Antonio** Amazonas, S Venezuela 03°31′N 66°47′W
54 I7 **San Antonio** Barinas, C Venezuela 07°24′N 71°28′W
55 O5 **San Antonio** Monagas, NE Venezuela 10°03′N 63°45′W
25 S12 **San Antonio** ✈ Texas, SW USA 29°31′N 98°11′W
San Antonio *see* San Antonio del Táchira
San Antonio Abad *see* Sant Antoni de Portmany
25 U13 **San Antonio Bay** *inlet* Texas, SW USA
61 E22 **San Antonio, Cabo** *headland* E Argentina 36°45′S 56°40′W
44 A5 **San Antonio, Cabo de** *headland* W Cuba 21°51′N 84°58′W
105 T11 **San Antonio, Cabo de** *headland* E Spain 38°50′N 00°09′E
54 H7 **San Antonio de Caparo** Táchira, W Venezuela 07°34′N 71°28′W
62 J5 **San Antonio de los Cobres** Salta, NE Argentina 24°10′S 66°17′W
54 H7 **San Antonio del Táchira** *var.* San Antonio. Táchira, W Venezuela 07°48′N 72°28′W
35 T15 **San Antonio, Mount** ▲ California, W USA 34°18′N 117°37′W
63 K16 **San Antonio Oeste** Río Negro, E Argentina 40°45′S 64°58′W
25 T13 **San Antonio River** ≈ Texas, SW USA
54 J5 **Sanare** Lara, N Venezuela 09°45′N 69°39′W
103 T16 **Sanary-sur-Mer** Var, SE France 43°07′N 05°48′E
104 G3 **Sanata Uxía de Ribeira** *var.* Ribeira. Galicia, NW Spain 42°33′N 09°01′W
25 X8 **San Augustine** Texas, SW USA 31°32′N 94°09′W
San Augustine *see* Minami-Iō-jima
141 T13 **Sanāw** *var.* Sanaw. NE Yemen 18°N 51°E
41 O11 **San Bartolo** San Luis Potosí, C Mexico 22°20′N 100°05′W
107 L16 **San Bartolomeo in Galdo** Campania, S Italy 41°24′N 15°01′E
106 K13 **San Benedetto del Tronto** Marche, C Italy 42°57′N 13°53′E
42 E3 **San Benito** Petén, N Guatemala 16°56′N 89°53′W
25 T17 **San Benito** Texas, SW USA 26°07′N 97°37′W
54 E6 **San Benito Abad** Sucre, N Colombia 08°56′N 75°02′W
35 P11 **San Benito Mountain** ▲ California, W USA 36°21′N 120°37′W
35 O10 **San Benito River** ≈ California, W USA
108 H10 **San Bernardino** Graubünden, S Switzerland 46°21′N 09°13′E
35 U15 **San Bernardino** California, W USA 34°06′N 117°15′W
35 U15 **San Bernardino Mountains** ⛰ California, W USA
62 H11 **San Bernardo** Santiago, C Chile 33°37′S 70°45′W
40 J8 **San Bernardo** Durango, C Mexico 25°58′N 105°27′W
164 G12 **Sanbe-san** ▲ Kyūshū, SW Japan 35°09′N 132°36′E
San Bizenti-Barakaldo *see* San Vicente de Barakaldo
40 J12 **San Blas** Nayarit, C Mexico 21°35′N 105°20′W
40 H8 **San Blas** Sinaloa, C Mexico 26°05′N 108°44′W
San Blas *see* Kuna Yala
43 U14 **San Blas, Archipiélago de** *island group* NE Panama
23 Q10 **San Blas, Cape** *headland* Florida, SE USA 29°39′N 85°21′W
43 V14 **San Blas, Cordillera de** ⛰ NE Panama
62 J8 **San Blas de los Sauces** Catamarca, NW Argentina 28°18′S 67°12′W
106 G8 **San Bonifacio** Veneto, NE Italy 45°22′N 11°14′E
29 S12 **Sanborn** Iowa, C USA 43°10′N 95°39′W
40 M7 **San Buenaventura** Coahuila, NE Mexico 27°04′N 101°32′W
105 S5 **San Caprasio** ▲ N Spain 41°45′N 00°26′W
62 G13 **San Carlos** Bío Bío, C Chile 36°25′S 71°58′W
40 E9 **San Carlos** Baja California Sur, NW Mexico 24°52′N 112°15′W
41 N5 **San Carlos** Coahuila, NE Mexico 29°00′N 100°51′W
41 P9 **San Carlos** Tamaulipas, C Mexico 24°36′N 98°42′W
42 L12 **San Carlos** Río San Juan, S Nicaragua 11°06′N 84°46′W
43 T16 **San Carlos** West Panamá, C Panama 08°29′N 79°58′W
171 N3 **San Carlos** *off.* San Carlos City. Luzon, N Philippines 15°57′N 120°18′E
61 G20 **San Carlos** Maldonado, S Uruguay 34°46′S 54°58′W
36 M14 **San Carlos** Arizona, SW USA 33°21′N 110°27′W
54 K5 **San Carlos** Cojedes, N Venezuela 09°39′N 68°35′W
San Carlos *see* Quesada, Costa Rica
San Carlos *see* Luba, Equatorial Guinea
61 B17 **San Carlos Centro** Santa Fe, C Argentina 31°45′S 61°05′W
171 P6 **San Carlos City** Negros, C Philippines 10°34′N 123°24′E
San Carlos City *see* San Carlos
San Carlos de Ancud *see* Ancud
63 H16 **San Carlos de Bariloche** Río Negro, SW Argentina 41°08′S 71°15′W
61 B21 **San Carlos de Bolívar** Buenos Aires, E Argentina 36°15′S 61°06′W
54 H6 **San Carlos del Zulia** Zulia, W Venezuela 09°01′N 71°58′W
54 L12 **San Carlos de Río Negro** Amazonas, S Venezuela 01°54′N 67°54′W
San Carlos, Estrecho de *see* Falkland Sound
36 M14 **San Carlos Reservoir** ⊠ Arizona, SW USA
42 M12 **San Carlos, Río** ≈ N Costa Rica
65 D24 **San Carlos Settlement** East Falkland, Falkland Islands
61 C23 **San Cayetano** Buenos Aires, E Argentina 38°20′S 59°37′W
103 O8 **Sancerre** Cher, C France 47°19′N 02°53′E
158 G7 **Sanchakou** Xinjiang Uygur Zizhiqu, NW China 39°56′N 78°28′E
Sanchoku *see* Samcheok
41 O12 **San Ciro** San Luis Potosí, C Mexico 21°40′N 99°50′W
105 P10 **San Clemente** Castilla-La Mancha, C Spain 39°24′N 02°25′W
35 T16 **San Clemente** California, W USA 33°25′N 117°36′W
61 E21 **San Clemente del Tuyú** Buenos Aires, E Argentina 36°22′S 56°43′W
35 S17 **San Clemente Island** *island* Channel Islands, California, W USA
103 O9 **Sancoins** Cher, C France 46°49′N 03°00′E
61 B16 **San Cristóbal** Santa Fe, C Argentina 30°20′S 61°14′W
44 B4 **San Cristóbal** Pinar del Río, W Cuba 22°43′N 83°03′W
45 O9 **San Cristóbal** *var.* Benemérita de San Cristóbal. S Dominican Republic 18°27′N 70°07′W
54 H7 **San Cristóbal** Táchira, W Venezuela 07°46′N 72°15′W
187 N10 **San Cristobal** *var.* Makira. *island* SE Solomon Islands
San Cristóbal *see* San Cristóbal de Las Casas
41 U16 **San Cristóbal de Las Casas** *var.* San Cristóbal. Chiapas, SE Mexico 16°44′N 92°40′W
187 N10 **San Cristóbal, Isla** *var.* Chatham Island. *island* Galapagos Islands, Ecuador, E Pacific Ocean
42 D5 **San Cristóbal Verapaz** Alta Verapaz, C Guatemala 15°21′N 90°22′W
44 F6 **Sancti Spíritus** Sancti Spíritus, C Cuba 21°54′N 79°27′W
103 O11 **Sancy, Puy de** ▲ C France 45°33′N 02°48′E
95 D15 **Sand** Rogaland, S Norway 59°28′N 06°16′E
169 W7 **Sandakan** Sabah, East Malaysia 05°52′N 118°04′E
182 K9 **Sandalwood** South Australia 34°51′S 140°13′E
Sandalwood Island *see* Sumba, Pulau
94 D11 **Sandane** Sogn Og Fjordane, S Norway 61°47′N 06°14′E
114 G12 **Sandanski** *prev.* Sveti Vrach. Blagoevgrad, SW Bulgaria 41°36′N 23°19′E
Sandaohezi *see* Shawan
76 J11 **Sandaré** Kayes, W Mali 14°36′N 10°22′W
95 J19 **Sandared** Västra Götaland, S Sweden 57°43′N 12°47′E
94 N12 **Sandarne** Gävleborg, C Sweden 61°15′N 17°10′E
Sandaun *see* West Sepik
96 K4 **Sanday** *island* NE Scotland, United Kingdom
31 P15 **Sand Creek** ≈ Indiana, N USA
95 H15 **Sande** Vestfold, S Norway 59°34′N 10°13′E
95 H16 **Sandefjord** Vestfold, S Norway 59°10′N 10°15′E
77 O15 **Sandégué** E Ivory Coast 07°59′N 03°33′W
77 P14 **Sandema** N Ghana 10°42′N 01°17′W
37 O11 **Sanders** Arizona, SW USA 35°13′N 109°21′W
24 M11 **Sanderson** Texas, SW USA 30°08′N 102°25′W
23 U4 **Sandersville** Georgia, SE USA 32°58′N 82°48′W
92 H4 **Sandgerði** Suðurnes, SW Iceland 64°01′N 22°42′W
28 K14 **Sand Hills** ⛰ Nebraska, C USA
25 S14 **Sandia** Texas, SW USA 27°59′N 97°52′W
35 T17 **San Diego** California, W USA 32°43′N 117°09′W
25 S14 **San Diego** Texas, SW USA 27°47′N 98°15′W
136 F14 **Sandıklı** Afyon, W Turkey 38°28′N 30°17′E
152 L12 **Sandila** Uttar Pradesh, N India 27°05′N 80°37′E
San Dimitri Point *see* Il-Ponta ta' San Dimitri
San Dimitri, Ras *see* Il-Ponta ta' San Dimitri
168 J13 **Sanding, Selat** *strait* W Indonesia
30 J3 **Sand Island** *island* Apostle Islands, Wisconsin, N USA
95 C16 **Sandnes** Rogaland, S Norway 58°51′N 05°45′E
92 F13 **Sandnessjøen** Nordland, C Norway 66°00′N 12°37′E
Sandø *see* Sandoy
79 L24 **Sandoa** Katanga, S Dem. Rep. Congo 09°41′S 22°56′E
111 N15 **Sandomierz** *Rus.* Sandomir. Świętokrzyskie, C Poland 50°42′N 21°45′E
Sandomir *see* Sandomierz
54 C13 **Sandoná** Nariño, SW Colombia 01°18′N 77°28′W
106 I7 **San Donà di Piave** Veneto, NE Italy 45°38′N 12°34′E
124 K14 **Sandovo** Tverskaya Oblast', W Russian Federation 58°26′N 36°30′E
Sandoway *see* Thandwe
97 M24 **Sandown** S England, United Kingdom 50°40′N 01°11′W
95 B19 **Sandoy** *Dan.* Sandø. *island* C Faroe Islands
39 N16 **Sand Point** Popof Island, Alaska, USA 55°20′N 160°30′W
32 M7 **Sandpoint** Idaho, NW USA 48°16′N 116°33′W
65 N24 **Sand Point** *headland* E Tristan da Cunha
31 R7 **Sand Point** *headland* Michigan, N USA 43°54′N 83°24′W
93 H14 **Sandsele** Västerbotten, N Sweden 65°16′N 17°40′E
10 I14 **Sandspit** Moresby Island, British Columbia, SW Canada 53°14′N 131°50′W
27 P9 **Sand Springs** Oklahoma, C USA 36°08′N 96°06′W
29 W7 **Sandstone** Minnesota, N USA 46°07′N 92°51′W
36 K15 **Sand Tank Mountains** ⛰ Arizona, SW USA
31 S8 **Sandusky** Michigan, N USA 43°26′N 82°50′W
31 S11 **Sandusky** Ohio, N USA 41°27′N 82°42′W
31 S12 **Sandusky River** ≈ Ohio, N USA
83 D22 **Sandverhaar** Karas, S Namibia 26°50′S 17°25′E
95 L24 **Sandvig** Bornholm, E Denmark 55°15′N 14°45′E
95 H15 **Sandvika** Akershus, S Norway 59°54′N 10°29′E
94 N13 **Sandviken** Gävleborg, C Sweden 60°38′N 16°50′E
30 M11 **Sandwich** Illinois, N USA 41°39′N 88°37′W
Sandwich Island *see* Efate
Sandwich Islands *see* Hawai'ian Islands
153 V16 **Sandwip** *island* SE Bangladesh
11 U12 **Sandy Bay** Saskatchewan, C Canada 55°31′N 102°14′W
183 N16 **Sandy Cape** *headland* Tasmania, SE Australia 41°27′S 144°43′E
36 L3 **Sandy City** Utah, W USA 40°36′N 111°53′W
31 U12 **Sandy Creek** ≈ Ohio, N USA
21 O5 **Sandy Hook** Kentucky, S USA 38°05′N 83°09′W
18 K15 **Sandy Hook** *headland* New Jersey, NE USA 40°27′N 73°59′W
Sandykachi/Sandykgachy *see* Sandykgaçy
146 J15 **Sandykaçy** *var.* Sandykgachy. Maryyskiy Velayat, S Turkmenistan 36°34′N 62°28′E
Sandykgachy *see* Sandykaçy
146 J15 **Sandykgaçy** *var.* Sandykgachy, *Rus.* Sandykachi. Mary Welaýaty, S Turkmenistan 36°34′N 62°28′E
146 L13 **Sandykly Gumy** *Rus.* Peski Sandykly. *desert* E Turkmenistan
Sandykly, Peski *see* Sandykly Gumy
11 Q13 **Sandy Lake** Alberta, W Canada 55°50′N 113°30′W
12 B8 **Sandy Lake** Ontario, C Canada 53°00′N 93°25′W
12 B8 **Sandy Lake** ◎ Ontario, C Canada
23 S3 **Sandy Springs** Georgia, SE USA 33°57′N 84°23′W
24 H8 **San Elizario** Texas, SW USA 31°35′N 106°16′W
99 L25 **Sanem** Luxembourg, SW Luxembourg 49°33′N 05°56′E
42 K5 **San Esteban** Olancho, C Honduras 15°19′N 85°52′W
105 O6 **San Esteban de Gormaz** Castilla y León, N Spain 41°34′N 03°13′W
40 E5 **San Esteban, Isla** *island* NW Mexico
San Eugenio/San Eugenio del Cuareim *see* Artigas
62 H11 **San Felipe** *var.* San Felipe de Aconcagua. Valparaíso, C Chile 32°45′S 70°42′W
40 D3 **San Felipe** Baja California Norte, NW Mexico 31°03′N 114°52′W
40 N12 **San Felipe** Guanajuato, C Mexico 21°30′N 101°15′W
54 K5 **San Felipe** Yaracuy, NW Venezuela 10°25′N 68°40′W
44 B5 **San Felipe, Cayos de** *island group* W Cuba
San Felipe de Aconcagua *see* San Felipe
San Felipe de Puerto Plata *see* Puerto Plata
37 R11 **San Felipe Pueblo** New Mexico, SW USA 35°25′N 106°27′W
San Feliú de Guixols *see* Sant Feliu de Guíxols
193 T10 **San Félix, Isla** *Eng.* San Felix Island. *island* W Chile
San Felix Island *see* San Félix, Isla
40 C4 **San Fernando** *var.* Misión San Fernando. Baja California Norte, NW Mexico 29°58′N 115°14′W
41 P9 **San Fernando** Tamaulipas, C Mexico 24°50′N 98°10′W
171 N2 **San Fernando** Luzon, N Philippines 16°45′N 120°21′E
171 O3 **San Fernando** Luzon, N Philippines 15°01′N 120°41′E
104 J16 **San Fernando** *prev.* Isla de León. Andalucía, S Spain 36°28′N 06°12′W
45 U14 **San Fernando** Trinidad, Trinidad and Tobago 10°17′N 61°27′W
35 S15 **San Fernando** California, W USA 34°16′N 118°26′W
54 L7 **San Fernando** *var.* San Fernando de Apure. Apure, C Venezuela 07°54′N 67°28′W
San Fernando de Apure *see* San Fernando
54 L11 **San Fernando de Atabapo** Amazonas, S Venezuela 04°00′N 67°42′W
62 L8 **San Fernando del Valle de Catamarca** *var.* Catamarca. Catamarca, NW Argentina 28°28′S 65°46′W
San Fernando de Monte Cristi *see* Monte Cristi
41 P9 **San Fernando, Río** ≈ C Mexico
23 X11 **Sanford** Florida, SE USA 28°48′N 81°16′W
19 P9 **Sanford** Maine, NE USA 43°26′N 70°46′W
21 T10 **Sanford** North Carolina, SE USA 35°29′N 79°10′W
25 N2 **Sanford** Texas, SW USA 35°42′N 101°31′W
39 T10 **Sanford, Mount** ▲ Alaska, USA 62°21′N 144°12′W
42 G8 **San Francisco** *var.* Gotera, San Francisco Gotera. Morazán, E El Salvador 13°41′N 88°06′W
43 R16 **San Francisco** Veraguas, C Panama 08°19′N 80°59′W
171 N2 **San Francisco** *var.* Aurora. Luzon, N Philippines 13°22′N 122°31′E
35 L8 **San Francisco** California, W USA 37°47′N 122°25′W
54 H5 **San Francisco** Zulia, NW Venezuela 10°36′N 71°39′W
34 M8 **San Francisco** ✈ California, W USA 37°37′N 122°23′W
35 N9 **San Francisco Bay** *bay* California, W USA
61 C24 **San Francisco de Bellocq** Buenos Aires, E Argentina 38°42′S 60°01′W
40 I6 **San Francisco de Borja** Chihuahua, N Mexico 27°57′N 106°42′W
42 J6 **San Francisco de la Paz** Olancho, C Honduras 14°55′N 86°14′W
40 J7 **San Francisco del Oro** Chihuahua, N Mexico 26°52′N 105°50′W
40 M12 **San Francisco del Rincón** Jalisco, SW Mexico 21°00′N 101°51′W
45 O8 **San Francisco de Macorís** C Dominican Republic 19°19′N 70°15′W
San Francisco de Satipo *see* Satipo
San Francisco Gotera *see* San Francisco
San Francisco Telixtlahuaca *see* Telixtlahuaca
107 K23 **San Fratello** Sicilia, Italy, C Mediterranean Sea 38°00′N 14°35′E
San Fructuoso *see* Tacuarembó
82 C12 **Sanga** Kwanza Sul, NW Angola 11°10′S 15°27′E
56 C5 **San Gabriel** Carchi, N Ecuador 0°35′N 77°48′W
159 S15 **Sa'ngain** Xizang Zizhiqu, W China 30°47′N 98°45′E
154 E13 **Sangamner** Mahārāshtra, W India 19°37′N 74°18′E
152 H12 **Sāngāner** Rājasthān, N India 26°48′N 75°48′E
Sangan, Koh-i- *see* Sangān, Kūh-e
149 N6 **Sangān, Kūh-e** *Pash.* Koh-i-Sangan. ▲ C Afghanistan
123 P10 **Sangar** Respublika Sakha (Yakutiya), NE Russian Federation 63°48′N 127°37′E
155 H15 **Sangāreddi** Telangana, India 17°37′N 78°04′E
169 V11 **Sangasanga** Borneo, C Indonesia 0°36′S 117°12′E
103 N1 **Sangatte** Pas-de-Calais, N France 50°56′N 01°41′E
107 B19 **San Gavino Monreale** Sardegna, Italy, C Mediterranean Sea 39°33′N 08°47′E
57 D16 **Sangayan, Isla** *island* W Peru
30 L14 **Sangchris Lake** ⊠ Illinois, N USA
171 N16 **Sangeang, Pulau** *island* S Indonesia
116 I10 **Sângeorgiu de Pădure** *prev.* Erdăt-Sângeorz, Sîngeorgiu de Pădure, *Hung.* Erdőszentgyörgy. Mureș, C Romania 46°27′N 24°50′E
116 I9 **Sângeorz-Băi** *var.* Singeroz Băi, *Ger.* Rumänisch-Sankt-Georgen, *Hung.* Oláhszentgyőrgy; *prev.* Sîngeorz-Băi. Bistrița-Năsăud, N Romania 47°24′N 24°40′E
35 R10 **Sanger** California, W USA 36°42′N 119°33′W
25 T5 **Sanger** Texas, SW USA 33°21′N 97°10′W
Sângerei *see* Sîngerei
101 L15 **Sangerhausen** Sachsen-Anhalt, C Germany 51°29′N 11°18′E
45 S6 **San Germán** W Puerto Rico 18°05′N 67°02′W
San Germano *see* Cassino
161 N2 **Sanggan He** ≈ E China
169 Q11 **Sanggau** Borneo, C Indonesia 0°08′N 110°35′E
79 G17 **Sangha** ◈ *province* N Congo
79 H16 **Sangha** ≈ Central African Republic/Congo
79 G16 **Sangha-Mbaéré** ◈ *prefecture* SW Central African Republic
149 Q15 **Sānghar** Sind, SE Pakistan 26°10′N 68°59′E
115 F22 **Sangiás** ▲ S Greece 36°39′N 22°24′E
Sangihe, Kepulauan *see* Sangir, Kepulauan
171 Q9 **Sangihe, Pulau** *var.* Sangir. *island* N Indonesia
54 G8 **San Gil** Santander, C Colombia 06°35′N 73°08′W
121 P16 **San Ġiljan** *var.* St Julian's. N Malta 35°55′N 14°29′E
106 F12 **San Gimignano** Toscana, C Italy 43°30′N 11°00′E
148 M8 **Sangīn** *var.* Sangin. Helmand, S Afghanistan 32°03′N 64°50′E
Sangin *see* Sangīn
107 O21 **San Giovanni in Fiore** Calabria, SW Italy 39°15′N 16°42′E
107 M16 **San Giovanni Rotondo** Puglia, SE Italy 41°43′N 15°44′E
106 G12 **San Giovanni Valdarno** Toscana, C Italy 43°34′N 11°31′E
Sangir *see* Sangihe, Pulau
171 Q10 **Sangir, Kepulauan** *var.* Kepulauan Sangihe. *island group* N Indonesia
Sangiyn Dalay *see* Erdenedalay, Dundgovĭ, Mongolia
Sangiyn Dalay *see* Erdene, Govĭ-Altay, Mongolia
Sangiyn Dalay *see* Nomgon, Ömnögovĭ, Mongolia
Sangiyn Dalay *see* Öldziyt, Övörhangay, Mongolia
163 Y15 **Sangju** *Jap.* Shōshū. C South Korea 36°26′N 128°09′E
167 R11 **Sangkha** Surin, E Thailand 14°36′N 103°43′E
169 W10 **Sangkulirang** Borneo, N Indonesia 01°00′N 117°56′E
169 W10 **Sangkulirang, Teluk** *bay* Borneo, N Indonesia
155 E16 **Sāngli** Mahārāshtra, W India 16°55′N 74°37′E
79 E16 **Sangmélima** Sud, S Cameroon 02°57′N 11°56′E
35 V15 **San Gorgonio Mountain** ▲ California, W USA 34°06′N 116°50′W
37 T8 **Sangre de Cristo Mountains** ⛰ Colorado/New Mexico, C USA
61 A20 **San Gregorio** Santa Fe, C Argentina 34°18′S 62°02′W
61 F18 **San Gregorio de Polanco** Tacuarembó, C Uruguay 32°37′S 55°50′W
45 V14 **Sangre Grande** Trinidad, Trinidad and Tobago 10°35′N 61°08′W
159 N16 **Sangri** Xizang Zizhiqu, W China 29°17′N 92°01′E
152 H9 **Sangrūr** Punjab, NW India 30°16′N 75°52′E
44 I11 **Sangster** *off.* Sir Donald Sangster International Airport, *var.* Montego Bay. ✈ (Montego Bay) W Jamaica 18°30′N 77°54′W
59 G17 **Sangue, Rio do** ≈ W Brazil
105 R4 **Sangüesa** *Bas.* Zangoza. Navarra, N Spain 42°34′N 01°17′W
61 C16 **San Gustavo** Entre Ríos, E Argentina 30°41′S 59°23′W
Sangyuan *see* Wuqiao
40 C6 **San Hipólito, Punta** *headland* NW Mexico 26°57′N 114°00′W
23 W15 **Sanibel** Sanibel Island, Florida, SE USA 26°27′N 82°01′W
23 V15 **Sanibel Island** *island* Florida, SE USA
60 F13 **San Ignacio** Misiones, NE Argentina 27°15′S 55°32′W
42 F2 **San Ignacio** *prev.* Cayo, El Cayo. Cayo, W Belize 17°09′N 89°02′W
57 L16 **San Ignacio** El Beni, N Bolivia 14°54′S 65°35′W
57 O18 **San Ignacio** Santa Cruz, E Bolivia 16°23′S 60°59′W
42 M14 **San Ignacio** *var.* San Ignacio de Acosta. San José, W Costa Rica 09°46′N 84°10′W
40 E6 **San Ignacio** Baja California Sur, NW Mexico 27°18′N 112°51′W
40 J10 **San Ignacio** Sinaloa, W Mexico 23°55′N 106°25′W
56 B9 **San Ignacio** Cajamarca, N Peru 05°09′S 79°00′W
San Ignacio de Acosta *see* San Ignacio
40 D7 **San Ignacio, Laguna** *lagoon* W Mexico
12 I6 **Sanikiluaq** Belcher Islands, Nunavut, C Canada 55°20′N 77°50′W
171 O3 **San Ildefonso Peninsula** *peninsula* Luzon, N Philippines
Saniquillie *see* Sanniquellie
Sanirajak *see* Hall Beach
61 D20 **San Isidro** Buenos Aires, E Argentina 34°28′S 58°31′W
43 N14 **San Isidro** *var.* San Isidro de El General. San José, SE Costa Rica 09°28′N 83°42′W
San Isidro de El General *see* San Isidro
54 E5 **San Jacinto** Bolívar, N Colombia 09°53′N 75°06′W
35 U16 **San Jacinto** California, W USA 33°47′N 116°58′W
35 V15 **San Jacinto Peak** ▲ California, W USA 33°48′N 116°40′W
61 F14 **San Javier** Misiones, NE Argentina 27°50′S 55°06′W
61 C16 **San Javier** Santa Fe, C Argentina 30°35′S 59°59′W
105 S13 **San Javier** Murcia, SE Spain 37°49′N 00°50′W
61 D18 **San Javier** Río Negro, W Uruguay 32°41′S 58°08′W
61 C16 **San Javier, Río** ≈ C Argentina
160 L12 **Sanjiang** *var.* Guyi, Sanjiang Dongzu Zizhixian. Guangxi Zhuangzu Zizhiqu, S China 25°46′N 109°26′E
Sanjiang *see* Jinping, Guizhou
Sanjiang Dongzu Zizhixian *see* Sanjiang
Sanjiaocheng *see* Haiyan
165 N11 **Sanjō** *var.* Sanzyô. Niigata, Honshū, C Japan 37°39′N 139°00′E
57 M15 **San Joaquín** El Beni, N Bolivia 13°06′S 64°46′W
55 O6 **San Joaquín** Anzoátegui, NE Venezuela 09°21′N 64°30′W
35 O9 **San Joaquin River** ≈ California, W USA
35 P10 **San Joaquin Valley** *valley* California, W USA
61 A18 **San Jorge** Santa Fe, C Argentina 31°50′S 61°50′W
40 D3 **San Jorge, Bahía de** *bay* NW Mexico
63 J19 **San Jorge, Golfo** *var.* Gulf of San Jorge. *gulf* S Argentina
San Jorge, Gulf of *see* San Jorge, Golfo
San Jorge, Isla de *see* Weddell Island
61 F14 **San José** Misiones, NE Argentina 27°46′S 55°47′W
57 P19 **San José** *var.* San José de Chiquitos. Santa Cruz, E Bolivia 14°13′S 68°05′W
42 M14 **San José** ● (Costa Rica) San José, C Costa Rica 09°55′N 84°05′W
42 C7 **San José** *var.* Puerto San José. Escuintla, S Guatemala 14°00′N 90°50′W
40 G6 **San José** Sonora, NW Mexico 27°32′N 110°09′W
188 K8 **San Jose** Tinian, S Northern Mariana Islands 15°00′S 145°38′E
105 U11 **San José** Eivissa, Spain, W Mediterranean Sea 38°55′N 01°18′E
35 N9 **San Jose** California, W USA 37°18′N 121°53′W
54 H5 **San José** Zulia, NW Venezuela 10°02′N 72°24′W
42 M14 **San José** *off.* Provincia de San José. ◈ *province* W Costa Rica
61 E19 **San José** ◈ *department* S Uruguay
42 M13 **San José** ✈ Alajuela, C Costa Rica 10°03′N 84°12′W

◆ Country
● Country Capital
◇ Dependent Territory
○ Dependent Territory Capital
◈ Administrative Regions
✈ International Airport
▲ Mountain
⛰ Mountain Range
Volcano
≈ River
◎ Lake
⊠ Reservoir

San José – Santana, Coxilha de

San José *see* San José del Guaviare, Colombia
San Jose *see* Oleai
San Jose *see* Sant Josep de sa Talaia, Ibiza, Spain
San José *see* San José de Mayo, Uruguay
171 O3 **San Jose City** Luzon, N Philippines 15°49′N 120°57′E
San José de Chiquitos *see* San José
San José de Cúcuta *see* Cúcuta
61 D16 **San José de Feliciano** Entre Ríos, E Argentina 30°26′S 58°46′W
55 O6 **San José de Guanipa** *var.* El Tigrito. Anzoátegui, NE Venezuela 08°54′N 64°10′W
62 I9 **San José de Jáchal** San Juan, W Argentina 30°15′S 68°46′W
40 G10 **San José del Cabo** Baja California Sur, NW Mexico 23°01′N 109°40′W
54 G12 **San José del Guaviare** *var.* San José. Guaviare, S Colombia 02°34′N 72°38′W
61 E20 **San José de Mayo** *var.* San José. San José, S Uruguay 34°20′S 56°42′W
54 I10 **San José de Ocuné** Vichada, E Colombia 04°10′N 70°21′W
41 O9 **San José de Raíces** Nuevo León, NE Mexico 24°32′N 100°15′W
63 K17 **San José, Golfo** *gulf* E Argentina
40 F9 **San José, Isla** *island* NW Mexico
43 U16 **San José, Isla** *island* SE Panama
25 U14 **San Jose Island** *island* Texas, SW USA
San José, Provincia de *see* San José
62 I10 **San Juan** San Juan, W Argentina 31°37′S 68°27′W
45 N9 **San Juan** *var.* San Juan de la Maguana. C Dominican Republic 18°49′N 71°12′W
57 E17 **San Juan** Ica, S Peru 15°22′S 75°07′W
45 U5 **San Juan** ○ (Puerto Rico) NE Puerto Rico 18°28′N 66°06′W
62 H10 **San Juan** *off.* Provincia de San Juan. ◈ *province* W Argentina
San Juan *see* San Juan de los Morros
62 O7 **San Juan Bautista** Misiones, S Paraguay 26°40′S 57°08′W
35 O10 **San Juan Bautista** California, W USA 36°50′N 121°34′W
San Juan Bautista *see* Villahermosa
San Juan Bautista Cuicatlán *see* Cuicatlán
San Juan Bautista Tuxtepec *see* Tuxtepec
79 C17 **San Juan, Cabo** *headland* S Equatorial Guinea 01°09′N 09°25′E
San Juan de Alicante *see* Sant Joan d'Alacant
54 H7 **San Juan de Colón** Táchira, NW Venezuela 08°02′N 72°17′W
40 L9 **San Juan de Guadalupe** Durango, C Mexico 25°12′N 100°50′W
San Juan de la Maguana *see* San Juan
54 G4 **San Juan del Cesar** La Guajira, N Colombia 10°45′N 73°00′W
40 L15 **San Juan de Lima, Punta** *headland* SW Mexico 18°34′N 103°40′W
42 I8 **San Juan de Limay** Estelí, NW Nicaragua 13°10′N 86°36′W
43 N12 **San Juan del Norte** *var.* Greytown. Río San Juan, SE Nicaragua 10°58′N 83°40′W
54 K4 **San Juan de los Cayos** Falcón, N Venezuela 11°11′N 68°27′W
40 M12 **San Juan de los Lagos** Jalisco, C Mexico 21°15′N 102°15′W
54 L5 **San Juan de los Morros** *var.* San Juan. Guárico, N Venezuela 09°53′N 67°23′W
40 K9 **San Juan del Río** Durango, C Mexico 25°12′N 100°50′W
41 O13 **San Juan del Río** Querétaro de Arteaga, C Mexico 20°24′N 100°00′W
42 J11 **San Juan del Sur** Rivas, SW Nicaragua 11°16′N 85°51′W
54 M9 **San Juan de Manapiare** Amazonas, S Venezuela 05°15′N 66°05′W
40 E7 **San Juanico** Baja California Sur, NW Mexico
40 D7 **San Juanico, Punta** *headland* NW Mexico 26°01′N 112°17′W
32 G6 **San Juan Islands** *island group* Washington, NW USA
40 I6 **San Juanito** Chihuahua, N Mexico
40 I12 **San Juanito, Isla** *island* C Mexico
37 R8 **San Juan Mountains** ▲ Colorado, C USA
54 E5 **San Juan Nepomuceno** Bolívar, NW Colombia 09°57′N 75°06′W
44 E5 **San Juan, Pico** ▲ C Cuba 21°58′N 80°10′W
San Juan, Provincia de *see* San Juan
191 W15 **San Juan, Punta** *headland* Easter Island, Chile, E Pacific Ocean 27°03′S 109°22′W
42 M12 **San Juan, Río** ≈ Costa Rica/Nicaragua
41 S15 **San Juan, Río** ≈ SE Mexico
37 O8 **San Juan River** ≈ Colorado/Utah, W USA
San Julián *see* Puerto San Julián
61 B17 **San Justo** Santa Fe, C Argentina 30°47′S 60°32′W
109 W5 **Sankt Aegyd am Neuwalde** Niederösterreich, E Austria 47°51′N 15°34′E
109 U9 **Sankt Andrä** *Slvn.* Šent Andraž. Kärnten, S Austria 46°46′N 14°49′E
Sankt Andrä *see* Szentendre
Sankt Anna *see* Sântana
108 K8 **Sankt Anton-am-Arlberg** Vorarlberg, W Austria 47°08′N 10°11′E
101 E16 **Sankt Augustin** Nordrhein-Westfalen, W Germany 50°46′N 07°10′E
Sankt-Bartholomäi *see* Palamuse
101 F24 **Sankt Blasien** Baden-Württemberg, SW Germany 47°43′N 08°09′E
109 R3 **Sankt Florian am Inn** Oberösterreich, N Austria 48°24′N 13°27′E
108 I7 **Sankt Gallen** *var.* St. Gallen, *Eng.* Saint Gall, *Fr.* St-Gall. Sankt Gallen, NE Switzerland 47°25′N 09°23′E
108 H8 **Sankt Gallen** *var.* St.Gallen, *Eng.* Saint Gall, *Fr.* St-Gall. ◈ *canton* NE Switzerland
108 J8 **Sankt Gallenkirch** Vorarlberg, W Austria 47°00′N 10°59′E
109 Q5 **Sankt Georgen** Salzburg, N Austria 47°59′N 12°57′E
Sankt Georgen *see* Đurđevac
Sankt-Georgen *see* Sfântu Gheorghe
109 R6 **Sankt Gilgen** Salzburg, NW Austria 47°46′N 13°21′E
Sankt Gotthard *see* Szentgotthárd
101 E20 **Sankt Ingbert** Saarland, SW Germany 49°17′N 07°07′E
Sankt-Jakobi *see* Viru-Jaagupi, Lääne-Virumaa, Estonia
Sankt-Jakobi *see* Pärnu-Jaagupi, Pärnumaa, Estonia
Sankt Johann *see* Sankt Johann in Tirol
109 T7 **Sankt Johann am Tauern** Steiermark, E Austria 47°20′N 14°27′E
109 Q7 **Sankt Johann im Pongau** Salzburg, NW Austria 47°22′N 13°13′E
109 P6 **Sankt Johann in Tirol** *var.* Sankt Johann. Tirol, W Austria 47°32′N 12°26′E
Sankt-Johannis *see* Järva-Jaani
108 L8 **Sankt Leonhard** Tirol, W Austria 47°05′N 10°53′E
Sankt Margarethen *see* Sankt Margarethen im Burgenland
109 Y5 **Sankt Margarethen im Burgenland** *var.* Sankt Margarethen. Burgenland, E Austria 47°49′N 16°38′E
Sankt Martin *see* Martin
109 X8 **Sankt Martin an der Raab** Burgenland, SE Austria 46°59′N 16°12′E
109 U7 **Sankt Michael in Obersteiermark** Steiermark, SE Austria 47°21′N 14°59′E
Sankt Michel *see* Mikkeli
Sankt Moritz *see* St. Moritz
108 E11 **Sankt Niklaus** Valais, S Switzerland 46°09′N 07°48′E
109 S7 **Sankt Nikolai** *var.* Sankt Nikolai im Sölktal. Steiermark, SE Austria 47°18′N 14°04′E
Sankt Nikolai im Sölktal *see* Sankt Nikolai
109 U9 **Sankt Paul** *var.* Sankt Paul im Lavanttal. Kärnten, S Austria 46°42′N 14°53′E
Sankt Paul im Lavanttal *see* Sankt Paul
Sankt Peter *see* Pivka
109 W9 **Sankt Peter am Ottersbach** Steiermark, SE Austria 46°49′N 15°48′E
124 J13 **Sankt-Peterburg** *prev.* Leningrad, Petrograd, *Eng.* Saint Petersburg, *Fin.* Pietari. Leningradskaya Oblast', NW Russian Federation 59°55′N 30°25′E
100 H8 **Sankt Peter-Ording** Schleswig-Holstein, N Germany 54°18′N 08°37′E
109 V4 **Sankt Pölten** Niederösterreich, N Austria 48°14′N 15°38′E
109 W7 **Sankt Ruprecht** *var.* Sankt Ruprecht an der Raab. Steiermark, SE Austria 47°10′N 15°41′E
Sankt Ruprecht an der Raab *see* Sankt Ruprecht
Sankt-Ulrich *see* Ortisei
109 T4 **Sankt Valentin** Niederösterreich, C Austria 48°11′N 14°33′E
Sankt Veit am Flaum *see* Rijeka
109 T9 **Sankt Veit an der Glan** *Slvn.* Št. Vid. Kärnten, S Austria 46°47′N 14°22′E
99 M21 **Sankt-Vith** *var.* Saint-Vith. Liège, E Belgium 50°17′N 06°07′E
101 E20 **Sankt Wendel** Saarland, SW Germany 49°28′N 07°10′E
109 R6 **Sankt Wolfgang** Salzburg, NW Austria 47°43′N 13°30′E
79 K21 **Sankuru** ≈ C Dem. Rep. Congo
40 D8 **San Lázaro, Cabo** *headland* NW Mexico 24°46′N 112°15′W
137 O16 **Şanlıurfa** *prev.* Sanli Urfa, Urfa; *anc.* Edessa. şanlıurfa, S Turkey 37°08′N 38°45′E
137 O16 **Şanlıurfa** *prev.* Urfa. ◈ *province* SE Turkey
Sanli Urfa *see* Şanlıurfa
137 O16 **Şanlıurfa Yaylası** *plateau* SE Turkey
61 B18 **San Lorenzo** Santa Fe, C Argentina 32°45′S 60°45′W
57 M21 **San Lorenzo** Tarija, S Bolivia 21°25′S 64°45′W
56 C5 **San Lorenzo** Esmeraldas, N Ecuador 01°15′N 78°51′W
42 H8 **San Lorenzo** Valle, S Honduras 13°24′N 87°27′W
56 A6 **San Lorenzo, Cabo** *headland* W Ecuador 0°57′S 80°49′W
105 N8 **San Lorenzo de El Escorial** *var.* El Escorial. Madrid, C Spain 40°36′N 04°07′W
40 E5 **San Lorenzo, Isla** *island* NW Mexico
57 C14 **San Lorenzo, Isla** *island* W Peru
63 G20 **San Lorenzo, Monte** ▲ S Argentina 47°40′S 72°12′W
40 I9 **San Lorenzo, Río** ≈ C Mexico
104 J15 **Sanlúcar de Barrameda** Andalucía, S Spain 36°46′N 06°21′W
104 J14 **Sanlúcar la Mayor** Andalucía, S Spain 37°24′N 06°13′W
40 E6 **San Lucas** *var.* Cabo San Lucas. Baja California Sur, NW Mexico 27°14′N 112°15′W
40 F11 **San Lucas** Baja California Sur, NW Mexico 22°50′N 109°52′W
40 G11 **San Lucas, Cabo** *var.* San Lucas Cape. *headland* NW Mexico 22°52′N 109°53′W
San Lucas Cape *see* San Lucas, Cabo
62 J11 **San Luis** San Luis, C Argentina 33°18′S 66°18′W
42 E4 **San Luis** Petén, NE Guatemala 16°16′N 89°27′W
42 M7 **San Luis** Región Autónoma Atlántico Norte, NE Nicaragua 13°59′N 84°10′W
36 H15 **San Luis** Arizona, SW USA 32°27′N 114°45′W
37 T8 **San Luis** Colorado, C USA 37°09′N 105°24′W
54 J4 **San Luis** Falcón, N Venezuela 11°09′N 69°39′W
62 J11 **San Luis** *off.* Provincia de San Luis. ◈ *province* C Argentina
41 N12 **San Luis de la Paz** Guanajuato, C Mexico 21°15′N 100°33′W
40 K8 **San Luis del Cordero** Durango, C Mexico 25°25′N 104°09′W
40 D4 **San Luis, Isla** *island* NW Mexico
42 E6 **San Luis Jilotepeque** Jalapa, SE Guatemala 14°40′N 89°42′W
57 M16 **San Luis, Laguna de** ◎ NW Bolivia
35 P13 **San Luis Obispo** California, W USA 35°17′N 120°40′W
37 R7 **San Luis Peak** ▲ Colorado, C USA 37°59′N 106°55′W
41 N11 **San Luis Potosí** San Luis Potosí, C Mexico 22°10′N 100°57′W
41 N11 **San Luis Potosí** ◈ *state* C Mexico
San Luis, Provincia de *see* San Luis
35 O10 **San Luis Reservoir** ⊠ California, W USA
40 D2 **San Luis Río Colorado** *var.* San Luis Río Colorado. Sonora, NW Mexico 32°26′N 114°48′W
San Luis Río Colorado *see* San Luis Río Colorado
37 S8 **San Luis Valley** *basin* Colorado, C USA
107 C19 **Sanluri** Sardegna, Italy, C Mediterranean Sea 39°34′N 08°54′E
61 D23 **San Manuel** Buenos Aires, E Argentina 37°47′S 58°50′W
36 M15 **San Manuel** Arizona, SW USA 32°36′N 110°37′W
106 F11 **San Marcello Pistoiese** Toscana, C Italy 44°03′N 10°46′E
107 N20 **San Marco Argentano** Calabria, SW Italy 39°31′N 16°07′E
54 E6 **San Marcos** Sucre, N Colombia 08°38′N 75°10′W
42 M14 **San Marcos** San José, C Costa Rica 09°39′N 84°00′W
42 B5 **San Marcos** San Marcos, W Guatemala 14°58′N 91°48′W
42 F6 **San Marcos** Ocotepeque, SW Honduras 14°23′N 88°57′W
41 O16 **San Marcos** Guerrero, S Mexico 16°45′N 99°22′W
25 S11 **San Marcos** Texas, SW USA 29°54′N 97°57′W
42 A5 **San Marcos** *off.* Departamento de San Marcos. ◈ *department* W Guatemala
San Marcos de Arica *see* Arica
San Marcos, Departamento de *see* San Marcos
40 E6 **San Marcos, Isla** *island* NW Mexico
106 H11 **San Marino** ● (San Marino) C San Marino 43°54′N 12°27′E
106 I11 **San Marino** *off.* Republic of San Marino. ◆ *republic* S Europe
San Marino, Republic of *see* San Marino
62 I11 **San Martín** Mendoza, C Argentina 33°05′S 68°28′W
54 F11 **San Martín** Meta, C Colombia 03°43′N 73°42′W
56 D11 **San Martín** *off.* Departamento de San Martín. ◈ *department* C Peru
194 I5 **San Martín** *Argentinian research station* Antarctica 68°18′S 67°03′W
63 H16 **San Martín de los Andes** Neuquén, W Argentina 40°11′S 71°22′W
San Martín, Departamento de *see* San Martín
104 M8 **San Martín de Valdeiglesias** Madrid, C Spain 40°21′N 04°24′W
63 G21 **San Martín, Lago** *var.* Lago O'Higgins. ◎ S Argentina
106 H6 **San Martino di Castrozza** Trentino-Alto Adige, N Italy 46°16′N 11°50′E
57 N16 **San Martín, Río** ≈ N Bolivia
San Martín Texmelucan *see* Texmelucan
35 N9 **San Mateo** California, W USA 37°33′N 122°19′W
55 O6 **San Mateo** Anzoátegui, NE Venezuela 09°48′N 64°36′W
42 B4 **San Mateo Ixtatán** Huehuetenango, W Guatemala 15°50′N 91°30′W
57 Q18 **San Matías** Santa Cruz, E Bolivia 16°20′S 58°24′W
63 K16 **San Matías, Golfo** *var.* Gulf of San Matías. *gulf* E Argentina
San Matías, Gulf of *see* San Matías, Golfo
15 O8 **Sanmaur** Québec, SE Canada 47°52′N 73°47′W
161 T10 **Sanmen Wan** *bay* E China
160 M6 **Sanmenxia** *var.* Shan Xian. Henan, C China 34°46′N 111°17′E
Sânmiclăuș Mare *see* Sânnicolau Mare
61 D14 **San Miguel** Corrientes, NE Argentina 28°02′S 57°41′W
57 L16 **San Miguel** El Beni, N Bolivia 16°43′S 61°06′W
42 G8 **San Miguel** San Miguel, SE El Salvador 13°27′N 88°11′W
40 L6 **San Miguel** Coahuila, N Mexico 29°10′N 101°28′W
40 J9 **San Miguel** *var.* San Miguel de Cruces. Durango, C Mexico 24°25′N 105°55′W
43 U16 **San Miguel** Panamá, SE Panama 08°27′N 78°51′W
35 P12 **San Miguel** California, W USA 35°45′N 120°42′W
42 B9 **San Miguel** ◈ *department* E El Salvador
41 N13 **San Miguel de Allende** Guanajuato, C Mexico 20°56′N 100°48′W
San Miguel de Cruces *see* San Miguel
San Miguel de Ibarra *see* Ibarra
61 D21 **San Miguel del Monte** Buenos Aires, E Argentina 35°26′S 58°50′W
62 J7 **San Miguel de Tucumán** *var.* Tucumán. Tucumán, N Argentina 26°47′S 65°15′W
43 V16 **San Miguel, Golfo de** *gulf* S Panama
35 P15 **San Miguel Island** *island* California, W USA
42 L11 **San Miguelito** Río San Juan, S Nicaragua 11°22′N 84°54′W
43 T15 **San Miguelito** Panamá, C Panama 08°58′N 79°31′W
57 N18 **San Miguel, Río** ≈ E Bolivia
56 D6 **San Miguel, Río** ≈ Colombia/Ecuador
40 I7 **San Miguel, Río** ≈ N Mexico
42 G8 **San Miguel, Volcán de** ※ SE El Salvador 13°27′N 88°18′W
161 Q12 **Sanming** Fujian, SE China 26°11′N 117°37′E
106 F11 **San Miniato** Toscana, C Italy 43°40′N 10°53′E
San Murezzan *see* St. Moritz
Sannār *see* Sennar
107 M15 **Sannicandro Garganico** Puglia, SE Italy 41°50′N 15°32′E
40 H6 **San Nicolás** Sonora, NW Mexico 28°31′N 109°24′W
61 C19 **San Nicolás de los Arroyos** Buenos Aires, E Argentina 33°20′S 60°13′W
35 R16 **San Nicolas Island** *island* Channel Islands, California, W USA
Sânnicolaul-Mare *see* Sânnicolau Mare
116 E11 **Sânnicolau Mare** *var.* Sânnicolaul-Mare, *Hung.* Nagyszentmiklós; *prev.* Sânmiclăuș Mare, Sînnicolau Mare. Timiș, W Romania 46°05′N 20°38′E
123 Q6 **Sannikova, Proliv** *strait* NE Russian Federation
76 K16 **Sanniquellie** *var.* Saniquillie. NE Liberia 07°24′N 08°45′W
165 R7 **Sannohe** Aomori, Honshū, C Japan 40°23′N 141°16′E
Sanntaler Alpen *see* Kamniško-Savinjske Alpe
111 O17 **Sanok** Podkarpackie, SE Poland 49°31′N 22°14′E
54 E5 **San Onofre** Sucre, NW Colombia 09°45′N 75°33′W
57 K21 **San Pablo** Potosí, S Bolivia 21°43′S 66°38′W
171 O4 **San Pablo** *off.* San Pablo City. Luzon, N Philippines 14°04′N 121°16′E
San Pablo Balleza *see* Balleza
35 N8 **San Pablo Bay** *bay* California, W USA
San Pablo City *see* San Pablo
40 C6 **San Pablo, Punta** *headland* NW Mexico 27°12′N 114°30′W
43 R16 **San Pablo, Río** ≈ C Panama
171 P4 **San Pascual** Burias Island, C Philippines 13°06′N 122°59′E
121 Q16 **San Pawl il Baħar** *Eng.* Saint Paul's Bay. E Malta 35°57′N 14°24′E
61 C19 **San Pedro** Buenos Aires, E Argentina 33°43′S 59°45′W
62 K5 **San Pedro** Jujuy, N Argentina 24°12′S 64°55′W
60 G13 **San Pedro** Misiones, NE Argentina 26°38′S 54°12′W
42 H1 **San Pedro** Corozal, NE Belize 17°58′N 87°55′W
76 M17 **San-Pédro** S Ivory Coast 04°45′N 06°37′W
40 L8 **San Pedro** *var.* San Pedro de las Colonias. Coahuila, NE Mexico 25°47′N 102°57′W
62 O5 **San Pedro** San Pedro, SE Paraguay 24°08′S 57°08′W
62 O6 **San Pedro** *off.* Departamento de San Pedro. ◈ *department* C Paraguay
44 G6 **San Pedro** ≈ C Cuba
77 N16 **San Pedro** ✕ (Yamoussoukro) C Ivory Coast 06°49′N 05°14′W
San Pedro *see* San Pedro del Pinatar
42 D5 **San Pedro Carchá** Alta Verapaz, C Guatemala 15°30′N 90°12′W
35 S16 **San Pedro Channel** *channel* California, W USA
62 I5 **San Pedro de Atacama** Antofagasta, N Chile 22°52′S 68°10′W
San Pedro de Durazno *see* Durazno
40 G5 **San Pedro de la Cueva** Sonora, NW Mexico 29°17′N 109°47′W
San Pedro de las Colonias *see* San Pedro
56 B11 **San Pedro de Lloc** La Libertad, NW Peru 07°26′S 79°31′W
105 S13 **San Pedro del Pinatar** *var.* San Pedro. Murcia, SE Spain 37°50′N 00°47′W
45 P9 **San Pedro de Macorís** SE Dominican Republic 18°30′N 69°18′W
San Pedro, Departamento de *see* San Pedro
40 C3 **San Pedro Mártir, Sierra** ▲ NW Mexico
San Pedro Pochutla *see* Pochutla
42 D2 **San Pedro, Río** ≈ Guatemala/Mexico
40 K10 **San Pedro, Río** ≈ C Mexico
104 J10 **San Pedro, Sierra de** ▲ W Spain
42 G5 **San Pedro Sula** Cortés, NW Honduras 15°26′N 88°01′W
San Pedro Tapanatepec *see* Tapanatepec
62 I4 **San Pedro, Volcán** ※ N Chile 21°46′S 68°13′W
106 E7 **San Pellegrino Terme** Lombardia, N Italy 45°53′N 09°42′E
25 T16 **San Perlita** Texas, SW USA 26°30′N 97°38′W
San Pietro *see* Supetar
San Pietro del Carso *see* Pivka
107 A20 **San Pietro, Isola di** *island* W Italy
32 K7 **Sanpoil River** ≈ Washington, NW USA
165 O9 **Sanpoku** *var.* Sampoku. Niigata, Honshū, C Japan 38°32′N 139°33′E
40 C3 **San Quintín** Baja California Norte, NW Mexico 30°28′N 115°58′W
40 B3 **San Quintín, Bahía de** *bay* NW Mexico
40 B3 **San Quintín, Cabo** *headland* NW Mexico 30°22′N 116°01′W
62 I12 **San Rafael** Mendoza, W Argentina 34°44′S 68°15′W
41 N9 **San Rafael** Nuevo León, NE Mexico 25°01′N 100°33′W
34 M8 **San Rafael** California, W USA 37°58′N 122°31′W
37 Q11 **San Rafael** New Mexico, SW USA 35°03′N 107°52′W
54 H4 **San Rafael** *var.* El Moján. Zulia, NW Venezuela 10°58′N 71°45′W
42 J8 **San Rafael del Norte** Jinotega, NW Nicaragua 13°12′N 86°06′W
42 J10 **San Rafael del Sur** Managua, SW Nicaragua 11°51′N 86°24′W
36 M5 **San Rafael Knob** ▲ Utah, W USA 38°46′N 110°45′W
35 Q14 **San Rafael Mountains** ▲ California, W USA
42 M13 **San Ramón** Alajuela, C Costa Rica 10°04′N 84°31′W
57 E14 **San Ramón** Junín, C Peru 11°08′S 75°18′W
61 F19 **San Ramón** Canelones, S Uruguay 34°18′S 55°55′W
62 K5 **San Ramón de la Nueva Orán** Salta, N Argentina 23°08′S 64°20′W
57 O16 **San Ramón, Río** ≈ E Bolivia
106 B11 **San Remo** Liguria, NW Italy 43°48′N 07°47′E
54 J3 **San Román, Cabo** *headland* NW Venezuela 12°10′N 70°01′W
61 C15 **San Roque** Corrientes, NE Argentina 28°35′S 58°45′W
188 I4 **San Roque** Saipan, S Northern Mariana Islands 15°15′S 145°47′E
104 K16 **San Roque** Andalucía, S Spain 36°13′N 05°23′W
25 R9 **San Saba** Texas, SW USA 31°13′N 98°44′W
25 Q9 **San Saba River** ≈ Texas, SW USA
61 D17 **San Salvador** Entre Ríos, E Argentina 31°38′S 58°30′W
42 F7 **San Salvador** ● (El Salvador) San Salvador, SW El Salvador 13°42′N 89°12′W
42 A10 **San Salvador** ◈ *department* C El Salvador
42 F8 **San Salvador** ✕ La Paz, S El Salvador 13°27′N 89°04′W
44 K4 **San Salvador** *prev.* Watlings Island. *island* E The Bahamas
62 J5 **San Salvador de Jujuy** *var.* Jujuy. Jujuy, N Argentina 24°10′S 65°20′W
42 F7 **San Salvador, Volcán de** ※ C El Salvador 13°58′N 89°14′W
77 Q14 **Sansanné-Mango** *var.* Mango. N Togo 10°21′N 00°28′E
45 S5 **San Sebastián** W Puerto Rico 18°21′N 67°00′W
63 J24 **San Sebastián, Bahía** *bay* S Argentina
Sansenhō *see* Sacheon
106 H12 **Sansepolcro** Toscana, C Italy 43°35′N 12°12′E
107 M16 **San Severo** Puglia, SE Italy 41°41′N 15°23′E
112 F11 **Sanski Most** ◈ Federacija Bosne I Hercegovine, NW Bosnia and Herzegovina
171 W12 **Sansundi** Papua, E Indonesia 0°42′S 135°48′E
162 K9 **Sant** *var.* Mayhan. Övörhangay, C Mongolia 46°02′N 104°00′E
104 K11 **Santa Amalia** Extremadura, W Spain 39°00′N 06°01′W
60 F13 **Santa Ana** Misiones, NE Argentina 27°22′S 55°34′W
57 L16 **Santa Ana** El Beni, N Bolivia 13°43′S 65°37′W
42 E7 **Santa Ana** Santa Ana, NW El Salvador 13°59′N 89°34′W
40 F4 **Santa Ana** Sonora, NW Mexico 30°31′N 111°08′W
35 T16 **Santa Ana** California, W USA 33°45′N 117°52′W
55 N6 **Santa Ana** Nueva Esparta, NE Venezuela 09°15′N 64°39′W
42 A9 **Santa Ana** ◈ *department* NW El Salvador
Santa Ana de Coro *see* Coro
35 U16 **Santa Ana Mountains** ▲ California, W USA
42 E7 **Santa Ana, Volcán de** *var.* La Matepec. ※ W El Salvador 13°49′N 89°36′W
42 G6 **Santa Bárbara** Santa Bárbara, W Honduras 14°56′N 88°11′W
40 J7 **Santa Barbara** Chihuahua, N Mexico 26°46′N 105°46′W
35 Q14 **Santa Barbara** California, W USA 34°24′N 119°40′W
54 L11 **Santa Bárbara** Amazonas, S Venezuela 03°55′N 67°06′W
54 I7 **Santa Bárbara** Barinas, W Venezuela 07°48′N 71°10′W
42 F5 **Santa Bárbara** ◈ *department* NW Honduras
Santa Bárbara *see* Iscuandé
35 Q15 **Santa Barbara Channel** *channel* California, W USA
Santa Bárbara de Samaná *see* Samaná
35 R16 **Santa Barbara Island** *island* Channel Islands, California, W USA
54 E5 **Santa Catalina** Bolívar, N Colombia 10°36′N 75°17′W
43 R15 **Santa Catalina** Ngöbe Bugle, W Panama 08°46′N 81°18′W
104 G2 **Santa Catalina de Armada** Galicia, NW Spain 43°02′N 08°49′W
35 T17 **Santa Catalina, Gulf of** *gulf* California, W USA
40 F8 **Santa Catalina, Isla** *island* NW Mexico
35 S16 **Santa Catalina Island** *island* Channel Islands, California, W USA
41 N8 **Santa Catarina** Nuevo León, NE Mexico 25°39′N 100°30′W
60 H13 **Santa Catarina** *off.* Estado de Santa Catarina. ◈ *state* S Brazil
Santa Catarina de Tepehuanes *see* Tepehuanes
Santa Catarina, Estado de *see* Santa Catarina
60 L13 **Santa Catarina, Ilha de** *island* S Brazil
45 Q16 **Santa Catherina** Curaçao 12°07′N 68°46′W
44 E5 **Santa Clara** Villa Clara, C Cuba 22°25′N 78°01′W
35 N9 **Santa Clara** California, W USA 37°20′N 121°57′W
36 J8 **Santa Clara** Utah, W USA 37°07′N 113°39′W
Santa Clara *see* Santa Clara de Olimar
61 F18 **Santa Clara de Olimar** *var.* Santa Clara. Cerro Largo, NE Uruguay 32°50′S 54°54′W
61 A17 **Santa Clara de Saguier** Santa Fe, C Argentina 31°21′S 61°50′W
Santa Coloma *see* Santa Coloma de Gramenet
105 X5 **Santa Coloma de Farners** *var.* Santa Coloma de Farnés. Cataluña, NE Spain 41°52′N 02°39′E
Santa Coloma de Farnés *see* Santa Coloma de Farners
Santa Coloma de Gramanet *see* Santa Coloma de Gramenet
105 W6 **Santa Coloma de Gramenet** *var.* Santa Coloma; *prev.* Santa Coloma de Gramanet. Cataluña, NE Spain 41°28′N 02°14′E
Santa Comba *see* Uaco Cungo
104 H8 **Santa Comba Dão** Viseu, N Portugal 40°23′N 08°07′W
82 C10 **Santa Cruz** Uíge, NW Angola 06°56′S 16°25′E
57 N19 **Santa Cruz** *var.* Santa Cruz de la Sierra. Santa Cruz, C Bolivia 17°49′S 63°11′W
62 G12 **Santa Cruz** Libertador, C Chile 34°38′S 71°27′W
42 K13 **Santa Cruz** Guanacaste, W Costa Rica 10°15′N 85°35′W
44 I12 **Santa Cruz** W Jamaica 18°03′N 77°43′W
64 P6 **Santa Cruz** Madeira, Portugal, NE Atlantic Ocean 32°43′N 16°47′W
35 N10 **Santa Cruz** California, W USA 36°58′N 122°01′W
63 H20 **Santa Cruz** *off.* Provincia de Santa Cruz. ◈ *province* S Argentina
57 O18 **Santa Cruz** ◈ *department* E Bolivia
Santa Cruz *see* Puerto Santa Cruz
Santa Cruz *see* Viru-Viru
Santa Cruz Barillas *see* Barillas
59 O18 **Santa Cruz Cabrália** Bahia, E Brazil 16°17′S 39°03′W
Santa Cruz de El Seibo *see* El Seibo
64 N11 **Santa Cruz de la Palma** La Palma, Islas Canarias, Spain, NE Atlantic Ocean 28°41′N 17°46′W
Santa Cruz de la Sierra *see* Santa Cruz
105 O9 **Santa Cruz de la Zarza** Castilla-La Mancha, C Spain 39°59′N 03°10′W
42 C5 **Santa Cruz del Quiché** Quiché, W Guatemala 15°02′N 91°06′W
105 N8 **Santa Cruz del Retamar** Castilla-La Mancha, C Spain 40°08′N 04°14′W
Santa Cruz del Seibo *see* El Seibo
44 G7 **Santa Cruz del Sur** Camagüey, C Cuba 20°44′N 78°00′W
105 O11 **Santa Cruz de Mudela** Castilla-La Mancha, C Spain 38°37′N 03°27′W
64 Q11 **Santa Cruz de Tenerife** Tenerife, Islas Canarias, Spain, NE Atlantic Ocean 28°28′N 16°15′W
64 P11 **Santa Cruz de Tenerife** ◈ *province* Islas Canarias, Spain, NE Atlantic Ocean
60 K9 **Santa Cruz do Rio Pardo** São Paulo, S Brazil 22°52′S 49°37′W
61 H15 **Santa Cruz do Sul** Rio Grande do Sul, S Brazil 29°42′S 52°25′W
57 C17 **Santa Cruz, Isla** *var.* Indefatigable Island, Isla Chávez. *island* Galapagos Islands, Ecuador, E Pacific Ocean
40 F8 **Santa Cruz, Isla** *island* NW Mexico
35 Q15 **Santa Cruz Island** *island* California, W USA
187 Q10 **Santa Cruz Islands** *island group* E Solomon Islands
Santa Cruz, Provincia de *see* Santa Cruz
63 I22 **Santa Cruz, Río** ≈ S Argentina
36 L15 **Santa Cruz River** ≈ Arizona, SW USA
61 C17 **Santa Elena** Entre Ríos, E Argentina 30°58′S 59°47′W
42 F2 **Santa Elena** Cayo, W Belize 17°08′N 89°04′W
25 R16 **Santa Elena** Texas, SW USA 26°43′N 98°30′W
56 A7 **Santa Elena, Bahía de** *bay* W Ecuador
55 R10 **Santa Elena de Uairén** Bolívar, E Venezuela 04°40′N 61°03′W
42 K12 **Santa Elena, Península** *peninsula* NW Costa Rica
56 A7 **Santa Elena, Punta** *headland* W Ecuador 02°11′S 81°00′W
104 L11 **Santa Eufemia** Andalucía, S Spain 38°36′N 04°54′W
107 N21 **Santa Eufemia, Golfo di** *gulf* S Italy
105 S4 **Santa Eulalia de Gállego** Aragón, NE Spain 42°16′N 00°46′W
105 V11 **Santa Eulalia del Río** Ibiza, Spain, W Mediterranean Sea 39°00′N 01°33′E
61 B17 **Santa Fe** Santa Fe, C Argentina 31°36′S 60°47′W
44 C6 **Santa Fé** *var.* La Fe. Isla de la Juventud, W Cuba 21°45′N 82°45′W
43 R16 **Santa Fé** Veraguas, C Panama 08°29′N 80°50′W
105 N14 **Santa Fe** Andalucía, S Spain 37°11′N 03°43′W
37 S10 **Santa Fe** *state capital* New Mexico, SW USA 35°41′N 105°56′W
61 B15 **Santa Fe** *off.* Provincia de Santa Fe. ◈ *province* C Argentina
Santa Fe *see* Bogotá
Santa Fe de Bogotá *see* Bogotá
60 J7 **Santa Fé do Sul** São Paulo, S Brazil 20°13′S 50°56′W
57 B18 **Santa Fe, Isla** *var.* Barrington Island. *island* Galapagos Islands, Ecuador, E Pacific Ocean
Santa Fe, Provincia de *see* Santa Fe
23 V9 **Santa Fe River** ≈ Florida, SE USA
59 M15 **Santa Filomena** Piauí, E Brazil 09°06′S 45°52′W
40 G10 **Santa Genoveva** ▲ NW Mexico 23°07′N 109°56′W
153 S14 **Santahar** Rajshahi, NW Bangladesh 24°45′N 89°03′E
60 G11 **Santa Helena** Paraná, S Brazil 24°53′S 54°19′W
54 J5 **Santa Inés** Lara, N Venezuela 10°37′N 69°18′W
63 G24 **Santa Inés, Isla** *island* S Chile
62 J13 **Santa Isabel** La Pampa, C Argentina 36°11′S 66°59′W
43 U14 **Santa Isabel** Colón, N Panama 09°31′N 79°12′W
186 L8 **Santa Isabel** *var.* Bughotu. *island* N Solomon Islands
Santa Isabel *see* Malabo
58 D11 **Santa Isabel do Rio Negro** Amazonas, NW Brazil 0°40′S 64°56′W
61 C15 **Santa Lucia** Corrientes, NE Argentina 28°58′S 59°05′W
57 I17 **Santa Lucía** Puno, S Peru 15°45′S 70°34′W
61 F20 **Santa Lucía** *var.* Santa Lucia. Canelones, S Uruguay 34°26′S 56°25′W
42 B6 **Santa Lucía Cotzumalguapa** Escuintla, SW Guatemala 14°20′N 91°00′W
107 L23 **Santa Lucia del Mela** Sicilia, Italy, C Mediterranean Sea 38°08′N 15°17′E
35 O11 **Santa Lucia Range** ▲ California, W USA
40 D9 **Santa Margarita, Isla** *island* NW Mexico
62 J7 **Santa María** Catamarca, N Argentina 26°51′S 66°02′W
61 G15 **Santa Maria** Rio Grande do Sul, S Brazil 29°41′S 53°48′W
35 P13 **Santa Maria** California, W USA 34°56′N 120°25′W
64 Q4 **Santa Maria** ✕ Santa Maria, Azores, Portugal, NE Atlantic Ocean
64 P3 **Santa Maria** *island* Azores, Portugal, NE Atlantic Ocean
Santa Maria *see* Gaua
Santa María Asunción Tlaxiaco *see* Tlaxiaco
40 G9 **Santa María, Bahía** *bay* W Mexico
83 L21 **Santa Maria, Cabo de** *headland* S Mozambique 26°05′S 32°58′E
104 G15 **Santa Maria, Cabo de** *headland* S Portugal 36°57′N 07°55′W
44 J4 **Santa Maria, Cape** *headland* Long Island, C The Bahamas 23°40′N 75°20′W
107 J17 **Santa Maria Capua Vetere** Campania, S Italy 41°05′N 14°15′E
104 G7 **Santa Maria da Feira** Aveiro, N Portugal 40°55′N 08°32′W
59 M17 **Santa Maria da Vitória** Bahia, E Brazil 13°26′S 44°09′W
55 N9 **Santa Maria de Erebato** Bolívar, SE Venezuela 05°09′N 64°50′W
55 N6 **Santa María de Ipire** Guárico, C Venezuela 08°51′N 65°21′W
Santa María del Buen Aire *see* Buenos Aires
40 J8 **Santa María del Oro** Durango, C Mexico 25°57′N 105°22′W
41 N12 **Santa María del Río** San Luis Potosí, C Mexico 21°48′N 100°42′W
Santa Maria di Castellabate *see* Castellabate
107 Q20 **Santa Maria di Leuca, Capo** *headland* SE Italy 39°48′N 18°21′E
108 K10 **Santa Maria-im-Munstertal** Graubünden, SE Switzerland 46°36′N 10°25′E
57 B18 **Santa María, Isla** *var.* Isla Floreana, Charles Island. *island* Galapagos Islands, Ecuador, E Pacific Ocean
40 J3 **Santa María, Laguna de** ◎ N Mexico
61 G16 **Santa Maria, Rio** ≈ S Brazil
43 R16 **Santa María, Río** ≈ C Panama
36 J12 **Santa Maria River** ≈ Arizona, SW USA
107 G15 **Santa Marinella** Lazio, C Italy 42°01′N 11°51′E
54 F4 **Santa Marta** Magdalena, N Colombia 11°14′N 74°13′W
104 J11 **Santa Marta** Extremadura, W Spain 38°37′N 06°39′W
54 F4 **Santa Marta, Sierra Nevada de** ▲ NE Colombia
Santa Maura *see* Lefkáda
35 S15 **Santa Monica** California, W USA 34°01′N 118°29′W
116 F10 **Sântana** *Ger.* Sankt Anna, *Hung.* Újszentanna; *prev.* Sîntana. Arad, W Romania 46°20′N 21°30′E
61 F16 **Santana, Coxilha de** *hill range* S Brazil

◆ Country ● Country Capital ◇ Dependent Territory ○ Dependent Territory Capital ◈ Administrative Regions ✕ International Airport ▲ Mountain ▲ Mountain Range ※ Volcano ≈ River ◎ Lake ⊠ Reservoir

Santana da Boa Vista – Sasmik, Cape

61 H16 **Santana da Boa Vista** Rio Grande do Sul, S Brazil 30°52′S 53°03′W
61 F16 **Santana do Livramento** *prev.* Livramento. Rio Grande do Sul, S Brazil 30°52′S 55°30′W
105 N2 **Santander** Cantabria, N Spain 43°28′N 03°48′W
54 F8 **Santander** *off.* Departamento de Santander. ⬖ *province* C Colombia
Santander, Departamento de *see* Santander
Santander Jiménez *see* Jiménez
Sant'Andrea *see* Svetac
107 B20 **Sant'Antioco** Sardegna, Italy, C Mediterranean Sea 39°03′N 08°28′E
105 V11 **Sant Antoni de Portmany** *Cas.* San Antonio Abad. Ibiza, Spain, W Mediterranean Sea 38°58′N 01°18′E
105 Y10 **Santanyí** Mallorca, Spain, W Mediterranean Sea 39°22′N 03°07′E
104 J13 **Santa Olalla del Cala** Andalucía, S Spain 37°54′N 06°13′W
35 R15 **Santa Paula** California, W USA 34°21′N 119°03′W
36 L4 **Santaquin** Utah, W USA 39°58′N 111°46′W
58 I12 **Santarém** Pará, N Brazil 02°26′S 54°41′W
104 G10 **Santarém** *anc.* Scalabis. Santarém, W Portugal 39°14′N 08°40′W
104 G10 **Santarém** ⬖ *district* C Portugal
44 F4 **Santaren Channel** *channel* W The Bahamas
54 K10 **Santa Rita** Vichada, E Colombia 04°51′N 68°27′W
188 B16 **Santa Rita** SW Guam
42 H5 **Santa Rita** Cortés, NW Honduras 15°10′N 87°54′W
40 E9 **Santa Rita** Baja California Sur, NW Mexico 27°29′N 100°33′W
54 H5 **Santa Rita** Zulia, NW Venezuela 10°35′N 71°30′W
59 I19 **Santa Rita de Araguaia** Goiás, S Brazil 17°17′S 53°13′W
59 M16 **Santa Rita de Cassia** *var.* Cássia. Bahia, E Brazil 11°03′S 44°16′W
61 D14 **Santa Rosa** Corrientes, NE Argentina 28°18′S 58°04′W
62 K13 **Santa Rosa** La Pampa, C Argentina 36°38′S 64°15′W
61 G14 **Santa Rosa** Rio Grande do Sul, S Brazil 27°50′S 54°29′W
58 E10 **Santa Rosa** Roraima, N Brazil 03°41′N 62°29′W
56 B8 **Santa Rosa** El Oro, SW Ecuador 03°29′S 79°57′W
57 I16 **Santa Rosa** Puno, S Peru 14°38′S 70°45′W
34 M7 **Santa Rosa** California, W USA 38°27′N 122°42′W
37 U11 **Santa Rosa** New Mexico, SW USA 34°54′N 104°43′W
55 O6 **Santa Rosa** Anzoátegui, NE Venezuela 09°37′N 64°20′W
42 A3 **Santa Rosa** *off.* Departamento de Santa Rosa. ⬖ *department* SE Guatemala
Santa Rosa *see* Santa Rosa de Copán
63 J15 **Santa Rosa, Bajo de** *basin* E Argentina
42 F6 **Santa Rosa de Copán** *var.* Santa Rosa. Copán, W Honduras 14°48′N 88°43′W
54 E8 **Santa Rosa de Osos** Antioquia, C Colombia 06°40′N 75°27′W
Santa Rosa, Departamento de *see* Santa Rosa
35 Q15 **Santa Rosa Island** *island* California, W USA
23 O9 **Santa Rosa Island** *island* Florida, SE USA
40 E6 **Santa Rosalía** Baja California Sur, NW Mexico 27°20′N 112°20′W
54 K6 **Santa Rosalía** Portuguesa, NW Venezuela 09°02′N 69°01′W
188 C15 **Santa Rosa, Mount** ▲ NE Guam
35 V16 **Santa Rosa Mountains** ▲▲ California, W USA
35 T2 **Santa Rosa Range** ▲▲ Nevada, W USA
62 M8 **Santa Sylvina** Chaco, N Argentina 27°49′S 61°09′W
Santa Tecla *see* Nueva San Salvador
61 B19 **Santa Teresa** Santa Fe, C Argentina 33°30′S 60°45′W
59 O20 **Santa Teresa** Espírito Santo, SE Brazil 19°51′S 40°49′W
61 E21 **Santa Teresita** Buenos Aires, E Argentina 36°32′S 56°41′W
61 H19 **Santa Vitória do Palmar** Rio Grande do Sul, S Brazil 33°32′S 53°25′W
35 Q14 **Santa Ynez River** ≈ California, W USA
Sant Carles de la Rápida *see* Sant Carles de la Ràpita
105 U7 **Sant Carles de la Ràpita** *var.* Sant Carles de la Rápida. Cataluña, NE Spain 40°37′N 00°36′E
105 W5 **Sant Celoni** Cataluña, NE Spain 41°39′N 02°25′E
35 U17 **Santee** California, W USA 32°50′N 116°58′W
21 T13 **Santee River** ≈ South Carolina, SE USA
40 K15 **San Telmo, Punta** *headland* SW Mexico 18°19′N 103°30′W
107 O17 **Santeramo in Colle** Puglia, SE Italy 40°47′N 16°45′E
107 M23 **San Teresa di Riva** Sicilia, Italy, C Mediterranean Sea 38°00′N 15°25′E
105 X5 **Sant Feliu de Guíxols** *var.* San Feliú de Guixols. Cataluña, NE Spain 41°47′N 03°02′E
105 W6 **Sant Feliu de Llobregat** Cataluña, NE Spain 41°22′N 02°00′E
106 C7 **Santhià** Piemonte, NE Italy 45°21′N 08°11′E
61 F15 **Santiago** Rio Grande do Sul, S Brazil 29°11′S 54°52′W
62 H11 **Santiago** *var.* Gran Santiago. ● (Chile) Santiago, C Chile 33°30′S 70°40′W
45 N8 **Santiago** *var.* Santiago de los Caballeros. N Dominican Republic 19°27′N 70°42′W
40 G10 **Santiago** Baja California Sur, NW Mexico 23°32′N 109°47′W
41 O8 **Santiago** Nuevo León, NE Mexico 25°22′N 100°09′W
43 R16 **Santiago** Veraguas, S Panama 08°06′N 80°59′W
57 E16 **Santiago** Ica, SW Peru 14°14′S 75°44′W
62 H11 **Santiago** *off.* Región Metropolitana de Santiago, *var.* Metropolitan. ⬖ *region* C Chile
76 D10 **Santiago** *var.* São Tiago. *island* Ilhas de Sotavento, S Cape Verde
62 H11 **Santiago** ✈ Santiago, C Chile 33°27′S 70°40′W
104 G3 **Santiago** ✈ Galicia, NW Spain
Santiago *see* Santiago de Cuba, Cuba
Santiago *see* Grande de Santiago, Río, Mexico
Santiago *see* Santiago de Compostela
42 B6 **Santiago Atitlán** Sololá, SW Guatemala 14°39′N 91°12′W
43 Q16 **Santiago, Cerro** ▲ W Panama 08°27′N 81°42′W
104 G3 **Santiago de Compostela** *var.* Santiago, *Eng.* Compostella; *anc.* Campus Stellae. Galicia, NW Spain 42°52′N 08°33′W
44 I8 **Santiago de Cuba** *var.* Santiago. Santiago de Cuba, E Cuba 20°01′N 75°51′W
Santiago de Guayaquil *see* Guayaquil
62 K8 **Santiago del Estero** Santiago del Estero, C Argentina 27°51′S 64°16′W
61 A15 **Santiago del Estero** *off.* Provincia de Santiago del Estero. ⬖ *province* N Argentina
Santiago del Estero, Provincia de *see* Santiago del Estero
40 I8 **Santiago de los Caballeros** Sinaloa, W Mexico 25°33′N 107°22′W
Santiago de los Caballeros *see* Santiago, Dominican Republic
Santiago de los Caballeros *see* Ciudad de Guatemala, Guatemala
42 F8 **Santiago de María** Usulután, SE El Salvador 13°28′N 88°28′W
104 F12 **Santiago do Cacém** Setúbal, S Portugal 38°01′N 08°42′W
40 J12 **Santiago Ixcuintla** Nayarit, C Mexico 21°50′N 105°11′W
Santiago Jamiltepec *see* Jamiltepec
24 L11 **Santiago Mountains** ▲▲ Texas, SW USA
40 J9 **Santiago Papasquiaro** Durango, C Mexico 25°00′N 105°27′W
Santiago Pinotepa Nacional *see* Pinotepa Nacional
Santiago, Región Metropolitana de *see* Santiago
56 C8 **Santiago, Río** ≈ N Peru
40 M10 **San Tiburcio** Zacatecas, C Mexico 24°08′N 101°29′W
105 N2 **Santillana** Cantabria, N Spain 43°24′N 04°06′W
54 I5 **San Timoteo** Zulia, NW Venezuela 09°50′N 71°05′W
Santi Quaranta *see* Sarandë
Santissima Trinidad *see* Jilong
105 O12 **Santisteban del Puerto** Andalucía, S Spain 38°15′N 03°11′W
105 S12 **Sant Joan d'Alacant** *Cast.* San Juan de Alicante. Valenciana, E Spain 38°26′N 00°27′W
105 U7 **Sant Jordi, Golf de** *gulf* NE Spain
105 U11 **Sant Josep de sa Talaia** *var.* San Jose. Ibiza, Spain, W Mediterranean Sea 38°55′N 1°18′E
162 G6 **Santmargats** *var.* Holboo. Dzavhan, W Mongolia 48°35′N 95°25′E
105 T8 **Sant Mateu** Valenciana, E Spain 40°28′N 00°10′E
25 S7 **Santo** Texas, SW USA 32°35′N 98°06′W
Santo *see* Espíritu Santo
60 M10 **Santo Amaro, Ilha de** *island* SE Brazil
61 G14 **Santo Ângelo** Rio Grande do Sul, S Brazil 28°17′S 54°15′W
76 C9 **Santo Antão** *island* Ilhas de Barlavento, N Cape Verde
60 J10 **Santo Antônio da Platina** Paraná, S Brazil 23°20′S 50°05′W
58 C13 **Santo Antônio do Içá** Amazonas, N Brazil 03°05′S 67°56′W
57 Q18 **Santo Corazón, Río** ≈ E Bolivia
44 E5 **Santo Domingo** Villa Clara, C Cuba 22°35′N 80°15′W
45 O9 **Santo Domingo** *prev.* Ciudad Trujillo. ● (Dominican Republic) SE Dominican Republic 18°30′N 69°57′W
40 E8 **Santo Domingo** Baja California Sur, NW Mexico 25°34′N 112°00′W
40 M10 **Santo Domingo** San Luis Potosí, C Mexico 23°18′N 101°42′W
42 L10 **Santo Domingo** Chontales, S Nicaragua 12°15′N 84°59′W
105 P4 **Santo Domingo de la Calzada** La Rioja, N Spain 42°26′N 02°57′W
56 B6 **Santo Domingo de los Colorados** Pichincha, NW Ecuador 0°13′S 79°09′W
Santo Domingo Tehuantepec *see* Tehuantepec
55 O6 **San Tomé** Anzoátegui, NE Venezuela 08°58′N 64°08′W
San Tomé de Guayana *see* Ciudad Guayana
105 R13 **Santomera** Murcia, SE Spain 38°03′N 01°05′W
105 O2 **Santoña** Cantabria, N Spain 43°27′N 03°28′W
Santorin *see* Santoríni
115 K22 **Santoríni** *var.* Santorin, *prev.* Thíra; *anc.* Thera. *island* Kykládes, Greece, Aegean Sea
60 M10 **Santos** São Paulo, S Brazil 23°56′S 46°22′W
65 J17 **Santos Plateau** *undersea feature* SW Atlantic Ocean 25°00′S 43°00′W
104 G6 **Santo Tirso** Porto, N Portugal 41°20′N 08°25′W
40 B2 **Santo Tomás** Baja California Norte, NW Mexico 31°32′N 116°26′W
42 L10 **Santo Tomás** Chontales, S Nicaragua 12°04′N 85°02′W
42 G5 **Santo Tomás de Castilla** Izabal, E Guatemala 15°40′N 88°36′W
40 B2 **Santo Tomás, Punta** *headland* NW Mexico 31°30′N 116°40′W
57 H16 **Santo Tomás, Río** ≈ C Peru
57 B18 **Santo Tomás, Volcán** 🌋 Galapagos Islands, Ecuador, E Pacific Ocean 0°46′S 91°01′W
61 F14 **Santo Tomé** Corrientes, NE Argentina 28°31′S 56°03′W
Santo Tomé de Guayana *see* Ciudad Guayana
98 H10 **Santpoort** Noord-Holland, W Netherlands 52°26′N 04°38′E
Santurce *see* Santurtzi
105 O2 **Santurtzi** *var.* Santurce, Santurzi. País Vasco, N Spain 43°20′N 03°03′W
Santurzi *see* Santurtzi
63 G20 **San Valentín, Cerro** ▲ S Chile 46°36′S 73°17′W
42 F8 **San Vicente** San Vicente, C El Salvador 13°38′N 88°42′W*
40 C2 **San Vicente** Baja California Norte, NW Mexico 31°20′N 116°15′W
188 H6 **San Vicente** Saipan, S Northern Mariana Islands
42 B9 **San Vicente** ⬖ *department* E El Salvador
104 I10 **San Vicente de Alcántara** Extremadura, W Spain 39°21′N 07°07′W
105 N2 **San Vicente de Barakaldo** *var.* Baracaldo, *Basq.* San Bizenti-Barakaldo. País Vasco, N Spain 43°17′N 02°59′W
57 E15 **San Vicente de Cañete** *var.* Cañete. Lima, W Peru 13°06′S 76°23′W
104 M2 **San Vicente de la Barquera** Cantabria, N Spain 43°23′N 04°24′W
54 E12 **San Vicente del Caguán** Caquetá, S Colombia 02°07′N 74°47′W
42 F8 **San Vincente, Volcán de** 🌋 C El Salvador 13°34′N 88°50′W
43 O15 **San Vito** Puntarenas, SE Costa Rica 08°49′N 82°58′W
106 I7 **San Vito al Tagliamento** Friuli-Venezia Giulia, NE Italy 45°54′N 12°55′E
107 H23 **San Vito, Capo** *headland* Sicilia, Italy, C Mediterranean Sea 38°11′N 12°41′E
107 P18 **San Vito dei Normanni** Puglia, SE Italy 40°40′N 17°42′E
160 L17 **Sanya** *var.* Ya Xian. Hainan, S China 18°25′N 109°27′E
83 J16 **Sanyati** ≈ N Zimbabwe
25 Q16 **San Ygnacio** Texas, SW USA 27°04′N 99°26′W
160 L6 **Sanyuan** Shaanxi, C China 34°40′N 108°56′E
123 P11 **Sanyyakhtakh** Respublika Sakha (Yakutiya), NE Russian Federation 60°34′N 124°09′E
146 J15 **S. A.Nyýazow Adyndaky** *Rus.* Imeni S. A. Niyazova. Maryyskiy Velayat, S Turkmenistan 36°44′N 62°23′E
82 C10 **Sanza Pombo** Uíge, NW Angola 07°20′S 16°00′E
Sanzyô *see* Sanjō
104 G14 **São Bartolomeu de Messines** Faro, S Portugal 37°12′N 08°16′W
60 M10 **São Bernardo do Campo** São Paulo, S Brazil 23°45′S 46°34′W
61 F15 **São Borja** Rio Grande do Sul, S Brazil 28°35′S 56°01′W
104 H14 **São Brás de Alportel** Faro, S Portugal 37°09′N 07°55′W
60 M10 **São Caetano do Sul** São Paulo, S Brazil 23°37′S 46°34′W
60 L9 **São Carlos** São Paulo, S Brazil 22°02′S 47°53′W
59 P16 **São Cristóvão** Sergipe, E Brazil 10°59′S 37°10′W
61 F15 **São Francisco de Assis** Rio Grande do Sul, S Brazil 29°32′S 55°07′W
58 K13 **São Félix** Pará, NE Brazil 06°43′S 51°56′W
São Félix *see* São Félix do Araguaia
59 J16 **São Félix do Araguaia** *var.* São Félix. Mato Grosso, W Brazil 11°36′S 50°40′W
59 J14 **São Félix do Xingu** Pará, NE Brazil 06°38′S 51°59′W
60 Q9 **São Fidélis** Rio de Janeiro, SE Brazil 21°37′S 41°40′W
76 D10 **São Filipe** Fogo, S Cape Verde 14°52′N 24°29′W
60 K12 **São Francisco do Sul** Santa Catarina, S Brazil 26°17′S 48°39′W
60 K12 **São Francisco, Ilha de** *island* S Brazil
59 P16 **São Francisco, Rio** ≈ E Brazil
61 G16 **São Gabriel** Rio Grande do Sul, S Brazil 30°17′S 54°17′W
60 P10 **São Gonçalo** Rio de Janeiro, SE Brazil 22°48′S 43°03′W
81 H23 **Sao Hill** Iringa, S Tanzania 08°19′S 35°11′E
60 R9 **São João da Barra** Rio de Janeiro, SE Brazil 21°39′S 41°04′W
104 G7 **São João da Madeira** Aveiro, N Portugal 40°52′N 08°28′W
58 M12 **São João de Cortes** Maranhão, E Brazil 02°30′S 44°27′W
59 M21 **São João del Rei** Minas Gerais, NE Brazil 21°08′S 44°15′W
59 N15 **São João do Piauí** Piauí, E Brazil 08°21′S 42°14′W
59 N14 **São João dos Patos** Maranhão, E Brazil 06°29′S 43°44′W
58 C11 **São Joaquim** Amazonas, NW Brazil 0°08′S 67°10′W
61 J14 **São Joaquim** Santa Catarina, S Brazil 28°20′S 49°55′W
60 L7 **São Joaquim da Barra** São Paulo, S Brazil 20°36′S 47°50′W
64 N2 **São Jorge** *island* Azores, Portugal, NE Atlantic Ocean
61 K14 **São José** Santa Catarina, S Brazil 27°34′S 48°39′W
60 M8 **São José do Rio Pardo** São Paulo, S Brazil 21°37′S 46°52′W
60 K8 **São José do Rio Preto** São Paulo, S Brazil 20°50′S 49°20′W
60 N10 **São Jose dos Campos** São Paulo, S Brazil 23°07′S 45°52′W
61 I17 **São Lourenço do Sul** Rio Grande do Sul, S Brazil 31°25′S 52°00′W
58 M12 **São Luís** *state capital* Maranhão, NE Brazil 02°34′S 44°16′W
58 F11 **São Luís** Roraima, N Brazil 01°11′N 60°15′W
58 M12 **São Luís, Ilha de** *island* NE Brazil
61 F14 **São Luiz Gonzaga** Rio Grande do Sul, S Brazil 28°24′S 54°58′W
São Mandol *see* São Manuel, Rio
47 U8 **São Manuel** ≈ C Brazil
59 H15 **São Manuel, Rio** *var.* São Mandol, Teles Pirés. ≈ C Brazil
58 C11 **São Marcelino** Amazonas, NW Brazil 0°53′N 67°16′W
58 N12 **São Marcos, Baía de** *bay* N Brazil
59 O20 **São Mateus** Espírito Santo, SE Brazil 18°44′S 39°53′W
60 J12 **São Mateus do Sul** Paraná, S Brazil 25°58′S 50°29′W
64 P3 **São Miguel** *island* Azores, Portugal, NE Atlantic Ocean
60 G13 **São Miguel d'Oeste** Santa Catarina, S Brazil 26°45′S 53°34′W
45 P9 **Saona, Isla** *island* SE Dominican Republic
172 H12 **Saondzou** ▲ Grande Comore, NW Comoros
103 R10 **Saône** ≈ E France
103 Q9 **Saône-et-Loire** ⬖ *department* C France
76 D9 **São Nicolau** *Eng.* Saint Nicholas. *island* Ilhas de Barlavento, N Cape Verde
60 M10 **São Paulo** *state capital* São Paulo, S Brazil 23°33′S 46°39′W
60 K9 **São Paulo** *off.* Estado de São Paulo. ⬖ *state* S Brazil
São Paulo de Loanda *see* Luanda
São Paulo, Estado de *see* São Paulo
São Pedro do Rio Grande do Sul *see* Rio Grande
104 H7 **São Pedro do Sul** Viseu, N Portugal 40°46′N 07°58′W
64 K13 **São Pedro e São Paulo** *undersea feature* C Atlantic Ocean 01°25′N 28°54′W
59 M14 **São Raimundo das Mangabeiras** Maranhão, E Brazil 07°00′S 45°30′W
59 Q14 **São Roque, Cabo de** *headland* E Brazil 05°29′S 35°16′W
São Salvador *see* Salvador, Brazil
São Salvador/São Salvador do Congo *see* M'Banza Congo, Angola
60 N10 **São Sebastião, Ilha de** *island* S Brazil
83 N19 **São Sebastião, Ponta** *headland* C Mozambique 22°09′S 35°33′E
104 F13 **São Teotónio** Beja, S Portugal 37°30′N 08°41′W
São Tiago *see* Santiago
79 B18 **São Tomé** ● (Sao Tome and Principe) São Tomé, S Sao Tome and Principe 0°22′N 06°41′E
79 B18 **São Tomé** ✈ São Tomé, S Sao Tome and Principe 0°24′N 06°39′E
79 B18 **São Tomé** *Eng.* Saint Thomas. *island* S Sao Tome and Principe
79 B17 **Sao Tome and Principe** *off.* Democratic Republic of Sao Tome and Principe. ◆ *republic* E Atlantic Ocean
Sao Tome and Principe, Democratic Republic of *see* Sao Tome and Principe
74 H9 **Saoura, Oued** ≈ NW Algeria
60 M10 **São Vicente** *Eng.* Saint Vincent. São Paulo, S Brazil 23°55′S 46°25′W
64 O5 **São Vicente** Madeira, Portugal, NE Atlantic Ocean 32°48′N 17°03′W
76 C9 **São Vicente** *Eng.* Saint Vincent. *island* Ilhas de Barlavento, N Cape Verde
104 F14 **São Vicente, Cabo de** *Eng.* Cape Saint Vincent, *Port.* Cabode São Vicente. *cape* S Portugal
São Vicente, Cabo de *see* São Vicente, Cabo de
Sápai *see* Sápes
Sapaleri, Cerro *see* Zapaleri, Cerro
Saparoea *see* Saparua
171 S13 **Saparua** *prev.* Saparoea. Pulau Saparau, C Indonesia 03°35′S 128°37′E
168 L11 **Sapat** Sumatera, W Indonesia 0°18′S 103°18′E
77 U17 **Sapele** Delta, S Nigeria 05°54′N 05°43′E
23 X7 **Sapelo Island** *island* Georgia, SE USA
23 X7 **Sapelo Sound** *sound* Georgia, SE USA
114 K13 **Sápes** *var.* Sápai. Anatolikí Makedonía kai Thráki, NE Greece 41°02′N 25°44′E
Sapiéntza *see* Sapiénza
115 D22 **Sapiénza** *var.* Sapiéntza. *island* S Greece
138 F12 **Sapir** *prev.* Sappir. Southern, S Israel 30°43′N 35°11′E
61 I15 **Sapiranga** Rio Grande do Sul, S Brazil 29°39′S 50°58′W
114 K13 **Sápka** ▲▲ NE Greece
105 X9 **Sa Pobla** Mallorca, Spain, W Mediterranean Sea 39°46′N 03°03′E
56 D11 **Saposoa** San Martín, N Peru 06°53′S 76°45′W
119 F16 **Sapotskin** *Pol.* Sopockinie, *Rus.* Sapotskino, Sopotskin. Hrodzyenskaya Voblasts', W Belarus 53°50′N 23°39′E
77 P13 **Sapoui** *var.* Sapouy. S Burkina Faso 11°34′N 01°44′W
Sapouy *see* Sapoui
Sappir *see* Sapir
165 S4 **Sapporo** Hokkaidō, NE Japan 43°05′N 141°21′E
107 M19 **Sapri** Campania, S Italy 40°05′N 15°36′E
169 T16 **Sapudi, Pulau** *island* S Indonesia
27 P9 **Sapulpa** Oklahoma, C USA 36°00′N 96°06′W
142 J4 **Saqqez** *var.* Saghez, Sakiz, Saqqiz. Kordestān, NW Iran 36°31′N 46°16′E
Saqqiz *see* Saqqez
139 U8 **Sarābādī** Wāsiț, E Iraq 33°00′N 44°52′E
167 P10 **Sara Buri** *var.* Saraburi. Saraburi, C Thailand 14°32′N 100°53′E
Saraburi *see* Sara Buri
Sarafjagān *see* Salafchegān
24 K9 **Saragosa** Texas, SW USA 31°03′N 103°39′W
Saragossa *see* Zaragoza
Saragt *see* Sarahs
56 B8 **Saraguro** Loja, S Ecuador 03°42′S 79°18′W
146 I15 **Sarahs** *var.* Saragt, *Rus.* Serakhs. Ahal Welaýaty, S Turkmenistan 36°33′N 61°10′E
126 M6 **Sarai** Ryazanskaya Oblast', W Russian Federation 53°43′N 39°59′E
Sarāi *see* Sarāy
154 M12 **Saraipāli** Chhattīsgarh, C India 21°21′N 83°01′E
149 T9 **Sarāi Sidhu** Punjab, E Pakistan 30°35′N 72°02′E
93 M15 **Säräisniemi** Kainuu, C Finland 64°25′N 26°50′E
113 I14 **Sarajevo** ● (Bosnia and Herzegovina) Federacija Bosne I Hercegovine, SE Bosnia and Herzegovina 43°53′N 18°24′E
112 I13 **Sarajevo** ✈ Federacija Bosne I Hercegovine, C Bosnia and Herzegovina 43°49′N 18°21′E
143 V4 **Sarakhs** Khorāsān-e Razavī, NE Iran 36°50′N 61°00′E
115 H17 **Sarakíniko, Akrotírio** *headland* Évvoia, C Greece 38°46′N 23°43′E
115 I18 **Sarakinó** *island* Vóreies Sporádes, Greece, Aegean Sea
127 V7 **Saraktash** Orenburgskaya Oblast', W Russian Federation 51°46′N 56°23′E
30 L15 **Sara, Lake** ▣ Illinois, N USA
23 N8 **Saraland** Alabama, S USA 30°49′N 88°04′W
55 V9 **Saramacca** ⬖ *district* N Suriname
55 V10 **Saramacca Rivier** ≈ C Suriname
166 M2 **Saramati** ▲ N Myanmar (Burma) 25°46′N 95°01′E
145 R10 **Saran'** *Kaz.* Saran. Karaganda, C Kazakhstan 49°47′N 73°02′E
18 K7 **Saranac Lake** New York, NE USA 44°18′N 74°06′W
18 K7 **Saranac River** ≈ New York, NE USA
Saranda *see* Sarandë
113 L23 **Sarandë** *var.* Saranda, *It.* Porto Edda; *prev.* Santi Quaranta. Vlorë, S Albania 39°53′N 20°E
61 H14 **Sarandi** Rio Grande do Sul, S Brazil 27°57′S 52°58′W
61 F19 **Sarandí del Yí** Durazno, C Uruguay 33°18′S 55°38′W
61 F19 **Sarandí Grande** Florida, S Uruguay 33°43′S 56°19′W
171 Q8 **Sarangani Islands** *island group* S Philippines
127 P5 **Saransk** Respublika Mordoviya, W Russian Federation 54°11′N 45°10′E
115 C14 **Sarantáporos** ≈ N Greece
114 H9 **Sarantsi** Sofia, W Bulgaria 42°43′N 23°46′E
127 T3 **Sarapul** Udmurtskaya Respublika, NW Russian Federation 56°26′N 53°52′E
Sarāqeb *see* Sarāqib
138 I3 **Sarāqib** *Fr.* Sarāqeb. Idlib, N Syria 35°52′N 36°48′E
54 J5 **Sarare** Lara, N Venezuela 09°47′N 69°10′W
55 O10 **Sarariña** Amazonas, S Venezuela 04°10′N 64°31′W
143 S10 **Sar Ashk** Kermān, C Iran
23 V13 **Sarasota** Florida, SE USA 27°20′N 82°31′W
117 O11 **Sarata** Odes'ka Oblast', SW Ukraine 46°01′N 29°40′E
116 I10 **Sărățel** *Hung.* Szeretfalva. Bistrița-Năsăud, N Romania 47°02′N 24°24′E
25 X10 **Saratoga** Texas, SW USA 30°15′N 94°31′W
18 K10 **Saratoga Springs** New York, NE USA 43°04′N 73°47′W
127 P8 **Saratov** Saratovskaya Oblast', W Russian Federation 51°33′N 45°58′E
127 P8 **Saratovskaya Oblast'** ⬖ *province* W Russian Federation
127 Q7 **Saratovskoye Vodokhranilishche** ▣ W Russian Federation
143 X13 **Sarāvān** Sīstān va Balūchestān, SE Iran 27°11′N 62°35′E
Saravan/Saravane *see* Salavan
169 S9 **Sarawak** ⬖ *state* East Malaysia
Sarawak *see* Kuching
139 U6 **Sarāy** *var.* Sarāi. Diyālá, E Iraq 34°06′N 45°06′E
136 D10 **Saray** Tekirdağ, NW Turkey 41°27′N 27°56′E
76 J12 **Saraya** SE Senegal 12°50′N 11°45′W
143 W14 **Sarbāz** Sīstān va Balūchestān, SE Iran 26°38′N 61°13′E
143 U8 **Sarbīsheh** Khorāsān-e Jonūbī, E Iran 32°35′N 59°50′E
111 J24 **Sárbogárd** Fejér, C Hungary 46°54′N 18°36′E
Şarcad *see* Sarkad
27 S7 **Sarcoxie** Missouri, C USA 37°04′N 94°07′W
152 L11 **Sārda** *Nep.* Kali. ≈ India/Nepal
152 G10 **Sardārshahr** Rājasthān, NW India 28°30′N 74°30′E
107 C18 **Sardegna** *Eng.* Sardinia. ⬖ *region* Italy, C Mediterranean Sea
107 A18 **Sardegna** *Eng.* Sardinia. *island* Italy, C Mediterranean Sea
42 K13 **Sardinal** Guanacaste, NW Costa Rica 10°30′N 85°38′W
54 G7 **Sardinata** Norte de Santander, N Colombia 08°07′N 72°47′W
Sardinia *see* Sardegna
120 K8 **Sardinia-Corsica Trough** *undersea feature* Tyrrhenian Sea, C Mediterranean Sea
22 L2 **Sardis** Mississippi, S USA 34°25′N 89°55′W
22 L2 **Sardis Lake** ▣ Mississippi, S USA
27 P12 **Sardis Lake** ▣ Oklahoma, C USA
92 H12 **Sarek** ▲▲ N Sweden
92 H11 **Sarektjåhkkå** ▲ N Sweden 67°28′N 17°56′E
Sar-e Pol *see* Sar-e Pol-e Žahāb
Sar-e Pol *see* Sar-e Pol
142 J6 **Sar-e Pol-e Žahāb** *var.* Sar-e Pol, Sar-i Pul. Kermānshāhān, W Iran 34°28′N 45°52′E
149 N3 **Sar-e Pul** *var.* Sar-i-Pul; *prev.* Sar-e Pol. Sar-e Pul, N Afghanistan 36°16′N 65°55′E
149 O3 **Sar-e Pul** ⬖ *province* N Afghanistan
Sarera, Teluk *see* Cenderawasih, Teluk
147 T13 **Sarez, Kŭli** *Rus.* Sarezskoye Ozero. ◎ SE Tajikistan
Sarezskoye Ozero *see* Sarez, Kŭli
64 G10 **Sargasso Sea** *sea* W Atlantic Ocean
149 U8 **Sargodha** Punjab, NE Pakistan 32°06′N 72°48′E
78 I13 **Sarh** *prev.* Fort-Archambault. Moyen-Chari, S Chad 09°08′N 18°22′E
143 P4 **Sārī** *var.* Sari, Sāri. Māzandarān, N Iran 36°37′N 53°05′E
115 N23 **Saría** *island* SE Greece
Sariasiya *see* Sariosiyo
40 F3 **Saric** Sonora, NW Mexico 31°08′N 111°22′W
188 K6 **Sarigan** *island* C Northern Mariana Islands
136 D14 **Sarıgöl** Manisa, SW Turkey 38°16′N 28°41′E
139 T6 **Sārihah** At Ta'mím, E Iraq 34°34′N 44°38′E
137 R12 **Sarıkamış** Kars, NE Turkey 40°18′N 42°36′E
169 R9 **Sarikei** Sarawak, East Malaysia 02°07′N 111°30′E
147 U12 **Sarikol Range** *Rus.* Sarykol'skiy Khrebet. ▲▲ China/Tajikistan
181 Y7 **Sarina** Queensland, NE Australia 21°34′S 149°12′E
Sarine *see* La Sarine
105 S5 **Sariñena** Aragón, NE Spain 41°47′N 00°10′W
147 O13 **Sariosiyo** *Rus.* Sariasiya. Surkhondaryo Viloyati, S Uzbekistan 38°25′N 67°51′E
Sar-i-Pul *see* Sar-e Pul, Afghanistan
Sar-i Pul *see* Sar-e Pol-e Žahāb, Iran
Sariqamish Kŭli *see* Sarygamyş Köli
149 V1 **Sarī Qūl** *Rus.* Ozero Zurkul', *Taj.* Zŭrkŭl. ◎ Afghanistan/Tajikistan *see also* Zŭrkŭl
Sarī Qūl *see* Zŭrkŭl
75 Q12 **Sarīr Tībistī** *var.* Serir Tibesti. *desert* S Libya
25 S15 **Sarita** Texas, SW USA 27°14′N 97°48′W
163 W14 **Sariwŏn** SW North Korea 38°30′N 125°52′E
114 P12 **Sarıyer** İstanbul, NW Turkey 41°11′N 29°03′E
97 L26 **Sark** *Fr.* Sercq. *island* Channel Islands
111 N24 **Sarkad** *Rom.* Şărcad. Békés, SE Hungary 46°44′N 21°25′E
145 W14 **Sarkand** *Kaz.* Sarqan. Almaty, SW Kazakhstan 45°24′N 79°55′E
152 D11 **Sarkāri Tala** Rājasthān, NW India 27°39′N 70°52′E
136 G15 **Şarkikaraağaç** *var.* Şarki Karaağaç. Isparta, SW Turkey 38°04′N 31°22′E
Şarki Karaağaç *see* Şarkikaraağaç
136 L13 **Şarkışla** Sivas, C Turkey 39°21′N 36°27′E
136 C11 **Şarköy** Tekirdağ, NW Turkey 40°37′N 27°07′E
Sárköz *see* Livada
Sarlat *see* Sarlat-la-Canéda
102 M13 **Sarlat-la-Canéda** *var.* Sarlat. Dordogne, SW France 44°54′N 01°12′E
109 S3 **Sarleinsbach** Oberösterreich, N Austria 48°33′N 13°55′E
Sarma *see* Ash Sharmah
171 Y12 **Sarmi** Papua, E Indonesia 01°51′S 138°45′E
63 I19 **Sarmiento** Chubut, S Argentina 45°38′S 69°07′W
63 H25 **Sarmiento, Monte** ▲ S Chile 54°28′S 70°49′W
94 J11 **Särna** Dalarna, C Sweden 61°40′N 13°10′E
108 F8 **Sarnen** Obwalden, C Switzerland 46°54′N 08°15′E
108 F9 **Sarner See** ◎ C Switzerland
14 D16 **Sarnia** Ontario, S Canada 42°58′N 82°23′W
116 L3 **Sarny** Rivnens'ka Oblast', NW Ukraine 51°20′N 26°35′E
171 O13 **Saroako** Sulawesi, C Indonesia 02°31′S 121°18′E
118 L13 **Sarochyna** *Rus.* Sorochino. Vitsyebskaya Voblasts', N Belarus 55°12′N 28°45′E
168 L12 **Sarolangun** Sumatera, W Indonesia 02°17′S 102°39′E
165 U3 **Saroma** Hokkaidō, NE Japan 44°01′N 143°43′E
165 V3 **Saroma-ko** ◎ Hokkaidō, NE Japan
Saronic Gulf *see* Saronikós Kólpos
115 H20 **Saronikós Kólpos** *Eng.* Saronic Gulf. *gulf* S Greece
106 D7 **Saronno** Lombardia, N Italy 45°38′N 09°02′E
136 B11 **Saros Körfezi** *gulf* NW Turkey
111 N20 **Sárospatak** Borsod-Abaúj-Zemplén, NE Hungary 48°18′N 21°30′E
127 O4 **Sarov** *prev.* Sarova. Respublika Mordoviya, SW Russian Federation 54°39′N 43°09′E
Sarova *see* Sarov
127 P12 **Sarpa** Respublika Kalmykiya, SW Russian Federation 47°00′N 45°42′E
127 P12 **Sarpa, Ozero** ◎ SW Russian Federation
Šar Planina *see* Mali i Sharrit
95 I16 **Sarpsborg** Østfold, S Norway 59°16′N 11°07′E
139 U5 **Sarqalā** At Ta'mím, N Iraq
Sarqan *see* Sarkand
103 U4 **Sarralbe** Moselle, NE France 49°02′N 07°01′E
Sarre *see* Saar, France/Germany
Sarre *see* Saarland, Germany
103 U5 **Sarrebourg** *Ger.* Saarburg. Moselle, NE France 48°43′N 07°03′E
Sarrebruck *see* Saarbrücken
103 U4 **Sarreguemines** *prev.* Saargemund. Moselle, NE France 49°06′N 07°04′E
104 I3 **Sarria** Galicia, NW Spain 42°47′N 07°25′W
105 S8 **Sarrión** Aragón, NE Spain 40°09′N 00°49′W
42 F4 **Sarstoon** *Sp.* Río Sarstún. ≈ Belize/Guatemala
Sarstún, Río *see* Sarstoon
123 Q9 **Sartang** ≈ NE Russian Federation
103 X16 **Sartène** Corse, France, C Mediterranean Sea 41°38′N 08°58′E
102 K7 **Sarthe** ⬖ *department* NW France
102 K7 **Sarthe** ≈ N France
115 H15 **Sárti** Kentrikí Makedonía, N Greece 40°05′N 24°E
Sartu *see* Daqing
165 T1 **Sarufutsu** Hokkaidō, NE Japan 45°20′N 142°03′E
Saruhan *see* Manisa
152 G9 **Sarūpsar** Rājasthān, NW India 29°25′N 73°50′E
137 U13 **Şärur** *prev.* Il'ichevsk. SW Azerbaijan 39°30′N 44°59′E
Sarvani *see* Marneuli
111 G23 **Sárvár** Vas, W Hungary 47°15′N 16°59′E
143 P11 **Sarvestān** Fārs, S Iran 29°16′N 53°13′E
171 W12 **Sarwon** Papua, E Indonesia 0°58′S 136°08′E
145 P17 **Saryagash** *Kaz.* Saryaghash. Yuzhnyy Kazakhstan, S Kazakhstan 41°29′N 69°10′E
Saryaghash *see* Saryagash
145 R9 **Saryarka** *Eng.* Kazakh Uplands, Kirghiz Steppe. *uplands* C Kazakhstan
147 W8 **Sary-Bulak** Narynskaya Oblast', C Kyrgyzstan 41°56′N 75°44′E
147 U10 **Sary-Bulak** Oshskaya Oblast', SW Kyrgyzstan 40°49′N 73°44′E
117 S14 **Sarych, Mys** *headland* S Ukraine 44°23′N 33°44′E
147 Z7 **Sary-Dzhaz** *var.* Aksu He. ≈ China/Kyrgyzstan *see also* Aksu He
Sary-Dzhaz *see* Aksu He
146 F8 **Sarygamyş Köli** *Rus.* Sarykamyshkoye Ozero, *Uzb.* Sariqamish Kŭli. *salt lake* Kazakhstan/Uzbekistan
144 G13 **Sarykamys** *Kaz.* Saryqamys. Mangistau, SW Kazakhstan 45°58′N 53°30′E
Sarykamyshkoye Ozero *see* Sarygamyş Köli
145 N7 **Sarykol'** *prev.* Uritskiy. Kustanay, N Kazakhstan 53°19′N 65°34′E
Sarykol'skiy Khrebet *see* Sarikol Range
144 M10 **Sarykopa, Ozero** ◎ C Kazakhstan
145 V15 **Saryozek** *Kaz.* Saryözek. Almaty, SE Kazakhstan 44°22′N 77°57′E
144 E10 **Saryozen** *Kaz.* Kishiözen; *prev.* Malyy Uzen'. ≈ Kazakhstan/Russian Federation
Saryqamys *see* Sarykamys
145 S13 **Saryshagan** *Kaz.* Saryshaghan. Karaganda, SE Kazakhstan 46°05′N 73°38′E
Saryshaghan *see* Saryshagan
145 O13 **Sarysu** ≈ S Kazakhstan
147 T11 **Sary-Tash** Oshskaya Oblast', SW Kyrgyzstan 39°44′N 73°14′E
145 T12 **Saryterek** Karaganda, C Kazakhstan 47°46′N 74°06′E
Saryyazynskoye Vodokhranilishche *see* Sarýazy Suw Howdany
146 J15 **Sarýazy Suw Howdany** *Rus.* Saryyazynskoye Vodokhranilishche. ▣ S Turkmenistan
145 T14 **Saryyesik-Atyrau, Peski** *desert* E Kazakhstan
106 E10 **Sarzana** Liguria, NW Italy 44°07′N 09°59′E
188 B17 **Sasalaguan, Mount** ▲ S Guam
153 O14 **Sasarām** Bihār, N India 24°58′N 84°01′E
186 M8 **Sasari, Mount** ▲ Santa Isabel, N Solomon Islands 08°09′S 159°32′E
164 C13 **Sasebo** Nagasaki, Kyūshū, SW Japan 33°10′N 129°42′E
14 I9 **Saseginaga, Lac** ◎ Québec, SE Canada
Saseno *see* Sazan
11 R13 **Saskatchewan** ⬖ *province* SW Canada
11 U14 **Saskatchewan** ≈ Manitoba/Saskatchewan, C Canada
11 T15 **Saskatoon** Saskatchewan, S Canada 52°10′N 106°40′W
11 T15 **Saskatoon** ✈ Saskatchewan, S Canada 52°15′N 107°05′W
123 N7 **Saskylakh** Respublika Sakha (Yakutiya), NE Russian Federation 71°56′N 114°07′E
42 L7 **Saslaya, Cerro** ▲ N Nicaragua 13°52′N 85°06′W
38 G17 **Sasmik, Cape** *headland* Tanaga Island, Alaska, USA 51°36′N 177°55′W

◆ Country
● Country Capital
◇ Dependent Territory
○ Dependent Territory Capital
⬖ Administrative Regions
✈ International Airport
▲ Mountain
▲▲ Mountain Range
🌋 Volcano
≈ River
◎ Lake
▣ Reservoir
319

SASNOVY BOR – SEGAF, KEPULAUAN

119 N19 **Sasnovy Bor** *Rus.* Sosnovyy Bor. Homyel'skaya Voblasts', SE Belarus *52°32′N 29°35′E*
127 N5 **Sasovo** Ryazanskaya Oblast', W Russian Federation *54°19′N 41°54′E*
25 S12 **Saspamco** Texas, SW USA *29°13′N 98°18′W*
109 W9 **Sass** *var.* Sassbach. ⛐ SE Austria
76 M17 **Sassandra** S Ivory Coast *04°58′N 06°08′W*
76 M17 **Sassandra** *var.* Ibo, Sassandra Fleuve. ⛐ S Ivory Coast
Sassandra Fleuve *see* Sassandra
107 B17 **Sassari** Sardegna, Italy, C Mediterranean Sea *40°44′N 08°33′E*
Sassbach *see* Sass
98 H11 **Sassenheim** Zuid-Holland, W Netherlands *52°14′N 04°31′E*
Sassmacken *see* Valdemārpils
100 O7 **Sassnitz** Mecklenburg-Vorpommern, NE Germany *54°32′N 13°39′E*
99 E16 **Sas van Gent** Zeeland, SW Netherlands *51°13′N 03°48′E*
145 W12 **Sasykkol', Ozero** ◎ E Kazakhstan
117 O12 **Sasyk, Ozero** *Rus.* Ozero Sasyk Kunduk, *var.* Ozero Kunduk. ◎ SW Ukraine
76 J12 **Satadougou** Kayes, SW Mali *12°40′N 11°25′W*
93 K18 **Satakunta** ◆ *region* W Finland
164 C17 **Sata-misaki** Kyūshū, SW Japan
26 I7 **Satanta** Kansas, C USA *37°23′N 102°00′W*
155 E15 **Sātāra** Mahārāshtra, W India *17°41′N 73°59′E*
192 G15 **Sātaua** Savai'i, NW Samoa *13°26′S 172°40′W*
188 M16 **Satawal** *island* Caroline Islands, C Micronesia
189 R17 **Satawan Atoll** *atoll* Mortlock Islands, C Micronesia
Satbaev *see* Satpayev
23 Y12 **Satellite Beach** Florida, SE USA *28°10′N 80°35′W*
95 M14 **Säter** Dalarna, C Sweden *60°21′N 15°45′E*
Sathmar *see* Satu Mare
23 V7 **Satilla River** ⛐ Georgia, SE USA
57 F14 **Satipo** *var.* San Francisco de Satipo. Junín, C Peru *11°19′S 74°37′W*
122 F11 **Satka** Chelyabinskaya Oblast', C Russian Federation *55°08′N 58°54′E*
153 T16 **Satkhira** Khulna, SW Bangladesh *22°43′N 89°06′E*
146 J13 **Şatlyk** *Rus.* Shatlyk. Mary Welaýaty, C Turkmenistan *37°55′N 61°00′E*
154 K9 **Satna** *prev.* Sutna. Madhya Pradesh, C India *24°33′N 80°50′E*
103 R11 **Satolas** ✈ (Lyon) Rhône, E France *45°44′N 05°01′E*
111 N20 **Sátoraljaújhely** Borsod-Abaúj-Zemplén, NE Hungary *48°24′N 21°39′E*
145 O12 **Satpayev** *Kaz.* Sätbaev; *prev.* Nikol'skiy. Karaganda, C Kazakhstan *47°59′N 67°27′E*
154 G11 **Sātpura Range** ▲▲ C India
165 Q10 **Satsuma-Sendai** Miyagi, Honshū, C Japan *38°16′N 140°52′E*
Satsuma-Sendai *see* Sendai
167 P12 **Sattahip** *var.* Ban Sattahip, Ban Sattahipp. Chon Buri, S Thailand *12°36′N 100°56′E*
92 L11 **Sattanen** Lappi, NE Finland *67°31′N 26°35′E*
Satul *see* Satun
116 H9 **Satulung** *Hung.* Kővárhosszúfalu. Maramureș, N Romania *47°34′N 23°26′E*
Satul-Vechi *see* Staro Selo
116 G8 **Satu Mare** *Ger.* Sathmar, *Hung.* Szatmárrnémeti. Satu Mare, NW Romania *47°46′N 22°55′E*
116 G8 **Satu Mare** ◆ *county* NW Romania
167 N16 **Satun** *var.* Satul, Setul. Satun, SW Thailand *06°40′N 100°01′E*
192 G16 **Satupa'itea** Savai'i, W Samoa *13°46′S 172°26′W*
Sau *see* Sava
14 F14 **Sauble** ⛐ Ontario, S Canada
14 F13 **Sauble Beach** Ontario, S Canada *44°36′N 81°15′W*
61 C16 **Sauce** Corrientes, NE Argentina *30°05′S 58°46′W*
Sauce *see* Juan L. Lacaze
36 K15 **Sauceda Mountains** ▲▲ Arizona, SW USA
61 C17 **Sauce de Luna** Entre Ríos, E Argentina *31°15′S 59°09′W*
63 L15 **Sauce Grande, Río** ⛐ E Argentina
40 K6 **Saucillo** Chihuahua, N Mexico *28°01′N 105°17′W*
95 D15 **Sauda** Rogaland, S Norway *59°38′N 06°23′E*
145 Q16 **Saudakent** *Kaz.* Saūdakent; *prev.* Baykadam, *Kaz.* Bayqadam. Zhambyl, S Kazakhstan *43°49′N 69°56′E*
92 J2 **Sauðárkrókur** Norðurland Vestra, N Iceland *65°45′N 19°39′W*
141 P9 **Saudi Arabia** *off.* Kingdom of Saudi Arabia, Al 'Arabīyah as Su'ūdīyah, *Ar.* Al Mamlakah al 'Arabīyah as Su'ūdīyah. ◆ *monarchy* SW Asia
Saudi Arabia, Kingdom of *see* Saudi Arabia
101 D19 **Sauer** *var.* Sûre. ⛐ NW Europe *see also* Sûre
Sauer *see* Sûre
101 F15 **Sauerland** *forest* W Germany
14 F14 **Saugeen** ⛐ Ontario, S Canada
18 K12 **Saugerties** New York, NE USA *42°04′N 73°55′W*
Saugor *see* Sāgar
10 K15 **Saugstad, Mount** ▲ British Columbia, SW Canada *52°12′N 126°35′W*
Sāūjbulāgh *see* Mahābād
102 J11 **Saujon** Charente-Maritime, W France *45°40′N 00°54′W*
29 T7 **Sauk Centre** Minnesota, N USA *45°44′N 94°57′W*
30 L8 **Sauk City** Wisconsin, N USA *43°16′N 89°43′W*
29 U7 **Sauk Rapids** Minnesota, N USA *45°35′N 94°09′W*
55 Y11 **Saül** C French Guiana *03°37′N 53°12′W*
103 O7 **Sauldre** ⛐ C France
101 I23 **Saulgau** Baden-Württemberg, SW Germany *48°03′N 09°28′E*
103 Q8 **Saulieu** Côte d'Or, C France *47°15′N 04°15′E*
118 G8 **Saulkrasti** C Latvia *57°14′N 24°25′E*
15 S6 **Sault-aux-Cochons, Rivière du** ⛐ Québec, SE Canada
31 Q4 **Sault Sainte Marie** Michigan, N USA *46°29′N 84°22′W*
12 F14 **Sault Ste. Marie** Ontario, S Canada *46°30′N 84°17′W*
145 P7 **Saumalkol'** *prev.* Volodarskoye. Severnyy Kazakhstan, N Kazakhstan *53°19′N 68°05′E*
190 E13 **Sauma, Pointe** *headland* Île Alofi, W Wallis and Futuna *14°21′S 177°58′W*
171 T16 **Saumlaki** *var.* Saumlakki. Pulau Yamdena, E Indonesia *07°53′S 131°18′E*
Saumlakki *see* Saumlaki
15 R12 **Saumon, Rivière au** ⛐ Québec, SE Canada
102 K8 **Saumur** Maine-et-Loire, NW France *47°16′N 00°04′W*
185 F23 **Saunders, Cape** *headland* South Island, New Zealand *45°53′S 170°40′E*
195 N13 **Saunders Coast** *physical region* Antarctica
65 B23 **Saunders Island** *island* NW Falkland Islands
65 C24 **Saunders Island Settlement** Saunders Island, NW Falkland Islands *51°22′S 60°05′W*
82 F11 **Saurimo** *Port.* Henrique de Carvalho, Vila Henrique de Carvalho. Lunda Sul, NE Angola *09°39′S 20°24′E*
55 S11 **Sauriwaunawa** S Guyana *03°10′N 59°51′W*
82 D12 **Sautar** Malanje, NW Angola *11°10′S 18°26′E*
45 S13 **Sauteurs** N Grenada *12°14′N 61°38′W*
102 K13 **Sauveterre-de-Guyenne** Gironde, SW France *44°43′N 00°02′W*
119 O14 **Sava** Mahilyowskaya Voblasts', E Belarus *54°22′N 30°49′E*
42 J5 **Savá** Colón, N Honduras *15°30′N 86°16′W*
84 H11 **Sava** *Eng.* Save, *Ger.* Sau, *Hung.* Száva. ⛐ SE Europe
33 Y8 **Savage** Montana, NW USA *47°28′N 104°17′W*
183 N16 **Savage River** Tasmania, SE Australia *41°34′S 145°15′E*
77 R15 **Savalou** S Benin *07°59′N 01°58′E*
30 K10 **Savanna** Illinois, N USA *42°05′N 90°09′W*
23 X6 **Savannah** Georgia, SE USA *32°02′N 81°01′W*
27 R2 **Savannah** Missouri, C USA *39°57′N 94°49′W*
20 H10 **Savannah** Tennessee, S USA *35°12′N 88°15′W*
21 O12 **Savannah River** ⛐ Georgia/South Carolina, SE USA
167 S9 **Savannakhét** *var.* Khanthabouli. Savannakhét, S Laos *16°38′N 104°49′E*
44 H12 **Savanna-La-Mar** W Jamaica *18°13′N 78°08′W*
12 B10 **Savant Lake** ◎ Ontario, S Canada
155 F17 **Savanūr** Karnātaka, W India *14°58′N 75°19′E*
93 J16 **Sävar** Västerbotten, N Sweden *63°52′N 20°33′E*
Savaria *see* Szombathely
154 C11 **Sāvarkundla** *var.* Kundla. Gujarāt, W India *21°21′N 71°20′E*
116 F11 **Săvârșin** *Hung.* Soborsin; *prev.* Săvirșin. Arad, W Romania *46°00′N 22°15′E*
136 C13 **Savaştepe** Balıkesir, W Turkey *39°20′N 27°38′E*
147 P11 **Savat** *Rus.* Savat. Sirdaryo Viloyati, E Uzbekistan *40°03′N 68°35′E*
Savat *see* Savat
Sävdijäri *see* Skaulo
77 R15 **Savè** SE Benin *08°04′N 02°29′E*
83 N18 **Save** Inhambane, E Mozambique *21°07′S 34°35′E*
102 L16 **Save** ⛐ S France
83 L17 **Save** *var.* Sabi. ⛐ Mozambique/Zimbabwe *see also* Sabi
Save *see* Sava
Save *see* Sabi
142 M6 **Sāveh** Markazī, W Iran *35°00′N 50°22′E*
116 L8 **Săveni** Botoşani, NE Romania *47°57′N 26°52′E*
103 N16 **Saverdun** Ariège, S France *43°15′N 01°34′E*
103 U5 **Saverne** *var.* Zabern; *anc.* Tres Tabernae. Bas-Rhin, NE France *48°45′N 07°22′E*
106 B9 **Savigliano** Piemonte, NW Italy *44°39′N 07°39′E*
Savigsivik *see* Savissivik
109 U10 **Savinja** ⛐ N Slovenia
106 H11 **Savio** ⛐ C Italy
Săvirșin *see* Săvârșin
197 O11 **Savissivik** *var.* Savigsivik. ◆ Qaasuitsup, N Greenland
93 N18 **Savitaipale** Etelä-Karjala, SE Finland *61°12′N 27°43′E*
113 J15 **Šavnik** C Montenegro *42°57′N 19°04′E*
108 I9 **Savognin** Graubünden, S Switzerland *46°34′N 09°35′E*
103 T12 **Savoie** ◆ *department* E France
106 C10 **Savona** Liguria, NW Italy *44°18′N 08°29′E*
93 N17 **Savonlinna** *Swe.* Nyslott. Etelä-Savo, E Finland *61°51′N 28°56′E*
93 N17 **Savonranta** Etelä-Savo, E Finland *62°10′N 29°10′E*
38 K10 **Savoonga** Saint Lawrence Island, Alaska, USA *63°40′N 170°29′W*
30 M13 **Savoy** Illinois, N USA *40°03′N 88°15′W*
117 O8 **Savran'** Odes'ka Oblast', SW Ukraine *48°07′N 30°00′E*
137 R11 **Şavşat** Artvin, NE Turkey *41°15′N 42°30′E*
95 L19 **Sävsjö** Jönköping, S Sweden *57°25′N 14°40′E*
Savu, Kepulauan *see* Sawu, Kepulauan
92 M11 **Savukoski** Lappi, NE Finland *67°17′N 28°14′E*
Savu, Pulau *see* Sawu, Pulau
187 Y14 **Savusavu** Vanua Levu, N Fiji *16°48′S 179°20′E*
171 O17 **Savu Sea** *Ind.* Laut Sawu. *sea* S Indonesia
83 H17 **Savute** North-West, N Botswana *18°33′S 24°06′E*
139 N7 **Şawāb Uqlat** *well* W Iraq
138 M7 **Şawāb, Wādī aş** *dry watercourse* W Iraq
152 H13 **Sawai Mādhopur** Rājasthān, N India *26°00′N 76°22′E*
Sawakin *see* Suakin
167 R8 **Sawang Daen Din** Sakon Nakhon, E Thailand *17°28′N 103°27′E*
167 O8 **Sawankhalok** *var.* Swankalok. Sukhothai, NW Thailand *17°19′N 99°50′E*
165 P13 **Sawara** Chiba, Honshū, S Japan *35°52′N 140°31′E*
37 R5 **Sawatch Range** ▲▲ Colorado, C USA
141 N12 **Sawdā', Jabal** ▲ SW Saudi Arabia *18°15′N 42°26′E*
75 P9 **Sawdā', Jabal as** ▲▲ C Libya
Sawdirī *see* Sodiri
97 F14 **Sawel Mountain** ▲ C Northern Ireland, United Kingdom *54°49′N 07°04′W*
75 X10 **Sawhāj** *var.* Sawhâj, *var.* Sohâg, Suliag. C Egypt *26°28′N 31°44′E*
Sawhâj *see* Sawhāj
77 O15 **Sawla** N Ghana *09°14′N 02°26′W*
141 X12 **Şawqirah** *var.* Suqrah. S Oman *18°16′N 56°34′E*
141 X12 **Şawqirah, Dawhat** *var.* Ghubbat Sawqirah, Sukra Bay, Suqrah Bay. *bay* S Oman
Sawqirah, Ghubbat *see* Şawqirah, Dawhat
183 V5 **Sawtell** New South Wales, SE Australia *30°22′S 153°04′E*
138 K7 **Şawt, Wādī aş** *dry watercourse* S Syria
171 O17 **Sawu, Kepulauan** *var.* Kepulauan Savu. *island group* S Indonesia
Sawu, Laut *see* Savu Sea
171 O17 **Sawu, Pulau** *var.* Pulau Sawu. *island* Kepulauan Sawu, S Indonesia
105 S12 **Sax** Valenciana, E Spain *38°33′N 00°49′W*
Saxe *see* Sachsen
108 C11 **Saxon** Valais, SW Switzerland *46°07′N 07°09′E*
Saxony *see* Sachsen
Saxony-Anhalt *see* Sachsen-Anhalt
77 R12 **Say** Niamey, SW Niger *13°08′N 02°20′E*
15 V7 **Sayabec** Québec, SE Canada *48°33′N 67°42′W*
Sayaboury *see* Xaignabouli
145 U12 **Sayak** *Kaz.* Sayaq. Karaganda, E Kazakhstan *46°54′N 77°17′E*
57 D14 **Sayán** Lima, W Peru *11°10′S 77°08′W*
129 T6 **Sayanskiy Khrebet** ▲▲ S Russian Federation
Sayaq *see* Sayak
146 K13 **Saýat** *Rus.* Sayat. Lebap Welaýaty, E Turkmenistan *38°44′N 63°51′E*
42 D3 **Sayaxché** Petén, N Guatemala *16°34′N 90°14′W*
Şaydā/Sayida *see* Saïda
162 J7 **Sayhan** *var.* Hüremt. Bulgan, C Mongolia *48°40′N 102°33′E*
163 N10 **Sayhandulaan** *var.* Öldziyt. Dornogovĭ, SE Mongolia *44°42′N 109°10′E*
162 K9 **Sayhan-Ovoo** *var.* Ongi. Dundgovĭ, C Mongolia *45°27′N 103°58′E*
141 T15 **Sayhūt** E Yemen *15°18′N 51°16′E*
29 U14 **Saylorville Lake** ◎ Iowa, C USA
Saymenskiy Kanal *see* Saimaa Canal
163 N10 **Saynshand** Dornogovĭ, SE Mongolia *44°51′N 110°07′E*
Saynshand *see* Sevrey
Sayn-Ust *see* Hohmorit
Say-Ötesh *see* Otes
138 J7 **Şayqal, Bahr** ◎ S Syria
158 H4 **Sayram Hu** ◎ NW China
26 K11 **Sayre** Oklahoma, C USA *35°18′N 99°38′W*
18 H12 **Sayre** Pennsylvania, NE USA *41°57′N 76°30′W*
18 K15 **Sayreville** New Jersey, NE USA *40°27′N 74°19′W*
147 N13 **Sayrob** *Rus.* Sayrab. Surkhondaryo Viloyati, S Uzbekistan *38°03′N 66°54′E*
40 L13 **Sayula** Jalisco, SW Mexico *19°52′N 103°36′W*
141 R14 **Say'ūn** *var.* Saywūn. C Yemen *15°53′N 48°32′E*
Say-Utës *see* Otes
10 K16 **Sayward** Vancouver Island, British Columbia, SW Canada *50°20′N 126°01′W*
Saywūn *see* Say'ūn
Sayyāl *see* As Sayyāl
139 U8 **Sayyid 'Abīd** *var.* Saiyid Abid. Wāsiţ, E Iraq *32°51′N 45°07′E*
113 J22 **Sazan** *var.* Ishulli i Sazanit, *It.* Saseno. *island* SW Albania
Sazanit, Ishulli i *see* Sazan
Sazau/Sazawa *see* Sázava
111 E17 **Sázava** *var.* Sazau, *Ger.* Sazawa. ⛐ C Czech Republic
114 K10 **Sazliyka** ⛐ C Bulgaria
124 J14 **Sazonovo** Vologodskaya Oblast', NW Russian Federation *59°04′N 35°10′E*
102 G6 **Scaër** Finistère, NW France *48°00′N 03°40′W*
97 J15 **Scafell Pike** ▲ NW England, United Kingdom *54°26′N 03°10′W*
Scalabis *see* Santarém
96 M2 **Scalloway** N Scotland, United Kingdom *60°10′N 01°17′W*
38 M11 **Scammon Bay** Alaska, USA *61°50′N 165°34′W*
Scammon Lagoon/Scammon, Laguna *see* Ojo de Liebre, Laguna
84 F7 **Scandinavia** *geophysical region* NW Europe
Scania *see* Skåne
96 K5 **Scapa Flow** *sea basin* N Scotland, United Kingdom
107 K26 **Scaramia, Capo** *headland* Sicilia, Italy, C Mediterranean Sea *36°46′N 14°29′E*
14 H15 **Scarborough** Ontario, SE Canada *43°46′N 79°14′W*
45 Z16 **Scarborough** *prev.* Port Louis. Tobago, Trinidad and Tobago *11°11′N 60°45′W*
97 N16 **Scarborough** N England, United Kingdom *54°17′N 00°24′W*
185 I17 **Scargill** Canterbury, South Island, New Zealand *42°57′S 172°57′E*
96 E7 **Scarp** *island* NW Scotland, United Kingdom
Scarpanto *see* Kárpathos
Scarpanto Strait *see* Karpathou, Stenó
107 G25 **Scauri** Sicilia, Italy, C Mediterranean Sea *36°45′N 12°06′E*
Scealg, Bá na *see* Ballinskelligs Bay
Scebeli *see* Shebeli
100 K10 **Schaale** ⛐ N Germany
100 K9 **Schaalsee** ◎ N Germany
99 G18 **Schaerbeek** Brussels, C Belgium *50°52′N 04°21′E*
108 G6 **Schaffhausen** *Fr.* Schaffhouse. Schaffhausen, N Switzerland *47°42′N 08°38′E*
108 G6 **Schaffhausen** *Fr.* Schaffhouse. ◆ *canton* N Switzerland
Schaffhouse *see* Schaffhausen
98 I8 **Schagen** Noord-Holland, NW Netherlands *52°47′N 04°47′E*
Schaken *see* Šakiai
98 M10 **Schalkhaar** Overijssel, E Netherlands *52°16′N 06°10′E*
109 R3 **Schärding** Oberösterreich, N Austria *48°27′N 13°26′E*
100 G9 **Scharhörn** *island* NW Germany
Schässburg *see* Sighişoara
Schaulen *see* Šiauliai
30 M10 **Schaumburg** Illinois, N USA *42°01′N 88°04′W*
Schebschi Mountains *see* Shebshi Mountains
98 P6 **Scheemda** Groningen, NE Netherlands *53°10′N 06°58′E*
100 I10 **Scheessel** Niedersachsen, NW Germany *53°11′N 09°33′E*
13 N8 **Schefferville** Québec, E Canada *54°50′N 67°W*
Schelde *see* Scheldt
99 D18 **Scheldt** *Dut.* Schelde, *Fr.* Escaut. ⛐ W Europe
35 X5 **Schell Creek Range** ▲▲ Nevada, W USA
18 K10 **Schenectady** New York, NE USA *42°48′N 73°57′W*
99 I17 **Scherpenheuvel** *Fr.* Montaigu. Vlaams Brabant, C Belgium *51°00′N 04°57′E*
98 K11 **Scherpenzeel** Gelderland, C Netherlands *52°07′N 05°30′E*
25 S12 **Schertz** Texas, SW USA *29°33′N 98°16′W*
98 G11 **Scheveningen** Zuid-Holland, W Netherlands *52°07′N 04°18′E*
98 G12 **Schiedam** Zuid-Holland, SW Netherlands *51°55′N 04°25′E*
99 M24 **Schieren** Diekirch, NE Luxembourg *49°50′N 06°06′E*
98 M4 **Schiermonnikoog** *Fris.* Skiermûntseach. Fryslân, N Netherlands *53°28′N 06°09′E*
98 M4 **Schiermonnikoog** *Fris.* Skiermûntseach. *island* Waddeneilanden, N Netherlands
99 K14 **Schijndel** Noord-Brabant, S Netherlands *51°37′N 05°27′E*
Schil *see* Jiu
99 H16 **Schilde** Antwerpen, N Belgium *51°14′N 04°35′E*
Schillen *see* Zhilino
103 V5 **Schiltigheim** Bas-Rhin, NE France *48°38′N 07°47′E*
106 G7 **Schio** Veneto, NE Italy *45°42′N 11°21′E*
98 H10 **Schiphol** ✈ (Amsterdam) Noord-Holland, C Netherlands *52°18′N 04°48′E*
Schippenbeil *see* Sępopol
Schiria *see* Şiria
Schivelbein *see* Świdwin
115 D22 **Schíza** *island* S Greece
175 U3 **Schjetman Reef** *reef* Antarctica
Schlackenwerth *see* Ostrov
109 R7 **Schladming** Steiermark, SE Austria *47°24′N 13°42′E*
Schlan *see* Slaný
Schlanders *see* Silandro
100 I7 **Schlei** *inlet* N Germany
101 D17 **Schleiden** Nordrhein-Westfalen, W Germany *50°31′N 06°30′E*
Schlelau *see* Szydłowiec
100 I7 **Schleswig** Schleswig-Holstein, N Germany *54°31′N 09°34′E*
29 T13 **Schleswig** Iowa, C USA *42°10′N 95°27′W*
100 H8 **Schleswig-Holstein** ◆ *state* N Germany
Schlettstadt *see* Sélestat
108 F7 **Schlieren** Zürich, N Switzerland *47°23′N 08°27′E*
Schlochau *see* Człuchów
Schloppe *see* Człopa
101 I18 **Schlüchtern** Hessen, C Germany *50°19′N 09°27′E*
101 J17 **Schmalkalden** Thüringen, C Germany *50°42′N 10°26′E*
109 W2 **Schmida** ⛐ NE Austria
65 P19 **Schmidt-Ott Seamount** *var.* Schmitt-Ott Seamount, Schmitt-Ott Tablemount. *undersea feature* SW Indian Ocean *39°37′S 13°00′E*
Schmiegel *see* Śmigiel
Schmitt-Ott Seamount/Schmitt-Ott Tablemount *see* Schmidt-Ott Seamount
15 V3 **Schmon** ⛐ Québec, SE Canada
101 M18 **Schneeberg** ▲ W Germany *50°03′N 11°51′E*
Schneeberg *see* Veliki Snežnik
Schnee-Eifel *see* Schneifel
Schneekoppe *see* Sněžka
Schneidemühl *see* Piła
101 D18 **Schneifel** *var.* Schnee-Eifel. *plateau* W Germany
Schnelle Körös/Schnelle Kreisch *see* Crişul Repede
100 I11 **Schneverdingen** *var.* Schneverdingen (Wümme). Niedersachsen, NW Germany *53°07′N 09°48′E*
Schneverdingen (Wümme) *see* Schneverdingen
Schoden *see* Skuodas
45 Q12 **Schœlcher** W Martinique *14°37′N 61°08′W*
18 K10 **Schoharie** New York, NE USA *42°40′N 74°20′W*
18 K11 **Schoharie Creek** ⛐ New York, NE USA
115 J21 **Schoinoússa** *island* Kykládes, Greece, Aegean Sea
100 L13 **Schönebeck** Sachsen-Anhalt, C Germany *52°01′N 11°45′E*
Schöneck *see* Skarszewy
100 O12 **Schönefeld** ✈ (Berlin) Berlin, NE Germany *52°23′N 13°29′E*
101 K24 **Schongau** Bayern, S Germany *47°49′N 10°54′E*
100 K13 **Schöningen** Niedersachsen, C Germany *52°07′N 10°58′E*
Schönlanke *see* Trzcianka
Schönsee *see* Kowalewo Pomorskie
31 P10 **Schoolcraft** Michigan, N USA *42°05′N 85°39′W*
98 O8 **Schoonebeek** Drenthe, NE Netherlands *52°39′N 06°57′E*
98 I12 **Schoonhoven** Zuid-Holland, C Netherlands *51°57′N 04°51′E*
98 H8 **Schoorl** Noord-Holland, NW Netherlands *52°42′N 04°40′E*
Schooten *see* Schoten
101 F24 **Schopfheim** Baden-Württemberg, SW Germany *47°39′N 07°49′E*
101 I21 **Schorndorf** Baden-Württemberg, S Germany *48°48′N 09°31′E*
100 F10 **Schortens** Niedersachsen, NW Germany *53°31′N 07°57′E*
99 H16 **Schoten** *var.* Schooten. Antwerpen, N Belgium *51°15′N 04°30′E*
183 Q17 **Schouten Island** *island* Tasmania, SE Australia
186 C5 **Schouten Islands** *island group* NW Papua New Guinea
98 E13 **Schouwen** *island* SW Netherlands
Schreiberhau *see* Szklarska Poręba
109 U2 **Schrems** Niederösterreich, E Austria *48°48′N 15°05′E*
101 L22 **Schrobenhausen** Bayern, SE Germany *48°33′N 11°14′E*
18 L8 **Schroon Lake** ◎ New York, NE USA
108 J8 **Schruns** Vorarlberg, W Austria *47°04′N 09°54′E*
Schubin *see* Szubin
25 U11 **Schulenburg** Texas, SW USA *29°40′N 96°54′W*
Schuls *see* Scuol
108 E8 **Schüpfheim** Luzern, C Switzerland *47°02′N 07°23′E*
35 S6 **Schurz** Nevada, W USA *38°55′N 118°48′W*
101 I24 **Schussen** ⛐ S Germany
Schüttenhofen *see* Sušice
29 R15 **Schuyler** Nebraska, C USA *41°25′N 97°04′W*
18 L10 **Schuylerville** New York, NE USA *43°05′N 73°34′W*
101 K20 **Schwabach** Bayern, SE Germany *49°20′N 11°02′E*
Schwabenalb *see* Schwäbische Alb
101 I23 **Schwäbische Alb** *var.* Schwabenalb, *Eng.* Swabian Jura. ▲▲ S Germany
101 I22 **Schwäbisch Gmünd** *var.* Gmünd. Baden-Württemberg, SW Germany *48°49′N 09°48′E*
101 I21 **Schwäbisch Hall** *var.* Hall. Baden-Württemberg, SW Germany *49°07′N 09°45′E*
101 H16 **Schwalm** ⛐ C Germany
109 V9 **Schwanberg** Steiermark, SE Austria *46°46′N 15°12′E*
108 H8 **Schwanden** Glarus, E Switzerland *47°02′N 09°04′E*
101 M20 **Schwandorf** Bayern, SE Germany *49°20′N 12°07′E*
109 S5 **Schwanenstadt** Oberösterreich, NW Austria *48°03′N 13°47′E*
169 S11 **Schwaner, Pegunungan** ▲▲ Borneo, N Indonesia
109 W5 **Schwarza** ⛐ E Austria
109 P9 **Schwarzach** ⛐ S Austria
101 M20 **Schwarzach** *Cz.* Černice. ⛐ Czech Republic/Germany
Schwarzach *see* Schwarzach im Pongau
109 Q7 **Schwarzach im Pongau** *var.* Schwarzach. Salzburg, NW Austria *47°19′N 13°09′E*
Schwarzawa *see* Svratka
101 N14 **Schwarze Elster** ⛐ E Germany
Schwarze Körös *see* Crişul Negru
108 D9 **Schwarzenburg** Bern, W Switzerland *46°51′N 07°28′E*
83 D21 **Schwarzrand** ▲▲ S Namibia
101 G23 **Schwarzwald** *Eng.* Black Forest. ▲▲ SW Germany
Schwarzwasser *see* Wda
39 P7 **Schwatka Mountains** ▲▲ Alaska, USA
109 N7 **Schwaz** Tirol, W Austria *47°21′N 11°44′E*
109 Y4 **Schwechat** Niederösterreich, NE Austria *48°09′N 16°29′E*
109 Y4 **Schwechat** ✈ (Wien) Wien, E Austria *48°04′N 16°31′E*
100 P11 **Schwedt** Brandenburg, NE Germany *53°04′N 14°16′E*
101 D19 **Schweich** Rheinland-Pfalz, SW Germany *49°49′N 06°44′E*
Schweidnitz *see* Świdnica
101 J18 **Schweinfurt** Bayern, SE Germany *50°03′N 10°13′E*
Schweiz *see* Switzerland
100 L9 **Schwerin** Mecklenburg-Vorpommern, N Germany *53°38′N 11°25′E*
Schwerin *see* Skwierzyna
100 L9 **Schweriner See** ◎ N Germany
Schwertberg *see* Świecie
101 F15 **Schwerte** Nordrhein-Westfalen, W Germany *51°27′N 07°34′E*
Schwiebus *see* Świebodzin
100 P13 **Schwielochsee** ◎ NE Germany
Schwihau *see* Švihov
Schwiz *see* Schwyz
108 G8 **Schwyz** *var.* Schwiz. Schwyz, C Switzerland *47°02′N 08°39′E*
108 G8 **Schwyz** *var.* Schwiz. ◆ *canton* C Switzerland
14 J11 **Schyan** ⛐ Québec, SE Canada
Schyl *see* Jiu
107 I24 **Sciacca** Sicilia, Italy, C Mediterranean Sea *37°31′N 13°05′E*
Sciasciamana *see* Shashemenē
107 L26 **Scicli** Sicilia, Italy, C Mediterranean Sea *36°48′N 14°43′E*
97 F25 **Scilly, Isles of** *island group* SW England, United Kingdom
111 H17 **Šcinawa** *Ger.* Steinau an der Elbe. Dolnośląskie, SW Poland *51°22′N 16°27′E*
Scio *see* Chíos
31 S14 **Scioto River** ⛐ Ohio, N USA
36 L5 **Scipio** Utah, W USA *39°15′N 112°06′W*
33 X6 **Scobey** Montana, NW USA *48°47′N 105°25′W*
183 T7 **Scone** New South Wales, SE Australia *32°02′S 150°51′E*
Scoresby Sound/Scoresbysund *see* Ittoqqortoormiit
Scoresby Sund *see* Kangertittivaq
Scorno, Punta dello *see* Caprara, Punta
34 K3 **Scotia** California, W USA *40°34′N 124°07′W*
47 Y14 **Scotia Plate** *tectonic feature*
47 V15 **Scotia Ridge** *undersea feature* S Atlantic Ocean
194 H2 **Scotia Sea** *sea* SW Atlantic Ocean
29 Q12 **Scotland** South Dakota, N USA *43°09′N 97°43′W*
25 R5 **Scotland** Texas, SW USA *33°37′N 98°27′W*
96 H11 **Scotland** ◇ *national region* Scotland, U K
21 W8 **Scotland Neck** North Carolina, SE USA *36°07′N 77°25′W*
195 R13 **Scott Base** NZ *research station* Antarctica *77°52′S 167°18′E*
10 J16 **Scott, Cape** *headland* Vancouver Island, British Columbia, SW Canada *50°43′N 128°24′W*
26 I5 **Scott City** Kansas, C USA *38°28′N 100°55′W*
27 Y7 **Scott City** Missouri, C USA *37°13′N 89°31′W*
195 R14 **Scott Coast** *physical region* Antarctica
18 C15 **Scottdale** Pennsylvania, NE USA *40°05′N 79°35′W*
195 Y11 **Scott Glacier** *glacier* Antarctica
195 Q17 **Scott Island** *island* Antarctica
26 L11 **Scott, Mount** ▲ Oklahoma, USA *34°52′N 98°34′W*
32 G13 **Scott, Mount** ▲ Oregon, NW USA *42°53′N 122°06′W*
34 M1 **Scott River** ⛐ California, W USA
28 I13 **Scottsbluff** Nebraska, C USA *41°52′N 103°40′W*
23 Q2 **Scottsboro** Alabama, S USA *34°40′N 86°01′W*
31 P15 **Scottsburg** Indiana, N USA *38°42′N 85°47′W*
183 P16 **Scottsdale** Tasmania, SE Australia *41°13′S 147°30′E*
36 L13 **Scottsdale** Arizona, SW USA *33°31′N 111°54′W*
45 O12 **Scotts Head Village** *var.* Cachacrou. S Dominica *15°12′N 61°22′W*
192 L14 **Scott Shoal** *undersea feature* S Pacific Ocean
20 K7 **Scottsville** Kentucky, S USA *36°45′N 86°11′W*
29 U14 **Scranton** Iowa, C USA *42°01′N 94°33′W*
18 I13 **Scranton** Pennsylvania, NE USA *41°25′N 75°40′W*
186 B6 **Screw** ⛐ NW Papua New Guinea
29 R14 **Scribner** Nebraska, C USA *41°40′N 96°40′W*
Scrobesbyrig' *see* Shrewsbury
14 I14 **Scugog** ⛐ Ontario, SE Canada
14 I14 **Scugog, Lake** ◎ Ontario, SE Canada
97 N17 **Scunthorpe** E England, United Kingdom *53°35′N 00°39′W*
108 K9 **Scuol** *Ger.* Schuls. Graubünden, E Switzerland *46°51′N 10°21′E*
Scupi *see* Skopje
Scutari *see* Shkodër
113 K17 **Scutari, Lake** *Alb.* Liqeni i Shkodrës, *SCr.* Skadarsko Jezero. ◎ Albania/Montenegro
Scyros *see* Skýros
Scythopolis *see* Beit She'an
138 E11 **Sderot** *prev.* Sederot. Southern, S Israel *31°31′N 34°35′E*
25 U13 **Seadrift** Texas, SW USA *28°25′N 96°42′W*
21 Y4 **Seaford** *var.* Seaford City. Delaware, NE USA *38°39′N 75°35′W*
Seaford City *see* Seaford
14 E15 **Seaforth** Ontario, S Canada *43°33′N 81°25′W*
24 M6 **Seagraves** Texas, SW USA *32°56′N 102°33′W*
11 X9 **Seal** ⛐ Manitoba, C Canada
182 M10 **Sea Lake** Victoria, SE Australia *35°34′S 142°51′E*
83 G26 **Seal, Cape** *headland* South Africa *34°06′S 23°18′E*
65 D26 **Sea Lion Islands** *island group* SE Falkland Islands
19 S8 **Seal Island** *island* Maine, NE USA
25 V11 **Sealy** Texas, SW USA *29°46′N 96°09′W*
35 X12 **Searchlight** Nevada, W USA *35°27′N 114°54′W*
27 V11 **Searcy** Arkansas, C USA *35°14′N 91°43′W*
19 R7 **Searsport** Maine, NE USA *44°28′N 68°54′W*
35 N10 **Seaside** California, W USA *36°36′N 121°51′W*
32 F10 **Seaside** Oregon, NW USA *45°57′N 123°55′W*
18 K16 **Seaside Heights** New Jersey, NE USA *39°56′N 74°03′W*
32 H8 **Seattle** Washington, NW USA *47°35′N 122°20′W*
32 H9 **Seattle-Tacoma** ✈ Washington, NW USA *47°04′N 122°27′W*
185 J16 **Seaward Kaikoura Range** ▲▲ South Island, New Zealand
42 J9 **Sébaco** Matagalpa, W Nicaragua *12°51′N 86°08′W*
19 P8 **Sebago Lake** ◎ Maine, NE USA
169 S13 **Sebangan, Teluk** *bay* Borneo, C Indonesia
Sebaste/Sebastia *see* Sivas
23 Y12 **Sebastian** Florida, SE USA *27°55′N 80°31′W*
40 C5 **Sebastián Vizcaíno, Bahía** *bay* NW Mexico
19 R6 **Sebasticook Lake** ◎ Maine, NE USA
34 M7 **Sebastopol** California, W USA *38°22′N 122°50′W*
Sebastopol *see* Sevastopol'
169 W8 **Sebatik, Pulau** *island* N Indonesia
19 R5 **Sebec Lake** ◎ Maine, NE USA
76 K12 **Sébékoro** Kayes, W Mali *13°00′N 09°03′W*
Sebenico *see* Šibenik
40 G6 **Seberi, Cerro** ▲ NW Mexico *27°49′N 110°18′W*
116 H11 **Sebeş** *Ger.* Mühlbach, *Hung.* Szászsebes; *prev.* Sebeşu Săsesc. Alba, W Romania *45°58′N 23°34′E*
Sebes-Körös *see* Crişul Repede
Sebeşu Săsesc *see* Sebeş
31 R8 **Sebewaing** Michigan, N USA *43°43′N 83°27′W*
124 F16 **Sebezh** Pskovskaya Oblast', W Russian Federation *56°19′N 28°31′E*
137 N12 **Şebinkarahisar** Giresun, N Turkey *40°19′N 38°25′E*
116 F11 **Sebiş** *Hung.* Borossebes. Arad, W Romania *46°21′N 22°09′E*
Sebkra Azz el Matti *see* Azzel Matti, Sebkha
19 Q4 **Seboomook Lake** ◎ Maine, NE USA
74 G6 **Sebou** *var.* Sebu. ⛐ N Morocco
20 I6 **Sebree** Kentucky, S USA *37°34′N 87°30′W*
23 X13 **Sebring** Florida, SE USA *27°30′N 81°26′W*
Sebta *see* Ceuta
Sebu *see* Sebou
169 U13 **Sebuku, Pulau** *island* N Indonesia
169 W8 **Sebuku, Teluk** *bay* Borneo, N Indonesia
106 F10 **Secchia** ⛐ N Italy
10 L17 **Sechelt** British Columbia, SW Canada *49°25′N 123°37′W*
56 C12 **Sechín, Río** ⛐ W Peru
56 A10 **Sechura, Bahía de** *bay* NW Peru
185 A22 **Secretary Island** *island* SW New Zealand
155 I15 **Secunderābād** *var.* Sikandarabad. Telangana, C India *17°30′N 78°33′E*
57 L17 **Sécure, Río** ⛐ C Bolivia
118 D10 **Seda** Telšiai, NW Lithuania *56°10′N 22°04′E*
27 T5 **Sedalia** Missouri, C USA *38°42′N 93°15′W*
103 R3 **Sedan** Ardennes, N France *49°42′N 04°56′E*
27 P7 **Sedan** Kansas, C USA *37°07′N 96°11′W*
105 N3 **Sedano** Castilla y León, N Spain *42°43′N 03°43′W*
104 H10 **Seda, Ribeira de** *stream* C Portugal
185 K15 **Seddon** Marlborough, South Island, New Zealand *41°42′S 174°05′E*
185 H15 **Seddonville** West Coast, South Island, New Zealand *41°35′S 171°59′E*
143 U7 **Sedeh** Khorāsān-e Jonūbī, E Iran *33°18′N 59°12′E*
Sederot *see* Sderot
65 B23 **Sedge Island** *island* NW Falkland Islands
76 G12 **Sédhiou** SW Senegal *12°39′N 15°33′W*
11 U16 **Sedley** Saskatchewan, S Canada *50°06′N 103°51′W*
Sedlez *see* Siedlce
117 Q2 **Sedniv** Chernihivs'ka Oblast', N Ukraine *51°39′N 31°34′E*
36 L11 **Sedona** Arizona, SW USA *34°52′N 111°45′W*
Sedunum *see* Sion
118 F12 **Šeduva** Šiaulai, N Lithuania *55°45′N 23°46′E*
141 Y8 **Seeb** *var.* Muscat Sīb Airport. ✈ (Masqaţ) NE Oman *23°36′N 58°27′E*
Seeb *see* As Sīb
108 M7 **Seefeld-in-Tirol** Tirol, W Austria *47°19′N 11°16′E*
83 E22 **Seeheim Noord** Karas, S Namibia *26°50′S 17°45′E*
Seeland *see* Sjælland
195 N9 **Seelig, Mount** ▲ Antarctica *81°45′S 102°15′W*
Seenu Atoll *see* Addu Atoll
Seeonee *see* Seoni
Seer *see* Dörgön
102 L5 **Sées** Orne, N France *48°36′N 00°11′E*
101 J14 **Seesen** Niedersachsen, C Germany *51°54′N 10°11′E*
Seesker Höhe *see* Szeska Góra
100 I10 **Seevetal** Niedersachsen, N Germany *53°24′N 10°01′E*
109 V6 **Seewiesen** Steiermark, E Austria *47°37′N 15°16′E*
136 J13 **Şefaatli** *var.* Kızılkoca. Yozgat, C Turkey *39°32′N 34°45′E*
143 V9 **Sefīdābeh** Khorāsān-e Janūbī, E Iran *31°05′N 60°30′E*
149 N3 **Sefid, Darya-ye** *Pash.* Āb-i-safed. ⛐ N Afghanistan
148 K5 **Sefīd Kūh, Selseleh-ye** *Eng.* Paropamisus Range. ▲▲ W Afghanistan
148 K5 **Sefīd Kūh, Selseleh-ye** *Eng.* Paropamisus Range. ▲▲ W Afghanistan
142 M4 **Sefīd, Rūd-e** ⛐ NW Iran
74 G6 **Sefrou** N Morocco *33°51′N 04°49′W*
185 E19 **Sefton, Mount** ▲ South Island, New Zealand *43°43′S 169°58′E*
171 S13 **Segaf, Kepulauan** *island group* E Indonesia

◆ Country ● Country Capital ◇ Dependent Territory ○ Dependent Territory Capital ◆ Administrative Regions ✈ International Airport ▲ Mountain ▲▲ Mountain Range ⛰ Volcano ⛐ River ◎ Lake ◎ Reservoir

SEGAMA, SUNGAI – SHADEHILL RESERVOIR

169 W7 **Segama, Sungai** 🌊 East Malaysia
168 L9 **Segamat** Johor, Peninsular Malaysia *02°30′N 102°48′E*
77 S13 **Ségbana** NE Benin *10°56′N 03°42′E*
Segestica *see* Sisak
Segesvár *see* Sighişoara
171 T12 **Seget** Papua Barat, E Indonesia *01°21′S 131°04′E*
Segewold *see* Sigulda
124 J9 **Segezha** Respublika Kareliya, NW Russian Federation *63°39′N 34°24′E*
Seghedin *see* Szeged
Segna *see* Senj
107 I16 **Segni** Lazio, C Italy *41°41′N 13°02′E*
Segodunum *see* Rodez
105 S9 **Segorbe** Valenciana, E Spain *39°51′N 00°30′W*
76 M12 **Ségou** *var.* Segu. Ségou, C Mali *13°26′N 06°12′W*
76 M12 **Ségou** ◆ *region* SW Mali
54 E8 **Segovia** Antioquia, N Colombia *07°08′N 74°39′W*
105 N7 **Segovia** Castilla y León, C Spain *40°57′N 04°07′W*
104 M6 **Segovia** ◆ *province* Castilla y León, N Spain
Segoviao Wangkí *see* Coco, Río
124 J9 **Segozerskoye Vodokhranilishche** *prev.* Ozero Segozero. ◎ NW Russian Federation
102 J7 **Segré** Maine-et-Loire, NW France *47°41′N 00°51′W*
105 U5 **Segre** 🌊 NE Spain
Segu *see* Ségou
38 I17 **Seguam Island** *island* Aleutian Islands, Alaska, USA
38 I17 **Seguam Pass** *strait* Aleutian Islands, Alaska, USA
77 Y7 **Séguédine** Agadez, NE Niger *20°12′N 13°03′E*
76 M15 **Séguéla** W Ivory Coast *07°58′N 06°44′W*
25 S11 **Seguin** Texas, SW USA *29°34′N 97°58′W*
38 E17 **Segula Island** *island* Aleutian Islands, Alaska, USA
62 K10 **Segundo, Río** 🌊 C Argentina
105 Q12 **Segura** 🌊 S Spain
105 P13 **Segura , Sierra de** ▲ S Spain
83 G18 **Sehithwa** North-West, N Botswana *20°28′S 22°43′E*
154 H10 **Sehore** Madhya Pradesh, C India *23°12′N 77°08′E*
186 G9 **Sehulea** Normanby Island, S Papua New Guinea *09°55′S 151°10′E*
149 P15 **Sehwān** Sind, SE Pakistan *26°26′N 67°52′N*
109 V8 **Seiersberg** Steiermark, SE Austria *47°01′N 15°22′E*
26 L9 **Seiling** Oklahoma, C USA *36°09′N 98°55′W*
103 S9 **Seille** 🌊 E France
99 J20 **Seilles** Namur, SE Belgium *50°31′N 05°12′E*
93 K17 **Seinäjoki** *Swe.* Östermyra. Etelä-Pohjanmaa, W Finland *62°45′N 22°55′E*
12 B12 **Seine** 🌊 Ontario, S Canada
102 M4 **Seine** 🌊 N France
102 K4 **Seine, Baie de la** *bay* N France
Seine, Banc de la *see* Seine Seamount
103 O5 **Seine-et-Marne** ◆ *department* N France
102 L3 **Seine-Maritime** ◆ *department* N France
84 B14 **Seine Plain** *undersea feature* E Atlantic Ocean *34°00′N 12°15′W*
84 B15 **Seine Seamount** *var.* Banc de la Seine. *undersea feature* E Atlantic Ocean *33°45′N 14°25′W*
102 E6 **Sein, Île de** *island* NW France
171 Y14 **Seinma** Papua, E Indonesia *04°10′S 138°54′E*
Seisbierrum *see* Sexbierum
109 U5 **Seitenstetten Markt** Niederösterreich, C Austria *48°03′N 14°41′E*
Seiyo *see* Uwa
Seiyu *see* Chŏnju
95 H22 **Sejerø** *island* E Denmark
110 P7 **Sejny** Podlaskie, NE Poland *54°09′N 23°21′E*
163 X15 **Sejong City** ● (South Korea) E South Korea *36°29′N 127°16′E*
81 G20 **Seke** Simiyu, N Tanzania *03°16′S 33°31′E*
164 L13 **Seki** Gifu, Honshū, SW Japan *35°30′N 136°54′E*
161 U12 **Sekibi-sho** *Chin.* Chiwei Yu. *island* (disputed) China/Japan/Taiwan
165 U3 **Sekihoku-tōge** *pass* Hokkaidō, NE Japan
Sekondi *see* Sekondi-Takoradi
77 P17 **Sekondi-Takoradi** *var.* Sekondi. S Ghana *04°55′N 01°45′W*
80 J11 **Sek'ot'a** Āmara, N Ethiopia *12°41′N 39°05′E*
Sekseüil *see* Saksaul'skoye
32 I9 **Selah** Washington, NW USA *46°39′N 120°31′W*
168 J8 **Selangor** *var.* Negeri Selangor Darul Ehsan. ◆ *state* Peninsular Malaysia
Selānik *see* Thessaloníki
167 R10 **Selaphum** Roi Et, E Thailand *16°00′N 103°54′E*
171 T16 **Selaru, Pulau** *island* Kepulauan Tanimbar, E Indonesia
171 U13 **Selassi** Papua Barat, E Indonesia *03°16′S 132°50′E*
168 J7 **Selatan, Selat** *strait* Peninsular Malaysia
168 K10 **Selatpanjang** Pulau Rantau, W Indonesia *01°00′N 102°44′E*
39 N8 **Selawik** Alaska, USA *66°36′N 160°00′W*
39 N8 **Selawik Lake** ◎ Alaska, USA
171 N14 **Selayar, Selat** *strait* Sulawesi, C Indonesia
95 C14 **Selbjørnsfjorden** *fjord* S Norway
94 H8 **Selbusjøen** ◎ S Norway
97 M17 **Selby** N England, United Kingdom *53°49′N 01°06′W*
29 N8 **Selby** South Dakota, N USA *45°30′N 100°01′W*
21 Z4 **Selbyville** Delaware, NE USA *38°28′N 75°12′W*
136 B15 **Selçuk** *var.* Akıncılar. İzmir, SW Turkey *37°56′N 27°25′E*
39 Q13 **Seldovia** Alaska, USA *59°26′N 151°42′W*
107 M18 **Sele** *anc.* Silarius. 🌊 S Italy
83 J19 **Selebi-Phikwe** Central, E Botswana *21°58′S 27°48′E*
42 B5 **Selegua, Río** 🌊 W Guatemala
129 X7 **Selemdzha** 🌊 SE Russian Federation
129 U7 **Selenga** *Mong.* Selenge Mörön. 🌊 Mongolia/Russian Federation
79 I19 **Selenge** Bandundu, W Dem. Rep. Congo *01°58′S 18°11′E*
162 K6 **Selenge** *var.* Ingettolgoy. Bulgan, N Mongolia *49°27′N 103°59′E*
162 L6 **Selenge** ◆ *province* N Mongolia
Selenge *see* Hyalganat, Bulgan, Mongolia
Selenge *see* Ih-Uul, Hövsgöl, Mongolia
Selenge Mörön *see* Selenga
123 N14 **Selenginsk** Respublika Buryatiya, S Russian Federation *52°00′N 106°40′E*
Selenica *see* Selenicë
113 K22 **Selenicë** *var.* Selenica. Vlorë, SW Albania *40°32′N 19°38′E*
123 Q8 **Selennyakh** 🌊 NE Russian Federation
100 J8 **Selenter See** ◎ N Germany
Sele Sound *see* Soela Väin
103 U6 **Sélestat** *Ger.* Schlettstadt. Bas-Rhin, NE France *48°16′N 07°28′E*
Seleucia *see* Silifke
92 I4 **Selfoss** Suðurland, SW Iceland *63°56′N 20°59′W*
28 M7 **Selfridge** North Dakota, N USA *46°01′N 100°52′W*
76 I15 **Seli** 🌊 N Sierra Leone
76 I11 **Sélibabi** *var.* Sélibaby. Guidimaka, S Mauritania *15°14′N 12°11′W*
Sélibaby *see* Sélibabi
Selidovka/Selidovo *see* Selydove
124 I15 **Seliger, Ozero** ◎ W Russian Federation
36 J11 **Seligman** Arizona, SW USA *35°20′N 112°56′W*
27 S8 **Seligman** Missouri, C USA *36°31′N 93°56′W*
80 E6 **Selima Oasis** *oasis* N Sudan
76 L13 **Sélingué, Lac de** ⊠ S Mali
Selinoús *see* Kréstena
18 G14 **Selinsgrove** Pennsylvania, NE USA *40°47′N 76°51′W*
Selishche *see* Syelishcha
124 I16 **Selizharovo** Tverskaya Oblast', W Russian Federation *56°50′N 33°24′E*
94 C10 **Selje** Sogn Og Fjordane, S Norway *62°02′N 05°22′E*
11 X16 **Selkirk** Manitoba, S Canada *50°10′N 96°52′W*
96 K13 **Selkirk** SE Scotland, United Kingdom *55°36′N 02°48′W*
96 K13 **Selkirk** *cultural region* SE Scotland, United Kingdom
11 O16 **Selkirk Mountains** ▲ British Columbia, SW Canada
193 T11 **Selkirk Rise** *undersea feature* SE Pacific Ocean
115 F21 **Sellasía** Pelopónnisos, S Greece *37°14′N 22°24′E*
44 M9 **Selle, Pic de la** *var.* La Selle. ▲ SE Haiti *18°18′N 71°55′W*
102 M8 **Selles-sur-Cher** Loir-et-Cher, C France *47°16′N 01°31′E*
36 K16 **Sells** Arizona, SW USA *31°54′N 111°52′W*
Sellye *see* Sal'a
23 P5 **Selma** Alabama, S USA *32°24′N 87°01′W*
35 Q11 **Selma** California, W USA *36°33′N 119°37′W*
20 G10 **Selmer** Tennessee, S USA *35°10′N 88°34′W*
173 N17 **Sel, Pointe au** *headland* W Réunion
Selseleh-ye Kuhe Vākhān *see* Nicholas Range
127 S2 **Selty** Udmurtskaya Respublika, NW Russian Federation *57°19′N 52°09′E*
Selukwe *see* Shurugwi
62 L9 **Selva** Santiago del Estero, N Argentina *29°46′S 62°02′W*
11 T9 **Selwyn Lake** ◎ Northwest Territories/Saskatchewan, C Canada
10 K6 **Selwyn Mountains** ▲ Yukon, NW Canada
181 T6 **Selwyn Range** ▲ Queensland, C Australia
117 W8 **Selydove** *var.* Selidovka, *Rus.* Selidovo. Donets'ka Oblast', SE Ukraine *48°06′N 37°16′E*
Selzaete *see* Zelzate
Seman *see* Semanit, Lumi i
168 M15 **Semangka, Teluk** *bay* Sumatera, SW Indonesia
113 D22 **Semanit, Lumi i** *var.* Seman. 🌊 W Albania
169 Q16 **Semarang** *var.* Samarang. Jawa, C Indonesia *06°58′S 110°29′E*
169 Q10 **Sematan** Sarawak, East Malaysia *01°50′N 109°44′E*
171 P17 **Semau, Pulau** *island* S Indonesia
169 V8 **Sembakung, Sungai** 🌊 Borneo, N Indonesia
79 G17 **Sembé** Sangha, NW Congo *01°38′N 14°35′E*
169 S13 **Sembulu, Danau** ◎ Borneo, N Indonesia
139 Q2 **Sēmēl** *Ar.* Sumayl, *var.* Summēl. Dahūk, N Iraq *36°52′N 42°51′E*
Semendria *see* Smederevo
117 R1 **Semenivka** Chernihivs'ka Oblast', N Ukraine *52°10′N 32°37′E*
117 S6 **Semenivka** *Rus.* Semenovka. Poltavs'ka Oblast', NE Ukraine *49°36′N 33°10′E*
127 O3 **Semenov** Nizhegorodskaya Oblast', W Russian Federation *56°47′N 44°27′E*
Semenovka *see* Semenivka
169 S17 **Semeru, Gunung** *var.* Mahameru. ▲ Jawa, S Indonesia *08°01′S 112°53′E*
145 V9 **Semey** *prev.* Semipalatinsk. Vostochnyy Kazakhstan, E Kazakhstan *50°26′N 80°16′E*
126 L7 **Semiluki** Voronezhskaya Oblast', W Russian Federation *51°46′N 39°00′E*
33 W16 **Seminoe Reservoir** ⊠ Wyoming, C USA
27 O11 **Seminole** Oklahoma, C USA *35°13′N 96°40′W*
24 M6 **Seminole** Texas, SW USA *32°43′N 102°39′W*
23 S8 **Seminole, Lake** ⊠ Florida/Georgia, SE USA
Semiozernoye *see* Auliyekol'
Semipalatinsk *see* Semey
143 O9 **Semirom** *var.* Samirum. Eşfahān, C Iran *31°20′N 51°50′E*
38 F17 **Semisopochnoi Island** *island* Aleutian Islands, Alaska, USA
169 R11 **Semitau** Borneo, C Indonesia *0°30′N 111°59′E*
81 E18 **Semliki** 🌊 Uganda/Dem. Rep. Congo
143 P5 **Semnān** *var.* Samnān. Semnān, N Iran *35°37′N 53°21′E*
143 Q5 **Semnān** *off.* Ostān-e Semnān. ◆ *province* N Iran
Semnān, Ostān-e *see* Semnān
99 K24 **Semois** 🌊 SE Belgium
108 E8 **Sempacher See** ◎ C Switzerland
Sena *see* Vila de Sena
30 L12 **Senachwine Lake** ◎ Illinois, N USA
59 O14 **Senador Pompeu** Ceará, E Brazil *05°30′S 39°25′W*
Sena Gallica *see* Senigallia
59 C15 **Sena Madureira** Acre, W Brazil *09°05′S 68°41′W*
155 L25 **Senanayake Samudra** ◎ E Sri Lanka
83 G15 **Senanga** Western, SW Zambia *16°09′S 23°16′E*
27 Y9 **Senath** Missouri, C USA *36°07′N 90°09′W*
22 L2 **Senatobia** Mississippi, S USA *34°37′N 89°58′W*
164 C16 **Sendai** *var.* Satsuma-Sendai. Kagoshima, Kyūshū, SW Japan *31°49′N 130°17′E*
165 Q11 **Sendai-wan** *bay* E Japan
101 J23 **Senden** Bayern, S Germany *48°18′N 10°04′E*
154 F11 **Sendhwa** Madhya Pradesh, C India *21°38′N 75°04′E*
111 H21 **Senec** *Ger.* Wartberg, *Hung.* Szenc; *prev.* Szempcz. Bratislavský Kraj, W Slovakia *48°14′N 17°24′E*
27 P3 **Seneca** Kansas, C USA *39°50′N 96°04′W*
27 R8 **Seneca** Missouri, C USA *36°50′N 94°36′W*
32 K13 **Seneca** Oregon, NW USA *44°06′N 118°57′W*
21 O11 **Seneca** South Carolina, SE USA *34°41′N 82°57′W*
18 G11 **Seneca Lake** ◎ New York, NE USA
31 U13 **Senecaville Lake** ⊠ Ohio, N USA
76 G11 **Senegal** *off.* Republic of Senegal, *Fr.* Sénégal. ◆ *republic* W Africa
76 H9 **Senegal** *Fr.* Sénégal. 🌊 W Africa
Senegal, Republic of *see* Senegal
31 O4 **Seney Marsh** *wetland* Michigan, N USA
101 P14 **Senftenberg** Brandenburg, E Germany *51°31′N 14°01′E*
82 L11 **Senga Hill** Northern, NE Zambia *09°26′S 31°12′E*
158 G13 **Sênggê Zangbo** 🌊 W China
171 Z13 **Senggi** Papua, E Indonesia *03°26′S 140°46′E*
127 R5 **Sengiley** Ul'yanovskaya Oblast', W Russian Federation *53°54′N 48°51′E*
63 I19 **Senguerr, Río** 🌊 S Argentina
83 J16 **Sengwa** 🌊 C Zimbabwe
Senia *see* Senj
111 H19 **Senica** *Ger.* Senitz, *Hung.* Szenice. Trnavský Kraj, W Slovakia *48°40′N 17°22′E*
Senica *see* Sjenica
106 J11 **Senigallia** *anc.* Sena Gallica. Marche, C Italy *43°43′N 13°13′E*
136 F15 **Senirkent** Isparta, SW Turkey *38°07′N 30°34′E*
Senitz *see* Senica
112 C10 **Senj** *Ger.* Zengg, *It.* Segna; *anc.* Senia. Lika-Senj, NW Croatia *44°58′N 14°55′E*
Senj *see* Örgön
92 H9 **Senja** *prev.* Senjen. *island* N Norway
Senjen *see* Senja
161 U12 **Senkaku-shotō** *Chin.* Diaoyutai. *island group* (disputed) SW Japan
137 R12 **Şenkaya** Erzurum, NE Turkey *40°33′N 42°17′E*
83 I16 **Senkobo** Southern, S Zambia *17°38′S 25°58′E*
103 O4 **Senlis** Oise, N France *49°13′N 02°33′E*
Senmonorom *see* Sênmonoŭrôm
167 T12 **Sênmonoŭrôm** *var.* Senmonorom. Môndól Kiri, E Cambodia *12°27′N 107°12′E*
80 G10 **Sennar** *var.* Sannār. Sinnar, C Sudan *13°31′N 33°38′E*
Senno *see* Syanno
Senones *see* Sens
109 W11 **Senovo** E Slovenia *46°01′N 15°24′E*
103 P6 **Sens** *anc.* Agendicum, Senones. Yonne, C France *48°12′N 03°17′E*
Sensburg *see* Mrągowo
167 S11 **Sên, Stœng** 🌊 C Cambodia
42 F7 **Sensuntepeque** Cabañas, NE El Salvador *13°52′N 88°38′W*
112 L8 **Senta** *Hung.* Zenta. Vojvodina, N Serbia *45°57′N 20°04′E*
Šent Andraž *see* Sankt Andrä
171 Y13 **Sentani, Danau** ◎ Papua, E Indonesia
28 J5 **Sentinel Butte** ▲ North Dakota, N USA *46°52′N 103°50′W*
10 M13 **Sentinel Peak** ▲ British Columbia, W Canada *54°51′N 122°02′W*
59 N16 **Sento Sé** Bahia, E Brazil *09°51′S 41°56′W*
Šent Peter *see* Pivka
Št. Vid *see* Sankt Veit an der Glan
Seo de Urgel *see* La Seu d'Urgell
163 X17 **Seogwipo** *prev.* Sŏgwip'o. S South Korea *33°14′N 126°33′E*
154 I7 **Seondha** Madhya Pradesh, C India *26°09′N 78°47′E*
163 Y17 **Seongsan** *prev.* Sŏngsan. S South Korea
154 J11 **Seoni** *prev.* Seonee. Madhya Pradesh, C India *22°06′N 79°36′E*
163 X14 **Seoul** , *Jap.* Keijō; *prev.* Kyŏngsŏng, Sŏul. ● (South Korea) NW South Korea *37°30′N 126°58′E*
83 I17 **Sepako** Central, NE Botswana *19°50′S 26°29′E*
184 I13 **Separation Point** *headland* South Island, New Zealand *40°46′S 172°58′E*
169 V10 **Sepasu** Borneo, N Indonesia *0°44′N 117°38′E*
186 B6 **Sepik** 🌊 Indonesia/Papua New Guinea
Sepone *see* Muang Xépôn
110 M7 **Sępopol** *Ger.* Schippenbeil. Warmińsko-Mazurskie, N Poland *54°16′N 21°09′E*
116 F10 **Şepreuş** *Hung.* Seprős. Arad, W Romania *46°34′N 21°44′E*
Seprős *see* Şepreuş
Şepşi-Sângeorz/Sepsiszentgyörgy *see* Sfântu Gheorghe
15 W4 **Sept-Îles** Québec, SE Canada *50°11′N 66°19′W*
105 N6 **Sepúlveda** Castilla y León, N Spain *41°18′N 03°45′W*
104 K8 **Sequeros** Castilla y León, N Spain *40°31′N 06°04′W*
104 L5 **Sequillo** 🌊 NW Spain
32 G7 **Sequim** Washington, NW USA *48°04′N 123°06′W*
35 S11 **Sequoia National Park** *national park* California, W USA
137 Q14 **Şerafettin Dağları** ▲ E Turkey
127 N10 **Serafimovich** Volgogradskaya Oblast', SW Russian Federation *49°34′N 42°43′E*
171 Q10 **Serai** Sulawesi, N Indonesia *01°45′N 124°58′E*
99 K19 **Seraing** Liège, E Belgium *50°37′N 05°31′E*
Sêraitang *see* Baima
Serakhs *see* Sarahs
171 W13 **Serami** Papua, E Indonesia *02°11′S 136°46′E*
171 R13 **Seram, Laut** *Eng.* Ceram Sea. *sea* E Indonesia
Serampore/Serampur *see* Shrīrāmpur
171 S13 **Seram, Pulau** *var.* Serang, *Eng.* Ceram. *island* Maluku, E Indonesia
169 N15 **Serang** Jawa, C Indonesia *06°07′S 106°09′E*
Serang *see* Seram, Pulau
169 P9 **Serasan, Pulau** *island* Kepulauan Natuna, W Indonesia
169 P9 **Serasan, Selat** *strait* Indonesia/Malaysia
112 M13 **Serbia** *off.* Federal Republic of Serbia; *prev.* Yugoslavia, *SCr.* Jugoslavija. ◆ *federal republic* SE Europe
112 M12 **Serbia** *Ger.* Serbien, *Serb.* Srbija. ◆ *republic* Serbia
Serbia, Federal Republic of *see* Serbia
Serbien *see* Serbia
Sercq *see* Sark
146 D12 **Serdar** *prev. Rus.* Gyzyrlabat, Kizyl-Arvat. Balkan Welaýaty, W Turkmenistan *39°02′N 56°15′E*
Serdica *see* Sofia
Serdobol' *see* Sortavala
127 O7 **Serdobsk** Penzenskaya Oblast', W Russian Federation *52°30′N 44°16′E*
145 X9 **Serebryansk** Vostochnyy Kazakhstan, E Kazakhstan *49°44′N 83°16′E*
123 Q12 **Serebryanyy Bor** Respublika Sakha (Yakutiya), NE Russian Federation *56°40′N 124°46′E*
111 H20 **Sered'** *Hung.* Szered. Trnavský Kraj, W Slovakia *48°19′N 17°45′E*
117 S1 **Seredyna-Buda** Sums'ka Oblast', NE Ukraine *52°09′N 34°00′E*
118 E13 **Seredžius** Tauragė, C Lithuania *55°04′N 23°24′E*
136 I14 **Şereflikoçhisar** Ankara, C Turkey *38°56′N 33°31′E*
106 D7 **Seregno** Lombardia, N Italy *45°39′N 09°12′E*
103 P7 **Serein** 🌊 C France
168 K9 **Seremban** Negeri Sembilan, Peninsular Malaysia *02°42′N 101°54′E*
81 H20 **Serengeti Plain** *plain* N Tanzania
82 K13 **Serenje** Central, E Zambia *13°12′S 30°15′E*
Seres *see* Sérres
116 J5 **Seret** 🌊 W Ukraine
Seret/Sereth *see* Siret
115 I21 **Serfopoúla** *island* Kykládes, Greece, Aegean Sea
127 P4 **Sergach** Nizhegorodskaya Oblast', W Russian Federation *55°31′N 45°29′E*
29 S13 **Sergeant Bluff** Iowa, C USA *42°24′N 96°19′W*
163 P7 **Sergelen** Dornod, NE Mongolia *48°31′N 114°01′E*
Sergelen *see* Tuvshinshiree
168 H8 **Sergeulangit, Pegunungan** ▲ Sumatera, NW Indonesia
122 L5 **Sergeya Kirova, Ostrova** *island* N Russian Federation
Sergeyevichi *see* Syarhyeyevichy
145 O7 **Sergeyevka** Severnyy Kazakhstan, N Kazakhstan *53°53′N 67°25′E*
Sergiopol *see* Ayagoz
59 P16 **Sergipe** *off.* Estado de Sergipe. ◆ *state* E Brazil
Sergipe, Estado de *see* Sergipe
126 L3 **Sergiyev Posad** Moskovskaya Oblast', W Russian Federation *56°21′N 38°10′E*
124 K5 **Sergozero, Ozero** ◎ NW Russian Federation
146 J17 **Serhetabat** *prev. Rus.* Gushgy, Kushka. Mary Welaýaty, S Turkmenistan *35°19′N 62°17′E*
169 Q10 **Serian** Sarawak, East Malaysia *01°10′N 110°35′E*
115 I21 **Sérifos** *anc.* Seriphos. *island* Kykládes, Greece, Aegean Sea
115 I21 **Sérifou, Stenó** *strait* SE Greece
136 F16 **Serik** Antalya, SW Turkey *36°55′N 31°06′E*
106 E7 **Serio** 🌊 N Italy
Seriphos *see* Sérifos
Serir Tibesti *see* Sarīr Tibistī
Sêrkog *see* Sêrtar
197 O14 **Sermersooq** *off.* Kommuneqarfik Sermersooq. ◆ *municipality* S Greenland
Sermersooq, Kommuneqarfik *see* Sermersooq
127 S5 **Sernovodsk** Samarskaya Oblast', W Russian Federation *53°56′N 51°16′E*
127 R2 **Sernur** Respublika Mariy El, W Russian Federation *56°55′N 49°09′E*
110 M11 **Serock** Mazowieckie, C Poland *52°30′N 21°03′E*
61 B18 **Serodino** Santa Fe, C Argentina *32°33′S 60°52′W*
Seroei *see* Serui
105 P14 **Serón** Andalucía, S Spain *37°20′N 02°28′W*
99 E14 **Serooskerke** Zeeland, SW Netherlands *51°42′N 03°52′E*
105 T6 **Seròs** Cataluña, NE Spain *41°27′N 00°24′E*
122 G10 **Serov** Sverdlovskaya Oblast', C Russian Federation *59°42′N 60°32′E*
83 I19 **Serowe** Central, SE Botswana *22°26′S 26°44′E*
104 H13 **Serpa** Beja, S Portugal *37°56′N 07°36′W*
Serpa Pinto *see* Menongue
182 A4 **Serpentine Lakes** *salt lake* South Australia
45 T15 **Serpent's Mouth, The** *Sp.* Boca de la Serpiente. *strait* Trinidad and Tobago/Venezuela
Serpiente, Boca de la *see* Serpent's Mouth, The
126 K4 **Serpukhov** Moskovskaya Oblast', W Russian Federation *54°54′N 37°26′E*
104 I10 **Serra de São Mamede** ▲ C Portugal *39°18′N 07°19′W*
60 K13 **Serra do Mar** ▲ S Brazil
Sérrai *see* Sérres
107 N22 **Serra San Bruno** Calabria, SW Italy *38°32′N 16°18′E*
103 S14 **Serres** Hautes-Alpes, SE France *44°26′N 05°42′E*
114 H13 **Sérres** *var.* Seres; *prev.* Sérrai. Kentrikí Makedonía, NE Greece *41°03′N 23°33′E*
62 J9 **Serrezuela** Córdoba, C Argentina *30°38′S 65°26′W*
59 O16 **Serrinha** Bahia, E Brazil *11°38′S 38°56′W*
59 M19 **Sêrro** *var.* Sêrro. Minas Gerais, NE Brazil *18°38′S 43°22′W*
Sêrro *see* Serro
Sert *see* Siirt
104 H9 **Sertã** *var.* Sertá. Castelo Branco, C Portugal *39°48′N 08°05′W*
Sertá *see* Sertã
60 L8 **Sertãozinho** São Paulo, S Brazil *21°04′S 47°55′W*
160 F7 **Sêrtar** *var.* Sêrkog. Sichuan, C China *32°18′N 100°18′E*
124 G12 **Sertolovo** Leningradskaya Oblast', NW Russian Federation *60°08′N 30°06′E*
171 W13 **Serui** *prev.* Seroei. Papua, E Indonesia *01°53′S 136°15′E*
83 J19 **Serule** Central, E Botswana *21°58′S 27°20′E*
169 S12 **Seruyan, Sungai** *var.* Sungai Pembuang. 🌊 Borneo, N Indonesia
115 E14 **Sérvia** Dytikí Makedonía, N Greece *40°12′N 22°01′E*
160 E7 **Sêrxü** *var.* Jugar. Sichuan, C China *32°54′N 98°06′E*
123 R13 **Seryshevo** Amurskaya Oblast', SE Russian Federation *51°03′N 128°16′E*
Sesana *see* Sežana
169 V8 **Sesayap, Sungai** 🌊 Borneo, N Indonesia
Sesdlets *see* Siedlce
79 N17 **Sese** Orientale, N Dem. Rep. Congo *02°13′N 25°52′E*
81 F18 **Sese Islands** *island group* S Uganda
83 H16 **Sesheke** *var.* Sesheko. Western, SE Zambia *17°28′S 24°20′E*
Sesheko *see* Sesheke
106 C8 **Sesia** *anc.* Sessites. 🌊 NW Italy
104 F11 **Sesimbra** Setúbal, S Portugal *38°26′N 09°06′W*
115 N22 **Seskló** *island* Dodekánisa, Greece, Aegean Sea
30 L16 **Sesser** Illinois, N USA *38°05′N 89°03′W*
Sessites *see* Sesia
106 G11 **Sesto Fiorentino** Toscana, C Italy *43°50′N 11°12′E*
106 E7 **Sesto San Giovanni** Lombardia, N Italy *45°32′N 09°14′E*
106 A8 **Sestriere** Piemonte, NE Italy *45°00′N 06°54′E*
106 D10 **Sestri Levante** Liguria, NW Italy *44°16′N 09°22′E*
107 C20 **Sestu** Sardegna, Italy, C Mediterranean Sea *39°15′N 09°06′E*
112 E8 **Sesvete** Zagreb, N Croatia *45°50′N 16°03′E*
118 G12 **Šėta** Kaunas, C Lithuania *55°17′N 24°16′E*
Setabis *see* Xàtiva
165 Q4 **Setana** Hokkaidō, NE Japan *42°27′N 139°52′E*
103 Q16 **Sète** *prev.* Cette. Hérault, S France *43°24′N 03°42′E*
58 J11 **Sete Ilhas** Amapá, NE Brazil *01°06′N 52°06′W*
59 L20 **Sete Lagoas** Minas Gerais, NE Brazil *19°29′S 44°15′W*
60 G10 **Sete Quedas, Ilha das** *island* S Brazil
92 I10 **Setermoen** Troms, N Norway *68°51′N 18°20′E*
95 E17 **Setesdal** *valley* S Norway
43 W16 **Setetule, Cerro** ▲ SE Panama *07°51′N 77°37′W*
21 Q5 **Seth** West Virginia, NE USA *38°06′N 81°40′W*
Setia *see* Sezze
74 K5 **Sétif** *var.* Stif. N Algeria *36°11′N 05°24′E*
164 L5 **Seto** Aichi, Honshū, SW Japan *35°14′N 137°06′E*
164 G13 **Seto-naikai** *Eng.* Inland Sea. *sea* S Japan
165 V16 **Setouchi** *var.* Setoushi. Kagoshima, Amami-Ō-shima, SW Japan *44°19′N 142°58′E*
Setoushi *see* Setouchi
74 F6 **Settat** W Morocco *33°03′N 07°37′W*
79 D20 **Setté Cama** Ogooué-Maritime, SW Gabon *02°32′S 09°46′E*
11 W13 **Setting Lake** ◎ Manitoba, C Canada
97 L16 **Settle** N England, United Kingdom *54°04′N 02°17′W*
189 Y12 **Settlement** E Wake Island *19°17′N 166°38′E*
104 F11 **Setúbal** *Eng.* Saint Ubes, Saint Yves. Setúbal, W Portugal *38°31′N 08°54′W*
104 F11 **Setúbal** ◆ *district* S Portugal
104 F12 **Setúbal, Baía de** *bay* W Portugal
Setul *see* Satun
12 B10 **Seul, Lac** ◎ Ontario, S Canada
103 R8 **Seurre** Côte d'Or, C France *47°00′N 05°09′E*
137 U11 **Sevan** C Armenia *40°32′N 44°56′E*
137 V12 **Sevana Lich** *Eng.* Lake Sevan, *Rus.* Ozero Sevan. ◎ E Armenia
Sevan, Lake/Sevan, Ozero *see* Sevana Lich
77 N11 **Sévaré** Mopti, C Mali *14°32′N 04°06′W*
117 S14 **Sevastopol'** *Eng.* Sebastopol. Avtonomna Respublika Krym, S Ukraine *44°36′N 33°33′E*
25 R14 **Seven Sisters** Texas, SW USA *27°57′N 98°34′W*
10 K13 **Seven Sisters Peaks** ▲ British Columbia, SW Canada *54°57′N 128°10′W*
99 M15 **Sevenum** Limburg, SE Netherlands *51°25′N 06°01′E*
103 P14 **Séverac-le-Château** Aveyron, S France *44°18′N 03°03′E*
14 H13 **Severn** 🌊 Ontario, S Canada
97 L21 **Severn** *Wel.* Hafren. 🌊 England/Wales, United Kingdom
125 O11 **Severnaya Dvina** *var.* Northern Dvina. 🌊 NW Russian Federation
127 N16 **Severnaya Osetiya-Alaniya, Respublika** *Eng.* North Ossetia; *prev.* Respublika Severnaya Osetiya, Severo-Osetinskaya SSR. ◆ *autonomous republic* SW Russian Federation
Severnaya Osetiya, Respublika *see* Severnaya Osetiya-Alaniya, Respublika
122 M5 **Severnaya Zemlya** *var.* Nicholas II Land. *island group* N Russian Federation
127 T5 **Severnoye** Orenburgskaya Oblast', W Russian Federation *54°03′N 52°31′E*
35 S3 **Severn Troughs Range** ▲ Nevada, W USA
125 W3 **Severnyy** Respublika Komi, NW Russian Federation *67°38′N 64°13′E*
144 I13 **Severnyy Chink Ustyurta** ▲ W Kazakhstan
125 Q13 **Severnyye Uvaly** *var.* Northern Ural Hills. *hill range* NW Russian Federation
145 O6 **Severnyy Kazakhstan** *off.* Severo-Kazakhstanskaya Oblast', *var.* North Kazakhstan, *Kaz.* Soltüstik Qazaqstan Oblysy. ◆ *province* N Kazakhstan
122 I6 **Severnyy, Ostrov** NW Russian Federation
125 V9 **Severnyy Ural** ▲ NW Russian Federation
Severo-Alichurskiy Khrebet *see* Alichuri Shimolí, Qatorkŭhi
123 N12 **Severobaykal'sk** Respublika Buryatiya, S Russian Federation *55°39′N 109°17′E*
Severodonetsk *see* Syeverodonets'k
124 M8 **Severodvinsk** *prev.* Molotov, Sudostroy. Arkhangel'skaya Oblast', NW Russian Federation *64°32′N 39°50′E*
Severo-Kazakhstanskaya Oblast' *see* Severnyy Kazakhstan
123 U11 **Severo-Kuril'sk** Sakhalinskaya Oblast', SE Russian Federation *50°38′N 155°57′E*
124 J3 **Severomorsk** Murmanskaya Oblast', NW Russian Federation *69°00′N 33°16′E*
Severo-Osetinskaya SSR *see* Severnaya Osetiya-Alaniya, Respublika
122 M7 **Severo-Sibirskaya Nizmennost'** *var.* North Siberian Plain, *Eng.* North Siberian Lowland. *lowlands* N Russian Federation
122 G10 **Severoural'sk** Sverdlovskaya Oblast', C Russian Federation *60°09′N 59°58′E*
122 L11 **Severo-Yeniseyskiy** Krasnoyarskiy Kray, C Russian Federation *60°29′N 93°13′E*
122 J12 **Seversk** Tomskaya Oblast', C Russian Federation *53°37′N 84°47′E*
126 M11 **Severskiy Donets** *Ukr.* Sivers'kyy Donets'. 🌊 Russian Federation/Ukraine *see also* Sivers'kyy Donets'
Severskiy Donets *see* Sivers'kyy Donets'
92 M9 **Sevettijärvi** Lappi, N Finland *69°31′N 28°40′E*
36 M5 **Sevier Bridge Reservoir** ⊠ Utah, W USA
36 J4 **Sevier Desert** *plain* Utah, W USA
36 J5 **Sevier Lake** ◎ Utah, W USA
21 N9 **Sevierville** Tennessee, S USA *35°53′N 83°34′W*
104 J14 **Sevilla** *Eng.* Seville; *anc.* Hispalis. Andalucía, SW Spain *37°24′N 05°59′W*
104 J13 **Sevilla** ◆ *province* Andalucía, SW Spain
Sevilla de Niefang *see* Niefang
43 O16 **Sevilla, Isla** *island* SW Panama
Seville *see* Sevilla
114 J9 **Sevlievo** Gabrovo, N Bulgaria *43°01′N 25°06′E*
Sevluš/Sevlyush *see* Vynohradiv
109 V11 **Sevnica** *Ger.* Lichtenwald. E Slovenia *46°00′N 15°20′E*
162 J11 **Sevrey** *var.* Saynshand. Ömnögovĭ, S Mongolia *43°30′N 102°08′E*
126 I7 **Sevsk** Bryanskaya Oblast', W Russian Federation *52°03′N 34°31′E*
76 J15 **Sewa** 🌊 E Sierra Leone
39 R12 **Seward** Alaska, USA *60°06′N 149°26′W*
29 R15 **Seward** Nebraska, C USA *40°52′N 97°06′W*
197 Q3 **Seward Peninsula** *peninsula* Alaska, USA
Seward's Folly *see* Alaska
62 H12 **Sewell** Libertador, C Chile *34°05′S 70°25′W*
98 K5 **Sexbierum** *Fris.* Seisbierrum. Fryslân, N Netherlands *53°13′N 05°28′E*
11 O13 **Sexsmith** Alberta, W Canada *55°18′N 118°45′W*
41 W13 **Seybaplaya** Campeche, SE Mexico *19°39′N 90°36′W*
173 N6 **Seychelles** *off.* Republic of Seychelles. ◆ *republic* W Indian Ocean
67 Z9 **Seychelles** *island group* NE Seychelles
173 N6 **Seychelles Bank** *var.* Le Banc des Seychelles. *undersea feature* W Indian Ocean *04°45′S 55°30′E*
Seychelles, Le Banc des *see* Seychelles Bank
Seychelles, Republic of *see* Seychelles
172 H17 **Seychellois, Morne** ▲ Mahé, NE Seychelles
146 J12 **Seýdi** *Rus.* Seydi; *prev.* Neftezavodsk. Lebap Welaýaty, E Turkmenistan *39°31′N 62°53′E*
136 G16 **Seydişehir** Konya, SW Turkey *37°25′N 31°51′E*
92 L2 **Seyðisfjörður** Austurland, E Iceland *65°15′N 14°00′W*
136 J13 **Seyfe Gölü** ◎ C Turkey
Seyhan *see* Adana
136 K16 **Seyhan Barajı** ⊠ S Turkey
136 K17 **Seyhan Nehri** 🌊 S Turkey
136 F13 **Seyitgazi** Eskişehir, W Turkey *39°27′N 30°42′E*
126 J7 **Seym** 🌊 W Russian Federation
117 S3 **Seym** 🌊 N Ukraine
123 T9 **Seymchan** Magadanskaya Oblast', E Russian Federation *62°54′N 152°27′E*
114 N12 **Seymen** Tekirdağ, NW Turkey *41°06′N 27°56′E*
183 O11 **Seymour** Victoria, SE Australia *37°01′S 145°10′E*
83 I25 **Seymour** Eastern Cape, S South Africa *32°33′S 26°46′E*
29 W16 **Seymour** Iowa, C USA *40°40′N 93°07′W*
27 U7 **Seymour** Missouri, C USA *37°09′N 92°46′W*
25 Q5 **Seymour** Texas, SW USA *33°36′N 99°16′W*
114 M12 **Şeytan Deresi** 🌊 NW Turkey
109 S12 **Sežana** *It.* Sesana. SW Slovenia *45°42′N 13°52′E*
103 P5 **Sézanne** Marne, N France *48°43′N 03°41′E*
107 I16 **Sezze** *anc.* Setia. Lazio, C Italy *41°29′N 13°04′E*
Sfákia *see* Chóra Sfakíon
115 D21 **Sfaktiría** *island* S Greece
116 J11 **Sfântu Gheorghe** *Ger.* Sankt-Georgen, *Hung.* Sepsiszentgyörgy; *prev.* Şepşi-Sângeorz, Sfîntu Gheorghe. Covasna, C Romania *45°52′N 25°49′E*
117 N13 **Sfântu Gheorghe, Braţul** *var.* Gheorghe Braţul. 🌊 E Romania
75 N6 **Sfax** *Ar.* Şafāqis. E Tunisia *34°45′N 10°45′E*
75 N6 **Sfax** ✕ E Tunisia *34°43′N 10°37′E*
Sfîntu Gheorghe *see* Sfântu Gheorghe
98 H13 **'s-Gravendeel** Zuid-Holland, SW Netherlands *51°48′N 04°36′E*
98 F11 **'s-Gravenhage** *var.* Den Haag, *Eng.* The Hague, *Fr.* La Haye. ● (Netherlands-seat of government) Zuid-Holland, W Netherlands *52°07′N 04°17′E*
98 G12 **'s-Gravenzande** Zuid-Holland, W Netherlands *52°00′N 04°10′E*
Shaan/Shaanxi Sheng *see* Shaanxi
159 X11 **Shaanxi** *var.* Shaan, Shaanxi Sheng, Shan-hsi, Shenshi, Shensi. ◆ *province* C China
Shaartuz *see* Shahrtuz
Shaba *see* Katanga
Shabani *see* Zvishavane
81 N17 **Shabeellaha Dhexe** *off.* Gobolka Shabeellaha Dhexe. ◆ *region* E Somalia
Shabeellaha Dhexe, Gobolka *see* Shabeellaha Dhexe
81 L17 **Shabeellaha Hoose** *off.* Gobolka Shabeellaha Hoose. ◆ *region* S Somalia
Shabeellaha Hoose, Gobolka *see* Shabeellaha Hoose
Shabeelle, Webi *see* Shebeli
114 O7 **Shabla** Dobrich, NE Bulgaria *43°33′N 28°31′E*
114 O7 **Shabla, Nos** *headland* NE Bulgaria *43°30′N 28°36′E*
13 N9 **Shabogama Lake** ◎ Newfoundland and Labrador, E Canada
79 N20 **Shabunda** Sud-Kivu, E Dem. Rep. Congo *02°42′S 27°20′E*
141 Q15 **Shabwah** C Yemen *15°09′N 46°46′E*
158 F8 **Shache** *var.* Yarkant. Xinjiang Uygur Zizhiqu, NW China *38°27′N 77°16′E*
Shacheng *see* Huailai
195 R12 **Shackleton Coast** *physical region* Antarctica
195 Z10 **Shackleton Ice Shelf** *ice shelf* Antarctica
Shaddādī *see* Ash Shadādah
28 K7 **Shadehill Reservoir** ⊠ South Dakota, N USA

◆ Country ● Country Capital ◇ Dependent Territory ○ Dependent Territory Capital ◆ Administrative Regions ✕ International Airport ▲ Mountain ▲ Mountain Range ⛰ Volcano 🌊 River ◎ Lake ⊠ Reservoir
321

SHADRINSK – SHĪRĪN TAGĀB

122 G11 **Shadrinsk** Kurganskaya Oblast', C Russian Federation *56°08′N 63°18′E*
31 O12 **Shafer, Lake** ⊠ Indiana, N USA
35 R13 **Shafter** California, W USA *35°27′N 119°15′W*
24 J11 **Shafter** Texas, SW USA *29°49′N 104°18′W*
97 L23 **Shaftesbury** S England, United Kingdom *51°01′N 02°12′W*
185 F22 **Shag** ~ South Island, New Zealand
145 V9 **Shagan** ~ E Kazakhstan
39 O11 **Shageluk** Alaska, USA *62°40′N 159°33′W*
122 K14 **Shagonar** Respublika Tyva, S Russian Federation *51°31′N 93°06′E*
185 F22 **Shag Point** *headland* South Island, New Zealand *45°28′S 170°50′E*
144 J12 **Shagyray, Plato** *plain* SW Kazakhstan
Shāhābād *see* Eslāmābād-e Gharb
168 K9 **Shah Alam** Selangor, Peninsular Malaysia *03°02′N 101°31′E*
117 O12 **Shahany, Ozero** ◎ SW Ukraine
138 H9 **Shahbā'** *anc.* Philippopolis. As Suwaydā', S Syria *32°50′N 36°38′E*
Shahbān *see* Ad Dayr
149 P17 **Shāh Bandar** Sind, SE Pakistan *23°59′N 67°54′E*
149 P13 **Shahdād Kot** Sind, SW Pakistan *27°49′N 67°49′E*
143 T10 **Shahdād, Namakzār-e** *salt pan* E Iran
149 Q15 **Shāhdādpur** Sind, SE Pakistan *25°56′N 68°40′E*
154 K10 **Shahdol** Madhya Pradesh, C India *23°19′N 81°26′E*
161 N7 **Sha He** ~ C China
Shahe *see* Linze
Shahepu *see* Linze
153 N13 **Shāhganj** Uttar Pradesh, N India *26°03′N 82°41′E*
152 C11 **Shāhgarh** Rājasthān, NW India *27°08′N 69°56′E*
Sha Hi *see* Orūmīyeh, Daryācheh-ye
Shāhī *see* Qā'emshahr
Shahjahanabad *see* Delhi
152 L11 **Shāhjahānpur** Uttar Pradesh, N India *27°53′N 79°55′E*
149 U7 **Shāhpur** Punjab, E Pakistan *32°15′N 72°32′E*
Shāhpur *see* Shāhpur Chākar
152 G13 **Shāhpura** Rājasthān, N India *25°38′N 75°01′E*
149 Q15 **Shāhpur Chākar** *var.* Shāhpur. Sind, SE Pakistan *26°11′N 68°44′E*
148 M5 **Shahrak** Gōwr, C Afghanistan *34°09′N 64°16′E*
143 Q11 **Shahr-e Bābak** Kermān, C Iran *30°08′N 55°04′E*
143 N8 **Shahr-e Kord** *var.* Shahr Kord. Chahār Maḩall va Bakhtīārī, C Iran *32°20′N 50°52′E*
143 O9 **Shahrezā** *var.* Qomisheh, Qumisheh, Shahriza; *prev.* Qomsheh. Eşfahān, C Iran *32°01′N 51°51′E*
147 S10 **Shahrikhon** *Rus.* Shakhrikhan. Andijon Viloyati, E Uzbekistan *40°42′N 72°03′E*
147 P11 **Shahriston** *Rus.* Shakhristan. NW Tajikistan *39°45′N 68°47′E*
Shahriza *see* Shahrezā
Shahr-i-Zabul *see* Zābol
Shahr Kord *see* Shahr-e Kord
147 P14 **Shahrtuz** *Rus.* Shaartuz. SW Tajikistan *37°13′N 68°05′E*
143 Q4 **Shāhrūd** *prev.* Emāmrūd, Emāmshahr. Semnān, N Iran *36°30′N 55°E*
Shahsavār/Shahsawar *see* Tonekābon
Shaikh Ābid *see* Shaykh 'Ābid
Shaikh Fāris *see* Shaykh Fāris
Shaikh Najm *see* Shaykh Najm
138 K5 **Shā'īr, Jabal** ▲ C Syria *34°51′N 37°49′E*
154 G10 **Shājāpur** Madhya Pradesh, C India *23°27′N 76°21′E*
80 J8 **Shakal, Ras** *headland* NE Sudan *18°04′N 38°34′E*
83 G17 **Shakawe** North West, NW Botswana *18°25′S 21°53′E*
Shakhdarinskiy Khrebet *see* Shokhdara, Qatorkŭhi
Shakhrikhan *see* Shahrikhon
Shakhrisabz *see* Sharixon
Shakhristan *see* Shahriston
Shakhtërsk *see* Zuhres
145 R10 **Shakhtinsk** Karaganda, C Kazakhstan *49°40′N 72°37′E*
126 L11 **Shakhty** Rostovskaya Oblast', SW Russian Federation *47°45′N 40°14′E*
127 P2 **Shakhun'ya** Nizhegorodskaya Oblast', W Russian Federation *57°42′N 46°36′E*
77 S15 **Shaki** Oyo, W Nigeria *08°37′N 03°25′E*
81 J15 **Shakiso** Oromīya, C Ethiopia *05°33′N 38°48′E*
29 V9 **Shakopee** Minnesota, N USA *44°48′N 93°31′W*
165 R3 **Shakotan-misaki** *headland* Hokkaidō, NE Japan *43°22′N 140°28′E*
39 N9 **Shaktoolik** Alaska, USA *64°19′N 161°05′W*
81 J14 **Shala Hāyk'** ◎ C Ethiopia
124 M10 **Shalakusha** Arkhangel'skaya Oblast', NW Russian Federation *62°16′N 40°16′E*
145 U8 **Shalday** Pavlodar, NE Kazakhstan *51°57′N 78°51′E*
127 P16 **Shali** Chechenskaya Respublika, SW Russian Federation *43°03′N 45°55′E*
141 W12 **Shalīm** *var.* Shelim. S Oman *18°07′N 55°39′E*
Shalīr, Āveh-ye *see* Shilayr, Wādī
Shaliuhe *see* Gangca
144 K12 **Shalkar** *var.* Chelkar. Aktyubinsk, W Kazakhstan *47°50′N 59°29′E*
144 F9 **Shalkar, Ozero** *prev.* Chelkar Ozero. ◎ W Kazakhstan
21 V12 **Shallotte** North Carolina, SE USA *33°58′N 78°21′W*
25 N5 **Shallowater** Texas, SW USA *33°41′N 102°00′W*
124 K11 **Shal'skiy** Respublika Kareliya, NW Russian Federation *61°45′N 36°02′E*
160 F9 **Shaluli Shan** ▲▲ C China
81 F22 **Shama** ~ C Tanzania
11 Z11 **Shamattawa** Manitoba, C Canada *55°52′N 92°05′W*
12 F8 **Shamattawa** ~ Ontario, C Canada
Shām, Bādiyat ash *see* Syrian Desert
Shamiya *see* Ash Shāmīyah
141 X8 **Shām, Jabal ash** *var.* Jebel Sham. ▲ NW Oman *23°21′N 57°08′E*
Sham, Jebel *see* Shām, Jabal ash
Shamkhor *see* Şämkir
18 G14 **Shamokin** Pennsylvania, NE USA *40°47′N 76°33′W*
25 P2 **Shamrock** Texas, SW USA *35°12′N 100°15′W*
Shana *see* Kuril'sk
Sha'nabī, Jabal ash *see* Chambi, Jebel
139 Y12 **Shanāwah** Al Başrah, E Iraq *30°57′N 47°25′E*
Shancheng *see* Taining
159 T8 **Shandan** *var.* Qingyuan. Gansu, N China *38°50′N 101°08′E*
Shandī *see* Shendi
161 Q5 **Shandong** *var.* Lu, Shandong Sheng, Shantung. ◇ *province* E China
161 R4 **Shandong Bandao** *var.* Shantung Peninsula. *peninsula* E China
Shandong Sheng *see* Shandong
139 U8 **Shandrūkh** Diyālá, E Iraq *33°20′N 45°19′E*
83 J17 **Shangani** ~ W Zimbabwe
161 O15 **Shangchuan Dao** *island* S China
Shangchuankou *see* Minhe
163 P12 **Shangdu** Nei Mongol Zizhiqu, N China *41°32′N 113°33′E*
161 O11 **Shanggao** *var.* Aoyang. Jiangxi, S China *28°16′N 114°55′E*
Shangguan *see* Daixian
161 S8 **Shanghai** *var.* Shang-hai. Shanghai Shi, E China *31°14′N 121°28′E*
161 S8 **Shanghai Shi** *var.* Hu, Shanghai. ◆ *municipality* E China
161 P13 **Shanghang** *var.* Linjiang. Fujian, SE China *25°03′N 116°25′E*
160 K14 **Shanglin** *var.* Dafeng. Guangxi Zhuangzu Zizhiqu, S China *23°26′N 108°32′E*
160 L7 **Shangluo** *prev.* Shangxian, Shangzhou. Shaanxi, C China *33°51′N 109°55′E*
83 G15 **Shangombo** Western, W Zambia *16°28′S 22°10′E*
Shangpai/Shangpaihe *see* Feixi
161 O6 **Shangqiu** *var.* Zhuji. Henan, C China *34°24′N 115°37′E*
161 Q10 **Shangrao** Jiangxi, S China *28°27′N 117°57′E*
Shangxian *see* Shangluo
161 S9 **Shangyu** *var.* Baiguan. Zhejiang, SE China *30°03′N 120°52′E*
163 X9 **Shangzhi** Heilongjiang, NE China *45°13′N 127°59′E*
Shangzhou *see* Shangluo
Shanhe *see* Zhengning
163 W9 **Shanhetun** Heilongjiang, NE China *44°42′N 127°12′E*
Shan-hsi *see* Shaanxi, China
Shan-hsi *see* Shanxi, China
159 O6 **Shankou** Xinjiang Uygur Zizhiqu, W China *42°02′N 94°08′E*
184 M13 **Shannon** Manawatu-Wanganui, North Island, New Zealand *40°32′S 175°24′E*
97 C17 **Shannon** *Ir.* An tSionainn. ~ W Ireland
97 B19 **Shannon** ✈ W Ireland *52°42′N 08°57′W*
167 N6 **Shan Plateau** *plateau* E Myanmar (Burma)
158 M6 **Shanshan** *var.* Piqan. Xinjiang Uygur Zizhiqu, NW China *42°53′N 90°18′E*
Shansi *see* Shanxi
167 N5 **Shan State** ◇ *state* E Myanmar (Burma)
Shantar Islands *see* Shantarskiye Ostrova
123 S12 **Shantarskiye Ostrova** *Eng.* Shantar Islands. *island group* E Russian Federation
161 Q14 **Shantou** *var.* Shan-t'ou, Swatow. Guangdong, S China *23°23′N 116°39′E*
Shan-t'ou *see* Shantou
Shantung *see* Shandong
Shantung Peninsula *see* Shandong Bandao
163 O14 **Shanxi** *var.* Jin, Shan-hsi, Shansi, Shanxi Sheng. ◆ *province* C China
161 P6 **Shanxian** *var.* Shan Xian. Shandong, E China *34°51′N 116°09′E*
Shan Xian *see* Sanmenxia
Shan Xian *see* Shanxian
Shanxi Sheng *see* Shanxi
160 L7 **Shanyang** Shaanxi, C China *33°35′N 109°48′E*
161 N13 **Shanyin** *var.* Daiyue. Shanxi, C China E Asia *39°30′N 112°56′E*
161 O13 **Shaoguan** *var.* Shao-kuan, *Cant.* Kukong; *prev.* Ch'u-chiang. Guangdong, S China *24°57′N 113°38′E*
Shao-kuan *see* Shaoguan
161 Q11 **Shaowu** Fujian, SE China *27°24′N 117°26′E*
161 S9 **Shaoxing** Zhejiang, SE China *30°02′N 120°35′E*
160 M12 **Shaoyang** *var.* Tangdukou. Hunan, S China *26°54′N 111°14′E*
160 M11 **Shaoyang** *var.* Baoqing, Shao-yang; *prev.* Pao-king. Hunan, S China *27°13′N 111°31′E*
Shao-yang *see* Shaoyang
96 K5 **Shapinsay** *island* NE Scotland, United Kingdom
125 S4 **Shapkina** ~ NW Russian Federation
Shāpūr *see* Salmās
158 M4 **Shaqiuhe** Xinjiang Uygur Zizhiqu, W China *45°00′N 88°52′E*
Shaqlāwa *see* Sheqlawe
Shaqlāwah *see* Sheqlawe
138 I8 **Shaqqā** As Suwaydā', S Syria *32°53′N 36°42′E*
141 P7 **Shaqrā'** Ar Riyāḑ, C Saudi Arabia *25°11′N 45°08′E*
Shaqrā *see* Shuqrah
145 W10 **Shar** *var.* Charsk. Vostochnyy Kazakhstan, E Kazakhstan *49°33′N 81°03′E*
149 O6 **Sharan** Dāykundī, SE Afghanistan *33°28′N 66°19′E*
149 Q7 **Sharan** *var.* Zareh Sharan. Paktīkā, E Afghanistan *33°08′N 68°47′E*
Sharaqpur *see* Sharqpur
145 U8 **Sharbakty** *Kaz.* Sharbaqty; *prev.* Shcherbakty. Pavlodar, E Kazakhstan *52°28′N 78°00′E*
Sharbaqty *see* Sharbakty
141 X12 **Sharbatāt** S Oman *17°57′N 56°14′E*
Sharbatāt, Ra's *see* Sharbithāt, Ras
141 X12 **Sharbithāt, Ras** *var.* Ra's Sharbatāt. *headland* S Oman *17°55′N 56°30′E*
14 K14 **Sharbot Lake** Ontario, SE Canada *44°45′N 76°46′W*
145 P17 **Shardara** *var.* Chardara. Yuzhnyy Kazakhstan, S Kazakhstan *41°15′N 68°01′E*
Shardara Dalasy *see* Step' Shardara
145 P17 **Shardarinskoye Vodokhranilishche** *prev.* Chardarinskoye Vodokhranilishche. ⊠ S Kazakhstan
162 F8 **Sharga** Govĭ-Altay, W Mongolia *46°16′N 95°32′E*
Sharga *see* Tsagaan-Uul
116 M7 **Sharhorod** Vinnyts'ka Oblast', C Ukraine *48°46′N 28°05′E*
Sharhulsan *see* Mandal-Ovoo
165 V3 **Shari** Hokkaidō, NE Japan *43°54′N 144°42′E*
Shari *see* Chari
139 T6 **Shārī, Buḩayrat** ◎ C Iraq
147 N12 **Sharixon** *Rus.* Shakhrisabz. Qashqadaryo Viloyati, S Uzbekistan *39°01′N 66°45′E*
Sharjah *see* Ash Shāriqah
118 K12 **Sharkawshchyna** *var.* Sharkowshchyna, *Pol.* Szarkowszczyzna, *Rus.* Sharkovshchina. Vitsyebskaya Voblasts', NW Belarus *55°27′N 27°28′E*
180 G9 **Shark Bay** *bay* Western Australia
141 Y9 **Sharkh** E Oman *21°20′N 59°04′E*
Sharkovshchina/ Sharkowshchyna *see* Sharkawshchyna
127 U6 **Sharlyk** Orenburgskaya Oblast', W Russian Federation *52°52′N 54°45′E*
75 Y9 **Sharm ash Shaykh** *var.* Ofiral, Sharm el Sheikh. E Egypt *27°51′N 34°16′E*
Sharm el Sheikh *see* Sharm ash Shaykh
18 B13 **Sharon** Pennsylvania, NE USA *41°12′N 80°28′W*
26 H4 **Sharon Springs** Kansas, C USA *38°54′N 101°46′W*
31 Q14 **Sharonville** Ohio, N USA *39°16′N 84°24′W*
Sharourah *see* Sharūrah
29 O10 **Sharpe, Lake** ⊠ South Dakota, N USA
Sharqī, Al Jabal ash/Sharqi, Jebel esh *see* Anti-Lebanon
Sharqiyah, Al Minṭaqah ash *see* Ash Sharqīyah
138 I6 **Sharqiyat an Nabk, Jabal** ▲ W Syria
149 W8 **Sharqpur** *var.* Sharaqpur. Punjab, E Pakistan *31°29′N 74°08′E*
141 Q13 **Sharūrah** *var.* Sharourah. Najrān, S Saudi Arabia *17°29′N 47°05′E*
125 O14 **Shar'ya** Kostromskaya Oblast', NW Russian Federation *58°22′N 45°30′E*
145 W15 **Sharyn** *prev.* Charyn. Almaty, SE Kazakhstan *43°48′N 79°22′E*
145 V15 **Sharyn** *var.* Charyn. ~ SE Kazakhstan
122 K13 **Sharypovo** Krasnoyarskiy Kray, C Russian Federation *55°33′N 89°12′E*
83 J18 **Shashe** Central, NE Botswana *21°25′S 27°28′E*
83 J18 **Shashe** *var.* Shashi. ~ Botswana/Zimbabwe
81 J14 **Shashemenē** *var.* Shashemenne, Shashhamana, *It.* Sciasciamana. Oromīya, C Ethiopia *07°16′N 38°38′E*
Shashemenne/ Shashhamana *see* Shashemenē
Shashi *see* Shashe
Shashi/Sha-shih/Shasi *see* Jingzhou, Hubei
35 N3 **Shasta Lake** ⊠ California, W USA
35 N2 **Shasta, Mount** ▲ California, W USA *41°24′N 122°11′W*
127 O4 **Shatki** Nizhegorodskaya Oblast', W Russian Federation *55°09′N 44°04′E*
Shatlyk *see* Şatlyk
Shatra *see* Ash Shatrah
119 K17 **Shatsk** Minskaya Voblasts', C Belarus *53°25′N 27°41′E*
127 N5 **Shatsk** Ryazanskaya Oblast', W Russian Federation *54°02′N 41°38′E*
26 J9 **Shattuck** Oklahoma, C USA *36°16′N 99°52′W*
145 P16 **Shauil'dir** *prev.* Shaul'der. Yuzhnyy Kazakhstan, S Kazakhstan *42°45′N 68°21′E*
Shaul'der *see* Shauil'dir
11 S17 **Shaunavon** Saskatchewan, S Canada *49°49′N 108°25′W*
Shavat *see* Shovot
Shaviyani Atoll *see* North Miladhunmadulu Atoll
158 K4 **Shawan** *var.* Sandaohezi. Xinjiang Uygur Zizhiqu, NW China *44°21′N 85°37′E*
14 G12 **Shawanaga** Ontario, S Canada *45°29′N 80°16′W*
30 M6 **Shawano** Wisconsin, N USA *44°46′N 88°38′W*
30 M6 **Shawano Lake** ◎ Wisconsin, N USA
15 P10 **Shawinigan** *prev.* Shawinigan Falls. Québec, SE Canada *46°33′N 72°45′W*
Shawinigan Falls *see* Shawinigan
15 P10 **Shawinigan-Sud** Québec, SE Canada *46°30′N 72°43′W*
138 J5 **Shawmarīyah, Jabal ash** ▲ C Syria
27 O11 **Shawnee** Oklahoma, C USA *35°20′N 96°55′W*
14 K12 **Shawville** Québec, SE Canada *45°37′N 76°31′W*
145 Q16 **Shayan** *var.* Chayan. Yuzhnyy Kazakhstan, S Kazakhstan *42°59′N 69°22′E*
Shaykh *see* Ash Shakk
139 W9 **Shaykh 'Ābid** *var.* Shaikh Ābid. Wāsiţ, E Iraq *32°40′N 46°09′E*
139 Y10 **Shaykh Fāris** *var.* Shaikh Fāris. Maysān, E Iraq *32°06′N 47°39′E*
139 T7 **Shaykh Ḩātim** Baghdād, E Iraq *33°29′N 44°15′E*
Shaykh, Jabal ash *see* Hermon, Mount
139 X10 **Shaykh Najm** *var.* Shaikh Najm. Maysān, E Iraq *32°04′N 46°54′E*
139 W9 **Shaykh Sa'd** Maysān, E Iraq *32°35′N 46°16′E*
147 T14 **Shazud** SE Tajikistan *37°45′N 72°22′E*
119 N18 **Shchadryn** *Rus.* Shchedrin. Homyel'skaya Voblasts', SE Belarus *52°53′N 29°33′E*
119 H18 **Shchara** ~ SW Belarus
Shchedrin *see* Shchadryn
Shcheglovsk *see* Kemerovo
126 K5 **Shchëkino** Tul'skaya Oblast', W Russian Federation *53°57′N 37°33′E*
125 S7 **Shchel'yayur** Respublika Komi, NW Russian Federation *65°19′N 53°27′E*
Shcherbakty *see* Sharbakty
126 K7 **Shchigry** Kurskaya Oblast', W Russian Federation *51°53′N 36°49′E*
Shchitkovichi *see* Shchytkavichy
117 Q2 **Shchors** Chernihivs'ka Oblast', N Ukraine *51°49′N 31°58′E*
117 T8 **Shchors'k** Dnipropetrovs'ka Oblast', E Ukraine *48°20′N 34°07′E*
Shchuchin *see* Shchuchyn
145 Q7 **Shchuchinsk** *prev.* Shchuchye. Akmola, N Kazakhstan *52°57′N 70°10′E*
Shchuchye *see* Shchuchinsk
119 G16 **Shchuchyn** *Pol.* Szczuczyn Nowogródzki, *Rus.* Shchuchin. Hrodzyenskaya Voblasts', W Belarus *53°36′N 24°45′E*
119 K17 **Shchytkavichy** *Rus.* Shchitkovichi. Minskaya Voblasts', C Belarus *53°13′N 27°59′E*
122 J13 **Shebalino** Respublika Altay, S Russian Federation *51°16′N 85°41′E*
126 J9 **Shebekino** Belgorodskaya Oblast', W Russian Federation *50°25′N 36°55′E*
Shebelē Wenz, Wabē *see* Shebeli
81 L14 **Shebeli** *Amh.* Wabē Shebelē Wenz, *It.* Scebeli, *Som.* Webi Shabeelle. ~ Ethiopia/ Somalia
113 M20 **Shebenikut, Maja e** ▲ E Albania *41°13′N 20°27′E*
Sheberghān *see* Shibirghān
144 F14 **Shebir** Mangistau, SW Kazakhstan *44°52′N 52°01′E*
31 N8 **Sheboygan** Wisconsin, N USA *43°46′N 87°44′W*
77 X15 **Shebshi Mountains** *var.* Schebschi Mountains. ▲▲ E Nigeria
Shechem *see* Nablus
Shedadi *see* Ash Shadādah
13 P14 **Shediac** New Brunswick, SE Canada *46°13′N 64°35′W*
126 L15 **Shedok** Krasnodarskiy Kray, SW Russian Federation *44°12′N 40°49′E*
Sheekh *see* Shiikh
38 M11 **Sheenjek River** ~ Alaska, USA
96 D13 **Sheep Haven** *Ir.* Cuan na gCaorach. *inlet* N Ireland
35 X10 **Sheep Range** ▲▲ Nevada, W USA
98 M13 **'s-Heerenberg** Gelderland, E Netherlands *51°52′N 06°15′E*
97 P22 **Sheerness** SE England, United Kingdom *51°27′N 00°45′E*
13 Q15 **Sheet Harbour** Nova Scotia, SE Canada *44°56′N 62°31′W*
185 H18 **Sheffield** Canterbury, South Island, New Zealand *43°22′S 172°01′E*
97 M18 **Sheffield** N England, United Kingdom *53°23′N 01°30′W*
23 O2 **Sheffield** Alabama, S USA *34°46′N 87°42′W*
29 V12 **Sheffield** Iowa, C USA *42°53′N 93°13′W*
25 N10 **Sheffield** Texas, SW USA *30°42′N 101°49′W*
63 H22 **Shehuen, Río** ~ S Argentina
Shekhem *see* Nablus
149 V8 **Shekhūpura** Punjab, NE Pakistan *31°42′N 74°08′E*
Sheki *see* Şäki
124 L14 **Sheksna** Vologodskaya Oblast', NW Russian Federation *59°11′N 38°32′E*
123 T5 **Shelagskiy, Mys** *headland* NE Russian Federation *70°04′N 170°39′E*
27 V3 **Shelbina** Missouri, C USA *39°41′N 92°02′W*
13 P16 **Shelburne** Nova Scotia, SE Canada *43°47′N 65°20′W*
14 G14 **Shelburne** Ontario, S Canada *44°04′N 80°12′W*
33 R7 **Shelby** Montana, NW USA *48°30′N 111°52′W*
21 Q10 **Shelby** North Carolina, SE USA *35°15′N 81°34′W*
31 S12 **Shelby** Ohio, N USA *40°52′N 82°39′W*
30 L14 **Shelbyville** Illinois, N USA *39°24′N 88°47′W*
31 P14 **Shelbyville** Indiana, N USA *39°31′N 85°46′W*
20 L5 **Shelbyville** Kentucky, S USA *38°13′N 85°12′W*
27 V2 **Shelbyville** Missouri, C USA *39°48′N 92°01′W*
20 J10 **Shelbyville** Tennessee, S USA *35°29′N 86°30′W*
25 X8 **Shelbyville** Texas, SW USA *31°42′N 94°03′W*
30 L14 **Shelbyville, Lake** ⊠ Illinois, N USA
29 S12 **Sheldon** Iowa, C USA *43°10′N 95°51′W*
38 M11 **Sheldons Point** Alaska, USA *62°31′N 165°03′W*
145 V15 **Shelek** *prev.* Chilik. Almaty, SE Kazakhstan *43°35′N 78°12′E*
145 V15 **Shelek** *prev.* Chilik. ~ SE Kazakhstan
Shelekhov Gulf *see* Shelikhova, Zaliv
123 U9 **Shelikhova, Zaliv** *Eng.* Shelekhov Gulf. *gulf* E Russian Federation
39 P14 **Shelikof Strait** *strait* Alaska, USA
Shelim *see* Shalīm
11 T14 **Shellbrook** Saskatchewan, S Canada *53°14′N 106°24′W*
28 L3 **Shell Creek** ~ North Dakota, N USA
Shellif *see* Chelif, Oued
22 I10 **Shell Keys** *island group* Louisiana, S USA
30 I4 **Shell Lake** Wisconsin, N USA *45°44′N 91°56′W*
29 W12 **Shell Rock** Iowa, C USA *42°42′N 92°34′W*
185 C26 **Shelter Point** *headland* Stewart Island, New Zealand *47°04′S 168°13′E*
18 L13 **Shelton** Connecticut, NE USA *41°19′N 73°06′W*
32 G8 **Shelton** Washington, NW USA *47°13′N 123°06′W*
Shemakha *see* Şamaxı
145 W9 **Shemonaikha** Vostochnyy Kazakhstan, E Kazakhstan *50°38′N 81°55′E*
127 Q4 **Shemursha** Chuvashskaya Respublika, W Russian Federation *54°57′N 47°27′E*
38 D16 **Shemya Island** *island* Aleutian Islands, Alaska, USA
29 T16 **Shenandoah** Iowa, C USA *40°46′N 95°23′W*
21 U4 **Shenandoah** Virginia, NE USA *38°26′N 78°34′W*
21 U4 **Shenandoah Mountains** *ridge* West Virginia, NE USA
21 V3 **Shenandoah River** ~ West Virginia, NE USA
77 W15 **Shendam** Plateau, C Nigeria *08°52′N 09°30′E*
80 G8 **Shendi** *var.* Shandī. River Nile, NE Sudan *16°41′N 33°22′E*
76 I15 **Shenge** SW Sierra Leone *07°54′N 12°54′W*
146 L10 **Shengeldi** *Rus.* Chingildi. Navoiy Viloyati, N Uzbekistan *40°59′N 64°13′E*
145 U15 **Shengel'dy** Almaty, SE Kazakhstan *44°04′N 77°31′E*
113 K18 **Shëngjin** *var.* Shëngjini. Lezhë, NW Albania *41°49′N 19°34′E*
Shëngjini *see* Shëngjin
Shengking *see* Liaoning
Sheng Xian/Shengxian *see* Shengzhou
161 S9 **Shengzhou** *var.* Shengxian, Sheng Xian. Zhejiang, SE China *29°36′N 120°42′E*
Shenking *see* Liaoning
125 N11 **Shenkursk** Arkhangel'skaya Oblast', NW Russian Federation *62°10′N 42°58′E*
160 L3 **Shenmu** Shaanxi, C China *38°49′N 110°27′E*
113 L19 **Shën Noj i Madh** ▲ C Albania *41°23′N 20°07′E*
160 L8 **Shennong Ding** ▲ C China *31°24′N 110°16′E*
Shenshi/Shensi *see* Shaanxi
163 V12 **Shenyang** *Chin.* Shen-yang, *Eng.* Moukden, Mukden; *prev.* Fengtien. *province capital* Liaoning, NE China *41°50′N 123°26′E*
Shen-yang *see* Shenyang
161 O15 **Shenzhen** Guangdong, S China *22°39′N 114°02′E*
154 G8 **Sheopur** Madhya Pradesh, C India *25°41′N 76°42′E*
116 L5 **Shepetivka** *Rus.* Shepetovka. Khmel'nyts'ka Oblast', NW Ukraine *50°12′N 27°01′E*
Shepetovka *see* Shepetivka
25 W10 **Shepherd** Texas, SW USA *30°30′N 95°00′W*
187 R14 **Shepherd Islands** *island group* C Vanuatu
20 K5 **Shepherdsville** Kentucky, S USA *38°00′N 85°42′W*
183 O11 **Shepparton** Victoria, SE Australia *36°25′S 145°26′E*
97 P22 **Sheppey, Isle of** *island* SE England, United Kingdom
139 T2 **Sheqlawe** *Ar.* Shaqlāwah, *var.* Shaqlāwa. Arbil, E Iraq *36°24′N 44°21′E*
Sherabad *see* Sherobod
9 O4 **Sherard, Cape** *headland* Nunavut, N Canada *74°36′N 80°10′W*
97 L23 **Sherborne** S England, United Kingdom *50°58′N 02°30′W*
76 H16 **Sherbro Island** *island* SW Sierra Leone
15 Q12 **Sherbrooke** Québec, SE Canada *45°23′N 71°55′W*
29 T11 **Sherburn** Minnesota, N USA *43°39′N 94°43′W*
78 H6 **Sherda** Tibesti, N Chad *20°04′N 16°48′E*
80 G7 **Shereik** River Nile, N Sudan *18°44′N 33°37′E*
126 K3 **Sheremet'yevo** ✈ (Moskva) Moskovskaya Oblast', W Russian Federation *56°05′N 37°10′E*
153 P14 **Sherghāti** Bihār, N India *24°35′N 84°51′E*
27 U12 **Sheridan** Arkansas, C USA *34°18′N 92°22′W*
33 W12 **Sheridan** Wyoming, C USA *44°47′N 106°59′W*
182 G8 **Sheringa** South Australia *33°51′S 135°13′E*
25 U5 **Sherman** Texas, SW USA *33°39′N 96°35′W*
194 J10 **Sherman Island** *island* Antarctica
19 S4 **Sherman Mills** Maine, NE USA *45°51′N 68°23′W*
29 O15 **Sherman Reservoir** ⊠ Nebraska, C USA
147 N14 **Sherobod** *Rus.* Sherabad. Surkhondaryo Viloyati, S Uzbekistan *37°43′N 66°59′E*
147 O13 **Sherobod** *Rus.* Sherabad. ~ S Uzbekistan
153 T14 **Sherpur** Dhaka, N Bangladesh *25°00′N 90°01′E*
37 T4 **Sherrelwood** Colorado, C USA *39°49′N 105°00′W*
99 J14 **'s-Hertogenbosch** *Fr.* Bois-le-Duc, *Ger.* Herzogenbusch. Noord-Brabant, S Netherlands *51°41′N 05°19′E*
28 M2 **Sherwood** North Dakota, N USA *48°55′N 101°36′W*
11 Q14 **Sherwood Park** Alberta, SW Canada *53°34′N 113°04′W*
56 F13 **Sheshea, Río** ~ E Peru
143 T5 **Sheshtamad** Khorāsān-e Razavī, NE Iran *36°03′N 57°45′E*
29 S10 **Shetek, Lake** ◎ Minnesota, N USA
96 M2 **Shetland Islands** *island group* NE Scotland, United Kingdom
144 F14 **Shetpe** Mangistau, SW Kazakhstan *44°06′N 52°03′E*
154 C11 **Shetrunji** ~ W India
Shevchenko *see* Aktau
117 W5 **Shevchenkove** Kharkivs'ka Oblast', E Ukraine *49°40′N 37°13′E*
81 H14 **Shewa Gīmīra** Southern Nationalities, S Ethiopia *07°12′N 35°49′E*
161 Q9 **Shexian** *var.* Huicheng, She Xian. Anhui, E China *29°53′N 118°27′E*
She Xian *see* Shexian
161 R6 **Sheyang** *prev.* Hede. Jiangsu, E China *33°49′N 120°13′E*
29 O4 **Sheyenne** North Dakota, N USA *47°49′N 99°08′W*
29 P4 **Sheyenne River** ~ North Dakota, N USA
96 G7 **Shiant Islands** *island group* NW Scotland, United Kingdom
123 U12 **Shiashkotan, Ostrov** *island* Kuril'skiye Ostrova, SE Russian Federation
31 R9 **Shiawassee River** ~ Michigan, N USA
141 R14 **Shibām** C Yemen *15°49′N 48°24′E*
Shibarghān *see* Shibirghān
165 O10 **Shibata** *var.* Sibata. Niigata, Honshū, C Japan *37°57′N 139°20′E*
Shiberghan/Shiberghān *see* Shibirghān
Shibh Jazīrat Sīnā' *see* Sinai
75 W8 **Shibīn al Kawm** *var.* Shibîn el Kôm. N Egypt *30°33′N 31°00′E*
Shibîn el Kôm *see* Shibīn al Kawm
149 N2 **Shibirghān** *var.* Shibarghān, Shiberghan, Shiberghān; *prev.* Sheberghān. Jowzjān, N Afghanistan *36°41′N 65°45′E*
143 O13 **Shīb, Kūh-e** ▲ S Iran
12 D8 **Shibogama Lake** ◎ Ontario, C Canada
Shibotsu-jima *see* Zelënyy, Ostrov
164 B16 **Shibushi** Kagoshima, Kyūshū, SW Japan *31°27′N 131°05′E*
189 U13 **Shichiyo Islands** *island group* Chuuk, C Micronesia
Shickshock Mountains *see* Chic-Chocs, Monts
shiderti *see* Shiderty
Shiderti *see* Shiderty
145 S8 **Shiderty** *prev.* Shiderti. Pavlodar, NE Kazakhstan *51°40′N 74°50′E*
145 S9 **Shiderty** *prev.* Shiderti. ~ N Kazakhstan
96 G10 **Shiel, Loch** ◎ N Scotland, United Kingdom
164 J13 **Shiga** *off.* Shiga-ken, *var.* Siga. ◆ *prefecture* Honshū, SW Japan
Shiga-ken *see* Shiga
Shigatse *see* Xigazê
141 U13 **Shiḩan** *oasis* NE Yemen
Shih-chia-chuang/Shihmen *see* Shijiazhuang
158 K4 **Shihezi** Xinjiang Uygur Zizhiqu, NW China *44°21′N 85°59′E*
Shiichi *see* Shyichy
80 N12 **Shiikh** *prev.* Sheekh. Togdheer, N Somalia *10°01′N 45°21′E*
113 K19 **Shijak** *var.* Shijaku. Durrës, W Albania *41°21′N 19°34′E*
Shijaku *see* Shijak
161 O4 **Shijiazhuang** *var.* Shih-chia-chuang; *prev.* Shihmen. *province capital* Hebei, E China *38°04′N 114°28′E*
165 R5 **Shikabe** Hokkaidō, NE Japan *42°03′N 140°45′E*
149 Q13 **Shikārpur** Sind, S Pakistan *27°59′N 68°39′E*
127 Q7 **Shikhany** Saratovskaya Oblast', W Russian Federation *52°07′N 47°13′E*
189 V12 **Shiki Islands** *island group* Chuuk, C Micronesia
164 G14 **Shikoku** *var.* Sikoku. *island* SW Japan
192 H5 **Shikoku Basin** *var.* Sikoku Basin. *undersea feature* N Philippine Sea *28°00′N 137°00′E*
164 G14 **Shikoku-sanchi** ▲▲ Shikoku, SW Japan
165 X4 **Shikotan, Ostrov** *Jap.* Shikotan-tō. *island* NE Russian Federation
Shikotan-tō *see* Shikotan, Ostrov
165 R4 **Shikotsu-ko** *var.* Sikotu Ko. ◎ Hokkaidō, NE Japan
81 N15 **Shilabo** Sumalē, E Ethiopia *06°05′N 44°48′E*
139 V3 **Shilayr, Wādī** *var.* Āw-e Shilēr, Āveh-ye Shalīr. ~ E Iraq
127 X7 **Shil'da** Orenburgskaya Oblast', W Russian Federation *51°46′N 59°48′E*
Shilēr, Āw-e *see* Shilayr, Wādī
153 S12 **Shiliguri** *var.* Siliguri. West Bengal, NE India *26°46′N 88°24′E*
Shiliu *see* Changjiang
129 V7 **Shilka** ~ S Russian Federation
18 H15 **Shillington** Pennsylvania, NE USA *40°18′N 75°57′W*
153 V13 **Shillong** *state capital* Meghālaya, NE India *25°37′N 91°54′E*
126 M5 **Shilovo** Ryazanskaya Oblast', W Russian Federation *54°18′N 40°53′E*
164 C14 **Shimabara** *var.* Simabara. Nagasaki, Kyūshū, SW Japan *32°48′N 130°20′E*
164 C14 **Shimabara-wan** *bay* SW Japan
164 F12 **Shimane** *off.* Shimane-ken, *var.* Simane. ◆ *prefecture* Honshū, SW Japan
164 G11 **Shimane-hantō** *peninsula* Honshū, SW Japan
Shimane-ken *see* Shimane
123 Q13 **Shimanovsk** Amurskaya Oblast', SE Russian Federation *52°00′N 127°36′E*
Shimanto *see* Nakamura
Shimbir Berris *see* Shimbiris
80 O12 **Shimbiris** *var.* Shimbir Berris. ▲ N Somalia *10°43′N 47°10′E*
165 T4 **Shimizu** Hokkaidō, NE Japan *42°58′N 142°54′E*
164 M14 **Shimizu** *var.* Simizu. Shizuoka, Honshū, S Japan *35°01′N 138°29′E*
152 I8 **Shimla** *prev.* Simla. *state capital* Himāchal Pradesh, N India *31°07′N 77°09′E*
165 N14 **Shimoda** *var.* Simoda. Shizuoka, Honshū, S Japan *34°40′N 138°55′E*
165 O13 **Shimodate** *var.* Simodate. Ibaraki, Honshū, S Japan *36°20′N 140°00′E*
155 F18 **Shimoga** Karnātaka, W India *13°56′N 75°31′E*
164 C15 **Shimo-jima** *island* SW Japan
164 B15 **Shimo-Koshiki-jima** *island* SW Japan
81 J21 **Shimoni** Kwale, S Kenya *04°40′S 39°22′E*
164 D13 **Shimonoseki** *var.* Simonoseki, *hist.* Akamagaseki, Bakan. Yamaguchi, Honshū, SW Japan *33°57′N 130°54′E*
124 G14 **Shimsk** Novgorodskaya Oblast', NW Russian Federation *58°12′N 30°43′E*
141 W7 **Shināş** N Oman *24°45′N 56°24′E*
148 J6 **Shīndand** *prev.* Shīndand. Herāt, W Afghanistan *33°19′N 62°09′E*
Shīndand *see* Shīndand
Shinei *see* Xinying
162 H10 **Shinejinst** *var.* Dzalaa. Bayanhongor, C Mongolia *44°29′N 99°19′E*
25 T12 **Shiner** Texas, SW USA *29°25′N 97°10′W*
167 N1 **Shingbwiyang** Kachin State, N Myanmar (Burma) *26°40′N 96°14′E*
Shingozha *see* Shynkozha
164 J15 **Shingū** *var.* Singû. Wakayama, Honshū, SW Japan *33°41′N 135°57′E*
14 F8 **Shining Tree** Ontario, S Canada *47°36′N 81°12′W*
165 P9 **Shinjō** *var.* Sinzyô. Yamagata, Honshū, C Japan *38°47′N 140°17′E*
96 I7 **Shin, Loch** ◎ N Scotland, United Kingdom
21 S3 **Shinnston** West Virginia, NE USA *39°22′N 80°19′W*
138 I6 **Shinshār** *Fr.* Chinnchâr. Ḩimş, W Syria *34°36′N 36°45′E*
Shinshū *see* Jinju
165 T4 **Shintoku** Hokkaidō, NE Japan *43°03′N 142°50′E*
81 G20 **Shinyanga** Shinyanga, NW Tanzania *03°40′S 33°25′E*
81 G20 **Shinyanga** ◆ *region* N Tanzania
165 Q10 **Shiogama** *var.* Siogama. Miyagi, Honshū, C Japan *38°19′N 141°00′E*
164 M12 **Shiojiri** *var.* Sioziri. Nagano, Honshū, S Japan *36°08′N 137°58′E*
164 I15 **Shiono-misaki** *headland* Honshū, SW Japan *33°25′N 135°45′E*
165 Q12 **Shioya-zaki** *headland* Honshū, C Japan *37°00′N 140°57′E*
114 J9 **Shipchenski Prohod** *var.* Shipchenski Prokhod. *pass* C Bulgaria
Shipchenski Prokhod *see* Shipchenski Prohod
160 G14 **Shiping** Yunnan, SW China *23°45′N 102°23′E*
13 P13 **Shippagan** *var.* Shippegan. New Brunswick, SE Canada *47°45′N 64°44′W*
Shippegan *see* Shippagan
18 F15 **Shippensburg** Pennsylvania, NE USA *40°03′N 77°31′W*
37 P9 **Shiprock** New Mexico, SW USA *36°47′N 108°41′W*
37 O9 **Ship Rock** ▲ New Mexico, SW USA *36°41′N 108°50′W*
15 R6 **Shipshaw** ~ Québec, SE Canada
123 V10 **Shipunskiy, Mys** *headland* E Russian Federation *53°04′N 159°57′E*
160 K7 **Shiquan** Shaanxi, C China *33°05′N 108°15′E*
Shiquanhe *see* Gar
122 K13 **Shira** Respublika Khakasiya, S Russian Federation *54°35′N 89°58′E*
Shirajganj Ghat *see* Sirajganj
165 P12 **Shirakawa** *var.* Sirakawa. Fukushima, Honshū, C Japan *37°07′N 140°11′E*
164 M13 **Shirane-san** ▲ Honshū, S Japan *35°39′N 138°13′E*
165 U14 **Shiranuka** Hokkaidō, NE Japan *42°57′N 144°01′E*
195 N12 **Shirase Coast** *physical region* Antarctica
165 U3 **Shirataki** Hokkaidō, NE Japan *43°55′N 143°14′E*
143 O11 **Shīrāz** *var.* Shiraz. Fārs, S Iran *29°38′N 52°34′E*
83 N15 **Shire** *var.* Chire. ~ Malawi/ Mozambique
Shiree *see* Tsagaanhayrhan
Shireet *see* Bayandelger
165 W3 **Shiretoko-hantō** *headland* Hokkaidō, NE Japan *44°06′N 145°07′E*
165 W3 **Shiretoko-misaki** *headland* Hokkaidō, NE Japan *44°20′N 145°19′E*
127 N5 **Shiringushi** Respublika Mordoviya, W Russian Federation *53°50′N 42°49′E*
148 M3 **Shīrīn Tagāb** Fāryāb, N Afghanistan *36°49′N 65°01′E*
149 N2 **Shīrīn Tagāb** ~ N Afghanistan

◆ Country ● Country Capital ◇ Dependent Territory ○ Dependent Territory Capital ◆ Administrative Regions ✈ International Airport ▲ Mountain ▲▲ Mountain Range Volcano ~ River ◎ Lake ⊠ Reservoir

SHIRIYA-ZAKI – SINDRI

165 R6 **Shiriya-zaki** *headland* Honshū, C Japan 41°24′N 141°27′E
144 I12 **Shirkala, Gryada** *plain* W Kazakhstan
152 F11 **Shir Kolāyat** *var.* Kolāyat. Rājasthān, NW India 27°56′N 73°02′E
165 P10 **Shiroishi** *var.* Siroisi. Miyagi, Honshū, C Japan 38°00′N 140°38′E
Shirokoye *see* Shyroke
165 O10 **Shirone** *var.* Sirone. Niigata, Honshū, C Japan 37°46′N 139°00′E
164 L12 **Shirotori** Gifu, Honshū, SW Japan 35°53′N 136°52′E
197 T1 **Shirshov Ridge** *undersea feature* W Bering Sea
Shirshütür/Shirshyutyur, Peski *see* Şirşütür Gumy
143 T3 **Shirvān** *var.* Shīrwān. Khorāsān-e Shomālī, NE Iran 37°25′N 57°55′E
Shirwa, Lake *see* Chilwa, Lake
Shīrwān *see* Shirvān
159 N5 **Shisanjianfang** Xinjiang Uygur Zizhiqu, W China 43°10′N 91°15′E
38 M16 **Shishaldin Volcano** ▲ Unimak Island, Alaska, USA 54°45′N 163°58′W
Shishchitsy *see* Shyshchytsy
83 G16 **Shishikola** North West, N Botswana 18°09′S 23°08′E
38 M8 **Shishmaref** Alaska, USA 66°15′N 166°04′W
Shisur *see* Ash Shişar
164 L13 **Shitara** Aichi, Honshū, SW Japan 35°06′N 137°33′E
152 D12 **Shiv** Rājasthān, NW India 26°11′N 71°14′E
Shivāji Sāgar *see* Konya Reservoir
154 H8 **Shivpuri** Madhya Pradesh, C India 25°28′N 77°41′E
36 J9 **Shivwits Plateau** *plain* Arizona, SW USA
Shiwalik Range *see* Siwalik Range
160 M8 **Shiyan** Hubei, C China 32°31′N 110°45′E
145 O15 **Shiyeli** *prev.* Chiili. Kzylorda, S Kazakhstan 44°13′N 66°46′E
Shizilu *see* Junan
160 H13 **Shizong** *var.* Danfeng. Yunnan, SW China 24°53′N 104°E
165 R10 **Shizugawa** Miyagi, Honshū, NE Japan 38°40′N 141°26′E
165 T5 **Shizunai** Hokkaidō, NE Japan 42°20′N 142°24′E
165 M14 **Shizuoka** *var.* Sizuoka. Shizuoka, Honshū, S Japan 34°59′N 138°20′E
164 M13 **Shizuoka** *off.* Shizuoka-ken, *var.* Sizuoka. ◆ *prefecture* Honshū, S Japan
Shizuoka-ken *see* Shizuoka
Shklov *see* Shklow
119 N15 **Shklow** *Rus.* Shklov. Mahilyowskaya Voblasts', E Belarus 54°13′N 30°18′E
113 K18 **Shkodër** *var.* Shkodra, *It.* Scutari, *SCr.* Skadar. Shkodër, NW Albania 42°03′N 19°31′E
113 K17 **Shkodër** ◆ *district* NW Albania
Shkodra *see* Shkodër
Shkodrës, Liqeni i *see* Scutari, Lake
113 L20 **Shkumbinit, Lumi i** *var.* Shkumbi, Shkumbin. ∽ C Albania
Shkumbi/Shkumbin *see* Shkumbinit, Lumi i
Shligigh, Cuan *see* Sligo Bay
122 L4 **Shmidta, Ostrov** *island* Severnaya Zemlya, N Russian Federation
183 S10 **Shoalhaven River** ∽ New South Wales, SE Australia
11 W16 **Shoal Lake** Manitoba, S Canada 50°28′N 100°36′W
31 O15 **Shoals** Indiana, N USA 38°40′N 86°47′W
164 I13 **Shōdo-shima** *island* SW Japan
Shōka *see* Zhanghua
122 M5 **Shokal'skogo, Proliv** *strait* N Russian Federation
147 T14 **Shokhdara, Qatorkŭhi** *Rus.* Shakhdarinskiy Khrebet. ▲ SE Tajikistan
145 T15 **Shokpar** *Kaz.* Shoqpar; *prev.* Chokpar. Zhambyl, S Kazakhstan 43°49′N 74°25′E
145 P15 **Sholakkorgan** *var.* Chulakkurgan. Yuzhnyy Kazakhstan, S Kazakhstan 43°45′N 69°10′E
145 N9 **Sholaksay** Kostanay, N Kazakhstan 51°45′N 64°45′E
Sholāpur *see* Solāpur
Sholdaneshty *see* Şoldăneşti
145 W15 **Shonzhy** *prev.* Chundzha. Almaty, SE Kazakhstan 43°32′N 79°28′E
Shoqpar *see* Shokpar
155 G21 **Shoranūr** Kerala, SW India 10°53′N 76°06′E
155 G16 **Shorāpur** Karnātaka, C India 16°34′N 76°48′E
147 O14 **Sho'rchi** *Rus.* Shurchi. Surkhondaryo Viloyati, S Uzbekistan 37°58′N 67°40′E
30 M11 **Shorewood** Illinois, N USA 41°31′N 88°12′W
Shorkazakhly, Solonchak *see* Kazakhlyshor, Solonchak
145 Q9 **Shortandy** Akmola, C Kazakhstan 51°45′N 71°01′E
149 O2 **Shōr Tappeh** *var.* Shortepa, Shor Tepe; *prev.* Shūr Tappeh. Balkh, N Afghanistan 37°22′N 66°49′E
Shortepa/Shor Tepe *see* Shōr Tappeh
186 J7 **Shortland Island** *var.* Alu. *island* Shortland Islands, NW Solomon Islands
Shosambetsu *see* Shosanbetsu
165 S2 **Shosanbetsu** *var.* Shosambetsu. Hokkaidō, NE Japan 44°31′N 141°47′E
33 O15 **Shoshone** Idaho, NW USA 42°56′N 114°24′W
35 T6 **Shoshone Mountains** ▲ Nevada, W USA
33 U12 **Shoshone River** ∽ Wyoming, C USA
83 I19 **Shoshong** Central, SE Botswana 23°02′S 26°31′E
33 V14 **Shoshoni** Wyoming, C USA 43°13′N 108°06′W
Shōshū *see* Sangju
117 S2 **Shostka** Sums'ka Oblast', NE Ukraine 51°52′N 33°30′E
185 C21 **Shotover** ∽ South Island, New Zealand
146 H9 **Shovot** *Rus.* Shavat. Xorazm Viloyati, W Uzbekistan 41°41′N 60°13′E
37 N12 **Show Low** Arizona, SW USA 34°15′N 110°01′W
Show Me State *see* Missouri
125 O4 **Shoyna** Nenetskiy Avtonomnyy Okrug, NW Russian Federation 67°50′N 44°09′E
124 M11 **Shozhma** Arkhangel'skaya Oblast', NW Russian Federation 61°57′N 40°10′E
117 Q7 **Shpola** Cherkas'ka Oblast', N Ukraine 49°00′N 31°27′E
Shqipëria/Shqipërisë, Republika e *see* Albania
22 G5 **Shreveport** Louisiana, S USA 32°32′N 93°45′W
97 K19 **Shrewsbury** *hist.* Scrobesbyrig'. W England, United Kingdom 52°43′N 02°45′W
152 D11 **Shri Mohangarh** *prev.* Sri Mohangorh. Rājasthān, NW India 27°17′N 71°18′E
153 S16 **Shrīrāmpur** *prev.* Serampore, Serampur. West Bengal, NE India 22°44′N 88°20′E
97 K19 **Shropshire** *cultural region* W England, United Kingdom
113 N17 **Shtime** *Serb.* Štimlje. C Kosovo 42°27′N 21°03′E
145 S16 **Shu** *Kaz.* Shū. Zhambyl, SE Kazakhstan 43°34′N 73°41′E
129 Q7 **Shu** *Kaz.* Shū; *prev.* Chu. ∽ Kazakhstan/Kyrgyzstan
160 G13 **Shuangbai** *var.* Tuodian. Yunnan, SW China 24°45′N 101°38′E
163 W9 **Shuangcheng** Heilongjiang, NE China 45°20′N 126°21′E
Shuangcheng *see* Zherong
160 E14 **Shuangjiang** *var.* Weiyuan. Yunnan, SW China 23°28′N 99°43′E
Shuangjiang *see* Jiangkou
Shuangjiang *see* Tongdao
163 U10 **Shuangliao** *var.* Zhengjiatun. Jilin, NE China 43°31′N 123°32′E
Shuang-liao *see* Liaoyuan
Shuangshipu *see* Fengxian
163 Y7 **Shuangyashan** *var.* Shuang-ya-shan. Heilongjiang, NE China 46°37′N 131°10′E
Shuang-ya-shan *see* Shuangyashan
Shu'aymiah *see* Shu'aymīyah
141 W12 **Shu'aymīyah** *var.* Shu'aymiah. S Oman 17°55′N 55°39′E
Shubarkuduk *see* Shubarkudyk
144 I10 **Shubarkudyk** *prev.* Shubarkuduk, *Kaz.* Shubarqudyq. Aktyubinsk, W Kazakhstan 49°09′N 56°31′E
Shubarqudyq *see* Shubarkudyk
145 N12 **Shubar-Tengiz, Ozero** ◎ C Kazakhstan
39 S5 **Shublik Mountains** ▲ Alaska, USA
Shubrā al Khaymah *see* Shubrâ el Kheima
121 U13 **Shubrâ el Kheima** *var.* Shubrā al Khaymah. N Egypt 30°06′N 31°15′E
158 E8 **Shufu** *var.* Tuokezhake. Xinjiang Uygur Zizhiqu, NW China 39°18′N 75°43′E
147 S14 **Shughnon, Qatorkŭhi** *Rus.* Shugnanskiy Khrebet. ▲ SE Tajikistan
Shugnanskiy Khrebet *see* Shughnon, Qatorkŭhi
161 Q6 **Shu He** ∽ E China
Shuicheng *see* Lupanshui
Shuiding *see* Huocheng
Shuidong *see* Dianbai
Shuiji *see* Laixi
Shū-Ile Taūlary *see* Gory Shu-Ile
Shuilocheng *see* Zhuanglang
Shuiluo *see* Zhuanglang
149 T10 **Shujāābād** Punjab, E Pakistan 29°53′N 71°23′E
Shū, Kazakhstan *see* Shu
Shū, Kazakhstan/Kyrgyzstan *see* Shu
163 W9 **Shulan** Jilin, NE China 44°28′N 126°57′E
158 E8 **Shule** Xinjiang Uygur Zizhiqu, NW China 39°19′N 76°06′E
Shuleh *see* Shule He
159 Q8 **Shule He** *var.* Shuleh, Sulo. ∽ C China
30 K9 **Shullsburg** Wisconsin, N USA 42°37′N 90°12′W
Shulu *see* Xinji
39 N16 **Shumagin Islands** *island group* Alaska, USA
146 G7 **Shumanay** Qoraqalpog'iston Respublikasi, W Uzbekistan 42°42′N 58°56′E
114 M8 **Shumen** Shumen, NE Bulgaria 43°17′N 26°57′E
114 M8 **Shumen** ◆ *province* NE Bulgaria
127 P4 **Shumerlya** Chuvashskaya Respublika, W Russian Federation 55°31′N 46°24′E
122 G11 **Shumikha** Kurganskaya Oblast', C Russian Federation 55°12′N 63°09′E
118 M12 **Shumilina** *Rus.* Shumilino. Vitsyebskaya Voblasts', NE Belarus 55°18′N 29°37′E
Shumilino *see* Shumilina
123 V11 **Shumshu, Ostrov** *island* SE Russian Federation
116 K5 **Shums'k** Ternopil's'ka Oblast', W Ukraine 50°06′N 26°04′E
Shunan *see* Tokuyama
39 O7 **Shungnak** Alaska, USA 66°53′N 157°08′W
Shunsen *see* Chuncheon
Shuoxian *see* Shuozhou
161 N3 **Shuozhou** *var.* Shuoxian. Shanxi, C China 39°20′N 112°25′E
141 P16 **Shuqrah** *var.* Shaqrā. SW Yemen 13°26′N 45°44′E
Shurab *see* Shŭrob
Shurchi *see* Sho'rchi
147 R11 **Shŭrob** *Rus.* Shurab. NW Tajikistan 40°02′N 70°31′E
143 T10 **Shūr, Rūd-e** ∽ E Iran
Shūr Tappeh *see* Shōr Tappeh
83 K17 **Shurugwi** *prev.* Selukwe. Midlands, C Zimbabwe 19°40′S 30°00′E
142 L8 **Shūsh** *anc.* Susa, *Bibl.* Shushan. Khūzestān, SW Iran 32°12′N 48°20′E
Shushan *see* Shūsh
142 L9 **Shūshtar** *var.* Shustar, Shushter. Khūzestān, SW Iran 32°03′N 48°51′E
Shushter/Shustar *see* Shūshtar
141 T9 **Shuṭfah, Qalamat** *well* E Saudi Arabia
139 V9 **Shuwayjah, Hawr ash** *var.* Hawr as Suwayqiyah. ◎ E Iraq
124 M16 **Shuya** Ivanovskaya Oblast', W Russian Federation 56°51′N 41°24′E
39 Q14 **Shuyak Island** *island* Alaska, USA
166 M4 **Shwebo** Sagaing, C Myanmar (Burma) 22°35′N 95°42′E
166 L7 **Shwedaung** Bago, W Myanmar (Burma) 18°44′N 95°12′E
166 M7 **Shwegyin** Bago, SW Myanmar (Burma) 17°56′N 96°59′E
167 N4 **Shweli** *Chin.* Longchuan Jiang. ∽ Myanmar (Burma)/China
166 M6 **Shwemyo** Mandalay, C Myanmar (Burma) 20°04′N 96°13′E
145 S14 **Shyganak** *var.* Čiganak, Chiganak, *Kaz.* Shyghanaq. Zhambyl, SE Kazakhstan 45°10′N 73°55′E
Shyghanaq *see* Shyganak
Shyghys Qazagastan Oblysy *see* Vostochnyy Kazakhstan
Shyghys Qongyrat *see* Shygys Konyrat
145 T12 **Shygys Konyrat**, *Kaz.* Shyghys Qongyrat. Karaganda, C Kazakhstan 47°01′N 75°05′E
119 M19 **Shyichy** *Rus.* Shiichi. Homyel'skaya Voblasts', SE Belarus 52°15′N 29°14′E
145 Q17 **Shymkent** *prev.* Chimkent. Yuzhnyy Kazakhstan, S Kazakhstan 42°19′N 69°36′E
144 H9 **Shyngghyrlau** *prev.* Chingirlau. Zapadnyy Kazakhstan, W Kazakhstan 51°10′N 53°44′E
144 G9 **Shyngyrlau** *prev.* Utva. ∽ NW Kazakhstan
145 W11 **Shynkozha** *prev.* Shingozha. Vostochnyy Kazakhstan, E Kazakhstan 47°46′N 80°38′E
152 J5 **Shyok** Jammu and Kashmir, NW India 34°13′N 78°12′E
117 S9 **Shyroke** *Rus.* Shirokoye. Dnipropetrovs'ka Oblast', E Ukraine 47°41′N 33°16′E
117 O9 **Shyryayeve** Odes'ka Oblast', SW Ukraine 47°21′N 30°11′E
117 S5 **Shyshaky** Poltavs'ka Oblast', C Ukraine 49°54′N 34°00′E
119 K17 **Shyshchytsy** *Rus.* Shishchitsy. Minskaya Voblasts', C Belarus 53°13′N 27°33′E
149 Y3 **Siachen Muztāgh** ▲ NE Pakistan
Siadehan *see* Tākestān
148 M13 **Siāhān Range** ▲ W Pakistan
142 I1 **Siāh Chashmeh** *var.* Chālderān. Āžarbāyjān-e Gharbī, N Iran 39°02′N 44°22′E
149 W7 **Siālkot** Punjab, NE Pakistan 32°29′N 74°35′E
186 E7 **Sialum** Morobe, C Papua New Guinea 06°02′S 147°37′E
Siam *see* Thailand
Siam, Gulf of *see* Thailand, Gulf of
Sian *see* Xi'an
Siang *see* Brahmaputra
Siangtan *see* Xiangtan
169 N8 **Siantan, Pulau** *island* Kepulauan Anambas, W Indonesia
54 H11 **Siare, Río** ∽ C Colombia
171 R6 **Siargao Island** *island* S Philippines
186 F7 **Siassi** Umboi Island, C Papua New Guinea 05°34′S 147°50′E
115 D14 **Siátista** Dytikí Makedonía, N Greece 40°16′N 21°34′E
166 K4 **Siatlai** Chin State, W Myanmar (Burma) 22°05′N 93°36′E
171 P6 **Siaton** Negros, C Philippines 09°03′N 123°03′E
171 P6 **Siaton Point** *headland* Negros, C Philippines 09°03′N 123°00′E
118 F11 **Šiauliai** *Ger.* Schaulen. Šiauliai, N Lithuania 55°55′N 23°21′E
118 E11 **Šiauliai** ◆ *province* N Lithuania
171 Q10 **Siau, Pulau** *island* N Indonesia
83 J15 **Siavonga** Southern, SE Zambia 16°33′S 28°42′E
81 G18 **Siaya** ◆ *county* W Kenya
Siazan' *see* Siyəzən
107 N20 **Sibari** Calabria, S Italy 39°45′N 16°26′E
Sibata *see* Shibata
127 X6 **Sibay** Respublika Bashkortostan, W Russian Federation 52°40′N 58°39′E
93 M19 **Sibbo** *Fin.* Sipoo. Uusimaa, S Finland 60°22′N 25°20′E
112 D13 **Šibenik** *It.* Sebenico. Šibenik-Knin, S Croatia 43°43′N 15°54′E
Šibenik *see* Šibenik-Knin
112 E13 **Šibenik-Knin** *off.* Šibenska Županija, *var.* Sibenik. ◆ *province* S Croatia
Šibenik-Knin *see* Drniš
Šibenska Županija *see* Šibenik-Knin
Siberia *see* Sibir'
Siberoet *see* Siberut, Pulau
168 H12 **Siberut, Pulau** *prev.* Siberoet. *island* Kepulauan Mentawai, W Indonesia
168 I12 **Siberut, Selat** *strait* W Indonesia
149 P11 **Sibi** Baluchistān, SW Pakistan 29°31′N 67°54′E
186 B9 **Sibidiri** Western, SW Papua New Guinea 08°58′S 142°14′E
123 N10 **Sibir'** *var.* Siberia. *physical region* NE Russian Federation
79 F20 **Sibiti** Lékoumou, S Congo 03°41′S 13°20′E
81 G21 **Sibiti** ∽ C Tanzania
116 I12 **Sibiu** *Ger.* Hermannstadt, *Hung.* Nagyszeben. Sibiu, C Romania 45°48′N 24°09′E
116 I11 **Sibiu** ◆ *county* C Romania
29 S11 **Sibley** Iowa, C USA 43°24′N 95°45′W
153 Y11 **Sibsāgar** *var.* Sivasagar. Assam, NE India 26°59′N 94°38′E
169 R9 **Sibu** Sarawak, East Malaysia 02°18′N 111°49′E
42 G2 **Sibun** ∽ E Belize
79 I15 **Sibut** *prev.* Fort-Sibut. Kémo, S Central African Republic 05°44′N 19°07′E
171 P4 **Sibuyan Island** *island* C Philippines
189 U1 **Sibylla Island** *island* N Marshall Islands
11 N16 **Sicamous** British Columbia, SW Canada 50°49′N 118°52′W
Sichelburger Gerbirge *see* Gorjanci
167 N14 **Sichon** *var.* Ban Sichon, Si Chon. Nakhon Si Thammarat, SW Thailand 09°03′N 99°51′E
Si Chon *see* Sichon
160 H9 **Sichuan** *var.* Chuan, Sichuan Sheng, Ssu-ch'uan, Szechuan, Szechwan. ◆ *province* C China
160 I9 **Sichuan Pendi** *basin* C China
Sichuan Sheng *see* Sichuan
103 S16 **Sicie, Cap** *headland* SE France 43°03′N 05°50′E
107 J24 **Sicilia** *Eng.* Sicily; *anc.* Trinacria. ◆ *region* Italy, C Mediterranean Sea
107 M24 **Sicilia** *Eng.* Sicily; *anc.* Trinacria. *island* Italy, C Mediterranean Sea
Sicilian Channel *see* Sicily, Strait of
107 H24 **Sicily, Strait of** *var.* Sicilian Channel. *strait* C Mediterranean Sea
42 K5 **Sico Tinto, Río** *var.* Río Negro. ∽ NE Honduras
57 H16 **Sicuani** Cusco, S Peru 14°21′S 71°13′W
112 J10 **Šid** Vojvodina, NW Serbia 45°07′N 19°13′E
115 A15 **Sidári** Kérkyra, Iónia Nisiá, Greece, C Mediterranean Sea 39°47′N 19°43′E
169 Q11 **Sidas** Borneo, C Indonesia 0°24′N 109°46′E
98 O5 **Siddeburen** Groningen, NE Netherlands 53°15′N 06°52′E
154 D9 **Siddhapur** *prev.* Siddhpur, Sidhpur. Gujarāt, W India 23°57′N 72°28′E
Siddhpur *see* Siddhapur
155 I15 **Siddipet** Telangana, C India 18°10′N 78°54′E
77 N14 **Sidéradougou** SW Burkina Faso 10°39′N 04°16′W
107 N23 **Siderno** Calabria, SW Italy 38°18′N 16°19′E
Siders *see* Sierre
154 L9 **Sidhi** Madhya Pradesh, C India 24°24′N 81°54′E
Sidhirókastron *see* Sidirókastro
Sidhpur *see* Siddhapur
75 U7 **Sīdī Barrānī** NW Egypt 31°38′N 25°58′E
74 I6 **Sidi Bel Abbès** *var.* Sidi bel Abbès, Sidi-Bel-Abbès. NW Algeria 35°12′N 00°43′W
74 E7 **Sidi-Bennour** W Morocco 32°39′N 08°28′W
74 M6 **Sidi Bouzid** *var.* Gammouda, Sīdī Bu Zayd. C Tunisia 35°05′N 09°20′E
Sidi Bu Zayd *see* Sidi Bouzid
74 D8 **Sidi-Ifni** SW Morocco 29°33′N 10°04′W
74 G6 **Sidi-Kacem** *prev.* Petitjean. N Morocco 34°21′N 05°46′W
114 G12 **Sidirókastro** *prev.* Sidhirókastron. Kentrikí Makedonía, NE Greece 41°14′N 23°23′E
194 L12 **Sidley, Mount** ▲ Antarctica 76°39′S 124°48′W
29 S16 **Sidney** Iowa, C USA 40°45′N 95°39′W
33 Y7 **Sidney** Montana, NW USA 47°42′N 104°10′W
28 J15 **Sidney** Nebraska, C USA 41°09′N 102°57′W
18 I11 **Sidney** New York, NE USA 42°15′N 75°30′W
31 R13 **Sidney** Ohio, N USA 40°16′N 84°09′W
23 T2 **Sidney Lanier, Lake** ⊠ Georgia, SE USA
Sidon *see* Saïda
122 J9 **Sidorovsk** Yamalo-Nenetskiy Avtonomnyy Okrug, N Russian Federation 66°34′N 82°12′E
Sidra *see* Surt
Sidra/Sidra, Gulf of *see* Surt, Khalīj, N Libya
Siebenbürgen *see* Transylvania
Sieben Dörfer *see* Săcele
110 O12 **Siedlce** *Ger.* Sedlez, *Rus.* Sesdlets. Mazowieckie, C Poland 52°10′N 22°18′E
101 E16 **Sieg** ∽ W Germany
101 F16 **Siegen** Nordrhein-Westfalen, W Germany 50°53′N 08°02′E
109 X4 **Sieghartskirchen** Niederösterreich, E Austria 48°13′N 16°01′E
167 S12 **Siĕmbok** *prev.* Phumĭ Siĕmbok. Stœng Trêng, N Cambodia 13°28′N 105°59′E
110 O11 **Siemiatycze** Podlaskie, NE Poland 52°27′N 22°52′E
167 T11 **Siĕmpang** Stœng Trêng, NE Cambodia 14°07′N 106°24′E
167 R11 **Siĕmréab** *prev.* Siemreap. Siĕmréab, NW Cambodia 13°21′N 103°50′E
Siemreap *see* Siĕmréab
106 G12 **Siena** *Fr.* Sienne; *anc.* Saena Julia. Toscana, C Italy 43°20′N 11°20′E
Sienne *see* Siena
92 K12 **Sieppijärvi** Lappi, NW Finland 67°09′N 23°58′E
110 J13 **Sieradz** Sieradz, C Poland 51°36′N 18°42′E
110 K10 **Sierpc** Mazowieckie, C Poland 52°51′N 19°44′E
24 I9 **Sierra Blanca** Texas, SW USA 31°10′N 105°22′W
37 S14 **Sierra Blanca Peak** ▲ New Mexico, SW USA 33°22′N 105°48′W
35 P5 **Sierra City** California, W USA 39°34′N 120°35′W
63 I16 **Sierra Colorada** Río Negro, S Argentina 40°37′S 67°48′W
63 J16 **Sierra Grande** Río Negro, E Argentina 41°34′S 65°21′W
76 G15 **Sierra Leone** *off.* Republic of Sierra Leone. ◆ *republic* W Africa
64 M13 **Sierra Leone Basin** *undersea feature* E Atlantic Ocean 05°00′N 17°00′W
66 K8 **Sierra Leone Fracture Zone** *tectonic feature* E Atlantic Ocean
Sierra Leone, Republic of *see* Sierra Leone
Sierra Leone Ridge *see* Sierra Leone Rise
64 L13 **Sierra Leone Rise** *var.* Sierra Leone Ridge, Sierra Leone Schwelle. *undersea feature* E Atlantic Ocean 05°30′N 21°00′W
Sierra Leone Schwelle *see* Sierra Leone Rise
40 L7 **Sierra Mojada** Coahuila, NE Mexico 27°13′N 103°42′W
37 N16 **Sierra Vista** Arizona, SW USA 31°33′N 110°18′W
108 D10 **Sierre** *Ger.* Siders. Valais, SW Switzerland 46°18′N 07°33′E
36 L16 **Sierrita Mountains** ▲ Arizona, SW USA
Siete Moai *see* Ahu Akivi
76 M15 **Sifié** W Ivory Coast 07°59′N 06°55′W
115 I21 **Sífnos** *anc.* Siphnos. *island* Kykládes, Greece, Aegean Sea
115 I21 **Sífnou, Stenó** *strait* SE Greece
Siga *see* Shiga
103 P16 **Sigean** Aude, S France 43°02′N 02°58′E
Sighet *see* Sighetu Marmaţiei
Sighetul Marmaţiei *see* Sighetu Marmaţiei
116 I8 **Sighetu Marmaţiei** *var.* Sighet, Sighetul Marmaţiei, *Hung.* Máramarossziget. Maramureş, N Romania 47°56′N 23°53′E
116 I11 **Sighişoara** *Ger.* Schässburg, *Hung.* Segesvár. Mureş, C Romania 46°12′N 24°48′E
168 G7 **Sigli** Sumatera, W Indonesia 05°21′N 95°56′E
92 J1 **Siglufjörður** Norðurland Vestra, N Iceland 66°09′N 18°56′W
101 H23 **Sigmaringen** Baden-Württemberg, S Germany 48°04′N 09°12′E
101 N20 **Signalberg** ▲ SE Germany 49°30′N 12°34′E
36 I13 **Signal Peak** ▲ Arizona, SW USA 33°20′N 114°03′W
Signan *see* Xi'an
194 H1 **Signy** *UK research station* South Orkney Islands, Antarctica 60°27′S 45°35′W
29 X15 **Sigourney** Iowa, C USA 41°19′N 92°12′W
115 K17 **Sígri, Akrotírio** *headland* Lésvos, E Greece 39°12′N 25°49′E
Sigsbee Deep *see* Mexico Basin
47 N2 **Sigsbee Escarpment** *undersea feature* N Gulf of Mexico 26°00′N 92°30′W
56 C8 **Sigsig** Azuay, S Ecuador 03°04′S 78°50′W
95 O15 **Sigtuna** Stockholm, C Sweden 59°36′N 17°44′E
42 H6 **Siguatepeque** Comayagua, W Honduras 14°39′N 87°48′W
105 P7 **Sigüenza** Castilla-La Mancha, C Spain 41°04′N 02°38′W
105 R4 **Sigüés** Aragón, NE Spain 42°39′N 01°00′W
76 K13 **Siguiri** NE Guinea 11°26′N 09°08′W
118 G8 **Sigulda** *Ger.* Segewold. C Latvia 57°08′N 24°51′E
167 Q14 **Sihanoukville** *var.* Kâmpóng Saôm; *prev.* Kompong Som. Kâmpóng Saôm, SW Cambodia 10°38′N 103°30′E
108 G8 **Sihlsee** ◎ NW Switzerland
93 K18 **Siikainen** Satakunta, SW Finland 61°52′N 21°49′E
93 M16 **Siilinjärvi** Pohjois-Savo, C Finland 63°05′N 27°40′E
137 R15 **Siirt** *var.* Sert; *anc.* Tigranocerta. Siirt, SE Turkey 37°56′N 41°56′E
137 R15 **Siirt** *var.* Sert. ◆ *province* SE Turkey
187 N8 **Sikaiana** *var.* Stewart Islands. *island group* W Solomon Islands
Sikandarabad *see* Secunderābād
152 J11 **Sikandra Rao** Uttar Pradesh, N India 27°42′N 78°21′E
10 M11 **Sikanni Chief** British Columbia, W Canada 57°16′N 122°44′W
10 M11 **Sikanni Chief** ∽ British Columbia, W Canada
152 H11 **Sīkar** Rājasthān, N India 27°33′N 75°12′E
76 M13 **Sikasso** Sikasso, S Mali 11°21′N 05°43′W
76 L13 **Sikasso** ◆ *region* SW Mali
167 N3 **Sikaw** Kachin State, C Myanmar (Burma) 23°50′N 97°04′E
83 H14 **Sikelenge** Western, W Zambia 14°51′S 24°13′E
27 Y7 **Sikeston** Missouri, C USA 36°52′N 89°35′W
93 J14 **Sikfors** Norrbotten, N Sweden 65°29′N 21°17′E
123 T14 **Sikhote-Alin', Khrebet** ▲ SE Russian Federation
Siking *see* Xi'an
115 J22 **Síkinos** *island* Kykládes, Greece, Aegean Sea
153 S11 **Sikkim** *Tib.* Denjong. ◆ *state* N India
111 I26 **Siklós** Baranya, SW Hungary 45°51′N 18°18′E
Sikoku *see* Shikoku
Sikoku Basin *see* Shikoku Basin
83 G14 **Sikongo** Western, W Zambia 15°03′S 22°07′E
Sikotu Ko *see* Shikotsu-ko
Sikouri/Sikoúrion *see* Sykoúrio
123 P8 **Siktyakh** Respublika Sakha (Yakutiya), NE Russian Federation 69°45′N 124°42′E
78 K11 **Sila** *off.* Région du Sila. ◆ *region* E Chad
118 D12 **Šilalė** Tauragė, W Lithuania 55°29′N 22°10′E
106 G5 **Silandro** *Ger.* Schlanders. Trentino-Alto Adige, N Italy 46°39′N 10°55′E
41 N12 **Silao** Guanajuato, C Mexico 20°56′N 101°28′W
Sila, Région du *see* Sila
Silarius *see* Sele
153 W14 **Silchar** Assam, NE India 24°49′N 92°48′E
108 G9 **Silenen** Uri, C Switzerland 46°49′N 08°39′E
21 T9 **Siler City** North Carolina, SE USA 35°43′N 79°27′W
33 U11 **Silesia** Montana, NW USA 45°32′N 108°52′W
110 F13 **Silesia** *physical region* SW Poland
74 K12 **Silet** S Algeria 22°45′N 04°51′E
Sileti *see* Silety
Siletitengiz *see* Siletyteniz, Ozero
145 R8 **Silety** *prev.* Sileti. ∽ N Kazakhstan
145 R7 **Siletyteniz, Ozero** *Kaz.* Siletitengiz. ◎ N Kazakhstan
172 H16 **Silhouette** *island* Inner Islands, SE Seychelles
136 I17 **Silifke** *anc.* Seleucia. İçel, S Turkey 36°22′N 33°57′E
Siliguri *see* Shiliguri
156 J10 **Siling Co** ◎ W China
Silinhot *see* Xilinhot
192 G14 **Silisili, Mauga** ▲ Savai'i, C Samoa 13°37′S 172°26′W
114 M6 **Silistra** *var.* Silistria; *anc.* Durostorum. Silistra, NE Bulgaria 44°06′N 27°17′E
114 M7 **Silistra** ◆ *province* NE Bulgaria
Silistria *see* Silistra
136 D10 **Silivri** İstanbul, NW Turkey 41°05′N 28°15′E
94 L13 **Siljan** ◎ C Sweden
95 G22 **Silkeborg** Midtjylland, C Denmark 56°10′N 09°34′E
108 M8 **Sill** ∽ W Austria
105 S10 **Silla** Valenciana, E Spain 39°22′N 00°25′W
62 H3 **Sillajguay, Cordillera** ▲ N Chile 19°45′S 68°39′W
118 K3 **Sillamäe** *Ger.* Sillamäggi. Ida-Virumaa, NE Estonia 59°23′N 27°45′E
Sillamäggi *see* Sillamäe
Sillein *see* Žilina
109 P9 **Sillian** Tirol, W Austria 46°45′N 12°25′E
112 B10 **Šilo** Primorje-Gorski Kotar, NW Croatia 45°09′N 14°39′E
27 R9 **Siloam Springs** Arkansas, C USA 36°11′N 94°32′W
25 X10 **Silsbee** Texas, SW USA 30°21′N 94°10′W
143 W15 **Sīlūp, Rūd-e** ∽ SE Iran
118 C12 **Šilutė** *Ger.* Heydekrug. Klaipėda, W Lithuania 55°20′N 21°30′E
137 Q15 **Silvan** Diyarbakır, SE Turkey 38°08′N 41°E
108 J10 **Silvaplana** Graubünden, S Switzerland 46°27′N 09°45′E
Silva Porto *see* Kuito
58 M12 **Silva, Recife do** *reef* E Brazil
154 D12 **Silvassa** Dādra and Nagar Haveli, W India 20°13′N 73°03′E
29 X4 **Silver Bay** Minnesota, N USA 47°17′N 91°15′W
37 P15 **Silver City** New Mexico, SW USA 32°47′N 108°16′W
18 D10 **Silver Creek** New York, NE USA 42°32′N 79°10′W
37 N12 **Silver Creek** ∽ Arizona, SW USA
27 P4 **Silver Lake** Kansas, C USA 39°06′N 95°51′W
32 I14 **Silver Lake** Oregon, NW USA 43°07′N 121°04′W
35 T9 **Silver Peak Range** ▲ Nevada, W USA
21 W3 **Silver Spring** Maryland, NE USA 39°00′N 77°01′W
Silver State *see* Colorado
Silver State *see* Nevada
37 Q7 **Silverton** Colorado, C USA 37°48′N 107°39′W
18 K16 **Silverton** New Jersey, NE USA 40°00′N 74°09′W
32 G11 **Silverton** Oregon, NW USA 45°00′N 122°46′W
25 N4 **Silverton** Texas, SW USA 34°28′N 101°18′W
104 G14 **Silves** Faro, S Portugal 37°11′N 08°26′W
54 D12 **Silvia** Cauca, SW Colombia 02°37′N 76°24′W
108 J9 **Silvrettagruppe** ▲ Austria/Switzerland
Sily-Vajdej *see* Vulcan
108 L7 **Silz** Tirol, W Austria 47°17′N 11°00′E
172 I13 **Sima** Anjouan, SE Comoros 12°11′S 44°18′E
Simabara *see* Shimabara
83 H15 **Simakando** Western, W Zambia 16°43′S 24°46′E
Simane *see* Shimane
119 L20 **Simanichy** *Rus.* Simonichi. Homyel'skaya Voblasts', SE Belarus 51°53′N 28°05′E
160 F14 **Simao** Yunnan, SW China 22°30′N 101°06′E
153 P12 **Simarā** Central, C Nepal 27°14′N 85°00′E
14 I8 **Simard, Lac** ◎ Québec, SE Canada
136 D13 **Simav** Kütahya, W Turkey 39°05′N 28°59′E
136 D13 **Simav Çayı** ∽ NW Turkey
79 L18 **Simba** Orientale, N Dem. Rep. Congo 0°46′N 22°54′E
186 C7 **Simbai** Madang, N Papua New Guinea 05°12′S 144°33′E
Simbirsk *see* Ul'yanovsk
Simbu *see* Chimbu
14 F17 **Simcoe** Ontario, S Canada 42°50′N 80°19′W
14 H14 **Simcoe, Lake** ◎ Ontario, S Canada
80 J11 **Sīmēn** ▲ N Ethiopia
114 K11 **Simeonovgrad** *prev.* Maritsa. Haskovo, S Bulgaria 42°03′N 25°36′E
116 G11 **Simeria** *Ger.* Pischk, *Hung.* Piski. Hunedoara, W Romania 45°51′N 23°00′E
107 L24 **Simeto** ∽ Sicilia, Italy, C Mediterranean Sea
168 G9 **Simeulue, Pulau** *island* NW Indonesia
117 T13 **Simferopol'** Avtonomna Respublika Krym, S Ukraine 44°55′N 33°06′E
117 T13 **Simferopol'** ✈ Avtonomna Respublika Krym, S Ukraine 44°55′N 34°04′E
Simi *see* Sými
152 M9 **Simikot** Far Western, NW Nepal 30°02′N 81°49′E
54 F7 **Simití** Bolívar, N Colombia 07°57′N 73°57′W
114 G11 **Simitla** Blagoevgrad, SW Bulgaria 41°57′N 23°06′E
35 S15 **Simi Valley** California, W USA 34°16′N 118°47′W
81 G21 **Simiyu** *off.* Mkoa wa Simiyu. ◆ *region* N Tanzania
Simiyu, Mkoa wa *see* Simiyu
Simizu *see* Shimizu
Simla *see* Shimla
Şimlăul Silvaniei/Şimleul Silvaniei *see* Şimleu Silvaniei
116 G9 **Şimleu Silvaniei** *Hung.* Szilágysomlyó; *prev.* Şimlăul Silvaniei, Şimleul Silvaniei. Sălaj, NW Romania 47°12′N 22°49′E
Simmer *see* Simmerbach
101 E19 **Simmerbach** *var.* Simmer. ∽ W Germany
101 F18 **Simmern** Rheinland-Pfalz, W Germany 50°00′N 07°30′E
22 I7 **Simmesport** Louisiana, S USA 30°58′N 91°48′W
119 F14 **Simnas** Alytus, S Lithuania 54°23′N 23°40′E
92 L13 **Simo** Lappi, NW Finland 65°40′N 25°04′E
Simoda *see* Shimoda
Simodate *see* Shimodate
92 M13 **Simojärvi** ◎ N Finland
92 L13 **Simojoki** ∽ NW Finland
41 U15 **Simojovel** *var.* Simojovel de Allende. Chiapas, SE Mexico 17°14′N 92°40′W
Simojovel de Allende *see* Simojovel
56 B7 **Simon Bolivar** *var.* Guayaquil. ✈ (Quayaquil) Guayas, W Ecuador 02°16′S 79°54′W
54 L5 **Simon Bolívar** ✈ (Caracas) Vargas, N Venezuela 10°33′N 66°54′W
Simonichi *see* Simanichy
14 M12 **Simon, Lac** ◎ Québec, SE Canada
Simonoseki *see* Shimonoseki
Šimonovany *see* Partizánske
Simonstad *see* Simon's Town
83 E26 **Simon's Town** *var.* Simonstad. Western Cape, SW South Africa 34°12′S 18°26′E
Simony *see* Partizánske
Simpeln *see* Simplon
99 M18 **Simpelveld** Limburg, SE Netherlands 50°50′N 05°59′E
108 E11 **Simplon** *var.* Simpeln. Valais, SW Switzerland 46°13′N 08°01′E
108 E11 **Simplon Pass** *pass* S Switzerland
106 C6 **Simplon Tunnel** *tunnel* Italy/Switzerland
Simpson *see* Fort Simpson
182 G1 **Simpson Desert** *desert* Northern Territory/South Australia
10 J9 **Simpson Peak** ▲ British Columbia, W Canada 59°43′N 131°29′W
9 N7 **Simpson Peninsula** *peninsula* Nunavut, NE Canada
21 P11 **Simpsonville** South Carolina, SE USA 34°44′N 82°15′W
95 L23 **Simrishamn** Skåne, S Sweden 55°34′N 14°20′E
123 U13 **Simushir, Ostrov** *island* Kuril'skiye Ostrova, SE Russian Federation
168 G9 **Sinabang** Sumatera, W Indonesia 02°27′N 96°24′E
81 N15 **Sina Dhaqa** Galguduud, C Somalia 05°21′N 46°21′E
75 X8 **Sinai** *var.* Sinai Peninsula, *Ar.* Shibh Jazīrat Sīnā', Sīnā. *physical region* NE Egypt
116 J12 **Sinaia** Prahova, SE Romania 45°20′N 25°33′E
188 B16 **Sinajana** C Guam 13°28′N 144°45′E
40 H8 **Sinaloa** ◆ *state* C Mexico
54 H4 **Sinamaica** Zulia, NW Venezuela 11°06′N 71°52′W
163 X14 **Sinan-ni** SE North Korea 38°37′N 127°43′E
Sīnā/Sinai Peninsula *see* Sinai
Sīnāwan *see* Sīnāwin
75 N8 **Sīnāwin** *var.* Sīnāwan. NW Libya 31°00′N 10°37′E
83 J16 **Sinazongwe** Southern, S Zambia 17°14′S 27°27′E
166 L6 **Sinbaungwe** Magway, W Myanmar (Burma) 19°44′N 95°10′E
166 L5 **Sinbyugyun** Magway, W Myanmar (Burma) 20°38′N 94°40′E
54 E6 **Since** Sucre, NW Colombia 09°14′N 75°08′W
54 E6 **Sincelejo** Sucre, NW Colombia 09°17′N 75°23′W
23 U4 **Sinclair, Lake** ⊠ Georgia, SE USA
10 M14 **Sinclair Mills** British Columbia, SW Canada 54°03′N 121°37′W
154 I8 **Sind** ∽ N India
Sind *see* Sindh
95 H19 **Sindal** Nordjylland, N Denmark 57°29′N 10°13′E
171 P7 **Sindañgan** Mindanao, S Philippines 08°09′N 122°59′E
79 D19 **Sindara** Ngounié, W Gabon 01°07′S 10°41′E
152 E13 **Sindari** *prev.* Sindri. Rājasthān, N India 25°32′N 71°58′E
114 N8 **Sindel** Varna, E Bulgaria 43°07′N 27°35′E
101 H22 **Sindelfingen** Baden-Württemberg, SW Germany 48°43′N 09°E
155 G16 **Sindgi** Karnātaka, C India 17°01′N 76°22′E
149 Q14 **Sindh** *prev.* Sind. ◆ *province* SE Pakistan
118 G5 **Sindi** *Ger.* Zintenhof. Pärnumaa, SW Estonia 58°28′N 24°41′E
136 C13 **Sındırgı** Balıkesir, W Turkey 39°13′N 28°10′E
77 N14 **Sindou** SW Burkina Faso 10°35′N 05°04′W
Sindri *see* Sindari

◆ Country ◇ Dependent Territory ◆ Administrative Regions ▲ Mountain ⚲ Volcano ◎ Lake
• Country Capital ○ Dependent Territory Capital ✈ International Airport ▲ Mountain Range ∽ River ⊠ Reservoir

Sind Sāgar Doāb – Smith Bay

149 T9 Sind Sāgar Doāb desert E Pakistan
126 M11 Sinegorskiy Rostovskaya Oblast', SW Russian Federation 48°01′N 40°52′E
123 S9 Sinegor'ye Magadanskaya Oblast', E Russian Federation 62°04′N 150°33′E
114 O12 Sinekli İstanbul, NW Turkey 41°13′N 28°13′E
104 F12 Sines Setúbal, S Portugal 37°58′N 08°52′W
104 F12 Sines, Cabo de headland S Portugal 37°57′N 08°55′W
92 L12 Sinettä Lappi, NW Finland 66°39′N 25°25′E
186 H6 Sinewit, Mount ▲ New Britain, C Papua New Guinea 04°42′S 151°58′E
80 G11 Singa var. Sinja, Sinjah. Sinnar, E Sudan 13°11′N 33°55′E
78 J12 Singako Moyen-Chari, S Chad 09°52′N 19°31′E
Singan see Xi'an
168 K10 Singapore ● (Singapore) S Singapore 01°17′N 103°48′E
168 L10 Singapore off. Republic of Singapore. ◆ republic SE Asia
Singapore, Republic of see Singapore
169 U17 Singaraja Bali, C Indonesia 08°06′S 115°04′E
167 O10 Sing Buri var. Singhaburi. Sing Buri, C Thailand 14°56′N 100°21′E
101 H24 Singen Baden-Württemberg, S Germany 47°46′N 08°50′E
Sîngeorgiu de Pădure see Sângeorgiu de Pădure
Sîngeorz-Băi/Singeroz Băi see Sângeorz-Băi
116 M9 Sîngerei var. Sângerei; prev. Lazovsk. N Moldova 47°38′N 28°08′E
Singhaburi see Sing Buri
81 H21 Singida Singida, C Tanzania 04°45′S 34°48′E
81 G22 Singida ◈ region C Tanzania
Singidunum see Beograd
Singkaling Hkamti see Hkamti
171 N14 Singkang Sulawesi, C Indonesia 04°09′S 119°58′E
168 J11 Singkarak, Danau ◎ Sumatera, W Indonesia
169 N10 Singkawang Borneo, C Indonesia 00°57′N 108°57′E
168 M11 Singkep, Pulau island Kepulauan Lingga, W Indonesia
168 H9 Singkilbaru Sumatera, W Indonesia 02°18′N 97°47′E
183 T7 Singleton New South Wales, SE Australia 32°38′S 151°00′E
Singora see Songkhla
Singû see Shingū
Sining see Xining
107 D17 Siniscola Sardegna, Italy, C Mediterranean Sea 40°34′N 09°42′E
113 F14 Sinj Split-Dalmacija, SE Croatia 43°41′N 16°37′E
Sinjajevina see Sinjavina
139 P3 Sinjār Nīnawýa, NW Iraq 36°20′N 41°51′E
139 P2 Sinjār, Jabal ▲▲ N Iraq
Sinja/Sinjah see Singa
113 K15 Sinjavina var. Sinjajevina. ▲▲ C Montenegro
80 I7 Sinkat Red Sea, NE Sudan 18°52′N 36°51′E
Sinkiang/Sinkiang Uighur Autonomous Region see Xinjiang Uygur Zizhiqu
Sînmartin see Târnăveni
163 V13 Sinmi-do island NW North Korea
101 I18 Sinn ∼ C Germany
Sinnamarie see Sinnamary
55 Y9 Sinnamary var. Sinnamarie. N French Guiana 05°23′N 53°00′W
80 G11 Sinnar ◈ state E Sudan
Sinneh see Sanandaj
18 E13 Sinnemahoning Creek ∼ Pennsylvania, NE USA
Sînnicolau Mare see Sânnicolau Mare
Sinoe, Lacul see Sinoie, Lacul
Sinoia see Chinhoyi
117 N14 Sinoie, Lacul prev. Lacul Sinoe. lagoon SE Romania
59 H16 Sinop Mato Grosso, W Brazil 11°38′S 55°27′W
136 K10 Sinop anc. Sinope. Sinop, N Turkey 42°02′N 35°09′E
136 J10 Sinop ◈ province N Turkey
136 K10 Sinop Burnu headland N Turkey 42°02′N 35°12′E
Sinope see Sinop
Sino/Sinoe see Greenville
163 Y12 Sinp'o E North Korea 40°01′N 128°10′E
101 H20 Sinsheim Baden-Württemberg, SW Germany 49°15′N 08°53′E
Sîntana see Sântana
169 R11 Sintang Borneo, C Indonesia 0°03′N 111°31′E
99 F14 Sint Annaland Zeeland, SW Netherlands 51°36′N 04°07′E
98 L5 Sint Annaparochie Fris. Sint Anne. Fryslân, N Netherlands 53°20′N 05°46′E
Sint Anne see Sint Annaparochie
45 V9 Sint Eustatius var. Statia, Eng. Saint Eustatius. ◇ Dutch special municipality Sint Maarten
99 G19 Sint-Genesius-Rode Fr. Rhode-Saint-Genèse. Vlaams Brabant, C Belgium 50°45′N 04°21′E
99 F16 Sint-Gillis-Waas Oost-Vlaanderen, N Belgium 51°14′N 04°08′E
99 H17 Sint-Katelijne-Waver Antwerpen, C Belgium 51°05′N 04°31′E
99 E18 Sint-Lievens-Houtem Oost-Vlaanderen, NW Belgium 50°55′N 03°52′E
45 V9 Sint Maarten Eng. Saint Martin. ◇ Dutch self-governing territory NE Caribbean Sea
99 F14 Sint Maartensdijk Zeeland, SW Netherlands 51°33′N 04°05′E
99 L19 Sint-Martens-Voeren Fr. Fouron-Saint-Martin. Limburg, NE Belgium 50°46′N 05°49′E
99 J14 Sint-Michielsgestel Noord-Brabant, S Netherlands 51°38′N 05°21′E
Sîn-Miclăuş see Gheorgheni
45 O16 Sint Nicholaas S Aruba 12°25′N 69°52′W
99 F16 Sint-Niklaas Fr. Saint-Nicolas. Oost-Vlaanderen, N Belgium 51°10′N 04°09′E
99 K14 Sint-Oedenrode Noord-Brabant, S Netherlands 51°34′N 05°28′E
25 T14 Sinton Texas, SW USA 28°02′N 97°33′W
99 G14 Sint Philipsland Zeeland, SW Netherlands 51°37′N 04°11′E
99 G19 Sint-Pieters-Leeuw Vlaams Brabant, C Belgium 50°47′N 04°16′E
104 E11 Sintra prev. Cintra. Lisboa, W Portugal 38°48′N 09°22′W
99 J18 Sint-Truiden Fr. Saint-Trond. Limburg, NE Belgium 50°48′N 05°13′E
99 H14 Sint Willebrord Noord-Brabant, S Netherlands 51°33′N 04°35′E
163 V13 Sinŭiju W North Korea 40°08′N 124°33′E
80 P13 Sinujiif Nugaal, NE Somalia 08°33′N 49°05′E
Sinus Aelaniticus see Aqaba, Gulf of
Sinus Gallicus see Lion, Golfe du
Sinyang see Xinyang
Sinyavka see Sinyawka
119 I18 Sinyawka Rus. Sinyavka. Minskaya Voblasts', SW Belarus 52°57′N 26°29′E
Sinying see Xinying
Sinyukha see Synyukha
Sinzyô see Shinjō
111 I24 Sió ∼ W Hungary
171 O7 Siocon Mindanao, S Philippines 07°37′N 122°09′E
111 I24 Siófok Somogy, Hungary 46°54′N 18°03′E
Siogama see Shiogama
83 G15 Sioma Western, SW Zambia 16°39′S 23°36′E
108 D11 Sion Ger. Sitten; anc. Sedunum. Valais, SW Switzerland 46°15′N 07°23′E
103 O11 Sioule ∼ C France
29 S12 Sioux Center Iowa, C USA 43°04′N 96°10′W
29 R13 Sioux City Iowa, C USA 42°30′N 96°24′W
29 R11 Sioux Falls South Dakota, N USA 43°33′N 96°45′W
12 B11 Sioux Lookout Ontario, S Canada 49°27′N 94°06′W
29 T12 Sioux Rapids Iowa, C USA 42°53′N 95°09′W
Sioux State see North Dakota
Sioziri see Shiojiri
171 P6 Sipalay Negros, C Philippines 09°46′N 122°25′E
55 V11 Sipaliwini ◈ district S Suriname
45 U15 Siparia Trinidad, Trinidad and Tobago 10°08′N 61°31′W
Siphnos see Sífnos
163 V11 Siping var. Ssu-p'ing, Szeping; prev. Ssu-p'ing-chieh. Jilin, NE China 43°09′N 124°22′E
11 X12 Sipiwesk Manitoba, C Canada 55°28′N 97°16′W
11 W13 Sipiwesk Lake ◎ Manitoba, C Canada
195 O11 Siple Coast physical region Antarctica
194 K12 Siple Island island Antarctica
194 K13 Siple, Mount ▲ Siple Island, Antarctica 73°25′S 126°24′W
Sipoo see Sibbo
112 G12 Šipovo Republika Srpska, W Bosnia and Herzegovina 44°16′N 17°05′E
23 O4 Sipsey River ∼ Alabama, S USA
168 I13 Sipura, Pulau island W Indonesia
0 G16 Siqueiros Fracture Zone tectonic feature E Pacific Ocean
42 L10 Siquia, Río ∼ SE Nicaragua
43 N13 Siquirres Limón, E Costa Rica 10°05′N 83°30′W
54 J5 Siquisique Lara, N Venezuela 10°36′N 69°45′W
155 G19 Sira Karnātaka, W India 13°46′N 76°54′E
95 D16 Sira ∼ S Norway
167 P12 Si Racha var. Ban Si Racha, Si Racha. Chon Buri, S Thailand 13°10′N 100°57′E
Si Racha see Si Racha
107 L25 Siracusa Eng. Syracuse. Sicilia, Italy, C Mediterranean Sea 37°04′N 15°17′E
153 T14 Sirajganj var. Shirajganj Ghat. Rajshahi, C Bangladesh 24°27′N 89°42′E
Sirakawa see Shirakawa
11 N14 Sir Alexander, Mount ▲ British Columbia, W Canada 54°00′N 120°33′W
137 O12 Şiran Gümüşhane, NE Turkey 40°12′N 39°07′E
77 Q12 Sirba ∼ E Burkina Faso
143 O17 Şīr Banī Yās island W United Arab Emirates
95 D17 Sirdalsvatnet ◎ S Norway
Sir Darya/Sirdaryo see Syr Darya
147 P10 Sirdaryo Sirdaryo Viloyati, E Uzbekistan 40°46′N 68°34′E
147 O11 Sirdaryo Viloyati Rus. Syrdar'inskaya Oblast'. ◈ province E Uzbekistan
Sir Donald Sangster International Airport see Sangster
181 S3 Sir Edward Pellew Group island group Northern Territory, NE Australia
116 K8 Siret Ger. Sereth, Hung. Szeret. Suceava, N Romania 47°55′N 26°05′E
116 K8 Siret var. Siretul, Ger. Sereth, Rus. Seret. ∼ Romania/Ukraine
Siretul see Siret
140 K3 Sirhān, Wādī as dry watercourse Jordan/Saudi Arabia
152 I8 Sirhind Punjab, N India 30°39′N 76°28′E
116 F11 Şiria Ger. Schiria. Arad, W Romania 46°16′N 21°38′E
Şiria see Syria
143 S14 Sīrīk Hormozgān, SE Iran 26°32′N 57°07′E
167 P8 Sirikit Reservoir ⊠ N Thailand
58 K12 Sirituba, Ilha island NE Brazil
143 R11 Sīrjān prev. Sa'īdābād. Kermān, S Iran 29°29′N 55°39′E
182 H9 Sir Joseph Banks Group island group South Australia
92 K11 Sirkka Lappi, N Finland 67°49′N 24°48′E
Sirna see Sýrna
137 R16 Şırnak Şırnak, SE Turkey 37°33′N 42°27′E
137 S16 Şırnak ◈ province SE Turkey
Siroisi see Shiroishi
155 J14 Sironcha Mahārāshtra, C India 18°51′N 80°03′E
Sirone see Shirone
Síros see Sýros
Sirotino see Sirotsina
118 M12 Sirotsina Rus. Sirotino. Vitsyebskaya Voblasts', N Belarus 55°23′N 29°37′E
152 H9 Sirsa Haryāna, NW India 29°32′N 75°04′E
173 Y17 Sir Seewoosagur Ramgoolam ✈ (port Louis) SE Mauritius
155 E18 Sirsi Karnātaka, W India 14°46′N 74°49′E
146 K12 Şirşütür Gumy var. Shirshütür, Rus. Peski Shirshyutyur. desert E Turkmenistan
Sirte see Surt
182 A2 Sir Thomas, Mount ▲ South Australia 27°09′S 129°49′E
Sirti, Gulf of see Surt, Khalij
137 Y12 Şirvan prev. Äli-Bayramli. SE Azerbaijan 39°57′N 48°54′E
142 J5 Sīrvān, Rūdkhāneh-ye var. Nahr Diyālá, Sirwan. ∼ Iran/Iraq see also Diyālá, Nahr
Sīrvān, Rudkhaneh-ye see Diyālá, Sirwan Nahr
118 H13 Širvintos Vilnius, SE Lithuania 55°01′N 24°58′E
Sirwan see Diyālá, Nahr/Sīrvān, Rudkhaneh-ye
11 N15 Sir Wilfrid Laurier, Mount ▲ British Columbia, SW Canada 52°45′N 119°51′W
14 M10 Sir-Wilfrid, Mont ▲ Québec, SE Canada 46°57′N 75°33′W
Sisačko-Moslavačka Županija see Sisak-Moslavina
112 E9 Sisak var. Siscia, Ger. Sissek, Hung. Sziszek; anc. Segestica. Sisak-Moslavina, C Croatia 45°28′N 16°21′E
167 R10 Si Sa Ket var. Sisaket, Sri Saket. Si Sa Ket, E Thailand 15°08′N 104°18′E
Sisaket see Si Sa Ket
112 E9 Sisak-Moslavina off. Sisačko-Moslavačka Županija. ◈ province C Croatia
167 O8 Si Satchanalai Sukhothai, NW Thailand
Siscia see Sisak
83 G22 Sishen Northern Cape, NW South Africa 27°47′S 22°59′E
137 V13 Sisian SE Armenia 39°31′N 46°03′E
197 N13 Sisimiut var. Holsteinborg, Holsteinsborg, Holstenborg, Holstensborg. Qeqqata, S Greenland 67°07′N 53°42′W
30 M1 Siskiwit Bay lake bay Michigan, N USA
34 L1 Siskiyou Mountains ▲▲ California/Oregon, W USA
Sisŏphŏn see Bantéay Méan Choăy
108 E7 Sissach Basel Landschaft, NW Switzerland 47°28′N 07°48′E
186 B5 Sissano West Sepik, NW Papua New Guinea 03°02′S 142°01′E
Sissek see Sisak
29 R7 Sisseton South Dakota, N USA 45°39′N 97°03′W
143 W9 Sīstān, Daryācheh-ye var. Daryācheh-ye Hāmūn, Hāmūn-e Şāberī. ◎ Afghanistan/Iran see also Şāberī, Hāmūn-e
Sīstān, Daryācheh-ye see Şāberī, Hāmūn-e
143 V12 Sīstān va Balūchestān off. Ostān-e Sīstān va Balūchestān, var. Balūchestān va Sīstān. ◈ province SE Iran
Sīstān va Balūchestān, Ostān-e see Sīstān va Balūchestān
103 T14 Sisteron Alpes-de-Haute-Provence, SE France 44°12′N 05°55′E
32 H13 Sisters Oregon, NW USA 44°17′N 121°33′W
65 G15 Sisters Peak ▲ N Ascension Island 07°56′S 14°23′W
21 R3 Sistersville West Virginia, NE USA 39°33′N 81°00′W
Sistova see Svishtov
Sitakund see Sitakunda
153 V16 Sitakunda var. Sitakund. Chittagong, SE Bangladesh 22°35′N 91°40′E
153 P12 Sītāmarhi Bihār, N India 26°36′N 85°30′E
152 L11 Sītāpur Uttar Pradesh, N India 27°33′N 80°40′E
Sitaş Cristuru see Cristuru Secuiesc
115 L25 Siteía var. Sitía. Kríti, Greece, E Mediterranean Sea 35°13′N 26°06′E
105 V6 Sitges Cataluña, NE Spain 41°14′N 01°49′E
115 H15 Sithonía' Atavyros peninsula NE Greece
Sitía see Siteía
54 F4 Sitionuevo Magdalena, N Colombia 10°46′N 74°43′W
39 X13 Sitka Baranof Island, Alaska, USA 57°03′N 135°19′W
39 Q15 Sitkinak Island island Trinity Islands, Alaska, USA
Sittang see Sittoung
99 L17 Sittard Limburg, SE Netherlands 51°00′N 05°52′E
Sitten see Sion
108 H7 Sitter ∼ NW Switzerland
109 U10 Sittersdorf Kärnten, S Austria 46°31′N 14°34′E
166 M7 Sittoung var. Sittang. ∼ S Myanmar (Burma)
166 K6 Sittwe var. Akyab. Rakhine State, W Myanmar (Burma) 20°09′N 92°51′E
42 L8 Siuna Región Autónoma Atlántico Norte, NE Nicaragua 13°44′N 84°46′W
153 R15 Siuri West Bengal, NE India 23°54′N 87°32′E
Siut see Asyūt
123 Q13 Sivaki Amurskaya Oblast', SE Russian Federation 52°39′N 126°43′E
136 M13 Sivas anc. Sebastia, Sebaste. Sivas, C Turkey 39°44′N 37°01′E
136 M13 Sivas ◈ province C Turkey
Sivasagar see Sibsāgar
137 O15 Siverek Şanlıurfa, S Turkey 37°46′N 39°19′E
117 X6 Sivers'k Donets'ka Oblast', E Ukraine 48°52′N 38°07′E
124 G13 Siverskiy Leningradskaya Oblast', NW Russian Federation 59°21′N 30°01′E
117 X6 Sivers'kyy Donets' Rus. Severskiy Donets. ∼ Russian Federation/Ukraine see also Severskiy Donets
Sivers'kyy Donets' see Severskiy Donets
125 W5 Sivomaskinskiy Respublika Komi, NW Russian Federation 66°42′N 62°33′E
136 G13 Sivrihisar Eskişehir, W Turkey 39°29′N 31°32′E
99 F22 Sivry Hainaut, S Belgium 50°10′N 04°11′E
123 V9 Sivuchiy, Mys headland E Russian Federation 56°45′N 163°13′E
Siwa see Sīwah
75 U9 Sīwah var. Siwa. NW Egypt 29°11′N 25°31′E
152 J9 Siwalik Range var. Shiwalik Range. ▲▲ India/Nepal
153 O13 Siwān Bihār, N India 26°14′N 84°21′E
43 O14 Sixaola, Río ∼ Costa Rica/Panama
Six Counties, The see Northern Ireland
103 T16 Six-Fours-les-Plages Var, SE France 43°05′N 05°50′E
161 Q7 Sixian var. Si Xian. Anhui, E China 33°29′N 117°53′E
Si Xian see Sixian
22 J9 Six Mile Lake ◎ Louisiana, S USA
Siyāh Gūz see Syagwêz
155 L25 Siyambalanduwa Uva Province, SE Sri Lanka 06°54′N 81°32′E
137 Y10 Siyäzän Rus. Siazan'. NE Azerbaijan 41°05′N 49°05′E
Sizebolu see Sozopol
Sizuoka see Shizuoka
95 I24 Sjælland ◈ county SE Denmark
95 I24 Sjælland Eng. Zealand, Ger. Seeland. island E Denmark
Sjar see Sääre
113 L15 Sjenica Turk. Seniça. Serbia, SW Serbia 43°16′N 20°01′E
94 G11 Sjoa ∼ S Norway
95 K23 Sjöbo Skåne, S Sweden 55°37′N 13°45′E
94 E9 Sjøholt Møre og Romsdal, S Norway 62°29′N 06°50′E
92 O1 Sjuøyane island group N Svalbard
Skadar see Shkodër
Skadarsko Jezero see Scutari, Lake
117 R11 Skadovs'k Khersons'ka Oblast', S Ukraine 46°07′N 32°53′E
95 I24 Skælskør Sjælland, E Denmark 55°16′N 11°18′E
92 I2 Skagaströnd prev. Höfdhakaupstadhur. Nordurland Vestra, N Iceland 65°49′N 20°18′W
95 H19 Skagen Nordjylland, N Denmark 57°44′N 10°37′E
Skagerak see Skagerrak
95 L16 Skagern ◎ C Sweden
197 T17 Skagerrak var. Skagerak. channel N Europe
94 G12 Skaget ▲ S Norway 61°19′N 09°07′E
32 H7 Skagit River ∼ Washington, NW USA
39 W12 Skagway Alaska, USA 59°27′N 135°18′W
92 K8 Skaidi Finnmark, N Norway 70°26′N 24°31′E
115 F21 Skála Pelopónnisos, S Greece 36°51′N 22°39′E
116 K6 Skalat Pol. Skałat. Ternopil's'ka Oblast', W Ukraine 49°27′N 25°59′E
95 J22 Skälderviken inlet Denmark/Sweden
92 I12 Skalka Lapp. Skalkká. ◎ N Sweden
Skalkká see Skalka
114 I12 Skaloti Anatolikí Makedonía kai Thráki, NE Greece 41°24′N 24°16′E
95 G22 Skanderborg Midtjylland, C Denmark 56°02′N 09°57′E
95 K22 Skåne prev. Eng. Scania. ◈ county S Sweden
75 N6 Skanès ✈ (Sousse) E Tunisia 35°36′N 10°56′E
95 C15 Skånevik Hordaland, S Norway 59°43′N 06°35′E
95 M18 Skänninge Östergötland, S Sweden 58°24′N 15°05′E
95 J23 Skanör med Falsterbo Skåne, S Sweden 55°24′N 12°48′E
115 H17 Skantzoúra island Vóreies Sporádes, Greece, Aegean Sea
95 K18 Skara Västra Götaland, S Sweden 58°23′N 13°25′E
95 M17 Skärblacka Östergötland, S Sweden 58°34′N 15°54′E
95 I18 Skärhamn Västra Götaland, S Sweden 57°59′N 11°33′E
95 I14 Skarnes Hedmark, S Norway 60°14′N 11°41′E
119 M21 Skarodnaye Rus. Skorodnoye. Homyel'skaya Voblasts', SE Belarus 51°38′N 28°50′E
110 I8 Skarszewy Ger. Schöneck. Pomorskie, NW Poland 54°04′N 18°25′E
111 M14 Skarżysko-Kamienna Świętokrzyskie, C Poland 51°07′N 20°52′E
114 H8 Skat var. Skŭt. ∼ NW Bulgaria
95 K16 Skattkärr Värmland, C Sweden 59°25′N 13°42′E
118 D12 Skaudvilė Tauragė, SW Lithuania 55°25′N 22°33′E
92 J12 Skaulo Lapp. Sávdijári. Norrbotten, N Sweden 67°21′N 21°03′E
111 K17 Skawina Małopolskie, S Poland 49°59′N 19°49′E
10 K12 Skeena ∼ British Columbia, SW Canada
10 J11 Skeena Mountains ▲▲ British Columbia, W Canada
97 O18 Skegness E England, United Kingdom 53°10′N 00°21′E
92 J4 Skeiðarársandur coast S Iceland
93 J15 Skellefteå Västerbotten, N Sweden 64°45′N 20°58′E
93 I14 Skellefteälven ∼ N Sweden
93 J15 Skelleftehamn Västerbotten, N Sweden 64°41′N 21°13′E
25 O2 Skellytown Texas, SW USA 35°34′N 101°10′W
95 J19 Skene Västra Götaland, S Sweden 57°30′N 12°34′E
97 G17 Skerries Ir. Na Sceirí. Dublin, E Ireland 53°35′N 06°07′W
95 H15 Ski Akershus, S Norway 59°43′N 10°50′E
115 G17 Skíathos Skíathos, Vóreies Sporádes, Greece, Aegean Sea 39°10′N 23°30′E
115 G17 Skíathos island Vóreies Sporádes, Greece, Aegean Sea
27 P9 Skiatook Oklahoma, C USA 36°22′N 96°00′W
27 P9 Skiatook Lake ⊠ Oklahoma, C USA
97 B22 Skibbereen Ir. An Sciobairín. Cork, SW Ireland 51°33′N 09°15′W
92 I9 Skibotn Troms, N Norway 69°22′N 20°18′E
119 F16 Skidal' Rus. Skidel'. Hrodzyenskaya Voblasts', W Belarus 53°35′N 24°15′E
97 K15 Skiddaw ▲ NW England, United Kingdom 54°37′N 03°07′W
Skidel' see Skidal'
25 T14 Skidmore Texas, SW USA 28°13′N 97°40′W
95 G16 Skien Telemark, S Norway 59°14′N 09°37′E
Skiermûntseach see Schiermonnikoog
110 L12 Skierniewice Łódzkie, C Poland 51°58′N 20°10′E
74 L5 Skikda prev. Philippeville. NE Algeria 36°51′N 07°E
30 M16 Skillet Fork ∼ Illinois, N USA
95 L19 Skillingaryd Jönköping, S Sweden 57°27′N 14°05′E
115 B19 Skinári, Akrotírio headland Iónia Nisiá, Greece 37°55′N 20°57′E
95 M15 Skinnskatteberg Västmanland, C Sweden 59°50′N 15°41′E
182 M12 Skipton Victoria, SE Australia 37°44′S 143°21′E
97 L16 Skipton N England, United Kingdom 53°57′N 02°W
Skiropoula see Skyropoúla
Skíros see Skýros
95 F21 Skive Midtjylland, NW Denmark 56°34′N 09°02′E
94 F11 Skjåk Oppland, S Norway 61°52′N 08°21′E
95 F22 Skjern Midtjylland, W Denmark 55°57′N 08°30′E
95 F22 Skjern Å var. Skjern Aa. ∼ W Denmark
Skjern Aa see Skjern Å
92 G12 Skjerstad Nordland, C Norway 67°14′N 15°00′E
92 J8 Skjervøy Troms, N Norway 70°03′N 20°56′E
92 I10 Skjold Troms, N Norway 69°03′N 19°18′E
111 I17 Skoczów Śląskie, S Poland 49°49′N 18°45′E
109 T11 Škofja Loka Ger. Bischoflack. NW Slovenia 46°12′N 14°16′E
94 N12 Skog Gävleborg, C Sweden 61°10′N 16°49′E
95 K16 Skoghall Värmland, C Sweden 59°20′N 13°30′E
31 N10 Skokie Illinois, N USA 42°01′N 87°43′W
116 H6 Skole L'vivs'ka Oblast', W Ukraine 49°04′N 23°29′E
115 D19 Skóllis ▲ S Greece 37°58′N 21°33′E
167 S13 Skon Kâmpóng Cham, C Cambodia 12°56′N 104°36′E
115 H17 Skópelos Skópelos, Vóreies Sporádes, Greece, Aegean Sea 39°07′N 23°43′E
115 H17 Skópelos island Vóreies Sporádes, Greece, Aegean Sea
126 L5 Skopin Ryazanskaya Oblast', W Russian Federation 53°46′N 39°37′E
113 N18 Skopje var. Üsküb, Turk. Üsküp, prev. Skoplje; anc. Scupi. ● (FYR Macedonia) N FYR Macedonia 42°N 21°28′E
113 O18 Skopje ✈ N FYR Macedonia 41°58′N 21°35′E
Skoplje see Skopje
110 I8 Skórcz Ger. Skurz. Pomorskie, N Poland 53°46′N 18°43′E
Skorodnoye see Skarodnaye
93 H16 Skorped Västernorrland, C Sweden 63°23′N 17°55′E
95 G21 Skørping Nordjylland, N Denmark 56°50′N 09°55′E
95 K18 Skövde Västra Götaland, S Sweden 58°24′N 13°52′E
123 Q13 Skovorodino Amurskaya Oblast', SE Russian Federation 54°03′N 123°47′E
19 Q6 Skowhegan Maine, NE USA 44°46′N 69°41′W
11 W15 Skownan Manitoba, S Canada 51°55′N 99°34′W
94 H13 Skreia Oppland, S Norway 60°37′N 11°00′E
Skripón see Orchómenos
118 H9 Skrīveri S Latvia 56°39′N 25°08′E
118 J11 Skrudaliena SE Latvia 55°50′N 26°42′E
118 D9 Skrunda W Latvia 56°39′N 22°01′E
95 C16 Skudeneshavn Rogaland, S Norway 59°10′N 05°16′E
95 C16 Skudnesfjorden S Norway
83 L20 Skukuza Mpumalanga, NE South Africa 25°01′S 31°35′E
97 B22 Skull Ir. An Scoil. SW Ireland 51°32′N 09°34′W
22 L3 Skuna River ∼ Mississippi, S USA
29 X15 Skunk River ∼ Iowa, C USA
Skuø see Skúvoy
118 C10 Skuodas Ger. Schoden, Pol. Szkudy. Klaipėda, NW Lithuania 56°16′N 21°30′E
95 K23 Skurup Skåne, S Sweden 55°28′N 13°30′E
Skurz see Skórcz
Skŭt see Skat
94 O13 Skutskär Uppsala, C Sweden 60°39′N 17°25′E
95 B19 Skúvoy Dan. Skuø. island C Faroe Islands
Skvira see Skvyra
117 O5 Skvyra Rus. Skvira. Kyyivs'ka Oblast', N Ukraine 49°44′N 29°42′E
39 Q11 Skwentna Alaska, USA 61°56′N 151°03′W
110 E11 Skwierzyna Ger. Schwerin. Lubuskie, W Poland 52°37′N 15°27′E
96 G9 Skye, Isle of island NW Scotland, United Kingdom
36 K13 Sky Harbor ✈ (Phoenix) Arizona, SW USA 32°26′N 112°00′W
32 I8 Skykomish Washington, NW USA 47°40′N 121°20′W
Skylge see Terschelling
63 F19 Skyring, Peninsula peninsula S Chile
63 H24 Skyring, Seno inlet S Chile
115 H17 Skyropoúla var. Skiropoula. island Vóreies Sporádes, Greece, Aegean Sea
115 I17 Skýros var. Skiros. Skýros, Vóreies Sporádes, Greece, Aegean Sea 38°55′N 24°34′E
115 I17 Skýros var. Skíros; anc. Scyros. island Vóreies Sporádes, Greece, Aegean Sea
118 J12 Slabodka Rus. Slobodka. Vitsyebskaya Voblasts', NW Belarus 55°41′N 27°11′E
95 I23 Slagelse Sjælland, E Denmark 55°25′N 11°22′E
93 I14 Slagnäs Norrbotten, N Sweden 65°36′N 18°10′E
39 T10 Slana Alaska, USA 62°46′N 144°00′W
97 F20 Slaney Ir. An tSláine. ∼ SE Ireland
116 J13 Slănic Prahova, SE Romania 45°14′N 25°58′E
116 K11 Slănic Moldova Bacău, E Romania 46°12′N 26°23′E
113 H16 Slano Dubrovnik-Neretva, SE Croatia 42°47′N 17°54′E
124 F13 Slantsy Leningradskaya Oblast', NW Russian Federation 59°06′N 28°00′E
111 C16 Slaný Ger. Schlan. Střední Čechy, NW Czech Republic 50°14′N 14°05′E
111 K16 Śląskie ◈ province S Poland
12 C10 Slate Falls Ontario, S Canada 51°11′N 91°32′W
27 T4 Slater Missouri, C USA 39°13′N 93°04′W
112 H9 Slatina Hung. Szlatina; prev. Podravska Slatina. Virovitica-Podravina, NE Croatia 45°40′N 17°46′E
116 I14 Slatina Olt, S Romania 45°27′N 24°21′E
25 N5 Slaton Texas, SW USA 33°26′N 101°38′W
95 H14 Slattum Akershus, S Norway 60°00′N 10°55′E
11 R10 Slave ∼ Alberta/Northwest Territories, C Canada
68 E12 Slave Coast coastal region W Africa
11 P13 Slave Lake Alberta, W Canada 55°17′N 114°46′W
122 I13 Slavgorod Altayskiy Kray, S Russian Federation 52°55′N 78°46′E
Slavgorod see Slawharad
Slavonia see Slavonija
112 G9 Slavonija Eng. Slavonia, Ger. Slawonien, Hung. Szlavonia, Szlavonország. cultural region NE Croatia
112 H10 Slavonski Brod Ger. Brod, Hung. Bród; prev. Brod, Brod na Savi. Brod-Posavina, NE Croatia 45°09′N 18°00′E
112 G10 Slavonski Brod-Posavina off. Brodsko-Posavska Županija, var. Brod-Posavina. ◈ province NE Croatia
116 L4 Slavuta Khmel'nyts'ka Oblast', NW Ukraine 50°18′N 26°52′E
117 P2 Slavutych Chernihivs'ka Oblast', N Ukraine 51°31′N 30°47′E
123 R15 Slavyanka Primorskiy Kray, SE Russian Federation 42°46′N 131°19′E
114 J8 Slavyanovo Pleven, N Bulgaria 43°28′N 24°52′E
Slavyansk see Slov"yans'k
126 K14 Slavyansk-na-Kubani Krasnodarskiy Kray, SW Russian Federation 45°16′N 38°09′E
119 N20 Slavyechna Rus. Slovechna. ∼ Belarus/Ukraine
119 O16 Slawharad Rus. Slavgorod. Mahilyowskaya Voblasts', E Belarus 53°27′N 31°00′E
110 G7 Sławno Zachodnio-pomorskie, NW Poland 54°23′N 16°43′E
Slawonien see Slavonija
29 S10 Slayton Minnesota, N USA 43°59′N 95°45′W
97 N18 Sleaford E England, United Kingdom 52°59′N 00°28′W
97 A20 Slea Head Ir. Ceann Sléibhe. headland SW Ireland 52°05′N 10°25′W
96 G9 Sleat, Sound of strait NW Scotland, United Kingdom
Sledyuki see Slyedzyuki
12 I5 Sleeper Islands island group Nunavut, C Canada
31 O6 Sleeping Bear Point headland Michigan, N USA 44°54′N 86°02′W
29 T10 Sleepy Eye Minnesota, N USA 44°18′N 94°43′W
39 O11 Sleetmute Alaska, USA 61°42′N 157°10′W
Sléibhe, Ceann see Slea Head
Slēmānī see As Sulaymānīyah
Slêmânî see As Sulaymānīyah
Slêmânî, Parêzga-î see As Sulaymānīyah
195 O5 Slessor Glacier glacier Antarctica
22 L9 Slidell Louisiana, S USA 30°16′N 89°46′W
18 K12 Slide Mountain ▲ New York, NE USA 42°00′N 74°23′W
98 I13 Sliedrecht Zuid-Holland, C Netherlands 51°50′N 04°46′E
121 P16 Sliema N Malta 35°54′N 14°31′E
97 G16 Slieve Donard ▲ SE Northern Ireland, United Kingdom 54°10′N 05°57′W
Sligeach see Sligo
97 D16 Sligo Ir. Sligeach. NW Ireland 54°17′N 08°28′W
97 C16 Sligo Ir. Sligeach. cultural region NW Ireland
97 C15 Sligo Bay Ir. Cuan Shligigh. inlet NW Ireland
18 B13 Slippery Rock Pennsylvania, NE USA 41°02′N 80°02′W
95 P19 Slite Gotland, SE Sweden 57°37′N 18°46′E
114 M8 Slivak Shumen, NE Bulgaria 43°27′N 26°56′E
114 L9 Sliven var. Slivno. Sliven, C Bulgaria 42°42′N 26°21′E
114 L10 Sliven ◈ province C Bulgaria
114 G9 Slivnitsa Sofia, W Bulgaria 42°51′N 23°01′E
Slivno see Sliven
114 L7 Slivo Pole Ruse, N Bulgaria 43°57′N 26°15′E
29 S13 Sloan Iowa, C USA 42°13′N 96°13′W
35 X12 Sloan Nevada, W USA 35°56′N 115°13′W
Slobodka see Slabodka
125 R14 Slobodskoy Kirovskaya Oblast', NW Russian Federation 58°43′N 50°12′E
Slobodzeya see Slobozia
117 O10 Slobozia Rus. Slobodzeya. E Moldova 46°45′N 29°42′E
116 L14 Slobozia Ialomiţa, SE Romania 44°34′N 27°23′E
98 O5 Slochteren Groningen, NE Netherlands 53°13′N 06°48′E
119 H17 Slonim Pol. Słonim. Hrodzyenskaya Voblasts', W Belarus 53°06′N 25°19′E
Słonim see Slonim
98 K7 Sloter Meer ◎ N Netherlands
Slot, The see New Georgia Sound
97 N22 Slough S England, United Kingdom 51°31′N 00°36′W
111 J20 Slovakia off. Slovenská Republika, Ger. Slowakei, Hung. Szlovákia, Slvk. Slovensko. ◆ republic C Europe
Slovak Ore Mountains see Slovenské rudohorie
Slovechna see Slavyechna
109 S12 Slovenia off. Republic of Slovenia, Ger. Slowenien, Slvn. Slovenija. ◆ republic SE Europe
Slovenia, Republic of see Slovenia
Slovenija see Slovenia
109 V10 Slovenj Gradec Ger. Windischgraz. N Slovenia 46°29′N 15°05′E
109 W10 Slovenska Bistrica Ger. Windischfeistritz. NE Slovenia 46°21′N 15°27′E
Slovenská Republika see Slovakia
109 W10 Slovenske Konjice E Slovenia 46°21′N 15°28′E
111 K20 Slovenské rudohorie Eng. Slovak Ore Mountains, Ger. Slowakisches Erzgebirge, Ungarisches Erzgebirge. ▲▲ C Slovakia
Slovensko see Slovakia
117 Y7 Slov"yanoserbs'k Luhans'ka Oblast', E Ukraine 48°41′N 39°00′E
117 W6 Slov"yans'k Rus. Slavyansk. Donets'ka Oblast', E Ukraine 48°51′N 37°38′E
Slowakei see Slovakia
Slowakisches Erzgebirge see Slovenské rudohorie
Slowenien see Slovenia
110 D11 Słubice Ger. Frankfurt. Lubuskie, W Poland 52°20′N 14°35′E
119 K19 Sluch Rus. Sluch'. ∼ C Belarus
116 L4 Sluch ∼ NW Ukraine
99 D16 Sluis Zeeland, SW Netherlands 51°18′N 03°22′E
112 D10 Slunj Hung. Szluin. Karlovac, C Croatia 45°06′N 15°35′E
110 I11 Słupca Wielkopolskie, C Poland 52°17′N 17°52′E
110 G6 Słupia Ger. Stolpe. ∼ NW Poland
110 G6 Słupsk Ger. Stolp. Pomorskie, N Poland 54°28′N 17°01′E
119 K18 Slutsk Minskaya Voblasts', S Belarus 53°02′N 27°32′E
119 O16 Slyedzyuki Rus. Sledyuki. Mahilyowskaya Voblasts', E Belarus 53°35′N 30°22′E
97 A17 Slyne Head Ir. Ceann Léime. headland W Ireland 53°25′N 10°11′W
27 U14 Smackover Arkansas, C USA 33°21′N 92°43′W
95 L20 Småland cultural region S Sweden
95 K20 Smålandsstenar Jönköping, S Sweden 57°10′N 13°24′E
Small Malaita see Maramasike
13 O8 Smallwood Reservoir ◎ Newfoundland and Labrador, S Canada
119 N14 Smalyany Rus. Smolyany. Vitsyebskaya Voblasts', N Belarus 54°36′N 30°04′E
119 L15 Smalyavichy Rus. Smolevichi. Minskaya Voblasts', C Belarus 54°02′N 28°05′E
74 C9 Smara var. Es Semara. N Western Sahara 26°45′N 11°44′W
119 I14 Smarhon' Pol. Smorgonie, Rus. Smorgon'. Hrodzyenskaya Voblasts', W Belarus 54°29′N 26°24′E
112 M11 Smederevo Ger. Semendria. Serbia, N Serbia 44°41′N 20°56′E
112 M12 Smederevska Palanka Serbia, C Serbia 44°24′N 20°56′E
95 M14 Smedjebacken Dalarna, C Sweden 60°08′N 15°25′E
116 L13 Smeeni Buzău, SE Romania 45°00′N 26°52′E
Smela see Smila
107 D16 Smeralda, Costa cultural region Sardegna, Italy, C Mediterranean Sea
111 J22 Śmigiel Ger. Schmiegel. Wielkopolskie, C Poland 52°02′N 16°33′E
117 Q6 Smila Rus. Smela. Cherkas'ka Oblast', C Ukraine 49°15′N 31°54′E
98 N7 Smilde Drenthe, NE Netherlands 52°57′N 06°28′E
11 S16 Smiley Saskatchewan, S Canada 51°40′N 109°24′W
25 T12 Smiley Texas, SW USA 29°16′N 97°38′W
Smilten see Smiltene
118 I8 Smiltene Ger. Smilten. N Latvia 57°25′N 25°53′E
123 T13 Smirnykh Ostrov Sakhalin, Sakhalinskaya Oblast', SE Russian Federation 49°43′N 142°48′E
11 Q13 Smith Alberta, W Canada 55°06′N 113°57′W
39 P4 Smith Bay bay Alaska, NW USA

◆ Country ● Country Capital ◇ Dependent Territory ○ Dependent Territory Capital ◈ Administrative Regions ✈ International Airport ▲ Mountain ▲▲ Mountain Range Volcano ∼ River ◎ Lake ⊠ Reservoir

Smith, Cape – Sotavento, Ilhas de

12 I3 Smith, Cape headland Québec, NE Canada 60°50′N 78°06′W
26 L3 Smith Center Kansas, C USA 39°46′N 98°46′W
10 K13 Smithers British Columbia, SW Canada 54°45′N 127°10′W
21 V10 Smithfield North Carolina, SE USA 35°30′N 78°21′W
36 L1 Smithfield Utah, W USA 41°50′N 111°49′W
21 X7 Smithfield Virginia, NE USA 36°41′N 76°38′W
12 I3 Smith Island island Nunavut, C Canada
Smith Island see Sumisu-jima
20 H7 Smithland Kentucky, S USA 37°06′N 88°24′W
21 T7 Smith Mountain Lake var. Leesville Lake. ⊠ Virginia, NE USA
34 L1 Smith River California, W USA 41°54′N 124°09′W
33 R9 Smith River ≈ Montana, NW USA
14 L13 Smiths Falls Ontario, SE Canada 44°54′N 76°01′W
33 N13 Smiths Ferry Idaho, NW USA 44°19′N 116°04′W
20 K7 Smiths Grove Kentucky, S USA 37°01′N 86°14′W
183 N15 Smithton Tasmania, SE Australia 40°54′S 145°06′E
18 L14 Smithtown Long Island, New York, NE USA 40°52′N 73°13′W
20 K9 Smithville Tennessee, S USA 35°59′N 85°49′W
25 T11 Smithville Texas, SW USA 30°04′N 97°32′W
Šmohor see Hermagor
35 Q4 Smoke Creek Desert desert Nevada, W USA
11 O14 Smoky ≈ Alberta, W Canada
182 F7 Smoky Bay South Australia 32°22′S 133°57′E
183 V6 Smoky Cape headland New South Wales, SE Australia 30°54′S 153°06′E
26 L4 Smoky Hill River ≈ Kansas, C USA
26 L4 Smoky Hills hill range Kansas, C USA
11 Q14 Smoky Lake Alberta, SW Canada 54°08′N 112°26′W
94 E8 Smøla island W Norway
126 H4 Smolensk Smolenskaya Oblast', W Russian Federation 54°48′N 32°08′E
126 H4 Smolenskaya Oblast' ◐ province W Russian Federation
Smolensk-Moscow Upland see Smolensko-Moskovskaya Vozvyshennost'
126 J3 Smolensko-Moskovskaya Vozvyshennost' var. Smolensk-Moscow Upland. ⛰ W Russian Federation
Smolevichi see Smalyavichy
115 C15 Smólikas var. Smolikás. ▲ W Greece 40°06′N 20°54′E
114 I12 Smolyan prev. Pashmakli. Smolyan, S Bulgaria 41°34′N 24°42′E
114 I12 Smolyan ◐ province S Bulgaria
Smolyany see Smalyany
33 S15 Smoot Wyoming, C USA 42°37′N 110°55′W
12 G12 Smooth Rock Falls Ontario, S Canada 49°17′N 81°37′W
Smorgon'/Smorgonie see Smarhon'
95 K23 Smygehamn Skåne, S Sweden 55°19′N 13°25′E
194 I7 Smyley Island island Antarctica
21 Y3 Smyrna Delaware, NE USA 39°18′N 75°36′W
23 S3 Smyrna Georgia, SE USA 33°52′N 84°30′W
20 J9 Smyrna Tennessee, S USA 36°00′N 86°30′W
Smyrna see İzmir
97 I16 Snaefell ▲ C Isle of Man 54°15′N 04°29′W
92 H3 Snæfellsjökull ▲ W Iceland 64°51′N 23°51′W
92 J3 Snækollur ▲ C Iceland 64°38′N 19°18′W
10 J4 Snake ≈ Yukon, NW Canada
29 O8 Snake Creek ≈ South Dakota, N USA
183 P13 Snake Island island Victoria, SE Australia
35 Y6 Snake Range ⛰ Nevada, W USA
32 K10 Snake River ≈ NW USA
29 V6 Snake River ≈ Minnesota, N USA
28 L12 Snake River ≈ Nebraska, C USA
33 Q14 Snake River Plain plain Idaho, NW USA
93 F15 Snåsa Nord-Trøndelag, C Norway 64°16′N 12°25′E
21 O8 Sneedville Tennessee, S USA 36°31′N 83°13′W
98 K6 Sneek Fris. Snits. Fryslân, N Netherlands 53°02′N 05°40′E
Sneeuw-gebergte see Maoke, Pegunungan
95 F22 Snejbjerg Midtjylland, C Denmark 56°08′N 08°55′E
122 K9 Snezhnogorsk Krasnoyarskiy Kray, N Russian Federation 68°06′N 87°37′E
124 J3 Snezhnogorsk Murmanskaya Oblast', NW Russian Federation 69°12′N 33°20′E
Snezhnoye see Snizhne
111 G15 Sněžka Ger. Schneekoppe, Pol. Śnieżka. ▲ N Czech Republic/Poland 50°42′N 15°55′E
110 N8 Śniardwy, Jezioro Ger. Spirdingsee. ◎ NE Poland
Sniečkus see Visaginas
Śnieżka see Sněžka
117 R10 Snihurivka Mykolayivs'ka Oblast', S Ukraine 47°05′N 32°48′E
116 I5 Snilov ✈ (L'viv) L'vivs'ka Oblast', W Ukraine 49°45′N 23°59′E
111 O19 Snina Hung. Szinna. Prešovský Kraj, E Slovakia 49°N 22°10′E
Snits see Sneek
117 Y8 Snizhne Rus. Snezhnoye. Donets'ka Oblast', SE Ukraine 48°01′N 38°46′E
94 G10 Snøhetta var. Snohetta. ▲ S Norway 62°22′N 09°08′E
92 G12 Snøtinden ▲ C Norway 66°39′N 13°50′E
97 I18 Snowdon ▲ NW Wales, United Kingdom 53°04′N 04°04′W
97 I18 Snowdonia ⛰ NW Wales, United Kingdom
Snowdrift see Lutselk'e
Snowdrift see Lutselk'e
37 N12 Snowflake Arizona, SW USA 34°30′N 110°04′W
21 Y5 Snow Hill Maryland, NE USA 38°11′N 75°23′W
21 W10 Snow Hill North Carolina, SE USA 35°26′N 77°39′W
194 H3 Snowhill Island island Antarctica
11 V13 Snow Lake Manitoba, C Canada 54°56′N 100°02′W
37 R5 Snowmass Mountain ▲ Colorado, C USA 39°07′N 107°04′W
18 M10 Snow, Mount ▲ Vermont, NE USA 42°56′N 72°52′W
34 M5 Snow Mountain ▲ California, W USA 39°44′N 123°01′W
Snow Mountains see Maoke, Pegunungan
33 N7 Snowshoe Peak ▲ Montana, NW USA 48°15′N 115°44′W
182 I8 Snowtown South Australia 33°49′S 138°13′E
36 K1 Snowville Utah, W USA 41°59′N 112°42′W
35 X3 Snow Water Lake ◎ Nevada, W USA
183 Q11 Snowy Mountains ⛰ New South Wales/Victoria, SE Australia
183 Q12 Snowy River ≈ New South Wales/Victoria, SE Australia
44 K5 Snug Corner Acklins Island, SE The Bahamas 22°31′N 73°51′W
167 T13 Snuŏl Krâchéh, E Cambodia 12°04′N 106°26′E
116 J7 Snyatyn Ivano-Frankivs'ka Oblast', W Ukraine 48°30′N 25°50′E
26 L12 Snyder Oklahoma, C USA 34°37′N 98°56′W
25 O6 Snyder Texas, SW USA 32°43′N 100°54′W
172 H3 Soalala Mahajanga, W Madagascar 16°05′S 45°21′E
172 J4 Soanierana-Ivongo Toamasina, E Madagascar 16°53′S 49°35′E
171 R11 Soasiu var. Tidore. Pulau Tidore, E Indonesia 0°40′N 127°25′E
54 G8 Soatá Boyacá, C Colombia 06°23′N 72°40′W
172 I5 Soavinandriana Antananarivo, C Madagascar 19°09′S 46°43′E
77 V13 Soba Kaduna, C Nigeria 10°58′N 08°06′E
163 Y16 Sobaek-sanmaek ⛰ S South Korea
80 F13 Sobat ≈ NE South Sudan
171 Z14 Sobger, Sungai ≈ Papua, E Indonesia
171 V13 Sobiei Papua Barat, E Indonesia 02°31′S 134°30′E
126 M3 Sobinka Vladimirskaya Oblast', W Russian Federation 56°00′N 39°55′E
127 S7 Sobolevo Orenburgskaya Oblast', W Russian Federation 51°57′N 51°42′E
Soborsin see Săvârşin
164 D15 Sobo-san ▲ Kyūshū, SW Japan 32°50′N 131°16′E
111 G14 Sobótka Dolnośląskie, SW Poland 50°53′N 16°48′E
59 O15 Sobradinho Bahia, E Brazil 09°33′S 40°56′W
Sobradinho, Barragem de see Sobradinho, Represa de
59 O16 Sobradinho, Represa de var. Barragem de Sobradinho. ⊠ E Brazil
58 O13 Sobral Ceará, E Brazil 03°45′S 40°20′W
105 T4 Sobrarbe physical region NE Spain
109 R10 Soča It. Isonzo. ≈ Italy/Slovenia
111 L14 Sochaczew Mazowieckie, C Poland 52°15′N 20°15′E
126 L15 Sochi Krasnodarskiy Kray, SW Russian Federation 43°35′N 39°46′E
114 G13 Sochós var. Sohos, Sokhós. Kentrikí Makedonía, N Greece 40°49′N 23°23′E
191 R11 Société, Archipel de la var. Archipel de Tahiti, Îles de la Société, Eng. Society Islands. island group W French Polynesia
Société, Îles de la/Society Islands see Société, Archipel de la
21 T11 Society Hill South Carolina, SE USA 34°28′N 79°54′W
175 W9 Society Ridge undersea feature C Pacific Ocean
62 I5 Socompa, Volcán 🌋 N Chile 24°18′S 68°03′W
Soconusco, Sierra de see Madre, Sierra
54 G8 Socorro Santander, C Colombia 06°30′N 73°16′W
37 R13 Socorro New Mexico, SW USA 33°58′N 106°55′W
Socotra see Suquṭrā
167 S14 Soc Trăng var. Khanh Hung. Soc Trăng, S Vietnam 09°36′N 105°58′E
105 P10 Socuéllamos Castilla-La Mancha, C Spain 39°18′N 02°48′W
35 W13 Soda Lake salt flat California, W USA
92 L11 Sodankylä Lappi, N Finland 67°26′N 26°35′E
Sodari see Sodiri
33 R15 Soda Springs Idaho, NW USA 42°39′N 111°36′W
Soddo/Soddu see Sodo
20 L10 Soddy Daisy Tennessee, S USA 35°14′N 85°11′W
95 N14 Söderfors Uppsala, C Sweden 60°23′N 17°14′E
94 N12 Söderhamn Gävleborg, C Sweden 61°19′N 17°10′E
95 N17 Söderköping Östergötland, S Sweden 58°28′N 16°20′E
95 N17 Södermanland ◐ county C Sweden
95 O16 Södertälje Stockholm, C Sweden 59°11′N 17°39′E
80 D10 Sodiri var. Sawdirī, Sodari. Northern Kordofan, C Sudan 14°23′N 29°06′E
81 I14 Sodo var. Soddo, Soddu. Southern Nationalities, S Ethiopia 06°49′N 37°43′E
Södra Karelen see Etelä-Karjala
Södra Österbotten see Etelä-Pohjanmaa
Södra Savolax see Etelä-Savo
95 M19 Södra Vi Kalmar, S Sweden 57°45′N 15°45′E
18 G9 Sodus Point headland New York, NE USA 43°16′N 76°59′W
171 Q17 Soe prev. Soë. Timor, C Indonesia 09°51′S 124°29′E
Soebang see Subang
Soekaboemi see Sukabumi
169 N15 Soekarno-Hatta ✈ (Jakarta) Jawa, S Indonesia
Soëla-Sund see Soela Väin
118 E5 Soela Väin prev. Eng. Sele Sound, Ger. Dagden-Sund, Soëla-Sund. strait W Estonia
Soemba see Sumba, Pulau
Soembawa see Sumbawa
Soemenep see Sumenep
Soengaipenoeh see Sungaipenuh
Soerabaja see Surabaya
Soerakarta see Surakarta
101 G14 Soest Nordrhein-Westfalen, W Germany 51°34′N 08°06′E
98 J11 Soest Utrecht, C Netherlands 52°10′N 05°20′E
100 F11 Soeste ≈ NW Germany
98 J11 Soesterberg Utrecht, C Netherlands 52°07′N 05°17′E
115 E16 Sofádes var. Sofádhes. Thessalía, C Greece 39°20′N 22°06′E
Sofádhes see Sofádes
83 N18 Sofala Sofala, C Mozambique 20°04′S 34°43′E
83 N17 Sofala ◐ province C Mozambique
83 N18 Sofala, Baia de bay C Mozambique
114 G10 Sofia var. Sophia, Sofiya, Eng. Sofia, Lat. Serdica. ● (Bulgaria) Sofia Grad, W Bulgaria 42°42′N 23°20′E
114 H9 Sofia ◐ province W Bulgaria
114 G9 Sofia ✈ Sofia Grad, W Bulgaria 42°42′N 23°26′E
172 J3 Sofia seasonal river NW Madagascar
Sofia see Sofiya
114 G9 Sofia Grad ◐ municipality W Bulgaria
115 G19 Sofikó Pelopónnisos, S Greece 37°46′N 23°04′E
Sofi-Kurgan see Sopu-Korgon
Sofiya see Sofia
Sofiyevka see Sofiyivka
117 S8 Sofiyivka Rus. Sofiyevka. Dnipropetrovs'ka Oblast', E Ukraine 48°04′N 33°55′E
123 R12 Sofiysk Khabarovskiy Kray, SE Russian Federation 51°32′N 139°46′E
123 R13 Sofiysk Khabarovskiy Kray, SE Russian Federation 52°20′N 133°37′E
124 I6 Sofporog Respublika Kareliya, NW Russian Federation 65°48′N 31°30′E
115 L23 Sofraná prev. Záfora. island Kykládes, Greece, Aegean Sea
165 Y14 Sōfu-gan island Izu-shotō, SE Japan
156 K10 Sog Xizang Zizhiqu, W China 31°52′N 93°40′E
54 G9 Sogamoso Boyacá, C Colombia 05°43′N 72°56′W
136 I11 Soğanlı Çayı ≈ N Turkey
94 E12 Sogn physical region S Norway
Sogndal see Sogndalsfjøra
94 E12 Sogndalsfjøra var. Sogndal. Sogn Og Fjordane, S Norway 61°13′N 07°05′E
95 E18 Søgne Vest-Agder, S Norway 58°05′N 07°49′E
94 D12 Sognefjorden fjord NE North Sea
94 C12 Sogn Og Fjordane ◐ county S Norway
162 I11 Sogo Nur ◎ N China
159 T12 Sogruma Qinghai, W China 32°32′N 100°52′E
Sŏgwip'o see Seogwipo
Sohâg see Sawhāj
Sohar see Şuḩār
64 H9 Sohm Plain undersea feature NW Atlantic Ocean
100 H7 Soholmer Au ≈ N Germany
Sohos see Sochós
Sohrau see Żory
99 F20 Soignies Hainaut, SW Belgium 50°35′N 04°04′E
159 R15 Soila Xizang Zizhiqu, W China 30°40′N 97°07′E
103 P4 Soissons anc. Augusta Suessionum, Noviodunum. Aisne, N France 49°23′N 03°20′E
164 H13 Sōja Okayama, Honshū, SW Japan 34°40′N 133°42′E
152 F13 Sojat Rājasthān, N India 25°53′N 73°45′E
163 W13 Sŏjosŏn-man inlet W North Korea
116 I4 Sokal' Rus. Sokal. L'vivs'ka Oblast', NW Ukraine 50°29′N 24°17′E
163 Y14 Sokcho prev. Sokch'o. N South Korea 38°07′N 128°34′E
Sokch'o see Sokcho
136 B15 Söke Aydın, SW Turkey 37°46′N 27°24′E
189 N12 Sokehs Island island E Micronesia
79 M24 Sokele Katanga, SE Dem. Rep. Congo 09°54′S 24°38′E
147 R11 Sokh Uzb. Sūkh. ≈ Kyrgyzstan/Uzbekistan
Sokh see So'x
Sokhós see Sochós
137 Q8 Sokhumi Rus. Sukhumi. NW Georgia 43°02′N 41°01′E
113 O14 Sokobanja Serbia, E Serbia 43°39′N 21°51′E
77 R15 Sokodé C Togo 08°58′N 01°10′E
123 T10 Sokol Magadanskaya Oblast', E Russian Federation 59°51′N 150°56′E
124 M13 Sokol Vologodskaya Oblast', NW Russian Federation 59°26′N 40°09′E
110 P9 Sokółka Podlaskie, NE Poland 53°24′N 23°31′E
76 M11 Sokolo Ségou, W Mali 14°43′N 06°02′W
111 A16 Sokolov Ger. Falkenau an der Eger; prev. Falknov nad Ohří. Karlovarský Kraj, W Czech Republic 50°10′N 12°38′E
111 O16 Sokołów Małopolski Podkarpackie, SE Poland 50°12′N 22°07′E
110 O11 Sokołów Podlaski Mazowieckie, C Poland 52°26′N 22°14′E
76 G11 Sokone W Senegal 13°53′N 16°22′W
77 T12 Sokoto Sokoto, NW Nigeria 13°05′N 05°16′E
77 T12 Sokoto ◐ state NW Nigeria
77 S12 Sokoto ≈ NW Nigeria
Sokotra see Suquṭrā
147 U7 Sokuluk Chuyskaya Oblast', N Kyrgyzstan 42°53′N 74°19′E
116 L7 Sokyryany Chernivets'ka Oblast', W Ukraine 48°28′N 27°25′E
95 C16 Sola Rogaland, S Norway 58°53′N 05°36′E
187 R12 Sola Vanua Lava, N Vanuatu 13°51′S 167°34′E
95 C17 Sola ✈ (Stavanger) Rogaland, S Norway 58°54′N 05°36′E
81 H18 Solai Nakuru, W Kenya 0°02′N 36°03′E
152 I8 Solan Himāchal Pradesh, N India 30°54′N 77°06′E
185 A25 Solander Island island SW New Zealand
Solano see Bahía Solano
155 F15 Solāpur var. Sholāpur. Mahārāshtra, W India 17°43′N 75°54′E
93 H16 Solberg Västernorrland, C Sweden 63°48′N 17°40′E
116 K9 Solca Ger. Solka. Suceava, N Romania 47°40′N 25°50′E
105 O16 Sol, Costa del coastal region S Spain
106 F5 Solda Ger. Sulden. Trentino-Alto Adige, N Italy 46°33′N 10°35′E
117 N9 Şoldăneşti Rus. Sholdaneshty. N Moldova 47°49′N 28°45′E
Soldau see Wkra
108 L8 Sölden Tirol, W Austria 46°58′N 11°01′E
27 P3 Soldier Creek ≈ Kansas, C USA
39 R12 Soldotna Alaska, USA 60°29′N 151°03′W
110 I10 Solec Kujawski Kujawsko-pomorskie, C Poland 53°04′N 18°09′E
61 B16 Soledad Santa Fe, C Argentina 30°38′S 60°52′W
55 E4 Soledad Atlántico, N Colombia 10°54′N 74°48′W
35 O11 Soledad California, W USA 36°25′N 121°19′W
55 O7 Soledad Anzoátegui, NE Venezuela 08°10′N 63°36′W
61 H15 Soledade Rio Grande do Sul, S Brazil 28°50′S 52°30′W
Isla Soledad see East Falkland
103 Y15 Solenzara Corse, France, C Mediterranean Sea 41°55′N 09°24′E
Soleure see Solothurn
94 C12 Solheim Hordaland, S Norway 60°54′N 05°30′E
125 N14 Soligalich Kostromskaya Oblast', NW Russian Federation 59°05′N 42°15′E
Soligorsk see Salihorsk
Soligorskoye Vodokhranilishche see Salihorskaye Vodaskhovishcha
97 L20 Solihull C England, United Kingdom 52°25′N 01°45′W
125 U13 Solikamsk Permskiy Kray, NW Russian Federation 59°37′N 56°46′E
127 V8 Sol'-Iletsk Orenburgskaya Oblast', W Russian Federation 51°09′N 55°05′E
57 G17 Solimana, Nevado ▲ S Peru 15°24′S 72°49′W
58 E13 Solimões, Rio ≈ C Brazil
113 E14 Solin It. Salona; anc. Salonae. Split-Dalmacija, S Croatia 43°33′N 16°29′E
101 E15 Solingen Nordrhein-Westfalen, W Germany 51°10′N 07°05′E
Solka see Solca
93 H16 Sollefteå Västernorrland, C Sweden 63°09′N 17°15′E
95 O15 Sollentuna Stockholm, C Sweden 59°26′N 17°56′E
105 X9 Sóller Mallorca, Spain, W Mediterranean Sea 39°46′N 02°42′E
94 L13 Sollerön Dalarna, C Sweden 60°55′N 14°34′E
101 I14 Solling hill range C Germany
95 O16 Solna Stockholm, C Sweden 59°22′N 17°58′E
126 K3 Solnechnogorsk Moskovskaya Oblast', W Russian Federation 56°07′N 37°04′E
123 R10 Solnechnyy Khabarovskiy Kray, SE Russian Federation 50°41′N 136°42′E
123 S13 Solnechnyy Respublika Sakha (Yakutiya), NE Russian Federation 60°13′N 137°42′E
Solo see Surakarta
107 L17 Solofra Campania, S Italy 40°49′N 14°48′E
168 J11 Solok Sumatera, W Indonesia 0°45′S 100°42′E
42 C6 Sololá Sololá, W Guatemala 14°46′N 91°09′W
42 A2 Sololá off. Departamento de Sololá. ◐ department SW Guatemala
Sololá, Departamento de see Sololá
81 J16 Sololo Marsabit, N Kenya 03°31′N 38°39′E
42 C4 Soloma Huehuetenango, W Guatemala 15°38′N 91°25′W
38 M9 Solomon Alaska, USA 64°33′N 164°26′W
27 N4 Solomon Kansas, C USA 38°55′N 97°22′W
187 N9 Solomon Islands prev. British Solomon Islands Protectorate. ◆ commonwealth republic W Solomon Islands
N Melanesia W Pacific Ocean
186 L7 Solomon Islands island group Papua New Guinea/Solomon Islands
26 M3 Solomon River ≈ Kansas, C USA
186 H8 Solomon Sea sea W Pacific Ocean
31 U11 Solon Ohio, N USA 41°23′N 81°26′W
117 T8 Solone Dnipropetrovs'ka Oblast', E Ukraine 48°12′N 34°49′E
171 P16 Solor, Kepulauan island group S Indonesia
126 M4 Solotcha Ryazanskaya Oblast', W Russian Federation 54°43′N 39°50′E
108 D7 Solothurn Fr. Soleure. Solothurn, NW Switzerland 47°13′N 07°32′E
108 D7 Solothurn Fr. Soleure. ◐ canton NW Switzerland
124 J7 Solovetskiye Ostrova island group NW Russian Federation
105 V5 Solsona Cataluña, NE Spain 42°00′N 01°31′E
113 E14 Šolta It. Solta. island S Croatia
Solṭānābād see Kāshmar
142 L4 Solṭānīyeh Zanjān, NW Iran 36°24′N 48°50′E
100 I11 Soltau Niedersachsen, NW Germany 52°59′N 09°50′E
124 G14 Sol'tsy Novgorodskaya Oblast', W Russian Federation 58°09′N 30°23′E
Soltüstik Qazaqstan Oblysy see Severnyy Kazakhstan
Solun see Thessaloníki
113 O19 Solunska Glava ▲ C FYR Macedonia 41°43′N 21°24′E
95 L22 Sölvesborg Blekinge, S Sweden 56°04′N 14°35′E
97 J15 Solway Firth inlet England/Scotland, United Kingdom
82 I13 Solwezi North Western, NW Zambia 12°11′S 26°23′E
165 Q11 Sōma Fukushima, Honshū, C Japan 37°49′N 140°52′E
136 C13 Soma Manisa, W Turkey 39°10′N 27°36′E
Somali see Sumalē
81 O15 Somalia off. Somali Democratic Republic, Som. Jamuuriyada Demuqraadiga Soomaaliyeed, Soomaaliya; prev. Italian Somaliland, Somaliland Protectorate. ◆ republic E Africa
173 N6 Somali Basin undersea feature W Indian Ocean 0°00′N 52°00′E
Somali Democratic Republic see Somalia
80 N12 Somaliland ◐ disputed territory N Somalia
Somaliland Protectorate see Somalia
67 Y8 Somali Plain undersea feature W Indian Ocean 01°00′N 51°30′E
112 J8 Sombor Hung. Zombor. Vojvodina, NW Serbia 45°46′N 19°07′E
99 H20 Sombreffe Namur, S Belgium 50°32′N 04°39′E
40 L10 Sombrerete Zacatecas, C Mexico 23°38′N 103°40′W
45 V8 Sombrero island N Anguilla
151 Q21 Sombrero Channel channel Nicobar Islands, India
116 H9 Şomcuta Mare Hung. Nagysomkút; prev. Somcuţa Mare. Maramureş, N Romania 47°29′N 23°30′E
Şomcuţa Mare see Şomcuta Mare
167 R9 Somdet Kalasin, E Thailand 16°41′N 103°44′E
99 L15 Someren Noord-Brabant, SE Netherlands 51°23′N 05°42′E
93 L19 Somero Varsinais-Suomi, SW Finland 60°37′N 23°30′E
33 P7 Somers Montana, NW USA 48°04′N 114°16′W
64 A12 Somerset var. Somerset Village. W Bermuda 32°18′N 64°53′W
37 Q5 Somerset Colorado, C USA 38°55′N 107°27′W
20 M7 Somerset Kentucky, S USA 37°05′N 84°36′W
19 O12 Somerset Massachusetts, NE USA 41°46′N 71°07′W
97 K23 Somerset cultural region SW England, United Kingdom
Somerset East see Somerset-Oos
64 A12 Somerset Island island W Bermuda
197 N9 Somerset Island island Queen Elizabeth Islands, Nunavut, NW Canada
Somerset Nile see Victoria Nile
83 I25 Somerset-Oos var. Somerset East. Eastern Cape, S South Africa 32°44′S 25°35′E
Somerset Village see Somerset
83 E26 Somerset-Wes var. Somerset West. Western Cape, SW South Africa 34°05′S 18°51′E
Somerset West see Somerset-Wes
Somers Islands see Bermuda
18 J17 Somers Point New Jersey, NE USA 39°18′N 74°34′W
19 P9 Somersworth New Hampshire, NE USA 43°15′N 70°52′W
36 H15 Somerton Arizona, SW USA 32°36′N 114°42′W
18 J14 Somerville New Jersey, NE USA 40°34′N 74°36′W
20 F10 Somerville Tennessee, S USA 35°14′N 89°24′W
25 U10 Somerville Texas, SW USA 30°21′N 96°31′W
25 T10 Somerville Lake ⊠ Texas, SW USA
Someş/Somesch/Someşul see Szamos
103 N2 Somme ◐ department N France
103 N2 Somme ≈ N France
95 L18 Sommen Jönköping, S Sweden 58°07′N 14°58′E
95 M18 Sommen ◎ S Sweden
101 K16 Sömmerda Thüringen, C Germany 51°10′N 11°07′E
Sommerein see Šamorín
Sommerfeld see Lubsko
55 Y11 Sommet Tabulaire var. Mont Itoupé. ▲ S French Guiana
111 H25 Somogy off. Somogy Megye. ◐ county SW Hungary
Somogy Megye see Somogy
Somorja see Šamorín
105 N7 Somosierra, Puerto de pass N Spain
187 Y14 Somosomo Taveuni, N Fiji 16°46′S 179°57′W
42 I9 Somotillo Chinandega, NW Nicaragua 13°01′N 86°53′W
42 I8 Somoto Madriz, NW Nicaragua 13°29′N 86°36′W
110 H10 Sompolno Wielkopolskie, C Poland 52°24′N 18°30′E
102 J17 Somport, Col du var. Puerto de Somport, Sp. Somport; anc. Summus Portus. pass France/Spain see also Somport
99 K15 Son Noord-Brabant, S Netherlands 51°32′N 05°34′E
95 H15 Son Akershus, S Norway 59°32′N 10°42′E
154 L9 Son var. Sone. ≈ C India
43 R16 Soná Veraguas, W Panama 08°00′N 81°20′W
Song see Zêkog
Sonapur see Subarnapur
95 G24 Sønderborg Ger. Sonderburg. Syddanmark, SW Denmark 54°55′N 09°48′E
Sonderburg see Sønderborg
101 K15 Sondershausen Thüringen, C Germany 51°22′N 10°52′E
Sondre Strømfjord see Kangerlussuaq
106 E6 Sondrio Lombardia, N Italy 46°11′N 09°52′E
Sone see Son
Sonepur see Subarnapur
57 K22 Sonequera ▲ S Bolivia 22°06′S 67°10′W
167 V12 Sông Câu Phu Yên, C Vietnam 13°26′N 109°12′E
167 R15 Sông Đôc Minh Hai, S Vietnam 09°03′N 104°51′E
81 H25 Songea Ruvuma, S Tanzania 10°42′S 35°39′E
163 X10 Songhua Hu ◎ NE China
163 Y7 Songhua Jiang var. Sungari. ≈ NE China
161 S8 Songjiang Shanghai Shi, E China 31°01′N 121°14′E
Sŏngjin see Kimch'aek
167 O16 Songkhla var. Songkla, Mal. Singora. Songkhla, SW Thailand 07°12′N 100°35′E
Songkla see Songkhla
163 T13 Song Ling ⛰ NE China
129 U12 Sông Ma Laos Ma, Nam. ≈ Laos/Vietnam
163 W14 Songnim SW North Korea 38°43′N 125°40′E
82 B10 Songo Uíge, NW Angola 07°30′S 14°56′E
83 M15 Songo Tete, NW Mozambique 15°36′S 32°45′E
79 F21 Songololo Bas-Congo, SW Dem. Rep. Congo 05°40′S 14°05′E
160 H7 Songpan var. Jin'an, Tib. Sungpu. Sichuan, C China 32°49′N 103°39′E
Songsan see Seongsan
161 R11 Songxi Fujian, SE China 27°33′N 118°46′E
160 M6 Songxian var. Song Xian. Henan, C China 34°11′N 112°04′E
Song Xian see Songxian
161 R10 Songyang var. Xiping; prev. Songyin. Zhejiang, SE China 28°29′N 119°27′E
Songyin see Songyang
163 V9 Songyuan var. Fu-yü, Petuna; prev. Fuyu. Jilin, NE China 45°10′N 124°52′E
Sonid Youqi see SaihanTal
Sonid Zuoqi see Mandalt
152 I10 Sonīpat Haryāna, N India 29°00′N 77°01′E
93 M15 Sonkajärvi Pohjois-Savo, C Finland 63°40′N 27°30′E
167 R6 Sơn La Sơn La, N Vietnam 21°20′N 103°55′E
149 O16 Sonmiāni Balochistān, S Pakistan 25°24′N 66°37′E
149 O16 Sonmiāni Bay bay S Pakistan
101 K18 Sonneberg Thüringen, C Germany 50°22′N 11°10′E
101 N24 Sonntagshorn ▲ Austria/Germany 47°40′N 12°42′E
Sonoita see Sonoyta
40 E3 Sonoita, Rio var. Río Sonoyta. ≈ Mexico/USA
35 N7 Sonoma California, W USA 38°16′N 122°28′W
35 T3 Sonoma Peak ▲ Nevada, W USA 40°50′N 117°34′W
35 P8 Sonora California, W USA 37°58′N 120°22′W
25 O10 Sonora Texas, SW USA 30°34′N 100°39′W
40 F5 Sonora ◐ state NW Mexico
35 X17 Sonoran Desert var. Desierto de Altar. desert Mexico/USA see also Altar, Desierto de
40 G5 Sonora, Río ≈ NW Mexico
40 E2 Sonoyta var. Sonoita. Sonora, NW Mexico 31°49′N 112°50′W
Sonoyta, Río see Sonoita, Rio
142 K6 Sonqor var. Sunqur. Kermānshāhān, W Iran 34°45′N 47°39′E
105 N9 Sonseca var. Sonseca con Casalgordo. Castilla-La Mancha, C Spain 39°40′N 03°59′W
Sonseca con Casalgordo see Sonseca
54 E9 Sonsón Antioquia, W Colombia 05°45′N 75°18′W
42 E7 Sonsonate Sonsonate, W El Salvador 13°44′N 89°43′W
42 A9 Sonsonate ◐ department SW El Salvador
188 A10 Sonsorol Islands island group S Palau
112 J9 Sonta Hung. Szond; prev. Szonta. Vojvodina, NW Serbia 45°34′N 19°06′E
167 S6 Sơn Tây var. Sontay. Ha Tây, N Vietnam 21°06′N 105°32′E
Sontay see Sơn Tây
101 J25 Sonthofen Bayern, S Germany 47°31′N 10°16′E
Soochow see Suzhou
80 O13 Sool off. Gobolka Sool. ◐ region N Somalia
Soomaaliya/Soomaaliyeed, Jamuuriyada Demuqraadiga see Somalia
Soome Laht see Finland, Gulf of
23 V5 Soperton Georgia, SE USA 32°22′N 82°35′W
167 S6 Sop Hao Houaphan, N Laos 20°33′N 104°25′E
Sophia see Sofia
171 S10 Sopi Pulau Morotai, E Indonesia 02°36′N 128°32′E
Sopianae see Pécs
171 U13 Sopinusa Papua Barat, E Indonesia 03°31′S 132°55′E
81 B14 Sopo ≈ W South Sudan
Sopockinie/Sopotskin/Sopotskino see Sapotskin
114 I9 Sopot Plovdiv, C Bulgaria 42°39′N 24°43′E
110 I7 Sopot Ger. Zoppot. Pomorskie, N Poland 54°26′N 18°33′E
167 O8 Sop Prap var. Ban Sop Prap. Lampang, NW Thailand 17°55′N 99°20′E
111 G22 Sopron Ger. Ödenburg. Győr-Moson-Sopron, NW Hungary 47°40′N 16°35′E
147 U11 Sopu-Korgon var. Sofi-Kurgan. Oshskaya Oblast', SW Kyrgyzstan 40°03′N 73°30′E
152 H5 Sopur Jammu and Kashmir, NW India 34°19′N 74°30′E
107 J15 Sora Lazio, C Italy 41°43′N 13°37′E
154 N13 Sorada Odisha, E India 19°46′N 84°29′E
93 H17 Söråker Västernorrland, C Sweden 62°32′N 17°32′E
57 J17 Sorata La Paz, W Bolivia 15°47′S 68°38′W
Sorau/Sorau in der Niederlausitz see Żary
105 Q14 Sorbas Andalucía, S Spain 37°06′N 02°06′W
94 N11 Sördellen ◎ C Sweden
Sord/Sórd Choluim Chille see Swords
15 O11 Sorel Québec, SE Canada 46°03′N 73°06′W
183 P17 Sorell Tasmania, SE Australia 42°49′S 147°34′E
183 O17 Sorell, Lake ◎ Tasmania, SE Australia
106 E8 Soresina Lombardia, N Italy 45°17′N 09°51′E
95 D14 Sørfjorden fjord S Norway
94 N11 Sörforsa Gävleborg, C Sweden 61°45′N 17°00′E
103 R14 Sorgues Vaucluse, SE France 44°N 04°52′E
136 K13 Sorgun Yozgat, C Turkey 39°49′N 35°10′E
105 P5 Soria Castilla y León, N Spain 41°47′N 02°26′W
105 P6 Soria ◐ province Castilla y León, N Spain
61 D19 Soriano Soriano, SW Uruguay 33°25′S 58°21′W
61 D19 Soriano ◐ department SW Uruguay
92 O4 Sørkapp headland SW Svalbard 76°34′N 16°33′E
143 T5 Sorkh, Kūh-e ⛰ NE Iran
95 I23 Sorø Sjælland, E Denmark 55°26′N 11°34′E
Soro see Ghazal, Bahr el
116 M8 Soroca Rus. Soroki. N Moldova 48°10′N 28°18′E
60 L10 Sorocaba São Paulo, S Brazil 23°29′S 47°27′W
Sorochino see Sarochyna
127 T7 Sorochinsk Orenburgskaya Oblast', W Russian Federation 52°26′N 53°10′E
Soroki see Soroca
188 H15 Sorol atoll Caroline Islands, W Micronesia
171 T12 Sorong Papua Barat, E Indonesia 0°49′S 131°16′E
81 G17 Soroti C Uganda 01°42′N 33°37′E
Sørøy see Sørøya
92 J8 Sørøya var. Sørøy, Lapp. Sállan. island N Norway
104 G11 Sorraia, Rio ≈ C Portugal
92 I10 Sørreisa Troms, N Norway 69°08′N 18°09′E
107 K18 Sorrento anc. Surrentum. Campania, S Italy 40°37′N 14°23′E
104 H10 Sor, Ribeira de stream C Portugal
195 T3 Sør Rondane Eng. Sor Rondane Mountains. ⛰ Antarctica
Sor Rondane Mountains see Sør Rondane
93 H14 Sorsele Västerbotten, N Sweden 65°31′N 17°34′E
107 B17 Sorso Sardegna, Italy, C Mediterranean Sea 40°46′N 08°33′E
171 P4 Sorsogon Luzon, N Philippines 12°57′N 124°04′E
105 U4 Sort Cataluña, NE Spain 42°25′N 01°07′E
124 H11 Sortavala prev. Serdobol'. Respublika Kareliya, NW Russian Federation 61°45′N 30°37′E
107 L25 Sortino Sicilia, Italy, C Mediterranean Sea 37°10′N 15°02′E
92 G10 Sortland Nordland, C Norway 68°44′N 15°25′E
94 G9 Sør-Trøndelag ◐ county S Norway
95 I15 Sørumsand Akershus, S Norway 59°58′N 11°13′E
118 D6 Sõrve Säär headland SW Estonia 57°54′N 22°02′E
95 K22 Sösdala Skåne, S Sweden 56°00′N 13°36′E
105 R4 Sos del Rey Católico Aragón, NE Spain 42°30′N 01°13′W
93 F15 Sösjöfjällen ▲ C Sweden 63°51′N 13°15′E
126 K7 Sosna ≈ W Russian Federation
62 H12 Sosneado, Cerro ▲ W Argentina 34°44′S 69°52′W
125 S9 Sosnogorsk Respublika Komi, NW Russian Federation 63°33′N 53°55′E
124 J8 Sosnovets Respublika Kareliya, NW Russian Federation 64°25′N 34°23′E
Sosnovets see Sosnowiec
127 Q3 Sosnovka Chuvashskaya Respublika, W Russian Federation 56°18′N 47°14′E
125 S16 Sosnovka Kirovskaya Oblast', NW Russian Federation 56°15′N 51°20′E
124 M6 Sosnovka Murmanskaya Oblast', NW Russian Federation 66°28′N 40°31′E
126 M6 Sosnovka Tambovskaya Oblast', W Russian Federation 53°14′N 41°19′E
124 H12 Sosnovo Fin. Rautu. Leningradskaya Oblast', NW Russian Federation 60°30′N 30°13′E
127 V3 Sosnovyy Bor Respublika Bashkortostan, W Russian Federation 55°51′N 57°09′E
Sosnovyy Bor see Sasnovy Bor
111 J16 Sosnowiec Ger. Sosnowitz, Rus. Sosnovets. Śląskie, S Poland 50°16′N 19°07′E
Sosnowitz see Sosnowiec
117 R2 Sosnytsya Chernihivs'ka Oblast', N Ukraine 51°31′N 32°30′E
109 V10 Šoštanj N Slovenia 46°23′N 15°03′E
122 G10 Sos'va Sverdlovskaya Oblast', C Russian Federation 59°13′N 61°58′E
54 D12 Sotará, Volcán 🌋 S Colombia 02°04′N 76°40′W
76 D10 Sotavento, Ilhas de var. Leeward Islands. island group S Cape Verde

SOTKAMO – STAFFORD

93 N15 **Sotkamo** Kainuu, C Finland 64°06′N 28°30′E
109 W11 **Sotla** ≈ E Slovenia
41 P10 **Soto la Marina** Tamaulipas, C Mexico 23°44′N 98°10′W
41 P10 **Soto la Marina, Río** ≈ C Mexico
95 B14 **Sotra** *island* S Norway
41 X12 **Sotuta** Yucatán, SE Mexico 20°34′N 89°00′W
79 F17 **Souanké** Sangha, NW Congo 02°03′N 14°02′E
76 M17 **Soubré** S Ivory Coast 05°50′N 06°35′W
115 H24 **Soúda** *var.* Soúdha, *Eng.* Suda. Kríti, Greece, E Mediterranean Sea 35°29′N 24°04′E
Soúdha *see* Soúda
Soueida *see* As Suwaydā'
114 L12 **Souflí** *prev.* Souflíon. Anatolikí Makedonía kai Thráki, NE Greece 41°12′N 26°18′E
Souflíon *see* Souflí
45 S11 **Soufrière** W Saint Lucia 13°51′N 61°03′W
45 X6 **Soufrière** ⛰ Basse Terre, S Guadeloupe 16°03′N 61°39′W
102 M13 **Souillac** Lot, S France 44°53′N 01°29′E
173 Y17 **Souillac** S Mauritius 20°31′S 57°31′E
74 M5 **Souk Ahras** NE Algeria 36°14′N 08°00′E
Souk el Arba du Rharb/ Souk-el-Arba-du-Rharb/ Souk-el-Arba-el-Rhab *see* Souk-el-Arba-Rharb
74 E6 **Souk-el-Arba-Rharb** *var.* Souk el Arba du Rharb, Souk-el-Arba-du-Rharb, Souk-el-Arba-el-Rhab. NW Morocco 34°38′N 06°00′W
Soukhné *see* As Sukhnah
Sŏul *see* Seoul
102 J11 **Soulac-sur-Mer** Gironde, SW France 45°31′N 01°06′W
99 L19 **Soumagne** Liège, E Belgium 50°36′N 05°48′E
18 M14 **Sound Beach** Long Island, New York, NE USA 40°56′N 72°58′W
95 J22 **Sound, The** *Dan.* Øresund, *Swe.* Öresund. *strait* Denmark/Sweden
115 H20 **Soúnio, Akrotírio** *headland* C Greece 37°39′N 24°01′E
138 F8 **Soûr** *var.* Şūr; *anc.* Tyre. SW Lebanon 33°18′N 35°30′E
Sources, Mont-aux- *see* Phofung
104 G8 **Soure** Coimbra, N Portugal 40°04′N 08°38′W
11 W17 **Souris** Manitoba, S Canada 49°38′N 100°17′W
13 Q14 **Souris** Prince Edward Island, SE Canada 46°22′N 62°16′W
28 L2 **Souris River** *var.* Mouse River. ≈ Canada/USA
25 X10 **Sour Lake** Texas, SW USA 30°08′N 94°24′W
115 F17 **Soúrpi** Thessalía, C Greece 39°07′N 22°55′E
104 H11 **Sousel** Portalegre, C Portugal 38°57′N 07°40′W
75 N6 **Sousse** *var.* Sūsah. NE Tunisia 35°46′N 10°38′E
14 H11 **South** ≈ Ontario, S Canada
South *see* Sud
83 G23 **South Africa** *off.* Republic of South Africa, *Afr.* Suid-Afrika. ◆ *republic* S Africa
South Africa, Republic of *see* South Africa
46-47 **South America** *continent*
2 J17 **South American Plate** *tectonic feature*
97 M23 **Southampton** *hist.* Hamwih, *Lat.* Clausentum. S England, United Kingdom 50°54′N 01°23′W
19 N14 **Southampton** Long Island, New York, NE USA 40°52′N 72°22′W
9 P8 **Southampton Island** *island* Nunavut, NE Canada
151 P20 **South Andaman** *island* Andaman Islands, India, NE Indian Ocean
13 Q6 **South Aulatsivik Island** *island* Newfoundland and Labrador, E Canada
182 E4 **South Australia** ◈ *state* S Australia
South Australian Abyssal Plain *see* South Australian Plain
192 G11 **South Australian Basin** *undersea feature* SW Indian Ocean 38°00′S 126°00′E
173 X12 **South Australian Plain** *var.* South Australian Abyssal Plain. *undersea feature* SE Indian Ocean
37 R13 **South Baldy** ▲ New Mexico, SW USA 33°59′N 107°11′W
23 Y14 **South Bay** Florida, SE USA 26°39′N 80°43′W
14 E12 **South Baymouth** Manitoulin Island, Ontario, S Canada 45°33′N 82°01′W
30 L10 **South Beloit** Illinois, N USA 42°29′N 89°02′W
31 O11 **South Bend** Indiana, N USA 41°40′N 86°15′W
25 R6 **South Bend** Texas, SW USA 32°58′N 98°39′W
32 F9 **South Bend** Washington, NW USA 46°38′N 123°48′W
South Beveland *see* Zuid-Beveland
South Borneo *see* Kalimantan Selatan
21 U7 **South Boston** Virginia, NE USA 36°42′N 78°58′W
182 F2 **South Branch Neales** *seasonal river* South Australia
21 U3 **South Branch Potomac River** ≈ West Virginia, NE USA
185 H19 **Southbridge** Canterbury, South Island, New Zealand 43°49′S 172°17′E
19 N12 **Southbridge** Massachusetts, NE USA 42°03′N 72°00′W
183 P17 **South Bruny Island** *island* Tasmania, SE Australia
18 L7 **South Burlington** Vermont, NE USA 44°27′N 73°08′W
44 M6 **South Caicos** *island* S Turks and Caicos Islands
South Cape *see* Ka Lae
23 V3 **South Carolina** *off.* State of South Carolina, *also known as* The Palmetto State. ◈ *state* SE USA
South Carpathians *see* Carpații Meridionalii
South Celebes *see* Sulawesi Selatan
21 Q5 **South Charleston** West Virginia, NE USA 38°22′N 81°42′W
192 D7 **South China Basin** *undersea feature* SE South China Sea 15°00′N 115°00′E
169 R8 **South China Sea** *Chin.* Nan Hai, *Ind.* Laut Cina Selatan, *Vtn.* Biên Đông. *sea* SE Asia
33 Z10 **South Dakota** *off.* State of South Dakota, *also known as* The Coyote State, Sunshine State. ◈ *state* N USA
23 X10 **South Daytona** Florida, SE USA 29°09′N 81°01′W
37 R10 **South Domingo Pueblo** New Mexico, SW USA 35°28′N 106°24′W
97 N23 **South Downs** *hill range* SE England, United Kingdom
83 I21 **South East** ◈ *district* SE Botswana
65 H15 **South East Bay** *bay* Ascension Island, C Atlantic Ocean
183 O17 **South East Cape** *headland* Tasmania, SE Australia 43°36′S 146°52′E
38 K10 **Southeast Cape** *headland* Saint Lawrence Island, Alaska, USA 62°56′N 169°39′W
South-East Celebes *see* Sulawesi Tenggara
192 G12 **Southeast Indian Ridge** *undersea feature* Indian Ocean/Pacific Ocean 50°00′S 110°00′E
Southeast Island *see* Tagula Island
193 P13 **Southeast Pacific Basin** *var.* Belling Hausen Mulde. *undersea feature* SE Pacific Ocean 60°00′S 115°00′W
65 H15 **South East Point** *headland* SE Ascension Island
183 O14 **South East Point** *headland* Victoria, S Australia 39°10′S 146°21′E
191 Z3 **South East Point** *headland* Kiritimati, NE Kiribati 01°42′N 157°10′W
44 L5 **Southeast Point** *headland* Mayaguana, SE The Bahamas 22°15′N 72°44′W
South-East Sulawesi *see* Sulawesi Tenggara
11 U12 **Southend** Saskatchewan, C Canada 56°20′N 103°14′W
97 P22 **Southend-on-Sea** E England, United Kingdom 51°33′N 00°43′E
83 H20 **Southern** *var.* Bangwaketse, Ngwaketze. ◈ *district* SE Botswana
138 E13 **Southern** ◈ *district* S Israel
83 N15 **Southern** ◈ *region* S Malawi
155 J26 **Southern** ◈ *province* S Sri Lanka
83 I15 **Southern** ◈ *province* S Zambia
185 E19 **Southern Alps** ▲▲ South Island, New Zealand
190 K15 **Southern Cook Islands** *island group* S Cook Islands
180 K12 **Southern Cross** Western Australia 31°17′S 119°15′E
80 A12 **Southern Darfur** ◈ *state* W Sudan
186 B7 **Southern Highlands** ◈ *province* W Papua New Guinea
11 V11 **Southern Indian Lake** ◎ Manitoba, C Canada
80 E11 **Southern Kordofan** ◈ *state* C Sudan
187 Z15 **Southern Lau Group** *island group* Lau Group, SE Fiji
81 I15 **Southern Nationalities** ◈ *region* S Ethiopia
173 S13 **Southern Ocean** *ocean*
21 T10 **Southern Pines** North Carolina, SE USA 35°10′N 79°23′W
96 I13 **Southern Uplands** ▲▲ S Scotland, United Kingdom
Southern Urals *see* Yuzhnyy Ural
183 P16 **South Esk River** ≈ Tasmania, SE Australia
11 U16 **Southey** Saskatchewan, S Canada 50°53′N 104°27′W
27 V2 **South Fabius River** ≈ Missouri, C USA
31 S10 **Southfield** Michigan, N USA 42°28′N 83°12′W
192 K10 **South Fiji Basin** *undersea feature* S Pacific Ocean 26°00′S 175°00′E
97 Q22 **South Foreland** *headland* SE England, United Kingdom 51°08′N 01°22′E
35 P7 **South Fork American River** ≈ California, W USA
28 K7 **South Fork Grand River** ≈ South Dakota, N USA
35 T12 **South Fork Kern River** ≈ California, W USA
39 Q7 **South Fork Koyukuk River** ≈ Alaska, USA
39 Q11 **South Fork Kuskokwim River** ≈ Alaska, USA
26 H2 **South Fork Republican River** ≈ C USA
26 L3 **South Fork Solomon River** ≈ Kansas, C USA
31 P5 **South Fox Island** *island* Michigan, N USA
20 G8 **South Fulton** Tennessee, S USA 36°28′N 88°53′W
195 U10 **South Geomagnetic Pole** *pole* Antarctica
65 J20 **South Georgia** *island* South Georgia and the South Sandwich Islands, SW Atlantic Ocean
65 K21 **South Georgia and the South Sandwich Islands** ◇ *UK Dependent Territory* SW Atlantic Ocean
47 Y14 **South Georgia Ridge** *var.* North Scotia Ridge. *undersea feature* SW Atlantic Ocean 54°00′S 40°00′W
181 Q1 **South Goulburn Island** *island* Northern Territory, N Australia
153 U16 **South Hatia Island** *island* SE Bangladesh
31 O10 **South Haven** Michigan, N USA 42°24′N 86°16′W
21 V7 **South Hill** Virginia, NE USA 36°43′N 78°07′W
South Holland *see* Zuid-Holland
21 P8 **South Holston Lake** ⊠ Tennessee/Virginia, S USA
175 N1 **South Honshu Ridge** *undersea feature* W Pacific Ocean
26 M6 **South Hutchinson** Kansas, C USA 38°01′N 97°56′W
151 K21 **South Huvadhu Atoll** *atoll* S Maldives
173 U14 **South Indian Basin** *undersea feature* Indian Ocean/Pacific Ocean 60°00′S 120°00′E
11 W11 **South Indian Lake** Manitoba, C Canada 56°48′N 98°56′W
81 I17 **South Island** *island* NW Kenya
185 C20 **South Island** *island* S New Zealand
65 B23 **South Jason** *island* Jason Islands, NW Falkland Islands
South Kalimantan *see* Kalimantan Selatan
South Karelia *see* Etelä-Karjala
South Kazakhstan *see* Yuzhnyy Kazakhstan
163 X15 **South Korea** *off.* Republic of Korea, *Kor.* Taehan Min'guk. ◆ *republic* E Asia
35 Q6 **South Lake Tahoe** California, W USA 38°56′N 119°57′W
25 N6 **Southland** Texas, SW USA 33°16′N 101°31′W
185 B23 **Southland** *off.* Southland Region. ◈ *region* South Island, New Zealand
Southland Region *see* Southland
29 N15 **South Loup River** ≈ Nebraska, C USA
151 K19 **South Maalhosmadulu Atoll** *atoll* N Maldives
14 E15 **South Maitland** ≈ Ontario, S Canada
192 E8 **South Makassar Basin** *undersea feature* E Java Sea
31 O6 **South Manitou Island** *island* Michigan, N USA
151 K18 **South Miladhunmadulu Atoll** *var.* Noonu. *atoll* N Maldives
21 X8 **South Mills** North Carolina, SE USA 36°28′N 76°18′W
8 H9 **South Nahanni** ≈ Northwest Territories, NW Canada
39 P13 **South Naknek** Alaska, USA 58°39′N 157°01′W
14 M13 **South Nation** ≈ Ontario, SE Canada
44 F9 **South Negril Point** *headland* W Jamaica 18°14′N 78°21′W
151 K20 **South Nilandhe Atoll** *var.* Dhaalu Atoll. *atoll* C Maldives
36 L2 **South Ogden** Utah, W USA 41°09′N 111°58′W
18 M14 **Southold** Long Island, New York, NE USA 41°03′N 72°24′W
194 H1 **South Orkney Islands** *island group* Antarctica
137 S9 **South Ossetia** *former autonomous region* SW Georgia
South Ostrobothnia *see* Etelä-Pohjanmaa
South Pacific Basin *see* Southwest Pacific Basin
19 P7 **South Paris** Maine, NE USA 44°14′N 70°33′W
189 U13 **South Pass** *passage* Chuuk Islands, C Micronesia
33 U15 **South Pass** *pass* Wyoming, C USA
20 K10 **South Pittsburg** Tennessee, S USA 35°00′N 85°42′W
28 K15 **South Platte River** ≈ Colorado/Nebraska, C USA
31 T16 **South Point** Ohio, N USA 38°25′N 82°35′W
65 G15 **South Point** *headland* S Ascension Island
31 R6 **South Point** *headland* Michigan, N USA 44°51′N 83°17′W
South Point *see* Ka Lae
195 Q9 **South Pole** *pole* Antarctica
183 P17 **Southport** Tasmania, SE Australia 43°26′S 146°57′E
97 K17 **Southport** NW England, United Kingdom 53°39′N 03°01′W
21 V12 **Southport** North Carolina, SE USA 33°55′N 78°00′W
19 P8 **South Portland** Maine, NE USA 43°38′N 70°14′W
14 H12 **South River** Ontario, S Canada 45°50′N 79°23′W
21 U11 **South River** ≈ North Carolina, SE USA
96 K5 **South Ronaldsay** *island* NE Scotland, United Kingdom
36 L2 **South Salt Lake** Utah, W USA 40°42′N 111°52′W
65 L21 **South Sandwich Islands** *island group* SW Atlantic Ocean
65 K21 **South Sandwich Trench** *undersea feature* SW Atlantic Ocean 56°30′S 25°00′W
11 S16 **South Saskatchewan** ≈ Alberta/Saskatchewan, S Canada
65 I21 **South Scotia Ridge** *undersea feature* S Scotia Sea
11 V10 **South Seal** ≈ Manitoba, C Canada
194 G4 **South Shetland Islands** *island group* Antarctica
65 H22 **South Shetland Trough** *undersea feature* Atlantic Ocean/Pacific Ocean 61°00′S 59°30′W
97 M14 **South Shields** NE England, United Kingdom 55°N 01°25′W
29 R13 **South Sioux City** Nebraska, C USA 42°28′N 96°24′W
192 J9 **South Solomon Trench** *undersea feature* W Pacific Ocean
183 V3 **South Stradbroke Island** *island* Queensland, E Australia
81 E15 **South Sudan** *off.* Republic of South Sudan. ◆ E Africa
South Sulawesi *see* Sulawesi Selatan
South Sumatra *see* Sumatera Selatan
184 K11 **South Taranaki Bight** *bight* SE Tasman Sea
South Tasmania Plateau *see* Tasman Plateau
36 M15 **South Tucson** Arizona, SW USA 32°11′N 110°56′W
12 H9 **South Twin Island** *island* Nunavut, C Canada
South Tyrol *see* Trentino-Alto Adige
96 E9 **South Uist** *island* NW Scotland, United Kingdom
South-West *see* Sud-Ouest
South-West Africa/South West Africa *see* Namibia
65 F15 **South West Bay** *bay* Ascension Island, C Atlantic Ocean
183 N18 **South West Cape** *headland* Tasmania, SE Australia 43°34′S 146°01′E
185 B26 **South West Cape** *headland* Stewart Island, New Zealand 47°15′S 167°28′E
38 J10 **Southwest Cape** *headland* Saint Lawrence Island, Alaska, USA 63°19′N 171°27′W
Southwest Indian Ocean Ridge *see* Southwest Indian Ridge
173 N11 **Southwest Indian Ridge** *var.* Southwest Indian Ocean Ridge. *undersea feature* SW Indian Ocean 43°00′S 40°00′E
192 L10 **Southwest Pacific Basin** *var.* South Pacific Basin. *undersea feature* SE Pacific Ocean 40°00′S 150°00′W
191 X3 **South West Point** *headland* Kiritimati, NE Kiribati 01°53′N 157°34′E
65 G25 **South West Point** *headland* SW Saint Helena 16°00′S 05°48′W
44 H2 **Southwest Point** *headland* Great Abaco, N The Bahamas 25°50′N 77°12′W
25 P5 **South Wichita River** ≈ Texas, SW USA
97 Q20 **Southwold** E England, United Kingdom 52°15′N 01°36′E
19 Q12 **South Yarmouth** Massachusetts, NE USA 41°38′N 70°09′W
116 J10 **Sovata** *Hung.* Szováta. Mureș, C Romania 46°36′N 25°04′E
107 N22 **Soverato** Calabria, SW Italy 38°40′N 16°31′E
121 O4 **Sovereign Base Area** *uk military installation* S Cyprus
Sovetabad *see* Ghafurov
126 C2 **Sovetsk** *Ger.* Tilsit. Kaliningradskaya Oblast', W Russian Federation 53°04′N 21°52′E
125 Q15 **Sovetsk** Kirovskaya Oblast', NW Russian Federation 57°37′N 49°02′E
127 N10 **Sovetskaya** Rostovskaya Oblast', SW Russian Federation 49°00′N 42°09′E
Sovetskoye *see* Ketchenery
146 I15 **Sovet"yab** *prev.* Sovet"yap. Ahal Welaýaty, S Turkmenistan 36°29′N 61°13′E
Sovet"yap *see* Sovet"yab
117 U12 **Sovyets'kyy** Avtonomna Respublika Krym, S Ukraine 45°20′N 34°54′E
83 I18 **Sowa** *var.* Sua. Central, NE Botswana 20°33′S 26°18′E
83 I18 **Sowa Pan** *var.* Sua Pan. *salt lake* NE Botswana
83 J21 **Soweto** Gauteng, NE South Africa 26°08′S 27°54′E
147 R11 **So'x** *Rus.* Sokh. Farg'ona Viloyati, E Uzbekistan 39°56′N 71°10′E
Sōya-kaikyō *see* La Pérouse Strait
165 T1 **Sōya-misaki** *headland* Hokkaidō, NE Japan 45°31′N 141°55′E
125 N7 **Soyana** ≈ NW Russian Federation
146 A8 **Soye, Mys** *var.* Mys Suz. *headland* NW Turkmenistan 41°47′N 52°27′E
82 A10 **Soyo** Zaire Province, NW Angola 06°07′S 12°18′E
80 J10 **Soyra** ▲ C Eritrea 14°46′N 39°29′E
145 P15 **Sozak** *Kaz.* Sozaq; *prev.* Suzak. Yuzhnyy Kazakhstan, S Kazakhstan 44°09′N 68°28′E
Sozaq *see* Sozak
119 P16 **Sozh** ≈ NE Europe
114 N10 **Sozopol** *prev.* Sizebolu; *anc.* Apollonia. Burgas, E Bulgaria 42°25′N 27°42′E
99 L20 **Spa** Liège, E Belgium 50°29′N 05°52′E
194 I7 **Spaatz Island** *island* Antarctica
144 M14 **Space Launching Centre** *space station* Kzylorda, S Kazakhstan
105 O7 **Spain** *off.* Kingdom of Spain, *Sp.* España; *anc.* Hispania, Iberia, *Lat.* Hispana. ◆ *monarchy* SW Europe
Spain, Kingdom of *see* Spain
Spalato *see* Split
97 O19 **Spalding** E England, United Kingdom 52°49′N 00°06′W
14 D11 **Spanish** Ontario, S Canada 46°12′N 82°21′W
36 L3 **Spanish Fork** Utah, W USA 40°09′N 111°40′W
64 B12 **Spanish Point** *headland* C Bermuda 32°18′N 64°49′W
14 E9 **Spanish River** ≈ Ontario, S Canada
44 K13 **Spanish Town** *hist.* St.Iago de la Vega. C Jamaica 18°N 76°57′W
Spánta, Akrotírio *see* Spátha, Akrotírio
35 Q5 **Sparks** Nevada, W USA 39°32′N 119°45′W
Sparnacum *see* Épernay
95 N16 **Sparreholm** Södermanland, C Sweden 59°04′N 16°51′E
23 U4 **Sparta** Georgia, SE USA 33°16′N 82°58′W
30 K16 **Sparta** Illinois, N USA 38°07′N 89°42′W
31 P9 **Sparta** Michigan, N USA 43°09′N 85°42′W
21 R8 **Sparta** North Carolina, SE USA 36°30′N 81°07′W
20 L9 **Sparta** Tennessee, S USA 35°55′N 85°30′W
30 I7 **Sparta** Wisconsin, N USA 43°57′N 90°50′W
Sparta *see* Spárti
21 Q11 **Spartanburg** South Carolina, SE USA 34°56′N 81°57′W
115 F21 **Spárti** *Eng.* Sparta. Pelopónnisos, S Greece 37°05′N 22°25′E
107 B21 **Spartivento, Capo** *headland* Sardegna, Italy, C Mediterranean Sea 38°52′N 08°50′E
11 P17 **Sparwood** British Columbia, SW Canada 49°45′N 114°45′W
126 I4 **Spas-Demensk** Kaluzhskaya Oblast', W Russian Federation 54°22′N 34°16′E
126 M4 **Spas-Klepiki** Ryazanskaya Oblast', W Russian Federation 55°08′N 40°15′E
Spasovo *see* Kulen Vakuf
123 R15 **Spassk-Dal'niy** Primorskiy Kray, SE Russian Federation 44°34′N 132°52′E
126 M5 **Spassk-Ryazanskiy** Ryazanskaya Oblast', W Russian Federation 54°25′N 40°21′E
115 H19 **Spáta** Attikí, C Greece 37°58′N 23°55′E
121 Q11 **Spátha, Akrotírio** *var.* Akrotírio Spánta. *headland* Kríti, Greece, E Mediterranean Sea 35°42′N 23°44′E
Špatrjan *see* Paternion
28 I9 **Spearfish** South Dakota, N USA 44°29′N 103°51′W
25 O1 **Spearman** Texas, SW USA 36°12′N 101°13′W
65 C25 **Speedwell Island** *island* S Falkland Islands
65 C25 **Speedwell Island Settlement** S Falkland Islands 52°13′S 59°41′W
65 G25 **Speery Island** *island* S Saint Helena
45 N14 **Speightstown** NW Barbados 13°15′N 59°39′W
106 I13 **Spello** Umbria, C Italy 43°00′N 12°41′E
39 R12 **Spenard** Alaska, USA 61°09′N 150°03′W
Spence Bay *see* Taloyoak
31 O14 **Spencer** Indiana, N USA 39°18′N 86°46′W
29 T12 **Spencer** Iowa, C USA 43°09′N 95°07′W
29 P12 **Spencer** Nebraska, C USA 42°52′N 98°42′W
21 S9 **Spencer** North Carolina, SE USA 35°41′N 80°26′W
20 L9 **Spencer** Tennessee, S USA 35°46′N 85°27′W
21 Q4 **Spencer** West Virginia, NE USA 38°48′N 81°22′W
30 K6 **Spencer** Wisconsin, N USA 44°46′N 90°17′W
182 G10 **Spencer, Cape** *headland* South Australia 35°17′S 136°52′E
39 V13 **Spencer, Cape** *headland* Alaska, USA 58°12′N 136°39′W
182 H9 **Spencer Gulf** *gulf* South Australia
18 F9 **Spencerport** New York, NE USA 43°11′N 77°48′W
31 Q12 **Spencerville** Ohio, N USA 40°42′N 84°21′W
115 E17 **Spercheiáda** *var.* Sperhiada, Sperkhiás. Stereá Elláda, C Greece 38°54′N 22°07′E
115 E17 **Spercheiós** ≈ C Greece
Sperhiada *see* Spercheiáda
95 G14 **Sperillen** ◎ S Norway
Sperkhiás *see* Spercheiáda
101 I18 **Spessart** *hill range* C Germany
Spétsai *see* Spétses
115 G21 **Spétses** *prev.* Spétsai. Spétses, S Greece 37°16′N 23°09′E
115 G21 **Spétses** *island* S Greece
96 J8 **Spey** ≈ NE Scotland, United Kingdom
101 G20 **Speyer** *Eng.* Spires; *anc.* Civitas Nemetum, Spira. Rheinland-Pfalz, SW Germany 49°18′N 08°26′E
101 G20 **Speyerbach** ≈ W Germany
107 N20 **Spezzano Albanese** Calabria, SW Italy 39°40′N 16°17′E
Spice Islands *see* Maluku
100 F9 **Spiekeroog** *island* NW Germany
109 W9 **Spielfeld** Steiermark, SE Austria 46°43′N 15°36′E
65 N21 **Spiess Seamount** *undersea feature* S Atlantic Ocean 53°00′S 02°00′W
108 E9 **Spiez** Bern, W Switzerland 46°42′N 07°41′E
98 G13 **Spijkenisse** Zuid-Holland, SW Netherlands 51°52′N 04°19′E
39 T6 **Spike Mountain** ▲ Alaska, USA 67°42′N 141°39′W
115 I25 **Spíli** Kríti, Greece, E Mediterranean Sea 35°12′N 24°33′E
108 D10 **Spillgerten** ▲ W Switzerland 46°34′N 07°25′E
118 F9 **Spilve** ✈ (Rīga) C Latvia 56°55′N 24°03′E
107 N17 **Spinazzola** Puglia, SE Italy 40°58′N 16°06′E
149 O9 **Spīn Bōldak** *prev.* Spīn Būldak. Kandahār, S Afghanistan 31°01′N 66°23′E
Spīn Būldak *see* Spīn Bōldak
Spira *see* Speyer
Spirdingsee *see* Śniardwy, Jezioro
Spires *see* Speyer
29 T11 **Spirit Lake** Iowa, C USA 43°25′N 95°06′W
29 T11 **Spirit Lake** ◎ Iowa, C USA
11 N13 **Spirit River** Alberta, W Canada 55°46′N 118°51′W
11 S14 **Spiritwood** Saskatchewan, S Canada 53°18′N 107°33′W
27 R11 **Spiro** Oklahoma, C USA 35°14′N 94°37′W
111 L19 **Spišská Nová Ves** *Ger.* Neudorf, Zipser Neudorf, *Hung.* Igló. Košický Kraj, E Slovakia 48°58′N 20°35′E
137 T11 **Spitak** NW Armenia 40°51′N 44°17′E
92 O2 **Spitsbergen** *island* NW Svalbard
Spittal *see* Spittal an der Drau
109 R9 **Spittal an der Drau** *var.* Spittal. Kärnten, S Austria 46°48′N 13°30′E
109 V3 **Spitz** Niederösterreich, NE Austria 48°24′N 15°22′E
94 D9 **Spjelkavik** Møre og Romsdal, S Norway 62°28′N 06°22′E
25 W10 **Splendora** Texas, SW USA 30°13′N 95°09′W
113 E14 **Split** *It.* Spalato. Split-Dalmacija, S Croatia 43°31′N 16°27′E
113 E14 **Split** ✈ Split-Dalmacija, S Croatia 43°33′N 16°19′E
113 E14 **Split-Dalmacija** *off.* Splitsko-Dalmatinska Županija. ◈ *province* S Croatia
11 X12 **Split Lake** ◎ Manitoba, C Canada
Splitsko-Dalmatinska Županija *see* Split-Dalmacija
108 H10 **Splügen** Graubünden, S Switzerland 46°33′N 09°18′E
Spodnji Dravograd *see* Dravograd
25 P12 **Spofford** Texas, SW USA 29°10′N 100°24′W
118 J11 **Špoģi** SE Latvia 56°03′N 26°47′E
32 L8 **Spokane** Washington, NW USA 47°40′N 117°26′W
32 L8 **Spokane River** ≈ Washington, NW USA
106 I13 **Spoleto** Umbria, C Italy 42°44′N 12°44′E
30 I4 **Spooner** Wisconsin, N USA 45°51′N 91°49′W
30 K12 **Spoon River** ≈ Illinois, N USA
21 W5 **Spotsylvania** Virginia, NE USA 38°12′N 77°35′W
32 L8 **Sprague** Washington, NW USA 47°19′N 117°55′W
170 J5 **Spratly Island** *Chin.* Nanwei Dao, *Viet.* Đao Tr ng Sa L n. *island* SW Spratly Islands
192 E6 **Spratly Islands** *Chin.* Nansha Qundao, *Viet.* Quân Đao Tr ng Sa. ◇ *disputed territory* SE Asia
32 J12 **Spray** Oregon, NW USA 44°30′N 119°38′W
112 I11 **Spreča** ≈ N Bosnia and Herzegovina
100 P13 **Spree** ≈ E Germany
100 P13 **Spreewald** *wetland* NE Germany
101 P14 **Spremberg** Brandenburg, E Germany 51°34′N 14°22′E
25 W11 **Spring** Texas, SW USA 30°03′N 95°24′W
31 Q10 **Spring Arbor** Michigan, N USA 42°12′N 84°33′W
83 E23 **Springbok** Northern Cape, W South Africa 29°44′S 17°56′E
18 I15 **Spring City** Pennsylvania, NE USA 40°10′N 75°33′W
20 L9 **Spring City** Tennessee, S USA 35°41′N 84°51′W
36 L4 **Spring City** Utah, W USA 39°28′N 111°30′W
35 W3 **Spring Creek** Nevada, W USA 40°45′N 115°40′W
27 S9 **Springdale** Arkansas, C USA 36°11′N 94°07′W
31 Q14 **Springdale** Ohio, N USA 39°17′N 84°29′W
100 I13 **Springe** Niedersachsen, N Germany 52°13′N 09°33′E
37 U9 **Springer** New Mexico, SW USA 36°21′N 104°35′W
37 W7 **Springfield** Colorado, C USA 37°24′N 102°36′W
23 W5 **Springfield** Georgia, SE USA 32°21′N 81°20′W
30 K14 **Springfield** *state capital* Illinois, N USA 39°48′N 89°39′W
20 L6 **Springfield** Kentucky, S USA 37°42′N 85°18′W
18 M12 **Springfield** Massachusetts, NE USA 42°06′N 72°32′W
29 T10 **Springfield** Minnesota, N USA 44°15′N 94°58′W
27 T7 **Springfield** Missouri, C USA 37°13′N 93°18′W
31 R13 **Springfield** Ohio, N USA 39°55′N 83°49′W
32 G13 **Springfield** Oregon, NW USA 44°03′N 123°01′W
29 Q12 **Springfield** South Dakota, N USA 42°51′N 97°54′W
20 J8 **Springfield** Tennessee, S USA 36°30′N 86°54′W
18 M9 **Springfield** Vermont, NE USA 43°18′N 72°27′W
30 K14 **Springfield, Lake** ⊠ Illinois, N USA
55 T8 **Spring Garden** NE Guyana 06°58′N 58°34′W
30 K8 **Spring Green** Wisconsin, N USA 43°10′N 90°02′W
29 X11 **Spring Grove** Minnesota, N USA 43°33′N 91°38′W
13 P15 **Springhill** Nova Scotia, SE Canada 45°40′N 64°04′W
23 V12 **Spring Hill** Florida, SE USA 28°28′N 82°36′W
27 R4 **Spring Hill** Kansas, C USA 38°44′N 94°49′W
22 G4 **Springhill** Louisiana, S USA 33°01′N 93°27′W
20 I9 **Spring Hill** Tennessee, S USA 35°46′N 86°55′W
21 U10 **Spring Lake** North Carolina, SE USA 35°10′N 78°58′W
24 M4 **Springlake** Texas, SW USA 34°13′N 102°18′W
35 W11 **Spring Mountains** ▲▲ Nevada, W USA
65 B24 **Spring Point** West Falkland, Falkland Islands 51°49′S 60°27′W
27 W9 **Spring River** ≈ Arkansas/Missouri, C USA
27 S7 **Spring River** ≈ Missouri/Oklahoma, C USA
83 J21 **Springs** Gauteng, NE South Africa 26°16′S 28°26′E
185 H16 **Springs Junction** West Coast, South Island, New Zealand 42°21′S 172°11′E
181 X8 **Springsure** Queensland, E Australia 24°09′S 148°06′E
29 W11 **Spring Valley** Minnesota, N USA 43°41′N 92°23′W
18 K13 **Spring Valley** New York, NE USA 41°10′N 73°58′W
29 N12 **Springview** Nebraska, C USA 42°49′N 99°45′W
18 D11 **Springville** New York, NE USA 42°27′N 78°52′W
36 L3 **Springville** Utah, W USA 40°10′N 111°36′W
Sprottau *see* Szprotawa
15 V4 **Sproule, Pointe** *headland* Québec, SE Canada 49°47′N 67°02′W
11 Q14 **Spruce Grove** Alberta, SW Canada 53°36′N 113°55′W
21 T4 **Spruce Knob** ▲ West Virginia, NE USA 38°40′N 79°37′W
35 X3 **Spruce Mountain** ▲ Nevada, W USA 40°33′N 114°46′W
21 P9 **Spruce Pine** North Carolina, SE USA 35°55′N 82°03′W
98 G13 **Spui** ≈ SW Netherlands
107 O19 **Spulico, Capo** *headland* S Italy 39°57′N 16°38′E
25 O5 **Spur** Texas, SW USA 33°28′N 100°51′W
97 O17 **Spurn Head** *headland* E England, United Kingdom 53°34′N 00°06′E
99 H20 **Spy** Namur, S Belgium 50°29′N 04°43′E
95 I15 **Spydeberg** Østfold, S Norway 59°36′N 11°04′E
185 J17 **Spy Glass Point** *headland* South Island, New Zealand 42°33′S 173°31′E
10 L17 **Squamish** British Columbia, SW Canada 49°41′N 123°11′W
19 O8 **Squam Lake** ◎ New Hampshire, NE USA
19 S2 **Squa Pan Mountain** ▲ Maine, NE USA 46°36′N 68°09′W
39 N16 **Squaw Harbor** Unga Island, Alaska, USA 55°12′N 160°41′W
14 E11 **Squaw Island** *island* Ontario, S Canada
107 O22 **Squillace, Golfo di** *gulf* S Italy
107 Q18 **Squinzano** Puglia, SE Italy 40°26′N 18°03′E
Sráid na Cathrach *see* Milltown Malbay
167 S11 **Srâlau** Stœng Trêng, N Cambodia 14°03′N 105°46′E
Srath an Urláir *see* Stranorlar
112 G10 **Srbac** ◈ Republika Srpska, N Bosnia and Herzegovina
Srbija *see* Serbia
Srbinje *see* Foča
112 K9 **Srbobran** *var.* Bácsszenttamás, *Hung.* Szenttamás. Vojvodina, N Serbia 45°33′N 19°46′E
Srbobran *see* Donji Vakuf
167 R13 **Srê Âmběl** Kaôh Kông, SW Cambodia 11°07′N 103°46′E
112 K13 **Srebrenica** Republika Srpska, E Bosnia and Herzegovina 44°04′N 19°18′E
112 I11 **Srebrenik** Federacija Bosne I Hercegovine, NE Bosnia and Herzegovina 44°42′N 18°30′E
114 K10 **Sredets** *prev.* Syulemeshlii. Stara Zagora, C Bulgaria 42°16′N 25°40′E
114 M10 **Sredets** *prev.* Grudovo. ◈ Burgas, E Bulgaria
114 M10 **Sredetska Reka** ≈ SE Bulgaria
123 U9 **Sredinnyy Khrebet** ▲▲ E Russian Federation
114 N7 **Sredishte** *Rom.* Beibunar; *prev.* Knyazhevo. Dobrich, NE Bulgaria 43°51′N 27°30′E
114 I10 **Sredna Gora** ▲▲ C Bulgaria
123 R7 **Srednekolymsk** Respublika Sakha (Yakutiya), NE Russian Federation 67°28′N 153°52′E
126 K7 **Srednerusskaya Vozvyshennost'** *Eng.* Central Russian Upland. ▲▲ W Russian Federation
122 L9 **Srednesibirskoye Ploskogor'ye** *var.* Central Siberian Uplands, *Eng.* Central Siberian Plateau. ▲▲ N Russian Federation
125 V13 **Sredniy Ural** ▲▲ NW Russian Federation
167 T12 **Srê Khtŭm** Môndól Kiri, E Cambodia 12°10′N 106°52′E
110 G12 **Śrem** Wielkopolskie, C Poland 52°07′N 17°00′E
112 K10 **Sremska Mitrovica** *prev.* Mitrovica, *Ger.* Mitrowitz. Vojvodina, NW Serbia 44°58′N 19°37′E
167 R11 **Srêng, Stœng** ≈ NW Cambodia
167 R11 **Srê Noy** Siĕmréab, NW Cambodia 13°47′N 104°03′E
Srepok, Sông *see* Srêpôk, Tônle
167 T12 **Srêpôk, Tônle** *var.* Sông Srepok. ≈ Cambodia/Vietnam
123 P13 **Sretensk** Zabaykal'skiy Kray, S Russian Federation 52°14′N 117°33′E
169 R10 **Sri Aman** Sarawak, East Malaysia 01°13′N 111°25′E
117 R4 **Sribne** Chernihivs'ka Oblast', N Ukraine 50°40′N 32°55′E
Sri Jayawardanapura *see* Sri Jayewardenapura Kotte
155 I25 **Sri Jayewardenapura Kotte** *var.* Sri Jayawardanapura. ● (legislative) Western Province, W Sri Lanka 06°54′N 79°58′E
155 M14 **Srīkākulam** Andhra Pradesh, E India 18°18′N 83°54′E
155 I25 **Sri Lanka** *off.* Democratic Socialist Republic of Sri Lanka; *prev.* Ceylon. ◆ *republic* S Asia
130 F14 **Sri Lanka** *island* S Asia
Sri Lanka, Democratic Socialist Republic of *see* Sri Lanka
153 V14 **Srimangal** Sylhet, E Bangladesh 24°19′N 91°40′E
Sri Mohangorh *see* Shri Mohangarh
152 H5 **Srīnagar** *state capital* Jammu and Kashmir, N India 34°07′N 74°50′E
167 N10 **Srinagarind Reservoir** ◎ W Thailand
155 F19 **Sringeri** Karnātaka, W India 13°26′N 75°13′E
155 K25 **Sri Pada** *Eng.* Adam's Peak. ▲ S Sri Lanka 06°49′N 80°25′E
Sri Saket *see* Si Sa Ket
111 G14 **Sroda Śląska** *Ger.* Neumarkt. Dolnośląskie, SW Poland 51°10′N 16°30′E
110 H12 **Środa Wielkopolska** Wielkopolskie, C Poland 52°13′N 17°17′E
Srpska Kostajnica *see* Bosanska Kostajnica
113 G14 **Srpska, Republika** ◈ *republic* Bosnia and Herzegovina
Srpski Brod *see* Bosanski Brod
Ssu-ch'uan *see* Sichuan
Ssu-p'ing/Ssu-p'ing-chieh *see* Siping
Stablo *see* Stavelot
99 G15 **Stabroek** Antwerpen, N Belgium 51°21′N 04°22′E
Stackeln *see* Strenči
96 I5 **Stack Skerry** *island* N Scotland, United Kingdom
100 I9 **Stade** Niedersachsen, NW Germany 53°36′N 09°29′E
94 C10 **Stadlandet** *peninsula* S Norway
109 R5 **Stadl-Paura** Oberösterreich, NW Austria 48°05′N 13°52′E
119 L20 **Stadolichy** *Rus.* Stodolichi. Homyel'skaya Voblasts', SE Belarus 51°44′N 28°30′E
98 P7 **Stadskanaal** Groningen, NE Netherlands 53°N 06°55′E
101 H16 **Stadtallendorf** Hessen, C Germany 50°49′N 09°01′E
101 K23 **Stadtbergen** Bayern, S Germany 48°21′N 10°50′E
108 G7 **Stäfa** Zürich, NE Switzerland 47°14′N 08°45′E
95 K23 **Staffanstorp** Skåne, S Sweden 55°38′N 13°13′E
101 K18 **Staffelstein** Bayern, C Germany 50°05′N 11°00′E
97 L19 **Stafford** C England, United Kingdom 52°48′N 02°07′W
26 L6 **Stafford** Kansas, C USA 37°57′N 98°36′W
21 W4 **Stafford** Virginia, NE USA 38°26′N 77°27′W

326

◆ Country ● Country Capital ◇ Dependent Territory ○ Dependent Territory Capital ◈ Administrative Regions ✈ International Airport ▲ Mountain ▲▲ Mountain Range ⛰ Volcano ≈ River ◎ Lake ⊠ Reservoir

STAFFORDSHIRE – STRONGILÍ

97 L19 **Staffordshire** *cultural region* C England, United Kingdom
19 N12 **Stafford Springs** Connecticut, NE USA 41°57′N 72°18′W
115 H14 **Stágira** Kentrikí Makedonía, N Greece 40°31′N 23°46′E
118 G7 **Staicele** N Latvia 57°52′N 24°48′E
Ştaierdorf-Anina *see* Anina
109 V8 **Stainz** Steiermark, SE Austria 46°55′N 15°18′E
Stájerlakanina *see* Anina
117 Y7 **Stakhanov** Luhans'ka Oblast', E Ukraine 48°30′N 38°42′E
108 E11 **Stalden** Valais, SW Switzerland 46°12′N 07°55′E
15 S8 **St-Alexandre** Québec, SE Canada 47°39′N 69°36′W
Stalin *see* Varna
Stalinabad *see* Dushanbe
Stalingrad *see* Volgograd
Staliniri *see* Tskhinvali
Stalino *see* Donets'k
Stalinobod *see* Dushanbe
Stalinov Štít *see* Gerlachovský štít
Stalinsk *see* Novokuznetsk
Stalins'kaya Oblast' *see* Donets'ka Oblast'
Stalinski Zaliv *see* Varnenski Zaliv
Stalin, Yazovir *see* Iskar, Yazovir
111 N15 **Stalowa Wola** Podkarpackie, SE Poland 50°35′N 22°02′E
114 I11 **Stamboliyski** Plovdiv, C Bulgaria 42°09′N 24°32′E
15 Q7 **St-Ambroise** Québec, SE Canada 48°35′N 71°19′W
97 N19 **Stamford** E England, United Kingdom 52°39′N 00°32′W
18 L14 **Stamford** Connecticut, NE USA 41°03′N 73°32′W
25 P6 **Stamford** Texas, SW USA 32°55′N 99°49′W
25 Q6 **Stamford, Lake** ⊞ Texas, SW USA
108 I10 **Stampa** Graubünden, SE Switzerland 46°21′N 09°35′E
Stampalia *see* Astypálaia
27 T14 **Stamps** Arkansas, C USA 33°22′N 93°30′W
92 G11 **Stamsund** Nordland, C Norway 68°07′N 13°50′E
27 R2 **Stanberry** Missouri, C USA 40°12′N 94°33′W
195 O3 **Stancomb-Wills Glacier** *glacier* Antarctica
83 K21 **Standerton** Mpumalanga, E South Africa 26°57′S 29°14′E
31 R7 **Standish** Michigan, N USA 43°59′N 83°58′W
20 M6 **Stanford** Kentucky, S USA 37°30′N 84°40′W
33 S9 **Stanford** Montana, NW USA 47°08′N 110°15′W
95 P19 **Stånga** Gotland, SE Sweden 57°16′N 18°30′E
94 I13 **Stange** Hedmark, S Norway 60°40′N 11°05′E
83 L23 **Stanger** KwaZulu/Natal, E South Africa 29°20′S 31°18′E
Stanger *see* KwaDukuza
Stanimaka *see* Asenovgrad
Stanislau *see* Ivano-Frankivs'k
35 P8 **Stanislaus River** ≈ California, W USA
Stanislav *see* Ivano-Frankivs'k
Stanislavskaya Oblast' *see* Ivano-Frankivs'ka Oblast'
Stanisławów *see* Ivano-Frankivs'k
Stanke Dimitrov *see* Dupnitsa
183 O15 **Stanley** Tasmania, SE Australia 40°48′S 145°18′E
65 E24 **Stanley** *var.* Port Stanley, Puerto Argentino. ○ (Falkland Islands) East Falkland, Falkland Islands 51°45′S 57°56′W
33 O13 **Stanley** Idaho, NW USA 44°12′N 114°58′W
28 L3 **Stanley** North Dakota, N USA 48°19′N 102°23′W
21 U4 **Stanley** Virginia, NE USA 38°34′N 78°30′W
30 J6 **Stanley** Wisconsin, N USA 44°58′N 90°54′W
79 G21 **Stanley Pool** *var.* Pool Malebo. ◎ Congo/Dem. Rep. Congo
155 H20 **Stanley Reservoir** ⊞ S India
Stanleyville *see* Kisangani
42 G3 **Stann Creek** ◐ *district* SE Belize
Stann Creek *see* Dangriga
123 Q12 **Stanovoy Khrebet** ▲▲ SE Russian Federation
108 F8 **Stans** Nidwalden, C Switzerland 46°57′N 08°23′E
97 O21 **Stansted** ✈ (London) Essex, E England, United Kingdom 51°53′N 00°16′E
183 U4 **Stanthorpe** Queensland, E Australia 28°35′S 151°52′E
21 N6 **Stanton** Kentucky, S USA 37°51′N 83°51′W
31 Q8 **Stanton** Michigan, N USA 43°19′N 85°04′W
29 Q14 **Stanton** Nebraska, C USA 41°57′N 97°13′W
28 L5 **Stanton** North Dakota, N USA 47°19′N 101°22′W
25 N7 **Stanton** Texas, SW USA 32°07′N 101°47′W
32 H7 **Stanwood** Washington, NW USA 48°14′N 122°22′W
117 Y7 **Stanychno-Luhans'ke** Luhans'ka Oblast', E Ukraine 48°39′N 39°30′E
108 K7 **Stanzach** Tirol, W Austria 47°24′N 10°36′E
98 M9 **Staphorst** Overijssel, E Netherlands 52°38′N 06°12′E
14 D18 **Staples** Ontario, S Canada 42°09′N 82°34′W
29 T6 **Staples** Minnesota, N USA 46°21′N 94°47′W
28 M14 **Stapleton** Nebraska, C USA 41°29′N 100°40′W
25 S8 **Star** Texas, SW USA 31°27′N 98°16′W
111 M14 **Starachowice** Świętokrzyskie, C Poland 51°04′N 21°02′E
Stara Kanjiža *see* Kanjiža
111 M18 **Stará L'ubovňa** *Ger.* Altlublau, *Hung.* Ólubló. Prešovský Kraj, E Slovakia 49°19′N 20°40′E
112 L10 **Stara Pazova** *Ger.* Altpasua, *Hung.* Ópazova. Vojvodina, N Serbia 44°59′N 20°10′E
Stara Planina *see* Balkan Mountains
114 L9 **Stara Reka** ≈ C Bulgaria
116 M5 **Stara Synyava** Khmel'nyts'ka Oblast', W Ukraine 49°39′N 27°39′E
116 I2 **Stara Vyzhivka** Volyns'ka Oblast', NW Ukraine 51°27′N 24°25′E
Staraya Belitsa *see* Staraya Byelitsa
119 M14 **Staraya Byelitsa** *Rus.* Staraya Belitsa. Vitsyebskaya Voblasts', NE Belarus 54°42′N 29°38′E
127 R5 **Staraya Mayna** Ul'yanovskaya Oblast', W Russian Federation 54°36′N 48°57′E
119 O18 **Staraya Rudnya** Homyel'skaya Voblasts', SE Belarus 52°50′N 30°17′E
124 H14 **Staraya Russa** Novgorodskaya Oblast', W Russian Federation 57°59′N 31°18′E
114 K10 **Stara Zagora** *Lat.* Augusta Trajana. Stara Zagora, C Bulgaria 42°26′N 25°39′E
114 K10 **Stara Zagora** ◐ *province* C Bulgaria
29 S8 **Starbuck** Minnesota, N USA 45°36′N 95°31′W
191 W4 **Starbuck Island** *prev.* Volunteer Island. *island* E Kiribati
27 V13 **Star City** Arkansas, C USA 33°56′N 91°52′W
112 F13 **Staretina** ▲▲ W Bosnia and Herzegovina
Stargard in Pommern *see* Stargard Szczeciński
110 E9 **Stargard Szczeciński** *Ger.* Stargard in Pommern. Zachodnio-pomorskie, NW Poland 53°20′N 15°02′E
187 N10 **Star Harbour** *harbor* San Cristobal, SE Solomon Islands
Stari Bečej *see* Bečej
113 F15 **Stari Grad** *It.* Cittavecchia. Split-Dalmacija, S Croatia 43°11′N 16°36′E
124 J16 **Staritsa** Tverskaya Oblast', W Russian Federation 56°28′N 34°51′E
23 V9 **Starke** Florida, SE USA 29°56′N 82°07′W
22 M4 **Starkville** Mississippi, S USA 33°27′N 88°49′W
186 B7 **Star Mountains** *Ind.* Pegunungan Sterren. ▲▲ Indonesia/Papua New Guinea
101 L23 **Starnberg** Bayern, SE Germany 48°00′N 11°19′E
101 L24 **Starnberger See** ◎ SE Germany
Starobel'sk *see* Starobil's'k
117 X8 **Starobesheve** Donets'ka Oblast', E Ukraine 47°45′N 38°01′E
117 Y6 **Starobil's'k** *Rus.* Starobel'sk. Luhans'ka Oblast', E Ukraine 49°16′N 38°56′E
119 K18 **Starobin** *var.* Starobyn. Minskaya Voblasts', S Belarus 52°44′N 27°28′E
Starobyn *see* Starobin
126 H6 **Starodub** Bryanskaya Oblast', W Russian Federation 52°30′N 32°56′E
110 I8 **Starogard Gdański** *Ger.* Preussisch-Stargard. Pomorskie, N Poland 53°57′N 18°29′E
Staroikan *see* Ikan
Starokonstantinov *see* Starokostyantyniv
116 L5 **Starokostyantyniv** *Rus.* Starokonstantinov. Khmel'nyts'ka Oblast', NW Ukraine 49°43′N 27°13′E
126 K12 **Starominskaya** Krasnodarskiy Kray, SW Russian Federation 46°31′N 39°03′E
114 L7 **Staro Selo** *Rom.* Satul-Vechi; *prev.* Star-Smil. Silistra, NE Bulgaria 43°58′N 26°32′E
126 K12 **Staroshcherbinovskaya** Krasnodarskiy Kray, SW Russian Federation 46°36′N 38°42′E
127 V6 **Starosubkhangulovo** Respublika Bashkortostan, W Russian Federation 53°05′N 57°22′E
35 S4 **Star Peak** ▲ Nevada, W USA 40°31′N 118°09′W
15 T8 **St-Arsène** Québec, SE Canada 47°55′N 69°21′W
Star-Smil *see* Staro Selo
97 J25 **Start Point** *headland* SW England, United Kingdom 50°13′N 03°38′W
Startsy *see* Kirawsk
Starum *see* Stavoren
119 L18 **Staryya Darohi** *Rus.* Staryye Dorogi. Minskaya Voblasts', S Belarus 53°02′N 28°16′E
Staryye Dorogi *see* Staryya Darohi
127 T2 **Staryye Zyattsy** Udmurtskaya Respublika, NW Russian Federation 57°22′N 52°42′E
117 U13 **Staryy Krym** Avtonomna Respublika Krym, S Ukraine 45°03′N 35°06′E
126 K8 **Staryy Oskol** Belgorodskaya Oblast', W Russian Federation 51°21′N 37°52′E
116 H6 **Staryy Sambir** L'vivs'ka Oblast', W Ukraine 49°27′N 23°00′E
101 L14 **Stassfurt** *var.* Staßfurt. Sachsen-Anhalt, C Germany 51°51′N 11°35′E
Staßfurt *see* Stassfurt
111 M15 **Staszów** Świętokrzyskie, C Poland 50°33′N 21°07′E
29 W13 **State Center** Iowa, C USA 42°01′N 93°09′W
18 E14 **State College** Pennsylvania, NE USA 40°48′N 77°52′W
18 K15 **Staten Island** *island* New York, NE USA
Staten Island *see* Estados, Isla de los
23 U8 **Statenville** Georgia, SE USA 30°42′N 83°00′W
23 W5 **Statesboro** Georgia, SE USA 32°28′N 81°47′W
States, The *see* United States of America
21 R9 **Statesville** North Carolina, SE USA 35°46′N 80°54′W
95 G16 **Stathelle** Telemark, S Norway 59°01′N 09°40′E
Statia *see* Sint Eustatius
30 K15 **Staunton** Illinois, N USA 39°00′N 89°47′W
21 T5 **Staunton** Virginia, NE USA 38°10′N 79°05′W
95 C16 **Stavanger** Rogaland, S Norway 58°58′N 05°43′E
99 L21 **Stavelot** *Dut.* Stablo. Liège, E Belgium 50°24′N 05°56′E
95 G16 **Stavern** Vestfold, S Norway 58°58′N 10°01′E
Stavers Island *see* Vostok Island
98 J7 **Stavoren** *Fris.* Starum. Fryslân, N Netherlands 52°52′N 05°22′E
115 K21 **Stavrí, Akrotírio** *var.* Akrotírio Stavrós. *headland* Naxos, Kykládes, Greece, Aegean Sea 37°12′N 25°32′E
126 M14 **Stavropol'** *prev.* Voroshilovsk. Stavropol'skiy Kray, SW Russian Federation 45°02′N 41°58′E
Stavropol' *see* Tol'yatti
126 M14 **Stavropol'skaya Vozvyshennost'** ▲▲ SW Russian Federation
126 M14 **Stavropol'skiy Kray** ◐ *territory* SW Russian Federation
115 H14 **Stavrós** Kentrikí Makedonía, N Greece 40°39′N 23°43′E
115 J24 **Stavrós, Akrotírio** *headland* Kríti, Greece, E Mediterranean Sea 35°25′N 24°57′E
Stavrós, Akrotírio *see* Stavrí, Akrotírio
114 I12 **Stavroúpoli** *prev.* Stavroúpolis. Anatolikí Makedonía kai Thráki, NE Greece 41°12′N 24°45′E
Stavroúpolis *see* Stavroúpoli
117 O6 **Stavyshche** Kyyivs'ka Oblast', N Ukraine 49°23′N 30°10′E
182 M11 **Stawell** Victoria, SE Australia 37°06′S 142°52′E
110 N9 **Stawiski** Podlaskie, NE Poland 53°22′N 22°08′E
14 G14 **Stayner** Ontario, S Canada 44°25′N 80°05′W
14 D17 **St. Clair** ≈ Canada/USA
37 R3 **Steamboat Springs** Colorado, C USA 40°28′N 106°51′W
15 U4 **Ste-Anne, Lac** ◎ Québec, SE Canada
20 M8 **Stearns** Kentucky, S USA 36°39′N 84°27′W
39 N10 **Stebbins** Alaska, USA 63°30′N 162°15′W
15 U7 **Ste-Blandine** Québec, SE Canada 48°22′N 68°27′W
108 K7 **Steeg** Tirol, W Austria 47°15′N 10°18′E
27 Y9 **Steele** Missouri, C USA 36°04′N 89°49′W
29 N5 **Steele** North Dakota, N USA 46°51′N 99°55′W
194 J5 **Steele Island** *island* Antarctica
30 K16 **Steeleville** Illinois, N USA 38°00′N 89°39′W
27 W6 **Steelville** Missouri, C USA 37°57′N 91°21′W
99 G14 **Steenbergen** Noord-Brabant, S Netherlands 51°35′N 04°19′E
Steenkool *see* Bintuni
11 O10 **Steen River** Alberta, W Canada 59°37′N 117°17′W
98 M8 **Steenwijk** Overijssel, N Netherlands 52°47′N 06°07′E
65 A23 **Steeple Jason** *island* Jason Islands, NW Falkland Islands
174 J8 **Steep Point** *headland* Western Australia 26°09′S 113°11′E
116 L9 **Ştefăneşti** Botoşani, NE Romania 47°44′N 27°15′E
Stefanie, Lake *see* Ch'ew Bahir
8 L5 **Stefansson Island** *island* Nunavut, N Canada
117 O10 **Ştefan Vodă** *Rus.* Suvorovo. SE Moldova 46°33′N 29°39′E
63 H18 **Steffen, Cerro** ▲ S Chile 44°27′S 71°42′W
108 D9 **Steffisburg** Bern, C Switzerland 46°47′N 07°38′E
95 J24 **Stege** Sjælland, SE Denmark 54°59′N 12°18′E
116 G10 **Ştei** *Hung.* Vaskohsziklás. Bihor, W Romania 46°34′N 22°28′E
Steier *see* Steyr
Steierdorf/Steierdorf-Anina *see* Anina
109 T7 **Steiermark** *off.* Land Steiermark, *Eng.* Styria. ◐ *state* C Austria
Steiermark, Land *see* Steiermark
101 J19 **Steigerwald** *hill range* C Germany
99 L17 **Stein** Limburg, SE Netherlands 50°58′N 05°45′E
Stein *see* Stein an der Donau
Stein *see* Kamnik, Slovenia
108 M8 **Steinach** Tirol, W Austria 47°07′N 11°30′E
Steinamanger *see* Szombathely
109 W3 **Stein an der Donau** *var.* Stein. Niederösterreich, NE Austria 48°25′N 15°35′E
Steinau an der Elbe *see* Ścinawa
11 Y16 **Steinbach** Manitoba, S Canada 49°32′N 96°40′W
Steiner Alpen *see* Kamniško-Savinjske Alpe
99 L24 **Steinfort** Luxembourg, W Luxembourg 49°39′N 05°55′E
100 H12 **Steinhuder Meer** ◎ NW Germany
93 E15 **Steinkjer** Nord-Trøndelag, C Norway 64°01′N 11°29′E
Stejarul *see* Karapelit
99 F16 **Stekene** Oost-Vlaanderen, NW Belgium 51°13′N 04°04′E
83 E26 **Stellenbosch** Western Cape, SW South Africa 33°56′S 18°51′E
98 F13 **Stellendam** Zuid-Holland, SW Netherlands 51°48′N 04°01′E
39 T12 **Steller, Mount** ▲ Alaska, USA 60°36′N 142°49′W
103 Y14 **Stello, Monte** ▲ Corse, France, C Mediterranean Sea 42°49′N 09°24′E
106 F5 **Stelvio, Passo dello** *pass* Italy/Switzerland
15 S7 **Ste-Maguerite Nord-Est** ≈ Québec, SE Canada
15 V4 **Ste-Marguerite, Pointe** *headland* Québec, SE Canada 50°01′N 66°43′W
12 J14 **Ste-Marie, Lac** ⊞ Québec, S Canada
103 R3 **Stenay** Meuse, NE France 49°29′N 05°12′E
100 L12 **Stendal** Sachsen-Anhalt, C Germany 52°36′N 11°52′E
118 E8 **Stende** NW Latvia 57°09′N 22°33′E
182 H10 **Stenhouse Bay** South Australia 35°15′S 136°58′E
95 J23 **Stenløse** Hovedstaden, E Denmark 55°47′N 12°13′E
95 L19 **Stensjön** Jönköping, S Sweden 57°36′N 14°42′E
95 K18 **Stenstorp** Västra Götaland, S Sweden 58°15′N 13°45′E
95 I18 **Stenungsund** Västra Götaland, S Sweden 58°05′N 11°49′E
Stepanakert *see* Xankändi
137 T11 **Step'anavan** N Armenia 41°00′N 44°27′E
100 K9 **Stepenitz** ≈ N Germany
29 O10 **Stephan** South Dakota, N USA 44°12′N 99°25′W
29 R3 **Stephen** Minnesota, N USA 48°27′N 96°54′W
27 T14 **Stephens** Arkansas, C USA 33°25′N 93°04′W
184 J13 **Stephens, Cape** *headland* D'Urville Island, Marlborough, SW New Zealand 40°42′S 173°56′E
21 V3 **Stephens City** Virginia, NE USA 39°03′N 78°10′W
182 L6 **Stephens Creek** New South Wales, SE Australia 31°51′S 141°30′E
184 K13 **Stephens Island** *island* C New Zealand
31 N5 **Stephenson** Michigan, N USA 45°27′N 87°36′W
13 S12 **Stephenville** Newfoundland, Newfoundland and Labrador, SE Canada 48°33′N 58°34′W
25 S7 **Stephenville** Texas, SW USA 32°12′N 98°13′W
Step' Nardara *see* Step' Shardara
145 R8 **Stepnogorsk** Akmola, C Kazakhstan 52°04′N 72°18′E
127 O15 **Stepnoye** Stavropol'skiy Kray, SW Russian Federation 44°18′N 44°34′E
145 Q8 **Stepnyak** Akmola, N Kazakhstan 52°52′N 70°49′E
145 P17 **Step' Shardara** *Kaz.* Shardara Dalasy; *prev.* Step' Nardara. *grassland* S Kazakhstan
192 J17 **Steps Point** *headland* Tutuila, W American Samoa 14°23′S 170°46′W
115 F17 **Stereá Elláda** *Eng.* Greece Central *var.* Stereá Ellás. ◐ *region* C Greece
Stereá Ellás *see* Stereá Elláda
83 J24 **Sterkspruit** Eastern Cape, SE South Africa 30°31′S 27°22′E
127 U6 **Sterlibashevo** Respublika Bashkortostan, W Russian Federation 53°19′N 55°12′E
39 R12 **Sterling** Alaska, USA 60°32′N 150°51′W
37 V3 **Sterling** Colorado, C USA 40°37′N 103°12′W
30 K11 **Sterling** Illinois, N USA 41°47′N 89°42′W
26 M5 **Sterling** Kansas, C USA 38°12′N 98°12′W
25 O8 **Sterling City** Texas, SW USA 31°50′N 101°00′W
31 S9 **Sterling Heights** Michigan, N USA 42°34′N 83°01′W
21 W3 **Sterling Park** Virginia, NE USA 39°00′N 77°24′W
37 V2 **Sterling Reservoir** ⊞ Colorado, C USA
22 I5 **Sterlington** Louisiana, S USA 32°42′N 92°05′W
127 U6 **Sterlitamak** Respublika Bashkortostan, W Russian Federation 53°39′N 56°00′E
Sternberg *see* Šternberk
111 H17 **Šternberk** *Ger.* Sternberg. Olomoucký Kraj, E Czech Republic 49°45′N 17°20′E
141 V17 **Stēroh** Suquţrā, S Yemen 12°21′N 53°50′E
Sterren, Pegunungan *see* Star Mountains
110 G11 **Stęszew** Wielkopolskie, C Poland 52°16′N 16°41′E
Stettin *see* Szczecin
Stettiner Haff *see* Szczeciński, Zalew
11 Q15 **Stettler** Alberta, SW Canada 52°21′N 112°40′W
31 V13 **Steubenville** Ohio, N USA 40°21′N 80°37′W
97 O21 **Stevenage** E England, United Kingdom 51°55′N 00°14′W
23 Q1 **Stevenson** Alabama, S USA 34°52′N 85°50′W
32 H11 **Stevenson** Washington, NW USA 45°43′N 121°54′W
182 E1 **Stevenson Creek** *seasonal river* South Australia
39 Q13 **Stevenson Entrance** *strait* Alaska, USA
30 L6 **Stevens Point** Wisconsin, N USA 44°32′N 89°33′W
39 R8 **Stevens Village** Alaska, USA 66°01′N 149°02′W
33 P10 **Stevensville** Montana, NW USA 46°30′N 114°05′W
93 E25 **Stevns Klint** *headland* E Denmark 55°15′N 12°25′E
10 J12 **Stewart** British Columbia, W Canada 55°58′N 129°52′W
10 J6 **Stewart** ≈ Yukon, NW Canada
10 I6 **Stewart Crossing** Yukon, NW Canada 63°22′N 136°37′W
63 H25 **Stewart, Isla** *island* S Chile
185 B25 **Stewart Island** *island* S New Zealand
Stewart Islands *see* Sikaiana
181 W6 **Stewart, Mount** ▲ Queensland, E Australia 20°11′S 145°29′E
10 H6 **Stewart River** Yukon, NW Canada 63°17′N 139°24′W
27 R3 **Stewartsville** Missouri, C USA 39°45′N 94°30′W
11 S16 **Stewart Valley** Saskatchewan, S Canada 50°34′N 107°47′W
29 W10 **Stewartville** Minnesota, N USA 43°51′N 92°29′W
Steyerlak-Anina *see* Anina
109 T5 **Steyr** *var.* Steier. Oberösterreich, N Austria 48°02′N 14°26′E
109 T5 **Steyr** ≈ NW Austria
15 T7 **St-Fabien** Québec, SE Canada 48°19′N 68°51′W
15 R11 **St-François, Lac** ◎ Québec, SE Canada
83 E25 **St. Helena Bay** *bay* SW South Africa
15 T8 **St-Hubert** Québec, SE Canada 47°46′N 69°15′W
29 P11 **Stickney** South Dakota, N USA 43°24′N 98°23′W
98 L5 **Stiens** Fryslân, N Netherlands 53°15′N 05°45′E
Stif *see* Sétif
27 Q11 **Stigler** Oklahoma, C USA 35°16′N 95°08′W
107 N18 **Stigliano** Basilicata, S Italy 40°24′N 16°13′E
95 N17 **Stigtomta** Södermanland, C Sweden 58°48′N 16°47′E
10 I11 **Stikine** ≈ British Columbia, W Canada
Stilida/Stilís *see* Stylída
95 G22 **Stilling** Midtjylland, C Denmark 56°04′N 10°00′E
29 W8 **Stillwater** Minnesota, N USA 45°03′N 92°48′W
27 O9 **Stillwater** Oklahoma, C USA 36°07′N 97°03′W
35 S5 **Stillwater Range** ▲▲ Nevada, W USA
18 I8 **Stillwater Reservoir** ⊞ New York, NE USA
107 O22 **Stilo, Punta** *headland* S Italy 38°27′N 16°36′E
27 R10 **Stilwell** Oklahoma, C USA 35°48′N 94°37′W
Štimlje *see* Shtime
25 N1 **Stinnett** Texas, SW USA 35°49′N 101°27′W
113 P18 **Štip** E FYR Macedonia 41°45′N 22°10′E
Stira *see* Stýra
96 J12 **Stirling** C Scotland, United Kingdom 56°07′N 03°57′W
96 I12 **Stirling** *cultural region* C Scotland, United Kingdom
180 J14 **Stirling Range** ▲▲ Western Australia
15 R8 **St-Jean** ≈ Québec, SE Canada
93 E16 **Stjørdalshalsen** Nord-Trøndelag, C Norway 63°27′N 10°57′E
83 L22 **St. Lucia** KwaZulu/Natal, E South Africa 28°22′N 32°25′E
Stochód *see* Stokhid
101 H24 **Stockach** Baden-Württemberg, S Germany 47°51′N 09°01′E
25 S12 **Stockdale** Texas, SW USA 29°14′N 97°57′W
109 X3 **Stockerau** Niederösterreich, NE Austria 48°24′N 16°13′E
93 H20 **Stockholm** ● (Sweden) Stockholm, C Sweden 59°17′N 18°03′E
95 O15 **Stockholm** ◐ *county* C Sweden
Stockmannshof *see* Pļaviņas
97 L18 **Stockport** NW England, United Kingdom 53°25′N 02°10′W
65 K15 **Stocks Seamount** *undersea feature* C Atlantic Ocean 11°42′S 33°48′W
35 O8 **Stockton** California, W USA 37°56′N 121°19′W
26 L3 **Stockton** Kansas, C USA 39°27′N 99°17′W
27 S6 **Stockton** Missouri, C USA 37°43′N 93°49′W
30 K3 **Stockton Island** *island* Apostle Islands, Wisconsin, N USA
27 S7 **Stockton Lake** ⊞ Missouri, C USA
97 M15 **Stockton-on-Tees** *var.* Stockton on Tees. N England, United Kingdom 54°34′N 01°19′W
Stockton on Tees *see* Stockton-on-Tees
24 M10 **Stockton Plateau** *plain* Texas, SW USA
28 M16 **Stockville** Nebraska, C USA 40°33′N 100°20′W
93 H17 **Stöde** Västernorrland, C Sweden 62°27′N 16°34′E
Stodolichi *see* Stadolichy
113 M19 **Stogovo Karaorman** ▲▲ W FYR Macedonia
Stoke *see* Stoke-on-Trent
97 L19 **Stoke-on-Trent** *var.* Stoke. C England, United Kingdom 53°N 02°10′W
182 M15 **Stokes Point** *headland* Tasmania, SE Australia 40°09′S 143°55′E
116 J2 **Stokhid** *Pol.* Stochód, *Rus.* Stokhod. ≈ NW Ukraine
Stokhod *see* Stokhid
92 I4 **Stokkseyri** Suðurland, SW Iceland 63°49′N 21°00′W
92 G10 **Stokmarknes** Nordland, C Norway 68°34′N 14°55′E
Stol *see* Veliki Krš
113 H15 **Stolac** Federacija Bosne I Hercegovine, S Bosnia and Herzegovina 43°04′N 17°58′E
Stolbce *see* Stowbtsy
101 D16 **Stolberg** *var.* Stolberg im Rheinland. Nordrhein-Westfalen, W Germany 50°45′N 06°15′E
Stolberg im Rheinland *see* Stolberg
123 P6 **Stolbovoy, Ostrov** *island* NE Russian Federation
Stolbtsy *see* Stowbtsy
119 J20 **Stolin** Brestskaya Voblasts', SW Belarus 51°53′N 26°51′E
95 K14 **Stöllet** *var.* Norra Ny. Värmland, C Sweden 60°24′N 13°15′E
Stolp *see* Słupsk
Stolpe *see* Słupia
Stolpmünde *see* Ustka
115 F15 **Stómio** Thessalía, C Greece 39°51′N 22°45′E
14 J11 **Stonecliffe** Ontario, SE Canada 46°12′N 77°58′W
96 L10 **Stonehaven** NE Scotland, United Kingdom 56°59′N 02°14′W
97 M23 **Stonehenge** *ancient monument* Wiltshire, S England, United Kingdom
23 T3 **Stone Mountain** ▲ Georgia, SE USA 33°48′N 84°10′W
11 X16 **Stonewall** Manitoba, S Canada 50°08′N 97°20′W
21 S3 **Stonewood** West Virginia, NE USA 39°15′N 80°18′W
14 D17 **Stoney Point** Ontario, S Canada 42°18′N 82°32′W
92 H10 **Stonglandseidet** Troms, N Norway 69°03′N 17°06′E
65 N25 **Stonybeach Bay** *bay* Tristan da Cunha, SE Atlantic Ocean
35 N5 **Stony Creek** ≈ California, W USA
65 N25 **Stonyhill Point** *headland* S Tristan da Cunha
14 I14 **Stony Lake** ◎ Ontario, SE Canada
11 Q14 **Stony Plain** Alberta, SW Canada 53°31′N 114°04′W
21 R9 **Stony Point** North Carolina, SE USA 35°51′N 81°04′W
18 G8 **Stony Point** *headland* New York, NE USA 43°50′N 76°18′W
11 T10 **Stony Rapids** Saskatchewan, C Canada 59°14′N 105°48′W
39 P11 **Stony River** Alaska, USA 61°48′N 156°37′W
Stony Tunguska *see* Podkamennaya Tunguska
12 G10 **Stooping** ≈ Ontario, C Canada
100 I9 **Stör** ≈ N Germany
95 M15 **Storå** Örebro, S Sweden 59°44′N 15°10′E
95 J16 **Stora Gla** ◎ C Sweden
95 I16 **Stora Le** *Nor.* Store Le. ◎ Norway/Sweden
92 I12 **Stora Lulevatten** ◎ N Sweden
92 H13 **Storavan** ◎ N Sweden
93 I20 **Storby** Åland, SW Finland 60°12′N 19°33′E
94 E10 **Stordalen** Møre og Romsdal, S Norway 62°22′N 07°00′E
95 H23 **Storebælt** *var.* Store Bælt, *Eng.* Great Belt, Storebelt. *channel* Baltic Sea/Kattegat
Store Bælt *see* Storebælt
Storebelt *see* Storebælt
95 M19 **Storebro** Kalmar, S Sweden 57°36′N 15°30′E
95 J24 **Store Heddinge** Sjælland, SE Denmark 55°19′N 12°24′E
Store Le *see* Stora Le
93 E16 **Støren** Sør-Trøndelag, S Norway 63°02′N 10°16′E
92 O4 **Storfjorden** *fjord* S Norway
95 L15 **Storfors** Värmland, C Sweden 59°33′N 14°16′E
92 G13 **Storforshei** Nordland, C Norway 66°25′N 14°25′E
Storhammer *see* Hamar
93 F16 **Storlien** Jämtland, C Sweden 63°18′N 12°10′E
183 P17 **Storm Bay** *inlet* Tasmania, SE Australia
29 T12 **Storm Lake** Iowa, C USA 42°38′N 95°12′W
29 S13 **Storm Lake** ◎ Iowa, C USA
96 G7 **Stornoway** NW Scotland, United Kingdom 58°13′N 06°23′W
Storojineţ *see* Storozhynets'
92 P1 **Storøya** *island* NE Svalbard
125 S10 **Storozhevsk** Respublika Komi, NW Russian Federation 61°56′N 52°18′E
Storozhinets *see* Storozhynets'
116 K8 **Storozhynets'** *Ger.* Storozynetz, *Rom.* Storojineţ, *Rus.* Storozhinets. Chernivets'ka Oblast', W Ukraine 48°11′N 25°42′E
Storozynetz *see* Storozhynets'
92 H11 **Storriten** *Lapp.* Stuorrarijjda. ▲ C Norway 68°09′N 17°12′E
19 N12 **Storrs** Connecticut, NE USA 41°48′N 72°15′W
94 I11 **Storsjøen** ◎ S Norway
94 N13 **Storsjön** ◎ C Sweden
93 F16 **Storsjön** ◎ C Sweden
92 J9 **Storslett** Troms, N Norway 69°45′N 21°03′E
94 H11 **Storsølnkletten** ▲ S Norway 61°52′N 11°32′E
92 I9 **Storsteinnes** Troms, N Norway 69°13′N 19°14′E
93 J14 **Storsund** Norrbotten, N Sweden 65°36′N 20°40′E
94 J9 **Storsylen** *Swe.* Sylarna. ▲ Norway/Sweden 63°00′N 12°14′E
92 H11 **Stortoppen** ▲ N Sweden 67°33′N 17°27′E
93 H14 **Storuman** Västerbotten, N Sweden 65°05′N 17°10′E
93 H14 **Storuman** ◎ N Sweden
94 N13 **Storvik** Gävleborg, C Sweden 60°37′N 16°30′E
95 O14 **Storvreta** Uppsala, C Sweden 59°58′N 17°42′E
100 L10 **Störwasserstrasse** *canal* N Germany
29 V13 **Story City** Iowa, C USA 42°10′N 93°36′W
11 V17 **Stoughton** Saskatchewan, S Canada 49°40′N 103°01′W
19 O11 **Stoughton** Massachusetts, NE USA 42°07′N 71°06′W
30 L9 **Stoughton** Wisconsin, N USA 42°56′N 89°12′W
97 L23 **Stour** ≈ E England, United Kingdom
97 P21 **Stour** ≈ S England, United Kingdom
27 T5 **Stover** Missouri, C USA 38°26′N 92°59′W
95 G21 **Støvring** Nordjylland, N Denmark 56°53′N 09°52′E
119 J17 **Stowbtsy** *Pol.* Stolbce, *Rus.* Stolbtsy. Minskaya Voblasts', C Belarus 53°29′N 26°44′E
25 X11 **Stowell** Texas, SW USA 29°47′N 94°22′W
97 P20 **Stowmarket** E England, United Kingdom 52°05′N 00°54′E
114 N8 **Stozher** Dobrich, NE Bulgaria 43°27′N 27°49′E
97 E14 **Strabane** *Ir.* An Srath Bán. W Northern Ireland, United Kingdom 54°49′N 07°27′W
121 S11 **Strabo Trench** *undersea feature* C Mediterranean Sea
27 T7 **Strafford** Missouri, C USA 37°16′N 93°07′W
183 N17 **Strahan** Tasmania, SE Australia 42°10′S 145°18′E
111 C18 **Strakonice** *Ger.* Strakonitz. Jihočeský Kraj, S Czech Republic 49°14′N 13°55′E
Strakonitz *see* Strakonice
100 N8 **Stralsund** Mecklenburg-Vorpommern, NE Germany 54°18′N 13°06′E
99 L16 **Stramproy** Limburg, SE Netherlands 51°12′N 05°43′E
83 E26 **Strand** Western Cape, SW South Africa 34°06′S 18°50′E
94 E10 **Stranda** Møre og Romsdal, S Norway 62°18′N 06°56′E
97 G15 **Strangford Lough** *Ir.* Loch Cuan. *inlet* E Northern Ireland, United Kingdom
95 N16 **Strängnäs** Södermanland, C Sweden 59°22′N 17°02′E
97 E14 **Stranorlar** *Ir.* Srath an Urláir. NW Ireland 54°48′N 07°46′W
97 H14 **Stranraer** S Scotland, United Kingdom 54°54′N 05°02′W
11 U16 **Strasbourg** Saskatchewan, S Canada 51°05′N 104°58′W
103 V5 **Strasbourg** *Ger.* Strassburg; *anc.* Argentoratum. Bas-Rhin, NE France 48°35′N 07°45′E
109 T8 **Strassburg** Kärnten, S Austria 46°54′N 14°21′E
37 U4 **Strasburg** Colorado, C USA 39°42′N 104°13′W
29 N7 **Strasburg** North Dakota, N USA 46°07′N 100°10′W
31 U12 **Strasburg** Ohio, N USA 40°35′N 81°31′W
21 U3 **Strasburg** Virginia, NE USA 38°59′N 78°21′W
117 N10 **Străşeni** *var.* Strasheny. C Moldova 47°07′N 28°37′E
Strasheny *see* Străşeni
Strassburg *see* Strasbourg, France
Strassburg *see* Aiud, Romania
99 M25 **Strassen** Luxembourg, S Luxembourg 49°37′N 06°05′E
109 R5 **Strasswalchen** Salzburg, C Austria 47°59′N 13°19′E
14 F16 **Stratford** Ontario, S Canada 43°22′N 81°00′W
184 K10 **Stratford** Taranaki, North Island, New Zealand 39°20′S 174°16′E
35 Q11 **Stratford** California, W USA 36°10′N 119°47′W
29 V13 **Stratford** Iowa, C USA 42°16′N 93°55′W
27 O12 **Stratford** Oklahoma, C USA 34°48′N 96°57′W
25 N1 **Stratford** Texas, SW USA 36°21′N 102°05′W
30 K6 **Stratford** Wisconsin, N USA 44°53′N 90°13′W
Stratford *see* Stratford-upon-Avon
97 M20 **Stratford-upon-Avon** *var.* Stratford. C England, United Kingdom 52°12′N 01°41′W
183 O17 **Strathgordon** Tasmania, SE Australia 42°49′S 146°04′E
11 Q16 **Strathmore** Alberta, SW Canada 51°05′N 113°20′W
35 R11 **Strathmore** California, W USA 36°07′N 119°04′W
14 E16 **Strathroy** Ontario, S Canada 42°57′N 81°40′W
96 I6 **Strathy Point** *headland* N Scotland, United Kingdom 58°36′N 04°04′W
37 W4 **Stratton** Colorado, C USA 39°16′N 102°34′W
19 P6 **Stratton** Maine, NE USA 45°08′N 70°25′W
18 M10 **Stratton Mountain** ▲ Vermont, NE USA 43°05′N 72°55′W
101 N21 **Straubing** Bayern, SE Germany 48°53′N 12°35′E
100 O12 **Strausberg** Brandenburg, E Germany 52°34′N 13°52′E
32 K13 **Strawberry Mountain** ▲ Oregon, NW USA 44°18′N 118°43′W
29 X12 **Strawberry Point** Iowa, C USA 42°40′N 91°31′W
36 M3 **Strawberry Reservoir** ⊞ Utah, W USA
36 M4 **Strawberry River** ≈ Utah, W USA
25 R7 **Strawn** Texas, SW USA 32°33′N 98°30′W
113 P17 **Straža** ▲ Bulgaria/FYR Macedonia 42°16′N 22°13′E
111 I19 **Strážov** *Hung.* Sztrazsó. ▲ NW Slovakia 48°59′N 18°29′E
182 F7 **Streaky Bay** South Australia 32°49′S 134°13′E
182 E7 **Streaky Bay** *bay* South Australia
30 L12 **Streator** Illinois, N USA 41°07′N 88°50′W
Streckenbach *see* Świdnik
Strednogorie *see* Pirdop
111 C17 **Středočeský Kraj** ◐ *region* C Czech Republic
29 O6 **Streeter** North Dakota, N USA 46°37′N 99°23′W
25 U8 **Streetman** Texas, SW USA 31°52′N 96°19′W
116 G13 **Strehaia** Mehedinţi, SW Romania 44°37′N 23°10′E
Strehlen *see* Strzelin
114 I10 **Strelcha** Pazardzhik, C Bulgaria 42°28′N 24°21′E
122 L12 **Strelka** Krasnoyarskiy Kray, C Russian Federation 58°05′N 92°54′E
124 L6 **Strel'na** ≈ NW Russian Federation
118 H7 **Strenči** *Ger.* Stackeln. N Latvia 57°38′N 25°42′E
15 V6 **St-René-de-Matane** Québec, SE Canada 48°42′N 67°22′W
108 K8 **Strengen** Tirol, W Austria 47°07′N 10°25′E
106 C6 **Stresa** Piemonte, NE Italy 45°52′N 08°32′E
Streshin *see* Streshyn
119 N18 **Streshyn** *Rus.* Streshin. Homyel'skaya Voblasts', SE Belarus 52°43′N 30°07′E
95 B18 **Streymoy** *Dan.* Strømø. *island* N Faroe Islands
95 G23 **Strib** Syddtjylland, C Denmark 55°33′N 09°47′E
111 A17 **Stříbro** *Ger.* Mies. Plzeňský Kraj, W Czech Republic 49°44′N 12°55′E
186 B7 **Strickland** ≈ SW Papua New Guinea
Striegau *see* Strzegom
Strigonium *see* Esztergom
98 H13 **Strijen** Zuid-Holland, SW Netherlands 51°45′N 04°34′E
63 H21 **Strobel, Lago** ◎ S Argentina
61 B25 **Stroeder** Buenos Aires, E Argentina 40°11′S 62°35′W
115 C20 **Strofádes** *island* Iónia Nisiá, Greece, C Mediterranean Sea
Strofilia *see* Strofyliá
115 G17 **Strofyliá** *var.* Strofilia. Évvoia, C Greece 38°49′N 23°25′E
100 O10 **Strom** ≈ NE Germany
107 L22 **Stromboli** ✸ Isola Stromboli, SW Italy 38°48′N 15°13′E
107 L22 **Stromboli, Isola** *island* Isole Eolie, S Italy
96 H9 **Stromeferry** N Scotland, United Kingdom 57°20′N 05°35′W
96 J5 **Stromness** N Scotland, United Kingdom 58°57′N 03°18′W
Strømø *see* Streymoy
94 N11 **Strömsbruk** Gävleborg, C Sweden 61°52′N 17°19′E
29 Q15 **Stromsburg** Nebraska, C USA 41°06′N 97°36′W
95 K21 **Strömsnäsbruk** Kronoberg, S Sweden 56°35′N 13°45′E
95 I17 **Strömstad** Västra Götaland, S Sweden 58°56′N 11°11′E
93 G16 **Strömsund** Jämtland, C Sweden 63°51′N 15°35′E
93 G15 **Ströms Vattudal** *valley* N Sweden
27 V14 **Strong** Arkansas, C USA 33°06′N 92°19′W
Strongilí *see* Strongylí

◆ Country
● Country Capital
◇ Dependent Territory
○ Dependent Territory Capital
◐ Administrative Regions
✈ International Airport
▲ Mountain
▲▲ Mountain Range
✸ Volcano
≈ River
◎ Lake
⊞ Reservoir

STRONGOLI – SUROVIKINO

107 O21 **Strongoli** Calabria, SW Italy 39°17′N 17°03′E
31 T11 **Strongsville** Ohio, N USA 41°18′N 81°50′W
115 Q23 **Strongylí** *var.* Strongilí. *island* SE Greece
96 K5 **Stronsay** *island* NE Scotland, United Kingdom
97 L21 **Stroud** C England, United Kingdom 51°46′N 02°15′W
27 O10 **Stroud** Oklahoma, C USA 35°45′N 96°39′W
18 I14 **Stroudsburg** Pennsylvania, NE USA 40°59′N 75°12′W
95 F21 **Struer** Midtjylland, W Denmark 56°29′N 08°37′E
113 M20 **Struga** SW FYR Macedonia 41°11′N 20°40′E
Strugi-Kranyse *see* Strugi-Krasnyye
124 G14 **Strugi-Krasnyye** *var.* Strugi-Kranyse. Pskovskaya Oblast', W Russian Federation 58°19′N 29°09′E
114 G11 **Struma** *Gk.* Strymónas. ≈ Bulgaria/Greece *see also* Strymónas
Struma *see* Strymónas
97 G21 **Strumble Head** *headland* SW Wales, United Kingdom 52°01′N 05°05′W
Strumeshnitsa *see* Strumica
113 Q19 **Strumica** E FYR Macedonia 41°27′N 22°39′E
113 Q19 **Strumica** *Bulg.* Strumeshnitsa. ≈ Bulgaria/ FYR Macedonia
114 G11 **Strumyani** Blagoevgrad, SW Bulgaria 41°41′N 23°13′E
31 V12 **Struthers** Ohio, N USA 41°03′N 80°36′W
114 I10 **Stryama** ≈ C Bulgaria
114 G13 **Strymónas** *Bul.* Struma. ≈ Bulgaria/Greece *see also* Struma
Strymónas *see* Struma
115 H14 **Strymonikós Kólpos** *gulf* N Greece
116 I6 **Stryy** L'vivs'ka Oblast', NW Ukraine 49°16′N 23°51′E
116 H6 **Stryy** ≈ W Ukraine
111 F14 **Strzegom** *Ger.* Striegau. Wałbrzych, SW Poland 50°59′N 16°20′E
110 E10 **Strzelce Krajeńskie** *Ger.* Friedeberg Neumark. Lubuskie, W Poland 52°52′N 15°30′E
111 I15 **Strzelce Opolskie** *Ger.* Gross Strehlitz. Opolskie, SW Poland 50°31′N 18°19′E
182 K3 **Strzelecki Creek** *seasonal river* South Australia
182 J3 **Strzelecki Desert** *desert* South Australia
111 G15 **Strzelin** *Ger.* Strehlen. Dolnośląskie, SW Poland 50°48′N 17°03′E
110 I11 **Strzelno** Kujawsko-pomorski, C Poland 52°38′N 18°11′E
111 N17 **Strzyżów** Podkarpackie, SE Poland 49°52′N 21°46′E
15 S8 **St-Siméon** Québec, SE Canada 47°50′N 69°55′W
Stua Laighean *see* Leinster, Mount
23 Y13 **Stuart** Florida, SE USA 27°12′N 80°15′W
29 U14 **Stuart** Iowa, C USA 41°30′N 94°19′W
29 O13 **Stuart** Nebraska, C USA 42°36′N 99°08′W
21 S8 **Stuart** Virginia, NE USA 36°38′N 80°19′W
10 L13 **Stuart** ≈ British Columbia, SW Canada
39 N10 **Stuart Island** *island* Alaska, USA
10 L13 **Stuart Lake** ◎ British Columbia, SW Canada
185 B22 **Stuart Mountains** ▲▲ South Island, New Zealand
182 F3 **Stuart Range** *hill range* South Australia
Stubaital *see* Neustift im Stubaital
95 I24 **Stubbekøbing** Sjælland, SE Denmark 54°53′N 12°04′E
45 P14 **Stubbs** Saint Vincent, Saint Vincent and the Grenadines 13°08′N 61°09′W
109 V6 **Stübming** ≈ E Austria
114 J11 **Studen Kladenets, Yazovir** ⊠ S Bulgaria
185 G21 **Studholme** Canterbury, South Island, New Zealand 44°44′S 171°08′E
Stuhlweissenberg *see* Székesfehérvár
Stuhm *see* Sztum
12 C7 **Stull Lake** ◎ Ontario, C Canada
Stuorrarijjda *see* Storriten
126 L4 **Stupino** Moskovskaya Oblast', W Russian Federation 54°54′N 38°06′E
27 U4 **Sturgeon** Missouri, C USA 39°13′N 92°16′W
14 G10 **Sturgeon** ≈ Ontario, S Canada
31 N6 **Sturgeon Bay** Wisconsin, N USA 44°51′N 87°21′W
14 G11 **Sturgeon Falls** Ontario, S Canada 46°22′N 79°57′W
12 C11 **Sturgeon Lake** ◎ Ontario, S Canada
30 M3 **Sturgeon River** ≈ Michigan, N USA
20 H6 **Sturgis** Kentucky, S USA 37°33′N 87°58′W
31 P11 **Sturgis** Michigan, N USA 41°48′N 85°25′W
28 J9 **Sturgis** South Dakota, N USA 44°24′N 103°30′W
112 D10 **Šturlić** ◐ Federacija Bosne I Hercegovine, NW Bosnia and Herzegovina
111 J22 **Štúrovo** *Hung.* Párkány; *prev.* Parkan. Nitriansky Kraj, SW Slovakia 47°49′N 18°40′E
182 L4 **Sturt, Mount** *hill* New South Wales, SE Australia
181 P4 **Sturt Plain** *plain* Northern Territory, N Australia
181 T9 **Sturt Stony Desert** *desert* South Australia
83 J25 **Stutterheim** Eastern Cape, S South Africa 32°35′S 27°26′E
101 H21 **Stuttgart** Baden-Württemberg, SW Germany 48°47′N 09°12′E
27 W12 **Stuttgart** Arkansas, C USA 34°30′N 91°32′W
92 H2 **Stykkishólmur** Vesturland, W Iceland 65°04′N 22°43′W
115 F17 **Stylída** *var.* Stilida, Stilís. Stereá Elláda, C Greece 38°55′N 22°37′E
116 K2 **Styr** *Rus.* Styr'. ≈ Belarus/ Ukraine
115 I19 **Stýra** *var.* Stira. Évvoia, C Greece 38°10′N 24°13′E
Styria *see* Steiermark
15 Y5 **St-Yvon** Québec, SE Canada 49°09′N 64°51′W
Su *see* Jiangsu
Sua *see* Sowa
171 Q17 **Suai** W East Timor 09°19′S 125°16′E
54 G9 **Suaita** Santander, C Colombia 06°07′N 73°30′W
80 I7 **Suakin** *var.* Sawakin. Red Sea, NE Sudan 19°06′N 37°17′E
161 T13 **Su'ao** *Jap.* Suō. N Taiwan 24°33′N 121°48′E
Suao *see* Suau
Sua Pan *see* Sowa Pan
40 G6 **Suaqui Grande** Sonora, NW Mexico 28°22′N 109°52′W
61 A16 **Suardi** Santa Fe, C Argentina 30°32′S 61°58′W
54 D11 **Suárez** Cauca, SW Colombia 02°55′N 76°41′W
186 G10 **Suau** *var.* Suao. Suaul Island, SE Papua New Guinea 10°39′S 150°03′E
118 G12 **Subačius** Panevėžys, NE Lithuania 55°46′N 24°45′E
168 K9 **Subang** *prev.* Soebang. Jawa, C Indonesia 06°32′S 107°45′E
169 O16 **Subang** ✈ (Kuala Lumpur) Pahang, Peninsular Malaysia
129 S10 **Subansiri** ≈ NE India
154 M12 **Subarnapur** *prev.* Sonapur, Sonepur. Odisha, E India 20°50′N 83°58′E
118 I11 **Subate** SE Latvia 56°00′N 25°54′E
139 N5 **Subaykhān** Dayr az Zawr, E Syria 34°52′N 40°35′E
Subei/Subei Mongolzu Zizhixian *see* Dangchengwan
169 P9 **Subi Besar, Pulau** *island* Kepulauan Natuna, W Indonesia
Subiyah *see* Aş Şubayḩīyah
26 I7 **Sublette** Kansas, C USA 37°28′N 100°52′W
112 K8 **Subotica** *Ger.* Maria-Theresiopel, *Hung.* Szabadka. Vojvodina, N Serbia 46°06′N 19°41′E
116 K9 **Suceava** *Ger.* Suczawa, *Hung.* Szucsava. Suceava, NE Romania 47°41′N 26°16′E
116 J9 **Suceava** ◐ *county* NE Romania
116 K9 **Suceava** *Ger.* Suczawa. ≈ N Romania
112 E12 **Sučevići** Zadar, SW Croatia 44°13′N 16°04′E
111 K17 **Sucha Beskidzka** Małopolskie, S Poland 49°44′N 19°36′E
111 M14 **Suchedniów** Świętokrzyskie, C Poland 51°01′N 20°49′E
42 A2 **Suchitepéquez** *off.* Departamento de Suchitepéquez. ◐ *department* SW Guatemala
Suchitepéquez, Departamento de *see* Suchitepéquez
Su-chou *see* Suzhou
Suchow *see* Suzhou, Jiangsu, China
Suchow *see* Xuzhou, Jiangsu, China
97 D17 **Suck** ≈ C Ireland
Sucker State *see* Illinois
186 F9 **Suckling, Mount** ▲ S Papua New Guinea 09°36′S 149°00′E
57 L19 **Sucre** *hist.* Chuquisaca, La Plata. ● (Bolivia-legal capital) Chuquisaca, S Bolivia 18°53′S 65°25′W
54 E6 **Sucre** Santander, N Colombia 08°50′N 74°22′W
56 A7 **Sucre** Manabí, W Ecuador 01°21′S 80°27′W
54 E6 **Sucre** *off.* Departamento de Sucre. ◐ *province* N Colombia
55 O5 **Sucre** *off.* Estado Sucre. ◐ *state* NE Venezuela
Sucre, Departamento de *see* Sucre
Sucre, Estado *see* Sucre
56 D6 **Sucumbíos** ◐ *province* NE Ecuador
113 G15 **Sućuraj** Split-Dalmacija, S Croatia 43°07′N 17°10′E
58 K10 **Sucuriju** Amapá, NE Brazil 01°31′N 50°W
Suczawa *see* Suceava
79 E16 **Sud** *Eng.* South. ◐ *province* S Cameroon
124 K13 **Suda** ≈ NW Russian Federation
Suda *see* Soúda
117 U13 **Sudak** Avtonomna Respublika Krym, S Ukraine 44°52′N 34°57′E
24 M4 **Sudan** Texas, SW USA 34°04′N 102°32′W
80 C10 **Sudan** *off.* Republic of Sudan, *Ar.* Jumhuriyat as-Sudan; *prev.* Anglo-Egyptian Sudan. ◆ *republic* N Africa
Sudanese Republic *see* Mali
Sudan, Jumhuriyat as- *see* Sudan
Sudan, Republic of *see* Sudan
14 F10 **Sudbury** Ontario, S Canada 46°29′N 81°W
97 P20 **Sudbury** E England, United Kingdom 52°04′N 00°43′E
Sud, Canal de *see* Gonâve, Canal de la
80 E13 **Sudd** *swamp region* S South Sudan
100 K10 **Sude** ≈ N Germany
Suderø *see* Suduroy
Sudest Island *see* Tagula Island
111 E15 **Sudeten** *var.* Sudetes, Sudetic Mountains, *Cz./Pol.* Sudety. ▲▲ Czech Republic/ Poland
Sudetes/Sudetic Mountains/Sudety *see* Sudeten
95 B19 **Suðuroy** *Dan.* Suderø. *island* S Faroe Islands
124 M15 **Sudislavl'** Kostromskaya Oblast', NW Russian Federation 57°55′N 41°45′E
Südkarpaten *see* Carpații Meridionalii
79 N20 **Sud-Kivu** *off.* Région Sud Kivu. ◐ *region* E Dem. Rep. Congo
Sud Kivu, Région *see* Sud-Kivu
Südliche Morava *see* Južna Morava
100 E12 **Süd-Nord-Kanal** *canal* NW Germany
126 M3 **Sudogda** Vladimirskaya Oblast', W Russian Federation 55°58′N 40°57′E
Sudostroy *see* Severodvinsk
79 C15 **Sud-Ouest** *Eng.* South-West. ◐ *province* W Cameroon
173 X17 **Sud Ouest, Pointe** *headland* SW Mauritius 20°27′S 57°18′E
187 P17 **Sud, Province** ◐ *province* S New Caledonia
92 G1 **Suðureyri** Vestfirðir, NW Iceland 66°08′N 23°31′W
92 J4 **Suðurland** ◐ *region* S Iceland
92 H4 **Suðurnes** ◐ *region* SW Iceland
126 J8 **Sudzha** Kurskaya Oblast', W Russian Federation 51°12′N 35°19′E
81 D15 **Sue** ≈ W South Sudan
105 S10 **Sueca** Valenciana, E Spain 39°13′N 00°19′W
Süedinenie *see* Saedinenie
Suero *see* Alzira
75 X8 **Suez** *Ar.* As Suways, El Suweis. NE Egypt 29°59′N 32°33′E
75 W7 **Suez Canal** *Ar.* Qanāt as Suways. *canal* NE Egypt
Suez, Gulf of *see* Khalīj as Suways
11 R17 **Suffield** Alberta, SW Canada 50°15′N 111°05′W
21 X7 **Suffolk** Virginia, NE USA 36°44′N 76°37′W
97 P20 **Suffolk** *cultural region* E England, United Kingdom
142 J2 **Sūfīān** Āzarbāyjān-e Sharqī, N Iran 38°15′N 45°59′E
31 N12 **Sugar Creek** ≈ Illinois, N USA
30 L13 **Sugar Creek** ≈ Illinois, N USA
31 R3 **Sugar Island** *island* Michigan, N USA
25 V11 **Sugar Land** Texas, SW USA 29°37′N 95°37′W
19 P6 **Sugarloaf Mountain** ▲ Maine, NE USA 45°01′N 70°18′W
65 G24 **Sugar Loaf Point** *headland* N Saint Helena 15°54′S 05°43′W
136 G16 **Suğla Gölü** ◎ SW Turkey
123 T8 **Sugoy** ≈ E Russian Federation
158 F7 **Sugun** Xinjiang Uygur Zizhiqu, W China 39°46′N 76°45′E
147 U11 **Sugut, Gora** ▲ SW Kyrgyzstan 39°52′N 73°36′E
169 V6 **Sugut, Sungai** ≈ East Malaysia
159 O9 **Suhai Hu** ◎ C China
162 K14 **Suhait** Nei Mongol Zizhiqu, N China 39°29′N 105°11′E
141 X7 **Şuḩār** *var.* Sohar. N Oman 24°20′N 56°43′E
114 N8 **Suha Reka** ≈ NE Bulgaria
113 M17 **Suharekë** *Serb.* Suva Reka. S Kosovo 42°23′N 20°50′E
162 L6 **Sühbaatar** Selenge, N Mongolia 50°12′N 106°14′E
163 P8 **Sühbaatar** *var.* Haylaastay. Sühbaatar, E Mongolia 46°44′N 113°51′E
163 P9 **Sühbaatar** ◐ *province* E Mongolia
114 J8 **Suhindol** *var.* Sukhindol. Veliko Turnovo, N Bulgaria 43°11′N 24°10′E
101 K17 **Suhl** Thüringen, C Germany 50°37′N 10°43′E
108 F7 **Suhr** Aargau, N Switzerland 47°23′N 08°05′E
Sui'an *see* Zhangpu
Suicheng *see* Suixi
161 O12 **Suichuan** *var.* Quanjiang. Jiangxi, S China 26°26′N 114°34′E
Suid-Afrika *see* South Africa
160 L4 **Suide** *var.* Mingzhou. Shaanxi, C China 37°30′N 110°10′E
Suidwes-Afrika *see* Namibia
163 Y9 **Suifenhe** Heilongjiang, NE China 44°22′N 131°12′E
Suigen *see* Suwon
163 W8 **Suihua** Heilongjiang, NE China 46°40′N 127°00′E
Súilí, Loch *see* Swilly, Lough
161 Q6 **Suining** Jiangsu, E China 33°54′N 117°58′E
160 I9 **Suining** Sichuan, C China 30°31′N 105°33′E
103 Q4 **Suippes** Marne, N France 49°08′N 04°31′E
97 E20 **Suir** *Ir.* An tSiúir. ≈ S Ireland
165 J13 **Suita** Ōsaka, Honshū, SW Japan 34°39′N 135°27′E
160 L16 **Suixi** *var.* Suicheng. Guangdong, S China 21°23′N 110°14′E
Sui Xian *see* Suizhou
163 T13 **Suizhong** Liaoning, NE China 40°19′N 120°22′E
161 N8 **Suizhou** *prev.* Sui Xian. Hubei, C China 31°46′N 113°20′E
149 P17 **Sujāwal** Sind, SE Pakistan 24°36′N 68°06′E
169 O16 **Sukabumi** *prev.* Soekaboemi. Jawa, C Indonesia 06°55′S 106°56′E
169 Q12 **Sukadana, Teluk** *bay* Borneo, W Indonesia
165 P11 **Sukagawa** Fukushima, Honshū, C Japan 37°16′N 140°20′E
Sukarnapura *see* Jayapura
Sukarno, Puntjak *see* Jaya, Puncak
Sükh *see* Sokh
Sukhindol *see* Suhindol
126 J5 **Sukhinichi** Kaluzhskaya Oblast', W Russian Federation 54°06′N 35°22′E
Sukhne *see* As Sukhnah
129 Q4 **Sukhona** *var.* Tot'ma. ≈ NW Russian Federation
167 O8 **Sukhothai** *var.* Sukotai. Sukhothai, W Thailand 17°00′N 99°51′E
Sukhumi *see* Sokhumi
Sukkertoppen *see* Maniitsoq
149 Q13 **Sukkur** Sind, SE Pakistan 27°45′N 68°46′E
Sukotai *see* Sukhothai
Sukra Bay *see* Şawqirah, Dawḩat
125 V15 **Suksun** Permskiy Kray, NW Russian Federation 57°10′N 57°27′E
165 F15 **Sukumo** Kōchi, Shikoku, SW Japan 32°55′N 132°42′E
94 B12 **Sula** *island* S Norway
125 Q5 **Sula** ≈ NW Russian Federation
117 R5 **Sula** ≈ N Ukraine
42 H6 **Sulaco, Río** ≈ NW Honduras
Sulaimaniya *see* As Sulaymānīyah
149 S10 **Sulaimān Range** ▲▲ C Pakistan
127 Q16 **Sulak** Respublika Dagestan, SW Russian Federation 43°19′N 47°28′E
127 Q16 **Sulak** ≈ SW Russian Federation
171 Q13 **Sula, Kepulauan** *island group* C Indonesia
136 I12 **Sulakyurt** *var.* Konur. Kırıkkale, N Turkey 40°10′N 33°42′E
171 P17 **Sulamu** Timor, S Indonesia 09°57′S 123°33′E
96 F5 **Sula Sgeir** *island* NW Scotland, United Kingdom
171 N13 **Sulawesi** *Eng.* Celebes. *island* C Indonesia
171 N13 **Sulawesi Barat** *off.* Provinsi Sulawesi Barat, *var.* Sulbar. ◐ *province* C Indonesia
Sulawesi Barat, Provinsi *see* Sulawesi Barat
Sulawesi, Laut *see* Celebes Sea
171 N14 **Sulawesi Selatan** *off.* Propinsi Sulawesi Selatan, *var.* Sulsel, *Eng.* South Celebes, South Sulawesi. ◐ *province* C Indonesia
Sulawesi Selatan, Propinsi *see* Sulawesi Selatan
171 P12 **Sulawesi Tengah** *off.* Propinsi Sulawesi Tengah, *var.* Sulteng, *Eng.* Central Celebes, Central Sulawesi. ◐ *province* N Indonesia
Sulawesi Tengah, Propinsi *see* Sulawesi Tengah
171 O14 **Sulawesi Tenggara** *off.* Propinsi Sulawesi Tenggara, *var.* Sultenggara, *Eng.* South-East Celebes, South-East Sulawesi. ◐ *province* C Indonesia
Sulawesi Tenggara, Propinsi *see* Sulawesi Tenggara
171 P11 **Sulawesi Utara** *off.* Propinsi Sulawesi Utara, *var.* Sulut, *Eng.* North Celebes, North Sulawesi. ◐ *province* N Indonesia
Sulawesi Utara, Propinsi *see* Sulawesi Utara
139 T5 **Sulaymān Beg** At Ta'mím, N Iraq
Sulbar *see* Sulawesi Barat
95 D15 **Suldalsvatnet** ◎ S Norway
Sulden *see* Solda
110 E12 **Sulechów** *Ger.* Züllichau. Lubuskie, W Poland 52°05′N 15°37′E
110 E11 **Sulęcin** Lubuskie, W Poland 52°29′N 15°06′E
77 U14 **Suleja** Niger, C Nigeria 09°15′N 07°10′E
111 K14 **Sulejów** Lodzkie, S Poland 51°21′N 19°57′E
96 I5 **Sule Skerry** *island* N Scotland, United Kingdom
Suliag *see* Sawhāj
76 J16 **Sulima** S Sierra Leone 06°59′N 11°34′W
117 O13 **Sulina** Tulcea, SE Romania 45°07′N 29°40′E
117 N13 **Sulina, Braţul** ≈ SE Romania
100 H12 **Sulingen** Niedersachsen, NW Germany 52°40′N 08°48′E
Sulisjielmmá *see* Sulitjelma
92 H12 **Sulitjelma** ◎ C Norway 67°10′N 16°16′E
92 H12 **Sulitjelma** *Lapp.* Sulisjielmmá. Nordland, C Norway 67°10′N 16°05′E
56 A9 **Sullana** Piura, NW Peru 04°54′S 80°42′W
23 N3 **Sulligent** Alabama, S USA 33°54′N 88°07′W
30 M14 **Sullivan** Illinois, N USA 39°36′N 88°36′W
31 N15 **Sullivan** Indiana, N USA 39°05′N 87°24′W
27 W5 **Sullivan** Missouri, C USA 38°12′N 91°09′W
Sullivan Island *see* Lanbi Kyun
96 M1 **Sullom Voe** NE Scotland, United Kingdom 60°24′N 01°09′W
103 O7 **Sully-sur-Loire** Loiret, C France 47°46′N 02°21′E
Sulmo *see* Sulmona
107 K15 **Sulmona** *anc.* Sulmo. Abruzzo, C Italy 42°03′N 13°56′E
Sulo *see* Shule He
114 M11 **Süloğlu** Edirne, NW Turkey 41°46′N 26°55′E
22 G9 **Sulphur** Louisiana, S USA 30°14′N 93°22′W
27 O12 **Sulphur** Oklahoma, C USA 34°31′N 96°58′W
28 K9 **Sulphur Creek** ≈ South Dakota, N USA
24 M5 **Sulphur Draw** ≈ Texas, SW USA
25 W5 **Sulphur River** ≈ Arkansas/ Texas, SW USA
25 V6 **Sulphur Springs** Texas, SW USA 33°09′N 95°36′W
24 M6 **Sulphur Springs Draw** ≈ Texas, SW USA
Sulsel *see* Sulawesi Selatan
14 D8 **Sultan** Ontario, S Canada 47°34′N 82°45′W
Sultānābād *see* Arāk
Sultan Alonto, Lake *see* Lanao, Lake
136 G15 **Sultan Dağları** ▲▲ C Turkey
114 N13 **Sultanköy** Tekirdağ, NW Turkey 41°01′N 27°58′E
171 Q7 **Sultan Kudarat** *var.* Nuling. Mindanao, S Philippines 07°20′N 124°16′E
152 M13 **Sultānpur** Uttar Pradesh, N India 26°15′N 82°04′E
Sulteng *see* Sulawesi Tengah
Sultenggara *see* Sulawesi Tenggara
171 O9 **Sulu Archipelago** *island group* SW Philippines
192 F7 **Sulu Basin** *undersea feature* SE South China Sea 08°00′N 121°30′E
Sülüktü *see* Sulyukta
Sulu, Laut *see* Sulu Sea
169 X6 **Sulu Sea** *var.* Laut Sulu. *sea* SW Philippines
Sulut *see* Sulawesi Utara
145 O15 **Sulutobe** *Kaz.* Sulūtōbe. Kzylorda, S Kazakhstan 44°31′N 66°17′E
Sulūtōbe *see* Sulutobe
147 Q11 **Sulyukta** *Kir.* Sülüktü. Batkenskaya Oblast', SW Kyrgyzstan 39°57′N 69°31′E
Sulz *see* Sulz am Neckar
101 G22 **Sulz am Neckar** *var.* Sulz. Baden-Württemberg, SW Germany 48°22′N 08°37′E
101 L20 **Sulzbach-Rosenberg** Bayern, SE Germany 49°30′N 11°43′E
195 N13 **Sulzberger Bay** *bay* Antarctica
81 M14 **Sumalē** *var.* Somali. ◐ E Ethiopia
113 F15 **Šumartin** Split-Dalmacija, S Croatia 43°17′N 16°52′E
32 H6 **Sumas** Washington, NW USA 49°00′N 122°15′W
168 J10 **Sumatera** *Eng.* Sumatra. *island* W Indonesia
168 J12 **Sumatera Barat** *off.* Propinsi Sumatera Barat, *var.* Sumbar, *Eng.* West Sumatra. ◐ *province* W Indonesia
Sumatera Barat, Propinsi *see* Sumatera Barat
168 L13 **Sumatera Selatan** *off.* Propinsi Sumatera Selatan, *var.* Sumsel, *Eng.* South Sumatra. ◐ *province* W Indonesia
Sumatera Selatan, Propinsi *see* Sumatera Selatan
168 H10 **Sumatera Utara** *off.* Propinsi Sumatera Utara, *var.* Sumut, *Eng.* North Sumatra. ◐ *province* W Indonesia
Sumatera Utara, Propinsi *see* Sumatera Utara
Sumatra *see* Sumatera
Šumava *see* Bohemian Forest
Sumayl *see* Sēmēl
139 U7 **Sumayr al Muḩammad** Diyālá, E Iraq 33°34′N 45°06′E
171 N17 **Sumba, Pulau** *Eng.* Sandalwood Island; *prev.* Soemba. *island* Nusa Tenggara, C Indonesia
146 D12 **Sumbar** ≈ W Turkmenistan
Sumbar *see* Sumatera Barat
192 E9 **Sumbawa** *prev.* Soembawa. *island* Nusa Tenggara, C Indonesia
170 L16 **Sumbawabesar** Sumbawa, S Indonesia 08°30′S 117°25′E
81 F23 **Sumbawanga** Rukwa, W Tanzania 07°57′S 31°37′E
82 B12 **Sumbe** *var.* N'Gunza, *Port.* Novo Redondo. Kwanza Sul, W Angola 11°13′S 13°53′E
96 M3 **Sumburgh Head** *headland* NE Scotland, United Kingdom 59°51′N 01°16′W
111 H23 **Sümeg** Veszprém, W Hungary 47°01′N 17°13′E
80 C12 **Sumeih** Eastern Darfur, S Sudan 09°50′N 27°39′E
169 T16 **Sumenep** *prev.* Soemenep. Pulau Madura, C Indonesia 07°01′S 113°51′E
Sumgait *see* Sumqayıtçay, Azerbaijan
Sumgait *see* Sumqayıt, Azerbaijan
165 Y14 **Sumisu-jima** *Eng.* Smith Island. *island* SE Japan
Summēl *see* Sēmēl
31 O5 **Summer Island** *island* Michigan, N USA
32 H15 **Summer Lake** ◎ Oregon, NW USA
11 N17 **Summerland** British Columbia, SW Canada 49°35′N 119°45′W
13 P14 **Summerside** Prince Edward Island, SE Canada 46°24′N 63°46′W
21 R5 **Summersville** West Virginia, NE USA 38°17′N 80°52′W
21 R5 **Summersville Lake** ⊠ West Virginia, NE USA
21 S13 **Summerton** South Carolina, SE USA 33°36′N 80°21′W
23 R2 **Summerville** Georgia, SE USA 34°28′N 85°21′W
21 S14 **Summerville** South Carolina, SE USA 33°01′N 80°10′W
39 R10 **Summit** Alaska, USA 63°21′N 148°50′W
35 V6 **Summit Mountain** ▲ Nevada, W USA 39°23′N 116°25′W
37 R8 **Summit Peak** ▲ Colorado, C USA 37°21′N 106°42′W
Summus Portus *see* Somport, Col du
29 X12 **Sumner** Iowa, C USA 42°51′N 92°05′W
22 K3 **Sumner** Mississippi, S USA 33°58′N 90°22′W
185 H17 **Sumner, Lake** ◎ South Island, New Zealand
37 U12 **Sumner, Lake** ⊠ New Mexico, SW USA
111 G17 **Šumperk** *Ger.* Mährisch-Schönberg. Olomoucký Kraj, E Czech Republic 49°58′N 17°00′E
42 F7 **Sumpul, Río** ≈ El Salvador/ Honduras
137 Z11 **Sumqayıt** *Rus.* Sumgait. E Azerbaijan 40°33′N 49°41′E
137 Y11 **Sumqayıtçay** *Rus.* Sumgait. ≈ E Azerbaijan
147 R9 **Sumsar** Dzhalal-Abadskaya Oblast', W Kyrgyzstan 41°12′N 71°16′E
Sumsel *see* Sumatera Selatan
117 S3 **Sums'ka Oblast'** *var.* Sumy, *Rus.* Sumskaya Oblast'. ◐ *province* NE Ukraine
Sumskaya Oblast' *see* Sums'ka Oblast'
124 J8 **Sumskiy Posad** Respublika Kareliya, NW Russian Federation 64°12′N 35°22′E
21 S12 **Sumter** South Carolina, SE USA 33°54′N 80°22′W
Sumut *see* Sumatera Utara
117 T3 **Sumy** Sums'ka Oblast', NE Ukraine 50°54′N 34°49′E
Sumy *see* Sums'ka Oblast'
159 Q15 **Sumzom** Xizang Zizhiqu, W China 29°45′N 96°14′E
125 R15 **Suna** Kirovskaya Oblast', NW Russian Federation 57°53′N 50°04′E
124 I10 **Suna** ≈ NW Russian Federation
165 S3 **Sunagawa** Hokkaidō, NE Japan 43°30′N 141°55′E
153 V13 **Sunamganj** Sylhet, NE Bangladesh 25°04′N 91°24′E
163 W14 **Sunan** ✈ (P'yŏngyang) SW North Korea 39°12′N 125°40′E
Sunan/Sunan Yugurzu Zizhixian *see* Hongwansi
19 N9 **Sunapee Lake** ◎ New Hampshire, NE USA
139 P4 **Şunaysilah** *salt marsh* N Iraq
20 M8 **Sunbright** Tennessee, S USA 36°12′N 84°39′W
33 R6 **Sunburst** Montana, NW USA 48°51′N 111°54′W
183 N12 **Sunbury** Victoria, SE Australia 37°36′S 144°45′E
21 X8 **Sunbury** North Carolina, SE USA 36°27′N 76°34′W
18 G14 **Sunbury** Pennsylvania, NE USA 40°51′N 76°47′W
61 A17 **Sunchales** Santa Fe, C Argentina 30°58′S 61°35′W
163 Y16 **Suncheon** *prev.* Sunch'ŏn. S South Korea 34°56′N 127°29′E
Sunch'ŏn *see* Suncheon
36 K13 **Sun City** Arizona, SW USA 33°36′N 112°16′W
19 O9 **Suncook** New Hampshire, NE USA 43°07′N 71°25′W
161 P5 **Suncun** *prev.* Xinwen. Shandong, E China 35°49′N 117°36′E
Sunda Islands *see* Greater Sunda Islands
33 Z12 **Sundance** Wyoming, C USA 44°24′N 104°22′W
153 T17 **Sundarbans** *wetland* Bangladesh/India
154 M11 **Sundargarh** Odisha, E India 22°07′N 84°02′E
129 U15 **Sunda Shelf** *undersea feature* S South China Sea 05°00′N 107°00′E
Sunda Trench *see* Java Trench
129 U17 **Sunda Trough** *undersea feature* E Indian Ocean 08°50′S 109°30′E
95 O16 **Sundbyberg** Stockholm, C Sweden 59°22′N 17°58′E
97 M14 **Sunderland** *var.* Wearmouth. NE England, United Kingdom 54°55′N 01°23′W
101 F15 **Sundern** Nordrhein-Westfalen, W Germany 51°19′N 08°00′E
136 F12 **Sündiken Dağları** ▲▲ C Turkey
24 M5 **Sundown** Texas, SW USA 33°27′N 102°29′W
11 P16 **Sundre** Alberta, SW Canada 51°49′N 114°46′W
14 H12 **Sundridge** Ontario, S Canada 45°45′N 79°25′W
93 H17 **Sundsvall** Västernorrland, C Sweden 62°22′N 17°20′E
26 H4 **Sunflower, Mount** ▲ Kansas, C USA 39°01′N 102°02′W
Sunflower State *see* Kansas
169 N14 **Sungaibuntu** Sumatera, SW Indonesia 04°04′S 105°37′E
168 K12 **Sungaidareh** Sumatera, W Indonesia 0°58′S 101°30′E
167 P17 **Sungai Kolok** *var.* Sungai Ko-Lok. Narathiwat, SW Thailand 06°02′N 101°58′E
Sungai Ko-Lok *see* Sungai Kolok
168 K12 **Sungaipenuh** *prev.* Soengaipenoeh. Sumatera, W Indonesia 02°S 101°28′E
169 P11 **Sungaipinyuh** Borneo, C Indonesia 0°16′N 109°06′E
Sungari *see* Songhua Jiang
Sungaria *see* Dzungaria
Sungei Pahang *see* Pahang, Sungai
167 O8 **Sung Men** Phrae, NW Thailand 17°59′N 100°07′E
83 M15 **Sungo** Tete, NW Mozambique 16°31′S 33°58′E
Sungpu *see* Songpan
168 M13 **Sungsang** Sumatera, W Indonesia 02°22′S 104°50′E
114 M9 **Sungurlare** Burgas, E Bulgaria 42°47′N 26°46′E
136 J12 **Sungurlu** Çorum, N Turkey 40°10′N 34°23′E
112 F9 **Sunja** Sisak-Moslavina, C Croatia 45°21′N 16°33′E
153 Q12 **Sun Koshi** ≈ E Nepal
94 F9 **Sunndalen** *valley* S Norway
94 F9 **Sunndalsøra** Møre og Romsdal, S Norway 62°39′N 08°37′E
95 K15 **Sunne** Värmland, C Sweden 59°52′N 13°05′E
95 O15 **Sunnersta** Uppsala, C Sweden 59°46′N 17°40′E
94 C11 **Sunnfjord** *physical region* S Norway
95 C15 **Sunnhordland** *physical region* S Norway
94 D10 **Sunnmøre** *physical region* S Norway
37 N4 **Sunnyside** Utah, W USA 39°33′N 110°23′W
32 J10 **Sunnyside** Washington, NW USA 46°01′N 119°58′W
35 N9 **Sunnyvale** California, W USA 37°22′N 122°02′W
30 L8 **Sun Prairie** Wisconsin, N USA 43°12′N 89°12′W
Sunqur *see* Sonqor
25 N1 **Sunray** Texas, SW USA 36°01′N 101°49′W
22 I8 **Sunset** Louisiana, S USA 30°24′N 92°04′W
25 S5 **Sunset** Texas, SW USA 33°24′N 97°45′W
Sunset State *see* Oregon
181 Z10 **Sunshine Coast** *cultural region* Queensland, E Australia
Sunshine State *see* Florida
Sunshine State *see* New Mexico
Sunshine State *see* South Dakota
123 O10 **Suntar** Respublika Sakha (Yakutiya), NE Russian Federation 62°10′N 117°34′E
39 R10 **Suntrana** Alaska, USA 63°51′N 148°51′W
148 J15 **Suntsar** Baluchistān, SW Pakistan 25°30′N 62°03′E
163 W15 **Sunwi-do** *island* SW North Korea
163 W6 **Sunwu** Heilongjiang, NE China 49°29′N 127°15′E
77 O16 **Sunyani** W Ghana 07°22′N 02°18′W
Suō *see* Su'ao
93 M17 **Suolahti** Keski-Suomi, C Finland 62°32′N 25°51′E
Suoločielgi *see* Saariselkä
Suomenlahti *see* Finland, Gulf of
Suomen Tasavalta/Suomi *see* Finland
93 N14 **Suomussalmi** Kainuu, E Finland 64°54′N 29°05′E
165 E13 **Suō-nada** *sea* SW Japan
93 M17 **Suonenjoki** Pohjois-Savo, C Finland 62°36′N 27°07′E
167 S13 **Suŏng** Kâmpóng Cham, C Cambodia 11°53′N 105°41′E
124 I10 **Suoyarvi** Respublika Kareliya, NW Russian Federation 62°02′N 32°24′E
Supanburi *see* Suphan Buri
57 D14 **Supe** Lima, W Peru 10°49′S 77°40′W
15 V7 **Supérieur, Lac** ◎ Québec, SE Canada
Supérieur, Lac *see* Superior, Lake
36 M14 **Superior** Arizona, SW USA 33°17′N 111°06′W
33 O9 **Superior** Montana, NW USA 47°11′N 114°53′W
29 P17 **Superior** Nebraska, C USA 40°01′N 98°04′W
30 I3 **Superior** Wisconsin, N USA 46°42′N 92°04′W
41 S17 **Superior, Laguna** *lagoon* S Mexico
31 N2 **Superior, Lake** *Fr.* Lac Supérieur. ◎ Canada/USA
36 L13 **Superstition Mountains** ▲▲ Arizona, SW USA
113 F14 **Supetar** *It.* San Pietro. Split-Dalmacija, S Croatia 43°22′N 16°34′E
167 O10 **Suphan Buri** *var.* Supanburi. Suphan Buri, W Thailand 14°29′N 100°10′E
171 V12 **Supiori, Pulau** *island* E Indonesia
188 K2 **Supply Reef** *reef* N Northern Mariana Islands
195 O7 **Support Force Glacier** *glacier* Antarctica
137 R10 **Supsa** *prev.* Sup'sa. ≈ W Georgia
Sup'sa *see* Supsa
Sūq 'Abs *see* 'Abs
139 W12 **Sūq ash Shuyūkh** Dhī Qār, SE Iraq 30°53′N 46°28′E
138 H4 **Şuqaylibīyah** Ḩamāh, W Syria 35°21′N 36°24′E
161 Q6 **Suqian** Jiangsu, E China 33°57′N 118°18′E
Suqrah *see* Şawqirah
Suqrah Bay *see* Şawqirah, Dawḩat
141 V16 **Suquṭrā** *var.* Sokotra, *Eng.* Socotra. *island* SE Yemen
141 Z8 **Şūr** NE Oman 22°32′N 59°33′E
Şūr *see* Soûr
127 P5 **Sura** Penzenskaya Oblast', W Russian Federation 53°23′N 45°03′E
127 P4 **Sura** ≈ W Russian Federation
149 N12 **Sūrāb** Baluchistān, SW Pakistan 28°28′N 66°15′E
Surabaja *see* Surabaya
192 E8 **Surabaya** *prev.* Surabaja, Soerabaja. Jawa, C Indonesia 07°14′S 112°45′E
95 N15 **Surahammar** Västmanland, C Sweden 59°43′N 16°13′E
169 Q16 **Surakarta** *Eng.* Solo; *prev.* Soerakarta. Jawa, S Indonesia 07°32′S 110°50′E
Surakhany *see* Suraxanı
137 S10 **Surami** C Georgia 41°59′N 43°36′E
143 X13 **Sūrān** Sīstān va Balūchestān, SE Iran 27°18′N 61°58′E
111 I21 **Šurany** *Hung.* Nagysurány. Nitriansky Kraj, SW Slovakia 48°05′N 18°10′E
154 D12 **Sūrat** Gujarāt, W India 21°10′N 72°54′E
Suratdhani *see* Surat Thani
152 G9 **Sūratgarh** Rājasthān, NW India 29°20′N 73°59′E
167 N14 **Surat Thani** *var.* Suratdhani. Surat Thani, SW Thailand 09°09′N 99°20′E
119 Q16 **Suraw** *Rus.* Surov. ≈ E Belarus
137 Z11 **Suraxanı** *Rus.* Surakhany. E Azerbaijan 40°25′N 49°59′E
141 Y11 **Surayr** E Oman 19°56′N 57°47′E
138 K2 **Suraysāt** Ḩalab, N Syria 36°42′N 38°01′E
118 O12 **Surazh** Vitsyebskaya Voblasts', NE Belarus 55°25′N 30°44′E
126 H6 **Surazh** Bryanskaya Oblast', W Russian Federation 53°04′N 32°29′E
191 V17 **Sur, Cabo** *headland* Easter Island, Chile, E Pacific Ocean 27°11′S 109°26′W
112 L11 **Surčin** Serbia, N Serbia 44°48′N 20°19′E
116 H9 **Surduc** *Hung.* Szurduk. Sălaj, NW Romania 47°13′N 23°20′E
113 P16 **Surdulica** Serbia, SE Serbia 42°43′N 22°10′E
99 L24 **Sûre** *var.* Sauer. ≈ W Europe *see also* Sauer
Sûre *see* Sauer
154 C10 **Surendranagar** Gujarāt, W India 22°44′N 71°43′E
18 K16 **Surf City** New Jersey, NE USA 39°21′N 74°24′W
183 V3 **Surfers Paradise** Queensland, E Australia 27°54′S 153°18′E
21 U13 **Surfside Beach** South Carolina, SE USA 33°36′N 78°58′W
102 J10 **Surgères** Charente-Maritime, W France 46°07′N 00°44′W
122 H10 **Surgut** Khanty-Mansiyskiy Avtonomnyy Okrug-Yugra, C Russian Federation 61°13′N 73°28′E
122 K10 **Surgutikha** Krasnoyarskiy Kray, N Russian Federation 64°44′N 87°13′E
98 M6 **Surhuisterveen** *Fris.* Surhusterfean. Fryslân, N Netherlands 53°11′N 06°10′E
Surhusterfean *see* Surhuisterveen
105 V5 **Súria** Cataluña, NE Spain 41°50′N 01°45′E
143 P10 **Sūriān** Fārs, S Iran
155 J15 **Suriāpet** Telangana, C India 17°10′N 79°42′E
171 Q6 **Surigao** Mindanao, S Philippines 09°43′N 125°31′E
167 R10 **Surin** Surin, E Thailand 14°53′N 103°29′E
Surinam *see* Suriname
55 U11 **Suriname** *off.* Republic of Suriname, *var.* Surinam; *prev.* Dutch Guiana, Netherlands Guiana. ◆ *republic* N South America
Suriname, Republic of *see* Suriname
Sūriya/Sūrīyah, Al-Jumhūrīyah al-'Arabīyah as- *see* Syria
Surkhab, Darya-i- *see* Kahmard, Daryā-ye
Surkhandar'inskaya Oblast' *see* Surxondaryo Viloyati
Surkhandar'ya *see* Surxondaryo
Surkhet *see* Birendranagar
147 R12 **Surkhob** ≈ C Tajikistan
137 P11 **Sürmene** Trabzon, NE Turkey 40°56′N 40°03′E
Surov *see* Suraw
127 N11 **Surovikino** Volgogradskaya Oblast', SW Russian Federation 48°39′N 42°46′E

◆ Country
● Country Capital
◇ Dependent Territory
○ Dependent Territory Capital
◐ Administrative Regions
✈ International Airport
▲ Mountain
▲▲ Mountain Range
Volcano
≈ River
◎ Lake
⊠ Reservoir

SUR, POINT – TAHANEA

35 N11 **Sur, Point** headland California, W USA 36°18′N 121°54′W
187 N15 **Surprise, Île** island N New Caledonia
61 E22 **Sur, Punta** headland E Argentina 50°59′S 69°10′W
Surrentum see Sorrento
28 M3 **Surrey** North Dakota, N USA 48°13′N 101°05′W
97 O22 **Surrey** cultural region SE England, United Kingdom
21 X7 **Surry** Virginia, NE USA 37°08′N 81°34′W
108 F8 **Sursee** Luzern, W Switzerland 47°11′N 08°07′E
127 P6 **Sursk** Penzenskaya Oblast', W Russian Federation 53°06′N 45°46′E
127 P5 **Surskoye** Ul'yanovskaya Oblast', W Russian Federation 54°28′N 46°47′E
75 P8 **Surt** var. Sidra, Sirte. N Libya 31°13′N 16°35′E
95 I19 **Surte** Västra Götaland, S Sweden 57°49′N 12°01′E
75 Q8 **Surt, Khalīj** Eng. Gulf of Sidra, Gulf of Sirti, Sidra. gulf N Libya
92 I5 **Surtsey** island S Iceland
137 N17 **Suruç** Şanlıurfa, S Turkey 36°58′N 38°24′E
168 L13 **Surulangun** Sumatera, W Indonesia 02°35′S 102°47′E
147 P13 **Surxondaryo** Rus. Surkhandar'ya. ≈ Tajikistan/Uzbekistan
147 N13 **Surxondaryo Viloyati** Rus. Surkhandar'inskaya Oblast'. ◈ province S Uzbekistan
Süs see Susch
106 A8 **Susa** Piemonte, NE Italy 45°10′N 07°01′E
165 E12 **Susa** Yamaguchi, Honshū, SW Japan 34°35′N 131°34′E
Susa see Shūsh
113 E16 **Sušac** It. Cazza. island SW Croatia
Sūsah see Sousse
164 G14 **Susaki** Kōchi, Shikoku, SW Japan 33°22′N 133°13′E
165 I15 **Susami** Wakayama, Honshū, SW Japan 33°32′N 135°32′E
142 K9 **Sūsangerd** var. Susangird. Khūzestān, SW Iran 31°40′N 48°06′E
Susangird see Sūsangerd
35 P4 **Susanville** California, W USA 40°25′N 120°39′W
108 J9 **Susch** var. Süs. Graubünden, SE Switzerland 46°45′N 10°04′E
137 N12 **Suşehri** Sivas, N Turkey 40°11′N 38°06′E
Susiana see Khūzestān
111 B18 **Sušice** Ger. Schüttenhofen. Plzeňský Kraj, W Czech Republic 49°14′N 13°32′E
39 R11 **Susitna** Alaska, USA 61°32′N 150°30′W
39 R11 **Susitna River** ≈ Alaska, USA
127 Q3 **Suslonger** Respublika Mariy El, W Russian Federation 56°18′N 48°16′E
105 N14 **Suspiro del Moro, Puerto del** pass S Spain
18 H16 **Susquehanna River** ≈ New York/Pennsylvania, NE USA
13 O15 **Sussex** New Brunswick, SE Canada 45°43′N 65°32′W
18 J13 **Sussex** New Jersey, NE USA 41°12′N 74°34′W
21 W7 **Sussex** Virginia, NE USA 36°54′N 77°16′W
97 O23 **Sussex** cultural region S England, United Kingdom
183 S10 **Sussex Inlet** New South Wales, SE Australia 35°10′S 150°35′E
99 L17 **Susteren** Limburg, SE Netherlands 51°04′N 05°50′E
10 K12 **Sustut Peak** ▲ British Columbia, W Canada 56°25′N 126°34′W
123 S9 **Susuman** Magadanskaya Oblast', E Russian Federation 62°46′N 148°08′E
188 H6 **Susupe** ● (Northern Mariana Islands-judicial capital) Saipan, S Northern Mariana Islands
136 D12 **Susurluk** Balıkesir, NW Turkey 39°55′N 28°10′E
114 M13 **Susuzmüsellim** Tekirdağ, NW Turkey 41°04′N 27°03′E
136 F15 **Sütçüler** Isparta, SW Turkey 37°31′N 31°00′E
116 L13 **Şuteşti** Brăila, SE Romania 45°13′N 27°27′E
83 F25 **Sutherland** Western Cape, SW South Africa 32°24′S 20°40′E
28 L15 **Sutherland** Nebraska, C USA 41°09′N 101°07′W
96 I7 **Sutherland** cultural region N Scotland, United Kingdom
185 B21 **Sutherland Falls** waterfall South Island, New Zealand
32 F14 **Sutherlin** Oregon, NW USA 43°23′N 123°18′W
149 V10 **Sutlej** ≈ India/Pakistan
Sutna see Satna
35 P7 **Sutter Creek** California, W USA 38°22′N 120°49′W
39 R11 **Sutton** Alaska, USA 61°42′N 148°53′W
29 Q16 **Sutton** Nebraska, C USA 40°36′N 97°52′W
21 R4 **Sutton** West Virginia, NE USA 38°41′N 80°43′W
12 F8 **Sutton** ≈ Ontario, C Canada
97 M19 **Sutton Coldfield** C England, United Kingdom 52°34′N 01°48′W
21 R4 **Sutton Lake** ⊠ West Virginia, NE USA
15 P13 **Sutton, Monts** hill range Québec, SE Canada
12 F8 **Sutton Ridges** ▲▲ Ontario, C Canada
165 Q4 **Suttsu** Hokkaidō, NE Japan 42°46′N 140°12′E
39 P15 **Sutwik Island** island Alaska, USA
Süüj see Dashinchilen
118 H5 **Suure-Jaani** Ger. Gross-Sankt-Johannis. Viljandimaa, S Estonia 58°34′N 25°28′E
118 J7 **Suur Munamägi** var. Munamägi, Ger. Eier-Berg. ▲ SE Estonia 57°42′N 27°03′E
118 F5 **Suur Väin** Ger. Grosser Sund. strait W Estonia
147 U8 **Suusamyr** Chuyskaya Oblast', C Kyrgyzstan 42°07′N 73°55′E
187 X14 **Suva** ● (Fiji) Viti Levu, W Fiji 18°08′S 178°27′E
187 X15 **Suva** ✈ Viti Levu, C Fiji 18°01′S 178°30′E
113 N18 **Suva Gora** ▲▲ W FYR Macedonia
118 H11 **Suvainiškis** Panevėžys, NE Lithuania 56°09′N 25°15′E
Suvalkai/Suvalki see Suwałki
113 P15 **Suva Planina** ▲▲ SE Serbia
Suva Reka see Suharekë
126 K5 **Suvorov** Tul'skaya Oblast', W Russian Federation 54°08′N 36°33′E
117 N12 **Suvorove** Odes'ka Oblast', SW Ukraine 45°35′N 28°58′E
114 M8 **Suvorovo** Varna, E Bulgaria 43°19′N 27°26′E
Suvorovo see Ştefan Vodă
Suwaik see As Suwayq
Suwaira see Aş Şuwayrah
110 O7 **Suwałki** Lith. Suvalkai, Rus. Suvalki. Podlaskie, NE Poland 54°06′N 22°56′E
167 R10 **Suwannaphum** Roi Et, E Thailand 15°36′N 103°46′E
23 V8 **Suwannee River** ≈ Florida/Georgia, SE USA
Şuwār see Aş Şuwār
190 K14 **Suwarrow** atoll N Cook Islands
143 R16 **Suwaydān** var. Sweiham. Abū Ẓaby, E United Arab Emirates 24°30′N 55°19′E
Suwaydá/Suwaydā', Muḥāfaẓat as see As Suwaydā'
Suwayqiyah, Hawr as see Shuwayjah, Hawr ash
Suways, Qanāt as see Suez Canal
Suweida see As Suwaydā'
Suweon see Suwon
163 X15 **Suwon** var. Suweon; prev. Suwŏn, Jap. Suigen. NW South Korea 37°17′N 127°03′E
Suwŏn see Suwon
Su Xian see Suzhou
143 R14 **Sūzā** Hormozgān, S Iran 26°50′N 56°05′E
Suzak see Sozak
165 K14 **Suzaka** Mie, Honshū, SW Japan 34°52′N 136°37′E
165 N12 **Suzaka** Nagano, Honshū, S Japan 36°38′N 138°20′E
126 M3 **Suzdal'** Vladimirskaya Oblast', W Russian Federation 56°27′N 40°29′E
161 P7 **Suzhou** var. Su Xian. Anhui, E China 33°38′N 117°02′E
161 R8 **Suzhou** var. Soochow, Su-chou, Suchow; prev. Wuhsien. Jiangsu, E China 31°23′N 120°34′E
Suzhou see Jiuquan
163 V12 **Suzi He** ≈ NE China
Suz, Mys see Soye, Mys
165 M10 **Suzu** Ishikawa, Honshū, SW Japan 37°24′N 137°12′E
165 M10 **Suzu-misaki** headland Honshū, SW Japan 37°31′N 137°19′E
Svågälv see Svågan
94 M10 **Svågan** var. Svågälv. ≈ C Sweden
Svalava/Svaljava see Svalyava
92 O2 **Svalbard** ◇ constituent part of Norway Arctic Ocean
92 J2 **Svalbarðseyri** Norðurland Eystra, N Iceland 65°43′N 18°03′W
95 K22 **Svalöv** Skåne, S Sweden 55°55′N 13°06′E
116 H7 **Svalyava** Cz. Svalava, Svaljava, Hung. Szolyva. Zakarpats'ka Oblast', W Ukraine 48°33′N 23°00′E
92 O2 **Svanbergfjellet** ▲ C Svalbard 78°40′N 18°10′E
95 M24 **Svaneke** Bornholm, E Denmark 55°07′N 15°08′E
95 L22 **Svängsta** Blekinge, S Sweden 56°16′N 14°46′E
95 J16 **Svanskog** Värmland, C Sweden 59°10′N 12°34′E
95 L16 **Svartå** Örebro, C Sweden 59°13′N 14°07′E
95 L15 **Svartälven** ≈ C Sweden
92 G12 **Svartisen** glacier C Norway
117 X6 **Svatove** Rus. Svatovo. Luhans'ka Oblast', E Ukraine 49°24′N 38°11′E
Svatovo see Svatove
Sväty Kríž nad Hronom see Žiar nad Hronom
167 Q11 **Svay Chék, Stœ̆ng** ≈ Cambodia/Thailand
167 S13 **Svay Riĕng** Svay Riĕng, S Cambodia 11°05′N 105°48′E
92 O3 **Sveagruva** Spitsbergen, W Svalbard 77°53′N 16°42′E
95 K23 **Svedala** Skåne, S Sweden 55°30′N 13°15′E
118 H12 **Svėdesai** Utena, NE Lithuania 55°42′N 25°22′E
93 G18 **Sveg** Jämtland, C Sweden 62°02′N 14°20′E
118 C12 **Švėkšna** Klaipėda, W Lithuania 55°31′N 21°37′E
94 C11 **Svelgen** Sogn Og Fjordane, S Norway 61°47′N 05°18′E
95 H15 **Svelvik** Vestfold, S Norway 59°37′N 10°24′E
118 I13 **Švenčionėliai** Pol. Nowo-Święciany. Vilnius, SE Lithuania 55°10′N 26°00′E
118 I13 **Švenčionys** Pol. Święciany. Vilnius, SE Lithuania 55°08′N 26°08′E
95 H24 **Svendborg** Syddtjylland, C Denmark 55°04′N 10°38′E
95 K19 **Svenljunga** Västra Götaland, S Sweden 57°30′N 13°05′E
92 P2 **Svenskøya** island E Svalbard
93 G17 **Svenstavik** Jämtland, C Sweden 62°40′N 14°24′E
95 G20 **Svenstrup** Nordjylland, N Denmark 56°58′N 09°52′E
118 H12 **Šventoji** ≈ C Lithuania
117 Z8 **Sverdlovs'k** Rus. Sverdlovsk; prev. Imeni Sverdlova Rudnik. Luhans'ka Oblast', E Ukraine 48°05′N 39°37′E
Sverdlovsk see Yekaterinburg
127 W2 **Sverdlovskaya Oblast'** ◈ province C Russian Federation
122 K6 **Sverdrupa, Ostrov** island N Russian Federation
Sverige see Sweden
113 D15 **Svetac** prev. Sveti Andrea, It. Sant'Andrea. island SW Croatia
Sveti Andrea see Svetac
Sveti Nikola see Sveti Nikole
113 O18 **Sveti Nikole** prev. Sveti Nikola. C FYR Macedonia 41°54′N 21°55′E
Sveti Vrach see Sandanski
123 T14 **Svetlaya** Primorskiy Kray, SE Russian Federation 46°33′N 138°20′E
126 B2 **Svetlogorsk** Kaliningradskaya Oblast', W Russian Federation 54°56′N 20°09′E
122 K9 **Svetlogorsk** Krasnoyarskiy Kray, N Russian Federation 66°51′N 88°29′E
Svetlogorsk see Svyetlahorsk
127 N14 **Svetlograd** Stavropol'skiy Kray, SW Russian Federation 45°20′N 42°53′E
Svetlovodsk see Svitlovods'k
119 A14 **Svetlyy** Ger. Zimmerbude. Kaliningradskaya Oblast', W Russian Federation 54°42′N 20°07′E
127 Y8 **Svetlyy** Orenburgskaya Oblast', W Russian Federation 50°34′N 60°42′E
127 P7 **Svetlyy** Saratovskaya Oblast', W Russian Federation 51°42′N 45°40′E
124 G11 **Svetogorsk** Fin. Enso. Leningradskaya Oblast', NW Russian Federation 61°06′N 28°52′E
Svetozarevo see Jagodina
111 B18 **Švihov** Ger. Schwihau. Plzeňský Kraj, W Czech Republic 49°31′N 13°18′E
112 E13 **Svilaja** ▲▲ SE Croatia
112 N12 **Svilajnac** Serbia, C Serbia 44°15′N 21°12′E
114 L11 **Svilengrad** prev. Mustafa-Pasha. Haskovo, S Bulgaria 41°45′N 26°14′E
Svinecea Mare, Munte see Svinecea Mare, Vârful
116 F13 **Svinecea Mare, Vârful** var. Munte Svinecea Mare. ▲ SW Romania 44°47′N 22°10′E
Svinø see Svínoy
95 B18 **Svínoy** Dan. Svinø. island NE Faroe Islands
147 N14 **Svintsovyy Rudnik** Turkm. Swintsowyy Rudnik. Lebap Welaýaty, E Turkmenistan 37°54′N 66°25′E
118 I13 **Svir** Rus. Svir'. Minskaya Voblasts', NW Belarus 54°51′N 26°24′E
124 I12 **Svir'** canal NW Russian Federation
Svir', Ozero see Svir, Vozyera
119 I14 **Svir, Vozyera** Rus. Ozero Svir'. ◎ C Belarus
114 J7 **Svishtov** prev. Sistova. Veliko Tarnovo, N Bulgaria 43°37′N 25°20′E
119 F18 **Svislach** Pol. Świsłocz, Rus. Svisloch'. Hrodzyenskaya Voblasts', W Belarus 53°02′N 24°06′E
119 M17 **Svislach** Rus. Svisloch'. Mahilyowskaya Voblasts', E Belarus 53°26′N 28°59′E
119 L17 **Svislach** Rus. Svisloch'. ≈ E Belarus
Svisloch' see Svislach
111 F17 **Svitavy** Ger. Zwittau. Pardubický Kraj, C Czech Republic 49°45′N 16°27′E
117 S6 **Svitlovods'k** Rus. Svetlovodsk. Kirovohrads'ka Oblast', C Ukraine 49°05′N 33°15′E
Svizzera see Switzerland
123 Q13 **Svobodnyy** Amurskaya Oblast', SE Russian Federation 51°24′N 128°05′E
114 G9 **Svoge** Sofia, W Bulgaria 42°58′N 23°20′E
92 G11 **Svolvær** Nordland, C Norway 68°15′N 14°40′E
111 F18 **Svratka** Ger. Schwarzawa. ≈ SE Czech Republic
113 P14 **Svrljig** Serbia, E Serbia 43°26′N 22°07′E
197 U10 **Svyataya Anna Trough** var. Saint Anna Trough. undersea feature N Kara Sea
124 M4 **Svyatoy Nos, Mys** headland NW Russian Federation 68°07′N 39°49′E
119 N18 **Svyetlahorsk** Rus. Svetlogorsk. Homyel'skaya Voblasts', SE Belarus 52°38′N 29°46′E
Swabian Jura see Schwäbische Alb
97 P19 **Swaffham** E England, United Kingdom 52°39′N 00°40′E
23 V5 **Swainsboro** Georgia, SE USA 32°36′N 82°19′W
83 C19 **Swakop** ≈ W Namibia
83 C19 **Swakopmund** Erongo, W Namibia 22°40′S 14°34′E
97 M15 **Swale** ≈ N England, United Kingdom
Swallow Island see Nendö
99 M16 **Swalmen** Limburg, SE Netherlands 51°13′N 06°02′E
12 G8 **Swan** ≈ Ontario, C Canada
97 L24 **Swanage** S England, United Kingdom 50°37′N 01°59′W
182 M10 **Swan Hill** Victoria, SE Australia 35°23′S 143°37′E
11 P13 **Swan Hills** Alberta, W Canada 54°41′N 116°20′W
65 D24 **Swan Island** island C Falkland Islands
Swankalok see Sawankhalok
29 U10 **Swan Lake** ◎ Minnesota, N USA
21 Y10 **Swanquarter** North Carolina, SE USA 35°24′N 76°20′W
182 J9 **Swan Reach** South Australia 34°39′S 139°35′E
11 V15 **Swan River** Manitoba, S Canada 52°06′N 101°17′W
183 P17 **Swansea** Tasmania, SE Australia 42°09′S 148°03′E
97 J22 **Swansea** Wel. Abertawe. S Wales, United Kingdom 51°38′N 03°57′W
21 R13 **Swansea** South Carolina, SE USA 33°47′N 81°06′W
19 S7 **Swans Island** island Maine, NE USA
28 L17 **Swanson Lake** ⊠ Nebraska, C USA
31 R11 **Swanton** Ohio, N USA 41°35′N 83°53′W
110 G11 **Swarzędz** Poznań, W Poland 52°24′N 17°05′E
Swatow see Shantou
83 L22 **Swaziland** off. Kingdom of Swaziland. ◆ monarchy S Africa
Swaziland, Kingdom of see Swaziland
93 G18 **Sweden** off. Kingdom of Sweden, Swe. Sverige. ◆ monarchy N Europe
Sweden, Kingdom of see Sweden
Swedru see Agona Swedru
25 V12 **Sweeny** Texas, SW USA 29°02′N 95°42′W
33 R6 **Sweetgrass** Montana, NW USA 48°58′N 111°58′W
32 G12 **Sweet Home** Oregon, NW USA 44°24′N 122°44′W
25 T12 **Sweet Home** Texas, SW USA 29°21′N 97°04′W
27 T4 **Sweet Springs** Missouri, C USA 38°57′N 93°24′W
20 M10 **Sweetwater** Tennessee, S USA 35°36′N 84°27′W
25 P7 **Sweetwater** Texas, SW USA 32°27′N 100°25′W
33 V15 **Sweetwater River** ≈ Wyoming, C USA
Sweiham see Suwaydān
83 F26 **Swellendam** Western Cape, SW South Africa 34°01′S 20°26′E
111 G15 **Świdnica** Ger. Schweidnitz. Wałbrzych, SW Poland 50°51′N 16°29′E
111 O14 **Świdnik** Ger. Streckenbach. Lubelskie, E Poland 51°14′N 22°41′E
110 F8 **Świdwin** Ger. Schivelbein. Zachodnio-pomorskie, NW Poland 53°47′N 15°44′E
111 F15 **Świebodzice** Ger. Freiburg in Schlesien, Swiebodzice. Walbrzych, SW Poland 50°54′N 16°23′E
110 E11 **Świebodzin** Ger. Schwiebus. Lubuskie, W Poland 52°15′N 15°31′E
Święciany see Švenčionys
110 I9 **Świecie** Ger. Schwertberg. Kujawsko-pomorskie, C Poland 53°24′N 18°24′E
111 L15 **Świętokrzyskie** ◈ province S Poland
11 T16 **Swift Current** Saskatchewan, S Canada 50°17′N 107°49′W
98 K9 **Swifterbant** Flevoland, C Netherlands 52°36′N 05°33′E
183 Q12 **Swifts Creek** Victoria, SE Australia 37°17′S 147°41′E
96 E13 **Swilly, Lough** Ir. Loch Súilí. inlet N Ireland
97 M22 **Swindon** S England, United Kingdom 51°34′N 01°47′W
Swinemünde see Świnoujście
110 D8 **Świnoujście** Ger. Swinemünde. Zachodnio-pomorskie, NW Poland 53°54′N 14°13′E
Swintsowyy Rudnik see Svintsovyy Rudnik
Świsłocz see Svislach
Swiss Confederation see Switzerland
108 E9 **Switzerland** off. Swiss Confederation, Fr. La Suisse, Ger. Schweiz, It. Svizzera; anc. Helvetia. ◆ federal republic C Europe
97 F17 **Swords** Ir. Sord, Sórd Choluim Chille. Dublin, E Ireland 53°28′N 06°13′W
18 H13 **Swoyersville** Pennsylvania, NE USA 41°18′N 75°48′W
139 V3 **Syagwêz** Ar. Siyāh Gūz. As Sulaymānīyah, E Iraq 35°49′N 45°45′E
124 I10 **Syamozero, Ozero** ◎ NW Russian Federation
124 M13 **Syamzha** Vologodskaya Oblast', NW Russian Federation 60°02′N 41°09′E
118 N13 **Syanno** Rus. Senno. Vitsyebskaya Voblasts', NE Belarus 54°49′N 29°43′E
119 K16 **Syarhyeyevichy** Rus. Sergeyevichi. Minskaya Voblasts', C Belarus 53°30′N 27°45′E
124 I12 **Syas'stroy** Leningradskaya Oblast', NW Russian Federation 60°05′N 32°37′E
Sycaminum see Hefa
30 M10 **Sycamore** Illinois, N USA 41°59′N 88°41′W
126 J3 **Sychëvka** Smolenskaya Oblast', W Russian Federation 55°52′N 34°19′E
111 H14 **Syców** Ger. Gross Wartenberg. Dolnosląskie, SW Poland 51°18′N 17°42′E
95 F24 **Syddanmark** ◈ county SW Denmark
14 E17 **Sydenham** ≈ Ontario, S Canada
Sydenham Island see Nonouti
183 T9 **Sydney** state capital New South Wales, SE Australia 33°55′S 151°10′E
13 R14 **Sydney** Cape Breton Island, Nova Scotia, SE Canada 46°10′N 60°10′W
Sydney Island see Manra
13 R14 **Sydney Mines** Cape Breton Island, Nova Scotia, SE Canada 46°14′N 60°19′W
Syedpur see Saidpur
119 K18 **Syelishcha** Rus. Selishche. Minskaya Voblasts', C Belarus 53°01′N 27°25′E
119 J18 **Syemyezhava** Rus. Semezhevo. Minskaya Voblasts', C Belarus 52°58′N 27°00′E
Syene see Aswān
117 X6 **Syeverodonets'k** Rus. Severodonetsk. Luhans'ka Oblast', E Ukraine 48°59′N 38°28′E
161 T6 **Sÿiao Shan** island SE China
100 H11 **Syke** Niedersachsen, NW Germany 52°55′N 08°49′E
94 D10 **Sykkylven** Møre og Romsdal, S Norway 62°23′N 06°35′E
Sykoúri see Sykoúrio
115 F15 **Sykoúrio** var. Sikouri, Sykoúri; prev. Sikoúrion. Thessalía, C Greece 39°46′N 22°35′E
125 R11 **Syktyvkar** prev. Ust'-Sysol'sk. Respublika Komi, NW Russian Federation 61°42′N 50°45′E
23 Q4 **Sylacauga** Alabama, S USA 33°10′N 86°15′W
Sylarna see Storsylen
153 V14 **Sylhet** Sylhet, NE Bangladesh 24°53′N 91°51′E
153 V13 **Sylhet** ◈ division NE Bangladesh
100 G6 **Sylt** island NW Germany
21 O10 **Sylva** North Carolina, SE USA 35°23′N 83°13′W
125 V15 **Sylva** ≈ NW Russian Federation
23 W5 **Sylvania** Georgia, SE USA 32°45′N 81°38′W
31 R11 **Sylvania** Ohio, N USA 41°43′N 83°42′W
11 Q15 **Sylvan Lake** Alberta, SW Canada 52°18′N 114°02′W
33 T13 **Sylvan Pass** pass Wyoming, C USA
23 T7 **Sylvester** Georgia, SE USA 31°31′N 83°50′W
25 P6 **Sylvester** Texas, SW USA 32°42′N 100°15′W
10 L11 **Sylvia, Mount** ▲ British Columbia, W Canada 58°03′N 124°26′W
122 K11 **Sym** ≈ C Russian Federation
115 N22 **Sými** var. Simi. island Dodekánisa, Greece, Aegean Sea
117 U8 **Synel'nykove** Dnipropetrovs'ka Oblast', E Ukraine 48°19′N 35°32′E
125 U6 **Synya** Respublika Komi, NW Russian Federation 65°21′N 58°01′E
117 P7 **Synyukha** Rus. Sinyukha. ≈ S Ukraine
195 V2 **Syowa** Japanese research station Antarctica 68°58′S 40°07′E
26 H6 **Syracuse** Kansas, C USA 38°00′N 101°43′W
29 S16 **Syracuse** Nebraska, C USA 40°39′N 96°11′W
18 H10 **Syracuse** New York, NE USA 43°03′N 76°09′W
Syracuse see Siracusa
Syrdar'inskaya Oblast' see Sirdaryo Viloyati
Syrdariya see Syr Darya
144 L14 **Syr Darya** var. Sai Hun, Sir Darya, Syrdarya, Kaz. Syrdariya, Rus. Syrdar'ya, Uzb. Sirdaryo; anc. Jaxartes. ≈ C Asia
Syrdarya see Syr Darya
138 J6 **Syria** off. Syrian Arab Republic, var. Siria, Syrie, Ar. Al-Jumhūrīyah al-'Arabīyah as-Sūrīyah, Sūrīya. ◆ republic SW Asia
Syrian Arab Republic see Syria
138 L9 **Syrian Desert** Ar. Al Hamad, Bādiyat ash Shām. desert SW Asia
Syrie see Syria
115 L22 **Sýrna** var. Sirna. island Kykládes, Greece, Aegean Sea
115 I20 **Sýros** var. Síros. island Kykládes, Greece, Aegean Sea
93 M18 **Sysmä** Päijät-Häme, S Finland 61°28′N 25°37′E
125 R12 **Sysola** ≈ NW Russian Federation
Syulemeshlii see Sredets
127 S2 **Syumsi** Udmurtskaya Respublika, NW Russian Federation 57°07′N 51°35′E
Syvash, Zaliv see Syvash, Zatoka
117 U12 **Syvash, Zatoka** Rus. Zaliv Syvash. inlet S Ukraine
127 Q6 **Syzran'** Samarskaya Oblast', W Russian Federation 53°10′N 48°23′E
Szabadka see Subotica
111 N21 **Szabolcs-Szatmár-Bereg** off. Szabolcs-Szatmár-Bereg Megye. ◈ county E Hungary
Szabolcs-Szatmár-Bereg Megye see Szabolcs-Szatmár-Bereg
110 G10 **Szamocin** Ger. Samotschin. Wielkopolskie, C Poland 53°02′N 17°04′E
116 H8 **Szamos** var. Someş, Someşul, Ger. Samosch, Somesch. ≈ Hungary/Romania
Szamosújvár see Gherla
110 G11 **Szamotuły** Poznań, W Poland 52°35′N 16°36′E
Szarkowszczyzna see Sharkawshchyna
111 M24 **Szarvas** Békés, SE Hungary 46°51′N 20°35′E
Szászmagyarós see Măieruş
Szászrégen see Reghin
Szászsebes see Sebeş
Szászváros see Orăştie
Szatmárnémeti see Satu Mare
Száva see Sava
111 P15 **Szczebrzeszyn** Lubelskie, E Poland 50°43′N 23°00′E
110 D9 **Szczecin** Eng./Ger. Stettin. Zachodnio-pomorskie, NW Poland 53°25′N 14°32′E
110 G8 **Szczecinek** Ger. Neustettin. Zachodnio-pomorskie, NW Poland 53°43′N 16°40′E
110 D8 **Szczeciński, Zalew** var. Stettiner Haff, Ger. Oderhaff. bay Germany/Poland
111 K15 **Szczekociny** Śląskie, S Poland 50°38′N 19°46′E
110 N8 **Szczuczyn** Podlaskie, NE Poland 53°34′N 22°17′E
Szczuczyn Nowogródzki see Shchuchyn
110 M8 **Szczytno** Ger. Ortelsburg. Warmińsko-Mazurskie, NE Poland 53°34′N 21°E
Szechuan/Szechwan see Sichuan
111 K21 **Szécsény** Nógrád, N Hungary 48°07′N 19°30′E
111 L25 **Szeged** Ger. Szegedin, Rom. Seghedin. Csongrád, SE Hungary 46°17′N 20°06′E
Szegedin see Szeged
111 N23 **Szeghalom** Békés, SE Hungary 47°01′N 21°09′E
Székelyhid see Săcueni
Székelykeresztúr see Cristuru Secuiesc
111 I23 **Székesfehérvár** Ger. Stuhlweissenberg; anc. Alba Regia. Fejér, W Hungary 47°13′N 18°24′E
Szeklerburg see Miercurea-Ciuc
Szekler Neumarkt see Târgu Secuiesc
111 J25 **Szekszárd** Tolna, S Hungary 46°21′N 18°41′E
Szempcz/Szenc see Senec
Szenice see Senica
Szentágota see Agnita
111 J22 **Szentendre** Ger. Sankt Andrä. Pest, N Hungary 47°40′N 19°02′E
111 L24 **Szentes** Csongrád, SE Hungary 46°40′N 20°17′E
111 F23 **Szentgotthárd** Eng. Saint Gotthard, Ger. Sankt Gotthard. Vas, W Hungary 46°57′N 16°18′E
Szentgyörgy see Đurđevac
Szenttamás see Srbobran
Széphely see Jebel
Szeping see Siping
Szered see Sered'
111 N21 **Szerencs** Borsod-Abaúj-Zemplén, NE Hungary 48°10′N 21°11′E
Szeret see Siret
Szeretfalva see Sărăţel
110 N7 **Szeska Góra** var. Szeskie Wygórza, Ger. Seesker Höhe. hill NE Poland
Szeskie Wygórza see Szeska Góra
111 H25 **Szigetvár** Baranya, SW Hungary 46°01′N 17°50′E
Szilágysomlyó see Şimleu Silvaniei
Szinna see Snina
Sziszek see Sisak
Szitás-Keresztúr see Cristuru Secuiesc
111 E15 **Szklarska Poręba** Ger. Schreiberhau. Dolnośląskie, SW Poland 50°50′N 15°30′E
Szkudy see Skuodas
Szlatina see Slatina
Szlavonia/Szlavonország see Slavonija
Szlovákia see Slovakia
Szluin see Slunj
111 L23 **Szolnok** Jász-Nagykun-Szolnok, C Hungary 47°11′N 20°12′E
Szolyva see Svalyava
111 G23 **Szombathely** Ger. Steinamanger; anc. Sabaria, Savaria. Vas, W Hungary 47°14′N 16°38′E
Szond/Szonta see Sonta
Szováta see Sovata
110 F13 **Szprotawa** Ger. Sprottau. Lubuskie, W Poland 51°33′N 15°32′E
Sztálinváros see Dunaújváros
Sztrazsó see Strážov
110 J8 **Sztum** Ger. Stuhm. Pomorskie, N Poland 53°54′N 19°01′E
110 H10 **Szubin** Ger. Schubin. Kujawsko-pomorskie, C Poland 53°04′N 17°49′E
Szucsava see Suceava
Szurduk see Surduc
111 M14 **Szydłowiec** Ger. Schlelau. Mazowieckie, C Poland 51°14′N 20°50′E

T

Taalintehdas see Dalsbruk
171 O4 **Taal, Lake** ◎ Luzon, NW Philippines
95 J23 **Taastrup** var. Tåstrup. Sjælland, E Denmark 55°39′N 12°19′E
111 I24 **Tab** Somogy, W Hungary 46°45′N 18°01′E
171 P4 **Tabaco** Luzon, N Philippines 13°22′N 123°42′E
186 G4 **Tabalo** Mussau Island, NE Papua New Guinea 01°22′S 149°37′E
104 K5 **Tábara** Castilla y León, N Spain 41°49′N 05°57′W
186 H5 **Tabar Islands** island group NE Papua New Guinea
Tabariya, Bahrat see Tiberias, Lake
143 S7 **Ţabas** var. Golshan. Khorāsān-e Jonūbī, C Iran 33°37′N 56°54′E
43 P15 **Tabasará, Serranía de** ▲▲ W Panama
41 U15 **Tabasco** ◈ state SE Mexico
Tabasco see Grijalva, Río
127 Q2 **Tabashino** Respublika Mariy El, W Russian Federation 57°00′N 47°47′E
58 B13 **Tabatinga** Amazonas, N Brazil 04°14′S 69°44′W
74 G9 **Tabelbala** W Algeria 29°22′N 03°01′W
11 Q17 **Taber** Alberta, SW Canada 49°48′N 112°09′W
171 V15 **Taberfane** Pulau Trangan, E Indonesia 06°14′S 134°08′E
95 L19 **Taberg** Jönköping, S Sweden 57°42′N 14°05′E
191 O3 **Tabiteuea** prev. Drummond Island. atoll Tungaru, W Kiribati
171 O5 **Tablas Island** island C Philippines
184 Q10 **Table Cape** headland North Island, New Zealand 39°07′S 178°00′E
13 S13 **Table Mountain** ▲ Newfoundland, Newfoundland and Labrador, E Canada 47°39′N 59°15′W
173 P17 **Table, Pointe de la** headland SE Réunion 21°19′S 55°49′E
27 S8 **Table Rock Lake** ⊠ Arkansas/Missouri, C USA
36 K14 **Table Top** ▲ Arizona, SW USA 32°45′N 112°07′W
186 D8 **Tabletop, Mount** ▲ C Papua New Guinea 06°51′S 146°00′E
111 D18 **Tábor** Jihočeský Kraj, S Czech Republic 49°25′N 14°41′E
123 R7 **Tabor** Respublika Sakha (Yakutiya), NE Russian Federation 71°14′N 150°23′E
29 S15 **Tabor** Iowa, C USA 40°54′N 95°40′W
81 F21 **Tabora** Tabora, W Tanzania 05°04′S 32°49′E
81 F21 **Tabora** ◈ region C Tanzania
21 U12 **Tabor City** North Carolina, SE USA 34°09′N 78°52′W
147 Q10 **Taboshar** NW Tajikistan 40°37′N 69°33′E
76 L18 **Tabou** var. Tabu. S Ivory Coast 04°28′N 07°20′W
142 J2 **Tabrīz** var. Tebriz; anc. Tauris. Āẕarbāyjān-e Sharqi, NW Iran 38°05′N 46°18′E
Tabu see Tabou
191 W1 **Tabuaeran** prev. Fanning Island. atoll Line Islands, E Kiribati
171 O2 **Tabuk** Luzon, N Philippines 17°26′N 121°25′E
140 J4 **Tabūk** Tabūk, NW Saudi Arabia 28°25′N 36°34′E
140 J5 **Tabūk** off. Minţaqat Tabūk. ◈ province NW Saudi Arabia
Tabūk, Minţaqat see Tabūk
187 Q13 **Tabwemasana, Mount** ▲ Espiritu Santo, W Vanuatu 15°22′S 166°44′E
95 O15 **Täby** Stockholm, C Sweden 59°29′N 18°04′E
41 N14 **Tacámbaro** Michoacán, SW Mexico 19°12′N 101°27′W
42 A5 **Tacaná, Volcán** ⧍ Guatemala/Mexico 15°07′N 92°06′W
43 X16 **Tacarcuna, Cerro** ▲ SE Panama 08°08′N 77°15′W
Tachau see Tachov
158 J3 **Tacheng** var. Qoqek. Xinjiang Uygur Zizhiqu, NW China 46°45′N 82°55′E
54 H7 **Táchira** off. Estado Táchira. ◈ state W Venezuela
Táchira, Estado see Táchira
Tachoshui see Dazhuoshui
111 A17 **Tachov** Ger. Tachau. Plveňský Kraj, W Czech Republic 49°48′N 12°38′E
171 Q5 **Tacloban** off. Tacloban City. Leyte, C Philippines 11°15′N 125°E
Tacloban City see Tacloban
57 I19 **Tacna** Tacna, SE Peru 18°S 70°15′W
57 H18 **Tacna** off. Departamento de Tacna. ◈ department S Peru
Tacna, Departamento de see Tacna
32 H8 **Tacoma** Washington, NW USA 47°15′N 122°27′W
18 L11 **Taconic Range** ▲▲ NE USA
62 L6 **Taco Pozo** Formosa, N Argentina 25°35′S 63°15′W
57 M20 **Tacsara, Cordillera de** ▲▲ S Bolivia
61 F17 **Tacuarembó** prev. San Fructuoso. Tacuarembó, C Uruguay 31°42′S 56°W
61 E18 **Tacuarembó** ◈ department C Uruguay
61 F17 **Tacuarembó, Río** ≈ C Uruguay
83 I14 **Taculi** North Western, NW Zambia 14°17′S 26°51′E
171 Q8 **Tacurong** Mindanao, S Philippines 06°42′N 124°40′E
77 V8 **Tadek** ≈ NW Niger
74 J9 **Tademaït, Plateau du** plateau C Algeria
187 R17 **Tadine** Province des Îles Loyauté, E New Caledonia 21°33′S 167°54′E
80 M11 **Tadjoura, Golfe de** Eng. Gulf of Tajura. inlet E Djibouti
80 L11 **Tadjourah** E Djibouti 11°47′N 42°51′E
Tadmor/Tadmur see Tudmur
11 W10 **Tadoule Lake** ◎ Manitoba, C Canada
15 S8 **Tadoussac** Québec, SE Canada 48°09′N 69°43′W
155 H18 **Tādpatri** Andhra Pradesh, E India 14°55′N 77°59′E
Tadzhikabad see Tojikobod
Tadzhikistan see Tajikistan
163 Y14 **Taebaek-sanmaek** prev. T'aebaek-sanmaek. ▲▲ E South Korea
T'aebaek-sanmaek see Taebaek-sanmaek
Taechŏng-do see Daecheong-do
163 X13 **Taedong-gang** ≈ C North Korea
Taegu see Daegu
Taehan-haehyŏp see Korea Strait
Taehan Min'guk see South Korea
Taejŏn see Daejeon
193 Z13 **Tafahi** island N Tonga
105 Q4 **Tafalla** Navarra, N Spain 42°32′N 01°41′W
77 W7 **Tafassâsset, Ténéré du** desert N Niger
75 M12 **Tafassâsset, Oued** ≈ SE Algeria
55 U11 **Tafelberg** ▲ S Suriname 03°55′N 56°09′W
97 J21 **Taff** ≈ SE Wales, United Kingdom
Tafila/Ţafīlah, Muḥāfaẓat aţ see Aţ Ţafīlah
77 N15 **Tafiré** N Ivory Coast 09°04′N 05°10′W
142 M6 **Tafresh** Markazī, W Iran 34°40′N 50°00′E
143 Q9 **Taft** Yazd, C Iran 31°45′N 54°14′E
35 R13 **Taft** California, W USA 35°08′N 119°27′W
25 T14 **Taft** Texas, SW USA 27°58′N 97°24′W
143 W12 **Taftān, Kūh-e** ▲ SE Iran 28°38′N 61°06′E
35 R13 **Taft Heights** California, W USA 35°06′N 119°29′W
189 Y14 **Tafunsak** Kosrae, E Micronesia 05°21′N 162°58′E
192 G16 **Tāga** Savai'i, SW Samoa 13°46′S 172°31′W
149 O6 **Tagāb** Dāikondī, E Afghanistan 33°53′N 66°23′E
39 O8 **Tagagawik River** ≈ Alaska, USA
165 Q10 **Tagajō** var. Tagazyô. Miyagi, Honshū, C Japan 38°18′N 140°58′E
126 K12 **Taganrog** Rostovskaya Oblast', SW Russian Federation 47°10′N 38°55′E
126 K12 **Taganrog, Gulf of** Rus. Taganrogskiy Zaliv, Ukr. Tahanroz'ka Zatoka. gulf Russian Federation/Ukraine
Taganrogskiy Zaliv see Taganrog, Gulf of
76 J8 **Tagant** ◈ region C Mauritania
148 M14 **Tagas** Baluchistān, SW Pakistan 27°09′N 64°36′E
171 O4 **Tagaytay** Luzon, N Philippines 14°04′N 120°55′E
Tagazyô see Tagajō
171 P6 **Tagbilaran** var. Tagbilaran City. Bohol, C Philippines 09°41′N 123°54′E
Tagbilaran City see Tagbilaran
106 B10 **Taggia** Liguria, NW Italy 43°51′N 07°48′E
77 V9 **Taghouaji, Massif de** ▲ C Niger 17°13′N 08°37′E
107 J15 **Tagliacozzo** Lazio, C Italy 42°03′N 13°15′E
106 J7 **Tagliamento** ≈ NE Italy
149 N3 **Tagow Bāy** var. Bai. Sar-e Pul, N Afghanistan 35°41′N 66°01′E
146 H9 **Tagta** var. Tahta, Rus. Takhta. Daşoguz Welaýaty, N Turkmenistan 41°40′N 59°51′E
146 J16 **Tagtabazar** var. Takhtabazar. Mary Welaýaty, S Turkmenistan 35°57′N 62°49′E
59 L17 **Taguatinga** Tocantins, C Brazil 12°16′S 46°25′W
186 I10 **Tagula** Tagula Island, SE Papua New Guinea 11°21′S 153°11′E
186 I11 **Tagula Island** prev. Southeast Island, Sudest Island. island SE Papua New Guinea
171 Q7 **Tagum** Mindanao, S Philippines 07°22′N 125°51′E
54 C7 **Tagún, Cerro** elevation Colombia/Panama
105 P7 **Tagus** Port. Rio Tejo, Sp. Río Tajo. ≈ Portugal/Spain
64 M9 **Tagus Plain** undersea feature E Atlantic Ocean 37°30′N 12°00′W
191 S10 **Tahaa** island Îles Sous le Vent, W French Polynesia
191 U10 **Tahanea** atoll Îles Tuamotu, C French Polynesia

◆ Country ● Country Capital ◇ Dependent Territory ○ Dependent Territory Capital ◈ Administrative Regions ✈ International Airport ▲ Mountain ▲▲ Mountain Range ⧍ Volcano ≈ River ◎ Lake ⊠ Reservoir

329

TAHANROZ'KA ZATOKA – TARAKAN

Tahanroz'ka Zatoka *see* Taganrog, Gulf of
74 K12 **Tahat** ▲ SE Algeria 23°15′N 05°34′E
163 U4 **Tahe** Heilongjiang, NE China 52°21′N 124°42′E
Tahilt *see* Tsogt
191 T10 **Tahiti** *island* Îles du Vent, W French Polynesia
Tahiti, Archipel de *see* Société, Archipel de la
118 E4 **Tahkuna Nina** *headland* W Estonia 59°06′N 22°35′E
148 K12 **Tāhlāb** ≈ W Pakistan
148 K12 **Tāhlāb, Dasht-i** *desert* SW Pakistan
27 R10 **Tahlequah** Oklahoma, C USA 35°57′N 94°58′W
35 Q6 **Tahoe City** California, W USA 39°09′N 120°09′W
35 P6 **Tahoe, Lake** ◎ California/Nevada, W USA
25 N6 **Tahoka** Texas, SW USA 33°10′N 101°47′W
32 F8 **Taholah** Washington, NW USA 47°19′N 124°17′W
77 T11 **Tahoua** Tahoua, W Niger 14°53′N 05°18′E
77 T11 **Tahoua** ◈ *department* W Niger
31 P3 **Tahquamenon Falls** *waterfall* Michigan, N USA
31 P4 **Tahquamenon River** ≈ Michigan, N USA
139 V10 **Taḩrīr** Al Qādisīyah, S Iraq 31°58′N 45°34′E
10 K17 **Tahsis** Vancouver Island, British Columbia, SW Canada 49°42′N 126°31′W
75 W9 **Ṭahṭā** *var.* Ṭahta. C Egypt 26°47′N 31°31′E
Tahta *see* Tagta
136 L15 **Tahtalı Dağları** ▲▲ C Turkey
57 I14 **Tahuamanu, Río** ≈ Bolivia/Peru
56 F13 **Tahuanía, Río** ≈ E Peru
191 X7 **Tahuata** *island* Îles Marquises, NE French Polynesia
76 L17 **Taï** SW Ivory Coast 05°52′N 07°28′W
161 P5 **Tai'an** Shandong, E China 36°13′N 117°12′E
191 R8 **Taiarapu, Presqu'île de** *peninsula* Tahiti, W French Polynesia
Taibad *see* Tāybād
160 K7 **Taibai Shan** ▲ C China 33°57′N 107°31′E
161 T13 **Taibei** ● (Taiwan) N Taiwan 25°02′N 121°28′E
105 Q12 **Taibilla, Sierra de** ▲▲ S Spain
Taibus Qi *see* Baochang
Taichū *see* Taizhong
T'aichung *see* Taizhong
Taiden *see* Daejeon
161 T14 **Taidong** *Jap.* Taitō; *prev.* T'aitung. S Taiwan 22°43′N 121°10′E
185 E23 **Taieri** ≈ South Island, New Zealand
115 E21 **Taïgetos** ▲▲ S Greece
161 N4 **Taihang Shan** ▲▲ C China
184 M11 **Taihape** Manawatu-Wanganui, North Island, New Zealand 39°41′S 175°47′E
161 O7 **Taihe** Anhui, E China 33°14′N 115°35′E
161 O12 **Taihe** *var.* Chengjiang. Jiangxi, S China 26°47′N 114°52′E
Taihoku *see* Taibei
161 P9 **Taihu** Anhui, E China 30°22′N 116°20′E
161 R8 **Tai Hu** ◎ E China
159 O9 **Taikang** *var.* Dorbod, Dorbod Mongolzu Zizhixian. Heilongjiang, NE China 46°50′N 124°25′E
161 O6 **Taikang** Henan, C China 34°01′N 114°59′E
165 T5 **Taiki** Hokkaidō, NE Japan 42°29′N 143°15′E
166 L8 **Taikkyi** Yangon, SW Myanmar (Burma) 17°16′N 95°55′E
Taikyū *see* Daegu
163 U8 **Tailai** Heilongjiang, NE China 46°25′N 123°25′E
168 I12 **Taileleo** Pulau Siberut, W Indonesia 01°45′S 99°06′E
182 J10 **Tailem Bend** South Australia 35°20′S 139°34′E
96 I8 **Tain** N Scotland, United Kingdom 57°49′N 04°04′W
161 S14 **Tainan** *prev.* Dainan, T'ainan. S Taiwan 23°01′N 120°05′E
115 E22 **Taínaro, Akrotírio** *cape* S Greece
161 Q11 **Taining** *var.* Shancheng. Fujian, SE China 26°55′N 117°13′E
191 W7 **Taiohae** *prev.* Madisonville. Nuku Hiva, NE French Polynesia 08°55′S 140°04′W
Taipei *see* Taibei
T'aipei *see* Taibei
168 J7 **Taiping** Perak, Peninsular Malaysia 04°54′N 100°42′E
Taiping *see* Chongzuo
163 S8 **Taiping Ling** ▲ NE China 47°27′N 120°27′E
165 Q4 **Taisei** Hokkaidō, NE Japan 42°13′N 139°52′E
165 G12 **Taisha** Shimane, Honshū, SW Japan 35°23′N 132°40′E
Taishō-tō *see* Sekibi-sho
109 R4 **Taiskirchen** Oberösterreich, NW Austria 48°15′N 13°33′E
63 F20 **Taitao, Peninsula de** *peninsula* S Chile
81 J21 **Taita/Taveta** ◈ *county* S Kenya
Taitō *see* Taidong
T'aitung *see* Taidong
92 M13 **Taivalkoski** Pohjois-Pohjanmaa, E Finland 65°35′N 28°20′E
93 K19 **Taivassalo** Varsinais-Suomi, SW Finland 60°35′N 21°36′E
161 T14 **Taiwan** *off.* Republic of China, *var.* Formosa, Formo'sa. ◆ *republic* E Asia
192 F5 **Taiwan** *var.* Formosa. *island* E Asia
Taiwan *see* Taizhong
T'aiwan Haihsia/Taiwan Haixia *see* Taiwan Strait
Taiwan Shan *see* Chungyang Shanmo
161 R13 **Taiwan Strait** *var.* Formosa Strait, *Chin.* T'aiwan Haihsia, Taiwan Haixia. *strait* China/Taiwan
161 S12 **Taiwan Taoyuan** *prev.* Chiang Kai-shek. ✈ (T'aibei) N Taiwan 25°09′N 121°20′E
161 N4 **Taiyuan** *var.* T'ai-yuan, T'ai-yüan; *prev.* Yangku. *province capital* Shanxi, C China 37°48′N 112°33′E
T'ai-yuan/T'ai-yüan *see* Taiyuan
161 S13 **Taizhong** *Jap.* Taichū; *prev.* T'aichung, Taiwan. C Taiwan 24°09′N 120°40′E
161 R7 **Taizhou** Jiangsu, E China 32°36′N 119°52′E
161 S10 **Taizhou** *var.* Jiaojiang; *prev.* Haimen. Zhejiang, SE China 28°36′N 121°19′E
Taizhou *see* Linhai
141 O16 **Ta'izz** SW Yemen 13°36′N 44°04′E
141 O16 **Ta'izz** ✈ SW Yemen 13°40′N 44°10′E
75 P12 **Tajarhī** SW Libya 24°21′N 14°28′E
147 P13 **Tajikistan** *off.* Republic of Tajikistan, *Rus.* Tadzhikistan, *Taj.* Jumhurii Tojikiston; *prev.* Tajik S.S.R. ◆ *republic* C Asia
Tajikistan, Republic of *see* Tajikistan
Tajik S.S.R *see* Tajikistan
165 O11 **Tajima** Fukushima, Honshū, C Japan 37°10′N 139°46′E
Tajoe *see* Tayu
Tajo, Río *see* Tagus
42 B5 **Tajumulco, Volcán** ▲ W Guatemala 15°04′N 91°50′W
105 P7 **Tajuña** ≈ C Spain
Tajura, Gulf of *see* Tadjoura, Golfe de
167 O9 **Tak** *var.* Rahaeng. Tak, W Thailand 16°51′N 99°08′E
189 U4 **Taka Atoll** *var.* Tōke. *atoll* Ratak Chain, N Marshall Islands
165 P12 **Takahagi** Ibaraki, Honshū, S Japan 36°42′N 140°42′E
165 H13 **Takahashi** *var.* Takahasi. Okayama, Honshū, SW Japan 34°48′N 133°38′E
Takahasi *see* Takahashi
189 P12 **Takaieu Island** *island* E Micronesia
184 I13 **Takaka** Tasman, South Island, New Zealand 40°52′S 172°49′E
170 M14 **Takalar** Sulawesi, C Indonesia 05°28′S 119°24′E
165 H13 **Takamatsu** *var.* Takamatu. Kagawa, Shikoku, SW Japan 34°19′N 133°59′E
Takamatu *see* Takamatsu
165 D14 **Takamori** Kumamoto, Kyūshū, SW Japan 32°50′N 131°08′E
165 D16 **Takanabe** Miyazaki, Kyūshū, SW Japan 32°13′N 131°31′E
170 M16 **Takan, Gunung** ▲ Pulau Sumba, S Indonesia 08°52′S 117°32′E
165 Q7 **Takanosu** *var.* Kita-Akita. Akita, Honshū, C Japan 40°13′N 140°23′E
Takao *see* Gaoxiong
165 L11 **Takaoka** Toyama, Honshū, SW Japan 36°44′N 137°02′E
184 N12 **Takapau** Hawke's Bay, North Island, New Zealand 40°01′S 176°21′E
191 U9 **Takapoto** *atoll* Îles Tuamotu, C French Polynesia
184 L5 **Takapuna** Auckland, North Island, New Zealand 36°48′S 174°46′E
165 J3 **Takarazuka** Hyōgo, Honshū, SW Japan 34°49′N 135°21′E
191 U9 **Takaroa** *atoll* Îles Tuamotu, C French Polynesia
165 N12 **Takasaki** Gunma, Honshū, S Japan 36°20′N 139°00′E
164 L12 **Takayama** Gifu, Honshū, SW Japan 36°09′N 137°16′E
164 K12 **Takefu** *var.* Echizen, Takehu. Fukui, Honshū, SW Japan 35°55′N 136°11′E
Takehu *see* Takefu
164 C14 **Takeo** Saga, Kyūshū, SW Japan 33°13′N 130°00′E
Takeo *see* Takêv
164 C17 **Take-shima** *island* Nansei-shotō, SW Japan
142 M5 **Tākestān** *var.* Takistan; *prev.* Siadehan. Qazvin, N Iran 36°02′N 49°40′E
164 D14 **Taketa** Ōita, Kyūshū, SW Japan 33°00′N 131°23′E
167 R13 **Takêv** *prev.* Takeo. Takêv, S Cambodia 10°59′N 104°47′E
167 O10 **Tak Fah** Nakhon Sawan, C Thailand
139 T13 **Takhādīd** *well* S Iraq
149 R3 **Takhār** ◈ *province* NE Afghanistan
Takhiatash *see* Taxiatosh
Ta Khmau *see* Kândal
Takhta *see* Tagta
Takhtabazar *see* Tagtabazar
145 O8 **Takhtabrod** Severnyy Kazakhstan, N Kazakhstan 52°35′N 67°37′E
Takhtakupyr *see* Taxtako'pir
142 M8 **Takht-e Shāh, Kūh-e** ▲▲ C Iran
77 V12 **Takiéta** Zinder, S Niger 13°43′N 08°33′E
8 J8 **Takijuq Lake** ◎ Nunavut, NW Canada
165 S3 **Takikawa** Hokkaidō, NE Japan 43°35′N 141°54′E
165 U3 **Takinoue** Hokkaidō, NE Japan 44°10′N 143°09′E
Takistan *see* Tākestān
185 B23 **Takitimu Mountains** ▲▲ South Island, New Zealand
Takkaze *see* Tekezē
165 R7 **Takko** Aomori, Honshū, C Japan 40°19′N 141°11′E
10 L13 **Takla Lake** ◎ British Columbia, SW Canada
Takla Makan Desert *see* Taklimakan Shamo
158 H9 **Taklimakan Shamo** *Eng.* Takla Makan Desert. *desert* NW China
167 T12 **Takôk** Môndól Kiri, E Cambodia 12°37′N 106°30′E
39 P10 **Takotna** Alaska, USA 62°59′N 156°03′W
Takow *see* Gaoxiong
123 O12 **Taksimo** Respublika Buryatiya, S Russian Federation 56°18′N 114°53′E
164 C13 **Taku** Saga, Kyūshū, SW Japan 33°19′N 130°06′E
10 I10 **Taku** ≈ British Columbia, W Canada
166 M15 **Takua Pa** *var.* Ban Takua Pa. Phangnga, SW Thailand 08°55′N 98°20′E
77 W16 **Takum** Taraba, E Nigeria 07°16′N 10°00′E
191 V10 **Takume** *atoll* Îles Tuamotu, C French Polynesia
190 L16 **Takutea** *island* S Cook Islands
186 K6 **Takuu Islands** *prev.* Mortlock Group. *island group* NE Papua New Guinea
119 L18 **Tal'** Minskaya Voblasts', S Belarus 52°52′N 27°58′E
40 L13 **Tala** Jalisco, C Mexico 20°39′N 103°45′W
61 F19 **Tala** Canelones, S Uruguay 34°24′S 55°45′W
Talabriga *see* Aveiro, Portugal
Talabriga *see* Talavera de la Reina, Spain
119 N14 **Talachyn** *Rus.* Tolochin. Vitsyebskaya Voblasts', NE Belarus 54°25′N 29°42′E
149 U7 **Talagang** Punjab, E Pakistan 32°55′N 72°29′E
105 V11 **Talaiassa** ▲ Ibiza, Spain, W Mediterranean Sea 38°55′N 1°17′E
155 J23 **Talaimannar** Northern Province, NW Sri Lanka 09°05′N 79°43′E
117 R3 **Talalayivka** Chernihivs'ka Oblast', N Ukraine 50°51′N 33°09′E
43 O15 **Talamanca, Cordillera de** ▲▲ S Costa Rica
56 A9 **Talara** Piura, NW Peru 04°31′S 81°17′W
104 L11 **Talarrubias** Extremadura, W Spain 39°03′N 05°14′W
147 S8 **Talas** Talasskaya Oblast', NW Kyrgyzstan 42°29′N 72°21′E
147 S8 **Talas** ≈ NW Kyrgyzstan
186 G7 **Talasea** New Britain, E Papua New Guinea 05°20′S 150°01′E
Talas Oblasty *see* Talasskaya Oblast'
147 S8 **Talasskaya Oblast'** *Kir.* Talas Oblasty. ◈ *province* NW Kyrgyzstan
147 S8 **Talasskiy Alatau, Khrebet** ▲▲ Kazakhstan/Kyrgyzstan
77 U12 **Talata Mafara** Zamfara, NW Nigeria 12°33′N 06°01′E
171 R9 **Talaud, Kepulauan** *island group* E Indonesia
104 M9 **Talavera de la Reina** *anc.* Caesarobriga, Talabriga. Castilla-La Mancha, C Spain 39°58′N 04°50′W
104 J11 **Talavera la Real** Extremadura, W Spain 38°53′N 06°46′W
186 F7 **Talawe, Mount** ▲ New Britain, C Papua New Guinea 05°30′S 148°24′E
23 S5 **Talbotton** Georgia, SE USA 32°40′N 84°32′W
183 R7 **Talbragar River** ≈ New South Wales, SE Australia
62 G13 **Talca** Maule, C Chile 35°28′S 71°42′W
62 F13 **Talcahuano** Bío Bío, C Chile 36°43′S 73°07′W
154 N12 **Tālcher** Odisha, E India 20°57′N 85°13′E
25 W5 **Talco** Texas, SW USA 33°21′N 95°06′W
145 V14 **Taldykorgan** *Kaz.* Taldyqorghan; *prev.* Taldy-Kurgan. Taldykorgan, SE Kazakhstan 45°N 78°23′E
Taldy-Kurgan/Taldyqorghan *see* Taldykorgan
147 Y7 **Taldy-Suu** Issyk-Kul'skaya Oblast', E Kyrgyzstan 42°49′N 78°33′E
147 U10 **Taldy-Suu** Oshskaya Oblast', SW Kyrgyzstan 40°33′N 73°52′E
Tal-e Khosravī *see* Yāsūj
193 Y15 **Taleki Tonga** *island* Otu Tolu Group, C Tonga
193 Y15 **Taleki Vavu'u** *island* Otu Tolu Group, C Tonga
102 J13 **Talence** Gironde, SW France 44°49′N 00°35′W
145 U16 **Talgar** *Kaz.* Talghar. Almaty, SE Kazakhstan 43°17′N 77°15′E
Talghar *see* Talgar
171 Q12 **Taliabu, Pulau** *island* Kepulauan Sula, C Indonesia
115 L22 **Taliarós, Akrotírio** *headland* Astypálaia, Kykládes, Greece, Aegean Sea 36°31′N 26°18′E
Ta-lien *see* Dalian
27 Q12 **Talihina** Oklahoma, C USA 34°45′N 95°03′W
Talimardzhan *see* Tollimarjon
137 T12 **T'alin** *Rus.* Talin; *prev.* Verin T'alin. W Armenia 40°23′N 43°51′E
Talin *see* T'alin
81 E15 **Tali Post** Central Equatoria, S South Sudan 05°55′N 30°44′E
Taliq-an *see* Tāloqān
Talış Dağları *see* Talish Mountains
142 L2 **Talish Mountains** *Az.* Talış Dağları, *Per.* Kūhhā-ye Ţavālesh, *Rus.* Talyshskiye Gory. ▲▲ Azerbaijan/Iran
170 M16 **Taliwang** Sumbawa, C Indonesia 08°45′S 116°55′E
119 L17 **Tal'ka** Minskaya Voblasts', C Belarus 53°22′N 28°21′E
39 R11 **Talkeetna** Alaska, USA 62°19′N 150°06′W
39 R11 **Talkeetna Mountains** ▲▲ Alaska, USA
Talkhof *see* Puurmani
92 H2 **Tálknafjörður** Vestfirðir, W Iceland 65°38′N 23°51′W
139 Q3 **Tall 'Abṭah** Nīnawá, N Iraq 35°52′N 42°40′E
138 M2 **Tall Abyaḍ** *var.* Tell Abiad. Ar Raqqah, N Syria 36°42′N 38°56′E
23 Q4 **Talladega** Alabama, S USA 33°26′N 86°06′W
139 Q2 **Tall 'Afar** Nīnawýa, N Iraq 36°22′N 42°27′E
23 S8 **Tallahassee** *prev.* Muskogean. *state capital* Florida, SE USA 30°26′N 84°17′W
22 L2 **Tallahatchie River** ≈ Mississippi, S USA
Tall al Abyaḍ *see* At Tall al Abyaḍ
139 W12 **Tall al Laḩm** Dhī Qār, S Iraq 30°46′N 46°22′E
183 P11 **Tallangatta** Victoria, SE Australia 36°15′S 147°13′E
23 R4 **Tallapoosa River** ≈ Alabama/Georgia, S USA
103 T13 **Tallard** Hautes-Alpes, SE France 44°30′N 06°04′E
139 Q3 **Tall ash Sha'īr** Nīnawá, N Iraq 36°11′N 42°26′E
23 Q5 **Tallassee** Alabama, S USA 32°32′N 85°53′W
139 R4 **Tall 'Azbah** Nīnawýa, NW Iraq 35°47′N 43°13′E
138 I5 **Tall Bīsah** Ḩimş, W Syria 34°50′N 36°44′E
139 R3 **Tall Ḩassūnah** Al Anbār, N Iraq 36°05′N 43°10′E
139 Q2 **Tall Ḩuqnah** *var.* Tell Huqnah. Nīnawá, N Iraq 36°33′N 42°34′E
Tallin *see* Tallinn
118 G3 **Tallinn** *Ger.* Reval, *Rus.* Tallin; *prev.* Revel. ● (Estonia) Harjumaa, NW Estonia 59°26′N 24°42′E
118 H3 **Tallinn** ✈ Harjumaa, NW Estonia 59°23′N 24°52′E
138 H5 **Tall Kalakh** *var.* Tell Kalakh. Ḩimş, C Syria 34°40′N 36°18′E
139 R2 **Tall Kayf** Nīnawýa, NW Iraq 36°30′N 43°08′E
Tall Kūchak *see* Tall Kūshik
139 P2 **Tall Kūshik** *var.* Tall Kūchak. Al Ḩasakah, E Syria 36°48′N 42°01′E
31 U12 **Tallmadge** Ohio, N USA 41°06′N 81°26′W
22 J5 **Tallulah** Louisiana, S USA 32°22′N 91°12′W
139 Q2 **Tall 'Uwaynāt** Nīnawýa, NW Iraq 36°43′N 42°18′E
139 Q2 **Tall Ẕāhir** Nīnawá, N Iraq 36°51′N 42°29′E
122 J13 **Tal'menka** Altayskiy Kray, S Russian Federation 53°55′N 83°26′E
122 K8 **Talnakh** Krasnoyarskiy Kray, N Russian Federation 69°26′N 88°27′E
117 P7 **Tal'ne** *Rus.* Tal'noye. Cherkas'ka Oblast', C Ukraine 48°55′N 30°40′E
Tal'noye *see* Tal'ne
80 E12 **Talodi** Southern Kordofan, C Sudan 10°40′N 30°25′E
188 B16 **Talofofo** SE Guam 13°21′N 144°45′E
188 B16 **Talofofo Bay** *bay* SE Guam
26 L9 **Taloga** Oklahoma, C USA 36°02′N 98°58′W
123 T10 **Talon** Magadanskaya Oblast', E Russian Federation 59°47′N 148°46′E
14 H11 **Talon, Lake** ◎ Ontario, S Canada
149 R2 **Tāloqān** *var.* Taliq-an. Takhār, NE Afghanistan 36°44′N 69°33′E
126 M8 **Talovaya** Voronezhskaya Oblast', W Russian Federation 51°07′N 40°46′E
9 N6 **Taloyoak** *prev.* Spence Bay. Nunavut, N Canada 69°30′N 93°25′W
25 Q8 **Talpa** Texas, SW USA 31°46′N 99°42′W
40 K13 **Talpa de Allende** Jalisco, C Mexico 20°22′N 104°51′W
23 S9 **Talquin, Lake** ⊠ Florida, SE USA
Talsen *see* Talsi
Talshand *see* Chandmanĭ
118 E8 **Talsi** *Ger.* Talsen. NW Latvia 57°14′N 22°35′E
143 V11 **Tal Sīāh** Sīstān va Balūchestān, SE Iran 28°19′N 57°43′E
62 G6 **Taltal** Antofagasta, N Chile 25°22′S 70°27′W
8 K10 **Taltson** ≈ Northwest Territories, NW Canada
168 K11 **Taluk** Sumatera, W Indonesia 0°32′S 101°35′E
92 J8 **Talvik** Finnmark, N Norway 70°02′N 22°59′E
182 M7 **Talyawalka Creek** ≈ New South Wales, SE Australia
Talyshskiye Gory *see* Talish Mountains
29 W14 **Tama** Iowa, C USA 41°58′N 92°34′W
Tama Abu, Banjaran *see* Penambo, Banjaran
169 U9 **Tamabo, Banjaran** ▲▲ East Malaysia
190 B16 **Tamakautoga** SW Niue 19°05′S 169°55′W
127 N7 **Tamala** Penzenskaya Oblast', W Russian Federation 52°32′N 43°18′E
77 P15 **Tamale** C Ghana 09°21′N 00°54′W
191 P3 **Tamana** *prev.* Rotcher Island. *atoll* Tungaru, W Kiribati
74 K12 **Tamanrasset** *var.* Tamenghest. S Algeria 22°49′N 05°32′E
74 J13 **Tamanrasset** *wadi* Algeria/Mali
166 M2 **Tamanthi** Sagaing, N Myanmar (Burma) 25°17′N 95°18′E
97 I24 **Tamar** ≈ SW England, United Kingdom
Tamar *see* Tudmur
54 H9 **Támara** Casanare, C Colombia 05°51′N 72°09′W
54 F7 **Tamar, Alto de** ▲ C Colombia 07°25′N 74°28′W
173 X16 **Tamarin** E Mauritius 20°20′S 57°22′E
105 T5 **Tamarite de Litera** *var.* Tararite de Llitera. Aragón, NE Spain 41°52′N 00°25′E
111 I24 **Tamási** Tolna, S Hungary 46°39′N 18°17′E
Tamatave *see* Toamasina
41 O9 **Tamaulipas** ◈ *state* C Mexico
41 P10 **Tamaulipas, Sierra de** ▲▲ C Mexico
56 F12 **Tamaya, Río** ≈ E Peru
40 I9 **Tamazula** Durango, C Mexico 24°43′N 106°33′W
40 L14 **Tamazula** Jalisco, C Mexico 19°41′N 103°18′W
Tamazulápam *see* Tamazulapán
41 Q15 **Tamazulapán** *var.* Tamazulápam. Oaxaca, SE Mexico 17°41′N 97°33′W
41 P12 **Tamazunchale** San Luis Potosí, C Mexico 21°17′N 98°46′W
76 H11 **Tambacounda** SE Senegal 13°44′N 13°43′W
83 M16 **Tambara** Manica, C Mozambique 16°42′S 34°14′E
77 T13 **Tambawel** Sokoto, NW Nigeria 12°24′N 04°42′E
186 M9 **Tambea** Guadalcanal, C Solomon Islands 09°19′S 159°42′E
169 N10 **Tambelan, Kepulauan** *island group* W Indonesia
57 E15 **Tambo de Mora** Ica, W Peru 13°30′S 76°08′W
170 L16 **Tambora, Gunung** ⛰ Sumbawa, S Indonesia 08°16′S 117°59′E
61 E17 **Tambores** Paysandú, W Uruguay 31°50′S 56°17′W
57 F14 **Tambo, Río** ≈ C Peru
56 F7 **Tamboryacu, Río** ≈ N Peru
126 M7 **Tambov** Tambovskaya Oblast', W Russian Federation 52°43′N 41°28′E
126 L6 **Tambovskaya Oblast'** ◈ *province* W Russian Federation
104 H3 **Tambre** ≈ NW Spain
169 V7 **Tambunan** Sabah, East Malaysia 05°40′N 116°22′E
81 C15 **Tambura** Western Equatoria, SW South Sudan 05°38′N 27°30′E
Tamchaket *see* Tâmchekkeț
76 J9 **Tâmchekkeț** *var.* Tamchaket. Hodh el Gharbi, S Mauritania 17°23′N 10°37′W
167 T7 **Tam Điệp** Ninh Binh, N Vietnam 20°09′N 105°54′E
Tamdybulak *see* Tomdibuloq
54 H8 **Tame** Arauca, C Colombia 06°27′N 71°42′W
104 H6 **Tâmega, Rio** *Sp.* Río Támega. ≈ Portugal/Spain
Támega, Río *see* Tâmega, Rio
115 H20 **Tamélos, Akrotírio** *headland* Tziá, Kykládes, Greece, Aegean Sea 37°31′N 24°16′E
Tamenghest *see* Tamanrasset
77 V8 **Tamgak, Adrar** ▲ C Niger 19°10′N 08°39′E
76 I13 **Tamgue** ▲ NW Guinea 12°14′N 12°18′W
41 Q12 **Tamiahua** Veracruz-Llave, E Mexico 21°15′N 97°27′W
41 Q12 **Tamiahua, Laguna de** *lagoon* E Mexico
23 Y16 **Tamiami Canal** *canal* Florida, SE USA
188 F17 **Tamil Harbor** *harbor* Yap, W Micronesia
155 H21 **Tamil Nādu** *prev.* Madras. ◈ *state* SE India
99 H20 **Tamines** Namur, S Belgium 50°27′N 04°37′E
116 E12 **Tamiš** *Ger.* Temesch, *Hung.* Temes. ≈ Romania/Serbia
167 U10 **Tam Ky** Quang Nam-Đa Năng, C Vietnam 15°32′N 108°30′E
Tammerfors *see* Tampere
Tammisaari *see* Ekenäs
95 N14 **Tämnaren** ◎ C Sweden
191 Q7 **Tamotoe, Passe** *passage* Tahiti, W French Polynesia
23 V12 **Tampa** Florida, SE USA 27°57′N 82°27′W
23 V12 **Tampa** ✈ Florida, SE USA 27°57′N 82°29′W
23 V13 **Tampa Bay** *bay* Florida, SE USA
93 L18 **Tampere** *Swe.* Tammerfors. Pirkanmaa, W Finland 61°30′N 23°45′E
41 Q11 **Tampico** Tamaulipas, C Mexico 22°18′N 97°52′W
171 P14 **Tampo** Pulau Muna, C Indonesia 04°38′S 122°40′E
167 V11 **Tam Quan** Binh Đinh, C Vietnam 14°34′N 109°00′E
162 J13 **Tamsag Muchang** Nei Mongol Zizhiqu, N China 40°28′N 102°34′E
Tamsal *see* Tamsalu
118 I4 **Tamsalu** *Ger.* Tamsal. Lääne-Virumaa, NE Estonia 59°10′N 26°07′E
109 S8 **Tamsweg** Salzburg, SW Austria 47°08′N 13°49′E
188 C15 **Tamuning** NW Guam 13°29′N 144°47′E
183 T6 **Tamworth** New South Wales, SE Australia 31°07′S 150°54′E
97 M19 **Tamworth** C England, United Kingdom 52°39′N 01°40′W
81 K19 **Tana** *Finn.* Tenojoki, *Lapp.* Deatnu. ≈ SE Kenya *see also* Deatnu, Tenojoki
Tana *see* Deatnu/Tana
164 I15 **Tanabe** Wakayama, Honshū, SW Japan 33°43′N 135°22′E
92 L8 **Tana Bru** Finnmark, N Norway 70°11′N 28°06′E
39 T10 **Tanacross** Alaska, USA 63°30′N 143°21′W
92 L7 **Tanafjorden** *Lapp.* Deanuvuotna. *fjord* N Norway
38 G17 **Tanaga Island** *island* Aleutian Islands, Alaska, USA
38 G17 **Tanaga Volcano** ▲ Tanaga Island, Alaska, USA 51°53′N 178°08′W
107 M18 **Tanagro** ≈ S Italy
80 H11 **T'ana Hāyk'** *var.* Lake Tana. ◎ NW Ethiopia
168 H11 **Tanahbela, Pulau** *island* Kepulauan Batu, W Indonesia
171 H15 **Tanahjampea, Pulau** *island* W Indonesia
168 H11 **Tanahmasa, Pulau** *island* Kepulauan Batu, W Indonesia
Tanais *see* Don
152 L10 **Tanakpur** Uttarakhand, N India 29°04′N 80°06′E
Tana, Lake *see* T'ana Hāyk'
181 P5 **Tanami Desert** *desert* Northern Territory, N Australia
167 T14 **Tân An** Long An, S Vietnam 10°32′N 106°24′E
39 Q9 **Tanana** Alaska, USA 65°12′N 152°00′W
Tananarive *see* Antananarivo
39 Q9 **Tanana River** ≈ Alaska, USA
95 C16 **Tananger** Rogaland, S Norway 58°55′N 05°34′E
188 H5 **Tanapag** Saipan, S Northern Mariana Islands 15°14′S 145°45′E
188 H5 **Tanapag, Puetton** *bay* Saipan, S Northern Mariana Islands
81 J20 **Tana River** ◈ *county* SE Kenya
106 C9 **Tanaro** ≈ N Italy
163 Y12 **Tanch'ŏn** E North Korea 40°22′N 128°49′E
40 M14 **Tancitaro, Cerro** ⛰ C Mexico 19°16′N 102°25′W
153 N12 **Tānda** Uttar Pradesh, N India 26°33′N 82°39′E
77 O15 **Tanda** E Ivory Coast 07°48′N 03°10′W
116 L14 **Ţăndărei** Ialomiţa, SE Romania 44°39′N 27°40′E
63 N14 **Tandil** Buenos Aires, E Argentina 37°18′S 59°10′W
78 H12 **Tandjilé** *off.* Région du Tandjilé. ◈ *region* SW Chad
Tandjilé, Région du *see* Tandjilé
Tandjoeng *see* Tanjung
Tandjoengkarang *see* Bandar Lampung
Tandjoengpandan *see* Tanjungpandan
Tandjoengpinang *see* Tanjungpinang
Tandjoengredeb *see* Tanjungredeb
149 Q16 **Tando Allāhyār** Sind, SE Pakistan 25°28′N 68°44′E
149 Q17 **Tando Bāgo** Sind, SE Pakistan 24°48′N 68°59′E
149 Q16 **Tando Muhammad Khān** Sind, SE Pakistan 25°07′N 68°35′E
182 L7 **Tandou Lake** *seasonal lake* New South Wales, SE Australia
94 L11 **Tandsjöborg** Gävleborg, C Sweden 61°40′N 14°40′E
155 H15 **Tāndūr** Telangana, C India 17°16′N 77°37′E
164 C17 **Tanega-shima** *island* Nansei-shotō, SW Japan
165 R7 **Taneichi** Iwate, Honshū, C Japan 40°23′N 141°42′E
Tanen Taunggyi *see* Tane Range
167 N8 **Tane Range** *Bur.* Tanen Taunggyi. ▲▲ W Thailand
111 P15 **Tanew** ≈ SE Poland
21 W2 **Taneytown** Maryland, NE USA 39°39′N 77°10′W
74 H12 **Tanezrouft** *desert* Algeria/Mali
138 L7 **Ţanf, Jabal aţ** ▲ SE Syria 33°32′N 38°43′E
81 J21 **Tanga** Tanga, E Tanzania 05°07′S 39°05′E
81 I22 **Tanga** ◈ *region* E Tanzania
153 T14 **Tangail** Dhaka, C Bangladesh 24°15′N 89°55′E
186 I5 **Tanga Islands** *island group* NE Papua New Guinea
155 K26 **Tangalla** Southern Province, S Sri Lanka 06°02′N 80°47′E
Tanganyika and Zanzibar *see* Tanzania
68 J13 **Tanganyika, Lake** ◎ E Africa
56 E7 **Tangarana, Río** ≈ N Peru
191 V16 **Tangaroa, Maunga** ⛰ Easter Island, Chile, E Pacific Ocean
Tangdukou *see* Shaoyang
74 G5 **Tanger** *var.* Tangiers, Tangier, *Fr./Ger.* Tangerk, *Sp.* Tánger; *anc.* Tingis. NW Morocco 35°49′N 05°49′W
169 N15 **Tangerang** Jawa, C Indonesia 06°14′S 106°36′E
Tangerk *see* Tanger
100 M12 **Tangermünde** Sachsen-Anhalt, C Germany 52°35′N 11°57′E
159 O12 **Tanggulashan** *var.* Togton Heyan, Tuotuoheyan. Qinghai, C China 34°13′N 92°25′E
156 K10 **Tanggula Shan** *var.* Dangla, Tangla Range. ▲▲ W China
159 N13 **Tanggula Shan** ▲ W China 33°18′N 91°10′E
156 K10 **Tanggula Shankou** *Tib.* Dang La. *pass* W China
161 N7 **Tanghe** Henan, C China 32°40′N 112°53′E
149 T5 **Tāngi** Khyber Pakhtunkhwa, NW Pakistan 34°18′N 71°42′E
Tangier *see* Tanger
21 Y5 **Tangier Island** *island* Virginia, NE USA
Tangiers *see* Tanger
22 K8 **Tangipahoa River** ≈ Louisiana, S USA
Tangla Range *see* Tanggula Shan
156 I10 **Tangra Yumco** *var.* Tangro Tso. ◎ W China
Tangro Tso *see* Tangra Yumco
157 T7 **Tangshan** *var.* T'ang-shan. Hebei, E China 39°39′N 118°15′E
T'ang-shan *see* Tangshan
77 R14 **Tanguiéta** NW Benin 10°35′N 01°19′E
163 X7 **Tangwang He** ≈ NE China
163 X7 **Tangyuan** Heilongjiang, NE China 46°45′N 129°52′E
167 S14 **Tân Hiệp** *var.* Phung Hiệp. Cân Thơ, S Vietnam 09°50′N 105°48′E
92 M11 **Tanhua** Lappi, N Finland 67°31′N 27°30′E
171 U16 **Tanimbar, Kepulauan** *island group* Maluku, E Indonesia
Tanintharyi *see* Taninthayi
Tanintharyi *see* Taninthayi
167 N12 **Taninthayi** *var.* Tenasserim. Taninthayi, S Myanmar (Burma) 12°05′N 99°00′E
167 N11 **Taninthayi** *var.* Tenasserim; *prev.* Tanintharyi. ◈ *region* S Myanmar (Burma)
Tānjarō *see* Tancero
Tānjarū *see* Tancero
129 T15 **Tanjong Piai** *headland* Peninsular Malaysia
Tanjore *see* Thanjāvūr
169 U12 **Tanjung** *prev.* Tandjoeng. Borneo, C Indonesia 02°08′S 115°23′E
169 W9 **Tanjungbatu** Borneo, N Indonesia 02°19′N 118°03′E
Tanjungkarang/Tanjungkarang-Telukbetung *see* Bandar Lampung
169 N13 **Tanjungpandan** *prev.* Tandjoengpandan. Pulau Belitung, W Indonesia 02°44′S 107°36′E
168 M10 **Tanjungpinang** *prev.* Tandjoengpinang. Pulau Bintan, W Indonesia 0°55′N 104°28′E
169 V9 **Tanjungredeb** *var.* Tanjungredep; *prev.* Tandjoengredeb. Borneo, C Indonesia 02°09′N 117°29′E
Tanjungredep *see* Tanjungredeb
169 V9 **Tanjung Selor** Borneo, C Indonesia 2°50′N 117°22′E
149 S8 **Tānk** Khyber Pakhtunkhwa, NW Pakistan 32°14′N 70°29′E
187 S15 **Tanna** *island* S Vanuatu
93 F17 **Tännäs** Jämtland, C Sweden 62°27′N 12°40′E
Tannenhof *see* Krynica
108 K7 **Tannheim** Tirol, W Austria 47°30′N 10°32′E
Tannu-Tuva *see* Tyva, Respublika
171 Q12 **Tano** Pulau Taliabu, E Indonesia 01°51′S 124°55′E
77 O17 **Tano** ≈ S Ghana
152 D10 **Tanot** Rājasthān, NW India 27°44′N 70°17′E
77 V11 **Tanout** Zinder, C Niger 14°58′N 08°54′E
Tân Phu *see* Đinh Quan
41 P12 **Tanquián** San Luis Potosí, C Mexico 21°38′N 98°39′W
77 R13 **Tansarga** E Burkina Faso 11°51′N 01°51′E
167 T13 **Tan Son Nhat** ✈ (Hô Chi Minh) Tây Ninh, S Vietnam 10°52′N 106°38′E
75 V8 **Ţanţā** *var.* Tanta, Tantā, Tantâ. N Egypt 30°42′N 31°00′E
74 D9 **Tan-Tan** SW Morocco 28°30′N 11°11′W
41 P12 **Tantoyuca** Veracruz-Llave, E Mexico 21°21′N 98°12′W
152 J12 **Tāntpur** Uttar Pradesh, N India 26°51′N 77°29′E
Tan-tung *see* Dandong
38 M12 **Tanunak** Alaska, USA 60°35′N 165°15′W
166 L5 **Ta-nyaung** Magway, W Myanmar (Burma) 20°49′N 94°40′E
167 S5 **Tân Yên** Tuyên Quang, N Vietnam 22°08′N 104°58′E
81 F22 **Tanzania** *off.* United Republic of Tanzania, *Swa.* Jamhuri ya Muungano wa Tanzania; *prev.* German East Africa, Tanganyika and Zanzibar. ◆ *republic* E Africa
Tanzania, Jamhuri ya Muungano wa *see* Tanzania
Tanzania, United Republic of *see* Tanzania
Tao'an *see* Taonan
163 T8 **Tao'er He** ≈ NE China
159 U11 **Tao He** ≈ C China
163 U9 **Taonan** *var.* Tao'an. Jilin, NE China 45°20′N 122°46′E
T'aon-an *see* Baicheng
Taongi *see* Bokaak Atoll
107 M23 **Taormina** *anc.* Tauromenium. Sicilia, Italy, C Mediterranean Sea 37°54′N 15°18′E
37 S9 **Taos** New Mexico, SW USA 36°24′N 105°35′W
Taoudenit *see* Taoudenni
77 O6 **Taoudenni** *var.* Taoudenit. Tombouctou, N Mali 22°46′N 03°54′W
74 G6 **Taounate** N Morocco 34°33′N 04°39′N
Taoyang *see* Lintao
161 S13 **Taoyuan** *Jap.* Tōen; *prev.* T'aoyüan. N Taiwan 25°00′N 121°15′E
118 I3 **Tapa** *Ger.* Taps. Lääne-Virumaa, NE Estonia 59°15′N 26°E
41 V17 **Tapachula** Chiapas, SE Mexico 14°53′N 92°18′W
Tapaiu *see* Gvardeysk
59 H14 **Tapajós, Rio** *var.* Tapajóz. ≈ NW Brazil
Tapajóz *see* Tapajós, Rio
61 C21 **Tapalqué** *var.* Tapalquén. Buenos Aires, E Argentina 36°21′S 60°01′W
Tapalquén *see* Tapalqué
Tapanahoni *see* Tapanahony Rivier
55 W11 **Tapanahony Rivier** *var.* Tapanahoni. ≈ E Suriname
41 T16 **Tapanatepec** *var.* San Pedro Tapanatepec. Oaxaca, SE Mexico 16°23′N 94°09′W
185 D23 **Tapanui** Otago, South Island, New Zealand 45°55′S 169°16′E
59 E14 **Tapauá** Amazonas, N Brazil 05°42′S 64°15′W
47 R7 **Tapauá, Rio** ≈ W Brazil
185 I14 **Tapawera** Tasman, South Island, New Zealand 41°24′S 172°50′E
61 I16 **Tapes** Rio Grande do Sul, S Brazil 30°40′S 51°25′W
76 K16 **Tapeta** C Liberia 06°36′N 08°52′W
154 H11 **Tāpi** *prev.* Tāpti. ≈ W India
104 J2 **Tapia de Casariego** Asturias, N Spain 43°34′N 06°56′W
56 F10 **Tapiche, Río** ≈ N Peru
167 N15 **Tapi, Mae Nam** *var.* Luang. ≈ SW Thailand
186 E8 **Tapini** Central, S Papua New Guinea 08°16′S 146°09′E
Tapirapecó, Serra *see* Tapirapecó, Sierra
55 N13 **Tapirapecó, Sierra** *Port.* Serra Tapirapecó. ▲▲ Brazil/Venezuela
77 R13 **Tapoa** ≈ Benin/Niger
188 H5 **Tapochau, Mount** ▲ Saipan, S Northern Mariana Islands
111 H24 **Tapolca** Veszprém, W Hungary 46°54′N 17°29′E
21 X5 **Tappahannock** Virginia, NE USA 37°55′N 76°54′W
31 U13 **Tappan Lake** ⊠ Ohio, N USA
165 Q6 **Tappi-zaki** *headland* Honshū, C Japan 41°15′N 140°19′E
Taps *see* Tapa
Tāpti *see* Tāpi
Tapuaemanu *see* Maiao
185 J16 **Tapuaenuku** ▲ South Island, New Zealand 42°00′S 173°39′E
171 N8 **Tapul Group** *island group* Sulu Archipelago, SW Philippines
58 E11 **Tapurmcuará** *var.* Tapuruquara. Amazonas, NW Brazil 0°17′S 65°00′W
Tapuruquara *see* Tapurmcuará
192 J17 **Taputapu, Cape** *headland* Tutuila, W American Samoa 14°20′S 170°51′W
141 W13 **Ţāqah** S Oman 17°02′N 54°23′E
139 T3 **Taqtaq** *Ar.* Ţaqţaq. Arbīl, N Iraq 35°54′N 44°36′E
Ţaqţaq *see* Taqtaq
61 J15 **Taquara** Rio Grande do Sul, S Brazil 29°36′S 50°46′W
59 H19 **Taquari, Rio** ≈ C Brazil
60 L8 **Taquaritinga** São Paulo, S Brazil 21°22′S 48°29′W
122 I11 **Tara** Omskaya Oblast', C Russian Federation 56°54′N 74°17′E
83 I16 **Tara** Southern, S Zambia 16°56′S 26°50′E
113 J15 **Tara** ≈ Montenegro
112 K13 **Tara** ▲▲ W Serbia
77 W15 **Taraba** ◈ *state* E Nigeria
77 X15 **Taraba** ≈ E Nigeria
75 O7 **Ţarābulus** *var.* Ţarābulus al Gharb, *Eng.* Tripoli. ● (Libya) NW Libya 32°54′N 13°11′E
75 O7 **Tarābulus** ✈ NW Libya 32°37′N 13°07′E
Ţarābulus al Gharb *see* Ţarābulus
Tarābulus/Ţarābulus ash Shām *see* Tripoli
105 O7 **Taracena** Castilla-La Mancha, C Spain 40°39′N 03°08′W
117 N12 **Taraclia** *Rus.* Tarakilya. S Moldova 45°55′N 28°40′E
139 V10 **Tarād al Kahf** Dhī Qār, SE Iraq 31°58′N 45°58′E
183 R10 **Tarago** New South Wales, SE Australia 35°04′S 149°40′E
162 J8 **Taragt** *var.* Hüremt. Övörhangay, C Mongolia 46°18′N 102°27′E
169 V8 **Tarakan** Borneo, C Indonesia 03°20′N 117°38′E

◆ Country
● Country Capital
◇ Dependent Territory
○ Dependent Territory Capital
◈ Administrative Regions
✈ International Airport
▲ Mountain
▲▲ Mountain Range
⛰ Volcano
≈ River
◎ Lake
⊠ Reservoir

Tarakan, Pulau – Tekes

169 V9 **Tarakan, Pulau** island N Indonesia
Tarakilya see Taraclia
165 P16 **Tarama-jima** island Sakishima-shotō, SW Japan
184 K10 **Taranaki** off. Taranaki Region. ◈ region North Island, New Zealand
184 K10 **Taranaki, Mount** var. Egmont. ⛰ North Island, New Zealand 39°16′S 174°04′E
Taranaki Region see Taranaki
105 O9 **Tarancón** Castilla-La Mancha, C Spain 40°01′N 03°01′W
188 M15 **Tarang Reef** reef C Micronesia
96 E7 **Taransay** island NW Scotland, United Kingdom
107 P18 **Taranto** var. Tarentum. Puglia, SE Italy 40°30′N 17°11′E
107 O19 **Taranto, Golfo di** Eng. Gulf of Taranto. gulf S Italy
Taranto, Gulf of see Taranto, Golfo di
62 G3 **Tarapacá** off. Región de Tarapacá. ◈ region N Chile
Tarapacá, Región de see Tarapacá
187 N9 **Tarapaina** Maramasike Island, N Solomon Islands 09°28′S 161°24′E
56 D10 **Tarapoto** San Martín, N Peru 06°31′S 76°23′W
138 M6 **Ţaraq an Na'jah** hill range E Syria
138 M6 **Ţaraq Sidāwī** hill range E Syria
103 Q11 **Tarare** Rhône, E France 45°54′N 04°26′E
Tararite de Llitera see Tamarite de Litera
184 M13 **Tararua Range** ▲▲ North Island, New Zealand
151 Q22 **Tarāsa Dwīp** island Nicobar Islands, India, NE Indian Ocean
103 Q15 **Tarascon** Bouches-du-Rhône, SE France 43°48′N 04°39′E
102 M17 **Tarascon-sur-Ariège** Ariège, S France 42°51′N 01°35′E
117 P6 **Tarashcha** Kyyivs'ka Oblast', N Ukraine 49°34′N 30°31′E
57 L18 **Tarata** Cochabamba, C Bolivia 17°35′S 66°04′W
57 I18 **Tarata** Tacna, SW Peru 17°30′S 70°00′W
190 H2 **Taratai** atoll Tungaru, W Kiribati
59 B15 **Tarauacá** Acre, W Brazil 08°06′S 70°45′W
59 B15 **Tarauacá, Rio** ≈ NW Brazil
191 Q8 **Taravao** Tahiti, W French Polynesia 17°44′S 149°19′W
191 R8 **Taravao, Baie de** bay Tahiti, W French Polynesia
191 Q8 **Taravao, Isthme de** isthmus Tahiti, W French Polynesia
103 X16 **Taravo** ≈ Corse, France, C Mediterranean Sea
190 J3 **Tarawa** ✈ Tarawa, W Kiribati 0°53′S 169°32′E
190 H2 **Tarawa** atoll Tungaru, W Kiribati
184 N10 **Tarawera** Hawke's Bay, North Island, New Zealand 39°03′S 176°34′E
184 N8 **Tarawera, Lake** ◎ North Island, New Zealand
184 N8 **Tarawera, Mount** ▲ North Island, New Zealand 38°13′S 176°29′E
105 S8 **Tarayuela** ▲ N Spain 40°28′N 00°22′W
145 R16 **Taraz** prev. Aulie Ata, Auliye-Ata, Dzhambul, Zhambyl. Zhambyl, S Kazakhstan 42°55′N 71°27′E
105 Q5 **Tarazona** Aragón, NE Spain 41°54′N 01°44′W
105 Q10 **Tarazona de la Mancha** Castilla-La Mancha, C Spain 39°16′N 01°55′W
145 X12 **Tarbagatay, Khrebet** ▲▲ China/Kazakhstan
96 J8 **Tarbat Ness** headland N Scotland, United Kingdom 57°51′N 03°48′W
149 U5 **Tarbela Reservoir** ▣ N Pakistan
96 H12 **Tarbert** W Scotland, United Kingdom 55°52′N 05°26′W
96 F7 **Tarbert** NW Scotland, United Kingdom 57°54′N 06°48′W
102 K16 **Tarbes** anc. Bigorra. Hautes-Pyrénées, S France 43°14′N 00°04′E
21 W9 **Tarboro** North Carolina, SE USA 35°54′N 77°34′W
Tarca see Torysa
106 J6 **Tarcento** Friuli-Venezia Giulia, NE Italy 46°13′N 13°13′E
182 F5 **Tarcoola** South Australia 30°44′S 134°34′E
105 S5 **Tardienta** Aragón, NE Spain 41°58′N 00°31′W
102 L11 **Tardoire** ≈ W France
183 U7 **Taree** New South Wales, SE Australia 31°56′S 152°29′E
92 K12 **Tärendö** Lapp. Deargget. Norrbotten, N Sweden 67°10′N 22°40′E
Tarentum see Taranto
74 C9 **Tarfaya** SW Morocco 27°56′N 12°55′W
114 L8 **Targovishte** , Tărgovište; prev. Eski Dzhumaya. Targovishte, N Bulgaria 43°15′N 26°34′E
114 L8 **Targovishte** var. Tŭrgovishte. ◈ province N Bulgaria
116 J13 **Târgovişte** prev. Tîrgovişte. Dâmboviţa, S Romania 44°54′N 25°29′E
Tărgovište see Targovishte
116 M12 **Târgu Bujor** prev. Tîrgu Bujor. Galaţi, E Romania 45°52′N 27°55′E
116 H13 **Târgu Cărbuneşti** prev. Tîrgu. Gorj, SW Romania 44°57′N 23°32′E
116 L9 **Târgu Frumos** prev. Tîrgu Frumos. Iaşi, NE Romania 47°12′N 27°00′E
116 H13 **Targu Jiu** prev. Tîrgu Jiu. Gorj, W Romania 45°03′N 23°20′E
116 H9 **Târgu Lăpuş** prev. Tîrgu Lăpuş. Maramureş, N Romania 47°28′N 23°54′E
Târgul-Neamţ see Târgu-Neamţ
Târgul-Săcuiesc see Târgu Secuiesc
116 I10 **Târgu Mureş** prev. Oşorhei, Tîrgu Mures, Ger. Neumarkt, Hung. Marosvásárhely. Mureş, C Romania 46°33′N 24°36′E
116 K9 **Târgu-Neamţ** var. Târgul-Neamţ; prev. Tîrgu-Neamţ. Neamţ, NE Romania 47°12′N 26°25′E
116 K11 **Târgu Ocna** Hung. Aknavásár; prev. Tîrgu Ocna. Bacău, E Romania 46°17′N 26°37′E
116 K11 **Târgu Secuiesc** Ger. Neumarkt, Szekler Neumarkt, Hung. Kezdivásárhely; prev. Chezdi-Oşorheiu, Târgul-Săcuiesc, Tîrgu Secuiesc. Covasna, E Romania 46°00′N 26°08′E
145 X10 **Targyn** Vostochnyy Kazakhstan, E Kazakhstan 49°32′N 82°47′E
Tar Heel State see North Carolina
186 C7 **Tari** Hela, W Papua New Guinea 05°52′S 142°58′E
162 J6 **Tarialan** var. Badrah. Hövsgöl, N Mongolia 49°33′N 101°58′E
162 I7 **Tariat** var. Horgo. Arhangay, C Mongolia 48°06′N 99°52′E
143 P17 **Ţarīf** Abū Ẓaby, C United Arab Emirates 24°02′N 53°47′E
104 K16 **Tarifa** Andalucía, S Spain 36°01′N 05°36′W
84 C14 **Tarifa, Punta de** headland SW Spain 36°01′N 05°39′W
57 M21 **Tarija** Tarija, S Bolivia 21°33′S 64°42′W
57 M21 **Tarija** ◈ department S Bolivia
141 R14 **Tarīm** C Yemen 16°N 48°50′E
Tarim Basin see Tarim Pendi
81 G19 **Tarime** Mara, N Tanzania 01°20′S 34°24′E
129 S8 **Tarim He** ≈ NW China
159 H8 **Tarim Pendi** Eng. Tarim Basin. basin NW China
149 N7 **Tarīn Kōt** var. Terinkot; prev. Tarīn Kowt. Uruzgān, C Afghanistan 32°38′N 65°52′E
Tarīn Kowt see Tarīn Kōt
171 O12 **Taripa** Sulawesi, C Indonesia 01°51′S 120°46′E
117 Q12 **Tarkhankut, Mys** headland S Ukraine 45°20′N 32°32′E
27 Q1 **Tarkio** Missouri, C USA 40°25′N 95°24′W
122 J9 **Tarko-Sale** Yamalo-Nenetskiy Avtonomnyy Okrug, N Russian Federation 64°55′N 77°34′E
77 P17 **Tarkwa** S Ghana 05°16′N 01°59′W
171 O3 **Tarlac** Luzon, N Philippines 15°29′N 120°34′E
95 F22 **Tarm** Midtjylland, W Denmark 55°55′N 08°32′E
57 E14 **Tarma** Junín, C Peru 11°28′S 75°41′W
103 N15 **Tarn** ◈ department S France
102 M15 **Tarn** ≈ S France
111 L22 **Tarna** ≈ C Hungary
92 G13 **Tärnaby** Västerbotten, N Sweden 65°44′N 15°20′E
149 P8 **Tarnak Rūd** ≈ SE Afghanistan
116 J11 **Târnava Mare** Ger. Grosse Kokel, Hung. Nagy-Küküllő; prev. Tîrnava Mare. ≈ S Romania
116 I11 **Târnava Mică** Ger. Kleine Kokel, Hung. Kis-Küküllo; prev. Tîrnava Mică. ≈ C Romania
116 I11 **Târnăveni** Ger. Marteskirch, Martinskirch, Hung. Dicsőszentmárton; prev. Sînmartin, Tîrnăveni. Mureş, C Romania 46°20′N 24°17′E
102 L14 **Tarn-et-Garonne** ◈ department S France
111 P18 **Tarnica** ▲ SE Poland 49°05′N 22°43′E
111 N15 **Tarnobrzeg** Podkarpackie, SE Poland 50°35′N 21°40′E
125 N12 **Tarnogskiy Gorodok** Vologodskaya Oblast', NW Russian Federation 60°28′N 43°45′E
Tarnopol see Ternopil'
111 M16 **Tarnów** Małopolskie, S Poland 50°01′N 20°59′E
Tarnowice/Tarnowitz see Tarnowskie Góry
111 J16 **Tarnowskie Góry** var. Tarnowice, Tarnowskie Gory, Ger. Tarnowitz. Śląskie, S Poland 50°27′N 18°52′E
95 N14 **Tärnsjö** Västmanland, C Sweden 60°10′N 16°57′E
186 K7 **Taro** Choiseul, NW Solomon Islands 07°00′S 156°57′E
106 E9 **Taro** ≈ NW Italy
186 I6 **Taron** New Ireland, NE Papua New Guinea 04°22′S 153°04′E
74 E8 **Taroudannt** var. Taroudant. SW Morocco 30°31′N 08°50′W
Taroudant see Taroudannt
23 V12 **Tarpon, Lake** ◎ Florida, SE USA
23 V12 **Tarpon Springs** Florida, SE USA 28°09′N 82°45′W
107 G14 **Tarquinia** anc. Tarquinii, hist. Corneto. Lazio, C Italy 42°23′N 11°45′E
Tarquinii see Tarquinia
Tarraco see Tarragona
76 D10 **Tarrafal** Santiago, S Cape Verde 15°16′N 23°45′W
105 V6 **Tarragona** anc. Tarraco. Cataluña, E Spain 41°07′N 01°15′E
105 T7 **Tarragona** ◈ province Cataluña, NE Spain
183 O17 **Tarraleah** Tasmania, SE Australia 42°11′S 146°29′E
23 P3 **Tarrant City** Alabama, S USA 33°34′N 86°45′W
185 D21 **Tarras** Otago, South Island, New Zealand 44°48′S 169°25′E
Tarrasa see Terrassa
105 U5 **Tàrrega** var. Tarrega. Cataluña, NE Spain 41°39′N 01°09′E
21 W9 **Tar River** ≈ North Carolina, SE USA
Tarsatica see Rijeka
136 J17 **Tarsus** İçel, S Turkey 36°52′N 34°52′E
62 K4 **Tartagal** Salta, N Argentina 22°32′S 63°50′W
137 V12 **Tärtär** Rus. Terter. ≈ SW Azerbaijan
102 J15 **Tartas** Landes, SW France 43°52′N 00°45′W
Tartlau see Prejmer
Tartous/Tartouss see Ţarţūs
118 J5 **Tartu** Ger. Dorpat; prev. Rus. Yurev, Yury'ev. Tartumaa, SE Estonia 58°20′N 26°44′E
118 I5 **Tartumaa** off. Tartu Maakond. ◈ province E Estonia
Tartu Maakond see Tartumaa
138 H5 **Ţarţūs** Fr. Tartouss; anc. Tortosa. Ţarţūs, W Syria 34°55′N 35°52′E
138 H5 **Ţarţūs** off. Muḩāfaẓat Ţarţūs, var. Tartous, Tartus. ◈ governorate W Syria
Ţarţūs, Muḩāfaẓat see Ţarţūs
164 C16 **Tarumizu** Kagoshima, Kyūshū, SW Japan 31°30′N 130°40′E
126 K4 **Tarusa** Kaluzhskaya Oblast', W Russian Federation 54°45′N 37°10′E
117 N11 **Tarutyne** Odes'ka Oblast', SW Ukraine 46°11′N 29°09′E
162 I7 **Tarvagatyn Nuruu** ▲▲ N Mongolia
106 J6 **Tarvisio** Friuli-Venezia Giulia, NE Italy 46°31′N 13°33′E
Tarvisium see Treviso
57 O16 **Tarvo, Río** ≈ E Bolivia
14 G8 **Tarzwell** Ontario, S Canada 48°00′N 79°58′W
40 K5 **Tasajera, Sierra de la** ▲▲ N Mexico
145 S13 **Tasaral** Karaganda, C Kazakhstan 46°17′N 73°54′E
145 N15 **Tasboget** Kaz. Tasböget; prev. Tasbuget. Kzylorda, S Kazakhstan 44°46′N 65°38′E
Tasböget see Tasboget
Tasbuget see Tasboget
108 E11 **Tasch** Valais, SW Switzerland 46°04′N 07°43′E
Tasek Kenyir see Kenyir, Tasik
122 J14 **Tashanta** Respublika Altay, S Russian Federation 49°42′N 89°15′E
Tashauz see Daşoguz
Tashi Chho Dzong see Thimphu
153 U11 **Tashigang** E Bhutan 27°19′N 91°33′E
137 T11 **Tashir** prev. Kalinino. N Armenia 41°07′N 44°16′E
143 Q11 **Ţashk, Daryācheh-ye** ◎ C Iran
Tashkent see Toshkent
Tashkentskaya Oblast' see Toshkent Viloyati
Tashkepri see Daşköpri
Tash-Kömür see Tash-Kumyr
147 S9 **Tash-Kumyr** Kir. Tash-Kömür. Dzhalal-Abadskaya Oblast', W Kyrgyzstan 41°22′N 72°09′E
127 T7 **Tashla** Orenburgskaya Oblast', W Russian Federation 51°42′N 52°33′E
Tashqurghan see Khulm
122 J13 **Tashtagol** Kemerovskaya Oblast', S Russian Federation 52°49′N 88°00′E
95 H24 **Tåsinge** island C Denmark
12 M5 **Tasiujaq** Québec, E Canada 58°43′N 69°58′W
144 F8 **Taskala** prev. Kamenka. Zapadnyy Kazakhstan, NW Kazakhstan 51°06′N 51°16′E
77 W11 **Tasker** Zinder, C Niger 15°06′N 10°42′E
145 W12 **Taskesken** Vostochnyy Kazakhstan, E Kazakhstan 47°15′N 80°45′E
136 J10 **Taşköprü** Kastamonu, N Turkey 41°30′N 34°12′E
Taskuduk, Peski see Tosquduq Qumlari
186 G5 **Taskul** New Ireland, NE Papua New Guinea 02°34′S 150°25′E
137 S13 **Taşlıçay** Ağrı, E Turkey 39°37′N 43°23′E
185 H14 **Tasman** off. Tasman District. ◈ unitary authority South Island, New Zealand
192 J12 **Tasman Basin** var. East Australian Basin. undersea feature S Tasman Sea
185 I14 **Tasman Bay** inlet South Island, New Zealand
Tasman District see Tasman
192 I13 **Tasman Fracture Zone** tectonic feature S Indian Ocean
185 E19 **Tasman Glacier** glacier South Island, New Zealand
Tasman Group see Nukumanu Islands
183 N15 **Tasmania** prev. Van Diemen's Land. ◈ state SE Australia
183 Q16 **Tasmania** island SE Australia
185 H14 **Tasman Mountains** ▲▲ South Island, New Zealand
183 P17 **Tasman Peninsula** peninsula Tasmania, SE Australia
192 I11 **Tasman Plain** undersea feature W Tasman Sea
192 I12 **Tasman Plateau** var. South Tasmania Plateau. undersea feature SW Tasman Sea
192 I11 **Tasman Sea** sea SW Pacific Ocean
116 G9 **Tăşnad** Ger. Trestenberg, Trestendorf, Hung. Tasnád. Satu Mare, NW Romania 47°30′N 22°33′E
136 L11 **Taşova** Amasya, N Turkey 40°45′N 36°20′E
77 T10 **Tassara** Tahoua, W Niger 16°40′N 05°34′E
12 K4 **Tassialouc, Lac** ◎ Québec, C Canada
Tassili du Hoggar see Tassili Ta-n-Ahaggar
74 L11 **Tassili-n-Ajjer** plateau E Algeria
74 K14 **Tassili Ta-n-Ahaggar** var. Tassili du Hoggar, Tassili ta-n-Ahaggar. plateau S Algeria
Tassili ta-n-Ahaggar see Tassili Ta-n-Ahaggar
59 M15 **Tasso Fragoso** Maranhão, E Brazil 08°22′S 45°53′W
Tåstrup see Taastrup
145 O9 **Tasty-Taldy** Akmola, C Kazakhstan 50°47′N 66°31′E
143 W10 **Tāsūkī** Sīstān va Balūchestān, SE Iran
111 I22 **Tata** Ger. Totis. Komárom-Esztergom, NW Hungary 47°39′N 18°19′E
74 E8 **Tata** SW Morocco 29°38′N 08°04′W
Tataaihoa, Pointe see Vénus, Pointe
111 I22 **Tatabánya** Komárom-Esztergom, NW Hungary 47°33′N 18°23′E
191 X10 **Tatakoto** atoll Îles Tuamotu, E French Polynesia
75 N7 **Tataouine** var. Taţāwīn. SE Tunisia 32°48′N 10°27′E
55 O5 **Tataracual, Cerro** ▲ NE Venezuela 10°13′N 64°20′W
117 O12 **Tatarbunary** Odes'ka Oblast', SW Ukraine 45°50′N 29°37′E
119 M17 **Tatarka** Mahilyowskaya Voblasts', E Belarus 53°15′N 28°50′E
Tatar Pazardzhik see Pazardzhik
122 I12 **Tatarsk** Novosibirskaya Oblast', C Russian Federation 55°08′N 75°58′E
Tatarskaya ASSR see Tatarstan, Respublika
123 T13 **Tatarskiy Proliv** Eng. Tatar Strait. strait SE Russian Federation
127 R4 **Tatarstan, Respublika** prev. Tatarskaya ASSR. ◈ autonomous republic W Russian Federation
Tatar Strait see Tatarskiy Proliv
Taţāwīn see Tataouine
171 N12 **Tate** Sulawesi, N Indonesia 0°12′S 119°44′E
141 N11 **Tathlīth** 'Asīr, S Saudi Arabia 19°38′N 43°32′E
141 O11 **Tathlīth, Wādī** dry watercourse S Saudi Arabia
183 R11 **Tathra** New South Wales, SE Australia 36°46′S 149°58′E
127 P8 **Tatishchevo** Saratovskaya Oblast', W Russian Federation 51°43′N 45°35′E
39 S12 **Tatitlek** Alaska, USA 60°49′N 146°29′W
10 L15 **Tatla Lake** British Columbia, SW Canada 51°54′N 124°39′W
121 Q2 **Tatlısu** Gk. Akanthoú. N Cyprus 35°21′N 33°45′E
11 Z10 **Tatnam, Cape** headland Manitoba, C Canada 57°16′N 91°03′W
111 K18 **Tatra Mountains** Ger. Tatra, Hung. Tátra, Pol./Slvk. Tatry. ▲▲ Poland/Slovakia
Tatra/Tátra see Tatra Mountains
Tatry see Tatra Mountains
164 I13 **Tatsuno** var. Tatuno. Hyōgo, Honshū, SW Japan 34°54′N 134°30′E
Tatti see Tatty
145 S16 **Tatty** prev. Tatti. Zhambyl, S Kazakhstan 43°11′N 73°22′E
60 L10 **Tatuí** São Paulo, S Brazil 23°21′S 47°49′W
37 V14 **Tatum** New Mexico, SW USA 33°15′N 103°19′W
25 X7 **Tatum** Texas, SW USA 32°19′N 94°31′W
Ta-t'ung/Tatung see Datong
Tatuno see Tatsuno
137 R14 **Tatvan** Bitlis, SE Turkey 38°31′N 42°15′E
95 C16 **Tau** Rogaland, S Norway 59°04′N 05°55′E
192 L17 **Ta'ū** var. Tau. island Manua Islands, E American Samoa
193 W15 **Tau** island Tongatapu Group, N Tonga
59 O14 **Tauá** Ceará, E Brazil 06°04′S 40°26′W
60 N10 **Taubaté** São Paulo, S Brazil 23°S 45°36′W
101 I19 **Tauber** ≈ SW Germany
101 I19 **Tauberbiscofsheim** Baden-Württemberg, C Germany 49°37′N 09°39′E
Tauchik see Taushyk
191 W10 **Tauere** atoll Îles Tuamotu, C French Polynesia
101 H17 **Taufstein** ▲ C Germany 50°31′N 09°18′E
190 I17 **Taukoka** island SE Cook Islands
145 T15 **Taukum, Peski** desert SE Kazakhstan
184 L10 **Taumarunui** Manawatu-Wanganui, North Island, New Zealand 38°52′S 175°14′E
59 A15 **Taumaturgo** Acre, W Brazil 08°54′S 72°48′W
27 X6 **Taum Sauk Mountain** ▲ Missouri, C USA 37°34′N 90°43′W
83 H22 **Taung** North-West, N South Africa 27°32′S 24°48′E
166 L6 **Taungdwingyi** Magway, C Myanmar (Burma) 20°01′N 95°20′E
166 M6 **Taunggyi** Shan State, C Myanmar (Burma) 20°47′N 97°00′E
166 M7 **Taungoo** Bago, C Myanmar (Burma) 18°57′N 96°26′E
166 L5 **Taungtha** Mandalay, C Myanmar (Burma) 21°16′N 95°25′E
Taungup see Toungup
149 S9 **Taunsa** Punjab, E Pakistan 30°43′N 70°41′E
97 K23 **Taunton** SW England, United Kingdom 51°01′N 03°06′W
19 O12 **Taunton** Massachusetts, NE USA 41°54′N 71°03′W
101 F18 **Taunus** ▲▲ W Germany
101 G18 **Taunusstein** Hessen, W Germany 50°09′N 08°09′E
184 N9 **Taupo** Waikato, North Island, New Zealand 38°42′S 176°05′E
184 M9 **Taupo, Lake** ◎ North Island, New Zealand
109 R8 **Taurach** var. Taurachbach. ≈ E Austria
Taurachbach see Taurach
118 D12 **Tauragė** Ger. Tauroggen. Tauragė, SW Lithuania 55°16′N 22°17′E
118 D13 **Tauragė** ◈ province Lithuania
54 G10 **Tauramena** Casanare, C Colombia 05°02′N 72°43′W
184 N7 **Tauranga** Bay of Plenty, North Island, New Zealand 37°42′S 176°09′E
15 O10 **Taureau, Réservoir** ▣ Québec, SE Canada
107 N22 **Taurianova** Calabria, SW Italy 38°22′N 16°01′E
Tauris see Tabrīz
184 I2 **Tauroa Point** headland North Island, New Zealand 35°09′S 173°02′E
Tauroggen see Tauragė
Tauromenium see Taormina
Taurus Mountains see Toros Dağları
Taus see Domažlice
144 E14 **Taushyk** Kaz. Taūshyq; prev. Tauchik. Mangistau, SW Kazakhstan 44°17′N 51°22′E
Taūshyq see Taushyk
105 R5 **Tauste** Aragón, NE Spain 41°55′N 01°15′W
191 V16 **Tautara, Motu** island Easter Island, Chile, E Pacific Ocean
191 R8 **Tautira** Tahiti, W French Polynesia 17°45′S 149°10′W
Tauz see Tovuz
Ţavālesh, Kūhhā-ye see Talish Mountains
136 D15 **Tavas** Denizli, SW Turkey 37°33′N 29°04′E
Tavastehus see Hämeenlinna
Tavau see Davos
122 G10 **Tavda** Sverdlovskaya Oblast', C Russian Federation 58°01′N 65°07′E
122 G10 **Tavda** ≈ C Russian Federation
105 T11 **Tavernes de la Valldigna** Valenciana, E Spain 39°03′N 00°13′W
81 I20 **Taveta** Taita/Taveta, S Kenya 03°23′S 37°40′E
187 Y14 **Taveuni** island N Fiji
147 R13 **Tavildara** Rus. Tavil'dara, Tovil'-Dora. C Tajikistan 38°42′N 70°27′E
Tavin see Adaatsag
104 H14 **Tavira** Faro, S Portugal 37°07′N 07°39′W
97 I24 **Tavistock** SW England, United Kingdom 50°33′N 04°08′W
Tavoy see Dawei
Tavoy Island see Mali Kyun
115 E16 **Tavropoú, Techníti Límni** ▣ C Greece
136 E13 **Tavşanlı** Kütahya, NW Turkey 39°34′N 29°28′E
187 X14 **Tavua** Viti Levu, W Fiji 17°27′S 177°51′E
97 J23 **Taw** ≈ SW England, United Kingdom
185 L14 **Tawa** Wellington, North Island, New Zealand 41°10′S 174°50′E
25 V6 **Tawakoni, Lake** ▣ Texas, SW USA
153 V11 **Tawang** Arunāchal Pradesh, NE India 27°34′N 91°54′E
169 R17 **Tawang, Teluk** bay Jawa, S Indonesia
31 R7 **Tawas Bay** ◎ Michigan, N USA
31 R7 **Tawas City** Michigan, N USA 44°16′N 83°33′W
169 V8 **Tawau** Sabah, East Malaysia 04°16′N 117°54′E
141 U10 **Ţawīl, Qalamat aţ** well SE Saudi Arabia
171 N9 **Tawitawi** island Tawitawi Group, SW Philippines
Ţawkar see Tokar
Tāwūq see Dāqūq
Tawzar see Tozeur
41 O15 **Taxco** var. Taxco de Alarcón. Guerrero, S Mexico 18°32′N 99°37′W
Taxco de Alarcón see Taxco
146 H8 **Taxiatosh** Rus. Takhiatash. Qoraqalpog'iston Respublikasi, W Uzbekistan 42°27′N 59°27′E
158 D9 **Taxkorgan** var. Taxkorgan Tajik Zizhixian. Xinjiang Uygur Zizhiqu, NW China 37°43′N 75°13′E
Taxkorgan Tajik Zizhixian see Taxkorgan
146 H7 **Taxtako'pir** Rus. Takhtakupyr. Qoraqalpog'iston Respublikasi, NW Uzbekistan 43°04′N 60°23′E
96 J10 **Tay** ≈ C Scotland, United Kingdom
143 V6 **Tāybād** var. Taibad, Tāyybād, Tayyebāt. Khorāsān-e Raẓavī, NE Iran 34°48′N 60°46′E
Taybert at Turkz see Ţayyibat at Turkī
124 J3 **Taybola** Murmanskaya Oblast', NW Russian Federation 68°30′N 33°18′E
81 M16 **Tayeeglow** Bakool, C Somalia 04°01′N 44°25′E
96 K11 **Tay, Firth of** inlet E Scotland, United Kingdom
122 J12 **Tayga** Kemerovskaya Oblast', S Russian Federation 56°02′N 85°26′E
Taygan see Delger
123 T9 **Taygonos, Mys** headland E Russian Federation 60°36′N 160°09′E
96 I11 **Tay, Loch** ◎ C Scotland, United Kingdom
11 N12 **Taylor** British Columbia, W Canada 56°09′N 120°43′W
29 O14 **Taylor** Nebraska, C USA 41°47′N 99°23′W
18 I13 **Taylor** Pennsylvania, NE USA 41°22′N 75°41′W
25 T10 **Taylor** Texas, SW USA 30°34′N 97°24′W
37 Q11 **Taylor, Mount** ▲ New Mexico, SW USA 35°14′N 107°36′W
37 R5 **Taylor Park Reservoir** ▣ Colorado, C USA
37 R6 **Taylor River** ≈ Colorado, C USA
21 P11 **Taylors** South Carolina, SE USA 34°55′N 82°18′W
20 L5 **Taylorsville** Kentucky, S USA 38°01′N 85°21′W
21 R6 **Taylorsville** North Carolina, SE USA 35°56′N 81°10′W
30 L14 **Taylorville** Illinois, N USA 39°33′N 89°17′W
140 K5 **Taymā'** Tabūk, NW Saudi Arabia 27°39′N 38°32′E
122 M10 **Taymura** ≈ C Russian Federation
123 O7 **Taymylyr** Respublika Sakha (Yakutiya), NE Russian Federation 72°32′N 121°54′E
122 L7 **Taymyr, Ozero** ◎ N Russian Federation
122 M6 **Taymyr, Poluostrov** peninsula N Russian Federation
122 L8 **Taymyrskiy (Dolgano-Nenetskiy) Avtonomnyy Okrug** ◈ autonomous district Krasnoyarskiy Kray, N Russian Federation
167 S13 **Tây Ninh** Tây Ninh, S Vietnam 11°21′N 106°07′E
122 L12 **Tayshet** Irkutskaya Oblast', S Russian Federation 55°51′N 98°04′E
162 G8 **Tayshir** var. Tsagaan-Olom. Govĭ-Altay, C Mongolia 46°42′N 96°30′E
171 N5 **Taytay** Palawan, W Philippines 10°49′N 119°30′E
169 Q16 **Tayu** prev. Tajoe. Jawa, C Indonesia 06°32′S 111°02′E
Tāyybād/Tayyebāt see Tāybād
138 L5 **Ţayyibah** var. At Taybé. Ḩimş, C Syria 35°13′N 38°51′E
138 I4 **Ţayyibat at Turkī** var. Taybert at Turkz. Ḩamāh, W Syria 35°16′N 36°55′E
145 P7 **Tayynsha** prev. Krasnoarmeysk. Severnyy Kazakhstan, N Kazakhstan 53°52′N 69°51′E
122 J10 **Taz** ≈ N Russian Federation
74 G6 **Taza** NE Morocco 34°13′N 04°06′W
139 T4 **Tāza Khurmātū** Kirkūk, E Iraq 35°18′N 44°22′E
165 Q8 **Tazawa-ko** ◎ Honshū, C Japan
21 N8 **Tazewell** Tennessee, S USA 36°27′N 83°34′W
21 Q7 **Tazewell** Virginia, NE USA 37°08′N 81°34′W
75 S11 **Tāzirbū** SE Libya 25°43′N 21°16′E
39 S11 **Tazlina Lake** ◎ Alaska, USA
122 J8 **Tazovskiy** Yamalo-Nenetskiy Avtonomnyy Okrug, N Russian Federation 67°33′N 78°21′E
137 U10 **Tbilisi** Eng. Tiflis. ● (Georgia) SE Georgia 41°41′N 44°55′E
137 T10 **Tbilisi** ✈ S Georgia 41°43′N 44°49′E
79 E14 **Tchabal Mbabo** ▲ NW Cameroon 07°12′N 12°16′E
Tchad see Chad
Tchad, Lac see Chad, Lake
77 S15 **Tchaourou** E Benin 08°58′N 02°40′E
79 E20 **Tchibanga** Nyanga, S Gabon 02°49′S 11°00′E
Tchien see Zwedru
77 Z6 **Tchigaï, Plateau du** ▲▲ NE Niger
77 V9 **Tchighozérine** Agadez, C Niger 17°15′N 07°48′E
77 T10 **Tchin-Tabaradene** Tahoua, W Niger 15°57′N 05°49′E
78 G13 **Tcholliré** Nord, NE Cameroon 08°48′N 14°00′E
Tchongking see Chongqing
22 K4 **Tchula** Mississippi, S USA 33°10′N 90°13′W
110 I7 **Tczew** Ger. Dirschau. Pomorskie, N Poland 54°05′N 18°46′E
116 I10 **Teaca** Ger. Tekendorf, Hung. Teke; prev. Ger. Teckendorf. Bistriţa-Năsăud, N Romania 46°55′N 24°30′E
40 J11 **Teacapán** Sinaloa, C Mexico 22°33′N 105°44′W
190 A10 **Teafuafou** island Funafuti Atoll, C Tuvalu
25 U8 **Teague** Texas, SW USA 31°37′N 96°16′W
191 R9 **Teahupoo** Tahiti, W French Polynesia 17°51′S 149°15′W
190 H15 **Te Aiti Point** headland Rarotonga, S Cook Islands 21°11′S 59°47′W
65 D24 **Teal Inlet** East Falkland, Falkland Islands 51°34′S 58°25′W
185 B22 **Te Anau** Southland, South Island, New Zealand 45°25′S 167°45′E
185 B22 **Te Anau, Lake** ◎ South Island, New Zealand
41 U15 **Teapa** Tabasco, SE Mexico 17°36′N 92°57′W
184 Q7 **Te Araroa** Gisborne, North Island, New Zealand 37°37′S 178°21′E
184 M7 **Te Aroha** Waikato, North Island, New Zealand 37°32′S 175°58′E
Teate see Chieti
190 A9 **Te Ava Fuagea** channel Funafuti Atoll, SE Tuvalu
190 B8 **Te Ava I Te Lape** channel Funafuti Atoll, SE Tuvalu
190 B9 **Te Ava Pua Pua** channel Funafuti Atoll, SE Tuvalu
184 M8 **Te Awamutu** Waikato, North Island, New Zealand 38°00′S 177°18′E
171 X12 **Teba** Papua, E Indonesia 01°27′S 137°54′E
104 L15 **Teba** Andalucía, S Spain 36°59′N 04°54′W
126 M15 **Teberda** Karachayevo-Cherkesskaya Respublika, SW Russian Federation 43°28′N 41°45′E
74 M6 **Tébessa** NE Algeria 35°21′N 08°06′E
62 O7 **Tebicuary, Río** ≈ S Paraguay
168 L13 **Tebingtinggi** Sumatera, W Indonesia 03°33′S 103°00′E
168 I8 **Tebingtinggi** Sumatera, N Indonesia 03°20′N 99°08′E
Tebingtinggi, Pulau see Rantau, Pulau
Tebriz see Tabrīz
137 U9 **T'ebulos Mta** Rus. Gora Tebulosmta; prev. Tebulos Mta. ▲ Georgia/Russian Federation 42°33′N 45°21′E
Tebulos Mta see T'ebulos Mta
Tebulosmta, Gora see T'ebulos Mta
41 Q14 **Tecamachalco** Puebla, S Mexico 18°52′N 97°44′W
40 B1 **Tecate** Baja California Norte, NW Mexico 32°33′N 116°38′W
136 M13 **Tecer Dağları** ▲▲ C Turkey
103 O17 **Tech** ≈ S France
77 P16 **Techiman** W Ghana 07°35′N 01°56′W
117 N15 **Techirghiol** Constanţa, SE Romania 44°03′N 28°37′E
74 A12 **Techla** var. Techlé. SW Western Sahara 21°39′N 14°57′W
Techlé see Techla
63 H18 **Tecka, Sierra de** ▲▲ SW Argentina
Teckendorf see Teaca
40 K13 **Tecolotlán** Jalisco, SW Mexico 20°10′N 104°07′W
40 K14 **Tecomán** Colima, SW Mexico 18°53′N 103°54′W
35 V12 **Tecopa** California, W USA 35°51′N 116°14′W
40 G5 **Tecoripa** Sonora, NW Mexico 28°38′N 109°58′W
41 N16 **Tecpan** var. Tecpan de Galeana. Guerrero, S Mexico 17°12′N 100°39′W
Tecpan de Galeana see Tecpan
40 J11 **Tecuala** Nayarit, C Mexico 22°24′N 105°30′W
116 L12 **Tecuci** Galaţi, E Romania 45°50′N 27°27′E
31 R10 **Tecumseh** Michigan, N USA 42°00′N 83°57′W
29 S16 **Tecumseh** Nebraska, C USA 40°22′N 96°12′W
27 O11 **Tecumseh** Oklahoma, C USA 35°15′N 96°56′W
Tedzhen see Harīrūd/Tejen
Tedzhen see Tejen
146 H15 **Tedzhenstroy** Turkm. Tejenstroy. Ahal Welaýaty, S Turkmenistan 36°57′N 60°49′E
Teel see Öndör-Ulaan
97 L15 **Tees** ≈ N England, United Kingdom
14 E15 **Teeswater** Ontario, S Canada 44°00′N 81°17′W
190 A10 **Tefala** island Funafuti Atoll, C Tuvalu
58 D13 **Tefé** Amazonas, N Brazil 03°24′S 64°45′W
74 K11 **Tefedest** ▲▲ S Algeria
136 E16 **Tefenni** Burdur, SW Turkey 37°19′N 29°45′E
58 D13 **Tefé, Rio** ≈ NW Brazil
169 P16 **Tegal** Jawa, C Indonesia 06°52′S 109°07′E
100 O12 **Tegel** ✈ (Berlin) Berlin, NE Germany 52°33′N 13°16′E
99 M15 **Tegelen** Limburg, SE Netherlands 51°20′N 06°09′E
101 L24 **Tegernsee** ◎ SE Germany
107 M18 **Teggiano** Campania, S Italy 40°25′N 15°28′E
77 U14 **Tegina** Niger, C Nigeria 10°06′N 06°10′E
Tegucigalpa see Central District
Tegucigalpa see Toncontín
Tegucigalpa see Francisco Morazán
77 U9 **Teguidda-n-Tessoumt** Agadez, C Niger 17°27′N 06°40′E
64 Q11 **Teguise** Lanzarote, Islas Canarias, Spain, NE Atlantic Ocean 29°04′N 13°38′W
122 K12 **Tegul'det** Tomskaya Oblast', C Russian Federation 57°16′N 87°58′E
35 S13 **Tehachapi** California, W USA 35°07′N 118°27′W
35 S13 **Tehachapi Mountains** ▲▲ California, W USA
Tehama see Tihāmah
Teheran see Tehrān
77 O14 **Téhini** NE Ivory Coast 09°36′N 03°40′W
143 N5 **Tehrān** var. Teheran. ● (Iran) Tehrān, N Iran 35°44′N 51°27′E
143 N6 **Tehrān** off. Ostān-e Tehrān, var. Tehran. ◈ province N Iran
Tehrān, Ostān-e see Tehrān
Tehri see Tīkamgarh
Tehri see New Tehri
41 Q15 **Tehuacán** Puebla, S Mexico 18°29′N 97°24′W
41 S17 **Tehuantepec** var. Santo Domingo Tehuantepec. Oaxaca, SE Mexico 16°18′N 95°14′W
41 S17 **Tehuantepec, Golfo de** var. Gulf of Tehuantepec. gulf S Mexico
Tehuantepec, Gulf of see Tehuantepec, Golfo de
Tehuantepec, Isthmus of see Tehuantepec, Istmo de
41 T16 **Tehuantepec, Istmo de** var. Isthmus of Tehuantepec. isthmus SE Mexico
0 I16 **Tehuantepec Ridge** undersea feature E Pacific Ocean 13°30′N 98°00′W
41 S16 **Tehuantepec, Río** ≈ SE Mexico
191 W10 **Tehuata** atoll Îles Tuamotu, C French Polynesia
64 O11 **Teide, Pico del** ▲ Gran Canaria, Islas Canarias, Spain, NE Atlantic Ocean 28°16′N 16°39′W
97 I21 **Teifi** ≈ SW Wales, United Kingdom
80 B9 **Teiga Plateau** plateau W Sudan
97 J24 **Teignmouth** SW England, United Kingdom 50°34′N 03°29′W
Teisen see Jecheon
116 H1 **Teiuş** Ger. Dreikirchen, Hung. Tövis. Alba, C Romania 46°12′N 23°40′E
169 U17 **Tejakula** Bali, C Indonesia 08°09′S 115°19′E
146 H14 **Tejen** Rus. Tedzhen. Ahal Welaýaty, S Turkmenistan 37°24′N 60°29′E
146 I15 **Tejen** Per. Harīrūd, Rus. Tedzhen. ≈ Afghanistan/Iran see also Harīrūd
Tejen see Harīrūd
Tejenstroy see Tedzhenstroy
35 S14 **Tejon Pass** pass California, W USA
Tejo, Rio see Tagus
41 O14 **Tejupilco** var. Tejupilco de Hidalgo. México, S Mexico 18°55′N 100°10′W
Tejupilco de Hidalgo see Tejupilco
184 P7 **Te Kaha** Bay of Plenty, North Island, New Zealand 37°45′S 177°42′E
29 S14 **Tekamah** Nebraska, C USA 41°46′N 96°13′W
184 I1 **Te Kao** Northland, North Island, New Zealand 34°40′S 172°57′E
185 F20 **Tekapo** ≈ South Island, New Zealand
185 F19 **Tekapo, Lake** ◎ South Island, New Zealand
184 P9 **Te Karaka** Gisborne, North Island, New Zealand 38°30′S 177°52′E
184 L7 **Te Kauwhata** Waikato, North Island, New Zealand 37°22′S 175°07′E
41 X12 **Tekax** var. Tekax de Álvaro Obregón. Yucatán, SE Mexico 20°07′N 89°10′W
Tekax de Álvaro Obregón see Tekax
136 A14 **Teke Burnu** headland W Turkey 38°06′N 26°35′E
114 M12 **Teke Deresi** ≈ NW Turkey
146 D10 **Tekedzhik, Gory** hill range NW Turkmenistan
145 V14 **Tekeli** Almaty, SE Kazakhstan 44°50′N 78°47′E
145 R7 **Teke, Ozero** ◎ N Kazakhstan
158 I5 **Tekes** Xinjiang Uygur Zizhiqu, NW China 43°15′N 81°43′E
145 W16 **Tekes** Almaty, SE Kazakhstan 42°40′N 80°01′E

◆ Country
● Country Capital
◇ Dependent Territory
○ Dependent Territory Capital
◈ Administrative Regions
✈ International Airport
▲ Mountain
▲▲ Mountain Range
⛰ Volcano
≈ River
◎ Lake
▣ Reservoir

TEKES – THIEF RIVER FALLS

Tekes see Tekes He
158 H5 Tekes He Rus. Tekes. ≈ China/Kazakhstan
Teke/Tekendorf see Teaca
80 I10 Tekezē var. Takkaze. ≈ Eritrea/Ethiopia
Tekhtin see Tsyakhtsin
136 C10 Tekirdağ It. Rodosto; anc. Bisanthe, Raidestos, Rhaedestus. Tekirdağ, NW Turkey 40°59′N 27°31′E
136 C10 Tekirdağ ◐ province NW Turkey
155 N14 Tekkali Andhra Pradesh, E India 18°37′N 84°15′E
115 K15 Tekke Burnu Turk. İlyasbaba Burnu. headland NW Turkey 40°03′N 26°12′E
137 Q13 Tekman Erzurum, NE Turkey 39°39′N 41°31′E
32 M9 Tekoa Washington, NW USA 47°13′N 117°05′W
190 H16 Te Kou ▲ Rarotonga, S Cook Islands 21°14′S 159°46′W
Tekrit see Tikrīt
171 P12 Teku Sulawesi, N Indonesia 0°46′S 123°25′E
184 L9 Te Kuiti Waikato, North Island, New Zealand 38°21′S 175°10′E
42 H4 Tela Atlántida, NW Honduras 15°46′N 87°25′W
138 F12 Telalim Southern, S Israel 30°58′N 34°47′E
Telanaipura see Jambi
155 I15 Telangana off. State of Telangana. ◐ state E India
Telangana, State of see Telangana
137 U10 Telavi prev. T'elavi. E Georgia 41°55′N 45°29′E
T'elavi see Telavi
138 F10 Tel Aviv ◐ district W Israel
Tel Aviv-Jaffa see Tel Aviv-Yafo
138 F10 Tel Aviv-Yafo var. Tel Aviv-Jaffa. Tel Aviv, C Israel 32°05′N 34°46′E
111 E18 Telč Ger. Teltsch. Vysočina, C Czech Republic 49°10′N 15°28′E
186 B6 Telefomin West Sepik, NW Papua New Guinea 05°08′S 141°31′E
10 J10 Telegraph Creek British Columbia, W Canada 57°56′N 131°10′W
190 B10 Telele island Funafuti Atoll, C Tuvalu
60 J11 Telêmaco Borba Paraná, S Brazil 24°20′S 50°44′W
95 E15 Telemark ◐ county S Norway
62 J13 Telén La Pampa, C Argentina 36°20′S 65°31′W
Teleneshty see Telenești
116 M9 Telenești Rus. Teleneshty. C Moldova 47°35′N 28°20′E
104 J4 Teleno, El ▲ NW Spain 42°19′N 06°21′W
116 I15 Teleorman ◐ county S Romania
116 I14 Teleorman ≈ S Romania
25 V5 Telephone Texas, SW USA 33°48′N 96°00′W
35 U11 Telescope Peak ▲ California, W USA 36°09′N 117°03′W
Teles Pirés see São Manuel, Rio
97 L19 Telford W England, United Kingdom 52°42′N 02°28′W
108 L7 Telfs Tirol, W Austria 47°19′N 11°05′E
42 I9 Telica León, NW Nicaragua 12°30′N 86°52′W
42 J6 Telica, Río ≈ C Honduras
76 I13 Télimélé W Guinea 10°45′N 13°02′W
43 O14 Telire, Río ≈ Costa Rica/Panama
114 I8 Telish prev. Azizie. Pleven, N Bulgaria 43°20′N 24°15′E
41 R16 Telixtlahuaca var. San Francisco Telixtlahuaca. Oaxaca, SE Mexico 17°18′N 96°54′W
10 K13 Telkwa British Columbia, SW Canada 54°39′N 126°51′W
25 P4 Tell Texas, SW USA 34°18′N 100°20′W
Tell Abiad see Tall Abyaḍ
Tell Abiad/Tell Abyad see At Tall al Abyaḍ
31 O16 Tell City Indiana, N USA 37°56′N 86°47′W
38 M9 Teller Alaska, USA 65°15′N 166°21′W
Tell Huqnah see Tall Ḥuqnah
155 F20 Tellicherry var. Thalashsheri, Thalassery. Kerala, SW India 11°44′N 75°29′E see also Thalassery
20 M10 Tellico Plains Tennessee, S USA 35°19′N 84°18′W
Tell Kalakh see Tall Kalakh
Tell Mardikh see Ebla
54 E11 Tello Huila, C Colombia 03°06′N 75°08′W
Tell Shedadi see Ash Shadādah
37 Q7 Telluride Colorado, C USA 37°56′N 107°48′W
117 X9 Tel'manove Donets'ka Oblast', E Ukraine 47°24′N 38°03′E
Tel'man/Tel'mansk see Gubadag
162 H6 Telmen var. Övögdiy. Dzavhan, C Mongolia 48°38′N 97°39′E
162 H6 Telmen Nuur ◎ NW Mongolia
Teloekbetoeng see Bandar Lampung
41 O15 Teloloapán Guerrero, S Mexico 18°21′N 99°52′W
Telo Martius see Toulon
Telposiz, Gora see Telpoziz, Gora
125 V8 Telpoziz, Gora prev. Gora Telposiz. ▲ NW Russian Federation 63°52′N 59°15′E
Telschen see Telšiai
63 J17 Telsen Chubut, S Argentina 42°27′S 66°59′W
118 D11 Telšiai Ger. Telschen. Telšiai, NW Lithuania 55°59′N 22°21′E
118 D11 Telšiai ◐ province NW Lithuania
Teltsch see Telč
Telukbetung see Bandar Lampung
168 H10 Telukdalam Pulau Nias, W Indonesia 0°34′N 97°47′E
14 H9 Temagami Ontario, S Canada 47°03′N 79°47′W
14 G9 Temagami, Lake ◎ Ontario, S Canada
190 H16 Te Manga ▲ Rarotonga, S Cook Islands 21°13′S 159°45′W
191 W12 Tematagi prev. Tematangi. atoll Îles Tuamotu, S French Polynesia
Tematangi see Tematagi
41 X11 Temax Yucatán, SE Mexico 21°10′N 88°53′W
171 E14 Tembagapura Papua, E Indonesia 04°10′S 137°19′E
129 U5 Tembenchi ≈ N Russian Federation
55 P6 Temblador Monagas, NE Venezuela 08°59′N 62°44′W
105 N9 Tembleque Castilla-La Mancha, C Spain 39°41′N 03°30′W
Temboni see Mitemele, Río
35 U16 Temecula California, W USA 33°29′N 117°09′W
168 K7 Temengor, Tasik ◎ Peninsular Malaysia
112 L9 Temerin Vojvodina, N Serbia 45°25′N 19°54′E
Temeschburg/Temeschwar see Timișoara
Temes-Kubin see Kovin
Temes/Temesch see Tamiš
Temesvár/Temeswar see Timișoara
Teminaboean see Teminabuan
171 U12 Teminabuan prev. Teminaboean. Papua Barat, E Indonesia 01°30′S 131°59′E
145 P17 Temirlan prev. Temirlanovka. Yuzhnyy Kazakhstan, S Kazakhstan 42°36′N 69°17′E
Temirlanovka see Temirlan
145 R10 Temirtau prev. Samarkandski, Samarkandskoye. Karaganda, C Kazakhstan 50°05′N 72°55′E
14 H10 Témiscaming Québec, SE Canada 46°40′N 79°04′W
Témiscamingue, Lac see Timiskaming, Lake
15 T8 Témiscouata, Lac ◎ Québec, SE Canada
127 N5 Temnikov Respublika Mordoviya, W Russian Federation 54°39′N 43°09′E
191 Y13 Temoe island Îles Gambier, E French Polynesia
183 Q9 Temora New South Wales, SE Australia 34°28′S 147°33′E
40 H7 Témoris Chihuahua, W Mexico 27°16′N 108°15′W
40 I5 Temósachic Chihuahua, N Mexico 28°55′N 107°42′W
187 Q10 Temotu var. Temotu Province. ◐ province E Solomon Islands
Temotu Province see Temotu
36 L14 Tempe Arizona, SW USA 33°24′N 111°54′W
Tempelburg see Czaplinek
107 C17 Tempio Pausania Sardegna, Italy, C Mediterranean Sea 40°55′N 09°07′E
42 K12 Tempisque, Río ≈ NW Costa Rica
25 T9 Temple Texas, SW USA 31°06′N 97°22′W
100 O12 Tempelhof ✈ (Berlin) Berlin, NE Germany 52°28′N 13°24′E
97 D19 Templemore Ir. An Teampall Mór. Tipperary, C Ireland 52°48′N 07°50′W
100 O11 Templin Brandenburg, NE Germany 53°07′N 13°31′E
41 P12 Tempoal var. Tempoal de Sánchez. Veracruz-Llave, E Mexico 21°32′N 98°23′W
Tempoal de Sánchez see Tempoal
41 P13 Tempoal, Río ≈ C Mexico
83 E14 Tempué Moxico, C Angola 13°36′S 18°56′E
126 J14 Temryuk Krasnodarskiy Kray, SW Russian Federation 45°15′N 37°26′E
99 G17 Temse Oost-Vlaanderen, N Belgium 51°08′N 04°13′E
63 F15 Temuco Araucanía, C Chile 38°45′S 72°37′W
185 G20 Temuka Canterbury, South Island, New Zealand 44°14′S 171°17′E
189 P13 Temwen Island island E Micronesia
56 C6 Tena Napo, C Ecuador 01°00′S 77°48′W
41 W13 Tenabo Campeche, E Mexico 20°02′N 90°12′W
Tenaghau see Aola
25 X7 Tenaha Texas, SW USA 31°56′N 94°14′W
39 X13 Tenake Chichagof Island, Alaska, USA 57°46′N 135°13′W
155 K16 Tenāli Andhra Pradesh, E India 16°13′N 80°36′E
Tenan see Cheonan
41 O14 Tenancingo var. Tenencingo de Degollado. México, S Mexico 18°57′N 99°39′W
191 X12 Tenararo island Groupe Actéon, SE French Polynesia
Tenasserim see Taninthayi
Tenasserim see Taninthayi
98 O5 Ten Boer Groningen, NE Netherlands 53°16′N 06°42′E
97 I21 Tenby SW Wales, United Kingdom 51°41′N 04°43′W
80 K11 Tendaho Āfar, NE Ethiopia 11°39′N 40°59′E
103 V14 Tende Alpes Maritimes, SE France 44°04′N 07°34′E
151 Q20 Ten Degree Channel strait Andaman and Nicobar Islands, India, E Indian Ocean
80 F11 Tendelti White Nile, E Sudan 13°01′N 31°55′E
76 G8 Te-n-Dghâmcha, Sebkhet var. Sebkha de Ndrhamcha, Sebkra de Ndaghamcha. salt lake W Mauritania
165 P10 Tendō Yamagata, Honshū, C Japan 38°22′N 140°22′E
74 H7 Tendrara NE Morocco 33°06′N 01°58′W
117 Q11 Tendrivs'ka Kosa spit S Ukraine
117 Q11 Tendrivs'ka Zatoka gulf S Ukraine
Tenencingo de Degollado see Tenancingo
77 N11 Ténenkou Mopti, C Mali 14°28′N 04°55′W
77 W9 Ténéré physical region C Niger
77 W9 Ténére, Erg du desert C Niger
64 O11 Tenerife island Islas Canarias, Spain, NE Atlantic Ocean
74 J5 Ténès NW Algeria 36°35′N 01°18′E
170 M15 Tengah, Kepulauan island group C Indonesia
Tengcheng see Tengxian
169 V11 Tenggarong Borneo, C Indonesia 0°23′S 117°00′E
162 J15 Tengger Shamo desert N China
168 L8 Tenggul, Pulau island Peninsular Malaysia
Tengiz Köl see Teniz, Ozero
76 M14 Tengréla var. Tingréla. N Ivory Coast 10°26′N 06°20′W
160 M14 Tengxian var. Tengcheng, Tengxian, Teng Xian. Guangxi Zhuangzu Zizhiqu, S China 23°24′N 110°49′E
Teng Xian see Tengxian
Tengxian see Tengxian
194 H2 Teniente Rodolfo Marsh Chilean research station South Shetland Islands, Antarctica 61°57′S 58°23′W
32 G9 Tenino Washington, NW USA 46°51′N 122°51′W
145 P9 Teniz, Ozero Kaz. Tengiz Köl. salt lake C Kazakhstan
112 I9 Tenja Osijek-Baranja, E Croatia 45°30′N 18°45′E
188 B16 Tenjo, Mount ▲ W Guam
155 H23 Tenkāsi Tamil Nādu, SE India 08°58′N 77°22′E
79 N24 Tenke Katanga, SE Dem. Rep. Congo 10°34′S 26°12′E
Tenke see Tinca
123 Q7 Tenkeli Respublika Sakha (Yakutiya), NE Russian Federation 70°09′N 140°39′E
27 R10 Tenkiller Ferry Lake ▣ Oklahoma, C USA
77 Q13 Tenkodogo S Burkina Faso 11°54′N 00°19′W
181 Q5 Tennant Creek Northern Territory, C Australia 19°40′S 134°16′E
20 G9 Tennessee off. State of Tennessee, also known as The Volunteer State. ◐ state SE USA
37 R5 Tennessee Pass pass Colorado, C USA
20 H10 Tennessee River ≈ S USA
23 N2 Tennessee Tombigbee Waterway canal Alabama/Mississippi, S USA
99 K22 Tenneville Luxembourg, SE Belgium 50°05′N 05°31′E
92 M11 Tenniöjoki ≈ NE Finland
92 L9 Tenojoki Lapp. Deatnu, Nor. Tana. ≈ Finland/Norway see also Deatnu, Tana
Tenojoki see Tana
169 U7 Tenom Sabah, East Malaysia 05°07′N 115°57′E
Tenos see Tínos
41 V15 Tenosique var. Tenosique de Pino Suárez. Tabasco, SE Mexico 17°30′N 91°24′W
Tenosique de Pino Suárez see Tenosique
22 I6 Tensas River ≈ Louisiana, S USA
23 O8 Tensaw River ≈ Alabama, S USA
74 E7 Tensift seasonal river W Morocco
171 O12 Tentena var. Tenteno. Sulawesi, C Indonesia 01°46′S 120°40′E
Tenteno see Tentena
183 U4 Tenterfield New South Wales, SE Australia 29°04′S 152°02′E
23 X16 Ten Thousand Islands island group Florida, SE USA
60 H9 Teodoro Sampaio São Paulo, S Brazil 22°30′S 52°13′W
59 N19 Teófilo Otoni var. Theophilo Ottoni. Minas Gerais, NE Brazil 17°52′S 41°31′W
116 K5 Teofipol' Khmel'nyts'ka Oblast', W Ukraine 50°00′N 26°22′E
Teohatu see Toahotu
41 P14 Teotihuacán ruins México, S Mexico
Teotilán see Teotitlán del Camino
41 Q15 Teotitlán del Camino var. Teotitlán. Oaxaca, S Mexico 18°10′N 97°08′W
190 G12 Tepa Île Uvea, E Wallis and Futuna 13°19′S 176°09′W
191 P8 Tepaee, Récif reef Tahiti, W French Polynesia
40 L14 Tepalcatepec Michoacán, SW Mexico 19°11′N 102°50′W
190 A16 Tepa Point headland SW Niue 19°07′S 169°56′E
40 L13 Tepatitlán var. Tepatitlán de Morelos. Jalisco, SW Mexico 20°50′N 102°46′W
Tepatitlán de Morelos see Tepatitlán
40 J9 Tepehuanes var. Santa Catarina de Tepehuanes. Durango, C Mexico 25°22′N 105°42′W
Tepelena see Tepelenë
113 L22 Tepelenë var. Tepelena, It. Tepeleni. Gjirokastër, S Albania 40°18′N 20°00′E
Tepeleni see Tepelenë
40 K12 Tepic Nayarit, C Mexico 21°30′N 104°55′W
111 C15 Teplice Ger. Teplitz; prev. Teplice-Šanov, Teplitz-Schönau. Ústecký Kraj, NW Czech Republic 50°38′N 13°49′E
Teplice-Šanov/Teplitz/Teplitz-Schönau see Teplice
117 O7 Teplyk Vinnyts'ka Oblast', C Ukraine 48°40′N 29°46′E
123 R10 Teplyy Klyuch Respublika Sakha (Yakutiya), NE Russian Federation 62°46′N 137°01′E
40 E5 Tepoca, Cabo headland NW Mexico 29°19′N 112°24′W
191 W9 Tepoto island Îles du Désappointement, C French Polynesia
92 L11 Tepsa Lappi, N Finland 67°34′N 25°36′E
190 B8 Tepuka atoll Funafuti Atoll, C Tuvalu
184 N7 Te Puke Bay of Plenty, North Island, New Zealand 37°48′S 176°19′E
40 L13 Tequila Jalisco, SW Mexico 20°52′N 103°48′W
41 O13 Tequisquiapan Querétaro de Arteaga, C Mexico 20°34′N 99°52′W
77 Q12 Téra Tillabéri, W Niger 14°01′N 00°45′E
104 J5 Tera ≈ NW Spain
191 V1 Teraina prev. Washington Island. atoll Line Islands, E Kiribati
81 F15 Terakeka Central Equatoria, S South Sudan 05°26′N 31°45′E
107 J14 Teramo anc. Interamna. Abruzzi, C Italy 42°40′N 13°43′E
98 P7 Ter Apel Groningen, NE Netherlands 52°52′N 07°05′E
104 H11 Tera, Ribeira de ≈ S Portugal
185 K14 Terawhiti, Cape headland North Island, New Zealand 41°17′S 174°36′E
98 N12 Terborg Gelderland, E Netherlands 51°55′N 06°22′E
137 P13 Tercan Erzincan, NE Turkey 39°47′N 40°23′E
64 O2 Terceira ✈ Terceira, Azores, Portugal, NE Atlantic Ocean 38°43′N 27°13′W
64 O2 Terceira var. Ilha Terceira. island Azores, Portugal, NE Atlantic Ocean
Terceira, Ilha see Terceira
116 K6 Terebovlya Ternopil's'ka Oblast', W Ukraine 49°18′N 25°44′E
127 O15 Terek ≈ SW Russian Federation
Terekhkovka see Tsyerakhowka
147 R9 Terek-Say Dzhalal-Abadskaya Oblast', W Kyrgyzstan 41°28′N 71°06′E
145 Z10 Terekty prev. Alekseevka, Alekseyevka. Vostochnyy Kazakhstan, E Kazakhstan 48°25′N 85°38′E
168 L7 Terengganu var. Trengganu. ◐ state Peninsular Malaysia
127 X7 Terensay Orenburgskaya Oblast', W Russian Federation 51°35′N 59°28′E
58 N13 Teresina var. Therezina. state capital Piauí, NE Brazil 05°09′S 42°46′W
60 P9 Teresópolis Rio de Janeiro, SE Brazil 22°25′S 42°59′W
110 P12 Terespol Lubelskie, E Poland 52°05′N 23°37′E
191 V16 Terevaka, Maunga ⛰ Easter Island, Chile, E Pacific Ocean 27°05′S 109°23′W
103 P3 Tergnier Aisne, N France 49°39′N 03°18′E
43 O14 Teribe, Río ≈ NW Panama
124 K3 Teriberka Murmanskaya Oblast', NW Russian Federation 69°10′N 35°18′E
Terinkot see Tarīn Kōt
Terisaqqan see Tersakkan
24 K12 Terlingua Texas, SW USA 29°18′N 103°36′W
24 K11 Terlingua Creek ≈ Texas, SW USA
62 K7 Termas de Río Hondo Santiago del Estero, N Argentina 27°29′S 64°52′W
136 M11 Terme Samsun, N Turkey 41°12′N 37°00′E
Termez see Termiz
Termia see Kýthnos
107 J23 Termini Imerese anc. Thermae Himerenses. Sicilia, Italy, C Mediterranean Sea 37°59′N 13°42′E
41 V14 Términos, Laguna de lagoon SE Mexico
77 X10 Termit-Kaoboul Zinder, C Niger 15°34′N 11°31′E
147 O14 Termiz Rus. Termez. Surkhondaryo Viloyati, S Uzbekistan 37°17′N 67°12′E
107 L15 Termoli Molise, C Italy 42°00′N 14°58′E
Termonde see Dendermonde
98 P5 Termunten Groningen, NE Netherlands 53°18′N 07°02′E
171 R11 Ternate Pulau Ternate, E Indonesia 0°48′N 127°23′E
109 T5 Ternberg Oberösterreich, N Austria 47°57′N 14°22′E
99 E15 Terneuzen var. Neuzen. Zeeland, SW Netherlands 51°20′N 03°50′E
123 T14 Terney Primorskiy Kray, SE Russian Federation 45°03′N 136°43′E
107 I14 Terni anc. Interamna Nahars. Umbria, C Italy 42°34′N 12°38′E
109 X6 Ternitz Niederösterreich, E Austria 47°43′N 16°02′E
117 V7 Ternivka Dnipropetrovs'ka Oblast', E Ukraine 48°30′N 36°05′E
116 K6 Ternopil' Pol. Tarnopol, Rus. Ternopol'. Ternopil's'ka Oblast', W Ukraine 49°32′N 25°38′E
Ternopil' see Ternopil's'ka Oblast'
116 I6 Ternopil's'ka Oblast' var. Ternopil', Rus. Ternopol'skaya Oblast'. ◐ province NW Ukraine
Ternopol' see Ternopil'
Ternopol'skaya Oblast' see Ternopil's'ka Oblast'
123 U13 Terpeniya, Mys headland Ostrov Sakhalin, SE Russian Federation 48°37′N 144°40′E
Térraba, Río see Grande de Térraba, Río
10 J13 Terrace British Columbia, W Canada 54°31′N 128°32′W
12 D12 Terrace Bay Ontario, S Canada 48°47′N 87°06′W
107 I16 Terracina Lazio, C Italy 41°18′N 13°13′E
93 F14 Terråk Troms, N Norway 65°03′N 12°22′E
26 M13 Terral Oklahoma, C USA 33°55′N 97°54′W
107 B19 Terralba Sardegna, Italy, C Mediterranean Sea 39°47′N 08°35′E
Terranova di Sicilia see Gela
Terranova Pausania see Olbia
105 W5 Terrassa Cast. Tarrasa. Cataluña, E Spain 41°34′N 02°01′E
15 O12 Terrebonne Québec, SE Canada 45°42′N 73°37′W
22 J11 Terrebonne Bay bay Louisiana, SE USA
31 N14 Terre Haute Indiana, N USA 39°27′N 87°24′W
25 U6 Terrell Texas, SW USA 32°44′N 96°16′W
Terre Neuve see Newfoundland and Labrador
33 Q14 Terreton Idaho, NW USA 43°49′N 112°25′W
103 T7 Territoire-de-Belfort ◐ department E France
33 X9 Terry Montana, NW USA 46°46′N 105°16′W
28 I9 Terry Peak ▲ South Dakota, N USA 44°19′N 103°49′W
136 H14 Tersakan Gölü ◎ C Turkey
145 O10 Tersakkan Kaz. Terisaqqan. ≈ C Kazakhstan
98 J4 Terschelling Fris. Skylge. island Waddeneilanden, N Netherlands
78 H10 Tersef Hadjer-Lamis, C Chad 12°55′N 16°49′E
147 X8 Terskey Ala-Too, Khrebet ▲▲ Kazakhstan/Kyrgyzstan
Terter see Tärtär
105 R8 Teruel anc. Turba. Aragón, E Spain 40°21′N 01°06′W
105 R7 Teruel ◐ province Aragón, E Spain
114 M7 Tervel prev. Kurtbunar, Rom. Curtbunar. Dobrich, NE Bulgaria 43°45′N 27°25′E
93 M16 Tervo Pohjois-Savo, C Finland 62°57′N 26°48′E
92 L13 Tervola Lappi, NW Finland 66°04′N 24°49′E
Tervueren see Tervuren
99 H18 Tervuren var. Tervueren. Vlaams Brabant, C Belgium 50°48′N 04°28′E
162 G5 Tes var. Dzür. Dzavhan, W Mongolia 49°37′N 95°46′E
112 H11 Tešanj Federacija Bosne I Hercegovine, N Bosnia and Herzegovina 44°37′N 18°00′E
Teschen see Cieszyn
83 M19 Tesenane Inhambane, S Mozambique 22°48′S 34°02′E
80 I9 Teseney var. Tessenei. W Eritrea 15°05′N 36°42′E
39 P5 Teshekpuk Lake ◎ Alaska, USA
Teshevle see Tsyeshawlya
162 K6 Teshig Bulgan, N Mongolia 49°51′N 102°45′E
165 T2 Teshio Hokkaidō, NE Japan 44°49′N 141°46′E
165 T2 Teshio-sanchi ▲▲ Hokkaidō, NE Japan
14 L7 Tessier, Lac ◎ Québec, SE Canada
Těšín see Cieszyn
Tesiyn Gol see Tes-Khem
129 T7 Tes-Khem var. Tesiyn Gol. ≈ Mongolia/Russian Federation
112 H11 Teslić Republika Srpska, N Bosnia and Herzegovina 44°35′N 17°50′E
10 I9 Teslin Yukon, W Canada 60°12′N 132°44′W
10 I8 Teslin ≈ British Columbia/Yukon, W Canada
77 Q8 Tessalit Kidal, NE Mali 20°12′N 00°58′E
77 V12 Tessaoua Maradi, S Niger 13°46′N 07°55′E
99 J17 Tessenderlo Limburg, NE Belgium 51°05′N 05°04′E
Tessenei see Teseney
Tessin see Ticino
97 M23 Test ≈ S England, United Kingdom
Testama see Tõstamaa
55 P4 Testigos, Islas los island group N Venezuela
37 S10 Tesuque New Mexico, SW USA 35°45′N 105°55′W
103 O17 Têt var. Tet. ≈ S France
Tet see Têt
54 G5 Tetas, Cerro de las ▲ NW Venezuela 09°58′N 73°00′W
83 M15 Tete Tete, NW Mozambique 16°14′S 33°34′E
83 M15 Tete off. Província de Tete. ◐ province NW Mozambique
11 N15 Tête Jaune Cache British Columbia, SW Canada 52°52′N 119°22′W
184 O8 Te Teko Bay of Plenty, North Island, New Zealand 38°03′S 176°48′E
186 K9 Tetepare island New Georgia Islands, NW Solomon Islands
Tete, Província de see Tete
Teterev see Teteriv
116 M5 Teteriv Rus. Teterev. ≈ N Ukraine
100 M9 Teterow Mecklenburg-Vorpommern, NE Germany 53°47′N 12°34′E
114 I9 Teteven Lovech, N Bulgaria 42°54′N 24°18′E
191 T10 Tetiaroa atoll Îles du Vent, W French Polynesia
105 P14 Tetica de Bacares ▲ S Spain 37°15′N 02°24′W
Tetiyev see Tetiyiv
117 O6 Tetiyiv Rus. Tetiyev. Kyyivs'ka Oblast', N Ukraine 49°21′N 29°40′E
39 T10 Tetlin Alaska, USA 63°08′N 142°31′W
33 R8 Teton River ≈ Montana, NW USA
74 G5 Tétouan var. Tetouan, Tetuán. N Morocco 35°33′N 05°22′W
Tetova/Tetovë see Tetovo
114 L7 Tetovo Razgrad, N Bulgaria 43°49′N 26°21′E
113 N18 Tetovo Alb. Tetova, Tetovë, Turk. Kalkandelen. NW FYR Macedonia 42°01′N 20°58′E
115 E20 Tetrázio ▲ S Greece
Tetschen see Děčín
Tetuán see Tétouan
191 Q8 Tetufera, Mont ▲ Tahiti, W French Polynesia 17°40′S 149°26′W
127 R4 Tetyushi Respublika Tatarstan, W Russian Federation 54°55′N 48°46′E
108 I7 Teufen Ausser Rhoden, NE Switzerland 47°24′N 09°24′E
40 L12 Teul var. Teul de Gonzáles Ortega. Zacatecas, C Mexico 21°30′N 103°28′W
107 B21 Teulada Sardegna, Italy, C Mediterranean Sea 38°58′N 08°46′E
Teul de Gonzáles Ortega see Teul
11 X16 Teulon Manitoba, S Canada 50°20′N 97°14′W
42 I7 Teupasenti El Paraíso, S Honduras 14°13′N 86°42′W
165 S2 Teuri-tō island NE Japan
100 G13 Teutoburger Wald Eng. Teutoburg Forest. hill range NW Germany
Teutoburg Forest see Teutoburger Wald
93 K17 Teuva Swe. Östermark. Etelä-Pohjanmaa, W Finland 62°29′N 21°45′E
107 H15 Tevere Eng. Tiber. ≈ C Italy
Teverya see Tverya
96 K13 Teviot ≈ SE Scotland, United Kingdom
Tevli see Tewli
122 H11 Tevriz Omskaya Oblast', C Russian Federation 57°30′N 72°13′E
185 B24 Te Waewae Bay bay South Island, New Zealand
97 L21 Tewkesbury C England, United Kingdom 51°59′N 02°09′W
119 F19 Tewli Rus. Tevli. Brestskaya Voblasts', SW Belarus 52°20′N 24°15′E
159 U12 Têwo var. Dêngka; prev. Dêngkagoin. Gansu, C China 34°05′N 103°15′E
Tewulike see Hoxud
25 U12 Texana, Lake ◎ Texas, SW USA
27 S14 Texarkana Arkansas, C USA 33°26′N 94°02′W
25 X5 Texarkana Texas, SW USA 33°26′N 94°03′W
25 N9 Texas off. State of Texas, also known as Lone Star State. ◐ state S USA
25 W12 Texas City Texas, SW USA 29°23′N 94°55′W
41 P14 Texcoco México, C Mexico 19°32′N 98°52′W
98 I6 Texel island Waddeneilanden, NW Netherlands
26 H8 Texhoma Oklahoma, C USA 36°30′N 101°46′W
25 N1 Texhoma Texas, SW USA 36°30′N 101°46′W
37 W12 Texico New Mexico, SW USA 34°23′N 103°03′W
24 L1 Texline Texas, SW USA 36°22′N 103°01′W
41 P14 Texmelucan var. San Martín Texmelucan. Puebla, S Mexico 19°46′N 98°53′W
27 O13 Texoma, Lake ▣ Oklahoma/Texas, C USA
25 N9 Texon Texas, SW USA 31°13′N 101°42′W
83 J23 Teyateyaneng NW Lesotho 29°04′S 27°51′E
124 M16 Teykovo Ivanovskaya Oblast', W Russian Federation 56°49′N 40°31′E
124 M16 Teza ≈ W Russian Federation
41 Q13 Teziutlán Puebla, S Mexico 19°49′N 97°22′W
153 W12 Tezpur Assam, NE India 26°39′N 92°47′E
Thaa Atoll see Kolhumadulu
9 N10 Tha-Anne ≈ Nunavut, NE Canada
83 K23 Thabana Ntlenyana var. Thabantshonyana, Mount Ntlenyana. ▲ E Lesotho 29°26′S 29°16′E
Thabantshonyana see Thabana Ntlenyana
83 J23 Thaba Putsoa ▲ C Lesotho 29°48′S 27°46′E
167 Q8 Tha Bo Nong Khai, E Thailand 17°52′N 102°34′E
103 T12 Thabor, Pic du ▲ E France 45°07′N 06°34′E
Tha Chin see Samut Sakhon
166 M7 Thagaya Bago, C Myanmar (Burma) 19°19′N 96°16′E
Thai, Ao see Thailand, Gulf of
167 T6 Thai Binh Thai Binh, N Vietnam 20°27′N 106°20′E
167 S7 Thai Hoa var. Nghia Dan. Nghê An, N Vietnam 19°21′N 105°26′E
167 P9 Thailand off. Kingdom of Thailand, Th. Prathet Thai; prev. Siam. ◆ monarchy SE Asia
167 P13 Thailand, Gulf of var. Gulf of Siam, Th. Ao Thai, Vtn. Vinh Thai Lan. gulf SE Asia
Thailand, Kingdom of see Thailand
Thai Lan, Vinh see Thailand, Gulf of
167 T6 Thai Nguyên Bắc Thai, N Vietnam 21°36′N 105°50′E
167 S8 Thakhèk var. Muang Khammouan. Khammouan, C Laos 17°25′N 104°51′E
153 S13 Thakurgaon Rajshahi, NW Bangladesh 26°05′N 88°34′E
149 S6 Thal Khyber Pakhtunkhwa, NW Pakistan 33°24′N 70°32′E
167 S11 Thalabârívât prev. Phumĭ Thalabârĭvăt. Stœng Trêng, N Cambodia 13°34′N 105°57′E
166 M15 Thalang Phuket, SW Thailand 08°00′N 98°21′E
Thalashsheri see Tellicherry/Thalassery
167 Q10 Thalat Khae Nakhon Ratchasima, C Thailand 15°15′N 102°24′E
109 Q5 Thalgau Salzburg, NW Austria 47°49′N 13°19′E
108 G7 Thalwil Zürich, NW Switzerland 47°17′N 08°35′E
83 I20 Thamaga Kweneng, SE Botswana 24°41′S 25°31′E
Thamarīd see Thamarīt
141 V13 Thamarīt var. Thamarīd, Thumrayt. SW Oman 17°39′N 54°02′E
141 P16 Thamar, Jabal ▲ SW Yemen 13°46′N 45°32′E
184 M6 Thames Waikato, North Island, New Zealand 37°10′S 175°33′E
14 D17 Thames ≈ Ontario, S Canada
97 O22 Thames ≈ S England, United Kingdom
184 M6 Thames, Firth of gulf North Island, New Zealand
14 D17 Thamesville Ontario, S Canada 42°33′N 81°58′W
141 S13 Thamūd N Yemen 17°18′N 49°57′E
167 N9 Thanbyuzayat Mon State, S Myanmar (Burma) 15°58′N 97°44′E
166 K7 Thandwe var. Sandoway. Rakhine State, W Myanmar (Burma) 18°28′N 94°20′E
152 I9 Thānesar Haryāna, NW India 29°58′N 76°48′E
167 T7 Thanh Hoa Thanh Hoa, N Vietnam 19°49′N 105°48′E
Thanintari Taungdan see Bilauktaung Range
155 I21 Thanjāvūr prev. Tanjore. Tamil Nādu, SE India 10°46′N 79°09′E
Thanlwin see Salween
103 U7 Thann Haut-Rhin, NE France 47°51′N 07°04′E
167 O16 Tha Nong Phrom Phatthalung, SW Thailand 07°24′N 100°04′E
167 N13 Thap Sakae var. Thap Sakau. Prachuap Khiri Khan, SW Thailand 11°30′N 99°35′E
Thap Sakau see Thap Sakae
81 I19 Tharaka-Nithi ◐ county C Kenya
98 L10 't Harde Gelderland, E Netherlands 52°25′N 05°53′E
152 D11 Thar Desert var. Great Indian Desert, Indian Desert. desert India/Pakistan
181 V10 Thargomindah Queensland, C Australia 28°00′S 143°47′E
150 D11 Thar Pārkar desert SE Pakistan
139 S7 Tharthār al Furāt, Qanāt ath canal C Iraq
139 R7 Tharthār, Buḥayrat ath ◎ C Iraq
139 R5 Tharthār, Wādī ath dry watercourse N Iraq
167 N13 Tha Sae Chumphon, SW Thailand
167 N15 Tha Sala Nakhon Si Thammarat, SW Thailand 08°43′N 99°54′E
114 I13 Thásos Thásos, E Greece 40°47′N 24°43′E
115 I14 Thásos island E Greece
37 N14 Thatcher Arizona, SW USA 32°47′N 109°46′W
167 T5 Thất Khê var. Tràng Dinh. Lang Sơn, N Vietnam 22°15′N 106°26′E
166 M8 Thaton Mon State, S Myanmar (Burma) 16°56′N 97°20′E
167 S9 That Phanom Nakhon Phanom, E Thailand 16°52′N 104°41′E
167 R10 Tha Tum Surin, E Thailand 15°18′N 103°39′E
103 P16 Thau, Bassin de var. Étang de Thau. ◎ S France
Thau, Étang de see Thau, Bassin de
166 L3 Thaungdut Sagaing, N Myanmar (Burma) 24°26′N 94°45′E
167 O8 Thaungyin Th. Mae Nam Moei. ≈ Myanmar (Burma)/Thailand
167 R8 Tha Uthen Nakhon Phanom, E Thailand 17°32′N 104°34′E
109 W2 Thaya var. Dyje. ≈ Austria/Czech Republic see also Dyje
Thaya see Dyje
27 V8 Thayer Missouri, C USA 36°31′N 91°34′W
166 L6 Thayetmyo Magway, C Myanmar (Burma) 19°20′N 95°10′E
33 S15 Thayne Wyoming, C USA 42°54′N 111°01′W
166 M5 Thazi Mandalay, C Myanmar (Burma) 20°50′N 96°04′E
Thebes see Thíva
44 L5 The Carlton var. Abraham Bay. Mayaguana, SE The Bahamas 22°21′N 72°56′W
45 O14 The Crane var. Crane. S Barbados 13°06′N 59°27′W
32 I11 The Dalles Oregon, NW USA 45°36′N 121°10′W
28 M14 Thedford Nebraska, C USA 41°59′N 100°32′W
The Flatts Village see Flatts Village
The Hague see 's-Gravenhage
Theiss see Tisa/Tisza
8 M9 Thelon ≈ Northwest Territories, N Canada
11 V15 Theodore Saskatchewan, S Canada 51°25′N 103°01′W
23 N8 Theodore Alabama, S USA 30°33′N 88°10′W
36 L13 Theodore Roosevelt Lake ▣ Arizona, SW USA
Theodosia see Feodosiya
Theophilo Ottoni see Teófilo Otoni
11 V13 The Pas Manitoba, C Canada 53°49′N 101°09′W
31 T14 The Plains Ohio, N USA 39°22′N 82°07′W
Thera see Santoríni
172 H17 Thérèse, Île island Inner Islands, NE Seychelles
Therezina see Teresina
115 L20 Thérma Ikaría, Dodekánisa, Greece, Aegean Sea 37°37′N 26°18′E
Thermae Himerenses see Termini Imerese
Thermae Pannonicae see Baden
Thermaic Gulf/Thermaicus Sinus see Thermaïkós Kólpos
121 Q8 Thermaïkós Kólpos Eng. Thermaic Gulf; anc. Thermaicus Sinus. gulf N Greece
Thermiá see Kýthnos
115 L17 Thermís Lésvos, E Greece 39°08′N 26°32′E
115 E18 Thérmo Dytikí Elláda, C Greece 38°32′N 21°42′E
33 V14 Thermopolis Wyoming, C USA 43°39′N 108°12′W
183 P10 The Rock New South Wales, SE Australia 35°18′S 147°07′E
195 O5 Theron Mountains ▲▲ Antarctica
The Sooner State see Oklahoma
115 G18 Thespiés Stereá Elláda, C Greece 38°18′N 23°08′E
115 E16 Thessalía Eng. Thessaly. ◐ region C Greece
14 C10 Thessalon Ontario, S Canada 46°15′N 83°34′W
115 G14 Thessaloníki Eng. Salonica, Salonika, SCr. Solun, Turk. Selânik. Kentrikí Makedonía, N Greece 40°38′N 22°58′E
115 G14 Thessaloníki ✈ Kentrikí Makedonía, N Greece 40°30′N 22°58′E
Thessaly see Thessalía
84 B12 Theta Gap undersea feature E Atlantic Ocean 12°40′W 43°30′N
97 P20 Thetford E England, United Kingdom 52°25′N 00°45′E
15 R11 Thetford-Mines Québec, SE Canada 46°07′N 71°16′W
113 K17 Theth var. Thethi. Shkodër, N Albania 42°25′N 19°45′E
Thethi see Theth
99 L20 Theux Liège, E Belgium 50°33′N 05°48′E
45 V9 The Valley ○ (Anguilla) E Anguilla 18°13′N 63°00′W
27 N10 The Village Oklahoma, C USA 35°33′N 97°33′W
The Volunteer State see Tennessee
25 W10 The Woodlands Texas, SW USA 30°09′N 95°27′E
Thiamis see Kalamás
Thian Shan see Tien Shan
Thibet see Xizang Zizhiqu
22 J9 Thibodaux Louisiana, S USA 29°48′N 90°49′W
29 S3 Thief Lake ◎ Minnesota, N USA
29 S3 Thief River ≈ Minnesota, C USA
29 S3 Thief River Falls Minnesota, N USA 48°07′N 96°10′W

◆ Country
● Country Capital
◇ Dependent Territory
○ Dependent Territory Capital
◐ Administrative Regions
✈ International Airport
▲ Mountain
▲▲ Mountain Range
⛰ Volcano
≈ River
◎ Lake
▣ Reservoir

Thièle – Toamasina

Thièle see La Thielle
32 G14 Thielsen, Mount ▲ Oregon, NW USA 43°09′N 122°04′W
Thielt see Tielt
106 G7 Thiene Veneto, NE Italy 45°43′N 11°29′E
Thienen see Tienen
103 P11 Thiers Puy-de-Dôme, C France 45°51′N 03°33′E
76 F11 Thiès W Senegal 14°49′N 16°52′W
81 I19 Thika Kiambu, S Kenya 01°03′S 37°05′E
Thikombia see Cikobia
151 K18 Thiladhunmathi Atoll var. Tiladummati Atoll. atoll N Maldives
Thimbu see Thimphu
153 T11 Thimphu var. Thimbu; prev. Tashi Chho Dzong. ● (Bhutan) W Bhutan 27°28′N 89°37′E
92 H2 Þingeyri Vestfirðir, NW Iceland 65°52′N 23°28′W
92 I3 Þingvellir Suðurland, SW Iceland 64°15′N 21°06′W
187 Q17 Thio Province Sud, C New Caledonia 21°37′S 166°13′E
103 T4 Thionville Ger. Diedenhofen. Moselle, NE France 49°22′N 06°11′E
77 O12 Thiou NW Burkina Faso 13°42′N 02°34′W
115 K22 Thíra Santorini, Kykládes, Greece, Aegean Sea 36°25′N 25°26′E
Thíra see Santoríni
115 J22 Thirasía island Kykládes, Greece, Aegean Sea
97 M16 Thirsk N England, United Kingdom 54°07′N 01°17′W
14 F12 Thirty Thousand Islands island group Ontario, S Canada
95 F20 Thisted Midtjylland, NW Denmark 56°58′N 08°42′E
Thistil Fjord see Þistilfjörður
92 L1 Þistilfjörður var. Thistil Fjord. fjord NE Iceland
182 G9 Thistle Island island South Australia
Thithia see Cicia
Thiukhaoluang Phrahang see Luang Prabang Range
115 G18 Thíva Eng. Thebes; prev. Thívai. Stereá Elláda, C Greece 38°19′N 23°19′E
Thívai see Thíva
102 M12 Thiviers Dordogne, SW France 45°24′N 00°54′E
92 J4 Þjórsá ≈ C Iceland
9 N10 Thlewiaza ≈ Nunavut, NE Canada
8 L10 Thoa ≈ Northwest Territories, NW Canada
99 G14 Tholen Zeeland, SW Netherlands 51°31′N 04°13′E
99 F14 Tholen island SW Netherlands
26 L10 Thomas Oklahoma, C USA 35°44′N 98°45′W
21 T3 Thomas West Virginia, NE USA 39°09′N 79°28′W
27 U3 Thomas Hill Reservoir ⊠ Missouri, C USA
23 S5 Thomaston Georgia, SE USA 32°53′N 84°19′W
19 R7 Thomaston Maine, NE USA 44°06′N 69°10′W
25 T12 Thomaston Texas, SW USA 28°56′N 97°07′W
23 O6 Thomasville Alabama, S USA 31°54′N 87°42′W
23 T8 Thomasville Georgia, SE USA 30°49′N 83°57′W
21 S9 Thomasville North Carolina, SE USA 35°52′N 80°04′W
35 N5 Thomes Creek ≈ California, W USA
11 W12 Thompson Manitoba, C Canada 55°45′N 97°54′W
29 R4 Thompson North Dakota, N USA 47°45′N 97°07′W
0 F8 Thompson ≈ Alberta/British Columbia, SW Canada
33 O8 Thompson Falls Montana, NW USA 47°36′N 115°20′W
29 Q10 Thompson, Lake ◎ South Dakota, N USA
34 M3 Thompson Peak ▲ California, W USA 41°00′N 123°01′W
27 S2 Thompson River ≈ Missouri, C USA
185 A22 Thompson Sound sound South Island, New Zealand
8 J5 Thomsen ≈ Banks Island, Northwest Territories, NW Canada
23 V4 Thomson Georgia, SE USA 33°28′N 82°30′W
103 T10 Thonon-les-Bains Haute-Savoie, E France 46°22′N 06°30′E
103 O15 Thoré var. Thore. ≈ S France
Thore see Thoré
37 P11 Thoreau New Mexico, SW USA 35°24′N 108°13′W
Thorenburg see Turda
92 J3 Þórisvatn ◎ C Iceland
92 P4 Thor, Kapp headland S Svalbard 76°25′N 25°01′E
92 I4 Þorlákshöfn Suðurland, SW Iceland 63°52′N 21°24′W
Thorn see Toruń
25 T10 Thorndale Texas, SW USA 30°36′N 97°12′W
14 H10 Thorne Ontario, S Canada 46°38′N 79°04′W
97 J14 Thornhill S Scotland, United Kingdom 55°13′N 03°46′W
25 U8 Thornton Texas, SW USA 31°24′N 96°34′W
Thornton Island see Millennium Island
14 H16 Thorold Ontario, S Canada 43°07′N 79°15′W
32 I9 Thorp Washington, NW USA 47°03′N 120°40′W
Thorshavn see Tórshavn
195 S3 Thorshavnheiane physical region Antarctica
92 L1 Þórshöfn Norðurland Eystra, NE Iceland 66°09′N 15°18′W
Thospitis see Van Gölü
167 S14 Thôt Nôt Cân Thơ, S Vietnam 10°17′N 105°31′E
102 K8 Thouars Deux-Sèvres, W France 46°59′N 00°13′W
153 X14 Thoubal Manipur, NE India 24°40′N 94°00′E
102 K9 Thouet ≈ W France
Thoune see Thun
18 H7 Thousand Islands island Canada/USA
35 S15 Thousand Oaks California, W USA 34°10′N 118°50′W
114 L12 Thrace cultural region SE Europe
114 J13 Thracian Sea Gk. Thrakikó Pélagos; anc. Thracium Mare. sea Greece/Turkey
Thracium Mare/Thrakikó Pélagos see Thracian Sea
Thrá Lí, Bá see Tralee Bay
33 R11 Three Forks Montana, NW USA 45°53′N 111°34′W
162 M8 Three Gorges Dam dam Hubei, C China
160 L9 Three Gorges Reservoir ⊠ C China
11 Q16 Three Hills Alberta, SW Canada 51°43′N 113°15′W
183 N15 Three Hummock Island island Tasmania, SE Australia
184 H1 Three Kings Islands island group N New Zealand
175 P10 Three Kings Rise undersea feature W Pacific Ocean
77 O18 Three Points, Cape headland S Ghana 04°43′N 02°03′W
31 P10 Three Rivers Michigan, N USA 41°56′N 85°37′W
25 S13 Three Rivers Texas, SW USA 28°27′N 98°10′W
83 G24 Three Sisters Northern Cape, SW South Africa 31°51′S 23°04′E
32 H13 Three Sisters ▲ Oregon, NW USA 44°08′N 121°46′W
187 N10 Three Sisters Islands island group SE Solomon Islands
25 Q6 Throckmorton Texas, SW USA 33°11′N 99°12′W
180 M10 Throssell, Lake salt lake Western Australia
115 K25 Thryptís var. Thrýptis. ▲ Kríti, Greece, E Mediterranean Sea 35°06′N 25°51′E
167 U14 Thuận Nam prev. Ham Thuân Nam. Bình Thuận, S Vietnam 10°49′N 107°49′E
167 T13 Thu Dâu Một var. Phu Cũơng. Sông Be, S Vietnam 10°58′N 106°40′E
167 S6 Thu Do ✈ (Ha Nôi) Ha Nôi, N Vietnam 21°13′N 105°46′E
99 G21 Thuin Hainaut, S Belgium 50°21′N 04°18′E
149 Q12 Thul Sind, SE Pakistan 28°14′N 68°50′E
Thule see Qaanaaq
83 J18 Thuli var. Tuli. ≈ S Zimbabwe
Thumrayt see Thamarīt
108 D9 Thun Fr. Thoune. Bern, W Switzerland 46°46′N 07°38′E
12 C12 Thunder Bay Ontario, S Canada 48°27′N 89°12′W
30 M1 Thunder Bay lake bay S Canada
31 R6 Thunder Bay lake bay Michigan, N USA
31 R6 Thunder Bay River ≈ Michigan, N USA
27 N11 Thunderbird, Lake ◎ Oklahoma, C USA
28 L8 Thunder Butte Creek ≈ South Dakota, N USA
108 E9 Thuner See ◎ C Switzerland
167 N15 Thung Song var. Cha Mai. Nakhon Si Thammarat, SW Thailand 08°10′N 99°41′E
108 H7 Thur ≈ N Switzerland
108 G6 Thurgau Fr. Thurgovie. ❖ canton NE Switzerland
Thurgovie see Thurgau
Thuringe see Thüringen
108 J7 Thüringen Vorarlberg, W Austria 47°12′N 09°48′E
101 J17 Thüringen Eng. Thuringia, Fr. Thuringe. ❖ state C Germany
101 J17 Thüringer Wald Eng. Thuringian Forest. ▲▲ C Germany
Thuringia see Thüringen
Thuringian Forest see Thüringer Wald
97 D19 Thurles Ir. Durlas. S Ireland 52°41′N 07°49′W
21 W2 Thurmont Maryland, NE USA 39°36′N 77°22′W
Thurø see Thurø By
95 H24 Thurø By var. Thurø. Syddtjylland, C Denmark 55°03′N 10°43′E
14 M12 Thurso Québec, SE Canada 45°36′N 75°13′W
96 J6 Thurso N Scotland, United Kingdom 58°35′N 03°32′W
194 I10 Thurston Island island Antarctica
108 I9 Thusis Graubünden, S Switzerland 46°40′N 09°27′E
Thýamis see Kalamás
95 E21 Thyborøn var. Tyborøn. Midtjylland, W Denmark 56°40′N 08°12′E
195 U3 Thyer Glacier glacier Antarctica
115 L20 Thýmaina island Dodekánisa, Greece, Aegean Sea
83 N15 Thyolo var. Cholo. Southern, S Malawi 16°03′S 35°11′E
183 U6 Tia New South Wales, SE Australia 31°14′S 151°51′E
54 H5 Tía Juana Zulia, NW Venezuela 10°18′N 71°24′W
Tiancheng see Chongyang
160 J14 Tiandong var. Pingma. Guangxi Zhuangzu Zizhiqu, S China 23°37′N 107°06′E
161 O3 Tianjin var. Tientsin. Tianjin Shi, E China 39°13′N 117°06′E
Tianjin see Tianjin Shi
161 P3 Tianjin Shi var. Jin, Tianjin, T'ien-ching, Tientsin. ❖ municipality E China
159 S10 Tianjun var. Xinyuan. Qinghai, C China 37°16′N 99°03′E
160 J13 Tianlin var. Leli. Guangxi Zhuangzu Zizhiqu, S China 24°27′N 106°03′E
Tian Shan see Tien Shan
159 W11 Tianshui Gansu, C China 34°33′N 105°51′E
150 I7 Tianshuihai Xinjiang Uygur Zizhiqu, W China 35°17′N 79°30′E
161 S10 Tiantai Zhejiang, SE China 29°11′N 121°01′E
160 J14 Tianyang var. Tianzhou. Guangxi Zhuangzu Zizhiqu, S China 23°50′N 106°52′E
Tianzhou see Tianyang
159 U9 Tianzhu var. Huazangsi, Tianzhu Zangzu Zizhixian. Gansu, C China 37°01′N 103°04′E
Tianzhu Zangzu Zizhixian see Tianzhu
191 Q7 Tiarei Tahiti, W French Polynesia 17°32′S 149°20′W
74 J6 Tiaret var. Tihert. NW Algeria 35°20′N 01°20′E
77 N17 Tiassalé S Ivory Coast 05°54′N 04°50′W
192 I16 Ti'avea Upolu, SE Samoa 13°58′S 171°30′W
Tiba see Chiba
60 J11 Tibagi var. Tibají. Paraná, S Brazil 24°29′S 50°29′W
60 J10 Tibagi, Rio var. Rio Tibají. ≈ S Brazil
Tibají see Tibagi
Tibají, Rio see Tibagi, Rio
139 Q9 Tibal, Wādī dry watercourse S Iraq
54 G9 Tibaná Boyacá, C Colombia 05°19′N 73°25′W
79 F14 Tibati Adamaoua, N Cameroon 06°25′N 12°33′E
76 K15 Tibé, Pic de ▲ SE Guinea 08°39′N 08°58′W
Tiber see Tevere, Italy
Tiber see Tivoli, Italy
Tiberias see Tverya
138 G8 Tiberias, Lake var. Chinnereth, Sea of Bahr Tabariya, Sea of Galilee, Ar. Bahrat Tabariya, Heb. Yam Kinneret. ◎ N Israel
78 H5 Tibesti off. Région du Tibesti. ❖ region N Chad
78 H6 Tibesti var. Tibesti Massif, Ar. Tibisti. ▲▲ N Africa
Tibesti Massif see Tibesti
Tibesti, Région du see Tibesti
Tibet see Xizang Zizhiqu
Tibetan Autonomous Region see Xizang Zizhiqu
Tibet, Plateau of see Qingzang Gaoyuan
Tibistī see Tibesti
14 K7 Tiblemont, Lac ◎ Québec, SE Canada
139 X9 Ţīb, Nahr aţ ≈ S Iraq
Tibnī see At Tibnī
182 L4 Tibooburra New South Wales, SE Australia 29°24′S 142°01′E
95 L18 Tibro Västra Götaland, S Sweden 58°25′N 14°11′E
40 E5 Tiburón, Isla var. Isla del Tiburón. island NW Mexico
Tiburón, Isla del see Tiburón, Isla
23 W14 Tice Florida, SE USA 26°40′N 81°49′W
Tichau see Tychy
114 L8 Ticha, Yazovir ◎ NE Bulgaria
76 K9 Tichit var. Tichitt. Tagant, C Mauritania 18°26′N 09°31′W
Tichitt see Tichit
108 G11 Ticino Fr./Ger. Tessin. ❖ canton S Switzerland
106 D8 Ticino ≈ Italy/Switzerland
108 H11 Ticino Ger. Tessin. ≈ SW Switzerland
Ticinum see Pavia
41 X12 Ticul Yucatán, SE Mexico 20°22′N 89°30′W
95 K18 Tidaholm Västra Götaland, S Sweden 58°12′N 13°55′E
Tidjikdja see Tidjikja
76 J8 Tidjikja var. Tidjikdja; prev. Fort-Cappolani. Tagant, C Mauritania 18°31′N 11°24′W
Tidore see Soasiu
171 R11 Tidore, Pulau island E Indonesia
Tidra, Île see Et Tîdra
77 N16 Tiébissou var. Tiebissou. C Ivory Coast 07°10′N 05°10′W
Tiebissou see Tiébissou
Tiefa see Diaobingshan
108 I9 Tiefencastel Graubünden, S Switzerland 46°40′N 09°33′E
Tiegenhof see Nowy Dwór Gdański
T'ieh-ling see Tieling
98 K13 Tiel Gelderland, C Netherlands 51°53′N 05°26′E
163 W7 Tieli Heilongjiang, NE China 46°57′N 128°01′E
163 V11 Tieling var. T'ieh-ling. Liaoning, NE China 42°19′N 123°52′E
152 L4 Tielongtan China/India 35°10′N 79°32′E
99 D17 Tielt var. Thielt. West-Vlaanderen, W Belgium 51°00′N 03°20′E
T'ien-ching see Tianjin Shi
99 I18 Tienen var. Thienen, Fr. Tirlemont. Vlaams Brabant, C Belgium 50°48′N 04°56′E
Tiên Giang, Sông see Mekong
147 X9 Tien Shan Chin. Thian Shan, Tian Shan, T'ien Shan, Rus. Tyan'-Shan'. ▲▲ C Asia
Tientsin see Tianjin
167 U6 Tiên Yên Quang Ninh, N Vietnam 21°19′N 107°24′E
95 O14 Tierp Uppsala, C Sweden 60°20′N 17°30′E
62 H7 Tierra Amarilla Atacama, N Chile 27°28′S 70°17′W
37 R9 Tierra Amarilla New Mexico, SW USA 36°42′N 106°31′W
41 R15 Tierra Blanca Veracruz-Llave, E Mexico 18°28′N 96°21′W
41 O16 Tierra Colorada Guerrero, S Mexico 17°10′N 99°30′W
63 J17 Tierra Colorada, Bajo de la basin SE Argentina
63 I25 Tierra del Fuego off. Provincia de la Tierra del Fuego. ❖ province S Argentina
63 J24 Tierra del Fuego island Argentina/Chile
Tierra del Fuego, Provincia de la see Tierra del Fuego
54 D7 Tierralta Córdoba, NW Colombia 08°10′N 76°04′W
104 K9 Tiétar ≈ W Spain
60 L10 Tietê São Paulo, S Brazil 23°04′S 47°41′W
60 J8 Tietê, Rio ≈ S Brazil
32 I9 Tieton Washington, NW USA 46°41′N 120°43′W
32 J6 Tiffany Mountain ▲ Washington, NW USA 48°40′N 119°55′W
31 S12 Tiffin Ohio, N USA 41°06′N 83°10′W
31 Q11 Tiffin River ≈ Ohio, N USA
Tiflis see Tbilisi
23 U7 Tifton Georgia, SE USA 31°27′N 83°31′W
171 R13 Tifu Pulau Buru, E Indonesia 03°46′S 126°36′E
38 L17 Tigalda Island island Aleutian Islands, Alaska, USA
115 I15 Tigáni, Akrotírio headland Límnos, E Greece 39°50′N 25°03′E
169 V6 Tiga Tarok Sabah, East Malaysia 06°57′N 117°07′E
117 O10 Tighina Rus. Bendery; prev. Bender. E Moldova 46°51′N 29°28′E
145 X9 Tigiretskiy Khrebet ▲▲ E Kazakhstan
79 F14 Tignère Adamaoua, N Cameroon 07°24′N 12°35′E
13 P14 Tignish Prince Edward Island, SE Canada 46°58′N 64°03′W
186 M10 Tigoa var. Tinggoa. Rennell, S Solomon Islands 11°39′S 160°13′E
Tigranocerta see Siirt
80 I11 Tigray ❖ federal region N Ethiopia
41 O11 Tigre, Cerro del ▲ C Mexico 23°06′N 99°13′W
56 F8 Tigre, Río ≈ N Peru
139 X10 Tigris Ar. Dijlah, Turk. Dicle. ≈ Iraq/Turkey
76 G9 Tiguent Trarza, SW Mauritania 17°15′N 16°00′W
74 M10 Tiguentourine E Algeria 27°59′N 09°16′E
77 V10 Tiguidit, Falaise de ridge C Niger
141 N13 Tihāmah var. Tehama. plain Saudi Arabia/Yemen
Tihert see Tiaret
Ti-hua/Tihwa see Ürümqi
41 Q13 Tihuatlán Veracruz-Llave, E Mexico 20°44′N 97°30′W
40 B1 Tijuana Baja California Norte, NW Mexico 32°32′N 117°01′W
42 E2 Tikal Petén, N Guatemala 17°11′N 89°36′W
154 I9 Tīkamgarh prev. Tehri. Madhya Pradesh, C India 24°44′N 78°50′E
158 L7 Tikanlik Xinjiang Uygur Zizhiqu, NW China 40°34′N 87°37′E
77 P12 Tikaré N Burkina Faso 13°16′N 01°39′W
39 O12 Tikchik Lakes lakes Alaska, USA
191 T9 Tikehau atoll Îles Tuamotu, C French Polynesia
191 V9 Tikei island Îles Tuamotu, C French Polynesia
126 L13 Tikhoretsk Krasnodarskiy Kray, SW Russian Federation 45°51′N 40°07′E
124 I13 Tikhvin Leningradskaya Oblast', NW Russian Federation 59°37′N 33°30′E
193 P9 Tiki Basin undersea feature S Pacific Ocean
76 K13 Tikinsso ≈ NE Guinea
Tikirarjuaq see Whale Cove
184 Q8 Tikitiki Gisborne, North Island, New Zealand 37°49′S 178°23′E
79 D16 Tiko Sud-Ouest, SW Cameroon 04°02′N 09°19′E
139 S6 Tikrīt var. Tekrit. Şalāḩ ad Dīn, N Iraq 34°36′N 43°42′E
124 I8 Tiksha Respublika Kareliya, NW Russian Federation 64°07′N 32°31′E
124 I6 Tikshozero, Ozero ◎ NW Russian Federation
123 P7 Tiksi Respublika Sakha (Yakutiya), NE Russian Federation 71°40′N 128°47′E
Tiladummati Atoll see Thiladhunmathi Atoll
42 A6 Tilapa San Marcos, SW Guatemala 14°31′N 92°11′W
42 L13 Tilarán Guanacaste, NW Costa Rica 10°28′N 84°57′W
99 J14 Tilburg Noord-Brabant, S Netherlands 51°34′N 05°05′E
14 D17 Tilbury Ontario, S Canada 42°15′N 82°26′W
182 K4 Tilcha South Australia 29°37′S 140°52′E
Tilcha Creek see Callabonna Creek
29 Q14 Tilden Nebraska, C USA 42°03′N 97°49′W
25 R13 Tilden Texas, SW USA 28°27′N 98°43′W
14 H10 Tilden Lake Ontario, S Canada 46°35′N 79°36′W
116 G9 Tileagd Hung. Mezőtelegd. Bihor, W Romania 47°03′N 22°11′E
145 P9 Tilekey prev. Ladyzhenka. Akmola, C Kazakhstan 50°58′N 68°44′E
77 Q8 Tîlemsi, Vallée de ≈ C Mali
123 V8 Tilichiki Krasnoyarskiy Kray, E Russian Federation 60°25′N 165°55′E
Tiligul see Tilihul
Tiligul'skiy Liman see Tilihul's'kyy Lyman
117 P9 Tilihul Rus. Tiligul. ≈ SW Ukraine
117 P10 Tilihul's'kyy Lyman Rus. Tiligul'skiy Liman. ◎ S Ukraine
Tilimsen see Tlemcen
Tilio Martius see Toulon
77 R11 Tillabéri var. Tillabéry. Tillabéri, W Niger 14°13′N 01°27′E
77 R11 Tillabéri ❖ department SW Niger
Tillabéry see Tillabéri
32 F11 Tillamook Oregon, NW USA 45°28′N 123°50′W
32 E11 Tillamook Bay inlet Oregon, NW USA
151 Q22 Tillanchāng Dwīp island Nicobar Islands, India, NE Indian Ocean
95 N15 Tillberga Västmanland, C Sweden 59°41′N 16°37′E
Tillenberg see Dyleň
21 S10 Tillery, Lake ⊠ North Carolina, SE USA
77 T10 Tillia Tahoua, W Niger 16°13′N 04°51′E
23 N6 Tillmans Corner Alabama, S USA 30°35′N 88°10′W
14 F17 Tillsonburg Ontario, S Canada 42°53′N 80°44′W
115 N22 Tílos island Dodekánisa, Greece, Aegean Sea
183 N5 Tilpa New South Wales, SE Australia 30°56′S 144°24′E
Tilsit see Sovetsk
31 N13 Tilton Illinois, N USA 40°06′N 87°39′W
126 K7 Tim Kurskaya Oblast', W Russian Federation 51°39′N 37°11′E
54 D12 Timaná Huila, S Colombia 01°58′N 75°55′W
Timan Ridge see Timanskiy Kryazh
125 Q6 Timanskiy Kryazh Eng. Timan Ridge. ridge NW Russian Federation
185 G20 Timaru Canterbury, South Island, New Zealand 44°23′S 171°15′E
127 S6 Timashevo Samarskaya Oblast', W Russian Federation 53°22′N 51°13′E
126 K13 Timashevsk Krasnodarskiy Kray, SW Russian Federation 45°37′N 38°57′E
Timbaki/Timbákion see Tympáki
22 K10 Timbalier Bay bay Louisiana, S USA
22 K11 Timbalier Island island Louisiana, S USA
76 L10 Timbedgha var. Timbédra. Hodh ech Chargui, SE Mauritania 16°17′N 08°14′W
Timbédra see Timbedgha
32 G10 Timber Oregon, NW USA 45°42′N 123°19′W
181 O3 Timber Creek Northern Territory, N Australia 15°35′S 130°21′E
28 M8 Timber Lake South Dakota, N USA 45°25′N 101°01′W
54 D12 Timbío Cauca, SW Colombia 02°20′N 76°40′W
54 C12 Timbiquí Cauca, SW Colombia 02°43′N 77°45′W
83 O17 Timbue, Ponta headland C Mozambique 18°49′S 36°22′E
Timbuktu see Tombouctou
169 W8 Timbun Mata, Pulau island E Malaysia
77 P8 Timétrine var. Ti-n-Kâr. oasis C Mali
Timfi see Týmfi
Timfristos see Tymfristós
77 V9 Timia Agadez, C Niger 18°07′N 08°49′E
171 X14 Timika Papua, E Indonesia 04°39′S 137°15′E
74 I9 Timimoun C Algeria 29°18′N 00°21′E
Timiris, Cap see Timirist, Râs
76 F8 Timirist, Râs var. Cap Timiris. headland NW Mauritania 19°18′N 16°28′W
145 O7 Timiryazevo Severnyy Kazakhstan, N Kazakhstan 53°45′N 66°33′E
116 E11 Timiş ❖ county SW Romania
14 H9 Timiskaming, Lake Fr. Lac Témiscamingue. ◎ Ontario/Québec, SE Canada
116 E11 Timişoara Ger. Temeschwar, Temeswar, Hung. Temesvár; prev. Temeschburg. Timiş, W Romania 45°46′N 21°17′E
116 E11 Timişoara ✈ Timiş, SW Romania 45°50′N 21°21′E
Timkovichi see Tsimkavichy
77 U8 Ti-m-Meghsoï ≈ NW Niger
100 K8 Timmendorfer Strand Schleswig-Holstein, N Germany 53°59′N 10°50′E
14 F7 Timmins Ontario, S Canada 48°09′N 80°01′W
21 S12 Timmonsville South Carolina, SE USA 34°07′N 79°56′W
30 K5 Timms Hill ▲ Wisconsin, N USA 45°27′N 90°12′W
112 P12 Timok ≈ E Serbia
58 N13 Timon Maranhão, E Brazil 05°08′S 42°52′W
171 Q17 Timor Sea sea E Indian Ocean
Timor Timur see East Timor
Timor Trench see Timor Trough
192 G8 Timor Trough var. Timor Trench. undersea feature NE Timor Sea
61 A21 Timote Buenos Aires, E Argentina 35°22′S 62°13′W
54 I6 Timotes Mérida, NW Venezuela 08°57′N 70°46′W
25 X8 Timpson Texas, SW USA 31°54′N 94°24′W
123 Q11 Timpton ≈ NE Russian Federation
93 H17 Timrå Västernorrland, C Sweden 62°29′N 17°20′E
20 J10 Tims Ford Lake ⊠ Tennessee, S USA
168 L7 Timur, Banjaran ▲▲ Peninsular Malaysia
171 Q8 Tinaca Point headland Mindanao, S Philippines 05°35′N 125°18′E
54 K5 Tinaco Cojedes, N Venezuela 09°44′N 68°28′W
64 Q11 Tinajo Lanzarote, Islas Canarias, Spain, NE Atlantic Ocean 29°03′N 13°41′W
187 P10 Tinakula island Santa Cruz Islands, E Solomon Islands
54 K5 Tinaquillo Cojedes, N Venezuela 09°57′N 68°20′W
116 F10 Tinca Hung. Tenke. Bihor, W Romania 46°46′N 21°58′E
155 J20 Tindivanam Tamil Nādu, SE India 12°15′N 79°41′E
74 E9 Tindouf W Algeria 27°43′N 08°09′W
74 E9 Tindouf, Sebkha de salt lake W Algeria
104 J2 Tineo Asturias, N Spain 43°20′N 06°25′W
77 R9 Ti-n-Essako Kidal, E Mali 18°30′N 02°27′E
Tinggoa see Tigoa
183 T5 Tingha New South Wales, SE Australia 29°56′S 151°13′E
Tingis see Tanger
Tinglett see Tinglev
95 F24 Tinglev Ger. Tinglett. Syddanmark, SW Denmark 54°57′N 09°15′E
56 E12 Tingo María Huánuco, C Peru 09°10′S 76°00′W
Tingréla see Tengréla
158 K16 Tingri var. Xégar. Xizang Zizhiqu, W China 28°40′N 87°04′E
95 M21 Tingsryd Kronoberg, S Sweden 56°30′N 15°E
95 P19 Tingstäde Gotland, SE Sweden 57°45′N 18°36′E
62 H12 Tinguiririca, Volcán ⛰ C Chile 34°52′S 70°24′W
94 F9 Tingvoll Møre og Romsdal, S Norway 62°55′N 08°13′E
188 K8 Tinian island S Northern Mariana Islands
Ti-n-Kâr see Timétrine
Tinnevelly see Tirunelveli
95 G15 Tinnoset Telemark, S Norway 59°43′N 09°03′E
95 F15 Tinnsjå prev. Tinnsjø, var. Tinnsjö. ◎ S Norway see also Tinnsjø
Tinnsjø see Tinnsjå
Tinnsjö see Tinnsjå
Tino see Chino
115 J20 Tínos Tínos, Kykládes, Greece, Aegean Sea 37°33′N 25°08′E
115 J20 Tínos anc. Tenos. island Kykládes, Greece, Aegean Sea
153 R14 Tinpahar Jhārkhand, NE India 25°00′N 87°43′E
153 X11 Tinsukia Assam, NE India 27°28′N 95°20′E
76 K10 Tîntâne Hodh el Gharbi, S Mauritania 16°25′N 10°08′W
62 L7 Tintina Santiago del Estero, N Argentina 27°00′S 62°45′W
182 K10 Tintinara South Australia 35°54′S 140°04′E
104 I14 Tinto ≈ SW Spain
77 S8 Ti-n-Zaouâtene Kidal, NE Mali 19°56′N 02°45′E
Tiobraid Árann see Tipperary
28 K3 Tioga North Dakota, N USA 48°24′N 102°56′W
18 G12 Tioga Pennsylvania, NE USA 41°54′N 77°07′W
25 T5 Tioga Texas, SW USA 33°28′N 96°55′W
35 Q8 Tioga Pass pass California, W USA
18 G12 Tioga River ≈ New York/Pennsylvania, NE USA
Tioman Island see Tioman, Pulau
168 M9 Tioman, Pulau var. Tioman Island. island Peninsular Malaysia
18 C12 Tionesta Pennsylvania, NE USA 41°31′N 79°30′W
18 D12 Tionesta Creek ≈ Pennsylvania, NE USA
168 J13 Tiop Pulau Pagai Selatan, W Indonesia 03°12′S 100°21′E
18 H11 Tioughnioga River ≈ New York, NE USA
74 J5 Tipasa var. Tipaza. N Algeria 36°35′N 02°27′E
Tipaza see Tipasa
42 J10 Tipitapa Managua, W Nicaragua 12°08′N 86°04′W
31 R13 Tipp City Ohio, N USA 39°57′N 84°10′W
31 O12 Tippecanoe River ≈ Indiana, N USA
97 D20 Tipperary Ir. Tiobraid Árann. S Ireland 52°29′N 08°10′W
97 D19 Tipperary Ir. Tiobraid Árann. cultural region S Ireland
35 R12 Tipton California, W USA 36°02′N 119°19′W
31 P13 Tipton Indiana, N USA 40°19′N 86°00′W
29 Y14 Tipton Iowa, C USA 41°46′N 91°07′W
27 U5 Tipton Missouri, C USA 38°39′N 92°46′W
36 I10 Tipton, Mount ▲ Arizona, SW USA 35°32′N 114°11′W
20 F8 Tiptonville Tennessee, S USA 36°21′N 89°30′W
12 E12 Tip Top Mountain ▲ Ontario, S Canada 48°18′N 86°06′W
155 G19 Tiptūr Karnātaka, W India 13°17′N 76°31′E
Tiquisate see Pueblo Nuevo Tiquisate
58 L13 Tiracambu, Serra do ▲▲ E Brazil
Tirana see Tiranë
113 K19 Tirana Rinas ✈ Durrës, W Albania 41°25′N 19°41′E
113 L20 Tiranë var. Tirana. ● (Albania) Tiranë, C Albania 41°20′N 19°50′E
113 K20 Tiranë ❖ district W Albania
140 I5 Tīrān, Jazīrat island Egypt/Saudi Arabia
106 F6 Tirano Lombardia, N Italy 46°13′N 10°10′E
182 I2 Tirari Desert desert South Australia
117 O10 Tiraspol Rus. Tiraspol'. E Moldova 46°50′N 29°35′E
Tiraspol' see Tiraspol
184 M8 Tirau Waikato, North Island, New Zealand 37°59′S 175°44′E
136 C14 Tire İzmir, SW Turkey 38°04′N 27°45′E
137 O11 Tirebolu Giresun, N Turkey 41°01′N 38°49′E
96 F11 Tiree island W Scotland, United Kingdom
Tîrgovişte see Târgovişte
Tîrgu see Târgu Cărbuneşti
Tîrgu Bujor see Târgu Bujor
Tîrgu Frumos see Târgu Frumos
Tîrgu Jiu see Târgu Jiu
Tîrgu Lăpuş see Târgu Lăpuş
Tîrgu Mures see Târgu Mureş
Tîrgu-Neamţ see Târgu-Neamţ
Tîrgu Ocna see Târgu Ocna
Tîrgu Secuiesc see Târgu Secuiesc
149 T3 Tirich Mīr ▲ NW Pakistan 36°12′N 71°51′E
76 J5 Tiris Zemmour ❖ region N Mauritania
Tirlemont see Tienen
127 W5 Tirlyanskiy Respublika Bashkortostan, W Russian Federation 54°09′N 58°32′E
Tîrnava Mare see Târnava Mare
Tîrnava Mică see Târnava Mică
Tîrnăveni see Târnăveni
Tírnavos see Týrnavos
Tirnovo see Veliko Tarnovo
154 J11 Tirodi Madhya Pradesh, C India 21°40′N 79°44′E
108 K8 Tirol off. Land Tirol, var. Tyrol, It. Tirolo. ❖ state W Austria
Tirol, Land see Tirol
Tirolo see Tirol
Tirreno, Mare see Tyrrhenian Sea
107 B19 Tirso ≈ Sardegna, Italy, C Mediterranean Sea
95 H22 Tirstrup ✈ (Århus) Århus, C Denmark 56°17′N 10°36′E
155 I21 Tiruchchirāppalli prev. Trichinopoly. Tamil Nādu, SE India 10°50′N 78°43′E
155 H23 Tirunelveli var. Tinnevelly. Tamil Nādu, SE India 08°45′N 77°43′E
155 J19 Tirupati Andhra Pradesh, E India 13°39′N 79°25′E
155 I20 Tiruppattūr Tamil Nādu, SE India 12°28′N 78°31′E
155 H21 Tiruppur prev. Tiruppūr. Tamil Nādu, SW India 11°05′N 77°20′E
Tiruvantapuram see Thiruvanathapuram/Trivandrum
155 I20 Tiruvannāmalai Tamil Nādu, SE India 12°13′N 79°07′E
112 L10 Tisa Ger. Theiss, Hung. Tisza, Rus. Tissa, Ukr. Tysa. ≈ SE Europe see also Tisza
Tisa see Tisza
Tischnowitz see Tišnov
11 U14 Tisdale Saskatchewan, S Canada 52°51′N 104°01′W
27 O13 Tishomingo Oklahoma, C USA 34°15′N 96°41′W
95 M17 Tisnaren ◎ S Sweden
111 F18 Tišnov Ger. Tischnowitz. Jihomoravský Kraj, SE Czech Republic 49°22′N 16°24′E
Tissa see Tisa/Tisza
74 J6 Tissemsilt N Algeria 35°37′N 01°48′E
153 S12 Tista ≈ NE India
112 L8 Tisza Ger. Theiss, Rom./Slvn./SCr. Tisa, Rus. Tissa, Ukr. Tysa. ≈ SE Europe see also Tisa
Tisza see Tisa
111 L23 Tiszaföldvár Jász-Nagykun-Szolnok, E Hungary 47°00′N 20°16′E
111 M22 Tiszafüred Jász-Nagykun-Szolnok, E Hungary 47°38′N 20°45′E
111 L23 Tiszakécske Bács-Kiskun, C Hungary 46°56′N 20°04′E
111 M21 Tiszaújváros prev. Leninváros. Borsod-Abaúj-Zemplén, NE Hungary 47°56′N 21°03′E
111 N21 Tiszavasvári Szabolcs-Szatmár-Bereg, NE Hungary 47°56′N 21°21′E
57 I17 Titicaca, Lake ◎ Bolivia/Peru
190 H17 Titikaveka Rarotonga, S Cook Islands 21°16′S 159°45′W
154 M13 Titilāgarh var. Titlagarh. Odisha, E India 20°18′N 83°09′E
168 K8 Titiwangsa, Banjaran ▲▲ Peninsular Malaysia
Titlagarh see Titilāgarh
Titograd see Podgorica
Titose see Chitose
Titova Mitrovica see Mitrovicë
Titovo Užice see Užice
113 M18 Titov Vrv ▲ NW FYR Macedonia 41°58′N 20°49′E
94 F7 Titran Sør-Trøndelag, S Norway 63°40′N 08°20′E
31 Q8 Tittabawassee River ≈ Michigan, N USA
116 J13 Titu Dâmboviţa, S Romania 44°40′N 25°32′E
79 M16 Titule Orientale, N Dem. Rep. Congo 03°20′N 25°23′E
23 X11 Titusville Florida, SE USA 28°37′N 80°50′W
18 C12 Titusville Pennsylvania, NE USA 41°36′N 79°39′W
76 G11 Tivaouane W Senegal 14°59′N 16°50′W
113 I17 Tivat SW Montenegro 42°25′N 18°43′E
14 E14 Tiverton Ontario, S Canada 44°15′N 81°31′W
97 J23 Tiverton SW England, United Kingdom 50°54′N 03°30′W
19 O12 Tiverton Rhode Island, NE USA 41°38′N 71°10′W
107 I15 Tivoli anc. Tiber. Lazio, C Italy 41°58′N 12°45′E
25 U13 Tivoli Texas, SW USA 28°26′N 96°54′W
141 Z8 Ţīwī NE Oman 22°43′N 59°20′E
41 Y11 Tizimín Yucatán, SE Mexico 21°10′N 88°09′W
74 K5 Tizi Ouzou var. Tizi-Ouzou. N Algeria 36°44′N 04°06′E
Tizi-Ouzou see Tizi Ouzou
74 D8 Tiznit SW Morocco 29°43′N 09°39′W
95 F23 Tjæreborg Syddtjylland, W Denmark 55°28′N 08°35′E
113 I14 Tjentište Republika Srpska, SE Bosnia and Herzegovina 43°23′N 18°42′E
98 L7 Tjeukemeer ◎ N Netherlands
Tjiamis see Ciamis
Tjiandjoer see Cianjur
Tjilatjap see Cilacap
Tjirebon see Cirebon
95 J18 Tjörn island S Sweden
92 O3 Tjuvfjorden fjord S Svalbard
Tkvarcheli see T'q'varcheli
40 L8 Tlahualilo Durango, N Mexico 26°06′N 103°25′W
41 P14 Tlalnepantla México, C Mexico 19°34′N 99°12′W
41 Q13 Tlapacoyán Veracruz-Llave, E Mexico 19°58′N 97°13′W
41 P16 Tlapa de Comonfort Guerrero, S Mexico 17°33′N 98°33′W
40 L13 Tlaquepaque Jalisco, C Mexico 20°36′N 103°19′W
Tlascala see Tlaxcala
41 P14 Tlaxcala var. Tlascala, Tlaxcala de Xicohténcatl. Tlaxcala, C Mexico 19°17′N 98°16′W
41 P14 Tlaxcala ❖ state S Mexico
Tlaxcala de Xicohténcatl see Tlaxcala
41 P14 Tlaxco var. Tlaxco de Morelos. Tlaxcala, S Mexico 19°38′N 98°06′W
Tlaxco de Morelos see Tlaxco
41 Q16 Tlaxiaco var. Santa María Asunción Tlaxiaco. Oaxaca, S Mexico 17°18′N 97°42′W
74 I6 Tlemcen var. Tilimsen, Tlemsen. NW Algeria 34°53′N 01°21′W
Tlemsen see Tlemcen
138 L4 Tlété Ouâte Rharbi, Jebel ▲ N Syria
116 J7 Tlumach Ivano-Frankivs'ka Oblast', W Ukraine 48°53′N 25°00′E
127 P17 Tlyarata Respublika Dagestan, SW Russian Federation 42°10′N 46°30′E
116 K10 Toaca, Vârful prev. Vîrful Toaca. ▲ NE Romania 46°58′N 25°55′E
Toaca, Vîrful see Toaca, Vârful
191 Q8 Toahotu prev. Teohatu. Tahiti, W French Polynesia
187 R13 Toak Ambrym, C Vanuatu 16°21′S 168°16′E
172 J4 Toamasina prev./Fr. Tamatave. Toamasina, E Madagascar 18°10′S 49°23′E
172 J4 Toamasina ❖ province E Madagascar

◆ Country
● Country Capital
◇ Dependent Territory
○ Dependent Territory Capital
❖ Administrative Regions
✈ International Airport
▲ Mountain
▲▲ Mountain Range
⛰ Volcano
≈ River
◎ Lake
⊠ Reservoir

TOAMASINA – TOTTENHAM

172 J4 **Toamasina** ✈ Toamasina, E Madagascar 18°10′S 49°23′E
21 X6 **Toano** Virginia, NE USA 37°22′N 76°46′W
191 U10 **Toau** *atoll* Îles Tuamotu, C French Polynesia
45 T6 **Toa Vaca, Embalse** ⊠ C Puerto Rico
62 K13 **Toay** La Pampa, C Argentina 36°43′S 64°22′W
159 R14 **Toba** Xizang Zizhiqu, W China 31°17′N 97°37′E
164 K14 **Toba** Mie, Honshū, SW Japan 34°29′N 136°51′E
168 I9 **Toba, Danau** ◎ Sumatera, W Indonesia
45 Y16 **Tobago** *island* NE Trinidad and Tobago
149 Q9 **Toba Kākar Range** ▲▲ NW Pakistan
105 Q12 **Tobarra** Castilla-La Mancha, C Spain 38°36′N 01°41′W
149 U9 **Toba Tek Singh** Punjab, E Pakistan 30°54′N 72°30′E
171 R11 **Tobelo** Pulau Halmahera, E Indonesia 01°45′N 127°59′E
14 E12 **Tobermory** Ontario, S Canada 45°15′N 81°39′W
96 G10 **Tobermory** W Scotland, United Kingdom 56°37′N 06°12′W
165 S4 **Tōbetsu** Hokkaidō, NE Japan 43°12′N 141°28′E
180 M6 **Tobin Lake** ◎ Western Australia
11 U14 **Tobin Lake** ◎ Saskatchewan, C Canada
35 T4 **Tobin, Mount** ▲ Nevada, W USA 40°25′N 117°28′W
165 O9 **Tobi-shima** *island* C Japan
169 N13 **Toboali** Pulau Bangka, W Indonesia 03°00′S 106°30′E
Tobol *see* Tobyl
122 H11 **Tobol'sk** Tyumenskaya Oblast', C Russian Federation 58°15′N 68°12′E
Tobruch/Tobruk *see* Ţubruq
125 R3 **Tobseda** Nenetskiy Avtonomnyy Okrug, NW Russian Federation 68°37′N 52°24′E
144 M8 **Tobyl** *prev.* Tobol. Kustanay, N Kazakhstan 52°42′N 62°36′E
144 L8 **Tobyl** *prev.* Tobol. ≋ Kazakhstan/Russian Federation
125 Q6 **Tobysh** ≋ NW Russian Federation
54 F10 **Tocaima** Cundinamarca, C Colombia 04°30′N 74°38′W
59 K16 **Tocantins** *off.* Estado do Tocantins. ◈ *state* C Brazil
Tocantins, Estado do *see* Tocantins
59 K15 **Tocantins, Rio** ≋ N Brazil
23 T2 **Toccoa** Georgia, SE USA 34°34′N 83°19′W
165 O12 **Tochigi** *off.* Tochigi-ken, *var.* Totigi. ◈ *prefecture* Honshū, S Japan
Tochigi-ken *see* Tochigi
165 O11 **Tochio** *var.* Totio. Niigata, Honshū, C Japan 37°27′N 139°00′E
95 I15 **Töcksfors** Värmland, C Sweden 59°30′N 11°49′E
42 J5 **Tocoa** Colón, N Honduras 15°40′N 86°01′W
62 H4 **Tocopilla** Antofagasta, N Chile 22°06′S 70°08′W
62 I4 **Tocorpuri, Cerro de** ▲ Bolivia/Chile 22°26′S 67°53′W
183 O10 **Tocumwal** New South Wales, SE Australia 35°53′S 145°35′E
54 K4 **Tocuyo de la Costa** Falcón, NW Venezuela 11°04′N 68°23′W
152 H13 **Toda Rāisingh** Rājasthān, N India 26°02′N 75°35′E
106 H13 **Todi** Umbria, C Italy 42°47′N 12°25′E
108 G9 **Tödi** ▲ NE Switzerland 46°52′N 08°53′E
171 T12 **Todlo** Papua Barat, E Indonesia 0°46′S 130°50′E
165 S9 **Todoga-saki** *headland* Honshū, C Japan 39°33′N 142°02′E
59 P17 **Todos os Santos, Baía de** *bay* E Brazil
40 F10 **Todos Santos** Baja California Sur, NW Mexico 23°28′N 110°14′W
40 B2 **Todos Santos, Bahía de** *bay* NW Mexico
Toeban *see* Tuban
Toekang Besi Eilanden *see* Tukangbesi, Kepulauan
Toeloengagoeng *see* Tulungagung
Tōen *see* Taoyuan
185 D25 **Toetoes Bay** *bay* South Island, New Zealand
11 Q14 **Tofield** Alberta, SW Canada 53°22′N 112°39′W
10 K17 **Tofino** Vancouver Island, British Columbia, SW Canada 49°05′N 125°51′W
189 X17 **Tofol** Kosrae, E Micronesia
95 J20 **Tofta** Halland, S Sweden 57°10′N 12°19′E
95 H15 **Tofte** Buskerud, S Norway 59°31′N 10°33′E
95 F24 **Toftlund** Syddanmark, SW Denmark 55°12′N 09°04′E
193 X15 **Tofua** *island* Ha'apai Group, C Tonga
187 Q12 **Toga** *island* Torres Islands, N Vanuatu
80 N13 **Togdheer** *off.* Gobolka Togdheer. ◈ *region* NW Somalia
Togdheer, Gobolka *see* Togdheer
164 L11 **Togi** Ishikawa, Honshū, SW Japan 37°06′N 136°44′E
39 N13 **Togiak** Alaska, USA 59°03′N 160°31′W
171 O11 **Togian, Kepulauan** *island group* C Indonesia
77 Q15 **Togo** *off.* Togolese Republic; *prev.* French Togoland. ◆ *republic* W Africa
Togolese Republic *see* Togo
162 F8 **Tögrög** Govĭ-Altay, SW Mongolia 45°51′N 95°04′E
162 F8 **Tögrög** *var.* Hoolt. Övörhangay, C Mongolia 45°31′N 103°06′E
Tögrög *see* Manhan
159 N12 **Togton He** *var.* Tuotuo He. ≋ C China
Togton Heyan *see* Tanggulashan
Toguzak *see* Togyzak
144 L7 **Togyzak** *prev.* Toguzak. ≋ Kazakhstan/Russian Federation
37 P10 **Tohatchi** New Mexico, SW USA 35°51′N 108°45′W
191 O7 **Tohiea, Mont** ✵ Moorea, W French Polynesia 17°33′S 149°48′W
137 N14 **Tohma Çayı** ≋ C Turkey
93 O17 **Tohmajärvi** Pohjois-Karjala, SE Finland 62°12′N 30°19′E
93 L16 **Toholampi** Keski-Pohjanmaa, W Finland 63°46′N 24°15′E
Tōhōm *see* Mandah
23 X12 **Tohopekaliga, Lake** ◎ Florida, SE USA
164 M14 **Toi** Shizuoka, Honshū, S Japan 34°55′N 138°45′E
190 B15 **Toi** N Niue 18°57′S 169°51′W
93 L19 **Toijala** Pirkanmaa, SW Finland 61°09′N 23°51′E
171 P12 **Toima** Sulawesi, N Indonesia 0°48′S 122°21′E
164 D17 **Toi-misaki** Kyūshū, SW Japan
171 Q17 **Toineke** Timor, S Indonesia 10°06′S 124°22′E
Toirc, Inis *see* Inishturk
35 U6 **Toiyabe Range** ▲▲ Nevada, W USA
Tojikiston, Jumhurii *see* Tajikistan
147 R12 **Tojikobod** *Rus.* Tadzhikabad. C Tajikistan 39°08′N 70°54′E
164 G12 **Tōjō** Hiroshima, Honshū, SW Japan 34°54′N 133°15′E
39 T10 **Tok** Alaska, USA 63°20′N 142°59′W
164 K13 **Tōkai** Aichi, Honshū, SW Japan 35°01′N 136°51′E
111 N21 **Tokaj** Borsod-Abaúj-Zemplén, NE Hungary 48°08′N 21°25′E
165 N11 **Tōkamachi** Niigata, Honshū, C Japan 37°08′N 138°44′E
185 D25 **Tokanui** Southland, South Island, New Zealand 46°33′S 169°02′E
80 I7 **Tokar** *var.* Ţawkar. Red Sea, NE Sudan 18°27′N 37°41′E
136 L12 **Tokat** Tokat, N Turkey 40°20′N 36°35′E
136 L12 **Tokat** ◈ *province* N Turkey
Tŏkchŏk-kundo *see* Deokjeok-gundo
Tōke *see* Taka Atoll
190 J9 **Tokelau** ◇ *NZ overseas territory* W Polynesia
Tőketerebes *see* Trebišov
Tokhtamyshbek *see* Tŭkhtamish
24 M6 **Tokio** Texas, SW USA 33°09′N 102°31′W
Tokio *see* Tōkyō
189 W11 **Toki Point** *point* NW Wake Island
Tokkuztara *see* Gongliu
117 V9 **Tokmak** *var.* Velykyy Tokmak. Zaporiz'ka Oblast', SE Ukraine 47°13′N 35°43′E
Tokmak *see* Tokmok
184 Q8 **Tokomaru Bay** Gisborne, North Island, New Zealand 38°10′S 178°18′E
184 M8 **Tokoroa** Waikato, North Island, New Zealand 38°14′S 175°52′E
76 K14 **Tokounou** C Guinea 09°43′N 09°46′W
38 M12 **Toksook Bay** Alaska, USA 60°33′N 165°01′W
Toksu *see* Xinhe
Toksum *see* Toksun
158 L6 **Toksun** *var.* Toksum. Xinjiang Uygur Zizhiqu, NW China 42°47′N 88°38′E
147 T8 **Toktogul** Talasskaya Oblast', NW Kyrgyzstan 41°51′N 72°56′E
147 T9 **Toktogul'skoye Vodokhranilishche** ⊠ W Kyrgyzstan
Toktomush *see* Tŭkhtamish
193 Y14 **Toku** *island* Vava'u Group, N Tonga
165 U16 **Tokunoshima** Kagoshima, Tokuno-shima, SW Japan
165 U16 **Tokuno-shima** *island* Nansei-shotō, SW Japan
164 I14 **Tokushima** *var.* Tokusima. Tokushima, Shikoku, SW Japan 34°04′N 134°28′E
164 H14 **Tokushima** *off.* Tokushima-ken, *var.* Tokusima. ◈ *prefecture* Shikoku, SW Japan
Tokushima-ken *see* Tokushima
Tokusima *see* Tokushima
164 E13 **Tokuyama** *var.* Shūnan. Yamaguchi, Honshū, SW Japan 34°04′N 131°48′E
165 N13 **Tōkyō** *var.* Tokio. ● (Japan) Tōkyō, Honshū, S Japan 35°40′N 139°45′E
165 O13 **Tōkyō** *off.* Tōkyō-to. ◈ *capital district* Honshū, S Japan
Tōkyō-to *see* Tōkyō
145 T12 **Tokyrau** ≋ C Kazakhstan
149 O3 **Tokzār** *Pash.* Tukzār. Sar-e Pul, N Afghanistan 35°47′N 66°28′E
145 W13 **Tokzhaylau** *prev.* Dzerzhinskoye. Almaty, SE Kazakhstan 45°49′N 81°04′E
145 W13 **Tokzhaylau** *var.* Dzerzhinskoye. Taldykorgan, SE Kazakhstan 45°49′N 81°04′E
189 U12 **Tol** *atoll* Chuuk Islands, C Micronesia
184 Q9 **Tolaga Bay** Gisborne, North Island, New Zealand 38°22′S 178°17′E
172 I7 **Tōlañaro** *prev.* Faradofay, Fort-Dauphin. Toliara, SE Madagascar
162 D6 **Tolbo** Bayan-Ölgiy, W Mongolia 48°22′N 90°22′E
Tolbukhin *see* Dobrich
60 G11 **Toledo** Paraná, S Brazil 24°45′S 53°41′W
54 G8 **Toledo** Norte de Santander, N Colombia 07°16′N 72°28′W
105 N9 **Toledo** *anc.* Toletum. Castilla-La Mancha, C Spain 39°52′N 04°02′W
30 M14 **Toledo** Illinois, N USA 39°16′N 88°15′W
29 W13 **Toledo** Iowa, C USA 42°00′N 92°34′W
31 R11 **Toledo** Ohio, N USA 41°40′N 83°33′W
32 F12 **Toledo** Oregon, NW USA 44°37′N 123°56′W
32 G9 **Toledo** Washington, NW USA 46°27′N 122°49′W
42 F3 **Toledo** ◈ *district* S Belize
104 M9 **Toledo** ◈ *province* Castilla-La Mancha, C Spain
25 Y7 **Toledo Bend Reservoir** ⊠ Louisiana/Texas, SW USA
104 M10 **Toledo, Montes de** ▲▲ C Spain
106 J12 **Tolentino** Marche, C Italy 43°08′N 13°17′E
Toletum *see* Toledo
94 H10 **Tolga** Hedmark, S Norway 62°25′N 11°00′E
158 J3 **Toli** Xinjiang Uygur Zizhiqu, NW China 45°55′N 83°33′E
172 H7 **Toliara** *var.* Toliary; *prev.* Tuléar. Toliara, SW Madagascar 23°20′S 43°41′E
172 H7 **Toliara** ◈ *province* SW Madagascar
Toliary *see* Toliara
118 H11 **Toliejai** *prev.* Kamajai. Panevėžys, NE Lithuania 55°16′N 25°30′E
54 D11 **Tolima** *off.* Departamento del Tolima. ◈ *province* C Colombia
Tolima, Departamento del *see* Tolima
171 N11 **Tolitoli** Sulawesi, C Indonesia 01°05′N 120°50′E
95 K22 **Tollarp** Skåne, S Sweden 55°55′N 14°00′E
100 N9 **Tollense** ≋ NE Germany
100 N10 **Tollensesee** ◎ NE Germany
36 K13 **Tolleson** Arizona, SW USA 33°25′N 112°15′W
146 M13 **Tollimarjon** *Rus.* Talimardzhan. Qashqadaryo Viloyati, S Uzbekistan 38°22′N 65°31′E
Tolmein *see* Tolmin
106 J6 **Tolmezzo** Friuli-Venezia Giulia, NE Italy 46°27′N 13°01′E
109 S11 **Tolmin** *Ger.* Tolmein, *It.* Tolmino. W Slovenia 46°12′N 13°39′E
Tolmino *see* Tolmin
111 J25 **Tolna** *Ger.* Tolnau. Tolna, S Hungary 46°26′N 18°47′E
111 I24 **Tolna** *off.* Tolna Megye. ◈ *county* SW Hungary
Tolna Megye *see* Tolna
Tolnau *see* Tolna
79 I20 **Tolo** Bandundu, W Dem. Rep. Congo 02°57′S 18°35′E
Tolochin *see* Talachyn
190 D12 **Toloke** Île Futuna, W Wallis and Futuna
30 M13 **Tolono** Illinois, N USA 39°59′N 88°16′W
105 Q3 **Tolosa** País Vasco, N Spain 43°09′N 02°04′W
Tolosa *see* Toulouse
171 O13 **Tolo, Teluk** *bay* Sulawesi, C Indonesia
39 R9 **Tolovana River** ≋ Alaska, USA
123 U10 **Tolstoy, Mys** *headland* E Russian Federation 59°12′N 155°04′E
63 G15 **Toltén** Araucanía, C Chile 39°13′S 73°15′W
63 G15 **Toltén, Río** ≋ S Chile
54 E6 **Tolú** Sucre, NW Colombia 09°32′N 75°34′W
41 O14 **Toluca** *var.* Toluca de Lerdo. México, S Mexico 19°20′N 99°40′W
Toluca de Lerdo *see* Toluca
41 O14 **Toluca, Nevado de** ✵ C Mexico 19°05′N 99°45′W
127 R6 **Tol'yatti** *prev.* Stavropol'. Samarskaya Oblast', W Russian Federation 53°32′N 49°27′E
77 O12 **Toma** NW Burkina Faso 12°46′N 02°51′W
30 K7 **Tomah** Wisconsin, N USA 43°59′N 90°31′W
30 L5 **Tomahawk** Wisconsin, N USA 45°27′N 89°40′W
117 T8 **Tomakivka** Dnipropetrovs'ka Oblast', E Ukraine 47°47′N 34°45′E
165 S4 **Tomakomai** Hokkaidō, NE Japan 42°38′N 141°32′E
165 S2 **Tomamae** Hokkaidō, NE Japan 44°18′N 141°38′E
104 G9 **Tomar** Santarém, W Portugal 39°36′N 08°25′W
123 T13 **Tomari** Ostrov Sakhalin, Sakhalinskaya Oblast', SE Russian Federation 47°47′N 142°09′E
115 C16 **Tómaros** ▲ W Greece 39°31′N 20°45′E
Tomaschow *see* Tomaszów Mazowiecki
Tomaschow *see* Tomaszów Lubelski
61 E16 **Tomás Gomensoro** Artigas, N Uruguay 30°28′S 57°28′W
117 N7 **Tomashpil'** Vinnyts'ka Oblast', C Ukraine 48°32′N 28°31′E
Tomaszów *see* Tomaszów Mazowiecki
111 P15 **Tomaszów Lubelski** *Ger.* Tomaschow. Lubelskie, E Poland 50°29′N 23°23′E
Tomaszów Mazowiecka *see* Tomaszów Mazowiecki
110 L13 **Tomaszów Mazowiecki** *var.* Tomaszów Mazowiecka; *prev.* Tomaszów, *Ger.* Tomaschow. Łódzkie, C Poland 51°33′N 20°E
40 J13 **Tomatlán** Jalisco, C Mexico 19°53′N 105°18′W
81 F15 **Tombe** Jonglei, E South Sudan 05°52′N 31°40′E
23 N4 **Tombigbee River** ≋ Alabama/Mississippi, S USA
82 A10 **Tomboco** Zaire Province, NW Angola 06°50′S 13°20′E
77 O10 **Tombouctou** *Eng.* Timbuktu. Tombouctou, N Mali 16°47′N 03°03′W
77 N9 **Tombouctou** ◈ *region* W Mali
37 N16 **Tombstone** Arizona, SW USA 31°42′N 110°04′W
83 A15 **Tombua** *Port.* Porto Alexandre. Namibe, SW Angola 15°49′S 11°53′E
83 J19 **Tom Burke** Limpopo, NE South Africa 23°07′S 28°01′E
146 L9 **Tomdibuloq** *Rus.* Tamdybulak. Navoiy Viloyati, N Uzbekistan 41°48′N 64°33′E
146 L9 **Tomditov-Togʻlari** ▲ N Uzbekistan
62 G13 **Tomé** Bío Bío, C Chile 36°38′S 72°57′W
58 L12 **Tomé-Açu** Pará, NE Brazil 02°25′S 48°09′W
95 L23 **Tomelilla** Skåne, S Sweden 55°33′N 14°00′E
105 O10 **Tomelloso** Castilla-La Mancha, C Spain 39°09′N 03°01′W
14 H10 **Tomiko Lake** ◎ Ontario, S Canada
77 N12 **Tominian** Ségou, C Mali 13°18′N 04°39′W
Tomini, Gulf of *see* Tomini, Teluk
171 N12 **Tomini, Teluk** *var.* Gulf of Tomini; *prev.* Teluk Gorontalo. *bay* Sulawesi, C Indonesia
165 Q11 **Tomioka** Fukushima, Honshū, C Japan 37°19′N 140°57′E
113 G14 **Tomislavgrad** Federacija Bosne I Hercegovine, SW Bosnia and Herzegovina 43°43′N 17°15′E
181 O9 **Tomkinson Ranges** ▲▲ South Australia/Western Australia
123 Q11 **Tommot** Respublika Sakha (Yakutiya), NE Russian Federation 58°57′N 126°24′E
171 Q11 **Tomohon** Sulawesi, N Indonesia 01°19′N 124°49′E
147 V7 **Tomok** *prev.* Tokmak. Chuyskaya Oblast', N Kyrgyzstan 42°50′N 75°18′E
54 K9 **Tomo, Río** ≋ E Colombia
113 L21 **Tomorrit, Mali i** ▲ S Albania 40°43′N 20°12′E
11 S17 **Tompkins** Saskatchewan, S Canada 50°03′N 108°49′W
20 K8 **Tompkinsville** Kentucky, S USA 36°43′N 85°41′W
171 N11 **Tompo** Sulawesi, N Indonesia 0°56′N 120°16′E
180 I8 **Tom Price** Western Australia 22°48′S 117°49′E
122 J12 **Tomsk** Tomskaya Oblast', C Russian Federation 56°30′N 85°05′E
122 I11 **Tomskaya Oblast'** ◈ *province* C Russian Federation
18 K16 **Toms River** New Jersey, NE USA 39°56′N 74°09′W
Tom Steed Lake *see* Tom Steed Reservoir
26 L12 **Tom Steed Reservoir** *var.* Tom Steed Lake. ⊠ Oklahoma, C USA
171 U13 **Tomu** Papua Barat, E Indonesia 02°07′S 133°01′E
158 H6 **Tomür Feng** *var.* Pobeda Peak, *Rus.* Pik Pobedy. ▲ China/Kyrgyzstan 42°02′N 80°07′E *see also* Pobedy, Pik
Tomür Feng *see* Pobedy, Pik
189 N13 **Tomworoahlang** Pohnpei, E Micronesia
41 U17 **Tonalá** Chiapas, SE Mexico 16°08′N 93°41′W
106 F6 **Tonale, Passo del** *pass* N Italy
164 I11 **Tonami** Toyama, Honshū, SW Japan 36°40′N 136°55′E
58 C12 **Tonantins** Amazonas, W Brazil 02°58′S 67°30′W
32 K6 **Tonasket** Washington, NW USA 48°41′N 119°27′W
55 Y9 **Tonate** *var.* Macouria. N French Guiana 05°00′N 52°28′W
18 D10 **Tonawanda** New York, NE USA 43°00′N 78°51′W
42 I7 **Toncontín** *prev.* Tegucigalpa. ● (Honduras) Francisco Morazán, SW Honduras 14°04′N 87°11′W
42 H7 **Toncontín** ✈ Central District, C Honduras 14°03′N 87°20′W
171 Q11 **Tondano** Sulawesi, C Indonesia 01°19′N 124°56′E
104 H7 **Tondela** Viseu, N Portugal 40°31′N 08°05′W
95 F24 **Tønder** *Ger.* Tondern. Syddanmark, SW Denmark 54°57′N 08°53′E
Tondern *see* Tønder
143 N4 **Tonekābon** *var.* Shahsawar, Tonkābon; *prev.* Shahsavār. Māzandarān, N Iran 36°40′N 51°25′E
Tonezh *see* Tonyezh
193 Y14 **Tonga** *off.* Kingdom of Tonga, *var.* Friendly Islands. ◆ *monarchy* SW Pacific Ocean
175 R9 **Tonga** *island group* SW Pacific Ocean
Tongaat *see* oThongathi
Tonga, Kingdom of *see* Tonga
161 Q13 **Tong'an** *var.* Datong, Tong an. Fujian, SE China 24°43′N 118°07′E
Tong an *see* Tong'an
27 Q4 **Tonganoxie** Kansas, C USA 39°06′N 95°05′W
39 Y13 **Tongass National Forest** *reserve* Alaska, USA
193 Y16 **Tongatapu** ✈ Tongatapu, S Tonga 21°10′S 175°10′W
193 Y16 **Tongatapu** *island* Tongatapu Group, S Tonga
193 Y16 **Tongatapu Group** *island group* S Tonga
175 S9 **Tonga Trench** *undersea feature* S Pacific Ocean
161 N8 **Tongbai Shan** ▲▲ C China
161 P8 **Tongcheng** Anhui, E China 31°16′N 117°00′E
160 L6 **Tongchuan** Shaanxi, C China 35°10′N 109°03′E
160 L12 **Tongdao** *var.* Tongdao Dongzu Zizhixian; *prev.* Shuangjiang. Hunan, S China 26°06′N 109°46′E
Tongdao Dongzu Zizhixian *see* Tongdao
159 T11 **Tongde** *var.* Gabasumdo. Qinghai, C China 35°13′N 100°39′E
99 K19 **Tongeren** *Fr.* Tongres. Limburg, NE Belgium 50°47′N 05°28′E
Tonghae *see* Donghae
160 G13 **Tonghai** *var.* Xiushan. Yunnan, SW China 24°06′N 102°45′E
163 X8 **Tonghe** Heilongjiang, NE China 46°00′N 128°45′E
163 W11 **Tonghua** Jilin, NE China 41°45′N 125°50′E
163 Z6 **Tongjiang** Heilongjiang, NE China 47°39′N 132°29′E
163 Y13 **Tongjosŏn-man** *prev.* Broughton Bay. *bay* E North Korea
163 V7 **Tongken He** ≋ NE China
Tongking, Gulf of *see* Tonkin, Gulf of
163 U10 **Tongliao** Nei Mongol Zizhiqu, N China 43°37′N 122°15′E
161 Q9 **Tongling** Anhui, E China 30°55′N 117°50′E
161 R9 **Tonglu** Zhejiang, SE China 29°50′N 119°38′E
187 R14 **Tongoa** *island* Shepherd Islands, S Vanuatu
62 G9 **Tongoy** Coquimbo, C Chile 30°16′S 71°31′W
160 L11 **Tongren** *var.* Rongwo. Guizhou, S China 27°44′N 109°10′E
159 T11 **Tongren** *var.* Rongwo. Qinghai, C China 35°31′N 101°58′E
Tongres *see* Tongeren
153 U11 **Tongsa** *var.* Tongsa Dzong. C Bhutan 27°33′N 90°30′E
Tongsa Dzong *see* Tongsa
Tongshan *see* Fuding, Fujian, China
Tongshan *see* Xuzhou, Jiangsu, China
Tongshi *see* Wuzhishan
159 P12 **Tongtian He** *var.* Zhi Qu. ≋ C China
96 I6 **Tongue** N Scotland, United Kingdom 58°30′N 04°25′W
44 H3 **Tongue of the Ocean** *strait* C The Bahamas
33 X10 **Tongue River** ≋ Montana, NW USA
33 W11 **Tongue River Reservoir** ⊠ Montana, NW USA
159 V11 **Tongwei** *var.* Pingxiang. Gansu, C China 35°09′N 105°15′E
159 W9 **Tongxin** Ningxia, N China 37°00′N 105°41′E
163 U9 **Tongyu** *var.* Kaitong. Jilin, NE China 44°49′N 123°08′E
160 J11 **Tongzi** *var.* Loushanguan. Guizhou, S China 28°08′N 106°49′E
162 F8 **Tonhil** *var.* Dzüyl. Govĭ-Altay, SW Mongolia 46°09′N 93°55′E
40 G5 **Tónichi** Sonora, NW Mexico 28°37′N 109°34′W
81 D14 **Tonj** Warap, W South Sudan 07°18′N 28°41′E
152 H13 **Tonk** Rājasthān, N India 26°10′N 75°50′E
Tonkābon *see* Tonekābon
27 N8 **Tonkawa** Oklahoma, C USA 36°40′N 97°18′W
167 T7 **Tonkin, Gulf of** *var.* Gulf of Tongking, *Chin.* Beibu Wan, *Vtn.* Vinh Bắc Bộ. *gulf* China/Vietnam
167 Q12 **Tônlé Sap** *Eng.* Great Lake. ◎ W Cambodia
102 L14 **Tonneins** Lot-et-Garonne, SW France 44°21′N 00°21′E
103 Q7 **Tonnerre** Yonne, C France 47°50′N 04°00′E
Tonoas *see* Dublon
35 U8 **Tonopah** Nevada, W USA 38°04′N 117°13′W
164 H13 **Tonoshō** Okayama, Shōdo-shima, SW Japan 34°29′N 134°10′E
43 S17 **Tonosí** Los Santos, S Panama 07°23′N 80°26′W
95 H16 **Tønsberg** Vestfold, S Norway 59°16′N 10°25′E
39 T11 **Tonsina** Alaska, USA 61°39′N 145°10′W
95 D17 **Tonstad** Vest-Agder, S Norway 58°40′N 06°42′E
193 X15 **Tonumea** *island* Nomuka Group, W Tonga
137 O11 **Tonya** Trabzon, NE Turkey 40°52′N 39°17′E
119 K20 **Tonyezh** *Rus.* Tonezh. Homyel'skaya Voblasts', SE Belarus 51°50′N 27°48′E
36 L3 **Tooele** Utah, W USA 40°32′N 112°18′W
122 L13 **Toora-Khem** Respublika Tyva, S Russian Federation 52°25′N 96°01′E
183 O5 **Toorale East** New South Wales, SE Australia 30°29′S 145°25′E
83 H25 **Toorberg** ▲ S South Africa 32°02′S 24°02′E
118 G5 **Tootsi** Pärnumaa, SW Estonia 58°34′N 24°43′E
183 U3 **Toowoomba** Queensland, E Australia 27°35′S 151°54′E
27 Q4 **Topeka** *state capital* Kansas, C USA 39°03′N 95°41′W
122 J12 **Topki** Kemerovskaya Oblast', S Russian Federation 55°12′N 85°40′E
111 M18 **Topla** *Hung.* Toplya. ≋ NE Slovakia
Toplicza *see* Toplița
116 J10 **Toplița** *Ger.* Töplitz, *Hung.* Marosheviz; *prev.* Toplița Română, *Hung.* Oláh-Toplicza, Toplicza. Harghita, C Romania 45°56′N 25°20′E
Toplița Română/Töplitz *see* Toplița
Toplya *see* Topla
111 I20 **Topoľčany** *Hung.* Nagytapolcsány. Nitriansky Kraj, W Slovakia 48°33′N 18°10′E
40 G8 **Topolobampo** Sinaloa, C Mexico 25°36′N 109°04′W
116 I13 **Topoloveni** Argeș, S Romania 44°49′N 25°02′E
114 L11 **Topolovgrad** *prev.* Kavakli. Haskovo, S Bulgaria 42°06′N 26°20′E
Topolya *see* Bačka Topola
32 J10 **Toppenish** Washington, NW USA 46°22′N 120°18′W
181 P4 **Top Springs Roadhouse** Northern Territory, N Australia 16°37′S 131°49′E
189 U11 **Tora** *island* Chuuk, C Micronesia
Toraigh *see* Tory Island
189 U11 **Tora Island Pass** *passage* Chuuk Islands, C Micronesia
143 U5 **Torbat-e Ḩeydarīyeh** *var.* Turbat-i-Haidari. Khorāsān-e Razavī, NE Iran 35°18′N 59°12′E
143 V5 **Torbat-e Jām** *var.* Turbat-i-Jam. Khorāsān-e Razavī, NE Iran 35°16′N 60°36′E
39 Q11 **Torbert, Mount** ▲ Alaska, USA 61°30′N 152°15′W
31 P6 **Torch Lake** ◎ Michigan, N USA
Törcsvár *see* Bran
Torda *see* Turda
104 L6 **Tordesillas** Castilla y León, N Spain 41°30′N 05°00′W
92 K13 **Töre** Norrbotten, N Sweden 65°55′N 22°40′E
95 L17 **Töreboda** Västra Götaland, S Sweden 58°41′N 14°07′E
95 J21 **Torekov** Skåne, S Sweden 56°25′N 12°39′E
92 O3 **Torell Land** *physical region* SW Svalbard
105 U4 **Toretta de l'Orrí** *var.* Llorri; *prev.* Tossal de l'Orrí. ▲ NE Spain 42°24′N 01°15′E
144 M14 **Toretum** Kyzylorda, S Kazakhstan 45°38′N 63°20′E
117 Y8 **Torez** Donets'ka Oblast', SE Ukraine 48°00′N 38°38′E
101 N14 **Torgau** Sachsen, E Germany 51°34′N 13°01′E
145 R8 **Torgay** *Kaz.* Torghay; *prev.* Turgay. Akmola, W Kazakhstan 51°46′N 72°45′E
145 N10 **Torgay** *prev.* Turgay. ≋ C Kazakhstan
Torgay Üstirti *see* Turgayskaya Stolovaya Strana
Torghay *see* Torgay
95 N22 **Torhamn** Blekinge, S Sweden 56°04′N 15°49′E
99 C17 **Torhout** West-Vlaanderen, W Belgium 51°04′N 03°06′E
106 B8 **Torino** *Eng.* Turin. Piemonte, NW Italy 45°03′N 07°39′E
Tori-shima *see* Io-Tori-shima
81 F16 **Torit** Eastern Equatoria, S South Sudan 04°27′N 32°31′E
186 H6 **Toriu** New Britain, E Papua New Guinea 04°39′S 151°42′E
148 M4 **Torkestān, Selseleh-ye Band-e** *var.* Bandi-i Turkistan. ▲▲ NW Afghanistan
104 L7 **Tormes** ≋ W Spain
Tornacum *see* Tournai
Torneå *see* Tornio
92 K12 **Torneälven** *var.* Torniojoki, *Fin.* Tornionjoki. ≋ Finland/Sweden
92 I11 **Torneträsk** ◎ N Sweden
13 O4 **Torngat Mountains** ▲▲ Newfoundland and Labrador, NE Canada
24 H8 **Tornillo** Texas, SW USA 31°26′N 106°06′W
92 K13 **Tornio** *Swe.* Torneå. Lappi, NW Finland 65°50′N 24°18′E
Torniojoki/Tornionjoki *see* Torneälven
61 B23 **Tornquist** Buenos Aires, E Argentina 38°08′S 62°15′W
104 L6 **Toro** Castilla y León, N Spain 41°31′N 05°24′W
62 H9 **Toro, Cerro del** ▲ N Chile 29°10′S 69°43′W
77 R12 **Torodi** Tillabéri, SW Niger 13°05′N 01°46′E
Törökbecse *see* Novi Bečej
186 J7 **Torokina** Bougainville, NE Papua New Guinea 06°12′S 155°04′E
111 L23 **Törökszentmiklós** Jász-Nagykun-Szolnok, E Hungary 47°11′N 20°26′E
42 G7 **Torola, Río** ≋ El Salvador/Honduras
Toronaíos, Kólpos *see* Kassándras, Kólpos
14 H15 **Toronto** *province capital* Ontario, S Canada 43°42′N 79°25′W
31 V12 **Toronto** Ohio, N USA 40°27′N 80°36′W
Toronto *see* Lester B. Pearson
27 P6 **Toronto Lake** ⊠ Kansas, C USA
35 V16 **Toro Peak** ▲ California, W USA 33°31′N 116°25′W
124 H16 **Toropets** Tverskaya Oblast', W Russian Federation 56°29′N 31°37′E
81 G18 **Tororo** E Uganda 0°42′N 34°12′E
136 H16 **Toros Dağları** *Eng.* Taurus Mountains. ▲▲ S Turkey
183 N13 **Torquay** Victoria, SE Australia 38°21′S 144°18′E
97 J24 **Torquay** SW England, United Kingdom 50°28′N 03°30′W
104 M5 **Torquemada** Castilla y León, N Spain 42°02′N 04°17′W
35 S16 **Torrance** California, W USA 33°50′N 118°20′W
104 G12 **Torrão** Setúbal, S Portugal 38°18′N 08°13′W
104 H8 **Torre, Alto da** ▲ C Portugal 40°21′N 07°31′W
107 K18 **Torre Annunziata** Campania, S Italy 40°45′N 14°27′E
105 T8 **Torreblanca** Valenciana, E Spain 40°14′N 00°12′E
104 L15 **Torrecilla** ▲ S Spain 36°38′N 04°54′W
105 P4 **Torrecilla en Cameros** La Rioja, N Spain 42°18′N 02°33′W
105 N13 **Torredelcampo** Andalucía, S Spain 37°45′N 03°52′W
107 K17 **Torre del Greco** Campania, S Italy 40°46′N 14°22′E
104 I6 **Torre de Moncorvo** *var.* Moncorvo, Tôrre de Moncorvo. Bragança, N Portugal 41°10′N 07°03′W
104 J9 **Torrejoncillo** Extremadura, W Spain 39°54′N 06°28′W
105 O8 **Torrejón de Ardoz** Madrid, C Spain 40°27′N 03°29′W
105 N7 **Torrelaguna** Madrid, C Spain 40°50′N 03°33′W
105 N2 **Torrelavega** Cantabria, N Spain 43°21′N 04°03′W
107 M16 **Torremaggiore** Puglia, SE Italy 41°42′N 15°17′E
104 M15 **Torremolinos** Andalucía, S Spain 36°38′N 04°30′W
182 I6 **Torrens, Lake** *salt lake* South Australia
105 S10 **Torrent** *Cas.* Torrente, *var.* Torrent de l'Horta. Valenciana, E Spain 39°27′N 00°28′W
Torrent de l'Horta/Torrente *see* Torrent
40 L8 **Torreón** Coahuila, NE Mexico 25°47′N 103°21′W
105 R13 **Torre-Pacheco** Murcia, SE Spain 37°43′N 00°57′W
106 A8 **Torre Pellice** Piemonte, NE Italy 44°49′N 07°12′E
105 O13 **Torreperogil** Andalucía, S Spain 38°02′N 03°17′W
61 J15 **Torres** Rio Grande do Sul, S Brazil 29°20′S 49°43′W
Torrès, Îles *see* Torres Islands
187 Q11 **Torres Islands** *Fr.* Îles Torrès. *island group* N Vanuatu
104 G9 **Torres Novas** Santarém, C Portugal 39°28′N 08°32′W
181 V1 **Torres Strait** *strait* Australia/Papua New Guinea
104 F10 **Torres Vedras** Lisboa, C Portugal 39°05′N 09°15′W
105 S13 **Torrevieja** Valenciana, E Spain 37°59′N 00°40′W
186 B6 **Torricelli Mountains** ▲▲ NW Papua New Guinea
96 G8 **Torridon, Loch** *inlet* NW Scotland, United Kingdom
106 D9 **Torriglia** Liguria, NW Italy 44°31′N 09°08′E
104 M9 **Torrijos** Castilla-La Mancha, C Spain 39°59′N 04°18′W
18 L12 **Torrington** Connecticut, NE USA 41°48′N 73°07′W
33 Z15 **Torrington** Wyoming, C USA 42°04′N 104°10′W
Torröjen *see* Torrön
94 F16 **Torrön** *prev.* Torröjen. ◎ C Sweden
105 N15 **Torrox** Andalucía, S Spain 36°45′N 03°58′W
94 N13 **Torsåker** Gävleborg, C Sweden 60°31′N 16°30′E
95 N21 **Torsås** Kalmar, S Sweden 56°24′N 16°00′E
95 J14 **Torsby** Värmland, C Sweden 60°07′N 13°E
95 N16 **Torshälla** Södermanland, C Sweden 59°25′N 16°28′E
95 B19 **Tórshavn** *Dan.* Thorshavn. ○ Faroe Islands 62°02′N 06°47′W
Torshiz *see* Kāshmar
146 I9 **To'rtkok'l** *var.* Türtkül, *Rus.* Turtkul'; *prev.* Petroaleksandrovsk. Qoraqalpog'iston Respublikasi, W Uzbekistan 41°35′N 61°E
Tortoise Islands *see* Colón, Archipiélago de
45 T9 **Tortola** *island* C British Virgin Islands
106 D9 **Tortona** *anc.* Dertona. Piemonte, NW Italy 44°54′N 08°52′E
107 L23 **Tortorici** Sicilia, Italy, C Mediterranean Sea 38°02′N 14°49′E
105 U7 **Tortosa** *anc.* Dertosa. Cataluña, E Spain 40°49′N 00°31′E
Tortosa *see* Ţarţūs
105 U7 **Tortosa, Cap** *cape* E Spain
44 L8 **Tortue, Île de la** *var.* Tortuga Island. *island* N Haiti
55 Y10 **Tortue, Montagne** ▲ C French Guiana
Tortuga, Isla *see* La Tortuga, Isla
Tortuga Island *see* Tortue, Île de la
54 C11 **Tortugas, Golfo** *gulf* W Colombia
45 T5 **Tortuguero, Laguna** *lagoon* N Puerto Rico
137 Q12 **Tortum** Erzurum, NE Turkey 40°20′N 41°36′E
Torugart, Pereval *see* Turugart Shankou
137 O12 **Torul** Gümüşhane, NE Turkey 40°35′N 39°18′E
110 J10 **Toruń** *Ger.* Thorn. Toruń, Kujawsko-pomorskie, C Poland 53°02′N 18°36′E
95 K20 **Torup** Halland, S Sweden 56°57′N 13°04′E
118 I6 **Tõrva** *Ger.* Törwa. Valgamaa, S Estonia 58°00′N 25°54′E
Törwa *see* Tõrva
96 D13 **Tory Island** *Ir.* Toraigh. *island* NW Ireland
111 N19 **Torysa** *Hung.* Tarca. ≋ NE Slovakia
Törzburg *see* Bran
124 J16 **Torzhok** Tverskaya Oblast', W Russian Federation 57°04′N 34°55′E
164 F15 **Tosa-Shimizu** *var.* Tosasimizu. Kōchi, Shikoku, SW Japan 32°47′N 132°58′E
Tosasimizu *see* Tosa-Shimizu
164 G15 **Tosa-wan** *bay* SW Japan
83 H21 **Tosca** North-West, N South Africa 25°51′S 23°56′E
106 F12 **Toscana** *Eng.* Tuscany. ◈ *region* C Italy
107 E14 **Toscano, Archipelago** *Eng.* Tuscan Archipelago. *island group* C Italy
106 G10 **Tosco-Emiliano, Appennino** *Eng.* Tuscan-Emilian Mountains. ▲ C Italy
Tösei *see* Dongshi
165 N15 **To-shima** *island* Izu-shotō, SE Japan
147 Q9 **Toshkent** *Eng./Rus.* Tashkent. ● Toshkent Viloyati, E Uzbekistan 41°19′N 69°17′E
147 Q9 **Toshkent** ✈ Toshkent Viloyati, E Uzbekistan 41°13′N 69°15′E
147 P9 **Toshkent Viloyati** *Rus.* Tashkentskaya Oblast'. ◈ *province* E Uzbekistan
124 H13 **Tosno** Leningradskaya Oblast', NW Russian Federation 59°34′N 30°48′E
159 Q10 **Toson Hu** ◎ C China
162 H6 **Tosontsengel** Dzavhan, NW Mongolia 48°42′N 98°14′E
162 J6 **Tosontsengel** *var.* Tsengel. Hövsgöl, N Mongolia 49°29′N 101°09′E
146 I8 **Tosquduq Qumlari** *var.* Goshquduq Qum, Taskuduk, Peski. *desert* W Uzbekistan
Tossal de l'Orrí *see* Toretta de l'Orrí
61 A15 **Tostado** Santa Fe, C Argentina 29°15′S 61°45′W
118 F6 **Tõstamaa** *Ger.* Testama. Pärnumaa, SW Estonia 58°20′N 23°59′E
100 I10 **Tostedt** Niedersachsen, NW Germany 53°16′N 09°42′E
136 J11 **Tosya** Kastamonu, N Turkey 41°02′N 34°02′E
95 F15 **Totak** ◎ S Norway
105 R13 **Totana** Murcia, SE Spain 37°45′N 01°30′W
94 H13 **Toten** *physical region* S Norway
83 G18 **Toteng** North-West, C Botswana 20°25′S 23°00′E
102 M3 **Tôtes** Seine-Maritime, N France 49°40′N 01°02′E
Totigi *see* Tochigi
Totio *see* Tochio
Totis *see* Tata
189 U13 **Totiw** *island* Chuuk, C Micronesia
125 N13 **Tot'ma** *var.* Totma. Vologodskaya Oblast', NW Russian Federation 59°58′N 42°42′E
Tot'ma *see* Sukhona
55 V9 **Totness** Coronie, N Suriname 05°53′N 56°19′W
42 C5 **Totonicapán** Totonicapán, W Guatemala 14°58′N 91°12′W
42 A2 **Totonicapán** *off.* Departamento de Totonicapán. ◈ *department* W Guatemala
Totonicapán, Departamento de *see* Totonicapán
61 B18 **Totoras** Santa Fe, C Argentina 32°35′S 61°11′W
187 Y15 **Totoya** *island* S Fiji
183 Q7 **Tottenham** New South Wales, SE Australia 32°16′S 147°23′E

334

◆ Country ● Country Capital ◇ Dependent Territory ○ Dependent Territory Capital ◈ Administrative Regions ✈ International Airport ▲ Mountain ▲▲ Mountain Range ✵ Volcano ≋ River ◎ Lake ⊠ Reservoir

- 164 I12 **Tottori** Tottori, Honshū, SW Japan 35°29´N 134°14´E
- 164 H12 **Tottori** off. Tottori-ken. ◆ prefecture Honshū, SW Japan
- **Tottori-ken** see Tottori
- 76 I6 **Touâjîl** Tiris Zemmour, N Mauritania 22°03´N 12°40´W
- 76 L15 **Touba** W Ivory Coast 08°17´N 07°41´W
- 76 G11 **Touba** N Senegal 14°53´N 15°53´W
- 74 E7 **Toubkal, Jbel** ▲ W Morocco 31°00´N 07°50´W
- 32 K10 **Touchet** Washington, NW USA 46°03´N 118°40´W
- 103 P7 **Toucy** Yonne, C France 47°45´N 03°18´E
- 77 O12 **Tougan** W Burkina Faso 13°06´N 03°03´W
- 74 L7 **Touggourt** NE Algeria 33°08´N 06°04´E
- 77 Q12 **Tougouri** N Burkina Faso 13°22´N 00°25´W
- 76 J13 **Tougué** NW Guinea 11°29´N 11°48´W
- 76 K12 **Toukoto** Kayes, W Mali 13°27´N 09°52´W
- 103 S5 **Toul** Meurthe-et-Moselle, NE France 48°41´N 05°54´E
- 76 L16 **Toulépleu** var. Touloblí. W Ivory Coast 06°37´N 08°27´W
- **Touliu** see Douliu
- 15 U3 **Touhustouc** ♠ Québec, SE Canada
- **Touloblí** see Toulépleu
- 103 T16 **Toulon** anc. Telo Martius, Tilio Martius. Var, SE France 43°07´N 05°56´E
- 30 K12 **Toulon** Illinois, N USA 41°05´N 89°54´W
- 102 M15 **Toulouse** anc. Tolosa. Haute-Garonne, S France 43°37´N 01°25´E
- 102 M15 **Toulouse** ✕ Haute-Garonne, S France 43°38´N 01°19´E
- 77 N16 **Toumodi** C Ivory Coast 06°34´N 05°01´W
- 74 G9 **Tounassine, Hamada** hill range W Algeria
- **Toungoo** see Taungoo
- 166 K7 **Toungup** var. Taungup. Rakhine State, W Myanmar (Burma) 18°51´N 94°14´E
- 102 L8 **Touraine** cultural region C France
- **Tourane** see Đa Nang
- 103 P1 **Tourcoing** Nord, N France 50°44´N 03°07´E
- 104 F2 **Touriñán, Cabo** headland NW Spain 43°02´N 09°20´W
- 76 J6 **Tourine** Tiris Zemmour, N Mauritania 21°11´50´W
- 102 J3 **Tourlaville** Manche, N France 49°38´N 01°34´W
- 99 D19 **Tournai** var. Tournay, Dut. Doornik; anc. Tornacum. Hainaut, SW Belgium 50°36´N 03°24´E
- 102 L16 **Tournay** Hautes-Pyrénées, S France 43°10´N 00°01´E
- **Tournay** see Tournai
- 103 R12 **Tournon** Ardèche, E France 45°05´N 04°49´E
- 103 R9 **Tournus** Saône-et-Loire, C France 46°33´N 04°53´E
- 59 Q14 **Touros** Rio Grande do Norte, E Brazil 05°10´S 35°29´W
- 102 L8 **Tours** anc. Caesarodunum, Turoni. Indre-et-Loire, C France 47°22´N 00°40´E
- 183 Q17 **Tourville, Cape** headland Tasmania, SE Australia 42°09´S 148°20´E
- 44 M9 **Toussaint Louverture** ✕ E Haiti 18°38´N 72°13´W
- 162 L8 **Töv** ◇ province C Mongolia
- 54 H7 **Tovar** Mérida, NW Venezuela 08°22´N 71°50´W
- 126 L5 **Tovarkovskiy** Tul'skaya Oblast', W Russian Federation 53°41´N 38°18´E
- **Tovil'-Dora** see Tavildara
- **Tovis** see Teiuș
- 137 V11 **Tovuz** Rus. Tauz. W Azerbaijan 40°58´N 45°41´E
- 165 R11 **Towada** Aomori, Honshū, C Japan 40°35´N 141°13´E
- 184 K3 **Towai** Northland, North Island, New Zealand 35°29´S 174°06´E
- 18 H12 **Towanda** Pennsylvania, NE USA 41°45´N 76°25´W
- 29 W4 **Tower** Minnesota, N USA 47°48´N 92°16´W
- 171 N12 **Towera** Sulawesi, N Indonesia 0°29´S 120°01´E
- **Tower Island** see Genovesa, Isla
- 180 M13 **Tower Peak** ▲ Western Australia 33°23´S 123°27´E
- 35 U11 **Towne Pass** pass California, W USA
- 29 N3 **Towner** North Dakota, N USA 48°20´N 100°27´W
- 33 R10 **Townsend** Montana, NW USA 46°19´N 111°31´W
- 181 X6 **Townsville** Queensland, NE Australia 19°24´S 146°53´E
- **Towoeti Meer** see Towuti, Danau
- 148 K4 **Towraghoudi** Herāt, NW Afghanistan 35°13´N 62°19´E
- 21 X3 **Towson** Maryland, NE USA 39°25´N 76°36´W
- 171 O13 **Towuti, Danau** Dut. Towoeti Meer. ◎ Sulawesi, C Indonesia
- **Toxkan He** see Ak-say
- 24 K9 **Toyah** Texas, SW USA
- 165 R4 **Tōya-ko** ◎ Hokkaidō, NE Japan
- 164 L11 **Toyama** Toyama, Honshū, SW Japan 36°41´N 137°13´E
- 164 L11 **Toyama** off. Toyama-ken. ◆ prefecture Honshū, SW Japan
- 164 L11 **Toyama-wan** bay W Japan
- 164 H15 **Tōyo** Kōchi, Shikoku, SW Japan 33°22´N 134°18´E
- **Toyohara** see Yuzhno-Sakhalinsk
- 164 L14 **Toyohashi** var. Toyohasi. Aichi, Honshū, SW Japan 34°46´N 137°22´E
- **Toyohasi** see Toyohashi
- 164 L14 **Toyokawa** Aichi, Honshū, SW Japan 34°47´N 137°24´E
- 164 I14 **Toyooka** Hyōgo, Honshū, SW Japan 35°35´N 134°48´E
- 164 L13 **Toyota** Aichi, Honshū, SW Japan 35°04´N 137°09´E
- 165 T1 **Toyotomi** Hokkaidō, NE Japan 45°07´N 141°45´E
- 147 Q10 **Toýtepa** Rus. To'ytepa.
- **Toytepa** see To'ytepa

- 74 M6 **Tozeur** var. Tawzar. W Tunisia 34°00´N 08°09´E
- 39 Q8 **Tozi, Mount** ▲ Alaska, USA 65°43´N 151°01´W
- 137 Q9 **T'q'varcheli** Rus. Tkvarcheli; prev. Tqvarch'eli. NW Georgia 42°51´N 41°42´E
- **Tqvarch'eli** see T'q'varcheli
- **Tráblous** see Tripoli
- 137 O11 **Trabzon** Eng. Trebizond; anc. Trapezus. Trabzon, NE Turkey 41°N 39°43´E
- 137 O11 **Trabzon** Eng. Trebizond. ◆ province NE Turkey
- 13 P13 **Tracadie** New Brunswick, SE Canada 47°32´N 64°57´W
- 15 O11 **Tracy** Québec, SE Canada 45°59´N 73°07´W
- 35 O8 **Tracy** California, W USA 37°43´N 121°27´W
- 29 S10 **Tracy** Minnesota, N USA 44°14´N 95°37´W
- 20 K10 **Tracy City** Tennessee, S USA 35°15´N 85°44´W
- 106 D7 **Tradate** Lombardia, N Italy 45°43´N 08°57´E
- 84 F6 **Traena Bank** undersea feature N Norwegian Sea 66°15´N 09°45´E
- 29 W13 **Traer** Iowa, C USA 42°11´N 92°28´W
- 104 J16 **Trafalgar, Cabo de** headland SW Spain 36°10´N 06°03´W
- **Traiectum ad Mosam/ Traiectum Tungorum** see Maastricht
- **Traiectum Mosae** see Maastricht
- **Tráigh Mhór** see Tramore
- 58 B11 **Traíra, Serra do** ▲ NW Brazil
- 109 V5 **Traisen** Niederösterreich, NE Austria 48°03´N 15°37´E
- 109 W4 **Traisen** ≈ NE Austria
- 109 X4 **Traiskirchen** Niederösterreich, NE Austria 48°01´N 16°18´E
- **Trajani Portus** see Civitavecchia
- **Trajectum ad Rhenum** see Utrecht
- 119 H14 **Trakai** Ger. Traken, Pol. Troki. Vilnius, SE Lithuania 54°39´N 24°58´E
- **Traken** see Trakai
- 97 B20 **Tralee** Ir. Trá Lí. SW Ireland 52°16´N 09°42´W
- 97 A20 **Tralee Bay** Ir. Bá Thrá Lí. bay SW Ireland
- **Trá Lí** see Tralee
- **Trälleborg** see Trelleborg
- **Tralles Aydin** see Aydin
- 61 J16 **Tramandaí** Rio Grande do Sul, S Brazil 30°01´S 50°11´W
- 108 C7 **Tramelan** Bern, W Switzerland 47°13´N 07°07´E
- **Trá Mhór** see Tramore
- 97 E20 **Tramore** Ir. Tráigh Mhór, Trá Mhór. Waterford, S Ireland 52°10´N 07°10´W
- 114 F9 **Tran** var. Trŭn. Pernik, W Bulgaria 42°51´N 22°37´E
- 95 L18 **Tranås** Jönköping, S Sweden 58°03´N 15°00´E
- 62 J7 **Trancas** Tucumán, N Argentina 26°11´S 65°20´W
- 104 I7 **Trancoso** Guarda, N Portugal 40°46´N 07°21´W
- 95 H22 **Tranebjerg** Midtjylland, C Denmark 55°51´N 10°36´E
- 95 K19 **Tranemo** Västra Götaland, S Sweden 57°30´N 13°21´E
- 167 N16 **Trang** Trang, S Thailand 07°33´N 99°36´E
- 171 V15 **Trangan, Pulau** island Kepulauan Aru, E Indonesia
- **Tràng Định** see Thất Khê
- 183 Q7 **Trangie** New South Wales, SE Australia 32°01´S 147°58´E
- 94 K12 **Trängslet** Dalarna, C Sweden 61°22´N 13°43´E
- 107 N16 **Trani** Puglia, SE Italy 41°17´N 16°25´E
- 61 F17 **Tranqueras** Rivera, ... Uruguay 31°13´S 55°45´W
- 63 G17 **Tranquí, Isla** island S Chile
- 39 V6 **Trans-Alaska pipeline** oil pipeline Alaska, USA
- 195 Q10 **Transantarctic Mountains** ▲▲ Antarctica
- **Transcarpathian Oblast** see Zakarpats'ka Oblast'
- 122 E9 **Trans-Siberian Railway** railroad Russian Federation
- **Transilvania** see Transylvania
- **Transilvaniei, Alpi** see Carpații Meridionali
- **Transjordan** see Jordan
- 172 L11 **Transkei Basin** undersea feature SW Indian Ocean 35°30´S 29°00´E
- 117 O10 **Transnistria** cultural region E Moldavia
- 81 H18 **Trans Nzoia** ◆ county W Kenya
- **Transylvanische Alpen/ Transylvanian Alps** see Carpații Meridionali
- 94 K12 **Transtrand** Dalarna, C Sweden 61°06´N 13°19´E
- 116 I10 **Transylvania** Eng. Ardeal, Transilvania, Ger. Siebenbürgen, Hung. Erdély. cultural region NW Romania
- 167 S14 **Tra Ôn** Vinh Long, S Vietnam 09°58´N 105°58´E
- 107 H23 **Trapani** anc. Drepanum. Sicilia, Italy, C Mediterranean Sea 38°02´N 12°30´E
- **Trapang Véng** see Kâmpóng Thum
- **Trapezus** see Trabzon
- 113 L19 **Trapoklovo** Sliven, C Bulgaria 42°00´N 26°34´E
- 183 P13 **Traralgon** Victoria, SE Australia 38°15´S 146°36´E
- 76 H9 **Trarza** ◆ region SW Mauritania
- **Trasimenischersee** see Trasimeno, Lago
- 106 H12 **Trasimeno, Lago** Eng. Lake of Perugia, Ger. Trasimenischersee. ◎ C Italy
- 95 J20 **Träslövsläge** Halland, S Sweden 57°02´N 12°18´E
- **Trás-os-Montes** see Cucumbi
- 104 I6 **Trás-os-Montes e Alto Douro** former province N Portugal
- 167 Q12 **Trat** var. Bang Phra. Trat, S Thailand 12°16´N 102°30´E
- **Trá Tholl, Inis** see Inishtrahull
- 109 T4 **Traun** Oberösterreich, N Austria 48°13´N 14°15´E
- 109 S5 **Traun** ≈ N Austria
- **Traun, Lake** see Traunsee

- 101 N23 **Traunreut** Bayern, SE Germany 47°58´N 12°36´E
- 109 S5 **Traunsee** var. Gmundner See, Eng. Lake Traun. ◎ N Austria
- **Trautenau** see Trutnov
- 21 P11 **Travelers Rest** South Carolina, SE USA 34°58´N 82°26´W
- 182 L8 **Travellers Lake** seasonal lake New South Wales, SE Australia
- 31 P6 **Traverse City** Michigan, N USA 44°45´N 85°37´W
- 29 R7 **Traverse, Lake** ◎ Minnesota/South Dakota, N USA
- 185 I16 **Travers, Mount** ▲ South Island, New Zealand 42°01´S 172°46´E
- 11 P17 **Travers Reservoir** ▣ Alberta, SW Canada
- 167 T14 **Tra Vinh** var. Phu Vinh. Tra Vinh, S Vietnam 09°57´N 106°20´E
- 25 T7 **Travis, Lake** ▣ Texas, SW USA
- 112 H12 **Travnik** Federacija Bosne I Hercegovine, C Bosnia and Herzegovina 44°14´N 17°40´E
- 109 V11 **Trbovlje** Ger. Trifail. C Slovenia 46°10´N 15°03´E
- 23 V13 **Treasure Island** Florida, SE USA 27°46´N 82°46´W
- 186 I8 **Treasure State** see Montana
- 106 D9 **Trebbia** anc. Trebia. ≈ NW Italy
- 100 N8 **Trebel** ≈ NE Germany
- 103 O16 **Trèbes** Aude, S France 43°12´N 02°26´E
- **Trebia** see Trebbia
- 111 F18 **Třebíč** Ger. Trebitsch. Vysočina, C Czech Republic 49°13´N 15°52´E
- 113 I16 **Trebinje** Republika Srpska, S Bosnia and Herzegovina 42°42´N 18°19´E
- 113 H16 **Trebišnjica** var. Trebišnica. ≈ S Bosnia and Herzegovina
- 111 N20 **Trebišov** Hung. Tőketerebes. Košický Kraj, E Slovakia 48°37´N 21°44´E
- **Trebitsch** see Třebíč
- **Trebizond** see Trabzon
- 109 V12 **Trebnje** SE Slovenia 45°54´N 15°01´E
- 111 D19 **Třeboň** Ger. Wittingau. Jihočeský Kraj, S Czech Republic 49°00´N 14°46´E
- 104 J15 **Trebujena** Andalucía, S Spain 36°52´N 06°11´W
- 108 I7 **Treene** ≈ N Germany
- **Tree Planters State** see Nebraska
- 101 S9 **Treffen** Kärnten, S Austria 46°40´N 13°51´E
- **Trefynwy** see Monmouth
- 102 G5 **Tréguier** Côtes d'Armor, NW France 48°50´N 03°12´W
- 61 G18 **Treinta y Tres** Treinta y Tres, E Uruguay 33°16´S 54°17´W
- 61 F18 **Treinta y Tres** ◆ department E Uruguay
- 122 F11 **Trëkhgornyy** Chelyabinskaya Oblast', C Russian Federation 54°42´N 58°25´E
- 114 F9 **Treklyanska Reka** ≈ W Bulgaria
- 102 K8 **Trélazé** Maine-et-Loire, NW France 47°27´N 00°28´E
- 63 K17 **Trelew** Chubut, SE Argentina 43°13´S 65°15´W
- 95 K23 **Trelleborg** var. Trälleborg. Skåne, S Sweden 55°22´N 13°10´E
- 113 P15 **Trem** ▲ SE Serbia 43°10´N 22°12´E
- 15 N12 **Tremblant, Mont** ▲ Québec, SE Canada 46°13´N 74°34´W
- 99 H17 **Tremelo** Vlaams Brabant, C Belgium 51°N 04°54´E
- 107 M15 **Tremiti, Isole** island group SE Italy
- 30 K12 **Tremont** Illinois, N USA 40°30´N 89°31´W
- 36 L1 **Tremonton** Utah, W USA 41°42´N 112°09´W
- 105 U4 **Tremp** Cataluña, NE Spain 42°09´N 00°53´E
- 30 J7 **Trempealeau** Wisconsin, N USA 44°00´N 91°25´W
- 15 P8 **Trenche, Lac** ◎ Québec, SE Canada
- 111 I20 **Trenčiansky Kraj** ◆ region W Slovakia
- 111 I19 **Trenčín** Ger. Trentschin, Hung. Trencsén. Trenčiansky Kraj, W Slovakia 48°54´N 18°03´E
- **Trencsén** see Trenčín
- **Trengganu** see Terengganu
- 61 A21 **Trenque Lauquen** Buenos Aires, E Argentina 36°01´S 62°47´W
- 14 J14 **Trent** ♠ Ontario, SE Canada
- 97 N18 **Trent** ≈ C England, United Kingdom
- **Trent** see Trento
- 106 F5 **Trentino-Alto Adige** Eng. Trentino-High Adige; prev. Venezia Tridentina. ◆ region N Italy
- **Trentino-Südtirol** see Trentino-Alto Adige
- 106 G6 **Trento** Eng. Trent, Ger. Trient; anc. Tridentum. Trentino-Alto Adige, N Italy 46°05´N 11°08´E
- 14 J14 **Trenton** Ontario, SE Canada 44°07´N 77°34´W
- 23 R1 **Trenton** Georgia, SE USA 34°52´N 85°27´W
- 31 S10 **Trenton** Michigan, N USA 42°08´N 83°10´W
- 21 S2 **Trenton** Missouri, C USA 40°N 93°37´W
- 28 M17 **Trenton** Nebraska, C USA 40°10´N 101°00´W
- 18 J15 **Trenton** state capital New Jersey, NE USA 40°13´N 74°45´W
- 21 W10 **Trenton** North Carolina, SE USA 35°05´N 77°20´W
- 20 G9 **Trenton** Tennessee, S USA 35°59´N 88°59´W
- 36 L1 **Trenton** Utah, W USA 41°55´N 111°57´W
- **Trentschin** see Trenčín

- 61 C23 **Tres Arroyos** Buenos Aires, E Argentina 38°22´S 60°17´W
- 61 J15 **Três Cachoeiras** Rio Grande do Sul, S Brazil 29°21´S 49°49´W
- 106 E7 **Trescore Balneario** Lombardia, N Italy 45°43´N 09°52´E
- 41 V17 **Tres Cruces, Cerro** ▲ SE Mexico 15°25´N 92°27´W
- 57 K18 **Tres Cruces, Cordillera** ▲▲ W Bolivia
- 113 N18 **Treska** ≈ N FYR Macedonia
- 113 I14 **Treskavica** ▲ SE Bosnia and Herzegovina
- 59 J20 **Três Lagoas** Mato Grosso do Sul, SW Brazil 20°46´S 51°43´W
- 40 H12 **Tres Marías, Islas** island group C Mexico
- 59 M19 **Três Marias, Represa** ▣ SE Brazil
- 63 F20 **Tres Montes, Península** headland S Chile 46°49´S 75°29´W
- 105 O3 **Trespaderne** Castilla y León, N Spain 42°47´N 03°24´W
- 60 G13 **Três Passos** Rio Grande do Sul, S Brazil 27°33´S 53°55´W
- 61 A23 **Tres Picos, Cerro** ▲ E Argentina 38°10´S 61°54´W
- 63 G17 **Tres Picos, Cerro** ▲ SW Argentina
- 60 I12 **Três Pinheiros** Paraná, S Brazil 25°25´S 51°51´W
- 59 M21 **Três Pontas** Minas Gerais, SE Brazil 21°33´S 45°18´W
- **Três Puntas, Punta** see Manabíque, Punta
- 60 P9 **Três Rios** Rio de Janeiro, SE Brazil 22°06´S 43°15´W
- **Tres Tabernae** see Saverne
- **Trestenberg/Trestendorf** see Tășnad
- 41 R15 **Tres Valles** Veracruz-Llave, SE Mexico 18°14´N 96°03´W
- 94 H12 **Tretten** Oppland, S Norway 61°19´N 10°19´E
- 101 K21 **Treuchtlingen** Bayern, S Germany 48°57´N 10°55´E
- 100 N13 **Treuenbrietzen** Brandenburg, E Germany 52°06´N 12°52´E
- 63 H17 **Trevelín** Chubut, S Argentina 43°02´S 71°27´W
- **Treves/Trèves** see Trier
- 106 E7 **Trevi** Umbria, C Italy 42°52´N 12°46´E
- 106 E7 **Treviglio** Lombardia, N Italy 45°32´N 09°35´E
- 104 J4 **Trevinca, Peña** ▲ NW Spain 42°10´N 06°49´W
- 105 P3 **Treviño** Castilla y León, N Spain 42°45´N 02°43´W
- 106 H7 **Treviso** anc. Tarvisium. Veneto, NE Italy 45°40´N 12°15´E
- 97 G24 **Trevose Head** headland SW England, United Kingdom 50°33´N 05°03´W
- 115 F20 **Trípoli** prev. Trípolis. Peloponnisos, S Greece 37°31´N 22°22´E
- 21 W4 **Triangle** Virginia, NE USA 38°30´N 77°17´W
- 83 L18 **Triangle** Masvingo, SE Zimbabwe 21°00´S 31°28´E
- 115 L23 **Tría Nisiá** island Kykládes, Greece, Aegean Sea
- 101 G23 **Triberg im Schwarzwald** Baden-Württemberg, SW Germany 48°07´N 08°13´E
- 153 V15 **Tripura** var. Hill Tippera. ◆ state NE India
- 108 K8 **Trisanna** ≈ W Austria
- 100 H8 **Trischen** island NW Germany
- 89 I14 **Tristan da Cunha** ◇ dependency of Saint Helena SE Atlantic Ocean
- 67 P15 **Tristan da Cunha** island SW Atlantic Ocean
- 89 I14 **Tristan da Cunha** see Saint Helena, Ascension and Tristan da Cunha
- 65 L18 **Tristan da Cunha Fracture Zone** tectonic feature S Atlantic Ocean
- 167 S14 **Tri Tôn** An Giang, S Vietnam 10°26´N 105°01´E
- 167 S14 **Tri Tôn, Đao** see Triton Island
- 167 W10 **Triton Island** Chin. Zhongjian Dao, Viet. Đao Tri Tôn. island S Paracel Islands
- 155 G24 **Trivandrum** Thiruvananthapuram, Tiruvantapuram. state capital Kerala, SW India see also Thiruvananthapuram
- 101 H20 **Trnava** Ger. Tyrnau, Hung. Nagyszombat. Trnavský Kraj, W Slovakia 48°23´N 17°36´E
- 101 H20 **Trnavský Kraj** ◆ region W Slovakia
- **Trnovo** see Veliko Tarnovo
- 101 D19 **Trier** Eng./Fr. Trèves; anc. Augusta Treverorum. Rheinland-Pfalz, SW Germany 49°45´N 06°39´E
- 106 K7 **Trieste** Slvn. Trst. Friuli-Venezia Giulia, NE Italy 45°39´N 13°45´E
- 106 J8 **Trieste, Golfo di/Triest, Golf von** see Trieste, Gulf of
- 112 F13 **Trieste, Gulf of** Cro. Tršćanski Zaljev, Ger. Golf von Triest, It. Golfo di Trieste, Slvn. Tržaški Zaliv. gulf S Europe
- **Trieste** see Trieste
- 109 W4 **Triesting** ≈ W Austria
- **Trièu Hai** see Quang Tri
- **Trifail** see Trbovlje
- 116 L9 **Trifești** Iași, NE Romania 47°27´N 27°44´E
- 109 S10 **Triglav** ▲ NW Slovenia 46°23´N 13°50´E
- 109 S10 **Triglav** see Triglav
- 104 I14 **Trigueros** Andalucía, S Spain 37°24´N 06°50´W
- 74 H5 **Trois Fourches, Cap des** headland NE Morocco
- 115 E16 **Trikala** prev. Trikkala. Thessalía, C Greece 39°33´N 21°46´E
- 115 E17 **Trikeriótis** ≈ C Greece
- **Trikkala** see Trikala
- **Trikomo/Trikomon** see Iskele
- 97 F17 **Trim** Ir. Baile Átha Troim. Meath, E Ireland 53°34´N 06°47´W
- 122 F11 **Troitsk** Chelyabinskaya Oblast', C Russian Federation 54°04´N 61°31´E

- 29 U11 **Trimont** Minnesota, N USA 43°45´N 94°42´W
- **Trimontium** see Plovdiv
- **Trinacria** see Sicilia
- 155 K24 **Trincomalee** var. Trinkomali. Eastern Province, NE Sri Lanka 08°34´N 81°13´E
- 65 K16 **Trindade, Ilha da** island Brazil, W Atlantic Ocean
- 47 Y9 **Trindade Spur** undersea feature SW Atlantic Ocean 21°00´S 35°00´W
- 111 J17 **Třinec** Ger. Trzynietz. Moravskoslezský Kraj, E Czech Republic 49°41´N 18°39´E
- 57 M16 **Trinidad** El Beni, N Bolivia 14°52´S 64°54´W
- 54 H9 **Trinidad** Casanare, E Colombia 05°25´N 71°39´W
- 44 E6 **Trinidad** Sancti Spíritus, C Cuba 21°48´N 80°00´W
- 61 E19 **Trinidad** Flores, S Uruguay 33°35´S 56°54´W
- 37 U8 **Trinidad** Colorado, C USA 37°11´N 104°31´W
- 45 Y17 **Trinidad** island C Trinidad and Tobago
- **Trinidad** see Jose Abad Santos
- 45 Y16 **Trinidad and Tobago** off. Republic of Trinidad and Tobago. ◆ republic SE West Indies
- **Trinidad and Tobago, Republic of** see Trinidad and Tobago
- 63 F22 **Trinidad, Golfo** gulf S Chile
- 61 B24 **Trinidad, Isla** island E Argentina
- 107 N16 **Trinitapoli** Puglia, SE Italy 41°22´N 16°06´E
- 55 X10 **Trinité, Montagnes de la** ▲ C French Guiana
- 55 W9 **Trinity** Texas, SW USA 30°57´N 95°22´W
- 35 N5 **Trinity Bay** inlet Newfoundland, Newfoundland and Labrador, E Canada
- 39 P15 **Trinity Islands** island group Alaska, USA
- 35 N2 **Trinity Mountains** ▲ California, W USA
- 35 S4 **Trinity Peak** ▲ Nevada, W USA 40°13´N 118°43´W
- 35 S5 **Trinity Range** ▲ Nevada, W USA
- 35 N2 **Trinity River** ≈ California, W USA
- 25 V8 **Trinity River** ≈ Texas, SW USA
- **Trinkomali** see Trincomalee
- 173 Y15 **Triolet** NW Mauritius
- 107 O20 **Trionto, Capo** headland S Italy 39°37´N 16°46´E
- 117 T4 **Trostyanets'** Rus. Trostyanets. Sums'ka Oblast', NE Ukraine 50°30´N 34°59´E
- 117 N7 **Trostyanets'** Rus. Trostyanets. Vinnyts'ka Oblast', C Ukraine 48°35´N 29°10´E
- **Trostyanets** see Trostyanets'
- 116 L11 **Trotuș** ≈ E Romania
- 44 M8 **Trou-du-Nord** N Haiti 19°34´N 71°57´W
- 25 W7 **Troup** Texas, SW USA
- 162 G7 **Tsagaanchuluut** Dzavhan, C Mongolia 47°06´N 96°04´E
- 162 M8 **Tsagaandelger** var. Haraat. Dundgovĭ, C Mongolia 46°30´N 107°12´E
- **Tsagaanders** see Bayantümen
- 162 J5 **Tsagaan-Üür** var. Bulgan. Hövsgöl, N Mongolia 50°30´N 101°28´E
- 127 P12 **Tsagan Aman** Respublika Kalmykiya, SW Russian Federation 47°37´N 46°43´E
- 23 V11 **Tsala Apopka Lake** ◎ Florida, SE USA
- **Tsamkong** see Zhanjiang
- **Tsangpo** see Brahmaputra
- **Tsant** see Deren
- **Tsao** see Tsau
- 172 J4 **Tsaratanana** Mahajanga, C Madagascar 16°46´S 47°40´E
- 114 N10 **Tsarevo** prev. Michurin. Burgas, E Bulgaria 42°10´N 27°51´E
- **Tsaribrod** see Dimitrovgrad
- **Tsarigrad** see Istanbul
- 114 K7 **Tsar Kaloyan** Ruse, N Bulgaria 43°36´N 26°14´E
- 117 T7 **Tsarychanka** Dnipropetrovs'ka Oblast', E Ukraine 48°57´N 34°29´E
- 83 H21 **Tsau** var. Tsao. North-West, NW Botswana 20°08´S 22°29´E
- 81 J20 **Tsavo** Taita/Taveta, S Kenya 02°59´S 38°28´E
- 81 J19 **Tsawisis** Karas, S Namibia 26°18´S 18°09´E
- **Tschakathurn** see Čakovec
- **Tschaslau** see Čáslav
- **Tschenstochau** see Częstochowa
- **Tschernembl** see Črnomelj
- 28 K6 **Tschida, Lake** ◎ North Dakota, N USA
- 162 G8 **Tseel** Govĭ-Altay, SW Mongolia 45°35´N 95°54´E
- 138 G8 **Tsefat** var. Safed, Ar. Safad; prev. Zefat. N Israel 32°57´N 35°27´E
- 126 J13 **Tselina** Rostovskaya Oblast', SW Russian Federation 46°31´N 41°01´E
- **Tselinograd** see Astana
- **Tselinogradskaya Oblast** see Akmola
- **Tsengel** see Tosontsengel
- 163 N8 **Tsenher** var. Altan-Ovoo. Arhangay, C Mongolia
- **Tsenher** see Mönhhayrhan
- **Tsenhermandal** var. Modon. Hentiy, C Mongolia 47°45´N 109°23´E
- **Tsentral'nyye Nizmennyye Garagumy** see Merkezi Garagumy

- 54 I6 **Trujillo**, NW Venezuela 09°20´N 70°38´W
- 54 I6 **Trujillo** off. Estado Trujillo. ◇ state W Venezuela
- **Trujillo, Estado** see Trujillo
- **Truk** see Chuuk
- **Truk Islands** see Chuuk
- 29 U10 **Truman** Minnesota, N USA 43°49´N 94°26´W
- 27 X10 **Trumann** Arkansas, C USA 35°40´N 90°30´W
- 36 J9 **Trumbull, Mount** ▲ Arizona, SW USA 36°22´N 113°09´W
- **Trŭn** see Tran
- 183 Q8 **Trundle** New South Wales, SE Australia 32°55´S 147°42´E
- 129 U13 **Trung Phân** physical region S Vietnam
- **Trṇg Sa L n, Đao** see Spratly Island
- **Trṇg Sa, Quần Đao** see Spratly Islands
- **Trupčilar** see Orlyak
- 13 Q15 **Truro** Nova Scotia, SE Canada 45°24´N 63°18´W
- 97 H25 **Truro** SW England, United Kingdom 50°16´N 05°03´W
- 25 P5 **Truscott** Texas, SW USA 33°43´N 99°48´W
- 116 K9 **Trușești** Botoșani, NE Romania 47°45´N 27°01´E
- 116 H6 **Truskavets'** L'vivs'ka Oblast', W Ukraine 49°17´N 23°30´E
- 10 M11 **Trutch** British Columbia, W Canada 57°42´N 123°00´W
- 37 Q14 **Truth or Consequences** New Mexico, SW USA 33°07´N 107°15´W
- 111 F15 **Trutnov** Ger. Trautenau. Královéhradecký Kraj, N Czech Republic 50°34´N 15°55´E
- 103 P3 **Truyère** ≈ C France
- 114 K9 **Tryavna** Lovech, N Bulgaria 42°52´N 25°30´E
- 28 M11 **Tryon** Nebraska, C USA 41°33´N 100°57´W
- 115 J16 **Trypití, Akrotírio** var. Ákra Tripití. headland Ágios Efstrátios, E Greece
- 94 J12 **Trysil** Hedmark, S Norway 61°18´N 12°16´E
- 94 I11 **Trysilelva** ≈ S Norway
- 112 D10 **Tržac** Federacija Bosni I Hercegovine, NW Bosnia and Herzegovina 44°58´N 15°48´E
- **Tržaški Zaliv** see Trieste, Gulf of
- 110 G10 **Trzcianka** Ger. Schönlanke. Pila, Wielkopolskie, C Poland 53°02´N 16°24´E
- 110 E7 **Trzebiatów** Ger. Treptow an der Rega. Zachodnio-pomorskie, NW Poland 54°04´N 15°14´E
- 111 G14 **Trzebnica** Ger. Trebnitz. Dolnośląskie, SW Poland 51°19´N 17°03´E
- 109 T10 **Tržič** Ger. Neumarktl. NW Slovenia 46°22´N 14°17´E
- 83 G21 **Tsabong** see Tshabong.
- 162 G7 **Tsagaanchuluut** Dzavhan, C Mongolia 47°06´N 96°04´E

◆ Country	◇ Dependent Territory	◆ Administrative Regions	▲ Mountain	▲ Volcano	◎ Lake
● Country Capital	○ Dependent Territory Capital	✕ International Airport	▲▲ Mountain Range	≈ River	▣ Reservoir

TSES – TUZLA

83 E21 **Tses** Karas, S Namibia *25°58′S 18°08′E*
Tseshevlya *see* Tsyeshawlya
162 E7 **Tsetseg** *var.* Tsetsegnuur. Hovd, W Mongolia *46°30′N 93°16′E*
Tsetsegnuur *see* Tsetseg
Tsetsen Khan *see* Öndörhaan
162 J8 **Tsetserleg** Arhangay, C Mongolia *47°29′N 101°19′E*
162 H6 **Tsetserleg** *var.* Halban. Hövsgöl, N Mongolia *49°30′N 97°33′E*
162 J8 **Tsetserleg** *var.* Hujirt. Övörhangay, C Mongolia *46°50′N 102°38′E*
77 R16 **Tsévié** S Togo *06°25′N 01°13′E*
Tshabong *see* Tsabong
83 G20 **Tshane** Kgalagadi, SW Botswana *24°05′S 21°54′E*
Tshangalele, Lac *see* Lufira, Lac de Retenue de la
83 H17 **Tshauxaba** Central, C Botswana *19°56′S 25°09′E*
79 F21 **Tshela** Bas-Congo, W Dem. Rep. Congo *04°56′S 13°02′E*
79 K22 **Tshibala** Kasai-Occidental, S Dem. Rep. Congo *06°53′S 22°01′E*
79 J22 **Tshikapa** Kasai-Occidental, SW Dem. Rep. Congo *06°23′S 20°47′E*
79 L22 **Tshilenge** Kasai Oriental, S Dem. Rep. Congo *06°17′S 23°48′E*
79 L24 **Tshimbalanga** Katanga, S Dem. Rep. Congo *09°42′S 23°04′E*
79 L22 **Tshimbulu** Kasai-Occidental, S Dem. Rep. Congo *06°27′S 22°54′E*
Tshiumbe *see* Chiumbe
79 M21 **Tshofa** Kasai-Oriental, C Dem. Rep. Congo *05°13′S 25°13′E*
79 K18 **Tshuapa** ≈ C Dem. Rep. Congo
114 G7 **Tsibritsa** ≈ NW Bulgaria
Tsien Tang *see* Fuchun Jiang
114 I12 **Tsigansko Gradishte** *Gr.* Giftokastro. ▲ Bulgaria/Greece *41°24′N 24°41′E*
Tsihombe *see* Tsiombe
8 H7 **Tsiigehtchic** *prev.* Arctic Red River. Northwest Territories, NW Canada *67°24′N 133°40′W*
125 Q7 **Tsil'ma** ≈ NW Russian Federation
119 J17 **Tsimkavichy** *Rus.* Timkovichi. Minskaya Voblasts', C Belarus *53°04′N 26°59′E*
126 M11 **Tsimlyansk** Rostovskaya Oblast', SW Russian Federation *47°39′N 42°05′E*
127 N11 **Tsimlyanskoye Vodokhranilishche** *var.* Tsimlyansk Vodoskhovshche, *Eng.* Tsimlyansk Reservoir. ⊠ SW Russian Federation
Tsimlyansk Reservoir *see* Tsimlyanskoye Vodokhranilishche
Tsimlyansk Vodoskhovshche *see* Tsimlyanskoye Vodokhranilishche
Tsinan *see* Jinan
Tsing Hai *see* Qinghai Hu, China
Tsinghai *see* Qinghai, China
Tsingtao/Tsingtau *see* Qingdao
Tsingyuan *see* Baoding
Tsinkiang *see* Quanzhou
Tsintao *see* Qingdao
83 D17 **Tsintsabis** Oshikoto, N Namibia *18°45′S 17°51′E*
172 H8 **Tsiombe** *var.* Tsihombe. Toliara, S Madagascar
123 O13 **Tsipa** ≈ S Russian Federation
172 H5 **Tsiribihina** ≈ W Madagascar
172 I5 **Tsiroanomandidy** Antananarivo, C Madagascar *18°44′S 46°02′E*
189 U13 **Tsis** *island* Chuuk, C Micronesia
Tsitsihar *see* Qiqihar
127 Q3 **Tsivil'sk** Chuvashskaya Respublika, W Russian Federation *55°51′N 47°30′E*
137 T9 **Tskhinvali** *prev.* Staliniri, Ts'khinvali. C Georgia *42°12′N 43°58′E*
119 J19 **Tsna** ≈ SW Belarus
124 I15 **Tsna** *var.* Zna. ≈ W Russian Federation
162 G9 **Tsogt** *var.* Tahilt. Govĭ-Altay, W Mongolia *45°20′N 96°42′E*
162 K10 **Tsogt-Ovoo** *var.* Doloon. Ömnögovĭ, S Mongolia *44°28′N 105°22′E*
162 L10 **Tsogttsetsiy** *var.* Baruunsuu. Ömnögovĭ, S Mongolia *43°46′N 105°28′E*
114 M9 **Tsonevo, Yazovir** *prev.* Yazovir Georgi Traykov. ⊠ NE Bulgaria
Tsoohor *see* Hürmen
164 K14 **Tsu** *var.* Tu. Mie, Honshū, SW Japan *34°41′N 136°30′E*
165 O10 **Tsubame** *var.* Tubame. Niigata, Honshū, C Japan *37°40′N 138°56′E*
165 V3 **Tsubetsu** Hokkaidō, NE Japan *43°43′N 144°01′E*
165 O13 **Tsuchiura** *var.* Tutiura. Ibaraki, Honshū, S Japan *36°05′N 140°11′E*
165 Q6 **Tsugaru-kaikyō** *strait* N Japan
164 E14 **Tsukumi** *var.* Tukumi. Ōita, Kyūshū, SW Japan *33°00′N 131°51′E*
Tsul-Ulaan *see* Bayannuur
Tsul-Ulaan *see* Bayannuur
83 D17 **Tsumeb** Oshikoto, N Namibia *19°13′S 17°42′E*
83 F17 **Tsumkwe** Otjozondjupa, NE Namibia *19°37′S 20°30′E*
164 D15 **Tsuno** Miyazaki, Kyūshū, SW Japan *32°43′N 131°32′E*
164 D12 **Tsuno-shima** *island* SW Japan
164 K12 **Tsuruga** *var.* Turuga. Fukui, Honshū, SW Japan *35°38′N 136°01′E*
164 H12 **Tsurugi-san** ▲ Shikoku, SW Japan *33°50′N 134°04′E*
165 P9 **Tsuruoka** *var.* Turuoka. Yamagata, Honshū, C Japan *38°44′N 139°48′E*
164 C12 **Tsushima** *prev.* Izuhara. Nagasaki, Tsushima, SW Japan *34°11′N 129°16′E*
164 C12 **Tsushima** *var.* Tsushima-tō, Tusima. *island group* SW Japan
Tsushima-tō *see* Tsushima
164 H12 **Tsuyama** *var.* Tuyama. Okayama, Honshū, SW Japan *35°04′N 134°01′E*
83 G19 **Tswaane** Ghanzi, W Botswana *22°21′S 21°52′E*
119 N16 **Tsyakhtsin** *Rus.* Tekhtin. Mahilyowskaya Voblasts', E Belarus *53°51′N 29°44′E*
119 P19 **Tsyerakhowka** *Rus.* Terekhovka. Homyel'skaya Voblasts', SE Belarus *52°13′N 31°24′E*
119 I17 **Tsyeshawlya** *prev.* Cheshevlya, Tseshevlya, *Rus.* Tesheyle. Brestskaya Voblasts', SW Belarus *53°14′N 25°49′E*
Tsyurupinsk *see* Tsyurupyns'k
117 R10 **Tsyurupyns'k** *Rus.* Tsyurupinsk. Khersons'ka Oblast', S Ukraine *46°35′N 32°43′E*
Tu *see* Tsu
186 C7 **Tua** ≈ C Papua New Guinea
Tuaim *see* Tuam
184 L6 **Tuakau** Waikato, North Island, New Zealand *37°16′S 174°56′E*
97 C17 **Tuam** *Ir.* Tuaim. Galway, W Ireland *53°31′N 08°50′W*
185 K14 **Tuamarina** Marlborough, South Island, New Zealand *41°27′S 174°00′E*
Tuamotu, Archipel des *see* Tuamotu, Îles
193 Q9 **Tuamotu Fracture Zone** *tectonic feature* E Pacific Ocean
191 W9 **Tuamotu, Îles** *var.* Archipel des Tuamotu, Dangerous Archipelago, Tuamotu Islands. *island group* N French Polynesia
Tuamotu Islands *see* Tuamotu, Îles
175 X10 **Tuamotu Ridge** *undersea feature* C Pacific Ocean
167 R5 **Tuân Giao** Lai Châu, N Vietnam *21°34′N 103°24′E*
171 O2 **Tuao** Luzon, N Philippines *17°42′N 121°25′E*
190 B15 **Tuapa** NW Niue *18°57′S 169°59′W*
43 N7 **Tuapí** Región Autónoma Atlántico Norte, NE Nicaragua *14°10′N 83°20′W*
126 K15 **Tuapse** Krasnodarskiy Kray, SW Russian Federation *44°08′N 39°07′E*
169 U6 **Tuaran** Sabah, East Malaysia *06°12′N 116°12′E*
106 I7 **Tua, Rio** ≈ N Portugal
192 H15 **Tuasivi** Savai'i, C Samoa *13°38′S 172°08′W*
185 B24 **Tuatapere** Southland, South Island, New Zealand *46°09′S 167°43′E*
36 M9 **Tuba City** Arizona, SW USA *36°08′N 111°14′W*
138 H11 **Ţubah, Qaşr aţ** *castle* 'Ammān, C Jordan
Tubame *see* Tsubame
169 R16 **Tuban** *prev.* Toeban. Jawa, C Indonesia *06°55′S 112°01′E*
141 O16 **Tuban, Wādī** *dry watercourse* SW Yemen
61 K14 **Tubarão** Santa Catarina, S Brazil *28°29′S 49°00′W*
98 O10 **Tubbergen** Overijssel, E Netherlands *52°25′N 06°46′E*
Tubeke *see* Tubize
101 H22 **Tübingen** *var.* Tuebingen. Baden-Württemberg, SW Germany *48°32′N 09°04′E*
127 W6 **Tubinskiy** Respublika Bashkortostan, W Russian Federation *52°48′N 58°18′E*
99 G19 **Tubize** *Dut.* Tubeke. Walloon Brabant, C Belgium *50°43′N 04°14′E*
76 J16 **Tubmanburg** NW Liberia *06°50′N 10°53′W*
75 T7 **Ţubruq** *Eng.* Tobruk, *It.* Tobruch. NE Libya *32°05′N 23°59′E*
191 T13 **Tubuai** *island* Îles Australes, SW French Polynesia
Tubuai, Îles/Tubuai Islands *see* Australes, Îles
Tubuai-Manu *see* Maiao
40 F3 **Tubutama** Sonora, NW Mexico *30°51′N 111°31′W*
54 K4 **Tucacas** Falcón, N Venezuela *10°50′N 68°22′W*
59 P16 **Tucano** Bahia, E Brazil *10°52′S 38°48′W*
57 P19 **Tucavaca, Río** ≈ E Bolivia
110 H8 **Tuchola** Kujawsko-pomorskie, C Poland *53°36′N 17°50′E*
111 M17 **Tuchów** Małopolskie, S Poland *49°53′N 21°04′E*
23 S3 **Tucker** Georgia, SE USA *33°53′N 84°10′W*
27 W10 **Tuckerman** Arkansas, C USA *35°43′N 91°12′W*
64 B12 **Tucker's Town** E Bermuda *32°20′N 64°42′W*
Tuckum *see* Tukums
36 M15 **Tucson** Arizona, SW USA *32°14′N 111°01′W*
62 J7 **Tucumán** *off.* Provincia de Tucumán. ◈ *province* N Argentina
Tucumán *see* San Miguel de Tucumán
Tucumán, Provincia de *see* Tucumán
37 V11 **Tucumcari** New Mexico, SW USA *35°10′N 103°43′W*
58 H13 **Tucunaré** Pará, N Brazil *05°15′S 55°49′W*
55 Q6 **Tucupita** Delta Amacuro, NE Venezuela *09°02′N 62°04′W*
58 K13 **Tucuruí, Represa de** ⊠ NE Brazil
110 F9 **Tuczno** Zachodnio-pomorskie, NW Poland *53°12′N 16°08′E*
Tuddo *see* Tudu
105 Q5 **Tudela** *Basq.* Tutera; *anc.* Tutela. Navarra, N Spain *42°04′N 01°37′W*
104 M6 **Tudela de Duero** Castilla y León, N Spain *41°35′N 04°34′W*
162 G6 **Tüdevtey** *var.* Oygon. Dzavhan, N Mongolia *48°57′N 96°33′E*
138 K6 **Tudmur** *var.* Tadmur, Tamar, *Gk.* Palmyra, *Bibl.* Tadmor. Ḥimş, C Syria *34°36′N 38°15′E*
118 J4 **Tudu** *Ger.* Tuddo. Lääne-Virumaa, NE Estonia *59°12′N 26°52′E*
Tuebingen *see* Tübingen
122 J14 **Tuekta** Respublika Altay, S Russian Federation *50°51′N 85°52′E*
104 I5 **Tuela, Rio** ≈ N Portugal
153 X12 **Tuensang** Nāgāland, NE India *26°16′N 94°45′E*
136 L15 **Tufanbeyli** Adana, C Turkey *38°15′N 36°13′E*
Tüffer *see* Laško
186 F9 **Tufi** Northern, S Papua New Guinea *09°08′S 149°20′E*
193 O3 **Tufts Plain** *undersea feature* N Pacific Ocean
Tugalan *see* Kolkhozobod
67 V14 **Tugela** ≈ SE South Africa
21 P6 **Tug Fork** ≈ S USA
39 P15 **Tugidak Island** *island* Trinity Islands, Alaska, USA
171 O2 **Tuguegarao** Luzon, N Philippines *17°37′N 121°48′E*
123 S12 **Tugur** Khabarovskiy Kray, SE Russian Federation *53°43′N 137°00′E*
161 P4 **Tuhai He** ≈ E China
104 G4 **Tui** Galicia, NW Spain *42°02′N 08°37′W*
77 O13 **Tui** *var.* Grand Balé. ≈ W Burkina Faso
57 J16 **Tuichi, Río** ≈ W Bolivia
64 Q11 **Tuineje** Fuerteventura, Islas Canarias, Spain, NE Atlantic Ocean *28°18′N 14°03′W*
43 X16 **Tuira, Río** ≈ SE Panama
Tuisarkan *see* Tūysarkān
Tujiabu *see* Yongxiu
127 W5 **Tukan** Respublika Bashkortostan, W Russian Federation *53°58′N 57°29′E*
171 P14 **Tukangbesi, Kepulauan** *Dut.* Toekang Besi Eilanden. *island group* C Indonesia
147 V13 **Tŭkhtamish** *Rus.* Toktomush; *prev.* Tokhtamyshbek. SE Tajikistan *37°51′N 74°41′E*
184 O12 **Tukituki** ≈ North Island, New Zealand
Tu-k'ou *see* Panzhihua
121 P12 **Tūkrah** NE Libya *32°32′N 20°35′E*
8 H6 **Tuktoyaktuk** Northwest Territories, NW Canada *69°27′N 133°W*
168 I9 **Tuktuk** Pulau Samosir, W Indonesia *02°32′N 98°43′E*
Tukumi *see* Tsukumi
118 E9 **Tukums** *Ger.* Tuckum. W Latvia *56°58′N 23°12′E*
81 G24 **Tukuyu** *prev.* Neu-Langenburg. Mbeya, S Tanzania *09°14′S 33°39′E*
Tukzār *see* Tokzār
41 O13 **Tula** *var.* Tula de Allende. Hidalgo, C Mexico *20°01′N 99°21′W*
41 O11 **Tula** Tamaulipas, C Mexico *22°59′N 99°43′W*
126 K5 **Tula** Tul'skaya Oblast', W Russian Federation *54°11′N 37°39′E*
Tulach Mhór *see* Tullamore
Tula de Allende *see* Tula
186 M9 **Tulaghi** *var.* Tulagi. Florida Islands, C Solomon Islands *09°04′S 160°09′E*
Tulagi *see* Tulaghi
159 N10 **Tulagt Ar Gol** ≈ W China
41 P13 **Tulancingo** Hidalgo, C Mexico *20°34′N 98°25′W*
35 R11 **Tulare** California, W USA *36°12′N 119°21′W*
29 P9 **Tulare** South Dakota, N USA *44°43′N 98°29′W*
35 Q12 **Tulare Lake Bed** *salt flat* California, W USA
37 S14 **Tularosa** New Mexico, SW USA *33°04′N 106°01′W*
37 P13 **Tularosa Mountains** ▲ New Mexico, SW USA
37 S15 **Tularosa Valley** *basin* New Mexico, SW USA
83 E25 **Tulbagh** Western Cape, SW South Africa *33°17′S 19°09′E*
56 C5 **Tulcán** Carchi, N Ecuador *00°44′N 77°43′W*
117 N13 **Tulcea** Tulcea, E Romania *45°11′N 28°49′E*
117 N13 **Tulcea** ◈ *county* SE Romania
Tul'chin *see* Tul'chyn
117 N7 **Tul'chyn** *Rus.* Tul'chin. Vinnyts'ka Oblast', C Ukraine *48°40′N 28°49′E*
Tuléar *see* Toliara
35 O1 **Tulelake** California, W USA *41°57′N 121°30′W*
116 J10 **Tulgheş** *Hung.* Gyergyótölgyes. Harghita, C Romania *46°57′N 25°46′E*
Tuli *see* Thuli
25 N4 **Tulia** Texas, SW USA *34°32′N 101°46′W*
8 I9 **Tulita** *prev.* Fort Norman, Norman. Northwest Territories, NW Canada *64°55′N 125°25′W*
20 J10 **Tullahoma** Tennessee, S USA *35°21′N 86°12′W*
183 N12 **Tullamarine** ✈ (Melbourne) Victoria, SE Australia *37°40′S 144°46′E*
183 Q7 **Tullamore** New South Wales, SE Australia *32°39′S 147°35′E*
97 E18 **Tullamore** *Ir.* Tulach Mhór. Offaly, C Ireland *53°16′N 07°30′W*
103 N12 **Tulle** *anc.* Tutela. Corrèze, C France *45°16′N 01°46′E*
109 X3 **Tulln** *var.* Oberhollabrunn. Niederösterreich, NE Austria *48°20′N 16°02′E*
109 W4 **Tulln** ≈ NE Austria
22 H6 **Tullos** Louisiana, S USA *31°48′N 92°19′W*
97 F19 **Tullow** *Ir.* An Tullach. Carlow, SE Ireland *52°48′N 06°44′W*
181 W5 **Tully** Queensland, NE Australia *18°03′S 145°56′E*
124 J3 **Tuloma** ≈ NW Russian Federation
27 P9 **Tulsa** Oklahoma, C USA *36°09′N 96°W*
153 N11 **Tulsipur** Mid Western, W Nepal *28°01′N 82°22′E*
126 K6 **Tul'skaya Oblast'** ◈ *province* W Russian Federation
126 L14 **Tul'skiy** Respublika Adygeya, SW Russian Federation *44°26′N 40°12′E*
186 E5 **Tulu** Manus Island, N Papua New Guinea *01°58′S 146°50′E*
54 D10 **Tuluá** Valle del Cauca, W Colombia *04°01′N 76°16′W*
116 M12 **Tulucești** Galați, E Romania *45°35′N 28°01′E*
39 N12 **Tuluksak** Alaska, USA *61°06′N 160°57′W*
41 Z12 **Tulum, Ruinas de** *ruins* Quintana Roo, SE Mexico
169 R17 **Tulungagung** *prev.* Toeloengagoeng. Jawa, C Indonesia *08°03′S 111°54′E*
186 J6 **Tulun Islands** *var.* Kilinailau Islands; *prev.* Carteret Islands. *island group* NE Papua New Guinea
126 M4 **Tuma** Ryazanskaya Oblast', W Russian Federation *55°09′N 40°27′E*
54 B12 **Tumaco** Nariño, SW Colombia *01°51′N 78°46′W*
54 B12 **Tumaco, Bahía de** *bay* SW Colombia
Tuman-gang *see* Tumen
42 L8 **Tuma, Río** ≈ N Nicaragua
95 O16 **Tumba** Stockholm, C Sweden *59°12′N 17°49′E*
Tumba, Lac *see* Ntomba, Lac
169 S12 **Tumbangsenamang** Borneo, C Indonesia *01°17′S 112°21′E*
183 Q10 **Tumbarumba** New South Wales, SE Australia *35°47′S 148°03′E*
56 A8 **Tumbes** Tumbes, NW Peru *03°33′S 80°27′W*
56 A9 **Tumbes** *off.* Departamento de Tumbes. ◈ *department* NW Peru
Tumbes, Departamento de *see* Tumbes
19 P5 **Tumbledown Mountain** ▲ Maine, NE USA *45°27′N 70°28′W*
11 N13 **Tumbler Ridge** British Columbia, W Canada *55°06′N 120°51′W*
95 N16 **Tumbo** *prev.* Rekarne. Västmanland, C Sweden *59°25′N 16°04′E*
167 Q12 **Tumbŏt, Phnum** ▲ W Cambodia *12°23′N 102°57′E*
182 G9 **Tumby Bay** South Australia *34°22′S 136°05′E*
163 Y10 **Tumen** Jilin, NE China *42°56′N 129°47′E*
163 Y11 **Tumen** *Chin.* Tumen Jiang, *Kor.* Tuman-gang, *Rus.* Tumyn'tszyan. ≈ E Asia
Tumen Jiang *see* Tumen
55 Q8 **Tumeremo** Bolívar, E Venezuela *07°17′N 61°30′W*
155 G19 **Tumkūr** Karnātaka, W India *13°20′N 77°06′E*
96 I10 **Tummel** ≈ C Scotland, United Kingdom
188 B15 **Tumon Bay** *bay* W Guam
77 P14 **Tumu** NW Ghana *10°55′N 01°59′W*
58 I10 **Tumuc-Humac Mountains** *var.* Serra Tumucumaque. ▲ N South America
Tumucumaque, Serra *see* Tumuc-Humac Mountains
183 Q10 **Tumut** New South Wales, SE Australia *35°20′S 148°14′E*
158 F7 **Tumxuk** *var.* Urad Qianqi. Xinjiang Uygur Zizhiqu, NW China *78°40′N 39°54′E*
Tumyn'tszyan *see* Tumen
Tün *see* Ferdows
45 U14 **Tunapuna** Trinidad, Trinidad and Tobago *10°38′N 61°23′W*
60 K11 **Tunas** Paraná, S Brazil *24°57′S 49°05′W*
Tunbridge Wells *see* Royal Tunbridge Wells
114 L11 **Tunca Nehri** *Bul.* Tundzha. ≈ Bulgaria/Turkey *see also* Tundzha
Tunca Nehri *see* Tundzha
137 O14 **Tunceli** *var.* Kalan. Tunceli, E Turkey *39°07′N 39°34′E*
137 O14 **Tunceli** ◈ *province* C Turkey
152 J12 **Tūndla** Uttar Pradesh, N India *27°13′N 78°14′E*
81 I25 **Tunduru** Ruvuma, S Tanzania *11°08′S 37°21′E*
114 L10 **Tundzha** *Turk.* Tunca Nehri. ≈ Bulgaria/Turkey *see also* Tunca Nehri
Tundzha *see* Tunca Nehri
162 I6 **Tünel** *var.* Bulag. Hövsgöl, N Mongolia *49°51′N 100°41′E*
155 H17 **Tungabhadra** ≈ S India
155 F17 **Tungabhadra Reservoir** ⊠ S India
191 P2 **Tungaru** *prev.* Gilbert Islands. *island group* W Kiribati
171 P7 **Tungawan** Mindanao, S Philippines *07°33′N 122°22′E*
Tungdor *see* Mainling
T'ung-shan *see* Xuzhou
161 Q16 **Tungsha Tao** *Chin.* Dongsha Qundao, *Eng.* Pratas Island. *island* S Taiwan
Tungshih *see* Dongshi
8 H9 **Tungsten** Northwest Territories, W Canada *62°N 128°09′W*
Tung-t'ing Hu *see* Dongting Hu
56 A13 **Tungurahua** ◈ *province* C Ecuador
95 F14 **Tunhovdfjorden** ◎ S Norway
22 K2 **Tunica** Mississippi, S USA *34°40′N 90°22′W*
75 N5 **Tunis** *var.* Tūnis. ● (Tunisia) NE Tunisia *36°53′N 10°10′E*
75 N5 **Tunis, Golfe de** *Ar.* Khalīj Tūnis. *gulf* NE Tunisia
75 N6 **Tunisia** *off.* Tunisian Republic, *Ar.* Al Jumhūrīyah at Tūnisīyah, *Fr.* République Tunisienne. ◆ *republic* N Africa
Tunisian Republic *see* Tunisia
Tunisienne, République *see* Tunisia
Tūnisīyah, Al Jumhūrīyah at *see* Tunisia
Tūnis, Khalīj *see* Tunis, Golfe de
54 G9 **Tunja** Boyacá, C Colombia *05°33′N 73°23′W*
93 F14 **Tunnsjøen** *Lapp.* Dåtnejaevrie. ◎ C Norway
39 N12 **Tuntutuliak** Alaska, USA *60°21′N 162°40′W*
197 P14 **Tunu** ◈ *province* E Greenland
147 U8 **Tunuk** Chuyskaya Oblast', C Kyrgyzstan *42°11′N 73°55′E*
13 Q6 **Tunungayualok Island** *island* Newfoundland and Labrador, E Canada
62 H11 **Tunuyán** Mendoza, W Argentina *33°35′S 69°00′W*
62 I11 **Tunuyán, Río** ≈ W Argentina
Tunxi *see* Huangshan
Tuodian *see* Shuangbai
Tuoji *see* Zhongba
Tuokezhake *see* Shufu
35 P9 **Tuolumne River** ≈ California, W USA
Tuong Buong *see* Tương Đương
167 R7 **Tương Đương** *var.* Tuong Buong. Nghê An, N Vietnam *19°15′N 104°30′E*
160 I13 **Tuoniang Jiang** ≈ S China
Tuotiereke *see* Jeminay
Tuotuo He *see* Togton He
Tuotuoheyan *see* Tanggulashan
Tüp *see* Tyup
60 J9 **Tupã** São Paulo, S Brazil *21°57′S 50°28′W*
191 S10 **Tupai** *var.* Motu Iti. *atoll* Îles Sous le Vent, W French Polynesia
61 G15 **Tupanciretã** Rio Grande do Sul, S Brazil *29°06′S 53°48′W*
22 M2 **Tupelo** Mississippi, S USA *34°15′N 88°42′W*
59 K18 **Tupiraçaba** Goiás, S Brazil *14°33′S 48°40′W*
57 L21 **Tupiza** Potosí, S Bolivia *21°27′S 65°45′W*
144 D14 **Tupkaragan, Mys** *prev.* Mys Tyub-Karagan. *headland* SW Kazakhstan *44°40′N 50°19′E*
11 N13 **Tupper** British Columbia, W Canada *55°30′N 119°59′W*
18 J8 **Tupper Lake** ◎ New York, NE USA
146 J10 **Tuproqqal'a** Khorazm Viloyati, W Uzbekistan *40°52′N 62°00′E*
146 J10 **Tuproqqal'a** *Rus.* Turpakkla. Xorazm Viloyati, W Uzbekistan *40°52′N 62°00′E*
62 H11 **Tupungato, Volcán** ▲ W Argentina *33°27′S 69°42′W*
163 T9 **Tuquan** Nei Mongol Zizhiqu, N China *45°21′N 121°36′E*
54 C13 **Túquerres** Nariño, SW Colombia *01°06′N 77°37′W*
153 U13 **Tura** Meghālaya, NE India *25°33′N 90°14′E*
122 M10 **Tura** Krasnoyarskiy Kray, N Russian Federation *64°20′N 100°17′E*
122 G10 **Tura** ≈ C Russian Federation
140 M10 **Turabah** Makkah, W Saudi Arabia *22°00′N 42°00′E*
55 O8 **Turagua, Cerro** ▲ C Venezuela *06°59′N 64°34′W*
184 L12 **Turakina** Manawatu-Wanganui, North Island, New Zealand *40°03′S 175°13′E*
185 K15 **Turakirae Head** *headland* North Island, New Zealand *41°26′S 174°54′E*
186 B8 **Turama** ≈ S Papua New Guinea
122 K13 **Turan** Respublika Tyva, S Russian Federation *52°11′N 93°40′E*
184 M10 **Turangi** Waikato, North Island, New Zealand *39°01′S 175°47′E*
146 F11 **Turan Lowland** *var.* Turan Plain, *Kaz.* Turan Oypaty, *Rus.* Turanskaya Nizmennost', *Turk.* Turan Pesligi, *Uzb.* Turan Pasttekisligi. *plain* C Asia
Turan Oypaty/Turan Pesligi/Turan Plain/Turanskaya Nizmennost' *see* Turan Lowland
Turan Pasttekisligi *see* Turan Lowland
138 K7 **Ţuraq al 'Ilab** *hill range* S Syria
119 K20 **Turaw** *Rus.* Turov. Homyel'skaya Voblasts', SE Belarus *52°04′N 27°44′E*
140 L2 **Turayf** Al Ḥudūd ash Shamālīyah, NW Saudi Arabia *31°43′N 38°40′E*
Turba *see* Teruel
54 E5 **Turbaco** Bolívar, N Colombia *10°20′N 75°25′W*
148 K15 **Turbat** Baluchistān, SW Pakistan *26°02′N 62°56′E*
Turbat-i-Haidari *see* Torbat-e Ḥeydarīyeh
Turbat-i-Jam *see* Torbat-e Jām
54 D7 **Turbo** Antioquia, NW Colombia *08°06′N 76°44′W*
Turčianský Svätý Martin *see* Martin
116 H10 **Turda** *Ger.* Thorenburg, *Hung.* Torda. Cluj, NW Romania *46°35′N 23°50′E*
142 M7 **Ţūreh** Markazī, W Iran
191 X12 **Tureia** *atoll* Îles Tuamotu, SE French Polynesia
110 I12 **Turek** Wielkopolskie, C Poland *52°01′N 18°30′E*
93 L19 **Turenki** Kanta-Häme, SW Finland *60°55′N 24°38′E*
Turfan *see* Turpan
Turgay *see* Torgay
Turgay *see* Torgay
144 M8 **Turgayskaya Stolovaya Strana** *Kaz.* Torgay Üstirti. *plateau* Kazakhstan/Russian Federation
Turgel *see* Türi
Türgovishte *see* Targovishte
Türgovishte *see* Targovishte
136 C14 **Turgutlu** Manisa, W Turkey *38°30′N 27°43′E*
136 L12 **Turhal** Tokat, N Turkey *40°23′N 36°05′E*
118 H4 **Türi** *Ger.* Turgel. Järvamaa, N Estonia *58°48′N 25°28′E*
105 S9 **Turia** ≈ E Spain
58 M12 **Turiaçu** Maranhão, E Brazil *01°40′S 45°22′W*
Turin *see* Torino
116 J2 **Turiya** *Pol.* Turja, *Rus.* Tur'ya; *prev.* Tur"ya. ≈ NW Ukraine
116 I3 **Turiys'k** Volyns'ka Oblast', NW Ukraine *51°05′N 24°31′E*
Turja *see* Turiya
116 H6 **Turka** L'vivs'ka Oblast', W Ukraine *49°07′N 23°01′E*
81 H16 **Turkana** ◈ *county* Kenya
81 H16 **Turkana, Lake** *var.* Lake Rudolf. ◎ N Kenya
Turkestan *see* Turkistan
147 Q12 **Turkestan Range** *Rus.* Turkestanskiy Khrebet. ▲ C Asia
Turkestanskiy Khrebet *see* Turkestan Range
111 M23 **Túrkeve** Jász-Nagykun-Szolnok, E Hungary *47°06′N 20°42′E*
25 O4 **Turkey** Texas, SW USA *34°23′N 100°54′W*
136 H14 **Turkey** *off.* Republic of Turkey, *Turk.* Türkiye Cumhuriyeti. ◆ *republic* SW Asia
181 N4 **Turkey Creek** Western Australia *16°54′S 128°12′E*
26 M9 **Turkey Creek** ≈ Oklahoma, C USA
37 T9 **Turkey Mountains** ▲ New Mexico, SW USA
Turkey, Republic of *see* Turkey
29 X11 **Turkey River** ≈ Iowa, C USA
127 N7 **Turki** Saratovskaya Oblast', W Russian Federation *52°00′N 43°16′E*
121 O1 **Turkish Republic of Northern Cyprus** ◇ *disputed territory* Cyprus
145 P16 **Turkistan** *prev.* Turkestan. Yuzhnyy Kazakhstan, S Kazakhstan *43°18′N 68°18′E*
Turkistan, Bandi-i *see* Torkestān, Selseleh-ye Band-e
Türkiye Cumhuriyeti *see* Turkey
146 K12 **Türkmenabat** *prev. Rus.* Chardzhev, Chardzhou, Chardzhui, Lenin-Turkmenski, *Turkm.* Chärjew. Lebap Welaýaty, E Turkmenistan *39°07′N 63°30′E*
146 A11 **Türkmen Aylagy** *Rus.* Turkmenskiy Zaliv. *lake gulf* W Turkmenistan
Turkmenbashi *see* Türkmenbaşy
146 A10 **Türkmenbaşy** *Rus.* Turkmenbashi; *prev.* Krasnovodsk. Balkan Welaýaty, W Turkmenistan *40°N 53°04′E*
146 A10 **Türkmenbasy Aylagy** *prev. Rus.* Krasnovodskiy Zaliv, *Turkm.* Krasnowodsk Aylagy. *lake Gulf* W Turkmenistan
146 J14 **Türkmengala** *Rus.* Turkmen-kala; *prev.* Turkmen-Kala. Mary Welaýaty, S Turkmenistan *37°25′N 62°19′E*
146 G13 **Turkmenistan**; *prev.* Turkmenskaya Soviet Socialist Republic. ◆ *republic* C Asia
Turkmen-kala/Turkmen-Kala *see* Türkmengala
Turkmenskaya Soviet Socialist Republic *see* Turkmenistan
Turkmenskiy Zaliv *see* Türkmen Aylagy
136 L16 **Türkoğlu** Kahramanmaraş, S Turkey *37°24′N 36°49′E*
44 L6 **Turks and Caicos Islands** ◇ *UK dependent territory* N West Indies
64 G10 **Turks and Caicos Islands** *UK dependent territory* N West Indies
45 N6 **Turks Islands** *island group* SE Turks and Caicos Islands
93 K19 **Turku** *Swe.* Åbo. Varsinais-Suomi, SW Finland *60°27′N 22°17′E*
81 H17 **Turkwel** *seasonal river* NW Kenya
27 P9 **Turley** Oklahoma, C USA *36°14′N 95°58′W*
35 P9 **Turlock** California, W USA *37°29′N 120°52′W*
118 I12 **Turmantas** Utena, NE Lithuania *55°41′N 26°27′E*
Turmberg *see* Wieżyca
54 L5 **Turmero** Aragua, N Venezuela *10°14′N 66°40′W*
184 N13 **Turnagain, Cape** *headland* North Island, New Zealand *40°30′S 176°36′E*
Turnau *see* Turnov
42 H2 **Turneffe Islands** *island group* E Belize
18 M11 **Turners Falls** Massachusetts, NE USA *42°36′N 72°31′W*
11 P16 **Turner Valley** Alberta, SW Canada *50°43′N 114°19′W*
99 I16 **Turnhout** Antwerpen, N Belgium *51°19′N 04°57′E*
109 V5 **Türnitz** Niederösterreich, E Austria *47°56′N 15°26′E*
11 S12 **Turnor Lake** ◎ Saskatchewan, C Canada
111 E15 **Turnov** *Ger.* Turnau. Liberecký Kraj, N Czech Republic *50°36′N 15°10′E*
Tŭrnovo *see* Veliko Tarnovo
116 I15 **Turnu Măgurele** *var.* Turnu-Măgurele. Teleorman, S Romania *43°44′N 24°53′E*
Turnu Severin *see* Drobeta-Turnu Severin
Turócszentmárton *see* Martin
Turoni *see* Tours
Turov *see* Turaw
Turpakkla *see* Tuproqqal'a
158 M6 **Turpan** *var.* Turfan. Xinjiang Uygur Zizhiqu, NW China *42°55′N 89°06′E*
Turpan Depression *see* Turpan Pendi
158 M6 **Turpan Pendi** *Eng.* Turpan Depression. *depression* NW China
158 M5 **Turpan Zhan** Xinjiang Uygur Zizhiqu, W China *43°10′N 89°06′E*
Turpentine State *see* North Carolina
44 H8 **Turquino, Pico** ▲ E Cuba *19°54′N 76°55′W*
27 Y10 **Turrell** Arkansas, C USA *35°22′N 90°13′W*
43 N14 **Turrialba** Cartago, E Costa Rica *09°56′N 83°40′W*
96 K8 **Turriff** NE Scotland, United Kingdom *57°32′N 02°28′W*
139 V7 **Tursāq** Diyālá, E Iraq *33°27′N 45°47′E*
Turshiz *see* Kāshmar
Tursunzade *see* Tursunzoda
147 P13 **Tursunzoda** *Rus.* Tursunzade; *prev.* Regar. W Tajikistan *38°30′N 68°10′E*
Turt *see* Hanh
Türtkŭl/Turtkul' *see* To'rtko'l
29 O9 **Turtle Creek** ≈ South Dakota, N USA
30 K4 **Turtle Flambeau Flowage** ⊠ Wisconsin, N USA
11 S14 **Turtleford** Saskatchewan, S Canada *53°21′N 108°48′W*
28 M4 **Turtle Lake** North Dakota, N USA *47°31′N 100°53′W*
92 K12 **Turtola** Lappi, NW Finland *66°39′N 23°55′E*
122 M10 **Turu** ≈ N Russian Federation
Turuga *see* Tsuruga
147 V10 **Turugart Pass** *pass* China/Kyrgyzstan
158 E7 **Turugart Shankou** *var.* Pereval Torugart. *pass* China/Kyrgyzstan
122 K9 **Turukhan** ≈ N Russian Federation
122 K9 **Turukhansk** Krasnoyarskiy Kray, N Russian Federation *65°50′N 87°48′E*
139 N3 **Ţurumbah** *well* NE Syria
Turuoka *see* Tsuruoka
Turush *see* Turysh
60 K7 **Turvo, Rio** ≈ S Brazil
Tur"ya *see* Turiya
144 H14 **Turysh** *prev.* Turush. Mangistau, SW Kazakhstan *45°24′N 56°02′E*
23 O4 **Tuscaloosa** Alabama, S USA *33°13′N 87°34′W*
23 O4 **Tuscaloosa, Lake** ◎ Alabama, S USA
Tuscan Archipelago *see* Toscano, Archipelago
Tuscan-Emilian Mountains *see* Tosco-Emiliano, Appennino
Tuscany *see* Toscana
35 V2 **Tuscarora** Nevada, W USA *41°16′N 116°13′W*
18 F15 **Tuscarora Mountain** *ridge* Pennsylvania, NE USA
30 M14 **Tuscola** Illinois, N USA *39°46′N 88°19′W*
25 P7 **Tuscola** Texas, SW USA *32°12′N 99°48′W*
23 O2 **Tuscumbia** Alabama, S USA *34°43′N 87°42′W*
92 O4 **Tusenøyane** *island group* S Svalbard
144 K13 **Tushchybas, Zaliv** *prev.* Zaliv Paskevicha. *lake gulf* SW Kazakhstan
Tusima *see* Tsushima
171 Y15 **Tusirah** Papua, E Indonesia *06°46′S 140°19′E*
23 Q5 **Tuskegee** Alabama, S USA *32°25′N 85°41′W*
94 E8 **Tustna** *island* S Norway
39 R12 **Tustumena Lake** ◎ Alaska, USA
110 K13 **Tuszyn** Łódzkie, C Poland *51°36′N 19°31′E*
137 S13 **Tutak** Ağrı, E Turkey *39°34′N 42°48′E*
185 C20 **Tutamoe Range** ▲ North Island, New Zealand
Tutasev *see* Tutayev
124 L15 **Tutayev** *var.* Tutasev. Yaroslavskaya Oblast', W Russian Federation *57°51′N 39°29′E*
Tutela *see* Tulle, France
Tutela *see* Tudela, Spain
Tutera *see* Tudela
155 H23 **Tuticorin** Tamil Nādu, SE India *08°48′N 78°10′E*
113 L15 **Tutin** Serbia, S Serbia *43°00′N 20°20′E*
184 O10 **Tutira** Hawke's Bay, North Island, New Zealand *39°14′S 176°53′E*
Tutiura *see* Tsuchiura
122 K10 **Tutonchny** Krasnoyarskiy Kray, N Russian Federation *64°12′N 93°52′E*
114 L6 **Tutrakan** Silistra, NE Bulgaria *44°03′N 26°38′E*
29 N5 **Tuttle** North Dakota, N USA *47°07′N 99°58′W*
26 M11 **Tuttle** Oklahoma, C USA *35°17′N 97°48′W*
27 O3 **Tuttle Creek Lake** ⊠ Kansas, C USA
101 H23 **Tuttlingen** Baden-Württemberg, S Germany *47°59′N 08°49′E*
171 R16 **Tutuala** East Timor *08°23′S 127°12′E*
192 K17 **Tutuila** *island* W American Samoa
83 I18 **Tutume** Central, E Botswana *20°26′S 27°02′E*
39 N7 **Tututalak Mountain** ▲ Alaska, USA *67°51′N 161°27′W*
22 K3 **Tutwiler** Mississippi, S USA *34°00′N 90°25′W*
162 L8 **Tuul Gol** ≈ N Mongolia
93 O16 **Tuupovaara** Pohjois-Karjala, E Finland *62°30′N 30°40′E*
Tuva *see* Tyva, Respublika
190 E7 **Tuvalu** *prev.* Ellice Islands. ◆ *commonwealth republic* SW Pacific Ocean
Tuvinskaya ASSR *see* Tyva, Respublika
163 O9 **Tüvshinshiree** *var.* Sergelen. Sühbaatar, E Mongolia *46°12′N 111°48′E*
141 P9 **Ţuwayq, Jabal** ▲ C Saudi Arabia
138 H13 **Ţuwayyil ash Shihāq** *desert* S Jordan
11 U16 **Tuxford** Saskatchewan, S Canada *50°33′N 105°32′W*
167 U12 **Tu Xoay** Đặc Lắc, S Vietnam *12°18′N 107°33′E*
40 L14 **Tuxpan** Jalisco, C Mexico *19°33′N 103°21′W*
40 J12 **Tuxpan** Nayarit, C Mexico *21°57′N 105°12′W*
41 Q12 **Tuxpán** *var.* Tuxpán de Rodríguez Cano. Veracruz-Llave, E Mexico *20°58′N 97°23′W*
Tuxpán de Rodríguez Cano *see* Tuxpán
41 R15 **Tuxtepec** *var.* San Juan Bautista Tuxtepec. Oaxaca, S Mexico *18°02′N 96°05′W*
41 U16 **Tuxtla** *var.* Tuxtla Gutiérrez. Chiapas, SE Mexico *16°44′N 93°03′W*
Tuxtla *see* San Andrés Tuxtla
Tuxtla Gutiérrez *see* Tuxtla
Tuyama *see* Tsuyama
167 T5 **Tuyên Quang** Tuyên Quang, N Vietnam *21°48′N 105°18′E*
167 U13 **Tuy Hoa** Binh Thuân, S Vietnam *11°03′N 108°12′E*
167 V12 **Tuy Hoa** Phu Yên, S Vietnam *13°02′N 109°15′E*
127 U5 **Tuymazy** Respublika Bashkortostan, W Russian Federation *54°36′N 53°40′E*
Tuy Phong *see* Liên Hương
142 L6 **Tūysarkān** *var.* Tuisarkan, Tuyserkān. Hamadān, W Iran *34°31′N 48°30′E*
Tuyserkān *see* Tūysarkān
Tuyuk *see* Tuyyk
145 W16 **Tuyyk** *Kaz.* Tuyyq; *prev.* Tuyuk. Taldykorgan, SE Kazakhstan *43°07′N 79°24′E*
Tuyyq *see* Tuyyk
136 I14 **Tuz Gölü** ◎ C Turkey
125 Q15 **Tuzha** Kirovskaya Oblast', NW Russian Federation *57°37′N 48°02′E*
113 K17 **Tuzi** S Montenegro *42°22′N 19°21′E*
139 T5 **Ţūz Khurmātū** At Ta'mím, N Iraq *34°56′N 44°38′E*
112 I11 **Tuzla** Federacija Bosne I Hercegovine, NE Bosnia and Herzegovina *44°33′N 18°40′E*
117 N15 **Tuzla** Constanța, SE Romania *43°58′N 28°38′E*

◆ Country
● Country Capital
◇ Dependent Territory
○ Dependent Territory Capital
◈ Administrative Regions
✈ International Airport
▲ Mountain
▲ Mountain Range
Volcano
≈ River
◎ Lake
⊠ Reservoir

◆ Country / ● Country Capital
◇ Dependent Territory / ○ Dependent Territory Capital
✘ Administrative Regions / ✈ International Airport
▲ Mountain / ▲ Mountain Range
℞ Volcano / ~ River
◎ Lake / ☐ Reservoir

337

UNTURÁN, SIERRA DE – VALLE DE GUANAPE

55 N12 **Unturán, Sierra de** ▲▲ Brazil/Venezuela
159 N11 **Unuli Horog** Qinghai, W China 35°10′N 91°50′E
136 M11 **Ünye** Ordu, W Turkey 41°08′N 37°14′E
Unza *see* Unzha
125 O14 **Unzha** *var.* Unza. ≈ NW Russian Federation
79 E17 **Uolo, Río** *var.* Eyo (lower course), Mbini, Uele (upper course), Woleu; *prev.* Benito. ≈ Equatorial Guinea/Gabon
55 Q10 **Uonán** Bolívar, SE Venezuela 04°33′N 62°10′W
161 T12 **Uotsuri-shima** *Chin.* Diaoyu Dao. *island* (disputed) China/Japan/Taiwan
165 M11 **Uozu** Toyama, Honshū, SW Japan 36°50′N 137°25′E
42 L12 **Upala** Alajuela, NW Costa Rica 10°52′N 85°W
55 P7 **Upata** Bolívar, E Venezuela 08°02′N 62°25′W
79 M23 **Upemba, Lac** ◎ SE Dem. Rep. Congo
145 R11 **Upenskoye** *prev.* Uspenskiy. Karaganda, C Kazakhstan 48°45′N 72°46′E
197 O12 **Upernavik** *var.* Upernivik. Qaasuitsup, C Greenland 73°06′N 55°42′W
Upernivik *see* Upernavik
83 F22 **Upington** Northern Cape, W South Africa 28°28′S 21°14′E
Uplands *see* Ottawa
192 I16 **'Upolu** *island* SE Samoa
38 G11 **'Upolu Point** *headland* Hawai'i, USA, C Pacific Ocean 20°15′N 155°51′W
Upper Austria *see* Oberösterreich
Upper Bann *see* Bann
14 M13 **Upper Canada Village** *tourist site* Ontario, SE Canada
18 I16 **Upper Darby** Pennsylvania, NE USA 39°57′N 75°15′W
28 L2 **Upper Des Lacs Lake** ⊠ North Dakota, N USA
185 L14 **Upper Hutt** Wellington, North Island, New Zealand 41°06′S 175°06′E
29 X11 **Upper Iowa River** ≈ Iowa, C USA
32 H15 **Upper Klamath Lake** ◎ Oregon, NW USA
34 M6 **Upper Lake** California, W USA 39°07′N 122°53′W
35 Q1 **Upper Lake** ◎ California, W USA
10 K9 **Upper Liard** Yukon, W Canada 60°01′N 128°59′W
97 E16 **Upper Lough Erne** ◎ SW Northern Ireland, United Kingdom
80 F12 **Upper Nile** ◆ *state* NE South Sudan
29 T3 **Upper Red Lake** ◎ Minnesota, N USA
31 S12 **Upper Sandusky** Ohio, N USA 40°49′N 83°16′W
Upper Volta *see* Burkina Faso
95 O15 **Upplands Väsby** *var.* Upplandsväsby. Stockholm, C Sweden 59°29′N 18°04′E
Upplandsväsby *see* Upplands Väsby
95 O15 **Uppsala** Uppsala, C Sweden 59°52′N 17°38′E
95 O14 **Uppsala** ◆ *county* C Sweden
38 J12 **Upright Cape** *headland* Saint Matthew Island, Alaska, USA 60°19′N 172°15′W
20 K6 **Upton** Kentucky, S USA 37°25′N 85°53′W
33 Y13 **Upton** Wyoming, C USA 44°06′N 104°37′W
141 N7 **'Uqlat aş Şuqūr** Al Qaşīm, W Saudi Arabia 25°51′N 42°13′E
Uqsuqtuuq *see* Gjoa Haven
Uqturpan *see* Wushi
54 C7 **Urabá, Golfo de** *gulf* NW Colombia
Uracas *see* Farallon de Pajaros
uradqianqi *see* Wulashan, N China
Uradar'ya *see* O'radaryo
Urad Qianqi *see* Xishanzui, N China
165 U5 **Urahoro** Hokkaidō, NE Japan 42°47′N 143°41′E
165 T5 **Urakawa** Hokkaidō, NE Japan 42°11′N 142°42′E
Ural *see* Zhayyk
183 T6 **Uralla** New South Wales, SE Australia 30°39′S 151°30′E
Ural Mountains *see* Ural'skiy Gory
144 F8 **Ural'sk** *Kaz.* Oral. Zapadnyy Kazakhstan, NW Kazakhstan 51°12′N 51°17′E
Ural'skaya Oblast' *see* Zapadnyy Kazakhstan
127 W5 **Ural'skiy Gory** *var.* Ural'skiy Khrebet, *Eng.* Ural Mountains. ▲▲ Kazakhstan/Russian Federation
Ural'skiy Khrebet *see* Ural'skiy Gory
138 I3 **Urām aş Şughrá** Ḩalab, N Syria 36°10′N 36°55′E
183 P10 **Urana** New South Wales, SE Australia 35°22′S 146°16′E
11 S10 **Uranium City** Saskatchewan, C Canada 59°30′N 108°46′W
58 F10 **Uraricoera** Roraima, N Brazil 03°26′N 60°54′W
47 S5 **Uraricoera, Rio** ≈ N Brazil
Ura-Tyube *see* Üroteppa
165 O13 **Urawa** *var.* Saitama. Saitama, Honshū, S Japan 35°52′N 139°40′E
122 H10 **Uray** Khanty-Mansiyskiy Avtonomnyy Okrug-Yugra, C Russian Federation 60°07′N 64°38′E
141 R7 **'Uray'irah** Ash Sharqīyah, E Saudi Arabia 25°59′N 48°52′E
30 M13 **Urbana** Illinois, N USA 40°06′N 88°12′W
31 R13 **Urbana** Ohio, N USA 40°04′N 83°46′W
29 V14 **Urbandale** Iowa, C USA 41°37′N 93°42′W
106 I11 **Urbania** Marche, C Italy 43°40′N 12°33′E
106 I11 **Urbino** Marche, C Italy 43°45′N 12°38′E
57 H16 **Urcos** Cusco, S Peru 13°40′S 71°38′W
105 N10 **Urda** Castilla-La Mancha, C Spain 39°25′N 03°43′W
Urda *see* Khan Ordasy
Urdgol *see* Chandmani
105 O3 **Urduña** *var.* Orduña. País Vasco, N Spain 43°00′N 03°00′W
Urdunn *see* Jordan
Urdzhar *see* Urzhar
97 L16 **Ure** ≈ N England, United Kingdom
119 K18 **Urechcha** *Rus.* Urech'ye. Minskaya Voblasts', S Belarus 52°57′N 27°54′E
Urech'ye *see* Urechcha
127 P2 **Uren'** Nizhegorodskaya Oblast', W Russian Federation 57°30′N 45°48′E
122 J9 **Urengoy** Yamalo-Nenetskiy Avtonomnyy Okrug, N Russian Federation 65°52′N 78°42′E
184 K10 **Urenui** Taranaki, North Island, New Zealand 38°59′S 174°25′E
187 Q12 **Ureparapara** *island* Banks Islands, N Vanuatu
40 G5 **Ures** Sonora, NW Mexico 29°26′N 110°24′W
Urfa *see* Şanlıurfa
Urga *see* Ulaanbaatar
162 F6 **Urgamal** *var.* Hungiy. Dzavhan, W Mongolia 48°31′N 94°15′E
146 H9 **Urganch** *Rus.* Urgench; *prev.* Novo-Urgench. Xorazm Viloyati, W Uzbekistan 41°40′N 60°32′E
Urgench *see* Urganch
136 J14 **Ürgüp** Nevşehir, C Turkey 38°39′N 34°55′E
147 O12 **Urgut** Samarqand Viloyati, C Uzbekistan 39°26′N 67°15′E
158 K3 **Urho** Xinjiang Uygur Zizhiqu, W China 46°05′N 84°51′E
152 G5 **Uri** Jammu and Kashmir, NW India 34°05′N 74°03′E
108 G9 **Uri** ◆ *canton* C Switzerland
54 F11 **Uribe** Meta, C Colombia 03°01′N 74°33′W
54 H4 **Uribia** La Guajira, N Colombia 11°45′N 72°19′W
116 G12 **Uricani** *Hung.* Hobicaurikány. Hunedoara, SW Romania 45°18′N 23°03′E
57 M21 **Uriondo** Tarija, S Bolivia 21°43′S 64°40′W
40 I7 **Urique** Chihuahua, N Mexico 27°16′N 107°51′W
40 I7 **Urique, Río** ≈ N Mexico
56 E9 **Urituyacu, Río** ≈ N Peru
Uritskiy *see* Sarykol'
98 K8 **Urk** Flevoland, N Netherlands 52°40′N 05°35′E
136 B14 **Urla** İzmir, W Turkey 38°19′N 26°47′E
116 K13 **Urlaţi** Prahova, SE Romania 44°59′N 26°15′E
127 V4 **Urman** Respublika Bashkortostan, W Russian Federation 54°53′N 56°52′E
147 P12 **Urmetan** W Tajikistan 39°27′N 68°13′E
Urmia *see* Orūmīyeh
Urmia, Lake *see* Orūmīyeh, Daryācheh-ye
Urmiyeh *see* Orūmīyeh
Uroševac *see* Ferizaj
147 P11 **Üroteppa** *Rus.* Ura-Tyube. NW Tajikistan 39°55′N 68°57′E
54 D8 **Urrao** Antioquia, W Colombia 06°16′N 76°10′W
Ursat'yevskaya *see* Xovos
Urt *see* Gurvantes
127 X7 **Urtazym** Orenburgskaya Oblast', W Russian Federation 52°12′N 58°48′E
59 K18 **Uruaçu** Goiás, C Brazil 14°38′S 49°06′W
40 M14 **Uruapan** *var.* Uruapan del Progreso. Michoacán, SW Mexico 19°26′N 102°04′W
Uruapan del Progreso *see* Uruapan
57 G15 **Urubamba, Cordillera** ▲▲ C Peru
57 G14 **Urubamba, Río** ≈ C Peru
58 G12 **Urucará** Amazonas, N Brazil 02°30′S 57°45′W
61 E16 **Uruguaiana** Rio Grande do Sul, S Brazil 29°45′S 57°05′W
Uruguai, Rio *see* Uruguay
61 E18 **Uruguay** *off.* Oriental Republic of Uruguay; *prev.* La Banda Oriental. ◆ *republic* E South America
61 E15 **Uruguay** *var.* Rio Uruguai, Río Uruguay. ≈ E South America
Uruguay, Oriental Republic of *see* Uruguay
Uruguay, Río *see* Uruguay
Urukthapel *see* Ngeruktabel
Urumchi *see* Ürümqi
Urumi Yeh *see* Orūmīyeh, Daryācheh-ye
158 L5 **Ürümqi** *var.* Tihwa, Urumchi, Urumqi, Urumtsi, Wu-lu-k'o-mu-shi, Wu-lu-mu-ch'i; *prev.* Ti-hua. Xinjiang Uygur Zizhiqu, NW China 43°52′N 87°31′E
Urumtsi *see* Ürümqi
Urundi *see* Burundi
183 V6 **Urunga** New South Wales, SE Australia 30°33′S 152°58′E
188 C15 **Uruno Point** *headland* NW Guam 13°37′N 144°50′E
123 U13 **Urup, Ostrov** *island* Kuril'skiye Ostrova, SE Russian Federation
141 P11 **'Uruq al Mawārid** *desert* S Saudi Arabia
Urusan *see* Ulsan
127 T5 **Urussu** Respublika Tatarstan, W Russian Federation 54°34′N 53°23′E
184 K10 **Uruti** Taranaki, North Island, New Zealand 38°57′S 174°32′E
57 K19 **Uru Uru, Lago** ◎ W Bolivia
55 P9 **Uruyén** Bolívar, SE Venezuela 05°40′N 62°26′W
149 O7 **Uruzgān** ; *prev.* Orūzgān. Uruzgān, C Afghanistan 32°58′N 66°39′E
149 N6 **Urūzgān** *prev.* Orūzgān. ◆ *province* C Afghanistan
165 T3 **Uryū-gawa** ≈ Hokkaidō, NE Japan
165 T2 **Uryū-ko** ◎ Hokkaidō, NE Japan
127 N8 **Uryupinsk** Volgogradskaya Oblast', SW Russian Federation 50°51′N 41°59′E
145 X12 **Urzhar** *prev.* Urdzhar. Vostochnyy Kazakhstan, E Kazakhstan 47°06′N 81°33′E
125 R16 **Urzhum** Kirovskaya Oblast', NW Russian Federation 57°09′N 49°56′E
116 K13 **Urziceni** Ialomiţa, SE Romania 44°43′N 26°39′E
164 E14 **Usa** Ōita, Kyūshū, SW Japan 33°31′N 131°22′E
125 T6 **Usa** ≈ NW Russian Federation
136 E14 **Uşak** *prev.* Ushak. Uşak, W Turkey 38°42′N 29°25′E
136 D14 **Uşak** *var.* Ushak. ◆ *province* W Turkey
83 C19 **Usakos** Erongo, W Namibia 22°01′S 15°32′E
81 J21 **Usambara Mountains** ▲▲ NE Tanzania
81 G23 **Usangu Flats** *wetland* SW Tanzania
65 D24 **Usborne, Mount** ▲ East Falkland, Falkland Islands 51°35′S 58°57′W
100 O8 **Usedom** *island* NE Germany
99 M24 **Useldange** Diekirch, C Luxembourg 49°47′N 05°59′E
119 L16 **Usha** ≈ C Belarus
Ushachi *see* Ushachy
118 L13 **Ushachy** *Rus.* Ushachi. Vitsyebskaya Voblasts', N Belarus 55°11′N 28°37′E
Ushak *see* Uşak
122 L4 **Ushakova, Ostrov** *island* Severnaya Zemlya, N Russian Federation
Ushant *see* Ouessant, Île d'
145 W13 **Usharal** *prev.* Ucharal. Almaty, E Kazakhstan 46°08′N 80°55′E
164 B15 **Ushibuka** *var.* Usibuka. Kumamoto, Shimo-jima, SW Japan 32°12′N 130°00′E
Ushi Point *see* Sabaneta, Puntan
145 V14 **Ushtobe** *Kaz.* Üshtöbe. Almaty, SE Kazakhstan 45°15′N 77°59′E
Üshtöbe *see* Ushtobe
63 I25 **Ushuaia** Tierra del Fuego, S Argentina 54°48′S 68°19′W
39 R10 **Usibelli** Alaska, USA 63°54′N 148°41′W
Usibuka *see* Ushibuka
186 D7 **Usino** Madang, N Papua New Guinea 05°40′S 145°31′E
125 U6 **Usinsk** Respublika Komi, NW Russian Federation 66°01′N 57°37′E
97 K22 **Usk** *Wel.* Wysg. ≈ SE Wales, United Kingdom
Uskočke Planine/Uskokengebirge *see* Gorjanci
Uskoplje *see* Gornji Vakuf
Üsküb/Üsküp *see* Skopje
114 M11 **Üsküpdere** Kırklareli, NW Turkey 41°41′N 27°21′E
126 L7 **Usman'** Lipetskaya Oblast', W Russian Federation 52°04′N 39°41′E
118 D8 **Usmas Ezers** ◎ NW Latvia
125 U13 **Usol'ye** Permskiy Kray, NW Russian Federation 59°27′N 56°33′E
41 T16 **Uspanapa, Río** ≈ SE Mexico
Uspenskiy *see* Upenskoye
103 O11 **Ussel** Corrèze, C France 45°33′N 02°18′E
163 Z6 **Ussuri** *var.* Usuri, Wusuri, *Chin.* Wusuli Jiang. ≈ China/Russian Federation
123 S15 **Ussuriysk** *prev.* Nikol'sk, Nikol'sk-Ussuriyskiy, Voroshilov. Primorskiy Kray, SE Russian Federation 43°48′N 131°59′E
136 J10 **Usta Burnu** *headland* N Turkey 41°58′N 34°30′E
149 P13 **Usta Muhammad** Baluchistān, SW Pakistan 28°07′N 68°00′E
123 V11 **Ust'-Bol'sheretsk** Kamchatskiy Kray, E Russian Federation 52°48′N 156°12′E
127 N9 **Ust'-Buzulukskaya** Volgogradskaya Oblast', SW Russian Federation 50°12′N 42°06′E
111 C16 **Ústecký Kraj** ◆ *region* NW Czech Republic
108 G7 **Uster** Zürich, NE Switzerland 47°21′N 08°49′E
107 J22 **Ustica, Isola d'** *island* S Italy
122 M11 **Ust'-Ilimsk** Irkutskaya Oblast', C Russian Federation 57°57′N 102°30′E
111 C15 **Ústí nad Labem** *Ger.* Aussig. Ústecký Kraj, NW Czech Republic 50°41′N 14°04′E
111 F17 **Ústí nad Orlicí** *Ger.* Wildenschwert. Pardubický Kraj, C Czech Republic 49°58′N 16°24′E
Ustinov *see* Izhevsk
113 J14 **Ustiprača** ◆ Republika Srpska, SE Bosnia and Herzegovina
122 H11 **Ust'-Ishim** Omskaya Oblast', C Russian Federation 57°42′N 70°58′E
110 G6 **Ustka** *Ger.* Stolpmünde. Pomorskie, N Poland 54°35′N 16°50′E
123 V9 **Ust'-Kamchatsk** Kamchatskiy Kray, E Russian Federation 56°14′N 162°28′E
145 X9 **Ust'-Kamenogorsk** *Kaz.* Öskemen. Vostochnyy Kazakhstan, E Kazakhstan 49°58′N 82°36′E
123 T10 **Ust'-Khayryuzovo** Krasnoyarskiy Kray, E Russian Federation 57°07′N 156°37′E
122 I14 **Ust'-Koksa** Respublika Altay, S Russian Federation 50°15′N 85°45′E
125 S11 **Ust'-Kulom** Respublika Komi, NW Russian Federation 61°42′N 53°42′E
123 Q8 **Ust'-Kuyga** Respublika Sakha (Yakutiya), NE Russian Federation 69°59′N 135°27′E
126 L14 **Ust'-Labinsk** Krasnodarskiy Kray, SW Russian Federation 44°40′N 40°46′E
123 R10 **Ust'-Maya** Respublika Sakha (Yakutiya), NE Russian Federation 60°27′N 134°28′E
123 R9 **Ust'-Nera** Respublika Sakha (Yakutiya), NE Russian Federation 64°28′N 143°01′E
123 P12 **Ust'-Nyukzha** Amurskaya Oblast', S Russian Federation 56°30′N 121°32′E
123 O7 **Ust'-Olenëk** Respublika Sakha (Yakutiya), NE Russian Federation 73°03′N 119°34′E
123 T9 **Ust'-Omchug** Magadanskaya Oblast', E Russian Federation 61°07′N 149°17′E
122 M13 **Ust'-Ordynskiy** Irkutskaya Oblast', S Russian Federation 52°50′N 104°42′E
125 N8 **Ust'-Pinega** Arkhangel'skaya Oblast', NW Russian Federation 64°09′N 41°55′E
122 K8 **Ust'-Port** Krasnoyarskiy Kray, N Russian Federation 69°42′N 84°25′E
114 L11 **Ustrem** *prev.* Vakav. Yambol, E Bulgaria 42°01′N 26°28′E
111 O18 **Ustrzyki Dolne** Podkarpackie, SE Poland 49°26′N 22°36′E
Ust'-Sysol'sk *see* Syktyvkar
125 R7 **Ust'-Tsil'ma** Respublika Komi, NW Russian Federation 65°25′N 52°09′E
Ust Urt *see* Ustyurt Plateau
125 O11 **Ust'ya** ≈ NW Russian Federation
124 K6 **Ust'ye Varzugi** Murmanskaya Oblast', NW Russian Federation 66°16′N 36°47′E
123 V10 **Ust'yevoye** *prev.* Kirovskiy. Kamchatskiy Kray, E Russian Federation 54°06′N 155°48′E
117 R8 **Ustynivka** Kirovohrads'ka Oblast', C Ukraine 47°58′N 32°32′E
144 H15 **Ustyurt Plateau** *var.* Ust Urt, *Uzb.* Ustyurt Platosi. *plateau* Kazakhstan/Uzbekistan
Ustyurt Platosi *see* Ustyurt Plateau
124 K14 **Ustyuzhna** Vologodskaya Oblast', NW Russian Federation 58°50′N 36°25′E
158 J4 **Usu** Xinjiang Uygur Zizhiqu, NW China 44°27′N 84°37′E
171 O13 **Usu** Sulawesi, C Indonesia 02°34′S 120°58′E
164 E14 **Usuki** Ōita, Kyūshū, SW Japan 33°07′N 131°48′E
42 G8 **Usulután** Usulután, SE El Salvador 13°20′N 88°26′W
42 B9 **Usulután** ◆ *department* SE El Salvador
41 W16 **Usumacinta, Río** ≈ Guatemala/Mexico
Usumbura *see* Bujumbura
Usuri *see* Ussuri
U.S./USA *see* United States of America
171 W14 **Uta** Papua, E Indonesia 04°28′S 136°03′E
36 K5 **Utah** *off.* State of Utah, *also known as* Beehive State, Mormon State. ◆ *state* W USA
36 L3 **Utah Lake** ◎ Utah, W USA
Utaidhani *see* Uthai Thani
93 M14 **Utajärvi** Pohjois-Pohjanmaa, C Finland 64°45′N 26°25′E
Utamboni *see* Mitemele, Río
Utaradit *see* Uttaradit
165 T3 **Utashinai** *var.* Utasinai. Hokkaidō, NE Japan 43°32′N 142°03′E
Utasinai *see* Utashinai
193 Y14 **'Uta Vava'u** *island* Vava'u Group, N Tonga
37 V9 **Ute Creek** ≈ New Mexico, SW USA
118 H12 **Utena** Utena, E Lithuania 55°30′N 25°34′E
118 H12 **Utena** ◆ *province* Lithuania
37 V10 **Ute Reservoir** ⊠ New Mexico, SW USA
167 O10 **Uthai Thani** *var.* Muang Uthai Thani, Udayadhani, Utaidhani. Uthai Thani, W Thailand 15°22′N 100°03′E
149 O15 **Uthal** Baluchistān, SW Pakistan 25°53′N 66°37′E
uThongathi *see* oThongathi
18 I10 **Utica** New York, NE USA 43°06′N 75°15′W
105 R10 **Utiel** Valenciana, E Spain 39°33′N 01°13′W
11 O13 **Utikuma Lake** ◎ Alberta, W Canada
42 J4 **Utila, Isla de** *island* Islas de la Bahía, N Honduras
Utina *see* Udine
59 O17 **Utinga** Bahia, E Brazil 12°05′S 41°07′W
Utirik *see* Utrik Atoll
95 M22 **Utlängan** *island* S Sweden
117 U11 **Utlyuts'kyy Lyman** *bay* S Ukraine
95 P16 **Utö** Stockholm, C Sweden 58°55′N 18°19′E
25 Q12 **Utopia** Texas, SW USA 29°30′N 99°31′W
98 J11 **Utrecht** *Lat.* Trajectum ad Rhenum. Utrecht, C Netherlands 52°06′N 05°07′E
83 K22 **Utrecht** KwaZulu/Natal, E South Africa 27°40′S 30°20′E
98 I11 **Utrecht** ◆ *province* C Netherlands
104 K14 **Utrera** Andalucía, S Spain 37°10′N 05°47′W
189 V4 **Utrik Atoll** *var.* Utirik, Utrõk, Utrõnk. *atoll* Ratak Chain, N Marshall Islands
Utrõk/Utrõnk *see* Utrik Atoll
95 B16 **Utsira** *island* SW Norway
92 L8 **Utsjoki** *var.* Ohcejohka. Lappi, N Finland 69°54′N 27°01′E
165 O12 **Utsunomiya** *var.* Utunomiya. Tochigi, Honshū, S Japan 36°36′N 139°53′E
127 P13 **Utta** Respublika Kalmykiya, SW Russian Federation 46°22′N 46°03′E
167 O8 **Uttaradit** *var.* Utaradit. Uttaradit, N Thailand 17°38′N 100°05′E
152 J9 **Uttarakhand** ◆ *state* N India
152 J8 **Uttarkāshi** Uttarakhand, N India 30°45′N 78°19′E
152 K11 **Uttar Pradesh** *prev.* United Provinces, United Provinces of Agra and Oudh. ◆ *state* N India
45 T5 **Utuado** C Puerto Rico 18°17′N 66°41′W
158 K3 **Utubulak** Xinjiang Uygur Zizhiqu, W China 46°50′N 86°15′E
39 N5 **Utukok River** ≈ Alaska, USA
Utunomiya *see* Utsunomiya
187 P10 **Utupua** *island* Santa Cruz Islands, E Solomon Islands
Utva *see* Shyngyrlau
189 Y15 **Utwe** Kosrae, E Micronesia
189 X15 **Utwe Harbor** *harbor* Kosrae, E Micronesia
Uubulan *see* Hayrhan
163 O8 **Uulbayan** *var.* Dzüünbulag. Sühbaatar, E Mongolia 46°30′N 112°22′E
118 G6 **Uulu** Pärnumaa, SW Estonia 58°15′N 24°32′E
197 N13 **Uummannaq** *var.* Umanak, Umanaq. ◆ Qaasuitsup, C Greenland
Uummannarsuaq *see* Nunap Isua
162 E4 **Üüreg Nuur** ◎ NW Mongolia
Uusikaarlepyy *see* Nykarleby
93 J19 **Uusikaupunki** *Swe.* Nystad. Varsinais-Suomi, SW Finland 60°48′N 21°25′E
93 M20 **Uusimaa** *Swe.* Nyland. ◆ *region* S Finland
127 S2 **Uva** Udmurtskaya Respublika, NW Russian Federation 56°41′N 52°15′E
155 K25 **Uva** ◆ *province* SE Sri Lanka
113 L14 **Uvac** ≈ W Serbia
25 Q12 **Uvalde** Texas, SW USA 29°14′N 99°49′W
119 O18 **Uvaravichy** *Rus.* Uvarovichi. Homyel'skaya Voblasts', SE Belarus 52°36′N 30°44′E
54 J11 **Uvá, Río** ≈ E Colombia
Uvarovichi *see* Uvaravichy
127 N7 **Uvarovo** Tambovskaya Oblast', W Russian Federation 51°58′N 42°13′E
122 H10 **Uvat** Tyumenskaya Oblast', C Russian Federation 59°11′N 68°37′E
190 G12 **Uvea, Île** *island* N Wallis and Futuna
81 E21 **Uvinza** Kigoma, W Tanzania 05°08′S 30°23′E
79 O20 **Uvira** Sud-Kivu, E Dem. Rep. Congo 03°24′S 29°05′E
162 E5 **Uvs** ◆ *province* NW Mongolia
162 F5 **Uvs Nuur** *var.* Ozero Ubsu-Nur. ◎ Mongolia/Russian Federation
164 F14 **Uwa** *var.* Seiyo. Ehime, Shikoku, SW Japan 33°22′N 132°29′E
164 F14 **Uwajima** *var.* Uwazima. Ehime, Shikoku, SW Japan 33°13′N 132°32′E
80 B5 **'Uwaynāt, Jabal al** *var.* Jebel Uweinat. ▲ Libya/Sudan 21°51′N 25°01′E
Uwazima *see* Uwajima
Uweinat, Jebel *see* 'Uwaynāt, Jabal al
14 H14 **Uxbridge** Ontario, S Canada 44°07′N 79°07′W
Uxellodunum *see* Issoudun
Uxin Qi *see* Dabqig, N China
41 X12 **Uxmal, Ruinas** *ruins* Yucatán, SE Mexico
129 Q5 **Uy** ≈ Kazakhstan/Russian Federation
144 K15 **Uyaly** Kzylorda, S Kazakhstan 44°22′N 61°16′E
123 R8 **Uyandina** ≈ NE Russian Federation
162 J8 **Uyanga** *var.* Ongi. Övörhangay, C Mongolia 46°30′N 102°18′E
Üydzen *see* Manlay
Uyeda *see* Ueda
122 K5 **Uyedineniya, Ostrov** *island* N Russian Federation
77 V17 **Uyo** Akwa Ibom, S Nigeria 05°00′N 07°57′E
162 D8 **Üyönch** Hovd, W Mongolia 46°04′N 92°05′E
Uyuk *see* Oyyk
141 V13 **'Uyūn** SW Oman 17°19′N 53°50′E
57 K20 **Uyuni** Potosí, W Bolivia 20°27′S 66°48′W
57 J20 **Uyuni, Salar de** *wetland* SW Bolivia
146 I9 **Uzbekistan** *off.* Republic of Uzbekistan. ◆ *republic* C Asia
Uzbekistan, Republic of *see* Uzbekistan
158 D8 **Uzbel Shankou** *Rus.* Pereval Kyzyl-Dzhiik. *pass* China/Tajikistan
146 B11 **Uzboý** *prev. Rus.* Imeni 26 Bakinskikh Komissarov, *Turkm.* 26 Baku Komissarlary Adyndaky. Balkan Welaýaty, W Turkmenistan 39°24′N 54°04′E
119 J17 **Uzda** Minskaya Voblasts', C Belarus 53°29′N 27°10′E
103 N12 **Uzerche** Corrèze, C France 45°24′N 01°35′E
103 R14 **Uzès** Gard, S France 44°00′N 04°25′E
147 T10 **Uzgen** *Kir.* Özgön. Oshskaya Oblast', SW Kyrgyzstan 40°42′N 73°17′E
117 O3 **Uzh** ≈ N Ukraine
Uzhgorod *see* Uzhhorod
116 G7 **Uzhhorod** *Rus.* Uzhgorod; *prev.* Ungvár. Zakarpats'ka Oblast', W Ukraine 48°36′N 22°19′E
Uzi *see* Uji
112 K13 **Užice** *prev.* Titovo Užice. Serbia, W Serbia 43°52′N 19°51′E
Uzin *see* Uzyn
126 L5 **Uzlovaya** Tul'skaya Oblast', W Russian Federation 54°01′N 38°15′E
108 H7 **Uznach** Sankt Gallen, NE Switzerland 47°12′N 09°00′E
Uzunagach *see* Uzynagash
136 B10 **Uzunköprü** Edirne, NW Turkey 41°18′N 26°40′E
118 D11 **Užventis** Šiauliai, C Lithuania 55°49′N 22°38′E
117 P5 **Uzyn** *Rus.* Uzin. Kyyivs'ka Oblast', N Ukraine 49°52′N 30°27′E
145 U16 **Uzynagash** *prev.* Uzunagach. Almaty, SE Kazakhstan 43°08′N 76°20′E
145 N7 **Uzynkol'** *prev.* Lenin, Leninskoye. Kustanay, N Kazakhstan 54°05′N 65°23′E

V

Vääksy *see* Asikkala
83 H23 **Vaal** ≈ C South Africa
93 M14 **Vaala** Kainuu, C Finland 64°34′N 26°49′E
93 N19 **Vaalimaa** Etelä-Karjala, SE Finland 60°34′N 27°49′E
99 M19 **Vaals** Limburg, SE Netherlands 50°46′N 06°01′E
93 J16 **Vaasa** *Swe.* Vasa; *prev.* Nikolainkaupunki. Österbotten, W Finland 63°07′N 21°39′E
98 L10 **Vaassen** Gelderland, E Netherlands 52°18′N 05°59′E
118 G11 **Vabalninkas** Panevėžys, NE Lithuania 55°59′N 24°45′E
Vabkent *see* Vobkent
111 J22 **Vác** *Ger.* Waitzen. Pest, N Hungary 47°46′N 19°08′E
61 I14 **Vacaria** Rio Grande do Sul, S Brazil 28°31′S 50°52′W
35 N7 **Vacaville** California, W USA 38°21′N 121°59′W
103 R15 **Vaccarès, Étang de** ◎ SE France
114 I11 **Vacha** *var.* Vŭcha. ≈ SW Bulgaria
44 L10 **Vache, Île à** *island* SW Haiti
173 Y16 **Vacoas** W Mauritius 20°18′S 57°29′E
32 G10 **Vader** Washington, NW USA 46°23′N 122°58′W
94 D12 **Vadheim** Sogn Og Fjordane, S Norway 61°12′N 05°48′E
154 D11 **Vadodara** *prev.* Baroda. Gujarāt, W India 22°19′N 73°14′E
92 M8 **Vadsø** *Fin.* Vesisaari. Finnmark, N Norway 70°07′N 29°47′E
95 L17 **Vadstena** Östergötland, S Sweden 58°26′N 14°55′E
108 I8 **Vaduz** ● (Liechtenstein) W Liechtenstein 47°08′N 09°32′E
Våg *see* Váh
125 N12 **Vaga** ≈ NW Russian Federation
94 G11 **Vågåmo** Oppland, S Norway 61°52′N 09°06′E
112 D12 **Vaganski Vrh** ▲ W Croatia 44°24′N 15°32′E
95 A19 **Vágar** *Dan.* Vågø. *island* W Faroe Islands
Vágbeszterce *see* Považská Bystrica
95 L19 **Vaggeryd** Jönköping, S Sweden 57°30′N 14°10′E
137 T12 **Vagharshapat** *var.* Ejmiadzin, Ejmiatsin, Etchmiadzin, *Rus.* Echmiadzin. W Armenia 40°10′N 44°17′E
95 O16 **Vagnhärad** Södermanland, C Sweden 58°57′N 17°32′E
Vågø *see* Vágar
104 G7 **Vagos** Aveiro, N Portugal 40°33′N 08°42′W
Vágselye *see* Šal'a
92 H10 **Vågsfjorden** *fjord* N Norway
94 C10 **Vågsøy** *island* S Norway
Vágújhely *see* Nové Mesto nad Váhom
111 I21 **Váh** *Ger.* Waag, *Hung.* Vág. ≈ W Slovakia
93 K16 **Vähäkyrö** Österbotten, W Finland 63°04′N 22°05′E
191 X11 **Vahitahi** *atoll* Îles Tuamotu, E French Polynesia
Váhtjer *see* Gällivare
190 C9 **Vaiaku** *var.* Funafuti. Funafuti Atoll, SE Tuvalu 08°31′N 179°11′E
Vaidei *see* Vulcan
22 L4 **Vaiden** Mississippi, S USA 33°19′N 89°42′W
155 I23 **Vaigai** ≈ SE India
191 V16 **Vaihu** Easter Island, Chile, E Pacific Ocean 27°10′S 109°22′W
118 I6 **Väike Emajõgi** ≈ S Estonia
118 I4 **Väike-Maarja** *Ger.* Klein-Marien. Lääne-Virumaa, NE Estonia 59°07′N 26°16′E
Väike-Salatsi *see* Mazsalaca
37 R4 **Vail** Colorado, C USA 39°36′N 106°20′W
193 V15 **Vaina** Tongatapu, S Tonga 21°12′S 175°10′W
118 E5 **Väinameri** *prev.* Muhu Väin, *Ger.* Moon-Sund. *sea* E Baltic Sea
93 N18 **Vainikkala** Etelä-Karjala, SE Finland 60°54′N 28°18′E
118 D10 **Vainode** SW Latvia 56°25′N 21°52′E
155 H23 **Vaippār** ≈ SE India
191 W11 **Vairaatea** *atoll* Îles Tuamotu, C French Polynesia
191 R8 **Vairao** Tahiti, W French Polynesia 17°48′S 149°17′W
103 R14 **Vaison-la-Romaine** Vaucluse, SE France 44°15′N 05°04′E
190 G11 **Vaitupu** Île Uvea, E Wallis and Futuna 13°14′S 176°09′W
190 F7 **Vaitupu** *atoll* C Tuvalu
Vajdahunyad *see* Hunedoara
Vajdej *see* Vulcan
78 K12 **Vakaga** ◆ *prefecture* NE Central African Republic
114 H10 **Vakarel** Sofia, W Bulgaria 42°35′N 23°40′E
Vakav *see* Ustrem
137 O11 **Vakfıkebir** Trabzon, NE Turkey 41°03′N 39°19′E
122 J10 **Vakh** ≈ C Russian Federation
Vakhon, Qatorkŭhi *see* Nicholas Range
147 P14 **Vakhsh** SW Tajikistan 37°46′N 68°48′E
147 Q12 **Vakhsh** ≈ SW Tajikistan
127 P1 **Vakhtan** Nizhegorodskaya Oblast', W Russian Federation 58°00′N 46°43′E
94 C13 **Vaksdal** Hordaland, S Norway 60°29′N 05°45′E
Valachia *see* Wallachia
108 D11 **Valais** *Ger.* Wallis. ◆ *canton* SW Switzerland
113 M21 **Valamarës, Mali i** ▲ SE Albania 40°48′N 20°31′E
127 S2 **Valamaz** Udmurtskaya Respublika, NW Russian Federation 57°36′N 52°07′E
113 Q19 **Valandovo** SE FYR Macedonia 41°20′N 22°33′E
111 I18 **Valašské Meziříčí** *Ger.* Wallachisch-Meseritsch, *Pol.* Wałeckie Międzyrzecze. Zlínský Kraj, E Czech Republic 49°29′N 17°57′E
115 I17 **Valáxa** *island* Vóreies Sporádes, Greece, Aegean Sea
95 K16 **Vålberg** Värmland, C Sweden 59°24′N 13°12′E
116 H12 **Vâlcea** *prev.* Vîlcea. ◆ *county* SW Romania
Vâlcedrăm *see* Valchedram
114 G7 **Valchedram** *var.* Vâlcedrăm, Vŭlchedrŭm. Montana, NW Bulgaria 43°42′N 23°25′E
63 J16 **Valcheta** Río Negro, E Argentina 40°42′S 66°08′W
114 N8 **Valchi Dol** *var.* Vŭlchidol; *prev.* Kurt-Dere. Varna, E Bulgaria 43°25′N 27°33′E
15 P12 **Valcourt** Québec, SE Canada 45°28′N 72°18′W
Valdai Hills *see* Valdayskaya Vozvyshennost'
104 M3 **Valdavia** ≈ N Spain
124 I15 **Valday** Novgorodskaya Oblast', W Russian Federation 57°57′N 33°20′E
124 I15 **Valdayskaya Vozvyshennost'** *var.* Valdai Hills. *hill range* W Russian Federation
104 L9 **Valdecañas, Embalse de** ⊠ W Spain
118 E8 **Valdemārpils** *Ger.* Sassmacken. NW Latvia 57°23′N 22°36′E
95 N18 **Valdemarsvik** Östergötland, S Sweden 58°13′N 16°35′E
105 N8 **Valdemoro** Madrid, C Spain 40°12′N 03°40′W
105 O11 **Valdepeñas** Castilla-La Mancha, C Spain 38°46′N 03°24′W
104 L5 **Valderaduey** ≈ NE Spain
104 L5 **Valderas** Castilla y León, N Spain 42°05′N 05°27′W
105 T7 **Valderrobres** *var.* Vall-de-roures. Aragón, NE Spain 40°53′N 00°08′E
63 K17 **Valdés, Península** *peninsula* SE Argentina
56 C5 **Valdéz** *var.* Limones. Esmeraldas, NW Ecuador 01°13′N 79°00′W
39 S11 **Valdez** Alaska, USA 61°08′N 146°21′W
Valdia *see* Weldiya
103 U11 **Val d'Isère** Savoie, E France 45°27′N 07°03′E
63 G15 **Valdivia** Los Ríos, C Chile 39°50′S 73°13′W
Valdivia Bank *see* Valdivia Seamount
65 P17 **Valdivia Seamount** *var.* Valdivia Bank. *undersea feature* E Atlantic Ocean 26°15′S 06°25′E
103 N4 **Val-d'Oise** ◆ *department* N France
14 J8 **Val-d'Or** Québec, SE Canada 48°06′N 77°42′W
23 U8 **Valdosta** Georgia, SE USA 30°49′N 83°16′W
94 G13 **Valdres** *physical region* S Norway
32 L13 **Vale** Oregon, NW USA 43°59′N 117°15′W
116 F9 **Valea lui Mihai** *Hung.* Érmihályfalva. Bihor, NW Romania 47°31′N 22°08′E
11 N15 **Valemount** British Columbia, SW Canada 52°46′N 119°17′W
59 O17 **Valença** Bahia, E Brazil 13°22′S 39°06′W
104 F4 **Valença do Minho** Viana do Castelo, N Portugal 42°02′N 08°38′W
59 N14 **Valença do Piauí** Piauí, E Brazil 06°26′S 41°46′W
103 N8 **Valençay** Indre, C France 47°10′N 01°31′E
103 R13 **Valence** *anc.* Valentia, Valentia Julia, Ventia. Drôme, E France 44°56′N 04°54′E
105 S10 **Valencia** Valenciana, E Spain 39°29′N 00°24′W
54 K5 **Valencia** Carabobo, N Venezuela 10°12′N 68°02′W
105 R10 **Valencia** *Cat.* València. ◆ *province* Valenciana, E Spain
105 S10 **Valencia** ✈ Valencia, E Spain
104 I10 **Valencia de Alcántara** Extremadura, W Spain 39°25′N 07°14′W
104 L4 **Valencia de Don Juan** Castilla y León, N Spain 42°17′N 05°31′W
105 U9 **Valencia, Golfo de** *var.* Gulf of Valencia. *gulf* E Spain
Valencia, Gulf of *see* Valencia, Golfo de
97 A21 **Valencia Island** *Ir.* Dairbhre. *island* SW Ireland
105 R10 **Valenciana** *var.* Valencia, *Cat.* València; *anc.* Valentia. ◆ *autonomous community* NE Spain
Valencia/València *see* Valenciana
103 P2 **Valenciennes** Nord, N France 50°21′N 03°32′E
116 K13 **Vălenii de Munte** Prahova, SE Romania 45°11′N 26°02′E
Valentia *see* Valence, France
Valentia *see* Valenciana
Valentia Julia *see* Valence
103 T8 **Valentigney** Doubs, E France 47°27′N 06°49′E
28 M12 **Valentine** Nebraska, C USA 42°53′N 100°31′W
24 J10 **Valentine** Texas, SW USA 30°35′N 104°30′W
Valentine State *see* Oregon
106 C8 **Valenza** Piemonte, NW Italy 45°01′N 08°37′E
94 I13 **Våler** Hedmark, S Norway 60°39′N 11°52′E
54 I6 **Valera** Trujillo, NW Venezuela 09°21′N 70°38′W
192 M11 **Valerie Guyot** S Pacific Ocean 33°00′S 164°00′W
Valetta *see* Valletta
118 I7 **Valga** *Ger.* Walk, *Latv.* Valka. Valgamaa, S Estonia 57°48′N 26°04′E
118 I7 **Valgamaa** *var.* Valga Maakond. ◆ *province* S Estonia
Valga Maakond *see* Valgamaa
43 Q15 **Valiente, Península** *peninsula* NW Panama
103 X16 **Valinco, Golfe de** *gulf* Corse, France, C Mediterranean Sea
112 L12 **Valjevo** Serbia, W Serbia 44°17′N 19°54′E
Valjok *see* Válljohka
118 I7 **Valka** *Ger.* Walk. N Latvia 57°48′N 26°01′E
Valka *see* Valga
93 L18 **Valkeakoski** Pirkanmaa, W Finland 61°17′N 24°05′E
93 M19 **Valkeala** Kymenlaakso, S Finland 60°55′N 26°49′E
99 L18 **Valkenburg** Limburg, SE Netherlands 50°52′N 05°50′E
99 K15 **Valkenswaard** Noord-Brabant, S Netherlands 51°21′N 05°29′E
119 G15 **Valkininkai** Alytus, S Lithuania 54°22′N 24°51′E
117 U5 **Valky** Kharkivs'ka Oblast', E Ukraine 49°51′N 35°40′E
41 Y12 **Valladolid** Yucatán, SE Mexico 20°39′N 88°13′W
104 M5 **Valladolid** Castilla y León, NW Spain 41°39′N 04°45′W
104 L5 **Valladolid** ◆ *province* Castilla y León, N Spain
103 U15 **Vallauris** Alpes-Maritimes, SE France 43°34′N 07°03′E
Vall-de-roures *see* Valderrobres
Vall D'Uxó *see* La Vall d'Uixó
95 E16 **Valle** Aust-Agder, S Norway 59°13′N 07°33′E
105 N2 **Valle** Cantabria, N Spain 43°14′N 04°16′W
42 H8 **Valle** ◆ *department* S Honduras
105 N8 **Vallecas** Madrid, C Spain 40°23′N 03°37′W
37 Q8 **Vallecito Reservoir** ⊠ Colorado, C USA
106 A7 **Valle d'Aosta** *Fr.* Vallée d'Aoste. ◆ *region* NW Italy
41 O14 **Valle de Bravo** México, S Mexico 19°19′N 100°08′W
55 N5 **Valle de Guanape** Anzoátegui, N Venezuela 09°54′N 65°41′W

◆ Country
● Country Capital
◇ Dependent Territory
○ Dependent Territory Capital
◆ Administrative Regions
✈ International Airport
▲ Mountain
▲▲ Mountain Range
Volcano
≈ River
◎ Lake
⊠ Reservoir

VALLE DE LA PASCUA – VERDE, CAPE

54 M6 **Valle de La Pascua** Guárico, N Venezuela 09°15′N 66°00′W
54 B11 **Valle del Cauca** *off.* Departamento del Valle del Cauca. ◈ *province* W Colombia
Valle del Cauca, Departamento del *see* Valle del Cauca
41 N13 **Valle de Santiago** Guanajuato, C Mexico 20°25′N 101°15′W
40 J7 **Valle de Zaragoza** Chihuahua, N Mexico 27°25′N 105°50′W
54 G5 **Valledupar** Cesar, N Colombia 10°31′N 73°16′W
Vallée d'Aoste *see* Valle d'Aosta
76 G10 **Vallée de Ferlo** ≈ NW Senegal
57 M19 **Vallegrande** Santa Cruz, C Bolivia 18°30′S 64°06′W
41 P8 **Valle Hermoso** Tamaulipas, C Mexico 25°39′N 97°49′W
35 N8 **Vallejo** California, W USA 38°08′N 122°16′W
62 G8 **Vallenar** Atacama, N Chile 28°35′S 70°44′W
95 O15 **Vallentuna** Stockholm, C Sweden 59°32′N 18°05′E
121 P16 **Valletta** *prev.* Valetta. ● (Malta) E Malta 35°54′N 14°31′E
27 N6 **Valley Center** Kansas, C USA 37°49′N 97°22′W
29 Q5 **Valley City** North Dakota, N USA 46°57′N 97°58′W
32 I15 **Valley Falls** Oregon, NW USA 42°28′N 120°16′W
Valleyfield *see* Salaberry-de-Valleyfield
21 S4 **Valley Head** West Virginia, NE USA 38°33′N 80°01′W
25 T8 **Valley Mills** Texas, SW USA 31°36′N 97°27′W
75 W10 **Valley of the Kings** *ancient monument* E Egypt
29 R11 **Valley Springs** South Dakota, N USA 43°34′N 96°28′W
20 K5 **Valley Station** Kentucky, S USA 38°06′N 85°52′W
11 O13 **Valleyview** Alberta, W Canada 55°02′N 117°17′W
25 T5 **Valley View** Texas, SW USA 33°27′N 97°08′W
61 C21 **Vallimanca, Arroyo** ≈ E Argentina
92 L9 **Válljohka** *var.* Valjok. Finnmark, N Norway 69°40′N 25°52′E
107 M19 **Vallo della Lucania** Campania, S Italy 40°13′N 15°15′E
108 B9 **Vallorbe** Vaud, W Switzerland 46°43′N 06°21′E
105 V6 **Valls** Cataluña, NE Spain 41°18′N 01°15′E
94 N11 **Vallsta** Gävleborg, C Sweden 61°30′N 16°25′E
94 N12 **Vallvik** Gävleborg, C Sweden 61°10′N 17°15′E
11 T17 **Val Marie** Saskatchewan, S Canada 49°15′N 107°44′W
118 H7 **Valmiera** *Est.* Volmari, *Ger.* Wolmar. N Latvia 57°34′N 25°26′E
105 N3 **Valnera** ▲ N Spain 43°08′N 03°39′W
102 J3 **Valognes** Manche, N France 49°31′N 01°28′W
Valona *see* Vlorë
Valona Bay *see* Vlorës, Gjiri i
104 G6 **Valongo** *var.* Valongo de Gaia. Porto, N Portugal 41°11′N 08°30′W
Valongo de Gaia *see* Valongo
104 M5 **Valoria la Buena** Castilla y León, N Spain 41°48′N 04°33′W
119 J15 **Valozhyn** *Pol.* Wołożyn, *Rus.* Volozhin. Minskaya Voblasts', C Belarus 54°05′N 26°32′E
104 I5 **Valpaços** Vila Real, N Portugal 41°36′N 07°17′W
62 G11 **Valparaíso** Valparaíso, C Chile 33°05′S 71°18′W
40 L11 **Valparaíso** Zacatecas, C Mexico 22°49′N 103°28′W
23 P8 **Valparaiso** Florida, SE USA 30°30′N 86°28′W
31 N11 **Valparaiso** Indiana, N USA 41°28′N 87°04′W
62 G11 **Valparaíso** *off.* Región de Valparaíso. ◈ *region* C Chile
Valparaíso, Región de *see* Valparaíso
Valpo *see* Valpovo
112 I9 **Valpovo** *Hung.* Valpo. Osijek-Baranja, E Croatia 45°40′N 18°25′E
103 R14 **Valréas** Vaucluse, SE France 44°22′N 05°00′E
Vals *see* Vals-Platz
154 D12 **Valsād** *prev.* Bulsar. Gujarāt, W India 20°40′N 72°55′E
Valsbaai *see* False Bay
171 T12 **Valse Pisang, Kepulauan** *island group* E Indonesia
108 H9 **Vals-Platz** *var.* Vals. Graubünden, S Switzerland 46°39′N 09°09′E
171 X16 **Vals, Tanjung** *headland* Papua, SE Indonesia 08°26′S 137°35′E
93 N15 **Valtimo** Pohjois-Karjala, E Finland 63°39′N 28°49′E
115 D17 **Váltou** ▲▲ C Greece
127 O12 **Valuyevka** Rostovskaya Oblast', SW Russian Federation 46°48′N 43°49′E
126 K9 **Valuyki** Belgorodskaya Oblast', W Russian Federation 50°11′N 38°07′E
36 L2 **Val Verda** Utah, W USA 40°51′N 111°53′W
64 N12 **Valverde** Hierro, Islas Canarias, Spain, NE Atlantic Ocean 27°48′N 17°55′W
104 I13 **Valverde del Camino** Andalucía, S Spain 37°35′N 06°45′W
95 G23 **Vamdrup** Syddanmark, C Denmark 55°26′N 09°18′E
94 L12 **Våmhus** Dalarna, C Sweden 61°07′N 14°30′E
93 K18 **Vammala** Pirkanmaa, SW Finland 61°20′N 22°55′E
Vámosudvarhely *see* Odorheiu Secuiesc
137 S14 **Van** Van, E Turkey 38°30′N 43°23′E
25 V7 **Van** Texas, SW USA 32°31′N 95°38′W
137 T14 **Van** ◈ *province* E Turkey
137 T11 **Vanadzor** *prev.* Kirovakan. N Armenia 40°49′N 44°29′E
25 U5 **Van Alstyne** Texas, SW USA 33°25′N 96°34′W
33 W10 **Vananda** Montana, NW USA 46°22′N 106°58′W
116 I11 **Vânători** *Hung.* Héjjasfalva; *prev.* Vînători. Mureş, C Romania 46°14′N 24°56′E
191 W12 **Vanavana** *atoll* Îles Tuamotu, SE French Polynesia
Vana-Vändra *see* Vändra
122 M11 **Vanavara** Krasnoyarskiy Kray, C Russian Federation 60°19′N 102°19′E
15 Q8 **Van Bruyssel** Québec, SE Canada 47°56′N 72°08′W
27 R10 **Van Buren** Arkansas, C USA 35°28′N 94°25′W
19 S1 **Van Buren** Maine, NE USA 47°07′N 67°57′W
27 W7 **Van Buren** Missouri, C USA 37°00′N 91°00′W
19 T5 **Vanceboro** Maine, NE USA 45°31′N 67°25′W
21 W10 **Vanceboro** North Carolina, SE USA 35°16′N 77°06′W
21 O4 **Vanceburg** Kentucky, S USA 38°36′N 84°40′W
45 W10 **Vance W. Amory** ✈ Nevis, Saint Kitts and Nevis 17°08′N 62°36′W
Vanch *see* Vanj
10 L17 **Vancouver** British Columbia, SW Canada 49°13′N 123°06′W
32 G11 **Vancouver** Washington, NW USA 45°38′N 122°39′W
10 L17 **Vancouver** ✈ British Columbia, SW Canada 49°03′N 123°00′W
10 K16 **Vancouver Island** *island* British Columbia, SW Canada
Vanda *see* Vantaa
171 X13 **Van Daalen** ≈ Papua, E Indonesia
30 L15 **Vandalia** Illinois, N USA 38°57′N 89°05′W
27 V3 **Vandalia** Missouri, C USA 39°18′N 91°29′W
31 R13 **Vandalia** Ohio, N USA 39°53′N 84°12′W
25 U13 **Vanderbilt** Texas, SW USA 28°45′N 96°37′W
31 Q10 **Vandercook Lake** Michigan, N USA 42°11′N 84°23′W
10 L14 **Vanderhoof** British Columbia, SW Canada 53°54′N 124°00′W
18 K8 **Vanderwhacker Mountain** ▲ New York, NE USA 43°54′N 74°06′W
181 P1 **Van Diemen Gulf** *gulf* Northern Territory, N Australia
Van Diemen's Land *see* Tasmania
118 H5 **Vändra** *Ger.* Fennern; *prev.* Vana-Vändra. Pärnumaa, SW Estonia 58°39′N 25°00′E
Vandsburg *see* Więcbork
34 L4 **Van Duzen River** ≈ California, W USA
118 F13 **Vandžiogala** Kaunas, C Lithuania 55°07′N 23°55′E
41 N10 **Vanegas** San Luis Potosí, C Mexico 23°53′N 100°55′W
Vaner, Lake *see* Vänern
95 K17 **Vänern** *Eng.* Lake Vaner; *prev.* Lake Vener. ◎ S Sweden
95 J18 **Vänersborg** Västra Götaland, S Sweden 58°16′N 12°22′E
94 F12 **Vang** Oppland, S Norway 61°07′N 08°34′E
172 I7 **Vangaindrano** Fianarantsoa, SE Madagascar 23°21′S 47°35′E
137 S14 **Van Gölü** *Eng.* Lake Van; *anc.* Thospitis. *salt lake* E Turkey
186 L9 **Vangunu** *island* New Georgia Islands, NW Solomon Islands
24 J9 **Van Horn** Texas, SW USA 31°03′N 104°51′W
187 Q11 **Vanikolo** *var.* Vanikoro. *island* Santa Cruz Islands, E Solomon Islands
Vanikoro *see* Vanikolo
186 A5 **Vanimo** West Sepik, NW Papua New Guinea 02°40′S 141°17′E
123 T13 **Vanino** Khabarovskiy Kray, SE Russian Federation 49°10′N 140°18′E
155 G19 **Vānīvilāsa Sāgara** ◎ SW India
147 S13 **Vanj** *Rus.* Vanch. S Tajikistan 38°22′N 71°27′E
116 G14 **Vânju Mare** *prev.* Vînju Mare. Mehedinţi, SW Romania 44°25′N 22°52′E
15 N12 **Vankleek Hill** Ontario, SE Canada 45°32′N 74°39′W
Van, Lake *see* Van Gölü
93 I16 **Vännäs** Västerbotten, N Sweden 63°54′N 19°43′E
93 I15 **Vännäsby** Västerbotten, N Sweden 63°55′N 19°53′E
102 H7 **Vannes** *anc.* Dariorigum. Morbihan, NW France 47°40′N 02°45′W
92 I8 **Vannøya** *island* N Norway
103 T12 **Vanoise, Massif de la** ▲▲ E France
83 E24 **Vanrhynsdorp** Western Cape, SW South Africa 31°36′S 18°45′E
21 P7 **Vansant** Virginia, NE USA 37°13′N 82°03′W
94 L13 **Vansbro** Dalarna, C Sweden 60°32′N 14°15′E
95 D18 **Vanse** Vest-Agder, S Norway 58°04′N 06°40′E
9 P7 **Vansittart Island** *island* Nunavut, NE Canada
93 M20 **Vantaa** *Swe.* Vanda. Uusimaa, S Finland 60°18′N 25°01′E
93 L19 **Vantaa** ✈ (Helsinki) Uusimaa, S Finland 60°18′N 25°01′E
32 J9 **Vantage** Washington, NW USA 46°55′N 119°55′W
187 Z14 **Vanua Balavu** *prev.* Vanua Mbalavu. *island* Lau Group, E Fiji
187 R12 **Vanua Lava** *island* Banks Islands, N Vanuatu
187 Y13 **Vanua Levu** *island* N Fiji
Vanua Mbalavu *see* Vanua Balavu
187 R12 **Vanuatu** *off.* Republic of Vanuatu; *prev.* New Hebrides. ◆ *republic* SW Pacific Ocean
175 P8 **Vanuatu** *island group* SW Pacific Ocean
Vanuatu, Republic of *see* Vanuatu
31 Q12 **Van Wert** Ohio, N USA 40°52′N 84°34′W
187 Q17 **Vao** Province Sud, S New Caledonia 22°40′S 167°29′E
Vapincum *see* Gap
117 N7 **Vapnyarka** Vinnyts'ka Oblast', C Ukraine 48°31′S 28°44′E
103 T15 **Var** ◈ *department* SE France
103 U14 **Var** ≈ SE France
95 J18 **Vara** Västra Götaland, S Sweden 58°16′N 12°57′E
Varađinska Županija *see* Varaždin
118 J10 **Varakļāni** C Latvia 56°36′N 26°40′E
106 C7 **Varallo** Piemonte, NE Italy 45°51′N 08°16′E
143 O5 **Varāmīn** *var.* Veramin. Tehrān, N Iran 35°19′N 51°40′E
153 N14 **Vārānasi** *prev.* Banaras, Benares, *hist.* Kasi. Uttar Pradesh, N India 25°20′N 83°E
125 T3 **Varandey** Nenetskiy Avtonomnyy Okrug, NW Russian Federation 68°48′N 57°54′E
92 M8 **Varangerbotn** *Lapp.* Vuonnabahta. Finnmark, N Norway 70°09′N 28°28′E
92 M8 **Varangerfjorden** *Lapp.* Várjjatvuotna. *fjord* N Norway
92 M8 **Varangerhalvøya** *Lapp.* Várnjárga. *peninsula* N Norway
Varannó *see* Vranov nad Topl'ou
107 M15 **Varano, Lago di** ◎ SE Italy
118 J13 **Varapayeva** *Rus.* Voropayevo. Vitsyebskaya Voblasts', NW Belarus 55°09′N 27°13′E
Varasd *see* Varaždin
112 E7 **Varaždin** *Ger.* Warasdin, *Hung.* Varasd. Varaždin, N Croatia 46°18′N 16°21′E
112 E7 **Varaždin** *off.* Varađinska Županija. ◈ *province* N Croatia
106 C10 **Varazze** Liguria, NW Italy 44°21′N 08°35′E
95 J20 **Varberg** Halland, S Sweden 57°06′N 12°15′E
114 J11 **Varbitsa** *var,* Vŭrbitsa; *prev.* Filevo. Haskovo, S Bulgaria 42°02′N 25°25′E
114 J12 **Varbitsa** ≈ S Bulgaria
113 Q19 **Vardar** *Gk.* Axiós. ≈ FYR Macedonia/Greece *see also* Axiós
Vardar *see* Axiós
95 F23 **Varde** Syddtjylland, W Denmark 55°38′N 08°31′E
137 V12 **Vardenis** E Armenia 40°11′N 45°43′E
92 N8 **Vardø** *Fin.* Vuoreija. Finnmark, N Norway 70°22′N 31°06′E
115 E18 **Vardoúsia** ▲▲ C Greece
Vareia *see* Logroño
100 G10 **Varel** Niedersachsen, NW Germany 53°24′N 08°07′E
119 G15 **Varėna** *Pol.* Orany. Alytus, S Lithuania 54°13′N 24°35′E
15 O12 **Varennes** Québec, SE Canada 45°42′N 73°25′W
103 P10 **Varennes-sur-Allier** Allier, C France 46°17′N 03°24′E
112 I12 **Vareš** Federacija Bosne I Hercegovine, E Bosnia and Herzegovina 44°12′N 18°19′E
106 D7 **Varese** Lombardia, N Italy 45°49′N 08°50′E
116 J12 **Vârful Moldoveanu** *var.* Moldoveanul; *prev.* Vîrful Moldoveanu. ▲ C Romania 45°35′N 24°48′E
Varganzi *see* Warganza
95 J18 **Vårgårda** Västra Götaland, S Sweden 58°00′N 12°49′E
54 L4 **Vargas** *off.* Estado Vargas. ◈ *state* N Venezuela
95 J18 **Vargön** Västra Götaland, S Sweden 58°21′N 12°22′E
95 C17 **Varhaug** Rogaland, S Norway 58°37′N 05°39′E
Várjjatvuotna *see* Varangerfjorden
93 N17 **Varkaus** Pohjois-Savo, C Finland 62°20′N 27°50′E
92 J2 **Varmahlið** Norðurland Vestra, N Iceland 65°32′N 19°33′W
95 J15 **Värmland** ◈ *county* C Sweden
95 K16 **Värmlandsnäs** *peninsula* S Sweden
114 N8 **Varna** *prev.* Stalin; *anc.* Odessus. Varna, E Bulgaria 43°14′N 27°56′E
114 N8 **Varna** ◈ *province* E Bulgaria
114 N8 **Varna** ✈ Varna, E Bulgaria 43°16′N 27°52′E
95 L20 **Värnamo** Jönköping, S Sweden 57°11′N 14°03′E
114 N8 **Varnenski Zaliv** *prev.* Stalinski Zaliv. *bay* E Bulgaria
114 N8 **Varnensko Ezero** *estuary* E Bulgaria
118 D11 **Varniai** Telšiai, W Lithuania 55°45′N 22°22′E
Várnjárga *see* Varangerhalvøya
Varnoús *see* Baba
111 D14 **Varnsdorf** *Ger.* Warnsdorf. Ústecký Kraj, NW Czech Republic 50°57′N 14°35′E
111 I23 **Várpalota** Veszprém, W Hungary 47°12′N 18°08′E
Varshava *see* Warszawa
114 G8 **Varshets** *var.* Vŭrshets. Montana, NW Bulgaria 43°14′N 23°20′E
93 K20 **Varsinais-Suomi** *Swe.* Egentliga Finland. ◈ *region* W Finland
118 K6 **Värska** Põlvamaa, SE Estonia 57°58′N 27°37′E
98 N12 **Varsseveld** Gelderland, E Netherlands 51°55′N 06°28′E
115 D19 **Vartholomió** *prev.* Vartholomión. Dytikí Elláda, S Greece 37°52′N 21°12′E
Vartholomión *see* Vartholomió
137 Q14 **Varto** Muş, E Turkey 39°10′N 41°28′E
95 K18 **Vartofta** Västra Götaland, S Sweden 58°06′N 13°40′E
93 O17 **Värtsilä** Pohjois-Karjala, E Finland 62°10′N 30°35′E
Värtsilä *see* Vyartsilya
117 R4 **Varva** Chernihivs'ka Oblast', NE Ukraine 50°31′N 32°43′E
59 H18 **Várzea Grande** Mato Grosso, SW Brazil 15°39′S 56°08′W
106 D9 **Varzi** Lombardia, N Italy 44°51′N 09°13′E
Varzimanor Ayni *see* Ayní
124 K5 **Varzuga** ≈ NW Russian Federation
103 P8 **Varzy** Nièvre, C France 47°22′N 03°22′E
111 G23 **Vas** *off.* Vas Megye. ◈ *county* W Hungary
Vasa *see* Vaasa
190 A9 **Vasafua** *island* Funafuti Atoll, C Tuvalu
111 O21 **Vásárosnamény** Szabolcs-Szatmár-Bereg, E Hungary 48°10′N 22°18′E
104 H13 **Vascão, Ribeira de** ≈ S Portugal
116 G10 **Vaşcău** *Hung.* Vaskoh. Bihor, NE Romania 46°28′N 22°30′E
Vascongadas, Provincias *see* País Vasco
Vashess Bay *see* Vaskess Bay
125 O8 **Vashka** ≈ NW Russian Federation
Vāsht *see* Khāsh
Vasilevichi *see* Vasilyevichy
115 G14 **Vasiliká** Kentrikí Makedonía, NE Greece 40°28′N 23°08′E
115 C18 **Vasilikí** Lefkáda, Iónia Nisiá, Greece, C Mediterranean Sea 38°36′N 20°37′E
115 K25 **Vasilikí** Kríti, Greece, E Mediterranean Sea 35°04′N 25°49′E
119 G16 **Vasilishki** *Pol.* Wasiliszki. Hrodzyenskaya Voblasts', W Belarus 53°47′N 24°51′E
Vasil Kolarov *see* Pamporovo
Vasil'kov *see* Vasyl'kiv
119 N19 **Vasilyevichy** *Rus.* Vasilevichi. Homyel'skaya Voblasts', SE Belarus 52°15′N 29°50′E
191 Y3 **Vaskess Bay** *var.* Vashess Bay. *bay* Kiritimati, E Kiribati
Vaskoh *see* Vaşcău
Vaskohsziklás *see* Ştei
116 M10 **Vaslui** Vaslui, C Romania 46°38′N 27°44′E
116 L11 **Vaslui** ◈ *county* NE Romania
Vas Megye *see* Vas
31 R8 **Vassar** Michigan, N USA 43°22′N 83°34′W
95 E15 **Vassdalsegga** ▲ S Norway 59°47′N 07°07′E
60 P9 **Vassouras** Rio de Janeiro, SE Brazil 22°24′S 43°38′W
95 N15 **Västerås** Västmanland, C Sweden 59°37′N 16°33′E
93 G15 **Västerbotten** ◈ *county* N Sweden
94 K12 **Västerdalälven** ≈ C Sweden
95 O16 **Västerhaninge** Stockholm, C Sweden 59°07′N 18°06′E
94 M10 **Västernorrland** ◈ *county* C Sweden
95 N19 **Västervik** Kalmar, S Sweden 57°44′N 16°40′E
95 M15 **Västmanland** ◈ *county* C Sweden
107 L15 **Vasto** *anc.* Histonium. Abruzzo, C Italy 42°07′N 14°43′E
95 J19 **Västra Götaland** ◈ *county* S Sweden
95 J16 **Västra Silen** ◎ S Sweden
111 G23 **Vasvár** *Ger.* Eisenburg. Vas, W Hungary 47°03′N 16°48′E
117 U9 **Vasylivka** Zaporiz'ka Oblast', SE Ukraine 47°26′N 35°18′E
117 O5 **Vasyl'kiv** *var.* Vasil'kov. Kyyivs'ka Oblast', N Ukraine 50°12′N 30°18′E
117 V8 **Vasyl'kivka** Dnipropetrovs'ka Oblast', E Ukraine 48°12′N 36°00′E
122 I11 **Vasyugan** ≈ C Russian Federation
103 N8 **Vatan** Indre, C France 47°06′N 01°49′E
Vaté *see* Efate
115 C18 **Vathy** *prev.* Itháki. Itháki, Iónia Nisiá, Greece, C Mediterranean Sea 38°22′N 20°43′E
107 G15 **Vatican City** *off.* Vatican City. ◆ *papal state* S Europe
Vatican City *see* Vatican City
107 M22 **Vaticano, Capo** *headland* S Italy 38°37′N 15°49′E
92 K3 **Vatnajökull** *glacier* SE Iceland
95 P15 **Vätö** Stockholm, C Sweden 59°48′N 18°55′E
187 Z16 **Vatoa** *island* Lau Group, SE Fiji
172 J5 **Vatomandry** Toamasina, E Madagascar 19°20′S 48°58′E
116 J9 **Vatra Dornei** *Ger.* Dorna Watra. Suceava, NE Romania 47°20′N 25°21′E
116 J9 **Vatra Moldoviţei** Suceava, NE Romania 47°37′N 25°36′E
Vatter, Lake *see* Vättern
95 L18 **Vättern** *Eng.* Lake Vatter; *prev.* Lake Vetter. ◎ S Sweden
187 X5 **Vatulele** *island* SW Fiji
117 P7 **Vatutine** Cherkas'ka Oblast', C Ukraine 49°01′N 31°04′E
187 W15 **Vatu Vara** *island* Lau Group, E Fiji
103 R14 **Vaucluse** ◈ *department* SE France
103 S5 **Vaucouleurs** Meuse, NE France 48°37′N 05°38′E
108 B9 **Vaud** *Ger.* Waadt. ◈ *canton* SW Switzerland
15 N12 **Vaudreuil** Québec, SE Canada 45°24′N 74°01′W
37 T12 **Vaughn** New Mexico, SW USA 34°36′N 105°12′W
54 I14 **Vaupés** *off.* Comisaría del Vaupés. ◈ *province* SE Colombia
Vaupés, Comisaría del *see* Vaupés
54 J13 **Vaupés, Río** *var.* Rio Uaupés. ≈ Brazil/Colombia *see also* Uaupés, Rio
Vaupés, Río *see* Uaupés, Rio
103 Q15 **Vauvert** Gard, S France 43°42′N 04°16′E
11 R17 **Vauxhall** Alberta, SW Canada 50°05′N 112°09′W
99 K23 **Vaux-sur-Sûre** Luxembourg, SE Belgium 49°55′N 05°34′E
172 J4 **Vavatenina** Toamasina, E Madagascar 17°25′S 49°11′E
193 Y14 **Vava'u Group** *island group* N Tonga
76 M16 **Vavoua** W Ivory Coast 07°23′N 06°29′W
127 S2 **Vavozh** Udmurtskaya Respublika, NW Russian Federation 56°48′N 51°53′E
155 K23 **Vavuniya** Northern Province, N Sri Lanka 08°45′N 80°30′E
119 G17 **Vawkavysk** *Pol.* Wołkowysk, *Rus.* Volkovysk. Hrodzyenskaya Voblasts', W Belarus 53°10′N 24°28′E
119 F17 **Vawkavyskaye Wzvyshsha** *Rus.* Volkovyskiye Vysoty. *hill range* W Belarus
95 P15 **Vaxholm** Stockholm, C Sweden 59°25′N 18°21′E
95 L21 **Växjö** *var.* Vexiö. Kronoberg, S Sweden 56°52′N 14°50′E
125 T1 **Vaygach, Ostrov** *island* NW Russian Federation
137 V13 **Vayk'** *prev.* Azizbekov. SE Armenia 39°41′N 45°28′E
Vazáš *see* Vittangi
125 P8 **Vazhgort** *prev.* Chasovo. Respublika Komi, NW Russian Federation 64°06′N 46°44′E
45 V10 **V. C. Bird** ✈ (St. John's) Antigua, Antigua and Barbuda 17°07′N 61°49′W
167 R13 **Veal Renh** *prev.* Phumĭ Veal Renh. Kâmpôt, SW Cambodia 10°43′N 103°49′E
29 Q7 **Veblen** South Dakota, N USA 45°50′N 97°17′W
98 N9 **Vecht** *Ger.* Vechte. ≈ Germany/Netherlands *see also* Vechte
Vecht *see* Vechte
100 G12 **Vechta** Niedersachsen, NW Germany 52°44′N 08°16′E
100 E12 **Vechte** *Dut.* Vecht. ≈ Germany/Netherlands *see also* Vecht
Vechte *see* Vecht
118 I8 **Vecpiebalga** C Latvia 57°03′N 25°47′E
118 G9 **Vecumnieki** C Latvia 56°36′N 24°30′E
95 C16 **Vedavågen** Rogaland, S Norway 59°18′N 5°13′E
Vedāvati *see* Hagari
95 J20 **Veddige** Halland, S Sweden 57°16′N 12°19′E
116 J15 **Vedea** ≈ S Romania
127 P16 **Vedeno** Chechenskaya Respublika, SW Russian Federation 42°57′N 46°02′E
95 C16 **Vedvågen** Rogaland, S Norway 59°18′N 05°13′E
98 O6 **Veendam** Groningen, NE Netherlands 53°05′N 06°53′E
98 K12 **Veenendaal** Utrecht, C Netherlands 52°03′N 05°33′E
99 E14 **Veere** Zeeland, SW Netherlands 51°33′N 03°40′E
24 M2 **Vega** Texas, SW USA 35°14′N 102°26′W
92 E13 **Vega** *island* C Norway
45 T5 **Vega Baja** C Puerto Rico 18°27′N 66°23′W
38 D17 **Vega Point** *headland* Kiska Island, Alaska, USA 51°49′N 177°19′E
95 F17 **Vegår** ◎ S Norway
99 K14 **Veghel** Noord-Brabant, S Netherlands 51°37′N 05°33′E
Veglia *see* Krk
114 E13 **Vegorítida, Límni** *var.* Límni Vegorítis. ◎ N Greece
Vegorítis, Límni *see* Vegorítida, Límni
11 Q14 **Vegreville** Alberta, SW Canada 53°30′N 112°02′W
95 K21 **Veinge** Halland, S Sweden 56°33′N 13°04′E
61 B21 **Veinticinco de Mayo** *var.* 25 de Mayo. Buenos Aires, E Argentina 35°27′S 60°11′W
63 I14 **Veinticinco de Mayo** La Pampa, C Argentina 37°45′S 67°40′W
119 F15 **Veisiejai** Alytus, S Lithuania 54°06′N 23°42′E
95 F23 **Vejen** Syddtjylland, W Denmark 55°29′N 09°13′E
104 K16 **Vejer de la Frontera** Andalucía, S Spain 36°15′N 05°58′W
95 G23 **Vejle** Syddanmark, C Denmark 55°43′N 09°33′E
114 M7 **Vekilski** Shumen, NE Bulgaria 43°33′N 27°19′E
54 G3 **Vela, Cabo de la** *headland* NE Colombia 12°14′N 72°13′W
Vela Goa *see* Goa
113 F15 **Vela Luka** Dubrovnik-Neretva, S Croatia 42°57′N 16°43′E
61 G19 **Velázquez** Rocha, E Uruguay 34°05′S 54°16′W
101 E15 **Velbert** Nordrhein-Westfalen, W Germany 51°22′N 07°03′E
109 S9 **Velden** Kärnten, S Austria 46°37′N 13°59′E
Veldes *see* Bled
99 K15 **Veldhoven** Noord-Brabant, S Netherlands 51°24′N 05°24′E
112 C11 **Velebit** ▲▲ C Croatia
114 N11 **Veleka** ≈ SE Bulgaria
109 V10 **Velenje** *Ger.* Wöllan. N Slovenia 46°22′N 15°07′E
190 E12 **Vele, Pointe** *headland* Île Futuna, S Wallis and Futuna
113 O18 **Veles** *Turk.* Köprülü. C FYR Macedonia 41°43′N 21°49′E
113 M20 **Velešta** SW FYR Macedonia 41°16′N 20°37′E
115 F16 **Velestíno** *prev.* Velestínon. Thessalía, C Greece 39°23′N 22°45′E
Velestínon *see* Velestíno
Velevshchina *see* Vyelyewshchyna
54 F9 **Vélez** Santander, C Colombia 06°02′N 73°43′W
105 Q13 **Vélez Blanco** Andalucía, S Spain 37°43′N 02°07′W
104 M17 **Vélez de la Gomera, Peñon de** *island group* S Spain
105 N15 **Vélez-Málaga** Andalucía, S Spain 36°47′N 04°06′W
105 Q13 **Vélez Rubio** Andalucía, S Spain 37°39′N 02°04′W
Velha Goa *see* Goa
Velho *see* Porto Velho
112 E8 **Velika Gorica** Zagreb, N Croatia 45°43′N 16°03′E
112 C9 **Velika Kapela** ▲▲ NW Croatia
Velika Kikinda *see* Kikinda
112 D10 **Velika Kladuša** Federacija Bosne I Hercegovine, NW Bosnia and Herzegovina 45°10′N 15°48′E
112 N11 **Velika Morava** *var.* Glavn'a Morava, Morava, *Ger.* Grosse Morava. ≈ C Serbia
112 N12 **Velika Plana** Serbia, C Serbia 44°20′N 21°01′E
109 U10 **Velika Raduha** ▲ N Slovenia 46°24′N 14°46′E
123 V7 **Velikaya** ≈ NE Russian Federation
124 F15 **Velikaya** ≈ W Russian Federation
Velikaya Berestovitsa *see* Vyalikaya Byerastavitsa
Velikaya Lepetikha *see* Velyka Lepetykha
Veliki Bečkerek *see* Zrenjanin
112 P12 **Veliki Krš** *var.* Stol. ▲ E Serbia 44°10′N 22°09′E
114 L8 **Veliki Preslav** *prev.* Preslav. Shumen, NE Bulgaria 43°09′N 26°50′E
112 B9 **Veliki Risnjak** ▲ NW Croatia 45°30′N 14°31′E
109 T13 **Veliki Snežnik** *Ger.* Schneeberg, *It.* Monte Nevoso. ▲ SW Slovenia 45°36′N 14°25′E
112 J13 **Veliki Stolac** ▲ E Bosnia and Herzegovina 43°55′N 19°15′E
Velikiy Bor *see* Vyaliki Bor
124 G16 **Velikiye Luki** Pskovskaya Oblast', W Russian Federation 56°20′N 30°27′E
124 H14 **Velikiy Novgorod** *prev.* Novgorod. Novgorodskaya Oblast', W Russian Federation 58°32′N 31°15′E
125 P12 **Velikiy Ustyug** Vologodskaya Oblast', NW Russian Federation 60°46′N 46°18′E
112 N11 **Veliko Gradište** Serbia, NE Serbia 44°46′N 21°28′E
155 I18 **Velikonda Range** ▲▲ SE India
114 K9 **Veliko Tarnovo** *prev.* Tirnovo, Trnovo, Tŭrnovo, *var.* Veliko Tărnovo, Veliko Tŭrnovo. Veliko Tarnovo, N Bulgaria 43°05′N 25°40′E
114 K8 **Veliko Tarnovo** *var.* Veliko Tŭrnovo. ◈ *province* N Bulgaria
Veliko Tărnovo *see* Veliko Tarnovo
Veliko Tŭrnovo *see* Veliko Tarnovo
Veliko Tŭrnovo *see* Veliko Tarnovo
Velikovec *see* Völkermarkt
125 R5 **Velikovisochnoye** Nenetskiy Avtonomnyy Okrug, NW Russian Federation 67°13′N 52°00′E
76 H12 **Vélingara** C Senegal 15°00′N 14°39′W
76 H11 **Vélingara** S Senegal 13°12′N 14°05′W
114 H11 **Velingrad** Pazardzhik, C Bulgaria 42°01′N 24°00′E
126 H3 **Velizh** Smolenskaya Oblast', W Russian Federation 55°30′N 31°06′E
111 F16 **Velká Deštná** *var.* Deštná, Grosskoppe, *Ger.* Deschnaer Koppe. ▲ NE Czech Republic 50°18′N 16°25′E
111 F18 **Velké Meziříčí** *Ger.* Grossmeseritsch. Vysočina, C Czech Republic 49°22′N 16°02′E
92 N1 **Velkomstpynten** *headland* NW Svalbard 79°51′N 11°37′E
111 K21 **Veľký Krtíš** Banskobystrický Kraj, C Slovakia 48°13′N 19°21′E
186 J8 **Vella Lavella** *var.* Mbilua. *island* New Georgia Islands, NW Solomon Islands
107 I15 **Velletri** Lazio, C Italy 41°41′N 12°47′E
95 K23 **Vellinge** Skåne, S Sweden 55°29′N 13°00′E
155 I19 **Vellore** Tamil Nādu, SE India 12°56′N 79°09′E
Velobriga *see* Viana do Castelo
115 G21 **Velopoúla** *island* S Greece
98 M12 **Velp** Gelderland, SE Netherlands 52°00′N 05°59′E
Velsen *see* Velsen-Noord
98 H9 **Velsen-Noord** *var.* Velsen. Noord-Holland, W Netherlands 52°27′N 04°40′E
125 N12 **Vel'sk** *var.* Velsk. Arkhangel'skaya Oblast', NW Russian Federation 61°03′N 42°01′E
Velsuna *see* Orvieto
98 K10 **Veluwemeer** *lake channel* C Netherlands
28 M3 **Velva** North Dakota, N USA 48°03′N 100°55′W
Velvendos/Velvendós *see* Velventós
115 E14 **Velventós** *var.* Velvendos, Velvendós. Dytikí Makedonía, N Greece 40°15′N 22°04′E
117 S5 **Velyka Bahachka** Poltavs'ka Oblast', C Ukraine 49°46′N 33°44′E
117 S9 **Velyka Lepetykha** *Rus.* Velikaya Lepetikha. Khersons'ka Oblast', S Ukraine 47°09′N 33°59′E
117 O10 **Velyka Mykhaylivka** Odes'ka Oblast', SW Ukraine 47°07′N 29°49′E
117 W8 **Velyka Novosilka** Donets'ka Oblast', E Ukraine 47°49′N 36°49′E
117 S9 **Velyka Oleksandrivka** Khersons'ka Oblast', S Ukraine 47°17′N 33°16′E
117 T4 **Velyka Pysarivka** Sums'ka Oblast', NE Ukraine 50°25′N 35°28′E
116 G6 **Velykyy Bereznyy** Zakarpats'ka Oblast', W Ukraine 48°54′N 22°27′E
117 W4 **Velykyy Burluk** Kharkivs'ka Oblast', E Ukraine 50°04′N 37°25′E
Velykyy Tokmak *see* Tokmak
173 P7 **Vema Fracture Zone** *tectonic feature* W Indian Ocean
65 P18 **Vema Seamount** *undersea feature* SW Indian Ocean 31°38′S 08°19′E
93 F17 **Vemdalen** Jämtland, C Sweden 62°26′N 13°50′E
95 N19 **Vena** Kalmar, S Sweden 57°31′N 16°00′E
41 N11 **Venado** San Luis Potosí, C Mexico 22°56′N 101°05′W
62 L11 **Venado Tuerto** Entre Ríos, E Argentina 33°45′S 61°56′W
61 A19 **Venado Tuerto** Santa Fe, C Argentina 33°46′S 61°57′W
107 K16 **Venafro** Molise, C Italy 41°28′N 14°03′E
55 Q9 **Venamo, Cerro** ▲ E Venezuela 05°56′N 61°25′W
106 B8 **Venaria** Piemonte, NW Italy 45°09′N 07°40′E
103 U15 **Vence** Alpes-Maritimes, SE France 43°45′N 07°07′E
104 H5 **Venda Nova** Vila Real, N Portugal 41°40′N 07°58′W
104 G11 **Vendas Novas** Évora, S Portugal 38°41′N 08°27′W
102 J9 **Vendée** ◈ *department* NW France
103 Q6 **Vendeuvre-sur-Barse** Aube, NE France 48°08′N 04°17′E
102 M7 **Vendôme** Loir-et-Cher, C France 47°48′N 01°04′E
Venedig *see* Venezia
Vener, Lake *see* Vänern
106 I8 **Veneta, Laguna** *lagoon* NE Italy
Venetia *see* Venezia
39 S7 **Venetie** Alaska, USA 67°00′N 146°25′W
106 H8 **Veneto** *var.* Venezia Euganea. ◈ *region* NE Italy
114 M7 **Venets** Shumen, NE Bulgaria 43°33′N 26°56′E
126 L5 **Venev** Tul'skaya Oblast', W Russian Federation 54°18′N 38°16′E
106 I8 **Venezia** *Eng.* Venice, *Fr.* Venise, *Ger.* Venedig; *anc.* Venetia. Veneto, NE Italy 45°26′N 12°20′E
Venezia Euganea *see* Veneto
Venezia, Golfo di *see* Venice, Gulf of
Venezia Tridentina *see* Trentino-Alto Adige
54 K8 **Venezuela** *off.* Republic of Venezuela; *prev.* Estados Unidos de Venezuela, United States of Venezuela. ◆ *republic* N South America
Venezuela, Cordillera de *see* Costa, Cordillera de la
Venezuela, Estados Unidos de *see* Venezuela
54 I4 **Venezuela, Golfo de** *Eng.* Gulf of Maracaibo, Gulf of Venezuela. *gulf* NW Venezuela
Venezuela, Gulf of *see* Venezuela, Golfo de
64 F11 **Venezuelan Basin** *undersea feature* E Caribbean Sea
Venezuela, Republic of *see* Venezuela
Venezuela, United States of *see* Venezuela
155 D16 **Vengurla** Mahārāshtra, W India 15°55′N 73°39′E
39 O15 **Veniaminof, Mount** ▲ Alaska, USA 56°12′N 159°24′W
23 V14 **Venice** Florida, SE USA 27°06′N 82°27′W
22 L10 **Venice** Louisiana, S USA 29°15′N 89°20′W
Venice *see* Venezia
106 J8 **Venice, Gulf of** *It.* Golfo di Venezia, *Slvn.* Beneški Zaliv. *gulf* N Adriatic Sea
Venise *see* Venezia
94 K13 **Venjan** Dalarna, C Sweden 60°58′N 13°55′E
94 K13 **Venjanssjön** ◎ C Sweden
155 J18 **Venkatagiri** Andhra Pradesh, E India 14°00′N 79°39′E
99 M15 **Venlo** *prev.* Venloo. Limburg, SE Netherlands 51°22′N 06°11′E
Venloo *see* Venlo
95 E18 **Vennesla** Vest-Agder, S Norway 58°15′N 08°00′E
107 M17 **Venosa** *anc.* Venusia. Basilicata, S Italy 40°57′N 15°49′E
Venoste, Alpi *see* Ötztaler Alpen
Venraij *see* Venray
99 M14 **Venray** *var.* Venraij. Limburg, SE Netherlands 51°32′N 05°59′E
118 C8 **Venta** *Ger.* Windau. ≈ Latvia/Lithuania
Venta Belgarum *see* Winchester
40 G9 **Ventana, Punta Arena de la** *var.* Punta de la Ventana. *headland* NW Mexico 24°03′N 109°49′W
Ventana, Punta de la *see* Ventana, Punta Arena de la
61 B23 **Ventana, Sierra de la** *hill range* E Argentina
Ventia *see* Valence
191 S11 **Vent, Îles du** *var.* Windward Islands. *island group* Archipel de la Société, W French Polynesia
191 R10 **Vent, Îles Sous le** *var.* Leeward Islands. *island group* Archipel de la Société, W French Polynesia
106 B11 **Ventimiglia** Liguria, NW Italy 43°47′N 07°37′E
97 M24 **Ventnor** S England, United Kingdom 50°36′N 01°11′W
18 J17 **Ventnor City** New Jersey, NE USA 39°19′N 74°27′W
103 S14 **Ventoux, Mont** ▲ SE France 44°12′N 05°21′E
118 C8 **Ventspils** *Ger.* Windau. NW Latvia 57°22′N 21°34′E
54 M10 **Ventuari, Río** ≈ S Venezuela
35 R15 **Ventura** California, W USA 34°15′N 119°18′W
182 F8 **Venus Bay** South Australia 33°15′S 134°42′E
Venusia *see* Venosa
191 P7 **Vénus, Pointe** *var.* Pointe Tataaihoa. *headland* Tahiti, W French Polynesia 17°28′S 149°29′W
41 V16 **Venustiano Carranza** Chiapas, SE Mexico 16°21′N 92°33′W
41 N7 **Venustiano Carranza, Presa** ⊠ NE Mexico
61 B15 **Vera** Santa Fe, C Argentina 29°28′S 60°10′W
105 Q14 **Vera** Andalucía, S Spain 37°15′N 01°51′W
63 K18 **Vera, Bahía** *bay* E Argentina
41 R14 **Veracruz** *var.* Veracruz Llave. Veracruz-Llave, E Mexico 19°10′N 96°09′W
Veracruz *see* Veracruz-Llave
41 Q13 **Veracruz-Llave** *var.* Veracruz. ◈ *state* E Mexico
Veracruz Llave *see* Veracruz
43 Q16 **Veraguas** *off.* Provincia de Veraguas. ◈ *province* W Panama
Veraguas, Provincia de *see* Veraguas
Veramin *see* Varāmīn
154 B12 **Verāval** Gujarāt, W India 20°54′N 70°22′E
106 C6 **Verbania** Piemonte, NW Italy 45°56′N 08°34′E
107 N20 **Verbicaro** Calabria, SW Italy 39°44′N 15°51′E
108 D11 **Verbier** Valais, SW Switzerland 46°06′N 07°14′E
Vercellae *see* Vercelli
106 C8 **Vercelli** *anc.* Vercellae. Piemonte, NW Italy 45°19′N 08°25′E
103 S13 **Vercors** *physical region* E France
Verdal *see* Verdalsøra
93 E16 **Verdalsøra** *var.* Verdal. Nord-Trøndelag, C Norway 63°47′N 11°27′E
Verde, Cabo *see* Cape Verde
44 J5 **Verde, Cape** *headland* Long Island, C The Bahamas 22°51′N 75°50′W

◆ Country
● Country Capital
◇ Dependent Territory
○ Dependent Territory Capital
◈ Administrative Regions
✈ International Airport
▲ Mountain
▲▲ Mountain Range
⛰ Volcano
≈ River
◎ Lake
⊠ Reservoir

VERDE, COSTA – VILLEFRANCHE

104 M2 **Verde, Costa** *coastal region* N Spain
Verde Grande, Río/Verde Grande y de Belem, Río *see* Verde, Río
100 H11 **Verden** Niedersachsen, NW Germany *52°55′N 09°14′E*
57 P16 **Verde, Río** ≈ Bolivia/Brazil
59 J19 **Verde, Río** ≈ SE Brazil
40 M12 **Verde, Río** *var.* Río Verde Grande, Río Verde Grande y de Belem. ≈ C Mexico
41 Q16 **Verde, Río** ≈ SE Mexico
36 L13 **Verde River** ≈ Arizona, SW USA
Verdhikoúsa/Verdhikoússa *see* Verdikoússa
27 Q8 **Verdigris River** ≈ Kansas/Oklahoma, C USA
115 E15 **Verdikoússa** *var.* Verdhikoúsa, Verdhikoússa. Thessalía, C Greece *39°47′N 21°59′E*
103 S15 **Verdon** ≈ SE France
15 O12 **Verdun** Québec, SE Canada *45°27′N 73°36′W*
103 S4 **Verdun** *var.* Verdun-sur-Meuse; *anc.* Verodunum. Meuse, NE France *49°09′N 05°25′E*
Verdun-sur-Meuse *see* Verdun
83 J21 **Vereeniging** Gauteng, NE South Africa *26°41′S 27°56′E*
Veremeyki *see* Vyeramyeyki
125 T14 **Vereshchagino** Permskiy Kray, NW Russian Federation *58°06′N 54°38′E*
76 G14 **Verga, Cap** *headland* W Guinea *10°12′N 14°27′W*
61 G18 **Vergara** Treinta y Tres, E Uruguay *32°58′S 53°54′W*
108 G11 **Vergeletto** Ticino, S Switzerland *46°13′N 08°34′E*
18 L8 **Vergennes** Vermont, NE USA *44°09′N 73°13′W*
Veria *see* Véroia
104 I5 **Verín** Galicia, NW Spain *41°55′N 07°26′W*
Verin T'alin *see* T'alin
118 K6 **Veriora** Põlvamaa, SE Estonia *57°57′N 27°23′E*
117 T7 **Verkhivtseve** Dnipropetrovs'ka Oblast', E Ukraine *48°27′N 34°15′E*
Verkhnedvinsk *see* Vyerkhnyadzvinsk
122 K10 **Verkhneimbatsk** Krasnoyarskiy Kray, N Russian Federation *63°06′N 88°03′E*
124 I3 **Verkhnetulomskiy** Murmanskaya Oblast', NW Russian Federation *68°37′N 31°46′E*
124 I3 **Verkhnetulomskoye Vodokhranilishche** ⊠ NW Russian Federation
Verkhneudinsk *see* Ulan-Ude
123 P10 **Verkhnevilyuysk** Respublika Sakha (Yakutiya), NE Russian Federation *63°44′N 119°59′E*
127 W5 **Verkhniy Avzyan** Respublika Bashkortostan, W Russian Federation *53°31′N 57°26′E*
127 Q11 **Verkhniy Baskunchak** Astrakhanskaya Oblast', SW Russian Federation *48°14′N 46°43′E*
127 W3 **Verkhniye Kigi** Respublika Bashkortostan, W Russian Federation *55°25′N 58°40′E*
117 T9 **Verkhniy Rohachyk** Khersons'ka Oblast', S Ukraine *47°16′N 34°16′E*
123 Q11 **Verkhnyaya Amga** Respublika Sakha (Yakutiya), NE Russian Federation *59°34′N 127°07′E*
125 V6 **Verkhnyaya Inta** Respublika Komi, NW Russian Federation *65°55′N 60°07′E*
125 O10 **Verkhnyaya Toyma** Arkhangel'skaya Oblast', NW Russian Federation *62°12′N 44°57′E*
126 K6 **Verkhov'ye** Orlovskaya Oblast', W Russian Federation *52°49′N 37°20′E*
116 I8 **Verkhovyna** Ivano-Frankivs'ka Oblast', W Ukraine *48°09′N 24°48′E*
123 P8 **Verkhoyanskiy Khrebet** ▲▲ NE Russian Federation
117 T7 **Verkn'odniprovs'k** Dnipropetrovs'ka Oblast', E Ukraine *48°40′N 34°17′E*
101 G14 **Verl** Nordrhein-Westfalen, NW Germany *51°52′N 08°30′E*
92 N1 **Verlegenhuken** *headland* N Svalbard *80°03′N 16°15′E*
82 A9 **Vermelha, Ponta** *headland* NW Angola *05°40′S 12°09′E*
103 P7 **Vermenton** Yonne, C France *47°40′N 03°43′E*
11 R14 **Vermilion** Alberta, SW Canada *53°21′N 110°52′W*
31 T11 **Vermilion** Ohio, N USA *41°25′N 82°21′W*
22 I10 **Vermilion Bay** *bay* Louisiana, S USA
29 V4 **Vermilion Lake** ◎ Minnesota, N USA
14 F9 **Vermilion River** ≈ Ontario, S Canada
30 L12 **Vermilion River** ≈ Illinois, N USA
29 R12 **Vermillion** South Dakota, N USA *42°46′N 96°55′W*
29 R12 **Vermillion River** ≈ South Dakota, N USA
15 O9 **Vermillon, Rivière** ≈ Québec, SE Canada
115 E14 **Vérmio** ▲▲ N Greece
18 L8 **Vermont** *off.* State of Vermont, *also known as* Green Mountain State. ◈ *state* NE USA
113 K16 **Vermosh** *var.* Vermoshi. Shkodër, N Albania *42°37′N 19°42′E*
Vermoshi *see* Vermosh
37 O3 **Vernal** Utah, W USA *40°27′N 109°31′W*
14 G11 **Verner** Ontario, S Canada *46°24′N 80°04′W*
102 M5 **Verneuil-sur-Avre** Eure, N France *48°44′N 00°55′E*
114 D13 **Vérno** ▲ N Greece
11 N17 **Vernon** British Columbia, SW Canada *50°17′N 119°19′W*
102 M4 **Vernon** Eure, N France *49°04′N 01°28′E*
23 N3 **Vernon** Alabama, S USA *33°45′N 88°06′W*
31 P15 **Vernon** Indiana, N USA *38°59′N 85°39′W*
25 Q4 **Vernon** Texas, SW USA *34°11′N 99°17′W*
32 G10 **Vernonia** Oregon, NW USA *45°51′N 123°11′W*
14 G12 **Vernon, Lake** ◎ Ontario, S Canada
22 G7 **Vernon Lake** ⊠ Louisiana, S USA
23 Y13 **Vero Beach** Florida, SE USA *27°38′N 80°24′W*
Verőcze *see* Virovitica
Verodunum *see* Verdun
115 E14 **Véroia** *var.* Veria, Vérroia, *Turk.* Karaferiye. Kentrikí Makedonía, N Greece *40°32′N 22°11′E*
106 E8 **Verolanuova** Lombardia, N Italy *45°20′N 10°06′E*
14 K14 **Verona** Ontario, SE Canada *44°30′N 76°42′W*
106 G8 **Verona** Veneto, NE Italy *45°27′N 11°E*
29 P6 **Verona** North Dakota, N USA *46°19′N 98°03′W*
30 L9 **Verona** Wisconsin, N USA *42°59′N 89°33′W*
61 E20 **Verónica** Buenos Aires, E Argentina *35°25′S 57°16′W*
22 J9 **Verret, Lake** ◎ Louisiana, S USA
Vérroia *see* Véroia
103 N5 **Versailles** Yvelines, N France *48°48′N 02°08′E*
31 P15 **Versailles** Indiana, N USA *39°04′N 85°16′W*
20 M5 **Versailles** Kentucky, S USA *38°02′N 84°45′W*
27 U5 **Versailles** Missouri, C USA *38°25′N 92°51′W*
31 Q13 **Versailles** Ohio, N USA *40°13′N 84°28′W*
Versecz *see* Vršac
108 A10 **Versoix** Genève, SW Switzerland *46°17′N 06°10′E*
15 Z6 **Verte, Pointe** *headland* Québec, SE Canada *48°36′N 64°10′W*
111 I22 **Vértes** ▲▲ NW Hungary
44 G6 **Vertientes** Camagüey, C Cuba *21°18′N 78°11′W*
115 G13 **Vertískos** ▲▲ N Greece
102 I8 **Vertou** Loire-Atlantique, NW France *47°10′N 01°28′W*
Verulamium *see* St Albans
99 L19 **Verviers** Liège, E Belgium *50°36′N 05°52′E*
103 Y14 **Vescovato** Corse, France, C Mediterranean Sea *42°30′N 09°27′E*
99 L20 **Vesdre** ≈ E Belgium
117 U10 **Vesele** *Rus.* Veseloye. Zaporiz'ka Oblast', S Ukraine *47°00′N 34°52′E*
111 D18 **Veselí nad Lužnicí** *var.* Weseli an der Lainsitz, *Ger.* Frohenbruck. Jihočeský Kraj, S Czech Republic *49°11′N 14°40′E*
114 M9 **Veselinovo** Shumen, NE Bulgaria *43°01′N 27°02′E*
126 L12 **Veselovskoye Vodokhranilishche** ⊠ SW Russian Federation
Veseloye *see* Vesele
117 Q9 **Veselynove** Mykolayivs'ka Oblast', S Ukraine *47°21′N 31°15′E*
Veseya *see* Vyasyeya
126 M10 **Veshenskaya** Rostovskaya Oblast', SW Russian Federation *49°37′N 41°43′E*
127 Q5 **Veshkayma** Ul'yanovskaya Oblast', W Russian Federation *54°04′N 47°06′E*
Vesisaari *see* Vadsø
Vesontio *see* Besançon
103 T7 **Vesoul** *anc.* Vesulium, Vesulum. Haute-Saône, E France *47°37′N 06°09′E*
95 J20 **Vessigebro** Halland, S Sweden *56°58′N 12°40′E*
95 D17 **Vest-Agder** ◈ *county* S Norway
23 P4 **Vestavia Hills** Alabama, S USA *33°27′N 86°47′W*
84 F6 **Vesterålen** *island* NW Norway
92 G10 **Vesterålen** *island group* N Norway
87 V3 **Vestervig** Midtjylland, NW Denmark *56°46′N 08°20′E*
92 H2 **Vestfirðir** ◈ *region* NW Iceland
92 G11 **Vestfjorden** *fjord* C Norway
95 G16 **Vestfold** ◈ *county* S Norway
Vestmanhavn *see* Vestmanna
95 B18 **Vestmanna** *Dan.* Vestmanhavn. Streymoy, N Faroe Islands *62°09′N 07°11′W*
92 I4 **Vestmannaeyjar** Suðurland, S Iceland *63°26′N 20°14′W*
94 E9 **Vestnes** Møre og Romsdal, S Norway *62°39′N 07°00′E*
92 H3 **Vesturland** ◈ *region* W Iceland
92 G11 **Vestvågøya** *island* C Norway
Vesulium/Vesulum *see* Vesoul
107 K17 **Vesuvio** *Eng.* Vesuvius. ✕ S Italy *40°48′N 14°29′E*
Vesuvius *see* Vesuvio
124 K14 **Ves'yegonsk** Tverskaya Oblast', W Russian Federation *58°40′N 37°13′E*
111 I23 **Veszprém** *Ger.* Veszprim. Veszprém, W Hungary *47°06′N 17°54′E*
111 H23 **Veszprém** *off.* Veszprém Megye. ◈ *county* W Hungary
Veszprém Megye *see* Veszprém
Veszprim *see* Veszprém
Vetka *see* Vyetka
95 M19 **Vetlanda** Jönköping, S Sweden *57°26′N 15°05′E*
127 P1 **Vetluga** Nizhegorodskaya Oblast', W Russian Federation *57°51′N 45°45′E*
125 P14 **Vetluga** ≈ NW Russian Federation
125 O14 **Vetluzhskiy** Kostromskaya Oblast', NW Russian Federation *58°21′N 45°25′E*
127 P2 **Vetluzhskiy** Nizhegorodskaya Oblast', W Russian Federation *57°10′N 45°07′E*
114 K7 **Vetovo** Ruse, N Bulgaria *43°42′N 26°16′E*
107 H14 **Vetralla** Lazio, C Italy *42°18′N 12°03′E*
Vetrino *see* Vyetryna
122 L7 **Vetrovaya, Gora** ▲ N Russian Federation *73°54′N 95°00′E*
Vetter, Lake *see* Vättern
106 J13 **Vettore, Monte** ▲ C Italy *42°49′N 13°15′E*
99 A17 **Veurne** *var.* Furnes. West-Vlaanderen, W Belgium *51°04′N 02°40′E*
31 Q15 **Vevay** Indiana, N USA *38°45′N 85°08′W*
108 C10 **Vevey** *Ger.* Vivis; *anc.* Vibiscum. Vaud, SW Switzerland *46°28′N 06°51′E*
Vexiö *see* Växjö
103 S13 **Veynes** Hautes-Alpes, SE France *44°33′N 05°51′E*
103 N11 **Vézère** ≈ W France
114 I9 **Vezhen** ▲ C Bulgaria *42°45′N 24°22′E*
136 K11 **Vezirköprü** Samsun, N Turkey *41°09′N 35°27′E*
57 J18 **Viacha** La Paz, W Bolivia *16°40′S 68°17′W*
27 R10 **Vian** Oklahoma, C USA *35°30′N 94°56′W*
Viana de Castelo *see* Viana do Castelo
104 H12 **Viana do Alentejo** Évora, S Portugal *38°20′N 08°00′W*
104 I4 **Viana do Bolo** Galicia, NW Spain *42°10′N 07°06′W*
104 G5 **Viana do Castelo** *var.* Viana de Castelo; *anc.* Velobriga. Viana do Castelo, NW Portugal *41°41′N 08°50′W*
104 G5 **Viana do Castelo** *var.* Viana de Castelo. ◈ *district* N Portugal
98 J12 **Vianen** Utrecht, C Netherlands *52°N 05°06′E*
167 Q8 **Viangchan** *Eng./Fr.* Vientiane. ● (Laos) C Laos *17°58′N 102°38′E*
167 P6 **Viangphoukha** *var.* Vieng Pou Kha. Louang Namtha, N Laos *20°41′N 101°03′E*
104 K13 **Viar** ≈ SW Spain
106 E11 **Viareggio** Toscana, C Italy *43°52′N 10°15′E*
103 O14 **Viaur** ≈ S France
Vibiscum *see* Vevey
95 G21 **Viborg** Midtjylland, NW Denmark *56°28′N 09°25′E*
29 R12 **Viborg** South Dakota, N USA *43°10′N 97°04′W*
107 N22 **Vibo Valentia** *prev.* Monteleone di Calabria; *anc.* Hipponium. Calabria, SW Italy *38°40′N 16°06′E*
105 W5 **Vic** *var.* Vich; *anc.* Ausa, Vicus Ausonensis. Cataluña, NE Spain *41°56′N 02°16′E*
102 K16 **Vic-en-Bigorre** Hautes-Pyrénées, S France *43°23′N 00°04′E*
40 K10 **Vicente Guerrero** Durango, C Mexico *23°30′N 104°24′W*
41 P10 **Vicente Guerrero, Presa** *var.* Presa de las Adjuntas. ⊠ NE Mexico
Vicentia *see* Vicenza
106 G8 **Vicenza** *anc.* Vicentia. Veneto, NE Italy *45°32′N 11°31′E*
Vich *see* Vic
54 J10 **Vichada** *off.* Comisaría del Vichada. ◈ *province* E Colombia
Vichada, Comisaría del *see* Vichada
54 K10 **Vichada, Río** ≈ E Colombia
61 G17 **Vichadero** Rivera, NE Uruguay *31°45′S 54°41′W*
Vichegda *see* Vychegda
124 M16 **Vichuga** Ivanovskaya Oblast', W Russian Federation *57°13′N 41°51′E*
103 P10 **Vichy** Allier, C France *46°08′N 03°26′E*
26 K9 **Vici** Oklahoma, C USA *36°09′N 99°18′W*
95 I19 **Vickan** Halland, S Sweden *57°25′N 12°00′E*
31 P10 **Vicksburg** Michigan, N USA *42°07′N 85°31′W*
22 J5 **Vicksburg** Mississippi, S USA *32°21′N 90°52′W*
103 O12 **Vic-sur-Cère** Cantal, C France *45°00′N 02°36′E*
59 I21 **Victor** Mato Grosso do Sul, SW Brazil *21°39′S 53°21′W*
29 X14 **Victor** Iowa, C USA *41°45′N 92°18′W*
182 I10 **Victor Harbor** South Australia *35°33′S 138°37′E*
61 C18 **Victoria** Entre Ríos, E Argentina *32°40′S 60°10′W*
10 L17 **Victoria** *province capital* Vancouver Island, British Columbia, SW Canada *48°25′N 123°22′W*
45 R14 **Victoria** NW Grenada *12°12′N 61°42′W*
42 H6 **Victoria** Yoro, NW Honduras *15°01′N 87°28′W*
121 O15 **Victoria** *var.* Rabat. Gozo, NW Malta *36°02′N 14°14′E*
116 I12 **Victoria** *Ger.* Viktoriastadt. Brașov, C Romania *45°44′N 24°41′E*
172 H17 **Victoria** ● (Seychelles) Mahé, SW Seychelles *04°38′S 28°28′E*
25 U13 **Victoria** Texas, SW USA *28°47′N 96°59′W*
183 N12 **Victoria** ◈ *state* SE Australia
174 K7 **Victoria** ≈ Western Australia
Victoria *see* Labuan, East Malaysia
Victoria *see* Masvingo, Zimbabwe
Victoria Bank *see* Vitória Seamount
11 Y15 **Victoria Beach** Manitoba, S Canada *50°40′N 96°30′W*
Victoria de Durango *see* Durango
Victoria de las Tunas *see* Las Tunas
83 I16 **Victoria Falls** Matabeleland North, W Zimbabwe *17°55′S 25°51′E*
83 I16 **Victoria Falls** *waterfall* Zambia/Zimbabwe
83 I16 **Victoria Falls** ✈ Matabeleland North, W Zimbabwe *18°03′S 25°48′E*
Victoria Falls *see* Iguaçu, Salto do
63 F19 **Victoria, Isla** *island* Archipiélago de los Chonos, S Chile
8 K6 **Victoria Island** *island* Northwest Territories/Nunavut, NW Canada
182 L8 **Victoria, Lake** ◎ New South Wales, SE Australia
68 I12 **Victoria, Lake** *var.* Victoria Nyanza. ◎ E Africa
195 S13 **Victoria Land** *physical region* Antarctica
187 X14 **Victoria, Mount** ▲ Viti Levu, W Fiji *17°37′S 178°00′E*
166 L5 **Victoria, Mount** ▲ W Myanmar (Burma) *21°13′N 93°53′E*
186 E9 **Victoria, Mount** ▲ S Papua New Guinea *08°51′S 147°36′E*
81 F17 **Victoria Nile** *var.* Somerset Nile. ≈ C Uganda
Victoria Nyanza *see* Victoria, Lake
42 G3 **Victoria Peak** ▲ SE Belize *16°50′N 88°38′W*
185 H16 **Victoria Range** ▲▲ South Island, New Zealand
181 O3 **Victoria River** ≈ Northern Territory, N Australia
181 P3 **Victoria River Roadhouse** Northern Territory, N Australia *15°37′S 131°07′E*
15 Q11 **Victoriaville** Québec, SE Canada *46°04′N 71°57′W*
Victoria-Wes *see* Victoria West
83 G24 **Victoria West** *Afr.* Victoria-Wes. Northern Cape, W South Africa *31°25′S 23°08′E*
62 J13 **Victorica** La Pampa, C Argentina *36°15′S 65°25′W*
Victor, Mount *see* Dronning Fabiolafjella
35 U14 **Victorville** California, W USA *34°32′N 117°17′W*
62 G9 **Vicuña** Coquimbo, N Chile *30°00′S 70°44′W*
62 K11 **Vicuña Mackenna** Córdoba, C Argentina *33°53′S 64°25′W*
Vicus Ausonensis *see* Vic
Vicus Elbii *see* Viterbo
33 X7 **Vida** Montana, NW USA *47°52′N 105°30′W*
23 V6 **Vidalia** Georgia, SE USA *32°13′N 82°24′W*
22 J7 **Vidalia** Louisiana, S USA *31°34′N 91°25′W*
95 F22 **Videbæk** Midtjylland, C Denmark *56°08′N 08°38′E*
60 I13 **Videira** Santa Catarina, S Brazil *27°00′S 51°08′W*
116 J14 **Videle** Teleorman, S Romania *44°15′N 25°27′E*
Videm-Krško *see* Krško
Vídeň *see* Wien
104 H12 **Vidigueira** Beja, S Portugal *38°12′N 07°48′W*
114 J9 **Vidima** ≈ N Bulgaria
114 G7 **Vidin** *anc.* Bononia. Vidin, NW Bulgaria *44°00′N 22°52′E*
114 F8 **Vidin** ◈ *province* NW Bulgaria
154 H10 **Vidisha** Madhya Pradesh, C India *23°30′N 77°50′E*
25 Y10 **Vidor** Texas, SW USA *30°07′N 94°01′W*
95 L20 **Vidöstern** ◎ S Sweden
92 J13 **Vidsel** Norrbotten, N Sweden *65°49′N 20°31′E*
118 H9 **Vidzemes Augstiene** ▲ C Latvia
118 J12 **Vidzy** Vitsyebskaya Voblasts', NW Belarus *55°24′N 26°38′E*
63 L16 **Viedma** Río Negro, E Argentina *40°50′S 62°58′W*
63 H22 **Viedma, Lago** ◎ S Argentina
45 O11 **Vieille Case** *var.* Itassi. N Dominica *15°36′N 61°24′W*
104 M2 **Vieja, Peña** ▲ N Spain *43°09′N 04°47′W*
24 J10 **Vieja, Sierra** ▲▲ Texas, SW USA
40 E4 **Viejo, Cerro** ▲ NW Mexico *30°16′N 112°18′W*
56 B9 **Viejo, Cerro** ▲ N Peru *04°54′S 79°24′W*
118 E10 **Viekšniai** Telšiai, NW Lithuania *56°14′N 22°33′E*
105 U3 **Vielha** *var.* Viella. Cataluña, NE Spain *42°41′N 00°47′E*
Viella *see* Vielha
99 L21 **Vielsalm** Luxembourg, E Belgium *50°17′N 05°55′E*
Vieng Pou Kha *see* Viangphoukha
23 T6 **Vienna** Georgia, SE USA *32°05′N 83°48′W*
30 L17 **Vienna** Illinois, N USA *37°24′N 88°55′W*
27 V5 **Vienna** Missouri, C USA *38°12′N 91°59′W*
21 Q3 **Vienna** West Virginia, NE USA *39°19′N 81°33′W*
Vienna *see* Wien, Austria
Vienna *see* Vienne, France
103 R11 **Vienne** *anc.* Vienna. Isère, E France *45°32′N 04°53′E*
102 L10 **Vienne** ◈ *department* W France
102 L9 **Vienne** ≈ W France
Vientiane *see* Viangchan
Vientos, Paso de los *see* Windward Passage
45 V6 **Vieques** *var.* Isabel Segunda. E Puerto Rico *18°08′N 65°25′W*
45 V6 **Vieques, Isla de** *island* E Puerto Rico
45 V6 **Vieques, Pasaje de** *passage* E Puerto Rico
45 V5 **Vieques, Sonda de** *sound* E Puerto Rico
Vierdörfer *see* Săcele
93 M15 **Vieremä** Pohjois-Savo, C Finland *63°42′N 27°02′E*
99 M14 **Vierlingsbeek** Noord-Brabant, SE Netherlands *51°36′N 06°01′E*
101 G20 **Viernheim** Hessen, W Germany *49°32′N 08°35′E*
101 D15 **Viersen** Nordrhein-Westfalen, W Germany *51°16′N 06°24′E*
108 G8 **Vierwaldstätter See** *Eng.* Lake of Lucerne. ◎ C Switzerland
103 N8 **Vierzon** Cher, C France *47°13′N 02°04′E*
40 L8 **Viesca** Coahuila, NE Mexico *25°25′N 102°45′W*
118 H10 **Viesīte** *Ger.* Eckengraf. S Latvia *56°21′N 25°30′E*
107 N15 **Vieste** Puglia, SE Italy *41°52′N 16°11′E*
167 T8 **Vietnam** *off.* Socialist Republic of Vietnam, *Vtn.* Công Hoa Xa Hôi Chu Nghia Viêt Nam. ◆ *republic* SE Asia
Vietnam, Socialist Republic of *see* Vietnam
167 S5 **Viêt Quang** Ha Giang, N Vietnam *22°24′N 104°48′E*
Vietri *see* Viêt Tri
167 S6 **Viêt Tri** *var.* Vietri. Vinh Phu, N Vietnam *21°20′N 105°26′E*
30 L4 **Vieux Desert, Lac** ◎ Michigan/Wisconsin, N USA
45 Y13 **Vieux Fort** S Saint Lucia *13°43′N 60°57′W*
45 X6 **Vieux-Habitants** Basse Terre, SW Guadeloupe *16°04′N 61°45′W*
119 G14 **Vievis** Vilnius, S Lithuania *54°46′N 24°51′E*
171 N2 **Vigan** Luzon, N Philippines *17°34′N 102°21′E*
106 D8 **Vigevano** Lombardia, N Italy *45°19′N 08°51′E*
107 N18 **Viggiano** Basilicata, S Italy *40°21′N 15°54′E*
58 L12 **Vigia** Pará, NE Brazil *0°50′S 48°07′W*
41 Y12 **Vigía Chico** Quintana Roo, SE Mexico *19°49′N 87°31′W*
45 T11 **Vigie** *prev.* George F L Charles. ✈ (Castries) NE Saint Lucia *14°01′N 60°59′W*
102 K17 **Vignemale** *var.* Pic de Vignemale. ▲ France/Spain *42°48′N 00°06′W*
Vignemale, Pic de *see* Vignemale
106 G10 **Vignola** Emilia-Romagna, C Italy *44°28′N 11°00′E*
104 G4 **Vigo** Galicia, NW Spain *42°15′N 08°44′W*
104 G4 **Vigo, Ría de** *estuary* NW Spain
94 D9 **Vigra** *island* S Norway
95 C17 **Vigrestad** Rogaland, S Norway *58°34′N 05°42′E*
93 L15 **Vihanti** Pohjois-Pohjanmaa, C Finland *64°29′N 25°E*
149 U10 **Vihāri** Punjab, E Pakistan *30°03′N 72°32′E*
102 K8 **Vihiers** Maine-et-Loire, NW France *47°09′N 00°37′W*
81 H18 **Vihiga** ◈ *county* W Kenya
111 O19 **Vihorlat** ▲ E Slovakia *48°54′N 22°09′E*
114 G11 **Vihren** *var.* Vikhren. ▲ SW Bulgaria *41°45′N 23°24′E*
93 L19 **Vihti** Uusimaa, S Finland *60°25′N 24°16′E*
Viipuri *see* Vyborg
93 M16 **Viitasaari** Keski-Suomi, C Finland *63°05′N 25°52′E*
118 K3 **Viivikonna** Ida-Virumaa, NE Estonia *59°19′N 27°41′E*
155 K16 **Vijayawāda** *prev.* Bezwada. Andhra Pradesh, SE India *16°34′N 80°40′E*
Vijosa/Vijosë *see* Aóos, Albania/Greece
Vijosa/Vijosë *see* Vjosës, Lumi i, Albania/Greece
92 J4 **Vík** Suðurland, S Iceland *63°25′N 18°58′W*
Vik *see* Vikøyri
94 L13 **Vika** Dalarna, C Sweden *60°55′N 14°30′E*
92 L12 **Vikajärvi** Lappi, N Finland *66°37′N 26°10′E*
94 L13 **Vikarbyn** Dalarna, C Sweden *60°57′N 15°00′E*
95 J22 **Viken** Skåne, S Sweden *56°09′N 12°36′E*
95 L17 **Viken** ◎ C Sweden
95 G15 **Vikersund** Buskerud, S Norway *59°58′N 09°59′E*
Vikhren *see* Vihren
11 R15 **Viking** Alberta, SW Canada *53°07′N 111°50′W*
84 E7 **Viking Bank** *undersea feature* N North Sea *60°35′N 02°35′E*
95 M14 **Vikmanshyttan** Dalarna, C Sweden *60°19′N 15°55′E*
94 D12 **Vikøyri** *var.* Vik. Sogn Og Fjordane, S Norway *61°04′N 06°34′E*
93 H17 **Viksjö** Västernorrland, C Sweden *62°45′N 17°30′E*
Viktoriastadt *see* Victoria
Vila *see* Port-Vila
Vila Arriaga *see* Bibala
Vila Artur de Paiva *see* Cubango
Vila Baleira *see* Porto Santo
Vila Bela da Santissima Trindade *see* Mato Grosso
58 B12 **Vila Bittencourt** Amazonas, NW Brazil *01°25′S 69°24′W*
Vila da Ponte *see* Cubango
64 O2 **Vila da Praia da Vitória** Terceira, Azores, Portugal, NE Atlantic Ocean *38°44′N 27°04′W*
Vila de Aljustrel *see* Cangamba
Vila de Almoster *see* Chiange
Vila de João Belo *see* Xai-Xai
Vila de Macia *see* Macia
Vila de Manhiça *see* Manhiça
Vila de Manica *see* Manica
Vila de Mocímboa da Praia *see* Mocímboa da Praia
83 N16 **Vila de Sena** *var.* Sena. Sofala, C Mozambique *17°25′S 34°59′E*
104 F14 **Vila do Bispo** Faro, S Portugal *37°05′N 08°53′W*
104 G6 **Vila do Conde** Porto, NW Portugal *41°21′N 08°45′W*
Vila do Maio *see* Maio
64 P3 **Vila do Porto** Santa Maria, Azores, Portugal, NE Atlantic Ocean *36°57′N 25°10′W*
83 K15 **Vila do Zumbo** *prev.* Vila do Zumbu, Zumbo. Tete, NW Mozambique *15°36′S 30°30′E*
Vila do Zumbu *see* Vila do Zumbo
105 T8 **Vilafamés** *prev.* Villafamés. Valenciana, E Spain *40°07′N 00°03′W*
104 I6 **Vila Flor** *var.* Vila Flôr. Bragança, N Portugal *41°18′N 07°09′W*
105 V6 **Vilafranca del Penedès** *var.* Villafranca del Panadés. Cataluña, NE Spain *41°21′N 01°42′E*
104 F10 **Vila Franca de Xira** *var.* Vilafranca de Xira. Lisboa, C Portugal *38°57′N 08°59′W*
Vila Gago Coutinho *see* Lumbala N'Guimbo
104 G3 **Vilagarcía** *var.* Villagarcía de Arosa. Galicia, NW Spain *42°35′N 08°45′W*
Vila General Machado *see* Camacupa
Vila Henrique de Carvalho *see* Saurimo
102 I7 **Vilaine** ≈ NW France
Vila João de Almeida *see* Chibia
118 K8 **Viļaka** *Ger.* Marienhausen. NE Latvia *57°12′N 27°43′E*
104 I2 **Vilalba** Galicia, NW Spain *43°17′N 07°41′W*
Vila Marechal Carmona *see* Uíge
Vila Mariano Machado *see* Ganda
172 G3 **Vilanandro, Tanjona** *prev./Fr.* Cap Saint-André. *headland* W Madagascar *16°10′S 44°27′E*
Vilanculos *see* Vilankulo
118 J10 **Viļāni** E Latvia *56°33′N 26°55′E*
83 N19 **Vilankulo** *var.* Vilanculos. Inhambane, E Mozambique *22°01′S 35°19′E*
Vila Norton de Matos *see* Balombo
104 G6 **Vila Nova de Famalicão** *var.* Vila Nova de Famalicao. Braga, N Portugal *41°24′N 08°31′W*
104 I6 **Vila Nova de Foz Côa** *var.* Vila Nova de Fozcôa. Guarda, N Portugal *41°05′N 07°09′W*
Vila Nova de Fozcôa *see* Vila Nova de Foz Côa
104 F6 **Vila Nova de Gaia** Porto, NW Portugal *41°08′N 08°37′W*
Vila Nova de Portimão *see* Portimão
105 V6 **Vilanova i La Geltru** Cataluña, NE Spain *41°15′N 01°42′E*
Vila Pereira de Eça *see* Ondjiva
104 H6 **Vila Pouca de Aguiar** Vila Real, N Portugal *41°30′N 07°38′W*
104 H6 **Vila Real** *var.* Vila Rial. Vila Real, N Portugal *41°17′N 07°45′W*
105 T9 **Vila-real** *var.* Vila-real de los Infantes; *prev.* Villarreal. Valenciana, E Spain *39°56′N 00°08′W*
104 H6 **Vila Real** ◈ *district* N Portugal
Vila-real de los Infantes *see* Vila-real
104 H14 **Vila Real de Santo António** Faro, S Portugal *37°12′N 07°25′W*
104 J7 **Vilar Formoso** Guarda, N Portugal *40°37′N 06°50′W*
Vila Rial *see* Vila Real
59 J15 **Vila Rica** Mato Grosso, W Brazil *09°52′S 50°44′W*
Vila Robert Williams *see* Caála
Vila Salazar *see* N'Dalatando
Vila Serpa Pinto *see* Menongue
Vila Teixeira da Silva *see* Bailundo
Vila Teixeira de Sousa *see* Luau
104 H9 **Vila Velha de Ródão** Castelo Branco, C Portugal *39°39′N 07°40′W*
104 G5 **Vila Verde** Braga, N Portugal *41°39′N 08°27′W*
104 H11 **Vila Viçosa** Évora, S Portugal *38°46′N 07°25′W*
57 G15 **Vilcabamba, Cordillera de** ▲▲ C Peru
Vîlcea *see* Vâlcea
122 J4 **Vil'cheka, Zemlya** *Eng.* Wilczek Land. *island* Zemlya Frantsa-Iosifa, NW Russian Federation
95 F22 **Vildbjerg** Midtjylland, C Denmark *56°12′N 08°47′E*
Vileyka *see* Vilyeyka
93 H15 **Vilhelmina** Västerbotten, N Sweden *64°38′N 16°40′E*
59 F17 **Vilhena** Rondônia, W Brazil *12°40′S 60°08′W*
115 G19 **Vília** Attikí, C Greece *38°09′N 23°21′E*
119 I14 **Viliya** *Lith.* Neris. ≈ W Belarus
Viliya *see* Neris
118 H5 **Viljandi** *Ger.* Fellin. Viljandimaa, S Estonia *58°22′N 25°30′E*
118 H5 **Viljandimaa** *var.* Viljandi Maakond. ◈ *province* SW Estonia
Viljandi Maakond *see* Viljandimaa
119 E14 **Vilkaviškis** *Pol.* Wyłkowyszki. Marijampolė, SW Lithuania *54°39′N 23°03′E*
118 F13 **Vilkija** Kaunas, C Lithuania *55°02′N 23°36′E*
197 V9 **Vil'kitskogo, Proliv** *strait* N Russian Federation
Vilkovo *see* Vylkove
57 L21 **Villa Abecia** Chuquisaca, S Bolivia *21°00′S 65°18′W*
41 N5 **Villa Acuña** *var.* Ciudad Acuña. Coahuila, NE Mexico *29°18′N 100°58′W*
40 J4 **Villa Ahumada** Chihuahua, N Mexico *30°38′N 106°30′W*
45 O9 **Villa Altagracia** C Dominican Republic *18°43′N 70°13′W*
56 L13 **Villa Bella** El Beni, N Bolivia *10°21′S 65°25′W*
104 J3 **Villablino** Castilla y León, N Spain *42°55′N 06°21′W*
54 K6 **Villa Bruzual** Portuguesa, N Venezuela *09°20′N 69°06′W*
105 O9 **Villacañas** Castilla-La Mancha, C Spain *39°38′N 03°20′W*
105 O12 **Villacarrillo** Andalucía, S Spain *38°07′N 03°05′W*
104 M7 **Villacastín** Castilla y León, N Spain *40°46′N 04°25′W*
Villa Cecilia *see* Ciudad Madero
109 S9 **Villach** *Slvn.* Beljak. Kärnten, S Austria *46°36′N 13°49′E*
107 B20 **Villacidro** Sardegna, Italy, C Mediterranean Sea *39°28′N 08°43′E*
Villa Concepción *see* Concepción
104 L4 **Villada** Castilla y León, N Spain *42°15′N 04°59′W*
40 M10 **Villa de Cos** Zacatecas, C Mexico *23°20′N 102°20′W*
54 L5 **Villa de Cura** *var.* Cura. Aragua, N Venezuela *10°00′N 67°30′W*
Villa del Nevoso *see* Ilirska Bistrica
Villa del Pilar *see* Pilar
104 M13 **Villa del Río** Andalucía, S Spain *37°59′N 04°17′W*
Villa de Méndez *see* Méndez
42 H6 **Villa de San Antonio** Comayagua, W Honduras *14°24′N 87°37′W*
105 N4 **Villadiego** Castilla y León, N Spain *42°31′N 04°01′W*
Villafamés *see* Vilafamés
41 U16 **Villa Flores** Chiapas, SE Mexico *16°12′N 93°16′W*
104 J3 **Villafranca del Bierzo** Castilla y León, N Spain *42°36′N 06°49′W*
105 S8 **Villafranca del Cid** Valenciana, E Spain *40°25′N 00°15′W*
104 J11 **Villafranca de los Barros** Extremadura, W Spain *38°34′N 06°20′W*
105 N10 **Villafranca de los Caballeros** Castilla-La Mancha, C Spain *39°26′N 03°21′W*
Villafranca del Panadés *see* Vilafranca del Penedès
106 F8 **Villafranca di Verona** Veneto, NE Italy *45°22′N 10°51′E*
107 J23 **Villafrati** Sicilia, Italy, C Mediterranean Sea *37°53′N 13°30′E*
Villagarcía de Arosa *see* Vilagarcía
41 O9 **Villagrán** Tamaulipas, C Mexico *24°29′N 99°30′W*
61 C17 **Villaguay** Entre Ríos, E Argentina *31°55′S 59°01′W*
62 O6 **Villa Hayes** Presidente Hayes, S Paraguay *25°05′S 57°25′W*
41 U15 **Villahermosa** *prev.* San Juan Bautista. Tabasco, SE Mexico *17°56′N 92°50′W*
105 O11 **Villahermosa** Castilla-La Mancha, C Spain *38°46′N 02°52′W*
64 O11 **Villahermoso** Gomera, Islas Canarias, Spain, NE Atlantic Ocean *38°46′N 02°52′W*
Villa Hidalgo *see* Hidalgo
105 T12 **Villajoyosa** *Cat.* La Vila Joíosa. Valenciana, E Spain *38°31′N 00°14′W*
Villa Juárez *see* Juárez
Villalba *see* Collado Villalba
41 N8 **Villaldama** Nuevo León, NE Mexico *26°29′N 100°27′W*
104 L5 **Villalón de Campos** Castilla y León, N Spain *42°05′N 05°03′W*
61 A25 **Villalonga** Buenos Aires, E Argentina *39°55′S 62°35′W*
104 L5 **Villalpando** Castilla y León, N Spain *41°51′N 05°25′W*
40 K9 **Villa Madero** *var.* Francisco I. Madero. Durango, C Mexico *24°28′N 104°20′W*
41 O9 **Villa Mainero** Tamaulipas, C Mexico *24°32′N 99°39′W*
Villamaña *see* Villamañán
104 L4 **Villamañán** *var.* Villamaña. Castilla y León, N Spain *42°19′N 05°35′W*
62 L10 **Villa María** Córdoba, C Argentina *32°23′S 63°15′W*
61 C17 **Villa María Grande** Entre Ríos, E Argentina *31°39′S 59°54′W*
57 K21 **Villa Martín** Potosí, SW Bolivia *20°46′S 67°45′W*
104 K15 **Villamartin** Andalucía, S Spain *36°52′N 05°38′W*
62 J8 **Villa Mazán** La Rioja, NW Argentina *28°43′S 66°25′W*
62 J11 **Villa Mercedes** *var.* Mercedes. San Luis, C Argentina *33°40′S 65°25′W*
Villamil *see* Puerto Villamil
Villa Nador *see* Nador
54 G5 **Villanueva** La Guajira, N Colombia *10°37′N 72°58′W*
42 H5 **Villanueva** Cortés, NW Honduras *15°14′N 88°00′W*
40 L11 **Villanueva** Zacatecas, C Mexico *22°24′N 102°53′W*
42 I9 **Villa Nueva** Chinandega, NW Nicaragua *12°58′N 86°46′W*
37 T11 **Villanueva** New Mexico, SW USA *35°18′N 105°20′W*
104 M12 **Villanueva de Córdoba** Andalucía, S Spain *38°20′N 04°38′W*
105 O12 **Villanueva del Arzobispo** Andalucía, S Spain *38°10′N 03°00′W*
104 K11 **Villanueva de la Serena** Extremadura, W Spain *38°58′N 05°48′W*
104 L5 **Villanueva del Campo** Castilla y León, N Spain *41°59′N 05°25′W*
105 O11 **Villanueva de los Infantes** Castilla-La Mancha, C Spain *38°45′N 03°01′W*
61 C14 **Villa Ocampo** Santa Fe, C Argentina *28°28′S 59°22′W*
40 J8 **Villa Ocampo** Durango, C Mexico *26°29′N 105°30′W*
40 J7 **Villa Orestes Pereyra** Durango, C Mexico *26°30′N 105°38′W*
105 N3 **Villarcayo** Castilla y León, N Spain *42°56′N 03°34′W*
104 L5 **Villardefrades** Castilla y León, N Spain *41°43′N 05°15′W*
105 S9 **Villar del Arzobispo** Valenciana, E Spain *39°44′N 00°50′W*
105 Q6 **Villaroya de la Sierra** Aragón, NE Spain *41°28′N 01°46′W*
Villarreal *see* Vila-real
62 P6 **Villarrica** Guairá, SE Paraguay *25°45′S 56°28′W*
63 G15 **Villarrica, Volcán** ✕ S Chile *39°28′S 71°57′W*
105 P10 **Villarrobledo** Castilla-La Mancha, C Spain *39°16′N 02°36′W*
105 N10 **Villarrubia de los Ojos** Castilla-La Mancha, C Spain *39°14′N 03°36′W*
18 J17 **Villas** New Jersey, NE USA *39°01′N 74°54′W*
105 O3 **Villasana de Mena** Castilla y León, N Spain *43°05′N 03°16′W*
107 M23 **Villa San Giovanni** Calabria, S Italy *38°12′N 15°39′E*
61 D18 **Villa San José** Entre Ríos, E Argentina *32°01′S 58°20′W*
Villa Sanjurjo *see* Al-Hoceïma
105 P6 **Villasayas** Castilla y León, N Spain *41°19′N 02°36′W*
107 C20 **Villasimius** Sardegna, Italy, C Mediterranean Sea *39°10′N 09°30′E*
41 N6 **Villa Unión** Coahuila, NE Mexico *28°18′N 100°43′W*
40 K10 **Villa Unión** Durango, C Mexico *23°59′N 104°01′W*
40 J10 **Villa Unión** Sinaloa, C Mexico *23°10′N 106°12′W*
62 K12 **Villa Valeria** Córdoba, C Argentina *34°21′S 64°56′W*
105 N8 **Villaverde** Madrid, C Spain *40°21′N 03°43′W*
54 F10 **Villavicencio** Meta, C Colombia *04°09′N 73°38′W*
104 L2 **Villaviciosa** Asturias, N Spain *43°29′N 05°26′W*
104 L12 **Villaviciosa de Córdoba** Andalucía, S Spain *38°04′N 05°00′W*
57 L22 **Villazón** Potosí, S Bolivia *22°05′S 65°35′W*
14 J8 **Villebon, Lac** ◎ Québec, SE Canada
Ville de Kinshasa *see* Kinshasa
102 J5 **Villedieu-les-Poêles** Manche, N France *48°51′N 01°12′W*
Villefranche *see* Villefranche-sur-Saône

◆ Country
● Country Capital
◇ Dependent Territory
○ Dependent Territory Capital
◈ Administrative Regions
✈ International Airport
▲ Mountain
▲▲ Mountain Range
✕ Volcano
≈ River
◎ Lake
⊠ Reservoir

Villefranche-de-Lauragais – Vučitrn

103 N16 **Villefranche-de-Lauragais** Haute-Garonne, S France 43°24′N 01°42′E
103 N14 **Villefranche-de-Rouergue** Aveyron, S France 44°21′N 02°02′E
103 R10 **Villefranche-sur-Saône** *var.* Villefranche. Rhône, E France 46°00′N 04°40′E
14 H9 **Ville-Marie** Québec, SE Canada 47°21′N 79°26′W
102 M15 **Villemur-sur-Tarn** Haute-Garonne, S France 43°50′N 01°32′E
105 S11 **Villena** Valenciana, E Spain 38°39′N 00°52′W
Villeneuve-d'Agen *see* Villeneuve-sur-Lot
102 L13 **Villeneuve-sur-Lot** *var.* Villeneuve-d'Agen, *hist.* Gajac. Lot-et-Garonne, SW France 44°24′N 00°43′E
103 P6 **Villeneuve-sur-Yonne** Yonne, C France 48°04′N 03°21′E
22 H8 **Ville Platte** Louisiana, S USA 30°41′N 92°16′W
103 R11 **Villeurbanne** Rhône, E France 45°46′N 04°54′E
101 G23 **Villingen-Schwenningen** Baden-Württemberg, S Germany 48°04′N 08°27′E
29 T15 **Villisca** Iowa, C USA 40°55′N 94°58′W
Villmanstrand *see* Lappeenranta
Vilna *see* Vilnius
119 H14 **Vilnius** *Pol.* Wilno, *Ger.* Wilna; *prev. Rus.* Vilna. ● (Lithuania) Vilnius, SE Lithuania 54°41′N 25°20′E
119 H14 **Vilnius** ✈ Vilnius, SE Lithuania 54°33′N 25°17′E
117 S7 **Vil'nohirs'k** Dnipropetrovs'ka Oblast', E Ukraine 48°31′N 34°01′E
117 U8 **Vil'nyans'k** Zaporiz'ka Oblast', SE Ukraine 47°56′N 35°22′E
93 L17 **Vilppula** Pirkanmaa, W Finland 62°02′N 24°30′E
101 M20 **Vils** ≈ SE Germany
118 C5 **Vilsandi** *island* W Estonia
117 P8 **Vil'shanka** *Rus.* Olshanka. Kirovohrads'ka Oblast', C Ukraine 48°12′N 30°54′E
101 O22 **Vilshofen** Bayern, SE Germany 48°36′N 13°10′E
155 J20 **Viluppuram** Tamil Nādu, SE India 12°54′N 79°40′E
113 I16 **Vilusi** W Montenegro 42°44′N 18°34′E
99 G18 **Vilvoorde** *Fr.* Vilvorde. Vlaams Brabant, C Belgium 50°56′N 04°25′E
Vilvorde *see* Vilvoorde
119 J14 **Vilyeyka** *Pol.* Wilejka, *Rus.* Vileyka. Minskaya Voblasts', NW Belarus 54°30′N 26°55′E
122 V11 **Vilyuchinsk** Kamchatskiy Kray, E Russian Federation 52°55′N 158°28′E
123 P10 **Vilyuy** ≈ NE Russian Federation
123 P10 **Vilyuysk** Respublika Sakha (Yakutiya), NE Russian Federation 63°42′N 121°20′E
123 N10 **Vilyuyskoye Vodokhranilishche** ⊠ NE Russian Federation
104 G2 **Vimianzo** Galicia, NW Spain 43°06′N 09°03′W
95 M19 **Vimmerby** Kalmar, S Sweden 57°40′N 15°50′E
102 L5 **Vimoutiers** Orne, N France 48°56′N 00°10′E
93 L16 **Vimpeli** Etelä-Pohjanmaa, W Finland 63°10′N 23°50′E
79 G14 **Vina** ≈ Cameroon/Chad
62 G11 **Viña del Mar** Valparaíso, C Chile 33°02′S 71°35′W
19 R8 **Vinalhaven Island** *island* Maine, NE USA
105 T8 **Vinaròs** Valenciana, E Spain 40°29′N 00°28′E
Vînători *see* Vânători
31 N15 **Vincennes** Indiana, N USA 38°42′N 87°30′W
195 Y12 **Vincennes Bay** *bay* Antarctica
25 O7 **Vincent** Texas, SW USA 32°30′N 101°10′W
95 H24 **Vindeby** Syddtjylland, C Denmark 54°55′N 11°09′E
93 I15 **Vindeln** Västerbotten, N Sweden 64°11′N 19°45′E
95 F21 **Vinderup** Midtjylland, C Denmark 56°29′N 08°48′E
Vindhya Mountains *see* Vindhya Range
153 N14 **Vindhya Range** *var.* Vindhya Mountains. ▲▲ N India
Vindobona *see* Wien
20 K6 **Vine Grove** Kentucky, S USA 37°48′N 85°58′W
18 J17 **Vineland** New Jersey, NE USA 39°29′N 75°02′W
116 E11 **Vinga** Arad, W Romania 46°00′N 21°14′E
95 M16 **Vingåker** Södermanland, C Sweden 59°02′N 15°52′E
167 S8 **Vinh** Nghê An, N Vietnam 18°42′N 105°41′E
104 I5 **Vinhais** Bragança, N Portugal 41°50′N 07°00′W
Vinh Linh *see* Hô Xa
Vinh Loi *see* Bac Liêu
167 S14 **Vinh Long** *var.* Vinhlong. Vinh Long, S Vietnam 10°15′N 105°59′E
Vinhlong *see* Vinh Long
113 Q18 **Vinica** NE FYR Macedonia 41°53′N 22°30′E
109 V13 **Vinica** SE Slovenia 45°28′N 15°12′E
114 G8 **Vinishte** Montana, NW Bulgaria 43°30′N 23°04′E
27 Q8 **Vinita** Oklahoma, C USA 36°38′N 95°09′W
Vînju Mare *see* Vânju Mare
98 I11 **Vinkeveen** Utrecht, C Netherlands 52°13′N 04°55′E
116 L6 **Vin'kivtsi** Khmel'nyts'ka Oblast', W Ukraine 49°02′N 27°13′E
112 I10 **Vinkovci** *Ger.* Winkowitz, *Hung.* Vinkovcze. Vukovar-Srijem, E Croatia 45°18′N 18°45′E
Vinkovcze *see* Vinkovci
Vinnitsa *see* Vinnytsya
Vinnitskaya Oblast'/Vinnytsya *see* Vinnyts'ka Oblast'
116 M7 **Vinnyts'ka Oblast'** *var.* Vinnytsya, *Rus.* Vinnitskaya Oblast'. ◈ *province* C Ukraine
117 N6 **Vinnytsya** *Rus.* Vinnitsa. Vinnyts'ka Oblast', C Ukraine 49°14′N 28°30′E
117 N6 **Vinnytsya** ✈ Vinnyts'ka Oblast', N Ukraine 49°13′N 28°40′E
Vinogradov *see* Vynohradiv
194 L8 **Vinson Massif** ▲ Antarctica 78°45′S 85°19′W
94 G11 **Vinstra** Oppland, S Norway 61°36′N 09°45′E
116 K12 **Vintilă Vodă** Buzău, SE Romania 45°28′N 26°43′E
29 X13 **Vinton** Iowa, C USA 42°10′N 92°01′W
22 F9 **Vinton** Louisiana, S USA 30°10′N 93°37′W
155 J17 **Vinukonda** Andhra Pradesh, E India 16°03′N 79°41′E
Vioara *see* Ocnele Mari
83 E23 **Vioolsdrif** Northern Cape, SW South Africa 28°50′S 17°38′E
82 M13 **Viphya Mountains** ▲▲ C Malawi
171 Q4 **Virac** Catanduanes Island, N Philippines 13°39′N 124°17′E
124 K8 **Virandozero** Respublika Kareliya, NW Russian Federation 63°59′N 36°00′E
137 P16 **Viranşehir** Şanlıurfa, SE Turkey 37°13′N 39°32′E
154 D13 **Virār** Mahārāshtra, W India 19°30′N 72°48′E
11 W16 **Virden** Manitoba, S Canada 49°50′N 100°57′W
30 K14 **Virden** Illinois, N USA 39°30′N 89°46′W
Virdois *see* Virrat
102 J5 **Vire** Calvados, N France 48°50′N 00°53′W
102 J4 **Vire** ≈ N France
83 A15 **Virei** Namibe, SW Angola 15°43′S 12°54′E
Virful Moldoveanu *see* Vârful Moldoveanu
35 R5 **Virgina Peak** ▲ Nevada, W USA 39°46′N 119°26′W
45 U9 **Virgin Gorda** *island* C British Virgin Islands
83 I22 **Virginia** Free State, C South Africa 28°06′S 26°53′E
30 K13 **Virginia** Illinois, N USA 39°57′N 90°12′W
29 W4 **Virginia** Minnesota, N USA 47°31′N 92°32′W
21 T6 **Virginia** *off.* Commonwealth of Virginia, *also known as* Mother of Presidents, Mother of States, Old Dominion. ◈ *state* NE USA
21 Y7 **Virginia Beach** Virginia, NE USA 36°51′N 75°59′W
33 R11 **Virginia City** Montana, NW USA 45°17′N 111°54′W
35 Q6 **Virginia City** Nevada, W USA 39°19′N 119°39′W
14 H8 **Virginiatown** Ontario, S Canada 48°09′N 79°35′W
Virgin Islands *see* British Virgin Islands
45 T9 **Virgin Islands (US)** *var.* Virgin Islands of the United States; *prev.* Danish West Indies. ◇ *US unincorporated territory* E West Indies
Virgin Islands of the United States *see* Virgin Islands (US)
45 T9 **Virgin Passage** *passage* Puerto Rico/Virgin Islands (US)
35 Y10 **Virgin River** ≈ Nevada/Utah, W USA
Virihaur *see* Virihaure
92 H12 **Virihaure** *Lapp.* Virihávrre, *var.* Virihaur. ◎ N Sweden
Virihávrre *see* Virihaure
167 T11 **Virôchey** Rôtânôkiri, NE Cambodia 13°59′N 106°49′E
93 N19 **Virolahti** Kymenlaakso, S Finland 60°33′N 27°37′E
30 J8 **Viroqua** Wisconsin, N USA 43°33′N 90°54′W
112 G8 **Virovitica** *Ger.* Virovititz, *Hung.* Verőcze; *prev. Ger.* Werowitz. Virovitica-Podravina, NE Croatia 45°49′N 17°25′E
112 G8 **Virovitica-Podravina** *off.* Virovitičko-Podravska Županija. ◈ *province* NE Croatia
Virovitičko-Podravska Županija *see* Virovitica-Podravina
Virovititz *see* Virovitica
113 J17 **Virpazar** S Montenegro 42°15′N 19°06′E
93 L17 **Virrat** *Swe.* Virdois. Pirkanmaa, W Finland 62°14′N 23°50′E
95 M20 **Virserum** Kalmar, S Sweden 57°17′N 15°18′E
99 K25 **Virton** Luxembourg, SE Belgium 49°34′N 05°32′E
118 F5 **Virtsu** *Ger.* Werder. Läänemaa, W Estonia 58°35′N 23°33′E
56 C12 **Virú** La Libertad, C Peru 08°24′S 78°40′W
Virudhunagar *see* Virudunagar
155 H23 **Virudunagar** *var.* Virudhunagar; *prev.* Virudupatti. Tamil Nādu, SE India 09°35′N 77°57′E
Virudupatti *see* Virudunagar
118 I3 **Viru-Jaagupi** *Ger.* Sankt-Jakobi. Lääne-Virumaa, NE Estonia 59°14′N 26°29′E
57 N19 **Viru-Viru** *var.* Santa Cruz. ✈ (Santa Cruz) Santa Cruz, C Bolivia 17°49′S 63°12′W
113 E15 **Vis** *It.* Lissa; *anc.* Issa. *island* S Croatia
Vis *see* Fish
118 I12 **Visaginas** *prev.* Sniečkus. Utena, E Lithuania 55°36′N 26°22′E
155 M15 **Visākhapatnam** *var.* Vishakhapatnam. Andhra Pradesh, SE India 17°45′N 83°19′E
35 R11 **Visalia** California, W USA 36°19′N 119°19′W
Vişău *see* Vişeu
95 P19 **Visby** *Ger.* Wisby. Gotland, SE Sweden 57°37′N 18°20′E
197 N9 **Viscount Melville Sound** *prev.* Melville Sound. *sound* Northwest Territories, N Canada
99 L19 **Visé** Liège, E Belgium 50°44′N 05°42′E
112 K13 **Višegrad** Republika Srpska, SE Bosnia and Herzegovina 43°46′N 19°18′E
58 L12 **Viseu** Pará, NE Brazil 01°10′S 46°09′W
104 H7 **Viseu** *prev.* Vizeu. Viseu, N Portugal 40°40′N 07°55′W
104 H7 **Viseu** *var.* Vizeu. ◈ *district* N Portugal
116 I8 **Vişeu** *Hung.* Visó; *prev.* Vişău. ≈ NW Romania
116 I8 **Vişeu de Sus** *var.* Vişeul de Sus, *Ger.* Oberwischau, *Hung.* Felsővisó. Maramureş, N Romania 47°43′N 23°24′E
Vişeul de Sus *see* Vişeu de Sus
Vishakhapatnam *see* Visākhapatnam
125 R10 **Vishera** ≈ NW Russian Federation
95 J19 **Viskafors** Västra Götaland, S Sweden 57°37′N 12°50′E
95 J20 **Viskan** ≈ S Sweden
95 L21 **Vislanda** Kronoberg, S Sweden 56°46′N 14°30′E
Vislinskiy Zaliv *see* Vistula Lagoon
Visó *see* Vişeu
112 H13 **Visoko** ◈ Federacija Bosne I Hercegovine, C Bosnia and Herzegovina
106 A9 **Viso, Monte** ▲ NW Italy 44°42′N 07°04′E
108 E10 **Visp** Valais, SW Switzerland 46°18′N 07°53′E
108 E10 **Vispa** ≈ S Switzerland
95 M21 **Vissefjärda** Kalmar, S Sweden 56°31′N 15°34′E
100 I11 **Visselhövede** Niedersachsen, NW Germany 52°58′N 09°36′E
95 G23 **Vissenbjerg** Syddtjylland, C Denmark 55°23′N 10°08′E
35 U17 **Vista** California, W USA 33°12′N 117°14′W
58 C11 **Vista Alegre** Amazonas, NW Brazil 01°23′N 68°13′W
114 J13 **Vistonída, Límni** ◎ NE Greece
Vistula *see* Wisła
119 A14 **Vistula Lagoon** *Ger.* Frisches Haff, *Pol.* Zalew Wiślany, *Rus.* Vislinskiy Zaliv. *lagoon* Poland/Russian Federation
114 I8 **Vit** ≈ NW Bulgaria
Vitebsk *see* Vitsyebsk
Vitebskaya Oblast' *see* Vitsyebskaya Voblasts'
107 H14 **Viterbo** *anc.* Vicus Elbii. Lazio, C Italy 42°25′N 12°08′E
112 H12 **Vitez** Federacija Bosne I Hercegovine, C Bosnia and Herzegovina 44°08′N 17°47′E
167 S14 **Vi Thanh** Cân Thơ, S Vietnam 09°45′N 105°28′E
Viti *see* Fiji
186 E7 **Vitiaz Strait** *strait* NE Papua New Guinea
104 J7 **Vitigudino** Castilla y León, N Spain 41°00′N 06°26′W
175 Q9 **Viti Levu** *island* W Fiji
187 W15 **Viti Levu** *island* W Fiji
123 O11 **Vitim** ≈ C Russian Federation
123 O12 **Vitimskiy** Irkutskaya Oblast', C Russian Federation 58°12′N 113°10′E
109 V2 **Vitis** Niederösterreich, N Austria 48°45′N 15°09′E
59 O20 **Vitória** *state capital* Espírito Santo, SE Brazil 20°19′S 40°21′W
Vitoria *see* Vitoria-Gasteiz
Vitoria Bank *see* Vitória Seamount
59 N18 **Vitória da Conquista** Bahia, E Brazil 14°53′S 40°52′W
105 P3 **Vitoria-Gasteiz** *var.* Vitoria, *Eng.* Vittoria. País Vasco, N Spain 42°51′N 02°40′W
65 J16 **Vitória Seamount** *var.* Victoria Bank, Vitoria Bank. *undersea feature* C Atlantic Ocean 18°48′S 37°24′W
112 F13 **Vitorog** ▲ SW Bosnia and Herzegovina 44°06′N 17°03′E
102 J6 **Vitré** Ille-et-Vilaine, NW France 48°07′N 01°12′W
103 R5 **Vitry-le-François** Marne, N France 48°43′N 04°36′E
114 D13 **Vítsi** *var.* Vítsoi. ▲ N Greece 40°39′N 21°23′E
Vítsoi *see* Vítsi
118 N13 **Vitsyebsk** *Rus.* Vitebsk. Vitsyebskaya Voblasts', NE Belarus 55°11′N 30°10′E
118 K13 **Vitsyebskaya Voblasts'** *Rus.* Vitebskaya Oblast'. ◈ *province* N Belarus
92 J11 **Vittangi** *Lapp.* Vazáš. Norrbotten, N Sweden 67°40′N 21°39′E
103 R8 **Vitteaux** Côte d'Or, C France 47°24′N 04°31′E
103 S6 **Vittel** Vosges, NE France 48°13′N 05°57′E
95 N15 **Vittinge** Västmanland, C Sweden 59°52′N 17°04′E
107 K25 **Vittoria** Sicilia, Italy, C Mediterranean Sea 36°56′N 14°30′E
Vittoria *see* Vitoria-Gasteiz
106 I7 **Vittorio Veneto** Veneto, NE Italy 45°59′N 12°18′E
175 Q7 **Vityaz Trench** *undersea feature* W Pacific Ocean
108 G8 **Vitznau** Luzern, W Switzerland 47°01′N 08°28′E
104 I1 **Viveiro** Galicia, NW Spain 43°39′N 07°35′W
105 S9 **Viver** Valenciana, E Spain 39°55′N 00°36′W
103 Q13 **Viverais, Monts du** ▲▲ C France
122 L9 **Vivi** ≈ N Russian Federation
22 F4 **Vivian** Louisiana, S USA 32°52′N 93°59′W
29 N10 **Vivian** South Dakota, N USA 43°53′N 100°16′W
103 R13 **Viviers** Ardèche, E France 44°31′N 04°40′E
Vivis *see* Vevey
83 K19 **Vivo** Limpopo, NE South Africa 22°58′S 29°13′E
102 L10 **Vivonne** Vienne, W France 46°25′N 00°15′E
Vizakna *see* Ocna Sibiului
Vizcaya *see* Bizkaia
Vizcaya, Golfo de *see* Biscay, Bay of
136 C10 **Vize** Kırklareli, NW Turkey 41°34′N 27°45′E
122 K4 **Vize, Ostrov** *island* Severnaya Zemlya, N Russian Federation
Vizeu *see* Viseu
155 M15 **Vizianagaram** *var.* Vizianagram. Andhra Pradesh, E India 18°07′N 83°25′E
Vizianagram *see* Vizianagaram
103 S12 **Vizille** Isère, E France 45°05′N 05°46′E
125 R11 **Vizinga** Respublika Komi, NW Russian Federation 61°06′N 50°09′E
116 M13 **Viziru** Brăila, SE Romania 45°00′N 27°43′E
113 K21 **Vjosës, Lumi i** *var.* Vijosa, Vijosë, *Gk.* Aóos. ≈ Albania/Greece *see also* Aóos
Vjosës, Lumi i *see* Aóos
99 H18 **Vlaams Brabant** ◈ *province* C Belgium
99 G18 **Vlaanderen** *Eng.* Flanders, *Fr.* Flandre. ◈ Belgium/France
98 G12 **Vlaardingen** Zuid-Holland, SW Netherlands 51°55′N 04°21′E
116 F10 **Vlădeasa, Vârful** *prev.* Vîrful Vlădeasa. ▲ NW Romania 46°45′N 22°46′E
Vlădeasa, Vîrful *see* Vlădeasa, Vârful
113 P16 **Vladičin Han** Serbia, SE Serbia 42°44′N 22°04′E
127 O16 **Vladikavkaz** *prev.* Dzaudzhikau, Ordzhonikidze. Respublika Severnaya Osetiya, SW Russian Federation 42°58′N 44°41′E
126 M3 **Vladimir** Vladimirskaya Oblast', W Russian Federation 56°09′N 40°21′E
Vladimirets *see* Voldymyerets'
144 M7 **Vladimirovka** Kostanay, N Kazakhstan 53°30′N 64°02′E
Vladimirovka *see* Yuzhno-Sakhalinsk
126 L3 **Vladimirskaya Oblast'** ◈ *province* W Russian Federation
126 I3 **Vladimirskiy Tupik** Smolenskaya Oblast', W Russian Federation 55°45′N 33°25′E
Vladimir-Volynskiy *see* Volodymyr-Volyns'kyy
123 Q7 **Vladivostok** Primorskiy Kray, SE Russian Federation 43°09′N 131°53′E
117 U13 **Vladyslavivka** Avtonomna Respublika Krym, S Ukraine 45°09′N 35°25′E
98 P6 **Vlagtwedde** Groningen, NE Netherlands 53°02′N 07°07′E
Vlajna *see* Kukavica
112 J12 **Vlasenica** ◈ Republika Srpska, E Bosnia and Herzegovina
112 G12 **Vlašić** ▲ C Bosnia and Herzegovina 44°18′N 17°40′E
111 D17 **Vlašim** *Ger.* Wlaschim. Středočesky Kraj, C Czech Republic 49°42′N 14°54′E
113 P15 **Vlasotince** Serbia, SE Serbia 42°58′N 22°07′E
123 Q7 **Vlasovo** Respublika Sakha (Yakutiya), NE Russian Federation 70°41′N 134°49′E
98 I11 **Vleuten** Utrecht, C Netherlands 52°07′N 05°01′E
98 I5 **Vlieland** *Fris.* Flylân. *island* Waddeneilanden, N Netherlands
98 I5 **Vliestroom** *strait* NW Netherlands
99 J14 **Vlijmen** Noord-Brabant, S Netherlands 51°42′N 05°14′E
99 E15 **Vlissingen** *Eng.* Flushing, *Fr.* Flessingue. Zeeland, SW Netherlands 51°26′N 03°34′E
Vlodava *see* Włodawa
Vlonë/Vlora *see* Vlorë
113 K22 **Vlorë** *prev.* Vlonë, *It.* Valona, Vlora. Vlorë, SW Albania 40°28′N 19°31′E
113 K22 **Vlorë** ◈ *district* SW Albania
113 K22 **Vlorës, Gjiri i** *var.* Valona Bay. *bay* SW Albania
Vlotslavsk *see* Włocławek
111 C16 **Vltava** *Ger.* Moldau. ≈ W Czech Republic
126 K3 **Vnukovo** ✈ (Moskva) Gorod Moskva, W Russian Federation 55°30′N 36°52′E
146 L11 **Vobkent** *Rus.* Vabkent. Buxoro Viloyati, C Uzbekistan 40°01′N 64°25′E
25 Q9 **Voca** Texas, SW USA 30°58′N 99°09′W
109 R5 **Vöcklabruck** Oberösterreich, NW Austria 48°01′N 13°38′E
112 D13 **Vodice** Šibenik-Knin, S Croatia 43°46′N 15°46′E
124 K10 **Vodlozero, Ozero** ◎ NW Russian Federation
112 A10 **Vodnjan** *It.* Dignano d'Istria. Istra, NW Croatia 44°57′N 13°51′E
125 S9 **Vodnyy** Respublika Komi, NW Russian Federation 63°31′N 53°21′E
145 V15 **Vodokhranilishche Kapshagay** *Kaz.* Qapshagay Böyeni; *prev.* Kapchagayskoye Vodokhranilishche. ⊠ SE Kazakhstan
124 I6 **Vodokhranilishche, Kumskoye** ◎ NW Russian Federation
95 G20 **Vodskov** Nordjylland, N Denmark 57°07′N 10°02′E
92 H4 **Vogar** Suðurnes, SW Iceland 63°58′N 22°20′W
Vogelkop *see* Doberai, Jazirah
77 X15 **Vogel Peak** *prev.* Dimlang. ▲ E Nigeria 08°16′N 11°44′E
101 H17 **Vogelsberg** ▲▲ C Germany
106 D8 **Voghera** Lombardia, N Italy 44°59′N 09°01′E
112 I13 **Vogošća** Federacija Bosne I Hercegovine, SE Bosnia and Herzegovina 43°55′N 18°20′E
101 M17 **Vogtland** *historical region* E Germany
125 V12 **Vogul'skiy Kamen', Gora** ▲ NW Russian Federation 60°10′N 58°41′E
187 P16 **Voh** Province Nord, C New Caledonia 20°57′S 164°41′E
Vohémar *see* Iharaña
172 H8 **Vohimena, Tanjona** *Fr.* Cap Sainte Marie. *headland* S Madagascar 25°20′S 45°06′E
172 J6 **Vohipeno** Fianarantsoa, SE Madagascar 22°21′S 47°51′E
118 H5 **Võhma** *Ger.* Wöchma. Viljandimaa, S Estonia 58°37′N 25°34′E
81 J20 **Voi** Taita/Taveta, S Kenya 03°23′S 38°35′E
76 K15 **Voinjama** N Liberia 08°25′N 09°42′W
103 S12 **Voiron** Isère, E France 45°22′N 05°35′E
109 V8 **Voitsberg** Steiermark, SE Austria 47°04′N 15°09′E
95 F24 **Vojens** *Ger.* Woyens. Syddanmark, SW Denmark 55°15′N 09°19′E
112 K9 **Vojvodina** *Ger.* Wojwodina. Vojvodina, N Serbia
15 S6 **Volant** ≈ Québec, SE Canada
Volaterrae *see* Volterra
43 P15 **Volcán** *var.* Hato del Volcán. Chiriquí, W Panama 08°45′N 82°38′W
Volcano Islands *see* Kazan-rettō
Volchansk *see* Vovchans'k
Volchya *see* Vovcha
94 D10 **Volda** Møre og Romsdal, S Norway 62°07′N 06°04′E
116 K3 **Voldymyerets'** *Russ.* Vladimirets. Rivnens'ka Oblast', NW Ukraine 51°24′N 25°52′E
98 J9 **Volendam** Noord-Holland, C Netherlands 52°30′N 05°04′E
124 L15 **Volga** Yaroslavskaya Oblast', W Russian Federation 57°56′N 38°23′E
29 R10 **Volga** South Dakota, N USA 44°19′N 96°55′W
122 C11 **Volga** ≈ NW Russian Federation
Volga-Baltic Waterway *see* Volgo-Baltiyskiy Kanal
Volga Uplands *see* Privolzhskaya Vozvyshennost'
124 L13 **Volgo-Baltiyskiy Kanal** *var.* Volga-Baltic Waterway. *canal* NW Russian Federation
126 M12 **Volgodonsk** Rostovskaya Oblast', SW Russian Federation 47°35′N 42°03′E
127 O10 **Volgograd** *prev.* Stalingrad, Tsaritsyn. Volgogradskaya Oblast', SW Russian Federation 48°42′N 44°29′E
127 N9 **Volgogradskaya Oblast'** ◈ *province* SW Russian Federation
127 P10 **Volgogradskoye Vodokhranilishche** ⊠ SW Russian Federation
101 J19 **Volkach** Bayern, C Germany 49°51′N 10°15′E
109 U9 **Völkermarkt** *Slvn.* Velikovec. Kärnten, S Austria 46°40′N 14°38′E
124 I12 **Volkhov** Leningradskaya Oblast', NW Russian Federation 59°56′N 32°19′E
101 D20 **Völklingen** Saarland, SW Germany 49°15′N 06°51′E
Volkovysk *see* Vawkavysk
Volkovyskiye Vysoty *see* Vawkavyskaye Wzvyshsha
83 K22 **Volksrust** Mpumalanga, E South Africa 27°22′S 29°54′E
98 L8 **Vollenhove** Overijssel, N Netherlands 52°40′N 05°58′E
119 L16 **Volma** ≈ C Belarus
Volmari *see* Valmiera
117 W9 **Volnovakha** Donets'ka Oblast', SE Ukraine 47°36′N 37°32′E
116 K6 **Volochys'k** Khmel'nyts'ka Oblast', W Ukraine 49°32′N 26°14′E
117 O6 **Volodarka** Kyyivs'ka Oblast', N Ukraine 49°31′N 29°55′E
117 W9 **Volodars'ke** Donets'ka Oblast', E Ukraine 47°11′N 37°19′E
127 R13 **Volodarskiy** Astrakhanskaya Oblast', SW Russian Federation 46°23′N 48°39′E
Volodarskoye *see* Saumalkol'
117 N8 **Volodars'k-Volyns'kyy** Zhytomyrs'ka Oblast', N Ukraine 50°37′N 28°28′E
116 I3 **Volodymyr-Volyns'kyy** *Pol.* Włodzimierz, *Rus.* Vladimir-Volynskiy. Volyns'ka Oblast', NW Ukraine 50°51′N 24°19′E
124 L14 **Vologda** Vologodskaya Oblast', W Russian Federation 59°10′N 39°55′E
124 L12 **Vologodskaya Oblast'** ◈ *province* NW Russian Federation
126 K3 **Volokolamsk** Moskovskaya Oblast', W Russian Federation 56°03′N 35°57′E
126 K9 **Volokonovka** Belgorodskaya Oblast', W Russian Federation 50°30′N 37°54′E
115 G16 **Vólos** Thessalía, C Greece 39°21′N 22°58′E
124 M11 **Voloshka** Arkhangel'skaya Oblast', NW Russian Federation 61°19′N 40°06′E
Vološinovo *see* Novi Bečej
116 H7 **Volovets'** Zakarpats'ka Oblast', W Ukraine 48°42′N 23°12′E
Volozhin *see* Valozhyn
127 Q7 **Vol'sk** Saratovskaya Oblast', W Russian Federation 52°04′N 47°20′E
77 Q17 **Volta** ≈ SE Ghana
Volta Blanche *see* White Volta
77 P16 **Volta, Lake** ⊠ SE Ghana
Volta Noire *see* Black Volta
60 O9 **Volta Redonda** Rio de Janeiro, SE Brazil 22°31′S 44°05′W
Volta Rouge *see* Red Volta
106 F12 **Volterra** *anc.* Volaterrae. Toscana, C Italy 43°23′N 10°52′E
107 K17 **Volturno** ≈ S Italy
113 I15 **Volujak** ▲▲ NW Montenegro
Volunteer Island *see* Starbuck Island
65 F24 **Volunteer Point** *headland* East Falkland, Falkland Islands 51°32′S 57°44′W
114 H13 **Vólvi, Límni** ◎ N Greece
Volyn *see* Volyns'ka Oblast'
116 I3 **Volyns'ka Oblast'** *var.* Volyn, *Rus.* Volynskaya Oblast'. ◈ *province* NW Ukraine
Volynskaya Oblast' *see* Volyns'ka Oblast'
127 Q3 **Volzhsk** Respublika Mariy El, W Russian Federation 55°53′N 48°21′E
127 O10 **Volzhskiy** Volgogradskaya Oblast', SW Russian Federation 48°49′N 44°40′E
172 I7 **Vondrozo** Fianarantsoa, SE Madagascar 22°50′S 47°20′E
39 P10 **Von Frank Mountain** ▲ Alaska, USA 63°36′N 154°29′W
115 C17 **Vónitsa** Dytikí Elláda, W Greece 38°55′N 20°53′E
118 J6 **Võnnu** *Ger.* Wendau. Tartumaa, SE Estonia 58°17′N 27°06′E
98 G12 **Voorburg** Zuid-Holland, W Netherlands 52°04′N 04°22′E
98 H11 **Voorschoten** Zuid-Holland, W Netherlands 52°08′N 04°26′E
98 M11 **Voorst** Gelderland, E Netherlands 52°10′N 06°10′E
98 K11 **Voorthuizen** Gelderland, C Netherlands 52°12′N 05°36′E
92 L2 **Vopnafjördhur** *bay* E Iceland
92 L2 **Vopnafjörður** Austurland, E Iceland 65°45′N 14°51′W
Vora *see* Vorë
119 H15 **Voranava** *Pol.* Werenów, *Rus.* Voronovo. Hrodzyenskaya Voblasts', W Belarus 54°09′N 25°19′E
108 I8 **Vorarlberg** *off.* Land Vorarlberg. ◈ *state* W Austria
Vorarlberg, Land *see* Vorarlberg
109 X7 **Vorau** Steiermark, E Austria 47°22′N 15°55′E
98 N11 **Vorden** Gelderland, E Netherlands 52°07′N 06°18′E
108 H9 **Vorderrhein** ≈ SE Switzerland
95 I24 **Vordingborg** Sjælland, SE Denmark 55°01′N 11°55′E
92 J2 **Vorðufell** ▲ N Iceland 65°42′N 18°45′W
113 K19 **Vorë** *var.* Vora. Tiranë, W Albania 41°23′N 19°37′E
115 H17 **Vóreies Sporádes** *var.* Vóreioi Sporádes, Vórioi Sporádhes, *Eng.* Northern Sporades. *island group* E Greece
Vóreioi Sporádes *see* Vóreies Sporádes
115 J17 **Vóreion Aigaíon** *Eng.* Aegean North. ◈ *region* SE Greece
115 G18 **Vóreios Evvoïkós Kólpos** *var.* Voreiós Evvoïkós Kólpos. *gulf* E Greece
197 S16 **Voring Plateau** *undersea feature* N Norwegian Sea 67°00′N 04°00′E
Vórioi Sporádhes *see* Vóreies Sporádes
125 W4 **Vorkuta** Respublika Komi, NW Russian Federation 67°27′N 64°E
95 I14 **Vorma** ≈ S Norway
118 E4 **Vormsi** *var.* Vormsi Saar, *Ger.* Worms, *Swed.* Ormsö. *island* W Estonia
Vormsi Saar *see* Vormsi
127 N7 **Vorona** ≈ W Russian Federation
126 L7 **Voronezh** Voronezhskaya Oblast', W Russian Federation 51°40′N 39°13′E
126 L7 **Voronezh** ≈ W Russian Federation
126 K8 **Voronezhskaya Oblast'** ◈ *province* W Russian Federation
Voronovitsya *see* Voronovytsya
Voronovo *see* Voranava
117 N6 **Voronovytsya** *Rus.* Voronovitsya. Vinnyts'ka Oblast', C Ukraine 49°06′N 28°49′E
122 K7 **Vorontsovo** Krasnoyarskiy Kray, N Russian Federation 71°45′N 83°31′E
124 K3 **Voron'ya** ≈ NW Russian Federation
Voropayevo *see* Varapayeva
Voroshilov *see* Ussuriysk
Voroshilovgrad *see* Luhans'ka Oblast', Ukraine
Voroshilovgrad *see* Luhans'k, Ukraine
Voroshilovgradskaya Oblast' *see* Luhans'ka Oblast'
Voroshilovsk *see* Stavropol', Russian Federation
Voroshilovsk *see* Alchevs'k
137 V13 **Vorotan** *Az.* Bärguşad. ≈ Armenia/Azerbaijan
127 P3 **Vorotynets** Nizhegorodskaya Oblast', W Russian Federation 56°06′N 46°06′E
117 S3 **Vorozhba** Sums'ka Oblast', NE Ukraine 51°10′N 34°15′E
117 T5 **Vorskla** ≈ Russian Federation/Ukraine
99 I17 **Vorst** Antwerpen, N Belgium 51°06′N 05°01′E
83 G21 **Vorstershoop** North-West, N South Africa 25°46′S 22°57′E
118 H6 **Võrtsjärv** *Ger.* Wirz-See. ◎ SE Estonia
118 J7 **Võru** *Ger.* Werro. Võrumaa, SE Estonia 57°51′N 27°01′E
147 R11 **Vorukh** N Tajikistan 39°51′N 70°34′E
118 I7 **Võrumaa** *off.* Võru Maakond. ◈ *province* SE Estonia
Võru Maakond *see* Võrumaa
83 G24 **Vosburg** Northern Cape, W South Africa 30°35′S 22°52′E
147 Q14 **Vose'** *Rus.* Vose; *prev.* Aral. SW Tajikistan 37°51′N 69°31′E
103 S6 **Vosges** ◈ *department* NE France
103 U6 **Vosges** ▲▲ NE France
124 K13 **Voskresenkoye** Vologodskaya Oblast', NW Russian Federation 59°25′N 37°56′E
126 L4 **Voskresensk** Moskovskaya Oblast', W Russian Federation 55°19′N 38°42′E
127 P2 **Voskresenskoye** Nizhegorodskaya Oblast', W Russian Federation 57°00′N 45°33′E
127 V6 **Voskresenskoye** Respublika Bashkortostan, W Russian Federation 53°07′N 56°07′E
94 D13 **Voss** Hordaland, S Norway 60°38′N 06°25′E
94 D13 **Voss** *physical region* S Norway
99 I16 **Vosselaar** Antwerpen, N Belgium 51°19′N 04°55′E
94 D13 **Vosso** ≈ S Norway
Vostochno-Kazakhstanskaya Oblast' *see* Vostochnyy Kazakhstan
123 S5 **Vostochno-Sibirskoye More** *Eng.* East Siberian Sea. *sea* Arctic Ocean
145 X10 **Vostochnyy Kazakhstan** *off.* Vostochno-Kazakhstanskaya Oblast', *var.* East Kazakhstan, *Kaz.* Shyghys Qazagastan Oblysy. ◈ *province* E Kazakhstan
122 L13 **Vostochnyy Sayan** *Eng.* Eastern Sayans, *Mong.* Dzüün Soyonï Nuruu. ▲▲ Mongolia/Russian Federation
Vostock Island *see* Vostok Island
195 U10 **Vostok** *Russian research station* Antarctica 77°18′S 105°32′E
191 X5 **Vostok Island** *var.* Vostock Island; *prev.* Stavers Island. *island* Line Islands, SE Kiribati
127 T2 **Votkinsk** Udmurtskaya Respublika, NW Russian Federation 57°04′N 54°00′E
125 U15 **Votkinskoye Vodokhranilishche** *var.* Votkinsk Reservoir. ⊠ NW Russian Federation
Votkinsk Reservoir *see* Votkinskoye Vodokhranilishche
60 J7 **Votuporanga** São Paulo, S Brazil 20°26′S 49°53′W
104 H7 **Vouga, Rio** ≈ N Portugal
115 E14 **Voúrinos** ▲▲ N Greece
115 G24 **Voúxa, Akrotírio** *headland* Kríti, Greece, E Mediterranean Sea 35°37′N 23°34′E
103 R4 **Vouziers** Ardennes, N France 49°24′N 04°42′E
117 V7 **Vovcha** *Rus.* Volchya. ≈ E Ukraine
117 V4 **Vovchans'k** *Rus.* Volchansk. Kharkivs'ka Oblast', E Ukraine 50°19′N 36°55′E
103 N6 **Voves** Eure-et-Loir, C France 48°18′N 01°39′E
79 M14 **Vovodo** ≈ S Central African Republic
94 M12 **Voxna** Gävleborg, C Sweden 61°21′N 15°35′E
94 L11 **Voxnan** ≈ C Sweden
114 F7 **Voynishka Reka** ≈ NW Bulgaria
125 T9 **Voyvozh** Respublika Komi, NW Russian Federation 62°54′N 54°52′E
124 M12 **Vozhega** Vologodskaya Oblast', NW Russian Federation 60°27′N 40°11′E
124 L12 **Vozhe, Ozero** ◎ NW Russian Federation
117 Q9 **Voznesens'k** *Rus.* Voznesensk. Mykolayivs'ka Oblast', S Ukraine 47°34′N 31°21′E
124 J12 **Voznesen'ye** Leningradskaya Oblast', NW Russian Federation 61°00′N 35°24′E
144 J14 **Vozrozhdeniya, Ostrov** *Uzb.* Wozrojdeniye Oroli. *island* Kazakhstan/Uzbekistan
95 G20 **Vrå** *var.* Vraa. Nordjylland, N Denmark 57°21′N 09°57′E
Vraa *see* Vrå
114 H9 **Vrachesh** Sofia, W Bulgaria 42°52′N 23°45′E
115 C19 **Vrachíonas** ▲ Zákynthos, Iónia Nisiá, Greece, C Mediterranean Sea 37°49′N 20°43′E
117 P8 **Vradiyivka** Mykolayivs'ka Oblast', S Ukraine 47°51′N 30°37′E
113 G14 **Vran** ▲ SW Bosnia and Herzegovina 43°35′N 17°30′E
116 K12 **Vrancea** ◈ *county* E Romania
147 T14 **Vrang** SE Tajikistan 37°03′N 72°26′E
123 T4 **Vrangelya, Ostrov** *Eng.* Wrangel Island. *island* NE Russian Federation
112 H13 **Vranica** ▲ C Bosnia and Herzegovina 43°57′N 17°43′E
113 O16 **Vranje** Serbia, SE Serbia 42°33′N 21°55′E
Vranov *see* Vranov nad Topl'ou
111 N19 **Vranov nad Topl'ou** *var.* Vranov, *Hung.* Varannó. Prešovský Kraj, E Slovakia 48°54′N 21°41′E
114 H8 **Vratsa** Vratsa, NW Bulgaria 43°13′N 23°34′E
114 H8 **Vratsa** ◈ *province* NW Bulgaria
114 F10 **Vrattsa** *prev.* Mirovo. Kyustendil, W Bulgaria 42°15′N 22°33′E
112 G11 **Vrbanja** ≈ NW Bosnia and Herzegovina
112 K9 **Vrbas** Vojvodina, NW Serbia 45°34′N 19°39′E
112 G13 **Vrbas** ≈ N Bosnia and Herzegovina
112 E8 **Vrbovec** Zagreb, N Croatia 45°53′N 16°24′E
112 C9 **Vrbovsko** Primorje-Gorski Kotar, NW Croatia 45°22′N 15°06′E
111 E15 **Vrchlabí** *Ger.* Hohenelbe. Královéhradecký Kraj, N Czech Republic 50°38′N 15°35′E
83 J22 **Vrede** Free State, E South Africa 27°25′S 29°10′E
100 E13 **Vreden** Nordrhein-Westfalen, NW Germany 52°01′N 06°50′E
83 E25 **Vredenburg** Western Cape, SW South Africa 32°55′S 18°00′E
99 I23 **Vresse-sur-Semois** Namur, SE Belgium 49°52′N 04°56′E
95 L16 **Vretstorp** Örebro, C Sweden 59°03′N 14°51′E
113 G15 **Vrgorac** *prev.* Vrhgorac. Split-Dalmacija, SE Croatia 43°10′N 17°24′E
Vrhgorac *see* Vrgorac
109 T12 **Vrhnika** *Ger.* Oberlaibach. W Slovenia 45°57′N 14°18′E
155 I21 **Vriddhāchalam** Tamil Nādu, SE India 11°33′N 79°18′E
98 N6 **Vries** Drenthe, NE Netherlands 53°04′N 06°34′E
98 O10 **Vriezenveen** Overijssel, E Netherlands 52°25′N 06°37′E
95 L20 **Vrigstad** Jönköping, S Sweden 57°19′N 14°30′E
108 H9 **Vrin** Graubünden, S Switzerland 46°40′N 09°06′E
112 E13 **Vrlika** Split-Dalmacija, S Croatia 43°54′N 16°24′E
113 M14 **Vrnjačka Banja** Serbia, C Serbia 43°36′N 20°55′E
Vrondádhes/Vrondados *see* Vrontádos
115 L18 **Vrontádos** *var.* Vrondados; *prev.* Vrondádhes. Chíos, E Greece 38°25′N 26°08′E
98 N9 **Vroomshoop** Overijssel, E Netherlands 52°28′N 06°35′E
112 N10 **Vršac** *Ger.* Werschetz, *Hung.* Versecz. Vojvodina, NE Serbia 45°08′N 21°18′E
112 M10 **Vršački Kanal** *canal* N Serbia
83 H21 **Vryburg** North-West, N South Africa 26°57′S 24°44′E
83 K22 **Vryheid** KwaZulu/Natal, E South Africa 27°45′S 30°48′E
111 I18 **Vsetín** *Ger.* Wsetin. Zlínský Kraj, E Czech Republic 49°21′N 17°57′E
111 J20 **Vtáčnik** *Hung.* Madaras, Ptacsnik; *prev.* Ptačnik. ▲ W Slovakia 48°38′N 18°38′E
Vuadil' *see* Wodil
Vŭcha *see* Vacha
Vučitrn *see* Vushtrri

◆ Country
● Country Capital
◇ Dependent Territory
○ Dependent Territory Capital
◈ Administrative Regions
✈ International Airport
▲ Mountain
▲▲ Mountain Range
Volcano
≈ River
◎ Lake
⊠ Reservoir

VUGHT – WARRAP

99 J14 **Vught** Noord-Brabant, S Netherlands 51°37′N 05°19′E
117 W8 **Vuhledar** Donets'ka Oblast', E Ukraine 47°48′N 37°11′E
112 I9 **Vuka** ≈ E Croatia
113 K17 **Vukël** *var.* Vukli. Shkodër, N Albania 42°29′N 19°39′E
Vukli *see* Vukël
112 J9 **Vukovar** *Hung.* Vukovár. Vukovar-Srijem, E Croatia 45°18′N 18°45′E
Vukovarsko-Srijemska Županija *see* Vukovar-Srijem
112 I10 **Vukovar-Srijem** *off.* Vukovarsko-Srijemska Županija. ◈ *province* E Croatia
125 U8 **Vuktyl** Respublika Komi, NW Russian Federation 63°49′N 57°07′E
11 Q17 **Vulcan** Alberta, SW Canada 50°27′N 113°12′W
116 G12 **Vulcan** *Ger.* Wulkan, *Hung.* Zsilyvajdevulkán; *prev.* Crivadia Vulcanului, Vaidei, *Hung.* Sily-Vajdej, Vajdej. Hunedoara, W Romania 45°22′N 23°16′E
116 M12 **Vulcănești** *Rus.* Vulkaneshty. S Moldova 45°41′N 28°25′E
107 L22 **Vulcano, Isola** *island* Isole Eolie, S Italy
Vŭlchedrŭm *see* Valchedram
Vŭlchidol *see* Valchi Dol
Vulkaneshty *see* Vulcănești
123 V11 **Vulkannyy** Kamchatskiy Kray, E Russian Federation 53°01′N 158°26′E
36 J13 **Vulture Mountains** ▲▲ Arizona, SW USA
167 T14 **Vung Tau** *prev. Fr.* Cape Saint Jacques, Cap Saint-Jacques. Ba Ria-Vung Tau, S Vietnam 10°21′N 107°04′E
187 X15 **Vunisea** Kadavu, SE Fiji 19°04′S 178°10′E
Vuohčču *see* Vuotso
93 N15 **Vuokatti** Kainuu, C Finland 64°08′N 28°16′E
93 M15 **Vuolijoki** Kainuu, C Finland 64°09′N 27°00′E
Vuolleriebme *see* Vuollerim
92 J13 **Vuollerim** *Lapp.* Vuolleriebme. Norrbotten, N Sweden 66°24′N 20°36′E
Vuonnabahta *see* Varangerbotn
Vuoreija *see* Vardø
92 L10 **Vuotso** *Lapp.* Vuohčču. Lappi, N Finland 68°04′N 27°05′E
Vŭrbitsa *see* Varbitsa
127 Q4 **Vurnary** Chuvashskaya Respublika, W Russian Federation 55°30′N 46°59′E
Vŭrshets *see* Varshets
Vusan *see* Busan
113 N16 **Vushtrri** *Serb.* Vučitrn. N Kosovo 42°49′N 21°00′E
119 F17 **Vyalikaya Byerastavitsa** *Pol.* Brzostowica Wielka, *Rus.* Bol'shaya Berëstovitsa; *prev.* Velikaya Berestovitsa. Hrodzyenskaya Voblasts', SW Belarus 53°12′N 24°03′E
119 N20 **Vyaliki Bor** *Rus.* Velikiy Bor. Homyel'skaya Voblasts', SE Belarus 52°02′N 29°56′E
119 J18 **Vyaliki Rozhan** *Rus.* Bol'shoy Rozhan. Minskaya Voblasts', S Belarus 52°46′N 27°07′E
124 H10 **Vyartsilya** *Fin.* Värtsilä. Respublika Kareliya, NW Russian Federation 62°07′N 30°43′E
119 K17 **Vyasyeya** *Rus.* Veseya. Minskaya Voblasts', C Belarus 53°04′N 27°41′E
125 R15 **Vyatka** ≈ NW Russian Federation
Vyatka *see* Kirov
125 S16 **Vyatskiye Polyany** Kirovskaya Oblast', NW Russian Federation 56°15′N 51°06′E
123 S14 **Vyazemskiy** Khabarovskiy Kray, SE Russian Federation 47°28′N 134°39′E
126 I4 **Vyaz'ma** Smolenskaya Oblast', W Russian Federation 55°09′N 34°20′E
127 N3 **Vyazniki** Vladimirskaya Oblast', W Russian Federation 56°15′N 42°06′E
127 O8 **Vyazovka** Volgogradskaya Oblast', SW Russian Federation 50°57′N 43°57′E
119 J14 **Vyazyn'** Minskaya Voblasts', NW Belarus 54°25′N 27°10′E
124 G11 **Vyborg** *Fin.* Viipuri. Leningradskaya Oblast', NW Russian Federation 60°44′N 28°47′E
125 P11 **Vychegda** *var.* Vichegda. ≈ NW Russian Federation
119 L14 **Vyelyewshchyna** *Rus.* Velevshchina. Vitsyebskaya Voblasts', N Belarus 54°44′N 28°35′E
119 P16 **Vyeramyeyki** *Rus.* Veremeyki. Mahilyowskaya Voblasts', E Belarus 53°46′N 31°17′E
118 K11 **Vyerkhnyadzvinsk** *Rus.* Verkhnedvinsk. Vitsyebskaya Voblasts', N Belarus 55°47′N 27°56′E
119 P18 **Vyetka** *Rus.* Vetka. Homyel'skaya Voblasts', SE Belarus 52°33′N 31°10′E
118 L12 **Vyetryna** *Rus.* Vetrino. Vitsyebskaya Voblasts', N Belarus 55°25′N 28°28′E
Vygonovskoye, Ozero *see* Vyhanashchanskaye, Vozyera
124 J9 **Vygozero, Ozero** ◎ NW Russian Federation
119 I18 **Vyhanashchanskaye, Vozyera** *prev.* Vozyera Vyhanawskaye, *Rus.* Ozero Vygonovskoye. ◎ SW Belarus
Vyhanawskaye, Vozyera *see* Vyhanashchanskaye, Vozyera
127 N4 **Vyksa** Nizhegorodskaya Oblast', W Russian Federation 55°21′N 42°10′E
117 O12 **Vylkove** *Rus.* Vilkovo. Odes'ka Oblast', SW Ukraine 45°24′N 29°37′E
125 R9 **Vym'** ≈ NW Russian Federation
116 H8 **Vynohradiv** *Cz.* Sevluš, *Hung.* Nagyszöllös, *Rus.* Vinogradov; *prev.* Sevlyush. Zakarpats'ka Oblast', W Ukraine 48°09′N 23°01′E
124 G13 **Vyritsa** Leningradskaya Oblast', NW Russian Federation 59°25′N 30°20′E
97 J19 **Vyrnwy** *Wel.* Afon Efyrnwy. ≈ E Wales, United Kingdom
145 X9 **Vyshe Ivanovskiy Belak, Gora** ▲ E Kazakhstan 50°16′N 83°46′E
117 P4 **Vyshhorod** Kyyivs'ka Oblast', N Ukraine 50°36′N 30°28′E
124 I15 **Vyshniy Volochek** Tverskaya Oblast', W Russian Federation 57°37′N 34°33′E
111 G18 **Vyškov** *Ger.* Wischau. Jihomoravský Kraj, SE Czech Republic 49°17′N 17°01′E
111 E18 **Vysočina** *prev.* Jihlavský Kraj. ◈ *region* N Czech Republic
119 E19 **Vysokaye** *Rus.* Vysokoye. Brestskaya Voblasts', SW Belarus 52°20′N 23°18′E
111 F17 **Vysoké Mýto** *Ger.* Hohenmauth. Pardubický Kraj, C Czech Republic 49°57′N 16°10′E
117 S9 **Vysokopillya** Khersons'ka Oblast', S Ukraine 47°28′N 33°30′E
126 K3 **Vysokovsk** Moskovskaya Oblast', W Russian Federation 56°12′N 36°42′E
Vysokoye *see* Vysokaye
124 K12 **Vytegra** Vologodskaya Oblast', NW Russian Federation 60°59′N 36°27′E
116 J8 **Vyzhnytsya** Chernivets'ka Oblast', W Ukraine 48°14′N 25°10′E

W

77 O14 **Wa** NW Ghana 10°07′N 02°28′W
Waadt *see* Vaud
Waag *see* Váh
Waagbistritz *see* Považská Bystrica
Waagneustadtl *see* Nové Mesto nad Váhom
81 M16 **Waajid** Gedo, SW Somalia 03°37′N 43°19′E
98 L13 **Waal** ≈ S Netherlands
187 O16 **Waala** Province Nord, W New Caledonia 19°46′S 163°41′E
99 I14 **Waalwijk** Noord-Brabant, S Netherlands 51°42′N 05°04′E
99 E16 **Waarschoot** Oost-Vlaanderen, NW Belgium 51°09′N 03°35′E
186 C7 **Wabag** Enga, W Papua New Guinea 05°28′S 143°40′E
15 N7 **Wabano** ≈ Québec, SE Canada
11 P11 **Wabasca** ≈ Alberta, SW Canada
31 P12 **Wabash** Indiana, N USA 40°47′N 85°48′W
29 X9 **Wabasha** Minnesota, N USA 44°22′N 92°01′W
31 N13 **Wabash River** ≈ N USA
14 C7 **Wabatongushi Lake** ◎ Ontario, S Canada
81 L15 **Wabē Gestro Wenz** ≈ SE Ethiopia
14 B9 **Wabos** Ontario, S Canada 46°48′N 84°06′W
11 W13 **Wabowden** Manitoba, C Canada 54°57′N 98°38′W
110 J9 **Wąbrzeźno** Kujawsko-pomorskie, C Poland 53°18′N 18°55′E
21 U12 **Waccamaw River** ≈ South Carolina, SE USA
23 U11 **Waccasassa Bay** *bay* Florida, SE USA
99 F16 **Wachtebeke** Oost-Vlaanderen, NW Belgium 51°10′N 03°52′E
25 T8 **Waco** Texas, SW USA 31°33′N 97°10′W
26 M3 **Waconda Lake** *var.* Great Elder Reservoir. ▣ Kansas, C USA
Wadai *see* Ouaddaï
Wad Al-Hajarah *see* Guadalajara
164 I12 **Wadayama** Hyōgo, Honshū, SW Japan 35°19′N 134°51′E
80 D10 **Wad Banda** Western Kordofan, C Sudan 13°08′N 27°56′E
75 P9 **Waddān** NW Libya 29°10′N 16°08′E
98 J4 **Waddeneilanden** *Eng.* West Frisian Islands. *island group* N Netherlands
98 J6 **Waddenzee** *var.* Wadden Zee. *sea* SE North Sea
Wadden Zee *see* Waddenzee
10 L16 **Waddington, Mount** ▲ British Columbia, SW Canada 51°17′N 125°16′W
98 H12 **Waddinxveen** Zuid-Holland, C Netherlands 52°03′N 04°38′E
11 U15 **Wadena** Saskatchewan, S Canada 51°57′N 103°48′W
29 T6 **Wadena** Minnesota, N USA 46°27′N 95°07′W
108 G7 **Wädenswil** Zürich, N Switzerland 47°14′N 08°41′E
21 S11 **Wadesboro** North Carolina, SE USA 34°59′N 80°03′W
155 G16 **Wādi** Karnātaka, C India 17°00′N 76°58′E
138 G10 **Wādī as Sīr** *var.* Wadi es Sir. ʻAmmān, NW Jordan 31°57′N 35°49′E
Wadi es Sir *see* Wādī as Sīr
78 J9 **Wadi Fira** *off.* Région du Wadi Fira; *prev.* Préfecture de Biltine. ◈ *region* E Chad
Wadi Fira, Région du *see* Wadi Fira
80 F5 **Wadi Halfa** *var.* Wādī Ḥalfā'. Northern, N Sudan 21°46′N 31°17′E
138 G13 **Wādī Mūsā** *var.* Petra. Ma'ān, S Jordan 30°19′N 35°29′E
23 V4 **Wadley** Georgia, SE USA 32°52′N 82°24′W
Wad Madanī *see* Wad Medani
80 G10 **Wad Medani** *var.* Wad Madanī. Gezira, C Sudan 14°24′N 33°30′E
80 F10 **Wad Nimr** White Nile, C Sudan 14°32′N 32°10′E
165 U16 **Wadomari** Kagoshima, Okinoerabu-jima, SW Japan 27°25′N 128°40′E
111 K17 **Wadowice** Małopolskie, S Poland 49°54′N 19°29′E
35 R5 **Wadsworth** Nevada, W USA 39°39′N 119°16′W
31 T12 **Wadsworth** Ohio, N USA 41°01′N 81°43′W
25 T11 **Waelder** Texas, SW USA 29°42′N 97°16′W
Waereghem *see* Waregem
163 U13 **Wafangdian** *var.* Fuxian, Fu Xian. Liaoning, NE China 39°36′N 122°00′E
171 R13 **Waflia** Pulau Buru, E Indonesia 03°10′S 126°05′E
Wagadugu *see* Ouagadougou
98 K12 **Wageningen** Gelderland, SE Netherlands 51°58′N 05°40′E
55 V9 **Wageningen** Nickerie, NW Suriname 05°44′N 56°45′W
9 O8 **Wager Bay** *inlet* Nunavut, N Canada
183 P10 **Wagga Wagga** New South Wales, SE Australia 35°11′S 147°22′E
180 J13 **Wagin** Western Australia 33°16′S 117°26′E
108 H8 **Wägitaler See** ◎ SW Switzerland
29 P12 **Wagner** South Dakota, N USA 43°04′N 98°17′W
27 Q9 **Wagoner** Oklahoma, C USA 35°58′N 92°23′W
37 U10 **Wagon Mound** New Mexico, SW USA 36°00′N 104°42′W
32 J14 **Wagontire** Oregon, NW USA 43°15′N 119°51′W
110 H10 **Wągrowiec** Wielkopolskie, C Poland 52°49′N 17°11′E
149 U6 **Wāh** Punjab, NE Pakistan 33°50′N 72°44′E
171 S13 **Wahai** Pulau Seram, E Indonesia 02°48′S 129°29′E
169 V10 **Wahau, Sungai** ≈ Borneo, C Indonesia
Wahaybah, Ramlat Al *see* Wahībah, Ramlat Āl
Wahda *see* Unity
38 D9 **Wahiawā** *var.* Wahiawa. O'ahu, Hawaii, USA, C Pacific Ocean 21°30′N 158°01′W
Wahībah, Ramlat Ahl *see* Wahībah, Ramlat Āl
141 Y9 **Wahībah, Ramlat Āl** *var.* Ramlat Ahl Wahībah, Ramlat Al Wahaybah, *Eng.* Wahībah Sands. *desert* N Oman
Wahībah Sands *see* Wahībah, Ramlat Āl
101 E16 **Wahn** ✈ (Köln) Nordrhein-Westfalen, W Germany 50°51′N 07°09′E
29 R15 **Wahoo** Nebraska, C USA 41°12′N 96°37′W
29 R6 **Wahpeton** North Dakota, N USA 46°16′N 96°36′W
Wahran *see* Oran
36 J6 **Wah Wah Mountains** ▲▲ Utah, W USA
38 D9 **Waialua** O'ahu, Hawaii, USA, C Pacific Ocean 21°34′N 158°07′W
38 D9 **Wai'anae** *var.* Waianae. O'ahu, Hawaii, USA, C Pacific Ocean 21°26′N 158°11′W
184 Q8 **Waiapu** ≈ North Island, New Zealand
185 I17 **Waiau** Canterbury, South Island, New Zealand 42°39′S 173°03′E
185 I17 **Waiau** ≈ South Island, New Zealand
185 B23 **Waiau** ≈ South Island, New Zealand
101 H21 **Waiblingen** Baden-Württemberg, S Germany 48°49′N 09°19′E
Waidhofen *see* Waidhofen an der Ybbs, Niederösterreich, Austria
Waidhofen *see* Waidhofen an der Thaya, Niederösterreich, Austria
109 V2 **Waidhofen an der Thaya** *var.* Waidhofen. Niederösterreich, NE Austria 48°49′N 15°17′E
109 U5 **Waidhofen an der Ybbs** *var.* Waidhofen. Niederösterreich, E Austria 47°58′N 14°47′E
171 T11 **Waigeo, Pulau** *island* Papua Barat, E Indonesia
184 L5 **Waiheke Island** *island* N New Zealand
184 M7 **Waihi** Waikato, North Island, New Zealand 37°22′S 175°51′E
185 C20 **Waihou** ≈ North Island, New Zealand
Waikaboebak *see* Waikabubak
171 N17 **Waikabubak** *prev.* Waikaboebak. Pulau Sumba, C Indonesia 09°40′S 119°25′E
185 D23 **Waikaia** ≈ South Island, New Zealand
185 D23 **Waikaka** Southland, South Island, New Zealand 45°55′S 168°59′E
184 L13 **Waikanae** Wellington, North Island, New Zealand 40°52′S 175°03′E
184 M7 **Waikare, Lake** ◎ North Island, New Zealand
184 O9 **Waikaremoana, Lake** ◎ North Island, New Zealand
185 I17 **Waikari** Canterbury, South Island, New Zealand 42°50′S 172°41′E
184 L8 **Waikato** *off.* Waikato Region. ◈ *region* North Island, New Zealand
184 M8 **Waikato** ≈ North Island, New Zealand
Waikato Region *see* Waikato
182 J9 **Waikerie** South Australia 34°12′S 139°57′E
185 F23 **Waikouaiti** Otago, South Island, New Zealand 45°36′S 170°39′E
38 H11 **Wailea** Hawaii, USA, C Pacific Ocean 19°53′N 155°07′W
38 F10 **Wailuku** Maui, Hawaii, USA, C Pacific Ocean 20°53′N 156°30′W
185 H18 **Waimakariri** ≈ South Island, New Zealand
38 D9 **Waimānalo Beach** *var.* Waimanalo Beach. O'ahu, Hawaii, USA, C Pacific Ocean 21°20′N 157°42′W
185 G15 **Waimangaroa** West Coast, South Island, New Zealand 41°41′S 171°49′E
185 G21 **Waimate** Canterbury, South Island, New Zealand 44°44′S 171°03′E
38 B8 **Waimea** Kaua'i, Hawaii, USA, C Pacific Ocean 21°57′N 159°40′W
38 G11 **Waimea** *var.* Kamuela. Hawaii, USA, C Pacific Ocean 20°02′N 155°20′W
38 D9 **Waimea** *var.* Maunawai. O'ahu, Hawaii, USA, C Pacific Ocean 21°39′N 158°04′W
99 M20 **Waimes** Liège, E Belgium 50°25′N 06°10′E
154 J11 **Wainganga** *var.* Wain River. ≈ C India
Waingapoe *see* Waingapu
171 N17 **Waingapu** *prev.* Waingapoe. Pulau Sumba, C Indonesia 09°40′S 120°16′E
55 S7 **Waini** ≈ N Guyana
55 S7 **Waini Point** *headland* NW Guyana 08°24′N 59°48′W
Wain River *see* Wainganga
11 R15 **Wainwright** Alberta, SW Canada 52°50′N 110°51′W
39 O5 **Wainwright** Alaska, USA 70°38′N 160°02′W
184 K4 **Waiotira** Northland, North Island, New Zealand 35°56′S 174°11′E
184 M11 **Waiouru** Manawatu-Wanganui, North Island, New Zealand 39°28′S 175°41′E
171 W14 **Waipa** Papua, E Indonesia 03°47′S 136°16′E
184 L8 **Waipa** ≈ North Island, New Zealand
184 P9 **Waipaoa** ≈ North Island, New Zealand
185 D25 **Waipapa Point** *headland* South Island, New Zealand 46°39′S 168°51′E
185 I18 **Waipara** Canterbury, South Island, New Zealand 43°04′S 172°45′E
184 N12 **Waipawa** Hawke's Bay, North Island, New Zealand 39°57′S 176°36′E
184 K4 **Waipu** Northland, North Island, New Zealand 35°58′S 174°25′E
184 N12 **Waipukurau** Hawke's Bay, North Island, New Zealand 40°01′S 176°34′E
171 U14 **Wair** Pulau Kai Besar, E Indonesia 05°16′S 133°09′E
Wairakai *see* Wairakei
184 N9 **Wairakei** *var.* Wairakai. Waikato, North Island, New Zealand 38°37′S 176°05′E
185 M14 **Wairarapa, Lake** ◎ North Island, New Zealand
185 J15 **Wairau** ≈ South Island, New Zealand
184 P10 **Wairoa** Hawke's Bay, North Island, New Zealand 39°03′S 177°26′E
184 P10 **Wairoa** ≈ North Island, New Zealand
184 J4 **Wairoa** ≈ North Island, New Zealand
184 N9 **Waitahanui** Waikato, North Island, New Zealand 38°48′S 176°04′E
184 M6 **Waitakaruru** Waikato, North Island, New Zealand 37°14′S 175°22′E
185 F21 **Waitaki** ≈ South Island, New Zealand
184 K10 **Waitara** Taranaki, North Island, New Zealand 39°01′S 174°14′E
184 M7 **Waitoa** Waikato, North Island, New Zealand 37°36′S 175°37′E
184 L8 **Waitomo Caves** Waikato, North Island, New Zealand 38°17′S 175°06′E
184 L11 **Waitotara** Taranaki, North Island, New Zealand 39°49′S 174°43′E
184 L11 **Waitotara** ≈ North Island, New Zealand
32 L10 **Waitsburg** Washington, NW USA 46°16′N 118°09′W
Waitzen *see* Vác
184 L6 **Waiuku** Auckland, North Island, New Zealand 37°15′S 174°45′E
164 L10 **Wajima** *var.* Wazima. Ishikawa, Honshū, SW Japan 37°23′N 136°53′E
81 K17 **Wajir** Wajir, NE Kenya 01°46′N 40°05′E
81 K18 **Wajir** ◈ *county* NE Kenya
79 J17 **Waka** Equateur, NW Dem. Rep. Congo 01°04′N 20°11′E
81 I14 **Waka** Southern Nationalities, S Ethiopia 07°12′N 37°19′E
14 D9 **Wakami Lake** ◎ Ontario, S Canada
164 I12 **Wakasa** Tottori, Honshū, SW Japan 35°18′N 134°25′E
164 J12 **Wakasa-wan** *bay* C Japan
185 C22 **Wakatipu, Lake** ◎ South Island, New Zealand
11 T15 **Wakaw** Saskatchewan, S Canada 52°40′N 105°45′W
164 I14 **Wakayama** Wakayama, Honshū, SW Japan 34°12′N 135°09′E
164 I15 **Wakayama** *off.* Wakayama-ken. ◈ *prefecture* Honshū, SW Japan
Wakayama-ken *see* Wakayama
26 K4 **Wa Keeney** Kansas, C USA 39°02′N 99°53′W
185 I14 **Wakefield** Tasman, South Island, New Zealand 41°24′S 173°03′E
97 M17 **Wakefield** N England, United Kingdom 53°42′N 01°29′W
27 O4 **Wakefield** Kansas, C USA 39°12′N 97°00′W
30 L4 **Wakefield** Michigan, N USA 46°27′N 89°55′W
21 U9 **Wake Forest** North Carolina, SE USA 35°58′N 78°30′W
Wakeham Bay *see* Kangiqsujuaq
189 Y11 **Wake Island** ◇ *US unincorporated territory* NW Pacific Ocean
189 Y12 **Wake Island** ✈ NW Pacific Ocean
189 Y12 **Wake Island** *atoll* NW Pacific Ocean
189 X12 **Wake Lagoon** *lagoon* Wake Island, NW Pacific Ocean
166 L8 **Wakema** Ayeyawady, SW Myanmar (Burma) 16°36′N 95°11′E
Wakhan *see* Khandūd
164 H14 **Waki** Tokushima, Shikoku, SW Japan 34°04′N 134°10′E
165 T1 **Wakkanai** Hokkaidō, NE Japan 45°25′N 141°39′E
83 K22 **Wakkerstroom** Mpumalanga, E South Africa 27°21′S 30°10′E
14 C10 **Wakomata Lake** ◎ Ontario, S Canada
183 N10 **Wakool** New South Wales, SE Australia 35°30′S 144°22′E
Wakra *see* Al Wakrah
Waku Kungo *see* Uaco Cungo
186 J7 **Wakunai** Bougainville, NE Papua New Guinea 05°52′S 155°10′E
Walachei/Walachia *see* Wallachia
155 K26 **Walawe Ganga** ≈ S Sri Lanka
111 F15 **Wałbrzych** *Ger.* Waldenburg, Waldenburg in Schlesien. Dolnośląskie, SW Poland 50°45′N 16°20′E
183 T6 **Walcha** New South Wales, SE Australia 31°01′S 151°38′E
101 K24 **Walchensee** ◎ SE Germany
99 D14 **Walcheren** *island* SW Netherlands
29 Z14 **Walcott** Iowa, C USA 41°34′N 90°46′W
33 W16 **Walcott** Wyoming, C USA 41°46′N 106°46′W
99 G21 **Walcourt** Namur, S Belgium 50°16′N 04°26′E
110 G9 **Wałcz** *Ger.* Deutsch Krone. Zachodnio-pomorskie, NW Poland 53°17′N 16°29′E
108 H7 **Wald** Zürich, N Switzerland 47°17′N 08°56′E
109 U3 **Waldaist** ≈ N Austria
180 I9 **Waldburg Range** ▲▲ Western Australia
37 R3 **Walden** Colorado, C USA 40°43′N 106°16′W
18 K13 **Walden** New York, NE USA 41°35′N 74°09′W
Waldenburg/Waldenburg in Schlesien *see* Wałbrzych
11 T15 **Waldheim** Saskatchewan, S Canada 52°38′N 106°35′W
Waldia *see* Weldiya
101 M23 **Waldkraiburg** Bayern, SE Germany 48°10′N 12°23′E
27 T14 **Waldo** Arkansas, C USA 33°21′N 93°18′W
23 V9 **Waldo** Florida, SE USA 29°47′N 82°07′W
19 R7 **Waldoboro** Maine, NE USA 44°06′N 69°22′W
21 W4 **Waldorf** Maryland, NE USA 38°36′N 76°54′W
32 F12 **Waldport** Oregon, NW USA 44°25′N 124°04′W
27 S11 **Waldron** Arkansas, C USA 34°54′N 94°09′W
195 Y13 **Waldron, Cape** *headland* Antarctica 66°08′S 116°00′E
101 F24 **Waldshut-Tiengen** Baden-Württemberg, S Germany 47°37′N 08°13′E
171 P12 **Walea, Selat** *strait* Sulawesi, C Indonesia
Wałeckie Międzyrzecze *see* Valašské Meziříčí
108 H8 **Walensee** ◎ NW Switzerland
38 L8 **Wales** Alaska, USA 65°36′N 168°03′W
97 J20 **Wales** *Wel.* Cymru. ◈ *national region* Wales, United Kingdom
9 O7 **Wales Island** *island* Nunavut, NE Canada
77 P14 **Walewale** N Ghana 10°21′N 00°48′W
99 M24 **Walferdange** Luxembourg, C Luxembourg 49°39′N 06°08′E
183 Q5 **Walgett** New South Wales, SE Australia 30°02′S 148°14′E
194 K10 **Walgreen Coast** *physical region* Antarctica
29 Q2 **Walhalla** North Dakota, N USA 48°55′N 97°55′W
21 O11 **Walhalla** South Carolina, SE USA 34°46′N 83°05′W
79 O19 **Walikale** Nord-Kivu, E Dem. Rep. Congo 01°29′S 28°05′E
Walk *see* Valga, Estonia
Walk *see* Valka, Latvia
29 U5 **Walker** Minnesota, N USA 47°06′N 94°35′W
15 V4 **Walker, Lac** ◎ Québec, SE Canada
35 S7 **Walker Lake** ◎ Nevada, W USA
35 R6 **Walker River** ≈ Nevada, W USA
28 K10 **Wall** South Dakota, N USA 43°58′N 102°12′W
173 U9 **Wallaby Plateau** *undersea feature* E Indian Ocean
33 N8 **Wallace** Idaho, NW USA 47°28′N 115°55′W
21 V11 **Wallace** North Carolina, SE USA 34°42′N 77°59′W
14 D17 **Wallaceburg** Ontario, S Canada 42°34′N 82°22′W
22 F5 **Wallace Lake** ▣ Louisiana, S USA
11 P13 **Wallace Mountain** ▲ Alberta, W Canada 54°50′N 115°57′W
116 J14 **Wallachia** *var.* Walachia, *Ger.* Walachei, *Rom.* Valachia. *cultural region* S Romania
Wallachisch-Meseritsch *see* Valašské Meziříčí
183 U4 **Wallangarra** New South Wales, SE Australia 28°56′S 151°57′E
182 I8 **Wallaroo** South Australia 33°56′S 137°38′E
32 L10 **Walla Walla** Washington, NW USA 46°03′N 118°20′W
101 H19 **Walldürn** Baden-Württemberg, SW Germany 49°34′N 09°22′E
100 F12 **Wallenhorst** Niedersachsen, NW Germany 52°21′N 08°01′E
Wallenthal *see* Hațeg
109 S4 **Wallern** Oberösterreich, N Austria 48°13′N 13°58′E
Wallern *see* Wallern im Burgenland
109 Z5 **Wallern im Burgenland** *var.* Wallern. Burgenland, E Austria 47°44′N 16°57′E
18 M9 **Wallingford** Vermont, NE USA 43°27′N 72°56′W
25 V11 **Wallis** Texas, SW USA 29°38′N 96°05′W
Wallis *see* Valais
192 K9 **Wallis and Futuna** *Fr.* Territoire de Wallis et Futuna. ◇ *French overseas collectivity* C Pacific Ocean
108 G7 **Wallisellen** Zürich, N Switzerland 47°25′N 08°36′E
Wallis et Futuna, Territoire de *see* Wallis and Futuna
190 H11 **Wallis, Îles** *island group* N Wallis and Futuna
99 G20 **Wallonia** ◈ *cultural region* SW Belgium
31 Q5 **Walloon Lake** ◎ Michigan, N USA
32 K10 **Wallula** Washington, NW USA 46°03′N 118°54′W
32 K10 **Wallula, Lake** ◎ Washington, NW USA
21 S8 **Walnut Cove** North Carolina, SE USA 36°18′N 80°08′W
35 N8 **Walnut Creek** California, W USA 37°52′N 122°04′W
26 K5 **Walnut Creek** ≈ Kansas, C USA
27 W9 **Walnut Ridge** Arkansas, C USA 36°06′N 90°56′W
25 S7 **Walnut Springs** Texas, SW USA 32°05′N 97°42′W
182 L10 **Walpeup** Victoria, SE Australia 35°09′S 142°01′E
187 R17 **Walpole, Île** *island* SE New Caledonia
39 N13 **Walrus Islands** *island group* Alaska, USA
97 L19 **Walsall** C England, United Kingdom 52°35′N 01°58′W
37 T7 **Walsenburg** Colorado, C USA 37°37′N 104°46′W
11 S17 **Walsh** Alberta, SW Canada 49°58′N 110°03′W
37 W7 **Walsh** Colorado, C USA 37°20′N 102°17′W
100 I11 **Walsrode** Niedersachsen, NW Germany 52°52′N 09°36′E
Waltenberg *see* Zalău
21 R14 **Walterboro** South Carolina, SE USA 32°54′N 80°21′W
Walter F.George Lake *see* Walter F. George Reservoir
23 R6 **Walter F. George Reservoir** *var.* Walter F.George Lake. ▣ Alabama/Georgia, SE USA
26 M12 **Walters** Oklahoma, C USA 34°22′N 98°18′W
101 J16 **Waltershausen** Thüringen, C Germany 50°53′N 10°33′E
173 N10 **Walters Shoal** *var.* Walters Shoals. *reef* S Madagascar
Walters Shoals *see* Walters Shoal
22 M3 **Walthall** Mississippi, S USA 33°36′N 89°16′W
20 M4 **Walton** Kentucky, S USA 38°52′N 84°36′W
18 J11 **Walton** New York, NE USA 42°10′N 75°07′W
79 O20 **Walungu** Sud-Kivu, E Dem. Rep. Congo 02°40′S 28°37′E
Walvisbaai *see* Walvis Bay
83 C19 **Walvis Bay** *Afr.* Walvisbaai. Erongo, NW Namibia 22°59′S 14°34′E
83 B19 **Walvis Bay** *bay* NW Namibia
Walvish Ridge *see* Walvis Ridge
65 O17 **Walvis Ridge** *var.* Walvish Ridge. *undersea feature* E Atlantic Ocean 28°00′S 03°00′E
171 X16 **Wamal** Papua, E Indonesia 08°00′S 139°06′E
171 U15 **Wamar, Pulau** *island* Kepulauan Aru, E Indonesia
79 O17 **Wamba** Orientale, NE Dem. Rep. Congo 02°10′N 27°59′E
77 V15 **Wamba** Nassarawa, C Nigeria 08°57′N 08°35′E
79 H22 **Wamba** *var.* Uamba. ≈ Angola/Dem. Rep. Congo
27 P4 **Wamego** Kansas, C USA 39°12′N 96°18′W
18 I10 **Wampsville** New York, NE USA 43°03′N 75°40′W
42 K6 **Wampú, Río** ≈ E Honduras
171 X16 **Wan** Papua, E Indonesia 08°15′S 138°00′E
Wan *see* Anhui
183 N4 **Wanaaring** New South Wales, SE Australia 29°42′S 144°07′E
185 D21 **Wanaka** Otago, South Island, New Zealand 44°42′S 169°09′E
185 D20 **Wanaka, Lake** ◎ South Island, New Zealand
171 W14 **Wanapiri** Papua, E Indonesia 04°21′S 135°52′E
14 F9 **Wanapitei** ≈ Ontario, S Canada
14 F10 **Wanapitei Lake** ◎ Ontario, S Canada
18 K14 **Wanaque** New Jersey, NE USA 41°02′N 74°17′W
171 U12 **Wanau** Papua Barat, E Indonesia 01°20′S 132°40′E
185 F22 **Wanbrow, Cape** *headland* South Island, New Zealand 45°07′S 170°59′E
Wancheng *see* Wanning
Wanchuan *see* Zhangjiakou
171 W13 **Wandai** *var.* Komeyo. Papua, E Indonesia 03°35′S 136°15′E
163 Z8 **Wanda Shan** ▲▲ NE China
197 R11 **Wandel Sea** *sea* Arctic Ocean
160 D13 **Wanding** *var.* Wandingzhen. Yunnan, SW China 24°01′N 98°00′E
Wandingzhen *see* Wanding
99 H20 **Wanfercée-Baulet** Hainaut, S Belgium 50°27′N 04°37′E
184 L12 **Wanganui** Manawatu-Wanganui, North Island, New Zealand 39°56′S 175°02′E
184 L11 **Wanganui** ≈ North Island, New Zealand
183 P11 **Wangaratta** Victoria, SE Australia 36°22′S 146°17′E
160 J8 **Wangcang** *var.* Donghe; *prev.* Fengjiaba, Hongjiang. Sichuan, C China 32°15′N 106°16′E
Wangda *see* Zogang
101 I24 **Wangen im Allgäu** Baden-Württemberg, S Germany 47°40′N 09°49′E
Wangerin *see* Węgorzyno
100 F9 **Wangerooge** *island* NW Germany
171 W13 **Wanggar** Papua, E Indonesia 03°22′S 135°15′E
160 J13 **Wangmo** *var.* Fuxing. Guizhou, S China 25°08′N 106°08′E
Wangolodougou *see* Ouangolodougou
161 S9 **Wangpan Yang** *sea* E China
163 Y10 **Wangqing** Jilin, NE China 43°19′N 129°42′E
167 P8 **Wang Saphung** Loei, C Thailand 17°18′N 101°45′E
167 O6 **Wan Hsa-la** Shan State, E Myanmar (Burma) 20°27′N 98°39′E
55 W9 **Wanica** ◈ *district* N Suriname
79 M18 **Wanie-Rukula** Orientale, C Dem. Rep. Congo 0°13′N 25°34′E
Wankie *see* Hwange
Wanki, Río *see* Coco, Río
81 N17 **Wanlaweyn** *var.* Wanle Weyn, *It.* Uanle Uen. Shabeellaha Hoose, SW Somalia 02°36′N 44°47′E
Wanle Weyn *see* Wanlaweyn
180 I12 **Wanneroo** Western Australia 31°40′S 115°35′E
160 L17 **Wanning** *var.* Wancheng. Hainan, S China 18°55′N 110°27′E
167 Q8 **Wanon Niwat** Sakon Nakhon, E Thailand 17°39′N 103°45′E
155 H16 **Wanparti** Telangana, C India 16°19′N 78°06′E
Wansen *see* Wiązów
160 L11 **Wanshan** Guizhou, S China 27°45′N 109°12′E
99 M14 **Wanssum** Limburg, SE Netherlands 51°31′N 06°04′E
184 N12 **Wanstead** Hawke's Bay, North Island, New Zealand 40°09′S 176°31′E
Wanxian *see* Wanzhou
188 F16 **Wanyaan** Yap, Micronesia
160 K8 **Wanyuan** Sichuan, C China 32°05′N 108°08′E
161 O11 **Wanzai** *var.* Kangle. Jiangxi, S China 28°06′N 114°27′E
99 J20 **Wanze** Liège, E Belgium 50°32′N 05°16′E
160 K9 **Wanzhou** *var.* Wanxian. Chongqing Shi, C China 30°48′N 108°21′E
31 R12 **Wapakoneta** Ohio, N USA 40°34′N 84°11′W
12 D7 **Wapaseese** ≈ Ontario, C Canada
32 I10 **Wapato** Washington, NW USA 46°27′N 120°25′W
29 Y15 **Wapello** Iowa, C USA 41°10′N 91°13′W
11 N13 **Wapiti** ≈ Alberta/British Columbia, SW Canada
27 X7 **Wappapello Lake** ▣ Missouri, C USA
18 K13 **Wappingers Falls** New York, NE USA 41°36′N 73°54′W
29 X13 **Wapsipinicon River** ≈ Iowa, C USA
14 L9 **Wapus** ≈ Québec, SE Canada
160 H7 **Waqên** Sichuan, C China 33°05′N 102°34′E
21 Q7 **War** West Virginia, NE USA 37°18′N 81°39′W
Warab *see* Warrap
155 J15 **Warangal** Telangana, C India 18°N 79°35′E
Warasdin *see* Varaždin
183 O16 **Waratah** Tasmania, SE Australia 41°28′S 145°34′E
183 O14 **Waratah Bay** *bay* Victoria, SE Australia
101 H15 **Warburg** Nordrhein-Westfalen, W Germany 51°30′N 09°11′E
182 I1 **Warburton Creek** *seasonal river* South Australia
180 M9 **Warbuton** Western Australia 26°17′S 126°18′E
99 M20 **Warche** ≈ E Belgium
Wardag/Wardak *see* Wardak
149 P5 **Wardak** *prev.* Vardak, *Pash.* Wardag. ◈ *province* E Afghanistan
32 K9 **Warden** Washington, NW USA 46°58′N 119°02′W
154 I12 **Wardha** Mahārāshtra, W India 20°41′N 78°40′E
Wardija Point *see* Wardija, Ras il-
121 N15 **Wardija, Ras il-** *var.* Ras il- Wardija, Wardija Point. *headland* Gozo, NW Malta 36°03′N 14°11′E
Wardija, Ras il- *see* Wardija, Ras il-
139 P3 **Wardīyah** Nīnawá, N Iraq 36°18′N 41°45′E
185 E19 **Ward, Mount** ▲ South Island, New Zealand 43°49′S 169°54′E
10 L11 **Ware** British Columbia, W Canada 57°26′N 125°41′W
99 D18 **Waregem** *var.* Waereghem. West-Vlaanderen, W Belgium 50°53′N 03°26′E
99 J19 **Waremme** Liège, E Belgium 50°41′N 05°15′E
100 N10 **Waren** Mecklenburg-Vorpommern, NE Germany 53°32′N 12°42′E
171 W13 **Waren** Papua, E Indonesia 02°13′S 136°21′E
101 F14 **Warendorf** Nordrhein-Westfalen, W Germany 51°57′N 08°00′E
21 P12 **Ware Shoals** South Carolina, SE USA 34°24′N 82°15′W
98 N4 **Warffum** Groningen, NE Netherlands 53°22′N 06°34′E
81 O15 **Wargalo** Mudug, E Somalia 06°06′N 47°40′E
146 M12 **Warganza** *Rus.* Varganzi. Qashqadaryo Viloyati, S Uzbekistan 39°18′N 66°00′E
Wargla *see* Ouargla
183 T4 **Warialda** New South Wales, SE Australia 29°34′S 150°35′E
154 F13 **Wāri Godri** Mahārāshtra, C India 19°28′N 75°43′E
167 R10 **Warin Chamrap** Ubon Ratchathani, E Thailand 15°11′N 104°51′E
25 R11 **Waring** Texas, SW USA 29°56′N 98°48′W
39 O8 **Waring Mountains** ▲▲ Alaska, USA
110 M12 **Warka** Mazowieckie, E Poland 51°45′N 21°12′E
184 L5 **Warkworth** Auckland, North Island, New Zealand 36°23′S 174°42′E
171 U12 **Warmandi** Papua Barat, E Indonesia 0°21′S 132°38′E
83 E22 **Warmbad** Karas, S Namibia 28°29′S 18°41′E
98 H8 **Warmenhuizen** Noord-Holland, NW Netherlands 52°43′N 04°45′E
110 M8 **Warmińsko-Mazurskie** ◈ *province* C Poland
97 L22 **Warminster** S England, United Kingdom 51°13′N 02°12′W
18 I15 **Warminster** Pennsylvania, NE USA 40°11′N 75°04′W
35 V8 **Warm Springs** Nevada, W USA 38°10′N 116°21′W
32 H12 **Warm Springs** Oregon, NW USA 44°51′N 121°24′W
21 S5 **Warm Springs** Virginia, NE USA 38°03′N 79°48′W
100 M8 **Warnemünde** Mecklenburg-Vorpommern, NE Germany 54°10′N 12°03′E
27 Q10 **Warner** Oklahoma, C USA 35°29′N 95°18′W
35 Q2 **Warner Mountains** ▲▲ California, W USA
23 T5 **Warner Robins** Georgia, SE USA 32°38′N 83°38′W
57 N18 **Warnes** Santa Cruz, C Bolivia 17°30′S 63°11′W
100 M9 **Warnow** ≈ NE Germany
Warnsdorf *see* Varnsdorf
98 M11 **Warnsveld** Gelderland, E Netherlands 52°08′N 06°14′E
154 I13 **Warora** Mahārāshtra, C India 20°12′N 79°01′E
182 L11 **Warracknabeal** Victoria, SE Australia 36°17′S 142°26′E
183 O13 **Warragul** Victoria, SE Australia 38°11′S 145°55′E
80 D13 **Warrap** Warrap, W South Sudan 08°08′N 28°37′E
81 D14 **Warrap** *var.* Warab. ◈ *state* W South Sudan

◆ Country
● Country Capital
◇ Dependent Territory
○ Dependent Territory Capital
◈ Administrative Regions
✈ International Airport
▲ Mountain
▲▲ Mountain Range
Volcano
≈ River
◎ Lake
▣ Reservoir

Warrego River – West Kalimantan

183 *O4* **Warrego River** *seasonal river* New South Wales/ Queensland, E Australia
183 *Q6* **Warren** New South Wales, SE Australia *31°41′S 147°51′E*
11 *X16* **Warren** Manitoba, S Canada *50°05′N 97°33′W*
27 *V14* **Warren** Arkansas, C USA *33°38′N 92°05′W*
31 *S10* **Warren** Michigan, N USA *42°29′N 83°02′W*
29 *R3* **Warren** Minnesota, N USA *48°12′N 96°46′W*
31 *U11* **Warren** Ohio, N USA *41°14′N 80°49′W*
18 *D12* **Warren** Pennsylvania, NE USA *41°52′N 79°09′W*
25 *X10* **Warren** Texas, SW USA *30°33′N 94°24′W*
97 *G16* **Warrenpoint** ***Ir.*** An Pointe. SE Northern Ireland, United Kingdom *54°07′N 06°16′W*
27 *S4* **Warrensburg** Missouri, C USA *38°46′N 93°44′W*
83 *H22* **Warrenton** Northern Cape, N South Africa *28°07′S 24°51′E*
23 *U4* **Warrenton** Georgia, SE USA *33°24′N 82°39′W*
27 *W4* **Warrenton** Missouri, C USA *38°48′N 91°08′W*
21 *V8* **Warrenton** North Carolina, SE USA *36°24′N 78°11′W*
21 *V4* **Warrenton** Virginia, NE USA *38°43′N 77°48′W*
77 *U17* **Warri** Delta, S Nigeria *05°26′N 05°34′E*
97 *L18* **Warrington** C England, United Kingdom *53°24′N 02°37′W*
23 *O9* **Warrington** Florida, SE USA *30°22′N 87°16′W*
23 *P3* **Warrior** Alabama, S USA *33°49′N 86°49′W*
182 *L13* **Warrnambool** Victoria, SE Australia *38°23′S 142°30′E*
29 *T2* **Warroad** Minnesota, N USA *48°55′N 95°18′W*
183 *S6* **Warrumbungle Range** ▲▲ New South Wales, SE Australia
154 *J12* **Wārsa** Mahārāshtra, C India *20°42′N 79°58′E*
31 *P11* **Warsaw** Indiana, N USA *41°13′N 85°52′W*
20 *L4* **Warsaw** Kentucky, S USA *38°47′N 84°55′W*
27 *T5* **Warsaw** Missouri, C USA *38°14′N 93°23′W*
18 *E10* **Warsaw** New York, NE USA *42°44′N 78°06′W*
21 *V10* **Warsaw** North Carolina, SE USA *35°00′N 78°05′W*
21 *X5* **Warsaw** Virginia, NE USA *37°57′N 76°46′W*
Warsaw/Warschau *see* Warszawa
81 *N17* **Warshiikh** Shabeellaha Dhexe, C Somalia *02°22′N 45°52′E*
101 *G15* **Warstein** Nordrhein-Westfalen, W Germany *51°27′N 08°21′E*
110 *M11* **Warszawa** ***Eng.*** Warsaw, ***Ger.*** Warschau, ***Rus.*** Varshava. ● (Poland) Mazowieckie, C Poland *52°15′N 21°E*
110 *J13* **Warta** Sieradz, C Poland *51°43′N 18°37′E*
110 *D11* **Warta** ***Ger.*** Warthe. ≈ W Poland
Wartberg *see* Senec
20 *M9* **Wartburg** Tennessee, S USA *36°08′N 84°37′W*
108 *J7* **Warth** Vorarlberg, NW Austria *47°16′N 10°11′E*
Warthe *see* Warta
169 *U12* **Waru** Borneo, C Indonesia *01°24′S 116°37′E*
171 *T13* **Waru** Pulau Seram, E Indonesia *03°24′S 130°38′E*
139 *N6* **Wa'r, Wādī al** *dry watercourse* E Syria
183 *U3* **Warwick** Queensland, E Australia *28°12′S 152°E*
15 *Q11* **Warwick** Québec, SE Canada *45°55′N 72°00′W*
97 *M20* **Warwick** C England, United Kingdom *52°17′N 01°34′W*
18 *K13* **Warwick** New York, NE USA *41°15′N 74°21′W*
29 *P4* **Warwick** North Dakota, N USA *47°49′N 98°42′W*
19 *O12* **Warwick** Rhode Island, NE USA *41°40′N 71°21′W*
97 *L20* **Warwickshire** *cultural region* C England, United Kingdom
14 *G14* **Wasaga Beach** Ontario, S Canada *44°30′N 80°00′W*
77 *U13* **Wasagu** Kebbi, NW Nigeria *11°25′N 05°48′E*
36 *M2* **Wasatch Range** ▲▲ W USA
35 *R12* **Wasco** California, W USA *35°34′N 119°20′W*
29 *V10* **Waseca** Minnesota, N USA *44°04′N 93°30′W*
14 *H13* **Washago** Ontario, S Canada *44°46′N 78°48′W*
19 *S2* **Washburn** Maine, NE USA *46°46′N 68°08′W*
28 *M5* **Washburn** North Dakota, N USA *47°15′N 101°02′W*
30 *K3* **Washburn** Wisconsin, N USA *46°41′N 90°53′W*
31 *S14* **Washburn Hill** *hill* Ohio, N USA
154 *H13* **Wāshīm** Mahārāshtra, C India *20°06′N 77°08′E*
97 *M14* **Washington** NE England, United Kingdom *54°54′N 01°31′W*
23 *U3* **Washington** Georgia, SE USA *33°44′N 82°44′W*
30 *L12* **Washington** Illinois, N USA *40°42′N 89°24′W*
31 *N15* **Washington** Indiana, N USA *38°40′N 87°10′W*
29 *X15* **Washington** Iowa, C USA *41°18′N 91°41′W*
27 *O3* **Washington** Kansas, C USA *39°49′N 97°03′W*
27 *W5* **Washington** Missouri, C USA *38°31′N 91°01′W*
21 *X9* **Washington** North Carolina, SE USA *35°33′N 77°04′W*
18 *B15* **Washington** Pennsylvania, NE USA *40°11′N 80°16′W*
25 *V10* **Washington** Texas, SW USA *30°18′N 96°08′W*
36 *J8* **Washington** Utah, W USA *37°07′N 113°30′W*
21 *V4* **Washington** Virginia, NE USA *38°43′N 78°11′W*
32 *I9* **Washington** ***off.*** State of Washington, ***also known as*** Chinook State, Evergreen State. ⬖ *state* NW USA
Washington *see* Washington Court House
31 *S14* **Washington Court House** ***var.*** Washington. Ohio, NE USA *39°32′N 83°29′W*
21 *W4* **Washington DC** ● (USA) District of Columbia, NE USA *38°54′N 77°02′W*
31 *O5* **Washington Island** *island* Wisconsin, N USA
Washington Island *see* Teraina
19 *O7* **Washington, Mount** ▲ New Hampshire, NE USA *44°16′N 71°18′W*
26 *M11* **Washita River** ≈ Oklahoma/Texas, C USA
97 *O18* **Wash, The** *inlet* E England, United Kingdom
32 *L9* **Washtucna** Washington, NW USA *46°44′N 118°19′W*
Wasiliszki *see* Vasilishki
110 *P9* **Wasilków** Podlaskie, NE Poland *53°12′N 23°15′E*
39 *R11* **Wasilla** Alaska, USA *61°34′N 149°26′W*
139 *V9* **Wāsiț** ***off.*** Muḥāfa at Wāsiț, ***var.*** Al Kūt. ⬖ *governorate* E Iraq
Wāsiț, Muḥāfa at *see* Wāsiț
55 *U9* **Wasjabo** Sipaliwini, NW Suriname *05°09′N 57°09′W*
12 *I10* **Waskaganish** ***prev.*** Fort Rupert, Rupert House. Québec, SE Canada *51°30′N 79°45′W*
11 *X11* **Waskaiowaka Lake** ◎ Manitoba, C Canada
11 *T14* **Waskesiu Lake** Saskatchewan, C Canada *53°56′N 106°05′W*
25 *X7* **Waskom** Texas, SW USA *32°28′N 94°03′W*
110 *G13* **Wąsosz** Dolnośląskie, SW Poland *51°36′N 16°30′E*
42 *M6* **Waspam** ***var.*** Waspán. Región Autónoma Atlántico Norte, NE Nicaragua *14°41′N 84°04′W*
Waspán *see* Waspam
165 *T3* **Wassamu** Hokkaidō, NE Japan *44°01′N 142°25′E*
108 *G9* **Wassen** Uri, C Switzerland *46°42′N 08°34′E*
98 *G11* **Wassenaar** Zuid-Holland, W Netherlands *52°09′N 04°23′E*
99 *N24* **Wasserbillig** Grevenmacher, E Luxembourg *49°43′N 06°30′E*
Wasserburg *see* Wasserburg am Inn
101 *M23* **Wasserburg am Inn** ***var.*** Wasserburg. Bayern, SE Germany *48°02′N 12°12′E*
101 *I17* **Wasserkuppe** ▲ C Germany *50°30′N 09°55′E*
103 *R5* **Wassy** Haute-Marne, N France *48°32′N 04°54′E*
171 *N14* **Watampone** ***var.*** Bone. Sulawesi, C Indonesia *04°33′S 120°20′E*
171 *R13* **Watawa** Pulau Buru, E Indonesia *03°36′S 127°13′E*
Watenstedt-Salzgitter *see* Salzgitter
18 *M13* **Waterbury** Connecticut, NE USA *41°33′N 73°01′W*
21 *R11* **Wateree Lake** ▣ South Carolina, SE USA
21 *R12* **Wateree River** ≈ South Carolina, SE USA
97 *E20* **Waterford** ***Ir.*** Port Láirge. Waterford, S Ireland *52°15′N 07°08′W*
31 *S9* **Waterford** Michigan, N USA *42°42′N 83°24′W*
97 *E20* **Waterford** ***Ir.*** Port Láirge. *cultural region* S Ireland
97 *E21* **Waterford Harbour** ***Ir.*** Cuan Phort Láirge. *inlet* S Ireland
98 *G12* **Wateringen** Zuid-Holland, W Netherlands *52°02′N 04°16′E*
99 *G19* **Waterloo** Walloon Brabant, C Belgium *50°43′N 04°24′E*
14 *F16* **Waterloo** Ontario, S Canada *43°28′N 80°32′W*
15 *P12* **Waterloo** Québec, SE Canada *45°20′N 72°28′W*
30 *K16* **Waterloo** Illinois, N USA *38°20′N 90°09′W*
29 *X13* **Waterloo** Iowa, C USA *42°31′N 92°16′W*
18 *G10* **Waterloo** New York, NE USA *42°54′N 76°51′W*
30 *L4* **Watersmeet** Michigan, N USA *46°16′N 89°10′W*
23 *V9* **Watertown** Florida, SE USA *30°11′N 82°36′W*
18 *I8* **Watertown** New York, NE USA *43°37′N 75°56′W*
29 *R9* **Watertown** South Dakota, N USA *44°54′N 97°07′W*
30 *M8* **Watertown** Wisconsin, N USA *43°12′N 88°44′W*
22 *L3* **Water Valley** Mississippi, S USA *34°09′N 89°37′W*
27 *O3* **Waterville** Kansas, C USA *39°41′N 96°45′W*
17 *V6* **Waterville** Maine, NE USA *44°34′N 69°41′W*
29 *V10* **Waterville** Minnesota, N USA *44°13′N 93°34′W*
18 *I10* **Waterville** New York, NE USA *42°55′N 75°18′W*
14 *E16* **Watford** Ontario, S Canada *42°57′N 81°51′W*
97 *N21* **Watford** E England, United Kingdom *51°39′N 00°24′W*
28 *K4* **Watford City** North Dakota, N USA *47°48′N 103°16′W*
141 *X12* **Wāṭif** S Oman *18°34′N 56°31′E*
18 *G11* **Watkins Glen** New York, NE USA *42°23′N 76°53′W*
Watlings Island *see* San Salvador
171 *U15* **Watnil** Pulau Kai Kecil, E Indonesia *05°45′S 132°39′E*
26 *M10* **Watonga** Oklahoma, C USA *35°52′N 98°26′W*
11 *T16* **Watrous** Saskatchewan, S Canada *51°40′N 105°29′W*
37 *T10* **Watrous** New Mexico, SW USA *35°48′N 104°58′W*
79 *P16* **Watsa** Orientale, NE Dem. Rep. Congo *03°00′N 29°31′E*
31 *N12* **Watseka** Illinois, N USA *40°46′N 87°44′W*
79 *J19* **Watsikengo** Equateur, C Dem. Rep. Congo *0°49′S 20°34′E*
182 *C5* **Watson** South Australia *30°32′S 131°29′E*
11 *U15* **Watson** Saskatchewan, S Canada *52°13′N 104°30′W*
195 *O10* **Watson Escarpment** ▲▲ Antarctica
10 *K9* **Watson Lake** Yukon, W Canada *60°05′N 128°47′W*
35 *N10* **Watsonville** California, W USA *36°53′N 121°43′W*
167 *Q8* **Wattay** ✈ (Viangchan) Viangchan, C Laos *18°03′N 102°36′E*
109 *N7* **Wattens** Tirol, W Austria *47°18′N 11°37′E*
20 *M9* **Watts Bar Lake** ▣ Tennessee, S USA
108 *H7* **Wattwil** Sankt Gallen, NE Switzerland *47°18′N 09°06′E*
171 *T14* **Watubela, Kepulauan** *island group* E Indonesia
101 *N24* **Watzmann** ▲ SE Germany *47°32′N 12°56′E*
186 *E8* **Wau** Morobe, C Papua New Guinea *07°22′S 146°40′E*
81 *D14* **Wau** ***var.*** Wāw. Western Bahr el Ghazal, W South Sudan *07°43′N 28°01′E*
29 *Q8* **Waubay** South Dakota, N USA *45°19′N 97°18′W*
29 *Q8* **Waubay Lake** ◎ South Dakota, N USA
183 *U7* **Wauchope** New South Wales, SE Australia *31°30′S 152°46′E*
23 *W13* **Wauchula** Florida, SE USA *27°33′N 81°48′W*
30 *M10* **Wauconda** Illinois, N USA *42°15′N 88°08′W*
182 *J7* **Waukaringa** South Australia *32°19′S 139°27′E*
31 *N10* **Waukegan** Illinois, N USA *42°21′N 87°50′W*
30 *M9* **Waukesha** Wisconsin, N USA *43°01′N 88°14′W*
29 *X11* **Waukon** Iowa, C USA *43°16′N 91°28′W*
30 *L8* **Waunakee** Wisconsin, N USA *43°13′N 89°28′W*
30 *L7* **Waupaca** Wisconsin, N USA *44°23′N 89°04′W*
30 *M8* **Waupun** Wisconsin, N USA *43°40′N 88°43′W*
26 *M13* **Waurika** Oklahoma, C USA *34°10′N 98°00′W*
26 *M12* **Waurika Lake** ▣ Oklahoma, C USA
30 *L6* **Wausau** Wisconsin, N USA *44°58′N 89°40′W*
31 *R11* **Wauseon** Ohio, N USA *41°33′N 84°08′W*
30 *L7* **Wautoma** Wisconsin, N USA *44°05′N 89°17′W*
30 *M9* **Wauwatosa** Wisconsin, N USA *43°03′N 88°03′W*
22 *L9* **Waveland** Mississippi, S USA *30°17′N 89°22′W*
97 *Q20* **Waveney** ≈ E England, United Kingdom
184 *L11* **Waverley** Taranaki, North Island, New Zealand *39°45′S 174°35′E*
29 *W12* **Waverly** Iowa, C USA *42°43′N 92°28′W*
27 *T4* **Waverly** Missouri, C USA *39°12′N 93°31′W*
29 *R15* **Waverly** Nebraska, C USA *40°56′N 96°27′W*
18 *G12* **Waverly** New York, NE USA *42°00′N 76°33′W*
20 *H8* **Waverly** Tennessee, S USA *36°04′N 87°49′W*
21 *W7* **Waverly** Virginia, NE USA *37°02′N 77°06′W*
99 *H19* **Wavre** Walloon Brabant, C Belgium *50°43′N 04°37′E*
166 *M8* **Waw** Bago, SW Myanmar (Burma) *17°26′N 96°40′E*
Wāw *see* Wau
14 *B7* **Wawa** Ontario, S Canada *47°59′N 84°43′W*
77 *T14* **Wawa** Niger, W Nigeria *09°52′N 04°33′E*
75 *Q11* **Wāw al Kabīr** S Libya *25°21′N 16°41′E*
43 *N7* **Wawa, Río** ***var.*** Rio Huahua. ≈ NE Nicaragua
186 *B8* **Wawoi** ≈ SW Papua New Guinea
25 *T7* **Waxahachie** Texas, SW USA *32°23′N 96°52′W*
158 *L9* **Waxxari** Xinjiang Uygur Zizhiqu, NW China *38°43′N 87°11′E*
Wayaobu *see* Zichang
23 *V7* **Waycross** Georgia, SE USA *31°13′N 82°21′W*
180 *K10* **Way, Lake** ◎ Western Australia
31 *P9* **Wayland** Michigan, N USA *42°40′N 85°38′W*
29 *R13* **Wayne** Nebraska, C USA *42°13′N 97°01′W*
18 *K14* **Wayne** New Jersey, NE USA *40°57′N 74°16′W*
21 *P5* **Wayne** West Virginia, NE USA *38°14′N 82°27′W*
23 *V4* **Waynesboro** Georgia, SE USA *33°04′N 82°01′W*
22 *M7* **Waynesboro** Mississippi, S USA *31°40′N 88°39′W*
20 *H10* **Waynesboro** Tennessee, S USA *35°20′N 87°49′W*
21 *U5* **Waynesboro** Virginia, NE USA *38°04′N 78°54′W*
18 *B16* **Waynesburg** Pennsylvania, NE USA *39°51′N 80°10′W*
27 *U6* **Waynesville** Missouri, C USA *37°48′N 92°11′W*
21 *O10* **Waynesville** North Carolina, SE USA *35°29′N 82°59′W*
26 *L8* **Waynoka** Oklahoma, C USA *36°36′N 98°53′W*
Wazan *see* Ouazzane
Wazima *see* Wajima
149 *V7* **Wazīrābād** Punjab, NE Pakistan *32°28′N 74°04′E*
Wazzan *see* Ouazzane
110 *I8* **Wda** ***var.*** Czarna Woda, ***Ger.*** Schwarzwasser. ≈ N Poland
187 *Q16* **Wé** Province des Îles Loyauté, E New Caledonia *20°55′S 167°15′E*
97 *O23* **Weald, The** *lowlands* SE England, United Kingdom
186 *A9* **Weam** Western, SW Papua New Guinea *08°33′S 141°10′E*
97 *L15* **Wear** ≈ N England, United Kingdom
Wearmouth *see* Sunderland
26 *L10* **Weatherford** Oklahoma, C USA *35°31′N 98°42′W*
25 *S6* **Weatherford** Texas, SW USA *32°47′N 97°48′W*
34 *M3* **Weaverville** California, W USA *40°42′N 122°57′W*
27 *R7* **Webb City** Missouri, C USA *37°07′N 94°28′W*
192 *G8* **Weber Basin** *undersea feature* S Ceram Sea
Webfoot State *see* Oregon
18 *F9* **Webster** New York, NE USA *43°12′N 77°25′W*
29 *Q8* **Webster** South Dakota, N USA *45°19′N 97°31′W*
29 *V13* **Webster City** Iowa, C USA *42°28′N 93°49′W*
27 *X5* **Webster Groves** Missouri, C USA *38°32′N 90°20′W*
21 *S4* **Webster Springs** ***var.*** Addison. West Virginia, NE USA *38°29′N 80°52′W*
171 *S11* **Weda, Teluk** *bay* Pulau Halmahera, E Indonesia
65 *B25* **Weddell Island** ***var.*** Isla de San Jorge. *island* W Falkland Islands
65 *K22* **Weddell Plain** *undersea feature* SW Atlantic Ocean *65°00′S 40°00′W*
65 *K23* **Weddell Sea** *sea* SW Atlantic Ocean
65 *B25* **Weddell Settlement** Weddell Island, W Falkland Islands *52°53′S 60°54′W*
182 *M11* **Wedderburn** Victoria, SE Australia *36°26′S 143°37′E*
100 *I9* **Wedel** Schleswig-Holstein, N Germany *53°35′N 09°42′E*
92 *N3* **Wedel Jarlsberg Land** *physical region* SW Svalbard
100 *I12* **Wedemark** Niedersachsen, NW Germany *52°33′N 09°43′E*
10 *M17* **Wedge Mountain** ▲ British Columbia, SW Canada *50°10′N 122°43′W*
23 *R4* **Wedowee** Alabama, S USA *33°16′N 85°28′W*
171 *U15* **Weduar** Pulau Kai Besar, E Indonesia *05°55′S 132°51′E*
35 *N2* **Weed** California, W USA *41°26′N 122°24′W*
15 *Q12* **Weedon Centre** Québec, SE Canada *45°40′N 71°28′W*
18 *E13* **Weedville** Pennsylvania, NE USA *41°15′N 78°28′W*
100 *F10* **Weener** Niedersachsen, NW Germany *53°09′N 07°19′E*
29 *S16* **Weeping Water** Nebraska, C USA *40°52′N 96°08′W*
99 *L16* **Weert** Limburg, SE Netherlands *51°15′N 05°43′E*
98 *I10* **Weesp** Noord-Holland, C Netherlands *52°18′N 05°03′E*
183 *S5* **Wee Waa** New South Wales, SE Australia *30°16′S 149°27′E*
110 *N7* **Węgorzewo** ***Ger.*** Angerburg. Warmińsko-Mazurskie, NE Poland *54°12′N 21°49′E*
110 *E9* **Węgorzyno** ***Ger.*** Wangerin. Zachodnio-pomorskie, NW Poland *53°34′N 15°35′E*
110 *N11* **Węgrów** ***Ger.*** Bingerau. Mazowieckie, C Poland *52°22′N 22°00′E*
98 *N5* **Wehe-Den Hoorn** Groningen, NE Netherlands *53°20′N 06°29′E*
98 *M12* **Wehl** Gelderland, E Netherlands *51°58′N 06°13′E*
Wehlau *see* Znamensk
168 *F7* **Weh, Pulau** *island* NW Indonesia
Wei *see* Weifang
161 *P1* **Weichang** ***prev.*** Zhuizishan. Hebei, E China *41°55′N 117°45′E*
Weichang *see* Weishan
Weichsel *see* Wisła
101 *M16* **Weida** Thüringen, C Germany *50°46′N 12°05′E*
Weiden *see* Weiden in der Oberpfalz
101 *M19* **Weiden in der Oberpfalz** ***var.*** Weiden. Bayern, SE Germany *49°40′N 12°10′E*
161 *Q4* **Weifang** ***var.*** Wei, Wei-fang; ***prev.*** Weihsien. Shandong, E China *36°44′N 119°10′E*
161 *S4* **Weihai** Shandong, E China *37°30′N 122°04′E*
160 *K6* **Wei He** ≈ C China
Weihsien *see* Weifang
101 *G17* **Weilburg** Hessen, W Germany *50°31′N 08°18′E*
101 *K24* **Weilheim in Oberbayern** Bayern, SE Germany *47°50′N 11°09′E*
183 *P4* **Weilmoringle** New South Wales, SE Australia *29°13′S 146°51′E*
101 *L16* **Weimar** Thüringen, C Germany *50°59′N 11°20′E*
25 *U11* **Weimar** Texas, SW USA *29°42′N 96°46′W*
160 *L6* **Weinan** Shaanxi, C China *34°30′N 109°30′E*
108 *H6* **Weinfelden** Thurgau, NE Switzerland *47°33′N 09°09′E*
101 *I24* **Weingarten** Baden-Württemberg, S Germany *47°49′N 09°37′E*
101 *G20* **Weinheim** Baden-Württemberg, SW Germany *49°33′N 08°40′E*
160 *H11* **Weining** ***var.*** Caohai, Weining Yizu Huizu Miaozu Zizhixian. Guizhou, S China *26°51′N 104°16′E*
Weining Yizu Huizu Miaozu Zizhixian *see* Weining
181 *V2* **Weipa** Queensland, NE Australia *12°43′S 142°01′E*
11 *Y11* **Weir River** Manitoba, C Canada *56°44′N 94°06′W*
21 *R1* **Weirton** West Virginia, NE USA *40°24′N 80°37′W*
32 *M13* **Weiser** Idaho, NW USA *44°15′N 116°58′W*
160 *F12* **Weishan** ***var.*** Weichang. Yunnan, SW China *25°22′N 100°19′E*
161 *P6* **Weishan Hu** ◎ E China
101 *M15* **Weisse Elster** ***Eng.*** White Elster. ≈ Czech Republic/ Germany
Weisse Körös/Weisse Kreisch *see* Crişul Alb
108 *L7* **Weissenbach am Lech** Tirol, W Austria *47°27′N 10°39′E*
Weissenburg *see* Wissembourg, France
Weissenburg *see* Alba Iulia, Romania
101 *K21* **Weissenburg in Bayern** Bayern, SE Germany *49°02′N 10°59′E*
101 *M15* **Weissenfels** ***var.*** Weißenfels. Sachsen-Anhalt, C Germany *51°12′N 11°58′E*
109 *R9* **Weissensee** ◎ S Austria
Weissenstein *see* Paide
108 *E11* **Weisshorn** ***var.*** Flüela Wisshorn. ▲ SW Switzerland *46°06′N 07°43′E*
Weisskirchen *see* Bela Crkva
23 *R3* **Weiss Lake** ▣ Alabama, S USA
101 *Q14* **Weisswasser** ***Lus.*** Běla Woda. Sachsen, E Germany *51°30′N 14°37′E*
99 *M22* **Weiswampach** Diekirch, N Luxembourg *50°08′N 06°05′E*
109 *U2* **Weitra** Niederösterreich, N Austria *48°41′N 14°54′E*
161 *O4* **Weixian** ***var.*** Wei Xian. Hebei, E China *36°59′N 115°15′E*
Wei Xian *see* Weixian
159 *V11* **Weiyuan** ***var.*** Qingyuan. Gansu, C China *35°07′N 104°12′E*
Weiyuan *see* Shuangjiang
160 *F14* **Weiyuan Jiang** ≈ SW China
109 *W7* **Weiz** Steiermark, SE Austria *47°13′N 15°38′E*
Weizhou *see* Wenchuan
160 *K16* **Weizhou Dao** *island* S China
110 *I6* **Wejherowo** Pomorskie, NW Poland *54°36′N 18°12′E*
27 *Q8* **Welch** Oklahoma, C USA *36°52′N 95°06′W*
24 *M6* **Welch** Texas, SW USA *32°52′N 102°06′W*
21 *Q6* **Welch** West Virginia, NE USA *37°26′N 81°36′W*
45 *O14* **Welchman Hall** C Barbados *13°10′N 59°34′W*
80 *J11* **Weldiya** ***var.*** Waldia, ***It.*** Valdia. Āmara, N Ethiopia *11°45′N 39°39′E*
21 *W8* **Weldon** North Carolina, SE USA *36°25′N 77°36′W*
25 *V9* **Weldon** Texas, SW USA *31°00′N 95°33′W*
99 *M19* **Welkenraedt** Liège, E Belgium *50°40′N 05°58′E*
193 *O2* **Welker Seamount** *undersea feature* N Pacific Ocean *55°07′N 140°18′W*
83 *I22* **Welkom** Free State, C South Africa *27°59′S 26°44′E*
14 *H16* **Welland** Ontario, S Canada *45°59′N 79°14′W*
14 *G16* **Welland** ≈ Ontario, S Canada
97 *O19* **Welland** ≈ C England, United Kingdom
14 *H17* **Welland Canal** *canal* Ontario, S Canada
155 *K25* **Wellawaya** Uva Province, SE Sri Lanka *06°44′N 81°07′E*
Welle *see* Uele
181 *T4* **Wellesley Islands** *island group* Queensland, N Australia
99 *J22* **Wellin** Luxembourg, SE Belgium *50°06′N 05°05′E*
97 *N20* **Wellingborough** C England, United Kingdom *52°19′N 00°42′W*
183 *R7* **Wellington** New South Wales, SE Australia *32°33′S 148°59′E*
14 *J15* **Wellington** Ontario, SE Canada *43°59′N 77°21′W*
185 *L14* **Wellington** ● Wellington, North Island, New Zealand *41°17′S 174°47′E*
83 *E26* **Wellington** Western Cape, SW South Africa *33°39′S 19°00′E*
37 *T2* **Wellington** Colorado, C USA *40°42′N 105°00′W*
27 *N7* **Wellington** Kansas, C USA *37°17′N 97°25′W*
35 *R7* **Wellington** Nevada, W USA *38°45′N 119°22′W*
31 *T11* **Wellington** Ohio, N USA *41°10′N 82°13′W*
25 *P3* **Wellington** Texas, SW USA *34°52′N 100°13′W*
36 *M4* **Wellington** Utah, W USA *39°31′N 110°45′W*
185 *M14* **Wellington** ***off.*** Wellington Region. ⬖ *region* (New Zealand) North Island, New Zealand
185 *L14* **Wellington** ✈ Wellington, North Island, New Zealand *41°19′S 174°48′E*
Wellington *see* Wellington, Isla
63 *F22* **Wellington, Isla** ***var.*** Wellington. *island* S Chile
183 *P12* **Wellington, Lake** ◎ Victoria, SE Australia
Wellington Region *see* Wellington
29 *X14* **Wellman** Iowa, C USA *41°27′N 91°50′W*
24 *M6* **Wellman** Texas, SW USA *33°03′N 102°25′W*
97 *K22* **Wells** SW England, United Kingdom *51°13′N 02°39′W*
29 *V11* **Wells** Minnesota, N USA *43°45′N 93°43′W*
35 *X2* **Wells** Nevada, W USA *41°07′N 114°58′W*
25 *W8* **Wells** Texas, SW USA *31°28′N 94°54′W*
18 *F12* **Wellsboro** Pennsylvania, NE USA *41°43′N 77°39′W*
21 *R1* **Wellsburg** West Virginia, NE USA *40°15′N 80°37′W*
184 *K4* **Wellsford** Auckland, North Island, New Zealand *36°17′S 174°30′E*
180 *L9* **Wells, Lake** ◎ Western Australia
181 *N4* **Wells, Mount** ▲ Western Australia *17°39′S 127°08′E*
97 *P18* **Wells-next-the-Sea** E England, United Kingdom *52°58′N 00°48′E*
31 *T15* **Wellston** Ohio, N USA *39°07′N 82°31′W*
27 *O10* **Wellston** Oklahoma, C USA *35°41′N 97°03′W*
18 *E11* **Wellsville** New York, NE USA *42°06′N 77°55′W*
31 *V12* **Wellsville** Ohio, N USA *40°36′N 80°39′W*
36 *L1* **Wellsville** Utah, W USA *41°38′N 111°55′W*
36 *I14* **Wellton** Arizona, SW USA *32°40′N 114°09′W*
109 *S4* **Wels** ***anc.*** Ovilava. Oberösterreich, N Austria *48°10′N 14°02′E*
99 *K15* **Welschap** ✈ (Eindhoven) Noord-Brabant, S Netherlands *51°27′N 05°22′E*
100 *P10* **Welse** ≈ NE Germany
22 *H9* **Welsh** Louisiana, S USA *30°12′N 92°49′W*
97 *K19* **Welshpool** ***Wel.*** Y Trallwng. E Wales, United Kingdom *52°38′N 03°06′W*
97 *O21* **Welwyn Garden City** E England, United Kingdom *51°48′N 00°13′W*
79 *K18* **Wema** Equateur, NW Dem. Rep. Congo *0°25′S 21°33′E*
81 *G21* **Wembere** ≈ C Tanzania
11 *N13* **Wembley** Alberta, W Canada *55°07′N 119°12′W*
12 *I9* **Wemindji** ***prev.*** Nouveau-Comptoir, Paint Hills. Québec, C Canada *53°00′N 78°42′W*
99 *G18* **Wemmel** Vlaams Brabant, C Belgium *50°54′N 04°18′E*
32 *J8* **Wenatchee** Washington, NW USA *47°50′N 120°48′W*
160 *M17* **Wenchang** Hainan, S China *19°34′N 110°46′E*
161 *R11* **Wencheng** ***var.*** Daxue. Zhejiang, SE China *27°48′N 120°01′E*
77 *P16* **Wenchi** W Ghana *07°45′N 02°02′W*
Wen-chou/Wenchow *see* Wenzhou
160 *H8* **Wenchuan** ***var.*** Weizhou. Sichuan, C China *31°29′N 103°39′E*
Wendau *see* Võnnu
Wenden *see* Cēsis
161 *S4* **Wendeng** Shandong, E China *37°10′N 122°00′E*
81 *J14* **Wendo** Southern Nationalities, S Ethiopia *06°34′N 38°28′E*
36 *J2* **Wendover** Utah, W USA *40°41′N 114°00′W*
14 *D9* **Wenebegon** ≈ Ontario, S Canada
14 *D8* **Wenebegon Lake** ◎ Ontario, S Canada
108 *E9* **Wengen** Bern, W Switzerland *46°38′N 07°57′E*
161 *O13* **Wengyuan** ***var.*** Longxian. Guangdong, S China *24°22′N 114°06′E*
189 *P15* **Weno** ***prev.*** Moen. Chuuk, C Micronesia
189 *V12* **Weno** ***prev.*** Moen. *atoll* Chuuk Islands, C Micronesia
158 *N13* **Wenquan** Qinghai, C China *33°16′N 91°44′E*
159 *H4* **Wenquan** ***var.*** Arixang, Bogeda'er. Xinjiang Uygur Zizhiqu, NW China *45°00′N 81°02′E*
Wenquan *see* Yingshan
160 *H14* **Wenshan** ***var.*** Kaihua. Yunnan, SW China *23°22′N 104°21′E*
158 *H6* **Wensu** Xinjiang Uygur Zizhiqu, W China *41°15′N 80°11′E*
182 *L8* **Wentworth** New South Wales, SE Australia *34°04′S 141°53′E*
27 *W4* **Wentzville** Missouri, C USA *38°48′N 90°51′W*
159 *V12* **Wenxian** ***var.*** Wen Xian. Gansu, C China *32°57′N 104°42′E*
Wen Xian *see* Wenxian
161 *S10* **Wenzhou** ***var.*** Wen-chou, Wenchow. Zhejiang, SE China *28°02′N 120°36′E*
34 *L4* **Weott** California, W USA *40°19′N 123°57′W*
99 *I20* **Wépion** Namur, SE Belgium *50°24′N 04°53′E*
100 *O11* **Werbellinsee** ◎ NE Germany
99 *L21* **Werbomont** Liège, E Belgium *50°22′N 05°43′E*
83 *G20* **Werda** Kgalagadi, S Botswana *25°13′S 23°16′E*
81 *N14* **Werdēr** Sumalē, E Ethiopia *06°59′N 45°20′E*
Werder *see* Virtsu
Werenów *see* Voranava
171 *U13* **Weri** Papua Barat, E Indonesia *03°10′S 132°39′E*
98 *I13* **Werkendam** Noord-Brabant, S Netherlands *51°48′N 04°54′E*
101 *M20* **Wernberg-Köblitz** Bayern, SE Germany *49°31′N 12°10′E*
101 *J18* **Werneck** Bayern, C Germany *50°00′N 10°06′E*
101 *K14* **Wernigerode** Sachsen-Anhalt, C Germany *51°51′N 10°48′E*
Werowitz *see* Virovitica
101 *J16* **Werra** ≈ C Germany
183 *N12* **Werribee** Victoria, SE Australia *37°55′S 144°39′E*
183 *T6* **Werris Creek** New South Wales, SE Australia *31°22′S 150°40′E*
Werro *see* Võru
Werschetz *see* Vršac
101 *K23* **Wertach** ≈ S Germany
101 *I19* **Wertheim** Baden-Württemberg, SW Germany *49°45′N 09°31′E*
98 *J8* **Wervershoof** Noord-Holland, NW Netherlands *52°43′N 05°09′E*
Wervicq *see* Wervik
99 *C18* **Wervik** ***var.*** Wervicq, Werwick. West-Vlaanderen, W Belgium *50°47′N 03°03′E*
Werwick *see* Wervik
101 *D14* **Wesel** Nordrhein-Westfalen, W Germany *51°39′N 06°37′E*
Weseli an der Lainsitz *see* Veselí nad Lužnicí
Wesenberg *see* Rakvere
100 *H12* **Weser** ≈ NW Germany
Wes-Kaap *see* Western Cape
25 *S17* **Weslaco** Texas, SW USA *26°09′N 97°59′W*
14 *J13* **Weslemkoon Lake** ◎ Ontario, SE Canada
181 *R1* **Wessel Islands** *island group* Northern Territory, N Australia
29 *P9* **Wessington** South Dakota, N USA *44°27′N 98°40′W*
29 *P10* **Wessington Springs** South Dakota, N USA *44°02′N 98°33′W*
25 *T8* **West** Texas, SW USA *31°48′N 97°05′W*
West *see* Ouest
30 *M9* **West Allis** Wisconsin, N USA *43°01′N 88°00′W*
182 *E8* **Westall, Point** *headland* South Australia *32°54′S 134°04′E*
194 *M10* **West Antarctica** ***prev.*** Lesser Antarctica. *physical region* Antarctica
14 *G11* **West Arm** Ontario, S Canada *46°16′N 80°25′W*
West Australian Basin *see* Wharton Basin
West Azerbaijan *see* Āžarbāyjān-e Gharbī
11 *N17* **Westbank** British Columbia, SW Canada *49°50′N 119°37′W*
138 *F10* **West Bank** *disputed region* SW Asia
14 *E11* **West Bay** Manitoulin Island, Ontario, S Canada *45°48′N 82°09′W*
22 *L11* **West Bay** *bay* Louisiana, S USA
30 *M8* **West Bend** Wisconsin, N USA *43°26′N 88°13′W*
153 *R16* **West Bengal** ⬖ *state* NE India
West Borneo *see* Kalimantan Barat
29 *Y14* **West Branch** Iowa, C USA *41°40′N 91°21′W*
31 *R7* **West Branch** Michigan, N USA *44°16′N 84°14′W*
18 *F13* **West Branch Susquehanna River** ≈ Pennsylvania, NE USA
97 *L20* **West Bromwich** C England, United Kingdom *52°29′N 01°59′W*
19 *P8* **Westbrook** Maine, NE USA *43°42′N 70°21′W*
29 *T10* **Westbrook** Minnesota, N USA *44°02′N 95°26′W*
29 *Y15* **West Burlington** Iowa, C USA *40°49′N 91°09′W*
96 *L2* **West Burra** *island* NE Scotland, United Kingdom
30 *J8* **Westby** Wisconsin, N USA *43°39′N 90°52′W*
44 *L6* **West Caicos** *island* W Turks and Caicos Islands
185 *A24* **West Cape** *headland* South Island, New Zealand *45°51′S 166°26′E*
174 *L4* **West Caroline Basin** *undersea feature* SW Pacific Ocean *04°00′N 138°00′E*
18 *I16* **West Chester** Pennsylvania, NE USA *39°56′N 75°35′W*
185 *E18* **West Coast** ***off.*** West Coast Region. ⬖ *region* South Island, New Zealand
West Coast Region *see* West Coast
25 *V12* **West Columbia** Texas, SW USA *29°08′N 95°39′W*
29 *W10* **West Concord** Minnesota, N USA *44°09′N 92°54′W*
29 *V14* **West Des Moines** Iowa, C USA *41°33′N 93°42′W*
37 *Q6* **West Elk Peak** ▲ Colorado, C USA *38°43′N 107°12′W*
44 *F1* **West End** Grand Bahama Island, N The Bahamas *26°36′N 78°55′W*
44 *F1* **West End Point** *headland* Grand Bahama Island, N The Bahamas *26°40′N 78°58′W*
98 *O7* **Westerbork** Drenthe, NE Netherlands *52°49′N 06°36′E*
98 *N3* **Westereems** *strait* Germany/ Netherlands
98 *O9* **Westerhaar-Vriezenveensewijk** Overijssel, E Netherlands *52°28′N 06°38′E*
100 *G6* **Westerland** Schleswig-Holstein, N Germany *54°54′N 08°19′E*
99 *I17* **Westerlo** Antwerpen, N Belgium *51°05′N 04°55′E*
19 *N13* **Westerly** Rhode Island, NE USA *41°22′N 71°45′W*
153 *N11* **Western** ⬖ *zone* C Nepal
186 *A8* **Western** ***var.*** Fly River. ⬖ *province* SW Papua New Guinea
186 *K9* **Western** ⬖ *province* S Solomon Islands
186 *J8* **Western** ***off.*** Western Province. ⬖ *province* NW Solomon Islands
155 *J26* **Western** ⬖ *province* SW Sri Lanka
83 *G15* **Western** ⬖ *province* SW Zambia
180 *K8* **Western Australia** ⬖ *state* W Australia
80 *A13* **Western Bahr el Ghazal** ⬖ *state* W South Sudan
Western Bug *see* Bug
83 *F25* **Western Cape** ***off.*** Western Cape Province, ***Afr.*** Wes-Kaap. ⬖ *province* SW South Africa
Western Cape Province *see* Western Cape
80 *A11* **Western Darfur** ⬖ *state* W Sudan
Western Desert *see* Şaḥrā' al Gharbīyah
118 *G9* **Western Dvina** ***Bel.*** Dzvina, ***Ger.*** Düna, ***Latv.*** Daugava, ***Rus.*** Zapadnaya Dvina. ≈ W Europe
81 *D15* **Western Equatoria** ⬖ *state* SW South Sudan
155 *E16* **Western Ghats** ▲▲ SW India
186 *C7* **Western Highlands** ⬖ *province* C Papua New Guinea
Western Isles *see* Outer Hebrides
80 *D11* **Western Kordofan** ⬖ *state* S Sudan
21 *T3* **Westernport** Maryland, NE USA *39°29′N 79°03′W*
Western Province *see* Western
Western Punjab *see* Punjab
74 *B10* **Western Sahara** ◇ *disputed territory* N Africa
Western Samoa *see* Samoa
Western Sayans *see* Zapadnyy Sayan
Western Scheldt *see* Westerschelde
Western Sierra Madre *see* Madre Occidental, Sierra
99 *E15* **Westerschelde** ***Eng.*** Western Scheldt; ***prev.*** Honte. *inlet* S North Sea
31 *S13* **Westerville** Ohio, N USA *40°07′N 82°55′W*
101 *F17* **Westerwald** ▲▲ W Germany
65 *C25* **West Falkland** ***var.*** Gran Malvina, Isla Gran Malvina. *island* W Falkland Islands
29 *R5* **West Fargo** North Dakota, N USA *46°49′N 96°51′W*
188 *M15* **West Fayu Atoll** *atoll* Caroline Islands, C Micronesia
18 *C11* **Westfield** New York, NE USA *42°18′N 79°34′W*
30 *L7* **Westfield** Wisconsin, N USA *43°56′N 89°31′W*
West Flanders *see* West-Vlaanderen
27 *S10* **West Fork** Arkansas, C USA *35°55′N 94°11′W*
29 *P16* **West Fork Big Blue River** ≈ Nebraska, C USA
29 *U12* **West Fork Des Moines River** ≈ Iowa/Minnesota, C USA
25 *S5* **West Fork Trinity River** ≈ Texas, SW USA
30 *L16* **West Frankfort** Illinois, N USA *37°54′N 88°55′W*
98 *I8* **West-Friesland** *physical region* NW Netherlands
West Frisian Islands *see* Waddeneilanden
19 *T5* **West Grand Lake** ◎ Maine, NE USA
18 *M12* **West Hartford** Connecticut, NE USA *41°44′N 72°45′W*
18 *M13* **West Haven** Connecticut, NE USA *41°16′N 72°57′W*
27 *X12* **West Helena** Arkansas, C USA *34°33′N 90°38′W*
28 *M2* **Westhope** North Dakota, N USA *48°54′N 101°01′W*
195 *Y8* **West Ice Shelf** *ice shelf* Antarctica
47 *R2* **West Indies** *island group* SE North America
West Irian *see* Papua
West Java *see* Jawa Barat
36 *L3* **West Jordan** Utah, W USA *40°37′N 111°55′W*
West Kalimantan *see* Kalimantan Barat

◆ Country
● Country Capital
◇ Dependent Territory
○ Dependent Territory Capital
⬖ Administrative Regions
✈ International Airport
▲ Mountain
▲▲ Mountain Range
Volcano
≈ River
◎ Lake
▣ Reservoir

WESTKAPELLE – WINNIPEGOSIS

99 D14 **Westkapelle** Zeeland, SW Netherlands *51°32′N 03°26′E*
West Kazakhstan *see* Zapadnyy Kazakhstan
31 O13 **West Lafayette** Indiana, N USA *40°24′N 86°54′W*
31 T13 **West Lafayette** Ohio, N USA *40°16′N 81°45′W*
West Lake *see* Kagera
29 Y14 **West Liberty** Iowa, C USA *41°34′N 91°15′W*
21 O5 **West Liberty** Kentucky, S USA *37°56′N 83°16′W*
Westliche Morava *see* Zapadna Morava
10 J13 **Westlock** Alberta, SW Canada *54°12′N 113°50′W*
14 E17 **West Lorne** Ontario, S Canada *42°36′N 81°35′W*
96 J12 **West Lothian** *cultural region* S Scotland, United Kingdom
99 H16 **Westmalle** Antwerpen, N Belgium *51°18′N 04°40′E*
192 G6 **West Mariana Basin** *var.* Perece Vela Basin. *undersea feature* W Pacific Ocean *15°00′N 137°00′E*
97 E17 **Westmeath** *Ir.* An Iarmhí, Na H-Iarmhidhe. *cultural region* C Ireland
27 Y11 **West Memphis** Arkansas, C USA *35°09′N 90°11′W*
21 W2 **Westminster** Maryland, NE USA *39°34′N 77°00′W*
21 O11 **Westminster** South Carolina, SE USA *34°39′N 83°06′W*
22 I5 **West Monroe** Louisiana, S USA *32°31′N 92°09′W*
18 D15 **Westmont** Pennsylvania, NE USA *40°16′N 78°55′W*
27 O3 **Westmoreland** Kansas, C USA *39°23′N 96°30′W*
35 W17 **Westmorland** California, W USA *33°02′N 115°37′W*
186 E6 **West New Britain** ◆ *province* E Papua New Guinea
West New Guinea *see* Papua
83 K18 **West Nicholson** Matabeleland South, S Zimbabwe *21°06′S 29°25′E*
29 T14 **West Nishnabotna River** ≈ Iowa, C USA
175 P11 **West Norfolk Ridge** *undersea feature* W Pacific Ocean
25 P12 **West Nueces River** ≈ Texas, SW USA
West Nusa Tenggara *see* Nusa Tenggara Barat
29 T11 **West Okoboji Lake** ◎ Iowa, C USA
33 R16 **Weston** Idaho, NW USA *42°01′N 119°29′W*
21 R4 **Weston** West Virginia, NE USA *39°03′N 80°28′W*
97 J22 **Weston-super-Mare** SW England, United Kingdom *51°21′N 02°59′W*
23 Z14 **West Palm Beach** Florida, SE USA *26°43′N 80°03′W*
43 T15 **West Panamá** ◆ *province* C Panamá
West Papua *see* Papua Barat
West Papua *see* Papua
188 E9 **West Passage** *passage* Babeldaob, N Palau
23 O9 **West Pensacola** Florida, SE USA *30°25′N 87°16′W*
27 V8 **West Plains** Missouri, C USA *36°44′N 91°51′W*
35 P7 **West Point** California, W USA *38°21′N 120°33′W*
23 R5 **West Point** Georgia, SE USA *32°52′N 85°10′W*
22 M3 **West Point** Mississippi, S USA *33°36′N 88°39′W*
29 R14 **West Point** Nebraska, C USA *41°50′N 96°42′W*
21 X6 **West Point** Virginia, NE USA *37°31′N 76°48′W*
182 G10 **West Point** *headland* South Australia *35°01′S 135°58′E*
65 B24 **Westpoint Island Settlement** Westpoint Island, NW Falkland Islands *51°21′S 60°41′W*
23 R4 **West Point Lake** ⊠ Alabama/Georgia, SE USA
81 H18 **West Pokot** ◆ *county* W Kenya
97 B16 **Westport** *Ir.* Cathair na Mart. Mayo, W Ireland *53°48′N 09°32′W*
185 G15 **Westport** West Coast, South Island, New Zealand *41°46′S 171°37′E*
32 F10 **Westport** Oregon, NW USA *46°07′N 123°22′W*
32 F9 **Westport** Washington, NW USA *46°53′N 124°06′W*
31 S15 **West Portsmouth** Ohio, N USA *38°45′N 83°01′W*
West Punjab *see* Punjab
11 V14 **Westray** Manitoba, C Canada *53°30′N 101°19′W*
96 J4 **Westray** *island* NE Scotland, United Kingdom
14 F9 **Westree** Ontario, S Canada *47°25′N 81°32′W*
97 L16 **West Riding** *cultural region* N England, United Kingdom
West River *see* Xi Jiang
30 J7 **West Salem** Wisconsin, N USA *43°54′N 91°04′W*
65 H21 **West Scotia Ridge** *undersea feature* W Scotia Sea
186 B5 **West Sepik** *prev.* Sandaun. ◆ *province* NW Papua New Guinea
173 N4 **West Sheba Ridge** *undersea feature* W Indian Ocean *12°45′N 48°15′E*
West Siberian Plain *see* Zapadno-Sibirskaya Ravnina
31 S11 **West Sister Island** *island* Ohio, N USA
West-Skylge *see* West-Terschelling
West Sumatra *see* Sumatera Barat
98 J5 **West-Terschelling** *Fris.* West-Skylge. Fryslân, N Netherlands *53°23′N 05°15′E*
64 J7 **West Thulean Rise** *undersea feature* N Atlantic Ocean
29 X12 **West Union** Iowa, C USA *42°57′N 91°48′W*
31 R15 **West Union** Ohio, N USA *38°47′N 83°33′W*
21 R3 **West Union** West Virginia, NE USA *39°18′N 80°47′W*
31 N13 **Westville** Illinois, N USA *40°02′N 87°38′W*
21 R3 **West Virginia** *off.* State of West Virginia, *also known as* Mountain State. ◆ *state* NE USA
99 A17 **West-Vlaanderen** *Eng.* West Flanders. ◆ *province* W Belgium
35 R7 **West Walker River** ≈ California/Nevada, W USA
35 P4 **Westwood** California, W USA *40°18′N 121°02′W*
183 P9 **West Wyalong** New South Wales, SE Australia *33°56′S 147°10′E*
171 Q16 **Wetar, Pulau** *island* Kepulauan Damar, E Indonesia
Wetar, Selat *see* Wetar Strait
171 R16 **Wetar Strait** *var.* Selat Wetar. *strait* Nusa Tenggara, S Indonesia
11 Q15 **Wetaskiwin** Alberta, SW Canada *52°57′N 113°20′W*
81 K21 **Wete** Pemba, E Tanzania *05°03′S 39°41′E*
166 M4 **Wetlet** Sagaing, C Myanmar (Burma) *22°43′N 95°22′E*
37 T6 **Wet Mountains** ▲▲ Colorado, C USA
101 E15 **Wetter** Nordrhein-Westfalen, W Germany *51°22′N 07°24′E*
101 H17 **Wetter** ≈ W Germany
99 F17 **Wetteren** Oost-Vlaanderen, NW Belgium *51°06′N 03°59′E*
108 F7 **Wettingen** Aargau, N Switzerland *47°28′N 08°20′E*
27 P11 **Wetumka** Oklahoma, C USA *35°14′N 96°14′W*
23 Q5 **Wetumpka** Alabama, S USA *32°32′N 86°12′W*
108 G7 **Wetzikon** Zürich, N Switzerland *47°19′N 08°48′E*
101 G17 **Wetzlar** Hessen, W Germany *50°33′N 08°30′E*
99 C18 **Wevelgem** West-Vlaanderen, W Belgium *50°48′N 03°12′E*
38 M6 **Wevok** *var.* Wewuk. Alaska, USA *68°52′N 166°05′W*
23 R9 **Wewahitchka** Florida, SE USA *30°06′N 85°12′W*
186 C6 **Wewak** East Sepik, NW Papua New Guinea *03°35′S 143°35′E*
27 O11 **Wewoka** Oklahoma, C USA *35°09′N 96°30′W*
Wewuk *see* Wevok
97 F20 **Wexford** *Ir.* Loch Garman. SE Ireland *52°21′N 06°31′W*
97 F20 **Wexford** *Ir.* Loch Garman. *cultural region* SE Ireland
30 L7 **Weyauwega** Wisconsin, N USA *44°16′N 88°54′W*
11 U17 **Weyburn** Saskatchewan, S Canada *49°39′N 103°51′W*
Weyer *see* Weyer Markt
109 U5 **Weyer Markt** *var.* Weyer. Oberösterreich, N Austria *47°52′N 14°39′E*
100 H11 **Weyhe** Niedersachsen, NW Germany *53°00′N 08°52′E*
97 L24 **Weymouth** S England, United Kingdom *50°36′N 02°28′W*
19 P11 **Weymouth** Massachusetts, NE USA *42°13′N 70°56′W*
99 H18 **Wezembeek-Oppem** Vlaams Brabant, C Belgium *50°51′N 04°28′E*
98 M9 **Wezep** Gelderland, E Netherlands *52°28′N 06°E*
184 M9 **Whakamaru** Waikato, North Island, New Zealand *38°27′S 175°48′E*
184 O8 **Whakatane** Bay of Plenty, North Island, New Zealand *37°58′S 177°E*
184 O8 **Whakatane** ≈ North Island, New Zealand
9 O9 **Whale Cove** *var.* Tikirarjuaq. Nunavut, C Canada *62°14′N 92°10′W*
96 M2 **Whalsay** *island* NE Scotland, United Kingdom
184 L11 **Whangaehu** ≈ North Island, New Zealand
184 M6 **Whangamata** Waikato, North Island, New Zealand *37°13′S 175°54′E*
184 Q9 **Whangara** Gisborne, North Island, New Zealand *38°34′S 178°12′E*
184 K3 **Whangarei** Northland, North Island, New Zealand *35°44′S 174°18′E*
184 K3 **Whangaruru Harbour** *inlet* North Island, New Zealand
25 V12 **Wharton** Texas, SW USA *29°19′N 96°08′W*
173 U8 **Wharton Basin** *var.* West Australian Basin. *undersea feature* E Indian Ocean
185 E18 **Whataroa** West Coast, South Island, New Zealand *43°17′S 170°20′E*
8 K10 **Wha Ti** *prev.* Lac la Martre. Northwest Territories, W Canada *63°10′N 117°12′W*
8 J9 **Wha Ti** Northwest Territories, W Canada *63°10′N 117°12′W*
184 K6 **Whatipu** Auckland, North Island, New Zealand *37°17′S 174°44′E*
33 Y16 **Wheatland** Wyoming, C USA *42°03′N 104°57′W*
14 D18 **Wheatley** Ontario, S Canada *42°06′N 82°27′W*
30 M10 **Wheaton** Illinois, N USA *41°52′N 88°06′W*
29 R7 **Wheaton** Minnesota, N USA *45°48′N 96°30′W*
37 T4 **Wheat Ridge** Colorado, C USA *39°44′N 105°06′W*
25 P2 **Wheeler** Texas, SW USA *35°26′N 100°17′W*
23 O2 **Wheeler Lake** ⊠ Alabama, S USA
35 Y6 **Wheeler Peak** ▲ Nevada, W USA *39°00′N 114°17′W*
37 T9 **Wheeler Peak** ▲ New Mexico, SW USA *36°34′N 105°25′W*
31 S15 **Wheelersburg** Ohio, N USA *38°43′N 82°51′W*
21 R2 **Wheeling** West Virginia, NE USA *40°05′N 80°43′W*
97 L16 **Whernside** ▲ N England, United Kingdom *54°13′N 02°27′W*
182 F9 **Whidbey, Point** *headland* South Australia *34°36′S 135°08′E*
180 I7 **Whim Creek** Western Australia *20°51′S 117°54′E*
10 L17 **Whistler** British Columbia, SW Canada *50°07′N 122°57′W*
21 W8 **Whitakers** North Carolina, SE USA *36°06′N 77°43′W*
14 H15 **Whitby** Ontario, S Canada *43°52′N 78°56′W*
97 N15 **Whitby** N England, United Kingdom *54°29′N 00°37′W*
10 G6 **White** ≈ Yukon, W Canada
13 T11 **White Bay** *bay* Newfoundland, Newfoundland and Labrador, E Canada
20 I8 **White Bluff** Tennessee, S USA *36°06′N 87°13′W*
28 J6 **White Butte** ▲ North Dakota, N USA *46°23′N 103°18′W*
19 R5 **White Cap Mountain** ▲ Maine, NE USA *45°33′N 69°15′W*
22 J9 **White Castle** Louisiana, S USA *30°10′N 91°09′W*
182 M5 **White Cliffs** New South Wales, SE Australia *30°52′S 143°04′E*
31 P8 **White Cloud** Michigan, N USA *43°34′N 85°47′W*
11 P14 **Whitecourt** Alberta, SW Canada *54°10′N 115°38′W*
25 O2 **White Deer** Texas, SW USA *35°26′N 101°10′W*
White Elster *see* Weisse Elster
24 M5 **Whiteface** Texas, SW USA *33°36′N 102°36′W*
18 K7 **Whiteface Mountain** ▲ New York, NE USA *44°22′N 73°54′W*
29 W5 **Whiteface Reservoir** ⊠ Minnesota, N USA
33 O7 **Whitefish** Montana, NW USA *48°24′N 114°20′W*
31 N9 **Whitefish Bay** Wisconsin, N USA *43°09′N 87°54′W*
31 Q3 **Whitefish Bay** *lake bay* Canada/USA
14 E11 **Whitefish Falls** Ontario, S Canada *46°06′N 81°42′W*
14 B7 **Whitefish Lake** ◎ Ontario, S Canada
29 U6 **Whitefish Lake** ◎ Minnesota, C USA
31 Q3 **Whitefish Point** *headland* Michigan, N USA *46°46′N 84°57′W*
31 O4 **Whitefish River** ≈ Michigan, N USA
25 O4 **Whiteflat** Texas, SW USA *34°06′N 100°55′W*
27 V12 **White Hall** Arkansas, C USA *34°18′N 92°05′W*
30 K14 **White Hall** Illinois, N USA *39°26′N 90°24′W*
31 O8 **Whitehall** Michigan, N USA *43°24′N 86°21′W*
18 L9 **Whitehall** New York, NE USA *43°33′N 73°24′W*
31 S13 **Whitehall** Ohio, N USA *39°58′N 82°53′W*
30 J7 **Whitehall** Wisconsin, N USA *44°22′N 91°20′W*
97 J15 **Whitehaven** NW England, United Kingdom *54°33′N 03°35′W*
10 I8 **Whitehorse** *territory capital* Yukon, W Canada *60°41′N 135°08′W*
184 O7 **White Island** *island* NE New Zealand
14 K13 **White Lake** ◎ Ontario, SE Canada
22 H10 **White Lake** ◎ Louisiana, S USA
186 G7 **Whiteman Range** ▲▲ New Britain, E Papua New Guinea
183 Q15 **Whitemark** Tasmania, SE Australia *40°10′S 148°01′E*
35 S9 **White Mountains** ▲▲ California/Nevada, W USA
19 N7 **White Mountains** ▲▲ Maine/New Hampshire, NE USA
80 F11 **White Nile** ◆ *state* C Sudan
81 E14 **White Nile** *Ar.* Al Bahr al Abyaḑ, An Nīl al Abyaḑ, Bahr el Jebel. ≈ NE South Sudan
67 U7 **White Nile** *var.* Bahr el Jebel. ≈ S Sudan
25 W5 **White Oak Creek** ≈ Texas, SW USA
10 H9 **White Pass** *pass* Canada/USA
32 I9 **White Pass** *pass* Washington, NW USA
21 O9 **White Pine** Tennessee, S USA *36°06′N 83°17′W*
18 K14 **White Plains** New York, NE USA *41°01′N 73°45′W*
37 N13 **Whiteriver** Arizona, SW USA *33°50′N 109°57′W*
28 M11 **White River** South Dakota, N USA *43°34′N 100°45′W*
27 W12 **White River** ≈ Arkansas, SE USA
37 P3 **White River** ≈ Colorado/Utah, C USA
31 N15 **White River** ≈ Indiana, N USA
31 O8 **White River** ≈ Michigan, N USA
28 K11 **White River** ≈ South Dakota, N USA
25 O5 **White River** ≈ Texas, SW USA
18 M8 **White River** ≈ Vermont, NE USA
25 O5 **White River Lake** ⊠ Texas, SW USA
32 H11 **White Salmon** Washington, NW USA *45°43′N 121°29′W*
18 I10 **Whitesboro** New York, NE USA *43°07′N 75°17′W*
25 T5 **Whitesboro** Texas, SW USA *33°39′N 96°54′W*
21 O7 **Whitesburg** Kentucky, S USA *37°07′N 82°52′W*
White Sea *see* Beloye More
White Sea-Baltic Canal/ White Sea Canal *see* Belomorsko-Baltiyskiy Kanal
63 I25 **Whiteside, Canal** *channel* S Chile
33 S10 **White Sulphur Springs** Montana, NW USA *46°33′N 110°54′W*
21 R6 **White Sulphur Springs** West Virginia, NE USA *37°48′N 80°18′W*
20 J6 **Whitesville** Kentucky, S USA *37°40′N 86°48′W*
32 I10 **White Swan** Washington, NW USA *46°22′N 120°46′W*
21 U12 **Whiteville** North Carolina, SE USA *34°20′N 78°42′W*
20 F10 **Whiteville** Tennessee, S USA *35°19′N 89°09′W*
77 Q13 **White Volta** *var.* Nakambé, *Fr.* Volta Blanche. ≈ Burkina Faso/Ghana
30 M9 **Whitewater** Wisconsin, N USA *42°51′N 88°43′W*
37 P14 **Whitewater Baldy** ▲ New Mexico, SW USA *33°19′N 108°38′W*
23 X17 **Whitewater Bay** *bay* Florida, SE USA
31 Q14 **Whitewater River** ≈ Indiana/Ohio, N USA
11 V16 **Whitewood** Saskatchewan, S Canada *50°19′N 102°16′W*
28 J9 **Whitewood** South Dakota, N USA *44°27′N 103°38′W*
25 U5 **Whitewright** Texas, SW USA *33°30′N 96°23′W*
97 I15 **Whithorn** S Scotland, United Kingdom *54°44′N 04°26′W*
184 M6 **Whitianga** Waikato, North Island, New Zealand *36°50′S 175°42′E*
19 N11 **Whitinsville** Massachusetts, NE USA *42°06′N 71°40′W*
20 M8 **Whitley City** Kentucky, S USA *36°45′N 84°29′W*
21 Q11 **Whitmire** South Carolina, SE USA *34°30′N 81°36′W*
31 R10 **Whitmore Lake** Michigan, N USA *42°26′N 83°44′W*
195 N9 **Whitmore Mountains** ▲▲ Antarctica
14 I12 **Whitney** Ontario, SE Canada *45°29′N 78°11′W*
25 T8 **Whitney** Texas, SW USA *31°56′N 97°20′W*
25 S8 **Whitney, Lake** ⊠ Texas, SW USA
35 S11 **Whitney, Mount** ▲ California, W USA *36°35′N 118°17′W*
181 Y6 **Whitsunday Group** *island group* Queensland, E Australia
25 S6 **Whitt** Texas, SW USA *32°55′N 98°01′W*
29 U12 **Whittemore** Iowa, C USA *43°03′N 94°25′W*
39 R12 **Whittier** Alaska, USA *60°46′N 148°40′W*
35 T15 **Whittier** California, W USA *33°58′N 118°01′W*
83 I25 **Whittlesea** Eastern Cape, S South Africa *32°08′S 26°51′E*
20 K10 **Whitwell** Tennessee, S USA *35°12′N 85°31′W*
8 L10 **Wholdaia Lake** ◎ Northwest Territories, NW Canada
182 H7 **Whyalla** South Australia *33°04′S 137°34′E*
Whydah *see* Ouidah
14 F13 **Wiarton** Ontario, S Canada *44°44′N 81°10′W*
171 O13 **Wiau** Sulawesi, C Indonesia *03°08′S 121°22′E*
111 H15 **Wiązów** *Ger.* Wansen. Dolnośląskie, SW Poland *50°49′N 17°13′E*
33 Y8 **Wibaux** Montana, NW USA *46°57′N 104°11′W*
27 N6 **Wichita** Kansas, C USA *37°42′N 97°20′W*
25 R5 **Wichita Falls** Texas, SW USA *33°55′N 98°30′W*
26 L11 **Wichita Mountains** ▲▲ Oklahoma, C USA
25 R5 **Wichita River** ≈ Texas, SW USA
96 K6 **Wick** N Scotland, United Kingdom *58°26′N 03°06′W*
36 K13 **Wickenburg** Arizona, SW USA *33°57′N 112°41′W*
24 L8 **Wickett** Texas, SW USA *31°34′N 103°00′W*
180 I7 **Wickham** Western Australia *20°40′S 117°11′E*
182 M14 **Wickham, Cape** *headland* Tasmania, SE Australia *39°36′S 143°55′E*
20 G7 **Wickliffe** Kentucky, S USA *36°58′N 89°04′W*
97 G19 **Wicklow** *Ir.* Cill Mhantáin. E Ireland *52°59′N 06°03′W*
97 F19 **Wicklow** *Ir.* Cill Mhantáin. *cultural region* E Ireland
97 G19 **Wicklow Head** *Ir.* Ceann Chill Mhantáin. *headland* E Ireland *52°57′N 06°00′W*
97 F18 **Wicklow Mountains** *Ir.* Sléibhte Chill Mhantáin. ▲▲ E Ireland
14 H10 **Wicksteed Lake** ◎ Ontario, S Canada
Wida *see* Ouidah
65 G15 **Wideawake Airfield** ✈ (Georgetown) SW Ascension Island
97 K18 **Widnes** NW England, United Kingdom *53°22′N 02°44′W*
110 H9 **Więcbork** *Ger.* Vandsburg. Kujawsko-pomorskie, C Poland *53°21′N 17°31′E*
101 E17 **Wied** ≈ W Germany
101 F16 **Wiehl** Nordrhein-Westfalen, W Germany *50°57′N 07°33′E*
111 L17 **Wieliczka** Małopolskie, S Poland *50°N 20°02′E*
110 H12 **Wielkopolskie** ◆ *province* SW Poland
111 J14 **Wieluń** Sieradz, C Poland *51°14′N 18°33′E*
109 X4 **Wien** *Eng.* Vienna, *Hung.* Bécs, *Slvk.* Viedeň, *Slvn.* Dunaj; *anc.* Vindobona. ● (Austria) Wien, NE Austria *48°13′N 16°22′E*
109 X4 **Wien** *off.* Land Wien, *Eng.* Vienna. ◆ *state* NE Austria
109 X5 **Wiener Neustadt** Niederösterreich, E Austria *47°49′N 16°08′E*
Wien, Land *see* Wien
110 G7 **Wieprza** *Ger.* Wipper. ≈ NW Poland
98 O10 **Wierden** Overijssel, E Netherlands *52°22′N 06°35′E*
98 I7 **Wieringerwerf** Noord-Holland, NW Netherlands *52°51′N 05°01′E*
Wieruschow *see* Wieruszów
111 I14 **Wieruszów** *Ger.* Wieruschow. Łódzkie, C Poland *51°18′N 18°09′E*
109 V9 **Wies** Steiermark, SE Austria *46°40′N 15°16′E*
Wiesbachhorn *see* Grosses Wiesbachhorn
101 G18 **Wiesbaden** Hessen, W Germany *50°06′N 08°14′E*
Wieselburg and Ungarisch-Altenburg/Wieselburg-Ungarisch-Altenburg *see* Mosonmagyaróvár
Wiesenhof *see* Ostrołęka
101 G20 **Wiesloch** Baden-Württemberg, SW Germany *49°18′N 08°42′E*
100 F10 **Wiesmoor** Niedersachsen, NW Germany *53°22′N 07°46′E*
110 I7 **Wieżyca** *Ger.* Turmberg. *hill* Pomorskie, N Poland
97 L17 **Wigan** NW England, United Kingdom *53°33′N 02°38′W*
37 U3 **Wiggins** Colorado, C USA *40°11′N 104°03′W*
22 M8 **Wiggins** Mississippi, S USA *30°50′N 89°09′W*
Wigorna Ceaster *see* Worcester
97 I14 **Wigtown** S Scotland, United Kingdom *54°53′N 04°27′W*
97 H14 **Wigtown** *cultural region* SW Scotland, United Kingdom
97 I15 **Wigtown Bay** *bay* SW Scotland, United Kingdom
98 L13 **Wijchen** Gelderland, SE Netherlands *51°48′N 05°44′E*
92 N1 **Wijdefjorden** *fjord* NW Svalbard
98 M10 **Wijhe** Overijssel, E Netherlands *52°22′N 06°07′E*
98 J12 **Wijk bij Duurstede** Utrecht, C Netherlands *51°58′N 05°21′E*
98 J13 **Wijk en Aalburg** Noord-Brabant, S Netherlands *51°46′N 05°06′E*
99 H16 **Wijnegem** Antwerpen, N Belgium *51°13′N 04°32′E*
14 E11 **Wikwemikong** Manitoulin Island, Ontario, S Canada *45°46′N 81°43′W*
108 H7 **Wil** Sankt Gallen, NE Switzerland *47°28′N 09°03′E*
29 R16 **Wilber** Nebraska, C USA *40°28′N 96°57′W*
32 K8 **Wilbur** Washington, NW USA *47°45′N 118°42′W*
27 Q11 **Wilburton** Oklahoma, C USA *34°57′N 95°20′W*
182 M6 **Wilcannia** New South Wales, SE Australia *31°34′S 143°23′E*
18 D12 **Wilcox** Pennsylvania, NE USA *41°34′N 78°40′W*
Wilczek Land *see* Vil'cheka, Zemlya
109 U6 **Wildalpen** Steiermark, E Austria *47°40′N 14°54′E*
31 O13 **Wildcat Creek** ≈ Indiana, N USA
108 L9 **Wilde Kreuzspitze** *It.* Picco di Croce. ▲ Austria/Italy *46°53′N 10°51′E*
Wildenschwert *see* Ústí nad Orlicí
98 O6 **Wildervank** Groningen, NE Netherlands *53°04′N 06°52′E*
100 G11 **Wildeshausen** Niedersachsen, NW Germany *52°54′N 08°26′E*
108 D10 **Wildhorn** ▲ SW Switzerland *46°21′N 07°22′E*
11 R17 **Wild Horse** Alberta, SW Canada *49°00′N 110°19′W*
27 N12 **Wildhorse Creek** ≈ Oklahoma, C USA
28 L14 **Wild Horse Hill** ▲ Nebraska, C USA *41°52′N 101°56′W*
109 W8 **Wildon** Steiermark, SE Austria *46°53′N 15°29′E*
24 M2 **Wildorado** Texas, SW USA *35°12′N 102°10′W*
29 R6 **Wild Rice River** ≈ Minnesota/North Dakota, N USA
Wilejka *see* Vilyeyka
195 Y9 **Wilhelm II Coast** *physical region* Antarctica
195 X9 **Wilhelm II Land** *physical region* Antarctica
55 U11 **Wilhelmina Gebergte** ▲▲ C Suriname
18 B13 **Wilhelm, Lake** ⊠ Pennsylvania, NE USA
92 O2 **Wilhelmøya** *island* C Svalbard
Wilhelm-Pieck-Stadt *see* Guben
109 W4 **Wilhelmsburg** Niederösterreich, E Austria *48°07′N 15°37′E*
100 G10 **Wilhelmshaven** Niedersachsen, NW Germany *53°31′N 08°07′E*
Wilia/Wilja *see* Neris
18 H13 **Wilkes Barre** Pennsylvania, NE USA *41°15′N 75°50′W*
21 R9 **Wilkesboro** North Carolina, SE USA *36°08′N 81°09′W*
195 W15 **Wilkes Coast** *physical region* Antarctica
189 W12 **Wilkes Island** *island* N Wake Island
195 X12 **Wilkes Land** *physical region* Antarctica
11 S15 **Wilkie** Saskatchewan, S Canada *52°27′N 108°42′W*
194 I6 **Wilkins Ice Shelf** *ice shelf* Antarctica
182 D4 **Wilkinsons Lakes** *salt lake* South Australia
Wiłkomierz *see* Ukmergė
182 K11 **Willalooka** South Australia *36°24′S 140°20′E*
32 G11 **Willamette River** ≈ Oregon, NW USA
183 O8 **Willandra Billabong Creek** *seasonal river* New South Wales, SE Australia
32 F9 **Willapa Bay** *inlet* Washington, NW USA
27 T7 **Willard** Missouri, C USA *37°18′N 93°25′W*
37 S12 **Willard** New Mexico, SW USA *34°36′N 106°01′W*
31 S12 **Willard** Ohio, N USA *41°03′N 82°43′W*
36 L1 **Willard** Utah, W USA *41°23′N 112°01′W*
186 G6 **Willaumez Peninsula** *headland* New Britain, E Papua New Guinea *05°03′S 150°04′E*
37 N15 **Willcox** Arizona, SW USA *32°13′N 109°49′W*
37 N16 **Willcox Playa** *salt flat* Arizona, SW USA
99 G17 **Willebroek** Antwerpen, C Belgium *51°04′N 04°22′E*
45 P16 **Willemstad** ○ Curaçao *12°07′N 68°54′W*
99 G14 **Willemstad** Noord-Brabant, S Netherlands *51°40′N 04°27′E*
11 S11 **William** ≈ Saskatchewan, C Canada
23 O6 **William "Bill" Dannelly Reservoir** ⊠ Alabama, S USA
182 G3 **William Creek** South Australia *28°55′S 136°23′E*
181 T15 **William, Mount** ▲ South Australia *37°20′S 142°41′E*
36 K11 **Williams** Arizona, SW USA *35°15′N 112°11′W*
29 X14 **Williamsburg** Iowa, C USA *41°39′N 92°00′W*
20 M8 **Williamsburg** Kentucky, S USA *36°44′N 84°10′W*
31 R15 **Williamsburg** Ohio, N USA *39°00′N 84°02′W*
21 X6 **Williamsburg** Virginia, NE USA *37°17′N 76°43′W*
10 M15 **Williams Lake** British Columbia, SW Canada *52°08′N 122°09′W*
21 P6 **Williamson** West Virginia, NE USA *37°42′N 82°16′W*
31 N13 **Williamsport** Indiana, N USA *40°18′N 87°18′W*
18 G13 **Williamsport** Pennsylvania, NE USA *41°16′N 77°03′W*
21 W9 **Williamston** North Carolina, SE USA *35°55′N 77°05′W*
21 P11 **Williamston** South Carolina, SE USA *34°37′N 82°28′W*
20 M4 **Williamstown** Kentucky, S USA *38°39′N 84°32′W*
18 L10 **Williamstown** Massachusetts, NE USA *42°41′N 73°11′W*
18 J16 **Willingboro** New Jersey, NE USA *40°01′N 74°52′W*
11 Q14 **Willingdon** Alberta, SW Canada *53°49′N 112°08′W*
25 W10 **Willis** Texas, SW USA *30°25′N 95°28′W*
108 F8 **Willisau** Luzern, W Switzerland *47°07′N 08°00′E*
83 F24 **Williston** Northern Cape, W South Africa *31°20′S 20°52′E*
23 V10 **Williston** Florida, SE USA *29°23′N 82°27′W*
28 J3 **Williston** North Dakota, N USA *48°07′N 103°37′W*
21 Q13 **Williston** South Carolina, SE USA *33°24′N 81°25′W*
10 L12 **Williston Lake** ⊠ British Columbia, W Canada
34 L5 **Willits** California, W USA *39°24′N 123°22′W*
29 T8 **Willmar** Minnesota, N USA *45°07′N 95°02′W*
10 K11 **Will, Mount** ▲ British Columbia, W Canada *57°31′N 128°48′W*
31 T11 **Willoughby** Ohio, N USA *41°38′N 81°24′W*
11 U17 **Willow Bunch** Saskatchewan, S Canada *49°30′N 105°41′W*
32 J11 **Willow Creek** ≈ Oregon, NW USA
39 R11 **Willow Lake** Alaska, USA *61°44′N 150°02′W*
8 I9 **Willowlake** ≈ Northwest Territories, NW Canada
83 H25 **Willowmore** Eastern Cape, S South Africa *33°18′S 23°30′E*
30 L5 **Willow Reservoir** ⊠ Wisconsin, N USA
35 N5 **Willows** California, W USA *39°28′N 122°12′W*
27 V7 **Willow Springs** Missouri, C USA *36°59′N 91°58′W*
182 I7 **Wilmington** South Australia *32°42′S 138°08′E*
21 Y2 **Wilmington** Delaware, NE USA *39°45′N 75°33′W*
21 V12 **Wilmington** North Carolina, SE USA *34°14′N 77°55′W*
31 R14 **Wilmington** Ohio, N USA *39°27′N 83°49′W*
20 M6 **Wilmore** Kentucky, S USA *37°51′N 84°39′W*
29 R8 **Wilmot** South Dakota, N USA *45°24′N 96°51′W*
Wilna/Wilno *see* Vilnius
101 G16 **Wilnsdorf** Nordrhein-Westfalen, W Germany *50°49′N 08°06′E*
99 G16 **Wilrijk** Antwerpen, N Belgium *51°11′N 04°25′E*
100 I10 **Wilseder Berg** *hill* NW Germany
67 Z12 **Wilshaw Ridge** *undersea feature* W Indian Ocean *17°30′S 56°30′E*
21 V9 **Wilson** North Carolina, SE USA *35°43′N 77°56′W*
25 N5 **Wilson** Texas, SW USA *33°21′N 101°44′W*
182 A7 **Wilson Bluff** *headland* South Australia/Western Australia *31°41′S 129°01′E*
35 Y7 **Wilson Creek Range** ▲▲ Nevada, W USA
23 O1 **Wilson Lake** ⊠ Alabama, S USA
26 M4 **Wilson Lake** ⊠ Kansas, SE USA
37 P7 **Wilson, Mount** ▲ Colorado, C USA *37°50′N 107°59′W*
183 P13 **Wilsons Promontory** *peninsula* Victoria, SE Australia
29 Y14 **Wilton** Iowa, C USA *41°35′N 91°01′W*
19 P7 **Wilton** Maine, NE USA *44°35′N 70°15′W*
28 M5 **Wilton** North Dakota, N USA *47°09′N 100°46′W*
97 L22 **Wiltshire** *cultural region* S England, United Kingdom
99 M23 **Wiltz** Diekirch, NW Luxembourg *49°58′N 05°56′E*
180 K9 **Wiluna** Western Australia *26°34′S 120°14′E*
99 M23 **Wilwerwiltz** Diekirch, NE Luxembourg *49°59′N 06°00′E*
29 P5 **Wimbledon** North Dakota, N USA *47°08′N 98°25′W*
42 K7 **Wina** *var.* Güina. Jinotega, N Nicaragua *14°00′N 85°14′W*
31 O12 **Winamac** Indiana, N USA *41°03′N 86°37′W*
81 G19 **Winam Gulf** *var.* Kavirondo Gulf. *gulf* SW Kenya
83 I22 **Winburg** Free State, C South Africa *28°31′S 27°01′E*
19 N10 **Winchendon** Massachusetts, NE USA *42°41′N 72°01′W*
14 M13 **Winchester** Ontario, SE Canada *45°07′N 75°19′W*
97 M23 **Winchester** *hist.* Wintanceaster, *Lat.* Venta Belgarum. S England, United Kingdom *51°04′N 01°19′W*
32 M10 **Winchester** Idaho, NW USA *46°13′N 116°35′W*
30 J14 **Winchester** Illinois, N USA *39°38′N 90°28′W*
31 Q13 **Winchester** Indiana, N USA *40°11′N 84°57′W*
20 M5 **Winchester** Kentucky, S USA *38°00′N 84°10′W*
18 M10 **Winchester** New Hampshire, NE USA *42°46′N 72°21′W*
20 K10 **Winchester** Tennessee, S USA *35°11′N 86°06′W*
21 V3 **Winchester** Virginia, NE USA *39°11′N 78°12′W*
99 L22 **Wincrange** Diekirch, NW Luxembourg *50°03′N 05°55′E*
10 I5 **Wind** ≈ Yukon, NW Canada
183 S8 **Windamere, Lake** ◎ New South Wales, SE Australia
Windau *see* Ventspils, Latvia
Windau *see* Venta, Latvia/Lithuania
18 D15 **Windber** Pennsylvania, NE USA *40°12′N 78°47′W*
23 T3 **Winder** Georgia, SE USA *33°59′N 83°43′W*
97 K15 **Windermere** NW England, United Kingdom *54°24′N 02°54′W*
14 C7 **Windermere Lake** ◎ Ontario, S Canada
31 U11 **Windham** Ohio, N USA *41°14′N 81°03′W*
83 D19 **Windhoek** *Ger.* Windhuk. ● (Namibia) Khomas, C Namibia *22°34′S 17°06′E*
83 D20 **Windhoek** ✈ Khomas, C Namibia *22°31′S 17°04′E*
Windhuk *see* Windhoek
15 O8 **Windigo** Québec, SE Canada *47°45′N 73°19′W*
15 O8 **Windigo** ≈ Québec, SE Canada
Windischfeistritz *see* Slovenska Bistrica
109 T6 **Windischgarsten** Oberösterreich, W Austria *47°42′N 14°21′E*
Windischgraz *see* Slovenj Gradec
37 T16 **Wind Mountain** ▲ New Mexico, SW USA *32°01′N 105°35′W*
29 T10 **Windom** Minnesota, N USA *43°52′N 95°07′W*
37 Q7 **Windom Peak** ▲ Colorado, C USA *37°37′N 107°35′W*
181 U9 **Windorah** Queensland, C Australia *25°25′S 142°41′E*
37 O10 **Window Rock** Arizona, SW USA *35°40′N 109°03′W*
31 N9 **Wind Point** *headland* Wisconsin, N USA *42°46′N 87°46′W*
33 U14 **Wind River** ≈ Wyoming, C USA
13 P15 **Windsor** Nova Scotia, SE Canada *45°00′N 64°09′W*
14 C17 **Windsor** Ontario, S Canada *42°18′N 83°W*
15 Q12 **Windsor** Québec, SE Canada *45°34′N 72°00′W*
97 N22 **Windsor** S England, United Kingdom *51°29′N 00°39′W*
37 T3 **Windsor** Colorado, C USA *40°28′N 104°54′W*
18 M12 **Windsor** Connecticut, NE USA *41°51′N 72°38′W*
27 T5 **Windsor** Missouri, C USA *38°31′N 93°31′W*
21 X9 **Windsor** North Carolina, SE USA *36°00′N 76°57′W*
18 M12 **Windsor Locks** Connecticut, NE USA *41°55′N 72°37′W*
25 R5 **Windthorst** Texas, SW USA *33°34′N 98°26′W*
45 Z14 **Windward Islands** *island group* E West Indies
Windward Islands *see* Barlavento, Ilhas de, Cape Verde
Windward Islands *see* Vent, Îles du, Archipel de la Société, French Polynesia
44 K8 **Windward Passage** *Sp.* Paso de los Vientos. *channel* Cuba/Haiti
55 T9 **Wineperu** C Guyana *06°10′N 58°34′W*
23 O3 **Winfield** Alabama, S USA *33°55′N 87°49′W*
29 Y15 **Winfield** Iowa, C USA *41°07′N 91°26′W*
27 O7 **Winfield** Kansas, C USA *37°14′N 97°00′W*
25 W6 **Winfield** Texas, SW USA *33°10′N 95°06′W*
21 Q4 **Winfield** West Virginia, NE USA *38°30′N 81°54′W*
29 N5 **Wing** North Dakota, N USA *47°06′N 100°16′W*
183 U7 **Wingham** New South Wales, SE Australia *31°52′S 152°24′E*
12 G16 **Wingham** Ontario, S Canada *43°54′N 81°19′W*
33 T8 **Winifred** Montana, NW USA *47°33′N 109°26′W*
12 E9 **Winisk Lake** ◎ Ontario, C Canada
24 L8 **Wink** Texas, SW USA *31°45′N 103°09′W*
36 M14 **Winkelman** Arizona, SW USA *32°59′N 110°46′W*
11 X17 **Winkler** Manitoba, S Canada *49°12′N 97°55′W*
109 Q9 **Winklern** Tirol, W Austria *46°54′N 12°54′E*
Winkowitz *see* Vinkovci
32 G9 **Winlock** Washington, NW USA *46°29′N 122°56′W*
77 P17 **Winneba** SE Ghana *05°22′N 00°38′W*
29 U11 **Winnebago** Minnesota, N USA *43°46′N 94°10′W*
29 R13 **Winnebago** Nebraska, C USA *42°14′N 96°28′W*
30 M7 **Winnebago, Lake** ◎ Wisconsin, N USA
30 M7 **Winneconne** Wisconsin, N USA *44°07′N 88°44′W*
35 T3 **Winnemucca** Nevada, W USA *40°59′N 117°44′W*
35 R4 **Winnemucca Lake** ◎ Nevada, W USA
101 H21 **Winnenden** Baden-Württemberg, SW Germany *48°52′N 09°22′E*
29 N11 **Winner** South Dakota, N USA *43°22′N 99°51′W*
33 U9 **Winnett** Montana, NW USA *47°00′N 108°18′W*
14 I9 **Winneway** Québec, SE Canada *47°35′N 78°33′W*
22 H6 **Winnfield** Louisiana, S USA *31°55′N 92°38′W*
29 U4 **Winnibigoshish, Lake** ◎ Minnesota, N USA
25 X11 **Winnie** Texas, SW USA *29°49′N 94°22′W*
11 Y16 **Winnipeg** *province capital* Manitoba, S Canada *49°53′N 97°10′W*
11 X16 **Winnipeg** ✈ Manitoba, S Canada *49°56′N 97°16′W*
0 J8 **Winnipeg** ≈ Manitoba, S Canada
11 X16 **Winnipeg Beach** Manitoba, S Canada *50°25′N 96°59′W*
11 W14 **Winnipeg, Lake** ◎ Manitoba, C Canada
11 W15 **Winnipegosis** Manitoba, S Canada *51°36′N 99°59′W*

◆ Country
● Country Capital
◇ Dependent Territory
○ Dependent Territory Capital
◆ Administrative Regions
✈ International Airport
▲ Mountain
▲▲ Mountain Range
Volcano
≈ River
◎ Lake
⊠ Reservoir

◆ Country
◉ Country Capital
◇ Dependent Territory
○ Dependent Territory Capital
✕ Administrative Regions
✕ International Airport
▲ Mountain
▲ Mountain Range
℞ Volcano
♒ River
◉ Lake
▣ Reservoir

163 Q10 **Xin Hot** Nei Mongol Zizhiqu, N China 43°58′N 114°59′E
Xinhua see Funing
163 T12 **Xinhui** var. Aohan Qi. Nei Mongol Zizhiqu, N China 42°12′N 119°57′E
159 T10 **Xining** var. Hsining, Hsi-ning, Sining. province capital Qinghai, C China 36°37′N 101°46′E
161 O4 **Xinji** prev. Shulu. Hebei, E China 37°55′N 115°14′E
161 P10 **Xinjian** Jiangxi, S China 28°37′N 115°46′E
Xinjiang see Xinjiang Uygur Zizhiqu
162 D8 **Xinjiang Uygur Zizhiqu** var. Sinkiang, Sinkiang Uighur Autonomous Region, Xin, Xinjiang. ◆ autonomous region NW China
160 H9 **Xinjin** var. Meixing, Tib. Zainlha. Sichuan, C China 30°27′N 103°46′E
Xinjin see Pulandian
Xinjing see Jingxi
163 U12 **Xinmin** Liaoning, NE China 41°58′N 122°51′E
160 M12 **Xinning** var. Jinshi. Hunan, S China 26°34′N 110°57′E
Xinning see Ningxian
Xinning see Fusui
Xinpu see Lianyungang
Xinshan see Anyuan
161 P5 **Xintai** Shandong, E China 35°54′N 117°44′E
Xinwen see Suncun
Xin Xian see Xinzhou
161 N6 **Xinxiang** Henan, C China 35°13′N 113°48′E
161 O8 **Xinyang** var. Hsin-yang, Sinyang. Henan, C China 32°09′N 114°04′E
161 Q6 **Xinyi** var. Xin'anzhen. Jiangsu, E China 34°17′N 118°14′E
161 Q6 **Xinyi He** ≈ E China
161 S14 **Xinying** var. Sinying, Jap. Shinei; prev. Hsinying. C Taiwan 23°12′N 120°15′E
161 O11 **Xinyu** Jiangxi, S China 27°51′N 115°00′E
158 I5 **Xinyuan** var. Künes. Xinjiang Uygur Zizhiqu, NW China 43°25′N 83°12′E
Xinyuan see Tianjun
162 M13 **Xinzhao Shan** ▲ N China 39°37′N 107°51′E
161 N3 **Xinzhou** var. Xin Xian. Shanxi, C China 38°24′N 112°43′E
Xinzhou see Longlin
161 S13 **Xinzhu** var. Hsinchu. N Taiwan 24°48′N 120°59′E
104 H4 **Xinzo de Limia** Galicia, NW Spain 42°05′N 07°45′W
Xions see Książ Wielkopolski
161 O7 **Xiping** Henan, C China 33°22′N 114°00′E
Xiping see Songyang
159 T11 **Xiqing Shan** ▲▲ C China
59 N16 **Xique-Xique** Bahia, E Brazil 10°47′S 42°44′W
Xireg see Ulan
115 E14 **Xirovoúni** ▲ N Greece 40°31′N 21°58′E
162 M13 **Xishanzui** prev. Urad Qianqi. Nei Mongol Zizhiqu, N China 40°43′N 108°41′E
Xisha Qundao see Paracel Islands
160 J11 **Xishui** var. Donghuang. Guizhou, S China 28°24′N 106°09′E
Xi Ujimqin Qi see Bayan Ul
160 K11 **Xiushan** var. Zhonghe. Chongqing Shi, C China 28°23′N 108°52′E
Xiushan see Tonghai
161 O10 **Xiu Shui** ≈ S China
Xiuyan see Qingjian
146 H9 **Xiva** Rus. Khiva, Khiwa. Xorazm Viloyati, W Uzbekistan 41°22′N 60°22′E
158 J16 **Xixabangma Feng** ▲ W China 28°25′N 85°47′E
160 M7 **Xixia** Henan, C China 33°30′N 111°25′E
Xixón see Gijón
Xixona see Jijona
Xizang see Xizang Zizhiqu
Xizang Gaoyuan see Qingzang Gaoyuan
160 E9 **Xizang Zizhiqu** var. Thibet, Tibetan Autonomous Region, Xizang, Eng. Tibet. ◆ autonomous region W China
163 U14 **Xizhong Dao** island N China
Xoi see Qüxü
146 H8 **Xo'jayli** Rus. Khodzheyli. Qoraqalpog'iston Respublikasi, W Uzbekistan 42°23′N 59°27′E
Xolotlán see Managua, Lago de
147 I9 **Xonqa** var. Khonqa, Rus. Khanka. Xorazm Viloyati, W Uzbekistan 41°31′N 60°39′E
146 H9 **Xorazm Viloyati** Rus. Khorezmskaya Oblast'. ◆ province W Uzbekistan
159 N9 **Xorkol** Xinjiang Uygur Zizhiqu, NW China 38°45′N 91°07′E
147 P11 **Xovos** var. Ursat'yevskaya, Rus. Khavast. Sirdaryo Viloyati, E Uzbekistan 40°14′N 68°46′E
41 X14 **Xpujil** Quintana Roo, E Mexico 18°30′N 89°24′W
161 Q8 **Xuancheng** var. Xuanzhou. Anhui, E China 30°57′N 118°53′E
Xuande Qundao see Amphitrite Group
167 T9 **Xuân Đuc** Quang Binh, C Vietnam 17°19′N 106°38′E
160 L9 **Xuan'en** var. Zhushan. Hubei, C China 30°03′N 109°26′E
160 K8 **Xuanhan** Sichuan, C China 31°25′N 107°41′E
161 O2 **Xuanhua** Hebei, E China 40°36′N 115°01′E
161 P4 **Xuanhui He** ≈ E China
167 T8 **Xuân Sơn** Quang Binh, C Vietnam 17°42′N 105°58′E
H12 **Xuanwei** Yunnan, China 26°08′N 104°04′E
Xuanzhou see Xuancheng
161 N7 **Xuchang** Henan, C China 34°03′N 113°48′E
Xucheng see Xuwen
137 X10 **Xudat** Rus. Khudat. NE Azerbaijan 41°37′N 48°39′E
81 M16 **Xuddur** var. Hudur, It. Oddur. Bakool, SW Somalia 04°07′N 43°47′E
80 O13 **Xudun** Sool, N Somalia 09°12′N 47°34′E
160 L11 **Xuefeng Shan** ▲▲ S China
161 S13 **Xue Shan** prev. Hsüeh Shan. ▲ N Taiwan
147 O13 **Xufar** Surkhondaryo Viloyati, S Uzbekistan 38°31′N 67°45′E
Xulun Hobot Qagan see Qagan Nur
42 F2 **Xunantunich** ruins Cayo, W Belize
163 W6 **Xun He** ≈ NE China
160 L7 **Xun He** ≈ C China
160 L14 **Xun Jiang** ≈ S China
163 W5 **Xunke** var. Bianjiang; prev. Qike. Heilongjiang, NE China 49°35′N 128°27′E
161 P13 **Xunwu** var. Changning. Jiangxi, S China 24°59′N 115°33′E
139 V4 **Xurmal** Ar. Khūrmāl, var. Khormal. As Sulaymānīyah, NE Iraq 35°19′N 46°06′E
161 O3 **Xushui** Hebei, E China 39°01′N 115°38′E
160 L16 **Xuwen** var. Xucheng. Guangdong, S China 20°21′N 110°09′E
160 I11 **Xuyong** var. Yongning. Sichuan, C China 28°17′N 105°21′E
161 P6 **Xuzhou** var. Hsu-chou, Suchow, Tongshan; prev. T'ung-shan. Jiangsu, E China 34°17′N 117°09′E
114 K13 **Xylaganí** var. Xilaganí. Anatolikí Makedonía kai Thráki, NE Greece 40°58′N 25°27′E
115 F19 **Xylókastro** var. Xilokastro. Pelopónnisos, S Greece 38°04′N 22°36′E

Y

160 H9 **Ya'an** var. Yaan. Sichuan, C China 30°N 102°57′E
182 L10 **Yaapeet** Victoria, SE Australia 35°48′S 142°03′E
79 D15 **Yabassi** Littoral, W Cameroon 04°30′N 09°59′E
81 J15 **Yabēlo** Oromīya, C Ethiopia 04°53′N 38°01′E
114 H9 **Yablanitsa** Lovech, N Bulgaria 43°01′N 24°06′E
43 N7 **Yablis** Región Autónoma Atlántico Norte, NE Nicaragua 14°08′N 83°44′W
123 O14 **Yablonovyy Khrebet** ▲▲ S Russian Federation
162 J14 **Yabrai Shan** ▲▲ NE China
45 U6 **Yabucoa** E Puerto Rico 18°02′N 65°53′W
160 J11 **Yachi He** ≈ S China
32 H10 **Yacolt** Washington, NW USA 45°49′N 122°22′W
54 M10 **Yacuaray** Amazonas, S Venezuela 04°12′N 66°30′W
57 M22 **Yacuiba** Tarija, S Bolivia 22°00′S 63°43′W
57 K16 **Yacuma, Río** ≈ C Bolivia
155 H16 **Yādgīr** Karnātaka, C India 16°46′N 77°09′E
21 R8 **Yadkin River** ≈ North Carolina, SE USA
21 R9 **Yadkinville** North Carolina, SE USA 36°07′N 80°40′W
127 P3 **Yadrin** Chuvashskaya Respublika, W Russian Federation 55°55′N 46°10′E
Yaegama-shotō see Yaeyama-shotō
Yaeme-saki see Paimi-saki
165 O16 **Yaeyama-shotō** var. Yaegama-shotō. island group SW Japan
75 O8 **Yafran** NW Libya 32°04′N 12°31′E
165 S2 **Yagashiri-tō** island NE Japan
65 H21 **Yaghan Basin** undersea feature SE Pacific Ocean
123 S9 **Yagodnoye** Magadanskaya Oblast', E Russian Federation 62°37′N 149°18′E
Yagotin see Yahotyn
78 G12 **Yagoua** Extrême-Nord, NE Cameroon 10°23′N 15°13′E
159 Q11 **Yagradagzê Shan** ▲ C China 35°06′N 95°41′E
Yaguachi see Yaguachi Nuevo
56 B7 **Yaguachi Nuevo** var. Yaguachi. Guayas, W Ecuador 02°06′S 79°43′W
Yaguarón, Río see Jaguarão, Rio
117 Q11 **Yahorlyts'kyy Lyman** bay S Ukraine
117 Q5 **Yahotyn** Rus. Yagotin. Kyyivs'ka Oblast', N Ukraine 50°15′N 31°48′E
40 L12 **Yahualica** Jalisco, SW Mexico 21°11′N 102°29′W
79 L17 **Yahuma** Orientale, N Dem. Rep. Congo 01°12′N 23°00′E
136 K15 **Yahyalı** Kayseri, C Turkey 38°08′N 35°23′E
167 N15 **Yai, Khao** ▲ SW Thailand 08°45′N 99°32′E
164 M14 **Yaizu** Shizuoka, Honshū, S Japan 34°52′N 138°20′E
160 G9 **Yajiang** var. Hekou, Tib. Nyagquka. Sichuan, C China 30°05′N 100°57′E
119 O14 **Yakawlyevichi** Rus. Yakovlevichi. Vitsyebskaya Voblasts', NE Belarus 54°20′N 30°31′E
163 S6 **Yakeshi** Nei Mongol Zizhiqu, N China 49°16′N 120°42′E
32 I9 **Yakima** Washington, NW USA 46°36′N 120°30′W
32 J10 **Yakima River** ≈ Washington, NW USA
Yakkabag see Yakkabog'
147 N12 **Yakkabog'** Rus. Yakkabag. Qashqadaryo Viloyati, S Uzbekistan 38°57′N 66°35′E
148 L12 **Yakmach** Baluchistān, SW Pakistan 28°48′N 63°48′E
77 O12 **Yako** W Burkina Faso 12°59′N 02°15′W
39 W13 **Yakobi Island** island Alexander Archipelago, Alaska, USA
79 K16 **Yakoma** Equateur, N Dem. Rep. Congo 04°04′N 22°23′E
114 H11 **Yakoruda** Blagoevgrad, SW Bulgaria 42°01′N 23°40′E
Yakovlevichi see Yakawlyevichi
127 T2 **Yakshur-Bod'ya** Udmurtskaya Respublika, NW Russian Federation 57°10′N 53°10′E
165 Q5 **Yakumo** Hokkaidō, NE Japan 42°18′N 140°15′E
164 B17 **Yaku-shima** island Nansei-shotō, SW Japan
39 V12 **Yakutat** Alaska, USA 59°33′N 139°44′W
39 U12 **Yakutat Bay** inlet Alaska, USA
Yakutia/Yakutiya/Yakutiya, Respublika see Sakha (Yakutiya), Respublika
123 Q10 **Yakutsk** Respublika Sakha (Yakutiya), NE Russian Federation 62°10′N 129°50′E
167 O17 **Yala** Yala, SW Thailand 06°32′N 101°19′E
182 D6 **Yalata** South Australia 31°30′S 131°53′E
31 S9 **Yale** Michigan, N USA 43°07′N 82°45′W
180 I11 **Yalgoo** Western Australia 28°23′S 116°43′E
114 O12 **Yalıköy** İstanbul, NW Turkey 41°29′N 28°19′E
79 L14 **Yalinga** Haute-Kotto, C Central African Republic 06°47′N 23°09′E
119 M17 **Yalizava** Rus. Yelizovo. Mahilyowskaya Voblasts', E Belarus 53°24′N 29°01′E
44 L13 **Yallahs Hill** ▲ E Jamaica 17°53′N 76°31′W
22 L3 **Yalobusha River** ≈ Mississippi, S USA
79 H15 **Yaloké** Ombella-Mpoko, W Central African Republic 05°15′N 17°12′E
160 E7 **Yalong Jiang** ≈ C China
136 E11 **Yalova** Yalova, NW Turkey 40°40′N 29°17′E
136 E11 **Yalova** ◆ province NW Turkey
Yaloveny see Ialoveni
Yalpug see Ialpug
Yalpug, Ozero see Yalpuh, Ozero
117 N12 **Yalpuh, Ozero** Rus. Ozero Yalpug. ◎ SW Ukraine
117 T14 **Yalta** Avtonomna Respublika Krym, S Ukraine 44°30′N 34°09′E
163 W12 **Yalu** Chin. Yalu Jiang, Jap. Oryokko, Kor. Amnok-kang. ≈ China/North Korea
Yalu Jiang see Yalu
136 F14 **Yalvaç** Isparta, SW Turkey 38°16′N 31°09′E
165 R9 **Yamada** Iwate, Honshū, C Japan 39°27′N 141°56′E
165 D14 **Yamaga** Kumamoto, Kyūshū, SW Japan 33°02′N 130°41′E
165 P10 **Yamagata** Yamagata, Honshū, C Japan 38°15′N 140°19′E
165 P9 **Yamagata** off. Yamagata-ken. ◆ prefecture Honshū, C Japan
Yamagata-ken see Yamagata
164 C16 **Yamagawa** Kagoshima, Kyūshū, SW Japan 31°12′N 130°37′E
164 E13 **Yamaguchi** var. Yamaguti. Yamaguchi, Honshū, SW Japan 34°11′N 131°26′E
164 E13 **Yamaguchi** off. Yamaguchi-ken, var. Yamaguti. ◆ prefecture Honshū, SW Japan
Yamaguchi-ken see Yamaguchi
Yamaguti see Yamaguchi
125 X5 **Yamalo-Nenetskiy Avtonomnyy Okrug** ◆ autonomous district N Russian Federation
122 J7 **Yamal, Poluostrov** peninsula N Russian Federation
165 N13 **Yamanashi** off. Yamanashi-ken, var. Yamanasi. ◆ prefecture Honshū, S Japan
Yamanashi-ken see Yamanashi
Yamanasi see Yamanashi
Yamanīyah, Al Jumhūrīyah al see Yemen
127 W5 **Yamantau** ▲ W Russian Federation 53°11′N 57°30′E
Yamasaki see Yamazaki
15 P12 **Yamaska** ≈ Québec, SE Canada
192 G4 **Yamato Ridge** undersea feature S Sea of Japan 39°20′N 135°00′E
164 I13 **Yamazaki** var. Yamasaki. Hyōgo, Honshū, SW Japan 35°00′N 134°31′E
183 V5 **Yamba** New South Wales, SE Australia 29°28′S 153°22′E
81 D16 **Yambio** var. Yambiyo. Western Equatoria, S South Sudan 04°34′N 28°21′E
Yambiyo see Yambio
114 L10 **Yambol** Turk. Yanboli. Yambol, E Bulgaria 42°29′N 26°30′E
114 M11 **Yambol** ◆ province E Bulgaria
79 M17 **Yambuya** Orientale, N Dem. Rep. Congo 01°22′N 24°21′E
171 T15 **Yamdena, Pulau** prev. Jamdena. island Kepulauan Tanimbar, E Indonesia
165 O14 **Yame** Fukuoka, Kyūshū, SW Japan 33°14′N 130°32′E
166 M6 **Yamethin** Mandalay, C Myanmar (Burma) 20°24′N 96°08′E
186 C6 **Yaminbot** East Sepik, NW Papua New Guinea 04°34′S 143°56′E
181 U9 **Yamma Yamma, Lake** ◎ Queensland, C Australia
76 M16 **Yamoussoukro** ● (Ivory Coast) C Ivory Coast 06°51′N 05°21′W
37 P3 **Yampa River** ≈ Colorado, C USA
117 S2 **Yampil'** Sums'ka Oblast', NE Ukraine 51°57′N 33°49′E
116 M8 **Yampil'** Vinnyts'ka Oblast', C Ukraine 48°15′N 28°18′E
123 T9 **Yamsk** Magadanskaya Oblast', E Russian Federation 59°33′N 154°04′E
152 J8 **Yamuna** prev. Jumna. ≈ N India
152 I9 **Yamunānagar** Haryāna, N India 30°07′N 77°17′E
Yamundá see Nhamundá, Rio
145 U8 **Yamyshevo** Pavlodar, NE Kazakhstan 51°49′N 77°28′E
159 N16 **Yamzho Yumco** ◎ W China
123 Q8 **Yana** ≈ NE Russian Federation
186 H9 **Yanaba Island** island SE Papua New Guinea
155 L16 **Yanam** var. Yanaon. Puducherry, E India 16°45′N 82°16′E
160 L5 **Yan'an** var. Yanan. Shaanxi, C China 36°35′N 109°27′E
Yanaon see Yanam
127 U3 **Yanaul** Respublika Bashkortostan, W Russian Federation 56°15′N 54°57′E
118 O12 **Yanavichy** Rus. Yanovichi. Vitsyebskaya Voblasts', NE Belarus 55°17′N 30°42′E
Yanboli see Yambol
140 K8 **Yanbu' al Baḩr** Al Madīnah, W Saudi Arabia 24°07′N 38°03′E
21 T8 **Yanceyville** North Carolina, SE USA 36°25′N 79°22′W
161 R7 **Yancheng** Jiangsu, E China 33°28′N 120°10′E
159 W8 **Yanchi** Ningxia, N China 37°49′N 107°24′E
160 L5 **Yanchuan** Shaanxi, C China 36°54′N 110°04′E
183 O10 **Yanco Creek** seasonal river New South Wales, SE Australia
183 O6 **Yanda Creek** seasonal river New South Wales, SE Australia
182 K4 **Yandama Creek** seasonal river New South Wales/South Australia
161 S11 **Yandang Shan** ▲▲ SE China
159 O6 **Yandun** Xinjiang Uygur Zizhiqu, W China 42°24′N 94°08′E
76 L13 **Yanfolila** Sikasso, SW Mali 11°08′N 08°12′W
79 M18 **Yangambi** Orientale, N Dem. Rep. Congo 0°46′N 24°24′E
158 M15 **Yangbajain** Xizang Zizhiqu, W China 30°05′N 90°35′E
Yangcheng see Yangshan, Guangdong, S China
Yangchow see Yangzhou
160 M15 **Yangchun** var. Chuncheng. Guangdong, S China 22°16′N 111°49′E
161 N2 **Yanggao** var. Longquan. Shanxi, C China 40°24′N 113°51′E
Yanggeta see Yaqeta
Yangiabad see Yangiobod
Yangibazar see Dzhany-Bazar, Kyrgyzstan
Yangi-Bazar see Kofarnihon, Tajikistan
Yangiklshak see Yangiqishloq
147 M13 **Yangi-Nishon** Rus. Yang-Nishan. Qashqadaryo Viloyati, S Uzbekistan 38°37′N 65°39′E
147 Q9 **Yangiobod** Rus. Yangiabad. Toshkent Viloyati, E Uzbekistan 41°10′N 70°10′E
147 O10 **Yangiqishloq** Rus. Yangiklshak. Jizzax Viloyati, C Uzbekistan 40°27′N 67°06′E
147 P11 **Yangiyer** Sirdaryo Viloyati, E Uzbekistan 40°19′N 68°48′E
147 P9 **Yangiyo'l** Rus. Yangiyul'. Toshkent Viloyati, E Uzbekistan 41°12′N 69°05′E
Yangiyul' see Yangiyo'l
160 M15 **Yangjiang** Guangdong, S China 21°50′N 112°02′E
Yangku see Taiyuan
Yang-Nishan see Yangi-Nishon
166 L8 **Yangon** Eng. Rangoon. Yangon, S Myanmar (Burma) 16°50′N 96°11′E
166 M8 **Yangon** Eng. Rangoon. ◆ region SW Myanmar (Burma)
161 N4 **Yangquan** Shanxi, C China 37°52′N 113°29′E
161 N13 **Yangshan** var. Yangcheng. Guangdong, S China 24°32′N 112°36′E
167 U12 **Yang Sin, Chu** ▲ S Vietnam 12°23′N 108°25′E
Yangtze see Chang Jiang/Jinsha Jiang
Yangtze see Chang Jiang
Yangtze Kiang see Chang Jiang
161 R7 **Yangzhou** var. Yangchow. Jiangsu, E China 32°22′N 119°22′E
160 L5 **Yan He** ≈ C China
163 Y10 **Yanji** Jilin, NE China 42°54′N 129°31′E
Yanji see Longjing
Yanjing see Yanyuan
29 Q12 **Yankton** South Dakota, N USA 42°52′N 97°24′W
161 O12 **Yanling** prev. Lingxian, Ling Xian. Hunan, S China 26°32′N 113°48′E
Yannina see Ioánnina
123 Q7 **Yano-Indigirskaya Nizmennost'** plain NE Russian Federation
Yanovichi see Yanavichy
155 K24 **Yan Oya** ≈ N Sri Lanka
158 K6 **Yanqi** var. Yanqi Huizu Zizhixian. Xinjiang Uygur Zizhiqu, NW China 42°04′N 86°32′E
Yanqi Huizu Zizhixian see Yanqi
161 Q10 **Yanshan** var. Hekou. Jiangxi, S China 28°18′N 117°43′E
160 H14 **Yanshan** var. Jiangna. Yunnan, SW China 23°36′N 104°20′E
161 P2 **Yan Shan** ▲▲ E China
163 X8 **Yanshou** Heilongjiang, NE China 45°27′N 128°19′E
123 Q7 **Yanskiy Zaliv** bay N Russian Federation
183 O4 **Yantabulla** New South Wales, SE Australia 29°22′S 145°00′E
161 R4 **Yantai** var. Yan-t'ai; prev. Chefoo, Chih-fu. Shandong, E China 37°30′N 121°22′E
118 A13 **Yantarnyy** Ger. Palmnicken. Kaliningradskaya Oblast', W Russian Federation 54°53′N 19°59′E
114 J9 **Yantra** Gabrovo, N Bulgaria 42°58′N 25°19′E
114 K9 **Yantra** ≈ N Bulgaria
160 G11 **Yanyuan** var. Yanjing. Sichuan, C China 27°30′N 101°22′E
161 P5 **Yanzhou** Shandong, E China 35°35′N 116°53′E
79 E16 **Yaoundé** var. Yaunde. ● (Cameroon) Centre, S Cameroon 03°51′N 11°31′E
188 I14 **Yap** ◇ state W Micronesia
188 F16 **Yap** island Caroline Islands, W Micronesia
57 M18 **Yapacani, Río** ≈ C Bolivia
171 W14 **Yapa Kopra** Papua, E Indonesia 04°18′S 135°05′E
Yapan see Yapen, Selat
Yapanskoye More see East Sea/Japan, Sea of
77 P15 **Yapei** N Ghana 09°10′N 01°08′W
12 M10 **Yapeitso, Mont** ▲ Québec, E Canada 52°18′N 70°24′W
171 W12 **Yapen, Pulau** prev. Japen. island E Indonesia
171 W12 **Yapen, Selat** var. Yapan. strait Papua, E Indonesia
61 E15 **Yapeyú** Corrientes, NE Argentina 29°28′S 56°50′W
136 I11 **Yapraklı** Çankırı, N Turkey 40°45′N 33°46′E
174 M3 **Yap Trench** var. Yap Trough. undersea feature SE Philippine Sea 08°30′N 138°00′E
Yap Trough see Yap Trench
Yapurá see Caquetá, Río, Brazil/Colombia
Yapurá see Japurá, Rio, Brazil/Colombia
197 I12 **Yaqaga** island N Fiji
197 H12 **Yaqeta** prev. Yanggeta. island Yasawa Group, NW Fiji
40 G6 **Yaqui** Sonora, NW Mexico 27°21′N 109°59′W
32 E12 **Yaquina Bay** bay Oregon, NW USA
40 G6 **Yaqui, Río** ≈ NW Mexico
54 K5 **Yaracuy** off. Estado Yaracuy. ◆ state NW Venezuela
Yaracuy, Estado see Yaracuy
146 E13 **Yarajy** Rus. Yaradzhi. Ahal Welaýaty, C Turkmenistan 38°12′N 57°40′E
Yaradzhi see Yarajy
125 Q15 **Yaransk** Kirovskaya Oblast', NW Russian Federation 57°18′N 47°52′E
136 F17 **Yardımcı Burnu** headland SW Turkey 36°10′N 30°25′E
97 Q19 **Yare** ≈ E England, United Kingdom
125 S9 **Yarega** Respublika Komi, NW Russian Federation 63°27′N 53°28′E
116 I7 **Yaremcha** Ivano-Frankivs'ka Oblast', W Ukraine 48°27′N 24°34′E
189 Q9 **Yaren (district)** ● (Nauru) SW Nauru 0°33′S 166°54′E
125 Q10 **Yarensk** Arkhangel'skaya Oblast', NW Russian Federation 62°09′N 49°03′E
155 F16 **Yargatti** Karnātaka, W India 16°07′N 75°11′E
164 M12 **Yariga-take** ▲ Honshū, S Japan 36°20′N 137°38′E
141 O15 **Yarim** W Yemen 14°15′N 44°23′E
54 F14 **Yarí, Río** ≈ SW Colombia
54 K5 **Yaritagua** Yaracuy, N Venezuela 10°05′N 69°07′W
Yarkand see Yarkant He
Yarkant see Shache
158 E9 **Yarkant He** var. Yarkand. ≈ NW China
149 U3 **Yarkhūn** ≈ NW Pakistan
Yarlung Zangbo Jiang see Brahmaputra
116 L6 **Yarmolyntsi** Khmel'nyts'ka Oblast', W Ukraine 49°13′N 26°53′E
13 O16 **Yarmouth** Nova Scotia, SE Canada 43°53′N 66°09′W
Yarmouth see Great Yarmouth
Yaroslav see Jarosław
124 L15 **Yaroslavl'** Yaroslavskaya Oblast', W Russian Federation 57°38′N 39°53′E
124 K14 **Yaroslavskaya Oblast'** ◆ province W Russian Federation
123 N11 **Yaroslavskiy** Respublika Sakha (Yakutiya), NE Russian Federation 60°10′N 114°12′E
183 P13 **Yarram** Victoria, SE Australia 38°36′S 146°40′E
183 O11 **Yarrawonga** Victoria, SE Australia 36°04′S 145°58′E
182 L4 **Yarriarraburra Swamp** wetland New South Wales, SE Australia
122 I8 **Yar-Sale** Yamalo-Nenetskiy Avtonomnyy Okrug, N Russian Federation 66°52′N 70°42′E
122 K11 **Yartsevo** Krasnoyarskiy Kray, C Russian Federation 60°15′N 90°09′E
126 I4 **Yartsevo** Smolenskaya Oblast', W Russian Federation 55°03′N 32°46′E
54 E8 **Yarumal** Antioquia, NW Colombia 06°59′N 75°25′W
187 W14 **Yasawa Group** island group NW Fiji
Yasel'da see Yasyel'da
77 V12 **Yashi** Katsina, N Nigeria 12°21′N 07°56′E
77 S14 **Yashikera** Kwara, W Nigeria 09°40′N 03°19′E
147 T14 **Yashilkŭl** Rus. Ozero Yashil'kul'. ◎ SE Tajikistan
Yashil'kul', Ozero see Yashilkŭl
165 P9 **Yashima** Akita, Honshū, C Japan 39°10′N 140°10′E
127 P13 **Yashkul'** Respublika Kalmykiya, SW Russian Federation 46°09′N 45°22′E
146 F13 **Yashlyk** Ahal Welaýaty, C Turkmenistan 37°46′N 58°51′E
Yasinovataya see Yasynuvata
114 N10 **Yasna Polyana** Burgas, E Bulgaria 42°18′N 27°35′E
167 R10 **Yasothon** Yasothon, E Thailand 15°46′N 104°12′E
183 R10 **Yass** New South Wales, SE Australia 34°52′S 148°55′E
Yassy see Iaşi
164 H12 **Yasugi** Shimane, Honshū, SW Japan 35°25′N 133°12′E
143 N10 **Yāsūj** var. Yesuj; prev. Tal-e Khosravi. Kohkīlūyeh va Būyer AHmad, C Iran 30°40′N 51°34′E
136 M11 **Yasun Burnu** headland N Turkey 41°07′N 37°40′E
119 I20 **Yasyel'da** Rus. Yasel'da. ≈ Brestskaya Voblasts', SW Belarus Europe
117 X8 **Yasynuvata** Rus. Yasinovataya. Donets'ka Oblast', SE Ukraine 48°05′N 37°57′E
136 C15 **Yatağan** Muğla, SW Turkey 37°22′N 28°08′E
165 Q7 **Yatate-tōge** pass Honshū, C Japan
187 Q17 **Yaté** Province Sud, S New Caledonia 22°10′S 166°56′E
27 P6 **Yates Center** Kansas, C USA 37°54′N 95°44′W
185 B21 **Yates Point** headland South Island, New Zealand 44°30′S 167°49′E
9 N9 **Yathkyed Lake** ◎ Nunavut, NE Canada
171 T16 **Yatoke** Pulau Babar, E Indonesia 07°51′S 129°49′E
79 M18 **Yatolema** Orientale, N Dem. Rep. Congo 0°25′N 24°35′E
164 C15 **Yatsushiro** var. Yatusiro. Kumamoto, Kyūshū, SW Japan 32°30′N 130°34′E
164 C15 **Yatsushiro-kai** bay SW Japan
138 F11 **Yatta** var. Yuta. S West Bank 31°29′N 35°10′E
81 J20 **Yatta Plateau** plateau SE Kenya
Yatusiro see Yatsushiro
57 F17 **Yauca, Río** ≈ SW Peru
45 S6 **Yauco** W Puerto Rico 18°02′N 66°51′W
Yaunde see Yaoundé
Yavan see Yovon
Yavarí see Javari, Rio
56 G9 **Yavari Mirim, Río** ≈ NE Peru
40 G7 **Yavaros** Sonora, NW Mexico 26°40′N 109°32′W
154 I13 **Yavatmāl** Mahārāshtra, C India 20°22′N 78°11′E
54 M9 **Yaví, Cerro** ▲ C Venezuela 05°43′N 65°51′W
43 W16 **Yaviza** Darién, SE Panama 08°09′N 77°41′W
138 F10 **Yavne** Central, W Israel 31°52′N 34°45′E
116 H5 **Yavoriv** Pol. Jaworów, Rus. Yavorov. L'vivs'ka Oblast', NW Ukraine 49°57′N 23°22′E
Yavorov see Yavoriv
164 F14 **Yawatahama** Ehime, Shikoku, SW Japan 33°27′N 132°24′E
Ya Xian see Sanya
136 L17 **Yayladaği** Hatay, S Turkey 35°55′N 36°00′E
125 V13 **Yayva** Permskiy Kray, NW Russian Federation 59°19′N 57°15′E
125 V12 **Yayva** ≈ NW Russian Federation
143 Q9 **Yazd** var. Yezd. Yazd, C Iran 31°55′N 54°22′E
143 Q8 **Yazd** off. Ostān-e Yazd, var. Yezd. ◆ province C Iran
Yazd, Ostān-e see Yazd
Yazgulemskiy Khrebet see Yazgulom, Qatorkŭhi
147 S13 **Yazgulom, Qatorkŭhi** Rus. Yazgulemskiy Khrebet. ▲▲ S Tajikistan
22 K5 **Yazoo City** Mississippi, S USA 32°51′N 90°24′W
22 K5 **Yazoo River** ≈ Mississippi, S USA
Yazovir Georgi Traykov see Tsonevo, Yazovir
127 Q5 **Yazykovka** Ul'yanovskaya Oblast', W Russian Federation 54°19′N 47°22′E
109 U4 **Ybbs** Niederösterreich, NE Austria 48°10′N 15°03′E
109 U4 **Ybbs** ≈ C Austria
95 G22 **Yding Skovhøj** hill C Denmark
115 G20 **Ýdra** var. Ídhra, Idra. Ýdra, S Greece 37°20′N 23°28′E
115 G21 **Ýdra** var. Ídhra. island Ýdra, S Greece
115 G20 **Ýdras, Kólpos** strait S Greece
167 N10 **Ye** Mon State, S Myanmar (Burma) 15°15′N 97°50′E
183 O12 **Yea** Victoria, SE Australia 37°15′S 145°27′E
167 P13 **Yeay Sên** prev. Phumĭ Yeay Sên. Kaôh Kŏng, SW Cambodia 11°09′N 103°09′E
Yebaishou see Jianping
78 I5 **Yebbi-Bou** Tibesti, N Chad 21°12′N 17°55′E
158 F9 **Yecheng** var. Kargilik. Xinjiang Uygur Zizhiqu, NW China 37°54′N 77°26′E
105 R11 **Yecla** Murcia, SE Spain 38°36′N 01°07′W
40 H6 **Yécora** Sonora, NW Mexico 28°23′N 108°56′W
Yedintsy see Edineţ
124 J13 **Yefimovskiy** Leningradskaya Oblast', NW Russian Federation 59°32′N 34°34′E
126 K6 **Yefremov** Tul'skaya Oblast', W Russian Federation 53°10′N 38°02′E
159 T11 **Yêgainnyin** var. Henan Mongolzu Zizhixian. Qinghai, C China 34°42′N 101°36′E
137 U12 **Yeghegis** Rus. Yekhegis. ≈ C Armenia
137 U12 **Yeghegnadzor** C Armenia 39°45′N 45°20′E
145 T10 **Yegindybulak** Kaz. Egindibulaq. Karaganda, C Kazakhstan 49°45′N 75°45′E
126 L4 **Yegor'yevsk** Moskovskaya Oblast', W Russian Federation 55°29′N 39°03′E
Yehuda, Haré see Judaean Hills
81 E15 **Yei** ≈ S South Sudan
161 P8 **Yeji** var. Yejiaji. Anhui, E China 31°52′N 115°58′E
Yejiaji see Yeji
122 G10 **Yekaterinburg** prev. Sverdlovsk. Sverdlovskaya Oblast', C Russian Federation 56°52′N 60°35′E
Yekaterinodar see Krasnodar
Yekaterinoslav see Dnipropetrovs'k
123 R13 **Yekaterinoslavka** Amurskaya Oblast', SE Russian Federation 50°23′N 129°03′E
127 O7 **Yekaterinovka** Saratovskaya Oblast', W Russian Federation 52°01′N 44°11′E
76 K16 **Yekepa** NE Liberia 07°35′N 08°32′W
Yekhegis see Yeghegis
145 T8 **Yekibastuz** prev. Ekibastuz. Pavlodar, NE Kazakhstan 51°42′N 75°22′E
127 T3 **Yelabuga** Respublika Tatarstan, W Russian Federation 55°46′N 52°07′E
Yela Island see Rossel Island
127 O8 **Yelan'** Volgogradskaya Oblast', SW Russian Federation 51°00′N 43°40′E
117 Q9 **Yelanets'** Rus. Yelanets. Mykolayivs'ka Oblast', S Ukraine 47°40′N 31°51′E
144 I9 **Yelek** Kaz. Elek; prev. Ilek. ≈ Kazakhstan/Russian Federation
126 L7 **Yelets** Lipetskaya Oblast', W Russian Federation 52°37′N 38°29′E
125 W4 **Yeletskiy** Respublika Komi, NW Russian Federation 67°03′N 64°05′E
76 J11 **Yélimané** Kayes, W Mali 15°06′N 10°43′W
Yelisavetpol see Gäncä
Yelizavetgrad see Kirovohrad
123 T12 **Yelizavety, Mys** headland SE Russian Federation 54°20′N 142°39′E
Yelizovo see Yalizava
127 S5 **Yelkhovka** Samarskaya Oblast', W Russian Federation 53°51′N 50°16′E
96 M1 **Yell** island NE Scotland, United Kingdom
155 E17 **Yellāpur** Karnātaka, W India 15°06′N 74°50′E
11 U17 **Yellow Grass** Saskatchewan, S Canada 49°51′N 104°09′W
Yellowhammer State see Alabama
11 O15 **Yellowhead Pass** pass Alberta/British Columbia, SW Canada
8 K10 **Yellowknife** territory capital Northwest Territories, W Canada 62°30′N 114°29′W
8 K9 **Yellowknife** ≈ Northwest Territories, NW Canada
23 P8 **Yellow River** ≈ Alabama/Florida, S USA
30 K7 **Yellow River** ≈ Wisconsin, N USA
30 I4 **Yellow River** ≈ Wisconsin, N USA
30 J6 **Yellow River** ≈ Wisconsin, N USA
Yellow River see Huang He
157 V8 **Yellow Sea** Chin. Huang Hai, Kor. Hwang-Hae. sea E Asia
33 S13 **Yellowstone Lake** ◎ Wyoming, C USA
33 T13 **Yellowstone National Park** national park Wyoming, NW USA
33 Y8 **Yellowstone River** ≈ Montana/Wyoming, NW USA
96 L1 **Yell Sound** strait N Scotland, United Kingdom
27 U9 **Yellville** Arkansas, C USA 36°12′N 92°41′W
122 K10 **Yeloguy** ≈ C Russian Federation
Yëloten see Yolöten
119 M20 **Yel'sk** Homyel'skaya Voblasts', SE Belarus 51°48′N 29°09′E
77 T13 **Yelwa** Kebbi, W Nigeria 10°52′N 04°46′E
21 R15 **Yemassee** South Carolina, SE USA 32°41′N 80°51′W
141 O15 **Yemen** off. Republic of Yemen, Ar. Al Jumhuriyah al Yamaniyah, Al Yaman. ◆ republic SW Asia
Yemen, Republic of see Yemen
116 M4 **Yemil'chyne** Zhytomyrs'ka Oblast', N Ukraine 50°51′N 27°49′E
124 M10 **Yemtsa** Arkhangel'skaya Oblast', NW Russian Federation 63°04′N 40°18′E
124 M10 **Yemtsa** ≈ NW Russian Federation
125 R10 **Yemva** prev. Zheleznodorozhnyy. Respublika Komi, NW Russian Federation 62°38′N 50°59′E
77 U17 **Yenagoa** Bayelsa, S Nigeria 4°58′N 6°16′E
117 X7 **Yenakiyeve** Rus. Yenakiyevo; prev. Ordzhonikidze, Rykovo. Donets'ka Oblast', E Ukraine 48°13′N 38°13′E
Yenakiyevo see Yenakiyeve
166 L6 **Yenangyaung** Magway, W Myanmar (Burma) 20°28′N 94°54′E
167 S5 **Yên Bai** Yên Bai, N Vietnam 21°43′N 104°54′E
183 P9 **Yenda** New South Wales, SE Australia 34°16′S 146°15′E
77 Q14 **Yendi** NE Ghana 09°30′N 00°01′W
Yêndum see Zhag'yab
158 E8 **Yengisar** Xinjiang Uygur Zizhiqu, NW China 38°50′N 76°11′E
136 H11 **Yenice Çayı** var. Filyos Çayı. ≈ N Turkey
121 R1 **Yenierenköy** var. Yialousa, Gk. Agialoúsa. NE Cyprus 35°33′N 34°13′E
Yenipazar see Novi Pazar
136 E12 **Yenişehir** Bursa, NW Turkey 40°17′N 29°38′E
Yenisei Bay see Yeniseyskiy Zaliv
122 K12 **Yeniseysk** Krasnoyarskiy Kray, C Russian Federation 58°23′N 92°06′E
197 W10 **Yeniseyskiy Zaliv** var. Yenisei Bay. bay N Russian Federation
127 Q12 **Yenotayevka** Astrakhanskaya Oblast', SW Russian Federation 47°16′N 47°01′E
124 L4 **Yenozero, Ozero** ◎ NW Russian Federation
Yenping see Nanping
39 Q11 **Yentna River** ≈ Alaska, USA
180 M10 **Yeo, Lake** salt lake Western Australia
163 Z15 **Yeongcheon** Jap. Eisen; prev. Yŏngch'ŏn. SE South Korea 35°56′N 128°55′E

◆ Country
● Country Capital
◇ Dependent Territory
○ Dependent Territory Capital
◆ Administrative Regions
✈ International Airport
▲ Mountain
▲▲ Mountain Range
Volcano
≈ River
◎ Lake
⊠ Reservoir

Yeongju – Zagreb

163 Y15 Yeongju Jap. Eishū; prev. Yŏngju. C South Korea 36°48′N 128°37′E
163 Y17 Yeosu Jap. Reisui; prev. Yŏsu. S South Korea 34°45′N 127°41′E
183 R7 Yeoval New South Wales, SE Australia 32°45′S 148°39′E
97 K23 Yeovil SW England, United Kingdom 50°57′N 02°39′W
40 H6 Yepachic Chihuahua, N Mexico 28°27′N 108°25′W
181 Y8 Yeppoon Queensland, E Australia 23°05′S 150°42′E
126 M5 Yeraktur Ryazanskaya Oblast', W Russian Federation 54°45′N 41°09′E
Yeraliyev see Kuryk
146 F12 Yerbent Ahal Welaýaty, C Turkmenistan 39°19′N 58°34′E
123 N11 Yerbogachën Irkutskaya Oblast', C Russian Federation 61°07′N 108°03′E
137 T12 Yerevan Eng. Erivan. ● (Armenia) C Armenia 40°12′N 44°31′E
137 U12 Yerevan ✈ C Armenia 40°07′N 44°34′E
145 R9 Yereymentau var. Jermentau, Kaz. Ereymentaū. Akmola, C Kazakhstan 51°38′N 73°10′E
145 R9 Yereymentau, Gory prev. Gory Yermentau. ▲ C Kazakhstan
127 O12 Yergeni hill range SW Russian Federation
Yeriho see Jericho
35 R6 Yerington Nevada, W USA 38°58′N 119°10′W
136 J13 Yerköy Yozgat, C Turkey 39°39′N 34°28′E
114 L13 Yerlisu Edirne, NW Turkey 40°45′N 26°38′E
Yermak see Aksu
Yermentau, Gory see Yereymentau, Gory
125 R5 Yërmitsa Respublika Komi, NW Russian Federation 66°57′N 52°15′E
35 V14 Yermo California, W USA 34°54′N 116°49′W
123 P13 Yerofey Pavlovich Amurskaya Oblast', SE Russian Federation 53°58′N 121°49′E
99 F15 Yerseke Zeeland, SW Netherlands 51°30′N 04°03′E
127 Q8 Yershov Saratovskaya Oblast', W Russian Federation 51°18′N 48°16′E
145 S7 Yertis Kaz. Ertis; prev. Irtyshsk. Pavlodar, NE Kazakhstan 53°21′N 75°27′E
129 R5 Yertis var. Irtish, Kaz. Ertis; prev. Irtysh. ≈ C Asia
125 P9 Yërtom Respublika Komi, NW Russian Federation 63°27′N 47°52′E
56 D13 Yerupaja, Nevado ▲ C Peru 10°23′S 76°58′W
Yerushalayim see Jerusalem
105 R4 Yesa, Embalse de ⊠ NE Spain
144 F11 Yesbol prev. Kulagino. Atyrau, W Kazakhstan 48°30′N 51°33′E
144 F9 Yesensay Zapadnyy Kazakhstan, NW Kazakhstan 49°59′N 51°19′E
144 F9 Yesensay Zapadnyy Kazakhstan, NW Kazakhstan 49°58′N 51°19′E
145 V15 Yesik Kaz. Esik; prev. Issyk. Almaty, SE Kazakhstan 42°23′N 77°25′E
145 O8 Yesil' Kaz. Esil. Akmola, C Kazakhstan 51°58′N 66°24′E
129 R6 Yesil' Kaz. Esil. ≈ Kazakhstan/Russian Federation
136 K15 Yeşilhisar Kayseri, C Turkey 38°22′N 35°08′E
136 L11 Yeşilırmak var. Iris. ≈ N Turkey
37 U12 Yeso New Mexico, SW USA 34°25′N 104°36′W
Yeso see Hokkaidō
127 N15 Yessentuki Stavropol'skiy Kray, SW Russian Federation 44°06′N 42°51′E
122 M9 Yessey Krasnoyarskiy Kray, N Russian Federation 68°18′N 101°49′E
105 P12 Yeste Castilla-La Mancha, C Spain 38°21′N 02°18′W
Yesuj see Yāsūj
183 T4 Yetman New South Wales, SE Australia 28°56′S 150°47′E
76 L4 Yetti physical region N Mauritania
166 M4 Ye-u Sagaing, C Myanmar (Burma) 22°49′N 95°26′E
102 H9 Yeu, Île d' island NW France
Yevlakh see Yevlax
137 W11 Yevlax Rus. Yevlakh. C Azerbaijan 40°36′N 47°10′E
117 S13 Yevpatoriya Avtonomna Respublika Krym, S Ukraine 45°12′N 33°23′E
Ye Xian see Laizhou
126 K12 Yeya ≈ SW Russian Federation
158 I10 Yeyik Xinjiang Uygur Zizhiqu, W China 36°44′N 83°14′E
126 K12 Yeysk Krasnodarskiy Kray, SW Russian Federation 46°41′N 38°15′E
Yezd see Yazd
Yezerishche see Yezyaryshcha
Yezhou see Jianshi
Yezo see Hokkaidō
118 N11 Yezyaryshcha Rus. Yezerishche. Vitsyebskaya Voblasts', NE Belarus 55°50′N 29°59′E
Yialí see Gyalí
Yialousa see Yenierenköy
163 V7 Yi'an Heilongjiang, NE China 47°52′N 125°13′E
Yiannitsá see Giannitsá
160 I10 Yibin Sichuan, C China 28°50′N 104°35′E
158 K13 Yibug Caka ◎ W China
160 M9 Yichang Hubei, C China 30°37′N 111°02′E
160 L5 Yichuan var. Danzhou. Shaanxi, C China 36°05′N 110°02′E
157 W3 Yichun Heilongjiang, NE China 47°41′N 129°10′E
161 O11 Yichun Jiangxi, S China 27°45′N 114°22′E
160 M9 Yidu prev. Zhicheng. Hubei, C China 30°21′N 111°27′E
Yidu see Qingzhou
188 C15 Yigo NE Guam 13°33′N 144°53′E
161 Q5 Yi He ≈ E China
136 X8 Yilan Heilongjiang, NE China 46°18′N 129°36′E
136 C9 Yıldız Dağları ▲▲ NW Turkey
136 L13 Yıldızeli Sivas, N Turkey 39°52′N 36°37′E
163 U4 Yilehuli Shan ▲▲ NE China
163 S7 Yimin He ≈ NE China
159 W8 Yinchuan var. Yinch'uan, Yin-ch'uan, Yinchwan. province capital Ningxia, N China 38°30′N 106°19′E
Yinchwan see Yinchuan
Yindu He see Indus
Yingcheng see Yingde
161 N14 Yingde var. Yingcheng. Guangdong, S China 24°08′N 113°21′E
163 U13 Yingkou var. Ying-k'ou, Yingkow; prev. Newchwang, Niuchwang. Liaoning, NE China 40°40′N 122°17′E
Yingkow see Yingkou
161 P9 Yingshan var. Wenquan. Hubei, C China 30°45′N 115°41′E
161 Q10 Yingtan Jiangxi, S China 28°17′N 117°03′E
Yin-hsien see Ningbo
158 H5 Yining var. I-ning, Uigh. Gulja, Kuldja. Xinjiang Uygur Zizhiqu, NW China 43°53′N 81°18′E
160 K11 Yinjiang var. Yinjiang Tujiazu Miaozu Zizhixian. Guizhou, S China 28°22′N 108°07′E
Yinjiang Tujiazu Miaozu Zizhixian see Yinjiang
166 L4 Yinmabin Sagaing, C Myanmar (Burma) 22°05′N 94°57′E
163 N13 Yin Shan ▲▲ N China
Yinshan see Guangshui
Yin-tu Ho see Indus
159 P15 Yi'ong Zangbo ≈ W China
Yioúra see Gyáros
81 J14 Yirga 'Alem It. Irgalem. Southern Nationalities, S Ethiopia 06°43′N 38°24′E
61 E19 Yí, Río ≈ C Uruguay
81 E14 Yirol Lakes, C South Sudan 06°34′N 30°33′E
163 S8 Yirshi var. Yirxie. Nei Mongol Zizhiqu, N China 47°16′N 119°51′E
Yirxie see Yirshi
Yishan see Guanyun
Yishi see Linyi
161 Q5 Yishui Shandong, E China 35°50′N 118°39′E
Yisrael/Yisra'el see Israel
Yíthion see Gýtheio
Yitiaoshan see Jingtai
163 W10 Yitong var. Yitong Manzu Zizhixian. Jilin, NE China 43°23′N 125°19′E
Yitong Manzu Zizhixian see Yitong
159 P5 Yiwu var. Aratürük. Xinjiang Uygur Zizhiqu, NW China 43°16′N 94°38′E
163 U12 Yiwulü Shan ▲ N China
163 T12 Yixian var. Yizhou. Liaoning, NE China 41°29′N 121°21′E
159 R15 Yixing Jiangsu, China 31°14′N 119°18′E
161 N10 Yiyang Hunan, S China 28°39′N 112°10′E
161 Q10 Yiyang Jiangxi, S China 28°21′N 117°23′E
161 N13 Yizhang Hunan, S China 25°24′N 112°51′E
Yizhou see Yixian
93 K19 Yläne Varsinais-Suomi, SW Finland 60°51′N 22°26′E
93 L14 Yli-Ii Pohjois-Pohjanmaa, C Finland 65°21′N 25°55′E
93 L14 Ylikiiminki Pohjois-Pohjanmaa, C Finland 65°00′N 26°10′E
92 N13 Yli-Kitka ◎ NE Finland
93 K17 Ylistaro Etelä-Pohjanmaa, W Finland 62°58′N 22°30′E
92 K13 Ylitornio Lappi, NW Finland 66°19′N 23°40′E
93 L15 Ylivieska Pohjois-Pohjanmaa, W Finland 64°05′N 24°30′E
93 L18 Ylöjärvi Pirkanmaa, W Finland 61°31′N 23°37′E
95 N17 Yngaren ◎ C Sweden
25 T12 Yoakum Texas, SW USA 29°17′N 97°09′W
77 X13 Yobe ◐ state NE Nigeria
165 R3 Yobetsu-dake ⅄ Hokkaidō, NE Japan 43°15′N 140°27′E
80 L11 Yoboki C Djibouti 11°30′N 42°04′E
22 M4 Yockanookany River ≈ Mississippi, S USA
22 L2 Yocona River ≈ Mississippi, S USA
171 Y15 Yodom Papua, E Indonesia 07°12′S 139°24′E
169 Q16 Yogyakarta prev. Djokjakarta, Jogjakarta, Jokyakarta. Jawa, C Indonesia 07°48′S 110°24′E
169 P17 Yogyakarta off. Daerah Istimewa Yogyakarta, var. Djokjakarta, Jogjakarta, Jokyakarta. ◐ autonomous district S Indonesia
Yogyakarta, Daerah Istimewa see Yogyakarta
165 Q3 Yoichi Hokkaidō, NE Japan 43°11′N 140°45′E
42 G6 Yojoa, Lago de ◎ NW Honduras
79 G16 Yokadouma Est, SE Cameroon 03°26′N 15°06′E
164 K13 Yokkaichi var. Yokkaiti. Mie, Honshū, SW Japan 34°58′N 136°38′E
Yokkaiti see Yokkaichi
79 E15 Yoko Centre, C Cameroon 05°29′N 12°19′E
165 V15 Yokoate-jima island Nansei-shotō, SW Japan
165 R6 Yokohama Aomori, Honshū, C Japan 41°04′N 141°14′E
165 O14 Yokosuka Kanagawa, Honshū, S Japan 35°18′N 139°39′E
164 G12 Yokota Shimane, Honshū, SW Japan 35°10′N 133°03′E
165 Q9 Yokote Akita, Honshū, C Japan 39°20′N 140°33′E
77 Y14 Yola Adamawa, E Nigeria 09°08′N 12°24′E
79 L19 Yolombo Equateur, C Dem. Rep. Congo 01°36′S 23°13′E
146 J14 Ýolöten Rus. Yëloten; prev. Iolotan'. Mary Welaýaty, S Turkmenistan 37°15′N 62°18′E
165 Y15 Yome-jima island Ogasawara-shotō, SE Japan
76 K16 Yomou SE Guinea 07°30′N 09°13′W
171 Y15 Yomuka Papua, E Indonesia 07°25′S 138°36′E
188 C16 Yona E Guam 13°24′N 144°46′E
164 H12 Yonago Tottori, Honshū, SW Japan 35°30′N 134°15′E
165 N16 Yonaguni Okinawa, SW Japan 24°29′N 123°00′E
165 N16 Yonaguni-jima island Nansei-shotō, SW Japan
165 T16 Yonaha-dake ▲ Okinawa, SW Japan 26°43′N 128°13′E
163 X14 Yŏnan SW North Korea 37°50′N 126°15′E
165 P10 Yonezawa Yamagata, Honshū, C Japan 37°56′N 140°06′E
161 Q12 Yong'an var. Yongan. Fujian, SE China 25°58′N 117°26′E
Yong'an see Fengjie
159 T9 Yongchang Gansu, N China 38°15′N 101°56′E
161 P7 Yongcheng Henan, C China 33°56′N 116°21′E
Yŏngch'ŏn see Yeongcheon
160 J10 Yongchuan Chongqing Shi, C China 29°27′N 105°56′E
159 U10 Yongdeng Gansu, C China 35°58′N 103°27′E
Yongding see Yongren
129 W9 Yongding He ≈ E China
161 P11 Yongfeng var. Enjiang. Jiangxi, S China 27°19′N 115°23′E
158 L5 Yongfeng var. Yongfengqu. Xinjiang Uygur Zizhiqu, W China 43°28′N 87°09′E
Yongfengqu see Yongfeng
160 L13 Yongfu Guangxi Zhuangzu Zizhiqu, S China 24°57′N 109°59′E
163 X13 Yŏnghŭng E North Korea 39°31′N 127°14′E
159 U10 Yongjing var. Liujiaxia. Gansu, C China 36°00′N 103°30′E
Yongjing see Xifeng
Yŏngju see Yeongju
Yongle Qundao see Crescent Group
Yongning see Xuyong
160 E12 Yongping Yunnan, SW China 25°30′N 99°28′E
160 G12 Yongren var. Yongding. Yunnan, SW China 26°09′N 101°40′E
160 L10 Yongshun var. Lingxi. Hunan, S China 29°02′N 109°46′E
161 P10 Yongxiu var. Tujiabu. Jiangxi, S China 29°09′N 115°47′E
Yongzhou see Lingling
Yongzhou see Zhishan
18 K14 Yonkers New York, NE USA 40°56′N 73°52′W
103 Q7 Yonne ◐ department C France
103 P6 Yonne ≈ C France
54 H9 Yopal var. El Yopal. Casanare, C Colombia 05°20′N 72°19′W
158 E8 Yopurga var. Yukuriawat. Xinjiang Uygur Zizhiqu, NW China 39°13′N 76°44′E
147 S11 Yordon var. Iordan, Rus. Jardan. Farg'ona Viloyati, E Uzbekistan 39°59′N 71°44′E
180 J12 York Western Australia 31°55′S 116°52′E
97 M16 York anc. Eboracum, Eburacum. N England, United Kingdom 53°58′N 01°05′W
23 N5 York Alabama, S USA 32°29′N 88°18′W
29 Q15 York Nebraska, C USA 40°52′N 97°35′W
18 G16 York Pennsylvania, NE USA 39°57′N 76°44′W
21 R11 York South Carolina, SE USA 34°59′N 81°14′W
14 J13 York ≈ Ontario, SE Canada
15 X6 York ≈ Québec, SE Canada
181 V1 York, Cape headland Queensland, NE Australia 10°40′S 142°36′E
182 I9 Yorke Peninsula peninsula South Australia
182 I9 Yorketown South Australia 35°01′S 137°38′E
19 P9 York Harbor Maine, NE USA 43°10′N 70°37′W
York, Kap see Innaanganeq
21 X6 York River ≈ Virginia, NE USA
97 M16 Yorkshire cultural region N England, United Kingdom
97 L16 Yorkshire Dales physical region N England, United Kingdom
11 V16 Yorkton Saskatchewan, S Canada 51°12′N 102°29′W
25 T12 Yorktown Texas, SW USA 28°58′N 97°30′W
21 X6 Yorktown Virginia, NE USA 37°14′N 76°32′W
30 M11 Yorkville Illinois, N USA 41°38′N 88°27′W
42 I5 Yoro Yoro, C Honduras 15°08′N 87°10′W
42 H5 Yoro ◐ department N Honduras
165 T16 Yoron-jima island Nansei-shotō, SW Japan
77 N13 Yorosso Sikasso, S Mali 12°21′N 04°47′W
35 R8 Yosemite National Park national park California, W USA
127 Q3 Yoshkar-Ola Respublika Mariy El, W Russian Federation 56°38′N 47°54′E
Yösönbulag see Altay
162 K8 Yösöndzüyl var. Mönhbulag. Övörhangay, C Mongolia 46°48′N 103°25′E
171 Y16 Yos Sudarso, Pulau var. Pulau Dolak, Pulau Kolepom; prev. Jos Sudarso. island E Indonesia
Yosu see Yeosu
165 R4 Yotei-zan ▲ Hokkaidō, NE Japan 42°50′N 140°46′E
97 D21 Youghal Ir. Eochaill. Cork, S Ireland 51°57′N 07°50′W
97 D21 Youghal Bay Ir. Cuan Eochaille. inlet S Ireland
18 C15 Youghiogheny River ≈ Pennsylvania, NE USA
160 K14 You Jiang ≈ S China
183 Q9 Young New South Wales, SE Australia 34°19′S 148°20′E
11 T15 Young Saskatchewan, S Canada 51°44′N 105°44′W
61 E18 Young Río Negro, W Uruguay 32°44′S 57°36′W
182 G5 Younghusband, Lake salt lake South Australia
182 J10 Younghusband Peninsula peninsula South Australia
184 Q10 Young Nicks Head headland North Island, New Zealand 39°38′S 177°03′E
185 D20 Young Range ▲▲ South Island, New Zealand
191 Q15 Young's Rock island Pitcairn Island, Pitcairn Islands
11 R16 Youngstown Alberta, SW Canada 51°32′N 111°12′W
31 V12 Youngstown Ohio, N USA 41°06′N 80°39′W
159 N9 Youshashan Qinghai, C China 38°12′N 90°58′E
Youth, Isle of see Juventud, Isla de la
77 N11 Youvarou Mopti, C Mali 15°19′N 04°15′W
160 K10 Youyang var. Zhongduo. Chongqing Shi, C China 28°48′N 108°48′E
163 Y7 Youyi Heilongjiang, NE China 46°51′N 131°54′E
147 P13 Yovon Rus. Yavan. SW Tajikistan 38°19′N 69°02′E
136 J13 Yozgat Yozgat, C Turkey 39°49′N 34°48′E
136 K13 Yozgat ◐ province C Turkey
62 O6 Ypacaraí var. Ypacaray. Central, S Paraguay 25°23′S 57°16′W
Ypacaray see Ypacaraí
62 P5 Ypané, Río ≈ C Paraguay
Ypres see Ieper
114 I13 Ypsário var. Ipsario. ▲ Thásos, E Greece 40°43′N 24°39′E
31 R10 Ypsilanti Michigan, N USA 42°12′N 83°36′W
34 M1 Yreka California, W USA 41°43′N 122°39′W
Yrendagüé see General Eugenio A. Garay
144 L11 Yrghyz prev. Irgiz. Aktyubinsk, C Kazakhstan 48°36′N 61°14′E
186 G5 Ysabel Channel channel N Papua New Guinea
14 K8 Yser, Lac ◎ Québec, SE Canada
147 Y8 Yshtyk Issyk-Kul'skaya Oblast', E Kyrgyzstan 41°34′N 78°21′E
Yssel see IJssel
103 Q12 Yssingeaux Haute-Loire, C France 45°09′N 04°07′E
95 K23 Ystad Skåne, S Sweden 55°25′N 13°51′E
Ysyk-Köl see Issyk-Kul', Ozero
Ysyk-Köl see Balykchy
Ysyk-Köl Oblasty see Issyk-Kul'skaya Oblast'
96 L8 Ythan ≈ NE Scotland, United Kingdom
Y Trallwng see Welshpool
94 C13 Ytre Arna Hordaland, S Norway 60°28′N 05°25′E
94 B12 Ytre Sula island S Norway
93 G17 Ytterhogdal Jämtland, C Sweden 62°10′N 14°55′E
Yu see Henan
Yuan see Red River
Yuancheng see Heyuan, Guangdong, S China
Yuan Jiang see Red River
161 S13 Yuanlin Jap. Inrin; prev. Yüanlin. C Taiwan 23°57′N 120°33′E
161 N3 Yuanping Shanxi, C China 38°26′N 112°42′E
Yuanquan see Anxi
Yuanshan see Lianping
161 O11 Yuan Shui ≈ S China
35 O6 Yuba City California, W USA 39°07′N 121°40′W
35 O6 Yuba River ≈ California, W USA
80 H13 Yubdo Oromīya, C Ethiopia 09°05′N 35°28′E
41 X12 Yucatán ◐ state SE Mexico
47 O3 Yucatan Basin var. Yucatan Deep. undersea feature N Caribbean Sea 20°00′N 84°00′W
Yucatán, Canal de see Yucatan Channel
41 Y10 Yucatan Channel Sp. Canal de Yucatán. channel Cuba/Mexico
Yucatan Deep see Yucatan Basin
Yucatan Peninsula see Yucatán, Península de
41 X13 Yucatán, Península de Eng. Yucatan Peninsula. peninsula Guatemala/Mexico
36 I11 Yucca Arizona, SW USA 34°49′N 114°06′W
35 V15 Yucca Valley California, W USA 34°06′N 116°30′W
161 P4 Yucheng Shandong, E China 37°01′N 116°37′E
Yuci see Jinzhong
129 X5 Yudoma ≈ E Russian Federation
161 P12 Yudu var. Gongjiang. Jiangxi, C China 26°02′N 115°24′E
Yue see Guangdong
Yuecheng see Yuexi
160 M12 Yuecheng Ling ▲▲ S China
Yuegai see Qumarlêb
Yuegaitan see Qumarlêb
181 P7 Yuendumu Northern Territory, N Australia 22°19′S 131°51′E
Yue Shan, Tai see Lantau Island
160 H10 Yuexi var. Yuecheng. Sichuan, C China 28°50′N 102°36′E
161 N10 Yueyang Hunan, S China 29°24′N 113°08′E
125 U14 Yug Permskiy Kray, NW Russian Federation 57°49′N 56°08′E
125 P13 Yug ≈ NW Russian Federation
123 R10 Yugorënok Respublika Sakha (Yakutiya), NE Russian Federation 59°46′N 137°36′E
122 H9 Yugorsk Khanty-Mansiyskiy Avtonomnyy Okrug-Yugra, C Russian Federation 61°17′N 63°25′E
122 H7 Yugorskiy Poluostrov peninsula NW Russian Federation
Yugoslavia see Serbia
146 K14 Yugo-Vostochnyye Garagumy prev. Yugo-Vostochnyye Karakumy. desert E Turkmenistan
Yugo-Vostochnyye Karakumy see Yugo-Vostochnyye Garagumy
Yuhu see Eryuan
161 S10 Yuhuan Dao island SE China
160 L14 Yu Jiang ≈ S China
Yujin see Qianwei
159 P9 Yuka Qinghai, W China 38°03′N 94°45′E
123 S7 Yukagirskoye Ploskogor'ye plateau NE Russian Federation
159 P9 Yuke He ≈ C China
118 L11 Yukhavichy Rus. Yukhovichi. Vitsyebskaya Voblasts', N Belarus 56°02′N 28°39′E
126 J4 Yukhnov Kaluzhskaya Oblast', W Russian Federation 54°43′N 35°15′E
Yukhovichi see Yukhavichy
79 J20 Yuki var. Yuki Kengunda. Bandundu, W Dem. Rep. Congo 03°57′S 19°30′E
Yuki Kengunda see Yuki
26 M10 Yukon Oklahoma, C USA 35°30′N 97°45′W
10 I5 Yukon prev. Yukon Territory, Fr. Territoire du Yukon. ◐ territory NW Canada
0 F4 Yukon ≈ Canada/USA
39 S7 Yukon Flats salt flat Alaska, USA
Yukon, Territoire du see Yukon
Yukon Territory see Yukon
137 T16 Yüksekova Hakkâri, SE Turkey 37°35′N 44°17′E
123 N10 Yukta Krasnoyarskiy Kray, C Russian Federation 63°16′N 106°04′E
165 O13 Yukuhashi var. Yukuhasi. Fukuoka, Kyūshū, SW Japan 33°41′N 131°00′E
Yukuhasi see Yukuhashi
Yukuriawat see Yopurga
125 O9 Yula ≈ NW Russian Federation
181 P8 Yulara Northern Territory, N Australia 25°15′S 130°57′E
127 W6 Yuldybayevo Respublika Bashkortostan, W Russian Federation 52°22′N 57°55′E
23 W8 Yulee Florida, SE USA 30°37′N 81°36′W
158 K7 Yuli var. Lopnur. Xinjiang Uygur Zizhiqu, NW China 41°24′N 86°12′E
161 T14 Yuli prev. Yüli. C Taiwan 23°23′N 121°18′E
160 L15 Yulin Guangxi Zhuangzu Zizhiqu, S China 22°37′N 110°08′E
160 L4 Yulin Shaanxi, C China 38°14′N 109°48′E
161 T14 Yuli Shan prev. Yüli Shan. ▲ E Taiwan 23°23′N 121°14′E
160 F11 Yulong Xueshan ▲ SW China 27°09′N 100°10′E
36 H14 Yuma Arizona, SW USA 32°40′N 114°38′W
37 W3 Yuma Colorado, C USA 40°07′N 102°43′W
54 K5 Yumare Yaracuy, N Venezuela 10°37′N 68°41′W
63 G14 Yumbel Bío Bío, C Chile 37°05′S 72°40′W
79 N19 Yumbi Maniema, E Dem. Rep. Congo 01°14′S 26°14′E
159 Q7 Yumen prev. Yumenzhen. Gansu, N China 40°19′N 97°12′E
159 R8 Yumendong prev. Laojunmiao. Gansu, N China 39°49′N 97°47′E
Yumenzhen see Yumendong
158 J3 Yumin var. Karabura. Xinjiang Uygur Zizhiqu, NW China 46°14′N 82°52′E
Yun see Yunnan
136 G14 Yunak Konya, W Turkey 38°50′N 31°42′E
45 O8 Yuna, Río ≈ E Dominican Republic
38 I17 Yunaska Island island Aleutian Islands, Alaska, USA
160 M6 Yuncheng Shanxi, C China 35°07′N 110°45′E
Yuncheng see Yunfu
161 N14 Yunfu var. Yuncheng. Guangdong, S China 22°56′N 112°02′E
57 L18 Yungas physical region E Bolivia
Yungki see Jilin
Yung-ning see Nanning
160 I12 Yungui Gaoyuan plateau SW China
Yunjinghong see Jinghong
160 M15 Yunkai Dashan ▲▲ S China
Yunki see Jilin
160 E11 Yun Ling ▲▲ SW China
Yunling see Yunxiao
161 N9 Yunmeng Hubei, C China 31°04′N 113°45′E
157 N14 Yunnan var. Yun, Yunnan Sheng, Yünnan, Yun-nan. ◐ province SW China
Yunnan see Kunming
Yunnan Sheng see Yunnan
Yünnan/Yun-nan see Yunnan
165 P15 Yunomae Kumamoto, Kyūshū, SW Japan 32°16′N 131°00′E
161 N8 Yun Shui ≈ C China
182 J7 Yunta South Australia 32°37′S 139°33′E
161 Q14 Yunxiao var. Yunling. Fujian, SE China 23°56′N 117°16′E
160 K9 Yunyang Sichuan, C China 31°03′N 109°43′E
Yunzhong see Huairen
193 S9 Yupanqui Basin undersea feature E Pacific Ocean
Yuping see Libo, Guizhou, China
Yuping see Pingbian, Yunnan, China
Yuratishki see Yuratsishki
119 I15 Yuratsishki Pol. Juraciszki, Rus. Yuratishki. Hrodzyenskaya Voblasts', W Belarus 54°02′N 25°56′E
Yurev see Tartu
122 J12 Yurga Kemerovskaya Oblast', S Russian Federation 55°42′N 84°59′E
56 E10 Yurimaguas Loreto, N Peru 05°54′S 76°07′W
Yurihonjō see Honjō
127 P3 Yurino Respublika Mariy El, W Russian Federation 56°19′N 46°15′E
41 N13 Yuriria Guanajuato, C Mexico 20°12′N 101°09′W
125 T13 Yurla Komi-Permyatskiy Okrug, NW Russian Federation 59°18′N 54°19′E
Yuruá, Río see Juruá, Rio
114 M13 Yürük Tekirdağ, NW Turkey 40°58′N 27°09′E
158 G10 Yurungkax He ≈ W China
125 Q14 Yur'ya var. Jarja. Kirovskaya Oblast', NW Russian Federation 59°01′N 49°22′E
Yury'ev see Tartu
125 N16 Yur'yevets Ivanovskaya Oblast', W Russian Federation 57°19′N 43°01′E
126 M3 Yur'yev-Pol'skiy Vladimirskaya Oblast', W Russian Federation 56°28′N 39°39′E
117 V7 Yur"yivka Dnipropetrovs'ka Oblast', E Ukraine 48°45′N 36°01′E
42 I7 Yuscarán El Paraíso, S Honduras 13°55′N 86°51′W
161 P2 Yu Shan ▲ S China
124 I7 Yushkozero Respublika Kareliya, NW Russian Federation 64°46′N 32°13′E
124 I7 Yushkozerskoye Vodokhranilishche var. Ozero Kujto. ◎ NW Russian Federation
169 W9 Yushu Jilin, China E Asia 44°48′N 126°55′E
159 R13 Yushu var. Gyêgu. Qinghai, C China 33°04′N 97°E
127 P12 Yusta Respublika Kalmykiya, SW Russian Federation 47°06′N 46°16′E
124 I10 Yustozero Respublika Kareliya, NW Russian Federation 62°44′N 33°31′E
137 Q11 Yusufeli Artvin, NE Turkey 40°50′N 41°31′E
164 F14 Yusuhara Kōchi, Shikoku, SW Japan 33°22′N 132°52′E
125 T14 Yus'va Permskiy Kray, NW Russian Federation 58°48′N 54°59′E
Yuta see Yatta
161 P2 Yutian Hebei, E China 39°52′N 117°44′E
158 H10 Yutian var. Keriya, Mugalla. Xinjiang Uygur Zizhiqu, NW China 36°49′N 81°31′E
62 K5 Yuto Jujuy, NW Argentina 23°35′S 64°28′W
62 P7 Yuty Caazapá, S Paraguay 26°31′S 56°20′W
160 G13 Yuxi Yunnan, SW China 24°22′N 102°28′E
161 O2 Yuxian prev. Yu Xian. Hebei, E China 39°50′N 114°33′E
Yu Xian see Yuxian
165 Q9 Yuzawa Akita, Honshū, C Japan 39°11′N 140°29′E
125 N16 Yuzha Ivanovskaya Oblast', W Russian Federation 56°34′N 42°00′E
Yuzhno-Alichurskiy Khrebet see Alichuri Janubí, Qatorkŭhi
Yuzhno-Kazakhstanskaya Oblast' see Yuzhnyy Kazakhstan
123 T13 Yuzhno-Sakhalinsk Jap. Toyohara; prev. Vladimirovka. Ostrov Sakhalin, Sakhalinskaya Oblast', SE Russian Federation 46°58′N 142°45′E
127 P14 Yuzhno-Sukhokumsk Respublika Dagestan, SW Russian Federation 44°43′N 45°32′E
145 Z10 Yuzhnyy Altay, Khrebet ▲▲ E Kazakhstan
Yuzhnyy Bug see Pivdennyy Buh
145 O15 Yuzhnyy Kazakhstan off. Yuzhno-Kazakhstanskaya Oblast', Eng. South Kazakhstan, Kaz. Ongtüstik Qazaqstan Oblysy; prev. Chimkentskaya Oblast'. ◐ province S Kazakhstan
123 U10 Yuzhnyy, Mys headland E Russian Federation 57°44′N 156°49′E
122 H6 Yuzhnyy, Ostrov island NW Russian Federation
127 W6 Yuzhnyy Ural var. Southern Urals. ▲▲ W Russian Federation
159 V10 Yuzhong Gansu, C China 35°52′N 104°09′E
Yuzhou see Chongqing
103 N5 Yvelines ◐ department N France
108 B9 Yverdon var. Yverdon-les-Bains, Ger. Iferten; anc. Eborodunum. Vaud, W Switzerland 46°47′N 06°38′E
Yverdon-les-Bains see Yverdon
102 M3 Yvetot Seine-Maritime, N France 49°37′N 00°48′E
Yylanly see Gurbansoltan Eje

Z

147 T12 Zaalayskiy Khrebet Taj. Qatorkŭhi Pasi Oloy. ▲▲ Kyrgyzstan/Tajikistan
Zaamin see Zomin
Zaandam see Zaanstad
98 I10 Zaanstad prev. Zaandam. Noord-Holland, C Netherlands 52°27′N 04°49′E
Zabadani see Az Zabdānī
Zabalatstsye see Zabalotstsye
112 L9 Žabalj Ger. Josefsdorf, Hung. Zsablya; prev. Józseffalva. Vojvodina, N Serbia 45°22′N 20°01′E
119 L18 Zabalotstsye prev. Zabalatstsye, Rus. Zabolot'ye. Homyel'skaya Voblasts', SE Belarus 52°40′N 28°34′E
Zāb aş Şaghīr, Nahraz see Little Zab
123 P14 Zabaykal'sk Zabaykal'skiy Kray, S Russian Federation 49°37′N 117°20′E
123 O12 Zabaykal'skiy Kray ◐ province S Russian Federation
Zāb-e Kūchek, Rūdkhāneh-ye see Little Zab
Zabeln see Sabile
Zabéré see Zabré
Zabern see Saverne
141 N16 Zabīd W Yemen 14°N 43°E
141 O16 Zabīd, Wādī dry watercourse SW Yemen
Żabinka see Zhabinka
Ząbkowice see Ząbkowice Śląskie
111 G15 Ząbkowice Śląskie var. Ząbkowice, Ger. Frankenstein, Frankenstein in Schlesien. Dolnośląskie, SW Poland 50°35′N 16°48′E
110 P10 Zabłudów Podlaskie, NE Poland 53°00′N 23°21′E
112 D8 Zabok Krapina-Zagorje, N Croatia 46°00′N 15°48′E
143 W9 Zābol var. Shahr-i-Zabul, Zabul; prev. Nasratabad. Sīstān va Balūchestān, E Iran 31°N 61°32′E
Zābol see Zābul
143 W13 Zābolī Sīstān va Balūchestān, SE Iran 27°09′N 61°32′E
Zabolot'ye see Zabalotstsye
77 Q13 Zabré var. Zabéré. S Burkina Faso 11°13′N 00°34′W
111 G17 Zábřeh Ger. Hohenstadt. Olomoucký Kraj, E Czech Republic 49°52′N 16°53′E
111 J16 Zabrze Ger. Hindenburg, Hindenburg in Oberschlesien. Śląskie, S Poland 50°18′N 18°47′E
149 O7 Zābul prev. Zābol. ◐ province SE Afghanistan
Zabul/Zābul see Zābol
42 E6 Zacapa Zacapa, E Guatemala 14°59′N 89°33′W
42 A3 Zacapa off. Departamento de Zacapa. ◐ department E Guatemala
Zacapa, Departamento de see Zacapa
40 M14 Zacapú Michoacán, SW Mexico 19°49′N 101°48′W
41 V14 Zacatal Campeche, SE Mexico 18°40′N 91°52′W
40 M11 Zacatecas Zacatecas, C Mexico 22°46′N 102°33′W
40 L10 Zacatecas ◐ state C Mexico
42 F8 Zacatecoluca La Paz, S El Salvador 13°29′N 88°51′W
41 P15 Zacatepec Morelos, S Mexico 18°40′N 99°11′W
41 Q13 Zacatlán Puebla, S Mexico 19°56′N 97°58′W
144 F8 Zachagansk Kaz. Zashaghan. Zapadnyy Kazakhstan, NW Kazakhstan 51°04′N 51°13′E
115 D20 Zacháro var. Zaharo, Zakháro. Dytikí Elláda, S Greece 37°29′N 21°40′E
22 J8 Zachary Louisiana, S USA 30°39′N 91°09′W
117 U6 Zachepylivka Kharkivs'ka Oblast', E Ukraine 49°13′N 35°15′E
Zachist'ye see Zachystsye
110 E9 Zachodnio-pomorskie ◐ province NW Poland
119 L14 Zachystsye Rus. Zachist'ye. Minskaya Voblasts', NW Belarus 54°24′N 28°45′E
40 L13 Zacoalco var. Zacoalco de Torres. Jalisco, SW Mexico 20°14′N 103°33′W
Zacoalco de Torres see Zacoalco
41 P13 Zacualtipán Hidalgo, C Mexico 20°39′N 98°42′W
112 C12 Zadar It. Zara; anc. Iader. Zadar, SW Croatia 44°07′N 15°15′E
112 C12 Zadar off. Zadarsko-Kninska Županija, Zadar-Knin. ◐ province SW Croatia
Zadar-Knin see Zadar
Zadarsko-Kninska Županija see Zadar
166 M14 Zadetkyi Kyun var. St.Matthew's Island. island Mergui Archipelago, S Myanmar (Burma)
67 Q9 Zadié var. Djadié. ≈ NE Gabon
159 Q13 Zadoi var. Qapugtang. Qinghai, C China 32°56′N 95°21′E
126 L7 Zadonsk Lipetskaya Oblast', W Russian Federation 52°25′N 38°55′E
Za'farâna see Za'farānah
75 X8 Za'farānah var. Za'farâna. E Egypt 29°06′N 32°34′E
149 W7 Zafarwāl Punjab, E Pakistan 32°20′N 74°53′E
121 Q1 Zafer Burnu var. Cape Andreas, Cape Apostolas Andreas, Gk. Akrotíri Apostólou Andréa. cape NE Cyprus
107 J23 Zafferano, Capo headland Sicilia, Italy, C Mediterranean Sea 38°06′N 13°31′E
114 M7 Zafirovo Silistra, NE Bulgaria 44°00′N 26°51′E
Záfora see Sofraná
104 J12 Zafra Extremadura, W Spain 38°25′N 06°27′W
110 E13 Żagań var. Zagań, Żegań, Ger. Sagan. Lubuskie, W Poland 51°37′N 15°20′E
118 F10 Žagarė Pol. Žagory. Šiaulai, N Lithuania 56°22′N 23°16′E
Zagazig see Az Zaqāzīq
74 M5 Zaghouan var. Zaghwān. NE Tunisia 36°26′N 10°05′E
Zaghwān see Zaghouan
115 G16 Zagorá Thessalía, C Greece 39°27′N 23°06′E
Zagorod'ye see Zaharoddzye
Žagory see Žagarė
Zágráb see Zagreb
112 E8 Zagreb Ger. Agram, Hung. Zágráb. ● (Croatia) Zagreb, N Croatia 45°48′N 15°58′E
112 E8 Zagreb prev. Grad Zagreb. ◐ province N Croatia

◆ Country
● Country Capital
◇ Dependent Territory
○ Dependent Territory Capital
◐ Administrative Regions
✈ International Airport
▲ Mountain
▲▲ Mountain Range
⅄ Volcano
≈ River
◎ Lake
⊠ Reservoir

ZĀGROS, KŪHHĀ-YE – ZHUJI

142 L7 Zāgros, Kūhhā-ye Eng. Zagros Mountains. ▲▲ W Iran
Zagros Mountains see Zāgros, Kūhhā-ye
112 O12 Žagubica Serbia, E Serbia 44°13′N 21°47′E
Zagunao see Lixian
111 L22 Zagyva ∿ N Hungary
Zaharo see Zacháro
119 G19 Zaharoddzye Rus. Zagorod'ye. physical region SW Belarus
143 W11 Zāhedān var. Zahidan; prev. Duzdab. Sīstān va Balūchestān, SE Iran 29°31′N 60°51′E
Zahidan see Zāhedān
Zaḩlah see Zahlé
138 H7 Zahlé var. Zaḩlah. C Lebanon 33°51′N 35°54′E
146 J14 Zähmet Rus. Zakhmet. Mary Welaýaty, C Turkmenistan 37°48′N 62°33′E
111 O20 Záhony Szabolcs-Szatmár-Bereg, NE Hungary 48°26′N 22°11′E
141 N13 Zahrān 'Asīr, S Saudi Arabia 17°48′N 43°28′E
139 R12 Zahrat al Baṭn hill range S Iraq
120 H11 Zahrez Chergui var. Zahrez Chergüi. marsh N Algeria
Zainlha see Xinjin
127 S4 Zainsk Respublika Tatarstan, W Russian Federation 55°12′N 52°01′E
82 A10 Zaire prev. Congo. ◈ province NW Angola
Zaire see Congo (river)
Zaire see Congo (Democratic Republic of)
112 P13 Zaječar Serbia, E Serbia 43°54′N 22°16′E
83 L18 Zaka Masvingo, E Zimbabwe 20°20′S 31°29′E
122 M14 Zakamensk Respublika Buryatiya, S Russian Federation 50°18′N 102°57′E
116 G7 Zakarpats'ka Oblast' Eng. Transcarpathian Oblast, Rus. Zakarpatskaya Oblast'. ◈ province W Ukraine
Zakarpatskaya Oblast' see Zakarpats'ka Oblast'
Zakataly see Zaqatala
Zakháro see Zacháro
Zakhidnyy Buh/Zakhodni Buh see Bug
Zakhmet see Zähmet
Zākhō see Zaxo
Zākhū see Zaxo
Zákinthos see Zákynthos
111 L18 Zakopane Małopolskie, S Poland 49°17′N 19°57′E
78 J12 Zakouma Salamat, S Chad 10°47′N 19°51′E
115 L25 Zákros Kríti, Greece, E Mediterranean Sea 35°06′N 26°12′E
115 C19 Zákynthos var. Zákinthos. Zákynthos, W Greece 37°47′N 20°54′E
115 C20 Zákynthos var. Zákinthos, It. Zante. island Iónia Nísoi, Greece, C Mediterranean Sea
115 C19 Zakýnthou, Porthmós strait SW Greece
111 G24 Zala off. Zala Megye. ◈ county W Hungary
111 G24 Zala ∿ W Hungary
138 M4 Zalābiyah Dayr az Zawr, C Syria 35°39′N 39°51′E
111 G24 Zalaegerszeg Zala, W Hungary 46°51′N 16°49′E
104 K11 Zalamea de la Serena Extremadura, W Spain 38°38′N 05°37′W
104 J13 Zalamea la Real Andalucía, S Spain 37°41′N 06°40′W
Zala Megye see Zala
163 U7 Zalantun var. Butha Qi. Nei Mongol Zizhiqu, N China 47°58′N 122°44′E
111 G23 Zalaszentgrót Zala, SW Hungary 46°57′N 17°05′E
Zalatna see Zlatna
116 G9 Zalău Ger. Waltenberg, Hung. Zilah; prev. Ger. Zillenmarkt. Sălaj, NW Romania 47°11′N 23°03′E
109 V10 Žalec Ger. Sachsenfeld. C Slovenia 46°15′N 15°08′E
110 K8 Zalewo Ger. Saalfeld. Warmińsko-Mazurskie, NE Poland 53°54′N 19°39′E
141 N9 Zalim Makkah, W Saudi Arabia 22°46′N 42°12′E
80 A11 Zalingei var. Zalinje. Central Darfur, W Sudan 12°51′N 23°29′E
Zalinje see Zalingei
116 K7 Zalishchyky Ternopil's'ka Oblast', W Ukraine 48°40′N 25°43′E
Zallah see Zillah
`Zaltni Pjašaci see Zlatni Pyasatsi
98 J13 Zaltbommel Gelderland, C Netherlands 51°49′N 05°15′E
124 H15 Zaluch'ye Novgorodskaya Oblast', NW Russian Federation 57°40′N 31°45′E
Zamak see Zamakh
141 Q14 Zamakh var. Zamak. N Yemen 16°26′N 47°35′E
136 K15 Zamantı Irmağı ∿ C Turkey
Zambesi/Zambeze see Zambezi
83 G14 Zambezi North Western, W Zambia 13°34′S 23°08′E
83 K15 Zambezi var. Zambesi, Port. Zambeze. ∿ S Africa
83 O15 Zambézia off. Província da Zambézia. ◈ province C Mozambique
Zambézia, Província da see Zambézia
83 I14 Zambia off. Republic of Zambia; prev. Northern Rhodesia. ◆ republic S Africa
Zambia, Republic of see Zambia
171 O8 Zamboanga off. Zamboanga City. Mindanao, S Philippines 06°56′N 122°03′E
Zamboanga City see Zamboanga
54 E5 Zambrano Bolívar, N Colombia 09°45′N 74°50′W
110 N10 Zambrów Łomża, E Poland 52°59′N 22°14′E
83 L14 Zambuè Tete, NW Mozambique 15°03′S 30°49′E
77 T13 Zamfara ∿ NW Nigeria
Zamkog see Zamtang
56 C9 Zamora Zamora Chinchipe, S Ecuador 04°04′S 78°52′W
104 K6 Zamora Castilla y León, NW Spain 41°30′N 05°45′W
104 K5 Zamora ◈ province Castilla y León, NW Spain
Zamora see Barinas
56 A13 Zamora Chinchipe ◈ province S Ecuador
40 M13 Zamora de Hidalgo Michoacán, SW Mexico 20°N 102°18′W
111 P15 Zamość Rus. Zamoste. Lubelskie, E Poland 50°44′N 23°16′E
Zamoste see Zamość
160 G7 Zamtang var. Zamkog; prev. Gamba. Sichuan, C China 32°19′N 100°55′E
75 O8 Zamzam, Wādī dry watercourse NW Libya
79 F20 Zanaga Lékoumou, S Congo 02°50′S 13°53′E
41 T16 Zanatepec Oaxaca, SE Mexico 16°28′N 94°24′W
105 P9 Záncara ∿ C Spain
Zancle see Messina
158 G14 Zanda Xizang Zizhiqu, W China 31°29′N 79°50′E
98 H10 Zandvoort Noord-Holland, W Netherlands 52°22′N 04°31′E
39 P8 Zane Hills hill range Alaska, USA
31 T13 Zanesville Ohio, N USA 39°55′N 82°02′W
Zanga see Hrazdan
Zangoza see Sangüesa
142 L4 Zanjān var. Zenjan, Zinjan. Zanjān, NW Iran 36°40′N 48°30′E
142 L4 Zanjān off. Ostān-e Zanjān, var. Zenjan, Zinjan. ◈ province NW Iran
Zanjān, Ostān-e see Zanjān
Zante see Zákynthos
81 J22 Zanzibar Zanzibar, E Tanzania 06°10′S 39°12′E
81 J22 Zanzibar ◈ region E Tanzania
81 J22 Zanzibar Swa. Unguja. island E Tanzania
81 J22 Zanzibar Channel channel E Tanzania
161 N8 Zaoyang Hubei, C China 32°10′N 112°45′E
165 P10 Zaō-zan ▲ Honshū, C Japan 38°06′N 140°27′E
124 J2 Zaozërsk Murmanskaya Oblast', NW Russian Federation 69°25′N 32°25′E
161 Q6 Zaozhuang Shandong, E China 34°53′N 117°38′E
28 L4 Zap North Dakota, N USA 47°18′N 101°55′W
112 L13 Zapadna Morava Ger. Westliche Morava. ∿ C Serbia
124 H16 Zapadnaya Dvina Tverskaya Oblast', W Russian Federation 56°17′N 32°03′E
Zapadnaya Dvina see Western Dvina
Zapadno-Kazakhstanskaya Oblast' see Zapadnyy Kazakhstan
122 J9 Zapadno-Sibirskaya Ravnina Eng. West Siberian Plain. plain C Russian Federation
Zapadnyy Bug see Bug
144 E9 Zapadnyy Kazakhstan off. Zapadno-Kazakhstanskaya Oblast', Eng. West Kazakhstan, Kaz. Batys Qazaqstan Oblysy; prev. Ural'skaya Oblast'. ◈ province NW Kazakhstan
122 K13 Zapadnyy Sayan Eng. Western Sayans. ▲▲ S Russian Federation
63 H15 Zapala Neuquén, W Argentina 38°54′S 70°06′W
62 I4 Zapaleri, Cerro var. Cerro Sapaleri. ▲ N Chile 22°51′S 67°10′W
25 Q16 Zapata Texas, SW USA 26°57′N 99°17′W
44 D5 Zapata, Península de peninsula W Cuba
61 G19 Zapicán Lavalleja, S Uruguay 33°31′S 54°55′W
65 J19 Zapiola Ridge undersea feature SW Atlantic Ocean
65 L19 Zapiola Seamount undersea feature S Atlantic Ocean 38°15′S 26°15′W
124 I2 Zapolyarnyy Murmanskaya Oblast', NW Russian Federation 69°24′N 30°53′E
117 U8 Zaporizhzhya Rus. Zaporozh'ye; prev. Aleksandrovsk. Zaporiz'ka Oblast', SE Ukraine 47°47′N 35°12′E
Zaporizhzhya see Zaporiz'ka Oblast'
117 U9 Zaporiz'ka Oblast' var. Zaporizhzhya, Rus. Zaporozhskaya Oblast'. ◈ province SE Ukraine
Zaporozhskaya Oblast' see Zaporiz'ka Oblast'
Zaporozh'ye see Zaporizhzhya
40 L14 Zapotiltic Jalisco, SW Mexico 19°40′N 103°29′W
158 G13 Zapug Xizang Zizhiqu, W China
137 V10 Zaqatala Rus. Zakataly. NW Azerbaijan 41°38′N 46°38′E
159 P13 Zaqên Qinghai, W China 33°23′N 94°31′E
159 Q13 Za Qu ∿ C China
136 M13 Zara Sivas, C Turkey 39°55′N 37°44′E
Zara see Zadar
Zarafshan see Zarafshon
147 P12 Zarafshon Rus. Zeravshan. W Tajikistan 39°12′N 68°36′E
146 L9 Zarafshon Rus. Zarafshan. Navoiy Viloyati, N Uzbekistan 41°33′N 64°09′E
Zarafshon see Zeravshan
147 O12 Zarafshon, Qatorkŭhi Rus. Zeravshanskiy Khrebet, Uzb. Zarafshon Tizmasi. ▲▲ Tajikistan/Uzbekistan
Zarafshon Tizmasi see Zarafshon, Qatorkŭhi
54 E7 Zaragoza Antioquia, N Colombia 07°30′N 74°52′W
40 I5 Zaragoza Chihuahua, N Mexico 29°36′N 107°41′W
41 N6 Zaragoza Coahuila, NE Mexico 28°31′N 100°54′W
41 O10 Zaragoza Nuevo León, NE Mexico 23°59′N 99°49′W
105 R5 Zaragoza Eng. Saragossa; anc. Caesaraugusta, Salduba. Aragón, NE Spain 41°39′N 00°54′W
105 R6 Zaragoza ◈ province Aragón, NE Spain
105 R5 Zaragoza ✈ Aragón, NE Spain 41°38′N 00°53′W
143 S10 Zarand Kermān, C Iran 30°50′N 56°35′E
148 J9 Zaranj Nīmrōz, SW Afghanistan 30°59′N 61°54′E
118 I11 Zarasai Utena, E Lithuania 55°44′N 26°17′E
62 N12 Zárate prev. General José F.Uriburu. Buenos Aires, E Argentina 34°06′S 59°03′W
105 Q2 Zarautz var. Zarauz. País Vasco, N Spain 43°17′N 02°10′W
Zarauz see Zarautz
Zaravecchia see Biograd na Moru
Zarāyīn see Zarēn
126 L4 Zaraysk Moskovskaya Oblast', W Russian Federation 54°48′N 38°54′E
55 N6 Zaraza Guárico, N Venezuela 09°23′N 65°20′W
Zarbdar see Zarbdor
147 P11 Zarbdor Rus. Zarbdar. Jizzax Viloyati, C Uzbekistan 40°04′N 68°10′E
142 M8 Zard Kūh ▲ SW Iran 32°19′N 50°03′E
124 I5 Zarechensk Murmanskaya Oblast', NW Russian Federation 66°39′N 31°27′E
127 P6 Zarechnyy Penzenskaya Oblast', W Russian Federation 53°12′N 45°12′E
Zareh Sharan see Sharan
39 Y14 Zarembo Island island Alexander Archipelago, Alaska, USA
139 V4 Zarēn var. Zarāyīn. As Sulaymānīyah, E Iraq 35°16′N 45°43′E
149 Q7 Zarghūn Shahr var. Katawaz. Paktīkā, SE Afghanistan 32°40′N 68°20′E
77 V13 Zaria Kaduna, C Nigeria 11°06′N 07°42′E
116 K2 Zarichne Rivnens'ka Oblast', NW Ukraine 51°49′N 26°09′E
122 J13 Zarinsk Altayskiy Kray, S Russian Federation 53°34′N 85°22′E
116 J12 Zărneşti Hung. Zernest. Braşov, C Romania 45°34′N 25°18′E
115 J25 Zarós Kríti, Greece, E Mediterranean Sea 35°08′N 24°54′E
100 O9 Zarow ∿ NE Germany
Zarqa see Az Zarqā'
Zarqa/Muḩāfaẓat az Zarqā' see Az Zarqā'
111 G20 Záruby ▲ W Slovakia 48°30′N 17°24′E
56 B8 Zaruma El Oro, SW Ecuador 03°46′S 79°38′W
110 E13 Żary Ger. Sorau, Sorau in der Niederlausitz. Lubuskie, W Poland 51°44′N 15°09′E
54 D10 Zarzal Valle del Cauca, W Colombia 04°24′N 76°01′W
42 I7 Zarzalar, Cerro ▲ S Honduras 14°15′N 86°49′W
Zashaghan see Zachagansk
152 I5 Zāskār ∿ NE India
152 I5 Zāskār Range ▲▲ NE India
Zaslavl' see Zaslawye
119 K15 Zaslawye Rus. Zaslavl'. Minskaya Voblasts', C Belarus 54°01′N 27°16′E
116 K7 Zastavna Chernivets'ka Oblast', W Ukraine 48°30′N 25°51′E
111 B16 Žatec Ger. Saaz. Ústecký Kraj, NW Czech Republic 50°20′N 13°35′E
Zaumgarten see Chrzanów
Zaunguzskiye Garagumy see Üngüz Angyrsyndaky Garagum
25 X9 Zavalla Texas, SW USA 31°09′N 94°25′W
99 H18 Zaventem Vlaams Brabant, C Belgium 50°53′N 04°28′E
99 H18 Zaventem ✈ (Brussel/Bruxelles) Vlaams Brabant, C Belgium 50°55′N 04°28′E
Zavertse see Zawiercie
114 L7 Zavet Razgrad, NE Bulgaria 43°46′N 26°40′E
127 O12 Zavetnoye Rostovskaya Oblast', SW Russian Federation 47°10′N 43°54′E
156 M3 Zavhan Gol ∿ W Mongolia
112 H12 Zavidovići Federacija Bosne I Hercegovine, N Bosnia and Herzegovina 44°26′N 18°07′E
123 R13 Zavitinsk Amurskaya Oblast', SE Russian Federation 50°23′N 128°57′E
Zawia see Az Zāwiyah
111 K15 Zawiercie Rus. Zavertse. Śląskie, S Poland 50°29′N 19°24′E
75 P11 Zawīlah var. Zuwaylah, It. Zueila. C Libya 26°10′N 15°07′E
138 I4 Zāwīyah, Jabal az ▲▲ NW Syria
139 Q1 Zaxo Ar. Zākhū, var. Zākhō. Dahūk, N Iraq 37°09′N 42°40′E
109 Y3 Zaya ∿ NE Austria
166 M8 Zayatkyi Bago, C Myanmar (Burma) 17°48′N 96°27′E
145 Y11 Zaysan Vostochnyy Kazakhstan, E Kazakhstan 47°30′N 84°55′E
Zaysan Köl see Zaysan, Ozero
145 Y11 Zaysan, Ozero Kaz. Zaysan Köl. ◎ E Kazakhstan
159 R16 Zayü var. Gyigang. Xizang Zizhiqu, W China 28°36′N 97°25′E
Zayyq see Zhayyk
44 F6 Zaza ∿ C Cuba
116 K5 Zbarazh Ternopil's'ka Oblast', W Ukraine 49°40′N 25°47′E
116 J5 Zboriv Ternopil's'ka Oblast', W Ukraine 49°40′N 25°07′E
111 F18 Zbraslav Jihomoravský Kraj, SE Czech Republic 49°13′N 16°19′E
116 K6 Zbruch ∿ W Ukraine
Žďár see Žďár nad Sázavou
111 F17 Žďár nad Sázavou Ger. Saar in Mähren; prev. Žďár. Vysočina, C Czech Republic 49°34′N 16°00′E
116 K4 Zdolbuniv Pol. Zdolbunów, Rus. Zdolbunov. Rivnens'ka Oblast', NW Ukraine 50°33′N 26°15′E
Zdolbunov/Zdolbunów see Zdolbuniv
110 J13 Zduńska Wola Sieradz, C Poland 51°37′N 18°57′E
117 O4 Zdvizh ∿ N Ukraine
Zdzięciół see Dzyatlava
111 I16 Zdzieszowice Ger. Odertal. Opolskie, SW Poland 50°24′N 18°06′E
Zealand see Sjælland
188 K6 Zealandia Bank undersea feature C Pacific Ocean
63 H20 Zeballos, Monte ▲ S Argentina 47°04′S 71°32′W
83 K20 Zebediela Limpopo, NE South Africa 24°16′S 29°21′E
113 L18 Zebës, Mali i var. Mali i Zebës. ▲ NE Albania 41°57′N 20°16′E
Zebës, Mali i see Zebës, Mali i
21 V9 Zebulon North Carolina, SE USA 35°49′N 78°19′W
112 K8 Žednik Hung. Bácsjózseffalva. Vojvodina, N Serbia 45°58′N 19°40′E
99 C15 Zeebrugge West-Vlaanderen, NW Belgium 51°20′N 03°13′E
183 N16 Zeehan Tasmania, SE Australia 41°54′S 145°19′E
99 L14 Zeeland Noord-Brabant, SE Netherlands 51°42′N 05°40′E
29 N7 Zeeland North Dakota, N USA 45°57′N 99°49′W
99 E14 Zeeland ◈ province SW Netherlands
83 I21 Zeerust North-West, N South Africa 25°33′S 26°06′E
98 K10 Zeewolde Flevoland, C Netherlands 52°20′N 05°32′E
Zefat see Tsefat
Żegań see Żagań
Zehden see Cedynia
100 O11 Zehdenick Brandenburg, NE Germany 52°58′N 13°19′E
Zē-i Bādīnān see Great Zab
Zeiden see Codlea
146 M14 Zeidskoye Vodokhranilishche ⊠ E Turkmenistan
Zē-i Kōya see Little Zab
181 P7 Zeil, Mount ▲ Northern Territory, C Australia 23°31′S 132°41′E
98 J12 Zeist Utrecht, C Netherlands 52°05′N 05°15′E
101 M16 Zeitz Sachsen-Anhalt, E Germany 51°03′N 12°08′E
159 T11 Zêkog var. Zequ; prev. Sonag. Qinghai, C China 35°03′N 101°30′E
Zelaya Norte see Atlántico Norte, Región Autónoma
Zelaya Sur see Atlántico Sur, Región Autónoma
99 F17 Zele Oost-Vlaanderen, NW Belgium 51°04′N 04°02′E
110 N12 Żelechów Lubelskie, E Poland 51°49′N 21°57′E
113 H14 Zelena Glava ▲ SE Bosnia and Herzegovina 43°32′N 17°55′E
113 P20 Zelen Breg ▲ S Macedonia 41°10′N 22°14′E
113 I14 Zelengora ▲▲ S Bosnia and Herzegovina
124 I5 Zelenoborskiy Murmanskaya Oblast', NW Russian Federation 66°52′N 32°25′E
127 R3 Zelenodol'sk Respublika Tatarstan, W Russian Federation 55°52′N 48°49′E
117 S9 Zelenodols'k Dnipropetrovs'ka Oblast', E Ukraine 47°38′N 33°41′E
122 J12 Zelenogorsk Krasnoyarskiy Kray, C Russian Federation 56°08′N 94°29′E
126 K3 Zelenograd Moskovskaya Oblast', W Russian Federation 56°02′N 37°08′E
118 B13 Zelenogradsk Ger. Cranz, Kranz. Kaliningradskaya Oblast', W Russian Federation 54°58′N 20°30′E
127 O15 Zelenokumsk Stavropol'skiy Kray, SW Russian Federation 44°22′N 43°48′E
165 X4 Zelënyy, Ostrov var. Shibotsu-jima. island NE Russian Federation
Železna Kapela see Eisenkappel
Železna Vrata see Demir Kapija
112 L11 Železniki Serbia, N Serbia 44°45′N 20°23′E
98 N12 Zelhem Gelderland, E Netherlands 52°00′N 06°21′E
113 N18 Želino NW FYR Macedonia 42°00′N 21°06′E
113 M14 Željin ▲ C Serbia
101 K17 Zella-Mehlis Thüringen, C Germany 50°40′N 10°40′E
109 P7 Zell am See var. Zell-am-See. Salzburg, S Austria 47°19′N 12°47′E
Zell-am-See see Zell am See
109 N7 Zell am Ziller Tirol, W Austria 47°13′N 11°52′E
Zelle see Celle
109 W2 Zellerndorf Niederösterreich, NE Austria 48°40′N 15°57′E
109 U7 Zeltweg Steiermark, S Austria 47°12′N 14°46′E
119 G17 Zel'va Pol. Zelwa. Hrodzyenskaya Voblasts', W Belarus 53°09′N 24°49′E
118 H13 Zelva Vilnius, C Lithuania 55°13′N 25°07′E
Zelwa see Zel'va
99 E16 Zelzate var. Selzaete. Oost-Vlaanderen, NW Belgium 51°12′N 03°49′E
118 E11 Žemaičių Aukštumas physical region W Lithuania
118 C12 Žemaičių Naumiestis Klaipėda, SW Lithuania 55°22′N 21°39′E
119 L14 Zembin var. Zyembin. Minskaya Voblasts', C Belarus 54°22′N 28°13′E
127 N6 Zemetchino Penzenskaya Oblast', W Russian Federation 53°31′N 42°35′E
79 M15 Zémio Haut-Mbomou, E Central African Republic 05°04′N 25°07′E
41 R16 Zempoaltepec, Cerro ▲ SE Mexico 17°04′N 95°54′W
99 G17 Zemst Vlaams Brabant, C Belgium 50°59′N 04°28′E
112 L11 Zemun Serbia, N Serbia 44°52′N 20°25′E
Zendajan see Zindah Jān
Zendeh Jān see Zindah Jān
Zengg see Senj
112 H12 Zenica Federacija Bosne I Hercegovine, C Bosnia and Herzegovina 44°12′N 17°53′E
Zenjan see Zanjān
Zen'kov see Zin'kiv
Zenshū see Jeonju
Zenta see Senta
82 B11 Zenza do Itombe Kwanza Norte, NW Angola 09°22′S 14°10′E
112 H12 Žepče Federacija Bosne I Hercegovine, N Bosnia and Herzegovina 44°26′N 18°00′E
23 W12 Zephyrhills Florida, SE USA 28°13′N 82°10′W
158 F9 Zepu var. Poskam. Xinjiang Uygur Zizhiqu, NW China 38°10′N 77°18′E
Zequ see Zêkog
147 Q12 Zeravshan Taj./Uzb. Zarafshon. ∿ Tajikistan/Uzbekistan
Zeravshan see Zarafshon
Zeravshanskiy Khrebet see Zarafshon, Qatorkŭhi
101 M14 Zerbst Sachsen-Anhalt, E Germany 51°57′N 12°05′E
Zerenda see Zerendy
145 P8 Zerendy prev. Zerenda. Akmola, N Kazakhstan 52°56′N 69°09′E
110 H12 Żerków Wielkopolskie, C Poland 52°03′N 17°33′E
108 E11 Zermatt Valais, SW Switzerland 46°00′N 07°45′E
Zernest see Zărneşti
108 J9 Zernez Graubünden, SE Switzerland 46°42′N 10°06′E
126 L12 Zernograd Rostovskaya Oblast', SW Russian Federation 46°52′N 40°13′E
Zestafoni see Zest'aponi
137 S9 Zest'aponi Rus. Zestafoni; prev. Zestap'oni. C Georgia 42°09′N 43°00′E
Zestap'oni see Zest'aponi
98 H12 Zestienhoven ✈ (Rotterdam) Zuid-Holland, SW Netherlands 51°57′N 04°30′E
113 J16 Zeta ∿ C Montenegro
8 L6 Zeta Lake ◎ Victoria Island, Northwest Territories, N Canada
98 L12 Zetten Gelderland, SE Netherlands 51°55′N 05°43′E
101 M17 Zeulenroda Thüringen, C Germany 50°40′N 11°58′E
100 H10 Zeven Niedersachsen, NW Germany 53°17′N 09°16′E
98 M12 Zevenaar Gelderland, SE Netherlands 51°55′N 06°05′E
99 H14 Zevenbergen Noord-Brabant, S Netherlands 51°39′N 04°36′E
129 X6 Zeya ∿ SE Russian Federation
Zeya Reservoir see Zeyskoye Vodokhranilishche
143 T11 Zeynalābād Kermān, C Iran 29°56′N 57°29′E
123 R12 Zeyskoye Vodokhranilishche Eng. Zeya Reservoir. ⊠ SE Russian Federation
104 H8 Zêzere, Rio ∿ C Portugal
Zgerzh see Zgierz
138 H6 Zgharta N Lebanon 34°24′N 35°54′E
110 K12 Zgierz Ger. Neuhof, Rus. Zgerzh. Łódź, C Poland 51°55′N 19°20′E
111 E14 Zgorzelec Ger. Görlitz. Dolnośląskie, SW Poland 51°10′N 15°E
119 F19 Zhabinka Pol. Żabinka. Brestskaya Voblasts', SW Belarus 52°12′N 24°01′E
Zhaggo see Luhuo
159 R15 Zhag'yab var. Yêndum. Xizang Zizhiqu, W China 30°42′N 97°33′E
Zhailma see Zhayylma
145 N14 Zhalagash prev. Dzhalagash. Kzylorda, S Kazakhstan 45°06′N 64°40′E
145 V16 Zhalanash Almaty, SE Kazakhstan 43°04′N 78°08′E
145 S7 Zhalauly, Ozero ◎ NE Kazakhstan
145 W10 Zhalgyztobe prev. Zhangiztobe. Vostochnyy Kazakhstan, E Kazakhstan 49°15′N 81°16′E
144 E9 Zhalpaktal Kaz. Zhalpaqtal; prev. Furmanovo. Zapadnyy Kazakhstan, W Kazakhstan 49°43′N 49°28′E
Zhalpaqtal see Zhalpaktal
119 G16 Zhaludok Rus. Zheludok. Hrodzyenskaya Voblasts', W Belarus 53°36′N 24°59′E
Zhaman-Akkol', Ozero see Akkol', Ozero
145 Q14 Zhambyl off. Zhambylskaya Oblast', Kaz. Zhambyl Oblysy; prev. Dzhambulskaya Oblast'. ◈ province S Kazakhstan
Zhambyl see Taraz
Zhambyl Oblysy/Zhambylskaya Oblast' see Zhambyl
Zhamo see Bomi
145 S12 Zhamshy ∿ C Kazakhstan
144 M15 Zhanadariya prev. Zhanadar'ya. Kzylorda, S Kazakhstan 44°41′N 64°39′E
Zhanadar'ya see Zhanadariya
144 E10 Zhanakazan prev. Novaya Kazanka. Zapadnyy Kazakhstan, W Kazakhstan 48°57′N 49°34′E
145 O15 Zhanakorgan Kaz. Zhangaqorghan. Kzylorda, S Kazakhstan 43°57′N 67°14′E
159 N16 Zhanang var. Chatang. Xizang Zizhiqu, W China 29°15′N 91°20′E
145 T12 Zhanaortalyk Karaganda, C Kazakhstan 47°31′N 75°42′E
144 F15 Zhanaozen Kaz. Zhangaözen; prev. Novyy Uzen'. Mangistau, W Kazakhstan 43°22′N 52°50′E
145 Q16 Zhanatas Zhambyl, S Kazakhstan 43°36′N 69°43′E
Zhangaözen see Zhanaozen
Zhangaqazaly see Ayteke Bi
Zhangaqorghan see Zhanakorgan
161 O2 Zhangbei Hebei, E China 41°13′N 114°43′E
Zhang-chia-k'ou see Zhangjiakou
Zhangdian see Zibo
Zhanggu see Danba
163 X9 Zhangguangcai Ling ▲▲ NE China
161 S13 Zhanghua Jap. Shōka; prev. Changhua. C Taiwan 24°06′N 120°31′E
Zhangiztobe see Zhalgyztobe
159 W11 Zhangjiachuan Gansu, N China 34°55′N 106°26′E
160 L10 Zhangjiajie var. Dayong. Hunan, S China 29°10′N 110°22′E
161 O2 Zhangjiakou var. Changkiakow, Zhang-chia-k'ou, Eng. Kalgan; prev. Wanchuan. Hebei, E China 40°48′N 114°51′E
161 Q13 Zhangping Fujian, SE China 25°21′N 117°29′E
161 Q13 Zhangpu var. Sui'an. Fujian, SE China 24°08′N 117°36′E
163 U11 Zhangwu Liaoning, NE China 42°21′N 122°32′E
159 S8 Zhangye var. Ganzhou. Gansu, N China 38°58′N 100°30′E
161 Q13 Zhangzhou Fujian, SE China 24°31′N 117°40′E
163 W6 Zhan He ∿ NE China
Zhänibek see Dzhanibek
160 L16 Zhanjiang var. Chanchiang, Chan-chiang, Cant. Tsamkong, Fr. Fort-Bayard. Guangdong, S China 21°10′N 110°20′E
145 V14 Zhansugirov prev. Dzhansugurov. Almaty, SE Kazakhstan 45°23′N 79°29′E
163 V8 Zhaodong Heilongjiang, NE China 46°03′N 125°58′E
Zhaoge see Qixian
160 H11 Zhaojue var. Xincheng. Sichuan, C China 28°03′N 102°50′E
161 N14 Zhaoqing Guangdong, S China 23°08′N 112°26′E
Zhaoren see Changwu
158 H5 Zhaosu var. Mongolküre. Xinjiang Uygur Zizhiqu, NW China 43°09′N 81°07′E
160 H11 Zhaotong Yunnan, SW China 27°20′N 103°29′E
163 V9 Zhaoyuan Heilongjiang, NE China 45°30′N 125°05′E
163 V9 Zhaozhou Heilongjiang, NE China 45°42′N 125°11′E
145 X13 Zharbulak Vostochnyy Kazakhstan, E Kazakhstan 46°04′N 82°05′E
158 J15 Zhari Namco ◎ W China
144 I12 Zharkamys Kaz. Zharqamys. Aktyubinsk, W Kazakhstan 47°58′N 56°33′E
145 W15 Zharkent prev. Panfilov. Taldykorgan, SE Kazakhstan 44°10′N 80°01′E
124 H17 Zharkovskiy Tverskaya Oblast', W Russian Federation 55°51′N 32°19′E
145 W11 Zharma Vostochnyy Kazakhstan, E Kazakhstan 48°48′N 80°55′E
144 F14 Zharmysh Mangistau, SW Kazakhstan 44°12′N 52°27′E
Zharqamys see Zharkamys
118 L13 Zhary Vitsyebskaya Voblasts', N Belarus 55°05′N 28°40′E
Zhaslyk see Jasliq
158 J14 Zhaxi Co ◎ W China
127 X6 Zhayyk Kaz. Zayyq, var. Ural. ∿ Kazakhstan/Russian Federation
144 L9 Zhayylma prev. Zhailma. Kostanay, N Kazakhstan 51°34′N 61°39′E
Zhdanov see Beyläqan
Zhdanov see Mariupol'
Zhe see Zhejiang
161 R10 Zhejiang var. Che-chiang, Chekiang, Zhe, Zhejiang Sheng. ◈ province SE China
Zhejiang Sheng see Zhejiang
145 S7 Zhelezinka Pavlodar, N Kazakhstan 53°35′N 75°16′E
119 C14 Zheleznodorozhnyy Ger. Gerdauen. Kaliningradskaya Oblast', W Russian Federation 54°21′N 21°17′E
Zheleznodorozhnyy see Yemva
122 K12 Zheleznogorsk Krasnoyarskiy Kray, C Russian Federation 56°20′N 93°36′E
126 J7 Zheleznogorsk Kurskaya Oblast', W Russian Federation 52°22′N 35°21′E
127 N15 Zheleznovodsk Stavropol'skiy Kray, SW Russian Federation 44°12′N 43°01′E
Zhëltyye Vody see Zhovti Vody
Zheludok see Zhaludok
144 H12 Zhem prev. Emba. ∿ W Kazakhstan
160 K7 Zhenba Shaanxi, C China 32°42′N 107°55′E
160 I13 Zhenfeng var. Mingu. Guizhou, S China 25°27′N 105°38′E
Zhengjiatun see Shuangliao
159 X10 Zhengning var. Shanhe. Gansu, N China 35°29′N 108°21′E
Zhengxiangbai Qi see Qagan Nur
161 N6 Zhengzhou var. Ch'eng-chou, Chengchow; prev. Chenghsien. province capital Henan, C China 34°45′N 113°38′E
161 R8 Zhenjiang var. Chenkiang. Jiangsu, E China 32°08′N 119°30′E
163 U9 Zhenlai Jilin, NE China 45°52′N 123°11′E
160 I11 Zhenxiong Yunnan, SW China 27°31′N 104°52′E
160 K11 Zhenyuan var. Wuyang. Guizhou, S China 27°07′N 108°33′E
161 R11 Zherong var. Shuangcheng. Fujian, SE China 27°16′N 119°54′E
145 U15 Zhetigen prev. Nikolayevka. Almaty, SE Kazakhstan 43°39′N 77°10′E
Zhetiqara see Zhitikara
144 F15 Zhetybay Mangistau, SW Kazakhstan 43°35′N 52°05′E
145 P17 Zhetysay var. Dzhetysay. Yuzhnyy Kazakhstan, S Kazakhstan 40°45′N 68°18′E
145 W14 Zhetysuskiy Alatau prev. Dzhungarskiy Alatau. ▲▲ China/Kazakhstan
160 M11 Zhexi Shuiku ⊠ C China
145 O12 Zhezdy Karaganda, C Kazakhstan 48°06′N 67°01′E
145 O12 Zhezkazgan Kaz. Zhezqazghan; prev. Dzhezkazgan. Karaganda, C Kazakhstan 47°49′N 67°44′E
Zhezqazghan see Zhezkazgan
Zhicheng see Yidu
Zhidachov see Zhydachiv
159 Q12 Zhidoi var. Gyaijêpozhanggê. Qinghai, C China 33°55′N 95°39′E
122 M13 Zhigalovo Irkutskaya Oblast', S Russian Federation 54°47′N 105°00′E
127 R6 Zhigulevsk Samarskaya Oblast', W Russian Federation 53°24′N 49°30′E
118 D13 Zhilino Ger. Schillen. Kaliningradskaya Oblast', W Russian Federation 54°55′N 21°54′E
Zhiloy, Ostrov see Çilov Adası
127 O8 Zhirnovsk Volgogradskaya Oblast', SW Russian Federation 51°01′N 44°49′E
160 M12 Zhishan prev. Yongzhou. Hunan, S China 26°12′N 111°36′E
Zhishan see Lingling
144 L8 Zhitikara Kaz. Zhetiqara; prev. Džetygara. Kostanay, NW Kazakhstan 52°14′N 61°12′E
Zhitkovichi see Zhytkavichy
Zhitomir see Zhytomyr
Zhitomirskaya Oblast' see Zhytomyrs'ka Oblast'
126 J5 Zhizdra Kaluzhskaya Oblast', W Russian Federation 53°38′N 34°39′E
119 N18 Zhlobin Homyel'skaya Voblasts', SE Belarus 52°53′N 30°01′E
Zhmerinka see Zhmerynka
116 M7 Zhmerynka Rus. Zhmerinka. Vinnyts'ka Oblast', C Ukraine 49°00′N 28°02′E
149 R9 Zhob var. Fort Sandeman. Baluchistān, SW Pakistan 31°21′N 69°31′E
149 R8 Zhob ∿ C Pakistan
Zhodino see Zhodzina
119 L15 Zhodzina Rus. Zhodino. Minskaya Voblasts', C Belarus 54°06′N 28°21′E
123 Q5 Zhokhova, Ostrov island Novosibirskiye Ostrova, NE Russian Federation
Zholkev/Zholkva see Zhovkva
Zhondor see Jondor
158 I15 Zhongba var. Tuoji. Xizang Zizhiqu, W China 29°37′N 84°11′E
Zhongba see Jiangyou
Zhongdian see Xamgyi'nyilha
Zhongduo see Youyang
Zhonghe see Xiushan
Zhonghua Renmin Gongheguo see China
Zhongjian Dao see Triton Island
159 V9 Zhongning Ningxia, N China 37°26′N 105°40′E
Zhongping see Huize
161 N15 Zhongshan Guangdong, S China 22°30′N 113°20′E
195 X7 Zhongshan Chinese research station Antarctica 69°23′S 76°34′E
160 M6 Zhongtiao Shan ▲▲ C China
159 V9 Zhongwei Ningxia, N China 37°31′N 105°10′E
160 K9 Zhongxian var. Zhongzhou. Chongqing Shi, C China 30°16′N 108°03′E
161 N9 Zhongxiang Hubei, C China 31°12′N 112°35′E
Zhongzhou see Zhongxian
144 M14 Zhosaly prev. Dzhusaly. Kzylorda, SW Kazakhstan 45°29′N 64°04′E
161 O7 Zhoukou var. Zhoukouzhen. Henan, C China 33°32′N 114°40′E
Zhoukouzhen see Zhoukou
161 S9 Zhoushan Zhejiang, S China 30°02′N 122°06′E
Zhoushan Islands see Zhoushan Qundao
161 S9 Zhoushan Qundao Eng. Zhoushan Islands. island group SE China
116 I5 Zhovkva Pol. Żółkiew, Rus. Zholkev, Zholkva; prev. Nesterov. L'vivs'ka Oblast', NW Ukraine 50°04′N 24°E
117 S7 Zhovti Vody Rus. Zhëltyye Vody. Dnipropetrovs'ka Oblast', E Ukraine 48°24′N 33°30′E
117 Q10 Zhovtneve Rus. Zhovtnevoye. Mykolayivs'ka Oblast', S Ukraine 46°50′N 32°00′E
Zhovtnevoye see Zhovtneve
Zhi Qu see Tongtian He
114 K9 Zhrebchevo, Yazovir ⊠ C Bulgaria
163 V13 Zhuanghe Liaoning, NE China 39°42′N 123°00′E
159 W11 Zhuanglang var. Shuiluo; prev. Shuilocheng. Gansu, C China 35°06′N 106°21′E
145 P15 Zhuantobe Kaz. Zhūantōbe. Yuzhnyy Kazakhstan, S Kazakhstan 44°45′N 68°50′E
161 Q5 Zhucheng Shandong, E China 35°58′N 119°24′E
159 V12 Zhugqu Gansu, C China 33°51′N 104°14′E
161 N15 Zhuhai Guangdong, S China 22°16′N 113°30′E
Zhuizishan see Weichang
Zhuji see Shangqiu

◆ Country ● Country Capital ◇ Dependent Territory ○ Dependent Territory Capital ◈ Administrative Regions ✈ International Airport ▲ Mountain ▲▲ Mountain Range Volcano ∿ River ◎ Lake ⊠ Reservoir

ZHUKOVKA – ZYRYANOVSK

126 *I5* **Zhukovka** Bryanskaya Oblast’, W Russian Federation *53°33′N 33°48′E*
161 *N7* **Zhumadian** Henan, C China *32°58′N 114°03′E*
161 *S13* **Zhunan** ***prev.*** Chunan. N Taiwan *24°44′N 120°51′E*
Zhuo Xian *see* Zhuozhou
161 *O3* **Zhuozhou** ***prev.*** Zhuo Xian. Hebei, E China *39°22′N 115°40′E*
162 *L14* **Zhuozi Shan** ▲ N China *39°28′N 106°58′E*
113 *M17* **Zhur** ***Serb.*** Žur. S Kosovo *42°10′N 20°37′E*
Zhuravichi *see* Zhuravichy
119 *O17* **Zhuravichy** ***Rus.*** Zhuravichi. Homyel’skaya Voblasts’, SE Belarus *53°15′N 30°33′E*
145 *Q8* **Zhuravlevka** Akmola, N Kazakhstan *52°00′N 69°59′E*
117 *Q4* **Zhurivka** Kyyivs’ka Oblast’, N Ukraine *50°28′N 31°48′E*
144 *J11* **Zhuryn** Aktyubinsk, W Kazakhstan *49°13′N 57°36′E*
145 *T15* **Zhusandala, Step’** *grassland* SE Kazakhstan
160 *L8* **Zhushan** Hubei, C China *32°11′N 110°05′E*
Zhushan *see* Xuan’en
Zhuyang *see* Dazhu
161 *N11* **Zhuzhou** Hunan, S China *27°52′N 112°52′E*
116 *I6* **Zhydachiv** ***Pol.*** Żydaczów, ***Rus.*** Zhidachov. L’vivs’ka Oblast’, NW Ukraine *49°20′N 24°08′E*
144 *G9* **Zhympity** ***Kaz.*** Zhympīty; ***prev.*** Dzhambeyty. Zapadnyy, W Kazakhstan *50°16′N 52°34′E*
119 *K19* **Zhytkavichy** ***Rus.*** Zhitkovichi. Homyel’skaya Voblasts’, SE Belarus *52°14′N 27°52′E*
117 *N4* **Zhytomyr** ***Rus.*** Zhitomir. Zhytomyrs’ka Oblast’, NW Ukraine *50°17′N 28°40′E*
Zhytomyr *see* Zhytomyrs’ka Oblast’
116 *M4* **Zhytomyrs’ka Oblast’** ***var.*** Zhytomyr, ***Rus.*** Zhitomirskaya Oblast’. ⬖ *province* N Ukraine
153 *U15* **Zia** ✈ (Dhaka) Dhaka, C Bangladesh
111 *J20* **Žiar nad Hronom** ***var.*** Sväty Kríž nad Hronom, ***Ger.*** Heiligenkreuz, ***Hung.*** Garamszentkereszt. Banskobystrický Kraj, C Slovakia *48°36′N 18°52′E*
161 *Q4* **Zibo** ***var.*** Zhangdian. Shandong, E China *36°51′N 118°01′E*
160 *L4* **Zichang** ***prev.*** Wayaobu. Shaanxi, C China *37°08′N 109°40′E*
Zichenau *see* Ciechanów
111 *G15* **Ziębice** ***Ger.*** Münsterberg in Schlesien. Dolnośląskie, SW Poland *50°37′N 17°01′E*
Ziebingen *see* Cybinka
Ziegenhais *see* Głuchołazy
110 *E12* **Zielona Góra** ***Ger.*** Grünberg, Grünberg in Schlesien, Grüneberg. Lubuskie, W Poland *51°56′N 15°31′E*
99 *F14* **Zierikzee** Zeeland, SW Netherlands *51°39′N 03°55′E*
160 *I10* **Zigong** ***var.*** Tzekung. Sichuan, C China *29°20′N 104°48′E*
76 *G12* **Ziguinchor** SW Senegal *12°34′N 16°20′W*
41 *N16* **Zihuatanejo** Guerrero, S Mexico *17°39′N 101°33′W*
Ziketan *see* Xinghai
Zilah *see* Zalău
127 *W7* **Zilair** Respublika Bashkortostan, W Russian Federation *52°12′N 57°15′E*
136 *L12* **Zile** Tokat, N Turkey *40°18′N 35°52′E*
111 *J18* **Žilina** ***Ger.*** Sillein, ***Hung.*** Zsolna. Zlínský Kraj, N Slovakia *49°13′N 18°44′E*
111 *J19* **Žilinský Kraj** ⬖ *region* N Slovakia
75 *Q9* **Zillah** ***var.*** Zallah. C Libya *28°30′N 17°33′E*
Zillenmarkt *see* Zalău
109 *N7* **Ziller** ∼ W Austria
Zillertal Alps *see* Zillertaler Alpen
109 *N8* **Zillertaler Alpen** ***Eng.*** Zillertal Alps, ***It.*** Alpi Aurine. ⛰ Austria/Italy
118 *K10* **Zilupe** ***Ger.*** Rosenhof. E Latvia *56°10′N 28°06′E*
41 *O13* **Zimapán** Hidalgo, C Mexico *20°45′N 99°21′W*
83 *I16* **Zimba** Southern, S Zambia *17°20′S 26°11′E*
83 *J17* **Zimbabwe** ***off.*** Republic of Zimbabwe; ***prev.*** Rhodesia. ◆ *republic* S Africa
Zimbabwe, Republic of *see* Zimbabwe
116 *H10* **Zimbor** ***Hung.*** Magyarzsombor. Sălaj, NW Romania *47°00′N 23°16′E*
Zimmerbude *see* Svetlyy
116 *J15* **Zimnicea** Teleorman, S Romania *43°39′N 25°21′E*
114 *L9* **Zimnitsa** Yambol, E Bulgaria *42°34′N 26°37′E*
127 *N12* **Zimovniki** Rostovskaya Oblast’, SW Russian Federation *47°07′N 42°29′E*
148 *J5* **Zindah Jān** ***var.*** Zendajan, Zindajān; ***prev.*** Zendeh Jān. Herāt, NW Afghanistan *34°55′N 61°53′E*
Zindajān *see* Zindah Jān
77 *V12* **Zinder** Zinder, S Niger *13°47′N 09°02′E*
77 *W11* **Zinder** ⬖ *department* S Niger
77 *P12* **Ziniaré** C Burkina Faso *12°35′N 01°18′W*
Zinjan *see* Zanjān
141 *P16* **Zinjibār** SW Yemen *13°08′N 45°23′E*
117 *T4* **Zin’kiv** ***var.*** Zen’kov. Poltavs’ka Oblast’, NE Ukraine *50°11′N 34°22′E*
Zinov’yevsk *see* Kirovohrad
Zintenhof *see* Sindi
31 *N10* **Zion** Illinois, N USA *42°27′N 87°49′W*
54 *F10* **Zipaquirá** Cundinamarca, C Colombia *05°03′N 74°01′W*
Zipser Neudorf *see* Spišská Nová Ves
111 *H23* **Zirc** Veszprém, W Hungary *47°16′N 17°52′E*
113 *D14* **Žirje** ***It.*** Zuri. *island* S Croatia
Zirknitz *see* Cerknica
108 *M7* **Zirl** Tirol, W Austria *47°17′N 11°16′E*
101 *K20* **Zirndorf** Bayern, SE Germany *49°27′N 10°57′E*
160 *M11* **Zi Shui** ∼ C China
109 *Y3* **Zistersdorf** Niederösterreich, NE Austria *48°32′N 16°45′E*
41 *O14* **Zitácuaro** Michoacán, SW Mexico *19°28′N 100°21′W*
Zito *see* Lhorong
101 *Q16* **Zittau** Sachsen, E Germany *50°53′N 14°48′E*
112 *I12* **Živinice** Federacija Bosne I Hercegovine, E Bosnia and Herzegovina *44°26′N 18°39′E*
Ziwa Magharibi *see* Kagera
81 *J14* **Ziway Hāyk’** ◎ C Ethiopia
161 *N12* **Zixing** Hunan, S China *26°01′N 113°25′E*
127 *W7* **Ziyanchurino** Orenburgskaya Oblast’, W Russian Federation *51°36′N 56°58′E*
160 *K8* **Ziyang** Shaanxi, C China *32°33′N 108°27′E*
111 *I20* **Zlaté Moravce** ***Hung.*** Aranyosmarót. Nitriansky Kraj, SW Slovakia *48°24′N 18°20′E*
112 *K13* **Zlatibor** ⛰ W Serbia
114 *L9* **Zlati Voyvoda** Sliven, C Bulgaria *42°36′N 26°13′E*
116 *G11* **Zlatna** ***Ger.*** Kleinschlatten, ***Hung.*** Zalatna; ***prev. Ger.*** Goldmarkt. Alba, C Romania *46°08′N 23°11′E*
114 *I8* **Zlatna Panega** Lovech, N Bulgaria *43°07′N 24°09′E*
114 *N8* **Zlatni Pyasatsi** ***var.*** `Załni Pjašaci, Zlatni Pyasŭtsi, Golden Sands. Varna, NE Bulgaria *43°19′N 28°03′E*
Zlatni Pyasŭtsi *see* Zlatni Pyasatsi
122 *F11* **Zlatoust** Chelyabinskaya Oblast’, C Russian Federation *55°12′N 59°33′E*
111 *M19* **Zlatý Stôl** ***Ger.*** Goldener Tisch, ***Hung.*** Aranyosasztal. ▲ C Slovakia *48°45′N 20°39′E*
113 *P18* **Zletovo** NE FYR Macedonia *42°00′N 22°14′E*
111 *H18* **Zlín** ***prev.*** Gottwaldov. Zlínský Kraj, E Czech Republic *49°14′N 17°40′E*
111 *H19* **Zlínský Kraj** ⬖ *region* E Czech Republic
75 *O7* **Zlīṭan** W Libya *32°28′N 14°34′E*
110 *F9* **Złocieniec** ***Ger.*** Falkenburg in Pommern. Zachodnio-pomorskie, NW Poland *53°31′N 16°01′E*
110 *J13* **Złoczew** Sieradz, S Poland *51°24′N 18°36′E*
Złoczów *see* Zolochiv
111 *F14* **Złotoryja** ***Ger.*** Goldberg. Dolnośląskie, W Poland *51°08′N 15°57′E*
110 *G9* **Złotów** Wielkopolskie, C Poland *53°22′N 17°02′E*
110 *G13* **Żmigród** ***Ger.*** Trachenberg. Dolnośląskie, SW Poland *51°31′N 16°55′E*
126 *J6* **Zmiyevka** Orlovskaya Oblast’, W Russian Federation *52°39′N 36°20′E*
117 *V5* **Zmiyiv** Kharkivs’ka Oblast’, E Ukraine *49°40′N 36°22′E*
Zna *see* Tsna
Znaim *see* Znojmo
126 *M7* **Znamenka** Tambovskaya Oblast’, W Russian Federation *52°24′N 42°28′E*
Znamenka *see* Znam”yanka
119 *C14* **Znamens** Astrakhanskaya Oblast’, W Russian Federation *54°37′N 21°13′E*
127 *P10* **Znamensk** ***Ger.*** Wehlau. Kaliningradskaya Oblast’, W Russian Federation *48°33′N 46°18′E*
117 *R7* **Znam”yanka** ***Rus.*** Znamenka. Kirovohrads’ka Oblast’, C Ukraine *48°41′N 32°40′E*
110 *H10* **Żnin** Kujawsko-pomorskie, C Poland *52°50′N 17°41′E*
111 *F19* **Znojmo** ***Ger.*** Znaim. Jihomoravský Kraj, SE Czech Republic *48°52′N 16°04′E*
79 *N16* **Zobia** Orientale, N Dem. Rep. Congo *02°57′N 25°55′E*
83 *N15* **Zóbuè** Tete, NW Mozambique *15°36′S 34°26′E*
98 *G12* **Zoetermeer** Zuid-Holland, W Netherlands *52°04′N 04°30′E*
108 *E7* **Zofingen** Aargau, N Switzerland *47°18′N 07°57′E*
159 *R15* **Zogang** ***var.*** Wangda. Xizang Zizhiqu, W China *29°41′N 97°54′E*
106 *E7* **Zogno** Lombardia, N Italy *45°49′N 09°42′E*
142 *M10* **Zohreh, Rūd-e** ∼ SW Iran
160 *H7* **Zoigê** ***var.*** Dagcagoin. Sichuan, C China *33°44′N 102°57′E*
Zółkiew *see* Zhovkva
108 *D8* **Zollikofen** Bern, W Switzerland *47°00′N 07°24′E*
Zolochev *see* Zolochiv
117 *U4* **Zolochiv** ***Rus.*** Zolochev. Kharkivs’ka Oblast’, E Ukraine *50°16′N 35°58′E*
116 *J5* **Zolochiv** ***Pol.*** Złoczów, ***var.*** Zolochev. L’vivs’ka Oblast’, W Ukraine *49°48′N 24°51′E*
117 *X7* **Zolote** ***Rus.*** Zolotoye. Luhans’ka Oblast’, E Ukraine *48°42′N 38°33′E*
117 *Q6* **Zolotonosha** Cherkas’ka Oblast’, C Ukraine *49°39′N 32°05′E*
Zolotoye *see* Zolote
Zólyom *see* Zvolen
83 *N15* **Zomba** Southern, S Malawi *15°22′S 35°23′E*
Zombor *see* Sombor
99 *D17* **Zomergem** Oost-Vlaanderen, NW Belgium *51°07′N 03°31′E*
147 *P11* **Zomin** ***Rus.*** Zaamin. Jizzax Viloyati, C Uzbekistan *39°56′N 68°16′E*
79 *I15* **Zongo** Equateur, N Dem. Rep. Congo *04°18′N 18°42′E*
136 *G10* **Zonguldak** Zonguldak, NW Turkey *41°26′N 31°47′E*
136 *H10* **Zonguldak** ⬖ *province* NW Turkey
99 *K17* **Zonhoven** Limburg, NE Belgium *50°59′N 05°22′E*
142 *J2* **Zonūz** Āẕarbāyjān-e Khāvarī, NW Iran *38°32′N 45°54′E*
103 *Y16* **Zonza** Corse, France, C Mediterranean Sea *41°49′N 09°13′E*
Zoppot *see* Sopot
Zorgho *see* Zorgo
77 *Q13* **Zorgo** ***var.*** Zorgho. C Burkina Faso *12°15′N 00°37′W*
104 *K10* **Zorita** Extremadura, W Spain *39°18′N 05°42′W*
147 *U14* **Zorkŭl** ***Rus.*** Ozero Zorkul’. ◎ SE Tajikistan
Zorkul’, Ozero *see* Zorkŭl
56 *A8* **Zorritos** Tumbes, N Peru *03°43′S 80°42′W*
111 *J16* **Żory** ***var.*** Zory, ***Ger.*** Sohrau. Śląskie, S Poland *50°04′N 18°42′E*
76 *K15* **Zorzor** N Liberia *07°46′N 09°28′W*
99 *E18* **Zottegem** Oost-Vlaanderen, NW Belgium *50°52′N 03°49′E*
77 *R15* **Zou** ∼ S Benin
78 *H6* **Zouar** Tibesti, N Chad *20°25′N 16°28′E*
76 *J6* **Zouérat** ***var.*** Zouérate, Zouîrât. Tiris Zemmour, N Mauritania *22°44′N 12°29′W*
Zouérate *see* Zouérat
Zoug *see* Zug
Zouîrât *see* Zouérat
76 *M16* **Zoukougbeu** C Ivory Coast *09°47′N 06°50′W*
98 *M5* **Zoutkamp** Groningen, NE Netherlands *53°22′N 06°17′E*
99 *J18* **Zoutleeuw** ***Fr.*** Leau. Vlaams Brabant, C Belgium *50°49′N 05°06′E*
112 *L9* **Zrenjanin** ***prev.*** Petrovgrad, Veliki Bečkerek, ***Ger.*** Grossbetschkerek, ***Hung.*** Nagybecskerek. Vojvodina, N Serbia *45°23′N 20°24′E*
112 *E10* **Zrinska Gora** ⛰ C Croatia
Zsablya *see* Žabalj
101 *N16* **Zschopau** ∼ E Germany
Zsebely *see* Jebel
Zsibó *see* Jibou
Zsilyvajdevulkán *see* Vulcan
Zsil/Zsily *see* Jiu
Zsolna *see* Žilina
Zsombolya *see* Jimbolia
Zsupanya *see* Županja
55 *N7* **Zuata** Anzoátegui, NE Venezuela *08°24′N 65°13′W*
105 *N14* **Zubia** Andalucía, S Spain *37°10′N 03°36′W*
65 *P16* **Zubov Seamount** *undersea feature* E Atlantic Ocean *20°45′S 08°45′E*
124 *I16* **Zubtsov** Tverskaya Oblast’, W Russian Federation *56°10′N 34°34′E*
108 *M8* **Zuckerhütl** ▲ SW Austria *46°57′N 11°07′E*
Zueila *see* Zawīlah
76 *M16* **Zuénoula** C Ivory Coast *07°26′N 06°03′W*
105 *S5* **Zuera** Aragón, NE Spain *41°52′N 00°47′W*
141 *V13* **Ẓufār** ***Eng.*** Dhofar. *physical region* SW Oman
108 *G8* **Zug** ***Fr.*** Zoug. Zug, C Switzerland *47°11′N 08°31′E*
108 *G8* **Zug** ***Fr.*** Zoug. ⬖ *canton* C Switzerland
137 *R9* **Zugdidi** W Georgia *42°30′N 41°52′E*
108 *G8* **Zuger See** ◎ NW Switzerland
101 *K25* **Zugspitze** ▲ S Germany *47°25′N 10°58′E*
117 *X8* **Zuhres** ***Rus.*** Shakhtërsk. Donets’ka Oblast’, SE Ukraine *48°01′N 38°16′E*
99 *E15* **Zuid-Beveland** ***var.*** South Beveland. *island* SW Netherlands
98 *K10* **Zuidelijk-Flevoland** *polder* C Netherlands
Zuider Zee *see* IJsselmeer
98 *G12* **Zuid-Holland** ***Eng.*** South Holland. ⬖ *province* W Netherlands
98 *N5* **Zuidhorn** Groningen, NE Netherlands *53°15′N 06°25′E*
98 *O6* **Zuidlaardermeer** ◎ NE Netherlands
98 *O6* **Zuidlaren** Drenthe, NE Netherlands *53°06′N 06°41′E*
99 *K14* **Zuid-Willemsvaart Kanaal** *canal* S Netherlands
98 *N8* **Zuidwolde** Drenthe, NE Netherlands *52°40′N 06°25′E*
Zuitai/Zuitaizi *see* Kangxian
105 *O14* **Zújar** Andalucía, S Spain *37°33′N 02°52′W*
104 *L11* **Zújar** ∼ W Spain
104 *L11* **Zújar, Embalse del** ⊠ W Spain
80 *J9* **Zula** E Eritrea *15°19′N 39°40′E*
54 *G6* **Zulia** ***off.*** Estado Zulia. ⬖ *state* NW Venezuela
Zulia, Estado *see* Zulia
Zullapara *see* Maungdaw
Züllichau *see* Sulechów
105 *P3* **Zumárraga** País Vasco, N Spain *43°05′N 02°19′W*
112 *D8* **Žumberačko Gorje** ***var.*** Gorjanči, Uskocke Planine, Žumberak, ***Ger.*** Uskokengebirge; ***prev.*** Sichelburger Gerbirge. ▲ Croatia/Slovenia ***see also*** Gorjanci
Žumberak *see* Gorjanci/Žumberačko Gorje
194 *K7* **Zumberge Coast** *coastal feature* Antarctica
Zumbo *see* Vila do Zumbo
29 *W10* **Zumbro Falls** Minnesota, N USA *44°15′N 92°25′W*
29 *W10* **Zumbro River** ∼ Minnesota, N USA
29 *W10* **Zumbrota** Minnesota, N USA *44°18′N 92°37′W*
99 *H15* **Zundert** Noord-Brabant, S Netherlands *51°28′N 04°40′E*
Zungaria *see* Dzungaria
77 *U14* **Zungeru** Niger, C Nigeria *09°49′N 06°10′E*
161 *P2* **Zunhua** Hebei, E China *40°10′N 117°58′E*
37 *O11* **Zuni** New Mexico, SW USA *35°03′N 108°52′W*
37 *P11* **Zuni Mountains** ⛰ New Mexico, SW USA
160 *J11* **Zunyi** Guizhou, S China *27°40′N 106°56′E*
160 *J15* **Zuo Jiang** ∼ China/Vietnam
Zuoqi *see* Gegan Gol
108 *J9* **Zuoz** Graubünden, SE Switzerland *46°37′N 09°58′E*
112 *I10* **Županja** ***Hung.*** Zsupanya. Vukovar-Srijem, E Croatia *45°03′N 18°42′E*
Žur *see* Zhur
127 *T2* **Zura** Udmurtskaya Respublika, NW Russian Federation *57°36′N 53°19′E*
139 *V8* **Zurbāṭīyah** Wāsiṭ, E Iraq *33°13′N 46°07′E*
Zuri *see* Žirje
108 *F7* **Zürich** ***Eng./Fr.*** Zurich, ***It.*** Zurigo. Zürich, N Switzerland *47°23′N 08°33′E*
108 *G6* **Zürich** ***Eng./Fr.*** Zurich. ⬖ *canton* N Switzerland
Zurich, Lake *see* Zürichsee
108 *G7* **Zürichsee** ***Eng.*** Lake Zurich. ◎ NE Switzerland
Zurigo *see* Zürich
149 *V1* **Zŭrkŭl** ***Pash.*** Sarī Qūl, ***Rus.*** Ozero Zurkul’. ◎ Afghanistan/Tajikistan ***see also*** Sarī Qūl
Zŭrkŭl *see* Sarī Qūl
Zurkul’, Ozero *see* Sarī Qūl/Zŭrkŭl
110 *K10* **Żuromin** Mazowieckie, C Poland *53°00′N 19°54′E*
108 *J8* **Zürs** Vorarlberg, W Austria *47°11′N 10°11′E*
77 *T13* **Zuru** Kebbi, W Nigeria *11°28′N 05°13′E*
108 *F6* **Zurzach** Aargau, N Switzerland *47°33′N 08°21′E*
101 *J22* **Zusam** ∼ S Germany
98 *M11* **Zutphen** Gelderland, E Netherlands *52°09′N 06°12′E*
75 *N7* **Zuwārah** NW Libya *32°56′N 12°06′E*
Zuwaylah *see* Zawīlah
125 *R14* **Zuyevka** Kirovskaya Oblast’, NW Russian Federation *58°24′N 51°08′E*
161 *N10* **Zuzhou** Hunan, S China *27°52′N 113°00′E*
Zvenigorodka *see* Zvenyhorodka
117 *P6* **Zvenyhorodka** ***Rus.*** Zvenigorodka. Cherkas’ka Oblast’, C Ukraine *49°05′N 30°58′E*
123 *N12* **Zvezdnyy** Irkutskaya Oblast’, C Russian Federation *56°43′N 106°22′E*
125 *U14* **Zvëzdnyy** Permskiy Kray, NW Russian Federation *57°45′N 56°20′E*
83 *K18* **Zvishavane** ***prev.*** Shabani. Matabeleland South, S Zimbabwe *20°20′S 30°02′E*
111 *J20* **Zvolen** ***Ger.*** Altsohl, ***Hung.*** Zólyom. Banskobystrický Kraj, C Slovakia *48°35′N 19°06′E*
112 *J12* **Zvornik** E Bosnia and Herzegovina *44°24′N 19°07′E*
98 *M5* **Zwaagwesteinde** ***Fris.*** De Westerein. Fryslân, N Netherlands *53°16′N 06°08′E*
98 *H10* **Zwanenburg** Noord-Holland, C Netherlands *52°22′N 04°44′E*
98 *L8* **Zwarte Meer** ◎ N Netherlands
98 *M9* **Zwarte Water** ∼ N Netherlands
98 *M8* **Zwartsluis** Overijssel, E Netherlands *52°39′N 06°04′E*
76 *L17* **Zwedru** ***var.*** Tchien. E Liberia *06°04′N 08°07′W*
98 *O8* **Zweeloo** Drenthe, NE Netherlands *52°48′N 06°45′E*
101 *E20* **Zweibrücken** ***Fr.*** Deux-Ponts, ***Lat.*** Bipontium. Rheinland-Pfalz, SW Germany *49°15′N 07°22′E*
108 *D9* **Zweisimmen** Fribourg, W Switzerland *46°33′N 07°22′E*
101 *M15* **Zwenkau** Sachsen, E Germany *51°11′N 12°19′E*
109 *V3* **Zwettl** Wien, NE Austria *48°28′N 14°17′E*
109 *T3* **Zwettl an der Rodl** Oberösterreich, N Austria *48°28′N 14°17′E*
99 *D18* **Zwevegem** West-Vlaanderen, W Belgium *50°48′N 03°20′E*
101 *M17* **Zwickau** Sachsen, E Germany *50°43′N 12°31′E*
101 *N16* **Zwickauer Mulde** ∼ E Germany
101 *O21* **Zwiesel** Bayern, SE Germany *49°02′N 13°14′E*
98 *H13* **Zwijndrecht** Zuid-Holland, SW Netherlands *51°49′N 04°39′E*
Zwischenwässern *see* Medvode
Zwittau *see* Svitavy
110 *N13* **Zwoleń** Mazowieckie, SE Poland *51°21′N 21°37′E*
98 *M9* **Zwolle** Overijssel, E Netherlands *52°31′N 06°06′E*
22 *G6* **Zwolle** Louisiana, S USA *31°37′N 93°38′W*
110 *K12* **Żychlin** Łódzkie, C Poland *52°15′N 19°38′E*
Żydaczów *see* Zhydachiv
Zyembin *see* Zembin
Zyôetu *see* Jōetsu
110 *L12* **Żyrardów** Mazowieckie, C Poland *52°02′N 20°28′E*
123 *S8* **Zyryanka** Respublika Sakha (Yakutiya), NE Russian Federation *65°45′N 150°43′E*
145 *Y9* **Zyryanovsk** Vostochnyy Kazakhstan, E Kazakhstan *49°45′N 84°16′E*

◆ Country
● Country Capital
◇ Dependent Territory
○ Dependent Territory Capital
⬖ Administrative Regions
✈ International Airport
▲ Mountain
⛰ Mountain Range
⩘ Volcano
∼ River
◎ Lake
⊠ Reservoir

Picture Credits

Dorling Kindersley would like to express their thanks to the following individuals, companies, and institutions for their help in preparing this atlas.

Earth Resource Mapping Ltd., Egham, Surrey
Brian Groombridge, World Conservation Monitoring Centre, Cambridge
The British Library, London
British Library of Political and Economic Science, London
The British Museum, London
The City Business Library, London
King's College, London
National Meteorological Library and Archive, Bracknell
The Printed Word, London
The Royal Geographical Society, London
University of London Library
Paul Beardmore
Philip Boyes
Hayley Crockford
Alistair Dougal
Reg Grant
Louise Keane
Zoe Livesley
Laura Porter
Jeff Eidenshink
Chris Hornby
Rachelle Smith
Ray Pinchard
Robert Meisner
Fiona Strawbridge

Every effort has been made to trace the copyright holders and we apologize in advance for any unintentional omissions. We would be pleased to insert the appropriate acknowledgment in any subsequent edition of this publication.

Adams Picture Library: 86CLA; **G Andrews:** 186CR; **Ardea London Ltd:** K Ghana 150C; M Iljima 132TC; R Waller 148TR; Art Directors **Aspect Picture Library:** P Carmichael 160TR; 131CR(below); G Tompkinson 190TRB; **Axiom:** C Bradley 148CA, 158CA; J Holmes xivCRA, xxivBCR, xxviiCRB, 150TCR, 166TL, J Morris 75TL, 77CRB, J Spaull 134BL; **Bridgeman Art Library, London / New York:** Collection of the Earl of Pembroke, Wilton House xxBC; **The J. Allan Cash Photolibrary:** xlBR, xliiCLA, xlivCL, 10BC, 60CL, 69CLB, 70CL, 72CLB, 75BR, 76BC, 87BL, 109BR, 138BCL, 141TL, 154CR, 178BR, 181TR; **Bruce Coleman Ltd:** 86BC, 98CL, 100TC; S Alden 192BC(below); Atlantide xxviTCR, 138BR; E Bjurstrom 141BR; S Bond 96CRB; T Buchholz xvCL, 92TR, 123TCL; J Burton xxiiiC; J Cancalosi 181TRB; B J Coates xxvBL, 192CL; B Coleman 63TL; B & C Colhoun 2TR, 36CB; A Compost xxiiiCBR; Dr S Coyne 45TL; G Cubitt xviTCL, 169BR, 178TR, 184TR; P Davey xxviiCLB, 121TL(below); N Devore 189CBL; S J Doylee xxiiCRR; H Flygare xviiCRA; M P L Fogden 17C(above) ; Jeff Foott Productions xxiiiCRB, 11CRA; M Freeman 91BRA; P van Gaalen 86TR; G Gualco 140C; B Henderson 194CR; Dr C Henneghien 69C; HPH Photography, H Van den Berg 69CR; C Hughes 69BCL; C James xxxixTC; J Johnson 39CR, 197TR; J Jurka 91CA; S C Kaufman 28C; S J Krasemann 33TR; H Lange 10TRB, 68CA; C Lockwood 32BC; L C Marigo xxiiBC, xxviiCLA, 49CRA, 59BR; M McCoy 187TR; D Meredith 3CR; J Murray xvCR, 179BR; Orion Press 165CR(above); Orion Services & Trading Co. Inc. 164CR; C Ott 17BL; Dr E Pott 9TR, 40CL, 87C, 93TL, 194CLB; F Prenzel 186BC, 193BC; M Read 42BR, 43CRB; H Reinhard xxiiCR, xxviiTR, 194BR; L Lee Rue III 151BCL; J Shaw xixTL; K N Swenson 194BC; P Terry 115CR; N Tomalin 54BCL; P Ward 78TC; S Widstrand 57TR; K Wothe 91C, 173TCL; J T Wright 127BR; **Colorific:** Black Star / L Mulvehil 156CL; Black Star / R Rogers 57BR; Black Star / J Rupp 161BCR; Camera Tres / C. Meyer 59BRA; R Caputo / Matrix 78CL; J. Hill 117CLB; M Koene 55TR; G Satterley xliiCLAR; M Yamashita 156BL, 167CR(above); **Comstock:** 108CRB; Corbis UK Ltd: 170TR, 170BL; **D Cousens:** 147 CRA; **Corbis:** Bob Daemmrich 6BL; Bryan Denton xxxCBL; Julie Dermansky / Julie Dermansky xxviiiTC; Everett Kennedy Brown / Epa 165CB; Kimimasa Mayama / Reuters 168CL(above); mosaaberizing / Demotix xxxCBR; Ocean 60BL; Ocean 135CL; Sucheta DAS / Reuters xxviBCR; Rob Widdis / epa 30CA; **Sue Cunningham Photographic:** 51CR; S Alden 192BC(below) **James Davis Travel Photography:** xxxviTCB, xxxviTR, xxxviCL, 13CA, 19BC, 49TLB, 56BCR, 57CLA, 61BCL, 93BC, 94TC, 102TR, 120CB, 158BC, 179CRA, 191BR; **Dorling Kindersley:** Paul Harris xxiiTR; Nigel Hicks xxiiBM; Jamie Marshall 181TR; Bharath Ramamrutham 155BR; Colin Sinclair 133BMR; George Dunnet: 124CA; **Environmental Picture Library:** Chris Westwood 126C; **Eye Ubiquitous:** xlCA; L. Fordyce 12CLA; L Johnstone 6CRA, 28BLA, 30CB; S. Miller xxiCA; M Southern 73BLA; **Chris Fairclough Colour Library:** xliBR; **Ffotograff:** N. Tapsell 158CL; **FLPA -Images of nature:** 123TR; **Geoscience Features:** xviBCR, xviBR, 102CL, 108BC, 122BR; Solar Film 64TC; **Getty Images:** Kim Steele 161BCL; **gettyone stone**: 131BC, 133BR, 164CR(above); G Johnson 130BL; R Passmore 120TR; D Austen 187CL; G Allison 186CL; L Ulrich 17TL; M Vines 17BL; R Wells 193BL; **Robert Harding Picture Library:** xviiTC, xxivCR, xxxC, xxxvTC, 2TLB, 3CA, 15CRB, 15CR, 37BC, 38CRA, 50BL, 95BR, 99CR, 114CR, 122BL, 131CLA, 142CB, 143TL, 147TR, 168TR, 168CA, 166BR; P G. Adam 13TCB; D Atchison-Jones 70BLA; J Bayne 72BCL; B Schuster 80CR; C Bowman 50BR, 53CA, 62CL, 70CRL; C Campbell xxiBC; G Corrigan 159CRB, 161CRB; P Craven xxxvBL; R Cundy 69BR; Delu 79BC; A Durand 111BR; Financial Times 142BR; R Frerck 51BL; T Gervis 3BCL, 7CR; I Griffiths xxxCL, 77TL; T Hall 166CRA; D Harney 142CA; S Harris xliiiBCL; G Hellier xvCRB, 135BL; F Jackson 137BCR; Jacobs xxxviiTL; P Koch 139TR; F Joseph Land 122TR; Y Marcoux 9BR; S Massif xvBC; A Mills 88CLB; L Murray 114TR; R Rainford xlivBL; G Renner 74CB, 194C; C Rennie 48CL, 116BR; R Richardson 118CL; P Van Riel 48BR; E Rooney 124TR; Sassoon xxivCL, 148CLB; Jochen Schlenker 193CL; P Scholey 176TR; M Short 137TL; E Simanor xxviiCR; V Southwell 139CR; J Strachan 42TR, 111BL, 132BCR; C Tokeley 131CLA; A C Waltham 161C; T Waltham xviiBL, xxiiCLLL, 138CRB; Westlight 37CR; N Wheeler 139BL; A Williams xxxviiiBR, xlTR; A Woolfitt 95BRA; Paul Harris: 168TC; **Hutchison Library:** 131CR (above) 6BL; P. Collomb 137CR; C. Dodwell 130TR; S Errington 70BCL; P. Hellyer 142BC; J. Horner xxxiTC; R. Ian Lloyd 134CRA; J.Nowell 135CLB, 143TC; A Zvoznikov xxiiCL; **Image Bank:** 87BR; J Banagan 190BCA; A Becker xxivBCL; M Khansa 121CR, M Isy-Schwart 193CR(above), 191CL; Khansa K Forest 163TR; Lomeo xxivTCR; T Madison 170TL(below); C Molyneux xxiiCRRR; C Navajas xviiiTR; Ocean Images Inc. 192CLB; J van Os xviiTCR; S Proehl 6CL; T Rakke xixTC, 64CL; M Reitz 196CA; M Romanelli 166CL(below); G A Rossi 151BCR, 176BLA; B Roussel 109TL; S Satushek xviiBCR; Stock Photos / J M Spielman xxivTRL; **Images Colour Library:** xxiiCLL, xxxixTR, xliCR, xliiiBL, 3BR, 19BR, 37TL, 44TL, 62TC, 91BR, 102CLB, 103CR, 150CL, 180CA; 164BC, 165TL; **Impact Photos:** J & G Andrews 186BL; C. Bluntzer 156BR; Cosmos / G. Buthaud 65BC; S Franklin 126BL; A. le Garsmeur 131C; C Jones xxxiCB, 70BL; V. Nemirousky 137BR; J Nicholl 76TCR; C. Penn 187C(below); G Sweeney xviiBR, 196CB, 196TR, J & G Andrews 186TR; **JVZ Picture Library:** T Nilson 135TC; **Frank Lane Picture Agency:** xxiTCR, xxiiiBL, 93TR; A Christiansen 58CRA; J Holmes xivBL; S. McCutcheon 3C; Silvestris 173TCR; D Smith xxiiBCL; W Wisniewsli 195BR; **Leeds Castle Foundation:** xxxviiBC; **Magnum:** Abbas 83CR, 136CA; S Franklin 134CRB; D Hurn 4BCL; P. Jones-Griffiths 191BL; H Kubota xviBCL, 156CLB; F Maver xviBL; S McCurry 73CL, 133BCR; G. Rodger 74TR; C Steele Perkins 72BL; **Mountain Camera / John Cleare:** 153TR; C Monteath 153CR; **Nature Photographers:** E.A. Janes 112CL; **Natural Science Photos:** M Andera 110C; **Network Photographers Ltd.:** C Sappa / Rapho 119BL; **N.H.P.A.:** N. J. Dennis xxiiiCL; D Heuchlin xxiiiCLA; S Krasemann 15BL, 25BR, 38TC; K Schafer 49CB; R Tidman 160CLB; D Tomlinson 145CR; M Wendler 48TR; **Nottingham Trent University:** T Waltham xivCL, xvBR; **Novosti:** 144BLA; **Oxford Scientific Films:** D Allan xxiiTR; H R Bardarson xviiiBC; D Bown xxiiiCBLL; M Brown 140BL; M Colbeck 147CAR; W Faidley 3TL; L Gould xxiiiTRB; D Guravich xxiiiTR; P Hammerschmidy / Okapia 87CLA; M Hill 57TL, 195TR; C Menteath ; J Netherton 2CRB; S Osolinski 82CA; R Packwood 72CA; M Pitts 179TC; N Rosing xxiiiCBL, 9TR, 197BR; D Simonson 57C; Survival Anglia / C Catton 137TR; R Toms xxiiiBR; K Wothe xxiBL, xviiCLA; **Panos Pictures:** B Aris 133C; P Barker xxivBR; T Bolstao 153BR; N Cooper 82CB, 153TC; J-L Dugast 166C(below), 167BR; J Hartley 73CA, 90CL; J Holmes 149BC; J Morris 76CLB; M Rose 146TR; D Sansoni 155CL; C Stowers 163TL; **Edward Parker:** 49TL, 49CLB; **Pictor International:** xivBR, xvBRA, xixTCL, xxCL, 3CLA, 17BR, 20TR, 20CRB, 23BCA, 23CL, 26CB, 27BC, 33TRB, 34BC, 34BR, 34CR, 38CB, 38CL, 43CL, 63BR, 65TC, 82CL, 83CLB, 99BR, 107CLA, 166TR, 171CL(above), 180CLB, 185TL; **Pictures Colour Library:** xxiBCL, xxiiBR, xxviBCL, 6BR, 15TR, 8TR, 16CL(above), 19TL, 20BL, 24C, 24CLA, 27TR, 32TRB, 36BC, 41CA, 43CRA, 68BL, 90TCB, 94BL, 99BL, 106CA, 107CLB, 107CR, 107BR, 117BL, 164BC, 192BL, K Forest 165TL(below); **Planet Earth Pictures:** 193CR(below); D Barrett 148CB, 184CA; R Coomber 16BL; G Douwma 172BR; E Edmonds 173BR; J Lythgoe 196BL; A Mounter 172CR; M Potts 6CA; P Scoones xxTR; J Walencik 110TR; J Waters 53BCL; **Popperfoto:** Reuters / J Drake xxxiiCLA; **Rex Features:** 165CR; Antelope xxxiiCLB; M Friedel xxiCR; J Shelley xxxCR; Sipa Press xxxCR; Sipa Press / Chamussy 176BL; **Robert Harding Picture Library:** C. Tokeley 131TL; J Strachan 132BL; Franz Joseph Land 122TR; Franz Joseph Land 364/7088 123BL, 169C(above), 170C(above), Tony Waltham 186CR(below), Y Marcoux 9BR; **Russia & Republics Photolibrary:** M Wadlow 118CR, 119CL, 124BC, 124CL, 125TL, 125BR, 126TCR; **Science Photo Library:** Earth Satellite Corporation xixTRB, xxxiCR, 49BCL; F Gohier xiCR; J Heseltine xviTCB; K Kent xvBLA; P Menzell xvBL; N.A.S.A. xBC; D Parker xivBC; University of Cambridge Collection Air Pictures 87CLB; RJ Wainscoat / P Arnold, Inc. xiBC; D Weintraub xiBL; **South American Pictures:** 57BL, 62TR; R Francis 52BL; Guyana Space Centre 50TR; T Morrison 49CRB, 49BL, 50CR, 52TR, 54TR, 61C; **Southampton Oceanography:** xviiiBL; **Sovofoto / Eastfoto:** xxxiiCBR; **Spectrum Colour Library:** 50BC, 160BC; J King 145BR; **Frank Spooner Pictures:** Gamma-Liason/Vogel 131CL(above); 26CRB; E. Baitel xxxiiBC; Bernstein xxxiCL; Contrast 112CR; Diard / Photo News 113CL; Liaison / C. Hires xxxiiTCB; Liaison / Nickelsberg xxxiiTR; Marleen 113TL; Novosti 116CA; P. Piel xxxCA; H Stucke 188CLB, 190CA; Torrengo / Figaro 78BR; A Zamur 113BL; **Still Pictures:** C Caldicott 77TC; A Crump 189CL; M & C Denis-Huot xxiiBL, 78CR, 81BL; M Edwards xxiCRL, 53BL, 64CR, 69BLA, 155BR; J Frebet 53CLB; H Giradet 53TC; E Parker 52CL; M Gunther 121BC; **Tony Stone Images:** xxviTR, 4CA, 7BL, 7CL, 13CRB, 39BR, 58C, 97BC, 101BR, 106TR, 109CL, 109CRB, 164CLB, 165C,180CB, 181BR, 188BC, 192TR; G Allison 18TR, 31CRB, 187CRB; D Armand 14TCB; D Austen 180TR, 186CL, 187CL; J Beatty 74CL; O Benn xxviBR; K Biggs xxiTL; R Bradbury 44BR; R A Butcher xxviTL; J Callahan xxviiCRA; P Chesley 185BCL, 188C; W Clay 30BL, 31CRA; J Cornish 96BL, 107TL; C Condina 41CB; T Craddock xxivTR; P Degginger 36CLB; Demetrio 5BR; N DeVore xxivBC; A Diesendruck 60BR; S Egan 87CRA, 96BR; R Elliot xxiiBCR; S Elmore 19C; J Garrett 73CR; S Grandadam 14BR; R Grosskopf 28BL; D Hanson 104BC; C Harvey 69TL; G Hellier 110BL, 165CR; S Huber 103CRB; D Hughs xxxiBR; A Husmo 91TR; G Irvine 31BC; J Jangoux 58CL; D Johnston xviiTR; A Kehr 113C; R Koskas xviTR; J Lamb 96CRA; J Lawrence 75CRA; L Lefkowitz 7CA; M Lewis 45CLA; S Mayman 55BR; Murray & Associates 45CR; G Norways 104CA; N Parfitt xxviiCL, 68TCR, 81TL; R Passmore 121TR; N Press xviBCA; E Pritchard 88CA, 90CLR; T Raymond 21BL, 29TR; L Resnick 74BR; M Rogers 80BR; A Sacks 28TCB; C Saule 90CR; S Schulhof xxivTC; P Seaward 34CL; M Segal 32BL; V Shenai 152CL; R Sherman 26CL; H Sitton 136CR; R Smith xxvBLA, 56C; S Studd 108CLA; H Strand 49BR, 63TR; P Tweedie 177CR; L Ulrich 17BL; M Vines 17TC; A B Wadham 60CR; J Warden 63CLB; R Wells 23CRA, 193BL; G Yeowell 34BL; **Telegraph Colour Library:** 61CRB, 61TCR, 157TL; R Antrobus xxxixBR; J Sims 26BR; **Topham Picturepoint:** xxxiCBL, 162BR, 168TR, 168BC; **Travel Ink:** A Cowin 88TR; **Trip:** 140BR, 144CA, 155CRA; B Ashe 159TR; D Cole 190BCL, 190CR; D Davis 89BL; I Deineko xxxiTR; J Dennis 22BL; Dinodia 154CL; Eye Ubiquitous / L Fordyce 2CLB; A Gasson 149CR; W Jacobs 43TL, 54BL, 177BC, 178CLA, 185BCR, 186BL; P Kingsbury 112C; K Knight 177BR; V Kolpakov 147BL; T Noorits 87TL, 119BR, 146CL; R Power 41TR; N Ray 166BL, 168TC; C Rennie 116CLB; V Sidoropolev 145TR; E Smith 183BC, 183TL; **Woodfin Camp & Associates:** 92BLR; **World Pictures:** xvCRA, xviiCRA, 9CRB, 22CL, 23BC, 24BL, 35BL, 40TR, 51TR, 71BR, 80TCR, 82TR, 83BL, 86BCR, 96TC, 98BL, 100CR, 101CR, 103BC, 105TC, 157BL, 161BCL, 162CLB, 172CLB, 172BC, 179BL, 182CB, 183C, 184CL, 185CR; 121BR, 121TT; **Zefa Picture Library:** xviBLR, xviiBCL, xviiiCL, 3CL, 8BC, 8CT, 9CR, 13BC, 14TC, 16TR, 21TL, 22CRB, 25BL, 32TCR, 36BCR, 59BCL, 65TCL, 69CLA, 79TL, 81BR, 87CRB, 92C, 98C, 99TL, 100BL, 107TR, 118CRB, 120BL; 122C(below), 124CLA, 164BR, 183TR; Anatol 113BR; Barone 114BL; Brandenburg 5C; A J Brown 44TR; H J Clauss 55CLB; Damm 71BC; Evert 92BL; W Felger 3BL; J Fields 189CRA; R Frerck 4BL; G Heil 56BR; K Heibig 115BR; Heilman 28BC; Hunter 8C; Kitchen 10TR, 8CL, 8BL, 9TR; Dr H Kramarz 7BLA, 123CR(below); Mehlio 155BL; J F Raga 24TR; Rossenbach 105BR; Streichan 89TL; T Stewart 13TR, 19CR; Sunak 54BR, 162TR; D H Teuffen 95TL; B Zaunders 40BC.
Additional Photography: Geoff Dann; Rob Reichenfeld; H Taylor; Jerry Young.

Map Credits

World Population Density map, page xxiv:

Source:LandScanTM Global Population Database. Oak Ridge, TN; Oak Ridge National Laboratory. Available at http://www.ornl.gov/landscan/.

350
Country · Dependent Territory · Administrative Region · Mountain · Volcano · Lake
Country Capital · Dependent Territory Capital · International Airport · Mountain Range · River · Reservoir

NORTH AMERICA

CANADA
Pages 8–15

UNITED STATES OF AMERICA
Pages 16–39

MEXICO
Pages 40–41

BELIZE
Pages 42–43

COSTA RICA
Pages 42–43

EL SALVADOR
Pages 42–43

GUATEMALA
Pages 42–43

HONDURAS
Pages 42–43

SOUTH AMERICA

GRENADA
Pages 44–45

HAITI
Pages 44–45

JAMAICA
Pages 44–45

ST KITTS & NEVIS
Pages 44–45

ST LUCIA
Pages 44–45

ST VINCENT & THE GRENADINES
Pages 44–45

TRINIDAD & TOBAGO
Pages 44–45

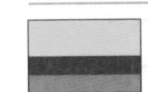
COLOMBIA
Pages 54–55

AFRICA

URUGUAY
Pages 60–61

CHILE
Pages 62–63

PARAGUAY
Pages 62–63

ALGERIA
Pages 74–75

EGYPT
Pages 74–75

LIBYA
Pages 74–75

MOROCCO
Pages 74–75

TUNISIA
Pages 74–75

LIBERIA
Pages 76–77

MALI
Pages 76–77

MAURITANIA
Pages 76–77

NIGER
Pages 76–77

NIGERIA
Pages 76–77

SENEGAL
Pages 76–77

SIERRA LEONE
Pages 76–77

TOGO
Pages 76–77

BURUNDI
Pages 80–81

DJIBOUTI
Pages 80–81

ERITREA
Pages 80–81

ETHIOPIA
Pages 80–81

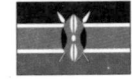
KENYA
Pages 80–81

RWANDA
Pages 80–81

SOMALIA
Pages 80–81

SUDAN
Pages 80–81

NAMIBIA
Pages 82–83

SOUTH AFRICA
Pages 82–83

SWAZILAND
Pages 82–83

ZAMBIA
Pages 82–83

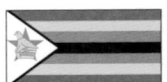
ZIMBABWE
Pages 82–83

COMOROS
Pages 172–173

MADAGASCAR
Pages 172–173

MAURITIUS
Pages 172–173

LUXEMBOURG
Pages 98–99

NETHERLANDS
Pages 98–99

GERMANY
Pages 100–101

FRANCE
Pages 102–103

MONACO
Pages 102–103

ANDORRA
Pages 104–105

PORTUGAL
Pages 104–105

SPAIN
Pages 104–105

POLAND
Pages 110–111

SLOVAKIA
Pages 110–111

ALBANIA
Pages 112–113

BOSNIA & HERZEGOVINA
Pages 112–113

CROATIA
Pages 112–113

KOSOVO (disputed)
Pages 112–113

MACEDONIA
Pages 112–113

MONTENEGRO
Pages 112–113

ASIA

LATVIA
Pages 118–119

LITHUANIA
Pages 118–119

CYPRUS
Pages 120–121

MALTA
Pages 120–121

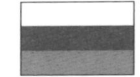
RUSSIAN FEDERATION
Pages 122–127

ARMENIA
Pages 136–137

AZERBAIJAN
Pages 136–137

GEORGIA
Pages 136–137

TURKEY
Pages 136–137/114–115

QATAR
Pages 140–143

SAUDI ARABIA
Pages 140–141

UNITED ARAB EMIRATES
Pages 140–143

YEMEN
Pages 140–141

IRAN
Pages 142–143

KAZAKHSTAN
Pages 144–145

KYRGYZSTAN
Pages 146–147

TAJIKISTAN
Pages 146–147

CHINA
Pages 156–163

MONGOLIA
Pages 156–157/162–163

NORTH KOREA
Pages 156–157/162–163

SOUTH KOREA
Pages 156–157/162–163

TAIWAN
Pages 156–157/160–161

JAPAN
Pages 164–165

MYANMAR (BURMA)
Pages 166–167

CAMBODIA
Pages 166–167

AUSTRALASIA & OCEANIA

SINGAPORE
Pages 168–169

MALDIVES
Pages 172–173

AUSTRALIA
Pages 180–183

NEW ZEALAND
Pages 184–185

PAPUA NEW GUINEA
Pages 186–187

FIJI
Pages 186–187

SOLOMON ISLANDS
Pages 186–187

VANUATU
Pages 186–187